McGRAW-HILL
DICTIONARY OF
SCIENTIFIC AND
TECHNICAL
TERMS

Sixth
Edition

McGRAW-HILL
DICTIONARY OF
SCIENTIFIC AND
TECHNICAL
TERMS

Sixth
Edition

McGraw-Hill

New York   Chicago   San Francisco
Lisbon   London   Madrid   Mexico City
Milan   New Delhi   San Juan   Seoul   Singapore   Sydney   Toronto

On the cover: Representation of a fullerene molecule with a noble gas atom trapped inside. At the Permian-Triassic sedimentary boundary the noble gases helium and argon have been found trapped inside fullerenes. They exhibit isotope ratios quite similar to those found in meteorites, suggesting that a fireball meteorite or asteroid exploded when it hit the Earth, causing major changes in the environment. *(Image copyright © Dr. Luann Becker. Reproduced with permission.)*

Over the six editions of the Dictionary, material has been drawn from the following references: G. M. Garrity et al., *Taxonomic Outline of the Procaryotes,* Release 2, Springer-Verlag, January 2002; D. W. Linzey, *Vertebrate Biology,* McGraw-Hill, 2001; J. A. Pechenik, *Biology of the Invertebrates,* 4th ed., McGraw-Hill, 2000; *U.S. Air Force Glossary of Standardized Terms,* AF Manual 11-1, vol. 1, 1972; F. Casey, ed., *Compilation of Terms in Information Sciences Technology,* Federal Council for Science and Technology, 1970; *Communications-Electronics Terminology,* AF Manual 11-1, vol. 3, 1970; P. W. Thrush, comp. and ed., *A Dictionary of Mining, Mineral, and Related Terms,* Bureau of Mines, 1968; *A DOD Glossary of Mapping, Charting and Geodetic Terms,* Department of Defense, 1967; J. M. Gilliland, *Solar-Terrestrial Physics: A Glossary of Terms and Abbreviations,* Royal Aircraft Establishment Technical Report 67158, 1967; W. H. Allen, ed., *Dictionary of Technical Terms for Aerospace Use,* National Aeronautics and Space Administration, 1965; *Glossary of Stinfo Terminology,* Office of Aerospace Research, U.S. Air Force, 1963; *Naval Dictionary of Electronic, Technical, and Imperative Terms,* Bureau of Naval Personnel, 1962; R. E. Huschke, *Glossary of Meteorology,* American Meteorological Society, 1959; *ADP Glossary,* Department of the Navy, NAVSO P-3097; *Glossary of Air Traffic Control Terms,* Federal Aviation Agency; *A Glossary of Range Terminology,* White Sands Missile Range, New Mexico, National Bureau of Standards, AD 467-424; *Nuclear Terms: A Glossary,* 2d ed., Atomic Energy Commission.

ISBN 0-07-042313-X

Library of Congress Cataloging-in-Publication Data

McGraw-Hill dictionary of scientific and technical terms--6th ed.
  p.   cm.
  ISBN 0-07-042313-X (alk. paper)
  1. Science--Dictionaries.   2. Technology--Dictionaries.   I. Title: Dictionary of scientific and technical terms.

Q123.M15   2002
503—dc21   2002026436

Contents

Contents

Preface

The first edition of the *McGraw-Hill Dictionary of Scientific and Technical Terms,* published in 1974, was a response to the inadequate representation of scientific terminology in general English language dictionaries. It was a time when the fruits of research and development in the era following World War II were transforming everyday life in areas ranging from medicine, to transportation, to telecommunications and computing. Humans had already landed on the Moon; the creation of the Internet had begun; the first microprocessors were in operation; and the revolution in molecular biology was underway. English was becoming more and more the common language of scientific research. The first edition of the *Dictionary* was intended primarily for those involved in these developments: the communities of scientific and engineering specialists and their students. However, over the span of subsequent editions, an understanding of the language of science and technology became important if not essential in many areas of commerce and culture, and even in everyday life as we try to make informed decisions about our environment, medical issues, and even the foods we eat. Thus, the audience for this, the sixth edition of the *Dictionary,* has expanded to the nonspecialist needing a comprehensive yet accessible resource for scientific terminology.

The language of science and technology is expanding not only in its role in our culture; it is growing in its breadth and depth as scientific disciplines mature and whole new technologies, such as nanotechnology and genomics, arise. The effects of the ready availability of powerful, networked computers and broadband communications have been felt in all areas of science; the pace of scientific discovery and dissemination of information has increased dramatically. The sequencing of the human genome well ahead of the original schedule is a prime example of the accelerating pace of discovery enabled by powerful technologies. Often, the economic and other benefits of scientific and technical advances must be weighed against potential or real deleterious consequences, for example in relation to biotechnology, environmental protection, and human health. The need to understand these issues has grown outward from the scientific specialists to educators, journalists, political leaders, and informed citizens.

To keep pace with the expanding language of science as well as the growing circle of persons concerned with it, some 5000 new terms have been added to this edition of the *Dictionary,* and many other terms have been revised as their usage evolves. The classification of terminology into fields has also seen changes reflecting more modern usage. For example, forensic science and neuroscience are now separate categories, whereas cytology and molecular biology have been merged as cell and molecular biology. Other fields have been kept, but their definitions updated as these sciences evolve. The reader will also appreciate that the multidisciplinary approach of modern science often makes neat categorization of terms difficult; for example, a term might fit in biochemistry, genetics, or microbiology as well as in cell and molecular biology. The *Dictionary* now has some 110,000 terms with 125,000 definitions. Many definitions are complemented by illustrations—about 3000 in all. In addition to new illustrations, many older ones were replaced with modern examples. Synonyms, acronyms, and abbreviations are given within definitions as well as in the alphabetical sequence as separate entries, where cross references to principal terms are provided. Every term is accompanied by its pronunciation, and a detailed guide to pronunciation follows this Preface. Where units of measurement are essential to the definition of a term, U.S. Customary units are used with International System (SI) or metric equivalents.

The editorial staff of the *Dictionary* endeavored to provide definitions that the nonspecialist reader could understand without losing the scientific meaning and context of the term. Each definition is identified by its field of use. There are 104 fields, ranging from general categorizations such as astronomy [ASTRON] and physics [PHYS] to specialized ones such as engineering acoustics [ENG ACOUS] and naval architecture [NAV ARCH]. A definition is identified as belonging to the vocabulary of a specific field; where it is used in more than one field, a more general field is designated. For example, if a definition is used in the field of analytical chemistry and inorganic chemistry, it is assigned to the field of chemistry. An alphabetical list of field abbreviations and an explanation of the scope of each field begins on page x.

The Appendix contains a full explanation of the International System of units, including units of temperature, with conversion tables for the U.S. Customary and the metric systems. It also includes a table of the chemical elements, along with an explanation of chemical nomenclature; a periodic table; lists of mathematical notation; mathematical signs and symbols, and other symbols used in scientific writing; tables of fundamental constants and elementary particles; a short chart of schematic electronic symbols; a geological time scale; a biographical listing of more than 1600 noted scientists, both historical and modern, many of whose names appear in dictionary terms; and an outline of the classification of living organisms.

An explanation of how to use the *Dictionary,* describing alphabetization, format, cross referencing, and more, can be found on page ix. The Notes on Pronunciation, in which the transcription system is explained, begins on page xv. A Pronunciation Key appears on page xvii.

This sixth edition of the *McGraw-Hill Dictionary of Scientific and Technical Terms* continues to serve the needs of both the scientific community and the general reader for high-quality information, and to contribute to scientific education and technological literacy.

Mark D. Licker
Publisher

Staff

Mark D. Licker, Publisher—Science

Elizabeth Geller, Managing Editor
Jonathan Weil, Senior Staff Editor
David Blumel, Staff Editor
Alyssa Rappaport, Staff Editor
Charles Wagner, Digital Content Manager
Renee Taylor, Editorial Assistant

Roger Kasunic, Vice President—Editing, Design, and Production

Joe Faulk, Editing Manager
Frank Kotowski, Jr., Senior Editing Supervisor

Ron Lane, Art Director
Vincent Piazza, Assistant Art Director

Thomas G. Kowalczyk, Production Manager
Pamela A. Pelton, Senior Production Supervisor

Henry F. Beechhold, Pronunciation Editor
Professor Emeritus of English
Former Chairman, Linguistics Program
The College of New Jersey
Trenton, New Jersey

This dictionary was set in Times Roman and Helvetica Bold by the
Clarinda Company, Clarinda, Iowa. It was printed and bound by
RR Donnelley.

Consulting Editors

The participation of the following individuals in the preparation of the
first five editions of the Dictionary is acknowledged with gratitude:

Prof. Eugene A. Avallone
Dr. Patrick Barry
Prof. George S. Bonn
Waldo G. Bowman
Dr. John M. Carroll
Dr. John F. Clark
Dr. Richard B. Couch
Dr. Charles B. Curtin
Robert L. Davidson
Prof. Roland H. Good, Jr.
Dr. J. Allen Hynek
Philip B. Jordain
Dr. Gary Judd
Alvin W. Knoerr

John Markus
Dr. Nathaniel Martin
Dr. Edward C. Monahan
Dr. N. Karle Mottet
Dr. Charles Oviatt
Dr. Guido Pontecorvo
Dr. John Quick
Prof. Alan Saleski
Brig. Gen. Peter C. Sandretto
Prof. Frederic Schwab
Dr. W. R. Sistrom
Dr. Leonard Spero
Dr. C. N. Touart
Dr. Joachim Weindling

Additional Consulting Editors for the Sixth Edition are:

Dr. Milton B. Adesnik
Robert D. Briskman

Dr. Orlando J. Miller
Dr. Kenneth P. H. Pritzker

How to Use the Dictionary

I. ALPHABETIZATION

The terms in the *McGraw-Hill Dictionary of Scientific and Technical Terms* are alphabetized on a letter-by-letter basis: word spacing, hyphen, comma, solidus, and apostrophe in a term are ignored in the sequencing. For example, an ordering of terms would be:

air-earth current
air ejector
airfield
air filter
AKF diagram

Also ignored in the sequencing of terms (usually, chemistry terms) are italic elements, numbers, small capitals, and Greek letters. For example, the following terms appear within alphabet letter "A":

N-acetylethanolamine
α-aminohydrocinnamic acid
***ortho*-aminophenol**
2-aminopropane

II. FORMAT

The basic format for a defining entry provides the term in boldface, the field in small capitals, and the single definition in lightface:

term [FIELD] Definition.

A field may be followed by multiple definitions, each introduced by a boldface number.

term [FIELD] **1.** Definition. **2.** Definition. **3.** Definition.

A term may have definitions in two or more fields:

term [BOT] Definition. [GEOL] Definition.

A simple cross-reference entry appears as:

term *See* another term.

A cross-reference may also appear in combination with definitions:

term [BOT] Definition. [GEOL] *See* another term.

III. CROSS-REFERENCING

A cross-reference entry directs the user to the defining entry. For example, the user looking up "average life" finds:

average life *See* mean life.

The user then turns to the "M" terms for the definition.

Cross-references are also made from variant spellings, acronyms, abbreviations, and symbols.

aesthacyte *See* esthacyte.
ASROC *See* antisubmarine rocket.
at. wt *See* atomic weight.
Au *See* gold.

The user should observe that an element ignored in alphabetizing may appear in a cross-reference entry. For example, the following directs the user to another term in alphabet letter "A," not in "N":

ASC *See* N-acetylsulfanilyl chloride.

IV. ALSO KNOWN AS . . . , etc.

A definition may conclude with a mention of a synonym of the term, a variant spelling, an abbreviation for the term, or other such information, introduced by "Also known as . . . ," "Also spelled . . . ," "Abbreviated . . . ," "Symbolized . . . ," "Derived from . . . ." When a term has more than one definition, the positioning of any of these phrases conveys the extent of applicability. For example:

term [BOT] **1.** Definition. Also known as synonym. **2.** Definition. Symbolized T.

In the above arrangement, "Also known as . . ." applies only to the first definition; "Symbolized . . ." applies only to the second definition.

term [BOT] **1.** Definition. **2.** Definition. [GEOL] Definition. Also known as synonym.

In the above arrangement, "Also known as . . ." applies only to the second field.

term [BOT] Also known as synonym. **1.** Definition. **2.** Definition. [GEOL] Definition.

In the above arrangement, "Also known as . . ." applies to both definitions in the first field.

term Also known as synonym. [BOT] **1.** Definition. **2.** Definition. [GEOL] Definition.

In the above arrangement, "Also known as . . ." applies to all definitions in both fields.

V. CHEMICAL FORMULAS

Chemistry definitions may include either an empirical formula (say, for acetaldehyde, C_2H_4O) or a line formula (for acrylic acid, $CH_2CHCOOH$), whichever is appropriate.

VI. PRONUNCIATION

All terms include pronunciations based on the Pronunciation Key on page xvii. Generally a single, selected pronunciation is given, without variants. Abbreviations, and acronyms spoken letter by letter, are not pronounced.

Field Abbreviations

ACOUS	acoustics	INV ZOO	invertebrate zoology
AERO ENG	aerospace engineering	LAP	lapidary
AGR	agriculture	LING	linguistics
ANALY CHEM	analytical chemistry	MAP	mapping
ANAT	anatomy	MATER	materials
ANTHRO	anthropology	MATH	mathematics
ARCH	architecture	MECH	mechanics
ARCHEO	archeology	MECH ENG	mechanical engineering
ASTRON	astronomy	MED	medicine
ASTROPHYS	astrophysics	MET	metallurgy
ATOM PHYS	atomic physics	METEOROL	meteorology
BIOCHEM	biochemistry	MICROBIO	microbiology
BIOL	biology	MIN ENG	mining engineering
BIOPHYS	biophysics	MINERAL	mineralogy
BOT	botany	MYCOL	mycology
BUILD	building construction	NAV	navigation
CELL MOL	cell and molecular biology	NAV ARCH	naval architecture
CHEM	chemisty	NEUROSCI	neuroscience
CHEM ENG	chemical engineering	NUCLEO	nucleonics
CIV ENG	civil engineering	NUC PHYS	nuclear physics
CLIMATOL	climatology	OCEANOGR	oceanography
COMMUN	communications	OPTICS	optics
COMPUT SCI	computer science	ORD	ordnance
CONT SYS	control systems	ORG CHEM	organic chemistry
CRYO	cryogenics	PALEOBOT	paleobotany
CRYSTAL	crystallography	PALEON	paleontology
DES ENG	design engineering	PARTIC PHYS	particle physics
ECOL	ecology	PATH	pathology
ELEC	electricity	PETR	petrology
ELECTR	electronics	PETRO ENG	petroleum engineering
ELECTROMAG	electromagnetism	PHARM	pharmacology
EMBRYO	embryology	PHYS	physics
ENG	engineering	PHYS CHEM	physical chemistry
ENG ACOUS	engineering acoustics	PHYSIO	physiology
EVOL	evolution	PL PATH	plant pathology
FL MECH	fluid mechanics	PL PHYS	plasma physics
FOOD ENG	food engineering	PSYCH	psychology
FOR	forestry	QUANT MECH	quantum mechanics
FOREN SCI	forensic science	RELAT	relativity
GEN	genetics	SCI TECH	science and technology
GEOCHEM	geochemistry	SOLID STATE	solid-state physics
GEOD	geodesy	SPECT	spectroscopy
GEOGR	geography	STAT	statistics
GEOL	geology	STAT MECH	statistical mechanics
GEOPHYS	geophysics	SYS ENG	systems engineering
GRAPHICS	graphic arts	SYST	systematics
HISTOL	histology	TEXT	textiles
HOROL	horology	THERMO	thermodynamics
HYD	hydrology	VERT ZOO	vertebrate zoology
IMMUNOL	immunology	VET MED	veterinary medicine
IND ENG	industrial engineering	VIROL	virology
INORG CHEM	inorganic chemistry	ZOO	zoology

Scope of Fields

acoustics—The science of the production, transmission, and effects of sound.

aerospace engineering—The branch of engineering pertaining to the design and construction of aircraft and space vehicles and of power units, and dealing with the special problems of flight in both the earth's atmosphere and space, such as in the flight of air vehicles and the launching, guidance, and control of missiles, earth satellites, and space vehicles and probes.

agriculture—The production of plants and animals useful to humans, involving soil cultivation and the breeding and management of crops and livestock.

analytical chemistry—The science of the characterization and measurement of chemicals; qualitative analysis is concerned with the description of chemical composition in terms of elements, compounds, or structural units, whereas quantitative analysis is concerned with the measurement of amount.

anatomy—The branch of morphology concerned with the gross and microscopic structure of animals, especially humans.

anthropology—The study of the interrelations of biological, cultural, geographical, and historical aspects of the human race.

archeology—The scientific study of the material remains of the cultures of historical and prehistorical peoples.

architecture—The art or practice of designing structures, especially habitable structures in accordance with principles determined by esthetic and practical or material considerations.

astronomy—The science concerned with celestial bodies and with the observation and interpretation of radiation received from the component parts of the universe.

astrophysics—The branch of astronomy that treats of the physical properties of celestial bodies, such as luminosity, size, mass, density, temperature, and chemical composition, and their origin and evolution.

atomic physics—The branch of physics concerned with the structures of the atom, the characteristics of the electrons and other elementary particles of which the atom is composed, the arrangement of the atom's energy states, and the processes involved in the radiation of light and x-rays.

biochemistry—The study of the chemical substances that occur in living organisms, the processes by which these substances enter into or are formed in the organisms and react with each other and the environment, and the methods by which the substances and processes are identified, characterized, and measured.

biology—The science of living organisms.

biophysics—The science that uses the experimental and theoretical approaches of physics to study the mechanisms of biological processes.

botany—That branch of biology dealing with the structure, function, diversity, evolution, reproduction, and utilization of plants and their interactions within the environment.

building construction—The technology of assembling materials into a structure, especially one designated for occupany.

cell and molecular biology—The study of the structures, functions, and molecular aspects (proteins, enzymes, nucleic acids) of the living cell.

chemical engineering—A branch of engineering which involves the design and operation of chemical plants.

chemistry—The scientific study of the properties, composition, and structure of matter, the changes in structure and composition of matter, and accompanying energy changes.

civil engineering—The planning, design, construction, and maintenance of fixed structures and ground facilities for industry, for transportation, for use and control of water, for occupancy, and for harbor facilities.

climatology—That branch of meteorology concerned with the mean physical state of the atmosphere together with its statistical variations in both space and time as reflected in the weather behavior over a period of many years.

communications—The science and technology by which information is collected from an originating source; converted into a form suitable for transmission; transmitted over a pathway such as a satellite channel, underwater acoustic channel, telephone cable, or fiber-optic link; and reconverted into a form suitable for interpretation by a receiver.

computer science—The study of computing, including computer hardware, software, programming, networking, database systems, information technology, interactive systems, and security.

control systems—The study of those systems in which one or more outputs are forced to change in a desired manner as time progresses.

cryogenics—The science of producing and maintaining very low temperatures, of phenomena at those temperatures, and of technical operations performed at very low temperatures.

crystallography—The branch of science that deals with the geometric description of crystals, their internal arrangement, and their properties.

design engineering—The branch of engineering concerned with the design of a product or facility according to generally accepted uniform standards and procedures, such as the specification of a linear dimension, or a manufacturing practice, such as the consistent use of a particular size of screw to fasten covers.

ecology—The study of the interrelationships between organisms and their environment.

electricity—The science of physical phenomena involving electric charges and their effects when at rest and when in motion.

electromagnetism—The branch of physics dealing with the observations and laws relating electricity to magnetism, and with magnetism produced by an electric current.

electronics—The technological area involving the manipulation of voltages and electric currents through the use of various devices for the purpose of performing some useful action with the currents and voltages; this field is generally divided into analog electronics, in which the signals to be manipulated take the form of continuous currents or voltages, and digital electronics, in which signals are represented by a finite set of states.

embryology—The study of the development of the organism from the zygote, or fertilized egg.

engineering—The science by which the properties of matter and the sources of power in nature are made useful to humans in structures, machines, and products.

engineering acoustics—The field of acoustics that deals with the production, detection, and control of sound by electrical devices, including the study, design, and construction of such things as microphones, loudspeakers, sound recorders and reproducers, and public address sytems.

evolution—The processes of biological and organic change in organisms by which descendants come to differ from their ancestors, and a history of the sequence of such change.

fluid mechanics—The science concerned with fluids, either at rest or in motion, and dealing with pressures, velocities, and accelerations in the fluid, including fluid deformation and compression or expansion.

food engineering—The technical discipline involved in food manufacturing and processing.

forensic science—The recognition, collection, identification, individualization, and interpretation of physical evidence, and the application of science and medicine for criminal and civil law or regulatory purposes.

forestry—The science of developing, cultivating, and managing forest lands for wood, forage, water, wildlife, and recreation; the management of growing timber.

genetics—The science concerned with biological inheritance, that is, with the causes of the resemblances and differences among related individuals.

geochemistry—The field that encompasses the investigation of the chemical composition of the earth, other planets, and the solar system and universe as a whole, as well as the chemical processes that occur within them.

geodesy—The subdividision of geophysics which includes determinations of the size and shape of the earth, the earth's gravitational field, and the location of point fixed to the earth's crust in an earth-referred coordinate system.

geography—The science that deals with the description of land, sea, and air and the distribution of plant and animal life, including humans.

geology—The study or science of earth, its history, and its life as recorded in the rocks; includes the study of the geologic features of an area, such as the geometry of rock formations, weathering and erosion, and sedimentation.

geophysics—The branch of geology in which the principles and practices of physics are used to study the earth and its environment, that is, earth, air, and (by extension) space.

graphic arts—The fine and applied arts of representation, decoration, and writing or printing on flat surfaces together with the tech-niques and crafts associated with each: includes painting, drawing, engraving, etching, lithography, photography, and printing arts.

histology—The study of the structure and chemical composition of animal tissues as related to their function.

horology—The science of time measurement and the principles and technology of constructing time-measuring instruments.

hydrology—The science dealing with all aspects of the waters on earth, including their occurrence, circulation, and distribution; their chemical and physical properties; and their reaction with the environment, including their relation to living things.

immunology—The division of biological science concerned with the native or acquired resistance of higher animal forms and humans to infection with microorganisms.

industrial engineering—A branch of engineering dealing with the design, development, and implementation of integrated systems of humans, machines, and information resources to provide products and services.

inorganic chemistry—The branch of chemistry that deals with reactions and properties of all chemical elements and their compounds, excluding hydrocarbons but usually including carbides and other simple carbon compounds (such as CO_2, CO, and HCN).

invertebrate zoology—The branch of zoology concerned with the taxonomy, behavior, and morphology of invertebrate animals.

lapidary—The study relating to precious stones or the art of cutting them.

linguistics—The study of the structure, meaning, and development of language and the production of speech, encompassing the subfields of grammatical theory, semantics, anthropological and sociological linguistics, psycholinguistics, neurolinguistics, and phonetics (the sounds associated with language), among others.

mapping—The art and practice of making a drawing or other representation, usually on a flat surface, of the whole or part of an area (as the surface of the earth or some other planet), indicating relative position and size according to a specified scale or projection of selected features, such as countries, cities, rock formations, or bodies of water.

materials—A multidisciplinary field concerned with the properties and uses of materials in terms of composition, structure, and processing.

mathematics—The deductive study of shape, quantity, and dependence; the two main areas are applied mathematics and pure mathematics, the former arising from the study of physical phenomena, the latter involving the intrinsic study of mathematical structures.

mechanical engineering—The branch of engineering concerned with energy conversion, mechanics, and mechanisms and devices for diverse applications, ranging from automotive parts through nanomachines.

mechanics—The branch of physics which seeks to formulate general rules for predicting the behavior of a physical system under the influence of any type of interaction with its environment.

medicine—The study of cause and treatment of human disease, including the healing arts dealing with diseases which are treated by a physician or a surgeon.

metallurgy—The branch of engineering concerned with the production of metals and alloys, their adaptation to use, and their performance in service; and the study of chemical reactions involved in the processes by which metals are produced, and of the laws governing the physical, chemical, and mechanical behavior of metallic materials.

meteorology—The science concerned primarily with the observation of the atmosphere and its phenomena, including temperature, density, winds, clouds, and precipitation.

microbiology—The science and study of microorganisms and of antibiotic substances.

mineralogy—The study of naturally occurring inorganic substances, called minerals, whether of terrestrial or extraterrestrial origin.

mining engineering—The branch of engineering concerned with the location and evaluation of coal and mineral deposits, the survey of mining areas, the layout and equipment of mines, the supervision of mining operations, and the cleaning, sizing, and dressing of the product.

mycology—The branch of biological science concerned with the study of fungi.

naval architecture—The study of the physical characteristics and the design and construction of buoyant structures which operate in water, and of the construction and operation of the power plant and other mechanical equipment of these structures.

navigation—The science or art of directing the movement of a craft, such as a ship, small marine craft, underwater vehicle, land vehicle, aircraft, missile, or spacecraft, from one place to another with the assistance of onboard equipment, objects, or devices, or of systems external to the craft.

neuroscience—The study of the brain and nervous system, including the anatomy and histology of the nervous system, development, sensation and perception, learning, memory, motor control, behavior, aging, and neurological and psychiatric disorders. Studies range from the molecular basis of nervous system development and function to attempts to understand the basis of consciousness and behavior.

nuclear physics—The study of the characteristics, behavior, and internal structure of the atomic nucleus.

nucleonics—The technology based on phenomena of the atomic nucleus such as radioactivity, fission, and fusion; includes nuclear reactors, various applications of radioisotopes and radiation, particle accelerators, and radiation detection devices.

oceanography—The science of the sea, including physical oceanography (the study of the physical properties of seawater and its motion in waves, tides, and currents), marine chemistry, marine geology, and marine biology.

optics—The study of phenomena associated with the generation, transmission, and detection of electromagnetic radiation in the spectral range extending from the long-wave edge of the x-ray region to the short-wave edge of the radio region; and the science of light.

ordnance—That military area concerned with supplies, including weapons, ammunition, combat vehicles, and the necessary repair equipment; and with heavy firearms discharged from mounts, including cannons and artillery.

organic chemistry—The study of the structure, preparation, properties, and reactions of carbon compounds.

paleobotany—The study of fossil plants and vegetation of the geologic past.

paleontology—The study of life in the geologic past as recorded by fossil remains.

particle physics—The branch of physics concerned with understanding the properties, behavior, and structure of elementary particles, especially through study of collisions or decays involving energies of hundreds of megaelectronvolts or more.

pathology—The branch of biological science which deals with the nature of disease, through study of its causes, its processes, and its effects, together with the associated alterations of structure and function; and the laboratory findings of disease, as distinguished from clinical signs and symptoms.

petroleum engineering—The branch of engineering concerned with the search for and extraction of oil, gas, and liquefiable hydrocarbons; usually subdivided into petrophysical, geological, reservoir drilling, production, and construction engineering.

petrology—The branch of geology dealing with the origin, occurrence, structure, and history of rocks, especially igneous and metamorphic rocks.

pharmacology—The science of detecting and measuring the therapeutic and toxic effects of drugs or other chemicals on biological systems, as well as the development and testing of new drugs and alternative uses of existing drugs.

physical chemistry—The branch of chemistry that deals with the interpretation of chemical phenomena and properties in terms of the underlying physical processes, and with the development of techniques for their investigation.

physics—The science concerned with those aspects of nature which can be understood in terms of elementary principles and laws.

physiology—The branch of biological science concerned with the basic activities that occur in cells and tissues of living organisms and involving physical and chemical studies of these organisms.

plant pathology—The branch of botany concerned with diseases of plants.

plasma physics—The study of highly ionized gases.

psychology—The science of the function of the mind and the behavior of an organism, both animal and human, in relation to its environment.

quantum mechanics—The modern theory of matter, of electromagnetic radiation, and of the interaction between matter and radiation; it differs from classical physics, which it generalizes and supersedes, mainly in the realm of atomic and subatomic phenomena.

relativity—The study of the physical theory which recognizes the universal character of the propagation speed of light and the consequent dependence of space, time, and other mechanical measurements on the motion of the observer performing the measurements; the two main divisions are special theory and general theory.

science and technology—The study of the natural sciences and the application of this knowledge for practical purposes.

solid-state physics—The branch of physics centering on the physical properties of solid materials; it is usually concerned with the properties of crystalline materials only, but it is sometimes extended to include the properties of glasses or polymers.

spectroscopy—The branch of physics concerned with the production, measurement, and interpretation of electromagnetic spectra arising from either emission or absorption of radiant energy by various substances.

statistical mechanics—That branch of physics which endeavors to explain and predict the macroscopic properties and behavior of a system on the basis of the known characteristics and interactions of the microscopic constituents of the system, usually when the number of such constituents is very large.

statistics—The science dealing with the collection, analysis, interpretation, and presentation of masses of numerical data.

systematics—The science of animal and plant classification.

systems engineering—The branch of engineering dealing with the design of a complex interconnection of many elements (a system) to maximize an agreed-upon measure of system performance.

textiles—The area of industry involving the production of fibers, filaments, or yarn, and the cloth made from these materials.

thermodynamics—The branch of physics which seeks to derive, from a few basic postulates, relations between properties of substances, especially those which are affected by changes in temperature, and a description of the conversion of energy from one form to another.

vertebrate zoology—The branch of zoology concerned with the taxonomy, behavior, and morphology of vertebrate animals.

veterinary medicine—The branch of medical practice which treats the diseases and injuries of animals.

virology—The science that deals with the study of viruses.

zoology—The science that deals with the taxonomy, behavior, and morphology of animal life.

Notes on Pronunciation

All graphic representations of speech, that is, transcription systems or phonetic alphabets, encounter common problems in communicating with the reader. Phonation shares nothing directly with visualization. Of course, since speaking and writing are both manifestations of language, language is a unifying reality. Nevertheless, symbols on a page cannot wholly convey the sounds of speech. Further, everyone filters written text through his or her own idiosyncratic dialect (idiolect, in linguistic terms). The idiolect of the writer will never exactly match that of the reader. When the Pronunciation Key in this volume says, for example, that the symbol "a" should be pronounced to rhyme with the vowel in *bat*, that vowel will be the version used in your dialect, which may be quite different from mine.

In a typical dictionary, each word is transcribed as a citation form. However, we do not speak as though reciting lists of words; we speak in contexts, that is, in macrosegments of speech (phrases, sentences). Thus, the dictionary pronunciation of a word may not capture the pronunciation of that word as it is uttered in a normal context. The most obvious difference arises from the intonational (sing-song) patterns that are so basic to connected speech. The stress (loudness) indications for individual words may not obtain when those words occur in phrases. For example, in the *Dictionary of Scientific and Technical Terms*, the term *formula* is shown with a primary stress (') on the first syllable. But when *formula* occurs as the second member of many phrases, the primary stress is reduced to a secondary stress (ˌ), as in *Helmert's formula*, where the first syllable of the scientist's name takes the primary stress. Actually, one of the unique features of this dictionary is a phrasal approach to pronunciation to give the reader a sense of the sound of a phrase rather than the sound of the individual words. Wherever appropriate, the stress indications are distributed as though the term were spoken in response to the question "What _____ is that?" or "What kind of _____ is that?" Thus: "What formula is that?" "HELmert's formula." "What kind of pipe is that?" "A LIGHT pipe."

The basic rule in sounding out phrases is to speak them as connected speech, with an alternation of stress values from lighter to heavier (or vice versa). There are some constructions in which both elements carry roughly the same stress, but this is less common than stress alternation. In general, a phrase consisting of an adjective modifier followed by a noun (for example, *hot air*) calls for a secondary stress on the adjective and a primary on the noun ("hot AIR"). But in a phrase in which a noun is being modified by another noun (such as peat moss), the stress values reverse, with the modifying noun receiving the greater stress ("PEAT moss"). These examples illustrate the difference between a noun with attribution and a phrasal noun. In the former, the main word (the noun) is being modified. In the second, the two words form a compound word. In phrases containing multiple modifiers, of which there are many in this dictionary, the patterns vary.

Strictly speaking, the English intonational system uses four levels of stress (amplitude) in concert with four levels of pitch (frequency). The pitch refers to tonal differences, noted particularly at the ends of sentences. The full set of stresses can be detected in the example "blue BIRD" versus "BLUEbird." In the former, we are talking about a bird of a certain color; in the latter, about a species. In the first example the adjective carries secondary stress,

the noun primary. In the second example the modifier is given primary stress, the modificand, tertiary stress. An article in front of either construction will carry fourth-level or weak stress. In this dictionary we have ignored pitch entirely, for it will sound approximately right when you speak the phrases out loud. The subtleties of stress are not fully expressed, but we feel that there is enough marking for practical utility. The indeterminate primary/secondary marker (ˌ) is used to help the reader achieve a reasonably natural intonation. It is especially valuable in the long chemical terms where the tendency is to drop into a monotone, stressing every part of the compound equally.

In addition to stress and pitch, English intonation includes junctural components. Internal junctures commonly, but not invariably, occur between words, though occasionally within words. Terminal junctures mark the close of a structural macrosegment (phrase, clause/sentence). In both cases, juncture serves as a boundary indicator and is signaled by various phonetic markers. Contrast, for example, the character of your voice at the end of an introductory phrase with its character at the end of a declarative sentence (and many questions as well): "Driving to the laboratory this morning. . . ." versus "I drove to the laboratory this morning." At the end of the phrase, the voice hangs suspended; at the end of the sentence, it drops away. Of course, we cannot mark the occurrence of these terminals in the Pronunciation Key, but when you speak they do occur and do affect pronunciation.

Another problem in representing speech graphically is phonological conditioning, that is, the effect of sounds on other sounds in proximity. For example, many dictionaries offer "t" as the pronunciation of that letter in words like *appendicitis* and *capacitor*. But it sounds stilted to pronounce such words with a frank "t" instead of a quickly flapped "d." In fact, there are about ten distinctive ways to pronounce the letter "t." (The term "letter" always refers to the alphabetic symbol, not the speech sound, which is properly a phone.) The phonological rule at work here is that when a voiceless sound (like a "t") occurs between vowels, it becomes voiced. A voiced "t" is effectively a "d." This rule operates at the phrase level as well. We say "cut" (with a "t"), but "CUD out" or even "KUH dout" for the phrase "cut out."

The niceties of pronunciation are many, far more than we can either address in these notes or represent in our transcriptions. And beyond absolute phonetic (including intonational) complexities lies the matter of dialect. In the United States there are numerous regional dialects, each with its own set of subdialects. For many utterances there may be little or no difference; for others the differences are appreciable. Level of education of the speaker plays an important role.

For each entry, the *Dictionary of Scientific and Technical Terms* provides a single pronunciation that represents a broadly acceptable rendering of the word or phrase. We have opted for the inclusion of preconsonantal and final "r." Thus, "mark" not "mahk"; "fiber" not "fibuh." Our vowel choices include a rendering of "o" in *forest* as "ah" (rhymes with the vowel of *what*), not "aw" (rhymes with the vowel of *door*), though the latter is widely distributed in American English. Likewise, we have chosen "ah," not "o" (rhymes with *go*), in most contexts in which both are common. "Sahlstice," then, not "solstice." But if your inclination is to use the "o" in preference to the "ah," by all means do so.

Representations of pronounciation of non-English words and phrases are close enough to make it clear to a listener what word or phrase is intended. Since English lacks certain sounds altogether and pronounces certain other non-English sounds differently from their counterparts, our transcription seeks to achieve an acceptable approximation.

Commonly, but not inevitably, vowels in unstressed syllables are reduced to the schwa sound, represented graphically by ə. Hence the "o" in *centrosphere* is spoken as a schwa, that is, with a sound like "uh" (rhymes with the vowel of *but*). In some contexts, unstressed vowels vanish altogether, as in *memory*, spoken "memry," though "mem-uh-ry" is certainly acceptable.

Even the phonetic division of syllables is arguable, for syllables do not necessarily divide according to spelling convention. Does one say "di-spersion" or "dis-persion"? We have generally followed *Webster's Third New International Dictionary*, but "instinct" has also been involved in our syllable division. This is an academic issue, for in normal speech these graphic variants cannot be detected.

Since the *Dictionary of Scientific and Technical Terms* is not a general-purpose dictionary, we did not feel that a listing of variant pronunciations was generally necessary. However, we do offer alternatives in selected instances, as for the terms *data* and *angina*.

Indeed, given dialectal variability, especially with respect to vowels, one can argue that for any entry there are as many correct pronunciations as there are dialects of educated speech in the English-speaking world.

The transcription system presented in the accompanying Pronunciation Key will be familiar to users of dictionaries published by G. & C. Merriam Company, such as *Webster's Third New International* and *Webster's New Collegiate*, for ours is essentially the same, though somewhat simplified. That there are numerous orthoepic discrepancies among dictionaries reminds us that orthoepy is an art, not a science, and that the purpose of a dictionary affects the contents. The *Dictionary of Scientific and Technical Terms* goes its own way, agreeing or disagreeing with other sources as deemed appropriate. A spoken language changes constantly. Vocabulary, grammar, meaning, and pronunciation are not immutable except in dead languages; and English, spoken in one form or another by perhaps half the population of the planet, is most assuredly alive.

Henry F. Beechhold, Ph.D.
Professor Emeritus of English
Former Chairman, Linguistics Program
The College of New Jersey
Trenton, New Jersey

Pronunciation Key

Vowels

a   as in **b**a**t**, **th**a**t**
ā   as in **b**ai**t**, **cr**a**te**
ä   as in **b**o**ther**, **f**a**ther**
e   as in **b**e**t**, **n**e**t**
ē   as in **b**ee**t**, **tr**ea**t**
i   as in **b**i**t**, **sk**i**t**
ī   as in **b**i**te**, **l**igh**t**
ō   as in **b**oa**t**, **n**o**te**
ȯ   as in **b**ough**t**, **t**au**t**
u̇   as in **b**oo**k**, **p**u**ll**
ü   as in **b**oo**t**, **p**oo**l**
ə   as in **b**u**t**, **s**o**fa**
au̇ as in **cr**o**w**d, **p**o**wer**
ȯi as in **b**oi**l**, **sp**oi**l**
yə as in **form**u**la**, **spectac**u**lar**
yü as in **f**ue**l**, **m**u**le**

Semivowels/Semiconsonants

w   as in **w**ind, **tw**in
y   as in **y**et, on**i**on

Stress (Accent)

ˈ   precedes syllable with primary stress

ˌ   precedes syllable with secondary stress

ˌ   precedes syllable with variable or indeterminate
    primary/secondary stress

Consonants

b   as in **b**i**b**, **d**ri**b**le
ch as in **ch**arge, stre**tch**
d   as in **d**og, ba**d**
f   as in **f**ix, sa**f**e
g   as in **g**ood, si**g**nal
h   as in **h**and, be**h**ind
j   as in **j**oint, di**g**it
k   as in **c**ast, bri**ck**
k̲   as in Ba**ch** (used rarely)
l   as in **l**oud, be**ll**
m   as in **m**ild, su**mm**er
n   as in **n**ew, de**n**t
n̲   indicates nasalization of preceding vowel
ŋ   as in ri**ng**, si**ng**le
p   as in **p**ier, sli**p**
r   as in **r**ed, sca**r**
s   as in **s**ign, po**s**t
sh as in **s**ugar, **sh**oe
t   as in **t**imid, ca**t**
th as in **th**in, brea**th**
th̲ as in **th**en, brea**the**
v   as in **v**eil, wea**v**e
z   as in **z**oo, crui**s**e
zh as in bei**g**e, trea**s**ure

Syllabication

·   indicates syllable boundary when following syllable
    is unstressed

A

a *See* ampere; atto-.

aΩ *See* abohm.

(aΩ)⁻¹ *See* abmho.

A *See* ampere; angstrom.

Å *See* angstrom.

A+ *See* A positive.

aA *See* abampere.

AA *See* antiaircraft.

AAA *See* antiaircraft artillery.

aa channel [GEOL] A narrow, sinuous channel in which a lava river moves down and away from a central vent to feed an aa lava flow. { 'ä·'ä 'chan·əl }

aAcm² *See* abampere centimeter squared.

aA/cm² *See* abampere per square centimeter.

aa lava *See* block lava. { 'ä·'ä 'lä·və }

Aalenian [GEOL] Lowermost Middle or uppermost Lower Jurassic geologic time. { ö'lēn·ē,ən }

AAM *See* air-to-air missile; antiaircraft missile.

aapamoor [ECOL] A moor with elevated areas or mounds supporting dwarf shrubs and sphagnum, interspersed with low areas containing sedges and sphagnum, thus forming a mosaic. { 'äp·ə,mür }

aardvark [VERT ZOO] A nocturnal, burrowing, insectivorous mammal of the genus *Orycteropus* in the order Tubulidentata. Also known as earth pig. { 'ärd,värk }

aardwolf [VERT ZOO] *Proteles cristatus.* A hyenalike African mammal of the family Hyaenidae. { 'ärd,wulf }

Aaron's rod [ARCH] A decorative rounded molding on which are entwined a single serpent and sometimes vines and leaves. { 'ar·ənz 'räd }

a axis [CRYSTAL] One of the crystallographic axes used as reference in crystal description, usually oriented horizontally, front to back. [GEOL] The direction of movement or transport in a tectonite. [MECH ENG] The angle that specifies the rotation of a machine tool about the *x* axis. { 'ā 'ak,sis }

ab- [ELECTROMAG] A prefix used to identify centimeter-gram-second electromagnetic units, as in abampere, abcoulomb, abfarad, abhenry, abmho, abohm, and abvolt. { ab }

ABA *See* abscisic acid.

abac *See* nomograph. { ə'bak }

abaca [BOT] *Musa textilis.* A plant of the banana family native to Borneo and the Philippines, valuable for its hard fiber. Also known as Manila hemp. { 'ä·bä,kä *or* 'ä·bə,kä }

abactinal [INV ZOO] In radially symmetrical animals, pertaining to the surface opposite the side where the mouth is located. { a'bak·tin·əl }

abacus [ARCH] A slab forming the topmost division of the capital of a column. [MATH] An instrument for performing arithmetical calculations manually by sliding markers on rods or in grooves. { 'ab·ə,kəs }

abaft [NAV ARCH] In a direction farther aft in a ship than a specified reference position, such as abaft the mast. { ə'baft }

abalienation [PSYCH] Mental deterioration or derangement. { ab,āl·yə'nā,shən }

abalone [INV ZOO] A gastropod mollusk composing the single genus *Haliotis* of the family Haliotidae. Also known as ear shell; ormer; paua. { ,ab·ə'lō·nē }

abalyn [ORG CHEM] A liquid rosin that is a methyl ester of abietic acid; prepared by treating rosin with methyl alcohol; used as a plasticizer. { 'ab·ə,lin }

abambulacral [INV ZOO] Pertaining to that part of the surface of an echinoderm that lacks tube feet. { ab,am·byə 'lak·rəl }

abampere [ELEC] The unit of electric current in the electromagnetic centimeter-gram-second system; 1 abampere equals 10 amperes in the absolute meter-kilogram-second-ampere system. Abbreviated aA. Also known as Bi; biot. { ab'am· pēr }

abampere centimeter squared [ELECTROMAG] The unit of magnetic moment in the electromagnetic centimeter-gram-second system. Abbreviated aAcm². { ab'am·pēr 'sen·tə,mē· dər 'skwerd }

abampere per square centimeter [ELEC] The unit of current density in the electromagnetic centimeter-gram-second system. Abbreviated aA/cm². { ab'am·pēr pər 'skwer 'sen· tə,med·ər }

abamurus [ARCH] A masonry block, in the form of a buttress, used to support a structure. { ,a·bə'myur·əs }

A band [HISTOL] The region between two adjacent I bands in a sarcomere; characterized by partial overlapping of actin and myosin filaments. { 'ā band }

abandon [ENG] To stop drilling and remove the drill rig from the site of a borehole before the intended depth or target is reached. [PETRO ENG] To terminate oil and gas production from a well when it becomes unprofitable. { ə'ban·dən }

abandoned channel *See* oxbow. { ə'ban·dənd 'chan·əl }

abandoned mine *See* abandoned workings. { ə'ban·dənd 'min }

abandoned workings [MIN ENG] Deserted excavations, either caved or sealed, in which further mining is not intended, and opening workings which are not ventilated and inspected regularly. Also known as abandoned mine. { ə'ban·dənd 'wər·kinz }

abandonment [MIN ENG] Failure to perform work, by conveyance, by absence, and by lapse of time, on a mining claim. [PETRO ENG] *See* abandonment contour. { ə'ban·dən· mənt }

abandonment contour [PETRO ENG] A graph of actual cumulative yield of an oil well compared with its estimated ultimate yield; useful in determining the most economic time to abandon an oil well. Also known as abandonment. { ə'ban· dən·mənt 'kän,tür }

abapertural [INV ZOO] Away from the shell aperture, referring to mollusks. { ab'ap·ər,chür·əl }

abapical [BIOL] On the opposite side to, or directed away from, the apex. { ab'ap·i·kəl }

abarognosis [MED] Lack of ability to estimate the weight of an object one is holding. { ā,bar·əg'nō·sis }

abasia [MED] Lack of muscular coordination in walking. { ā'bā·zhə }

abate [ENG] 1. To remove material, for example, in carving stone. 2. In metalwork, to excise or beat down the surface in order to create a pattern or figure in low relief. { ə'bāt }

abatement [ENG] 1. The waste produced in cutting a timber, stone, or metal piece to a desired size and shape. 2. A decrease in the amount of a substance or other quantity, such as atmospheric pollution. { ə'bāt·mənt }

abat-jour [BUILD] A device that is used to deflect daylight downward as it streams through a window. { ä·bä'zhür }

A battery [ELECTR] The battery that supplies power for filaments or heaters of electron tubes in battery-operated equipment. { 'ā ,bat·ə,rē }

AARDVARK

The aardvark (*Orycteropus afer*), a nocturnal, burrowing animal ranging from Ethiopia to southern Africa.

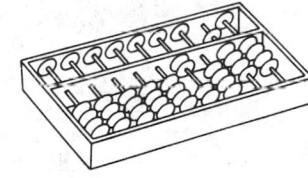

ABACUS

Drawing of an abacus.

ABALONE

Typical abalone ear-shaped shell perforated by pores.

abattoir [IND ENG] A building in which cattle or other animals are slaughtered. { ‚ab·ə'twär }

abat-vent [BUILD] A series of sloping boards or metal strips, or some similar contrivance, to break the force of wind without being an obstruction to the passage of air or sound, as in a louver or chimney cowl. { ‚ä‚bä'vän }

abaxial [BIOL] On the opposite side to, or facing away from, the axis of an organ or organism. { ab'ak·sē·əl }

abb [TEXT] Yarn made of abb wool. [VERT ZOO] A coarse wool from the fleece areas of lesser quality. { ab }

abbazzo [GRAPHICS] A rough sketch, draft, or model. { a'bat·sō }

Abbe condenser [OPTICS] A variable large-aperture lens system arranged substage to image a light source into the focal plane of a microscope objective. { 'ä·bə kən'dens·ər }

Abbe number [OPTICS] A number which expresses the deviating effect of an optical glass on light of different wavelengths. { 'ä·bə ‚nəm·bər }

Abbe prism [OPTICS] A system used for image erection which is composed of two double right-angle prisms and involves four reflections. { 'ä·bə 'priz·əm }

Abbe refractometer [OPTICS] An optical instrument for the measurement of the refractive index of liquids. { 'ä·bə ‚rē·frak'täm·əd·ər }

Abbe's sine condition [OPTICS] A relationship which must hold to prevent aberration of a mirror or lens from producing a coma. { 'ä·bəz 'sīn kən‚dish·ən }

Abbe's theory [OPTICS] The theory that for a lens to produce a true image, it must be large enough to transmit the entire diffraction pattern of the object. { 'ä·bəz 'thē·ə·rē }

abbreviated dialing [COMMUN] A feature that requires less than the usual number of dialing operations to connect two or more subscribers. { ə'brē·vē·ād·əd 'dī·liŋ }

ABC See automatic brightness control.

abcoulomb [ELEC] The unit of electric charge in the electromagnetic centimeter-gram-second system, equal to 10 coulombs. Abbreviated aC. { ab'kü·lōm }

abcoulomb centimeter [ELEC] In the electromagnetic centimeter-gram-second system of units, the unit of electric dipole moment. Abbreviated aCcm. { ab'kü·lōm 'sen·tə‚mēd·ər }

abcoulomb per cubic centimeter [ELEC] The electromagnetic centimeter-gram-second unit of volume density of charge. Abbreviated aC/cm³. { ab'kü·lōm pər 'kyü·bik 'sen·tə ‚mēd·ər }

abcoulomb per square centimeter [ELEC] The electromagnetic centimeter-gram-second unit of surface density of charge, electric polarization, and displacement. Abbreviated aC/cm². { ab'kü·lōm pər skwer 'sen·tə‚mēd·ər }

ABC system [GEOD] See airborne control system. [GEOPHYS] A procedure in seismic surveying to determine the effect of irregular weathering thickness. [ORD] An atomic, biological, or chemical weapons system. { 'ā‚bē'sē 'sis·təm }

Abderhalden reaction [PATH] A chemical blood test for the identification of certain enzymes associated with pregnancy and a few diseases. { 'äp·dər‚häl·dən rē'ak·shən }

abdomen [ANAT] The portion of the vertebrate body between the thorax and the pelvis.The cavity of this part of the body. [INV ZOO] The elongate region posterior to the thorax in arthropods. { ab'dōm·ən or 'ab·də‚mən }

abdominal depth [ANTHRO] Maximum horizontal contact dimension, measured front to back. { ab'däm·ə·nəl depth }

abdominal gestation [MED] Development of a fetus outside the uterus in the abdominal cavity. { ab'däm·ə·nəl je'stä·shən }

abdominal gills [INV ZOO] Paired, segmental, leaflike, filamentous expansions of the abdominal cuticle for respiration in the aquatic larvae of many insects. { ab'däm·ə·nəl 'gilz }

abdominal hernia See ventral hernia. { ab'däm·ə·nəl 'hər·nē·ə }

abdominal hysterectomy [MED] Surgical removal of all or part of the uterus through an incision in the abdomen. { ab'däm·ə·nəl ‚his·tə'rek·tə·mē }

abdominal limb [INV ZOO] In most crustaceans, any of the segmented abdominal appendages. { ab'däm·ə·nəl 'lim }

abdominal pregnancy See abdominocyesis. { ab'däm·ə·nəl 'preg·nən·sē }

abdominal regions [ANAT] Nine theoretical areas delineated on the abdomen by two horizontal and two parasagittal lines: above, the right hypochondriac, epigastric, and left hypochondriac; in the middle, the right lateral, umbilical, and left lateral; and below, the right inguinal, hypogastric, and left inguinal. { ab'däm·ə·nəl 'rē·jənz }

abdominal vascular accident [MED] Vascular occlusion and hemorrhage in an abdominal organ, usually the small intestine, or in the peritoneal cavity. { ab'däm·ə·nəl 'vas·kyə·lər 'ak·sə·dənt }

abdominocyesis [MED] Implantation and development of the fertilized ovum in the peritoneal cavity. Also known as abdominal pregnancy. { ab‚däm·ə·nō·sī'ē·səs }

abducens [NEUROSCI] The sixth cranial nerve in vertebrates; a paired, somatic motor nerve arising from the floor of the fourth ventricle of the brain and supplying the lateral rectus eye muscles. { ab'dyü·sənz }

abduction [PHYSIO] Movement of an extremity or other body part away from the axis of the body. { ab'dək·shən }

abductor [PHYSIO] Any muscle that draws a part of the body or an extremity away from the body axis. { ab'dək·tər }

abeam See on the beam. { a'bēm }

Abegg's rule [CHEM] An empirical rule, holding for a large number of elements, that the sum of the maximum positive and negative valencies of an element equals eight. { 'ä·begz 'rül }

Abelian domain See Abelian field. { ə'bēl·yən dō'mān }

Abelian extension [MATH] A Galois extension whose Galois group is Abelian. { ə'bēl·yən ik'sten·chən }

Abelian field [MATH] A set of elements $a, b, c, \ldots$ forming Abelian groups with addition and multiplication as group operations where $a(b + c) = ab + ac$. Also known as Abelian domain; domain. { ə'bēl·yən 'fēld }

Abelian group [MATH] A group whose binary operation is commutative; that is, $ab = ba$ for each a and b in the group. Also known as commutative group. { ə'bēl·yən 'grüp }

Abelian operation See commutative operation. { ə'bēl·yən ‚äp·ə'rā·shən }

Abelian quantum Hall state [CRYO] A quantum Hall state that contains two or more components of incompressible fluid, has a filling factor equal to p/q, where q is not divisible by p, and has a topological order that can be described as a pattern of dancing steps of the electrons and can be characterized by a symmetric matrix and a charge vector, both with integer entries. { ə'bē·lē·ən ‚kwän·təm ‚hȯl ‚stät }

Abelian ring See commutative ring. { ə'bēl·yən 'riŋ }

Abelian theorems [MATH] A class of theorems which assert that if a sequence or function behaves regularly, then some average of the sequence or function behaves regularly; examples include the Abel theorem (second definition) and the statement that if a sequence converges to s, then its Cesaro summation exists and is equal to s. { ə'bēl·yən 'thir·əmz }

abelite [MATER] A substance made of ammonium nitrate and a nitrated aromatic hydrocarbon and used as an explosive. { 'a·bə·līt }

Abell richness classes [ASTRON] A scale of six categories of richness into which clusters of galaxies are classified, based on the number of galaxies observed in the cluster that are not more than 2 magnitudes fainter than the third-brightest member. { 'ā·bəl 'rich·nəs ‚klas·əz }

Abel's inequality [MATH] An inequality which states that the absolute value of the sum of n terms, each in the form ab, where the b's are positive numbers, is not greater than the product of the largest b with the largest absolute value of a partial sum of the a's. { 'ä·bəlz ‚in·ē'kwäl·i·dē }

Abel's integral equation [MATH] The equation

$$f(x) = \int_a^x u(z)(x - z)^{-a}dz \quad (0 < a < 1, \; x \geq a)$$

where $f(x)$ is a known function and $u(z)$ is the function to be determined; when $a = \frac{1}{2}$, this equation has application to Abel's problem. { 'ä·bəlz 'in·tə·grəl i'kwā·zhən }

Abel's problem [MATH] The problem which asks what path a particle will follow if it moves under the influence of gravity alone and its altitude-time function is to follow a specific law. { 'ä·bəlz 'präb·ləm }

Abel's summation method [MATH] A method of attributing a sum to an infinite series whose nth term is a_n by taking the limit on the left at $x = 1$ of the sum of the series whose nth term is $a_n x^n$. { 'ä·bəlz sə'mā·shən ‚meth·əd }

Abel tester [PHYS CHEM] A laboratory instrument used in testing the flash point of kerosine and other volatile oils having flash points below 120°F (49°C); the oil is contained in a closed cup which is heated by a fixed flame below and a movable flame above. { 'ä·bəl 'tes·tər }

Abel theorem [MATH] **1.** A theorem stating that if a power series in z converges for $z = a$, it converges absolutely for $|z| < |a|$. **2.** A theorem stating that if a power series in z converges to $f(z)$ for $|z| < 1$ and to a for $z = 1$, then the limit of $f(z)$ as z approaches 1 equals a. **3.** A theorem stating that if the three series with nth term a_n, b_n, and $c_n = a_0 b_n + a_1 b_{n-1} + \cdots + a_n b_0$, respectively, converge, then the third series equals the product of the first two series. { 'ä·bəl 'thir·əm }

abend [COMPUT SCI] An unplanned program termination that occurs when a computer is directed to execute an instruction or to process information that it cannot recognize. Also known as blow up; bomb; crash. { 'ab·end }

abenteric [MED] Involving abdominal organs and structures outside the intestine. { ,ab·en'ter·ik }

aberrant [BIOL] An atypical group, individual, or structure, especially one with an aberrant chromosome number. { ə'ber·ənt }

aberration [ASTRON] The apparent angular displacement of the position of a celestial body in the direction of motion of the observer, caused by the combination of the velocity of the observer and the velocity of light. [OPTICS] *See* optical aberration. { ,ab·ə'rä·shən }

abfarad [ELEC] A unit of capacitance in the electromagnetic centimeter-gram-second system equal to 10^9 farads. Abbreviated aF. { ab'far·ad }

abhenry [ELEC] A unit of inductance in the electromagnetic centimeter-gram-second system of units which is equal to 10^{-9} henry. Abbreviated aH. { ab'hen·rē }

abherent [MATER] A substance that inhibits a material from adhering to itself or another material. Also known as abhesive. { ab'her·ənt }

abhesive *See* abherent. { ab'hē·ziv }

Abies [BOT] The firs, a genus of trees in the pine family characterized by erect cones, absence of resin canals in the wood, and flattened needlelike leaves. { 'ā·bē,ēz }

abietic acid [ORG CHEM] $C_{20}H_{30}O_2$ A tricyclic, crystalline acid obtained from rosin; used in making esters for plasticizers. { ,a·bē'et·ik 'as·əd }

abietine [MATER] The distillate of the gums of the Jeffrey and digger pines; comprises 96% heptane; used as a cleaning agent, insecticide, and constituent of standard gasolines to measure detonation of engines. { 'a·bē·ə,tēn }

ab initio computation [PHYS CHEM] Computation of the geometry of a molecule solely from a knowledge of its composition and molecular structure as derived from the solution of the Schrödinger equation for the given molecule. { ,ab ə'nish·ē·ō ,käm·pyə'tā·shən }

abiocoen [ECOL] A nonbiotic habitat. { 'ā,bī·ō,sēn }

abiogenesis [BIOL] The obsolete concept that plant and animal life arise from nonliving organic matter. Also known as autogenesis; spontaneous generation. { ,ā,bī·ō'jen·ə·sis }

abioseston [OCEANOGR] A general term for dead organic matter floating in ocean water. { ,ā,bī·ō'ses·tən }

abiotic [BIOL] Referring to the absence of living organisms. { ,ā,bī'äd·ik }

abiotic environment [ECOL] All physical and nonliving chemical factors, such as soil, water, and atmosphere, which influence living organisms. { ,ā,bī'äd·ik in'vī·rən,mənt }

abiotic substance [ECOL] Any fundamental chemical element or compound in the environment. { ,ā,bī'äd·ik 'səb·stəns }

abiotrophy [MED] Disordered functioning of an organ or system, as in Huntington's chorea, due to an inherited pathologic trait, which trait, however, may remain latent in the individual rather than becoming apparent; this mechanism is still conceptual. { ,ā·bī'ä·trə·fē }

abjection [MYCOL] The discharge or casting off of spores by the spore-bearing structure of a fungus. { ab'jek·shən }

ablastin [IMMUNOL] An antibodylike substance elicited by *Trypanosoma lewisi* in the blood serum of infected rats that inhibits reproduction of the parasite. { ə'blas·tən }

ablating material *See* ablative agent. { ə'blād·iŋ mə,tir·ē·əl }

ablation [AERO ENG] The intentional removal of material from a nose cone or spacecraft during high-speed movement through a planetary atmosphere to provide thermal protection to the underlying structure. [GEOL] The wearing away of rocks, as by erosion or weathering. [HYD] The reduction in volume of a glacier due to melting and evaporation. [MED] The removal of tissue or a part of the body by surgery, such as by excision or amputation. { ə'blā·shən }

ablation area [HYD] The section in a glacier or snowfield where ablation exceeds accumulation. { ə'blā·shən 'er·ē·ə }

ablation cone [HYD] A debris-covered cone of ice, firn, or snow formed by differential ablation. { ə'blā·shən kōn }

ablation factor [HYD] The rate at which a snow or ice surface wastes away. { ə'blā·shən 'fak·tər }

ablation form [HYD] A feature on a snow or ice surface caused by melting or evaporation. { ə'blā·shən fòrm }

ablation moraine [GEOL] **1.** A layer of rock particles overlying ice in the ablation of a glacier. **2.** Drift deposited from a superglacial position through the melting of underlying stagnant ice. { ə'blā·shən mə'rān }

ablative agent [MATER] A material from which the surface layer is to be removed, often for the purpose of dissipating extreme heat energy, as in space vehicles reentering the earth's atmosphere. Also known as ablating material; ablative material; ablator. { 'a·blə·div 'ā·jənt }

ablative cooling [AERO ENG] The carrying away of heat, generated by aerodynamic heating, from a vital part by arranging for its absorption by a nonvital part. { 'a·blə·div 'kül·iŋ }

ablative material *See* ablative agent. { 'a·blə·div mə'tir·ē·əl }

ablative shielding [AERO ENG] A covering of material designed to reduce heat transfer to the internal structure through sublimation and loss of mass. { 'a·blə·div 'shēld·iŋ }

ablatograph [ENG] An instrument that records ablation by measuring the distance a snow or ice surface falls during the observation period. { ə'blā·də,graf }

ablator *See* ablative agent. { 'ab,lād·ər }

able [COMPUT SCI] A name for the hexadecimal digit whose decimal equivalent is 10. { 'ā·bəl }

A block [CIV ENG] A hollow concrete masonry block with one end closed and the other open and with a web between, so that when the block is laid in a wall two cells are produced. { 'ā ,bläk }

ABM *See* antiballistic missile.

abmho [ELEC] A unit of conductance in the electromagnetic centimeter-gram-second system of units equal to 10^9 mhos. Abbreviated $(a\Omega)^{-1}$. Also known as absiemens (aS). { 'ab,mō }

Abney effect [OPTICS] A shift in the apparent hue of a light which occurs as colored light is desaturated by the addition of white light. { 'ab·nē ə'fekt }

Abney law [OPTICS] The shift in apparent hue of spectral color that is desaturated by addition of white light is toward the red end of the spectrum if the wavelength is below 570 nanometers and toward the blue if it is above. { 'ab·nē ,lò }

Abney level *See* clinometer. { 'ab·nē 'lev·əl }

Abney mounting [SPECT] A modification of the Rowland mounting in which only the slit is moved to observe different parts of the spectrum. { 'ab·nē ,maùnt·iŋ }

abnormal anticlinorium [GEOL] An anticlinorium with axial planes of subsidiary folds diverging upward. { ab'nòr·məl ,an·tə·kli'nò·rē·əm }

abnormal behavior [PSYCH] Personality functioning that is socially undesirable or that renders the individual unable to cope with day-to-day living. Also known as behavior disorder. { ab'nòr·məl be'hāv·yər }

abnormal fold [GEOL] An anticlinorium in which there is an upward convergence of the axial surfaces of the subsidiary folds. { ab'nòr·məl 'fōld }

abnormal glow discharge [ELECTR] A discharge of electricity in a gas tube at currents somewhat higher than those of an ordinary glow discharge, at which point the glow covers the entire cathode and the voltage drop decreases with increasing current. { ab'nòr·məl 'glō 'dis·chärj }

abnormality [SCI TECH] Any deviation from normal characteristics. { ab·nòr'mal·ə·tē }

abnormal magnetic variation [GEOPHYS] The anomalous value in magnetic compass readings made in some local areas containing unknown sources that deflect the compass needle from the magnetic meridian. { ab'nòr·məl mag'ned·ik ve·rē'ā·shən }

abnormal place [MIN ENG] An area in a coal mine where the

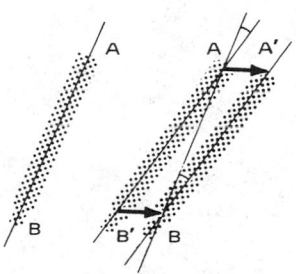

ABERRATION

The aberration of light as seen in astronomy. Starlight arriving along AB and seen in this direction by a stationary observer (left) appears to the observer in transverse motion AA′ (right) to come from the direction AB′ (or A′B). *(From G. de Vaucouleurs, Discovery of the Universe, 1957; reprinted by permission of Faber and Faber Ltd.)*

geological conditions render mining uneconomical. { ab'nòr·məl 'plās }

abnormal pressure [PETRO ENG] A pressure beyond the range of pressure values that would normally be expected at a given depth. { ab'nòr·məl 'presh·ər }

abnormal propagation [COMMUN] Phenomena of unstable or changing atmospheric or ionospheric conditions acting upon transmitted radio waves, preventing such waves from following their normal path, thereby causing difficulties and disruptions of communications. { ab'nòr·məl ,präp·ə'gā·shən }

abnormal psychology [PSYCH] A branch of psychology that deals with behavior disorders and internal psychic conflict in addition to certain normal phenomena such as dreams, motivations, and anxiety. { ab'nòr·məl sī·'käl·ə·jē }

abnormal reading See abnormal time. { ab'nòr·məl 'rēd·iŋ }

abnormal reflections [ELECTROMAG] Sharply defined reflections of substantial intensity at frequencies greater than the critical frequency of the ionized layer of the ionosphere. { ab'nòr·məl re'flek·shənz }

abnormal series See anomalous series. { ab'nòr·məl 'sir,ēz }

abnormal statement [COMPUT SCI] An element of a FORTRAN V (UNIVAC) program which specifies that certain function subroutines must be called every time they are referred to. { ab'nòr·məl 'stāt·mənt }

abnormal synclinorium [GEOL] A synclinorium with axial planes of subsidiary folds converging downward. { ab'nòr·məl ¦sin·kli'nòr·ē·əm }

abnormal time [IND ENG] During a time study, an elapsed time for any element which is excessively longer or shorter than the median of the elapsed times. Also known as abnormal reading. { 'ab,nòr·məl 'tīm }

ABO blood group [IMMUNOL] An immunologically distinct, genetically determined group of human erythrocyte antigens represented by two blood factors (A and B) and four blood types (A, B, AB, and O). { ā·bē'ō 'bləd ,grüp }

ABO blood group system [IMMUNOL] A set of multiple alleles found on a single locus on human chromosome 9 that specifies the presence or absence of certain red cell antigens, which determines the ABO blood group. { ,ā,bē'ō 'bləd ,grüp ,sis·təm }

abohm [ELEC] The unit of electrical resistance in the centimeter-gram-second system; 1 abohm equals 10^{-9} ohm in the meter-kilogram-second system. Abbreviated aΩ. { a'bōm }

abohm centimeter [ELEC] The centimeter-gram-second unit of resistivity. Abbreviated aΩcm. { a'bōm 'sen·tə,mē·dər }

abomasitis [VET MED] Inflammation of the abomasum in ruminants. { ,a·bō·mə'sīd·əs }

abomasum [VERT ZOO] The final chamber of the complex stomach of ruminants; has a glandular wall and corresponds to a true stomach. { ,ab·ō'mā·səm }

A bomb See atomic bomb. { 'ā ,bäm }

aboral [INV ZOO] Opposite to the mouth. { a'bòr·əl }

abort [AERO ENG] **1.** To cut short or break off an action, operation, or procedure with an aircraft, space vehicle, or the like, especially because of equipment failure. **2.** An aircraft, space vehicle, or the like which aborts. **3.** An act or instance of aborting. [COMPUT SCI] To terminate a procedure, such as the running of a computer program or the printing of a document, while it is still in progress. { ə'bòrt }

abort branch [CONT SYS] A branching instruction in the program controlling a robot that causes a test to be performed on whether the tool-center point is properly positioned, and to reposition it if it drifts out of the acceptable range. { ə'bòrt ,branch }

aborted firing [ORD] A firing of a gun or launching of a missile which is cut off either manually or automatically after the firing command has been given but before ignition has been initiated. { ə'bòrd·əd 'fīr·iŋ }

abortifacient [MED] Any agent that induces abortion. { ə,bòrd·ə'fā·shənt }

abortion [MED] The spontaneous or induced expulsion of the fetus prior to the time of viability, most often during the first 20 weeks of the human gestation period. { ə'bòr·shən }

abortive [BIOL] Imperfectly formed or developed. { ə'bòrd·iv }

abortive infection [VIROL] The viral infection of a cell in which viral components may be synthesized without the production of infective viruses. Also known as nonproductive infection. { ə'bòrd·iv in'fek·shən }

abortive transduction [MICROBIO] Failure of exogenous fragments that were introduced into a bacterial cell by viruses to become inserted into the bacterial chromosome. { ə'bòrd·iv tranz'dək·shən }

abortus [MED] An aborted fetus. { ə'bòrd·əs }

abort zone [AERO ENG] The area surrounding the launch within which malperforming missiles will be contained with known and acceptable probability. { ə'bòrt 'zōn }

aboundikro See Sapele mahogany. { ə'baùn·dē,krō }

about-sledge [MET] A large hammer that is utilized in blacksmithing. { ə'baùt ,slej }

a-b plane [GEOL] The surface along which differential movement takes place. { ā¦bē ,plān }

AB power pack [ELEC] **1.** Assembly in a single unit of the A battery and B battery for a battery-operated vacuum-tube circuit. **2.** Unit that supplies the necessary A and B direct-current voltages from an alternating-current source of power. { ā¦bē 'paù·ər ,pak }

abrachiocephalia [MED] Congenital lack of arms and head. Also known as acephalobrachia. { ¦ā,brāk·ē·ə·sə'fal·yə }

abradant See abrasive. { ə'brād·ant }

abrade [GEOL] To wear away by abrasion or friction. { ə'brād }

Abraham's tree [METEOROL] The popular name given to a form of cirrus radiatus clouds, consisting of an assemblage of long feathers and plumes of cirrus that seems to radiate from a single point on the horizon. { 'ā·brə,hamz 'trē }

Abrams' law [CIV ENG] In concrete materials, for a mixture of workable consistency the strength of concrete is determined by the ratio of water to cement. { 'ā·brəmz 'lò }

abranchiate [ZOO] Without gills. { ā'braŋk·ē·ət }

abrasion [ENG] **1.** The removal of surface material from any solid through the frictional action of another solid, a liquid, or a gas or combination thereof. **2.** A surface discontinuity brought about by roughening or scratching. [GEOL] Wearing away of sedimentary rock chiefly by currents of water laden with sand and other rock debris and by glaciers. [MED] A spot denuded of skin, mucous membrane, or superficial epithelium by rubbing or scraping. { ə'brā·zhən }

abrasion drilling [PETRO ENG] An oil-drilling method in which abrasive material under pressure replaces the conventional drill string and bit. { ə'brā·zhən ,dril·iŋ }

abrasion mark [GRAPHICS] **1.** A very fine mark on a print or film surface that does not penetrate to the base. **2.** A defect occurring in a print or a film as a result of moving contact with another surface. { ə'brā·zhən ,märk }

abrasion platform [GEOL] An uplifted marine peneplain or plain, according to the smoothness of the surface produced by wave erosion, which is of large area. { ə'brā·zhən 'plat·fòrm }

abrasion resistance [MATER] The ability of a surface to resist wearing due to contact with another surface moving with respect to it. { ə'brā·zhən rə'zis·təns }

abrasion-resistance index [MATER] In vulcanized material or synthetic rubber compounds, a measure of abrasion resistance relative to a standard rubber compound under defined conditions. { ə'brā·zhən rə'zis·təns 'in·deks }

abrasion test [MECH ENG] The measurement of abrasion resistance, usually by the weighing of a material sample before and after subjecting it to a known abrasive stress throughout a known time period, or by reflectance or surface finish comparisons, or by dimensional comparisons. { ə'brā·zhən test }

abrasive [GEOL] A small, hard, sharp-cornered rock fragment, used by natural agents in abrading rock material or land surfaces. Also known as abrasive ground. [MATER] **1.** A material used, usually as a grit sieved by a specified mesh but also as a solid shape or as a paste or slurry or air suspension, for grinding, honing, lapping, superfinishing, polishing, pressure blasting, or barrel tumbling. **2.** A material sintered or formed into a solid mass such as a hone or a wheel disk, cone, or burr for grinding or polishing other materials. **3.** Having qualities conducive to or derived from abrasion. Also known as abradant. { ə'brās·əv }

abrasive belt [MECH ENG] A cloth, leather, or paper band impregnated with grit and rotated as an endless loop to abrade materials through continuous friction. { ə'brās·əv belt }

abrasive blasting [MECH ENG] The cleaning or finishing of surfaces by the use of an abrasive entrained in a blast of air. { ə'brās·əv 'blast·iŋ }

abrasive cloth [MECH ENG] Tough cloth to whose surface an abrasive such as sand or emery has been bonded for use in grinding or polishing. { ə'brās·əv 'klòth }

abrasive cone [MECH ENG] An abrasive sintered or shaped into a solid cone to be rotated by an arbor for abrasive machining. { ə'brās·əv 'kōn }

abrasive disk [MECH ENG] An abrasive sintered or shaped into a disk to be rotated by an arbor for abrasive machining. { ə'brās·əv 'disk }

abrasive drilling [MIN ENG] A rotary drilling method in which drilling is effected by the abrasive action of the drill steel or drilling medium which rotates while being pressed against the rock. { ə'brās·əv 'dril·iŋ }

abrasive ground See abrasive. { ə'brās·əv 'graùnd }

abrasive jet cleaning [ENG] The removal of dirt from a solid by a gas or liquid jet carrying abrasives to ablate the surface. { ə'brās·əv 'jet 'klēn·iŋ }

abrasive machining [MECH ENG] Grinding, drilling, shaping, or polishing by abrasion. { ə'brās·əv mə'shēn·iŋ }

abrasiveness [MATER] **1.** The property of a material causing wear of a surface by friction. **2.** The quality or characteristic of being able to scratch, abrade, or wear away another material. { ə'brās·əv·nəs }

abrasive paper [MATER] Tough paper to whose surface an abrasive, such as sand or emery, has been bonded for use in grinding or polishing. { ə'brās·əv 'pā·pər }

abrasive sand [MATER] Grit used as abrasive, usually graded as to which sieve mesh it will pass through. { ə'brās·əv 'sand }

abreaction [PSYCH] In psychoanalytic theory, the weakening or elimination of anxiety by reexperiencing, either through imagination or in reality, the original anxiety-provoking experience. { ,ab·rē'ak·shən }

abreast milling [MECH ENG] A milling method in which parts are placed in a row parallel to the axis of the cutting tool and are milled simultaneously. { ə'brest 'mil·iŋ }

abreuvoir [CIV ENG] A space between stones in masonry to be filled with mortar. { ab·rü'vwär }

Abridged Nautical Almanac See Nautical Almanac. { ə'brijd 'nòt·ə·kəl 'òl·mə·nak }

Abrikosov-Suhl resonance [SOLID STATE] For materials that display the Kondo effect, a long-lived scattering resonance of electronic states near the Fermi level that forms at temperatures less than the Kondo temperature; accounts for the qualitative behavior of resistivity and magnetic susceptibility as functions of temperature. Also known as Kondo resonance. { ,ab·rə'kä·sòf 'sül 'rez·ən·əns }

abrin [BIOCHEM] A highly poisonous protein found in the seeds of Abrus precatorius, the rosary pea. { 'a·brin }

abrupt [BOT] Ending suddenly, as though broken off. { ə'brəpt }

abruptio placenta [MED] A pregnancy disorder in which the placenta separates prematurely from the uterus. { ə,brəp·shē·o plə'sent·ə }

abrupt junction [ELECTR] A pn junction in which the concentration of impurities changes suddenly from acceptors to donors. { ə'brəpt 'jəŋk·shən }

abs [COMPUT SCI] A special function occurring in ALGOL, which yields the absolute value, or modulus, of its argument. [METEOROL] See absolute.

ABS See acrylonitrile butadiene styrene resin; antilock braking system.

absarokite [PETR] An alkalic basalt of about equal portions of olivine, augite, labradorite, and sanidine with accessory biotite, apatite, and opaque oxides; leucite is occasionally present in small amounts. { ab'sä·rə·kīt }

abscess [MED] A localized collection of pus surrounded by inflamed tissue. { 'ab·ses }

abscisic acid [BIOCHEM] $C_{16}H_{20}O_4$ A plant hormone produced by fruits and leaves that promotes abscission and dormancy and retards vegetative growth. Abbreviated ABA. Formerly known as abscisin. { ab'sis·ik 'as·əd }

abscisin See abscisic acid. { ab'sis·ən }

abscissa [MATH] One of the coordinates of a two-dimensional coordinate system, usually the horizontal coordinate, denoted by x. { ab'sis·ə }

abscission [BOT] A physiological process promoted by abscisic acid whereby plants shed a part, such as a leaf, flower, seed, or fruit. { ab'sizh·ən }

abscission layer [BOT] A zone of cells whose breakdown causes separation of a leaf or other structure from the stem. { ab'sizh·ən ,lā·ər }

absiemens See abmho. { ab'sē·mənz }

absinthe [FOOD ENG] A green liqueur having a bitter licorice flavor and a high alcohol content. { ab'santh }

absinthe oil [MATER] A toxic essential oil obtained from the dried leaves of Artemisia absinthium; soluble in alcohol; formerly used in medicine. Also known as wormwood oil. { ab'santh ,òil }

absolute [METEOROL] Referring to the highest or lowest recorded value of a meteorological element, whether at a single station or over an area, during a given period. Abbreviated abs. { ,ab·sə'lüt }

absolute address [COMPUT SCI] The numerical identification of each storage location which is wired permanently into a computer by the manufacturer. { 'ab·sə,lüt ə'dres }

absolute addressing [COMPUT SCI] The identification of storage locations in a computer program by their physical addresses. { 'ab·sə,lüt ə'dres·iŋ }

absolute age [GEOL] The geologic age of a fossil, or a geologic event or structure expressed in units of time, usually years. Also known as actual age. { 'ab·sə,lüt 'āj }

absolute alcohol [ORG CHEM] Ethyl alcohol that contains no more than 1% water. Also known as anhydrous alcohol. { 'ab·sə,lüt 'al·kə·hòl }

absolute altimeter [ENG] An instrument which employs radio, sonic, or capacitive technology to produce on its indicator the measurement of distance from the aircraft to the terrain below. Also known as terrain-clearance indicator. { 'ab·sə,lüt al'tim·ə·dər }

absolute altitude [ENG] Altitude above the actual surface, either land or water, of a planet or natural satellite. { 'ab·sə,lüt 'al·tə·tüd }

absolute angle of attack [AERO ENG] The acute angle between the chord of an airfoil at any instant in flight and the chord of that airfoil at zero lift. { 'ab·sə,lüt 'aŋ·gəl əv ə'tak }

absolute blocking [CIV ENG] A control arrangement for rail traffic in which a track is divided into sections or blocks upon which a train may not enter until the preceding train has left. { 'ab·sə,lüt 'bläk·iŋ }

absolute block system [CIV ENG] A block system in which only a single railroad train is permitted within a block section during a given period of time. { 'ab·sə,lüt 'bläk ,sis·təm }

absolute boiling point [CHEM] The boiling point of a substance expressed in the unit of an absolute temperature scale. { 'ab·sə,lüt 'bòil·iŋ ,pòint }

absolute ceiling [AERO ENG] The greatest altitude at which an aircraft can maintain level flight in a standard atmosphere and under specified conditions. { 'ab·sə,lüt 'sēl·iŋ }

absolute cell reference [COMPUT SCI] A cell reference used in a formula in a spreadsheet program that does not change when the formula is copied or moved. { 'ab·sə,lüt 'sel ,ref·rəns }

absolute code [COMPUT SCI] A code used when the addresses in a program are to be written in machine language exactly as they will appear when the instructions are executed by the control circuits. { 'ab·sə,lüt 'kōd }

absolute configuration [ORG CHEM] The three-dimensional arrangement of substituents around a chiral center in a molecule. Also known as absolute stereochemistry. { 'ab·sə,lüt kən,fig·yə'rā·shən }

absolute convergence [MATH] That property of an infinite series (or infinite product) of real or complex numbers if the series (product) of absolute values converges; absolute convergence implies convergence. { 'ab·sə,lüt kən'vərj·əns }

absolute coordinates [MATH] Coordinates given with reference to a fixed point of origin. { 'ab·sə,lüt kō'òrd·ən·ats }

absolute coordinate system [NAV] The inertial coordinate system which has its origin on the axis of the earth and is fixed with respect to the stars. Also known as absolute reference frame. { 'ab·sə,lüt kō'òrd·ə·nət 'sis·təm }

absolute delay [NAV] In loran, the time interval between transmission of a signal from the A station and transmission of the next signal from the B station. { 'ab·sə,lüt də'lā }

absolute density See absolute gravity. { 'ab·sə,lüt 'dens·ə·dē }

ABSCISIC ACID

Structural formula for + -abscisic acid, the naturally occurring form.

absolute detection limit [ANALY CHEM] The smallest amount of an element or compound that is detectable in or on a given sample; expressed in terms of mass units or numbers of atoms or molecules. { 'ab·sə¦lüt di'tek·shən ¦lim·ət }

absolute deviation [ORD] The shortest distance between the center of the target and the point where a projectile hits or bursts. [STAT] The difference, without regard to sign, between a variate value and a given value. { 'ab·sə¦lüt dēv·ē'ā·shən }

absolute drought [METEOROL] In Britain, a period of at least 15 consecutive days during which no measurable daily precipitation has fallen. { 'ab·sə¦lüt ¦draut }

absolute efficiency [ENG ACOUS] The ratio of the power output of an electroacoustic transducer, under specified conditions, to the power output of an ideal electroacoustic transducer. { 'ab·sə¦lüt ə'fish·ən·sē }

absolute electrometer [ELEC] A very precise type of attracted disk electrometer in which the attraction between two disks is balanced against the force of gravity. { 'ab·sə¦lüt ə,lek'träm·əd·ər }

absolute error [MATH] In an approximate number, the numerical difference between the number and a number considered exact. [ORD] **1.** Shortest distance between the center of impact or the center of burst of a group of shots and the point of impact or burst of a single shot within the group. **2.** Error of a sight consisting of its error in relation to a master service sight with which it is tested and of the known error of the master service sight. { 'ab·sə¦lüt 'er·ər }

absolute expansion [THERMO] The true expansion of a liquid with temperature, as calculated when the expansion of the container in which the volume of the liquid is measured is taken into account; in contrast with apparent expansion. { 'ab·sə¦lüt ik'span·shən }

absolute gain of an antenna [ELECTROMAG] Gain in a given direction when the reference antenna is an isotropic antenna isolated in space. Also known as isotropic gain of an antenna. { 'ab·sə¦lüt ¦gān əv ən an'ten·ə }

absolute geopotential topography *See* geopotential topography. { 'ab·sə¦lüt jē·ō·pə'ten·shəl tə'päg·rə·fē }

absolute gravity [CHEM] Density or specific gravity of a fluid reduced to standard conditions; for example, with gases, to 760 mmHg pressure and 0°C temperature. Also known as absolute density. { 'ab·sə¦lüt 'grav·ə·dē }

absolute humidity [PHYS] The ratio of the mass of water vapor in a sample of air to the volume of the sample. { 'ab·sə¦lüt hyü'mid·ə·dē }

absolute index of refraction *See* index of refraction. { 'ab·sə¦lüt 'in,deks əv ri'frak·shən }

absolute inequality *See* unconditional inequality. { 'ab·sə¦lüt ,in·ē'kwäl·ə·dē }

absolute instability [METEOROL] The state of a column of air in the atmosphere when it has a superadiabatic lapse rate of temperature, that is, greater than the dry-adiabatic lapse rate. Also known as autoconvective instability; mechanical instability. { 'ab·sə¦lüt ,in·stə'bil·ə·dē }

absolute instruction [COMPUT SCI] A computer instruction in its final form, in which it can be executed. { 'ab·sə¦lüt in'strək·shən }

absolute instrument [ENG] An instrument which measures a quantity (such as pressure or temperature) in absolute units by means of simple physical measurements on the instrument. { 'ab·sə¦lüt 'in·strə·mənt }

absolute isohypse [METEOROL] A line that has the properties of both constant pressure and constant height above mean sea level. { 'ab·sə¦lüt 'ī·sō,hīps }

absolute linear momentum *See* absolute momentum. { 'ab·sə¦lüt 'lin·ē·ər mə'ment·əm }

absolute luminosity [OPTICS] The luminosity of an object expressed in units of fundamental quantities. { 'ab·sə¦lüt lü·mə'näs·ə·dē }

absolutely continuous function [MATH] A function defined on a closed interval with the property that for any positive number ϵ there is another positive number η such that, for any finite set of nonoverlapping intervals, $(a_1, b_1), (a_2, b_2), \ldots, (a_n, b_n)$, whose lengths have a sum less than η, the sum over the intervals of the absolute values of the differences in the values of the function at the ends of the intervals is less than ϵ. { ¦ab·sə¦lüt·lē kən¦tin·yə·wəs 'fəŋk·shən }

absolutely continuous measure [MATH] A sigma finite measure m on a sigma algebra is absolutely continuous with respect to another sigma finite measure n on the same sigma algebra if every element of the sigma algebra whose measure n is zero also has measure m equal to zero. { ab·sə¦lüt·lē kən,tin·yə·wəs 'mezh·ər }

absolute magnetometer [ENG] An instrument used to measure the intensity of a magnetic field without reference to other magnetic instruments. { 'ab·sə¦lüt mag·nə'täm·ə·dər }

absolute magnitude [ASTROPHYS] **1.** A measure of the brightness of a star equal to the magnitude the star would have at a distance of 10 parsecs from the observer. **2.** The stellar magnitude any meteor would have if placed in the observer's zenith at a height of 100 kilometers. [MATH] The absolute value of a number or quantity. { 'ab·sə¦lüt 'mag·nə·tüd }

absolute manometer [ENG] **1.** A gas manometer whose calibration, which is the same for all ideal gases, can be calculated from the measurable physical constants of the instrument. **2.** A manometer that measures absolute pressure. { 'ab·sə¦lüt mə'näm·ə·dər }

absolute mean deviation [STAT] The arithmetic mean of the absolute values of the deviations of a variable from its expected value. { 'ab·sə¦lüt ,mēn dē·vē'ā·shən }

absolute method [ANALY CHEM] A method of chemical analysis that bases characterization completely on standards defined in terms of physical properties. { 'ab·sə¦lüt 'meth·əd }

absolute moment [MATH] The nth absolute moment of a distribution $f(x)$ about a point x_0 is the expected value of the nth power of the absolute value of $x - x_0$. { ¦ab·sə¦lüt 'mō·mənt }

absolute momentum [METEOROL] The sum of the (vector) momentum of a particle relative to the earth and the (vector) momentum of the particle due to the earth's rotation. Also known as absolute linear momentum. { 'ab·sə¦lüt mə'ment·əm }

absolute motion [NAV] Motion relative to a point fixed on the earth's surface or to an apparently fixed celestial point. [PHYS] Motion of an object described by its measurement in a frame of reference that is preferred over all other frames. { 'ab·sə¦lüt 'mō·shən }

absolute number [MATH] A number represented by numerals rather than by letters. { 'ab·sə¦lüt 'nəm·bər }

absolute orientation [NAV] The adjusting to proper scale, orientating of the model datum parallel to sea level or another given vertical datum, and positioning of the model with reference to the horizontal datum of a stereoscopic model or group of models. { 'ab·sə¦lüt ȯr·ē·ən'tā·shən }

absolute parallax *See* absolute stereoscopic parallax. { 'ab·sə¦lüt 'par·ə·laks }

absolute permeability [ELECTROMAG] The ratio of the magnetic flux density to the intensity of the magnetic field in a medium; measurement is in webers per square meter in the meter-kilogram-second system. Also known as induced capacity. [PETRO ENG] A measurement of the capacity for flow of a single fluid (water, gas, or oil) through a rock formation when the formation is completely saturated with that fluid. { 'ab·sə¦lüt pər·mē·ə'bil·ə·dē }

absolute pitch [ACOUS] The pitch of a musical tone expressed as the frequency of the sound wave of that tone. { 'ab·sə¦lüt 'pich }

absolute plating efficiency [CYTOL] The percentage of individual cells that give rise to colonies when the cells are inoculated into culture media. { 'ab·sə¦lüt 'plād·iŋ i,fish·ən·sē }

absolute porosity [PETRO ENG] The ratio of the volume of the pore spaces or voids in a rock to the total bulk volume of the rock. { 'ab·sə¦lüt pə'räs·əd·ē }

absolute potential vorticity *See* potential vorticity. { 'ab·sə¦lüt pə'ten·shəl vȯr'tis·ə·dē }

absolute pressure [PHYS] The pressure above the absolute zero value of pressure that theoretically obtains in empty space or at the absolute zero of temperature, as distinguished from gage pressure. { 'ab·sə¦lüt 'presh·ər }

absolute pressure gage [ENG] A device that measures the pressure exerted by a fluid relative to a perfect vacuum; used to measure pressures very close to a perfect vacuum. { 'ab·sə¦lüt 'presh·ər ,gāj }

absolute pressure transducer [ENG] A device that

responds to absolute pressure as the input and provides a measurable output of a nature different than but proportional to absolute pressure. { 'ab·sə̇lüt 'presh·ər tranz'dü·sər }

absolute programming [COMPUT SCI] Programming with the use of absolute code. { 'ab·sə̇lüt 'prō·gram·iŋ }

absolute reaction rate [PHYS CHEM] The rate of a chemical reaction as calculated by means of the (statistical-mechanics) theory of absolute reaction rates. { 'ab·sə̇lüt rē'ak·shən ̧rāt }

absolute reference frame See absolute coordinate system. { 'ab·sə̇lüt 'ref·ə·rəns ̧frām }

absolute refractive constant See index of refraction. { 'ab·sə̇lüt ri'frak·tiv 'kän·stənt }

absolute refractory period [NEURO] A period ranging from 0.5 to 2 milliseconds during which neural tissue is totally unresponsive. { 'ab·sə̇lüt ri'frak·trē ̧pir·ē·əd }

absolute retract [MATH] A topological space, A, such that, if B is a closed subset of another topological space, C, and if A is homeomorphic to B, then B is a retract of C. { 'ab·sə̇lüt ri'trakt }

absolute roof [MIN ENG] The entire mass of strata overlying a subsurface point of reference. { 'ab·sə̇lüt 'rüf }

absolute scale See absolute temperature scale. { 'ab·sə̇lüt ̧skāl }

absolute space-time [PHYS] A concept underlying Newtonian mechanics which postulates the existence of a preferred reference system of time and spatial coordinates; replaced in relativistic mechanics by Einstein's equivalency principle. Also known as absolute time. { 'ab·sə̇lüt 'spās ̧tīm }

absolute specific gravity [MECH] The ratio of the weight of a given volume of a substance in a vacuum at a given temperature to the weight of an equal volume of water in a vacuum at a given temperature. { 'ab·sə̇lüt spə'sif·ək 'grav·əd·ē }

absolute stability [METEOROL] The state of a column of air in the atmosphere when its lapse rate of temperature is less than the saturation-adiabatic lapse rate. { 'ab·sə̇lüt stə'bil·ə·dē }

absolute standard [PHYS] A particle or object designated as a standard by assigning to it a mass of one unit; used in defining quantities in Newton's second law of motion. { 'ab·sə̇lüt 'stan·dərd }

absolute stereochemistry See absolute configuration. { 'ab·sə̇lüt ̧ster·ē·ō'kem·ə·strē }

absolute stereoscopic parallax [GRAPHICS] Considering a pair of aerial photographs of equal principal distance, the absolute stereoscopic parallax of a point is the algebraic difference of the distances of the two images from their respective photograph nadirs, measured in a horizontal plane and parallel to the air base. Also known as absolute parallax; horizontal parallax; linear parallax; parallax; stereoscopic parallax; x-parallax. { 'ab·sə̇lüt ster·ē·ō'skäp·ik 'par·ə̧laks }

absolute stop [CIV ENG] A railway signal which indicates that the train must make a full stop and not proceed until there is a change in the signal. Also known as stop and stay. { 'ab·sə̇lüt 'stäp }

absolute system of units [PHYS] A set of units for measuring physical quantities, defined by interrelated equations in terms of arbitrary fundamental quantities of length, mass, time, and charge or current. { 'ab·sə̇lüt 'sis·təm əv 'yü·nəts }

absolute temperature [THERMO] **1.** The temperature measurable in theory on the thermodynamic temperature scale. **2.** The temperature in Celsius degrees relative to the absolute zero at −273.16°C (the Kelvin scale) or in Fahrenheit degrees relative to the absolute zero at −459.69°F (the Rankine scale). { 'ab·sə̇lüt 'tem·prə·chür }

absolute temperature scale [THERMO] A scale with which temperatures are measured relative to absolute zero. Also known as absolute scale. { 'ab·sə̇lüt 'tem·prə·chür ̧skāl }

absolute term See constant term. { 'ab·sə̇lüt 'tərm }

absolute threshold [PHYSIO] The minimum stimulus energy that an organism can detect. { 'ab·sə̇lüt 'thresh·hōld }

absolute time [GEOL] Geologic time measured in years, as determined by radioactive decay of elements. [PHYS] See absolute space-time. { 'ab·sə̇lüt 'tīm }

absolute unit [PHYS] A unit defined in terms of units of fundamental quantities such as length, time, mass, and charge or current. { 'ab·sə̇lüt 'yü·nət }

absolute vacuum [PHYS] A void completely empty of matter. Also known as perfect vacuum. { 'ab·sə̇lüt 'vak·yüm }

absolute value Also known as magnitude. [MATH] **1.** For a real number, the number if it is nonnegative, and the negative of the number if it is negative. Also known as numerical value. **2.** For a complex number, the square root of the sum of the squares of its real and imaginary parts. Also known as modulus. **3.** The length of a vector, disregarding its direction; the square root of the sum of the squares of its orthogonal components. { 'ab·sə̇lüt 'val·yü }

absolute-value computer [COMPUT SCI] A computer that processes the values of the variables rather than their increments. { 'ab·sə̇lüt 'val·yü kəm'pyüd·ər }

absolute vector [COMPUT SCI] In computer graphics, a vector whose end points are given in absolute coordinates. { 'ab·sə̇lüt 'vek·tər }

absolute velocity [PHYS] The vector sum of the velocity of a fluid parcel relative to the earth and the velocity of the parcel due to the earth's rotation; the east-west component is the only one affected. { 'ab·sə̇lüt və'läs·ə·dē }

absolute viscosity [FL MECH] The tangential force per unit area of two parallel planes at unit distance apart when the space between them is filled with a fluid and one plane moves with unit velocity in its own plane relative to the other. Also known as coefficient of viscosity. { 'ab·sə̇lüt vis'käs·ə·dē }

absolute volume [ENG] The total volume of the particles in a granular material, including both permeable and impermeable voids but excluding spaces between particles. { 'ab·sə̇lüt 'väl·yüm }

absolute vorticity [FL MECH] The vorticity of a fluid relative to an absolute coordinate system; especially, the vorticity of the atmosphere relative to axes not rotating with the earth. { 'ab·sə̇lüt vȯr'tis·ə·dē }

absolute wavemeter [ELECTROMAG] A type of wavemeter in which the frequency of an injected radio-frequency voltage is determined by measuring the length of a resonant line. { 'ab·sə̇lüt 'wāv̧mēd·ər }

absolute weighing [ENG] Determination of the mass of a sample and expressing its value in units, fractions, and multiples of the mass of the prototype of the international kilogram. { 'ab·sə̇lüt 'wā·iŋ }

absolute zero [THERMO] The temperature of −273.16°C, or −459.69°F, or 0 K, thought to be the temperature at which molecular motion vanishes and a body would have no heat energy. { 'ab·sə̇lüt 'zir·ō }

absorb [CHEM] To take up a substance in bulk. [ELECTROMAG] To take up energy from radiation. [PHYS] To take up matter or radiation. { əb'sȯrb }

absorbance [PHYS CHEM] The common logarithm of the reciprocal of the transmittance of a pure solvent. Also known as absorbancy; extinction. { əb'sȯr·bəns }

absorbancy See absorbance. { əb'sȯr·bən·sē }

absorbed charge [ELEC] Charge on a capacitor which arises only gradually when the potential difference across the capacitor is maintained, due to gradual orientation of permanent dipolar molecules. { əb'sȯrbd 'chärj }

absorbed dose [MED] The part of an administered medication which is not excreted by the recipient's body. [NUCLEO] The amount of energy imparted by ionizing particles to a unit mass of irradiated material at a place of interest. Also known as dosage; dose. { əb'sȯrbd 'dōs }

absorbed-dose rate [NUCLEO] The absorbed dose of ionizing radiation imparted at a given location per unit of time (second, minute, hour, or day). { əb'sȯrbd 'dōs ̧rāt }

absorbency [CHEM] Penetration of one substance into another. { əb'sȯr·bən·sē }

absorbency index See absorptivity. { əb'sȯr·bən·sē 'in·deks }

absorbent [MATER] A material which, in contact with a liquid or gas, extracts one or more substances for which it has an affinity, and is altered physically or chemically during the process. { əb'sȯr·bənt }

absorbent cotton [MATER] A cotton fiber that absorbs water because its natural waxes have been removed. { əb'sȯr·bənt 'kät·ən }

absorbent paper [MATER] Paper capable of absorbing and holding liquids by the capillarity of the pores between or within the closely matted cellulosic fibers. { əb'sȯr·bənt 'pā·pər }

absorber [CHEM ENG] Equipment in which a gas is absorbed by contact with a liquid. [ELECTR] A material or device that takes up and dissipates radiated energy; may be used to shield an object from the energy, prevent reflection of the energy, determine the nature of the radiation, or selectively transmit one or more components of the radiation.　[ENG]　The surface on a solar collector that absorbs the solar radiation.　[MECH ENG]　**1.** A device which holds liquid for the absorption of refrigerant vapor or other vapors.　**2.** That part of the low-pressure side of an absorption system used for absorbing refrigerant vapor.　[NUCLEO]　A material that absorbs neutrons or other ionizing radiation.　{ əb'sȯr·bər }

absorber capacity　[CHEM ENG]　During natural gas processing, the maximum volume of the gas that can be processed through an absorber without alteration of specified operating conditions.　{ əb'sȯr·bər kə,pas·əd·ē }

absorber control　*See* absorption control.　{ əb'sȯr·bər kən'trōl }

absorber oil　*See* absorption oil.　{ əb'sȯr·bər ,ȯil }

absorber plate　[ENG]　A part of a flat-plate solar collector that provides a surface for absorbing incident solar radiation.　{ əb'sȯr·bər ,plāt }

absorbing boom　[CIV ENG]　A device that floats on the water and is used to stop the spread of an oil spill and aid in its removal.　{ əb'sȯrb·iŋ ,büm }

absorbing rod　*See* control rod.　{ əb'sȯrb·iŋ ,räd }

absorbing state　[MATH]　A special case of recurrent state in a Markov process in which the transition probability, P_{ii}, equals 1; a process will never leave an absorbing state once it enters.　{ əb'sȯrb·iŋ ,stāt }

absorbing subset　[MATH]　A subset, A, of a vector space such that, for any point, x, there exists a number, b, greater than zero such that ax is a member of A whenever the absolute value of a is greater than zero and less than b.　{ əb,sȯrb·iŋ 'səb,set }

absorbing well　[CIV ENG]　A shaft that permits water to drain through an impermeable stratum to a permeable stratum.　{ əb'sȯrb·iŋ ,wel }

absorptance　[PHYS]　The ratio of the total unabsorbed radiation to the total incident radiation; equal to one (unity) minus the transmittance.　{ əb'sȯrp·təns }

absorptiometer　[ANALY CHEM]　**1.** An instrument equipped with a filter system or other simple dispersing system to measure the absorption of nearly monochromatic radiation in the visible range by a gas or a liquid, and so determine the concentration of the absorbing constituents in the gas or liquid.　**2.** A device for regulating the thickness of a liquid in spectrophotometry.　{ əb,sȯrp·tē'ä·məd·ər }

absorptiometric analysis　[ANALY CHEM]　Chemical analysis of a gas or a liquid by measurement of the peak electromagnetic absorption wavelengths that are unique to a specific material or element.　{ əb,sȯrp·tē·ə'met·rik ə'nal·ə·sis }

absorption　[BIOL]　The net movement (transport) of water and solutes into a cell or an organism to the interior.　[CHEM]　The taking up of matter in bulk by other matter, as in dissolving of a gas by a liquid.　[ELEC]　The property of a dielectric in a capacitor which causes a small charging current to flow after the plates have been brought up to the final potential, and a small discharging current to flow after the plates have been short-circuited, allowed to stand for a few minutes, and short-circuited again.　Also known as dielectric soak.　[ELECTROMAG]　Taking up of energy from radiation by the medium through which the radiation is passing.　[HYD]　Entrance of surface water into the lithosphere.　[IMMUNOL]　**1.** Removal of antibodies from an antiserum by addition of antigen.　**2.** Removal of antigens from a mixture by addition of antibodies.　[NUCLEO]　The process by which the quantity of particles entering a body is reduced by their interaction with the matter.　[PHYSIO]　Passage of a chemical substance, a pathogen, or radiant energy through a body membrane.　{ əb'sȯrp·shən }

absorption atelectasis　*See* obstructive atelectasis.　{ əb'sȯrp·shən ,ad·ə'lek·tə·sis }

absorption band　[PHYS]　A range of wavelengths or frequencies in the electromagnetic spectrum within which radiant energy is absorbed by a substance.　{ əb'sȯrp·shən ,band }

absorption bed　[CIV ENG]　A sizable pit containing coarse aggregate about a distribution pipe system; absorbs the effluent of a septic tank.　{ əb'sȯrp·shən ,bed }

absorption cell　[OPTICS]　A vessel with transparent walls for holding a gas or liquid whose absorptivity or absorption spectrum is to be measured.　{ əb'sȯrp·shən ,sel }

absorption circuit　[ELECTR]　A series-resonant circuit used to absorb power at an unwanted signal frequency by providing a low impedance to ground at this frequency.　{ əb'sȯrp·shən 'sər·kət }

absorption coefficient　Also known as absorption factor; absorption ratio; coefficient of absorption.　[ACOUS]　The ratio of the sound energy absorbed by a surface of a medium or material to the sound energy incident on the surface.　[PHYS]　If a flux through a material decreases with distance x in proportion to e^{-ax}, then a is called the absorption coefficient.　{ əb'sȯrp·shən ,kō·ə'fish·ənt }

absorption column　*See* absorption tower.　{ əb'sȯrp·shən ,käl·əm }

absorption constant　*See* absorptivity.　{ əb'sȯrp·shən ,käns·tənt }

absorption control　[ELECTR]　*See* absorption modulation.　[NUCLEO]　Control of a nuclear reactor by a material that absorbs neutrons, such as cadmium or boron steel.　Also known as absorber control.　{ əb'sȯrp·shən kən'trōl }

absorption cross section　[ELECTROMAG]　In radar, the ratio of the amount of power removed from a beam by absorption of radio energy by a target to the power in the beam incident upon the target.　{ əb'sȯrp·shən ¦krȯs ¦sek·shən }

absorption current　[ELEC]　The component of a dielectric current that is proportional to the rate of accumulation of electric charges within the dielectric.　{ əb'sȯrp·shən 'kər·ənt }

absorption curve　[PHYS]　A graph showing the curvilinear relationship of the variation in absorbed radiation as a function of wavelength.　{ əb'sȯrp·shən ,kərv }

absorption cycle　[MECH ENG]　In refrigeration, the process whereby a circulating refrigerant, for example, ammonia, is evaporated by heat from an aqueous solution at elevated pressure and subsequently reabsorbed at low pressure, displacing the need for a compressor.　{ əb'sȯrp·shən ,sī·kəl }

absorption dynamometer　[ENG]　A device for measuring mechanical forces or power in which the mechanical energy input is absorbed by friction or electrical resistance.　{ əb'sȯrp·shən dīn·ə'mäm·əd·ər }

absorption edge　[SPECT]　The wavelength corresponding to a discontinuity in the variation of the absorption coefficient of a substance with the wavelength of the radiation.　Also known as absorption limit.　{ əb'sȯrp·shən ,ej }

absorption-emission pyrometer　[ENG]　A thermometer for determining gas temperature from measurement of the radiation emitted by a calibrated reference source before and after this radiation has passed through and been partially absorbed by the gas.　{ əb'sȯrp·shən ə'mish·ən pī'räm·əd·ər }

absorption factor　*See* absorption coefficient.　{ əb'sȯrp·shən ,fak·tər }

absorption fading　[COMMUN]　Slow type of fading, primarily caused by variations in the absorption rate along the radio path.　{ əb'sȯrp·shən 'fād·iŋ }

absorption field　[CIV ENG]　Trenches containing coarse aggregate about distribution pipes permitting septic-tank effluent to seep into surrounding soil.　Also known as disposal field.　{ əb'sȯrp·shən ,fēld }

absorption gasoline　[MATER]　A gasoline obtained by using an oil to absorb the natural or refinery gas containing the gasoline and then distilling it from the oil.　{ əb'sȯrp·shən gas·ə'lēn }

absorption hygrometer　[ENG]　An instrument with which the water vapor content of the atmosphere is measured by means of the absorption of vapor by a hygroscopic chemical.　{ əb'sȯrp·shən hī'gräm·əd·ər }

absorption index　[OPTICS]　The complex index of refraction may be written as $n(1 + ik)$; the coefficient k is the absorption index.　Also known as index of absorption.　{ əb'sȯrp·shən 'in·deks }

absorption lens　[OPTICS]　Glass which prevents selected wavelengths from passing through it; used in eyeglasses.　{ əb'sȯrp·shən ,lenz }

absorption limit　*See* absorption edge.　{ əb'sȯrp·shən 'lim·ət }

absorption line　[SPECT]　A minute range of wavelength or frequency in the electromagnetic spectrum within which radiant energy is absorbed by the medium through which it is passing.　{ əb'sȯrp·shən ,līn }

absorption loss　[CIV ENG]　The quantity of water that is lost during the initial filling of a reservoir because of absorption

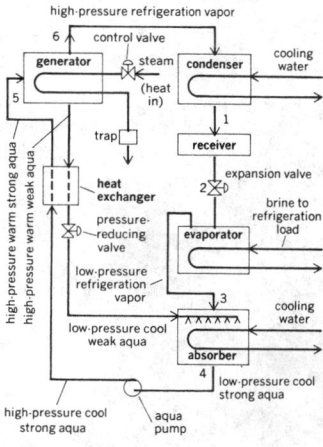

ABSORPTION CYCLE

high-pressure refrigeration vapor

6 — control valve

generator — steam (heat in) — condenser — cooling water

5

trap — 1

receiver

heat exchanger — 2 — expansion valve — brine to refrigeration load

pressure-reducing valve — evaporator

low-pressure refrigeration vapor — 3 — cooling water

low-pressure cool weak aqua — absorber

high-pressure warm strong aqua
high-pressure warm weak aqua

low-pressure cool strong aqua — 4

high-pressure cool strong aqua — aqua pump

Basic absorption cycle for an air-conditioning system.

by soil and rocks. [COMMUN] That part of the transmission loss due to the dissipation or conversion of either sound energy or electromagnetic energy into other forms of energy, either within the medium or attendant upon a reflection. { əb'sȯrp·shən ‚lȯs }

absorption meter [ENG] An instrument designed to measure the amount of light transmitted through a transparent substance, using a photocell or other light detector. { əb'sȯrp·shən 'mēd·ər }

absorption modulation [ELECTR] A system of amplitude modulation in which a variable-impedance device is inserted in or coupled to the output circuit of the transmitter. Also known as absorption control; loss modulation. { əb'sȯrp·shən ‚mäd·yü'lā·shən }

absorption nebula *See* dark nebula. { əb'sȯrp·shən 'neb·yə·lə }

absorption number [ENG] A dimensionless group used in the field of gas absorption in a wetted-wall column; represents the liquid side mass-transfer coefficient. { əb'sȯrp·shən ‚nəm·bər }

absorption oil [MATER] A petroleum or coal tar oil that is contacted with a vapor or gas mixture to remove heavy components, as in the recovery of natural gasoline from wet natural gas. Also known as absorber oil; scrubbing oil; wash oil. { əb'sȯrp·shən ‚ȯil }

absorption peak [SPECT] A wavelength of maximum electromagnetic absorption by a chemical sample; used to identify specific elements, radicals, or compounds. { əb'sȯrp·shən ‚pēk }

absorption plant [CHEM ENG] A facility to recover the condensable portion of natural or refinery gas. { əb'sȯrp·shən ‚plant }

absorption process [CHEM ENG] A method in which light oil is introduced into an absorption tower so that it absorbs the gasoline in the rising wet gas; the light oil is then distilled to separate the gasoline. { əb'sȯrp·shən ‚präs·əs }

absorption property [MATH] For set theory or for a Boolean algebra, the property that the union of a set, *A*, with the intersection of *A* and any set is equal to *A*, or the property that the intersection of *A* with the union of *A* and any set is also equal to *A*. { əb'sȯrp·shən ‚präp·ərd·ē }

absorption ratio *See* absorption coefficient. { əb'sȯrp·shən ‚rā·shō }

absorption refrigeration [MECH ENG] Refrigeration in which cooling is effected by the expansion of liquid ammonia into gas and absorption of the gas by water; the ammonia is reused after the water evaporates. { əb'sȯrp·shən rə‚frij·ə'rā·shən }

absorption spectrophotometer [SPECT] An instrument used to measure the relative intensity of absorption spectral lines and bands. Also known as difference spectrophotometer. { əb'sȯrp·shən ‚spek·trə·fə'täm·ə·dər }

absorption spectroscopy [SPECT] An instrumental technique for determining the concentration and structure of a substance by measuring the intensity of electromagnetic radiation it absorbs at various wavelengths. { əb'sȯrp·shən ‚spek'träs·kə·pē }

absorption spectrum [SPECT] A plot of how much radiation a sample absorbs over a range of wavelengths; the spectrum can be a plot of either absorbance or transmittance versus wavelength, frequency, or wavenumber. { əb'sȯrp·shən ‚spek·trəm }

absorption system [MECH ENG] A refrigeration system in which the refrigerant gas in the evaporator is taken up by an absorber and is then, with the application of heat, released in a generator. { əb'sȯrp·shən ‚sis·təm }

absorption test [IMMUNOL] Analysis of the antigenic components of bacterial cells and large macromolecules by a series of precipitation or agglutination reactions with specific antibodies. { əb'sȯrp·shən ‚test }

absorption tower [ENG] A vertical tube in which a rising gas is partially absorbed by a liquid in the form of falling droplets. Also known as absorption column. { əb'sȯrp·shən ‚tau·ər }

absorption trench [CIV ENG] A trench containing coarse aggregate about a distribution tile pipe through which septic-tank effluent may move beneath earth. { əb'sȯrp·shən ‚trench }

absorption tube [CHEM] A tube filled with a solid absorbent and used to absorb gases and vapors. { əb'sȯrp·shən ‚tüb }

absorption unit *See* sabin. { əb'sȯrp·shən ‚yü·nət }

absorption wavemeter [ELECTR] A frequency- or wavelength-measuring instrument consisting of a calibrated tunable circuit and a resonance indicator. { əb'sȯrp·shən 'wāv‚mēd·ər }

absorptive laws [MATH] Either of two laws satisfied by the operations, usually denoted ∪ and ∩, on a Boolean algebra, namely $a \cup (a \cap b) = a$ and $a \cap (a \cup b) = a$, where a and b are any two elements of the algebra; if the elements of the algebra are sets, then ∪ and ∩ represent union and intersection of sets. { əb'sȯrp·tiv ‚lȯz }

absorptive power *See* absorptivity. { əb'sȯrp·tiv ‚pau·ər }

absorptivity [ANALY CHEM] The constant a in the Beer's law relation $A = abc$, where A is the absorbance, b the path length, and c the concentration of solution. Also known as absorptive power. Formerly known as absorbency index; absorption constant; extinction coefficient. [THERMO] The ratio of the radiation absorbed by a surface to the total radiation incident on the surface. { əb‚sȯrp'tiv·əd·ē }

absorptivity-emissivity ratio [ASTROPHYS] In space applications, the ratio of absorptivity for solar radiation of a material to its infrared emissivity. Also known as A/E ratio. { əb‚sȯrp'tiv·əd·ē ‚ē·mə'siv·ə·tē ‚rā·shō }

abstinence syndrome [MED] A disturbance of metabolic equilibrium that occurs when a narcotic drug is withdrawn from the user. { 'abz·tə·nəns 'sin‚drōm }

abstract algebra [MATH] The study of mathematical systems consisting of a set of elements, one or more binary operations by which two elements may be combined to yield a third, and several rules (axioms) for the interaction of the elements and the operations; includes group theory, ring theory, and number theory. { 'abz·trakt 'al·jə·brə }

abstract automata theory [COMPUT SCI] The mathematical theory which characterizes automata by three sets: input signals, internal states, and output signals; and two functions: input functions and output functions. { 'abz·trakt ȯ'täm·ə·tə 'thē·ə·rē }

abstract data type [COMPUT SCI] A mathematical model which may be used to capture the essentials of a problem domain in order to translate it into a computer program; examples include queues, lists, stacks, trees, graphs, and sets. Abbreviated ADT. { 'abz·trakt 'dad·ə ‚tīp }

abstraction [HYD] **1.** The draining of water from a stream by another having more rapid corroding action. **2.** The part of precipitation that does not become direct runoff. { ab'strak·shən }

abstraction reaction [CHEM] A bimolecular chemical reaction in which an atom that is either neutral or charged is removed from a molecular entity. { ab'strak·shən re‚ak·shən }

abstract theory [SCI TECH] A theory in which a system is described without specifying a structure. { abz'trakt 'thē·ə·rē }

abstriction [MYCOL] In fungi, the cutting off of spores in hyphae by formation of septa followed by abscission of the spores, especially by constriction. { ab'strik·shən }

abT *See* gauss.

abterminal [BIOL] Referring to movement from the end toward the middle; specifically, describing the mode of electric current flow in a muscle. { ab'tərm·ən·əl }

abtesla *See* gauss. { ab'tes·lə }

A-B toxin [BIOCHEM] A toxin found in some bacteria and plants that is composed of two functionally distinct parts termed A, the enzymatically active portion, and B, the receptor binding portion; it can catalyze chemical reactions inside animal cells. { ¦ā¦bē 'täk·sən }

Abt track [CIV ENG] One of the cogged rails used for railroad tracking in mountains and so arranged that the cogs are not opposite one another on any pair of rails. { 'apt ‚trak }

Abukuma-type facies [PETR] A type of dynathermal regional metamorphism characterized by low pressure. { ab·ə'kü·mə ‚tīp 'fā·shēz }

abulia [PSYCH] Loss of ability to make decisions. { ā'bü·lē·ə }

abundance [GEOCHEM] The relative amount of a given element among other elements. [NUCLEO] *See* abundance ratio. { ə'bən·dəns }

abundance ratio [NUCLEO] The ratio of the number of

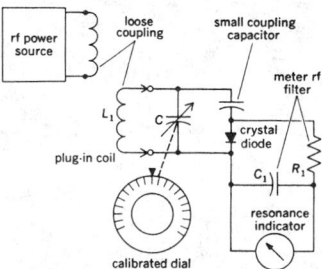

ABSORPTION WAVEMETER

Schematic diagram of inductance-capacitance type of absorption wavemeter (for frequencies between approximately 50 kilohertz and 1000 megahertz).

atoms of one isotope to the total number of atoms in a mixture of isotopes. Also known as abundance. { ə'bən·dəns 'rā·shō }

abundant number [MATH] A positive integer that is greater than the sum of all its divisors, including unity. Also known as redundant number. { ə'bən·dənt 'nəm·bər }

aburton [NAV ARCH] Of an object, having its length directed across a ship from side to side. { ə'bərt·ən }

abutment [CIV ENG] A surface or mass provided to withstand thrust; for example, end supports of an arch or a bridge. [MIN ENG] In a coal mine, the place where the natural ground material makes contact with the ends of an embankment. { ə'bət·mənt }

abutting joint [DES ENG] A joint which connects two pieces of wood in such a way that the direction of the grain in one piece is angled (usually at 90°) with respect to the grain in the other. { ə'bət·iŋ ˌjȯint }

abutting tenons [DES ENG] Two tenons inserted into a common mortise from opposite sides so that they contact. { ə'bət·iŋ 'ten·ənz }

abvolt [ELEC] The unit of electromotive force in the electromagnetic centimeter-gram-second system; 1 abvolt equals 10^{-8} volt in the absolute meter-kilogram-second system. Abbreviated aV. { 'ab,vōlt }

abvolt per centimeter [ELEC] In the electromagnetic centimeter-gram-second system of units, the unit of electric field strength. Abbreviated aV/cm. { 'ab,vōlt pər 'sen·tə,mēd·ər }

abwatt [ELEC] The unit of electrical power in the centimeter-gram-second system; 1 abwatt equals 1 watt in the absolute meter-kilogram-second system. { 'ab,wät }

abWb See maxwell.

abweber See maxwell. { 'ab,web·ər }

abyssal [GEOL] See plutonic. [OCEANOGR] Pertaining to the abyssal zone. { ə'bis·əl }

abyssal-benthic [OCEANOGR] Pertaining to the bottom of the abyssal zone. { ə'bis·əl 'ben·thik }

abyssal cave See submarine fan. { ə'bis·əl 'kāv }

abyssal fan See submarine fan. { ə'bis·əl 'fan }

abyssal floor [GEOL] The ocean floor, or bottom of the abyssal zone. { ə'bis·əl 'flȯr }

abyssal gap [GEOL] A gap in a sill, ridge, or rise that lies between two abyssal plains. { ə'bis·əl 'gap }

abyssal hill [GEOL] A hill 2000 to 3000 feet (600 to 900 meters) high and a few miles wide within the deep ocean. { ə'bis·əl 'hil }

abyssal injection [GEOL] The process of driving magmas, originating at considerable depths, up through deep-seated contraction fissures in the earth's crust. { ə'bis·əl in'jek·shən }

abyssal plain [GEOL] A flat, almost level area occupying the deepest parts of many of the ocean basins. { ə'bis·əl 'plān }

abyssal rock [GEOL] Plutonic, or deep-seated, igneous rocks. { ə'bis·əl 'räk }

abyssal theory [GEOL] A theory of the origin of ores involving the separation of ore silicates from the liquid stage during the cooling of the earth. { ə'bis·əl 'thē·ə·rē }

abyssal zone [OCEANOGR] The biogeographic realm of the great depths of the ocean beyond the limits of the continental shelf, generally below 1000 meters. { ə'bis·əl 'zōn }

abyssolith [GEOL] A molten mass of eruptive material passing up without a break from the zone of permanently molten rock within the earth. { ə'bis·ō,lith }

abyssopelagic [OCEANOGR] Pertaining to the open waters of the abyssal zone. { ə'bis·ō·pə'la·jik }

abzyme See catalytic antibody. { 'ab,zīm }

ac See alternating current.

aC See abcoulomb.

Ac See actinium; altocumulus cloud.

Ac_0 [MET] The temperature at which a magnetic change occurs in cementite; the Curie point of cementite.

Ac_1 [MET] The temperature at which austenite begins to be formed upon heating a steel.

Ac_2 [MET] The Curie point of ferrite.

Ac_3 [MET] The temperature at which the transformation of ferrite to austenite is completed upon heating a steel.

Ac_4 [MET] The temperature at which delta iron is formed from gamma iron upon heating a steel.

Ac_{cm} [MET] The temperature at which the solution of cementite in austenite is completed upon heating a hypereutectoid steel.

acacia gum See gum arabic. { ə'kā·shyə ˌgəm }

Acadian orogeny [GEOL] The period of formation accompanied by igneous intrusion that took place during the Middle and Late Devonian in the Appalachian Mountains. { ə'kād·ē·ən ȯr'äj·ə·nē }

Acala [BOT] A type of cotton indigenous to Mexico and cultivated in Texas, Oklahoma, and Arkansas. { ə'kal·ə }

acalculia [MED] A form of aphasia characterized by inability to do mathematical calculations. { ˌā·kal'kyü·lē·ə }

acalyculate [BOT] Lacking a calyx. { ¦ā·kə¦lik·yü,lāt }

Acalyptratae See Acalyptreatae. { ¦ā·kə'lip·trəd,ē }

Acalyptreatae [INV ZOO] A large group of small, two-winged flies in the suborder Cyclorrhapha characterized by small or rudimentary calypters. Also spelled Acalyptratae. { ¦ā·kə·lip'trē·ə,dē }

acantha [BIOL] A sharp spine; a spiny process, as on vertebrae. { ə'kan·thə }

Acanthaceae [BOT] A family of dicotyledonous plants in the order Scrophulariales distinguished by their usually herbaceous habit, irregular flowers, axile placentation, and dry, dehiscent fruits. { ə,kan'thās·ē,ē }

acanthaceous [BOT] Having sharp points or prickles; prickly. { ə,kan'thā·shəs }

Acantharia [INV ZOO] A subclass of essentially pelagic protozoans in the class Actinopodea characterized by skeletal rods constructed of strontium sulfate (celestite). { ə,kan'tha·rē·ə }

Acanthaster [INV ZOO] A genus of Indo-Pacific starfishes, including the crown-of-thorns, of the family Asteriidae; economically important as a destroyer of oysters in fisheries. { ə,kan'thas·tər }

acanthella [INV ZOO] A transitional larva of the phylum Acanthocephala in which rudiments of reproductive organs, lemnisci, a proboscis, and a proboscis receptacle are formed. { ə,kan'thel·ə }

acanthite [MINERAL] Ag_2S A blackish to lead-gray silver sulfide mineral, crystallizing in the orthorhombic system. { ə'kan·thīt }

acanthocarpous [BOT] Having spiny fruit. { ə,kan·thə'kär·pəs }

Acanthocephala [INV ZOO] The spiny-headed worms, a phylum of helminths; adults are parasitic in the alimentary canal of vertebrates. { ə,kan·thō'sef·ə·lə }

Acanthocheilonema perstans [INV ZOO] A tropical filarial worm, parasitic in humans. { ə,kan·thə,kī·lə'nē·mə 'pərs·tənz }

acanthocheilonemiasis [MED] A parasitic infection of humans caused by the filarial nematode *Acanthocheilonema perstans*. { ə,kan·thə,kī·lə·ne'mī·ə·səs }

acanthocladous [BOT] Having spiny branches. { ə,kan·thə'klad·əs }

acanthocyte [PATH] A crenated red cell which has a distinctive spiky outline. { ə'kan·thō,sīt }

acanthocytosis [MED] A disorder of erythrocytes in which spiny projections appear on the blood cells. { ə,kan·thə,sī'tō·səs }

Acanthodes [PALEON] A genus of Carboniferous and Lower Permian eellike acanthodian fishes of the family Acanthodidae. { ə,kan'thō·dēz }

Acanthodidae [PALEON] A family of extinct acanthodian fishes in the order Acanthodiformes. { ə,kan'thō·də,dē }

Acanthodiformes [PALEON] An order of extinct fishes in the class Acanthodii having scales of acellular bone and dentine, one dorsal fin, and no teeth. { ə,kan·thō·də'fȯr,mēz }

Acanthodii [PALEON] A class of extinct fusiform fishes, the first jaw-bearing vertebrates in the fossil record. { ə,kan'thō·dē,ī }

acanthoid [BIOL] Shaped like a spine. { ə'kan·thȯid }

acanthoma [MED] A tumor composed of epidermal cells. { ,ak·ən'thō·mə }

Acanthometrida [INV ZOO] An order of marine protozoans in the subclass Acantharia with 20 or less skeletal rods. { ə,kan·thə'met·rə·də }

Acanthophis antarcticus [VERT ZOO] The death adder, a venomous snake found in Australia and New Guinea; venom is neurotoxic. { ə'kan·thə·fəs ant'ärk·tə·kəs }

Acanthophractida [INV ZOO] An order of marine protozoans in the subclass Acantharia; skeleton includes a latticework shell and skeletal rods. { ə,kan·thə'frak·tə·də }

acanthopodia [INV ZOO] The long subpseudopodia of

ACANTHELLA

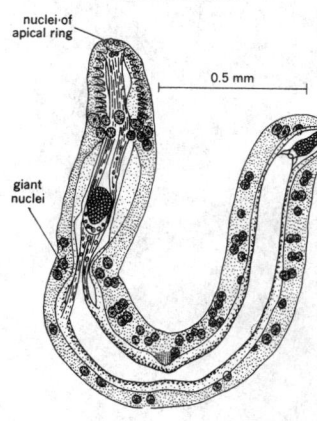

A stage in the life history of *Moniliformis dubius*, a helminth, with the acanthella dissected from its enveloping sheath.

ACANTHODES

A lateral view of *Acanthodes* species, about 30 centimeters long. (*After D. M. S. Watson*)

amoebas of the suborder Acanthopodina, order Amoebida. { ə¸kan·thə'pōd·ē·ə }

acanthopodous [BOT] Having a spiny or prickly petiole or peduncle. { ¦ā¸kan¦thä·pə·dəs }

acanthopore [PALEON] A tubular spine in some fossil bryozoans. { ə'kan·thə¸pór }

Acanthopteri [VERT ZOO] An equivalent name for the Perciformes. { a¸kan'thäp·tə·rī }

Acanthopterygii [VERT ZOO] An equivalent name for the Perciformes. { a¸kan¸thäp·tə'rē·jē·ī }

acanthosis [MED] Any thickening of the prickle-cell layer of the epidermis; associated with many skin diseases. { a¸kan'thō·səs }

acanthosoma [INV ZOO] The last primitive larval stage, the mysis, in the family Sergestidae. { ə¸kan·thə'sō·mə }

Acanthosomatidae [INV ZOO] A small family of insects in the order Hemiptera. { ə¸kan·thə·sə'mad·ə¸dē }

acanthosphere [BOT] A specialized ciliated body in *Nitella* cells. { ə'kan·thə¸sfir }

acanthostegous [INV ZOO] Being overlaid with two series of spines, as the ovicell or ooecium of certain bryozoans. { ə¸kan·thə'steg·əs }

acanthozooid [INV ZOO] A specialized individual in a bryozoan colony that secretes tubules which project as spines above the colony's outer surface. { ə¸kan·thə'zō·óid }

Acanthuridae [VERT ZOO] The surgeonfishes, a family of perciform fishes in the suborder Acanthuroidei. { ə¸kan'thù·rə·dē }

Acanthuroidei [VERT ZOO] A suborder of chiefly herbivorous fishes in the order Perciformes. { ə¸kan·thə'róid·ē¸ī }

acanthus [ARCH] A sculptured ornamentation representing the leaves of an *Acanthus*, a Mediterranean prickly herb. { ə'kan·thəs }

acapau [MATER] A type of ebony obtained from *Voucapapoua americana* in the Amazon valley; used for fine inlay work. Also known as partridge wood. Properly spelled acapu. { ¦äk·ə'paù }

acapnia [MED] Absence of carbon dioxide in the blood and tissues. { ə'kap·nē·ə }

acapu *See* acapau. { ak·ə'pü }

Acari [INV ZOO] The equivalent name for Acarina. { 'a·kə·rē }

acariasis [MED] Any skin disease resulting from infestation with acarids or mites. { a·kə'rī·ə·səs }

acaricide [MATER] A pesticide used to destroy mites on domestic animals, crops, and humans. Also known as miticide. { ə'kar·ə¸sīd }

Acaridiae [INV ZOO] A group of pale, weakly sclerotized mites in the suborder Sarcoptiformes, including serious pests of stored food products and skin parasites of warm-blooded vertebrates. { ¸a·kə'rid·ē¸ē }

Acarina [INV ZOO] The ticks and mites, a large order of the class Arachnida, characterized by lack of body demarcation into cephalothorax and abdomen. { ¸a·kə'rēn·ə }

acaroid resin [ORG CHEM] A gum resin from aloelike trees of the genus *Xanthorrhoea* in Australia and Tasmania; used in varnishes and inks. Also known as gum accroides; yacca gum. { 'a·kə¸róid 'rez·ən }

acarology [INV ZOO] A branch of zoology dealing with the mites and ticks. { ¸a·kə'rä·lə·je }

acarophily [ECOL] A symbiotic relationship between plants and mites. { ¦a·kə·rō¦fil·ē }

acarophobia [PSYCH] Abnormal fear of mites. { ¸a·kə·rə'fōb·ē·ə }

acarpellous [BOT] Lacking carpels. { ā'kär·pə·ləs }

acarpous [BOT] Not producing fruit. { ā'kär·pəs }

ACAS *See* airborne collision avoidance system.

acatalasemia [MED] Lack of catalase in the blood. { ¦ā·ka·tə·lə'sē·mē·ə }

acatalasia [MED] Congenital absence of the enzyme catalase. { ¦ā·ka·tə'lā·zhē·ə *or* zhə }

acatamathesia [MED] 1. Inability to understand conversation. 2. Morbid blunting or deterioration of the senses, as in mental deafness and blindness. { ¦ā·ka·tə·mə'thē·zhē·ə *or* zhə }

acaulous [BOT] 1. Lacking a stem. 2. Being apparently stemless but having a short underground stem. { ¦ā'kól·əs }

acaustobiolith [PETR] A noncombustible organic rock, or one formed by organic accumulation of minerals. { ¦ā¸kós·tə'bī·ə·lith }

acaustophytolith [PETR] An acaustobiolith resulting from plant activity, such as a pelagic ooze that contains diatoms. { ¦ā¸kós·tə'fīd·ə¸lith }

accelerated aging [ENG] Hastening the deterioration of a product by a laboratory procedure in order to determine long-range storage and use characteristics. { ak'sel·ə¸rād·əd 'āj·iŋ }

accelerated erosion [GEOL] Soil erosion that occurs more rapidly than soil horizons can form from the parent regolith. { ak'sel·ər¸ā·dəd i'rō·zhən }

accelerated graphics port [COMPUT SCI] A personal computer graphics bus that transfers data at a greater rate than a PCI bus. { ak¸sel·ə¸rād·əd 'graf·iks ¸pórt }

accelerated hypertension *See* malignant hypertension. { ak'sel·ər¸ā·dəd 'hī·pər¸ten·shən }

accelerated life test [ENG] Operation of a device, circuit, or system above maximum ratings to produce premature failure; used to estimate normal operating life. { ak'sel·ər¸ā·dəd 'līf ¸test }

accelerated test [ELEC] A test of the serviceability of an electric cable in use for some time by applying twice the voltage normally carried. { ak'sel·ər¸ā·dəd 'test }

accelerated weathering [ENG] A laboratory test used to determine, in a short period of time, the resistance of a paint film or other exposed surface to weathering. { ak'sel·ər¸ā·dəd 'weth·ər·iŋ }

accelerating agent [MATER] 1. A substance which increases the speed of a chemical reaction. 2. A compound which hastens and improves the curing of rubber. { ak'sel·ər¸ād·iŋ 'ā'jənt }

accelerating electrode [ELECTR] An electrode used in cathode-ray tubes and other electron tubes to increase the velocity of the electrons that contribute the space current or form a beam. { ak'sel·ər¸ād·iŋ i'lek¸trōd }

accelerating incentive *See* differential piece-rate system. { ak'sel·ər¸ād·iŋ in'sen·tiv }

accelerating potential [ELECTR] The energy potential in electron-beam equipment that imparts additional speed and energy to the electrons. { ak'sel·ər¸ād·iŋ pə'ten·shəl }

accelerating relay [ELEC] Any relay that is used to assist in starting a motor or increasing its speed. { ak'sel·ə¸rād·iŋ 'rē¸lā }

acceleration [MECH] The rate of change of velocity with respect to time. { ak¸sel·ə'rā·shən }

acceleration analysis [MECH ENG] A mathematical technique, often done graphically, by which accelerations of parts of a mechanism are determined. { ak¸sel·ə'rā·shən ə¸nal·ə·səs }

acceleration error [NAV] That error resulting from change in a craft's velocity vector: specifically, either the deviation of an aircraft magnetic compass caused by the action of the vertical component of the earth's magnetic field on the compass magnets when the compass card is thrown off level by accelerations of the aircraft; or the deflection of the apparent vertical, as indicated by an artificial horizon, due to acceleration. { ak¸sel·ə'rā·shən ¸er·ər }

acceleration-error constant [CONT SYS] The ratio of the acceleration of a controlled variable of a servomechanism to the actuating error when the actuating error is constant. { ak¸sel·ə'rā·shən 'er·ər 'kän·stənt }

acceleration feedback [AERO ENG] The use of accelerometers strategically located within the body of a missile so that they sense body accelerations during flight and interact with another device on board the missile or with a control center on the ground or in an airplane to keep the missile's speed within design limits. { ak¸sel·ə'rā·shən 'fēd¸bak }

acceleration globulin [BIOCHEM] A globulin that acts to accelerate the conversion of prothrombin to thrombin in blood clotting; found in blood plasma in an inactive form. { ak¸sel·ə'rā·shən 'gläb·yə·lən }

acceleration measurement [MECH] The technique of determining the magnitude and direction of acceleration, including translational and angular acceleration. [NAV] A fundamental measurement required for the operation of the inertial navigator. { ak¸sel·ə'rā·shən 'mezh·ər·mənt }

acceleration mechanisms [ASTROPHYS] The ways in which cosmic-ray and solar-flare particles may have acquired their high energies. { ak¸sel·ə'rā·shən 'mek·ə·niz·əmz }

acceleration of free fall See acceleration of gravity. { ak,sel·ə'rā·shən əv 'frē ,fȯl }

acceleration of gravity [MECH] The acceleration imparted to bodies by the attractive force of the earth; has an international standard value of 980.665 cm/s² but varies with latitude and elevation. Also known as acceleration of free fall; apparent gravity. { ak,sel·ə'rā·shən əv 'grav·ə·dē }

acceleration phase [MICROBIO] A period of increasing growth which is followed by the log phase in a culture of microbes. { ak·sel·ə'rā·shən ,fāz }

acceleration potential [FL MECH] The sum of the potential of the force field acting on a fluid and the ratio of the pressure to the fluid density; the negative of its gradient gives the acceleration of a point in the fluid. { ak,sel·ə'rā·shən pə'ten·shəl }

acceleration signature [IND ENG] A printed record that shows the pattern of acceleration and deceleration of an anatomical reference point in the performance of a task. { ak,sel·ə'rā·shən 'sig·nə·chər }

acceleration stress [MED] The effect of an increase in gravitational force upon human physiology and behavior, particularly during takeoff and reentry in space flight. { ak,sel·ə'rā·shən ,stres }

acceleration switch [ELEC] A switch that opens or closes in the presence of acceleration that exceeds a certain value. { ak,sel·ə'rā·shən ,swich }

acceleration time [COMPUT SCI] The time required for a magnetic tape transport or any other mechanical device to attain its operating speed. { ak,sel·ə'rā·shən ,tīm }

acceleration tolerance [ENG] The degree to which personnel or equipment withstands acceleration. [PHYSIO] The maximum g forces an individual can withstand without losing control or consciousness. { ak,sel·ə'rā·shən 'täl·ər·əns }

acceleration voltage [ELECTR] The voltage between a cathode and accelerating electrode of an electron tube. { ak,sel·ə'rā·shən 'vōl·təj }

accelerator [GRAPHICS] The constituent of a photographic developer that speeds up development rate. Also known as activator. [MATER] **1.** Any substance added to stucco, plaster, mortar, concrete, cement, and so on to hasten the set. **2.** In the vulcanization process, a substance, added with a curing agent, to speed processing and enhance physical characteristics of a vulcanized material. [MECH ENG] A device for varying the speed of an automotive vehicle by varying the supply of fuel. [PHYS] See particle accelerator. { ak'sel·ə,rād·ər }

accelerator catalyst [MATER] A catalyst that increases the rate of a chemical reaction. { ak'sel·ə,rād·ər 'kad·ə·list }

accelerator jet [MECH ENG] The jet through which the fuel is injected into the incoming air in the carburetor of an automotive vehicle with rapid demand for increased power output. { ak'sel·ə,rād·ər ,jet }

accelerator linkage [MECH ENG] The linkage connecting the accelerator pedal of an automotive vehicle to the carburetor throttle valve or fuel injection control. { ak'sel·ə,rād·ər ,liŋ·kij }

accelerator mass spectrometer [SPECT] A combination of a mass spectrometer and an accelerator that can be used to measure the natural abundances of very rare radioactive isotopes. { ak'sel·ə,rād·ər ¦mas spek'träm·əd·ər }

accelerator pedal [MECH ENG] A pedal that operates the carburetor throttle valve or fuel injection control of an automotive vehicle. { ak'sel·ə,rād·ər ,ped·əl }

accelerator pump [MECH ENG] A small cylinder and piston controlled by the throttle of an automotive vehicle so as to provide an enriched air-fuel mixture during acceleration. { ak'sel·ə,rād·ər ,pəmp }

acceleratory reflex [PHYSIO] Any reflex originating in the labyrinth of the inner ear in response to a change in the rate of movement of the head. { ak'sel·ə·rə,tȯr·ē 'rē·fleks }

accelerogram [ENG] A record made by an accelerograph. { ak'sel·ə·rə,gram }

accelerograph [ENG] An accelerometer having provisions for recording the acceleration of a point on the earth during an earthquake or for recording any other type of acceleration. { ak'sel·ə·rə,graf }

accelerometer [ENG] An instrument which measures acceleration or gravitational force capable of imparting acceleration. { ak,sel·ə'räm·əd·ər }

accelerometry [IND ENG] The quantitative determination of acceleration and deceleration in the entire human body or a part of the body in the performance of a task. { ak,sel·ə'räm·ə·drē }

accelofilter [CHEM] A filtration device that uses a vacuum or pressure to draw or force the liquid through the filter to increase the rate of filtration. { ak'sel·ō,fil·tər }

accent lighting [CIV ENG] Directional lighting which highlights an object or attracts attention to a particular area. { 'ak·sent ,līd·iŋ }

accentuation [ELECTR] The enhancement of signal amplitudes in selected frequency bands with respect to other signals. { ak,sen·chə'wā·shən }

accentuator [ELECTR] A circuit that provides for the first part of a process for increasing the strength of certain audio frequencies with respect to others, to help these frequencies override noise or to reduce distortion. Also known as accentuator circuit. [MATER] A material that acts to increase the selectivity or intensity of a stain. { ak'sen·chə,wād·ər }

accentuator circuit See accentuator. { ak'sen·chə,wād·ər 'sər·kət }

accept [COMPUT SCI] A data transmission statement which is used in FORTRAN when the computer is in conversational mode, and which enables the programmer to input, through the teletypewriter, data the programmer wishes stored in memory. { ak'sept }

acceptability [ENG] State or condition of meeting minimum standards for use, as applied to methods, equipment, or consumable products. { ak,sep·tə'bil·ə·dē }

acceptable quality level [IND ENG] The maximum percentage of defects that has been determined tolerable as a process average for a sampling plan during inspection or test of a product with respect to economic and functional requirements of the item. Abbreviated AQL. { ak¦sep·tə·bəl 'kwäl·ə·dē ,lev·əl }

acceptable reliability level [IND ENG] The required level of reliability for a part, system, device, and so forth; may be expressed in a variety of terms, for example, number of failures allowable in 1000 hours of operating life. Abbreviated ARL. { ak¦sep·tə·bəl rə,lī·ə'bil·ə·dē ,lev·əl }

acceptable risk [GEOPHYS] In seismology, that level of earthquake effects which is judged to be of sufficiently low social and economic consequence, and which is useful for determining design requirements in structures or for taking certain actions. { ak¦sep·tə·bəl 'risk }

acceptance criteria [IND ENG] Standards of judging the acceptability of manufactured items. { ak'sep·təns krī'tēr·ē·ə }

acceptance number [IND ENG] The maximum allowable number of defective pieces in a sample of specified size. { ak'sep·təns ,nəm·bər }

acceptance sampling [IND ENG] Taking a sample from a batch of material to inspect for determining whether the entire lot will be accepted or rejected. { ak'sep·təns ,sam·pliŋ }

acceptance test [IND ENG] A test used to determine conformance of a product to design specifications, as a basis for its acceptance. { ak'sep·təns ,test }

accepted indicator [NAV] An airborne indicator which has been proven capable of accurate and reliable measurement. { ak'sep·təd 'in·də,kād·ər }

acceptor [CHEM] **1.** A chemical whose reaction rate with another chemical increases because the other substance undergoes another reaction. **2.** A species that accepts electrons, protons, electron pairs, or molecules such as dyes. [CHEM ENG] A calcined carbonate used to absorb the carbon dioxide evolved during a coal gasification process. [SOLID STATE] An impurity element that increases the number of holes in a semiconductor crystal such as germanium or silicon; aluminum, gallium, and indium are examples. Also known as acceptor impurity; acceptor material. { ak'sep·tər }

acceptor atom [SOLID STATE] An atom of a substance added to a semiconductor crystal to increase the number of holes in the conduction band. { ak'sep·tər 'ad·əm }

acceptor circuit [ELECTR] A series-resonant circuit that has a low impedance at the frequency to which it is tuned and a higher impedance at all other frequencies. { ak'sep·tər 'sər·kət }

acceptor impurity See acceptor. { ak'sep·tər im'pyùr·ə·dē }

acceptor level [SOLID STATE] An energy level in a semiconductor that results from the presence of acceptor atoms. { ak'sep·tər ,lev·əl }

acceptor material *See* acceptor. { ak'sep·tər mə'tir·ē·əl }

access [CIV ENG] Freedom, ability, or the legal right to pass without obstruction from a given point on earth to some other objective, such as the sea or a public highway. [COMPUT SCI] The reading of data from storage or the writing of data into storage. { 'ak,ses }

access arm [COMPUT SCI] The mechanical device which positions the read/write head on a magnetic storage unit. { 'ak,ses ,ärm }

access code [COMMUN] **1.** Numeric identification for internetwork or facility switching. **2.** The preliminary digits that a user must dial to be connected through an automatic PBX to the serving switching center. [COMPUT SCI] A sequence of characters which a user must enter into a terminal in order to use a computer system. { 'ak,ses ,kōd }

access control [COMPUT SCI] A restriction on the operations that a user of a computer system may perform on files and other resources of the system. { 'ak,ses kən,trōl }

access-control list [COMPUT SCI] A column of an access matrix, containing the access rights of various users of a computer system to a given file or other resource of the system. { 'ak,ses kən,trōl ,list }

access-control mechanism *See* reference monitor. { 'ak,ses kən'trōl ,me·kə·ni·zəm }

access-control register [COMPUT SCI] A storage device which controls the word-by-word transmission over a given channel. { 'ak,ses kən'trōl ,rej·ə·stər }

access-control words [COMPUT SCI] Permanently wired instructions channeling transmitted words into reserved locations. { 'ak,ses kən'trōl ,wərdz }

access door [BUILD] A provision for access to concealed plumbing or other equipment without disturbing the wall or fixtures. { 'ak,ses ,dór }

access eye [CIV ENG] A threaded plug fitted into bends and junctions of drain, waste, or soil pipes to provide access when a blockage occurs. [ENG] *See* cleanout. { 'ak,ses ,ī }

access flooring *See* raised flooring. { 'ak,ses ,flor·iŋ }

access gap *See* memory gap. { 'ak,ses ,gap }

access hole *See* manhole. { 'ak,ses ,hōl }

accessibility condition [MATH] The condition that any state of a finite Markov chain can be reached from any other state. { ak,ses·ə'bil·əd·ē kən,dish·ən }

access line [COMMUN] Four-wire circuit between a subscriber or a local PBX to the serving switching center. { 'ak,ses ,līn }

access management [COMPUT SCI] The use of techniques to allow various components of a computer's operating system to be used only by authorized personnel. { 'ak,ses ,man·ij·mənt }

access matrix [COMPUT SCI] A method of representing discretionary authorization information, with rows representing subjects or users of the system, columns corresponding to objects or resources of the system, and cells (intersections of rows and columns) composed of allowable operations that a subject may apply to an object. { 'ak,ses ,mā·triks }

access mechanism [COMPUT SCI] The mechanism of positioning reading or writing heads onto the required tracks of a magnetic disk. { 'ak,ses 'mek·ə,niz·əm }

access method [COMMUN] The procedures required to obtain access to a communications network. [COMPUT SCI] A set of programming routines which links programs and the data that these programs transfer into and out of memory. { 'ak,ses ,meth·əd }

access mode [COMPUT SCI] A programming clause in COBOL which is required when using a random-access device so that a specific record may be read out of or written into a mass storage bin. { 'ak,ses ,mōd }

accessorius [ANAT] Any muscle that reinforces the action of another. { ak·sə'sòr·ē·əs }

accessory [MECH ENG] A part, subassembly, or assembly that contributes to the effectiveness of a piece of equipment without changing its basic function; may be used for testing, adjusting, calibrating, recording, or other purposes. { ak'ses·ə·rē }

accessory bud [BOT] An embryonic shoot occurring above or to the side of an axillary bud. Also known as supernumerary bud. { ak'ses·ə·rē ,bəd }

accessory cell [BOT] A morphologically distinct epidermal cell adjacent to, and apparently functionally associated with, guard cells on the leaves of many plants. [IMMUNOL] Any nonlymphocytic cell that helps in the induction of the immune response by presenting antigen to a helper T lymphocyte. { ak'ses·ə·rē ,sel }

accessory chromosome *See* supernumerary chromosome. { ak'ses·ə·rē 'krōm·ə,sōm }

accessory cloud [METEOROL] A cloud form that is dependent, for its formation and continuation, upon the existence of one of the major cloud genera; may be an appendage of the parent cloud or an immediately adjacent cloudy mass. { ak'ses·ə·rē ,klaůd }

accessory ejecta [GEOL] Pyroclastic material formed from solidified volcanic rocks that are from the same volcano as the ejecta. { ak'ses·ə·rē i'jek·tə }

accessory element *See* trace element. { ak'ses·ə·rē 'el·ə·mənt }

accessory fruit [BOT] A fruit derived from the ovary and its contents as well as other parts of the flower; it is usually derived from an inferior (inserted below other floral parts) ovary. Also known as inferior fruit. { ak'ses·ə·rē ,früt }

accessory gland [ANAT] A mass of glandular tissue separate from the main body of a gland. [INV ZOO] A gland associated with the male reproductive organs in insects. { ak'ses·ə·rē ,gland }

accessory mineral [MINERAL] A minor mineral in an igneous rock that does not affect its general character. { ak'ses·ə·rē ,min·rəl }

accessory movement *See* synkinesia. { ak'ses·ə·rē ,müv·mənt }

accessory nerve [NEURO] The eleventh cranial nerve in tetrapods, a paired visceral motor nerve; the bulbar part innervates the larynx and pharynx, and the spinal part innervates the trapezius and sternocleidomastoid muscles. { ak'ses·ə·rē ,nərv }

accessory pigments [BIOCHEM] Light-absorbing pigments, including carotenoids and phycobilins, which complement chlorophyll in plants, algae, and bacteria by trapping light energy for photosynthesis. { ak'ses·ə·rē 'pig·məns }

accessory plate [OPTICS] Thin plate of quartz, gypsum, or mica used with a petrological microscope to modify the effects of polarized light and intensify qualities in translucent minerals. { ak'ses·ə·rē ,plāt }

accessory sexual characters [ANAT] Those structures and organs (excluding the gonads) composing the genital tract and including accessory glands and external genitalia. { ak'ses·ə·rē 'seksh·ə·wəl 'kar·ik·tərz }

accessory species [ECOL] A species comprising 25-50% of a community. { ak'ses·ə·rē 'spē·shēz }

access privileges [COMPUT SCI] The extent to which a user of a computer in a network is allowed to use and read, write to, and execute files in other computers in the network. { 'ak,ses ,priv·ə·ləj·əs }

access protocol [COMMUN] A set of rules observed by all nodes in a local-area network so that one node can get the attention of another and its data packet can be transferred, and so that no two data packets can be simultaneously transmitted over the same medium. { 'ak,ses ,prōd·ə,kól }

access provider *See* service provider. { 'ak,ses prə,vīd·ər }

access road [CIV ENG] A route, usually paved, that enables vehicles to reach a designated facility expeditiously. { 'ak·ses ,rōd }

access time [COMPUT SCI] The time period required for reading out of or writing into the computer memory. { 'ak·ses ,tīm }

access tunnel [CIV ENG] A tunnel provided for an access road. { 'ak·ses ,tən·əl }

access type [COMPUT SCI] One of the allowable operations that a given user of a computer system governed by access controls may perform on a file or other resource of the system, such as own, read, write, or execute. { 'ak·ses ,tīp }

accident [HYD] An interruption in a river that interferes with, or sometimes stops, the normal development of the river system. { 'ak·sə,dent }

accidental ejecta [GEOL] Pyroclastic rock formed from preexisting nonvolcanic rocks or from volcanic rocks unrelated to the erupting volcano. { 'ak·sə'den·təl i'jek·tə }

accidental error [SCI TECH] In experimental observations, an error which does not always recur when an observation is repeated under the same conditions. { 'ak·sə'den·təl 'er·ər }

accidental inclusion See xenolith. { ¦ak·sə¦den·təl in'klü·zhən }

accidental point [GRAPHICS] The vanishing point of a group of lines in a perspective drawing that are parallel neither to the line of sight nor to the horizon. { ¦ak·sə¦den·təl 'pȯint }

accidental species [ECOL] A species which constitutes less than one-fourth of the population of a stand. { ¦ak·sə¦den·təl 'spē·shēz }

accidental whorl [FOREN SCI] A type of whorl fingerprint pattern which is a combination of two different types of pattern, with the exception of the plain arch, with two or more deltas; or a pattern which possesses some of the requirements for two or more different types; or a pattern which conforms to none of the definitions; in accidental whorl tracing three types appear: an outer (O), inner (I), or meeting (M). { ¦ak·sə¦den·təl 'wərl }

accident block [GEOL] A solid chip of rock broken off from the subvolcanic basement and ejected from a volcano. { 'ak·sə,dent ,bläk }

accident-cause code [IND ENG] Sponsored by the American Standards Association, the code that classifies accidents under eight defective working conditions and nine improper working practices. { 'ak·sə,dent ¦kȯz ,kōd }

accident frequency rate [IND ENG] The number of all disabling injuries per million worker-hours of exposure. { 'ak·sə,dent 'frē·kwən·sē ,rāt }

accident-prone [MED] Predisposed to sustain more accidents than others exposed to the same hazard. { 'ak·sə,dent ,prōn }

accident severity rate [IND ENG] The number of worker-days lost as a result of disabling injuries per thousand worker-hours of exposure. { 'ak·sə,dent sə'ver·əd·ē ,rāt }

Accipitridae [VERT ZOO] The diurnal birds of prey, the largest and most diverse family of the order Falconiformes, including hawks, eagles, and kites. { ,ak·sə'pi·trə,dē }

acclimated microorganism [ECOL] Any microorganism that is able to adapt to environmental changes such as a change in temperature, or a change in the quantity of oxygen or other gases. { ə'klīm·əd·əd ,mī·krō'ȯr·gə·niz·əm }

acclimation See acclimatization. { ,ak·lə'mā·shən }

acclimatization [BIOL] Physiological, emotional, and behavioral adjustment by an individual to changes in the environment. [EVOL] Adaptation of a species or population to a changed environment over several generations. Also known as acclimation. { ə,klī·mə·tə'zā·shən }

acclivity [GEOL] A slope that is ascending from a reference point. { ə'kliv·əd·ē }

aCcm See abcoulomb centimeter.

aC/cm² See abcoulomb per square centimeter.

aC/cm³ See abcoulomb per cubic centimeter.

accolade [ARCH] Decorative molding in which two ogee curves meet centrally over the top of a window or door. { 'ak·ə,läd or läd }

accommodation [CONT SYS] Any alteration in a robot's motion in response to the robot's environment; it may be active or passive. [ECOL] A population's location within a habitat. [MAP] The limits or range within which a stereo-plotting instrument is capable of operating. [PHYSIO] A process in most vertebrates whereby the focal length of the eye is changed by automatic adjustment of the lens to bring images of objects from various distances into focus on the retina. { ə,käm·ə'dā·shən }

accommodation coefficient [STAT MECH] The ratio of the average energy actually transferred between a surface and impinging gas molecules scattered by the surface, to the average energy which would theoretically be transferred if the impinging molecules reached complete thermal equilibrium with the surface. { ə,käm·ə'dā·shən ,kō·ə'fish·ənt }

accommodation ladder [NAV ARCH] A light ladder or similar structure, usually portable, hung over a ship's side at the gangway to permit access to small boats. { ə,käm·ə'dā·shən ,lad·ər }

accommodation reflex [PHYSIO] Changes occurring in the eyes when vision is focused from a distant to a near object; involves pupil contraction, increased lens convexity, and convergence of the eyes. { ə,käm·ə'dā·shən 'rē·fleks }

accommodation rig [MIN ENG] A part of an offshore drilling complex containing areas for sleeping, storing supplies, and recreation. { ə,käm·ə'dā·shən ,rig }

accommodation time [ELECTR] The time from the production of the first electron to the production of a steady electric discharge in a gas. { ə,käm·ə'dā·shən ,tīm }

accordant [GEOL] Pertaining to topographic features that have nearly the same elevation. { ə'kȯrd·ənt }

accordant fold [GEOL] One of several folds that are similarly oriented. { ə'kȯrd·ənt ,fōld }

accordant drainage [HYD] Flow of surface water that follows the dip of the strata over which it flows. Also known as concordant drainage. { ə'kȯrd·ənt 'drān·ij }

accordant summit level [GEOL] A hypothetical horizontal plane that can be drawn over a broad region connecting mountain summits of similar elevation. { ə'kȯrd·ənt 'səm·ət ,lev·əl }

accordion cable [ELEC] A flat, multiconductor cable prefolded into a zigzag shape and used to make connections to movable equipment such as a chassis mounted on pullout slides. { ə'kȯrd·ē·ən 'kā·bəl }

accordion door [BUILD] A door that folds and unfolds like an accordion when it is opened and closed. { ə'kȯrd·ē·ən ,dȯr }

accordion fold [GRAPHICS] In a binding operation, two or more parallel folds of printed sheets. { ə'kȯrd·ē·ən ,fōld }

accordion partition [BUILD] A movable, fabric-faced partition which is fitted into an overhead track and folds like an accordion. { ə'kȯrd·ē·ən pər'tish·ən }

accordion roller conveyor [MECH ENG] A conveyor with a flexible latticed frame which permits variation in length. { ə'kȯrd·ē·ən 'rōl·ər kən'vā·ər }

accounting package [COMPUT SCI] A set of special routines that allow collection of information about the usage level of various components of a computer system by each production program. { ə'kaunt·iŋ 'pak·ij }

accouplement [ARCH] A pair of elements of a structure that are very close or touching, such as two columns. { ə'kəp·lə·mənt }

accrescent [BOT] Growing continuously with age, especially after flowering. { ə'krēs·ənt }

accretion [ASTRON] A process in which a star gathers molecules of interstellar gas to itself by gravitational attraction. [CIV ENG] Artificial buildup of land due to the construction of a groin, breakwater, dam, or beach fill. [GEOL] **1.** Gradual buildup of land on a shore due to wave action, tides, currents, airborne material, or alluvial deposits. **2.** The process whereby stones or other inorganic masses add to their bulk by adding particles to their surfaces. Also known as aggradation. **3.** See accretion tectonics. [METEOROL] The growth of a precipitation particle by the collision of a frozen particle (ice crystal or snowflake) with a supercooled liquid droplet which freezes upon contact. { ə'krē·shən }

accretionary lapilli See mud ball. { ə'krē·shən,er·ē lə'pi·lē }

accretionary lava ball [GEOL] A rounded ball of lava that occurs on the surface of an aa lava flow. { ə'krē·shən,er·ē 'lä·və ,bȯl }

accretionary limestone [PETR] A type of limestone formed by the slow accumulation of organic remains. { ə'krē·shən,er·ē 'līm·stōn }

accretionary ridge [GEOL] A beach ridge located inland from the modern beach, indicating that the coast has been built seaward. { ə'krē·shən,er·ē ,rij }

accretion disk [ASTRON] A viscous structure consisting of gas lost by a red giant or supergiant flowing around a companion main-sequence star or compact object (white dwarf, neutron star, or black hole). { ə'krē·shən ,disk }

accretion hypothesis [ASTRON] Any hypothesis which assumes that the earth originated by the gradual addition of solid bodies, such as meteorites, that were formerly revolving about the sun but were drawn by gravitation to the earth. { ə'krē·shən hī'päth·ə·səs }

accretion line [HISTOL] A microscopic line on a tooth, marking the addition of a layer of enamel or dentin. { ə'krē·shən ,līn }

accretion tectonics [GEOL] The bringing together, or suturing, of terranes; regarded by many geologists as an important mechanism of continental growth. Also known as accretion. { ə'krē·shən tek'tän·iks }

accretion theory [ASTRON] A theory that the solar system originated from vortices in a disk-shaped mass. { ə'krē·shən 'thē·ə·rē }

ACCIDENTAL WHORL

A reproduction of an I tracing of an accidental whorl. *(Federal Bureau of Investigation)*

accretion topography [GEOL] Topographic features built by accumulation of sediment. { ə'krē·shən tä'päg·rə·fē }

accretion vein [GEOL] A type of vein formed by the repeated filling of channels followed by their opening because of the development of fractures in the zone undergoing mineralization. { ə'krē·shən ˌvān }

accretion zone [GEOL] Any beach area undergoing accretion. { ə'krē·shən ˌzōn }

accretive operator [MATH] A linear operator *T* defined on a subspace *D* of a Hilbert space which satisfies the following condition: the real part of the inner product of *Tu* with *u* is nonnegative for all *u* belonging to *D*. { ə¦krēd·iv 'äp·ə,rād·ər }

accumbent [BOT] Describing an organ that leans against another; specifically referring to cotyledons having their edges folded against the hypocotyl. { ə'kəm·bənt }

accumulated discrepancy [ENG] The sum of the separate discrepancies which occur in the various steps of making a survey. { ə'kyü·myə,lād·əd də'skrep·ən·sē }

accumulated divergence [MAP] In making a map, the algebraic sum of the divergences for the sections of a line of levels, from the beginning of the line to any section end at which it is desired to compute the total divergence. { ə'kyü·myə,lād·əd də'vər·jəns }

accumulated dose [PHYSIO] The total amount of radiation absorbed by an organism as a result of exposure to radiation. { ə'kyü·myə,lād·əd 'dōs }

accumulated temperature [METEOROL] A value based on the integrated product of the number of degrees that air temperature rises above a given threshold value and the number of days in the period during which this excess is maintained. { ə'kyü·myə,lād·əd 'tem·prə·chər }

accumulation [HYD] The quantity of snow or other solid form of water added to a glacier or snowfield by alimentation. [MIN ENG] **1.** In coal mining, firedamp that collects in higher parts of mine workings and at the edge of wastes. **2.** Oil or gas in some form of trap. { ə·kyü·myə'lā·shən }

accumulation area [HYD] The portion of a glacier above the firn line, where the accumulation exceeds ablation. Also known as firn field; zone of accumulation. { ə·kyü·myə'lā·shən ˌer·ē·ə }

accumulation factor [MATH] The quantity $(1 + r)$ in the formula for compound interest, where *r* is the rate of interest; measures the rate at which the principal grows. { ə·kyü·myə'lā·shən 'fak·tər }

accumulation point *See* cluster point. { ə·kyü·myə'lā·shən ˌpóint }

accumulation zone [GEOL] The area where the bulk of the snow contributing to an avalanche was originally deposited. { ə·kyü·myə'lā·shən ˌzōn }

accumulative error *See* cumulative error. { ə'kyü·myə,lād·iv 'er·ər }

accumulative timing [IND ENG] A time-study method that allows direct reading of the time for each element of an operation by the use of two stopwatches which operate alternately. { ə'kyü·myə,lād·iv 'tīm·iŋ }

accumulator [AERO ENG] A device sometimes incorporated in the fuel system of a gas-turbine engine to store fuel and release it under pressure as an aid in starting. [CHEM ENG] An auxiliary ram extruder on blow-molding equipment used to store melted material between deliveries. [COMPUT SCI] A specific register, in the arithmetic unit of a computer, in which the result of an arithmetic or logical operation is formed; here numbers are added or subtracted, and certain operations such as sensing, shifting, and complementing are performed. Also known as accumulator register; counter. [ELEC] *See* storage battery. [ENG] *See* air vessel. [MECH ENG] **1.** A device, such as a bag containing pressurized gas, which acts upon hydraulic fluid in a vessel, discharging it rapidly to give high hydraulic power, after which the fluid is returned to the vessel with the use of low hydraulic power. **2.** A device connected to a steam boiler to enable a uniform boiler output to meet an irregular steam demand. **3.** A chamber for storing low-side liquid refrigerant in a refrigeration system. Also known as surge drum; surge header. [PETRO ENG] A tank or chamber for receiving and temporarily storing a liquid used in a gas processing plant during a continuous process. { ə'kyü·myə,lād·ər }

accumulator battery *See* storage battery. { ə'kyü·myə,lād·ər 'bad·ə·rē }

accumulator jump instruction [COMPUT SCI] An instruction which programs a computer to ignore the previously established program sequence depending on the status of the accumulator. Also known as accumulator transfer instruction. { ə'kyü·myə,lād·ər ˌjəmp in'strək·shən }

accumulator plant [BOT] A plant or tree that grows in a metal-bearing soil and accumulates an abnormal content of the metal. { ə'kyü·myə,lād·ər ,plant }

accumulator register *See* accumulator. { ə'kyü·myə,lād·ər 'rej·ə,stər }

accumulator shift instruction [COMPUT SCI] A computer instruction which causes the word in a register to be displaced a specified number of bit positions to the left or right. { ə'kyü·myə,lād·ər 'shift in'strək·shən }

accumulator transfer instruction *See* accumulator jump instruction. { ə'kyü·myə,lād·ər 'trans·fər in'strək·shən }

accuracy [SCI TECH] The extent to which the results of a calculation or the readings of an instrument approach the true values of the calculated or measured quantities, and are free from error. { 'ak·yə·rə·sē }

accuracy checking [MAP] The procurement of presumptive evidence of a map's compliance with specified accuracy standards; indicates the relative (rather than the absolute) accuracy of map features. { 'ak·yə·rə·sē ,chek·iŋ }

accuracy control system [COMPUT SCI] Any method which attempts error detection and control, such as random sampling and squaring. { 'ak·yə·rə·sē kən'trōl ,sis·təm }

accuracy life [ORD] The estimated average number of rounds that a particular weapon can fire before its tube becomes so worn that its accuracy tolerance is exceeded. { 'ak·yə·rə·sē ,līf }

accuracy of fire [ORD] The measurement of the precision of fire expressed as the distance of the center of impact from the center of the target. { 'ak·yə·rə·sē əv 'fīr }

accuracy testing [MAP] The procurement of confirmed evidence, on a sampling basis, of a map's compliance with specified accuracy standards; indicates both the relative and absolute accuracy of map features. { 'ak·yə·rə·sē ,test·iŋ }

accurate contour [MAP] A contour line whose accuracy lies within one-half of the basic vertical interval. Also known as normal contour. { 'ak·yə,rət 'kän·tür }

accustomization [ENG] The process of learning the techniques of living with a minimum of discomfort in an extreme or new environment. { ə,kəs·tə·mə'zā·shən }

ac/dc motor *See* universal motor. { ,ā·sē,dē·sē 'mōd·ər }

ac/dc receiver [ELECTR] A radio receiver designed to operate from either an alternating- or direct-current power line. Also known as universal receiver. { ,ā·sē,dē·sē ri'sēv·ər }

acellular [BIOL] Not composed of cells. { 'ā·sel·yə·lər }

acellular gland [PHYSIO] A gland, such as intestinal glands, the pancreas, and parotid glands, that secretes a noncellular product. { 'ā·sel·yə·lər ˌgland }

acellular slime mold [MYCOL] The common name for members of the Myxomycetes. { 'ā·sel·yə·lər 'slīm ˌmōld }

acellular vaccine [IMMUNOL] A vaccine consisting of one or more parts of an infectious agent, rather than the whole cell. { ¦ā¦sel·yü·lər ,vak'sēn }

Ac-Em *See* actinon.

acenaphthene [ORG CHEM] $C_{12}H_{10}$ An unsaturated hydrocarbon whose colorless crystals melt at 92°C; insoluble in water; used as a dye intermediate and as an agent for inducing polyploidy. { ,as·ə'naf·thēn }

acenaphthequinone [ORG CHEM] $C_{10}H_6(CO)_2$ A three-ring hydrocarbon in the form of yellow needles melting at 261-263°C; insoluble in water and soluble in alcohol; used in dye synthesis. { ,as·ə¦naf·thə·kwa¦nōn }

acene [ORG CHEM] Any condensed polycyclic compound with fused rings in a linear arrangement; for example, anthracene. { ə'sēn }

acenocoumarin *See* acenocoumarol. { ə,sēn·ə'kü·mə·rən }

acenocoumarol [ORG CHEM] $C_{19}H_{15}NO_6$ A tasteless, odorless, white, crystalline powder with a melting point of 197°C; slightly soluble in water and organic solvents; used as an anticoagulant. Also known as acenocoumarin. { ə,sēn·ə'kü·mə·rəl }

acentric [BIOL] Not oriented around a middle point. [GEN]

A chromosome or chromosome fragment lacking a centromere. { ,ā'sen·trik }

acentrous [VERT ZOO] Lacking vertebral centra and having the notochord persistent throughout life, as in certain primitive fishes. { ,ā'sen·trəs }

Acephalina [INV ZOO] A suborder of invertebrate parasites in the protozoan order Eugregarinida characterized by nonseptate trophozoites. { ā,sef·ə'līn·ə }

acephalobrachia *See* abrachiocephalia. { ā,sef·ə· lə'brāk· ē·ə }

acephalocardia [MED] Congenital lack of a head and a heart. { ā,sef·ə·lə'kärd·ē·ə }

acephalochiria [MED] Congenital lack of a head and hands. { ā,sef·ə·lə'kir·ē·ə }

acephalocyst [INV ZOO] An abnormal cyst of the *Echinococcus granulosus* larva, lacking a head and brood capsules, found in human organs. { ā'sef·ə·lə,sist }

acephalopodia [MED] Congenital lack of a head and feet. { ā,sef·ə·lə'pōd·ē·ə }

acephalorrhachia [MED] Congenital lack of a head and vertebral column. { ā¦sef·ə·lə'räk·ē·ə }

acephalostomia [MED] Congenital lack of a head, with a mouthlike orifice in the neck or chest. { ā¦sef·ə·lə'stōm·ē·ə }

acephalothoracica [MED] Congenital lack of a head and thorax. { ā¦sef·ə·lə·thə'ras·ə·kə }

acephalous [BOT] Having the style originate at the base instead of at the apex of the ovary. [ZOO] Lacking a head. { ā'sef·ə·ləs }

acephate [ORG CHEM] $C_4H_{10}NO_3PS$ A white solid with a melting point of 72-80°C; very soluble in water; used as an insecticide for a wide range of aphids and foliage pests. { 'as· ə·fāt }

acephatemet [ORG CHEM] $CH_3OCH_3SPONH_2$ A white, crystalline solid with a melting point of 39-41°C; limited solubility in water; used as an insecticide to control cutworms and borers on vegetables. { as·ə'fāt·mət }

Acer [BOT] A genus of broad-leaved, deciduous trees of the order Sapindales, commonly known as the maples; the sugar or rock maple (*A. saccharum*) is the most important commercial species. { 'ā·sər *or* 'ā,kər }

acerate [BOT] Needle-shaped, specifically referring to leaves. { 'as·ə,rāt }

acerbophobia [PSYCH] Abnormal fear of sour taste sensations. Also known as acerophobia. { ə,sər·bə'fōb·ē·ə }

Acerentomidae [INV ZOO] A family of wingless insects belonging to the order Protura; the body lacks tracheae and spiracles. { ,a·sə·rən'tōm·ə·dē }

acerophobia *See* acerbophobia. { ə,sər·ə'fōb·ē·a }

acervate [BIOL] Growing in heaps or dense clusters. { ə'sər,vāt }

acervulus [MYCOL] A cushion- or disk-shaped mass of hyphae, peculiar to the Melanconiales, on which there are dense aggregates of conidiophores. { ə'sər·vyə·ləs }

acetabulectomy [MED] Excision of the acetabulum. { ə,sēd·ə·bə'lek·tə·mē }

acetabuloplasty [MED] Plastic surgery involving repair or enlargement of the cavity of the acetabulum to restore its normal state. { ə,sēd·ə'būl·ə,plas·tē }

acetabulum [ANAT] A cup-shaped socket on the hipbone that receives the head of the femur. [INV ZOO] **1.** A cavity on an insect body into which a leg inserts for articulation. **2.** The sucker of certain invertebrates such as trematodes and tapeworms. { ,a·sə'tab·yə·ləm }

acetal [ORG CHEM] **1.** $CH_3CH(OC_2H_5)_2$ A colorless, flammable, volatile liquid used as a solvent and in manufacture of perfumes. Also known as 1,1-diethoxyethane. **2.** Any one of a class of compounds formed by the addition of alcohols to aldehydes. { 'as·ə,tal }

acetaldehydase [BIOCHEM] An enzyme that catalyzes the oxidation of acetaldehyde to acetic acid. { ¦as·əd,al· də'hī,dās }

acetaldehyde [ORG CHEM] C_2H_4O A colorless, flammable liquid used chiefly to manufacture acetic acid. { ,as· əd'al·də,hīd }

acetaldehyde cyanohydrin *See* lactonitrile. { ,as·əd'al· də,hīd ,sī·ə·nō'hīd·rən }

acetal resins [ORG CHEM] Linear, synthetic resins produced by the polymerization of formaldehyde (acetal homopolymers) or of formaldehyde with trioxane (acetal copolymers); hard,

tough plastics used as substitutes for metals. Also known as polyacetals. { 'as·ə,tal 'rez·ənz }

acetamide [ORG CHEM] CH_3CONH_2 The crystalline, colorless amide of acetic acid, used in organic synthesis and as a solvent. { ə'sed·ə,mīd }

acetamidine hydrochloride [ORG CHEM] $C_2H_6N_2 \cdot HCl$ Deliquescent crystals that are long prisms with a melting point reported as either 174°C or 164–166°C; soluble in water and alcohol; used in the synthesis of imidazoles, pyrimidines, and triazines. { ə'sed·am·ə,dēn hī·drə'klō,rīd }

3-acetamido-4-hydroxybenzenearsonic acid [PHARM] $(HO)(CH_3CONH)C_6H_3AsO(OH)_2$ An odorless, white to slightly yellow powder with a slightly acid taste; soluble in alkali and alkali carbonate solutions; used in veterinary medicine. Also known as acetarsone. { ,thrē ,as·əd'am·ə·dō ,fòr ,hī,dräk·sē,ben,zēn·är'sän·ik 'as·əd }

acetaminophen [ORG CHEM] $C_8H_9O_2N$ Large monoclinic prisms with a melting point of 169-170°C; soluble in organic solvents such as methanol and ethanol; used in the manufacture of azo dyes and photographic chemicals, and as an analgesic and antipyretic. { ə,sēd·ə'mēn·ə·fən }

acetanilide [ORG CHEM] An odorless compound in the form of white, shining, crystalline leaflets or a white, crystalline powder with a melting point of 114-116°C; soluble in hot water, alcohol, ether, chloroform, acetone, glycerol, and benzene; used as a rubber accelerator, in the manufacture of dyestuffs and intermediates, as a precursor in penicillin manufacture, and as a painkiller. { ,a·səd'an·ə,līd }

acetarsone *See* 3-acetamido-4-hydroxybenzenearsonic acid. { ,a·səd'är,sòn }

acetate [ORG CHEM] One of two species derived from acetic acid, CH_3COOH; one type is the acetate ion, CH_3COO^-; the second type is a compound whose structure contains the acetate ion, such as ethyl acetate. [TEXT] The official name for the textile fiber produced from partially hydrolyzed cellulose acetate. Formerly known as acetate rayon. { 'as·ə,tāt }

acetate dye [CHEM] **1.** Any of a group of water-insoluble azo or anthroquinone dyes used for dyeing acetate fibers. **2.** Any of a group of water-insoluble amino azo dyes that are treated with formaldehyde and bisulfate to make them water-soluble. { 'as·ə,tāt ,dī }

acetate film [MATER] A cellulose acetate resin sheet that is transparent, airproof, hygienic, and resistant to grease, oil, and dust; used for photographic film, magnetic tapes, and packaging. { 'as·ə,tāt ,film }

acetate of lime [ORG CHEM] Calcium acetate made from pyroligneous acid and a water suspension of calcium hydroxide. { 'as·ə,tāt əv 'līm }

acetate process [CHEM ENG] Acetylation of cellulose (wood pulp or cotton linters) with acetic acid or acetic anhydride and sulfuric acid catalyst to make cellulose acetate resin or fiber. { 'as·ə,tāt 'präs·əs }

acetate rayon *See* acetate. { 'as·ə,tāt 'rā,än }

acetazolamide [PHARM] $(CH_3CONH)C_2N_2S(SO_2NH_2)$ An odorless, white to faintly yellowish-white, crystalline powder with a melting point of 258°C; slightly soluble in water; used as a diuretic. { ,as·ə·tə'zäl·ə,mīd }

acetenyl *See* ethinyl. { ə'sed·ə,nil }

acetic acid [ORG CHEM] CH_3COOH **1.** A clear, colorless liquid or crystalline mass with a pungent odor, miscible with water or alcohol; crystallizes in deliquescent needles; a component of vinegar. Also known as ethanoic acid. **2.** A mixture of the normal and acetic salts; used as a mordant in the dyeing of wool. { ə'sēd·ik 'as·əd }

acetic acid bacteria *See* Acetobacter. { ə'sēd·ik 'as·əd ,bak'tir·ē·ə }

acetic anhydride [ORG CHEM] $(CH_3CO)_2O$ A liquid with a pungent odor that combines with water to form acetic acid; used as an acetylating agent. { ə'sēd·ik an'hīd,rīd }

acetic ester *See* ethyl acetate. { ə'sēd·ik 'es·tər }

acetic ether *See* ethyl acetate. { ə'sēd·ik 'ē·thər }

acetic fermentation [MICROBIO] Oxidation of alcohol to produce acetic acid by the action of bacteria of the genus *Acetobacter*. { ə'sēd·ik fər·mən'tā·shən }

acetic thiokinase [BIOCHEM] An enzyme that catalyzes the formation of acetyl coenzyme A from acetate and adenosine triphosphate. { ə'sēd·ik ,thī·ə'kīn,ās }

acetidin *See* ethyl acetate. { ə'sed·ə·din }

acetin [ORG CHEM] $C_3H_5(OH)_2OOCCH_3$ A thick, colorless, hygroscopic liquid with a boiling point of 158°C, made by heating glycerol and strong acetic acid; soluble in water and alcohol; used in tanning, as a dye solvent and food additive, and in explosives. Also spelled acetine. { 'as·ə·tin }

acetine See acetin. { 'as·ə,tēn }

acetoacetate [ORG CHEM] A salt which contains the CH_3COCH_2COO radical; derived from acetoacetic acid. { ¦as·ə,tō·'as·ə,tāt }

acetoacetic acid [ORG CHEM] CH_3COCH_2COOH A colorless liquid miscible with water; derived from β-hydroxybutyric acid in the body. { ¦as·ə,tō,ə'sēd·ik 'as·əd }

acetoacetic ester See ethyl acetoacetate. { ¦as·ə,tō,ə'sēd·ik 'es·tər }

acetoacetyl coenzyme A [BIOCHEM] $C_{25}H_{41}O_{18}N_7P_3S$ An intermediate product in the oxidation of fatty acids. { ¦as·ə,tō·ə'sēd·əl ¸kō'en·zīm 'ā }

acetoamidoacetic acid See aceturic acid. { ¦as·ə,tō¦am·ə,dō·ə'sēd·ik 'as·əd }

Acetobacter [MICROBIO] A genus of gram-negative, aerobic bacteria of uncertain affiliation comprising ellipsoidal to rod-shaped cells as singles, pairs, or chains; they oxidize ethanol to acetic acid. Also known as acetic acid bacteria; vinegar bacteria. { ə'sēd·ō,bak·tər }

Acetobacter aceti [MICROBIO] An aerobic, rod-shaped bacterium capable of efficient oxidation of glucose, ethyl alcohol, and acetic acid; found in vinegar, beer, and souring fruits and vegetables. { ə'sēd·ō,bak·tər ə'sēd·ē }

Acetobacter suboxydans [MICROBIO] A short, nonmotile vinegar bacterium that can oxidize ethanol to acetic acid; useful for industrial production of ascorbic and tartaric acids. { ə'sēd·ō,bak·tər səb'äks·ə·dəns }

acetoclastic bacteria [MICROBIO] Bacteria that utilize acetic acid only and produce methane during anaerobic fermentation. { ¸a·sə·tō¦klas·tik bak'tir·ē·ə }

acetoclastis [MICROBIO] The process, carried out by some methanogens, of splitting acetate into methane and carbon dioxide. { ¸a·sə·tō'klas·təs }

acetogenic bacteria [BIOCHEM] Anaerobic bacteria capable of reducing carbon dioxide to acetic acid or converting sugars into acetate. { ¸a·sə·tō¦jen·ik bak'tir·ē·ə }

acetoin [ORG CHEM] $CH_3COCHOHCH_3$ A slightly yellow liquid, melting point 15°C, used as an aroma carrier in the preparation of flavors and essences; produced by fermentation or from diacetyl by partial reduction with zinc and acid. { ə'sed·ə·wən }

acetol [ORG CHEM] CH_3COCH_2OH A colorless liquid soluble in water; a reducing agent. { 'as·ə·tol }

acetolactic acid [BIOCHEM] $C_5H_8O_4$ A monocarboxylic acid formed as an intermediate in the synthesis of the amino acid valine. { ¸as·ə·tə'lak·tik 'as·əd }

acetolysis [ORG CHEM] Decomposition of an organic molecule through the action of acetic acid or acetic anhydride. { ¸as·ə'täl·ə·səs }

acetomeroctol [PHARM] $CH_3COOHgC_6H_3(OH)C(CH_3)_2$- $CH_3C(CH_3)_3$ A white solid with a melting point of 155-157°C; soluble in alcohol, ether, and chloroform; used as an antiseptic. { ¸as·ə,tō·mə'räk·tol }

Acetomonas [MICROBIO] A genus of aerobic, polarly flagellated vinegar bacteria in the family Pseudomonadaceae; used industrially to produce vinegar, gluconic acid, and L-sorbose. { ə¸sed·ə'mon·əs }

acetone [ORG CHEM] CH_3COCH_3 A colorless, volatile, extremely flammable liquid, miscible with water; used as a solvent and reagent. Also known as 2-propanone. { 'as·ə,tōn }

acetone-benzol process [CHEM ENG] A dewaxing process in petroleum refining, with acetone and benzol used as solvents. { 'as·ə,tōn 'ben·zól ¸präs·əs }

acetone body See ketone body. { 'as·ə,tōn ¸bäd·ē }

acetone cyanohydrin [ORG CHEM] $(CH_3)_2COHCN$ A colorless liquid obtained from condensation of acetone with hydrocyanic acid; used as an insecticide or as an organic chemical intermediate. { 'as·ə,tōn sī,ə·nō'hīd·rən }

acetone fermentation [MICROBIO] Formation of acetone by the metabolic action of certain anaerobic bacteria on carbohydrates. { 'as·ə,tōn fər·mən'tā·shən }

acetone glucose See acetone sugar. { 'as·ə,tōn 'glü·kōs }

acetonemia [MED] A condition characterized by large amounts of acetone bodies in the blood. Also known as ketonemia. { ¸as·ə·tə'nēm·ē·ə }

acetone number [CHEM] A ratio used to estimate the degree of polymerization of materials such as drying oils; it is the weight in grams of acetone added to 100 grams of a drying oil to cause an insoluble phase to form. { 'as·ə·tōn 'nəm·bər }

acetone pyrolysis [ORG CHEM] Thermal decomposition of acetone into ketene. { 'as·ə·tōn pī'räl·ə·səs }

acetone-sodium bisulfite [ORG CHEM] $(CH_3)_2C(OH)$-SO_3Na Crystals that have a slight sulfur dioxide odor and slightly fatty feel; freely soluble in water, decomposed by acids; used in photography and in textile dyeing and printing. { 'as·ə·tōn 'sōd·ē·əm ¸bī'səl,fāt }

acetone sugar [ORG CHEM] Any reducing sugar that contains acetone; examples are 1,2-monoacetone-D-glucofuranose and 1,2-5,6-diacetone-D-glucofuranose. Also known as acetone glucose. { 'as·ə·tōn 'shùg·ər }

acetonitrile [ORG CHEM] CH_3CN A colorless liquid soluble in water; used in organic synthesis. { ¸as·ə·tō'nī,tril }

acetonylacetone [ORG CHEM] $CH_3COCH_2CH_2COCH_3$ A colorless liquid with a boiling point of 192.2°C; soluble in water; used as a solvent and as an intermediate for pharmaceuticals and photographic chemicals. { ¸as·ə,tän·əl'as·ə,tōn }

acetophenetidin [PHARM] $CH_3CONHC_6H_4OC_2H_5$ Odorless, white crystals or powder with a melting point of 135°C; soluble in alcohol, chloroform, and ether; used as an analgesic. { ¸as·ə,tä·fə'net·ə,din }

acetophenone [ORG CHEM] $C_6H_5COCH_3$ Colorless crystals with a melting point of 19.6°C and a specific gravity of 1.028; used as a chemical intermediate. { ¸as·ə,tä·fə'nōn }

acetostearin [ORG CHEM] A general term for monoglycerides of stearic acid acetylated with acetic anhydride; used as a protective food coating and as plasticizers for waxes and synthetic resins to improve low-temperature characteristics. { ə,sē·dō'stēr·ən }

acetovanillon [PHARM] $HOC_6H_3(OCH_3)COCH_3$ Fine, needlelike crystals with a melting point of 115°C and a faint vanilla odor; freely soluble in hot water, alcohol, benzene, chloroform, and ether; used as a cardiotonic drug. { ə,sē·dō·və'nil·ən }

acetoxime [ORG CHEM] $(CH_3)_2CNOH$ Colorless crystals with a chlorallike odor and a melting point of 61°C; soluble in alcohol, ethers, and water; used in organic synthesis and as a solvent for cellulose ethers. { ¸as·ò'täk,sēm }

21-acetoxypregnenolone [PHARM] $C_{23}H_{34}O_4$ Needlelike crystals which become opaque at about 80°C on prolonged standing; melting point is 184-185°C; soluble in chloroform and toluene; used as an antiarthritic drug. { ¸twen·tē'wən ¸as·ə,täk·sē,preg'nen·ə,lōn }

aceturic acid [ORG CHEM] $CH_3CONHCHCH_2COOH$ Long, needlelike crystals with a melting point of 206–208°C; soluble in water and alcohol; forms stable salts with organic bases; used in MED. { ¸as·ə¦tùr·ik 'as·əd }

acetyl [ORG CHEM] CH_3CO- A two-carbon organic radical containing a methyl group and a carbonyl group. { ə'sēd·əl }

α-acetylacetanilide See acetoacetic acid. { 'al·fə ə¦sed·əl¦a·səd'an·ə,līd }

acetylacetone [ORG CHEM] $CH_3COCH_2OCCH_3$ A colorless liquid with a pleasant odor and a boiling point of 140.5°C; soluble in water; used as a solvent, lubricant additive, paint drier, and pesticide. { ə¦sed·əl'as·ə,tōn }

***para*-acetylaminophenyl salicylate** [PHARM] C_6H_4-$(NHCCH_3)OOCC_6H_4OH$ Odorless and tasteless, fine, white, crystalline scales with a melting point of 187–188°C; soluble in alcohol, ether, and hot water; used as an analgesic. Also known as phenetsal. { ¦par·ə ¦as·ə·təl¦am·ə,nō'fēn·əl sə'lis·ə,lāt }

acetylase [BIOCHEM] Any enzyme that catalyzes the formation of acetyl esters. { ə'sed·əl,ās }

acetylated cotton [MATER] Mildewproof cotton made by the chemical conversion of part of the raw cotton fiber to cellulose acetate. { ə'sed·əl,āt·əd 'kät·ən }

acetylated lanolin [MATER] A clear, nongreasy liquid made by reacting lanolin with polyoxyethylenes; soluble in water, oils, and alcohol; used in cosmetics. { ə'sed·əl,āt·əd 'lan·əl·ən }

acetylating agent [ORG CHEM] A reagent, such as acetic

anhydride, capable of bonding an acetyl group onto an organic molecule. { ə'sed·əl,āt·iŋ ,ā·jənt }

acetylation [ORG CHEM] The process of bonding an acetyl group onto an organic molecule. { ə,sed·əl'ā·shən }

acetyl benzoyl peroxide [ORG CHEM] $C_6H_5CO·O_2·OCCH_3$ White crystals with a melting point of 36.6°C; moderately soluble in ether, chloroform, carbon tetrachloride, and water; used as a germicide and disinfectant. { ə'sed·əl 'ben·zȯil pə'räk,sīd }

acetyl bromide [ORG CHEM] CH_3COBr A colorless, fuming liquid with a boiling point of 81°C; soluble in ether, chloroform, and benzene; used in organic synthesis and dye manufacture. { ə'sed·əl 'brō,mīd }

α-acetylbutyrolactone [ORG CHEM] $C_6H_8O_3$ A liquid with an esterlike odor; soluble in water; used in the synthesis of 3,4-disubstituted pyridines. { ¦al·fə ə,sed·əl¦byüd·ə·rō'lak·tōn }

acetylcarbromal [PHARM] $(C_2H_5)_2CBrCONHCONHCO·CH_3$ Crystals with a slightly bitter taste and a melting point of 109°C; soluble in alcohol and ethyl acetate; used as a sedative. { ə,sed·əl'kär·bə,mal }

acetyl chloride [ORG CHEM] CH_3COCl A colorless, fuming liquid with a boiling point of 51–52°C; soluble in ether, acetone, and acetic acid; used in organic synthesis, and in the manufacture of dyestuffs and pharmaceuticals. { ə'sed·əl 'klȯ,rīd }

acetylcholine [NEURO] $C_7H_{17}O_3N$ A compound released from certain autonomic nerve endings which acts in the transmission of nerve impulses to excitable membranes. { ə,sed·əl'kō,lēn }

acetylcholine receptor [CELL MOL] A receptor in the membranes of certain cell structures, such as synapses or the neuromuscular junction, to which the transmitter substance acetylcholine binds. Nicotinic acetylcholine receptors are gated ion channels that open in response to acetylcholine, leading to an increase in membrane conductance; muscarinic acetylcholine receptors are G-protein-linked receptors inducing membrane ion channel changes or intracellular processes such as smooth muscle contraction. { ə,sed·əl 'kō,lēn ri,sep·tər }

acetylcholinesterase [BIOCHEM] An enzyme found in excitable membranes that inactivates acetylcholine. { ə,sed·əl·kō·lən'es·tər,ās }

acetyl-CoA pathway [CELL BIOL] A pathway of autotrophic carbon dioxide fixation. { a¦sed·əl ,kō¦ā 'path,wā }

acetyl coenzyme A [BIOCHEM] $C_{23}H_{39}O_{17}N_7P_3S$ A coenzyme, derived principally from the metabolism of glucose and fatty acids, that takes part in many biological acetylation reactions; oxidized in the Krebs cycle. { ə,sed·əl ,kō'en,zīm 'ā }

acetylcysteine [PHARM] $HSCH_2CH(NHCOCH_3)COOH$ Crystals with a melting point of 109-110°C; used as a mucolytic drug. { ə,sed·əl'sis·tē,ēn or ,tēn }

acetylene [ORG CHEM] C_2H_2 A colorless, highly flammable gas that is explosive when compressed; the simplest compound containing a triple bond; used in organic synthesis and as a welding fuel. Also known as ethyne. { ə'sed·əl,ēn }

acetylene black [ORG CHEM] A form of carbon with high electrical conductivity; made by decomposing acetylene by heat. { ə'sed·əl,ēn 'blak }

acetylene cutting See oxyacetylene cutting. { ə'sed·əl,ēn 'kət·iŋ }

acetylene generator [ENG] A steel cylinder or tank that provides for controlled mixing of calcium carbide and water to generate acetylene. { ə'sed·əl,ēn 'jen·ə,rād·ər }

acetylene series [ORG CHEM] A series of unsaturated aliphatic hydrocarbons, each containing at least one triple bond and having the general formula C_nH_{2n-2}. { ə'sed·əl,ēn 'sir·ēz }

acetylene snow [MATER] A solid form of acetylene that does not react to shock or flame. { ə¦set·əl·ən 'snō }

acetylene tetrabromide [ORG CHEM] $CHBr_2CHBr_2$ A yellowish liquid with a boiling point of 239–242°C; soluble in alcohol and ether; used for separating minerals and as a solvent. { ə'sed·əl,ēn ,te·trə'brō,mīd }

acetylene tetrachloride See sym-tetrachloroethane. { ə'sed·əl,ēn ,te·trə'klȯr,īd }

acetylene torch See oxyacetylene torch. { ə'sed·əl,ēn ,tȯrch }

acetylene welding See oxyacetylene welding. { ə'sed·əl,ēn 'weld·iŋ }

acetylenic [ORG CHEM] Pertaining to acetylene or being like acetylene, such as having a triple bond. { ə,sed·ə'len·ik }

acetylenyl See ethinyl. { ə,sed·ə'len·əl }

***N*-acetylethanolamine** [ORG CHEM] $CH_3CONHC_2H_4OH$ A brown, viscous liquid with a boiling range of 150–152°C; soluble in alcohol, ether, and water; used as a plasticizer, humectant, high-boiling solvent, and textile conditioner. { ¦en ə,sed·əl,eth·ə'näl·ə,mēn }

acetylide [ORG CHEM] A compound formed from acetylene with the H atoms replaced by metals, as in cuprous acetylide (Cu_2C_2). { ə'sed·əl,īd }

acetyl iodide [ORG CHEM] CH_3COI A colorless, transparent, fuming liquid with a boiling point of 105–108°C; soluble in ether and benzene; used in organic synthesis. { ə'sed·əl 'ī·ə,dīd }

acetylisoeugenol [ORG CHEM] $C_6H_3(CHCHCH_3)(OCH_3)$ (OCOCH$_3$) White crystals with a clovelike odor and a congealing point of 77°C; used in perfumery and flavoring. { ə,sed·əl,ī·sō'yü·jə,nȯl }

acetyl number [ANALY CHEM] A measure of free hydroxyl groups in fats or oils determined by the amount of potassium hydroxide used to neutralize the acetic acid formed by saponification of acetylated fat or oil. { ə'sed·əl ,nəm·bər }

acetyl peroxide [ORG CHEM] $(CH_3CO)_2O_2$ Colorless crystals with a melting point of 30°C; soluble in alcohol and ether; used as an initiator and catalyst for resins. { ə'sed·əl pə'räk,sīd }

acetyl phosphate [BIOCHEM] $C_2H_5O_5P$ The anhydride of acetic and phosphoric acids occurring in the metabolism of pyruvic acid by some bacteria; phosphate is used by some microorganisms, in place of adenosine triphosphate, for the phosphorylation of hexose sugars. { ə'sed·əl 'fäs,fāt }

acetyl propionyl [ORG CHEM] $CH_3COCOCH_2CH_3$ A yellow liquid with a boiling point of 106-110°C; used in butterscotch- and chocolate-type flavors. { ə'sed·əl 'prō·pē·ə,nil }

acetylsalicylic acid [ORG CHEM] $CH_3COOC_6H_4COOH$ A white, crystalline, weakly acidic substance, with melting point 137°C; slightly soluble in water; used medicinally as an antipyretic. Also known by trade name aspirin. { ə¦sed·əl¦sal·ə¦sil·ik 'as·əd }

***N*-acetylsulfanilyl chloride** [ORG CHEM] $C_8H_8ClNO_3S$ Thick, light tan prisms ranging to brown powder or fine crystals with a melting point of 149°C; soluble in benzene, chloroform, and ether; used as an intermediate in the preparation of sulfanilamide and its derivatives. Abbreviated ASC. { ¦en ə¦sed·əl·səl'fan·ə·lil 'klȯr,īd }

acetylurea [ORG CHEM] $CH_3CONHCONH_2$ Crystals that are colorless and are slightly soluble in water. { ə,sēd·əl,yü'rē·ə }

acetyl valeryl [ORG CHEM] $CH_3COCOC_4H_9$ A yellow liquid used for cheese, butter, and other flavors. Also known as heptadione-2,3. { ə'sēd·əl ,val·ə,ril }

ACF diagram [PETR] A triangular diagram showing the chemical character of a metamorphic rock; the three components plotted are A $= Al_2O_3 + Fe_2O_3 - (Na_2O + K_2O)$, C$= CaO$, F $= FeO + MgO + MnO$. { ,ā,sē'ef 'dī·ə,gram }

acfm See actual cubic feet per minute.

a-c fracture [CRYSTAL] A type of tension fracture lying parallel to the a-c fabric plane and normal to plane b in a crystal. { 'a'sē 'frak·chər }

a-c girdle [GEOL] A girdle of points in a petrofabric diagram that have a tread parallel with the plane of the a and c fabric axes. { 'a'sē 'gərd·əl }

Achaenodontidae [PALEON] A family of Eocene dichobunoids, piglike mammals belonging to the suborder Palaeodonta. { ə,kēn·ə'dän·tə·dē }

achaetous [INV ZOO] Without setae. Also known as asetigerous. { ə'kēd·əs }

A chain See heavy chain. { 'ā ,chān }

achalasia [MED] Inability of a hollow muscular organ or ring of muscle (sphincter) to relax. { ,ak·ə'lāzh·ē·ə }

achalasia of the cardia [MED] Enlargement of the esophagus as a result of cardiospasm. { ,ak·ə'lāzh·ē·ə əv thə 'kärd·ē·ə }

ache [MED] A constant dull or throbbing pain. { āk }

acheb [ECOL] Short-lived vegetation regions of the Sahara composed principally of mustards (Cruciferae) and grasses (Gramineae). { ə'cheb }

achene [BOT] A small, dry, indehiscent fruit formed from a simple ovary bearing a single seed. { ə'kēn }

Acheulean [ARCHEO] Lower Paleolithic archeological time, characterized by biface tools having cutting edges all around. { ə'shül·ē·ən }

achiasmate [CYTOL] Pertaining to meiosis that lacks chiasma. { ˌā·kī'az͟ˌmāt }

achievement age [PSYCH] Accomplishment, or actual level of scholastic performance, expressed as equivalent to the age in years of the average child showing similar attainments. { ə'chēv·mənt 'āj }

achievement quotient [PSYCH] The ratio, usually multiplied by 100, between the achievement age, or actual scholastic level, and the mental age. { ə'chēv·mənt kwō·shənt }

achilary [BOT] In flowers, having the lip (labellum) undeveloped or lacking. { ə'kil·ə·rē }

Achilles [ASTRON] An asteroid; member of the group known as the Trojan planets. { ə'kil·ēz }

Achilles group See Greek group. { ə'kil·ēz ˌgrüp }

Achilles jerk [PHYSIO] A reflex action seen as plantar flection in response to a blow to the Achilles tendon. Also known as Achilles tendon reflex. { ə'kil·ēz ˌjərk }

Achilles tendon [ANAT] The tendon formed by union of the tendons of the calf muscles, the soleus and gastrocnemius, and inserted into the heel bone. { ə'kil·ēz 'ten·dən }

Achilles tendon reflex See Achilles jerk. { ə'kil·ēz 'ten·dən 'rē·fleks }

achiral molecules [ORG CHEM] Molecules which are superposable to their mirror images. { ¦ā·kī·rəl 'mäl·ə·kyülz }

achlamydeous [BOT] Lacking a perianth. { ¦ā·klə'mid·ē·əs }

achlorhydria [MED] Absence of hydrochloric acid in the stomach. { ¦ā·klór'hīd·rē·ə }

achluophobia [PSYCH] Abnormal fear of darkness. { ¦ak·lü·ə'fōb·ē·ə }

Acholeplasma [MICROBIO] The single genus of the family Acholeplasmataceae, comprising spherical and filamentous cells. { ə,kōl·ə'plaz·mə }

Acholeplasmataceae [MICROBIO] A family of the order Mycoplasmatales; characters same as for the order and class (Mollicutes); members do not require sterol for growth. { ə,kōl·ə,plaz·mə'tās·ē,ē }

acholia [MED] Suppression or absence of bile secretion into the small intestine. { ¦ā'kōl·ē·ə }

achondrite [GEOL] A stony meteorite that contains no chondrules. { ¦ā'kän,drīt }

achondroplasia [MED] A hereditary deforming disease of the skeletal system, inherited in humans as an autosomal dominant trait and characterized by insufficient growth of the long bones, resulting in reduced length. Also known as chondrodystrophy fetalis. { ¦ā,kän·drə'plāzh·ē·ə }

achondroplastic dwarf [MED] A human with short legs and arms due to achondroplasia. { ¦ā,kän·drə'plas·tik 'dwórf }

achordate [VERT ZOO] Lacking a notochord. { ¦ā'kór,dāt }

achreocythemia [MED] An anemia characterized by pale erythrocytes due to hemoglobin deficiency. { ¦ā,krē·ə,sī'thēm·ē·ə }

achroglobin [BIOCHEM] A colorless respiratory pigment present in some mollusks and urochordates. { ¦ak·rə'glōb·ən }

achroite [MINERAL] A colorless variety of tourmalines found in Malagasy. { 'ak·rō,īt }

achromasia [MED] Absence of normal skin pigment. { ¦ā·krō'māzh·ē·ə }

achromat See achromatic lens. { 'ak·rə,mat }

Achromatiaceae [MICROBIO] A family of gliding bacteria of uncertain affiliation; cells are spherical to ovoid or cylindrical, movements are slow and jerky, and microcysts are not known. { a,krō·mə·dē'ās·ē,ē }

achromatic [OPTICS] Capable of transmitting light without decomposing it into its constituent colors. { ¦a·krə¦mad·ik }

achromatic color [OPTICS] A color that has no hue or saturation but only brightness, such as white, black, and various shades of gray. { ¦a·krə¦mad·ik 'kəl·ər }

achromatic condenser [OPTICS] A condenser designed to eliminate chromatic and spherical aberrations, usually through the use of four elements, two of which are achromatic lenses; used in microscopes having high magnification. { ¦a·krə¦mad·ik kən'den·sər }

achromatic fringe [OPTICS] An interference fringe of light whose position is independent of the wavelength of the light used; the first fringe of a Lloyd's mirror system and the central fringe of a Fresnel biprism system are examples. { ¦a·krə¦mad·ik 'frinj }

achromatic interval [PHYSIO] The difference between the achromatic threshold and the smallest light stimulus at which the hue is detectable. { ¦a·krə¦mad·ik 'int·ər·vəl }

achromatic lens [OPTICS] A combination of two or more lenses having a focal length that is the same for two quite different wavelengths, thereby removing a major portion of chromatic aberration. Also known as achromat. { ¦a·krə¦mad·ik 'lenz }

achromatic locus [OPTICS] A region that includes those points in the chromaticity diagram which represent acceptable reference standards of illumination. Commonly referred to as white light. Also known as achromatic region. { ¦ak·rə¦mad·ik 'lō·kəs }

achromatic prism [OPTICS] A prism consisting of two or more prisms with different refractive indices combined so that light passing through the device is deviated but not dispersed. { ¦a·krə¦mad·ik 'priz·əm }

achromatic threshold [PHYSIO] The smallest light stimulus that can be detected by a dark-adapted eye, so called because all colors lose their hue at this illumination. { ¦a·krə¦mad·ik 'thresh,hōld }

achromatin [CYTOL] The portion of the cell nucleus which does not stain easily with basic dyes. { ¦ā'krō·mə·tən }

Achromatium [MICROBIO] The type genus of the family Achromatiaceae. { a·krə'māsh·ē·əm }

achromatophilia [BIOL] The property of not staining readily. { ¦ā·krō,mad·ə'fil·ē·ə }

achromic [BIOL] Colorless; lacking normal pigmentation. { ¦ā'krō·mik }

Achromobacter [MICROBIO] A genus of motile and nonmotile, gram-negative, rod-shaped bacteria in the family Achromobacteraceae. { ˌā'krō·mə,bak·tər }

Achromobacteraceae [MICROBIO] Formerly a family of true bacteria, order Eubacteriales, characterized by aerobic metabolism. { ˌā,krō·mə,bak·tər'ās·ē,ē }

achronal set [RELAT] A set of points in a space-time with no two points of the set having timelike separation. { ¦ā,krón·əl 'set }

Achroonema [MICROBIO] A genus of bacteria in the family Pelonemataceae; cells have smooth, delicate, porous walls. { ˌak·rō'ōn·ə·mə }

achylia [MED] Absence of chyle. { ¦ā'kil·ē·ə }

achylia gastrica [MED] Lack of secretion of hydrochloric acid and proteolytic enzymes by the stomach. { ˌā'kil·ē·ə 'gas·trik·ə }

ACI See acoustic comfort index.

acicula [BOT] A needle-shaped part, for example, of a plant or crystal. { ə'sik·yə·lə }

acicular [SCI TECH] Needlelike; slender and pointed. { ə'sik·yə·lər }

acicular ice [HYD] Fresh-water ice composed of many long crystals and layered hollow tubes of varying shape containing air bubbles. Also known as fibrous ice; satin ice. { ə'sik·yə·lər 'īs }

acicular powder [MET] A metal powder whose grains are needle-shaped. { ə'sik·yə·lər 'paúd·ər }

aciculate [BOT] Finely scored on the surface. { ə'sik·yə·lət }

aciculignosa [ECOL] Narrow sclerophyll or coniferous vegetation that is mostly subalpine, subarctic, or continental. { ə,sik·yə·lig'nōs·ə }

acid [CHEM] **1.** Any of a class of chemical compounds whose aqueous solutions turn blue litmus paper red, react with and dissolve certain metals to form salts, and react with bases to form salts. **2.** A compound capable of transferring a hydrogen ion in solution. **3.** A substance that ionizes in solution to yield the positive ion of the solvent. **4.** A molecule or ion that combines with another molecule or ion by forming a covalent bond with two electrons from the other species. { 'as·əd }

π-acid [ORG CHEM] An acid that readily forms stable complexes with aromatic systems. { 'pī 'as·əd }

acid acceptor [ORG CHEM] A stabilizer compound added to

plastic and resin polymers to combine with trace amounts of acids formed by decomposition of the polymers. { 'as·əd ək'sep·tər }

acid alcohol [ORG CHEM] A compound containing both a carboxyl group ($-COOH$) and an alcohol group ($-CH_2OH$, $=CHOH$, or $\equiv COH$). { 'as·əd 'al·kə·hȯl }

acid amide [ORG CHEM] A compound derived from an acid in which the hydroxyl group ($-OH$) of the carboxyl group ($-COOH$) has been replaced by an amino group ($-NH_2$) or a substituted amino group ($-NHR$ or $-NHR_2$). { 'as·əd 'a‚mīd }

Acidaminococcus [MICROBIO] A genus of bacteria in the family Veillonellaceae; cells are often oval or kidney-shaped and occur in pairs; amino acids can supply the single energy source. { ‚as·əd‚a·mə·nō'käk·əs }

acid anhydride [CHEM] An acid with one or more molecules of water removed; for example, SO_3 is the acid anhydride of H_2SO_4, sulfuric acid. { 'as·əd ‚an'hīd‚rīd }

acid azide [ORG CHEM] A compound in which the hydroxy group of a carboxylic acid is replaced by the azido group ($-NH_3$).An acyl or aroyl derivative of hydrazoic acid. Also known as acyl azide. { 'as·əd 'ā‚zīd }

acid-base balance [PHYSIO] Physiologically maintained equilibrium of acids and bases in the body. { 'as·əd 'bās 'bal·əns }

acid-base catalysis [CHEM] The increase in speed of certain chemical reactions due to the presence of acids and bases. { 'as·əd 'bās kə'tal·ə·sis }

acid-base equilibrium [CHEM] The condition when acidic and basic ions in a solution exactly neutralize each other; that is, the pH is 7. { 'as·əd 'bās ‚ik·wə'lib·rē·əm }

acid-base indicator [ANALY CHEM] A substance that reveals, through characteristic color changes, the degree of acidity or basicity of solutions. { 'as·əd 'bās 'in·də‚kād·ər }

acid-base pair [CHEM] A concept in the Braunsted theory of acids and bases; the pair consists of the source of the proton (acid) and the base generated by the transfer of the proton. { 'as·əd 'bās 'pār }

acid-base titration [ANALY CHEM] A titration in which an acid of known concentration is added to a solution of base of unknown concentration, or the converse. { 'as·əd 'bās tī'trā·shən }

acid blowcase See blowcase. { 'as·əd 'blō·kās }

acid bottom and lining [MET] A melting furnace's inner bottom and lining composed of materials that at operating temperatures of the furnace react with the melt and slag to give an acid reaction; examples of materials are sand, siliceous rock, and silica brick. { 'as·əd 'bät·əm an 'līn·iŋ }

acid brittleness [MET] Low ductility of a metal due to its absorption of hydrogen gas, which may occur during an electrolytic process or during cleaning. Also known as hydrogen embrittlement. { 'as·əd 'brit·əl·nəs }

acid bronze [MET] A copper-tin alloy containing lead and nickel; used in pumping equipment. { 'as·əd 'bränz }

acid calcium phosphate See calcium phosphate. { 'as·əd 'kal·sē·əm 'fäs‚fāt }

acid cell [HISTOL] A parietal cell of the stomach. [PHYS CHEM] An electrolytic cell whose electrolyte is an acid. { 'as·əd ‚sel }

acid chloride [ORG CHEM] A compound containing the radical $-COCl$; an example is benzoyl chloride. { 'as·əd 'klȯr‚īd }

acid clay [GEOL] A type of clay that gives off hydrogen ions when it dissolves in water. { 'as·əd 'klā }

acid cleaning [ENG] The use of circulating acid to remove dirt, scale, or other foreign matter from the interior of a pipe. { 'as·əd 'klēn·iŋ }

acid conductor [CHEM ENG] A vessel designed for refortification of hydrolyzed acid by heating and evaporation of water, or sometimes by distillation of water under partial vacuum. { 'as·əd kən'dək·tər }

acid cure [MET] The removal of some gangue carbonates from uranium ore by agitation with sulfuric acid prior to the leaching process. { 'as·əd 'kyu̇r }

acid detergent fiber [AGR] The fraction of undigestible plant material in forage, usually cellulose fiber coated with lignin. Abbreviated ADF. { 'as·əd di‚tər·jənt ‚fī·bər }

acid dilution [PETRO ENG] Dilution of concentrated hydrochloric acid with water prior to oil-well acidizing. { 'as·əd də'lü·shən }

acid disproportionation [CHEM] The self-oxidation of a sample of an oxidized element to the next higher oxidation state and then a corresponding reduction to lower oxidation states. { 'as·əd ‚dis·prə‚pȯr·shə'nā·shən }

acid dye [ORG CHEM] Any of a group of sodium salts of sulfonic and carboxylic acids used to dye natural and synthetic fibers, leather, and paper. { 'as·əd ‚dī }

acid egg See blowcase. { 'as·əd ‚eg }

acid electrolyte [INORG CHEM] A compound, such as sulfuric acid, that dissociates into ions when dissolved, forming an acidic solution that conducts an electric current. { 'as·əd ə'lek·trə‚līt }

acidemia [MED] A condition in which the pH of the blood falls below normal. { ‚as·ə'dēm·ē·ə }

acid-fast bacteria [MICROBIO] Bacteria, especially mycobacteria, that stain with basic dyes and fluorochromes and resist decoloration by acid solutions. { 'as·əd ‚fast bak'tir·ē·ə }

acid-fast stain [MICROBIO] A differential stain used in identifying species of *Mycobacterium* and one species of *Nocardia*. { 'as·əd ‚fast 'stān }

acid-fracture [PETRO ENG] To open or enlarge a fracture in a productive, hard limestone formation by using a mixture of oil and acid or of water and acid under high pressure. { 'as·əd ‚frak·chər }

acid gases [CHEM ENG] The hydrogen sulfide and carbon dioxide found in natural and refinery gases which, when combined with moisture, form corrosive acids; known as sour gases when hydrogen sulfide and mercaptans are present. { 'as·əd 'gas·əz }

acid halide [ORG CHEM] A compound of the type RCOX, where R is an alkyl or aryl radical and X is a halogen. { 'as·əd 'hā‚līd }

acid heat test [ANALY CHEM] The determination of degree of unsaturation of organic compounds by reacting with sulfuric acid and measuring the heat of reaction. { 'as·əd 'hēt ‚test }

acid hydrolase [BIOCHEM] Any of a group lysosomal digestive enzymes that function optimally in an acidic environment and can sever (by hydrolysis) particular chemical bonds found in natural materials. { ‚as·əd 'hī·drə‚lās }

acidic [CHEM] 1. Pertaining to an acid or to its properties. 2. Forming an acid during a chemical process. { ə'sid·ik }

acidic dye [ORG CHEM] An organic anion that binds to and stains positively charged macromolecules. { ə‚sid·ik 'dī }

acidic group [ORG CHEM] The radical COOH, present in organic acids. { ə‚sid·ik 'grüp }

acidic lava [GEOL] Extruded felsic igneous magma which is rich in silica (SiO_2 content exceeds 65). { ə‚sid·ik 'lä·və }

acidic oxide [INORG CHEM] An oxygen compound of a nonmetal, for example, SO_2 or P_2O_5, which yields an oxyacid with water. { ə‚sid·ik 'äk‚sīd }

acidic rock [PETR] Igneous rock containing more than 66% SiO_2, making it silicic. { ə'sid·ik 'räk }

acidic titrant [ANALY CHEM] An acid solution of known concentration used to determine the basicity of another solution by titration. { ə'sid·ik 'tī·trənt }

acidification [CHEM] Addition of an acid to a solution until the pH falls below 7. { ə‚sid·ə·fə'kā·shən }

acidimeter [ANALY CHEM] An apparatus or a standard solution used to determine the amount of acid in a sample. { ‚as·ə'dim·ə·tər }

acidimetry [ANALY CHEM] The titration of an acid with a standard solution of base. { ‚as·ə'dim·ə·trē }

aciding [ENG] A light etching of a building surface of cast stone. { 'as·əd·iŋ }

acidity [CHEM] The state of being acid. { ə'sid·ə·tē }

acidity coefficient [GEOCHEM] The ratio of the oxygen content of the bases in a rock to the oxygen content in the silica. Also known as oxygen ratio. { ə'sid·ə·tē ‚kō·ə'fish·ənt }

acidity function [CHEM] A quantitative scale for measuring the acidity of a solvent system; usually established over a range of compositions. { ə'sid·əd·ē ‚fəŋk·shən }

acidizing [PETRO ENG] Well-stimulation method to increase oil production by injecting hydrochloric acid into the oil-bearing formation; the acid dissolves rock to enlarge the porous passages through which the oil must flow. { ‚as·ə'dīz·iŋ }

acid jetting [PETRO ENG] The jetting, from a device lowered through oil-well tubing, of an acid spray onto bottom-hole rock to clean away mud and scale interfering with oil flow. { 'as·əd 'jed·iŋ }

acid lead [MET] A 99.9% pure commercial lead made by adding copper to fully refined lead. { 'as·əd 'led }

acid lining [ENG] In steel production, a silica-brick lining used in furnaces. { 'as·əd 'līn·iŋ }

acid mine drainage [MIN ENG] Drainage from bituminous coal mines containing a large concentration of acidic sulfates, especially ferrous sulfate. { 'as·əd 'mīn ‚drā·nij }

acid mine water [MIN ENG] Mine water with free sulfuric acid, due to the weathering of iron pyrites. { 'as·əd 'mīn ‚wȯd·ər }

acid number [CHEM] See acid value. [ENG] A number derived from a standard test indicating the acid or base composition of lubricating oils; it in no way indicates the corrosive attack of the used oil in service. Also known as corrosion number. { 'as·əd ¦näm·bər }

acidolysis [ORG CHEM] A chemical reaction involving the decomposition of a molecule, with the addition of the elements of an acid to the molecule; the reaction is comparable to hydrolysis or alcoholysis, in which water or alcohol, respectively, is used in place of the acid. Also known as acyl exchange. { ‚as·ə'däl·ə·səs }

acid open-hearth process [MET] A steelmaking process employing an open-hearth furnace lined with siliceous-type refractories. { 'as·əd ‚ō·pən 'härth ‚prä·səs }

acidophil [BIOL] **1.** Any substance, tissue, or organism having an affinity for acid stains. **2.** An organism having a preference for an acid environment. [HISTOL] **1.** An alpha cell of the adenohypophysis. **2.** See eosinophil. { ə'sid·ə‚fil }

acidophilia See eosinophilia. { ə‚sid·ə'fil·ē·ə }

acidophilic erythroblast See normoblast. { ə'sid·ə‚fil·ik ə'rith·rə‚blast }

acidosis [MED] A condition of decreased alkali reserve of the blood and other body fluids. { ‚as·ə'dō·səs }

acidotrophic [BIOL] Having an acid nutrient requirement. { ə‚sid·ə'trōf·ik }

acid phosphatase [BIOCHEM] An enzyme in blood which catalyzes the release of phosphate from phosphate esters; optimum activity at pH 5. { 'as·əd 'fäs·fə‚tās }

acid phosphate [INORG CHEM] A mono- or dihydric phosphate; for example, M_2HPO_4 or MH_2PO_4, where M represents a metal atom. { 'as·əd 'fäs‚fāt }

acid pickle [MATER] Industrial waste water that is the spent liquor from a chemical process used to clean metal surfaces. { 'as·əd 'pik·əl }

acid polishing [ENG] The use of acids to polish a glass surface. { 'as·əd 'päl·ish·iŋ }

acid potassium phthalate See potassium biphthalate. { 'as·əd pə'tas·ē·əm 'tha‚lāt }

acid potassium sulfate See potassium bisulfate. { 'as·əd pə'tas·ē·əm 'səl‚fāt }

acid precipitation [METEOROL] Rain or snow with a pH of less than 5.6. { 'as·əd prə‚sip·ə'tā·shən }

acid process [CHEM ENG] In paper manufacture, a pulp digestion process that uses an acidic reagent, for example, a bisulfite solution containing free sulfur dioxide. [MET] A melting process carried out in a furnace lined with acidic materials which combine readily with the oxides in the ore. { 'as·əd ‚prä·səs }

acid-producing material [MIN ENG] In coal mining, rock strata containing sufficient amounts of pyrite to cause acid formation when the strata are weathered by air and water. { 'as·əd prə¦düs·iŋ mə'tir·ē·əl }

acid-proof coating [MATER] Material in liquid form suitable for application, by spraying, to the wall of projectile or bomb cavities to protect the metal from attack by explosives or other shell fillers. { 'as·əd ¦prüf ¦kōt·iŋ }

acid rain [METEOROL] Precipitation in the form of water drops that incorporates anthropogenic acids and acid materials. { ¦as·əd 'rān }

acid reaction [CHEM] A chemical reaction produced by an acid. { 'as·əd rē'ak·shən }

acid recovery plant [CHEM ENG] In some refineries, a facility for separating sludge acid into acid oil, tar, and weak sulfuric acid, with provision for later reconcentration. { 'as·əd rə'kəv·ə·rē ‚plant }

acid-refined oils [MATER] A class of linseed oils with the mucilaginous component removed by treatment of raw oil with sulfuric acid. { 'as·əd rə¦fīnd 'ȯilz }

acid refractory [MATER] A refractory that is composed principally of silica and reacts at high temperatures with bases such as lime, alkalies, and basic oxides. { 'as·əd rə'frak·tə·rē }

acid-resistant [MATER] Able to withstand chemical reaction with or degeneration by acids. { 'as·əd rə¦zis·tənt }

acid salt [CHEM] A compound derived from an acid and base in which only a part of the hydrogen is replaced by a basic radical; for example, the acid sulfate $NaHSO_4$. { 'as·əd ‚sȯlt }

acid slag [MET] Furnace slag in which there is more silica and silicates than lime and magnesia. { 'as·əd ‚slag }

acid sludge [CHEM ENG] The residue left after treating petroleum oil with sulfuric acid for the removal of impurities. { 'as·əd ‚sləj }

acid sodium tartrate See sodium bitartrate. { 'as·əd sōd·ē·əm 'tär‚trāt }

acid soil [GEOL] A soil with pH less than 7; results from presence of exchangeable hydrogen and aluminum ions. { 'as·əd 'sȯil }

acid solution [CHEM] An aqueous solution containing more hydrogen ions than hydroxyl ions. { 'as·əd sə'lü·shən }

acid soot [ENG] Carbon particles that have absorbed acid fumes as a by-product of combustion; hydrochloric acid absorbed on carbon particulates is frequently the cause of metal corrosion in incineration. { 'as·əd 'süt }

acid spar [MINERAL] A grade of fluorspar containing over 98% CaF_2 and no more than 1% SiO_2; produced by flotation; used for the production of hydrofluoric acid. { 'as·əd 'spär }

acid spoil [MIN ENG] Spoil material having pyrite content sufficiently high to produce acidic water as a result of weathering. { 'as·əd ‚spȯil }

acid steel [MET] Steel produced in a melting furnace employing siliceous-type refractories. { 'as·əd 'stēl }

acid tartrate See bitartrate. { 'as·əd 'tär‚trāt }

acid tide [MED] A period of increased acidity of urine and body fluids. { 'as·əd ‚tīd }

acid treatment [CHEM ENG] A refining process in which unfinished petroleum products, such as gasoline, kerosine, and diesel oil, are contacted with sulfuric acid to improve their color, odor, and other properties. [PETRO ENG] In petroleum-bearing formations, use of acid to enlarge the pore spaces and passages through which the reservoir fluids flow. { 'as·əd 'trēt·mənt }

acidulant [FOOD ENG] One of a class of chemicals added to food to increase either tartness or acidity, such as malic or citric acids for tartness and phosphoric acid for acidity. { ə'sij·ə·lənt }

acidulous water [HYD] Mineral water either with dissolved carbonic acid or dissolved sulfur compounds such as sulfates. { ə'sij·ə·ləs 'wȯd·ər }

acid value Also known as acid number. [CHEM] The acidity of a solution expressed in terms of normality. [ORG CHEM] A number indicating the amount of nonesterified fatty acid present in a sample of fat or fatty oil as determined by alkaline titration. { 'as·əd 'val·yü }

acid wash [MATER] A solution of phosphoric acid applied to steel parts that removes and neutralizes the alkaline solutions used for grease removal after machining; also leaves a metallic phosphate coating which accepts paint well and, of itself, provides a degree of protection against rust. { 'as·əd 'wash }

acid-water pollution [ENG] Industrial wastewaters that are acidic; usually appears in effluent from the manufacture of chemicals, batteries, artificial and natural fiber, fermentation processes (beer), and mining. { 'as·əd 'wȯd·ər pə'lü·shən }

acieration [MET] Electrolytic coating of a thin metal plate with iron; the iron hardens to steellike strength. { ‚a‚sī·ə'rā·shən }

acinar [ANAT] Pertaining to an acinus. { 'as·ə·nər }

acinar cell [ANAT] Any of the cells lining an acinous gland. { 'as·ə·nər 'sel }

Acinetobacter [MICROBIO] A genus of nonmotile, short, plump, almost spherical rods in the family Neisseriaceae; strictly aerobic; resistant to penicillin. { ‚as·ə‚nēd·ō'bak·tər }

aciniform [ZOO] Shaped like a berry or a bunch of grapes. { ə'sin·ə‚fȯrm }

acinotubular gland See tubuloalveolar gland. { ‚as·ə‚nō'tü·byə·lər 'gland }

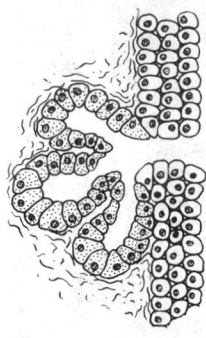

ACINOUS GLAND

An example of a typical compound type of acinous gland.

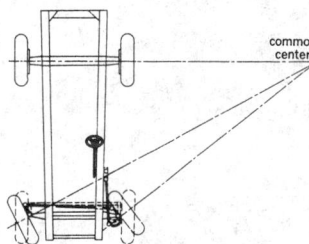

ACKERMAN STEERING GEAR

common center

All wheels turn about a common center.

acinous [BIOL] Of or pertaining to acini. { 'as·ə,nəs }

acinous gland [ANAT] A multicellular gland with sac-shaped secreting units. Also known as alveolar gland. { 'as·ə,nəs ,gland }

acinus [ANAT] The small terminal sac of an acinous gland, lined with secreting cells. [BOT] An individual drupelet of a multiple fruit. { 'as·ə·nəs }

Acipenser [VERT ZOO] A genus of actinopterygian fishes in the sturgeon family, Acipenseridae. { 'as·ə,pen·sər }

Acipenseridae [VERT ZOO] The sturgeons, a family of actinopterygian fishes in the order Acipenseriformes. { ,as·ə,pen'ser·ə·dē }

Acipenseriformes [VERT ZOO] An order of the subclass Actinopterygii represented by the sturgeons and paddlefishes. { as·ə,pen,ser·ə'fȯr,mēz }

ACK *See* acknowledge character.

Ackeret method [FL MECH] A method of studying the behavior of an airfoil in a supersonic airstream based on the hypothesis that the disturbance caused by the airfoil consists of two plane waves, at the leading and trailing edges, which propagate outward like sound waves and each makes an angle equal to the Mach angle with the direction of flow. { 'ak·ə·rət ,meth·əd }

Ackerman linkage *See* Ackerman steering gear. { 'ak·ər·mən ,liŋ·kij }

Ackerman steering gear [MECH ENG] Differential gear or linkage that turns the two steered road wheels of a self-propelled vehicle so that all wheels roll on circles with a common center. Also known as Ackerman linkage. { 'ak·ər·mən 'stēr·iŋ ,gir }

acknowledge character [COMPUT SCI] A signal that a receiving station transmits in order to indicate that a block of information has been received and that its validity has been checked. Also known as acknowledgement. Abbreviated ACK. { ak'nä·lij 'kar·ək·tər }

acknowledgement *See* acknowledge character. { ak'nä·lij·mənt }

aclastic [OPTICS] Having the property of not refracting light. { |ā'klas·tik }

aclinal [GEOL] Without dip; horizontal. { |ā'klīn·əl }

aclinic [GEOPHYS] Referring to a situation where a freely suspended magnetic needle remains in a horizontal position. { a'klin·ik }

aclinic line *See* magnetic equator. { a'klin·ik 'līn }

aΩcm *See* abohm centimeter.

Acmaeidae [INV ZOO] A family of gastropod mollusks in the order Archaeogastropoda; includes many limpets. { ak'mē·ə,dē }

acme [PALEON] The time of largest abundance or variety of a fossil taxon; the taxon may be either general or local. { 'ak·mē }

acme harrow [AGR] A type of harrow having a transverse horizontal frame with stiff curved blades. Also known as blade harrow; curved knife-tooth harrow; pulverizer. { 'ak·mē 'ha·rō }

acme screw thread [DES ENG] A standard thread having a profile angle of 29° and a flat crest; used on power screws in such devices as automobile jacks, presses, and lead screws on lathes. Also known as acme thread. { 'ak·mē 'skrü ,thred }

acme thread *See* acme screw thread. { 'ak·mē ,thred }

acmic [ECOL] A phase or period in which an aquatic population undergoes seasonal changes. { 'ak·mik }

acmite [MINERAL] NaFeSi$_2$O$_6$ A brown or green silicate mineral of the pyroxene group, often in long, pointed prismatic crystals; hardness is 6–6.5 on Mohs scale, and specific gravity is 3.50–3.55; found in igneous and metamorphic rocks. { 'ak,mīt }

acne [MED] A pleomorphic, inflammatory skin disease involving sebaceous follicles of the face, back, and chest and characterized by blackheads, whiteheads, papules, pustules, and nodules. { 'ak·nē }

acne rosacea [MED] A form of acne occurring in older persons and seen as reddened inflamed areas on the forehead, nose, and cheeks. { 'ak·nē rō'zā·shə }

Acnidosporidia [INV ZOO] An equivalent name for the Haplospora. { ak,nī·də,spō·'rid·ē·ə }

acnode *See* isolated point. { 'ak·nōd }

Acoela [INV ZOO] An order of marine flatworms in the class Turbellaria characterized by the lack of a digestive tract and coelomic cavity. { ā'sēl·ə }

Acoelea [INV ZOO] An order of gastropod mollusks in the subclass Opistobranchia; includes many sea slugs. { ,ā·sə'lē·ə }

Acoelomata [INV ZOO] A subdivision of the animal kingdom; individuals are characterized by lack of a true body cavity. { ā,sēl·ə'mäd·ə }

acoelomate [ZOO] Pertaining to an animal that lacks a coelom. { ā'sēl·ə,māt }

acoelous [ZOO] **1.** Lacking a true body cavity or coelom. **2.** Lacking a true stomach or digestive tract. { ,ā'sēl·əs }

acolpate [BOT] Of pollen grains, lacking furrows or grooves. { ,ā'kōl,pāt }

Aconchulinida [INV ZOO] An order of protozoans in the subclass Filosia comprising a small group of naked amebas having filopodia. { ə,kän·kə'lin·ə·də }

aconitase [BIOCHEM] An enzyme involved in the Krebs citric acid cycle that catalyzes the breakdown of citric acid to *cis*-aconitic and isocitric acids. { ə'kän·ə,tās }

aconite [BOT] Any plant of the genus *Aconitum*. Also known as friar's cowl; monkshood; mousebane; wolfsbane. [PHARM] A toxic drug obtained from the dried tuberous root of *Aconitum napellus;* the principal alkaloid is aconitine. { 'ak·ə,nīt }

aconitic acid [ORG CHEM] C$_6$H$_6$O$_6$ A white, crystalline organic acid found in sugarcane and sugarbeet; obtained during manufacture of sugar. { ,ak·ə'nid·ik 'as·əd }

aconitine [PHARM] C$_{34}$H$_{47}$O$_{11}$N A poisonous, white, crystalline alkaloid compound obtained from aconites such as monkshood (*Aconitum napellus*). { ə'kän·ə,tēn }

acorn [BOT] The nut of the oak tree, usually surrounded at the base by a woody involucre. { 'ā,kȯrn }

acorn barnacle [INV ZOO] Any of the sessile barnacles that are enclosed in conical, flat-bottomed shells and attach to ships and near-shore rocks and piles. { 'ā,kȯrn ,bär·nə·kəl }

acorn disease [PL PATH] A virus disease of citrus plants characterized by malformation of the fruit, which is somewhat acorn-shaped. { 'ā,kȯrn diz,ēz }

acorn sugar *See* quercitol. { 'ā,kȯrn ,shug·ər }

acorn tube [ELECTR] An ultra-high-frequency electron tube resembling an acorn in shape and size. { 'ā,kȯrn ,tüb }

acorn worm [INV ZOO] Any member of the class Enteropneusta, free-living animals that usually burrow in sand or mud. Also known as tongue worm. { 'ā,kȯrn ,wȯrm }

acotyledon [BOT] A plant without cotyledons. { ā,käd·əl'ēd·ən }

acoubuoy [ENG] An acoustic listening device similar to a sonobuoy, used on land to form an electronic fence that will pick up sounds of enemy movements and transmit them to orbiting aircraft or land stations. { ə'kü,bȯi }

acouchi [VERT ZOO] A hystricomorph rodent represented by two species in the family Dasyproctidae; believed to be a dwarf variety of the agouti. { ə'kü·shē }

acoustic [ACOUS] Relating to, containing, producing, arising from, actuated by, or carrying sound. { ə'küs·tik }

acoustic absorption *See* sound absorption. { ə'küs·tik əb'sȯrp·shən }

acoustic absorption coefficient *See* sound absorption coefficient. { ə'küs·tik əb'sȯrp·shən ,kō·ə,fish·ənt }

acoustic absorptivity *See* sound absorption coefficient. { ə'küs·tik ab,sorp'tiv·ə·tē }

acoustical [ACOUS] Having a characteristic concerning sound, of an object or quantity that in and of itself does not have properties associated with sound, such as a device, measurement, or symbol. { ə'küs·tə·kəl }

acoustical ceiling [BUILD] A ceiling covered with or built of material with special acoustical properties. { ə'küs·tə·kəl 'sēl·iŋ }

acoustical ceiling system [BUILD] A system for the structural support of an acoustical ceiling; lighting and air diffusers may be included as part of the system. { ə'küs·tə·kəl 'sēl·iŋ 'sis·təm }

acoustical door [BUILD] A solid door with gasketing along the top and sides, and usually an automatic door bottom, designed to reduce noise transmission. { ə'küs·tə·kəl 'dȯr }

acoustical Doppler effect [ACOUS] The change in pitch of a sound observed when there is relative motion between source and observer. { ə'küs·tə·kəl 'däp·lər ə'fekt }

acoustical holography [PHYS] A technique for using sound to form visible images, in which acoustic beams form an interference pattern of an object and a beam of light interacts with this pattern and is focused to form an optical image. { ə'küs·tə·kəl hō'läg·rə·fē }

acoustical insulation board [MATER] A porous board designed or used for acoustical applications or for sound-insulating construction. { ə'küs·tə·kəl in·sə'lā·shən ,bórd }

acoustical material [MATER] Any natural or synthetic material that absorbs sound; acoustical tile is an example. { ə'küs·tə·kəl mə'tir·ē·əl }

acoustical model [CIV ENG] A model used to investigate certain acoustical properties of an auditorium or room such as sound pressure distribution, sound-ray paths, and focusing effects. { ə'küs·tə·kəl 'mäd·əl }

acoustical plaster [MATER] A low-density sound-absorbing plaster applied as a finish coat to provide a uniform finished surface. { ə'küs·tə·kəl 'plas·tər }

acoustical scintillation [COMMUN] Irregular fluctuation in the received intensity of sounds propagated through the atmosphere from a source of uniform output; produced by nonhomogeneous structure of the atmosphere along the path of the sound. { ə'küs·tə·kəl sin·tə'lā·shən }

acoustical tile [MATER] A sound-absorptive material, usually having unit dimensions of 24 × 24 inches (approximately 61 × 61 centimeters) or less, used to cover an acoustical ceiling. { ə'küs·tə·kəl 'tīl }

acoustical treatment [CIV ENG] That part of building planning that is designed to provide a proper acoustical environment; includes the use of acoustical material. { ə'küs·tə·kəl 'trēt·mənt }

acoustic amplifier [ELECTR] A device that amplifies mechanical vibrations directly at audio and ultrasonic frequencies. Also known as acoustoelectric amplifier. { ə'küs·tik 'am·plə,fī·ər }

acoustic approximation [FL MECH] The approximation that leads from the nonlinear hydrodynamic equations of a gas to the linear wave equation for sound wave propagation. { ə'küs·tik ə,präk·sə'mā·shən }

acoustic array [ENG ACOUS] A sound-transmitting or sound-receiving system whose elements are arranged to give desired directional characteristics. { ə'küs·tik ə'rā }

acoustic axis See axis of acoustic symmetry. { ə'küs·tik 'ak·səs }

acoustic bearing See sonic bearing. { ə'küs·tik 'ber·iŋ }

acoustic Bessel bullet [ACOUS] One of a class of localized wave solutions to the three-dimensional wave equation that maintain their shape and amplitude as they propagate in space. { ə'küs·tik 'be·səl ,bül·ət }

acoustic branch [SOLID STATE] One of the parts of the dispersion relation, frequency as a function of wave number, for crystal lattice vibrations, representing vibration at low (acoustic) frequencies. { ə'küs·tik ,branch }

acoustic bridge [ELECTR] A device, based on the principle of the electrical Wheatstone bridge, used for analysis of deafness. { ə'küs·tik 'brij }

acoustic capacitance See acoustic compliance. { ə'küs·tik kə'pas·ə·təns }

acoustic cavitation [FL MECH] The formation of vapor-filled bubbles in a liquid during the short periodic intervals of negative pressure, or tensile stress, that accompany the passage of a sound wave. { ə,kü·stik ,kav·ə'tā·shən }

acoustic center [ENG ACOUS] The center of the spherical sound waves radiating outward from an acoustic transducer. { ə'küs·tik 'sen·tər }

acoustic clarifier [ENG ACOUS] System of cones loosely attached to the baffle of a loudspeaker and designed to vibrate and absorb energy during sudden loud sounds to suppress these sounds. { ə'küs·tik 'klar·ə,fī·ər }

acoustic comfort index [ACOUS] An arbitrarily designed scale to indicate the noise inside the passenger cabin of an aircraft; on this scale +100 represents ideal conditions or zero noise, 0 represents barely tolerable conditions, and −100 represents intolerable conditions. Abbreviated ACI. { ə'küs·tik 'käm·fərt 'in·deks }

acoustic compliance [ACOUS] The reciprocal of acoustic stiffness. Also known as acoustic capacitance. { ə'küs·tik kəm'plī·əns }

acoustic convolver See convolver. { ə'küs·tik kən'välv·ər }

acoustic coupler [ENG ACOUS] A device used between the modem of a computer terminal and a standard telephone line to permit transmission of digital data in either direction without making direct connections. { ə'küs·tik 'kəp·lər }

acoustic delay [ENG ACOUS] A delay which is deliberately introduced in sound reproduction by having the sound travel a certain distance along a pipe before conversion into electric signals. { ə'küs·tik di'lā }

acoustic delay line [ELECTR] A device in which acoustic signals are propagated in a medium to make use of the sonic propagation time to obtain a time delay for the signals. Also known as sonic delay line. { ə'küs·tik di'lā ,līn }

acoustic detection [ENG] Determination of the profile of a geologic formation, an ocean layer, or some object in the ocean by measuring the reflection of sound waves off the object. { ə'küs·tik di'tek·shən }

acoustic detector [ELECTR] The stage in a receiver at which demodulation of a modulated radio wave into its audio component takes place. { ə'küs·tik di'tek·tər }

acoustic dispersion [ACOUS] A complex sound wave's separation into its frequency components as it passes through a medium; usually measured by the rate of change of velocity with frequency. { ə'küs·tik dis'pər·zhən }

acoustic domain [ACOUS] A concentration of crystal lattice vibrations traveling at the speed of sound; used to generate light from an array of *pn* junctions. { ə'küs·tik də'mān }

acoustic emission [ACOUS] The phenomenon of transient elastic-wave generation due to a rapid release of strain energy caused by a structural alteration in a solid material. Also known as stress-wave emission. { ə'küs·tik ē'mish·ən }

acoustic energy See sound energy. { ə'küs·tik 'en·ər·jē }

acoustic fatigue [MECH] The tendency of a material, such as a metal, to lose strength after acoustic stress. { ə'küs·tik fə'tēg }

acoustic feedback [ENG ACOUS] The reverberation of sound waves from a loudspeaker to a preceding part of an audio system, such as to the microphone, in such a manner as to reinforce, and distort, the original input. Also known as acoustic regeneration. { ə'küs·tik 'fēd,bak }

acoustic filter See filter. { ə'küs·tik 'fil·tər }

acoustic fix See sonic fix. { ə'küs·tik ,fiks }

acoustic generator [ENG ACOUS] A transducer which converts electrical, mechanical, or other forms of energy into sound. { ə'küs·tik 'jen·ə,rād·ər }

acoustic grating [ACOUS] A series of rods or other suitable objects of equal size placed in a row a fixed distance apart; causes sounds with different wavelengths to be diffracted in different directions. { ə'küs·tik 'grāt·iŋ }

acoustic heat engine [ENG] A device that transforms heat energy first into sound energy and then into electrical power, without the use of moving mechanical parts. { ə'küs·tik 'hēt ,en·jən }

acoustic hologram [ENG] The phase interference pattern, formed by acoustic beams, that is used in acoustical holography; when light is made to interact with this pattern, it forms an image of an object placed in one of the beams. { ə'küs·tik 'häl·ə,gram }

acoustic homing [NAV] The following of a path of acoustic energy to or toward its source or point of reflection. { ə'küs·tik 'hōm·iŋ }

acoustic horn See horn. { ə'küs·tik 'hórn }

acoustic image [ACOUS] The geometric space figure that is made up of the acoustic foci of an acoustic lens, mirror, or other acoustic optical system and is the acoustic counterpart of an extended source of sound. Also known as image. { ə'küs·tik 'im·ij }

acoustic imaging [ACOUS] The use of ultrasound to produce real-time images of the internal structure of a metallic or biological object that is opaque to light. Also known as sonography; ultrasonic imaging; ultrasonography. { ə'küs·tik 'im·ij·iŋ }

acoustic impedance [ACOUS] The complex ratio of the sound pressure on a given surface to the sound flux through that surface, expressed in acoustic ohms. { ə'küs·tik im'pēd·əns }

acoustic inertance See acoustic mass. { ə'küs·tik i'nərt·əns }

acoustic insulation [MATER] A material used to diminish sound energy that passes through it or strikes its surface. { ə'küs·tik in·sə'lā·shən }

acoustic intensity [ACOUS] The limit approached by the ratio of the acoustic power in a given area to the magnitude

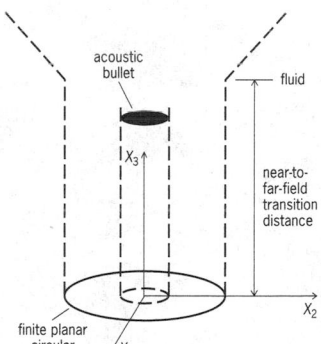

ACOUSTIC BESSEL BULLET

Acoustic Bessel bullet launched from a finite planar circular aperture.

of this area as the magnitude approaches zero. { ə'küs·tik in'ten·səd·ē }

acoustic interferometer [ACOUS] A device for measuring the velocity and attenuation of sound waves in a gas or liquid by an interference method. { ə'küs·tik in·tər·fə'rä·məd·ər }

acoustic jamming [ENG ACOUS] The deliberate radiation or reradiation of mechanical or electroacoustic signals with the objectives of obliterating or obscuring signals which the enemy is attempting to receive and of deterring enemy weapons systems. { ə'küs·tik 'jam·iŋ }

acoustic labyrinth [ENG ACOUS] Special baffle arrangement used with a loudspeaker to prevent cavity resonance and to reinforce bass response. { ə'küs·tik 'lab·ə,rinth }

acoustic lens [MATER] Selected materials shaped to refract sound waves in accordance with the principles of geometrical optics, as is done for light. Also known as lens. { ə'küs·tik 'lenz }

acoustic levitation [ACOUS] The use of a very intense sound wave to keep a body suspended above the device producing the sound wave. { ə'küs·tik lev·ə'tā·shən }

acoustic line [ENG ACOUS] The acoustic equivalent of an electrical transmission line, involving baffles, labyrinths, or resonators placed at the rear of a loudspeaker and arranged to help reproduce the very low audio frequencies. { ə'küs·tik 'līn }

acoustic line of position See sonic line of position. { ə'küs·tik 'līn əv pə'zish·ən }

acoustic logging [PETRO ENG] A porosity measurement technique for drill holes; depth is compared to the travel time of a sonic impulse through a given portion of the formation—a rate dependent upon rock composition and fluids in the formation. { ə'küs·tik 'läg·iŋ }

acoustic Mach meter [AERO ENG] A device which registers data on sound propagation for the calculation of Mach number. { ə'küs·tik 'mäk ,mēd·ər }

acoustic mass [ACOUS] The quantity which, after multiplication by 2π times the frequency, results in the acoustic reactance associated with the kinetic energy of the sound medium. Also known as acoustic inertance. { ə'küs·tik 'mas }

acoustic mass reactance [ACOUS] The part of the acoustic reactance associated with the kinetic energy of a medium. Also known as mass reactance. { ə'küs·tik 'mas rē'ak·təns }

acoustic measurement [ACOUS] The process of quantitatively determining one or more properties of sound. { ə'küs·tik 'mezh·ər·mənt }

acoustic memory [COMPUT SCI] A computer memory that uses an acoustic delay line, in which a train of pulses travels through a medium such as mercury or quartz. { ə'küs·tik 'mem·ə·rē }

acoustic microscope [OPTICS] An instrument which employs acoustic radiation at microwave frequencies to allow visualization of the microscopic detail exhibited in elastic properties of an object. { ə'küs·tik 'mīk·rə,skōp }

acoustic mine [ORD] A naval mine which is activated by acoustic means. { ə'küs·tik 'mīn }

acoustic mode [SOLID STATE] The type of crystal lattice vibrations which for long wavelengths act like an acoustic wave in a continuous medium, but which for shorter wavelengths approach the Debye frequency, showing a dispersive decrease in phase velocity. { ə'küs·tik 'mōd }

acoustic navigation See sonic navigation. { ə'küs·tik nav·ə'gā·shən }

acoustic nerve [NEURO] See auditory nerve. { ə'küs·tik ,nərv }

acoustic noise [ACOUS] Noise in the acoustic spectrum; usually measured in decibels. { ə'küs·tik ,nóiz }

acoustic ocean-current meter [ENG] An instrument that measures current flow in rivers and oceans by transmitting acoustic pulses in opposite directions parallel to the flow and measuring the difference in pulse travel times between transmitter-receiver pairs. { ə'küs·tik 'ō·shən ,kər·ənt 'mēd·ər }

acoustic ohm [ACOUS] The unit of acoustic impedance. Also known as acoustic reactance unit; acoustic resistance unit. { ə'küs·tik 'ōm }

acousticophobia [PSYCH] Abnormal fear of sounds. { a,küs·tə·kə'fōb·ē·ə }

acoustic particle detection [PARTIC PHYS] A technique for detecting charged particles traversing a medium by recording

the impulsive acoustic signals that result from rapid thermal expansion of the medium. { ə'küs·tik 'pärd·ə·kəl di,tek·shən }

acoustic phonon [SOLID STATE] A quantum of excitation of an acoustic mode of vibration. { ə'küs·tik 'fōn,än }

acoustic plaster [MATER] Plaster having good acoustic absorbing properties; it contains metal which, upon contact with water, evolves gas to aerate the mass. { ə'küs·tik 'plas·tər }

acoustic position reference system [ENG] An acoustic system used in offshore oil drilling to provide continuous information on ship position with respect to an ocean-floor acoustic beacon transmitting an ultrasonic signal to three hydrophones on the bottom of the drilling ship. { ə'küs·tik pə'zish·ən ¦ref·rəns ,sis·təm }

acoustic power See sound power. { ə'küs·tik 'paú·ər }

acoustic radar [ENG] Use of sound waves with radar techniques for remote probing of the lower atmosphere, up to heights of about 5000 feet (1500 meters), for measuring wind speed and direction, humidity, temperature inversions, and turbulence. { ə'küs·tik 'rā,där }

acoustic radiation [ACOUS] Infrasonic, sonic, or ultrasonic waves propagating through a solid, liquid, or gaseous medium. { ə'küs·tik ,räd·ē'ā·shən }

acoustic radiation pressure [ACOUS] A unidirectional, steady-state pressure exerted upon a surface exposed to a sound wave. { ə'küs·tik ,räd·ē'ā·shən ,presh·ər }

acoustic radiator [ENG ACOUS] A vibrating surface that produces sound waves, such as a loudspeaker cone or a headphone diaphragm. { ə'küs·tik 'räd·ē,äd·ər }

acoustic radiometer [ENG] An instrument for measuring sound intensity by determining the unidirectional steady-state pressure caused by the reflection or absorption of a sound wave at a boundary. { ə'küs·tik ,räd·ē·ə'ä·məd·ər }

acoustic ratio [ENG ACOUS] The ratio of the intensity of sound radiated directly from a source to the intensity of sound reverberating from the walls of an enclosure, at a given point in the enclosure. { ə'küs·tik 'rā·shō }

acoustic reactance [ACOUS] The imaginary component of the acoustic impedance. { ə'küs·tik rē'ak·təns }

acoustic reactance unit See acoustic ohm. { ə'küs·tik rē'ak·təns ,yü·nət }

acoustic receiver [ELECTR] The complete equipment required for receiving modulated radio waves and converting them into sound. { ə'küs·tik rə'sēv·ər }

acoustic reciprocity theorem [ACOUS] A theorem which states that in the acoustic field due to a sound source at point A, the sound pressure received at any other point B is the same as that which would be produced at A if the source were placed at B, and that this can be generalized for multiple sources and receivers. { ə'küs·tik ,res·ə'präs·əd·ē ,thir·əm }

acoustic reflection coefficient See acoustic reflectivity. { ə'küs·tik ri'flek·shən ,kō·ə,fish·ənt }

acoustic reflectivity [ACOUS] Ratio of the rate of flow of sound energy reflected from a surface, on the side of incidence, to the incident rate of flow. Also known as acoustic reflection coefficient; sound reflection coefficient. { ə'küs·tik ,rē·flek'tiv·əd·ē }

acoustic reflex [NEURO] Brief, involuntary closure of the eyes due to stimulation of the acoustic nerve by a sudden sound. { ə'küs·tik 'rē,fleks }

acoustic reflex enclosure [ENG ACOUS] A loudspeaker cabinet designed with a port to allow a low-frequency contribution from the rear of the speaker cone to be radiated forward. { ə'küs·tik 'rē,fleks in,klō·zhər }

acoustic refraction [ACOUS] Variation of the direction of sound transmission due to spatial variation of the wave velocity in the medium. { ə'küs·tik ri'frak·shən }

acoustic regeneration See acoustic feedback. { ə'küs·tik rē,jen·ə'rā·shən }

acoustic resistance [ACOUS] The real component of the acoustic impedance. Also known as resistance. { ə'küs·tik ri'zis·təns }

acoustic resistance unit See acoustic ohm. { ə'küs·tik ri'zis·təns ,yü·nət }

acoustic resonance [ACOUS] A phenomenon exhibited by an acoustic system, such as an organ pipe or Helmholtz resonator, in which the response of the system to sound waves becomes very large when the frequency of the sound approaches a natural vibration frequency of the air in the system. { ə'küs·tik 'rez·ə·nəns }

acoustic resonator [ACOUS] An enclosure that produces sound-wave resonance at a particular frequency. { ə'küs·tik 'rez·ə‚nād·ər }

acoustics [PHYS] **1.** The science of the production, transmission, and effects of sound. **2.** The characteristics of a room that determine the qualities of sound in it relevant to hearing. { ə'küs·tiks }

acoustic scattering [ACOUS] The irregular reflection, refraction, and diffraction of sound in many directions. { ə'küs·tik 'skad·ər·iŋ }

acoustic seal [ENG ACOUS] A joint between two parts to provide acoustical coupling with low losses of energy, such as between an earphone and the human ear. { ə'küs·tik 'sēl }

acoustic shadow [ACOUS] A region immediately behind an object placed in the path of a sound wave whose wavelength is much smaller than the object, in which the initial sound wave is cut off by the object and the sound intensity is determined by the diffraction and interference of sound waves bent around the obstacle. { ə'küs·tik 'shad·ō }

acoustic shielding [ACOUS] A sound barrier that prevents the transmission of acoustic energy. { ə'küs·tik 'shēld·iŋ }

acoustic signal processing [ACOUS] The extraction of information from signals propagated undersea, in the atmosphere, or in the solid earth in the presence of acoustic noise. { ə'küs·tik 'sig‚nəl ‚prä‚ses·iŋ }

acoustic signature [ENG] In acoustic detection, the profile characteristic of a particular object or class of objects, such as a school of fish or a specific ocean-bottom formation. { ə'küs·tik 'sig·nə·chər }

acoustic spectrograph [ENG] A spectrograph used with sound waves of various frequencies to study the transmission and reflection properties of ocean thermal layers and marine life. { ə'küs·tik 'spek·trə‚graf }

acoustic spectrometer [ENG ACOUS] An instrument that measures the intensities of the various frequency components of a complex sound wave. Also known as audio spectrometer. { ə'küs·tik spek'träm·əd·ər }

acoustic spectrum [ACOUS] The range of acoustic frequencies, extending from subsonic to ultrasonic frequencies, that is, approximately from zero to at least 1 megahertz. { ə'küs·tik 'spek·trəm }

acoustic stiffness [ACOUS] The product of the angular frequency and the acoustic stiffness reactance. { ə'küs·tik 'stif·nəs }

acoustic stiffness reactance [ACOUS] The part of acoustic reactance associated with the potential energy of a medium or its boundaries. Also known as stiffness reactance. { ə'küs·tik 'stif·nəs rē'ak·təns }

acoustic strain gage [ENG] An instrument used for measuring structural strains; consists of a length of fine wire mounted so its tension varies with strain; the wire is plucked with an electromagnetic device, and the resulting frequency of vibration is measured to determine the amount of strain. { ə'küs·tik ‚strān ‚gāj }

acoustic streaming [FL MECH] Unidirectional flow currents in a fluid that are due to the presence of sound waves. { ə'küs·tik 'strēm·iŋ }

acoustic survey *See* sonic logging. { ə'küs·tik 'sər‚vā }

acoustic theodolite [ENG] An instrument that uses sound waves to provide a continuous vertical profile of ocean currents at a specific location. { ə'küs·tik thē'äd·əl‚īt }

acoustic theory [AERO ENG] The linearized small-disturbance theory used to predict the approximate airflow past an airfoil when the disturbance velocities caused by the flow are small compared to the flight speed and to the speed of sound. { ə'küs·tik 'thē·ə·rē }

acoustic tile [MATER] A thin, often decorative tile with sound-absorbing properties, used to cover ceilings and walls. { ə'küs·tik 'tīl }

acoustic tomography [ACOUS] An imaging or remote sensing technique in which information is collected from beams of acoustic radiation which have passed through an object, generally in the form of an image or other representation of a two-dimensional slice through the object. { ə'küs·tik tə'mäg·rə·fē }

acoustic torpedo [ORD] A naval torpedo which is directed toward its target either by the noise emitted by the target or by sonar. { ə'küs·tik tor'pēd·ō }

acoustic transducer [ENG ACOUS] A device that converts acoustic energy to electrical or mechanical energy, such as a microphone or phonograph pickup. { ə'küs·tik tranz'dü·sər }

acoustic transformer [ENG ACOUS] A device, such as a horn or megaphone, for increasing the efficiency of sound radiation. { ə'küs·tik tranz'for·mər }

acoustic transmission [ACOUS] The transfer of energy in the form of regular mechanical vibration through a gaseous, liquid, or solid medium. { ə'küs·tik tranz'mish·ən }

acoustic transmission coefficient *See* sound transmission coefficient. { ə'küs·tik tranz'mish·ən ‚kō·ə‚fish·ənt }

acoustic transmissivity *See* sound transmission coefficient. { ə'küs·tik ‚tranz·mis'iv·ə·dē }

acoustic transponder [NAV] A device used in underwater NAV which, on being interrogated by coded acoustic signals, emits acoustic reply. { ə'küs·tik tranz'pän·dər }

acoustic treatment [BUILD] The use of sound-absorbing materials to give a room a desired degree of freedom from echo and reverberation. { ə'küs·tik 'trēt·mənt }

acoustic velocity *See* speed of sound. { ə'küs·tik və'läs·ə·dē }

acoustic wave [ACOUS] **1.** An elastic nonelectromagnetic wave that has a frequency which may extend into the gigahertz range; one type is a surface acoustic wave, and the other type is a bulk or volume acoustic wave. Also known as elastic wave. **2.** *See* sound. { ə'küs·tik 'wāv }

acoustic-wave amplifier [ELECTR] An amplifier in which the charge carriers in a semiconductor are coupled to an acoustic wave that is propagated in a piezoelectric material, to produce amplification. { ə'küs·tik wāv 'am·plə‚fī·ər }

acoustic-wave-based sensor [ENG] A device that employs a surface acoustic wave, a thickness-shear-mode resonance (a resonant oscillation of a thin plate of material), or other type of acoustic wave to measure the physical properties of a thin film or liquid layer or, in combination with chemically sensitive thin films, to detect the presence and concentration of chemical analytes. { ə‚kü·stik 'wāv‚bāst ‚sen·sər }

acoustic well logging [ENG] A ground exploration method that uses a high-energy sound source and a receiver, both underground. { ə'küs·tik 'wel ‚läg·iŋ }

acoustoelectric amplifier *See* acoustic amplifier. { ə‚küs·tō·ə‚lek·trik 'am·plə‚fi·ər }

acoustoelectric effect [ELECTR] **1.** The development of a direct-current voltage in a semiconductor or metal by an acoustic wave traveling parallel to the surface of the material. Also known as electroacoustic effect. **2.** The amplification of a sound wave propagating in a piezoelectric semiconductor subject to a steady electric field that is strong enough that the resulting electron drift velocity exceeds the speed of sound. { ə‚küs·tō·ə'lek·trik i‚fekt }

acoustoelectronics [ENG ACOUS] The branch of electronics that involves use of acoustic waves at microwave frequencies (above 500 megahertz), traveling on or in piezoelectric or other solid substrates. Also known as pretersonics. { ə‚küs·tō·ə‚lek'trän·iks }

acoustooptical cell [ELEC] An electric-to-optical transducer in which an acoustic or ultrasonic electric input signal modulates or otherwise acts on a beam of light. { ə‚küs·tō‚äp·tə·kəl 'sel }

acoustooptical filter [OPTICS] An optical filter that is tuned across the visible spectrum by acoustic waves in the frequency range of 40 to 68 megahertz. { ə‚küs·tō‚äp·tə·kəl 'fil·tər }

acoustooptical material [MATER] A material in which the refractive index or some other optical property can be changed by an acoustic wave. { ə‚küs·tō‚äp·tə·kəl mə'tir·ē·əl }

acoustooptic interaction [OPTICS] A way to influence the propagation characteristics of an optical wave by applying a low-frequency acoustical field to the medium through which the wave passes. { ə‚küs·tō‚äp·tik ‚in·tə'rak·shən }

acoustooptic modulator [OPTICS] A device utilizing acoustooptic interaction ultrasonically to vary the amplitude or the phase of a light beam. Also known as Bragg cell. { ə‚küs·tō‚äp·tik 'mäd·yə‚lād·ər }

acoustooptics [OPTICS] The science that deals with interactions between acoustic waves and light. { ə‚küs·tō‚äp·tiks }

ACP *See* acyl carrier protein.

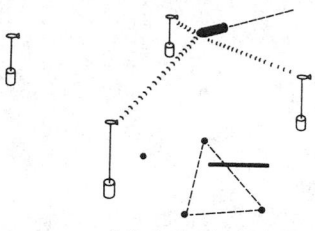

ACOUSTIC TRANSPONDER

pilot's dead reckoning tracer (DRT) presentation

Submersible navigating from underwater acoustic transponders.

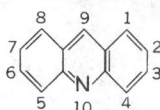

ACRIDINE

Structural formula for acridine.

a-c plane [CRYSTAL] A plane at right angles to the surface of movement in a crystal. { 'ā'sē 'plān }

acquire [ELECTR] **1.** Of acquisition radars, the process of detecting the presence and location of a target in sufficient detail to permit identification. **2.** Of tracking radars, the process of positioning a radar beam so that a target is in that beam to permit the effective employment of weapons. Also known as target acquisition. { ə'kwīr }

acquired [BIOL] Not present at birth, but developed by an individual in response to the environment and not subject to hereditary transmission. { ə'kwīrd }

acquired drive [PSYCH] Any aroused state of behavior that originates from and is nurtured by experience rather than inheritance. { ə'kwīrd 'drīv }

acquired immune deficiency syndrome [MED] A disease that is caused by the human immunodeficiency virus (HIV) and compromises the competency of the immune system; characterized by persistent lymphadenopathy and opportunistic infections such as *Pneumocystis carinii* pneumonia, cytomegalovirus, disseminated histoplasmosis, candidiasis, isosporiasis, and malignancies such as Kaposi's sarcoma. Abbreviated AIDS. { ə'kwīrd ə¦myün də¦fish·ən·sē 'sin‚drōm }

acquired immunity [IMMUNOL] Resistance to a microbial or other antigenic substance taken on by a naturally susceptible individual; may be either active or passive. { ə'kwīrd ə'myün·ə·dē }

acquired immunological tolerance [IMMUNOL] Failure of immunological responsiveness, that is, inability of antigen-sensitive cells to synthesize antibodies; induced by exposure to large amounts of an antigen. Also known as immunological paralysis. { ə'kwīrd ‚im·yü·nə¦laj·ə·kəl 'täl·ə·rəns }

acquisition [ENG] The process of pointing an antenna or a telescope so that it is properly oriented to allow gathering of tracking or telemetry data from a satellite or space probe. { ‚ak·wə'zish·ən }

acquisition and tracking radar [ENG] A radar set capable of locking onto a received signal and tracking the object emitting the signal; the radar may be airborne or on the ground. { ‚ak·wə'zish·ən ən 'trak·iŋ ‚rā‚där }

acquisition tone [COMPUT SCI] An audible tone that verifies entry into a minicomputer, microcomputer, or calculator. { ‚ak·wə'zish·ən ‚tōn }

acrania [MED] Partial or complete absence of the cranium at birth. { ā'krān·ē·ə }

Acrania [ZOO] A group of lower chordates with no cranium, jaws, vertebrae, or paired appendages; includes the Tunicata and Cephalochordata. { ā'krān·ē·ə }

acrasia [PSYCH] Lack of self-control. { ə'krā·zē·ə }

Acrasiales [BIOL] A group of microorganisms that have plant and animal characteristics; included in the phylum Myxomycophyta by botanists and Mycetozoia by zoologists. { ə'krazh·ē¦ā·lēz }

Acrasida [MYCOL] An order of Mycetozoia containing cellular slime molds. { ə'kras·ə·də }

Acrasieae [BIOL] An equivalent name for the Acrasiales. { ə‚krāz·ē¦ē‚ē }

acrasin [BIOCHEM] The chemotactic substance thought to be secreted by, and to effect aggregation of, myxamebas during their fruiting phase. { ə'krāz·ən }

Acrasiomycota [MYCOL] The phylum containing the cellular slime molds. { ə¦krā·zē·ō·mī'käd·ə }

acraspedote [INV ZOO] Describing tapeworm segments which are not overlapping. { ə'kras·pə‚dōt }

acre [MECH] A unit of area, equal to 43,560 square feet, or to 4046.8564224 square meters. { 'ā·kər }

Acree's reaction [ANALY CHEM] A test for protein in which a violet ring appears when concentrated sulfuric acid is introduced below a mixture of the unknown solution and a formaldehyde solution containing a trace of ferric chloride. { 'ak·rēz rē'ak·shən }

acre-foot [HYD] The volume of water required to cover 1 acre to a depth of 1 foot, hence 43,560 cubic feet; a convenient unit for measuring irrigation water, runoff volume, and reservoir capacity. { 'ā·kər 'fůt }

acre-foot per day [HYD] The United States unit of volume rate of water flow. Abbreviated acre-ft/d. { 'ā·kər 'fůt pər 'dā }

acre-ft/d See acre-foot per day.

acre-in. See acre-inch.

acre-inch [HYD] A unit of volume used in the United States for water flow, equal to 3630 cubic feet. Abbreviated acre-in. { 'ā·kər 'inch }

acre-yield [GEOL] The average amount of oil, gas, or water taken from one acre of a reservoir. { 'ā·kər ¦yēld }

acridine [ORG CHEM] $(C_6H_4)_2NCH$ A typical member of a group of organic heterocyclic compounds containing benzene rings fused to the 2,3 and 5,6 positions of pyridine; derivatives include dyes and medicines. { 'ak·rə‚dēn }

acridine dye [ORG CHEM] Any of a class of basic dyes containing the acridine nucleus that bind to deoxyribonucleic acid. { 'ak·rə‚dēn ¦dī }

acridine orange [ORG CHEM] A dye with an affinity for nucleic acids; the complexes of nucleic acid and dye fluoresce orange with RNA and green with DNA when observed in the fluorescence microscope. { 'ak·rə‚dēn 'är·inj }

acriflavine [ORG CHEM] $C_{14}H_{14}N_3Cl$ A yellow acridine dye obtained from proflavine by methylation in the form of red crystals; used as an antiseptic in solution. { ‚ak·rə'flā‚vēn }

acritarch [PALEON] A unicellular microfossil of unknown or uncertain biological origin that occurs abundantly in strata from the Precambrian and Paleozoic. { 'ak·rə‚tark }

acroagnosis [MED] Loss or absence of sense perception in a limb. { ‚ak·rō·ag'nō·səs }

acrobatholithic [GEOL] A stage in batholithic erosion where summits of cupolas and stocks are exposed without any exposure of the surface separating the barren interior of the batholith from the mineralized upper part. { ‚ak·rə¦bath·ə¦lith·ik }

acroblast [CYTOL] A vesicular structure in the spermatid formed from Golgi material. { 'ak·rə‚blast }

acrocarpous [BOT] In some mosses of the subclass Eubrya, having the sporophyte at the end of a stem and therefore exhibiting the erect habit. { ‚ak·rə'kär·pəs }

acrocentric chromosome [CYTOL] A chromosome having the centromere close to one end. { ¦ak·rə¦sen·trik 'krō·mə‚sōm }

acrocephalosyndactylism [MED] A congenital malformation consisting of an enlarged, pointed skull and defective separation of fingers and toes. Also known as Apert's syndrome. { ‚ak·rə‚sef·ə·lō‚sin'dak·tə‚liz·əm }

acrocephaly See oxycephaly. { ‚ak·rə'sef·ə·lē }

Acroceridae [INV ZOO] The humpbacked flies, a family of orthorrhaphous dipteran insects in the series Brachycera. { a‚krä'ser·ə·de }

acrodermatitis enteropathica [MED] An often fatal inherited disease involving inefficient intestinal absorption of zinc; readily treated by adding zinc to the diet. { ‚a·krō‚dər·mə¦tī·təs ‚en·tə·rə¦pa·thə·kə }

acrodomatia [ECOL] Specialized structures on certain plants adapted to shelter mites; relationship is presumably symbiotic. { ‚ak·rə·də'māsh·ē·ə }

acrodont [ANAT] Having teeth fused to the edge of the supporting bone. { 'ak·rə‚dänt }

acrodynia [MED] A childhood syndrome associated with mercury ingestion and characterized by periods of irritability alternating with apathy, anorexia, pink itching hands and feet, photophobia, sweating, tachycardia, hypertension, and hypotonia. { ‚ak·rō'din·ē·ə }

acrolein [ORG CHEM] $CH_2=CHCHO$ A colorless to yellow liquid with a pungent odor and a boiling point of 52.7°C; soluble in water, alcohol, and ether; used in organic synthesis, pharmaceuticals manufacture, and as an herbicide and tear gas. { ə'krōl·ē·ən }

acrolein cyanohydrin [ORG CHEM] $CH_2:CHCH(OH)CN$ A liquid soluble in water and boiling at 165°C; copolymerizes with ethylene and acrylonitrile; used to modify synthetic resins. { ə'krōl·ē·ən ‚sī·ə·nō'hī·drən }

acrolein dimer [ORG CHEM] $C_6H_8O_2$ A flammable, water-soluble liquid used as an intermediate for resins, dyestuffs, and pharmaceuticals. { ə'krōl·ē·ən 'dī·mər }

acrolein test [ANALY CHEM] A test for the presence of glycerin or fats; a sample is heated with potassium bisulfate, and acrolein is released if the test is positive. { ə'krōl·ē·ən ‚test }

acromegaly [MED] A chronic condition in adults caused by hypersecretion of the growth hormone and marked by enlarged jaws, extremities, and viscera, accompanied by certain physiological changes. { ‚ak·rō'meg·ə·lē }

acromelalgia See erythromelalgia. { ‚ak·rō·mə'läl·jē·ə }

acromere [HISTOL] The distal portion of a rod or cone in the retina. { 'ak·rō,mēr }

acrometer [ENG] An instrument to measure the density of oils. { ə'kräm·əd·ər }

acromion [ANAT] The flat process on the outer end of the scapular spine that articulates with the clavicle and forms the outer angle of the shoulder. { ə'krō·me,än }

acromorph [GEOL] A salt dome. { 'ak·rō,mórf }

acron [EVOL] Unsegmented head of the ancestral arthropod. [INV ZOO] **1.** The preoral, nonsegmented portion of an arthropod embryo. **2.** The prostomial region of the trochophore larva of some mollusks. { 'ak,rän }

acronematic [BIOL] Referring to a flagellum without hairs. { ,ak·rō·nə'mad·ik }

acroparesthesia [MED] A chronic self-limited symptom complex associated with a variety of systemic diseases, characterized by tingling, pins-and-needles sensations, numbness or stiffness, and occasionally pains in the hands and feet. { ,ak·rō,par·ə'thēzh·ē·ə }

acropetal [BOT] From the base toward the apex, as seen in the formation of certain organs or the spread of a pathogen. { ə'krä·pəd·əl }

acrophobia [PSYCH] Abnormal fear of great heights. { ,ak·rə'fōb·ē·ə }

Acrosaleniidae [PALEON] A family of Jurassic and Cretaceous echinoderms in the order Salenoida. { ¦ak·rō,sal·ə'nī·ə·dē }

acroscopic [BOT] Facing, or on the side toward, the apex. { ,ak·rə'skäp·ik }

acrosin [BIOCHEM] A proteolytic enzyme located in the acrosome of a spermatozoon; thought to be involved in penetration of the egg. { 'ak·rə·sin }

acrosome [CYTOL] The anterior, crescent-shaped body of spermatozoon, formed from Golgi material of the spermatid. Also known as perforatorium. { 'ak·rə,sōm }

acrosome reaction [CELL MOL] A form of cellular exocytosis that allows sperm to penetrate the zona pellucida of ovulated eggs. { 'ak·rə,sōm rē,ak·shən }

acrospore [MYCOL] In fungi, a spore formed at the outer tip of a hypha. { 'ak·rə,spór }

acrotarsium [ANAT] Instep of the foot. { ,ak·rō'tär·sē·əm }

acroterion [ARCH] Also known as acroterium. **1.** A pedestal on a pediment to support an ornamental, such as a statue. **2.** An ornamental placed on such a pedestal. { ,ak·rə'tir·ē,än }

acroterium See acroterion. { ,ak·rə'tir·ē·əm }

Acrothoracica [INV ZOO] A small order of burrowing barnacles in the subclass Cirripedia that inhabit corals and the shells of mollusks and barnacles. { ,ak·rə·thə'ras·ik·ə }

Acrotretacea [PALEON] A family of Cambrian and Ordovician inarticulate brachiopods of the suborder Acrotretidina. { ,ak·rō·tre'tās·ē·ə }

Acrotretida [INV ZOO] An order of brachiopods in the class Inarticulata; representatives are known from Lower Cambrian to the present. { ,ak·rō'tred·ə·də }

Acrotretidina [INV ZOO] A suborder of inarticulate brachiopods of the order Acrotretida; includes only species with shells composed of calcium phosphate. { ,ak·rō·tre'tī·də·nə }

acrozone See range zone. { 'ak·rō,zōn }

Acrux See α Crucis. { 'ā,crəks }

acrylamide [ORG CHEM] $CH_2CHCONH_2$ Colorless, odorless crystals with a melting point of 84.5°C; soluble in water, alcohol, and acetone; used in organic synthesis, polymerization, sewage treatment, ore processing, and permanent press fabrics. { ə'kril·ə,mīd }

acrylamide copolymer [ORG CHEM] A thermosetting resin formed of acrylamide with other resins, such as the acrylic resins. { ə'kril·ə,mīd kō'päl·ə·mər }

acrylate [ORG CHEM] **1.** A salt or ester of acrylic acid. **2.** See acrylate resin. { 'ak·rə,lāt }

acrylate resin [ORG CHEM] Acrylic acid or ester polymer with a −CH_2−$CH(COOR)$− structure; used in paints, sizings and finishes for paper and textiles, adhesives, and plastics. Also known as acrylate. { 'ak·rə,lāt 'rez·ən }

acrylate rubber [MATER] A member of a class of elastomers based on acrylate esters. { 'ak·rə,lāt 'rəb·ər }

acrylic acid [ORG CHEM] $CH_2CHCOOH$ An easily polymerized, colorless, corrosive liquid used as a monomer for acrylate resins. { ə'kril·ik 'as·əd }

acrylic ester [ORG CHEM] An ester of acrylic acid. { ə'kril·ik 'es·tər }

acrylic fiber [TEXT] Any of numerous synthetic textile fibers made by polymerization of acrylonitrile. { ə'kril·ik 'fī·bər }

acrylic resin [ORG CHEM] A thermoplastic synthetic organic polymer made by the polymerization of acrylic derivatives such as acrylic acid, methacrylic acid, ethyl acrylate, and methyl acrylate; used for adhesives, protective coatings, and finishes. { ə'kril·ik 'rez·ən }

acrylic rubber [ORG CHEM] Synthetic rubber containing acrylonitrile; for example, nitrile rubber. { ə'kril·ik 'rəb·ər }

acrylic syrup [MATER] Lucite in a liquid form; used as a low-pressure laminating resin; produces stiff, strong, tough laminates that can be adapted to bright or translucent colors. { ə'kril·ik 'sir·əp }

acrylonitrile [ORG CHEM] CH_2CHCN A colorless liquid compound used in the manufacture of acrylic rubber and fibers. Also known as vinylcyanide. { ,ak·rə,lō'nī·trəl }

acrylonitrile-butadiene rubber See nitrile rubber. { ,ak·rə,lō'nī·trəl ,byüd·ə'dī,ēn 'rəb·ər }

acrylonitrile butadiene styrene resin [ORG CHEM] A polymer made by blending acrylonitrile-styrene copolymer with a butadiene-acrylonitrile rubber or by interpolymerizing polybutadiene with styrene and acrylonitrile; combines the advantages of hardness and strength of the vinyl resin component with the toughness and impact resistance of the rubbery component. Abbreviated ABS. { ,ak·rə,lō'nī·trəl ,byüd·ə'dī,ēn 'stī·rēn 'rez·ən }

acrylonitrile copolymer [ORG CHEM] Oil-resistant synthetic rubber made by polymerization of acrylonitrile with compounds such as butadiene or acrylic acid. { ,ak·rə,lō'nī·trəl kō'päl·ə·mər }

acrylonitrile rubber See nitrile rubber. { ,ak·rə,lō'nī·trəl 'rəb·ər }

ACSR See aluminum cable steel-reinforced.

Actaeonidae [INV ZOO] A family of gastropod mollusks in the order Tectibranchia. { ,ak·tē'än·ə·dē }

Actaletidae [INV ZOO] A family of insects belonging to the order Collembola characterized by simple tracheal systems. { ,ak·tə'led·ə·dē }

ACTH See adrenocorticotropic hormone.

Actidione [MICROBIO] Trade name for the antibiotic cyclohexamide. { ,ak·tə'dī,ōn }

actin [BIOCHEM] A muscle protein that is the chief constituent of the Z-band myofilaments of each sarcomere. { 'ak·tən }

actinal [INV ZOO] In radially symmetrical animals, referring to the part from which the tentacles or arms radiate or to the side where the mouth is located. { 'ak·tə·nəl }

Aotiniaria [INV ZOO] The sea anemones, an order of cnidarians in the subclass Zoantharia. { ak,tin·ē'a·rē·ə }

actinic [PHYS] Pertaining to electromagnetic radiation capable of initiating photochemical reactions, as in photography or the fading of pigments. { ,ak'tin·ik }

actinic achromatism [OPTICS] **1.** The design of a photographic lens system so that light sources at the wavelength of the Fraunhofer D line near 589 nanometers and the G line at 430.8 nanometers are focused at the same point and produce images of the same size. **2.** The design of an astronomical lens system so that light sources at the wavelength of the Fraunhofer F line at 486.1 nanometers and the G line at 430.8 nanometers are focused at the same point and produce images of the same size. Also known as FG achromatism. { ,ak'tin·ik ,ā'krōm·ə,tiz·əm }

actinic focus [OPTICS] The point in an optical system at which the chemically most effective rays (usually those in the ultraviolet) converge. Also known as chemical focus. { ,ak'tin·ik 'fō·kəs }

actinic glass [OPTICS] Glass that transmits more of the visible components of incident radiation and less of the infrared and ultraviolet components. { ,ak'tin·ik 'glas }

actinide series [CHEM] The group of elements of atomic number 89 through 103. Also known as actinoid elements. { 'ak·tə,nīd 'sir,ēz }

actinism [CHEM] The production of chemical changes in a substance upon which electromagnetic radiation is incident. { 'ak·tə'niz·əm }

actinium [CHEM] A radioactive element, symbol Ac, atomic number 89; its longest-lived isotope is ^{227}Ac with a half-life of 21.7 years; the element is trivalent; chief use is, in equilibrium with its decay products, as a source of alpha rays. { ak'tin·ē·əm }

actinium decay series [NUCLEO] A series of radioactive disintegration products derived from uranium-235. { ‚ak'tin·ē·əm di'kā ‚sir·ēz }

actinium emanation *See* actinon. { ‚ak'tin·ē·əm ‚em·ə'nā·shən }

actinobacillosis [VET MED] A bacterial disease of domestic animals caused by *Actinobacillus lignieresii*. { ‚ak·tə·nō‚bas·ə'lō·səs }

Actinobacillus [MICROBIO] A species of gram-negative, oval, spherical, or rod-shaped bacteria that are of uncertain affiliation; coccal and bacillary cells are often interspersed, giving a "Morse code" form; species are pathogens of animals, occasionally of humans. { ‚ak·tə·nō·bə'sil·əs }

Actinobacillus lignieresii [MICROBIO] The causative agent of actinobacillosis. { ‚ak·tə·nō·bə‚sil·əs ‚lin·yir'ās·ē‚ē }

Actinobacillus pleuropneumoniae [MICROBIO] The etiologic agent of pleuropneumoniae in swine. { ‚ac·tə·nō·bə‚sil·əs ‚plùr·ə·nə'mō·nē‚ī }

Actinobacillus suis [MICROBIO] The etiologic agent of various lesions in piglets. { ‚ak·tə·nō‚bə‚sil·əl 'sü·is }

Actinobifida [MICROBIO] A genus of bacteria in the family Micromonosporaceae with a dichotomously branched substrate; an aerial mycelium is formed which produces single spores. { ‚ak·tə·nō'bī·fə·də }

actinocarpous [BOT] Having flowers and fruit radiating from one point. { ‚ak·tə·nō'kär·pəs }

actinochemistry [CHEM] A branch of chemistry concerned with chemical reactions produced by light or other radiation. { ‚ak·tə·nō'kem·ə·strē }

actinochitin [BIOCHEM] A form of birefringent or anisotropic chitin found in the seta of certain mites. { ‚ak·tə·nō'kī·tən }

Actinochitinosi [INV ZOO] A group name for two closely related suborders of mites, the Trombidiformes and the Sarcoptiformes. { ‚ak·tə·nō‚kī·tə'nō·sē }

actinodielectric [ELEC] Of a substance, exhibiting an increase in electrical conductivity when electromagnetic radiation is incident upon it. { ‚ak·tə·nō‚dī·ə'lek·trik }

actinoelectricity [ELEC] The electromotive force produced in a substance by electromagnetic radiation incident upon it. { ‚ak·tə·nō·i‚lek'tris·ə·dē }

actinogram [ENG] The record of heat from a source, such as the sun, as detected by a recording actinometer. { ‚ak'tin·ə‚gram }

actinograph [ENG] A recording actinometer. { ‚ak'tin·ə‚graf }

actinoid elements *See* actinide series. { 'ak·tə‚nóid 'el·ə·məns }

Actinolaimoidea [INV ZOO] A superfamily of nematodes in the order Dorylaimida, containing some species with remarkable elaborations of the stoma and the characteristic axial spear. { ‚ak·tə·nō·lə'móid·ē·ə }

actinolite [MINERAL] $Ca_2(Mg,Fe)_5Si_8O_{22}(OH)_2$ A green, monoclinic rock-forming amphibole; a variety of asbestos occurring in needlelike crystals and in fibrous or columnar forms; specific gravity 3–3.2. { ‚ak'tin·ə‚līt }

actinology [PHYS] The branch of physics dealing with electromagnetic radiation and its chemical effects. { ‚ak·tə'näl·ə·jē }

actinomere [INV ZOO] One of the segments composing the body of a radially symmetrical animal. { ‚ak'tin·ə‚mir }

actinometer [ENG] Any instrument used to measure the intensity of radiant energy, particularly that of the sun. { ‚ak·tə'näm·əd·ər }

actinometry [ASTROPHYS] The science of measurement of radiant energy, particularly that of the sun, in its thermal, chemical, and luminous aspects. { ‚ak·tə'näm·ə·trē }

actinomorphic [BIOL] Descriptive of an organism, organ, or part that is radially symmetrical. { ‚ak·tə·nō'mòr·fik }

Actinomyces [MICROBIO] The type genus of the family Actinomycetaceae; anaerobic to facultatively anaerobic; includes human and animal pathogens. { ‚ak·tə·nō'mī·sēs }

Actinomycetaceae [MICROBIO] A family of bacteria in the order Actinomycetales; gram-positive, diphtheroid cells which form filaments but not mycelia; chemoorganotrophs that ferment carbohydrates. { ‚ak·tə·nō‚mī·sə'tās·ē‚ē }

Actinomycetales [MICROBIO] An order of bacteria; cells form branching filaments which develop into mycelia in some families. { ‚ak·tə·nō‚mī·sə'tā·lēz }

actinomycete [MICROBIO] Any member of the bacterial family Actinomycetaceae. { ‚ak·tə·nō'mī‚sēt }

actinomycin [MICROBIO] The collective name for a large number of red chromoprotein antibiotics elaborated by various strains of *Streptomyces*. { ‚ak·tə·nō'mī·sən }

actinomycosis [MED] An infectious bacterial disease caused by *Actinomyces bovis* in cattle, hogs, and occasionally in humans. Also known as lumpy jaw. { ‚ak·tə·nō‚mī'kō·səs }

actinomyosin [BIOCHEM] A protein complex formed by the combination of actin and myosin during muscle contraction. { ‚ak·tə·nō'mī·əs·ən }

Actinomyxida [INV ZOO] An order of protozoan invertebrate parasites of the class Myxosporidea characterized by trivalved spores with three polar capsules. { ‚ak·tə·nō'mik·sə·də }

actinon [NUC PHYS] A radioactive isotope of radon, symbol An, atomic number 86, atomic weight 219, belonging to the actinium series. Also known as actinium emanation (Ac-Em). { 'ak·tə‚nän }

actinophage [MICROBIO] A bacteriophage that infects and lyses members of the order Actinomycetales. { ak'tin·ə‚fāj }

Actinophryida [INV ZOO] An order of protozoans in the subclass Heliozoia; individuals lack an organized test, a centroplast, and a capsule. { ‚ak·tə·nō'frī·ə·də }

Actinoplanaceae [MICROBIO] A family of bacteria in the order Actinomycetales with well-developed mycelia and spores formed on sporangia. { ‚ak·tə·nō·plə'nās·ē‚ē }

Actinoplanes [MICROBIO] A genus of bacteria in the family Actinoplanaceae having aerial mycelia and spherical to subspherical sporangia; spores are spherical and motile by means of a tuft of polar flagella. { ‚ak·tə·nō'plā‚nēz }

Actinopodea [INV ZOO] A class of protozoans belonging to the superclass Sarcodina; most are free-floating, with highly specialized pseudopodia. { ‚ak·tə·nō'pōd·ē·ə }

Actinopteri [VERT ZOO] An equivalent name for the Actinopterygii. { ‚ak·tə'näp·tə‚rī }

Actinopterygii [VERT ZOO] The ray-fin fishes, a subclass of the Osteichthyes distinguished by the structure of the paired fins, which are supported by dermal rays. { ‚ak·tə‚näp·tə'rij·ē‚ī }

actinostele [BOT] A protostele characterized by xylem that is either star-shaped in cross section or has ribs radiating from the center. { ak'tin·ə‚stēl }

actinostome [BIOL] **1.** The mouth of a radiate animal. **2.** The peristome of an echinoderm. { ak'tin·ə‚sōm }

Actinostromariidae [PALEON] A sphaeractinoid family of extinct marine hydrozoans. { ‚ak·tə·nō‚strō·mə'rī·ə‚dē }

actinotherapy *See* radiation therapy. { ‚ak·tə·nō'the·rə·pē }

actinotrocha [INV ZOO] The free-swimming larva of *Phoronis*, a genus of small, marine, tubicolous worms. { ‚ak·tə·nō'trō·kə }

actinouranium [NUC PHYS] A naturally occurring radioactive isotope of the actinium series, emitting only alpha decay; symbol AcU; atomic number 92; mass number 235; half-life 7.1×10^8 years; isotopic symbol ^{235}U. { ‚ak·tə·nō‚yù'rā·nē·əm }

actinula [INV ZOO] A larval stage of some hydrozoans that has tentacles and a mouth; attaches and develops into a hydroid in some species, or metamorphoses into a medusa. { ak'tin·yə·lə }

action [MECH] An integral associated with the trajectory of a system in configuration space, equal to the sum of the integrals of the generalized momenta of the system over their canonically conjugate coordinates. Also known as phase integral. [ORD] The mechanism of a gun, usually breechloading, by which it is loaded, fired, and unloaded. { 'ak·shən }

action at a distance theory [PHYS] A theory of the interaction of two bodies separated in space, without concern for a detailed mechanism of the propagation of effects between bodies. { 'ak·shən at ə 'dis·təns ‚thē·ə·rē }

action current [PHYSIO] The electric current accompanying membrane depolarization and repolarization in an excitable cell. { 'ak·shən ‚kə·rənt }

action entries [COMPUT SCI] The lower right-hand portion of a decision table, indicating which of the various possible actions result from each of the various possible conditions. { 'ak·shən ‚en·trēz }

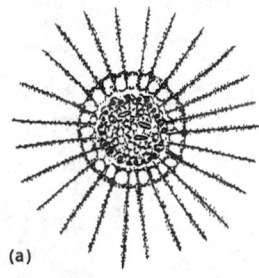
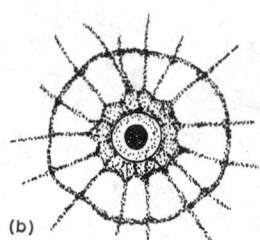

ACTINOPHRYIDA

(a)

(b)

Examples of Actinophryida. *(a)* Single specimen of *Actinosphaerium eichorni (after Pernard). (b)* Single specimen of *Actinophrys pontica. (From R. P. Hall, Protozoology, Prentice-Hall, 1953)*

action integral *See* action variable. { 'ak·shən 'int·ə·grəl }

action period [ELECTR] The period of time during which data in a Williams tube storage device can be read or new data can be written into this storage. { 'ak·shən ,pir·ē·əd }

action portion [COMPUT SCI] The lower portion of a decision table, comprising the action stub and action entries. { 'ak·shən ,pȯr·shən }

action potential [NEURO] A transient change in electric potential at the surface of a nerve or muscle cell occurring at the moment of excitation. { 'ak·shən pə,ten·chəl }

action-reaction law [PHYS] The law that when one body exerts force on another, the second body exerts a collinear force on the first equal in magnitude but oppositely directed. { 'ak·shən 'rē,ak·shən ,lȯ }

action spectrum [PHYSIO] Graphic representation of the comparative effects of different wavelengths of light on living systems or their components. { 'ak·shən ,spek·trəm }

action stub [COMPUT SCI] The lower left-hand portion of a decision table, consisting of a single column listing the various possible actions (transformations to be done on data and materials). { 'ak·shən ,stəb }

action variable [PHYS] The integral ∫ *pdq* over a cycle of a dynamical system; *q* is some coordinate, and *p* the conjugate momentum. Also known as action integral. { 'ak·shən ,ver·ē·ə·bəl }

actium [ECOL] A rocky seashore community. { 'ak·tē·əm }

activate [ELEC] To make a cell or battery operative by addition of a liquid. [ELECTR] To treat the filament, cathode, or target of a vacuum tube to increase electron emission. [ENG] To set up conditions so that the object will function as designed or required. [NUCLEO] To induce radioactivity through bombardment by neutrons or by other types of radiation. [ORD] **1.** To bring into existence by official order a unit, post, camp, station, base, or shore activity which has previously been constituted and designated by name or number, or both, so that it can be organized to function in its assigned capacity. **2.** To prepare for active service a naval ship or craft which has been in an inactive or reserve status. [PHYS] To start activity or motion in a device or material. { 'ak·tə,vāt }

activated alumina [MATER] Highly porous, granular aluminum oxide that preferentially absorbs liquids from gases and vapors, and moisture from some liquids; also used as a catalyst or catalyst carrier, as an absorbent to remove fluorides from drinking water, and in chromatography. { 'ak·tə,vād·əd ə'lüm·ə·nə }

activated bauxite *See* filter bauxite. { 'ak·tə,vād·əd 'bȯk·sīt }

activated carbon [MATER] A powdered, granular, or pelleted form of amorphous carbon characterized by very large surface area per unit volume because of an enormous number of fine pores. Also known as activated charcoal. { 'ak·tə,vād·əd 'kär·bən }

activated cathode [ELECTR] A thermionic cathode consisting of a tungsten filament to which thorium has been added, and then brought to the surface, by a process such as heating in the absence of an electric field in order to increase thermionic emission. { 'ak·tə,vād·əd 'kath,ōd }

activated charcoal *See* activated carbon. { 'ak·tə,vād·əd 'char,kōl }

activated clay [MATER] Bentonite, or other clay, treated with acid to enhance its ability to absorb or bleach. { 'ak·tə,vād·əd klā }

activated coal plough [MIN ENG] A type of power-operated cutting blade used for coal seams too hard to be sheared by a normal blade. { 'ak·tə,vād·əd 'kōl ,plau̇ }

activated complex [PHYS CHEM] An energetically excited state which is intermediate between reactants and products in a chemical reaction. Also known as transition state. { 'ak·tə,vād·əd 'käm·pleks }

activated diffusion [SOLID STATE] Movement of atoms, ions, or lattice defects across a potential barrier in a solid. { 'ak·də,vād·əd di'fyü·zhən }

activated macrophage [IMMUNOL] A macrophage whose ability to destroy microbes or other cells has been enhanced because of stimulation by a lymphokine. { 'ak·tə,vād·əd 'mak·rə,fāj }

activated rosin flux [MATER] Soldering flux containing activating agents which promote wetting by the solder. { 'ak·tə,vād·əd 'räz·ən 'fləks }

activated sintering [MET] Sintering of a metal powder compact in contact with a gaseous atmosphere which reacts with the metal surfaces and enhances the joining of metal particles. { 'ak·tə,vād·əd 'sin·tər·iŋ }

activated sludge [CIV ENG] A semiliquid mass removed from the liquid flow of sewage and subjected to aeration and aerobic microbial action; the end product is dark to golden brown, partially decomposed, granular, and flocculent, and has an earthy odor when fresh. { 'ak·tə,vād·əd 'sləj }

activated-sludge effluent [CIV ENG] The liquid from the activated-sludge treatment that is further processed by chlorination or by oxidation. { 'ak·tə,vād·əd ,sləj 'ef,lü·ənt }

activated-sludge process [CIV ENG] A sewage treatment process in which the sludge in the secondary stage is put into aeration tanks to facilitate aerobic decomposition by microorganisms; the sludge and supernatant liquor are separated in a settling tank; the supernatant liquor or effluent is further treated by chlorination or oxidation. { 'ak·tə,vād·əd ,sləj 'prä,səs }

activating enzyme [BIOCHEM] An enzyme that catalyzes a reaction involving adenosine triphosphate and a specific amino acid to give a product that subsequently reacts with a specific transfer ribonucleic acid. { 'ak·tə,vād·iŋ 'en,zīm }

activating reagent [MATER] Material added to another material or mixture so that a physical or chemical change will take place more rapidly or completely. { 'ak·tə,vād·iŋ ,rē'ā·jənt }

activating receptor [NEURO] A sense organ at the end of a nerve that triggers a specific response when it is stimulated. { 'ak·tə,vād·iŋ rə'sep·tər }

activation [CHEM] Treatment of a substance by heat, radiation, or activating reagent to produce a more complete or rapid chemical or physical change. [ELEC] The process of adding liquid to a manufactured cell or battery to make it operative. [ELECTR] The process of treating the cathode or target of an electron tube to increase its emission. Also known as sensitization. [MET] **1.** A process of facilitating the separation and collection of ore powders by the use of substances which change the response of the particle surfaces to a flotation fluid. **2.** A process that increases the rate of pressing and heating a metal powder into cohesion. [MOL BIO] A change that is induced in an amino acid before it is utilized for protein synthesis. [NUCLEO] The process of inducing radioactivity by bombardment with neutrons or with other types of radiation. [PHYSIO] The designation for all changes in the ovum during fertilization, from sperm contact to the dissolution of nuclear membranes. { ,ak·tə'vā·shən }

activation analysis [NUCLEO] A method of chemical analysis based on the detection of characteristic radionuclides following a nuclear bombardment. Also known as radioactivity analysis. { ,ak·tə'vā·shən ə'nal·ə·səs }

activation cross section [NUC PHYS] The cross section for formation of a radionuclide by a particular interaction. { ,ak·tə|vā·shən 'krȯs ,sek·shən }

activation energy [PHYS CHEM] The energy, in excess over the ground state, which must be added to an atomic or molecular system to allow a particular process to take place. { ,ak·tə'vā·shən 'en·ər·jē }

activation record [COMPUT SCI] A variable part of a program module, such as data and control information, that may vary with different instances of execution. { ,ak·tə'vā·shən 'rek·ərd }

activator [CHEM] **1.** A substance that increases the effectiveness of a rubber vulcanization accelerator; for example, zinc oxide or litharge. **2.** A trace quantity of a substance that imparts luminescence to crystals; for example, silver or copper in zinc sulfide or cadmium sulfide pigments. [GEN] A molecule that modifies a repressor in a way that enables it to stimulate operon transcription. [GRAPHICS] *See* accelerator. { 'ak·tə,vād·ər }

activator ribonucleic acid [GEN] Ribonucleic acid molecules which form a sequence-specific complex with receptor genes linked to producer genes. { 'ak·tə,vā·tər ¦rī·bō¦nü¦klē·ik 'as·əd }

active accommodation [CONT SYS] The alteration of preprogrammed robotic motions by the integrated effects of sensors, controllers, and the robotic motion itself. { 'ak·tiv ə,käm·ə'dā·shən }

active anaphylaxis [IMMUNOL] The allergic response following reintroduction of an antigen into a hypersensitive individual. { 'ak·tiv ¦an·ə·fə'lak·səs }

active antiroll system [NAV ARCH] A system of antiroll tanks in a ship in which pumps are used to transfer the liquid, through a connecting channel, from one tank in a pair to the other. { 'ak·tiv ¦an·tē'rōl 'sis·təm }

active area [ELECTR] The area of a metallic rectifier that acts as the rectifying junction and conducts current in the forward direction. { 'ak·tiv 'er·ē·ə }

active balance [COMMUN] Summation of all return currents, in telephone repeater operation, at a terminal network balanced against the impedance of the local circuit or drop. { 'ak·tiv 'bal·əns }

active biomass [MICROBIO] The amount of a culture that is actively growing. { ¦ak·tiv 'bī·ō,mas }

active cell [COMPUT SCI] The cell that continues the value being used or modified in a spreadsheet program, and that is highlighted by the cell pointer. Also known as current cell. { ¦ak·tiv 'sel }

active center [ASTRON] A localized, transient region of the solar atmosphere in which sunspots, faculae, plages, prominences, solar flares, and so forth are observed. [BIOCHEM] **1.** A flexible portion of an enzyme that binds to the substrate and converts it into the reaction product. **2.** In carrier and receptor proteins, the portion of the molecule that interacts with the specific target compounds. [CHEM] **1.** Any one of the points on the surface of a catalyst at which the chemical reaction is initiated or takes place. **2.** *See* active site. { 'ak·tiv 'sen·tər }

active chaff [ORD] An expendable battery-powered jammer, usually supported by parachute or balloon, dropped by aircraft to saturate enemy radars or produce delayed false returns when triggered by enemy radars. { 'ak·tiv 'chaf }

active communications satellite [AERO ENG] Satellite which receives, regenerates, and retransmits signals between stations. { 'ak·tiv kə,myü·nə'kā·shənz 'sad·ə,līt }

active component [ELEC] In the phasor representation of quantities in an alternating-current circuit, the component of current, voltage, or apparent power which contributes power, namely, the active current, active voltage, or active power. Also known as power component. [ELECTR] *See* active element. { 'ak·tiv kəm'pō·nənt }

active computer [COMPUT SCI] When two or more computers are installed, the one that is on-line and processing data. { 'ak·tiv kəm'pyüd·ər }

active controls technology [AERO ENG] The development of special forms of augmentation systems to stabilize airplane configurations and to limit, or tailor, the design loads that the airplane structure must support. { 'ak·tiv kən'trōlz ,tek'näl·ə·jē }

active-cord mechanism [MECH ENG] A slender, chainlike grouping of joints and links that makes active and flexible winding motions under the control of actuators attached along its body. { 'ak·tiv ¦kòrd 'mek·ə,niz·əm }

active current [ELEC] The component of an electric current in a branch of an alternating-current circuit that is in phase with the voltage. Also known as watt current. { 'ak·tiv 'kə·rənt }

active detection system [ENG] A guidance system which emits energy as a means of detection; for example, sonar and radar. { 'ak·tiv di'tek·shən ,sis·təm }

active device [ELECTR] A component, such as an electron tube or transistor, that is capable of amplifying the current or voltage in a circuit. { 'ak·tiv di'vīs }

active door *See* active leaf. { 'ak·tiv 'dòr }

active earth pressure [CIV ENG] The horizontal pressure that an earth mass exerts on a wall. { 'ak·tiv 'ərth 'presh·ər }

active electric network [ELEC] Electric network containing one or more sources of energy. { 'ak·tiv ə'lek·trik 'net,wərk }

active electronic countermeasures [ELECTR] The major subdivision of electronic countermeasures that concerns electronic jamming and electronic deceptions. { 'ak·tiv ə,lek'trän·ik 'kaùnt·ər,mezh·ərz }

active element [ELECTR] Any generator of voltage or current in an impedance network. Also known as active component. [NUC PHYS] A chemical element which has one or more radioactive isotopes. { 'ak·tiv 'el·ə·mənt }

active entry [MIN ENG] An entry in which coal is being mined from a portion or from connected sections. { 'ak·tiv 'en·trē }

active file [COMPUT SCI] A collection of records that is currently being used or is available for use. { 'ak·tiv 'fīl }

active filter [ELECTR] A filter that uses an amplifier with conventional passive filter elements to provide a desired fixed or tunable pass or rejection characteristic. { 'ak·tiv 'fil·tər }

active front [METEOROL] A front, or portion thereof, which produces appreciable cloudiness and, usually, precipitation. { 'ak·tiv frənt }

active galactic nucleus [ASTRON] A central region of a galaxy, a light-year or less in diameter, where violent and apparently explosive behavior is observed which is manifested in many ways, including the high-velocity outflow of gas, strong nonthermal radio emission, intense and often polarized and highly variable radiation over a wide range of wavelength bands, and ejection of jets of relativistic material. { ¦ak·tiv gə,lak·tik 'nü·klē·əs }

active galaxy [ASTRON] A galaxy whose central region exhibits strong emission activity, from radio to x-ray frequencies, probably as a result of gravitational collapse; this category includes M82 galaxies, Seyfert galaxies, N galaxies, and possibly quasars. { 'ak·tiv 'gal·ək·sē }

active glacier [HYD] A glacier in which some of the ice is flowing. { 'ak·tiv 'glā·shər }

active homing [NAV] **1.** The homing of an aerodynamic missile by radar, in which radio signals are transmitted from the missile to the target and reflected to the missile to direct it toward the target. **2.** Homing in which the homing device on the missile reveals the presence of the missile to the target. { 'ak·tiv 'hōm·iŋ }

active illumination [ENG] Lighting whose direction, intensity, and pattern are controlled by commands or signals. { 'ak·tiv ə,lüm·ə'nā·shən }

active immunity [IMMUNOL] Disease resistance in an individual due to antibody production after exposure to a microbial antigen following disease, inapparent infection, or inoculation. { 'ak·tiv im'yü·nət·ē }

active immuno-gene therapy [MED] Immuno-gene therapy in which the tumor is directly altered with cytokine genes in order to induce an endogenous reaction against it. { ¦ak·tiv ,im·yə·nō'jēn ,ther·ə·pē }

active immunotherapy [IMMUNOL] A type of immunotherapy that attempts to stimulate the host's intrinsic immune response to the tumor, either nonspecifically or specifically. { ¦ak·tiv ¦im·yə·nō'ther·ə·pē }

active infrared detection system [ENG] An infrared detection system in which a beam of infrared rays is transmitted toward possible targets, and rays reflected from a target are detected. { 'ak·tiv 'in·frə,red di'tek·shən ,sis·təm }

active jamming *See* jamming. { 'ak·tiv 'jam·iŋ }

active layer [GEOL] That part of the soil which is within the suprapermafrost layer and which usually freezes in winter and thaws in summer. Also known as frost zone. { 'ak·tiv 'lā·ər }

active leaf [BUILD] In a door with two leaves, the leaf which carries the latching or locking mechanism. Also known as active door. { 'ak·tiv 'lēf }

active leg [ELECTR] An electrical element within a transducer which changes its electrical characteristics as a function of the application of a stimulus. { 'ak·tiv 'leg }

active location system [NAV] A navigation system in which the navigation satellite interrogates the craft, and the craft responds; useful for surveillance by a ground station, or for automated navigation if the satellite subsequently transmits data. { 'ak·tiv lō'kā·shən ,sis·təm }

active logic [ELECTR] Logic that incorporates active components which provide such functions as level restoration, pulse shaping, pulse inversion, and power gain. { 'ak·tiv 'läj·ik }

active margin [GEOL] A continental margin that is characterized by earthquakes, volcanic activity, and orogeny resulting from movement of tectonic plates. { 'ak·təv 'mär·jən }

active master file [COMPUT SCI] A relatively active computer master file, as determined by usage data. { 'ak·tiv 'mas·tər 'fīl }

active master item [COMPUT SCI] A relatively active item in a computer master file, as determined by usage data. { 'ak·tiv 'mas·tər 'ī·təm }

active material [ELEC] **1.** A fluorescent material used in screens for cathode-ray tubes. **2.** An energy-storing material, such as lead oxide, used in the plates of a storage battery. **3.** A material, such as the iron of a core or the copper of a winding, that is involved in energy conversion in a circuit. **4.** In a battery, the chemically reactive material in either of the electrodes that participates in the charge and discharge reactions. [ELECTR] The material of the cathode of an electron tube that emits electrons when heated. [NUCLEO] A material capable of releasing substantial quantities of nuclear energy during fission. { 'ak·tiv mə'tir·ē·əl }

active-matrix liquid-crystal display [ELEC] A liquid-crystal display that has an active element, such as a transistor or diode, on every picture element. Abbreviated AMLCD. { ¦ak·tiv ¦mā·triks ¦lik·wid 'kris·təl di₁splā }

active mirror [OPTICS] A mirror whose position and shape are continually adjusted in response to changing environmental conditions in order to obtain optimum performance. { 'ak·tiv 'mir·ər }

active oxygen method [FOOD ENG] An accelerated test used to determine the potential stability of fats toward oxidation (rancidity). { 'ak·tiv 'äk·sə·jən ₁meth·əd }

active permafrost [GEOL] Permanently frozen ground (permafrost) which, after thawing by artificial or unusual natural means, reverts to permafrost under normal climatic conditions. { 'ak·tiv 'pər·mə₁fröst }

active power [ELEC] The product of the voltage across a branch of an alternating-current circuit and the component of the electric current that is in phase with the voltage. { 'ak·tiv 'pau·ər }

active prominence [ASTRON] A classification of prominences of the sun; such a prominence is rapidly moving, and is the most frequent type. { 'ak·tiv 'präm·ə·nəns }

active prominence region [ASTRON] Portions of the solar limb that display active prominences, characterized by downflowing knots and streamers, sprays, frequent surges, and curved loops. Abbreviated APR. { 'ak·tiv 'präm·ə·nəns ₁rē·jən }

active-RC filter [ELEC] An active filter whose frequency-sensitive mechanism is the charging of a capacitor (C) through a resistor (R), giving a characteristic frequency at which the impedances of the resistor and the capacitor are equal. { ¦ak·tiv ¦är¦sē 'fil·tər }

active region [ASTRON] A localized, transient, nonuniform region on the sun's surface, penetrating well down into the lower chromosphere. [ELECTR] The region in which amplifying, rectifying, light emitting, or other dynamic action occurs in a semiconductor device. { 'ak·tiv 'rē·jən }

active-RLC filter [ELEC] An integrated-circuit filter that uses both inductors (L), made as spirals of metallization on the top layer, and amplifiers, connected to simulate negative resistors (R), that enhance the performance of the inductors as well as capacitors (C). { ak·tiv ¦är¦el¦sē ₁fil·tər }

active satellite [AERO ENG] A satellite which transmits a signal. { 'ak·tiv 'sad·ə₁līt }

active site [CHEM] The effective site at which a given heterogeneous catalytic reaction can take place. Also known as active center. [MOL BIO] The region of an enzyme molecule at which binding with the substrate occurs. Also known as binding site; catalytic site. { 'ak·tiv 'sīt }

active sludge [CIV ENG] A sludge rich in destructive bacteria used to break down raw sewage. { 'ak·tiv 'sləj }

active solar system [MECH ENG] A solar heating or cooling system that operates by mechanical means, such as motors, pumps, or valves. { 'ak·tiv 'sō·lər ₁sis·təm }

active solid [CHEM] A porous solid possessing adsorptive properties and used for chromatographic separations. { 'ak·tiv 'säl·əd }

active sonar [ENG] A system consisting of one or more transducers to send and receive sound, equipment for the generation and detection of the electrical impulses to and from the transducer, and a display or recorder system for the observation of the received signals. { 'ak·tiv 'sō₁när }

active sound cancellation [ACOUS] Any technique in which a control sound source creates sound in a selected region equal in amplitude and opposite in phase to sound that would otherwise exist, but this sound cancellation cannot be maintained in the presence of system changes unless there is also a feedback mechanism. { ¦ak·tiv 'saund ₁kan·sə₁lā·shən }

active sound control [ACOUS] Any modification of sound fields by loudspeakers, controlled, for example, through the use of a feedforward mechanism, for reduction, equalization, or cancellation of sound. { ¦ak·tiv 'saund kən₁tröl }

active substrate [SOLID STATE] A semiconductor or ferrite material in which active elements are formed; also a mechanical support for the other elements of a semiconductor device or integrated circuit. { 'ak·tiv 'səb₁strāt }

active Sun [ASTRON] The Sun during the portion of its 11-year cycle in which sunspots, flares, prominences, and variations in radio-frequency emission reach their maximum. { 'ak·tiv 'sən }

active system [ENG] In radio and radar, a system that requires transmitting equipment, such as a beacon or transponder. { 'ak·tiv 'sis·təm }

active termination [COMPUT SCI] A means of ending a chain of peripheral devices connected to a small computer system interface (SCSI) port, suitable for longer chains, where it can reduce electrical interference. { ¦ak·tiv ₁tər·mə'nā·shən }

active tracking system [NAV] A system with a transponder or transmitter on board the vehicle to repeat, transmit, or retransmit information to the tracking equipment; for example, Dovap, Secor, Azusa, Miran, and Minitrack. { 'ak·tiv 'trak·iŋ ₁sis·təm }

active transducer [ELECTR] A transducer whose output is dependent upon sources of power, apart from that supplied by any of the actuating signals, which power is controlled by one or more of these signals. { 'ak·tiv tranz'düs·ər }

active transport [PHYSIO] The pumping of ions or other substances across a cell membrane against an osmotic gradient, that is, from a lower to a higher concentration. { 'ak·tiv 'tranz₁pört }

active vibration suppression [MECH ENG] The prevention of undesirable vibration by techniques involving feedback control of the vibratory motion, whereby the forces designed to reduce the vibration depend on the system displacements and velocities. { 'ak·tiv vī'brā·shən sə₁presh·ən }

active volcano [GEOL] A volcano capable of venting lava, pyroclastic material, or gases. { 'ak·tiv ₁väl'kā·nō }

active voltage [ELEC] In an alternating-current circuit, the component of voltage which is in phase with the current. { 'ak·tiv 'vōl·tij }

active window [COMPUT SCI] In a windowing environment, the window in which the user is currently working and which receives keyboard input. { ¦ak·tiv 'win₁dō }

active workings [MIN ENG] All places in a mine that are ventilated and inspected regularly. { 'ak·tiv 'wərk·iŋz }

activity [COMPUT SCI] The use or modification of information contained in a file. [NUC PHYS] The intensity of a radioactive source. Also known as radioactivity. [PHYS CHEM] A thermodynamic function that correlates changes in the chemical potential with changes in experimentally measurable quantities, such as concentrations or partial pressures, through relations formally equivalent to those for ideal systems. [SYS ENG] The representation in a PERT or critical-path-method network of a task that takes up both time and resources and whose performance is necessary for the system to move from one event to the next. { ₁ak'tiv·əd·ē }

activity chart [IND ENG] A tabular presentation of a series of operations of a process plotted against a time scale. { ₁ak'tiv·əd·ē ₁chärt }

activity coefficient [PHYS CHEM] A characteristic of a quantity expressing the deviation of a solution from ideal thermodynamic behavior; often used in connection with electrolytes. { ₁ak'tiv·əd·ē ₁kō·ə'fish·ənt }

activity duration [SYS ENG] In critical-path-method terminology, the estimated amount of time required to complete an activity. { ₁ak'tiv·əd·ē də'rā·shən }

activity level [COMPUT SCI] **1.** The value assumed by a structural variable during the solution of a programming problem. **2.** A measure of the number of times that use or modification is made of the information contained in a file. { ₁ak'tiv·əd·ē 'lev·əl }

activity ratio [COMPUT SCI] The ratio between used or modified records and the total number of records in a file. [GEOL] The ratio of plasticity index to percentage of clay-sized minerals in sediment. { ₁ak'tiv·əd·ē 'rā·shō }

activity sampling *See* work sampling. { ₁ak'tiv·əd·ē ₁sam·pliŋ }

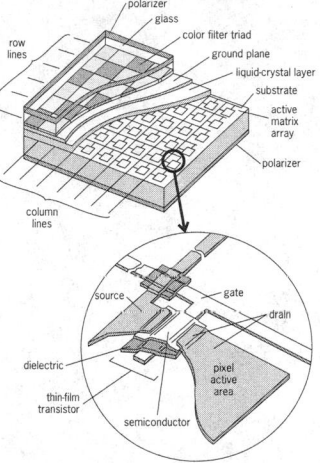

ACTIVE-MATRIX LIQUID-CRYSTAL DISPLAY

polarizer
glass
color filter triad
ground plane
liquid-crystal layer
substrate
active matrix array
polarizer
row lines
column lines
source
gate
drain
dielectric
pixel active area
thin-film transistor
semiconductor

Exploded section of active-matrix liquid-crystal display.

activity sequence method [COMPUT SCI] A method of organizing records in a file so that the records most frequently used are located where they can be found most quickly. { ak'tiv·əd·ē 'sē·kwəns ,meth·əd }

activity series [CHEM] A series of elements that have similar properties—for example, metals—arranged in descending order of chemical activity. { ak'tiv·əd·ē ,sir·ēz }

actol *See* silver lactate. { 'ak,tȯl }

actomyosin [BIOCHEM] A protein complex consisting of myosin and actin; the major constituent of a contracting muscle fibril. { ,ak·tə'mī·ə·sən }

actophilous [ECOL] Having a seashore growing habit. { ,ak'tä·fə·ləs }

actual age *See* absolute age. { 'ak·chə·wəl āj }

actual argument [COMPUT SCI] The variable which replaces a dummy argument when a procedure or macroinstruction is called up. { 'ak·chə·wəl 'är·gyə·mənt }

actual cost [IND ENG] Cost determined by an allocation of cost factors recorded during production. { 'ak·chə·wəl 'kȯst }

actual cubic feet per minute [CHEM ENG] A measure of the volume of gas at operating temperature and pressure, as distinct from volume of gas at standard temperature and pressure. Abbreviated acfm. { 'ak·chə·wəl 'kyü·bik ,fēt pər 'min·ət }

actual decimal point [COMPUT SCI] The period appearing on a printed report as opposed to the virtual point defined only by the data structure within the computer. { 'ak·chə·wəl 'des·məl 'pȯint }

actual elevation [METEOROL] The vertical distance above mean sea level of the ground at the meteorological station. { 'ak·chə·wəl ,el·ə'vā·shən }

actual exhaust velocity [AERO ENG] **1.** The real velocity of the exhaust gas leaving a nozzle as determined by accurately measuring at a specified point in the nozzle exit plane. **2.** The velocity obtained when the kinetic energy of the gas flow produces actual thrust. { 'ak·chə·wəl ig'zȯst və,läs·əd·ē }

actual height [ELECTROMAG] Highest altitude at which refraction of radio waves actually occurs. { 'ak·chə·wəl 'hīt }

actual horsepower *See* actual power. { 'ak·chə·wəl 'hȯrs,pau̇·ər }

actual instruction *See* effective instruction. { 'ak·chə·wəl in'strək·shən }

actualism *See* uniformitarianism. { 'ak·chü·ə,liz·əm }

actual key [COMPUT SCI] A data item in COBOL computer language which can be used as an address. { 'ak·chə·wəl 'kē }

actual motion [NAV] Motion of a craft relative to the earth. { 'ak·chə·wəl 'mō·shən }

actual power [MECH ENG] The power delivered at the output shaft of a source of power. Also known as actual horsepower. { 'ak·chə·wəl 'pau̇·ər }

actual pressure [METEOROL] The atmospheric pressure at the level of the barometer (elevation of ivory point), as obtained from the observed reading after applying the necessary corrections for temperature, gravity, and instrumental errors. { 'ak·chə·wəl 'presh·ər }

actual relative movement *See* slip. { 'ak·chə·wəl 'rel·ə·tiv 'müv·mənt }

actual time [IND ENG] Time taken by a worker to perform a given task. { 'ak·chə·wəl tīm }

actual time of arrival [NAV] The time at which a craft arrives at a specified point at a destination. { 'ak·chə·wəl tīm əv ə'rī·vəl }

actual time of departure [NAV] **1.** The time of leaving a specified point at a place. **2.** The actual time an aircraft becomes airborne. { 'ak·chə·wəl tīm əv də'pär·chər }

actual time of interception [NAV] The time of intercepting a craft by another craft. { 'ak·chə·wəl tīm əv in·tər'sep·shən }

actuate [MECH ENG] To put into motion or mechanical action, as by an actuator. { 'ak·chə·wāt }

actuated roller switch [MECH ENG] A centrifugal sequence-control switch that is placed in contact with a belt conveyor, immediately preceding the conveyor which it controls. { 'ak·chə,wād·əd 'rō·lər 'swich }

actuating system [CONT SYS] An electric, hydraulic, or other system that supplies and transmits energy for the operation of other mechanisms or systems. { 'ak·chə,wād·iŋ ,sis·təm }

actuator [CONT SYS] A mechanism to activate process control equipment by use of pneumatic, hydraulic, or electronic signals; for example, a valve actuator for opening or closing a valve to control the rate of fluid flow. [ENG ACOUS] An auxiliary external electrode used to apply a known electrostatic force to the diaphragm of a microphone for calibration purposes. Also known as electrostatic actuator. [MECH ENG] A device that produces mechanical force by means of pressurized fluid. [ORD] Part of the receiver mechanism in certain types of automatic weapons. { 'ak·chə,wād·ər }

AcU *See* actinouranium.

acuate [BIOL] **1.** Having a sharp point. **2.** Needle-shaped. { 'ak·yə,wāt }

acuity [BIOL] Sharpness of sense perception, as of vision or hearing. { ə'kyü·ə·dē }

Aculeata [INV ZOO] A group of seven superfamilies that constitute the stinging forms of hymenopteran insects in the suborder Apocrita. { ə,kyü·lē'ä·də }

aculeate [BIOL] Pertaining to something that is prickly. { ə'kyü·lē·ət }

aculeus [INV ZOO] **1.** A sharp, hairlike spine, as on the wings of certain lepidopterans. **2.** An insect stinger modified from an ovipositor. { ə'kyü·lē·əs }

Aculognathidae [INV ZOO] The ant-sucking beetles, a family of coleopteran insects in the superfamily Cucujoidea. { ə,kyü·ləg'nath·ə,dē }

acuminate [BOT] Tapered to a slender point, especially referring to leaves. { ə'kyüm·ə·nət }

acupuncture [MED] The ancient Chinese art of puncturing the body with long, fine gold or silver needles to relieve pain and cure disease. { 'ak·yü,pəŋk·chər }

acutance [OPTICS] An objective measure of the ability of a photographic system to show a sharp edge between contiguous areas of low and high illuminance. { ə'kyü·təns }

acute [BIOL] Ending in a sharp point. [MED] Referring to a disease or disorder of rapid onset, short duration, and pronounced symptoms. { ə'kyüt }

acute alcoholism [MED] Drunkenness accompanied by an acute, transient disturbance of physiological and mental functions. { ə'kyüt 'al·kə·hȯ,liz·əm }

acute angle [MATH] An angle of less than 90°. { ə'kyüt 'aŋ·gəl }

acute angle block [GEOL] A fault block in which the strike of strata on the down-dip side meets a diagonal fault at an acute angle. { ə'kyüt 'aŋ·gəl 'bläk }

acute appendicitis [MED] A sudden, severe attack of appendicitis characterized by abdominal pain, usually localized in the lower right quadrant, with nausea, vomiting, and constipation. { ə'kyüt ə,pen·də'sīd·əs }

acute arch [ARCH] An arch with a sharply pointed apex and narrow width. Also known as lancet arch. { ə'kyüt 'ärch }

acute arthritis [MED] A severe joint inflammation with a short course. { ə'kyüt ärth'rīd·əs }

acute ascending myelitis [MED] Severe inflammation of the spinal cord beginning in the lower segments and progressing toward the head. { ə'kyüt ə'send·iŋ ,mi·ə'līd·əs }

acute bacterial endocarditis [MED] Fulminant, rapidly progressive endocarditis, usually associated with a significant systemic infection. { ə'kyüt bak'tir·ē·əl ,en·dō,kär'dīd·əs }

acute benign lymphoblastosis [MED] *See* infectious mononucleosis. { ə'kyüt bə'nīn ,lim·fə,blas'tōs·əs }

acute berylliosis [MED] Severe chemical pneumoconiosis that is caused by inhalation of beryllium salts. { ə'kyüt bə,ril·ē'os·əs }

acute bisectrix [MINERAL] A bisecting line of the acute angle of the optic axes of biaxial minerals. { ə'kyüt ,bī'sek·triks }

acute cerebellar ataxia [MED] A severe childhood syndrome of sudden onset characterized by muscular incoordination, impaired articulation, oscillations of the eyeballs, and decreased intraocular pressure. { ə'kyüt ,ser·ə'bel·ər ə'tak·sē·ə }

acute dermatitis [MED] Any severe inflammation of the skin. { ə'kyüt ,dər·mə'tīd·əs }

acute glomerulonephritis [MED] Severe kidney inflammation, usually following infection with group A hemolytic streptococci, particularly type 12. { ə'kyüt glä,mer·yə,lō·nə'frīd·əs }

acute granulocytic leukemia [MED] A severe blood disorder in which the abnormal white cells are immature forms of granulocytes. Also known as myeloblastic leukemia. { ə'kyüt gran·yə·lə'sid·ik lü'kē·mē·ə }

acute infective encephalomyelitis *See* epidemic neuromyasthenia. { ə'kyüt in'fek·təv en,sef·ə·lō,mī·ə'līd·əs }

acute inflammation [MED] Severe inflammation with rapid progress and pronounced symptoms. { ə'kyüt in·flə'mā·shən }

acute leukemia [MED] A severe blood disorder characterized by rapid onset and progress, with anemia and hemorrhagic manifestations; immature forms of leukocytes are predominant. { ə'kyüt lü'kē·mē·ə }

acute lymphocytic leukemia [MED] A severe blood disorder in which abnormal leukocytes are identified as immature forms of lymphocytes. Also known as lymphoblastic leukemia. { ə'kyüt lim·fə'sid·ik lü'kē·mē·ə }

acute monocytic leukemia [MED] A severe blood disorder in which abnormal leukocytes are identified as immature forms of monocytes. Also known as monoblastic leukemia. { ə'kyüt ,män·ə'sid·ik lü'kē·mē·ə }

acute necrotizing hemorrhagic encephalomyelitis [MED] A sudden, severe central nervous system disease with variable symptoms; pathology includes hemorrhages and necrosis of the white matter. { ə'kyüt ,nek·rə'tīz·iŋ hem·ə'raj·ik en,sef·ə·lō,mī·ə'līd·əs }

acute nonsuppurative hepatitis *See* interstitial hepatitis. { ə'kyüt ,nän'səp·yə,rād·iv hep·ə'tīd·əs }

acute pancreatitis [MED] A disease of unknown etiology that causes sudden liberation of activated pancreatic enzymes that digest the pancreatic parenchyma, leading to dissolution of fat and production of calcium soaps, and rupture of pancreatic vessels with resultant hemorrhage and shock. { ə'kyüt 'pan·krē·ə'tīd·əs }

acute-phase protein [IMMUNOL] Any of a group of proteins that are produced by the liver and appear in the blood in increased amounts shortly after the onset of infection or tissue damage; they include C-reactive protein, fibrinogen, proteolytic enzyme inhibitors, and transferrin. { ə'kyüt ¦fāz 'prō,tēn }

acute-phase reaction [IMMUNOL] During inflammation, change in the rates of synthesis of certain serum proteins that are important in nonspecific defense reactions. { ə'kyüt ¦fāz rē'ak·shən }

acute radiation syndrome [MED] A complex of symptoms involving the intestinal tract, blood-forming organs, and skin following whole-body irradiation. { ə'kyüt 'rād·ē·a·shən 'sin,drōm }

acute respiratory disease [MED] Severe adenovirus infection of the respiratory tract characterized by fever, sore throat, and cough. { ə'kyüt 'res·prə,tòr·ē di,zēz }

acute rheumatic fever [MED] A severe infectious process caused by beta hemolytic streptococci; characterized by fever and frequently accompanied by painful inflamed joints, endocarditis, chorea, or glomerulonephritis. { ə'kyüt rù'mad·ik 'fē·vər }

acute rhinitis [MED] Inflammation of the nasal mucous membrane due to either infection or allergy. { ə'kyüt rī 'nīd·əs }

acute toxic encephalopathy [MED] A severe childhood syndrome characterized by sudden onset of coma or stupor, fever, convulsions, and impaired respiratory and cardiovascular functioning. { ə'kyüt 'täk·sik en,sef·ə'läp·ə·thē }

acute transfection [GEN] Short-term deoxyribonucleic acid infection of cells. { ə'kyüt tranz'fek·shən }

acute triangle [MATH] A triangle each of whose angles is less than 90°. { ə'kyüt 'trī,aŋ·gəl }

acute tubular necrosis *See* lower nephron nephrosis. { ə'kyüt 'tüb·yə·lər nə'krō·səs }

acute yellow atrophy [MED] Rapid liver destruction following viral hepatitis, toxic chemicals, or other agents. { ə'kyüt 'yel·ō 'a·trə·fē }

acutifoliate [BOT] Possessing sharply pointed leaves. { ə,kyüd·ə'fō·lē·āt }

acutilobate [BOT] Possessing sharply pointed lobes. { ə,kyüd·ə'lō,bāt }

acyclic [BOT] Having flowers arranged in a spiral instead of a whorl. [MATH] **1.** A transformation on a set to itself for which no nonzero power leaves an element fixed. **2.** A chain complex all of whose homology groups are trivial. [PHYS] Continually varying without a regularly repeated pattern. { ā'sik·lik }

acyclic compound [ORG CHEM] A chemical compound with an open-chain molecular structure rather than a ring-shaped structure; for example, the alkane series. { ā'sik·lik 'kam,paund }

acyclic digraph [MATH] A directed graph with no directed cycles. { ā¦sik·lik 'dī,graf }

acyclic feeding [COMPUT SCI] A method employed by alphanumeric readers in which the trailing edge or some other document characteristic is used to activate the feeding of the succeeding document. { ā¦sik·lik 'fēd·iŋ }

acyclic graph [MATH] A graph with no cycles. Also known as forest. { ā¦sik·lik 'graf }

acyclic machine *See* homopolar generator. { ā'sik·lik mə'shēn }

acyclic motion *See* irrotational flow. { ā'sik·lik 'mō·shən }

acyl [ORG CHEM] A radical formed from an organic acid by removal of a hydroxyl group; the general formula is RCO, where R may be aliphatic, alicyclic, or aromatic. { 'a·səl }

acylation [ORG CHEM] Any process whereby the acyl group is incorporated into a molecule by substitution. { ,as·ə'lā·shən }

acyl azide *See* acid azide. { 'a·səl 'a,zīd }

acylcarbene [ORG CHEM] A carbene radical in which at least one of the groups attached to the divalent carbon is an acyl group; for example, acetylcarbene. { ,a·səl'kär,bēn }

acyl carnitine *See* fatty acyl carnitine. { 'a·səl 'kär·nə,tēn }

acyl carrier protein [BIOCHEM] A protein in fatty acid synthesis that picks up acetyl and malonyl groups from acetyl coenzyme A and malonyl coenzyme A and links them by condensation to form β-keto acid acyl carrier protein, releasing carbon dioxide and the sulfhydryl form of acyl carrier protein. Abbreviated ACP. { 'a·səl 'kar·ē·ər 'prō,tēn }

acyl-coenzyme A *See* fatty acyl-coenzyme A. { 'a·səl kō 'en,zim 'ā }

acyl exchange *See* acidolysis. { 'a·səl iks'chānj }

acyl halide [ORG CHEM] One of a large group of organic substances containing the halocarbonyl group; for example, acyl fluoride. { 'a·səl 'hal,īd }

acylnitrene [ORG CHEM] A nitrene in which the nitrogen is covalently bonded to an acyl group. { ,a·səl'nī,trēn }

acyloin [ORG CHEM] An organic compound that may be synthesized by condensation of aldehydes; an example is benzoin, $C_6H_5COCHOHC_6H_5$. { ə'sil·ə·wən }

acyloin condensation [ORG CHEM] The reaction of an aliphatic ester with metallic sodium to form intermediates converted by hydrolysis into aliphatic α-hydroxyketones called acyloins. { ə'sil·ə·wən ,kän,den'sā·shən }

AD *See* Alzheimer's disease; average deviation.

Ada [COMPUT SCI] A computer language that was chosen by the United States Department of Defense to support the development of embedded systems, and uses the language Pascal as a base to meet the reliablity and efficiency requirements imposed by these systems. { 'ā·də }

ADA *See* air-defense artillery.

adakites [GEOL] Rocks formed from lavas that melted from subducting slabs associated with either volcanic arcs or arc/continent collision zones; they were first described from Adak Island in the Aleutians. { 'a·də,kīts }

adalert [GEOPHYS] An advance alert issued by a regional warning center to give prompt warning of a change in solar activity. { 'ad·ə,lərt }

adamantane [ORG CHEM] A $C_{10}H_{16}$ alicyclic hydrocarbon whose structure has the same arrangement of carbon atoms as does the basic unit of the diamond lattice. { ,ad·ə'man,tān }

adamantine drill [MECH ENG] A core drill with hardened steel shot pellets that revolve under the rim of the rotating tube; employed in rotary drilling in very hard ground. { ,ad·ə'man,tēn 'dril }

adamantine spar [MINERAL] A silky brown variety of corundum. { ,ad·ə'man,tēn 'spär }

adamantinoma *See* ameloblastoma. { ad·ə,man·tə'nōm·ə }

adambulacral [INV ZOO] Lying adjacent to the ambulacrum. { ¦ad·am·byə'lāk·rəl }

adamellite *See* quartz monzonite. { ə'dam·ə,līt }

adamite [MINERAL] $Zn_2(AsO_4)(OH)$ A colorless, white, or yellow mineral consisting of basic zinc arsenate, crystallizing in the orthorhombic system; hardness is 3.5 on Mohs scale, and specific gravity is 4.34–4.35. { 'ad·ə,mīt }

Adams-Bashforth process [MATH] A method of numerically integrating a differential equation of the form

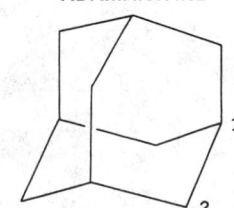

ADAMANTANE

Structure of adamantane showing bridgehead carbon labeled 1 and alternate position for substitution at carbon 2.

$(dy/dx) = f(x,y)$ that uses one of Gregory's interpolation formulas to expand f. { 'a·dəmz 'bash,förth ,präs·əs }

Adam's catalyst [CHEM ENG] Finely divided plantinum(IV) oxide, made by fusing hexachloroplatinic(IV) acid with $NaNO_3$. { 'a·dəmz 'kad·əl·əst }

adamsite [MINERAL] Greenish-black mica. [ORG CHEM] $C_6H_4 \cdot NH \cdot C_6H_4 \cdot AsCl$ A yellow crystalline arsenical; used in leather tanning and in warfare and riot control to produce skin and eye irritation, chest distress, and nausea; U.S. Army code is DM. Also known as diphenylaminechloroarsine; phenarsazine chloride. { 'a·dəm,zīt }

ada mud [ENG] A conditioning material added to drilling mud to obtain satisfactory cores and samples of formations. { 'ā·də ,məd }

adapertural [INV ZOO] Near the aperture, specifically of a conch. { ,ad'ap·ə,chər·əl }

adapical [BOT] Near or toward the apex or tip. { ,ad'a·pi·kəl }

adaptability test [ORD] A test to ascertain the adaptability of a standardized item or equipment to a particular unit or organization. { ə,dap·tə'bil·ə·dē ,test }

adaptation [GEN] Adjustment to new or altered environmental conditions by changes in genotype (natural selection) or phenotype. [PHYSIO] The occurrence of physiological changes in an individual exposed to changed conditions; for example, tanning of the skin in sunshine, or increased red blood cell counts at high altitudes. { ,a,dap'tā·shən }

adaptation brightness *See* adaptation luminance. { ,a,dap'tā·shən ,brīt·nəs }

adaptation illuminance *See* adaptation luminance. { ,a,dap 'tā·shən ə'lü·mə·nəns }

adaptation level *See* adaptation luminance. { ,a,dap'tā·shən ,lev·əl }

adaptation luminance [OPTICS] The average luminance, or brightness, of objects and surfaces in the immediate vicinity of an observer estimating the visual range. Also known as adaptation brightness; adaptation illuminance; adaptation level; brightness level; field brightness; field luminance. { ,a,dap'tā·shən 'lü·mə·nəns }

adaptation syndrome [MED] Endocrine-mediated stress reaction of the body in response to systemic injury; involves an initial stage of shock, followed by resistance or adaptation and then healing or exhaustion. { ,a,dap'tā·shən ,sin,drōm }

adapter [COMPUT SCI] A device which converts bits of information received serially into parallel bit form for use in the inquiry buffer unit. [ENG] A device used to make electrical or mechanical connections between items not originally intended for use together. [MET] A connecting piece, usually made of fireclay, between a horizontal zinc retort and the condenser in which the molten zinc collects. [OPTICS] An attachment to a camera that permits its use in a manner for which it was not designed. { ə'dap·tər }

adapter skirt [AERO ENG] A flange or extension of a space vehicle that provides a ready means for fitting some object to a stage or section. { ə'dap·tər ,skərt }

adapter spool [PETRO ENG] A device for connecting blowout preventers of different sizes or capacities to the casinghead. { ə'dap·tər ,spül }

adapter transformer [ELEC] A transformer designed to supply a single electric lamp; its primary terminals are designed to fit into an ordinary lampholder, its secondary terminals into a lampholder of a low-voltage lamp. { ə'dap·tər tranz,förmər }

adaptins [CELL MOL] Peripheral membrane proteins that play an important role in the assembly of clathrin-coated vesicles in the trans-Golgi network and plasma membrane during receptor-mediated endocytosis. { ə'dap·tins }

adaptive behavior [PSYCH] Any behavior that helps the organism adjust to its environment. { ə'dap·tiv bə'hāv·yər }

adaptive branch [CONT SYS] A branch instruction in the computer program controlling a robot that may lead the robot to execute a series of instructions, depending on external conditions. { ə'dap·tiv 'branch }

adaptive colitis *See* irritable colon. { ə'dap·tiv kə'līd·əs }

adaptive communications [COMMUN] A communications system capable of automatic change to meet changing inputs

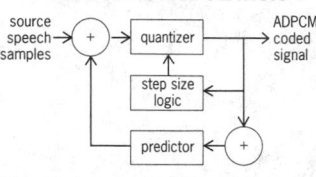

ADAPTIVE DIFFERENTIAL PULSE-CODE MODULATION

Diagram of an ADPCM encoder.

or changing characteristics of the device or process being controlled. Also known as self-adjusting communications; self-optimizing communications. { ə'dap·tiv kə,myü·nə'kā·shənz }

adaptive control [CONT SYS] A control method in which one or more parameters are sensed and used to vary the feedback control signals in order to satisfy the performance criteria. { ə'dap·tiv kən'trōl }

adaptive-control function [CONT SYS] That level in the functional decomposition of a large-scale control system which updates parameters of the optimizing control function to achieve a best fit to current plant behavior, and updates parameters of the direct control function to achieve good dynamic response of the closed-loop system. { ə'dap·tiv kən'trōl ,fəŋk·shən }

adaptive differential pulse-code modulation [COMMUN] A method of compressing speech and music signals in which the transmitted signals represent differences between input signals and predicted signals, and these predicted signals are synthesized by predictors with response functions representative of the short- and long-term correlation inherent in the signal. Abbreviated ADPCM. { ə'dap·tiv ,dif·ə'ren·chəl 'pəls,cōd ,mäj·ə,lā·shən }

adaptive disease [PHYSIO] The physiologic changes impairing an organism's health as the result of exposure to an unfamiliar environment. { ə'dap·tiv di,zēz }

adaptive divergence [EVOL] Divergence of new forms from a common ancestral form due to adaptation to different environmental conditions. { ə'dap·tiv də'vər·jəns }

adaptive enzyme [MICROBIO] Any bacterial enzyme formed in response to the presence of a substrate specific for that enzyme. { ə'dap·tiv 'en,zīm }

adaptive equalization [COMMUN] A signal-processing technique designed to compensate for impairments in received signals over a communications channel resulting from imperfect transmission characteristics. { ə'dap·tiv ,ē·kwə·lə,zā·shən }

adaptive filter [ELECTR] An electric filter whose frequency response varies with time, as a function of the input signal. { ə,dap·tiv 'fil·tər }

adaptive immune response [IMMUNOL] An immune response based on the principle of clonal recognition, such that upon first exposure to an antigen, primed lymphocytes either differentiate into immune effector cells or form an expanded pool of memory cells that respond to secondary exposure to the same antigen by mounting an amplified and more rapid response. { ə'dap·tiv i'myün ri,späns }

adaptive integration [MATH] A numerical technique for obtaining the definite integral of a function whose smoothness, or lack thereof, is unknown, to a desired degree of accuracy, while doing only as much work as necessary on each subinterval of the interval in question. { ə'dap·tiv ,int·ə'grā·shən }

adaptive mutations [GEN] Mutations conferring an advantage in a selective environment which arise after nongrowing or slowly growing cells are exposed to the selective environment. { ə'dap·tiv myü'tā·shənz }

adaptive norm [GEN] The mix of genotypes of a well-adapted species or population. { ə'dap·tiv 'nörm }

adaptive optics [OPTICS] The theory and design of optical systems that measure and correct wavefront aberrations in real time, that is, simultaneously with the operation of the system. { ə'dap·tiv 'äp·tiks }

adaptive radiation [EVOL] Diversification of a dominant evolutionary group into a large number of subsidiary types adapted to more restrictive modes of life (different adaptive zones) within the range of the larger group. { ə'dap·tiv ,rād·ē'ā·shən }

adaptive robot [CONT SYS] A robot that can alter its responses according to changes in the environment. { ə'dap·tiv 'rō,bät }

adaptive signal processing [COMMUN] The design of adaptive systems for signal-processing applications. { ə'dap·tiv 'sig·nəl 'prä·sə·siŋ }

adaptive sound cancellation [ACOUS] A form of active sound cancellation that is maintained in the presence of system changes through a feedback mechanism. { ə'dap·tiv 'saund ,kan·sə,lā·shən }

adaptive sound control [ACOUS] Any modification of sound fields by loudspeakers, controlled through the use of a feedback mechanism, for reduction, equalization, or cancellation of sound.　{ ə¦dap·tiv 'saund kən¸trōl }

adaptive structure [ENG] A structure whose geometric and inherent structural characteristics can be changed beneficially in response to external stimulation by either remote commands or automatic means.　{ ə¸dap·tiv 'strək·chər }

adaptive system [SYS ENG] A system that can change itself in response to changes in its environment in such a way that its performance improves through a continuing interaction with its surroundings.　{ ə'dap·tiv 'sis·təm }

adaptive system theory [COMPUT SCI] The branch of automata theory dealing with adaptive, or self-organizing, systems.　{ ə'dap·tiv 'sis·təm ¸the·ə·rē }

adaptive value [GEN] The property of a given genotype that confers fitness to an organism in a given environment.　{ ə'dap·tiv 'val·yü }

adaptometer [ENG] An instrument that measures the lowest brightness of an extended area that can barely be detected by the eye.　{ ¸a¸dap'tä·məd·ər }

adaptor [COMPUT SCI] A printed circuit board that is plugged into an expansion slot in a computer to communicate with an external peripheral device. [GEN] Any of the short synthetic oligonucleotide strands that have one sticky end and one blunt end; the blunt end joins to the blunt end of a deoxyribonucleic acid fragment, forming a new fragment with two sticky ends that can be more easily spliced into a vector.　{ ə'dap·tər }

adaptor protein [CELL MOL] A specialized protein that links protein components of the signaling pathway, thereby aiding intracellular signal transduction.　{ ə'dap·tər ¸prō·tēn }

ADAR See advanced-design array radar.　{ 'ā¸där }

adatom [PHYS CHEM] An atom adsorbed on a surface so that it will migrate over the surface.　{ 'ad¸ad·əm }

adaxial [BIOL] On the same side as or facing toward the axis of an organ or organism.　{ ¸ad'ak·sē·əl }

Adcock antenna [ELECTROMAG] A pair of vertical antennas separated by a distance of one-half wavelength or less and connected in phase opposition to produce a radiation pattern having the shape of a figure eight.　{ 'ad¸käk ¸an'ten·ə }

Adcock direction finder [NAV] A radio direction finder utilizing one or more pairs of Adcock antennas.　{ 'ad¸käk də'rek·shən ¸fīn·dər }

ADCON See address constant.　{ 'ad¸kän }

adconductor cathode [ELECTR] A cathode in which adsorbed alkali metal atoms provide electron emission in a glow or arc discharge.　{ ¦ad·kən¦dək·tər 'kath¸ōd }

adcumulus [PETR] Pertaining to the growth of a cumulus crystal so as to exclude the growth of other phases; results in a monomineralic rock.　{ ad'kyü·myə·ləs }

add [COMPUT SCI] See add operation. [MATH] To perform addition. [ORD] A fire correction term used by observers in adjusting fire to indicate that an increase in range (of so many yards) will follow and is desired.　{ ad }

addend [MATH] One of a collection of numbers to be added.　{ 'a¸dend }

addendum [DES ENG] The radial distance between two concentric circles on a gear, one being that whose radius extends to the top of a gear tooth (addendum circle) and the other being that which will roll without slipping on a circle on a mating gear (pitch line).　{ ə'den·dəm }

addendum circle [DES ENG] The circle on a gear passing through the tops of the teeth.　{ ə'den·dəm ¸sər·kəl }

adder [COMPUT SCI] A computer device that can form the sum of two or more numbers or quantities. [ELECTR] A circuit in which two or more signals are combined to give an output-signal amplitude that is proportional to the sum of the input-signal amplitudes. Also known as adder circuit. [VERT ZOO] Any of the venomous viperine snakes included in the family Viperidae.　{ 'ad·ər }

adder circuit See adder.　{ 'ad·ər ¸sər·kət }

addiction [MED] Habituation to a specific practice, such as drinking alcoholic beverages or using drugs.　{ ə'dik·shən }

addictive disorder [PSYCH] A disorder characterized by the chronic use of an agent and resulting in the development of tolerance, physical dependence, and finally drug-seeking behavior.　{ ə'dik·tiv dis'ōr·dər }

add-in [COMPUT SCI] An electronic component that can be placed on a printed circuit board already installed in a computer to enhance the computer's capability.　{ 'ad¸in }

adding circuit [ELECTR] A circuit that performs the mathematical operation of addition.　{ 'ad·iŋ 'sər·kət }

adding machine [COMPUT SCI] A device which performs the arithmetical operation of addition and subtraction.　{ 'ad·iŋ mə¸shēn }

adding tape [ENG] A surveyor's tape that is calibrated from 0 to 100 by full feet (or meters) in one direction, and has 1 additional foot (or meter) beyond the zero end which is subdivided in tenths or hundredths.　{ 'ad·iŋ ¸tāp }

add-in program [COMPUT SCI] A computer program that enhances the capabilities of a particular application.　{ 'ad¸in ¸prō·grəm }

Addis count [PATH] A renal function test which estimates the blood cell count in a 12-hour urine specimen.　{ 'ad·əs ¸kaůnt }

Addison's disease [MED] A primary failure or insufficiency of the adrenal cortex to secrete hormones.　{ 'ad·ə·sənz di¸zēz }

addition [MATH] **1.** An operation by which two elements of a set are combined to yield a third; denoted +; usually reserved for the operation in an Abelian group or the group operation in a ring or vector space. **2.** The combining of complex quantities in which the individual real parts and the individual imaginary parts are separately added. **3.** The combining of vectors in a prescribed way; for example, by algebraically adding corresponding components of vectors or by forming the third side of the triangle whose other sides each represent a vector. Also known as composition.　{ ə'di·shən }

addition agent [PHYS CHEM] A substance added to a plating solution to change characteristics of the deposited substances.　{ ə'di·shən ¸ā·jənt }

addition formula [MATH] An equation expressing a function of the sum of two quantities in terms of functions of the quantities themselves.　{ ə'dish·ən ¸fȯr·myə·lə }

addition item [COMPUT SCI] An item which is to be filed in its proper place in a computer.　{ ə'di·shən 'īd·əm }

addition polymer [ORG CHEM] A polymer formed by the chain addition of unsaturated monomer molecules, such as olefins, with one another without the formation of a by-product, as water; examples are polyethylene, polypropylene, and polystyrene. Also known as addition resin.　{ ə'di·shən 'päl·ə·mər }

addition polymerization [ORG CHEM] A reaction initiated by an anion, cation, or radical in which a large number of monomer units are added rapidly (a chain reaction) until terminated by some mechanism, forming a high-molecular-weight polymer in a very short time; an example is the free-radical polymerization of propylene to polypropylene.　{ ə'dish·ən pə¸lim·ə·rə'zā·shən }

addition reaction [ORG CHEM] A type of reaction of unsaturated hydrocarbons with hydrogen, halogens, halogen acids, and other reagents, so that no change in valency is observed and the organic compound forms a more complex one.　{ ə'di·shən rē'ak·shən }

addition record [COMPUT SCI] A new record inserted into an updated master file.　{ ə'di·shən ¸rek·ərd }

addition resin See addition polymer.　{ ə'di·shən 'rez·ən }

addition sign [MATH] The symbol +, used to indicate addition. Also known as plus sign.　{ ə'di·shən ¸sīn }

addition solid solution [CRYSTAL] Random addition of atoms or ions in the interstices within a crystal structure.　{ ə'di·shən 'säl·əd sə'lü·shən }

addition table [COMPUT SCI] The part of memory that holds the table of numbers used in addition in a computer employing table look-up techniques to carry out this operation.　{ ə'di·shən ¸tā·bəl }

additive [MATER] **1.** A substance added to another to strengthen or otherwise alter it for the purpose of improving the performance of the finished product. **2.** See admixture. [MATH] Pertaining to addition. [STAT] That property of a process in which increments of the dependent variable are independent for nonoverlapping intervals of the independent variable.　{ 'ad·əd·iv }

additive factor [GEN] Any of a group of nonallelic genes that affect the same phenotypic characteristics.　{ 'ad·ə·div 'fak·tər }

additive function [MATH] Any function f that preserves addition; that is, $f(x + y) = f(x) + f(y)$.　{ 'ad·əd·iv 'fəŋ·shən }

additive gene action [GEN] **1.** A form of allelic interaction

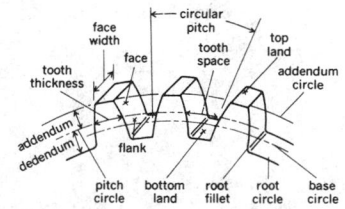

ADDENDUM

Drawing of the principal features of a gear tooth showing the addendum in relation to the other parts.

in which dominance is absent, resulting in a heterozygote that is intermediate in phenotype between homozygotes for the alternative alleles. **2.** The cumulative contribution made by all loci in a group of nonallelic genes to a polygenic trait. { ¦ad·ə·div ¦jēn ˌak·shən }

additive genetic variance [GEN] That part of the genetic variance of a quantitative character attributed to the average effects of substituting one allele for another at a given locus or at the multiple loci governing a polygenic trait. { ¦ad·ə·div jə¦ned·ik 'ver·ē·əns }

additive identity [MATH] In a mathematical system with an operation of addition denoted +, an element 0 such that $0 + e = e + 0 = e$ for any element e in the system. { 'ad·ə·div ī'den·ə·dē }

additive inverse [MATH] In a mathematical system with an operation of addition denoted +, an additive inverse of an element e is an element $-e$ such that $e + (-e) = (-e) + e = 0$, where 0 is the additive identity. { 'ad·ə·div 'in,vərs }

additive level [MATER] The total percentage of all the additives in an oil sample. { 'ad·əd·iv 'lev·əl }

additive primary colors [OPTICS] The three colors, usually red, green, and blue, which are mixed together in an additive process. { 'ad·əd·iv 'prīm·ə·rē 'kəl·ərz }

additive process [OPTICS] The process of producing colors by mixing lights of additive primary colors in various proportions. { 'ad·əd·iv ˌprä·səs }

additive set function [MATH] A set function with the properties that (1) the union of any two sets in the range of the function is also in this range and (2) the value of the function at a finite union of disjoint sets in the range of the set function is equal to the sum of the values at each set in the union. Also known as finitely additive set function. { ¦ad·əd·iv ¦set ˌfəŋk·shən }

additive synthesis [ENG ACOUS] A method of synthesizing complex tones by adding together an appropriate number of simple sine waves at harmonically related frequencies. { ¦ad·ə·div 'sin·thə·səs }

add-on [COMPUT SCI] A peripheral device, such as a printer or disk drive, that is added to a basic computer. { 'ad,ȯn }

add-on memory [COMPUT SCI] Computer storage that is added to the original main storage to enhance the computer's processing capability. { 'ad,ȯn 'mem·rē }

add operation [COMPUT SCI] An operation in computer processing in which the sum of two or more numbers is placed in a storage location previously occupied by one of the original numbers. Also known as add. { 'ad ˌäp·ə,rā·shən }

address [COMPUT SCI] The number or name that uniquely identifies a register, memory location, or storage device in a computer. { 'ad·res }

addressable [COMPUT SCI] Capable of being located by a computer through an addressing technique. { ə'dres·ə·bəl }

addressable cursor [COMPUT SCI] A cursor that can be moved by software or keyboard controls to any point on the screen. { ə'dres·ə·bəl 'kər·sər }

address book [COMPUT SCI] A feature in an e-mail program for storing e-mail addresses. { 'ad·rəs ˌbuk }

address bus [COMPUT SCI] An internal computer communications channel that carries addresses from the central processing unit to components under the unit's control. { 'ad·res ˌbəs }

address computation [COMPUT SCI] The modification by a computer of an address within an instruction, or of an instruction based on results obtained so far. Also known as address modification. { 'ad·res ˌkäm·pyə'tā·shən }

address constant [COMPUT SCI] A value, or its expression, used in the calculation of storage addresses from relative addresses for computers. Abbreviated ADCON. Also known as base address; presumptive address; reference address. { 'ad·res ˌkän·stənt }

address conversion [COMPUT SCI] The use of an assembly program to translate symbolic or relative computer addresses. { 'ad·res kən,vər·zhən }

address counter [COMPUT SCI] A counter which increments an initial memory address as a block of data is being transferred into the memory locations indicated by the counter. { 'ad·res ˌkaunt·ər }

address field [COMPUT SCI] The portion of a computer program instruction which specifies where a particular piece of

information is located in the computer memory. { 'ad·res ˌfēld }

address format [COMPUT SCI] A description of the number of addresses included in a computer instruction. { 'ad·res ˌfȯr·mat }

address-free program [COMPUT SCI] A computer program in which all addresses are represented as displacements from the expected contents of a base register. { 'ad·res ¦frē 'prō·grəm }

address generation [COMPUT SCI] An addressing technique which facilitates addressing large storages and implementing dynamic program relocation; the effective main storage address is obtained by adding together the contents of the base register of the index register and of the displacement field. { 'ad·res ˌjen·ə'rā·shən }

addressing [COMPUT SCI] **1.** The methods of locating and gaining access to information in a computer's storage. **2.** The methods of selecting a particular peripheral device from several that are available at a given time. { ə'dres·iŋ }

addressing mode [COMPUT SCI] The specific technique by means of which a memory reference instruction will be spelled out if the computer word is too small to contain the memory address. { ə'dres·iŋ ˌmōd }

addressing system [COMPUT SCI] A labeling technique used to identify storage locations within a computer system. { ə'dres·iŋ ˌsis·təm }

address interleaving [COMPUT SCI] The assignment of consecutive addresses to physically separate modules of a computer memory, making possible the very-high-speed access of a sequence of contiguously addressed words, since all modules operate nearly simultaneously. { 'ad·res ˌin·tər'lēv·iŋ }

addressless instruction format See zero-address instruction format. { ə'dres·ləs ˌin'strək·shən 'fȯr·mat }

address modification [COMPUT SCI] See address computation. { 'ad·res ˌmäd·ə·fə'kā·shən }

address part [COMPUT SCI] That part of a computer instruction which contains the address of the operand, of the result, or of the next instruction. { 'ad·res ˌpärt }

address register [COMPUT SCI] A register wherein the address part of an instruction is stored by a computer. { 'ad·res ˌrej·ə·stər }

address resolution [COMPUT SCI] **1.** The process of obtaining the actual machine address needed to perform an operation. **2.** The process by which the address used to identify a workstation on a local-area network is translated to an address that can be handled on the Internet. { 'ad·res ˌrez·ə,lü·shən }

address sort routine [COMPUT SCI] A debugging routine which scans all instructions of the program being checked for a given address. { 'ad·res 'sȯrt ˌrü'tēn }

address space [COMPUT SCI] The number of storage locations available to a computer program. { 'ad·rəs ˌspās }

address track [COMPUT SCI] A path on a magnetic tape, drum, or disk on which are recorded addresses used in the retrieval of data stored on other tracks. { 'ad·res ˌtrak }

address translation [COMPUT SCI] The assignment of actual locations in a computer memory to virtual addresses in a computer program. { 'ad·res tranz'lā·shən }

add-subtract time [COMPUT SCI] The time required to perform an addition or subtraction, exclusive of the time required to obtain the quantities from storage and put the sum or difference back into storage. { 'ad səb'trakt ˌtīm }

add time [COMPUT SCI] The time required by a computer to perform an addition, not including the time needed to obtain the addends from storage and put the sum back into storage. { 'ad ˌtīme }

add-to-memory technique [COMPUT SCI] In direct-memory-access systems, a technique which adds a data word to a memory location; permits linear operations such as data averaging on process data. { 'ad tə ¦mem·rē 'tek·nēk }

adduct [CHEM] **1.** A chemical compound that forms from chemical addition of two species; for example, reaction of butadiene with styrene forms an adduct, 4-phenyl-1-cyclohexene. **2.** The complex compound formed by association of an inclusion complex. { 'a,dəkt }

adduction [PHYSIO] Movement of one part of the body toward another or toward the median axis of the body. { ə'dək·shən }

adductor [ANAT] Any muscle that draws a part of the body toward the median axis. { ə'dək·tər }

Adeleina [INV ZOO] A suborder of protozoan invertebrate parasites in the order Eucoccida in which the sexual and asexual stages are in different hosts. { ,ad·ə'līn·ə }

adelite [MINERAL] CaMg(AsO₄)(OH,F) A colorless to gray, bluish-gray, yellowish-gray, yellow, or light green orthorhombic mineral consisting of a basic arsenate of calcium and magnesium; usually occurs in massive form. { 'ad·əl,īt }

adelphous [BOT] Having stamens fused together by their filaments. { ə'del·fəs }

adenase [BIOCHEM] An enzyme that catalyzes the hydrolysis of adenine to hypoxanthine and ammonia. { 'ad·ən,ās }

adenine [BIOCHEM] C₅H₅N₅ A purine base, 6-aminopurine, occurring in ribonucleic acid and deoxyribonucleic acid and as a component of adenosine triphosphate. { 'ad·ən,ēn }

adenitis [MED] Inflammation of a gland or lymph node. { ,ad·ən'īd·əs }

adenoacanthoma [MED] An adenocarcinoma, common in the endometrium, in which squamous cells replace the cylindrical epithelium. { |ad·ən,ō·ə,kan'thō·mə }

adeno-associated satellite virus [VIROL] A defective virus that is unable to reproduce without the help of an adenovirus. { |ad·ən,ō·ə'sō·shē,ād·əd 'sad·ə,līt |vī·rəs }

adenocarcinoma [MED] A malignant tumor originating in glandular or ductal epithelium and tending to produce acinic structures. { |ad·ən,ō,kär·sən'ō·mə }

adenohypophysis [ANAT] The glandular part of the pituitary gland, composing the anterior and intermediate lobes. { |ad·ən,ō,hī'pä·fə·səs }

adenoid [ANAT] **1.** A mass of lymphoid tissue. **2.** Lymphoid tissue of the nasopharynx. Also known as pharyngeal tonsil. { 'ad,nóid }

adenoma [MED] A benign tumor of glandular origin and structure. { ,ad·ən'ō·mə }

adenomatoid tumor [MED] A benign genital-tract tumor composed of stroma whose spaces are lined by cells that resemble epithelium, endothelium, and mesothelium. { ,ad·ən'ä·mə,tóid 'tü·mər }

adenomatosis [MED] A condition characterized by multiple adenomas within an organ or in several related organs. { ,ad·ən,ō·mə'tō·səs }

adenomatous goiter [MED] An asymmetric goiter due to isolated nodular masses of thyroid tissue. Also known as multiple colloid goiter; nodular goiter. { ,ad·ən,ō'ma·təs 'góit·ər }

adenomere [EMBRYO] The embryonic structure which will become the functional portion of a gland. { ,ad·ən'ō·mir }

adenomyoma [MED] A benign tumor of glandular and muscular elements occurring principally in the uterus and rectum. { ,ad·ən,ō,mī'ō·mə }

adenomyosis [MED] **1.** The invasion of muscular tissue, such as of the uterine wall or Fallopian tubes, by endometrial tissue. **2.** Any abnormal growth of muscle or glandular tissues. { ,ad·ən,ō,mī'ō·səs }

adenopathy [MED] Any glandular disease; common usage limits the term to any abnormal swelling or enlargement of lymph nodes. { ,ad·ən'ä·pə·thē }

Adenophorea [INV ZOO] A class of unsegmented worms in the phylum Nematoda. { ,ad·ən·ə'fór·ē·ə }

adenophyllous [BOT] Having leaves with glands. { ,ad·ən'ä·fə·ləs }

adenosine [BIOCHEM] C₁₀H₁₃N₅O₄ A nucleoside composed of adenine and D-ribose. { ə'den·ə,sēn }

adenosine 3′,5′-cyclic monophosphate See cyclic adenylic acid. { ə|den·ə·sēn |thrē|prīm |fīv|prīm 'sīk·lik 'mä·nō'fäs·fāt }

adenosine 3′,5′-cyclic phosphate See cyclic adenylic acid. { ə|den·ə·sēn |thrē|prīm |fīv|prīm 'sīk·lik 'fäs·fāt }

adenosine diphosphatase [BIOCHEM] An enzyme that catalyzes the hydrolysis of adenosine diphosphate. Abbreviated ADPase. { ə|den·ə,sēn ,dī'fäs·fə,tās }

adenosine diphosphate [BIOCHEM] C₁₀H₁₅N₅O₁₀P₂ A coenzyme composed of adenosine and two molecules of phosphoric acid that is important in intermediate cellular metabolism. Abbreviated ADP. { ə|den·ə,sēn ,dī'fäs·fāt }

adenosine monophosphate See adenylic acid. { ə|den·ə,sēn 'mä·nō'fäs·fāt }

adenosine 3′,5′-monophosphate See cyclic adenylic acid. { ə|den·ə,sēn |thrē|prīm |fīv|prīm 'mä·nō'fäs·fāt }

adenosine triphosphatase [BIOCHEM] An enzyme that catalyzes the hydrolysis of adenosine triphosphate. Abbreviated ATPase. { ə|den·ə,sēn ,trī'fäs·fə,tās }

adenosine triphosphate [BIOCHEM] C₁₀H₁₆N₅O₁₂P₃ A coenzyme composed of adenosine diphosphate with an additional phosphate group; an important energy compound in metabolism. Abbreviated ATP. { ə|den·ə,sēn,tri'fäs,fāt }

adenosis [MED] Any nonneoplastic glandular disease, especially one involving the lymph nodes. { ,ad·ən'ō·səs }

adeno-SV40 hybrid virus [VIROL] A defective virus particle in which part of the genetic material of papovavirus SV40 is encased in an adenovirus protein coat. { |ad·ən,ō |es\|vē|fór·tē 'hī·brəd 'vī·rəs }

Adenoviridae [VIROL] A family of double-stranded DNA viruses with icosahedral symmetry; usually found in the respiratory tract of the host species and often associated with respiratory diseases. Also known as adenovirus. { ,ad·ən·ō'vīr·ə,dē }

adenovirus See Adenoviridae. { |ad·ən,o'vī·rəs }

adenylcyclase [BIOCHEM] The catalyzing enzyme in the conversion of adenosine triphosphate to cyclic adenosine monophosphate during metabolism. { |ad·ən,il'sī,klās }

adenylic acid [BIOCHEM] **1.** A generic term for a group of isomeric nucleotides. **2.** The phosphoric acid ester of adenosine. Also known as adenosine monophosphate (AMP). { |ad·ən|il·ik 'as·əd }

adeoniform [INV ZOO] **1.** A lobate, bilamellar zooarium. **2.** Resembling the fossil bryozoan *Adeona*. { ,ad·ē'ä·nə,fórm }

Adephaga [INV ZOO] A suborder of insects in the order Coleoptera characterized by fused hind coxae that are immovable. { ə'def·ə·gə }

adequacy [ELEC] The existence of sufficient facilities within an electric power system to satisfy the customer load requirement under static system conditions. { 'ad·ə·kwə·sē }

adequate contact [MED] The degree of contact required between an infectious and a susceptible individual to cause infection of the latter. { 'ad·ə·kwət 'kän,takt }

adequate stimulus [NEURO] The energy of any specific mode that is sufficient to elicit a response in an excitable tissue. { 'ad·ə·kwət 'stim·yə·ləs }

ader wax See ozocerite. { 'äd·ər ,waks }

ADF See acid detergent fiber; automatic direction finder.

ADF bearing indicator [NAV] An instrument used with an airborne radio direction finder to indicate automatically the relative, magnetic, or true bearing (or reciprocal) of a transmitter. { |ā|dē|ef 'ber·iŋ ,in·də,kāt·ər }

adfluvial [BIOL] Migrating between lakes and rivers or streams. { ad'flü·vē·əl }

adfreezing [HYD] The process by which one object adheres to another by the binding action of ice; applied to permafrost studies. { ,ad'frēz·iŋ }

ADF reversal [NAV] The swinging of the needle on the direction indicator of an airborne automatic direction finder (ADF) through 180°, indicating that the station to which the direction finder is tuned has been passed. { |ā|dē|ef ri'vər·səl }

ADH See vasopressin.

Adhara [ASTRON] A star of spectral type B2II. Also known as ε Canis Majoris. { ə'där·ə }

adherend [MATER] **1.** A body attached to another by means of an adhesive substance. **2.** The surface to which an adhesive adheres. { ,ad'hir·ənd }

adherens junction [CELL MOL] An integrin-mediated anchoring junction that connects the cytoskeleton (actin filaments) of a cell to the extracellular matrix or to the cytoskeleton of surrounding cells. { ad'hir·enz jəŋk·shən }

adherent point [MATH] For a set in a topological space, a point that is either a member of the set or an accumulation point of the set. { ad|hir·ənt 'póint }

adhering junction [CYTOL] An intercellular junction that promotes adhesion between cells. Also known as desmosome. { ad|hir·iŋ 'jəŋk·shən }

adhesion [BOT] Growing together of members of different and distinct whorls. [ELECTROMAG] Any mutually attractive force holding together two magnetic bodies, or two oppositely charged nonconducting bodies. [ENG] Intimate sticking together of metal surfaces under compressive stresses by formation of metallic bonds. [MECH] The force of static friction between two bodies, or the effects of this force. [MED] The abnormal union of an organ or part with some other part by

ADENINE

Structural formula of adenine.

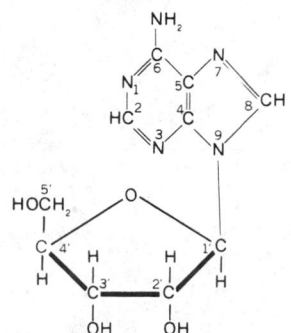

ADENOSINE

Structural formula of adenosine.

formation of fibrous tissue. [PHYS] The tendency, due to intermolecular forces, for matter to cling to other matter. { ad'hē·zhən }

adhesional work [THERMO] The work required to separate a unit area of a surface at which two substances are in contact. Also known as work of adhesion. { ad'hē·zhən·əl ‚wərk }

adhesive [MATER] A substance used to bond two or more solids so that they act or can be used as a single piece; examples are resins, formaldehydes, glue, paste, cement, putty, and polyvinyl resin emulsions. { ad'hēz·iv }

adhesive bond [MECH] The forces such as dipole bonds which attract adhesives and base materials to each other. { ad'hēz·iv 'bänd }

adhesive bonding [ENG] The fastening together of two or more solids by the use of glue, cement, or other adhesive. { ad'hēz·iv 'bänd·iŋ }

adhesive cell [INV ZOO] Any of various glandular cells in ctenophores, turbellarians, and hydras used for adhesion to a substrate and for capture of prey. Also known as colloblast; glue cell; lasso cell. { ad'hēz·iv 'sel }

adhesive strength [ENG] The strength of an adhesive bond, usually measured as a force required to separate two objects of standard bonded area, by either shear or tensile stress. { ad'hēz·iv 'streŋkth }

adhesive tape [MATER] Tape coated with a substance that binds or sticks to a surface. { ad'hēz·iv ‚tāp }

ad hoc inquiry [COMPUT SCI] A single request for a piece of information, such as a report. { ad ‚häk in'kwī·rē }

adiabat [METEOROL] The relatively constant rate (5.5°F/100 feet or 10°C/kilometer) at which a mass of air cools as it rises. { 'ad·ē·ə‚bat }

adiabatic [THERMO] Referring to any change in which there is no gain or loss of heat. { ¦ad·ē·ə¦bad·ik }

adiabatic approximation [ASTROPHYS] The approximation that the pressure and density of gas in a star are related by the adiabatic law. [PHYS CHEM] *See* Born-Oppenheimer approximation. { ¦ad·ē·ə¦bad·ik ə‚präk·sə'mā·shən }

adiabatic atmosphere [METEOROL] A model atmosphere characterized by a dry-adiabatic lapse rate throughout its vertical extent. { ¦ad·ē·ə¦bad·ik 'at·mə‚sfir }

adiabatic calorimeter [PHYS CHEM] An instrument used to study chemical reactions which have a minimum loss of heat. { ¦ad·ē·ə¦bad·ik ‚kal·ə'rim·əd·ər }

adiabatic chart *See* Stuve chart. { ¦ad·ē·ə¦bad·ik 'chärt }

adiabatic compression [THERMO] A reduction in volume of a substance without heat flow, in or out. { ¦ad·ē·ə¦bad·ik kəm'presh·ən }

adiabatic condensation pressure *See* condensation pressure. { ¦ad·ē·ə¦bad·ik ‚kän‚den'sā·shən ‚presh·ər }

adiabatic condensation temperature *See* condensation temperature. { ¦ad·ē·ə¦bad·ik ‚kän‚den'sā·shən 'tem·prə·chər }

adiabatic cooling [THERMO] A process in which the temperature of a system is reduced without any heat being exchanged between the system and its surroundings. { ¦ad·ē·ə¦bad·ik 'kül·iŋ }

adiabatic curing [ENG] The curing of concrete or mortar under conditions in which there is no loss or gain of heat. { ¦ad·ē·ə¦bad·ik 'kyŭr·iŋ }

adiabatic demagnetization [CRYO] A method of cooling paramagnetic salts to temperatures of 10^{-3} K; the sample is cooled to the boiling point of helium in a strong magnetic field, thermally isolated, and then removed from the field to demagnetize it. Also known as Giaque-Debye method; magnetic cooling; paramagnetic cooling. { ¦ad·ē·ə¦bad·ik ‚de‚mag·nəd·ə'zā·shən }

adiabatic ellipse [FL MECH] A plot of the speed of sound as a function of the speed of flow for the adiabatic flow of a gas, which forms one quadrant of an ellipse. { ¦ad·ē·ə¦bad·ik i'lips }

adiabatic engine [MECH ENG] A heat engine or thermodynamic system in which there is no gain or loss of heat. { ¦ad·ē·ə¦bad·ik 'en·jən }

adiabatic envelope [THERMO] A surface enclosing a thermodynamic system in an equilibrium which can be disturbed only by long-range forces or by motion of part of the envelope; intuitively, this means that no heat can flow through the surface. { ¦ad·ē·ə¦bad·ik 'en·və‚lōp }

adiabatic equilibrium [METEOROL] A vertical distribution of temperature and pressure in an atmosphere in hydrostatic equilibrium such that an air parcel displaced adiabatically will continue to possess the same temperature and pressure as its surroundings, so that no restoring force acts on a parcel displaced vertically. Also known as convective equilibrium. { ¦ad·ē·ə¦bad·ik ‚ē·kwə'lib·rē·əm }

adiabatic equivalent temperature *See* equivalent temperature. { ¦ad·ē·ə¦bad·ik i'kwiv·ə·lənt 'tem·prə‚chər }

adiabatic expansion [THERMO] Increase in volume without heat flow, in or out. { ¦ad·ē·ə¦bad·ik ik'span·chən }

adiabatic extrusion [ENG] Forming plastic objects by energy produced by driving the plastic mass through an extruder without heat flow. { ¦ad·ē·ə¦bad·ik ik'strü·zhən }

adiabatic flame temperature [PHYS CHEM] The highest possible temperature of combustion obtained under the conditions that the burning occurs in an adiabatic vessel, that it is complete, and that dissociation does not occur. { ¦ad·ē·ə¦bad·ik ‚flām 'tem·prə·chər }

adiabatic flow [FL MECH] Movement of a fluid without heat transfer. { ¦ad·ē·ə¦bad·ik 'flō }

adiabatic invariant [PHYS] A physical quantity which may be quantized and which, to a certain degree of approximation, remains unchanged under the slow variation of any parameter. { ¦ad·ē·ə¦bad·ik in'ver·ē·ənt }

adiabatic lapse rate *See* dry-adiabatic lapse rate. { ¦ad·ē·ə¦bad·ik 'laps ‚rāt }

adiabatic law [PHYS] The relationship which states that, for adiabatic expansion of gases, $P\rho^{-\gamma}$ = constant, where P = pressure, ρ = density, and γ ratio of specific heats C_P/C_V. { ¦ad·ē·ə¦bad·ik 'lò }

adiabatic process [THERMO] Any thermodynamic procedure which takes place in a system without the exchange of heat with the surroundings. { ¦ad·ē·ə¦bad·ik prä·səs }

adiabatic rate *See* dry adiabatic lapse rate. { ¦ad·ē·ə¦bad·ik 'rāt }

adiabatic recovery temperature [FL MECH] **1.** The temperature reached by a moving fluid when brought to rest through an adiabatic process. Also known as recovery temperature; stagnation temperature. **2.** The final and initial temperature in an adiabatic, Carnot cycle. { ¦ad·ē·ə¦bad·ik ri'kəv·ə·rē ‚tem·prə·chər }

adiabatic saturation pressure *See* condensation pressure. { ¦ad·ē·ə¦bad·ik ‚sach·ə'rā·shən ‚presh·ər }

adiabatic saturation temperature *See* condensation temperature. { ¦ad·ē·ə¦bad·ik ‚sach·ə'rā·shən ‚tem·prə·chər }

adiabatic system [SCI TECH] A body or system whose condition is altered without gaining heat from or losing heat to the surroundings. { ¦ad·ē·ə¦bad·ik 'sis·təm }

adiabatic vaporization [THERMO] Vaporization of a liquid with virtually no heat exchange between it and its surroundings. { ¦ad·ē·ə¦bad·ik ‚vā·pər·ə'zā·shən }

adiabatic wall temperature [FL MECH] The temperature assumed by a wall in a moving fluid stream when there is no heat transfer between the wall and the stream. { ¦ad·ē·ə¦bad·ik 'wòl ‚tem·prə·chər }

adiadochokinesis [MED] A type of motor incoordination associated with cerebellar damage in which repetitive movements controlled by antagonistic muscles cannot be performed without severe muscular incoordination. { ə‚dē·ə‚dō·kō·kə'nē·səs }

adiagnostic [PETR] Pertaining to a rock texture in which identification of individual components is not possible macroscopically or microscopically; applied especially to igneous rock. { ‚ā‚dī·əg'näs·tik }

adiathermanous [PHYS] Not capable of transmitting radiant heat. Also known as adiathermic. { ‚ā‚dī·ə'thərm·ə·nəs }

adiathermic *See* adiathermanous. { ‚ā‚dī·ə'thərm·ik }

Adie's syndrome [MED] Impaired pupillary reaction to light and absent tendon reflexes. { 'a‚dēz ‚sin‚drōm }

Adimeridae [INV ZOO] An equivalent name for the Colydiidae. { ‚ad·ə'mer·ə·dē }

adinole [GEOL] An argillaceous sediment that has undergone albitization at the margin of a basic intrusion. { 'ad·ən‚ōl }

adipate [ORG CHEM] Salt produced by reaction of adipic acid with a basic compound. { 'ad·ə‚pāt }

adiphenine [PHARM] $C_{20}H_{25}NO_2 \cdot HCl$ The compound 2-diethylaminoethyl diphenylacetate hydrochloride; a cholinergic blocking agent. { ə'dif·ə‚nēn }

adipic acid [ORG CHEM] $HOOC(CH_2)_4COOH$ A colorless

crystalline dicarboxylic acid, sparingly soluble in water; used in nylon manufacture. { ə'dip·ik 'as·əd }

adipocellulose [BIOCHEM] A type of cellulose found in the cell walls of cork tissue. { ¦ad·ə·pō'sel·yə,lōs }

adipocere [MED] A light-colored, waxy material formed by postmortem conversion of body fats to higher fatty acids. { 'ad·ə·pə,sir }

adipocerite See hatchettite. { ¦ad·ə'päs·ə,rīt }

adipocire See hatchettite. { ¦ad·ə'pä,sir }

adipocyte [CELL BIOL] An animal cell that stores fat. { 'a·də·pə,sīt }

adipogenesis [PHYSIO] The formation of fat or fatty tissue. { ¦ad·ə·pō'jen·ə·səs }

adiponecrosis neonatorum [MED] Localized fatty-tissue necrosis occurring in large, healthy infants born after difficult labor. { ¦ad·ə·pō·nə'krō·səs ,nē·ō·nə'tȯr·əm }

adiponitrile [ORG CHEM] $NC(CH_2)_4CN$ The high-boiling liquid dinitrile of adipic acid; used to make nylon intermediates. { ¦ad·ə·pō'nī·trəl }

adipose [BIOL] Fatty; of or relating to fat. { 'ad·ə,pōs }

adipose fin [VERT ZOO] A modified posterior dorsal fin that is fleshy and lacks rays; found in salmon and typical catfishes. { 'ad·ə,pōs ,fin }

adipose tissue [HISTOL] A type of connective tissue specialized for lipid storage. { 'ad·ə,pōs 'tish·ü }

adiposis dolorosa [MED] An uncommon type of obesity in which the excess fat deposits are tender and painful. { ad·ə'pō·səs ,dō·lə'rōs·ə }

adiposogenital dystrophy [MED] A syndrome involving obesity, retarded gonad development, and sometimes diabetes insipidus resulting from impaired functioning of the pituitary and hypothalamus. Also known as Froehlich's syndrome. { ¦ad·ə,pō·sō'jen·əd·əl 'dis·trə·fē }

adipsia [MED] Absence of thirst or avoidance of drinking. { ā'dip·sē·ə }

A display [ELECTR] A radar oscilloscope display in cartesian coordinates; the targets appear as vertical deflection lines; their Y coordinates are proportional to signal intensity; their X coordinates are proportional to distance to targets. { 'ā di,splā }

adit [CIV ENG] An access tunnel used for excavation of the main tunnel. [MIN ENG] A nearly horizontal tunnel for access, drainage, or ventilation of a mine. Also known as side drift. { 'ad·ət }

adjab butter [MATER] A semisolid oil extracted from the nuts of the tree *Bassia toxisperma*; used to make soap. Also known as djave butter. { 'a,jab ,bəd·ər }

adjacency [COMPUT SCI] A condition in character recognition in which two consecutive graphic characters are separated by less than a specified distance. { ə'jās·ən·sē }

adjacency matrix [MATH] **1.** For a graph with n vertices, the $n \times n$ matrix $A = a_{ij}$, where the nondiagonal entry a_{ij} is the number of edges joining vertex i and vertex j, and the diagonal entry a_{ii} is twice the number of loops at vertex i. **2.** For a diagraph with no loops and not more than one arc joining any two vertices, an $n \times n$ matrix $A = [a_{ij}]$, in which $a_{ij} = 1$ if there is an arc directed from vertex i to vertex j, and otherwise $a_{ij} = 0$. { ə'jās·ən·sē ,mā·triks }

adjacency structure [MATH] A listing, for each vertex of a graph, of all the other vertices adjacent to it. { ə'jās·ən·sē ,strək·chər }

adjacent angle [MATH] One of a pair of angles with a common side formed by two intersecting straight lines. { ə'jās·ənt 'aŋ·gəl }

adjacent-channel interference [COMMUN] Interference that is caused by a transmitter operating in an adjacent channel. { ə'jās·ənt 'chan·əl in·tər'fir·əns }

adjacent-channel selectivity [ELECTR] The ability of a radio receiver to respond to the desired signal and to reject signals in adjacent frequency channels. { ə'jās·ənt 'chan·əl sə,lek'tiv·əd·ē }

adjacent sea [GEOGR] A sea connected with the oceans but semienclosed by land; examples are the Caribbean Sea and North Polar Sea. { ə'jās·ənt 'sē }

adjacent side [MATH] For a given vertex of a polygon, one of the sides of the polygon that terminates at the vertex. { ə'jās·ənt 'sīd }

adjective dye [CHEM] Any dye that needs a mordant. { ə'jek·tiv ,dī }

adjoined number [MATH] A number z that is added to a number field F to form a new field consisting of all numbers that can be derived from z and the numbers in F by the operations of addition, subtraction, multiplication, and division. { ə¦jȯind 'nəm·bər }

adjoining sheets [MAP] Those maps that are contiguous to one or more sides and corners of a map series. { ə'jȯin·iŋ 'shēts }

adjoint of a matrix See adjugate; Hermitian conjugate. { 'aj,ȯint əv ə 'mā·triks }

adjoint operator [MATH] An operator B such that the inner products (Ax,y) and (x,By) are equal for a given operator A and for all elements x and y of a Hilbert space. Also known as associate operator; Hermitian conjugate operator. { 'aj,ȯint 'äp·ə,rād·ər }

adjoint variable [PHYS] In classical dynamics, the canonically conjugate p_i interpreted as generalized momenta. { 'aj,ȯint 'ver·ē·ə·bəl }

adjoint vector space [MATH] The complete normed vector space constituted by a class of bounded, linear, homogeneous scalar functions defined on a normed vector space. { 'aj,ȯint 'vek·tər ,spās }

adjoint wave functions [QUANT MECH] Functions in the Dirac electron theory which are formed by applying the Dirac matrix B to the Hermitian conjugates of the original wave functions. { 'aj,ȯint 'wāv ,fəŋk·shənz }

adjugate [MATH] For a matrix A, the matrix obtained by replacing each element of A with the cofactor of the transposed element. Also known as adjoint of a matrix. { 'aj·ə,gāt }

adjustable base anchor [BUILD] An item which holds a doorframe above a finished floor. { ə'jəs·tə·bəl ¦bās 'aŋ·kər }

adjustable choke [PETRO ENG] A choke fitted with a movable conical needle of sleeve that can be changed with respect to its seat so that the rate of flow can be varied. { ə'jəs·tə·bəl 'chōk }

adjustable parallels [ENG] Wedge-shaped iron bars placed with the thin end of one on the thick end of the other, so that the top face of the upper and the bottom face of the lower remain parallel, but the distance between the two faces is adjustable; the bars can be locked in position by a screw to prevent shifting. { ə'jəs·tə·bəl 'par·ə,lelz }

adjustable propeller [NAV ARCH] A screw propeller with blades that can be rotated around their axes to change the pitch. Also known as controllable-pitch propeller. { ə'jəs·tə·bəl prə'pel·ər }

adjustable resistor [ELEC] A resistor having one or more sliding contacts whose position may be changed. { ə'jəs·tə·bəl ri'zis·tər }

adjustable square [ENG] A try square with an arm that is at right angles to the ruler; the position of the arm can be changed to form an L or a T. Also known as double square. { ə'jəs·tə·bəl 'skwer }

adjustable transformer See variable transformer. { ə'jəs·tə·bəl tranz'fȯr·mər }

adjustable wrench [ENG] A wrench with one jaw which is fixed and another which is adjustable; the size is adjusted by a knurled screw. { ə'jəs·tə·bəl 'rench }

adjusted decibel [ELECTR] A unit used to show the relationship between the interfering effect of a noise frequency, or band of noise frequencies, and a reference noise power level of −85 dBm. Abbreviated dBa. Also known as decibel adjusted. { ə'jəs·təd 'des·ə,bel }

adjusted elevation [GEOD] **1.** The elevation resulting from the application of an adjustment correction to an orthometric elevation. **2.** The elevation resulting from the application of both an orthometric correction and an adjustment correction to a preliminary elevation. { ə'jəs·təd ,el·ə'vā·shən }

adjusted position [MAP] An adjusted value of the coordinate position of a point. { ə'jəs·təd pə'zi·shən }

adjusted stream [HYD] A stream which flows mostly parallel to the strike and as little as necessary in other courses. { ə'jəs·təd 'strēm }

adjusted value [SCI TECH] A value of a quantity derived from observed data by some orderly process which eliminates discrepancies arising from errors in those data. { ə'jəs·təd 'val·yü }

adjusting [ENG] In measurement technology, setting or compensating a measuring instrument or a weight in such a way that the indicated value deviates as little as possible from the actual value. { ə'jəst·iŋ }

A DISPLAY

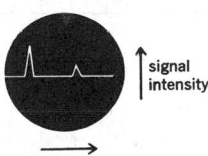

signal intensity

range

Drawing of a frequently employed display for presentation of radar outputs.

adjusting point [ORD] A distinctive terrain feature or some portion of the target at or near the center of the area upon which the observer wishes to place fire. { ə'jəst·iŋ póint }

adjusting ring [ORD] Part of a fuse setter which is used to adjust the setting element on the fuse of an explosive projectile. { ə'jəst·iŋ riŋ }

adjustment [GEOD] **1.** The determination and application of corrections to orthometric differences of elevation or to orthometric elevations to make the elevation of all bench marks consistent and independent of the circuit closures. **2.** The placing of detail or control stations in their positions relative to other detail or control stations. [PSYCH] The process of mental change to allow an individual to function harmoniously with his environment. { ə'jəst·mənt }

adjustment correction See arbitrary correction. { ə'jəst·mənt kə'rek·shən }

adjustment disorder [PSYCH] A category of emotional disorder in which an individual exhibits maladaptive reactions to identifiable life events or circumstances. { ə'jəst·mənt dis ,órd·ər }

adjustment of fire [ORD] The determining and applying of corrections to firing data to bring the center of impact or of burst, or the cone of fire of automatic weapons, to the adjusting point and to keep it there. { ə'jəst·mənt əv 'fīr }

adjustment reaction [PSYCH] A transient, situational personality disorder occurring in reaction to some significant person, immediate event, or internal emotional conflict. { ə'jəst· mənt rē'ak·shən }

adjustor neuron [NEURO] Any of the interconnecting nerve cells between sensory and motor neurons of the central nervous system. { ə'jəs·tər 'nü,rän }

adjutage [ENG] A tube attached to a container of liquid at an orifice to facilitate or regulate outflow. { 'aj·ə,tazh }

adjuvant [PHARM] A material that enhances the action of a drug or antigen. { 'aj·ə·vənt }

adjuvant chemotherapy [MED] Chemotherapy that is used to destroy suspected undetectable residual tumor after surgery or radiation treatment has eradicated all detectable tumor; effective in the treatment of breast and colon cancer. { 'a·jə·vənt ,kē·mō'ther·ə·pē }

ad lib [BIOL] Shortened form for ad libitum; without limit or restraint. { ,ad 'lib }

adlittoral [OCEANOGR] Of, pertaining to, or occurring in shallow waters adjacent to a shore. { ,ad'lid·ə·rəl }

administrative map [MAP] **1.** A map with graphically recorded information pertaining to administrative matters, such as supply and evacuation installations, medical facilities, service and maintenance areas, main supply roads, boundaries, and other details necessary to show the administrative situation in relation to the tactical situation. **2.** Any map on which are delineated political subdivisions and boundaries of a country or countries. { ad'min·ə,strād·iv map }

admiralty brass [MET] An alloy containing 71% copper, 28% zinc, and 0.75-1% tin for additional corrosion resistance. { 'ad·mrəl·tē 'bras }

admiralty coal [MIN ENG] A high-quality, smokeless steam coal. { 'ad·mrəl·tē 'kōl }

Admiralty constant [NAV ARCH] Any of the numbers that give displacement and midship section. { 'ad·mrəl·tē 'kän·stənt }

admittance [ELEC] A measure of how readily alternating current will flow in a circuit; the reciprocal of impedance, it is expressed in siemens. { əd'mit·əns }

admittance matrix [ELEC] A matrix Y whose elements are the mutual admittances between the various meshes of an electrical network, it satisfies the matrix equation $I = YV$, where I and V are column vectors whose elements are the currents and voltages in the meshes. { əd'mit·əns 'mā·triks }

admixture [GEOL] One of the lesser or subordinate grades of sediment. [MATER] A material (other than aggregate, cement, or water) added in small quantities to concrete to produce some desired change in properties. Also known as additive. { ,ad'miks·chər }

aDNA See ancient DNA.

adnate [BIOL] United through growth; used especially for unlike parts. [BOT] Pertaining to growth with one side adherent to a stem. { 'ad,nāt }

adnexa [BIOL] Subordinate or accessory parts, such as eyelids, Fallopian tubes, and extraembryonic membranes. { ad'neks·ə }

adobe [GEOL] Heavy-textured clay soil found in the southwestern United States and in Mexico. { ə'dō·bē }

adobe brick [MATER] An earth or clay, straw brick of large but varying dimensions, roughly molded and sun-dried. { ə'dō·bē 'brik }

adobe construction [BUILD] Wall construction with sun-dried blocks of adobe soil. { ə'dō·bē kən'strək·shən }

adobe flats [GEOL] Broad flats that are floored with sandy clay and have been formed from sheet floods. { ə'dō·bē 'flats }

adolescence [GEOL] Stage in the cycle of erosion following youth and preceding maturity. [PSYCH] The period of life from puberty to maturity. { ,ad·əl'es·əns }

adolescent coast [GEOL] A type of shoreline characterized by low but nearly continuous sea cliffs. { ,ad·əl'es·ənt ,kōst }

adolescent river [HYD] A river with a graded bed and a well-cut channel that reaches base level at its mouth, its waterfalls and lakes of the youthful stage having been destroyed. { ,ad·əl'es·ənt 'riv·ər }

adolescent stream [HYD] A stream characterized by a well-cut, smoothly graded channel that may reach base level at its mouth. { ,ad·əl'es·ənt 'strēm }

Adonis [ASTRON] An asteroid with an orbital eccentricity of 0.779 and a perihelion well inside the orbit of Venus that passed about 1×10^6 miles (1.6×10^6 kilometers) from earth in 1936. { ə'dän·əs }

adonite See adonitol. { 'ad·ə,nīt }

adonitol [BIOCHEM] $C_5H_{12}O_5$ A pentitol from the dicotyledenous plant Adonis vernalis; large crystals that are optically inactive and melt at 102°C; it does not reduce Fehling's solution, and is freely soluble in water and hot alcohol. Also known as adonite; ribitol. { ə'dän·ə,tōl }

adont hinge [INV ZOO] A type of ostracod hinge articulation which either lacks teeth and has overlapping valves or has a ridge and groove. { 'ā,dänt ,hinj }

adoptive immunity [IMMUNOL] Immunity resulting from the transfer of an immune function from one organism to another through the transfer of immunologically competent cells. Also known as transfer immunity. { ə'däp·təv ə'myü·nəd·ē }

adoptive immunotherapy [IMMUNOL] The transfer of immunologically competent white blood cells or their precursors into the host. { ə'däp·tiv ,im·yə·nō'ther·ə·pē }

adoral [ZOO] Near the mouth. { ,a'dór·əl }

ADP See adenosine diphosphate; automatic data processing.

ADPase See adenosine diphosphatase.

ADPCM See adaptive differential pulse-code modulation.

ADPE See automatic data-processing equipment.

Adrastea [ASTRON] A small satellite of Jupiter, having an orbital radius of 80,140 miles (128,980 kilometers) and radial dimensions of 7, 6, and 5 miles (12, 10, and 8 kilometers). Also known as Jupiter XV. { ə'dras·tē·ə }

adrenal cortex [ANAT] The cortical moietie of the suprarenal glands which secretes glucocorticoids, mineralocorticoids, androgens, estrogens, and progestagens. { ə'drēn·əl 'kór·teks }

adrenal cortex hormone [BIOCHEM] Any of the steroids produced by the adrenal cortex. Also known as adrenocortical hormone; corticoid. { ə'drēn·əl 'kór·teks 'hór,mōn }

adrenal cortical insufficiency [MED] Failure of the adrenal cortex to secrete adequate hormones. { ə'drēn·əl 'kórt·i·kəl ,in·sə'fish·ən·sē }

adrenalectomy [MED] Surgical removal of an adrenal gland. { ə,drēn·əl'ek·tə·mē }

adrenal gland [ANAT] An endocrine organ located close to the kidneys of vertebrates and consisting of two morphologically distinct components, the cortex and medulla. Also known as suprarenal gland. { ə'drēn·əl ,gland }

adrenaline See epinephrine. { ə'dren·əl·ən }

adrenal medulla [ANAT] The hormone-secreting chromaffin cells of the adrenal gland that produce epinephrine and norepinephrine. { ə'drēn·əl mə'dəl·ə }

adrenal virilism [MED] **1.** The development of male characteristics in the female resulting from excessive production of adrenal hormones with androgenic activity. **2.** A rare form of pseudohermaphroditism. { ə'drēn·əl 'vir·ə,liz·əm }

adrenergic [PHYSIO] Describing the chemical activity of epinephrine or epinephrine-like substances. { ˌad·rə'nər·jik }

adrenergic blocking agent [BIOCHEM] Any substance that blocks the action of epinephrine or an epinephrine-like substance. { ˌad·rə·nər·jik 'bläk·iŋ ˌā·jənt }

adrenochrome [BIOCHEM] $C_9H_9O_3N$ A brick-red oxidation product of epinephrine which can convert hemoglobin into methemoglobin. { ə'dren·ə‚krōm }

adrenocortical hormone See adrenal cortex hormone. { əˌdrēn·ō'kȯrd·ə·kəl 'hȯr‚mōn }

adrenocorticosteroid [BIOCHEM] **1.** A steroid that is obtained from the adrenal cortex. **2.** A steroid that resembles adrenal cortex steroids or has physiological effects like them. { əˌdrē·nō‚kȯrd·ə·kō'stir‚ȯid }

adrenocorticotropic hormone [BIOCHEM] The chemical secretion of the adenohypophysis that stimulates the adrenal cortex. Abbreviated ACTH. Also known as adrenotropic hormone. { əˌdrēn·ō‚kȯrd·ə·kō'träp·ik 'hȯr‚mōn }

adrenogenital syndrome [MED] A group of symptoms associated with hypersecretion of adrenal cortex hormones; effects vary with sex and time of development. { əˌdrēn·ō'jen·ə·təl 'sin‚drōm }

adrenomedullary [PHYSIO] Pertaining to the adrenal gland medulla. { əˌdrē·nō·mə'dəl·ə·rē }

adrenotropic [PHYSIO] Of or pertaining to an effect on the adrenal cortex. { əˌdrēn·ə'träp·ik }

adrenotropic hormone See adrenocorticotropic hormone. { əˌdrēn·ə'träp·ik 'hȯr‚mōn }

adret [ECOL] The sunny (usually south) face of a mountain featuring high timber and snow lines. { 'ad·rət }

ADR studio [ENG ACOUS] A sound-recording studio used in motion-picture and television production to allow an actor who did not intelligibly record his or her speech during the original filming or video recording to do so by watching himself or herself on the screen and repeating the original speech with lip synchronism; it is equipped with facilities for recreating the acoustical liveness and background sound of the environment of the original dialog. Derived from automatic dialog replacement studio. Also known as postsynchronizing studio. { ˌāˌdē'är ˌstüd·ē·ō }

ADSEL See Mode S.

ADSL See asymmetric digital subscriber line; asynchronous digital subscriber loop. { a·dē·es'el or 'ad·səl }

adsorbate [CHEM] A solid, liquid, or gas which is adsorbed as molecules, atoms, or ions by such substances as charcoal, silica, metals, water, and mercury. { ad'sȯr‚bāt }

adsorbent [CHEM] A solid or liquid that adsorbs other substances; for example, charcoal, silica, metals, water, and mercury. { ad'sȯr·bənt }

adsorption [CHEM] The surface retention of solid, liquid, or gas molecules, atoms, or ions by a solid or liquid, as opposed to absorbtion, the penetration of substances into the bulk of the solid or liquid. { ad'sȯrp·shən }

adsorption catalysis [PHYS CHEM] A catalytic reaction in which the catalyst is an adsorbent. { ad'sȯrp·shən kə'tal·ə·səs }

adsorption chromatography [ANALY CHEM] Separation of a chemical mixture (gas or liquid) by passing it over an adsorbent bed which adsorbs different compounds at different rates. { ad'sȯrp·shən ‚krō·mə'täg·rə·fē }

adsorption complex [CHEM] An entity consisting of an adsorbate and that portion of the adsorbent to which it is bound. { ad'sȯrp·shən ‚käm‚pleks }

adsorption gasoline [MATER] Gasoline extracted from natural gas or refinery gas. { ad'sȯrp·shən ‚gas·ə'lēn }

adsorption indicator [ANALY CHEM] An indicator used in solutions to detect slight excess of a substance or ion; precipitate becomes colored when the indicator is adsorbed. An example is fluorescein. { ad'sȯrp·shən ‚in·də‚kād·ər }

adsorption isobar [PHYS CHEM] A graph showing how adsorption varies with some parameter, such as temperature, while holding pressure constant. { ad'sȯrp·shən 'ī·sō‚bär }

adsorption isotherm [PHYS CHEM] The relationship between the gas pressure p and the amount w, in grams, of a gas or vapor taken up per gram of solid at a constant temperature. { ad'sȯrp·shən 'ī·sō‚thərm }

adsorption potential [PHYS CHEM] A change in the chemical potential that occurs as an ion moves from a gas or solution phase to the surface of an adsorbent. { ad'sȯrp·shən pə‚ten·chəl }

adsorption system [MECH ENG] A device that dehumidifies air by bringing it into contact with a solid adsorbing substance. { ad'sȯrp·shən ‚sis·təm }

ADT See abstract data type.

adularescence [OPTICS] A certain type of white or bluish light seen in a gemstone (usually adularia) as it is turned. { ˌaj·ə·lə'res·əns }

adularia [MINERAL] A weakly triclinic form of the mineral orthoclase occurring in transparent, colorless to milky-white pseudo-orthorhombic crystals. { ˌaj·ə'la·rē·ə }

adularization [GEOL] Replacement by or introduction of the mineral adularia. { əˌjül·ə·rə'zā·shən }

adult polycystic kidney disease [MED] An autosomal dominant disease that is characterized by the formation of cysts along the length of the nephron that causes the kidneys to enlarge, resulting in kidney failure in midadulthood. { əˌdəlt ‚päl·ə‚sis·tik 'kid·nē diz‚ēz }

adult rickets See osteomalacia. { ə'dəlt'rik·əts }

ad valorem tax [PETRO ENG] Property tax for oil-producing properties, assessed at a flat rate for each net barrel of oil produced. { ˌad və‚lȯr·əm 'taks }

advance [CIV ENG] In railway engineering, a length of track that extends beyond the signal that controls it. [GEOL] **1.** A continuing movement of a shoreline toward the sea. **2.** A net movement over a specified period of time of a shoreline toward the sea. [HYD] The forward movement of a glacier. [MECH ENG] To effect the earlier occurrence of an event, for example, spark advance or injection advance. [NAV] **1.** In making a turn, the distance a vessel moves in its initial direction from the point where the rudder is started over until the heading has changed 90°. **2.** The distance a vessel moves in the initial direction for heading changes of less than 90°. { əd'vans }

advanced [EVOL] Denoting a later stage within a lineage that demonstrates evolutionary progression. { əd'vanst }

advanced battery [ELEC] A large battery storage system designed to harness solar or wind energy or to store excess electricity during low-demand periods for use during higher-demand periods. { əd'vanst 'bad·ə·rē }

advanced-design array radar [ORD] A radar system that uses two antennas and a data-processing center to locate and identify enemy intercontinental ballistic missiles. Abbreviated ADAR. { əd'vanst də‚zīn ə'rā 'rā‚där }

advanced fuel fusion [NUCLEO] All energy-producing reactions of light nuclei other than the reaction of a deuteron and a triton to produce a helium-4 nucleus and a neutron. { əd‚vanst ‚fyül 'fyü·zhən }

advanced gallery [MIN ENG] A small heading driven in advance of the main tunnel in tunnel excavation. { əd'vanst 'gal‚rē }

advanced gas-cooled reactor [NUCLEO] A power-generating nuclear reactor which has steel-clad uranium dioxide fuel elements and is cooled by carbon dioxide gas. { əd'vanst 'gas ‚küld rē'ak·tər }

advanced line of position [NAV] A line of position which has been moved forward along the course line to allow for the run since the line was established; the opposite is a retired line of position. { əd'vanst ‚līn əv pə'zish·ən }

advanced potential [ELECTROMAG] Any electromagnetic potential arising as a solution of the classical Maxwell field equations, analogous to a retarded potential solution, but lying on the future light cone of space-time; the potential appears, at present, to have no physical interpretation. { əd'vanst pə'ten·chəl }

advanced programmatic risk analysis [IND ENG] A method for managing engineering programs with multiple projects and strict resource constraints which balances both technical and management risks. { əd'vanst ‚prō·grə‚mad·ik 'risk ə‚nal·ə·səs }

Advanced Research Projects Agency Network [COMPUT SCI] The computer network developed by the U.S. Department of Defense in 1969 from which the Internet originated. Abbreviated ARPANET. { əd‚vanst ri'sərch ‚präjeks ‚ā·jən·sē ‚net‚wərk }

advanced sewage treatment See tertiary sewage treatment. { əd'vanst 'sü·ij ‚trēt·mənt }

advanced signal-processing system [COMPUT SCI] A portable data-processing system for military use; its complete

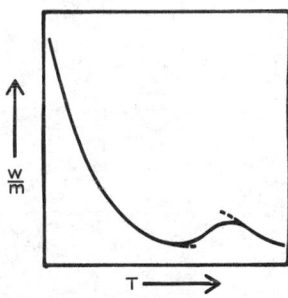

ADSORPTION ISOBAR

A typical adsorption isobar; w/m is weight of material adsorbed per unit weight of adsorbent, and T is absolute temperature.

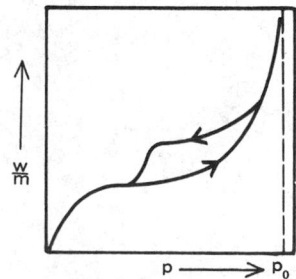

ADSORPTION ISOTHERM

A typical adsorption isotherm; w/m is weight of material adsorbed per weight of adsorbent, and p is pressure.

configuration may consist of the analyzer unit, a postprocessing unit (for data-processing and control tasks), and an advanced signal-processing display unit. Also known as Proteus. { əd'vanst 'sig·nəl 'präs·əs·iŋ ‚sis·təm }

advance of the perihelion [ASTRON] The slow rotation of the major axis of a planet's orbit in the direction of the planet's revolution, due to gravitational interactions with other planets and other effects such as those of general relativity. { əd'vans əv thə ‚per·ə¦hēl·yən }

advance overburden [MIN ENG] Overburden in excess of the average overburden-to-ore ratio that must be removed in opencut mining. { əd'vans 'ō·vər‚bərd·ən }

advance signal [CIV ENG] A signal in a block system up to which a train may proceed within a block that is not completely cleared. { əd'vans 'sig·nəl }

advance slope grouting [ENG] A grouting technique in which the front of the mass of grout is forced to move horizontally through preplaced aggregate. { əd'vans 'slōp 'graüd·iŋ }

advance slope method [ENG] A method of concrete placement in which the face of the fresh concrete, which is not vertical, moves forward as the concrete is placed. { əd'vans 'slōp ‚meth·əd }

advance stripping [MIN ENG] The removal of barren or sub-ore-grade earthy or rock materials to expose the minable grade of ore. { əd'vans 'strip·iŋ }

advance wave [MIN ENG] The air-pressure wave preceding the flame in a coal dust explosion. { əd'vans ‚wāv }

advancing [MIN ENG] Mining outward from the shaft toward the boundary. { əd'vans·iŋ }

advancing fire See assault fire. { əd'vans·iŋ ‚fīr }

advancing longwall [MIN ENG] Mining coal outward from the shaft pillar, with roadways maintained through the worked-out portion of the mine. { əd'vans·iŋ 'lȯŋ‚wȯl }

advantage factor [NUCLEO] The ratio of the radiation dose received in a specified time interval at a position in a nuclear reactor where some enhanced effect is produced, to the radiation dose in the same time interval at a reference position in the reactor. { əd'van·tij ‚fak·tər }

advection [METEOROL] The process of transport of an atmospheric property solely by the mass motion of the atmosphere. [OCEANOGR] The process of transport of water, or of an aqueous property, solely by the mass motion of the oceans, most typically via horizontal currents. { ‚ad'vek·shən }

advectional inversion [METEOROL] An inverted temperature gradient in the air resulting from a horizontal inflow of colder air into an area. { ad'vek·shən·əl in'vər·zhən }

advection fog [METEOROL] A type of fog caused by the horizontal movement of moist air over a cold surface and the consequent cooling of that air to below its dew point. { ‚ad'vek·shən ‚fäg }

advective hypothesis [METEOROL] The assumption that local temperature changes are the result only of horizontal or isobaric advection. { ‚ad'vek·tiv hī'päth·ə·səs }

advective model [METEOROL] A mathematical or dynamic model of fluid flow which is characterized by the advective hypothesis. { ‚ad'vek·tiv 'mäd·əl }

advective thunderstorm [METEOROL] A thunderstorm resulting from static instability produced by advection of relatively colder air at high levels or relatively warmer air at low levels or by a combination of both conditions. { ‚ad'vek·tiv 'thən·dər‚stȯrm }

adventitia [ANAT] The external, connective-tissue covering of an organ or blood vessel. Also known as tunica adventitia. { ‚ad·ven'tish·ə }

adventitious [BIOL] **1.** Also known as adventive. Acquired spontaneously or accidentally, not by heredity. **2.** Arising, as a tissue or organ, in an unusual or abnormal place. { ‚ad·ven'tish·əs }

adventitious bud [BOT] A bud that arises at points on the plant other than at the stem apex or a leaf axil. { ‚ad·ven'tish·əs 'bəd }

adventitious deafness [MED] A type of deafness that occurs at any point during a lifetime and may have a course either of gradual, progressive development or of sudden onset. { ‚ad·ven'tish·əs 'def·nəs }

adventitious embryo [MED] An embryo developing outside the uterus. { ‚ad·ven'tish·əs 'em·brē·ō }

adventitious root [BOT] A root that arises from any plant

part other than the primary root (radicle) or its branches. { ‚ad·ven'tish·əs 'rüt }

adventitious vein [INV ZOO] The vessel between the intercalary and accessory veins on certain insect wings. { ‚ad·ven'tish·əs 'vān }

adventitious virus [VIROL] A contaminant virus present by chance in a virus preparation. { ‚ad·ven'tish·əs 'vī·rəs }

adventive [BIOL] **1.** An organism that is introduced accidentally and is imperfectly naturalized; not native. **2.** See adventitious. { ad'ven·tiv }

adventive cone [GEOL] A volcanic cone that is on the flank of and subsidiary to a larger volcano. Also known as lateral cone; parasitic cone. { ad'ven·tiv 'kōn }

adventive crater [GEOL] A crater opened on the flank of a large volcanic cone. { ad'ven·tiv 'krāt·ər }

advisory area [NAV] A designated area within a flight information region to which air-traffic advisory notices apply. { əd'vīz·ə·rē ‚er·ē·ə }

advolution [BIOL] Development or growth with increasing similarities; growth toward; the opposite of evolution. { ‚ad·və'lü·shən }

adz [DES ENG] A cutting tool with a thin arched blade, sharpened on the concave side, at right angles on the handle; used for rough dressing of timber. { adz }

adz block [MECH ENG] The part of a machine for wood planing that carries the cutters. { 'adz ‚bläk }

Ae [MET] The temperature of equilibrium for corresponding phase change Ac; Ae_{cm}, Ae_3, and Ae_4 are similarly related to Ac_{cm}, Ac_3, and Ac_4.

Ae₁ [MET] The temperature, attained without thermal lag, at which cementite-austenite conversion takes place in a hypoeutectoid steel or ferrite-austenite conversion takes place in a hypereutectoid steel.

aebi [BIOL] A unit for the standardization of a phosphatase. { 'ā'ē·bē }

Aechminidae [PALEON] A family of extinct ostracodes in the order Paleocopa in which the hollow central spine is larger than the valve. { ēk'min·ə‚dē }

aeciospore [MYCOL] A spore produced by an aecium. { 'ēsh·ē·ə‚spȯr }

aecium [MYCOL] The fruiting body or sporocarp of rust fungi. { 'ēsh·ē·əm }

aedeagus [INV ZOO] The copulatory organ of a male insect. { ¦ē·dē'ā·gəs }

Aedes [INV ZOO] A genus of the dipterous subfamily Culicinae in the family Culicidae, with species that are vectors for many diseases of humans. { ā'ē·dēz }

Aeduellidae [PALEON] A family of Lower Permian palaeoniscoid fishes in the order Palaeonisciformes. { ē·dü'el·ə‚dī }

Aegeriidae [INV ZOO] The clearwing moths, a family of lepidopteran insects in the suborder Heteroneura characterized by the lack of wing scales. { ‚ē·jə'rē·ə‚dē }

Aegialitidae [INV ZOO] An equivalent name for the Salpingidae. { ‚ē·jyə'lid·ə‚dē }

Aegidae [INV ZOO] A family of isopod crustaceans in the suborder Flabellifera whose members are economically important as fish parasites. { 'ē·jə·dē }

aegirine [MINERAL] $NaFe(SiO_3)_2$ A brown or green clinopyroxene occurring in alkali-rich igneous rocks. Also known as aegirite. { 'ā·gə‚rēn }

aegirite See aegirine. { 'ā·gə‚rīt }

aegithognathous [VERT ZOO] Referring to a bird palate in which the vomers are completely fused and truncate in appearance. { ‚ē·gə‚thäg'nā·thəs }

Aegothelidae [VERT ZOO] A family of small Australo-Papuan owlet-nightjars in the avian order Caprimulgiformes. { ‚ē·gə'thel·ə‚dē }

Aegypiinae [VERT ZOO] The Old World vultures, a subfamily of diurnal carrion feeders of the family Accipitridae. { ‚ē·jə'pī·ə‚nē }

Aegyptianella [MICROBIO] A genus of the family Anaplasmataceae; organisms from inclusions in red blood cells of birds. { ə‚jip·shə'nel·ə }

Aegyptopithecus [PALEON] A primitive primate that is thought to represent the common ancestor of both the human and ape families. { ə‚jip·tō'pith·ə‚kəs }

aelophilous [BOT] Describing a plant whose disseminules are dispersed by wind. { ē'lä·fə·ləs }

Aelosomatidae [INV ZOO] A family of microscopic freshwater annelid worms in the class Oligochaeta characterized by a ventrally ciliated prostomium. { ‚e‚lä·sə'mad·ə‚dē }

aenigmatite *See* enigmatite. { ə'nig·mə‚tīt }

aeolian *See* eolian. { ē'ōl·ē·ən }

aeolotropic *See* anisotropic. { ‚ē·ə·lō'träp·ik }

aeolotropy *See* anisotropy. { ‚ē·ə'lä·trə·pē }

aeon [ASTRON] A billion (10^9) years. Also spelled eon. { 'ē‚än }

Aepophilidae [INV ZOO] A family of bugs in the hemipteran superfamily Saldoidea. { ‚ē·pō'fil·ə‚dē }

Aepyornis [PALEON] A genus of extinct ratite birds representing the family Aepyornithidae. { ‚ē·pē'ȯrn·əs }

Aepyornithidae [PALEON] The single family of the extinct avian order Aepyornithiformes. { ‚ē·pē‚ȯr'nith·ə‚dē }

Aepyornithiformes [PALEON] The elephant birds, an extinct order of ratite birds in the superorder Neognathae. { ‚ē·pē‚ȯr‚nith·ə'fȯr‚mēz }

aequorin [BIOCHEM] A bioluminescent protein that is produced by jellyfish of the genus *Aequorea* and emits light in the presence of calcium or strontium. { ē·kwə‚rin }

aerated concrete [MATER] Concrete made by adding substances which will liberate gases by chemical reaction; entrapped gases reduce its density and enhance insulation properties. { 'e‚rād·əd ‚kän'krēt }

aerated flow [ENG] Flowing liquid in which gas is dispersed as fine bubbles throughout the liquid. { 'e‚rād·əd 'flō }

aerating agent [FOOD ENG] A gas such as carbon dioxide, nitrous oxide, or food-grade Freon used to flavor beverages or to dispense liquids from charged containers. { 'e‚rād·iŋ ‚ā·jənt }

A/E ratio *See* absorptivity-emissivity ratio. { 'ā‚ē ‚rā·shō }

aeration [ENG] **1.** Exposing to the action of air. **2.** Causing air to bubble through. **3.** Introducing air into a solution by spraying, stirring, or similar method. **4.** Supplying or infusing with air, as in sand or soil. [FOOD ENG] Charging a liquid with some gas, such as water with carbon dioxide (soda water). [MIN ENG] The introduction of air into the pulp in a flotation cell to form air bubbles. { e'rā·shən }

aeration cell [PHYS CHEM] An electrolytic cell whose electromotive force is due to electrodes of the same material located in different concentrations of dissolved air. Also known as oxygen cell. { e'rā·shən ‚sel }

aeration tank [ENG] A fluid-holding tank with provisions to aerate its contents by bubbling air or another gas through the liquid or by spraying the liquid into the air. { e'rā·shən ‚taŋk }

aerator [DES ENG] A tool having a roller equipped with hollow fins; used to remove cores of soil from turf. [ENG] **1.** One who aerates. **2.** Equipment used for aeration. **3.** Any device for supplying air or gas under pressure, as for fumigating, welding, or ventilating. [MECH ENG] Equipment used to inject compressed air into sewage in the treatment process. [MET] A device which decreases the density of sand by mixing it with air, thus facilitating the movement of sand particles in packing. { 'e‚rād·ər }

aerenchyma [BOT] A specialized tissue in some water plants characterized by thin-walled cells and large intercellular air spaces. { ‚a'reŋk·ə·mə }

aerial [BIOL] Of, in, or belonging to the air or atmosphere. [ELECTR] *See* antenna. [ORD] **1.** Of or pertaining to operations in the air or to aircraft. **2.** Of weapons or missiles used in aircraft or launched, dropped, or shot from aircraft. { 'e·rē·əl }

aerial archeology [ARCHEO] Location of ancient earthworks, walls, village sites, ditches, and other features through the use of aerial surveys. { 'e·rē·əl ‚är·kē'äl·ə·jē }

aerial bomb [ORD] A bomb designed to be dropped from an aircraft, carrying either a high explosive or another agent, and normally detonated on contact or by a timing device. { 'e·rē·əl 'bäm }

aerial burst *See* airburst. { 'e·rē·əl 'bərst }

aerial cableway *See* aerial tramway. { 'e·rē·əl 'kā·bəl‚wā }

aerial camera [OPTICS] A camera designed for use in aircraft and containing a mechanism to expose the film in continuous sequence at a steady rate. Also known as aerocamera. { 'e·rē·əl 'kam·rə }

aerial cannon *See* aircraft cannon. { 'e·rē·əl 'kan·ən }

aerial dart [ORD] A metal dart designed to be dropped as a missile from an aircraft. Also known as flechette. { 'e·rē·əl 'därt }

aerial film [GRAPHICS] Specially designed roll film supplied in many lengths and widths, with various emulsion types, for use in aerial cameras. { 'e·rē·əl 'film }

aerial mapping [MAP] The making of planimetric and contoured maps and charts on the basis of photographs of the ground surface from an aircraft, spacecraft, or rocket. Also known as aerocartography. { 'e·rē·əl 'map·iŋ }

aerial mine [ORD] **1.** A mine designed to be dropped from an aircraft, especially into water. **2.** An early World War II light case bomb, the predecessor of the blockbuster, that was normally dropped by parachute. { 'e·rē·əl 'mīn }

aerial mycelium [MYCOL] A mass of hyphae that occurs above the surface of a substrate. { 'e·rē·əl mī'sē·lē·əm }

aerial observation [ORD] Observation from aircraft of artillery fire. { 'e·rē·əl ‚äb·zər'vā·shən }

aerial perspective [OPTICS] The effect produced by diffusion of light in the atmosphere whereby more distant objects have less clarity of outline and are lighter in tone. { 'e·rē·əl pər'spek·tiv }

aerial photogrammetry [ENG] Use of aerial photographs to make accurate measurements in surveying and mapmaking. { 'e·rē·əl ‚fōt·ə'gram·ə·trē }

aerial photographic reconnaissance *See* aerial photoreconnaissance. { 'e·rē·əl ‚fōd·ə‚graf·ik ri'kän·ə·səns }

aerial photography [ENG] The making of photographs of the ground surface from an aircraft, spacecraft, or rocket. Also known as aerophotography. { 'e·rē·əl fə'täg·rə·fē }

aerial photoreconnaissance [ENG] The obtaining of information by air photography; the three types are strategic, tactical, and survey-cartographic photoreconnaissance. Also known as aerial photographic reconnaissance. { 'e·rē·əl ‚fōd·ō‚ri'kän·ə·səns }

aerial reconnaissance [ENG] The collection of information by visual, electronic, or photographic means while aloft. { 'e·rē·əl ‚ri'kän·ə·səns }

aerial root [BOT] A root exposed to the air, usually anchoring the plant to a tree, and often functioning in photosynthesis. { 'e·rē·əl 'rüt }

aerial ropeway *See* aerial tramway. { 'e·rē·əl 'rōp‚wā }

aerial sound ranging [AERO ENG] The process of locating an aircraft by means of the sounds it emits. { 'e·rē·əl 'saund ‚rānj·iŋ }

aerial spud [MECH ENG] A cable for moving and anchoring a dredge. { 'e·rē·əl 'spəd }

aerial stem [BOT] A stem with an erect or vertical growth habit above the ground. { 'e·rē·əl 'stem }

aerial survey [ENG] A survey utilizing photographic, electronic, or other data obtained from an airborne station. Also known as aerosurvey; air survey. { 'e·rē·əl 'sər·vā }

aerial torpedo [ORD] **1.** A torpedo designed or adapted to be launched from a low-flying aircraft into water. **2.** Formerly, the explosive projectile thrown by a trench mortar and designed so as to fall point down. { 'e·rē·əl tȯr'pēd·ō }

aerial tramway [MECH ENG] A system for transporting bulk materials that consists of one or more cables supported by steel towers and is capable of carrying a traveling carriage from which loaded buckets can be lowered or raised. Also known as aerial cableway; aerial ropeway. { 'e·rē·əl 'tram‚wā }

aeriform [PHYS] Having the form or nature of air. { 'e·rə‚fȯrm }

aeroacoustics [ACOUS] The science of aerodynamically generated sounds, encompassing the study of how such sounds are produced, how they propagate through various media (including propagation through ducts carrying fluid flows), how they interact with obstacles, or barriers or wave-bearing surfaces, how they radiate away, how they can be managed or controlled, and how they are measured or predicted. { ‚e·rō·ə'küs·tiks }

aeroallergen [MED] Any airborne particulate matter that can induce allergic responses in sensitive persons. { ‚e·rō'al·ər·jən }

aeroballistics [MECH] The study of the interaction of projectiles or high-speed vehicles with the atmosphere. { ‚e·rō·bə'lis·tiks }

aerobe [BIOL] An organism that requires air or free oxygen to maintain its life processes. { 'e‚rōb }

aerobic adhesive [MATER] A two-compartment structural

adhesive that contains an acrylic resin and retains its adhesiveness in the presence of oxygen. { e¦rō₀bik ad'hēz·iv }

aerobic-anaerobic interface [CIV ENG] That point in bacterial action in the body of a sewage sludge or compost heap where both aerobic and anaerobic microorganisms participate, and the decomposition of the material goes no further. { e'rōb·ik 'an·ə₀rōb·ik 'in·tər₀fās }

aerobic-anaerobic lagoon [CIV ENG] A pond in which the solids from a sewage plant are placed in the lower layer; the solids are partially decomposed by anaerobic bacteria, while air or oxygen is bubbled through the upper layer to create an aerobic condition. { e'rōb·ik 'an·ə₀rōb·ik lə'gün }

aerobic bacteria [MICROBIO] Any bacteria requiring free oxygen for the metabolic breakdown of materials. { e'rōb·ik ₀bak'tir·ē·ə }

aerobic digestion [CHEM ENG] Digestion of matter suspended or dissolved in waste by microorganisms under favorable conditions of oxygenation. { e'rōb·ik də'jes·chən }

aerobic lagoon [CIV ENG] An aerated pond in which sewage solids are placed, and are decomposed by aerobic bacteria. Also known as aerobic pond. { e'rō·bik lə'gün }

aerobic pond See aerobic lagoon. { e¦rō·bik 'pand }

aerobic process [BIOL] A process requiring the presence of oxygen. { e'rōb·ik 'präs·əs }

aerobiology [BIOL] The study of the atmospheric dispersal of airborne fungus spores, pollen grains, and microorganisms; and, more broadly, of airborne propagules of algae and protozoans, minute insects such as aphids, and pollution gases and particles which exert specific biologic effects. { ₀e·rō₀bī'äl·ə·jē }

aerobioscope [MICROBIO] An apparatus for collecting and determining the bacterial content of a sample of air. { ₀e·rō'bi·ə₀skōp }

aerobiosis [BIOL] Life existing in air or oxygen. { ₀e·rō₀bi'ō·səs }

aerocamera See aerial camera. { 'e·rō₀kam·rə }

aerocartography See aerial mapping. { ₀e·rō₀kär'täg·rə·fē }

aerochlorination [CIV ENG] Treatment of sewage with compressed air and chlorine gas to remove fatty substances. { ₀e·rō₀klȯr·ə'nā·shən }

Aerococcus [MICROBIO] A genus of bacteria in the family Streptococcaceae; spherical cells have the tendency to form tetrads; they ferment glucose with production of dextrorotatory lactic acid (homofermentative). { 'e·rō₀käk·əs }

AERO code [METEOROL] An international code used to encode for transmission, in words five numerical digits long, synoptic weather observations of particular interest to aviation operations. { 'e·rō 'kōd }

aerocyst [BOT] An air vesicle in certain species of algae. { 'e·rō₀sist }

aerodiscone antenna [ELECTROMAG] Electrically small antenna for airborne applications in the very-high-frequency and ultra-high-frequency bands; it is derived from, and preserves, the desirable electrical characteristics of the discone antenna and can be designed in various physical shapes. { ₀e·rō'dis₀kōn an'ten·ə }

aerodontalgia [MED] A toothache brought on by atmospheric decompression. { ₀e₀rō₀dän'tal·jē·ə }

aerodrome See airport. { 'e·rō₀drōm }

aeroduct [AERO ENG] A ramjet type of engine designed to scoop up ions and electrons freely available in the outer reaches of the atmosphere or in the atmospheres of other spatial bodies and, by a metachemical process within the engine duct, to expel particles derived from the ions and electrons as a propulsive jetstream. { 'e·rō₀dəkt }

aerodynamic [FL MECH] Pertaining to forces acting upon any solid or liquid body moving relative to a gas (especially air). { ₀e·rō·dī'nam·ik }

aerodynamically generated sound See aerodynamic sound. { ₀e·rō·dī'nam·ik·le ¦jen·ə₀rād·əd 'saund }

aerodynamically rough surface [FL MECH] A surface whose irregularities are sufficiently high that the turbulent boundary layer reaches right down to the surface. { ₀e·rō·dī'nam·ik·lē 'rəf 'sər·fəs }

aerodynamically smooth surface [FL MECH] A surface whose irregularities are sufficiently small to be entirely embedded in the laminar sublayer. { ₀e·rō·dī'nam·ik·lē 'smüth 'sər·fəs }

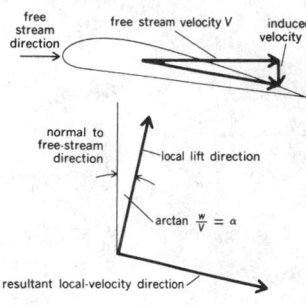

AERODYNAMIC FORCE

Effect of the induced velocity on the direction of local velocity and lift.

aerodynamic balance [ENG] A balance used for the measurement of the forces exerted on the surfaces of instruments exposed to flowing air; frequently used in tests made on models in wind tunnels. { ₀e·ro·dī'nam·ik 'bal·əns }

aerodynamic center [AERO ENG] A point on a cross section of a wing or rotor blade through which the forces of drag and lift are acting and about which the pitching moment coefficient is practically constant. { ₀e·ro·dī'nam·ik 'sent·ər }

aerodynamic characteristics [AERO ENG] The performance of a given airfoil profile as related to lift and drag, to angle of attack, and to velocity, density, viscosity, compressibility, and so on. { ₀e·ro·dī'nam·ik ₀kar·ik·tə'ris·tiks }

aerodynamic chord [AERO ENG] A straight line intersecting or touching an airfoil profile at two points; specifically, that part of such a line between two points of intersection. { ₀e·ro·dī'nam·ik 'kȯrd }

aerodynamic coefficient [FL MECH] Any nondimensional coefficient relating to aerodynamic forces or moments, such as a coefficient of drag or a coefficient of lift. { ₀e·ro·dī'nam·ik ₀kō·ə'fish·ənt }

aerodynamic configuration [AERO ENG] The form of an aircraft, incorporating desirable aerodynamic qualities. { ₀e·ro·dī'nam·ik kən₀fig·yə'rā·shən }

aerodynamic control [AERO ENG] A control surface whose use causes local aerodynamic forces. { ₀e·ro·dī'nam·ik kən'trōl }

aerodynamic decelerator [AERO ENG] Any device made from textiles, such as a parachute, that is designed to produce drag. { ₀e·ro·dī'nam·ik ₀de'sel·ə₀rād·ər }

aerodynamic drag [FL MECH] A retarding force that acts upon a body moving through a gaseous fluid and that is parallel to the direction of motion of the body; it is a component of the total fluid force acting on the body. Also known as aerodynamic resistance. { ₀e·ro·dī'nam·ik 'drag }

aerodynamic force [FL MECH] The force between a body and a gaseous fluid caused by their relative motion. Also known as aerodynamic load. { ₀e·ro·dī'nam·ik 'fȯrs }

aerodynamic heating [FL MECH] The heating of a body produced by passage of air or other gases over its surface; caused by friction and by compression processes and significant chiefly at high speeds. { ₀e·ro·dī'nam·ik 'hēt·iŋ }

aerodynamic instability [AERO ENG] An unstable state caused by oscillations of a structure that are generated by spontaneous and more or less periodic fluctuations in the flow, particularly in the wake of the structure. { ₀e·ro·dī'nam·ik ₀in·stə'bil·əd·ē }

aerodynamic lift [FL MECH] That component of the total aerodynamic force acting on a body perpendicular to the undisturbed airflow relative to the body. Also known as lift. { ₀e·ro·dī'nam·ik 'lift }

aerodynamic load See aerodynamic force. { ₀e·ro·dī'nam·ik 'lōd }

aerodynamic missile [AERO ENG] A missile with surfaces which produce lift during flight. { ₀e·ro·dī'nam·ik 'mis·əl }

aerodynamic moment [AERO ENG] The torque about the center of gravity of a missile or projectile moving through the atmosphere, produced by any aerodynamic force which does not act through the center of gravity. { ₀e·ro·dī'nam·ik 'mō·mənt }

aerodynamic noise See aerodynamic sound. { ₀e·ro·dī'nam·ik 'nȯiz }

aerodynamic phenomena [FL MECH] Acoustic, thermal, electrical, and mechanical effects, among others, that result from the flow of air over a body. { ₀e·ro·dī'nam·ik fə'näm·ə·nə }

aerodynamic resistance See aerodynamic drag. { ₀e·rō·dī'nam·ik ri'zis·təns }

aerodynamics [FL MECH] The science that deals with the motion of air and other gaseous fluids and with the forces acting on bodies when they move through such fluids or when such fluids move against or around the bodies. { ₀e·rō·dī'nam·iks }

aerodynamic size [PHYS] Particle size determined from inertia or settling velocity, assuming Stokes' law for the resistance to a sphere moving through a fluid. Also known as inertial size. { ₀e·ro·dī'nam·ik 'sīz }

aerodynamic sound [ACOUS] Sound that is generated by

the unsteady motion of a gas and its interaction with surrounding surfaces. Also known as aerodynamically generated sound; aerodynamic noise. { ‚e·rō·dī¦nam·ik 'saůnd }

aerodynamic stability [AERO ENG] The property of a body in the air, such as an aircraft or rocket, to maintain its attitude, or to resist displacement, and if displaced, to develop aerodynamic forces and moments tending to restore the original condition. { ‚e·ro·dī'nam·ik stə'bil·əd·ē }

aerodynamic time [AERO ENG] A characteristic time equal to the mass of an aircraft divided by the product of the gross wing area, the density of air, and the air speed. { ‚e·ro·dī'nam·ik 'tīm }

aerodynamic trail [FL MECH] A condensation trail formed by adiabatic cooling to saturation (or slight supersaturation) of air passing over the surfaces of high-speed aircraft. { ‚e·ro·dī'nam·ik 'trāl }

aerodynamic trajectory [MECH] A trajectory or part of a trajectory in which the missile or vehicle encounters sufficient air resistance to stabilize its flight or to modify its course significantly. { ‚e·ro·dī'nam·ik trə'jek·trē }

aerodynamic turbulence [FL MECH] A state of fluid flow in which the instantaneous velocities exhibit irregular and apparently random fluctuations. { ‚e·ro·dī'nam·ik 'tərb·yə·ləns }

aerodynamic vehicle [AERO ENG] A device, such as an airplane or glider, capable of flight only within a sensible atmosphere and relying on aerodynamic forces to maintain flight. { ‚e·ro·dī'nam·ik 'vē·ə·kəl }

aerodynamic wave drag [FL MECH] The force retarding an airplane, especially in supersonic flight, as a consequence of the formation of shock waves ahead of it. { ‚e·ro·dī'nam·ik 'wāv ‚drag }

aerodyne [AERO ENG] Any heavier-than-air craft that derives its lift in flight chiefly from aerodynamic forces, such as the conventional airplane, glider, or helicopter. { 'e·rō‚dīn }

aeroelasticity [MECH] The deformation of structurally elastic bodies in response to aerodynamic loads. { ‚e·rō·i‚las'tis·əd·ē }

aeroembolism [MED] A condition marked by the presence of nitrogen bubbles in the blood and other body tissues resulting from a sudden fall in atmospheric pressure. Also known as air embolism. { ‚e·rō'em·bə‚liz·əm }

aerofall mill [MECH ENG] A grinding mill of large diameter with either lumps of ore, pebbles, or steel balls as crushing bodies; the dry load is airswept to remove mesh material. { 'e·rō‚fól ‚mil }

aerofilter [CIV ENG] A filter bed for sewage treatment consisting of coarse material and operated at high speed, often with recirculation. { 'e·rō‚fil·tər }

aerofoil See airfoil. { 'e·rō‚fóil }

aerogel [CHEM] A porous solid formed from a gel by replacing the liquid with a gas with little change in volume so that the solid is highly porous. { 'e·rō‚jel }

aerogenerator [ELEC] A generator that is driven by the wind, designed to utilize wind power on a commercial scale. { ‚e·rō'jen·ə‚rād·ər }

aerogeography [GEOGR] The geographic study of earth features by means of aerial observations and aerial photography. { ‚e·rō·jē'äg·rə·fē }

aerogeology [GEOL] The geologic study of earth features by means of aerial observations and aerial photography. { ‚e·rō·jē'äl·ə·jē }

aerograph [ENG] Any self-recording instrument carried aloft by any means to obtain meteorological data. { 'e·rō‚graf }

aerography [METEOROL] **1.** The study of the air or atmosphere. **2.** The practice of weather observation, map plotting, and maintaining records. **3.** See descriptive meteorology. { e'räg·rə·fē }

aerohydrous mineral [MINERAL] A mineral containing water in small cavities. { ¦e·rō¦hī·drəs 'min·rəl }

aerolite See stony meteorite. { 'e·rō‚līt }

aerologation [NAV] The science of long-range navigation by altimetry. Also known as minimal flight path; pressure pattern flying; single-heading flight. { e‚räl·ə'gā·shən }

aerological days [METEOROL] Specified days on which additional upper-air observations are made; an outgrowth of the International Polar Year. { ‚e·rə'lä·jə·kəl 'dāz }

aerological diagram [METEOROL] A diagram of atmospheric thermodynamics plotted from upper-atmospheric soundings; usually contains various reference lines such as isobars and isotherms. { ‚e·rə¦lä·jə·kəl 'dī·ə‚gram }

aerology [METEOROL] **1.** Synonym for meteorology, according to official usage in the U.S. Navy until 1957. **2.** The study of the free atmosphere throughout its vertical extent, as distinguished from studies confined to the layer of the atmosphere near the earth's surface. { e'rä·lə·jē }

aeromagnetic surveying [GEOPHYS] The mapping of the magnetic field of the earth through the use of electronic magnetometers suspended from aircraft. { ‚e·rō·mag'ned·ik sər'vā·iŋ }

aeromechanics [FL MECH] The science of air and other gases in motion or equilibrium; has two branches, aerostatics and aerodynamics. { ‚e·rō·mi'kan·iks }

aeromedicine See aerospace medicine. { ¦e·rō¦med·ə·sin }

aerometeorograph [ENG] A self-recording instrument used on aircraft for the simultaneous recording of atmospheric pressure, temperature, and humidity. { ‚e·rō‚mēd· ē'ór·ə‚graf }

aerometer [ENG] An instrument to ascertain the weight or density of air or other gases. { e'rä·məd·ər }

Aeromonas [MICROBIO] A genus of bacteria in the family Vibrionaceae; straight, motile rods with rounded ends; most species are pathogenic to marine and fresh-water animals. { e·rō'mōn·əs }

aeromotor [AERO ENG] An engine designed to provide motive power for an aircraft. { 'e·rō‚mōd·ər }

aeronaut [AERO ENG] A person who operates or travels in an airship or balloon. { 'e·rō‚nót }

aeronautical beacon [NAV] A visual aid to navigation, displaying flashes of white or colored light or both, used to indicate the location of airports, landmarks, and certain points of the Federal airways in mountainous terrain and to mark hazards. { e·rə'nód·ə·kəl 'bē·kən }

aeronautical chart [MAP] A basic map of countries or lands made primarily on Lambert conformal ionic projection and layer-tinted; air navigational data are overprinted. { e·rə'nód·ə·kəl 'chärt }

aeronautical climatology [METEOROL] The application of the data and techniques of climatology to aviation meteorological problems. { e·rə'nód·ə·kəl ‚klī·mə'täl·ə·je }

aeronautical engineering [AERO ENG] The branch of engineering concerned primarily with the design and construction of aircraft structures and power units, and with the special problems of flight in the atmosphere. { e·rə'nód·ə·kəl en·jə'nir·iŋ }

aeronautical flutter [FL MECH] An aeroelastic, self-excited vibration in which the external source of energy is the airstream and which depends on the elastic, inertial, and dissipative forces of the system in addition to the aerodynamic forces. Also known as flutter. { e·rə'nód·ə·kəl 'fləd·ər }

aeronautical information overprint [NAV] Additional information which is printed or stamped on a map or chart for the specific purpose of air navigation. { e·rə'nód·ə·kəl in'fər·ma·shən 'ō·vər‚print }

aeronautical light [NAV] A luminous or lighted aid to navigation intended primarily for air navigation. { e·rə'nód·ə·kəl 'līt }

aeronautical meteorology [METEOROL] The study of the effects of weather upon aviation. { e·rə'nód·ə·kəl ‚mēd·ē·ə'räl·ə·je }

aeronautical mile See air mile. { e·rə'nód·ə·kəl 'mīl }

aeronautical mobile satellite service [COMMUN] A mobile satellite service in which the mobile earth stations are located on board aircraft. Abbreviated AMSS. { ‚er·ə¦nód·ə·kəl ‚mō·bəl 'sad·əl‚īt ‚sər·vəs }

aeronautical mobile service [COMMUN] A mobile service between aircraft stations and land stations, or between aircraft stations, in which survival craft stations may also participate. { ‚er·ə¦nód·ə·kəl ¦mō·bəl 'sər·vəs }

aeronautical pilotage chart [NAV] An aeronautical chart designed primarily for air navigation. Also known as contact chart. { e·rə'nód·ə·kəl 'pī·lət·ij ‚chärt }

aeronautical planning chart [NAV] An aeronautical chart of small scale designed to satisfy long-range air navigation and mission-planning requirements. { e·rə'nód·ə·kəl 'plan·iŋ chärt }

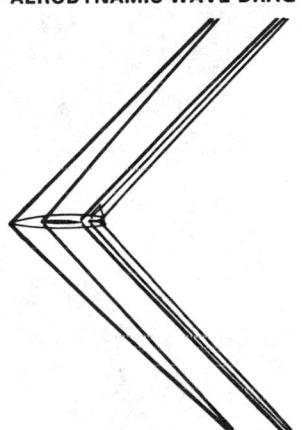

AERODYNAMIC WAVE DRAG

Shock waves generated by an airplane at supersonic speeds act to retard the plane.

aeronautics [FL MECH] The science that deals with flight through the air. { ,e·rə'nȯd·iks }

aeronomy [GEOPHYS] The study of the atmosphere of the earth or other bodies, particularly in relation to composition, properties, relative motion, and radiation from outer space or other bodies. { e'rän·ə·mē }

aerootitis *See* barotitis. { ,e·rō,ō'tīd·əs }

aeropalynology [PALEOBOT] A branch of palynology that focuses on the study of pollen grains and spores that are dispersed into the atmosphere. { ,er·ō,pal·ə'näl·ə·jē }

aeropause [GEOPHYS] A region of indeterminate limits in the upper atmosphere, considered as a transition region between the denser portion of the atmosphere and interplanetary space. { 'e·rə,pȯz }

aerophare *See* radio beacon. { 'e·rə'fer }

aerophotography *See* aerial photography. { ,e·rō·fə'täg·rə'fē }

aerophysics [AERO ENG] The physics dealing with the design, construction, and operation of aerodynamic devices. { ¦e·rō¦fiz·iks }

aerophyte *See* epiphyte. { 'e·rō,fīt }

aeroplankton [ECOL] Small airborne organisms such as insects. { ¦e·rō¦plaŋk·tən }

aeroponics [AGR] The practice of growing plants without soil while suspended in air; a nutrient and water solution is sprayed on the roots and allowed to drain off to be discarded or recycled. { ,er·ə'pän·iks }

aeropulse engine *See* pulsejet engine. { 'e·rō,pȯls 'en·jən }

aeroservoelasticity [AERO ENG] The study of the interaction of automatic flight controls on aircraft and aeroelastic response and stability. { ¦e·rō,sər·vō,i·las'tis·əd·ē }

aerosiderite [GEOL] A meteorite composed principally of iron. { ,e·rō'sīd·ə,rīt }

aerosinusitis *See* barosinusitis. { ,e·rō,sī·nə'sīd·əs }

aerosol [CHEM] A suspension of small particles in a gas; the particles may be solid or liquid or a mixture of both; aerosols are formed by the conversion of gases to particles, the disintegration of liquids or solids, or the suspension of powdered material. [METEOROL] A small droplet or particle suspended in the atmosphere and formed from both natural and anthropogenic sources. { 'e·rə,sȯl }

aerosol generator [MECH ENG] A mechanical means of producing a system of dispersed phase and dispersing medium, that is, an aerosol. { 'e·rə,sȯl 'jen·ə,rād·ər }

aerosol lidar [OPTICS] A type of lidar that is designed to measure the scattering of laser light from atmospheric dust and aerosols. { 'e·rə,sȯl 'lī,där }

aerosol propellant [MATER] Compressed gas or vapor in a container which, upon release of pressure and expansion through a valve, carries another substance from the container; used for cosmetics, household cleaners, and so on; examples are butanes, propane, nitrogen, fluorocarbons, and carbon dioxide. { 'e·rə,sȯl prə'pel·ənt }

aerospace *See* airspace. { ¦e·rō¦spās }

aerospace electronics [ELECTR] The field of electronics as applied to aircraft and spacecraft. { ¦e·rō¦spās i,lek'trän·iks }

aerospace engineering [ENG] Engineering pertaining to the design and construction of aircraft and space vehicles and of power units, and to the special problems of flight in both the earth's atmosphere and space, as in the flight of air vehicles and in the launching, guidance, and control of missiles, earth satellites, and space vehicles and probes. { ¦e·rō¦spās ,en·jə'nir·iŋ }

aerospace environment [GEOPHYS] **1.** The conditions, influences, and forces that are encountered by vehicles, missiles, and so on in the earth's atmosphere or in space. **2.** External conditions which resemble those of atmosphere and space, and in which a piece of equipment, a living organism, or a system operates. { ¦e·rō¦spās in'vī·rən·mənt }

aerospace ground equipment [AERO ENG] Support equipment for air and space vehicles. Abbreviated AGE. { ¦e·rō¦spās 'graůnd i,kwip·mənt }

aerospace industry [ENG] Industry concerned with the use of vehicles in both the earth's atmosphere and space. { ¦e·rō¦spās 'in·dəs·trē }

aerospace medicine [MED] The branch of medicine dealing with the effects of flight in the atmosphere or space upon the human body and with the prevention or cure of physiological or psychological malfunctions arising from these effects. Also known as aeromedicine; aviation medicine. { ¦e·rō¦spās 'med·ə·sin }

aerospace vehicle [AERO ENG] A vehicle capable of flight both within and outside the sensible atmosphere. { ¦e·rō¦spās 'vē·ə·kəl }

aerospike engine [AERO ENG] An advanced liquid-propellant rocket engine that uses an axisymmetric plug nozzle, in combination with a torus-shaped combustion chamber and a turbine exhaust system that injects the turbine drive gases into the nozzle base, to achieve a geometry that is only one-quarter the length of a conventional rocket engine, as well as automatic altitude compensation, resulting in superior low-altitude performance. { ¦e·rō¦spīk 'en·jən }

Aerosporin [MICROBIO] Trade name for the antibiotic polymyxin B. { ¦e·rō¦spȯr·ən }

aerostat [AERO ENG] Any aircraft that derives its buoyancy or lift from a lighter-than-air gas contained within its envelope or one of its compartments; for example, ships and balloons. { 'e·rō,stat }

aerostatic balance [ENG] An instrument for weighing air. { ¦e·rō¦stad·ik 'bal·əns }

aerostatics [FL MECH] The science of the equilibrium of gases and of solid bodies immersed in them when under the influence only of natural gravitational forces. { ¦e·rō¦stad·iks }

aerosurvey *See* aerial survey. { ¦e·rō¦sər,vā }

aerotaxis [BIOL] The movement of an organism, especially aerobic and anaerobic bacteria, with reference to the direction of oxygen or air. { ,e·rō'tak·səs }

aerothermochemistry [FL MECH] The study of gases which takes into account the effect of motion, heat, and chemical changes. { ¦e·rō,thər·mō'kem·ə·strē }

aerothermodynamic border [GEOPHYS] An altitude of about 100 miles (160 kilometers), above which the atmosphere is so rarefied that the skin of an object moving through it at high speeds generates no significant heat. { ¦e·rō,thər·mō·dī'nam·ik 'bȯrd·ər }

aerothermodynamics [FL MECH] The study of aerodynamic phenomena at sufficiently high gas velocities that thermodynamic properties of the gas are important. { ,e·rō,thər·mō·dī'nam·iks }

aerothermoelasticity [FL MECH] The study of the response of elastic structures to the combined effects of aerodynamic heating and loading. { ¦e·rō,thər·mō,i,las'tis·əd·ē }

aerotolerant [MICROBIO] Able to survive in the presence of oxygen. { ,e·rō'täl·ə·rənt }

aerotrain [ENG] A train that is propelled by a fan jet engine and floats on a cushion of low-pressure air, traveling at speeds up to 267 miles (430 kilometers) per hour. { 'e·rō,trān }

aerotropism [BOT] A response in which the growth direction of a plant component changes due to modifications in oxygen tension. { ,e·rō'trō,piz·əm }

aerozine [MATER] Hydrazine mixed 50:50 with dimethylhydrazine; the most-used bipropellant rocket fuel. { 'e·rō,zēn }

AES *See* Auger electron spectroscopy.

aeschynomenous [BOT] Having sensitive leaves that droop when touched, such as members of the Leguminosae. { ,es·kə'näm·ə·nəs }

Aesculus [BOT] A genus of deciduous trees or shrubs belonging to the order Sapindales. Commonly known as buckeye. { 'es·kyə·ləs }

Aeshnidae [INV ZOO] A family of odonatan insects in the suborder Anisoptera distinguished by partially fused eyes. { 'esh·nə·dē }

aesthacyte *See* esthacyte. { 'es·thə,sīt }

aesthesia *See* esthesia. { es'thē·zhə }

aesthesiometer *See* esthesiometer. { es,thē·zē'äm·əd·ər }

aesthete [BOT] A plant organ with the capacity to respond to definite physical stimuli. { 'es,thēt }

aestidurilignosa [ECOL] A mixed woodland of evergreen and deciduous hardwoods. { ,es·tə·də,ril·əg'nōs·ə }

aestilignosa [ECOL] A woodland of tropophytic vegetation in temperate regions. { ,es·tə·lig'nōs·ə }

aestival *See* estival. { 'es·tə·vəl }

aestivation [BOT] The arrangement of floral parts in a bud. [PHYSIO] The condition of dormancy or torpidity. { ,es·tə'vā·shən }

Aetosauria [PALEON] A suborder of Triassic archosaurian quadrupedal reptiles in the order Thecodontia armored by rings of thick, bony plates. { ā¦et·ə'sór·ē·ə }

af *See* audio frequency.

aF *See* abfarad.

AFC *See* automatic frequency control.

afebrile [MED] Without fever. { ¦ā¦fēb,ril }

affect [PSYCH] Conscious awareness of feelings; mood. { 'af,ekt }

affection [MED] Any pathology or diseased state of the body. [PSYCH] The feeling aspect of consciousness. { ə'fek·shən }

affective disorder [PSYCH] Any of a group of disorders in which there is a prominent and persistent disturbance of mood and a full syndrome of associated symptoms, such as depressive disorders or bipolar disorder. Also known as mood disorder. { ə¦fek·tiv dis'órd·ər }

affectivity [PSYCH] The state of being susceptible to emotional stimuli. { a,fek'tiv·əd·ē }

afferent [PHYSIO] Conducting or conveying inward or toward the center, specifically in reference to nerves and blood vessels. { 'af·ə·rənt }

afferent neuron [NEURO] A nerve cell that conducts impulses toward a nerve center, such as the central nervous system. { 'af·ə·rənt 'nù,rän }

affination [FOOD ENG] Removing the film of adhering molasses from raw sugar crystals by treating with a heavy sugar syrup. { ,af·ə'nā·shən }

affine connection [MATH] A structure on an *n*-dimensional space that, for any pair of neighboring points *P* and *Q*, specifies a rule whereby a definite vector at *Q* is associated with each vector at *P*; the two vectors are said to be parallel. { ə'fīn kə'nek·shən }

affine deformation [GEOL] A type of deformation in which very thin layers slip against each other so that each moves equally with respect to its neighbors; generally does not result in folding. [MAP] A deformation in which the scale along one axis, or reference plane, is different from the scale along the other axis, or plane, normal to the first. { ə'fīn ,dē·fòr'mā·shən }

affine geometry [MATH] The study of geometry using the methods of linear algebra. { ə'fīn jē'äm·ə·trē }

affine Hjelmslev plane [MATH] A generalization of an affine plane in which more than one line may pass through two distinct points. Also known as Hjelmslev plane. { ə¦fīn 'hyelm,slev ,plān }

affine plane [MATH] In projective geometry, a plane in which (1) every two points lie on exactly one line, (2) if *p* and *L* are a given point and line such that *p* is not on *L*, then there exists exactly one line that passes through *p* and does not intersect *L*, and (3) there exist three noncollinear points. { ə'fīn ,plān }

affine space [MATH] An *n* dimensional vector space which has an affine connection defined on it. { ə'fīn ,spās }

affine strain [GEOPHYS] A strain in the earth that does not differ from place to place. { ə'fīn ,strān }

affine transformation [MATH] A function on a linear space to itself, which is the sum of a linear transformation and a fixed vector. { ə'fīn ,tranz·fər'mā·shən }

affinity [CHEM] The extent to which a substance or functional group can enter into a chemical reaction with a given agent. Also known as chemical affinity. [COMPUT SCI] A specific relationship between data processing elements that requires one to be used with the other, where a choice might otherwise exist. [IMMUNOL] The strength of the attractive forces between an antigen and an antibody. { ə'fin·əd·ē }

affinity chromatography [ANALY CHEM] A chromatographic technique that utilizes the ability of biological molecules to bend to certain ligands specifically and reversibly; used in protein biochemistry. { ə'fin·əd·ē ,krō·mə'täg·rə·fē }

affinity labeling [BIOCHEM] A method for introducing a label into the active site of an enzyme by relying on the tight binding between the enzyme and its substrate (or cofactors). { ə'fin·əd·ē 'lā·bə·liŋ }

afforestation [FOR] Establishment of a new forest by seeding or planting on nonforested land. { a,fär·ə'stā·shən }

affreightment [IND ENG] The lease of a vessel for the transportation of goods. { ə'frāt·mənt }

afibrinogenemia [MED] Complete absence of fibrinogen in the blood. { ,ā·fī¦brin·ə·jə¦nē·mē·ə }

A15 compound *See* A15 phase. { ¦ā'fif¦tēn 'käm¦paünd }

A15 phase [SOLID STATE] An intermetallic compound having the chemical formula A_3B, where A represents a transition element, and a crystal structure in which the B atoms are located at the corners and in the center of a cubic unit cell, while the A atoms are arranged in pairs on the cube faces. Also known as A15 compound. { ¦ā·fif'tēn ,fāz }

aflatoxicosis [MED] Aflatoxin poisoning. { ə¦flād·ō,täk·sə'kō·səs }

aflatoxin [BIOCHEM] The toxin produced by some strains of the fungus *Aspergillus flavus*, the most potent carcinogen yet discovered. { ,af·lə'täk·sin }

afocal lens [OPTICS] A lens of zero convergent power, whose focal points are infinitely distant. { ¦ā·fō·kəl 'lenz }

afocal system [OPTICS] An optical system of zero convergent power, for example, a telescope. { 'ā·fō·kəl 'sis·təm }

a format [COMPUT SCI] A nonexecutable statement in FORTRAN which permits alphanumeric characters to be transmitted in a manner similar to numeric data. { 'ā 'fór,mat }

A frame [BUILD] A dwelling whose main frames are in the shape of the letter A. [ENG] Two poles supported in an upright position by braces or guys and used for lifting equipment. Also known as double mast. [OCEANOGR] An A-shaped frame used for outboard suspension of oceanographic gear on a research vessel. { 'ā ,frām }

Africa [GEOGR] The second largest continent, with an area of 11,700,000 square miles (30,420,000 square kilometers); bisected midway by the Equator, above and below which it shows symmetry of climate and vegetation zones. { 'af·ri·kə }

African greenheart [MATER] The yellowish-brown wood of *Piptadena africana;* used for shipbuilding and dock timbers. Also known as dahoma. { 'af·ri·kən 'grēn,härt }

African horse sickness [VET MED] An infectious, mosquito-borne virus disease of equines characterized by fever and edematous swelling. { 'af·ri·kən hórs 'sik·nəs }

African sleeping sickness [MED] A disease of humans confined to tropical Africa, caused by the protozoans *Trypanosoma gambiense* or *T. rhodesiense;* symptoms include local reaction at the site of the bite, fever, enlargement of adjacent lymph nodes, skin rash, edema, and during the late phase, somnolence and emaciation. Also known as African trypanosomiasis; maladie du sommeil; sleeping sickness. { 'af·ri·kən 'slēp·iŋ 'sik·nəs }

African superplume [GEOPHYS] A large, discrete, slowly rising plume of heated material in the earth's mantle, beneath southern Africa, believed by some to contribute to the movement of tectonic plates. { ¦af·ri·kən 'sü·pər,plüm }

African swine fever *See* hog cholera. { 'af·ri·kən 'swīn ,fēv·ər }

African trypanosomiasis *See* African sleeping sickness. { 'af·ri·kən trə,pan·ə·sə¦mī·ə·səs }

African violet [BOT] *Saintpaulia ionantha.* A flowering plant typical of the family Gesneriaceae. { 'af·ri·kən 'vī·ə·lət }

aft [NAV ARCH] Near, toward, or at the stern of a ship. { aft }

afterbirth [EMBRYO] The placenta and fetal membranes expelled from the uterus following birth of offspring in viviparous mammals. { 'af·tər,bərth }

afterblow [MET] A blow in a Bessemer process, occurring after the flame for the removal of carbon has dropped, to remove phosphorus. { 'af·tər,blō }

afterbody [AERO ENG] **1.** A companion body that trails a satellite. **2.** A section or piece of a rocket or spacecraft that enters the atmosphere unprotected behind the nose cone or other body that is protected for entry. **3.** The afterpart of a vehicle. { 'af·tər,bäd·ē }

afterboil [MECH ENG] In an automotive engine, coolant boiling after the engine has stopped because of the inability of the engine at rest to dissipate excess heat. { 'af·tər,bóil }

afterbreak [MIN ENG] A phenomenon occurring during mine subsidence; material slides inward after the main break, assumed at right angles to the plane of the seam. { 'af·tər,brāk }

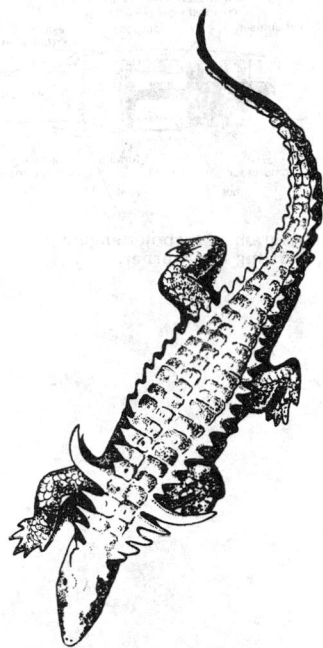

Restoration of *Desmatosuchus*, a quadrupedal thecodont, length to 3 meters. *(From E. H. Colbert, Evolution of the Vertebrates, 2d ed., copyright © 1969 by John Wiley & Sons, Inc.; reprinted by permission)*

AFTERBURNER

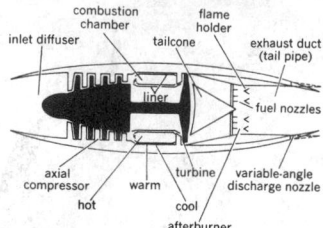

Diagram of turbojet engine showing afterburner.

afterburner [AERO ENG] A device for augmenting the thrust of a jet engine by burning additional fuel in the uncombined oxygen in the gases from the turbine. { 'af·tər‚bər·nər }

afterburning [AERO ENG] The function of an afterburner. [MECH ENG] Combustion in an internal combustion engine following the maximum pressure of explosion. { 'af·tər‚bərn·iŋ }

afterburnt [AERO ENG] Descriptive of the condition following the complete transformation of the solid propellant to gaseous form. { 'af·tər‚bərnt }

afterburst [MIN ENG] **1.** A tremor that sometimes follows a rock blast as the ground adjusts to the new stress distribution. **2.** A sudden collapse of rock in an underground mine subsequent to a rock burst. { 'af·tər‚bərst }

afterchromed dye [MATER] A dye that is improved in color quality or fastness after the textile is dyed by treatment with sodium dichromate, copper sulfate, or similar materials. { 'af·tər‚krōmd 'dī }

aftercondenser [MECH ENG] A condenser in the second stage of a two-stage ejector; used in steam power plants, refrigeration systems, and air conditioning systems. { 'af·tər‚kən'dens·ər }

aftercooler [MECH ENG] A heat exchanger which cools air that has been compressed; used on turbocharged engines. { 'af·tər‚kül·ər }

aftercooling [MECH ENG] The cooling of a gas after its compression. [NUCLEO] The cooling of a reactor after it has been shut down. { 'af·tər‚kül·iŋ }

afterdamp [MIN ENG] The mixture of gases which remains in a mine after a mine fire or an explosion of firedamp. { 'af·tər‚damp }

afterfilter [MECH ENG] In an air-conditioning system, a high-efficiency filter located near a terminal unit. Also known as final filter. { 'af·tər‚fil·tər }

afterflaming [AERO ENG] With liquid- or solid-propellant rocket thrust chambers, a characteristic low-grade combustion that takes place in the thrust chamber assembly and around its nozzle exit after the main propellant flow has been stopped. { 'af·tər‚flām·iŋ }

aftergases [MIN ENG] Gases produced by mine explosions or mine fires. { 'af·tər‚gas·əs }

afterglow [ATOM PHYS] See phosphorescence. [METEOROL] A broad, high arch of radiance or glow seen occasionally in the western sky above the highest clouds in deepening twilight, caused by the scattering effect of very fine particles of dust suspended in the upper atmosphere. [PL PHYS] The transient decay of a plasma after the power has been turned off. { 'af·tər‚glō }

afterheat [NUCLEO] Heat derived from residual radioactivity after a reactor has been shut down. { 'af·tər‚hēt }

afterimage [NEURO] A visual sensation occurring after the stimulus to which it is a response has been removed. { 'af·tər‚im·əj }

afterloading [MED] Placing nonradioactive holders in a patient during an operative procedure to provide radiographic information used for dosimetric evaluation, and to receive the radioactive sources during the course of postoperative treatment. { 'af·tər‚lōd·iŋ }

afterpain [MED] Lower abdominal pain after passage of the placenta secondary to uterine contractions. { 'af·tər‚pān }

afterpeak [NAV ARCH] The compartment closest to the stern in the hold of a ship. { 'af·tər‚pēk }

after-perpendicular [NAV ARCH] A vertical line through the intersection of the design waterline with the after-side of the straight portion of the rudder post of a ship. Abbreviated AP. { 'af·tər‚pər·pən¦dik·yə·lər }

afterpotential [NEURO] A small positive or negative wave that follows and is dependent on the main spike potential, seen in the oscillograph tracing of an action potential passing along a nerve. { 'af·tərpə¦ten·chəl }

afterripening [BOT] A period of dormancy after a seed is shed during which the synthetic machinery of the seed is prepared for germination and growth. { 'af·tər‚rī·pən·iŋ }

afterrunning [MECH ENG] In an automotive engine, continued operation of the engine after the ignition switch is turned off. Also known as dieseling; run-on. { 'af·tər‚rən·iŋ }

afterscour water [TEXT] Wash water that has been discharged from the dyeing vat in the chemical treatment of fabric. { 'af·tər‚skaúr 'wòd·ər }

aftershaft [VERT ZOO] An accessory, plumelike feather near the upper umbilicus on the feathers of some birds. { 'af·tər‚shaft }

aftershock [GEOPHYS] A small earthquake following a larger earthquake and originating at or near the larger earthquake's epicenter. { 'af·tər‚shäk }

aftertack [MATER] Of a paint film, tackiness or stickiness which remains over a long period of time. Also known as residual tack. { 'af·tər‚tak }

after top dead center [MECH ENG] The position of the piston after reaching the top of its stroke in an automotive engine. { 'af·tər 'täp 'ded 'sen·tər }

afterwind [NUCLEO] A wind produced by the updraft accompanying the rise of the fireball of a nuclear explosion and directed toward the burst center. { 'af·tər‚wind }

Aftonian interglacial [GEOL] Post-Nebraska interglacial geologic time. { ‚af'ton·ē·ən ‚in·tər'glā·shəl }

afwillite [MINERAL] $Ca_3Si_2O_4(OH)_6$ A colorless mineral consisting of a hydrous calcium silicate and occurring in monoclinic crystals; specific gravity is 2.6. { 'af·wə‚līt }

Ag See silver.

agalactia [MED] Nonsecretion or imperfect secretion of milk after childbirth. { ‚ā·gə'lak·shə }

agalite [MINERAL] A mineral with the same composition as talc but with a less soapy feel; used as a filler in writing paper. { 'a·gə‚līt }

agalmatolite [GEOL] A soft, waxy, gray, green, yellow, or brown mineral or stone, such as pinite and steatite; used by the Chinese for carving images. Also known as figure stone; lardite; pagodite. { ‚a·gəl'mad·əl‚īt }

Agamemnon [ASTRON] An asteroid, one of a group of Trojan planets whose periods of revolution are approximately equal to that of Jupiter, or about 12 years. { ‚ag·ə'mem·nən }

agameon [BIOL] An organism which reproduces only by asexual means. Also known as agamospecies. { ā'gam·ē·ən }

agamete [BIOL] An asexual reproductive cell that develops into an adult individual. { ā'ga‚mēt }

agamic [BIOL] Referring to a species or generation which does not reproduce sexually. { ā'gam·ik }

Agamidae [VERT ZOO] A family of Old World lizards in the suborder Sauria that have acrodont dentition. { ə'gam·ə‚dē }

agammaglobulinemia [MED] The condition characterized by lack of or extremely low levels of gamma globulin in the blood, together with defective antibody production and frequent infections; primary agammaglobulinemia occurs in three clinical forms: congenital, acquired, and transient. { ā‚gam·ə'gläb·yə·lən·ē·mē·ə }

agamogony [BIOL] Asexual reproduction, specifically schizogony. { ‚ā·gə'mäg·ə·nē }

agamospecies See agameon. { ‚a·gə·mō'spē·shēz }

agamospermy [BOT] Apogamy in which sexual union is incomplete because of abnormal development of the pollen and the embryo sac. { 'ā·gam·ə‚spərm·ē }

Agaontidae [INV ZOO] A family of small hymenopteran insects in the superfamily Chalcidoidea; commonly called fig insects for their role in cross-pollination of figs. { ‚a·gā'än·tə‚dē }

agar [MATER] A gelatinous product extracted from certain red algae and used chiefly as a gelling agent in culture media. { 'äg·ər }

agar attar See oriental linaloe. { 'äg·ər 'at·ər }

agar-gel reaction [IMMUNOL] A precipitin type of antigen-antibody reaction in which the reactants are introduced into different regions of an agar gel and allowed to diffuse toward each other. { 'äg·ər ‚jel ri'ak·shən }

agaric acid [ORG CHEM] $C_{19}H_{36}(OH)(COOH)_3$ An acid with melting point 141°C; soluble in water, insoluble in benzene; used as an irritant. Also known as agaricin. { ə'gar·ik 'as·əd }

Agaricales [MYCOL] An order of fungi in the class Basidiomycetes containing all forms of fleshy, gilled mushrooms. { ə‚gar·ə'kā·lēz }

agaricin See agaric acid. { ə'gar·ə·sən }

agaric mineral See rock milk. { ə'gar·ik 'min·rəl }

agarophyte [BOT] Any seaweed that yields agar. { ə'gar·ə‚fīt }

agarose [BIOCHEM] The gelling component of agar; possesses a double-helical structure which forms a three-dimensional framework capable of holding water molecules in the interstices. { 'ag·ə,rōs }

agar plate count See plate count. { 'äg·ər 'plāt ,kau̇nt }

Agassiz orogeny [GEOL] A phase of diastrophism confined to North America Cordillera occurring at the boundary between the Middle and Late Jurassic. { 'ag·ə·sē ȯ'räj·ə·nē }

Agassiz trawl [OCEANOGR] A dredge consisting of a net attached to an iron frame with a hoop at each end that is used to collect organisms, particularly invertebrates, living on the ocean bottom. { 'ag·ə·sē 'trȯl }

Agassiz Valleys [GEOL] Undersea valleys in the Gulf of Mexico between Cuba and Key West. { 'ag·ə·sē 'val·ēz }

agate [GRAPHICS] A type size in printing of about 5½ points, where 72 points equals 1 inch. [MINERAL] SiO₂ A fine-grained, fibrous variety of chalcedony with color banding or irregular clouding. { 'ag·ət }

agate glass [MATER] Multicolor glass made by blending glasses of two or more colors or by rolling transparent glass into powdered glass of various colors. { 'ag·ət ,glas }

agate jasper [MINERAL] An impure variety of quartz consisting of jasper and agate. Also known as jaspagate. { 'ag·ət 'jas·pər }

agatized wood See silicified wood. { 'ag·ə·tīzd 'wu̇d }

Agavaceae [BOT] A family of flowering plants in the order Liliales characterized by parallel, narrow-veined leaves, a more or less corolloid perianth, and an agavaceous habit. { 'ag·ə'vās·ē,ē }

agavose [ORG CHEM] C₁₂H₂₂O₁₁ A sugar found in the juice of the agave tree; used in medicine as a diuretic and laxative. { 'ag·ə,vōs }

AGB See asymptotic giant branch.

AGC See automatic gain control.

age [BIOL] Period of time from origin or birth to a later time designated or understood; length of existence. [GEOL] **1.** Any one of the named epochs in the history of the earth marked by specific phases of physical conditions or organic evolution, such as the Age of Mammals. **2.** One of the smaller subdivisions of the epoch as geologic time, corresponding to the stage or the formation, such as the Lockport Age in the Niagara Epoch. { āj }

AGE See aerospace ground equipment.

age coating [ELEC] The black deposit that is formed on the inner surface of an electric lamp by material evaporated from the filament. { āj 'kōd·iŋ }

aged [GEOL] Of a ground configuration, having been reduced to base level. { 'ā·jəd }

age determination [GEOL] Identification of the geologic age of a biological or geological specimen by using the methods of dendrochronology or radiometric dating. { 'āj di,tər·mə'nā·shən }

age distribution [ECOL] The distribution of different age groups in a population. { 'āj dis·trə'byü·shən }

aged shore [GEOL] A shore long established at a constant level and adjusted to the waves and currents of the sea. { 'ā·jəd 'shȯr }

age hardening [MET] Increasing the hardness of an alloy by a relatively low-temperature heat treatment that causes precipitation of components or phases of the alloy from the super-saturated solid solution. Also known as precipitation hardening. { ,āj 'härd·ən·iŋ }

agel fiber [MATER] Fiber from the leaves and stem of the gebang palm; used for rope, sailcloth, and fishnets. { 'ā·jəl 'fī·bər }

Agena See β Centauri. { ə'jen·ə }

agenda [COMPUT SCI] **1.** The sequence of control statements required to carry out the solution of a computer problem. **2.** A collection of programs used for manipulating a matrix in the solution of a problem in linear programming. { ə'jen·də }

agenesis [BIOL] Absence of a tissue or organ due to lack of development. { 'ā·jen·ə·səs }

age of diurnal inequality [GEOPHYS] The time interval between the maximum semimonthly north or south declination of the moon and the time that the maximum effect of the declination upon the range of tide or speed of the tidal current occurs. Also known as age of diurnal tide; diurnal age. { 'āj əv dī'ərn·əl ,in·ē'kwäl·əd·ē }

age of diurnal tide See age of diurnal inequality. { 'āj əv dī'ərn·əl ,tīd }

Age of Fishes [GEOL] An informal designation of the Silurian and Devonian periods of geologic time. { 'āj əv 'fish·əz }

Age of Mammals [GEOL] An informal designation of the Cenozoic era of geologic time. { 'āj əv 'mam·əlz }

Age of Man [GEOL] An informal designation of the Quaternary period of geologic time. { 'āj əv 'man }

age of parallax inequality [GEOPHYS] The time interval between the perigee of the moon and the maximum effect of the parallax (distance of the moon) upon the range of tide or speed of tidal current. Also known as parallax age. { 'āj əv 'par·ə,laks ,in·ē'kwäl·əd·ē }

age of phase inequality [GEOPHYS] The time interval between the new or full moon and the maximum effect of these phases upon the range of tide or speed of tidal current. Also known as age of tide; phase age. { 'āj əv 'fāz ,in·ē'kwäl·əd·ē }

age of the moon [ASTRON] The elapsed time, usually expressed in days, since the last new moon. { 'āj əv thə 'mün }

age of tide See age of phase inequality. { 'āj əv 'tīd }

ageostrophic wind See geostrophic departure. { ə'jē·ə¦sträf·ik 'wind }

age ratio [GEOL] The ratio of the amount of daughter to parent isotope in a mineral being dated radiometrically. { 'āj ,rā·shō }

age softening [MET] The loss of strength and hardness which takes place at room temperature in some alloys because of the spontaneous decrease of residual stresses in the strain-hardened structure. { 'āj 'sȯf·niŋ }

age structure [ANTHRO] Categorization of the population of communities or countries by age groups, allowing demographers to make projections of the growth or decline of the particular population. { 'āj ,strək·chər }

agger [CIV ENG] A material used for road fill over low ground. { 'a·jər }

agglomerate [GEOL] A pyroclastic rock composed of angular rock fragments in a matrix of volcanic ash; typically occurs in volcanic vents. [MET] A rigid mass of metallic particles that have been joined together by a powder metallurgical technique, such as sintering. { ə'gläm·ə·rət }

agglomeration [FOOD ENG] A technique that combines powdered material to form larger, more soluble particles by intermingling in a humid atmosphere. [MET] Conversion of small pieces of low-grade iron ore into larger lumps by application of heat. [METEOROL] The process in which particles grow by collision with and assimilation of cloud particles or other precipitation particles. Also known as coagulation. [SCI TECH] An indiscriminately formed cluster of particles. { ə,gläm·ə'rā·shən }

agglomeration test [MIN ENG] A test of a button of coke whose results are used as a measure of the binding qualities of the coal. { ə,gläm·ə'rā·shən ,test }

agglutinate cone See spatter cone. { ə'glüt·ən,āt ,kōn }

agglutination [CELL MOLEC] The joining of two organisms of the same species for the purpose of sexual reproduction. { ə,glü·tə'nā·shən }

agglutination reaction [IMMUNOL] Clumping of a particulate suspension of antigen by a reagent, usually an antibody. { ə,glüt·ən'ā·shən rē'ak·shən }

agglutinin [IMMUNOL] An antibody from normal or immune serum that causes clumping of its complementary particulate antigen, such as bacteria or erythrocytes. { ə'glüt·ən·ən }

agglutinogen [IMMUNOL] An antigen that stimulates production of a specific antibody (agglutinin) when introduced into an animal body. { ə,glü·tin·ə·jən }

agglutinoid [IMMUNOL] An agglutin that lacks the power to agglutinate but has the ability to unite with its agglutinogen. { ə'glüt·ən,ȯid }

aggradation [GEOL] See accretion. [HYD] A process of shifting equilibrium of stream deposition, with upbuilding approximately at grade. { ,ag·rə'dā·shən }

aggradation recrystallization [GEOL] Recrystallization resulting in the enlargement of crystals. { ,ag·rə'dā·shən rē,kris·tə·lə'zā·shən }

aggraded valley floor [GEOL] The surface of a flat deposit of alluvium which is thicker than the stream channel's depth and is formed where a stream has aggraded its valley. { ə'grād·əd 'val·ē 'flȯr }

aggraded valley plain See alluvial plain. { ə'grād·əd 'val· ē 'plān }

aggregate [BOT] Referring to fruit formed in a cluster, from a single flower, such as raspberry, or from several flowers, such as pineapple. [CHEM] A group of atoms or molecules that are held together in any way, for example, a micelle. [GEOL] A collection of soil grains or particles gathered into a mass. [MATER] The natural sands, gravels, and crushed stone used for mixing with cementing material in making mortars and concretes. { 'ag·rə·gət }

aggregate bin [ENG] A structure designed for storing and dispensing dry granular construction materials such as sand, crushed stone, and gravel; usually has a hopperlike bottom that funnels the material to a gate under the structure. { 'ag·rə· gət ‚bin }

aggregate data type See scalar data type. { 'ag·rə·gət 'da· də ‚tīp }

aggregate fruit [BOT] A type of fruit composed of a number of small fruitlets all derived from the ovaries of a single flower. { 'ag·rə·gət 'früt }

aggregate function [COMPUT SCI] A command in a database management program that performs an arithmetic operation on the values in a specified column or field in all the records in the database, such as computing their sum or average or counting the number of records that satisfy particular criteria. { ‚ag·rə·gət 'fəŋk·shən }

aggregate interlock [ENG] The projection of aggregate particles or portions thereof from one side of a joint or crack in concrete into recesses in the other side so as to effect load transfer in compression and shear, and to maintain mutual alignment. { 'ag·rə·gət 'in·tər‚läk }

aggregate production scheduling [IND ENG] A type of planning at a broad level without consideration of individual products and activities in order to develop a program of output that will meet future demand under given constraints. { ‚ag· ri‚gət prə‚dək·shən ‚skej·ə·liŋ }

aggregate recoil [NUC PHYS] The ejection of atoms from the surface of a sample as a result of their being attached to one atom that is recoiling as the result of alpha-particle emission. { 'ag·rə·gət 'rē‚cȯil }

aggregate structure [GEOL] A mass composed of separate small crystals, scales, and grains that, under a microscope, extinguish at different intervals during the rotation of the stage. { 'ag·rə·gət 'strək·chər }

aggregation [BIOL] A grouping or clustering of separate organisms. [CHEM] A process that results in the formation of aggregates. [FL MECH] See axisymmetrization. { ‚ag· rə'gā·shən }

aggressin [BIOCHEM] A protein produced by a pathogenic microbe which inhibits the host's immune system. { ə'gre· sən }

aggression [PSYCH] Feelings and behavior of anger and hostility usually manifested by punitive or destructive actions; often associated with frustration. { ə'gresh·ən }

aggressive carbon dioxide [CHEM ENG] The carbon dioxide dissolved in water in excess of the amount required to precipitate a specified concentration of calcium ions as calcium carbonate; used as a measure of the corrosivity and scaling properties of water. { ə'gres·iv 'kär·bən dī'äk‚sīd }

aggressive device [COMPUT SCI] A unit of a computer that can initiate a request for communication with another device. { ə'gres·iv di'vīs }

aggressive magma [GEOL] A magma that forces itself into place. { ə'gres·iv 'mag·mə }

aggressive mimicry [ZOO] Mimicry used to attract or deceive a species in order to prey upon it. { ə'gres·iv 'mim· ə·krē }

aggressive water [HYD] Any of the waters which force their way into place. { ə'gres·iv 'wȯd·ər }

agile manufacturing [IND ENG] Operations that can be rapidly reconfigured to satisfy changing market demands. { ‚a· jəl ‚man·yü'fak·chə·riŋ }

aging [ACOUS] The process by which the pressure disturbance from a passing aircraft is distorted as it propagates away from the aircraft, causing the signature to stretch out in duration and length, lose detail, and form shock waves. [BIOL] Growing older. [CHEM] All irreversible structural changes that occur in a precipitate after it has formed. [ELEC] Allowing a permanent magnet, capacitor, meter, or other device to remain in storage for a period of time, sometimes with a voltage applied, until the characteristics of the device become essentially constant. [ELECTROMAG] Change in the magnetic properties of iron with passage of time, for example, increase in the hysteresis. [ENG] **1.** The changing of the characteristics of a device due to its use. **2.** Operation of a product before shipment to stabilize characteristics or detect early failures. [MATER] **1.** Change in the properties of any substance with time. **2.** Change occurring in powders or slips with the passage of time. **3.** Curing of ceramic materials, such as clays and glazes, by a definite period of time under controlled storage conditions. [MET] **1.** Change in properties of an alloy or metal which generally proceeds slowly at room temperatures and faster at elevated temperatures. **2.** Strain relief, occurring through long storage outdoors under varying temperatures, of iron castings intended for use as toolroom plates or lathe-bed supports. **3.** A second heat treatment of an alloy at a lower temperature, causing precipitation of the unstable phase and increasing hardness, strength, and electrical conductivity. [NUCLEO] The slowing down of neutrons. { 'āj·iŋ }

aging blemish See redox blemish. { 'āj·iŋ ‚blem·ish }

aging-lung emphysema [MED] An asymptomatic pulmonary disease associated with aging, characterized by alveolar dilation due to loss of tissue elasticity. { ‚āj·iŋ ‚ləŋ em· fə'zē·mə }

agitated depression [PSYCH] A depressive disorder that includes loss of pleasurable feelings, physical and emotional overactivity, weight loss, and insomnia. { ‚a·jə‚tād·əd də'presh·ən }

agitating speed [MECH ENG] The rate of rotation of the drum or blades of a truck mixer or other device used for agitation of mixed concrete. { 'aj·ə‚tād·iŋ ‚spēd }

agitating truck [MECH ENG] A vehicle carrying a drum or agitator body, in which freshly mixed concrete can be conveyed from the point of mixing to that of placing, the drum being rotated continuously to agitate the contents. { 'aj·ə‚tād·iŋ ‚trək }

agitator [MECH ENG] A device for keeping liquids and solids in liquids in motion by mixing, stirring, or shaking. { 'aj· ə‚tād·ər }

agitator body [MECH ENG] A truck-mounted drum for transporting freshly mixed concrete; rotation of internal paddles or of the drum prevents the setting of the mixture prior to delivery. { 'aj·ə‚tād·ər 'bäd·ē }

Aglaspida [PALEON] An order of Cambrian and Ordovician merostome arthropods in the subclass Xiphosurida characterized by a phosphatic exoskeleton and vaguely trilobed body form. { ə'glas·pə·də }

aglomerular [HISTOL] Lacking glomeruli. { ‚ā·glə'mər· yə·lər }

Aglossa [VERT ZOO] A suborder of anuran amphibians represented by the single family Pipidea and characterized by the absence of a tongue. { ā'gläs·ə }

aglycon [BIOCHEM] The nonsugar compound resulting from the hydrolysis of glycosides; an example is 3,5,7,3',4'-pentahydroxyflavylium, or cyanidin. { ə'glī‚kän }

aglyphous [VERT ZOO] Having solid teeth. { 'a·glə·fəs }

agmatine [BIOCHEM] $C_5H_{14}N_4$ Needlelike crystals with a melting point of 231°C; soluble in water; a product of the enzymatic decarboxylation of arginine. { 'ag·mə‚tēn }

agmatite [PETR] **1.** A migmatite that contains xenoliths. **2.** Fragmental plutonic rock with granitic cement. { 'ag·mə‚tīt }

agnate [BIOL] Related exclusively through male descent. { 'ag‚nāt }

Agnatha [VERT ZOO] The most primitive class of vertebrates, characterized by the lack of true jaws. { 'ag·nə·thə }

agnathia [MED] Lack of the jaws. { ag'nath·ē·ə }

agnosia [MED] Loss of the ability to recognize persons or objects and their meaning. { ag'nozh·ē·ə }

agonic line [GEOPHYS] The imaginary line through all points on the earth's surface at which the magnetic declination is zero; that is, the locus of all points at which magnetic north and true north coincide. { ā'gän·ik līn }

Agonidae [VERT ZOO] The poachers, a small family of marine perciform fishes in the suborder Cottoidei. { ə'gän· ə·dē }

agonist [BIOCHEM] A chemical substance that can combine with a cell receptor and cause a reaction or create an active site. [PHYSIO] A contracting muscle that is resisted or counteracted by another muscle, called an antagonist, with which it is paired. { 'ag·ə‚nist }

agonistic behavior [PSYCH] In social animals, fighting and escape behavior common in males during the rutting season. { ¦ag·ə¦nis·tik bi'hāv·yər }

agoraphobia [PSYCH] Abnormal fear of open places. { ə‚gȯr·ə'fōb·ē·ə }

agostic [ORG CHEM] A three-center, two-electron bonding interaction in which a hydrogen atom is bonded to both a carbon atom and a metal atom, such as the interaction of a CH bond and an unsaturated transition-metal compound. { ə'gäs·tik }

agouti [VERT ZOO] A hystricomorph rodent, *Dasyprocta*, in the family Dasyproctidae, with 13 species. { ə'güd·ē }

AGP *See* accelerated graphics port.

agpaite [PETR] A group of igneous rocks containing feldspathoids; includes naujaite, lujavrite, and kakortokite. { 'ag·pə‚īt }

agranular leukocyte [HISTOL] A type of white blood cell, including lymphocytes and monocytes, characterized by the absence of cytoplasmic granules and by a relatively large spherical or indented nucleus. { ‚ā'gran·yə·lər 'lü·kə‚sīt }

agranular reticulum [CYTOL] Endoplasmic reticulum lacking ribosomes. { ā'gran·yə·lər ri'tik·yə·ləm }

agranulocytosis [MED] An acute febrile illness, usually resulting from drug hypersensitivity, manifested as severe leukopenia, often with complete disappearance of granulocytes. { ā‚gran·yə·lō‚sī'tō·səs }

agraphia [MED] Loss of the ability to write. { ‚ā'graf·ē·ə }

agravic [GEOPHYS] Of or pertaining to a condition of no gravitation. { ‚ā'grav·ik }

agravic illusion [MED] An apparent movement of a target in the visual field due to otolith response in zero gravity. Also known as oculoagravic illusion. { ‚ā'grav·ik il'ü·zhən }

agreement residual [SOLID STATE] The sum of the differences between the observed and calculated structure amplitudes of a crystal, for all observed reflections, divided by the sum of the observed amplitudes. { ə‚grē·mənt rə'zij·ə·wəl }

agrestal [ECOL] Growing wild in the fields. { ə'grest·əl }

agretope [IMMUNOL] In antigen presentation, the part of an antigen that interacts with a class II histocompatibility molecule. { 'ag·rə‚tōp }

agribiotechnology [AGR] Biotechnology applied to agriculture. { ‚ag·rə¦bī·ō‚tek'näl·ə·jē }

agricere [GEOL] A waxy or resinous organic coating on soil particles. { 'ag·rə‚sir }

agricolite *See* eulytite. { ə'grik·ə‚līt }

agricultural aircraft [AERO ENG] Aircraft adapted or designed for use in agriculture and forestry. { ¦ag·rə¦kəl·chə·rəl 'er‚kraft }

agricultural chemicals [MATER] Fertilizers, soil conditioners, fungicides, insecticides, weed killers, and other chemicals used to increase farm crop productivity and quality. { ¦ag·rə¦kəl·chə·rəl 'kem·ə·kəls }

agricultural chemistry [AGR] The science of chemical compositions and changes involved in the production, protection, and use of crops and livestock; includes all the life processes through which food and fiber are obtained for humans and their animals, and control of these processes to increase yields, improve quality, and reduce costs. { ¦ag·rə¦kəl·chə·rəl 'kem·ə·strē }

agricultural climatology [AGR] In general, the study of climate as to its effect on crops; it includes, for example, the relation of growth rate and crop yields to the various climatic factors and hence the optimum and limiting climates for any given crop. Also known as agroclimatology. { ¦ag·rə¦kəl·chə·rəl ‚klī·mə'täl·ə·jē }

agricultural engineering [AGR] A discipline concerned with developing and improving the means for providing food and fiber for human needs. { ¦ag·rə¦kəl·chə·rəl ‚en·jə'nir·iŋ }

agricultural geography [GEOGR] A branch of geography that deals with areas of land cultivation and the effect of such cultivation on the physical landscape. { ¦ag·ri¦kəl·chə·rəl jē'ag·rə·fē }

agricultural geology [GEOL] A branch of geology that deals with the nature and distribution of soils, the occurrence of mineral fertilizers, and the behavior of underground water. { ¦ag·rə¦kəl·chə·rəl jē'äl·ə·jē }

agricultural lime [MATER] A hydrated lime which is used to condition soil. { ¦ag·rə¦kəl·chə·rəl 'līm }

agricultural machinery [AGR] Machines utilized for tillage, planting, cultivation, and harvesting of crops. { ¦ag·rə¦kəl·chə·rəl mə'shēn·rē }

agricultural meteorology [AGR] The study and application of relationships between meteorology and agriculture, involving problems such as timing the planting of crops. Also known as agrometeorology. { ¦ag·rə¦kəl·chə·rəl ‚mēd·ē·ə'räl·ə·jē }

agricultural pipe drain [CIV ENG] A system of porous or perforated pipes laid in a trench filled with gravel or the like; used for draining subsoil. { ¦ag·rə¦kəl·chə·rəl ‚pīp ‚drān }

agricultural robot [CONT SYS] A robot used to pick and harvest farm products and fruits. { ¦ag·rə¦kəl·chə·rəl 'rō‚bät }

agricultural science [AGR] A discipline dealing with the selection, breeding, and management of crops and domestic animals for more economical production. { ¦ag·rə¦kəl·chə·rəl 'sī·əns }

agricultural wastes [AGR] Those liquid or solid wastes that result from agricultural practices, such as cattle manure, crop residue (for example, corn stalks), pesticides, and fertilizers. { ¦ag·rə¦kəl·chə·rəl 'wāsts }

agriculture [BIOL] The production of plants and animals useful to humans, involving soil cultivation and the breeding and management of crops and animals. { 'ag·rə‚kəl·chər }

Agriochoeridae [PALEON] A family of extinct tylopod ruminants in the superfamily Merycoidodontoidea. { ‚ag·rē·ō'kir·ə‚dē }

agrioecology [ECOL] The ecology of cultivated plants. { ¦ag·rē·ō‚ē'käl·ə·jē }

Agrionidae [INV ZOO] A family of odonatan insects in the suborder Zygoptera characterized by black or red markings on the wings. { ‚ag·rē'an·ə‚dē }

Agrobacterium [MICROBIO] A genus of bacteria in the family Rhizobiaceae; cells do not fix free nitrogen, and three of the four species are plant pathogens, producing galls and hairy root. { ‚ag·rō‚bak'tir·e·əm }

agroecology [AGR] The application of ecological knowledge about organism distribution and abundance to enhance crop yields and protect crops from insect or disease attack. { ‚ag·rō‚ē'käl·ə·jē }

agroecosystem [ECOL] A model for the functionings of an agricultural system with all its inputs and outputs. { ¦ag·rō'ek·ō‚sis·təm }

agroenvironment [AGR] The soil and climate of a region as they affect agriculture. { ¦ag·rō‚en'vī·rən·mənt }

agroforestry [AGR] The practice of growing trees in association with agricultural crops or animals to provide both ecological and economic benefits. { ¦ag·rō'fär·əs‚trē }

agroinfection [AGR] A method of using the t-DNA of plasmids to infect plant cells with DNA from a plant virus. { ¦ag·rō‚in'fek·shən }

Agromyzidae [INV ZOO] A family of myodarian cyclorrhaphous dipteran insects of the subsection Acalypteratae; commonly called leaf-miner flies because the larvae cut channels in leaves. { ¦ag·rō'mīz·ə‚dē }

agronomy [AGR] The principles and procedures of soil management and of field crop and special-purpose plant improvement, management, and production. { ə'grän·ə·mē }

agrophilous [ECOL] Having a natural habitat in grain fields. { ə'gräf·ə·ləs }

agrostology [BOT] A division of systematic botany concerned with the study of grasses. { ‚ag·rə‚stä·lə·jē }

agrotechnology [AGR] An innovative technology designed to render agricultural production more efficient and profitable. { ¦ag·rō‚tek'näl·ə·jē }

aguilarite [MINERAL] Ag₄SeS An iron-black mineral associated with argentite and silver in Mexico. { ‚äg·ə'lä‚rīt }

Agulhas Current [OCEANOGR] A fast current flowing in a southwestward direction along the southeastern coast of Africa. { ə'gəl·əs 'kər·ənt }

AGV *See* automated guided vehicle.

agyiophobia [PSYCH] Abnormal fear of crossing streets. { ‚a·jē·ə'fōb·ē·ə }

aH *See* abhenry.

Ah *See* ampere-hour.

ahaptoglobinemia [MED] An inherited lack of haptoglobin, a blood serum protein. { ā¦hap·tō‚glō·bə'nēm·ē·ə }

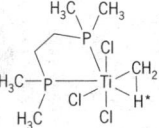

AGOSTIC

Agostic-bridged hydrogen indicated by *.

AGOUTI

The agouti (*Dasyprocta aguti*), a rodent found in Mexico, South America, and the West Indies.

Aharonov-Bohm effect [QUANT MECH] An effect manifested when a beam of electrons is split into two beams that travel in opposite directions around a region containing magnetic flux and are then recombined, whereby the intensity of the resulting beam oscillates periodically as the enclosed magnetic field is changed.   { ä·hä′rō,nóf ′bäm i,fekt }

ahermatypic [INV ZOO] Non-reef-building, as applied to corals.   { ¦ā,hər·mə¦tip·ik }

ahlfeldite [MINERAL] (Ni,Co)SeO$_3$·2H$_2$O A triclinic mineral identified as green to yellow crystals with a reddish-brown coating, consisting of a hydrous selenite of nickel.   { äl′fel,dīt }

AIA See anti-immunoglobulin antibody.

aichmophobia [PSYCH] Abnormal fear of sharp or pointed objects.   { ,äk·mō′fōb·ē·ə }

aided matching [ORD] Mechanical system for transferring firing or other data from a data-transmission line to a gun data computer or other device.   { ′ād·əd ′mach·iŋ }

aided tracking [ENG] A system of radar-tracking a target signal in bearing, elevation, or range, or any combination of these variables, in which the rate of motion of the tracking equipment is machine-controlled in collaboration with an operator so as to minimize tracking error.   { ′ād·əd ′trak·iŋ }

aided-tracking mechanism [ENG] A device consisting of a motor and variable-speed drive which provides a means of setting a desired tracking rate into a director or other fire-control instrument, so that the process of tracking is carried out automatically at the set rate until it is changed manually.   { ′ād·əd ′trak·iŋ ,mek·ə,niz·əm }

aided-tracking ratio [ENG] The ratio between the constant velocity of the aided-tracking mechanism and the velocity of the moving target.   { ′ād·əd ′trak·iŋ ,rā·shō }

AIDS See acquired immune deficiency syndrome.   { ādz }

AIDS-related complex [MED] A set of symptoms, such as lymph node enlargement, fever, loss of weight, diarrhea, and minor opportunistic diseases, associated with a weakened immune system, indicating a less severe form of infection by the HIV virus than AIDS itself. Abbreviated ARC.   { ′ādz rə,lād·əd ¦käm,pleks }

aid to navigation [NAV] A device external to a craft, designed to assist in determination of position of the craft, or of a safe course, or to warn of dangers or obstructions.   { ād tə ,nav·ə′gā·shən }

aiguille [ENG] A slender form of drill used for boring or drilling a blasthole in rock.   [GEOL] The needle-top of the summit of certain glaciated mountains, such as near Mont Blanc.   { ,ā′gwēl }

aikinite [MINERAL] PbCuBiS$_3$ A mineral crystallizing in the orthorhombic system and occurring massive and in gray needle-shaped crystals; hardness is 2 on Mohs scale, and specific gravity is 7.07. Also known as needle ore.   { ′ā·kə,nīt }

aileron [AERO ENG] The hinged rear portion of an aircraft wing moved differentially on each side of the aircraft to obtain lateral or roll control moments.   [ARCH] A half gable, such as that which closes the end of a penthouse roof or of a church aisle.   { ′āl·ə,rän }

ailsyte [PETR] An alkalic microgranite containing a considerable amount of riebeckite. Also known as paisanite.   { ′āl,sīt }

ailurophobia [PSYCH] Abnormal fear of cats.   { i,lü·rə ′fōb·ē·ə }

aiming circle [ENG] An instrument for measuring angles in azimuth and elevation in connection with artillery firing and general topographic work; equipped with fine and coarse azimuth micrometers and a magnetic needle.   { ′ām·iŋ ,sər·kəl }

aiming point [ORD] An object or point on which the sight of a weapon is laid for direction, or on which an observer orients his observing instrument. **2.** The point used by an aerial bombardier or pilot to determine where to release bombs, rockets, mines, torpedoes, and so forth.   { ′ām·iŋ ,póint }

aiming post [ORD] In mortar firing, a wooden or metal post having contrasting painted transverse bands and a metal point or stake for driving into the ground; used as a sighting point in direct fire.   { ′ām·iŋ ,pōst }

aiming screws [MECH ENG] On an automotive vehicle, spring-loaded screws designed to secure headlights to a support frame and permit aiming of the headlights in horizontal and vertical planes.   { ′aim·iŋ ,skrüz }

aimless drainage [HYD] Drainage without a well-developed system, as in areas of glacial drift or karst topography.   { ′ām·ləs ′drān·ij }

A/in.^{2} See ampere per square inch.

A indicator See A scope.   { ′ā ,in·də,kād·ər }

ainhum [MED] A tropical disease of unknown etiology that is peculiar to black males, in which a toe is slowly and spontaneously amputated by a fibrous ring.   { ′ī·nyüm }

aiophyllous See evergreen.   { ,ī·ō′fil·əs }

air [CHEM] A predominantly mechanical mixture of a variety of individual gases forming the earth's enveloping atmosphere.   [MECH ENG] See air-injection reactor.   { er }

air-acetylene welding [MET] A gas-welding process in which the heat is obtained from the combustion of acetylene and air.   { er ə′sed·əl·ən ′weld·iŋ }

air-actuated [ENG] Powered by compressed air.   { ′er ′ak·chə,wād·əd }

AIRAD [NAV] Airmen advisory on NOTAM which is given only local dissemination during preflight or inflight briefings.   { ′er,ad }

air adit [MIN ENG] An adit driven for the purpose of ventilating a mine.   { ′er ′ad·ət }

Air Almanac [NAV] A periodical publication of astronomical data useful to and designed primarily for air navigation. A joint publication of the U.S. Naval Observatory and the Royal Greenwich Observatory, listing the Greenwich hour angle and declination of various celestial bodies at 10-minute intervals, the time of sunrise, sunset, moonrise, and moonset, and other astronomical information arranged in a form convenient for navigators; each publication covers 4 months.   { ′er ′ól·mə,nak }

air-arc furnace [ENG] An arc furnace designed to power wind tunnels, the air being superheated to 20,000 K and expanded to emerge at supersonic speeds.   { ′er ,ärk ′fər·nəs }

air armament [ORD] All equipment through which a combat aircraft can release destructive power on a target.   { ′er ′är·mə·mənt }

air aspirator valve [MECH ENG] On certain automotive engines, a one-way valve installed on the exhaust manifold to allow air to enter the exhaust system; provides extra oxygen to convert carbon monoxide to carbon dioxide. Also known as gulp valve.   { ′er ′as·pə,rād·ər ,valv }

air-assist forming [ENG] A plastics thermoforming method in which air pressure is used to partially preform a sheet before it enters the mold.   { ′er ə′sist ′fórm·iŋ }

air-atomizing oil burner [ENG] An oil burner in which a stream of fuel oil is broken into very fine droplets through the action of compressed air.   { ′er ′at·ə′mīz·iŋ ,óil ′bərn·ər }

air bag [MECH ENG] An automotive vehicle passenger safety device consisting of a passive restraint in the form of a bag which is automatically inflated with gas to provide cushioned protection against the impact of a collision.   { ′er ,bag }

air barrage [MIN ENG] An airtight wall dividing a ventilation gallery in a mine into two parts, so that air is led in through one part and out through the other.   [ORD] **1.** Bombs dropped in a predetermined pattern from a fleet of aircraft to reduce the effectiveness of ground military personnel and equipment. **2.** The shells fired by an antiaircraft battery of guns at attacking airplanes.   { ′er bə′räzh }

air base [AERO ENG] **1.** In the U.S. Air Force, an establishment, comprising an airfield, its installations, facilities, personnel, and activities, for the flight operation, maintenance, and supply of aircraft and air organizations. **2.** A similar establishment belonging to any other air force. **3.** In a restricted sense, only the physical installation.   [MAP] **1.** The line joining two air stations, or the length of this line. **2.** The distance, at the scale of the stereoscopic model, between adjacent perspective centers as reconstructed in the plotting instrument.   { ′er ,bās }

air battery [ELEC] A connected group of two or more air cells; also, a single air cell.   { ′er ′bad·ə·rē }

air belt [MECH ENG] The chamber which equalizes the pressure that is blasted into the cupola at the tuyeres.   { ′er ,belt }

air-bend die [MET] A device for forming metals in which only the two edges of the lower section are in contact with the metal.   { ′er ,bend ′dī }

air bind [ENG] The presence of air in a conduit or pump which impedes passage of the liquid.   { ′er ,bīnd }

airblast [MIN ENG] A disturbance in underground workings accompanied by a strong rush of air.   { ′er,blast }

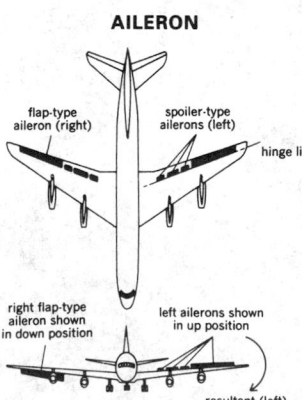

AILERON

Flap- and spoiler-type ailerons on jet transport airplane.

flap-type aileron (right)

spoiler-type ailerons (left)

hinge line

right flap-type aileron shown in down position

left ailerons shown in up position

resultant (left) rolling moment

airblast circuit breaker [ELEC] An electric switch which, on opening, utilizes a high-pressure gas blast (air or sulfur hexafluoride) to break the arc. { 'er,blast 'sər·kət 'brāk·ər }

airblasting [ENG] A blasting technique in which air at very high pressure is piped to a steel shell in a shot hole and discharged. Also known as air breaking. { 'er,blast·iŋ }

air bleeder [MECH ENG] A device, such as a needle valve, for removing air from a hydraulic system. { 'er 'blēd·ər }

air-blower noise [ACOUS] Noise in blowers in heater and air-conditioning systems due to air turbulence. { 'er 'blō·ər }

air-blown asphalt [MATER] A bituminous product made by reacting the residual oil of petroleum distillation with air at 400–600°F (204–316°C). { 'er ¦blōn 'as,fȯlt }

airboat [AERO ENG] *See* seaplane. [NAV ARCH] A shallow boat that is propelled by an airplane propeller and steered by an airplane rudder. { 'er,bōt }

airborne [AERO ENG] Of equipment and material, carried or transported by aircraft. [ORD] **1.** Of a force or organization, transported, or designed to be transported, by aircraft. **2.** Of an action or operation, carried out with transport aircraft. { 'er,bȯrn }

airborne alert [ORD] A state of aircraft readiness wherein combat-equipped aircraft are rapidly airborne and ready for immediate action; it is designed to reduce reaction time and to increase the survival factor. { 'er,bȯrn ə'lərt }

airborne assault weapon [ORD] An unarmored, mobile, full-tracked gun providing a mobile antitank capability for airborne troops; it can be air-dropped. { 'er,bȯrn ə'sȯlt 'wep·ən }

airborne collision avoidance system [NAV] A navigation system for preventing collisions between aircraft that relies primarily on equipment carried on the aircraft itself, but which may make use of equipment already employed in the ground-based air-traffic control system. Abbreviated ACAS. { 'er,bȯrn kə'lizh·ən ə'vȯid·əns ,sis·təm }

airborne collision warning system [ENG] A system such as a radar set or radio receiver carried by an aircraft to warn of the danger of possible collision. { 'er,bȯrn kə'lizh·ən 'wȯrn·iŋ ,sis·təm }

airborne command post [ORD] A suitably equipped aircraft used by the commander for the control of his forces. { 'er,bȯrn kə'mand ,pōst }

airborne control system [GEOD] A survey system for fourth-order horizontal and vertical control surveys involving electromagnetic distance measurements and horizontal and vertical measurements from two or more known positions to a helicopter hovering over the unknown position. Also known as ABC system. { 'er,bȯrn kən'trōl ,sis·təm }

airborne detector [ENG] A device, transported by an aircraft, whose function is to locate or identify an air or surface object. { 'er,bȯrn di'tek·tər }

airborne early warning [ORD] The detection of enemy air or surface units by radar or other equipment carried in an airborne vehicle, and the transmitting of a warning to friendly units. { 'er,bȯrn 'ər·lē 'wȯrn·iŋ }

airborne early warning and control [ORD] Air surveillance and control provided by airborne early-warning vehicles which are equipped with search and height-finding radar and communications equipment for controlling weapons. { 'er,bȯrn 'ər·lē 'wȯrn·iŋ ən kən'trōl }

airborne early-warning station [ORD] An airborne radar station for detecting aircraft, which are incapable of being detected by a ground radar station because of line-of-sight propagation limitations. { 'er,bȯrn 'ər·lē 'wȯrn·iŋ ,stā·shən }

airborne electronic survey control [ENG] The airborne portion of very accurate positioning systems used in controlling surveys from aircraft. { 'er,bȯrn i,lek'trän·ik 'sər·vā kən'trōl }

airborne intercept radar [ENG] Airborne radar used to track and "lock on" to another aircraft to be intercepted or followed. { 'er,bȯrn 'in·tər,sept ,rā,där }

airborne magnetometer [ENG] An airborne instrument used to measure the magnetic field of the earth. { 'er,bȯrn ,mag·nə'täm·əd·ər }

airborne operation [ORD] An operation involving the movement and delivery by air, into an objective area, of combat forces and their logistic support for execution of a tactical or a strategic mission; the means may be any combination of airborne units, air-transportable units, and types of transport aircraft, depending on the mission and the overall situation. { 'er,bȯrn ,äp·ə'rā·shən }

airborne profile [GEOD] Continuous terrain-profile data produced by an absolute altimeter in an aircraft which is making an altimeter-controlled flight along a prescribed course. { 'er,bȯrn 'prō,fīl }

airborne profile recorder [ENG] An electronic instrument that emits a pulsed-type radar signal from an aircraft to measure vertical distances between the aircraft and the earth's surface. Abbreviated APR. Also known as terrain profile recorder (TPR). { 'er,bȯrn 'prō,fīl ri,kȯrd·ər }

airborne radar [ENG] Radar equipment carried by aircraft to assist in navigation by pilotage, to determine drift, and to locate weather disturbances; a very important use is locating other aircraft either for avoidance or attack. { 'er,bȯrn 'rā,där }

airborne self-protection jammer [ELECTR] An electronic system carried by an aircraft to prevent detection by enemy radar, by both emitting signals that obscure radar returns and disguising the jamming signal itself. { 'er,bȯrn ¦self·prə'tek·shən ,jam·ər }

airborne troops [ORD] Ground units whose primary mission is to make assault landings from the air; may refer specifically to troops landed by aircraft as distinguished from parachute troops, or may refer only to those troops landed by parachute or glider as distinguished from those landed in powered aircraft. { 'er,bȯrn 'trüps }

airborne unit [ORD] A ground unit organized, trained, and equipped for airborne assault. { 'er,bȯrn 'yü·nət }

airborne warning and control system [ORD] A U.S. Air Force program that uses airborne radar to detect low-flying enemy aircraft and high-altitude attacking bombers and to provide associated command and control functions for strategic interceptor forces. Abbreviated AWACS. { 'er,bȯrn 'wȯrn·iŋ ən kən'trōl ,sis·təm }

airborne waste [ENG] Vapors, gases, or particulates introduced into the atmosphere by evaporation, chemical, or combustion processes; a frequent cause of smog and an irritant to eyes and breathing passages. { 'er,bȯrn 'wāst }

air-bound [ENG] Of a pipe or apparatus, containing a pocket of air that prevents or reduces the desired liquid flow. { 'er ,baȯnd }

air brake [MECH ENG] An energy-conversion mechanism activated by air pressure and used to retard, stop, or hold a vehicle or, generally, any moving element. { 'er ,brāk }

air breaking *See* airblasting. { 'er ,brāk·iŋ }

air-break switch *See* air switch. { 'er ¦brāk ,swich }

air breakup [AERO ENG] The breakup of a test reentry body within the atmosphere. { 'er 'brāk,əp }

air-breathing [MECH ENG] Of an engine or aerodynamic vehicle, required to take in air for the purpose of combustion. { 'er 'brēth·iŋ }

air-breathing missile [ORD] A missile having an engine that requires air intake for combustion of its fuel, such as a ramjet or turbojet; it cannot operate beyond the atmosphere. { 'er 'brēth·iŋ 'mis·əl }

air brick [MATER] A brick or brick-sized metal box that is hollow or perforated and used for ventilation. { 'er ,brik }

airbrush [GRAPHICS] A pencil-shaped air gun that fine-sprays paint; used in retouching photographs and in shading drawings. { 'er,brəsh }

airburst [ORD] Any burst in the air, but usually the bursting of a projectile or bomb above the ground with resulting spray of fragments. Also known as aerial burst. { 'er,bərst }

air bypass valve [MECH ENG] Also known as diverter valve. { 'er 'bī,pas ,valv }

air cap [MECH ENG] A device used in thermal spraying which directs the air pattern for purposes of atomization. { 'er ,kap }

air capacitor [ELEC] A capacitor having only air as the dielectric material between its plates. Also known as air condenser. { 'er kə'pas·əd·ər }

air carbon arc cutting [MET] A carbon arc cutting process in which the molten metal in the cut is removed by an airblast. { 'er 'kär·bən ,ärk 'kəd·iŋ }

air casing [ENG] A metal casing surrounding a pipe or reservoir and having a space between to prevent heat transmission. { 'er ,kās·iŋ }

air cell [ELECTR] A cell in which depolarization at the positive electrode is accomplished chemically by reduction of the oxygen in the air. [MECH ENG] A small auxiliary combustion chamber used to promote turbulence and improve combustion in certain types of diesel engines. [ZOO] A cavity or receptacle for air such as an alveolus, an air sac in birds, or a dilation of the trachea in insects. { 'er ˌsel }

air chamber [MECH ENG] A pressure vessel, partially filled with air, for converting pulsating flow to steady flow of water in a pipeline, as with a reciprocating pump. { 'er ˌchām·bər }

air change [ENG] A measure of the movement of a given volume of air in or out of a building or room in a specified time period; usually expressed in cubic feet per minute. { 'er ˌchānj }

air check [ENG ACOUS] A recording made of a live radio broadcast for filing purposes at the broadcasting facility. { 'er ˌchek }

air classifier [MECH ENG] A device to separate particles by size through the action of a stream of air. Also known as air elutriator. { 'er 'klas·ə,fī·ər }

air cleaner [ENG] Any of various devices designed to remove particles and aerosols of specific sizes from air; examples are screens, settling chambers, filters, wet collectors, and electrostatic precipitators. { 'er ˌklēn·ər }

Airco-Hoover sweetening [CHEM ENG] Removal of mercaptans from gasoline by caustic and water washes, then heating the dried gasoline and passing it with some oxygen through a reactor containing a slurry of diatomaceous earth impregnated with copper chloride; the oxygen regenerates the catalyst. { 'er,kō 'hüv·ər 'swēt·niŋ }

air composition [METEOROL] The kinds and amounts of the constituent substances of air, the amounts being expressed as percentages of the total volume or mass. { 'er ˌkäm·pə'zish·ən }

air compression [PHYS] The decrease of volume of a quantity of air as a result of an increase in pressure, as is accomplished by a piston moving in a cylinder. { 'er ˌkəm'presh·ən }

air compressor [MECH ENG] A machine that increases the pressure of air by increasing its density and delivering the fluid against the connected system resistance on the discharge side. { 'er ˌkəm'pres·ər }

air-compressor unloader [MECH ENG] A device for control of air volume flowing through an air compressor. { 'er ˌkəm'pres·ər ən'lōd·ər }

air-compressor valve [MECH ENG] A device for controlling the flow into or out of the cylinder of a compressor. { 'er ˌkəm'pres·ər ˌvalv }

air condenser [ELEC] See air capacitor. [MECH ENG] **1.** A steam condenser in which the heat exchange occurs through metal walls separating the steam from cooling air. Also known as air-cooled condenser. **2.** A device that removes vapors, such as of oil or water, from the airstream in a compressed-air line. { 'er ˌkən'dens·ər }

air conditioner [MECH ENG] A mechanism primarily for comfort cooling that lowers the temperature and reduces the humidity of air in buildings. { 'er ˌkən'dish·ən·ər }

air conditioning [MECH ENG] The maintenance of certain aspects of the environment within a defined space to facilitate the function of that space; aspects controlled include air temperature and motion, radiant heat level, moisture, and concentration of pollutants such as dust, microorganisms, and gases. Also known as climate control. [TEXT] A chemical process by which small fibers are sealed into yarn. { 'er ˌkən'dish·ən·iŋ }

air content [MATER] The volume of air voids in cement paste, mortar, or concrete, exclusive of pore space in aggregate particles; usually expressed as a percentage of total volume of the mixture. { 'er ˌkän,tent }

air-control center [COMMUN] An area set aside in a submarine for the control of aircraft; it is the equivalent of a combat information center on an aircraft or a ship. { 'er kən'trōl ˌsent·ər }

air conveyor See pneumatic conveyor. { 'er kən,vā·ər }

air-cooled blast-furnace slag [MET] The material resulting from solidification of molten blast-furnace slag under atmospheric conditions. { 'er ˌküld 'blast ¦fər·nəs ˌslag }

air-cooled condenser See air condenser. { 'er ˌküld kən'dens·ər }

air-cooled engine [MECH ENG] An engine cooled directly by a stream of air without the interposition of a liquid medium. { 'er ˌküld 'en·jən }

air-cooled heat exchanger [MECH ENG] A finned-tube (extended-surface) heat exchanger with hot fluids inside the tubes, and cooling air that is fan-blown (forced draft) or fan-pulled (induced draft) across the tube bank. { 'er ˌküld ˌhēt ˌiks'chānj·ər }

air cooling [MECH ENG] Lowering of air temperature for comfort, process control, or food preservation. { 'er ˌkül·iŋ }

air coordinates [ORD] A system of coordinates used in determining the forces acting on a missile. { 'er kō'órd·ən,ots }

air-core coil [ELECTR] An inductor without a magnetic core. { 'er ˌkór ˌkóil }

air-core transformer [ELECTROMAG] Transformer (usually radio-frequency) having a nonmetallic core. { 'er ˌkór tranz 'fórm·ər }

air course See airway. { 'er ˌkórs }

air cover [ORD] **1.** The protection against attack, especially air attack, given by airplanes to surface or airborne forces. **2.** The airplanes giving, or designated to give, this protection. { 'er ˌkəv·ər }

aircraft [AERO ENG] Any structure, machine, or contrivance, especially a vehicle, designed to be supported by the air, either by the dynamic action of the air upon the surfaces of the structure or object or by its own buoyancy. Also known as air vehicle. { 'er,kraft }

aircraft ammunition [ORD] Ammunition designed to be shot, launched, or dropped from aircraft. { 'er,kraft am·yə'nish·ən }

aircraft antenna [ELECTR] An airborne device used to detect or radiate electromagnetic waves. { 'er,kraft an'ten·ə }

aircraft axes See axes of an aircraft. { 'er,kraft 'ak,sēz }

aircraft bonding [AERO ENG] Electrically connecting together all of the metal structure of the aircraft, including the engine and metal covering on the wiring. { 'er,kraft 'bänd·iŋ }

aircraft cannon [ORD] A cannon designed or modified for use in an aircraft. Also known as aerial cannon. { 'er,kraft 'kan·ən }

aircraft carrier [NAV ARCH] A ship that carries aircraft, has a takeoff and landing deck, and is otherwise designed and equipped to serve as a base of operations for the aircraft. Also known as carrier. { 'er,kraft 'kar·ē·ər }

aircraft ceiling [METEOROL] After United States weather observing practice, the ceiling classification applied when the reported ceiling value has been determined by a pilot while in flight within 1.5 nautical miles (2.8 kilometers) of any runway of the airport. { 'er,kraft 'sēl·iŋ }

aircraft compass system [NAV] A wholly airborne direction-giving system consisting of means for sensing the earth's magnetic field, stabilizing means, and usually remote indicating equipment. { 'er,kraft 'käm·pəs ˌsis·təm }

aircraft control and warning system [ORD] A system established to control and report the movement of aircraft; consists of observation facilities (radar, passive electronic, visual, or other means), control centers, and necessary communication. { 'er,kraft kən'trōl an 'wórn·iŋ ˌsis·təm }

aircraft decibel rating [ELECTROMAG] The ratio of the radar reflectivity of a specific type of aircraft to that of a selected reference aircraft, measured in decibels. { 'er,kraft 'des·ə,bəl ˌrād·iŋ }

aircraft detection [ENG] The sensing and discovery of the presence of aircraft; major techniques include radar, acoustical, and optical methods. { 'er,kraft di'tek·shən }

aircraft dispersal area [ORD] An area on a military installation designed primarily for the dispersal of parked aircraft, to make such aircraft less vulnerable in the event of an enemy air raid. { 'er,kraft dis'pər·səl ˌer·ē·ə }

aircraft early-warning station [NAV] A station with long-range radar for detecting the approach of aircraft at a distance. { 'er,kraft 'ər·lē 'wórn·iŋ ˌstā·shən }

aircraft electrification [METEOROL] **1.** The accumulation of a net electric charge on the surface of an aircraft. **2.** The separation of electric charge into two concentrations of opposite sign on distinct portions of an aircraft surface. { 'er,kraft i,lek·trə·fə'kā·shən }

aircraft engine [AERO ENG] A component of an aircraft that develops either shaft horsepower or thrust and incorporates design features advantageous for aircraft propulsion. { 'er,kraft ˌen·jən }

aircraft fuel [MATER] The material used as a source of energy by an aircraft engine to provide propulsion. { 'er ˌkraft ˌfyül }

aircraft gun [ORD] A gun of any type mounted in a combat aircraft, such as a machine gun or cannon. { 'er ˌkraft ˌgən }

aircraft icing [METEOROL] The accumulation of ice on the exposed surfaces of aircraft when flying through supercooled water drops (cloud or precipitation). { 'er ˌkraft 'īs·iŋ }

aircraft impactor [ENG] An instrument carried by an aircraft for the purpose of obtaining samples of airborne particles. { 'er ˌkraft im'pak·tər }

aircraft instrumentation [AERO ENG] Electronic, gyroscopic, and other instruments for detecting, measuring, recording, telemetering, processing, or analyzing different values or quantities in the flight of an aircraft. { 'er ˌkraft ˌin·strə·mən'tā·shən }

aircraft instrument panel [AERO ENG] A coordinated instrument display arranged to provide the pilot and flight crew with information about the aircraft's speed, altitude, attitude, heading, and condition; also advises the pilot of the aircraft's response to his control efforts. { 'er ˌkraft 'in·strə·mənt ˌpan·əl }

aircraft low-approach system [NAV] Any of the various means for furnishing guidance in the vertical and horizontal planes to aircraft during descent from an initial approach altitude to a point near the ground. { 'er ˌkraft ¦lō ə'prōch ˌsis·təm }

aircraft machine-gun turret [ORD] An armored enclosure installed in an aircraft for housing the armament and related accessories; it is designed to rotate about one or more axes, thus permitting positioning and firing of a machine gun in a number of directions or angles. { 'er ˌkraft mə'shēn ˌgən ˌtə·rət }

aircraft noise [ACOUS] Effective sound output of the various sources of noise associated with aircraft operation, such as propeller and engine exhaust, jet noise, and sonic boom. { 'er ˌkraft ˌnóiz }

aircraft propeller [AERO ENG] A hub-and-multiblade device for transforming the rotational power of an aircraft engine into thrust power for the purpose of moving an aircraft through the air. { 'er ˌkraft prə·pel·ər }

aircraft propulsion [AERO ENG] The means, other than gliding, whereby an aircraft moves through the air; effected by the rearward acceleration of matter through the use of a jet engine or by the reactive thrust of air on a propeller. { 'er ˌkraft prə·pəl·shən }

aircraft pylon [AERO ENG] A suspension device externally installed under the wing or fuselage of an aircraft; it is aerodynamically designed to fit the configuration of specific aircraft so as to create the least amount of drag; it provides a means of attaching fuel tanks, bombs, rockets, torpedoes, rocket motors, or machine-gun pods. { 'er ˌkraft 'pī ˌlän }

aircraft report See pilot report. { 'er ˌkraft ri ˌpórt }

aircraft rocket [ORD] A rocket missile designed to be launched from an aircraft and employing warheads that are armor-piercing, incendiary, or explosive; it is carried beneath the wings of fighter or ground-attack airplanes. { 'er ˌkraft ˌräk·ət }

aircraft testing [AERO ENG] The subjecting of an aircraft or its components to simulated or actual flight conditions while measuring and recording pertinent physical phenomena that indicate operating characteristics. { 'er ˌkraft ˌtest·iŋ }

aircraft thermometry [METEOROL] The science of temperature measurement from aircraft. { 'er ˌkraft thər'mäm·ə·trē }

aircraft vectoring [NAV] The directional control of in-flight aircraft through transmission of bearing and altitude instructions from the ground. { 'er ˌkraft 'vek·tə·riŋ }

aircraft weather reconnaissance [METEOROL] The making of detailed weather observations or investigations from aircraft in flight. { 'er ˌkraft ˌweth·ər ri ˌkän·ə·səns }

air crossing [MIN ENG] A mine passage in which two airways cross each other. { 'er ˌkrós·iŋ }

air-cure [CHEM ENG] To vulcanize at ordinary room temperatures, or without the aid of heat. { 'er ˌkyür }

air current [FL MECH] Very generally, any moving stream of air. [GEOPHYS] See air-earth conduction current. [MIN ENG] The flow of air ventilating the workings of a mine. Also known as airflow. { 'er ˌkər·ənt }

air curtain [MECH ENG] A stream of high-velocity temperature-controlled air which is directed downward across an opening; it excludes insects, exterior drafts, and so forth, prevents the transfer of heat across it, and permits air-conditioning of a space with an open entrance. { 'er ˌkərt·ən }

air cushion [MECH ENG] A mechanical device using trapped air to arrest motion without shock. { 'er ˌkush·ən }

air-cushion vehicle [MECH ENG] A transportation device supported by low-pressure, low-velocity air capable of traveling equally well over water, ice, marsh, or relatively level land. Also known as ground-effect machine (GEM); hovercraft. { 'er ˌkush·ən ˌvē·ə·kəl }

air-cut [ENG] Referring to the inadvertent mechanical incorporation of air into a liquid system. { 'er ˌkət }

air cycle [MECH ENG] A refrigeration cycle characterized by the working fluid, air, remaining as a gas throughout the cycle rather than being condensed to a liquid; used primarily in airplane air conditioning. { 'er ˌsī·kəl }

air cylinder [MECH ENG] A cylinder in which air is compressed by a piston, compressed air is stored, or air drives a piston. { 'er ˌsil·ən·dər }

air defense [ORD] All measures designed to destroy, nullify, or reduce the effectiveness of enemy attack by aircraft or missiles after they are airborne. { 'er di ˌfens }

air-defense artillery [ORD] Weapons and equipment for actively combating air targets from the ground. Abbreviated ADA. { 'er di ˌfens ˌär'til·ə·rē }

air deficiency [CHEM] Insufficient air in an air-fuel mixture causing either incomplete fuel oxidation or lack of ignition. { 'er di ˌfish·ən·sē }

air delivery [ORD] A method of air movement wherein personnel, supplies, and equipment are unloaded from aircraft in flight. Also known as air drop. { 'er di ˌliv·ə·rē }

air density [MECH] The mass per unit volume of air. { 'er ˌden·səd·ē }

air-depolarized battery [ELEC] A primary battery which is kept depolarized by atmospheric oxygen rather than chemical compounds. Also known as metal-air battery. { 'er dē'pōl·ə ˌrīzd 'bad·ə·rē }

air diffuser [BUILD] An air distribution outlet, usually located in the ceiling and consisting of deflecting vanes discharging supply air in various directions and planes, and arranged to promote mixing of the supplied air with the air already in the room. { 'er di ˌfyüz·ər }

air discharge [GEOPHYS] **1.** A form of lightning discharge, intermediate in character between a cloud discharge and a cloud-to-ground discharge, in which the multibranching lightning channel descending from a cloud base does not reach the ground, but succeeds only in neutralizing the space charge distributed in the subcloud layer. **2.** A type of diffuse electrical discharge occasionally reported as occurring in the region above an active thunderstorm. { 'er 'dis ˌchärj }

air-distributing acoustical ceiling [BUILD] A suspended acoustical ceiling in which the board or tile is provided with small, evenly distributed mechanical perforations; designed to provide a desired flow of air from a pressurized plenum above. { 'er di'strib·yəd·iŋ ə'kü·sti·kəl 'sēl·iŋ }

air diving [ENG] A type of diving in which the diver's breathing medium is a normal atmospheric mixture of oxygen and nitrogen; limited to depths of 190 feet (58 meters). { 'er ˌdīv·iŋ }

air door [MIN ENG] A door placed in a mine roadway to prevent the passage of air. { 'er ˌdór }

air dose [NUCLEO] The total absorbed dose at a point in free air, resulting from both the radiation of a primary beam of x-rays or gamma rays and the radiation scattered by the surrounding air. { 'er ˌdōs }

air drain [CIV ENG] An empty space left around the external foundation wall of a building to prevent the earth from lying against it and causing dampness. { 'er ˌdrān }

air drainage [METEOROL] General term for gravity-induced, downslope flow of relatively cold air. { 'er ˌdrān·ij }

airdraulic [MECH ENG] Combining pneumatic and hydraulic action for operation. { ˌer'dról·ik }

air-dried lumber [MATER] Wood dried by exposure to air

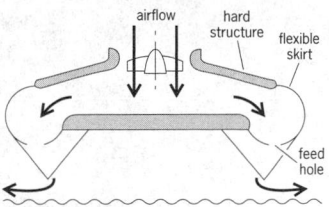

AIR-CUSHION VEHICLE

Basic construction of an air-cushion vehicle.

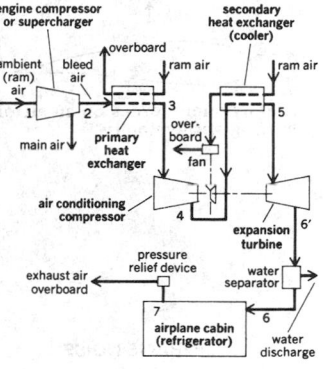

AIR CYCLE

Open air-cycle bootstrap system for airplanes.

under natural conditions; usually has a moisture content not greater than 24. Also known as air-seasoned lumber; natural-seasoned lumber. { ¦er ¦drīd 'ləm·bər }

air-dried strength [MET] Tenacity of a sand mixture after a core or mold has been dried in air without application of heat. { ¦er ¦drīd 'streŋkth }

air drift [MIN ENG] A roadway, generally inclined, driven in stone for ventilation purposes. { ¦er ˌdrift }

air drill [MECH ENG] A drill powered by compressed air. { ¦er ˌdril }

air drilling [PETRO ENG] A form of rotary drilling in which compressed air is used as the circulation medium. { ¦er ˌdril·iŋ }

air drop *See* air delivery. { ¦er ˌdräp }

air drying [ENG] Removing moisture from a material by exposure to air to the extent that no further moisture is released on contact with air; important in lumber manufacture. { ¦er 'drī·iŋ }

air duct *See* airflow duct. { ¦er ˌdəkt }

air-earth conduction current [GEOPHYS] That part of the air-earth current contributed by the electrical conduction of the atmosphere itself; represented as a downward movement of positive space charge in storm-free regions all over the world. Also known as air current. { ¦er ¦ərth kən'dək·shən ˌkər·ənt }

air-earth current [GEOPHYS] The transfer of electric charge from the positively charged atmosphere to the negatively charged earth; made up of the air-earth conduction current, a precipitation current, a convection current, and miscellaneous smaller contributions. { ¦er ¦ərth 'kər·ənt }

air ejector [MECH ENG] A device that uses a fluid jet to remove air or other gases, as from a steam condenser. { ¦er i'jek·tər }

air eliminator [MECH ENG] In a piping system, a device used to remove air from water, steam, or refrigerant. { ¦er i'lim·ə,nād·ər }

air elutriator *See* air classifier. { ¦er ē'lü·trē,ād·ər }

air embolism *See* aeroembolism. { ¦er 'em·bə,liz·əm }

air endway [MIN ENG] A narrow roadway driven in a coal seam parallel and close to a winning headway for ventilation. { ¦er 'end,wā }

air engine [MECH ENG] An engine in which compressed air is the actuating fluid. { ¦er 'en·jən }

air-entrained cement [MATER] Cement with improved qualities due to the introduction of air bubbles in its preparation. { ¦er in'trānd sə'ment }

air-entraining agent [MATER] An admixture, usually a resin, soap, or grease, for portland cement or concrete to effect air entrainment and, thus, superior properties. { ¦er in¦trān·iŋ 'ā·jənt }

air entrainment [ENG] The inclusion of minute bubbles of air in cement or concrete through the addition of some material during grinding or mixing to reduce the surface tension of the water, giving improved properties for the end product. { ¦er in'trān·mənt }

air equivalent [NUCLEO] A measure of the effectiveness of an absorber of nuclear radiation, equal to the thickness of a layer of air at standard pressure and temperature that absorbs the same fraction of radiation or results in the same energy loss as does the absorber. { ¦er i'kwiv·ə·lənt }

air escape [DES ENG] A device that is fitted to a pipe carrying a liquid for releasing excess air; it contains a valve that controls air release while preventing loss of liquid. { ¦er ə,skāp }

air-exhaust ventilator [MECH ENG] Any air-exhaust unit used to carry away dirt particles, odors, or fumes. { ¦er ig'zöst 'ven·tə,lād·ər }

air feed [MET] In thermal spraying, transmittal of powdered material by air pressure through the gun into the heat source. { ¦er ˌfēd }

airfield [CIV ENG] The area of an airport for the takeoff and landing of airplanes. { 'er,fēld }

air filter [ENG] A device that reduces the concentration of solid particles in an airstream to a level that can be tolerated in a process or space occupancy; a component of most systems in which air is used for industrial processes, ventilation, or comfort air conditioning. { 'er ,fil·tər }

airfloat clay [MIN ENG] Fine particles of clay obtained by air separation from coarser particles following a grinding operation. { 'er,flōt 'klā }

air flotation *See* dissolved air flotation. { 'er flō'tā·shən }

airflow [FL MECH] **1.** A flow or stream of air which may take lace in a wind tunnel or, as a relative airflow, past the wing or other parts of a moving craft. Also known as airstream. **2.** A rate of flow, measured by mass or volume per unit of time. [MIN ENG] *See* air current. { 'er,flō }

airflow duct [ENG] A pipe, tube, or channel through which air moves into or out of an enclosed space. Also known as air duct. { 'er,flō ,dəkt }

airflow orifice [ENG] An opening through which air moves out of an enclosed space. { 'er,flō 'ör·ə·fəs }

airflow pipe [ENG] A tube through which air is conveyed from one location to another. { 'er,flō ,pīp }

airflow stack effect [FL MECH] The variation of pressure with height in air flowing in a vertical duct due to a difference in temperature between the flowing air and the air outside the duct. { 'er,flō 'stak i,fekt }

airfoil [AERO ENG] A body of such shape that the force exerted on it by its motion through a fluid has a larger component normal to the direction of motion than along the direction of motion; examples are the wing of an airplane and the blade of a propeller. Also known as aerofoil. { 'er,föil }

airfoil profile [AERO ENG] The closed curve defining the external shape of the cross section of an airfoil. Also known as airfoil section; airfoil shape; wing section. { 'er,föil 'prō,fīl }

airfoil section *See* airfoil profile. { 'er,föil ,sek·shən }

airfoil shape *See* airfoil profile. { 'er,föil ,shāp }

airfoil-vane fan [MIN ENG] A centrifugal-type mine fan; the vanes are curved backward from the direction of rotation. { 'er,föil 'vān 'fan }

airframe [AERO ENG] The basic assembled structure of any aircraft or rocket vehicle, except lighter-than-air craft, necessary to support aerodynamic forces and inertia loads imposed by the weight of the vehicle and contents. { 'er,frām }

airframe noise [ACOUS] Noise generated by the nonpropulsive components of an aircraft. { 'er,frām ,nöiz }

air-fuel mixture [MECH ENG] In a carbureted gasoline engine, the charge of air and fuel that is mixed in the appropriate ratio in the carburetor and subsequently fed into the combustion chamber. { 'er 'fyül ,miks·chər }

air-fuel ratio [CHEM] The ratio of air to fuel by weight or volume which is significant for proper oxidative combustion of the fuel. { 'er 'fyül ,rā·shō }

air furnace [MET] **1.** A furnace using a natural air draft. **2.** A furnace in which the metal is melted by a flame originating from fuel burned at one end, passing over the hearth in the middle, and exiting at the other end. { 'er ,fər·nəs }

air gage [ENG] **1.** A device that measures air pressure. **2.** A device that compares the shape of a machined surface to that of a reference surface by measuring the rate of passage of air between the surfaces. { 'er ,gāj }

air gap [ELECTR] **1.** A gap or an equivalent filler of nonmagnetic material across the core of a choke, transformer, or other magnetic device. **2.** A spark gap consisting of two electrodes separated by air. **3.** The space between the stator and rotor in a motor or generator. [ENG] **1.** The distance between two components or parts. **2.** In plastic extrusion coating, the distance from the opening of the extrusion die to the nip formed by the pressure and chill rolls. **3.** The unobstructed vertical distance between the lowest opening of a faucet (or the like) which supplies a plumbing fixture (such as a tank or washbowl) and the level at which the fixture will overflow. [GEOL] *See* wind gap. [PETRO ENG] In an offshore drilling operation, the distance from the normal sea surface level to the bottom of the base of the drilling platform. { 'er ,gap }

air gas [MATER] A gaseous fuel made by blowing air through a coal or coke bed so that CO_2 is reduced to CO. { 'er ,gas }

airglow [GEOPHYS] The quasi-steady radiant emission from the upper atmosphere over middle and low latitudes, as distinguished from the sporadic emission of auroras which occur over high latitudes. Also known as light-of-the-night-sky; night-sky light; night-sky luminescence; permanent aurora. { 'er,glō }

air grating [BUILD] A fixed metal grille on the exterior of a building through which air is brought into or discharged from the building for purposes of ventilation. { 'er ,grād·iŋ }

air-ground communication [COMMUN] Two-way communication between aircraft and stations on the ground. { ¦er ¦graund kə,myü·nə'kā·shən }

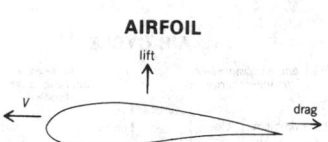

AIRFOIL

Aerodynamic forces on an airfoil moving with velocity *V*.

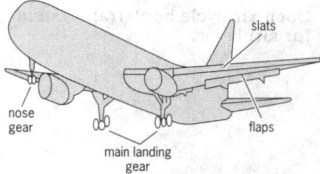

AIRFRAME NOISE

Sources of airframe noise.

air gun *See* air rifle. { 'er ‚gən }

air hammer *See* pneumatic hammer. { 'er ‚ham·ər }

air-handling system [MECH ENG] An air-conditioning system in which an air-handling unit provides part of the treatment of the air. { 'er ¦hand·liŋ ‚sis·təm }

air-handling unit [MECH ENG] A packaged assembly of air-conditioning components (coils, filters, fan humidifier, and so forth) which provides for the treatment of air before it is distributed. { 'er ¦hand·liŋ ‚yü·nət }

air-hardening steel [MET] A steel whose content of carbon and other alloying elements is sufficient for the steel to harden fully by cooling in air or any other atmosphere from a temperature above its transformation range. Also known as self-hardening steel. { 'er ¦härd·niŋ ‚stēl }

airhead [ORD] A designated geographical area in an area of operations used as a base of supply and evacuation by air. { 'er‚hed }

air heater *See* air preheater. { 'er ‚hēd·ər }

air-heating system *See* air preheater. { 'er ‚hēd·iŋ ‚sis·təm }

air heave [GEOL] Deformation of plastic sediments on a tidal flat as a result of the growth of air pockets in them; the growth occurs by accretion of smaller air bubbles oozing through the sediment. { 'er ‚hēv }

air hoar [HYD] A hoar growing on objects above the ground or snow. { 'er ‚hór }

air hoist [MECH ENG] A lifting tackle or tugger constructed with cylinders and pistons for reciprocating motion and air motors for rotary motion, all powered by compressed air. Also known as pneumatic hoist. { 'er ‚hóist }

airhole [MIN ENG] A small excavation or hole made to improve ventilation by communication with other workings or with the surface. { 'er‚hōl }

air horn [MECH ENG] In an automotive engine, the upper portion of the carburetor barrel through which entering air passes in quantities controlled by the choke plate and the throttle plate. { 'er ‚hórn }

air horsepower [MECH ENG] The theoretical (minimum) power required to deliver the specified quantity of air under the specified pressure conditions in a fan, blower, compressor, or vacuum pump. Abbreviated air hp. { 'er ‚hórs‚pau̇·ər }

air hp *See* air horsepower.

air hunger [MED] The deep, gasping respiration characteristic of severe diabetic acidosis and coma. { 'er ‚həŋ·gər }

air-injection reactor [MECH ENG] A unit installed in an automotive engine which mixes fresh air with hot exhaust gases in the exhaust manifold to react with any gasoline that has escaped unburned from the cylinders. Abbreviated AIR. { 'er in¦jek·shən rē'ak·tər }

air-injection system [MECH ENG] A device that uses compressed air to inject the fuel into the cylinder of an internal combustion engine. Also known as thermactor. { 'er in¦jek·shən ‚sis·təm }

air inlet [MECH ENG] In an air-conditioning system, a device through which air is exhausted from a room or building. { 'er ¦in‚let }

air-inlet valve [MECH ENG] In a heating/air-conditioning system of a motor vehicle, a valve in the plenum blower assembly that permits selection of either inside or outside air. { 'er ¦in‚let ‚valv }

air-insulated substation [ELEC] An electric power substation that has the busbars and equipment terminations generally open to air and utilizes insulation properties of ambient air for insulation to ground. { 'er 'in·sə'lād·əd 'səb‚stā·shən }

air intake [AERO ENG] An open end of an air duct or similar projecting structure so that the motion of the aircraft is utilized in capturing air to be conducted to an engine or ventilator. [MIN ENG] A device for supplying a compressor with clean air at the lowest possible temperature. { 'er ¦in‚tāk }

air-jet process [TEXT] A technique by which filament yarns are exposed to a forceful stream of air in order to increase their bulk. { 'er ¦jet ‚präs·əs }

air knife [ENG] A device that uses a thin, flat jet of air to remove the excess coating from freshly coated paper. [MET] A device that employs a high-velocity airstream to separate metallic particles with different densities by shifting the trajectory of lighter particles more than that of heavier particles. { 'er ‚nīf }

air-knife coating [ENG] An even film of coating left on paper after treatment with an air knife. { 'er ¦nīf ‚kōd·iŋ }

air-lance [ENG] To direct a pressurized-air stream to remove unwanted accumulations, as in boiler-wall cleaning. { 'er ‚lans }

air launch [AERO ENG] Launching from an aircraft in the air. { 'er ‚lónch }

air-launched ballistic missile [ORD] A ballistic missile launched from an airborne vehicle. { 'er ‚lóncht bə'lis·tik 'mis·əl }

air layering [BOT] A method of vegetative propagation, usually of a wounded part, in which the branch or shoot is enclosed in a moist medium until roots develop, and then it is severed and cultivated as an independent plant. { 'er ‚lā·ə·riŋ }

air leakage [MECH ENG] **1.** In ductwork, air which escapes from a joint, coupling, and such. **2.** The undesired leakage or uncontrolled passage of air from a ventilation system. { 'er ‚lēk·əj }

airless spraying [ENG] The spraying of paint by means of high fluid pressure and special equipment. Also known as hydraulic spraying. { 'er·ləs 'sprā·iŋ }

airlift [AERO ENG] **1.** To transport passengers and cargo by the use of aircraft. **2.** The total weight of personnel or cargo carried by air. { 'er‚lift }

air lift [MECH ENG] **1.** Equipment for lifting slurry or dry powder through pipes by means of compressed air. **2.** *See* air-lift pump. { 'er¦lift }

air-lift hammer [MECH ENG] A gravity drop hammer used in closed die forging in which the ram is raised to its starting point by means of an air cylinder. { 'er‚lift 'ham·ər }

air-lift pump [MECH ENG] A device composed of two pipes, one inside the other, used to extract water from a well; the lower end of the pipes is submerged, and air is delivered through the inner pipe to form a mixture of air and water which rises in the outer pipe above the water in the well; also used to move corrosive liquids, mill tailings, and sand. Also known as air lift. { 'er‚lift 'pəmp }

airlight [METEOROL] In determinations of visual range, light from sun and sky which is scattered into the eyes of an observer by atmospheric suspensoids (and, to slight extent, by air molecules) lying in the observer's cone of vision. { 'er‚līt }

airlight formula [OPTICS] A fundamental equation of visual-range theory, relating the apparent luminance of a distant black object, the apparent luminance of the background sky above the horizon, and the extinction coefficient of the air layer near the ground. { 'er‚līt 'fór·myə·lə }

air line [ENG] A fault, in the form of an elongated bubble, in glass tubing. Also known as hairline. [MECH ENG] A duct, hose, or pipe that supplies compressed air to a pneumatic tool or piece of equipment. [SPECT] Lines in a spectrum due to the excitation of air molecules by spark discharges, and not ordinarily present in arc discharges. { 'er ‚līn }

air-line lubricator *See* line oiler. { 'er ‚līn 'lü·brə‚kād·ər }

air-line main [MIN ENG] The pipe column supplying air from the compressors to the quarry face. { 'er ‚līn ‚mān }

air lock [ENG] **1.** A chamber capable of being hermetically sealed that provides for passage between two places of different pressure, such as between an altitude chamber and the outside atmosphere, or between the outside atmosphere and the work area in a tunnel or shaft being excavated through soil subjected to water pressure higher than atmospheric. Also known as lock. **2.** An air bubble in a pipeline which impedes liquid flow. **3.** A depression on the surface of a molded plastic part that results from air trapped between the surface of the mold and the plastic. [MIN ENG] A casing atop an upcast mine shaft to minimize surface air leakage into the fan. { 'er ‚läk }

air-lock strip [BUILD] The weather stripping which is fastened to the edges of each wing of a revolving door. { 'er ‚läk ‚strip }

air log [AERO ENG] A distance-measuring device used especially in certain guided missiles to control range. { 'er ‚läg }

airman [MIN ENG] A worker who constructs brattices. [ORD] An enlisted man in the military air force. { 'er·mən }

air mass [METEOROL] An extensive body of the atmosphere which approximates horizontal homogeneity in its weather characteristics, particularly with reference to temperature and moisture distribution. { 'er ‚mas }

air-mass analysis [METEOROL] In general, the theory and practice of synoptic surface-chart analysis by the so-called Norwegian methods, which involve the concepts of the polar

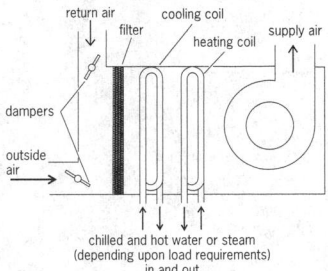

AIR-HANDLING UNIT

return air / filter / cooling coil / heating coil / supply air / dampers / outside air

chilled and hot water or steam (depending upon load requirements) in and out

Schematic of central air-handling unit.

front and of the broad-scale air masses which it separates. { 'er ˌmas ə'nal·ə·səs }

air-mass climatology [CLIMATOL] The representation of the climate of a region by the frequency and characteristics of the air masses under which it lies; basically, a type of synoptic climatology. { 'er ˌmas klīm·ə'täl·ə·jē }

air-mass precipitation [METEOROL] Any precipitation that can be attributed only to moisture and temperature distribution within an air mass when that air mass is not, at that location, being influenced by a front or by orographic lifting. { 'er ˌmas pri,sip·ə'tā·shən }

air-mass shower [METEOROL] A shower that is produced by local convection within an unstable air mass; the most common type of air-mass precipitation. { 'er ˌmas 'shaů·ər }

air-mass source region [METEOROL] An extensive area of the earth's surface over which bodies of air frequently remain for a sufficient time to acquire characteristic temperature and moisture properties imparted by that surface. { 'er ˌmas 'sórs ˌrē·jən }

air-measuring station [MIN ENG] A place in a mine airway where the volume of air passing is measured periodically. { 'er ˌmezh·ər·iŋ ˌstā·shən }

air meter [ENG] A device that measures the flow of air, or gas, expressed in volumetric or weight units per unit time. Also known as airometer. { 'er ˌmēd·ər }

air mile [NAV] A unit of length used in air navigation and equal, since 1954, to 1 international nautical mile (1852 meters). Also known as aeronautical mile. { 'er ˌmīl }

air mileage [NAV] The number of miles flown relative to the air; true air speed multiplied by time. { 'er ˌmī·lij }

air mileage indicator [ENG] An instrument on an airplane which continuously indicates mileage through the air. { ¦er ˌmī·lij 'in·də'kād·ər }

air mileage unit [ENG] A device which derives continuously and automatically the air distance flown, and feeds this information into other units, such as an air mileage indicator. { 'er ˌmī·lij ˌyü·nət }

air-mixing plenum [MECH ENG] In an air-conditioning system, a chamber in which the recirculating air is mixed with air from outdoors. { 'er ˌmiks·iŋ 'plēn·əm }

air monitoring [CIV ENG] A practice of continuous air sampling by various levels of government or particular industries. { 'er ˌmän·ə·triŋ }

air motor [MECH ENG] A device in which the pressure of confined air causes the rotation of a rotor or the movement of a piston. { 'er ˌmōd·ər }

air mover [MIN ENG] A portable compressed-air appliance, used as a blower or exhauster. { 'er ˌmüv·ər }

air navigation [NAV] The process of directing and monitoring the progress of an aircraft between selected geographic points or with respect to some predetermined plan. Also known as avigation. { ¦er ˌnav·ə'gā·shən }

air nozzle [MECH ENG] In an automotive engine, a device for supplying air to the air-injection reactor. { 'er ˌnäz·əl }

airometer [ENG] **1.** An apparatus for both holding air and measuring the quantity of air admitted into it. **2.** See air meter. { ˌer'ä·məd·ər }

air outlet [MECH ENG] In an air-conditioning system, a device at the end of a duct through which air is supplied to a space. { 'er ˌaůt·lət }

air parcel [METEOROL] An imaginary body of air to which may be assigned any or all of the basic dynamic and thermodynamic properties of atmospheric air. { 'er ˌpär·səl }

air permeability [TEXT] The quality of porosity of a fabric; indicates the wind resistance, and therefore warmth or coolness, of the fabric. { ¦er ˌpər·mē·ə'bil·ə·dē }

air-permeability test [ENG] A test for the measurement of the fineness of powdered materials, such as portland cement. { ¦er ˌpər·mē·ə'bil·ə·dē ˌtest }

air pickets [ORD] Airborne early-warning aircraft disposed around a position, area, or formation primarily to detect, report, and track approaching enemy aircraft and to control intercepts. { 'er ˌpik·əts }

airplane [AERO ENG] A heavier-than-air vehicle designed to use the pressures created by its motion through the air to lift and transport useful loads. { 'er,plān }

airplane flare [ENG] A flare, often magnesium, that is dropped from an airplane to illuminate a ground area; a small parachute decreases the rate of descent. { 'er,plān ˌfler }

air plot [NAV] A continuous plot of the position of an airborne object represented graphically to show true headings steered and distances flown. { 'er ˌplät }

air-plot wind velocity [NAV] The wind velocity calculated from a knowledge of an air position and a fix. { ¦er ˌplät ¦wind və,läs·ə·dē }

air pocket [ENG] An air-filled space that is normally occupied by a liquid. Also known as air trap. [METEOROL] An expression used in the early days of aviation for a downdraft; such downdrafts were thought to be pockets in which there was insufficient air to support the plane. { 'er ˌpäk·ət }

air pollution [ECOL] The presence in the outdoor atmosphere of one or more contaminants such as dust, fumes, gas, mist, odor, smoke, or vapor in quantities and of characteristics and duration such as to be injurious to human, plant, or animal life or to property, or to interfere unreasonably with the comfortable enjoyment of life and property. { ¦er pə'lü·shən }

air-pollution control [ENG] A practical means of treating polluting sources to maintain a desired degree of air cleanliness. { ¦er pə'lü·shən kən,trōl }

airport [CIV ENG] A terminal facility used for aircraft takeoff and landing and including facilities for handling passengers and cargo and for servicing aircraft. Also known as aerodrome. { 'er,pórt }

airport engineering [CIV ENG] The planning, design, construction, and operation and maintenance of facilities providing for the landing and takeoff, loading and unloading, servicing, maintenance, and storage of aircraft. { 'er,pórt en·jə'nir·iŋ }

airport light beacon [NAV] A visual navigational aid, usually a rotating beacon displaying flashes of white or colored light or both, located at or near an airport to indicate its specific or general location. { 'er,pórt 'līt ˌbē·kən }

airport surface detection equipment [NAV] A short-range radar using millimeter waves and giving a panoramic presentation of all aircraft and vehicles, moving or stationary, on the surface of an aerodrome; used by air traffic controllers for expeditious movement of surface aircraft on the ramp, taxiway, and runway. { 'er,pórt 'sər·fəs di'tek·shən i,kwip·mənt }

airport traffic [NAV] Aircraft both in the air and on the active runways in the airport zone. { 'er,pórt ˌtraf·ik }

airport traffic control tower [NAV] The terminal control point at an airport for all takeoff and landing operations, departure and approach operations, and ground movements of aircraft and airport vehicles. Abbreviated ATCT. { 'er,pórt 'traf·ik kən,trōl ˌtaů·ər }

air position [NAV] The theoretical position of an aircraft or missile at a given moment, assuming it to have been unaffected in flight by wind. Also known as no-wind position. { 'er pə'zish·ən }

air-position indicator [NAV] An airborne computing system which presents a continuous indication of the aircraft position on the basis of aircraft heading, airspeed, and elapsed time. Abbreviated API. { 'er pə'zish·ən ˌin·də'kād·ər }

air preheater [MECH ENG] A device used in steam boilers to transfer heat from the flue gases to the combustion air before the latter enters the furnace. Also known as air heater; air-heating system. { 'er ˌprē'hēd·ər }

air pressure [PHYS] The force per unit area that the air exerts on any surface in contact with it, arising from the collisions of the air molecules with the surface. { 'er ˌpresh·ər }

air-pressure drop [FL MECH] The pressure lost in overcoming friction along an airway. { 'er ˌpresh·ər ˌdräp }

airproof See airtight. { 'er,prüf }

air propeller [AERO ENG] A hub-and-multiblade device for changing rotational power of an aircraft engine into thrust power for the purpose of propelling an aircraft through the air. [MECH ENG] A rotating fan for moving air. { 'er prə,pel·ər }

air properties [PHYS] Characteristics of air as a gas, such as density, molecular weight, specific heats, boiling point, critical temperature, and critical pressure. { 'er ˌprä·pər,tēz }

air pump [MECH ENG] A device for removing air from an enclosed space or for adding air to an enclosed space. { 'er ˌpəmp }

Air Pump See Antlia. { 'er ˌpəmp }

air puncher [ENG] A machine consisting essentially of a reciprocating chisel or pick, driven by air. { 'er ˌpən·chər }

air purge [MECH ENG] Removal of particulate matter from air within an enclosed vessel by means of air displacement. { 'er ˌpərj }

air quench [MET] In a heat-treatment process, the rapid cooling of a metal by a blast of cold air. { 'er ˌkwench }

air raid [ORD] An attack of a place by aircraft; may consist of a bombing or strafing attack against a position or a city. { 'er ˌrād }

air-raid shelter [CIV ENG] A chamber, often underground, provided with living facilities and food, for sheltering people against air attacks. { ¦er ˌrād ¦shel·tər }

air receiver [MECH ENG] A vessel designed for compressed-air installations that is used both to store the compressed air and to permit pressure to be equalized in the system. { 'er ri,sē·vər }

air register [ENG] A device attached to an air-distributing duct for the purpose of controlling the discharge of air into the space to be heated, cooled, or ventilated. { 'er ¦rej·ə·stər }

air regulator [MECH ENG] A device for regulating airflow, as in the burner of a furnace. [MIN ENG] An adjustable door installed in permanent air stoppings to control ventilating current. { ¦er ¦reg·yə,lād·ər }

air reheater [MECH ENG] In a heating system, any device used to add heat to the air circulating in the system. { ¦er ˌrē¦hēd·ər }

air release valve [MECH ENG] A valve, usually manually operated, which is used to release air from a water pipe or fitting. { 'er ri¦lēs ,valv }

air resistance [MECH] Wind drag giving rise to forces and wear on buildings and other structures. { 'er ri'zis·təns }

air rifle [ORD] A low-powered rifle that shoots small metal pellets (BBs) by the action of compressed air. Also known as air gun; BB gun. { 'er ˌrīf·əl }

air ring [ENG] In plastics forming, a circular manifold which distributes an even flow of cool air into a hollow tubular form passing through the manifold. { 'er ˌriŋ }

air route [NAV] The navigable airspace between two points, identified to the extent necessary for the application of flight rules. { 'er ˌrüt }

air-route surveillance radar [NAV] A long-range (approximately 150-mile or 240-kilometer) radar used by the Federal Aviation Agency to control air traffic between terminals. Abbreviated ARSR. { ¦er ˌrüt sər¦vā·ləns ˌrā,där }

air-route traffic control [NAV] Traffic control service provided to aircraft which are operating under instrument flight rules within controlled airspace, and principally en route. { 'er ˌrüt 'traf·ik kən,trōl }

air-route traffic control center [NAV] A place from which traffic is directed along a controlled air route. { 'er ˌrüt ¦traf·ik kən'trōl ,sen·tər }

air sac [GEOL] See vesicle. [INV ZOO] One of the large, thin-walled structures associated with the tracheal system of some insects. [VERT ZOO] In birds, any of the small vesicles that are connected with the respiratory system and located in bones and muscles to increase buoyancy. { 'er ,sak }

air sampling [ENG] The collection and analysis of samples of air to measure the amounts of various pollutants or other substances in the air, or the air's radioactivity. { 'er ,sam·pliŋ }

air scoop [DES ENG] An air-duct cowl projecting from the outer surface of an aircraft or automobile, which is designed to utilize the dynamic pressure of the airstream to maintain a flow of air. { 'er ,küp }

air screw [MECH ENG] A screw propeller that operates in air. { 'er ,skrü }

air-seasoned [ENG] Treated by exposure to air to give a desired quality. { 'er ,sēz·ənd }

air-seasoned lumber See air-dried lumber. { 'er ¦sēz·ənd 'ləm·bər }

air-sensitive crystal [CHEM] A crystal that decomposes when exposed to air. { 'er ,sen·səd·iv 'krist·əl }

air separator [MECH ENG] A device that uses an air current to separate a material from another of greater density or particles from others of greater size. { 'er ,sep·ə,rād·ər }

air-setting mortar [MATER] Mortar that sets in air at atmospheric pressure and ordinary temperatures. { ¦er ,sed·iŋ ¦mord·ər }

air sextant [NAV] A sextant designed primarily for air navigation; generally provided with some mechanical devices for furnishing horizon indications. { 'er ,seks·tənt }

air shaft [BUILD] An open space surrounded by the walls of a building or buildings to provide ventilation for windows.

Also known as air well. [MIN ENG] A usually vertical earth bore or shaft to supply surface air to an underground facility such as a mine. { 'er ,shaft }

airshed [GEOGR] The geographical area associated with a given air supply. [METEOROL] The air supply in a given region. { 'er,shed }

airship [AERO ENG] A propelled and steered aerial vehicle, dependent on gases for flotation. { 'er,ship }

air shooting [GEOPHYS] In seismic prospecting, the technique of applying a seismic pulse to the earth by detonating a charge or charges in the air. { 'er ,shüd·iŋ }

air shot [ENG] A shot prepared by loading (charging) so that an air space is left in contact with the explosive for the purpose of lessening its shattering effect. { 'er ,shät }

air shower See cosmic-ray shower. { 'er ,shaù·ər }

airsickness [MED] Motion sickness associated with flying due to the effects of acceleration. { 'er,sik·nəs }

air-slaked [CHEM] Having the property of a substance, such as lime, that has been at least partially converted to a carbonate by exposure to air. { 'er ,slākt }

Airslide conveyor [MECH ENG] An air-activated gravity-type conveyor, of the Fuller Company, using low-pressure air to aerate or fluidize pulverized material to a degree which will permit it to flow on a slight incline by the force of gravity. { ¦er,slīd kən¦vā·ər }

air sounding [METEOROL] The act of measuring atmospheric phenomena or determining atmospheric conditions at altitude, especially by means of apparatus carried by balloons or rockets. { 'er ,saùnd·iŋ }

air space [ENG] An enclosed space containing air in a wall for thermal insulation. [ORD] The space in a firearm between the powder charge and projectile. { ¦er ¦spās }

airspace [AERO ENG] **1.** The space occupied by an aircraft formation or used in a maneuver. **2.** The area around an airplane in flight that is considered an integral part of the plane in order to prevent collision with another plane; the space depends on the speed of the plane. [METEOROL] **1.** Of or pertaining to both the earth's atmosphere and space. Also known as aerospace. **2.** The portion of the atmosphere above a particular land area, especially a nation or other political subdivision. { 'er,spās }

air-spaced coax [ELECTROMAG] Coaxial cable in which air is basically the dielectric material; the conductor may be centered by means of a spirally wound synthetic filament, beads, or braided filaments. { 'er ,spāst 'kō,aks }

airspace reservation [NAV] An airspace designated by government authority in which flight is prohibited or restricted. { 'er,spās ,rez·ər'vā·shən }

airspace warning area [NAV] An airspace which is outside the territorial limits of any country and in which exists an actual or potential hazard to aircraft in flight; such an area has no legal status. { ¦er,spās ¦worn·iŋ ,er·ē·ə }

airspeed [AERO ENG] The speed of an airborne object relative to the atmosphere; in a calm atmosphere, airspeed equals ground speed. { 'er,spēd }

airspeed head [ENG] Any instrument or device, usually a pitot tube, mounted on an aircraft for receiving the static and dynamic pressures of the air used by the airspeed indicator. { 'er,spēd ,hed }

airspeed indicator [ENG] A device that computes and displays the speed of an aircraft relative to the air mass in which the aircraft is flying. { 'er,spēd ,in·də,kād·ər }

air spora [BIOL] Airborne fungus spores, pollen grains, and microorganisms. { 'er ,spór·ə }

air spring [MECH ENG] A spring in which the energy storage element is air confined in a container that includes an elastomeric bellows or diaphragm. { 'er ,spriŋ }

air stack [AERO ENG] A group of planes flying at prescribed heights while waiting to land at an airport. [MIN ENG] A chimney for ventilating a coal mine. { 'er ,stak }

air-standard cycle [THERMO] A thermodynamic cycle in which the working fluid is considered to be a perfect gas with such properties of air as a volume of 12.4 cubic feet per pound at 14.7 pounds per square inch (approximately 0.7756 cubic meter per kilogram at 101.36 kilopascals) and 492°R and a ratio of specific heats of 1:4. { ¦er ¦stan·dərd 'sī·kəl }

air-standard engine [MECH ENG] A heat engine operated in an air-standard cycle. { ¦er ¦stan·dərd 'en·jən }

AIRSHIP

(a)

(b)

Two types of airships, (a) rigid and (b) nonrigid.

airstart [AERO ENG] An act or instance of starting an aircraft's engine while in flight, especially a jet engine after flameout. { 'er‚stärt }

air starting valve [MECH ENG] A device that admits compressed air to an air starter. { 'er ¦stärd·iŋ ‚valv }

air station *See* camera station. { 'er ‚stā·shən }

airstream *See* airflow. { 'er‚strēm }

air strike [ORD] An attack on specific objectives by fighter, bomber, or attack air craft on an offensive mission; may consist of several air organizations under a single command in the air. { 'er ‚strīk }

air strip *See* landing strip. { 'er ‚strip }

air stripping [CHEM ENG] The process of bubbling air through water to remove volatile organic substances from the water. { 'er ‚strip·iŋ }

air-supply mask [ENG] *See* air-tube breathing apparatus. { 'er sə¦plī ‚mask }

air surveillance [ENG] Systematic observation of the airspace by visual, electronic, or other means, primarily for identifying all aircraft in that airspace, and determining their movements. { 'er sər'vā·ləns }

air surveillance radar [ENG] Radar of moderate range providing position of aircraft by azimuth and range data without elevation data; used for air-traffic control. { 'er sər¦vā·ləns ¦ra‚där }

air survey *See* aerial survey. { 'er ¦sər‚vā }

air-suspension encapsulation [CHEM ENG] A technique for microencapsulation of various types of solid particles; the particles undergo a series of cycles in which they are first suspended by a vertical current of air while they are sprayed with a solution of coating material, and are then moved by the airstream into a region where they undergo a drying treatment. Also known as Wurster process. { 'ər sə¦spen·shən in‚kap·sə'lā·shən }

air-suspension system [MECH ENG] Parts of an automotive vehicle that are intermediate between the wheels and the frame, and support the car body and frame by means of a cushion of air to absorb road shock caused by passage of the wheels over irregularities. { 'er sə¦spen·shən ¦sis·təm }

air sweetening [CHEM ENG] A process in which air or oxygen is used to oxidize lead mercaptides to disulfides instead of using elemental sulfur. { 'er ‚swēt·ən·iŋ }

air switch [ELEC] A switch in which the breaking of the electric circuit takes place in air. Also known as air-break switch. { 'er ‚swich }

air system [MECH ENG] A mechanical refrigeration system in which air serves as the refrigerant in a cycle comprising compressor, heat exchanger, expander, and refrigerating core. { 'er ‚sis·təm }

air taxi [AERO ENG] A carrier of passengers and cargo engaged in charter flights, feeder air services to large airline facilities, or contract airmail transportation. { 'er ‚tak·sē }

air temperature [METEOROL] **1.** The temperature of the atmosphere which represents the average kinetic energy of the molecular motion in a small region and is defined in terms of a standard or calibrated thermometer in thermal equilibrium with the air. **2.** The temperature that the air outside of the aircraft is assumed to have as indicated on a cockpit instrument. { 'er ¦tem·prə·chər }

air-temperature correction [NAV] **1.** A correction to a sextant altitude reading, required because of the air temperature. **2.** A correction required in the airspeed as read by a pitot tube to arrive at true airspeed. { 'er ¦tem·prə·chər kə'rek·shən }

air terminal [CIV ENG] A facility providing a place of assembly and amenities for airline passengers and space for administrative functions. [ELEC] A structure, such as a tower, that serves as a lightning arrester. { 'er ‚tərm·ən·əl }

air thermometer [ENG] A device that measures the temperature of an enclosed space by means of variations in the pressure or volume of air contained in a bulb placed in the space. { 'er thə¦mäm·əd·ər }

airtight [ENG] Not permitting the passage of air. Also known as airproof. { 'er‚tīt }

air-to-air missile [ORD] A missile launched from an aircraft at an airborne target. Abbreviated AAM. { ¦er tü ¦er 'mis·əl }

air-to-air resistance [CIV ENG] The resistance provided by the wall of a building to the flow of heat. { ¦er tü ¦er ri'sis·təns }

air-to-ground missile [ORD] A missile launched from an aircraft at a ground target. Abbreviated AGM. { ¦er tü ¦graùnd 'mis·əl }

air-to-space missile [ORD] A missile launched from an aircraft at a space target, such as an earth satellite. { ¦er tü ¦spās 'mis·əl }

air-to-surface missile [ORD] A missile launched from aircraft at a ground or sea target. Abbreviated ASM. { ¦er tü ¦sər·fəs 'mis·əl }

air-to-underwater missile [ORD] A missile launched from an aircraft at an underwater target. { ¦er tü ‚ən·dər¦wòd·ər 'mis·əl }

air toxics *See* hazardous air pollutants. { 'er ¦täk·siks }

air-track drill [MIN ENG] A heavy drilling machine for quarry or opencast blasting, equipped with caterpillar tracks and operated by independent air motors. { 'er ¦trak ‚dril }

air traffic [NAV] Aircraft operating in the air or on an airport surface. { 'er ‚traf·ik }

air-traffic area [NAV] The airspace within a circular limit defined by a 5-statute-mile (approximately 8-kilometer) horizontal radius from the geographical center of an airport at which an operative airport traffic control tower is located, and an upward extent to 2000 feet (609.6 meters) above the surface. { 'er ¦traf·ik ‚er·ē·ə }

air-traffic control [NAV] **1.** A service which promotes the safe and fast movement of aircraft operating in the air or on an airport surface by providing rules, procedures, and information and advisory services for pilots. Abbreviated ATC. **2.** A system comprising enabling legislation, operating procedures, and navigation and communication equipment which is intended to make for the safe and expeditious movement of aircraft from the time that they leave the departure gates to arrival at the terminal gates. { 'er ¦traf·ik kən‚trōl }

air-traffic control center [NAV] A place for receiving information regarding aircraft movement, interpreting this information, and issuing instructions to aircraft to promote safe, orderly, expeditious flow of traffic; it directs traffic along controlled airspace. { 'er ¦traf·ik kən'trōl ‚sen·tər }

air-traffic controller [NAV] A person responsible for controlling air traffic. { 'er ¦traf·ik kən'trōl·ər }

air-traffic control radar beacon system [NAV] A system adopted by the Federal Aviation Agency for use in controlling air traffic over the United States; the aircraft carry identification transponders designed to transmit an airplane identity code, altitude, and additional message when interrogated by an air-traffic controller's equipment. Abbreviated ATCRBS. { 'er ¦traf·ik kən'trōl ¦rā‚där 'bē·kən ‚sis·təm }

air-traffic control zone *See* control zone. { 'er ¦traf·ik kən'trōl ‚zōn }

air transport [MIN ENG] Movement from one place to another of the filling material in a mine through pneumatic pipelines. Also known as air transportation. { 'er ¦tranz ‚pórt }

air transportation [AERO ENG] The use of aircraft, predominantly airplanes, to move passengers and cargo. [MIN ENG] *See* air transport. { 'er ‚tranz·pər'tā·shən }

air trap [CIV ENG] A U-shaped pipe filled with water that prevents the escape of foul air or gas from such systems as drains and sewers. [ENG] *See* air pocket. { 'er ‚trap }

air-tube breathing apparatus [ENG] A device consisting of a smoke helmet, mask, or mouthpiece supplied with fresh air by means of a flexible tube. Also known as air-supply mask. { 'er ‚tüb 'brēth·iŋ ‚a·pə¦rad·əs }

air-tube clutch [MECH ENG] A clutch fitted with a tube whose inflation causes the clutch to engage, and deflation, to disengage. { 'er ¦tüb ‚kləch }

air turbulence [METEOROL] Highly irregular atmospheric motion characterized by rapid changes in wind speed and direction and by the presence, usually, of up and down currents. { 'er ¦tər·byə·ləns }

air valve [MECH ENG] A valve that automatically lets air out of or into a liquid-carrying pipe when the internal pressure drops below atmospheric. { 'er ‚valv }

air vane [AERO ENG] A vane that acts in the air, as contrasted to a jet vane which acts within a jetstream. { 'er ‚vān }

air-variable capacitor [ELEC] A device with one rotating and one fixed set of metal plates positioned in meshed fashion and separated by air; capacitance is varied by rotating one set of plates to vary the overlap with the fixed plates. { 'er ¦ver·ē·ə·bəl kə'pas·əd·ər }

air vehicle *See* aircraft. { 'er ¦vē·ə·kəl }

air-velocity measurement [FL MECH] The measurement of the rate of displacement of air or gas at a specific location, as when ascertaining wind speed or airspeed of an aircraft. { 'er və'läs·əd·ē 'mezh·ər·mənt }

air vessel [ENG] **1.** An enclosed volume of air which uses the compressibility of air to minimize water hammer. Also known as accumulator. **2.** An enclosed chamber using the compressibility of air to promote a more uniform flow of water in a piping system. { 'er ,ves·əl }

air void [MATER] A space which is filled with air in cement paste, mortar, or concrete. { 'er ,vȯid }

air volcano [GEOL] An eruptive opening in the earth from which large volumes of gas emanate, in addition to mud and stones; a variety of mud volcano. { 'er ¦väl¦kā·nō }

air wall [NUCLEO] A wall of an ionization chamber designed so that its effect on ionizing radiation approximates that of air. { 'er ,wȯl }

air washer [MECH ENG] **1.** A device for cooling and cleaning air in which the entering warm, moist air is cooled below its dew point by refrigerated water so that although the air leaves close to saturation with water, it has less moisture per unit volume than when it entered. **2.** Apparatus to wash particulates and soluble impurities from air by passing the airstream through a liquid bath or spray. { 'er ,wash·ər }

air-water jet [ENG] A jet of mixed air and water which leaves a nozzle at high velocity; used in cleaning the surfaces of concrete or rock. { 'er ¦wȯd·ər 'jet }

air-water storage tank [ENG] A water storage tank in which the air above the water is compressed. { 'er ¦wȯd·ər 'stȯr·ij ,taŋk }

air-water vapor mixture [PHYS] A mixture of dry air and water vapor, such as the atmosphere. { 'er ¦wȯd·ər 'vā·pər ,miks·chər }

airwave [ELECTR] A radio wave used in radio and television broadcasting. [METEOROL] A wavelike oscillation in the pattern of wind flow aloft, usually with reference to the stronger portion of the westerly current. { 'er,wāv }

airway [BUILD] A passage for ventilation between thermal insulation and roof boards. [MIN ENG] A passage for air in a mine. Also known as air course. [NAV] A designated route of passage for aircraft. { 'er,wā }

airway beacon [NAV] A revolving light indicating the course of an airway. { 'er,wā ,bē·kən }

airways code *See* United States airways code. { 'er,wāz ,kōd }

airways forecast *See* aviation weather forecast. { 'er,wāz ,fȯr,kast }

airways observation *See* aviation weather observation. { 'er,wāz ,äb·zər'vā·shən }

air wedge [OPTICS] A wedge-shaped film of air between two flat reflecting surfaces that produces an interference pattern consisting of a series of light and dark bands parallel to the thin edge of the wedge. { 'er ,wej }

air well *See* air shaft. { 'er ,wel }

Airy differential equation [MATH] The differential equation $(d^2f/dz^2) - zf = 0$, where z is the independent variable and f is the value of the function; used in studying the diffraction of light near caustic surface. { ¦er·ē ,dif·ə¦ren·chəl i'kwā·zhən }

Airy disk [OPTICS] The bright, diffuse central spot of light formed by an optical system imaging a point source of light. { ¦er·ē ,disk }

Airy function [MATH] Either of the solutions of the Airy differential equation. { ¦er·ē ¦fəŋk·shən }

Airy isostasy [GEOPHYS] A theory of hydrostatic equilibrium of the earth's surface which contends that mountains are floating on a fluid lava of higher density, and that higher mountains have a greater mass and deeper roots. { ¦er·ē ī'säs·tə·sē }

Airy phase [ACOUS] An acoustic wave formed by an explosion in shallow water over a flat bottom. { ¦er·ē ,fāz }

Airy points [ENG] The points at which a horizontal rod is optionally supported to avoid its bending. { ¦er·ē ,pȯins }

Airy spirals [OPTICS] Spiral interference patterns formed by quartz cut perpendicularly to the axis in convergent circularly polarized light. { ¦er·ē 'spī·rəlz }

Airy stress function [MECH] A biharmonic function of two

variables whose second partial derivatives give the stress components of a body subject to a plane strain. { ¦er·ē 'stres ,fəŋk·shən }

aisle [ARCH] **1.** A passageway between or alongside blocks of seats, as in an auditorium. **2.** One of the parts of a basilica which are located at the sides of the nave, with each aisle separated from it by a row of columns. { īl }

aisleway [CIV ENG] A passage or walkway within a factory, storage building, or shop permitting the flow of inside traffic. { 'īl,wā }

Aistopoda [PALEON] An order of Upper Carboniferous amphibians in the subclass Lepospondyli characterized by reduced or absent limbs and an elongate, snakelike body. { ,ā·ə'stäp·ə·də }

Aitken dust counter [ENG] An instrument for determining the dust content of the atmosphere. Also known as Aitken nucleus counter. { ¦āt·kən 'dəst ,kaunt·ər }

Aitken nuclei [METEOROL] The microscopic particles in the atmosphere which serve as condensation nuclei for droplet growth during the rapid adiabatic expansion produced by an Aitken dust counter. { ¦āt·kən 'nü·klē,ī }

Aitken nucleus counter *See* Aitken dust counter. { ¦āt·kən 'nü·klē·əs ,kaunt·ər }

Aitken's formula [ASTRON] The expression used to determine the separation limit for true binary stars: $\log p'' = 2.5 - 0.2m$, where p'' = limit, m = magnitude. { ¦āt·kənz ¦fȯrm·yə·lə }

Aitoff equal-area map projection [MAP] A Lambert equal-area azimuthal projection of a hemisphere converted into a map projection of the entire sphere by a manipulation suggested by Aitoff. { ¦ī,tȯf ¦ē·kwəl ¦er·ē·ə 'map prə,jek·shən }

Al Velorum stars *See* dwarf Cepheids. { ¦al¦ī və'lȯr·əm ,stärz }

Aizoaceae [BOT] A family of flowering plants in the order Caryophyllales; members are unarmed leaf-succulents, chiefly of Africa. { ā,īz·ə'wā·sē,ē }

Ajax powder [MATER] A high-strength, high-density, gelatinous permitted explosive having good water resistance; contains nitroglycerine, potassium perchlorate, ammonium oxalate, and wood flour. { ¦ā,jaks ¦paud·ər }

ajmaline [ORG CHEM] $C_{20}H_{26}N_2O_2$ An amber, crystalline alkaloid obtained from *Rauwolfia* plants, especially *R. serpentina*. { 'aj·mə,lēn }

AJO breathing apparatus [MIN ENG] A breathing device consisting of a Siebe-Gorman mining gas mask with a small oxygen cylinder and a canister which neutralizes mining gases, such as carbon monoxide, sulfureted hydrogen, and nitrous fumes. { ¦a¦jā¦ō 'brēth·iŋ ,ap·ə'rad·əs }

ajowan oil [MATER] A yellow essential oil distilled from seeds of the herbaceous plant *Carum copticum* (*Ptychotis ajowan*) and used in pharmaceuticals. Also known as ptychotis oil. { 'aj·ə,wän ,ȯil }

akaganeite [MINERAL] β-FeO(OH) A mineral found in meteorites and considered to be formed in flight or by alteration. { ,a·kə'gan·ē,īt }

akaryote [CYTOL] A cell that lacks a nucleus. { ,ā'ka·rē,ōt }

akathisia [MED] Motor restlessness ranging from a feeling of inner disquiet, often localized in the muscles, to an inability to sit still or lie quietly. { ,a·kə'thiz·ē·ə }

akenobeite [PETR] A form of aplite composed of orthoclase and oligoclase with quartz in the interstices. { ,a·kə'nōb·ē,it }

akerite [PETR] A rock composed of quartz syenite containing soda microcline, oligoclase, and augite. { 'ȯ·kə,rīt }

akermanite [MINERAL] $Ca_2MgSi_2O_7$ Anhydrous calcium-magnesium silicate found in igneous rocks; a melilite. { 'ȯ·kər·mə,nīt }

AKF diagram [PETR] A triangular diagram showing the chemical character of a metamorphic rock in which the three components plotted are A = Al_2O_3 + Fe_2O_3 + (CaO + Na_2O), K = K_2O, and F = FeO + MgO + MnO. { ¦ā¦kā¦ef 'dī·ə,gram }

akimbo span [ANTHRO] The distance measured between elbow points when the subject stands with arms flexed in a horizontal plane, with the wrists straight, palms down, fingers straight and together, and the thumbs touching the chest; or the distance measured with upper arms horizontal, forearms vertical. { ə'kim,bō ,span }

akinesia [MED] **1.** Loss or impairment of motor function. **2.** Immobility from any cause. { ¦a·ki'nēzh·ə }

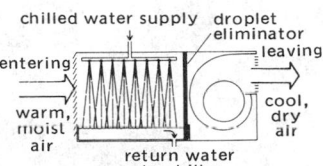

AIR WASHER

chilled water supply   droplet eliminator leaving

entering

warm, moist air   cool, dry air

return water to chiller

Schematic of air washer.

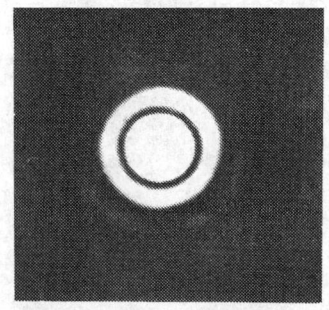

AIRY DISK

Picture of an Airy disk, one type of diffraction pattern. (*R. W. Ditchburn*)

akinete [BOT] A thick-walled resting cell of unicellular and filamentous green algae. { ˌaˈkīˌnēt }

Akins' classifier [MIN ENG] A device for separating fine-size solids from coarser solids in a wet pulp; consists of an interrupted-flight screw conveyor, operating in an inclined trough. { ˈāˌkənz ˈklasˌəˌfī·ər }

akrochordite [MINERAL] $Mn_4Mg(AsO_4)_2(OH)_4·4H_2O$ Mineral consisting of a hydrous basic manganese magnesium arsenate and occurring in reddish-brown rounded aggregates; hardness is 3 on Mohs scale, and specific gravity is 3.2. { ˌak·rōˈkȯrˌdit }

aktological [GEOL] Nearshore shallow-water areas, conditions, sediments, or life. { ˌak·təˈläj·ə·kəl }

akund [MATER] A silky cotton fiber from the shrub *Calotropis gigantea;* used in combination with kapok fiber for insulation. { ˈäˌkünd }

akureyri disease *See* epidemic neuromyasthenia. { ˌaˈkyüˈrā·ri diˌzēz }

Al *See* aluminum.

ala [BIOL] A wing or winglike structure. { ˈāˌlə }

alabandite [MINERAL] MnS A complex sulfide mineral that is a component of meteorites and usually occurs in iron-black massive or granular form. Also known as manganblende. { ˌal·əˈbanˌdīt }

alabaster [MINERAL] **1.** $CaSO_4·2H_2O$ A fine-grained, colorless gypsum. **2.** *See* onyx marble. { ˈal·əˌbas·tər }

alabaster glass [MATER] A glass that contains small inclusions of different diffractive indexes and shows no color reaction to light. { ˈal·ə·bas·tər ˌglas }

alalia [MED] Loss of speech. { ˌāˈlāl·yə }

alamalt [FOOD ENG] A powder made from cooked and toasted sweet potato and used to make candy. { ˈal·əˌmȯlt }

alamosite [MINERAL] $PbSiO_3$ A white or colorless monoclinic mineral consisting of lead silicate and occurring in radiating fibers; hardness is 4.5 on Mohs scale, and specific gravity is 6.5. { ˌal·əˈmōˌsīt }

alang-alang *See* cogon. { ˈäˌläŋˈäˌläŋ }

alanine [BIOCHEM] $C_3H_7NO_2$ A white, crystalline, nonessential amino acid of the pyruvic acid family. { ˈal·əˌnēn }

alanyl [ORG CHEM] The radical CH_3CHNH_2CO-; occurs in, for example, alanyl alanine, a dipeptide. { ˈal·əˌnil }

alar [BIOL] Winglike or pertaining to a wing. { ˈāˌlər }

alarm gage [ENG] A device that actuates a signal either when the steam pressure in a boiler is too high or when the water level in a boiler is too low. { əˈlärm ˌgaj }

alarm reaction [BIOL] The sum of all nonspecific phenomena which are elicited by sudden exposure to stimuli, which affect large portions of the body, and to which the organism is quantitatively or qualitatively not adapted. { əˈlärm rēˈak·shən }

alarm signal [ELECTR] The international radiotelegraph alarm signal transmitted to actuate automatic devices that sound an alarm indicating that a distress message is about to be broadcast. { əˈlärm ˌsig·nəl }

alarm song [INV ZOO] A stress signal occurring in many families of beetles. { əˈlärm ˌsȯŋ }

alarm system [ENG] A system which operates a warning device after the occurrence of a dangerous or undesirable condition. { əˈlärm ˌsis·təm }

alarm valve [ENG] A device that sounds an alarm when water flows in an automatic sprinkler system. { əˈlärm ˌvalv }

Alaska cedar [MATER] Wood from *Chamaecyparis nootkaensis* or *Cupressus sitkaensis*; has a fine, uniform, straight grain and is light and moderately hard; used for furniture and boat building. Also known as Sitka cypress; yellow cedar; yellow cypress. { əˈlas·kə ˈsēd·ər }

Alaska Current [OCEANOGR] A current that flows northwestward and westward along the coasts of Canada and Alaska to the Aleutian Islands. { əˈlas·kə ˈkər·ənt }

alaskaite [MINERAL] A light lead-gray sulfide mineral consisting of a mixture of lead, silver, copper, and bismuth. { əˈlas·kəˌīt }

alaskite [PETR] A granitic rock composed mainly of quartz and alkali feldspar, with few dark mineral components. { əˈlasˌkīt }

alate [BIOL] Possessing wings or winglike structures. { ˈāˌlāt }

Alaudidae [VERT ZOO] The larks, a family of Oscine birds in the order Passeriformes. { əˈlau̇·dəˌdē }

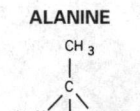

ALANINE

Structural formula of alanine.

ALBATROSS

The laysan albatross, with the characteristic hooked bill and long tubular nostrils of oceanic birds.

Albada finder [OPTICS] A viewfinder used with a camera held at eye level; the field of view is enclosed by a white frame that is made to appear very distant by reflection from the rear surface of the objective lens. { alˈbä·də ˌfīn·dər }

albafite [MINERAL] Greenish to brownish bitumen which becomes white when exposed to air; contains up to 15% oxygen; fusible; insoluble in organic solvents; varies from soft to hard, porous to compact; atomic ratio H/C 1.75–2.25. { ˈal·bəˌfīt }

albanite [PETR] A melanocratic leucitite found near Rome, Italy. { ˈal·bəˌnīt }

albarium [MATER] A white lime used for stucco; made by burning marble. { alˈbar·ē·əm }

albatross [VERT ZOO] Any of the large, long-winged oceanic birds composing the family Diomedeidae of the order Procellariformes. { ˈal·bəˌtrȯs }

albedo [NUCLEO] The reflection factor a surface, such as paraffin, has for neutrons. [OPTICS] That fraction of the total light incident on a reflecting surface, especially a celestial body, which is reflected back in all directions. { alˈbēˌdō }

albedometer [ENG] An instrument used for the measurement of the reflecting power, that is, the albedo, of a surface. { al·bəˈdä·məd·ər }

albedo neutrons *See* albedo particles. { alˈbēˌdō ˈnüˌtränz }

albedo particles [GEOPHYS] Neutrons or other particles, such as electrons or protons, which leave the earth's atmosphere, having been produced by nuclear interactions of energetic particles within the atmosphere. Also known as albedo neutrons. { alˈbēˌdō ˈpärd·ə·kəlz }

Alberger process [CHEM ENG] A method of manufacturing salt by heating brine at high pressure and passing it to a graveler which removes calcium sulfate; the salt crystallizes as the pressure is reduced and thus is separated from the brine. { ˈälˌbər·gər ˈpräs·əs }

Albers projection [MAP] An equal-area projection of the conical type, on which the meridians are straight lines that meet in a common point beyond the limits of the map, and the parallels are concentric circles whose center is at the point of intersection of the meridians. { ˈälˌbərz ˈprȯjek·shən }

Alberta low [METEOROL] A low centered on the eastern slope of the Canadian Rockies in the province of Alberta, Canada. { alˈbərt·ə ˈlō }

albertite [MINERAL] Jet-black, brittle natural hydrocarbon with conchoidal fracture, hardness of 1-2, and specific gravity of approximately 1.1. Also known as asphaltite coal. { ˈal·bərˌtīt }

Albertosaurus [PALEON] A carnivorous therapod dinosaur, 30 feet (9 meters) long, from the Late Cretaceous Period that had long muscular hindlimbs, comparatively weak forelimbs (with two-fingered hands), and powerful jaws lined with sharp teeth; related to Tyrannosaurus. { alˌber·dəˈsȯr·əs }

albertype *See* photogelatin printing plate. { ˈal·bərˌtīp }

Albian [GEOL] Uppermost Lower Cretaceous geologic time. { ˈal·bē·ən }

albic horizon [GEOL] A soil horizon from which clay and free iron oxides have been removed or in which the iron oxides have been segregated. { ˈal·bik həˈrīz·ən }

albinism [BIOL] The state of having colorless chromatophores, which results in the absence of pigmentation in animals that are normally pigmented. [MED] A hereditary, metabolic disorder transmitted as an autosomal recessive and characterized by the inability to form melanin in the skin, hair, and eyes due to tyrosinase deficiency. { ˈal·bəˌniz·əm }

albino [BIOL] A human or animal with a congenital deficiency of pigment in the skin, hair, and eyes. [BOT] An abnormal plant with colorless chromatophores. { alˈbī·nō }

Albionian [GEOL] Lower Silurian geologic time. { ˌalˈbē·ȯn·ē·ən }

albite [MINERAL] $NaAlSi_3O_8$ A colorless or milky-white variety of plagioclase of the feldspar group found in granite and various igneous and metamorphic rocks. Also known as sodaclase; sodium feldspar; white feldspar; white schorl. { ˈalˌbīt }

albite-epidote-amphibolite facies [PETR] Rocks of metamorphic type formed under intermediate temperature and pressure conditions by regional metamorphism or in the outer contact metamorphic zone. { ˈalˌbīt ˈep·əˌdōt ˌam·fibˈə·līt ˈfāˌshēz }

albite law [CRYSTAL] A rule specifying the orientation of alternating lamellae in multiple twin feldspar crystals; the twinning plane is brachypinacoid and is common in albite. { 'al‚bīt ‚lō }

albitite [PETR] A porphyritic dike rock that is coarse-grained and composed almost wholly of albite; common accessory minerals are muscovite, garnet, apatite, quartz, and opaque oxides. { 'al‚bə‚tīt }

albitization [PETR] The formation of albite in a rock as a secondary mineral. { ‚al‚bəd‚ə'zā‚shən }

albitophyre [PETR] A porphyritic rock that contains albite phenocrysts in a groundmass composed mostly of albite. { al'bid‚ə‚fīr }

albolite [MATER] A plastic cement composed principally of magnesia and silica. { 'al‚bə‚līt }

Alboll [GEOL] A suborder of the soil order Mollisol with distinct horizons, wet for some part of the year; occurs mostly on upland flats and in shallow depressions. { 'al‚bȯl }

albomaculatus [BOT] A variegation consisting of irregularly distributed white and green regions on plants due to the mitotic segregation of genes or plastids. { ‚al‚bō‚ma‚kyə'läd‚əs }

albomycin [MICROBIO] An antibiotic produced by *Actinomyces subtropicus*; effective against penicillin-resistant pneumococci and staphylococci. { ‚al‚bō'mīs‚ən }

alboranite [PETR] Olivine-free hypersthene basalt. { ‚al‚bə'ra‚nīt }

albuginea [HISTOL] A layer of white, fibrous connective tissue investing an organ or other body part. { ‚al‚byü'jin‚ē‚ə }

albumen [CYTOL] The white of an egg, composed principally of albumin. { ‚al'byü‚mən }

albumin [BIOCHEM] Any of a group of plant and animal proteins which are soluble in water, dilute salt solutions, and 50% saturated ammonium sulfate. { ‚al'byü‚mən }

albumin-globulin ratio [BIOCHEM] The ratio of the concentrations of albumin to globulin in blood serum. { ‚al'byü‚mən 'gläb‚yə‚lən ‚rā‚shō }

albumin glue [MATER] A bonding agent composed of soluble dried blood with minor additives and giving strong, durable bonds when coagulated in plywood joints at temperatures of 160–180°F (71–82°C). { ‚al'byü‚mən 'glü }

albuminoid [BIOCHEM] *See* scleroprotein. [BIOL] Having the characteristics of albumin. { ‚al'byü‚mə‚nȯid }

albumin suspension test [PATH] A blood-grouping test in which the determination is made by suspending the red cells in diluted bovine albumin instead of in a saline solution. { ‚al'byü‚mən sə'spen‚chən ‚test }

albuminuria [MED] The presence of albumin in the urine; usually indicating renal disease. { ‚al‚byü‚mə'nūr‚ē‚ə }

albumose [BIOCHEM] A protein derivative formed by the action of a hydrolytic enzyme, such as pepsin. { 'al‚byə‚mōs }

alburnum *See* sapwood. { al'bər‚nəm }

ALC *See* automatic level control.

Alcaligenes [MICROBIO] A genus of gram-negative, aerobic rods and cocci of uncertain affiliation; cells are motile, and species are commonly found in the intestinal tract of vertebrates. { ‚al‚kə'lij‚ə‚nēz }

alcaptonuria *See* alkaptonuria. { al‚kap‚tə'nūr‚ē‚ə }

Alcator [PL PHYS] A type of tokamak with a high toroidal field (up to 8.7 teslas) and high-density plasma (up to 3×10^{15} particles per cubic centimeter). { 'al‚kād‚ər }

Alcedinidae [VERT ZOO] The kingfishers, a worldwide family of colorful birds in the order Coraciiformes; characterized by large heads, short necks, and heavy, pointed bills. { ‚al‚sə'din‚ə‚dē }

alchemy [CHEM] A speculative chemical system having as its central aims the transmutation of base metals to gold and the discovery of the philosopher's stone. { 'al‚kə‚mē }

alcian blue [MATER] A copper phthalocyanin derivative used as a stain for connective tissue mucins and a number of epithelial mucins and as a gelling agent for lubricating fluids. { 'al‚shən 'blü }

Alcidae [VERT ZOO] A family of shorebirds, predominantly of northern coasts, in the order Charadriiformes, including auks, puffins, murres, and guillemots. { 'al‚sə‚dē }

Alciopidae [INV ZOO] A pelagic family of errantian annelid worms in the class Polychaeta. { ‚al‚sē'äp‚ə‚dē }

alcogel [CHEM] A gel formed by an alcosol. { 'al‚kə‚jel }

alcohol [ORG CHEM] Any member of a class of organic compounds in which a hydrogen atom of a hydrocarbon has been replaced by a hydroxy (−OH) group. { 'al‚kə‚hȯl }

alcoholate [ORG CHEM] A compound formed by the reaction of an alcohol with an alkali metal. Also known as alkoxide. { ‚al‚kə'hȯ‚lāt }

alcohol dehydrogenase [BIOCHEM] The enzyme that catalyzes the oxidation of ethanol to acetaldehyde. { 'al‚kə‚hȯl ‚dē‚hī'drä‚jə‚nās }

alcohol fuel [MATER] A motor fuel of gasoline blended with 5–25% of anhydrous ethyl alcohol; used particularly in Europe. { 'al‚kə‚hȯl ‚fyül }

alcoholic [MED] An individual who consumes excess amounts of alcoholic beverages to the extent of being addicted, habituated, or dependent. { ‚al‚kə'hȯl‚ik }

alcoholic beverage [FOOD ENG] A potable preparation containing ethyl alcohol. { ‚al‚kə'hȯl‚ik 'bev‚rij }

alcoholic fermentation [MICROBIO] The process by which certain yeasts decompose sugars in the absence of oxygen to form alcohol and carbon dioxide; method for production of ethanol, wine, and beer. { ‚al‚kə'hȯl‚ik ‚fər‚mən'tā‚shən }

alcoholic hepatitis [MED] A frequently occurring form of hepatitis that is caused by excessive ethyl alcohol intake and is characterized by fever, high white blood cell count, and jaundice. { ‚al‚kə'hȯl‚ik ‚hep‚ə'tīd‚əs }

alcoholic hyaline *See* Mallory bodies. { ‚al‚kə'hȯl‚ik 'hī‚ə‚lən }

alcoholimeter *See* alcoholometer. { ‚al‚kə'hȯ'lim‚ əd‚ər }

alcoholism [MED] Compulsive consumption of and dependence on alcoholic beverages, usually leading to pathology of the digestive and nervous systems. { 'al‚kə‚hȯ‚liz‚əm }

alcoholmeter *See* alcoholometer. { 'al‚kə‚hȯl ‚mēd‚ər }

alcoholometer [ENG] A device, such as a form of hydrometer, that measures the quantity of an alcohol contained in a liquid. Also known as alcoholimeter; alcoholmeter. { ‚al‚kə‚hȯ'lä‚məd‚ər }

alcohol thermometer [ENG] A liquid-in-glass thermometer that uses ethyl alcohol as its working substance. { 'al‚kə‚hȯl thər'mäm‚əd‚ər }

alcoholysis [ORG CHEM] The breaking of a carbon-to-carbon bond by addition of an alcohol. { ‚al‚kə'hȯl‚ə‚səs }

Alcor [ASTRON] The star 80 Ursae Majoris. { 'al‚kȯr }

alcosol [CHEM] Mixture of an alcohol and a colloid. { 'al‚kə‚sȯl }

alcove [ARCH] **1.** A recessed part of a room. **2.** A small room that opens into a larger one. **3.** An arched opening in a wall. [GEOL] A large niche formed by a stream in a face of horizontal strata. { 'al‚kōv }

alcove hologram [OPTICS] A type of hologram whose surface is bent into an arc that is concave as seen by the viewer, allowing the formation of real images that are viewable over a 180° angle. { 'al‚kōv 'häl‚ə‚gram }

alcove lands [GEOL] Terrain where the mud rocks or sandy clays and shales that compose the hills (badlands) are interstratified by occasional harder beds; the slopes are terraced. { 'al‚kōv ‚lanz }

Alcyonacea [INV ZOO] The soft corals, an order of littoral anthozoans of the subclass Alcyonaria. { ‚al‚sī‚ə'nās‚ē‚ə }

Alcyonaria [INV ZOO] A subclass of the Anthozoa; members are colonial cnidarians, most of which are sedentary and littoral. { ‚al‚sī‚ə'ner‚ē‚ə }

Aldebaran [ASTRON] A red giant star of visual magnitude 1.06, spectral classification K5-III, in the constellation Taurus; the star α Tauri. { al'deb‚ə‚rən }

aldehyde [ORG CHEM] One of a class of organic compounds containing the CHO radical. { 'al‚də‚hīd }

aldehyde ammonia [ORG CHEM] $CH_3CHOHNH_2$ A white, crystalline solid with a melting point of 97°C; soluble in water and alcohol; used in organic synthesis and as a vulcanization accelerator. { 'al‚də‚hīd ə'mō‚nyə }

aldehyde dehydrogenase [BIOCHEM] An enzyme that catalyzes the conversion of an aldehyde to its corresponding acid. { 'al‚də‚hīd ‚dē'hī‚drə‚jə‚nās }

aldehyde lyase [BIOCHEM] Any enzyme that catalyzes the nonhydrolytic cleavage of an aldehyde. { 'al‚də‚hīd 'lī‚ās }

aldehyde polymer [ORG CHEM] Any of the plastics based on aldehydes, such as formaldehyde, acetaldehyde, butyraldehyde, or acrylic aldehyde (acrolein). { 'al‚də‚hīd 'päl‚ə‚mər }

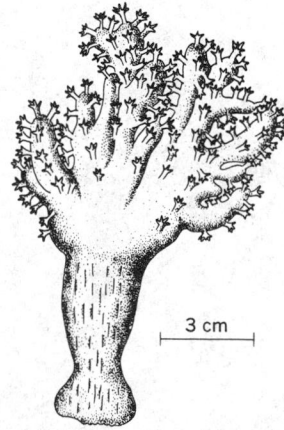

ALCYONACEA

3 cm

Representative alcyonacean *Alcyonium palmatum.* (*After Y. Delage*)

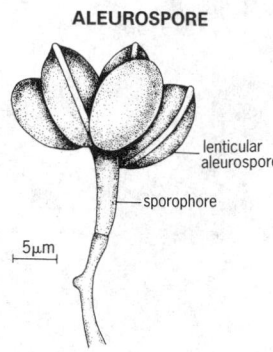

ALEUROSPORE

lenticular
aleurospore

sporophore

5μm

*Papularia sphaerosperma
(Coniosporium arundinis)
with aleurospores. (After
G. Goldanich, 1938)*

alder [BOT] The common name for several trees of the genus *Alnus*. { 'ȯl·dər }

aldicarb [ORG CHEM] $C_7H_{14}N_2O_2S$ A colorless, crystalline compound with a melting point of 100°C; used as an insecticide, miticide, and nematicide to treat soil for cotton, sugarbeets, potatoes, peanuts, and ornamentals. { 'al·də,kärb }

Aldis signaling lamp [COMMUN] A hand-carried signaling lamp used to transmit messages from ships and aircraft. { 'ȯl·dəs 'sig·nəl·iŋ ,lamp }

aldohexose [ORG CHEM] A hexose, such as glucose or mannose, containing the aldehyde group. { ,al·dō'hek,sōs }

aldol [ORG CHEM] $CH_3CH(OH)CH_2CHO$ A colorless, thick liquid with a boiling point of 83°C; used in manufacturing rubber age resistors, accelerators, and vulcanizers. { 'al,dȯl }

aldolase [BIOCHEM] An enzyme in anaerobic glycolysis that catalyzes the cleavage of fructose 1,6-diphosphate to glyceraldehyde 3-phosphate; used also in the reverse reaction. { 'al·də,lās }

aldol condensation [ORG CHEM] Formation of a β-hydroxycarbonyl compound by the condensation of an aldehyde or a ketone in the presence of an acid or base catalyst. Also known as aldol reaction. { 'al,dȯl ,kän·dən'sā·shən }

aldol reaction *See* aldol condensation. { 'al,dȯl rē'ak·shən }

aldose [ORG CHEM] A class of monosaccharide sugars; the molecule contains an aldehyde group. { 'al,dōs }

aldosterone [BIOCHEM] $C_{21}H_{28}O_5$ A steroid hormone extracted from the adrenal cortex that functions chiefly in regulating sodium and potassium metabolism. { al'däs·tə,rōn }

aldosteronism [MED] Hypertension induced by excessive secretion of aldosterone. { al'däs·tə·rə,niz·əm }

Aldrich syndrome [MED] A recessive, sex-linked disease characterized by a complex of symptoms, including eczematoid dermatitis, thrombocytopenia, black stool, and a deficiency of immune globulins. { 'ȯl·drich ,sin,drōm }

Aldrin [ORG CHEM] $C_{12}H_8Cl_6$ Trade name for a water-insoluble, white, crystalline compound, consisting mainly of chlorinated dimethanonaphthalene; used as a pesticide. { 'al·drən }

ale [FOOD ENG] A fermented malt beverage, differing from beer in containing up to 8% alcohol by volume and being hopped more heavily. { āl }

alecithal [CYTOL] Referring to an egg without yolk, such as the eggs of placental mammals. { ā'les·ə·thəl }

alee basin [GEOL] A basin formed in the deep sea by turbidity currents aggrading courses where the currents were deflected around a submarine ridge. { ə'lē ,bās·ən }

aleishtite [GEOL] A bluish or greenish mixture of dickite and other clay minerals. { ə'lē·ish,tīt }

Alembert *See* d'Alembert entries. { 'a·ləm,ber }

alençon [TEXT] A type of point lace characterized by a fine mesh ground and a floral design that is outlined with heavy thread. { ə'len,sän }

aleph null [MATH] The cardinal number of any set which can be put in one-to-one correspondence with the set of positive integers. Also known as aleph zero. { 'ä,lef ,nəl }

aleph one [MATH] The smallest cardinal number that is larger than aleph zero. { 'äl,ef 'wən }

aleph zero *See* aleph null. { 'äl,ef 'zir·ō }

Alepocephaloidei [VERT ZOO] The slickheads, a suborder of deap-sea teleostean fishes of the order Salmoniformes. { ə,lep·ō·sə·fə'lȯi·de,ī }

alert [ORD] A state of readiness against impending danger or for going into action. { ə'lərt }

alert box [COMPUT SCI] A dialog box that warns of an existing condition or the consequences of a command that has been given, or explains why a command cannot be executed. { ə'lərt ,bäks }

alerting signal [COMMUN] Specific signal that is applied to subscriber access lines to indicate an incoming call. { ə'lərt·iŋ 'sig·nəl }

alert limit [NAV] The maximum position error allowable for a given phase of flight without an alert being raised. { ə·'lərt ,lim·it }

aletophyte [ECOL] A weedy plant growing on the roadside or in fields where natural vegetation has been disrupted by humans. { ə'lēd·ə,fīt }

alette [ARCH] **1.** A minor wing of a building. **2.** A door jamb. **3.** A rear pilaster, partially visible within a cluster of columnar elements. **4.** The wing of the pier on both sides of an engaged column. { ə'let }

aleukemia [MED] Leukemia in which the white blood cell count is normal or low. { ¦ā·lü'kē·mē·ə }

aleuron [BOT] Protein in the form of grains stored in the embryo, endosperm, or perisperm of many seeds. { 'al·yə,rän }

aleurospore [MYCOL] A simple terminal or lateral, thick-walled, nondeciduous spore produced by some fungi of the order Moniliales. { ə'lyür·ə,spȯr }

Aleutian Current [OCEANOGR] A current setting southwestward along the southern coasts of the Aleutian Islands. { ə'lü·shən ,kər·ənt }

Aleutian disease [VET MED] A disease of mink characterized by accumulations of plasma cells in several organs, hyaline changes in the walls of small arteries, and interstitial fibrosis of the kidneys. { ə'lü·shən diz,ēz }

Aleutian low [METEOROL] The low-pressure center located near the Aleutian Islands on mean charts of sea-level pressure; represents one of the main centers of action in the atmospheric circulation of the Northern Hemisphere. { ə'lü·shən ,lō }

alewife [VERT ZOO] *Pomolobus pseudoharengus.* A food fish of the herring family that is very abundant on the Atlantic coast. { 'āl,wīf }

Alexanderson antenna [ELECTROMAG] An antenna, used at low or very low frequencies, consisting of several base-loaded vertical radiators connected together at the top and fed at the bottom of one radiator. { ,al·ig'zan·dər·sən an,ten·ə }

Alexander's subbase theorem [MATH] The theorem that a topological space is compact if and only if its topology has a subbase with the property that any set that is contained in the union of a collection of members of the subbase is contained in the union of a finite number of members of this collection. { ,al·ig'zan·dərz ¦səb,bās ,thir·əm }

Alexandrian [GEOL] Lower Silurian geologic time. { ,al·ig'zan·dre·ən }

alexandrite [MINERAL] A gem variety of chrysoberyl; emerald green in natural light but red in transmitted or artificial light. { ,al·ig'zan,drīt }

Alexandroff compactification *See* one-point compactification. { al·ik¦san,drȯf kəm,pak·tə·fə'kā·shən }

alexia [MED] Loss of the ability to read. { ə'lek·sē·ə }

Alexinic unit [BIOL] A unit for the standardization of blood serum. { ¦a,lek¦sin·ik 'yü·nət }

alexithymia [PSYCH] An affective or cognitive disorder in which individuals are preoccupied with mundane, chronologically oriented details and emotions are experienced in undifferentiated form. { ə,lek·sə'thī·mē·ə }

Aleyrodidae [INV ZOO] The whiteflies, a family of homopteran insects included in the series Sternorrhyncha; economically important as plant pests. { ,al·ə'räd·ə,dē }

alfalfa [BOT] *Medicago sativa.* A herbaceous perennial legume in the order Rosales, characterized by a deep taproot. Also known as lucerne. { al'fal·fə }

Alfalfa mosaic virus [VIROL] The type species of the genus *Alfamovirus* of the family Bromoviridae. Abbreviated AMV. { al¦fal·fə mō¦zā·ik 'vī·rəs }

Alfalfa mosaic virus group *See* Alfamovirus. { al¦fal·fə mō¦zā·ik 'vī·rəs ,grüp }

Alfamovirus [VIROL] A genus of plant viruses in the family Bromoviridae that is characterized by virions which are either bacilliform or ellipsoidal and contain single-stranded ribonucleic acid genomes; alfalfa mosaic virus is the type species. Also known as Alfalfa mosaic virus group. { al'fam·ə,vī·rəs }

alfa process [FOOD ENG] A German process for butter manufacture in which high-fat cream (78%) is run through a three-cylinder cooler, inside of which cooled spiral-ribbed rollers turn. { 'al·fə 'präs·əs }

alfenol [MET] A permeability alloy that has 16% aluminum and 84% iron; it is brittle and at 572°F (300°C) can be rolled into thin sheets; used for transformer cores and tape recorder heads. { 'al·fə,nȯl }

alfin catalyst [ORG CHEM] A catalyst derived from reaction of an alkali alcoholate with an olefin halide; used to convert olefins (for example, ethylene, propylene, or butylenes) into polyolefin polymers. { 'al·fin 'kad·ə,list }

Alfisol [GEOL] An order of soils with gray to brown surface horizons, a medium-to-high base supply, and horizons of clay accumulation. { 'al·fə₁sȯl }

Alford loop [ELECTROMAG] An antenna utilizing multielements which usually are contained in the same horizontal plane and adjusted so that the antenna has approximately equal and in-phase currents uniformly distributed along each of its peripheral elements and produces a substantially circular radiation pattern in the plane of polarization; it is known for its purity of polarization. { 'ȯl·fərd ₁lüp }

Alfvén number [PHYS] The ratio of the speed of the Alfvén wave to the speed of the fluid at a point in the fluid. { äl'vän ₁nəm·bər }

Alfvén speed [PHYS] The speed of motion of the Alfvén wave, which is $v_a = B_0/\sqrt{\rho\mu}$, where B_0 is the magnetic field strength, ρ the fluid density, and μ the magnetic permeability (in meter-kilogram-second units). { äl'vän ₁spēd }

Alfvén wave [PHYS] A hydromagnetic shear wave which moves along magnetic field lines; a major accelerative mechanism of charged particles in plasma physics and astrophysics. { äl'vän ₁wāv }

algae [BOT] General name for the chlorophyll-bearing organisms in the plant subkingdom Thallobionta. { 'al·jē }

algae bloom [ECOL] A heavy growth of algae in and on a body of water as a result of high phosphate concentration from farm fertilizers and detergents. { 'al·jē ₁blüm }

algae wash [ECOL] A shoreline drift consisting almost entirely of filamentous algae. { 'al·jē ₁wash }

algal [BOT] Of or pertaining to algae. [GEOL] Formed from or by algae. { 'al·gəl }

algal biscuit [GEOL] A disk-shaped or spherical mass, up to 20 centimeters in diameter, made up of carbonate that is probably the result of precipitation by algae. { 'al·gəl ₁bis·kət }

algal coal [GEOL] Coal formed mainly from algal remains. { 'al·gəl ₁kōl }

algal limestone [PETR] A type of limestone either formed from the remains of calcium-secreting algae or formed when algae bind together the fragments of other lime-secreting organisms. { 'al·gəl ₁līm₁stōn }

algal pit [GEOL] An ablation depression that is small and contains algae. { 'al·gəl ₁pit }

algal reef [GEOL] An organic reef which has been formed largely of algal remains and in which algae are or were the main lime-secreting organisms. { 'al·gəl ₁rēf }

algal ridge [GEOL] Elevated margin of a windward coral reef built by actively growing calcareous algae. { 'al·gəl ₁rij }

algal rim [GEOL] Low rim built by actively growing calcareous algae on the lagoonal side of a leeward reef or on the windward side of a patch reef in a lagoon. { 'al·gəl ₁rim }

algal structure [GEOL] A deposit, most frequently calcareous, with banding, irregular concentric structures, crusts, and pseudo-pisolites or pseudo-concretionary forms resulting from organic, colonial secretion and precipitation. { 'al·gəl ₁strək·chər }

algebra [MATH] **1** A method of solving practical problems by using symbols, usually letters, for unknown quantities. **2.** The study of the formal manipulations of equations involving symbols and numbers. **3.** An abstract mathematical system consisting of a vector space together with a multiplication by which two vectors may be combined to yield a third, and some axioms relating this multiplication to vector addition and scalar multiplication. Also known as hypercomplex system. { 'al·jə·brə }

algebraic addition [MATH] The addition of algebraic quantities in the sense that adding a negative quantity is the same as subtracting a positive one. { ₁al·jə₁brā·ik ə'dish·ən }

algebraically closed field [MATH] **1.** A field F such that every polynomial of degree equal to or greater than 1 with coefficients in F has a root in F. **2.** A field F is said to be algebraically closed in an extension field K if any root in K of a polynominal with coefficients in F also lies in F. Also known as algebraically complete field. { ₁al·jə₁brā·ik·lē ₁klōzd 'fēld }

algebraically complete field See algebraically closed field. { ₁al·jə₁brā·ik·lē kəm₁plēt 'fēld }

algebraically independent [MATH] A subset S of a commutative ring B is said to be algebraically independent over a subring A of B (or the elements of S are said to be algebraically independent over A) if, whenever a polynominal in elements

of S, with coefficients in A, is equal to 0, then all the coefficients in the polynomial equal 0. { ₁al·jə₁brā·ik·lē ₁in·də'pen·dənt }

algebraic closure of a field [MATH] An algebraic extension field which has no algebraic extensions but itself. { ₁al·jə₁brā·ik 'klō·zhər əv ə 'fēld }

algebraic computation system See symbolic system. { ₁al·jə₁brā·ik ₁käm·pyə'tā·shən ₁sis·təm }

algebraic curve [MATH] **1.** The set of points in the plane satisfying a polynomial equation in two variables. **2.** More generally, the set of points in n-space satisfying a polynomial equation in n variables. { ₁al·jə₁brā·ik 'kərv }

algebraic deviation [STAT] The difference between a variate and a given value, which is counted positive if the variate is greater than the given value, and negative if less. { ₁al·jə₁brā·ik ₁dē·vē'ā·shən }

algebraic equation [MATH] An equation in which zero is set equal to an algebraic expression. { ₁al·jə₁brā·ik i'kwā·zhən }

algebraic expression [MATH] An expression which is obtained by performing a finite number of the following operations on symbols representing numbers: addition, subtraction, multiplication, division, raising to a power. { ₁al·jə₁brā·ik ik'spresh·ən }

algebraic extension of a field [MATH] A field which contains both the given field and all roots of polynomials with coefficients in the given field. { ₁al·jə₁brā·ik ik'sten·shən əv ə 'fēld }

algebraic function [MATH] A function whose value is obtained by performing only the following operations to its argument: addition, subtraction, multiplication, division, raising to a rational power. { ₁al·jə₁brā·ik 'fəŋk·shən }

algebraic geometry [MATH] The study of geometric properties of figures using methods of abstract algebra. { ₁al·jə₁brā·ik jē'äm·ə·trē }

algebraic hypersurface [MATH] For an n-dimensional Euclidean space with coordinates $x_1, x_2, \ldots, x_n$, the set of points that satisfy an equation of the form $f(x_1, x_2, \ldots, x_n) = 0$, where f is a polynomial in the coordinates. { ₁al·jə₁brā·ik 'hī·pər₁sər·fəs }

algebraic identity [MATH] A relation which holds true for all possible values of the literal symbols occurring in it; for example, $(x + y)(x - y) = x^2 - y^2$. { ₁al·jə₁brā·ik i'den·ə·tē }

algebraic integer [MATH] The root of a polynomial whose coefficients are integers and whose leading coefficient is equal to 1. { ₁al·jə₁brā·ik 'in·tə·jər }

algebraic invariant [MATH] A polynomial in coefficients of a quadratic or higher form in a collection of variables whose value is unchanged by a specified class of linear transformations of the variables. { ₁al·jə₁brā·ik in'ver·ē·ənt }

algebraic K theory [MATH] The study of the mathematical structure resulting from associating with each ring A the group $K(A)$, the Grothendieck group of A. { ₁al·jə₁brā·ik 'kā ₁thē·ə·rē }

algebraic language [MATH] The conventional method of writing the symbols, parentheses, and other signs of formulas and mathematical expressions. { ₁al·jə₁brā·ik 'laŋ·gwij }

algebraic manipulation language [COMPUT SCI] A programming language used in the solution of analytic problems by symbolic computation. { ₁al·jə₁brā·ik mə·ni·pyə'lā·shən ₁laŋ·gwij }

algebraic number [MATH] Any root of a polynomial with rational coefficients. { ₁al·jə₁brā·ik 'nəm·bər }

algebraic number field [MATH] A finite extension field of the field of rational numbers. { ₁al·jə₁brā·ik 'nəm·bər ₁fēld }

algebraic number theory [MATH] The study of properties of real numbers, especially integers, using the methods of abstract algebra. { ₁al·jə₁brā·ik 'nəm·bər ₁thē·ə·rē }

algebraic object [MATH] Either an algebraic structure, such as a group, ring, or field, or an element of such an algebraic structure. { ₁al·jə₁brā·ik 'äb₁jekt }

algebraic operation [MATH] Any of the operations of addition, subtraction, multiplication, division, raising to a power, or extraction of roots. { ₁al·jə₁brā·ik ₁äp·ə'rā·shən }

algebraic scattering theory [PHYS] An approach to the analysis of reactions between composite particles in which the fundamental role is played by the scattering matrix, which is obtained algebraically, without the use of a wave equation, by using the concept of dynamic symmetry. { ₁al·jə₁brā·ik 'skad·ə·riŋ ₁thē·ə·rē }

algebraic set [MATH] A set made up of all zeros of some

specified set of polynomials in n variables with coefficients in a specified field F, in a specified extension field of F. { ¦al·jə¦brā·ik 'set }

algebraic subtraction [MATH] The subtraction of signed numbers, equivalent to reversing the sign of the subtrahend and adding it to the minuend. { ¦al·jə¦brā·ik səb'trak·shən }

algebraic sum [MATH] **1.** The result of the addition of two or more quantities, with the addition of a negative quantity equivalent to subtraction of the corresponding positive quantity. **2.** For two fuzzy sets A and B, with membership functions m_A and m_B, that fuzzy set whose membership function m_{A+B} satisfies the equation $m_{A+B}(x) = m_A(x) + m_B(x) - [m_A(x) \cdot m_B(x)]$ for every element x. { ¦al·jə¦brā·ik 'səm }

algebraic surface [MATH] A subset S of a complex n-space which consists of the set of complex solutions of a system of polynomial equations in n variables such that S is a complex two-manifold in the neighborhood of most of its points. { ¦al·jə¦brā·ik 'sər·fəs }

algebraic symbol [MATH] A letter that represents a number or a symbol indicating an algebraic operation. { ¦al·jə¦brā·ik 'sim·bəl }

algebraic term [MATH] In an expression, a term that contains only numbers and algebraic symbols. { ¦al·jə¦brā·ik 'tərm }

algebraic topology [MATH] The study of topological properties of figures using the methods of abstract algebra; includes homotopy theory, homology theory, and cohomology theory. { ¦al·jə¦brā·ik tə'päl·ə·jē }

algebraic variety [MATH] A set of points in a vector space that satisfy each of a set of polynomial equations with coefficients in the underlying field of the vector space. { ¦al·jə¦brā·ik və'rī·əd·ē }

algebra of subsets [MATH] An algebra of subsets of a set S is a family of subsets of S that contains the null set, the complement (relative to S) of each of its members, and the union of any two of its members. { ¦al·jə·brə əv 'səb,sets }

algebra with identity [MATH] An algebra which has an element, not equal to 0 and denoted by 1, such that, for any element x in the algebra, $x1 = 1x = x$. { ¦al·jə·brə with i'den·ə·tē }

alged malaria See falciparum malaria. { 'al·jəd mə'ler·ē·ə }

Algenib [ASTRON] A star in the constellation Pegasus. { ,al'jen·əb }

Algerian onyx See onyx marble. { al'jer·ē·ən 'än·iks }

algesia [PHYSIO] Sensitivity to pain. { al'jēz·ē·ə }

algesimeter [PHYSIO] A device used to determine pain thresholds. { ,al·jə'sim·əd·ər }

algesiroreceptor [PHYSIO] A pain-sensitive cutaneous sense organ. { ,al·jə¦si·rō·ri¦sep·tər }

algicide [MATER] A chemical used to kill algae. { 'al·jə,sīd }

algin [MATER] A hydrophilic polysaccharide extracted from brown algae, such as giant kelp. [ORG CHEM] See sodium alginate. { 'al·jən }

alginate [BOT] An algal polysaccharide that is a major constituent of the cell walls of brown algae. { 'al·jə,nāt }

alginic acid [ORG CHEM] $(C_6H_8O_6)_n$ An insoluble colloidal acid obtained from brown marine algae; it is hard when dry and absorbent when moist. Also known as algin. { al'jin·ik 'as·əd }

alginic acid sodium salt See sodium alginate. { al'jin·ik 'as·əd 'sōd·ē·əm 'sȯlt }

alginite See algite. { 'al·jə,nīt }

algite [PETR] The petrological unit that constitutes algal material present in considerable amounts in algal or boghead coal. Also known as alginite. { 'al,jīt }

algodonite [MINERAL] Cu_6As A steel gray to silver white mineral consisting of copper arsenide and occurring as minute hexagonal crystals or in massive and granular form. { al'gäd·ə,nīt }

Algol [ASTRON] An eclipsing variable star of spectral classification B8 in the constellation Perseus; the star β Persei. Also known as Demon Star. [COMPUT SCI] An algorithmic and procedure-oriented computer language used principally in the programming of scientific problems. { 'al,gȯl }

algology [BOT] The study of algae. Also known as phycology. [MED] The science and study of phenomena associated with pain. { al'gäl·ə·jē }

Algol symbiotic [ASTRON] A symbiotic star consisting of a red giant, a main-sequence star, and an accretion disk of gas from the red giant that forms around the main-sequence star and is heated by it. { 'al,gȯl ,sim·bē'äd·ik }

Algoman orogeny [GEOL] Orogenic episode affecting Archean rocks of Canada about 2.4 billion years ago. Also known as Kenoran orogeny. { al'gōm·ən ȯ'räj·ə·nē }

algometer [MED] An instrument for measuring pressure stimuli which produce pain. { al'gä·məd·ər }

Algonkian See Proterozoic. { al'gän·kē·ən }

algophage See cyanophage. { 'al·gə,fāj }

algophobia [PSYCH] Abnormal fear of pain. { al·gə'fōb·ē·ə }

algorithm [MATH] A set of well-defined rules for the solution of a problem in a finite number of steps. { 'al·gə,rith·əm }

algorithmic error [COMPUT SCI] An error in computer processing resulting from imprecision in the method used to carry out mathematical computations, usually associated with either rounding or truncation of numbers. { ¦al·gə¦rith·mik 'er·ər }

algorithmic language [COMPUT SCI] A language in which a procedure or scheme of calculations can be expressed accurately. { ¦al·gə¦rith·mik 'laŋ·gwij }

algorithm translation [COMPUT SCI] A step-by-step computerized method of translating one programming language into another programming language. { 'al·gə,rith·əm tranz'lā·shən }

algor mortis [PATH] Postmortem cooling of the body. { ¦al·gər ¦mȯr·təs }

alias [COMPUT SCI] **1.** An alternative entry point in a computer subroutine at which its execution may begin, if so instructed by another routine. **2.** An alternative name for a file or device. [STAT] Either of two effects in a factorial experiment which cannot be differentiated from each other on the basis of the experiment. { 'ā·lē·əs }

aliasing [COMPUT SCI] In computer graphics, the jagged appearance of diagonal lines on printouts and on video monitors. [MATH] Introduction of error into the computed amplitudes of the lower frequencies in a Fourier analysis of a function carried out using discrete time samplings whose interval does not allow the proper analysis of the higher frequencies present in the analyzed function. { 'āl·yəs·iŋ }

alicyclic [ORG CHEM] **1.** Having the properties of both aliphatic and cyclic substances. **2.** Referring to a class of saturated hydrocarbon compounds whose structures contain one ring. Also known as cycloaliphatic; cycloalkane. **3.** Any one of the compounds of the alicyclic class. Also known as cyclane. { al·ə¦sī·klik }

alidade [ENG] **1.** An instrument for topographic surveying and MAP by the plane-table method. **2.** Any sighting device employed for angular measurement. { 'al·ə,dād }

alien substitution [GEN] The replacement of one or more chromosomes by those from a different species. { ¦āl·ē·ən ,səb·stə'tü·shən }

aliesterase [BIOCHEM] Any one of the lipases or nonspecific esterases. { al·ē'es·tə,rās }

aligning drift [MECH ENG] A rod or bar that is used for aligning parts during assembly. { ə'līn·iŋ ,drift }

alignment [ARCHEO] An arrangement of a single row or of multiple rows of standing stones at a sites formerly occupied by humans. [CIV ENG] In a survey for a highway, railroad, or similar installation, a ground plan that shows the horizontal direction of the route. [ELECTR] The process of adjusting components of a system for proper interrelationship, including the adjustment of tuned circuits for proper frequency response and the time synchronization of the components of a system. [ENG] Placing of surveying points along a straight line. [MAP] Representing of the correct direction, character, and relationships of a line or feature on a map. [MIN ENG] The act of laying out a tunnel or regulating by line; adjusting to a line. [NUC PHYS] A population $p(m)$ of the $2I + 1$ orientational substates of a nucleus; $m = -I$ to $+I$, such that $p(m) = p(-m)$. { ə'līn·mənt }

alignment chart See nomograph. { ə'līn·mənt ,chärt }

alignment correction [ENG] A correction applied to the measured length of a line to allow for not holding the tape exactly in a vertical plane of the line. { ə'līn·mənt kə'rek·shən }

alignment pin [DES ENG] Pin in the center of the base of an octal, loctal, or other tube having a single vertical projecting rib that aids in correctly inserting the tube in its socket. { ə'līn·mənt ˌpin }

alignment wire *See* ground wire. { ə'līn·mənt ˌwīr }

alimentary [BIOL] Of or relating to food, nutrition, or diet. { ˌal·ə¦men·trē }

alimentary canal [ANAT] The tube through which food passes; in humans, includes the mouth, pharynx, esophagus, stomach, and intestine. { ¦al·ə¦men·trē kə'nal }

alimentation [BIOL] Providing nourishment by feeding. [HYD] *See* accumulation. { ˌal·ə·mən'tā·shən }

Alioth [ASTRON] Traditional name for a second-magnitude star in the Big Dipper; the star ε Ursae Majoris. { 'al·ē·äth }

aliphatic [ORG CHEM] Of or pertaining to any organic compound of hydrogen and carbon characterized by a straight chain of the carbon atoms; three subgroups of such compounds are alkanes, alkenes, and alkynes. { ¦al·ə¦fad·ik }

aliphatic acid [ORG CHEM] Any organic acid derived from aliphatic hydrocarbons. { ¦al·ə¦fad·ik 'as·əd }

aliphatic acid ester [ORG CHEM] Any organic ester derived from aliphatic acids. { ¦al·ə¦fad·ik 'as·əd 'es·tər }

aliphatic polycyclic hydrocarbon [ORG CHEM] A hydrocarbon compound in which at least two of the aliphatic structures are cyclic or closed. { ¦al·ə¦fad·ik ˌpä·lə'sī·klik ˌhī·drə'kär·bən }

aliphatic polyene compound [ORG CHEM] Any unsaturated aliphatic or alicyclic compound with more than four carbons in the chain and with at least two double bonds; for example, hexadiene. { ¦al·ə¦fad·ik 'päl·ē·ˌēn ˌkäm·paund }

aliphatic series [ORG CHEM] A series of open-chained carbon-hydrogen compounds; the two major classes are the series with saturated bonds and with the unsaturated. { ¦al·ə¦fad·ik 'sir·ēz }

aliquant [CHEM] A part of a sample that has been divided into a set of equal parts plus a smaller remainder part. [MATH] A divisor that does not divide a quantity into equal parts. { 'al·ə·ˌkwänt }

aliquot [CHEM] A part of a sample that has been divided into exactly equal parts with no remainder. [MATH] A divisor that divides a quantity into equal parts with no remainder. [MED] A representative sample of a larger quantity. { 'al·ə·ˌkwät }

Alismataceae [BOT] A family of flowering plants belonging to the order Alismatales characterized by schizogenous secretory cells, a horseshoe-shaped embryo, and one or two ovules. { ə,liz·mə'tās·ē,ē }

Alismatales [BOT] A small order of flowering plants in the subclass Alismatidae, including aquatic and semiaquatic herbs. { ə,liz·mə'tā·lēz }

Alismatidae [BOT] A relatively primitive subclass of aquatic or semiaquatic herbaceous flowering plants in the class Liliopsida, generally having apocarpous flowers, and trinucleate pollen and lacking endosperm. { ə,liz'mad·ə,dē }

alisphenoid [ANAT] 1. The bone forming the greater wing of the sphenoid in adults. 2. Of or pertaining to the sphenoid wing. { ¦al·ə¦sfe,noid }

alite [MATER] A constituent of portland cement clinker consisting mostly of calcium silicate. { 'ā,līt }

alive [ELEC] *See* energized. [MIN ENG] That portion of a lode that is productive. { ə'līv }

alivincular [INV ZOO] In some bivalves, having the long axis of the short ligament transverse to the hinge line. { ¦al·ə¦viŋ·kyə·lər }

alizarin [ORG CHEM] $C_{14}H_6O_2(OH)_2$ An orange crystalline compound, insoluble in cold water; made synthetically from anthraquinone; used in the manufacture of dyes and red pigments. { ə'liz·ə·rən }

alizarin dye [ORG CHEM] Sodium salts of sulfonic acids derived from alizarin. { ə'liz·ə·rən 'dī }

alizarin red [ORG CHEM] Any of several red dyes derived from anthraquinone. { ə'liz·ə·rən 'red }

alizarin yellow [MATER] A dye useful as an acid-base indicator; solutions change color from yellow (acid) to purple (basic) in the pH range 10.1 to 12.0. { ə'liz·ə·rən 'yel·ō }

alkadiene *See* diene. { ˌal·kə'dī,ēn }

alkalemia [MED] An increase in blood pH above normal levels. { ˌal·kə'lēm·ē·ə }

alkalescence [CHEM] The property of a substance that is alkaline, that is, having a pH greater than 7. { ˌal·kə'les·əns }

alkali [CHEM] Any compound having highly basic qualities. [PETR] *See* alkalic. { 'al·kə,lī }

alkali-aggregate reaction [CHEM] The chemical reaction of an aggregate with the alkali in a cement, resulting in a weakening of the concrete. { 'al·kə,lī 'ag·rə·gət rē'ak·shən }

alkali alcoholate [ORG CHEM] A compound formed from an alcohol and an alkali metal base; the alkali metal replaces the hydrogen in the hydroxyl group. { 'al·kə,lī ,al·kə'hò,lāt }

alkali blue [ORG CHEM] The sodium salt of triphenylrosanilinesulfonic acid; used as an indicator. { 'al·kə,lī 'blü }

alkalic Also known as alkali. [PETR] 1. Of igneous rock, containing more than average alkali (K_2O and Na_2O) for that clan in which they are found. 2. Of igneous rock, having feldspathoids or other minerals, such as acmite, so that the molecular ratio of alkali to silica is greater than 1:6. 3. Of igneous rock, having a low alkali-lime index (51 or less). { ,al'kal·ik }

alkali-calcic series [PETR] The series of igneous rocks with weight percentage of silica in the range 51–55, and weight percentages of CaO and $K_2O + Na_2O$ equal. { ¦al·kə,lī ¦kal,sik ,sir·ēz }

alkali cellulose [MATER] Product of wood pulp steeped with sodium hydroxide; first step in manufacture of viscose rayon and other cellulosics. { 'al·kə,lī 'sel·yə,lōs }

alkali chlorosis [PL PATH] Yellowing of plant foliage due to excess amounts of soluble salts in the soil. { 'al·kə,lī klə'rō·səs }

alkalide [INORG CHEM] A member of a class of crystalline salts with an alkali metal atom. { 'al·kə,līd }

alkali denaturation test [PATH] A blood test for the measurement of fetal hemoglobin in terms of its resistance to alkali denaturation. { 'al·kə,lī də,nach·ə'rā·shən ,test }

alkali disease [MED] Selenium poisoning. [VET MED] 1. Botulism of ducks. 2. Trembles of cattle. { 'al·kə,lī diz,ēz }

alkali emission [GEOPHYS] Light emission from free lithium, potassium, and especially sodium in the upper atmosphere. { 'al·kə,lī i'mish·ən }

alkali feldspar [MINERAL] A feldspar composed of potassium feldspar and sodium feldspar, such as orthoclase, microcline, albite, and anorthoclase; all are considered alkali-rich. { 'al·kə,lī 'feld,spar }

alkali flat [GEOL] A level lakelike plain formed by the evaporation of water in a depression and deposition of its fine sediment and dissolved minerals. { 'al·kə,lī ,flat }

alkali ion diode [ENG] In testing for leaks, a device which senses the presence of halogen gases by the use of positive ions of alkali metal on the heated diode surfaces. { 'al·kə,lī 'ī·ən ,dī,ōd }

alkali lake [HYD] A lake with large quantities of dissolved sodium and potassium carbonates as well as sodium chloride. { 'al·kə,lī 'lāk }

alkali lead [MET] An alloy of lead hardened with small quantities of alkali metals; used as bearing metals. { 'al·kə,lī 'led }

alkali lignin [MATER] A type of lignin produced by treating the black liquor from the soda process with acid; used as an extender in the negative plates of storage batteries, in asphalt, and in paperboard products. { 'al·kə,lī 'lig·nən }

alkali-lime index [PETR] The percentage by weight of silica in a sequence of igneous rocks on a variation diagram where the weight percentages of CaO and of K_2O and Na_2O are equal. { 'al·kə,lī 'līm ,in·deks }

alkali metal [CHEM] Any of the elements of group I in the periodic table: lithium, sodium, potassium, rubidium, cesium, and francium. { 'al·kə,lī ,med·əl }

alkalimeter [ANALY CHEM] 1. An apparatus for measuring the quantity of alkali in a solid or liquid. 2. An apparatus for measuring the quantity of carbon dioxide formed in a reaction. { ,al·kə'lim·əd·ər }

alkalimetry [ANALY CHEM] Quantitative measurement of the concentration of bases or the quantity of one free base in a solution; techniques include titration and other analytical methods. { ,al·kə'lim·ə·trē }

alkaline [CHEM] 1. Having properties of an alkali. 2. Having a pH greater than 7. { 'al·kə,līn }

alkaline cell [ELEC] A primary cell that uses an alkaline electrolyte, usually potassium hydroxide, and delivers about 1.5 volts at much higher current rates than the common carbon-zinc cell. Also known as alkaline-manganese cell. { 'al·kə,līn ,sel }

alkaline cleaner [MET] An aqueous solution of an alkali used for metal cleaning. { 'al·kə,līn 'klēn·ər }

alkaline earth [INORG CHEM] An oxide of an element of group 2 in the periodic table, such as barium, calcium, and strontium. Also known as alkaline-earth oxide. { ¦al·kə‚līn 'ərth }

alkaline-earth metals [CHEM] The heaviest members of group 2 in the periodic table; usually calcium, strontium, magnesium, and barium. { ¦al·kə‚līn 'ərth 'med·əlz }

alkaline-earth oxide See alkaline earth. { ¦al·kə‚līn 'ərth 'äk‚sīd }

alkaline flooding [PETRO ENG] A type of enhanced oil recovery in which alkaline chemicals are injected during a water flooding or are combined with polymer flooding; the chemicals react with acids in the crude oil to form surfactants. { 'al·kə‚līn 'fləd·iŋ }

alkaline-manganese cell See alkaline cell. { ¦al·kə‚līn ¦maŋ·gə‚nēs ‚sel }

alkaline phosphatase [BIOCHEM] A phosphatase active in alkaline media. { 'al·kə‚līn 'fäs·fə‚tās }

alkaline soil [GEOL] Soil containing soluble salts of magnesium, sodium, or the like, and having a pH value between 7.3 and 8.5. { 'al·kə‚līn 'sȯil }

alkaline storage battery [ELEC] A storage battery in which the electrolyte consists of an alkaline solution, usually potassium hydroxide. { 'al·kə‚līn 'stȯr·ij ‚bad·ə·rē }

alkaline tide [PHYSIO] The temporary decrease in acidity of urine and body fluids after eating, attributed by some to the withdrawal of acid from the body due to gastric digestion. { 'al·kə‚līn 'tīd }

alkaline wash [CHEM ENG] The removal of impurities from kerosine, used for illuminating purposes, by caustic soda solution. { 'al·kə‚līn ‚wäsh }

alkalinity [CHEM] The property of having excess hydroxide ions in solution. { ‚al·kə'lin·ə·dē }

alkaliphile [PHYSIO] An organism that prefers or is able to withstand an alkaline environment (pH value above 9). { 'al·kə·lə‚fīl }

alkali reactivity [MATER] Susceptibility of a concrete aggregate to alkali-aggregate reaction { 'al·kə‚lī ‚rē·ak'tiv·əd·ē }

alkali-resisting paint [MATER] A paint, such as one made with a synthetic resin, that does not undergo saponification when used in such places as bathrooms or on such materials as new concretes. { ¦al·kə‚lī rə‚zist·iŋ 'pānt }

alkali soil [GEOL] A soil, with salts injurious to plant life, having a pH value of 8.5 or higher. { 'al·kə‚lī ‚sȯil }

alkaloid [ORG CHEM] One of a group of nitrogenous bases of plant origin, such as nicotine, cocaine, and morphine. { 'al·kə‚lȯid }

alkalometry [ANALY CHEM] The measurement of the quantity of alkaloids present in a substance. { ‚al·kə'läm·ə·trē }

alkalosis [MED] A condition of high blood alkalinity caused either by high intake of sodium bicarbonate or by loss of hydrochloric acid or blood carbon dioxide. { ‚al·kə'lō·səs }

alkamine [ORG CHEM] A compound that has both the alcohol and amino groups. Also known as amino alcohol. { 'al·kə‚mēn }

alkane [ORG CHEM] A member of a series of saturated aliphatic hydrocarbons having the empirical formula C_nH_{2n+2}. Also known as paraffin; paraffinic hydrocarbon. { 'al‚kān }

alkanet [MATER] A chemical indicator made from the root of *Alkanna tinctoria*. { 'al·kə‚net }

alkannin [ORG CHEM] $C_{16}H_{16}O_5$ A red powder, the coloring ingredient of alkanet; soluble in alcohol, benzene, ether, and oils; used as a coloring agent for fats and oils, wines, and wax. { al'ka·nən }

alkanolamine [ORG CHEM] One of a group of viscous, water-soluble amino alcohols of the aliphatic series. { ‚al·kə'näl·ə‚mēn }

alkaptonuria [MED] A hereditary metabolic disorder transmitted as an autosomal recessive in humans in which large amounts of homogentisic acid (alkapton) are excreted in the urine due to a deficiency of homogentisic acid oxidase. Also spelled alcaptonuria. { al‚kap·tə'nŭr·ē·ə }

Alkar process [CHEM ENG] Catalytic alkylation of aromatic hydrocarbons with olefins to produce alkylaromatics; for example, production of ethylbenzene from benzene and ethylene. { 'al‚kar 'präs·əs }

alkene [ORG CHEM] One of a class of unsaturated aliphatic hydrocarbons containing one or more carbon-to-carbon double bonds. { 'al‚kēn }

alkenones [GEOL] Long-chain (37–39 carbon atoms) di-, tri-, and tetraunsaturated methyl and ethyl ketones produced by certain phytoplankton (coccolithophorids), which biosynthetically control the degree of unsaturation (number of carbon-carbon double bonds) in response to the water temperature; the survival of this temperature signal in marine sediment sequences provides a temporal record of sea surface temperatures that reflect past climates. { 'al·kə‚nōnz }

alkoxide See alcoholate. { al'käk‚sīd }

alkoxy [ORG CHEM] An alkyl radical attached to a molecule by oxygen, such as the ethoxy radical. { al'käk·sē }

alkyd paint [MATER] A paint using an alkyd resin as the vehicle for the pigment. { 'al·kəd ‚pānt }

alkyd resin [ORG CHEM] A class of adhesive resins made from unsaturated acids and glycerol. { 'al·kəd 'rez·ən }

alkyl [ORG CHEM] An organic group that results from removal of a hydrogen atom from an acyclic, saturated hydrocarbon; may be represented in a chemical formula by R−. { 'al‚kil }

alkylamine [ORG CHEM] A compound consisting of an alkyl group attached to the nitrogen of an amine; an example is ethylamine, $C_2H_5NH_2$. { ¦al·kəl·ə‚mēn }

alkylaryl sulfonates [ORG CHEM] General name for alkylbenzene sulfonates. { ¦al·kəl·ə¦rəl 'səl·fə‚nāts }

alkylate [ORG CHEM] A product of the alkylation process in petroleum refining. { 'al·kə‚lāt }

alkylate bottom [CHEM ENG] Residue from fractionation of total alkylate which boils at a higher temperature than aviation gasolines. { 'al·kə‚lāt 'bäd·əm }

alkylated gasoline [MATER] A cleaning-burning gasoline with a high-octane rating; prepared by adding neohexane or some other alkylate. { 'al·kə‚lād·əd ‚gas·ə'lēn }

alkylation [CHEM ENG] A refinery process for chemically combining isoparaffin with olefin hydrocarbons. [ORG CHEM] A chemical process in which an alkyl radical is introduced into an organic compound by substitution or addition. { ‚al·kə'lā·shən }

alkylbenzene sulfonates [ORG CHEM] Widely used nonbiodegradable detergents, commonly dodecylbenzene or tridecylbenzene sulfonates. { ¦al·kəl¦ben‚zēn 'səl·fə‚nāts }

alkylene [ORG CHEM] An organic radical formed from an unsaturated aliphatic hydrocarbon; for example, the ethylene radical $C_2H_3−$. { 'al·kə‚lēn }

alkyl halide [ORG CHEM] A compound consisting of an alkyl group and a halogen; an example is ethylbromide. { 'al·kəl 'hal‚īd }

alkyloxonium ion [ORG CHEM] $(ROH_2)^+$ An oxonium ion containing one alkyl group. { ¦al·kil‚äk¦sō·nē·əm 'ī‚än }

alkyne [ORG CHEM] One of a group of organic compounds containing a carbon-to-carbon triple bond. { 'al‚kīn }

allachesthesia [MED] A tactile sensation experienced remote from the point of stimulation but on the same side of the body. { 'al·ək·əs'thēzh·ə }

allactite [MINERAL] $Mn_7(AsO_4)_2(OH)_8$ A brownish-red mineral consisting of a basic manganese arsenate. { ə'lak‚tīt }

allalinite [PETR] An altered gabbro with original texture and euhedral pseudomorphs. { ə'lal·ə‚nīt }

allanite [MINERAL] $(Ca,Ce,La,Y)_2(Al,Fe)_3Si_3O_{12}(OH)$ Monoclinic mineral distinguished from all other members of the epidote group of silicates by a relatively high content of rare earths. Also known as bucklandite; cerine; orthite; treanorite. { 'al·ə‚nīt }

allantoic acid [BIOCHEM] $C_4H_8N_4O_4$ A crystalline acid obtained by hydrolysis of allantoin; intermediate product in nucleic acid metabolism. { ¦al·ən¦tō·ik 'as·əd }

allantoin [BIOCHEM] $C_4H_6N_4O_3$ A crystallizable oxidation product of uric acid found in allantoic and amniotic fluids and in fetal urine. { ə'lan·tə‚wən }

allantoinase [BIOCHEM] An enzyme, occurring in nonmammalian vertebrates, that catalyzes the hydrolysis of allantoin. { ə'lan·tə·wə‚nās }

allantois [EMBRYO] A fluid-filled, saclike, extraembryonic membrane lying between the chorion and amnion of reptilian, bird, and mammalian embryos. { ə'lan·tə'wəs }

allantoxanic acid [BIOCHEM] $C_4H_3N_3O_4$ An acid formed by oxidation of uric acid or allantoin. { ‚a‚lan‚täk'san·ik 'as·əd }

allanturic acid [BIOCHEM] $C_3H_4N_2O_3$ An acid formed

principally by the oxidation of allantoin. { ¦al·ən¦tür·ik 'as·əd }

Allard's law [OPTICS] A mathematical formula defining the relationship between the intensity of a light, atmospheric conditions, and the amount of light received at any given distance. { 'al·ərdz ˌlȯ }

all-around traverse [ORD] A turn in a complete circle in a horizontal plane; a weapon has this capability when it can be turned 360° by its traversing mechanism. { ¦ȯl ə¦raůnd trə'vərs }

all-burnt time [AERO ENG] The point in time at which a rocket has consumed its propellants. { ¦ȯl ¦bərnt ˌtīm }

all-burnt velocity See burnout velocity. { ¦ȯl ¦bərnt və'läs·əd·ē }

all-channel tuning [COMMUN] The ability of a television set to receive ultra-high-frequency as well as very-high-frequency channels. { ¦ȯl ˌchan·əl 'tün·iŋ }

allcharite [MINERAL] A lead gray mineral, supposed to be a lead arsenic sulfide and known only crystallographically as orthorhombic crystals. { 'ȯl·kəˌrīt }

all-diffused monolithic integrated circuit [ELECTR] Microcircuit consisting of a silicon substrate into which all of the circuit parts (both active and passive elements) are fabricated by diffusion and related processes. { ¦ȯl də¦fyüzd ˌmän·ə'lith·ik 'in·tə ˌgrād·əd 'sər·kət }

Alleculidae [INV ZOO] The comb claw beetles, a family of mostly tropical coleopteran insects in the superfamily Tenebrionoidea. { ˌȯl·ə'kyü·lə ˌdē }

Allee's principle [GEN] The concept of an intermediate optimal population density by which groups of organisms often flourish best if neither too few nor too many individuals are present. { a'lēz ˌprin·sə·bəl }

allège [BUILD] A part of a wall which is thinner than the rest, especially the spandrel under a window. { a'lezh }

alleghanyite [MINERAL] $Mn_5(SiO_4)_2(OH)_2$ A pink mineral consisting of basic manganese silicate. { 'al·ə¦gä·nēˌīt }

Alleghenian [GEOL] Lower Middle Pennsylvanian geologic time. { 'al·ə¦gān·ē·ən }

Alleghenian life zone [ECOL] A biome that includes the eastern mixed coniferous and deciduous forests of New England. { ¦al·ə¦gān·ən 'līf ˌzōn }

Alleghenian orogeny [GEOL] Pennsylvanian and Early Permian orogenic episode which deformed the rocks of the Appalachian Valley and the Ridge and Plateau provinces. { ¦al·ə¦gān·ē·ən ȯ'räj·ə·nē }

allele [GEN] One of the alternate forms of a gene at a gene locus on a chromosome. Also known as allelomorph. { ə'lēl }

allele frequency [GEN] The fraction of all alleles at a given locus constituted by a particular allele in a population. Also known as gene frequency. { ə'lēl ¦frē·kwən·sē }

allelic exclusion [GEN] Expression of a single immunoglobulin allele by a B lymphocyte. { ə¦lē·lik iks'klü·zhən }

allelic mutant [GEN] A cell or organism with characters different from those of the parent due to alterations of one or more alleles. { ə¦lēl·ik ¦myüt·ənt }

allelochemic [PHYSIO] Pertaining to a semiochemical that acts as an interspecific agent of communication. { ə¦lē·lō¦kem·ik }

allelochemistry [CHEM] The science of compounds synthesized by one organism that stimulate or inhibit other organisms. { ə¦lē·lō¦kem·ə·strē }

allelomimetic behavior [PSYCH] Behavior in social animals in which each animal does the same thing as those nearby. { ə¦lē·lō·mə¦med·ik bi'hāv·yər }

allelomorph See allele. { ə'lē·ləˌmȯrf }

allelopathy [PL PHYS] The harmful effect of one plant or microorganism on another owing to the release of secondary metabolic products into the environment. { ˌa·lə'läp·ə·thē }

allelotoxin [PL PHYS] A toxic compound released in an allelopathic process. { ə¦lē·lō¦täk·sən }

allelotropism [BIOL] A mutual attraction between two cells or organisms. { ə¦lē·lō¦trä ˌpiz·əm }

allemontite [MINERAL] AsSb Rhombohedric, gray or reddish, native antimony arsenide occurring in reniform masses. Also known as arsenical antimony. { ˌal·ə'mänˌtīt }

Allende meteorite [GEOL] A meteorite that fell in Mexico in 1969 and contains inclusions that have been radiometrically dated at 4.56×10^9 years, the oldest found so far, presumably indicating the time of formation of the first solid bodies in the solar system. { aiˌyen·de 'mēd·ē·əˌrīt }

Allen-Doisy unit [BIOL] A unit for the standardization of estrogens. { ¦al·ən ¦dȯiz·ē ˌyü·nət }

allene [ORG CHEM] C_3H_4 An unsaturated aliphatic hydrocarbon with two double bonds. Also known as propadiene. { ˌa'lēn }

Allen red metal [MET] An alloy of copper and lead containing 50% lead and a small quantity of sulfur to hold the lead in solution. { ¦al·ən ¦red ˌmed·əl }

Allen screw [DES ENG] A screw or bolt which has an axial hexagonal socket in its head. { 'al·ən ˌskrü }

Allen's rule [VERT ZOO] The generalization that the protruding parts of a warm-blooded animal's body, such as the tail, ears, and limbs, are shorter in animals from cold parts of the species range than from warm parts. { 'al·ənz ˌrül }

Allen wrench [DES ENG] A wrench made from a straight or bent hexagonal rod, used to turn an Allen screw. { 'al·ən ˌrench }

allergen [IMMUNOL] Any antigen, such as pollen, a drug, or food, that induces an allergic state in humans or animals. { 'al·ərˌjen }

allergic arteritis [MED] Inflammation of the arterial walls resulting from an allergic state. { ə'lərj·ik ˌärt·ə'rīd·əs }

allergic dermatitis [MED] Inflammation of the skin following contact of an allergen with sensitized tissue. { ə'lərj·ik 'dər·mə'tīd·əs }

allergic granulomatosis [MED] A form of pulmonary vasculitis, frequently associated with asthma. { ə¦lər·jik ¦gran·yə·lō·mə'tō·səs }

allergic reaction See allergy. { ə'lərj·ik rē'ak·shən }

allergic rhinitis See hay fever. { ə'lərj·ik rī'nīd·əs }

allergic vasculitis syndrome [MED] A skin disease, possibly immunologic, characterized by ulcers which result from destructive inflammation of underlying blood vessels. { ə'lərj·ik vas·kyü'līd·əs ˌsin·drōm }

allergy [MED] A type of antigen-antibody reaction marked by an exaggerated physiologic response to a substance that causes no symptoms in nonsensitive individuals. Also known as allergic reaction. { 'al·ər·jē }

Allerod oscillation [CLIMATOL] A temporary increase in temperature during the closing stages of the Pleistocene ice age, dated in Europe about 9850-8850 B.C. { 'al·əˌräd ˌäs·ə'lā·shən }

allethrin [ORG CHEM] An insecticide, a synthetic pyrethroid, more effective than pyrethrin. { 'al·ə·thrən }

allevardite See rectorite. { ˌal·ə'värˌdīt }

alliaceous [SCI TECH] Having a garlic- or onionlike smell. { ¦al·ē¦ā·shəs }

alliance [SYST] A group of related families ranking between an order and a class. { ə'lī·əns }

allicin [MATER] An oily liquid extracted from garlic which has a sharp garlic odor; used in medicine as an antibacterial agent. { 'al·ə·sən }

allidochlor [ORG CHEM] $C_8H_{12}NOCl$ An amber liquid having slight solubility in water; used as a preemergence herbicide for vegetable crops, soybeans, sorghum, and ornamentals. { ə'lid·əˌklȯr }

alligator [VERT ZOO] Either of two species of archosaurian reptiles in the genus *Alligator* of the family Alligatoridae. { 'al·əˌgād·ər }

alligator clip [ELEC] A long, narrow spring clip with meshing jaws; used with test leads to make temporary connections quickly. Also known as crocodile clip. { 'al·əˌgād·ər ˌklip }

alligator effect See orange peel. { 'al·əˌgād·ər ə'fekt }

Alligatorinae [VERT ZOO] A subgroup of the crocodilian family Crocodylidae that includes alligators, caimans, *Melanosuchus*, and *Paleosuchus*. { ˌal·ə·gə'tȯr·əˌnē }

alligatoring [MATER] Cracking of a film of paint or varnish, with broad, deep cracks through one or more coats. Also known as crocodiling. [MET] **1.** A splitting of an end of a rolled steel slab in which the plane of the split is parallel to the rolled surface. Also known as fishmouthing. **2.** The roughening of a sheet-metal surface during forming due to the coarse grain of the metal used. { 'al·əˌgād·ər·iŋ }

alligator pear oil See avocado oil. { ¦al·əˌgād·ər ¦per ˌȯil }

ALLIGATOR

The American alligator (*Alligator mississipiensis*).

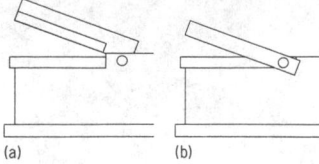

(a)   (b)

Alligator shears. (a) High-knife.
(b) Low-knife.

alligator shears [ENG] A cutting tool with a fixed lower blade and a movable upper blade (shearing arm) that moves in an arc around a fulcrum pin; used mainly for shearing applications that do not require great accuracy. { 'al·ə₁gād·ər ₁shirz }

alligator squeezer [MET] A tool with a fixed upper jaw and a movable lower jaw used to squeeze a ball of iron produced by the paddling process into a bloom or billet. { 'al·ə₁gād·ər ₁skwēz·ər }

alligator wrench [DES ENG] A wrench having fixed jaws forming a V, with teeth on one or both jaws. { 'al·ə₁gād·ər ₁rench }

all-inertial guidance [NAV] **1.** The guidance of a vehicle entirely by use of inertial devices. **2.** A self-contained system which can automatically determine the position, velocity, and attitude of a moving vehicle for the purpose of directing its future course. { ¦ȯl ə¦nər·shəl 'gīd·əns }

Allium [BOT] A genus of bulbous herbs in the family Liliaceae including leeks, onions, and chives. { 'al·ē·əm }

allivalite [PETR] A form of gabbro composed of anorthite and olivine; accessories are augite, apatite, and opaque iron oxides. { 'al·ə·və₁līt }

allo- [CHEM] Prefix applied to the stabler form of two isomers. { 'a·lō }

alloantibody [IMMUNOL] Antibody that reacts with an antigen occurring in a genetically different member of the same species. { ¦a·lō¦ant·i₁bäd·ē }

alloantigen See isoantigen. { ¦a·lō¦ant·i·jən }

allobar [NUC PHYS] A form of an element differing in its atomic weight from the naturally occurring form and hence being of different isotopic composition. [PHYS] A barometric pressure change. { 'a·lo₁bär }

allocate [COMPUT SCI] To place a portion of a computer memory or a peripheral unit under control of a computer program, through the action of an operator, program instruction, or executive program. [IND ENG] To assign a portion of a resource to an activity. { 'a·lō₁kāt }

allocation [PETRO ENG] The amount of oil or gas distributed from a well per unit time. { ₁al·ə'kā·shən }

allocheiria [MED] A form of allachesthesia in which the tactile sensation is experienced on the side opposite the one to which the stimulus was applied; seen in tabes dorsalis and other conditions. { ₁a·lō'kī·rē·ə }

allochem [GEOL] Sediment formed by chemical or biochemical precipitation within a depositional basin; includes intraclasts, oolites, fossils, and pellets. { 'a·lō₁kem }

allochemical metamorphism [PETR] Metamorphism that is accompanied by addition or removal of material so that the bulk chemical composition of the rock is changed. { ₁a·lō¦kem·ə·kəl ₁med·ə'mȯr₁fiz·əm }

allochetite [PETR] A porphyritic igneous rock composed of phenocrysts of labradorite, orthoclase, titanaugite, nepheline, magnetite, and apatite in a groundmass of augite, biotite, magnetite, hornblende, nepheline, and orthoclase. { ₁a·lə'ked₁īt }

allochoric [BOT] Describing a species that inhabits two or more closely related communities, such as forest and grassland, in the same region. { ₁a·lə'kȯr·ik }

allochromatic crystal [CRYSTAL] A crystal having photoconductive properties due to the presence of small particles within it. { ₁a·lə·krə'mad·ik 'kris·təl }

allochromy [PHYS] Emission of electromagnetic radiation that results from incident radiation at a different wavelength, as occurs in fluorescence or the Raman effect. { ¦a·lə ¦krȯm·ē }

allochthon [GEOL] A rock that was transported a great distance from its original deposition by some tectonic process, generally related to overthrusting, recumbent folding, or gravity sliding. { ə'läk·thən }

allochthonous [ECOL] Pertaining to organisms or organic sediments in a given ecosystem that originated in another system. [PETR] Of rocks whose primary constituents have not been formed in situ. { ə'läk·thə·nəs }

allochthonous coal [GEOL] A type of coal arising from accumulations of plant debris moved from their place of growth and deposited elsewhere. { ə'läk·thə·nəs ₁kōl }

allochthonous stream [HYD] A stream flowing in a channel that it did not form. { ə'läk·thə·nəs ₁strēm }

allodynia [MED] Painful response to a stimulus that does not normally elicit pain. { ₁a·lə'dī·nē·ə }

Alloeocoela [INV ZOO] An order of platyhelminthic worms of the class Turbellaria distinguished by a simple pharynx and a diverticulated intestine. { ə₁lē·ə'sēl·ə }

allogene [GEOL] A mineral or rock that has been moved to the site of deposition. Also known as allothigene; allothogene. { 'a·lə₁jēn }

allogeneic [IMMUNOL] Referring to a transplant made to a different genotype within the same species. Also spelled allogenic. { ¦al·ə·jə¦nē·ik }

allogeneic graft See allograft. { ¦al·ə·jə¦nē·ik 'graft }

allogenic [ECOL] Caused by external factors, as in reference to the change in habitat of a natural community resulting from drought. [GEOL] See allothogenic. [IMMUNOL] See allogeneic. { ¦a·lə¦jen·ik }

allograft [BIOL] Graft from a donor transplanted to a genetically dissimilar recipient of the same species. Also known as allogeneic graft. { 'a·lō₁graft }

Allogromiidae [INV ZOO] A little-known family of protozoans in the order Foraminiferida; adults are characterized by a chitinous test. { ₁a·lə·grə'mī·ə₁dē }

Allogromiina [INV ZOO] A suborder of marine and freshwater protozoans in the order Foraminiferida characterized by an organic test of protein and acid mucopolysaccharide. { ₁a·lə·grə'mī·ə·nə }

allogyric birefringence [OPTICS] The phenomenon in active optical media whereby circularly polarized light is transmitted unchanged but the velocity of right-handed circularly polarized light is different from that of left-handed. { ₁a·lō'jī·rik ₁bī·ri'frin·jəns }

Alloionematoidea [INV ZOO] A superfamily of parasitic nematodes belonging to the order Rhabditida, having either no lips or six small amalgamated lips, and a rhabditiform esophagus with a weakly developed valve in the posterior bulb. { ə₁lȯi·ō₁nem·ə'tȯid·ē·ə }

allomerism [CRYSTAL] A constancy in crystal form in spite of a variation in chemical composition. { ə'läm·ə₁riz·əm }

allometry [BIOL] **1.** The quantitative relation between a part and the whole or another part as the organism increases in size. Also known as heterauxesis; heterogony. **2.** The quantitative relation between the size of a part and the whole or another part, in a series of related organisms that differ in size. [MATH] A relation between two variables x and y that can be written in the form $y = ax^n$, where a and n are constants. { ə'läm·ə·trē }

allomone [PHYSIO] A chemical produced by an organism which induces in a member of another species a behavioral or physiological reaction favorable to the emitter; may be mutualistic or antagonistic. { 'a·lə₁mōn }

allomorphism See paramorphism. { ₁a·lə'mȯr₁fiz·əm }

allomorphite [MINERAL] A mineral consisting of barite that is pseudomorphous after anhydrite. { ₁a·lə'mȯr₁fīt }

allomorphosis [EVOL] Allometry in phylogenetic development. { ₁al·ō·mȯr'fō·səs }

Allomyces [MYCOL] A genus of aquatic phycomycetous fungi in the order Blastocladiales characterized by basal rhizoids, terminal hyphae, and zoospores with a single posterior flagellum. { a·lō'mī₁sēz }

alloparapatric speciation [EVOL] A mode of gradual speciation in which new species originate through populations that are initially allopatric but eventually become parapatric before effective reproductive isolation has evolved. { ¦al·ə₁pa·rə¦pa·trik ₁spē·sē'ā·shən }

allopathy [MED] A system of medicine that employs remedies whose effects are unlike those of the disease, in contrast to homeopathy. { ə'läp·ə·thē }

allopatric [ECOL] Referring to populations or species that occupy naturally exclusive, but usually adjacent, geographical areas. { ¦a·lō¦pa·trik }

allopatric speciation [ECOL] Differentiation of populations in geographical isolation to the point where they are recognized as separate species. { ¦al·ō¦pa·trik ₁spē·sē'ā·shən }

allopelagic [ECOL] Relating to organisms living at various depths in the sea in response to influences other than temperature. { ¦a·lō·pə'laj·ik }

allophane [GEOL] $Al_2O_3 \cdot SiO_2 \cdot nH_2O$ A clay mineral composed of hydrated aluminosilicate gel of variable composition; P_2O_5 may be present in appreciable quantity. { 'a·lə₁fān }

allophene [GEN] A mutant phenotype that can revert to a normal phenotype if it is transplanted to a wild-type host. { 'al·ə₁fēn }

allophore [HISTOL] A chromatophore which contains a red pigment that is soluble in alcohol; found in the skin of fishes, amphibians, and reptiles. { 'a·lō,fòr }

alloploid See allopolyploid. { 'al·ə,plóid }

allopolyploid [GEN] An organism or strain arising from a combination of chromosome sets from two diploid organisms belonging to different species. Also known as alloploid. { ,a·lə'päl·ə,plóid }

allopregnancy [IMMUNOL] A pregnancy in which the male partner is allogeneic with respect to the female. { ¦al·ō'preg·nən·sē }

allopurinol [PHARM] $C_5H_4ON_4$ The compound 4-hydroxypyrazolo-3,4-d-pyrimidine; inhibits xanthine oxidase, an enzyme required for uric acid formation. { ,a·lə'pyür·ə,nól }

all-or-none law [NEURO] The principle that transmission of a nerve impulse is either at full strength or not at all. { ¦ól ər ¦nən ,lò }

Allosaurus [PALEON] A carnivorous therapod dinosaur, 40 feet (12 meters) long, and weighing 1.5 tons, from the Late Jurassic Period that had muscular hindlimbs, small forelimbs (with three-fingered hands), and sharp teeth; similar to but smaller than Tyrannosaurus. { ,al·ə'sòr·əs }

allosome [GEN] **1.** Sex chromosome. **2.** Any atypical chromosome. { 'a·lō,sōm }

allostasis [PSYCH] The ongoing adaptive efforts of the body to maintain stability (homeostasis) in response to stressors. { ,al·ə'stā·səs }

allostatic load [PSYCH] The physiological wear and tear on the body that results from ongoing adaptive efforts to maintain stability (homeostasis) in response to stressors. { ,al·ə¦stad·ik 'lòd }

allosteric activation [BIOCHEM] The increase in an enzyme's activity that occurs when an allosteric effector binds to its specific regulatory site on the enzyme. Also known as positive allosteric control. { ¦a·lə¦stir·ik ,ak·tə'vā·shən }

allosteric control See feedback inhibition. { ¦a·lə¦stir·ik kən'tról }

allosteric effector [BIOCHEM] A small molecule that reacts either with a nonbinding site of an enzyme molecule, or with a protein molecule, and causes a change in the function of the molecule. Also known as allosteric modulator. { ¦a·lə¦stir·ik ə'fek·tər }

allosteric enzyme [BIOCHEM] Any of the regulatory bacterial enzymes, such as those involved in end-product inhibition. { ¦a·lə¦stir·ik 'en,zīm }

allosteric modulator See allosteric effector. { ¦a·lə¦stir·ik 'mäd·yə,lād·ər }

allosteric site [BIOCHEM] The inactive (or less active) region of an enzyme molecule. { ¦a·lə¦ster·ik 'sīt }

allosteric transition [BIOCHEM] A reversible exchange of one base pair for another on a protein molecule that alters the properties of the active site and changes the biological activity of the protein. { ¦a·lə¦stir·ik tranz'ish·ən }

allostery [BIOCHEM] The property of an enzyme able to shift reversibly between an active and an inactive configuration { 'a·lō,stir·ē }

allotetraploid See amphidiploid. { ,a·lō'te·trə,plóid }

Allotheria [PALEON] A subclass of Mammalia that appeared in the Upper Jurassic and became extinct in the Cenozoic. { ,a·lō'thir·ē·ə }

allothigene See allogene. { ə'läth·ə,jēn }

allothimorph [GEOL] A metamorphic rock constituent which retains its original crystal outlines in the new rock. { ə'läth·ə,mórf }

allothogene See allogene. { ə'läth·ə,jēn }

allothogenic [GEOL] Formed from preexisting rocks which have been transported from another location. Also known as allogenic. { ə¦läth·ə¦jen·ik }

allotrioblast See xenoblast. { ,a¦lə'trē·ə,blast }

Allotriognathi [VERT ZOO] An equivalent name for the Lampridiformes. { ə'lä·trē'äg·nə,thī }

allotriomorphic [MINERAL] Of minerals in igneous rock not bounded by their own crystal faces but having their outlines impressed on them by the adjacent minerals. Also known as anhedral; xenomorphic. { ə¦lä·trē·ə¦mór·fik }

allotriomorphism See allotropy. { ə¦lä·trē·ə¦mor,fiz·əm }

allotrope [CHEM] A form of an element showing allotropy. { 'a·lə,trōp }

allotropism See allotropy. { ,a·lə'trä,piz·əm }

allotropy [CHEM] The assumption by an element of two or more different forms or structures which are most frequently stable in different temperature ranges, such as different crystalline forms of carbon as charcoal, graphite, or diamond. Also known as allotriomorphism; allotropism. { ə'lä·trə,pē }

allotter [COMMUN] A telephone term referring to a distributor, associated with the finder control group relay assembly, which allots an idle line-finder in preparation for an additional call. { ə'läd·ər }

allotter relay [COMMUN] A telephone term referring to a relay of the line-finder circuit whose function is to preallot an idle line-finder to the next incoming call from the line, and to guard relays. { ə'läd·ər 'rē,lā }

allotype [IMMUNOL] Inherited variations in the genes coding for certain amino acid sequences in the constant region of the immunoglobulin molecules. [SYST] A paratype of the opposite sex to the holotype. { 'a·lə,tīp }

allowable bearing value [CIV ENG] The maximum permissible pressure on foundation soil that provides adequate safety against rupture of the soil mass or movement of the foundation of such magnitude as to impair the structure imposing the pressure. Also known as allowable soil pressure. { ə'laù·ə·bəl 'ber·iŋ ,val·yü }

allowable load [MECH] The maximum force that may be safely applied to a solid, or is permitted by applicable regulators. { ə'laù·ə·bəl 'lòd }

allowable oil [PETRO ENG] The oil an operator is permitted by law to remove from a well in one day. { ə'laù·ə·bəl 'oil }

allowable soil pressure See allowable bearing value. { ə'laù·ə·bəl 'soil ,presh·ər }

allowable stress [MECH] The maximum force per unit area that may be safely applied to a solid. { ə'laù·ə·bəl 'stres }

allowance [DES ENG] An intentional difference in sizes of two mating parts, allowing clearance usually for a film of oil, for running or sliding fits. { ə'laù·əns }

allowed energy bands [SOLID STATE] The restricted regions of possible electron energy levels in a solid. { ə'laùd 'en·ər·jē ,banz }

allowed hours See standard hour. { ə'laùd 'aù·ərz }

allowed time [IND ENG] Amount of time allowed each employee for personal needs during a work cycle. { ə'laùd 'tīm }

allowed transition [QUANT MECH] A transition between two states which is permitted by the selection rules and which consequently has a relatively high priority. { ə'laùd tranz'ish·ən }

alloxan [BIOCHEM] $C_4H_2N_2O_4$ Crystalline oxidation product of uric acid; induces diabetes experimentally by selective destruction of pancreatic beta cells. Also known as mesoxalyurea. { ə'läk·sən }

alloy [MATER] A composite plastic produced by blending and melting together different polymers. [MET] A metal product containing two or more elements as a solid solution, as an intermetallic compound, or as a mixture of metallic phases. { 'a,lói }

Alloy 750 [MET] A bearing alloy, containing 6.5% tin, 2.5% silicon, 1% copper, 0.5% nickel, and the remainder aluminum; used for automobile engine bearings. { 'a,lói ,sev·ən 'fif·tē }

alloy adhesive [MATER] An adhesive compounded from resins of two or more differential chemical classes, such as thermosetting and elastomeric, in order to combine performance characteristics. { 'a,lói ad,hē·ziv }

alloying [MET] The addition of a metal or alloy to another metal or alloy. { ə'lói·iŋ }

alloy junction [ELECTR] A junction produced by alloying one or more impurity metals to a semiconductor to form a p or n region, depending on the impurity used. Also known as fused junction. { 'a,lói jəŋk·shən }

alloy-junction diode [ELECTR] A junction diode made by placing a pill of doped alloying material on a semiconductor material and heating until the molten alloy melts a portion of the semiconductor, resulting in a pn junction when the dissolved semiconductor recrystallizes. Also known as fused-junction diode. { 'a,lói ¦jəŋk·shən 'dī,ōd }

alloy-junction transistor [ELECTR] A junction transistor made by placing pellets of a p-type impurity such as indium above and below an n-type wafer of germanium, then heating until the impurity alloys with the germanium to give a pnp

transistor. Also known as fused-junction transistor. { 'a,lȯi ¦jəŋk·shən tranz'is·tər }

alloy nuclear fuel [NUCLEO] A material used in nuclear reactors that is an alloy of a fissionable substance and a nonfissionable metal or metals. { 'a,lȯi 'nü·klē·ər 'fyül }

alloy plating [MET] The codeposition of two or more metals on an electrode by electrolysis. { 'a,lȯi ¦plāt·iŋ }

alloy steel [MET] A steel whose distinctive properties are due to the presence of one or more elements other than carbon. { ¦a,lȯi ¦stēl }

allozygote [GEN] An individual who has different mutations in the two alleles at a given locus. { ,al·ə'zī,gōt }

allozyme [GEN] One of two or more forms of an enzyme that are specified by allelic genes. { 'al·ə,zīm }

all-pass network [ELECTR] A network designed to introduce a phase shift in a signal without introducing an appreciable reduction in energy of the signal at any frequency. { ¦ȯl ¦pas 'net,wərk }

all-sky camera [OPTICS] A camera directed vertically downward toward a horizontal convex mirror so as to photograph the entire sky simultaneously. { ¦ȯl ,skī 'kam·rə }

allspice [BOT] The dried, unripe berries of a small, tropical evergreen tree, *Pimenta officinalis*, of the myrtle family; yields a pungent, aromatic spice. { 'ȯl,spīs }

allspice oil *See* pimenta oil. { 'ȯl,spīs ,ȯil }

all-translational system [CONT SYS] A simple robotic system in which there is no rotation of the robot or its components during movements of the robot's body. { ¦ȯl ,tranz'lā·shən·əl 'sis·təm }

allulose [ORG CHEM] $CH_2OHCO(CHOH)_3CH_2OH$ A constituent of cane sugar molasses; it is nonfermentable. { 'al·yə,lōs }

alluvial [GEOL] **1.** Of a placer, or its associated valuable mineral, formed by the action of running water. **2.** Pertaining to or consisting of alluvium, or deposited by running water. { ə'lüv·ē·əl }

alluvial cone [GEOL] An alluvial fan with steep slopes formed of loose material washed down the slopes of mountains by ephemeral streams and deposited as a conical mass of low slope at the mouth of a gorge. Also known as cone delta; cone of dejection; cone of detritus; debris cone; dry delta; hemicone; wash. { ə'lüv·ē·əl 'kōn }

alluvial dam [GEOL] A sedimentary deposit which is built by an overloaded stream and dams its channel; especially characteristic of distributaries on alluvial fans. { ə'lüv·ē·əl 'dam }

alluvial deposit *See* alluvium. { ə'lüv·ē·əl di'päz·ət }

alluvial fan [GEOL] A fan-shaped deposit formed by a stream either where it issues from a narrow mouutain valley onto a plain or broad valley, or where a tributary stream joins a main stream. { ə'lüv·ē·əl 'fan }

alluvial flat [GEOL] A small alluvial plain having a slope of about 5 to 20 feet per mile (1.5 to 6 meters per 1600 meters) and built of fine sandy clay or adobe deposited during flood. { ə'lüv·ē·əl 'flat }

alluvial mining [MIN ENG] The exploitation of alluvial deposits by dredging, hydraulicking, or drift mining. { ə'lüv·ē·əl 'mīn·iŋ }

alluvial ore deposit [GEOL] A deposit in which the valuable mineral particles have been transported and left by a stream. { ə'lüv·ē·əl ȯr di¦päz·ət }

alluvial plain [GEOL] A plain formed from the deposition of alluvium usually adjacent to a river that periodically overflows. Also known as aggraded valley plain; river plain; wash plain; waste plain. { ə'lüv·ē·əl 'plān }

alluvial slope [GEOL] A surface of alluvium which slopes down from mountainsides and merges with the plain or broad valley floor. { ə'lüv·ē·əl 'slōp }

alluvial soil [GEOL] A soil deposit developed on floodplain and delta deposits. { ə'lüv·ē·əl 'sȯil }

alluvial terrace [GEOL] A terraced embankment of loose material adjacent to the sides of a river valley. Also known as built terrace; drift terrace; fill terrace; stream-built terrace; wave-built platform; wave-built terrace. { ə'lüv·ē·əl 'ter·əs }

alluvial valley [GEOL] A valley filled with a stream deposit. { ə'lüv·ē·əl 'val·ē }

alluviation [GEOL] The deposition of sediment by a river. { ə,lüv·ē'ā·shən }

alluvion *See* alluvium. { ə'lüv·ē·ən }

ALLSPICE

A branch of *Pimenta officinalis* with the berry fruit.

alluvium [GEOL] The detrital materials that are eroded, transported, and deposited by streams; an important constituent of shelf deposits. Also known as alluvial deposit; alluvion. { ə'lüv·ē·əm }

all-wave receiver [ELECTR] A radio receiver capable of being tuned from about 535 kilohertz to at least 20 megahertz; some go above 100 megahertz and thus cover the FM band also. { ¦ȯl ¦wāv ri'sē·vər }

all-weather aircraft [AERO ENG] Aircraft that are designed or equipped to perform by day or night under any weather conditions. { ¦ȯl ¦weth·ər 'er,kraft }

all-weather airport [CIV ENG] An airport with facilities to permit the landing of qualified aircraft and aircrewmen without regard to operational weather limits. { ¦ȯl ¦weth·ər 'er,pȯrt }

all-weather fighter [AERO ENG] A fighter aircraft equipped with radar and other special devices which enable it to intercept its target in the dark, or in daylight weather conditions that do not permit visual interception; it is usually a multiplace (pilot plus navigator-observer) airplane. { ¦ȯl ¦weth·ər 'fīd·ər }

all-weather landing system [NAV] An instrument landing system having optimum operational capability in low and zero visibility. { ¦ȯl ¦weth·ər 'land·iŋ ,sis·təm }

all-weld-metal test specimen [MET] A specimen composed entirely of weld metal used in a weld tension test wherein the axis of the weld from which it is derived is parallel to the axis of the test bar. { ¦ȯl ¦weld ¦med·əl 'test ,spes·ə·mən }

allyl- [ORG CHEM] A prefix used in names of compounds whose structure contains an allyl cation. { 'al·əl }

allylacetone [ORG CHEM] $CH_2CHCH_2CH_2COCH_3$ A colorless liquid, soluble in water and organic solvents; used in pharmaceutical synthesis, perfumes, fungicides, and insecticides. { ,al·əl'as,tōn }

allyl alcohol [ORG CHEM] CH_2CHCH_2OH Colorless, pungent liquid, boiling at 96°C; soluble in water; made from allyl chloride by hydrolysis. { 'al·əl 'al·kə,hȯl }

allylamine [ORG CHEM] $CH_2CHCH_2NH_2$ A yellow oil that is miscible with water; boils at 58°C; prepared from mustard oil. { ¦al·əl·ə¦mēn }

allyl bromide [ORG CHEM] C_3H_5Br A colorless to light yellow, irritating toxic liquid with a boiling point of 71.3°C; soluble in organic solvents; used in organic synthesis and for the manufacture of synthetic perfumes. { 'al·əl 'brō,mīd }

allyl cation [ORG CHEM] A carbonium cation with a structure usually represented as $CH_2{=}CH{-}CH_2{}^+$; attachment site is the saturated carbon atom. { 'al·əl 'kat,ī·ən }

allyl chloride [ORG CHEM] CH_2CHCH_2Cl A volatile, pungent, toxic, flammable, colorless liquid, boiling at 46°C; insoluble in water; made by chlorination of propylene at high temperatures. { 'al·əl 'klȯr,īd }

allyl cyanide [ORG CHEM] C_4H_5N A liquid with an onionlike odor and a boiling point of 119°C; slightly soluble in water; used as a cross-linking agent in polymerization. { 'al·əl 'sī·ə,nīd }

allylene [ORG CHEM] $CH_3C{:}CH$ An acetylenic, three-carbon hydrocarbon; a colorless gas boiling at −24°C; soluble in ether. Also known as propyne. { 'al·ə,lēn }

allylic hydrogen [ORG CHEM] In an organic molecule, a hydrogen attached to a carbon atom that is adjacent to a double bond. { ə'lil·ik 'hī·drə·jən }

allylic substitution [ORG CHEM] A reaction occurring at position 1 of an allylic system (with the double bond between positions 2 and 3) in which the incoming group is attached to the same atom (position 1) as the leaving group, or the incoming group is attached at position 3, with the double bond moving from positions 2 and 3 to positions 1 and 2. { ə'lil·ik ,səb·stə'tü·shən }

allyl isothiocyanate [ORG CHEM] $CH_2CH{:}CH_2NCS$ A pungent, colorless to pale-yellow liquid; soluble in alcohol, slightly soluble in water; irritating odor; boiling point 152°C; used as a fumigant and as a poison gas. Also known as mustard oil. { 'al·əl ¦ī·sō,thī·ō'sī·ə,nāt }

allyl mercaptan [ORG CHEM] CH_2CHCH_2SH A colorless liquid with a boiling point of 67–68°C; soluble in ether and alcohol; used as intermediate in pharmaceutical manufacture. { 'al·əl mər'kap,tan }

allyl plastic *See* allyl resin. { 'al·əl 'plas·tik }

allyl resin [ORG CHEM] Any of a class of thermosetting synthetic resins derived from esters of allyl alcohol or allyl chloride; used in making cast and laminated products. Also known as allyl plastic. { 'al·əl 'rez·ən }

allyl sulfide [ORG CHEM] $(CH_2CHCH_2)_2S$ A colorless liquid with a garliclike odor and a boiling point of 139°C; used in synthetic oil of garlic. { 'al·əl 'səl,fīd }

allylthiourea [ORG CHEM] $C_3H_5NHCSNH_2$ A white, crystalline solid that melts at 78°C; soluble in water; used as a corrosion inhibitor. { ,al·əl,thī·ō,yu'rē·ə }

allyltrichlorosilane [ORG CHEM] $CH_2CHCH_2SiCl_3$ A pungent, colorless liquid with a boiling point of 117.5°C; used as an intermediate for silicones. { ,al·əl,trī,klȯr'äs·ə,lān }

allylurea [ORG CHEM] $C_4H_8N_2O$ Crystals with a melting point of 85°C; freely soluble in water and alcohol; used to manufacture allylthiourea and other corrosion inhibitors. { ,al·əl·yu'rē·ə }

allyxycarb [ORG CHEM] $C_{16}H_{22}N_2O_2$ A yellow, crystalline compound used as an insecticide for fruit orchards, vegetable crops, rice, and citrus. { ə'liks·ə,karb }

alm [ECOL] A meadow in alpine or subalpine mountain regions. { älm }

almanac [SCI TECH] A book that contains astronomical or meteorological data, arranged according to days, weeks, and months of a given year, and may also contain diverse information of a nonastronomical character. { 'ȯl·mə,nak }

almandine [MINERAL] $Fe_3Al_2(SiO_4)_3$ A variety of garnet, deep red to brownish red, found in igneous and metamorphic rocks in many parts of world; used as a gemstone and an abrasive. Also known as almandite. { 'al·mən,dēn }

almandite *See* almandine. { 'al·mən,dīt }

almendro [MATER] The wood of tonka bean tree species in Panama *(Coumarouna panamensis);* resistant to marine borers and used as a substitute for greenheart in marine construction. { äl'men·drō }

almeriite *See* natroalunite. { ,al·mə'rē,īt }

almon [MATER] A type of lauan from the tree *Shorea eximia;* the wood is fairly strong and hard with a coarse texture. { 'al,mōn }

almond [BOT] *Prunus amygdalus.* A small deciduous tree of the order Rosales; it produces a drupaceous edible fruit with an ellipsoidal, slightly compressed nutlike seed. { 'ä·mənd }

almost every [MATH] A proposition concerning the points of a measure space is said to be true at almost every point, or to be true almost everywhere, if it is true for every point in the space, with the exception at most of a set of points which form a measurable set of measure zero. { 'ȯl,mōst 'ev·rē }

almost-perfect number [MATH] An integer that is 1 greater than the sum of all its factors other than itself. { 'ȯl,mōst 'pər·fik 'nəm·bər }

almost-periodic function [MATH] A continuous function $f(x)$ such that for any positive number ϵ there is a number M so that for any real number x, any interval of length M contains a nonzero number t such that $|f(x + t) - f(x)| < \epsilon$. { 'ȯl,mōst ,pir·ē'äd·ik 'fəŋk·shən }

Almquist unit [BIOL] A unit for the standardization of vitamin K. { 'äm,kwist ,yü·nət }

almucantar *See* parallel of altitude. { ¦al·myü¦kan·tər }

alnico [MET] One of a series of ferrous alloys containing aluminum, nickel, and cobalt, valued because of their highly retentive magnetic properties; usually designated with a roman-numeral number, such as alnico VII. Also known as aluminum-nickel-cobalt alloy. { 'al·ni,kō }

alnico magnet [ELECTROMAG] A permanent magnet made of alnico. { 'al·ni,kō 'mag·nət }

Alnilam [ASTRON] A star in the constellation Orion. { al'nil·əm }

alnoite [PETR] A variety of biotite lamprophyres characterized by lepidomelane phenocrysts; it is feldspar-free but contains melitite, perovskite, olivine, and carbonate in the matrix. { 'al·nə,wit }

aloe [PHARM] The dried resinous juice extracted from the leaves of the genus *Aloe,* especially *A. vulgaris* of the West Indies and *A. perryi* of Africa; used in purgative mixtures. { 'a,lō }

aloe lace [TEXT] A fragile lace fabric made from the fibers of an aloe plant. { 'a,lō 'lās }

aloe wood oil *See* oriental linaloe. { 'a,lō ¦wud ,oil }

aloha [COMMUN] A radio-channel random-access technique that depends on positive acknowledgement of correct receipt for error control. { ə'lō·ə }

aloisite [MINERAL] A brown to violet mineral consisting of a hydrous subsilicate of calcium, iron, magnesium, and sodium, and occurring in amorphous masses. { ,a·lə'wis·ē,īt }

alongshore current *See* littoral current. { ə'lȯŋ,shȯr 'kər·ənt }

alopecia [MED] Loss of hair; baldness. { ,a·lə'pē·shə }

alopecia areata [MED] A type of alopecia that is characterized as an autoimmune disorder and usually presents with one or many oval, slightly erythematous, asymptomatic patches of hair loss. { ,a·lə'pē·shə ,a·rē'ad·ə }

Alopiidae [VERT ZOO] A family of pelagic isurid elasmobranchs commonly known as thresher sharks because of their long, whiplike tail. { ə'pī·ə,dē }

alpaca [VERT ZOO] *Lama pacos.* An artiodactyl of the camel family (Camelidae); economically important for its long, fine wool. { al'pak·ə }

alpage [ECOL] A summer grazing area composed of natural plant pasturage in upland or mountainous regions. { 'al·pəj }

alpenglow [METEOROL] A reappearance of sunset colors on a mountain summit after the original mountain colors have faded into shadow; also, a similar phenomenon preceding the regular coloration at sunrise. { 'al·pən,glō }

alpestrine [ECOL] Referring to organisms that live at high elevation but below the timberline. Also known as subalpine. { ,al'pes·trən }

alpha [ELECTR] The ratio between the change in collector current and the change in emitter current of a transistor. [SCI TECH] The first letter in the Greek alphabet: α, A. { 'al·fə }

alpha-adrenergic receptor [CYTOL] *See* alpha receptor. { ¦al·fə ,ad·rə¦nər·jik ri'sep·tər }

alphabet [SCI TECH] Any ordered set of unique graphics called characters, such as the 26 letters of the Roman alphabet. { 'al·fə,bet }

alphabetic character [COMPUT SCI] A letter or other symbol used to form data, other than a digit. { ¦al·fə¦bed·ik 'kar·ik·tər }

alphabetic coding [COMPUT SCI] **1.** Abbreviation of words for computer input. **2.** A system of coding with a number system of base 26, the letters of the alphabet being used instead of the cardinal numbers. { ¦al·fə¦bed·ik 'kōd·iŋ }

alphabetic string *See* character string. { ¦al·fə¦bed·ik 'striŋ }

alphabet length [GRAPHICS] The measure in points of a complete lowercase alphabet in a specified typeface and size that is used to determine the average number of characters per pica. { 'al·fə·bet ,leŋkth }

alpha brass [MET] An alloy of copper and zinc containing up to 36% zinc dissolved, rather than chemically combined, with the copper; ductile, easily cold-worked, and corrosion resistant; used for hot-water pipes. { 'al·fə ,bras }

alpha bronze [MET] A copper-tin alloy; commercial forms containing 4–5% tin are used in coinage, springs, and turbine blades. { 'al·fə ,bränz }

alpha cell [HISTOL] Any of the acidophilic chromophiles in the anterior lobe of the adenohypophysis. { 'al·fə ,sel }

alpha cellulose [ORG CHEM] A highly refined, insoluble cellulose from which sugars, pectin, and other soluble materials have been removed. Also known as chemical cellulose. { 'al·fə 'sel·yə,lōs }

Alpha Centauri [ASTRON] A double star, the brightest in the constellation Centaurus; apart from the sun, it is the nearest bright star to earth, about 4.3 light-years away; spectral classification G2. Also known as Rigil Kent. { ¦al·fə sen'tȯ·rē }

alpha cross section [NUCLEO] Total cross section for interaction with alpha particles. { 'al·fə 'krȯs ,sek·shən }

alpha cutoff frequency [ELECTR] The frequency at the high end of a transistor's range at which current amplification drops 3 decibels below its low-frequency value. { 'al·fə 'kəd,ȯf ,frē·kwən·sē }

alpha decay [NUC PHYS] A radioactive transformation in which an alpha particle is emitted by a nuclide. { 'al·fə di'kā }

alpha emission [NUC PHYS] Ejection of alpha particles from the atom's nucleus. { 'al·fə i'mish·ən }

alpha fetoprotein [IMMUNOL] A serum protein that has become associated with the detection of certain types of cancer and fetal abnormalities. Abbreviated AFP. { 'al·fə ¦fēd·ō'prō,tēn }

ALMOND

Almond twig with leaves and fruit.

ALPACA

The alpaca (*Lama pacos*), sheared every 2 years to yield 18–24 pounds (8–11 kilograms) of wool in its lifetime.

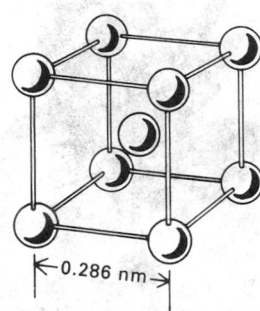

ALPHA IRON

← 0.286 nm →

The body-centered cubic crystal structure of alpha iron.

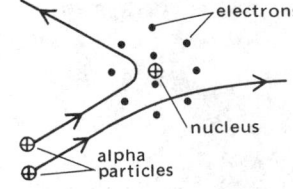

ALPHA–PARTICLE SCATTERING

electrons

nucleus

alpha particles

Scattering by an atom of alpha particles.

alphageometric technique *See* alphamosaic technique. { ¦al·fə‚jē·ə'me·trik ‚tek‚nēk }

alpha globulin [BIOCHEM] A heterogeneous fraction of serum globulins containing the proteins of greatest electrophoretic mobility. { 'al·fə 'gläb·yə·lən }

alpha gypsum [MATER] A specially processed gypsum having low consistency and high compressive strength, often exceeding 5000 pounds per square inch (34.5 megapascals). { 'al·fə 'jip·səm }

alpha helix [MOL BIO] A spatial configuration of the polypeptide chains of proteins in which the chain assumes a helical form, 0.54 nanometer in pitch, 3.6 amino acids per turn, presenting the appearance of a hollow cylinder with radiating side groups. { 'al·fə 'hē·liks }

alpha hemolysis [MICROBIO] Partial hemolysis of red blood cells with green discoloration in a blood agar medium by certain hemolytic streptococci. { 'al·fə hi'mäl·ə·səs }

alpha iron [MET] Iron with a body-centered cubic structure which is stable below 1670°F (910°C). { 'al·fə ‚ī·ərn }

alpha irradiation [NUCLEO] Subjection of a substance to a flux of alpha particles. { 'al·fə ir‚ād·ē'ā·shən }

alphameric characters *See* alphanumeric characters. { ¦al·fə¦mer·ik 'kar·ik·tərz }

alphameric typebar [COMPUT SCI] A metal bar containing the alphabet, the ten numerical characters, and the ampersand, in use in electromechanical accounting machines. { ¦al·fə¦mer·ik 'tīp‚bär }

alphamosaic technique [COMPUT SCI] In computer graphics, a technique for displaying very-low-resolution images by constructing them from a set of elementary graphics characters. Also known as alphageometric technique. { ¦al·fə·mō'zā·ik ‚tek‚nēk }

alphanumeric characters [COMPUT SCI] All characters used by a computer, including letters, numerals, punctuation marks, and such signs as $, @, and #. Also known as alphameric characters. { ¦al·fə·nü¦mer·ik 'kar·ik·tərz }

alphanumeric display device [ELECTR] A device which visibly represents alphanumeric output information from some signal source. { ¦al·fə·nü¦mer·ik dis'plā di‚vīs }

alphanumeric grid *See* atlas grid. { ¦al·fə·nü¦mer·ik 'grid }

alphanumeric instruction [COMPUT SCI] The name given to instructions which can be read equally well with alphabetic or numeric kinds of fields of data. { ¦al·fə·nü¦mer·ik in'strək·shən }

alphanumeric pager [COMMUN] A receiver in a radio paging system that contains a liquid-crystal display which can display text or numeric messages. { ¦al·fə·nü¦mer·ik 'pā·jər }

alphanumeric reader [ELECTR] A device capable of reading alphabetic, numeric, and special characters and punctuation marks. { ¦al·fə·nü¦mer·ik 'rēd·ər }

alpha olefin [ORG CHEM] An olefin where the unsaturation (double bond) is at the alpha position, that is, between the two end carbons of the carbon chain. { 'al·fə 'ō·lə‚fən }

alpha particle [ATOM PHYS] A positively charged particle consisting of two protons and two neutrons, identical with the nucleus of the helium atom; emitted by several radioactive substances. { 'al·fə ‚pärd·ə·kəl }

alpha-particle detector [NUCLEO] A device used to indicate the presence of alpha particles. { 'al·fə 'pärd·ə·kəl di‚tek·tər }

alpha-particle scattering [ATOM PHYS] Deviation at various angles of a stream of alpha particles passing through a foil of material. { 'al·fə 'pärd·ə·kəl 'skad·ər·iŋ }

alpha position [ORG CHEM] In chemical nomenclature, the position of a substituting group of atoms in the main group of a molecule; for example, in a straight-chain compound such as α-hydroxypropionic acid ($CH_3CHOHCOOH$), the hydroxyl radical is in the alpha position. { 'al·fə pə‚zish·ən }

alpha ray [NUCLEO] A stream of alpha particles. { 'al·fə ‚rā }

alpha-ray vacuum gage [ENG] An ionization gage in which the ionization is produced by alpha particles emitted by a radioactive source, instead of by electrons emitted from a hot filament; used chiefly for pressures from 10^{-3} to 10 torrs. Also known as alphatron. { 'al·fə ‚rā 'vak·yüm ‚gāj }

alpha receptor [CYTOL] Any of a group of receptors on cell membranes that are thought to be associated with vasoconstriction, relaxation of intestinal muscle, and contraction of the nictitating membrane, iris dilator muscle, smooth muscle of the spleen, and muscular layer of the uterine wall. Also called alpha-adrenergic receptor. { ¦al·fə ri¦sep·tər }

alpha rhythm [PHYSIO] An electric current from the occipital region of the brain cortex having a pulse frequency of 8 to 13 per second; associated with a relaxed state in normal human adults. { 'al·fə ‚rith·əm }

alpha rule *See* renaming rule. { 'al·fə ‚rül }

alpha system [COMMUN] A signaling system in which alphabetic characters designate the signaling code to be used. { 'al·fə ‚sis·təm }

alpha taxonomy [SYST] The level of taxonomic study concerned with characterizing and naming species. { 'al·fə tak'sän·ə·mē }

alpha test [COMPUT SCI] A test of software carried out at the user's location and using actual data. { 'al·fə ‚test }

alpha test site [COMPUT SCI] A place where a complete computer system is tested with actual data and transactions. { 'al·fə ‚test ‚sīt }

alphatopic [NUCLEO] Pertaining to the relationship between two nuclides that differ in composition or in mass by an alpha particle. { ¦al·fə¦täp·ik }

alphatron *See* alpha-ray vacuum gage. { 'al·fə‚trän }

Alpheidae [INV ZOO] The snapping shrimp, a family of decapod crustaceans that is included in the section Caridea. { ‚al'fē·ə‚dē }

Alphonsus [ASTRON] A moon crater. { al'fän·səs }

Alpides [GEOL] Great east-west structural belt including the Alps of Europe and the Himalayas and related mountains of Asia; mostly folded in Tertiary times. { 'al·pə‚dēz }

alpine [ECOL] Any plant native to mountain peaks or boreal regions. [GEOL] Similar to or characteristic of a lofty mountain or mountain system. { 'al‚pīn }

alpine glacier [HYD] A glacier lying on or occupying a depression in mountainous terrain. Also known as mountain glacier. { 'al‚pīn 'glā·shər }

Alpine orogeny [GEOL] Jurassic through Tertiary orogeny which affected the Alpides. { 'al‚pīn ȯ'räj·ə·nē }

alpine tundra [ECOL] Large, flat or gently sloping, treeless tracts of land above the timberline. { 'al‚pīn 'tən‚drə }

alpine-type facies [PETR] High-pressure, low-temperature (150–400°C) dynamothermal metamorphism characterized by the presence of the pumpellyite and glaucophane schist facies. { ¦al‚pīn¦tīp 'fā‚shez }

alpinotype tectonics [GEOL] Tectonics of the alpine-type geosynclinal mountain belts characterized by deep-seated plastic folding, plutonism, and lateral thrusting. { al'pē·nō‚tīp ‚tek'tän·iks }

Alport's syndrome [MED] A very rare genetic disease of the glomeruli that results in glomerular scarring and eventual renal failure within the second or third decade of life. { 'al‚pȯrts ‚sin‚drōm }

alsbachite [PETR] A plutonic rock of sodic plagioclase, quartz, and subordinate orthoclase and accessory garnet, biotite, and muscovite; a variety of porphyritic granodiorite. { 'ȯlz‚bä‚kīt }

alsifilm [MATER] A bentonite gel in the form of sheets used primarily for electrical insulation, because of its properties of heat and oil resistance. { 'alz·ə‚film }

alsike clover [BOT] *Trifolium hybridium.* A species of clover with pink or white flowers that is grown for forage; native to Alsike, Sweden. { 'al‚sik 'klō·ver }

alstonite *See* bromlite. { 'ȯlz·tə‚nīt }

alt *See* altitude.

Altaid orogeny [GEOL] Mountain building in Central Europe and Asia that occurred from the late Carboniferous to the Permian. { ¦al‚tād ȯ'räj·ə·nē }

Altair [ASTRON] A star that is 16.5 light-years from the sun; spectral type A7IV–V. Also known as α Aquilae. { al'tīr *or* 'al‚ter }

altaite [MINERAL] PbTe A tin-white lead-tellurium mineral occurring as isometric crystals with tin ores in central Asia. { al'tā‚īt }

Altar *See* Ara. { 'al‚tär }

altazimuth [ENG] An instrument equipped with both horizontal and vertical graduated circles, for the simultaneous observation of horizontal and vertical directions or angles. Also known as astronomical theodolite; universal instrument. { al'taz·ə·məth }

alt-azimuth mounting *See* altitude-azimuth mounting. { ˌalt 'az·ə·məth ˌmaunt·iŋ }

altazimuth telescope [OPTICS] A telescope equipped with an altazimuth mounting. { al'taz·ə·məth 'tel·ə‚skōp }

alteration [GRAPHICS] In printing composition, a change that is made in typeset copy. [PETR] A change in a rock's mineral composition. { ˌȯl·tə'rā·shən }

alteration enzyme [IMMUNOL] A protein of bacteriophage T4 that is injected into a host bacterium along with the deoxyribonucleic acid of the bacteriophage and that modifies the ribonucleic acid polymerase of the host so that the ribonucleic acid is unable to initiate transcription at host promoters. { ȯl·tə'rā·shən ‚en‚zīm }

alter ego [PSYCH] A close friend who represents a second self to an individual. { 'ȯl·tər 'ē‚gō }

alternant hydrocarbon [ORG CHEM] A member of a class of conjugated molecules whose carbon atoms can be divided into two sets so that members of one set are formally bonded only to members of the other set. { 'ȯl·tər·nənt ‚hī·drə'kär·bən }

alternate [BOT] **1.** Of the arrangement of leaves on opposite sides of the stem at different levels. **2.** Of the arrangement of the parts of one whorl between members of another whorl. { 'ȯl·tər·nət }

alternate airport [NAV] An airport designated in a pilot's flight plan at which an aircraft will have the capability to land if a landing at the intended airport becomes inadvisable. { 'ȯl·tər·nət 'er‚pȯrt }

alternate angles [MATH] A pair of nonadjacent angles that a transversal forms with each of two lines; they lie on opposite sides of the transversal, and are both interior, or both exterior, to the two lines. { 'ȯl·tər·nət 'aŋ·gəlz }

alternate-channel interference [COMMUN] Interference that is caused in one communications channel by a transmitter operating in the next channel beyond an adjacent channel. Also known as second-channel interference. { 'ȯl·tər·nət ‚chan·əl in·tər'fir·əns }

alternate energy [ENG] Any source of energy other than fossil fuels that is used for constructive purposes. { 'ȯl·tər·nət 'en·ər·jē }

alternate immersion test [MET] A corrosion test in which a specimen is repeatedly carried through a cycle of immersion in and removal from a corrosive medium over definite time intervals. { 'ȯl·tər·nət im'ər·zhən ‚test }

alternate index *See* secondary index. { 'ȯl·tər·nət 'in‚deks }

alternate key [COMPUT SCI] A key on a computer keyboard that does not itself generate a character but changes the nature of the character generated by another key when depressed simultaneously with it; similar to the control and shift keys. Abbreviated ALT key. { 'ȯl·tər·nət ‚kē }

alternate phyllotaxy [BOT] **1.** An arrangement of leaves that occur individually at nodes and on opposite sides of the stem. **2.** A spiral arrangement of leaves on a stem, with one leaf at a node. { 'ȯl·tər·nət 'fil·ə‚tak·sē }

alternate polarity operation [MET] A resistance welding method which utilizes alternating polarity for the progression of weld pulses. { 'ȯl·tər·nət pə'lar·əd·ē ‚äp·ə'rā·shən }

alternate routing [COMMUN] The operation of a switching center when all circuits are found busy in a programmed route to the destination, and the call is offered to another programmed route. { 'ȯl·tər·nət 'rüt·iŋ }

alternate track [COMPUT SCI] The disk track used if, after a disk volume is initialized, a defective track is sensed by the system. { 'ȯl·tər·nət 'trak }

alternate traversing fire [ORD] Method of covering a target that has both width and depth by firing a succession of traversing groups whose normal range dispersion will provide for distribution in depth. { 'ȯl·tər·nət trə‚vərs·iŋ 'fīr }

alternating copolymer [ORG CHEM] A polymer formed of two different monomer molecules that alternate in sequence in the polymer chain. { 'ȯl·tər‚nād·iŋ kō'päl·ə·mər }

alternating current [ELEC] Electric current that reverses direction periodically, usually many times per second. Abbreviated ac. { 'ȯl·tər‚nād·iŋ ‚kər·ənt }

alternating-current circuit theory [ELEC] The mathematical description of conditions in an electric circuit driven by an alternating source or sources. { 'ȯl·tər‚nād·iŋ ‚kər·ənt 'sər·kət ‚thē·ə·rē }

alternating-current coupling [ELECTR] A coupling which passes alternating-current signals but blocks direct-current signals. { 'ȯl·tər‚nād·iŋ ‚kər·ənt 'kəp·liŋ }

alternating-current/direct-current [ELECTR] Pertaining to electronic equipment capable of operation from either an alternating-current or direct-current primary power source. { 'ȯl·tər‚nād·iŋ ‚kər·ənt di‚rekt ‚kər·ənt }

alternating-current dump [ELECTR] The removal of all alternating-current power from a computer intentionally, accidentally, or conditionally. { 'ȯl·tər‚nād·iŋ ‚kər·ənt 'dəmp }

alternating-current erase [ELECTR] The use of an alternating current to energize a tape recorder erase head in order to remove previously recorded signals from a tape. { 'ȯl·tər‚nād·iŋ ‚kər·ənt ə'rās }

alternating-current erasing head [ELECTR] In magnetic recording, an erasing head which uses alternating current to produce the magnetic field necessary for erasing. { 'ȯl·tər‚nād·iŋ ‚kər·ənt ə'rās·iŋ ‚hed }

alternating-current generator [ELEC] A machine, usually rotary, which converts mechanical power into alternating-current electric power. { 'ȯl·tər‚nād·iŋ ‚kər·ənt 'jen·ə‚rād·ər }

alternating-current Josephson effect [CRYO] The oscillating current flow resulting from the tunneling of electron pairs through a thin insulating barrier between two superconductors when a steady voltage, V, is maintained across the barrier, with a frequency equal to $2eV/h$, where e is the magnitude of the charge of the electron and h is Planck's constant. { 'ȯl·tər‚nād·iŋ ‚kər·ənt 'jō·səf·sən i‚fekt }

alternating-current Kerr effect [OPTICS] Birefringence of light passing through a crystal that is simultaneously pumped with an intense laser beam. { 'ȯl·tər‚nād·iŋ ‚kər·ənt 'kər i‚fekt }

alternating-current magnetic biasing [ELECTR] Biasing with alternating current, usually well above the signal frequency range, in magnetic tape recording. { 'ȯl·tər‚nād·iŋ ‚kər·ənt mag'ned·ik 'bī·əs·iŋ }

alternating-current motor [ELEC] A machine that converts alternating-current electrical energy into mechanical energy by utilizing forces exerted by magnetic fields produced by the current flow through conductors. { 'ȯl·tər‚nād·iŋ ‚kər·ənt 'mōd·ər }

alternating-current network [ELEC] An electrical network that has elements with both resistance and reactance. { 'ȯl·tər‚nād·iŋ ‚kər·ənt 'net‚wərk }

alternating-current power supply [ELEC] A power supply that provides one or more alternating-current output voltages, such as an ac generator, dynamotor, inverter, or transformer. { 'ȯl·tər‚nād·iŋ ‚kər·ənt 'pau·ər sə‚plī }

alternating-current resistance *See* high-frequency resistance. { 'ȯl·tər‚nād·iŋ ‚kər·ənt ri'zis·təns }

alternating-current transmission [ELECTR] In television, that form of transmission in which a fixed setting of the controls makes any instantaneous value of signal correspond to the same value of brightness for only a short time. { 'ȯl·tər‚nād·iŋ ‚kər·ənt tranz'mish·ən }

alternating-current welder [ENG] A welding machine utilizing alternating current for welding purposes. { 'ȯl·tər‚nād·iŋ ‚kər·ənt 'weld·ər }

alternating flashing light [NAV] A light showing one or more flashes with color variations at regular intervals, the duration of light being less than that of darkness. { 'ȯl·tər‚nād·iŋ ‚flash·iŋ 'līt }

alternating form [MATH] A bilinear form f which changes sign under interchange of its independent variables; that is, $f(x,y) = -f(y, x)$ for all values of the independent variables x and y. { 'ȯl·tər‚nād·iŋ 'fȯrm }

alternating function [MATH] A function in which the interchange of two independent variables causes the dependent variable to change sign. { 'ȯl·tər‚nād·iŋ 'fəŋk·shən }

alternating gradient [ELECTROMAG] A magnetic field in which successive magnets have gradients of opposite sign, so that the field increases with radius in one magnet and decreases with radius in the next; used in synchrotrons and cyclotrons. { 'ȯl·tər‚nād·iŋ 'grād·ē·ənt }

alternating-gradient focusing [ELECTROMAG] A configuration of transverse electric or magnetic fields suitable for focusing or confining a charged-particle beam in which successive magnets or electrodes (in time or along the beam direction) have opposite polarity. { 'ȯl·tər‚nād·iŋ 'grād·ē·ənt 'fō·kəs·iŋ }

alternating-gradient synchrotron [NUCLEO] A proton synchrotron using an alternating magnetic-field gradient for focusing; beams of protons having extremely high energy (above 25 GeV) are produced. { 'ȯl·tər·nād·iŋ ¦grād·ē·ənt 'siŋk·rō‚trän }

alternating group [MATH] A group made up of all the even permutations of *n* objects. { 'ȯl·tər·nād·iŋ 'grüp }

alternating-group flashing light [NAV] A navigation light having groups of total eclipses at regular intervals and having color variations, the duration of light being equal to or greater than that of darkness. Also known as alternating-group occulting light. { 'ȯl·tər·nād·iŋ ¦grüp 'flash·iŋ ‚līt }

alternating-group occulting light See alternating-group flashing light. { 'ȯl·tər·nād·iŋ ¦grüp ə'kȯlt·iŋ ‚līt }

alternating occulting light [NAV] A navigation light having one or more total eclipses at regular intervals and having color variations, the duration of light being equal to or greater than that of darkness. { 'ȯl·tər·nād·iŋ ə'kȯlt·iŋ ‚līt }

alternating series [MATH] Any series of real numbers in which consecutive terms have opposite signs. { 'ȯl·tər·nād·iŋ 'sir·ēz }

alternating stress [MECH] A stress produced in a material by forces which are such that each force alternately acts in opposite directions. { 'ȯl·tər·nād·iŋ 'stres }

alternating voltage [ELEC] Periodic voltage, the average value of which over a period is zero. { 'ȯl·tər·nād·iŋ 'vōl·tij }

alternation [MATH] See disjunction. [PHYS] Variation, either positive or negative, of a waveform from zero to maximum and back to zero, equaling one-half of a cycle. { ‚ȯl·tər'nā·shən }

alternation of generations See metagenesis. { ‚ȯl·tər'nā·shən əv ‚jen·ə'rā·shənz }

alternation of multiplicities law [CHEM] The law that the periodic table arranges the elements in such a sequence that their number of orbital electrons, and hence their multiplicities, alternates between even and odd numbers. { ‚ȯl·tər'nā·shən əv ‚məl·tə'plis·əd·ēz ‚lȯ }

alternative algebra [MATH] A nonassociative algebra in which any two elements generate an associative algebra. { ȯl¦tər·nəd·iv 'al·jə·brə }

alternative hypothesis [STAT] Value of the parameter of a population other than the value hypothesized or believed to be true by the investigator. { ȯl¦tər·nət·iv hī'päth·ə·səs }

alternative RNA splicing [CELL MOL] A process in gene expression that enables the production of multiple forms of messenger ribonucleic acid (mRNA) from a single RNA transcript, thus enabling the production of multiple forms of protein from one gene. { ȯl¦tər·nəd·iv ¦är¦en'ā ‚splīs·iŋ }

alternator [ELEC] A mechanical, electrical, or electromechanical device which supplies alternating current. { 'ȯl·tər‚nād·ər }

alterne [ECOL] A community exhibiting alternating dominance with other communities in the same area. { 'ȯl‚tərn }

altherbosa [ECOL] Communities of tall herbs, usually succeeding where forests have been destroyed. { ¦al·thər¦bōs·ə }

altigraph [ENG] A pressure altimeter that has a recording mechanism to show the changes in altitude. { 'al·tə‚graf }

altimeter [ENG] An instrument which determines the altitude of an object with respect to a fixed level, such as sea level; there are two common types: the aneroid altimeter and the radio altimeter. { al'tim·əd·ər }

altimeter corrections [ENG] Corrections which must be made to the readings of a pressure altimeter to obtain true altitudes; involve horizontal pressure gradient error and air temperature error. { al'tim·əd·ər kə'rek·shənz }

altimeter setting [ENG] The value of atmospheric pressure to which the scale of an aneroid altimeter is set; after United States practice, the pressure that will indicate airport elevation when the altimeter is 10 feet (3 meters) above the runway (approximately cockpit height). { al'tim·əd·ər ‚sed·iŋ }

altimeter-setting indicator [ENG] A precision aneroid barometer calibrated to indicate directly the local altimeter setting. { al'tim·əd·ər ‚sed·iŋ 'in·də‚kād·ər }

altimetry [ENG] The measurement of heights in the atmosphere (altitude), generally by an altimeter. { al'tim·ə·trē }

altiplanation [GEOL] A phase of solifluction that may be seen as terracelike forms, flattened summits, and passes that are mainly accumulations of loose rock. { ‚al·tə·plā'nā·shən }

altiplanation surface [GEOL] A flat area fronted by scarps a few to hundreds of feet in height; the area ranges from several square rods to hundreds of acres. Also known as altiplanation terrace. { ‚al·tə·plā'nā·shən ‚sər·fəs }

altiplanation terrace See altiplanation surface. { ‚al·tə·plā'nā·shən ‚ter·əs }

altithermal [GEOPHYS] Period of high temperature, particularly the postglacial thermal optimum. Also known as hypsithermal. { ¦al·tə¦thər·məl }

Altithermal [GEOL] A dry postglacial interval centered about 5500 years ago during which temperatures were warmer than at present. { ¦al·tə¦thər·məl }

altithermal soil [GEOL] Soil recording a period of rising or high temperature. { ¦al·tə¦thər·məl 'sȯil }

altitude Abbreviated alt. [ENG] **1.** Height, measured as distance along the extended earth's radius above a given datum, such as average sea level. **2.** Angular displacement above the horizon measured by an altitude curve. [MATH] The perpendicular distance from the base to the top (a vertex or parallel line) of a geometric figure such as a triangle or parallelogram. { 'al·tə‚tüd }

altitude acclimatization [PHYSIO] A physiological adaptation to reduced atmospheric and oxygen pressure. { 'al·tə‚tüd ə‚klī·məd·ə'zā·shən }

altitude azimuth [ENG] An azimuth determined by solution of the navigational triangle with altitude, declination, and latitude given. { 'al·tə‚tüd 'az·ə·məth }

altitude-azimuth mounting [ENG] A two-axis telescope mounting in which the azimuth of the direction in which the telescope is pointed is determined by rotation about a vertical axis and the corresponding altitude is determined by rotation about a horizontal axis; computer-controlled motors must move the telescope in both altitude and azimuth to compensate for the earth's rotation. Also known as alt-azimuth mounting. { ¦al·tə‚tüd 'az·ə·məth ‚maunt·iŋ }

altitude chamber [ENG] A chamber within which the air pressure, temperature, and so on can be adjusted to simulate conditions at different altitudes; used for experimentation and testing. { 'al·tə‚tüd ‚chām·bər }

altitude circle [ASTRON] See parallel of altitude. [ELECTROMAG] A bright circle which surrounds the central dark portion of a plan position indicator display or photograph, and results from ground clutter. { 'al·tə‚tüd ‚sər·kəl }

altitude-contour See C factor. { 'al·tə‚tüd 'kän‚tür }

altitude curve [ENG] The arc of a vertical circle between the horizon and a point on the celestial sphere, measured upward from the horizon. [NAV] A graphical representation of the altitude of a celestial body as it would appear from a single assumed position or a series of assumed positions over a period of time; such curves are precomputed. { 'al·tə‚tüd ‚kərv }

altitude datum [ENG] The arbitrary level from which heights are reckoned. { 'al·tə‚tüd ‚dad·əm }

altitude delay [ELECTR] Synchronization delay introduced between the time of transmission of the radar pulse and the start of the trace on the indicator to eliminate the altitude/height hole on the plan position indicator-type display. { 'al·tə‚tüd di'lā }

altitude difference [ENG] The difference between computed and observed altitudes, or between precomputed and sextant altitudes. Also known as altitude intercept; intercept. { 'al·tə‚tüd 'dif·rəns }

altitude hole [ELECTR] The blank area in the center of a plan position indicator-type radarscope display caused by the time interval between transmission of a pulse and the receipt of the first ground return. { 'al·tə‚tüd ‚hōl }

altitude intercept See altitude difference. { 'al·tə‚tüd 'in·tər‚sept }

altitude reservation [NAV] **1.** The prior approval by the appropriate air traffic control agencies of flight plans, requesting use of certain airspace for the purpose of expediting mass movement of aircraft or other special air operations. **2.** A flight altitude assigned to an aircraft by an air-traffic control agency. { 'al·tə‚tüd ‚rez·ər'vā·shən }

altitude sickness [MED] In general, any sickness brought on by exposure to reduced oxygen tension and barometric pressure. { 'al·tə‚tüd ‚sik·nəs }

altitude signal [ELECTR] The radio signals returned to an airborne electricity device by the ground or sea surface directly beneath the aircraft. { 'al·tə‚tüd ‚sig·nəl }

altitude tints See gradient tints. { 'al·tə‚tüd ‚tints }

altitude valve [AERO ENG] A valve that adjusts the composition of the air-fuel mixture admitted into an airplane carburetor as the air density varies with altitude. { 'al·tə‚tüd ‚valv }

altitude wind tunnel [AERO ENG] A wind tunnel in which the air pressure, temperature, and humidity can be varied to simulate conditions at different altitudes. { 'al'tə‚tüd 'wind ‚tən·əl }

altitudinal vegetation zone [ECOL] A geographical band of physiognomically similar vegetation correlated with vertical and horizontal gradients of environmental conditions. { ¦al·tə¦tüd·ən·əl ‚vej·ə'tā·shən ‚zōn }

ALT key See alternate key. { 'ȯlt 'kē }

alto [GRAPHICS] The negative which is electrodeposited during the preparation of plates for printing currency. { 'al·tō }

altocumulus cloud [METEOROL] A principal cloud type, white or gray or both white and gray in color; occurs as a layer or patch with a waved aspect, the elements of which appear as laminae, rounded masses, or rolls; frequently appears at different levels in a given sky. Abbreviated Ac. { ¦al·tō¦kyüm·yə·ləs 'klau̇d }

altostratus cloud [METEOROL] A principal cloud type in the form of a gray or bluish (never white) sheet or layer of striated, fibrous, or uniform appearance; very often totally covers the sky and may cover an area of several thousand square miles; vertical extent may be from several hundred to thousands of meters. Abbreviated As. { ¦al·tō¦strat·əs 'klau̇d }

altricial [VERT ZOO] Pertaining to young that are born or hatched immature and helpless, thus requiring extended development and parental care. { ‚al'trish·əl }

ALU See arithmetical unit.

alula [ZOO] **1.** Digit of a bird wing homologous to the thumb. **2.** See calypter. { 'al·yə·lə }

alum [INORG CHEM] **1.** Any of a group of double sulfates of trivalent metals such as aluminum, chromium, or iron and a univalent metal such as potassium or sodium. **2.** See aluminum sulfate; ammonium aluminum sulfate; potassium aluminum sulfate. [MINERAL] $KAl(SO_4)_2 \cdot 12H_2O$ A colorless, white, astringent-tasting evaporite mineral. { 'al·əm }

alum cake [MATER] A material composed of silica and aluminum sulfate produced by the action of sulfuric acid on clay. { 'al·əm ‚kāk }

alum coal [GEOL] Argillaceous brown coal rich in pyrite in which alum is formed on weathering. { 'al·əm ‚kōl }

alumel [MET] An alloy containing 94% nickel, 2% aluminum, 3% chromium, and 1% silicon; used to form the chromel-alumel thermocouple. { 'al·yü·mel }

alumetized steel See aluminized steel. { ə'lüm·ə‚tīzd 'stēl }

alumina [INORG CHEM] Al_2O_3 The native form of aluminum oxide occurring as corundum or in hydrated forms, as a powder or crystalline substance. { ə'lüm·ə·nə }

alumina balls [MATER] Alumina in the form of balls $1/4$ to $3/4$ inch (6.4 to 19 millimeters) in diameter; usually composed of 99% alumina and having high resistance to chemicals and heat; used in reactor and catalytic beds. { ə'lüm·ə·nə ‚bȯlz }

alumina brick [MATER] A group of fireclay bricks containing 50, 60, or 70% alumina, used in high temperature applications. { ə'lüm·ə·nə ‚brik }

alumina bubble brick [MATER] A lightweight refractory brick used to line kiln walls; manufactured by passing an air jet over molten alumina to produce small hollow bubbles which are pressed into bricks and other shapes. { ə'lüm·ə·nə ¦bəb·əl ‚brik }

alumina cement [MATER] A cement made with bauxite and containing a high percentage of aluminate, having the property of setting to high strength in 24 hours. { ə'lüm·ə·nə si'ment }

alumina fibers [MATER] Short, linear crystals of alumina which have a strength of up to 200,000 pounds per square inch (1.38 gigapascals); used in plastics as a filler to improve heat resistance and dielectric properties. Also known as sapphire whiskers. { ə'lüm·ə·nə 'fī·bərz }

alumina porcelain [MATER] Porcelain composed principally of alumina; used to make spark plugs. { ə'lüm·ə·nə 'pȯr·slən }

aluminate [INORG CHEM] A negative ion usually assigned the formula AlO_2^- and derived from aluminum hydroxide. { ə'lüm·ə‚nāt }

aluminate cement [MATER] Cement made with bauxite, with high percentage of alumina; sets to a high strength in 24 hours and is used for constructing bank walls and laying roads.

Also known as aluminous cement; high-alumina cement; high-speed cement. { ə'lüm·ə‚nāt si'ment }

alumina trihydrate [INORG CHEM] $Al_2O_3 \cdot 3H_2O$, or $Al(OH)_3$ A white powder; insoluble in water, soluble in hydrochloric or sulfuric acid or sodium hydroxide; used in the manufacture of ceramic glasses and in paper coating. Also known as aluminum hydrate; aluminum hydroxide; hydrated alumina; hydrated aluminum oxide. { ə'lüm·ə·nə ‚trī'hī‚drāt }

aluminide [MAT SCI] An intermetallic alloy containing aluminum plus another element, such as nickel, iron, or titanium. { ə'lüm·ə‚nīd }

aluminite [MINERAL] $Al_2(SO_4)(OH)_4 \cdot 7H_2O$ Native monoclinic hydrous aluminum sulfate; used in tanning, papermaking, and water purification. Also known as websterite. { ə'lüm·ə‚nīt }

aluminium See aluminum. { ‚al·yü'min·ē·əm }

aluminize [ENG] To apply a film of aluminum to a material, such as glass. [MET] To form a protective surface alloy on a metal by treatment at elevated temperature with aluminum or an aluminum compound. { ə'lüm·ə‚nīz }

aluminized explosive [MATER] An explosive to which aluminum has been added. { ə'lüm·ə‚nīzd ik'splō·siv }

aluminized steel [MET] A steel coated with an aluminum-iron alloy coating; prepared by dip-coating and diffusing aluminum into steel at 1600°F (870°C); resists scaling and oxidation up to 1650°F (900°C). Also known as alumetized steel; calorized steel. { ə'lüm·ə‚nīzd 'stēl }

aluminon [ORG CHEM] $C_{22}H_{23}N_3O_9$ A yellowish-brown, glassy powder that is freely soluble in water; used for the detection and colorimetric estimation of aluminum in foods, water, and tissues, and as a pharyngeal aerosol spray. { ə'lüm·ə‚nän }

aluminosilicate [INORG CHEM] $3Al_2O_3 \cdot 2SiO_2$ A colorless, crystalline combination of silicate and aluminate in the form of rhombic crystals. { ə¦lüm·ə·nō¦sil·ə‚kāt }

aluminosis [MED] A lung disorder caused by inhalation of alumina dust. { ə‚lüm·ə'nō·səs }

aluminothermy [MET] The process of reducing a metallic oxide to the metal and producing great heat by mixing finely divided aluminum with the oxide, which is reduced as the aluminum is oxidized. { ə'lüm·ə·nō‚thər·mē }

aluminotype [GRAPHICS] A printing plate of the relief type; the raised surface is made of aluminum. { ə'lüm·ə·nō‚tīp }

aluminous cement See aluminate cement. { ə'lüm·ə·nəs si'ment }

aluminum [CHEM] A chemical element, symbol Al, atomic number 13, and atomic weight 26.9815. Also spelled aluminium. { ə'lüm·ə·nəm }

aluminum acetate [ORG CHEM] $Al(CH_3COO)_3$ A white, amorphous powder that is soluble in water; used in aqueous solution as an antiseptic. { ə'lüm·ə·nəm 'as·ə‚tāt }

aluminum acetylsalicylate [PHARM] $[C_6H_4(OCOCH_3)(COO)]_2AlOH$ White, odorless granules or powder; decomposes on melting and in alkali, hydroxides, and carbonates; used in MED. Also known as aluminum aspirin. { ə'lüm·ə·nəm ə¦sēd·əl·sə¦lis·ə‚lāt }

aluminum alloy [MET] An alloy of aluminum and relatively small amounts of other metals, such as copper, magnesium, or manganese. { ə'lüm·ə·nəm 'a‚lȯi }

aluminum ammonium sulfate See ammonium aluminum sulfate. { ə'lüm·ə·nəm ə'mon·ē·əm 'səl‚fāt }

aluminum arrester See aluminum-cell arrester. { ə'lüm·ə·nəm ə'res·tər }

aluminum-base grease See aluminum-soap grease. { ə'lüm·ə·nəm ‚bās 'grēs }

aluminum borohydride [ORG CHEM] $Al(BH_4)_3$ A volatile liquid with a boiling point of 44.5°C; used in organic synthesis and as a jet fuel additive. { ə'lüm·ə·nəm bȯr·ō'hī‚drīd }

aluminum brass [MET] **1.** A casting brass to which aluminum has been added as a flux to improve the casting qualities and, with the addition of lead, the machining qualities. **2.** A wrought brass to which aluminum has been added to improve the extruding and forging qualities and the oxidation resistance. { ə'lüm·ə·nəm 'bras }

aluminum bronze [MET] A copper-aluminum alloy which may also contain iron, manganese, nickel, or zinc. { ə'lüm·ə·nəm 'brānz }

aluminum cable steel-reinforced [ELEC] A type of power transmission line made of an aluminum conductor provided with a core of steel. Abbreviated ACSR. { ə'lüm·ə·nəm 'kā·bəl 'stēl ˌrē·in'fȯrst }

aluminum-cell arrester [ELEC] A lightning arrester consisting of a number of electrolytic cells in series formed from aluminum trays containing electrolyte. Also known as aluminum arrester; electrolytic arrester. { ə'lüm·ə·nəm ˌsel ə'res·tər }

aluminum chloride [INORG CHEM] $AlCl_3$ or Al_2Cl_6 A deliquescent compound in the form of white to colorless hexagonal crystals; fumes in air and reacts explosively with water; used as a catalyst. { ə'lüm·ə·nəm 'klȯr‚īd }

aluminum coating [MET] A film of aluminum applied to a metallic surface by, for example, spraying, electrolysis, or hot dipping. { ə'lüm·ə·nəm 'kōd·iŋ }

aluminum conductor [ELEC] Any of several aluminum alloys employed for conducting electric current; because its weight is one-half that of copper for the same conductance, it is used in high-voltage transmission lines. { ə'lüm·ə·nəm kən'dək·tər }

aluminum fluoride [INORG CHEM] $AlF_3·3^1/_2H_2O$ A white, crystalline powder, insoluble in cold water. { ə'lüm·ə·nəm 'flu̇r‚īd }

aluminum fluosilicate [INORG CHEM] $Al_2(SiF_6)_3$ A white powder that is soluble in hot water; used for artificial gems, enamels, and glass. Also known as aluminum silicofluoride. { ə'lüm·ə·nəm ‚flü·ə'sil·i·kət }

aluminum foil [MET] Aluminum in the form of a sheet of thickness not exceeding 0.005 inch (0.127 millimeter). { ə'lüm·ə·nəm 'fȯil }

aluminum grease See aluminum-soap grease. { ə'lüm·ə·nəm 'grēs }

aluminum halide [INORG CHEM] A compound of aluminum with a halogen element, such as aluminum chloride. { ə'lüm·ə·nəm 'ha‚līd }

aluminum hydrate See alumina trihydrate. { ə'lüm·ə·nam 'hī·drāt }

aluminum hydroxide See alumina trihydrate. { ə'lüm·ə·nam hī'dräk‚sīd }

aluminum-magnesium alloy [MET] An alloy of aluminum, 5–10% magnesium, and sometimes small amounts of other metals, characterized by high resistance to corrosion and high machinability. { ə'lüm·ə·nəm mag'nēz·ē·əm 'a‚lȯi }

aluminum monostearate [ORG CHEM] $Al(OH)_2$-[OOC-$(CH_2)_{16}CH_3$] A white to yellowish-white powder with a melting point of 155°C; used in the manufacture of medicine, paint, and ink, in waterproofing, and as a plastics stabilizer. { ə'lüm·ə·nəm ‚män·ō'stir‚āt }

aluminum-nickel-cobalt alloy See alnico. { ə'lüm·ə·nəm ‚nik·əl ‚kō‚bȯlt 'a‚lȯi }

aluminum nitrate [INORG CHEM] $Al(NO_3)_3·9H_2O$ White, deliquescent crystals with a melting point of 73°C; soluble in alcohol and acetone; used as a mordant for textiles, in leather tanning, and as a catalyst in petroleum refining. { ə'lüm·ə·nəm 'nī‚trāt }

aluminum oleate [ORG CHEM] A soaplike compound of aluminum and oleic acid, used in lubricating oils and greases to improve their viscosity. { ə'lüm·ə·nəm 'ō·lē‚āt }

aluminum ore [GEOL] A natural material from which aluminum may be economically extracted. { ə'lüm·ə·nəm 'ȯr }

aluminum orthophosphate [INORG CHEM] $AlPO_4$ White crystals, melting above 1500°C; insoluble in water, soluble in acids and bases; useful in ceramics, paints, pulp, and paper. Also known as aluminum phosphate. { ə'lüm·ə·nəm ‚ȯr·thō'fäs‚fāt }

aluminum oxide [INORG CHEM] Al_2O_3 A compound in the form of a white powder or colorless hexagonal crystals; melts at 2020°C; insoluble in water; used in aluminum production, paper, spark plugs, absorbing gases, light bulbs, artificial gems, and manufacture of abrasives, refractories, ceramics, and electrical insulators. { ə'lüm·ə·nəm 'äk‚sīd }

aluminum paint [MATER] A mixture of oil varnish and aluminum pigment in the form of thin flakes which overlap in the paint film; reflects the sun's radiation well and retains the heat in hot-air or hot-water pipes or tanks. { ə'lüm·ə·nəm 'pānt }

aluminum palmitate [ORG CHEM] $Al(C_{16}H_{31}O_2)·H_2O$ An aluminum soap used in waterproofing fabrics, paper, and leather and as a drier in paints. { ə'lüm·ə·nəm 'päm·ə‚tāt }

aluminum paste [MATER] Aluminum powder finely ground in oil; used in aluminum paints. { ə'lüm·ə·nəm 'pāst }

aluminum phosphate See aluminum orthophosphate. { ə'lüm·ə·nəm 'fäs‚fāt }

aluminum plate [GRAPHICS] A polymer-coated, anodized aluminum printing plate used in offset-lithographic printing. { ə'lüm·ə·nəm 'plāt }

aluminum potassium sulfate See potassium aluminum sulfate. { ə'lüm·ə·nəm pə'tas·ē·əm 'səl‚fāt }

aluminum powder [MATER] Small flakes of aluminum metal obtained by stamping or ball-milling foil in the presence of a fatty lubricant, such as stearic acid, which causes the flakes to orient in a pattern to give high brilliance. { ə'lüm·ə·nəm 'pau̇d·ər }

aluminum silicate [INORG CHEM] $Al_2(SiO_3)_3$ A white solid that is insoluble in water; used as a refractory in glassmaking. { ə'lüm·ə·nəm 'sil·ə‚kāt }

aluminum silicofluoride See aluminum fluosilicate. { ə'lüm·ə·nəm ‚sil·ə·kō‚flu̇r‚īd }

aluminum-silicon alloy [MET] An alloy of aluminum, 5-22% silicon, and sometimes small amounts of other metals, characterized by ease of casting and welding, light weight, and high resistance to corrosion. { ə'lüm·ə·nəm 'sil·i·kən 'a‚lȯi }

aluminum-silicon bronze [MET] An alloy consisting chiefly of copper, with aluminum and silicon added to give greater strength and hardness. { ə'lüm·ə·nəm 'sil·i·kən 'bränz }

aluminum soap [ORG CHEM] Any of various salts of higher carboxylic acids and aluminum that are insoluble in water and soluble in oils; used in lubricating greases, paints, varnishes, and waterproofing substances. { ə'lüm·ə·nəm 'sōp }

aluminum-soap grease [MATER] A lubricating grease consisting of a petroleum oil thickened with aluminum soap. Also known as aluminum-base grease; aluminum grease. { ə'lüm·ə·nəm 'sōp 'grēs }

aluminum sodium sulfate [INORG CHEM] $AlNa(SO_4)_2·12H_2O$ Colorless crystals with an astringent taste and a melting point of 61°C; soluble in water; used as a mordant and for waterproofing textiles, as a food additive, and for matches, tanning, ceramics, engraving, and water purification. Abbreviated SAS. Also known as porous alum; soda alum; sodium aluminum sulfate. { ə'lüm·ə·nəm 'sōd·ē·əm 'səl‚fāt }

aluminum solder [MET] A solder containing up to 15% aluminum, having a melting point above that of the tin-lead solders, and applied with a brazing torch. { ə'lüm·ə·nəm 'säd·ər }

aluminum stearate [ORG CHEM] $Al(C_{17}H_{35}COO)_3$ An aluminum soap in the form of a white powder that is insoluble in water and soluble in oils; used for waterproofing fabrics and concrete and as a drier in paints and varnishes. { ə'lüm·ə·nəm 'stir‚āt }

aluminum sulfate [INORG CHEM] $Al_2(SO_4)_3·18H_2O$ A colorless salt in the form of monoclinic crystals that decompose in heat and are soluble in water; used in papermaking, water purification, and tanning, and as a mordant in dyeing. Also known as alum. { ə'lüm·ə·nəm 'səl‚fāt }

aluminum therapy [MED] Therapy intended mainly for prevention rather than treatment of silicosis; provides for inhalation of powdered aluminum and alumina (Al_2O_3) dust by miners in the change house. { ə'lüm·ə·nəm 'ther·ə·pē }

aluminum triacetate [ORG CHEM] $Al(C_2H_3O_2)_3$ A white solid that is very slightly soluble in cold water. { ə'lüm·ə·nəm ‚trī'as·ə‚tāt }

alumite See alunite. { 'al·ə‚mīt }

alum rock See alunite. { 'al·əm ‚räk }

alum schist See alum shale. { 'al·əm ‚shist }

alum shale [PETR] A shale containing pyrite that is decomposed by weathering to form sulfuric acid, which acts on potash and alumina constituents to form alum. Also known as alum schist; alum slate. { 'al·əm ‚shāl }

alum slate See alum shale. { 'al·əm ‚slāt }

alumstone See alunite. { 'al·əm‚stōn }

alunite [MINERAL] $KAl_3(SO_4)_2(OH)_6$ A mineral composed of a basic potassium aluminum sulfate; it occurs as a hydrothermal-alteration product in feldspathic igneous rocks and is used in the manufacture of alum. Also known as alumite; alum rock; alumstone. { 'al·yə‚nīt }

alunitization [GEOL] Introduction of or replacement by alunite. { ‚al·yə·nə·tə'zā·shən }

alunogen [MINERAL] $Al_2(SO_4)_3·18H_2O$ A white mineral occurring as a fibrous incrustation of hydrated aluminum sulfate by volcanic action or decomposition of pyrite. Also known as feather alum; hair salt. { ə'lün·ə·jən }

alure [ARCH] A gallery or passage, such as along the parapets of a castle, around the roof of a church, or along a cloister. { 'al·yur }

alurgite [MINERAL] A purple manganiferous variety of muscovite mica. { 'a·lur,jīt }

alvar [ECOL] Dwarfed vegetation characteristic of certain Scandinavian steppelike communities with a limestone base. { 'al,vär }

alveator [INV ZOO] A type of pedicellaria in echinoderms. { 'al·vē,ād·ər }

alveolar [BIOL] Of or relating to an alveolus. { al'vē·ə·lər }

alveolar-capillary block syndrome [MED] Arterial oxygen deficiency due to improper functioning of the membranes between the alveoli and capillaries. { al'vē·ə·lər ¦kap·ə,ler·ē ¦bläk ,sin,drōm }

alveolar gland See acinous gland. { al'vē·ə·lər ,gland }

alveolar oxygen pressure [PHYSIO] The oxygen pressure in the alveoli; the value is about 105 mmHg. { al'vē·ə·lər 'äk·sə·jən ,presh·ər }

alveolar process [ANAT] The ridge of bone surrounding the alveoli of the teeth. { al¦vē·ə·lər ¦prä,ses }

alveolar ridge [ANAT] The bony remains of the alveolar process of the maxilla or mandible. { al¦vē·ə·lər 'rij }

alveolated cell See epithelioid cell. { al'vē·ə,lād·əd ,sel }

alveolitoid [INV ZOO] A type of tabulate coral having a vaulted upper wall and a lower wall parallel to the surface of attachment. { al'vē·ə·lə,tóid }

alveoloplasty [MED] Surgical shaping of the alveolar ridges in preparation for dentures or after the removal of several teeth. { al'vē·ə,lō,plas·tē }

alveolus [ANAT] **1.** A tiny air sac of the lung. **2.** A tooth socket. **3.** A sac of a compound gland. { al'vē·ə·ləs }

Alydidae [INV ZOO] A family of hemipteran insects in the superfamily Coreoidea. { ə'lid·ə,dē }

alymphocytic agammaglobulinemia [IMMUNOL] A type of immune globulin deficiency usually transmitted as an autosomal recessive that is characterized by a complete absence of lymphocytes; affected infants cannot produce a humoral or a cell-mediated immune response and are severely predisposed to infections, and usually die within the first year of life. { ,ā·lim·fə¦sid·ik ,ā,gam·ə,gläb·yü,li'nē·mē·ə }

alymphocytosis [MED] Absence or deficiency of blood lymphocytes. { ¦ā,lim·fō,sī'tō·səs }

alyphite [GEOL] Bitumen that yields a high percentage of open-chain aliphatic hydrocarbons upon distillation. { 'al·ə,fīt }

Alysiella [MICROBIO] A genus of bacteria in the family Simonsiellaceae; cells are arranged in pairs within the filaments. { ə,lis·ē'el·ə }

alysoid See catenary. { 'al·ə,sóid }

Alzheimer's disease [MED] A progressive neurodegenerative disease characterized by loss of function and death of nerve cells in several areas of the brain, leading to loss of cognitive function such as memory and language; the most common cause of senile dementia. Abbreviated AD. { 'älts,hī·mərz di,zēz }

Am See americium.

AM See amplitude modulation.

A/m See ampere per meter.

Am² See ampere meter squared.

A/m² See ampere per square meter.

amacratic lens See amasthenic lens. { ¦ā·mə¦krad·ik 'lenz }

amacrine cell [ANAT] An interneuron located in the inner plexiform layer of the vertebrate retina that influences retinal signal processing in response to visual stimuli at the level of contact between the bipolar and ganglion cells. { 'am·ə,krēn ,sel }

Amagat density unit [PHYS] A unit of density in the Amagat system, used in the study of the behavior of gases under pressure; it is equal to the density of a gas at a pressure of 1 atmosphere and a temperature of 0°C; for an ideal gas this is 44.6148 ± 0.0004 moles per cubic meter. { 'ä·mä·gä 'den·səd·ē ,yü·nət }

Amagat diagram [PHYS] A diagram that plots a series of isothermal curves for a gas pressure versus the gas pressure-volume product. { 'ä·mä·gä 'dī·ə,gram }

Amagat law See Amagat-Leduc rule. { 'ä·mä·gä ,ló }

Amagat-Leduc rule [PHYS] The rule which states that the volume taken up by a gas mixture equals the sum of the volumes each gas would occupy at the temperature and pressure of the mixture. Also known as Amagat law; Leduc law. { 'ä·mä·gä lə'dúk ,rül }

Amagat system [PHYS] A system of units in which the unit of pressure is the atmosphere and the unit of volume is the gram-molecular volume (22.4 liters at standard conditions). { 'ä·mä·gä ,sis·təm }

Amagat volume unit [PHYS] A unit of volume in the Amagat system, used in the study of the behavior of gases under pressure; it is equal to the volume occupied by 1 mole of a gas at a pressure of 1 atmosphere and a temperature of 0°C; for an ideal gas this is 0.0224141 ± 0.0000002 cubic meter. { 'ä·mä·gä 'väl·yəm ,yü·nət }

amalgam [MET] An alloy of mercury. [MINERAL] A silver mercury alloy occurring in nature. { ə'mal·gəm }

amalgamate [MET] **1.** To unite a metal in an alloy with mercury. **2.** To unite two dissimilar metals. **3.** To cover the zinc elements of a galvanic battery with mercury. { ə'mal·gə,māt }

amalgamating table [MET] A sloping wooden table covered with a copper plate on which mercury is spread to amalgamate with precious-metal particles. { ə'mal·gə,mād·iŋ ,tā·bəl }

amalgamation [MET] Also known as amalgam treatment. **1.** The process of separating metal from ore by alloying the metal with mercury; formerly used for gold and silver recovery, where it has been superseded by the cyanide process. **2.** The formation of an alloy of a metal with mercury. { ə,mal·gə'mā·shən }

amalgamation pan [MET] A circular cast-iron pan in which gold or silver ore is ground and the precious metal particles are amalgamated with mercury added to the pan. { ə,mal·gə'mā·shən ,pan }

amalgamator [MET] A device for bringing pulverized ore into contact with mercury to form an amalgam from which the metal is subsequently recovered. { ə'mal·gə,mād·ər }

amalgam barrel [MET] A small batching mill used to grind auriferous concentrates with mercury. { ə'mal·gəm ,bar·əl }

amalgam retort [MET] A retort in which mercury is distilled off from gold, or silver amalgam is obtained in amalgamation. { ə'mal·gəm ri,tórt }

amalgam treatment See amalgamation. { ə'mal·gəm ,trēt·mənt }

Amalthea [ASTRON] The innermost known satellite of Jupiter, orbiting at a mean distance of 1.13×10^5 miles (1.82×10^5 kilometers); it has a diameter of about 150 miles (240 kilometers). Also known as Jupiter V. { ,äm·äl'thē·ə }

amantadine [PHARM] $C_{10}H_{17}N$ A symmetrical amine used as a viral chemoprophylactic because it selectively inhibits certain myxoviruses; also of value in the treatment of parkinsonism. { ə'man·tə,dēn }

amanthophilous [BOT] Of plants having a habitat in sandy plains or hills. { ,a·mən'thä·fə·ləs }

Amaranthaceae [BOT] The characteristic family of flowering plants in the order Caryophyllales; they have a syncarpous gynoecium, a monochlamydeous perianth that is more or less scarious, and mostly perfect flowers. { ,a·mə·rə'thā·sē,ē }

amaranth [BOT] An annual plant (seldom perennial) of the genus Amaranthus that is distributed worldwide in warm and humid regions and is distinguished by small chaffy flowers (arranged in dense, green or red, monoecious or dioecious inflorescences) and by dry, membranous, indehiscent, one-seeded fruit. { 'am·ə·ranth }

amarantite [MINERAL] $Fe(SO_4)(OH)·3H_2O$ An amaranth red to brownish- or orange-red triclinic mineral consisting of a hydrated basic sulfate of ferric iron. { ,a·mə'ran,tīt }

amarillite [MINERAL] $NaFe(SO_4)_2·6H_2O$ A pale greenish-yellow mineral consisting of a hydrous sodium ferric sulfate. { ,a·mə'ri,līt }

Amaryllidaceae [BOT] The former designation for a family of plants now included in the Liliaceae. { ,a·mə,ri·lə'dā·sē,ē }

A-mast [ENG] An A-shaped arrangement of upright poles for supporting a mechanism designed to lift heavy loads. { 'ā ,mast }

amasthenic lens [OPTICS] A lens that refracts the rays of

AMANTADINE

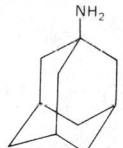

Structural formula of amantadine.

AMARANTH

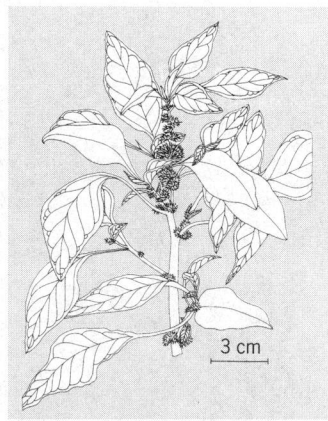

Flowers on branch of Amaranthus tricolor.

light into one focus. Also known as amacratic lens. { ¦ā·mǝs¦thēn·ik 'lenz }

amateur bands [COMMUN] Bands of frequencies assigned to licensed radio amateurs. { 'a·mǝ·chǝr ˌbanz }

amateur radio [ELECTR] A radio used for two-way radio communications by private individuals as leisure-time activity. Also known as ham radio. { 'a·mǝ·chǝr 'rād·ē·ō }

amathophobia [PSYCH] Abnormal fear of dust. { ǝˌmath·ǝ'fō·bē·ǝ }

amatol [MATER] An explosive mixture composed of ammonium nitrate and trinitrotoluene; mixtures with 50% and 80% ammonium nitrate are used for small and large shells, respectively. { 'am·ǝ·tȯl }

amatoxin [BIOCHEM] Any of a group of toxic peptides that selectively inhibit ribonucleic acid polymerase in mammalian cells; produced by the mushroom *Amanita phalloides*. { 'am·ǝ¦täk·sǝn }

amaurosis [MED] Total or partial blindness. { ˌa·mȯ'rō·sǝs }

amaurotic familial idiocy [MED] A hereditary condition, transmitted as an autosomal recessive, predominantly in Jewish children, characterized by blindness, muscular weakness, and subnormal mental development; when the onset is in infancy, the disease is known commonly as Tay-Sachs disease. { ¦a,mȯ¦räd·ik fǝ'mil·yǝl 'id·ē·ǝ·sē }

amavadin [BIOCHEM] A vanadium coordination complex and natural product obtained from the poisonous mushroom *Amanita muscaria*. { ǝ'mav·ǝˌdin }

amazonite [MINERAL] An apple-green, bright-green, or blue-green variety of microcline found in the United States and the former Soviet Union; sometimes used as a gemstone. Also known as amazon stone. { ¦a·mǝˌzō₂nīt }

amazon stone *See* amazonite. { 'a·mǝ·zän ˌstōn }

ambatoarinite [MINERAL] A mineral consisting of a carbonate of cerium metals and strontium. { ˌam·bǝˌtō'ä·rǝˌnīt }

amber [MINERAL] A transparent yellow, orange, or reddish-brown fossil resin derived from a coniferous tree; used for ornamental purposes; it is amorphous, has a specific gravity of 1.05–1.10, and a hardness of 2–2.5 on Mohs scale. { 'am·bǝr }

amber codon [VIROL] The polypeptide chain-termination messenger-RNA codon UAG, which brings about the termination of protein translation. { 'am·bǝr 'kō₂dän }

amber glass [MATER] A tinted glass made by using different mixtures of sulfur and iron oxide; the color can vary from pale yellow to ruby amber. { 'am·bǝr 'glas }

ambergris [PHYSIO] A fatty substance formed in the intestinal tract of the sperm whale; used in the manufacture of perfume. { 'am·bǝˌgris }

amberite [MATER] A smokeless powder composed of guncotton, barium nitrate, and paraffin. { 'am·bǝˌrīt }

amber mutation [GEN] Alteration of a codon to UAG, one of the three codons that result in premature polypeptide chain termination in all living organisms. { 'am·bǝr myü'tā·shǝn }

amberoid [MINERAL] A gem-quality mineral composed of small fragments of amber that have been reunited by heat or pressure. { 'am·bǝˌrȯid }

amber oil [MATER] **1.** A yellowish to brown essential oil made by destructive distillation of amber; has an acrid taste. **2.** A light essential oil prepared by destructive distillation of rosin. { 'am·bǝr ˌȯil }

ambident [ORG CHEM] A chemical species or molecule that possesses two alternative reactive sites, either of which can bond in a reaction; examples include cyanate ions, thiosulfate ions, oxime anions, and enolate ions. Also known as ambidentate. { 'am·bǝ·dǝnt }

ambidentate *See* ambident. { ˌam·bǝ'den₂tāt }

ambidextrous [PHYSIO] Capable of using both hands with equal skill. { ¦am·bǝ¦dek·strǝs }

ambient [ENG] Surrounding; especially, of or pertaining to the environment about a flying aircraft or other body but undisturbed or unaffected by it, as in ambient air or ambient temperature. { 'am·bē·ǝnt }

ambient light [OPTICS] The surrounding light, such as that reaching a television picture-tube screen from light sources in a room. { 'am·bē·ǝnt 'līt }

ambient noise [ACOUS] The pervasive noise associated with a given environment, being usually a composite of sounds from sources both near and distant. { 'am·bē·ǝnt 'nȯiz }

ambient pressure [FL MECH] The pressure of the surrounding medium, such as a gas or liquid, which comes into contact with an apparatus or with a reaction. { 'am·bē·ǝnt 'presh·ǝr }

ambient stress field [GEOPHYS] The distribution and numerical value of the stresses present in a rock environment prior to its disturbance by man. Also known as in-place stress field; primary stress field; residual stress field. { 'am·bē·ǝnt 'stres ˌfēld }

ambient temperature [PHYS] The temperature of the surrounding medium, such as gas or liquid, which comes into contact with the apparatus. { 'am·bē·ǝnt 'tem·prǝ·chǝr }

ambigenous [BOT] Of a perianth whose outer leaves resemble the calyx while the inner leaves resemble the corolla. { ˌam·bi·jǝn·ǝs }

ambiguity [ELECTR] The condition in which a synchro system or servosystem seeks more than one null position. [NAV] The condition in which navigation coordinates derived from a navigational instrument define more than one point, direction, line of position, or surface of position. { ˌam·bǝ'gyü·ǝd·ē }

ambiguity error [COMPUT SCI] An error in reading a number represented in a digital display that can occur when this representation is changing; for example, the number 699 changing to 700 might be read as 799 because of imprecise synchronization in the changing of digits. { ˌam·bǝ'gyü·ǝd·ē ˌer·ǝr }

ambiguous case [MATH] **1.** For the solution of a plane triangle, the case in which two sides and the angle opposite one of them is given, and there are two distinct solutions. **2.** For the solution of a spherical triangle, the case in which two sides and the angle opposite one of them is given, or two angles and the side opposite one of them is given, and there are two distinct solutions. { am'big·yǝ·wǝs 'kās }

ambiguous codon [GEN] A codon capable of coding for more than one amino acid sequence. { am'big·yǝ·wǝs 'kō₂dän }

ambiguous name [COMPUT SCI] A name of a file or other item which is only partially specified; it is useful in conducting a search of all the items to which it might apply. { am'big·yǝ·wǝs 'nām }

ambipolar [SCI TECH] Simultaneously operating in two opposite directions; for example, an electric current arising from the movement of positive and negative ions. { ¦am·bē'pōl·ǝr }

ambipolar diffusion [PHYS] The diffusion in a plasma of charged particles, such as electrons or ions, as a result of the almost exact local charge neutrality required. { ¦am·bē'pōl·ǝr dif'yü·zhǝn }

ambisexual [MED] An individual having undifferentiated primordia of both sexes. [PSYCH] Having feelings and exhibiting behavior common to both sexes. { ¦am·bi¦sek·shǝ·wǝl }

ambitus [BIOL] The periphery or external edge, as of a mollusk shell or leaf. { 'am·bǝ·tǝs }

ambivalence [PSYCH] The coexistence of contradictory emotions, attitudes, ideas, or desires with respect to a particular person, object, or situation. { am'bi·vǝ·lǝns }

amblygonite [MINERAL] $(Li,Na)AlPO_4(F,OH)$ A mineral occurring in white or greenish cleavable masses and found in the United States and Europe; important ore of lithium. { am'bli·gǝˌnīt }

amblyopia [MED] Dimness of vision, especially that not due to refractive errors or organic disease of the eye; may be congenital or acquired. { am'blē'ōp·ē·ǝ }

Amblyopsidae [VERT ZOO] The cave fishes, a family of actinopterygian fishes in the order Percopsiformes. { am·blē'äp·sǝˌdē }

Amblyopsiformes [VERT ZOO] An equivalent name for the Percopsiformes. { am·blē¦äp·sǝ'fȯr·mēz }

Amblypygi [INV ZOO] An order of chelicerate arthropods in the class Arachnida, commonly known as the tailless whip scorpions. { am'blip·iˌjī }

amboceptor [IMMUNOL] According to P. Ehrlich, an antibody present in the blood of immunized animals which contains two specialized elements: a cytophil group that unites with a cellular antigen, and a complementophil group that joins with the complement. { 'am·bōˌsep·tǝr }

amboceptor unit [BIOL] A unit for the standardization of blood serum. { 'am·bōˌsep·tǝr ˌyü·nǝt }

ambonite [PETR] Any of a group of hornblende-biotite andesites and dacites containing cordierite. { 'am·bǝˌnīt }

ambrette oil [MATER] A fixative for perfume having a strong musklike odor and distilled from the musk seed of *Hibiscus abelmoschus*. { 'am,bret ,oil }

ambrite [MINERAL] A yellow-gray, semitransparent fossil resin resembling amber; found in large masses in New Zealand coal fields and regarded as a semiprecious stone. { 'am,brīt }

ambrosine [MINERAL] A yellowish to clove-brown variety of amber rich in succinic acid; occurs as rounded masses in phosphate beds near Charleston, South Carolina. { 'am·brə,zēn }

ambrotype [GRAPHICS] An obsolete method of photography in which a negative is formed in a collodion emulsion on glass; when backed with black velvet or black varnish, the collodion surface reflects positive highlights and the resulting effect is that of a positive. { 'am·brə,tīp }

ambulacrum [INV ZOO] In echinoderms, any of the radial series of plates along which the tube feet are arranged. { ,am·byə'lak·rəm }

ambulatorial [ZOO] **1.** Capable of walking. **2.** In reference to a forest animal, having adapted to walking, as opposed to running, crawling, or leaping. { ,am·byə·lə'tȯr·ē·əl }

ambulatory [ARCH] **1.** A passageway around the apse of a church, or around a shrine. **2.** A covered walk of a cloister. { 'am·byə·lə,tȯr·ē }

ambulatory schizophrenia [PSYCH] A condition in which a person exhibits symptoms of both manic-depressive and schizophrenic psychosis but is not considered to require institutionalization. { 'am·byə·lə,tȯr·ē ,skit·sə'frē·nē·ə }

Ambystoma [VERT ZOO] A genus of common salamanders; the type genus of the family Ambystomatidae. { am'bis·tə·mə }

Ambystomatidae [VERT ZOO] A family of urodele amphibians in the suborder Salamandroidea; neoteny occurs frequently in this group. { am,bis·tə'mad·ə,dē }

Ambystomoidea [VERT ZOO] A suborder to which the family Ambystomatidae is sometimes elevated. { am,bis·tə'mȯid·ē·ə }

AMC *See* automatic modulation control.

AMCHA *See* tranexamic acid.

AM CVn star [ASTRON] A binary system consisting of two orbiting white dwarf stars which fill their respective Roche lobes, exchange mass at the point of contact, and vary in brightness due to eclipses of one white dwarf by the other. { ¦ā¦em ¦sē¦vē¦en ,stär }

Amdahl's law [COMPUT SCI] A law stating that the speedup that can be achieved by distributing a computer program over *p* processors cannot exceed $1/\{f + [1 - f)/p]\}$, where *f* is the fraction of the work of the program that must be done in serial mode. { 'am,dälz ,lȯ }

AME *See* angle-measuring equipment.

ameba [INV ZOO] The common name for a number of species of naked unicellular protozoans of the order Amoebida; an example is a member of the genus *Amoeba*. { ə'mē·bə }

Amebelodontinae [PALEON] A subfamily of extinct elephantoid proboscideans in the family Gomphotheriidae. { ,a·mə,bel·ə'dän·tə,nē }

amebiasis [MED] A parasitic disease of humans caused by the ameba *Entamoeba histolytica*, characterized by clinical-pathological intestinal manifestations, including an acute dysentery phase. Also known as amebic dysentery. { ,am·ē'bī·ə·səs }

amebic abscess [MED] Liquefactive necrosis of the brain and liver, without suppuration, caused by amebas, usually *Entamoeba histolytica*. { ə'mē·bik 'ab,ses }

amebic dysentery *See* amebiasis. { ə'mē·bik 'dis·ən,ter·ē }

amebicide [MATER] A chemical used to kill amebas, especially parasitic species. { ə'mēb·ə,sīd }

amebocyte [INV ZOO] One of the wandering ameboid cells in the tissues and fluids of many invertebrates that function in assimilation and excretion. { ə'mēb·ə,sīt }

ameboid movement [CYTOL] A type of cellular locomotion involving the formation of pseudopodia. { ə'mēb·ȯid 'mūv·mənt }

ameiosis [GEN] Nonreduction of chromosome number due to suppression failure of one of the two meiotic divisions, resulting in failure to reduce the chromosome complement from diploid to haploid, as in parthenogenesis. { ,ā·mī'ō·səs }

Ameiuridae [VERT ZOO] A family of North American catfishes belonging to the suborder Siluroidei. { ,a·mī'yür·ə,dē }

ameloblast [EMBRYO] One of the columnar cells of the enamel organ that form dental enamel in developing teeth. { 'am·ə·lō,blast }

ameloblastic odontoma [MED] A neoplasm of epithelial and mesenchymal odontogenic tissue. Also known as odontoblastoma. { ¦am·ə·lō¦blas·tik ,ō,dän'tō·mə }

ameloblastoma [MED] An epithelial tumor associated with the enamel organ; cells of basal layers resemble the ameloblast. Also known as adamantinoma. { ,am·ə·lō·bla'stō·mə }

amelogenesis imperfecta [MED] An inherited dental disorder that causes defective formation of tooth enamel. { ,am·ə·lō,jen·ə·səs ,im·pər'fek·tə }

amemolite [GEOL] A stalactite with one or more changes in its axis of growth. { ə'mem·ə,līt }

amendment record *See* change record. { ə'mend·mənt ,rek·ərd }

amenorrhea [MED] Absence of menstruation due to either normal or abnormal conditions. { ¦ā,men·ə'rē·ə }

amensalism [ECOL] A type of interaction that is neutral to one species but harmful to a second species. { ā'men·sə,liz·əm }

ament [BOT] A catkin. [MED] A person with congenital mental deficiency; an idiot. { 'a,ment }

amentia [MED] Congenital subnormal intellectual development. { ,ā'mensh·ə }

Amera [INV ZOO] One of the three divisions of the phylum Vermes proposed by O. Bütschli in 1910 and given the rank of a subphylum. { 'am·ə·rə }

American basement [BUILD] A basement located above ground level and containing the building's main entrance. { ə'mer·ə·kən 'bās·mənt }

American bond [CIV ENG] A bond in which every fifth, sixth, or seventh course of a wall consists of headers and the other courses consist of stretchers. Also known as common bond; Scotch bond. { ə'mer·ə·kən 'bänd }

American boreal faunal region [ECOL] A zoogeographic region comprising marine littoral animal communities of the coastal waters off east-central North America. { ə'mer·ə·kən 'bȯr·ē·əl 'fȯn·əl ,rē·jən }

American boring system [MIN ENG] A rope system of percussive boring, with a derrick which enables the complete set of boring tools to be raised clear of the hole. Also known as American system. { ə'mer·ə·kən 'bȯr·iŋ ,sis·təm }

American caisson *See* box caisson. { ə'mer·ə·kən 'kā,sän }

American-Egyptian cotton [BOT] A type of cotton developed by hybridization of Egyptian and American plants. { ə'mer·ə·kən i¦jip·shən 'kät·ən }

American Ephemeris and Nautical Almanac [ASTRON] An annual publication of the U.S. Naval Observatory containing tables of the predicted positions of various celestial bodies and other data of use to astronomers and navigators. { ə'mer·ə·kən i'fem·ə·rəs an 'nȯd·ə·kəl 'ȯl·mə,nak }

American explosive [MATER] One of many explosives that have passed U.S. Bureau of Mines tests and are used under certain conditions. { ə'mer·ə·kən ik'splō·siv }

American filter *See* disk filter. { ə'mer·ə·kən 'fil·tər }

American holly [FOR] *Ilex opaca*. A type of holly, widely valued as a Christmas decoration, that grows naturally in the eastern and southeastern United States close to the Atlantic and Gulf coasts, in the Mississippi Valley, and westward to Oklahoma and Missouri; it attains a maximum height of 40–50 ft (12–15 m) and has red berries on dark evergreen leaves. { ə,mer·ə·kən 'häl·ē }

American hophornbeam [FOR] *Ostrya virginiana*. A species widely distributed in the eastern half of the United States and in the highlands of southern Mexico and Guatemala; it may reach a height of 60 feet (18 meters) and can be recognized by its fruit, which closely resembles that of the hop vine, and by its very scaly bark. { ə,mer·ə·kən ,häp'hȯrn,bēm }

American hornbeam [FOR] *Carpinus caroliniana*. A tree sometimes attaining a height of 35 feet (10.7 meters) that is characterized by a smooth, steel-gray, fluted bark; it grows throughout the eastern half of the United States, especially in moist soil along banks of streams. Also known as water beech. { ə,mer·ə·kən 'hȯrn,bēm }

American jade *See* californite. { ə'mer·ə·kən 'jād }

American lion *See* puma. { ə'mer·ə·kən 'lī·ən }

American melting point [CHEM ENG] A temperature 3°F

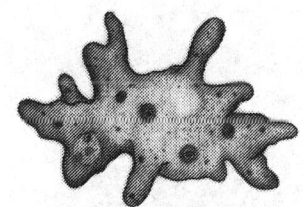

AMEBA

Typical species of ameba showing numerous pseudopodia, which are used for locomotion.

AMERICAN HOLLY

American holly (*Ilex opaca*).

(1.7°C) higher than the American Society for Testing and Materials Method D87 paraffin-wax melting point. { ə'mer·ə·kən 'melt·iŋ ‚póint }

American mucocutaneous leishmaniasis [MED] A form of leishmaniasis caused by *Leishmania brasiliensis,* transmitted by sandflies of the genus *Phlebotomus,* and characterized by skin ulcers and ulceration and necrosis of the mucosa of the mouth and nose. Also known as South American leishmaniasis. { ə'mer·ə·kən ‚myü·kō‚kyü'tān·e·əs ‚lēsh·mən'ī·ə·səs }

American pennyroyal oil *See* hedeoma oil. { ə'mer·ə·kən ¦pen·e‚róil ‚óil }

American run [TEXT] A system of numbering woolen yarns of 100 yards each in 1 ounce; thus if 1400 yards weigh 1 ounce, it would be a 14-run yarn. { ə'mer·ə·kən 'rən }

American spotted fever *See* Rocky Mountain spotted fever. { ə'mer·ə·kən ‚späd·əd 'fēv·ər }

American standard beam [CIV ENG] A type of I beam made of hot-rolled structural steel. { ə'mer·ə·kən 'stan·dərd 'bēm }

American standard channel [CIV ENG] A C-shaped structural member made of hot-rolled structural steel. { ə'mer·ə·kən 'stan·dərd 'chan·əl }

American Standard Code for Information Interchange [COMMUN] Coded character set to be used for the general interchange of information among information-processing systems, communications systems, and associated equipment; the standard code, comprising characters 0 through 127, includes control codes, upper- and lower-case letters, numerals, punctuation marks, and commonly used symbols; an additional set is known as extended ASCII. Abbreviated ASCII. { ə'mer·ə·kən 'stan·dərd 'kōd fər in·fər'mā·shən 'in·tər‚chānj }

American standard pipe thread [DES ENG] Taper, straight, or dryseal pipe thread whose dimensions conform to those of a particular series of specified sizes established as a standard in the United States. Also known as Briggs pipe thread. { ə'mer·ə·kən 'stan·dərd 'pīp ‚thred }

American standard screw thread [DES ENG] Screw thread whose dimensions conform to those of a particular series of specified sizes established as a standard in the United States; used for bolts, nuts, and machine screws. { ə'mer·ə·kən 'stan·dərd 'skrü ‚thred }

American system *See* American boring system. { ə'mer·ə·kən 'sis·təm }

American system drill *See* churn drill. { ə'mer·ə·kən ¦sis·təm ‚dril }

American Table of Distances [ENG] Published data concerning the safe storage of explosives and ammunition. { ə'mer·ə·kən ‚tā·bəl əv 'dis·təns·əz }

American wire gage [MET] A particular series of specified diameters and thicknesses established as a standard in the United States and used for nonferrous sheets, rods, and wires. Abbreviated AWG. Also known as Brown and Sharp gage (B and S gage). { ə'mer·ə·kən 'wīr ‚gāj }

American wormseed oil *See* chenopodium oil. { ə'mer·ə·kən 'wərm‚sēd ‚óil }

americium [CHEM] A chemical element, symbol Am, atomic number 95; the mass number of the isotope with the longest half-life is 243. { ‚am·ə'ris·ē·əm }

americyl ion [INORG CHEM] A dioxo monocation of americium, with the formula $(AmO_2)^-$. { ə'mer·ə·səl 'ī‚än }

Amerosporae [MYCOL] A spore group of the Fungi Imperfecti characterized by one-celled or threadlike spores. { ‚am·ə'räs·pə‚rē }

amesite [MINERAL] $(Mg,Fe)_4Al_4Si_2O_{10}(OH)_8$ An apple-green phyllosilicate mineral occurring in foliated hexagonal plates. { 'äm‚zīt }

Ames test [ANALY CHEM] A bioassay that uses a set of histidine auxotrophic mutants of *Salmonella typhimurium* for detecting mutagenic and possibly carcinogenic compounds. { 'āmz ‚test }

ametabolous metamorphosis [INV ZOO] A growth stage of certain insects characterized by an increase in size without distinct external changes. { ¦ā·mə¦tab·ə·ləs ‚med·ə'mór·fə·səs }

A-metal [MET] A type of permeability alloy containing 44% nickel and a small amount of copper; used to give nondistortion characteristics upon magnetization in transformers and loudspeakers. { 'ā ‚med·əl }

amethopterin [PHARM] $C_{20}H_{22}N_8O_5 \cdot H_2O$ An antimetabolite effective as a folic acid antagonist and used for treatment of acute and subacute leukemia. Also known as methotrexate. { ‚am·ə'thäp·tə‚rin }

amethyst [MINERAL] The transparent purple to violet variety of the mineral quartz; used as a jeweler's stone. { 'am·ə‚thist }

ametoecious [ECOL] Of a parasite that remains with the same host. { ‚am·ə'tēsh·əs }

ametropia [MED] Any deficiency in the refractive ability of the eye that causes an unfocused image to fall on the retina. { ‚a·mə'trōp·ē·ə }

AM field signature [ELECTR] The characteristic pattern of an alternating magnetic field, as displayed by detection and classification equipment. { ¦ā¦em 'fēld ‚sig·nə·chər }

AM Herculis star *See* polar. { ¦ā¦em 'hər·kyə·ləs ¦stär }

amherstite [PETR] A syenodiorite containing andesine and antiperthite. { 'a·mər‚stīt }

amianthus [MINERAL] A fine, silky variety of asbestos, such as chrysotile. { ‚a·mē'an·thəs }

amicable numbers [MATH] Two numbers such that the exact divisors of each number (except the number itself) add up to the other number. { 'am·ə·kə·bəl 'nəm·bərz }

Amici prism [OPTICS] A compound prism, used in direct-vision spectroscopes, that disperses a beam of light into a spectrum without causing the beam as a whole to undergo any net deviation; it is made up of alternate crown and flint glass components, refracting in opposite directions. Also known as direct-vision prism. { ə'mēch·ē ‚priz·əm }

amicron [PHYS CHEM] A particle having a size of 10^{-7} centimeter or less, which is a size in a system of classification of particle sizes in colloid chemistry. { ā'mī‚krän }

amictic [INV ZOO] **1.** In rotifers, producing diploid eggs that are incapable of being fertilized. **2.** Pertaining to the egg produced by the amictic female. { ə'mik·tik }

amictic lake [HYD] A lake that is perennially frozen. { ə'mik·tik 'lāk }

amidase [BIOCHEM] Any enzyme that catalyzes the hydrolysis of nonpeptide C=N linkages. { 'am·ə‚dās }

amidation [ORG CHEM] The process of forming an amide; for example, in the laboratory benzyl reacts with methyl amine to form *N*-methylbenzamide. { ‚am·ə‚dā·shən }

amide [ORG CHEM] One of a class of organic compounds containing the $CONH_2$ radical. { 'am‚īd }

amide hydrolysis [ORG CHEM] The cleavage of an amide into its constitutive acid and amine fragments by a net addition of water. { 'am‚īd hī'dräl·ə·səs }

amidine [ORG CHEM] A compound which contains the radical $CNHNH_2$. { 'am·ə‚dēn }

amido [ORG CHEM] Indicating the NH_2 radical when it is present in a molecule with the CO radical. { ə'mē‚dō }

amidohydrolase [BIOCHEM] An enzyme that catalyzes deamination. { ə¦mē·dō¦nī·drə‚lās }

amidol [ORG CHEM] $C_6H_3(NH_2)_2OH \cdot HCl$ A grayish-white crystalline salt; soluble in water, slightly soluble in alcohol; used as a developer in photography and as an analytical reagent. { 'am·i‚dól }

amidourea hydrochloride *See* semicarbazide hydrochloride. { ¦am·ə·dō·yú'rē·ə 'hī·drə‚klór‚īd }

amidships [NAV ARCH] At or toward the middle of a ship. { ə'mid‚ships }

Amiidae [VERT ZOO] A family of actinopterygian fishes in the order Amiiformes represented by a single living species, the bowfin *(Amia calva).* { ə'mī·ə‚dē }

Amiiformes [VERT ZOO] An order of actinopterygian fishes characterized by an abbreviate heterocercal tail, fusiform body, and median fin rays. { ə‚mī·ə'fór‚mēz }

amimia [PSYCH] The loss of the ability to express ideas by gestures or to understand the significance of gestures. { ‚ā'mim·ē·ə }

A min *See* ampere-minute.

amination [ORG CHEM] **1.** The preparation of amines. **2.** A process in which the amino group $(=NH_2)$ is introduced into organic molecules. { ‚am·ə'nā·shən }

amine [ORG CHEM] One of a class of organic compounds which can be considered to be derived from ammonia by replacement of one or more hydrogens by functional groups. { ə'mēn }

amine oxidase [BIOCHEM] An enzyme that catalyzes the oxidation of tyramine and tryptamine to aldehyde. { ə'mēn 'äk·sə‚dās }

amino-, amin- [CHEM] Having the property of a compound in which the group NH_2 is attached to a radical other than an acid radical. { ə'mē,nō }

aminoacetic acid *See* glycine. { ə'mē,no,ə'sēd·ik 'as·əd }

amino acid [BIOCHEM] Any of the organic compounds that contain one or more basic amino groups and one or more acidic carboxyl groups and that are polymerized to form peptides and proteins; only 20 of the more than 80 amino acids found in nature serve as building blocks for proteins; examples are tyrosine and lysine. { ə'mē,nō 'as·əd }

amino acid dating [GEOCHEM] Relative or absolute age determination of materials by measuring the degree of racemization of certain amino acids, which generally increases with geologic age. { ə,mē·nō 'as·əd 'dā·diŋ }

aminoaciduria [MED] A group of disorders in which excess amounts of amino acids are excreted in the urine; caused by abnormal protein metabolism. { ə'mē·nō,as·ə'dùr·ē·ə }

aminoacyl tRNA [CELL MOL] A molecular complex formed during protein synthesis by the linkage of a transfer ribonucleic acid molecule (tRNA) with its corresponding amino acid. { ə'mē·nō'as·əl 'tē,är,en'ā }

aminoacyl-tRNA synthetase [CELL MOL] An enzyme catalyzing the linkage of a transfer ribonucleic acid (tRNA) molecule to its corresponding amino acid during protein synthesis. { ə'mē·nō'as·əl 'tē,är,en'ā 'sin·thə,tās }

amino alcohol *See* alkaminc. { ə'mē,nō 'al·kə,hól }

1-aminoanthraquinone [ORG CHEM] $C_{14}H_9NO_2$ Ruby-red crystals with a melting point of 250°C; freely soluble in alcohol, benzene, chloroform, ether, glacial acetic acid, and hydrochloric acid; used in the manufacture of dyes and pharmaceuticals. { 'wən ə'mē·nō,an·thrə·kwē'nōn }

4-aminoantipyrine [PHARM] $C_{11}H_{13}N_3O$ Pale yellow crystals with a melting point of 109°C; soluble in water, benzene, and ethanol; used as an antipyretic and analgesic. { 'fór ə'mē·nō,an·tē'pī,rēn }

***para*-aminobenzoic acid** [BIOCHEM] $C_7H_7O_2N$ A yellow-red, crystalline compound that is part of the folic acid molecule; essential in metabolism of certain bacteria. Abbreviated PABA. { 'par·ə ə'mē·nō,ben'zō·ik 'as·əd }

2-amino-1-butanol [ORG CHEM] $CH_3CH_2CH(NH_2)CH_2OH$ A liquid miscible with water, soluble in alcohols; used in the synthesis of surface-active agents, vulcanizing accelerators, and pharmaceuticals, and as an emulsifying agent for such products as cosmetic creams and lotions. { 'tü ə'mē·nō 'wən 'byüt·ən,ól }

γ-aminobutyric acid [ORG CHEM] $H_2NCH_2CH_2CH_2COOH$ Crystals which are either leaflets or needles, with a melting point of 202°C; thought to be a central nervous system postsynaptic inhibitory transmitter. Abbreviated GABA. { 'gam·ə ə'mē·no,byu'tir·ik 'as·əd }

ε-aminocaproic acid [ORG CHEM] $C_6H_{13}NO_2$ Crystals with a melting point of 204–206°C; freely soluble in water; used as an antifibrinolytic agent and a spacer for affinity chromatography. { 'ep·sə,lən ə'mē·nō·kə,prō·ik 'as·əd }

aminocarb [ORG CHEM] $C_{11}H_{16}N_2O_2$ A tan, crystalline compound with a melting point of 93–94°C; slightly soluble in water; used as an insecticide for control of forest insects and pests of cotton, tomatoes, tobacco, and fruit crops. { ə'mē·nō,kärb }

aminocide *See* succinic acid 2,2-dimethylhydrazide. { ə 'mē·nō,sīd }

amino diabetes [MED] A congenital disorder characterized by excessive quantities of amino acids, glucose, and phosphate in the urine, resulting from deficient resorption in the proximal convoluted tubules of the kidney. Also known as Fanconi's syndrome. { ə'mē·nō ,dī·ə'bēd·ēz }

aminodiborane [INORG CHEM] Any compound derived from diborane (B_2H_6) in which one H of the bridge has been replaced by NH_2. { ə'mē·nō,dī'bór,ān }

3-amino-2,5-dichlorobenzoic acid [ORG CHEM] $C_7H_5O_2$-NCl_2 A white solid with a melting point of 200–201°C; solubility in water is 700 parts per million at 20°C; used as a preemergence herbicide for soybeans, corn, and sweet potatoes. { 'thrē ə'mē·nō 'tü 'fīv dī,klòr·ə,ben'zō·ik 'as·əd }

aminoethane *See* ethyl amine. { ə'mē·nō,eth·ən }

2-aminoethanethiol [PHARM] C_2H_7NS Crystals having a disagreeable odor and a melting point of 97–98.5°C; used in radiation sickness therapy. { 'tü ə'mē·nō,eth·ən,eth·ē,ól }

amino group [ORG CHEM] A functional group (-NH_2)

formed by the loss of a hydrogen atom from ammonia. { ə'mē·nō ,grüp }

α-aminohydrocinnamic acid *See* phenylalanine. { 'al·fə ə,mē·nō,hī·drō·sə,nam·ik 'as·əd }

4-amino-3-hydroxybutyric acid [PHARM] H_2NCH_2CH-(OH)CH_2COOH Crystals that decompose at 218°C; soluble in water; used as an anticonvulsant. { 'fòr ə,me·nō 'thrē ,hī'dräk·sē,byü'tir·ik 'as·əd }

***para*-(aminomethyl)benzenesulfonamide** [PHARM] C_7H_{10}-N_2O_2S Crystals with a melting point of 151-152°C; soluble in dilute alkali and acid; used as an antibacterial agent. { 'par·ə ə,mē·nō,meth·əl 'ben,zēn,səl,fän·ə,mīd }

2-amino-2-methyl-1,3-propanediol [ORG CHEM] $HOCH_2$-$C(CH_3)(NH_2)CH_2OH$ Crystals with a melting point of 109-111°C; soluble in water and alcohol; used in the synthesis of surface-active agents, pharmaceuticals, and vulcanizers, and as an emulsifying agent for cosmetics, leather dressings, polishes, and cleaning compounds. { 'tü ə'mē·nō 'tü 'meth·əl 'wən 'thrē 'prō,pān'dī,ól }

3-amino-2-naphthoic acid [ORG CHEM] $H_2NC_{10}H_6COOH$ Yellow crystals in the shape of scales with a melting point of 214°C; soluble in alcohol and ether; used in the determination of copper, nickel, and cobalt. { 'thrē ə'mē·nō 'tü naf'thō·ik 'as·əd }

1-amino-2-naphthol-4-sulfonic acid [ORG CHEM] H_2NC_{10}-$H_5(OH)SO_3H$ White or gray, needlelike crystals; soluble in hot sodium bisulfite solutions; used in the manufacture of azo dyes. { 'wən ə'mē·nō 'tü 'naf,thòl 'fòr səl'fän·ik 'as·əd }

2-amino-5-naphthol-7-sulfonic acid [ORG CHEM] $C_{10}H_5$-NH_2OHSO_3H Gray or white needles that are soluble in hot water; used as a dye intermediate. { 'tü ə'mē·nō 'fīv 'naf,thòl 'sev·ən səl'fän·ik 'as·əd }

amino nitrogen [CHEM] Nitrogen combined with hydrogen in the amino group. Also known as ammonia nitrogen. { ə'mē·nō 'nī·trə·jən }

2-amino-5-nitrothiazole [PHARM] $C_3H_3N_3O_2S$ A fluffy, greenish to orange-yellow powder with a slightly bitter taste; decomposes at 202°C; used as a veterinary drug in turkeys and chickens for histomonads and for trichomoniasis in pigeons. { 'tü ə'mē·nō 'fīv ,nī·trō'thī·ə,zōl }

6-aminopenicillanic acid [PHARM] $C_8H_{12}N_2O_3S$ Nonhygroscopic crystals that decompose at 209°C; used in the manufacture of synthetic penicillins. Abbreviated 6-APA. { 'siks ə'mē·nō,pen·ə·sil'an·ik 'as·əd }

aminopeptidase [BIOCHEM] An enzyme which catalyzes the liberation of an amino acid from the end of a peptide having a free amino group. { ə,mē·nō'pep·tə,dās }

aminophenol [ORG CHEM] A type of compound containing the NH_2 and OH groups joined to the benzene ring; examples are *para*-aminophenol and *ortho*-hydroxylaniline. { ə,mē·nō'fē,nól }

***ortho*-aminophenol** *See* *ortho*-hydroxylaniline. { 'or·thō ə,mē·nō·fē·nól }

***para*-aminophenol** [ORG CHEM] p $HOC_6H_4NO_2$ A phenol in which an amino ($-NH_2$) group is located on the benzene ring of carbon atoms para (p) to the hydroxyl ($-OH$) group; used as a photographic developer and as an intermediate in dye manufacture. { 'par·ə ə,mē·nō'fē·nól }

α-amino-β-phenylpropionic acid *See* phenylalanine. { 'al·fe ə'mē·nō 'bäd·ə 'fen·əl,prō·pē'añ·ik 'as·əd }

3-aminophthalic hydrazide *See* luminol. { 'thrē ə'mē·nō,thal·ik 'hī·drə·zīd }

aminophylline [PHARM] $C_{16}H_{24}N_{10}O_4$ A drug in the form of white or slightly yellowish, water-soluble granules or powder, used as a smooth-muscle relaxant, myocardial stimulant, and diuretic. { ə,mē·nō'fī,lēn }

amino plastic [MATER] Any plastic made of compounds derived from ammonia. { ə'mē·nō 'plas·tik }

aminopolypeptidase [BIOCHEM] A proteolytic enzyme that cleaves polypeptides containing either a free amino group or a basic nitrogen atom having at least one hydrogen atom. { ə'mē·nō,pä·lə'pep·tə,dās }

2-aminopropane *See* isopropylamine. { 'tü ə,mē·nō'prō·pän }

aminoprotease [BIOCHEM] An enzyme that hydrolyzes a protein and unites with its free amino group. { ə'mē·nō'prō·tē,ās }

aminopterin [PHARM] $C_{19}H_{20}N_8O_5 \cdot 2H_2O$ A yellow crystalline acid which is similar to folic acid and is used clinically as an antagonist of folic acid. { ,a·mə'näp·tə·rən }

3-aminopyridine See β-aminopyridine. { ¦thrē ə,me·nō'pī·rə,dēn }

4-aminopyridine [ORG CHEM] $C_5H_6N_2$ White crystals with a melting point of 158.9°C; soluble in water; used as a repellent for birds. Abbreviated 4-AP. { ¦fȯr ə,me·nō'pī·rə,dēn }

β-aminopyridine [ORG CHEM] $C_5H_6N_2$ Crystals with a melting point of 64°C; soluble in water, alcohol, and benzene; used in drug and dye manufacture. Also known as 3-aminopyridine. { ¦bād·ə ə,me·nō'pī·rə,dēn }

aminopyrine [PHARM] $C_{13}H_{17}N_3O$ Leafletlike crystals with a melting point of 107–109°C; soluble in alcohol, benzene, and chloroform; used as an antipyretic and analgesic in humans and animals. { ə'me·nō'pī,rēn }

amino resin [ORG CHEM] A type of resin prepared by condensation polymerization, with an aldehyde, of a compound containing an amino group. { ə'me·nō 'rez·ən }

para-aminosalicylic acid [PHARM] $C_7H_7NO_3$ White, crystalline drug used with other drugs in the treatment of tuberculosis. Abbreviated PAS. { ¦par·ə ə¦me·nō¦sal·ə¦sil·ik 'as·əd }

amino sugar [BIOCHEM] A monosaccharide in which a nonglycosidic hydroxyl group is replaced by an amino or substituted amino group; an example is D-glucosamine. { ə¦me·nō ¦shůg·ər }

amino terminal [BIOCHEM] The end part of a polypeptide chain which contains a free alpha-amino group. { ə¦me·nō ¦tərm·ən·əl }

2-aminothiazole [ORG CHEM] $C_3H_4N_2S$ Pale-yellow crystals that melt at 92°C; soluble in cold water, slightly soluble in ethyl alcohol; used as an intermediate in the synthesis of sulfathiazole. { ¦tü ə,me·nō'thī·ə,zōl }

aminotransferase See transaminase. { ə,me·nō'tranz·fə,rās }

aminotriazole [ORG CHEM] $C_2H_4N_4$ Crystals with a melting point of 159°C; soluble in water, methanol, chloroform, and ethanol; used as a herbicide, cotton plant defoliant, and growth regulator for annual grasses and broadleaf and aquatic weeds. Abbreviated ATA. { ə¦me·nō'trī·ə,zōl }

amitosis [CYTOL] Cell division by simple fission of the nucleus and cytoplasm without chromosome differentiation. { ¦ā,mī'tō·səs }

Am²/Js See ampere square meter per joule second.

AML See automatic modulation limiting.

AMLCD See active-matrix liquid-crystal display.

Ammanian [GEOL] Middle Upper Cretaceous geologic time. { ,ä'man·ē·ən }

ammeter [ENG] An instrument for measuring the magnitude of electric current flow. Also known as electric current meter. { 'a,mēd·ər }

ammine [INORG CHEM] One of a group of complex compounds formed by coordination of ammonia molecules with metal ions. { 'a,mēn }

ammocoete [ZOO] A protracted larval stage of lampreys. { 'a·mə,sēt }

ammocolous [ECOL] Describing plants having a habitat in dry sand. { ə'mä·kə·ləs }

Ammodiscacea [INV ZOO] A superfamily of foraminiferal protozoans in the suborder Textulariina, characterized by a simple to labyrinthic test wall. { ,a·mə,dis'kāsh·ə }

Ammodytoidei [VERT ZOO] The sand lances, a suborder of marine actinopterygian fishes in the order Perciformes, characterized by slender, eel-shaped bodies. { ə,mäd·i'tȯi·dē,ī }

ammonal [MATER] A high-explosive mixture, made of ammonium nitrate, trinitrotoluene (TNT), and flaked or powdered aluminum. { 'a·mə,nal }

ammonation [INORG CHEM] A reaction in which ammonia is added to other molecules or ions by covalent bond formation utilizing the unshared pair of electrons on the nitrogen atom, or through ion-dipole electrostatic interactions. { ,a·mə'nā·shən }

ammonia [INORG CHEM] NH_3 A colorless gaseous alkaline compound that is very soluble in water, has a characteristic pungent odor, is lighter than air, and is formed as a result of the decomposition of most nitrogenous organic material; used as a fertilizer and as a chemical intermediate. { ə'mōn·yə }

ammonia absorption refrigerator [MECH ENG] An absorption-cycle refrigerator which uses ammonia as the circulating refrigerant. { ə'mōn·yə əb¦sorp·shən ri'frij·ə,rād·ər }

ammonia alum See ammonium aluminum sulfate. { ə'mōn·yə 'al·əm }

ammonia-beam maser [PHYS] A gas maser using ammonia as the paramagnetic material. { ə'mōn·yə ¦bēm 'māz·ər }

ammoniac [INORG CHEM] See ammoniacal. [MATER] A gum resin obtained from the stems of the ammoniac plant; used in medicine, perfume, plaster, concrete, and adhesive. { ə'mōn·ē,ak }

ammoniacal [INORG CHEM] Pertaining to ammonia or its properties. Also known as ammoniac. { ¦a·mə¦nī·ə·kəl }

ammonia clock [HOROL] A time-measuring device dependent on the pyramidal ammonia molecule's property of turning inside out readily and oscillating between the two extreme positions at the precise frequency of 2.387013×10^{10} hertz. Also known as ammonia maser clock. { ə'mōn·yə 'klak }

ammonia compressor [MECH ENG] A device that decreases the volume of a quantity of gaseous ammonia by the amplification of pressure; used in refrigeration systems. { ə'mōn·yə kəm'pres·ər }

ammonia condenser [MECH ENG] A device in an ammonia refrigerating system that raises the pressure of the ammonia gas in the evaporating coil, conditions the ammonia, and delivers it to the condensing system. { ə'mōn·yə kən'dens·ər }

ammonia dynamite [CHEM] Dynamite with part of the nitroglycerin replaced by ammonium nitrate. { ə'mōn·yə 'dī·nə,mīt }

ammonia gelatin [MATER] An explosive of the gelatin dynamite class containing ammonium nitrate. { ə'mōn·yə 'jel·ət·ən }

ammonia liquor [CHEM ENG] Water solution of ammonia, ammonium compounds, and impurities, obtained from destructive distillation of bituminous coal. { ə'mōn·yə 'lik·ər }

ammonia maser clock See ammonia clock. { ə'mōn·yə ¦māz·ər ,kläk }

ammonia meter [ENG] A hydrometer designed specifically to determine the density of aqueous ammonia solutions. { ə'mō·nyə ,mēd·ər }

ammonia oil [MATER] Low-pour-test lubricating oil for use in an ammonia compressor in mechanical refrigeration. { ə'mōn·yə ,ȯil }

ammonia permissible [MATER] A permissible explosive that is an ammonia dynamite. { ə'mōn·yə pər'mis·ə·bəl }

ammonia synthesis [CHEM ENG] Chemical combination of nitrogen and hydrogen gases at high temperature and pressure in the presence of a catalyst to form ammonia. { ə'mōn·yə 'sin·thə·səs }

ammoniated mercuric chloride See ammoniated mercury. { ə¦mōn·ē·ād·əd mər¦kyür·ik 'klȯr,īd }

ammoniated mercury [INORG CHEM] $HgNH_2Cl$ A white powder that darkens on light exposure; insoluble in water and alcohol, soluble in ammonium carbonate solutions and in warm acids; used in pharmaceuticals and as a local anti-infective in medicine. { ə'mōn·ē·ād·əd 'mər·kyə·rē }

ammoniated superphosphate [INORG CHEM] A fertilizer containing 5 parts of ammonia to 100 parts of superphosphate. { ə'mōn·ē·ād·əd ,sü·pər'fäs,fāt }

ammoniation [CHEM] Treating or combining with ammonia. { ə,mōn·ē'ā·shən }

ammonia valve [ENG] A valve that is resistant to corrosion by ammonia. { ə'mōn·yə ,valv }

ammonia water [CHEM] A water solution of ammonia; a clear colorless liquid that is basic because of dissociation of NH_4OH to produce hydroxide ions; used as a reagent, solvent, and neutralizing agent. { ə'mōn·yə ,wȯd·ər }

ammonification [CHEM] Addition of ammonia or ammonia compounds, especially to the soil. { ə,män·ə·fə'kā·shən }

ammonifiers [ECOL] Fungi, or actinomycetous bacteria, that participate in the ammonification part of the nitrogen cycle and release ammonia (NH_3) by decomposition of organic matter. { ə'män·ə,fī·ərz }

ammonioborite [MINERAL] $(NH_4)_2B_{10}O_{16} \cdot 5H_2O$ A white mineral consisting of a hydrous ammonium borite and occurring as aggregates of minute plates. { ə,mōn·ē·ō'bȯr,īt }

ammoniojarosite [MINERAL] $(NH_4)Fe_3(SO_4)_2(OH)_6$ Pale-yellow mineral consisting of basic ferric ammonium sulfate. { ə,mōn·ē·ō·jə'rō,sīt }

PARA-AMINOSALICYLIC ACID

Structural formula of para-aminosalicylic acid.

AMINO SUGAR

The Haworth formula of D-glucosamine.

ammonite [MATER] An explosive containing 70-95% ammonium nitrate. [PALEON] A fossil shell of the cephalopod order Ammonoidea. { 'a·mə,nīt }

ammonium acetate [ORG CHEM] **1.** CH_3COONH_4 A normal salt formed by the neutralization of acetic acid with ammonium hydroxide; a white, crystalline, deliquescent material used in solution for the standardization of electrodes for hydrogen ions. **2.** $CH_3COONH_4 \cdot CH_3COOH$ An acid salt resulting from the distillation of the neutral salt or from its solution in hot acetic acid; crystallizes in deliquescent needles. **3.** A mixture of the normal and acetic salts; used as a mordant in the dyeing of wool. { ə'mōn·yəm 'as·ə,tāt }

ammonium alginate [ORG CHEM] $(C_6H_7O_6 \cdot NH_4)_n$ A high-molecular-weight, hydrophilic colloid; used as a thickening agent/stabilizer in ice cream, cheese, canned fruits, and other food products. { ə'mōn·yəm 'al·jə,nāt }

ammonium alum See ammonium aluminum sulfate. { ə'mōn·yəm 'al·əm }

ammonium aluminum sulfate [INORG CHEM] $NH_4Al(SO_4)_2 \cdot 12H_2O$ Colorless, odorless crystals that are soluble in water; used in manufacturing medicines and baking powder, dyeing, papermaking, and tanning. Also known as alum; aluminum ammonium sulfate; ammonia alum; ammonium alum. { ə'mōn·yəm ə'lü·mə·nəm 'səl,fāt }

ammonium benzoate [ORG CHEM] $NH_4C_7H_5O_2$ A salt of benzoic acid prepared as a coarse, white powder; used as a preservative in certain adhesives and rubber latex. { ə'mōn·yəm 'ben·zə,wāt }

ammonium bicarbonate [INORG CHEM] NH_4HCO_3 White, crystalline, water-soluble salt; used in baking powders and in fire-extinguishing mixtures. Also known as ammonium hydrogen carbonate. { ə'mōn·yəm bī'kär·bə,nāt }

ammonium bichromate See ammonium dichromate. { ə'mōn·yəm bī'krō,māt }

ammonium bifluoride [INORG CHEM] $NH_4F \cdot HF$ A salt that crystallizes in the orthorhombic system and is soluble in water; prepared in the form of white flakes from ammonia treated with hydrogen fluoride; used in solution as a fungicide and wood preservative. Also known as ammonium acid fluoride; ammonium hydrogen fluoride. { ə'mōn·yəm bī'flúr,īd }

ammonium bitartrate [ORG CHEM] $NH_4HC_4H_4O_6$ Colorless crystals that are soluble in water; used to make baking powder and to detect calcium. Also known as monoammonium tartrate. { ə'mōn·yəm bī'tär,trāt }

ammonium borate [INORG CHEM] NH_4BO_3 A white, crystalline, water-soluble salt which decomposes at 198°C; used as a fire retardant on fabrics. { ə'mōn·yəm 'bòr,āt }

ammonium bromide [INORG CHEM] NH_4Br An ammonium halide that crystallizes in the cubic system; made by the reaction of ammonia with hydrobromic acid or bromine; used in photography and for pharmaceutical preparations (sedatives). { ə'mōn·yəm 'brō,mīd }

ammonium carbamate [INORG CHEM] $NH_4NH_2CO_2$ A salt that forms colorless, rhombic crystals, which are very soluble in cold water; an important, unstable intermediate in the manufacture of urea; found in commercial ammonium carbonate. { ə'mōn·yəm 'kär·bə,māt }

ammonium carbonate [INORG CHEM] **1.** $(NH_4)_2CO_3$ The normal ammonium salt of carbonic acid, prepared by passing gaseous carbon dioxide into an aqueous solution of ammonia and allowing the vapors (ammonia, carbon dioxide, water) to crystallize. **2.** $NH_4HCO_3 \cdot NH_2COONH_4$ A white, crystalline double salt of ammonium bicarbonate and ammonium carbamate obtained commercially; the principal ingredient of smelling salts. { ə'mōn·yəm 'kär·bə,nāt }

ammonium chloride [INORG CHEM] NH_4Cl A white crystalline salt that occurs naturally as a sublimation product of volcanic action or is manufactured; used as an electrolyte in dry cells, as a flux for soldering, tinning, and galvanizing, and as an expectorant. { ə'mōn·yəm 'klòr,īd }

ammonium chromate [INORG CHEM] $(NH_4)_2CrO_4$ A salt that forms yellow, monoclinic crystals; made from ammonium hydroxide and ammonium dichromate; used in photography as a sensitizer for gelatin coatings. { ə'mōn·yəm 'krō,māt }

ammonium citrate [ORG CHEM] $(NH_4)_2HC_6H_5O_7$ White, granular material; used as a reagent. { ə'mōn·yəm 'sī,trāt }

ammonium dichromate [INORG CHEM] $(NH_4)_2Cr_2O_7$ A salt that forms orange, monoclinic crystals; made from ammonium sulfate and sodium dichromate; soluble in water and alcohol; ignites readily; used in photography, lithography, pyrotechnics, and dyeing. Also known as ammonium bichromate. { ə'mōn·yəm dī'krō,māt }

ammonium fluoride [INORG CHEM] NH_4F A white, unstable, crystalline salt with a strong odor of ammonia; soluble in cold water; used in analytical chemistry, glass etching, and wood preservation, and as a textile mordant. { ə'mōn·yəm 'flúr,īd }

ammonium fluosilicate [INORG CHEM] $(NH_4)_2SiF_6$ A toxic, white, crystalline powder; soluble in alcohol and water; used for mothproofing, glass etching, and electroplating. Also known as ammonium silicofluoride. { ə'mōn·yəm ,flü·ə'sil·ə,kāt }

ammonium formate [ORG CHEM] HCO_2NH_4 Deliquescent crystals or granules with a melting point of 116°C; soluble in water and alcohol; used in analytical chemistry to precipitate base metals from salts of the noble metals. { ə'mōn·yəm 'fòr·,māt }

ammonium gluconate [ORG CHEM] $NH_4C_6H_{11}O_7$ A white, crystalline powder made from gluconic acid and ammonia; soluble in water; used as an emulsifier for cheese and salad dressing and as a catalyst in textile printing. { ə'mōn·yəm 'glü·kə,nāt }

ammonium halide [INORG CHEM] A compound with the ammonium ion bonded to an ion formed from one of the halogen elements. { ə'mōn·yəm 'hal,īd }

ammonium hydrogen carbonate See ammonium bicarbonate. { ə'mōn·yəm 'hi·drə·jən 'kär·bə,nāt }

ammonium hydrogen fluoride See ammonium bifluoride. { ə'mōn·yəm 'hi·drə·jən 'flúr,īd }

ammonium hydroxide [INORG CHEM] NH_4OH A hydrate of ammonia, crystalline below −79°C; it is a weak base known only in solution as ammonia water. Also known as aqua ammonia. { ə'mōn·yəm ,hī'dräk,sīd }

ammonium iodide [INORG CHEM] NH_4I A salt prepared from ammonia and hydrogen iodide or iodine; it forms colorless, regular crystals which sublime when heated; used in photography and for pharmaceutical preparations. { ə'mōn·yəm 'ī·ə,dīd }

ammonium lactate [ORG CHEM] $NH_4C_3H_5O_3$ A yellow, syrupy liquid used in finishing leather. { ə'mōn·yəm 'lak,tāt }

ammonium lineolate [ORG CHEM] $C_{17}H_{31}COONH_4$ A soft, pasty material used as an emulsifying agent in various industrial applications. { ə'mōn·yəm lə'nē·ə,lāt }

ammonium metatungstate [ORG CHEM] $(NH_4)_6H_2W_{12}O_{40}$ A white powder, soluble in water, used for electroplating. { ə'mōn·yəm ,med·ə'təŋ,stāt }

ammonium molybdate [INORG CHEM] $(NH_4)_2MoO_4$ White, crystalline salt used as an analytic reagent, as a precipitant of phosphoric acid, and in pigments. { ə'mōn·yəm mə'lib,dāt }

ammonium nickel sulfate See nickel ammonium sulfate. { ə'mōn·yəm ¦nik·əl 'səl,fāt }

ammonium nitrate [INORG CHEM] NH_4NO_3 A colorless crystalline salt; very insensitive and stable high explosive; also used as a fertilizer. { ə'mōn·yəm 'nī,trāt }

ammonium oxalate [ORG CHEM] $(NH_4)_2C_2O_4 \cdot H_2O$ A salt in the form of colorless, rhombic crystals. { ə'mōn·yəm 'äk·sə,lāt }

ammonium perchlorate [INORG CHEM] NH_4ClO_4 A salt that forms colorless or white rhombic and regular crystals, which are soluble in water; it decomposes at 150°C, and the reaction is explosive at higher temperatures. { ə'mōn·yəm pər'klòr,āt }

ammonium perchlorate explosive [MATER] Any of several compositions consisting of ammonium perchlorate in combination with a high explosive and powdered metal; produces powerful blast. { ə'mōn·yəm pər'klòr,āt ik'splō·siv }

ammonium persulfate [INORG CHEM] $(NH_4)_2S_2O_8$ White crystals which decompose on melting; soluble in water; used as an oxidizing agent and bleaching agent, and in etching, electroplating, food preservation, and aniline dyes. { ə'mōn·yəm pər'səl,fāt }

ammonium phosphate [INORG CHEM] $(NH_4)_2HPO_4$ A salt of ammonia and phosphoric acid that forms white monoclinic crystals, which are soluble in water; used as a fertilizer and fire retardant. { ə'mōn·yəm 'fäs,fāt }

ammonium picrate [ORG CHEM] $NH_4C_6H_2O(NO_2)_3$ Compound with stable yellow and metastable red forms of orthorhombic crystals; used as a military explosive for armor-piercing shells. { ə'mōn·yəm 'pik,rāt }

ammonium salt [INORG CHEM] A product of a reaction between ammonia and various acids; examples are ammonium chloride and ammonium nitrate. { ə'mōn·yəm 'sôlt }

ammonium silicofluoride *See* ammonium fluosilicate. { ə'mōn·yəm ¦sil·ə·kō'flùr,īd }

ammonium soap [ORG CHEM] A product from reaction of a fatty acid with ammonium hydroxide; used in toiletry preparations such as soaps and in emulsions. { ə'mōn·yəm 'sōp }

ammonium stearate [ORG CHEM] $C_{17}H_{35}COONH_4$ A tan, waxlike substance with a melting point of 73–75°C; used in cosmetics and for waterproofing cements, paper, textiles, and other materials. { ə'mōn·yəm 'stir,āt }

ammonium sulfamate [INORG CHEM] $NH_4OSO_2NH_2$ White crystals with a melting point of 130°C; soluble in water; used for flameproofing textiles, in electroplating, and as an herbicide to control woody plant species. { ə'mōn·yəm 'səl·fə,māt }

ammonium sulfate [INORG CHEM] $(NH_4)_2SO_4$ Colorless, rhombic crystals which melt at 140°C and are soluble in water. { ə'mōn·yəm 'səl,fāt }

ammonium sulfate fractionation [BIOCHEM] A protein purification technique that utilizes the varying precipitation rates of different proteins in ammonium sulfate to obtain considerable separation. { ə¦mōn·ē·əm ¦səl,fāt ,frak·shə'nā·shən }

ammonium sulfide [INORG CHEM] $(NH_4)_2S$ Yellow crystals, stable only when dry and below 0°C; decomposes on melting; soluble in water and alcohol; used in photographic developers and for coloring brasses and bronzes. { ə'mōn·yəm 'səl,fīd }

ammonium tartrate [ORG CHEM] $C_4H_{12}N_2O_6$ Colorless, monoclinic crystals; used in textiles and in medicine. { ə'mōn·yəm 'tär,trāt }

ammonium thiocyanate [ORG CHEM] NH_4SCN Colorless, deliquescent crystals with a melting point of 149.6°C; soluble in water, acetone, alcohol, and ammonia; used in analytical chemistry, freezing solutions, fabric dyeing, electroplating, photography, and steel pickling. { ə'mōn·yəm ,thī·ō'sī·ə,nāt }

ammonium vanadate [INORG CHEM] NH_4VO_3 A white to yellow, water-soluble, crystalline powder; used in inks and as a paint drier and textile mordant. { ə'mōn·yəm 'van·ə,dāt }

ammonoid [PALEON] A cephalopod of the order Ammonoidea. { 'a·mə,nóid }

Ammonoidea [PALEON] An order of extinct cephalopod mollusks in the subclass Tetrabranchia; important as index fossils. { ,a·mə'nóid·ē·ə }

ammonolysis [CHEM] **1.** A dissociation reaction of the ammonia molecule producing H^+ and NH_2^- species. **2.** Breaking of a bond by addition of ammonia. { ,a·mə'näl·ə·səs }

ammonotelic [BIOL] Pertaining to the excretion of nitrogen primarily as ammonium ion, $[NH_4^+]$. { ə¦mä·nō'tēl·ik }

ammonoxidation *See* ammoxidation. { ,a·mən,äk·sə'dā·shən }

Ammon's law [ANTHRO] The law stating that cephalic index and stature vary inversely. { 'ä·mənz ,lò }

Ammotheidae [INV ZOO] A family of marine arthropods in the subphylum Pycnogonida. { ,a·mə'thē·ə,dē }

ammoxidation [CHEM ENG] A process in which mixtures of propylene, ammonia, and oxygen are converted in the presence of a catalyst, with acrylonitrile as the primary product. Also known as ammonoxidation; oxyamination. { ,am,äk·sə'dā·shən }

ammunition [ORD] **1.** All kinds of missiles to be thrown against an enemy. **2.** Missiles not for direct use against an enemy, with such purposes as illumination, signaling, and decelerating. **3.** A complete round and all its components, that is, the material required for firing a weapon such as a pistol. { ,am·yə'ni·shən }

ammunition belt [ORD] A fabric or metal band with loops for carrying cartridges that are fed from it into a machine gun or other automatic weapon; a feed belt. { ,am·yə'ni·shən ,belt }

ammunition box [ORD] A box in which linked ammunition is folded and from which it can be fed into a machine gun. { ,am·yə'ni·shən ,bäks }

ammunition carrier [ORD] **1.** A vehicle that accompanies guns and carries ammunition for them. **2.** A member of a gun or mortar squad who carries ammunition and helps load in actual firing. { ,am·yə'ni·shən ,kar·ē·ər }

ammunition chest [ORD] A receptacle, as on a caisson, limber, or gun carriage, in which ammunition is kept. { ,am·yə'ni·shən ,chest }

ammunition day of supply [ORD] The estimated quantity of ammunition required per day to sustain operations in an active theater. { ,am·yə'ni·shən ¦dā əv sə'plī }

ammunition depot [ORD] A storage and supply depot for projectiles, bombs, or other ammunition. { ,am·yə'ni·shən 'de,pō }

ammunition dump [ORD] An ammunition storage point, usually established in the field for temporary use. { ,am·yə'ni·shən ,dəmp }

amnesia [MED] The pathological loss or impairment of memory brought about by psychogenic or physiological disturbances. { am'nēzh·ə }

amnesic aphasia [MED] Loss of memory for the appropriate names of objects, conditions, or relations, accompanied by fragmented or hesitant speech. { am'nēz·ik ə'fāzh·ə }

amnicolous [ECOL] Describing plants having a habitat on sandy riverbanks. { ,am·nə'kä·ləs }

amniocentesis [MED] A procedure during pregnancy by which the abdominal wall and fetal membranes are punctured with a cannula to withdraw amniotic fluid. { ¦am·nē·ō,sen'tē·səs }

amniochorionic [EMBRYO] Relating to both amnion and chorion. { ¦am·nē·ō,kór·ē'än·ik }

amniogenesis [EMBRYO] The development or formation of the amnion. { ¦am·nē·ō'jen·ə·səs }

amniography [MED] Radiography of the fetus after injection of radiopaque material into the amniotic sac. { ,am·nē 'äg·rə·fē }

amnioma [MED] A broad flat tumor of the skin resulting from adhesion of the amnion after birth. { ,am·nē'ō·mə }

amnion [EMBRYO] A thin extraembryonic membrane forming a closed sac around the embryo in birds, reptiles, and mammals. { 'am·nē,än }

amnionitis [MED] Inflammation of the amnion resulting from infection. { ,am·nē·ō'nīd·əs }

amniorrhea [MED] The premature escape or discharge of the amniotic fluid. { ,am·ne·ō'rē·ə }

amniorrhexis [MED] Rupture of the amnion. { ¦am·nē·ō'rek·səs }

amnioscope [MED] An endoscope for studying amniotic fluid in the intact amniotic sac. { 'am·ne·ō,skōp }

amnioscopy [MED] Visual examination of the amniotic fluid in the lower part of the amniotic sac by means of an amnioscope. { 'am·ne·ō,skōp·ē }

Amniota [VERT ZOO] A collective term for the Reptilia, Aves, and Mammalia, all of which have an amnion during development. { ,am·nē'äd·ə }

amniote [ZOO] An animal that develops an amnion during its embryonic stage; includes birds, reptiles, and mammals. { 'am·nē,ōt }

amniotic fluid [PHYSIO] A substance that fills the amnion to protect the embryo from desiccation and shock. { ¦am·nē¦äd·ik 'flü·əd }

amniotome [MED] An instrument used to puncture the fetal membranes. { 'am·nē·ə,tōm }

amniotomy [MED] Artificial rupture of the fetal membranes by means of an amniotome to induce labor. { ,am·nē'äd·ə·mē }

amobarbital [PHARM] $C_{11}H_{18}N_2O_3$ A white, crystalline powder with a bitter taste and a melting point of 156–161°C; soluble in alcohol; used in medicine. { ,am·ō'bär·bə·təl }

A mode [ACOUS] A form of ultrasonic medical tomography that uses acoustic pulse emissions and echo reception along a single line-of-sight axial propagation path and usually displays the information on a cathode-ray oscilloscope in which the horizontal axis of the display is a linear time base, triggered at the time of the transmitted pulse, and the received echoes are manifested as vertical deflections, with vertical displacement a measure of the amplitude of the strength of the returning echo. { 'ā 'mōd }

amodiaquine [PHARM] $C_{20}H_{22}ClN_3O$ A crystalline compound that melts at 208°C; used as an antimalarial. { ,am·ō'dī·ə,kwēn }

Amoeba [INV ZOO] A genus of naked, rhizopod protozoans in the order Amoebida characterized by a thin pellicle and thick, irregular pseudopodia. { ə'mē·bə }

Amoebida [INV ZOO] An order of rhizopod protozoans in the subclass Lobosia characterized by the absence of a protective covering (test). { ˌam·ə'bī·də }

Amoebobacter [MICROBIO] A genus of bacteria in the family Chromatiaceae; cells are spherical and nonmotile, have gas vacuoles, and contain bacteriochlorophyll *a* on vesicular photosynthetic membranes. { əˌmē·bə'bak·tər }

amoeboid fold [GEOL] A fold or structure, such as an anticline, having no prevailing trend or definite shape. { ə'mē‚bȯid 'fōld }

amoeboid glacier [HYD] A glacier connected with its snowfield for a portion of the year only. { ə'mē‚bȯid 'glā·shər }

Amor [ASTRON] An asteroid with an orbital eccentricity of 0.448 that approached to about 1×10^7 miles (1.6×10^7 kilometers) from earth. { 'ä‚mȯr }

Amor object [ASTRON] Any asteroid which crosses the orbit of Mars. { 'ä‚mȯr 'äb·jekt }

amorphic allele [GEN] An allele that lacks gene activity. { ə'mȯr·fik ə'lēl }

Amorphosporangium [MICROBIO] A genus of bacteria in the family Actinoplanaceae with irregular sporangia and rod-shaped spores; they are motile by means of polar flagella. { ə¦mȯr·fə·spə'ran·jəm }

amorphous [PHYS] Pertaining to a solid which is noncrystalline, having neither definite form nor structure. { ə'mȯr·fəs }

amorphous film [MATER] A magnetically ordered metallic film that can be deposited on a semiconductor chip or on almost any other material without need for a crystal substrate, for use in magnetic bubble memories. { ə'mȯr·fəs 'film }

amorphous frost [HYD] Hoar frost which possesses no apparent simple crystalline structure; opposite of crystalline frost. { ə'mȯr·fəs 'frȯst }

amorphous laser *See* glass laser. { ə'mȯr·fəs 'lāz·ər }

amorphous memory array [COMPUT SCI] An array of memory switches made of amorphous material. { ə'mȯr·fəs 'mem·rē ə‚rā }

amorphous mineral [MINERAL] A mineral without definite crystalline structure. { ə'mȯr·fəs 'min·rəl }

amorphous peat [GEOL] Peat composed of fine grains of organic matter; it is plastic like wet, heavy soil, with all original plant structures destroyed by decomposition of cellulosic matter. { ə'mȯr·fəs 'pēt }

amorphous ribbon [MET] A metallic alloy that has an amorphous structure and is formed into a strip 25 to 50 micrometers thick and 1 to 150 millimeters (0.04 to 6 inches) wide by a process in which a melt of the required composition is ejected through an orifice onto a copper drum where it is instantly quenched and formed into a ribbon by rotation of the drum. { ə'mȯr·fəs 'rib·ən }

amorphous semiconductor [SOLID STATE] A semiconductor material which is not entirely crystalline, having only short-range order in its structure. { ə'mȯr·fəs ¦sem·ē·kən¦dək·tər }

amorphous sky [METEOROL] A sky characterized by an abundance of fractus clouds, usually accompanied by precipitation falling from a higher, overcast cloud layer. { ə'mȯr·fəs 'skī }

amorphous snow [HYD] A type of snow with irregular crystalline structure. { ə'mȯr·fəs 'snō }

amorphous solid [SOLID STATE] A rigid material whose structure lacks crystalline periodicity; that is, the pattern of its constituent atoms or molecules does not repeat periodically in three dimensions. { ə'mȯr·fəs 'säl·əd }

amortisseur winding *See* damper winding. { a¦mȯrd·ə'sər 'wīnd·iŋ }

amortize [IND ENG] To reduce gradually an obligation, such as a mortgage, by periodically paying a part of the principal as well as the interest. { 'am·ər‚tīz }

amortizement [ARCH] The sloping top of a buttress or projecting pier. { ¦am·ər¦tīz·mənt }

amosite [MINERAL] A monoclinic amphibole form of asbestos having long fibers and a high iron content; used in insulation. { 'am·ə‚zīt }

amount limit [IND ENG] In a test for a fixed quantity of work, the time required to complete the work or the total amount of work that can be completed in an unlimited time. { ə'maůnt ‚lim·ət }

amount of substance [CHEM] A measure of the number of elementary entities present in a substance or system; usually measured in moles. { ə'maůnt əv 'səb·stəns }

amp *See* amperage. *See* ampere. { amp }

AMP *See* adenylic acid.

3′,5′-AMP *See* cyclic adenylic acid.

ampacity [ELEC] Current-carrying capacity in amperes; used as a rating for power cables. { am'pas·əd·ē }

ampangabeite *See* samarskite. { ˌäm‚päŋ'gä·bē‚īt }

Ampeliscidae [INV ZOO] A family of tube-dwelling amphipod crustaceans in the suborder Gammaridea. { ˌam·pə'lis·ə‚dē }

ampelite [PETR] A graphite schist containing silica, alumina, and sulfur; used as a refractory. { 'am·pə‚līt }

amperage [ELEC] The amount of electric current in amperes. Abbreviated amp. { 'am·prij }

ampere [ELEC] The unit of electric current in the rationalized meter-kilogram-second system of units; defined in terms of the force of attraction between two parallel current-carrying conductors. Abbreviated a; A; amp. { 'am‚pir }

Ampère balance *See* current balance. { 'äm‚per ‚bal·əns }

Ampère currents [ELECTROMAG] Postulated "molecular-ring" currents to explain the phenomena of magnetism as well as the apparent nonexistence of isolated magnetic poles. { 'äm‚per 'kər·əns }

ampere-hour [ELEC] A unit for the quantity of electricity, obtained by integrating current flow in amperes over the time in hours for its flow; used as a measure of battery capacity. Abbreviated Ah; amp-hr. { 'am‚pir ¦aů·ər }

ampere-hour capacity [ELEC] The charge, measured in ampere-hours, that can be delivered by a storage battery up to the limit to which the battery may be safely discharged. { 'am‚pir ¦aů·ər kə'pas·əd·ē }

ampere-hour meter [ENG] A device that measures the total electric charge that passes a given point during a given period of time. { 'am‚pir ¦aů·ər ‚mēd·ər }

Ampère law [ELECTROMAG] **1.** A law giving the magnetic induction at a point due to given currents in terms of the current elements and their positions relative to the point. Also known as Laplace law. **2.** A law giving the line integral over a closed path of the magnetic induction due to given currents in terms of the total current linking the path. { 'äm‚per ‚lȯ }

ampere meter squared [ELECTROMAG] The SI unit of electromagnetic moment. Abbreviated Am². { ¦am‚pir ¦mēd·ər 'skwerd }

ampere-minute [ELEC] A unit of electrical charge, equal to the charge transported in 1 minute by a current of 1 ampere, or to 60 coulombs. Abbreviated A min. { ¦am‚pir ¦min·ət }

ampere per meter [ELECTROMAG] The SI unit of magnetic field strength and magnetization. Abbreviated A/m. { 'am‚pir pər 'mēd·ər }

ampere per square inch [ELEC] A unit of current density, equal to the uniform current density of a current of 1 ampere flowing through an area of 1 square inch. Abbreviated A/in². { 'am‚pir pər ‚skwer 'inch }

ampere per square meter [ELEC] The SI unit of current density. Abbreviated A/m². { 'am‚pir pər ‚skwer 'mēd·ər }

Ampère rule [ELECTROMAG] The rule which states that the direction of the magnetic field surrounding a conductor will be clockwise when viewed from the conductor if the direction of current flow is away from the observer. { 'äm‚per ‚rül }

ampere square meter per joule second [ELECTROMAG] The SI unit of gyromagnetic ratio. Abbreviated Am²/Js. { ¦am‚pir ¦skwer ¦mēd·ər pər ¦jül 'sek·ənd }

Ampère theorem [ELECTROMAG] The theorem which states that an electric current flowing in a circuit produces a magnetic field at external points equivalent to that due to a magnetic shell whose bounding edge is the conductor and whose strength is equal to the strength of the current. { 'äm‚per ‚thir·əm }

ampere-turn [ELECTROMAG] A unit of magnetomotive force in the meter-kilogram-second system defined as the force of a closed loop of one turn when there is a current of 1 ampere flowing in the loop. Abbreviated amp-turn. { 'am‚pir ‚tərn }

amperometric titration [PHYS CHEM] A titration that involves measuring an electric current or changes in current during the course of the titration. { ¦am·pə·rə¦me·trik tī'trā·shən }

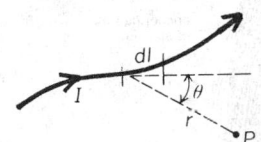

AMPÈRE LAW

A graphic representation of the Ampère law (def. 1). Contribution of current element to magnetic induction at *P* is proportional to $I\,dl\,\sin\theta/r^2$. I = current, dl = length of current element, r = distance of point *P* from current elements, angle θ is between current element and the line forming the element to point *P*. The contribution points perpendicularly into the page.

amperometric transducer [ENG] A transducer in which the concentration of a dissolved substance is determined from the electric current produced between two electrodes immersed in the test solution when one of the electrodes is kept at a selected electric potential with respect to the solution. { am,pir·ə¦me·trik tranz'dü·sər }

amperometry [PHYS CHEM] Chemical analysis by techniques which involve measuring electric currents. { ,am·pə'rä·mə·trē }

Ampharetidae [INV ZOO] A large, deep-water family of polychaete annelids belonging to the Sedentaria. { ,am·fə'red·ə,dē }

Ampharetinae [INV ZOO] A subfamily of annelids belonging to the family Ampharetidae. { ,am·fə'ret·ə,nē }

amphetamine [PHARM] $C_6H_5CH_2CHNHCH_3$ A volatile, colorless liquid used as a central nervous system stimulant. { am'fed·ə,mēn }

amphiarthrosis [ANAT] An articulation of limited movement in which bones are connected by fibrocartilage, such as that between vertebrae or that at the tibiofibular junction. { ,am·fī·är'thrō·səs }

amphiaster [INV ZOO] Type of spicule found in some sponges. { 'am·fē,as·tər }

Amphibia [VERT ZOO] A class of vertebrate animals in the superclass Tetrapoda characterized by a moist, glandular skin, gills at some stage of development, and no amnion during the embryonic stage. { am'fib·ē·ə }

Amphibicorisae [INV ZOO] A subdivision of the insect order Hemiptera containing surface water bugs with exposed antennae. { am,fib·ə'kòr·ə,sē }

Amphibioidei [INV ZOO] A family of tapeworms in the order Cyclophyllidea. { am,fə,bī'òid·ē,ī }

amphibiotic [ZOO] Being aquatic during the larval stage and terrestrial in the adult stage. { ¦am·fə,bī¦äd·ik }

amphibious [BIOL] Capable of living both on dry or moist land and in water. [MECH ENG] Said of vehicles or equipment designed to be operated or used on either land or water. [ORD] A military operation conducted by coordinated action of land, sea, and air forces. { ,am'fib·ē·əs }

amphibious assault landing model See assault-landing model. { ,am'fib·ē·əs ə'sólt 'land·iŋ ,mäd·əl }

amphibious assault ship [NAV ARCH] A ship designed to transport and land troops, equipment, and supplies by means of embarked helicopters. { ,am'fib·ē·əs ə'sólt ,ship }

amphibious command ship [NAV ARCH] A naval ship from which a commander exercises control in amphibious operations. { ,am'fib·ē·əs kə'mand ,ship }

amphibious mine [ORD] A mine designed especially to hinder beach landing and river-crossing operations by damaging or destroying landing craft, small boats, water-fording vehicles, and floating bridges; it may be of various types, such as contact, controlled, or drifting. { am'fib·ē·əs 'mīn }

amphibious tank [ORD] A vehicle mounting a howitzer or cannon, capable of delivering direct fire from the water as well as from land; used in providing early artillery support in amphibious operations. { am'fib·ē·əs 'taŋk }

amphibious tractor [ORD] A vehicle used for the movement of troops and cargo from ship to shore in the assault phase of amphibious operations, or for limited movement of troops and cargo over land or water. Abbreviated amtrac. { am'fib·ē·əs 'trak·tər }

amphiblastic cleavage [EMBRYO] The unequal but complete cleavage of telolecithal eggs. { ¦am·fə¦blas·tik 'klēv·ij }

amphiblastula [EMBRYO] A blastula resulting from amphiblastic cleavage. [INV ZOO] The free-swimming flagellated larva of many sponges. { ¦am·fə¦blas·chə·lə }

amphibole [MINERAL] Any of a group of rock-forming, ferromagnesian silicate minerals commonly found in igneous and metamorphic rocks; includes hornblende, anthophyllite, tremolite, and actinolite (asbestos minerals). { 'am·fə,bōl }

amphibolic [MED] Uncertain; wavering; refers to the stage of a disease when prognosis is uncertain. [ZOO] Possessing the ability to turn either backward or forward, as the outer toe of certain birds. { ¦am·fə¦bäl·ik }

amphibolic pathway [BIOCHEM] A microbial biosynthetic and energy-producing pathway, such as the glycolytic pathway. { ¦am·fə¦bäl·ik 'path,wā }

Amphibolidae [INV ZOO] A family of gastropod mollusks in the order Basommatophora. { ,am·fə'bäl·ə,dē }

amphibolite [PETR] A crystalloblastic metamorphic rock composed mainly of amphibole and plagioclase; quartz may be present in small quantities. { am'fib·ə,līt }

amphibolite facies [PETR] Rocks produced by medium- to high-grade regional metamorphism. { am'fib·ə,līt 'fā,shēz }

amphibolization [PETR] Formation of amphibole in a rock as a secondary mineral. { am,fib·ə·lə'zā·shən }

amphicarpic [BOT] Having two types of fruit, differing either in form or ripening time. { ¦am·fə¦kär·pik }

Amphichelydia [PALEON] A suborder of Triassic to Eocene anapsid reptiles in the order Chelonia; these turtles did not have a retractable neck. { ,am·fə·kə'lid·ē·ə }

amphichrome [BOT] A plant that produces flowers of different colors on the same stalk. { 'am·fə,krōm }

Amphicoela [VERT ZOO] A small suborder of amphibians in the order Anura characterized by amphicoelous vertebrae. { ¦am·fə¦sēl·ə }

amphicoelous [VERT ZOO] Describing vertebrae that have biconcave centra. { ¦am·fə¦sēl·əs }

amphicribral [BOT] Having the phloem surrounded by the xylem, as seen in certain vascular bundles. { ¦am·fə¦krib·rəl }

amphicryptophyte [BOT] A marsh plant with amphibious vegetative organs. { ¦am·fə¦krip·tə,fīt }

Amphicyonidae [PALEON] A family of extinct giant predatory carnivores placed in the infraorder Miacoidea by some authorities. { ¦am·fə·sī¦än·ə,dē }

amphicytula [EMBRYO] A zygote that is capable of holoblastic unequal cleavage. { ,am·fə'sich·ə·lə }

amphid [INV ZOO] Either of a pair of sensory receptors in nematodes, believed to be chemoreceptors and situated laterally on the anterior end of the body. { 'am·fəd }

amphidetic [INV ZOO] Of a bivalve ligament, extending both before and behind the beak. { ¦am·fə¦ded·ik }

amphidiploid [GEN] A tetraploid organism or species produced when chromosomes of a hybrid between two species double, yielding a diploid set of chromosomes from each parent. Also known as allotetraploid. { ¦am·fə¦di,plòid }

Amphidiscophora [INV ZOO] A subclass of sponges in the class Hexactinellida characterized by an anchoring tuft of spicules and no hexasters. { ¦am·fə·di'skäf·ə·rə }

Amphidiscosa [INV ZOO] An order of hexactinellid sponges in the subclass Amphidiscophora characterized by amphidisc spicules, that is, spicules having a stellate disk at each end. { ¦am·fə·di'skō·sə }

amphidromic [OCEANOGR] Of or pertaining to progression of a tide wave or bulge around a point or center of little or no tide. { ¦am·fə¦dräm·ik }

amphidromic point [MAP] On a chart of cotidal lines, a no-tide or nodal point from which the cotidal lines radiate. { ¦am·fə¦dräm·ik 'pòint }

amphidromic region [MAP] An area surrounding an amphidromic point in which the cotidal lines radiate from the no-tide point and progress through all hours of the tide cycle. { ¦am·fə¦dräm·ik 'rē·jən }

amphigastrium [BOT] Any of the small appendages located ventrally on the stem of some liverworts. { ¦am·fə¦gas·trē·əm }

amphigean [ECOL] An organism that is native to both Old and New Worlds. { ¦am·fə¦jē·ən }

amphigene See leucite. { 'am·fə,jēn }

Amphilestidae [PALEON] A family of Jurassic triconodont mammals whose subclass is uncertain. { ,am·fə'les·tə,dē }

Amphilinidea [INV ZOO] An order of tapeworms in the subclass Cestodaria characterized by a protrusible proboscis, anterior frontal glands, and no holdfast organ; they inhabit the coelom of sturgeon and other fishes. { ,am·fə·lə'nid·ē·ə }

Amphimerycidae [PALEON] A family of late Eocene to early Oligocene tylopod ruminants in the superfamily Amphimerycoidea. { ,am·fə·mə'ris·ə,dē }

Amphimerycoidea [PALEON] A superfamily of extinct ruminant artiodactyls in the infraorder Tylopoda. { ,am·fə,mir·ə'kòid·ē·ə }

amphimixis [PHYSIO] The union of egg and sperm in sexual reproduction. { ,am·fə'mik·səs }

Amphimonadidae [INV ZOO] A family of zoomastigophorean protozoans in the order Kinetoplastida. { ,am·fə·mə'näd·ə,dē }

amphimorphic [GEOL] A rock or mineral formed by two geologic processes. { ,am·fə'mòr·fik }

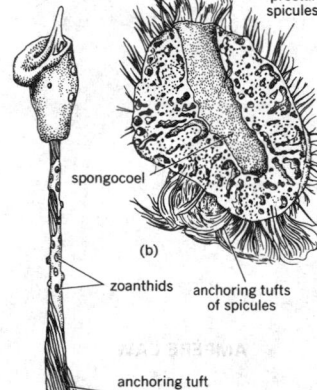

AMPHIDISCOPORA

prostal spicules

spongocoel

(b)

zoanthids   anchoring tufts of spicules

anchoring tuft of spicules

(a)

Representative examples. (a) *Hyalonema*, with zoanthids encrusting a root tuft of spicules (*after Hyman, 1940*). (b) *Pheronema*, sectioned longitudinally (*from Hyman, 1940, after Schulze, 1887*).

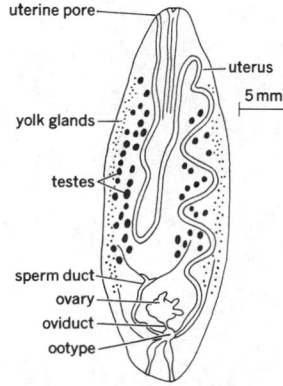

AMPHILINIDEA

uterine pore

uterus

5 mm

yolk glands

testes

sperm duct

ovary

oviduct

ootype

Amphilina, a tapeworm in the order Amphilinidea.

Amphineura [INV ZOO] A class of the phylum Mollusca; members are bilaterally symmetrical, elongate marine animals, such as the chitons. { ¸am·fə'nùr·ə }

Amphinomidae [INV ZOO] The stinging or fire worms, a family of amphinomorphan polychaetes belonging to the Errantia. { ¸am·fə'näm·ə¸dē }

Amphinomorpha [INV ZOO] Group name for three families of errantian polychaetes: Amphenomidae, Euphrosinidae, and Spintheridae. { ¦am·fə·nə¦mòr·fə }

amphioxus [ZOO] Former designation for the lancelet, *Branchiostoma*. { ¸am·fē'äk·səs }

amphipathic molecule [ORG CHEM] A molecule having both hydrophilic and hydrophobic groups; examples are wetting agents and membrane lipids such as phosphoglycerides. { ¸am·fə'path·ik 'mäl·ə¸kyül }

amphiphile [CHEM] A molecule which has a polar head attached to a long hydrophobic tail. { 'am·fə¸fīl }

amphiphilic [BIOCHEM] Describing a molecule having a polar region that is separated from the nonpolar region. { ¸am·fə'fil·ək }

amphiphloic [BOT] Pertaining to the central vascular cylinder of stems having phloem on both sides of the xylem. { ¸am·fə'flō·ik }

amphiphyte [ECOL] A plant growing on the boundary zone of wet land. { 'am·fə¸fīt }

amphiplatyan [ANAT] Describing vertebrae having centra that are flat both anteriorly and posteriorly. { ¦am·fə¦plad·ē·ən }

amphipneustic [VERT ZOO] Having both gills and lungs through all life stages, as in some amphibians. { ¦am·fə¦nüs·tik }

Amphipoda [INV ZOO] An order of crustaceans in the subclass Malacostraca; individuals lack a carapace, bear unstalked eyes, and respire through thoracic branchiae or gills. { am'fip·ə·də }

amphipodous [INV ZOO] Having both walking and swimming legs. { am'fip·ə·dəs }

amphiprotic See amphoteric. { ¦am·fə¦präd·ik }

amphisapropel [GEOL] Cellulosic ooze containing coarse plant debris. { ¸am¦fīz·ə¦prō¸pel }

amphisarca [BOT] An indehiscent fruit characterized by many cells and seeds, pulpy flesh, and a hard rind; melon is an example. { ¸am·fə'sär·kə }

Amphisbaenidae [VERT ZOO] A family of tropical snakelike lizards in the suborder Sauria. { ¦am·fəs¦bēn·ə¸dē }

Amphisopidae [INV ZOO] A family of isopod crustaceans in the suborder Phreactoicoidea. { ¦am·fə¦säp·ə¸dē }

amphispore [MYCOL] A specialized urediospore with a thick, colorful wall; a resting spore. { 'am·fə¸spór }

Amphissitidae [PALEON] A family of extinct ostracods in the suborder Beyrichicopina. { ¦am·fə¦sid·ə¸dē }

Amphistaenidae [VERT ZOO] The worm lizards, a family of reptiles in the suborder Sauria; structural features are greatly reduced, particularly the limbs. { ¦am·fə¦stēn·ə¸dē }

amphistome [INV ZOO] An adult type of digenetic trematode having a well-developed ventral sucker (acetabulum) on the posterior end. { 'am·fə¸stōm }

amphistylar [ARCH] Having free columns in porticoes either at both ends or at both sides and across the full ends of sides. { ¦am·fə¦stīl·ər }

amphistylic [VERT ZOO] Having the jaw suspended from the brain case and the hyomandibular cartilage, as in some sharks. { ¦am·fə¦stīl·ik }

amphitene See zygotene. { 'am·fə¸tēn }

amphitheater [ARCH] A structure or large room containing oval, circular, or semicircular tiers of seats facing an open space. [GEOGR] A valley or gulch having an oval or circular floor and formed by glacial action. { 'am·fə¸thē·ə·tər }

amphithecium [BOT] The external cell layer during development of the sporangium in mosses. { ¸am·fə'thē·shē·əm }

Amphitheriidae [PALEON] A family of Jurassic therian mammals in the infraclass Pantotheria. { ¸am·fə·thə'rī·ə¸dē }

amphitriaene [INV ZOO] A poriferan spicule having three divergent rays at each end. { ¸am·fə'trī¸ēn }

amphitrichous [BIOL] Having flagella at both ends, as in certain bacteria. { am'fi·trə·kəs }

Amphitritinae [INV ZOO] A subfamily of sedentary polychaete worms in the family Terebellidae. { ¸am·fə'trī·tə¸nē }

Amphitrite [ASTRON] An asteroid with a diameter of about 120 miles (200 kilometers), mean distance from the sun of 2.554 astronomical units, and S-type surface composition. { 'am·fə'trī¸dē }

amphitropical distribution [ECOL] Distribution of mostly temperate organisms which are discontinuous between the Northern and Southern hemispheres. { ¦am·fə¦träp·ə·kəl ¸dis·trə'byü·shən }

amphitropous [BOT] Having a half-inverted ovule with the funiculus attached near the middle. { ¸am'fi·trə·pəs }

Amphiumidae [VERT ZOO] A small family of urodele amphibians in the suborder Salamandroidea composed of three species of large, eellike salamanders with tiny limbs. { ¸am·fē'yü·mə¸dē }

amphivasal [BOT] Having the xylem surrounding the phloem, as seen in certain vascular bundles. { ¦am·fə¦vā·zəl }

Amphizoidae [INV ZOO] The trout stream beetles, a small family of coleopteran insects in the suborder Adephaga. { ¸am·fə'zō·ə¸dē }

ampholyte [CHEM] An amphoteric electrolyte. { 'am·fə¸līt }

ampholytic detergent [CHEM] A detergent that is cationic in acidic solutions and anionic in basic solutions. { ¦am·fə¦lid·ik di'tər·jənt }

Amphoriscidae [INV ZOO] A family of calcareous sponges in the order Sycettida. { ¦am·fə¦ris·ə¸dē }

amphoteric [CHEM] Having both acidic and basic characteristics. Also known as amphiprotic. { ¦am·fə¦ter·ik }

amphotericin [MICROBIO] An amphoteric antifungal antibiotic produced by *Streptomyces nodosus* and having of two components, A and B. { ¸am·fə'ter·ə·sən }

amphotericin A [MICROBIO] The relatively inactive component of amphotericin. { ¸am·fə'ter·ə·sən 'ā }

amphotericin B [MICROBIO] $C_{46}H_{73}O_{20}N$ The active component of amphotericin, suitable for systemic therapy of deep or superficial mycotic infections. { ¸am·fə'ter·ə·sən 'bē }

amphoterism [CHEM] The property of being able to react either as an acid or a base. { am'fäd·ə¸riz·əm }

amphoterite [GEOL] A stony meteorite containing bronzite and olivine with some oligoclase and nickel-rich iron. { am'fäd·ə¸rīt }

amp-hr See ampere-hour.

amplexicaul [BOT] Pertaining to a sessile leaf with the base or stipules embracing the stem. { am'plek·sə¸kòl }

amplexus [BOT] Having the edges of a leaf overlap the edges of a leaf above it in vernation. [VERT ZOO] The copulatory embrace of frogs and toads. { am'plek·səs }

ampliate [BIOL] Widened or enlarged. { 'am·plē¸āt }

amplidyne [ELEC] A rotating magnetic amplifier having special windings and brush connections so that small changes in power input to the field coils produce large changes in power output. { 'am·plə¸dīn }

amplification [GEN] **1.** Treatment with an antibiotic or other agent to increase the relative proportion of plasmid to bacterial deoxyribonucleic acid. **2.** Bulk replication of a gene library. [SCI TECH] The production of an output of greater magnitude than the input. { ¸am·plə·fə'kā·shən }

amplification factor [ELECTR] In a vacuum tube, the ratio of the incremental change in plate voltage to a given small change in grid voltage, under the conditions that the plate current and all other electrode voltages are held constant. { ¸am·plə·fə'kā·shən ¸fak·tər }

amplification noise [ELECTR] Noise generated in the vacuum tubes, transistors, or integrated circuits of an amplifier. { ¸am·plə·fə'kā·shən ¸nòiz }

amplified back bias [ELECTR] Degenerative voltage developed across a fast time-constant circuit within a stage of an amplifier and fed back into a preceding stage. { 'am·plə¸fīd 'bak ¸bī·əs }

amplifier [ENG] A device capable of increasing the magnitude or power level of a physical quantity, such as an electric current or a hydraulic mechanical force, that is varying with time, without distorting the wave shape of the quantity. { 'am·plə¸fī·ər }

amplifier-type meter [ENG] An electric meter whose characteristics have been enhanced by the use of preamplification for the signal input eventually used to actuate the meter. { 'am·plə¸fī·ər ¦tīp ¦mēd·ər }

amplify [ENG ACOUS] To strengthen a signal by increasing its amplitude or by raising its level. { 'am·plə¸fī }

amplifying delay line [ELECTR] Delay line used in pulse-compression systems to amplify delayed signals in the super-high-frequency region. { 'am·plə₁fī·iŋ di'lā ₁līn }

amplitron [ELECTR] Crossed-field continuous cathode reentrant beam backward-wave amplifier for microwave frequencies. { 'am·plə₁trän }

amplitude [ASTRON] The range in brightness of a variable star, usually expressed in magnitudes. [MATH] The angle between a vector representing a specified complex number on an Argand diagram and the positive real axis. Also argument. [NAV] Angular distance north or south of the prime vertical; the arc of the horizon, or the angle at the zenith between the prime vertical and a vertical circle, measured north or south from the prime vertical to the vertical circle. [PHYS] The maximum absolute value attained by the disturbance of a wave or by any quantity that varies periodically. { 'am·plə₁tüd }

amplitude discriminator See pulse-height discriminator. { 'am·plə₁tüd dis'krim·ə₁nād·ər }

amplitude distortion See frequency distortion. { 'am·plə₁tüd di'stòr·shən }

amplitude factor See crest factor. { 'am·plə₁tüd ₁fak·tər }

amplitude fading [COMMUN] Fading in which the amplitudes of frequency components of a modulated carrier wave are uniformly attenuated. { 'am·plə₁tüd ₁fād·iŋ }

amplitude-frequency distortion See frequency distortion. { 'am·plə₁tüd 'frē·kwən·sē di'stòr·shən }

amplitude-frequency response See frequency response. { 'am·plə₁tüd 'frē·kwən·sē ri'späns }

amplitude gate [ELECTR] A circuit which transmits only those portions of an input signal which lie between two amplitude boundary level values. Also known as slicer; slicer amplifier. { 'am·plə₁tüd ₁gāt }

amplitude level [PHYS] The natural logarithm of the ratio of two amplitudes, each measured in the same units. { 'am·plə₁tüd ₁lev·əl }

amplitude limiter See limiter. { 'am·plə₁tüd ₁lim·əd·ər }

amplitude-limiting circuit See limiter. { 'am·plə₁tüd ¦lim·əd·iŋ ₁sər·kət }

amplitude-modulated indicator [ENG] A general class of radar indicators, in which the sweep of the electron beam is deflected vertically or horizontally from a base line to indicate the existence of an echo from a target. Also known as deflection-modulated indicator; intensity-modulated indicator. { 'am·plə₁tüd ¦mäj·ə₁lād·əd ¦in·də₁kād·ər }

amplitude modulation [ELECTR] Abbreviated AM. **1.** Modulation in which the aplitude of a wave is the characteristic varied in accordance with the intelligence to be transmitted. **2.** In telemetry, those systems of modulation in which each component frequency f of the transmitted intelligence produces a pair of sideband frequencies at carrier frequency plus f and carrier minus f. { 'am·plə₁tüd ₁maj·ə'lā·shən }

amplitude-modulation noise [COMMUN] Noise produced by undesirable amplitude variations of a radio-frequency signal. { 'am·plə₁tüd ₁maj·ə'lā·shən ₁nòiz }

amplitude-modulation radio [COMMUN] Also known as AM radio.The system of radio communication employing amplitude modulation of a radio-frequency carrier to convey the intelligence.A receiver used in such a system. { 'am·plə₁tüd ₁maj·ə'lā·shən ₁rād·ē₁ō }

amplitude modulator [PHYS] Any device which imposes amplitude modulation upon a carrier wave in accordance with a desired program. { 'am·plə₁tüd ₁maj·ə₁lād·ər }

amplitude noise [ELECTROMAG] Effect on radar accuracy of the fluctuations in the amplitude of the signal returned by the target; these fluctuations are caused by any change in aspect if the target is not a point source. { 'am·plə₁tüd ₁nòiz }

amplitude of accommodation [PHYSIO] The range of focal powers of which the eye is capable, expressed in diopters. { 'am·plə₁tüd əv ə₁käm·ə'dā·shən }

amplitude resonance [PHYS] The frequency at which a given sinusoidal excitation produces the maximum amplitude of oscillation in a resonant system. { 'am·plə₁tüd 'rez·ə·nəns }

amplitude response [ELECTR] The maximum output amplitude obtainable at various points over the frequency range of an instrument operating under rated conditions. { 'am·plə₁tüd ri'späns }

amplitude selector See pulse-height selector. { 'am·plə₁tüd si₁lek·tər }

amplitude separator [ELECTR] A circuit used to isolate the portion of a waveform with amplitudes above or below a given value or between two given values. { 'am·plə₁tüd 'sep·ə₁rād·ər }

amplitude shift keying [COMMUN] A method of transmitting binary coded messages in which a sinusoidal carrier is pulsed so that one of the binary states is represented by the presence of the carrier while the other is represented by its absence. Abbreviated ASK. { 'am·plə₁tüd ¦shift ₁kē·iŋ }

amplitude splitting [OPTICS] A technique in which light falls on a partially reflecting surface; part of the light is transmitted, part reflected, and after further manipulation, these parts are recombined to give interference. { 'am·plə₁tüd ₁splid·iŋ }

amplitude suppression ratio [ELECTR] Ratio, in frequency modulation, of the undesired output to the desired output of a frequency-modulated receiver when the applied signal has simultaneous amplitude and frequency modulation. { 'am·plə₁tüd sə'presh·ən ₁rā·shō }

amplitude-versus-frequency distortion [ELECTR] The distortion caused by the nonuniform attenuation or gain of the system, with respect to frequency under specified terminal conditions. { 'am·plə₁tüd ¦vər·səs ¦frē·kwən·sē di'stòr·shən }

amp-turn See ampere-turn. { ¦amp¦tərn }

ampule [MED] A small, hermetically sealed flask that usually contains medicine for parenteral administration. { 'am₁pyül }

Ampulicidae [INV ZOO] A small family of hymenopteran insects in the superfamily Sphecoidea. { ₁am·pyə'lis·ə₁dē }

ampulla [ANAT] A dilated segment of a gland or tubule. [BOT] A small air bladder in some aquatic plants. [INV ZOO] The sac at the base of a tube foot in certain echinoderms. { am'pùl·ə }

ampulla of Lorenzini [VERT ZOO] Any of the cutaneous receptors in the head region of elasmobranchs; thought to have a thermosensory function. { am'pùl·ə əv ₁lò·rent'zē₁nē }

ampulla of Vater [ANAT] Dilation at the junction of the bile and pancreatic ducts and the duodenum in humans. Also known as papilla of Vater. { am'pùl·ə əv 'fät·ər }

Ampullariella [MICROBIO] A genus of bacteria in the family Actinoplanaceae having cylindrical or bottle-shaped sporangia and rod-shaped spores arranged in parallel chains; motile by means of a polar tuft of flagella. { ₁am·pyə₁lar·ē'el·ə }

ampullary organ [PHYSIO] An electroreceptor most sensitive to direct-current and low-frequency electric stimuli; found over the body surface of electric fish and also in certain other fish such as sharks, rays, and catfish. { 'am·pyə₁ler·ē 'òr·gən }

amputation [MED] The surgical, congenital, or spontaneous removal of a limb or projecting body part. { ₁am·pyə'tā·shən }

AM radio See amplitude-modulation radio. { ¦ā¦em 'rād·ē₁ō }

AM signature [COMMUN] A graphic representation of the significant identifying characteristics of an amplitude-modulated signal. { ¦ā¦em 'sig·nə·chər }

AMSS See aeronautical mobile satellite service.

amtrac See amphibious tractor. { 'am₁trak }

amu See atomic mass unit.

AMV See Alfalfa mosaic virus..

amychophobia [PSYCH] Abnormal fear of being scratched or clawed. { ₁am·ə·kə'fōb·ē·ə }

amygdalin [BIOCHEM] $C_6H_5CH(CN)OC_{12}H_{21}O_{10}$ A glucoside occurring in the kernels of certain plants of the genus *Prunus*. { ə'mig·də₁lən }

amygdaloid [GEOL] Lava rock containing amygdules. Also known as amygdaloidal lava. { ə'mig·də₁lòid }

amygdaloidal lava See amygdaloid. { ə'mig·də₁lòid·əl ₁läv·ə }

amygdule [GEOL] **1.** A mineral filling formed in vesicles (cavities) of lava flows; it may be chalcedony, opal, calcite, chlorite, or prehnite. **2.** An agate pebble. { ə'mig₁dyül }

amyl [ORG CHEM] Any of the eight isomeric arrangements of the radical C_5H_{11} or a mixture of them. Also known as pentyl. { 'am·əl }

amyl acetate [ORG CHEM] $CH_3COO(CH_2)_2CH(CH_3)_2$ A colorless liquid, boiling at 142°C; soluble in alcohol and ether, slightly soluble in water; used in flavors and perfumes. Also known as banana oil; isoamyl acetate. { 'am·əl 'as·ə₁tāt }

amyl alcohol [ORG CHEM] **1.** A colorless liquid that is a mixture of isomeric alcohols. **2.** An optically active liquid composed of isopentyl alcohol and active amyl alcohol. { 'am·əl 'al·kə,hól }

n-amylamine [ORG CHEM] $C_5H_{11}NH_2$ A colorless liquid with a boiling point of 104.4°C; soluble in water, alcohol, and ether; used in dyestuffs, insecticides, synthetic detergents, corrosion inhibitors, and pharmaceuticals, and as a gasoline additive. { ¦en ə'mil·ə,men }

amylase [BIOCHEM] An enzyme that hydrolyzes reserve carbohydrates, starch in plants and glycogen in animals. { 'am·ə,lās }

amyl benzoate See isoamyl benzoate. { 'am·əl 'ben·zə,wāt }

amylene [ORG CHEM] C_5H_{10} A highly flammable liquid with a low boiling point, 37.5–38.5°C; often a component of petroleum. Also known as 2-methyl-2-butene. { 'am·ə ,lēn }

amyl ether [ORG CHEM] **1.** Either of two isomeric compounds, *n*-amyl ether or isoamyl ether; both may be represented by the formula $(C_5H_{11})_2O$. **2.** A mixture mainly of isoamyl ether and *n*-amyl ether formed in preparation of amyl alcohols from amyl chloride; very slightly soluble in water; used mainly as a solvent. { 'am·əl 'ē·thər }

amyl mercaptan [ORG CHEM] $C_5H_{11}SH$ A colorless to light yellow liquid with a boiling range of 104–130°C; soluble in alcohol; used in odorant for detecting gas line leaks. { 'am· əl mər'kap,tan }

amyl nitrate [ORG CHEM] $C_5H_{11}ONO_2$ An ester of amyl alcohol added to diesel fuel to raise the cetane number. { 'am· əl 'nī,trāt }

amyl nitrite [ORG CHEM] $(CH_3)_2CH(CH_2)_2NO_2$ A yellow liquid; soluble in alcohol, very slightly soluble in water; fruity odor; it is flammable and the vapor is explosive; used in medicine and perfumes. Also known as isoamyl nitrite. { 'am· əl 'nī,trīt }

amyloglucosidase [FOOD ENG] An enzyme of microbial origin that breaks glucoside bonds in starch and dextrins to form glucose; used in the manufacturing of glucose and for converting carbohydrates to fermentable sugars (as in beer-brewing). Also known as glucoamylase. { ,am·ə·lō ,glü'kō·sə,dās }

amylograph [ENG] An instrument used to measure and record the viscosity of starch and flour pastes and the temperature at which they gelatinize. { ə'mīl·ə,graf }

amyloid [PATH] An abnormal protein deposited in tissues, formed from the infiltration of an unknown substance, probably a carbohydrate. { 'am·ə,lóid }

amyloid beta protein [BIOCHEM] A 42-amino-acid proteolytic product of the amyloid precursor protein that accumulates in the brains (frontal cortex and hippocampus) of persons with Alzheimer's disease. { 'am·ə,lóid ¦bād·ə 'prō,tēn }

amyloid body [PATH] Any of the microscopic hyaline bodies that stain like amyloid with metachromatic aniline dyes. { 'am·ə,lóid ,bäd·ē }

amyloidosis [MED] Deposition of amyloid in one or more organs of the body. { ,am·ə·loi'dō·səs }

amylolysis [BIOCHEM] The enzyme-catalyzed hydrolysis of starch to soluble products. { ,am·ə'läl·ə·səs }

amylolytic enzyme [BIOCHEM] A type of enzyme capable of denaturing starch molecules; used in textile manufacture to remove starch added to slash sizing agents. { ,am·ə'läd· ik 'en,zīm }

amylopectin [BIOCHEM] A highly branched, high-molecular-weight carbohydrate polymer composed of about 80% corn starch. { ,am·ə·lō'pek·tən }

amylopectinosis [MED] A hereditary disease arising from an enzyme deficiency and characterized by abnormal accumulation of glycogen in tissues. Also known as Andersen's disease. { ,am·ə·lō,pek·tə'nō·səs }

amyloplast [BOT] A colorless cell plastid packed with starch grains and occurring in cells of plant storage tissue. { 'am· ə·lō,plast }

amylopsin [BIOCHEM] An enzyme in pancreatic juice that acts to hydrolyze starch into maltose. { ,am·ə'läp·sən }

amylose [BIOCHEM] A linear starch polymer. { 'am·ə ,lōs }

amyl propionate [ORG CHEM] $CH_3CH_2COOC_5H_{11}$ A colorless liquid with an applelike odor and a distillation range of 135–175°C; used in perfumes, lacquers, and flavors. { 'am· əl 'prō·pē·ə,nāt }

amyl salicylate [ORG CHEM] $C_6H_4OHCOOC_5H_{11}$ A clear liquid that occasionally has a yellow tinge; boils at 280°C;

soluble in alcohol, insoluble in water; used in soap and perfumes. Also known as isoamyl salicylate. { 'am·əl sə'lis· ə,lāt }

amyl xanthate [ORG CHEM] A salt formed by replacing the hydrogen attached to the sulfur in amylxanthic acid by a metal; used as collector agent in the flotation of certain minerals. { 'am·əl 'zan,thāt }

Amynodontidae [PALEON] A family of extinct hippopotamuslike perissodactyl mammals in the superfamily Rhinoceratoidea. { ,a·mə·nə'dän·tə,dē }

amyotonia [MED] Absence of muscle tone. { ¦ā¦mī·ə'tō· nē·ə }

amyotonia congenita [MED] A congenital disease of the central nervous system characterized by absence of voluntary muscle tone and reflexes. Also known as Oppenheim's disease. { ¦ā¦mī·ə'tō·nē·ə kən'jen·əd·ə }

amyotrophic lateral sclerosis [MED] A progressive neurological disorder characterized by loss of connection and death of motor neurons in the cortex and spinal cord. { ¦a¦mī·ə¦träf· ik 'la·trəl sklə'rō·səs }

amyriotic field [QUANT MECH] A quantized field that has creation and annihilation operators satisfying specified commutation rules and a vacuum state. { ə¦mir·ē,äd·ik 'fēld }

An See actinon.

ANA See antinuclear antibody.

Anabaena [BOT] A genus of blue-green algae in the class Cyanophyceae; members fix atmospheric nitrogen. { ,an· ə'bēn· ə }

Anabantidae [VERT ZOO] A fresh-water family of actinopterygian fishes in the order Perciformes, including climbing perches and gourami. { ,an·ə'ban·tə,dē }

Anabantoidei [VERT ZOO] A suborder of fresh-water labyrinth fishes in the order Perciformes. { ,an·ə,ban'tói·dē,ī }

anabasine [ORG CHEM] A colorless, liquid alkaloid extracted from the plants *Anabasis aphylla* and *Nicotiana glauca*; boiling point is 105°C; soluble in alcohol and ether; used as an insecticide. { ə'na·bə,sēn }

anabatic wind [METEOROL] An upslope wind; usually applied only when the wind is blowing up a hill or mountain as the result of a local surface heating, and apart from the effects of the larger-scale circulation. { ¦an·ə¦bad·ik 'wind }

anabiosis [BIOL] State of suspended animation induced by desiccation and reversed by addition of moisture; can be achieved in rotifers. { ,an·ə,bī'ō·səs }

anabohitsite [PETR] A variety of olivine-pyroxenite containing hornblende and hypersthene and a high proportion (about 30%) of magnetite and ilmenite. { ,an·ə·bō'hit,sīt }

anabolic [BIOCHEM] Pertaining to anabolism. [EVOL] Pertaining to anaboly. { ,an·ə'bäl·ik }

anabolic steroid [BIOCHEM] Any of a group of steroid hormones that increase anabolism. { ,an·ə¦bäl·ik 'sti,róid }

anabolism [BIOCHEM] A part of metabolism involving the union of smaller molecules into larger molecules; the method of synthesis of tissue structure. { ,an'ab·ə,liz·əm }

anaboly [EVOL] The addition, through evolutionary differentiation, of a new terminal stage to the morphogenetic pattern. { ə'nab·ə·lī }

anabranch [HYD] A diverging branch of a stream or river that loses itself in sandy soil or rejoins the main flow downstream. { 'an·ə,branch }

Anacanthini [VERT ZOO] An equivalent name for the Gadiformes. { ,an·ə'kan·thə,nī }

Anacardiaceae [BOT] A family of flowering plants, the sumacs, in the order Sapindales; many species are allergenic to humans. { ,an·ə,kärd·ē'ās·ē,ē }

anacardium gum See cashew gum. { ,an·ə'kärd·ē·əm 'gəm }

anaclinal [GEOL] Having a downward inclination opposite to that of a stratum. { ¦an·ə'klīn·əl }

anaclitic [PSYCH] Pertaining to the dependence of the infant on the mother or mother substitute for a sense of well-being. { ¦an·ə'klid·ik }

anaclitic depression [PSYCH] Seriously impaired physical, social, and intellectual development in some infants following sudden separation from a loving mother or mother figure. { ¦an·ə¦klid·ik di'presh·ən }

anaclitic object choice [PSYCH] The choosing of a love object resembling the person upon whom the individual was emotionally dependent during infancy. { ¦an·ə¦klid·ik ¦ob,jekt ¦chói·s }

anaconda [VERT ZOO] *Eunectes murinus.* The largest living snake, an arboreal-aquatic member of the boa family (Boidae). { ˌan·ə'kän·də }

anacoustic zone [GEOPHYS] The zone of silence in space, starting at about 100 miles (160 kilometers) altitude, where the distance between air molecules is greater than the wavelength of sound, and sound waves can no longer be propagated. { ¦an·ə¦kü·stik ˌzōn }

Anactinochitinosi [INV ZOO] A group name for three closely related suborders of mites and ticks: Onychopalpida, Mesostigmata, and Ixodides. { ə¦nak·tə·nə¦kīt·ən'ō,sī }

Anacystis [BOT] A genus of blue-green algae in the class Cyanophycea. { ˌan·ə'sis·təs }

anadromous [VERT ZOO] Said of a fish, such as the salmon and shad, that ascends fresh-water streams from the sea to spawn. { ə'na·drə·məs }

Anadyomenaceae [BOT] A family of green marine algae in the order Siphonocladales characterized by the expanded blades of the thallus. { ə,na·dyə,men'ās·ē,ē }

anaerobe [BIOL] An organism that does not require air or free oxygen to maintain its life processes. { 'an·ə,rōb }

anaerobic adhesive [MATER] A single-component adhesive that hardens rapidly to form a strong bond between surfaces from which air is excluded. { ¦an·ə¦rōb·ik əd'hēz·iv }

anaerobic bacteria [MICROBIO] Any bacteria that can survive in the partial or complete absence of air; two types are facultative and obligate. { ¦an·ə¦rōb·ik ,bak'tir·ē·ə }

anaerobic condition [BIOL] The absence of oxygen, preventing normal life for organisms that depend on oxygen. { ¦an·ə¦rōb·ik kən'dish·ən }

anaerobic glycolysis [BIOCHEM] A metabolic pathway in plants by which, in the absence of oxygen, hexose is broken down to lactic acid and ethanol with some adenosinetriphosphate synthesis. { ¦an·ə¦rōb·ik glī'käl·ə·səs }

anaerobic petri dish [MICROBIO] A glass laboratory dish for plate cultures of anaerobic bacteria; a thioglycollate agar medium and restricted air space give proper conditions. { ¦an·ə¦rōb·ik 'pē·trē ,dish }

anaerobic process [SCI TECH] A process from which air or oxygen not in chemical combination is excluded. { ¦an·ə¦rōb·ik 'präs·əs }

anaerobic sediment [GEOL] A highly organic sediment formed in the absence or near absence of oxygen in water that is rich in hydrogen sulfide. { ¦an·ə¦rōb·ik 'sed·ə·mənt }

anaerobiosis [BIOL] A mode of life carried on in the absence of molecular oxygen. { ,an·ə,rō'bī·ə·səs }

anaerophyte [ECOL] A plant that does not need free oxygen for respiration. { ə'ner·ə,fīt }

anafront [METEOROL] A front at which the warm air is ascending the frontal surface up to high altitudes. { 'an·ə,frənt }

anagen effluvium [MED] Acute hair loss that usually follows chemotherapy or radiotherapy. { ¦an·ə·jən ə¦flü·vē·əm }

anaglyph [GRAPHICS] **1.** A stereogram in which the two views are printed or projected superimposed in complementary colors, usually red and blue; by viewing through filter spectacles of corresponding complementary colors, a stereoscopic image is formed. **2.** A surface worked in low relief. { 'an·ə,glif }

anagyrine [ORG CHEM] $C_{15}H_{20}N_2O$ A toxic alkaloid found in several species of *Lupinus* in the western United States; acute poisoning produces nervousness, depression, loss of muscular control, convulsions, and coma. { ,an·ə'jī,rēn }

anakinesis [BIOCHEM] A process in living organisms by which energy-rich molecules, such as adenosine triphosphate, are formed. { ,an·ə·kə'nē·səs }

anal [ANAT] Relating to or located near the anus. { 'ān·əl }

analbite [MINERAL] A triclinic albite which is not stable and becomes monoclinic at about 700°C. { ə'nal,bīt }

analbuminemia [MED] A disorder transmitted as an autosomal recessive, characterized by drastic reduction or absence of serum albumin. { ¦an,al,byü·mə¦nēm·ē·ə }

anal character [PSYCH] A personality type that exhibits excessive orderliness, miserliness, and obstinancy. { 'ān·əl 'kar·ik·tər }

analcime [MINERAL] $NaAlSi_2O_6·H_2O$ A white or slightly colored isometric zeolite found in diabase and in alkali-rich basalts. Also known as analcite. { ə'nal,sēm }

analcimite [PETR] An extrusive or hypabyssal rock that consists primarily of pyroxene and analcime. { ə'nal·sə,mīt }

analcimization [GEOL] The replacement in igneous rock of feldspars or feldspathoids by analcime. { ə¦nal·sə·mə¦zā·shən }

analcite See analcime. { ə'nal,sīt }

analemma [ASTRON] A figure-eight-shaped diagram on a globe showing the declination of the sun throughout the year and also the equation of time. [CIV ENG] Any raised construction which serves as a support or rest. { ,an·ə'lem·ə }

analeptic [PHARM] Any drug used to restore respiration and a wakeful state. { ,an·ə¦lep·tik }

anal fin [VERT ZOO] An unpaired fin located medially on the posterior ventral part of the fish body. { 'ān·əl ,fin }

analgesia [PHYSIO] Insensibility to pain with no loss of consciousness. { ,an·əl'jēzh·ə }

analgesic [PHARM] Any drug, such as salicylates, morphine, or opiates, used primarily for the relief of pain. { ,an·əl'jēz·ik }

anal gland [INV ZOO] A gland in certain mollusks that secretes a purple substance. [VERT ZOO] A gland located near the anus or opening into the rectum in many vertebrates. { 'ān·əl ,gland }

anallagmatic curve [MATH] A curve that is its own inverse curve with respect to some circle. { ə¦nal·ig¦mad·ik 'kərv }

anallobaric center See pressure-rise center. { ə¦nal·ə¦bär·ik 'sen·tər }

analog [CHEM] A compound whose structure is similar to that of another compound but whose composition differs by one element. [FOOD ENG] A meat-substitute food manufactured from vegetable ingredients, such as soybeans. [ELECTR] **1.** A physical variable which remains similar to another variable insofar as the proportional relationships are the same over some specified range; for example, a temperature may be represented by a voltage which is its analog. **2.** Pertaining to devices, data, circuits, or systems that operate with variables which are represented by continuously measured voltages or other quantities. [METEOROL] A past large-scale synoptic weather pattern which resembles a given (usually current) situation in its essential characteristics. { 'an·əl,äg }

analog adder [ELECTR] A device with one output voltage which is a weighted sum of two input voltages. { 'an·əl,äg 'ad·ər }

analog channel [ELECTR] A channel on which the information transmitted can have any value between the channel limits, such as a voice channel. { 'an·əl,äg 'chan·əl }

analog communications [COMMUN] System of telecommunications employing a nominally continuous electric signal that varies in frequency, amplitude, or other characteristic, in some direct correlation to nonelectrical information (sound, light, and so on) impressed on a transducer. { 'an·əl,äg kə,myü·nə'kā·shənz }

analog comparator [ELECTR] **1.** A comparator that checks digital values to determine whether they are within predetermined upper and lower limits. **2.** A comparator that produces high and low digital output signals when the sum of two analog voltages is positive and negative, respectively. { 'an·əl,äg kəm'par·əd·ər }

analog computer [COMPUT SCI] A computer is which quantities are represented by physical variables; problem parameters are translated into equivalent mechanical or electrical circuits as an analog for the physical phenomenon being investigated. { 'an·əl,äg kəm'pyüd·ər }

analog data [COMPUT SCI] Data represented in a continuous form, as contrasted with digital data having discrete values. { 'an·əl,äg 'dad·ə }

analog-digital computer See hybrid computer. { 'an·əl,äg 'dij·ə·təl kəm,pyüd·ər }

analog indicator [ELECTR] A device in which the result of a measurement is indicated by a pointer deflection or other visual quantity. { 'an·əl,äg 'in·də,kād·ər }

analog monitor [ELECTR] A display unit that accepts only analog signals, which must be converted from digital signals by the computer's video display board. { 'an·əl·äg ,män·əd·ər }

analog multiplexer [ELECTR] A multiplexer that provides switching of analog input signals to allow use of a common analog-to-digital converter. { 'an·əl,äg 'məl·tə,plek·sər }

analog multiplier [ELECTR] A device that accepts two or more inputs in analog form and then produces an output proportional to the product of the input quantities. { 'an·əl,äg 'məl·tə,plī·ər }

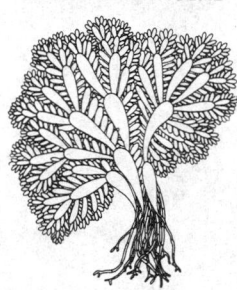

ANADYOMENACEAE

Anadyomene, a genus in Anadyomenaceae, with expanded blades.

ANAEROBIC PETRI DISH

dish cover

air space

anaerobic agar

Brewer anaerobic petri dish. *(Courtesy BioQuest, Division of Becton, Dickinson and Co.)*

analog network [ELECTR] A circuit designed so that circuit variables such as voltages are proportional to the values of variables in a system under study. { 'an·əl‚äg 'net‚wərk }

analogous [BIOL] Referring to structures that are similar in function and general appearance but not in origin, such as the wing of an insect and the wing of a bird. { ə'nal·ə·gəs }

analogous pole [SOLID STATE] The pole of a crystal that acquires a positive charge when the crystal is heated. { ə'nal·ə·gəs ‚pōl }

analog output [CONT SYS] Transducer output in which the amplitude is continuously proportional to a function of the stimulus. { 'an·əl‚äg 'aut‚put }

analog radio system *See* AR system. { 'an·əl‚äg 'rā·dē·ō 'sis·təm }

analog readout [ENG] A scale on a balance that continuously indicates measurement values by the position of an index mark, either a line or a pointer, opposite a graduated scale which is usually marked with numbers. { 'an·əl‚äg 'rēd‚aut }

analog recording [ELECTR] Any method of recording in which some characteristic of the recording signal, such as amplitude or frequency, is continuously varied in a manner analogous to the time variations of the original signal. { 'an·əl‚äg ri'kord·iŋ }

analog signal [ELECTR] A nominally continuous electrical signal that varies in amplitude or frequency in response to changes in sound, light, heat, position, or pressure. { 'an·əl‚äg 'sig·nəl }

analog simulation [COMPUT SCI] The representation of physical systems and phenomena by variables such as translation, rotation, resistance, and voltage. { 'an·əl‚äg ‚sim·yə'lā·shən }

analog states [NUC PHYS] Certain nuclear states belonging to neighboring nuclear isobars and possessing identical structure except for the transformation of one or more neutrons into the same number of protons. Also known as isobaric analog states. { 'an·əl‚äg ‚stāts }

analog switch [ELECTR] **1.** A device that either transmits an analog signal without distortion or completely blocks it. **2.** Any solid-state device, with or without a driver, capable of bilaterally switching voltages or current. { 'an·əl‚äg ‚swich }

analog-to-digital converter [ELECTR] A device which translates continuous analog signals into proportional discrete digital signals. { 'an·əl‚äg tə 'dij·ət·əl kən'vərd·ər }

analog-to-frequency converter [ELECTR] A converter in which an analog input in some form other than frequency is converted to a proportional change in frequency. { 'an·əl‚äg tə 'frē·kwən·sē kən'vərd·ər }

analog voltage [ELECTR] A voltage that varies in a continuous fashion in accordance with the magnitude of a measured variable. { 'an·əl‚äg 'vōl·tij }

anal phase [PSYCH] In psychoanalytic theory, the period of pregenital psychosexual development, usually from 1 to 3 years of age, in which the child has particular interest in the process of defecation and the sensations connected with the anus. { 'ān·əl ‚fāz }

anal plate [EMBRYO] An embryonic plate formed of endoderm and ectoderm through which the anus later ruptures. [VERT ZOO] **1.** One of the plates on the posterior portion of the plastron in turtles. **2.** A large scale anterior to the anus of most snakes. { 'ān·əl ‚plāt }

anal sphincter [ANAT] Either of two muscles, one voluntary and the other involuntary, controlling closing of the anus in vertebrates. { 'ān·əl 'sfiŋk·tər }

analysand [PSYCH] An individual in psychoanalytic treatment. { ə'nal·ə‚zand }

analysis [ANALY CHEM] The determination of the composition of a substance. [MATH] The branch of mathematics most explicitly concerned with the limit process or the concept of convergence; includes the theories of differentiation, integration and measure, infinite series, and analytic functions. Also known as mathematical analysis. [METEOROL] A detailed study in synoptic meteorology of the state of the atmosphere based on actual observations, usually including a separation of the entity into its component patterns and involving the drawing of families of isopleths for various elements. { ə'nal·ə·səs }

analysis by synthesis [COMMUN] A method of determining the parameters of a speech coder in which the consequence of choosing a particular value of a coder parameter is evaluated

by locally decoding the signal and comparing it to the original input signal. { ə'nal·ə·sis ‚bī 'sin·thə·səs }

analysis line [SPECT] The spectral line used in determining the concentration of an element in spectrographic analysis. { ə'nal·ə·səs ‚līn }

analysis of variance [STAT] A method for partitioning the total variance in experimental data into components assignable to specific sources. { ə'nal·ə·səs əv 'ver·ē·əns }

analyte [ANALY CHEM] **1.** The sample being analyzed. **2.** The specific component that is being measured in a chemical analysis. { 'an·ə‚līt }

analytical aerotriangulation [ENG] Analytical phototriangulation, performed with aerial photographs. { ‚an·əl'id·ə·kəl ‚er·ō‚trī‚aŋ·gyə'lā·shən }

analytical balance [ENG] A balance with a sensitivity of 0.1–0.01 milligram. { ‚an·əl'id·ə·kəl 'bal·əns }

analytical blank *See* blank. { ‚an·əl'id·ə·kəl 'blaŋk }

analytical centrifugation [ENG] Centrifugation following precipitation to separate solids from solid-liquid suspensions; faster than filtration. { ‚an·əl'id·ə·kəl sen‚trif·ə'gā·shən }

analytical chemistry [CHEM] The science of chemical characterization and measurement; qualitative analysis is concerned with the description of chemical composition in terms of elements, compounds, or structural units, whereas quantitative analysis is concerned with the measurement of amount. { ‚an·əl'id·ə·kəl 'kem·ə·strē }

analytical distillation [ANALY CHEM] Precise resolution of a volatile liquid mixture into its components; the mixture is vaporized by heat or vacuum, and the vaporized components are recondensed into liquids at their respective boiling points. { ‚an·əl'id·ə·kəl ‚dis·tə'lā·shən }

analytical engine [COMPUT SCI] An early-19th-century form of mechanically operated digital computer. { ‚an·əl'id·ə·kəl 'en·jən }

analytical extraction [ANALY CHEM] Precise transfer of one or more components of a mixture (liquid to liquid, gas to liquid, solid to liquid) by contacting the mixture with a solvent in which the component of interest is preferentially soluble. { ‚an·əl'id·ə·kəl ik'strak·shən }

analytical function generator [ELECTR] An analog computer device in which the dependence of an output variable on one or more input variables is given by a function that also appears in a physical law. Also known as natural function generator; natural law function generator. { ‚an·əl'id·ə·kəl 'fəŋk·shən ‚jen·ə‚rād·ər }

analytical geomorphology *See* dynamic geomorphology. { ‚an·əl'id·ə·kəl jē·ō‚mor'fäl·ə·jē }

analytical nadir-point triangulation [ENG] Radial triangulation performed by computational routines in which nadir points are utilized as radial centers. { ‚an·əl'id·ə·kəl ‚nā‚dir 'point ‚trī‚aŋ·gyə'lā·shən }

analytical orientation [ENG] The computational steps required to determine tilt, direction of principal line, flight height, angular elements, and linear elements in preparing aerial photographs for rectification. { ‚an·əl'id·ə·kəl ‚or·ē·ən'tā·shən }

analytical photogrammetry [ENG] A method of photogrammetry in which solutions are obtained by mathematical methods. { ‚an·əl'id·ə·kəl ‚fōd·ə'gram·ə·trē }

analytical photography [ENG] Photography, either motion picture or still, accomplished to determine (by qualitative, quantitative, or any other means) whether a particular phenomenon does or does not occur. { ‚an·əl'id·ə·kəl fə'täg·rə·fē }

analytical phototriangulation [ENG] A phototriangulation procedure in which the spatial solution is obtained by computational routines. { ‚an·əl'id·ə·kəl ‚fōd·ō‚trī‚aŋ·gyə'lā·shən }

analytical radar prediction [ENG] Prediction based on proven formulas, power tables, or graphs; considers surface height, structural and terrain information, and criteria for radar reflectivity together with the aspect angle and range to the target. { ‚an·əl'id·ə·kəl 'rā‚där prə'dik·shən }

analytical radial triangulation [ENG] Radial triangulation performed by computational routines. { ‚an·əl'id·ə·kəl 'rād·ē·əl ‚trī‚aŋ·gyə'lā·shən }

analytical three-point resection radial triangulation [ENG] A method of computing the coordinates of the ground principal points of overlapping aerial photographs by resecting on three horizontal control points appearing in the overlap area.

{ ,an·əl'id·ə·kəl ¦thrē ¦point rē'sek·shən 'rād·ē·əl ,tri,aŋ·gyə'lā·shən }

analytical transmission electron microscopy [BIOPHYS] A type of electron microscopy that allows the simultaneous characterization of crystal morphology, structure, and composition at high spatial resolution. { ,an·ə¦lid·ə·kəl ,tranz¦mish·ən i¦lek,trän mī'kräs·kə·pē }

analytical ultracentrifuge [ENG] An ultracentrifuge that uses one of three optical systems (schlieren, Rayleigh, or absorption) for the accurate determination of sedimentation velocity or equilibrium. { ,an·əl'id·ə·kəl ¦əl·trə¦sen·trə,fyüj }

analytic continuation [MATH] The process of extending an analytic function to a domain larger than the one on which it was originally defined. { ,an·əl'id·ik kən·tin·yü'ā·shən }

analytic curve [MATH] A curve whose parametric equations are real analytic functions of the same real variable. { ,an·əl'id·ik 'kərv }

analytic extension [RELAT] An extension, in a real analytic manner, past a coordinate singularity of a solution to Einstein's equations of general relativity. { ,an·əl'id·ik ik'sten·chən }

analytic function [MATH] A function which can be represented by a convergent Taylor series. Also known as holomorphic function. { ,an·əl'id·ik 'fuŋk·shən }

analytic geometry [MATH] The study of geometric figures and curves using a coordinate system and the methods of algebra. Also known as cartesian geometry. { ,an·əl'id·ik jē'äm·ə·trē }

analytic hierarchy [MATH] A systematic procedure for representing the elements of any problem which breaks down the problem into its smaller constituents and then calls for only simple pairwise comparison judgments to develop priorities at each level. { ,an·əl'id·ik 'hī·ər,är·kē }

analytic inertial navigation [NAV] Inertial navigation in which outputs of accelerometers that have inertia-maintained orientations are converted to geographic navigational data by automatic computers. { ,an·əl'id·ik in'ər·shəl ,nav·ə'gā·shən }

analytic mechanics [MECH] The application of differential and integral calculus to classical (nonquantum) mechanics. { ,an·əl'id·ik mi'kan·iks }

analytic number theory [MATH] The study of problems concerning the discrete domain of integers by means of the mathematics of continuity. { ,an·əl'id·ik 'nəm·bər ,thē·ə·rē }

analytic psychology [PSYCH] A theoretical system attributed to Carl Jung that minimizes the influence of sexual factors in emotional disorders and stresses mystical religious influences. Also known as Jungian psychology. { ,an·əl'id·ik sī'käl·ə·jē }

analytic regularization [QUANT MECH] A method of extracting a finite piece from an infinite result in quantum field theory, based on analytically continuing the propagators that appear in typically divergent integrals. { ,an·əl'id·ik ,reg·yə·lə·rə'zā·shən }

analytic set [MATH] A subset of a separable, complete metric space that is a continuous image of a Borel set in this metric space. { ,an·ə¦lid·ik 'set }

analytic structure [MATH] A covering of a locally Euclidean topological space by open sets, each of which is homeomorphic to an open set in Euclidean space, such that the coordinate transformation (in both directions) between the overlap of any two of these sets is given by analytic functions. { an·əl'id·ik 'strək·chər }

analytic trigonometry [MATH] The study of the properties and relations of the trigonometric functions. { ,an·əl'id·ik ,trig·ə'näm·ə·trē }

analyzer [COMPUT SCI] **1.** A routine for the checking of a program. **2.** One of several types of computers used to solve differential equations. [ENG] A multifunction test meter, measuring volts, ohms, and amperes. Also known as set analyzer. [MECH ENG] The component of an absorption refrigeration system where the mixture of water vapor and ammonia vapor leaving the generator meets the relatively cool solution of ammonia in water entering the generator and loses some of its vapor content. [OPTICS] A device, such as a Nicol prism, which passes only plane polarized light; used in the eyepiece of instruments such as the polariscope. { 'an·ə,līz·ər }

analyzing power [NUC PHYS] In a nuclear scattering process, a measure of the effect on scattering cross sections of

changes in the polarization of the beam or target nuclei. { 'an·ə,liz·iŋ ,paùr }

anamigmatism [GEOL] A process of high-temperature, high-pressure remelting of sediment to yield magma. { ,an·ə'mig·mə,tiz·əm }

anamnesis [MED] Information gained from the patient and others regarding the patient's medical history. [PSYCH] The faculty of memory. { ,an·əm'nē·səs }

anamnestic response [IMMUNOL] A rapidly increased antibody level following renewed contact with a specific antigen, even after several years. Also known as booster response. { ,an·əm'nes·tik ri'späns }

Anamnia [VERT ZOO] Vertebrate animals which lack an amnion in development, including Agnatha, Chondrichthyes, Osteichthyes, and Amphibia. { a'nam·nē·ə }

Anamniota [VERT ZOO] The equivalent name for Anamnia. { ¦a,nam·nē'ōd·ə }

anamniote [ZOO] An animal that does not develop an amnion during its embryonic stage. { an'am·nē,ōt }

anamorphic lens [OPTICS] A lens that produces different magnifications along lines in different directions in the image plane. { ¦an·ə¦mòr·fik 'lenz }

anamorphic system [OPTICS] An optical system incorporating a cylindrical surface in which the image is distorted so that the angle of coverage in a direction perpendicular to the cylinder is different for the image than for the object. { ¦an·ə¦mòr·fik 'sis·təm }

anamorphic zone [GEOL] The zone of rock flow, as indicated by reactions that may involve decarbonation, dehydration, and deoxidation; silicates are built up, and the formation of denser minerals and of compact crystalline structure takes place. { ¦an·ə¦mòr·fik 'zōn }

anamorphism [EVOL] *See* anamorphosis. [GEOL] A kind of metamorphism at considerable depth in the earth's crust and under great pressure, resulting in the formation of complex minerals from simple ones. { ,an·ə'mòr·fiz·əm }

anamorphoscope [OPTICS] An optical instrument, usually consisting of a cylindrical lens or mirror, that restores an image distorted by anamorphosis to its normal proportions. { ,an·ə'mòr·fə,skōp }

anamorphosis [EVOL] Gradual increase in complexity of form and function during evolution of a group of animals or plants. Also known as anamorphism. [GRAPHICS] A drawing which appears to be distorted unless viewed from a particular angle or with a special device. [OPTICS] The production of a distorted image by an optical system. { ,an·ə'mòr·fə·səs }

anamorphote lens [OPTICS] A lens designed to produce anamorphosis. { ¦an·ə¦mòr,fōt ,lenz }

Anancinae [PALEON] A subfamily of extinct proboscidean placental mammals in the family Gomphotheriidae. { ə'nan·sə,nē }

Ananke [ASTRON] A small satellite of Jupiter with a diameter of about 14 miles (23 kilometers), orbiting with retrograde motion at a mean distance of 1.3×10^7 miles (2.1×10^7 kilometers). Also known as Jupiter XII. { ə'naŋ·kē }

anapaite [MINERAL] $Ca_2Fe(PO_4)_2 \cdot 4H_2O$ A pale-green or greenish-white triclinic mineral consisting of a ferrous iron hydrous phosphate and occurring in crystals and massive forms; hardness is 3–4 on Mohs scale, and specific gravity is 3.81. { ə'nap·ə,īt }

anapeirean *See* Pacific suite. { ,an·ə'pir·ē·ən }

anaphase [CYTOL] **1.** The stage in mitosis and in the second meiotic division when the centromere splits and the chromatids separate and move to opposite poles. **2.** The stage of the first meiotic division when the two halves of a bivalent chromosome separate and move to opposite poles. { 'an·ə,fāz }

anaphase-promoting complex [CELL MOL] A large protein complex activated during mitosis that promotes the metaphase to anaphase transition by effecting the proteolytic destruction of several key mitotic regulators. Abbreviated APC. { ¦an·ə·fāz prə,mōd·iŋ 'käm,pleks }

anaphoresis [MED] Deficient functioning of sweat glands. [PHYS CHEM] Upon application of an electric field, the movement of positively charged colloidal particles or macromolecules suspended in a liquid toward the anode. [PHYSIO] Movement of positively charged ions into tissues under the influence of an electric current. { ¦an·ə·fə'rē·səs }

anaphylactic shock [MED] A syndrome seen as one of the

clinical manifestations of anaphylaxis. { ¦an·ə·fə¦lak·tik 'shäk }

anaphylactoid reaction [MED] A nonallergic reaction resembling anaphylaxis and depending on the toxicity of the inductant. { ¦an·ə·fə¦lak,tóid rē'ak·shən }

anaphylatoxin [IMMUNOL] The vasodilator principal, a toxic substance released by tissues of sensitized animals when antigen and antibody react. { ¦an·ə,fil·ə'täk·sən }

anaphylaxis [MED] Hypersensitivity following parenteral injection of an antigen; local or systemic allergenic reaction occurs when the antigen is reintroduced after a time lapse. { ¦an·ə·fə¦lak·səs }

anaplasia [MED] Reversion of cells to an embryonic, immature, or undifferentiated state; degree usually corresponds to malignancy of a tumor. { ¸an·ə'plāzh·ə }

Anaplasma [MICROBIO] A genus of the family Anaplasmataceae; organisms form inclusions in red blood cells of ruminants. { ¸an·ə'plaz·mə }

Anaplasmataceae [MICROBIO] A family of the order Rickettsiales; obligate parasites, either in or on red blood cells or in the plasma of various vertebrates. { ¸an·ə,plaz·mə'tās·ē,ē }

Anaplotheriidae [PALEON] A family of extinct tylopod ruminants in the superfamily Anaplotherioidea. { ¸an·ə,pläth·ə'rī·ə,dē }

Anaplotherioidea [PALEON] A superfamily of extinct ruminant artiodactyls in the infraorder Tylopoda. { ¸an·ə,pläth·ə,rē'óid·ē·ə }

anapolysis [INV ZOO] Lifetime retention of ripe proglottids in some tapeworms. { ¸an·ə'päl·ə·səs }

anapophysis [ANAT] An accessory process on the dorsal side of the transverse process of the lumbar vertebrae in humans and other mammals. { ¸an·ə'päf·ə·səs }

Anapsida [VERT ZOO] A subclass of reptiles characterized by a roofed temporal region in which there are no temporal openings. { ə'nap·sə·də }

anasarca [MED] Generalized edema of the subcutaneous connective tissue and serous body cavities. { ¸an·ə'sär·kə }

Anasca [PALEON] A suborder of extinct bryozoans in the order Cheilostomata. { ə'nas·kə }

anaseism [GEOPHYS] Movement of the earth in a direction away from the focus of an earthquake. { ¦an·ə,sīz·əm }

Anaspida [PALEON] An order of extinct fresh- or brackish-water vertebrates in the class Agnatha. { ə'nas·pə·də }

Anaspidacea [INV ZOO] An order of the crustacean superorder Syncarida. { ¸ə¦nas·pə'dās·ē·ə }

Anaspididae [INV ZOO] A family of crustaceans in the order Anaspidacea. { ¸ə¦nas·pə'dī,dē }

anastatic process [GRAPHICS] Reproduction of a printed page, either type or pictures, by moistening it with dilute acid and pressing it against a zinc plate; the acid etches the zinc wherever in contact with unprinted portions; the plate can then be inked and printed. { ¦an·ə¦stad·ik 'präs·əs }

anastatic water [HYD] That part of the subterranean water in the capillary fringe between the zone of aeration and the zone of saturation in the soil. { ¦an·ə¦stad·ik 'wód·ər }

anastigmat See anastigmatic lens. { a'nas·tig,mat }

anastigmatic lens [OPTICS] A compound lens corrected for astigmatism and curvature of field. Also known as anastigmat. { ¦an·ə·stig¦mad·ik 'lenz }

anastomosis [MED] **1.** A surgical communication made between blood vessels, for example, between the portal vein and the inferior vena cava. **2.** An opening created by surgery, trauma, or disease between two or more normally separate spaces or organs. [SCI TECH] **1.** The union or intercommunication of parts or branches, such as blood vessels, streams, or leaf veins. Also known as inosculation. **2.** A network of parts or branches created by the process of anastomosis. { ə,nas·tə'mō·səs }

anastomotic operation [MED] A surgical procedure to create an anastomosis. { ¸an·ə·stə¦mäd·ik ¸äp·ə'rā·shən }

anastral [CYTOL] Lacking asters. { a'nas·trəl }

anastral mitosis [CYTOL] Mitosis in which a spindle forms but no centrioles or asters are observed; typically occurs in plants. { ā¦nas·trəl mī'tō·səs }

anatabine [ORG CHEM] $C_{10}H_{12}N_2$ An alkaloid found in tobacco. { ə'nad·ə,bēn }

anatase [MINERAL] The brown, dark-blue, or black tetragonal crystalline form of titanium dioxide, TiO_2; used to make a white pigment. Also known as octahedrite. { 'an·ə,tās }

anatexis [GEOL] A high-temperature process of metamorphosis by which plutonic rock in the lowest levels of the crust is melted and regenerated as a magma. { ¸an·ə'tek·səs }

anathermal [GEOL] A period of time between the age of other strata or units of reference in which the temperature is increasing. { ¸an·ə'thər·məl }

Anatidae [VERT ZOO] A family of waterfowl, including ducks, geese, mergansers, pochards, and swans, in the order Anseriformes. { ə'nad·ə,dē }

anatomical dead space See dead space. { ¸an·ə'täm·ə·kəl 'ded ,spās }

anatomical landmark See anatomical reference point. { ¸an·ə¦täm·ə·kəl 'lan,märk }

anatomical position [ANAT] A reference posture used in anatomical description in which the subject stands erect against a wall with feet parallel and touching, and arms adducted and supinated, with palms facing forward. { ¸an·ə¦täm·ə·kəl pə'zi·shən }

anatomical reference point [ANAT] A prominent structure or feature of the human body that can be located and described by visual inspection or palpation at the body's surface; used to define movements and postures. Also known as anatomical landmark. { ¸an·ə¦täm·ə·kəl 'ref·rəns ,póint }

anatomy [BIOL] A branch of morphology dealing with the structure of animals and plants. { ə'nad·ə·mē }

anatomy of function [PHYSIO] A description of the changes in the configuration of limbs during the performance of a task; considered a subdiscipline of kinesiology. { ə¦nad·ə·mē əv 'fəŋk·shən }

anatropous [BOT] Having the ovule fully inverted so that the micropyle adjoins the funiculus. { ə'na·trə·pəs }

anautogenous insect [INV ZOO] Any insect in which the adult female must feed before producing eggs. { ¦ā,nó¦täj·ə·nəs 'in,sekt }

anauxite [MINERAL] $Al_2(SiO_7)(OH)_4$ A clay mineral that is a mixture of kaolinite and quartz. Also known as ionite. { ə'nók,sīt }

anaxial [BIOL] Lacking an axis, therefore being irregular in form. { a'nak·sē·əl }

Anbauhobel [MIN ENG] A rapid plough, traveling at a speed of 75 feet (22.5 meters) per minute, for use on longwall faces. { ¦an,baù¦hō·bəl }

Ancalomicrobium [MICROBIO] A genus of prosthecate bacteria; nonmotile, unicellular forms with two to eight prosthecae per cell; reproduction is by budding of cells. { ¸an¦kal·ō,mī¦krob·ē·əm }

ancestroecium [INV ZOO] The tube that encloses an ancestrula. { ¸an·səs'trēsh·əm }

ancestrula [INV ZOO] The first polyp of a bryozoan colony. { ¸an'ses·trə·lə }

anchieutectic [GEOL] A type of magma which is incapable of undergoing further notable main-stage differentiation because its mineral composition is practically in eutectic proportions. { ¦aŋ·kē·yü¦tek·tik }

anchimeric assistance [ORG CHEM] The participation by a neighboring group in the rate-determining step of a reaction; most often encountered in reactions of carbocation intermediates. Also known as neighboring-group participation. { ¦aŋ·kə¦mer·ik ə'sis·təns }

anchimonomineralic [PETR] Of rock composed mostly of one kind of mineral. { ¦aŋ·kē,män·ō,min·ə¦ral·ik }

anchor [CIV ENG] A device connecting a structure to a heavy masonry or concrete object to a metal plate or to the ground to hold the structure in place. [COMPUT SCI] A tag that indicates either the source or destination of a hyperlink; for example, HTML anchors are used to create links within a document or to another document. [ENG] A device, such as a metal rod, wire, or strap, for fixing one object to another, such as specially formed metal connectors used to fasten together timbers, masonry, or trusses. [INV ZOO] **1.** An anchor-shaped spicule in the integument of sea cucumbers. **2.** An anchor-shaped ossicle in echinoderms. [MECH ENG] **1.** In steam plowing, a vehicle located on the side of the field opposite that of the engine and maintaining the tension on the endless wire by means of a pulley. **2.** A device for a piping system that maintains the correct position and direction of the pipes and controls pipe movement occurring as a result of thermal expansion. [MET] A device that prevents the movement of sand cores in molds. [NAV ARCH] A device attached by cable to a ship and dropped overboard so that its hooks or flukes engage the bottom and hold the ship at that location. { 'aŋ·kər }

anchorage [ARCH] A permanent placement or foundation to which the lower members of a structure can be attached in order to provide stability for the entire structure. [CIV ENG] **1.** An area where a vessel anchors or may anchor because of either suitability or designation. Also known as anchor station. **2.** A device which anchors tendons to the posttensioned concrete member. **3.** In pretensioning, a device used to anchor tendons temporarily during the hardening of the concrete. **4.** See deadman. { 'aŋ·kə·rij }

anchorage chart [MAP] A nautical chart showing prescribed or recommended anchorages. { 'aŋ·kə·rij ‚chärt }

anchorage deformation [CIV ENG] The shortening of tendons due to their modification or slippage when the prestressing force is transferred to the anchorage device. Also known as anchorage slip. { 'aŋ·kə·rij dē‚fôr'mā·shən }

anchorage-dependent cell [CYTOL] A cell that grows, survives, or maintains function only when attached to an inert surface, such as glass or plastic. { ¦aŋ·kə‚rij di¦pen·dənt 'sel }

anchorage slip See anchorage deformation. { 'aŋ·kə·rij ‚slip }

anchorage zone [CIV ENG] **1.** In posttensioning, the region adjacent to the anchorage for the tendon which is subjected to secondary stresses as a result of the distribution of the prestressing force. **2.** In pretensioning, the region in which transfer bond stresses are developed. { 'aŋ·kə·rij ‚zōn }

anchor and collar [DES ENG] A door or gate hinge whose socket is attached to an anchor embedded in the masonry. { 'aŋ·kər ən 'käl·ər }

anchor ball [NAV ARCH] **1.** A projectile with grappling hooks which is fired into the rigging of a wrecked vessel for lifesaving purposes. **2.** A black, circular shape hoisted between the bow and foremast of a vessel to indicate that it is anchored in or near a channel. { 'aŋ·kər ‚bȯl }

anchor block [BUILD] A block of wood, replacing a brick in a wall to provide a nailing or fastening surface. [CIV ENG] See deadman. { 'aŋ·kər ‚bläk }

anchor bolster See hawse bolster. { 'aŋ·kər ‚bōl·stər }

anchor bolt [CIV ENG] A bolt used with its head embedded in masonry or concrete and its threaded part protruding to hold a structure or machinery in place. Also known as anchor rod. { 'aŋ·kər ‚bōlt }

anchor buoy [ENG] One of a series of buoys marking the limits of an anchorage. { 'aŋ·kər ‚bȯi }

anchor charge [ENG] A procedure that allows several charges to be preloaded in a seismic shot hole; the bottom charges are fired first, and the upper charges are held down by anchors. { 'aŋ·kər ‚chärj }

anchored bulkhead [CIV ENG] A bulkhead secured to anchor piles. { 'aŋ·kərd 'bəlk‚hed }

anchored catalyst See immobilized catalyst. { 'aŋ·kərd 'kad·ə‚list }

anchored dune [GEOL] A sand dune stabilized by growth of vegetation. { 'aŋ·kərd 'dün }

anchored graphic [COMPUT SCI] A picture or graph that remains at a fixed position on a page of a document rather than being attached to the text. { ¦aŋ·kərd 'graf·ik }

anchored-type ceramic veneer [MATER] Ceramic veneer which is attached to a backing by grout and nonferrous metal anchors. { 'aŋ·kərd ‚tīp sə'ram·ik və'nēr }

anchor escapement [HOROL] A clock escapement in which pallets of an anchor-shaped component cause the escape wheel to recoil slightly as the wheel is arrested. Also known as recoil escapement. { 'aŋ·kər is‚kāp·mənt }

anchor gear [NAV ARCH] Shipboard apparatus consisting of anchor windlass, chain stoppers, and hawsepipes; more generally includes anchor and chain. { 'aŋ·kər ‚gēr }

anchor ice [HYD] Ice formed beneath the surface of water, as in a lake or stream, and attached to the bottom or to submerged objects. Also known as bottom ice; ground ice. { 'aŋ·kər ‚īs }

anchorite [PETR] A variety of diorite having nodules of mafic minerals and veins of felsic minerals. { 'aŋ·kə‚rīt }

anchor light [NAV ARCH] A mast light shown by a vessel at anchor during the night. { 'aŋ·kər ‚līt }

anchor log [CIV ENG] A log, beam, or concrete block buried in the earth and used to hold a guy rope firmly. Also known as deadman; ground anchor. { 'aŋ·kər ‚läg }

anchor nut [DES ENG] A nut in the form of a tapped insert forced under steady pressure into a hole in sheet metal. { 'aŋ·kər ‚nət }

anchor packer [PETRO ENG] A device used in oil wells to seal the annular space between the tubing and its surrounding casing to help control the oil-producing gas lift. { 'aŋ·kər ‚pak·ər }

anchor pattern [MET] The pattern of minute projections formed on a metal surface by sandblasting, shot blasting, or chemical etching; used to enhance the adhesiveness of a surface coating. { 'aŋ·kər ‚pad·ərn }

anchor pile [CIV ENG] A pile that is located on the land side of a bulkhead or pier and anchors it through such devices as rods, cables, and chains. { 'aŋ·kər ‚pīl }

anchor plate [CIV ENG] A metal or wooden plate fastened to or embedded in a support, such as a floor, and used to hold a supporting cable firmly. { 'aŋ·kər ‚plāt }

anchor point [MATH] Either of the two end points of a Bézier curve. { 'aŋ·kər ‚pȯint }

anchor rod See anchor bolt. { 'aŋ·kər ‚räd }

anchor shackle [NAV ARCH] A shackle by which a chain is joined to the ring of an anchor. Also known as bending shackle. { 'aŋ·kər ‚shak·əl }

anchor station [CIV ENG] See anchorage. [OCEANOGR] An anchoring site by a research vessel for the purpose of making a set of scientific observations. { 'aŋ·kər ‚stā·shən }

anchor stone [GEOL] A rock or pebble that has marine plants attached to it. { 'aŋ·kər ‚stōn }

anchor tower [CIV ENG] **1.** A tower which is a part of a crane staging or stiffleg derrick and serves as an anchor. **2.** A tower that supports and anchors an overhead transmission line. { 'aŋ·kər ‚tau̇·ər }

anchor washpipe spear [PETRO ENG] A special type of fishing tool; attached to the washover pipe with slips and released from the pipe upon engagement of the fish by the tool, it allows the washover and retrieval of the fish to be accomplished in one return trip. { 'aŋ·kər 'wäsh‚pīp ‚spir }

anchor wall See deadman. { 'aŋ·kər ‚wȯl }

anchor well [NAV ARCH] A well in a ship's forward overhang for holding anchors. { 'aŋ·kər ‚wel }

anchor windlass [NAV ARCH] A machine, generally located on the forecastle head of a ship, designed to raise or lower an anchor; it consists of a horizontal barrel that is fitted with gearlike projections that engage the links of the anchor chain, and is turned by steam or electrical power. Also known as windlass. { 'aŋ·kər ‚wind·ləs }

anchovy [VERT ZOO] Any member of the Engraulidae, a family of herringlike fishes harvested commercially for human consumption. { 'an‚chō·vē }

anchylosis See ankylosis. { aŋ·ki'lō·səs }

ancient DNA [ANTHRO] Deoxyribonucleic acid that has been copied and sequenced from individuals representing extinct populations or species. Abbreviated aDNA. { ¦ān·chənt ¦dē¦en'ā }

ancipital [BOT] Having two edges, specifically referring to flattened stems, as of certain grasses. { an'sip·əd·əl }

ancon [ARCH] A bracket, elbow, or console at the top of a wall or window jamb to support a cornice. { 'aŋ‚kän }

ancora [INV ZOO] The initial, anchor-shaped growth stage of graptolithinids. { 'aŋ·kə·rə }

ancylite [MINERAL] $SrCe(CO_3)_2(OH)·H_2O$ A mineral consisting of hydrous basic carbonate of cerium and strontium. { 'an·sə‚līt }

ancyloid [INV ZOO] A limpet-shaped or patelliform shell with the apex directed anteriorly. { 'an·sə‚lȯid }

ancylopoda [PALEON] A suborder of extinct herbivorous mammals in the order Perissodactyla. { ‚an·sə'lä·pə·də }

Ancylostoma [INV ZOO] A genus of roundworms, commonly known as hookworms, in the order Ancylostomidae; parasites of humans, dogs, and cats. { ‚aŋ·kə'läs·tə·mə }

Ancylostoma duodenale [INV ZOO] The Old World hookworm, a human intestinal parasite that causes microcytic hypochromic anemia. { ‚aŋ·kə'läs·tə·mə ‚dü·ə·di¦näl }

Ancylostomidae [INV ZOO] A family of nematodes belonging to the group Strongyloidea. { ¦aŋ·kə·lō·stä'mad·ə‚dē }

And See Andromeda.

andalusite [MINERAL] Al_2SiO_5 A brown, yellow, green,

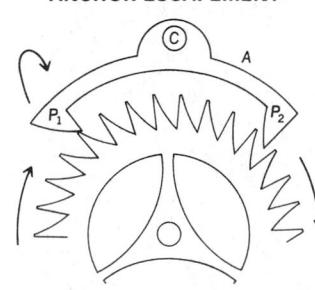

ANCHOR ESCAPEMENT

Common in domestic pendulum clocks, this escapement tends to compensate for changes in amplitude of the swing because of irregularly cut gears and varying lubrication; A = anchor, P_1, P_2 = pallets, C = point at which anchor swings about.

red, or gray neosilicate mineral crystallizing in the orthorhombic system, usually found in metamorphic rocks. { ¦an·də'lü‚sīt }

AND circuit *See* AND gate. { 'and ‚sər·kət }

Andean-type continental margin [GEOL] A continental margin, as along the Pacific coast of South America, where oceanic lithosphere descends beneath an adjacent continent producing andesitic continental margin volcanism. { 'an·dē·ən ‚tīp ‚känt·ən'ent·əl 'mär·jən }

Andept [GEOL] A suborder of the soil order Inceptisol, formed chiefly in volcanic ash or in regoliths with high components of ash. { ¦an¦dept }

Andersen's disease *See* amylopectinosis. { 'an·dər·sənz di‚zēz }

Anderson bridge [ELECTR] A six-branch modification of the Maxwell-Wien bridge, used to measure self-inductance in terms of capacitance and resistance; bridge balance is independent of frequency. { 'an·dər·sən ‚brij }

Anderson-Dayem bridge [CRYO] A Josephson junction in a superconducting film,formed by a constriction with length and width on the order of a few micrometers or less. { ¦an·dər·sən ¦dā·əm ‚brij }

andersonite [MINERAL] $Na_2Ca(UO_2)(CO_3)_3 \cdot 6H_2O$ Bright yellow-green secondary mineral consisting of a hydrous sodium calcium uranium carbonate. { 'an·dər·sən‚īt }

Andes glow *See* Andes lightning. { 'an‚dēz ‚glō }

andesine [MINERAL] A plagioclase feldspar with a composition ranging from $Ab_{70}An_{30}$ to $Ab_{50}An_{50}$, where $Ab = NaAl-Si_3O_8$ and $An = CaAl_2Si_2O_8$; it is a primary constituent of intermediate igneous rocks, such as andesites. { 'an·də‚zēn }

andesite [PETR] Very finely crystalline extrusive rock of volcanic origin composed largely of plagioclase feldspar (oligoclase or andesine) with smaller amounts of dark-colored mineral (hornblende, biotite, or pyroxene), the extrusive equivalent of diorite. { 'an·də‚zīt }

andesite line [GEOL] The postulated geographic and petrographic boundary between the andesite-dacite-rhyolite rock association of the margin of the Pacific Ocean and the olivine-basalt-trachyte rock association of the Pacific Ocean basin. { 'an·də‚zīt ‚līn }

andesitic glass [GEOL] A natural glass that is chemically equivalent to andesite. { 'an·də‚zīt·ik ‚glas }

Andes lightning [GEOPHYS] Electrical coronal discharges observable often as far as several hundred miles away, generally over any of the mountainous areas of the world when under disturbed electrical conditions. Also known as Andes glow. { 'an‚dēz 'līt·niŋ }

AND function [MATH] An operation in logical algebra on statements P, Q, R, such that the operation is true if all the statements P, Q, R, ... are true, and the operation is false if at least one statement is false. { 'and ‚fuŋk·shən }

AND gate [ELECTR] A circuit which has two or more input-signal ports and which delivers an output only if and when every input signal port is simultaneously energized. Also known as AND circuit; passive AND gate. { 'and ‚gāt }

AND/NOR gate [ELECTR] A single logic element whose operation is equivalent to that of two AND gates with outputs feeding into a NOR gate. { ¦and ¦nȯr ‚gāt }

AND NOT gate [ELECTR] A coincidence circuit that performs the logic operation AND NOT, under which a result is true only if statement A is true and statement B is not. Also known as A AND NOT B gate. { ¦and ¦nät ‚gāt }

AND-OR circuit [ELECTR] Gating circuit that produces a prescribed output condition when several possible combined input signals are applied; exhibits the characteristics of the AND gate and the OR gate. { ¦and ¦ȯr ‚sər·kət }

AND-OR-INVERT gate [ELECTR] A logic circuit with four inputs, a_1, a_2, b_1, and b_2, whose output is 0 only if either a_1 and a_2 or b_1 and b_2 are 1. Abbreviated A-O-I gate. { ¦and ¦ȯr in'vərt ‚gāt }

andorite [MINERAL] $AgPbSb_3S_6$ A dark-gray or black orthorhombic mineral. Also known as sundtite. { 'an·də‚rīt }

Andr *See* Andromeda.

Andrade's creep law [MECH] A law which states that creep exhibits a transient state in which strain is proportional to the cube root of time and then a steady state in which strain is proportional to time. { 'an‚drädz 'krēp ‚lȯ }

andradite [MINERAL] The calcium-iron end member of the garnet group. { an'drä‚dīt }

Andreaeales [BOT] The single order of mosses of the subclass Andreaeobrya. { ‚an·drē·ē'ā·lēz }

Andreaeceae [BOT] The single family of the Andreaeales, an order of mosses. { ‚an·drē'ē·sē‚ē }

Andreaeobrya [BOT] The granite mosses, a subclass of the class Bryopsida. { ‚an·drē·ē'äb·rē·ə }

Andreaeopsida [BOT] A class of the plant division Bryophyta distinguished by longitudinal splitting of the mature capsule into four valves; commonly known as granite mosses. { ‚an·drē·ē'äp·səd·ə }

Andrenidae [INV ZOO] The mining or burrower bees, a family of hymenopteran insects in the superfamily Apoidea. { ‚an'dren·ə‚dē }

Andrews's curves [THERMO] A series of isotherms for carbon dioxide, showing the dependence of pressure on volume at various temperatures. { 'an‚drüz ‚kərvz }

andrewsite [MINERAL] $(Cu,Fe^{2+})Fe_3{}^{3+}(PO_4)_3(OH)_2$ A bluish-green mineral consisting of a basic phosphate of iron and copper. { 'an·drü‚zīt }

andrite [GEOL] A meteorite composed principally of augite with some olivine and troilite. { 'an‚drīt }

androecious [BOT] Pertaining to plants that have only male flowers. { an'drē·shəs }

androecium [BOT] The aggregate of stamens in a flower. { ‚an'drēsh·ē·əm }

androgen [BIOCHEM] A class of steroid hormones produced in the testis and adrenal cortex which act to regulate masculine secondary sexual characteristics. { 'an·drə·jən }

androgenesis [EMBRYO] Development of an embryo from a fertilized irradiated egg, involving only the male nucleus. { ‚an·drə'jen·ə·səs }

androgenetic alopecia [MED] The most common cause of hair loss, characterized by gradual progression, with miniaturization of genetically programmed hair follicles. { ‚an·drō·jə¦ned·ik ‚al·ə'pē·shə }

androgenetic merogony [EMBRYO] The fertilization of egg fragments that lack a nucleus. { ¦an·drə·jə¦ned·ik mə'rä·gə·nē }

androgenic gland [INV ZOO] A gland found in most malacostracan crustaceans and producing hormones that control the development of the testes and male sexual characteristics. { ¦an·drə¦jen·ik 'gland }

androgen insensitivity syndrome *See* testicular feminization. { ¦an·drə·jən in‚sen·sə'tiv·ə·dē ‚sin‚drōm }

androgenital syndrome [MED] An inherited syndrome of humans in which there is masculinization of genitals in females with XX chromosomal constitution. { ¦an·drə¦gen·ə·dəl 'sin‚drōm }

androgen unit [BIOL] A unit for the standardization of male sex hormones. { 'an·drə·jən ‚yü·nət }

androgynous docking system [AERO ENG] A docking system that uses the principle of reverse symmetry to allow the mechanical linkage of otherwise dissimilar spacecraft. { an‚dräj·ə·nəs 'däk·iŋ ‚sis·təm }

androgyny [MED] A form of pseudohermaphroditism in humans in which the individual has female external sexual characteristics, but has undescended testes. Also known as male pseudohermaphroditism. { an'dräj·ə·nē }

android pelvis *See* masculine pelvis. { 'an‚drȯid 'pel·vəs }

Andromeda [ASTRON] A constellation with a right ascension of 1 hour and a declination of 40°N. Abbreviated And; Andr. { ‚an'dräm·ə·də }

Andromeda Galaxy [ASTRON] The spiral galaxy of type Sb nearest to the Milky Way. Also known as Andromeda Nebula. { ‚an'dräm·ə·də 'gal·ək‚sē }

Andromeda Nebula *See* Andromeda Galaxy. { ‚an'dräm·ə·də 'ne·byə·lə }

Andromedids [ASTRON] A meteor shower whose radiant is located near the star γ Andromedae, and which reaches its peak about November 27; associated with the Biela comet. Also known as Bielids. { ‚an'dräm·ə‚didz }

andromerogony [EMBRYO] Development of an egg fragment following cutting, shaking, or centrifugation of a fertilized or unfertilized egg. { ‚an‚drō·mə'räg·ə·nē }

Andronikashvili experiment [CRYO] An experiment to determine the fractional densities of the superfluid and normal fluid components of liquid helium by measuring the period and

ANEMOMETER

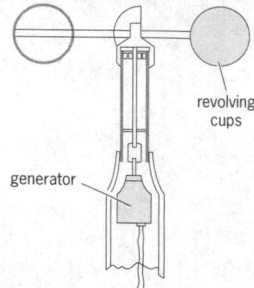

revolving cups

generator

Cutaway diagram of a rotating cup anemometer. *(After G. K. McMillan and D. M. Considine, eds., Process/Industrial Instruments and Controls Handbook, 5th ed., McGraw-Hill, 1999)*

decrement of a torsional pendulum immersed in the helium. { ˌanˈdrəˌniˈkəshˌvilˈē ikˈsperˈəˌmənt }

androphile [ECOL] An organism, such as a mosquito, showing a preference for humans as opposed to animals. { ˈanˈdrōˌfīl }

androphobia [PSYCH] Abnormal fear of men or of the male sex. { ˌanˈdrōˈfōbˈēˈə }

androphore [BOT] A stalk that supports stamens or antheridia. [INV ZOO] A gonophore in cnidarians in which only male elements develop. { ˈanˈdrōˌfōr }

androsin [BIOCHEM] $C_{15}H_{20}O_8$ A glucoside found in the herb *Apocynum androsaemifolium;* yields glucose and acetovanillone on hydrolysis. { ˈanˈdrəˌsən }

androsperm [BIOL] A sperm cell carrying a Y chromosome. { ˈanˈdrəˌspərm }

androstane [BIOCHEM] $C_{19}H_{32}$ The parent steroid hydrocarbon for all androgen hormones. Also known as etioallocholane. { ˈanˈdrəˌstān }

androstenedione [BIOCHEM] $C_{19}H_{26}O_2$ Any one of three isomeric androgens produced by the adrenal cortex. { ˌanˈdrōˈstəˌnedˈēˌōn }

androsterone [BIOCHEM] $C_{19}H_{30}O_2$ An androgenic hormone occurring as a hydroxy ketone in the urine of men and women. { ˌanˈdrästˈəˌrōn }

anechoic chamber [ENG] **1.** A test room in which all surfaces are lined with a sound-absorbing material to reduce reflections of sound to a minimum. Also known as dead room; free-field room. **2.** A room completely lined with a material that absorbs radio waves at a particular frequency or over a range of frequencies; used principally at microwave frequencies, such as for measuring radar beam cross sections. { ˈanˈəˈkōˈik ˈchāmˈbər }

anelastic creep [MATER] Time-dependent elastic (nonpermanent) creep. Also known as transient creep. { ˌanˈəˈlasˈtik ˈkrēp }

anelasticity [MECH] Deviation from a proportional relationship between stress and strain. { ˈanˈəˈlasˈtisˈədˈē }

anelectric [PHYS] Not becoming charged by friction. { ˈanˈəˈlekˈtrik }

Anelytropsidae [VERT ZOO] A family of lizards represented by a single Mexican species. { əˌnelˈəˈträpˈsəˌdē }

anemia [MED] A condition marked by significant decreases in hemoglobin concentration and in the number of circulating red blood cells. Also known as oligochromemia. { əˈnēmˈēˈə }

anemic galaxy [ASTRON] A spiral galaxy with unusually low surface brightness and inconspicuous spiral arms. { əˌnēmˈik ˈgalˈikˈsē }

anemic necrosis [MED] Tissue death following a critical decrease in blood flow or oxygen levels. { əˈnēmˈik nəˈkrōˈsəs }

anemobiagraph [ENG] A recording pressure-tube anemometer in which the wind scale of the float manometer is linear through the use of springs; an example is the Dines anemometer. { ˈaˈnəˈməˈbīˈəˌgraf }

anemochory [ECOL] Wind dispersal of plant and animal disseminules. { əˈnēmˈəˌkȯrˈē }

anemoclast [GEOL] A clastic rock that was fragmented and rounded by wind. { ˈaˈnəˈmōˌklast }

anemoclastic [GEOL] Referring to rock that was broken by wind erosion and rounded by wind action. { ˈaˈnəˈmōˌklasˈtik }

anemoclinometer [ENG] A type of instrument which measures the inclination of the wind to the horizontal plane. { ˈaˈnəˌmōˈkləˈnämˈədˈər }

anemogenic [MED] Causing anemia. { əˈnēmˈəˌjenˈik }

anemogram [ENG] A record made by an anemograph. { əˈnēmˈəˌgram }

anemograph [ENG] **1.** An instrument which records wind velocities. **2.** A recording anemometer. { əˈnēmˈəˌgraf }

anemology [METEOROL] Scientific investigation of winds. { ˌanˈəˈmälˈəˈjē }

anemometer [ENG] A device which measures air speed. { ˌanˈəˈmämˈədˈər }

anemometry [METEOROL] The study of measuring and recording the direction and speed (or force) of the wind, including its vertical component. { ˌanˈəˈmämˈəˈtrē }

anemophilous [BOT] Pollinated by wind-carried pollen. { ˈanˈəˌmäfˈəˈləs }

anemophobia [PSYCH] Abnormal fear of drafts or winds. { ˌanˈəˈməˈfōbˈēˈə }

anemoscope [ENG] An instrument for indicating the direction of the wind. { əˈnēmˈəˌskōp }

anemotaxis [BIOL] Orientation movement of a free-living organism in response to wind. { ˌanˈəˈmōˈtakˈsəs }

anemotropism [BIOL] Orientation response of a sessile organism to air currents and wind. { ˌanˈəˈmäˈtrəˌpizˈəm }

anemovane [ENG] A combined contact anemometer and wind vane used in the Canadian Meteorological Service. { ˈanˈəˈmōˌvān }

anencephalia [MED] A congenital anomaly marked by severely defective development of the brain, together with absence of the bones of the cranial vault and the cerebral and cerebellar hemispheres, and with only a rudimentary brainstem and some traces of basal ganglia present. { ˌanˌenˈsəˈfālˈyə }

anencephaly [MED] A neural tube defect resulting in gross underdevelopment of the brain in fetal life. { ˌanˌenˈsefˈəˈlē }

anenterous [ZOO] Having no intestine, as a tapeworm. { aˈnenˈtəˈrəs }

Anepitheliocystidia [INV ZOO] A superorder of digenetic trematodes proposed by G. LaRue. { ˈanˈəˈpəˌthēˈlīˈōˌsisˈtidˈēˈə }

anergy [IMMUNOL] The condition of exhibiting no response to an antigen or antibody. [MED] The condition of exhibiting a lack of energy. { ˈaˌnərˈjē }

aneroid [ENG] **1.** Containing no liquid or using no liquid. **2.** *See* aneroid barometer. { ˈanˈəˌrȯid }

aneroid altimeter [ENG] An altimeter containing an aneroid barometer that actuates the indicator. { ˈanˈəˌrȯid alˈtimˈədˈər }

aneroid barograph [ENG] An aneroid barometer arranged so that the deflection of the aneroid capsule actuates a pen which graphs a record on a rotating drum. Also known as aneroidograph; barograph; barometrograph. { ˈanˈəˌrȯid ˈbarˈəˌgraf }

aneroid barometer [ENG] A barometer which utilizes an aneroid capsule. Also known as aneroid. { ˈanˈəˌrȯid bəˈrämˈədˈər }

aneroid calorimeter [ENG] A calorimeter that uses a metal of high thermal conductivity as a heat reservoir. { ˈanˈəˌrȯid ˌkalˈəˈrimˈədˈər }

aneroid capsule [ENG] A thin, disk-shaped box or capsule, usually metallic, partially evacuated and sealed, held extended by a spring, which expands and contracts with changes in atmospheric or gas pressure. Also known as bellows. { ˈanˈəˌrȯid ˈkapˈsəl }

aneroid diaphragm [ENG] A thin plate, usually metal, covering the end of an aneroid capsule and moving axially as the ambient gas pressure increases or decreases. { ˈanˈəˌrȯid ˈdiˈəˌfram }

aneroid flowmeter [ENG] A mechanism to measure fluid flow rate by pressure of the fluid against a bellows counterbalanced by a calibrated spring. { ˈanˈəˌrȯid ˈflōˌmēdˈər }

aneroid liquid-level meter [ENG] A mechanism to measure fluid depth by pressure of the fluid against a bellows which in turn acts on a manometer or signal transmitter. { ˈanˈəˌrȯid ˈlikˈwəd ˈlevˈəl ˌmedˈər }

aneroidograph *See* aneroid barograph. { ˈanˈəˌrȯidˈəˈgraf }

aneroid valve [MECH ENG] A valve actuated or controlled by an aneroid capsule. { ˈanˈəˌrȯid ˈvalv }

anesthesia [PHYSIO] **1.** Insensibility, general or local, induced by anesthetic agents. **2.** Loss of sensation, of neurogenic or psychogenic origin. { ˌanˈəsˈthēzhˈə }

anesthesiology [MED] A branch of medicine dealing with the administration of anesthetics. { ˌanˈəsˌthēzˈēˈälˈəˈjē }

anesthetic [PHARM] A drug, such as ether, that produces loss of sensibility. { ˈanˈəsˈthedˈik }

anestrus [VERT ZOO] A prolonged period of inactivity between two periods of heat in cyclically breeding female mammals. { aˈnesˈtrəs }

anethole [ORG CHEM] $C_{10}H_{12}O$ White crystals that melt at

22.5°C; very slightly soluble in water; affected by light; odor resembles oil of anise; used in perfumes and flavors, and as a sensitizer in color-bleaching processes in color photography. { 'an·ə,thȯl }

aneucentric aberration [GEN] An aberration that results in a chromosome with more than one centromere. { ¦an·yə¦sen·trik ab·ə'rā·shən }

aneuploidy [GEN] Deviation from a normal haploid, diploid, or polyploid chromosome complement by the presence in excess of, or in defect of, one or more individual chromosomes. { 'a·nyü,plȯid·ē }

aneurine *See* thiamine. { 'an·yə,rēn }

aneurysm [MED] Localized abnormal dilation of an artery due to weakening of the vessel wall. { 'an·yə,riz·əm }

aneusomatic organism [GEN] An organism whose cells contain variable numbers of individual chromosomes. { ¦an·yə·sō¦mad·ik 'ȯr·gə,niz·əm }

aneutronic power [NUCLEO] The generation of energy by an aneutronic reactor. { ¦ā·nü¦trän·ik 'pau̇·ər }

aneutronic reaction [NUC PHYS] A nuclear reaction generating so few neutrons that its neutronicism is less than 0.01. { ¦ā·nü¦trän·ik rē'ak·shən }

aneutronic reactor [NUCLEO] A fusion reactor in which all reactions are aneutronic. { ¦ā·nü¦trän·ik rē'ak·tər }

angaralite [MINERAL] $Mg_2(Al,Fe)_{10}Si_6O_{29}$ A mineral of the chlorite group, occurring in thin black plates. { an'gar·ə,līt }

Angara Shield [GEOL] A shield area of crystalline rock in Siberia. { äŋ·gə'rä ,shēld }

angel dust *See* phencyclidine. { 'ān·jəl ,dəst }

angel echo [ENG] A radar echo from a region where there are no visible targets; may be caused by insects, birds, or refractive index variations in the atmosphere. { 'ān·jəl ,ek·ō }

angelica [FOOD ENG] **1.** A spice from the perennial herb *Angelica archangelica* of the ginger family. **2.** An amber or a yellow sweet wine without muscat flavor. { an'jel·ə·kə }

angelica oil [MATER] An essential oil with an odor that is strongly aromatic; soluble in alcohol; main ingredients are phellandrene and valeric acid; distilled from the seeds and roots of *Angelica archangelica;* used in medicines, liqueurs, and perfumes. { an'jel·ə·kə ,oil }

Angelman syndrome [MED] A genetic disorder that is caused by defects on the maternally derived chromosome 15, causing severe mental retardation, absence of speech, microcephaly, facial dysmorphism, seizures, neonatal hypotonia, ataxic movements, and inappropriate laughter. { 'aŋ·gəl·mən ,sin,drōm }

angiectasia [MED] Abnormal blood vessel dilation. { ,an·jē·ek'tā·zē·ə }

angiitis *See* vasculitis. { ,an·jē'īd·əs }

angina [MED] **1.** A sore throat. **2.** Any intense, constricting pain. { 'an·jə·nə *or* an'jī·nə }

angina pectoris [MED] Constricting chest pain which may be accompanied by pain radiating down the arms, up into the jaw, or to other sites. { 'an·jə·nə 'pek·tə·rəs }

angioblast [EMBRYO] A mesenchyme cell derived from extraembryonic endoderm that differentiates into embryonic blood cells and endothelium. { 'an·jē·ə,blast }

angiocardiography [MED] Roentgenographic visualization of the heart chambers and thoracic vessels following injection of a radiopaque material. { ¦an·jē·ə,kärd·ē¦äg·rə·fē }

angioedema [MED] The development of giant hives beneath the surface of the skin, especially around the eyes and lips; usually due to an allergic response. { ,an·jē·ō·i'dē·mə }

angiogenesis [EMBRYO] The origin and development of blood vessels. { ¦an·jē·ō¦jen·ə·səs }

angiogram [MED] An x-ray photograph of blood vessels following injection of a radiopaque material. { 'an·jē·ə,gram }

angiography [MED] Roentgenographic visualization of blood vessels following injection of a radiopaque material. { ,an·jē'äg·rə·fē }

angiokinesis *See* vasomotion. { ,an·jē·ō· kə'nē·səs }

angiology [MED] The branch of medicine concerned with the blood vessels and the lymphatic system. { ,an·jē'äl·ə·jē }

angioma [MED] A tumor composed of lymphatic vessels or blood. { ,an·jē'ō·mə }

angioneurotic edema [MED] Acute, localized accumulations of tissue fluid causing swellings around the face; condition

may be due either to heredity or a food allergy. { ¦an·jē·ō·nə¦räd·ik i'dē·mə }

angiopathy [MED] Any disease of the vascular system. { ,an·jē'äp·ə·thē }

angioplacentography [MED] Radiography of the blood vessels of placentas through the injection of radiopaque dye. { ¦an·jē·ō·plə,sen'täg·rə·fē }

angioplasty [MED] A procedure for alleviating blockage of an artery in which a balloon-tipped catheter is threaded into an artery to a point of obstruction and inflated to push the vessel open. { 'an·jē·ə,plas·tē }

angiosarcoma [MED] A malignant soft-tissue tumor arising from vascular elements. { ,an·jē·ō,sär'kō·mə }

angioscotoma [MED] A visual-field disturbance caused by dilated blood-vessels in the retina. { ¦an·jē·ō·skə'tō·mə }

angiospasm *See* vasospasm. { 'an·jē·ə,spaz·əm }

angiosperm [BOT] The common name for members of the plant division Magnoliophyta. { 'an·jē·ō,spərm }

Angiospermae [BOT] An equivalent name for the Magnoliophyta. { ¦an·jē·ə¦spər·mē }

angiotensin [BIOCHEM] A decapeptide hormone that influences blood vessel constriction and aldosterone secretion by the adrenal cortex. Also known as hypertensin. { ,an·jē·ə'ten·sən }

angitis [MED] An inflammatory condition of the walls of blood or lymph vessels. { an'jīd·əs }

angle [MATH] The geometric figure, arithmetic quantity, or algebraic signed quantity determined by two rays emanating from a common point or by two planes emanating from a common line. { 'aŋ·gəl }

angle back-pressure valve [MECH ENG] A back-pressure valve with its outlet opening at right angles to its inlet opening. { 'aŋ·gəl 'bak ,presh·ər ,valv }

angle bar [BUILD] An upright bar at the meeting of two faces of a polygonal window, bay window, or bow window. { 'aŋ·gəl ,bär }

angle bead [BUILD] A strip, usually of metal or wood, set at the corner of a plaster wall to protect the corner or serve as a guide to float the plaster flush with it. { 'aŋ·gəl ,bēd }

angle beam [ENG] Ultrasonic waves transmitted for the inspection of a metallic structure at an angle measured from the beam center line to a normal to the test surface. { 'aŋ·gəl ,bēm }

angle bisection [MATH] The division of an angle by a line or plane into two equal angles. { 'aŋ·gəl bī'sek·shən }

angle blasting [ENG] Sandblasting, or the like, at an angle of less that 90°. { 'aŋ·gəl ,blast·iŋ }

angle block [ENG] A small block of wood used to fasten adjacent pieces, usually at right angles, or glued into the corner of a wooden frame to stiffen it. Also known as glue block. { 'aŋ·gəl ,bläk }

angle board [DES ENG] A board whose surface is cut at a desired angle; serves as a guide for cutting or planing other boards at the same angle. { 'aŋ·gəl ,bȯrd }

angle bond [CIV ENG] A tie used to bond masonry work at wall corners. { 'aŋ·gəl ,bänd }

angle brace [ENG] A brace across the interior angle of two members that meet at an angle. Also known as angle tie. { 'aŋ·gəl ,brās }

angle bracket [ARCH] A bracket used in an angle or corner of a molded cornice. [GRAPHICS] Either of a pair of marks ⟨⟩ enclosing a mutilated passage or the explanation of an abbreviation in a text, or to enclose quotations or illustrations in a reference work. Also known as broken bracket; pointed bracket. { 'aŋ·gəl ,brak·ət }

angle brick [ENG] Any brick having an oblique shape to fit an oblique, salient corner. { 'aŋ·gəl ,brik }

angle buttress [ARCH] One of two buttresses at right angles to each other, forming the corner of a structure. { 'aŋ·gəl ,bə·trəs }

angle capital [ARCH] A capital at a corner column. { 'aŋ·gəl ,kap·ə·təl }

angle clip [CIV ENG] A short strip of angle iron used to secure structural elements at right angles. { 'aŋ·gəl ,klip }

angle closer [ENG] A specially shaped brick used to close the bond at the corner of a wall. { 'aŋ·gəl ,klōz·ər }

angle collar [DES ENG] A cast-iron pipe fitting which has a socket at each end for joining with the spigot ends of two pipes that are not in alignment. { 'aŋ·gəl ,käl·ər }

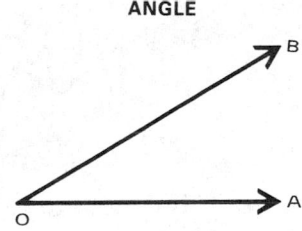

ANGLE

Angle formed by rays *OA* and *OB* emanating from common point *O*. *O* is vertex of angle; *OA* and *OB* are sides of angle.

angle-control section See crossover. { 'aŋ·gəl kən¦trōl ‚sek·shən }

angle cut [MIN ENG] A drilling pattern in which drill holes converge, so that a core can be blasted out and an open or relieved cavity or free face be left for the following shots, which are timed to ensue with a fractional delay. { 'aŋ·gəl ‚kət }

angle diversity [COMMUN] Diversity reception in which beyond-the-horizon tropospheric scatter signals are received at slightly different angles, equivalent to paths through different scatter volumes in the troposphere. { 'aŋ·gəl də'vər·səd·ē }

angle divider [DES ENG] A square for setting or bisecting angles; one side is an adjustable hinged blade. { 'aŋ·gəl də'vīd·ər }

angle dozer [MECH ENG] A power-operated machine fitted with a blade, adjustable in height and angle, for pushing, sidecasting, and spreading loose excavated material as for opencast pits, clearing land, or leveling runways. Also known as angling dozer. { 'aŋ·gəl‚dōz·ər }

angled stair [ARCH] A stair in which successive flights are at an angle of other than 180° to each other (often at 90°), with an intermediate platform between them. { 'aŋ·gəld 'ster }

angle equation [ENG] A condition equation which expresses the relationship between the sum of the measured angles of a closed figure and the theoretical value of that sum, the unknowns being the corrections to the observed directions or angles, depending on which are used in the adjustment. Also known as triangle equation. { 'aŋ·gəl i‚kwā·zhən }

angle fillet [ENG] A wooden strip, triangular in cross section, which is used to cover the internal joint between two surfaces meeting at an angle of less than 180°. { 'aŋ·gəl ‚fil·ət }

angle fishplates [CIV ENG] Plates which join the rails and prevent the rail joint from sagging where heavy cars and locomotives are used. Also known as angle; angle bar. { 'aŋ·gəl 'fish‚plāts }

angle float [ENG] A trowel having two edge surfaces bent at 90°; used to finish corners in freshly poured concrete and in plastering. { 'aŋ·gəl ‚flōt }

angle gauge [CIV ENG] A template used to set or check angles in building construction. { 'aŋ·gəl ‚gāj }

angle gear See angular gear. { 'aŋ·gəl ‚gēr }

angle globe valve [ENG] A globe valve having an angular configuration that permits it to be fitted at bends in pipework. { ¦aŋ·gəl ¦glōb ¦valv }

angle hip tile See arris hip tile. { 'aŋ·gəl 'hip ‚tīl }

angle iron [CIV ENG] **1.** An L-shaped cleat or brace. **2.** A length of steel having a cross section resembling the letter L. { 'aŋ·gəl ‚ī·ərn }

angle jamming [ELECTR] An electronic countermeasure in which azimuth and elevation information, from a scanning fire control radar present in the modulation components on the returning echo pulse, is jammed by transmitting a pulse similar to the radar pulse but with modulation information out of phase with the returning target angle modulation information. { 'aŋ·gəl jam·iŋ }

angle joint [ENG] A joint between two pieces of lumber which results in a change in direction. { 'aŋ·gəl ‚jȯint }

angle lacing [CIV ENG] A system of lacing in which angle irons are used in place of bars. { 'aŋ·gəl ‚lās·iŋ }

angle-lighting luminaire [OPTICS] A luminaire whose light distribution is asymmetric with respect to a direction of specific interest. { 'aŋ·gəl ‚līt·iŋ ‚lü·mə'ner }

angle-measuring equipment [NAV] A radio aid to navigation for measuring the vertical angle to a ground-based or seabased transmitter by determining the phase difference between signals received by antennas at different positions on the aircraft. Abbreviated AME. { 'aŋ·gəl ¦mezh·ər·iŋ i‚kwip·mənt }

angle method of adjustment [ENG] A method of adjustment of observations which determines corrections to observed angles. { 'aŋ·gəl ‚meth·əd əv ə'jəs·mənt }

angle modillion [ARCH] A modillion at the corner of a cornice. { 'aŋ·gəl mō'dil·yən }

angle modulation [ELECTR] The variation in the angle of a sine-wave carrier; particular forms are phase modulation and frequency modulation. Also known as sinusoidal angular modulation. { 'aŋ·gəl mäj·ə'lā·shən }

angle noise [ELECTROMAG] Tracking error introduced into radar by variations in the apparent angle of arrival of the echo from a target, because of finite target size. { 'aŋ·gəl ‚nȯiz }

angle of action [MECH ENG] The angle of revolution of either of two wheels in gear during which any particular tooth remains in contact. { 'aŋ·gəl əv 'ak·shən }

angle of advance See angular advance. { 'aŋ·gəl əv əd'vans }

angle of approach [CIV ENG] The maximum angle of an incline onto which a vehicle can move from a horizontal plane without interference. [MECH ENG] The angle that is turned through by either of paired wheels in gear from the first contact between a pair of teeth until the pitch points of these teeth fall together. [ORD] Angle between the line along which a moving target is traveling and the line along which the gun is pointed. { 'aŋ·gəl əv ə'prōch }

angle of arrival [ELECTROMAG] A measure of the direction of propagation of electromagnetic radiation upon arrival at a receiver (the term is most commonly used in radio); it is the angle between the plane of the phase front and some plane of reference, usually the horizontal, at the receiving antenna. { 'aŋ·gəl əv ə'rīv·əl }

angle of attack [AERO ENG] The angle between a reference line which is fixed with respect to an airframe (usually the longitudinal axis) and the direction of movement of the body. [MIN ENG] The angle in a mine fan made by the direction of air approach and the chord of the aerofoil section. { 'aŋ·gəl əv ə'tak }

angle of bite See angle of nip. { 'aŋ·gəl əv 'bīt }

angle of cant [AERO ENG] In a spin-stabilized rocket, the angle formed by the axis of a venturi tube and the longitudinal axis of the rocket. { 'aŋ·gəl əv 'kant }

angle of clearance [ORD] The angle between the line along which a gun or launcher is pointed at the target and the line along which the weapon must be pointed for a projectile or missile fired from it to clear any obstruction between the weapon and the target. { 'aŋ·gəl əv 'klir·əns }

angle of climb [AERO ENG] The angle between the flight path of a climbing vehicle and the local horizontal. { 'aŋ·gəl əv 'klīm }

angle of commutation [ASTRON] The difference between the celestial longitudes of the sun and a planet, as observed from the earth. { 'aŋ·gəl əv ‚käm·yə'tā·shən }

angle of contact [FL MECH] The angle between the surface of a liquid and the surface of a partially submerged object or of the container at the line of contact. Also known as contact angle. { 'aŋ·gəl əv 'kän‚takt }

angle of contingence [MATH] For two points on a plane curve, the angle between the tangents to the curve at those points. { 'aŋ·gəl əv kən'tin·jəns }

angle of current [HYD] In stream gaging, the angular difference between 90° and the angle made by the current with a measuring section. { 'aŋ·gəl əv 'kər·ənt }

angle of cut [NAV] The smaller angular difference of two bearings or lines of position. { 'aŋ·gəl əv 'kət }

angle of deflection [ELECTR] The angle through which the electron beam in a cathode-ray tube is diverted from a straight path. [ORD] The horizontal clockwise angle between the axis of the bore of a gun and the line of sighting when a gun is laid for direction. [PETRO ENG] In directional drilling with a whipstock or other deflecting tool, the angle at which a well is deflected from the vertical. { 'aŋ·gəl əv di'flek·shən }

angle of departure [AERO ENG] The vertical angle, at the origin, between the line of site and the line of departure. [CIV ENG] The maximum angle of an incline from which a vehicle can move onto a horizontal plane without interference, such as from rear bumpers. [ELECTR] See angle of radiation. { 'aŋ·gəl əv di'pär·chər }

angle of depression [ENG] The angle in a vertical plane between the horizontal and a descending line. Also known as depression angle; descending vertical angle; minus angle. { 'aŋ·gəl əv di'presh·ən }

angle of descent [AERO ENG] The angle between the flight path of a descending vehicle and the local horizontal. { 'aŋ·gəl əv di'sent }

angle of deviation See deviation. { 'aŋ·gəl əv ‚dē·vē'ā·shən }

angle of dip See dip. { 'aŋ·gəl əv 'dip }

angle of divergence [ELECTR] The angular spread of an electron beam in an oscilloscope. [NAV] In terminal instrument procedures, the smaller of the angles formed by the intersection of two courses, radials, bearings, or combinations thereof. [OPTICS] The angular spread of a light beam from a collimating device or laser. { 'aŋ·gəl əv də'vərj·əns }

angle of elevation [ENG] The angle in a vertical plane between the local horizontal and an ascending line, as from an observer to an object; used in astronomy, surveying, and so on. Also known as ascending vertical angle; elevation angle. [ORD] **1.** The vertical angle above the line of sight through which the axis of the gun bore must be raised so that the bullet or projectile will carry to the target. **2.** In aerial gunnery, an acute angle between the bore axis of a gun and the horizontal; called quadrant elevation in ground gunnery. { 'aŋ·gəl əv ‚el·ə'vā·shən }

angle of entry [ORD] The acute angle between the tangent to the trajectory at the point of impact of a bomb or projectile and the perpendicular to the surface of the ground or target at the point of impact. { 'aŋ·gəl əv 'en·trē }

angle of external friction [ENG] The angle between the abscissa and the tangent of the curve representing the relationship of shearing resistance to normal stress acting between soil and the surface of another material. Also known as angle of wall friction. { 'aŋ·gəl əv ek'stərn·əl 'frik·shən }

angle of fall [MECH] The vertical angle at the level point, between the line of fall and the base of the trajectory. { 'aŋ·gəl əv 'fol }

angle of friction See angle of repose. { 'aŋ·gəl əv 'frik·shən }

angle of geodesic contingence [MATH] For two points on a curve on a surface, the angle of intersection of the geodesics tangent to the curve at those points. { 'aŋ·gəl əv jē·ə‚des·ik kən'tin·jəns }

angle of glide [AERO ENG] Angle of descent for an airplane or missile in a glide. { 'aŋ·gəl əv 'glīd }

angle of impact [MECH] The acute angle between the tangent to the trajectory at the point of impact of a projectile and the plane tangent to the surface of the ground or target at the point of impact. { 'aŋ·gəl əv 'im‚pakt }

angle of incidence [OPTICS] The angle formed by a ray arriving at a surface and the perpendicular to that surface at the point of arrival. Also known as incidence angle. { 'aŋ·gəl əv 'in·sə·dəns }

angle of lag See lag angle. { 'aŋ·gəl əv 'lag }

angle of lead See lead angle. { 'aŋ·gəl əv 'lēd }

angle of nip [MECH ENG] The largest angle that will just grip a lump between the jaws, rolls, or mantle and ring of a crusher. Also known as angle of bite; nip. [MIN ENG] In a rock-crushing machine, the maximum angle subtended by its approaching jaws or roll surfaces at which a specified piece of ore can be gripped. { 'aŋ·gəl əv 'nip }

angle of obliquity See angle of pressure. { 'aŋ·gəl əv ō'blik·wəd·ē }

angle of orientation [MECH] Of a projectile in flight, the angle between the plane determined by the axis of the projectile and the tangent to the trajectory (direction of motion), and the vertical plane including the tangent to the trajectory. { 'aŋ·gəl əv ‚or·ē·ən'tā·shən }

angle of pitch [AERO ENG] The angle, as seen from the side, between the longitudinal body axis of an aircraft or similar body and a chosen reference line or plane, usually the horizontal plane. { 'aŋ·gəl əv 'pich }

angle of pressure [DES ENG] The angle between the profile of a gear tooth and a radial line at its pitch point. Also known as angle of obliquity. { 'aŋ·gəl əv 'presh·ər }

angle of radiation [ELECTROMAG] Angle between the surface of the earth and the center of the beam of energy radiated upward into the sky from a transmitting antenna. Also known as angle of departure. { 'aŋ·gəl əv rād·ē'ā·shən }

angle of recess [MECH ENG] The angle that is turned through by either of two wheels in gear, from the coincidence of the pitch points of a pair of teeth until the last point of contact of the teeth. { 'aŋ·gəl əv 'rē‚ses }

angle of reflection [PHYS] The angle between the direction of propagation of a wave reflected by a surface and the line perpendicular to the surface at the point of reflection. Also known as reflection angle. { 'aŋ·gəl əv ri'flek·shən }

angle of refraction [PHYS] The angle between the direction of propagation of a wave that is refracted by a surface and the line that is perpendicular to the surface at the point of refraction. { 'aŋ·gəl əv ri'frak·shən }

angle of repose [ENG] See angle of rest. [MECH] The angle between the horizontal and the plane of contact between two bodies when the upper body is just about to slide over the lower. Also known as angle of friction. { 'aŋ·gəl əv ri'pōz }

angle of rest [ENG] The maximum slope at which a heap of any loose or fragmented solid material will stand without sliding, or will come to rest when poured or dumped in a pile or on a slope. Also known as angle of repose. { 'aŋ·gəl əv 'rest }

angle of roll [AERO ENG] The angle that the lateral body axis of an aircraft or similar body makes with a chosen reference plane in rolling; usually, the angle between the lateral axis and a horizontal plane. { 'aŋ·gəl əv 'rōl }

angle of safety [ORD] Minimum permissible angular clearance, at the weapon, of the path of a projectile or missile above friendly troops. { 'aŋ·gəl əv 'sāf·tē }

angle of shear [GEOL] The angle between the planes of maximum shear which is bisected by the axis of greatest compression. { 'aŋ·gəl əv 'shēr }

angle of site [ORD] The vertical angle that is formed by the line of site and the horizontal. { 'aŋ·gəl əv 'sīt }

angle of slide [MIN ENG] The slope, measured in degrees of deviation from the horizontal, on which loose or fragmented solid materials will start to slide. { 'aŋ·gəl əv 'slīd }

angle of stall [AERO ENG] The angle of attack at which the flow of air begins to break away from the airfoil, the lift begins to decrease, and the drag begins to increase. Also known as stalling angle. { 'aŋ·gəl əv 'stol }

angle of thread [DES ENG] The angle occurring between the sides of a screw thread, measured in an axial plane. { 'aŋ·gəl əv 'thred }

angle of torsion [MECH] The angle through which a part of an object such as a shaft or wire is rotated from its normal position when a torque is applied. Also known as angle of twist. { 'aŋ·gəl əv 'tor·shən }

angle of train [ORD] The azimuth element of firing data furnished by a remote-control system. { 'aŋ·gəl əv 'trān }

angle of traverse [ORD] **1.** Horizontal angle through which a gun or launcher can be turned on its mount. **2.** Angle between the lines from a gun or launcher to the right and left limits of the front which is covered by its fire, that is, the angle through which the weapon is traversed. { 'aŋ·gəl əv trə'vərs }

angle of twist See angle of torsion. { 'aŋ·gəl əv 'twist }

angle of vertical [ASTRON] The angle on the celestial sphere between a given vertical circle and the prime vertical circle. { 'aŋ·gəl əv 'vərd·ə·kəl }

angle of view [OPTICS] The angle subtended by an image at the second nodal point of a lens. { 'aŋ·gəl əv 'vyü }

angle of wall friction See angle of external friction. { 'aŋ·gəl əv ‚wol ‚frik·shən }

angle of wrap [DES ENG] On a band brake mechanism, the distance, expressed in degrees, that the brake band wraps around the brake flange. { 'aŋ·gəl əv 'rap }

angle of yaw [AERO ENG] The angle, as seen from above, between the longitudinal body axis of an aircraft, a rocket, or the like and a chosen reference direction. Also known as yaw angle. { 'aŋ·gəl əv 'yo }

angle paddle [ENG] A hand tool used to finish a plastered surface. { 'aŋ·gəl ‚pad·əl }

angle plate [DES ENG] An L-shaped plate or a plate having an angular section. { 'aŋ·gəl ‚plāt }

angle post [BUILD] A railing support used at a landing or other break in the stairs. { 'aŋ·gəl ‚pōst }

angle press [MECH ENG] A hydraulic plastics-molding press with both horizontal and vertical rams; used to produce complex moldings with deep undercuts. { 'aŋ·gəl ‚press }

angle rafter [BUILD] A rafter, such as a hip rafter, at the angle of the roof. { 'aŋ·gəl ‚raf·tər }

angle-resolved photoelectron spectroscopy [SPECT] A type of photoelectron spectroscopy which measures the kinetic energies of photoelectrons emitted from a solid surface and the angles at which they are emitted relative to the surface. Abbreviated ARPES. { 'aŋ·gəl ri'zälvd ‚fōd·ō·ə'lek‚trän ‚spek'träs·kə·pē }

anglerfish [VERT ZOO] Any of several species of the order Lophiiformes characterized by remnants of a dorsal fin seen as a few rays on top of the head that are modified to bear a terminal bulb. { 'aŋ·glər‚fish }

angle rib [ARCH] **1.** One of the diagonal ribs dividing each rectangle of a gothic vaulting; it is a structural member. **2.** Molding ornamenting an angle in decorative work. { 'aŋ·gəl ‚rib }

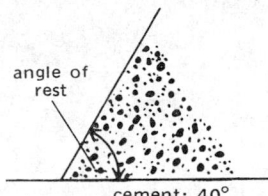

ANGLE OF REST

angle of rest

cement: 40°
round gravel: 30°

Angle of rest, one of the critical angles for bulk material.

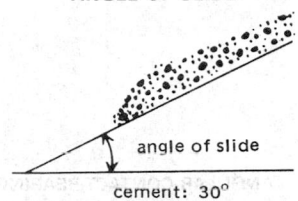

ANGLE OF SLIDE

angle of slide

cement: 30°
round gravel: 18°

Angle of slide, one of the critical angles for bulk material.

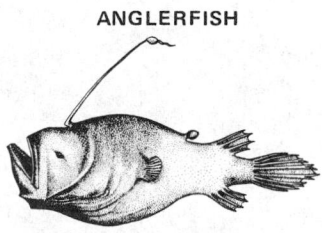

ANGLERFISH

Cryptosaras couesi, the anglerfish, length up to 18 inches (45 centimeters). *(After G. B. Goode and T. H. Bean, Oceanic Ichthyology, U.S. Nat. Mus. Spec. Bull. no. 2, 1895)*

angle section [CIV ENG] A structural steel member having an L-shaped cross section. { 'aŋ·gəl ,sek·shən }

angle set [MIN ENG] **1.** A timber set using an angle brace. **2.** One of a series of sets placed at angles to each other. { 'aŋ·gəl ,set }

angle shot [GRAPHICS] **1.** A photograph taken with the camera tilted at an angle from the horizontal. **2.** In cinematography, a motion picture that is taken from a different angle; it may repeat the action in a previous shot or continue the action. { 'aŋ·gəl ,shät }

anglesite [MINERAL] $PbSO_4$ A mineral occurring in white or gray, tabular or prismatic orthorhombic crystals or compact masses. Also known as lead spar; lead vitriol. { 'aŋ·glə,sīt }

angle-stem thermometer [ENG] A device used to measure temperatures in oil-custody tanks; the angle of the calibrated stem may be 90° or greater to the sensitive portion of the thermometer, as needed to fit the tank shell contour. { 'aŋ·gəl ¦stem thər'mäm·əd·ər }

angle stile [BUILD] A narrow strip of wood used to conceal the joint between a wall and a vertical wood surface which makes an angle with the wall, as at the edge of a corner cabinet. { 'aŋ·gəl ,stīl }

angle structure [CIV ENG] A method of building a tower for mechanical strength in which braces are placed at angles with respect to the vertical support rods. { 'aŋ·gəl ,strək·chər }

angle strut [CIV ENG] An angle-shaped structural member which is designed to carry a compression load. { 'aŋ·gəl ,strət }

angle sub See bent sub. { 'aŋ·gəl ,səb }

angle tracking noise [ELECTR] Any deviation of the tracking axis from the center of reflectivity of a radar target; it is the resultant of servo noise, receiver noise, angle noise, and amplitude noise. { 'aŋ·gəl ¦trak·iŋ ,nȯiz }

angle valve [DES ENG] A manually operated valve with its outlet opening oriented at right angles to its inlet opening; used for regulating the flow of a fluid in a pipe. { 'aŋ·gəl ,valv }

angle variable [MECH] The dynamical variable w conjugate to the action variable J, defined only for periodic motion. { 'aŋ·gəl ,ver·ē·ə·bəl }

angling dozer See angle dozer. { 'aŋ·gliŋ ,dōz·ər }

Angoumian [GEOL] Upper middle Upper Cretaceous (Upper Turonian) geologic time. { ,än'güm·ē·ən }

angrite [GEOL] An achondrite stony meteorite composed principally of augite with a little olivine and troilite. { 'aŋ,grīt }

angstrom [MECH] A unit of length, 10^{-10} meter, used primarily to express wavelengths of optical spectra. Abbreviated A; Å. Also known as tenthmeter. { 'aŋ·strəm }

Ångström coefficient [PHYS] The multiplying amplitude parameter inserted in Ångström's formula for the scattering of electromagnetic radiation by atmospheric dust. { 'ȯŋ·strəm ,kō·ə'fish·ənt }

Ångström compensation pyrheliometer [ENG] A pyrheliometer consisting of two identical Manganin strips, one shaded, the other exposed to sunlight; an electrical current is passed through the shaded strip to raise its temperature to that of the exposed strip, and the electric power required to accomplish this is a measure of the solar radiation. { 'ȯŋ·strəm käm·pən'sā·shən ¦pīr,hē,lē'äm·əd·ər }

Ångström's formula [PHYS] A formula stating that the scattering coefficient for dust in the atmosphere is inversely proportional to a positive power of the wavelength of the radiation, with the power depending on the size of the dust particles. { 'ȯŋ·strəmz 'fȯrm·yə·lə }

anguclast [GEOL] An angular phenoclast. { 'aŋ·gyu̇ ,klast }

Anguidae [VERT ZOO] A family of limbless, snakelike lizards in the suborder Sauria, commonly known as slowworms or glass snakes. { 'aŋ·gwə,dē }

Anguilliformes [VERT ZOO] A large order of actinopterygian fishes containing the true eels. { aŋ,gwil·ə'fȯr,mēz }

anguilliform motion [VERT ZOO] A type of locomotion in which a fish such as an eel moves its entire body, from head to tail, with considerable amplitude. { aŋ¦gwil·ə,form 'mō·shən }

Anguilloidei [VERT ZOO] The typical eels, a suborder of actinopterygian fishes in the order Anguilliformes. { aŋ·gwə'lȯid·ē,ī }

ANGULAR-CONTACT BEARING

Standard angular-contact ball bearing. (*Marlin-Rockwell*)

angular [SCI TECH] **1.** Having sharp corners. **2.** Having or forming angles. { 'aŋ·gyə·lər }

angular acceleration [MECH] The time rate of change of angular velocity. { 'aŋ·gyə·lər ak,sel·ə'rā·shən }

angular accelerometer [ENG] An accelerometer that measures the rate of change of angular velocity between two objects under observation. { 'aŋ·gyə·lər ak,sel·ə'räm·əd·ər }

angular advance [MECH ENG] The amount by which the angle between the crank of a steam engine and the virtual crank radius of the eccentric exceeds a right angle. Also known as angle of advance; angular lead. { 'aŋ·gyə·lər əd'vans }

angular aggregate [MATER] An aggregate whose particles possess well-defined edges formed at the intersection of roughly planar faces. { 'aŋ·gyə·lər 'ag·ri·gət }

angular aperture [OPTICS] The angle subtended at an axial object point of an optical instrument by the radius of the entrance pupil. { 'aŋ·gyə·lər 'ap·ə·chər }

angular bitstalk See angular bitstock. { 'aŋ·gyə·lər 'bit,stök }

angular bitstock [MECH ENG] A bitstock whose handles are positioned to permit its use in corners and other cramped areas. Also known as angular bitstalk. { 'aŋ·gyə·lər 'bit,stäk }

angular clearance [DES ENG] The relieved space located below the straight of a die, to permit passage of blanks or slugs. { 'aŋ·gyə·lər 'klir·əns }

angular-contact bearing [MECH ENG] A rolling-contact antifriction bearing designed to carry heavy thrust loads and also radial loads. { 'aŋ·gyə·lər 'kän,takt ,ber·iŋ }

angular correlations [NUC PHYS] A technique of nuclear experimentation for determining spins of nuclear states, the angular momentum mixtures of incoming or outgoing particles, and the multipole mixtures of emitted gamma rays, by measuring the dependence of the intensity or the cross section of a nuclear reaction on the directions of two or more radiations. { 'aŋ·gyə·lər ,kär·ə'lā·shənz }

angular cutter [MECH ENG] A tool-steel cutter used for finishing surfaces at angles greater or less than 90° with its axis of rotation. { 'an·gyə·lər 'kəd·ər }

angular diameter [ASTRON] The angle subtended at the observer by a diameter of a distant spherical body which is perpendicular to the line between the observer and the center of the body. { 'an·gyə·lər ,dī'am·əd·ər }

angular displacement [PHYS] A vector measure of the rotation of an object about an axis; the vector points along the axis according to the right-hand rule; the length of the vector is the rotation angle, in degrees or radians. { 'an·gyə·lər dis'plās·mənt }

angular distance [MATH] **1.** For two points, the angle between the lines from a point of observation to the points. **2.** The angular difference between two directions, numerically equal to the angle between two lines extending in the given directions. **3.** The arc of the great circle joining two points, expressed in angular units. [PHYS] The distance between two points, expressed in wavelengths at a specified frequency. { 'an·gyə·lər 'dis·təns }

angular distortion [MAP] Distortion in a map projection because of nonconformality. { 'an·gyə·lər di'stȯr·shən }

angular distribution [NUCLEO] The distribution in angle, relative to an experimentally specified direction, of the intensity of photons or particles resulting from a nuclear or extranuclear process. { 'an·gyə·lər dis·trə'byü·shən }

angular error of closure See error of closure. { 'an·gyə·lər 'er·ər əv 'klōzh·ər }

angular frequency [PHYS] For any oscillation, the number of vibrations per unit time, multiplied by 2π. Also known as angular velocity; radian frequency. { 'an·gyə·lər 'frē·kwən·sē }

angular gear [MECH ENG] A gear that transmits motion between two rotating shafts that are not parallel. Also known as angle gear. { 'an·gyə·lər 'gēr }

angular height [ORD] The vertical angle between the line of site and the horizontal. { 'an·gyə·lər 'hīt }

angular impulse [MECH] The integral of the torque applied to a body over time. { 'an·gyə·lər 'im,pəls }

angularity [SCI TECH] The sharpness of corners and edges. { 'aŋ·gyə'lar·əd·ē }

angular lead See angular advance. { 'aŋ·gyə·lər 'lēd }

angular leaf spot [PL PATH] A bacterial disease of plants characterized by angular spots on leaves and caused by *Pseudomonas lachrymans* in cucumbers and *Xanthomonas malvacearum* in cotton. { 'aŋ·gyə·lər 'lēf ˌspät }

angular length [MECH] A length expressed in the unit of the length per radian or degree of a specified wave. { 'aŋ·gyə·lər 'leŋkth }

angular magnification [OPTICS] For an optical system, the ratio of the angle subtended by the image at the eye to the angle subtended by the object at the eye. { 'aŋ·gyə·lər ˌmag·nə·fə'kā·shən }

angular milling [MECH ENG] Milling surfaces that are flat and at an angle to the axis of the spindle of the milling machine. { 'aŋ·gyə·lər 'mil·iŋ }

angular momentum [MECH] **1.** The cross product of a vector from a specified reference point to a particle, with the particle's linear momentum. Also known as moment of momentum. **2.** For a system of particles, the vector sum of the angular momenta (first definition) of the particles. { 'aŋ·gyə·lər mə'ment·əm }

angular momentum operator [QUANT MECH] Any vector operator satisfying communication rules of the type $[J_x, J_y] = iJ_z$. { 'aŋ·gyə·lər mə'ment·əm 'äp·ə,rād·ər }

angular parallax [NAV] The angle between grid north and true north; the angle increases the farther the grid departs from the standard meridian. { 'aŋ·gyə·lər 'par·ə,laks }

angular perspective [GRAPHICS] A form of plane linear perspective in which some of the principal lines of the picture are either parallel or perpendicular to the picture plane and some are oblique. { 'aŋ·gyə·lər pər'spek·tiv }

angular pitch [DES ENG] The angle determined by the length along the pitch circle of a gear between successive teeth. { 'aŋ·gyə·lər 'pich }

angular radius [MATH] For a circle drawn on a sphere, the smaller of the angular distances from one of the two poles of the circle to any point on the circle. { 'aŋ·gyə·lər 'rād·ē·əs }

angular rate *See* angular speed. { 'aŋ·gyə·lər ˌrāt }

angular resolution [ELECTROMAG] A measure of the ability of a radar to distinguish between two targets solely by the measurement of angles. { 'aŋ·gyə·lər ˌrez·ə'lü·shən }

angular resolver *See* resolver. { 'aŋ·gyə·lər ri'zälv·ər }

angular shear [MECH ENG] A shear effected by two cutting edges inclined to each other to reduce the force needed for shearing. { 'aŋ·gyə·lər 'shēr }

angular speed [MECH] Change of direction per unit time, as of a target on a radar screen, without regard to the direction of the rotation axis; in other words, the magnitude of the angular velocity vector. Also known as angular rate. { 'aŋ·gyə·lər 'spēd }

angular spreading [OCEANOGR] The lateral extension of ocean waves as they move out of the wave-generating area as swell. { 'aŋ·gyə·lər 'spred·iŋ }

angular-spreading factor [OCEANOGR] The ratio of the actual wave energy present at a point to that which would have been present in the absence of angular spreading. { 'aŋ·gyə·lər 'spred·iŋ ˌfak·tər }

angular travel [ORD] Angular distance covered by a moving target in a given time. { 'aŋ·gyə·lər 'trav·əl }

angular travel error [MECH] The error which is introduced into a predicted angle obtained by multiplying an instantaneous angular velocity by a time of flight. { 'aŋ·gyə·lər 'trav·əl ˌer·ər }

angular travel method [ORD] Method of calculating firing data based on the rate of angular travel of the target in direction and elevation. { 'aŋ·gyə·lər 'trav·əl ˌmeth·əd }

angular unconformity [GEOL] An unconformity in which the older strata dip at a different angle (usually steeper) than the younger strata. { 'aŋ·gyə·lər ˌən·kən'fòrm·əd·ē }

angular unit method [ORD] A method of adjusting antiaircraft artillery gunfire in which range deviations in miles obtained by a distant observer are converted into altitude corrections in yards for application at the data computer. { 'aŋ·gyə·lər ¦yü·nət ˌmeth·əd }

angular velocity [MECH] The time rate of change of angular displacement. [PHYS] *See* angular frequency. { 'aŋ·gyə·lər və'läs·əd·ē }

angular wave number [METEOROL] The number of waves of a given wavelength required to encircle the earth at the latitude of the disturbance. Also known as hemispheric wave number. { 'aŋ·gyə·lər ¦wāv 'nəm·bər }

angulator [ENG] An instrument for converting angles measured on an oblique plane to their corresponding projections on a horizontal plane; the rectoblique plotter and the photoangulator are types. { 'aŋ·gyə,lād·ər }

anhalonium alkaloid [PHARM] Any of the alkaloids found in mescal buttons, including hordenine and mescaline; used as a cerebral stimulant and motor depressant. Also known as cactus alkaloid. { ˌan·ə'lōn·ē·əm 'al·kə,lòid }

anharmonicity [PHYS] **1.** Mechanical vibration where the restoring force acting on a system does not vary linearly with displacement from equilibrium position. **2.** Variation from a linear relationship of dipole moment with internuclear distance in the infrared portion of the electromagnetic spectrum. { ¦an,här·mə'nis·əd·ē }

anharmonic oscillator [PHYS] An oscillating system in which the restoring force opposing a displacement from the position of equilibrium is a nonlinear function of the displacement. { ¦an,här¦män·ik 'äs·ə,lād·ər }

anharmonic oscillator spectrum [SPECT] A molecular spectrum which is significantly affected by anharmonicity of the forces between atoms in the molecule. { ˌan·här¦män·ik ¦äs·ə,lād·ər ¦spek·trəm }

anhedonia [PSYCH] Inability to experience pleasure from activities that ordinarily produce pleasurable feelings. { ˌan·ha'dōn·ē·ə }

anhedral *See* allotriomorphic. { an'hēd·rəl }

anhedron [PETR] Rock that has the organized internal structure of a crystal without the external geometric form of a crystal. { an'hēd·rən }

anhidrosis [MED] Absent or deficient secretion of sweat. { ˌan·hī'drō·səs }

Anhimidae [VERT ZOO] The screamers, a family of birds in the order Anseriformes characterized by stout bills, webbed feet, and spurred wings. { an'him·ə,dē }

Anhingidae [VERT ZOO] The anhingas or snakebirds, a family of swimming birds in the order Pelecaniformes. { an'hin·jə,dē }

anhydrase [BIOCHEM] Any enzyme that catalyzes the removal of water from a substrate. { an'hī,drās }

anhydremia [MED] A decreased amount of water in the plasma. { ˌan,hī'drēm·yə }

anhydride [CHEM] A compound formed from an acid by removal of water. { an'hī,drīd }

anhydrite [MINERAL] $CaSO_4$ A mineral that represents gypsum without its water of crystallization, occurring commonly in white and grayish granular to compact masses; the hardness is 3–3.5 on Mohs scale, and specific gravity is 2.90–2.99. Also known as cube spar. { an'hī,drīt }

anhydrite evaporite [PETR] $CuSO_4$ A sedimentary rock composed chiefly of copper sulfate in compact granular form deposited by evaporation of water; resembles marble and differs from gypsum in lack of water of hydration and hardness. { an'hī,drīt i'vap·ə,rīt }

anhydrobiosis [PHYSIO] A type of cryptobiosis induced by dehydration. { ¦an,hī·drō ˌbī'ō·səs }

anhydrock [PETR] A sedimentary rock chiefly made of anhydrite. { an'hi,dräk }

anhydrous [CHEM] Being without water, especially water of hydration. { an'hī·drəs }

anhydrous alcohol *See* absolute alcohol. { an'hī·drəs 'al·kə,hòl }

anhydrous ammonia [INORG CHEM] Liquid ammonia, a colorless liquid boiling at −33.3°C. { an'hī·drəs ə'mōn·yə }

anhydrous calcium sulfate [MATER] Gypsum from which all the water of crystallization has been removed. Also known as dead-burnt gypsum. { an'hī·drəs ¦kal·sē·əm 'səl,fāt }

anhydrous ferric chloride *See* ferric chloride. { an'hī·drəs ¦fer·ik 'klòr,īd }

anhydrous gypsum plaster [MATER] Plaster which has a greater percentage of the water of crystallization removed than normal gypsum plasters; used as a finish plaster. { an'hī·drəs ¦jip·səm 'plas·tər }

anhydrous hydrogen chloride [INORG CHEM] HCl Hazardous, toxic, colorless gas used in polymerization, isomerization, alkylation, nitration, and chlorination reactions; becomes

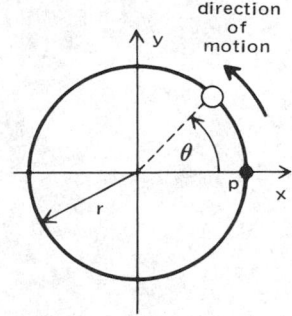

ANGULAR VELOCITY

Particle p moves on circular path with radius r. θ is angular displacement of p. Angular velocity is rate of change of θ with respect to time.

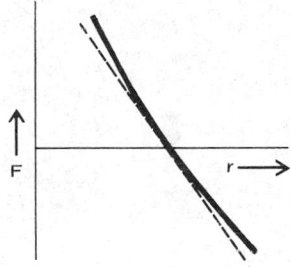

ANHARMONIC OSCILLATOR

Interatomic force F as a function of the atomic separation r. Anharmonicity is produced by the departure of the actual force (solid curve) from the dashed line.

hydrochloric acid in aqueous solutions. { an'hī·drəs ,hī·drə·jən 'klȯr,īd }

anhydrous phosphoric acid See phosphoric anhydride. { an'hī·dres fä'sfȯrik 'as·əd }

anhydrous plumbic acid See lead dioxide. { an'hī·dres 'pləmb·ik 'as·əd }

anhydrous sodium carbonate See soda ash. { an'hi·dres ,sōd·ē·əm 'kärb·ə,nāt }

anhydrous sodium sulfate [INORG CHEM] Na_2SO_4 Water-soluble, white crystals with bitter, salty taste; melts at 888°C; used in the manufacture of glass, paper, pharmaceuticals, and textiles, and as an analytical reagent. { an'hī·drəs ,sōd·ē·əm 'səl,fāt }

anhysteretic remanence [ELECTROMAG] The remanence in a magnetic recording medium that results from adding an alternating current to the signal current; at low fields it is linear and thus does not exhibit hysteresis. { ,an,his·tə¦red·ik 'rem·ə·nəns }

anilazine [ORG CHEM] $C_9H_5Cl_3N_4$ A tan solid with a melting point of 159-160°C; used for fungal diseases of lawns, turf, and vegetable crops. { ə'nil·ə,zēn }

anileridine [PHARM] $C_{22}H_{28}N_2O_2$ A white, crystalline powder; soluble in alcohol and chloroform; used as a narcotic. { ,an·əl'er·ə,dēn }

anilide [ORG CHEM] A compound that has the $C_6H_5NH_2-$ group; an example is benzanilide, $C_6H_5NHCOC_6H_5$. { 'an·əl,īd }

Aniliidae [VERT ZOO] A small family of nonvenomous, burrowing snakes in the order Squamata. { ,an·əl'ī·ə,dē }

aniline [ORG CHEM] $C_6H_5NH_2$ An aromatic amine compound that is a pale brown liquid at room temperature; used in the dye, pharmaceutical, and rubber industries. { 'an·əl·ən }

aniline black [ORG CHEM] A black dye produced on certain textiles, such as cotton, by oxidizing aniline or aniline hydrochloride. { 'an·əl·ən 'blak }

aniline N,N-dimethyl See N,N-dimethylaniline. { 'an·əl·ən ¦en ¦en di'meth·əl }

aniline dye [ORG CHEM] A dye derived from aniline. { 'an·əl·ən 'dī }

aniline-formaldehyde resin [CHEM] A thermoplastic resin made by polymerizing aniline and formaldehyde. { 'an·əl·ən ,fȯr'mal·də,hīd ,rez·ən }

aniline hydrochloride [ORG CHEM] $C_6H_5NH_2 \cdot HCl$ White crystals, although sometimes the commercial variety has a greenish tinge; melting point 198°C; soluble in water and ethanol; used in dye manufacture, dyeing, and printing. { 'an·əl·ən ,hī·drə'klȯr,īd }

aniline ink [MATER] A fast-drying printing ink that is a solution of a coal-tar dye in an organic solvent or a solution of a pigment in an organic solvent or water. { 'an·əl·ən ,iŋk }

aniline leather [MATER] Leather whose grain pattern has been accentuated by impregnation with aniline dye. { 'an·əl·ən 'leth·ər }

aniline point [CHEM ENG] The minimum temperature for a complete mixing of aniline and materials such as gasoline; used in some specifications to indicate the aromatic content of oils and to calculate approximate heat of combustion. { 'an·əl·ən ,pȯint }

aniline printing See flexography. { 'an·əl·ən ,print·iŋ }

aniline process See flexography. { 'an·əl·ən ,präs·əs }

anilol [MATER] An aniline-alcohol mixture used as a blending compound in petroleum products. { 'an·ə,lȯl }

anima [PSYCH] In analytic psychology, a person's inner being as opposed to the persona presented to a world. { 'an·ə·mə }

animal [ZOO] Any living organism distinguished from plants by the lack of chlorophyll, the requirement for complex organic nutrients, the lack of a cell wall, limited growth, mobility, and greater irritability. { 'an·ə·məl }

animal balance [ENG] A balance designed to weigh living animals, with a readout or display relatively unaffected by the pulse or movements of the animal. { 'an·ə·məl 'bal·əns }

animal black [CHEM] Finely divided carbon made by calcination of animal bones or ivory; used for pigments, decolorizers, and purifying agents; varieties include bone black and ivory black. { 'an·ə·məl ,blak }

animal charcoal [CHEM] Charcoal obtained by the destructive distillation of animal matter at high temperatures; used to adsorb organic coloring matter. { 'an·ə·məl 'chär,kōl }

animal communication [PSYCH] The discipline within the field of animal behavior that deals with the receipt and use of signals by animals. { 'an·ə·məl kə,myü·nə'kā·shən }

animal community [ECOL] An aggregation of animal species held together in a continuous or discontinuous geographic area by ties to the same physical environment, mainly vegetation. { 'an·ə·məl kə'myü·nəd·ē }

animal ecology [ECOL] A study of the relationships of animals to their environment. { 'an·ə·məl i'käl·ə·jē }

animal fiber [TEXT] A natural textile fiber of animal origin; wool and silk are the most important. { 'an·ə·məl ,fīb·ər }

animal glue [MATER] A glue made from the bones, hide, horns, and connective tissues of animals. { 'an·ə·məl ,glü }

animal husbandry [AGR] A branch of agriculture concerned with the breeding and feeding of domestic animals. { 'an·ə·məl 'həz·bən·drē }

Animalia [SYST] The animal kingdom. { ,an·ə'māl·yə }

animal kingdom [SYST] One of the two generally accepted major divisions of living organisms which live or have lived on earth (the other division being the plant kingdom). { 'an·ə·məl ,kiŋ·dəm }

animal locomotion [ZOO] Progressive movement of an animal body from one point to another. { 'an·ə·məl ,lō·kə'mō·shən }

animal oil See bone oil. { 'an·ə·məl ,ȯil }

animal pole [CYTOL] The region of an ovum which contains the least yolk and where the nucleus gives off polar bodies during meiosis. { 'an·ə·məl ,pōl }

animal power [MECH ENG] The time rate at which muscular work is done by a work animal, such as a horse, bullock, or elephant. { 'an·ə·məl ,pau̇·ər }

animal virus [VIROL] A small infectious agent able to propagate only within living animal cells. { 'an·ə·məl 'vī·rəs }

Animikean [GEOL] The middle subdivision of Proterozoic geologic time. Also known as Penokean; Upper Huronian. { ə¦nim·ə¦kē·ən }

animikite [GEOL] An ore of silver, composed of a mixture of sulfides, arsenides, and antimonides, and containing nickel and lead; occurs in white or gray granular masses. { ə'nim·ə,kīt }

anion [CHEM] An ion that is negatively charged. { 'a,nī·ən }

anion exchange [CHEM] A type of ion exchange in which the immobilized functional groups on the solid resin are positive. { 'a,nī·ən iks'chānj }

anionic detergent [MATER] A class of detergents having a negatively charged surface-active ion, such as sodium alkylbenzene sulfonate. { ¦a,nī¦än·ik di'tər·jənt }

anionic polymerization [ORG CHEM] A type of polymerization in which Lewis bases, such as alkali metals and metallic alkyls, act as catalysts. { ¦a,nī¦än·ik pə,lim·ə·rə'zā·shən }

anionotropy [CHEM] The breaking off of an ion such as hydroxyl or bromide from a molecule so that a positive ion remains in a state of dynamic equilibrium. { ¦a,nī·ə'nä·trə·pē }

Anisakidae [INV ZOO] A family of parasitic roundworms in the superfamily Ascaridoidea. { ,an·ə'säk·ə,dē }

anisaldehyde [ORG CHEM] $C_6H_4(OCH_3)CHO$ A compound with melting point 2.5°C, boiling point 249.5°C; insoluble in water, soluble in alcohol and ether; used in perfumery and flavoring, and as an intermediate in production of antihistamines. { ¦a·nəs'al·də,hīd }

anise [BOT] The small fruit of the annual herb *Pimpinella anisum* in the family Umbelliferae; fruit is used for food flavoring, and oil is used in medicines, soaps, and cosmetics. { 'an·əs }

Anisian [GEOL] Lower Middle Triassic geologic time. { ə'nis·ē·ən }

anisic acid [ORG CHEM] $CH_3OC_6H_4COOH$ White crystals or powder with a melting point of 184°C; soluble in alcohol and ether; used in medicine and as an insect repellent and ovicide. { ə'nis·ik 'as·əd }

anisic alcohol [ORG CHEM] $C_8H_{10}O_2$ A colorless liquid that boils in the range 255–265°C; it is obtained by reduction of anisic aldehyde; used in perfumery, and as an intermediate in the manufacture of pharmaceuticals. { ə'nis·ik 'al·kə,hȯl }

anisocarpous [BOT] Referring to a flower whose number of carpels is different from the number of stamens, petals, and sepals. { ¦a,nis·ə'kär·pəs }

anisochela [INV ZOO] A chelate sponge spicule with dissimilar ends. { ¦a,nis·ə'kēl·ə }

anisocytosis [MED] A condition in which the erythrocytes show a considerable variation in size due to excessive quantities of hemoglobin. { ¦a,nis·ə,sī'tō·səs }

anisodactylous [VERT ZOO] Having unequal digits, especially referring to birds with three toes forward and one backward. { ¦a,nis·ə'dak·tə·ləs }

anisodesmic [MINERAL] Pertaining to crystals or compounds in which the ionic bonds are unequal in strength. { ¦a,nis·ə'dez·mik }

anisogamete See heterogamete. { ¦a,nis·ə'ga ,mēt }

anisogamy See heterogamy. { ¦a,nis'äg·ə·mē }

anisole [ORG CHEM] $C_6H_5OCH_3$ A colorless liquid that is soluble in ether and alcohol, insoluble in water; boiling point is 155°C; vapors are highly toxic; used as a solvent and in perfumery. { 'an·ə,sōl }

anisomerous [BOT] Referring to flowers that do not have the same number of parts in each whorl. { ¦a,nī'säm·ə·rəs }

anisometric particle [VIROL] Any unsymmetrical, rod-shaped plant virus. { ¦a,nī·sə¦me·trik 'pärd·ə·kəl }

Anisomyaria [INV ZOO] An order of mollusks in the class Bivalvia containing the oysters, scallops, and mussels. { ¦a,nī·sə,mī'a·rē·ə }

anisophyllous [BOT] Having leaves of two or more shapes and sizes. { ¦a,nī·sə¦fil·əs }

Anisoptera [INV ZOO] The true dragonflies, a suborder of insects in the order Odonata. { ¦a,nī'säp·tə·rə }

anisostemonous [BOT] Referring to a flower whose number of stamens is different from the number of carpels, petals, and sepals. { ¦a,nī·sä'stem·ə·nəs }

Anisotomidae [INV ZOO] An equivalent name for Leiodidae. { ¦a,nī·sə'täm·ə,dē }

anisotropic [PHYS] Showing different properties as to velocity of light transmission, conductivity of heat or electricity, compressibility, and so on, in different directions. Also known as aeolotropic. { ¦a,nī·sə¦träp·ik }

anisotropic magnetoresistance [SOLID STATE] A type of magnetoresistance displayed by all metallic magnetic materials, which arises because conduction electrons have more frequent collisions when they move parallel to the magnetization in the material than when they move perpendicular to it. { ¦an·ə sə¦trō·pik ,mag,ned·ō·ri'sis·təns }

anisotropic membrane [CHEM ENG] An ultrafiltration membrane which has a thin skin at the separating surface and is supported by a spongy sublayer of membrane material. { ¦a,nī sə¦träp·ik 'mcm,brān }

anisotropy [ASTRON] The departure of the cosmic microwave radiation from equal intensity in all directions. [BOT] The property of a plant that assumes a certain position in response to an external stimulus. [PHYS] The characteristic of a substance for which a physical property, such as index of refraction, varies in value with the direction in or along which the measurement is made. Also known as aeolotropy; eolotropy. [ZOO] The property of an egg that has a definite axis or axes. { ¦a,nī'sä·trə·pē }

anisotropy constant [ELECTROMAG] In a ferromagnetic material, temperature-dependent parameters relating the magnetization in various directions to the anisotropy energy. { ¦a,nī'sä·trə·pē ,kän·stənt }

anisotropy energy [ELECTROMAG] Energy stored in a ferromagnetic crystal by virtue of the work done in rotating the magnetization of a domain away from the direction of easy magnetization. { ¦a,nī'sä·trə·pē ,en·ər·jē }

anisotropy factor See dissymmetry factor. { ¦a,nī'sä·trə·pē ,fak·tər }

ankaramite [PETR] A mafic olivine basalt primarily composed of pyroxene with smaller amounts of olivine and plagioclase and accessory biotite, apatite, and opaque oxides. { 'aŋ·kə·rä,mīt }

ankaratrite See olivine nephelinite. { 'aŋ·kə·rä,trīt }

anker [MECH] A unit of capacity equal to 10 U.S. gallons (37.854 liters); used to measure liquids, especially honey, oil, vinegar, spirits, and wine. { 'aŋ·kər }

ankerite [MINERAL] $Ca(Fe,Mg,Mn)(CO_3)_2$ A white, red, or gray iron-rich carbonate mineral associated with iron ores and found in thin veins in coal seams; specific gravity is 2.95-3.1. Also known as cleat spar. { 'aŋ·kə,rīt }

ankle [ANAT] The joint formed by the articulation of the leg bones with the talus, one of the tarsal bones. { 'aŋ·kəl }

ankle breadth [ANTHRO] The distance measured between projections at lower ends of the tibia and fibula. { 'aŋ·kəl ,bredth }

ankle thickness [ANTHRO] Distance measured perpendicular to ankle breadth. { 'aŋ·kəl ,thik·nəs }

Ankylosauria [PALEON] A suborder of Cretaceous dinosaurs in the reptilian order Ornithischia characterized by short legs and flattened, heavily armored bodies. { 'aŋ·kə·lə'sòr·ē·ə }

ankylosing spondylitis See spondylitis. { 'aŋ·kə,lōz·iŋ ,spän·də'līd·əs }

ankylosis Also spelled anchylosis. [MED] Stiffness or immobilization of a joint due to a surgical or pathologic process. [PHYS] The loss by a system of one or more degrees of freedom through development of one or more frictional constraints. { ,aŋ·kə'lō·səs }

ankyrin [CELL MOL] A protein found in the cell membrane of erythrocytes that attaches the membrane to the cytoskeleton protein spectrin. { aŋ'kī·rən }

ANL See automatic noise limiter.

anlage [EMBRYO] Any group of embryonic cells when first identifiable as a future organ or body part. Also known as blastema; primordium. { 'än,läg·ə }

annabergite [MINERAL] $(Ni,Co)_3(AsO_4)_2{\cdot}8H_2O$ A monoclinic mineral usually found as apple-green incrustations as an alteration product of nickel arsenides; it is isomorphous with erythrite. Also known as nickel bloom; nickel ocher. { 'a·nə,bər,gīt }

annatto [BOT] *Bixa orellana.* A tree found in tropical America, characterized by cordate leaves and spinose, seed-filled capsules; a yellowish-red dye obtained from the pulp around the seeds is used as a food coloring. { ə'näd·ō }

anneal [ENG] To treat a metal, alloy, or glass with heat and then cool to remove internal stresses and to make the material less brittle. Also known as temper. [GEN] To recombine complementary strands of deoxyribonucleic acid that were separated by heating or other means of denaturation. { ə'nēl }

annealing furnace [ENG] A furnace for annealing metals or glass. Also known as annealing oven. { ə'nēl·iŋ ,fər·nəs }

annealing oven See annealing furnace. { ə'nēl·iŋ ,əv·ən }

annealing point [THERMO] The temperature at which the viscosity of a glass is $10^{13.0}$ poises. Also known as annealing temperature; 13.0 temperature. { ə'nēl·iŋ ,pòint }

annealing temperature See annealing point. { ə'nēl·iŋ ,tem·prə·chər }

annealing twin [MET] A twinned crystal that is formed as molten metal is cooled and solidified. { ə'nēl·iŋ ,twin }

Annedidae [VERT ZOO] A small family of limbless, snake-like, burrowing lizards of the suborder Sauria. { ə'ned·ə,dē }

Annelida [INV ZOO] A diverse phylum comprising the multisegmented wormlike animals. { ə'nel·ə·də }

annex point [MAP] A point used to assist in the relative orientation of vertical and oblique photographs; selected in the overlap area between the vertical and its corresponding oblique photograph, about midway between the pass points. { 'an,eks ,pòint }

annidation [ECOL] The phenomenon whereby a mutant is maintained in a population because it can flourish in an available ecological niche that the parent organisms cannot utilize. { ,an·ə'dā·shən }

Anniellidae [VERT ZOO] A family of limbless, snakelike lizards in the order Squamata. { ,an·ē'el·ə,dē }

annihilation [PARTIC PHYS] A process in which an antiparticle and a particle combine and release their rest energies in other particles. { ə,nī·ə'lā·shən }

annihilation operator [QUANT MECH] An operator which reduces the occupation number of a single state by unity; for example, an annihilation operator applied to a state of one particle yields the vacuum. Also known as destruction operator. { ə,nī·ə'lā·shən ¦äp·ə,rād·ər }

annihilation radiation [PARTIC PHYS] Electromagnetic radiation arising from the collision, and resulting annihilation, of an electron and a positron, or of any particle and its antiparticle. { ə,nī·ə'lā·shən ,rād·ē·ā·shən }

annihilator [MATH] For a set S, the class of all functions of specified type whose value is zero at each point of S. { ə'nī·ə,läd·ər }

anniversary clock [HOROL] A clock that can run as long as

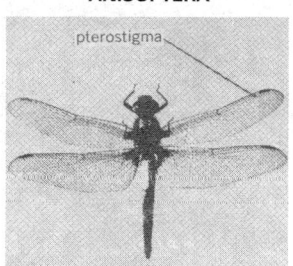

ANISOPTERA

pterostigma

An adult dragonfly, showing the thickened spot, pterostigma, on the costal margin of the wing.

ANKYLOSAURIA

Restoration of the armored Cretaceous dinosaur *Ankylosaurus* (about 20 feet or 6 meters long).

400 days on a single winding because of a slow torsion pendulum. Also known as four-hundred-day clock. { ‚an·ə'vərs·ə·rē ‚kläk }

Annonaceae [BOT] A large family of woody flowering plants in the order Magnoliales, characterized by hypogynous flowers, exstipulate leaves, a trimerous perianth, and distinct stamens with a short, thick filament. { ‚a·nə'näs·ē‚ē }

annotated photograph [MAP] A photograph on which planimetric, geologic, cultural, hydrographic, or vegetation information has been added to identify, classify, outline, clarify,or describe features that would not otherwise be apparent in examination of an unmarked photograph. { 'an·ə‚tād·əd 'fōd·ə‚graf }

annotation [COMPUT SCI] Any comment or note included in a program or flow chart in order to clarify some point at issue. { ‚an·ə'tā·shən }

annotation overprint [ORD] **1.** The outline delimiting a target or installation. **2.** A symbol which locates the position of a target together with an identifying reference number as depicted on a target graphic. { ‚an·ə'tā·shən 'ō·vər‚print }

annotation text [ORD] A descriptive text containing the identification, function, location, physical characteristics, and other information concerning a target or installation. { ‚an·ə'tā·shən ‚tekst }

annual aberration [ASTRON] Aberration caused by the velocity of the earth's revolution about the sun. { 'an·yə·wəl ab·ə'rā·shən }

annual cost comparison [IND ENG] A method of selecting from among several alternative projects or courses of action on the basis of their annual costs, including depreciation. { 'an·yə·wəl 'kȯst kəm‚par·ə·sən }

annual equation [ASTRON] A variation in the moon's apparent motion caused by variations in the distance of the earth from the sun during the course of the year. { 'an·yə·wəl i'kwā·zhən }

annual flood [HYD] The highest flow at a point on a stream during any particular calendar year or water year. { 'an·yə·wəl 'fləd }

annual growth ring *See* annual ring. { 'an·yə·wəl 'grōth ‚riŋ }

annual inequality [OCEANOGR] Seasonal variation in water level or tidal current speed, more or less periodic, due chiefly to meteorological causes. { 'an·yə·wəl ‚in·i'kwäl·əd·ē }

annual labor *See* assessment drilling. { 'an·yə·wəl 'lā·bər }

annual layer [GEOL] **1.** A sedimentary layer deposited, or presumed to have been deposited, during the course of a year; for example, a glacial varve. **2.** A dark layer in a stratified salt deposit containing disseminated anhydrite. { 'an·yə·wəl 'lā·ər }

annual magnetic change *See* magnetic annual change. { 'an·yə·wəl ‚mag'ned·ik 'chānj }

annual magnetic variation *See* magnetic annual variation. { 'an·yə·wəl ‚mag'ned·ik ver·ē'ā·shən }

annual parallax [ASTRON] The apparent displacement of a celestial body viewed from two separated observation points whose base line is the radius of the earth's orbit. Also known as heliocentric parallax. { 'an·yə·wəl 'par·ə‚laks }

annual plant [BOT] A plant that completes its growth in one growing season and therefore must be planted annually. { 'an·yə·wəl 'plant }

annual ring [BOT] A line appearing on tree cross sections marking the end of a growing season and showing the volume of wood added during the year. Also known as annual growth ring. { 'an·yə·wəl 'riŋ }

annual service availability index [ELEC] The ratio of customer-hours of service supplied by an electrical utility during one year to the customer-hours requested, expressed as a percentage. { ¦an·yə·wəl ¦sər·vəs ə‚vāl·ə'bil·əd·ē ‚in‚deks }

annual storage [HYD] The capacity of a reservoir that can handle a watershed's annual runoff but cannot carry over any portion of the water for longer than the year. { 'an·yə·wəl 'stȯr·ij }

annual variation [ASTRON] The change in the right ascension and declination of a star during one year, due to the combined effect of the star's proper motion and the precession of the equinoxes. [GEOPHYS] A component in the change with time in the earth's magnetic field at a specified location that has a period of 1 year. { 'an·yə·wəl ver·ē'ā·shən }

annuation [ECOL] The annual variation in the presence, absence, or abundance of a member of a plant community. { ‚an·yə'wā·shən }

annular [MIN ENG] The space between the casing and wall of a hole or between a drill pipe and casing. { 'an·yə·lər }

annular atoms [ORG CHEM] The atoms in a cyclic compound that are members of the ring. { 'an·yə·lər 'ad·əmz }

annular auger [DES ENG] A ring-shaped boring tool which cuts an annular channel, leaving the core intact. { 'an·yə·lər 'ȯg·ər }

annular blowout preventer [PETRO ENG] A sealing device used in a drilling operation that consists of a large valve, which provides a seal in the annular space between the pipe and the wellbore, or on the wellbore itself if no pipe is present. { 'an·yə·lər 'blō‚aut pri‚ven·tər }

annular budding [BOT] Budding by replacement of a ring of bark on a stock with a ring bearing a bud from a selected species or variety. { 'an·yə·lər 'bəd·iŋ }

annular conductor [ELEC] A number of wires stranded in three reversed concentric layers around a saturated hemp core. { 'an·yə·lər kən'dək·tər }

annular drainage pattern [HYD] A ringlike pattern subsequent in origin and associated with maturely dissected dome or basin structures. { 'an·yə·lər 'drān·ij ‚pad·ərn }

annular eclipse [ASTRON] An eclipse in which a thin ring of the source of light appears around the obscuring body. { 'an·yə·lər i'klips }

annular effect [FL MECH] A phenomenon observed in the flow of fluid in a tube when its motion is alternating rapidly, as in the propagation of sound waves, in which the mean velocity rises progressing from the center of the tube toward the walls and then falls within a thin laminar boundary layer to zero at the wall itself. { 'an·yə·lər i'fekt }

annular gear [DES ENG] A gear having a cylindrical form. { 'an·yə·lər 'gir }

annular hernia *See* umbilical hernia. { 'an·yə·lər 'hərn·ē·ə }

annular nozzle [DES ENG] A nozzle with a ring-shaped orifice. { 'an·yə·lər 'näz·əl }

annular section [ENG] The open space between two concentric tubes, pipes, or vessels. { 'an·yə·lər 'sek·shən }

annular solid [MATH] A solid generated by rotating a closed plane curve about a line which lies in the plane of the curve and does not intersect the curve. { 'an·yə·lər 'säl·əd }

annular transistor [ELECTR] Mesa transistor in which the semiconductor regions are arranged in concentric circles about the emitter. { 'an·yə·lər tran'zis·tər }

annular vault [ARCH] A vault arising from circular walls. { 'an·yə·lər 'vȯlt }

annular velocity [PETRO ENG] The rate of movement of the mud in the annular space of a drilling well. { 'an·yə·lər və'läs·əd·ē }

annular vessel [BOT] A xylem tube or duct with internal lignified rings. { 'an·yə·lər 'ves·əl }

annulated shaft [ARCH] A column consisting of a cluster of shafts which seem to be held together at intervals by an annular band. { 'an·yə‚lād·əd 'shaft }

annulene [ORG CHEM] One of a group of monocyclic conjugated hydrocarbons which have the general formula $[-CH=CH-]_n$. { 'an·yə‚lēn }

annulus [ANAT] Any ringlike anatomical part. [BOT] **1.** An elastic ring of cells between the operculum and the mouth of the capsule in mosses. **2.** A line of cells, partly or entirely surrounding the sporangium in ferns, which constricts, thus causing rupture of the sporangium to release spores. **3.** A whorl resembling a calyx at the base of the strobilus in certain horsetails. [MATH] The ringlike figure that lies between two concentric circles. [MYCOL] A ring of tissue representing the remnant of the veil around the stipe of some agarics. { 'an·yə·ləs }

annulus conjecture [MATH] For dimension n, the assertion that if f and g are locally flat embeddings of the $(n-1)$ sphere, S^{n-1}, in real n space, R^n, with $f(S^{n-1})$ in the bounded component of $R^n - g(S^{n-1})$, then the closed region in R^n bounded by $f(S^{n-1})$ and $g(S^{n-1})$ is homeomorphic to the direct product of S^{n-1} and the closed interval [0,1]; it is established for $n \neq 4$. { 'an·yə·ləs kən'jek·chər }

annunciator [ENG] A signaling apparatus which operates electromagnetically and serves to indicate visually, or visually and audibly, whether a current is flowing, has flowed, or has changed direction of flow in one or more circuits. { ə'nən·sē·ād·ər }

Anobiidae [INV ZOO] The deathwatch beetles, a family of coleopteran insects of the superfamily Bostrichoidea. { ,an·ə'bī·ə,dē }

anode [ELEC] The terminal at which current enters a primary cell or storage battery; it is positive with respect to the device, and negative with respect to the external circuit. [ELECTR] **1.** The collector of electrons in an electron tube. Also known as plate; positive electrode. **2.** In a semiconductor diode, the terminal toward which forward current flows from the external circuit. [PHYS CHEM] The positive terminal of an electrolytic cell. { 'a,nōd }

anode balancing coil [ELEC] A set of mutually coupled windings used to maintain approximately equal currents in anodes operating in parallel from the same transformer terminal. { 'a,nōd ¦bal·əns·iŋ ,kȯil }

anode characteristic [ELECTR] Relationship of anode current to anode voltage in a vacuum tube. { 'a,nōd ,kar·ik·tə'ris·tik }

anode circuit [ELECTR] Complete external electrical circuit connected between the anode and the cathode of an electron tube. Also known as plate circuit. { 'a,nōd ,sər·kət }

anode-circuit detector [ELECTR] Detector functioning by virtue of a nonlinearity in its anode-circuit characteristic. Also known as plate-circuit detector. { 'a,nod ,sər·kət di,tek·tər }

anode copper [MET] Slabs of refined blister copper used as anodes in the electrolytic refining of copper. { 'a,nōd ,käp·ər }

anode corrosion [MET] The disintegration of a metal acting as an anode. { 'a,nōd kə'rō·zhən }

anode-corrosion efficiency [PHYS CHEM] The ratio of actual weight loss of an anode due to corrosion to the theoretical loss as calculated by Faraday's law. { 'a,nōd kə¦rō·zhən i,fish·ən·sē }

anode current [ELECTR] The electron current flowing through an electron tube from the cathode to the anode. Also known as plate current. { 'a,nōd ,kər·ənt }

anode dark space [ELECTR] A thin, dark region next to the anode glow in a glow-discharge tube. { 'a,nōd 'därk ,spās }

anode detector [ELECTR] A detector in which rectification of radio-frequency signals takes place in the anode circuit of an electron tube. Also known as plate detector. { 'a,nōd di,tek·tər }

anode dissipation [ELECTR] Power dissipated as heat in the anode of an electron tube because of bombardment by electrons and ions. { 'a,nōd dis·ə'pā·shən }

anode drop See anode fall. { 'a,nōd ,dräp }

anode effect [PHYS CHEM] A condition produced by polarization of the anode in the electrolysis of fused salts and characterized by a sudden increase in voltage and a corresponding decrease in amperage. { 'a,nōd i,fekt }

anode efficiency [ELECTR] The ratio of the ac load circuit power to the dc anode power input for an electron tube. Also known as plate efficiency. { 'a,nōd i,fish·ən·sē }

anode fall [ELECTR] **1.** A very thin space-charge region in front of an anode surface, characterized by a steep potential gradient through the region. **2.** The voltage across this region. Also known as anode drop. { 'a,nōd ,fȯl }

anode film [CHEM] The portion of solution in immediate contact with the anode. { 'a,nōd ,film }

anode furnace [MET] A furnace in which blister copper or impure nickel is refined. { 'a,nōd ,fər·nəs }

anode glow [ELECTR] A thin, luminous layer on the surface of the anode in a glow-discharge tube. { 'a,nōd ,glō }

anode impedance [ELECTR] Total impedance between anode and cathode exclusive of the electron stream. Also known as plate impedance; plate-load impedance. { 'a,nōd im,pēd·əns }

anode input power [ELECTR] Direct-current power delivered to the plate (anode) of a vacuum tube by the source of supply. Also known as plate input power. { 'a,nōd 'in,pu̇t ,pau̇·ər }

anode metal [MET] The metal used as anode in an electroplating process. { 'a,nōd ,med·əl }

anode modulation [ELECTR] Modulation produced by introducing the modulating signal into the anode circuit of any tube in which the carrier is present. Also known as plate modulation. { 'a,nōd ,mäj·ə'lā·shən }

anode mud [MET] An insoluble substance or mixture that collects at the anode in an electrolytic refining or plating process. Also known as anode slime. { 'a,nōd ,məd }

anode neutralization [ELECTR] Method of neutralizing an amplifier in which the necessary 180° phase shift is obtained by an inverting network in the plate circuit. Also known as plate neutralization. { 'a,nōd ,nü·trə·lə'zā·shən }

anode pulse modulation [ELECTR] Modulation produced in an amplifier or oscillator by application of externally generated pulses to the plate circuit. Also known as plate-pulse modulation. { 'a,nōd 'pəls ,mäj·ə'lā·shən }

anode rays [ELECTR] Positive ions coming from the anode of an electron tube; generally due to impurities in the metal of the anode. { 'a,nōd ,rāz }

anode resistance [ELECTR] The resistance value obtained when a small change in the anode voltage of an electron tube is divided by the resulting small change in anode current. Also known as plate resistance. { 'a,nōd ri,zis·təns }

anode saturation [ELECTR] The condition in which the anode current of an electron tube cannot be further increased by increasing the anode voltage; the electrons are then being drawn to the anode at the same rate as they are emitted from the cathode. Also known as current saturation; plate saturation; saturation; voltage saturation. { 'a,nōd ,sach·ə'rā·shən }

anode scrap [MET] Portions of anode copper retrieved from electrolytic refining of the metal. { 'a,nōd ,skrap }

anode sheath [ELECTR] The electron boundary which exists in a gas-discharge tube between the plasma and the anode when the current demanded by the anode circuit exceeds the random electron current at the anode surface. { 'a,nōd ,shēth }

anode slime See anode mud. { 'a,nōd ,slīm }

anodic [PHYS] Pertaining to the anode. { ə'näd·ik }

anodic cleaning [MET] The removal of a foreign substance from a metallic surface by electrolysis with the metal as the anode. Also known as anodic pickling; reverse-current cleaning. { ə'näd·ik 'klēn·iŋ }

anodic coating [MET] A film of oxide produced on a metal by electrolysis with the metal as the anode. { ə'näd·ik 'kōd·iŋ }

anodic pickling See anodic cleaning. { ə'näd·ik 'pik·liŋ }

anodic polarization [PHYS CHEM] The change in potential of an anode caused by current flow. { ə'näd·ik pō·lə·rə'zā·shən }

anodic protection [MET] Reduction of the corrosion rate in an anode by polarizing it into a potential region where the dissolution rates low. { ə'näd·ik prə'tek·shən }

anodic reaction [MET] The reaction in the mechanism of electrochemical corrosion in which the metal forming the anode dissolves in the electrolyte in the form of positively charged ions. { ə'näd·ik rē'ak·shən }

anodize [MET] To form a decorative or protective passive film on a metal part by making it the anode of a cell and applying electric current. { 'an·ə,dīz }

anodized aluminum [MET] Aluminum coated with a layer of aluminum oxide by an anodic process in a suitable electrolyte such as chromic acid or sulfuric acid solution. { 'an·ə,dīzd ə'lüm·ə·nəm }

anodized dielectric film [ELEC] An insulating film produced on a conducting surface by anodizing; used for producing thin-film capacitors, trimming resistor values, and passivation in the manufacture of integrated circuits. { 'an·ə,dīzd dī·ə¦lek·trik 'film }

anodized magnesium [MET] An anodic coating on magnesium produced in one of various electrolytes, mainly of fluorides, phosphates, or chromates. { 'an·ə,dīzd ,mag'nēz·ē·əm }

anodontia See hypodontia. { ,an·ə'dän·chə }

anole [VERT ZOO] Any arboreal lizard of the genus Anolis, characterized by flattened adhesive digits and a prehensile outer toe. { ə'nō·lē }

anolyte [CHEM] The part of the electrolyte at or near the anode that is changed in composition by the reactions at the anode. { 'an·ə,līt }

ANOMALINACEA

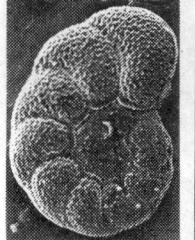 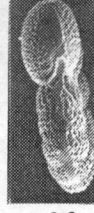

(a)   (b)

Scanning electron micrographs of the foraminiferan *Holmanella,* from the Miocene of California. *(a)* Spiral view and *(b)* edge view of bievolute planispiral test, with a coarsely perforate granular margin, and a slitlike aperture extending up the terminal face. *(R. B. MacAdam, Chevron Oil Field Research Co.)*

Anomalinacea [INV ZOO] A superfamily of marine and benthic sarcodinian protozoans in the order Foraminiferida. { ə͵näm·ə·lə′näs·ē·ə }

anomalistic month [ASTRON] The average period of revolution of the moon from perigee to perigee, a period of 27 days 13 hours 18 minutes 33.2 seconds. { ə¦näm·ə¦lis·tik ′mənth }

anomalistic period [ASTRON] The interval between two successive perigee passages of a satellite in orbit about a primary. Also known as perigee-to-perigee period. { ə¦näm·ə¦lis·tik ′pir·ē·əd }

anomalistic year [ASTRON] The period of one revolution of the earth about the sun from perihelion to perihelion; 365 days 6 hours 13 minutes 53.0 seconds in 1900 and increasing at the rate of 0.26 second per century. { ə¦näm·ə¦lis·tik ′yēr }

anomalon [NUC PHYS] A nuclear fragment, produced in the collision of a projectile nucleus at relativistic energy with a target nucleus at rest, that has an anomalously short mean free path, comparable to that of a uranium nucleus. { ə′näm·ə͵län }

anomaloscope [OPTICS] An optical instrument for testing color vision, in which a yellow light whose intensity may be varied is matched against red and green lights whose intensity is fixed. { ə′näm·ə·lə͵skōp }

anomalous [SCI TECH] Deviating from the normal; irregular. { ə′näm·ə·ləs }

anomalous Barkhausen effect [ELECTROMAG] The occurrence of large steps in the magnetization of an iron-aluminum alloy at temperature above about 400°C (750°F). { ə¦näm·ə·ləs ′bark͵haủz·ən i͵fekt }

anomalous dispersion [OPTICS] Extraordinary behavior in the curve of refractive index versus wavelength which occurs in the vicinity of absorption lines or bands in the absorption spectrum of a medium. { ə′näm·ə·ləs dis′pər·zhən }

anomalous expansion [THERMO] An increase in the volume of a substance that results from a decrease in its temperature, such as is displayed by water at temperatures between 0 and 4°C (32 and 39°F). { ə′näm·ə·ləs ik′span·shən }

anomalous Funkel effect [ELECTR] Current fluctuations in an electron tube resulting from positive ions entering the space-charge region in front of the cathode. { ə¦näm·ə·ləs ′fəŋ·kəl i͵fekt }

anomalous Hall effect [ELECTROMAG] **1.** In a current-carrying conductor in a magnetic field, development of a transverse voltage resulting from the deflection of positive charge carriers (hole states) by the Lorentz force. **2.** The Hall effect in ferromagnetic metals, which arises from the unsymmetrical scattering of conduction electrons at magnetic moments. { ə¦näm·ə·ləs ′hȯl i͵fekt }

anomalous magma [GEOL] Magma formed or obviously changed by assimilation. { ə′näm·ə·ləs ′mag·mə }

anomalous magnetic moment [PARTIC PHYS] The difference between the observed magnetic moment and the value predicted by Dirac's theory. { ə′näm·ə·ləs mag′ned·ik ′mō·mənt }

anomalous series [ATOM PHYS] A series of spectral lines associated with atomic energy levels whose Rydberg corrections do not vary smoothly with total quantum number, generally because they involve excitation of two electrons. Also known as abnormal series. { ə′näm·ə·ləs ′sir·ēz }

anomalous skin effect [ELEC] The skin effect at very low temperatures and high frequencies at which the thickness of the conducting skin layer is less than the electron mean free path, so that the classical theory of electrical conductivity breaks down. { ə¦näm·ə·ləs ′skin i͵fekt }

anomalous trichromatism [PHYSIO] A mild defect in red-green color vision in which the subject, when asked to mix red and green light to match a certain shade of yellow, produces a different shade than does someone with normal color vision. { ə′näm·ə·ləs trī′krōm·ə͵tiz·əm }

anomalous viscosity *See* non-Newtonian viscosity. { ə′näm·ə·ləs vis′käs·əd·ē }

anomalous Zeeman effect [SPECT] A type of splitting of spectral lines of a light source in a magnetic field which occurs for any line arising from a combination of terms of multiplicity greater than one; due to a nonclassical magnetic behavior of the electron spin. { ə′näm·ə·ləs ′zā͵män i͵fekt }

Anomaluridae [VERT ZOO] The African flying squirrels, a small family in the order Rodentia characterized by the climbing organ, a series of scales at the root of the tail. { ə′näm·ə͵lủr·ə͵dē }

anomaly [ASTRON] In celestial mechanics, the angle between the radius vector to an orbiting body from its primary (the focus of the orbital ellipse) and the line of apsides of the orbit, measured in the direction of travel, from the point of closest approach to the primary (perifocus). Also known as true anomaly. [BIOL] An abnormal deviation from the characteristic form of a group. [GEOL] A local deviation from the general geological properties of a region. [MED] Any part of the body that is abnormal in its position, form, or structure. [METEOROL] The deviation of the value of an element (especially temperature) from its mean value over some specified interval. [OCEANOGR] The difference between conditions actually observed at a serial station and those that would have existed had the water all been of a given arbitrary temperature and salinity. [SCI TECH] A deviation beyond normal variations. { ə′näm·ə·lē }

anomaly detection [COMPUT SCI] The technology that seeks to identify an attack on a computer system by looking for behavior that is out of the norm. { ə′näm·ə·lē di͵tek·shən }

anomaly finder [ENG] A computer-controlled data-plotting system used on ships to measure and record seismic, gravity, magnetic, and other geophysical data and water depth, time, course, and speed. { ə′näm·ə·lē ͵fīn·dər }

anomaly of geopotential difference *See* dynamic-height anomaly. { ə′näm·ə·lē əv ͵jē·ō·pə͵ten·shəl ′dif·rəns }

anomer [ORG CHEM] One of a pair of isomers of cyclic carbohydrates; resulting from creation of a new point of symmetry when a rearrangement of the atoms occurs at the aldehyde or ketone position. { ′an·ə·mər }

anomeric carbon [BIOCHEM] The carbon about which anomers rotate. { ͵an·ə͵mir·ik ′kär·bən }

anomic aphasia [PSYCH] A subtype of fluent aphasia in which principal nouns and verbs cannot be recalled by the individual; typically caused by injury at the temporo-parietal junction. { ə′näm·ik ə′fāzh·ə }

anomie [PSYCH] Apathy, alienation, and personal distress resulting from a lack of purpose or ideals. { ′an·ə·mē }

anomite [MINERAL] A variety of biotite different only in optical orientation. { ′an·ə͵mīt }

Anomocoela [VERT ZOO] A suborder of toadlike amphibians in the order Anura characterized by a lack of free ribs. { ¦an·ə·mō¦sē·lə }

anomocoelous [ANAT] Describing a vertebra with a centrum that is concave anteriorly and flat or convex posteriorly. { ¦an·ə·mō¦sē·ləs }

Anomphalacea [PALEON] A superfamily of extinct gastropod mollusks in the order Aspidobranchia. { ə͵näm·fə′läsh·ə }

Anomura [VERT ZOO] A section of the crustacean order Decapoda that includes lobsterlike and crablike forms. { ͵an·ə′mủr·ə }

anonymous dimensionless group 1-4 [CHEM ENG] Four of the dimensionless groups, used to solve problems in transfer processes, gas absorption in wetted-wall columns, and laminar boundary-layer flow. { ə′nän·ə·məs di¦men·shən·ləs ′grüp ¦wən tə ¦fȯr }

anonymous FTP [COMPUT SCI] A public FTP (file transfer protocol) site at which users can log in and download documents by entering "anonymous" as their user ID, and their e-mail address as password. { ə͵nän·ə·məs ¦ef¦tē′pē }

anoopsia [MED] Strabismus in which the eye is turned upward. { ͵an·ō′äp·sē·ə }

Anopheles [INV ZOO] A genus of mosquitoes in the family Culicidae; members are vectors of malaria, dengue, and filariasis. { ə′näf·ə͵lēz }

anopheline [INV ZOO] Pertaining to mosquitoes of the genus *Anopheles* or a closely related genus. { ə′näf·ə·lən }

Anopla [INV ZOO] A class or subclass of the phylum Rhynchocoela characterized by a simple tubular proboscis and by having the mouth opening posterior to the brain. { ′an·ə·plə }

Anoplocephalidae [INV ZOO] A family of tapeworms in the order Cyclophyllidea. { ¦an·ə͵plä·sə′fal·ə͵dē }

Anoplura [INV ZOO] The sucking lice, a small group of mammalian ectoparasites usually considered to constitute an order in the class Insecta. { ͵an·ə′plủr·ə }

anorexia [MED] Loss of appetite. { ͵an·ə′rek·sē·ə }

anorexia nervosa [PSYCH] A disorder in which dramatic

reduction in caloric intake consequent to excessive dieting leads to significant physiological, emotional, psychological, and behavioral disturbances.   { ‚an·ə'rek·sē·ə nər'vō·sə }

anorgasmy [MED] Inability, usually psychic, to reach a climax during coitus.   { ‚an·ȯr'gaz·mē }

anorogenic [GEOL] Of a feature, forming during tectonic quiescence between orogenic periods, that is, lacking in tectonic disturbance.   { ¦a‚nȯ·rō¦jen·ik }

anorogenic time [GEOL] Geologic time when no significant deformation of the crust occurred.   { ¦a‚nȯ·rō¦jen·ik 'tīm }

anorthic crystal *See* triclinic crystal.   { ə'nȯr·thik 'kris·təl }

anorthite [MINERAL] The white, grayish, or reddish calcium-rich end member of the plagioclase feldspar series; composition ranges from $Ab_{10}An_{90}$ to Ab_0An_{100}, where $Ab = NaAlSi_3O_8$ and $An = CaAl_2Si_2O_8$. Also known as calciclase; calcium feldspar.   { ə'nȯr‚thīt }

anorthite-basalt [PETR] A rock composed of a basic variety of basalt with anorthite instead of labradorite.   { ə'nȯr‚thīt bə'sȯlt }

anorthoclase [MINERAL] A triclinic alkali feldspar having a chemical composition ranging from $Or_{40}Ab_{60}$ to $Or_{10}Ab_{90}$ to about 20 mole % An, where $Or = KAlSi_3O_8$, $Ab = NaAlSi_3O_8$, and $An = CaAl_2Si_2O_8$. Also known as anorthose; soda microcline.   { ə'nȯr·thə‚klās }

anorthopia [MED] A defect of vision in which straight lines do not seem straight, and parallelism or symmetry is not properly perceived.   { ‚a·nȯr'thōp·ē·ə }

anorthose *See* anorthoclase.   { ‚ə'nȯr‚thōs }

anorthosite [PETR] A visibly crystalline plutonic rock composed almost entirely of plagioclase feldspar (andesine to anorthite) with minor amounts of pyroxene and olivine.   { ə'nȯr·thə‚sīt }

anorthositization [GEOL] A process of anorthosite formation by replacement or metasomatism.   { ə¦nȯr·thə‚sid·ə'zā·shən }

anoscope [MED] An instrument for examining the lower rectum and anal canal.   { 'ā·nə‚skōp }

anosmia [MED] Absence of the sense of smell.   { ə'näz·mē·ə }

Anostraca [INV ZOO] An order of shrimplike crustaceans generally referred to the subclass Branchiopoda.   { ə'näs·trə·kə }

anotron [ELECTR] A cold-cathode glow-discharge diode having a copper anode and a large cathode of sodium or other material.   { 'an·ə‚trän }

anovulation [MED] The absence of ovulation.   { ‚an‚äv·yə'lā·shən }

anoxemia [MED] Condition of having an insufficient supply of oxygen in the bloodstream.   { ‚a‚näk'sem·ē·ə }

anoxia [MED] The failure of oxygen to gain access to, or to be utilized by, the body tissues.   { ə'näk·sē·ə }

anoxic zone [OCEANOGR] An oxygen-depleted region in a marine environment.   { ə'näk·sik ‚zōn }

anoxybiosis [PHYSIO] A type of cryptobiosis induced by lack of oxygen.   { ə‚näk·sə‚bī'ō·səs }

A-N radio range [NAV] A type of radio beacon station whose signals provide definite track guidance for aircraft by establishing four radial lines of position which can be identified by a continuous-tone signal made up of keyed pulses of equal amplitude representing the Morse code letters A and N.   { 'ā ‚en 'rād·ē‚ō 'rānj }

ansae [ASTRON] **1.** The ends of the rings of Saturn, as seen from the earth. **2.** Opposing extension or knots of a celestial object, such as a planetary nebula or lenticular galaxy.   { 'an·sē }

Ansbacher unit [BIOL] A unit for the standardization of vitamin K.   { 'änz‚bäk·ər ‚yü·nət }

Anser [VERT ZOO] A genus of birds in the family Anatidae comprising the typical geese.   { 'an·sər }

Anseranatini [VERT ZOO] A subfamily of aquatic birds in the family Anatidae represented by a single species, the magpie goose.   { ‚an·sər·ə'nat·ə‚nī }

Anseriformes [VERT ZOO] An order of birds, including ducks, geese, swans, and screamers, characterized by a broad, flat bill and webbed feet.   { ‚an·sər·ə'fȯr‚mēz }

Anson unit [BIOL] A unit for the standardization of trypsin and proteinases.   { 'an·sən ‚yü·nət }

answer back [COMPUT SCI] The ability of a device such as a computer or terminal to automatically identify itself when it is contacted by another communicating device.   { 'an·sər ‚bak }

answering cord [ELEC] Cord nearest the face of the switchboard which is used for answering subscribers' calls and incoming trunks.   { 'an·sər·iŋ ‚kȯrd }

answering jack [ELEC] Jack on which a station calls in and is answered by an operator.   { 'an·sər·iŋ ‚jak }

answer lamp [ELEC] Telephone switchboard lamp that lights when an answer cord is plugged into a line jack; the lamp goes out when the call is completed.   { 'an·sər ‚lamp }

answer-only modem [COMMUN] A modem that can answer but not initiate a call.   { ¦an·sər ‚ōn·lē 'mō‚dem }

ant [INV ZOO] The common name for insects in the hymenopteran family Formicidae; all are social, and colonies exhibit a highly complex organization.   { ant }

Ant *See* Antlia.

anta [ARCH] A pier or pilaster formed by a thickening at the end of a wall.   { 'an·tə }

antacid [CHEM] Any substance that counteracts or neutralizes acidity.   { 'ant'as·əd }

antagonism [BIOL] **1.** Mutual opposition as seen between organisms, muscles, physiologic actions, and drugs. **2.** Opposing action between drugs and disease or drugs and functions.   { an'tag·ə‚niz·əm }

antagonist [BIOCHEM] A molecule that bears sufficient structural similarity to a second molecule to compete with that molecule for binding sites on a third molecule. [PHARM] A drug or other chemical substance capable of reducing the physiological activity of another chemical substance; refers especially to a drug that opposes the action of a drug or other chemical substance on the nervous system by combining with and blocking the nerve receptor. [PHYSIO] A muscle that contracts with, and limits the action of, another muscle, called an agonist, with which it is paired.   { an'tag·ə‚nist }

antarafacial [ORG CHEM] The stereochemistry when, simultaneously, two sigma bonds are formed or broken on the opposite faces of the component pi systems, such as in a cycloaddition reaction.   { ‚an·tə·rə'fā·shəl }

Antarctica [GEOGR] A continent roughly centered on the South Pole and surrounded by an ocean consisting of the southern parts of the Atlantic, Pacific, and Indian oceans.   { ‚ant'ärd·ik·ə }

antarctic air [METEOR] A type of air whose characteristics are developed in an antarctic region.   { ‚ant'ärd·ik 'er }

antarctic anticyclone [METEOR] The glacial anticyclone which has been said to overlie the continent of Antarctica; analogous to the Greenland anticyclone.   { ‚ant'ärd·ik ‚ant·i'sī‚klōn }

Antarctic Circle [GEOD] The parallel of latitude approximately 66°32' south of the Equator.   { ‚ant'ärd·ik 'sər·kəl }

Antarctic Circumpolar Current [OCEANOGR] The ocean current flowing from west to east through all the oceans around the Antarctic Continent. Also known as West Wind Drift.   { ‚ant'ärd·ik ‚sər·kəm¦pōl·ər 'kər·ənt }

Antarctic Convergence [OCEANOGR] The oceanic polar front indicating the boundary between the subantarctic and subtropical waters. Also known as Southern Polar Front.   { ‚ant'ärd·ik kən'vər·jəns }

antarctic faunal region [ECOL] A zoogeographic region describing both the marine littoral and terrestrial animal communities on and around Antarctica.   { ‚ant'ärd·ik 'fȯn·əl ‚rē·jən }

antarctic front [METEOR] The semipermanent, semicontinuous front between the antarctic air of the Antarctic continent and the polar air of the southern oceans; generally comparable to the arctic front of the Northern Hemisphere.   { ‚ant'ärd·ik 'frənt }

Antarctic Intermediate Water [OCEANOGR] A water mass in the Southern Hemisphere, formed at the surface near the Antarctic Convergence between 45° and 55°S; it can be traced in the North Atlantic to about 25°N.   { ‚ant'ärd·ik in·tər'mēd·ē·ət 'wȯd·ər }

Antarctic Ocean [GEOGR] A circumpolar ocean belt including those portions of the Atlantic, Pacific, and Indian oceans which reach the Antarctic continent and are bounded on the north by the Subtropical Convergence; not recognized as a separate ocean.   { ‚ant'ärd·ik ‚ō·shən }

Antarctic ozone hole [METEOR] In the spring, the depletion of stratospheric ozone over the Antarctic region, typically

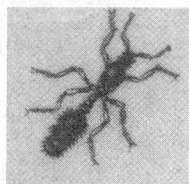

ANT

The female black ant
(*Monororium minimum*),
about $1/16$ inch (1.6 millimeters).

south of 55° latitude; the formation of the hole is explained by the activation of chlorine and the catalytic destruction of O_3; it occurs during September, when the polar regions are sunlit but the air is still cold and isolated from midlatitude air by a strong polar vortex.   Also known as ozone hole.   { ant'ärt·ik 'ō,zōn ,hōl }

Antarctic vortex *See* polar vortex.   { ant'ärt·ik 'vȯr,teks }

Antarctic Zone [GEOGR] The region between the Antarctic Circle (66°32'S) and the South Pole.   { ,ant'ärd·ik ,zōn }

Antares [ASTRON] A red supergiant variable binary star of stellar magnitude 0.9, 520 light-years from the sun, spectral classification M1-Ib, in the constellation Scorpius; the star α Scorpii.   { an'tar·ēz }

antazoline [PHARM] $C_{17}H_{19}N_3$ A bitter, white, crystalline powder with a melting point of 237-241°C; used as an antihistamine.   { an'taz·ə,lēn }

anteater [VERT ZOO] Any of several mammals, in five orders, which live on a diet of ants and termites.   { 'ant,ēd·ər }

antebrachium *See* forearm.   { ,an·tə'brāk·ē·əm }

antecedent [MATH] **1.** The numerator of a ratio. **2.** The first of the two statements in an implication. **3.** For an integer, *n*, that is greater than 1, the preceding integer, $n - 1$.   { 'an·tə,sēd·ənt }

antecedent platform [GEOL] A submarine platform 165 feet (50 meters) or more below sea level from which barrier reefs and atolls are postulated to grow toward the water's surface.   { ,ant·ə'sēd·ənt 'plat,fȯrm }

antecedent precipitation index [METEOROL] A weighted summation of daily precipitation amounts; used as an index of soil moisture.   { ,ant·ə'sēd·ənt pri,sip·ə'tā·shən 'in,deks }

antecedent stream [HYD] A stream that has retained its early course in spite of geologic changes since its course was assumed.   { ,ant·ə'sēd·ənt 'strēm }

antecedent valley [GEOL] A stream valley that existed before uplift, faulting, or folding occurred and which has maintained itself during and after these events.   { ,ant·ə'sēd·ənt 'val·ē }

anteconsequent stream [HYD] A stream consequent to the form assumed by the earth's surface as the result of early movement of the earth but antecedent to later movement.   { ,an·tē'kän·sə·kwənt 'strēm }

antecosta [INV ZOO] The internal, anterior ridge of the tergum or sternum of many insects that provides a surface for attachment of the longitudinal muscles.   { 'an·tə,käs·tə }

antediluvial [GEOL] Formerly referred to time or deposits antedating Noah's flood.   { ,an·tē·də'lüv·ē·əl }

antefix [ARCH] **1.** A decorated upright slab used in classical architecture and derivatives to close or conceal the open end of a row of tiles which cover the joints of roof tiles. **2.** A similar ornament on the ridge of a roof.   { 'an·tə,fiks }

anteflexion [MED] Forward bending of an organ on itself.   { ,an·tē'flek·shən }

antelope [VERT ZOO] Any of the hollow-horned, hoofed ruminants assigned to the artiodactyl subfamily Antilopinae; confined to Africa and Asia.   { 'an·təl,ōp }

antemedium [MATER] A substance, such as wax or resin solvent, used in tissue processing prior to infiltration for histological examination.   { ,an·tē'mēd·ē·əm }

ante meridian [ASTRON] **1.** A section of the celestial meridian; it lies below the horizon, and the nadir is included. **2.** Before noon, or the period of time between midnight (0000) and noon (1200).   { ¦an·tē mə¦rid·ē·ən }

antemortem [MED] Before death.   { ¦an·tē¦mȯr·dəm }

antenatal [MED] Occurring or existing before birth.   { ¦an·te¦nād·əl }

antenna [ELECTROMAG] A device used for radiating or receiving radio waves.   Also known as aerial; radio antenna. [INV ZOO] Any one of the paired, segmented, and movable sensory appendages occurring on the heads of many arthropods.   { an'ten·ə }

antenna amplifier [ELECTROMAG] One or more stages of wide-band electronic amplification placed within or physically close to a receiving antenna to improve signal-to-noise ratio and mutually isolate various devices receiving their feed from the antenna.   { an'ten·ə 'am·plə,fī·ər }

antenna bearing [NAV] The generated bearing of the antenna of a radar set, as delivered to the indicator.   { an'ten·ə ,ber·iŋ }

antenna chlorophyll [BIOCHEM] Chlorophyll molecules which collect light quanta.   { an'ten·ə 'klȯr·ə,fil }

antenna circuit [ELECTR] A complete electric circuit which includes an antenna.   { an'ten·ə ,sər·kət }

antenna coil [ELECTROMAG] Coil through which antenna current flows.   { an'ten·ə ,kȯil }

antenna coincidence [ELECTROMAG] That instance when two rotating, highly directional antennas are pointed toward each other.   { an'ten·ə kō'in·səd·əns }

antenna complex [CELL MOL] A complex of protein molecules found in the thylakoid membrane of chloroplasts that captures and transfers light energy to the photochemical reaction center.   { an'ten·ə ,käm,pleks }

antenna counterpoise *See* counterpoise.   { an'ten·ə 'kaunt·ər,pȯiz }

antenna coupler [ELECTROMAG] A radio-frequency transformer, tuned line, or other device used to transfer energy efficiently from a transmitter to a transmission line or from a transmission line to a receiver.   { an'ten·ə ,kəp·lər }

antenna crosstalk [ELECTROMAG] The ratio or the logarithm of the ratio of the undesired power received by one antenna from another to the power transmitted by the other.   { an'ten·ə 'krȯs,tȯk }

antenna detector [ELECTROMAG] Device consisting of an antenna and electronic equipment to warn aircraft crew members that they are being observed by radar sets.   { an'ten·ə di'tek·tər }

antenna directivity diagram [ELECTROMAG] Curve representing, in polar or cartesian coordinates, a quantity proportional to the gain of an antenna in the various directions in a particular plane or cone.   { an'ten·ə di·rek'tiv·əd·ē 'dī·ə,gram }

antenna effect [ELECTROMAG] A distortion of the directional properties of a loop antenna caused by an input to the direction-finding receiver which is generated between the loop and ground, in contrast to that which is generated between the two terminals of the loop.   Also known as electrostatic error; vertical component effect.   { an'ten·ə i'fekt }

antenna effective area [ELECTROMAG] In any specified direction, the square of the wavelength multiplied by the power gain (or directive gain) in that direction, and divided by 4π.   { an'ten·ə i'fek·tiv 'er·ē·ə }

antenna efficiency [ELECTROMAG] The ratio of the amount of power radiated into space by an antenna to the total energy received by the antenna.   { an'ten·ə i,fish·ən·sē }

antenna field [ELECTROMAG] A group of antennas placed in a geometric configuration.   { an'ten·ə ,fēld }

antenna gain [ELECTROMAG] A measure of the effectiveness of a directional antenna as compared to a standard nondirectional antenna.   Also known as gain.   { an'ten·ə ,gān }

antennal gland [INV ZOO] An excretory organ in the cephalon of adult crustaceans and best developed in the Malacostraca.   Also known as green gland.   { an'ten·ə ,gland }

antenna loading [ELECTR] **1.** The amount of inductance or capacitance in series with an antenna, which determines the antenna's electrical length. **2.** The practice of loading an antenna in order to increase its electrical length.   { an'ten·ə ,lōd·iŋ }

antenna matching [ELECTROMAG] Process of adjusting impedances so that the impedance of an antenna equals the characteristic impedance of its transmission line.   { an'ten·ə ,mach·iŋ }

antenna pair [ELECTROMAG] Two antennas located on a base line of accurately surveyed length, sometimes arranged so that the array may be rotated around an axis at the center of the base line; used to produce directional patterns and in direction finding.   { an'ten·ə ,per }

antenna pattern *See* radiation pattern.   { an'ten·ə ,pad·ərn }

antenna polarization [ELECTROMAG] The orientation of the electric field lines in the electromagnetic field radiated or received by the antenna.   { an'ten·ə ,pō·lə·rə'zā·shən }

antenna power [ELECTROMAG] Radio-frequency power delivered to an antenna.   { an'ten·ə ,pau·ər }

antenna power gain [ELECTROMAG] The power gain of an antenna in a given direction is 4π times the ratio of the radiation intensity in that direction to the total power delivered to the antenna.   { an'ten·ə 'pau·ər ,gān }

antenna resistance [ELECTROMAG] The power supplied to an entire antenna divided by the square of the effective antenna current measured at the point where power is supplied to the antenna. { an'ten·ə ri‚zis·təns }

antenna scanner [ELECTROMAG] A microwave feed horn which moves in such a way as to illuminate sequentially different reflecting elements of an antenna array and thus produce the desired field pattern. { an'ten·ə ‚skan·ər }

Antennata [INV ZOO] An equivalent name for the Mandibulata. { ‚an·tə'näd·ə }

antenna temperature [ELECTROMAG] The temperature of a blackbody enclosure which would produce the same amount of noise as the antenna if it completely surrounded the antenna and was in thermal equilibrium with it. { an'ten·ə ‚tem·prə·chər }

antenna tilt error [ENG] Angular difference between the tilt angle of a radar antenna shown on a mechanical indicator, and the electrical center of the radar beam. { an'ten·ə 'tilt ‚er·ər }

antenodal [INV ZOO] Before or in front of the nodus, a cross vein near the middle of the costal border of the wing of dragonflies. { ‚an·tē'nōd·əl }

antepartum [MED] Pertaining to the period before delivery or birth. { ‚an·tē'pard·əm }

anteport [ARCH] A preliminary portal; an outer gate or door. { 'an·tē‚pört }

anter [INV ZOO] Part of a bryozoan operculum which serves to close off a portion of the operculum. { 'an·tər }

anteriad [ZOO] Toward the anterior portion of the body. { an'tir·ē‚ad }

anterior [ZOO] Situated near or toward the front or head of an animal body. { an'tir·ē·ər }

anterior arm reach [ANTHRO] The distance measured from the wall to the tip of the right middle finger when the subject makes the maximum forward reach with both arms, standing with heels, buttocks, middle of back (in lateral sense), and occiput against the wall. { an'tir·ē·ər 'ärm ‚rēch }

anterior commissure [NEURO] A bundle of nerve fibers that cross the midline of the brain in front of the third ventricle and serve to connect parts of the cerebral hemispheres. { an'tir·ē·ər 'käm·ə·shùr }

anterior horn [NEURO] The ventral column of gray matter in the spinal cord containing the cell bodies of motor (efferent) neurons. { an'tir·ē·ər 'hórn }

anterior horn cell [NEURO] A motor neuron in the anterior horn gray matter of the spinal cord; directly innervates skeletal muscle. { an'tir·ē·ər 'hórn ‚sel }

anterograde amnesia [MED] Loss of memory for the period subsequent to a sudden trauma or a seizure. { 'an·tə·rə‚grād am'nēzhə }

anteromedial [ZOO] Anterior and toward the middle of the body. { ‚an·tə·rə‚mēd·ē·əl }

antetheca [PALEON] The last or exposed septum at any stage of fusulinid growth. { ‚an·tē‚thek·ə }

anthelate [BOT] An open, paniculate cyme. { 'an·thə‚lāt }

anthelic arc [ASTRON] A rare type of halo phenomenon appearing in an area 180° from the sun's azimuth and at the sun's elevation. { ant'hē·lik 'ärk }

anthelion [ASTRON] A luminous white spot which occasionally appears on the parhelic circle 180° in azimuth away from the sun. Also known as counter sun. { ant'hēl·yən }

anthelminthic [PHARM] A chemical substance used to destroy tapeworms in domestic animals. Also spelled anthelmintic. { ‚an·thel‚min·thik }

anthelmintic See anthelminthic. { ‚an·thel‚min·tik }

anther [BOT] The pollen-producing structure of a flower. { 'an·thər }

antheraxanthin [BIOCHEM] A neutral yellow plant pigment unique to the Euglenophyta. { ‚an·thər‚aks‚an·thən }

anther culture [BOT] A haploid tissue culture derived from anthers or pollen cells. { 'an·thər ‚kəl·chər }

antheridiophore [BOT] A specialized stemlike structure that supports an antheridium in some mosses and liverworts. { ‚an·thə'rid·ē·ə·fòr }

antheridium [BOT] **1.** The sex organ that produces male gametes in cryptogams. **2.** A minute structure within the pollen grain of seed plants. { ‚an·thə'rid·ē·əm }

antheriferous [BOT] Anther-bearing. { ‚an·thə'rif·ə·rəs }

antherozoid [BOT] A motile male gamete produced by plants. { ‚an·thə·rə'zō·əd }

anther smut [MYCOL] *Ustilago violacea.* A smut fungus that

attacks certain plants and forms spores in the anthers. { 'an·thər ‚smət }

anthesis [BOT] The flowering period in plants. { an'thē·səs }

Anthicidae [INV ZOO] The antlike flower beetles, a family of coleopteran insects in the superfamily Tenebrionoidea. { an'this·ə‚dē }

anthill [INV ZOO] A mound of earth deposited around the entrance to ant and termite nests in the ground. [MIN ENG] The cuttings around the hole collar in blasthole drilling. { 'ant‚hil }

anthoblast [INV ZOO] A developmental stage of some corals; produced by budding. { 'an·thə‚blast }

anthocarpous [BOT] Describing fruits having accessory parts. { ‚an·thə‚kär·pəs }

anthocaulis [INV ZOO] The stemlike basal portion of some solitary corals; the oral portion becomes a new zooid. { ‚an·thə‚kòl·əs }

Anthocerotae [BOT] A small class of the plant division Bryophyta, commonly known as hornworts or horned liverworts. { ‚an·thə'ser·ə‚tē }

anthocodium [INV ZOO] The free, oral end of an anthozoan polyp. { ‚an·thə'kōd·ē·əm }

Anthocoridae [INV ZOO] The flower bugs, a family of hemipteran insects in the superfamily Cimicimorpha. { ‚an·thə'kòr·ə‚dē }

anthocyanidin [BIOCHEM] Any of the colored aglycone plant pigments obtained by hydrolysis of anthocyanins. { ‚an·thə‚sī'an·ə·dən }

anthocyanin [BIOCHEM] Any of the intensely colored, sap-soluble glycoside plant pigments responsible for most scarlet, purple, mauve, and blue coloring in higher plants. { ‚an·thə'sī·ə·nən }

Anthocyathea [PALEON] A class of extinct marine organisms in the phylum Archaeocyatha characterized by skeletal tissue in the central cavity. { ‚an·thə‚sī'ā·thē·ə }

anthocyathus [INV ZOO] The oral portion of the anthocaulis of some solitary corals that becomes a new zooid. { ‚an·thə‚sī'ā·thəs }

anthodite [GEOL] Gypsum or aragonite growing in clumps of long needle- or hairlike crystals on the roof or wall of a cave. { 'an·thə‚dīt }

anthoinite [MINERAL] $Al_2W_2O_9·3H_2O$ A white mineral consisting of a hydrous basic aluminum tungstate. { ‚an'thói‚nīt }

Anthomedusae [INV ZOO] A suborder of hydrozoan cnidarians in the order Hydroida characterized by athecate polyps. { ‚an·thō·mi'dü·se }

Anthomyzidae [INV ZOO] A family of cyclorrhaphous myodarian dipteran insects belonging to the subsection Acalypteratae. { ‚an·thō'mīz·ə‚dē }

anthophagous [ZOO] Feeding on flowers. { an'thä·fə·gəs }

anthophobia [PSYCH] Abnormal fear of flowers. { ‚an·thə'fōb·ē·ə }

anthophore [BOT] A stalklike extension of the receptacle bearing the pistil and corolla in certain plants. { 'an·thə‚fòr }

anthophyllite [MINERAL] A clove-brown orthorhombic mineral of the amphibole group; a variety of asbestos occurring as lamellae, radiations, fibers, or massive in metamorphic rocks. Also known as bidalotite. { ‚an·thō'fi‚līt }

Anthosomidae [INV ZOO] A family of fish ectoparasites in the crustacean suborder Caligoida. { ‚an·thə'säm·ə‚dē }

anthostele [INV ZOO] A thick-walled, nonretractile aboral region of certain cnidarians. { 'an·thə‚stēl }

Anthozoa [INV ZOO] A class of marine organisms in the phylum Cnidaria including the soft, horny, stony, and black corals, the sea pens, and the sea anemones. { ‚an·thō'zō·ə }

anthozooid [INV ZOO] Any of the individual zooids of a compound anthozoan. { ‚an·thō'zō‚óid }

anthracene [ORG CHEM] $C_{14}H_{10}$ A crystalline tricyclic aromatic hydrocarbon, colorless when pure, melting at 218°C and boiling at 342°C; obtained in the distillation of coal tar; used as an important source of dyestuffs, and in coating applications. { 'an·thrə‚sēn }

anthracene oil [MATER] A heavy, green oil that is a coal-tar fraction boiling above 270°C; used as a source of anthracene, phenanthrene, and carbazole. { 'an·thrə‚sēn ‚óil }

anthracene violet See gallein. { 'an·thrə‚sēn 'vī·lət }

anthraciferous coal [ORG CHEM] Anthracite-hard coal containing or yielding anthracene. { 'an·thrə,sif·ə·rəs 'kōl }

anthracite [MINERAL] A high-grade metamorphic coal having a semimetallic luster, high content of fixed carbon, and high density, and burning with a short blue flame and little smoke or odor. Also known as hard coal; Kilkenny coal; stone coal. { 'an·thrə,sīt }

anthracite duff [MATER] In Wales, fine anthracite screenings used in briquets or mixed with bituminous coal to fuel cement kilns on chain grate stokers. { 'an·thrə,sīt ,dəf }

anthracite fines [MIN ENG] Small pieces of material from an anthracite coal preparation plant, usually below $^1/_8$-inch (3-millimeter) diameter. { 'an·thrə,sīt ,fīnz }

anthracitization [GEOCHEM] The natural process by which bituminous coal is transformed into anthracite coal. { ,an·thrə,sīd·ə'zā·shən }

anthracnose [PL PATH] A fungus disease of plants caused by members of the Melanconiales and characterized by dark or black limited stem lesions. { an'thrak,nōs }

Anthracosauria [PALEON] An order of Carboniferous and Permian labyrinthodont amphibians that includes the ancestors of living reptiles. { ,an·thrə·kə'sór·ē·ə }

anthracosilicosis [MED] Chronic lung inflammation caused by inhalation of carbon and silicon particles. { ,an·thrə·kə,sil·ə'kō·səs }

anthracosis [MED] The accumulation of inhaled black coal dust particles in the lung accompanied by chronic inflammation. Also known as blacklung. { 'an·thrə'kō·səs }

Anthracotheriidae [PALEON] A family of middle Eocene and early Pleistocene artiodactyl mammals in the superfamily Anthracotherioidea. { ,an·thrə·kə·thə'rī·ə,dē }

Anthracotherioidea [PALEON] A superfamily of extinct artiodactyl mammals in the suborder Paleodonta. { 'an·thrə·kə·thə,rī'oid·ē·ə }

anthracoxene [GEOL] A brownish resin that occurs in brown coal; in ether it dissolves into an insoluble portion, anthrocoxenite, and a soluble portion, schlanite. { ,an·thrə'käk,sēn }

anthralin [PHARM] $C_{14}H_{10}O_3$ A yellow powder with a melting point of 176–181°C; soluble in chloroform, acetone, and benzene; used in treatment of psoriasis. { 'an·thrə·lən }

anthranilic acid [ORG CHEM] o-NH$_2$C$_6$H$_4$COOH A white or pale yellow, crystalline acid melting at 146°C; used as an intermediate in the manufacture of dyes, pharmaceuticals, and perfumes. { ¦an·thrə¦lin·ik 'as·əd }

anthrapurpurin [ORG CHEM] C$_6$H$_3$OH(CO)$_2$C$_6$H$_2$(OH)$_2$ Orange-yellow, crystalline needles with a melting point of 369°C; soluble in alcohol and alkalies; used in dyeing. { ,an·thrə'pər·pə,rin }

anthraquinone [ORG CHEM] C$_6$H$_4$(CO)$_2$C$_6$H$_4$ Yellow crystalline diketone that is insoluble in water; used in the manufacture of dyes. { ,an·thrə·kwi'nōn }

anthraquinone pigments [BIOCHEM] Coloring materials which occur in plants, fungi, lichens, and insects; consists of about 50 derivatives of the parent compound, anthraquinone. { ,an·thrə·kwi'nōn 'pig·məns }

anthrax [VET MED] An acute, infectious bacterial disease of sheep and cattle caused by *Bacillus anthracis;* transmissible to humans. Also known as splenic fever; wool-sorter's disease. { 'an,thraks }

anthraxolite [GEOL] Anthracite-like asphaltic material occurring in veins in Precambrian slate of Sudbury District, Ontario. { an'thrak·sə,līt }

anthraxylon [GEOL] The vitreous-appearing components of coal that are derived from the woody tissues of plants. { an'thrak·sə,län }

Anthribidae [INV ZOO] The fungus weevils, a family of coleopteran insects in the superfamily Curculionoidea. { an'thrib·ə,dē }

anthrone [ORG CHEM] C$_{14}$H$_{10}$O Colorless needles with a melting point of 156°C; soluble in alcohol, benzene, and hot sodium hydroxide; used as a reagent for carbohydrates. { 'an,thrōn }

anthropic [ANTHRO] Pertaining to humans or the period of their existence on earth. { an'thräp·ik }

anthropic principle [ASTRON] The assertion that the presence of intelligent life on earth places limits on the many ways the universe could have developed and could have caused the conditions of temperature that prevail today. { an'thräp·ik 'prin·sə·pəl }

anthropocentric [PSYCH] **1.** Regarding humankind as the most important factor in the universe. **2.** Evaluating all occurrences solely by human values. { ¦an·thrə,pō¦sen·trik }

anthropochory [ECOL] Dispersal of plant and animal disseminules by humans. { ¦an·thrə·pə¦kòr·ē }

anthropodesoxycholic acid *See* chenodeoxycholic acid. { ¦an·thrə,pō·de¦zäk·sə'käl·ik 'as·əd }

anthropogenic [ECOL] Referring to environmental alterations resulting from the presence or activities of humans. { ¦an·thrə·pə¦jen·ik }

anthropogeography *See* human geography. { ¦an·thrə·pō·jē'äg·rə·fē }

anthropogeomorphology [ANTHRO] The study of the effects of humans on the physical landscape such as in the development and operation of an open-pit mine. { ¦an·thrə·pō,jē·ə·mòr'fäl·ə·jē }

anthropography [ANTHRO] A branch of anthropology that deals with the geographic distribution of divisions of humans based on physical character, language, customs, and institutions. Also known as cultural geography; human geography. { ,an·thrə'päg·rə·fē }

anthropoid [VERT ZOO] Pertaining to or resembling the Anthropoidea. { 'an·thrə,pòid }

Anthropoidea [VERT ZOO] A suborder of mammals in the order Primates including New and Old World monkeys. { ,an·thrə'pòid·ē·ə }

anthropology [BIOL] The study of the interrelations of biological, cultural, geographical, and historical aspects of humankind. { ,an·thrə'päl·ə·jē }

anthropometry [ANTHRO] Description of the physical variation in humankind by measurement; a basic technique of physical anthropology. { ,an·thrə'päm·ə·trē }

anthroponoses [MED] Diseases transmitted from humans to animals. { ,an·thrə·pə'nō,sēz }

anthropophobia [PSYCH] Abnormal fear of people. { ,an·thrə·pə'fōb·ē·ə }

anthroposcopy [ANTHRO] The description of physical variation in humankind by visual inspection; a basic technique of physical anthropology. { ,an·thrə'päs·kə·pē }

anthroposphere [ECOL] The biosphere of the great geological activities of humankind. Also known as noosphere. { an'thrä·pə,sfir }

Anthuridea [INV ZOO] A suborder of crustaceans in the order Isopoda characterized by slender, elongate, subcylindrical bodies, and by the fact that the outer branch of the paired tail appendage (uropod) arches over the base of the terminal abdominal segment, the telson. { ,an·thə'rīd·ē·ə }

anti [ORG CHEM] In stereochemistry, on the opposite side of a reference plane; for example, the stereochemical outcome of an addition reaction where the new bonds are on the opposite side of the original pi bond is called anti addition. { 'an,tē }

antiacid additive [MATER] A substance that prevents or retards corrosive acid formation in crankcase oils during use. { ¦an·tē¦as·əd 'ad·ə,tiv }

antiacid bronze [MET] A high-lead, acid-resistant bronze used for casting chemical machine parts. { ¦an·tē¦as·əd 'bränz }

antiagglutinin [IMMUNOL] A substance that neutralizes a corresponding agglutinin. { ,an·tē·ə'glüt·ən·ən }

antiaggressin [IMMUNOL] An antibody that neutralizes aggressin, a substance produced by pathogenic microorganisms to enhance virulence. { ,an·tē·ə'gres·ən }

antiaircraft [ORD] Used, or designed to be used, against airborne aircraft or missiles. Abbreviated AA. { 'an·tē'er,kraft }

antiaircraft artillery [ORD] Projectile weapons with related equipment, such as searchlights or radar, employed on the ground or on ships to strike at airborne aircraft or missiles. Abbreviated AAA. { 'an·tē'er,kraft är'til·ə·rē }

antiaircraft barrage [ORD] **1.** A concentration of antiaircraft fire covering an area through which enemy aircraft are likely to fly. Also known as predicted barrage. **2.** Less precisely, any concentration of antiaircraft fire. { 'an·tē'er,kraft bə'räzh }

antiaircraft gun [ORD] A gun designed for use against aircraft, easily shifted in direction and elevation, having great

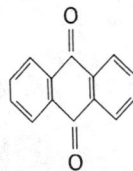

ANTHRAQUINONE

Structural formula of anthraquinone.

range, and capable of firing at high angles of elevation. { 'an·tē'er,kraft ,gən }

antiaircraft missile [ORD] A guided missile that is intended to be launched from the surface against an airborne target. Abbreviated AAM. { 'an·tē'er,kraft ,mis·əl }

antialiasing technique [COMPUT SCI] In computer graphics, a technique for smoothing the jagged appearance of diagonal lines on printouts and on video monitors. { ,an·tē'āl·ē·əs·iŋ ,tek,nēk }

antianaphylaxis [IMMUNOL] A condition in which a sensitized animal resists anaphylaxis. { ¦an·tē,an·ə·fə'lak·səs }

Antiarchi [PALEON] A division of highly specialized placoderms restricted to fresh-water sediments of the Middle and Upper Devonian. { ,an·tē'är,kī }

antiaromatic [CHEM] A cyclic compound with delocalized electrons that does not obey Hückel's rule, and is much less stable than similar nonaromatic compounds. { ,an·tē,ar·ə'mad·ik }

antiarrhythmic [MED] An agent that prevents or alleviates cardiac arrhythmia. { ,an·tē,ā'rith·mik }

antiatom [ATOM PHYS] An atom made up of antiprotons, antineutrons, and positrons in the same way that an ordinary atom is made up of protons, neutrons, and electrons. { 'an·tē,ad·əm }

antiautomorphism [MATH] An antiisomorphism of a ring, field, or integral domain with itself. { ,an·tē,ód·ə'mór,fiz·əm }

antiauxin [BIOCHEM] A molecule that competes with an auxin for receptor sites. { ¦an·tē'ók·sən }

antibacterial agent [MICROBIO] A synthetic or natural compound which inhibits the growth and division of bacteria. { ¦an·tē,bak'tir·ē·əl 'ā·jənt }

antiballistic missile [ORD] Any object thrown, dropped, fired, or otherwise projected with the purpose of intercepting a ballistic missile. Abbreviated ABM. { ,an·tē bə'lis·tik 'mis·əl }

antibaryon [ATOM PHYS] One of a class of antiparticles, including the antinucleons and the antihyperons, with strong interactions, baryon number −1, and hypercharge and charge opposite to those for the particles. { 'an·tē¦bar·ē·än }

antibiosis [ECOL] Antagonistic association between two organisms in which one is adversely affected. { ¦an·tē,hī¦ō·səs }

antibiotic [MICROBIO] A chemical substance, produced by microorganisms and synthetically, that has the capacity in dilute solutions to inhibit the growth of, and even to destroy, bacteria and other microorganisms. { ¦an·tē,bī'äd·ik }

antibiotic assay [MICROBIO] A method for quantitatively determining the concentration of an antibiotic by its effect in inhibiting the growth of a susceptible microorganism. { ¦an·tē,bī¦äd·ik 'a,sā }

antibody [IMMUNOL] A protein, found principally in blood serum, originating either normally or in response to an antigen and characterized by a specific reactivity with its complementary antigen. Also known as immune body. { 'an·tə,bäd·ē }

antibody binding site [CELL MOLEC] See antibody combining site. { ¦an·tə,bäd·ē 'bīnd·iŋ ,sīt }

antibody combining site [CELL MOLEC] **1.** The portion of an antibody molecule that makes physical contact with the corresponding antigenic determinant. **2.** The portion of an antigen that makes physical contact with the corresponding antibody. Also known as antibody binding site. { 'an·tə,bäd·ē kəm'bīn·iŋ ,sīt }

antibody-deficiency syndrome [MED] Any of the human defects of antibody production, such as hypogammaglobulinemia, agammaglobulinemia, and dysgammaglobulinemia, usually associated with reduced serum concentrations of immunoglobulins. { 'an·tə,bäd·ē di'fish·ən·sē ,sin,dróm }

antibody-dependent cell-mediated cytotoxicity [IMMUNOL] An immunologic response in which an immunologic effector cell binds to a target cell coated with antibodies, triggering a series of metabolic events that leads to lysis of the target cell. { ,ant·i,bäd·ē di¦pen·dənt ,sel¦mēd·ē,ād·əd ,sī·tō·täk'sis·əd·ē }

antibody-mediated immunity See humoral immunity. { 'an·tə,bäd·ē ¦mēd·ē,ād·əd i'myün·əd·ē }

antibonding orbital [PHYS] An atomic or molecular orbital whose energy increases as atoms are brought closer together,

indicating a net repulsion rather than a net attraction and chemical bonding. { 'an·tē'bänd·iŋ 'ór·bə·təl }

antiboreal faunal region [ECOL] A zoogeographic region including marine littoral faunal communities at the southern end of South America. { 'an·tē,bór·ē·əl 'fón·əl ,rē·jən }

anticapacitance switch [ELECTR] A switch designed to have low capacitance between its terminals when open. { ¦an·tē·kə'pas·ə·təns ,swich }

anticarcinogen [PHARM] Any substance which is antagonistic to the action of a carcinogen. { 'an·tē,kär'sin·ə·jən }

anticatalyst [CHEM] A material that slows down the action of a catalyst; an example is lead, which inhibits the action of platinum. { 'an·tē¦kad·əl·ist }

anticathode [ELECTR] The anode or target of an x-ray tube, on which the stream of electrons from the cathode is focused and from which x-rays are emitted. { 'an·tē'kath,ōd }

anticenter [ASTRON] The direction in the sky opposite to that of the center of the Galaxy, located in the constellation Auriga. [GEOL] The point on the surface of the earth that is diametrically opposite the epicenter of an earthquake. Also known as antiepicenter. { 'an·tē'sent·ər }

antichain [MATH] **1.** A subset of a partially ordered set in which no pair is a comparable pair. **2.** See Sperner set. { 'an·tē,chān }

antichlor [CHEM ENG] A chemical used in the manufacture of paper or textiles to remove excess chlorine or bleaching solution. { ¦an·ti'klór }

anticholinesterase [BIOCHEM] Any agent, such as a nerve gas, that inhibits the action of cholinesterase and thereby destroys or interferes with nerve conduction. { ,an·ti,kō·lə'nes·tə,rās }

anticipation [GEN] The occurrence of a phenotype at a younger age or in a more severe form in succeeding generations of a family. { an,tis·ə'pā·shən }

anticipatory staging [COMPUT SCI] Moving blocks of data from one storage device to another prior to the actual request for them by the program. { 'an'tis·ə·pə,tór·ē 'stāj·iŋ }

anticlastic [MATH] Having the property of a surface or portion of a surface whose two principal curvatures at each point have opposite signs, so that one normal section is concave and the other convex. { 'an·tē¦klas·tik }

anticlinal [BOT] Pertaining to a cell layer that runs at right angles across the circumference of a plant part. [GEOL] Folded as in an anticline. { ¦an·tē¦klīn·əl }

anticlinal axis [GEOL] The median line of a folded structure from which the strata dip on either side. { ¦an·tē¦klīn·əl 'ak·səs }

anticlinal bend [GEOL] An upwardly convex flexure of rock strata in which one limb dips gently toward the apex of the strata and the other dips steeply away from it. { ¦an·tē¦klīn·əl 'bend }

anticlinal mountain [GEOL] Ridges formed by a convex flexure of the strata. { ¦an·tē¦klīn·əl 'maún·tən }

anticlinal theory [GEOL] A theory relating trapped underground oil accumulation to anticlinal structures. { ¦an·tē¦klīn·əl 'thē·ə·rē }

anticlinal trap [GEOL] A formation in the top of an anticline in which petroleum has accumulated. { ¦ant·i¦klīn·əl 'trap }

anticlinal valley [GEOL] A valley that follows an anticlinal axis. { ¦an·tē¦klīn·əl 'val·ē }

anticline [GEOL] A fold in which layered strata are inclined down and away from the axes. { 'an·ti,klīn }

anticlinorium [GEOL] A series of anticlines and synclines that form a general arch or anticline. { ,an·ti,klī'nor·ē·əm }

anticlutter gain control [ELECTR] Device which automatically and smoothly increases the gain of a radar receiver from a low level to the maximum, within a specified period after each transmitter pulse, so that short-range echoes producing clutter are amplified less than long-range echoes. { ,an·tē'kləd·ər 'gān kən,trōl }

anticoagulant [PHARM] An agent, such as sodium citrate, that prevents coagulation of a colloid, especially blood. { ¦an·tē,kō'ag·yə·lənt }

anticoding strand See antisense strand. { ,an·tē'kōd·iŋ ,strand }

anticodon [GEN] A three-nucleotide sequence in transfer RNA that complements the codon in messenger RNA. { ¦an·tē'kō,dän }

anticoincidence [NUC PHYS] The occurrence of an event at

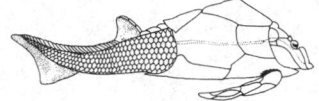

ANTIARCHI

Pterichthyodes (Pterichthys), a Middle Devonian antiarch. Lateral view showing scale-covered tail; about 15 centimeters in length.

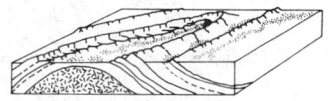

ANTICLINE

Diagram relating anticlinal structure to topography.

one place without a simultaneous event at another place. { ‚an·tē‚kō'in·sə·dəns }

anticoincidence circuit [ELECTR] Circuit that produces a specified output pulse when one (frequently predesignated) of two inputs receives a pulse and the other receives no pulse within an assigned time interval. { ‚an·tē‚kō'in·sə·dəns ‚sər·kət }

anticollision radar [ENG] A radar set designed to give warning of possible collisions during movements of ships or aircraft. { ‚an·tē·kə'li·zhən ‚rā‚där }

anticommutative operation [MATH] A method of combining two objects, *a · b*, such that *a · b* = −*b · a*. { ‚an·tē‚käm·yə‚tād·iv ‚äp·ə'rā·shən }

anticommutator [MATH] The anticommutator of two operators, *A* and *B*, is the operator *AB* + *BA*. { ‚an·tē'käm·yə‚täd·ər }

anticommute [MATH] Two operators anticommute if their anticommutator is equal to zero. { ‚an·tē·kə'myüt }

anticonvulsant [PHARM] An agent, such as Dilantin, that prevents or arrests a convulsion. { ‚an·tē·kən'vəl·sənt }

anticorona [OPTICS] A diffraction phenomenon appearing at a point before an observer with the sun or moon directly behind the observer; consists of rings of colored lights complementary to the coronal rings. Also known as Brocken bow. { ‚an·tē·kə'rō·nə }

anticorrosive paint [MATER] A paint formulated with a corrosive-resistant pigment (such as lead chromate, zinc chromate, or red lead) and a chemical- and moisture-resistant binder; used to protect iron and steel surfaces. { ‚an·tē·kə'rō·siv 'pānt }

anticosecant See arc cosecant. { ‚an·tē·kō'sē‚kant }

anticosine See arc cosine. { ‚an·tē'kō‚sīn }

anticotangent See arc contangent. { ‚an·tē·kō'tan·jənt }

anticreeper [CIV ENG] A device attached to a railroad rail to prevent it from moving in the direction of its length. { 'an·tē‚krēp·ər }

anticrepuscular rays [ASTRON] Extensions of crepuscular rays, converging toward a point 180° from the sun. { ‚an·tē·kri'pəs·kyə·lər 'rāz }

anticryptic [ECOL] Pertaining to protective coloration that makes an animal resemble its surroundings so that it is inconspicuous to its prey. { ‚an·tē'krip·tik }

anticurl coating [GRAPHICS] A thin layer generally applied to or on the back of photographic material to prevent front warping. { 'an·tē‚kərl 'kōd·iŋ }

anticusp [INV ZOO] An anterior, downward projection in conodonts. { 'an·tē‚kəsp }

anticyclogenesis [METEOROL] The process which creates an anticyclone or intensifies an existing one. { ‚an·tē‚sī·klō'jen·ə·səs }

anticyclolysis [METEOROL] Any weakening of anticyclonic circulation in the atmosphere. { ‚an·tē‚sī'kläl·ə·səs }

anticyclone [METEOROL] High-pressure atmospheric closed circulation whose relative direction of rotation is clockwise in the Northern Hemisphere, counterclockwise in the Southern Hemisphere, and undefined at the Equator. Also known as high-pressure area. { ‚an·tē'sī‚klōn }

anticyclonic [METEOROL] Referring to a rotation about the local vertical that is clockwise in the Northern Hemisphere, counterclockwise in the Southern Hemisphere, undefined at the Equator. { ‚an·tē‚sī'klän·ik }

anticyclonic shear [METEOROL] Horizontal wind shear of such a nature that it tends to produce anticyclonic rotation of the individual air particles along the line of flow. { ‚an·tē‚sī'klän·ik 'shēr }

anticyclonic winds [METEOROL] The winds associated with a high pressure area and constituting part of an anticyclone. { ‚an·tē‚sī'klän·ik 'winz }

antidepressant [PHARM] A drug, such as imipramine and tranylcypromine, that relieves depression by increasing central sympathetic activity. { ‚an·tē·di'pres·ənt }

antiderivative See indefinite integral. { ‚an·tē·di'riv·əd·iv }

antidesiccant [MATER] Material applied to plants prior to transplanting to reduce the amount of moisture lost by transpiration. { ‚an·tē'des·ə·kənt }

antidetonant See antiknock. { ‚an·tē‚det·ən·ənt }

antideuteron [ATOM PHYS] The antiparticle to the deuteron, composed of an antineutron and an antiproton. { ‚an·tē‚düt·ə‚rän }

antidiabetic [PHARM] An agent, such as insulin, that is effective in controlling diabetes. { ‚an·tē‚dī·ə‚bed·ik }

antidiarrheal [PHARM] An agent, such as Kaopectate, that prevents or arrests diarrhea. { ‚an·tē‚dī·ə‚rē·əl }

antidieseling solenoid See idle-stop solenoid. { ‚ant·i‚dēz·əl·iŋ 'sō·lə‚nȯid }

antidip stream [HYD] A stream that flows in a direction opposite to the general dip of the strata. { ‚ant·ē'dip ‚strēm }

antidiuretic [PHARM] An agent, such as vasopressin, that prevents the excretion of urine. { ‚an·tē‚dī·yə‚red·ik }

antidiuretic hormone See vasopressin. { ‚an·tē‚dī·yə‚red·ik 'hȯr‚mōn }

antidote [PHARM] An agent that relieves or counteracts the action of a poison. { 'an·tə‚dōt }

antidrag [AERO ENG] **1.** Describing structural members in an aircraft or missile that are designed or built to resist the effects of drag. **2.** Referring to a force acting against the force of drag. { ‚an·tē'drag }

antidune [GEOL] A temporary form of ripple on a stream bed analogous to a sand dune but migrating upcurrent. { 'an·tē‚dün }

antienzyme [BIOCHEM] An agent that selectively inhibits the action of an enzyme. { ‚an·tē'en‚zīm }

antiepicenter [ASTRON] See anticenter. { ‚an·tē'ep·i‚sent·ər }

antiestuarine circulation [OCEANOGR] In an estuary, the inflow of low-salinity surface water over a deeper outflowing (seaward), dense, high-salinity water layer. { ‚an·tē‚es·chə·wə‚rēn ‚ser·kyü'lā·shən }

antifading antenna [ELECTR] An antenna designed to confine radiation mainly to small angles of elevation to minimize the fading of radiation directed at larger angles of elevation. { ‚an·tē‚fād·iŋ an'ten·ə }

antiferroelectric crystal [SOLID STATE] A crystalline substance characterized by a state of lower symmetry consisting of two interpenetrating sublattices with equal but opposite electric polarization, and a state of higher symmetry in which the sublattices are unpolarized and indistinguishable. { ‚an·tē‚fer·ō·i'lek·trik 'kris·təl }

antiferromagnetic domain [SOLID STATE] A region in a solid within which equal groups of elementary atomic or molecular magnetic moments are aligned antiparallel. { ‚an·tē‚fer·ō‚mag'ned·ik dō'mān }

antiferromagnetic resonance [ELECTROMAG] Magnetic resonance in antiferromagnetic materials which may be observed by rotating magnetic fields in either of two opposite directions. { ‚an·tē‚fer·ō‚mag'ned·ik 'rez·ə·nəns }

antiferromagnetic substance [ELECTROMAG] A substance that is composed of antiferromagnetic domains. { ‚an·tē‚fer·ō‚mag'ned·ik 'səb·stəns }

antiferromagnetic susceptibility [ELECTROMAG] The magnetic response to an applied magnetic field of a substance whose atomic magnetic moments are aligned in antiparallel fashion. { ‚an·tē‚fer·ō‚mag'ned·ik sə‚sep·tə'bil·əd·ē }

antiferromagnetism [SOLID STATE] A property possessed by some metals, alloys, and salts of transition elements by which the atomic magnetic moments form an ordered array which alternates or spirals so as to give no net total moment in zero applied magnetic field. { ‚an·tē‚fer·ō'mag·nə‚tiz·əm }

antifertility agent [PHARM] A drug that prevents the formation of a fertilized ovum upon sexual intercourse. { ‚an·tē·fər'til·əd·ē 'ā·jənt }

antifertilizin [BIOCHEM] An immunologically specific substance produced by animal sperm to implement attraction by the egg before fertilization. { ‚an·tē·fər'til·ə·zən }

antifibrinolysin [BIOCHEM] Any substance that inhibits the proteolytic action of fibrinolysin. { ‚an·tē‚fī·brə'näl·ə·sən }

antifix [ARCH] In classical architecture, either an ornament at the eaves to hide the ends of the joint tiles of the roof or an ornament of the cymatium of a classic cornice, sometimes pierced so that water can flow through. { 'an·tē‚fiks }

antifoaming agent [ORG CHEM] A substance, such as silicones, organic phosphates, and alcohols, that inhibits the formation of bubbles in a liquid during its agitation by reducing its surface tension. { ‚an·tē‚fōm·iŋ ‚ā·jənt }

antifogging compound [MATER] A compound of one or more basic chemicals with filler or extenders for preventing condensation of moisture on glass and other transparent material, such as lenses or windshields. { 'an·tē¦fäg·iŋ ¦käm·paůnd }

antiform [GEOL] An anticline-like structure whose stratigraphic sequence is not known. { 'an·tē¦fȯrm }

antifouling coating [MATER] A special paint containing copper used on ships' bottoms to prevent marine organisms from attaching themselves. { ¦an·tē¦faůl·iŋ ¦kōd·iŋ }

antifreeze [CHEM] A substance added to a liquid to lower its freezing point; the principal automotive antifreeze component is ethylene glycol. { 'an·tē¦frēz }

antifreeze proteins [BIOCHEM] Proteins that decrease the nonequilibrium freezing point of water without significantly affecting the melting point by directly binding to the surface of an ice crystal, thereby disrupting its normal structure and growth pattern and inhibiting further ice growth; found in a number of fish, insects, and plants. { 'an·ti¦frēz ¦prō¦tēnz }

antifriction [MECH] Making friction smaller in magnitude. [MECH ENG] Employing a rolling contact instead of a sliding contact. { ¦an·tē¦frik·shən }

antifriction alloy [MET] An alloy generally having more than 50% tin as a base and cast as facings of machinery bearings, domestic equipment, and small parts in rolling contact. { ¦an·tē¦frik·shən 'al¸ȯi }

antifriction bearing [MECH ENG] Any bearing having the capability of reducing friction effectively. { ¦an·tē¦frik·shən ¸ber·iŋ }

antifriction material [ENG] A machine element made of Babbitt metal, lignum vitae, rubber, or a combination of a soft, easily deformable metal overlaid on a hard, resistant one. { ¦an·tē¦frik·shən mə'tir·ē·əl }

antifungal agent [MATER] A chemical compound that either destroys or inhibits the growth of fungi. { ¦an·tē¦fəŋ·gəl ¸ā·jənt }

antigen [IMMUNOL] A substance which reacts with the products of specific humoral or cellular immunity, even those induced by related heterologous immunogens. { 'an·tə·jən }

antigenaemia [IMMUNOL] A condition in which viral antigen is present in the blood; occurs in viral hepatitis and may occur in smallpox, myxomatosis, and yellow fever. { ¸an·tə·jə'nē·mē·ə }

antigen-antibody reaction [IMMUNOL] The combination of an antigen with its antibody. { 'an·tə·jən ¦an·tə¦bäd·ē rē'ak·shən }

antigenic competition [IMMUNOL] A decrease in immune response to one antigenic peptide due to a concurrent immune response to a different antigenic peptide. { ¦an·tə¦jen·ik ¸käm·pə'tish·ən }

antigenic determinant [IMMUNOL] The portion of an antigen molecule that determines the specificity of the antigen-antibody reaction. { ¦an·tə¦jen·ik di'tər·mə·nənt }

antigenic drift [IMMUNOL] Minor change of an antigen on the surface of a pathogenic microorganism. { ¦an·tə¦jen·ik 'drift }

antigenicity [IMMUNOL] Ability of an antigen to induce an immune response and combine with specific antibodies or T-cell receptors. { ¸an·tə·jə'nis·əd·ē }

antigenic mimicry [IMMUNOL] Acquisition or production of host antigens by a parasite, enabling it to avoid detection by the host's immune system. { 'an·tə¦jen·ik 'mim·ə·krē }

antigenic modulation [IMMUNOL] Loss of detectable antigen from the surface of a cell after incubation with antibodies. { ¦an·tə¦jen·ik ¸mäj·ə'lā·shən }

antigenic shift [VIROL] An abrupt major change in the antigenicity of a virus; believed to result from recombination of genes. { ¦an·tə¦jen·ik 'shift }

antigenic variation [IMMUNOL] Alteration of an antigen on the surface of a microorganism; may enable a pathogenic mocroorganism to evade destruction by the host's immune system. { ¦an·tə¦jen·ik ¸ver·ē'ā·shən }

antigen presentation [IMMUNOL] The process whereby a cell expresses antigen on its surface in a form that can be recognized by a T lymphocyte. { ¦an·tə·jən ¸prē·zən'tā·shən }

antiglare shield [COMPUT SCI] A sheet of nonreflective material placed over the screen of an electronic display to reduce the amount of light reflected from the screen. { 'an·tē¸gler 'shēld }

antigorite [MINERAL] Mg₃Si₂O₅(OH)₄ Brownish-green variety of the mineral serpentine. Also known as baltimorite; picrolite. { an'tig·ə¸rīt }

antigravity [PHYS] The repulsion of one body by another by means of a gravitational type of force; this has never been observed. { ¦an·tē¦grav·əd·ē }

anti-g suit See g suit. { ¦an·tē¦jē ¸süt }

antihalation backing [GRAPHICS] A coating on the back of film to minimize reflection of light from the base into the emulsion. { ¦an·tē·hə'lā·shən 'bak·iŋ }

antihelium [ATOM PHYS] The antimatter counterpart of helium, whose atoms each consist of two orbiting positrons and a nucleus composed of two antiprotons and either one or two antineutrons. { ¦an·tē'hē·lē·əm }

antihemophilic factor [BIOCHEM] A soluble protein clotting factor in mammalian blood. Also known as factor VIII; thromboplastinogen. { ¦an·tē¸hē·mə'fil·ik ¸fak·tər }

antihemorrhagic vitamin See vitamin K. { ¦an·tē¸hem·ə'raj·ik 'vīd·ə·mən }

antihistamine [PHARM] A drug that prevents or diminishes the effect of histamine; used in treating allergic reactions and common-cold symptoms. { ¸an·tē'hist·ə¸mēn }

antihunt circuit [ELECTR] A stabilizing circuit used in a closed-loop feedback system to prevent self-oscillations. { 'an·tē¸hənt ¸sər·kət }

antihydrogen [ATOM PHYS] The antimatter counterpart of hydrogen, whose atoms each consist of an orbiting positron and a nucleus that is an antiproton, antideuteron, or antitriton. { ¦an·tē'hī·drə·jən }

antihyperbolic function See inverse hyperbolic function. { ¸an·tē¸hī·pər¦bäl·ik 'fəŋk·shən }

antihyperon [PARTIC PHYS] An antiparticle to a hyperon, having the same mass, lifetime, and spin as the hyperon, but with charge and magnetic moment reversed in sign. { ¦an·tē¦hī·pə¸rän }

antihypertensive agent [PHARM] A substance, such as reserpine, that reduces hypertension. { ¸an·tē¸hī·pər'ten·siv 'ā·jənt }

anti-icing [AERO ENG] The prevention of the formation of ice upon any object, especially aircraft, by means of windshield sprayer, addition of antifreeze materials to the carburetor, or heating the wings and tail. { ¸an·tē'īs·iŋ }

anti-idiotype antibody [IMMUNOL] An antibody that is the mirror image of the original antibody formed against a specific surface antigen. { ¸an·tē¦id·ē·ə¸tīp 'an·tə¸bäd·ē }

anti-immunoglobulin antibody [IMMUNOL] An antibody produced in response to a foreign antibody introduced into an experimental animal. Abbreviated AIA. { ¦an·tē¸im·yə·nō¦gläb·yə·lən 'an·tē¸bäd·ē }

anti-infective vitamin See vitamin A. { ¸an·tē·in'fek·div ¸vīd·ə·mən }

anti-inflammatory agent [PHARM] A substance, such as cortisone, that counteracts inflammation. { ¸an·tē·in'flam·ə¸tȯr·ē ¸ā·jənt }

anti-intrusion technology [COMPUT SCI] One of the different ways in which an attack on a computer system can be detected and countered, including prevention, deterrence, detection, deflection, and diminution. { ¸an·tē¸in'trü·zhən ¸tek¦näl·ə·jē }

anti-isomorphism [MATH] A one-to-one correspondence between two rings, fields, or integral domains such that, if x′ corresponds to x and y′ corresponds to y, then x′ + y corresponds to x + y, but y′x′ corresponds to xy. { ¸an·tē¸ī·sə'mȯr¸fiz·əm }

antijamming [ELECTR] Any system or technique used to counteract the jamming of communications or of radar operation. { ¸an·tē'jam·iŋ }

antiknock [MATER] **1.** Resisting detonation or pinging in spark-ignited engines. **2.** A substance, such as tetraethyllead, added to motor and aviation gasolines to increase the resistance of the fuel to knock in spark-ignited engines. Also known as antidetonant. { 'an·tē¸näk }

antiknock blending value [ENG] The numerical improvement by an antiknock additive to gasoline octane, often a greater amount than the additive's own octane value. { 'an·tē¸näk 'blend·iŋ ¸val·yü }

antiknock gasoline [MATER] Gasoline containing, for example, tetraethyllead, which in small amounts prevents or lessens knocking in a spark-ignited engine. { 'an·tē¸näk ¸gas·ə'lēn }

antiknock rating [ENG] Measurement of the ability of an

automotive gasoline to resist detonation or pinging in spark-ignited engines. { 'an·tē,näk 'räd·iŋ }

antilepton [ATOM PHYS] An antiparticle of a lepton, such as an antineutrino or a positron. { ¦an·tē'lep·tän }

Antilles Current [OCEANOGR] A current formed by part of the North Equatorial Current that flows along the northern side of the Greater Antilles. { an'til·ēz ¦kər·ənt }

Antilocapridae [VERT ZOO] A family of artiodactyl mammals in the superfamily Bovoidea; the pronghorn is the single living species. { ,an·tə,lō'kap·rə,dē }

antilock braking system [MECH ENG] For vehicles, a sensor-control system found in braking systems which prevents wheel lockup while allowing the brakes to continue slowing the wheel. Abbreviated ABS. { ¦an·tē,läk 'brāk·iŋ ,sis·təm }

antilog See antilogarithm. { 'an·ti,läg }

antilogarithm [MATH] For a number x, a second number whose logarithm equals x. Abbreviated antilog. Also known as inverse logarithm. { ¦an·ti'läg·ə,rith·əm }

antilogous pole [SOLID STATE] That crystal pole which becomes electrically negative when the crystal is heated or is expanded by decompression. { an'til·ə·gəs 'pōl }

Antilopinae [VERT ZOO] The antelopes, a subfamily of artiodactyl mammals in the family Bovidae. { ,an·tə'lōp·ə,nē }

antilymphocyte serum [IMMUNOL] An immunosuppressive agent effective in prolonging the lives of homografts in experimental animals by reducing the circulating lymphocytes. { ¦an·tē'lim·fə,sīt ,sir·əm }

antilysin [IMMUNOL] A substance antagonistic to the action of a lysin. { 'an·tē¦lī·sən }

antimagnetic [ENG] Constructed so as to avoid the influence of magnetic fields, usually by the use of nonmagnetic materials and by magnetic shielding. { ,an·tē,mag'ned·ik }

antimalarial [PHARM] **1.** A drug, such as quinacrine, that prevents or suppresses malaria. **2.** Acting against malaria. { ,an·tē·mə'ler·ē·əl }

antimatter [PHYS] Material consisting of atoms which are composed of positrons, antiprotons, and antineutrons. { 'an·tē,mad·ər }

antimere [INV ZOO] Any one of the equivalent parts into which a radially symmetrical animal may be divided. { 'an·tē,mir }

antimetabolite [PHARM] A substance, such as sulfanilamide or amethopterin, that inhibits utilization of an essential metabolite because it is an analog of the metabolite. { ¦an·tē·mə'tab·ə,līt }

antimicrobial agent [MICROBIO] A chemical compound that either destroys or inhibits the growth of microscopic and submicroscopic organisms. { ,an·tē,mī'krōb·ē·əl ,ā·jənt }

antimissile missile [ORD] An explosive missile designed to intercept and destroy another missile in flight. Also known as auntie. { ,an·tē¦mis·əl ¦mis·əl }

antimitotic drug [PHARM] A substance, such as colchicine, vincristine, or vinblastine, that interferes with mitotic cellular division; used in the chemotherapy of leukemia. { ,an·tē,mī'täd·ik ,drəg }

antimolecule [ATOM PHYS] A molecule made up of antiprotons, antineutrons, and positrons in the same way that an ordinary molecule is made up of protons, neutrons, and electrons. { ¦an·tē'mäl·ə,kyül }

antimonate [CHEM] The radical $[Sb(OH)_6]^-$ in salts derived from antimony pentoxide, Sb_4O_{10}, and bases. { 'an·tə·mə,nāt }

antimonial lead [MET] A lead alloy containing up to 25% antimony and possessing greater hardness and tensile strength than lead; used for storage-battery plates, pipes, cable coverings, and roofing. { ,an·tə¦mōn·ē·əl 'led }

antimonic [CHEM] Derived from or pertaining to pentavalent antimony. { ¦an·tə¦män·ik }

antimonide [INORG CHEM] A binary compound of antimony with a more positive compound, for example, H_5Sb. Also known as stibide. { 'an·tə·mə,nīd }

antimonite [MINERAL] Sb_2S_3 A lead-gray antimony sulfide mineral, the primary source of antimony; sometimes contains gold or silver; has a brilliant metallic luster, and occurs as prismatic orthorhombic crystals in massive forms. Also known as antimony glance; gray antimony; stibium; stibnite. { 'an·tə·mə,nīt }

antimonous [CHEM] Pertaining to antimony, especially trivalent antimony. { 'an·tə·mə·nəs }

antimony [CHEM] A chemical element, symbol Sb, atomic number 51, atomic weight 121.75. [MINERAL] A very brittle, tin-white, hexagonal mineral, the native form of the element. { 'an·tə,mō·nē }

antimony-124 [NUC PHYS] Radioactive antimony with mass number of 124; 60-day half-life; used as tracer in solid-state and pipeline flow studies. { 'an·tə,mō·nē ,wən,twen·tē'fór }

antimony blende See kermesite. { 'an·tə,mō·nē 'blend }

antimony glance See antimonite. { 'an·tə,mō·nē 'glans }

antimonyl [CHEM] The inorganic radical SbO^-. { 'an·tə·mə,nil }

antimonyl potassium tartrate See tartar emetic. { 'an·tə·mə,nil pə'tas·ē·əm 'tär,trāt }

antimony(III) oxide [INORG CHEM] Sb_2O_3 Colorless, rhombic crystals, melting at 656°C; insoluble in water; powerful reducing agent. { 'an·tə,mō·nē ,thrē 'äk,sīd }

antimony pentachloride [INORG CHEM] $SbCl_5$ A reddish-yellow, oily liquid; hygroscopic, it solidifies after moisture is absorbed and decomposes in excess water; soluble in hydrochloric acid and chloroform; used in analytical testing for cesium and alkaloids, for dyeing, and as an intermediary in synthesis. Also known as antimony perchloride. { 'an·tə,mō·nē ,pent·ə'klór,īd }

antimony pentafluoride [INORG CHEM] SbF_5 A corrosive, hygroscopic, moderately viscous fluid; reacts violently with water; forms a clear solution with glacial acetic acid; used in the fluorination of organic compounds. { 'an·tə,mō·nē ,pent·ə'flúr,īd }

antimony pentasulfide [INORG CHEM] Sb_2S_5 An orange-yellow powder; soluble in alkali, soluble in concentrated hydrochloric acid, with hydrogen sulfide as a by-product, and insoluble in water; used as a red pigment. Also known as antimony persulfide; antimony red; golden antimony sulfide. { 'an·tə,mō·nē ,pent·ə'səl,fīd }

antimony perchloride See antimony pentachloride. { 'an·tə,mō·nē ,per'klór,īd }

antimony persulfide See antimony pentasulfide. { 'an·tə,mō·nē ,per'səl,fīd }

antimony red See antimony pentasulfide. { 'an·tə,mō·nē 'red }

antimony sodiate See sodium antimonate. { 'an·tə,mō·nē 'sō·dē·āt }

antimony sulfate [INORG CHEM] $Sb_2(SO_4)_3$ Antimony(III) sulfate, a white, deliquescent powder; soluble in acids. { 'an·tə,mō·nē 'səl,fāt }

antimony trichloride [INORG CHEM] $SbCl_3$ Hygroscopic, colorless, crystalline mass; fumes slightly in air, is soluble in alcohol and acetone, and forms antimony oxychloride in water; used as a mordant, as a chlorinating agent, and in fireproofing textiles. { 'an·tə,mō·nē ,trī'klór,īd }

antimony trisulfide [INORG CHEM] Sb_2S_3 Black and orange-red rhombic crystals; soluble in concentrated hydrochloric acid and sulfide solutions, insoluble in water; melting point 546°C; used as a pigment, and in matches and pyrotechnics. { 'an·tə,mō·nē ,trī'səl,fīd }

antimony yellow See lead antimonite. { 'an·tə,mō·nē 'ye·lō }

antimutagen [GEN] A compound that is antagonistic to the action of mutagenic agents on bacteria. { ¦an·tē'myüd·ə·jən }

antineoplastic drug [PHARM] An agent, such as mercapto-purine compounds, that is antagonistic to the growth of a neoplasm. { ¦an·tē,nē·ō¦plas·tik 'drəg }

antineutrino [PARTIC PHYS] The antiparticle to the neutrino; it has zero mass, spin $^1/_2$, and positive helicity; there are two antineutrinos, one associated with electrons and one with muons. { 'an·tē·nü¦trē·nō }

antineutron [PARTIC PHYS] The antiparticle to the neutron; a strongly interacting baryon which has no charge, mass of 939.6 MeV, spin $^1/_2$, and mean life of about 10^3 seconds. { 'an·tē¦nü,trän }

antinodal points See negative nodal points. { ,an·tē'nōd·əl ,póins }

antinode [ASTRON] Either of the two points on an orbit where a line in the orbit plane, perpendicular to the line of nodes and passing through the focus, intersects the orbit. [PHYS] A point, line, or surface in a standing-wave system at which some characteristic of the wave has maximum amplitude. Also known as loop. { 'an·tə,nōd }

antinoise [ACOUS] Noise that is deliberately created to mimic an existing noise field in antiphase so that the two fields cancel each other, resulting in silence. Also known as antisound. { ˌan·tēˈnȯiz }

antinoise microphone [ENG ACOUS] Microphone with characteristics which discriminate against acoustic noise. { ¦an·tē¦nȯiz ˈmi·krəˌfän }

antinuclear antibody [IMMUNOL] Antibody to deoxyribonucleic acid, ribonucleic acid, histone, or nonhistone proteins found in the serum of individuals with certain autoimmune diseases. Abbreviated ANA. { ¦an·tē¦nü·klē·ər ˈan·təˌbäd·ē }

antinucleon [PARTIC PHYS] An antineutron or antiproton, that is, particles having the same mass as their nucleon counterparts but opposite charge or opposite magnetic moment. { ˈan·tē¦nü·klē¸än }

antinucleus [NUC PHYS] A nucleus made up of antineutrons and antiprotons in the same way that an ordinary nucleus is made up of neutrons and protons. { ¦an·tē¦nü·klē·əs }

Antioch process [MET] A method of plaster molding in which a plaster-water mixture is poured over a pattern, after which the mold is steam-treated, allowed to set in air, dried in an oven, and cooled for use in casting certain alloys. { ˈan·tē¸äk ˈpräs·əs }

antioncogene [GEN] Any of a class of genes that are involved in the negative regulation of normal growth; the loss or mutation of these genes leads to malignant growth. More generally called tumor suppressor gene. { ¦an·tē¦aŋ·kəˌjēn }

antioxidant [PHYS CHEM] A substance that, when present at a lower concentration than that of the oxidizable substrate, significantly inhibits or delays oxidative processes, while being itself oxidized. In primary antioxidants, antioxidative activity is implemented by the donation of an electron or hydrogen atom to a radical derivative, and in secondary antioxidants by the removal of an oxidative catalyst and the consequent prevention of the initiation of oxidation. Antioxidants are used in polymers to prevent degradation, and in foods, beverages, and cosmetic products to inhibit deterioration and spoilage. { ˌan·tēˈäk·sə·dənt }

antiozonant [CHEM ENG] A protective agent which can be added to rubber during processing to diminish the deteriorating effects of ozone. { ˌan·tēˈō·zə·nənt }

antiparallel [GEN] The opposite orientation of the two complementary strands of deoxyribonucleic acid, 5′ to 3′ and 3′ to 5′. [MATH] Property of two nonzero vectors in a vector space over the real numbers such that one vector equals the product of the other vector and a negative number. [PHYS] Property of two displacements or other vectors which lie along parallel lines but point in opposite directions. { ¦an·tē¦par·əˌlel }

antiparallel lines [MATH] Two lines that make equal angles in opposite order with two specified lines. { ˌan·tē¸par·əˌlel ˈlīnz }

antiparasitic agent [PHARM] An agent, such as emetine or quinine, that destroys or suppresses human and animal parasites. { ˌan·tē¦par·əˌsid·ik ˈā·jənt }

antiparticle [PARTIC PHYS] A counterpart to a particle having mass, lifetime, and spin identical to the particle but with charge and magnetic moment reversed in sign. { ˈan·tē¦pärd·ə·kəl }

Antipatharia [INV ZOO] The black or horny corals, an order of tropical and subtropical cnidarians in the subclass Zoantharia. { ˌan·tē·pəˈthar·ē·ə }

antipercolator [MECH ENG] In an automotive engine, a valve in the carburetor that is designed to vent vapor when the throttle plate is closed; prevents fuel from dropping into the carburetor due to unvented pressure. { ˌan·tē¦pər·kəˌlād·ər }

antipersonnel [ORD] Applied to bombs, mines, and missiles designed to kill, wound, or obstruct people. Abbreviated apers. { ˌan·tē¸pərs·ənˈel }

antipersonnel weapon [ORD] All instruments of combat, either offensive or defensive, used to destroy, wound, obstruct, or threaten enemy personnel. { ˌan·tē¸pərs·ənˈel ˌwep·ən }

antiperthite [GEOL] Natural intergrowth of feldspars formed by separation of sodium feldspar (albite) and potassium feldspar (orthoclase) during slow cooling of molten mixtures; the potassium-rich phase is evolved in a plagioclase host, exactly the inverse of perthite. { ˌan·tiˈpərˌthīt }

antipetalous [BOT] Having stamens positioned opposite to, rather than alternating with, the petals. { ¦an·tiˈped·ə·ləs }

antipitching fin [NAV ARCH] A control surface protruding underwater at the bow or stern to generate lift opposed to the pitching of the ship. { ˌan·tēˈpich·iŋ ˌfin }

antipodal [BOT] Any of three cells grouped at the base of the embryo sac, that is, at the end farthest from the micropyle, in most angiosperms. { anˈtip·əd·əl }

antipodal points [MATH] The points at opposite ends of a diameter of a sphere. { anˈtip·əd·əl ˈpȯins }

antipodes [GEOD] Diametrically opposite points on the earth. { anˈtip·əˌdēz }

antiporter [CELL MOL] A channel protein that simultaneously or sequentially transports two different types of substrates across a cell membrane, one into the cell (for example sodium ion) and one out of the cell (for example calcium ion). { anˈtē′pȯrd·ər }

antiprincipal planes See negative principal planes. { ¦an·tē¦prin·sə·pəl ˈplānz }

antiprincipal point See negative principal point. { ¦an·tē¦prin·sə·pəl ˈpȯint }

antiproton [PARTIC PHYS] The antiparticle to the proton; a strongly interacting baryon which is stable, carries unit negative charge, has the same mass as the proton (938.3 MeV), and has spin $1/2$. { ¦an·tē¦prōˌtän }

antiprotonic atom [ATOM PHYS] An atom consisting of an ordinary nucleus with an orbiting antiproton. { ˌan·te·proˈtän·ik ˈad·əm }

antipruritic [PHARM] An agent, such as camphor, that relieves itching. { ˌan·tē·prüˈrid·ik }

antipyretic [PHARM] Any agent, such as aspirin, that reduces or prevents fever. { ˌan·tē·pīˈred·ik }

antipyrine [PHARM] $C_{11}H_{12}ON_2$ A compound used as an antipyretic, analgesic, and antirheumatic. { ˌan·tē′pīˌrēn }

antiquark [PARTIC PHYS] The hypothetical antiparticle of a quark, having electric charge, baryon number, and strangeness opposite in sign to that of the corresponding quark. { ˈan·tē¦kwärk }

antique finish [MATER] A paper finish, somewhat rougher than eggshell, obtained by operating wet presses and calender stacks at reduced pressures. { an¦tēk ˈfin·ish }

antiquing [ENG] **1.** Producing a rich glow on the surface of a leather by applying stain, wax, or oil, allowing it to set, and rubbing or brushing the leather. **2.** A technique of handling wet paint to expose parts of the undercoat, by combing, graining, or marbling. Also known as broken-color work. { anˈtēk·iŋ }

antirachitic vitamin See vitamin D. { ˌan·tē·rəˈkid·ik ˈvid·ə·mən }

antirad [CHEM ENG] An inhibitor incorporated into rubber during manufacturing to reduce the degrading effects of radiation. { ¦an·te¦rad }

antiradar coating [ENG] A surface treatment used to reduce the reflection of electromagnetic waves so as to avoid radar detection. { ˌan·tē′rāˌdär ¦kōd·iŋ }

antirattle spring [MECH ENG] In an automotive vehicle, a spring installed to hold parts in the clutches and the disk brakes together; prevents rattling. { ˈan·tē¦rad·əl ˈspriŋ }

anti-redeposition agent [CHEM ENG] An additive used in a detergent to help prevent soil from resettling on a fabric after it has been removed during washing. { ¦an·tē¸rē¸dep·əˈzish·ən ¸āˌjənt }

antireflection coating [ENG] The application of a thin film of dielectric material to a surface to reduce its reflection and to increase its transmission of light or other electromagnetic radiation. { ˌan·tē·riˈflek·shən ¸kōd·iŋ }

antiresonance [ELEC] See parallel resonance. [ENG] The condition for which the impedance of a given electric, acoustic, or dynamic system is very high, approaching infinity. { ˌan·tēˈrez·ən·əns }

antiresonant circuit See parallel resonant circuit. { ˌan·tēˈrez·ən·ənt ˈsər·kət }

anti-Rh agglutinin [IMMUNOL] An antibody against any Rh antigen; it must be acquired and is never natural. { ˈan·tēˌärˌāch əˈglüt·ən·ən }

anti-Rh immunoglobulin [IMMUNOL] A serum protein that destroys Rh-positive fetal erythrocytes in an Rh-negative mother when administered after delivery. { ¦an·tēˌärˌäch ˌim·yə·nəˈgläb·yə·lən }

anti-Rh serum [IMMUNOL] A blood serum containing anti-Rh antibodies. { ¦an·tēˌärˌäch ˈsir·əm }

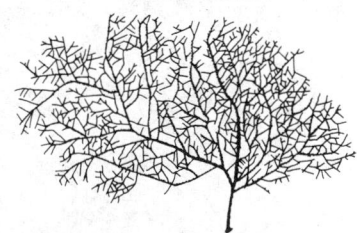

ANTIPATHARIA

Irregularly branching colony, *Antipathes rhipidion*.

antiricochet device [ORD] Usually a parachute unit, fuse adapter, and fuse attached to the tail of a bomb to prevent ricochet, with consequent loss of effectiveness and possible danger to the dropping plane. { ¦an·tē¦rik·ə₁shā di₁vīs }

antiroll fin [NAV ARCH] A control surface protruding from either of the lower sides of a ship to generate lift opposed to the rolling of the ship. { ¦an·tē¦rōl ₁fin }

antirolling gyroscope [NAV ARCH] A very large gyroscope having a vertical axis and mounted in a ship so that the axis can be tipped fore and aft by an electric motor, resulting in the gyroscope's exerting a large torque on the ship about the fore-and-aft axis. { ¦an·tē¦rōl·iŋ ¦jī·rə₁skōp }

antiroll tank [NAV ARCH] One of paired tanks at opposite sides within a ship which are partially filled with liquid whose flow from one tank to the other helps to offset ship rolls. { ¦an·tē¦rōl 'taŋk }

antirostrum [INV ZOO] The terminal segment of the appendages of certain mites. { ₁an·tē¦räs·trəm }

antisatellite missile [ORD] A missile whose target is an orbiting satellite. { an·tē¦sad·ə₁līt ₁mis·əl }

antisecant See arc secant. { ₁an·tē¦sē₁kant }

antisense [GEN] A strand of deoxyribonucleic acid having a sequence identical to messenger ribonucleic acid. { ¦an·tē¦sens }

antisense drug [MED] A gene-based drug containing material that inhibits the synthesis of abnormal protein (which is typically caused by a specific disease state) by specifically binding to the ribonucleic acid responsible for its formation. { ₁an·tē¦sens 'drəg }

antisense ribonucleic acid [MOL BIO] A ribonucleic acid (RNA) transcript (or portion of one) that is complementary to another nucleic acid, usually another RNA molecule. { ₁an·tē¦sens ₁rī·bō¦nü¦klē·ik 'as·əd }

antisense strand [MOL BIO] The strand of a double-stranded deoxyribonucleic acid molecule from which ribonucleic acid is transcribed. Also known as anticoding strand. { ¦an·tē¦sens ₁strand }

antiseptic [MICROBIO] A substance used to destroy or prevent the growth of infectious microorganisms on or in the human or animal body. { ¦an·tə¦sep·tik }

antiserum [IMMUNOL] The serum component of blood that contains antibodies specific to one or more antigens. { 'an·tē₁sir·əm }

antisetoff powder [GRAPHICS] Fine-powdered starch or limestone applied to a freshly printed sheet of paper to reduce unintentional transfer of ink to neighboring sheets in a stack. { ₁an·tē¦sed₁óf ₁paùd·ər }

antishock agent [PHARM] A substance, such as a cesium salt, that relieves a state of shock. { ₁an·tē¦shäk ₁ā·jənt }

antisideric [PHARM] A pharmaceutical that counteracts the physiological effects of iron. { ¦an·tē·sə¦der·ik }

anti-sidetone circuit [ELEC] Telephone circuit which prevents sound, introduced in the local transmitter, from being reproduced in the local receiver. { ¦an·tē¦sīd₁tōn ₁sər·kət }

antisine See arc sine. { ₁an·tē¦sīn }

antiskid plate [ENG] A sheet of metal roughed on both sides and placed between piled objects, such as boxes in a freight car, to prevent sliding. { ¦an·tē¦skid 'plāt }

antiskinning agents [GRAPHICS] Antioxidants that prevent printing inks from drying before use. { ₁an·tē¦skin·iŋ ₁ā·jəns }

antislip metal [MET] Metal, usually iron, bronze, or aluminum, containing abrasive grains cast with the metal or rolled out to the surface; used for car steps, stair treads, or floor plates. { ¦an·tē¦slip ₁med·əl }

antislip paint [MATER] A paint with a high coefficient of friction, caused by addition of sand, wood flour, or cork dust; used on steps, porches, and walkways to prevent slipping. { ¦an·tē¦slip ₁pānt }

antismallpox vaccine See smallpox vaccine. { ¦an·tē¦smól ₁päks ₁vak·sēn }

antismudge ring [BUILD] A frame attached around a ceiling-mounted air diffuser, to minimize the formation of rings of dirt on the ceiling. { ¦an·tē¦sməj 'riŋ }

antisolar point [ASTRON] The point on the celestial sphere which lies directly opposite the sun from the observer, that is, on the line from the sun through the observer. { ¦an·tē¦sō·lər ₁póint }

antisound See antinoise. { ₁an·tē¦saùnd }

antispasmodic [PHARM] An agent, such as benzyl benzoate, that relieves convulsions and the pain of muscular spasms. { ¦an·tē₁spaz¦mäd·ik }

antistat See antistatic agent. { 'an·tē₁stat }

antistatic agent [MATER] A material used with textiles, plastics, paper products, or wax polishes to reduce static-electrical charges by allowing the charge to leak off. Also known as antistat. { ¦an·tē¦stad·ik 'ā·jənt }

antistatic brush [GRAPHICS] A brush whose bristles do not retain electrostatic charge. { ¦an·tē¦stad·ik 'brəsh }

antistatic mat [COMPUT SCI] A floor mat placed in front of a device such as a tape drive that is sensitive to discharges of static electricity to safeguard against loss of data from such discharges during human handling of the device. { ¦an·tē¦stad·ik 'mat }

anti-Stokes lines [SPECT] Lines of radiated frequencies which are higher than the frequency of the exciting incident light. { ¦an·tē¦stōks ₁līnz }

antistreptolysin [IMMUNOL] The antibody that neutralizes the streptolysin of group A hemolytic streptococci. { ₁an·tē₁strep·tə¦līs·ən }

antistress mineral [MINERAL] Minerals such as leucite, nepheline, alkalic feldspar, andalusite, and cordierite which cannot form or are unstable in an environment of high shearing stress, and hence are not found in highly deformed rocks. { ¦an·tē¦stres ₁min·ə·rəl }

antistripping agent [MATER] An additive used in an asphaltic binder to overcome the affinity of an aggregate for water instead of asphalt; assists the asphalt to adhere to wet surfaces. { ₁an·tē¦strip·iŋ 'ā·jənt }

antisubmarine [ORD] Descriptive of equipment, mines, or missiles designed to attack or destroy submarines. { ¦an·tē₁səb·mə'rēn }

antisubmarine missile [ORD] A missile whose target is a submarine. { ¦an·tē₁səb·mə'rēn 'mis·əl }

antisubmarine rocket [ORD] A solid-propellant-rocket-driven torpedo that is surface-launched, with final track underwater. Abbreviated ASROC. { ¦an·tē₁səb·mə'rēn 'räk·ət }

antisubmarine weapon [ORD] Any naval device designed to destroy submarines. { ¦an·tē₁səb·mə'rēn 'wep·ən }

antisymmetric determinant [MATH] The determinant of an antisymmetric matrix. Also known as skew-symmetric determinant. { ₁an·tē·sə₁me·trik di'tər·mə·nənt }

antisymmetric dyadic [MATH] A dyadic equal to the negative of its conjugate. { ¦an·tē·si¦me·trik dī'ad·ik }

antisymmetric matrix [MATH] A matrix which is equal to the negative of its transpose. Also known as skew matrix; skew-symmetric matrix. { ¦an·tē·si¦me·trik 'mā·triks }

antisymmetric relation [MATH] A relation, which may be denoted ∈, among the elements of a set such that if $a \in b$ and $b \in a$ then $a = b$. { ₁ant·i·si¦me·trik ri'lā·shən }

antisymmetric tensor [MATH] A tensor in which interchanging two indices of an element changes the sign of the element. { ¦an·tē·si¦me·trik 'ten·sər }

antisymmetric wave function [PHYS] A many-particle wave function which changes its sign when the coordinates of two of the particles are interchanged. { ¦an·tē·si¦me·trik 'wāv ₁fəŋk·shən }

antisymmetrized wave function [QUANT MECH] A wave function of several identical fermions (such as electrons) which changes sign but in all other respects remains unaltered if two of the fermions are interchanged. { ₁an·tē₁sim·ə₁trīzd 'wāv ₁fəŋk·shən }

antitail [ASTRON] A structure occasionally observed in comets that appears to extend from the coma toward the sun, and usually has the appearance of a spike. { ¦an·tē¦tāl }

antitangent See arc tangent. { ₁an·tē¦tan·jənt }

antitank ditch [ORD] A deep ditch prepared as an obstacle to enemy tanks. Also known as tank ditch. { ¦an·tē¦taŋk ₁dich }

antitank grenade [ORD] A rifle grenade designed to be used against tanks or other armored vehicles. { ¦an·tē¦taŋk grə'nād }

antitank guided missile [ORD] An antitank missile whose flight path is controlled by a combination of optical sighting and command signals from an automatic computer through multiple wire command links; it may also contain an infrared homing device for final range correction; it is designed to be launched from any type of vehicle or ground emplacement;

this definition excludes items whose trajectory cannot be altered in flight. { ¦an·tē¦taŋk ¸gīd·əd 'mis·əl }

antitank gun [ORD] A gun suitable for use against tanks or other armored vehicles. { ¦an·tē¦taŋk ¸gən }

antitank mine activator [ORD] A nonmetallic item designed to adapt a firing device to an antitank mine; may be empty, inert-filled, or explosive-filled. { ¦an·tē¦taŋk 'mīn ¸ak·tə¸vād·ər }

antitank mine field [ORD] **1.** Area in which antitank mines are planted to stop or slow down enemy tanks or other armored vehicles. **2.** Pattern of antitank mines. { ¦an·tē¦taŋk 'mīn ¸fēld }

antitank missile [ORD] A missile designed to destroy or immobilize enemy tanks. { ¦an·tē¦taŋk ¦mis·əl }

antitank obstacle [ORD] Natural or artificial obstruction or barrier which will stop, slow down, or cause tanks or other armored vehicles to maneuver or change direction. Also known as tank obstacle. { ¦an·tē¦taŋk ¸äb·sti·kəl }

antitank weapon [ORD] Any weapon suitable for use against tanks or other armored vehicles. { ¦an·tē¦taŋk ¸wep·ən }

antitermination factor [BIOCHEM] Protein that interferes with normal termination of ribonucleic acid synthesis. { ¸an·tē¸tər·mə¦nā·shən 'fak·tər }

antitheft device [MECH ENG] A piece of equipment installed on an automotive vehicle in order to prevent or slow down theft; designs include mechanical locks on the steering wheel and ignition switch as well as other means of shutting off the ignition system, shutting off fuel flow, or sounding an alarm. { ¸an·tē'theft di¸vīs }

antithetic variable [STAT] One of two random variables having high negative correlation, used in the antithetic variate method of estimating the mean of a series of observations. { ¦an·tē¦thed·ik 'ver·ē·ə·bəl }

antithrombin [BIOCHEM] A substance in blood plasma that inactivates thrombin. { ¸an·tē¦thräm·bən }

antitorpedo [ORD] Descriptive of equipment, missiles, and the like designed to combat or destroy torpedoes. { ¸an·tē·tȯr'pēd·ō }

antitoxin [IMMUNOL] An antibody elaborated by the body in response to a bacterial toxin that will combine with and generally neutralize the toxin. { ¸an·tē'täk·sən }

antitrades [METEOROL] A deep layer of westerly winds in the troposphere above the surface trade winds of the tropics. { 'an·tē¸trādz }

anti-transmit-receive tube [ELECTR] A switching tube that prevents the received echo signal from being dissipated in the transmitter. { ¦an·tē·tranz¦mit ri¦sēv ¸tüb }

antitrigonometric function See inverse trigonometric function. { ¸an·tē¸trig·ə·nə¸me·trik 'fəŋk·shən }

antitriptic wind [METEOROL] A wind for which the pressure force exactly balances the viscous force, in which the vertical transfers of momentum predominate. { ¦an·tē¦trip·tik ¸wind }

antitriton [ATOM PHYS] The antiparticle to the triton, composed of an antiproton and two antineutrons. { ¦an·tē¦trīt·ən }

antitubercular agent [PHARM] A substance, such as streptomycin or isoniazid, used in the treatment of tuberculosis. { ¦an·tē·tə¦bərk·yə·lər 'ā·jənt }

antitumor antibiotic [MICROBIO] A substance, such as actinomycin, luteomycin, or mitomycin C, which is produced by microorganisms and is effective against some forms of cancer. { ¸an·tē¦tüm·ər ¸an·tē¸bī'äd·ik }

antitussive [PHARM] An agent, such as benylin expectorants, that relieves coughing. { ¦an·tē¦təs·iv }

antitwilight arch [METEOROL] The pink or purplish band of about 3° vertical angular width which lies just above the antisolar point at twilight; it rises with the antisolar point at sunset and sets with the antisolar point at sunrise. { ¦an·tē¦twī¸līt 'arch }

antivenin [IMMUNOL] An immune serum that neutralizes the venoms of certain poisonous snakes and black widow spiders. { ¦an·tē¦ven·ən }

antivernalization [BOT] Delayed flowering in plants due to treatment with heat. { ¦an·tē¸vərn·əl·ə'zā·shən }

antiviral agent [PHARM] A substance, such as interferon or amantadine, that decreases virus multiplication in the body. { ¦an·tē¦vī·rəl 'ā·jənt }

antivirus program See antivirus software. { ¸an·tē'vī·rəs ¸prō·grəm }

antivirus software [COMPUT SCI] Software that is designed to protect against computer viruses. { ¦an·tē¸vī·rəs 'sȯf¸wer }

antivitamin [BIOCHEM] Any substance that prevents a vitamin from normal metabolic functioning. { ¸an·tē¦vīd·ə·mən }

antixerophthalmic vitamin See vitamin A. { ¦an·tē¸zir¸äf'thal·mik 'vīd·ə·mən }

antler [VERT ZOO] One of a pair of solid bony, usually branched outgrowths on the head of members of the deer family (Cervidae); shed annually. { 'ant·lər }

antlerite [MINERAL] Cu$_3$SO$_4$(OH)$_4$ Emerald- to blackish-green mineral occurring in aggregates of needlelike crystals; an ore of copper. Also known as vernadskite. { 'ant·lə¸rīt }

Antler orogeny [GEOL] Late Devonian and Early Mississippian orogeny in Nevada, resulting in the structural emplacement of eugeosynclinal rocks over microgeosynclinal rocks. { 'ant·lər ȯ'räj·ə·nē }

Antlia [ASTRON] A constellation with a right ascension of 10 hours and declination of 35°S. Abbreviated Ant. Also known as Air Pump. { 'ant·lē·ə }

ant lion [INV ZOO] The common name for insects of the family Myrmeleontidae in the order Neuroptera; larvae are commonly called doodlebugs. { 'ant ¸lī·ən }

antlophobia [PSYCH] Abnormal fear of floods. { ¸ant·lə'fōb·ē·ə }

AN-TNT slurry [MATER] A mixture of ammonium nitrate and trinitrotoluene; used as an explosive. { ¸ā¸en ¸tē¸en¦tē 'slər·ē }

Antoine equation [PHYS] The empirical relationship between temperature and vapor pressure of liquids; log $P = B - A/(C + T)$, where A, B, C are experimental constants, T is absolute temperature, and P is vapor pressure. { 'an¸twän i¸kwā·zhən }

Antonadi scale [ASTRON] A scale for measuring seeing conditions, ranging from I for perfect conditions to V for very bad conditions. { ¸än·tō'näd·ē ¸skāl }

Antonoff's rule [PHYS] The rule which states that the surface tension at the interface between two saturated liquid layers in equilibrium is equal to the difference between the individual surface tensions of similar layers when exposed to air. { an'tä¸nȯfs ¸rül }

antrorse [BIOL] Turned or directed forward or upward. { 'an¸trȯrs }

antrum [ANAT] A cavity of a hollow organ or a sinus. { 'an·trəm }

ANTU See 1-(1-naphthyl)-2-thiourea.

Anura [VERT ZOO] An order of the class Amphibia comprising the frogs and toads. { ə'nur·ə }

anuresis [MED] Retention of urine in the urinary bladder due to inability to void. { ¸an·yə'rē·səs }

anuria [MED] Complete absence of urinary output. { ə'nyur·ē·ə }

anus [ANAT] The posterior orifice of the alimentary canal. { 'ā·nəs }

anvil [ANAT] See incus. [ENG] **1.** The part of a machine that absorbs the energy delivered by a sharp force or blow. **2.** The stationary end of a micrometer caliper. [MET] **1.** A heavy wrought-iron, cast-iron, or steel block upon which metal is hammered in smith forging. **2.** The base of the hammer, holding the die bed and lower die part in drop forging. [METEOROL] See incus. { 'an·vəl }

anvil cloud [METEOROL] The popular name given to a cumulonimbus capillatus cloud, a thunderhead whose upper portion spreads in the form of an anvil with a fibrous or smooth aspect; it also refers to such an upper portion alone when it persists beyond the parent cloud. { 'an·vəl ¸klaud }

anxiety [PSYCH] A physiological and mental state of apprehension and fear of something unknown to the conscious. { aŋ'zī·əd·ē }

anxiety neurosis [PSYCH] A psychoneurotic disorder characterized by diffuse anxious expectation not restricted to definite situations, persons, or objects, and by emotional instability, irritability, apprehensiveness, and a sense of fatigue; caused by incomplete resolution of repressed emotional problems, and frequently associated with somatic symptoms. { ¸aŋ'zī·əd·ē nu'rō·səs }

anxiolytic agent [PHARM] A drug that relieves anxiety. { ¸aŋk·sē·ō¦lid·ik 'ā·jənt }

anyon [QUANT MECH] A particle obeying an unconventional form of quantum statistics, which is characterized by a parameter that can take on any of a continuum of values, just two of which represent Bose-Einstein and Fermi-Dirac statistics. { 'an·ē¸än }

ANT LION

The ant lion (*Myrmeleon immaculatus*).

anyris oil [MATER] Sandalwood oil from West India sandalwood species. { ə'nī·rəs ,óil }

Ao horizon [GEOL] That portion of the A horizon of a soil profile which is composed of pure humus. { ¦ā¦ō hə'rīz·ən }

A-O-I gate See AND-OR-INVERT gate. { ,ā,ō'ī ,gāt }

A-1 time [ASTRON] A particular atomic time scale, established by the U.S. Naval Observatory, with the origin on January 1, 1958, at zero hours Universal Time and with the unit (second) equal to 9,192,631,770 cycles of cesium at zero field. { 'ā ¦wən ,tīm }

Aoo horizon [GEOL] Uppermost portion of the A horizon of a soil profile which consists of undecomposed vegetable litter. { ¦ā¦ō¦ō hə'rīz·ən }

AOQL See average outgoing quality limit.

aorta [ANAT] The main vessel of systemic arterial circulation arising from the heart in vertebrates. [INV ZOO] The large dorsal or anterior vessel in many invertebrates. { ā'órd·ə }

aortic aneurysm [MED] Dilation of the wall of the aorta, usually the ascending portion. { ā'órd·ik 'an·yə,riz·əm }

aortic arch [ANAT] The portion of the aorta extending from the heart to the third thoracic vertebra; single in warm-blooded vertebrates and paired in fishes, amphibians, and reptiles. { ā'órd·ik 'ärch }

aortic body See aortic paraganglion. { ā'órd·ik 'bäd·ē }

aortic incompetence [MED] A condition in which blood from the aorta flows back into the left ventricle because of the incapacity of the aortic valve. { ā'órd·ik in'käm·pə·təns }

aortic paraganglion [ANAT] A structure in vertebrates belonging to the chromaffin system and found on the front of the abdominal aorta near the mesenteric arteries. Also known as aortic body; organs of Zuckerkandl. { ā'órd·ik ,pa·rə'gaŋ·glē,än }

aortic stenosis [MED] Abnormal narrowing of the aortic valve orifice; may be either congenital or acquired. { ā'órd·ik stə'nō·səs }

aortic valve [ANAT] A heart valve comprising three flaps which guards the passage from the left ventricle to the aorta and prevents the backward flow of blood. { ā'órd·ik 'valv }

aortitis [MED] Inflammation of the aorta. { ,ā,ór'tīd·əs }

aortography [MED] Radiography of the aorta through a radiopaque dye injection. { ,ā,ór'tä·grə·fē }

AP See after-perpendicular.

4-AP See 4-aminopyridine.

6-APA See 6-aminopenicillanic acid.

apachite [PETR] A phonolite consisting of enigmatite and hornblende in about the same quantity as the pyroxene, but of a later crystallization phase. { ə'pa,chīt }

apandrous [BOT] Lacking male organs or having nonfunctional male organs. { ,a'pan·drəs }

APAP See acetaminophen.

apareon [ASTRON] The point on a Mars-centered orbit where a satellite is at its greatest distance from Mars. { ,a'par·ē·ən }

apastron [ASTRON] That point of the orbit of one member of a binary star system at which the stars are farthest apart. { ,a'pas·trən }

Apatemyidae [PALEON] A family of extinct rodentlike insectivorous mammals belonging to the Proteutheria. { ə,pad·ə'mī·ə,dē }

apatetic [ECOL] Pertaining to the imitative protective coloration of an animal subject to being preyed upon. { ,a·pə¦ted·ik }

Apathornithidae [PALEON] A family of Cretaceous birds, with two species, belonging to the order Ichthyornithiformes. { ,a·pə,thór'nith·ə,dē }

apatite [MINERAL] A group of phosphate minerals that includes 10 mineral species and has the general formula $X_5(YO_4)_3Z$, where X is usually Ca^{2+} or Pb^{3+}, Y is P^{5+} or As^{5+}, and Z is F^-, Cl^-, or OH^-. { 'ap·ə,tīt }

Apatosaurus [PALEON] A herbivorous sauropod dinosaur, approximately 70 feet (21 meters) long and weighing 30 tons, from the Jurassic Period that had much longer hindlimbs than forelimbs. Also known as Brontosaurus. { ə,pad·ə'sór·əs }

APC See anaphase-promoting complex; automatic phase control.

ape [VERT ZOO] Any of the tailless primates of the families Hylobatidae and Pongidae in the same superfamily as humans. { āp }

apeirophobia [PSYCH] Abnormal fear of infinity. { ə,pīr·ə'fōb·ē·ə }

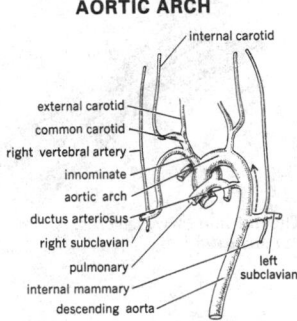

AORTIC ARCH

Ventral view of the aortic arch in the human, showing the blood vessels that arise from it. *(Modified from B. M. Patten)*

aperiodic [PHYS] Of irregular occurrence; not periodic; not displaying resonant response. { ¦a,pir·ē¦äd·ik }

aperiodic antenna [ELECTROMAG] Antenna designed to have constant impedance over a wide range of frequencies because of the suppression of reflections within the antenna system; includes terminated wave and rhombic antennas. { ¦a,pir·ē¦äd·ik an'ten·ə }

aperiodic compass [NAV] Literally, a compass without a period, that is, a compass that, after being deflected, returns by one direct movement to its proper reading, without oscillation. Also known as deadbeat compass. { ¦a,pir·ē¦äd·ik 'kəm·pəs }

aperiodic damping [PHYS] Condition of a system in which the amount of damping is so large that, when the system is subjected to a single disturbance, either constant or instantaneous, the system comes to a position of rest without passing through that position; while an aperiodically damped system is not strictly an oscillating system, it has such properties that it should become an oscillating system if the damping were sufficiently reduced. { ¦a,pir·ē¦äd·ik 'damp·iŋ }

aperiodic waves [ELEC] The transient current wave in a series circuit with resistance R, inductance L, and capacitance C when $R^2C = 4L$. [PHYS] Waves without a definite repetitive pattern; for example, transient waves. { ¦a,pir·ē¦äd·ik 'wāvz }

apers See antipersonnel. { 'ā,pərs }

apertometer [OPTICS] An instrument designed to measure the numerical aperture of microscope objectives. { ,a·pər'täm·əd·ər }

Apert's syndrome See acrocephalosyndactylism. { 'a,pərts ,sin,drōm }

aperture [ELECTR] An opening through which electrons, light, radio waves, or other radiation can pass. [GRAPHICS] A rectangular cutout on an aperture card. [OPTICS] The diameter of the objective of a telescope or other optical instrument, usually expressed in inches, but sometimes as the angle between lines from the principal focus to opposite ends of a diameter of the objective. { 'ap·ə,chər }

aperture aberration [OPTICS] Errors in optical imaging which occur because rays of different distances from the axis do not come to the same focus. { 'ap·ə,chər ,ab·ə'rā·shən }

aperture angle [OPTICS] The angle subtended by the radius of the entrance pupil of an optical instrument at the object. { 'ap·ə,chər ,aŋ·gəl }

aperture antenna [ELECTROMAG] Antenna in which the beam width is determined by the dimensions of a horn, lens, or reflector. { 'ap·ə,chər an'ten·ə }

aperture card [GRAPHICS] A card designed with apertures for mounting or inserting frames of microfilm. { 'ap·ə,chər ,kärd }

aperture conductivity [ACOUS] The ratio of the density of a medium to the acoustic mass at an aperture. { 'ap·ə,chər ,kän,dək'tiv·əd·ē }

aperture disk [ENG] A disk with a small round opening used in a densitometer to vary the amount of light or the area to be measured. { 'ap·ə,chər ,disk }

aperture grill picture tube [ELECTR] An in-line gun-type picture tube in which the shadow mask is perforated by long, vertical stripes and the screen is painted with vertical phosphor stripes. { 'ap·ə,chər ,gril 'pik·chər ,tüb }

aperture illumination [ELECTROMAG] Field distribution in amplitude and phase over an aperture. { 'ap·ə,chər i,lüm·ə'nā·shən }

aperture mask See shadow mask. { 'ap·ə,chər ,mask }

aperture plate [ELECTR] A small part of a piece of perforated ferromagnetic material that forms a magnetic cell. { 'ap·ə,chər ,plāt }

aperture ratio [OPTICS] The ratio of the effective diameter of a lens to its focal length. { 'ap·ə,chər 'rā·shō }

aperture sight [ORD] An irregularly shaped adjustable mechanical item usually integral to a rear sight; it functions as a peephole through which the sight at the opposite end of a gun is brought into view in aiming at a target or object. { 'ap·ə,chər ,sīt }

aperture slit See aperture slot. { 'ap·ə,chər ,slit }

aperture slot [OPTICS] A narrow rectangular opening in the optical system of a rotary camera through which light from a continuously moving document is transmitted to a film whose movement is synchronized to that of the document. Also known as aperture slit. { 'ap·ə,chər ,slät }

aperture splitting [OPTICS] A technique in which light from a single slit is divided by passing it through two other slits and is combined by a lens. { 'ap·ə·chər ,splid·iŋ }

aperture stop [OPTICS] That opening in an optical system that determines the size of the bundle of rays which traverse the system from a given point of the object to the corresponding point of the image. { 'ap·ə,chər ,stäp }

aperture synthesis [ELECTROMAG] The use of one or more pairs of instruments of relatively small aperture, acting as interferometers, to obtain the information-gathering capability of a telescope of much larger aperture. { 'ap·ə,chər ,sin·thə·səs }

apetalous [BOT] Lacking petals. { ,ā'ped·əl·əs }

apex [ANAT] **1.** The upper portion of a lung extending into the root. **2.** The pointed end of the heart. **3.** The tip of the root of a tooth. [BOT] The pointed tip of a leaf. [ENG] In architecture or construction, the highest point, peak, or tip of any structure. [GEOL] The part of a mineral vein nearest the surface of the earth. [MATH] **1.** The vertex of a triangle opposite the side which is regarded as the base. **2.** The vertex of a cone or pyramid. { 'ā,peks }

apex impulse [PHYSIO] The point of maximum outward movement of the left ventricle of the heart during systole, normally localized in the fifth left intercostal space in the midclavicular line. Also known as left ventricular thrust. { 'ā,peks ,im,pəls }

apex stone [ARCH] The uppermost stone in a gable, pediment, vault, or dome; usually triangular, often highly decorated. Also known as saddle stone. { 'ā,peks ,stōn }

Apgar score [MED] An index used to evaluate a newborn infant's physical condition based on a rating of 0–2 for each of five criteria: heart rate, respiratory effort, muscle tone, response to stimulation, and skin color. { 'ap·gär ,skȯr }

aphagia [MED] Inability to swallow; may be organic or psychic in origin. { ə'fāj·ə }

aphakia [MED] Absence of the lens of the eye. { ə'fāk·ē·ə }

aphaniphyric [PETR] Denoting a texture of porphyritic rocks with microaphanitic groundmasses. Also known as felsophyric. { ,af·ə·nə'fir·ik }

aphanite [PETR] **1.** A general term applied to dense, homogeneous rocks whose constituents are too small to be distinguished by the unaided eye. **2.** A rock having aphanitic texture. { 'af·ə,nīt }

aphanitic [PETR] Referring to the texture of an igneous rock in which the crystalline components are not distinguishable by the unaided eye. { ,af·ə'nid·ik }

Aphanomyces [MYCOL] A genus of fungi in the phycomycetous order Saprolegniales; species cause root rot in plants. { 'af·ə·nə'mī,sēz }

A phase See liquid A. { 'ā ,fāz }

A₁ phase See liquid A. { 'ā·səb¦wən ,fāz }

aphasia [MED] Impairment in the use or comprehension of language that is caused by lesions of the cerebral cortex. { ə'fāzh·ə }

aphasic seizure [MED] A transient inability to speak due to an abnormal electrical discharge from the speech areas of the brain. { ə'fāz·ik 'sēzh·ər }

Aphasmidea [INV ZOO] An equivalent name for the Adenophorea. { ¦ā,faz'mid·ē·ə }

Aphelenchoidea [INV ZOO] A superfamily of plant and insect-associated nematodes in the order Tylenchida. { ,af·ə,leŋ'kȯid·ē·ə }

Aphelenchoidoidea [INV ZOO] A superfamily of parasitic nematodes containing only one family, characterized by the lack of an isthmus in the esophagus and, in males, thorn-shaped spicules. { ,af·ə,leŋ,kȯi'dȯid·ē·ə }

aphelion [ASTRON] The point on a planetary orbit farthest from the sun. { ə'fēl·yən }

Aphelocheiridae [INV ZOO] A family of hemipteran insects belonging to the superfamily Naucoroidea. { ¦af·ə,läk·ə'rī,dē }

aphesperian [ASTRON] The farthest point of a satellite in its orbit about Venus. { ,ā·fə'spir·ē·ən }

aphid [INV ZOO] The common name applied to the soft-bodied insects of the family Aphididae; they are phytophagous plant pests and vectors for plant viruses and fungal parasites. { 'ā·fəd }

Aphididae [INV ZOO] The true aphids, a family of homopteran insects in the superfamily Aphidoidea. { ə'fid·ə,dē }

Aphidoidea [INV ZOO] A superfamily of sternorrhynchan insects in the order Homoptera. { ,a·fə'dȯid·ē·ə }

Aphis [INV ZOO] A genus of aphid, the type genus of the family Aphididae. { 'ā·fəs }

aphonia [MED] Loss of voice and power of speech. { ā'fōn·ē·ə }

aphotic zone [OCEANOGR] The deeper part of the ocean where sunlight is absent. { ā'fäd·ik ,zōn }

Aphredoderidae [VERT ZOO] A family of actinopterygian fishes in the order Percopsiformes containing one species, the pirate perch. { ,a·frə·də'der·ə,dē }

aphrodisiac [PHYSIO] Any chemical agent or odor that stimulates sexual desires. { ,af·rə'dē·zē,ak }

Aphroditidae [INV ZOO] A family of scale-bearing polychaete worms belonging to the Errantia. { ,af·rə'did·ə,dē }

Aphrosalpingoidea [PALEON] A group of middle Paleozoic invertebrates classified with the calcareous sponges. { ¦af·rō,sal,piŋ'gȯid·ē·ə }

aphrosiderite See ripidolite. { ,af·rō'sid·ə,rīt }

aphtha [MED] White, painful oral ulcer of unknown cause. { 'af·thə }

aphthitalite [MINERAL] (K,Na)₃Na(SO₄)₂ A white mineral crystallizing in the rhombohedral system and occurring massively or in crystals. { ,af'thid·əl,īt }

aphylactic map projection [MAP] A map projection which is neither conformal nor equal-area. { ,af·ə'lak·dik 'map prə'jek·shən }

Aphylidae [INV ZOO] An Australian family of hemipteran insects composed of two species; not placed in any higher taxonomic group. { ə'fil·ə,dē }

aphyllous [BOT] Lacking foliage leaves. { ā'fil·əs }

aphyric [PETR] Of the texture of fine-grained igneous rocks, showing two generations of the same mineral but without phenocrysts. { ā'fir·ik }

aphytic zone [ECOL] The part of a lake floor that lacks plants because it is too deep for adequate light penetration. { ā'fid·ik ,zōn }

API See air-position indicator; application program interface.

apiary [AGR] A place where bees are kept, especially for breeding and honey making. { 'ā·pē,er·ē }

apical [BOT] Relating to the apex or tip. { 'ap·i·kəl }

apical angle [MECH] The angle between the tangents to the curve outlining the contour of a projectile at its tip. [OPTICS] The dihedral angle between the refracting faces of a prism. Also known as refracting angle. { 'ap·i·kəl 'aŋ·gəl }

apical bud See terminal bud. { 'ap·i·kəl ,bəd }

apical dominance [BOT] Inhibition of lateral bud growth by the apical bud of a shoot, believed to be a response to auxins produced by the apical bud. { 'ap·i·kəl 'däm·ə·nəns }

apicalia [INV ZOO] Paired sensory cilia on the head of gnathostomulids. { ¦ap·ə¦kal·yə }

apical meristem [BOT] A region of embryonic tissue occurring at the tips of roots and stems. Also known as promeristem. { 'ap·i·kəl 'mer·ə,stem }

apical plate [INV ZOO] A group of cells at the anterior end of certain trochophore larvae; believed to have nervous and sensory functions. { 'ap·i·kəl 'plāt }

apiculate [BOT] Ending abruptly in a short, sharp point. { ə'pik·yə·lət }

apiculture [AGR] Large-scale commercial beekeeping. { 'ā·pə,kəl·chər }

Apidae [INV ZOO] A family of hymenopteran insects in the superfamily Apoidea including the honeybees, bumblebees, and carpenter bees. { 'a·pə,dē }

Apioceridae [INV ZOO] A family of orthorrhaphous dipteran insects in the series Brachycera. { ,ap·ē·ō'ser·ə,dē }

apioid [PHYS] A pear-shaped form taken by a rapidly revolving mass of liquid due to the force of gravity. { 'ap·ē,ȯid }

apiology [INV ZOO] The scientific study of bees, particularly honeybees. { ,ā·pē'äl·ə·jē }

Apis [INV ZOO] A genus of bees, the type genus of the Apidae. { 'ā·pəs }

API scale [CHEM ENG] The American Petroleum Institute hydrometer scale for the measurement of the specific gravity of liquids; used primarily in the American petroleum industry. { ¦ā,pē¦ī ,skāl }

Apistobranchidae [INV ZOO] A family of spioniform annelid worms belonging to the Sedentaria. { ə¦pis·tə¦braŋk·ə,dē }

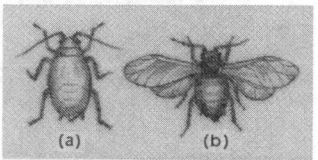

APHID

Two Forms during the life cycle of an aphid: *(a)* wingless female, *(b)* winged female.

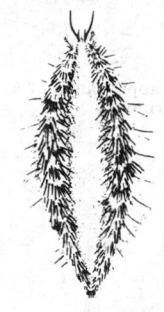

APHRODITIDAE

The sea mouse, *Aphrodita*, of the Aphroditidae.

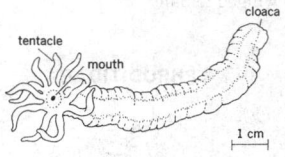

APODIDA

Typical appearance of an apodous holothurian.

apitong [MATER] A wood from the Philippine tree *Dipterocarpus grandiflorus;* sold as mahogany although it is not a true mahogany. { ə'pē,tȯŋ }

apjohnite [MINERAL] MnAl$_2$(SO$_4$)$_4$·22H$_2$O A white, rose-green, or yellow mineral containing water and occurring in crusts, fibrous masses, or efflorescences. { 'ap,jä,nīt }

APL [COMPUT SCI] An interactive computer language whose operators accept and produce arrays with homogeneous elements of type number or character.

Aplacophora [INV ZOO] A subclass of vermiform mollusks in the class Amphineura characterized by no shell and calcareous integumentary spicules. { ¦ä,plä'käf·ə·rə }

aplanatic lens [OPTICS] A lens corrected for spherical aberration. { ¦a·plə¦nad·ik 'lenz }

aplanatic points [OPTICS] Two points on the axis of an optical system which are located so that all the rays emanating from one converge to, or appear to diverge from, the other. { ¦a·plə¦nad·ik 'pȯins }

aplanogamete [BIOL] A gamete that lacks motility. { ā'plan·ə·gə,mēt }

aplanospore [MYCOL] A nonmotile, asexual spore, usually a sporangiospore, common in the Phycomycetes. { ā'plan·ə,spȯr }

aplasia [MED] Defective development which results in the virtual absence of a tissue or an organ; only a remnant appears. { ā'plāzh·ə }

aplastic anemia [MED] A blood disorder in which lymphocytes predominate while there is a deficiency of erythrocytes, hemoglobin, and granulocytes. { ā'plas·tik ə'nēm·yə }

aplite [PETR] Fine-grained granitic dike rock made up of light-colored mineral constituents, mostly quartz and feldspar; used to manufacture glass and enamel. { 'a,plīt }

aplysiatoxin [BIOCHEM] A bislactone toxin produced by the blue-green alga *Lyngbya majuscula.* { ə¦plīz·ē¦tak·sən }

apnea [MED] A transient cessation of respiration. { 'ap·nē·ə }

Apneumonomorphae [INV ZOO] A suborder of arachnid arthropods in the order Araneida characterized by the lack of book lungs. { ā,nü·mə,nō'mȯr,fē }

apneusis [PHYSIO] In certain lower vertebrates, sustained tonic contraction of the respiratory muscles to allow prolonged inspiration. { ap'nü·səs }

apo- [CHEM] A prefix that denotes formation from or relationship to another chemical compound. { 'ap·ō *or* 'ap·ə }

apoapsis [ASTRON] The point in an orbit farthest from the center of attraction. { ¦ap·ō¦ap·səs }

apoatropine [ORG CHEM] C$_{17}$H$_{21}$NO$_2$ An alkaloid melting at 61°C with decomposition of the compound; highly toxic; obtained by dehydrating atropine. { ,ap·ō'a·trə,pēn }

apob [METEOROL] An observation of pressure, temperature, and relative humidity taken aloft by means of an aerometeorograph; a type of aircraft sounding. { 'ā,päb }

apocarpous [BOT] Having carpels separate from each other. { ¦ap·ə¦kär·pəs }

apocenter *See* apofocus. { 'ap·ə,sen·tər }

apochromat *See* apochromatic lens. { ,ap·ə'krō·mat }

apochromatic lens [OPTICS] A lens with corrections for chromatic and spherical aberration. { ¦ap·ə·krō¦mad·ik 'lenz }

apochromatic system [OPTICS] An optical system which is free from both spherical and chromatic aberration for two or more colors. { ¦ap·ə·krō¦mad·ik 'sis·təm }

apocodeine [PHARM] C$_{18}$H$_{19}$NO$_2$ White crystals with a melting point of 124°C; decomposes on melting; soluble in alcohol and ether; used in medicine. { ¦ap·ə¦kō,dēn }

apocrine gland [PHYSIO] A multicellular gland, such as a mammary gland or an axillary sweat gland, that extrudes part of the cytoplasm with the secretory product. { 'ap·ə·krən ,gland }

apocronus [ASTRON] The farthest point of a satellite in its orbit about Saturn. Also known as aposaturnium. { 'ap·ə¦krō·nəs }

Apocynaceae [BOT] A family of tropical and subtropical flowering trees, shrubs, and vines in the order Gentianales, characterized by a well-developed latex system, granular pollen, a poorly developed corona, and the carpels often united by the style and stigma; well-known members are oleander and periwinkle. { ə,päs·ə'nās·ē,ē }

Apoda [VERT ZOO] The caecilians, a small order of wormlike, legless animals in the class Amphibia. { 'a·pəd·ə }

Apodacea [INV ZOO] A subclass of echinoderms in the class Holothuroidea characterized by simple or pinnate tentacles and reduced or absent tube feet. { ,a·pə'dās·ē·ə }

apodeme [INV ZOO] An internal ridge or process on an arthropod exoskeleton to which organs and muscles attach. { 'ap·ə,dēm }

Apodes [VERT ZOO] An equivalent name for the Anguilliformes. { 'ap·ə,dēz }

Apodi [VERT ZOO] The swifts, a suborder of birds in the order Apodiformes. { 'ap·ə,dī }

Apodida [INV ZOO] An order of worm-shaped holothurian echinoderms in the subclass Apodacea. { ə'päd·ə·də }

Apodidae [VERT ZOO] The true swifts, a family of apodiform birds belonging to the suborder Apodi. { ə'päd·ə,dē }

Apodiformes [VERT ZOO] An order of birds containing the hummingbirds and swifts. { ə,päd·ə'fȯr,mēz }

apodization [ELECTR] A technique for modifying the response of a surface acoustic wave filter by varying the overlap between adjacent electrodes of the interdigital transducer. [OPTICS] The modification of the amplitude transmittance of the aperture of an optical system so as to reduce or suppress the energy of the diffraction rings relative to that of the central Airy disk. [SPECT] A mathematical transformation carried out on data received from an interferometer to alter the instrument's response function before the Fourier transformation is calculated to obtain the spectrum. { ,a·pə·də'zā·shən }

apoenzyme [BIOCHEM] The protein moiety of an enzyme; determines the specificity of the enzyme reaction. { ¦a·pō¦en,zīm }

apoferritin [BIOCHEM] A protein found in intestinal mucosa cells that has the ability to combine with ferric ion. { ¦ap·ə¦fer·ət·ən }

apofocus [ASTRON] The point on an elliptic orbit at the greatest distance from the principal focus. Also known as apocenter. { ¦ap·ə¦fō·kəs }

apogalacteum [ASTRON] The point at which a celestial body is farthest from the center of the Milky Way. { ¦ap·ə·gə¦lak·tē·əm }

apogamy [BIOL] Asexual, parthenogenetic development of diploid cells, such as the development of a sporophyte from a gametophyte without fertilization. { ə'päg·ə·mē }

apogean tidal currents [OCEANOGR] Tidal currents of decreased speed occurring at the time of apogean tides. { ¦ap·ə¦jē·ən ¦tīd·əl ¦kər·əns }

apogean tides [OCEANOGR] Tides of decreased range occurring when the moon is near apogee. { ¦ap·ə¦jē·ən 'tīdz }

apogee [ASTRON] That point in an orbit at which the moon or an artificial satellite is most distant from the earth; the term is sometimes loosely applied to positions of satellites of other planets. { 'ap·ə,jē }

apogeny [BOT] Loss of the function of reproduction. { ə'päj·ə·nē }

apogeotropism [BOT] Negative geotropism; growth up or away from the soil. { ¦a·pō,jē·ō'trä,piz·əm }

Apogonidae [VERT ZOO] The cardinal fishes, a family of tropical marine fishes in the order Perciformes; males incubate eggs in the mouth. { ,ap·ə'gän·ə,dē }

Apoidea [INV ZOO] The bees, a superfamily of hymenopteran insects in the suborder Apocrita. { ə'pȯid·ē·ə }

apoinducer [BIOCHEM] A protein that, when bound to deoxyribonucleic acid, activates transcription by ribonucleic acid polymerase. { ¦a·pō·in¦dü·sər }

apojove [ASTRON] The farthest point of a satellite in its orbit about Jupiter. { 'ap·ə¦jōv }

apolipoprotein [BIOCHEM] A protein that combines with a lipid to form a lipoprotein. { ¦a·pō¦li·pō'prō,tēn }

Apollo [ASTRON] **1.** To the Greeks, the planet Mercury when it was a morning star. **2.** An asteroid with a very eccentric orbit and perihelion inside the orbit of Venus that passed about 1.8 × 10^6 miles (3 × 10^6 kilometers) from earth in 1932. { ə'päl·ō }

Apollonius' problem [MATH] The problem of constructing a circle that is tangent to three given circles. { ,ap·ə¦lōn·ē·əs ¦präb·ləm }

Apollo object [ASTRON] Any asteroid which crosses the earth's orbit. { ə'päl·ō 'äb·jikt }

Apollo program [AERO ENG] The scientific and technical program of the United States that involved placing men on the

moon and returning them safely to earth. { ə'päl·ō ¦prō·
grəm }

apolune [ASTRON] Farthest point of a satellite in an elliptic
orbit about the moon. Also known as aposelene. { ¦ap·
ə¦lün }

apolysis [INV ZOO] In most tapeworms, the shedding of ripe
proglottids. { ə'päl·ə·səs }

apomeiosis [CYTOL] Meiosis that is either suppressed or
imperfect. { ¦ap·ə,mī¦ō·səs }

apomercurian [ASTRON] The farthest point of a satellite in
its orbit about Mercury. { ¦ap·ə,mər'kyür·ē·ən }

apomixis [EMBRYO] Parthenogenetic development of sex
cells without fertilization. { ,ap·ə'mik·səs }

apomorph [SYST] Any derived character occurring at a
branching point and carried through one descending group in
a phyletic lineage. { 'ap·ə,mórf }

apomorphine [PHARM] $C_{17}H_{17}NO_2$ A crystalline alkaloid
obtained by dehydration of morphine; acts as a powerful emetic.
{ ¦ap·ə¦mór,fēn }

apomyoglobin [BIOCHEM] Myoglobin that lacks its heme
group. { ¦ap·ə¦mī·e,glōb·ən }

aponeurosis [ANAT] A broad sheet of regularly arranged
connective tissue that covers a muscle or serves to connect a
flat muscle to a bone. { ¦ap·ə,nü'rō·səs }

apophorometer [ENG] An apparatus used to identify minerals by sublimation. { ,ap·ə·fə'räm·əd·ər }

apophyge [ARCH] A small curvature applied to the top or
bottom of a column shaft where it expands to meet the fillet.
{ ə'päf·ə,jē }

apophyllite [MINERAL] A hydrous calcium potassium silicate containing fluorine and occurring as a secondary mineral
with zeolites with geodes and other igneous rocks; the composition is variable but approximates $KFCa_4(Si_2O_5)_4·8H_2O$. Also
known as fish-eye stone. { 'päf·ə,līt }

apophyllous [BOT] Having the parts of the perianth distinct.
{ ə'päf·ə·ləs }

apophysis [ANAT] An outgrowth or process on an organ
or bone. [MYCOL] A swollen filament in fungi. { ə'päf·
ə·səs }

apoplexy [MED] **1.** A symptom complex caused by an acute
vascular lesion of the brain and characterized by unconsciousness with various degrees of paralysis and sensory impairment.
2. Sudden, severe hemorrhage into any organ. { 'ap·ə,plek·
sē }

apoplutonian [ASTRON] The farthest point of a satellite in
its orbit about Pluto. { ¦ap·ə·plü'tōn·ē·ən }

apoposeidon [ASTRON] The farthest point of a satellite in
its orbit about Neptune. { ¦ap·ə·pə'sīd·ən }

apoprotein [BIOCHEM] The protein portion of a conjugated
protein exclusive of the prosthetic group. { ¦ap·ə¦prō,tēn }

apoptosis [CYTOL] Death of cells triggered by extracellular
signals or genetically programmed events, carried out by processes within the cell, and characterized by systemic breakdown
of cellular constituents, in particular chromosomal deoxyribonucleic acid; may be involved in normal development and aging,
or may serve to eliminate defective or damaged cells. Also
known as programmed cell death. { ,ā·pō'tō·səs }

apopyle [INV ZOO] Any one of the large pores in a sponge by
which water leaves a flagellated chamber to enter the exhalant
system. { 'ap·ə,pīl }

Aporidea [INV ZOO] An order of tapeworms of uncertain
composition and affinities; parasites of anseriform birds.
{ ,ap·ə'rīd·ē·ə }

aporogamy [BOT] Entry of the pollen tube into the embryo
sac through an opening other than the micropyle. { ,ap·ə'räg·
ə·mē }

aposaturnium See apocronus. { ¦ap·ə·sə'tər·nē·əm }

aposelene See apolune. { ¦ap·ə·sə¦lēn }

aposematic [ECOL] Pertaining to colors or structures on an
organism that provide a special means of defense against enemies. Also known as sematic. { ¦ap·ə·sə¦mad·ik }

A positive [ELEC] Symbolized A+. **1.** Positive terminal of
an A battery or positive polarity of other sources of filament
voltage. **2.** Denoting the terminal to which the positive side
of the filament voltage source should be connected. { ¦ā 'päz·
əd·iv }

apospory [MYCOL] Suppression of spore formation with
development of the haploid (sexual) generation directly from
the diploid (asexual) generation. { 'ap·ə,spór·ē }

apostatic selection [ECOL] Predation on the most abundant
forms in a population, leading to balanced distribution of a
variety of forms. { ¦ap·ə¦stad·ik sə'lek·shən }

a posteriori probability See empirical probability. { ¦ā
,pä,stir·ē'ór,ē ,präb·ə'bil·əd·ē }

apostilb [OPTICS] A luminance unit equal to one ten-thou-
sandth of a lambert. Also known as blondel. { 'ap·ə,stilb }

Apostomatida [INV ZOO] An order of ciliated protozoans in
the subclass Holotrichia; majority are commensals on marine
crustaceans. { ə,päs·tə'mad·ə·də }

apotele [ANAT] A scalloped ridge around the edge of an
otolith. { 'ap·ə,tēl }

apothecaries' dram See dram. { ə'päth·ə,ker·ēz 'dram }

apothecaries' measure [PHARM] A system of units of volume, usually of liquid drugs, in which 16 fluid ounces equals
1 pint. { ə'päth·ə,ker·ēz 'mezh·ər }

apothecaries' ounce See ounce. { ə'päth·ə,ker·ēz 'aúns }

apothecaries' pound See pound. { ə'päth·ə,ker·ēz 'paúnd }

apothecaries' weight [PHARM] A system of units of mass,
usually of drugs, in which 1 pound equals 5760 grains or 1
troy pound. { ə'päth·ə,ker·ēz 'wāt }

apothecium [MYCOL] A spore-bearing structure in some
Ascomycetes and lichens in which the fruiting surface or hymenium is exposed during spore maturation. { ¦ap·ə¦thēsh·əm }

apothem [MATH] The perpendicular distance from the center of a regular polygon to one of its sides. Also known as
short radius. { 'ap·ə,them }

apouranian [ASTRON] The farthest point of a satellite in its
orbit about Uranus. { ¦ap·ō,yü'rän·ē·ən }

A power supply See A supply. { ¦ā 'paú·ər sə,plī }

apozymase [BIOCHEM] The protein component of a zymase.
{ ,ap·ə'zī,mās }

Appalachia [GEOL] Proposed borderland along the southeastern side of North America, seaward of the Appalachian
geosyncline in Paleozoic time. { ,ap·ə¦lā·chə }

Appalachian orogeny [GEOL] An obsolete term referring
to Late Paleozoic diastrophism beginning perhaps in the Late
Devonian and continuing until the end of the Permian; now
replaced by Alleghenian orogeny. { ,ap·ə¦lā·chən ō'räj·ə·nē }

apparatus [SCI TECH] A compound instrument designed to
carry out a specific function. { ,ap·ə'rad·əs }

apparent [ASTRON] A term used to designate certain measured or measurable astronomic quantities to refer them to real
or visible objects, such as the sun or a star. { ə'pa·rənt }

apparent additional mass [FL MECH] A fictitious mass of
fluid added to the mass of the body to represent the force
required to accelerate the body through the fluid. { ə'pa·rənt
ə'dish·ən·əl 'mas }

apparent candlepower [OPTICS] For an extended source of
light, at a specified distance, the candlepower of a point source
that would produce the same illumination as the extended
source at the same distance. { ə'pa·rənt 'kan·dəl,paúr }

apparent cohesion [GEOL] In soil mechanics, the resistance
of particles to being pulled apart due to the surface tension of
the moisture film surrounding each particle. Also known as
film cohesion. { ə'pa·rənt ,kō'hē·zhən }

apparent concentration [ANALY CHEM] The value of analyte concentration obtained when the interference is not considered. { ə'par·ənt ,kän·sən'trā·shən }

apparent density [MET] The weight per unit volume of a
metal powder, in contrast to the weight per unit volume of the
individual particles. { ə'pa·rənt 'dens·əd·ē }

apparent depth [OPTICS] The depth of the image of an
object submerged in a transparent medium; it is reduced from
the real depth of the object by a factor equal to the relative
index of refraction of the medium with respect to air. { ə'pa·
rənt 'depth }

apparent dip [GEOL] Dip of a rock layer as it is exposed in
any section not at a right angle to the strike. { ə'pa·rənt 'dip }

apparent expansion [THERMO] The expansion of a liquid
with temperature, as measured in a graduated container without
taking into account the container's expansion. { ə'pa·rənt
ik'span·shən }

apparent force [MECH] A force introduced in a relative
coordinate system in order that Newton's laws be satisfied in
the system; examples are the Coriolis force and the centrifugal
force incorporated in gravity. { ə'pa·rənt 'fórs }

apparent gravity See acceleration of gravity. { ə'pa·rənt
'grav·əd·ē }

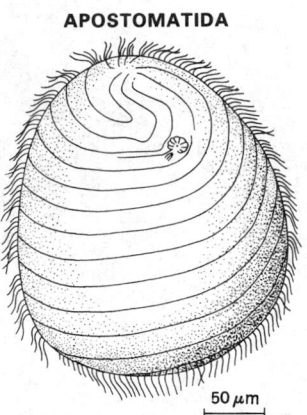

APOSTOMATIDA

Foettingeria, an example of an
apostomatid.

apparent horizon [ASTRON] *See* horizon. [RELAT] The boundary of a region in space-time in which the gravitational field is so strong that the cross-sectional area of an outgoing light pulse decreases. { ə'pa·rənt hə'rīz·ən }

apparent libration in longitude *See* lunar libration. { ə'pa·rənt lī'brā·shən in 'län·jə,tüd }

apparent luminance [OPTICS] Luminance, created by air light, of that portion of the visual field subtended by a dark, distant object; that is, the light scattered into the eye by particles, including air molecules, lying along the optic path from eye to object. { ə'pa·rənt 'lü·mə·nəns }

apparent magnitude [ASTRON] An index of a star's brightness relative to that of the other stars; it does not take into account the difference in distance between the stars and is not an indication of the star's true luminosity. { ə'pa·rənt 'mag·nə,tüd }

apparent motion *See* relative motion. { ə'pa·rənt 'mō·shən }

apparent movement of faults [GEOL] The apparent motion observed to have occurred in any chance section across a fault. { ə'pa·rənt ¦müv·mənt əv ¦fòlts }

apparent noon [ASTRON] Twelve o'clock apparent time, or the instant the apparent sun is over the upper branch of the meridian. { ə'pa·rənt 'nün }

apparent place *See* apparent position. { ə'pa·rənt 'plās }

apparent plunge [GEOL] Inclination of a normal projection of lineation in the plane of a vertical cross section. { ə'pa·rənt 'plənj }

apparent position [ASTRON] The position on the celestial sphere at which a heavenly body (or a space vehicle) would be seen from the center of the earth at a particular time. Also known as apparent place. { ə'pa·rənt pə'sish·ən }

apparent power [ELEC] The product of the root-mean-square voltage and the root-mean-square current delivered in an alternating-current circuit, no account being taken of the phase difference between voltage and current. { ə'pa·rənt 'paù·ər }

apparent precession *See* apparent wander. { ə'pa·rənt pri'sesh·ən }

apparent shoreline [MAP] The outer edge of marine vegetation (marsh, mangrove, cypress) delineated on photogrammetric surveys where the actual shoreline is obscured. { ə'pa·rənt 'shór,līn }

apparent slope [GRAPHICS] A true slope that appears vertically distorted or exaggerated in air photographs viewed with a stereoscope. { ə'pa·rənt 'slōp }

apparent solar day [ASTRON] The duration of one rotation of the earth on its axis with respect to the apparent sun. Also known as true solar day. { ə'pa·rənt ¦sō·lər 'dā }

apparent solar time [ASTRON] Time measured by the apparent diurnal motion of the sun. Also known as apparent time; true solar time. { ə'pa·rənt ¦so·lər 'tīm }

apparent source *See* effective center. { ə'pa·rənt 'sórs }

apparent sun [ASTRON] The sun as it appears to an observer. Also known as true sun. { ə'pa·rənt 'sən }

apparent time *See* apparent solar time. { ə'pa·rənt 'tīm }

apparent vertical [GEOPHYS] The direction of the resultant of gravitational and all other accelerations. Also known as dynamic vertical. { ə'pa·rənt 'verd·ə·kəl }

apparent viscosity [FL MECH] The value obtained by applying the instrumental equations used in obtaining the viscosity of a Newtonian fluid to viscometer measurements of a non-Newtonian fluid. { ə¦par·ənt vi'skäs·əd·ē }

apparent visual angle [OPTICS] The angle subtended by a source at the observer's eye as calculated from the source size and distance from the eye. { ə'pa·rənt 'vizh·ə·wəl 'aŋ·gəl }

apparent volume [PHYS] The difference between the volume of a binary solution and the volume of the pure solvent at the same temperature. { ə'pa·rənt 'väl·yəm }

apparent wander [GEOPHYS] Apparent change in the direction of the axis of rotation of a spinning body, such as a gyroscope, due to rotation of the earth. Also known as apparent precession; wander. { ə'pa·rənt 'wän·dər }

apparent water table *See* perched water table. { ə'pa·rənt 'wód·ər ,tā·bəl }

apparent weight [MECH] For a body immersed in a fluid (such as air), the resultant of the gravitational force and the buoyant force of the fluid acting on the body; equal in magnitude to the true weight minus the weight of the displaced fluid. { ə'pa·rənt 'wāt }

apparent wind *See* relative wind. { ə'pa·rənt 'wind }

apparition [ASTRON] A period during which a planet, asteroid, or comet is observable, generally between two successive conjunctions of the body with the sun. { ,ap·ə'rish·ən }

appearance potential [PHYS] The minimal potential which the electron beam in the ion source of a mass spectrometer must traverse in order to acquire enough energy to produce ions of a specified nuclide or molecular fragment. { ə'pir·əns pə'ten·chəl }

appearance ratio *See* hyperstereoscopy. { ə'pir·əns ,rā·shō }

appendage [BIOL] Any subordinate or nonessential structure associated with a major body part. [NAV ARCH] Any fitting installed outside the underwater hull proper, such as a rudder, shaft, shaft strut, and bilge keel. [ZOO] Any jointed, peripheral extension, especially limbs, of arthropod and vertebrate bodies. { ə'pen·dij }

appendectomy [MED] Surgical removal of the vermiform appendix. { ,ap·ən'dek·tə·mē }

appendicitis [MED] Inflammation of the vermiform appendix. { ə,pen·də'sīd·əs }

appendicular skeleton [ANAT] The bones of the pectoral and pelvic girdles and the paired appendages in vertebrates. { ,ap·ən'dik·yə·lər 'skel·ə·tən }

appendiculate [BIOL] Having or forming appendages. { ,ap·ən'dik·yə,lāt }

appendix [ANAT] **1.** Any appendage. **2.** *See* vermiform appendix. { ə'pen·diks }

appendix testis [MED] A remnant of the cranial part of the paramesonephric or Müllers duct, attached to the testis. Also known as hydatid of Morgagni. { ə'pen·diks 'tes·təs }

apperception [PSYCH] Perception as modified and enhanced by one's own emotions, memories, and biases. { ,ap·ər'sep·shən }

appersonification [PSYCH] Unconscious identification with and acquisition of another person's characteristics. { ,a·pər,sän·ə·fə'kā·shən }

appestat [PHYSIO] The center for appetite regulation in the hypothalamus. { 'ap·ə,stat }

appetitive behavior [ZOO] Any behavior that increases the probability that an animal will be able to satisfy a need; for example, a hungry animal will move around to find food. { ə'ped·ə·tiv bi'hāv·yər }

appinite [PETR] Hornblende-rich plutonic rock with high feldspar content. { 'ap·ə,nīt }

apple [BOT] *Malus domestica.* A deciduous tree in the order Rosales which produces an edible, simple, fleshy, pome-type fruit. { 'ap·əl }

apple-cedar rust [PL PATH] A disease of apples and Eastern red cedars that is caused by the fungus *Gymnosporangium juniperi-virginianae*; on cedar branches, it manifests itself as brown round galls that do not cause injury, and on apple leaves, as yellow spots that later turn brown and result in cupping and curling of the leaf. { 'ap·əl ¦sēd·ər ,rəst }

apple coal [GEOL] Easily mined soft coal that breaks into small pieces the size of apples. { 'ap·əl ,kōl }

apple essence *See* isoamyl valerate. { 'ap·əl 'es·əns }

Applegate diagram [ELECTR] A graph of the electron paths in a two-cavity klystron tube, showing how electron bunching occurs. { 'ap·əl,gāt 'dī·ə,gram }

apple honey *See* apple syrup. { 'ap·əl ,hən·ē }

applejack [FOOD ENG] An alcoholic beverage produced from hard cider by lowering its temperature so that much of the water freezes and precipitates as ice and the remaining liquid contains a higher percentage of alcohol. { 'ap·əl,jak }

apple of Peru *See* jimsonweed. { 'ap·əl əv pə'rü }

apple oil *See* isoamyl valerate. { 'ap·əl ,oil }

apple pox *See* blister canker. { 'ap·əl ,päks }

apple scab disease [PL PATH] A plant disease caused by the fungus *Venturia inaequalis* that may cause premature defoliation, June drop of young fruits, and unsightly blemishes on ripe apples. { 'ap·əl ¦skab diz,ēz }

apple syrup [FOOD ENG] A sweetening agent made from cull apples and composed mainly of levulose, sucrose, and dextrose; used in the food industry. Also known as apple honey. { 'ap·əl ,sir·əp }

applet [COMPUT SCI] A small program, typically written in Java. { 'ap·lət }

Appleton layer *See* F_2 layer. { 'ap·əl·tən ,lā·ər }

appliance [ENG] A piece of equipment that draws electric or other energy and produces a desired work-saving or other result, such as an electric heater, a radio, or an electronic range. { ə'plī·əns }

appliance panel [ENG] In electric systems, a metal housing containing two or more devices (such as fuses) for protection against excessive current in circuits which supply portable electric appliances. { ə'plī·əns ‚pan·əl }

applicable surfaces [MATH] Surfaces such that there is a length-preserving map of one onto the other. { ¦ap·lə·kə·bəl 'sər·fəs·əz }

application [COMPUT SCI] A computer program that performs a specific task, for example, a word processor, a Web browser, or a spread sheet. { ‚ap·lə'kā·shən }

application development language [COMPUT SCI] A very-high-level programming language that generates coding in a conventional programming language or provides the user of a data-base management system with a programming language that is easier to implement than conventional programming languages. { ‚ap·lə'kā·shən di'vel·əp·mənt ‚laŋ·gwij }

application development system [COMPUT SCI] An integrated group of software products used to assist in the efficient development of computer programs and systems. { ‚ap·lə'kā·shən di'vel·əp·mənt ‚sis·təm }

application generator [COMPUT SCI] A commercially prepared software package used to create applications programs or parts of such programs. { ‚ap·lə'kā·shən ‚jen·ə‚rād·ər }

application package [COMPUT SCI] A combination of required hardware, including remote inputs and outputs, plus programming of the computer memory to produce the specified results. { ‚ap·lə'kā·shən ‚pak·ij }

application processor [COMPUT SCI] A computer that processes data. { ‚ap·lə'kā·shən 'prä‚ses·ər }

application program [COMPUT SCI] A program written to solve a specific problem, produce a specific report, or update a specific file. { ‚ap·lə'kā·shən ‚prō·grəm }

application program interface [COMPUT SCI] A language that enables communication between computer programs, in particular between application programs and control programs. Abbreviated API. { ‚ap·lə'kā·shən ‚prō·grəm 'in·tər‚fās }

application server [COMPUT SCI] A computer that executes commands requested by a Web server to fetch data from data-bases. Also known as app server. { ‚ap·lə'kā·shən ‚ser·vər }

application-specific integrated circuit [ELECTR] An integrated circuit that is designed for a particular application by integrating standard cells from a library, making possible short design times and rapid production cycles. Abbreviated ASIC. { ‚ap·lə‚kā·shən spi¦sif·ik ‚int·i‚grād·əd 'sər·kət }

applications technology satellite [AERO ENG] Any artificial satellite in the National Aeronautics and Space Administration program for the evaluation of advanced techniques and equipment for communications, meteorological, and navigation satellites. Abbreviated ATS. { ‚ap·lə'kā·shənz ¦tek'näl·ə·jē ¦sad·ə‚līt }

application study [COMPUT SCI] The detailed process of determining a system or set of procedures for using a computer for definite functions of operations, and establishing specifications to be used as a base for the selection of equipment suitable to the specific needs. { ‚ap·lə'kā·shən ‚stəd·ē }

application system [COMPUT SCI] A group of related applications programs designed to perform a specific function. { ‚ap·lə'kā·shən ‚sis·təm }

application window [COMPUT SCI] In a graphical user interface, the chief window of an application program, with a title bar, a menu bar, and a work area. { ‚ap·lə'kā·shən ‚win‚dō }

applicative language [COMPUT SCI] A programming language in which functions are repeatedly applied to the results of other functions and, in its pure form, there are no statements, only expressions without side effects. { 'ap·lə‚kad·iv 'laŋ·gwij }

applied anatomy [ANAT] **1.** A discipline that considers problems involving the biomechanical functions of a body. **2.** The application of anatomical principles to specific fields of human endeavor, for example, surgical anatomy. { ə'plīd ə'nad·ə·mē }

applied climatology [CLIMATOL] The scientific analysis of climatic data in the light of a useful application for an operational purpose. { ə'plīd ‚klīm·ə'täl·ə·jē }

applied ecology [ECOL] Activities involved in the management of natural resources. { ə'plīd i'käl·ə·jē }

applied epistemology [COMPUT SCI] The use of machines or other models to simulate processes such as perception, recognition, learning, and selective recall, or the application of principles assumed to hold for human categorization, perception, storage, search, and so on, to the design of machines, machine programs, scanning, storage, and retrieval systems. { ə'plīd i¦pis·tə¦mäl·ə·jē }

applied inverse scattering theory [PHYS] The branch of inverse scattering theory that treats the case in which the data provided are incomplete or corrupted by noise. { ə'plīd 'in‚vərs ¦skad·ər·iŋ ‚thē·ə·rē }

applied meteorology [METEOROL] The application of current weather data, analyses, or forecasts to specific practical problems. { ə'plīd ‚mēd·ē·ə'räl·ə·jē }

applied potential tomography [MED] A method of producing images of the electrical impedance of tissues, in which potentials are applied to the body through skin electrodes, and the resulting currents give rise to measurable potentials elsewhere on the body from which the impedance of organs and tissues can be determined. { ə'plīd pə'ten·chəl tə'mäg·rə·fē }

applied research [ENG] Research directed toward using knowledge gained by basic research to make things or to create situations that will serve a practical or utilitarian purpose. { ə'plīd ri‚sərch }

applied strategic research [ENG] Research done to provide a basic understanding of a current applied project. { ə'plīd strə'tē·jik ri'sərch }

applied trim [BUILD] Supplementary and separate decorative strips of wood or moldings applied to the face or sides of a frame, such as a doorframe. { ə'plīd 'trim }

appliqué [GRAPHICS] A decoration or ornament made by cutting out and attaching one piece of material to the surface of another. [OPTICS] A combination of lenses that provides for the same focal length at three or more wavelengths. { ¦ap·lə¦kā }

appliqué armor [ORD] Material or attachment which can be installed on a tank to give it additional protection against kinetic- or nonkinetic-energy projectiles. { ¦ap·lə¦kā ¦är·mər }

appliqué circuit [ELEC] Special circuit which is provided to modify existing equipment to allow for special usage; for example, some carrier telephone equipment designed for ringdown manual operation can be modified through the use of an appliqué circuit to allow for use between points having dial equipment. { ¦ap·lə¦kā ‚sər·kət }

apposition beach [GEOL] One of a series of parallel beaches formed on the seaward side of an older beach. { ‚ap·ə'zish·ən ‚bēch }

apposition eye [INV ZOO] A compound eye found in diurnal insects and crustaceans in which each ommatidium focuses on a small part of the whole field of light, producing a mosaic image. { ‚ap·ə'zish·ən ‚ī }

apposition fabric [PETR] A primary orientation of the elements of a sedimentary rock that is developed or formed at time of deposition of the material; fabrics of most sedimentary rocks belong to this type. Also known as primary fabric. { ‚ap·ə'zish·ən ‚fab·rik }

appressed [BIOL] Pressed close to or lying flat against something. { ə'prest }

approach [MECH ENG] The difference between the temperature of the water leaving a cooling tower and the wet-bulb temperature of the surrounding air. [NAV] In air operations, a maneuver executed by an aircraft in making its transit from high-altitude enroute flight to the point where it begins the landing approach; includes maneuvers (such as flying racetrack pattern) required for traffic control. { ə'prōch }

approach and landing area [NAV] An airspace of defined dimensions together with its runways and water channels, used by aircraft arriving at or departing from or operating within the vicinity of the airport. { ə'prōch ən 'land·iŋ ‚er·ē·ə }

approach-approach conflict [PSYCH] Psychological conflict resulting from the necessity of choosing between two desirable alternatives. { a¦prōch a¦prōch ¦kän·flikt }

approach-avoidance conflict [PSYCH] Psychological conflict that results when a goal has both desirable and undesirable aspects. { ə¦prōch ə¦vòid·əns ¦kän‚flikt }

approach chart [NAV] An aeronautical chart which provides information about navigational facilities, flight patterns, radio aids and their frequencies, and so on, for use in making an approach to an airfield under either visual or instrument flight conditions. { ə'prōch ,chärt }

approach control [NAV] A radio guidance service for aircraft within 10 to 20 miles (16 to 32 kilometers) of their destination airport that provides steering vectors and speed values to properly place the aircraft for entry into the airport traffic pattern and to provide safe spacing from neighboring aircraft. { ə'prōch kən,trōl }

approach course [NAV] A flight track to runway or approach fix defined by a visual or electronic aid. Also known as approach lane. { ə'prōch ,kórs }

approach fix [NAV] A geographical position on a defined approach course. { ə'prōch ,fiks }

approach gate [NAV] That point on the final approach course which is 1 mile (1.6 kilometers) from the approach fix on the side away from the airport or 5 miles (8 kilometers) from the landing threshold, whichever is farthest from the landing threshold. { ə'prōch ,gāt }

approach lane See approach course. { ə'prōch ,lān }

approach lights [NAV] A group of aeronautical lights indicating a desirable line of approach to a landing area. { ə'prōch ,līts }

approach navigation [NAV] The navigation operation which must be performed in the proximity of an airport or harbor by an aircraft or ship when proceeding to terminal or dock. { ə'prōch ,nav·ə'gā·shən }

approach path [NAV] That portion of the flight path which extends from the point where a descent is started to the point where the aircraft touches down on the runway. { ə'prōch ,path }

approach segment [NAV] The basic functional division of an instrument approach procedure. Also known as approachway. { ə'prōch ,seg·mənt }

approach sequence [NAV] The position of an aircraft while awaiting approach clearance or while on approach. { ə'prōch ,sē·kwəns }

approach signal [CIV ENG] A railway signal warning an engineer of a signal ahead that displays a restrictive indication. { ə'prōch ,sig·nəl }

approach surface base line [NAV] In terminal instrument procedures, an imaginary horizontal line at threshold elevation. { ə'prōch ¦sər·fəs 'bās ,līn }

approach vector [CONT SYS] A vector that describes the orientation of a robot gripper and points in the direction from which the gripper approaches a workpiece. { ə'prōch ,vek·tər }

approach visibility [NAV] The distance from which a pilot on the instrument approach glide path can see landing aids at the runway threshold. Also known as slant visibility. { ə'prōch ,viz·ə'bil·əd·ē }

approachway See approach segment. { ə'prōch,wā }

approved flame safety lamp [MIN ENG] A flame safety lamp which has been approved for use in gaseous coal mines. { ə¦prüvd ¦flām 'saf·tē ,lamp }

approx See approximate. { ə'präks }

approximate [MATH] 1. To obtain a result that is not exact but is near enough to the correct result for some specified purpose. 2. To obtain a series of results approaching the correct result. [SCI TECH] 1. Close to the correct value. Abbreviated approx. 2. To be close to. { ə'präk·sə·mət (adjective) or ə'präk·sə,māt (verb) }

approximate absolute temperature [PHYS] A temperature scale with the ice point at 273° and boiling point of water at 373°; it is intended to approximate the Kelvin temperature scale with sufficient accuracy for many sciences, notably meteorology, and is widely used in the meteorological literature. Also known as tercentesimal thermometric scale. { ə'präk·sə·mət 'ab·sə,lüt 'tem·prə·chər }

approximate altitude [NAV] Of a star, an altitude determined by estimation, by a star finder or star chart. { ə'präk·sə·mət 'al·tə·tüd }

approximate contour [MAP] A contour substituted for a normal contour whenever there is a question as to its reliability; reliability usually is evaluated as exceeding one-half the contour interval. { ə'präk·sə·mət 'kän,tur }

approximate reasoning [MATH] The process by which a possibly imprecise conclusion is deduced from a collection of imprecise premises. { ə¦präks·ə·mət 'rēz·ən·iŋ }

approximation [MATH] 1. A result that is not exact but is near enough to the correct result for some specified purpose. 2. A procedure for obtaining such a result. { ə¦prak·sə¦mā·shən }

approximation property [MATH] The property of a Barach space, B, in which compact sets are approxiately finite-dimensional in the sense that, for any compact set, K, continuous linear transformations, L, from K to finite-dimensional subspaces of B can be found with arbitrarily small upper bounds on the norm of $L(x) - x$ for all points x in K. { ə,präk·sə'mā·shən ,präp·ərd·ē }

app server See application server. { 'ap ,sər·vər }

appulse [ASTRON] 1. The near approach of one celestial body to another on the celestial sphere, as in occultation or conjunction. 2. A penumbral eclipse of the moon. { ə'pəls }

APR See active-prominence region. See airborne profile recorder.

apraxia [MED] The inability to perform purposeful acts as a result of brain lesions; characteristically, paralysis is absent and kinesthesia is unimpaired. { ā'prak·sē·ə }

apricot [BOT] Prunus armeniaca. A deciduous tree in the order Rosales which produces a simple fleshy stone fruit. { 'ap·rə,kät }

apricot kernel oil See persic oil. { 'ap·rə,kät ¦kər·nəl ,óil }

a priori [MATH] Pertaining to deductive reasoning from assumed axioms or supposedly self-evident principles, supposedly without reference to experience. { ¦ā prē¦ór·ē }

a priori probability See mathematical probability. { ¦ā prē¦ór·ē ,präb·ə'bil·əd·ē }

aproctous [MED] Having an imperforate anus. [ZOO] Lacking an anus. { ā'präk·təs }

apron [AERO ENG] A protective device specially designed to cover an area surrounding the fuel inlet on a rocket or spacecraft. [BUILD] 1. A board on an interior wall beneath a windowsill. 2. The vertical rear panel of a sink attached to a wall. 3. A section of a concrete slab extending beyond the face of a building on adjacent ground. Also known as skirt; skirting. 4. A vertical panel installed behind a sink or lavatory. [CIV ENG] 1. A hard-surfaced area, usually paved, adjacent to a ship or the like, used to park, load, unload, or service vehicles. 2. A covering of a material such as concrete or timber over soil to prevent erosion by flowing water, as at the bottom of a dam. 3. A concrete or wooden shield that is situated along the bank of a river, along a sea wall, or below a dam. 4. In a railroad system, a bridge structure that carries tracks and is hinged to land for connecting the deck of a railroad-car ferry to the shore. [GEOL] See outwash plain. [HYD] See ram. [MECH ENG] A plate serving to protect or cover a machine. [MIN ENG] A canvas-covered frame set at such an angle in the miner's rocker that the gravel and water passing over it are carried to the head of the machine. [ORD] 1. That portion of the superior slope of a parapet or the interior slope of a pit designed to protect the slopes against blast. 2. The hinged portion of a shield. 3. A removal screen of camouflage material placed over or in front of artillery guns. { 'ā·prən }

apron conveyor [MECH ENG] A conveyor used for carrying granular or lumpy material and consisting of two strands of roller chain separated by overlapping plates, forming the carrying surface, with sides 2–6 inches (5–15 centimeters) high. { 'ā·prən kən,vā·ər }

apron feeder [MECH ENG] A limited-length version of apron conveyor used for controlled-rate feeding of pulverized material to a process or packaging unit. Also known as plate-belt feeder; plate feeder. { 'ā·prən ,fēd·ər }

apron flashing [BUILD] 1. The flashing that covers the joint between a vertical surface and a sloping roof, as at the lower edge of a chimney. 2. The flashing that diverts water from a vertical surface into a gutter. { 'ā·prən ,flash·iŋ }

apron lining [BUILD] The piece of boarding which covers the rough apron piece of a staircase. { 'ā·prən ,līn·iŋ }

apron piece [BUILD] A beam that supports a landing or a series of winders in a staircase. { 'ā·prən ,pēs }

apron rail [BUILD] A lock rail having a raised ornamental molding. { 'ā·prən ,rāl }

apron wall [BUILD] In an exterior wall, a panel which extends downward from a windowsill to the top of a window below. { 'ā·prən ,wól }

aprotic solvent [CHEM] A solvent that does not yield or accept a proton. { ā'präd·ik 'säl·vənt }

Aps *See* Apus.

apse [ARCH] A semicircular (or nearly semicircular) or semipolygonal space, usually in a church, terminating an axis and intended to house an altar. [ASTRON] *See* apsis. { 'aps }

apsidal motion [ASTRON] The precession of the periastron of a binary system in the orbital plane of the two stars, resulting from tidal gravitational moments. { 'ap·sə·dəl 'mō·shən }

Apsidospondyli [VERT ZOO] A term used to include, as a subclass, amphibians in which the vertebral centra are formed from cartilaginous arches. { ¦ap·sə·də'spän·də‚lī }

apsis [ASTRON] In celestial mechanics, either of the two orbital points nearest or farthest from the center of attraction. Also known as apse. { 'ap·səs }

APT *See* Automatic Programming Tool.

apterium [VERT ZOO] A bare space between feathers on a bird's skin. { ap'tir·ē·əm }

apterous [BIOL] Lacking wings, as in certain insects, or winglike expansions, as in certain seeds. { 'ap·tə·rəs }

Apterygidae [VERT ZOO] The kiwis, a family of nocturnal ratite birds in the order Apterygiformes. { ¦ap·tə'rij·ə‚dē }

Apterygiformes [VERT ZOO] An order of ratite birds containing three living species, the kiwis, characterized by small eyes, limited eyesight, and nostrils at the tip of the bill. { ‚ap·tə‚rij·ə'fȯr‚mēz }

Apterygota [INV ZOO] A subclass of the Insecta characterized by being primitively wingless. { ‚ap·tə·rə'god·ə }

Aptian [GEOL] Lower Cretaceous geologic time, between Barremian and Albian. Also known as Vectian. { 'ap·tē·ən }

aptitude [PSYCH] The natural inclination or capacity for skillful performance of an as yet unlearned task. { 'ap·tə‚tüd }

aptitude test [PSYCH] Any standardized examination used to evaluate a person'a ability to learn a particular skill. { 'ap·tə‚tüd ‚test }

APT system *See* automatic picture-transmission system. { ‚ā‚pē'tē ‚sis·təm }

aptyalism [MED] Deficiency or absence of saliva. { ā'tī·ə‚liz·əm }

APU *See* auxiliary power unit.

Apus [ASTRON] A constellation with a right ascension of 16 hours and declination of 75°S. Abbreviated Aps. [VERT ZOO] A genus of birds comprising the Old World swifts. { 'a·pəs }

APW method *See* augmented plane-wave method. { ¦ā¦pē'dəb·əl‚yü ‚meth·əd }

apyrase [BIOCHEM] Any enzyme that hydrolyzes adenosinetriphosphate, with liberation of phosphate and energy, and that is believed to be associated with actomyosin activity. { 'ap·ə‚rās }

apyrexia [MED] Absence of fever. { ‚ā‚pī'rek·sē·ə }

Aqil *See* Aquila.

Aql *See* Aquila.

AQL *See* acceptable quality level.

Aqr *See* Aquarius.

aqua [CHEM] Latin for water. { 'äk·wə }

aqua ammonia *See* ammonium hydroxide. { 'äk·wə ə'mōn·ē·ə }

aquaboard *See* aquaplane. { 'ak·wə‚bȯrd }

aquaculture *See* aquiculture. { 'ak·wə‚kəl·chər }

aquadag [ELECTR] Graphite coating on the inside of certain cathode-ray tubes for collecting secondary electrons emitted by the face of the tube. [MATER] A colloidal suspension of graphite in water. { 'ak·wə‚dag }

aquafortis *See* nitric acid. { ¦äk·wə'fȯrd·əs }

aquagene tuff *See* hyaloclastite. { ¦ak·wə‚jēn 'təf }

aqualf [GEOL] A suborder of the soil order Alfisol, seasonally wet and marked by gray or mottled colors; occurs in depressions on or wide flats in local landscapes. { 'ak·wəlf }

aqualung [ENG] A self-contained underwater breathing apparatus (scuba) of the demand or open-circuit type developed by J.Y. Cousteau. { 'ak·wə‚ləŋ }

aquamarine [MINERAL] A pale-blue or greenish-blue transparent gem variety of the mineral beryl. { ‚ak·wə·mə'rēn }

aquametry [ANALY CHEM] Analytical processes to measure the water present in materials; methods include Karl Fischer titration, reactions with acid chlorides and anhydrides, oven drying, distillation, and chromatography. { ə'kwäm·ə·trē }

aquaplane [NAV ARCH] A board on which a person rides while being towed by a motorboat at such speed that the front part of the board rises out of the water. Also known as aquaboard. { 'ak·wə‚plān }

aqua regia [INORG CHEM] A fuming, highly corrosive, volatile liquid with a suffocating odor made by mixing 1 part concentrated nitric acid and 3 parts concentrated hydrochloric acid; reacts with all metals, including silver and gold. { ¦äk·wə 'rē·jə }

δ Aquarids [ASTRON] A meteor shower consisting of relatively long, slow-moving meteors that has its maximum around July 30 and radiant near the star δ Aquarii. { ¦del·tə 'ak·wə‚ridz }

η Aquarids [ASTRON] A meteor shower associated with Halley's Comet that occurs in the first week of May with radiant near the star η Aquarii. { ¦¦ā·də 'ak·wə‚ridz }

Aquarius [ASTRON] A constellation with a right ascension of 23 hours and declination of 15°S. Abbreviated Aqr. Also known as Water Bearer. { ə'kwer·ē·əs }

aquasol *See* hydrosol. { 'ak·wə‚sȯl }

aquatic [BIOL] Living or growing in, on, or near water; having a water habitat. { ə'kwäd·ik }

aquatic weed cutter [NAV ARCH] A device attached to the bow of a boat that can cut and clear a wide path through a weed-choked lake. { ə'kwäd·ik 'wēd ‚kəd·ər }

aquatint [GRAPHICS] An etching process that produces several tones by varying the etching time of different areas of a copper plate; the resulting print resembles an ink or wash drawing. { 'ak·wə‚tint }

aquation [CHEM] Formation of a complex that contains water by replacement of other coordinated groups in the complex. { ə'kwā·shən }

aquatone [GRAPHICS] An offset printing process utilizing a zinc plate that is gelatin-coated and hardened and sensitized to print type, line drawings, and fine-screen halftones. { 'ak·wə‚tōn }

aqueduct [CIV ENG] An artificial tube or channel for conveying water. { 'ak·wə‚dəkt }

Aquent [GEOL] A suborder of the soil order Entisol, bluish gray or greenish gray in color; under water until very recent times; located at the margins of oceans, lakes, or seas. { 'ā·kwənt }

aqueous [SCI TECH] Relating to or made with water. { 'āk·wē·əs }

aqueous desert [ECOL] A marine bottom environment with little or no macroscopic invertebrate shelled life. { 'āk·wē·əs 'dez·ərt }

aqueous electron *See* hydrated electron. { 'āk·wē·əs i'lek‚trän }

aqueous humor [PHYSIO] The transparent fluid filling the anterior chamber of the eye. { 'āk·wē·əs 'yü·mər }

aqueous lava [GEOL] Mud lava produced by the mixing of volcanic ash with condensing volcanic vapor or other water. { 'āk·wē·əs 'läv·ə }

aqueous micelle [BIOCHEM] A spherical aggregate, 4-8 nanometers in diameter, formed dynamically from surfactants in water above a characteristic concentration, the critical micelle concentration. { 'āk·wē·əs mi'sel }

aqueous rock [PETR] A sedimentary rock deposited by or in water. Also known as hydrogenic rock. { 'āk·wē·əs 'räk }

aqueous solution [CHEM] A solution with the solvent as water. { 'āk·wē·əs sə'lü·shən }

Aquept [GEOL] A suborder of the soil order Inceptisol, wet or drained, which lacks silicate clay accumulation in the soil profiles; surface horizon varies in thickness. { 'ak·wəpt }

aquiclude [GEOL] A porous formation that absorbs water slowly but will not transmit it fast enough to furnish an appreciable supply for a well or spring. { 'ak·wə‚klüd }

aquiculture [BIOL] Cultivation of natural faunal resources of water. Also spelled aquaculture. { 'ak·wə‚kəl·chər }

aquifer [GEOL] A permeable body of rock capable of yielding quantities of groundwater to wells and springs. [HYD] A subsurface zone that yields economically important amounts of water to wells. { 'ak·wə·fər }

Aquifoliaceae [BOT] A family of woody flowering plants in the order Celastrales characterized by pendulous ovules, alternate leaves, imbricate petals, and drupaceous fruit; common members include various species of holly (*Ilex*). { ‚ak·wə‚fōl·ē'ās·ē‚ē }

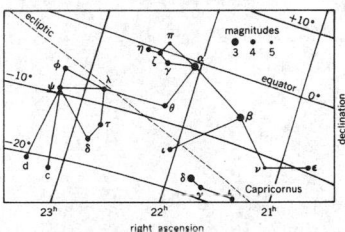

AQUARIUS

Line pattern of the constellation Aquarius. The grid lines represent the coordinates of the sky. The apparent brightness, or magnitude, of the stars is shown by the sizes of the dots, which are graded by appropriate numbers as indicated.

aquifuge [GEOL] An impermeable body of rock which contains no interconnected openings or interstices and therefore neither absorbs nor transmits water. { 'ak·wə,fyüj }

aquiherbosa [ECOL] Herbaceous plant communities in wet areas, such as swamps and ponds. { ,ak·wē,hər'bōs·ə }

Aquila [ASTRON] A constellation with a right ascension of 20 hours and declination of 5°N. Abbreviated Aqil; Aql. { 'ak·wə·lə }

α Aquilae *See* Altair. { ¦al·fə 'ak·wə,lē }

aquiprata [ECOL] Communities of plants which are found in areas such as wet meadows where groundwater is a factor. { ə'kwip·rəd·ə }

Aquitanian [GEOL] Lower lower Miocene or uppermost Oligocene geologic time. { ,ak·wə'tān·ē·ən }

aquitard [GEOL] A bed of low permeability adjacent to an aquifer; may serve as a storage unit for groundwater, although it does not yield water readily. { 'ak·wə,tärd }

Aquod [GEOL] A suborder of the soil order Spodosol, with a black or dark brown horizon just below the surface horizon; seasonally wet, it occupies depressions or wide flats from which water cannot escape easily. { 'ak·wəd }

aquo ion [CHEM] Any ion containing one or more water molecules. { 'a·kwō 'ī,än }

Aquoll [GEOL] A suborder of the soil order Mollisol, with thick surface horizons; formed under wet conditions; it may be under water at times, but is seasonally rather than continually wet. { 'ak·wȯl }

Aquox [GEOL] A suborder of the soil order Oxisol, seasonally wet, found chiefly in shallow depressions; deeper soil profiles are predominantly gray, sometimes mottled, and contain nodules or sheets of iron and aluminum oxides. { 'ak·wäks }

Aquult [GEOL] A suborder of the soil order Ultisol; seasonally wet, it is saturated with water a significant part of the year unless drained; surface horizon of the soil profile is dark and varies in thickness, grading to gray in the deeper portions; it occurs in depressions or on wide upland flats from which water drains very slowly. { 'ak·wəlt }

Ar *See* argon.

Ar₁ [MET] The temperature at which conversion of austenite to ferrite or to ferrite plus cementite is completed upon cooling a steel.

Ar₃ [MET] The temperature at which austenite begins to convert to ferrite upon cooling a steel.

Ar₄ [MET] The temperature at which delta ferrite is converted to gamma iron (austenite) upon cooling a steel.

Ar_cm [MET] The temperature at which austenite is converted to cementite upon cooling a hypereutectoid steel.

Ara [ASTRON] A constellation with a right ascension of 17 hours and declination of 55°S. Also known as Altar. { 'ä·rə }

araban [BIOCHEM] A polysaccharide composed of the pentose sugar L-arabinose. { 'ar·ə,ban }

Arabellidae [INV ZOO] A family of polychaete worms belonging to the Errantia. { ,ar·ə'bel·ə,dē }

arabic numerals [MATH] The numerals 0, 1, 2, 3, 4, 5, 6, 7, 8, and 9. Also known as Hindu-Arabic numerals. { 'ar·ə·bik 'nüm·rəlz }

arabine *See* harman. { 'ar·ə,bēn }

arabinose [BIOCHEM] $C_5H_{10}O_5$ A pentose sugar obtained in crystalline form from plant polysaccharides such as gums, hemicelluloses, and some glycosides. { ə'rab·ə,nōs }

arabite *See* arabitol. { 'ar·ə,bīt }

arabitol [ORG CHEM] $CH_2OH(CHOH)_3CH_2OH$ An alcohol that is derived from arabinose; a sweet, colorless crystalline material present in d and l forms; soluble in water; melts at 103°C. Also known as arabite. { ə'rab·ə,tȯl }

Araceae [BOT] A family of herbaceous flowering plants in the order Arales; plants have stems, roots, and leaves, the inflorescence is a spadix, and the growth habit is terrestrial or sometimes more or less aquatic; well-known members include dumb cane (*Dieffenbachia*), jack-in-the-pulpit (*Arisaema*), and *Philodendron*. { ə'rās·ē,ē }

arachic acid *See* eicosanoic acid. { ə'rak·ik 'as·əd }

arachidic acid *See* eicosanoic acid. { ,a·rə'kid·ik 'as·əd }

arachidonate [BIOCHEM] A salt or ester of arachidonic acid. { ə¦rak·ə¦dän,āt }

arachidonic acid [BIOCHEM] $C_{20}H_{32}O_2$ An essential unsaturated fatty acid that is a precursor in the biosynthesis

ARABINOSE

The structural formula of β-L-arabinose.

of prostaglandins, thromboxanes, and leukotrienes. { ə¦rak·ə¦dan·ik 'as·əd }

arachis oil *See* peanut oil. { 'ar·ə·kəs ,ȯil }

Arachnia [MICROBIO] A genus of bacteria in the family Actinomycetaceae; branched diphtheroid rods and branched filaments form filamentous microcolonies; facultatively anaerobic; the single species is a human pathogen. { ə'rak·nē·ə }

Arachnida [INV ZOO] A class of arthropods in the subphylum Chelicerata characterized by four pairs of thoracic appendages. { ə'rak·nəd·ə }

arachnodactyly [MED] A rare congenital defect of the skeletal system marked by abnormally long hand and foot bones. { ə¦rak·nə¦dak·tə·lē }

arachnoid [ANAT] A membrane that covers the brain and spinal cord and lies between the pia mater and dura mater. [BOT] Of cobweblike appearance, caused by fine white hairs. Also known as araneose. [INV ZOO] Any invertebrate related to or resembling the Arachnida. { ə'rak,nȯid }

arachnoidal granulations [ANAT] Projections of the arachnoid layer of the cerebral meninges through the dura mater. Also known as arachnoid villi; Pacchionian bodies. { ¦a,rak¦nȯid·əl ,gran·yə'lā·shənz }

Arachnoidea [INV ZOO] The name used in some classification schemes to describe a class of primitive arthropods. { ¦a,rak'nȯid·ē·ə }

arachnoid villi *See* arachnoidal granulations. { ə'rak,nȯid'·vil·ē }

arachnology [INV ZOO] The study of arachnids. { ,a,rak'näl·ə·jē }

arachnophobia [PSYCH] Abnormal fear of spiders. { ə¦rak·nō'fōb·ē·ə }

Aradidae [INV ZOO] The flat bugs, a family of hemipteran insects in the superfamily Aradoidea. { ə'rad·ə,dē }

Aradoidea [INV ZOO] A small superfamily of hemipteran insects belonging to the subdivision Geocorisae. { ,a·rə'dȯid·ē·ə }

Araeoscelidia [PALEON] A provisional order of extinct reptiles in the subclass Euryapsida. { ə¦rē·ə·sə'lid·ē·ə }

araeostyle [ARCH] Having an intercolumnation alternately of two and four column diameters. Also spelled areostyle. { ə'rē·ə,stīl }

Arago distance [ASTRON] The angular distance from the antisolar point to the Arago point. { 'a·rə,gō 'dis·təns }

aragonite [MINERAL] $CaCO_3$ A white, yellowish, or gray orthorhombic mineral species of calcium carbonate but with a crystal structure different from those of vaterite and calcite, the other two polymorphs of the same composition. Also known as Aragon spar. { ə'räg·ə,nīt }

Aragon spar *See* aragonite. { 'ar·ə,gän ,spär }

Arago point [OPTICS] A neutral point located about 20° directly above the antisolar point in relatively clear air and at higher elevations in turbid air. { 'a·rə,gō ,pȯint }

Arago's disk [ELECTROMAG] A device consisting of a horizontal disk of copper that can rotate about a vertical axis in an airtight box, and a horizontal bar magnet suspended above the disk but outside the box; upon rapid rotation of the disk, the bar magnet is deflected and eventually rotates in the same direction with smaller velocity. { 'a·rə,gōz ,disk }

Arales [BOT] An order of monocotyledonous plants in the subclass Arecidae. { ə'rā,lēz }

Araliaceae [BOT] A family of dicotyledonous trees and shrubs in the order Umbellales; there are typically five carpels and the fruit, usually a berry, is fleshy or dry; well-known members are ginseng (*Panax*) and English ivy (*Hedera helix*). { ə,rāl·ē'ās·ē,ē }

aralkyl [ORG CHEM] A radical in which an aryl group is substituted for an alkyl H atom. Derived from arylated alkyl. { a'ral,kil }

aramayoite [MINERAL] $Ag(Sb,Bi)S_2$ An iron-black mineral consisting of silver antimony bismuth sulfide. { ,ar·ə'mī·ə,wīt }

Aramidae [VERT ZOO] The limpkins, a family of birds in the order Gruiformes. { ə'ram·ə,dē }

Aran-Duchenne atrophy [MED] A muscular system disorder of adults involving progressive spinal muscular atrophy. { ¦ä·rän ,dyü¦shen 'a·trə·fē }

Araneae [INV ZOO] An equivalent name for Araneida. { ə'rän·ē,ē }

Araneida [INV ZOO] The spiders, an order of arthropods in the class Arachnida. { ,a·rə'nē·ə·də }

araneology [INV ZOO] The study of spiders. { ə,rān·ē'äl·ə·jē }

araneose *See* arachnoid. { ə'rān·ē,ōs }

arapahite [PETR] A dark-colored, porous, fine-grained basic basalt consisting of magnetite, bytownite, and augite. { ə'rap·ə,hīt }

araucaria [BOT] A primitive conifer of the genus *Araucaria* with broad leathery leaves, large cones, and edible seeds that is indigenous to South America and Australia. { ,ar,ō'kä·rē·ə }

Arbacioida [INV ZOO] An order of echinoderms in the superorder Echinacea. { är,bäs·ē'óid·ə }

arbilos [MATH] A plane figure bounded by a semicircle and two smaller semicircles which lie inside the larger semicircle, have diameters along the diameter of the larger semicircle, and are tangent to the larger semicircle and to each other. Also known as shoemaker's knife. { 'är·bī,lōs }

arbiter [COMPUT SCI] A computer unit that determines the priority sequence in which two or more processor inputs are connected to a single functional unit such as a multiplier or memory. { 'är·bəd·ər }

arbitrarily primed polymerase chain reaction [MOL BIO] A deoxyribonucleic acid (DNA) fingerprinting technique in which one short arbitrary primer is used to amplify multiple DNA fragments of different length, which yield a fingerprint after separation in gel electrophoresis. Also known as random amplification. { ,är·bə'trer·ə·lē 'prīmd pə'lim·ə,rās 'chān rē,ak·shən }

arbitrary correction [ORD] The correction of firing data or sound-locator data, applied to correct for observed errors after allowance has been made for all known causes of deviation. Also known as adjustment correction. { 'är·bə,trer·ē kə'rek·shən }

arbitrary course computer *See* course-line computer. { 'är·bə,trer·ē ¦kȯrs kəm,pyüd·ər }

arbitrary function generator *See* general-purpose function generator. { 'är·bə,trer·ē 'fəŋk·shən ,jen·ə,rād·ər }

arbitrary grid [MAP] Any reference system developed for use where no grid is available or practical, or where military security for the reference is desired. { 'är·bə,trer·ē 'grid }

arbitration [COMPUT SCI] The set of rules in a computer's operating system for allocating the resources of the computer, such as its peripheral devices or memory, to more than one program or user. [IND ENG] A semijudicial means of settling labor-management disputes in which both sides agree to be bound by the decision of one or more neutral persons selected by some method mutually agreed upon. { ,är·bə'trā·shən }

arbitration bar [MET] A cast-iron specimen, in the form of a standard-sized bar, to be tested for conformity to specifications of the American Society for Testing and Materials. { ,är·bə'trā·shən ,bär }

arbor [HOROL] The axle of a wheel in a watch or clock. [MECH ENG] **1.** A cylindrical device positioned between the spindle and outer bearing of a milling machine and designed to hold a milling cutter. **2.** A shaft or spindle used to hold a revolving cutting tool or the work to be cut. [MET] A device which supports sand cores in molds. { 'är·bər }

arbor collar [ENG] A cylindrical spacer that positions and secures a revolving cutter on an arbor. { 'är·bər ,käl·ər }

arboreal Also known as arboreous. [BOT] Relating to or resembling a tree. [ZOO] Living in trees. { är'bór·ē·əl }

arboreous [BOT] **1.** Wooded. **2.** *See* arboreal. { är'bór·ē·əs }

arborescence [BIOL] The state of being treelike in form and appearance. { 'är·bə¦res·əns }

arborescent powder *See* dendritic powder. { ¦är·bə¦res·ənt 'paúd·ər }

arboretum [BOT] An area where trees and shrubs are cultivated for educational and scientific purposes. { ,är·bə'rēd·əm }

arbor hole [DES ENG] A hole in a revolving cutter or grinding wheel for mounting it on an arbor. { 'är·bər ,hōl }

arboriculture [BOT] The cultivation of ornamental trees and shrubs. { 'är·bə·rə¦kəl·chər }

arborization [BIOL] A treelike arrangement, such as a branched dendrite or axon. { ,är·bə·rə'zā·shən }

arborization block *See* intraventricular heart block. { ,är·bə·rə'zā·shən ,bläk }

arborol *See* dendrimer. { 'är·bə,rōl }

arbor press [MECH ENG] A machine used for forcing an arbor or a mandrel into drilled or bored parts preparatory to turning or grinding. Also known as mandrel press. { 'är·bər ,pres }

arbor support [ENG] A device to support the outer end or intermediate point of an arbor. { 'är·bər sə,pȯrt }

arbor vitae [NEURO] The treelike arrangement of white nerve tissue seen in a median section of the cerebellum. { 'är·bər 'vīd·ē }

arborvitae [BOT] Any of the ornamental trees, sometimes called the tree of life, in the genus *Thuja* of the order Pinales. { ¦är·bər¦vīd·ē }

arbor vitae oil *See* thuja oil. { ¦är·bər ¦vīd·ē ,ȯil }

arboviral encephalitides [MED] Diseases which are caused by arthropod-borne viruses (arboviruses), such as the encephalitis infections. { är·bə¦vī·rəl ,en·sef·ə'līd·ə,dēz }

arbovirus [VIROL] Small, arthropod-borne animal viruses that are unstable at room temperature and inactivated by sodium deoxycholate; cause several types of encephalitis. Also known as arthropod-borne virus. { 'är·bə,vī·rəs }

Arbuckle orogeny [GEOL] Mid-Pennsylvanian episode of diastrophism in the Wichita and Arbuckle Mountains of Oklahoma. { 'är·bək·əl ȯ'räj·ə·nē }

arbuscule [MYCOL] A treelike haustorial organ in certain mycorrhizal fungi. { är'bə·skyül }

arbutin [ORG CHEM] $C_{12}H_{16}O_7$ A bitter glycoside from the bearberry and certain other plants; sometimes used as a urinary antiseptic. { är'byüt·ən }

arc [ELEC] *See* electric arc. [ENG] The graduated scale of an instrument for measuring angles, as a marine sextant; readings obtained on that part of the arc beginning at zero and extending in the direction usually considered positive are popularly said to be on the arc, and those beginning at zero and extending in the opposite direction are said to be off the arc. [GEOL] A geologic or topographic feature that is repeated along a curved line on the surface of the earth. [MATH] **1.** A continuous piece of the circumference of a circle. **2.** *See* edge. { ärk }

Arc [ASTRON] A radio source consisting of two bundles of parallel filaments adjoining the source Sagittarius A near the center of the Milky Way Galaxy. { ärk }

ARC *See* AIDS-related complex.

arcade [ARCH] **1.** A series of arches supported on columns. **2.** An arched passageway. [INV ZOO] A type of cell associated with the pharyngeal region of nematodes and united with like cells by an arch. { är'kād }

arcanite [MINERAL] K_2SO_4 A colorless, vitreous orthorhombic sulfate mineral. Also known as glaserite. { 'är·kə,nīt }

arcature [ARCH] **1.** A small arcade. **2.** A blind, usually decorative arcade. { 'är·kə,chər }

arcback [ELECTR] The flow of a principal electron stream in the reverse direction in a mercury-vapor rectifier tube because of formation of a cathode spot on an anode; this results in failure of the rectifying action. Also known as backfire. { 'ärk,bak }

arc blow [MET] The shifting of the arc in various directions in electric-arc welding due to the magnetic fields at the arc. { 'ärk ,blō }

arc brazing [MET] Brazing with the use of an electric arc. { 'ärk ,brāz·iŋ }

arc chute [ELEC] A collection of insulating barriers in a circuit breaker for confining the arc and preventing it from causing damage. { 'ärk ,shüt }

arc converter [ELECTR] A form of oscillator using an electric arc as the generator of alternating or pulsating current. { 'ärk kən,vər·dər }

arc cosecant [MATH] Also known as anticosecant; inverse cosecant. **1.** For a number x, any angle whose cosecant equals x. **2.** For a number x, the angle between $-\pi/2$ radians and $\pi/2$ radians whose cosecant equals x; it is the value at x of the inverse of the restriction of the cosecant function to the interval between $-\pi/2$ and $\pi/2$. { 'ärk kō'sē,kant }

arc cosine [MATH] Also known as anticosine; inverse cosine. **1.** For a number x, any angle whose cosine equals x. **2.** For a number x, the angle between 0 radians and π radians whose cosine equals x; it is the value at x of the inverse of the restriction of the cosine function to the interval between 0 and π. { 'ärk 'kō,sīn }

arc cotangent [MATH] Also known as anticotangent; inverse cotangent. **1.** For a number x, any angle whose cotangent equals x. **2.** For a number x, the angle between 0 radians and π radians whose cotangent equals x; it is the value at x of the inverse of the restriction of the cotangent function to the interval between 0 and π. { 'ärk kō'tan·jənt }

arc cutting [MET] A type of thermal cutting of metal using the temperature generated by an electric arc. { 'ärk ,kəd·iŋ }

arc discharge [ELEC] A direct-current electrical current between electrodes in a gas or vapor, having high current density and relatively low voltage drop. { 'ärk 'dis,chärj }

arc-disjoint paths [MATH] In a graph, two paths with common end points that have no arcs in common. { ,ärk'dis ,jóint pathz }

arc doubleau [ARCH] An arch, usually very massive, carried across a wide space, to support a groined vault or to stiffen a barrel vault. { 'ärk 'dü,blō }

Arcellinida [INV ZOO] An order of rhizopodous protozoans in the subclass Lobosia characterized by lobopodia and a well-defined aperture in the test. { ,är·sə'lin·ə·də }

arc excitation [ATOM PHYS] Use of electric-arc energy to move electrons into higher energy orbits. { 'ärk ,ek,sī'tā·shən }

arc force [MECH] The force of a plasma arc through a nozzle or opening. { 'ärk ,fórs }

arc furnace [MET] A furnace used to heat materials by the energy from an electric arc. Also known as electric-arc furnace. { 'ärk 'fər·nəs }

arc gouging [MET] The formation of a bevel or groove by an arc cutting process. { 'ärk ,gaúj·iŋ }

arch [CIV ENG] A structure curved and so designed that when it is subjected to vertical loads, its two end supports exert reaction forces with inwardly directed horizontal components; common uses for the arch are as a bridge, support for a roadway or railroad track, or part of a building. { ärch }

archaebacteria [MICROBIO] A group of unusual prokaryotic organisms that microscopically resemble true bacteria but differ biochemically and genetically, and form a distinct evolutionary group; some occur widely in oxygen-free environments and produce methane, while others are found in extreme salty or acidic conditions or grow at high temperatures. { ,ar·kē·bak'tir·ē·ə }

Archaeoceti [PALEON] The zeuglodonts, a suborder of aquatic Eocene mammals in the order Cetacea; the oldest known cetaceans. { ,ärk·ē·ə'sē,tī }

Archaeocidaridae [PALEON] A family of Carboniferous echinoderms in the order Cidaroida characterized by a flexible test and more than two columns of interambulacral plates. { ,ärk·ē·ə,sə'dar·ə,dē }

Archaeocopida [PALEON] An order of Cambrian crustaceans in the subclass Ostracoda characterized by only slight calcification of the carapace. { ,ärk·ē·ə'käp·ə·də }

Archaeogastropoda [INV ZOO] An order of gastropod mollusks that includes the most primitive snails. { ,ärk·ē·ə,gas'-träp·ə·də }

Archaeopteridales [PALEOBOT] An order of Upper Devonian sporebearing plants in the class Polypodiopsida characterized by woody trunks and simple leaves. { ,ärk·ē,äp·tə'rīd·ə·lēz }

Archaeopteris [PALEOBOT] A genus of fossil plants in the order Archaeopteridales; used sometimes as an index fossil of the Upper Devonian. { ,ärk·ē'äp·tə·rəs }

Archaeopterygiformes [PALEON] The single order of the extinct avian subclass Archaeornithes. { ,ärk·ē,äp·tə,rij·ə'fór,mēz }

Archaeopteryx [PALEON] The earliest known bird; a genus of fossil birds in the order Archaeopterygiformes characterized by flight feathers like those of modern birds. { ,ärk·ē'äp·tə·riks }

Archaeornithes [PALEON] A subclass of Upper Jurassic birds comprising the oldest fossil birds. { ,ärk·ē'ór·nə,thēz }

archaic [PSYCH] Designating elements, largely unconscious, in the psyche which are remnants of humankind's prehistoric past and which reappear in dreams and other symbolic manifestations. { är'kā·ik }

archallaxis [BIOL] Deviation from an ancestral pattern early

ARCHAEOPTERYX

Archaeopteryx lithographica. (*From W. E. Swinton, Fossil Birds, British Museum-Natural History, 1958*)

ARCH DAM

East Canyon Dam, a thin-arch concrete structure on the East Canyon River, Utah. (*U.S. Bureau of Reclamation*)

in development, eliminating duplication of the phylogenetic history. { ,ärk·ə'lak·səs }

Archangiaceae [MICROBIO] A family of bacteria in the order Myxobacterales; microcysts are rod-shaped, ovoid, or spherical and are not enclosed in sporangia, and fruiting bodies are irregular masses. { ,ärk,an·jē'ās·ē,ē }

Archangium [MICROBIO] The single genus of the family Archangiaceae; sporangia are lacking, and there is no definite slime wall. { ärk'an·jē·əm }

Archanthropinae [PALEON] A subfamily of the Hominidae, set up by F. Weidenreich, which is no longer used. { ,ärk·ən'thräp·ə,nē }

arch band [CIV ENG] Any narrow elongated surface forming part of or connected with an arch. { 'ärch ,band }

arch bar [BUILD] **1.** A curved chimney bar. **2.** A curved bar in a window sash. { 'ärch ,bar }

arch beam [CIV ENG] A curved beam, used in construction, with a longitudinal section bounded by two arcs having different radii and centers of curvature so that the beam cross section is larger at either end than at the center. { 'ärch ,bēm }

arch blocks [MIN ENG] Blocks applied to the wooden voussoirs used in framing a timber support for the roof when driving a tunnel. { 'ärch ,bläks }

arch brace [BUILD] A curved brace, usually used in pairs to support a roof frame and give the effect of an arch. { 'ärch ,brās }

arch brick [MATER] **1.** A wedge-shaped brick used in arches. **2.** An overburned brick, resulting from contact with the fire in the arch of a kiln. { 'ärch ,brik }

arch bridge [CIV ENG] A bridge having arches as the main supports. { 'ärch ,brij }

arch center [CIV ENG] A temporary structure for support of the parts of a masonry or concrete arch during its construction. { 'ärch ,sen·tər }

arch corner bead [BUILD] A corner bead which is cut on the job; used to form and reinforce the curved portion of arch openings. { 'ärch ,kòr·nər ,bēd }

arch dam [CIV ENG] A dam having a curved face on the downstream side, the curve being roughly a portion of a cylinder whose axis is vertical. { 'ärch ,dam }

Archean [GEOL] A term, meaning ancient, which has been applied to the oldest rocks of the Precambrian; as more physical measurements of geologic time are made, the usage is changing; the term Early Precambrian is preferred. { är'kē·ən }

arc heating [MET] The heating of a material by the heat energy from an electric arc, which has a very high temperature and very high concentration of heat energy. Also known as electric-arc heating. { 'ärk ,hēd·iŋ }

arched construction [BUILD] A method of construction relying on arches and vaults to support walls and floors. { 'ärcht kən'strək·shən }

archegoniophore [BOT] The stalk supporting the archegonium in liverworts and ferns. { ,ärk·ə'gōn·ē·ə,fór }

archegonium [BOT] The multicellular female sex organ in all plants of the Embryobionta except the Pinophyta and Magnoliophyta. { ,ark·ə'gōn·ē·əm }

archencephalon [EMBRYO] The primitive embryonic forebrain from which the forebrain and midbrain develop. { ,ärk,in'sef·ə·län }

archenteron [EMBRYO] The cavity of the gastrula formed by ingrowth of cells in vertebrate embryos. Also known as gastrocoele; primordial gut. { ,ärk'en·tə,rän }

archeoastronomy [ASTRON] The study which attempts to reconstruct the astronomical knowledge and activity of prehistoric people and its influence on their cultures and societies. { ,är·kē·ō·ə'strän·ə·mē }

archeocyte [INV ZOO] A type of ovoid amebocyte in sponges, characterized by large nucleolate nuclei and blunt pseudopodia; gives rise to germ cells. { 'är·kē·ə,sīt }

archeological chemistry [ARCHEO] The application of chemical techniques to the study of the material remains of the cultures of historical or prehistorical peoples, for example, to ascertain the age or composition of such remains. { ,är·kē·ə,läj·i·kəl 'kem·ə·strē }

archeological chronology [ARCHEO] The establishment of the temporal sequence of human cultures by applying a variety of dating methods to cultural remains. { ,är·kē·ə,läj·i·kəl krə'näl·ə·jē }

archeology [SCI TECH] The scientific study of the material remains of the cultures of historical or prehistorical peoples. { ¦är·kē¦äl·ə·jē }

archeomagnetic dating [GEOPHYS] An absolute dating method based on the earth's shifting magnetic poles. When clays and other rock and soil materials are fired to approximately 1300°F (700°C) and allowed to cool in the earth's magnetic field, they retain a weak magnetism which is aligned with the position of the poles at the time of firing. This allows for dating, for example, of when a fire pit was used, based on the reconstruction of pole position for earlier times. { ¦är·kē·ō,mag¦ned·ik 'dā·diŋ }

archeometric [ARCHEO] Referring to application of the techniques of physics and chemistry to the analysis of archeological and historical art objects. { är·kē·ə¦me·trik }

Archeozoic [GEOL] **1.** The era during which, or during the latter part of which, the oldest system of rocks was made. **2.** The last of three subdivisions of Archean time, when the lowest forms of life probably existed; as more physical measurements of geologic time are made, the usage is changing; it is now considered part of the Early Precambrian. { ¦är·kē·ə¦zō·ik }

Archer See Sagittarius. { 'är·chər }

archerfish [VERT ZOO] The common name for any member of the fresh-water family Toxotidae in the order Perciformes; individuals eject a stream of water from the mouth to capture insects. { 'är·chər,fish }

Archeria [PALEON] Genus of amphibians, order Embolomeri, in early Permian in Texas; fish eaters. { ¸är'kir·ē·ə }

archespore [BOT] A cell from which the spore mother cell develops in either the pollen sac or the ovule of an angiosperm. { 'är·kə,spór }

archetype [EVOL] A hypothetical ancestral type conceptualized by eliminating all specialized character traits. { 'är·ki,tīp }

arch girder [CIV ENG] A normal H-section steel girder bent to a circular shape. { 'ärch ¸gər·dər }

arch-gravity dam [CIV ENG] An arch dam stabilized by gravity due to great mass and breadth of the base. { 'ärch ¦grav·əd·ē ¸dam }

Archiacanthocephala [INV ZOO] An order of worms in the phylum Acanthocephala; adults are endoparasites of terrestrial vertebrates. { ¦är·kē·ə,kan·thə'sef·ə·lə }

Archiannelida [INV ZOO] A group name applied to three families of unrelated annelid worms: Nerillidae, Protodrilidae, and Dinophilidae. { ¦är·kē·ə'nel·ə·də }

archibenthic zone [OCEANOGR] The biogeographic realm of the ocean extending from a depth of about 665 feet to 2625-3610 feet (200 meters to 800-1100 meters). { ¦är·kē¦ben·thik ¸zōn }

archibole See positive element. { 'är·kē,bōl }

Archichlamydeae [BOT] An artificial group of flowering plants, in the Englerian system of classification, consisting of those families of dicotyledons that lack petals or have petals separate from each other. { ¦är·kē·klə'mid·ē,ē }

archicoel [ZOO] The segmentation cavity persisting between the ectoderm and endoderm as a body cavity in certain lower forms. { 'är·kē,sēl }

Archidiidae [BOT] A subclass of the plant class Bryopsida; consists of a single genus, *Archidium*, unique in having spores scattered in a single layer of the endothecium and having no quadrant stage in the early ontogeny of the capsule. { ¸är·kə'dī·ə,dē }

Archie [COMPUT SCI] A system of file servers that searches for specific files that are publicly available in File Transfer Protocol archives on the Internet. { 'är·chē }

archigastrula [EMBRYO] A gastrula formed by invagination, as opposed to ingrowth of cells. { ¸är·kē'gas·trə·lə }

Archigregarinida [INV ZOO] An order of telosporean protozoans in the subclass Gregarinia; endoparasites of invertebrates and lower chordates. { ¦är·kē,greg·ə'rin·ə·də }

Archimedean ordered field [MATH] A field with a linear order that satisfies the axiom of Archimedes. { ¸ärk·ə¦mē·dē·ən ¦órd·ərd 'fēld }

Archimedean principle [PHYS] The principle that a body immersed in a fluid undergoes an apparent loss in weight equal to the weight of the fluid it displaces. { ¦är·kə¦mēd·ē·ən 'prin·sə·pəl }

Archimedean solid [MATH] One of 13 possible solids whose faces are all regular polygons, though not necessarily all of the same type, and whose polyhedral angles are all equal. Also known as semiregular solid. { ¦är·kə¦mēd·ē·ən 'säl·əd }

Archimedean spiral [MATH] A plane curve whose equation in polar coordinates (r, θ) is $r^m = a^m\theta$, where a and m are a positive or negative integer. { ¦är·kə¦mēd·ē·ən 'spī·rəl }

Archimedes' axiom See axiom of Archimedes. { ¦är·kə¦mēd,ēz 'ak·sē·əm }

Archimedes number [FL MECH] One of a dimensionless group of numbers denoting the ratio of gravitational force to viscous force. { ¦är·kə¦mēd,ēz 'nəm·bər }

Archimedes' problem [MATH] The problem of dividing a hemisphere into two parts of equal volume with a plane parallel to the base of the hemisphere; it cannot be solved by Euclidean methods. { ¦är·kə¦mēd,ēz 'präb·ləm }

Archimedes' screw [MECH ENG] A device for raising water by means of a rotating broad-threaded screw or spirally bent tube within an inclined hollow cylinder. { ¦är·kə¦mēd,ēz 'skrü }

Archimedes' spiral See spiral of Archimedes. { ¦är·kə¦mēd'ēz 'spī·rəl }

archinephridium [INV ZOO] One of a pair of primitive nephridia found in each segment of some annelid larvae. { ¦är·kē·nə¦frid·ē·əm }

archinephros [VERT ZOO] The paired excretory organ of primitive vertebrates and the larvae of hagfishes and caecilians. { ¦är·kē¦ne,frōs }

arching [CIV ENG] **1.** The transfer of stress from a yielding part of a soil mass to adjoining less-yielding or restrained parts of the mass. **2.** A system of arches. **3.** The arched part of a structure. [GEOL] The folding of schists, gneisses, or sediments into anticlines. [MIN ENG] Curved support for roofs of openings in mines. { 'ärch·iŋ }

archipallium [PHYSIO] The olfactory pallium or the olfactory cerebral cortex; phylogenetically, the oldest part of the cerebral cortex. { ¦är·ki¦pal·ē·əm }

archipelagic apron [GEOL] A fan-shaped slope around an oceanic island differing from deep-sea fans in having little, if any, sediment cover. { ¦är·kə·pə¦laj·ik 'a·prən }

archipelago [GEOGR] **1.** A large group of islands. **2.** A sea that has a large group of islands within it. { ¦är·kə¦pel·ə,gō }

architect [ARCH] A person who is skilled and knowledgeable in the design of buildings and whose qualifications are recognized by a college degree and licensing by an appropriate professional organization. { 'är·kə,tekt }

architectonic [ARCH] **1.** Pertaining to or in accordance with the principles of architecture. **2.** Having the qualities of ARCH in terms of structure and concept. [GEOL] Of forces that determine structure. { ¦är·kə,tek¦tän·ik }

architect's scale [GRAPHICS] A rule with a scale on it so chosen that by placing the rule's edge on a reduced-scale drawing the scale of the drawing (say, in inches) may be converted directly into the dimensions of the object (say, in feet). { 'är·kə,teks ,skāl }

architectural acoustics [CIV ENG] The science of planning and building a structure to ensure the most advantageous flow of sound to all listeners. { ¦är·kə¦tek·chər·əl ə'kü·stiks }

architectural bronze [MET] An alloy containing 57% copper, 40% zinc, 2.75% lead, 0.25% tin; used for extruded moldings and forgings. { ¦är·kə¦tek·chər·əl 'bränz }

architectural concrete [MATER] Concrete used for ornamentation or finish on exterior or interior surfaces of a building or other structure. { ¦är·kə¦tek·chər·əl 'kän,krēt }

architectural engineering [CIV ENG] The branch of engineering dealing primarily with building materials and components and with the design of structural systems for buildings, in contrast to heavy construction such as bridges. { ¦är·kə¦tek·chər·əl ,en·jə'nir·iŋ }

architectural millwork [CIV ENG] Ready-made millwork especially fabricated to meet the specifications for a particular job, as distinguished from standard or stock items or sizes. Also known as custom millwork. { ¦är·kə¦tek·chər·əl 'mil,wərk }

architectural terra-cotta [MATER] A hard-burnt, glazed or unglazed clay unit used in building construction. { ¦är·kə¦tek·chər·əl ,ter·ə'käd·ə }

architectural volume [CIV ENG] The cubic content of a building calculated by multiplying the floor area by the height. { ¦är·kə¦tek·chər·əl 'väl·yəm }

architecture [ENG] **1.** The art and science of designing

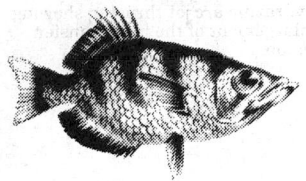

ARCHERFISH

The archerfish (*Toxotes jaculator*), maximum length 7 inches (18 centimeters).

buildings. **2.** The product of this art and science. { 'är·kə,tek·chər }

architrave [ARCH] **1.** The lowest division of an entablature that rests on the column capital. **2.** The molded band, group of moldings, or other architectural member around an opening, such as a door, especially if rectangular. { 'är·kə,trāv }

architrave cornice [ARCH] An entablature in which the cornice rests directly on the architrave, the frieze being omitted. { 'är·kə,trāv 'kór·nəs }

archival storage [COMPUT SCI] Storage of infrequently used or backup information that cannot be readily or immediately accessed by a computer system. { är,kīv·əl 'stòr·ij }

archiving [COMPUT SCI] The storage of files, in the form of punch cards, microfilm, or magnetic tape, for very long periods of time, in case it is necessary to regenerate the file due to subsequent introduction of errors. { 'är,kīv·iŋ }

archivolt [ARCH] **1.** The band or molding surrounding an arch. **2.** The architectural members of the inner surface of an arch. { 'är·kə,vōlt }

Archosauria [VERT ZOO] A subclass of reptiles composed of five orders: Thecodontia, Saurischia, Ornithschia, Pterosauria, and Crocodilia. { ,är·kə'sòr·ē·ə }

Archostemata [INV ZOO] A suborder of insects in the order Coleoptera. { ,är·kə·stə'mäd·ə }

arch pattern [FOREN SCI] A fingerprint pattern in which ridges enter on one side of the impression, form a wave or angular upthrust, and flow out the other side. { 'ärch ,pad·ərn }

arch press [MECH ENG] A punch press having an arch-shaped frame to permit operations on wide work. { 'ärch ,pres }

arch rib [CIV ENG] One of a set of projecting molded members subdividing the undersurface of an arch. { 'ärch ,rib }

arch ring [CIV ENG] A curved member that provides the main support of an arched structure. { 'ärch ,riŋ }

arch truss [CIV ENG] A truss having the form of an arch or arches. { 'ärch ,trəs }

archway [ARCH] A way or passage over which an arch extends. { 'ärch,wā }

arc-hyperbolic cosecant [MATH] For a number, x, not equal to zero, the number whose hyperbolic cosecant equals x; it is the value at x of the inverse of the hyperbolic cosecant function. Also known as inverse hyperbolic cosecant. { 'ärk ,hī·pər,bäl·ik kō'sē,kant }

arc-hyperbolic cosine [MATH] Also known as inverse hyperbolic cosine. **1.** For a number, x, equal to or greater than 1, either of the two numbers whose hyperbolic cosine equals x. **2.** For a number, x, equal to or greater than 1, the positive number whose hyperbolic cosine equals x; it is the value at x of the restriction of the inverse of the hyperbolic cosine function to the positive numbers. { ,ärk ,hī·pər,bäl·ik 'kō,sīn }

arc-hyperbolic cotangent [MATH] For a number, x, with absolute value greater than 1, the number whose hyperbolic cotangent equals x; it is the value at x of the inverse of the hyperbolic cotangent function. Also known as inverse hyperbolic cotangent. { ,ärk ,hī·pər,bäl·ik kō'tan·jənt }

arc-hyperbolic function See inverse hyperbolic function. { ärk ,hī·pər'bäl·ik 'fəŋk·shən }

arc-hyperbolic secant [MATH] Also known as inverse hyperbolic secant. **1.** For a number, x, equal to or greater than 0 and equal to or less than 1, either of the two numbers whose hyperbolic secant equals x. **2.** For a number, x, equal to or greater than 0, and equal to or less than 1, the positive number whose hyperbolic cosecant equals x; it is the value at x of the restriction of the hyperbolic secant function to the positive numbers. { ,ärk ,hī·pər,bäl·ik 'sē,kant }

arc-hyperbolic sine [MATH] For a number, x, the number whose hyperbolic sine equals x; it is the value at x of the inverse of the hyperbolic sine function. Also known as inverse hyperbolic sine. { ,ärk ,hī·pər,bäl·ik 'sīn }

arc-hyperbolic tangent [MATH] For a number, x, with absolute value less than 1, the number whose hyperbolic tangent equals x; it is the value at x of the inverse of the hyperbolic tangent function. Also known as inverse hyperbolic tangent. { ,ärk ,hī·pər,bäl·ik 'tan·jənt }

arcing contacts [ELEC] Special contacts on which the arc is drawn after the main contacts of a switch or circuit breaker have opened. { 'ärk·iŋ 'kän,taks }

arcing ring [ELEC] A metal ring attached to an insulator to protect it from damage by a power arc. { 'ärk·iŋ ,riŋ }

arcing time [ELEC] **1.** Interval between the parting, in a switch or circuit breaker, of the arcing contacts and the extension of the arc. **2.** Time elapsing, in a fuse, from the severance of the fuse link to the final interruption of the circuit under a specified condition. { 'ärk·iŋ ,tīm }

arc jet engine [AERO ENG] An electromagnetic propulsion engine used to supply motive power for flight; hydrogen and ammonia are used as the propellant, and some plasma is formed as the result of electric-arc heating. { 'ärk 'jet 'en·jən }

arc lamp [ELEC] An electric lamp in which the light is produced by an arc made when current flows through ionized gas between two electrodes. Also known as electric-arc lamp. { 'ärk ,lamp }

arc measurement [GEOD] A survey method used to determine the size of the earth. { 'ärk 'mezh·ər·mənt }

arc melting [MET] Melting and purification of metal in an electric-arc furnace. { 'ärk 'melt·iŋ }

arcmin See minute.

arc navigation [NAV] A navigation system in which the position of an airplane or ship is maintained along an arc of a circle which has a radius measured from a control station by means of electronic distance-measuring equipment, such as shoran or oboe. { 'ärk ,nav·ə'gā·shən }

arcocentrum [ANAT] A centrum formed of modified, fused mesial parts of the neural or hemal arches. { ,är·kō'sen·trəm }

arc of action See arc of contact. { 'ärk əv 'ak·shən }

arc of approach [DES ENG] In toothed gearing, the part of the arc of contact along which the flank of the driving wheel contacts the face of the driven wheel. { 'ärk əv ə'prōch }

arc of contact [MECH ENG] **1.** The angular distance over which a gear tooth travels while it is in contact with its mating tooth. Also known as arc of action. **2.** The angular distance a pulley travels while in contact with a belt or rope. { 'ärk əv 'kän,takt }

arc of parallel [GEOD] A part of an astronomic or a geodetic parallel. { 'ärk əv 'par·ə,lel }

arc of recess [DES ENG] In toothed gearing, the part of the arc of contact wherein the face of the driving wheel touches the flank of the driven wheel. { 'ärk əv 'rē,ses }

arc of visibility [NAV] The arc of a light sector, designated by its limiting bearings as observed from seaward. { 'ärk əv ,viz·ə'bil·əd·ē }

arcometer [ENG] A device for determining the density of a liquid by measuring the apparent weight loss of a solid of known mass and volume when it is immersed in the liquid. { är'käm·əd·ər }

arc-over [ELEC] An unwanted arc resulting from the opening of a switch or the breakdown of insulation. { 'ärk ,ō·vər }

arc process [CHEM ENG] A former process that used electric arcs for fixation (oxidation) of atmospheric nitrogen to manufacture nitric acid. { 'ärk 'präs·əs }

arc resistance [ELEC] **1.** A measure of the durability of an insulating or dielectric material against the formation of conductive paths along the surface by arc discharges. **2.** The ratio of the voltage that gives rise to an arc discharge to the current in the arc. { 'ärk ri,zis·təns }

arc seam weld [MET] A linear weld or overlapping spot welds made by an arc welding process. { 'ärk 'sēm ,weld }

arc secant [MATH] Also known as antisecant; inverse secant. **1.** For a number x, any angle whose secant equals x. **2.** For a number x, the angle between 0 radians and π radians whose secant equals x; it is the value at x of the inverse of the restriction of the secant function to the interval between 0 and π. { 'ärk 'sē,kant }

arc sine [MATH] Also known as antisine; inverse sine. **1.** For a number x, any angle whose sine equals x. **2.** For a number x, the angle between $-\pi/2$ radians and $\pi/2$ radians whose sine equals x; it is the value at x of the inverse of the restriction of the sine function to the interval between $-\pi/2$ and $\pi/2$. { 'ärk 'sīn }

arc sine transformation [STAT] A technique used to convert data made up of frequencies or proportions into a form that can be analyzed by analysis of variance or by regression analysis. { 'ärk 'sīn ,tranz·fər'mā·shən }

arc spectrum [SPECT] The spectrum of a neutral atom, as opposed to that of a molecule or an ion; it is usually produced by vaporizing the substance in an electric arc; designated by

---

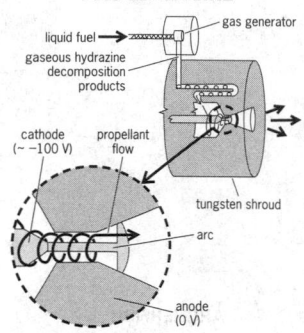

ARC JET ENGINE

liquid fuel → gas generator
gaseous hydrazine decomposition products
cathode (~ −100 V)
propellant flow
tungsten shroud
arc
anode (0 V)

Hydrazine arc jet thruster, showing enlargement of the heat transfer region.

the roman numeral I following the symbol for the element, for example, HeI.   { 'ärk ,spek·trəm }

arc-spot weld   [MET]   A weld that covers a very small area of the surface in contact and was made by an arc welding process.   { 'ärk 'spät ,weld }

arc spraying   [MET]   Depositing on a surface a metal melted by an electric arc and blown at high speed in an atomized state.   { 'ärk ,sprā·iŋ }

arc-suppression coil   [ELEC]   A grounding reactor, used in alternating-current power transmission systems, which is designed to limit the current flowing to ground at the location of a fault almost to zero by setting up a reactive current to ground that balances the capacitive current to ground flowing from the lines.   Also known as Petersen coil.   { ¦ärk sə'presh·ən ¦kȯil }

arc tangent   [MATH]   Also known as antitangent; inverse tangent.   **1.** For a number x, any angle whose tangent equals x.   **2.** For a number x, the angle between $-\pi/2$ radians and $\pi/2$ radians whose tangent equals x; it is the value at x of the inverse of the restriction of the tangent function to the interval between $-\pi/2$ and $\pi/2$.   { ¦ärk 'tan·jənt }

arc-through   [ELECTR]   Of a gas tube, a loss of control resulting in the flow of a principal electron stream in the normal direction during a scheduled nonconducting period.   { 'ärk ,thrü }

arctic air   [METEOROL]   An air mass whose characteristics are developed mostly in winter over arctic surfaces of ice and snow.   { ¦ärd·ik or 'ärk·tik 'er }

arctic-alpine   [ECOL]   Of or pertaining to areas above the timberline in mountainous regions.   { ¦ärd·ik 'al,pīn }

arctic anticyclone   See arctic high.   { ¦ärd·ik ,an·tē'sī,klōn }

Arctic Circle   [GEOD]   The parallel of latitude 66°32'N (often taken as 66$^1/_2$°N).   { ¦ärd·ik 'sər·kəl }

arctic climate   See polar climate.   { ¦ärd·ik 'klī·mət }

arctic desert   See polar desert.   { ¦ärd·ik 'dez·ərt }

arctic front   [METEOROL]   The semipermanent, semicontinuous front between the deep, cold arctic air and the shallower, basically less cold polar air of northern latitudes.   { ¦ärd·ik 'frənt }

arctic haze   [METEOROL]   A condition of reduced horizontal and slant visibility (but unimpeded vertical visibility) encountered by aircraft in flight (up to more than 30,000 feet, or 9140 meters) over arctic regions.   { ¦ärd·ik 'hāz }

arctic high   [METEOROL]   A weak high that appears on mean charts of sea-level pressure over the Arctic Basin during late spring, summer, and early autumn.   Also known as arctic anticyclone; polar anticyclone; polar high.   { ¦ärd·ik 'hī }

arcticization   [ENG]   The preparation of equipment for operation in an environment of extremely low temperatures.   { ¦ärd·ik,ī'zā·shən }

arctic mist   [METEOROL]   A mist of ice crystals; a very light ice fog.   { ¦ärd·ik 'mist }

Arctic Ocean   [GEOGR]   The north polar ocean lying between North America, Greenland, and Asia.   { ¦ärd·ik 'ō·shən }

Arctic Oscillation   [METEOROL]   Atmospheric pressure fluctuations (positive and negative phases) between the polar and middle latitudes (above 45° North) that strengthen and weaken the winds circulating counterclockwise from the surface to the lower stratosphere around the Arctic and, as a result, modulate the severity of the winter weather over most Northern Hemisphere middle and high latitudes.   Also known as the Northern Hemisphere annular mode.   { ¦ärd·ik ,äs·ə'lā·shən }

arctic sea smoke   [METEOROL]   Steam fog; but often specifically applied to steam fog rising from small areas of open water within sea ice.   { ¦ärd·ik ¦sē ¦smōk }

Arctic suite   [PETR]   A group of basic igneous rocks intermediate in composition between Atlantic and Pacific suites.   { ¦ärd·ik 'swēt }

arctic tree line   [ECOL]   The northern limit of tree growth; the sinuous boundary between tundra and boreal forest.   { 'ärd·ik ¦trē ¦līn }

Arctic Zone   [GEOGR]   The area north of the Arctic Circle (66°32'N).   { ¦ärd·ik ¦zōn }

Arctiidae   [INV ZOO]   The tiger moths, a family of lepidopteran insects in the suborder Heteroneura.   { ärk'tī·ə,dē }

arc time   [MET]   The time interval during which an arc is maintained in the making of an arc weld.   { 'ärk ,tīm }

arc-to-chord correction   See conversion angle.   { ¦ärk tə ¦kȯrd kə'rek·shən }

Arctocyonidae   [PALEON]   A family of extinct carnivore-like mammals in the order Condylarthra.   { ,ärk·tō,sī'än·ə,dē }

Arctolepiformes   [PALEON]   A group of the extinct joint-necked fishes belonging to the Arthrodira.   { ,ärk·tō,lep·ə'fȯr,mēz }

arc triangulation   [ENG]   A system of triangulation in which an arc of a great circle on the surface of the earth is followed in order to tie in two distant points.   { 'ärk ,trī,aŋ·gyə'lā·shən }

Arcturidae   [INV ZOO]   A family of isopod crustaceans in the suborder Valvifera characterized by an almost cylindrical body and extremely long antennae.   { ,ärk'tūr·ə,dē }

Arcturus   [ASTRON]   A star that is 36 light-years from the sun; spectral classification K2IIIp.   Also known as α Boötes.   { ,ärk'tūr·əs }

arcuale   [EMBRYO]   Any of the four pairs of primitive cartilages from which a vertebra is formed.   { ,ärk·yə'wä·lē }

arcuate   [ANAT]   Arched or curved; bow-shaped.   { 'ärk·yə·wət }

arcuated   [ARCH]   Based on or characterized by arches or archlike curves or vaults.   { 'ärk·yə,wād·əd }

arcuate delta   [GEOL]   A bowed or curved delta with the convex margin facing the body of water.   Also known as fan-shaped delta.   { 'ärk·yə·wət 'del·tə }

arcuation   [GEOL]   Production of an arc, as in rock flowage where movement proceeded in a fanlike manner.   { ,ärk·yə'wā·shən }

arcus   [METEOROL]   A dense and horizontal roll-shaped accessory cloud, with more or less tattered edges, situated on the lower front part of the main cloud.   { 'ar·kəs }

arc voltage   [MET]   The voltage across a welding arc.   { 'ärk ,vōl·tij }

arcwall coal cutter   [MIN ENG]   A type of electric or compressed-air cutter for under- or overcutting a coal seam in narrow work.   { 'ärk,wȯl 'kōl ,kəd·ər }

arc-welder's disease   See siderosis.   { 'ärk¦weld·ərz di,zēz }

arc welding   See electric-arc welding.   { 'ärk ,weld·iŋ }

arcwise-connected set   [MATH]   A set in which each pair of points can be joined by a simple arc whose points are all in the set.   Also known as path-connected set; pathwise-connected set.   { 'ärk,wīz kə,nek·təd 'set }

arc welding   See electric-arc welding..   { 'ärk ,weld·iŋ }

Arcyzonidae   [PALEON]   A family of Devonian paleocopan ostracods in the superfamily Kirkbyacea characterized by valves with a large central pit.   { ,är,sī'zän·ə,dē }

ARDC model atmosphere   See standard atmosphere.   { ¦ā¦är¦dē¦sē ,mäd·əl 'at·mə,sfer }

ardealite   [MINERAL] $Ca_2(HPO_4)(SO_4)\cdot 4H_2O$ A white or light-yellow mineral consisting of a hydrous acid calcium phosphate-sulfate.   { ,är·dē'ə,līt }

Ardeidae   [VERT ZOO]   The herons, a family of wading birds in the order Ciconiiformes.   { är'dē·ə,dē }

Ardennian orogeny   [GEOL]   A short-lived orogeny during the Ludlovian stage of the Silurian period of geologic time.   { är'den·ē·ən o'raj·ə·nē }

ardennite   [MINERAL] $Mn_5Al_5(VO_4)(SiO_4)_5(OH)_2\cdot 2H_2O$ A yellow to yellowish-brown mineral consisting of a hydrous silicate vanadate and arsenate of manganese and aluminum.   { är'den,īt }

arduinite   See mordenite.   { är'dwin,īt }

are   [MECH]   A unit of area, used mainly in agriculture, equal to 100 square meters.   { är }

area   [COMPUT SCI]   A section of a computer memory assigned by a computer program or by the hardware to hold data of a particular type.   [MATH]   A measure of the size of a two-dimensional surface, or of a region on such a surface.   { 'er·ē·ə }

area amniotica   [EMBRYO]   The transparent part of the blastodisc in mammals.   { 'er·ē·ə ,am·nē'äd·ə·kə }

area bombing   [ORD]   Dropping of bombs on a general locality rather than on a specific target or in a pattern.   { 'er·ē·ə ,bäm·iŋ }

area code   [COMMUN]   A three-digit prefix used in dialing long-distance telephone calls in the United States and Canada.   { 'er·ē·ə ,kōd }

area control   [NAV]   A form of air-traffic control in which aircraft flying in designated areas are furnished flight advisory service.   { 'er·ē·ə kən'trōl }

area coverage   [ENG]   Complete coverage of an area by aerial photography having parallel overlapping flight lines and

ARCTURIDAE

1 mm

Arcturella, a member of the family Arcturidae. (*From G. O. Sars, An Account of the Crustacea of Norway, vol. 2, 1899*)

stereoscopic overlap between exposures in the line of flight. { 'er·ē·ə ¦kəv·rij }

area defense [ORD] A defense against air attack organized to protect an area, as distinguished from a point defense or line defense. { 'er·ē·ə di'fens }

area delimiting line [MAP] A line fixing the boundary of an area. { 'er·ē·ə di'lim·əd·iŋ ¦līn }

area drain [CIV ENG] A receptacle designed to collect surface or rain water from an open area. { 'er·ē·ə ¦drān }

area effect [ELECTR] In general, the condition of the dielectric strength of a liquid or vacuum separating two electrodes being higher for electrodes of smaller area. { 'er·ē·ə i'fekt }

area fire [ORD] Fire delivered on a prescribed area. { 'er·ē·ə ¦fīr }

area forecast [METEOROL] A weather forecast for a specified geographic area; usually applied to a form of aviation weather forecast. Also known as regional forecast. { 'er·ē·ə 'fȯr,kast }

area landfill [CIV ENG] A sanitary landfill operation that takes care of the solid waste of more than one municipality in a region. { 'er·ē·ə 'land,fil }

areal density [COMPUT SCI] The amount of data that can be stored on a unit area of the surface of a hard disk, floppy disk, or other storage device. { ¸er·ē·əl 'den·səd·ē }

areal eruption [GEOL] Volcanic eruption resulting from collapse of the roof of a batholith; the volcanic rocks grade into parent plutonic rocks. { 'er·ē·əl i'rəp·shən }

areal geology [GEOL] Distribution and form of rocks or geologic units of any relatively large area of the earth's surface. { 'er·ē·əl jē'äl·ə·jē }

area light [CIV ENG] **1.** A source of light with significant dimensions in two directions, such as a window or luminous ceiling. **2.** A light used to illuminate large areas. { 'er·ē·ə ¸līt }

areal pattern [PETRO ENG] Distribution pattern of oil-production wells and water- or gas-injection wells over a given oil reservoir. { 'er·ē·əl 'pad·ərn }

areal sweep efficiency [PETRO ENG] Percentage of the total oil reservoir or pore volume which is within the area being swept of oil by a displacing fluid, as in a natural or artificial gas drive or water injection. { 'er·ē·əl ¦swēp i'fish·ən·sē }

areal velocity [ASTROPHYS] In celestial mechanics, the area swept out by the radius vector per unit time. { 'er·ē·əl və'läs·əd·ē }

area meter [ENG] A mechanism to measure fluid flow rate through a fixed-area conduit by the movement of a weighted piston or float supported by the flowing fluid; includes rotameters and piston-type meters. { 'er·ē·ə ¸mēd·ər }

area navigation [NAV] An aircraft navigation system in which radial and distance information from a Vortac station or distance information from two or more Vortacs is used to fix aircraft position, so that navigation is not restricted to airways or direct routes between stations. Abbreviated RNAV. { 'er·ē·ə ¸nav·ə'gā·shən }

area of error [NAV] An area (elliptical, circular, parallelogram-shaped, and so on) containing the true position, within which, for a stated level of probability, the true position is considered to lie. { 'er·ē·ə əv 'er·ər }

area of use [ENG] For a balance depending on gravitational acceleration, an area that includes a sufficient number of locations providing a mean value for the gravitational acceleration of the given balance. { 'er·ē·ə əv 'yüs }

area opaca [EMBRYO] The opaque peripheral area of the blastoderm of birds and reptiles, continuous with the yolk. { 'er·ē·ə ō'päk·ə }

area pellucida [EMBRYO] The central transparent area of the blastoderm of birds and reptiles, overlying the subgerminal cavity. { 'er·ē·ə pə'lü·səd·ə }

area placentalis [EMBRYO] The part of the trophoblast in immediate contact with the uterine mucosa in the embryos of early placental vertebrates. { 'er·ē·ə pla·sən'tāl·əs }

area redistribution [PHYS] A method of measuring the effective duration of an irregular pulse by constructing a rectangular pulse that has the same peak amplitude and the same area on a graph of amplitude versus time. { 'er·ē·ə ¸rē,dis·trə'byü·shən }

area rule [AERO ENG] A prescribed method of design for obtaining minimum zero-lift drag for a given aerodynamic

configuration, such as a wing-body configuration, at a given speed. { 'er·ē·ə ¸rül }

area sampling [STAT] A method in which the area to be sampled is subdivided into smaller blocks which are selected at random and then subsampled or fully surveyed; method is used when a complete frame of reference is not available. { ¦er·ē·ə ¦samp·liŋ }

area search [COMPUT SCI] A computer search that examines only those records which satisfy some broad criteria. { 'er·ē·ə ¸sərch }

area survey [ENG] A survey of areas large enough to require loops of control. { 'er·ē·ə ¦sər,vā }

area target [ORD] A target attacked by area bombing, consisting of an area such as an entire munitions factory, rather than a single building or similar point target. { 'er·ē·ə ¦tär·gət }

area triangulation [ENG] A system of triangulation designed to progress in every direction from a control point. { 'er·ē·ə ¸trī,aŋ·gyə'lā·shən }

area vitellina [EMBRYO] The outer nonvascular zone of the area opaca; consists of ectoderm and endoderm. { 'er·ē·ə ¸vid·ə'līn·ə }

area wall [CIV ENG] A retaining wall around an areaway. { 'er·ē·ə ¸wȯl }

areaway [CIV ENG] An open space at subsurface level adjacent to a building, providing access to and utilities for a basement. { 'er·ē·ə,wā }

Arecaceae [BOT] The palms, the single family of the order Arecales. { ¸ar·ə'ka·sē,ē }

Arecales [BOT] An order of flowering plants in the subclass Arecidae composed of the palms. { ¸ar·ə'kā,lēz }

Arecidae [BOT] A subclass of flowering plants in the class Liliopsida characterized by numerous, small flowers subtended by a prominent spathe and often aggregated into a spadix, and broad, petiolate leaves without typical parallel venation. { ə'res·ə,dē }

arecoline [ORG CHEM] $C_8H_{13}O_2N$ An alkaloid from the betel nut; an oily, colorless liquid with a boiling point of 209°C; soluble in water, ethanol, and ether; combustible; used as a medicine. { ə'rek·ə,lēn }

areg [ECOL] A sand desert. { 'a,reg }

A register See arithmetic register. { 'ā ¸rej·ə·stər }

arena See lek. { ə'rēn·ə }

arenaceous [GEOL] Of sediment or sedimentary rocks that have been derived from sand or that contain sand. Also known as arenarious; psammitic; sabulous. { ¦a·rə¦nāsh·əs }

arenarious See arenaceous. { ¦a·rə¦ner·ē·əs }

Arenaviridae [VIROL] A family of ribonucleic acid animal viruses consisting of a single genus, Arenavirus, having an enveloped, spherical pleomorphic form. { ə,rēn·ə'vī·rə,dē }

arendalite [MINERAL] A dark-green variety of epidote found in Arendal, Norway. { ə'rend·əl,īt }

arene See aromatic hydrocarbon. { 'a,rēn }

Arenicolidae [INV ZOO] The lugworms, a family of mud-swallowing worms belonging to the Sedentaria. { ə,ren·ə'käl·ə,dē }

arenicolite [GEOL] A hole, groove, or other mark in a sedimentary rock, generally sandstone, interpreted as a burrow made by an arenicolous marine worm or a trail of a mollusk or crustacean. { ¸a·rə'nik·ə,līt }

arenicolous [ZOO] Living or burrowing in sand. { ¸a·rə'nik·ə·ləs }

Arenigian [GEOL] A European stage including Lower Ordovician geologic time (above Tremadocian, below Llanvirnian). Also known as Skiddavian. { ¸a·rə'nij·ē·ən }

arenite [PETR] Consolidated sand-texture sedimentary rock of any composition. Also known as arenyte; psammite. { 'a·rə,nīt }

Arent [GEOL] A suborder of the soil order Entisol, consisting of soils formerly of other classifications that have been severely disturbed, completely disrupting the sequence of horizons. { 'a·rənt }

arenyte See arenite. { 'a·rə,nīt }

areocentric [ASTRON] With Mars as a center. { ¸ar·ē·ō'sen·trik }

areodesy [ASTRON] Determination, by observation and measurement, of the exact positions of points on, and the figures and areas of large portions of, the surface of the planet Mars, or the shape and size of the planet Mars. { ¸ar·ē·ō,des·ē }

areographic [ASTRON] Referring to positions on Mars

measured in latitude from the planet's equator and in longitude from a reference meridian. { ‚ar·ē·ō'graf·ik }

areography [ASTRON] The study of the surface features of Mars, or its geography. [ECOL] Descriptive biogeography. { ‚ar·ē'äg·rə·fē }

areola [ANAT] **1.** The portion of the iris bordering the pupil of the eye. **2.** A pigmented ring surrounding a nipple, vesicle, or pustule. **3.** A small space, interval, or pore in a tissue. { ə'rē·ə·lə }

areola mammae [ANAT] The circular pigmented area surrounding the nipple of the breast. Also known as areola papillaris; mammary areola. { ə'rē·ə·lə 'mam·ē }

areola papillaris See areola mammae. { ə'rē·ə·lə pap·ə'lär· əs }

areolar tissue [HISTOL] A loose network of fibrous tissue and elastic fiber that connects the skin to the underlying structures. { ə'rē·ə·lər 'tish·ü }

areology [ASTRON] The scientific study related to the properties of Mars. { ‚ar·ē'äl·ə·jē }

areostyle See araeostyle. { 'ar·ē·ō‚stīl }

Ares [ASTRON] The planet Mars. { 'er‚ēz }

arête [GEOL] Narrow, jagged ridge produced by the merging of glacial cirques. Also known as arris; crib; serrate ridge. { a'rāt }

Arethusa [ASTRON] An asteroid with a diameter of about 126 miles (210 kilometers), mean distance from the sun of 3.069 astronomical units, and C-type surface composition. { ‚ar·ə'thü·zə }

ARFOR [METEOROL] A code word used internationally to indicate an area forecast; usually applied to an aviation weather forecast. { 'är‚fòr }

ARFOT [METEOROL] A code word used internationally to indicate an area forecast with units in the English system; usually applied to an aviation weather forecast. { 'är‚fòt }

arfvedsonite [MINERAL] A black monoclinic amphibole, containing sodium and silicon trioxide with occluded water and some calcium. Also known as soda hornblende. { 'är· vəd·sə‚nīt }

Arg See Argo.

Argand diagram [MATH] A two-dimensional cartesian coordinate system for representing the complex numbers, the number $x + iy$ being represented by the point whose coordinates are x and y. { 'är‚gän 'di·ə‚gram }

Argand lamp [ENG] A gas lamp having a tube-shaped wick, allowing a current of air inside as well as outside the flame. { 'är‚gän 'lamp }

Argasidae [INV ZOO] The soft ticks, a family of arachnids in the suborder Ixodides; several species are important as ectoparasites and disease vectors for humans and domestic animals. { är'gas·ə‚dē }

Argelander method [ASTRON] A technique to estimate the brightness of variable stars; it involves estimating the difference in magnitude between the variable stars as compared to one or more stars that are invariable. { 'är·gə‚land·ər ‚meth·əd }

argentaffin cell [HISTOL] Any of the cells of the gastrointestinal tract that are thought to secrete serotonin. { är'jen·tə· fən ‚sel }

argentaffin fiber See reticular fiber. { är'jen·tə·fən ‚fī·bər }

argentic [CHEM] Relating to or containing silver. { är'jen·tik }

argentic oxide See silver suboxide. { är'jen·tik 'äk‚sīd }

Argentinoidei [VERT ZOO] A family of marine deepwater teleostean fishes, including deep-sea smelts, in the order Salmoniformes. { ‚är‚jen·tə'nóid·ē‚ī }

argentite [MINERAL] Ag_2S A lustrous, lead-gray ore of silver; it is a monoclinic mineral and is dimorphous with acanthite. Also known as argyrite; silver glance; vitreous silver. { 'är· jən‚tīt }

argentocyanides [INORG CHEM] Complexes formed, for example, in the cyanidation of silver ores and in electroplating, when silver cyanide reacts with solutions of soluble metal cyanides. Also known as dicyanoargentates. { är‚jen·tō'sī· ə‚nīdz }

argentojarosite [MINERAL] $AgFe_3(SO_4)_2(OH)_6$ A yellow or brownish mineral consisting of basic silver ferric sulfate. { är‚jen·tō'jär·ə‚sīt }

argentometer [ENG] A hydrometer used to find the amount of silver salt in a solution. { ‚är·jən'täm·əd·ər }

argentometry [ANALY CHEM] A volumetric analysis that employs precipitation of insoluble silver salts; the salts may be chromates or chlorides. { ‚är·jən'täm·ə·trē }

argentophil [BIOL] Of cells, tissues, or other structures, having an affinity for silver. { är'jen·tə‚fil }

argentum [CHEM] Latin for silver. { är'jen·təm }

Argid [GEOL] A suborder of the soil order Aridisol, well drained, having a characteristically brown or red color and a silicate accumulation below the surface horizon; occupies older land surfaces in deserts. { 'är·jəd }

Argidae [INV ZOO] A small family of hymenopteran insects in the superfamily Tenthredinoidea. { 'är·jə‚dē }

argil See potter's clay. { 'är·jəl }

argillaceous [GEOL] Of rocks or sediments made of or largely composed of clay-size particles or clay minerals. { ‚är· jə'lā·shəs }

argillation [GEOL] Development of clay minerals by weathering of aluminum silicates. { ‚är·jə'lā·shən }

argillic alteration [GEOL] A rock alteration in which certain minerals are converted to minerals of the clay group. { är'jil· ik ‚ól·tə'rā·shən }

argilliferous [GEOL] Abounding in or producing clay. { ‚är·jə‚lif·ə·rəs }

argillite [PETR] A compact rock formed from siltstone, shale, or claystone but intermediate in degree of induration and structure between them and slate; argillite is more indurated than mudstone but lacks the fissility of shale. { 'är·jə‚līt }

arginase [BIOCHEM] An enzyme that catalyzes the splitting of urea from the amino acid arginine. { 'är·jə‚nās }

arginine [BIOCHEM] $C_6H_{14}N_4O_2$ A colorless, crystalline, water-soluble, essential amino acid of the α-ketoglutaric acid family. { 'är·jə‚nēn }

Argo [ASTRON] The large Ptolemy constellation; a southern constellation, now divided into four groups (Carina, Pupis, Vela, and Pyxis Nautica). Abbreviated Arg. Also known as Ship. { 'är·gō }

argol [FOOD ENG] A deposit formed in casks during aging of wine. [MATER] Any of several manures used as fuel in parts of Asia. { 'är‚gól }

argon [CHEM] A chemical element, symbol Ar, atomic number 18, atomic weight 39.998. { 'är‚gän }

argon ionization detector [NUCLEO] An ionization chamber that is filled with argon gas. { 'är‚gän ‚ī ə nə'zā·shən di‚tek·tər }

argon laser [OPTICS] A gas laser using ionized argon; emits a 4880-angstrom line as well as infrared radiation. { 'är‚gän 'lā·zər }

ARGOS system [COMMUN] An operational international satellite system to collect, locate, and disseminate environmental data. { 'är‚gos ‚sis·təm }

Argovian [GEOL] Upper Jurassic (lower Lusitanian), a substage of geologic time in Great Britain. { är'gōv·ē·ən }

Arguesian plane See Desarguesian plane. { är‚gesh·ən 'plän }

Arguloida [INV ZOO] A group of crustaceans known as the fish lice; taxonomic status is uncertain. { ‚är·gə'lóid·ə }

argument [ASTRON] An angle or arc, as seen from a focus. [COMPUT SCI] A value applied to a procedure, subroutine, or macroinstruction which is required in order to evaluate any of these. [MATH] See amplitude. See independent variable. { 'är·gyə·mənt }

argument of latitude [ASTRON] The angular distance measured in the orbit plane from the ascending node to the orbiting object; the sum of the argument of perigee and the true anomaly. { 'är·gyə·mənt əv 'lad·ə‚tüd }

argument of perigee [ASTRON] The angle or arc, as seen from a focus of an elliptical orbit, from the ascending node to the closest approach of the orbiting body to the focus; the angle is measured in the orbital plane in the direction of motion of the orbiting body. Also known as argument of perihelion. { 'är·gyə·mənt əv 'per·ə‚jē }

argument of perihelion See argument of perigee. { 'är·gyə· mənt əv ‚per·ə‚hēl·yən }

argument separator [COMPUT SCI] A comma or other punctuation mark that separates successive arguments in a command or statement in a computer program. { 'är·gyü·mənt ‚sep· ə‚rād·ər }

ε Argus [ASTRON] The former name of ε Carinae, a star of visual magnitude 1.74, spectral type K0. { 'ep·sə‚län 'är·gəs }

argyria [MED] A dusky-gray or bluish discoloration of the

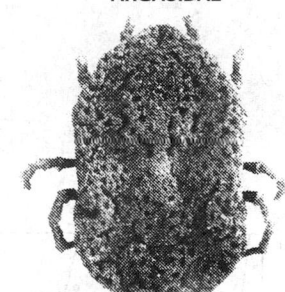

ARGASIDAE

Argasid tick, *Ornithodoros coriaceus,* enlarged.

skin and mucous membranes produced by the prolonged administration or application of silver preparations. { är'jir·ē·ə }

argyrite See argentite. { är'jir‚īt }

argyrodite [MINERAL] Ag_8GeS_6 A steel-gray mineral, one of two germanium minerals and a source for germanium; crystallizes in the isometric system and is isomorphous with canfieldite. { är'jir·ə‚dīt }

argyrophil lattice fiber See reticular fiber. { 'är·jə·rō‚fil 'lad·əs 'fī·bər }

Arhynchobdellae [INV ZOO] An order of annelids in the class Hirudinea characterized by the lack of an eversible proboscis; includes most of the important leech parasites of human and warm-blooded animals. { ¦ā‚riŋ'käb·də‚lē }

Arhynchodina [INV ZOO] A suborder of ciliophoran protozoans in the order Thigmotrichida. { ¦ā‚riŋ'kä·də·nə }

arhythmia See arrhythmia. { ā'rith·mē·ə }

arhythmicity [BIOL] A condition that is characterized by the absence of an expected behavioral or physiologic rhythm. { ¦ā‚rith'mis·əd·ē }

Ari See Aries.

ariboflavinosis [MED] Dietary deficiency of riboflavin, associated with the syndrome of angular cheilosis and stomatitis, corneal vascularity, nasolabial seborrhea, and genitorectal dermatitis. { ¦ā‚rīb·ə‚flav·ə'nō·səs }

arid biogeographic zone [ECOL] Any region of the world that supports relatively little vegetation due to lack of water. { 'ar·əd ¦bī·ō‚gē·ō'graf·ik ‚zōn }

arid climate [CLIMATOL] Any extremely dry climate. { 'ar·əd 'klī·mət }

arid erosion [GEOL] Erosion or wearing away of rock that occurs in arid regions, due largely to the wind. { 'ar·əd i'rō·zhən }

Aridisol [GEOL] A soil order characterized by pedogenic horizons; low in organic matter and nitrogen and high in calcium, magnesium, and more soluble elements; usually dry. { a'rid·ə‚sòl }

aridity [CLIMATOL] The degree to which a climate lacks effective, life-promoting moisture. { ə'rid·əd·ē }

aridity coefficient [CLIMATOL] A function of precipitation and temperature designed by W. Gorczynski to represent the relative lack of effective moisture (the aridity) of a place. { ə'rid·əd·ē ‚kō·ə'fish·ənt }

aridity index [CLIMATOL] An index of the degree of water deficiency below water need at any given station; a measure of aridity. { ə'rid·əd·ē ‚in‚deks }

Arid Transition life zone [ECOL] The zone of climate and biotic communities occurring in the chaparrals and steppes from the Rocky Mountain forest margin to California. { 'ar·əd trans'ish·ən 'līf ‚zōn }

arid zone See equatorial dry zone. { 'ar·əd ‚zōn }

ariegite [PETR] A group of pyroxenites composed principally of clinopyroxene, orthopyroxene, and spinel. { ‚ar·ē'ā‚zhīt }

Ariel [ASTRON] A satellite of the planet Uranus orbiting at a mean distance of 119,000 miles (192,000 kilometers). { 'ar·ē·əl }

Aries [ASTRON] A constellation with a right ascension of 3 hours and declination of 20°N. Abbreviated Ari. Also known as Ram. { 'er‚ēz }

arietiform [VERT ZOO] Shaped like a ram's horns; specifically, describing the dark facial marking that extends across the nose of kangaroo rats. { ‚ar·ē'ed·ə‚fòrm }

Ariidae [VERT ZOO] A family of tropical salt-water catfishes in the order Siluriformes. { ə'rī·ə‚dē }

Arikareean [GEOL] Lower Miocene geologic time. { ə‚rik·ə'rē·ən }

aril [BOT] An outgrowth of the funiculus in certain seeds that either remains as an appendage or envelops the seed. { 'ar·əl }

arilode [BOT] An aril originating from tissues in the micropyle region; a false aril. { 'ar·ə‚lōd }

Arionidae [INV ZOO] A family of mollusks in the order Stylommatophora, including some of the pulmonate slugs. { ‚ar·ē'än·ə‚dē }

arista [INV ZOO] The bristlelike or hairlike structure in many organisms, especially at or near the tip of the antenna of many Diptera. { ə'ris·tə }

Aristarchus [ASTRON] A crater on the moon. { ‚ar·ə'stär·kəs }

Aristolochiaceae [BOT] The single family of the plant order Aristolochiales. { ə‚ris·tə‚lō·kē'ās·ē‚ē }

Aristolochiales [BOT] An order of dicotyledonous plants in the subclass Magnoliidae; species are herbaceous to woody, often climbing, with perigynous to epigynous, apetalous flowers, uniaperturate or nonaperturate pollen, and seeds with a small embryo and copious endosperm. { ə‚ris·tə‚lō·kē'ā‚lēz }

aristolochic acid [ORG CHEM] $C_{17}H_{11}NO_7$ Crystals in the form of shiny brown leaflets that decompose at 281–286°C; soluble in alcohol, chloroform, acetone, ether, acetic acid, and aniline; used as an aromatic bitter. Also known as aristolochine. { ə¦ris·tə¦läk·ik 'as·əd }

aristolochine See aristolochic acid. { a‚ris'täl·ə‚kēn }

aristopedia [INV ZOO] Replacement of the arista by a nearly perfect leg. { ə‚ris·tə'pēd·ē·ə }

Aristotle's lantern [INV ZOO] A five-sided feeding and locomotor apparatus surrounding the esophagus of most sea urchins. { 'ar·ə‚städ·əlz 'lant·ərn }

arithlog paper [MATH] Graph paper marked with a semilogarithmic coordinate system. { ə'rith‚läg ‚pā·pər }

arithmetic [MATH] Addition, subtraction, multiplication, and division, usually of integers, rational numbers, real numbers, or complex numbers. { ə'rith·mə‚tik }

arithmetic address [COMPUT SCI] An address in a computer program that results from performing an arithmetic operation on another address. { ¦a·rith¦med·ik ə'dres }

arithmetical addition [MATH] The addition of positive numbers or of the absolute values of signed numbers. { ¦a·rith¦med·ə·kəl ə'dish·ən }

arithmetical element See arithmetical unit. { ¦a·rith¦med·ə·kəl 'el·ə·mənt }

arithmetical instruction [COMPUT SCI] An instruction in a computer program that directs the computer to perform an arithmetical operation (addition, subtraction, multiplication, or division) upon specified items of data. { ¦a·rith¦med·ə·kəl ‚in'strək·shən }

arithmetical operation [COMPUT SCI] A digital computer operation in which numerical quantities are added, subtracted, multiplied, divided, or compared. { ¦a·rith¦med·ə·kəl ‚äp·ə'rā·shən }

arithmetical unit [COMPUT SCI] The section of the computer which carries out all arithmetic and logic operations. Also known as arithmetical element; arithmetic-logic unit (ALU); arithmetic section; logic-arithmetic unit; logic section. { ¦a·rith¦med·ə·kəl 'yü·nət }

arithmetic average See arithmetic mean. { ¦a·rith¦med·ik 'av·rij }

arithmetic check [COMPUT SCI] The verification of an arithmetical operation or series of operations by another such process; for example, the multiplication of 73 by 21 to check the result of multiplying 21 by 73. { ə'rith·mə‚tik ‚chek }

arithmetic circuitry [COMPUT SCI] The section of the computer circuitry which carries out the arithmetic operations. { ¦a·rith¦med·ik 'sər·kə·trē }

arithmetic coding [COMMUN] A method of data compression in which a long character string is represented by a single number whose value is obtained by repeatedly partitioning the range of possible values in proportion to the probabilities of the characters. { ¦a·rith¦med·ik 'cōd·iŋ }

arithmetic-geometric mean [MATH] For two positive numbers a_1 and b_1, the common limit of the sequences $\{a_n\}$ and $\{b_n\}$ defined recursively by the equations $a_{n+1} = \frac{1}{2}(a_n + b_n)$ and $b_{n+1} = (a_n b_n)^{1/2}$. { ¦a·rith¦med·ik jē·ə¦me·trik 'mēn }

arithmetic-logic unit See arithmetical unit. { ə'rith·mə‚tik 'läj·ik ‚yü·nət }

arithmetic mean [MATH] The average of a collection of numbers obtained by dividing the sum of the numbers by the quantity of numbers. Also known as arithmetic average; average (av). { ¦a·rith¦med·ik 'mēn }

arithmetic processor See numeric processor extension. { ə'rith·mə‚tik ‚präs‚es·ər }

arithmetic progression [MATH] A sequence of numbers for which there is a constant d such that the difference between any two successive terms is equal to d. Also known as arithmetic sequence. { ¦a·rith¦med·ik prə'gresh·ən }

arithmetic register [COMPUT SCI] A specific memory location reserved for intermediate results of arithmetic operations. Also known as A register. { ¦a·rith¦med·ik 'rej·ə·stər }

arithmetic scan [COMPUT SCI] The procedure for examining

ARIES

Line pattern of constellation Aries. Grid lines represent the coordinates of the sky. Apparent brightness, or magnitude, of stars is shown by sizes of the dots, which are graded by appropriate numbers as indicated.

arithmetic expressions and determining the order of execution of operators, in the process of compilation into machine-executable code of a program written in a higher-level language. { ¦a·rith¦med·ik ˌskan }

arithmetic section *See* arithmetical unit. { ¦a·rith¦med·ik ˌsek·shən }

arithmetic sequence *See* arithmetic progression. { ¦a·rith¦med·ik 'sē·kwəns }

arithmetic series [MATH] A series whose terms form an arithmetic progression. { ¦a·rith¦med·ik ˌsirˌēz }

arithmetic shift [COMPUT SCI] A shift of the digits of a number, expressed in a positional notation system, in the register without changing the sign of the number. { ¦a·rith¦med·ik 'shift }

arithmetic sum [MATH] **1.** The result of the addition of two or more positive quantities. **2.** The result of the addition of the absolute values of two or more quantities. { ¦a·rith¦med·ik 'səm }

arithmetic symmetry [ELECTR] Property of a band-pass or band-rejection filter whose graph of amplitude versus frequency is symmetrical around a center frequency; that is, the left-hand side of the response is a mirror image of the right-hand side. { ¦a·rith¦med·ik 'sim·ə·trē }

arithmetization [MATH] The study of various branches of higher mathematics by methods that make use of only the basic concepts and operations of arithmetic. Representation of the elements of a finite or denumerable set by nonnegative integers. Also known as Gödel numbering. { ə¦rith·məd·ə'zā·shən }

Arizona ruby [MINERAL] A ruby-red pyrope garnet of igneous origin found in the southwestern United States. { ¦ar·ə¦zōn·ə 'rü·bē }

arizonite [MINERAL] $Fe_2Ti_3O_9$ A steel-gray mineral containing iron and titanium and found in irregular masses in pegmatite. [PETR] A dike rock composed of mostly quartz, some orthoclase, and accessory mica and apatite. { ˌar·ə'zō,nīt }

Arkansas stone [ENG] A whetstone made of Arkansas stone, for sharpening edged tools. [PETR] A variety of novaculite quarried in Arkansas. { 'är·kən,só ,stōn }

arkite [PETR] A feldspathoid-rich rock consisting largely of pseudoleucite and nepheline, subordinate melanite and pyroxene, and accessory orthoclase, apatite, and sphene. { 'är,kīt }

arkose [PETR] A sedimentary rock composed of sand-size fragments that contain a high proportion of feldspar in addition to quartz and other detrital minerals. { 'är,kōs }

arkose quartzite *See* arkosite. { 'är,kōs 'kwórt,sīt }

arkosic [PETR] Having wholly or partly the character of arkose. { är'kōs·ik }

arkosic bentonite [PETR] Bentonite derived from volcanic ash which contains 25–75% sandy impurities and whose detrital crystalline grains remain essentially unaltered. Also known as sandy bentonite. { är'kōs·ik 'ben·tə,nīt }

arkosic limestone [PETR] An impure clastic limestone composed of a relatively high proportion of grains or crystals of feldspar. { är'kōs·ik 'līm,stōn }

arkosic sandstone [PETR] A sandstone in which much feldspar is present, ranging from unassorted products of granular disintegration of granite to partly sorted river-laid or even marine deposits. { är'kōs·ik 'san,stōn }

arkosic wacke *See* feldspathic graywacke. { är'kōs·ik 'wak·ə }

arkosite [PETR] A quartzite with a high proportion of feldspar. Also known as arkose quartzite. { är'kō,sīt }

arksutite *See* chiolite. { ärk'sü,tīt }

ARL *See* acceptable reliability level.

arm [ANAT] The upper or superior limb in humans which comprises the upper arm with one bone and the forearm with two bones. [CONT SYS] A robot component consisting of an interconnected set of links and powered joints that move and support the wrist socket and end effector. [ELEC] *See* branch. [ENG ACOUS] *See* tone arm. [GEOL] A ridge or a spur that extends from a mountain. [MATH] A side of an angle. [NAV ARCH] The part of an anchor extending from the crown to one of the flukes. [OCEANOGR] A long, narrow inlet of water extending from another body of water. [ORD] **1.** A combat branch of a military force; specifically, a branch of the U.S. Army, such as the Infantry Armored Cavalry, the primary function of which is combat. **2.** *(Often plural)* Weapons for use in war. **3.** To supply with arms. **4.** To ready

ammunition for detonation, as by removal of safety devices or alignment of the explosive elements in the explosive train of the fuse. [PHYS] The perpendicular distance from the line along which a force is applied to a reference point. { ärm }

armadillo [VERT ZOO] Any of 21 species of edentate mammals in the family Dasypodidae. { ˌär·mə'dil·ō }

armament [ORD] **1.** The weapons of an airplane, tank, ship, or the like or of a unit or organized force. **2.** *(Often plural)* War equipment, weapons, and supplies. { 'är·mə·mənt }

armament error [ORD] The dispersion of shots from a particular gun. The deviation of any shot from the center of impact of a series of shots from a gun after all errors of personnel and adjustment have been accounted for. { 'är·mə·mənt ,er·ər }

armangite [MINERAL] $Mn_3(AsO_3)_2$ A black mineral crystallizing in the rhombohedral system and consisting of manganese arsenite. { är'man,gīt }

armature [ARCH] Framing or bars fashioned of structural ironwork and used to reinforce various features, for example, slender columns or hanging members. [ELECTROMAG] **1.** That part of an electric rotating machine that includes the main current-carrying winding in which the electromotive force produced by magnetic flux rotation is induced; it may be rotating or stationary. **2.** The movable part of an electromagnetic device, such as the movable iron part of a relay, or the spring-mounted iron part of a vibrator or buzzer. { 'är·mə,chər }

armature chatter [ELECTROMAG] Vibration of the armature of a relay caused by pulsating coil current or by marginally low coil current. { 'är·mə,chər ,chad·ər }

armature contact *See* movable contact. { 'är·mə,chər 'kän,takt }

armature reactance [ELECTROMAG] The inductive reactance due to the flux produced by the armature current and enclosed by the conductors in the armature slots and the end connections. { 'är·mə,chər ,rē'ak·təns }

armature reaction [ELECTROMAG] Interaction between the magnetic flux produced by armature current and that of the main magnetic field in an electric motor or generator. { 'är·mə,chər ,rē'ak·shən }

armature resistance [ELEC] The ohmic resistance in the main current-carrying windings of an electric generator or motor. { 'är·mə,chər ri'zis·təns }

armchair nanotube [PHYS CHEM] A carbon nanotube formed from a graphite sheet that is rolled up so that the edge is in the shape of armchairs. { ¦ärm,chär 'nan·ō,tüb }

arm conveyor [MECH ENG] A conveyor in the form of an endless belt or chain to which are attached projecting arms or shelves which carry the materials. { ¦ärm kən'vā·ər }

armed merchantman [NAV ARCH] An armed merchant ship of a neutral state; under international law, a vessel may be armed without prejudice to its status as a merchant ship. { ¦ärmd 'mər·chənt·mən }

arm elevator [MECH ENG] A chain elevator with protruding arms to cradle fixed-shape objects, such as drums or barrels, as they are moved upward. { ¦ärm ,el·ə'vād·ər }

armenite [MINERAL] $BaCa_2Al_6Si_8O_{28}·2H_2O$ Mineral composed of a hydrous calcium barium aluminosilicate. { är'mē,nīt }

ARMET [METEOROL] An international code word used to indicate an area forecast with units in the metric system. { 'är,met }

Armillaria root rot [PL PATH] A fungus disease of forest and orchard trees initiated by invasion of the root system, then of the lower trunk, by *Armillaria mellea*. Also known as bark-splitting disease. { ˌär·mə¦lar·ē·ə 'rüt ,rät }

Armilliferidae [INV ZOO] A family of pentastomid arthropods belonging to the suborder Porocephaloidea. { ˌär·mə·lə'fer·ə,dē }

arming [ORD] The changing of a fuse from a safe condition to a state of readiness for functioning. { 'ärm·iŋ }

arming device [ORD] Device for arming of a fuse under controlled conditions. { 'ärm·iŋ di,vīs }

arming distance *See* arming range. { 'ärm·iŋ ,dis·təns }

arming range [ORD] The distance from a weapon or launching point at which a fuse is expected to become armed. Also known as arming distance. { 'ärm·iŋ ,ränj }

arming resistance [ORD] The resistance to the displacement of certain fuse components; must be overcome in order to arm a fuse. { 'ärm·iŋ ri,zis·təns }

arm length [ANTHRO] The length of the arm measured from

ARMADILLO

Nine-banded armadillo (*Dasypus novemcinctus*), the only edentate inhabiting the United States.

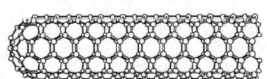

ARMCHAIR NANOTUBE

Model formed by rolling up a graphite sheet so that the edge is in the shape of armchairs.

top of the clavicle to the tip of the middle finger, with the arm straight down at the side of the body. { 'ärm ‚leŋkth }

armor [ELEC] Metal sheath enclosing a cable, primarily for mechanical protection. [ORD] **1.** Any physical protective covering, such as metal, used on vehicles or persons against projectiles or fragments. **2.** Armored units or forces. **3.** The component of a weapon system that gives protection to the vehicle or weapon on its way to the target. { 'är·mər }

armor castings [MET] A type of armor made of high-alloy steel frequently used when complicated shapes are involved. { 'är·mər ‚kast·iŋz }

armored artillery [ORD] Self-propelled artillery, with some armor, organic to the armored division or organized as separate battalions. { 'är·mərd är'til·ə·rē }

armored cable [ELEC] An electrical cable provided with a sheath of metal primarily for mechanical protection. { 'är·mərd 'kā·bəl }

armored cavalry [ORD] A unit organized and equipped to perform missions requiring great mobility, firepower, and shock action. { 'är·mərd 'kav·əl·rē }

armored faceplate [DES ENG] A tamper-proof faceplate or lock front, mortised in the edge of a door to cover the lock mechanism. { 'är·mərd 'fās‚plāt }

armored front [DES ENG] A lock front used on mortise locks that consists of two plates, the underplate and the finish plate. { 'är·mərd 'frənt }

armored infantry [ORD] A field army unit designed to close and destroy the enemy by fire and maneuver, to repel hostile assault in close combat, and to provide support for tanks. { 'är·mərd 'in·fən·trē }

armored mud ball [GEOL] A large (0.4–20 inches or 1–50 centimeters in diameter) subspherical mass of silt or clay coated with coarse sand and fine gravel. Also known as pudding ball. { 'är·mərd 'məd ‚bȯl }

armored personnel carrier [ORD] An armored vehicle which provides protection from small-arms fire and shell fragments; used to transport personnel both on and off the battlefield. { 'är·mərd ‚pərs·ən'el ‚kar·ē·ər }

armored vehicle [ORD] A wheeled or track-laying vehicle protected by armor plate that is used for combat security or cargo; for example, tanks, personnel carriers, armored cars, and self-propelled artillery. { 'är·mərd 'vē·ə·kəl }

armored wood [MATER] Wood which is faced on one or both sides with metal sheeting. { 'är·mərd 'wu̇d }

armorer [ORD] One who services and repairs small arms, fills ammunition belts, and performs similar duties necessary to keep small arms ready for use. { 'är·mər·ər }

Armorican orogeny [GEOL] Little-used term, now replaced by Hercynian or Variscan orogeny. { är'mȯr·ə·kən ȯ'räj·ə·nē }

armor-piercing [ORD] Of ammunition, bombs, bullets, and projectiles, designed to penetrate armor and other resistant targets. { 'är·mər ‚pirs·iŋ }

armor-piercing bomb [ORD] A missile, designed for dropping from aircraft, which is capable of penetrating the heaviest deck armor; also effective against reinforced-concrete structures; usually contains charge of explosive D, making up about 15% of the total weight. { 'är·mər ‚pirs·iŋ ‚bäm }

armor-piercing bullet [ORD] A bullet having a hard metal core, a soft metal envelope, and a jacket; when the bullet strikes armor, the envelope and jacket are stopped, but the armor-piercing core continues forward and penetrates the armor. { 'är·mər ‚pirs·iŋ ‚bu̇l·ət }

armor-piercing incendiary [ORD] Armor-piercing projectile designed to set fires after piercing armor. { 'är·mər ‚pirs·iŋ in'sen·dē‚er·ē }

armor-piercing incendiary tracer [ORD] An armor-piercing projectile designed to set fires after piercing armor, and also fitted with a tracer for spotting. { 'är·mər ‚pirs·iŋ in'sen·dē‚er·ē 'trā·sər }

armor-piercing sabot [ORD] A type of projectile which is armor-piercing and which incorporates a sabot. { 'är·mər ‚pirs·iŋ ‚sab·ō }

armor-piercing tracer [ORD] An armor-piercing projectile fitted with a tracer for spotting. { 'är·mər ‚pirs·iŋ 'trā·sər }

armor plate [BUILD] A metal plate which protects the lower part of a door from kicks and scratches, covering the door to a height usually 39 inches (1 meter) or more. [MET] Heavy, flat steel, either surface-hardened or hardened throughout, used as a sheathing for warships, tanks, and so forth to resist penetration and deformation from heavy gunfire. { 'är·mər ‚plāt }

Armour unit [BIOL] A unit for the standardization of adrenal cortical hormones and trypsin. { 'är·mər ‚yü·nət }

arm population *See* population I. { 'ärm ‚päp·yə‚lā·shən }

armrack [ORD] A frame with shelves, niches, hooks, or similar devices used to store small arms, to protect them, and to prevent unauthorized handling. { 'ärm‚rak }

arms locker [ORD] A chest, cupboard, or the like used for the safekeeping of small arms. { 'ärmz ‚läk·ər }

arm solution [CONT SYS] The computation performed by a robot controller to calculate the joint positions required to achieve desired tool positions. { 'ärm sə‚lü·shən }

Armstrong oscillator [ELECTR] Inductive feedback oscillator that consists of a tuned-grid circuit and an untuned-tickler coil in the plate circuit; control of feedback is accomplished by varying the coupling between the tickler and the grid circuit. { 'ärm‚strȯŋ 'äs·ə‚lād·ər }

Armstrong's acid *See* naphthalene-1,5-disulfonic acid. { 'ärm‚strȯŋz 'as·əd }

arm-tool aggregate [IND ENG] A biomechanical unit comprising the arm and the tool that it holds and manipulates. { ‚ärm ‚tül 'ag·rə·gət }

armure [TEXT] A plain, striped, ribbed, or woven fabric having small fancy designs that suggest chain armor. { 'ärm ‚myu̇r }

army [ORD] **1.** The land military forces of a nation. **2.** A unit of the U.S. Army made up of two or more army corps. { 'är·mē }

army artillery [ORD] Artillery assigned or attached to an army and retained under direct army command. { 'är·mē är'til·ə·rē }

army depot [ORD] A depot located within the area of an army and designated by the army commander where supplies from the communications zone or from local sources are received, classified, stored, and distributed. { 'är·mē 'dep·ō }

armyworm [INV ZOO] Any of the larvae of certain species of noctuid moths composing the family Phalaenidae; economically important pests of corn and other grasses. { 'är·mē‚wərm }

Arndt-Eistert synthesis [ORG CHEM] A method of increasing the length of an aliphatic acid by one carbon by reacting diazomethane with acid chloride. { 'ärnt 'ī·stərt ‚sin·thə·səs }

Arneth's classification *See* Arneth's index. { 'är‚nets ‚klas·ə·fə'kā·shən }

Arneth's count *See* Arneth's index. { 'är‚nets ‚kau̇nt }

Arneth's formula *See* Arneth's index. { 'är‚nets ‚fȯr·myə·lə }

Arneth's index [HISTOL] A system for dividing peripheral blood granulocytes into five classes according to the number of nuclear lobes, the least mature cells being tabulated on the left, giving rise to the terms "shift to left" and "shift to right" as an indication of granulocytic immaturity or hypermaturity respectively. Also known as Arneth's classification; Arneth's count; Arneth's formula. { 'är‚nets ‚in‚deks }

arnimite [MINERAL] $Cu_5(SO_4)_2(OH)_6 \cdot 3H_2O$ Mineral consisting of a hydrous copper sulfate. { 'ärn·ə‚mīt }

Arnold sterilizer [MICROBIO] An apparatus that employs steam under pressure at 212°F (100°C) for fractional sterilization of specialized bacteriological culture media. { 'ärn·əld 'ster·ə‚līz·ər }

Arodoidea [INV ZOO] A superfamily of hemipteran insects belonging to the subdivision Geocorisae. { ‚a·rə'dȯid·ē·ə }

arolium [INV ZOO] A pad projecting between the tarsal claws of some insects and arachnids. { ə'rōl·ē·əm }

aromatic [ORG CHEM] **1.** Pertaining to or characterized by the presence of at least one benzene ring. **2.** Describing those compounds having physical and chemical properties resembling those of benzene. { ‚ar·ə'mad·ik }

aromatic alcohol [ORG CHEM] Any of the compounds containing the hydroxyl group in a side chain to a benzene ring, such as benzyl alcohol. { ‚ar·ə'mad·ik 'al·kə‚hȯl }

aromatic aldehyde [ORG CHEM] An aromatic compound containing the CHO radical, such as benzaldehyde. { ‚ar·ə'mad·ik 'al·də‚hīd }

aromatic amine [ORG CHEM] An organic compound that contains one or more amino groups joined to an aromatic structure. { ‚ar·ə'mad·ik 'am‚ēn }

aromatic amino acid [BIOCHEM] An organic acid containing at least one amino group and one or more aromatic groups; for example, phenylalanine, one of the essential amino acids. { ¦ar·ə¦mad·ik ə¦mēn·ō ¹as·əd }

aromatic hydrocarbon [ORG CHEM] A member of the class of hydrocarbons, of which benzene is the first member, consisting of assemblages of cyclic conjugated carbon atoms and characterized by large resonance energies. Also known as arene. { ¦ar·ə¦mad·ik ¦hī·drə¹kär·bən }

aromatic ketone [ORG CHEM] An aromatic compound containing the −CO radical, such as acetophenone. { ¦ar·ə¦mad·ik ¹kē,tōn }

aromatic nucleus [ORG CHEM] The six-carbon ring characteristic of benzene and related series, or condensed six-carbon rings of naphthalene, anthracene, and so forth. { ¦ar·ə¦mad·ik ¹nü·klē·əs }

aromatic spirits of ammonia [PHARM] A flavored, hydroalcoholic solution of ammonia and ammonium carbonate having an aromatic, pungent odor; used as a reflex stimulant. [PHARM] A flavored, hydro-alcoholic solution of ammonia and ammonium carbonate having an aromatic, pungent odor; used as a reflex stimulant. Also known as hartshorn salts; smelling salts. { ¦ar·ə¦mad·ik ¦spir·ət əv ə¹mōn·yə }

aromatic sulfuric acid [PHARM] A preparation consisting of sulfuric acid, tincture of ginger, oil of cinnamon, and alcohol; formerly used as a tonic and astringent. { ¦ar·ə¦mad·ik ¸səl¹fyür·ik ¹as·əd }

aromatic vinegar [PHARM] A flavored solution of acetic acid used as smelling salts. { ¦ar·ə¦mad·ik ¹vin·ə·gər }

aromatization [CHEM ENG] Conversion of any nonaromatic hydrocarbon structure to aromatic hydrocarbon, particularly petroleum. { ə¸rō·məd·ə¹zā·shən }

arostat process [CHEM ENG] A process in which aromatic molecules are saturated by catalytic hydrogenation to produce high-quality jet fuels, low-aromatic-content solvents, and high-purity cyclohexane from benzene. { ¹ar·ə¸stat ¸präs·əs }

aroyl [ORG CHEM] The radical RCO, where R is an aromatic (benzoyl, napthoyl) group. { ¹ar·ə·wəl }

aroylation [ORG CHEM] A reaction in which the aroyl group is incorporated into a molecule by substitution. { ¸ar·ə·wə¹lā·shən }

ARPA See automated radar plotting aid. { ¹är·pə }

ARPANET See Advanced Research Projects Agency Network. { ¹är·pə¸net }

ARPES See angle-resolved photoelectron spectroscopy.

ARQ See automatic repeat request.

arquerite [MINERAL] A mineral consisting of a soft, malleable, silver-rich variety of amalgam, containing about 87% silver and 13% mercury. { är¹kē¸rīt }

arrastra See arrastre. { ə¸räs·trə }

arrastre [MIN ENG] A mill comprising a circular, rock-lined pit in which broken ore is pulverized by stones, attached to horizontal poles fastened in a central pillar, which stones are dragged around the pit. Also spelled arrastra. { ə¹rä·strə }

array [COMPUT SCI] A collection of data items with each identified by a subscript or key and arranged in such a way that a computer can examine the collection and retrieve data from these items associated with a particular subscript or key. [ELECTR] A group of components such as antennas, reflectors, or directors arranged to provide a desired variation of radiation transmission or reception with direction. [STAT] The arrangement of a sequence of items in statistics according to their values, such as from largest to smallest. { ə¹rā }

array element [COMPUT SCI] A single data item in an array. { ə¹rā ¸el·ə·mənt }

array processor [COMPUT SCI] A multiprocessor composed of a set of identical central processing units acting synchronously under the control of a common unit. { ə¹rā ¸präs·es·ər }

array radar [ENG] A radar incorporating a multiplicity of phased antenna elements. { ə¹rā ¹rā,där }

array sonar [ENG] A sonar system incorporating a phased array of radiating and receiving transducers. { ə¹rā ¹sō,när }

array-type microelectrode [NEURO] A type of microelectrode that monitors electrically active cells in culture and can potentially be used to explore electrical activity of neural networks during development and learning. { ə¹rā ¦tīp ¸mī·krō·i¹lek,trōd }

arrested decay [GEOL] A stage in coal formation where biochemical action ceases. { ə¹res·təd di¹kā }

arrested evolution [EVOL] Evolution that was extremely slow in comparison with that characteristic of most organic lineages. { ə¹res·təd ¸ev·ə¹lü·shən }

arrester [ELEC] See lightning arrester. [ENG] A wire screen at the top of an incinerator or chimney which prevents sparks or burning material from leaving the stack. { ə¹res·tər }

arrester hook [AERO ENG] A hook in the tail section of an airplane; used to engage the arrester wires on an aircraft carrier's deck. { ə¹res·tər ¸hük }

arrester wires [NAV ARCH] Cables that extend across the stern of an aircraft carrier's deck; they are raised above the deck on supports so that they may be engaged by the arrester hook of an aircraft when it lands on a carrier; the wires play out along the deck with increasing resistance until the aircraft stops moving. Also known as arresting gear. { ə¹res·tər ¸wīrz }

arresting gear See arrester wires. { ə¹rest·iŋ ¸gir }

arrestment device [ENG] A locking mechanism installed on a balance for holding one of several levers in place; serves to protect the balance. { ə¹rest·mənt di¸vīs }

arrhenite [MINERAL] A variety of fergusonite. { ə¹rä,nīt }

Arrhenius equation [PHYS CHEM] The relationship that the specific reaction rate constant k equals the frequency factor constant s times $\exp(-\Delta H_{act}/RT)$, where ΔH_{act} is the heat of activation, R the gas constant, and T the absolute temperature. { ar¹rä·nē·əs i¹kwä·zhən }

Arrhenius-Guzman equation [PHYS] The relation between the viscosity η of a liquid and the Kelvin temperature T at constant pressure: $\eta = A \exp(B/RT)$, where A and B are constants and R is the gas constant. { ar¹rä·nē·əs ¸güth·mən i,kwä·zhən }

Arrhenius viscosity formulas [PHYS] A series of three equations which relate the viscosity of a liquid to the temperature, the viscosity of a solution to its concentration and to the viscosity of the solvent, and the viscosity of a sol to the viscosity of the medium. { ar¹rä·nē·əs vis¹käs·əd·ē ¸för·myə·ləz }

arrhenoblastoma [MED] A solid, sometimes malignant, tumor of the ovary that usually produces male sex hormones, inducing virilism. { ¸a·rə,nō,bla¹stō·mə }

arrhenotoky [BIOL] Production of only male offspring by a parthenogenetic female. { ¸a·rə¹näd·ə·kē }

arrhinencephalia [MED] A congenital malformation in which part or all of the rhinencephalon is absent and the nose is malformed. { ə,rīn·en·sə¹fal·yə }

arrhythmia [MED] Absence of rhythm, especially of heart beat or respiration. Also spelled arhythmia. { ā¹rith·mē·ə }

Arridae [VERT ZOO] A family of catfishes in the suborder Siluroidei found from Cape Cod to Panama. { ¹a·rə¸dē }

arrière-voussure [BUILD] **1.** An arch or vault in a thick wall carrying the thickness of the wall, especially one over a door or window frame. **2.** A relieving arch behind the face of a wall. { ¹ar·ē,er,vü¹sùr }

arris [ARCH] A short edge or angle at the junction of two surfaces, especially moldings and raised edges. [GEOL] See arête. { ¹ar·əs }

arris fillet [BUILD] A triangular wooden piece that raises the slates of a roof against a chimney or wall so that rain runs off. { ¹ar·əs ¦fil·ət }

arris gutter [BUILD] A V-shaped wooden gutter fixed to the eaves of a building. { ¹ar·əs ¦gəd·ər }

arris hip tile [BUILD] A special roof tile having an L-shaped cross section, made to fit over the hip of a roof. Also known as hip tile. { ¹ar·əs ¹hip ¸tīl }

arris rail [CIV ENG] A rail of triangular section, usually formed by slitting diagonally a strip of square section. { ¹ar·əs ¸rāl }

arrissing tool [ENG] A tool similar to a float, but having a form suitable for rounding an edge of freshly placed concrete. { ¹ar·əs·iŋ ¸tül }

arris tile [BUILD] Any angularly shaped tile. { ¹ar·əs ¸tīl }

arrisways [CIV ENG] Diagonally, in respect to the manner of laying tiles, slates, bricks, or timber. Also known as arriswise. { ¹ar·əs,wāz }

arriswise See arrisways. { ¹ar·əs,wīz }

arrival rate [IND ENG] The mean number of new calling units arriving at a service facility per unit time. { ə¹rī·vəl ¸rāt }

arrival time [GEOPHYS] In seismological measurements, the time at which a given wave phase is detected by a seismic recorder. { ə¹rī·vəl ¸tīm }

ARROW OF TIME

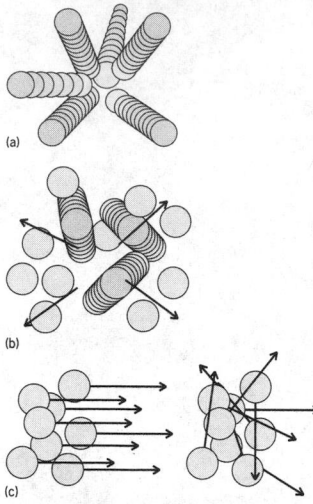

The three basic irreversibilities of nature which ensure that events are irreversible: (*a*) dispersion of matter, (*b*) dispersion of energy, and (*c*) disorganization of orderly motion.

ARROWWORM

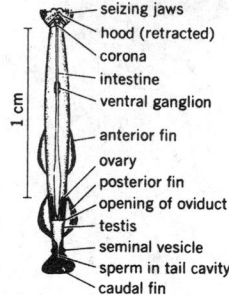

Dorsal view of *Sagitta enflata*, an arrowworm.

ARTACAMINAE

Dorsal view of *Artacamella*, a genus of the Artacaminae.

arrojadite [MINERAL] $Na_2(Fe,Mn)_5(PO_4)_4$ Dark-green mineral crystallizing in the monoclinic system, being isostructural with dickinsonite and occurring in masses. { ˌarə'jäˌdīt }

Arrow *See* Sagitta. { 'arˌō }

arrowhead [ARCHEO] The pointed or barbed tip (made of stone, bone, metal, or other material) of an arrow, often present at various sites of prehistoric peoples. Also known as arrowpoint. [BOT] Any aquatic plant of the genus *Sagitarria* (water plantain family) that has arrowhead-shaped leaves and white flowers. { 'aˌrōˌhed }

arrow of time [PHYSIO] The uniform and unique direction associated with the apparent inevitable flow of time into the future. { ˌarˌō əv 'tīm }

arrowroot [BOT] Any of the tropical American plants belonging to the genus *Maranta* in the family Marantaceae. { 'arˌōˌrüt }

arrowroot starch [FOOD ENG] A nutritive carbohydrate obtained from the underground stems of arrowroot plants and used as a food for infants and invalids. { 'arˌōˌrüt ˌstärch }

arrow wing [AERO ENG] An aircraft wing of V-shaped planform, either tapering or of constant chord, suggesting a stylized arrowhead. { 'arˌōˌrüt ˌwiŋ }

arrowworm [INV ZOO] Any member of the phylum Chaetognatha; useful indicator organism for identifying displaced masses of water. { 'arˌōˌwərm }

arroyo [GEOL] Small, deep gully produced by flash flooding in arid and semiarid regions of the southwestern United State. { ə'róiˌō }

arsenal [ORD] **1.** An installation whose primary mission is research, development, and manufacture pertaining to assigned items or components. **2.** An installation having coequal missions of maintenance and supply for assigned items or components. { 'ärsˌnəl }

arsenate [INORG CHEM] **1.** AsO_4^{3-} A negative ion derived from orthoarsenic acid, $H_3AsO_4 \cdot {}^1\!/_2H_2O$. **2.** A salt or ester of arsenic acid. { 'ärsˌnˌāt }

arsenic [CHEM] A chemical element, symbol As, atomic number 33, atomic weight 74.9216. [MINERAL] A brittle, steel-gray hexagonal mineral, the native form of the element. { 'ärsˌnˌik }

arsenic acid [INORG CHEM] $H_3AsO_4 \cdot {}^1\!/_2H_2O$ White, poisonous crystals, soluble in water and alcohol; used in manufacturing insecticides, glass, and arsenates and as a defoliant. Also known as orthoarsenic acid. { är'senˌik 'asˌəd }

arsenical [CHEM] **1.** Pertaining to arsenic. **2.** A compound that contains arsenic. { ar'senˌəˌkəl }

arsenical antimony *See* allemontite. { ar'senˌəˌkəl 'antˌəˌmōˌnē }

arsenical nickel *See* niccolite. { ar'senˌəˌkəl 'nikˌəl }

arsenic bloom *See* arsenolite. { 'ärsˌnˌik dī'sələˌfīd }

arsenic disulfide [INORG CHEM] As_2S_2 Red, orange, or black monoclinic crystals, insoluble in water; used in fireworks; occurs naturally as realgar. { 'ärsˌnˌik dī'sələˌfīd }

arsenic oxide [INORG CHEM] **1.** An oxide of arsenic. **2.** *See* arsenic pentoxide. **3.** *See* arsenic trioxide. { 'ärsˌnˌik 'äkˌsīd }

arsenic pentasulfide [INORG CHEM] As_2S_5 Yellow crystals that are insoluble in water and readily decompose to the trisulfide and sulfur; used as a pigment. { 'ärsˌnˌik ˌpentə'sələˌfīd }

arsenic pentoxide [INORG CHEM] As_2O_5 A white, deliquescent compound that decomposes by heat and is soluble in water. Also known as arsenic oxide. { 'ärsˌnˌik ˌpent'äkˌsīd }

arsenic trichloride [INORG CHEM] $AsCl_3$ An oily, colorless liquid that dissolves in water; used in ceramics, organic chemical syntheses, and in the preparation of pharmaceuticals. { 'ärsˌnˌik ˌtri'klórˌīd }

arsenic trioxide [INORG CHEM] As_2O_3 A toxic compound, slightly soluble in water; octahedral crystals change to the monoclinic form by heating at 200°C; occurs naturally as arsenolite and claudetite; used in small quantities in some medicinal preparations. Also known as arsenic oxide; arsenious acid. { 'ärsˌnˌik ˌtri'äkˌsīd }

arsenic trisulfide [INORG CHEM] As_2S_3 An acidic compound in the form of yellow or red monoclinic crystals with a melting point at 300°C; occurs as the mineral orpiment; used as a pigment. { 'ärsˌnˌik ˌtri'sələˌfīd }

arsenide [CHEM] A binary compound of negative, trivalent arsenic; for example, H_3As or $GaAs$. { 'ärsˌənˌīd }

arsenin [ORG CHEM] A heterocyclic organic compound composed of a six-membered ring system in which the carbon atoms are unsaturated and the unique heteroatom is arsenic, with no nitrogen atoms present. { 'ärsenˌən }

arseniopleite [MINERAL] A reddish-brown mineral consisting of a basic arsenate of manganese, calcium, iron, lead, and magnesium and occurring in cleavable masses. { ärˌsenˌēˌō'plēˌīt }

arseniosiderite [MINERAL] $Ca_3Fe_4(AsO_4)_4(OH)_4 \cdot 4H_2O$ A yellowish-brown mineral consisting of a basic iron calcium arsenate and occurring as concretions. { ärˌsenˌēˌōˌ'sidˌəˌrīt }

arsenious acid *See* arsenic trioxide. { är'senˌēˌəs 'asˌəd }

arsenite [INORG CHEM] **1.** AsO_3^{3-} A negative ion derived from aqueous solutions of As_4O_6. **2.** A salt or ester of arsenious acid. { 'ärsˌəˌnīt }

arsenobenzene [ORG CHEM] $C_6H_5As{:}AsC_6H_5$ White needles that melt at 212°C; insoluble in cold water, soluble in benzene; derivatives have some use in medicine. { ˌärsˌənˌō'benˌzēn }

arsenobismite [MINERAL] $Bi_2(AsO_4)(OH)_3$ A yellowish-green mineral consisting of a basic bismuth arsenate and occurring in aggregates. { ˌärsˌənˌō'bizˌmīt }

arsenoclasite [MINERAL] $Mn_5(AsO_4)_2(OH)_4$ A red mineral consisting of a basic manganese arsenate. Also spelled arsenoklasite. { ˌärsˌənˌō'klāˌsīt }

arseno compound [ORG CHEM] A compound containing an As-As bond with the general formula $(RAs)_n$, where R represents a functional group; structures are cyclic or long-chain polymers. { ˌärsˌənˌō 'kämˌpaùnd }

arsenoklasite *See* arsenoclasite. { ˌärsˌənˌō'klāˌsīt }

arsenolamprite [MINERAL] $FeAsS$ A lead gray mineral consisting of nearly pure arsenic; occurs in masses with a fibrous foliated structure. { ˌärsˌənˌō'lamˌprīt }

arsenolite [MINERAL] As_2O_3 A mineral crystallizing in the isometric system and usually occurring as a white bloom or crust. Also known as arsenic bloom. { är'senˌəlˌīt }

arsenopyrite [MINERAL] $FeAsS$ A white to steel-gray mineral crystallizing in the monoclinic system with pseudo-orthorhombic symmetry because of twinning; occurs in crystalline rock and is the principal ore of arsenic. Also known as mispickel. { ˌärsˌənˌō'pīˌrīt }

arsenotherapy [MED] Treatment of disease by means of arsenical drugs. { ˌärsˌənˌō'therˌəˌpē }

arsenous oxide *See* arsenic trioxide. { 'ärˌsəˌnəs 'äkˌsīd }

arsine [INORG CHEM] H_3As A colorless, highly poisonous gas with an unpleasant odor. { är'sēn }

arsinic acid [INORG CHEM] An acid of general formula R_2AsO_2H; derived from trivalent arsenic; an example is cacodylic acid, or dimethylarsinic acid, $(CH_3)_2AsO_2H$. { ärˌsin·ik 'asˌəd }

arsoite [PETR] An olivine-bearing diopside trachyte. { 'ärsōˌīt }

arsonic acid [INORG CHEM] An acid derived from orthoarsenic acid, $OAs(OH)_3$; the type formula is generally considered to be $RAsO(OH)_2$; an example is *para*-aminobenzenearsonic acid, $NH_2C_6H_4AsO(OH)_2$. { ärˌsän·ik 'asˌəd }

arsonium [INORG CHEM] $-AsH_4$ A radical which may be considered analogous to the ammonium radical in that a compound such as AsH_4OH may form. { är'sōnˌēˌəm }

Arsonval *See* d'Arsonval entries.

arsphenamine [PHARM] $C_{12}H_{12}As_2N_2O_2 \cdot 2HCl \cdot 2H_2O$ The antisyphilitic diaminodihydroxyarsenobenzene dihydrochloride, effective also on protozoan infections, first prepared by P. Ehrlich in 1909. Also known as Ehrlich's 606. { ärsˌfenˌəˌmēn }

ARSR *See* air-route surveillance radar.

Artacaminae [INV ZOO] A subfamily of polychaete annelids in the family Terebellidae of the Sedentaria. { ˌärˌtə'kamˌəˌnē }

arteriogram [MED] A roentgenogram of an artery after injection with radiopaque material. { är'tirˌēˌəˌgram }

arteriography [MED] **1.** Graphic presentation of the pulse. **2.** Roentgenography of the arteries after the intravascular injection of a radiopaque substance. { ärˌtirˌēˌ'ägˌrəˌfē }

arteriole [ANAT] An artery of small diameter that terminates in capillaries. { är'tirˌēˌōl }

arteriolopathy [MED] Disease of the arterioles. { är‚tir·ē·ə'läp·ə·thē }

arteriolosclerosis [MED] Thickening of the lining of arterioles, usually due to hyalinization or fibromuscular hyperplasia. { är‚tir·ē·ə‚lō·sklə'rō·səs }

arteriometer [MED] An instrument that measures arterial pulsations. { är‚tir·ē·'äm·əd·ər }

arteriosclerosis [MED] A degenerative arterial disease marked by hardening and thickening of the vessel walls. { är‚tir·ē·ō·sklə'rō·səs }

arteriosclerosis obliterans [MED] Hardening of the artery walls with obstruction of the lumen due to proliferation of the innermost vessel layer. { är‚tir·ē·ō·sklə'rō·səs ō'blid·ər‚änz }

arteriotomy [MED] Incision or opening of an artery. { är‚tir·ē'äd·ə·mē }

arteriovenous anastomosis [ANAT] A blood vessel that connects an arteriole directly to a venule without capillary intervention. { är‚tir·ē·ō've·nəs ə‚nas·tə'mō·səs }

arteriovenous aneurysm [MED] 1. Dilation of the walls of an artery and a vein via an abnormal canal (fistula) between the vessels. 2. Dilation of an arteriovenous fistula. { är‚tir·ē·ō've·nəs 'an·yə‚riz·əm }

arterite [PETR] 1. A migmatite produced as a result of regional contact metamorphism during which residual magmas were injected into the host rock. 2. Gneisses characterized by veins formed from the solution given off by deep-seated intrusions of molten granite. 3. A veined gneiss in which the vein material was injected from a magma. { är'tir‚īt }

arteritic migmatite [GEOL] Injection gneiss supposedly produced by introduction of pegmatite, granite, or aplite into schist parallel to the foliation. { ‚ärd·ə‚rid·ik 'mig·mə‚tīt }

arteritis [MED] Inflammation of an artery. { ‚ärd·ə'rīd·əs }

artery [ANAT] A vascular tube that carries blood away from the heart. { 'ärd·ə·rē }

artesian aquifer [HYD] An aquifer that is bounded above and below by impermeable beds and that contains artesian water. Also known as confined aquifer. { är'tē·zhən 'ak·wə·fər }

artesian basin [HYD] A geologic structural feature or combination of such features in which water is confined under artesian pressure. { är'tē·zhən 'bās·ən }

artesian leakage [HYD] The slow percolation of water from artesian formations into the confining materials of a less permeable, but not strictly impermeable, character. { är'tē·zhən 'lēk·ij }

artesian spring [HYD] A spring whose water issues under artesian pressure, generally through some fissure or other opening in the confining bed that overlies the aquifer. Also known as fissure spring. { är'tē·zhən 'spriŋ }

artesian water [HYD] Groundwater that is under sufficient pressure to rise above the level at which it encounters a well, but which does not necessarily rise to or above the surface of the ground. { är'tē·zhən 'wȯd·ər }

artesian well [HYD] A well in which the water rises above the top of the water-bearing bed. { är'tē·zhən 'wel }

arthochromatic erythroblast See normoblast. { ‚är·thrō‚krō‚mad·ik ə'rith·rə‚blast }

Arthoniaceae [BOT] A family of lichens in the order Hysteriales. { ‚är‚thän·ē'ās·ē‚ē }

arthritis [MED] Any inflammatory process affecting joints or their component tissues. { är'thrīd·əs }

arthritis urethritica See Reiter's syndrome. { är'thrīd·əs ‚yü·rē'thrid·ə·kə }

Arthrobacter [MICROBIO] A genus of gram-positive, aerobic rods in the coryneform group of bacteria; metabolism is respiratory, and cellulose is not attached. { 'är‚thrō‚bak·tər }

arthrobranch [INV ZOO] In Malacostraca, the gill attached to the joint between the body and the first leg segment. { 'är‚thrō‚braŋk }

arthrodesis [MED] Fusion of a joint by removing the articular surfaces and securing bony union. Also known as operative ankylosis. { är'thräd·ə·səs }

arthrodia [ANAT] A diarthrosis permitting only restricted motion between a concave and a convex surface, as in some wrist and ankle articulations. Also known as gliding joint. { är'thrōd·ē·ə }

Arthrodira [PALEON] The joint-necked fishes, an Upper Silurian and Devonian order of the Placodermi. { ‚är‚thrō'dī·rə }

Arthrodonteae [BOT] A family of mosses in the subclass Eubrya characterized by thin, membranous peristome teeth composed of cell walls. { ‚är‚thrō'dänt·ē‚ē }

arthrogram [MED] A roentgenogram of a joint space after injection of radiopaque material. { 'är·thrə‚gram }

arthrography [MED] Roentgenography of a joint space after the injection of radiopaque material. { ‚är'thräg·rə·fē }

arthrogryposis [MED] Permanent fixation of a joint in a flexed position. { ‚är‚thrō‚grī'pō·səs }

Arthromitaceae [MICROBIO] Formerly a family of nonmotile bacteria in the order Caryophanales found in the intestine of millipedes, cockroaches, and toads. { ‚är‚thräm·ə'tās·ē‚ē }

arthropathy [MED] 1. Any joint disease. 2. A neurotrophic disorder of a joint, usually due to lack of pain sensation, found in association with tabes dorsalis, leprosy, syringomyelia, diabetic polyneuropathy, and occasionally multiple sclerosis and myelodysplasias. { ‚är'thräp·ə·thē }

arthroplasty [MED] 1. The making of an artificial joint. 2. Reconstruction of a new and functioning joint from an ankylosed one; a plastic operation upon a joint. { 'är·thrō‚plas·tē }

arthropod [INV ZOO] Any invertebrate (of the phylum Arthropoda) with a hard exoskeleton, segmented body, and jointed legs (for example, insects, arachnids, myriapods, and crustaceans). { 'arth·rō‚päd }

Arthropoda [INV ZOO] The largest phylum in the animal kingdom; adults typically have a segmented body, a sclerotized integument, and many-jointed segmental limbs. { är'thräp·ə·də }

arthropod-borne virus See arbovirus. { ‚är·thrə‚päd ‚bȯrn 'vī·rəs }

arthropodin [BIOCHEM] A water-soluble protein which forms part of the endocuticle of insects. { är'thräp·ə·dən }

arthroscope [MED] An endoscope for examining the interior of a joint. { 'ärth·rə‚skōp }

arthroscopy [MED] Visual examination of the interior of a joint by means of an arthroscope. { är'thrä·skə·pē }

arthrosis [ANAT] An articulation or suture uniting two bones. [MED] Any degenerative joint disease. { är'thrō·səs }

arthrospore [BOT] A jointed, vegetative resting spore resulting from filament segmentation in some blue-green algae and hypha segmentation in many Basidiomycetes. { 'är‚thrō'spȯr }

Arthrotardigrada [INV ZOO] A suborder of microscopic invertebrates in the order Heterotardigrada characterized by toelike terminations on the legs. { ‚är·thrō‚tard·ə'gräd·ə }

arthrotomy [MED] Surgical incision into a joint. { är'thräd·ə·mē }

Arthur unit [BIOL] A unit for the standardization of splenin A. { 'är·thər ‚yü·nət }

Arthus reaction [IMMUNOL] An allergic reaction of the immediate hypersensitive type that results from the union of antigen and antibody, with complement present, in blood vessel walls. { 'är·thəs rē'ak·shən }

artichoke [BOT] *Cynara scolymus.* A herbaceous perennial plant belonging to the order Asterales; the flower head is edible. { 'ärd·ə‚chōk }

article [INV ZOO] A segment of an arthropod leg between two articulations. { 'ärd·ə·kəl }

articulamentum [INV ZOO] The innermost layer of a calcareous plate in a chiton. { är‚tik·yə·lə'men·təm }

articular cartilage [ANAT] Cartilage that covers the articular surfaces of bones. { är'tik·yə·lər 'kärt·lij }

articular disk [ANAT] A disk of fibrocartilage, dividing the cavity of certain joints. { är'tik·yə·lər 'disk }

articular membrane [INV ZOO] A flexible region of the cuticle between sclerotized areas of the exoskeleton of an arthropod; functions as a joint. { är'tik·yə·lər 'mem‚brän }

Articulata [INV ZOO] 1. A class of the Brachiopoda having hinged valves that usually bear teeth. 2. The only surviving subclass of the echinoderm class Crinoidea. { är‚tik·yə'läd·ə }

articulated drop chute [ENG] A drop chute, for a falling stream of concrete, which consists of a vertical succession of tapered metal cylinders, so designed that the lower end of each cylinder fits into the upper end of the one below. { är'tik·yə‚lād·əd 'dräp ‚shüt }

articulated leader [MECH ENG] A wheel-mounted transport

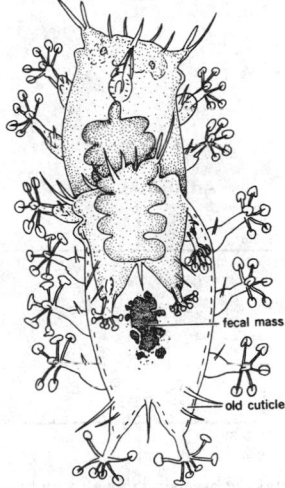

ARTHROTARDIGRADA

Batillipes, **a genus of the Arthrotardigrada; defecation during molt.**

ARTICHOKE

Edible artichoke flower heads. The bottom one is shown in cross section.

unit with a pivotal loading element used in earth moving. { är'tik·yə,lād·əd 'lēd·ər }

articulated structure [CIV ENG] A structure in which relative motion is allowed to occur between parts, usually by means of a hinged or sliding joint or joints. { är'tik·yə,lād·əd 'strək·chər }

articulated train [ENG] A railroad train whose cars are permanently or semipermanently connected. { är'tik·yə,lād·əd 'trān }

articulation [ANAT] *See* joint. [BOT] A joint between two parts of a plant that can separate spontaneously. [COMMUN] The percentage of speech units understood correctly by a listener in a communication system; it generally applies to unrelated words, as in code messages, in distinction to intelligibility. [CONT SYS] The manner and actions of joining components of a robot with connecting parts or links that allow motion. [INV ZOO] A joint between rigid parts of an animal body, such as the segments of an appendage in insects. [PHYSIO] The act of enunciating speech. { är,tik·yə'lā·shən }

articulation equivalent [COMMUN] Of a complete telephone connection, a measure of the articulation of speech reproduced over it, expressed numerically in terms of the trunk loss of a working reference system when the latter is adjusted to give equal articulation. { är,tik·yə'lā·shən i'kwiv·ə·lənt }

articulation index [PSYCH] An indication of the percentage of speech sounds that would be heard correctly when heard without a context. { är,tik·yə'lā·shən 'in,deks }

articulation point *See* cut point. { är,tik·yə'lā·shən ,pöint }

articulation test [PSYCH] A type of test used to assess hearing and loss of hearing for speech. { är,tik·yə'lā·shən ,test }

articulite *See* itacolumite. { är'tik·yə,līt }

artifact [ARCHEO] Any crafted object of common use that reflects the skills of humans in past cultures. [COMMUN] Any component of a signal that is extraneous to the variable represented by the signal. [HISTOL] A structure in a fixed cell or tissue formed by manipulation or by the reagent. [MED] Noise or spurious signals that occur during various radiological imaging techniques; can reach a level where they appear in the image with as much strength as the signals produced by real objects. { 'ärd·ə,fakt }

artificial aging [MET] The heat treatment of an alloy at moderately elevated temperatures to accelerate precipitation of a component from the supersaturated solid solution. { ärd·ə¦fish·əl 'āj·iŋ }

artificial antenna *See* dummy antenna. { ärd·ə¦fish·əl an'ten·ə }

artificial asteroid [AERO ENG] An object made by humans and placed in orbit about the sun. { ärd·ə¦fish·əl 'as·tə,röid }

artificial atmosphere [CHEM ENG] A mixture of gases used in industrial operations in place of air; classified as an active, or process, atmosphere, or an inactive, or protective, atmosphere. { ärd·ə¦fish·əl 'at·mə,sfir }

artificial atom [ELECTR] A structure, typically 50-100 nanometers in diameter, that is fabricated in a semiconductor crystal and holds a small number of electrons which are trapped in a bowllike potential well. { ,ärd·ə,fish·əl 'ad·əm }

artificial camphor *See* terpene hydrochloride. { ärd·ə¦fish·əl 'kam·fər }

artificial chromosome [GEN] A functional chromosome constructed by genetic engineering, having a centromere (and a telomere at each end, if linear rather than circular) and thus transmissable in cell division after introduction into a cell. { ,ärd·ə¦fish·əl 'krō·mə,sōm }

artificial crystal *See* superlattice. { ärd·ə¦fish·əl 'krist·əl }

artificial delay line *See* delay line. { ärd·ə¦fish·əl di'lā ,līn }

artificial ear [ENG ACOUS] A device designed to duplicate the frequency response, acoustic impedance, threshold sensitivity, and relative perception of loudness, consisting of a special microphone enclosed in a box with properties similar to those of the human ear. { ärd·ə¦fish·əl 'ir }

artificial echo [ELECTROMAG] 1. Received reflections of a transmitted pulse from an artificial target, such as an echo box, corner reflector, or other metallic reflecting surface. 2. Delayed signal from a pulsed radio-frequency signal generator. { ärd·ə¦fish·əl 'ek·ō }

artificial feel [AERO ENG] A type of force feedback incorporated in the control system of an aircraft or spacecraft whereby a portion of the forces acting on the control surfaces are transmitted to the cockpit controls. { ärd·ə¦fish·əl 'fēl }

artificial fiber [TEXT] A filament made from material such as glass, rayon, or nylon. Also known as synthetic fiber. { ärd·ə¦fish·əl 'fī·bər }

artificial gold *See* stannic sulfide. { ärd·ə¦fish·əl 'gōld }

artificial gravity [AERO ENG] A simulated gravity established within a space vehicle by rotation or acceleration. { ärd·ə¦fish·əl 'grav·əd·ē }

artificial ground [ELEC] A common correction for a radio-frequency electrical or electronic circuit that is not directly connected to the earth. { ärd·ə¦fish·əl 'graund }

artificial harbor [CIV ENG] 1. A harbor protected by breakwaters. 2. A harbor formed by dredging. { ärd·ə¦fish·əl 'här·bər }

artificial heart [MED] An endoprosthetic device used to replace or assist the heart. { ärd·ə¦fish·əl 'härt }

artificial horizon [NAV] 1. A gyro-operated flight instrument that shows the pitching and banking attitudes of an aircraft or spacecraft with respect to a reference line horizon, within limited degrees of movement, by means of the relative position of lines or marks on the face of the instrument representing the aircraft and the horizon. Also known as automatic horizon. 2. A device, such as a spirit level or pendulum, that establishes a horizontal reference in a navigation instrument. { ärd·ə¦fish·əl hə'rīz·ən }

artificial hypothermia [MED] A surgical technique used, for example, in heart surgery, in which blood is cooled by a heat exchanger; body temperature is lowered to approximately 85°F (29.4°C), reducing oxygen requirements of tissues, particularly brain cells, and permitting temporary cessation of circulation. { ärd·ə¦fish·əl ,hī·pō'thər·mē·ə }

artificial insemination [MED] A process by which spermatozoa are collected from males and deposited in female genitalia by instruments rather than by natural service. { ärd·ə¦fish·əl in,sem·ə'nā·shən }

artificial intelligence [COMPUT SCI] The property of a machine capable of reason by which it can learn functions normally associated with human intelligence. { ärd·ə¦fish·əl in'tel·ə·jəns }

artificial ionization [COMMUN] Introduction of an artificial reflecting or scattering layer into the atmosphere to permit beyond-the-horizon communications. { ärd·ə¦fish·əl ,ī·ə·nə'zā·shən }

artificial kidney [MED] An apparatus that performs the work of the kidney in purifying blood; used only in cases of renal failure or shutdown. { ärd·ə¦fish·əl 'kid·nē }

artificial language [COMPUT SCI] A computer language that is specifically designed to facilitate communication in a particular field, but is not yet natural to that field; opposite of a natural language, which evolves through long usage. { ärd·ə¦fish·əl 'laŋ·gwij }

artificial lift [PETRO ENG] Any method of lifting oil out of underground reservoirs, usually by injecting gas into the rock or sand formation to force fluids from wells; an example is a gas lift. { ärd·ə¦fish·əl 'lift }

artificial line [ELEC] Circuit made up of lumped constants, which is used to simulate various characteristics of a transmission line. { ärd·ə¦fish·əl 'līn }

artificial line duct [ELEC] Balancing network simulating the impedance of the real line and distant terminal apparatus, which is employed in a duplex circuit to make the receiving device unresponsive to outgoing signal currents. { ärd·ə¦fish·əl 'līn ,dəkt }

artificial load [ELEC] Dissipative but essentially nonradiating device having the impedance characteristics of an antenna, transmission line, or other practical utilization circuit. { ärd·ə¦fish·əl 'lōd }

artificially layered structure *See* superlattice. { ärd·ə¦fish·əl·ē ¦lā·ərd 'strək·chər }

artificial malachite *See* copper carbonate. { ärd·ə¦fish·əl 'mal·ə,kīt }

artificial monument [ENG] A relatively permanent object made by humans, such as an abutment or stone marker, used to identify the location of a survey station or corner. { ärd·ə¦fish·əl 'män·yə·mənt }

artificial neroli oil *See* methyl anthranilate. { ärd·ə¦fish·əl nə'rōl·ē ,öil }

artificial nerve graft [MED] Used to enhance peripheral nerve regeneration, a porous or resorbable tube containing

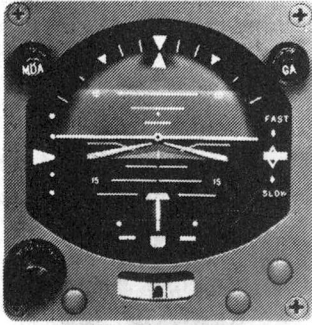

ARTIFICIAL ATOM

top plate

bottom plate

Making precision measurements on an artificial atom. The atom is positioned between two capacitor plates. Single electrons (*e*) can be induced to tunnel into or out of the atom by applying a voltage on the top plate.

ARTIFICIAL HORIZON

Artificial horizon indicator. (*Kollsman Instrument Co.*)

matrix material that may lead axons to grow in the desired direction. { ¦ärd·ə¦fish·əl 'nərv ˌgraft }

artificial nourishment [CIV ENG] The process of replenishing a beach by artificial means, such as the deposition of dredged material. { ¦ärd·ə¦fish·əl 'nər·ish·mənt }

artificial parthenogenesis [PHYSIO] Activation of an egg by chemical and physical stimuli in the absence of sperm. { ¦ärd·ə¦fish·əl ¦pär·thə·nō¦gen·ə·səs }

artificial radiation belt [GEOPHYS] High-energy electrons trapped in the earth's magnetic field as a result of high-altitude nuclear explosions. { ¦ärd·ə¦fish·əl ˌrād·ē¦ā·shən ˌbelt }

artificial radioactivity See induced radioactivity. { ¦ärd·ə¦fish·əl ˌrād·ē·ō·ak'tiv·əd·ē }

artificial radio aurora [COMMUN] Modification of the ionosphere by high-power high-frequency radio transmitters to improve scatter and auroral long-distance communication. Also known as radio aurora. { ¦ärd·ə¦fish·əl ¦rād·ē,ō ə'ror·ə }

artificial reality See virtual reality. { ¦ärd·ə¦fish·əl rē'al·əd·ē }

artificial recharge [CIV ENG] The recharge of an aquifer depleted by abnormally large withdrawals, by the use of injection wells and other techniques. { ¦ärd·ə¦fish·əl 'rē,chärj }

artificial respiration [MED] The maintenance of breathing by artificial ventilation, in the absence of normal spontaneous respiration; effective methods include mouth-to-mouth breathing and the use of a respirator. { ¦ärd·ə¦fish·əl ˌres·pə'rā·shən }

artificial satellite [AERO ENG] Any human-made object placed in a near-periodic orbit in which it moves mainly under the gravitational influence of one celestial body, such as the earth, sun, another planet, or a planet's moon. { ¦ärd·ə¦fish·əl 'sad·ə,līt }

artificial scheelite See calcium tungstate. { ¦ärd·ə¦fish·əl 'shā,līt }

artificial selection [GEN] A breeding method whereby particular genetic traits are selected by human manipulation. { ¦ärd·ə¦fish·əl si'lek·shən }

artificial sky [ARCH] A dome (usually hemispherical) illuminated by concealed light sources; used to illustrate and study daylighting techniques on architectural models placed near the center of the hemisphere. { ¦ärd·ə¦fish·əl 'skī }

artificial sweetener [FOOD ENG] A sugar substitute, such as saccharin. { ¦ärd·ə¦fish·əl 'swēt·nər }

artificial variable [IND ENG] One type of variable introduced in a linear program model in order to find an initial basic feasible solution; an artificial variable is used for equality constraints and for greater-than or equal inequality constraints. { ¦ärd·ə¦fish·əl 'ver·ē·ə·bəl }

artificial ventilation [MIN ENG] The inducing of a flow of air through a mine or part of a mine by mechanical or other means. { ¦ärd·ə¦fish·əl 'vent·əl'ā·shən }

artificial voice [ENG ACOUS] **1.** Small loudspeaker mounted in a shaped baffle which is proportioned to simulate the acoustical constants of the human head; used for calibrating and testing close-talking microphones. **2.** Synthetic speech produced by a multiple tone generator; used to produce a voice reply in some real-time computer applications. { ¦ärd·ə¦fish·əl 'vois }

artificial weathering [ENG] The controlled production of changes in materials under laboratory conditions to simulate actual outdoor exposure. { ¦ärd·ə¦fish·əl 'weth·ə·riŋ }

artillery [ORD] A gun or a rocket launcher with mounting too large or too heavy to be classed as a small arm. { är'til·ə·rē }

artillery bogie [ORD] The portion of an artillery weapon, consisting of wheels, axles, and various supporting appurtenances, which is the principal weight-bearing unit when the weapon is being transported. { är'til·ə·rē 'bō·gē }

artillery cart [ORD] A trailer that carries equipment used by artillery units for fire control, communications, and mapping; it is attached to a field gun for traveling. { är'til·ə·rē 'kärt }

artillery cartridge extractor [ORD] An extractor designed for pulling an empty cartridge case or unfired cartridge out of the chamber of an artillery weapon. { är'til·ə·rē 'kär·trij ik,strak·tər }

artillery cleaning staff [ORD] A round wooden or metal staff with or without a metal handle and with a head unit designed for holding a piece of fabric or cotton for swabbing the bore of a mortar or a short-barrel cannon. { är'til·ə·rē 'klēn·iŋ ,staf }

artillery sled [ORD] A flat-bottomed steel item usually curved up at one end; it usually has wheel welds and attaching facilities for fastening the wheels of artillery mounts; used primarily to transport weapons over snow, ice, swamps, or rough terrain. { är'til·ə·rē ,sled }

artillery survey [ORD] The process of determining, with sufficient exactness, the relative horizontal and vertical locations of the pieces and targets so that they may be plotted on the firing chart, and of providing accurate data for the pieces. { är'til·ə·rē 'sər,vā }

artillery train [ORD] A number of pieces of ordnance mounted on traveling carriages. { är'til·ə·rē ,trān }

Artinian ring [MATH] A ring is Artinian on left ideals (or right ideals) if every descending sequence of left ideals (or right ideals) has only a finite number of distinct members. { ar¦tin·ē·ən 'riŋ }

artinite [MINERAL] $Mg_2CO_3(OH)_2 \cdot 3H_2O$ A snow-white mineral crystallizing in the orthorhombic system and occurring in crystals or fibrous aggregates. { är'tē,nīt }

Artinskian [GEOL] A European stage of geologic time including Lower Permian (above Sakmarian, below Kungurian). { är'tin·skē·ən }

Artiodactyla [VERT ZOO] An order of terrestrial, herbivorous mammals characterized by having an even number of toes and by having the main limb axes pass between the third and fourth toes. { ¦ärd·ē·ō'dak·tə·lə }

artotype See photogelatin printing plate. { 'ärd·ə,tīp }

ARTS See automated radar terminal system.

artwork [GRAPHICS] **1.** Illustrative and decorative matter as distinguished from text. **2.** Illustrative matter and type proofs arranged on a mechanical. { 'ärt,wərk }

Arundel method [MAP] A combination of graphical and analytical methods, based on radial triangulation, for point-by-point topographic mapping from aerial photographs. { ə'rənd·əl ,meth·əd }

arviculture [AGR] The cultivation of field crops. { 'är·və,kəl·chər }

aryl [ORG CHEM] An organic group derived from an aromatic hydrocarbon by removal of one hydrogen. { 'ar·əl }

aryl acid [ORG CHEM] An organic acid that has an aryl group. { 'ar·əl 'as·əd }

arylamine [ORG CHEM] An organic compound formed from an aromatic hydrocarbon that has at least one amine group joined to it, such as aniline. { 'ar·əl·ə,mēn }

arylated alkyl See aralkyl. { ar·ə'lād·əd 'al·kəl }

aryl compound [ORG CHEM] Molecules with the six-carbon aromatic ring structure characteristic of benzene or compounds derived from aromatics. { 'ar·əl ,käm,paùnd }

aryl diazo compound [ORG CHEM] A diazo compound bonded to the ring structure characteristic of benzene or any other aromatic derivative. { 'ar·əl dī'āz·ō ,käm,paùnd }

arylene [ORG CHEM] A radical that is bivalent and formed by removal of hydrogen from two carbon sites on an aromatic nucleus. { 'ar·ə,lēn }

aryl halide [ORG CHEM] An aromatic derivative in which a ring hydrogen has been replaced by a halide atom. { 'ar·əl 'hal,īd }

arylide [ORG CHEM] A compound formed from a metal and an aryl group, for example, PbR_4, where R is the aryl group. { 'ar·ə,līd }

aryloxy compound [ORG CHEM] One of a group of compounds useful as organic weed killers, such as 2,4-dichlorophenoxyacetic acid (2,4-D). { 'ar·əl¦äk·sē ,käm,paùnd }

aryne [ORG CHEM] An aromatic species in which two adjacent atoms of a ring lack substituents, with two orbitals each missing an electron. Also known as benzyne. { 'a,rīn }

arytenoid [ANAT] Relating to either of the paired, pyramid-shaped, pivoting cartilages on the dorsal aspect of the larynx, in humans and most other mammals, to which the vocal cords and arytenoid muscles are attached. { ,ar·ə'tē,nòid }

arzrunite [MINERAL] A bluish-green mineral consisting of a basic copper sulfate with copper chloride and lead, and occurring as incrustations. { ärz'rü,nīt }

aS See abmho.

As See altostratus cloud; arsenic.

asar See esker. { 'a·sər }

asarone [ORG CHEM] $C_{12}H_{16}O_3$ A crystalline substance with melting point 67°C; insoluble in water, soluble in alcohol; found in plants of the genus Asarum; used as a constituent in essential oils such as calamus oil. { 'as·ə,rōn }

asbestine [MATER] A material with the properties of asbestos. { as'be,stēn }

asbestos [MINERAL] A general name for the useful, fibrous varieties of a number of rock-forming silicate minerals that are heat-resistant and chemically inert; two varieties exist: amphibole asbestos, the best grade of which approaches the composition $Ca_2Mg_5(OH)_2Si_8O_{22}$ (tremolite), and serpentine asbestos, usually chrysotile, $Mg_3Si_2(OH)_4O_5$. { as'bes·təs }

asbestos blanket [MATER] Asbestos fibers (alone or in combination with other fibers) stitched, bonded, or woven into flexible blanket form; used for high-temperature insulation or for fire barriers. { as'bes·təs ,blaŋ·kət }

asbestos board [MATER] A sheet of fire-resistant material made from asbestos fiber and portland cement. { as'bes·təs ,bȯrd }

asbestos cement [MATER] A building material composed of a mixture of asbestos fiber, portland cement, and water made into plain sheets, corrugated sheets, tiles, and piping. { as'bes·təs si,ment }

asbestos-cement cladding [BUILD] Asbestos board and component wall systems, directly supported by wall framing, forming a wall or wall facing. { as'bes·təs si,ment 'klad·iŋ }

asbestos-cement pipe [MATER] A concrete pipe made of a mixture of portland cement and asbestos fiber and highly resistant to corrosion; used in drainage systems, waterworks systems, and gas lines. { as'bes·təs si,ment 'pīp }

asbestos felt [MATER] A product made by saturating felted asbestos with asphalt or other suitable binder, such as a synthetic elastomer. { as'bes·təs 'felt }

asbestos insulation [MATER] A material composed of asbestos fibers bonded with mixtures of clay or sodium silicate; used as thermal insulation for temperatures above 1500°F (816°C). { as'bes·təs ,in·sə'lā·shən }

asbestosis [MED] A chronic lung inflammation caused by inhalation of asbestos dust. { as,be'stō·səs }

asbestos joint runner [MATER] An asbestos rope, wrapped around a pipe and then clamped in position; used to hold molten lead which is poured in a caulked joint. Also known as pouring rope. { as'bes·təs 'jȯint ,rən·ər }

asbestos plaster [MATER] A fireproof insulating material generally composed of asbestos with bentonite as the binder. { as'bes·təs 'plas·tər }

asbestos roofing [MATER] Roofing or wall cladding sheets made of asbestos cement. { as'bes·təs 'rüf·iŋ }

asbestos shingle [MATER] A shingle composed of asbestos cement formed under pressure; used on houses for roofing and siding that resist the destructive effects of time, weather, and fire. { as'bes·təs 'shiŋ·gəl }

asbolane See asbolite. { 'az·bə,lān }

asbolite [MINERAL] A black, earthy mineral aggregate containing hydrated oxides of manganese and cobalt. Also known as asbolane; black cobalt; earthy cobalt. { 'az·bə,līt }

as-brazed [MET] A brazement prior to any additional treatment such as thermal, chemical, or mechanical treatments. { ,az 'brāzd }

as-built drawing See as-fitted drawing. { ¦az 'bilt 'drȯ·iŋ }

as-built schedule [IND ENG] The final schedule for a project, reflecting the actual scope, actual completion dates, actual duration of the specified activities, and start dates. { ¦az ¦bilt ¦skej·əl }

ASC See N-acetylsulfanilyl chloride.

A scale [ACOUS] A system used to filter out sound below 55 decibels; its characteristics are equal to those of the human ear. { 'ā ,skāl }

A scan See A scope. { 'ā ,skan }

Ascaphidae [VERT ZOO] A family of amphicoelous frogs in the order Anura, represented by four living species. { ə'skaf·ə,dē }

ascariasis [MED] Any parasitic infection of humans or domestic mammals caused by species of *Ascaris*. { ,as·kə'rī·ə·səs }

ascarid [INV ZOO] The common name for any roundworm belonging to the superfamily Ascaridoidea. { 'as·kə·rəd }

Ascaridata [INV ZOO] An equivalent name for the Ascaridina. { ə,skar·ə'däd·ə }

Ascaridida [INV ZOO] An order of parasitic nematodes in the subclass Phasmidia. { ə,skar·ə'dī·də }

Ascarididae [INV ZOO] A family of parasitic nematodes in the superfamily Ascaridoidea. { ,as·kə'rid·ə,dē }

Ascaridina [INV ZOO] A suborder of parasitic nematodes in the order Ascaridida. { ə,skar·ə'dī·nə }

Ascaridoidea [INV ZOO] A large superfamily of parasitic nematodes of the suborder Ascaridina. { ə,skar·ə'dȯid·ē·ə }

ascaridole [ORG CHEM] $C_{10}H_{16}O_2$ A terpene peroxide, explosive when heated; used as an initiator in polymerization. { ə'skar·ə,dōl }

Ascaris [INV ZOO] A genus of roundworms that are intestinal parasites in mammals, including humans. { 'as·kə·rəs }

Ascaroidea [INV ZOO] An equivalent name for Ascaridoidea. { ,as·kə'rȯid·ē·ə }

ascender [GRAPHICS] That part of some lowercase letters which extends above the main body, as in b, d, and h. { ə'sen·dər }

ascending aorta [ANAT] The first part of the aorta, extending from its origin in the heart to the aortic arch. { ə'send·iŋ,ā'ȯrd·ə }

ascending branch [MECH] The portion of the trajectory between the origin and the summit on which a projectile climbs and its altitude constantly increases. { ə'send·iŋ 'branch }

ascending chain condition [MATH] The condition on a ring that every ascending sequence of left ideals (or right ideals) has only a finite number of distinct members. { ə,sen·diŋ 'chān kən,dish·ən }

ascending chromatography [ANALY CHEM] A technique for the analysis of mixtures of two or more compounds in which the mobile phase (sample and carrier) rises through the fixed phase. { ə'send·iŋ ,krō·mə'täg·rə·fē }

ascending colon [ANAT] The portion of the colon that extends from the cecum to the bend on the right side below the liver. { ə'send·iŋ 'kōl·ən }

ascending node [ASTRON] Also known as northbound node. **1.** The point at which a planet, planetoid, or comet crosses to the north side of the ecliptic. **2.** The point at which a satellite crosses to the north side of the equatorial plane of its primary. { ə'send·iŋ ¦nōd }

ascending sequence [MATH] **1.** A sequence of elements of a partially ordered set such that each member of the sequence is equal to or less than the following one. **2.** In particular, a sequence of sets such that each member of the sequence is a subset of the following one. { ə,sen·diŋ 'sē·kwəns }

ascending series [MATH] **1.** A series each of whose terms is greater than the preceding term. **2.** See power series. { ə'send·iŋ 'sir·ēz }

ascending sort [COMPUT SCI] The arrangement of records or other data into a sequence running from the lowest to the highest in a specified field. { ə'send·iŋ 'sȯrt }

ascending vertical angle See angle of elevation. { ə'send·iŋ ¦vərd·i·kəl 'aŋ·gəl }

ascent [AERO ENG] Motion of a craft in which the path is inclined upward with respect to the horizontal. { ə'sent }

Ascheim-Zondek test [PATH] A human pregnancy test that uses the reaction of ovaries in immature white mice to an injection of urine from a woman. { 'äsh,hīm 'tsän,dek ,test }

Aschelmintha [INV ZOO] A theoretical grouping erected by B. G. Chitwood as a series that includes the phylum Nematoda. { ,ask,hel'min·thə }

Aschelminthes [INV ZOO] A heterogeneous phylum of small to microscopic wormlike animals; individuals are pseudocoelomate and mostly unsegmented and are covered with a cuticle. { ,ask,hel'min,thēz }

aschistic [GEOL] Pertaining to rocks of minor igneous intrusions that have not been differentiated into light and dark portions but that have essentially the same composition as the larger intrusions with which they are associated. { ā'skis·tik }

Aschoff body [MED] The lesion of rheumatic fever found around blood vessels in the myocardium. { 'ä,shȯf ,bäd·ē }

Ascidiacea [INV ZOO] A large class of the phylum Tunicata; adults are sessile and may be solitary or colonial. { ə,sid·ē'äsh·ē·ə }

ascidiform [BOT] Pitcher-shaped, as certain leaves. { ə'sid·ə,fȯrm }

ascidium [BOT] A pitcher-shaped plant organ or part. { ə'sid·ē·əm }

ASCII See American Standard Code for Information Interchange. { 'as,kē }

ASCII file [COMPUT SCI] A data or text file that contains only codes that constitute the 128-character ASCII set. { ¦as,kē 'fīl }

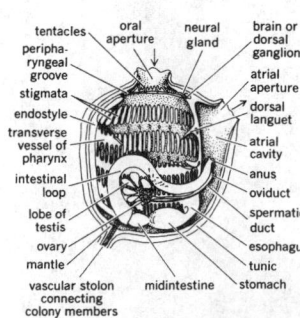

ASCIDIACEA

tentacles — oral aperture — neural gland — brain or dorsal ganglion — peripharyngeal groove — atrial aperture — stigmata — dorsal languet — endostyle — atrial cavity — transverse vessel of pharynx — anus — intestinal loop — oviduct — lobe of testis — spermatic duct — ovary — esophagus — mantle — tunic — vascular stolon connecting colony members — midintestine — stomach

Left side of a zooid of the colonial ascidian *Perophora*. The tunic, mantle, and anterior wall of the pharynx have been removed from the left. The two arrows indicate direction of water intake and expulsion.

ASCII protocol [COMMUN] A protocol for the simplest mode of transmitting ASCII data, with little or no error checking. { ¦as‚kē ¦prōd·ə‚kȯl }

ASCII sort order [COMPUT SCI] A sort order determined by the numbering of characters in the American Standard Code for Information Interchange. { ¦as‚kē ¦sȯrt ‚ȯrd·ər }

ascites [MED] An abnormal accumulation of serous fluid in the abdominal cavity. { ə'sīd·ēz }

Asclepiadaceae [BOT] A family of tropical and subtropical flowering plants in the order Gentianales characterized by a well-developed latex system; milkweed (*Asclepias*) is a well-known member. { ə‚sklēp·ē·ə'dās·ē‚ē }

ascocarp [MYCOL] The mature fruiting body bearing asci with ascospores in higher Ascomycetes. { 'as·kə‚kärp }

ascogenous [MYCOL] Pertaining to or producing asci. { a'skäj·ə·nəs }

ascogonium [MYCOL] The specialized female sexual organ in higher Ascomycetes. { ‚as·kə'gōn·ē·əm }

Ascolichenes [BOT] A class of the lichens characterized by the production of asci similar to those produced by Ascomycetes. { ‚as·kə‚lī'kē·nēz }

Ascoli's theorem [MATH] The theorem that a set of uniformly bounded, equicontinuous, real-valued functions on a closed set of a real Euclidean *n*-dimensional space contains a sequence of functions which converges uniformly on compact subsets. { as'kō‚lēz ‚thir·əm }

Ascomycetes [MYCOL] A class of fungi in the subdivision Eumycetes, distinguished by the ascus. { ‚as·kō‚mī'sēd·ēz }

ascon [INV ZOO] A sponge or sponge larva having incurrent canals leading directly to the paragaster. { 'a‚skän }

A scope [ELECTR] A radarscope on which the trace appears as a horizontal or vertical range scale and the signals appear as vertical or horizontal deflections. Also known as A indicator; A scan. { 'ā skōp }

ascorbic acid [BIOCHEM] $C_6H_8O_6$ A white, crystalline, water-soluble vitamin found in many plant materials, especially citrus fruit. Also known as vitamin C. { ə'skȯr·bik 'as·əd }

ascospore [MYCOL] An asexual spore representing the final product of the sexual process, borne on an ascus in Ascomycetes. { 'as·kə‚spȯr }

Ascothoracica [INV ZOO] An order of marine crustaceans in the subclass Cirripedia occurring as endo- and ectoparasites of echinoderms and cnidarians. { ‚as·kə·thə'ras·ə·kə }

ascus [MYCOL] An oval or tubular spore sac bearing ascospores in members of the Ascomycetes. { 'as·kəs }

asdic [ELECTR] Derived from British term for sonar and underwater listening devices. Anti-Submarine Detection Investigation Committee. { 'az‚dik }

aseismic [GEOPHYS] Not subject to the occurrence or destructive effects of earthquakes. { ā'sīz·mik }

A selection [ECOL] Selection that favors species adapted to consistently adverse environments. { 'ā si‚lek·shən }

Aselloidea [INV ZOO] A group of free-living, fresh-water isopod crustaceans in the suborder Asellota. { ‚a·sə'lȯid·ē·ə }

Asellota [INV ZOO] A suborder of morphologically and ecologically diverse aquatic crustaceans in the order Isopoda. { ə'sel·ə·də }

asepsis [MED] The state of being free from pathogenic microorganisms. { ā'sep·səs }

aseptic meningitis [MED] A type of meningitis in which the raised cell count in the cerebrospinal fluid is predominantly lymophocytic; usually caused by a viral infection, but can result from other causes. { ā'sep·tik ‚men·ən'jīd·əs }

asetigerous See achaetous. { ā·sə'tij·ə·rəs }

asexual [BIOL] **1.** Not involving sex. **2.** Exhibiting absence of sex or of functional sex organs. { ā'seksh·ə·wəl }

asexual reproduction [BIOL] Formation of new individuals from a single individual without the involvement of gametes. { ā'seksh·ə·wəl ‚rē·prə'dək·shən }

as-fitted drawing [ENG] A drawing as amended after completion of an industrial facility in order to provide an accurate record of the details of the entire installation in their final form. Also known as as-built drawing; as-made drawing. { ¦az‚fid·əd 'drȯ·iŋ }

ash [BOT] **1.** A tree of the genus *Fraxinus*, deciduous trees of the olive family (Oleaceae) characterized by opposite, pinnate leaflets. **2.** Any of various Australian trees having wood of great toughness and strength; used for tool handles and in work requiring flexibility. [CHEM] The incombustible matter remaining after a substance has been incinerated. [ENG] An undesirable constituent of diesel fuel whose quantitative measurement indicates degree of fuel cleanliness and freedom from abrasive material. [GEOL] Volcanic dust and particles less than 4 millimeters in diameter. { ash }

Ashby [GEOL] A North American stage of Middle Ordovician geologic time, forming the upper subdivision of Chazyan, and lying above Marmor and below Porterfield. { 'ash‚bē }

ash collector See dust chamber. { 'ash kə'lek·tər }

ash cone [GEOL] A volcanic cone built primarily of unconsolidated ash and generally shaped somewhat like a saucer, with a rim in the form of a wide circle and a broad central depression often nearly at the same elevation as the surrounding country. { 'ash ‚kōn }

ash content [MATER] The mass of incombustible material remaining after burning a given coal sample as a percentage of the original mass of the coal. { 'ash 'kän‚tent }

ash conveyor [MECH ENG] A device that transports refuse from a furnace by fluid or mechanical means. { 'ash kən'vā·ər }

ash dump [ENG] An opening in the floor of a fireplace or firebox through which ashes are swept to an ash pit below. { 'ash ‚dəmp }

ashen light [ASTRON] A faint, luminous glow sometimes observed over the right side of Venus when it is close to inferior conjunction, probably due to electrical disturbances in the ionosphere of Venus. { 'ash·ən ‚līt }

ash fall [GEOL] **1.** A fall of airborne volcanic ash from an eruption cloud; characteristic of Vulcanian eruptions. Also known as ash shower. **2.** Volcanic ash resulting from an ash fall and lying on the ground surface. { 'ash ‚fȯl }

ash field [GEOL] A thick, extensive deposit of volcanic ash. Also known as ash plain. { 'ash ‚fēld }

ash flow [GEOL] **1.** An avalanche of volcanic ash, generally a highly heated mixture of volcanic gases and ash, traveling down the flanks of a volcano or along the surface of the ground. Also known as glowing avalanche; incandescent tuff flow. **2.** A deposit of volcanic ash and other debris resulting from such a flow and lying on the surface of the ground. { 'ash ‚flō }

ash-flow tuff See ignimbrite. { 'ash‚flō ‚təf }

ash furnace [ENG] A furnace in which materials are fritted for glassmaking. { 'ash ‚fər·nəs }

ash fusibility [GEOL] The gradual softening and melting of coal ash that takes place with increase in temperature as a result of the melting of the constituents and chemical reactions. { 'ash ‚fyüz·ə'bil·əd·ē }

Ashgillian [GEOL] A European stage of geologic time in the Upper Orodovician (above Upper Caradocian, below Llandoverian of Silurian). { ash'gil·yən }

ashing [ANALY CHEM] An analytical process in which the chemical material being analyzed is oven-heated to leave only noncombustible ash. { 'ash·iŋ }

ashlar [CIV ENG] Masonry with an exposed side of square or rectangular stones. { 'ash·lər }

ashlar brick [MATER] A brick having rough-hackled faces resembling stone. { 'ash·lər ‚brik }

ashlar line [BUILD] The outer line of a wall above any projecting base. { 'ash·lər ‚līn }

ash pan [ENG] A metal receptacle beneath a fireplace or furnace grating for collection and removal of ashes. { 'ash ‚pan }

ash pit [BUILD] The ash-collecting area beneath a fireplace hearth. { 'ash ‚pit }

ash pit door [ENG] A cast-iron door providing access to an ash pit for ash removal. { 'ash ‚pit ‚dȯr }

ash plain See ash field. { 'ash ‚plān }

ash rock [GEOL] The material of arenaceous texture produced by volcanic explosions. { 'ash ‚räk }

ash shower See ash fall. { 'ash ‚shaù·ər }

ashstone [PETR] A rock composed of fine volcanic ash; particles are less than 0.06 millimeter in diameter. { 'ash‚stōn }

ashtonite See mordenite. { 'ash·tə‚nīt }

ash viscosity [GEOL] The ratio of shearing stress to velocity gradient of molten ash; indicates the suitability of a coal ash for use in a slag-tap-type boiler furnace. { 'ash vis'käs·əd·ē }

ashy grit [GEOL] **1.** Pyroclastic material of sand and smaller size. **2.** Mixture of ordinary sand and volcanic ash. { 'ash·ē 'grit }

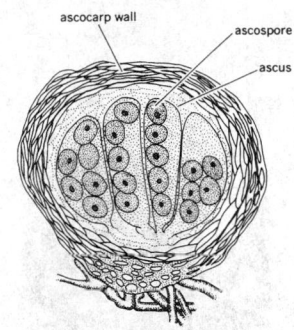

ASCOCARP

Longitudinal section showing several asci in ascocarp of *Erysiphe aggregata*.

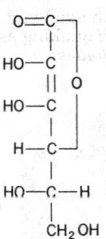

ASCORBIC ACID

Structural formula of ascorbic acid.

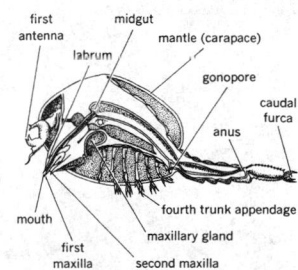

ASCOTHORACICA

Ascothorax ophioctenis, a parasite in the bursae of brittle stars. (*From R. D. Barnes, Invertebrate Zoology, 2d ed., Saunders, 1968*)

ASPEN

A leaf scar with axial bud, leaf, and twig of the quaking aspen (*Populus tremloides*).

Asia [GEOGR] The largest continent, comprising the major portion of the broad east-west extent of the Northern Hemisphere land masses. { 'āzh·ə }

Asian flu [MED] An acute viral respiratory infection of humans caused by influenza A-2 virus. { 'āzh·ən 'flü }

ASIC *See* application-specific integrated circuit. { 'ā,sik or 'ā¦es¦ī'sē }

asiderite *See* stony meteorite. { ə'sīd·ə,rīt }

Asilidae [INV ZOO] The robber flies, a family of predatory, orthorrhaphous, dipteran insects in the series Brachycera. { ə'sil·ə,dē }

A size [ENG] **1.** One of a series of sizes to which trimmed paper and board are manufactured; for size A*N*, with *N* equal to any integer from 0 to 10, the length of the longer side is $2^{-(2N-1)/4}$ meters, while the length of the shorter side is $2^{-(2N+1)/4}$ meters, with both lengths rounded off to the nearest millimeter. **2.** Of a sheet of paper, the dimensions 8.5 inches by 11 inches (216 millimeters by 279 millimeters). { 'ā ,sīz }

ASK *See* amplitude shift keying.

ASM *See* air-to-surface missile.

as-made drawing *See* as-fitted drawing. { ¦az ¦mād 'drȯ·iŋ }

Aso lava [GEOL] A type of indurated pyroclastic deposit produced during the explosive eruptions that formed the Aso Caldera of Kyushu, Japan. { 'äs·ō 'läv·ə }

asomatognosia [MED] Lacking awareness of paralysis because the brain is damaged. { ā,sōm·ə,täg'nōzh·ə }

Asopinae [INV ZOO] A family of hemipteran insects in the superfamily Pentatomoidea including some predators of caterpillars. { ə'sōp·ə,nē }

asparaginase [BIOCHEM] An enzyme that catalyzes the hydrolysis of asparagine to asparaginic acid and ammonia. { ,as·pə'raj·ə,nās }

asparagine [BIOCHEM] $C_4H_8N_2O_3$ A white, crystalline amino acid found in many plant seeds. { ə'spar·ə,jēn }

asparagolite *See* asparagus stone. { ,as·pə'rag·ə,līt }

asparagus [BOT] *Asparagus officinalis.* A dioecious, perennial monocot belonging to the order Liliales; the shoot of the plant is edible. { ə'spar·ə·gəs }

asparagus rust [PL PATH] A fungal disease of asparagus plants that is caused by *Puccinia asparagi*, characterized by uredinial lesions that appear as oval reddish-brown pustules. { ə'spar·ə·gəs ,rəst }

asparagus stone [MINERAL] A yellow-green variety of apatite occurring in crystals. Also known as asparagolite. { ə'spar·ə·gəs ,stōn }

aspartame [ORG CHEM] $C_{14}H_{18}N_2O_5$ A dipeptide ester about 160 times sweeter than sucrose in aqueous solution; used as a low-calorie sweetener. { ə'spär,tām }

aspartase [BIOCHEM] A bacterial enzyme that catalyzes the deamination of aspartic acid to fumaric acid and ammonia. { ə'spär,tās }

aspartate [BIOCHEM] A compound that is an ester or salt of aspartic acid. { ə'spär,tāt }

aspartic acid [BIOCHEM] $C_4H_7NO_4$ A nonessential, crystalline dicarboxylic amino acid found in plants and animals, especially in molasses from young sugarcane and sugarbeet. { ə'spärd·ik 'as·əd }

aspartokinase [BIOCHEM] An enzyme that catalyzes the reaction of aspartic acid with adenosinetriphosphate to give aspartyl phosphate. { ə,spärd·ō'kī,nās }

aspartoyl [BIOCHEM] $-COCH_2CH(NH_2)CO-$ A bivalent radical derived from aspartic acid. { ə'spärd·ə,wil }

aspartyl phosphate [BIOCHEM] $H_2O_3POOCH_2CHNH_2\text{-}COOH$ An intermediate in the biosynthesis of pyrimidines. { ə'spärd·əl 'fäs,fāt }

aspect [ARCH] The direction which a building faces with respect to the points of a compass. [ASTRON] The apparent position of a celestial body relative to another; particularly, the apparent position of the moon or a planet relative to the sun. [CIV ENG] Of railway signals, what the engineer sees when viewing the blades or lights in their relative positions or colors. [ECOL] Seasonal appearance. [GEOL] **1.** The general appearance of a specific geologic entity or fossil assemblage as considered more or less apart from relations in time and space. **2.** The direction toward which a valley side or slope faces with respect to the compass or rays of the sun. { 'a,spekt }

aspect angle [ENG] The angle formed between the longitudinal axis of a projectile in flight and the axis of a radar beam.

[GEOL] The angle between the aspect of a slope and the geographic south (Northern Hemisphere) or the geographic north (Southern Hemisphere). { 'a,spekt ,aŋ·gəl }

aspection [ECOL] Seasonal change in appearance or constitution of a plant community. { a'spek·shən }

aspect ratio [AERO ENG] The ratio of the square of the span of an airfoil to the total airfoil area, or the ratio of its span to its mean chord. [DES ENG] **1.** The ratio of frame width to frame height in television; it is 4:3 in the United States and Britain. **2.** In any rectangular configuration (such as the cross section of a rectangular duct), the ratio of the longer dimension to the shorter. [MECH ENG] In an automotive vehicle, the ratio of the height of a tire to its width. Also known as tire profile. [NUCLEO] The ratio of the plasma diameter of a toroidal controlled fusion device to the major diameter of the torus. { 'a,spekt ,rā·shō }

aspen [BOT] Any of several species of poplars (*Populus*) characterized by weak, flattened leaf stalks which cause the leaves to flutter in the slightest breeze. { 'as·pən }

Aspergillaceae [MYCOL] Former name for the Eurotiaceae. { ,as·pər·jə'lās·ē,ē }

Aspergillales [MYCOL] Former name for the Eurotiales. { ,as·pər·jə'lā·lēz }

aspergillic acid [BIOCHEM] $C_{12}H_{20}O_2N_2$ A diketopiperazine-like antifungal antibiotic produced by certain strains of *Aspergillus flavus*. { ¦as·pər¦jil·ik 'as·əd }

aspergillin [MYCOL] **1.** A black pigment found in spores of some molds of the genus *Aspergillus*. **2.** A broad-spectrum antibacterial antibiotic produced by the molds *Aspergillus flavus* and *A. fumigatus*. { ,as·pər¦jil·ən }

aspergillosis [MED] A rare fungus infection of humans and animals caused by several species of *Aspergillus*. { ,as·pər·jil'ō·səs }

Aspergillus [MYCOL] A genus of fungi including several species of common molds and some human and plant pathogens. { ,as·pər¦jil·əs }

asperifoliate [BOT] Rough-leaved. { ¦as·pər·ə¦fōl·ē,āt }

asperity [GEOL] A type of surface roughness appearing along the interface of two faults. { a'sper·ə·dē }

aspermatism [MED] **1.** Failure to ejaculate or secrete semen. **2.** Defective or absent sperm formation. { ā'spərm·ə,tiz·əm }

asperomagnetic state [SOLID STATE] The condition of a rare-earth glass in which the spins are oriented in fixed directions, with most nearest-neighbor spins parallel or nearly parallel, so that the spin directions are distributed in one hemisphere. { a¦sper·ō,mag'ned·ik 'stāt }

asperulate [BOT] Delicately roughened. { a'sper·ə,lāt }

asphalt [MATER] A brown to black, hard, brittle, or plastic bituminous material composed principally of hydrocarbons; formed in oil-bearing rocks near the Dead Sea, and in Trinidad; prepared by pyrolysis from coal tar, certain petroleums, and lignite tar; melts on heating; insoluble in water but soluble in gasoline; used for paving and roofing and in paints and varnishes. { 'a,sfȯlt }

asphalt base *See* naphthene base. { 'a,sfȯlt ¦bās }

asphalt base crude [MATER] A crude oil which yields napthenic asphalt residues when processed. { 'a,sfȯlt ¦bās 'krüd }

asphalt block [MATER] A paving block composed of a mixture of 88–92% crushed stone with the balance asphalt cement. { 'a,sfȯlt ,bläk }

asphalt cement [MATER] Fluxed or unfluxed asphalt especially prepared for direct use in making bituminous pavements. { 'a,sfȯlt si'ment }

asphalt cutter [MECH ENG] A powered machine having a rotating abrasive blade; used to saw through bituminous surfacing material. { 'a,sfȯlt ,kəd·ər }

asphalt emulsion [MATER] Asphalt cement in water containing a small amount of emulsifying agent. { 'a,sfȯlt i'məl·shən }

asphalt-emulsion slurry seal [MATER] A mixture of slow-setting emulsified asphalt, fine aggregate, and mineral filler, with water added to produce a slurry consistency. { 'a,sfȯlt i'məl·shən 'slə·rē ,sēl }

asphalt enamel [MATER] A surface coating composed of a mixture of asphalt, finely pulverized mica, clay, soapstone, or talc; applied to pipe that will be buried. { 'a,sfȯlt i'nam·əl }

asphaltene [MATER] Any of the dark, solid constituents of crude oils and other bitumens which are soluble in carbon disulfide but insoluble in paraffin naphthas; they hold most of the organic constituents of bitumens. { a'sfòl,tēn }

asphalt flux [MATER] An oil used to reduce the consistency or viscosity of hard asphalt to the point required for use. { 'a,sfòlt ,fläks }

asphalt fog seal [MATER] An asphalt surface treatment consisting of a light application of liquid asphalt without a mineral aggregate cover. { 'a,sfòlt 'fäg ,sēl }

asphalt heater [ENG] A piece of equipment for raising the temperature of bitumen used in paving. { 'a,sfòlt 'hēd·ər }

asphaltic base oil [MATER] A crude oil that has asphaltum as the predominating solid residual. { a'sfòlt·ik ¦bas ,óil }

asphaltic concrete [MATER] A special concrete consisting of a mixture of graded aggregate and heated asphalt; must be applied and spread while hot. { a'sfòlt·ik ¦kän,krēt }

asphaltic material [MATER] A solid, liquid, or semisolid mixture of heavy hydrocarbons and nonmetallic derivatives; obtained from naturally occurring bituminous deposits or from residues of petroleum refining. { a'sfòl·tik mə'tir·ē·əl }

asphaltic road oil [MATER] A thick, fluid solution of asphalt. { a'sfòlt·ik 'rōd ,óil }

asphaltic sand [GEOL] Deposits of sand grains cemented together with soft, natural asphalt. { a'sfòlt·ik 'sand }

asphaltite [GEOL] Any of the dark-colored, solid, naturally occurring bitumens that are insoluble in water, but more or less completely soluble in carbon disulfide, benzol, and so on, with melting points between 250 and 600°F (121 and 316°C); examples are gilsonite and grahamite. { a'sfòl,tīt }

asphaltite coal See albertite. { a'sfòl,tīt ,kōl }

asphalt lamination [MATER] A laminate of sheet material, such as paper or felt, which uses asphalt as the adhesive. { 'a,sfòlt ,lam·ə'nā·shən }

asphalt leveling course [CIV ENG] A layer of an asphalt-aggregate mixture of variable thickness, used to eliminate irregularities in contour of an existing surface, prior to the placement of a superimposed layer. { 'a,sfòlt 'lev·əl·iŋ ,kòrs }

asphalt macadam [MATER] Pavement made with an asphalt binder rather than tar. { 'a,sfòlt mə'kad·əm }

asphalt mastic [MATER] A mixture of asphalt with sand, asbestos, crushed rock, or similar material; used like cement. Also known as mastic asphalt. { 'a,sfòlt 'mas·tik }

asphalt overlay [CIV ENG] One or more layers of asphalt construction on an existing pavement. { 'a,sfòlt 'ov·ər,lā }

asphalt paint [MATER] Asphaltic material dissolved in volatile solvent with or without pigments, drying oils, resins, and so on. { 'a,sfòlt 'pānt }

asphalt paper [MATER] A paper that is coated or impregnated with asphalt. { 'a,sfòlt 'pā·pər }

asphalt pavement [CIV ENG] A pavement consisting of a surface layer of mineral aggregate, coated and cemented together with asphalt cement on supporting layers. { 'a,sfòlt 'pāv·mənt }

asphalt primer [MATER] Low-viscosity, liquid asphaltic material applied to and absorbed by nonbituminous surfaces, as in waterproofing. { 'a,sfòlt 'prīm·ər }

asphalt rock [GEOL] Natural asphalt-containing sandstone or dolomite. Also known as asphalt stone; bituminous rock; rock asphalt. { 'a,sfòlt 'räk }

asphalt roofing [MATER] A roofing material made by impregnating a dry roofing felt with a hot asphalt saturant, applying asphalt coatings to the weather and reverse sides, and embedding a mineral surfacing in the coating on the weather side. { 'a,sfòlt 'rüf·iŋ }

asphalt shingle [MATER] A roof shingle made of felt impregnated with asphalt and covered with mineral granules. { 'a,sfòlt 'shiŋ·gəl }

asphalt soil stabilization [CIV ENG] The treatment of naturally occurring nonplastic or moderately plastic soil with liquid asphalt at normal temperatures to improve the load-bearing qualities of the soil. { 'a,sfòlt ¦sóil ,stab·ə·lə'zā·shən }

asphalt stone See asphalt rock. { 'a,sfòlt 'stōn }

asphalt tile [MATER] Floor tile composed of asbestos fibers, mineral coloring pigments, and inert fillers bound together; used on rigid subfloors or hardwood floors. { 'a,sfòlt 'tīl }

asphaltum [MATER] Bituminous material in oil of turpentine; used in photomechanical work because of its ability to be rendered insoluble in light. { a'sfòl·təm }

aspheric surface [OPTICS] A lens or mirror surface which is altered slightly from a spherical surface in order to reduce aberrations. { ā'sfir·ik 'sər·fəs }

asphradium [INV ZOO] An organ, believed to be a chemoreceptor, in mollusks. Also spelled osphradium. { a'sfräd·ē·əm }

asphyxia [MED] Suffocation due to oxygen deprivation, resulting in anoxia and carbon dioxide accumulation in the body. { a'sfik·sē·ə }

asphyxiation [MED] Suffocation caused by lowering the oxygen supply through the blood to bodily organs to a level that is incapable of supporting life. { as,fik·sē'ā·shən }

aspiculate [INV ZOO] Lacking spicules, referring to Porifera. { a'spik·yə·lət }

Aspidiotinae [INV ZOO] A subfamily of homopteran insects in the superfamily Coccoidea. { a,spid·ē'ä·tə,nē }

Aspidiphoridae [INV ZOO] An equivalent name for the Sphindidae. { a,spid·ə'fòr·ə,dē }

Aspidobothria [INV ZOO] An equivalent name for the Aspidogastrea. { ¦as·pə,dō'bäth·rē·ə }

Aspidobothroidea [INV ZOO] A group of trematodes accorded class rank by W. J. Hargis. { ¦as·pə,dō·bə'thròid·ē·ə }

Aspidobranchia [INV ZOO] An equivalent name for the Archaeogastropoda. { as·pə,dō'braŋk·ē·ə }

Aspidochirotacea [INV ZOO] A subclass of bilaterally symmetrical echinoderms in the class Holothuroidea characterized by tube feet and 10-30 shield-shaped tentacles. { ¦as·pə,dō,kī·rə'tās·ē·ə }

Aspidochirotida [INV ZOO] An order of holothurioid echinoderms in the subclass Aspidochirotacea characterized by respiratory trees and dorsal tube feet converted into tactile warts. { ¦as·pə,dō,kī'räd·ə·də }

Aspidocotylea [INV ZOO] An equivalent name for the Aspidogastrea. { ¦as·pə,dō,käd·ə'lē·ə }

Aspidodiadematidae [INV ZOO] A small family of deep-sea echinoderms in the order Diadematoida. { ¦as·pə,dō,dī·ə·də'mad·ə,dē }

Aspidogastrea [INV ZOO] An order of endoparasitic worms in the class Trematoda having strongly developed ventral holdfasts. { ¦as·pə,dō'gas·trē·ə }

Aspidogastridae [INV ZOO] A family of trematode worms in the order Aspidogastrea occurring as endoparasites of mollusks. { ¦as·pə,dō'gas·trə,de }

Aspidorhynchidae [PALEON] The single family of the Aspidorhynchiformes, an extinct order of holostean fishes. { ¦as·pə,dō'riŋ·kə,dē }

Aspidorhynchiformes [PALEON] A small, extinct order of specialized holostean fishes. { ¦as·pə,dō,riŋk·ə'fór,mēz }

aspidospermine [PHARM] $C_{22}H_{30}O_2N_2$ White to brownish-yellow crystals with a melting point of 132-136°C; soluble in water, alcohol, chloroform, and ether; used in medicine. { ¦as·pə,dō'spər,mēn }

aspidospondyly [VERT ZOO] The condition in which the vertebral centra and spines are separate. { ,as·pi·dō'spän·də·lē }

Aspinothoracida [PALEON] The equivalent name for Brachythoraci. { a,spīn·ō·thə'ras·əd·ə }

aspirating See dedusting. { 'as·pə,rād·iŋ }

aspirating burner [ENG] A burner in which combustion air at high velocity is drawn over an orifice, creating a negative static pressure and thereby sucking fuel into the stream of air; the mixture of air and fuel is conducted into a combustion chamber, where the fuel is burned in suspension. { 'as·pə,rād·iŋ 'bər·nər }

aspirating screen [MIN ENG] A vibrating screen from which light, liberated particles are removed by suction. { 'as·pə,rād·iŋ 'skrēn }

aspiration [MED] The removal of fluids from a cavity by suction. [MICROBIO] The use of suction to draw up a sample in a pipette. [SCI TECH] Act or the result of removing, carrying along, or drawing by suction. { ,as·pə'rā·shən }

aspiration condenser [NUCLEO] An ion-counter collecting element consisting of a cylindrical condenser which when charged produces a radial field that collects ions from the aspirated air. { ,as·pə'rā·shən kən,den·sər }

aspiration meteorograph [ENG] An instrument for the continuous recording of two or more meteorological parameters, with the ventilation being provided by a suction fan. { ,as·pə'rā·shən ,mēd·ē'òr·ə,graf }

ASPIDORHYNCHIFORMES

Aspidorhynchus acutirostris (Blainville), Upper Jurassic, Bavaria, length to 3 feet (90 centimeters).

aspiration psychrometer [ENG] A psychrometer in which the ventilation is provided by a suction fan. { ˌas·pə'rā·shən ˌsi'kräm·əd·ər }

aspiration thermograph [ENG] A thermograph in which ventilation is provided by a suction fan. { ˌas·pə'rā·shən 'thərm·ə‚graf }

aspirator [ENG] Any instrument or apparatus that utilizes a vacuum to draw up gases or granular materials. [MIN ENG] A device made of wire gauze, of cloth, or of a fibrous mass held between pieces of meshed material and used to cover the mouth and nose to keep dusts from entering the lungs. { 'as·pə‚rād·ər }

aspirin See acetylsalicylic acid. { 'as·prən }

aspite [GEOL] A cratered volcano with the base wide in relation to the height; for example, Mauna Loa. { 'as‚pīt }

asporogenic mutant [MICROBIO] A bacillus that is unable to form spores due to alterations at any of several gene loci. { ¦ā‚spȯr·ə¦jen·ik 'myüt·ənt }

asporogenous [BOT] Not producing spores, especially of certain yeasts. { ¦ā·spə'räj·ə·nəs }

Aspredinidae [VERT ZOO] A family of salt-water catfishes in the order Siluriformes found off the coast of South America. { ˌa·sprə'din·ə·dē }

ASROC See antisubmarine rocket. { 'as‚räk }

ass [VERT ZOO] Any of several perissodactyl mammals in the family Equidae belonging to the genus *Equus,* especially *E. hemionus* and *E. asinus.* { as }

assault [ORD] **1.** Final phase of an attack; closing with the enemy in hand-to-hand fighting. **2.** The landing of troops for attack on the enemy's beach defenses. **3.** The landing of parachute and glider elements on unsecured and unprepared drop zones and landing zones to attack and seize an airhead. **4.** A short, violent, but well-ordered attack against a local objective, such as a gun emplacement, fort, or machine gun nest. { ə'sȯlt }

assault aircraft [AERO ENG] Powered aircraft, including helicopters, which move assault troops and cargo into an objective area and which provide for their resupply. { ə'sȯlt 'er‚kraft }

assault boat [NAV ARCH] A small boat that can easily be transported on land; used for amphibious military attacks or to cross lakes and rivers in land warfare. { ə'sȯlt ‚bōt }

assault fire [ORD] **1.** Fire delivered by attacking troops as they close with an enemy to engage him at close range or in hand-to-hand fighting, usually delivered from the hip or the standing position at a sustained rate. Also known as advancing fire. **2.** In artillery, extremely accurate, short-range destruction fire at point targets. { ə'sȯlt ‚fīr }

assault gun [ORD] Any of various sizes and types of guns that are self-propelled or mounted on tanks and are used for direct fire from close range against point targets. { ə'sȯlt ‚gən }

assault-landing model [ORD] A special form of assault model designed specifically for planning amphibious landings. Also known as amphibious-assault landing model. { ə'sȯlt ¦land·iŋ ‚mäd·əl }

assault model [ORD] Vehicle designed to provide direct fire in combat. { ə'sȯlt ‚mäd·əl }

assay [ANALY CHEM] Qualitative or quantitative determination of the components of a material, as an ore or a drug. { 'a‚sā }

assay balance [ENG] A sensitive balance used in the assaying of gold, silver, and other precious metals. { 'a‚sā ‚bal·əns }

assay bar [MET] A bar of pure or nearly pure gold and silver; used by a government as a standard. { 'a‚sā ‚bär }

assay plan [MIN ENG] A mine map showing the assay, stope, width, and so forth of samples taken from positions marked. { 'a‚sā ‚plan }

assay pound [MIN ENG] A weight which varies from time to time but is sometimes 0.5 gram, and is used by assayers to proportionately represent a pound. { 'a‚sā ‚paúnd }

assay ton [MIN ENG] A unit of weight of ore equal to 29,167 milligrams; the number of milligrams of precious metal in this measure equals the number of troy ounces in a short ton. { 'a‚sā ‚tən }

assay value [MIN ENG] The amount of gold or silver as shown by assay of any given sample and represented by ounces per ton of ore. { 'a‚sā ‚val·yü }

assay walls [MIN ENG] The planes to which an ore body can be profitably mined, the limiting factor being the metal content of the country rock as determined from assays. { 'a‚sā ‚wȯlz }

assemblage [ARCHEO] All related cultural traits and artifacts associated with one archeological manifestation. [ECOL] A group of organisms sharing a common habitat by chance. [GEOL] **1.** A group of fossils that, appearing together, characterize a particular stratum. **2.** A group of minerals that compose a rock. [ORD] A collection of items designed to accomplish one general function and identified and issued as a single item. [PALEON] A group of fossils occurring together at one stratigraphic level. { ə'sem·blij }

assemblage zone [PALEON] A biostratigraphic unit defined and identified by a group of associated fossils rather than by a single index fossil. { ə'sem·blij ‚zōn }

assembled stone [MATER] A stone made of two or more gem materials, whether genuine or imitation. { ə'sem·bəld 'stōn }

assembler [COMPUT SCI] A program designed to convert symbolic instruction into a form suitable for execution on a computer. Also known as assembly program; assembly routine. { ə'sem·blər }

assembler directive [COMPUT SCI] A statement in an assembly-language program that gives instructions to the assembler and does not generate machine language. { ə'sem·blər di‚rek·tiv }

assembler language See assembly language. { ə'sem·blər ‚laŋ·gwij }

assembler program [COMPUT SCI] A program that is written in assembly language. { ə'sem·blər ‚prō·grəm }

assembling bolt [CIV ENG] A threaded bolt for holding together temporarily the several parts of a structure during riveting. { ə'sem·bliŋ ‚bōlt }

assembly [COMPUT SCI] The automatic translation into machine language of a computer program written in symbolic language. [MECH ENG] A unit containing the component parts of a mechanism, machine, or similar device. { ə'sem·blē }

assembly drawing [GRAPHICS] A working-type engineering drawing depicting a complete unit, usually included with detail drawings of all parts in a set of working drawings. { ə'sem·blē ‚drȯ·iŋ }

assembly language [COMPUT SCI] A symbolic, nonbinary format for instructions (human-readable version of machine language) that allows mnemonic names to be used for instructions and data; for example, the instruction to add the number 39321 to the contents of register D1 in the central processing unit might be written as ADD#39321, D1 in assembly language, as opposed to a string of 0's and 1's in machine language. { ə'sem·blē ‚laŋ·gwij }

assembly line [IND ENG] A mass-production arrangement whereby the work in process is progressively transferred from one operation to the next until the product is assembled. { ə'sem·blē ‚līn }

assembly-line balancing [IND ENG] Assigning numbers of operators or machines to each operation of an assembly line so as to meet the required production rate with a minimum of idle time. { ə'sem·blē ‚līn 'bal·əns·iŋ }

assembly list [COMPUT SCI] A printed list which is the by-product of an assembly procedure; it lists in logical instruction sequence all details of a routine, showing the coded and symbolic notation next to the actual notations established by the assembly procedure; this listing is highly useful in the debugging of a routine. { ə'sem·blē ‚list }

assembly machine [MECH ENG] A machine in a manufacturing facility that produces a configuration of some practical value from discrete components. { ə'sem·blē mə‚shēn }

assembly method [IND ENG] The technique used to assemble a manufactured product, such as hand assembly, progressive line assembly, and automatic assembly. { ə'sem·blē ‚meth·əd }

assembly program See assembler. { ə'sem·blē 'prō·grəm }

assembly robot [COMPUT SCI] A robot that positions, mates, fits, and assembles components or parts and adjusts the finished product to function as intended. { ə'sem·blē ‚rō·bät }

assembly routine See assembler. { ə'sem·blē rü'tēn }

assembly system [COMPUT SCI] An automatic programming software system with a programming language and machine-language programs that aid the programmer by performing different functions such as checkout and updating. { ə'sem·blē ,sis·təm }

assembly time [ENG] **1.** The elapsed time after the application of an adhesive until its strength becomes effective. **2.** The time elapsed in performing an assembly or subassembly operation. { ə'sem·blē ,tīm }

assembly unit [COMPUT SCI] **1.** A device which performs the function of associating and joining several parts or piecing together a program. **2.** A portion of a program which is capable of being assembled into a larger program. { ə'sem·blē ,yü·nət }

assessment drilling [MIN ENG] Drilling to fulfill the requirement that a prescribed amount of work be done annually on an unpatented mining claim to retain title. Also known as annual labor. { ə'ses·mənt ,dril·iŋ }

assessment work [MIN ENG] Annual work at an unpatented mining claim in the public domain performed under law to maintain the claim title. { ə'ses·mənt ,wərk }

assets [IND ENG] All the resources, rights, and property owned by a person or a company; the book value of these items as shown on the balance sheet. { 'a,sets }

assign [COMPUT SCI] A control statement in FORTRAN which assigns a computed value i to a variable k, the latter representing the number of the statement to which control is then transferred. { ə'sīn }

assignable cause [IND ENG] Any identifiable factor which causes variation in a process outside the predicted limits, thereby altering quality. { ə'sīn·ə·bəl 'kóz }

assignment problem [COMPUT SCI] A special case of the transportation problem in a linear program, in which the number of sources (assignees) equals the number of designations (assignments) and each supply and each demand equals 1. { ə'sīn·mənt 'präb·ləm }

assignment statement [COMPUT SCI] A statement in a computer program that assigns a value to a variable. { ə'sīn·mənt ,stāt·mənt }

assili cotton [TEXT] A long-staple Egyptian cotton characterized by high tensile strength. { 'ä·sə·lē ,kät·ən }

assimilation [GEOL] Incorporation of solid or fluid material that was originally in the rock wall into a magma. [PHYSIO] The conversion of nutritive materials into protoplasm. { ə,sim·ə'lā·shən }

assimilative nitrate reduction [MICROBIO] The reduction of nitrates by some aerobic bacteria for purposes of assimilation. { ə,sim·ə'lād·iv 'nī,trāt ri,dək·shən }

assimilative sulfate reduction [MICROBIO] The reduction of sulfates by certain obligate anaerobic bacteria for purposes of assimilation. { ə,sim·ə'lād·iv 'səl,fāt ri,dək·shən }

assisted panel [COMPUT SCI] In an interactive system, a screen that explains a question the computer has asked, the available options, the expected format, and so forth. { ə'sis·təd 'pan·əl }

assisted takeoff [AERO ENG] A takeoff of an aircraft or a missile by using a supplementary source of power, usually rockets. { ə'sis·təd 'tāk,óf }

assize [CIV ENG] **1.** A cylindrical block of stone forming one unit in a column. **2.** A layer of stonework. { ə'sīz }

Assmann psychrometer [ENG] A special form of the aspiration psychrometer in which the thermometric elements are well shielded from radiation. { 'äs,män ,sī'kräm·əd·ər }

associate [PSYCH] An item or event that is linked to another in the mind of an individual. { ə'sō·sē,āt }

associate curve *See* Bertrand curve. { ə'sō·sē·ot ,kərv }

associated automatic movement *See* synkinesia. { ə'sō·sē,ād·əd |ód·ə'mad·ik 'müv·mənt }

associated corpuscular emission [GEOPHYS] The full complement of secondary charged particles associated with the passage of an x-ray or gamma-ray beam through air. { ə'sō·sē,ād·əd ,kór'pəs·kyə·lər i'mish·ən }

associated document [COMPUT SCI] A file that is linked to the application program in which it was created, so that the application can be started by choosing such a file. { ə,sō·sē,ād·əd 'däk·yə·mənt }

associated gas [PETRO ENG] Gaseous hydrocarbons occurring as a free-gas phase under original oil-reservoir conditions of temperature and pressure. Also known as gas-cap gas. { ə'sō·sē,ād·əd 'gas }

associated prime ideal [MATH] A prime ideal I in a commutative ring R is said to be associated with a module M over R if there exists an element x in M such that I is the annihilator of x. { ə'sō·sē,ād·əd 'prīm ,ī·dēl }

associated production [PARTIC PHYS] Production of strange particles invariably in twos, never one particle alone. { ə'sō·sē,ād·əd prə'dək·shən }

associated radii of convergence [MATH] For a power series in n variables, $z_1, \ldots, z_n$, any set of numbers, $r_1, \ldots, r_n$, such that the series converges when $|z_i| < r_i$, $i = 1, \ldots, n$, and diverges when $|z_i| > r_i$, $i = 1, \ldots, n$. { ə'sō·sē,ād·əd 'rād·dē,ī əv kən'vər·jəns }

associated tensor [MATH] A tensor obtained by taking the inner product of a given tensor with the metric tensor, or by performing a series of such operations. { ə'sō·sē,ād·əd 'ten·sər }

associate matrix *See* Hermitian conjugate. { ə'sō·sē·ət 'mā·triks }

associate operator *See* adjoint operator. { ə'sō·sē·ət 'äp·ə,rād·ər }

associates [MATH] Two elements x and y in a commutative ring with identity such that $x = ay$, where a is a unit. Also known as equivalent elements. { ə'sō·sē·ətz }

association [ASTRON] A sparsely populated grouping of very young stars that appear to have had a common origin and have not yet had time to disperse. [CHEM] Combination or correlation of substances or functions. [ECOL] Major segment of a biome formed by a climax community, such as an oak-hickory forest of the deciduous forest biome. [PSYCH] A connection formed through learning. { ə,sō·sē'ā·shən }

association area [PHYSIO] An area of the cerebral cortex that is thought to link and coordinate activities of the projection areas. { ə,sō·sē'ā·shən ,er·ē·ə }

association center [INV ZOO] In invertebrates, a nervous center coordinating and distributing stimuli from sensory receptors. { ə,sō·sē'ā·shən ,sen·tər }

association constant [BIOCHEM] A quantitative description of the affinity of a ligand for a protein that binds to it. { ə,sō·sē'ā·shən ,kän·stənt }

association fiber [NEURO] One of the white nerve fibers situated just beneath the cortical substance and connecting the adjacent cerebral gyri. { ə,sō·sē'ā·shən ,fi·bər }

association neuron [NEURO] A neuron, usually within the central nervous system, between sensory and motor neurons. { ə,sō·sē'ā·shən ,nü,rän }

association test [PSYCH] Any test designed to determine the nature of the mental or emotional link between a stimulus and a response. { ə,sō·sē'ā·shən ,test }

association trail [COMPUT SCI] A linkage between two or more documents or items of information, discerned during the process of their examination and recorded with the aid of an information retrieval system. { ə,sō·sē'ā·shən ,trāl }

associative algebra [MATH] An algebra in which the vector multiplication obeys the associative law. { ə'sō·sē,ād·iv 'al·jə·brə }

associative dimensioning system [COMPUT SCI] A system for making automatic changes in the dimensions of workpieces manufactured by machine tools. { ə'sō·sē,ād·iv di'men·shən·iŋ 'sis·təm }

associative facilitation [PSYCH] Ease in establishing a new association because of previous associations. { ə'sō·sē,ād·iv fə,sil·ə'tā·shən }

associative inhibition [PSYCH] Difficulty in establishing a new association because of previous associations. { ə'sō·sē,ād·iv ,in·ə'bish·ən }

associative key [COMPUT SCI] In a computer system with an associative memory, a field used to reference items through comparing the value of the field with corresponding fields in each memory cell and retrieving the contents of matching cells. { ə'sō·sē,ād·iv 'kē }

associative law [MATH] For a binary operation that is designated ∘, the relationship expressed by $a \circ (b \circ c) = (a \circ b) \circ c$. { ə'sō·sē,ād·iv 'ló }

associative learning [PSYCH] The principle that items experienced together are mentally linked so that they tend to reinforce one another. { ə'sō·sē,ād·iv 'lərn·iŋ }

associative memory [COMPUT SCI] A data-storage device in which a location is identified by its informational content rather than by names, addresses, or relative positions, and from which the data may be retrieved. Also known as associative storage. [PSYCH] Recalling a previously experienced item by thinking of something that is linked with it, thus invoking the association. { ə'sō·sē‚ād·iv 'mem·rē }

associative processor [COMPUT SCI] A digital computer that consists of a content-addressable memory and means for searching rapidly changing random digital data stored within, at speeds up to 1000 times faster than conventional digital computers. { ə'sō·sē‚ād·iv 'präs‚es·ər }

associative storage *See* associative memory. { ə'sō·sē‚ād·iv 'stór·ij }

associative thinking [PSYCH] **1.** The mental process of making associations between a given subject and all pertinent present factors without drawing on past experience. **2.** Free association. { ə'sō·sē‚ād·iv 'think·iŋ }

associator [COMPUT SCI] A device for bringing like entities into conjunction or juxtaposition. { ə'sō·sē‚ād·ər }

assortative mating [GEN] Nonrandom mating with respect to phenotypes. { ə'sòrd·əd·iv ‚mād·iŋ }

as-spun [TEXT] Pertaining to a spun fiber that has not undergone further processing. { ‚az 'spən }

assumed decimal point [COMPUT SCI] For a decimal number stored in a computer or appearing on a printout, a position in the number at which place values change from positive to negative powers of 10, but to which no location is assigned or at which no printed character appears, as opposed to an actual decimal point. Also known as virtual decimal point. { ə'sümd 'des·məl ‚póint }

assumed plane coordinates [ENG] A local plane-coordinate system set up at the convenience of the surveyor. { ə'sümd ‚plān ‚kō'órd·nəts }

assumed position [NAV] In celestial navigation, a point on the surface of the earth at which a craft is assumed to be located and for which the computed altitude is determined in the solution of a celestial observation. Also known as chosen position. { ə'sümd pə'zi·shən }

assured mineral *See* developed reserves. { ə'shúrd 'min·rəl }

assyntite [PETR] A plutonic rock consisting largely of orthoclase and pyroxene, lesser amounts of sodalite and nepheline, and accessory biotite, sphene, apatite, and opaque oxides. { ə'sin‚tīt }

astable circuit [ELECTR] A circuit that alternates automatically and continuously between two unstable states at a frequency dependent on circuit constants; for example, a blocking oscillator. { 'ā'stā·bəl 'sər·kət }

astable multivibrator [ELECTR] A multivibrator in which each active device alternately conducts and is cut off for intervals of time determined by circuit constants, without use of external triggers. Also known as free-running multivibrator. { ā'stā·bəl ‚məlt·i'vī‚brād·ər }

Astacidae [INV ZOO] A family of fresh-water crayfishes belonging to the section Macrura in the order Decapoda, occurring in the temperate regions of the Northern Hemisphere. { ‚as·tə'sī‚dē }

astacin [BIOCHEM] $C_{40}H_{48}O_4$ A red carotenoid ketone pigment found in crustaceans, as in the shell of a boiled lobster. { 'as·tə·sən }

Astacinae [INV ZOO] A subfamily of crayfishes in the family Astacidae including all North American species west of the Rocky Mountains. { ‚as·tə'sī‚nē }

A stage [ORG CHEM] An early stage in a thermosetting resin reaction characterized by linear structure, solubility, and fusibility of the material. { 'ā ‚stāj }

A star *See* A-type star. { 'ā ‚stär }

Astartian *See* Sequanian. { ə'stär·shən }

astasia [MED] Lack of muscular coordination in standing. { ə'stāzh·ə }

astatic [PHYS] Without orientation or directional characteristics; having no tendency to change position. { ā'stad·ik }

astatic coils [ELECTROMAG] Two identical coils, connected in series and suspended from the same axis, so that a uniform, external magnetic field exerts no net torque on the system. { ā'stad·ik 'kóilz }

astatic galvanometer [ENG] A sensitive galvanometer designed to be independent of the earth's magnetic field. { ā'stad·ik ‚gal·və'näm·əd·ər }

astatic governor *See* isochronous governor. { ā'stad·ik gəv·ə·nər }

astatic gravimeter [ENG] A sensitive gravimeter designed

to measure small changes in gravity. { ā'stad·ik grə'vim·əd·ər }

astatic magnetometer [ENG] A magnetometer for determining the gradient of a magnetic field by measuring the difference in reading from two magnetometers placed at different positions. { ā'stad·ik ‚mag·nə'täm·əd·ər }

astatic pair [ELECTROMAG] A pair of parallel magnets, equal in strength and having polarities in opposite directions, and perpendicular to an axis which bisects both of them; there is no net force or torque on the pair in a uniform field. { ā'stad·ik 'per }

astatic pendulum [PHYS] A pendulum which almost never takes a position of equilibrium. { ā'stad·ik 'pen·jə·ləm }

astatic system [ELECTROMAG] A system of magnets arranged so that the net force and torque exerted on the system by a uniform magnetic field equals 0. { ā¦stad·ik ¦sis·təm }

astatic wattmeter [ENG] An electrodynamic wattmeter designed to be insensitive to uniform external magnetic fields. { ā'stad·ik 'wät‚mēd·ər }

astatine [CHEM] A radioactive chemical element, symbol At, atomic number 85, the heaviest of the halogen elements. { 'as·tə‚tēn }

A station [NAV] In loran, the designation applied to one transmitting station of a pair, the signal of which always occurs less than half a repetition period after the preceding signal and more than half a repetition period before the succeeding signal of the other station, designated a B station. { 'ā ¦stā·shən }

astatized gravimeter [ENG] A gravimeter, sometimes referred to as unstable, where the force of gravity is maintained in an unstable equilibrium with the restoring force. { 'as·tə‚tīzd grə'vim·əd·ər }

astaxanthin [BIOCHEM] $C_{40}H_{52}O_4$ A violet carotenoid pigment found in combined form in certain crustacean shells and bird feathers. { ¦as·tə'zan·thən }

Asteidae [INV ZOO] A small, obscure family of cyclorrhaphous myodarian dipteran insects in the subsection Acalypteratae. { ‚as·tē'ī‚dē }

astel [MIN ENG] An overhead boarding or arching in a mine gallery. { 'as·təl }

astelic [BOT] Lacking a stele or having a discontinuous arrangement of vascular bundles. { ā'stēl·ik }

aster [BOT] Any of the herbaceous ornamental plants of the genus *Aster* belonging to the family Compositae. [CYTOL] The star-shaped structure that encloses the centrosome at the end of the spindle during mitosis. { 'as·tər }

Asteraceae [BOT] An equivalent name for the Compositae. { ‚as·tə'rās·ē‚ē }

Asterales [BOT] An order of dicotyledonous plants in the subclass Asteridae, including aster, sunflower, zinnia, lettuce, artichoke, and dandelion; the ovary is inferior, flowers are borne in involucrate, centripetally flowering heads, and the calyx, when present, is modified into a set of scale-, hair-, or bristlelike structures called the pappus. { ‚as·tə'rāl·ēz }

astereognosis [MED] Loss of recognition of objects by touch, although recognition occurs through another sense, usually vision. Also known as tactile agnosia. { ā‚ste·rē·äg'nō·səs }

asteria [LAP] A gemstone that displays a star or rayed figure when cut in the cabochon style in the proper crystallographic plane. { ə'stir·ē·ə }

Asteridae [BOT] A large subclass of dicotyledonous plants in the class Magnoliopsida; plants are sympetalous, with unitegmic, tenuinucellate ovules and with the stamens usually as many as, or fewer than, the corolla lobes and alternate with them. { ‚as·tə'rī‚dē }

Asteriidae [INV ZOO] A large family of echinoderms in the order Forcipulatida, including many predatory sea stars. { ‚as·tə'ri‚ə‚dē }

Asterinidae [INV ZOO] The starlets, a family of echinoderms in the order Spinulosida. { ‚as·tə'rin·ə‚dē }

asterism [ASTRON] A constellation or small group of stars. [OPTICS] A starlike optical phenomenon seen in gemstones called star stones; due to reflection of light by lustrous inclusions reduced to sharp lines of light by a domed cabochon style of cutting. [SPECT] A star-shaped pattern sometimes seen in x-ray spectrophotographs. { 'as·tə‚riz·əm }

astern [ENG] To the rear of an aircraft, vehicle, or vessel; behind; from the back. { ə'stərn }

ASTER

New England aster (*Aster novae-angliae*), showing ray and disk flowers. (*Courtesy of Alvin E. Staffan, from National Audubon Society*)

asternal [ANAT] **1.** Not attached to the sternum. **2.** Without a sternum. { ā'stərn·əl }

asteroid [ASTRON] One of the many small celestial bodies revolving around the sun, most of the orbits being between those of Mars and Jupiter. Also known as minor planet; planetoid. { 'as·tə,ròid }

asteroid belt [ASTRON] The region between 2.1 and 3.5 astronomical units from the sun where most of the asteroids are found. { 'as·tə,ròid ,belt }

Asteroidea [INV ZOO] The starfishes, a subclass of echinoderms in the subphylum Asterozoa characterized by five radial arms. { ,as·tə'ròid·ē·ə }

Asteroschematidae [INV ZOO] A family of ophiuroid echinoderms in the order Phrynophiurida with individuals having a small disk and stout arms. { ,as·tə·rō,skē'mad·ə,dē }

Asterozoa [INV ZOO] A subphylum of echinoderms characterized by a star-shaped body and radially divergent axes of symmetry. { ,as·tə·rə'zō·ə }

aster wilt [PL PATH] A fungus disease of asters caused by *Fusarium oxysporum* f. *callistephi*. { 'as·tər ,wilt }

aster yellows [PL PATH] A widespread virus disease of asters, other ornamental plants, and many vegetables, characterized by yellowing and dwarfing; leafhoppers are vectors. { 'as·tər 'yel,ōz }

asthenia [MED] Loss or lack of strength. { as'thē·nē·ə }

asthenolith [GEOL] A body of magma locally melted at any time within any solid portion of the earth. { as'then·ə,lith }

asthenopia [MED] Weakness of the eye muscles or of visual acuity, sometimes accompanied by pain and headache. { as·thə'nōp·ē·ə }

asthenosphere [GEOL] That portion of the upper mantle beneath the rigid lithosphere which is plastic enough for rock flowage to occur; extends from a depth of 30–60 miles (50–100 kilometers) to about 240 miles (400 kilometers) and is seismically equivalent to the low velocity zone. { as'then·ə,sfir }

asthma [MED] A pulmonary disease marked by labored breathing, wheezing, and coughing; cause may be emotional stress, chemical irritation, or exposure to an allergen. { 'az·mə }

Astian [GEOL] A European stage of geologic time: upper Pliocene, above Plaisancian, below the Pleistocene stage known as Villafranchian, Calabrian, or Günz. { 'as·tē·ən }

Asticcacaulis [MICROBIO] A genus of prosthecate bacteria; cells are rod-shaped with an appendage (pseudostalk), and reproduction is by binary fission of cells. { ə,stik·ə'kól·əs }

astichous [BOT] Not arranged in rows. { 'as·tə·kəs }

astigmat *See* astigmatic lens. { ə'stig,mat }

astigmatic difference [OPTICS] **1.** The distance between the primary and secondary foci of an astigmatic optical system. **2.** The difference between the reciprocals of the distances of the primary and secondary foci from an astigmatic thin lens or mirror. { ¦a·stig,mad·ik 'dif·rəns }

astigmatic foci [OPTICS] The two lines on which rays emanating from a point are focused by an astigmatic optical system. Also known as focal lines. { ¦a·stig,mad·ik 'fō,sī }

astigmatic interval [OPTICS] The portion of a pencil of rays in an astigmatic optical system that lies between the primary and secondary foci. Also known as conoid of Sturm; interval of Sturm. { ¦a·stig,mad·ik 'in·tər·vəl }

astigmatic lens [OPTICS] A planocylindrical, spherocylindrical, or spherotoric lens used in eyeglasses to correct astigmatism. Also known as astigmat. { ¦as·tig¦mad·ik 'lenz }

astigmatic mounting [SPECT] A mounting designed to minimize the astigmatism of a concave diffraction grating. { ¦a·stig,mad·ik 'mòunt·iŋ }

astigmatic surfaces [OPTICS] Two surfaces containing the astigmatic foci of points in a plane perpendicular to the optical axis of an astigmatic system. { ¦a·stig,mad·ik 'sər·fəs·əs }

astigmatism [ELECTR] In an electron-beam tube, a focus defect in which electrons in different axial planes come to focus at different points. [MED] A defect of vision due to irregular curvatures of the refractive surfaces of the eye so that focal points of light are distorted. [OPTICS] The failure of an optical system, such as a lens or a mirror, to image a point as a single point; the system images the point on two line segments separated by an interval. { ə'stig·mə,tiz·əm }

astigmatizer [OPTICS] A device, as attached to a rangefinder, for drawing out a point of light into a line or band. { ə'stig·mə,tīz·ər }

astigmometer [OPTICS] An instrument which measures the amount of astigmatism in an optical system. { ,as·tig'mäm·əd·ər }

astogeny [INV ZOO] Morphological and size changes associated with zooids of aging colonial animals. { a'stäj·ə·nē }

astomatal [BOT] Lacking stomata. Also known as astomous. { ā'stōm·əd·əl }

Astomatida [INV ZOO] An order of mouthless protozoans in the subclass Holotrichia; all species are invertebrate parasites, typically in oligochaete annelids. { as·tō'mad·ə·də }

astomatous [INV ZOO] Lacking a mouth, especially a cytostome, as in certain ciliates. { ā'stäm·əd·əs }

astomocnidae nematocyst [INV ZOO] A stinging cell whose thread has a closed end and either is adhesive or acts as a lasso to entangle prey. { ,as·tə'mäk·nə,dē ni'mad·ə,sist }

astomous [BOT] **1.** Having a capsule that bursts irregularly and is not dehiscent by an operculum. **2.** *See* astomatal. { 'as·tə·məs }

Aston dark space [ELECTR] A dark region in a glow-discharge tube which extends for a few millimeters from the cathode up to the cathode glow. { 'as·tən ¦därk ,spās }

Aston process [MET] A process for making controlled-quality wrought iron synthetically. { 'as·tən 'präs·əs }

Aston whole-number rule [PHYS] The rule which states that when expressed in atomic weight units, the atomic weights of isotopes are very nearly whole numbers, and the deviations found in samples of elements are due to the presence of several isotopes with different weights. { 'as·tən ,hōl 'nəm·bər ,rül }

astraeid [INV ZOO] Of a group of corals that are imperforate. { a'strē·əd }

astragal [BUILD] **1.** A small convex molding decorated with a string of beads or bead-and-reel shapes. **2.** A plain bead molding. **3.** A member, or combination of members, fixed to one of a pair of doors or casement windows to cover the joint between the meeting stiles and to close the clearance gap. { 'as·trə·gəl }

astragal front [DES ENG] A lock front which is shaped to fit the edge of a door with an astragal molding. { 'as·trə·gəl ¦frənt }

astragalus [ANAT] The bone of the ankle which articulates with the bones of the leg. Also known as talus. { ə'strag·ə·ləs }

astrakanite *See* bloedite. { 'as·trə·kə,nīt }

astral [ASTRON] Characteristic of a specific star or stars; stellar is the accepted term. { 'as·trəl }

astral dome *See* astrodome. { 'as·trəl ,dōm }

astral lamp [ENG] An Argand lamp designed so that its light is not prevented from reaching a table beneath it by the flattened annular reservoir holding the oil. { 'as·trəl ,lamp }

astraphobia [PSYCH] Abnormal fear of lightning and thunderstorms. Also known as astrapophobia. { ,as·trə'fōb·ē·ə }

astrapophobia *See* astraphobia. { ,as·trə·pō'fōb·ē·ə }

Astrapotheria [PALEON] A relatively small order of large, extinct South American mammals in the infraclass Eutheria. { ,as·trə·pə'thir·ē·ə }

Astrapotheroidea [PALEON] A suborder of extinct mammals in the order Astrapotheria, ranging from early Eocene to late Miocene. { ,as·trə·pə·thə'ròid·ē·ə }

Astrea [INV ZOO] A genus of mollusks in the class Gastropoda. { 'as·trē·ə }

astre fictif [ASTRON] Any of several fictitious stars assumed to move along the celestial equator at uniform rates corresponding to the speeds of the several harmonic constituents of the tide-producing force. { 'as·tər 'fik,tif }

astringent [MED] A substance applied to produce local contraction of blood vessels, to shrink mucous membranes, or to check discharges such as serum or mucus. { ə'strin·jənt }

astrionics [ELECTR] The science of adapting electricity to aerospace flight. { ,as·trē'än·iks }

astro- [ASTRON] A prefix meaning star or stars and, by extension, sometimes used as the equivalent of celestial, as in astronautics. { 'as·trō }

astroballistics [MECH] The study of phenomena arising out of the motion of a solid through a gas at speeds high enough to cause ablation; for example, the interaction of a meteoroid with the atmosphere. { ,as·trō·bə'lis·tiks }

astrobiology [BIOL] An approach to the scientific study of the living universe which seeks to understand the origin and

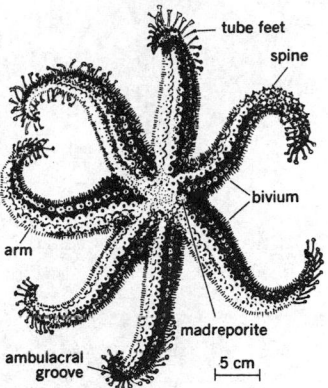

ASTEROIDEA

A representative asteroid, *Astrostole scabra*. Although five arms are common, species of this group may have 6–12 arms.

ASTRAPOTHERIA

Astrapotherium magnum skeleton.

evolution of life on earth, to determine if life exists elsewhere in the universe, and to predict the future of life on earth and in the rest of the universe. { ¦as·trō·bī'äl·ə·jē }

astrobleme [GEOL] A circular-shaped depression on the earth's surface produced by the impact of a cosmic body. { 'as·trō,blēm }

astrochanite *See* bloedite. { ə'sträk·ə,nīt }

astrochemistry [ASTRON] The science that applies the principles of chemistry to matter in space. { ¦as·trō'kem·ə·strē }

astrochronology [ASTRON] The use of stellar phenomena in chronology. { ¦as·trō·krə'näl·ə·jē }

astrocompass [NAV] A direction-determining instrument into which can be set the coordinates of any celestial body and the latitude of the observer and which will then give an indication of azimuth, true north, and heading. { 'as·trō,käm·pəs }

astrocyte [NEURO] A star-shaped cell; specifically, a neuroglial cell. { 'as·trə,sīt }

astrocytoma [MED] A slow-growing glial tumor made up of cells resembling astrocytes; often it will undergo malignant change and assume the appearance and growth characteristics of a glioblastoma. { ,as·trə,sī'tōm·ə }

astrodome [AERO ENG] A transparent dome in the fuselage or body of an aircraft or spacecraft intended primarily to permit taking celestial observations in navigating. Also known as astral dome; navigation dome. { 'as·trō,dōm }

astrodynamics [AERO ENG] The practical application of celestial mechanics, astroballistics, propulsion theory, and allied fields to the problem of planning and directing the trajectories of space vehicles. [ASTROPHYS] The dynamics of celestial objects. { ,as·trō·dī'nam·iks }

astrogation *See* astronavigation. { ,as·trə'gā·shən }

astrogeodetic [GEOD] Pertaining to direct measurements of the earth. { ,as·trō,jē·ə'ded·ik }

astrogeodetic datum orientation [GEOD] Adjustment of the ellipsoid of reference for a particular datum so that the sum of the squares of deflections of the vertical at selected points throughout the geodetic network is made as small as possible. { ,as·trō,jē·ə'ded·ik 'dad·əm ,ór·ē·ən'tā·shən }

astrogeodetic deflection [GEOD] The angle at a point between the normal to the geoid and the normal to the ellipsoid of an astrogeodetically oriented datum. Also known as relative deflection. { ,as·trō,jē·ə'ded·ik di'flek·shən }

astrogeodetic leveling [GEOD] A concept whereby the astrogeodetic deflections of the vertical are used to determine the separation of the ellipsoid and the geoid in studying the figure of the earth. Also known as astronomical leveling. { ,as·trō·,jē·ə'ded·ik 'lev·əl·iŋ }

astrogeodetic undulations [GEOD] Separations between the geoid and astrogeodetic ellipsoid. { ,as·trō,jē·ə'ded·ik ,ən·jə'lā·shənz }

astrogeology [ASTRON] The science that applies the principles of geology, geochemistry, and geophysics to the moon and planets other than the earth. { ,as·trō,jē'äl·ə·jē }

astroglia [NEURO] Neuroglia composed of astrocytes. { ə'sträg·lē·ə }

astrograph [ASTRON] A telescope designed to be used exclusively for astronomical photography. [MAP] A device for projecting a set of precomputed altitude curves onto a chart or plotting sheet, the curves moving with time such that if they are properly adjusted, they will remain in the correct position on the chart or plotting sheet; used in mapping the heavens. { 'as·trō,graf }

astrographic position *See* astrometric position. { ¦as·trō¦graf·ik pə'zish·ən }

astrograph mean time [ASTRON] A form of mean time, used in setting an astrograph; mean-time setting of 1200 occurs when the local hour angle of Aries is 0°. { 'as·trō,graf ¦mēn 'tīm }

astrogravimetric leveling [GEOD] A concept whereby a gravimetric map is used for the interpolation of the astrogeodetic deflections of the vertical to determine the separation of the ellipsoid and the geoid in studying the figure of the earth. { ,as·trō,grav·ə¦me·trik 'lev·əl·iŋ }

astrogravimetric points [GEOD] Astronomical positions corrected for the deflection of the vertical by gravimetric methods. { ,as·trō,grav·ə¦me·trik 'póins }

astroid [MATH] A hypocycloid for which the diameter of the fixed circle is four times the diameter of the rolling circle. { 'a,stróid }

astrolabe [ENG] An instrument designed to observe the positions and measure the altitudes of celestial bodies. { 'as·trə,lāb }

astrometric binary star [ASTRON] A binary star that may be distinguished from a single star only from the variable proper motion of one of its components. { ¦as·trə¦me·trik 'bī,ner·ē 'stär }

astrometric position [ASTRON] The position of a heavenly body or space vehicle on the celestial sphere corrected for aberration but not for planetary aberration. Also known as astrographic position. { ¦as·trə¦me·trik pə'zish·ən }

astrometry [ASTRON] The branch of astronomy dealing with the geometrical relations of the celestial bodies and their real and apparent motions. { ə'sträm·ə·trē }

astron [NUC PHYS] A proposed thermonuclear device in which a deuterium plasma is confined by an axial magnetic field produced by a shell of relativistic electrons. { 'a,strän }

Astron accelerator [NUCLEO] A linear induction accelerator in which ferromagnetic induction cores are used to generate the accelerating field. { 'a,strän ak'sel·ə,rād·ər }

astronaut [AERO ENG] In United States terminology, a person who rides in a space vehicle. { 'as·trə,nót }

astronautical enginering [AERO ENG] The engineering aspects of flight in space. { ¦as·trə¦nód·ə·kəl ,en·jə'nir·iŋ }

astronautics [AERO ENG] **1.** The art, skill, or activity of operating spacecraft. **2.** The science of space flight. { ,as·trə'nód·iks }

astronavigation [NAV] The plotting and directing of the movement of a vehicle from within by means of observations on celestial bodies. Also known as astrogation; celestial navigation. { ¦as·trō,nav·ə'gā·shən }

astronomic *See* astronomical. { ,as·trə'näm·ik }

astronomical [ASTRON] Of or pertaining to astronomy or to observations of the celestial bodies. Also known as astronomic. { ,as·trə'näm·ə·kəl }

astronomical almanac [ASTRON] A publication giving the tables of coordinates of a number of celestial bodies at a number of specific times during a given period. { ,as·trə'näm·ə·kəl 'ól·mə,nak }

astronomical atlas [ASTRON] A set of maps of celestial phenomena, often developed in conjunction with an astronomical catalog, and providing a clear picture of the spatial relations between the phenomena. { ,as·trə'näm·ə·kəl 'at·ləs }

astronomical azimuth [GEOD] The angle between the astronomical meridian plane of the observer and the plane containing the observed point and the true normal (vertical) of the observer, measured in the plane of the horizon, preferably clockwise from north. { ,as·trə'näm·ə·kəl 'az·ə·məth }

astronomical bearing *See* true bearing. { ,as·trə'näm·ə·kəl 'ber·iŋ }

astronomical camera [OPTICS] A camera designed to record either point sources (stars), extended sources (nebulae, galaxies, planets, or the sun and moon), or the spectra of celestial bodies. { ,as·trə'näm·ə·kəl 'kam·rə }

astronomical catalogue [ASTRON] A list or enumeration of astronomical data, generally ordered by increasing right ascension of the objects listed. { ,as·trə'näm·ə·kəl 'kad·əl,äg }

astronomical clock [HOROL] A clock that shows sidereal time. { ,as·trə'näm·ə·kəl 'kläk }

astronomical constants [ASTROPHYS] The elements of the orbits of the bodies of the solar system, their masses relative to the sun, their size, shape, orientation, rotation, and inner constitution, and the velocity of light. { ,as·trə'näm·ə·kəl 'kän·stəns }

astronomical coordinate system [ASTRON] Any system of spherical coordinates serving to locate astronomical objects on the celestial sphere. { ,as·trə'näm·ə·kəl ,kō'órd·ə·nət ,sis·təm }

astronomical date [ASTRON] Designation of epoch by year, month, day, and decimal fraction. { ,as·trə'näm·ə·kəl 'dāt }

astronomical day [ASTRON] A mean solar day beginning at mean noon, 12 hours later than the beginning of the civil day of the same date; astronomers now generally use the civil day. { ,as·trə'näm·ə·kəl 'dā }

astronomical distance [ASTRON] The distance of a celestial body expressed in units such as the light-year, astronomical unit, and parsec. { ,as·trə'näm·ə·kəl 'dis·təns }

astronomical eclipse *See* eclipse. { ‚as·trə'näm·ə·kəl i'klips }

astronomical ephemeris *See* ephemeris. { ‚as·trə'näm·ə·kəl i'fem·ə·rəs }

astronomical equator [GEOD] An imaginary line on the surface of the earth connecting points having 0° astronomical latitude. Also known as terrestrial equator. { ‚as·trə'näm·ə·kəl i'kwād·ər }

astronomical instruments [ENG] Specific kinds of telescopes and ancillary equipment used by astronomers to study the positions, motions, and composition of stars and members of the solar system. { ‚as·trə'näm·ə·kəl 'in·strə·məns }

astronomical latitude [GEOD] Angular distance between the direction of gravity (plumb line) and the plane of the celestial equator; applies only to positions on the earth and is reckoned from the astronomical equator. { ‚as·trə'näm·ə·kəl 'lad·ə‚tüd }

astronomical leveling *See* astrogeodetic leveling. { ‚as·trə'näm·ə·kəl 'lev·əl·iŋ }

astronomical longitude [GEOD] The angle between the plane of the reference meridian and the plane of the local celestial meridian. { ‚as·trə'näm·ə·kəl 'län·jə‚tüd }

astronomical meridian [GEOD] A line on the surface of the earth connecting points having the same astronomical longitude. Also known as terrestrial meridian. { ‚as·trə'näm·ə·kəl mə'rid·ē·ən }

astronomical meridian plane [GEOD] A plane that contains the vertical of the observer and is parallel to the instantaneous rotation axis of the earth. { ‚as·trə'näm·ə·kəl mə'rid·ē·ən ‚plān }

astronomical nutation [ASTRON] A small periodic motion of the celestial pole of celestial bodies, including the earth, with respect to the pole of the ecliptic. { ‚as·trə'näm·ə·kəl nü'tā·shən }

astronomical observatory [ASTRON] A building designed and equipped for making observations of astronomical phenomena. { ‚as·trə'näm·ə·kəl əb'zər·və‚tȯr·ē }

astronomical parallel [GEOD] A line connecting points having the same astronomical latitude. { ‚as·trə'näm·ə·kəl 'par·ə‚lel }

astronomical photography [OPTICS] The use of the photographic process to record surface features of celestial objects, their positions and motions (for measurement), and their radiation (photometry) and spectra (spectroscopy). Also known as astrophotography. { ‚as·trə'näm·ə·kəl fə'täg·rə·fē }

astronomical position [GEOD] **1.** A point on the earth whose coordinates have been determined as a result of observation of celestial bodies. Also known as astronomical station. **2.** A point on the earth defined in terms of astronomical latitude and longitude. { ‚as·trə'näm·ə·kəl pə'zish·ən }

astronomical refraction [GEOPHYS] The bending of a ray of celestial radiation as it passes through atmospheric layers of increasing density. { ‚as·trə'näm·ə·kəl ri'frak·shən }

astronomical scintillation [ASTROPHYS] Any scintillation phenomena, such as irregular oscillatory motion, variation of intensity, and color fluctuation, observed in the light emanating from an extraterrestrial source. Also known as stellar scintillation. { ‚as·trə'näm·ə·kəl sint·əl'ā·shən }

astronomical seeing *See* seeing. { ‚as·trə'näm·ə·kəl 'sē·iŋ }

astronomical spectrograph [SPECT] An instrument used to photograph spectra of stars. { ‚as·trə'näm·ə·kəl 'spek·trə‚graf }

astronomical spectroscopy [SPECT] The use of spectrographs in conjunction with telescopes to obtain observational data on the velocities and physical conditions of astronomical objects. { ‚as·trə'näm·ə·kəl ‚spek'träs·kə·pē }

astronomical station *See* astronomical position. { ‚as·trə'näm·ə·kəl 'stā·shən }

astronomical surveying [GEOD] The celestial determination of latitude and longitude; separations are calculated by computing distances corresponding to measured angular displacements along the reference spheroid. { ‚as·trə'näm·ə·kəl sər'vā·iŋ }

astronomical telescope [OPTICS] A telescope designed for viewing astronomical objects. { ‚as·trə'näm·ə·kəl 'tel·ə‚skōp }

astronomical theodolite *See* altazimuth. { ‚as·trə'näm·ə·kəl thē'äd·əl‚īt }

astronomical tide [OCEANOGR] An equilibrium tide due to attractions of the sun and moon. { ‚as·trə'näm·ə·kəl 'tīd }

astronomical time [ASTRON] Solar time in an astronomical day that begins at noon. { ‚as·trə'näm·ə·kəl 'tīm }

astronomical traverse [ENG] A survey traverse in which the geographic positions of the stations are obtained from astronomical observations, and lengths and azimuths of lines are obtained by computation. { ‚as·trə'näm·ə·kəl trə'vərs }

astronomical triangle [ASTRON] A spherical triangle on the celestial sphere. { ‚as·trə'näm·ə·kəl 'trī‚aŋ·gəl }

astronomical twilight [ASTRON] The period of incomplete darkness when the center of the sun is more than 6° but not more than 18° below the celestial horizon. { ‚as·trə'näm·ə·kəl 'twī‚līt }

astronomical unit [ASTRON] Abbreviated AU. **1.** A measure for distance within the solar system equal to the mean distance between earth and sun, that is, about 92,956,000 miles (149,598,000 kilometers). **2.** The semimajor axis of the elliptical orbit of earth. { ‚as·trə'näm·ə·kəl 'yü·nət }

astronomical year *See* tropical year. { ‚as·trə'näm·ə·kəl 'yir }

astronomy [SCI TECH] The science concerned with celestial bodies and the observation and interpretation of the radiation received in the vicinity of the earth from the component parts of the universe. { ə'strän·ə·mē }

Astropectinidae [INV ZOO] A family of echinoderms in the suborder Paxillosina occurring in all seas from tidal level downward. { ‚as·trō‚pek'tin·ə‚dē }

astrophobia [PSYCH] Abnormal fear of stars and celestial space. { ‚as·trō'fōb·ē·ə }

astrophotography *See* astronomical photography. { ‚as·trō·fə'täg·rə·fē }

astrophyllite [MINERAL] $(K,Na)_3(Fe,Mn)_7Ti_2Si_8O_{24}$-$(O,OH)_7$ A mineral composed of a basic silicate of potassium or sodium, iron or manganese, and titanium. { ‚as·trə'fī‚līt }

Astrophysics [ASTRON] A branch of astronomy that treats of the physical properties of celestial bodies, such as luminosity, size, mass, density, temperature, and chemical composition, and with their origin and evaluation. { ‚as·trō'fiz·iks }

astropyle [INV ZOO] A small, rounded projection from the central capsule of some radiolarians. { 'as·trō‚pīl }

astrosclereid [BOT] A type of sclereid cell that tends to be radiately branched but is otherwise quite variable, occurring in the leaves of many plants and in the petioles of *Camellia*. { ‚as·trə'skler·ē·əd }

astrosphere [CYTOL] The center of the aster exclusive of the rays. { 'as·trō‚sfir }

astrotracker [NAV] An automatic sextant which has the ability to sight on and continuously to track selected stars throughout the day and night, providing continuous heading and position data with no intervention on the part of the airman. Also known as star tracker. { 'as·trō‚trak·ər }

Asturian orogeny [GEOL] Mid-Upper Carboniferous diastrophism. { ə'stur·ē·ən ȯ'räj·ə·nē }

asty [INV ZOO] A bryozoan colony. { 'a‚stī }

astylar [ARCH] Not having columns or pilasters. { ¦ā'stīl·ər }

asulcal [BIOL] Without a sulcus. { ā'səl·kəl }

A supply [ELECTR] Battery, transformer filament winding, or other voltage source that supplies power for heating filaments of vacuum tubes. Also known as A power supply. { 'ā sə‚plī }

as-welded [MET] The condition of a weldment prior to any additional thermal, chemical, or mechanical treatment. { 'az‚weld·əd }

asymbolia [MED] An aphasia in which there is an inability to understand or use acquired symbols, such as speech, writing, or gestures, as a means of communication. { ‚ā‚sim'bōl·ē·ə }

asymmetrical bedding [GEOL] An order in which lithologic types or facies follow one another in a circuitous arrangement so that, for example, the sequence of types 1-2-3-1-2-3-1-2-3 indicates asymmetry (while the sequence 1-2-3-2-1-2-3-2-1 indicates symmetrical bedding). { ¦ā·sə¦me·tri·kəl 'bed·iŋ }

asymmetrical cell [ELECTR] A cell, such as a photoelectric cell, in which the impedance to the flow of current in one direction is greater than in the other direction. { ¦ā·sə¦me·tri·kəl 'sel }

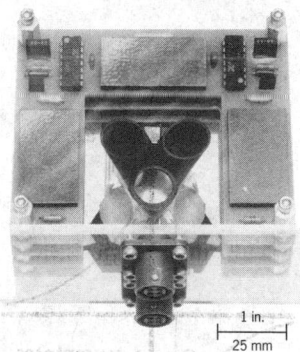

ASTROTRACKER

1 in.
├─────┤
25 mm

Strapdown astrotracker in housing. (*Northrop Grumman Corp.*)

asymmetrical conductivity [ELEC] A variation in the conductivity of a conductor over its cross section that is not symmetric about the conductor's central axis. { |ā·sə|me·tri·kəl ˌkän,dək'tiv·əd·ē }

asymmetrical deflection [ELECTR] A type of electrostatic deflection in which one deflector plate is maintained at a fixed potential and the deflecting voltage is supplied to the other plate. { |ā·sə|me·tri·kəl di'flek·shən }

asymmetrical fold [GEOL] A fold in which one limb dips more steeply than the other. { |ā·sə|me·tri·kəl 'fōld }

asymmetrical laccolith [GEOL] A laccolith in which the beds dip at conspicuously different angles in different sectors. { |ā·sə|me·tri·kəl 'lak·ə,lith }

asymmetrical modem [COMMUN] A modem that simultaneously transmits and receives data, but at different speeds. { ˌā·si|me·trə·kəl 'mō,dem }

asymmetrical ripple mark [GEOL] The normal form of ripple mark, with short downstream slopes and comparatively long, gentle upstream slopes. { |ā·sə|me·tri·kəl 'rip·əl ,märk }

asymmetrical-sideband transmission *See* vestigial-sideband transmission. { |ā·sə|me·tri·kəl 'sīd ,band ,tranz'mish·ən }

asymmetrical vein [GEOL] A crustified vein of geologic material with unlike layers on each side. { |ā·sə|me·tri·kəl 'vān }

asymmetric carbon atom [ORG CHEM] A carbon atom with four different atoms or groups of atoms bonded to it. Also known as chiral carbon atom; stereogenic center. { |ā·sə|me·trik |kär·bən 'ad·əm }

asymmetric cell division [CELL MOL] A phenomenon in which a mother cell divides into daughter cells that are unequal in size or cytoplasmic content, usually resulting in a different developmental fate for each. { |ā·sə,me·trik 'sel də,vizh·ən }

asymmetric digital subscriber line [COMMUN] A broadband communication technology designed for use on conventional telephone lines, which reserves more bandwidth for receiving data (1-8 megabits per second) than for sending data (100-800 kilobits per second). Abbreviated ADSL. { |ā·sə|me·trik 'dij·ə·dəl ,səb'skrī·bər ,līn }

asymmetric rotor [MECH ENG] A rotating element for which the axis (center of rotation) is not centered in the element. { |ā·sə|me·trik 'rōd·ər }

asymmetric synthesis [ORG CHEM] Chemical synthesis of a pure enantiomer, or of an enantiomorphic mixture in which one enantiomer predominates, without the use of resolution. { |ā·sə|me·trik 'sin·thə·səs }

asymmetric top [MECH] A system in which all three principal moments of inertia are different. { |ā·sə|me·trik 'täp }

asymmetry [PHYS CHEM] The geometrical design of a molecule, atom, or ion that cannot be divided into like portions by one or more hypothetical planes. Also known as molecular asymmetry. { |ā'sim·ə·trə }

asymmetry effect [PHYS CHEM] The asymmetrical distribution of the ion cloud around an ion that results from the finite relaxation time for the ion cloud when a voltage is applied; leads to a reduction in ion mobility. { ā'sim·ə·trē i,fekt }

asymptote [MATH] **1.** A line approached by a curve in the limit as the curve approaches infinity. **2.** The limit of the tangents to a curve as the point of contact approaches infinity. { 'as·əm,tōt }

asymptote of convergence *See* convergence line. { 'as·əm,tōt əv kən'vər·jəns }

asymptotically flat [RELAT] A space-time is asymptotically flat if it approaches Minkowski space-time at a prescribed rate at large spatial distances. { ˌā·sim|tät·ə·klē 'flat }

asymptotically simple [RELAT] A space-time is asymptotically simple if it satisfies certain mathematical requirements on the conformal structure of null infinity; these requirements are a definition of a type of asymptotic flatness. { ˌā·sim|tät·ə·klē 'sim·pəl }

asymptotic cone of acceptance [GEOPHYS] The solid angle in the celestial sphere from which particles have to come in order to contribute significantly to the counting rate of a given neutron monitor on the surface of the earth. { ā,sim'täd·ik |kōn əv ik'sep·təns }

asymptotic curve [MATH] A curve on a surface whose osculating plane at each point is the same as the tangent plane to the surface. { ā,sim'täd·ik 'kərv }

asymptotic direction of arrival [GEOPHYS] The direction at infinity of a positively charged particle, with given rigidity,

which impinges in a given direction at a given point on the surface of the earth, after passing through the geomagnetic field. { ā,sim'täd·ik di'rek·shən əv ə'rīv·əl }

asymptotic directions [MATH] For a hyperbolic point on a surface, the two directions in which the normal curvature vanishes; equivalently, the directions of the asymptotic curves passing through the point. { ā,sim'täd·ik də'rek·shənz }

asymptotic efficiency [STAT] The efficiency of an estimator within the limiting value as the size of the sample increases. { ā,sim'täd·ik ə'fish·ən·sē }

asymptotic expansion [MATH] A series of the form $a_0 + (a_1/x) + (a_2/x^2) + \cdots + (a_n/x_n) + \cdots$ is an asymptotic expansion of the function $f(x)$ if there exists a number N such that for all $n \neq N$ the quantity $x_n[f(x) - S_n(x)]$ approaches zero as x approaches infinity, where $S_n(x)$ is the sum of the first n terms in the series. Also known as asymptotic series. { ā,sim'täd·ik ik'span·shən }

asymptotic formula [MATH] A statement of equality between two functions which is not a true equality but which means the ratio of the two functions approaches 1 as the variable approaches some value, usually infinity. { ā,sim'täd·ik 'fȯr·myə·lə }

asymptotic freedom [PARTIC PHYS] In some gauge theories, the property of the strong interactions of growing steadily weaker at high energies. { ā,sim'täd·ik 'frēd·əm }

asymptotic giant branch [ASTRON] A grouping of stars on the Hertzsprung-Russell diagram that is roughly asymptotic to the giant branch; it represents a later stage in giant-star evolution in which hydrogen-fusing and helium-fusing shells surround a core in which both hydrogen fusion and helium fusion are exhausted. Abbreviated AGB. { ˌa,sim'täd·ik jī·ənt 'branch }

asymptotic series *See* asymptotic expansion. { ā,sim'täd·ik 'sir·ēz }

asymptotic stability [MATH] The property of a vector differential equation which satisfies the conditions that (1) whenever the magnitude of the initial condition is sufficiently small, small perturbations in the initial condition produce small perturbations in the solution; and (2) there is a domain of attraction such that whenever the initial condition belongs to this domain the solution approaches zero at large times. { ā,sim'täd·ik stə'bil·əd·ē }

asynapsis [CELL MOLEC] Absence of pairing of homologous chromosomes during meiosis. { ˌā·si'nap·səs }

asynchronous [COMPUT SCI] Operating at a speed determined by the circuit functions rather than by timing signals. [PHYS] Not synchronous. { ā'siŋ·krə·nəs }

asynchronous communications [COMMUN] The transmission and recognition of a single character at a time. { ā'siŋ·krə·nəs kə,myü·nə'kā·shənz }

asynchronous communication adaptor [COMPUT SCI] A device connected to a computer to allow it to carry out asynchronous communication over a telephone line. { ā'siŋ·krə·nəs kə,myü·nə'kā·shənz ə,dap·tər }

asynchronous computer [COMPUT SCI] A computer in which the performance of any operation starts as a result of a signal that the previous operation has been completed, rather than on a signal from a master clock. { ā'siŋ·krə·nəs kəm'pyüd·ər }

asynchronous control [CONT SYS] A method of control in which the time allotted for performing an operation depends on the time actually required for the operation, rather than on a predetermined fraction of a fixed machine cycle. { ā'siŋ·krə·nəs kən'trōl }

asynchronous data [COMPUT SCI] Information which is sampled at irregular intervals with respect to another operation. { ā'siŋ·krə·nəs 'dad·ə }

asynchronous device [CONT SYS] A device in which the speed of operation is not related to any frequency in the system to which it is connected. { ā'siŋ·krə·nəs di'vīs }

asynchronous digital subscriber loop *See* asymmetric digital subscriber line. { ˌā'siŋ·krə·nəs 'dij·ə·dəl ,səb'skrī·bər ,lüp }

asynchronous input/output [COMPUT SCI] The ability to receive input data while simultaneously outputting data. { ā'siŋ·krə·nəs 'in,pút 'aút,pút }

asynchronous inputs [ELECTR] The terminals in a flip-flop circuit which affect the output state of the flip-flop independently of the clock. { ā'siŋ·krə·nəs 'in,púts }

asynchronous logic [ELECTR] A logic network in which

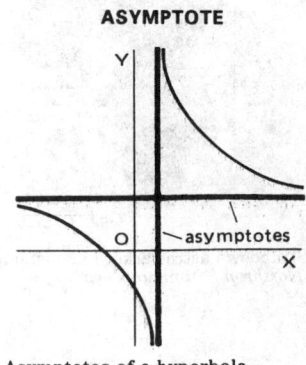

ASYMPTOTE

Asymptotes of a hyperbola.

the speed of operation depends only on the signal propagation through the network.   { ā'siŋ·krə·nəs 'läj·ik }

asynchronous machine [ELEC] An ac machine whose speed is not proportional to the frequency of the power line.   { ā'siŋ·krə·nəs mə'shēn }

asynchronous operation [ELECTR] An operation that is started by a completion signal from a previous operation, proceeds at the maximum speed of the circuits until finished, and then generates its own completion signal.   { ā'siŋ·krə·nəs ,äp·ə'rā·shən }

asynchronous tie [ELEC] An installation at which power is transmitted between two alternating-current power systems, operating at the same nominal frequency but with different frequency controls, by a direct-current link.   { ā'siŋ·krə·nəs 'tī }

asynchronous time-division multiplexing [COMMUN] A data-transmission technique in which several users utilize a single channel by means of a system which assigns time slots only to active channels.   { ā'siŋ·krə·nəs 'tīm də'vi·zhən 'məlt·i,pleks·iŋ }

asynchronous timing [IND ENG] A simulation method for queues in which the system model is updated at each arrival or departure, resulting in the master clock being increased by a variable amount.   { ā'siŋ·krə·nəs 'tīm·iŋ }

asynchronous transfer mode [COMMUN] A high-speed packet-switching technology based on cell-oriented switching and multiplexing that uses 53-byte packets to transfer different types of information, such as voice, video, and data, over the same communications network at different speeds.   Abbreviated ATM.   { ,a'siŋ·krə·nəs 'tranz·fər ,mōd }

asynchronous transmission [COMMUN] Data transmission in which each character contains its own start and stop pulses and there is no control over the time between characters.   { ā'siŋ·krə·nəs ,tranz'mish·ən }

asynchronous working [COMPUT SCI] The mode of operation of a computer in which an operation is performed only at the end of the preceding operation.   { ā'siŋ·krə·nəs 'wərk·iŋ }

asyndetic [COMPUT SCI] **1.** Omitting conjunctions or connectives.   **2.** Pertaining to a catalog without cross references.   { ¦as·ən¦ded·ik }

asynergia [MED] Faulty coordination of groups of organs or muscles normally acting in unison; particularly, the abnormal state of muscle antagonism in cerebellar disease.   { ‚ā·si'nərj·ē·ə }

asystole [MED] The absence of cardiac contraction; cardiac arrest.   { ā'sis·tə·lē }

at *See* technical atmosphere.

At *See* astatine.

ata [MECH] A unit of absolute pressure in the metric technical system equal to 1 technical atmosphere.   { 'a·tə }

ATA *See* aminotriazole.

atacamite [MINERAL] $Cu_2Cl(OH)_3$ Native, green hydrous-copper oxychloride crystallizing in the orthorhombic system.   { ,ad·ə'kam,īt }

atactic [ORG CHEM] Of the configuration for a polymer, having the opposite steric configurations for the carbon atoms of the polymer chain occur in equal frequency and more or less at random.   { ā'tak·tik }

atactostele [BOT] A type of monocotyledonous siphonostele in which the vascular bundles are dispersed irregularly throughout the center of the stem.   { ə'tak·tə,stēl }

atavism [EVOL] Appearance of a distant ancestral form of an organism or one of its parts due to reactivation of ancestral genes.   { 'ad·ə,viz·əm }

ataxia [MED] Lack of muscular coordination due to any of several nervous system diseases.   { ə'tak·sē·ə }

ataxic [GEOL] Pertaining to unstratified ore deposits.   { ə'tak·sik }

ataxiophobia [PSYCH] Abnormal fear of disorder.   { ə¦tak·sē·ə'fōb·ē·ə }

ataxite [GEOL] An iron meteorite that lacks the structure of either hexahedrite or octahedrite and contains more than 10% nickel.   [PETR] A taxitic rock whose components are arranged in a brecciaLike manner, that is, there is no specific arrangement.   { ə'tak,sīt }

ATC *See* air-traffic control.

ATCRBS *See* air-traffic control radar beacon system.

ATDM *See* asynchronous time-division multiplexing.

atectonic [GEOL] Of an event that occurs when orogeny is not taking place.   { ¦ā·tek'tän·ik }

atectonic pluton [GEOL] A pluton that is emplaced when orogeny is not occurring.   { ¦ā·tek'tän·ik 'plü,tän }

atelectasis [MED] **1.** Total or partial collapsed state of the lung.   **2.** Failure of the lung to expand at birth.   { ,ad·əl'ek·stə,səs }

Ateleopoidei [VERT ZOO] A family of oceanic fishes in the order Cetomimiformes characterized by an elongate body, lack of a dorsal fin, and an anal fin continuous with the caudal fin.   { ə¦tel·ē·ə'póid·ē,ī }

atelestite [MINERAL] $Bi_8(AsO_4)_3O_5(OH)_5$ A yellow mineral consisting of basic bismuth arsenate and occurring in minute crystals; specific gravity is 6.82.   { ,ad·əl'e,stīt }

ateliosis [MED] Infantilism or dwarfism characterized by general, but proportional, underdevelopment and normal intelligence; associated with anterior pituitary deficiencies.   { ə,tel·ē'ō·səs }

Atelopodidae [VERT ZOO] A family of small, brilliantly colored South and Central American frogs in the suborder Procoela.   { ə,tel·ə'päd·ə,dē }

Atelostomata [INV ZOO] A superorder of echinoderms in the subclass Euechinoidea characterized by a rigid, exocyclic test and lacking a lantern, or jaw, apparatus.   { ə,tel·ə'stöm·əd·ə }

Aten [ASTRON] The first asteroid found to have a period less than that of the earth, 346.93 days, with an orbital eccentricity of 0.19.   { 'ä,ten }

Aten asteroid [ASTRON] An asteroid whose period is less than that of the earth.   { 'ä,ten 'as·tə,róid }

Athalamida [INV ZOO] An order of naked amebas of the subclass Granuloreticulosia in which pseudopodia are branched and threadlike (reticulopodia).   { ,ath·ə'läm·əd·ə }

Athecanephria [INV ZOO] An order of tube-dwelling, tentaculate animals in the class Pogonophora characterized by a saclike anterior coelom.   { ,ath·ə·kə'nef·rē·ə }

athecate [INV ZOO] Lacking a theca.   { 'ath·ə,kāt }

Atherinidae [VERT ZOO] The silversides, a family of actinopterygian fishes of the order Atheriniformes.   { ,ath·ə'rin·ə,dē }

Atheriniformes [VERT ZOO] An order of actinopterygian fishes in the infraclass Teleostei, including flyingfishes, needlefishes, killifishes, silversides, and allied species.   { ,ath·ə,rin·ə'fòr,mēz }

athermalize [ENG] To make independent of temperature or of thermal effects.   { ¦ā'thər·mə,līz }

athermal transformation [PHYS] A chemical or physical change not requiring a change in the temperature of the substance, as in the formation of martensite.   { ¦ā'thər·məl ,tranz·fər'mā·shən }

athermancy [ELECTROMAG] Property of a substance which cannot transmit infrared radiation.   { ¦ā¦thər·mən·sē }

atheroma [MED] A lipid deposit in the inner wall of an artery; characteristic of atherosclerosis.   { ,ath·ə'rōm·ə }

atherosclerosis [MED] Deposition of lipid with proliferation of fibrous connective tissue cells in the inner walls of the arteries.   { ,ath·ə·rō,sklə'rō·səs }

athetosis [MED] Slow, recurrent, involuntary wormlike movements of various parts of the body associated with lesions of the basal ganglia.   { ,ath·ə'tō·səs }

athetotic speech [MED] Disorder of articulation rhythm involving a general jerkiness in speech production that interferes with the normal rate of speech; associated with athetosis.   { ¦ath·ə¦täd·ik 'spēch }

athey wagon [PETRO ENG] Equipment for fighting oil-field fires consisting of a track-mounted boom with a hook on the end that is used to remove debris, deliver explosives, or position a new wellhead.   { 'ath·ē ,wag·ən }

Athiorhodaceae [MICROBIO] Formerly the nonsulfur photosynthetic bacteria, a family of small, gram-negative, nonsporeforming, motile bacteria in the suborder Rhodobacteriineae.   { ā,thī·ə,rō'dās·ē,ē }

athlete's foot [MED] Dermatophytosis of the feet, usually affecting the skin between the toes.   { 'ath,lēts 'füt }

athodyd [AERO ENG] A type of jet engine, consisting essentially of a duct or tube of varying diameter and open at both ends, which admits air at one end, compresses it by the forward motion of the engine, adds heat to it by the combustion of fuel,

ATACTIC

Atactic type of spatially oriented polymers.

and discharges the resulting gases at the other end to produce thrust. { 'ath·ə‚did }

athrocyte [HISTOL] A cell that engulfs extraneous material and stores it as granules in the cytoplasm. { 'ath·rə‚sīt }

athrogenic [PETR] Of or pertaining to pyroclastics. { ¦ath·rə¦jen·ik }

athwartship [NAV ARCH] Perpendicular to the fore and aft centerline of a ship. { ə'thwȯrt‚ship }

Athyrididina [PALEON] A suborder of fossil articulate brachiopods in the order Spiriferida characterized by laterally or, more rarely, ventrally directed spires. { ‚ath·ə·rə'də'dī·nə }

Atlantacea [INV ZOO] A superfamily of mollusks in the subclass Prosobranchia. { ¦at‚lan¦tās·ē·ə }

Atlantic Ocean [GEOGR] The large body of water separating the continents of North and South America from Europe and Africa and extending from the Arctic Ocean to the continent of Antarctica. { ət'lan·tik 'ō·shən }

Atlantic salmon [INV ZOO] *Salmo salar.* A species of salmon that occurs throughout the North Atlantic Ocean and spawns in eastern North America and western Europe; it can complete more than one migratory cycle and thus can breed multiple times. { ət'lan·tik 'sa·mən }

Atlantic series [PETR] A great group of igneous rocks, based on tectonic setting, found in nonorogenic areas, often associated with block sinking and great crustal instability, and erupted along faults and fissures or through explosion vents. Also known as Atlantic suite. { ət'lan·tik 'sir·ēz }

Atlantic standard time *See* Atlantic time. { ət'lan·tik ¦stan·dərd ‚tīm }

Atlantic suite *See* Atlantic series. { ət'lan·tik 'swēt }

Atlantic time [ASTRON] A time zone; the fourth zone west of Greenwich. Also known as Atlantic standard time. { ət'lan·tik 'tīm }

Atlantic-type continental margin [GEOL] A continental margin typified by that of the Atlantic which is aseismic because oceanic and continental lithospheres are coupled. { ət'lan·tik ‚tīp ‚känt·ən'ent·əl 'mär·jən }

atlantite [PETR] An olivine-bearing nepheline tephrite. { ət'lan‚tīt }

atlas [ANAT] The first cervical vertebra. [MAP] A collection of charts or maps kept loose or bound in a volume. [MATH] An atlas for a manifold is a collection of coordinate patches that covers the manifold. { 'at·ləs }

Atlas [ASTRON] The innermost known satellite of Saturn, which orbits at a distance of 85 × 10³ miles (137 × 10³ kilometers), just outside the A ring, and has an irregular shape with an average diameter of 20 miles (30 kilometers). { 'at·ləs }

atlas grid [MAP] A reference system that permits the designation of the location of a point or an area on a map, photo, or other graphic in terms of numbers and letters. Also known as alphanumeric grid. { 'at·ləs ‚grid }

atm *See* atmosphere.

ATM *See* asynchronous transfer mode; automatic teller machine.

atmidometer *See* atmometer. { ‚at·mə'däm·əd· ər }

atmoclast [GEOL] A fragment of rock broken off in place by atmospheric weathering. { 'at·mə‚klast }

atmoclastic [PETR] Of a clastic rock, composed of atmoclasts that have been recemented without rearrangement. { ‚at·mə¦klas·tik }

atmogenic [GEOL] Of rocks, minerals, and other deposits derived directly from the atmosphere by condensation, wind action, or deposition from volcanic vapors; for example, snow. { ¦at·mə¦jen·ik }

atmolith [GEOL] A rock precipitated from the atmosphere, that is, an atmogenic rock. { 'at·mə‚lith }

atmolysis [FL MECH] The separation of gas mixtures by using their relative diffusibility through a porous partition. { ət'mäl·ə·səs }

atmo-meter *See* meter-atmosphere. { 'at·mō‚mēd·ər }

atmometer [ENG] The general name for an instrument which measures the evaporation rate of water into the atmosphere. Also known as atmidometer; evaporation gage; evaporimeter. { ət'mäm·əd·ər }

atmophile element [METEOROL] **1.** Any of the most typical elements of the atmosphere (hydrogen, carbon, nitrogen, oxygen, iodine, mercury, and inert gases). **2.** Any of the elements which either occur in the uncombined state or, as volatile compounds, concentrate in the gaseous primordial atmosphere. { 'at·mō‚fil 'el·ə·mənt }

atmosphere [MECH] A unit of pressure equal to 101.325 kilopascals, which is the air pressure measured at mean sea level. Abbreviated atm. Also known as standard atmosphere. [METEOROL] The gaseous envelope surrounding a planet or celestial body. { 'at·mə‚sfir }

atmospheric absorption [GEOPHYS] The reduction in energy of microwaves by gases and moisture in the atmosphere. { ¦at·mə¦sfir·ik əb'zȯrp·shən }

atmospheric acoustics [ACOUS] The science of sound waves in the open air. { at·mə¦sfir·ik ə'kü·stiks }

atmospheric attenuation [GEOPHYS] A process in which the flux density of a parallel beam of energy decreases with increasing distance from the source as a result of absorption or scattering by the atmosphere. { ¦at·mə¦sfir·ik ə‚ten·yə'wā·shən }

atmospheric boil *See* terrestrial scintillation. { ¦at·mə¦sfir·ik 'bȯil }

atmospheric boundary layer *See* surface boundary layer. { ¦at·mə¦sfir·ik 'baůn·drē ‚lā·ər }

atmospheric braking [AERO ENG] **1.** Slowing down an object entering the atmosphere of the earth or other planet from space by using the drag exerted by air or other gas particles in the atmosphere. **2.** The action of the drag so exerted. { ¦at·mə¦sfir·ik 'brāk·iŋ }

atmospheric cell [METEOROL] An air parcel that exhibits a specific type of motion within its boundaries, such as the vertical circular motion of the Hadley cell. { ¦at·mə‚sfir·ik 'sel }

atmospheric chemistry [METEOROL] The study of the production, transport, modification, and removal of atmospheric constituents in the troposphere and stratosphere. { ¦at·mə¦sfir·ik 'kem·ə·strē }

atmospheric composition [METEOROL] The chemical abundance in the earth's atmosphere of its constituents, including nitrogen, oxygen, argon, carbon dioxide, water vapor, ozone, neon, helium, krypton, methane, hydrogen, and nitrous oxide. { ¦at·mə¦sfir·ik ‚käm·pə'zish·ən }

atmospheric condensation [METEOROL] The transformation of water in the air from a vapor phase to dew, fog, or cloud. { ¦at·mə¦sfir·ik ‚kän·dən'sā·shən }

atmospheric control [ORD] **1.** Any device or system designed to operate movable aerodynamic control surfaces to direct a guided missile in an atmosphere dense enough for such controls to be effective. **2.** The control provided by such devices. { ¦at·mə¦sfir·ik kən'trōl }

atmospheric cooler [MECH ENG] A fluids cooler that utilizes the cooling effect of ambient air surrounding the hot, fluids-filled tubes. { ¦at·mə¦sfir·ik 'kül·ər }

atmospheric corrosion [MET] The gradual destruction or alteration of a metal or alloy by contact with substances present in the atmosphere, such as oxygen, carbon dioxide, water vapor, and sulfur and chlorine compounds. { ¦at·mə¦sfir·ik kə'rō·zhən }

atmospheric density [METEOROL] The ratio of the mass of a portion of the atmosphere to the volume it occupies. { ¦at·mə¦sfir·ik 'den·səd·ē }

atmospheric diffusion [METEOROL] The exchange of fluid parcels between regions in the atmosphere in the apparently random motions of a scale too small to be treated by equations of motion. { ¦at·mə¦sfir·ik di'fyü·zhən }

atmospheric distillation [CHEM ENG] Distillation operation conducted at atmospheric pressure, in contrast to vacuum distillation or pressure distillation. { ¦at·mə¦sfir·ik ‚dis·tə'lā·shən }

atmospheric disturbance [METEOROL] Any agitation or disruption of the atmospheric steady state. { ¦at·mə¦sfir·ik dis'tər·bəns }

atmospheric drag [FL MECH] A major perturbation of close artificial satellite orbits caused by the resistance of the atmosphere; the secular effects are decreasing eccentricity, semidiameter, and period. { ¦at·mə¦sfir·ik 'drag }

atmospheric duct [GEOPHYS] A stratum of the troposphere within which the refractive index varies so as to confine within the limits of the stratum the propagation of an abnormally large proportion of any radiation of sufficiently high frequency, as in a mirage. { ¦at·mə¦sfir·ik 'dəkt }

atmospheric electric field [GEOPHYS] The atmosphere's

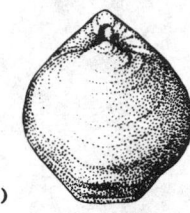

ATHYRIDIDINA

(a)

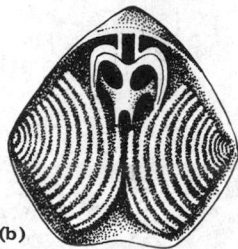

(b)

Composita, a genus in Athyrididina: *(a)* dorsal view, *(b)* ventral view with pedicle cut away to show spire. *(From R. C. Moore, ed., Treatise on Invertebrate Palentology, pt. H, Geological Society of America, Inc., and University of Kansas Press, 1965)*

electric field strength in volts per meter at any specified point in time and space; near the earth's surface, in fair-weather areas, a typical datum is about 100 and the field is directed vertically in such a way as to drive positive charges downward. { ¦at·mə¦sfir·ik i¦lek·trik 'fēld }

atmospheric electricity [GEOPHYS] The electrical processes occurring in the lower atmosphere, including both the intense local electrification accompanying storms and the much weaker fair-weather electrical activity over the entire globe produced by the electrified storms continuously in progress. { ¦at·mə¦sfir·ik i¦lek'tris·əd·ē }

atmospheric entry [AERO ENG] The penetration of any planetary atmosphere by any object from outer space; specifically, the penetration of the earth's atmosphere by a crewed or uncrewed capsule or spacecraft. { ¦at·mə¦sfir·ik 'en·trē }

atmospheric evaporation [HYD] The exchange of water between the earth's oceans, lakes, rivers, ice, snow, and soil and the atmosphere. { ¦at·mə¦sfir·ik i¦vap·ə'rā·shən }

atmospheric gas [METEOROL] One of the constituents of air, which is a gaseous mixture primarily of nitrogen, oxygen, argon, carbon dioxide, water vapor, ozone, neon, helium, krypton, methane, hydrogen, and nitrous oxide. { ¦at·mə¦sfir·ik 'gas }

atmospheric general circulation [METEOROL] The statistical mean global flow pattern of the atmosphere. { ¦at·mə¦sfir·ik ¦jen·rəl sərk·yə'lā·shən }

atmospheric impurity [ENG] An extraneous substance that is mixed as a contaminant with the air of the atmosphere. { ¦at·mə¦sfir·ik im¦pyür·əd·ē }

atmospheric interference [GEOPHYS] Electromagnetic radiation, caused by natural electrical disturbances in the atmosphere, which interferes with radio systems. Also known as atmospherics; sferics; strays. { ¦at·mə¦sfir·ik ‚in·tər'fir·əns }

atmospheric ionization [GEOPHYS] The process by which neutral atmospheric molecules or atoms are rendered electrically charged chiefly by collisions with high-energy particles. { ¦at·mə¦sfir·ik ‚ī·ə·nə'zā·shən }

atmospheric lapse rate See environmental lapse rate. { ¦at·mə¦sfir·ik 'laps ‚rāt }

atmospheric layer See atmospheric shell. { ¦at·mə¦sfir·ik 'lā·ər }

atmospheric noise [ELECTR] Noise heard during radio reception due to atmospheric interference. { ¦at·mə¦sfir·ik 'nóiz }

atmospheric optics See meteorological optics. { ¦at·mə¦sfir·ik 'óp·tiks }

atmospheric physics [GEOPHYS] The study of the physical phenomena of the atmosphere. { ¦at·mə¦sfir·ik 'fiz·iks }

atmospheric pressure [PHYS] The pressure at any point in an atmosphere due solely to the weight of the atmospheric gases above the point concerned. Also known as barometric pressure. { ¦at·mə¦sfir·ik 'presh·ər }

atmospheric-pressure cure [PETRO ENG] Preparation of test petroleum specimens by aging at normal atmospheric pressure for a given period of time at specified temperature and humidity. { ¦at·mə¦sfir·ik 'presh·ər 'kyúr }

atmospheric radiation [GEOPHYS] Infrared radiation emitted by or being propagated through the atmosphere. { ¦at·mə¦sfir·ik ‚rād·ē'ā·shən }

atmospheric radio wave [ELECTROMAG] Radio wave that is propagated by reflection in the atmosphere; may include either the ionospheric wave or the tropospheric wave, or both. { ¦at·mə¦sfir·ik 'rād·ē·ō ‚wāv }

atmospheric refraction [GEOPHYS] **1.** The angular difference between the apparent zenith distance of a celestial body and its true zenith distance, produced by refraction effects as the light from the body penetrates the atmosphere. **2.** Any refraction caused by the atmosphere's normal decrease in density with height. { ¦at·mə¦sfir·ik ri'frak·shən }

atmospheric region See atmospheric shell. { ¦at·mə¦sfir·ik 'rē·jən }

atmospherics See atmospheric interference. { ¦at·mə¦sfir·iks }

atmospheric scattering [GEOPHYS] A change in the direction of propagation, frequency, or polarization of electromagnetic radiation caused by interaction with the atoms of the atmosphere. { ¦at·mə¦sfir·ik 'skad·ər·iŋ }

atmospheric shell [METEOROL] Any one of a number of strata or layers of the earth's atmosphere; temperature distribution is the most common criterion used for denoting the various shells. Also known as atmospheric layer; atmospheric region. { ¦at·mə¦sfir·ik 'shel }

atmospheric shimmer See terrestrial scintillation. { ¦at·mə¦sfir·ik 'shim·ər }

atmospheric sounding [METEOROL] A measurement of atmospheric conditions aloft, above the effective range of surface weather observations. { ¦at·mə¦sfir·ik 'saund·iŋ }

atmospheric steam curing [ENG] The steam curing of concrete or cement products at atmospheric pressure, usually at a maximum ambient temperature between 100 and 200°F (40 and 95°C). { ¦at·mə¦sfir·ik 'stēm 'kyúr·iŋ }

atmospheric structure [METEOROL] Atmospheric characteristics, including wind direction and velocity, air density, and velocity of sound. { ¦at·mə¦sfir·ik 'strək·chər }

atmospheric suspensoids [METEOROL] Moderately finely divided particles suspended in the atmosphere; dust is an example. { ¦at·mə¦sfir·ik sə'spen‚soidz }

atmospheric tide [GEOPHYS] Periodic global motions of the earth's atmosphere, produced by gravitational action of the sun and moon; amplitudes are minute except in the upper atmosphere. { ¦at·mə¦sfir·ik 'tīd }

atmospheric turbulence [METEOROL] Apparently random fluctuations of the atmosphere that often constitute major deformations of its state of fluid flow. { ¦at·mə¦sfir·ik 'tər·byə·ləns }

Atokan [GEOL] A North American provincial series in lower Middle Pennsylvanian geologic time, above Morrowan, below Desmoinesian. { ə'tō·kən }

atoll [GEOGR] A ring-shaped coral reef that surrounds a lagoon without projecting land area and that is surrounded by open sea. { 'a‚tól }

atoll texture [GEOL] The surrounding of a ring of one mineral with another mineral, or minerals, within and without the ring. Also known as core texture. { 'a‚tól ‚teks·chər }

atom [COMPUT SCI] A primitive data element in a data structure. [CHEM] The individual structure which constitutes the basic unit of any chemical element. [MATH] An element, A, of a measure algebra, other than the zero element, which has the property that any element which is equal to or less than A is either equal to A or equal to the zero element. { 'ad·əm }

atom cluster [PHYS CHEM] An assembly of between three and a few thousand atoms or molecules that are weakly bound together and have properties intermediate between those of the isolated atom or molecule and the bulk or solid-state material. { 'ad·əm ‚kləs·tər }

atomic absorption coefficient [PHYS] The linear absorption coefficient divided by the number of atoms per unit volume. { ə'täm·ik əb'zórp·shən ‚kō·ə‚fish·ənt }

atomic absorption spectroscopy [SPECT] An instrumental technique for detecting concentrations of atoms to parts per million by measuring the amount of light absorbed by atoms or ions vaporized in a flame or an electrical furnace. { ə'tä·mik əb'sórp·shən ‚spek'träs·kə·pē }

atomic airburst [ORD] The explosion of an atomic weapon in the air at a height greater than the maximum radius of the fireball. { ə'täm·ik 'er‚bərst }

atomic ammunition [ORD] Formerly, ammunition deriving its explosive force from nuclear fission; now, all ammunition which derives its explosive force from a chain reaction of active nuclear material. { ə'täm·ik ‚am·yə'nish·ən }

atomic battery See nuclear battery. { ə'täm·ik 'bad·ə·rē }

atomic beam [PHYS] A stream of atoms, which may or may not be ionized. { ə'täm·ik 'bēm }

atomic-beam frequency standard [PHYS] A source of precisely timed signals which are derived from an atomic-beam resonance, such as a cesium-beam cell or a hydrogen maser. { ə'täm·ik ‚bēm 'frē·kwən·sē ‚stan·dərd }

atomic-beam resonance [PHYS] Phenomenon in which an oscillating magnetic field, superimposed on a uniform magnetic field at right angles to it, causes transitions between states with different magnetic quantum numbers of the nuclei of atoms in a beam passing through the field; the transitions occur only when the frequency of the oscillating field assumes certain characteristic values. { ə'täm·ik ‚bēm ‚rez·ən·əns }

atomic bomb [ORD] Also known as A bomb. **1.** A device for suddenly producing an explosively rapid neutron chain reaction in a fissile material such as uranium-235 or plutonium-239. Also known as fission bomb. **2.** Any explosive device which derives its energy from nuclear reactions, including a fusion bomb. Also known as nuclear bomb. { ə'täm·ik 'bäm }

atomic charge [ATOM PHYS] The electric charge of an ion, equal to the number of electrons the atom has gained or lost in its ionization multiplied by the charge on one electron. { ə'täm·ik 'chärj }

atomic clock [HOROL] An electronic clock whose frequency is supplied or governed by the natural resonance frequencies of atoms or molecules of suitable substances. { ə'täm·ik 'kläk }

atomic cloud [NUCLEO] The cloud of hot gases, smoke, dust, and other matter that is carried aloft after the explosion of a nuclear weapon in the air or near the surface; frequently has a mushroom shape. { ə'täm·ik 'klaud }

atomic connectivity [PHYS CHEM] The specific pattern of chemical bonds between atoms in a molecule. { ə'täm·ik kə,nek'tiv·əd·ē }

atomic constants *See* fundamental constants. { ə'täm·ik 'kän·stəns }

atomic core [ATOM PHYS] An atom stripped of its valence electrons, so that its remaining electrons are all in closed shells. { ə'täm·ik 'kór }

atomic crystal [OPTICS] A crystallike structure of atoms that occupy sites in an optical lattice. { ə'täm·ik 'krist·əl }

atomic device *See* atomic weapon. { ə'täm·ik di'vīs }

atomic diamagnetism [ATOM PHYS] Diamagnetic ionic susceptibility, important in providing correction factors for measured magnetic susceptibilities; calculated theoretically by considering electron density distributions summed for each electron shell. { ə'täm·ik ,dī·ə'mag·nə,tiz·əm }

atomic emission spectroscopy [SPECT] A form of atomic spectroscopy in which one observes the emission of light at discrete wavelengths by atoms which have been electronically excited by collisions with other atoms and molecules in a hot gas. { ə'täm·ik ə,mish·ən spek'träs·kə·pē }

atomic energy *See* nuclear energy. { ə'täm·ik 'en·ər·jē }

atomic energy level [ATOM PHYS] A definite value of energy possible for an atom, either in the ground state or an excited condition. { ə'täm·ik 'en·ər·jē ,lev·əl }

atomic fallout *See* fallout. { ə'täm·ik 'fȯl,aut }

atomic fission *See* fission. { ə'täm·ik 'fish·ən }

atomic fluorescence spectroscopy [SPECT] A form of atomic spectroscopy in which the sample atoms are first excited by absorbing radiation from an external source containing the element to be detected, and the intensity of radiation emitted at characteristic wavelengths during transitions of these atoms back to the ground state is observed. { ə'täm·ik flu̇'res·əns spek'träs·kə·pē }

atomic force microscope [ENG] A device for mapping surface atomic structure by measuring the force acting on the tip of a sharply pointed wire or other object that is moved over the surface. { ə'täm·ik 'fȯrs 'mī,krə,skōp }

atomic form factor *See* atomic scattering factor. { ə'täm·ik 'fȯrm ,fak·tər }

atomic fountain [ATOM PHYS] A device in which atoms in a magnetooptic trap from a thermal beam are pushed upward with a pulse of laser light, causing them to assume a ballistic trajectory; used for studying free-falling atoms. { ə'täm·ik 'faunt·ən }

atomic frequency [SOLID STATE] One of the vibrational frequencies of an atom in a crystal lattice. { ə'täm·ik 'frē·kwən·sē }

atomic funnel [ATOM PHYS] A device that uses a magnetic quadrupole field and trapping and cooling laser beams to form a slowed atomic beam into a highly localized and collimated beam with a peak phase-space density over 10,000 times that of the original beam. { ə'täm·ik 'fən·əl }

atomic fusion *See* fusion. { ə'täm·ik 'fyü·zhən }

atomic gas laser [OPTICS] A gas laser, such as the helium-neon laser, in which electrons and ions accelerated between electrodes by an electric field collide and excite atoms and ions to higher energy levels; laser action occurs during subsequent decay back to lower energy levels. { ə'täm·ik 'gas ,lā·zər }

atomic ground state [ATOM PHYS] The state of lowest energy in which an atom can exist. Also known as atomic unexcited state. { ə'täm·ik 'graund ,stāt }

atomic heat capacity [PHYS CHEM] The heat capacity of a gram-atomic weight of an element. { ə'täm·ik 'hēt kə'pas·əd·ē }

atomic hydrogen [CHEM] Gaseous hydrogen whose molecules are dissociated into atoms. { ə'täm·ik 'hī·drə·jən }

atomic hydrogen maser [PHYS] A maser in which dissociated hydrogen atoms from an electric discharge source are formed into a beam that undergoes selective magnetic processing; can be used as an atomic clock. { ə'täm·ik 'hī·drə·jən 'mā·zər }

atomic hydrogen welding [MET] An arc welding process in which hydrogen gas dissociated by the arc recombines outside the arc to provide intense heat and protection against oxidation for the weld. { ə'täm·ik 'hī·drə·jən 'weld·iŋ }

atomicity [CHEM] The number of atoms in a molecule of a compound. { ,ad·ə'mis·əd·ē }

atomic magnet [ATOM PHYS] An atom which possesses a magnetic moment either in the ground state or in an excited state. { ə'täm·ik 'mag·nət }

atomic magnetic moment [ATOM PHYS] A magnetic moment, permanent or temporary, associated with an atom, measured in magnetons. { ə'täm·ik ,mag¦ned·ik 'mō·mənt }

atomic mass [PHYS] The mass of a neutral atom usually expressed in atomic mass units. { ə'täm·ik 'mas }

atomic mass unit [PHYS] An arbitrarily defined unit in terms of which the masses of individual atoms are expressed; the standard is the unit of mass equal to one-twelfth the mass of the carbon atom, having as nucleus the isotope with mass number 12. Abbreviated amu. Also known as dalton. { ə'täm·ik 'mas 'yü·nət }

atomic moisture meter [ENG] An instrument that measures the moisture content of coal instantaneously and continuously by bombarding it with neutrons and measuring the neutrons which bounce back to a detector tube after striking hydrogen atoms of water. { ə'täm·ik 'mȯis·chər ,mēd·ər }

atomic nucleus *See* nucleus. { ə'täm·ik 'nü·klē·əs }

atomic number [NUC PHYS] The number of protons in an atomic nucleus. Also known as proton number. { ə'täm·ik 'nəm·bər }

atomic operation [COMPUT SCI] An operation that cannot be broken up into smaller parts that could be performed by different processors. { ə'täm·ik ,äp·ə'rā·shən }

atomic orbital [ATOM PHYS] The space-dependent part of a wave function describing an electron in an atom. { ə'täm·ik 'ȯr·bə·təl }

atomic paramagnetism [ELECTROMAG] The result of a permanent magnetic moment in an atom. { ə'täm·ik ,par·ə'mag·nə,tiz·əm }

atomic particle [ATOM PHYS] One of the particles of which an atom is constituted, as an electron, neutron, or proton. { ə'täm·ik 'pard·ə·kəl }

atomic percent [CHEM] The number of atoms of an element in 100 atoms representative of a substance. { ə'täm·ik pər'sent }

atomic photoelectric effect [PHYS CHEM] *See* photoionization. { ə'täm·ik ,fōd·ō·i'lek·trik i'fekt }

atomic physics [PHYS] The science concerned with the structure of the atom, the characteristics of the elementary particles of which the atom is composed, and the processes involved in the interactions of radiant energy with matter. { ə'täm·ik 'fiz·iks }

atomic pile *See* nuclear reactor. { ə'täm·ik 'pīl }

atomic polarization [PHYS CHEM] Polarization of a material arising from the change in dipole moment accompanying the stretching of chemical bonds between unlike atoms in molecules. { ə'täm·ik ,pōl·ə·rə'zā·shən }

atomic power plant *See* nuclear power plant. { ə'täm·ik 'pau̇·ər ,plant }

atomic radius [PHYS CHEM] Also known as covalent radius. **1.** Half the distance between the nuclei of two like atoms that are covalently bonded. **2.** The experimentally determined radius of an atom in a covalently bonded compound. Also known as covalent radius. { ə'täm·ik 'rād·ē·əs }

atomic reactor *See* nuclear reactor. { ə'täm·ik rē'ak·tər }

atomic rocket [AERO ENG] A rocket propelled by an engine in which the energy for the jetstream is to be generated by nuclear fission or fusion. Also known as nuclear rocket. { ə'täm·ik 'räk·ət }

atomic scattering factor [PHYS] A quantity which

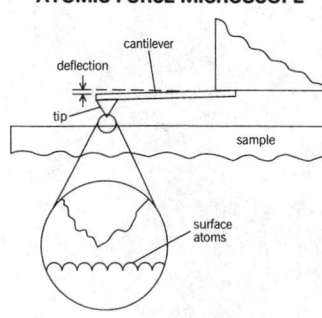

ATOMIC FORCE MICROSCOPE

cantilever

deflection

tip

sample

surface atoms

Diagram of atomic force microscope.

expresses the efficiency with which x-rays of a stated wavelength are scattered into a given direction by a particular atom, measured in terms of the corresponding scattering by a point electron. Also known as atomic form factor. { ə'täm·ik 'skad·ər·iŋ ,fak·tər }

atomic second [PHYS] As defined in 1967, the duration of 9,192,631,770 periods of the radiation corresponding to the two hyperfine levels of the fundamental state of the atom of cesium-133. { ə'täm·ik 'sek·ənd }

atomic spectroscopy [SPECT] The branch of physics concerned with the production, measurement, and interpretation of spectra arising from either emission or absorption of electromagnetic radiation by atoms. { ə'täm·ik ,spek'träs·kə·pē }

atomic spectrum [SPECT] The spectrum of radiations due to transitions between energy levels in an atom, either absorption or emission. { ə'täm·ik 'spek·trəm }

atomic standard [PHYS] Any supposedly immutable property of an atom, such as the wavelength or frequency of a characteristic spectral line, in terms of which a unit of a physical quantity is defined. { ə'täm·ik 'stan·dərd }

atomic stopping power [NUCLEO] For an ionizing particle passing through an element, the particle's energy loss per atom within a unit area normal to the particle's path; equal to the linear energy transfer (energy loss per unit path length) divided by the number of atoms per unit volume. { ə'täm·ik 'stäp·iŋ ,paὺ·ər }

atomic structure [ATOM PHYS] The arrangement of the parts of an atom, which consists of a massive, positively charged nucleus surrounded by a cloud of electrons arranged in orbits describable in terms of quantum mechanics. { ə'täm·ik 'strək·chər }

atomic surface burst [ORD] An atomic missile burst at an elevation such that the fireball touches the ground. { ə'täm·ik 'sər·fəs ,bərst }

atomic susceptibility [ELECTROMAG] The magnetization of a material per atom per unit of applied field; measured in ergs per oersted per atom. { ə'täm·ik sə,sep·tə'bil·əd·ē }

atomic theory [CHEM] The assumption that matter is composed of particles called atoms and that these are the limit to which matter can be subdivided. { ə'täm·ik 'thē·ə·rē }

atomic time [HOROL] Any time system standardized with reference to an atomic resonance, such as the international standard cesium-133 transition. { ə'täm·ik ,tīm }

atomic underground burst [ORD] The explosion of an atomic weapon with its center beneath the surface of the ground. { ə'täm·ik 'ən·dər,graὺnd 'bərst }

atomic underwater burst [ORD] The explosion of an atomic weapon with its center beneath the surface of the water. { ə'täm·ik 'ən·dər,wόd·ər 'bərst }

atomic unexcited state See atomic ground state. { ə'täm·ik ,ən·ek'sīd·əd 'stāt }

atomic units See Hartree units. { ə'täm·ik 'yü·nəts }

atomic vibration [ATOM PHYS] Periodic, nearly harmonic changes in position of the atoms in a molecule giving rise to many properties of matter, including molecular spectra, heat capacity, and heat conduction. { ə'täm·ik ,vī'brā·shən }

atomic volume [PHYS CHEM] The volume occupied by 1 gram-atom of an element in the solid state. { ə'täm·ik 'väl·yəm }

atomic weapon [ORD] Any bomb, warhead, or projectile using active nuclear material to cause a chain reaction upon detonation. Also known as atomic device; nuclear weapon. { ə'täm·ik 'wep·ən }

atomic weight [CHEM] The relative mass of an atom based on a scale in which a specific carbon atom (carbon-12) is assigned a mass value of 12. Abbreviated at. wt. Also known as relative atomic mass. { ə'täm·ik 'wāt }

atom interferometer [PHYS] A device which measures the interference effects that result when a beam of atoms is manipulated in such a way that the de Broglie waves of the atoms are split into two components and subsequently recombined. { ,ad·əm ,in·tə·fə'räm·əd·ər }

atomization [ANALY CHEM] In flame spectrometry, conversion of a volatilized sample into free atoms. [CHEM] A process in which the chemical bonds in a molecule are broken to yield separated (free) atoms. [MECH ENG] The mechanical subdivision of a bulk liquid or meltable solid, such as certain metals, to produce drops, which vary in diameter depending

on the process from under 10 to over 1000 micrometers. { ,ad·ə·mə'zā·shən }

atomizer [MECH ENG] A device that produces a mechanical subdivision of a bulk liquid, as by spraying, sprinkling, misting, or nebulizing. { 'ad·ə,mīz·ər }

atomizer burner [MECH ENG] A liquid-fuel burner that atomizes the unignited fuel into a fine spray as it enters the combustion zone. { 'ad·ə,mīz·ər 'bər·nər }

atomizer mill [MECH ENG] A solids grinder, the product from which is a fine powder. { 'ad·ə,mīz·ər ,mil }

atomizing humidifier [MECH ENG] A humidifier in which tiny particles of water are introduced into a stream of air. { 'ad·ə,mīz·iŋ ,hyü'mid·ə,fī·ər }

atom laser [PHYS] A device that generates intense coherent beams of atoms (coherent matter waves), analogous to coherent light waves emitted by a conventional laser, through a stimulated process that generally involves extraction of the beams from a Bose-Einstein condensate. { 'ad·əm 'lā·zər }

atom optics [PHYS] The use of laser light and nanofabricated structures to manipulate the motion of atoms in the same manner that rudimentary optical elements control light. { 'ad·əm 'äp·tiks }

atom probe [ENG] An instrument for identifying a single atom or molecule on a metal surface; it consists of a field ion microscope with a probe hole in its screen opening into a mass spectrometer; atoms that are removed from the specimen by pulsed field evaporation fly through the probe hole and are detected in the mass spectrometer. { 'ad·əm ,prōb }

atoms-in-molecules method [PHYS CHEM] The description of the electronic structure of a molecule as a perturbation of the isolated states of its constituent atoms. { 'ad·əmz in 'mäl·ə,kyülz ,meth·əd }

atom smasher See particle accelerator. { 'ad·əm ,smash·ər }

atom trap trace analysis [ANALY CHEM] An atom-counting method in which individual atoms of a chosen isotope are captured and detected with a laser trap. { 'ad·əm ,trap 'träs ə,nal·ə·səs }

atony [MED] Absence or extremely low degree of tonus. { 'at·ən·ē }

atopic allergy [MED] A type of immediate hypersensitivity in humans resulting from spontaneous sensitization, usually by inhaled or ingested antigens; for example, asthma, hay fever, or hives. { ā'täp·ik 'al·ər·jē }

atopic dermatitis [MED] A chronic eruption of red patches accompanied by intense itching that usually begins in infancy but may continue into adult life; the disease has a genetic predisposition, but its expression is modified by environmental factors. { ,ā,täp·ik ,dər·mə'tīd·əs }

atopite [MINERAL] A yellow or brown variety of romeite that contains fluorine. { 'ad·ə,pīt }

atopy [MED] Clinically evident hypersensitivity. { 'ad·ə·pē }

ATP See adenosine triphosphate.

ATPase [BIOCHEM] An enzyme that hydrolyzes adenosine triphosphate into adenosine diphosphate and phosphate. { ,ā,tē'pē,ās }

ATP synthase [BIOCHEM] An enzyme that catalyzes the conversion of phosphate and adenosine diphosphate into adenosine triphosphate during oxidative phosphorylation in mitochondria and bacteria or phosphorylation in chloroplasts. { ,ā,tē,pē 'sin,thās }

ATR See attenuated total reflectance.

A trace [ELECTR] The first trace of an oscilloscope, such as the upper trace of a loran indicator. { 'ā ,trās }

Atractidae [INV ZOO] A family of parasitic nematodes in the superfamily Oxyuroidea. { a'trak·tə,dē }

atran [NAV] An acronym for automatic terrain recognition and navigation, a system which depends upon the correlation of terrain images appearing on a radar cathode-ray tube with previously prepared maps or simulated radar images of the terrain. { 'ā,tran }

atrazine [CHEM] $C_8H_{14}ClN_5$ A white crystalline compound widely used as a photosynthesis-inhibiting herbicide for weeds. { 'a·trə,zēn }

atresia [MED] Imperforation or closure of a natural orifice or passage of the body. { ə'trē·zhə }

atrial flutter [MED] A cardiac arrhythmia characterized by rapid, irregular atrial impulses and ineffective atrial contractions; the heartbeat varies from 60 to 180 per minute and

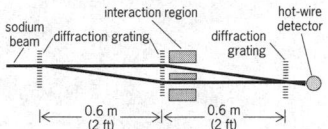

ATOM INTERFEROMETER

Diagram of a separated-beam atom interferometer. Diffraction gratings have a period of 200 nanometers. In the interaction region, the two components of the atom wave have centers separated by 55 micrometers.

ATOM LASER

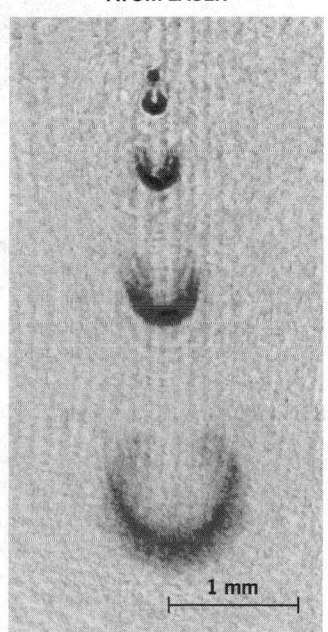

Pulsed atom laser in operation, with pulses of coherent sodium atoms coupled out from a Bose-Einstein condensate that is confined in a magnetic trap.

is grossly irregular in intensity and rhythm.   Also known as auricular flutter.   { 'ā·trē·əl 'fləd·ər }

atrial septum [ANAT]   The muscular wall between the atria of the heart.   Also known as interatrial septum.   { 'ā·trē·əl 'sep·təm }

atrichia [MED]   Any congenital or acquired condition in which hair is essentially absent.   { ā'trik·ē·ə }

Atrichornithidae [VERT ZOO]   The scrubbirds, a family of suboscine perching birds in the suborder Menurae.   { a,trī·kȯr'nith·ə,dē }

atrichous [CYTOL]   Lacking flagella.   { 'a·trə·kəs }

atriopore [ZOO]   The opening of an atrium as seen in lancelets and tunicates.   { 'ā·trē·ə,pȯr }

atrioventricular bundle [ANAT]   *See* bundle of His.   { ¦ā·trē·ō,ven'trik·yə·lər 'bən·dəl }

atrioventricular canal [EMBRYO]   The common passage between the atria and ventricles in the heart of the mammalian embryo before division of the organ into right and left sides.   { ¦ā·trē·ō,ven'trik·yə·lər kə'nal }

atrioventricularis communis [MED]   A congenital malformation of the heart in which partitioning has not occurred.   { ¦ā·trē·ō,ven,trik·yə'lär·əs kə'myü·nəs }

atrioventricular node [ANAT]   A group of slow-conducting fibers in the atrium of the vertebrate heart that are stimulated by impulses originating in the sinoatrial node and conduct impulses to the bundle of His.   { ¦ā·trē·o,ven¦trik·yə¦lər 'nōd }

atrioventricular valve [ANAT]   A structure located at the orifice between the atrium and ventricle which maintains a unidirectional blood flow through the heart.   { ¦a·trē·ō,ven'trik·yə·lər 'valv }

atrium [ANAT]   **1.** The heart chamber that receives blood from the veins.   **2.** The main part of the tympanic cavity, below the malleus.   **3.** The external chamber to receive water from the gills in lancelets and tunicates.   [ARCH]   An open court located within a building.   { 'ā·trē·əm }

atrophic arthritis *See* rheumatoid arthritis.   { ā'trōf·ik ar'thrīd·əs }

atrophic gastritis [MED]   Chronic inflammation of the stomach with atrophy of the mucosa.   { ā'trōf·ik gas'trīd·əs }

atrophy [MED]   Diminution in the size of a cell, tissue, or organ that was once fully developed and of normal size.   { 'a·trə·fē }

atropine [PHARM] $C_{17}H_{23}O_3N$ An alkaloid extracted from *Atropa belladonna* and related plants of the family Solanaceae; used to relieve muscle spasms and pain, and to dilate the pupil of the eye.   { 'a·trə,pēn }

atropine sulfate [PHARM] $C_{34}H_{48}N_2O_{10}S$ In the hydrate form the compound is either granules or powder; melting point is 190–194°C; used as an anticholinergic drug for humans and as an anticholinergic drug and smooth muscle relaxant drug for animals.   { 'a·trə,pēn 'səl,fāt }

atropinization [PHYSIO]   The physiological condition of being under the influence of atropine.   { ə¦trō·pə·nə'zā·shən }

atropisomer [ORG CHEM]   One of two conformations of a molecule whose interconversion is slow enough to allow separation and isolation under predetermined conditions.   { ¦a·trō¦pīz·ə·mər }

atropous *See* orthotropous.   { 'a·trə·pəs }

ATR tube *See* anti-transmit-receive tube.   { 'ā'tē,är ,tüb }

Atrypidina [PALEON]   A suborder of fossil articulate brachiopods in the order Spiriferida.   { a·trī'pid·ə·nə }

ATS *See* Administrative Terminal System; applications technology satellite.

attached dune [GEOL]   A dune that has formed around a rock or other geological feature in the path of windblown sand.   { ə'tacht 'dün }

attached file *See* attachment.   { ə'tacht 'fīl }

attached groundwater [HYD]   The portion of subsurface water adhering to pore walls in the soil.   { ə'tacht 'graünd,-wȯd·ər }

attached processing [COMPUT SCI]   A method of data processing in which several relatively inexpensive computers dedicated to specific tasks are connected together to provide a greater processing capability.   { ə'tacht 'präs,es·iŋ }

attached processor [COMPUT SCI]   A computer that is electronically connected to and operates under the control of another computer.   { ə'tacht 'präs,es·ər }

attached shock *See* attached shock wave.   { ə'tacht 'shäk }

attached shock wave [FL MECH]   An oblique or conical shock wave that appears to be in contact with the leading edge of an airfoil or the nose of a body in a supersonic flow field. Also known as attached shock.   { ə'tacht 'shäk ,wāv }

attached thermometer [ENG]   A thermometer which is attached to an instrument to determine its operating temperature.   { ə'tacht thər'mäm·əd·ər }

attaching gas [ELECTR]   A gas in which electron attachment takes place.   { ə'tach·iŋ ,gas }

attachment [COMPUT SCI]   An additional file sent with an e-mail message.   [ORG CHEM]   The conversion of a molecular entity into another molecular structure solely by formation of a single two-center bond with another molecular entity and no other changes in bonding.   [PSYCH]   The behavior of an individual who relates in an affiliative or dependent manner to another individual or object.   [VIROL]   The initial stage in the infection of a cell by a virus that follows a chance collision by the virus with a suitable receptor area on the cell.   { ə'tach·mənt }

attachment coefficient [ELECTR]   The probability that an electron drifting through a gas under the influence of a uniform electric field will undergo electron attachment in a unit distance of drift.   { ə'tach·mənt ,kō·ə,fish·ənt }

attachment plug [ELEC]   A device having an attached flexible cord containing conductors, and capable of being inserted in a receptacle so as to form an electrical connection between the conductors in the cord and conductors permanently connected to the receptacle.   { ə'tach·mənt ,pləg }

attachment unit interface [COMMUN]   A 15-pin connector on an Ethernet card for connecting a network cable.   Abbreviated AUI.   { ə¦tach·mənt ,yü·nət 'in·tər,fās }

attack director [COMPUT SCI]   An electromechanical analog computer which is designed for surface antisubmarine use and which computes continuous solution of several lines of submarine attack; it is part of several antisubmarine fire control systems.   { ə'tak di'rek·tər }

attack heading [NAV]   **1.** The interceptor heading during the attack phase which will achieve the desired track-crossing angle.   **2.** The assigned magnetic-compass heading to be flown by aircraft during the delivery phase of an air strike.   { ə'tak 'hed·iŋ }

attack plane [AERO ENG]   A multiweapon carrier aircraft which can carry bombs, torpedoes, and rockets.   { ə'tak ,plān }

attapulgite [MINERAL] $(Mg,Al)_2Si_4O_{10}(OH)\cdot4H_2O$ A clay mineral with a needlelike shape from Georgia and Florida; active ingredient in most fuller's earth, and used as a suspending agent, as an oil well drilling fluid, and as a thickener in latex paint.   { ,ad·ə'pəl,jīt }

attar of roses *See* rose oil.   { 'ad·ər əv 'rōz·əs }

attemperation [ENG]   The regulation of the temperature of a substance.   { ə,tem·pə'rā·shən }

attemperation of steam [MECH ENG]   The controlled cooling, in a steam boiler, of steam at the superheater outlet or between the primary and secondary stages of the superheater to regulate the final steam temperature.   { ə,tem·pə'rā·shən əv 'stēm }

attempt frequency [NUC PHYS]   The frequency with which an alpha particle attempts to cross the Gamow barrier in the Gamow-Condon-Gurney theory.   { ə'tempt ,frē·kwən·sē }

attendant's switchboard [COMMUN]   Switchboard of one or more positions in a central-office location which permits the central-office operator to receive, transmit, or cut in on a call to or from one of the lines which the office services.   { ə'ten·dəns 'swich,bȯrd }

attended time [COMPUT SCI]   The time in which a computer is either switched on and capable of normal operation (including time during which it is temporarily idle but still watched over by computer personnel) or out of service for maintenance work.   { ə'tend·əd ¦tīm }

attention deficit disorder [PSYCH]   A psychiatric disorder of childhood characterized by attention span problems and impulsivity.   { ə'ten·shən ¦def·ə·sət dis,ȯr·dər }

attention hypothesis [PSYCH]   The theory that a person's attention to objects and other persons is selectively drawn toward areas of great personal interest.   { ə'ten·shən 'hī'päth·ə·səs }

attention span [PSYCH]   The period of time a person is able to concentrate his attentions on a given item, usually with respect to learning.   { ə'ten·shən ,span }

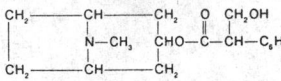

ATROPINE

Structural formula of atropine.

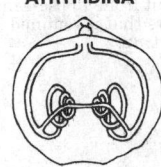

ATRYPIDINA

Spiralium of *Zygospira*.

attenuate [ENG ACOUS] To weaken a signal by reducing its level. { ə'ten·yə,wāt }

attenuated total reflectance [SPECT] A method of spectrophotometric analysis based on the reflection of energy at the interface of two media which have different refractive indices and are in optical contact with each other. Abbreviated ATR. Also known as frustrated internal reflectance; internal reflectance spectroscopy. { ə'ten·yə,wād·əd 'tōd·əl ri'flek·təns }

attenuated vaccine [IMMUNOL] A suspension of weakened bacteria, viruses, or fractions thereof used to produce active immunity. { ə'ten·yə,wād·əd ,vak'sēn }

attenuation [BOT] Tapering, sometimes to a long point. [ELEC] The exponential decrease with distance in the amplitude of an electrical signal traveling along a very long uniform transmission line, due to conductor and dielectric losses. [ENG] A process by which a material is fabricated into a thin, slender configuration, such as forming a fiber from molten glass. [MICROBIO] Weakening or reduction of the virulence of a microorganism. [PHYS] The reduction in level of a quantity, such as the intensity of a wave, over an interval of a variable, such as the distance from a source. { ə,ten·yə'wā·shən }

attenuation coefficient [ELECTROMAG] The space rate of attenuation of any transmitted electromagnetic radiation. { ə,ten·yə'wā·shən ,kō·ə'fish·ənt }

attenuation constant [PHYS] A rating for a line or medium through which a plane wave is being transmitted, equal to the relative rate of decrease of an amplitude of a field component, voltage, or current in the direction of propagation, in nepers per unit length. { ə,ten·yə'wā·shən ,kän·stənt }

attenuation distortion [COMMUN] 1. In a circuit or system, departure from uniform amplification or attenuation over the frequency range required for transmission. 2. The effect of such departure on a transmitted signal. { ə,ten·yə'wā·shən dis,tòr·shən }

attenuation equalizer [ELECTR] Corrective network which is designed to make the absolute value of the transfer impedance, with respect to two chosen pairs of terminals, substantially constant for all frequencies within a desired range. Also known as attenuation factor. { ə,ten·yə'wā·shən 'ē·kwə,līz·ər }

attenuation factor See attenuation constant. { ə,ten·yə'wā·shən ,fak·tər }

attenuation length [PHYS] The reciprocal of the attenuation coefficient. { ə,ten·yə'wā·shən ,leŋkth }

attenuation network [ELECTR] Arrangement of circuit elements, usually impedance elements, inserted in circuitry to introduce a known loss or to reduce the impedance level without reflections. { ə,ten·yə'wā·shən 'net,wərk }

attenuation ratio [PHYS] The magnitude of the propagation ratio. { ə,ten·yə'wā·shən ,rā·shō }

attenuator [ELECTR] An adjustable or fixed transducer for reducing the amplitude of a wave without introducing appreciable distortion. { ə'ten·yə,wād·ər }

Atterberg scale [GEOL] A geometric and decimal grade scale for classification of particles in sediments based on the unit value of 2 millimeters and involving a fixed ratio of 10 for each successive grade; subdivisions are geometric means of the limits of each grade. { 'at·ər,bərg ,skāl }

attic [BUILD] The part of a building immediately below the roof and entirely or partly within the roof framing. { 'ad·ik }

Attican orogeny [GEOL] Late Miocene diastrophism. { 'ad·ə·kən ō'räj·ə·nē }

attic tank [BUILD] An open tank which is installed above the highest plumbing fixture in a building and which supplies water to the fixtures by gravity. { 'ad·ik ,taŋk }

atticurge [BUILD] Of a doorway, having jambs which are inclined slightly inward, so that the opening is wider at the threshold than at the top. { 'ad·ə,kərj }

attic ventilator [BUILD] A mechanical fan located in the attic space of a residence; usually moves large quantities of air at a relatively low velocity. { 'ad·ik 'vent·əl,ād·ər }

attitude [AERO ENG] The position or orientation of an aircraft, spacecraft, and so on, either in motion or at rest, as determined by the relationship between its axes and some reference line or plane or some fixed system of reference axes. [GEOL] The position of a structural surface feature in relation to the horizontal. [GRAPHICS] The orientation of a camera or a photograph with respect to a given external reference system. { 'ad·ə,tüd }

attitude control [AERO ENG] 1. The regulation of the attitude of an aircraft, spacecraft, and so on. 2. A device or system that automatically regulates and corrects attitude, especially of a pilotless vehicle. { 'ad·ə,tüd kən,trōl }

attitude gyro [AERO ENG] Also known as attitude indicator. 1. A gyro-operated flight instrument that indicates the attitude of an aircraft or spacecraft with respect to a reference coordinate system. 2. Any gyro-operated instrument that indicates attitude. { 'ad·ə,tüd ,jī·rō }

attitude indicator See attitude gyro. { 'ad·ə,tüd 'in·də,kād·ər }

attitude jet [AERO ENG] 1. A stream of gas from a jet used to correct or alter the attitude of a flying body either in the atmosphere or in space. 2. The nozzle that directs this jet-stream. { 'ad·ə,tüd ,jet }

atto- [PHYS] A prefix representing 10^{-18}, which is 0.000 000 000 000 000 001, or one-millionth of a millionth of a millionth. Abbreviated a. { 'ad·ō }

attracted-disk electrometer [ELEC] A type of electrometer in which the attraction between two oppositely charged disks is measured. { ə'trak·təd ¦disk i,lek'träm·əd·ər }

attraction gripper [CONT SYS] A robot component that uses adhesion, suction, or magnetic forces to grasp a workpiece. { ə'trak·shən ,grip·ər }

attractor [PHYS] A geometrical object toward which the trajectory of a dynamical system, represented by a curve in phase space, converges in the course of time. { ə'trak·tər }

attribute [COMPUT SCI] 1. A data item containing information about a variable. 2. A characteristic of computer-generated characters, such as underline, boldface, or reverse image. { 'a·trə,byüt }

attribute sampling [IND ENG] A quality-control inspection method in which the sampled articles are classified only as defective or nondefective. { 'a·trə,byüt ,sam·pliŋ }

attributes testing [ENG] A reliability test procedure in which the items under test are classified according to qualitative characteristics. { 'a·trə,byüts ,test·iŋ }

attrital coal [GEOL] A bright coal composed of anthraxylon and of attritus in which the translucent cell-wall degradation matter or translucent humic matter predominates, with the ratio of anthraxylon to attritus being less than 1:3. { ə'trīd·əl 'kōl }

attrition [GEOL] The act of wearing and smoothing of rock surfaces by the flow of water charged with sand and gravel, by the passage of sand drifts, or by the movement of glaciers. [MATER] Wear caused by rubbing or friction. For metal surfaces, also known as scoring; scouring. { ə'trish·ən }

attrition mill [MECH ENG] A machine in which materials are pulverized between two toothed metal disks rotating in opposite directions. { ə'trish·ən ,mil }

attrition minesweeping [ORD] Minesweeping designed to minimize the number of live mines at any one time in a channel or area being subjected to heavy, continuous mining attack when clearance sweeping is impossible and port closure unacceptable. { ə'trish·ən 'mīn,swēp·iŋ }

attrition rate [ORD] A factor, normally expressed as a percentage, reflecting the degree of losses of personnel or nonconsumable supplies due to various causes within a specified period of time. { ə'trish·ən ,rāt }

attritor [MET] A high-energy stirred-ball mill used for mechanically alloying metal powder particles. { ə'trid·ər }

attritus [GEOL] 1. Visible-to-ultramicroscopic particles of vegetable matter produced by microscopic and other organisms in vegetable deposits, particularly in swamps and bogs. 2. The dull gray to nearly black, frequently striped portion of material that makes up the bulk of some coals and alternate bands of bright anthraxylon in well-banded coals. { ə'trīd·əs }

atu [PHYS] A unit of underpressure or pressure below atmospheric pressure in the metric technical system; equal to 1 technical atmosphere. { ¦at¦ü }

atü [PHYS] A unit of overpressure or gage pressure in the metric technical system; equal to 1 technical atmosphere. { ¦at¦yü }

Atwood machine [MECH ENG] A device comprising a pulley over which is passed a stretch-free cord with a weight hanging on each end. { 'at,wüd mə'shēn }

at. wt See atomic weight.

Atyidae [INV ZOO] A family of decapod crustaceans belonging to the section Caridea. { a'tī·ə,dē }

A-type star [ASTRON] In star classification based on spectral characteristics, the type of star in whose spectrum the hydrogen absorption lines are at a maximum. Also known as A star. { 'ā,tīp ,stär }

A-type virus particles [VIROL] A morphologically defined group of double-shelled spherical ribonucleic acid virus particles, often found in tumor cells. { 'ā ¦tīp 'vī·rəs ,pard·ə·kəlz }

Au *See* gold.

AU *See* astronomical unit.

Auberger blood group system [IMMUNOL] An immunologically distinct, genetically determined human erythrocyte antigen, demonstrated by reaction with anti-Au[2] antibody. { ¦ō·bər,zhā 'bləd ,grüp ,sis·təm }

Aubert phenomenon [PSYCH] The perception of a vertical line as oblique by an observer whose head is inclined to one side in a darkened room. { ō,ber fə'näm·ə·nən }

aubrite [GEOL] An enstatite achondrite (meteorite) consisting almost wholly of crystalline-granular enstatite (and clinoenstatite) poor in lime and practically free from ferrous oxide, with accessory oligoclase. Also known as bustite. { 'ō,brīt }

Auchenorrhyncha [INV ZOO] A group of homopteran families and one superfamily, in which the beak arises at the anteroventral extremity of the face and is not sheathed by the propleura. { ,ok·ə·nə'riŋ·kə }

audibility [ACOUS] **1.** The state or quality of being heard. **2.** The intensity of a received audio signal, usually expressed in decibels above or below 1 milliwatt using a stated single frequency sine wave. { ,od·ə'bil·əd·ē }

audibility curve [ACOUS] **1.** The limits of hearing represented graphically as an area by plotting the minimum audible intensity of a sine wave sound versus frequency. **2.** *See* equal loudness contour. { ,od·ə'bil·əd·ē ,kərv }

audibility threshold [ACOUS] The sound intensity at a given frequency which is the minimum perceptible by a normal human ear under specified standard conditions. { ,od·ə'bil·əd·ē ,thresh,hōld }

audible feedback [COMPUT SCI] A feature of a computer keyboard that generates sound each time a key is depressed sufficiently to generate a character on the screen. { ,od·ə·bəl 'fēd,bak }

audible frequency *See* audible tone. { ¦od·ə·bəl 'frē·kwən·sē }

audible leak detector [ENG] A device used as an auxiliary to the main leak detector for conversion of the output signal into audible sound. { ¦od·ə·bəl 'lēk di,tek·tər }

audible tone [ACOUS] Sound of a frequency which the average human can hear, ranging from 30 to 16,000 hertz. Also known as audible frequency. { ¦od·ə·bəl 'tōn }

audio [ACOUS] **1.** Of or pertaining to sound in the range of frequencies considered audible at reasonable listening intensities to the average young adult listener, approximately 15 to 20,000 hertz. **2.** Pertaining to equipment for the recording, transmission, reproduction, or amplification of such sound. { 'od·ē·ō }

audio adapter *See* sound board. { ,od·ē·ō ə'dap·tər }

audio amplifier *See* audio-frequency amplifier. { 'od·ē·ō 'am·plə,fī·ər }

audio frequency [ACOUS] A frequency that can be detected as a sound by the average young adult, approximately 15 to 20,000 hertz. Abbreviated af. Also known as sonic frequency; sound frequency. { ¦od·ē·ō 'frē·kwən·sē }

audio-frequency amplifier [ELECTR] An electronic circuit for amplification of signals within, and in some cases above, the audible range of frequencies in equipment used to record and reproduce sound. Also known as audio amplifier. { 'od·ē·ō ¦frē·kwən·sē ¦am·plə,fī·ər }

audio-frequency choke [ELECTROMAG] Choke used to impede the flow of audio-frequency currents; generally a coil wound on an iron core. { 'od·ē·ō ¦frē·kwən·sē ,chōk }

audio-frequency meter [ENG] One of a number of types of frequency meters usable in the audio range; for example, a resonant-reed frequency meter. { 'od·ē·ō ¦frē·kwən·sē ,mēd·ər }

audio-frequency oscillator [ELECTR] An oscillator circuit using an electron tube, transistor, or other nonrotating device to produce an audio-frequency alternating current. Also

known as audio oscillator. { 'od·ē·ō ¦frē·kwən·sē 'äs·ə,lād·ər }

audio-frequency peak limiter [ELEC] A circuit used in an audio-frequency system to cut off signal peaks that exceed a predetermined value. Also known as audio peak limiter. { 'od·ē·ō ¦frē·kwən·sē 'pēk ,lim·əd·ər }

audio-frequency range [ACOUS] The range of frequencies to which the human ear is sensitive, approximately 15 to 20,000 hertz. Also known as audio range. { 'od·ē·ō ¦frē·kwən·sē ,rānj }

audio-frequency shift modulation [COMMUN] System of facsimile transmission over radio, in which the frequency shift required is applied through an 800-hertz shift of an audio signal, rather than shifting the radio transmitter frequency; the radio signal is modulated by the shifting audio signal, usually at 1500 to 2300 hertz. { 'od·ē·ō ¦frē·kwən·sē ,shift mäj·ə'lā·shən }

audio-frequency transformer [ELEC] An iron-core transformer that is used for coupling audio-frequency circuits. Also known as audio transformer. { 'od·ē·ō ¦frē·kwən·sē tranz'för·mər }

audiogenic seizure [MED] A transient episode of muscular, sensory, or psychic dysfunction induced by sound. { ¦od·ē·ō¦jen·ik 'sē·zhər }

audiogram [ACOUS] A graph showing hearing loss, percent hearing loss, or percent hearing as a function of frequency. { 'od·ē·ō,gram }

audio image [ACOUS] A sound that originates, or appears to originate, at a certain point in space. { 'od·ē·ō ,im·ij }

audioimpedance measurement [ACOUS] The measurement of acoustic impedance, as in the direct assessment of the dynamic motor control of sound feedback of different parts of the ear. { ¦od·ē·ō,im'pēd·əns ,mezh·ər·mənt }

audiology [ACOUS] The science of hearing. { ,od·ē'äl·ə·jē }

audio masking *See* masking. { 'od·ē·ō ,mask·iŋ }

audiometer [ENG] An instrument composed of an oscillator, amplifier, and attenuator and used to measure hearing acuity for pure tones, speech, and bone conduction. { ,od·ē'äm·əd·ər }

audiometry [ACOUS] The study of hearing ability by means of audiometers. { ,od·ē'äm·ə·trē }

audio-modulated radiosonde [ENG] A radiosonde with a carrier wave modulated by audio-frequency signals whose frequency is controlled by the sensing elements of the instrument. { ¦od·ē·o¦mäj·ə·lād·əd ¦rād·ē·ō,sänd }

audio oscillator *See* audio-frequency oscillator. { 'od·ē·ō 'äs·ə·lād·ər }

audio patch bay [ENG ACOUS] Specific patch panels provided to terminate all audio circuits and equipment used in a channel and technical control facility; this equipment can also be found in transmitting and receiving stations. { 'od·ē·ō ¦pach ,bā }

audio peak limiter *See* audio-frequency peak limiter. { 'od·ē·ō 'pēk ,lim·ə·dər }

audio range *See* audio-frequency range. { 'od·ē·ō ,rānj }

audio response [COMMUN] A form of computer output in which prerecorded spoken syllables, words, or messages are selected and put together by a computer as the appropriate verbal response to a keyboarded inquiry on a time-shared on-line information system. { 'od·ē·ō ri'späns }

audio response unit [COMMUN] A magnetic recording system that provides voice response to an inquiry made from a typewriter or telephone-type terminal connected to a computer by a data transmission line. { 'od·ē·ō ri'späns ,yü·nət }

audio signal [ACOUS] An electric signal having the frequency of a mechanical wave that can be detected as a sound by the human ear. { 'od·ē·ō ,sig·nəl }

audio spectrometer *See* acoustic spectrometer. { 'od·ē·ō spek'träm·əd·ər }

audio system *See* sound-reproducing system. { 'od·ē·ō ,sis·təm }

audio taper [ENG ACOUS] A special type of potentiometer used in a volume-control apparatus to compensate for the nonlinearity of human hearing and give the impression of a linear increase in audibility as volume is raised. Also known as linear taper. { 'od·ē·ō ,tā·pər }

audio transformer *See* audio-frequency transformer. { 'od·ē·ō tranz'för·mər }

audiovisual [COMMUN] Pertaining to methods of education

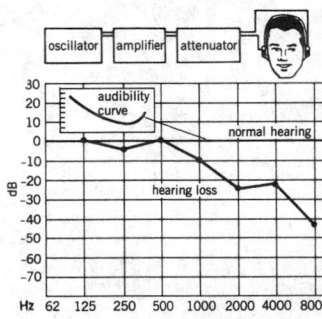

AUDIOGRAM

Audiogram for determining the audibility curve for pure-tone hearing loss at various frequency levels.

and training that make use of both hearing and sight. { ¦òd·ē·ō¦vizh·ə·wəl }

audiphone [ENG ACOUS] A device that enables persons with certain types of deafness to hear, consisting of a plate or diaphragm that is placed against the teeth and transmits sound vibrations to the inner ear. { 'òd·ə‚fōn }

audit [COMPUT SCI] The operations developed to corroborate the evidence as regards authenticity and validity of the data that are introduced into the data-processing problem or system. { 'òd·ət }

audition [PHYSIO] Ability to hear. { ò'dish·ən }

auditory [PHYSIO] Pertaining to the act or the organs of hearing. { 'òd·ə‚tōr·ē }

auditory association area [NEURO] The cortical association area in the brain just inferior to the auditory projection area, related to it anatomically and functionally by association fibers. { 'òd·ə‚tōr·ē ə‚sō·sē'ā·shən ‚er·ē·ə }

auditory impedance [PHYSIO] The acoustic impedance of the ear. { 'òd·ə‚tōr·ē im'pēd·əns }

auditory nerve [NEURO] The eighth cranial nerve in vertebrates; either of a pair of sensory nerves composed of two sets of nerve fibers, the cochlear nerve and the vestibular nerve. Also known as acoustic nerve; vestibulocochlear nerve. { 'òd·ə‚tōr·ē 'nərv }

auditory perspective [ACOUS] Three-dimensional realism of sound, as produced by an actual orchestra or by a stereophonic sound system. { 'òd·ə‚tōr·ē pər'spek·tiv }

auditory placode [EMBRYO] An ectodermal thickening from which the inner ear develops in vertebrates. { 'òd·ə‚tōr·ē 'pla‚kōd }

auditory-verbal agnosia [PSYCH] A disturbance in comprehension of spoken language in the presence of otherwise intact auditory functioning and essentially normal performance in other language modalities. Also known as pure word deafness. { ¦òd·ə‚tòr·ē ¦vər·bəl ‚ag'nō·zhə }

audit total [COMPUT SCI] A count or sum of a known quantity, calculated in order to verify data. { 'òd·ət ‚tōd·əl }

audit trail [COMPUT SCI] A system that provides a means for tracing items of data from processing step to step, particularly from a machine-produced report or other machine output back to the original source data. { 'òd·ət ‚trāl }

Auerbach's plexus See myenteric plexus. { 'aùr‚baks ¦plek·səs }

Aufbau principle [CHEM] A description of the building up of the elements in which the structure of each in sequence is obtained by simultaneously adding one positive charge (proton) to the nucleus of the atom and one negative charge (electron) to an atomic orbital. { 'aùf‚baù ¦prin·sə·pəl }

aufwuch [ECOL] A plant or animal organism which is attached or clings to surfaces of leaves or stems of rooted plants above the bottom stratum. { 'òf‚wək }

auganite [PETR] An olivine-free basalt (calcic plagioclase and augite are the essential mineral components) or an augite-bearing andesite. { 'òg·ə‚nīt }

augelite [MINERAL] Natural, basic aluminum phosphate. { 'òj·ə‚līt }

augen [PETR] Large, lenticular eye-shaped mineral grain or mineral aggregate visible in some metamorphic rocks. { 'òg·ən or 'aù·gən }

augend [MATH] A quantity to which another quantity is added. { 'ò‚jənd }

augen kohle See eye coal. { 'aù·gən ‚kōl·ə }

augen schist [PETR] A mylonitic rock characterized by the presence of recrystallization. { 'aù·gən ‚shist }

augen structure [PETR] A structure found in some gneisses and granites in which certain of the constituents are squeezed into elliptic or lens-shaped forms and, especially if surrounded by parallel flakes of mica, resemble eyes. { 'aù·gən ‚strək·chər }

auger [DES ENG] **1.** A wood-boring tool that consists of a shank with spiral channels ending in two spurs, a central tapered feed screw, and a pair of cutting lips. **2.** A large augerlike tool for boring into soil. { 'ò·gər }

auger bit [DES ENG] a A bit shaped like an auger but without a handle; used for wood boring and for earth drilling. [MIN ENG] Hard steel or tungsten carbide-tipped cutting teeth used in an auger running on a torque bar or in an auger-drill head running on a continuous-flight auger. { 'ò·gər ‚bit }

auger boring [ENG] **1.** The hole drilled by the use of auger equipment. **2.** See auger drilling. { 'ò·gər ‚bòr·iŋ }

Auger coefficient [ATOM PHYS] The ratio of the number of Auger electrons to the number of ejected x-ray photons. { ō'zhā ‚kō·ə'fish·ənt }

auger conveyor See screw conveyor. { 'ò·gər kən'vā·ər }

auger drilling [ENG] A method of drilling in which penetration is accomplished by the cutting or gouging action of chisel-type cutting edges forced into the substance by rotation of the auger bit. Also known as auger boring. { 'ò·gər ‚dril·iŋ }

Auger effect [ATOM PHYS] A two-electron process in which an electron makes a discrete transition from a less bound shell to a vacant electron shell and the energy gained in this process is tranferred via the electrostatic interaction to another bound electron which escapes the atom. Also known as Auger transition; internal absorption; internal photoionization. { ō'zhā i‚fekt }

Auger electron [ATOM PHYS] An electron that is expelled from an atom in the Auger effect. { ō'zhā i'lek‚trän }

Auger electron spectroscopy [SPECT] The energy analysis of Auger electrons produced when an excited atom relaxes by a radiationless process after ionization by a high-energy electron, ion, or x-ray beam. Abbreviated AES. { ō'zhā i'lek‚trän spek'träs·kə·pē }

auger mining [MIN ENG] A coal mining method that employs instruments to bore horizontally into coal seams. { 'ò·gər 'mīn·iŋ }

auger packer [MECH ENG] A feed mechanism that uses a continuous auger or screw inside a cylindrical sleeve to feed hard-to-flow granulated solids into shipping containers, such as bags or drums. { 'ò·gər 'pak·ər }

Auger recombination [ATOM PHYS] Recombination of an electron and a hole in which no electromagnetic radiation is emitted, and the excess energy and momentum of the recombining electron and hole are given up to another electron or hole. { ō'zhā ri‚käm·bə'nā·shən }

Auger shower [ASTRON] A very large cosmic-ray shower. Also known as extensive air shower. { ō'zhā ¦shaù·ər }

Auger transition See Auger effect. { ō'zhā tran‚zish·ən }

auget [ENG] A priming tube, used in blasting. Also spelled augette. { ō'zhet }

augette See auget. { ō'zhet }

augite [MINERAL] $(Ca,Mg,Fe)(Mg,Fe,Al)(Al,Si)_2O_6$ A general name for the monoclinic pyroxenes; occurs as dark green to black, short, stubby, prismatic crystals, often of octagonal outline. { 'ò‚jīt }

augitite [PETR] A volcanic rock consisting of abundant phenocrysts of augite in a glassy groundmass containing microlites of nepheline and plagioclase, with accessory biotite, apatite, and opaque oxides. { 'ò·jə‚tīt }

augitophyre [PETR] A porphyritic rock in which the phenocrysts are augite and the groundmass is potash feldspar. { ò'jid·ə‚fī·ər }

augmentation [ASTRON] The apparent increase in the semidiameter of a celestial body, as observed from the earth, as the body's altitude (angular distance above the horizon) increases, due to the reduced distance from the observer; used principally in reference to the moon. { ‚òg·mən'tā·shən }

augmentation correction [NAV] The sextant altitude correction due to the apparent increase in the semidiameter of a celestial body as its altitude increases. { ‚òg·mən'tā·shən kə‚rek·shən }

augmentation distance [NUC PHYS] The extrapolation distance, which is the distance between the time boundary of a nuclear reactor and its boundary calculated by extrapolation. { ‚òg·mən'tā·shən ‚dis·təns }

augmentation system [AERO ENG] An electronic servomechanism or feedback control system which provides improvements in aircraft performance or pilot handling characteristics over that of the basic unaugmented aircraft. { ‚òg·mən'tā·shən ‚sis·təm }

augmented matrix [MATH] The matrix of the coefficients, together with the constant terms, in a system of linear equations. { 'òg·men·təd 'mā·triks }

augmented operation code [COMPUT SCI] An operation code which is further defined by information from another portion of an instruction. { 'òg·men·təd äp·ə'rā·shən ‚kōd }

augmented plane-wave method [SOLID STATE] A method of approximating the energy states of electrons in a crystal lattice; the potential is assumed to be spherically symmetrical within spheres centered at each atomic nucleus and constant in the interstitial region, wave functions (the augmented plane waves) are constructed by matching solutions of the Schrödinger equation within each sphere with plane-wave solutions in the interstitial region, and linear combinations of these wave functions are then determined by the variational method. Abbreviated APW method. { $\text{òg}^{\shortmid}\text{ment·əd 'plān ,wāv ,meth·əd}$ }

augmenter tube [AERO ENG] A tube or pipe, usually one of several, through which the exhaust gases from an aircraft reciprocating engine are directed to provide additional thrust. { òg'men·tər ,tüb }

AUI See attachment unit interface.

auk [VERT ZOO] Any of several large, short-necked diving birds (*Alca*) of the family Alcidae found along North Atlantic coasts. { òk }

aulacogen [GEOL] A major fault-bounded trough considered to be one part of a three-rayed fault system on the domes above mantle hot spots; the other two rays open as proto-ocean basins. { ,aù'läk·ə·jən }

Aulodonta [INV ZOO] An order of echinoderms proposed by R. Jackson in 1921. { ,òl·ə'dän·tə }

aulodont dentition [INV ZOO] In echinoderms, grooved teeth with epiphyses that do not meet, so the foramen magnum of the jaw is open. { $\text{'òl·ə·,dänt ,den'tish·ən}$ }

Aulolepidae [PALEON] A family of marine fossil teleostean fishes in the order Ctenothrissiformes. { ,òl·ə'lep·ə,dē }

aulophobia [PSYCH] Abnormal fear of flutes. { ,òl·ə'fōb·ē·ə }

aulophyte [ECOL] A nonparasitic plant that lives in the cavity of another plant for shelter. { 'òl·ə,fīt }

Auloporidae [PALEON] A family of Paleozoic corals in the order Tabulata. { ,òl·ə'pór·ə,dē }

A* unit [PHYS] An atomic standard unit of length, based on the tungsten $K\alpha_1$ line, approximately 10^{-11} centimeter; used for measurements of x-ray wavelengths and of crystal dimensions. { 'ā,stär ,yü·nət }

auntie See antimissile missile. { 'an,tē }

Aur See Auriga.

aura [MED] An unusual sensation preceding the appearance of more definite symptoms; in epilepsy, auras frequently precede the convulsive seizure. { 'òr·ə }

aural [BIOL] Pertaining to the ear or the sense of hearing. { 'òr·əl }

auralization See virtual acoustics. { ,òr·əl·ə'zā·shən }

aural masking See masking. { 'òr·əl 'mask·iŋ }

aural null [NAV] A signal generated by the operator of a direction finder by the back-and-forth rotation of a loop antenna or the goniometer of an Adcock or other directional antenna system; the position of the rotating mechanism at which the null (minimum signal) is determined to be located indicates the azimuth or bearing of the radio station from which the emission is known to have originated. { 'òr·əl 'nəl }

aural radio range [ELECTR] A radio-range station providing lines of position by virtue of aural identification or comparison of signals at the output of a receiver. { $\text{'òr·əl 'rād·ē,ō ,rānj}$ }

aural signal [ACOUS] **1.** A signal that can be heard. **2.** The sound portion of a television signal. { 'òr·əl 'sig·nəl }

aural transmitter [COMMUN] Radio equipment used for transmitting aural (sound) signals from a television broadcast station. { $\text{'òr·əl ,tranz'mid·ər}$ }

auramine hydrochloride [ORG CHEM] $C_{17}H_{22}ClN_3\cdot H_2O$ A compound melting at 267°C; very soluble in water, soluble in ethanol; used as a dye and an antiseptic. Also known as yellow pyoktanin. { $\text{'òr·ə,mēn ,hī·drə'klór,īd}$ }

aurantia [ORG CHEM] $C_{12}H_8N_8O_{12}$ An orange aniline dye, used in stains in biology and in some photographic filters. { ò'ranch·ə }

aurantiin See naringin. { ò'ran·tē·ən }

aurelia [INV ZOO] A morphological grouping of paramecia, including the elongate, cigar-shaped species which appear to be nearly circular in cross section. { ò'rēl·yə }

Aurelia [INV ZOO] A genus of scyphozoans. { ò'rēl·yə }

aureofacin [MICROBIO] An antifungal antibiotic produced by a strain of *Streptomyces aureofaciens*. { ,òr·ē·ō'fās·ən }

aureole [GEOL] A ring-shaped contact zone surrounding an igneous intrusion. Also known as contact aureole; contact zone; exomorphic zone; metamorphic aureole; metamorphic

zone; thermal aureole. [METEOROL] A poorly developed corona in the atmosphere characterized by a bluish-white disk immediately around the luminous celestial body, as around the sun or moon in the fog. { 'òr·ē,ōl }

aureothricin [MICROBIO] $C_9H_{10}O_2N_2S_2$ An antibacterial antibiotic produced by a strain of *Actinomyces*. { ,òr·ē·ō'thrīs·ən }

aureusidin [BIOCHEM] $C_{15}H_{11}O_5$ A yellow flavonoid pigment found typically in the yellow snapdragon. Also known as 4,6,3',4'-tetrahydroxyaurone. { ,òr·e·ə'sīd·ən }

Auri See Auriga.

aurichalcite [MINERAL] $(Zn,Cu)_5(CO_3)_2(OH)_6$ Pale-green or pale-blue mineral consisting of a basic copper zinc carbonate and occurring in crystalline incrustations. Also known as brass ore. { ,òr·ə'kal,sīt }

auricle [ANAT] **1.** An ear-shaped appendage to an atrium of the heart. **2.** Any ear-shaped structure. **3.** See pinna. { 'òr·ə·kəl }

auric oxide See gold oxide. { 'òr·ik 'äk,sīd }

auricular fibrillation [MED] Arrhythmic contractions of the auricles. { $\text{ò'rik·yə·lər fib·rə'lā·shən}$ }

auricular flutter See atrial flutter. { $\text{ò'rik·yə·lər 'fləd·ər}$ }

auricular height [ANTHRO] Height of the cranium, measured from the auditory point to the vertex. { ò'rik·yə·lər 'hīt }

auricularia larva [INV ZOO] A barrel-shaped, food-gathering larval form with a winding ciliated band, common to holothurians and asteroids. { $\text{ò,rik·yə'lar·ē·ə 'lär·və}$ }

auricularis [ANAT] Any of the three muscles attached to the cartilage of the external ear. { $\text{ò}^{\shortmid}\text{rik·yə'lär·əs}$ }

auriferous [GEOL] Of a substance, especially a mineral deposit, bearing gold. { ò'rif·ə·rəs }

Auriga [ASTRON] A constellation with a right ascension of 6 hours and declination of 40°N. Abbreviated Aur; Auri. { ò'rī·gə }

aurin [ORG CHEM] $C_{19}H_{14}O_3$ A derivative of triphenylmethane; solid with red-brown color with green luster; melting point about 220°C; insoluble in water; used as a dye intermediate. { 'òr·ən }

aurophore [INV ZOO] A bell-shaped structure which is part of the float of certain cnidarians. { 'òr·ə,fór }

aurora [ELEC] See corona discharge. [GEOPHYS] The most intense of the several lights emitted by the earth's upper atmosphere, seen most often along the outer realms of the Arctic and Antarctic, where it is called the aurora borealis and aurora australis, respectively; excited by charged particles from space. { ə'rór·ə }

Aurora [ASTRON] An asteroid with a diameter of about 132 miles (220 kilometers), mean distance from the sun of 3.153 astronomical units, and C-type surface composition. { ə'rór·ə }

aurora australis [GEOPHYS] The aurora of southern latitudes. Also known as southern lights. { $\text{ə'rór·ə ò'strā·ləs}$ }

aurora borealis [GEOPHYS] The aurora of northern latitudes. Also known as northern lights. { $\text{ə'rór·ə ,bór·ē'al·əs}$ }

aurora gating [ELECTR] Operator-controlled gating to eliminate undesirable radar returns from aurora. { $\text{ə'rór·ə }^{\shortmid}\text{gād·iŋ}$ }

auroral [ASTRON] The period of dusk before sunrise. { ə'rór·əl }

auroral absorption event [GEOPHYS] A large increase in D-region electron density and associated radio-signal absorption, caused by electron-bombardment of the atmosphere during an aurora or a geomagnetic storm. { $\text{ə'rór·əl əb'zórp·shən i'vent}$ }

auroral caps [GEOPHYS] The regions surrounding the auroral poles, lying between the poles and the auroral zones. { ə'rór·əl ,kaps }

auroral electrojet [GEOPHYS] An intense electric current in the magnetosphere, flowing along the auroral zones during a polar substorm. { $\text{ə'rór·əl i'lek·trə,jet}$ }

auroral forms [GEOPHYS] Auroral display types, of which two are basic: ribbonlike bands and cloudlike surfaces. { ə'rór·əl 'fórmz }

auroral frequency [GEOPHYS] The percentage of nights on which an aurora is seen at a particular place, or on which one would be seen if clouds did not interfere. { $\text{ə'rór·əl 'frē·kwən,sē}$ }

auroral isochasm [GEOPHYS] A line connecting places of

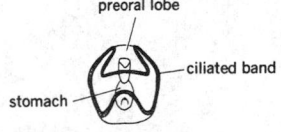

AULODONT DENTITION

tooth — epiphysis — foramen magnum — jaw — tooth — groove on inner surface of tooth — 1 cm

Lantern structure in aulodont dentition; a single interradial jaw is shown and below it a cross section of the tooth.

AURICULARIA LARVA

preoral lobe — ciliated band — stomach

Auricularia larva, showing the ciliated band.

equal auroral frequency, averaged over a number of years. { ə'rȯr·əl 'ī·sō,kaz·əm }

auroral line [SPECT] A prominent green line in the spectrum of the aurora at a wavelength of 5577 angstroms, resulting from a certain forbidden transition of oxygen. { ə'rȯr·əl ,līn }

auroral oval [GEOPHYS] An oval-shaped region centered on the earth's magnetic pole in which auroral emissions occur. { ə'rȯr·əl 'ō·vəl }

auroral poles [GEOPHYS] The points on the earth's surface on which the auroral isochasms are centered; coincide approximately with the magnetic-axis poles of the geomagnetic field. { ə'rȯr·əl 'pōlz }

auroral propagation [COMMUN] The propagation of radio waves that are reflected from the aurora in the presence of unusual solar activity. { ə'rȯr·əl ,präp·ə'gā·shən }

auroral region [GEOPHYS] The region within 30° geomagnetic latitude of each auroral pole. { ə'rȯr·əl ,rē·jən }

auroral storm [GEOPHYS] A rapid succession of auroral substorms, occurring in a short period, of the order of a day, during a geomagnetic storm. { ə'rȯr·əl ,stȯrm }

auroral substorm [GEOPHYS] A characteristic sequence of auroral intensifications and movements occurring around midnight, in which a rapid poleward movement of auroral arcs produces a bulge in the auroral oval. { ə'rȯr·əl 'səb,stȯrm }

auroral zone [GEOPHYS] A roughly circular band around either geomagnetic pole within which there is a maximum of auroral activity; lies about 10-15° geomagnetic latitude from the geomagnetic poles. { ə'rȯr·əl ,zōn }

auroral zone blackout [GEOPHYS] Communication fadeout in the auroral zone most often due to an increase of ionization in the lower atmosphere. { ə'rȯr·əl ,zōn 'blak,aut }

aurora polaris [GEOPHYS] A high-altitude aurora borealis or aurora australis. { ə'rȯr·əl pə'lar·əs }

aurosmiridium [MINERAL] A brittle, silver-white, isometric mineral consisting of a solid solution of gold and osmium in iridium. { ,ȯr·ō·smə'rid·ē·əm }

aurothioglucose [PHARM] $C_6H_{11}AuO_5S$ A compound of gold with thioglucose, used for the treatment of rheumatoid arthritis and nondisseminated lupus erythematosus; administered in oil suspension. { ,ȯr·ō,thī·ə'glü,kōs }

auscultation [MED] The act of listening to sounds from internal organs, especially the heart and lungs, to aid in diagnosing their physical state. { ,ȯs·kəl'tā·shən }

ausforging [MET] The forming of austenitic steel into required shapes by hammering or pressing after cooling. { ȯs'fȯrj·iŋ }

ausforming [MET] The processing of steel by subjecting it to deformation followed by quenching and tempering in order to increase strength, ductility, and resistance to fatigue. { 'aus,fȯrm·iŋ }

ausrolling [MET] The working of austenitic steel by passing it, after cooling, between oppositely revolving rollers. { ȯs'rōl·iŋ }

austausch coefficient See exchange coefficient. { 'aus,taush ,kō·ə'fish·ənt }

austempering [MET] A process for the heat treatment of austenitic steel. { ,ȯs'tem·pə·riŋ }

austenite [MET] Gamma iron with carbon in solution. { 'ȯs·tə,nīt }

austenitic [MET] Composed mainly of austenite. { ¦ȯs·tə¦nid·ik }

austenitic cast iron [MET] The product resulting from changing the basic crystalline structure of gray or ductile iron by alloying it with substantial amounts of nickel, manganese, silicon, or other elements. { ¦ȯs·tə¦nid·ik ¦kast 'ī·ərn }

austenitic manganese steel See Hadfield manganese steel. { ¦ȯs·tə¦nid·ik 'maŋ·ga,nēs 'stēl }

austenitic stainless steel [MET] Stainless steel composed principally of austenite made stable by alloying with nickel. { ¦ȯs·tə¦nid·ik 'stān·ləs 'stēl }

austenitic steel [MET] An alloy whose structure is typically that of austenite at room temperature. { ¦ȯs·tə¦nid·ik 'stēl }

austenitize [MET] To heat steel to the temperature range in which the crystalline form of the iron is austenite. { ¦ȯs·tə·nə,tīz }

auster See ostria. { 'ȯs·tər }

austinite [MINERAL] $CaZnAsO_4(OH)$ A colorless or yellowish mineral crystallizing in the orthorhombic system; consists of a basic calcium zinc arsenate; hardness is 4.5 on Mohs scale, and specific gravity is 4.13. { 'ȯs·tə,nīt }

austral [GEOD] Pertaining to south. { 'ȯs·trəl }

austral axis pole [GEOPHYS] The southern intersection of the geomagnetic axis with the earth's surface. { 'ȯs·trəl ¦ak·səs ,pōl }

Australia [GEOGR] An island continent of 2,941,526 square miles (7,618,517 square kilometers), with low elevation and moderate relief, situated in the southern Pacific. { ȯ'strāl·yə }

Australia antigen [IMMUNOL] An infectious agent that causes hepatitis in some people; similar to an inherited serum protein in being polymorphic. { ȯ'strāl·yə 'ant·i·jən }

Australian faunal region [ECOL] A zoogeographic region that includes the terrestrial animal communities of Australia and all surrounding islands except those of Asia. { ȯ'strāl·yən 'fȯn·əl ,rē·jən }

Australian X disease See Murray Valley encephalitis. { ȯ'strāl·yən 'eks di,zēz }

australite [GEOL] A tektite found in southern Australia, occurring as glass balls and spheroidal dumbbell forms of green and black, similar to obsidian and probably of cosmic origin. { 'ȯs·trə,līt }

Australopithecinae [PALEON] The near-men, a subfamily of the family Hominidae composed of the single genus *Australopithecus*. { ȯ,strā·lō,pith·ə'sī·nē }

Australopithecus [PALEON] A genus of near-men in the subfamily Australopithecinae representing a side branch of human evolution. { ȯ,strā·lō'pith·ə·kəs }

austral region [ECOL] A North American biogeographic region including the region between transitional and tropical zones. { 'ȯs·trəl ,rē·jən }

Austrian orogeny [GEOL] A short-lived orogeny during the end of the Early Cretaceous. { 'ȯs·trē·ən ȯ'räj·ə·nē }

Austroastacidae [INV ZOO] A family of crayfish in the order Decapoda found in temperate regions of the Southern Hemisphere. { ¦ȯs·trō·ə'stās·ə,dē }

Austrodecidae [INV ZOO] A monogeneric family of marine arthropods in the subphylum Pycnogonida. { ,ȯs·trə'des·ə,dē }

Austroriparian life zone [ECOL] The zone in which occurs the climate and biotic communities of the southeastern coniferous forests of North America. { ¦ȯs·trō,rī'per·ē·ən 'līf ,zōn }

autallotriomorphic [PETR] Pertaining to an aplitic texture in which all mineral constituents crystallized simultaneously, preventing the development of euhedral crystals. { ¦aud·ə¦lä·trē·ə¦mȯr·fik }

autecology See autoecology. { ,ȯd·i'käl·ə·jē }

authalic latitude [MAP] A latitude based on a sphere having the same area as the spheroid, and such that areas between successive parallels of latitude are exactly equal to the corresponding areas on the spheroid. Also known as equal-area latitude. { ȯ'thal·ik 'lad·ə·tüd }

authalic map projection See equal-area map projection. { ȯ'thal·ik 'map prə,jek·shən }

authentication [COMMUN] Security measure designed to protect a communications system against fraudulent transmissions and establish the authenticity of a message. { ə,thent·ə'kā·shən }

authenticator [COMMUN] Letter, numeral, or groups of letters or numerals attesting to the authenticity of a message or transmission. { ə'thent·ə,kād·ər }

authigene [MINERAL] A mineral which has not been transported but has been formed in place. Also known as authigenic mineral. { 'ȯ·thə,jēn }

authigenic [GEOL] Of constituents that came into existence with or after the formation of the rock of which they constitute a part; for example, the primary and secondary minerals of igneous rocks. { 'ȯ·thə¦jen·ik }

authigenic mineral See authigene. { 'ȯ·thə¦jen·ik 'min·rəl }

authigenic sediment [GEOL] Sediment occurring in the place where it was originally formed. { 'ȯ·thə¦jen·ik 'sed·ə·mənt }

authoring language [COMPUT SCI] A programming language designed to be convenient for authors of computer-based learning materials. { 'ȯ·thər·iŋ 'laŋ·gwij }

authoritarian character [PSYCH] A personality that asks unquestioning subordination and obedience but is intolerant of weakness in others, rejects members of groups other than his own, is very rigid, and requires all issues to be decided in black and white, yet may accept superior authority in a servile way. { ȯ‚thȯr·ə'ter·ē·ən 'kar·ik·tər }

authorization code [COMPUT SCI] A password or identifying number that is used to gain access to a computer system. { ‚ȯth·ə·rə'zā·shən ‚kōd }

authorized carrier frequency [COMMUN] A specific carrier frequency authorized for use, from which the actual carrier frequency is permitted to deviate, solely because of frequency instability, by an amount not to exceed the frequency tolerance. { 'ȯ·thə‚rīzd 'kar·ē·ər ‚frē·kwən·sē }

authorized library [COMPUT SCI] A group of authorized programs. { 'ȯ·thə‚rīzed 'li‚brer·ē }

authorized program [COMPUT SCI] A computer program that can alter the fundamental operation or status of a computer system. { 'ȯ·thə‚rizd 'prō·grəm }

autism [PSYCH] A schizophrenic symptom characterized by absorption in fantasy to the exclusion of perceptual reality. { 'ȯ‚tiz·əm }

autistic [PSYCH] **1.** Pertaining to or characterized by autism. **2.** Behavior, most commonly in children, characterized by aphonia, disregard for reality, self-manipulation, and sometimes a preoccupation with pointed or shiny objects. { ȯ'tis·tik }

auto-abstract [COMPUT SCI] **1.** To select key words from a document, commonly by an automatic or machine method, for the purpose of forming an abstract of the document. **2.** The material abstracted from a document by machine methods. { ‚ȯd·ō'ab‚strakt }

autoacceleration [ORG CHEM] The increase in polymerization rate and molecular weight of certain vinyl monomers during bulk polymerization. { ‚ȯd·ō·ik‚sel·ə'rā·shən }

autoaccelerator [NUCLEO] A linear induction accelerator in which air-core cavities excited by the electron beam's self-fields are used to generate the accelerating field. { ‚ȯd·ō·ik'sel·ə‚rād·ər }

autoadaptivity [CONT SYS] The ability of an advanced robot to sense the environment, accept commands, and analyze and execute operations. { ‚ȯd·ō·‚ə‚dap'tiv·əd·ē }

autoagglutination [IMMUNOL] Agglutination of an individual's erythrocytes by his own serum. Also known as autohemagglutination. { ‚ȯd·ō·ə‚glüt·ən'ā·sh; ppn }

autoagglutinin [IMMUNOL] An antibody in an individual's blood serum that causes agglutination of his own erythrocytes. { ‚ȯd·ō·ə'glüt·ən·ən }

autoalarm *See* automatic alarm receiver. { 'ȯd·ō·ə‚lärm }

autoallelopathy [PL PATH] Inhibition of a species by self-produced toxins. { ‚ȯd·ō‚a·lə'lä·pə·thē }

autoanalysis [PSYCH] Self-analysis, a technique of psychotherapy. { ‚ȯd·ō·ə'nal·ə·səs }

auto answer [COMMUN] The feature of a modem that receives the telephone ring for an incoming call and accepts the call to establish a connection. { ‚ȯd·ō 'an·sər }

autoantibody [IMMUNOL] An antibody formed by an individual against his own tissues; common in hemolytic anemias. { ‚ȯd·ō'ant·i‚bäd·ē }

autoantigen [IMMUNOL] A tissue within the body which acquires the ability to incite the formation of complementary antibodies. { ‚ȯd·ō'ant·i·jən }

autoasphyxiation [PHYSIO] Asphyxiation by the products of metabolic activity. { ‚ȯd·ō·as‚fik·sē'ā·shən }

autobarotropy [METEOROL] The state of a fluid which is characterized by both barotropy and piezotropy when the coefficients of barotropy and piezotropy are equal. { ‚ȯd·ō·bə'rä·trə·pē }

autobasidium [MYCOL] An undivided basidium typically found in higher Basidiomycetes. { ‚ȯd·ō·bə'sid·ē·əm }

autoboat *See* motorboat. { 'ȯd·ō‚bōt }

autobrecciation [GEOL] The process whereby portions of the first consolidated crust of a lava flow are incorporated into the still-fluid portion. { ‚ȯd·ō‚brech·ē'ā·shən }

auto bypass [COMPUT SCI] The ability of a computer network to bypass a terminal or other device if it fails, allowing other devices connected to the network to continue operation. { ‚ȯd·ō 'bī‚pas }

autocall [COMPUT SCI] The automatic placing of a telephone call by a computer or a computer-controlled modem. Also known as automatic call origination. { 'ȯd·ō‚kȯl }

autocarp [BOT] **1.** A fruit formed as the result of self-fertilization. **2.** A fruit consisting of the ripened pericarp without adnate parts. { 'ȯd·ō‚kärp }

autocarpy [BOT] Production of fruit by self-fertilization. { 'ȯd·ō‚kärp·ē }

autocatalysis [CHEM] A catalytic reaction started by the products of a reaction that was itself catalytic. { ‚ȯd·ō·kə'tal·ə·səs }

autochory [ECOL] Active self-dispersal of individuals or their disseminules. { 'ȯd·ō‚kȯr·ē }

autochrome plate [GRAPHICS] A photographic plate once used for direct color photography on coated glass; the glass must be viewed as a transparency, with the image formed in the film of dyed starch grains coating the glass. { 'ȯd·ō‚krōm ‚plāt }

autochthon [GEOL] A succession of rock beds that have been moved comparatively little from their original site of formation, although they may be folded and faulted extensively. [PALEON] A fossil occurring where the organism once lived. { ȯ'täk·thən }

autochthonous [ECOL] Pertaining to organisms or organic sediments that are indigenous to a given ecosystem. [GEOL] Having been formed or occurring in the place where found. { ȯ'täk·thə·nas }

autochthonous coal [GEOL] Coal believed to have originated from accumulations of plant debris at the place where the plants grew. Also known as indigenous coal. { ȯ'täk·thə·nas 'kōl }

autochthonous microorganism [MICROBIO] An indigenous form of soil microorganisms, responsible for chemical processes that occur in the soil under normal conditions. { ȯ'täk·thə·nas ‚mī·krō'ȯr·gə‚niz·əm }

autochthonous sediment [GEOL] A residual soil deposit formed in place through decomposition. { ȯ'täk·thə·nas 'sed·ə·mənt }

autochthonous stream [HYD] A stream flowing in its original channel. { ȯ'täk·thə·nas 'strēm }

autoclastic [GEOL] Of rock, fragmented in place by folding due to orogenic forces when the rock is not so heavily loaded as to render it plastic. { ‚ȯd·ō‚klas·tik }

autoclastic schist [GEOL] Schist formed in place from massive rocks by crushing and squeezing. { ‚ȯd·ō‚klas·tik 'shist }

autoclave [ENG] An airtight vessel for heating and sometimes agitating its contents under high steam pressure; used for industrial processing, sterilizing, and cooking with moist or dry heat at high temperatures. { 'ȯd·ō‚klāv }

autoclave curing [ENG] Steam curing of concrete products, sand-lime brick, asbestos cement products, hydrous calcium silicate insulation products, or cement in an autoclave at maximum ambient temperatures generally between 340 and 420°F (170 and 215°C). { 'ȯd·ō‚klāv 'kyūr·iŋ }

autoclave molding [ENG] A method of curing reinforced plastics that uses an autoclave with 50-100 pounds per square inch (345-690 kilopascals) steam pressure to set the resin. { 'ȯd·ō‚klāv 'mōld·iŋ }

autoclaving [SCI TECH] Heating of liquids or sterilizing of equipment at high steam pressure. { 'ȯd·ō‚klāv·iŋ }

autocode [COMPUT SCI] The process of using a computer to convert automatically a symbolic code into a machine code. Also known as automatic code. { 'ȯd·ō‚kōd }

autocoder [COMPUT SCI] A person or machine producing or using autocode as a part or the whole of a task. { 'ȯd·ō‚kōd·ər }

autocoenobium [INV ZOO] An asexually produced coenobium that is a miniature of the parent. { ‚ȯd·ō·sē'nō·bē·əm }

autocollimation [OPTICS] A procedure for collimating a telescope or other optical instrument with objective and crosshairs, in which the instrument is directed toward a plane mirror and the crosshairs and lens are adjusted so that the crosshairs coincide with their reflected image. { ‚ȯd·ō‚käl·ə'mā·shən }

autocollimator [OPTICS] **1.** A device by which a single lens collimates diverging light from a slit, and then focuses the light on an exit slit after it has passed through a prism to a mirror and been reflected back through the prism. **2.** A telescope which has a graduated reticle, enabling an observer to read off the angles subtended by distant objects. **3.** A convex mirror at the focus of the principal mirror of a reflecting telescope, which causes light to leave the telescope in a parallel beam.

4. A telescope equipped with an eyepiece designed for autocollimation. { ¦òd·ō'käl·ə¦mäd·ər }

autoconsequent falls [HYD] Waterfalls in streams carrying a heavy load of calcium carbonate in solution which develop at particular sites along the stream course where warming, evaporation, and other factors cause part of the solution load to be precipitated. { ¦òd·ō'kän·sə·kwənt 'fòlz }

autoconsequent stream [HYD] A stream in the process of building a fan or an alluvial plain, the course of which is guided by the slopes of the alluvium the stream itself has deposited. { ¦òd·ō'kän·sə·kwənt 'strēm }

autoconvection [METEOROL] The phenomenon of the spontaneous initiation of convection in an atmospheric layer in which the lapse rate is equal to or greater than the autoconvective lapse rate. { ¦òd·ō·kən'vek·shən }

autoconvection gradient See autoconvective lapse rate. { ¦òd·ō·kən'vek·shən 'grād·ē·ənt }

autoconvective instability See absolute instability. { ¦òd·ō·kən'vek·tiv ¸in·stə'bil·əd·ē }

autoconvective lapse rate [METEOROL] The largely hypothetical environmental lapse rate of temperature in an atmosphere in which the density is constant with height (homogeneous atmosphere); 3°C per 100 meters in dry air. Also known as autoconvection gradient. { ¦òd·ō·kən'vek·tiv 'laps ¸rāt }

autocopulation [INV ZOO] Self-copulation; sometimes occurs in certain hermaphroditic worms. { ¦òd·ō¸käp·yə'lā·shən }

autocorrelation [ELECTR] A technique used to detect cyclic activity in a complex signal. [STAT] In a time series, the relationship between values of a variable taken at certain times in the series and values of a variable taken at other, usually earlier times. { ¦òd·ō¸kär·ə'lā·shən }

autocorrelation function [MATH] For a specified function $f(t)$, the average value of the product $f(t)f(t - \tau)$, where τ is a time-delay parameter; more specifically, the limit as T approaches infinity of $1/(2T)$ times the integral from $-T$ to T of $F(t)f(t - \tau) dt$. { ¦òd·ō¸kär·ə'lā·shən ¸fuŋk·shən }

autocorrelator [ELECTR] A correlator in which the input signal is delayed and multiplied by the undelayed signal, the product of which is then smoothed in a low-pass filter to give an approximate computation of the autocorrelation function; used to detect a nonperiodic signal or a weak periodic signal hidden in noise. { ¸òd·ō'kär·ə¸lād·ər }

autocrine signalling [PHYSIO] Signaling in which cells respond to substances that they themselves release. { ¦òd·ə¸krīn 'sig·nəl·iŋ }

autodecrement addressing [COMPUT SCI] An addressing mode of minicomputers in which the register is first decremented and then used as a pointer. { ¸òd·ō'dek·rə·mənt ə'dres·iŋ }

autodeme [ECOL] A plant population in which most individuals are self-fertilized. { ¦òd·ō¸dēm }

autodetachment [ATOM PHYS] A process in which one of the excited electrons in a doubly excited state of a negative ion is ejected, while the other excited electron drops back to a lower energy level. { ¸òd·ō·di'tach·mənt }

auto dial [COMMUN] The feature of a modem that automatically opens a telephone line and dials the telephone of a receiving computer to establish a connection. { ¦òd·ō 'dīl }

autodyne circuit [ELECTR] A circuit in which the same tube elements serve as oscillator and detector simultaneously. { 'òd·ō¸dīn ¸sər·kət }

autodyne reception [COMMUN] System of heterodyne reception through the use of a device which is both an oscillator and a detector. { 'òd·ō¸dīn ri'sep·shən }

autoecious [BOT] See autoicous. [MYCOL] Referring to a parasitic fungus that completes its entire life cycle on a single host. { ò'tēsh·əs }

autoecology [ECOL] The study of how a particular species responds to the environment. Also spelled autecology. { ¸òd·ō·i'käl·ə·jē }

autoeroticism See autoerotism. { ¸òd·ō·ə'räd·ə¸siz·əm }

autoerotism [PSYCH] Sexual arousal or gratification using one's own body. Also known as autoeroticism. { ¸òd·ō'er·ə¸tiz·əm }

autofocus rectifier [OPTICS] A precise, vertical photoenlarger which permits the correction of distortion in an aerial negative caused by tilt. { 'òd·ō¸fō·kəs 'rek·tə¸fī·ər }

autofrettage [ENG] A process for manufacturing gun barrels; prestressing the metal increases the load at which its permanent deformation occurs. { 'òd·ō¸fred·ij }

autogamy [BIOL] A process of self-fertilization that results in homozygosis; occurs in some flowering plants, fungi, and protozoans. { 'òd·ō¸gam·ē }

autogenesis See abiogenesis. { ¦òd·ō·'jen·ə·səs }

autogenetic drainage [HYD] A self-established drainage system developed solely by headwater erosion. { ¦òd·ō·jə¦ned·ik 'drān·ij }

autogenetic topography [GEOL] Conformation of land due to the physical action of rain and streams. { ¦òd·ō·jə¦ned·ik tə'päg·rə·fē }

autogenous [BIOL] Originating or derived from sources within the same individual. [SCI TECH] Self-generated; produced without external influence. { ò'täj·ə·nəs }

autogenous control [MOL BIO] Regulation of gene expression by a product of the gene itself that either inhibits or enhances the gene's activity. { ò'täj·ə·nəs kən'trōl }

autogenous electrification [PHYS] The process by which net charge is built up on an object, such as an airplane, moving relative to air containing dust or ice crystals; produced by frictional effects (triboelectrification) accompanying contact between the object and the particulate matter. { ò'täj·ə·nəs i¸lek·trə·fə'kā·shən }

autogenous grinding [MECH ENG] The secondary grinding of material by tumbling the material in a revolving cylinder, without balls or bars taking part in the operation. { ò'täj·ə·nəs 'grīnd·iŋ }

autogenous healing [ENG] A natural process of closing and filling cracks in concrete or mortar while it is kept damp. { ò'täj·ə·nəs 'hēl·iŋ }

autogenous ignition temperature See ignition temperature. { ò'täj·ə·nəs ig'nish·ən ¸tem·prə·chər }

autogenous insect [INV ZOO] Any insect in which adult females can produce eggs without first feeding. { ò'täj·ə·nəs 'in¸sekt }

autogenous mill See autogenous tumbling mill. { ò'täj·ə·nəs 'mil }

autogenous tumbling mill [MECH ENG] A type of ball-mill grinder utilizing as the grinding medium the coarse feed (incoming) material. Also known as autogenous mill. { ò'täj·ə·nəs 'təm·bliŋ ¸mil }

autogenous vaccine [IMMUNOL] A vaccine prepared from a culture of microorganisms taken directly from the infected person. { ò'täj·ə·nəs ¸vak'sēn }

autogenous volume change [MATER] The change in volume produced by continued hydration of cement, exclusive of effects of external forces or change of water content or temperature. { ò'täj·ə·nəs 'väl·yəm ¸chānj }

autogenous welding [MET] A fusion welding process using heat without the addition of filler metal to join two pieces of the same metal. { ò'täj·ə·nəs 'weld·iŋ }

autogeosyncline [GEOL] A parageosyncline that subsides as an elliptical basin or trough nearly without associated highlands. Also known as intracratonic basin. { ¦òd·ō¦jē·ō'sin¸klīn }

autogiro [AERO ENG] A type of aircraft which utilizes a rotating wing (rotor) to provide lift and a conventional engine-propeller combination to propel the vehicle through the air. { ¸òd·ō'jī·rō }

autograft See autotransplant. { 'òd·ō¸graft }

autohemagglutination See autoagglutination. { ¦òd·ō¸hēm·ə¸glüt·ən'ā·shən }

autohemolysis [PHYSIO] The spontaneous lysis of blood which occurs during an incubation in a hematological procedure. { ¦òd·ō·hə'mäl·ə·səs }

autohemorrhage [INV ZOO] Voluntary exudation or ejection of nauseous or poisonous blood by certain insects as a defense against predators. { ¦òd·ō'hem·rij }

autohemotherapy [MED] Treatment of disease with the patient's own blood, withdrawn by venipuncture and then injected intramuscularly. { ¦òd·ō¸hēm·ə'ther·ə¸pē }

autoicous [BOT] Having male and female organs on the same plant but on different branches. Also spelled autoecious. { ò'tòi·kəs }

autoignition [MECH ENG] Spontaneous ignition of some or all of the fuel-air mixture in the combustion chamber of an

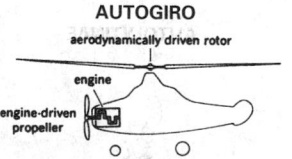

AUTOGIRO

aerodynamically driven rotor

engine

engine-driven propeller

The in-flight dynamic components of a typical autogiro.

internal combustion engine. Also known as spontaneous combustion. { ¦ȯd·ō·ig¦nish·ən }

autoignition temperature [CHEM] The temperature at which a material (solid, liquid, or gas) will self-ignite and sustain combustion in air without an external spark or flame. { ¦ȯd·ō·ig¦nish·ən 'tem·prə·chər }

autoimmune disease [IMMUNOL] An illness involving the formation of autoantibodies which appear to cause pathological damage to the host. { ¦ȯd·ō·ə'myün di‚zēz }

autoimmune hemolytic anemia [IMMUNOL] An autoimmune disease in which antibodies initiate complement lysis of red blood cells. { ¦ȯd·ō·i¦myün ‚hē·mə¦lid·ik ə'nēm·yə }

autoimmunity [IMMUNOL] An immune state in which antibodies are formed against the person's own body tissues. { ¦ȯd·ō·ə'myün·əd·ē }

autoimmunization [IMMUNOL] Immunization obtained by natural processes within the body. { ¦ȯd·ō‚i·myə·nə'zā·shən }

autoincrement addressing [COMPUT SCI] An addressing mode of minicomputers in which the operand address is gotten from the specified register which is then incremented. { ¦ȯd·ō'in‚krə·mənt ə'dres·iŋ }

autoindexing See automatic indexing. { ¦ȯd·ō'in‚deks·iŋ }

autoinfection [MED] Reinfection by an organism existing within the body or transferred from one part of the body to another. { ¦ȯd·ō‚in'fek·shən }

autoinjection See autointrusion. { ¦ȯd·ō‚in'jek·shən }

autoinoculation [MED] **1.** Spread of a disease from one part of the body to another. **2.** Injection of an autovaccine. { ¦ȯd·ō·in‚äk·yə'lā·shən }

autointoxication [MED] Poisoning by metabolic products elaborated within the body; generally, toxemia of pathologic states. { ¦ȯd·ō·in‚täk·sə'kā·shən }

autointrusion [GEOL] A process wherein the residual liquid of a differentiating magma is drawn into rifts formed in the crystal mesh at a late stage by deformation of unspecified origin. Also known as autoinjection. { ¦ȯd·ō·in'trü·zhən }

autoionization [ATOM PHYS] The radiationless transition of an electron in an atom from a discrete electronic level to an ionized continuum level of the same energy. Also known as preionization. { ¦ȯd·ō‚ī·ə·nə'zā·shən }

autolith [PETR] **1.** A fragment of igneous rock enclosed in another igneous rock of later consolidation, each being regarded as a derivative from a common parent magma. **2.** A round, oval, or elongated accumulation of iron-magnesium minerals of uncertain origin in granitoid rock. { 'ȯd·ō‚lith }

autolithography [GRAPHICS] A lithographic process in which the artist makes a drawing onto the printing surface directly. { ‚ȯd·ō·lith'äg·rə·fē }

autologous [BIOL] Derived from or produced by the individual in question, such as an autologous protein or an autologous graft. { ȯ'täl·ə·gəs }

autoluminescence [ATOM PHYS] Luminescence of a material (such as a radioactive substance) resulting from energy originating within the material itself. { ¦ȯd·ō‚lü·mə'nes·əns }

autolysis [GEOCHEM] Return of a substance to solution, as of phosphate removed from seawater by plankton and returned when these organisms die and decay. [PATH] Self-digestion by body cells following somatic or organ death or ischemic injury. { ȯ'täl·ə·səs }

autolysosome See autophagic vacuole. { ¦ȯd·ō'lī·sə‚sōm }

autolytic enzyme [BIOCHEM] A bacterial enzyme, located in the cell wall, that causes disintegration of the cell following injury or death. { ¦ȯd·əl¦id·ik 'en‚zīm }

Autolytinae [INV ZOO] A subfamily of errantian polychaetes in the family Syllidae. { ‚ȯd·ō'lid·ə·nē }

automanual system [CIV ENG] A railroad signal system in which signals are set manually but are activated automatically to return to the danger position by a passing train. { ¦ȯd·ō¦man· yə·wəl 'sis·təm }

automata theory [MATH] A theory concerned with models used to simulate objects and processes such as computers, digital circuits, nervous systems, cellular growth and reproduction. { ȯ'täm·əd·ə 'thē·ə·rē }

automated decision making [COMPUT SCI] The use of computers to carry out tasks requiring the generation or selection of options. { ¦ȯd·ə‚mād·əd di'sizh·ən ‚māk·iŋ }

automated fingerprint identification system [FOREN SCI] A searchable database of finger and palm print records used to verify the identity of users of systems or of criminals, or to link unsolved crimes. { ¦ȯd·ə‚mäd·əd ¦fiŋ·gər‚print ī‚den·tə·fə'kā·shən ‚sis·təm }

automated guided vehicle [IND ENG] In a flexible manufacturing system, a driverless computer-controlled vehicle equipped with guidance and collision-avoidance systems and used to transport workpieces and tools between work stations. Abbreviated AGV. { ‚ȯd·ə¦mäd·əd ¦gīd·əd 've·ə·kəl }

automated guided vehicle system [CONT SYS] A computer-controlled system that uses pallets and other interface equipment to transport workpieces to numerically controlled machine tools and other equipment in a flexible manufacturing system, moving in a predetermined pattern to ensure automatic, accurate, and rapid work-machine contact. { 'ȯd·ə‚mäd·əd ¦gīd·əd 've·ə·kəl ‚sis·təm }

automated identification system [COMPUT SCI] In a data processing system, the use of a technology such as bar coding, image recognition, or voice recognition instead of keyboarding for data entry. { ‚ȯd·ə¦mäd·əd ī‚den·tə·fə'ka·shən ‚sis·təm }

automated radar plotting aid [NAV] A marine computer-based anticollision system that automatically processes time coordinates of radar echo signals into space coordinates in digital form, determines consecutive coordinates and motion parameters of targets, calculates the predicted closest point of approach and time to closest point of approach and presents them in graphic or alphanumeric form on the radar display, and switches on alarms if there is a danger of collision. { 'ȯd· ə‚mäd·əd ¦rä‚där 'pläd·iŋ ‚ād }

automated radar terminal system [NAV] A system for carrying out air-traffic control in the vicinity of airports which uses both airport surveillance radar and the air-traffic radar beacon system; radar video, representing aircraft targets, is presented on the air-traffic controllers' displays, and the automation system automatically tracks controlled aircraft and presents alpha-numeric information adjacent to their targets. Abbreviated ARTS. { 'ȯd·ə‚mäd·əd 'rä‚där 'tərm·ən·əl ‚sis· təm }

automated tape library [COMPUT SCI] A computer storage system consisting of several thousand magnetic tapes and equipment under computer control which automatically brings the tapes from storage, mounts them on tape drives, dismounts the tapes when the job is completed, and returns them to storage. { 'ȯd·ə‚mäd·əd 'tāp ¦lī‚brer·ē }

automatic [ENG] Having a self-acting mechanism that performs a required act at a predetermined time or in response to certain conditions. [ORD] See automatic weapon. { ¦ȯd· ə¦mad·ik }

automatic abstracting [COMPUT SCI] Techniques whereby, on the basis of statistical properties, a subset of the sentences in a document is selected as representative of the general content of that document. { ¦ȯd·ə¦mad·ik 'ab‚strakt·iŋ }

automatic acceleration See dynamic resolution. { ¦ȯd·ə¦mad· ik ik‚sel·ə'rā·shən }

automatic alarm receiver [ELECTR] A complete receiving, selecting, and warning device capable of being actuated automatically by intercepted radio-frequency waves forming the international automatic alarm signal. Also known as autoalarm. { ¦ȯd·ə¦mad·ik ə'lärm ri‚sē·vər }

automatic-alarm-signal keying device [COMMUN] A device capable of automatically keying the radiotelegraph transmitter on board a vessel to transmit the international automatic-alarm signal, or to respond to receipt of an internationally agreed-upon distress signal and wake up the radio operator on ships not having a 24-hour radio watch. { ¦ȯd·ə¦mad·ik ə'lärm ‚sig·nəl 'kē·iŋ di‚vīs }

automatic back bias [ELECTR] Radar technique which consists of one or more automatic gain control loops to prevent overloading of a receiver by large signals, whether jamming or actual radar echoes. { ¦ȯd·ə¦mad·ik 'bak ‚bī·əs }

automatic background control See automatic brightness control. { ¦ȯd·ə¦mad·ik 'bak‚graúnd kən‚trōl }

automatic balance [ENG] A balance capable of performing weighing procedures without the intervention of an operator. { ¦ȯd·ə¦mad·ik 'ba·ləns }

automatic bass compensation [ELECTR] A circuit related to the volume control in some radio receivers and audio amplifiers to make bass notes sound properly balanced, in the audio spectrum, at low volume-control settings. { ¦ȯd·ə¦mad·ik 'bās ‚käm·pən'sā·shən }

automatic batcher [MECH ENG] A batcher for concrete

AUTOLYTINAE

Procerastea, dorsal view.

which is actuated by a single starter switch, opens automatically at the start of the weighing operations of each material, and closes automatically when the designated weight of each material has been reached.   { ¦öd·ə¦mad·ik 'bach·ər }

automatic bias   [ELECTR]   A method of obtaining the correct bias for a vacuum tube or transistor through use of a resistor, usually in the cathode or emitter circuit.   { ¦öd·ə¦mad·ik 'bī·əs }

automatic brazing   [MET]   Brazing by the use of either portable or stationary equipment which does not require constant supervision by the operator.   { ¦öd·ə¦mad·ik 'brāz·iŋ }

automatic breech mechanism   [ORD]   A device that utilizes the energy of recoil, or the pressure of the powder gases, to open the breech, withdraw the fired cartridge case, insert a new cartridge, and close the breech.   { ¦öd·ə¦mad·ik 'brēch ,mek·ə'niz·əm }

automatic brightness control   [ELECTR]   A circuit used in a television receiver to keep the average brightness of the reproduced image essentially constant. Abbreviated ABC. Also known as automatic background control.   { ¦öd·ə¦mad·ik 'brīt·nəs kən,trōl }

automatic calibration   [ENG]   A process in which an electronic device automatically performs the recalibration of a measuring range of a weighing instrument, for example an electronic balance.   { ¦öd·ə¦mad·ik ,kal·ə'brā·shən }

automatic calling unit   [COMPUT SCI]   A device that enables a business machine or computer to automatically dial calls over a communications network.   { ¦öd·ə¦mad·ik 'kȯl·iŋ ,yü·nət }

automatic call origination   See autocall.   { ¦öd·ə¦mad·ik 'kȯl·ə,rij·ə'nā·shən }

automatic carriage   [COMPUT SCI]   Any mechanism designed to feed continuous paper or plastic forms through a printing or writing device, often using sprockets to engage holes in the paper.   { ¦öd·ə¦mad·ik 'kar·ij }

automatic casing hanger   [PETRO ENG]   Unitized hanger-seal assembly latched at the lower end of an oil-well casing string to support the next smaller string and make a seal between the two strings.   { ¦öd·ə¦mad·ik 'kās·iŋ ,haŋ·ər }

automatic C bias   See self-bias.   { ¦öd·ə¦mad·ik 'sē ,bī·əs }

automatic celestial navigation   See celestial-inertial guidance.   { ¦öd·ə¦mad·ik sə'les·chəl ,nav·ə'gā·shən }

automatic character recognition   [COMPUT SCI]   The technology of using special machine systems to identify human-readable symbols, most often alphanumeric, and then to utilize this data.   { ¦öd·ə¦mad·ik 'kar·ik·tər ,rek·ig'nish·ən }

automatic check   [COMPUT SCI]   An error-detecting procedure performed by a computer as an integral part of the normal operation of a device, with no human attention required unless an error is actually detected.   { ¦öd·ə¦mad·ik 'chek }

automatic check-out system   [CONT SYS]   A system utilizing test equipment capable of automatically and simultaneously providing actions and information which will ultimately result in the efficient operation of tested equipment while keeping time to a minimum.   { ¦öd·ə¦mad·ik 'chek,aut ,sis·təm }

automatic choke   [MECH ENG]   A system for enriching the air-fuel mixture in a cold automotive engine when the accelerator is first depressed; the choke plate opens automatically when the engine achieves normal operating temperature.   { ¦öd·ə¦mad·ik 'chōk }

automatic chroma control   See automatic color control.   { ¦öd·ə¦mad·ik 'krōm·ə kən,trōl }

automatic chrominance control   See automatic color control.   { ¦öd·ə¦mad·ik 'krōm·ə·nəns kən,trōl }

automatic code   See autocode.   { ¦öd·ə¦mad·ik 'kōd }

automatic coding   [COMPUT SCI]   Any technique in which a computer is used to help bridge the gap between some intellectual and manual form of describing the steps to be followed in solving a given problem, and some final coding of the same problem for a given computer.   { ¦öd·ə¦mad·ik 'kōd·iŋ }

automatic color control   [ELECTR]   A circuit used in a color television receiver to keep color intensity levels essentially constant despite variations in the strength of the received color signal; control is usually achieved by varying the gain of the chrominance band-pass amplifier. Also known as automatic chroma control; automatic chrominance control.   { ¦öd·ə¦mad·ik 'kəl·ər kən,trōl }

automatic computer   [COMPUT SCI]   A computer which can carry out a special set of operations without human intervention.   { ¦öd·ə¦mad·ik kəm'pyüd·ər }

automatic connection   [ELECTR]   Ability of electronic switching equipment to make a connection between users without human intervention.   { ¦öd·ə¦mad·ik kə'nek·shən }

automatic contrast control   [ELECTR]   A circuit that varies the gain of the radio-frequency and video intermediate-frequency amplifiers in such a way that the contrast of the television picture is maintained at a constant average level.   { ¦öd·ə¦mad·ik 'kän,trast kən,trōl }

automatic control   [CONT SYS]   Control in which regulating and switching operations are performed automatically in response to predetermined conditions. Also known as automatic regulation.   { ¦öd·ə¦mad·ik kən,trōl }

automatic control balance   [ENG]   An automatic balance fitted with an accessory which determines whether a package has been filled within preselected limits. Also known as check-weigher.   { ¦öd·ə¦mad·ik kən,trōl ,bal·əns }

automatic-control block diagram   [CONT SYS]   A diagrammatic representation of the mathematical relationships defining the flow of information and energy through the automatic control system, in which the components of the control system are represented as functional blocks in series and parallel arrangements according to their position in the actual control system.   { ¦öd·ə¦mad·ik kən'trōl 'bläk ,dī·ə,gram }

automatic-control error coefficient   [CONT SYS]   Three numerical quantities that are used as a measure of the steady-state errors of an automatic control system when the system is subjected to constant, ramp, or parabolic inputs.   { ¦öd·ə¦mad·ik kən'trōl 'er·ər ,kō·ə'fish·ənt }

automatic-control frequency response   [CONT SYS]   The steady-state output of an automatic control system for sinusoidal inputs of varying frequency.   { ¦öd·ə¦mad·ik 'frē·kwən·sē ri,späns }

automatic controller   [CONT SYS]   An instrument that continuously measures the value of a variable quantity or condition and then automatically acts on the controlled equipment to correct any deviation from a desired preset value. Also known as automatic regulator; controller.   { ¦öd·ə¦mad·ik kən¦trōl·ər }

automatic-control servo valve   [CONT SYS]   A mechanically or electrically actuated servo valve controlling the direction and volume of fluid flow in a hydraulic automatic control system.   { ¦öd·ə¦mad·ik kən'trōl 'sər·vō ,valv }

automatic-control stability   [CONT SYS]   The property of an automatic control system whose performance is such that the amplitude of transient oscillations decreases with time and the system reaches a steady state.   { ¦öd·ə¦mad·ik kən'trōl stə,bil·ə·dē }

automatic-control system   [CONT SYS]   A control system having one or more automatic controllers connected in closed loops with one or more processes. Also known as regulating system.   { ¦öd·ə¦mad·ik kən'trōl ,sis·təm }

automatic-control transient analysis   [CONT SYS]   The analysis of the behavior of the output variable of an automatic control system as the system changes from one steady-state condition to another in terms of such quantities as maximum overshoot, rise time, and response time.   { ¦öd·ə¦mad·ik kən'trōl 'tran·zhənt ə,nal·ə·səs }

automatic coupling   [MECH ENG]   A device which couples rail cars when they are bumped together.   { ¦öd·ə¦mad·ik 'kəp·liŋ }

automatic custody transfer   [PETRO ENG]   Automatic system for measuring and sampling oil or petroleum products at points of receipt or delivery.   { ¦öd·ə¦mad·ik 'kəs·təd·ē ,tranz·fər }

automatic cutout   [ELEC]   A device, usually operated by centrifugal force or by an electromagnet, that automatically shorts part of a circuit at a particular time.   { ¦öd·ə¦mad·ik 'kəd,aut }

automatic dam   [MIN ENG]   In placer mining, a dam with a gate that automatically discharges the water when it reaches a certain height behind the dam.   { ¦öd·ə¦mad·ik 'dam }

automatic data processing   [ENG]   The machine performance, with little or no human assistance, of any of a variety of tasks involving informational data; examples include automatic and responsive reading, computation, writing, speaking, directing artillery, and the running of an entire factory. Abbreviated ADP.   { ¦öd·ə¦mad·ik 'dad·ə 'präs,əs·iŋ }

automatic degausser   [ELECTR]   An arrangement of degaussing coils mounted around a color television picture tube, combined with a special circuit that energizes these coils only while the set is warming up; demagnetizes any parts of the receiver that have been affected by the magnetic field of the earth or of any nearby home appliance. { ¦ȯd·ə¦mad·ik dē'gaús·ər }

automatic dialer [ELECTR] A device in which a telephone number up to a maximum of 14 digits can be stored in a memory and then activated, directly into the line, by the caller's pressing a button. Also known as mechanical dialer. { ¦ȯd·ə¦mad·ik 'dīl·ər }

automatic dialog replacement studio See ADR studio. { ¦ȯd·ə¦mad·ik ¸dī·ə¸läg ri'plās·mənt ¸stüd·ē¸ō }

automatic dictionary [COMPUT SCI] Any table within a computer memory which establishes a one-to-one correspondence between two sets of characters. { ¦ȯd·ə¦mad·ik 'dik·shə¸ner·ē }

automatic direction finder [ELECTR] A direction finder that without manual manipulation indicates the direction of arrival of a radio signal. Abbreviated ADF. Also known as radio compass. { ¦ȯd·ə¦mad·ik di'rek·shən ¸fīnd·ər }

automatic door bottom [ENG] A movable plunger, in the form of a horizontal bar at the bottom of a door, which drops automatically when the door is closed, sealing the threshold and reducing noise transmission. Also known as automatic threshold closer. { ¦ȯd·ə¦mad·ik 'dȯr ¸bäd·əm }

automatic drill [DES ENG] A straight brace for bits whose shank comprises a coarse-pitch screw sliding in a threaded tube with a handle at the end; the device is operated by pushing the handle. { ¦ȯd·ə¦mad·ik 'dril }

automatic error correction [COMMUN] A technique, usually requiring the use of special codes or automatic retransmission, which detects and corrects errors occurring in transmission; the degree of correction depends upon coding and equipment configuration. { ¦ȯd·ə¦mad·ik 'er·ər kə'rek·shən }

automatic exchange [ELECTR] A telephone, teletypewriter, or data-transmission exchange in which communication between subscribers is effected, without the intervention of an operator, by devices set in operation by the originating subscriber's instrument. Also known as automatic switching system; machine switching system. { ¦ȯd·ə¦mad·ik iks'chanj }

automatic exposure [OPTICS] Photoelectric exposure control by a special device that maintains an essentially constant exposure in the focal plane for a given range of field luminance. { ¦ȯd·ə¦mad·ik ik'spō·zhər }

automatic fillup shoe [PETRO ENG] A device, installed on the first joint of casing, that functions to regulate automatically the volume of mud in the casing. { ¦ȯd·ə¦mad·ik 'fil¸əp ¸shü }

automatic fine-tuning control [ELECTR] A circuit used in a color television receiver to maintain the correct oscillator frequency in the tuner for best color picture by compensating for drift and incorrect tuning. { ¦ȯd·ə¦mad·ik ¸fīn 'tün·iŋ kən¸trōl }

automatic fire [ORD] Continuous fire from an automatic gun until the pressure on the trigger is released. { ¦ȯd·ə¦mad·ik 'fīr }

automatic fire pump [MECH ENG] A pump which provides the required water pressure in a fire standpipe or sprinkler system; when the water pressure in the system drops below a preselected value, a sensor causes the pump to start. { ¦ȯd·ə¦mad·ik ¸fīr ¸pəmp }

automatic flushing system [CIV ENG] A water tank system which provides automatically for the periodic flushing of urinals or other plumbing fixtures, or of pipes having too small a slope to drain effectively. { ¦ȯd·ə¦mad·ik 'fləsh·iŋ ¸sis·təm }

automatic focus [OPTICS] A device in a camera or enlarger which automatically keeps the objective lens in focus through a range of magnification. { ¦ȯd·ə¦mad·ik 'fō·kəs }

automatic frequency control [ELECTR] Abbreviated AFC. **1.** A circuit used to maintain the frequency of an oscillator within specified limits, as in a transmitter. **2.** A circuit used to keep a superheterodyne receiver tuned accurately to a given frequency by controlling its local oscillator, as in an FM receiver. **3.** A circuit used in radar superheterodyne receivers to vary the local oscillator frequency so as to compensate for changes in the frequency of the received echo signal. **4.** A circuit used in television receivers to make the frequency of a sweep oscillator correspond to the frequency of the synchronizing pulses in the received signal. { ¦ȯd·ə¦mad·ik 'frē·kwən·sē kən¸trōl }

automatic gain control [ELECTR] A control circuit that automatically changes the gain (amplification) of a receiver or other piece of equipment so that the desired output signal remains essentially constant despite variations in input signal strength. Abbreviated AGC. { ¦ȯd·ə¦mad·ik 'gān kən¸trōl }

automatic grid bias See self-bias. { ¦ȯd·ə¦mad·ik 'grid ¸bī·əs }

automatic gun See automatic weapon. { ¦ȯd·ə¦mad·ik 'gən }

automatic gun charger [ORD] A gun charger that includes a mechanism for the clearance of gun stoppages and for the retention of the breech mechanism to the rear of the gun receiver. { ¦ȯd·ə¦mad·ik 'gən ¸chärj·ər }

automatic head parking [COMPUT SCI] A feature that moves the read/write head of a hard disk over the landing zone whenever power is shut off to ensure against a head crash. { ¦ȯd·ə¦mad·ik 'hed ¸pärk·iŋ }

automatic horizon See artificial horizon. { ¦ȯd·ə¦mad·ik hə'rīz·ən }

automatic ignition [ENG] A device that lights the fuel in a gas burner when the gas-control valve is turned on. { ¦ȯd·ə¦mad·ik ig'nish·ən }

automatic indexing [COMPUT SCI] Selection of key words from a document by computer for use as index entries. Also known as autoindexing. [CONT SYS] The procedure for determining the orientation and position of a workpiece with respect to an automatically controlled machine, such as a robot manipulator, that is to perform an operation on it. { ¦ȯd·ə¦mad·ik 'in¸deks·iŋ }

automatic intercept [COMMUN] Telephone service that automatically records messages a caller may leave when the called party is away from his telephone. { ¦ȯd·ə¦mad·ik 'in·tər¸sept }

automatic interrupt [COMPUT SCI] Interruption of a computer program brought about by a hardware device or executive program acting as a result of some event which has occurred independently of the interrupted program. { ¦ȯd·ə¦mad·ik 'in·tə¸rəpt }

automatic landing system [NAV] The means for automatically guiding and controlling aircraft from an initial approach altitude to a point where safe contact is made with the landing surface. { ¦ȯd·ə¦mad·ik 'land·iŋ ¸sis·təm }

automatic level compensation [COMMUN] System which automatically compensates for amplitude variations in a circuit. { ¦ȯd·ə¦mad·ik 'lev·əl ¸käm·pen'sā·shən }

automatic level control [ELECTR] A circuit that keeps the output of a radio transmitter, tape recorder, or other device essentially constant, even in the presence of large changes in the input amplitude. Abbreviated ALC. [MECH ENG] In an automotive vehicle, a system in which two air-chamber shock absorbers in the rear are fed compressed air by an electric compressor; pressure in the air chambers is determined automatically by sensors to maintain the vehicle at a predetermined height regardless of load. { ¦ȯd·ə¦mad·ik 'lev·əl kən¸trōl }

automatic light control [ELECTR] Automatic adjustment of illumination reaching a film, television camera, or other imaging device as a function of scene brightness. { ¦ȯd·ə¦mad·ik 'līt kən¸trōl }

automatic mathematical translator [COMPUT SCI] An automatic-programming computer capable of receiving a mathematical equation from a remote input and returning an immediate solution. { ¦ȯd·ə¦mad·ik ¸math·ə'mad·ə·kəl 'tranz¸lād·ər }

automatic message accounting [COMMUN] See automatic toll ticketing. { ¦ȯd·ə¦mad·ik 'mes·ij ə¸kaunt·iŋ }

automatic message-switching center [COMMUN] A center in which messages are automatically routed according to information in them. { ¦ȯd·ə¦mad·ik 'mes·ij ¸swich·iŋ ¸sen·tər }

automatic microfiller [ENG] A device used to place microfilm in jackets at relatively high speeds. { ¦ȯd·ə¦mad·ik 'mi·krō¸fil·ər }

automatic modulation control [ELECTR] A transmitter circuit that reduces the gain for excessively strong audio input signals without affecting the strength of normal signals, thereby permitting higher average modulation without overmodulation. Abbreviated AMC. { ¦ȯd·ə¦mad·ik ¸mäj·ə'lā·shən kən¸trōl }

automatic modulation limiting [COMMUN] A circuit that prevents overmodulation in some citizen-band radio transmitters by reducing the gain of one or more audio amplifier stages when the voice signal becomes stronger. Abbreviated AML. { ¦òd·ə¦mad·ik mäj·ə'lā·shən ¸lim·əd·iŋ }

automatic mold [ENG] A mold used in injection or compression molding of plastic objects so that repeated molding cycles are possible, including ejection, without manual assistance. { ¦òd·ə¦mad·ik 'mōld }

automatic noise limiter [ELECTR] A circuit that clips impulse and static noise peaks, and sets the level of limiting or clipping according to the strength of the incoming signal, so that the desired signal is not affected. Abbreviated ANL. { ¦òd·ə¦mad·ik 'nòiz ¸lim·əd·ər }

automatic peak limiter See limiter. { ¦òd·ə¦mad·ik 'pēk ¸lim·əd·ər }

automatic phase control [ELECTR] Abbreviated APC. **1.** A circuit used in color television receivers to reinsert a 3.58-megahertz carrier signal with exactly the correct phase and frequency by synchronizing it with the transmitted color-burst signal. **2.** An automatic frequency-control circuit in which the difference between two frequency sources is fed to a phase detector that produces the required control signal. { ¦òd·ə¦mad·ik 'fāz kən¸trōl }

automatic picture control [ELECTR] A multiple-contact switch used in some color television receivers to disconnect one or more of the regular controls and make connections to corresponding preset controls. { ¦òd·ə¦mad·ik 'pik·chər kən¸trōl }

automatic picture-transmission system [ELECTR] A system in which a meteorological satellite continuously scans and transmits a view of a transverse swath directly beneath it; transmissions can be recorded by simple ground equipment to reconstruct an image of the cloud patterns within a thousand kilometers of the ground station. Abbreviated APT system. { ¦òd·ə¦mad·ik 'pik·chər tranz'mish·ən ¸sis·təm }

automatic pilot [NAV] Also known as autopilot. **1.** Equipment which automatically stabilizes the attitude of an aircraft about its pitch, roll, and yaw axes and keeps it on a predetermined heading. **2.** Equipment for automatically steering ships. { ¦òd·ə¦mad·ik 'pī·lət }

automatic pistol [ORD] A pistol able to operate as an automatic weapon. { ¦òd·ə¦mad·ik 'pis·təl }

automatic press [MECH ENG] A press in which mechanical feeding of the work is synchronized with the press action. { ¦òd·ə¦mad·ik 'pres }

automatic programming [COMPUT SCI] The preparation of machine-language instructions by use of a computer. { ¦òd·ə¦mad·ik 'prō¸gram·iŋ }

Automatic Programming Tool [COMPUT SCI] A computer language used to program numerically controlled machine tools. Abbreviated APT. { ¦òd·ə¦mad·ik 'prō¸gram·iŋ ¸tül }

automatic pumping station [CHEM ENG] An installation on a pipeline that automatically provides the proper pressure when a fluid is being transported. { ¦òd·ə¦mad·ik 'pəmp·iŋ ¸stā·shən }

automatic ranging See autoranging. { ¦òd·ə¦mad·ik 'rānj·iŋ }

automatic record changer [ENG ACOUS] An electric phonograph that automatically plays a number of records one after another. { ¦òd·ə¦mad·ik 'rek·ord ¸chānj·ər }

automatic regulation See automatic control. { ¦òd·ə¦mad·ik ¸reg·yə'lā·shən }

automatic regulator See automatic controller. { ¦òd·ə¦mad·ik 'reg·yə¸lād·ər }

automatic relay [COMMUN] Means of selective switching which causes automatic equipment to record and retransmit communications. { ¦òd·ə¦mad·ik 'rē¸lā }

automatic repeat request [COMPUT SCI] A request from a receiving device to retransmit the most recent block of data. Abbreviated ARQ. { ¦òd·ə¦mad·ik ri'pēt ri¸kwest }

automatic rifle [ORD] A rifle capable of operating as an automatic weapon. { ¦òd·ə¦mad·ik 'rīf·əl }

automatic routine [COMPUT SCI] A routine that is executed independently of manual operations, but only if certain conditions occur within a program or record, or during some other process. { ¦òd·ə¦mad·ik rü'tēn }

automatic sampler [MECH ENG] A mechanical device to sample process streams (gas, liquid, or solid) either continuously or at preset time intervals. { ¦òd·ə¦mad·ik 'sam·plər }

automatic scanning receiver [ELECTR] A receiver which can automatically and continuously sweep across a preselected frequency, either to stop when a signal is found or to plot signal

occupancy within the frequency spectrum being swept. { ¦òd·ə¦mad·ik 'skan·iŋ ri¸sē·vər }

automatic screw machine [MECH ENG] A machine designed to automatically produce finished parts from bar stock at high production rates; the term is not an exact, specific machine-tool classification. { ¦òd·ə¦mad·ik 'skrü mə¸shēn }

automatic sensitivity control [ELECTR] Circuit used for automatically maintaining receiver sensitivity at a predetermined level; it is similar to automatic gain control, but it affects the receiver constantly rather than during the brief interval selected by the range gate. { ¦òd·ə¦mad·ik sen·sə'tiv·əd·ē kən¸trōl }

automatic sequences [COMPUT SCI] The characteristic of a computer that can perform successive operations without human intervention. { ¦òd·ə¦mad·ik 'sē·kwən·səs }

automatic short-circuiter [ELEC] Device designed to automatically short-circuit the commutator bars in some forms of single-phase commutator motors. { ¦òd·ə¦mad·ik ¸shòrt 'sər·kəd·ər }

automatic shutdown [COMPUT SCI] A procedure whereby a network or computer system stops work in an orderly fashion with as little data loss and other damage as possible when the system's software determines that it has encountered unacceptable conditions. { ¦òd·ə¦mad·ik 'shət¸daùn }

automatic shut-off [ENG ACOUS] A switch in some tape recorders which automatically stops the machine when the tape ends or breaks. { ¦òd·ə¦mad·ik 'shəd¸òf }

automatic slips [ENG] A pneumatic or hydraulic device for setting and removing slips automatically. Also known as power slips. { ¦òd·ə¦mad·ik 'slips }

automatic speed sensing [COMPUT SCI] The capability of a modem to automatically determine the maximum rate of data transfer over a connection. { ¦òd·ə¦mad·ik 'spēd ¸sen·siŋ }

automatic spider [MIN ENG] A foot or hydraulically actuated drill-rod clamping device similar to a Wommer safety clamp. { ¦òd·ə¦mad·ik 'spīd·ər }

automatic stability [AERO ENG] Stability achieved with the controls operated by automatic devices, as by an automatic pilot. { ¦òd·ə¦mad·ik stə'bil·əd·ē }

automatic stabilization equipment [AERO ENG] Apparatus which automatically operates control devices to maintain an aircraft in a stable condition. { ¦òd·ə¦mad·ik ¸stā·bə·lə'zā·shən i¸kwip·mənt }

automatic stoker [MECH ENG] A device that supplies fuel to a boiler furnace by mechanical means. Also known as mechanical stoker. { ¦òd·ə¦mad·ik 'stōk·ər }

automatic stop [COMPUT SCI] An automatic halting of a computer processing operation as the result of an error detected by built-in checking devices. { ¦òd·ə¦mad·ik 'stäp }

automatic switchboard [COMMUN] Telephone switchboard in which the connections are made by using remotely controlled switches. { ¦òd·ə¦mad·ik 'swich¸bòrd }

automatic switching system See automatic exchange. { ¦òd·ə¦mad·ik 'swich·iŋ ¸sis·təm }

automatic tank battery [PETRO ENG] Interconnected system of storage tanks with automatic controls to direct incoming oil to empty tanks in a desired sequence. { ¦òd·ə¦mad·ik 'tank 'bad·ə·rē }

automatic telegraph transmission [COMMUN] Form of telegraphy in which telegraph signals are transmitted mechanically from a perforated tape. { ¦òd·ə¦mad·ik 'tel·ə¸graf tranz·mish·ən }

automatic teller machine [COMPUT SCI] A banking terminal that is activated by inserting a magnetic card containing the user's account number, and that accepts deposits, dispenses cash, provides information about current balances, and may perform other services such as making payments and transfers and providing account statements. Abbreviated ATM. { ¦òd·ə¦mad·ik 'tel·ər mə¸shēn }

automatic test equipment [ENG] Test equipment that makes two or more tests in sequence without manual intervention; it usually stops when the first out-of-tolerance value is detected. { ¦òd·ə¦mad·ik 'test i¸kwip·mənt }

automatic threshold closer See automatic door bottom. { ¦òd·ə¦mad·ik 'thresh¸hōld ¸klōz·ər }

automatic threshold variation [ELECTR] Constant false-alarm rate scheme that is an open-loop of automatic gain control

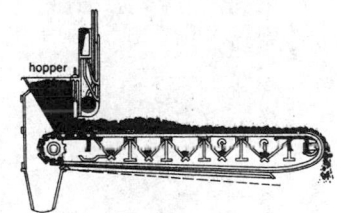

AUTOMATIC STOKER

A chain grate, one type of automatic stoker.

in which the decision threshold is varied continuously in proportion to the incoming intermediate frequency and video noise level. { ¦ȯd·ə¦mad·ik 'thresh‚hōld ‚yer·ē'ā·shən }

automatic time switch [ENG] Combination of a switch with an electric or spring-wound clock, arranged to turn an apparatus on and off at predetermined times. { ¦ȯd·ə¦mad·ik ‚tīm ‚swich }

automatic tint control [ELECTR] A circuit used in color television receivers to maintain correct flesh tones when a station changes cameras or switches to commercials, by correcting phase errors before the chroma signal is demodulated. { ¦ȯd·ə¦mad·ik 'tint kən‚trōl }

automatic titrator [ANALY CHEM] **1.** Titration with quantitative reaction and measured flow of reactant. **2.** Electrically generated reactant with potentiometric, amperometric, or colorimetric end-point or null-point determination. { ¦ȯd·ə¦mad·ik 'tī‚trād·ər }

automatic toll ticketing [COMMUN] System whereby toll calls are automatically recorded and timed, and toll tickets printed, under control of the calling telephone's dial pulses and without the intervention of an operator. Also known as automatic message accounting. { ¦ȯd·ə¦mad·ik 'tōl ‚tik· əd·iŋ }

automatic tracking [NAV] **1.** Tracking in which a servomechanism automatically follows some characteristic of the signal; specifically, a process by which tracking or data-acquisition systems are enabled to keep their antennas continuously directed at a moving target without manual operation. **2.** An instrument which displays the actual course made good through the use of navigation derived from several sources. { ¦ȯd·ə¦mad·ik 'trak·iŋ }

automatic track shift [ENG ACOUS] A system used with multiple-track magnetic tape recorders to index the tape head, after one track is played, to the correct position for the start of the next track. { ¦ȯd·ə¦mad·ik 'trak ‚shift }

automatic transfer equipment [ELEC] Equipment which automatically transfers a load so that a source of power may be selected from one of several incoming lines. { ¦ȯd·ə¦mad·ik 'tranz‚fər i‚kwip·mənt }

automatic traverse computer [NAV] An instrument which, by mechanical linkages and gears, converts courses into latitudes and departures; also totals the algebraic sum of the latitudes and departures. { ¦ȯd·ə¦mad·ik trə'vərs kəm‚pyüd·ər }

automatic tuning system [CONT SYS] An electrical, mechanical, or electromechanical system that tunes a radio receiver or transmitter automatically to a predetermined frequency when a button or lever is pressed, a knob turned, or a telephone-type dial operated. { ¦ȯd·ə¦mad·ik 'tün·iŋ ‚sis·təm }

automatic-type belt-tensioning device [MECH ENG] Any device which maintains a predetermined tension in a conveyor belt. { ¦ȯd·ə¦mad·ik ‚tīp 'belt ‚ten·shən·iŋ di‚vīs }

automatic video noise leveling [ELECTR] Constant false-alarm rate scheme in which the video noise level at the output of the receiver is sampled at the end of each range sweep and the receiver gain is readjusted accordingly to maintain a constant video noise level at the output. { ¦ȯd·ə¦mad·ik 'vid·ē·ō 'nȯiz ‚lev·əl·iŋ }

automatic voltage regulator See voltage regulator. { ¦ȯd·ə¦mad·ik 'vōl·tij ‚reg·yə‚lād·ər }

automatic volume compressor See volume compressor. { ¦ȯd·ə¦mad·ik 'väl·yəm kəm‚pres·ər }

automatic volume control [ELECTR] An automatic gain control that keeps the output volume of a radio receiver essentially constant despite variations in input-signal strength during fading or when tuning from station to station. Abbreviated AVC. { ¦ȯd·ə¦mad·ik 'väl·yəm kən‚trōl }

automatic volume expander See volume expander. { ¦ȯd·ə¦mad·ik 'väl·yəm ik‚spand·ər }

automatic wagon control [MIN ENG] A mechanism to keep the speed of wagons within certain designed limits; may consist of small hydraulic units fixed at intervals along the inside of the track. { ¦ȯd·ə¦mad·ik 'wag·ən kən‚trōl }

automatic weapon [ORD] A gun that fires, extracts, ejects, and reloads without application of power from an outside source, repeating the cycle as long as the firing mechanism is held in the proper position. Also known as automatic; automatic gun. { ¦ȯd·ə¦mad·ik 'wep·ən }

automatic weather station [METEOROL] A weather station at which the services of an observer are not required; usually equipped with telemetric apparatus. { ¦ȯd·ə¦mad·ik 'weth·ər ‚stā·shən }

automatic welding [MET] Electric-arc welding with automatic control of the arc movement along the welding line, the electrode feed, and the arc-gap length. { ¦ȯd·ə¦mad·ik 'weld·iŋ }

automatic wet-pipe sprinkler system [ENG] A sprinkler system, all of whose parts are filled with water at sufficient pressure to provide an immediate continuous discharge if the system is activated. { ¦ȯd·ə¦mad·ik ¦wet ¦pīp 'spriŋk·lər ‚sis·təm }

automatic zero setting [ENG] A system for automatic correction of zero-point drifts or for compensation of soiling of load receivers on a balance by means of a special accessory component. { ¦ȯd·ə¦mad·ik 'zir·ō ‚sed·iŋ }

automation [ENG] **1.** The use of technology to ease human labor or extend the mental or physical capabilities of humans. **2.** The mechanisms, machines, and systems that save or eliminate labor, or imitate actions typically associated with human beings. { ‚ȯd·ə'mā·shən }

automatism [BIOL] Spontaneous activity of tissues or cells. [MED] An act performed with no apparent exercise of will, as in sleepwalking and certain hysterical and epileptic states. { ȯ'täm·ə‚tiz·əm }

automaton [COMPUT SCI] A robot which functions without step-by-step guidance by a human operator. { ȯ'täm·ə‚tän }

automechanism [CONT SYS] A machine or other device that operates automatically or under control of a servomechanism. { ¦ȯd·ō'mek·ə‚niz·əm }

autometamorphism [PETR] Metamorphism of an igneous rock by the action of its own volatile fluids. Also known as autometasomatism. { ¦ȯd·ō‚med·ə'mȯr‚fiz·əm }

autometasomatism See autometamorphism. { ¦ȯd·ō‚med·ə'sō·mə‚tiz·əm }

automobile [MECH ENG] A four-wheeled, trackless, self-propelled vehicle for land transportation of as many as eight people. Also known as car. { ‚ȯd·ə·mə'bēl }

automobile chassis [MECH ENG] The automobile frame, together with the wheels, power train, brakes, engine, and steering system. { ‚ȯd·ə·mə'bēl 'chas·ē }

automonitor [COMPUT SCI] A computer program used in debugging which instructs a computer to make a record of its own operations. { ¦ȯd·ō'män·əd·ər }

automorphic [PETR] Of minerals in igneous rock bounded by their own crystal faces. Also known as euhedral; idiomorphic. { ¦ȯd·ō¦mȯr·fik }

automorphism [MATH] An isomorphism of an algebraic structure with itself. { ¦ȯd·ō'mȯr‚fiz·əm }

automorphosis [PETR] Metamorphosis of solidified igneous rock by solutions from its heated interior. { ‚ȯd·ə'mȯr·fə·səs }

automotive air conditioning [MECH ENG] A system for maintaining comfort of occupants of automobiles, buses, and trucks, limited to air cooling, air heating, ventilation, and occasionally dehumidification. { ¦ȯd·ə¦mōd·iv 'er kən‚dish·ən·iŋ }

automotive alternator [ELEC] An ac generator used in an automotive vehicle to provide current for the vehicle's electrical systems. { ¦ȯd·ə¦mōd·iv 'ȯl·tə‚nād·ər }

automotive body [ENG] An enclosure mounted on and attached to the frame of an automotive vehicle, to contain passengers and luggage, or in the case of commercial vehicles the commodities being carried. { ¦ȯd·ə¦mōd·iv 'bäd·ē }

automotive brake [MECH ENG] A friction mechanism that slows or stops the rotation of the wheels of an automotive vehicle, so that tire traction slows or stops the vehicle. { ¦ȯd·ə¦mōd·iv 'brāk }

automotive engine [MECH ENG] The fuel-consuming machine that provides the motive power for automobiles, airplanes, tractors, buses, and motorcycles and is carried in the vehicle. { ¦ȯd·ə¦mōd·iv 'en·jən }

automotive engineering [MECH ENG] The branch of mechanical engineering concerned primarily with the special problems of land transportation by a four-wheeled, trackless, automotive vehicle. { ¦ȯd·ə¦mōd·iv ‚en·jə'nir·iŋ }

automotive frame [ENG] The basic structure of all automotive vehicles, except tractors, which is supported by the suspension and upon which or attached to which are the power plant,

AUTOMOTIVE ALTERNATOR

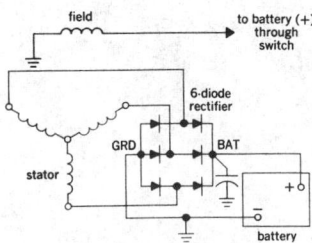

Wiring circuit of an automotive alternator. (Delco-Remy Division, General Motors Corp.)

transmission, clutch, and body or seat for the driver. { ˌȯd·ə¦mōd·iv ˈfrām }

automotive fuel [MATER] A material, generally a liquid fuel, gasoline, or distillate, whose combustion is used to supply chemical energy to provide the power for an automotive vehicle. { ˌȯd·ə¦mōd·iv ˈfyül }

automotive ignition system [MECH ENG] A device in an automotive vehicle which initiates the chemical reaction between fuel and air in the cylinder charge. { ˌȯd·ə¦mōd·iv igˈnish·ən ˌsis·təm }

automotive steering [MECH ENG] Mechanical means by which a driver controls the course of a moving automobile, bus, truck, or tractor. { ˌȯd·ə¦mōd·iv ˈstir·iŋ }

automotive suspension [MECH ENG] The springs and related parts intermediate between the wheels and frame of an automotive vehicle that support the frame on the wheels and absorb road shock caused by passage of the wheels over irregularities. { ˌȯd·ə¦mōd·iv səsˈpen·chən }

automotive transmission [MECH ENG] A device for providing different gear or drive ratios between the engine and drive wheels of an automotive vehicle, a principal function being to enable the vehicle to accelerate from rest through a wide speed range while the engine operates within its most effective range. { ˌȯd·ə¦mōd·iv tranzˈmish·ən }

automotive vehicle [MECH ENG] A self-propelled vehicle or machine for land transportation of people or commodities or for moving materials, such as a passenger car, bus, truck, motorcycle, tractor, airplane, motorboat, or earthmover. { ˌȯd·ə¦mōd·iv ˈvē·ə·kəl }

automotive voltage regulator [ELEC] A device in the automotive electrical system to prevent generator or alternator overvoltage. { ˌȯd·ə¦mōd·iv ˈvōl·tij ˌreg·yəˌlād·ər }

automutagen [GEN] Any mutagenic chemical formed as a product of metabolism. { ˌȯd·ōˈmyu·də·jən }

autonavigator [NAV] A navigation system that automatically measures absolute vehicle motions and computes distance and direction from the departure point. { ˌȯd·ōˈnav·əˌgād·er }

autonomic agent [NEURO] A compound that reduces or enhances nerve-impulse transmission across synaptic junctions, especially in the autonomic nervous system. { ˌȯd·əˈnäm·ik ˈā·jənt }

autonomic movement [BOT] A plant movement that results from internal growth changes and is independent of changes in the external environment. { ˌȯd·əˈnäm·ik ˈmüv·mənt }

autonomic nervous system [NEURO] The visceral or involuntary division of the nervous system in vertebrates, which enervates glands, viscera, and smooth, cardiac, and some striated muscles. { ˌȯd·əˈnäm·ik ˈnər·vəs ˌsis·təm }

autonomic reflex system [PHYSIO] An involuntary biological control system characterized by the uncontrolled functioning of smooth muscles and glands to maintain an acceptable internal environment. { ˌȯd·əˈnäm·ik ˈrēˌfleks ˌsis·təm }

autonomous channel operation [COMPUT SCI] The rapid transfer of data between computer peripherals and the main store in which an entire block of data is transferred, word by word; the cycles of storage time for the word transfer are stolen from those available to the central processing unit. { ȯˈtän·ə·məs ˈchan·əl ˌäp·əˈrā·shən }

autonomous robot [ENG] A robot that not only can maintain its own stability as it moves, but also can plan its movements. { ȯˈtän·ə·məs ˈrō·bät }

autonomous underwater vehicle [OCEANOGR] A crewless, untethered submersible which operates independent of direct human control. Abbreviated AUV. { ȯˈtän·ə·məs ˌən·dərˌwȯd·ər ˈvē·ə·kəl }

autonomous vehicle [ENG] A vehicle that is able to plan its path and to execute its plan without human intervention. { ȯˈtän·ə·məs ˈvē·ə·kəl }

autooxidation See autoxidation. { ˌȯd·ōˌäk·səˈdā·shən }

autopatch [ELECTR] A device for connecting radio transceivers to telephone lines by remote control, generally through the use of repeaters. { ˈȯd·ōˌpach }

autopatrol [MECH ENG] A self-powered blade grader. Also known as motor grader. { ˈȯd·ōˌpəˌtrōl }

autophagic vacuole [CYTOL] A membrane-bound cellular organelle that engulfs pieces of the substance of the cell itself. Also known as autolysosome. { ˌȯd·ōˈfā·jik ˈvak·yəˌwōl }

autophagocytosis [CYTOL] The cellular process of phagocytizing a portion of protoplasm by a vacuole within the cell. { ˌȯd·ōˌfag·ə·sīˈtō·səs }

autophagy [CYTOL] The cellular process of self-digestion. { ȯˈtäf·əˌfā·jē }

autophobia [PSYCH] Abnormal fear of one's self or of being alone. { ˌȯd·əˈfōb·ē·ə }

autophytograph [GEOL] An imprint on a rock surface made by chemical activity of a plant or plant part. { ˌȯd·əˈfīd·əˌgraf }

autopilot See automatic pilot. { ˈȯd·ōˌpī·lət }

autoplotter [COMPUT SCI] A machine which automatically draws a graph from input data. { ˈȯd·ōˌpläd·ər }

autopneumatolysis [GEOL] The occurrence of metamorphic changes at the pneumatolytic stage of a cooling magma when temperatures are approximately 400–600°C. { ˌȯd·ōˌnü·məˈtäl·ə·səs }

autopoisoning See self-poisoning. { ˌȯd·ōˈpȯiz·ən·iŋ }

autopolarity [ELECTR] Automatic interchanging of connections to a digital meter when polarity is wrong; a minus sign appears ahead of the value on the digital display if the reading is negative. { ˌȯd·ō·pəˈlär·əd·ē }

autopolyploid [GEN] A cell or organism having three or more sets of chromosomes all derived from the same species. { ˌȯd·ōˈpäl·iˌplȯid }

autopositive [GRAPHICS] A film or paper which produces a positive image when exposed to a positive (or a negative image from a negative) and which is processed in a single development stage. { ˌȯd·ōˈpäz·əd·iv }

autoprotolysis [CHEM] Transfer of a proton from one molecule to another of the same substance. { ˌȯd·ō·prəˈtäl·ə·səs }

autoprotolysis constant [CHEM] A constant denoting the equilibrium condition for the autoprotolysis reaction. { ˌȯd·ō·prəˈtäl·ə·səs ˈkän·stənt }

autopsy [PATH] A postmortem examination of the body to determine cause of death. { ˈȯˌtap·sē }

autoracemization [ORG CHEM] A racemization process that occurs spontaneously. { ˌȯd·ōˌrā·sə·məˈzā·shən }

autoradar plot See chart comparison unit. { ˈȯd·ōˌrāˌdär ˌplät }

autoradiography [ENG] A technique for detecting radioactivity in a specimen by producing an image on a photographic film or plate. Also known as radioautography. { ˌȯd·ōˌrād·ēˈäg·rə·fē }

autorail [MECH ENG] A self-propelled vehicle having both flange wheels and pneumatic tires to permit operation on both rails and roadways. { ˈȯd·ōˌrāl }

autoranging [ENG] Automatic switching of a multirange meter from its lowest to the next higher range, with the switching process repeated until a range is reached for which the full-scale value is not exceeded. Also known as automatic ranging. { ˈȯd·ōˌrānj·iŋ }

autoreducing tachymeter [ENG] A class of tachymeter by which horizontal and height distances are read simultaneously. { ˌȯd·ō·riˈdüs·iŋ təˈkim·əd·ər }

autoregressive series [MATH] A function of the form $f(t) = a_1 f(t-1) + a_2 f(t-2) + \cdots + a_m f(t-m) + k$, where k is any constant. { ˌȯd·ō·riˈgres·iv ˈsir·ēz }

autorotation [MECH] **1.** Rotation about any axis of a body that is symmetrical and exposed to a uniform airstream and maintained only by aerodynamic moments. **2.** Rotation of a stalled symmetrical airfoil parallel to the direction of the wind. { ˌȯd·ō·rōˈtā·shən }

autoserum [IMMUNOL] A serum obtained from a patient used for treatment of that patient. { ˈȯd·ōˌsir·əm }

autosexing [BIOL] Displaying differential sex characters at birth, noted particularly in fowl bred for sex-specific colors and patterns. { ˈȯd·ōˌseks·iŋ }

autoskeleton [INV ZOO] The endoskeleton of a sponge. { ˈȯd·ōˌskel·ə·tən }

autosled [MECH ENG] A propeller-driven machine equipped with runners and wheels and adaptable to use on snow, ice, or bare roads. { ˈȯd·ōˌsled }

autosomal dominant hearing loss [MED] Typically, a progressive form of hearing loss in which one of an individual's two copies of the autosomal dominant hearing loss gene is mutated. { ˌȯd·ə¦sōm·əl ¦däm·ə·nənt ˈhēr·iŋ ˌlȯs }

autosomal recessive hearing loss [MED] Typically, a congenital, severe loss of hearing (up to and including complete

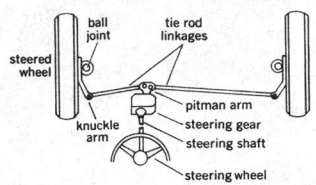

AUTOMOTIVE STEERING

Components of a typical steering system.

deafness) that occurs only if an individual inherits a mutant copy of an autosomal recessive hearing loss gene from each parent. { ‚òd·ə¦sōm·əl ri¦ses·iv 'hēr·iŋ ‚lòs }

autosomal trait [GEN] Any characteristic determined by autosomal genes. { ¦òd·ə¦sō·məl 'trāt }

autosome [GEN] Any chromosome other than a sex chromosome. { 'òd·ō‚sōm }

autospore [BOT] In algae, a nonmotile asexual reproductive cell or a nonmotile spore that is a miniature of the cell that produces it. { 'òd·ō‚spór }

autostability [CONT SYS] The ability of a device (such as a servomechanism) to hold a steady position, either by virtue of its shape and proportions, or by control by a servomechanism. { ¦òd·ō·stə¦bil·əd·ē }

autostarter [ELEC] **1.** Automatic starting and switchover generating system consisting of a standby generator coupled to the station load through an automatic power transfer control unit. **2.** See autotransformer starter. { 'òd·ō‚stärd·ər }

autostart routine [COMPUT SCI] A set of instructions that is permanently stored in a computer memory and activated when the computer is turned on, to perform diagnostic tests and then load the operating system. { 'òd·ō‚stärt rü‚tēn }

autostoper [MIN ENG] A stoper, or light compressed-air rock drill, mounted on an air-leg support which not only supports the drill but also exerts pressure on the drill bit. { 'òd·ō‚stōp·ər }

autostylic [VERT ZOO] Having the jaws attached directly to the cranium, as in chimeras, amphibians, and higher vertebrates. { ¦òd·ō¦stīl·ik }

autosyndesis [CYTOL] The act of pairing of homologous chromosomes from the same parent during meiosis in polyploids. { ¦òd·ō¦sin·də·səs }

autotest program [COMPUT SCI] A computer program within the operating system that aids in testing and debugging programs. { 'òd·ō‚test 'prō·grəm }

autotomy [MED] Surgical removal of a part of one's own body. [ZOO] The process of self-amputation of appendages in crabs and other crustaceans and tails in some salamanders and lizards under stress. { ò'täd·ə·mē }

autotrace [COMPUT SCI] A routine that locates outlines of raster graphics images and transforms them into vector graphics, usually at higher resolution. { 'òd·ō‚trās }

autotransductor [ELECTROMAG] A saturable reactor in which the same windings carry the main and control currents. { ¦òd·ō·tranz¦dək·tər }

autotransformer [ELEC] A power transformer having one continuous winding that is tapped; part of the winding serves as the primary and all of it serves as the secondary, or vice versa; small autotransformers are used to start motors. { ¦òd·ō·tranz¦fòr·mər }

autotransformer starter [ELEC] Motor starter having an autotransformer to furnish a reduced voltage for starting; includes the necessary switching mechanism. Also known as autostarter. { ¦òd·ō·tranz¦fòr·mər ‚stärd·ər }

autotransplant [BIOL] Tissue removed from an organism and grafted on another site of the same organism. Also known as autograft. { ¦òd·ō'tranz‚plant }

autotroph [BIOL] An organism capable of synthesizing organic nutrients directly from simple inorganic substances, such as carbon dioxide and inorganic nitrogen. { 'òd·ō‚träf }

autotrophic ecosystem [ECOL] An ecosystem that has primary producers as a principal component, and sunlight as the major initial energy source. { ¦òd·ə‚tròf·ik 'ek·ō‚sis·təm }

autotrophic succession [ECOL] A type of ecological succession that involves organisms that can utilize renewable resources. { ¦òd·ə‚trō·fik sək'sesh·ən }

autoxidation [CHEM] **1.** The slow, flameless combustion of materials by reaction with oxygen. **2.** An oxidation reaction that is self-catalyzed and spontaneous. **3.** An oxidation reaction begun only by an inductor. Also known as autooxidation. { ò¦täk·sə'dā·shən }

autozooecium [INV ZOO] The tube enclosing an autozooid. { ¦òd·ō‚zō¦esh·ē·əm }

autozooid [INV ZOO] An unspecialized feeding individual in a bryozoan colony, possessing fully developed organs and exoskeleton. { ¦òd·ō¦zō‚óid }

autumn [ASTRON] The season of the year which is the transition period from summer to winter, occurring as the sun approaches the winter solstice; beginning is marked by the autumnal equinox. Also known as fall. { 'òd·əm }

autumnal [ASTRON] Pertaining to the season autumn. { ò'təm·nəl }

autumnal equinox [ASTRON] The point on the celestial sphere at which the sun's rays at noon are 90° above the horizon at the Equator, or at an angle of 90° with the earth's axis, and neither North nor South Pole is inclined to the sun; occurs in the Northern Hemisphere on approximately September 23 and marks the beginning of autumn. Also known as first point of Libra. { ò'təm·nəl 'ē·kwə‚näks }

autumn ice [OCEANOGR] Sea ice in early stage of formation; comparatively salty, and crystalline in appearance. { ¦òd·əm 'īs }

Autunian [GEOL] A European stage of Lower Permian geologic time, above the Stephanian of the Carboniferous and below the Saxonian. { ‚ò'tün·ē·ən }

autunite [MINERAL] $Ca(UO_2)_2(PO_4)_2 \cdot 10H_2O$ A common fluorescent mineral that occurs as yellow tetragonal plates in uranium deposits; minor ore of uranium. { ō'tə‚nīt }

AUV See autonomous underwater vehicle.

Auversian See Ledian. { ‚ō'vərzh·ən }

auxanogram [MICROBIO] A plate culture provided with variable growth conditions to determine the effects of specific environmental factors. { òg'zan·ə‚gram }

auxanography [MICROBIO] The study of growth-inhibiting or growth-promoting agents by means of auxanograms. { ‚òg·zə'näg·rə·fē }

auxanometer [ENG] An instrument used to detect and measure plant growth rate. { ‚òg·zə'näm·əd·ər }

auxesis [PHYSIO] Growth resulting from increase in cell size. { ‚òg'zē·səs }

auxiliary anode [MET] A supplementary anode that alters the current distribution in electroplating to give a more uniform plating thickness. { òg¦zil·yə·rē 'an‚ōd }

auxiliary brake [PETRO ENG] A supplemental brake mechanism for assisting a mechanical brake; permits lowering of heavy hook loads safely at reduced rates without causing excessive wear on the mechanical brake. { òg¦zil·yə·rē 'brāk }

auxiliary channel [COMMUN] A secondary path for low-speed communication that uses the same circuit as a higher-speed stream of data. { òg¦zil·yə·rē 'chan·əl }

auxiliary circle [ASTRON] In celestial mechanics, a circumscribing circle to an orbital ellipse with radius a, the semimajor axis. { òg¦zil·yə·rē 'sər·kəl }

auxiliary contacts [ELEC] Contacts, in a switching device, in addition to the main circuit contacts, which function with the movement of the latter. { òg¦zil·yə·rē 'kän‚taks }

auxiliary dead latch [DES ENG] A supplementary latch in a lock which automatically deadlocks the main latch bolt when the door is closed. Also known as auxiliary latch bolt; deadlocking latch bolt; trigger bolt. { òg¦zil·yə·rē 'ded ‚lach }

auxiliary electrode [PHYS CHEM] An electrode in an electrochemical cell used for transfer of electric current to the test electrode. { òg¦zil·yə·rē i'lek‚trōd }

auxiliary equation [MATH] The equation that is obtained from a given linear differential equation by replacing with zero the term that involves only the independent variable. Also known as reduced equation. { òg¦zil·yə·re i'kwā·zhən }

auxiliary equipment See off-line equipment. { òg¦zil·yə·rē ə'kwip·mənt }

auxiliary fan [MIN ENG] A small fan installed underground for ventilating narrow coal drivages or hard headings which are not ventilated by the normal air current. { òg¦zil·yə·rē 'fan }

auxiliary fault [GEOL] A branch fault; a minor fault ending against a major one. { òg¦zil·yə·rē 'fòlt }

auxiliary fluid ignition [AERO ENG] A method of ignition of a liquid-propellant rocket engine in which a liquid that is hypergolic with either the fuel or the oxidizer is injected into the combustion chamber to initiate combustion. { òg¦zil·yə·rē 'flü·əd ig'nish·ən }

auxiliary instruction buffer [COMPUT SCI] A section of storage in the instruction unit, 16 bytes in length, used to hold prefetched instructions. { òg¦zil·yə·rē in'strək·shən ‚bəf·ər }

auxiliary landing gear [AERO ENG] The part or parts of a landing gear, such as an outboard wheel, which is intended to stabilize the craft on the surface but which bears no significant part of the weight. { òg¦zil·yə·rē 'land·iŋ ‚gir }

auxiliary latch bolt See auxiliary dead latch. { ȯg′zil·yə·rē ′lach ‚bōlt }

auxiliary memory [COMPUT SCI] **1.** A high-speed memory that is in a large main frame or supercomputer, is not directly addressable by the central processing unit, and is connected to the main memory by a high-speed data channel. **2.** See auxiliary storage. { ȯg′zil·yə·rē ′mem·rē }

auxiliary mineral [MINERAL] A light-colored, relatively rare or unimportant mineral in an igneous rock; examples are apatite, muscovite, corundrum, fluorite, and topaz. { ȯg′zil·yə·rē ′min·rəl }

auxiliary operation [COMPUT SCI] An operation performed by equipment not under continuous control of the central processing unit of a computer. { ȯg′zil·yə·rē ‚äp·ə′rā·shən }

auxiliary plane [GEOL] A plane at right angles to the net slip on a fault plane as determined from analysis of seismic data for an earthquake. { ȯg′zil·yə·rē ′plān }

auxiliary power plant [MECH ENG] Ancillary equipment, such as pumps, fans, and soot blowers, used with the main boiler, turbine, engine, waterwheel, or generator of a power-generating station. { ȯg′zil·yə·rē ′pau̇·ər ‚plant }

auxiliary power unit [AERO ENG] A power unit carried on an aircraft or spacecraft which can be used in addition to the main sources of power. Abbreviated APU. { ȯg′zil·yə·rē ′pau̇·ər ‚yü·nət }

auxiliary processor [COMPUT SCI] Any equipment which performs an auxiliary operation in a computer. { ȯg′zil·yə·rē ′präs‚es·ər }

auxiliary rafter [BUILD] A member strengthening the principal rafter in a truss. { ȯg′zil·yə·rē ′raf·tər }

auxiliary reinforcement [CIV ENG] In a prestressed structural member, any reinforcement in addition to that whose function is prestressing. { ȯg′zil·yə·rē ‚rē·ən′fȯrs·mənt }

auxiliary relay [ELEC] Relay that operates in response to the opening or closing of its operating circuit to assist another relay or device in performing a function. { ȯg′zil·yə·rē ′rē‚lā }

auxiliary rim lock [DES ENG] A secondary or extra lock that is surface-mounted on a door to provide additional security. { ȯg′zil·yə·rē ′rim ‚läk }

auxiliary rope-fastening device [MECH ENG] A device attached to an elevator car, to a counterweight, or to the overhead dead-end rope-hitch support, that automatically supports the car or counterweight in case the fastening for the wire rope (cable) fails. { ȯg′zil·yə·rē ′rōp ‚fas·ən·iŋ di‚vīs }

auxiliary routine [COMPUT SCI] A routine designed to assist in the operation of the computer and in debugging other routines. { ȯg′zil·yə·rē rü′tēn }

auxiliary ship [NAV ARCH] A naval vessel other than a combat ship, such as a troop ship, repair ship, and cargo ship. { ȯg′zil·yə·rē ′ship }

auxiliary storage [COMPUT SCI] Storage device in addition to the main storage of a computer; for example, magnetic tape, disk, or magnetic drum. Also known as auxiliary memory. { ȯg′zil·yə·rē ′stȯr·ij }

auxiliary switch [ELEC] A switch actuated by the main device (such as a circuit breaker) for signaling, interlocking, or other purposes. { ȯg′zil·yə·rē ′swich }

auxiliary thermometer [ENG] A mercury-in-glass thermometer attached to the stem of a reversing thermometer and read at the same time as the reversing thermometer so that the correction to the reading of the latter, resulting from change in temperature since reversal, can be computed. { ȯg′zil·yə·rē thər′mäm·əd·ər }

auximone [BIOCHEM] Any of certain growth-promoting substances occurring principally in sphagnum peat decomposed by nitrogen bacteria. { ′ȯk·sə‚mōn }

auxin [BIOCHEM] Any organic compound which promotes plant growth along the longitudinal axis when applied to shoots free from indigenous growth-promoting substances. { ′ȯk·sən }

auxoautotrophic [BIOL] Requiring no exogenous growth factors. { ‚ȯk·sō‚ȯd·ō‚ə′trä·fik }

auxochrome [CHEM] Any substituent group such as −NH₂ and −OH which, by affecting the spectral regions of strong absorption in chromophores, enhance the ability of the chromogen to act as a dye. { ′ȯk·sə‚krōm }

auxocyte [BIOL] A gamete-forming cell, such as an oocyte or spermatocyte, or a sporocyte during its growth period. Also known as gonotocont. { ′ȯk·sə‚sīt }

auxograph [ENG] An automatic device that records changes in the volume of a body. { ′ȯk·sə‚graf }

auxoheterotrophic [BIOL] Requiring exogenous growth factors. { ‚ȯk·sō‚hed·ə·rə‚trä·fik }

auxometer [ENG] An instrument that measures the magnification of a lens system. { ‚ȯk′säm·əd·ər }

auxospore [INV ZOO] A reproductive cell in diatoms formed in association with rejuvenescence by the union of two cells that have diminished in size through repeated divisions. { ′ȯk·sə‚spȯr }

auxotonic [BOT] Induced by growth rather than by exogenous stimuli. { ‚ȯk·sə′tän·ik }

auxotrophic mutant [GEN] An organism that requires a specific exogenous growth factor, such as an amino acid, for its growth. { ‚ȯk·sə‚trä·fik ′myüt·ənt }

av See arithmetic mean.

aV See abvolt.

availability [COMPUT SCI] Of data, data channels, and input-output devices in computers, the condition of being ready for use and not immediately committed to other tasks. [NAV] The probability that a navigational system will function and provide required levels of accuracy, integrity, and continuity. [PHYS] The difference between the enthalpy per unit mass of substance and the product of entropy per unit mass multiplied by the lowest temperature available to the substance for heat discard; used in determining the ratio of actual work performed during a process by a working substance to that which theoretically should have been performed. [SYS ENG] The probability that a system is operating satisfactorily at any point in time, excluding times when the system is under repair. { ə‚val·ə′bil·ə·dē }

availability ratio [IND ENG] The ratio of the amount of time a system is actually available for use to the amount of time it is supposed to be available. { ə‚val·ə′bil·əd·ē ′rā·shō }

available chlorine [CHEM] The quantity of chlorine released by a bleaching powder when treated with acid. { ə′val·ə·bəl ′klȯr‚ēn }

available-chlorine method [MICROBIO] A technique for the standardization of chlorine disinfectants intended for use as germicidal rinses on cleaned surfaces; increments of bacterial inoculum are added to different disinfectant concentrations, and after incubation the results indicate the capacity of the disinfectant to handle an increasing bacterial load before exhaustion of available chlorine, the germicidal principle. { ə′val·ə·bəl ′klȯr‚ēn ‚meth·əd }

available draft [MECH ENG] The usable differential pressure in the combustion air in a furnace, used to sustain combustion of fuel or to transport products of combustion. { ə′val·ə·bəl ′draft }

available energy [MECH ENG] Energy which can in principle be converted to mechanical work. { ə′val·ə·bəl ′en·ər·jē }

available heat [MECH ENG] The heat per unit mass of a working substance that could be transformed into work in an engine under ideal conditions for a given amount of heat per unit mass furnished to the working substance. { ə′val·ə·bəl ′hēt }

available line [ELECTR] Portion of the length of the scanning line which can be used specifically for picture signals in a facsimile system. { ə′val·ə·bəl ′līn }

available moisture [HYD] Moisture in soil that is available for use by plants. { ə′val·ə·bəl ′mȯis·chər }

available motions inventory [IND ENG] A list of all motions available to a human for performing a specific task. { ə‚val·ə·bəl ‚mō·shənz ′in·ven‚tȯr·ē }

available power [ELECTR] The power which a linear source of energy is capable of delivering into its conjugate impedance. { ə′val·ə·bəl ′pau̇·ər }

available-power gain [ELECTR] Ratio, in an electronic transducer, of the available power from the output terminals of the transducer, under specified input termination conditions, to the available power from the driving generator. { ə′val·ə·bəl ′pau̇·ər ‚gān }

available relief [GEOL] The vertical distance after uplift between the altitude of the original surface and the level at which grade is first attained. { ə′val·ə·bəl ri′lēf }

available space list [COMPUT SCI] A pool of inactive memory cells, available for use in a list-processing system, to which cells containing items deleted from data lists are added, and from which cells needed for newly inserted data items are removed. { ə'väl·ə·bəl 'spās ,list }

available time *See* up time. { ə'väl·ə·bəl 'tīm }

avalanche [ELECTR] **1.** The cumulative process in which an electron or other charged particle accelerated by a strong electric field collides with and ionizes gas molecules, thereby releasing new electrons which in turn have more collisions, so that the discharge is thus self-maintained. Also known as avalanche effect; cascade; cumulative ionization; electron avalanche; Townsend avalanche; Townsend ionization. **2.** Cumulative multiplication of carriers in a semiconductor as a result of avalanche breakdown. Also known as avalanche effect. [HYD] A mass of snow or ice moving rapidly down a mountain slope or cliff. { 'av·ə,lanch }

avalanche breakdown [ELECTR] Nondestructive breakdown in a semiconductor diode when the electric field across the barrier region is strong enough so that current carriers collide with valence electrons to produce ionization and cumulative multiplication of carriers. { 'av·ə,lanch 'brāk,daún }

avalanche conduction [NEURO] Conduction of a nerve impulse through several neurons which converge, increasing the discharge intensity by summation. { 'av·ə,lanch kən'dək·shən }

avalanche diode [ELECTR] A semiconductor breakdown diode, usually made of silicon, in which avalanche breakdown occurs across the entire *pn* junction and voltage drop is then essentially constant and independent of current; the two most important types are IMPATT and TRAPATT diodes. { 'av·ə,lanch 'dī,ōd }

avalanche effect *See* avalanche. { 'av·ə,lanch i,fekt }

avalanche impedance [ELECTR] The complex ratio of the reverse voltage of a device that undergoes avalanche breakdown to the reverse current. { 'av·ə,lanch im'pēd·əns }

avalanche-induced migration [ELECTR] A technique of forming interconnections in a field-programmable logic array by applying appropriate voltages for shorting selected base-emitter junctions. { 'av·ə,lanch in,düsd ,mī'grā·shən }

avalanche noise [ELECTR] **1.** A junction phenomenon in a semiconductor in which carriers in a high-voltage gradient develop sufficient energy to dislodge additional carriers through physical impact; this agitation creates ragged current flows which are indicated by noise. **2.** The noise produced when a junction diode is operated at the onset of avalanche breakdown. { 'av·ə,lanch ,nóiz }

avalanche oscillator [ELECTR] An oscillator that uses an avalanche diode as a negative resistance to achieve one-step conversion from direct-current to microwave outputs in the gigahertz range. { 'av·ə,lanch |äs·ə,lād·ər }

avalanche photodiode [ELECTR] A photodiode operated in the avalanche breakdown region to achieve internal photocurrent multiplication, thereby providing rapid light-controlled switching operation. { 'av·ə,lanch ,fōd·ō'dī,ōd }

avalanche protector [MECH ENG] Guard plates installed on an excavator to prevent loose material from sliding into the wheels or tracks. { 'av·ə,lanch prə,tek·tər }

avalanche transistor [ELECTR] A transistor that utilizes avalanche breakdown to produce chain generation of charge-carrying hole-electron pairs. { 'av·ə,lanch tran'zis·tər }

avalanche voltage [ELECTR] The reverse voltage required to cause avalanche breakdown in a *pn* semiconductor junction. { 'av·ə,lanch ,vōl·tij }

avalanche wind [METEOROL] The rush of air produced in front of an avalanche of dry snow or in front of a landslide. { 'av·ə,lanch ,wind }

avant-corps [ARCH] That part of a building which projects prominently from the main mass, for example, a pavilion. { ä,vänt'kór }

avatar [COMPUT SCI] A virtual representation of a person or a person's interactions with others in a virtual environment, conveying a sense of someone's presence (known as telepresence) by providing the location (position and orientation) and identity; examples include the graphical human figure model, the talking head, and the real-time reproduction of a three-dimensional human image. { 'av·ə,tär }

AVC *See* automatic volume control.

aV/cm *See* abvolt per centimeter.

aven [GEOL] *See* pothole. [MIN ENG] A vertical shaft leading upward from a cave passage, sometimes connecting with passages above. { 'av·ən }

Avena [BOT] A genus of grasses (family Gramineae), including oats, characterized by an inflorescence that is loosely paniculate, two-toothed lemmas, and deeply furrowed grains. { ə'vēn·ə }

avenin [BIOCHEM] The glutelin of oats. { ə'vēn·ən }

aventurine [MINERAL] **1.** A glass or mineral containing sparkling gold-colored particles, usually copper or chromic oxide. **2.** A shiny red or green translucent quartz having small, but microscopically visible, exsolved hematite or included mica particles. { ə'vench·ə,rēn }

average *See* arithmetic mean. { 'av·rij }

average acoustic output [ENG ACOUS] Vibratory energy output of a transducer measured by a radiation pressure balance; expressed in terms of watts per unit area of the transducer face. { 'av·rij ə'kü·stik 'aút,pút }

average assay value [MIN ENG] The weighted result of assays obtained from a number of samples by multiplying the assay value of each sample by the width or thickness of the ore face over which it is taken and dividing the sum of these products by the total width of cross section sampled. { 'av·rij 'a,sā ,val·yü }

average bisector [NAV] A line extending through a four-course radio range station into opposing quadrants and midway between the lines (and their extensions) bisecting these two quadrants; used when the courses are not symmetrical. { 'av·rij 'bī,sek·tər }

average bond dissociation energy [PHYS CHEM] The average value of the bond dissociation energies associated with the homolytic cleavage of several bonds of a set of equivalent bonds of a molecule. Also known as bond energy. { |av·rij |bänd di·sō·sē'ā·shən ,en·ər·jē }

average-calculating operation [COMPUT SCI] A common or typical calculating operation longer than an addition and shorter than a multiplication; often taken as the mean of nine additions and one multiplication. { 'av·rij |kal·kyə,lād·iŋ ,äp·ə,rā·shən }

average curvature [MATH] For a given arc of a plane curve, the ratio of the change in inclination of the tangent to the curve, over the arc, to the arc length. { |av·rij 'kərv·ə·chər }

average deviation [MATH] In statistics, the average or arithmetic mean of the deviation, taken without regard to sign, from some fixed value, usually the arithmetic mean of the data. Abbreviated AD. Also known as mean deviation. { 'av·rij ,dē·vē'ā·shən }

average discount factor *See* discount factor. { 'av·rij 'dis,-kaúnt ,fak·tər }

average-edge line [COMPUT SCI] The imaginary line which traces or smooths the shape of any written or printed character to be recognized by a computer through optical, magnetic, or other means. { 'av·rij |ej ,līn }

average effectiveness level *See* effectiveness level. { 'av·rij i'fek·tiv·nəs ,lev·əl }

average gradient [GRAPHICS] A measure of contrast in a photographic image, expressed as the slope of a straight line joining two density points on the sensitometric curve. { 'av·rij 'grād·ē·ənt }

average heading [NAV] The average heading flown for a given period; it should be the same value as desired heading if the drift was predicted accurately. { 'av·rij 'hed·iŋ }

average igneous rock [PETR] A hypothetical rock whose composition is thought to be similar to the average chemical composition of the outermost 10-mile (16-kilometer) shell of the earth. { 'av·rij 'ig·nē·əs 'räk }

average information content [COMMUN] The average of the information content per symbol emitted from a source. { 'av·rij ,in·fər'mā·shən ,kän·tent }

average life *See* mean life. { 'av·rij 'līf }

average limit of ice [OCEANOGR] The average seaward extent of ice formation during a normal winter. { 'av·rij 'lim·ət əv 'īs }

average molecular weight [ORG CHEM] The calculated number to average the molecular weights of the varying-length polymer chains present in a polymer mixture. { 'av·rij mə'lek·yə·lər 'wāt }

average noise figure [ELECTR] Ratio in a transducer of total output noise power to the portion thereof attributable to thermal noise in the input termination, the total noise being summed over frequencies from zero to infinity, and the noise temperature of the input termination being standard (290 K). { 'av·rij ,fig·yər }

average outgoing quality limit [IND ENG] The average quality of all lots that pass quality inspection, expressed in terms of percent defective. Abbreviated AOQL. { 'av·rij 'aút,gō·iŋ 'kwäl·əd·ē ,lim·ət }

average power output [ELECTR] Radio-frequency power, in an audio-modulation transmitter, delivered to the transmitter output terminals, averaged over a modulation cycle. { 'av·rij 'paú·ər 'aút,pút }

average sample number [IND ENG] An anticipated number of pieces that must be inspected to determine the acceptability of a particular lot. { 'av·rij ,sam·pəl ,nəm·bər }

average wind [NAV] In air navigation, the resultant wind which would produce, or has produced, the same wind effect during a given period as the summation of the actual winds which will affect, or have affected, the flight of an aircraft. { 'av·rij 'wind }

averaging [CONT SYS] The reduction of noise received by a robot sensor by screening it over a period of time. { 'av·rij·iŋ }

averaging device [ENG] A device for obtaining the arithmetic mean of a number of readings, as on a bubble sextant. { 'av·rij·iŋ di'vīs }

averaging pitot tube [ENG] A flowmeter that consists of a rod extending across a pipe with several interconnected upstream holes, which simulate an array of pitot tubes across the pipe, and a downstream hole for the static pressure reference. { 'av·rij·iŋ ,pē,tō 'tüb }

aversion therapy [PSYCH] A behavior therapy technique intended to suppress undesirable behavior by pairing a stimulus associated with an undesirable behavior together with a painful or unpleasant stimulus. { ə'vər·zhən ,ther·ə·pē }

aversive behavior [PSYCH] Avoidance behavior. { ə'vərs·iv bi'hāv·yər }

Aves [VERT ZOO] A class of animals composed of the birds, which are warm-blooded, egg-laying vertebrates primarily adapted for flying. { 'ā,vēz }

avgas See aviation gasoline. { 'av,gas }

avianize [VIROL] To attenuate a virus by repeated culture on chick embryos. { 'av·ē·ə,nīz }

avian leukosis [VET MED] A disease complex in fowl probably caused by viruses and characterized by autonomous proliferation of blood-forming cells. { 'av·ē·ən lü'kō·səs }

avian pneumoencephalitis [VET MED] See Newcastle disease. { 'av·ē·ən ¦nü·mō·in,sef·ə'līd·əs }

avian pseudoplague See Newcastle disease. { 'av·ē·ən 'süd·ō,plag }

avian tuberculosis [VET MED] A tuberculosis-like mycobacterial disease of fowl caused by *Mycobacterium avium*. { 'av·ē·ən tə,bər·kyə'lō·səs }

aviation [AERO ENG] **1.** The science and technology of flight through the air. **2.** The world of airplane business and its allied industries. { ,ā·vē'ā·shən }

aviation gasoline [MATER] Stable fuel with high volatility and high octane, especially suited for use in aircraft reciprocating engines. Abbreviated avgas. { ,ā·vē'ā·shən ,gas·ə'lēn }

aviation medicine See aerospace medicine. { ,ā·vē'ā·shən 'med·ə·sən }

aviation method [ENG] Determination of knock-limiting power, under lean-mixture conditions, of fuels used in spark-ignition aircraft engines. { ,ā·vē'ā·shən 'meth·əd }

aviation mix [MATER] Antiknock fluid containing tetraethyllead, ethylene dibromide, and dye; used in aviation gasoline. { ,ā·vē'ā·shən 'miks }

aviation weather forecast [METEOROL] A forecast of weather elements of particular interest to aviation, such as the ceiling, visibility, upper winds, icing, turbulence, and types of precipitation or storms. Also known as airways forecast. { ,ā·vē'ā·shən ¦weth·ər ,fȯr,kast }

aviation weather observation [METEOROL] An evaluation, according to set procedure, of those weather elements which are most important for aircraft operations. Also known as airways observation. { ,ā·vē'ā·shən ¦weth·ər ,äb·zər'vā·shən }

avicolous [ECOL] Living on birds, as of certain insects. { ā'vik·ə·ləs }

avicularium [INV ZOO] A specialized individual in a bryozoan colony with a beak that keeps other animals from settling on the colony. { ə,vik·yə'lar·ē·ən }

aviculture [VERT ZOO] Care and breeding of birds, especially wild birds, in captivity. { 'ā·və,kəl·chər }

avidin [BIOCHEM] A protein constituting 0.2% of the total protein in egg white; molecular weight is 70,000; combines firmly with biotin but loses this ability when subjected to heat. { 'av·əd·ən }

avidity [IMMUNOL] The total binding strength of a polyvalent antibody (an antibody that has multiple binding sites for the same antigen) with a polyvalent antigen (an antigen with multiple identical antibody-binding sites); equivalent to the sum of the affinities at each of the binding sites. { ə'vid·əd·ē }

avifauna [VERT ZOO] **1.** Birds, collectively. **2.** Birds characterizing a period, region, or environment. { ¦ā·və¦fȯn·ə }

avigation See air navigation. { ,a·və'gā·shən }

aviolite [PETR] A mica-cordierite-hornfels. { ā'vī·ə,līt }

avionics [ENG] The design and production of airborne electrical and electronic devices; term is derived from aviation electricity. { ,ā·vē'än·iks }

avitaminosis [MED] Any vitamin-deficiency disease. { ¦ā ,vīd·ə·mə'nō·səs }

avocado [BOT] *Persea americana*. A subtropical evergreen tree of the order Magnoliales that bears a pulpy pear-shaped edible fruit. { ,av·ə'käd·ō }

avocado oil [MATER] An oil extracted from ripe fruit of the avocado (*Persea americana*). Also known as alligator pear oil. { ,av·ə'käd·ō ,ȯil }

avogadrite [MINERAL] (K,Cs)BF$_4$ An orthorhombic fluoborate mineral occurring in small crystals on Vesuvian lava. { ,a·və'gäd,rīt }

Avogadro's hypothesis [PHYS] See Avogadro's law. { ¦a·və¦gäd·drōz hī'päth·ə·səs }

Avogadro's law [PHYS] The law which states that under the same conditions of pressure and temperature, equal volumes of all gases contain equal numbers of molecules; for example, 359 cubic feet at 32°F and 1 atmosphere for a perfect gas. Also known as Avogadro's hypothesis. { ¦a·və¦gäd·drōz ,lō }

Avogadro's number [PHYS] The number (6.02×10^{23}) of molecules in a gram-molecular weight of a substance. { ¦a·və¦gäd·drōz ,nəm·bər }

avogram [MECH] A unit of mass, equal to 1 gram divided by the Avogadro number. { 'a·və,gram }

avoidable delay [IND ENG] An interruption under the control of the operator during the normal operating time. { ə'vȯid·ə·bəl di'lā }

avoidance-avoidance conflict [PSYCH] Psychological conflict resulting from the necessity of choosing between two undesirable alternatives. { ə¦vȯid·əns ə¦vȯid·əns ¦kän,flikt }

avoirdupois pound See pound. { ,av·ərd·ə'pȯiz 'paúnd }

avoirdupois weight [MECH] The system of units which has been commonly used in English-speaking countries for measurement of the mass of any substance except precious stones, precious metals, and drugs; it is based on the pound (approximately 453.6 grams) and includes the short ton (2000 pounds), long ton (2240 pounds), ounce (one-sixteenth pound), and dram (one-sixteenth ounce). { ,av·ərd·ə'pȯiz 'wāt }

Avonian See Dinantian. { ə'vōn·ē·ən }

avulsion [HYD] A sudden change in the course of a stream by which a portion of land is cut off, as where a stream cuts across and forms an oxbow. [MED] Tearing one part away from the other, either by trauma or surgery. { ə'vəl·shən }

AWACS See airborne warning and control system. { 'ā·waks }

awaruite [MINERAL] Native nickel-iron alloy containing 57.7% nickel. { ,a·wä'rü,īt }

AWG See American wire gage.

awl [DES ENG] A point tool with a short wooden handle used to mark surfaces and to make small holes, as in leather or wood. { ȯl }

awn [BOT] Any of the bristles at the ends of glumes or bracts on the spikelets of oats, barley, and some wheat and grasses. Also known as beard. { ȯn }

awning deck [NAV ARCH] A light deck over the main deck, running the full length of a ship. Also known as hurricane deck. { 'ȯn·iŋ ,dek }

awning window [BUILD] A window consisting of a series of vertically arranged, top-hinged rectangular sections; designed to admit air while excluding rain. { 'ȯn·iŋ ,win·dō }

AVOCADO

Avocado foilage and fruit.

ax [DES ENG] An implement consisting of a heavy metal wedge-shaped head with one or two cutting edges and a relatively long wooden handle; used for chopping wood and felling trees. { aks }

axed brick [ENG] A brick, shaped with an ax, that has not been trimmed. Also known as rough-axed brick. { ¦akst ¦brik }

axenic culture [BIOL] The growth of organisms of a single species in the absence of cells or living organisms of any other species. { ā'zen·ik 'kəl·chər }

axes of an aircraft [AERO ENG] Three fixed lines of reference, usually centroidal and mutually perpendicular: the longitudinal axis, the normal or yaw axis, and the lateral or pitch axis. Also known as aircraft axes. { 'ak,sēz əv ən 'er,kraft }

axes of inertia [PHYS] The three principal axes of inertia, namely, one about which the moment of inertia is a maximum, one about which the moment of inertia is a minimum, and one perpendicular to both. { 'ak,sēz əv in'ər·shə }

axhammer [DES ENG] An ax having one cutting edge and one hammer face. { 'aks,ham·ər }

axial [SCI TECH] Of, pertaining to, or along an axis. { 'ak·sē·əl }

axial angle [CRYSTAL] 1. The acute angle between the two optic axes of a biaxial crystal. Also known as optic angle; optic-axial angle. 2. In air, the larger angle between the optic axes after refraction on leaving the crystal. { 'ak·sē·əl 'aŋ·gəl }

axial compression [GEOL] A compression applied parallel with the cylinder axis in experimental work involving rock cylinders. { 'ak·sē·əl kəm'presh·ən }

axial culmination [GEOL] Distortion of the fold axis upward in a form similar to an anticline. { 'ak·sē·əl ,kəl·mə'nā·shən }

axial dipole field [GEOPHYS] A postulated magnetic field for the earth, consisting of a dipolar field centered at the earth's center, with its axis coincident with the earth's rotational axis. { 'ak·sē·əl 'di,pōl ,fēld }

axial element [CRYSTAL] The lengths, length ratios, and angles which define a crystal's unit cell. { 'ak·sē·əl 'el·ə·mənt }

axial fan [MECH ENG] A fan whose housing confines the gas flow to the direction along the rotating shaft at both the inlet and outlet. { 'ak·sē·əl 'fan }

axial filament [CYTOL] The central microtubule elements of a cilium or flagellum. [INV ZOO] An organic fiber which serves as the core for deposition of mineral substance to form a ray of a sponge spicule. { 'ak·sē·əl 'fil·ə·mənt }

axial flow [FL MECH] Flow of fluid through an axially symmetric device such that the direction of the flow is along the axis of symmetry. Also known as axisymmetric flow. { 'ak·sē·əl 'flō }

axial-flow compressor [MECH ENG] A fluid compressor that accelerates the fluid in a direction generally parallel to the rotating shaft. { 'ak·sē·əl 'flō kəm'pres·ər }

axial-flow jet engine [AERO ENG] 1. A jet engine in which the general flow of air is along the longitudinal axis of the engine. 2. A turbojet engine that utilizes an axial-flow compressor and turbine. { 'ak·sē·əl 'flō ¦jet 'en·jən }

axial-flow pump [MECH ENG] A pump having an axial-flow or propeller-type impeller; used when maximum capacity and minimum head are desired. Also known as propeller pump. { 'ak·sē·əl 'flō ,pəmp }

axial force diagram [CIV ENG] In statics, a graphical representation of the axial load acting at each section of a structural member, plotted to scale and with proper sign as an ordinate at each point of the member and along a reference line representing the length of the member. { 'ak·sē·əl ¦fòrs ,di·ə,gram }

axial gland [INV ZOO] A structure enclosing the stone canal in certain echinoderms; its function is uncertain. { 'ak·sē·əl 'gland }

axial gradient [EMBRYO] In some invertebrates, a graded difference in metobolic activity along the anterior-posterior, dorsal-ventral, and medial-lateral embryonic axes. { 'ak·sē·əl 'grād·ē·ənt }

axial hydraulic thrust [MECH ENG] In single-stage and multistage pumps, the summation of unbalanced impeller forces acting in the axial direction. { 'ak·sē·əl hī'drò·lik 'thrəst }

axial jet [FL MECH] A flowing, turbulent stream which mixes with standing water in three dimensions. { 'ak·sē·əl 'jet }

axial lead [ELEC] A wire lead extending from the end along the axis of a resistor, capacitor, or other component. { 'ak·sē·əl 'lēd }

axial load [MECH] A force with its resultant passing through the centroid of a particular section and being perpendicular to the plane of the section. { 'ak·sē·əl 'lōd }

axial mining [ORD] Continuous or intermittent nuisance mining in great depth along the axis of enemy advance. { 'ak·sē·əl 'mīn·iŋ }

axial modulus [MECH] The ratio of a simple tension stress applied to a material to the resulting strain parallel to the tension when the sides of the sample are restricted so that there is no lateral deformation. Also known as modulus of simple longitudinal extension. { ¦ak·sē·əl 'mäj·ə·ləs }

axial moment of inertia [MECH] For any object rotating about an axis, the sum of its component masses times the square of the distance to the axis. { 'ak·sē·əl 'mō·mənt əv in'ər·shə }

axial musculature [ANAT] The muscles that lie along the longitudinal axis of the vertebrate body. { 'ak·sē·əl 'məs·kyə·lə·chər }

axial nozzle [MECH ENG] An inlet or outlet connection installed in the head of a shell-and-tube exchanger and aligned normal to the plane in which the tube lies. { 'ak·sē·əl 'näz·əl }

axial plane [CRYSTAL] 1. A plane that includes two of the crystallographic axes. 2. The plane of the optic axis of an optically biaxial crystal. [GEOL] A plane that intersects the crest or trough in such a manner that the limbs or sides of the fold are more or less symmetrically arranged with reference to it. Also known as axial surface. { 'ak·sē·əl 'plān }

axial-plane cleavage [GEOL] Rock cleavage essentially parallel to the axial plane of a fold. { 'ak·sē·əl ¦plān ,klē·vij }

axial-plane foliation [GEOL] Foliation developed in rocks parallel to the axial plane of a fold and perpendicular to the chief deformational pressure. { 'ak·sē·əl ¦plān ,fō·lē'ā·shən }

axial-plane schistosity [GEOL] Schistosity developed parallel to the axial planes of folds. { 'ak·sē·əl ¦plān ,shis'täs·əd·ē }

axial-plane separation [GEOL] The distance between axial planes of adjacent anticline and syncline. { 'ak·sē·əl ¦plān sep·ə'rā·shən }

axial quadrupole See longitudinal quadrupole. { 'ak·sē·əl 'kwäd·rə,pōl }

axial rake [MECH ENG] The angle between the face of a blade of a milling cutter or reamer and a line parallel to its axis of rotation. { 'ak·sē·əl 'rāk }

axial ratio [CRYSTAL] The ratio obtained by comparing the length of a crystallographic axis with one of the lateral axes taken as unity. [ELECTR] The ratio of the major axis to the minor axis of the polarization ellipse of a waveguide. Also known as ellipticity. { 'ak·sē·əl 'rā·shō }

axial relief [MECH ENG] The relief behind the end cutting edge of a milling cutter. { 'ak·sē·əl ri'lēf }

axial runout [MECH ENG] The total amount, along the axis of rotation, by which the rotation of a cutting tool deviates from a plane. { 'ak·sē·əl 'rən,aút }

axial skeleton [ANAT] The bones composing the skull, vertebral column, and associated structures of the vertebrate body. { 'ak·sē·əl 'skel·i·tən }

axial stream [HYD] 1. The chief stream of an intermontane valley, the course of which is along the deepest part of the valley and is parallel to its longer dimension. 2. A stream whose course is along the axis of an anticlinal or a synclinal fold. { 'ak·sē·əl 'strēm }

axial surface See axial plane. { 'ak·sē·əl 'sər·fəs }

axial symmetry [MATH] Property of a geometric configuration which is unchanged when rotated about a given line. { 'ak·sē·əl 'sim·ə·trē }

axial trace [GEOL] The intersection of the axial plane of a fold with the surface of the earth or any other specified surface; sometimes such a line is loosely and incorrectly called the axis. { 'ak·sē·əl 'trās }

axial trough [GEOL] Distortion of a fold axis downward into a form similar to a syncline. { 'ak·sē·əl 'tròf }

axial-type mass flowmeter [ENG] An instrument in which fluid in a pipe is made to rotate at a constant speed by a motor-driven impeller, and the torque required by a second, stationary impeller to straighten the flow again is a direct measurement of mass flow. { 'ak·sē·əl ¦tīp 'mas 'flō,med·ər }

axial vector See pseudovector. { 'ak·sē·əl 'vek·tər }

axial winding [MATER] A winding used in filament-wound

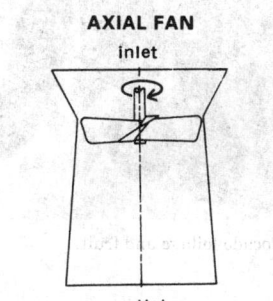

AXIAL FAN

inlet

outlet

Simplified diagram of an axial fan.

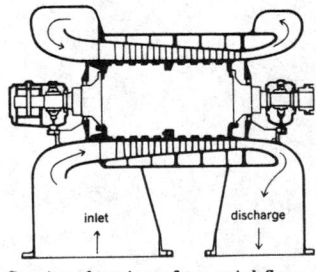

AXIAL-FLOW COMPRESSOR

inlet   discharge

Section drawing of an axial-flow compressor.

fiberglass-reinforced plastic construction in which the filaments run along the axis at a zero helix angle.   { 'ak·sē·əl 'wind·iŋ }

axiation [EMBRYO] The formation or development of axial structures, such as the neural tube.   { ˌak·sē'ā·shən }

Axiidae [INV ZOO] A family of decapod crustaceans, including the hermit crabs, in the suborder Reptantia.   { ˌak'sī·ə,dē }

axil [BIOL] The angle between a structure and the axis from which it arises, especially for branches and leaves.   { 'ak·səl }

axilla [ANAT] The depression between the arm and the thoracic wall; the armpit.   [BOT] An axil.   { ak'sil·ə }

axillary [ANAT] Of, pertaining to, or near the axilla or armpit. [BOT] Placed or growing in the axis of a branch or leaf.   { 'ak·sə,ler·ē }

axillary bud [BOT] A lateral bud borne in the axil of a leaf.   { 'ak·sə,ler·ē 'bəd }

axillary sweat gland [ANAT] An apocrine gland located in the axilla.   { 'ak·sə,ler·ē 'swet ,gland }

Axinellina [INV ZOO] A suborder of sponges in the order Clavaxinellida.   { ˌak·sə'nə'lī·nə }

axinite [MINERAL] $H_2(Ca,Fe,Mn)_4(BO)Al_2(SiO_4)_5$ Brown, blue, green, gray, or purplish gem mineral that commonly forms glassy triclinic crystals. Also known as glass schorl.   { 'ak·sə,nīt }

axinitization [GEOL] The replacement of rocks by axinite, as in the border zones of some granites.   { ak,zin·ə·tə'zā·shən }

axiolite [MINERAL] A variety of elongated spherulite in which there is an aggregation of minute acicular crystals arranged at right angles to a central axis.   { 'ak·sē·ə,līt }

axiom [MATH] Any of the assumptions upon which a mathematical theory (such as geometry, ring theory, and the real numbers) is based. Also known as postulate.   { 'ak·sē·əm }

axiomatic S-matrix theory [PARTIC PHYS] An approach to the study of elementary particles that seeks to formulate S-matrix theory in a rigorous manner based on a few fundamental axioms that include Lorentz invariance, unitarity, analyticity near the physical values of the energy and momentum variables, and singularities in the physical region that correspond to known particles and scattering thresholds.   { ˌak·sē·ə¦mad·ik 'es ˌmā·triks ˌthē·ə·rē }

axiom of Archimedes [MATH] The postulate that if x is any real number, there exists an integer n such that n is greater than x. Also known as Archimedes' axiom.   { ¦ak·sē·əm əv ˌärk·ə'mē,dēz }

axiom of choice [MATH] The axiom that for any family A of sets there is a function that assigns to each set S of the family A a member of S.   { ¦ak·sē·əm əv 'chóis }

axion [PARTIC PHYS] A hypothetical neutral pseudoscalar boson with mass roughly of order 100 keV to 1 MeV, postulated to preserve the parity and time-reversal invariance of strong interactions, despite the effects of instantons.   { 'ak·se,än }

axis [ANAT] **1.** The second cervical vertebra in higher vertebrates; the first vertebra of amphibians. **2.** The center line of an organism, organ, or other body part. [GEOL] **1.** A line where a folded bed has maximum curvature. **2.** The central portion of a mountain chain. [GRAPHICS] The locus of intersection of two pencils of lines in perspective position. [MATH] **1.** In a coordinate system, the line determining one of the coordinates, obtained by setting all other coordinates to zero. **2.** A line of symmetry for a geometric figure. **3.** For a cone whose base has a center, a line passing through this center and the vertex of the cone. [MECH] A line about which a body rotates.   { 'ak·səs }

axis cylinder [CYTOL] **1.** The central mass of a nerve fiber. **2.** The core of protoplasm in a medullated nerve fiber.   { 'ak·səs ¦sil·ən·dər }

axis of abscissas [MATH] The horizontal or x axis of a two-dimensional cartesian coordinate system, parallel to which abscissas are measured.   { 'ak·səs əv ab'sis·əz }

axis of acoustic symmetry [ACOUS] An axis such that the three-dimensional directivity pattern of a transducer may be generated by rotating a two-dimensional directivity pattern around it. Also known as acoustic axis.   { 'ak·səs əv ə'kü·stik 'sim·ə·trē }

axis of circulation [ELECTROMAG] The axis where the equiphase surfaces of a circulating electromagnetic wave converge.   { ¦ak·səs əv ˌsər·kyə'lā·shən }

axis of freedom [DES ENG] An axis in a gyro about which a gimbal provides a degree of freedom.   { 'ak·səs əv frēd·əm }

axis of homology [MAP] The intersection of the plane of

the photograph with the horizontal plane of the map or the plane of reference of the ground. Also known as axis of perspective; map parallel; perspective axis.   { 'ak·səs əv hə'mäl·ə·jē }

axis of ordinates [MATH] The vertical or y axis of a two-dimensional cartesian coordinate system, parallel to which ordinates are measured.   { 'ak·səs əv 'órd·nəts }

axis of pelvis [ANAT] A curved line which forms right angles to the pelvic-cavity planes.   { 'ak·səs əv 'pel·vəs }

axis of perspective *See* axis of homology.   { 'ak·səs əv pər'spek·tiv }

axis of rotation [MECH] A straight line passing through the points of a rotating rigid body that remain stationary, while the other points of the body move in circles about the axis.   { 'ak·səs əv rō'tā·shən }

axis of sighting [ENG] A line taken through the sights of a gun, or through the optical center and centers of curvature of lenses in any telescopic instrument.   { 'ak·səs əv 'sīd·iŋ }

axis of symmetry [MECH] An imaginary line about which a geometrical figure is symmetric. Also known as symmetry axis.   { 'ak·səs əv 'sim·ə·trē }

axis of the bore [ORD] The imaginary central line of the bore of a gun.   { 'ak·səs əv thə 'bòr }

axis of thrust *See* thrust axis.   { 'ak·səs əv ˌthrəst }

axis of tilt [GRAPHICS] A line through the perspective center perpendicular to the principal plane.   { 'ak·səs əv 'tilt }

axis of torsion [MECH] An axis parallel to the generators of a cylinder undergoing torsion, located so that the displacement of any point on the axis lies along the axis. Also known as axis of twist.   { ¦ak·səs əv 'tòr·shən }

axis of trunnions [ORD] The axis about which a gun is rotated in elevation to increase or decrease the range of fire.   { 'ak·səs əv 'trən·yənz }

axis of twist *See* axis of torsion.   { ¦ak·səs əv 'twist }

axis of weld [MET] A line along a weld used to describe the positions of the localized welds.   { 'ak·səs əv 'weld }

axisymmetric flow *See* axial flow.   { ¦ak·sə·sə¦me·trik 'flō }

axisymmetrization [FL MECH] The rounding of noncircular vortices in the absence of strong external deformation. Also known as aggregation.   { ¦ak·sə,sim·ə·trə'zā·shən }

axle [MECH ENG] A supporting member that carries a wheel and either rotates with the wheel to transmit mechanical power to or from it, or allows the wheel to rotate freely on it.   { 'ak·səl }

axle box [ENG] A bushing through which an axle passes in the hub of a wheel.   { 'ak·səl ,bäks }

axle grease [MATER] A lubricating grease containing suspended lime particles and thickened with rosin soap.   { 'ak·səl ,grēs }

axle ratio [MECH ENG] In an automotive vehicle, the ratio of the speed in revolutions per minute of the drive shaft to that of the drive wheels.   { 'ak·səl 'rā·shō }

axoblast [INV ZOO] **1.** The germ cell in mesozoans; cells are linearly arranged in the longitudinal axis and produce the primary nematogens. **2.** The individual scleroblast of the axis epithelium which produces spicules in octocorals.   { 'ak·sə,blast }

axocoel [INV ZOO] The anterior pair of coelomic sacs in the dipleurula larval ancestral stage of echinoderms.   { 'ak·sə,sēl }

axogamy [BOT] Having sex organs on a leafy stem.   { ak'säg·ə·mē }

axolemma [NEURO] The plasma membrane of an axon.   { ˌak·sə'lem·ə }

axolotl [VERT ZOO] The neotenous larva of some salamanders in the family Ambystomidae.   { ¦ak·sə¦läd·əl }

axolotl unit [BIOL] A unit for the standardization of thyroid extracts.   { ¦ak·sə¦läd·əl 'yü·nət }

axometer [ENG] An instrument that locates the optical axis of a lens, particularly a lens used in eyeglasses.   { ak'säm·əd·ər }

axon [NEURO] The process or nerve fiber of a neuron that carries the unidirectional nerve impulse away from the cell body. Also known as neuraxon; neurite.   { 'ak,sän }

axonal transport [NEUROSCI] The movement of organelles and molecules down a nerve cell's axon to its terminals along its cytoplasmic microtubule network.   { ak¦sän·əl 'tranz,pórt }

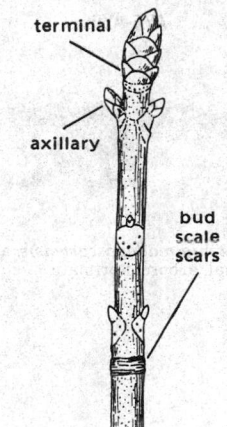

AXILLARY BUD

terminal

axillary

bud scale scars

Position of axillary bud in the buckeye.

axoneme [CYTOL] A bundle of fibrils enclosed by a membrane that is continuous with the plasma membrane. { 'ak·sə,nēm }

Axonolaimoidea [INV ZOO] A superfamily of free-living nematodes with species inhabiting marine and brackish-water environments. { ¦ak·sə·nō·lə'móid·ē·ə }

axonometric projection [GRAPHICS] A drawing that shows an object's inclined position with respect to the planes of projection. Also known as isometric projection. { ¦ak·sə·nō¦me·trik prə'jek·shən }

axoplasm [NEURO] The protoplasm of an axon. { 'ak·sə,plaz·əm }

axopodium [INV ZOO] A semipermanent pseudopodium composed of axial filaments surrounded by a cytoplasmic envelope. { ,ak·sə'pōd·ē·əm }

AYE-AYE

Daubentonia madagascariensis, a nocturnal, arboreal primate.

aye-aye [VERT ZOO] *Daubentonia madagascariensis*. A rare prosimian primate indigenous to eastern Madagascar; the single species of the family Daubentoniidae. { 'ī,ī }

Ayre method [NAV ARCH] A method of calculating the effective horsepower of ships based on a review of model tests. { 'er ,meth·əd }

Ayrton-Jones balance [ELEC] A type of balance with which force between current-carrying conductors is measured; uses single-layer solenoids as the fixed and movable coils. { ¦er·tən ¦jōnz 'bal·əns }

Ayrton-Perry winding [ELEC] Winding of two wires in parallel but opposite directions to give better cancellation of magnetic fields than is obtained with a single winding. { ¦er·tən ¦per·ē ,wind·iŋ }

Ayrton shunt [ELEC] A shunt used to increase the range of a galvanometer without changing the damping. Also known as universal shunt. { 'er·tən ,shənt }

azacrown ether [ORG CHEM] A crown ether that has nitrogen donor atoms as well as oxygen donor atoms to coordinate to the metal iron. { 'az·ə,kraún 'ē·thər }

9-azafluorene *See* carbazole. { ¦nīn ə'za·,flór·ēn }

azaserine [PHARM] $C_4H_7O_4N_3$ An antibiotic produced by a species of *Streptomyces* or by synthesis; used in treatment of acute leukemia. { ə'zas·ə,rēn }

6-azauradine [PHARM] $C_8H_{11}N_3O_6$ Crystals having a melting point of 160-161°C; the base is used as an antineoplastic agent, and the triacetate is used as a drug for psoriasis. { ¦siks ə'zór·ə,dēn }

azelaic acid [ORG CHEM] $HOOC(CH_2)_7CCOH$ Colorless leaflets; melting point 106.5°C; a dicarboxylic acid useful in lacquers, alkyd salts, organic synthesis, and formation of polyamides. { ¦az·ə¦lā·ik 'as·əd }

azelate [ORG CHEM] A salt of azelaic acid, for example, sodium azelate. { 'az·əl,āt }

azel display [ELECTR] Modified type of plan position indicator presentation showing two separate radar displays on one cathode-ray screen; one display presents bearing information and the other shows elevation. { 'az·el dis,plā }

azel mounting *See* altazimuth mounting. { 'az·əl ,maúnt·iŋ }

azeotrope *See* azeotropic mixture. { ā'zē·ə,trōp }

azeotropic distillation [CHEM ENG] A process by which a liquid mixture is separated into pure components with the help of an additional substance or solvent. { ¦āz·ē·ə,trō·pik ,dis·tə'lā·shən }

azeotropic mixture [CHEM] A solution of two or more liquids, the composition of which does not change upon distillation. Also known as azeotrope. { ¦a,zē·ə¦träp·ik 'miks·chər }

2-azetidinecarboxylic acid [BIOCHEM] $C_4H_7NO_2$ Crystals which discolor at 200°C and darken until 310°C; soluble in water; a specific antagonist of L-proline; used in the production of abnormally high-molecular-weight polypeptides. { ¦tü ə,zed·ə,dēn·ə,kär,bäk'sil·ik 'as·əd }

azide [ORG CHEM] One of several types of compounds containing the $-N_3$ group and derived from hydrazoic acid, HN_3. { 'ā,zīd }

azimino *See* diazoamine. { ā'zim·ē·nō }

azimuth [ASTRON] Horizontal direction of a celestial point from a terrestrial point, expressed as the angular distance from a reference direction, usually measured from 0° at the reference direction clockwise through 360°. [ENG] In directional drilling, the direction of the face of the deviation tool with respect to magnetic north. [GEOD] Horizontal direction on the earth's surface. { 'az·ə·məth }

azimuth-adjustment slide rule [ENG] A circular slide rule by which a known angular correction for fire at one elevation can be changed to the proper correction for any other elevation. { 'az·ə·məth ə¦jəs·mənt 'slīd ,rül }

azimuthal chart [MAP] A chart on an azimuthal projection. Also known as zenithal chart. { ,az·ə'məth·əl 'chärt }

azimuthal equidistant chart [MAP] A chart on the azimuthal equidistant map projection. { ,az·ə'məth·əl ,ē·kwə'dis·tənt ¦chärt }

azimuth alignment [ENG ACOUS] The condition whereby the center lines of the playback- and recording-head gaps are exactly perpendicular to the magnetic tape and parallel to each other. { 'az·ə·məth ə'līn·mənt }

azimuthal map projection [MAP] The transformation of a spherical representation of the earth's surface to a tangent or intersecting plane established perpendicular to a right line passing through the center of the spherical representation. { ,az·ə'məth·əl 'map prə'jek·shən }

azimuthal orthomorphic projection *See* stereographic projection. { ,az·ə'məth·əl ¦orth·ə¦mór·fik prə'jek·shən }

azimuthal quantum number [ATOM PHYS] The orbital angular momentum quantum number l, such that the eigenvalue of L^2 is $l(l + 1)$. { ,az·ə'məth·əl ¦kwän·təm ,nəm·bər }

azimuth angle [ENG] An angle in triangulation or in traverse through which the computation of azimuth is carried. { 'az·ə·məth 'aŋ·gəl }

azimuth-angle indicator *See* riser-angle indicator. { 'az·ə·məth' aŋ·gəl ,in·də,kād·ər }

azimuth bar *See* azimuth instrument. { 'az·ə·məth ,bär }

azimuth blanking [ELECTR] Blanking of the radar receiver as the scan traverses a selected azimuth region. { 'az·ə·məth ,blaŋk·iŋ }

azimuth circle [DES ENG] A ring calibrated from 0 to 360° over a compass, compass repeater, radar plan position indicator, direction finder, and so on, which provides means for observing compass bearings and azimuths. { 'az·ə·məth ,sər·kəl }

azimuth compass [NAV] A compass with vertical sights so that the magnetic azimuths of celestial bodies may be read. { 'az·ə·məth ,käm·pəs }

azimuth deviation [ORD] The angular difference in azimuth between the lines from the gun to the target and from the gun to the point where a projectile strikes or bursts. { 'az·ə·məth ,dē·vē'ā·shən }

azimuth dial [ENG] Any horizontal circle dial that reads azimuth. { 'az·ə·məth ,dīl }

azimuth equation [GEOD] A condition equation which expresses the relationship between the fixed azimuths of two lines that are connected by triangulation or traverse. { 'az·ə·məth i'kwā·zhən }

azimuth error [ASTRON] The angle by which the east-west axis of a transit telescope deviates from being perpendicular to the plane of the meridian. [ENG] An error in the indicated azimuth of a target detected by radar. { 'az·ə·məth ,er·ər }

azimuth gain reduction [ELECTR] Technique which allows control of the radar receiver system throughout any two azimuth sectors. { 'az·ə·məth 'gän ri,dək·shən }

azimuth gating [ELECTR] The practice of selectively brightening and enhancing the gain-desired sectors of a radar plan position indicator display, usually by applying a step waveform to the automatic gain control circuit. { 'az·ə·məth ,gād·iŋ }

azimuth indicator [ENG] An approach-radar scope which displays azimuth information. { 'az·ə·məth ,in·də,kād·ər }

azimuth instrument [ENG] An instrument for measuring azimuths, particularly a device which fits over a central pivot in the glass cover of a magnetic compass. Also known as azimuth bar; bearing bar. { 'az·ə·məth ,in·strə·mənt }

azimuth line [ENG] A radial line from the principal point, isocenter, or nadir point of a photograph, representing the direction to a similar point of an adjacent photograph in the same flight line; used extensively in radial triangulation. { 'az·ə·məth ,līn }

azimuth marker [ENG] **1.** A scale encircling the plan position indicator scope of a radar on which the azimuth of a target from the radar may be measured. **2.** Any of the reference limits inserted electronically at 10 or 15° intervals which extend radially from the relative position of the radar on an off-center plan position indicator scope. [NAV] *See* electronic azimuth marker. { 'az·ə·məth ,mär·kər }

azimuth rate [ORD] In gunnery, the rate of change in azimuth measured in mils or degrees per second. { 'az·ə·məth ‚rāt }

azimuth resolution [ELECTROMAG] Angle or distance by which two targets must be separated in azimuth to be distinguished by a radar set, when the targets are at the same range. { 'az·ə·məth ‚rez·ə'lü·shən }

azimuth scale [ENG] A graduated angle-measuring device on instruments, gun carriages, and so forth that indicates azimuth. { 'az·ə·məth ‚skāl }

azimuth-stabilized plan position indicator [ENG] A north-upward plan position indicator (PPI), a radarscope, which is stabilized by a gyrocompass so that either true or magnetic north is always at the top of the scope regardless of vehicle orientation. { 'az·ə·məth ¦sta·bə‚līzd 'plan pə'zish·ən 'in·də‚kād·ər }

azimuth tables [ASTRON] Publications providing tabulated azimuths or azimuth angles of celestial bodies for various combinations of declination, altitude, and hour angle; great-circle course angles can also be obtained by substitution of values. { 'az·ə·məth ‚ta·bəlz }

azimuth transfer [ENG] Connecting, with a straight line, the nadir points of two vertical photographs selected from overlapping flights. { 'az·ə·məth 'tranz‚fər }

azimuth traverse [ENG] A survey traverse in which the direction of the measured course is determined by azimuth and verified by back azimuth. { 'az·ə·məth trə'vərs }

azimuth versus amplitude [ELECTR] Electronic counter-countermeasures receiver with a plan position indicator type of display attached to the main antenna and used to display strobes due to jamming aircraft; it is useful in making passive fixes when two or more radar sites can operate together. { 'az·ə·məth ‚vər·səs 'am·plə‚tüd }

azine [ORG CHEM] A compound of six atoms in a ring; at least one of the atoms is nitrogen, and the ring structure resembles benzene; an example is pyridine. { 'ā‚zēn }

azine dyes [ORG CHEM] Benzene-type dyes derived from phenazine; members of the group, such as nigrosines and safranines, are quite varied in application. { 'ā‚zēn ‚dīz }

aziridine See ethyleneimine. { ə'zir‚ə‚dēn }

azlactone [ORG CHEM] A compound that is an anhydride of α-acylamino acid; the basic ring structure is the 5-oxazolone type. { az'lak‚tōn }

azlon [TEXT] Any textile fiber derived from protein, such as casein. { 'az‚län }

azo- [ORG CHEM] A prefix indicating the group −N=N−. { 'a‚zō }

azobenzene [ORG CHEM] $C_6H_5N_2C_6H_5$ A compound existing in cis and trans geometric isomers; the cis form melts at 71°C; the trans form comprises orange-red leaflets, melting at 68.5°C; used in manufacture of dyes and accelerators for rubbers. { ‚a·zō'ben‚zēn }

2,2′-azobisisobutyronitrile [ORG CHEM] $C_8H_{12}N_4$ Crystals that decompose at 107°C; soluble in methanol and in ethanol; used as an initiator of free radical reactions and as a blowing agent for plastics and elastomers. { ¦tü ¦tü‚prīm ‚a·zō·bī‚ī·sō‚byüd·ə·rə'nī·trəl }

azo compound [ORG CHEM] A compound having two organic groups separated by an azo group (−N=N−). { 'a‚zō ‚käm‚paünd }

azo dyes [ORG CHEM] Widely used commercial dyestuffs derived from amino compounds, with the −N− chromophore group; can be made as acid, basic, direct, or mordant dyes. { 'a·zō ‚dīz }

Azoic [GEOL] That portion of the earlier Precambrian time in which there is no trace of life. { ā'zō·ik }

azoic dye [ORG CHEM] A water-insoluble azo dye that is formed by coupling of the components on a fiber. Also known as ice color; ingrain color. { a'zō·ik 'dī }

azoic printing [GRAPHICS] The method whereby azoic compositions, that is, mixtures of naphtols and diazotized products, temporarily inhibited from color development are printed on cloth; when the printed material is passed through steam containing formic acid vapor, the coupling reaction occurs and development takes place. { a'zō·ik 'print·iŋ }

azole [ORG CHEM] One of a class of organic compounds with a five-membered N-heterocycle containing two double bonds; an example is 1,2,4-triazole. { 'ā‚zōl }

Azomonas [MICROBIO] A genus of large, motile, oval to spherical bacteria in the family Axotobacteraceae; members produce no cysts and secrete large quantities of capsular slime. { ‚az·ə'mō·nəs }

azomycin [MICROBIO] $C_3H_3O_2N_3$ An antimicrobial antibiotic produced by a strain of *Nocardia mesenterica*. { ā·zō'mis·ən }

azon [ORD] A glide bomb used in World War II; its control surfaces in the trail were adjusted by radio signals to control the bomb in azimuth only (hence, azon). { 'ā‚zän }

azonal soil [GEOL] Any group of soils without well-developed profile characteristics, owing to their youth, conditions of parent material, or relief that prevents development of normal soil-profile characteristics. Also known as immature soil. { 'ā‚zōn·əl 'sȯil }

azoospermia [PHYSIO] **1.** Absence of motile sperm in the semen. **2.** Failure of formation and development of sperm. { ¦ā‚zō·ō'spər·mē·ə }

Azores high [METEOROL] The semipermanent subtropical high over the North Atlantic Ocean, especially when it is located over the eastern part of the ocean; when in the western part of the Atlantic, it becomes the Bermuda high. { ‚ā‚zȯrz 'hī }

azotemia [MED] The presence of excessive amounts of nitrogenous compounds in the blood. { ‚az·ə'tem·ē·ə }

Azotobacter [MICROBIO] A genus of large, usually motile, rod-shaped, oval, or spherical bacteria in the family Azotobacteraceae; form thick-walled cysts, and may produce large quantities of capsular slime. { ə'zōd·ə‚bak·tər }

Azotobacteraceae [MICROBIO] A family of large, bluntly rod-shaped, gram-negative, aerobic bacteria capable of fixing molecular nitrogen. { ə¦zōd·ə‚bak·tə'rās·ē‚ē }

azotometer See nitrometer. { ‚az·ə'täm·əd·ər }

azoturia [MED] A condition characterized by excess amounts of urea or other nitrogenous substances in the urine. { ‚az·ə'tür·ē·ə }

azoxybenzene [ORG CHEM] $C_6H_5NO=N−C_6H_5$ A compound existing in cis and trans forms; the cis form melts at 87°C; the trans form comprises yellow crystals, melting at 36°C, insoluble in water, soluble in ethanol. { ə¦zäk·sē'ben‚zēn }

azoxy compound [ORG CHEM] A compound having an oxygen atom bonded to one of the nitrogen atoms of an azo compound. { ā¦zäk·sē ‚käm‚paünd }

azulene [ORG CHEM] $C_{16}H_{26}O$ The blue coloring matter of wormwood and other essential oils; an oily, blue liquid, boiling at 170°C; insoluble in water; used in cosmetics. { 'azh·ə‚lēn }

azulite [MINERAL] A translucent pale-blue variety of smithsonite found in large masses in Arizona and Greece. { 'azh·ə‚līt }

azure B [CYTOL] $C_{15}H_{16}ClN_3S$ A metachromatic basic dye that imparts a green color to chromosomes, a blue color to nucleoli and cytoplasmic ribosomes, and a red color to deposits containing mucopolysaccharides. { ¦azh·ər 'bē }

azurite [MINERAL] $Cu_3(CO_3)_2(OH)_2$ A blue monoclinic mineral consisting of a basic carbonate of copper; an ore of copper. Also known as blue copper ore; blue malachite; chessylite. { 'azh·ə‚rīt }

azurmalachite [MINERAL] A mixture of azurite and malachite, usually occurring massive with concentric banding; used as an ornamental stone. { ¦a·zhər'mal·ə‚kīt }

Azusa [ENG] A continuous-wave, high-accuracy, phase-comparison, single-station tracking system operating at C-band and giving two direction cosines and slant range which can be used to determine space position and velocity of a vehicle (usually a rocket or a missile). { ə'züs·ə }

azygospore [MYCOL] A spore which is morphologically similar to a zygospore but is formed parthenogenetically. Also known as parthenospore. { ā'zī·gə‚spȯr }

azygos vein [ANAT] A branch of the right precava which drains the intercostal muscles and empties into the superior vena cava. { ā'zī·gəs ‚vān }

azygote [BIOL] An individual produced by haploid parthenogenesis. { ā'zī‚gōt }

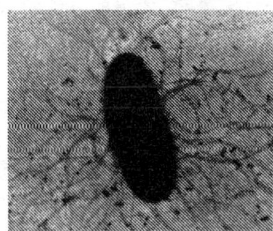

AZOTOBACTER

Azotobacter vinelandii with peritrichous flagella. (*From A. W. Hofer, Flagellation of Azotobacter, J. Bacteriol., 48:697-701, 1944*)

b *See* barn; bel.

B *See* bel; boron; brewster.

Ba *See* barium.

Baade's window [ASTRON] An unusually transparent region about 4° from the galactic center. { 'bä·dəz ,win·dō }

babassu oil [MATER] A nondrying oil obtained from the kernels of the babassu palm and composed principally of lauric, myristic, and oleic acids. { ¦bä·bə¦sü ,oil }

babbitt metal [MET] Any of the white alloys composed primarily of tin or lead and of lesser amounts of antimony, copper, and perhaps other metals, and used for bearings. { 'bab·ət ,med·əl }

babble [COMMUN] **1.** Aggregate crosstalk from a large number of channels. **2.** Unwanted disturbing sounds in a carrier or other multiple-channel system which result from the aggregate crosstalk or mutual interference from other channels. { 'bab·əl }

Babcock coefficient of friction [FL MECH] An approximation to the coefficient of friction for steam flowing in a circular pipe of diameter *d* inches, given by 0.0027[1 + (3.6/*d*)]. { 'bab,käk ,kō·ə¦fish·ənt əv 'frik·shən }

Babcock magnetograph [ASTRON] An instrument used to measure weak magnetic fields on the sun. { 'bab,käk mag'ned·ə,graf }

Babcock test [FOOD ENG] A test involving volumetric measurement of the fat content of a sample of milk; sulfuric acid is added, and the flask is heated to liquefy the fat, which is brought to the calibrated neck of the flask by centrifugal force. { 'bab,käk ,test }

Babesia [INV ZOO] The type genus of the Babesiidae, a protozoan family containing red blood cell parasites. { bə'bezh·ə }

babesiasis [VET MED] A tick-borne protozoan disease of mammals other than humans caused by species of *Babesia*. { ,bab·ə'zī·ə·səs }

Babesiidae [INV ZOO] A family of protozoans in the suborder Haemosporina containing parasites of vertebrate red blood cells. { ,bab·ə'zī·ə,dē }

Babinet compensator [OPTICS] A device for working with polarized light, made of two quartz prisms, assembled in a rhomb, to enable the optical retardation to be adjusted to positive or negative values. { bä·bi'nā ,käm·pen'sād·ər }

Babinet point [OPTICS] A neutral point located 15 to 20° directly above the sun. { bä·bi'nā ,point }

Babinet's principle [OPTICS] The principle that the diffraction patterns produced by complementary screens are identical; two screens are said to be complementary when the opaque parts of one correspond to the transparent parts of the other. { bä·bi'nāz ,prin·sə·pəl }

Babinski reflex [MED] An abnormal reflex after infancy associated with a disturbance of the pyramidal tract, characterized by extension of the great toe with fanning of the other toes on sharply stroking the lateral aspect of the sole. { bə'binz·kē 'ri,fleks }

baboon [VERT ZOO] Any of five species of large African and Asian terrestrial primates of the genus *Papio*, distinguished by a doglike muzzle, a short tail, and naked callosities on the buttocks. { ba'bün }

Babo's law [PHYS CHEM] A law stating that the relative lowering of a solvent's vapor pressure by a solute is the same at all temperatures. { 'bä,bōz ,lô }

babs *See* blind approach beacon system. { babz }

babuina [VERT ZOO] A female baboon. { ,bab·ə'wēn·ə }

baby-pig disease [VET MED] Acute hypoglycemia of newborn pigs; usually fatal if untreated. { ¦bā·bē 'pig di,zēz }

baby spot [ELEC] A small spotlight, usually equipped with a hood, used (as in the theater) to concentrate light on an area or an object a small distance from the spotlight. { ¦bā·bē 'spät }

baccate [BOT] **1.** Bearing berries. **2.** Having pulp like a berry. { 'bak,āt }

bacciferous [BOT] Bearing berries. { bak'sif·ə·rəs }

Bacillaceae [MICROBIO] The single family of endospore-forming rods and cocci. { ,bas·ə'lās·ē,ē }

Bacillariophyceae [BOT] The diatoms, a class of algae in the division Chrysophyta. { ,bas·ə,ler·ē·ə'fīs·ē,ē }

Bacillariophyta [BOT] An equivalent name for the Bacillariophyceae. { ,bas·ə,ler·ē'ä·fəd·ə }

bacillary [MICROBIO] **1.** Rod-shaped. **2.** Produced by, pertaining to, or resembling bacilli. { 'bas·ə,ler·ē }

bacillary dysentery [MED] A highly infectious bacterial disease of humans, localized in the bowels; caused by *Shigella*. { 'bas·ə,ler·ē 'dis·ən,ter·ē }

bacillary white diarrhea *See* pullorum disease. { 'bas·ə,ler·ē ¦wīt ,di·ə'rē·ə }

bacillophobia [PSYCH] An abnormal fear of bacilli. { bə,sil·ə'fōb·ē·ə }

bacilluria [MED] The presence of bacilli in the urine. { ,bas·ə'lùr·ē·ə }

bacillus [MICROBIO] Any rod-shaped bacterium. { bə'sil·əs }

Bacillus [MICROBIO] A genus of bacteria in the family Bacillaceae; rod-shaped cells are aerobes or facultative anaerobes and usually produce catalase. { bə'sil·əs }

Bacillus anthracis [MICROBIO] A gram-positive, rod-shaped, endospore-forming bacterium that is the causative agent of anthrax; its spores can remain viable for many years in soil, water, and animal hides and products. { bə¦sil·əs ,an'thrak·əs }

Bacillus Calmette-Guérin vaccine [IMMUNOL] A vaccine prepared from attenuated human tubercle bacilli and used to immunize humans against tuberculosis. Abbreviated BCG vaccine. { bə'sil·əs ,kal'met ,gā'raṇ ,vak,sēn }

Bacillus cereus [MICROBIO] A spore-forming bacterium that often survives cooking and grows to large numbers in improperly refrigerated foods; it produces both a diarrheal toxin and an emetic toxin in the gastrointestinal tract following its ingestion via contaminated meats, dried foods, and rice. { bə,sil·əs 'sir·ē·əs }

Bacillus sphaericus [MICROBIO] An aerobic, spore-producing bacterium that forms a protein complex during sporulation that is toxic to the larvae of certain mosquitoes but is apparently harmless to all other forms of life. { bə,sil·əs 'sfī·ri·kəs }

bacitracin [MICROBIO] A group of polypeptide antibiotics produced by *Bacillus licheniformis*. { ,bas·ə'trās·ən }

back [ANAT] The part of the human body extending from the neck to the base of the spine. [GRAPHICS] The part of a book where the binding and pages are stitched together. [MIN ENG] **1.** The upper part of any mining cavity. **2.** A joint,

BABOON

A baboon, a representative of Old World cercopithecoid monkeys.

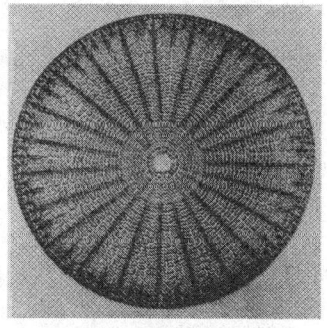

BACILLARIOPHYCEAE

Arachnoidiscus ehrenbergii, a concentric diatom with radial symmetry. *(From H. J. Fuller and O. Tippo, College Botany, rev. ed., Holt, 1954).*

usually a strike joint, perpendicular to the direction of working. { bak }

backacter *See* backhoe. { 'bak,ak·tər }

back-arc basin [GEOL] The region (small ocean basin) between an island arc and the continental mainland formed during oceanic plate subduction, containing sediment eroded from both. { 'bak,ärk ,bās·ən }

back arch [ARCH] A concealed arch which carries backing of a wall, where the exterior facing is carried by a lintel. { 'bak ,ärch }

back azimuth [NAV] An azimuth 180° from a given azimuth. { 'bak 'az·ə·məth }

back balance [MIN ENG] **1.** A kind of self-acting incline in a mine. **2.** The means of maintaining tension on a rope transmission or haulage system, consisting of the tension carriage, attached weight, and supporting structure. { 'bak ¦bal·əns }

backband [BUILD] A piece of millwork used around a rectangular window or door casing as a cover for the gap between the casing and the wall or as a decorative feature. Also known as backbend. { 'bak,band }

back beach *See* backshore. { 'bak ,bēch }

back bearing [NAV] A bearing along the reverse direction of a line. Also known as reciprocal bearing. { 'bak 'ber·iŋ }

backbend [BUILD] **1.** At the outer edge of a metal door or window frame, the face which returns to the wall surface. **2.** *See* backband. { 'bak,bend }

backbending [NUC PHYS] A discontinuity in the rotational levels of some rare-earth nuclei around spin $20\hbar$ (where $\hbar$ is Planck's constant divided by 2π), which appears as a backbend on a graph that plots the moment of inertia versus the square of the rotational frequency. { 'bak,bend·iŋ }

back-bent occlusion *See* bent-back occlusion. { 'bak ,bent ə'klü·zhən }

back bias [ELECTR] **1.** Degenerative or regenerative voltage which is fed back to circuits before its originating point; usually applied to a control anode of a tube or other device. **2.** Voltage applied to a grid of a tube (or tubes) or electrode of another device to reduce a condition which has been upset by some external cause. { 'bak ,bī·əs }

backblast area [ORD] Cone-shaped area to the rear of a recoilless weapon, rocket launcher, or rocket-assisted takeoff unit which is dangerous to personnel because of the expulsion of powder gases. { 'bak,blast 'er·ē·ə }

back bond [SOLID STATE] A chemical bond between an atom in the surface layer of a solid and an atom in the second layer. { 'bak ,bänd }

backbone [ANAT] *See* spine. [COMPUT SCI] The portion of a communication network that handles the largest volume of traffic, usually employing a high-speed, high-capacity medium designed to transmit data over long distances. [GEOL] **1.** A ridge forming the principal axis of a mountain. **2.** The principal mountain ridge, range, or system of a region. [GRAPHICS] *See* spine. { 'bak,bōn }

back boxing *See* backlining. { 'bak ¦bäk·siŋ }

backbreak *See* overbreak. { 'bak,brāk }

back bulb [BOT] A pseudobulb on certain orchid plants that remains on the plant after removal of the terminal growth, and that is used for propagation. { 'bak ,bəlb }

backcast stripping [MIN ENG] A stripping method using two draglines; one strips and casts the overburden, and the other recasts a portion of the overburden. { 'bak,kast 'strip·iŋ }

back check [DES ENG] In a hydraulic door closer, a mechanism that slows the speed with which a door may be opened. { 'bak ,chek }

back-coated mirror [OPTICS] Glass with a reflective coating applied against the rear surface. { 'bak ,kōd·əd 'mir·ər }

backcoating [GRAPHICS] A light-absorbing sensitized layer that is applied to the back of a film base material. { 'bak,kōd·iŋ }

back contact [ELEC] Normally closed stationary contact on a relay that is opened when the relay is energized. { 'bak ¦kän,takt }

back course [NAV] In the instrument low-approach system, the course which extends from the back of a localizer antenna system to furnish guidance in the horizontal plane at the rear of the localizer. { 'bak ,kȯrs }

backcross [GEN] A cross between an F_1 heterozygote and an individual of P_1 genotype. { 'bak,krȯs }

backcross parent *See* recurrent parent. { 'bak,krȯs ,per·ənt }

backdeep [GEOL] An epieugeosynclinal basin; a nonvolcanic postorogenic geosynclinal basin whose sediments are derived from an uplifted eugeosyncline. { 'bak,dēp }

backdigger *See* backhoe. { 'bak¦dig·ər }

back diode [ELECTR] A special type of tunnel diode operated at low levels of reverse bias at which the device has negative resistance. { 'bak ,dī,ōd }

back-door cold front [METEOROL] A front which leads a cold air mass toward the south and southwest along the Atlantic seaboard of the United States. { 'bak ,dȯr 'kōld ,frənt }

back draft [MET] A reversed taper given to a casting model or pattern to prevent its withdrawal from the mold. { 'bak ,draft }

back-draft damper [MECH ENG] A damper with blades actuated by gravity, permitting air to pass through them in one direction only. { 'bak ,draft 'dam·pər }

back echo [ELECTROMAG] An echo signal produced on a radar screen by one of the minor back lobes of a search radar beam. { 'bak ,ek·ō }

back-echo reflection [ELECTR] A radar echo produced by radiation reflected to the target by a large, fixed obstruction; that is, the ray path is from antenna to obstruction to target to antenna, instead of antenna to target to antenna. { 'bak ,ek·ō ri'flek·shən }

backed cloth [TEXT] Cloth made by weaving or knitting an extra weft or warp to the back to increase thickness or to obtain different color effects. { 'bakt 'klȯth }

back edging [ENG] Cutting through a glazed ceramic pipe by first chipping through the glaze around the outside and then chipping the pipe itself. { 'bak ,ej·iŋ }

back electromotive force *See* counterelectromotive force. { 'bak i¦lek·trō¦mōd·iv 'fȯrs }

back-emission electron radiography [ELECTR] A technique used in microradiography to visualize, among other things, the presence of material of different atomic numbers in the surface of the specimen being observed; the polished side of the specimen is facing and in close contact with the emulsion side of a fine-grain photographic plate; a light-tight cover holds the specimen and plate in place to be subjected to hardened x-rays. { 'bak i'mish·ən i'lek,trän ,rād·ē'äg·rə·fē }

back end *See* thrust yoke. { 'bak ,end }

back-end system [COMPUT SCI] A computer that operates on data which have been previously processed by another computer system. { 'bak ¦end ,sis·təm }

backfill [CIV ENG] Earth refilling a trench or an excavation around a building, bridge abutment, and the like. [MIN ENG] Waste sand or rock used to support the mine roof after removal of ore. { 'bak,fil }

back fillet [BUILD] The return of the margin of a groin, doorjamb, or window jamb when it projects beyond a wall. { 'bak ,fil·ət }

backfire [CIV ENG] A fire that is started in order to burn against and cut off a spreading fire. [ELECTR] *See* arcback. [ENG] Momentary backward burning of flame into the tip of a torch. Also known as flashback. [MECH ENG] In an internal combustion engine, an improperly timed explosion of the fuel mixture in a cylinder, especially one occurring during the period that the exhaust or intake valve is open and resulting in a loud detonation. [ORD] Rearward escapement of gases or cartridge fragments upon firing a gun. { 'bak,fīr }

backfire antenna [ELECTROMAG] An antenna which exhibits significant gain in a direction 180° from its principal lobe. { 'bak,fīr an'ten·ə }

backflap hinge [DES ENG] A hinge having a flat plate or strap which is screwed to the face of a shutter or door. Also known as flap hinge. { 'bak,flap ,hinj }

backflash [CHEM] Rapid combustion of a material occurring in an area that the reaction was not intended for. { 'bak,flash }

backflooding [HYD] A reversal of flow of water at the water table resulting from changes in precipitation. { 'bak,fləd·iŋ }

backflow [CIV ENG] The flow of water or other liquids, mixtures, or substances into the distributing pipes of a potable supply of water from any other than its intended source. [FL MECH] Any flow in a direction opposite to the natural or intended direction of flow. { 'bak,flō }

backflow connection [CIV ENG] Any arrangement of pipes,

plumbing fixtures, drains, and so forth, in which backflow can occur. { 'bak,flō kə'nek·shən }

backflow preventer *See* vacuum breaker. { 'bak,flō pri'ven·tər }

backflow valve *See* backwater valve. { 'bak,flō ,valv }

back focal length [OPTICS] The distance from the rear surface of a lens to its focal plane. { 'bak 'fō·kəl ,leŋkth }

backfolding [GEOL] Process in mountain forming in which the folds are overturned toward the interior of an orogenic belt. Also known as backward folding. { 'bak,fōld·iŋ }

backfurrow [AGR] To plow by throwing or turning together the first two furrows that were plowed, leaving clear furrows on the side. [CIV ENG] In an excavation procedure, the first cut made on undisturbed land. { 'bak,fər·ō }

back gearing [MECH ENG] The technique of using gears on machine tools to obtain an increase in the number of speed changes that can be gotten with cone belt drives. { 'bak ,gir·iŋ }

Back-Goudsmit effect [ATOM PHYS] Breakdown of the coupling between the nuclear-spin angular momentum and the total angular momentum of the electrons in an atom at relatively small magnetic fields. { 'bak 'gōd,smit i,fekt }

back gouging [MET] The elimination of excess material from both weld metal and base metal on the opposite side of a partly welded joint; a groove or bevel is formed in order to facilitate complete joint penetration. { 'bak 'gaúj·iŋ }

background [COMMUN] **1.** Picture white of the facsimile copy being scanned when the picture is black and white only. **2.** Undesired printing in the recorded facsimile copy of the picture being transmitted, resulting in shading of the background area. Noise heard during radio reception caused by atmospheric interference or the operation of the receiver at such high gain that inherent tube and circuit noises become noticeable. { 'bak,graúnd }

background count [PHYS] Responses of the radiation counting system to radiation coming from sources other than the source to be measured. { 'bak,graúnd ,kaúnt }

background discrimination [ENG] The ability of a measuring instrument, circuit, or other device to distinguish signal from background noise. { 'bak,graúnd dis,krim·ə'nā·shən }

background extinction [EVOL] Intervals of lower extinction intensity between mass extinctions. { 'bak,graúnd ik ,stiŋk·shən }

background genotype [GEN] The genotype of the organism at genetic loci other than those responsible for the phenotype. { 'bak,graúnd 'jē·nə,tīp }

background ink [COMPUT SCI] In optical character recognition, a highly reflective ink used to print the parts of a document that are to be ignored by the scanner. { 'bak,graúnd ,iŋk }

background luminance [OPTICS] In visual-range theory, the brightness of the background against which a target is viewed. { 'bak,graúnd 'lüm·i·nəns }

background mass spectrum [PHYS] The display of printed record obtained from a mass spectrometer or spectrograph before a sample has been inserted. { 'bak,graúnd ,mas 'spek·trəm }

background noise [ACOUS] The unwanted residual sound that is present whether or not the sound source being studied is in operation. [ENG] The undesired signals that are always present in an electronic or other system, independent of whether or not the desired signal is present. { 'bak,graúnd ,nóiz }

background processing [COMPUT SCI] **1.** The execution of lower-priority programs when higher-priority programs are not being handled by a data-processing system. **2.** Computer processing that is not interactive or visible on the display screen. { 'bak,graúnd 'prä·ses·iŋ }

background program [COMPUT SCI] A computer program that has low priority in a multiprogramming system. { 'bak,graúnd 'prō·grəm }

background radiation [NUCLEO] The radiation in humans' natural environment, including cosmic rays and radiation from the naturally radioactive elements. Also known as natural radiation. [PHYS] Radiation which is due to sources other than the source of interest in a measurement of radiation and which is detected by the measuring apparatus. { 'bak,graúnd ,rād·ē'ā·shən }

background reflectance [COMPUT SCI] The reflectance, relative to a standard, of the surface on which a printed or handwritten character has been inscribed in optical character recognition. { 'bak,graúnd ri'flek·təns }

background returns [ENG] **1.** Signals on a radar screen from objects which are of no interest. **2.** *See* clutter. { 'bak,-graúnd ri'tərnz }

background signal [ENG] The output of a leak detector caused by residual gas to which the detector element reacts. { 'bak,graúnd ,sig·nəl }

back gutter [BUILD] A gutter installed on the uphill side of a chimney on a sloping roof to divert water around the chimney. { 'bak ,gəd·ər }

backhand welding [MET] A welding technique in which the flame is directed back against the completed weld. Also known as backward welding. { 'bak,hand 'weld·iŋ }

backhaul [COMMUN] Point-to-point satellite transmission of video from a remote site to a network distribution center in real time. { 'bak,hól }

back hearth [BUILD] That part of the hearth (or floor) which is contained within the fireplace itself. Also known as inner hearth. { 'bak ,härth }

backhoe [MECH ENG] An excavator fitted with a hinged arm to which is rigidly attached a bucket that is drawn toward the machine in operation. Also known as backacter; backdigger; dragshovel; pullshovel. { 'bak ,hō }

back holes [MIN ENG] The holes which are shot last in mine shaft sinking. { 'bak ,hōlz }

backing [CIV ENG] **1.** The unexposed, rough masonry surface of a wall that is faced with finer work. **2.** The earth backfill of a retaining wall. [ELECTR] Flexible material, usually cellulose acetate or polyester, used on magnetic tape as the carrier for the oxide coating. [MET] *See* backing strip. [METEOROL] **1.** Internationally, a change in wind direction in a counterclockwise sense (for example, south to east) in either hemisphere of the earth. **2.** In United States usage, a change in wind direction in a counterclockwise sense in the Northern Hemisphere, clockwise in the Southern Hemisphere. [MIN ENG] Timbers across the top of a level, supported in notches cut in the rock. { 'bak·iŋ }

backing board [BUILD] In a suspended acoustical ceiling, a flat sheet of gypsum board to which acoustical tile is attached by adhesive or mechanical means. { 'bak·iŋ ,bórd }

backing brick [CIV ENG] A relatively low-quality brick used behind face brick or other masonry. { 'bak·iŋ ,brik }

backing cloth [GRAPHICS] Fiber material on the reverse side of photographic paper for strength. { 'bak·iŋ ,klóth }

backing deals [MIN ENG] Boards, 1–4 inches (2.5–10 centimeters) thick, of sufficient length to bridge the space between timber or steel sets or between rings in skeleton tubbing. { 'bak·iŋ ,dēlz }

backing off [ENG] Removing excessive body metal from badly worn bits. { 'bak·iŋ ,óf }

backing plate [ENG] A plate used to support the hardware for the cavity used in plastics injection molding. { 'bak·iŋ ,plāt }

backing pump [MECH ENG] A vacuum pump, in a vacuum system using two pumps in tandem, which works directly to the atmosphere and reduces the pressure to an intermediate value, usually between 100 and 0.1 pascals. Also known as fore pump. { 'bak·iŋ ,pəmp }

backing ring [ENG] A strip of metal attached at a pipe joint at the root of a weld to prevent spatter and to ensure the integrity of the weld. { 'bak·iŋ ,riŋ }

backing space [ENG] Space between a fore pump and a diffusion pump in a leak-testing system. { 'bak·iŋ ,spās }

backing-space technique [ENG] Testing for leaks by connecting a leak detector to the backing space. { 'bak·iŋ ,spās ,tek·nēk }

backing storage [COMPUT SCI] A computer storage device whose capacity is larger, but whose access time is slower, than that of the computer's main storage or immediate access storage; usually slower than main storage. Also known as bulk storage. { 'bak·iŋ ,stór·ij }

backing strip [MET] A piece of metal, asbestos, or other nonflammable material placed behind a joint to facilitate welding. Also known as backing. { 'bak·iŋ ,strip }

backing up [CIV ENG] In masonry, the laying of backing brick. [GRAPHICS] The process of printing on the reverse side of a printed sheet. { 'bak·iŋ ,əp }

back jamb *See* backlining. { 'bak ,jam }

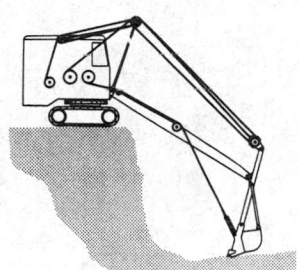

BACKHOE

Diagram of power crane fitted with backhoe. *(United States Steel Corp.)*

backjoint [CIV ENG] In masonry, a rabbet such as that made on the inner side of a chimneypiece to receive a slip. { 'bak,jȯint }

backlands [GEOL] A section of a river floodplain lying behind a natural levee. { 'bak,lanz }

backlash [DES ENG] The amount by which the tooth space of a gear exceeds the tooth thickness of the mating gear along the pitch circles. [ELECTR] A small reverse current in a rectifier tube caused by the motion of positive ions produced in the gas by the impact of thermoelectrons. [ENG] **1.** Relative motion of mechanical parts caused by looseness. **2.** The difference between the actual values of a quantity when a dial controlling this quantity is brought to a given position by a clockwise rotation and when it is brought to the same position by a counterclockwise rotation. { 'bak,lash }

backlight [GRAPHICS] A spotlight that illuminates from behind so that the subject is separated from the background; used in photography. { 'bak,līt }

backlimb [GEOL] Of the two limbs of an asymmetrical anticline, the one that is more gently dipping. { 'bak,lim }

backlining [BUILD] **1.** A thin strip which lines a window casing, next to the wall and opposite the pulley stile, and provides a smooth surface for the working of the weighted sash. Also known as back boxing; back jamb. **2.** That piece of framing forming the back recess for boxing shutters. [GRAPHICS] Paper strip that is cemented to a book's backbone to bind the signatures and permit space between the backbone and the cover. { 'bak,līn·iŋ }

back lintel [BUILD] A lintel which supports the backing of a masonry wall, as opposed to the lintel supporting the facing material. { 'bak ,lin·təl }

backlit display [ELECTR] An electronic display that incorporates a light source in back of a liquid-crystal or other electronic display to increase readability, especially in daylight. { 'bak,lit di'splā }

back lobe [ELECTROMAG] The three-dimensional portion of the radiation pattern of a directional antenna that is directed away from the intended direction. { 'bak ,lōb }

backlog [IND ENG] **1.** An accumulation of orders promising future work and profit. **2.** An accumulation of unprocessed materials or unperformed tasks. { 'bak,läg }

backmarsh [ECOL] Marshland formed in poorly drained areas of an alluvial floodplain. { 'bak,märsh }

back mixing [CHEM ENG] The tendency of reacted chemicals to intermingle with unreacted feed in reactors, such as stirred tanks, packed towers, and baffled tanks. { 'bak ,mik·siŋ }

back nailing [BUILD] Nailing the plies of a built-up roof to the substrate to prevent slippage. { 'bak ,nāl·iŋ }

back nut [DES ENG] **1.** A threaded nut, one side of which is dished to retain a grommet; used in forming a watertight pipe joint. **2.** A locking nut on the shank of a pipe fitting, tap, or valve. { 'bak ,nət }

back off [ENG] **1.** To unscrew or disconnect. **2.** To withdraw the drill bit from a borehole. **3.** To withdraw a cutting tool or grinding wheel from contact with the workpiece. { 'bak ,ȯf }

back order [IND ENG] **1.** An order held for future completion. **2.** A new order placed for previously unavailable materials of an old order. { 'bak ,ȯrd·ər }

backout [AERO ENG] An undoing of previous steps during a countdown, usually in reverse order. [COMPUT SCI] To remove a change that was previously made in a computer program. [MET] Process of nullifying the effect of positive electrical potentials occurring in an anodic area in a cathodic protection system. { 'bak,aut }

backplane [ELECTR] A wiring board, usually constructed as a printed circuit, used in microcomputers and minicomputers to provide the required connections between logic, memory, input/output modules, and other printed circuit boards which plug into it at right angles. { 'bak,plān }

backplastering [BUILD] A coat of plaster applied to the back side of lath, opposite the finished surface. { 'bak,plas·triŋ }

backplate [BUILD] A plate, usually metal or wood, which serves as a backing for a structural member. { 'bak,plāt }

backplate lamp holder [DES ENG] A lamp holder, integrally mounted on a plate, which is designed for screwing to a flat surface. { 'bak,plāt 'lamp ,hōl·dər }

back porch [ELECTR] The period of time in a television circuit immediately following a synchronizing pulse during which the signal is held at the instantaneous amplitude corresponding to a black area in the received picture. { 'bak ,pȯrch }

back pressure [MECH] Pressure due to a force that is operating in a direction opposite to that being considered, such as that of a fluid flow. [MECH ENG] Resistance transferred from rock into the drill stem when the bit is being fed at a faster rate than the bit can cut. { 'bak ,presh·ər }

back-pressure curve [PETRO ENG] A graph used to arrive at the capacity of a natural-gas well to deliver gas into a pipeline at a sustained rate; uses data from back-pressure testing. { 'bak ,presh·ər ,kərv }

back-pressure-relief port [ENG] In a plastics extrusion die, an opening for the release of excess material. { 'bak ,presh·ər ri'lēf ,pȯrt }

back-pressure testing [PETRO ENG] Method of estimating open-flow capacity of natural-gas wells by relating a series of gas-flow rates and their corresponding stabilized pressures at the bottom of the well bore. { 'bak ,presh·ər ,test·iŋ }

back-pressure valve [PETRO ENG] A check valve installed in a natural-gas well bore to shut off gas flow while replacing the blowout preventer (used during drilling) with a christmas tree piping arrangement, which controls gas flow out of the completed well. { 'bak ,presh·ər ,valv }

back putty [MATER] The bedding of glazing compound which is placed between the face of glass and the frame or sash containing it. Also known as bed glazing. { 'bak ,pəd·ē }

back radiation See backscattering; counterradiation. { 'bak ,rād·ē'ā·shən }

back rake [DES ENG] An angle on a single-point turning tool measured between the plane of the tool face and the reference plane. { 'bak ,rāk }

back range [NAV] A range (distance) measured astern, particularly one used as guidance for a craft moving away from the objects from which the distance information was deduced, forming the range. { 'bak ,ranj }

back reef [GEOGR] The area between a reef and the land. { 'bak ,rēf }

back-reflection photography [CRYSTAL] A method of studying crystalline structure by x-ray diffraction in which the photographic film is placed between the source of x-rays and the crystal specimen. { 'bak ri'flek·shən fə'täg·rə·fē }

back resistance [ELECTR] The resistance between the contacts opposing the inverse current of a metallic rectifier. { 'bak ri'sis·təns }

backrope [NAV ARCH] **1.** Either of two ropes or chains on a sailing ship, extending aft from the lower end of the dolphin striker to each side of the bows. **2.** See cat back. { 'bak,rōp }

back-run process [CHEM ENG] A process for manufacturing water gas in which part of the run is made down, by passing steam through the superheater, thence up through the carburetor, down through the generator, and direct to the scrubbers. { 'bak ,rən 'präs·əs }

back rush [OCEANOGR] Return of water seaward after the uprush of the waves. { 'bak ,rəsh }

backs [MIN ENG] Ore height available above a given working level. { baks }

backsaw [DES ENG] A fine-tooth saw with its upper edge stiffened by a metal rib to ensure straight cuts. { 'bak,sȯ }

backsawing [FOR] A method of converting timber so that the growth rings meet the face in any part of an angle of less than 45°. Also known as bastard-sawing; crown-cut; plainsawing; slash-sawing. { 'bak,sȯ·iŋ }

backscatter gage [ENG] A radar instrument used to measure the radiation scattered at 180° to the direction of the incident wave. { 'bak¦skad·ər ,gaj }

backscattering Also known as back radiation; backward scattering. [COMMUN] Propagation of extraneous signals by F- or E-region reflection in addition to the desired ionospheric scatter mode; the undesired signal enters the antenna through the back lobes. [ELECTROMAG] **1.** Radar echoes from a target. **2.** Undesired radiation of energy to the rear by a directional antenna. [PHYS] The deflection of radiation or nuclear particles by scattering processes through angles greater than 90° with respect to the original direction of travel. { 'bak¦skad·ə·riŋ }

backscattering thickness gage [ENG] A device that uses a radioactive source for measuring the thickness of materials, such as coatings, in which the source and the instrument measuring the radiation are mounted on the same side of the material, the backscattered radiation thus being measured.   { 'bak¦skad·ə·riŋ 'thik·nəs ,gāj }

backscatter radar   [ORD]   An over-the-horizon radar system designed primarily to detect an enemy missile attack from far beyond the horizon, by detecting radio energy reflected from the ionosphere disturbance created by the missiles.   { 'bak¦skad·ər 'rā,där }

backset   [BUILD]   The horizontal distance from the face of a lock or latch to the center of the keyhole, knob, or lock cylinder.   { 'bak,set }

back-set bed   [GEOL]   Cross bedding that dips in a direction against the flow of a depositing current.   { 'bak ,set ,bed }

backshore   [GEOL]   The upper shore zone that is beyond the advance of the usual waves and tides.   Also known as back beach; backshore beach.   { 'bak,shȯr }

backshore beach   See backshore.   { 'bak,shȯr ,bēch }

backshore terrace   See berm.   { 'bak,shȯr 'ter·əs }

back shot   [MIN ENG]   A shot used for widening an entry, placed at some distance from the head of an entry.   { 'bak ,shät }

backsight   [ENG]   1. A sight on a previously established survey point or line.   2. Reading a leveling rod in its unchanged position after moving the leveling instrument to a different location.   [NAV]   A marine sextant observation of a celestial body made by facing 180° from the azimuth of the body and using the visible horizon in the direction in which the observer is facing.   { 'bak,sīt }

backsight method   [ENG]   1. A plane-table traversing method in which the table orientation produces the alignment of the alidade on an established map line, the table being rotated until the line of sight is coincident with the corresponding ground line.   2. Sighting two pieces of equipment directly at each other in order to orient and synchronize one with the other in azimuth and elevation.   { 'bak,sīt 'meth·əd }

back siphonage   [CIV ENG]   The flowing back of used, contaminated, or polluted water from a plumbing fixture or vessel into the pipe which feeds it; caused by reduced pressure in the pipe.   { 'bak 'sī·fən·ij }

back slope   See dip slope.   { 'bak ,slōp }

back solution   [CONT SYS]   The calculation of the tool-coordinated positions that correspond to specified robotic joint positions.   { 'bak sə,lü·shən }

backspace   [COMPUT SCI]   To move a recording medium one unit in the reverse or background direction.   [MECH ENG]   To move a typewriter carriage back one space by depressing a backspace key.   { 'bak,spās }

backspring   [NAV ARCH]   A heavy line extending forward at an acute angle with a ship from the stern or midships to a wharf.   { 'bak,spriŋ }

backstay   [ENG]   1. A supporting cable that prevents a more or less vertical object from falling forward.   2. A spring used to keep together the cutting edges of purchase shears.   3. A rod that runs from either end of a carriage's rear axle to the reach.   4. A leather strip that covers and strengthens a shoe's back seam.   [GRAPHICS]   A rope or strap that keeps the carriage of a hand printing press from moving too far forward.   [NAV ARCH]   A rope, wire, or cable that runs from the top of a mast to the side of a ship and slants a little aft.   [TEXT]   A bar having a glass rod on its top that runs across a loom beneath the lowest motion of the warp yarns.   { 'bak,stā }

backstep sequence   [MET]   Sequential deposition of weld beads in the direction opposite to the direction of welding.   { 'bak,step ,sē·kwəns }

back stoping   See shrinkage stoping.   { 'bak ,stōp·iŋ }

back-surface field   [ELECTR]   A p^+ layer that is added to a silicon solar cell to reduce electron-hole recombination at the cell's back surface and thereby increase the cell's efficiency.   { 'bak ,sər·fəs ,fēld }

backswamp   [GEOL]   Swampy depressed area of a floodplain between the natural levees and the edge of the floodplain.   { 'bak,swamp }

backswamp depression   [ECOL]   A low swamp found adjacent to river levees.   { 'bak,swamp di'presh·ən }

back sweetening   [CHEM ENG]   The controlled addition of commercial-grade mercaptans to a petroleum stock having excess free sulfur in order to reduce free sulfur by forming a disulfide.   { 'bak ,swēt·ən·iŋ }

backtalk   [COMPUT SCI]   Passage of information from a standby computer to the active computer.   { 'bak,tȯk }

backthrusting   [GEOL]   The thrusting in the direction of the interior of an orogenic belt, opposite the general structural trend.   { 'bak,thrəst·iŋ }

back titration   [CHEM]   A titration to return to the end point which was passed.   { 'bak tī'trā·shən }

back-to-front ratio   [ELECTROMAG]   Ratio used in connection with an antenna, metal rectifier, or any device in which signal strength or resistance in one direction is compared with that in the opposite direction.   { ¦bak tə ¦frənt 'rā·shō }

backtracking   [COMPUT SCI]   A method of solving problems automatically by a systematic search of the possible solutions; the invalid solutions are eliminated and are not retried.   { 'bak,trak·iŋ }

backup   [BUILD]   That part of a masonry wall behind the exterior facing.   [CIV ENG]   Overflow in a drain or piping system, due to stoppage.   [COMPUT SCI]   1. Logical or physical facilities to aid the process of restarting a computer system and recovering the information in it following a failure.   2. The provision of such facilities.   [ENG]   1. An item under development intended to perform the same general functions that another item also under development performs.   2. A compressible material used behind a sealant to reduce its depth and to support the sealant against sag or indentation.   [GRAPHICS]   1. An image printed on the reverse side of a printed sheet.   2. The printing of such an image.   [MET]   A support used to balance the upsetting force in the workpieces during flash welding.   [PETRO ENG]   During drilling, the holding of one section of pipe while another is screwed out of it or into it.   { 'bak,əp }

backup arrangement   See cascade.   { 'bak,əp ,ə'rānj·mənt }

backup relay   [ELEC]   A relay designed to protect a power system in case a primary relay fails to operate as desired.   { 'bak,əp 'rē·lā }

backup strip   [BUILD]   A wood strip which is fixed at the corner of a partition or wall to provide a nailing surface for ends of lath.   Also known as lathing board.   { 'bak,əp ,strip }

backup system   [SYS ENG]   A system, normally redundant but kept available to replace a system which may fail in operation.   { 'bak,əp ,sis·təm }

backup tong   [ENG]   A heavy device used on a drill pipe to loosen the tool joints.   { 'bak,əp ,täng }

Backus-Naur form   [COMPUT SCI]   A metalanguage that specifies which sequences of symbols constitute a syntactically valid program language.   Abbreviated BNF.   { ¦bäk·əs ¦naúr ,fȯrm }

back veneer   [MATER]   In veneer plywood, the layer of veneer on the side of a plywood sheet which is opposite the face veneer; usually of lower quality.   { 'bak və,nir }

back vent   [CIV ENG]   An individual vent for a plumbing fixture located on the downstream (sewer) side of a trap to protect the trap against siphonage.   { 'bak ,vent }

backward-acting regulator   [ELECTR]   Transmission regulator in which the adjustment made by the regulator affects the quantity which caused the adjustment.   { 'bak·wərd 'ak·tiŋ 'reg·yə,lād·ər }

backward-bladed aerodynamic fan   [MECH ENG]   A fan that consists of several streamlined blades mounted in a revolving casing.   { 'bak·wərd ,blād·əd ,er·ō·dī'nam·ik ,fan }

backward chaining   [COMPUT SCI]   In artificial intelligence, a method of reasoning which starts with the problem to be solved and repeatedly breaks this goal into subgoals that are more readily solvable with the relevant data and the system's rules of inference.   { ¦bak·wərd 'chān·iŋ }

backward compatibility   See downward compatibility.   { ¦bak·wərd kəm,pad·ə'bil·əd·ē }

backward difference   [MATH]   One of a series of quantities obtained from a function whose values are known at a series of equally spaced points by repeatedly applying the backward difference operator to these values; used in interpolation and numerical calculation and integration of functions.   { ¦bak·wərd 'dif·rəns }

backward difference operator   [MATH]   A difference operator, denoted ∇, defined by the equation $\nabla f(x) = f(x) - f(x - h)$, where h is a constant denoting the difference between successive points of interpolation or calculation.   { ¦bak·wərd ¦dif·rəns 'äp·ə,rād·ər }

backward diode   [ELECTR]   A semiconductor diode similar to a tunnel diode except that it has no forward tunnel current; used as a low-voltage rectifier. { 'bak·wərd 'dī,ōd }

backward error analysis [COMPUT SCI] A form of error analysis which seeks to replace all errors made in the course of solving a problem by an equivalent perturbation of the original problem. { 'bak·wərd 'er·ər ə,nal·ə·səs }

backward folding See backfolding. { 'bak·wərd ¦fōld·iŋ }

backward pass [IND ENG] The calculation of late finish times (dates) for all uncompleted network activities for a specific project by subtracting durations of uncompleted activities from the scheduled finish time of the final activity. { 'bak·wərd 'pas }

backward read [COMPUT SCI] The transfer of data from a magnetic tape to computer storage when the tape is running in reverse. { 'bak·wərd 'rēd }

backward scattering See backscattering. { 'bak·wərd ¦skad·ə·riŋ }

backward search [COMPUT SCI] A search of a document or database that starts at the cursor's location and moves backwards toward the beginning of the document or database. { ¦bak·wərd 'sərch }

backward wave [ELECTROMAG] An electromagnetic wave traveling opposite to the direction of motion of some other physical quantity in an electronic device such as a traveling-wave tube or mismatched transmission line. { 'bak·wərd ,wāv }

backward-wave magnetron [ELECTR] A magnetron in which the electron beam travels in a direction opposite to the flow of the radio-frequency energy. { 'bak·wərd ,wāv 'mag·nə,trän }

backward-wave oscillator [ELECTR] An electronic device which amplifies microwave signals simultaneously over a wide band of frequencies and in which the traveling wave produced is reflected backward so as to sustain the wave oscillations. Abbreviated BWO. Also known as carcinotron. { 'bak·wərd ,wāv 'äs·ə,läd·ər }

backward-wave tube [ELECTR] A type of microwave traveling-wave electron tube in which electromagnetic energy on a slow-wave circuit flows opposite in direction to the travel of electrons in a beam. { 'bak·wərd ,wāv ,tüb }

backward welding See backhand welding. { 'bak·wərd 'weld·iŋ }

backwash [CHEM ENG] **1.** In an ion-exchange resin system, an upward flow of water through a resin bed that cleans and reclassifies the resin particles after exhaustion. **2.** See blowback. [OCEANOGR] **1.** Water or waves thrown back by an obstruction such as a ship or breakwater. **2.** The seaward return of water after a rush of waves onto the beach foreshore. { 'bak,wäsh }

backwash mark [GEOL] A crisscross ridge pattern in beach sand, caused by backwash. { 'bak,wäsh ,märk }

backwash ripple mark [GEOL] Ripple marks that are broad and flat and parallel to the shoreline, with narrow, shallow troughs and crests about 30 centimeters apart; formed by backwash above the maximum wave retreat level. { 'bak,wäsh 'rip·əl ,märk }

backwater [HYD] **1.** A series of connected lagoons, or a creek parallel to a coast, narrowly separated from the sea and connected to it by barred outlets. **2.** Accumulation of water resulting from and held back by an obstruction. **3.** Water reversed in its course by an obstruction. { 'bak,wòd·ər }

backwater valve [ENG] A type of check valve in a drainage pipe; reversal of flow causes the valve to close, thereby cutting off flow. Also known as backflow valve. { 'bak,wòd·ər ,valv }

backweld [MET] A weld placed behind a single groove weld. { 'bak,weld }

back work [MIN ENG] Any kind of operation in a mine not immediately concerned with production or transport; literally, work behind the face, such as repairs to roads. { 'bak ,wərk }

bacteremia [MED] Presence of bacteria in the blood. { 'bak·tə'rē·mē·ə }

bacteremic shock [MED] A state of shock occurring during the course of bacteremia, especially if caused by gram-negative bacteria. { ¦bak·tə¦rē·mik 'shäk }

bacteria [MICROBIO] Extremely small, relatively simple prokaryotic microorganisms traditionally classified with the fungi as Schizomycetes. { bak'tir·ē·ə }

Bacteriaceae [MICROBIO] A former designation for Brevibacteriaceae. { bak,tir·ē'ās·ē,ē }

bacterial blight [PL PATH] Any blight disease of plants caused by bacteria, including common bacterial blight, halo blight, and fuscous blight. { bak'tir·ē·əl 'blīt }

bacterial bronchopneumonia [MED] Bacterial infection of the lung which has spread from infected bronchi. { bak'tir·ē·əl ¦bräŋ·kō·nù'mō·nya }

bacterial brown spot [PL PATH] A bacterial blight disease of plants caused by Pseudomonas syringae; marked by water-soaked reddish-brown spots or cankers. Also known as bacterial canker. { bak'tir·ē·əl ¦braùn ,spät }

bacterial canker See bacterial brown spot. { bak'tir·ē·əl 'kaŋ·kər }

bacterial capsule [MICROBIO] A thick, mucous envelope, composed of polypeptides or carbohydrates, surrounding some bacteria. { bak'tir·ē·əl 'kap·səl }

bacterial coenzyme [MICROBIO] Organic molecules that participate directly in a bacterial enzymatic reaction and may be chemically altered during the reaction. { bak'tir·ē·əl kō'en,zīm }

bacterial competence [MICROBIO] The ability of cells in a bacterial culture to accept and be transformed by a molecule of transforming deoxyribonucleic acid. { bak'tir·ē·əl 'käm·pə·təns }

bacterial encephalitis [MED] Inflammation of the brain caused by primary or secondary bacterial infection. { bak'tir·ē·əl in,sef·ə'līd·əs }

bacterial endocarditis [MED] Inflammation of the endocardium due to bacterial invasion. Also known as subacute bacterial endocarditis. { bak'tir·ē·əl ,en·dō,kär'dīd·əs }

bacterial endoenzyme [MICROBIO] An enzyme produced and active within the bacterial cell. { bak'tir·ē·əl ,en·dō'en,zīm }

bacterial endospore [MICROBIO] A body, resistant to extremes of temperature and to dehydration, produced within the cells of gram-positive, sporeforming rods of Bacillus and Clostridium and by the coccus Sporosarcina. { bak'tir·ē·əl 'en·dō,spór }

bacterial enzyme [MICROBIO] Any of the metabolic catalysts produced by bacteria. { bak'tir·ē·əl 'en,zīm }

bacterial infection [MED] Establishment of an infective bacterial agent in or on the body of a host. { bak'tir·ē·əl in'fek·shən }

bacterial leaf spot [PL PATH] A bacterial disease of plants characterized by spotty discolorations on the leaves; examples are angular leaf spot and leaf blotch. { bak'tir·ē·əl 'lēf ,spät }

bacterial luminescence [MICROBIO] A light-producing phenomenon exhibited by certain bacteria. { bak'tir·ē·əl lü·mə'nes·əns }

bacterial metabolism [MICROBIO] Total chemical changes carried out by living bacteria. { bak'tir·ē·əl mə'tab·ə,liz·əm }

bacterial motility [MICROBIO] Self-propulsion in bacteria, either by gliding on a solid surface or by moving the flagella. { bak'tir·ē·əl mō'til·əd·ē }

bacterial photosynthesis [MICROBIO] Use of light energy to synthesize organic compounds in green and purple bacteria. { bak'tir·ē·əl ,fōd·ō'sin·thə·səs }

bacterial pigmentation [MICROBIO] The organic compounds produced by certain bacteria which give color to both liquid cultures and colonies. { bak'tir·ē·əl ,pig·mən'tā·shən }

bacterial pneumonia [MED] Consolidation of the lung caused by inflammatory exudation due to bacterial infection. { bak'tir·ē·əl nə'mō·nyə }

bacterial pustule [PL PATH] A bacterial blight of plants caused by Xanthomonas phaseoli; characterized by blisters on the leaves. { bak'tir·ē·əl 'pəs·chəl }

bacterial soft rot [PL PATH] A bacterial disease of plants marked by disintegration of tissues. { bak'tir·ē·əl ¦sóft ,rät }

bacterial speck [PL PATH] A bacterial disease of plants characterized by small lesions on plant parts. { bak'tir·ē·əl 'spek }

bacterial spot [PL PATH] Any bacterial disease of plants marked by spotting of the infected part. { bak'tir·ē·əl 'spät }

bacterial transformation See transformation. { bak'tir·ē·əl ,tranz·fər'mā·shən }

bacterial vaccine [IMMUNOL] A preparation of living, attenuated, or killed bacteria used to enhance the immune reaction in an individual already infected with the same bacteria. { bak'tir·ē·əl vak'sēn }

bacterial vaginosis [MED] Inflammation of the vagina that causes a nonirritating white or gray vaginal discharge, often with a distinctive fishy odor; it results from overgrowth of various normal vaginal bacteria and by depletion of *Lactobacillus* species, especially strains that produce hydrogen peroxide. { bak,tir·ē·əl ,vaj·ə'nō·səs }

bacterial wilt disease [PL PATH] A common bacterial disease of cucumber and muskmelon, caused by *Erwinia tracheiphila*, characterized by wilting and shriveling of the leaves and stems. { bak'tir·ē·əl 'wilt di,zēz }

bactericide [MATER] An agent that destroys bacteria. { bak'tir·ə,sīd }

bactericidin [IMMUNOL] An antibody that kills bacteria in the presence of complement. { bak'tir·ə¦sīd·ən }

bacterin [IMMUNOL] A suspension of killed or weakened bacteria used in artificial immunization. { 'bak·tə·rən }

bacteriochlorophyll [BIOCHEM] $C_{52}H_{70}O_6N_4Mg$ A tetrahydroporphyrin chlorophyll compound occurring in the forms *a* and *b* in photosynthetic bacteria; there is no evidence that *b* has the empirical formula given. { bak'tir·ē·ə'klȯr·ə,fil }

bacteriocin [MICROBIO] Any of a group of proteins produced by various strains of gram-negative bacteria that may inhibit the growth of other strains of the same or related species. { bak'tir·ē·ə,sīn }

bacteriocinogen [GEN] A plasmid deoxyribonucleic acid found in some strains of bacteria which specifies production of a bacteriocin, an antibiotic for some other bacteria. { bak,tir·ē·ə'sin·ə·jən }

bacteriocyte [INV ZOO] A modified fat cell found in certain insects that contains bacterium-shaped rods believed to be symbiotic bacteria. { bak'tir·ē·ə,sīt }

bacteriogenic [MICROBIO] Caused by bacteria. { bak¦tir·ē·ə'jen·ik }

bacteriological warfare [ORD] Warfare conducted with pathogenic microorganisms as offensive weapons; a type of biological warfare. { bak¦tir·ē·ə¦läj·ə·kəl 'wȯr,fer }

bacteriologist [MICROBIO] A specialist in the study of bacteria. { bak,tir·ē'äl·ə·jəst }

bacteriology [MICROBIO] The science and study of bacteria; a specialized branch of microbiology. { bak,tir·ē'äl·ə·jē }

bacteriolysin [MICROBIO] An antibody that is active against and causes lysis of specific bacterial cells. { bak,tir·ē·ə'līs·ən }

bacteriolysis [MICROBIO] Dissolution of bacterial cells. { bak,tir·ē'äl·ə·səs }

Bacterionema [MICROBIO] A genus of bacteria in the family Actinomycetaceae; characteristically cells are rods with filaments attached and produce filamentous microcolonies; facultative anaerobes; carbohydrates are fermented. { bak,tir·ē'än·ə·mə }

bacteriophage [VIROL] Any of the viruses that infect bacterial cells; each has a narrow host range. Also known as phage. { bak'tir·ē·ə,fāj }

bacteriophobia [PSYCH] An abnormal fear of bacteria and other microorganisms. { bak,tir·ē·ə'fōb·ē·ə }

bacteriorhodopsin [BIOCHEM] A purple substance in the cell membranes of halobacteria (found in extremely saline environments) during conditions of low oxygen, and consisting of the protein bacteriopsin and retinal, the same carotenoid found in the visual pigments of animals; in response to light, the purple membrane pumps protons out of the cell, providing the energy gradient for synthesis of adeniosine triphosphate. { bak,tir·ē·ō·rō'däp·sən }

bacterioruberin [MICROBIO] A carotenoid pigment found in some halophilic aerobic archaebacteria that gives them a striking red color and seems to protect them from strong sunlight in their natural environments. { bak,tir·ē·ə'rüb·ə·rən }

bacteriosis [PL PATH] Any bacterial disease of plants. { bak,tir·ē'ō·səs }

bacteriostasis [MICROBIO] Inhibition of bacterial growth and metabolism. { bak,tir·ē·ō'stā·səs }

bacteriostatic agent [MICROBIO] A substance that inhibits the growth of bacteria. { ,bak¦tir·ē·ō¦stad·ik 'ā·jənt }

bacteriotoxin [MICROBIO] **1.** Any toxin that destroys or inhibits growth of bacteria. **2.** A toxin produced by bacteria. { bak,tir·ē·ō'täk·sən }

bacteriotropin [IMMUNOL] An antibody that is increased in amount during specific immunization and that renders the corresponding bacterium more susceptible to phagocytosis. { bak,tir·ē'ä·trə·pən }

bacterioviridin *See* chlorobium chlorophyll. { bak,tir·ē·ə'vir·ə·dən }

bacteriuria [MED] The occurrence of bacteria in the urine. { bak,tir·ē'yü·rē·ə }

bacteroid [MICROBIO] A bacterial form of irregular shape, frequently associated with special conditions. { 'bak·tə,rȯid }

Bacteroidaceae [MICROBIO] The single family of gram-negative anaerobic bacteria; cells are nonsporeforming rods; some species are pathogenic. { ,bak·tə,rȯi'dās·ē,ē }

Bactrian camel [VERT ZOO] *Camelus bactrianus*. The two-humped camel. { 'bak·trē·ən 'kam·əl }

baculite [GEOL] A crystallite that looks like a dark rod. { 'bak·yə,līt }

Baculoviridae [VIROL] A family of invertebrate viruses comprising the genus *Baculovirus* that is characterized by enveloped rod-shaped virions containing a molecule of supercoiled double-stranded deoxyribonucleic. { ,bak·yə·lə'vir·ə,dī }

Baculovirus [VIROL] A genus of invertebrate viruses that contains enveloped, double-stranded, supercoiled deoxyribonucleic, includes the subgroups nuclear polyhedrosis virus and granulosis virus. { ,bak·yə·lə'vī·rəs }

baculum [VERT ZOO] The penis bone in lower mammals. Also known as os priapi. { 'bak·yə·ləm }

bad branch [COMPUT SCI] An error in which execution of a computer program jumps to an incorrect instruction, usually as a result of errors in the program. { ¦bad 'branch }

bad break [GRAPHICS] In type composition, the setting of the first line of a page beginning with only a portion of a word or containing only a single word. { ¦bad 'brāk }

baddeleyite [MINERAL] ZrO_2 A colorless, yellow, brown, or black monoclinic zirconium oxide mineral found in Brazil and Ceylon; used as heat- and corrosion-resistant linings for furnaces and muffles. { 'bad·əl·ē,īt }

badge meter *See* film badge. { 'baj ,mēd·ər }

badger [DES ENG] *See* badger plane. [ENG] A tool used inside a pipe or culvert to remove any excess mortar or deposits. [VERT ZOO] Any of eight species of carnivorous mammals in six genera comprising the subfamily Melinae of the weasel family (Mustelidae). { 'baj·ər }

badge reader [COMPUT SCI] A device that can read data appearing in the form of holes in plastic badges or prepunched cards. { 'baj ,rēd·ər }

badger plane [DES ENG] A hand plane whose mouth is cut obliquely from side to side, so that the plane can work close up to a corner. Also known as badger. { 'baj·ər ,plān }

Badger's rule [PHYS CHEM] An empirical relationship between the stretching force constant for a molecular bond and the bond length. { 'baj·ərz ,rül }

badlands [GEOGR] An erosive physiographic feature in semiarid regions characterized by sharp-edged, sinuous ridges separated by steep-sided, narrow, winding gullies. { 'bad,lanz }

bad page break [COMPUT SCI] A soft page break at an inappropriate location in a document, such as one that splits a table or leaves a single line of text at the top or bottom of a page. { ¦bad 'pāj ,brāk }

bad sector [COMPUT SCI] An area of disk storage that does not record data reliably and therefore is not used. { ,bad 'sek·tər }

bad top [MIN ENG] A weak roof in a coal mine; sometimes develops following a blast. { 'bad ,täp }

bad track [COMPUT SCI] A disk track that contains a bad sector. { ,bad 'trak }

bad track table [COMPUT SCI] A listing of the bad sectors on a disk, which is packaged with or attached to a disk. { ¦bad 'trak ,tā·bəl }

baeckeol [ORG CHEM] $C_{13}H_{18}O_4$ A phenolic ketone that is crystalline and pale yellow; found in oils from plants of species of the myrtle family. { 'bāk·ē,ȯl }

baeocyte [INV ZOO] A motile or nonmotile blue-green bacterial endospore. { 'bē·ə,sīt }

Baeyer strain theory [ORG CHEM] The theory that the relative stability of penta- and hexamethylene ring compounds is caused by a propitious bond angle between carbons and a lack of bond strain. { 'bā·ər 'strān ,thē·ə·rē }

baffle [ELEC] Device for deflecting oil or gas in a circuit

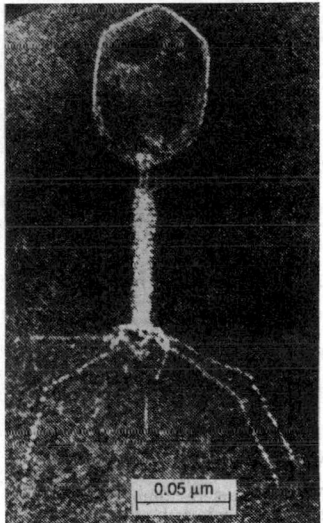

BACTERIOPHAGE

Electron micrograph of negatively stained T2 bacteriophage showing head and tail components. *(Courtesy of H. Fernandez-Moran)*

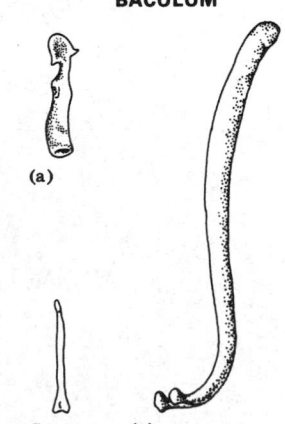

BACULUM

(a) (b) (c)

Bacula of *(a)* squirrel, *(b)* cotton mouse, *(c)* otter.

breaker. [ELECTR] An auxiliary member in a gas tube used, for example, to control the flow of mercury particles or deionize the mercury following conduction. [ENG] A plate that regulates the flow of a fluid, as in a steam-boiler flue or a gasoline muffler. [ENG ACOUS] A cabinet or partition used with a loudspeaker to reduce interaction between sound waves produced simultaneously by the two surfaces of the diaphragm. { 'baf·əl }

baffle plate [ELECTROMAG] Metal plate inserted in a waveguide to reduce the cross-sectional area for wave conversion purposes. { 'baf·əl ˌplāt }

baffling wind [METEOROL] A wind that is shifting so that nautical movement by sailing vessels is impeded. { ¦baf·liŋ 'wind }

bag [ENG] **1.** A flexible cover used in bag molding. **2.** A container made of paper, plastic, or cloth without rigid walls to transport or store material. { bag }

bagasse [FOOD ENG] Remains of sugarcane after the juice has been extracted by pressure between the rolls of a mill; used as a fuel and in applications requiring fibrous material. Also known as megass. { bə'gas }

bagasse disease *See* bagassosis. { bə'gas diˌzēz }

bagassosis [MED] A pneumoconiosis caused by the inhalation of bagasse dust, a dry sugarcane residue. Also known as bagasse disease. { ˌbag·ə'sō·səs }

bag filter [ENG] Filtering apparatus with porous cloth or felt bags through which dust-laden gases are sent, leaving the dust on the inner surfaces of the bags. { 'bag ˌfil·tər }

bagherra [TEXT] A woven or knitted velvet with uncut loop pile. { bə'ger·ə }

baghouse [ENG] The large chamber or room for holding bag filters used to filter gas streams from a furnace. { 'bagˌhaus }

bag molding [ENG] A method of molding plastic or plywood-plastic combinations into curved shapes, in which fluid pressure acting through a flexible cover, or bag, presses the material to be molded against a rigid die. { 'bag ˌmōld·iŋ }

Bagnold number [ENG] A dimensionless number used in saltation studies. { 'bag·nəld ˌnəm·bər }

bag plug [ENG] An inflatable drain stopper, located at the lowest point of a piping system, that acts to seal a pipe when inflated. { 'bag ˌpləg }

Bagridae [VERT ZOO] A family of semitropical catfishes in the suborder Siluroidei. { 'bag·rəˌdē }

bag trap [ENG] An S-shaped trap in which the vertical inlet and outlet pipes are in alignment. { 'bag ˌtrap }

baguette *See* bead molding. { ba'get }

baguio [METEOROL] A tropical cyclone that occurs in the Philippines. { bäg'yō }

bahada *See* bajada. { bə'häd·ə }

bahamite [PETR] A consolidated limestone formed of sediment similar to a type currently found accumulating in the Bahamas. { bə'hamˌīt }

bahiaite [PETR] Holocrystalline igneous rock formed mainly of hypersthene with subordinate hornblende and sometimes minor amounts of other minerals. { bə'hī·yəˌīt }

bahut [ARCH] **1.** In a masonry wall or parapet, the rounded upper course. **2.** A low wall surmounting a cornice to carry the roof structure. { 'bäˌhut }

bai [METEOROL] A yellow mist prevalent in China and Japan in spring and fall, when the loose surface of the interior of China is churned up by the wind, and clouds of sand rise to a great height and are carried eastward, where they collect moisture and fall as a yellow mist. { bī }

baikerite [MINERAL] A waxlike mineral from the vicinity of Lake Baikal, Siberia; apparently about 60% ozocerite with other tarry, waxy, and resinous hydrocarbons. { 'bī·kəˌrīt }

bail [ENG] A loop of heavy wire snap-fitted around two or more parts of a connector or other device to hold the parts together. { bāl }

bailer [ENG] A long, cylindrical vessel fitted with a bail at the upper end and a flap or tongue valve at the lower extremity; used to remove water, sand, and mud- or cuttings-laden fluids from a borehole. Also known as bailing bucket. { 'bāl·ər }

Bailey bridge [CIV ENG] A lattice bridge built of interchangeable panels connected at the corners with steel pins, permitting rapid construction; developed in Britain about 1942 as a military bridge. { 'bāl·ē ˌbrij }

Bailey meter [ENG] A flowmeter consisting of a helical quarter-turning vane which operates a counter to record the total weight of granular material flowing through vertical or near-vertical ducts, spouts, or pipes. { 'bāl·ē ˌmēd·ər }

bailing [ENG] Removal of the cuttings from a well during cable-tool drilling, or of the liquid from a well, by means of a bailer. { 'bāl·iŋ }

bailing bucket *See* bailer. { 'bāl·iŋ ˌbək·ət }

bailing drum [ENG] A reel for winding bailing line. { 'bāl·iŋ ˌdrəm }

bailing line [ENG] A cable attached to the bailer of a derrick; it is passed over a sheave at the top of the derrick and spooled on a reel. { 'bāl·iŋ ˌlīn }

bailout [AERO ENG] The exiting from a flying aircraft and descending by parachute in an emergency. { 'bālˌaut }

bailout bottle [AERO ENG] A personal supply of oxygen usually contained in a cylinder under pressure and utilized when the individual has left the central oxygen system, as in a parachute jump. { 'bālˌaut 'bäd·əl }

Baily's beads [ASTRON] Bright points of sunlight appearing around the edge of the moon just before and after the central phase of a total solar eclipse. { 'bāl·ēz ¦bēdz }

Bainbridge reflex [PHYSIO] A poorly understood reflex acceleration of the heart rate due to rise of pressure in the right atrium and vena cavae, possibly mediated through afferent vagal fibers. { 'bānˌbrij 'rē·flaks }

bainite [MET] Steel formed by austempering, having an acicular structure of ferrite and carbides, exhibiting considerable toughness, and combining high strength with high ductility. { 'bāˌnīt }

Bain-Ondarçuhu device [FL MECH] An inclined plane with a self-running droplet that coats its upper surface. { ¦bān ˌōn'dä·sə,hü diˌvīs }

Bairdiacea [INV ZOO] A superfamily of ostracod crustaceans in the suborder Podocopa. { ˌber·dē'ās·ē·ə }

Baire function [MATH] The smallest class of functions on a topological space which contains the continuous functions and is closed under pointwise limits. { 'ber ˌfaŋk·shən }

Baire measure [MATH] A measure defined on the class of all Baire sets such that the measure of any closed, compact set is finite. { 'ber ˌmezh·ər }

Baire's category theorem [MATH] The theorem that a complete metric space is of second category; equivalently, the intersection of any sequence of open dense sets in a complete metric space is dense. { ¦berz 'kad·əˌgór·ē ˌthir·əm }

Baire set [MATH] A member of the smallest sigma algebra containing all closed, compact subsets of a topological space. { 'ber ˌset }

Baire space [MATH] A topological space in which every countable intersection of dense, open subsets is dense in the space. { 'ber ˌspās }

Bairstow number [FL MECH] A term previously used for Mach number. { 'berˌstō ˌnəm·bər }

bajada [GEOL] An alluvial plain formed as a result of lateral growth of adjacent alluvial fans until they finally coalesce to form a continuous inclined deposit along a mountain front. Also spelled bahada. { bə'häd·ə }

bajada breccia [PETR] An imperfectly stratified accumulation of coarse, angular rock fragments mixed with mud that formed in arid climates and results from a mudflow containing considerable water. { bə'häd·ə 'brech·ə }

Bajocian [GEOL] A European stage: the middle Middle or lower Middle Jurassic geologic time; above Toarcian, below Bathonian. { bə'jō·shən }

Bakanae disease [PL PATH] A fungus disease of rice in Japan, caused by *Gibberella fujikurae*; a foot rot disease. { bə'kä·nē diˌzēz }

baked core [MET] In sand castings, a core which has been heated in core stoves or by dielectric heating to attain uniform physical properties. { 'bākt ¦kór }

baked finish [ENG] A paint or varnish finish obtained by baking, usually at temperatures above 150°F (65°C), thereby developing a tough, durable film. { 'bākt 'fin·ish }

baked permeability [MET] The property of a molded mass of sand which permits passage of gases through it when molten metal is poured in a mold, baked above 230°F (110°C), and then cooled. { 'bākt ˌper·mē·ə'bil·əd·ē }

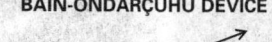

BAIN-ONDARÇUHU DEVICE

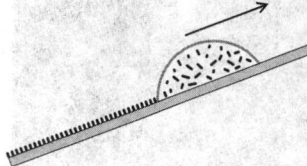

Motion of a self-running droplet in the Bain-Ondarçuhu device.

baked strength [MET] The strength of a molded sand mixture when baked above 230°F (110°C) and then cooled to ambient temperature. { 'bākt ¦streŋkth }

bakeout [ENG] The degassing of surfaces of a vacuum system by heating during the pumping process. { 'bāk‚aut }

baker bell dolphin [CIV ENG] A dolphin consisting of a heavy bell-shaped cap pivoted on a group of piles; a blow from a ship will tilt the bell, thus absorbing energy. { 'bāk·ər ¦bel ‚däl·fən }

bakerite [MINERAL] $8CaO \cdot 5B_2O_3 \cdot 6SiO_2 \cdot 6H_2O$ White mineral, occurring in fine-grained, nodular masses, resembling marble and unglazed porcelain, and consisting of hydrous calcium borosilicate. { 'bāk·ə‚rīt }

Baker-Nunn camera [OPTICS] A large camera with a Schmidt-type lens system used to track earth satellites. { ¦bāk·ər ¦nən 'kam·rə }

Baker-Schmidt telescope [OPTICS] A type of Schmidt telescope in which the light reflected from the near-spheroidal primary mirror is again reflected from a smaller, near-spheroidal secondary mirror, producing an image that is free of astigmatism and distortion. { ¦bāk·ər ¦shmit 'tel·ə‚skōp }

bakers' yeast [FOOD ENG] An industrial yeast used for baking purposes because of maximum growth and low alcohol production; composed of dry cells of one or more strains of *Saccharomyces cerevisiae*. { 'bāk·ərz ‚yēst }

baking [ENG] The use of heat on fresh paint films to speed the evaporation of thinners and to promote the reaction of binder components so as to form a hard polymeric film. Also known as stoving. [FOOD ENG] The use of heat in an oven to convert flour, water, yeast, and such, into baked goods. [MET] Heating metal at low temperatures to remove gases. { 'bāk·iŋ }

baking finish [MATER] Varnish or paint that must be baked at temperatures greater than 150°F (66°C) to develop desired final properties of strength and hardness. { 'bāk·iŋ ‚fin·ish }

baking powder [FOOD ENG] A yeast substitute of sodium bicarbonate plus potassium tartrate, tartaric acid, anhydrous sodium aluminum sulfate, monocalcium phosphate, or any combination of these acids so formulated as to release carbon dioxide from the sodium bicarbonate (baking soda) when moistened. { 'bāk·iŋ ‚paud·ər }

baking soda See sodium bicarbonate. { 'bāk·iŋ ‚sōd·ə }

baking varnish [MATER] A chemical-resistant varnish made of synthetic resins that requires baking to be dried. { 'bāk·iŋ ‚vär·nəsh }

BAL See dimercaprol.

Balaenicipitidae [VERT ZOO] A family of wading birds composed of a single species, the shoebill stork (*Balaeniceps rex*), in the order Ciconiiformes. { bə‚lēn·ə·sə'pid·ə‚dē }

Balaenidae [VERT ZOO] The right whales, a family of cetacean mammals composed of five species in the suborder Mysticeti. { bə'lēn·ə‚dē }

balance [ACOUS] The condition in a stereo system wherein both speakers produce the same average sound levels. [AERO ENG] **1.** The equilibrium attained by an aircraft, rocket, or the like when forces and moments are acting upon it so as to produce steady flight, especially without rotation about its axes. **2.** The equilibrium about any specified axis that counterbalances something, especially on an aircraft control surface, such as a weight installed forward of the hinge axis to counterbalance the surface aft of the hinge axis. [CHEM] To bring a chemical equation into balance so that reaction substances and reaction products obey the laws of conservation of mass and charge. [ELEC] The state of an electrical network when it is adjusted so that voltage in one branch induces or causes no current in another branch. [ENG] An instrument for measuring mass or weight. [MIN ENG] The counterpoise or weight attached by cable to the drum of a winding engine to balance the weight of the cage and hoisting cable and thus assist the engine in lifting the load out of the shaft. { 'bal·əns }

Balance See Libra. { 'bal·əns }

balance arm [BUILD] On a projected window, a side supporting arm which is constructed so that the center of gravity of the sash is not changed appreciably when the window is opened. { 'bal·əns ‚ärm }

balance bar See balance beam. { 'bal·əns ‚bär }

balance beam [CIV ENG] A long beam, attached to a gate (or drawbridge, and such) so as to counterbalance the weight of the gate during opening or closing. Also known as balance bar. { 'bal·əns ‚bēm }

balance car [MIN ENG] In quarrying, a car loaded with iron or stone and connected by means of a steel cable with a channeling machine operating on an inclined track; used to counteract the force of gravity and thus enable the channeling machine to operate with equal ease uphill and downhill. { 'bal·əns ‚kär }

balance coil [ELEC] An iron-core solenoid with adjustable taps near the center; used to convert a two-wire circuit to a three-wire circuit, the taps furnishing a neutral terminal for the latter. { 'bal·əns ‚kȯil }

balance control [ELECTR] A control used in a stereo sound system to vary the volume of one loudspeaker system relative to the other while maintaining their combined volume essentially constant. { 'bal·əns kən'trōl }

balanced amplifier [ELECTR] An electronic amplifier in which there are two identical signal branches connected so as to operate with the inputs in phase opposition and with the output connections in phase, each balanced to ground. { 'bal·ənst 'am·plə‚fī·ər }

balanced anesthesia [MED] Anesthesia produced by safe doses of two or more agents or methods of anesthesia, each of which contributes to the total desired effect. { 'bal·ənst an·əs'thē·zhə }

balanced armature unit [ENG ACOUS] Driving unit used in magnetic loudspeakers, consisting of an iron armature pivoted between the poles of a permanent magnet and surrounded by coils carrying the audio-frequency current; variations in audio-frequency current cause corresponding changes in armature magnetism and corresponding movements of the armature with respect to the poles of the permanent magnet. { 'bal·ənst 'ärm·ə·chər ‚yü·nət }

balanced bridge [ELEC] Wheatstone bridge circuit which, when in a quiescent state, has an output voltage of zero. { 'bal·ənst 'brij }

balanced circuit [ELEC] **1.** A circuit whose two sides are electrically alike and symmetrical with respect to a common reference point, usually ground. **2.** An electric circuit that has been adjusted to neutralize the mutual induction of an adjacent circuit. { 'bal·ənst 'sər·kət }

balanced cloth [TEXT] A fabric made up of equal numbers and sizes of warp and filling yarns. { 'bal·ənst 'klȯth }

balanced construction [BUILD] A plywood or sandwich-panel construction which has an odd number of plies laminated together so that the construction is identical on both sides of a plane through the center of the panel. { 'bal·ənst kən'strək·shən }

balanced converter See balun. { 'bal·ənst kən'vərd·ər }

balanced currents [ELEC] Currents flowing in the two conductors of a balanced line which, at every point along the line, are equal in magnitude and opposite in direction. Also known as push-pull currents. { 'bal·ənst 'kər·əns }

balanced design [ENG] A winding pattern used in fabricating filament-wound reinforced plastics that renders the stresses in all the filaments equal. { 'bal·ənst di'zīn }

balanced detector [ELECTR] A detector used in frequency-modulation receivers; in one form the audio output is the rectified difference between voltages produced across two resonant circuits, one being tuned slightly above the carrier frequency and one slightly below. { 'bal·ənst di'tek·tər }

balanced digit system [MATH] A number system in which the allowable digits in each position range in value from $-n$ to n, where n is some positive integer, and $n + 1$ is greater than one-half the base. { 'bal·ənst 'dij·ət ‚sis·təm }

balanced door [BUILD] A door equipped with double-pivoted hardware which is partially counterbalanced to provide easier operation. { 'bal·ənst 'dȯr }

balanced draft [ENG] The maintenance of a constant draft in a furnace by monitoring both the incoming air and products of combustion. { 'bal·ənst 'draft }

balanced earthwork [CIV ENG] Cut-and-fill work in which the amount of fill equals the amount of material excavated. { 'bal·ənst 'ərth‚wərk }

balanced fertilizer [MATER] A material of varying composition added to soil so as to provide essential mineral elements at required levels, improve soil structure, or enhance microbial activity. { 'bal·ənst 'fərd·ə‚līz·ər }

balanced gasoline [MATER] An automotive gasoline blended from petroleum hydrocarbons of varying volatilities to provide desired performance for engine starting, warm-up, acceleration, and mileage. { 'bal·ənst ,gas·ə'lēn }

balanced incomplete block design [MATH] For positive integers b, v, r, k, and λ, an arrangement of v elements into b subsets or blocks so that each block contains exactly k distinct elements, each element occurs in r blocks, and every combination of two elements occurs together in exactly λ blocks. Also known as (b,v,r,k,λ)-design. { 'bal·ənst ,iŋ·kəm,plēt 'blāk di,zīn }

balanced input [ELECTR] A symmetrical input circuit having equal impedance from both input terminals to reference. { 'bal·ənst ¦in,put }

balanced line [ELEC] A transmission line consisting of two conductors capable of being operated so that the voltages of the two conductors at any transverse plane are equal in magnitude and opposite in polarity with respect to ground. [IND ENG] A production line for which the time cycles of the operators are made approximately equal so that the work flows at a desired steady rate from one operator to the next. { 'bal·ənst ,līn }

balanced load [ELEC] A load that presents the same impedance, with respect to ground, at both ends or terminals. { 'bal·ənst 'lōd }

balanced merge [COMPUT SCI] A merge or sort operation in which the data involved are divided equally between the available storage devices. { 'bal·ənst 'mərj }

balanced method [ENG] Method of measurement in which the reading is taken at zero; it may be a visual or audible reading, and in the latter case the null is the no-sound setting. { 'bal·ənst ¦meth·əd }

balanced modulator [ELECTR] A modulator in which the carrier and modulating signal are introduced in such a way that the output contains the two sidebands without the carrier. { 'bal·ənst 'maj·ə,lād·ər }

balanced network [ELEC] Hybrid network in which the impedances of the opposite branches are equal. { 'bal·ənst ¦net,wərk }

balanced oscillator [ELECTR] Any oscillator in which, at the oscillator frequency, the impedance centers of the tank circuits are at ground potential, and the voltages between either end and their centers are equal in magnitude and opposite in phase. { 'bal·ənst 'äs·ə,lād·ər }

balanced output [ELECTR] A three-conductor output (as from an amplifier) in which the signal voltage alternates above and below a third, neutral wire. { 'bal·ənst ¦aut,put }

balanced polymorphism [GEN] Maintenance in a population of two or more alleles in equilibrium at frequencies too high to be explained, particularly for the rarer of them, by mutation; commonly due to the selective advantage of a heterozygote over both homozygotes. { 'bal·ənst ¦päl·i'mor,fiz·əm }

balanced range of error [STAT] A range of error in which the maximum and minimum possible errors are opposite in sign and equal in magnitude. { 'bal·ənst ¦rānj əv 'er·ər }

balanced reinforcement [CIV ENG] An amount and distribution of steel reinforcement in a flexural reinforced concrete member such that the allowable tensile stress in the steel and the allowable compressive stress in the concrete are attained simultaneously. { 'bal·ənst ,rē·ən'for·smənt }

balanced ring modulator [ELECTR] A modulator that uses tubes or diodes to suppress the carrier signal while providing double-sideband output. { 'bal·ənst ¦riŋ ,maj·ə,lād·ər }

balanced rock See perched block. { 'bal·ənst ,räk }

balanced rudder [NAV ARCH] A rudder, part of whose area is forward of the vertical axis, to counterbalance water pressure abaft of the axis. { 'bal·ənst ,rəd·ər }

balanced sash [BUILD] In a double-hung window, a sash which opens by being raised or lowered and which is balanced with counterweights or pretensioned springs so that little force is required to move the sash. { 'bal·ənst ,sash }

balanced set [ELECTR] Two or more components, such as tubes or transistors, connected in parallel or push-pull configuration, that have been chosen on the basis of identical, or nearly identical, gain and load characteristics. [MATH] A set S in a real or complex vector space X such that if x is in S and $|a| \leq 1$, then ax is in S. { 'bal·ənst ,set }

balanced step [BUILD] One of a series of winders arranged so that the width of each winder tread (at the narrow end) is almost equal to the tread width in the straight portion of the adjacent stair flight. Also known as dancing step; dancing winder. { 'bal·ənst ,step }

balanced stock [ORD] **1.** That condition in supply when availability and requirements are in equilibrium for specific items. **2.** Accumulation of supplies in quantities determined necessary to meet requirements for a fixed period of time. { 'bal·ənst ,stäk }

balanced surface [AERO ENG] A control surface that extends on both sides of the axis of the hinge or pivot, or that has auxiliary devices or extensions connected with it, in such a manner as to effect a small or zero resultant moment of the air forces about the hinge axis. { 'bal·ənst 'sər·fəs }

balanced translocation [GEN] Positional change of one or more chromosome segments in cells or gametes without alteration of the normal diploid or haploid complement of genetic material. { 'bal·ənst tranz·lō'kā·shən }

balanced transmission line [ELEC] Transmission line having equal conductor resistances per unit length and equal impedances from each conductor to earth and to other electrical circuits. { 'bal·ənst tranz'mish·ən ,līn }

balanced-tree [COMPUT SCI] A system of indexes that keeps track of stored data, and in which data keys are stored in a hierarchy that is continually modified in order to minimize access times. Abbreviated B-tree. { 'bal·ənst 'trē }

balanced valve [ENG] A valve having equal fluid pressure in both the opening and closing directions. { 'bal·ənst ,valv }

balanced voltages [ELEC] Voltages that are equal in magnitude and opposite in polarity with respect to ground. Also known as push-pull voltages. { 'bal·ənst ,vōl·tij·əz }

balanced wire circuit [ELEC] Circuit wherein the two sides are electrically alike and symmetrical with respect to ground and other conductors. { 'bal·ənst ¦wīr ,sər·kət }

balance equation [MATH] An equation expressing a balance of quantities in the sense that the local or individual rates of change are zero. [METEOROL] A diagnostic equation expressing a balance between the pressure field and the horizontal field of motion of the atmosphere. { 'bal·əns i'kwā·zhən }

balance error [COMPUT SCI] An error voltage that arises at the output of analog adders in an analog computer and is directly proportional to the drift error. { 'bal·əns ,er·ər }

balance method See null method. { 'bal·əns ,meth·əd }

balance pipe [ENG] A pipe in a compressed-air piping system that is used to displace trapped air so that the condensate can flow freely into the trap. { 'bal·əns ,pīpe }

balancer [ELEC] A mechanism for equalizing the loads on the outer lines of a three-wire system for electric power distribution, consisting of two similar shunt or compound machines which are coupled together with the armatures connected in series across the outer lines. [INV ZOO] See haltere. [VERT ZOO] Either of a pair of rodlike lateral appendages on the heads of some larval salamanders. { 'bal·ən·sər }

balancer set [ELEC] Two coupled direct-current generators or motors that are used to equalize the voltage on each side of a three-wire system. { 'bal·ən·sər ,set }

balance shot [MIN ENG] In coal mining, a shot for which the drill hole is parallel to the face of the coal that is to be broken by it. { 'bal·əns ,shät }

balance spring [HOROL] An oscillating spring of spiral or (in a chronometer) cylindrical shape which governs the movement of a balance wheel in a timepiece. Also known as hairspring. { 'bal·əns ,spriŋ }

balance tool [MECH ENG] A tool designed for taking the first cuts when the external surface of a piece in a lathe is being machined; it is supported in the tool holder at an unvarying angle. { 'bal·əns ,tül }

balance-to-unbalance transformer [ELEC] Device for matching a pair of lines, balanced with respect to earth, to a pair of lines not balanced with respect to earth. { 'bal·əns tü ¦ən,bal·əns tranz'for·mər }

balance wheel [MECH ENG] **1.** A wheel which governs or stabilizes the movement of a mechanism. **2.** See flywheel. { 'bal·əns ,wēl }

balancing [COMPUT SCI] The distribution of workload among computing resources to optimize performance. { 'bal·əns·iŋ }

balancing a survey [ENG] Distributing corrections through any traverse to eliminate the error of closure and to obtain an adjusted position for each traverse station. Also known as traverse adjustment. { 'bal·əns·iŋ ə 'sər,vā }

balancing band [NAV ARCH] A band at the balancing point of an anchor on either side of the shank, fitted with a ring. { 'bal·əns·iŋ ,band }

balancing capacitor [ELECTR] A variable capacitor used to improve the accuracy of a radio direction finder. Also known as compensating capacitor. { 'bal·əns·iŋ kə'pas·əd·ər }

balancing delay [IND ENG] In motion study, idleness of one hand while the other is active to catch up. { 'bal·əns·iŋ di,lā }

balancing plug cock See balancing valve. { 'bal·əns·iŋ 'pləg ,käk }

balancing unit [ELEC] **1.** Antenna-matching device used to permit efficient coupling of a transmitter or receiver having an unbalanced output circuit to an antenna having a balanced transmission line. **2.** Device for converting balanced to unbalanced transmission lines, and vice versa, by placing suitable discontinuities at the junction between the lines instead of using lumped components. { 'bal·əns·iŋ ,yü·nət }

balancing valve [ENG] A valve used in a pipe for controlling fluid flow; not usually used to shut off the flow. Also known as balancing plug cock. { 'bal·əns·iŋ ,valv }

Balanidae [INV ZOO] A family of littoral, sessile barnacles in the suborder Balanomorpha. { bə'lan·ə,dē }

balanitis [MED] Inflammation of the glans of the penis or of the clitoris. { bal·ə'nīd·əs }

Balanoglossus [INV ZOO] A cosmopolitan genus of tongue worms belonging to the class Enteropneusta. { ,bal·ə·nō'glä·səs }

Balanomorpha [INV ZOO] The symmetrical barnacles, a suborder of sessile crustaceans in the order Thoracica. { ,bal·ə·nō'mȯr·fə }

Balanopaceae [BOT] A small family of dioecious dicotyledonous plants in the order Fagales characterized by exstipulate leaves, seeds with endosperm, and the pistillate flower solitary in a multibracteate involucre. { ,bal·ə·nō'pās·ē,ē }

Balanopales [BOT] An ordinal name suggested for the Balanopaceae in some classifications. { ,bal·ə'näp·ə,lēz }

Balanophoraceae [BOT] A family of dicotyledonous terrestrial plants in the order Santalales characterized by dry nutlike fruit, one to five ovules, unisexual flowers, attachment to the stem of the host, and the lack of chlorophyll. { ,bal·ə,näf·ə'rās·ē,ē }

balanoposthitis [MED] Inflammation of the glans penis and of the prepuce. { ,bal·ə·nō·päs'thīd·əs }

Balanopsidales [BOT] An order in some systems of classification which includes only the Balanopaceae of the Fagales. { ,bal·ən,äp·sə'dā·lēz }

balantidiasis [MED] An intestinal infection of humans caused by the protozoan *Balantidium coli*. { ,bal·ənt·ə'dī·ə·səs }

Balantidium [INV ZOO] A genus of protozoans in the order Trichostomatida containing the only ciliated protozoan species parasitic in humans, *Balantidium coli*. { ,bal·ən'tid·ē·əm }

Balanus [INV ZOO] A genus of barnacles composed of sessile acorn barnacles; the type genus of the family Balanidae. { 'bal·ə·nəs }

balata [MATER] A hard substance, similar to gutta-percha, used mainly in golf balls and belting, which is made by drying the milky juice of the bully tree (*Manilkara bidentata*). Also known as gutta-balata. { bə'läd·ə }

Balbiani chromosome See polytene chromosome. { ,bäl·bē'än·ē 'krō·mə,sōm }

Balbiani rings [CYTOL] Localized swellings of a polytene chromosome. { ,bäl·bē'än·ē ,riŋz }

balbriggan [TEXT] Knitted yarn made of cotton, wool, rayon, or other fiber and which is characteristically lightweight and unbleached. { bal'brig·ən }

balconet [BUILD] A pseudobalcony; a low ornamental railing at a window, projecting only slightly beyond the threshold or sill. { ,bal·kə'net }

balcony [BUILD] A deck which projects from a building wall above ground level. { 'bal·kə·nē }

balcony outlet [BUILD] In a vertical rainwater pipe that passes through an exterior balcony, a fitting which provides an inlet for the drainage of rainwater from the balcony. { 'bal·kə·nē ,aut,let }

bald [GEOGR] An elevated grassy, treeless area, as on the top of a mountain. [MIN ENG] **1.** Without framing. **2.** A mine timber which has a flat end. { bȯld }

baldheaded anticline [GEOL] An upfold with a crest that has been deeply eroded before later deposition. { 'bȯld,hed·əd 'an·ti,klīn }

baldness [MED] Loss or absence of hair. { 'bȯld·nəs }

Baldwin spot See bitter pit. { 'bȯld·wən ,spät }

bale [IND ENG] **1.** A large package of material, pressed tightly together, tied with rope, wire, or hoops and usually covered with wrapping. **2.** The amount of material in a bale; sometimes used as a unit of measure, as 500 pounds (227 kilograms) of cotton in the United States. { bāl }

baleen [VERT ZOO] A horny substance, growing as fringed filter plates suspended from the upper jaws of whalebone whales. Also known as whalebone. { be'lēn }

baler [MECH ENG] A machine which takes large quantities of raw or finished materials and binds them with rope or metal straps or wires into a large package. { 'bāl·ər }

Balfour's law [EMBRYO] The law that the speed with which any part of the ovum segments is roughly proportional to the protoplasm's concentration in that area; the segment's size is inversely proportional to the protoplasm's concentration. { 'bal·fərz ,lȯ }

baling [CIV ENG] A technique used to convert loose refuse into heavy blocks by compaction; the blocks are then burned and are buried in sanitary landfill. { 'bāl·iŋ }

Bali wind [METEOROL] A strong east wind at the eastern end of Java. { 'bäl·ē 'wind }

balk [BUILD] A squared timber used in building construction. [CIV ENG] A low ridge of earth that marks a boundary line. [MIN ENG] A sudden thinning out, for a certain distance, of a bed of coal. { bȯk }

balking [IND ENG] The refusal of a customer to enter a queue for some reason, such as insufficient waiting room. { 'bȯk·iŋ }

ball [GEOL] **1.** A low sand ridge, underwater by high tide, which extends generally parallel with the shoreline; usually separated by an intervening trough from the beach. **2.** A spheroidal mass of sedimentary material. **3.** Common name for a nodule, especially of ironstone. [MECH ENG] In fine grinding, one of the crushing bodies used in a ball mill. [ORD] **1.** A bullet for general use, as distinguished from bullets for special uses such as armor-piercing, incendiary, or high explosive. **2.** A small-arms solid propellant which is oblate spheroidal in shape, generally a double-base propellant. { bȯl }

ball-and-race-type pulverizer [MECH ENG] A grinding machine in which balls rotate under an applied force between two races to crush materials, such as coal, to fine consistency. Also known as ball-bearing pulverizer. { 'bȯl ən 'rās ,tīp 'pəl·və,rīz·ər }

ball-and-ring method See ring-and-ball test. { 'bȯl ən 'riŋ ,meth·əd }

ball-and-socket joint [ANAT] See enarthrosis. [GEOL] See cup-and-ball joint. [MECH ENG] A joint in which a member ending in a ball is joined to a member ending in a socket so that relative movement is permitted within a certain angle in all planes passing through a line. Also known as ball joint. { 'bȯl ən 'säk·ət ,jȯint }

ball-and-trunnion joint [MECH ENG] A joint in which a universal joint and a slip joint are combined in a single assembly. { 'bȯl ən 'trən·yən ,jȯint }

ballas [MINERAL] A spherical aggregate of small diamond crystals; used in diamond drill bits and other diamond tools. { 'bal·əs }

ballast [AERO ENG] A relatively dense substance that is placed in the cab of a balloon and can be thrown out to reduce the load or can be shifted to change the center of gravity. [CIV ENG] Crushed stone used in a railroad bed to support the ties, hold the track in line, and help drainage. [ELEC] A circuit element that serves to limit an electric current or to provide a starting voltage, as in certain types of lamps, such as in fluorescent ceiling fixtures. [MATER] Coarse gravel used as an ingredient in concrete. [NAV ARCH] **1.** A relatively heavy material such as lead, iron, or water placed in a ship to ensure stability or to maintain the proper draft or trim. **2.** To pump seawater into empty fuel tanks of a ship to ensure its stability or suitable draft and trim for seaworthiness. { 'bal·əst }

ballast factor [ELEC] The ratio of the luminous output of a lamp when operated on a ballast to its luminous output when operated under standardized rating conditions. { 'bal·əst ,fak·tər }

ballast lamp [ELEC] A light-producing electrical resistance

BALL-AND-RACE-TYPE PULVERIZER

Coarse raw material is ground by crushing and attrition between balls and races and is then withdrawn from the pulverizer by an air stream.

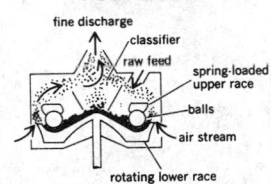

BALLAST

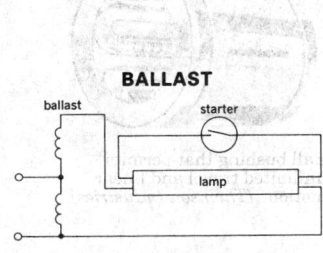

Position of the ballast in preheat lamp circuit.

device which maintains nearly constant current by increasing in resistance as the current increases. { 'bal·əst ,lamp }

ballast leg *See* ballast movement. { 'bal·əst ,leg }

ballast line [NAV ARCH] A ship's water line when the ship is ballasted. { 'bal·əst ,līn }

ballast movement [NAV] A voyage or leg of a voyage by a tanker during which no commercially valuable cargo is in the tanks; seawater is usually carried to maintain stability. Also known as ballast leg; ballast passage. { 'bal·əst ,müv·mənt }

ballast passage *See* ballast movement. { 'bal·əst ,pas·ij }

ballast pump [NAV ARCH] A pump designed to discharge water ballast. { 'bal·əst ,pəmp }

ballast reactor [ELEC] A coil wound on an iron core and connected in series with a fluorescent lamp to compensate for the negative-resistance characteristics of the lamp by providing an increased voltage drop as the current through the lamp is increased. { 'bal·əst rē'ak·tər }

ballast resistor [ELEC] A resistor that increases in resistance as current through it increases, and decreases in resistance as current decreases. Also known as barretter (British usage). { 'bal·əst ri'sis·tər }

ballast tank [NAV ARCH] **1.** One of several tanks in the hold of a ship which may be pumped full of water as ballast. **2.** One of the tanks in a submarine that are filled with water or air to submerge or surface. { 'bal·əst ,taŋk }

ballast tube [ELEC] A ballast resistor mounted in an evacuated glass or metal envelope, like that of a vacuum tube, to reduce radiation of heat from the resistance element and thereby improve the voltage-regulating action. { 'bal·əst ,tüb }

ball bearing [MECH ENG] An antifriction bearing permitting free motion between moving and fixed parts by means of balls confined between outer and inner rings. { 'bȯl 'ber·iŋ }

ball-bearing hinge [MECH ENG] A hinge which is equipped with ball bearings between the hinge knuckles in order to reduce friction. { ,bȯl 'ber·iŋ ,hinj }

ball-bearing pulverizer *See* ball-and-race-type pulverizer. { ,bȯl 'ber·iŋ ,pəl·və,rīz·ər }

ball bonding [ENG] The making of electrical connections in which a flame is used to cut a wire, the molten end of which solidifies as a ball, which is pressed against the bonding pad on an integrated circuit. { 'bȯl ,bänd·iŋ }

ball breaker [ENG] **1.** A steel or iron ball that is hoisted by a derrick and allowed to fall on blocks of waste stone to break them or to swing against old buildings to demolish them. Also known as skull cracker; wrecking ball. **2.** A coring and sampling device consisting of a hollow glass ball, 3 to 5 inches (7.5 to 12.5 centimeters) in diameter, held in a frame attached to the trigger line above the triggering weight of the corer; used to indicate contact between corer and bottom. { 'bȯl ¦brāk·ər }

ball burnishing [MET] A method of giving small stainless steel parts a lustrous finish by rotating them in a wood-lined barrel with water, burnishing soap, and hardened steel balls. { 'bȯl ,bər·nish·iŋ }

ball bushing [MECH ENG] A type of ball bearing that allows motion of the shaft in its axial direction. { 'bȯl ,bush·iŋ }

ball catch [DES ENG] A door fastener having a contained metal ball which is under pressure from a spring; the ball engages a striking plate and keeps the door from opening until force is applied. { 'bȯl ,kach }

ball check valve [ENG] A valve having a ball held by a spring against a seat; used to permit flow in one direction only. { 'bȯl 'chek ,valv }

ball clay [MATER] A clay used in ceramics that is characterized by strong binding properties, a tendency to ball, and excellent plasticity. { 'bȯl ,klā }

ball coal [GEOL] A variety of coal occurring in spheroidal masses. { 'bȯl ,kōl }

ball float [MECH ENG] A floating device, usually approximately spherical, which is used to operate a ball valve. { 'bȯl ,flōt }

ball-float liquid-level meter [ENG] A float which rises and falls with liquid level, actuating a pointer adjacent to a calibrated scale in order to measure the level of a liquid in a tank or other container. { 'bȯl ,flōt ¦lik·wəd ¦lev·əl ¦mēd·ər }

ball grinder *See* ball mill. { 'bȯl ,grind·ər }

ballhead [MECH ENG] That part of the governor which contains flyweights whose force is balanced, at least in part, by the force of compression of a speeder spring. { 'bȯl¦hed }

ball ice [OCEANOGR] Numerous floating spheres of sea ice having diameters of 1–2 inches (2.5–5 centimeters), generally in belts similar to slush which forms at the same time. { 'bȯl ,īs }

Balling hydrometer [ENG] A type of saccharometer used to determine the density of sugar solutions. { 'bȯl·iŋ hī'dräm·əd·ər }

balling-up [MET] Formation of balls of molten brazing filler metal when the base material has not been sufficiently wetted. { ¦bȯl·iŋ ¦əp }

ballistic area [ORD] Space lying between the centers of impact of two groups of shots, one consisting entirely of shots over the target and the other entirely of shots short of the target. { bə'lis·tik ,er·ē·ə }

ballistic body [ENG] A body free to move, behave, and be modified in appearance, contour, or texture by ambient conditions, substances, or forces, such as by the pressure of gases in a gun, by rifling in a barrel, by gravity, by temperature, or by air particles. { bə'lis·tik ,bäd·ē }

ballistic camera [OPTICS] A ground-based camera using multiple exposures on the same plate to record the trajectory of a rocket. { bə'lis·tik ,kam·rə }

ballistic case [ORD] Any shell or casing with efficient ballistic characteristics, used to enclose elements for delivery on a target; any bomb, projectile, or rocket case. { bə'lis·tik ,kās }

ballistic coefficient [MECH] The numerical measure of the ability of a missile to overcome air resistance; dependent upon the mass, diameter, and form factor. { bə'lis·tik ,kō·ə'fish·ənt }

ballistic conditions [MECH] Conditions which affect the motion of a projectile in the bore and through the atmosphere, including muzzle velocity, weight of projectile, size and shape of projectile, rotation of the earth, density of the air, temperature or elasticity of the air, and the wind. { bə'lis·tik kən'dish·əns }

ballistic correction [ORD] An adjustment in firing data that is based on ballistic conditions, affecting the flight of a projectile; does not include adjustment based on observation of fire. { bə'lis·tik kə'rek·shən }

ballistic curve [MECH] The curve described by the path of a bullet, a bomb, or other projectile as determined by the ballistic conditions, by the propulsive force, and by gravity. { bə'lis·tik 'kərv }

ballistic deflection [MECH] The deflection of a missile due to its ballistic characteristics. { bə'lis·tik di'flek·shən }

ballistic density [MECH] A representation of the atmospheric density encountered by a projectile in flight, expressed as a percentage of the density according to the standard artillery atmosphere. { bə'lis·tik 'den·səd·ē }

ballistic efficiency [MECH] **1.** The ability of a projectile to overcome the resistance of the air; depends chiefly on the weight, diameter, and shape of the projectile. **2.** The external efficiency of a rocket or other jet engine of a missile. { bə'lis·tik i'fish·ən·sē }

ballistic entry [MECH] Movement of a ballistic body from without to within a planetary atmosphere. { bə'lis·tik 'en·trē }

ballistic galvanometer [ELEC] A galvanometer having a long period of swing so that the deflection may measure the electric charge in a current pulse or the time integral of a voltage pulse. { bə'lis·tik ,gal·və'näm·əd·ər }

ballistic instrument [ENG] Any instrument, such as a ballistic galvanometer or a ballistic pendulum, that measures an impact or sudden pulse of energy. { bə'lis·tik 'in·strə·mənt }

ballistic lead [ORD] The correction or allowance for wind effects and gravity made when computing a lead angle. { bə'lis·tik 'lēd }

ballistic limit [MECH] The minimum velocity at which a particular armor-piercing projectile is expected to consistently and completely penetrate armor plate of given thickness and physical properties at a specified angle of obliquity. { bə'lis·tik 'lim·ət }

ballistic magnetometer [ENG] A magnetometer designed to employ the transient voltage induced in a coil when either the magnetized sample or coil are moved relative to each other. { bə'lis·tik ,mag·nə'täm·əd·ər }

ballistic measurement [MECH] Any measurement in which an impulse is applied to a device such as the bob of a ballistic pendulum, or the moving part of a ballistic galvanometer, and the subsequent motion of the device is used to determine the

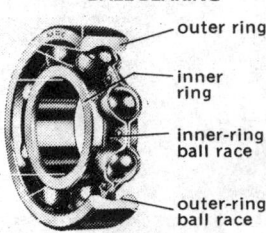

BALL BEARING

outer ring

inner ring

inner-ring ball race

outer-ring ball race

Deep-groove ball bearing. *(Marlin-Rockwell)*

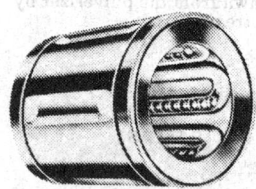

BALL BUSHING

Ball bushing that permits unlimited travel and linear motion. *(Thomson Industries)*

magnitude of the impulse, and, from this magnitude, the quantity to be measured.   { bə'lis·tik 'mezh·ər·mənt }

ballistic missile [ORD] A missile capable of guiding and propelling itself in a direction and to a velocity such that it will follow a ballistic trajectory to a desired point.   { bə'lis·tik 'mis·əl }

ballistic missile defense [ORD] A system utilizing laser beams and other esoteric techniques to detect, identify, and destroy ballistic missiles. Abbreviated BMD.   { bə'lis·tik 'mis·əl di'fens }

ballistic mortar [ORD] A heavy, short-barreled mortar, pendulum-mounted, for determining the relative power of explosives; a small sample of a test explosive is placed in the detonation chamber and a projectile is located forward of the charge; upon detonation the projectile is driven into a sand bank and the mortar swings through an arc; a marker records the maximum weight of the test explosive required to produce the same rise as 10 grams of TNT.   { bə'lis·tik 'mȯrd·ər }

ballistic pendulum [ENG] A device which uses the deflection of a suspended weight to determine the momentum of a projectile.   { bə'lis·tik 'pen·jə·ləm }

ballistic range [ORD] A long, instrumented enclosure wherein gun-launched projectiles are tested.   { bə'lis·tik ,rānj }

ballistics [MECH] Branch of applied mechanics which deals with the motion and behavior characteristics of missiles, that is, projectiles, bombs, rockets, guided missiles, and so forth, and of accompanying phenomena.   { bə'lis·tiks }

ballistic separator [CIV ENG] A device that takes out noncompostable material like stones, glass, metal, and rubber, from solid waste by passing the waste over a rotor that has impellers to fling the material in the air; the lighter organic (compostable) material travels a shorter distance than the heavier (noncompostable) material.   { bə'lis·tik 'sep·ə,rād·ər }

ballistics of penetration [MECH] That part of terminal ballistics which treats of the motion of a projectile as it forces its way into targets of solid or semisolid substances, such as earth, concrete, or steel.   { bə'lis·tiks əv pen·ə'trā·shən }

ballistic table [MECH] Compilation of ballistic data from which trajectory elements such as angle of fall, range to summit, time of flight, and ordinate at any time, can be obtained.   { bə'lis·tik 'tā·bəl }

ballistic temperature [MECH] That temperature (in °F) which, when regarded as a surface temperature and used in conjunction with the lapse rate of the standard artillery atmosphere, would produce the same effect on a projectile as the actual temperature distribution encountered by the projectile in flight.   { bə'lis·tik 'tem·prə·chər }

ballistic test [ORD] Proof test of weapons or ammunition to determine suitability.   { bə'lis·tik ,test }

ballistic tracking See dynamic resolution.   { bə,lis·tik 'trak·iŋ }

ballistic trajectory [MECH] The trajectory followed by a body being acted upon only by gravitational forces and resistance of the medium through which it passes.   { bə'lis·tik trə'jek·tə,rē }

ballistic transport [ELECTR] The passage of electrons through a semiconductor whose length is less than the mean free path of electrons in the semiconductor, so that most of the electrons pass through the semiconductor without scattering.   { bə'lis·tik 'tranz,pȯrt }

ballistic tube [ORD] A gun tube that is used for ballistic tests; usually selected because of certain demonstrated characteristics.   { bə'lis·tik 'tüb }

ballistic uniformity [MECH] The capability of a propellant, when fired under identical conditions from round to round, to impart uniform muzzle velocity and produce similar interior ballistic results.   { bə'lis·tik ,yü·nə'fȯr·məd·ē }

ballistic vehicle [ENG] A nonlifting vehicle; a vehicle that follows a ballistic trajectory.   { bə'lis·tik 've·ə·kəl }

ballistic wave [MECH] An audible disturbance caused by compression of air ahead of a missile in flight.   { bə'lis·tik ,wāv }

ballistic weapon [ORD] Any missile weapon, as a bomb, rocket, projectile, or bullet, affected by ballistic conditions.   { bə'lis·tik 'wep·ən }

ballistic wind [MECH] That constant wind which would produce the same effect upon the trajectory of a projectile as the actual wind encountered in flight.   { bə'lis·tik 'wind }

ballistite [MATER] A smokeless propellant containing nitrocellulose and nitroglycerin; used in some rocket, mortar, and small-arms ammunition.   { 'bal·ə,stīt }

ballistocardiogram [MED] The recording made by a ballistocardiograph.   { bə¦lis·tō'kärd·ē·ə,gram }

ballistocardiograph [MED] A device to measure the volume of blood passing through the heart in a given period of time.   { bə¦lis·tō'kärd·ē·ə,graf }

ballistophobia [PSYCH] An abnormal fear of projectiles or missiles.   { bə¦lis·tə'fōb·ē·ə }

ballistospore [MYCOL] A type of fungal spore that is forcibly discharged at maturity.   { bə'lis·tə,spȯr }

ball joint See ball-and-socket joint.   { 'bȯl ,jȯint }

ball lightning [GEOPHYS] A relatively rare form of lightning, consisting of a reddish, luminous ball, of the order of 1 foot (30 centimeters) in diameter, which may move rapidly along solid objects or remain floating in midair. Also known as globe lightning.   { 'bȯl ¦līt·niŋ }

ball mill [MECH ENG] A pulverizer that consists of a horizontal rotating cylinder, up to three diameters in length, containing a charge of tumbling or cascading steel balls, pebbles, or rods. Also known as ball grinder.   { 'bȯl ,mil }

ballonet [AERO ENG] One of the air cells in a blimp, fastened to the bottom or sides of the envelope, which are used to maintain the required pressure in the envelope without adding or valving gas as the ship ascends or descends. Also spelled ballonnet.   { ¦bal·ə¦nā }

ballonnet See ballonet.   { ¦bal·ə¦nā }

balloon [AERO ENG] A nonporous, flexible spherical bag, inflated with a gas such as helium that is lighter than air, so that it will rise and float in the atmosphere; a large-capacity balloon can be used to lift a payload suspended from it.   { bə'lün }

balloon apron [ORD] An antiaircraft device consisting of cables hanging perpendicularly from other cables extending between two or more balloons.   { bə'lün ,ā·prən }

balloon astronomy [ASTRON] The observation of celestial objects from instruments mounted on balloons and carried to altitudes up to 18 miles (30 kilometers), to detect electromagnetic radiation at wavelengths which do not penetrate to the earth's surface.   { bə'lün ə'strän·ə·mē }

balloon barrage [ORD] An antiaircraft defense consisting of a number of balloons, usually together and equipped with balloon aprons, held captive by steel cables and strategically moored near vital areas or installations.   { bə'lün bə'räzh }

balloon ceiling [METEOROL] The ceiling classification applied when the ceiling height is determined by timing the ascent and disappearance of a ceiling balloon or pilot balloon in United States weather observing practice.   { bə'lün ,sēl·iŋ }

balloon cover [AERO ENG] A cover which fits over a large, inflated balloon to facilitate handling in high or gusty winds. Also known as balloon shroud.   { bə'lün ,kəv·ər }

balloon drag [METEOROL] A small balloon, loaded with ballast and inflated so that it will explode at a predetermined altitude, which is attached to a larger balloon; frequently used to retard the ascent of a radiosonde during the early part of the flight so that more detailed measurements may be obtained.   { bə'lün ,drag }

balloon framing [CIV ENG] Framing for a building in which each stud is one piece from roof to foundation.   { bə'lün ,fram·iŋ }

balloon shroud See balloon cover.   { bə'lün ,shraúd }

balloon-type rocket [AERO ENG] A liquid-fuel rocket, such as the Atlas, that requires the pressure of its propellants (or other gases) within it to give it structural integrity.   { bə'lün ¦tīp 'räk·ət }

balloting [MECH] A tossing or bounding movement of a projectile, within the limits of the bore diameter, while moving through the bore under the influence of the propellant gases.   { 'bal·əd·iŋ }

ball-peen hammer [ENG] A hammer with a ball at one end of the head; used in riveting and forming metal.   { 'bȯl,pēn 'ham·ər }

ball pendulum test [ENG] A test for measuring the strength of explosives; consists of measuring the swing of a pendulum produced by the explosion of a weighed charge of material.   { 'bȯl 'pen·jə·ləm ,test }

ball race [DES ENG] A track, channel, or groove in which ball bearings turn.   { 'bȯl ,rās }

BALLISTIC PENDULUM

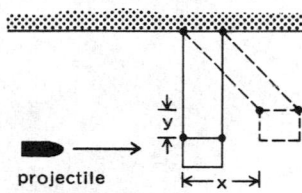

Ballistic pendulum before (solid lines) and after (broken lines) projectile impact. Measurements of x or y and knowledge of mass and length of the pendulum determine the initial momentum of the projectile.

BALL MILL

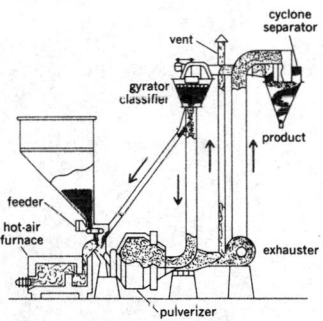

Conical ball mill pulverizer in closed circuit with classifier. *(Hardinge Co., Inc.)*

ball screw [MECH ENG] An element used to convert rotation to longitudinal motion, consisting of a threaded rod linked to a threaded nut by ball bearings constrained to roll in the space formed by the threads, in order to reduce friction. { 'bȯl ,skrü }

ball sealers [PETRO ENG] Balls of rubber, plastic, or metal that are dropped down the well bore to aid the acidizing of impermeable zones of an oil reservoir; they wedge into and plug the bottomhole tubing perforations that are adjacent to the more permeable reservoir zones. { 'bȯl ,sēl·ərz }

ball sizing [MET] Finishing a hole to a precise diameter and burnishing the surface by forcing a steel ball through it. { 'bȯl ,sīz·iŋ }

ballstone [GEOL] **1.** Large mass or concretion of fine, unstratified limestone resulting from growth of coral colonies. **2.** A nodule of rock, especially ironstone, in a stratified unit. { 'bȯl,stōn }

ball test [CIV ENG] In a drain, a test for freedom from obstruction and for circularity in which a ball (less than the diameter of the drain by a specified amount) is rolled through the drain. { 'bȯl ,test }

ball-up [ENG] **1.** During a drilling operation, collection by a portion of the drilling equipment of a mass of viscous consolidated material. **2.** Failure of an anchor to hold on a soft bottom, by pulling out with a large ball of mud attached. { 'bȯl ,əp }

balluster [GRAPHICS] One of the sections of wood used to make the frames on which silk, or some other material, is stretched in the silk-screen process. { 'bal·ə,stər }

ballute [AERO ENG] A cross between a balloon and a parachute, used to brake the free fall of sounding rockets. { 'ba,lüt }

ball valve [MECH ENG] A valve in which the fluid flow is regulated by a ball moving relative to a spherical socket as a result of fluid pressure and the weight of the ball. { 'bȯl ,valv }

balm [GEOL] A concave cliff or precipice that forms a shelter. [MED] A soothing or healing medication. { bäm }

Balmer continuum [SPECT] A continuous range of wavelengths (or wave numbers or frequencies) in the spectrum of hydrogen at wavelengths less than the Balmer limit, resulting from transitions between states with principal quantum number $n = 2$ and states in which the single electron is freed from the atom. { ¦bäl·mər kən'tin·yə·wəm }

Balmer discontinuity See Balmer jump. { 'bȯl·mər dis,känt·ən'ü·əd·ē }

Balmer formula [SPECT] An equation for the wavelengths of the spectral lines of hydrogen, $1/\lambda = R[(1/m^2) - (1/n^2)]$, where λ is the wavelength, R is the Rydberg constant, and m and n are positive integers (with n larger than m) that give the principal quantum numbers of the states between which occur the transition giving rise to the line. { 'bȯl·mər ,fȯr·myə·lə }

Balmer jump [SPECT] The sudden decrease in the intensity of the continuous spectrum of hydrogen at the Balmer limit. Also known as Balmer discontinuity. { 'bȯl·mər ,jəmp }

Balmer limit [SPECT] The limiting wavelength toward which the lines of the Balmer series crowd and beyond which they merge into a continuum, at approximately 365 nanometers. { 'bȯl·mər ,lim·ət }

Balmer lines [SPECT] Lines in the hydrogen spectrum, produced by transitions between $n = 2$ and $n > 2$ levels either in emission or in absorption; here n is the principal quantum number. { 'bȯl·mər ,līnz }

Balmer series [SPECT] The set of Balmer lines. { 'bȯl·mər ¦sir·ēz }

balsa [BOT] *Ochroma lagopus.* A tropical American tree in the order Malvales; its wood is strong and lighter than cork. { 'bȯl·sə }

balsam [MATER] An exudate of the balsam tree; a mixture of resins, essential oils, cinnamic acid, and benzoic acid. { 'bȯl·səm }

Balsaminaceae [BOT] A family of flowering plants in the order Geraniales, including touch-me-not (*Impatiens*); flowers are irregular with five stamens and five carpels, leaves are simple with pinnate venation, and the fruit is an elastically dehiscent capsule. { ,bȯl·sə·mə'nās·ē,ē }

balsam of Peru See Peru balsam. { 'bȯl·səm əv pə'rü }

Baltic Sea [GEOGR] An intracontinental, Mediterranean-type sea, connected with the North Sea and surrounded by Sweden, Denmark, Germany, Poland, the Baltic States, and Finland. { 'bȯl·tik 'sē }

Baltimore oil See wormseed oil. { 'bȯl·tə,mȯr ,ȯil }

balun [ELEC] A device used for matching an unbalanced coaxial transmission line or system to a balanced two-wire line or system. Also known as balanced converter; bazooka; line-balance converter. { 'ba,lən }

baluster [BUILD] A post which supports a handrail and encloses the open sections of a stairway. { 'bal·ə·stər }

balustrade [BUILD] The railing assembly of a stairway consisting of the handrail, balusters, and usually a bottom rail. { 'bal·ə,strād }

Bamberga [ASTRON] An asteroid with a diameter of about 129 miles (215 kilometers), mean distance from the sun of 2.686 astronomical units, and C-type surface composition. { 'bäm,bər·gə }

Bamberger's formula [ORG CHEM] A structural formula for naphthalene that shows the valencies of the benzene rings pointing toward the centers. { 'bäm,bər·gərz 'fȯr·myə·lə }

bamboo [BOT] The common name of various tropical and subtropical, perennial, ornamental grasses in five genera of the family Gramineae characterized by hollow woody stems up to 6 inches (15 centimeters) in diameter. { bam'bü }

Bambusoideae [BOT] A subfamily of grasses, composed of bamboo species, in the family Gramineae. { ,bam·bə'sȯid·ē,ē }

Banach algebra [MATH] An algebra which is a Banach space satisfying the property that for every pair of vectors, the norm of the product of those vectors does not exceed the product of their norms. { 'bä,näk 'al·jə·brə }

Banach's fixed-point theorem [MATH] A theorem stating that if a mapping f of a metric space E into itself is a contraction, then there exists a unique element x of E such that $fx = x$. Also known as Caccioppoli-Banach principle. { ¦bä,näks ,fikst ,pȯint 'thir·əm }

Banach space [MATH] A real or complex vector space in which each vector has a non-negative length, or norm, and in which every Cauchy sequence converges to a point of the space. Also known as complete normed linear space. { 'bä,näk ,spās }

Banach-Steinhaus theorem [MATH] If a sequence of bounded linear transformations of a Banach space is pointwise bounded, then it is uniformly bounded. { ¦bä,näk ¦stīn,haùs ,thir·əm }

Banach-Tarski paradox [MATH] A theorem stating that, for any two bounded sets, with interior points in a Euclidean space of dimension at least three, one of the sets can be disassembled into a finite number of pieces and reassembled to form the other set by moving the pieces with rigid motions (translations and rotations). { ¦bä,näk ¦tär·skē 'par·ə,däks }

banakite [PETR] An alkalic basalt made up of plagioclase, sanidine, and biotite, with small quantities of analcime, augite, and olivine; quartz or leucite may be present. { 'ban·ə,kīt }

banana [BOT] Any of the treelike, perennial plants of the genus *Musa* in the family Musaceae; fruit is a berry characterized by soft, pulpy flesh and a thin rind. { bə'nan·ə }

banana freckle [PL PATH] A fungus disease of the banana caused by *Macrophoma musae*, producing brown or black spots on the fruit and leaves. { bə'nan·ə ,frek·əl }

banana jack [ELEC] A jack that fits a banana plug; generally designed for panel mounting. { bə'nan·ə ,jak }

banana oil [ORG CHEM] **1.** A solution of nitrocellulose in amyl acetate having a bananalike odor. **2.** See amyl acetate. { bə'nan·ə ,ȯil }

banana plug [ELEC] A plug having a spring-metal tip shaped like a banana and used on test leads or as terminals for plug-in components. { bə'nan·ə ,pləg }

banco [HYD] A meander or oxbow lake separated from a river by a change in its course. { 'baŋ·kō }

Bancroft's filariasis See wuchereriasis. { 'baŋ,krȯfs ,fil·ə'rī·ə·səs }

band [ANALY CHEM] The position and spread of a solute within a series of tubes in a liquid-liquid extraction procedure. Also known as zone. [BUILD] Any horizontal flat member or molding or group of moldings projecting slightly from a wall plane and usually marking a division in the wall. Also known as band course; band molding. [COMMUN] A range of electromagnetic-wave frequencies between definite limits, such as that assigned to a particular type of radio service.

BANANA

Pseudostem of commercial banana plant *(Musa sapientum),* showing characteristic foliage and single stem of bananas. Plant grows to height of 30 feet (9 meters) or more.

[COMPUT SCI] A set of circular or cyclic recording tracks on a storage device such as a magnetic drum, disk, or tape loop. [CYTOL] Any of the characteristic transverse stripes exhibited by polytene or metaphase chromosomes that are stained. [DES ENG] A strip or cord crossing the back of a book to which the sections are sewn. [GEOD] Any latitudinal strip, designated by accepted units of linear or angular measurement, which circumscribes the earth. [GEOL] A thin layer or stratum of rock that is noticeable because its color is different from the colors of adjacent layers. [ORD] A metal sleeve joining together the barrel and stock of a gun. [SOLID STATE] A restricted range in which the energies of electrons in solids lie, or from which they are excluded, as understood in quantum-mechanical terms. Also known as energy bands. [SPECT] *See* band spectrum. { 'band }

bandage [BUILD] A strap, band, ring, or chain placed around a structure to secure and hold its parts together, as around the springing of a dome. [ELEC] Rubber ribbon about 4 inches (10 centimeters) wide for temporarily protecting a telephone or coaxial splice from moisture. [MED] A strip of gauze, muslin, flannel, or other material, usually in the form of a roll, but sometimes triangular or tailed, used to hold dressing in place, to apply pressure, to immobilize a part, to support a dependent or injured part, to obliterate tissue cavities, or to check hemorrhage. { 'ban·dij }

bandaite [PETR] A dacite type of extrusive rock composed of hypersthene and labradorite. { 'ban·də‚īt }

band brake [MECH ENG] A brake in which the frictional force is applied by increasing the tension in a flexible band to tighten it around the drum. { 'band ‚brāk }

band chain [ENG] A steel or Invar tape, graduated in feet and at least 100 feet (30.5 meters) long, used for accurate surveying. { 'band ‚chān }

band clamp [DES ENG] A two-piece metal clamp, secured by bolts at both ends; used to hold riser pipes. { 'band ‚klamp }

band clutch [MECH ENG] A friction clutch in which a steel band, lined with fabric, contracts onto the clutch rim. { 'band ‚kləch }

band course *See* band. { 'band ‚kȯrs }

banded [PETR] Pertaining to the appearance of rocks that have thin and nearly parallel bands of different textures, colors, and minerals. { 'ban·dəd }

banded anteater *See* marsupial anteater. { 'ban·dəd 'ant‚ēd·ər }

banded coal [GEOL] A variety of bituminous and subbituminous coal made up of a sequence of thin lenses of highly lustrous coalified wood or bark interspersed with layers of more or less striated bright or dull coal. { 'ban·dəd 'kōl }

banded differentiate [PETR] A type of igneous rock made up of bands of different composition, frequently alternating between two varieties as in a layered intrusion. { 'ban·dəd ‚dif·ə'ren·chē‚āt }

banded iron formation [GEOL] A sedimentary mineral deposit consisting of alternate silica-rich (chert or quartz) and iron-rich layers formed 2.5–3.5 billion years ago; the major source of iron ore. { 'band·əd 'ī‚ȯrn fȯr‚mā·shən }

banded ore [GEOL] Ore made up of layered bands composed either of the same minerals that differ from band to band in color or textures or proportion, or of different minerals. { 'ban·dəd 'ȯr }

banded peat [GEOL] Peat formed of alternate layers of vegetable debris. { 'ban·dəd 'pēt }

banded structure [MET] The appearance of a metal showing light and dark parallel bands in the direction of rolling or working. [METEOROL] The appearance of precipitation echoes in the form of long bands as presented on radar plan position indicator (PPI) scopes. [PETR] An outcrop feature in igneous and metamorphic rocks due to alternation of layers, stripes, flat lenses, or streaks that obviously differ in mineral composition or texture. { 'ban·dəd 'strək·chər }

banded vein [GEOL] A vein composed of layers of different minerals that lie parallel to the walls. Also known as ribbon vein. { 'ban·dəd 'vān }

band-elimination filter *See* band-stop filter. { 'band i‚lim·ə'nā·shən 'fil·tər }

band gap [SOLID STATE] An energy difference between two allowed bands of electron energy in a metal. { 'band ‚gap }

band groove [ORD] One of the channels cut into the rotating band of a projectile during the process of engraving. { 'band ‚grüv }

band head [SPECT] A location on the spectrogram of a molecule at which the lines of a band pile up. { 'band ‚hed }

bandicoot [VERT ZOO] **1.** Any of several large Indian rats of the genus *Nesokia* and related genera. **2.** Any of several small insectivorous and herbivorous marsupials comprising the family Peramelidae and found in Tasmania, Australia, and New Guinea. { 'ban·di‚küt }

banding [DES ENG] A strip of fabric which is used for bands. [HYD] In a glacier, a structure of alternate ice layers of different textures and appearance. [PETR] **1.** The series of layers occurring in a banded structure. **2.** In sedimentary rocks, the thin bedding of alternate layers of different materials. [SCI TECH] A pattern of bands. { 'band·iŋ }

band land [ORD] The raised portion of the rotating band of a projectile after engraving has taken place, produced by the rifling groove of the gun tube. { 'band ‚land }

band lightning *See* ribbon lightning. { 'band ‚līt·niŋ }

band molding *See* band. { 'band ‚mōld·iŋ }

band of position [NAV] An area which extends to either side of a line of position of imperfect accuracy and within which a craft is considered to be located. { 'band əv pə'zish·ən }

bandoleer [ORD] A closed loop of fabric, provided with pockets designed to accommodate small-arms ammunition; used by the individual soldier for carrying ammunition, by suspending one or more bandoleers over the shoulders. { ‚ban·də'lir }

band-pass [ELECTR] A range, in hertz or kilohertz, expressing the difference between the limiting frequencies at which a desired fraction (usually half power) of the maximum output is obtained. { 'band ‚pas }

band-pass amplifier [ELECTR] An amplifier designed to pass a definite band of frequencies with essentially uniform response. { 'band ‚pas ‚am·plə‚fī·ər }

band-pass filter [ELECTR] An electric filter which transmits more or less uniformly in a certain band, outside of which the frequency components are attenuated. [OPTICS] *See* Christiansen filter. { 'band ‚pas ‚fil·tər }

band-pass response [ELECTR] Response characteristics in which a definite band of frequencies is transmitted uniformly. Also known as flat top response. { 'band ‚pas ri'späns }

band-pass system [ENG ACOUS] A loudspeaker system, often used for subwoofers, in which the speaker is mounted inside an enclosure on a shelf that divides the enclosure into two parts, and one or both parts are coupled to the outside by a vent; the frequency response of the system is that of a fourth-order band-pass filter (one vent) or an asymmetrical sixth-order band-pass filter (two vents). { 'band‚pas ‚sis·təm }

band printer [COMPUT SCI] A line printer that uses a band of type characters as its printing mechanism. { 'band ‚print·ər }

band-rejection filter *See* band-stop filter. { 'band ri'jek·shən ‚fil·tər }

band saw [MECH ENG] A power-operated woodworking saw consisting basically of a flexible band of steel having teeth on one edge, running over two vertical pulleys, and operated under tension. { 'band ‚sȯ }

band scheme [SOLID STATE] The identification of energy bands of a solid with the levels of independent atoms from which they arise as the atoms are brought together to form the solid, together with the width and spacing of the bands. { 'band ‚skēm }

band selector [ELECTR] A switch that selects any of the bands in which a receiver, signal generator, or transmitter is designed to operate and usually has two or more sections to make the required changes in all tuning circuits simultaneously. Also known as band switch. { 'band sə'lek·tər }

B and S gage *See* American wire gage. { ‚bē ən ‚ēs ‚gāj }

band sound-pressure level [ACOUS] The sound-pressure level that results from the portion of sound within a specified frequency band. Abbreviated BSPL. { ‚band 'saund ‚presh·ər ‚lev·əl }

band spectrum [SPECT] A spectrum consisting of groups or bands of closely spaced lines in emission or absorption, characteristic of molecular gases and chemical compounds. Also known as band. { 'band ‚spek·trəm }

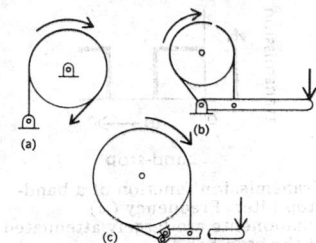

BAND BRAKE

Band brakes. *(a)* Basic structure. *(b)* With direct-action lever. *(c)* Differential self-assisting feature.

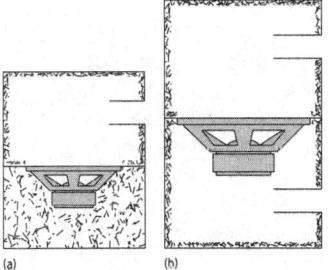

BAND-PASS SYSTEM

Side views of *(a)* fourth-order system and *(b)* sixth-order asymmetrical system.

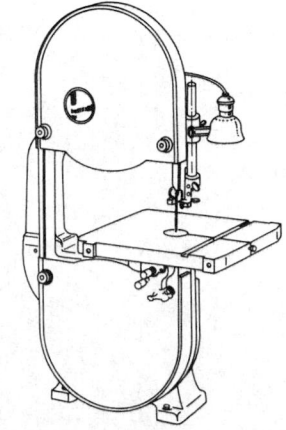

BAND SAW

The narrow band saw, a flexible band of steel, can make curved as well as straight cuts even in thick pieces. *(Delta)*

BAND-STOP FILTER

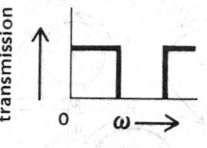

band-stop

Transmission function of a band-stop filter. Frequency (ω) components are largely attenuated at the stop band.

band spreading [COMMUN] Method of double-sideband transmission in which the frequency band of the modulating wave is shifted upward in frequency so that the sidebands produced by modulation are separated in frequency from the carrier by an amount at least equal to the bandwidth of the original modulating wave, and second order distortion products may be filtered from the demodulator output. { 'band ,spred·iŋ }

band-spread tuning control [ELECTR] A tuning control provided on some shortwave receivers to spread the stations in a single band of frequencies over an entire tuning dial. { 'band ,spred 'tün·iŋ kən'trōl }

band-stop filter [ELECTR] An electric filter which transmits more or less uniformly at all frequencies of interest except for a band within which frequency components are largely attenuated. Also known as band-elimination filter; band-rejection filter. { 'band ,stäp,fil·tər }

band switch See band selector. { 'band ,swich }

band theory of ferromagnetism [SOLID STATE] A theory according to which ferromagnetism is caused by electrons in the unfilled energy bands of a crystal. { ¦band ,thē·ə·rē əv ,fer·ō'mag·nə,tiz·əm }

band theory of solids [SOLID STATE] A quantum-mechanical theory of the motion of electrons in solids that predicts certain restricted ranges or bands for the energies of these electrons. Also known as energy-band theory of solids. { 'band ,thē·ə·rē əv ¦säl·ədz }

band wheel [MECH ENG] In a drilling operation, a large wheel that transmits power from the engine to the walking beam. { 'band ,wēl }

bandwidth [COMMUN] **1.** The difference between the frequency limits of a band containing the useful frequency components of a signal. **2.** A measure of the amount of data that can travel a communications path in a given time, usually expressed as thousands of bits per second (kbps) or millions of bits per second (Mbps). { 'band,width }

bandylite [MINERAL] CuB₂O₄·CuCl₂·4H₂O A tetragonal mineral that is deep blue with greenish lights and consists of a hydrated copper borate-chloride. { 'ban-də,līt }

Bangalore torpedo [ORD] A metal tube or pipe packed with a high-explosive charge; chiefly used to clear a path through barbed wire or minefields. { 'baŋ·gə,lòr tòr'pēd·ō }

bang-bang circuit [ELECTR] An operational amplifier with double feedback limiters that drive a high-speed relay (1–2 milliseconds) in an analog computer; involved in signal-controlled programming. { ¦baŋ ¦baŋ ,sər·kət }

bang-bang control [COMPUT SCI] Control of programming in an analog computer through a bang-bang circuit. [CONT SYS] A type of automatic control system in which the applied control signals assume either their maximum or minimum values. { ¦baŋ ¦baŋ kən'trōl }

bang-bang-off control See bang-zero-bang control. { ¦baŋ ¦baŋ 'òf kən,trōl }

bang-bang robot [CONT SYS] A simple robot that can make only two types of motions. { ¦baŋ ¦baŋ 'rō,bät }

Bangiophyceae [BOT] A class of red algae in the plant division Rhodophyta. { ,baŋ·ē·ə'fīs·ē,ē,ē }

Bang's disease See contagious abortion. { 'baŋz diz'ēz }

bang-zero-bang control [CONT SYS] A type of control in which the control values are at their maximum, zero, or minimum. Also known as bang-bang-off control. { ¦baŋ ,zir·ō 'baŋ kən,trōl }

banister [BUILD] A handrail for a staircase. { 'ban·ə·stər }

bank [AERO ENG] The lateral inward inclination of an airplane when it rounds a curve. [CIV ENG] See embankment. [ELEC] **1.** A number of similar electrical devices, such as resistors, connected together for use as a single device. **2.** An assemblage of fixed contacts over which one or more wipers or brushes move in order to establish electrical connections in automatic switching. [ENG] A pipework installation in which the pipes are set parallel to each other in proximity. [GEOL] **1.** The edge of a waterway. **2.** The rising ground bordering a body of water. **3.** A steep slope or face, generally consisting of unconsolidated material. [IND ENG] The amount of material allowed to accumulate at a point on a production line where it is not employed or worked upon, to permit reasonable fluctuations in line speed before and after the point. Also known as float. [MIN ENG] **1.** The top of the shaft. **2.** The surface around the mouth of a shaft. **3.**

The whole, or sometimes only one side or one end, of a working place underground. **4.** To manipulate materials such as coal, gravel, or sand on a bank. **5.** A terracelike bench in open-pit mining. [OCEANOGR] A relatively flat-topped raised portion of the sea floor occurring at shallow depth and characteristically on the continental shelf or near an island. { baŋk }

bank-and-turn indicator [AERO ENG] A device used to advise the pilot that the aircraft is turning at a certain rate, and that the wings are properly banked to preclude slipping or sliding of the aircraft as it continues in flight. Also known as bank indicator. { ¦baŋk ən 'tərn 'in·də,kād·ər }

bank-and-wiper switch [ELEC] Switch in which electromagnetic ratchets or other mechanisms are used, first, to move the wipers to a desired group of terminals, and second, to move the wipers over the terminals of the group to the desired bank contacts. { ¦baŋk ən 'wī·pər ,swich }

bank cushion [NAV] In nautical navigation, a force acting on the bow of a ship in a manner which forces the ship away from the bank in a restricted channel, especially where the banks are steep; it is a force which opposes bank suction. { ¦baŋk ,kùsh·ən }

bank deposit [GEOL] Mounds, ridges, and terraces of sediment rising above and about the surrounding sea bottom. { ¦baŋk di'päz·ət }

banked winding [ELECTR] A radio-frequency coil winding which proceeds from one end of the coil to the other without return by having, side by side, many flat spirals formed by winding single turns one over the other, thereby reducing the distributed capacitance of the coil. { 'baŋkt 'wīnd·iŋ }

banker [ENG] The bench or table upon which bricklayers and stonemasons prepare and shape their material. { 'baŋ·kər }

banket [GEOL] A conglomerate containing valuable metal to be exploited. { baŋ'ket }

bankfull stage [HYD] The flow stage of a river in which the stream completely fills its channel and the elevation of the water surface coincides with the bank margins. { ¦baŋk ¦fùl ,stāj }

bank gravel See bank-run gravel. { 'baŋk ,grav·əl }

bank height [MIN ENG] The vertical height of a bank as measured between its highest point or crest and its toe at the digging level or bench. Also known as bench height; digging height. { 'baŋk ,hīt }

bank indicator See bank-and-turn indicator. { 'baŋk 'in·də,kād·ər }

banking pin [HOROL] One of the erect pins in the bottom plate of a watch that restrict the movement of the lever. { 'baŋk·iŋ ,pin }

bank-inset reef [GEOL] A coral reef situated on island or continental shelves well inside the outer edges. { 'baŋk 'in,set ,rēf }

bank material [CIV ENG] Soil or rock in place before excavation or blasting. { 'baŋk mə'tir·ē·əl }

bank measure [CIV ENG] The volume of a given portion of soil or rock as measured in its original position before excavation. { 'baŋk ,mezh·ər }

bank reef [GEOL] A reef which rises at a distance back from the outer margin of rimless shoals. { 'baŋk ,rēf }

bank-run gravel [GEOL] A natural deposit comprising gravel or sand. [MATER] Aggregate taken directly from natural deposits; contains both large and small stones. Also known as bank gravel; run-of-bank gravel. { 'baŋk ,rən 'grav·əl }

bank sand [GEOL] Deposits occurring in banks or pits and containing a low percentage of clay; used in core making. { 'baŋk ,sand }

bank select [COMPUT SCI] To activate and deactivate blocks of memory or other internal system components using electronic control signals. Also known as bank switch. { 'baŋk si,lekt }

bank selected memory [COMPUT SCI] Auxiliary blocks of memory in a microcomputer that can be switched in to replace some or all of the internal memory by software-controlled switches located outside the microprocessor. { 'baŋk si,lek·təd 'mem·rē }

bank slope [MIN ENG] The angle, measured in degrees of deviation from the horizontal, at which the earthy or rock material will stand in an excavated, terracelike cut in an open-pit mine or quarry. Also known as bench slope. { 'baŋk ,slōp }

banksman *See* lander. { 'baŋks·mən }

banks oil *See* cod-liver oil. { 'baŋks ‚ȯil }

bank storage [HYD] Water absorbed in the permeable bed and banks of a lake, reservoir, or stream. { 'baŋk ‚stȯr·ij }

bank suction [NAV] The bodily movement of a ship toward the near bank due to a decrease in pressure as a result of increased velocity of flow of water past the hull in a restricted channel. { 'baŋk ‚sək·shən }

bank switch *See* bank select. { 'baŋk ‚swich }

banner [BOT] The fifth or posterior petal of a butterfly-shaped (papilionaceous) flower. { 'ban·ər }

banner cloud [METEOROL] A cloud plume often observed to extend downwind from isolated mountain peaks, even on otherwise cloud-free days. Also known as cloud banner. { 'ban·ər ‚klau̇d }

bantam tube [ELECTR] Vacuum tube having a standard octal base, but a considerably smaller glass tube than a standard glass tube. { 'ban·təm ‚tüb }

Banti's disease [MED] Portal hypertension, congestive splenomegaly, and hypersplenism due to an obstructive lesion in the splenic vein, portal vein, or intrahepatic veins. { 'bän·tēz di‚zēz }

bar [GEOL] **1.** Any of the various submerged or partially submerged ridges, banks, or mounds of sand, gravel, or other unconsolidated sediment built up by waves or currents within stream channels, at estuary mouths, and along coasts. **2.** Any band of hard rock, for example, a vein or dike, that extends across a lode. [MECH] A unit of pressure equal to 10^5 pascals, or 10^5 newtons per square meter, or 10^6 dynes per square centimeter. [MET] An elongated piece of metal of simple uniform cross-section dimensions, usually rectangular, circular, or hexagonal, produced by forging or hot rolling. Also known as barstock. [MIN ENG] *See* bar drill. { bär }

BAR *See* Browning automatic rifle.

baraboo [GEOL] A monadnock buried by a series of strata and then reexposed by the partial erosion of these younger strata. { 'bär·ə‚bü }

Bárány chair [ENG] A chair in which a person is revolved to test his susceptibility to vertigo. { bə'rän·ē ‚cher }

bararite [MINERAL] $(NH_4)_2SiF_6$ A white, hexagonal mineral consisting of ammonium silicon fluoride; occurs in tabular, arborescent, and mammillary forms. { bə'rä‚rīt }

barat [METEOROL] A heavy northwest squall in Manado Bay on the north coast of the island of Celebes, prevalent from December to February. { bə'rät }

barathea [TEXT] Material, such as silk, cotton, or rayon, made with a pebble weave. { ‚bar·ə'thē·ə }

barb [METEOROL] A means of representing wind speed in the plotting of a synoptic chart, being a short, straight line drawn obliquely toward lower pressure from the end of a wind-direction shaft. Also known as feather. [VERT ZOO] A side branch on the shaft of a bird's feather. { bärb }

Barbados earth [GEOL] A deposit of fossil radiolarians. { bär'bā·dəs ‚ərth }

barban [ORG CHEM] $C_{11}H_9O_2NCl_2$ A white, crystalline compound with a melting point of 75–76°C; used as a postemergence herbicide of wild oats in barley, flax, lentil, mustard, and peas. { 'bär‚ban }

barb bolt [DES ENG] A bolt having jagged edges to prevent its being withdrawn from the object into which it is driven. Also known as rag bolt. { 'bärb ‚bōlt }

bar beach [GEOL] A straight beach of offshore bars that are separated by shallow bodies of water from the mainland. { 'bär ‚bēch }

barbed tributary [HYD] A tributary that enters the main stream in an upstream direction instead of pointing downstream. { 'bärbd 'trib·yə‚ter·ē }

barbed wire [MATER] Two or more wires twisted together with addition of sharp hooks or points (or a single wire furnished with barbs); used for fences. Also known as barbwire. { 'bärb ‚dwī·ər }

barbel [VERT ZOO] **1.** A slender, tactile process near the mouth in certain fishes, such as catfishes. **2.** Any European fresh-water fish in the genus *Barbus*. { 'bär·bəl }

barbellate [BIOL] Having short, stiff, hooked bristles. { 'bär·bə‚lāt }

bar bending [CIV ENG] In reinforced concrete construction, the process of bending reinforcing bars to various shapes. { 'bär ‚ben·diŋ }

barber [METEOROL] A severe storm at sea during which spray and precipitation freeze onto the decks and rigging of ships. { 'bär·bər }

barberite [MET] A nonferrous alloy with good resistance to sulfuric acid, sea water, and mine waters; 88.5% copper, 5% nickel, 5% tin, 1.5% silicon. { 'bär·bə‚rīt }

barbertonite [MINERAL] $Mg_6Cr_2(OH)_{16}CO_3·4H_2O$ A lilac to rose pink, hexagonal mineral consisting of a hydrated carbonate-hydroxide of magnesium and chromium; occurs in massive form or in masses of fibers or plates. { 'bär·bər·tə‚nīt }

barbette [NAV ARCH] A nonrotating cylinder of armor that protects the rotating part of the turret of a warship below the gunhouse. [ORD] A mound of earth or a platform upon which guns are mounted to fire over a wall or parapet. { bär'bet }

barbicel [VERT ZOO] One of the small, hook-bearing processes on a barbule of the distal side of a barb or a feather. { 'bär·bə‚sel }

barbierite [MINERAL] $NaAlSi_3O_8$ A hypothetical soda feldspar thought to be isomorphous with orthoclase. { bar'bi‚rīt }

barbital [ORG CHEM] $C_8H_{12}N_2O_3$ A compound crystallizing in needlelike form from water; has a faintly bitter taste; melting point 188–192°C; used to make sodium barbital, a long-duration hypnotic and sedative. { 'bär·bə‚tȯl }

barbiturate [PHARM] Any of a group of ureides, such as phenobarbital, Amytal, or Seconal, that act as central nervous system depressants. { 'bär'bich·ə·rət }

barbituric acid [ORG CHEM] $C_4H_4O_3N_2$ 2,4,6-Trioxypyrimidine, the parent compound of the barbiturates; colorless crystals melting at 245°C, slightly soluble in water. { ‚bär·bə'tu̇r·ik 'as·əd }

barbiturism [MED] Intoxication following an overdose of barbiturates; characterized by delirium, coma, and sometimes death. { bär'bich·ə‚riz·əm }

bar buoy [NAV] A buoy marking the location of a bar at the mouth of a river or approach to a harbor. { 'bär ‚bȯi }

barbwire *See* barbed wire. { 'bärb'wī·ər }

bar chair *See* bar support. { 'bär ‚cher }

barchan [GEOL] A crescent-shaped dune or drift of wind-blown sand or snow, the arms of which point downwind; formed by winds of almost constant direction and of moderate speeds. Also known as barchane; barkhan; crescentic dune. { bär'kän }

barchane *See* barchan. { bär'kän }

bar chart *See* bar graph. { 'bär ‚chärt }

bar clamp [DES ENG] A clamping device consisting of a long bar with adjustable clamping jaws; used in carpentry. { 'bär ‚klamp }

bar code [COMPUT SCI] The representation of alphanumeric characters by series of adjacent stripes of various widths, for example, the universal product code. { 'bär ‚kōd }

bar-code reader *See* bar-code scanner. { 'bär ‚kōd 'rēd·ər }

bar-code scanner [COMPUT SCI] An optical scanning device that reads texts which have been converted into a special bar code. Also known as bar-code reader. { 'bär ‚kōd 'skan·ər }

Bardeen-Cooper-Schrieffer theory [SOLID STATE] A theory of superconductivity that describes quantum-mechanically those states of the system in which conduction electrons cooperate in their motion so as to reduce the total energy appreciably below that of other states by exploiting their effective mutual attraction; these states predominate in a superconducting material. Abbreviated BCS theory. { ‚bär‚dēn ‚kü·pər ‚shrē·fər ‚thē·ə·rē }

bar drawing [MET] An operation in which a metallic bar is pulled through a die so that the cross-sectional area of the bar is reduced. { 'bär ‚drȯ·iŋ }

bar drill [MIN ENG] A small diamond type or other type of rock drill mounted on a bar and used in an underground workplace. Also known as bar. { 'bär ‚dril }

bare board [ELECTR] A printed circuit board with conductors but no electronic components. { 'ber ‚bȯrd }

bareboat charter [IND ENG] An agreement to charter a ship without its crew or stores; the fee for its use for a predetermined period of time is based on the price per ton of cargo handled. { 'ber‚bōt ‚chärd·ər }

BAR DRAWING

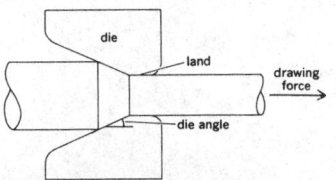

A cross section of a drawing die in operation showing the reduction in cross-sectional area of the bar being pulled through the die.

bare charge [ORD] An explosive charge without casing prepared for use in determining explosive blast characteristics. { ¦ber ¦chärj }

bare charm [PARTIC PHYS] Charm that is carried by a quark and is not canceled by the charm of the corresponding antiquark, so that the hadron of which the quark is a constituent has net charm different from zero. { ¦ber ¦chärm }

bare disk [ELECTR] A floppy-disk drive without electronic control circuits. { ¦ber ¦disk }

bare electrode [MET] An uncoated electrode used in submerged arc automatic welding with a gas-shielded arc or a granular flux deposited in an elongated mound over a joint. { ¦ber i'lek₁trōd }

barefaced tenon [ENG] A tenon having a shoulder cut on one side only. { ¦ber₁fās ¦ten·ən }

bareface fabric [TEXT] Any perfectly smooth fabric that has no nap. { ¦ber₁fās ¦fab·rik }

barefoot completion See open-hole completion. { ¦ber₁fût kəm'plē·shən }

baren [GRAPHICS] A pad for pressing paper against an inked wooden block to produce a print. { bä'ren }

bare rock [NAV] In U.S. Coast and Geodetic Survey terminology, a rock extending above the plane of mean high water. { ¦ber ¦räk }

bare tube [ENG] In a heat exchanger, a tube whose inner and outer surfaces are both smooth. { ¦ber 'tüb }

bare value [QUANT MECH] The value which some physical property of a particle, such as its mass or charge, is supposed to have in the absence of any interactions with fields. { ¦ber 'val·yü }

bar finger sand [GEOL] An elongated lenticular sand body that lies beneath a distributory in a birdfoot delta. { 'bär ¦fiŋ·gər ₁sand }

Barfoed's test [ANALY CHEM] A test for monosaccharides conducted in an acid solution; cupric acetate is reduced to cuprous oxide, a red precipitate. { 'bär·fûts ₁test }

bar folder [MET] A machine used to bend a metal sheet into a sharp, narrow, and accurate fold, or a rounded fold, along the edge. { 'bär ₁fōld·ər }

barge [NAV ARCH] A large cargo-carrying craft which is towed or pushed by a tug on both seagoing and inland waters. { bärj }

bargeboard See vergeboard. { 'bärj₁bord }

barge couple [BUILD] **1.** One of two rafters that support that part of a gable roof which projects beyond the gable wall. **2.** One of the rafters (under the barge course) which serve as grounds for the vergeboards and carry the plastering or boarding of the soffits. Also known as barge rafter. { 'bärj ₁kəp·əl }

barge course [BUILD] **1.** The coping of a wall, formed by a course of bricks set on edge. **2.** In a tiled roof, the part of the tiling which projects beyond the principal rafters where there is a gable. { 'bärj ₁kors }

bar generator [ELECTR] Generator of pulses or repeating waves which are equally separated in time; these pulses are synchronized by the synchronizing pulses of a television system, so that they can produce a stationary bar pattern on a television screen. { 'bär ¦jen·ə₁rād·er }

barge rafter See barge couple. { 'bärj ₁raf·tər }

barge spike See boat spike. { 'bärj ₁spīk }

barge stone [BUILD] One of the stones, generally projecting, which form the sloping top of a gable built of masonry. { 'bärj ₁stōn }

bar graph [STAT] A diagram of frequency-table data in which a rectangle with height proportional to the frequency is located at each value of a variate that takes only certain discrete values. Also known as bar chart; rectangular graph. { 'bär ₁graf }

bar hole [ENG] A small-diameter hole made in the ground along the route of a gas pipe in a bar test survey. { 'bär ₁hōl }

bariatrics [MED] A branch of medicine that deals with the treatment of obesity. { ₁ba·rē'a₁triks }

baric topography See height pattern. { 'bar·ik tə'päg·rə·fē }

baric wind law See Buys-Ballot's law. { 'bar·ik ¦wind ₁lo }

barines [METEOROL] Westerly winds in eastern Venezuela. { ba'rēnz }

baring See overburden. { 'ba·riŋ }

bar iron [MET] Wrought iron formed into bars. { 'bär ₁i·ərn }

Bari-Sol process [CHEM ENG] Removal of waxes from liquid hydrocarbons by extraction of the wax with a mixed ethylene dichloride-benzene solvent, followed by separation from the hydrocarbon in a centrifuge. { ¦bär·ē ¦säl ₁präs·əs }

barite [MINERAL] $BaSO_4$ A white, yellow, or colorless orthorhombic mineral occurring in tabular crystals, granules, or compact masses; specific gravity is 4.5; used in paints and drilling muds and as a source of barium chemicals; the principal ore of barium. Also known as baryte; barytine; cawk; heavy spar. { 'ba₁rīt }

barite dollar [MINERAL] Barite in the form of rounded disk-shaped masses; formed in a sandstone or sandy shale. { 'ba₁rīt ₁däl·ər }

BARITT diode See barrier injection transit-time diode. { 'bar·ət ¦dī₁ōd }

barium [CHEM] A chemical element, symbol Ba, with atomic number 56 and atomic weight of 137.34. { 'bar·ē·əm }

barium-140 [NUC PHYS] A radioactive isotope of barium with atomic mass 140; the half-life is 12.8 days, and the decay is by negative beta-particle emission. { 'bar·ē·əm ₁wən'förd·ē }

barium acetate [INORG CHEM] $Ba(C_2H_3O_2)_2 \cdot H_2O$ A barium salt made by treating barium sulfide or barium carbonate with acetic acids; it forms colorless, triclinic crystals that decompose upon heating; used as a reagent for sulfates and chromates. { 'bar·ē·əm 'as·ə₁tāt }

barium azide [INORG CHEM] $Ba(N_3)_2$ A crystalline compound soluble in water; used in high explosives. { 'bar·ē·əm 'ā₁zīd }

barium-base grease [MATER] A lubricating material made from lubricating oil and barium soap. { 'bar·ē·əm ₁bās 'grēs }

barium binoxide See barium peroxide. { 'bar·ē·əm bī'näk₁sīd }

barium bromate [INORG CHEM] $Ba(BrO_3)_2 \cdot H_2O$ A poisonous compound that forms colorless, monoclinic crystals, decomposing at 260°C; used for preparing other bromates. { 'bar·ē·əm 'brō₁māt }

barium bromide [INORG CHEM] $BaBr_2 \cdot 2H_2O$ Colorless crystals soluble in water and alcohol; used in photographic compounds. { 'bar·ē·əm 'brō₁mīd }

barium carbonate [INORG CHEM] $BaCO_3$ A white powder with a melting point of 174°C; soluble in acids (except sulfuric acid); used in rodenticides, ceramic flux, optical glass, and television picture tubes. { 'bar·ē·əm 'kär·bə₁nət }

barium chlorate [INORG CHEM] $Ba(ClO_3)_2 \cdot H_2O$ A salt prepared by the reaction of barium chloride and sodium chlorate; it forms colorless, monoclinic crystals, soluble in water; used in pyrotechnics. { 'bar·ē·əm 'klor₁āt }

barium chloride [INORG CHEM] $BaCl_2$ A toxic salt obtained as colorless, water-soluble cubic crystals, melting at 963°C; used as a rat poison, in metal surface treatment, and as a laboratory reagent. { 'bar·ē·əm 'klor₁īd }

barium chromate [INORG CHEM] $BaCrO_4$ A toxic salt that forms yellow, rhombic crystals, insoluble in water; used as a pigment in overglazes. { 'bar·ē·əm 'krō₁māt }

barium citrate [ORG CHEM] $Ba_3(C_6H_5O_7)_2 \cdot 2H_2O$ A grayish-white, toxic, crystalline powder; used as a stabilizer for latex paints. { 'bar·ē·əm 'sī₁trāt }

barium cyanide [ORG CHEM] $Ba(CN)_2$ A white, crystalline powder; soluble in water and alcohol; used in metallurgy and electroplating. { 'bar·ē·əm 'sī·ə₁nīd }

barium dioxide See barium peroxide. { 'bar·ē·əm dī'äk₁sīd }

barium enema [MED] A suspension of barium sulfate administered as an enema into the lower bowel to render it radiopaque. { 'bar·ē·əm 'en·ə·mə }

barium fluoride [INORG CHEM] BaF_2 Colorless, cubic crystals, slightly soluble in water; used in enamels. { 'bar·ē·əm flùr₁īd }

barium fluosilicate [INORG CHEM] $BaSiF_6H$ A white, crystalline powder; insoluble in water; used in ceramics and insecticides. Also known as barium silicofluoride. { 'bar·ē·əm ₁flü·ə'sil·ə₁kāt }

barium fuel cell [ELEC] A fuel cell in which barium is used with either oxygen or chlorine to convert chemical energy into electrical energy. { 'bar·ē·əm 'fyül ₁sel }

barium glass [MATER] Glass which differs from ordinary lime-soda glass in that barium oxide replaces part of the calcium oxide. { 'bar·ē·əm 'glas }

barium hydroxide [INORG CHEM] $Ba(OH)_2 \cdot 8H_2O$ Colorless, monoclinic crystals, melting at 78°C; soluble in water, insoluble in acetone; used for fat saponification and fusing of silicates. { 'bar·ē·əm hī'dräk,sīd }

barium hypophosphite [PHARM] $BaH_4(PO_2)_2$ A toxic, white, crystalline powder; soluble in water; used in medicine. { 'bar·ē·əm ,hī·pō'fäs,fāt }

barium hyposulfite See barium thiosulfate. { 'bar·ē·əm ,hī·pō'səl,fīt }

barium manganate [INORG CHEM] $BaMnO_4$ A toxic, emerald-green powder which is used as a paint pigment. Also known as Cassel green; manganese green. { 'bar·ē·əm 'maŋ·gə,nāt }

barium meal [MED] A suspension of barium sulfate taken orally to render the upper gastrointestinal tract radiopaque. { 'bar·ē·əm ,mēl }

barium mercury iodide See mercuric barium iodide. { 'bar·ē·əm 'mər·kyə·rē 'ī·ə,dīd }

barium molybdate [INORG CHEM] $BaMoO_4$ A toxic, white powder with a melting point of approximately $1600°C$; used in electronic and optical equipment and as a paint pigment. { 'bar·ē·əm mə'lib,dāt }

barium monosulfide [INORG CHEM] BaS A colorless, cubic crystal that is soluble in water; used in pigments. { 'bar·ē·əm ,män·ō'səl,fīd }

barium monoxide See barium oxide. { 'bar·ē·əm mə'näk,sīd }

barium nitrate [INORG CHEM] $Ba(NO_3)_2$ A toxic salt occurring as colorless, cubic crystals, melting at $592°C$, and soluble in water; used as a reagent, in explosives, and in pyrotechnics. Also known as nitrobarite. { 'bar·ē·əm 'nī,trāt }

barium oxide [INORG CHEM] BaO A white to yellow powder that melts at $1923°C$; it forms the hydroxide with water; may be used as a dehydrating agent. Also known as barium monoxide; barium protoxide. { 'bar·ē·əm 'äk,sīd }

barium perchlorate [INORG CHEM] $Ba(ClO_4)_2·4H_2O$ Tetrahydrate variety which forms colorless hexagons; used in pyrotechnics. { 'bar·ē·əm pər'klōr,āt }

barium permanganate [INORG CHEM] $Ba(MnO_4)_2$ Brownish-violet, toxic crystals; soluble in water; used as a disinfectant. { 'bar·ē·əm pər'maŋ·gə,nāt }

barium peroxide [INORG CHEM] BaO_2 A compound formed as white toxic powder, insoluble in water; used as a bleach and in the glass industry. Also known as barium binoxide; barium dioxide; barium superoxide. { 'bar·ē·əm pər'äk,sīd }

barium plaster [MATER] A special mill-mixed gypsum plaster containing barium salts; used to plaster walls of x-ray rooms. { 'bar·ē·əm ,plas·tər }

barium protoxide See barium oxide. { 'bar·ē·əm prō'täk,sīd }

barium silicide [INORG CHEM] $BaSi_2$ A compound that has the appearance of metal-gray lumps; melts at white heat; used in metallurgy to deoxidize steel. { 'bar·ē·əm 'sil·ə,sīd }

barium-sodium niobate [MATER] Synthetic electrooptical crystals used to produce coherent green light in lasers and to manufacture devices such as electrooptical modulators and optical polarimetric oscillators. { 'bar·ē·əm 'sōd·ē·əm 'nī·ə,bāt }

barium star [ASTRON] A peculiar, low-velocity, strong-lined red giant or subgiant star of spectral type G, K, or M, whose spectrum has anomalously strong lines of barium, sometimes with strong bands of methyldadyne (CH), molecular carbon (C_2), and cyanogen radical (CN). { 'bar·ē·əm ,stär }

barium stearate [ORG CHEM] $Ba(C_{18}H_{35}O_2)_2$ A white, crystalline solid; melting point $160°C$; used as a lubricant in manufacturing plastics and rubbers, in greases, and in plastics as a stabilizer against deterioration caused by heat and light. { 'bar·ē·əm 'stir,āt }

barium sulfate [INORG CHEM] $BaSO_4$ A salt occurring in the form of white, rhombic crystals, insoluble in water; used as a white pigment, as an opaque contrast medium for roentgenographic processes, and as an antidiarrheal. { 'bar·ē·əm 'səl,fāt }

barium sulfite [INORG CHEM] $BaSO_3$ A toxic, white powder; soluble in dilute hydrochloric acid; used in paper manufacturing. { 'bar·ē·əm 'səl,fīt }

barium superoxide See barium peroxide. { 'bar·ē·əm ,sü·pər'äk,sīd }

barium tetrasulfide [INORG CHEM] $BaS_4·H_2O$ Red or yellow, rhombic crystals, soluble in water. { 'bar·ē·əm ,te·trə'səl,fīd }

barium thiocyanate [INORG CHEM] $Ba(SCN)·2H_2O$ White crystals that deliquesce; used in dyeing and in photography. { 'bar·ē·əm ,thī·ō'sī·ə,nāt }

barium thiosulfate [INORG CHEM] $BaS_2O_3·H_2O$ A white powder that decomposes upon heating; used to make explosives and in matches. Also known as barium hyposulfite. { 'bar·ē·əm ,thī·ō'səl,fāt }

barium titanate [INORG CHEM] $BaTiO_3$ A grayish powder that is insoluble in water but soluble in concentrated sulfuric acid; used as a ferroelectric ceramic. { 'bar·ē·əm 'tī·tə,nāt }

barium tungstate [INORG CHEM] $BaWO_4$ A toxic, white powder used as a pigment and in x-ray photography. Also known as barium white; barium wolframate; tungstate white; wolfram white. { 'bar·ē·əm 'təŋ,stāt }

barium white See barium tungstate. { 'bar·ē·əm 'wīt }

barium wolframate See barium tungstate. { 'bar·ē·əm 'wùl·frə,māt }

bar joist [BUILD] A small steel truss with wire or rod web lacing used for roof and floor supports. { 'bär ,jöist }

bark [BOT] The tissues external to the cambium in a stem or root. [MET] The decarburized layer formed beneath the scale on the surface of steel heated in air. [NAV ARCH] A three-masted sailing ship whose foremast and mainmast are square-rigged and whose mizzenmast is fore-and-aft-rigged. { bärk }

bark cloth [TEXT] Fabric made from inner tree bark by beating it to a smooth, wearable thinness. { bärk ,klöth }

bark crepe [TEXT] A crepe fabric textured to simulate the appearance of tree bark. { bärk ,krāp }

bar keel [NAV ARCH] A solid keel with a rectangular cross section in an iron or steel ship. { 'bär ,kēl }

barker [DES ENG] See bark spud. [ENG] A machine, used mainly in pulp mills, which removes the bark from logs. [FOR] **1.** A worker who subjects logs and pulpwood to water pressure in a stream barker or tumbling in a drum barker, in order to free them from bark and dirt. Also known as power barker. **2.** A worker who prepares or shovels bark for tanning. { 'bär·kər }

Barker method [CRYSTAL] A method utilizing a number of convenient rules which allow two observers to choose the same reference system to describe the same noncubic crystal. { 'bär·kər ,meth·əd }

barkevikite [MINERAL] A brown or black member of the amphibole mineral group; looks like basaltic hornblende but differs from it in its iron concentration. { 'bär·kə,vi,kīt }

bark graft [BOT] A graft made by slipping the scion beneath a slit in the bark of the stock. { bärk ,graft }

barkhan See barchan. { bär'kän }

Barkhausen criterion [ELECTR] A criterion used to determine the stability of an oscillator circuit which states that, if the circuit is seen as a loop consisting of an amplifier with gain A and a linear circuit whose gain $\beta(j\omega)$ depends on frequency ω, then the loop will oscillate with a perfect sine wave at some frequency ω_0 if at that frequency $A\beta(j\omega_0) = 1$ exactly, that is, if the magnitude of $A\beta(j\omega_0)$ is exactly 1 and its phase is $0°$ or $360°$.

Barkhausen effect [ELECTROMAG] The succession of abrupt changes in magnetization occurring when the magnetizing force acting on a piece of iron or other magnetic material is varied. { 'bärk,haùz·ən in'fekt }

Barkhausen interference [COMMUN] Interference caused by Barkhausen oscillations. { 'bärk,haùz·ən in·tər'fir·əns }

Barkhausen-Kurz oscillator [ELECTR] An oscillator of the retarding-field type in which the frequency of oscillation depends solely on the transit time of electrons oscillating about a highly positive grid before reaching the less positive anode. Also known as Barkhausen oscillator; positive-grid oscillator. { 'bärk,haùz·ən 'kərts 'äs·ə,lād·ər }

Barkhausen oscillation [ELECTR] Undesired oscillation in the horizontal output tube of a television receiver, causing one or more ragged dark vertical lines on the left side of the picture. { 'bärk,haùz·ən ,äs·ə'lā·shən }

Barkhausen oscillator See Barkhausen-Kurz oscillator. { 'bärk,haùz·ən 'äs·ə,lād·ər }

barkometer [CHEM ENG] A hydrometer calibrated to test the strength of tanning liquors used in tanning leather. { bär'käm·əd·ər }

bark pocket [BOT] An opening between tree annual rings which contains bark. { bärk ,päk·ət }

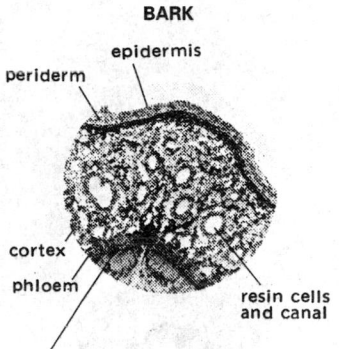

BARK

epidermis
periderm
cortex
phloem
resin cells and canal
cambium

Tranverse action of young twig of balsam fir (*Abies balsamae l.*) showing tissues often considered to compose the bark. (*Forest Products Laboratory, USDA*)

bark-splitting disease See Armillaria root rot. { 'bärk ˌsplit·iŋ di¦zēz }

bark spud [DES ENG] A tool which peels off bark. Also known as barker. { 'bärk ˌspəd }

barley [BOT] A plant of the genus *Hordeum* in the order Cyperales that is cultivated as a grain crop; the seed is used to manufacture malt beverages and as a cereal. { 'bär·lē }

barley coal [MIN ENG] A stream size of anthracite sized on a round punched plate; passes through 0.25-inch (6.4-millimeter) holes. Also known as buckwheat no. 3. { 'bär·lē ˌkōl }

barley scald [PL PATH] A fungus disease of barley caused by *Rhynchosporium secalis* and characterized by bluish-green to yellow blotches and blighting of the foliage. { 'bär·lē ˌskóld }

barley smut [PL PATH] **1.** A loose smut disease of barley caused by *Ustilago nuda*. **2.** A covered smut disease of barley caused by *U. hordei*. { 'bär·lē ˌsmət }

barley stripe [PL PATH] A fungus disease of barley characterized by light green or yellow stripes on the leaves; incited by the diffusible toxin of *Helminthosporium gramineum*. { 'bär·lē ˌstrīp }

barley yellow dwarf virus [VIROL] The type species of the genus *Luteovirus*. { 'bär·lē 'yel·ō ˌdwórf ˌvī·rəs }

barley yellow dwarf virus group See Luteovirus. { 'bär·lē 'yel·ō ˌdwórf ˌvī·rəs ˌgrüp }

bar linkage [MECH ENG] A set of bars joined together at pivots by means of pins or equivalent devices; used to transmit power and information. { 'bär ˌliŋ·kij }

Barlow lens [OPTICS] A lens with one plane surface and one concave surface that is placed between the objective and eyepiece of a telescope to decrease the convergence of the beam from the objective and thereby increase the effective focal length. { 'bär·lō ˌlenz }

Barlow's equation [MECH] A formula, $t = DP/2S$, used in computing the strength of cylinders subject to internal pressures, where t is the thickness of the cylinder in inches, D the outside diameter in inches, P the pressure in pounds per square inch, and S the allowable tensile strength in pounds per square inch. { 'bär·lōz i'kwā·zhən }

Barlow's rule [PHYS CHEM] The rule that the volume occupied by the atoms in a given molecule is proportional to the valences of the atoms, using the lowest valency values. { 'bär·lōz ˌrül }

bar magnet [ELECTROMAG] A bar of hard steel that has been strongly magnetized and holds its magnetism, thereby serving as a permanent magnet. { 'bär ˌmag·nət }

bar mining [MIN ENG] The mining of river bars, usually between low and high waters, although the stream is sometimes deflected and the bar worked below water level. { 'bär ˌmīn·iŋ }

barn [AGR] A farm building used for storage of agricultural products and equipment or for housing farm animals. [NUC PHYS] A unit of area equal to 10^{-24} square centimeter; used in specifying nuclear cross sections. Symbolized b. { 'bärn }

barnacle [ENG] A nodelike deposit that occurs on the surface of a heat exchanger tube or an evaporating device and has a semigranular outer shell bonded to the fouled surface, enclosing a slurry of putrefying organisms. [INV ZOO] The common name for a number of species of crustaceans which compose the subclass Cirripedia. { 'bär·nə·kəl }

Barnard's loop [ASTRON] A large emission nebula, about 10° by 140° in size, around the central portion of Orion, that consists of an expanding shell of gas that probably originated in a supernova. { 'bär·nərdz ˌlüp }

Barnard's star [ASTRON] A star 6.1 light-years away from earth, of visual magnitude 9.5 and proper motion of 10.31 seconds of arc annually. { 'bär·nərdz ˌstär }

Barnett effect [ELECTROMAG] The development of a slight magnetization in an initially unmagnetized iron rod when it is rotated at high speed about its axis. { 'bär·nit 'fekt }

Barnett method [ELECTROMAG] Use of the Barnett effect to determine the gyromagnetic moment of ferromagnetic material. { 'bär·nit ˌmeth·əd }

barney [MIN ENG] A small car or truck, attached to a rope or cable, used to push cars up a slope or an inclined plane. Also known as bullfrog; donkey; groundhog; larry; mule; ram; truck. { 'bär·nē }

baroclinic [PHYS] Of, pertaining to, or characterized by baroclinity. { ¦bar·ə¦klin·ik }

baroclinic disturbance [METEOROL] Any migratory cyclone associated with strong baroclinity of the atmosphere, evidenced on synoptic charts by temperature gradients in the constant-pressure surfaces, vertical wind shear, tilt of pressure troughs with height, and concentration of solenoids in the frontal surface near the ground. Also known as baroclinic wave. { ¦bar·ə¦klin·ik di'stər·bəns }

baroclinic field [METEOROL] A distribution of atmospheric pressure and mass such that the specific volume, or density, of air is a function not solely of pressure. { ¦bar·ə¦klin·ik 'fēld }

baroclinic instability [METEOROL] A hydrodynamic instability arising from the existence of a meridional temperature gradient (and hence of a thermal wind) in an atmosphere in quasi-geostrophic equilibrium and possessing static stability. { ¦bar·ə¦klin·ik in·stə'bil·əd·ē }

baroclinicity See baroclinity. { ˌbar·ə·klə'nis·əd·ē }

baroclinic model [METEOROL] A concept of stratification in the atmosphere, involving surfaces of constant pressure intersecting surfaces of constant density. { ¦bar·ə¦klin·ik 'mäd·əl }

baroclinic wave See baroclinic disturbance. { ¦bar·ə¦klin·ik 'wāv }

baroclinity [PHYS] The state of stratification in a fluid in which surfaces of constant pressure (isobaric surfaces) intersect surfaces of constant density (isosteric surfaces). Also known as baroclinicity; barocliny. { ˌbar·ə'klin·əd·ē }

barocliny See baroclinity. { ˌbar·ə'klin·ē }

baroduric bacteria [MICROBIO] Bacteria that can tolerate conditions of high hydrostatic pressure. { ¦bar·ə¦dùr·ik bak'tir·ē·ə }

barodynamics [MECH] The mechanics of heavy structures which may collapse under their own weight. { ˌbar·ə·dī'nam·iks }

barogram [ENG] The record of an aneroid barograph. { 'bar·əˌgram }

barograph See aneroid barograph. { 'bar·əˌgraf }

barometer [ENG] An absolute pressure gage specifically designed to measure atmospheric pressure. { bə'räm·əd·ər }

barometer elevation [METEOROL] The vertical distance above mean sea level of the ivory point (zero point) of a weather station's mercurial barometer; frequently the same as station elevation. Also known as elevation of ivory point. { bə'räm·əd·ər el·ə'vā·shən }

barometric [ENG] Pertaining to a barometer or to the results obtained by using a barometer. [PHYS] Loosely, pertaining to atmospheric pressure; for example, barometric gradient (meaning pressure gradient). { bar·ə'me·trik }

barometric altimeter See pressure altimeter. { bar·ə'met·rik al'tim·əd·ər }

barometric condenser [MECH ENG] A contact condenser that uses a long, vertical pipe into which the condensate and cooling liquid flow to accomplish their removal by the pressure created at the lower end of the pipe. { bar·ə'met·rik kən'den·sər }

barometric corrections [PHYS] The corrections which must be applied to the reading of a mercury barometer in order that the observed value may be rendered accurate. Also known as barometric errors. { bar·ə'met·rik kə'rek·shənz }

barometric draft regulator [MECH ENG] A damper usually installed in the breeching between a boiler and chimney; permits air to enter the breeching automatically as required, to maintain a constant overfire draft in the combustion chamber. { bar·ə'met·rik 'draft reg·yə'lād·ər }

barometric elevation [ENG] An elevation above mean sea level estimated from the difference in atmospheric pressure between the point in question and an elevation of known value. { bar·ə'met·rik el·ə'vā·shən }

barometric errors See barometric corrections. { bar·ə'met·rik 'er·ərz }

barometric fuel control [AERO ENG] A device that maintains the correct flow of fuel to an engine by adjusting to atmospheric pressure at different altitudes, as well as to impact pressure. { bar·ə'met·rik 'fyül kən'trōl }

barometric fuse [ENG] A fuse that functions as a result of change in the pressure exerted by the surrounding air. { bar·ə'met·rik 'fyüz }

barometric gradient See pressure gradient. { bar·ə'met·rik 'grād·ē·ənt }

BARNACLE

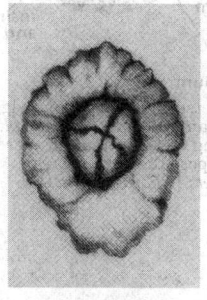

Top view of adult stage of *Balanus*, the acorn barnacle.

BAROMETRIC CONDENSER

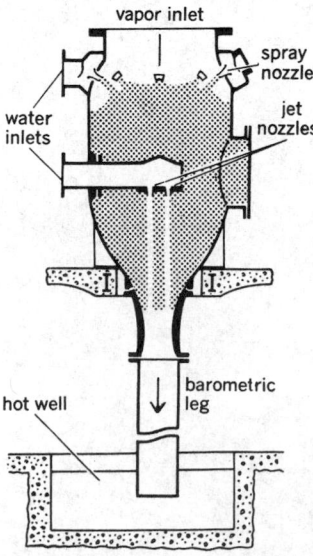

A multijet barometric condenser. (*Schutte and Koerting Co.*)

barometric hypsometry [ENG] The determination of elevations by means of either mercurial or aneroid barometers. { bar·ə'met·rik hip'säm·ə·trē }

barometric leveling [ENG] The measurement of approximate elevation differences in surveying with the aid of a barometer; used especially for large areas. { bar·ə'met·rik 'lev·əl·iŋ }

barometric pressure See atmospheric pressure. { bar·ə'met·rik 'presh·ər }

barometric surface [PHYS] A surface at each point of which the barometric pressure is the same. { bar·ə'met·rik 'sər·fəs }

barometric switch See baroswitch. { bar·ə'met·rik 'swich }

barometric tendency See pressure tendency. { bar·ə'met·rik 'ten·dən·sē }

barometric tide [GEOPHYS] A daily variation in atmospheric pressure due to the gravitational attraction of the sun and moon. { bar·ə'met·rik 'tīd }

barometric wave [METEOROL] Any wave in the atmospheric pressure field; the term is usually reserved for short-period variations not associated with cyclonic-scale motions or with atmospheric tides. { bar·ə'met·rik 'wāv }

barometrograph See aneroid barograph. { bar·ə'me·trə ,graf }

barometry [ENG] The study of the measurement of atmospheric pressure, with particular reference to ascertaining and correcting the errors of the different types of barometer. { bə'räm·ə·trē }

baromil [MECH] The unit of length used in graduating a mercury barometer in the centimeter-gram-second system. { 'bar·ə,mil }

barophile [MICROBIO] An organism that thrives under conditions of high hydrostatic pressure. { 'bar·ə,fīl }

barophobia [PSYCH] Abnormal fear of gravity. { 'bar·ə'fō·bē·ə }

baroscope [ENG] An apparatus which demonstrates the equality of the weight of air displaced by an object and its loss of weight in air. { 'bar·ə,skōp }

barosinusitis [MED] Inflammation of the sinuses, characterized by edema and hemorrhage, due to expansion of air within the sinuses at decreased barometric pressure. Also known as aerosinusitis. { ¦bar·ō,sī·nə'sīd·əs }

barostat [ENG] A mechanism which maintains constant pressure inside a chamber. { 'bar·ə,stat }

baroswitch [ENG] **1.** A pressure-operated switching device used in a radiosonde which determines whether temperature, humidity, or reference signals will be transmitted. **2.** Any switch operated by a change in barometric pressure. Also known as barometric switch. { 'bar·ə,swich }

barotaxis [BIOL] Orientation movement of an organism in response to pressure changes. { ¦bar·ə¦tak·səs }

barothermogram [ENG] The record made by a barothermograph. { ¦bar·ō'thər·mə,gram }

barothermograph [ENG] An instrument which automatically records pressure and temperature. { ¦bar·ō'thər·mə,graf }

barothermohygrogram [ENG] The record made by a barothermohygrograph. { ¦bar·ō¦thər·mō'hī·grə ,gram }

barothermohygrograph [ENG] An instrument that produces graphs of atmospheric pressure, temperature, and humidity on a single sheet of paper. { ¦bar·ō¦thər·mō'hī·grə,graf }

barotitis [MED] Inflammation of the ear, or a part of it, caused by changes in atmospheric pressure. Also known as aerootitis. { ,bar·ə'tīd·əs }

barotrauma [MED] Injury to air-containing structures, such as the middle ears, sinuses, lungs, and gastrointestinal tract, due to unequal pressure differences across their walls. { ,bar·ə'trau·mə }

barotropic [PHYS] Of, pertaining to, or characterized by a condition of barotropy. { ,bar·ə'träp·ik }

barotropic disturbance [METEOROL] Also known as barotropic wave. **1.** A wave disturbance in a two-dimensional nondivergent flow; the driving mechanism lies in the variation of either vorticity of the basic current or the variation of the vorticity of the earth about the local vertical. **2.** An atmospheric wave of cyclonic scale in which troughs and ridges are approximately vertical. { ,bar·ə'träp·ik dis'tər·bəns }

barotropic field [METEOROL] A distribution of atmospheric pressures and mass such that the specific volume, or density, of air is a function solely of pressure. { ,bar·ə'träp·ik 'fēld }

barotropic model [METEOROL] Any of a number of model atmospheres in which some of the following conditions exist throughout the motion: coincidence of pressure and temperature surfaces, absence of vertical wind shear, absence of vertical motions, absence of horizontal velocity divergence, and conservation of the vertical component of absolute vorticity. { ,bar·ə'träp·ik 'mäd·əl }

barotropic phenomenon [THERMO] The sinking of a vapor beneath the surface of a liquid when the vapor phase has the greater density. { ,bar·ə'träp·ik fə'näm·ə,nän }

barotropic wave See barotropic disturbance. { ,bar·ə'träp·ik ,wāv }

barotropy [PHYS] The state of a fluid in which surfaces of constant density (or temperature) are coincident with surfaces of constant pressure; it is the state of zero baroclinity. { bə'rä·trə·pē }

bar pattern [ELECTR] Pattern of repeating lines or bars on a television screen. { 'bär ,pad·ərn }

bar plain [GEOL] A plain formed by a stream without a low-water channel or an alluvial cover. { 'bär ,plān }

bar post [CIV ENG] One of the posts driven into the ground to form the sides of a field gate. { 'bär ,pōst }

bar printer [COMPUT SCI] An impact printer in which the character heads are mounted on type bars. { 'bär ,print·ər }

barracuda [VERT ZOO] The common name for about 20 species of fishes belonging to the genus *Sphyraena* in the order Perciformes. { ,bar·ə'küd·ə }

barrage [CIV ENG] An artificial dam which increases the depth of water of a river or watercourse, or diverts it into a channel for navigation or irrigation. [ORD] A prearranged barrier of fire designed to protect friendly troops and installations by impeding enemy movement across defensive lines or areas. { bə'räzh }

barrage balloon [ORD] A balloon restrained from free flight by means of a cable attaching it to the earth; used to support wires or nets as protection against air attacks. { bə'räzh bə'lün }

barrage jamming [COMMUN] The simultaneous jamming of a number of radio or radar bands of frequencies. { bə'räzh ,jam·iŋ }

barrage-type spillway [CIV ENG] A passage for surplus water with sluice gates across the width of the entrance. { bə'räzh ,tīp 'spil,wā }

barranca [GEOL] A hole or deep break made by heavy rain; a ravine. { bə'raŋ·kə }

Barr body [CYTOL] A condensed, inactivated X chromosome inside the nuclear membrane in interphase somatic cells of women and most female mammals. Also known as sex chromatin. { 'bär ,bäd·ē }

barre [TEXT] **1.** A pattern of bars or stripes parallel to the weft of a fabric. **2.** A defect in the production of a fabric that results in a streak parallel to the weft. { bär }

barred-and-braced gate [CIV ENG] A gate with a diagonal brace to reinforce the horizontal timbers. { ¦bärd ən ¦brāst 'gāt }

barred basin See restricted basin. { ¦bärd ¦bās·ən }

barred beach sequence [GEOL] A sequence comprising longshore bars, barrier beaches, and lagoons that develop when, under low-energy conditions, waves cross a broad continental shelf before impinging on a shoreline where sand-sized sediments are abundant. { ¦bärd ¦bēch 'sē·kwəns }

barred gate [CIV ENG] A gate with one or more horizontal timber rails. { ¦bärd 'gāt }

barred spiral galaxy [ASTRON] A spiral galaxy whose spiral arms originate at the ends of a bar-shaped structure centered at the nucleus of the galaxy. { ¦bärd ¦spī·rəl 'gal·ik·sē }

barrel [DES ENG] **1.** A container having a circular lateral cross section that is largest in the middle, and ends that are flat; often made of staves held together by hoops. **2.** A piece of small pipe inserted in the end of a cartridge to carry the squib to the powder. **3.** That portion of a pipe having a constant bore and wall thickness. [MECH] Abbreviated bbl. **1.** The unit of liquid volume equal to 31.5 gallons (approximately 119 liters). **2.** The unit of liquid volume for petroleum equal to 42 gallons (approximately 158 liters). **3.** The unit of dry volume equal to 105 quarts (approximately 116 liters). **4.** A unit of weight that varies in size according to the commodity being weighed. [NAV ARCH] The central part of a windlass or capstan about which the cable is wound. [OPTICS] A

BARRACUDA

The great or predatory barracuda (*Sphyraena barracuda*), with a long jaw and numerous sharp teeth.

tapering cylindrical housing which contains the lenses of a camera and the iris diaphragm. [ORD] The cylindrical metallic part of a gun which controls the initial direction of a projectile. { 'bar·əl }

barrel arch [ARCH] A plain arch with a barrellike cross section in which the length is greater than the span (diameter). { 'bar·əl ,ärch }

barrel assembly [ORD] The gun barrel together with the other parts necessary to attach it to the rest of the gun. { 'bar·əl ə'sem·blē }

barrel bolt [DES ENG] A door bolt which moves in a cylindrical casing; not driven by a key. Also known as tower bolt. { 'bar·əl ,bōlt }

barrel ceiling [ARCH] A ceiling of semicylindrical shape. { 'bar·əl ,sēl·iŋ }

barrel chest See emphysematous chest. { 'bar·əl ,chest }

barrel compressor [MECH ENG] A centrifugal compressor having a barrel-shaped housing. { 'bar·əl kəm,pres·ər }

barrel copper [MIN ENG] Copper in lumps small enough to be picked out of the mass of rock and put in the furnace without dressing. { 'bar·əl ,käp·ər }

barrel distortion [OPTICS] A defect in an optical system whereby lateral magnification decreases with object size; the image of a square then appears barrel-shaped. { 'bar·əl dis'tór·shən }

barrel drain [CIV ENG] Any drain which is cylindrical. { 'bar·əl ,drān }

barrel erosion [ORD] Wearing away of the interface of the bore due to the combined effects of gas washing, scoring, and mechanical abrasion; causes a reduction in muzzle velocity. { 'bar·əl i'rō·zhən }

barrel-etch reactor [ENG] A type of plasma reactor in which the specimens to be etched are placed in a quartz support stand and a plasma is generated that diffuses and contacts them. { ¦bar·əl ¦ech rē'ak·tər }

barrel extension [ORD] Metal projection fixed to the rear of the barrel in certain automatic guns; extends backward and holds the breech locked against the gas pressure in the chamber when the gun is fired. { 'bar·əl ik'sten·shən }

barrel fitting [DES ENG] A short length of threaded connecting pipe. { 'bar·əl ,fid·iŋ }

barrelhead [DES ENG] The flat end of a barrel. { 'bar·əl ,hed }

barreling [MET] See tumbling. [ORD] Expansion of the body of a cartridge case when the gun chamber recovers longitudinally following firing and the mouth of the case is not free to move in the chamber. { 'bar·əl·iŋ }

barrel life [ORD] As applied to small arms and automatic weapons, the number of rounds which may be fired through a barrel at a particular firing schedule before the barrel becomes unserviceable; varies with the firing schedule. { 'bar·əl ,līf }

barrel plating [MET] An electroplating process by which articles are brought into contact with an electrolyte while rotating in a perforated hardwood barrel. { 'bar·əl ,plād·iŋ }

barrel printer [COMPUT SCI] A computer printer in which the entire set of characters is placed around a rapidly rotating cylinder at each print position; computer-controlled print hammers opposite each print position strike the paper and press it against an inked ribbon between the paper and the cylinder when the appropriate character reaches a position opposite the print hammer. { 'bar·əl ,prin·tər }

barrel roof [BUILD] 1. A roof of semicylindrical section; capable of spanning long distances parallel to the axis of the cylinder. 2. See barrel vault. { 'bar·əl ,rüf }

barrels per calendar day [CHEM ENG] A unit measuring the average rate of oil processing in a petroleum refinery, with allowances for downtime over a period of time. Abbreviated BCD. { 'bar·əlz pər ¦kal·ən·dər ,dā }

barrels per day [CHEM ENG] A unit measuring the rate at which petroleum is produced at the refinery. Abbreviated BD; bpd. { 'bar·əlz pər 'dā }

barrels per month [CHEM ENG] A unit measuring the rate at which petroleum is produced at the refinery. Abbreviated BM; bpm. { 'bar·əlz pər 'mənth }

barrels per stream day [CHEM ENG] A measurement used to denote rate of oil or oil-product flow while a fluid-processing unit is in continuous operation. Abbreviated BSD. { 'bar·əlz pər ¦strēm ,dā }

barrel vault [ARCH] A masonry vault of plain, semicircular cross section, supported by parallel walls or arcades and adapted to longitudinal areas. Also known as barrel roof; cradle vault; tunnel vault; wagonhead vault; wagon vault. { 'bar·əl ,vólt }

barrel whip [ORD] The movement of a gun barrel in a plane normal to the longitudinal axis of the gun bore, as the gun operates through a complete firing cycle. { 'bar·əl ,wip }

Barremian [GEOL] Lower Cretaceous geologic age, between Hauterivian and Aptian. { bə'räm·ē·ən }

barren liquor [CHEM ENG] Liquid (liquor) from filter-cake washing in which there is little or no recovery value; for example, barren cyanide liquor from washing of gold cake slimes. { 'bar·ən ¦lik·ər }

barrens [GEOGR] An area that because of adverse environmental conditions is relatively devoid of vegetation compared with adjacent areas. { 'bar·ənz }

barretter [ELEC] 1. Bolometer that consists of a fine wire or metal film having a positive temperature coefficient of resistivity, so that resistance increases with temperature; used for making power measurements in microwave devices. 2. See ballast resistor. { bə'red·ər }

barricade [ENG] Structure composed essentially of concrete, earth, metal, or wood, or any combination thereof, and so constructed as to reduce or confine the blast effect and fragmentation of an explosion. { 'bar·ə,kād }

barricade shield [ENG] A type of movable shield made of a material designed to absorb ionizing radiation, for protection from radiation. { 'bar·ə,kād ,shēld }

barrier [ECOL] Any physical or biological factor that restricts the migration or free movement of individuals or populations. [NAV] Anything which obstructs or prevents passage of a craft. [PHYS] See potential barrier. { 'bar·ē·ər }

Barrier [ORD] A passive acoustic detection system for submarines, consisting of hydrophones positioned on the ocean floor and connected by undersea cable to a land-based computer center. { 'bar·ē·ər }

barrier bar [GEOL] Ridges whose crests are parallel to the shore, and which are usually made up of water-worn gravel put down by currents in shallow water at some distance from the shore. { 'bar·ē·ər ,bär }

barrier basin [GEOL] A basin formed by natural damming, for example, by landslides or moraines. { 'bar·ē·ər ,bās·ən }

barrier beach [GEOL] A single, long, narrow ridge of sand which rises slightly above the level of high tide and lies parallel to the shore, from which it is separated by a lagoon. Also known as offshore beach. { 'bar·ē·ər ,bēch }

barrier capacitance [ELECTR] The capacitance that exists between the *p*-type and *n*-type semiconductor materials in a semiconductor *pn* junction that is reverse-biased so that it does not conduct. Also known as depletion-layer capacitance; junction capacitance. { 'bar·ē·ər kə,pas·əd·əns }

barrier chain [GEOL] A series of barrier spits, barrier islands, and barrier beaches extending along a coastline. { 'bar·ē·ər ,chān }

barrier curb [CIV ENG] A curb with vertical sides high enough to keep vehicles from crossing it. { 'bar·ē·ər ,kərb }

barrier flat [GEOL] An area which is relatively flat and frequently occupied by pools of water that separate the seaward edge of the barrier from a lagoon on the landward side. { 'bar·ē·ər ,flat }

barrier-grid storage tube See radechon. { 'bar·ē·ər ,grid 'stór·ij ,tüb }

barrier ice See shelf ice. { 'bar·ē·ər ,īs }

barrier injection transit-time diode [ELECTR] A microwave diode in which the carriers that traverse the drift region are generated by minority carrier injection from a forward-biased junction instead of being extracted from the plasma of an avalanche region. Abbreviated BARITT diode. { 'bar·ē·ər in'jek·shən 'trans·ət ,tīm 'dī,ōd }

barrier island [GEOL] An elongate accumulation of sediment formed in the shallow coastal zone and separated from the mainland by some combination of coastal bays and their associated marshes and tidal flats; barrier islands are typically several times longer than their width and are interrupted by tidal inlets. { 'bar·ē·ər ,ī·lənd }

barrier lagoon [GEOGR] A shallow body of water that separates the shore and a barrier reef. { 'bar·ē·ər lə'gün }

barrier lake [HYD] A small body of water that lies in a basin, retained there by a natural dam or barrier. { 'bar·ē·ər ,lāk }

barrier layer See depletion layer. { 'bar·ē·ər ,lā·ər }

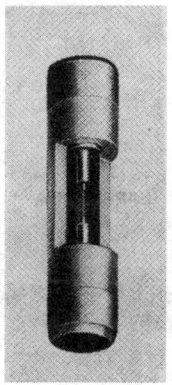

BARRETTER

Cutaway view of commercial barretter about 1 inch (2.5 centimeters) in length. *(Sperry Division, Sperry Rand Corp.)*

barrier-layer cell *See* photovoltaic cell. { 'bar·ē·ər ˌlā·ər ˌsel }

barrier-layer photocell *See* photovoltaic cell. { 'bar·ē·ər ˌlā·ər 'fōd·ō,sel }

barrier-layer rectification *See* depletion-layer rectification. { 'bar·ē·ər ˌlā·ər ˌrek·tə·fə'kā·shən }

barrier marsh [ECOL] A type of marsh that restricts or prevents invasion of the area beyond it by new species of animals. { 'bar·ē·ər ˌmärsh }

barrier material [MATER] Packing material impervious to moisture, vapor, or other liquids and gases. [ORD] An inert material placed in an explosive charge to shape the detonation wave. { 'bar·ē·ər mə'tir·ē·əl }

barrier penetration [QUANT MECH] The passage of a particle through a potential barrier, that is, through a region of finite extent in which the particle's potential energy is greater than its total energy. { 'bar·ē·ər ˌpen·ə'trā·shən }

barrier plastic [MATER] A polymer that is impermeable to gases and can be fabricated into an object such as a container that will not permit aromas or other gases to pass through it in either direction. { 'bar·ē·ər ˌplas·tik }

barrier reef [GEOL] A coral reef that runs parallel to the coast of an island or continent, from which it is separated by a lagoon. { 'bar·ē·ər ˌrēf }

barrier separation [CHEM ENG] The separation of a two-component gaseous mixture by selective diffusion of one component through a separative barrier (microporous metal or nonporous polymeric). { 'bar·ē·ər sep·ə'rā·shən }

barrier shield [ENG] A wall or enclosure made of a material designed to absorb ionizing radiation, shielding the operator from an area where radioactive material is being used or processed by remote-control equipment. { 'bar·ē·ər ˌshēld }

barrier spit [GEOL] A barrier of sand joined at one of its ends to the mainland. { 'bar·ē·ər ˌspit }

barrier strip [ELECTR] A device for connecting two cables without using lugs in which bare wires from one cable are connected to lugs of screws on one side of the strip and wires from the other cable are attached at corresponding points on the opposite side. { 'bar·ē·ər ˌstrip }

barrier theory of cyclones [METEOROL] A theory of cyclone development, proposed by F.M. Exner, which states that a slow-moving mass of cold air in the path of rapidly eastward-moving warmer air will bring about the formation of low pressure on the lee side of the cold air; analogous to the formation of a dynamic trough on the lee side of an orographic barrier. Also known as drop theory. { 'bar·ē·ər ˌthē·ə·rē əv 'sī,klōnz }

barrier voltage [ELECTR] The voltage necessary to cause electrical conduction in a junction of two dissimilar materials, such as *pn* junction diode. { 'bar·ē·ər ˌvōl·tij }

barrier zone [BOT] In a tree, new tissue formed by the cambium after it has been wounded; serves as both an anatomical and a chemical wall. { 'bar·ē·ər ˌzōn }

Barrovian metamorphism [GEOL] A regional metamorphism that can be zoned into facies that are metamorphic. { bə'rōv·ē·ən ˌmed·ə'mòr,fiz·əm }

barrow *See* handbarrow; wheelbarrow. { 'ba·rō }

barrow run [CIV ENG] A temporary pathway of wood planks or sheets to provide a smooth access for wheeled materials-handling carriers on a building site. { 'ba·rō ˌrən }

bar sash lift [BUILD] A type of handle, attached to the bottom rail of a sash, for raising or lowering it. { 'bär 'sash ˌlift }

bar scale *See* graphic scale. { 'bär ˌskāl }

bar screen [MECH ENG] A sieve with parallel steel bars for separating small from large pieces of crushed rock. { 'bär ˌskrēn }

bar sight [ORD] The rear sight of a firearm, consisting of a movable bar, usually with an open notch. { 'bär ˌsīt }

bar steel [MET] Steel formed into bars. { 'bär ˌstēl }

barstock *See* bar. { 'bär,stäk }

Barstovian [GEOL] Upper Miocene geologic time. { ˌbär 'stōv·ē·ən }

bar strainer [DES ENG] A screening device consisting of a bar or a number or parallel bars; used to prevent objects from entering a drain. { 'bär ˌstrān·ər }

bar support [CIV ENG] A device used to support or hold steel reinforcing bars in proper position before or during the placement of concrete. Also known as bar chair. { 'bär sə'pòrt }

bar test survey [ENG] A leakage survey in which bar holes are driven or bored at regular intervals along the way of an underground gas pipe and the atmosphere in the holes is tested with a combustible gas detector or such. { 'bär ˌtest 'sər,vā }

bar theory [GEOL] A theory that accounts for thick deposits of salt, gypsum, and other evaporites in terms of increased salinity of a solution in a lagoon caused by evaporation. { 'bär 'thē·ə·rē }

bartholinitis [MED] Inflammation of Bartholin's glands and/or their ducts which is caused by bacteria from feces or a sexually transmitted infection, such as gonorrhea. { ˌbär·tə·lə'nīd·əs }

Bartholin's glands [ANAT] Two pea-sized glands located on each side of the labia minora that secrete a lubricating fluid upon sexual arousal. { 'bärt·əl·ənz ˌglanz }

Barth plan [IND ENG] A wage incentive plan intended for a low task and for all efficiency points and defined as: earning = rate per hour × square root of the product (hours standard × hours actual). { 'bärth ˌplan }

Bartlett force [NUC PHYS] A force between nucleons in which spin is exchanged. { 'bärt·lət ˌfòrs }

Bartlett's test [STAT] A method to test for the equalities of variances from a number of independent normal samples by testing the hypothesis. { 'bärt·ləts ˌtest }

Bartonella [MICROBIO] A genus of the family Bartonellaceae; parasites in or on red blood cells and within fixed tissue cells; found in humans and in the arthropod genus *Phlebotomus*. { ˌbärt·ən'el·ə }

Bartonellaceae [MICROBIO] A family of the order Rickettsiales; rod-shaped, coccoid, ring- or disk-shaped cells; parasites of human and other vertebrate red blood cells. { ˌbärt·ən,e'lās·ē,ē }

bartonellosis *See* Carrion's disease. { ˌbärt·ən·e'lō·səs }

Bartonian [GEOL] A European stage: Eocene geologic time above Auversian, below Ludian. Also known as Marinesian. { bär'tōn·ē·ən }

Bart reaction [ORG CHEM] Formation of an aryl arsonic acid by treating the aryl diazo compound with trivalent arsenic compounds, such as sodium arsenite. { 'bärt rē'ak·shən }

bar turret lathe [MECH ENG] A turret lathe in which the bar stock is slid through the headstock and collet on line with the turning axis of the lathe and held firmly by the closed collet. { 'bär 'tər·ət ˌlāth }

bar-type grating [CIV ENG] An open grid assembly of metal bars in which the bearing bars (running in one direction) are spaced by rigid attachment to crossbars. { 'bär ˌtīp 'grād·iŋ }

bar winding [ELEC] An armature winding made up of a series of metallic bars connected at their ends. { 'bär ˌwīnd·iŋ }

barycenter [ASTRON] The center of gravity of the earth-moon system. [MATH] The center of mass of a system of finitely many equal point masses distributed in euclidean space in such a way that their position vectors are linearly independent. { 'bar·ə,sen·tər }

barycentric coordinates [MATH] The coefficients in the representation of a point in a simplex as a linear combination of the vertices of the simplex. { ˌbar·ə'sen·trik kō'órd·ən,əts }

barycentric element [ASTROPHYS] An orbital element referred to the center of mass of the solar system. { ˌbar·ə'sen·trik 'el·ə·mənt }

barycentric energy [MECH] The energy of a system in its center-of-mass frame. { ˌbar·ə'sen·trik 'en·ər·jē }

Barychilinidae [PALEON] A family of Paleozoic crustaceans in the suborder Platycopa. { ˌbar·ə·kə'lin·ə,dē }

barye [MECH] The pressure unit of the centimeter-gram-second system of physical units; equal to 1 dyne per square centimeter (0.001 millibar). Also known as microbar. { 'ba·rē }

Barylambdidae [PALEON] A family of late Paleocene and early Eocene aquatic mammals in the order Pantodonta. { ˌbar·ə'lam·də,dē }

baryon [PARTIC PHYS] Any elementary particle which can be transformed into a nucleon and some number of mesons and lighter particles. Also known as heavy particle. { 'bar·ē,än }

baryon number [PARTIC PHYS] A quantum number equal to the number of baryons minus the number of antibaryons in a system; it is conserved at the present level of detection, but may not be exactly conserved. { 'bar·ē,än ˌnəm·bər }

BARYON OCTET

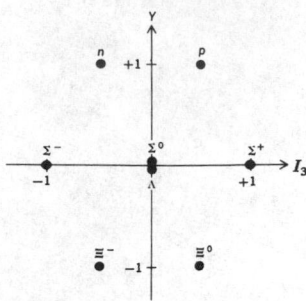

The baryon octet, arrayed with respect to I_3 as abscissa and Y as ordinate. The charge number Q is given by $Q = I_3 + Y/2$.

BASALT

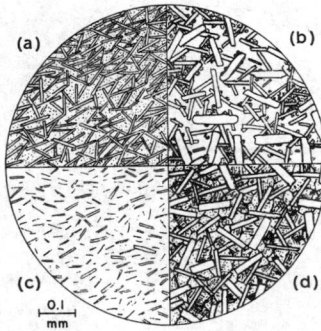

Textures of basalt. *(a)* Pilotaxitic texture with feldspar microlites and some interstitial, microcrystalline material. *(b)* Intersertal texture with feldspar microlites and some granular pyroxene and much interstitial glass. *(c)* Hyalopilitic texture with feldspar microlites in glass. *(d)* Intergranular texture with feldspar laths and some interstitial granular pyroxene.

baryon octet [PARTIC PHYS] The group of one lambda, three sigma, and two xi hyperons and two nucleons, all having spin $1/2$ and positive parity, and forming a symmetrical pattern as suggested by SU_3 symmetry. { 'bar·ē͵än äk'tet }

baryon resonance [PARTIC PHYS] A cross section anomaly indicating the existence of an unstable baryon. { 'bar·ē͵än 'rez·ən·əns }

baryon spectroscopy [PARTIC PHYS] The science of the energy levels and changes of state occurring among baryon particles. { 'bar·ē͵än spek'träs·kə·pē }

baryon-to-photon ratio [ASTRON] The estimated ratio of the number of baryons (mostly protons and neutrons) to photons (mostly in the cosmic microwave radiation) in the universe. { 'bar·ē͵än tə 'fō͵tän 'rā·shō }

barysphere *See* centrosphere. { 'bar·ə͵sfir }

baryta feldspar *See* hyalophane. { bə'rīd·ə 'fel͵spär }

baryta water [CHEM] A solution of barium hydroxide. { bə'rīd·ə 'wȯd·ər }

baryte *See* barite. { 'ba͵rīt }

Barytheriidae [PALEON] A family of extinct proboscidean mammals in the suborder Barytherioidea. { ͵bar·ə·thə'rī·ə͵dē }

Barytherioidea [PALEON] A suborder of extinct mammals of the order Proboscidea, in some systems of classification. { ͵bar·ə͵thir·ē'ȯid·ē·ə }

barytine *See* barite. { 'bar·ə͵tēn }

barytocalcite [MINERAL] $CaBa(CO_3)_2$ A colorless to white, grayish, greenish, or yellowish monoclinic mineral consisting of calcium and barium carbonate. { bə͵rīd·ə'kal͵sīt }

barytropic gas [PHYS] A gas whose pressure depends only on its density. { ͵bar·ə'träp·ik 'gas }

basal [BIOL] Of, pertaining to, or located at the base. [PHYSIO] Being the minimal level for, or essential for maintenance of, vital activities of an organism, such as basal metabolism. { 'bā·səl }

basal arkose [PETR] Partially reworked feldspathic residuum in the lower section of a sandstone that overlies granitic rock. { 'bā·səl 'är͵kōs }

basal body [CYTOL] A cellular organelle that induces the formation of cilia and flagella and is similar to and sometimes derived from a centriole. Also known as kinetosome. { 'bā·səl ͵bäd·ē }

basal-cell carcinoma [MED] A locally invasive, rarely metastatic nevoid tumor of the epidermis. Also known as basal-cell epithelioma. { 'bā·səl ͵sel ͵kärs·ən'ō·mə }

basal-cell epithelioma *See* basal-cell carcinoma. { 'bā·səl ͵sel ͵ep·ə͵thē·lē'ō·mə }

basal cleavage [CRYSTAL] Cleavage parallel to the base of the crystal structure or to the lattice plane which is normal to one of the lattice axes. { 'bā·səl 'klēv·ij }

basal complex *See* basement. { 'bā·səl 'käm͵pleks }

basal conglomerate [GEOL] A coarse gravelly sandstone or conglomerate forming the lowest member of a series of related strata which lie unconformably on older rocks; records the encroachment of the seabeach on dry land. { 'bā·səl kən'gläm·ə·rət }

basal coplane [GRAPHICS] The condition of exposure of a pair of photographs in which the two photographs lie in a common plane parallel to the air base. { 'bā·səl 'kō͵plān }

basal disc [BIOL] The expanded basal portion of the stalk of certain sessile organisms, used for attachment to the substrate. { 'bā·səl 'disk }

basal ganglia [NEURO] The corpus striatum, or the corpus striatum and the thalamus considered together as the important subcortical centers. { 'bā·səl 'gaŋ·glē·ə }

basal groundwater [HYD] A large body of groundwater that floats on and is in hydrodynamic equilibrium with sea water. { 'bā·səl 'graȯnd͵wȯd·ər }

basalia [VERT ZOO] The cartilaginous rods that support the base of the pectoral and pelvic fins in elasmobranchs. { bə'sal·ē·ə }

basalis [HISTOL] The basal portion of the endometrium; it is not shed during menstruation. { bə'sal·əs }

basal lamina [EMBRYO] The portion of the gray matter of the embryonic neural tube from which motor nerve roots develop. { 'bā·səl 'lam·ə·nə }

basal membrane [ANAT] The tissue beneath the pigment layer of the retina that forms the outer layer of the choroid. { 'bā·səl 'mem͵brän }

basal metabolic rate [PHYSIO] The amount of energy utilized per unit time under conditions of basal metabolism; expressed as calories per square meter of body surface or per kilogram of body weight per hour. Abbreviated BMR. { 'bā·səl med·ə'bäl·ik 'rāt }

basal metabolism [PHYSIO] The sum total of anabolic and catabolic activities of an organism in the resting state providing just enough energy to maintain vital functions. { 'bā·səl mə'tab·ə͵liz·əm }

basal orientation [CRYSTAL] A crystal orientation in which the surface is parallel to the base of the lattice or to the lattice plane which is normal to one of the lattice axes. { 'bā·səl ͵ȯr·ē·ən'tā·shən }

basal plane [CRYSTAL] The plane perpendicular to the long, or c, axis in all crystals except those of the isometric system. { 'bā·səl 'plān }

basal rot [PL PATH] Any rot that affects the basal parts of a plant, especially bulbs. { 'bā·səl 'rät }

basalt [PETR] An aphanitic crystalline rock of volcanic origin, composed largely of plagioclase feldspar (labradorite or bytownite) and dark minerals such as pyroxene and olivine; the extrusive equivalent of gabbro. { bə'sȯlt }

basalt glass *See* tachylite. { bə'sȯlt ͵glas }

basaltic dome *See* shield volcano. { bə'sȯl·tik 'dōm }

basaltic hornblende [PETR] A black or brown variety of hornblende rich in ferric iron and occurring in basalts and other iron-rich basic igneous rocks. Also known as basaltine; lamprobolite; oxyhornblende. { bə'sȯl·tik 'hȯrn͵blend }

basaltic lava [PETR] A volcanic fluid rock of basaltic composition. { bə'sȯl·tik 'lav·ə }

basaltic magma [GEOL] Mobile rock material of basaltic composition. { bə'sȯl·tik 'mag·mə }

basaltic rock [PETR] Igneous rock that is fine-grained and contains basalt, diabase, and dolerite; if andesite is included the rock is dark in color. { bə'sȯl·tik 'räk }

basaltic shell [GEOL] The lower crystal layer of basalt underlying the oceans and beneath the sialic layer of continents. { bə'sȯl·tik 'shel }

basaltiform [GEOL] Similar to basalt in form. { bə'sȯl·tə͵fȯrm }

basaltine *See* basaltic hornblende. { bə'sȯl͵tēn }

basalt obsidian *See* tachylite. { bə'sȯlt əb'sid·ē·ən }

basal tunnel [ENG] A water supply tunnel constructed along the basal water table. { 'bā·səl 'tən·əl }

basaluminite [MINERAL] $Al_4(SO_4)(OH)_{10}·5H_2O$ A white mineral consisting of hydrated basic aluminum sulfate; occurs in compact masses. { ͵bās·ə'lüm·ə͵nīt }

basal wall [BOT] The wall dividing the oospore into an anterior and a posterior half in plants bearing archegonia. { 'bā·səl 'wȯl }

basal water table [HYD] The water table of basal groundwater. { 'bā·səl 'wȯd·ər ͵tā·bəl }

basanite [PETR] A basaltic extrusive rock closely allied to chert, jasper, or flint. Also known as Lydian stone; lydite. { 'bas·ə͵nīt }

basculating fault *See* wrench fault. { 'ba·skyə͵lād·iŋ 'fȯlt }

bascule [ENG] A structure that rotates about an axis, as a seesaw, with a counterbalance (for the weight of the structure) at one end. { 'ba͵skül }

bascule bridge [CIV ENG] A movable bridge consisting primarily of a cantilever span extending across a channel; it rotates about a horizontal axis parallel with the waterway. { 'ba͵skül ͵brij }

bascule leaf [CIV ENG] The span of a bascule bridge. { 'ba͵skül ͵lēf }

base [CHEM] Any chemical species, ionic or molecular, capable of accepting or receiving a proton (hydrogen ion) from another substance; the other substance acts as an acid in giving of the proton. Also known as Brønsted base. [CHEM ENG] The primary substance in solution in crude oil, and remaining after distillation. [COMPUT SCI] *See* root. [ELECTR] **1.** The region that lies between an emitter and a collector of a transistor and into which minority carriers are injected. **2.** The part of an electron tube that has the pins, leads, or other terminals to which external connections are made either directly or through a socket. **3.** The plastic, ceramic, or other insulating board that supports a printed wiring pattern. [ENG] Foundation or part upon which an object or instrument rests. [GEN] *See* nitrogenous base. [GRAPHICS] A transparent plastic film on which a photographic emulsion is applied. [LAP] *See* pavilion. [MATH] **1.** A side or face upon which the altitude of a geometric configuration is thought of as being constructed. **2.** For a logarithm, the number of which the logarithm is the exponent. **3.** For a number system, the number whose powers determine place value. **4.** For a topological space, a collection of sets, unions of which form all the open sets of the space. [ORD] Station or installation from which military forces operate and from which supplies are obtained. { bās }

base address *See* address constant. { 'bās ə'dres }

base-altitude ratio [GRAPHICS] The ratio between the air-base length and the flight altitude of a stereoscopic pair of photographs. Also known as base-height ratio; K factor. { 'bās 'al·tə,tüd ,rā·shō }

base analog [MOL BIO] A molecule similar enough to a purine or pyrimidine base to substitute for the normal bases, resulting in abnormal base pairing. { 'bās 'an·ə,läg }

base anchor [BUILD] The metal piece attached to the base of a doorframe for the purpose of securing the frame to the floor. { 'bās ,aŋ·kər }

base angle [MATH] Either of the two angles of a triangle that have the base for a side. { 'bās ,aŋ·gəl }

base apparatus [ENG] Any apparatus designed for use in measuring with accuracy and precision the length of a base line in triangulation, or the length of a line in first- or second-order traverse. { 'bās ,ap·ə'rad·əs }

base area [GRAPHICS] A portion of the lower edge of a microfilm jacket that provides an area for notching. { 'bās 'er·ē·ə }

baseball [PL PHYS] A machine used in controlled fusion research to confine a plasma; consists of a linear magnetic bottle sealed by magnetic mirrors at both ends, and has current-carrying structures, which resemble the seams of a baseball in shape, to stabilize the plasma. { 'bās,bȯl }

baseband [COMMUN] The band of frequencies occupied by all transmitted signals used to modulate the radio wave. { 'bās,band }

baseband frequency response [COMMUN] Frequency response characteristics of the frequency band occupied by all of the signals used to modulate a transmitted carrier. { 'bās,band 'frē·kwən·sē ri'späns }

baseband system [COMMUN] A communications system in which information is transmitted over a single unmodulated band of frequencies. { 'bās,band ,sis·təm }

base bias [ELECTR] The direct voltage that is applied to the majority-carrier contact (base) of a transistor. { 'bās ,bī·əs }

base block [BUILD] **1.** A block of any material, generally with little or no ornament, forming the lowest member of a base, or itself fulfilling the functions of a base, as a member applied to the foot of a door or to window trim. **2.** A rectangular block at the base of a casing or column which the baseboard abuts. **3.** *See* skirting block. { 'bās ,bläk }

baseboard [BUILD] A finish board covering the interior wall at the junction of the wall and the floor. Also known as skirt; skirting. { 'bās,bȯrd }

baseboard heater [BUILD] Heating elements installed in panels along the baseboard of a wall. { 'bās,bȯrd 'hēd·ər }

baseboard radiator [CIV ENG] A heating unit which is located at the lower portion of a wall and to which heat is supplied by hot water, warm air, steam, or electricity. { 'bās,bȯrd 'rād·ē,ād·ər }

base box [MET] A unit of area used for tin-plated steel sheet; one base box is equivalent to 112 sheets, 14 by 20 inches (35.6 by 50.8 centimeters), or 62,720 square inches of surface, coated on two sides; 1 pound (0.454 kilogram) of tin per base box is equal to a coating of tin 0.000059 inch (0.0014986 millimeter) thick. { 'bās ,bäks }

base bullion [MET] Crude lead that has enough silver in it to make the extraction of silver worthwhile; gold may be present. { 'bās ,bùl·yən }

base cap *See* base molding. { 'bās ,kap }

base-centered lattice [CRYSTAL] A space lattice in which each unit cell has lattice points at the centers of each of two opposite faces as well as at the vertices; in a monoclinic crystal, they are the faces normal to one of the lattice axes. { 'bās ,sen·tərd 'lad·əs }

base circle [DES ENG] The circle on a gear such that each tooth-profile curve is an involute of it. { 'bās ,sər·kəl }

base conditions [PETRO ENG] Standard conditions of 14.65 psia pressure and 60°F (15.6°C) used to calculate the amount of gas contained in oil from a well (the gas-oil ratio). { 'bās kən'dish·ənz }

base correction [ENG] The adjustment made to reduce measurements taken in field exploration to express them with reference to the base station values. { 'bās kə'rek·shən }

base course [BUILD] The lowest course or first course of a wall. [CIV ENG] The first layer of material laid down in construction of a pavement. { 'bās ,kȯrs }

base density [OPTICS] The value of the inherent optical transmission density of a film base; does not include any contribution from the emulsion layer. { 'bās 'den·səd·ē }

Basedow's disease *See* exophthalmic goiter. { 'bäz·ə,dōz di,zēz }

base-displacement [COMPUT SCI] In machine-language programming, a technique in which addresses are specified relative to a base address where the beginning of the program is stored. { 'bās dis,plās·mənt }

base drag [FL MECH] Drag owing to a base pressure lower than the ambient pressure; it is a part of the pressure drag. { 'bās ,drag }

base elbow [DES ENG] A cast-iron pipe elbow having a baseplate or flange which is cast on it and by which it is supported. { 'bās ,el,bō }

base electrode [ELECTR] An ohmic or majority carrier contact to the base region of a transistor. { 'bās i'lek,trōd }

base exchange [GEOCHEM] Replacement of certain ions by others in clay. { 'bās iks'chānj }

base excision repair [CELL MOL] A deoxyribonucleic acid (DNA) repair system in which an altered base is removed from the sugar backbone by action of a specific DNA glycolase and then the abasic sugar is removed by apurinic/apyrimidic (AP) lyase and AP endonuclease, leaving a one-nucleotide gap that is then filled in and ligated. { 'bās ek'siz·zhən ri,per }

base flashing [BUILD] **1.** The flashing provided by upturned edges of a watertight membrane on a roof. **2.** Any metal or composition flashing at the joint between a roofing surface and a vertical surface, such as a wall or parapet. { 'bās ,flash·iŋ }

base flow [HYD] The flow of water entering stream channels from groundwater sources in the drainage of large lakes. { 'bās ,flō }

base font [COMPUT SCI] The font used in a document if none other is specified. { 'bās ,fänt }

base for the neighborhood system *See* local base. { 'bās fər thə 'nā·bər,hùd ,sis·təm }

base fracture [MIN ENG] In quarrying, the broken condition of the base after a blast; it may be a good or bad base fracture. { 'bās 'frak·chər }

base fuse [ORD] A fuse installed in the bottom of a projectile. { 'bās ,fyüz }

base-height ratio *See* base-altitude ratio. { 'bās ,hīt 'rā·shō }

base initiation [ORD] Detonation initiated at the base (rear) of the charge. { 'bās i,nish·ē'ā·shən }

base insulator [ELEC] Heavy-duty insulator used to support the weight of an antenna mast and insulate the mast from the ground or some other surface. { 'bās 'in·sə,lād·ər }

base isolators [CIV ENG] Components placed within a building (not always at the base) which are relatively flexible in the lateral direction, yet can sustain the vertical load. When an earthquake causes ground motions, base isolators allow the structure to respond much more slowly than it would without them, resulting in lower seismic demand on the structure. Isolators may be laminated steel with high-quality rubber pads, sometimes incorporating lead or other energy-absorbing materials. { 'bās ,ī·sə,lād·ərz }

base language [COMPUT SCI] The component of an extensible language which provides a complete but minimal set of primitive facilities, such as elementary data types, and simple operations and control constructs. { 'bās 'laŋ·gwij }

base level [GEOL] That critical plane of erosion and deposition represented by river level on continents and by wave or current base in the sea. { 'bās ,lev·əl }

base-leveled plain [GEOL] Any land surface changed almost to a plain by subaerial erosion. Also known as peneplain. { 'bās ,lev·əld 'plān }

base-leveling epoch *See* gradation period. { 'bās ,lev·əl·iŋ 'ep·ək }

base line Abbreviated BL. [ELECTR] The line traced on

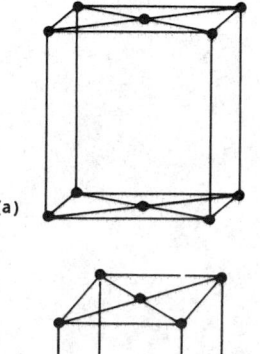

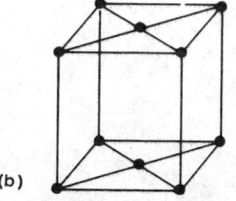

BASE-CENTERED LATTICE

(a)

(b)

Two of fourteen Bravais lattices.
(*a*) Base-centered orthorhombic.
(*b*) Base-centered monoclinic.

amplitude-modulated indicators which corresponds to the power level of the weakest echo detected by the radar; it is retraced with every pulse transmitted by the radar but appears as a nearly continuous display on the scope. [ENG] **1.** A surveyed line, established with more than usual care, to which surveys are referred for coordination and correlation. **2.** A cardinal line extending east and west along the astronomic parallel passing through the initial point, along which standard township, section, and quarter-section corners are established. [GRAPHICS] The bottom alignment of upper-case letters in a font. [NAV] The geodesic line joining the two stations between which electrical phase or time is compared in determining navigational coordinates. [SCI TECH] A line drawn in the graphical representation of a varying physical quantity, such as a voltage or current, to indicate a reference value, such as the voltage value of a bias. { 'bās ˌlīn }

base-line break [ELECTR] Technique in radar which uses the characteristic break in the base line on an A-scope display due to a pulse signal of significant strength in noise jamming. { 'bās ˌlīn ˌbrāk }

base-line check See ground check. { 'bās ˌlīn ˌchek }

base-line delay [NAV] The time interval which elapses while a signal travels from a loran master station to a slave station. { 'bās ˌlīn di'lā }

base-line extension [NAV] In the loran navigation system, the extension in both directions of a line passing through a slave station and its master; on this line the maximum time interval between the reception of signals from the master and slave stations may be received. { 'bās ˌlīn ik'sten·shən }

base-line technique [ANALY CHEM] A method for measurement of absorption peaks for quantitative analysis of chemical compounds in which a base line is drawn tangent to the spectrum background; the distance from the base line to the absorption peak is the absorbence due to the sample under study. { 'bās ˌlīn tek'nēk }

baseload [ELEC] Minimum load of a power generator over a given period of time. { 'bāsˌlōd }

base-loaded antenna [ELECTROMAG] Vertical antenna having an impedance in series at the base for loading the antenna to secure a desired electrical length. { 'bās ˌlōd·əd an'ten·ə }

base magnification [OPTICS] The ratio of the distance between the centers of the objectives of a pair of binoculars to the distance between the centers of the eyepieces. { 'bās ˌmag·nə·fə'kā·shən }

base map [MAP] A map having essential outlines and onto which additional geographical or topographical data may be placed for comparison or correlation. Also known as mother map. { 'bās ˌmap }

basement [BUILD] A building story which is wholly or less than half below ground; it is generally used for living space. [GEOL] **1.** A complex, usually of igneous and metamorphic rocks, that is overlain unconformably by sedimentary strata. Also known as basement rock. **2.** A crustal layer beneath a sedimentary one and above the Mohorovičić discontinuity. **3.** The ancient continental igneous rock base that lies beneath Precambrian rocks. Also known as basal complex; basement complex. { 'bāsˌmənt }

basement complex See basement. { 'bās·mənt 'käm,pleks }

basement membrane [HISTOL] A delicate connective-tissue layer underlying the epithelium of many organs. { 'bās·mənt 'mem,brān }

basement rock See basement. { 'bās·mənt ˌräk }

basement wall [BUILD] A foundation wall which encloses a usable area under a building. { 'bās·mənt ˌwol }

base metal [CHEM] Any of the metals on the lower end of the electrochemical series. [MET] **1.** The metal that is to be worked. **2.** The principal metal of an alloy. **3.** Any metal that will oxidize when heated in air. **4.** The metal of parts to be welded. Also known as parent metal. **5.** Metal to which cladding or plating is applied. Also known as basis metal. { 'bās ˌmed·əl }

base modulation [ELECTR] Amplitude modulation produced by applying the modulating voltage to the base of a transistor amplifier. { 'bās ˌmäj·ə'lā·shən }

base molding [BUILD] Molding used to trim the upper edge of interior baseboard. Also known as base cap. { 'bās ˌmōld·iŋ }

base mortar [ORD] Mortar in a platoon for which initial firing data are computed and with reference to which data for other mortars in the unit are computed. { 'bās ˌmord·ər }

basendite [INV ZOO] In crustaceans, either of a pair of lobes at the end of each specialized paired appendage. { 'bā 'sen,dīt }

base net [ENG] A system, in surveying, of quadrilaterals and triangles that include and are quite close to a base line in a triangulation system. { 'bās ˌnet }

base notation See radix notation. { 'bās nō'tā·shən }

base of projectile [ORD] The rearmost section of a projectile; for projectiles having a rotating band, it is the section located to the rear thereof. { 'bās əv prə'jek·tīl }

base oil [MATER] An oil to which other oils or substances are added to produce a lubricant. { 'bās ˌoil }

base ore [MIN ENG] Ore in which the gold is associated with sulfides, as contrasted with free-milling ores in which the sulfides have been removed by leaching. { 'bās ˌor }

base pair [MOL BIO] Two nitrogenous bases, one purine and one pyrimidine, that pair in double-stranded deoxyribonucleic acid. { 'bās ˌper }

base pairing [MOL BIO] The hydrogen bonding of complementary purine and pyrimidine bases-adenine with thymine, guanine with cytosine-in double-stranded deoxyribonucleic acids or ribonucleic acids or in DNA/RNA hybrid molecules. { 'bās ˌper·iŋ }

base peak [SPECT] The tallest peak in a mass spectrum; it is assigned a relative intensity value of 100, and lesser peaks are reported as a percentage of it. { 'bās ˌpēk }

base period [STAT] The period of a year, or other unit of time, used as a reference in constructing an index number. Also known as base year. { 'bās ˌpir·ē·əd }

base piece [ORD] The gun first selected after calibration fire whose center of burst or impact is taken as the reference point in determining calibration corrections for the remaining guns of the battery. { 'bās ˌpēs }

base pin See pin. { 'bās ˌpin }

base plate [DES ENG] The part of a theodolite which carries the lower ends of the three foot screws and attaches the theodolite to the tripod for surveying. [ENG] A metal plate that provides support or a foundation. { 'bās ˌplāt }

base-plus-fog density [OPTICS] The value of the inherent optical transmission density of a film base plus the nonimage density contributed by the developed emulsion. { 'bās ˌpləs 'fäg ˌden·səd·ē }

base point [ORD] A well-defined point in the target area used as a point of reference from which range and direction adjustments of artillery fire are made. { 'bās ˌpoint }

base pressure [FL MECH] The pressure exerted on the base or extreme aft end of a body, as of a cylindrical or boat-tailed body or of a blunt-trailing-edge wing in fluid flow. [MECH] A pressure used as a reference base, for example, atmospheric pressure. { 'bās ˌpresh·ər }

base quantity [PHYS] One of a small number of physical quantities in a system of measurement that are defined, independent of other physical quantities, by means of a physical standard and by procedures for comparing the quantity to be measured with the standard. Also known as fundamental quantity. { 'bās ˌkwän·ə·tē }

base rate area [COMMUN] Area within the telephone exchange in which all types of service are given without mileage charges. { 'bās ˌrāt ˌer·ē·ə }

base register See index register. { 'bās ˌrej·ə·stər }

base screed [ENG] A metal screed with expanded or short perforated flanges that serves as a dividing strip between plaster and cement and acts as a guide to indicate proper thickness of cement or plaster. { 'bās ˌskrēd }

base sequence [GEN] The specific order of purine and pyrimidine bases in a polynucleotide such as deoxyribonucleic acid or ribonucleic acid. { 'bās ˌsē·kwəns }

base sheet [BUILD] Saturated or coated felt sheeting which is laid as the first ply in a built-up roofing membrane. { 'bās ˌshēt }

base shoe [BUILD] A molding at the base of a baseboard. { 'bās ˌshü }

base shoe corner [BUILD] A molding piece or block applied in the corner of a room to eliminate the need for mitering the base shoe. { 'bās ˌshü ˌkor·nər }

base space of a bundle [MATH] The topological space B in the bundle (E,p,B). { ˌbās ˌspās əv ə 'bən·dəl; }

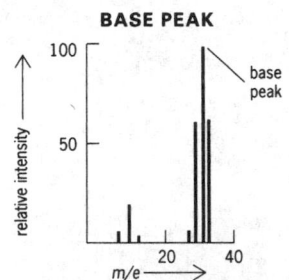

BASE PEAK

Base peak in the mass spectrum of methanol.

base spray [ORD] Fragments of a bursting projectile that are thrown to the rear in the line of flight, in contrast to nose spray and side spray. { 'bās ¦sprā }

base-spreading resistance [ELECTR] Resistance which is found in the base of any transistor and acts in series with it, generally a few ohms in value. { 'bās ¦spred·iŋ ri'zis·təns }

base stacking [MOL BIO] The orientation of adjacent base pairs such that their planes are parallel and their surfaces almost touch, as occurs in double-stranded deoxyribonucleic acid molecules. { 'bās ¦stak·iŋ }

base station [COMMUN] **1.** A land station, in the land mobile service, carrying on a service with land mobile stations (a base station may secondarily communicate with other base stations incident to communications with land mobile stations). **2.** A station in a land mobile system which remains in a fixed location and communicates with the mobile stations. [ENG] The point from which a survey begins. [GEOD] A geographic position whose absolute gravity value is known. { 'bās ¦stā·shən }

base system [COMPUT SCI] A computer system containing only program modules that carry out basic functions. { 'bās ¦sis·təm }

base tee [DES ENG] A pipe tee with a connected baseplate for supporting it. { 'bas ¦tē }

base tile [BUILD] The lowest course of tiles in a tiled wall. { 'bās ¦tīl }

base time *See* normal element time. *See* normal time. { 'bās ¦tīm }

base-timing sequencing [NAV] The control of the time sharing of a single transponder between several ground transmitters through the use of suitable coded timing signals. { 'bās ¦tīm·iŋ 'sē·kwəns·iŋ }

base unit [PHYS] One of a small number of units in a system of measurement that are defined, independent of other units, by means of a physical standard; equivalently, a unit of a base quantity. Also known as fundamental unit. { 'bās ¦yü·nət }

base vector [MATH] One of a set of linearly independent vectors in a vector space such that each vector in the space is a linear combination of vectors from the set; that is, a member of a basis. { 'bās ¦vek·tər }

base year *See* base period. { 'bās ¦yir }

base-year method *See* Laspeyre's index. { 'bās ¦yir 'meth·əd }

bashing [MIN ENG] **1.** The building of walls and nonporous stoppings for the complete isolation of a district of a mine in which a fire has occurred. **2.** The complete stowing of old mine workings or roadways after all equipment has been removed. { 'bash·iŋ }

basic [CHEM] Of a chemical species that has the properties of a base. [PETR] Of igneous rocks, having low silica content (generally less than 54%) and usually being rich in iron, magnesium, or calcium. { 'bā·sik }

BASIC [COMPUT SCI] A procedure-level computer language designed to be easily learned and used by nonprofessionals, and well suited for an interactive, conversational mode of operation. Derived from Beginners All-purpose Symbolic Instruction Code. { 'bā·sik }

basic batch [COMPUT SCI] The least complex level of computer processing, in which application systems are normally made up of small programs that are run through the computer one at a time and that can process transactions only from sequential files. { 'bā·sik 'bach }

basic converter *See* basic-lined converter. { 'bā·sik kən'vərd·ər }

basic copper carbonate *See* copper carbonate. { 'bā·sik 'käp·ər 'kär·bə¦nāt }

basic disk operating system [COMPUT SCI] The part of a computer's operating system that handles the transfer of data between programs and disk units and the control of files. Abbreviated BDOS. { 'bā·sik ¦disk ¦äp·ə·rād·iŋ 'sis·təm }

basic dye [MATER] Any of the dyes which are salts of the colored organic bases containing amino and imino groups, combined with a colorless acid, such as hydrochloric or sulfuric. { 'bā·sik ¦dī }

basic element *See* elemental motion. { 'bā·sik 'el·ə·mənt }

basic feasible solution [IND ENG] A basic solution to a linear program model in which all the variables are nonnegative. { 'bā·sik ¦fēz·ə·bəl sə'lü·shən }

basic frequency [PHYS] The frequency, in any wave, which is considered the most important; in a driven system, it would generally be the driving frequency, while in most periodic waves it would correspond to the fundamental frequency. { 'bā·sik ¦frē·kwən·sē }

basic front [GEOL] An advancing zone of granitization enriched in calcium, magnesium, and iron. { 'bā·sik ¦frənt }

basic grasp [IND ENG] Any one of the fundamental means of taking hold of an object. { 'bā·sik ¦grasp }

basic group [CHEM] A chemical group (for example, OH^-) which, when freed by ionization in solution, produces a pH greater than 7. { 'bā·sik ¦grüp }

basic hornfels [PETR] A type of hornfels derived from a basic igneous rock. { 'bā·sik ¦hórn¦felz }

basichromatic [BIOL] Staining readily with basic dyes. { ¦bā·si¦krō¦mad·ik }

basic input/output system [COMPUT SCI] The part of a computer's operating system that handles communications between a program and external devices such as printers and electronic displays. Abbreviated BIOS. { 'in¦put 'aut¦put ¦sis·təm }

basic instruction [COMPUT SCI] An instruction in a computer program which is systematically changed by the program to obtain the instructions which are actually carried out. Also known as presumptive instruction; unmodified instruction. { 'bā·sik in'strək·shən }

basic-lined [MET] Pertaining to the walls and bottom of a melting furnace made of refractory materials, such as lime or dolomite, that have a basic reaction in the melting process. { 'bā·sik ¦līnd }

basic-lined converter [MET] A converter, such as the Peir-Smith copper converter, which has a basic refractory lining. Also known as basic converter. { 'bā·sik ¦līnd kən'vərd·ər }

basic linkage [COMPUT SCI] Computer coding that provides a standard means of connecting a given routine or program with other routines and that can be used repeatedly according to the same rules. { 'bā·sik 'liŋ·kij }

basic motion [IND ENG] A single, complete movement of a body member; determined by motion studies. { 'bā·sik 'mō·shən }

basic motion-time study [IND ENG] A system of predetermined motion-time standards for basic motions. Abbreviated BMT study. { 'bā·sik 'mō·shən 'tīm ¦stəd·ē }

basic open-hearth process [MET] An open-hearth process for steelmaking under basic slag; used for pig iron and scrap with a phosphorus content too low for the Bessemer process and too high for the acid open-hearth process. { 'bā·sik ¦ō pən ¦härth 'präs·əs }

basic oxide [INORG CHEM] A metallic oxide that is a base, or that forms a hydroxide when combined with water, such as sodium oxide to sodium hydroxide. { 'bā·sik 'äk¦sīd }

basic processing unit [COMMUN] Principal controller and data processor within the communications system. { 'bā·sik 'präs¦es·iŋ ¦yü·nət }

basic pulse repetition rate [NAV] In the loran navigation system, the lowest pulse repetition rate used by a chain of stations; the chain has a number of stations, all operating on the same radio frequency; one of these stations uses the basic rate while the others use rates that differ by fractional numbers of pulses per second. { 'bā·sik 'pəls ¦rep·ə'tish·ən ¦rāt }

basic Q *See* nonloaded Q. { 'bā·sik 'kyü }

basic refractory [MATER] Any heat-resistant material used for basic linings; examples are dolomite and magnesite. { 'bā·sik ri'frak·trē }

basic research [SCI TECH] Fundamental theoretical or experimental investigation to advance scientific knowledge, immediate practical application not being a direct objective. Also known as pure research. { 'bā·sik ri'sərch }

basic rock [PETR] An igneous rock with a relatively low silica content, and rich in iron, magnesium, or calcium. { 'bā·sik 'räk }

basic salt [INORG CHEM] A compound that is a base and a salt because it contains elements of both, for example, copper carbonate hydroxide, $Cu_2(OH)_2CO_3$. { 'bā·sik 'sólt }

basic schist [PETR] A schistose rock that forms from the metamorphism of a basic igneous rock. { 'bā·sik 'shist }

basic sediment and water [PETRO ENG] Oil, water, and foreign matter that collects in the bottom of petroleum storage tanks. Abbreviated BS&W. Also known as bottoms; bottom sediment; bottom settlings; sediment and water. { 'bā·sik 'sed·ə·mənt ən 'wȯd·ər }

basic slag [MET] A slag resulting from the steelmaking process; rich in phosphorus, it is ground and used as a nutrient in grasslands. { 'bā·sik 'slag }

basic software [COMPUT SCI] Software requirements that are taken into account in the design of the data-processing hardware and usually are provided by the original equipment manufacturer. { 'bā·sik 'sȯft,wer }

basic solution [IND ENG] A solution to a linear program model, consisting of *m* equations in *n* variables, obtained by solving for *m* variables in terms of the remaining $(n - m)$ variables and setting the $(n - m)$ variables equal to zero. [MATH] In bifurcation theory, a simple, explicitly known solution of a nonlinear equation, in whose neighborhood other solutions are studied. { 'bā·sik sə'lü·shən }

basic steel [MET] Steel made by the basic process, in a furnace with a basic lining. { 'bā·sik 'stēl }

basic telecommunications access method [COMPUT SCI] A method of controlling data transmission between a computer's main storage and its terminals and of providing applications programs with the capability of communicating with printers, terminals, and other devices. Abbreviated BTAM. { 'bā·sik ,tel·ə·kə,myü·nə̇'kā·shənz 'ak,ses ,meth·əd }

basic titrant [CHEM] A standard solution of a base used for titration. { 'bā·sik 'tī·trənt }

basic truss [MECH] A framework of bars arranged so that for any given loading of the bars the forces on the bars are uniquely determined by the laws of statics. { 'bās·ik 'trəs }

basic variables [COMPUT SCI] The *m* variables in a basic feasible solution for a linear programming model. { 'bā·sik 'ver·ē·ə·bəlz }

basidiocarp [MYCOL] The fruiting body of a fungus in the class Basidiomycetes. { bə'sid·ē·ə,kärp }

Basidiolichenes [BOT] A class of the Lichenes characterized by the production of basidia. { bə,sid·ē·ō,lī'kē,nēz }

Basidiomycetes [MYCOL] A class of fungi in the subdivision Eumycetes; important as food and as causal agents of plant diseases. { bə,sid·ē·ō,mī'sēd,ēz }

basidiophore [MYCOL] A basidia-bearing sporophore. { bə'sid·ē·ə,fȯr }

basidiospore [MYCOL] A spore produced by a basidium. { bə'sid·ē·ə,spȯr }

basidium [MYCOL] A cell, usually terminal, occurring in Basidiomycetes and producing spores (basidiospores) by nuclear fusion followed by meiosis. { bə'sid·ē·əm }

basification [GEOL] Development of a more basic rock, usually with more hornblende, biotite, and oligoclase, by contamination of a granitic magma in the assimilation of country rock. { ,bās·ə·fə'kā·shən }

basifixed [BOT] Attached at or near the base. { ¦bās·ə ¦fikst }

basil [BOT] The common name for any of the aromatic plants in the genus *Ocimum* of the mint family; leaves of the plant are used for food flavoring. [MATER] Sheephide tanned with bark. { 'bāz·əl *or* 'baz·əl }

basilar [BIOL] Of, pertaining to, or situated at the base. { 'bas·ə·lər }

basilar groove [ANAT] The cavity which is located on the upper surface of the basilar process of the brain and upon which the medulla rests. { 'bas·ə·lər ¦grüv }

basilar index [ANTHRO] The ratio of the distance from the basion to the alveolar point to the total skull length multiplied by 100. { 'bas·ə·lər 'in,deks }

basilar membrane [ANAT] A membrane of the mammalian inner ear supporting the organ of Corti and separating two cochlear channels, the scala media and scala tympani. { 'bas·ə·lər 'mem,brān }

basilar meningitis [MED] Inflammation of the meninges which affects chiefly the base of the brain, or in which exudate collects predominantly at the basal cisterns. { 'bas·ə·lər ,men·ən'jīd·əs }

basilar papilla [ANAT] **1.** A sensory structure in the lagenar portion of an amphibian's membranous labyrinth between the oval and round windows. **2.** The organ of Corti in mammals. { 'bas·ə·lər pə'pil·ə }

basilar plate [EMBRYO] An embryonic cartilaginous plate in vertebrates that is formed from the parachordals and anterior

BASIDIUM

A drawing of the basidiospores and basidium in fungi.

notochord and gives rise to the ethmoid and other bones of the skull. { 'bas·ə·lər ¦plāt }

basilar process [ANAT] A strong, quadrilateral plate of bone forming the anterior portion of the occipital bone, in front of the foramen magnum. { 'bas·ə·lər 'präs·əs }

basilic vein [ANAT] The large superficial vein of the arm on the medial side of the biceps brachii muscle. { bə'sil·ik 'vān }

basil oil [MATER] Any of the yellow, aromatic essential oils derived from the leaves of sweet basil; used for flavors and perfumes and in medicines. Also known as sweet basil oil. { 'bāz·əl ,ȯil }

basimesostasis [GEOL] A process of the partial or entire enclosure of plagioclase crystals in a diabase by augite. { 'bā·zē,mez·ə'stā·səs }

basin [CIV ENG] **1.** A dock employing floodgates to keep water level constant during tidal variations. **2.** A harbor for small craft. [DES ENG] An open-top vessel with relatively low sloping sides for holding liquids. [GEOL] **1.** A low-lying area, wholly or largely surrounded by higher land, that varies from a small, nearly enclosed valley to an extensive, mountain-rimmed depression. **2.** An entire area drained by a given stream and its tributaries. **3.** An area in which the rock strata are inclined downward from all sides toward the center. **4.** An area in which sediments accumulate. [MET] The mouth of a sprue in a gating system of castings into which the molten metal is first poured. [OCEANOGR] Deep portion of sea surrounded by shallower regions. { 'bās·ən }

basin accounting *See* hydrologic accounting. { 'bās·ən ə'kaȯnt·iŋ }

basin-and-range structure [GEOL] Regional structure dominated by fault-block mountains separated by basins filled with sediment. { 'bās·ən ən 'ranj ,strək·chər }

basin cultivation [AGR] A type of cultivation in which small basins are enclosed by low earthen ridges to check runoff from heavy rains, thus conserving soil moisture and minimizing soil erosion. { 'bā·sən kəl·tə'vā·shən }

basin fold [GEOL] Synclinal and anticlinal folds in structural basins. { 'bās·ən ,fōld }

basining [GEOL] A settlement of earth in the form of basins due to the solution and transportation of underground deposits of salt and gypsum. { 'bās·ən·iŋ }

basin length [GEOL] Length in a straight line from the mouth of a stream to the farthest point on the drainage divide of its basin. { 'bās·ən ,leŋkth }

basin of attraction [PHYS] The collection of all possible initial conditions of a dynamical system for which the trajectories representing that system in phase space will converge to a particular attractor. { 'bās·ən əv ə'trak·shən }

basin order [GEOL] A classification of basins according to stream drainage; for example, a first-order basin contains all of the drainage area of a first-order stream. { 'bās·ən ,ȯrd·ər }

basin peat *See* local peat. { 'bās·ən ,pēt }

basin perimeter [MAP] The length of a map line that encloses the catchment area of a drainage basin. { 'bās·ən pə'rim·əd·ər }

basin range [GEOL] A mountain range characteristic of the Great Basin in the western United States and formed by a faulted and tilted block of strata. { 'bās·ən ,rānj }

basin swamp [ECOL] A fresh-water swamp at the margin of a small calm lake, or near a large lake protected by shallow water or a barrier. { 'bās·ən ,swämp }

basin valley [GEOL] The filled-in depression of large inter-mountain areas; an example is Salt Lake Valley in Utah. { 'bās·ən ,val·ē }

basion [ANAT] In craniometry, the point on the anterior margin of the foramen magnum where the midsagittal plane of the skull intersects the plane of the foramen magnum. { 'bās·ē,än }

basion-bregma height [ANTHRO] Distance between basion and bregma. { 'bās·ē,än ¦breg·mə ,hīt }

basipetal [BIOL] Movement or growth from the apex toward the base. { bā'sip·əd·əl }

basipodite [INV ZOO] The distal segment of the protopodite of a biramous appendage in arthropods. { bā'sip·ə,dīt }

basipterygium [VERT ZOO] A basal bone or cartilage supporting one of the paired fins in fishes. { bə,sip·tə'rij·ē·əm }

basis [MATH] A set of linearly independent vectors in a vector space such that each vector in the space is a linear combination of vectors from the set. { 'bā·səs }

basis metal *See* base metal. { 'bā·səs 'med·əl }

basisternum [INV ZOO] In insects, the anterior one of the two sternal skeletal plates. { ¦bās·ə¦stər·nəm }

basistyle [INV ZOO] Either of a pair of flexible processes on the hypopygium of certain male dipterans. { ¦bās·ə¦stīl }

basis weight [GRAPHICS] The weight in pounds of 500 sheets of standard-size paper; certain sizes for a given class of paper are accepted as standard. { 'bā·səs ‚wāt }

basitarsus [INV ZOO] The basal segment of the tarsus in arthropods. { ¦bās·ə¦tär·səs }

basket [DES ENG] A lightweight container with perforations. [MECH ENG] A type of single-tube core barrel made from thin-wall tubing with the lower end notched into points, which is intended to pick up a sample of granular or plastic rock material by bending in on striking the bottom of the borehole or solid layer; may be used to recover an article dropped into a borehole. Also known as basket barrel; basket tube; sawtooth barrel. { 'bas·kət }

basket barrel *See* basket. { 'bas·kət ‚bar·əl }

basket cell [HISTOL] A type of cell in the cerebellum whose axis-cylinder processes terminate in a basketlike network around the cells of Purkinje. { 'bas·kət ‚sel }

basket coil *See* basket winding. { 'bas·kət ‚kȯil }

basket-handle arch [ARCH] An arch whose width is much greater than its height and which resembles an ellipse and is drawn from three or more centers. Also known as multicenter arch. { 'bas·kət ‚hand·əl ‚ärch }

basket star [INV ZOO] The common name for ophiuroid echinoderms belonging to the family Gorgonocephalidae. { 'bas·kət ‚stär }

basket strainer [CHEM ENG] A porous-sided or screen-covered vessel used to screen solid particles out of liquid or gas streams. { 'bas·kət ‚strān·ər }

basket sub [ENG] A fishing tool run above a bit or a mill to recover small nondrillable pieces of metal or debris in the well. { 'bas·kət ‚səb }

basket tube *See* basket. { 'bas·kət ‚tüb }

basket-weave [BUILD] A checkerboard pattern of bricks, flat or on edge. [TEXT] Pertaining to fabric in which two or more yarns are worked in warp and weft. { 'bas·kət ‚wēv }

basket winding [ELECTR] A crisscross coil winding in which successive turns are far apart except at points of crossing, giving low distributed capacitance. Also known as basket coil. { 'bas·kət ‚wind·iŋ }

Basommatophora [INV ZOO] An order of mollusks in the subclass Pulmonata containing many aquatic snails. { bə‚säm·ə'täf·rə }

basonym [SYST] The original, validly published name of a taxon. { 'bās·ə‚nim }

basophil [HISTOL] A white blood cell with granules that stain with basic dyes and are water-soluble. { 'bās·ə‚fil }

basophilia [BIOL] An affinity for basic dyes. [MED] An increase in the number of basophils in the circulating blood. [PATH] Stippling of the red cells with basic staining granules, representing a degenerative condition as seen in severe anemia, leukemia, malaria, lead poisoning, and other toxic states. { ‚bās·ə'fil·ē·ə }

basophilous [BIOL] Staining readily with the basic dyes. [ECOL] Of plants, growing best in alkaline soils. { bə'säf·ə·ləs }

basophobia [PSYCH] Abnormal fear of walking or standing erect. { ‚bās·ō'fō·bē·ə }

bas-relief *See* low relief. { ‚ba·ri'lēf }

bass [ACOUS] Sounds having frequencies at the lower end of the audio range, below about 250 hertz. [VERT ZOO] The common name for a number of fishes assigned to two families, Centrarchidae and Serranidae, in the order Perciformes. { bās (sounds) *and* bas (fish) }

bassanite [MINERAL] A white mineral consisting of hydrated calcium sulfate; a pseudomorph of gypsum. { bə'sä‚nīt }

bass boost [ELECTR] A circuit that emphasizes the lower audio frequencies, generally by attenuating higher audio frequencies. { ¦bās ‚büst }

bass compensation [ELECTR] A circuit that emphasizes the low-frequency response of an audio amplifier at low volume levels to offset the lower sensitivity of the human ear to weak low frequencies. { 'bās ‚käm·pən'sā·shən }

bass control [ELECTR] A manual tone control that attenuates higher audio frequencies in an audio amplifier and thereby emphasizes bass frequencies. { 'bās kən'trōl }

basset [GEOL] The outcropping edge of a layer of rock exposed to the surface. [VERT ZOO] A French breed of short-legged hunting dogs with long ears and a typical hound coat. { 'bas·ət }

basse-taille [GRAPHICS] Having a background carved in low relief. { bä'stī }

bassetite [MINERAL] A transparent, yellow, monoclinic mineral presumably consisting of a hydrated uranium phosphate containing divalent iron; occurs in groups of thin tablets. { 'bas·əd‚īt }

basso [GRAPHICS] A printing plate used in electrotyping of currency, made by depositing a layer of nickel, followed by a thick layer of iron. { 'ba·sō }

bassora gum [MATER] A type of high-colored gum similar to tragacanth gum; includes Indian gum. { 'bas·ə·rə ‚gəm }

bass reflex baffle [ENG ACOUS] A loudspeaker baffle having an opening of such size that bass frequencies from the rear of the loudspeaker emerge to reinforce those radiated directly forward. { ¦bas 'rē‚fleks ‚baf·əl }

bass response [ELECTR] A measure of the output of an electronic device or system as a function of an input of low audio frequencies. { 'bās ri‚späns }

bass trap [ENG ACOUS] Any device used in a sound-recording studio to absorb sound at frequencies less than about 100 hertz. { 'bās ‚trap }

basswood [BOT] A common name for trees of the genus *Tilia* in the linden family of the order Malvales. Also known as linden. { 'bas‚wùd }

bassy [ENG ACOUS] Pertaining to sound reproduction that overemphasizes low-frequency notes. { 'bās·ē }

bast *See* phloem. { bast }

bastard-cut file [DES ENG] A file that has coarser teeth than a rough-cut file. { 'bas·tərd ¦kət ‚fīl }

bastard pointing *See* bastard tuck pointing. { 'bas·tərd ‚pȯint·iŋ }

bastard sawing *See* backsawing. { 'bas·tərd ‚sȯ·iŋ }

bastard thread [DES ENG] A screw thread that does not match any standard threads. { 'bas·tərd ‚thred }

bastard tuck pointing [BUILD] An imitation tuck pointing in which the external face is parallel to the wall, but projects slightly and casts a shadow. Also known as bastard pointing. { 'bas·tərd ¦tək ‚pȯint·iŋ }

bast fiber [BOT] Any fiber stripped from the inner bark of plants, such as flax, hemp, jute, and ramie; used in textile and paper manufacturing. { 'bast ‚fī·bər }

bastion [GEOL] A prominent aggregation of bedrock extending from the mouth of a hanging glacial trough and reaching well into the main glacial valley. { 'bas·chən }

bastite [MINERAL] A hydrated magnesium silicate, a variety of serpentine occurring from the alteration of orthorhombic pyroxenes such as enstatite. { 'ba‚stīt }

bastnaesite [MINERAL] (Ce,La)CO_3(F,OH) A greasy yellow to reddish-brown fluorocarbonate rare-earth metal mineral; source of rare earths, for example, cerium and lanthanum. { 'bast·nə‚sīt }

bat [VERT ZOO] The common name for all members of the mammalian order Chiroptera. { bat }

Batales [BOT] A small order of dicotyledonous plants in the subclass Caryophillidae of the class Magnoliopsida containing a single family with only one genus, *Batis*. { bə'tā‚lēz }

bat bolt [DES ENG] A bolt whose butt or tang is bashed or jagged. { 'bat ‚bȯlt }

batch [COMPUT SCI] A set of items, records, or documents to be processed as a single unit. [ENG] **1.** The quantity of material required for or produced by one operation. **2.** An amount of material subjected to some unit chemical process or physical mixing process to make the final product substantially uniform. [PETRO ENG] In a pipeline operation, a volume of a crude oil of a given density that is pumped next to one of a different density to help prevent mixing of deliveries. { bach }

batch-and-forward system [COMPUT SCI] A data-processing system in which data are collected for a time and then transmitted as a unit to a computer. { 'bach ən 'fȯr·wərd ‚sis·təm }

batch box [ENG] A container of known volume used to

The largemouth bass (*Micropterus salmoides*).

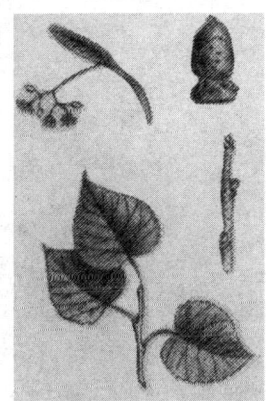

American basswood (*Tilia americana*).

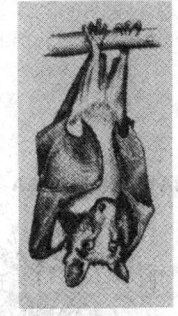

Epauletted fruit bat. (*Epomophorus wahlbergi*).

measure and mix the constituents of a batch of concrete, plaster, or mortar, to ensure proper proportions. { 'bach ¦bāks }

batch cementing [PETRO ENG] During an oil well cementing operation, the pumping of cement in a number of separate batches rather than pumping it in a single operation. { 'bach si¦ment·iŋ }

batch distillation [CHEM ENG] Distillation where the entire batch of liquid feed is placed into the still at the beginning of the operation, in contrast to continuous distillation, where liquid is fed continuously into the still. { 'bach dis·tə'lā·shən }

batched water [ENG] The mixing water added to a concrete or mortar mixture before or during the initial stages of mixing. { 'bacht ¦wȯd·ər }

batcher [MECH ENG] A machine in which the ingredients of concrete are measured and combined into batches before being discharged to the concrete mixer. { 'bach·ər }

batching [COMPUT SCI] Grouping records for the purpose of processing them in a computer. [ENG] Weighing or measuring the volume of the ingredients of a batch of concrete or mortar, and then introducing these ingredients into a mixer. { 'bach·iŋ }

batching sphere [PETRO ENG] A large rubber ball used in a pipeline to separate batches of crude oil to help prevent mixing of deliveries. { 'bach·iŋ ¦sfir }

Batchinsky relation [FL MECH] The relation stating that the fluidity of a liquid is proportional to the difference between the specific volume and a characteristic specific volume, approximately equal to the specific volume appearing in the van der Waals equation. { ba'chin·skē ri¦lā·shən }

batch job [COMPUT SCI] One of a group of jobs that are executed together by batch-processing techniques. { 'bach ¦jäb }

batch manufacturing [IND ENG] The manufacture of parts in discrete runs or lots, generally interspersed with other production procedures. { 'bach ¦man·ə'fak·chər·iŋ }

batch mixer [MECH ENG] A machine which mixes concrete or mortar in batches, as opposed to a continuous mixer. { 'bach ¦mik·sər }

batch oil [MATER] A pale, lemon-colored, low-viscosity mineral oil used particularly in the manufacture of cordage. { 'bach ¦ȯil }

batch-oriented applications [COMMUN] Applications of data communications that involve the transfer of thousands or even millions of bytes of data and are usually point-to-point and computer-to-computer. { 'bach ¦ȯr·ē¦ent·əd ¦ap·lə'kā·shənz }

batch plant [ENG] An operating installation of equipment including batchers and mixers as required for batching or for batching and mixing concrete materials. { 'bach ¦plant }

batch process [ENG] A process that is not in continuous or mass production; operations are carried out with discrete quantities of material or a limited number of items. { 'bach ¦präs·əs }

batch processing [COMPUT SCI] A technique that uses a single program loading to process many individual jobs, tasks, or requests for service. { 'bach ¦präs·es·iŋ }

batch production See series production. { 'bach prə'dək·shən }

batch reactor [CHEM ENG] A chemical reactor in which the reactants and catalyst are introduced in the desired quantities and the vessel is then closed to the delivery of additional material. { 'bach rē¦ak·tər }

batch rectification [CHEM ENG] Batch distillation in which the boiled-off vapor is re-condensed into liquid form and refluxed back into the still to make contact with the rising vapors. { 'bach ¦rek·tə·fə'kā·shən }

batch stream [COMPUT SCI] A group of batch processing programs that are scheduled to run on a computer. { 'bach ¦strēm }

batch system [COMPUT SCI] A computer system that uses batch processing. { 'bach ¦sis·təm }

batch total [COMPUT SCI] The total for a specified constituent quantity in a batch; used to verify the accuracy of operations on the batch. { 'bach ¦tōd·əl }

batch treatment [CHEM ENG] A corrosion control procedure in which chemical corrosion inhibitors are injected into the lines of a production system. [PETRO ENG] A process for separating an emulsion of crude oil and water into its components. { 'bach ¦trēt·mənt }

batch-type furnace [MECH ENG] A furnace used for heat treatment of materials, with or without direct firing; loading and unloading operations are carried out through a single door or slot. { 'bach ¦tīp 'fər·nəs }

bate [MATER] In tanning operations, any material used to remove lime from skins and to soften them prior to further processing. { bāt }

batea [MIN ENG] A conical-shaped wood unit (12.3 inches or 31 centimeters in diameter with about 150° apex angle) used to recover valuable metals from river channels and bars. { bə'tē·ə }

Bateman equations [NUC PHYS] A set of equations that give the number of atoms of each nuclide of a radioactive decay chain produced after a specified time, when a specified number of atoms of the parent nuclide are initially present. { 'bāt·mən i¦kwā·zhənz }

Batesian mimicry [ECOL] Resemblance of an innocuous species to one that is distasteful to predators. { 'bāt·sē·ən 'mim·ə·krē }

bat-handle switch [ELEC] A toggle switch having an actuating lever shaped like a baseball bat. { 'bat ¦hand·əl ¦swich }

bathochromatic shift [PHYS CHEM] The shift of the fluorescence of a compound toward the red part of the spectrum due to the presence of a bathochrome radical in the molecule. { ¦bath·ō¦krō¦mad·ik 'shift }

batholite [GEOL] An older massive protrusion of magma that solidifies as coarse crystalline rock in the deep horizons of the earth's crust. { 'bath·ə¦līt }

batholith [GEOL] A body of igneous rock, 40 square miles (100 square kilometers) or more in area, emplaced at great or intermediate depth in the earth's crust. { 'bath·ə¦lith }

bathometer [ENG] A mechanism which measures depths in water. { bə'thäm·əd·ər }

Bathonian [GEOL] A European stage of geologic time: Middle Jurassic, below Callovian, above Bajocian. Also known as Bathian. { bə'thōn·ē·ən }

bathophobia [PSYCH] Abnormal fear of depths. { ¦bath·ō'fōb·ē·ə }

Bathornithidae [PALEON] A family of Oligocene birds in the order Gruiformes. { ¦ba·thȯr'nith·ə¦dē }

bathtub capacitor [ELEC] A capacitor enclosed in a metal housing having broadly rounded corners like those on a bathtub. { 'bath¦təb kə'pas·əd·ər }

bathtub curve [IND ENG] An equipment failure-rate curve with an initial sharply declining failure rate, followed by a prolonged constant-average failure rate, after which the failure rate again increases sharply. { 'bath¦təb ¦kȯrv }

bathvillite [MINERAL] An oxygenated hydrocarbon mineral, found in Tortane Hill, Scotland, that is amorphous, fawn-brown, opaque, and quite friable. { 'bath·və¦līt }

bathyal zone [OCEANOGR] The biogeographic realm of the ocean depths between 100 and 1000 fathoms (180 and 1800 meters). { 'bath·ē·əl ¦zōn }

bathyclinograph [ENG] A mechanism which measures vertical currents in the deep sea. { ¦bath·ə¦klīn·ə¦graf }

bathyconductograph [ENG] A device to measure the electrical conductivity of sea water at various depths from a moving ship. { ¦bath·ə·kən'dək·tə¦graf }

Bathyctenidae [INV ZOO] A family of bathypelagic coelenterates in the phylum Ctenophora. { ¦ba·thik'ten·ə¦dē }

Bathyergidae [VERT ZOO] A family of mammals, including the South African mole rats, in the order Rodentia. { ¦bath·ē'ər·jə¦dē }

bathygram [ENG] A graph recording the measurements of sonic sounding instruments. { 'bath·ə¦gram }

Bathylaconoidei [VERT ZOO] A suborder of deep-sea fishes in the order Salmoniformes. { ¦bath·ə¦la·kə'nȯi·dē¦ī }

bathymetric biofacies [GEOL] The lateral distribution and character of underwater sedimentary strata. { ¦bath·ə'me·trik ¦bī·ō'fā·shēz }

bathymetric chart [MAP] A topographic map of the floor of the ocean. { ¦bath·ə'me·trik 'chärt }

bathymetry [ENG] The science of measuring ocean depths in order to determine the sea floor topography. { bə'thim·ə·trē }

Bathynellacea [INV ZOO] An order of crustaceans in the superorder Syncarida found in subterranean waters in England and central Europe. { ¦bath·ə·nel'ās·ē·ə }

Bathynellidae [INV ZOO] The single family of the crustacean order Bathynellacea. { ¦bath·ə'nel·ə¦dē }

BATHYNELLACEA

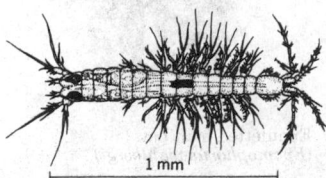

1 mm

Bathynella natans, female.

Bathyodontoidea [INV ZOO] A superfamily of nematodes of the order Mononchida containing the single family Bathyodontidae, characterized by a high, usually well-developed lip region, a hexaradiate oral opening with cuticularized walls, and a two-fold stoma; eight longitudinal rows of pores occur along the length of the body. { ¦bath·ē·ō·dän'tóid·ē·ə }

bathyorographical [GEOD] Concerned with depths of oceans and heights of mountains. { ¦bath·ē‚or·ə'graf·ə·kəl }

bathypelagic zone [OCEANOGR] The biogeographic realm of the ocean lying between depths of 500 and 2550 fathoms (900 and 3700 meters). { ¦bath·ə·pə'laj·ik 'zōn }

Bathypteroidae [VERT ZOO] A family of benthic, deep-sea fishes in the order Salmoniformes. { bə‚thip·tə'rói·dē }

bathyscaph [NAV ARCH] A free, crewed vehicle having a spherical cabin on the underside for exploring the deep ocean. { 'bath·ə‚scaf }

bathysphere [NAV ARCH] A spherical chamber in which persons are lowered for observation and study of ocean depths. { 'bath·ə‚sfir }

Bathysquillidae [INV ZOO] A family of mantis shrimps, with one genus (*Bathysquilla*) and two species, in the order Stomatopoda. { ¦bath·ə'skwil·ə‚dē }

bathythermogram [ENG] The record that is made by a bathythermograph. { ¦bath·ə'thər·mə‚gram }

bathythermograph [ENG] A device for obtaining a record of temperature against depth (actually, pressure) in the ocean from a ship underway. Also known as BT. Abbreviated BT. { ¦bath·ə'thər·mə‚graf }

bathythermosphere *See* bathythermograph. { ¦bath·ə'thər·mə‚sfir }

bathyvessel [NAV ARCH] A ship, such as a bathysphere or submarine, designed to operate far below the surface of the water. { 'bath·ə‚ves·əl }

batik [TEXT] **1.** A method of dyeing fabric in which parts of the cloth not intended to be dyed are covered with removable wax. **2.** The print so produced. **3.** The dyed cloth. { bə'tēk }

bating [CHEM ENG] Cleaning of depilated leather hides by the action of tryptic enzymes. { 'bād·iŋ }

batiste [TEXT] A plain-woven, sheer fabric made from fibers of relatively small diameter; texture is very soft. { bə'tēst }

Batoidea [VERT ZOO] The skates and rays, an order of the subclass Elasmobranchii. { bə'tóid·ē·ə }

batophobia [PSYCH] Abnormal fear of being near something of great height, such as a mountain or tall building. { ‚bad·ə'fōb·ē·ə }

Batrachoididae [VERT ZOO] The single family of the order Batrachoidiformes. { ‚ba·trə'kói·də‚dē }

Batrachoidiformes [VERT ZOO] The toadfishes, an order of teleostean fishes in the subclass Actinopterygii. { ‚ba·trə‚kói·də'för‚mēz }

batt [MATER] A blanket of insulating material usually 16 inches (41 centimeters) wide and 3 to 6 inches (8 to 15 centimeters) thick, used to insulate building walls and roofs. [MIN ENG] A thin layer of coal occurring in the lower part of shale strata that lie above and close to a coal bed. { bat }

batted work [ENG] A hand-dressed stone surface scored from top to bottom in narrow parallel strokes (usually 8-10 per inch or 20-25 per centimeter) by use of a batting tool. { 'bad·əd ‚wərk }

batten [AERO ENG] Metal, wood, or plastic panels laced to the envelope of a blimp in the nose cone to add rigidity to the nose and provide a good point of attachment for mooring. [BUILD] **1.** A sawed timber strip of specific dimension-usually 7 inches (18 centimeters) broad, less than 4 inches (10 centimeters) thick, and more than 6 feet (1.8 meters) long-used for outside walls of houses, flooring, and such. **2.** A strip of wood nailed across a door or other structure made of parallel boards to strengthen it and prevent warping. **3.** *See* furring. { 'bat·ən }

battenberg [TEXT] A coarse type of renaissance lace made from linen braid tape and linen thread. { 'bat·ən‚bərg }

batten door [BUILD] A wood door without stiles which is constructed of vertical boards held together by horizontal battens on the back side. Also known as ledged door. { 'bat·ən ‚dòr }

battened column [CIV ENG] A column consisting of two longitudinal shafts, rigidly connected to each other by batten plates. { 'bat·ənd 'käl·əm }

battened wall [BUILD] A wall to which battens have been affixed. Also known as strapped wall. { 'bat·ənd 'wòl }

batten plate [CIV ENG] A rectangular plate used to connect two parallel structural steel members by riveting or welding. { 'bat·ən ‚plāt }

batten roll [BUILD] In metal roofing, a roll joint formed over a triangular-shaped wood piece. Also known as conical roll. { 'bat·ən ‚ròl }

batten seam [BUILD] A seam in metal roofing which is formed around a wood strip. { 'bat·ən ‚sēm }

batter [CIV ENG] A uniformly steep slope in a retaining wall or pier; inclination is expressed as 1 foot horizontally per vertical unit (in feet). { 'bad·ər }

batter board [CIV ENG] Horizontal boards nailed to corner posts located just outside the corners of a proposed building to assist in the accurate layout of foundation and excavation lines. { 'bad·ər ‚bòrd }

batter brace [CIV ENG] A diagonal brace which reinforces one end of a truss. Also known as batter post. { 'bad·ər ‚brās }

battered-child syndrome [MED] A clinical condition in young children due to serious physical abuse, generally from a parent or foster parent. { ¦bad·ərd ¦chīld 'sin‚drōm }

batter level [ENG] A device for measuring the inclination of a slope. { 'bad·ər ‚lev·əl }

batter pile [CIV ENG] A pile driven at an inclination to the vertical to provide resistance to horizontal forces. Also known as brace pile; spur pile. { 'bad·ər ‚pīl }

batter post [CIV ENG] **1.** A post at one side of a gateway or at a corner of a building for protection against vehicles. **2.** *See* batter brace. { 'bad·ər ‚pōst }

batter stick [CIV ENG] A tapered board which is hung vertically and used to test the batter of a wall surface. { 'bad·ər ‚stik }

battery [CHEM ENG] A series of distillation columns or other processing equipment operated as a single unit. [ELEC] A direct-current voltage source made up of one or more units that convert chemical, thermal, nuclear, or solar energy into electrical energy. [ORD] A group of guns or other weapons, such as mortars, machine guns, artillery pieces, or of searchlights, set up under one tactical commander in a certain area. { 'bad·ə·rē }

battery amalgamation [MET] Amalgamation by means of mercury placed in the mortar box of a stamp battery. { 'bad·ə·rē ə‚mal·gə'mā·shən }

battery assay [MIN ENG] An assay of samples taken from ore as crushed in a stamp battery. { 'bad·ə·rē 'as‚ā }

battery charger [ELEC] A rectifier unit used to change alternating to direct power for charging a storage battery. Also known as charger. { 'bad·ə·rē ‚chär·jər }

battery clip [ELEC] A terminal of a connecting wire having spring jaws that can be quickly snapped on a terminal of a device, such as a battery, to which a temporary wire connection is desired. { 'bad·ə·rē ‚klip }

battery command periscope [OPTICS] An optical instrument consisting of dual telescope tubes positioned vertically on a common mounting; it provides periscopic vision for the observer, and may be used to observe artillery fire. { 'bad·ə·rē kə¦mand 'per·ə‚skōp }

battery depolarizer *See* depolarizer. { 'bad·ə·rē ‚dē'pōl·ə‚rīz·ər }

battery electrolyte [PHYS CHEM] A liquid, paste, or other conducting medium in a battery, in which the flow of electric current takes place by migration of ions. { 'bad·ə·rē i'lek·trə‚līt }

battery eliminator [ELECTR] A device which supplies electron tubes with voltage from electric power supply mains. { 'bad·ə·rē ə'lim·ə‚nād·ər }

battery limits [CHEM ENG] An area in a refinery or chemical plant encompassing a processing unit or battery of units along with their related utilities and services. { 'bad·ə·rē ‚lim·əts }

battery manganese *See* manganese dioxide. { 'bad·ə·rē ‚maŋ·gə‚nēs }

battery of genes [GEN] The set of producer genes which is activated when a particular sensor gene activates its set of integrator genes. { 'bad·ə·rē əv 'jēnz }

BATRACHOIDIFORMES

Atlantic midshipman (*Porichthys porosissimus*), about 8 inches (20 centimeters) long. (*After D. S. Jordan*)

battery, overvoltage, ringing, supervision, coding, hybrid and test access See BORSCHT. { 'bad·ə·rē ¦ō·vər¦vōl·tij 'riŋ·iŋ ,sü·pər'vizh·ən 'kōd·iŋ 'hī·brid ən 'test ,ak,ses }

battery reefs See Kimberley reefs. { 'bad·ə·rē ,rēfs }

battery separator [ELEC] An insulating plate inserted between the positive and negative plates of a battery to prevent them from touching. { 'bad·ə·rē ,sep·ə,rād·ər }

Battey disease [MED] A tuberculosislike disease of humans caused by *Mycobacterium intracellulare*. { 'bad·ē di'zēz }

batting tool [ENG] A mason's chisel usually 3–4 $\frac{1}{2}$ inches (7.6–11.4 centimeters) wide, used to dress stone to a striated surface. { 'bad·iŋ ,tül }

battlefield recovery [ORD] Removal of disabled or abandoned material, either enemy or friendly, from the battlefield and moving it to a recovery collecting point or to a maintenance or supply establishment. { 'bad·əl,fēld ri'kəv·rē }

battleship [NAV ARCH] One of a class of heaviest and most extensively armed and armored vessels, with at least 10-inch (25-centimeter) armor plating and guns of 12-inch (30-centimeter) or larger caliber. { 'bad·əl,ship }

battle sight [ORD] A predetermined sight setting, carried on a weapon, that enables the firer to engage targets effectively at battle ranges when conditions do not permit exact sight settings. { 'bad·əl ,sīt }

batture [GEOL] An elevation of the bed of a river under the surface of the water; sometimes used to signify the same elevation when it has risen above the surface. { ba'tùr }

baud [COMMUN] A unit of telegraph signaling speed equal to the number of code elements (pulses and spaces) per second or twice the number of pulses per second. { bòd }

Baudot code [COMMUN] A teleprinter code that uses a combination of five or six marking and spacing intervals of equal duration for each character. { bò'dō ,kōd }

baulk [ARCHEO] A strip of earth left in place between the trenches of an archeological dig to permit study of the stratigraphy of the area as long as possible. { bòk }

Baumé hydrometer scale [PHYS CHEM] A calibration scale for liquids that is reducible to specific gravity by the following formulas: for liquids heavier than water, specific gravity = $145 \div (145 - n)$ [at 60°F]; for liquids lighter than water, specific gravity = $140 \div (130 + n)$ [at 60°F]; n is the reading on the Baumé scale, in degrees Baumé; Baumé is abbreviated Bé. { bō'mā hī'dräm·əd·ər ,skāl }

baumhauerite [MINERAL] $Pb_4As_6S_{13}$ A lead to steel gray, monoclinic mineral consisting of lead arsenic sulfide. { baù'maù·ə,rīt }

Bauschinger effect [MET] A phenomenon by which the plastic deformation of a metal increases the tensile yield strength and decreases the compressive yield strength. { 'baù,shiŋ·ər i'fekt }

Bautz-Morgan classification [ASTRON] A classification of clusters of galaxies into three categories, ranging from type I in which the cluster contains a supergiant elliptical galaxy, to type II in which the cluster contains no member that is significantly brighter than the general bright population. { 'baùts 'mòr·gən ,klas·ə·fə,kā·shən }

bauxite [PETR] A whitish, grayish, brown, yellow, or reddish-brown rock composed of hydrous aluminum oxides and aluminum hydroxides and containing impurities such as free silica, silt, iron hydroxides, and clay minerals; the principal commercial source of aluminum. { 'bòk,sīt }

bauxite treating [CHEM ENG] A catalytic petroleum process in which a vaporized petroleum fraction is passed through beds of bauxite; conversion of many different sulfur compounds, particularly mercaptans into hydrogen sulfide, takes place. { 'bòk,sīt ,trēd·iŋ }

bauxitization [GEOL] Bauxite development from either primary aluminum silicates or secondary clay minerals. { ¦bòk·sə·də'zā·shən }

bavenite See duplexite. { bə've,nīt }

Baveno twin law [CRYSTAL] An uncommon twin law applicable in feldspar, in which the twin plane and composition surface are (021); a Baveno twin usually consists of two individuals. { bə've·nō ,twin ,lò }

b axis [CRYSTAL] A crystallographic axis that is oriented horizontally, right to left. [MECH ENG] The angle that specifies the rotation of a machine tool about the *y* axis. [PETR] A direction in the plane of movement that is at a right angle to the tectonic transport direction. { 'bē ,ak·səs }

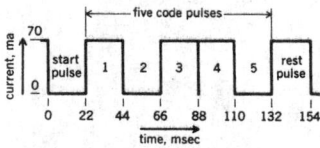

BAUDOT CODE

The five-unit stop-start telegraph code, seven-unit code pattern. This combination represents letter F.

bay [AERO ENG] A space formed by structural partitions on an aircraft. [ARCH] Division of a building between adjacent beams or columns. [BOT] *Laurus nobilis*. An evergreen tree of the laurel family. [COMPUT SCI] See drive bay. [ELECTROMAG] One segment of an antenna array. [ENG] A housing used for equipment. [GEOGR] **1.** A body of water, smaller than a gulf and larger than a cove in a recess in the shoreline. **2.** A narrow neck of water leading from the sea between two headlands. [GEOPHYS] A simple transient magnetic disturbance, usually an hour in duration, whose appearance on a magnetic record has the shape of a V or a bay of the sea. { bā }

Bayard-Alpert ionization gage [ELECTR] A type of ionization vacuum gage using a tube with an electrode structure designed to minimize x-ray-induced electron emission from the ion collector. { ¦bā·ərd ¦al,pərt ī·ən·ə'zā·shən ,gāj }

bay bar See baymouth bar. { 'bā ,bär }

bay barrier [GEOL] A narrow shoal or small point of land projecting from the shore across the mouth of a bay and severing the bay's connection with the main body of water. { 'bā ,bar·ē·ər }

bayberry [BOT] **1.** *Pimenta acris*. A West Indian tree related to the allspice; a source of bay oil. Also known as bay-rum tree; Jamaica bayberry; wild cinnamon. **2.** Any tree of the genus *Myrica*. { 'bā,ber·ē }

bayberry tallow See bayberry wax. { 'bā,ber·ē ,tal·ō }

bayberry wax [MATER] Green, bitter-tasting wax derived from boiling of wax myrtle (*Myrica*) berries; used for candles, soaps, medicine. Also known as bayberry tallow; laurel wax; myrtle wax. { 'bā,ber·ē ,waks }

bay delta [GEOL] A usually triangular alluvial deposit formed at the point where the mouth of a stream enters the head of a drowned valley. { 'bā ,del·tə }

Bayer letter [ASTRON] The Greek (or Roman) letter used in a Bayer name. { 'bī·ər ,led·ər }

Bayer name [ASTRON] The Greek (or Roman) letter and the possessive form of the Latin name of a constellation, used as a star name; examples are α Cygni (Deneb), β Orionis (Rigel), and η Ursae Majoris (Alkaid). { 'bī·ər ,nām }

Bayer process [MET] A method of producing alumina from bauxite by heating it in a sodium hydroxide solution. { 'bī·ər ,präs·əs }

Bayer's constellations [ASTRON] Thirteen constellations in the southern hemisphere named by J. Bayer. { 'bī·ərz ,kan·stə'lā·shənz }

Bayes decision rule [STAT] A decision rule under which the strategy chosen from among several available ones is the one for which the expected value of payoff is the greatest. { 'bāz di'sizh·ən ,rül }

Bayesian statistics [STAT] An approach to statistics in which estimates are based on a synthesis of a prior distribution and current sample data. { ¦bāz·ē·ən stə'tis·tiks }

Bayesian theory [STAT] A theory, as of statistical inference or decision making, in which probabilities are associated with individual events or statements rather than with sequences of events. { 'bāz·ē·ən ,thē·ə·rē }

Bayes rule [STAT] The rule that the probability $P(E_i|A)$ of some event E_i, given that another event A has been observed, is $P(E_i)P(A|E_i)/P(A)$, where $P(E_i)$ is the prior probability of E_i, determined either objectively or subjectively, and $P(A)$, the probability of A, is given by the sum over all possible events E_j of the quantity $P(E_j)P(A|E_j)$. { 'bāz ,rül }

Bayes' theorem [MATH] A theorem stating that the probability of a hypothesis, given the original data and some new data, is proportional to the probability of the hypothesis, given the original data only, and the probability of the new data, given the original data and the hypothesis. Also known as inverse probability principle. { 'bāz 'thir·əm }

bay head [GEOL] A swampy region at the head of a bay. { 'bā ¦hed }

bay-head bar [GEOL] A bar formed a short distance from the shore at the head of a bay. { 'bā ¦hed ,bär }

bay-head beach [GEOL] A beach formed around a bay head by storm waves; layers of sediment cover the bay floor and bare rock benches front the headland cliffs. { 'bā ¦hed ,bēch }

bay-head delta [GEOL] A delta at the head of an estuary or a bay into which a river discharges because of the margin of the land's late partial submergence. { 'bā ¦hed ,del·tə }

bay ice [OCEANOGR] Sea ice that is young and flat but sufficiently thick to impede navigation. { ¦bā ¦īs }

bayldonite [MINERAL] $Cu_3(AsO_4)_2(OH)_2$ An apple green to yellowish-green monoclinic mineral consisting of a basic arsenate of copper and lead; occurs in minute mammillary concretions, in massive form, and as crusts. { ′bāl·də,nīt }

bay leaf [FOOD ENG] A herb; the dried leaf of the bay tree (*Laurus nobilis*). { ′bā ,lēf }

bayleyite [MINERAL] $Mg_2(UO_2)(CO_3)_3 \cdot 18H_2O$ A sulfur yellow monoclinic mineral consisting of a hydrated carbonate of magnesium and uranium; occurs as minute, short-prismatic crystals. { ′bā·lē,īt }

baymouth bar [GEOL] A bar extending entirely or partially across the mouth of a bay. Also known as bay bar. { ′bā ,maůth ,bär }

bay oil [MATER] Yellow essential oil with clovelike odor and pungent taste; derived from distillation of West Indian bayberry (*Pimenta acris*) leaves used in flavors and perfumes. Also known as myrica oil. { ′bā ,ȯil }

bayonet [ORD] An edged steel blade with a tapered point and a formed handle with an underhand grip, designed to be attached to the muzzle end of a rifle, shotgun, or the like for use in hand-to-hand combat. { ′bā·ə′net }

bayonet base [ELEC] A tube base or lamp base having two projecting pins on opposite sides of a smooth cylindrical surface to engage in corresponding slots in a bayonet socket and hold the base firmly in the socket. { ′bā·ə′net ,bās }

bayonet coupling [DES ENG] A coupling in which two or more pins extend out from a plug and engage in grooves in the side of a socket. { ′bā·ə′net ,kəp·liŋ }

bayonet Neil-Concelman connector See BNC connector. { ,bā·ə′net ′nēl ′käns·əl·mən kə,nek·tər }

bayonet socket [DES ENG] A socket, having J-shaped slots on opposite sides, into which a bayonet base or coupling is inserted against a spring and rotated until its pins are seated firmly in the slots. { ′bā·ə′net ′säk·ət }

bayonet-tube exchanger [MECH ENG] A dual-tube apparatus with heating (or cooling) fluid flowing into the inner tube and out of the annular space between the inner and outer tubes; can be inserted into tanks or other process vessels to heat or cool the liquid contents. { ′bā·ə′net ,tüb iks′chānj·ər }

bayou [HYD] A small, sluggish secondary stream or lake that exists often in an abandoned channel or a river delta. { ′bī,yü }

bay rum [MATER] A product originally prepared from the distillation of rum and water mixed with bayberry (*Pimenta acris*) leaves but now made from a mixture of bay oil, orange oil, pimenta oil, alcohol, and water; used in shaving lotions and alcohol rubs. { ′bā ′rəm }

bay-rum tree See bayberry. { ′bā ′rəm ,trē }

bayside beach [GEOL] A beach formed at the side of a bay by materials eroded from nearby headlands and deposited by longshore currents. { ′bā,sīd ,bēch }

bay window [ARCH] A window that projects outward from the wall of a building and forms a small indoor alcove. { ′bā ′win·dō }

bazooka [ELECTR] See balun. [ORD] Popular name applied to the 2.36-inch (60-millimeter) rocket launcher. { bə′zü·kə }

bazzite [MINERAL] $Sc_2Be_3Si_6O_{18}$ An azure-blue mineral that crystallizes in the hexagonal system; the rare scandium analog of beryl. { ′ba,zīt }

B battery [ELECTR] The battery that furnishes required direct-current voltages to the plate and screen-grid electrodes of the electron tubes in a battery-operated circuit. { ′bē ,bad·ə·rē }

BBC See bromobenzylcyanide.

BBD See bucket brigade device.

B-B fraction [CHEM ENG] A mixture of butanes and butenes distilled from a solution of light liquid hydrocarbons. { ¦bē¦bē ′frak·shən }

BB gun See air rifle. { ¦bē¦bē ,gən }

B bit [MIN ENG] A nonstandard core bit no longer in common use except in drilling deep boreholes to sample gold-bearing deposits in South Africa; the set outside and inside diameters are about 21/16 and 13/8 inches (52 and 35 millimeters), respectively. { ′bē ,bit }

bbl See barrel.

B blasting powder [MATER] A mixture of nitrate of soda, charcoal, and sulfur; used in coal mines. Also known as soda blasting powder. { ′bē ′blas·tiŋ ,paůd·ər }

B box See index register. { ′bē ,bäks }

BBS See bulletin board system.

BCAS See beacon collision avoidance system.

BCD See barrels per calendar day.

BCD system See binary coded decimal system. { ′bē¦sē′dē ,sis·təm }

B cell [IMMUNOL] One of a heterogeneous population of bone-marrow-derived lymphocytes which participates in the immune responses. Also known as B lymphocyte. { ′bē ,sel }

b-c fracture [GEOL] A tension fracture parallel with the fabric plane and normal to the *a* axis. { ′bē¦sē ′frak·chər }

BCG vaccine See Bacillus Calmette-Guérin vaccine. { ′bē¦sē¦jē vak′sēn }

B chain See light chain. { ′bē ,chān }

bcl-2 [BIOCHEM] A family of proteins that operate in the effector phase of apoptosis and may either promote or inhibit apoptosis.

b-complete [RELAT] A criterion determining whether a space-time is free of singularities based on whether curves of finite length have an end point, where length is defined by a generalized affine parameter along the curve. { ′bē kəm′plēt }

b-c plane [GEOL] A plane that is perpendicular to the plane of movement and parallel to the *b* direction in that plane. { ′bē¦sē ,plān }

BCS theory See Bardeen-Cooper-Schrieffer theory. { ′bē¦sē¦es ,thē·ə·rē }

BD See barrels per day.

B damage [ORD] Combat damage such that the aircraft is unable to return to its base. { ′bē ,dam·ij }

BDC See bottom dead center.

Bdelloidea [INV ZOO] An order of the class Rotifera comprising animals which resemble leeches in body shape and manner of locomotion. { də′lȯid·ē·ə }

Bdellomorpha [INV ZOO] An order of ribbonlike worms in the class Enopla containing the single genus *Malacobdella*. { ,del·ə′mȯr·fə }

Bdellonemertini [INV ZOO] An equivalent name for the Bdellomorpha. { ,del·ə′ner·tə,nī }

Bdellovibrio [MICROBIO] A genus of bacteria of uncertain affiliation; parasites of other bacteria; curved, motile rods with a polar flagellum in parasitic state; in nonparasitic state, cells are helical. { ,del·ə′vī·brē,ō }

bdft See board-foot.

B display [ELECTR] The presentation of radar output data in rectangular coordinates in which range and azimuth are plotted on the coordinate axes. Also known as range-bearing display. { ′bē dis′plā }

BDOS See basic disk operating system. { ′bē,dȯs }

Be See beryllium.

Bé See Baumé hydrometer scale.

BE See binding energy.

beach [GEOL] The zone of unconsolidated material that extends landward from the low-water line to where there is marked change in material or physiographic form or to the line of permanent vegetation. [NAV] To intentionally run a craft ashore, as a landing ship. { bēch }

beach cusp See cusp. { ′bēch ,kəsp }

beach cycle [GEOL] Periodic retreat and outbuilding of beaches resulting from waves and tides. { ′bēch ,sī·kəl }

beach drift [GEOL] The material transported by drifting of beach. { ′bēch ,drift }

beach face See foreshore. { ′bēch ,fās }

beach gravel [GEOL] Gravels in which most of the particles cluster about one size. { ′bēch ,grav·əl }

beachhead [ORD] An area on an enemy shoreline which opposing troops have landed upon and occupied. { ′bēch,hed }

beach marker [NAV] Usually a light, buoy, panel, or electronic device, used to identify a beach, for incoming water traffic. { ′bēch ,mär·kər }

beach mining [MIN ENG] The mining of the heavy minerals, such as rutile, zircon, monazite, ilmenite, and sometimes gold, which occur in sand dunes, beaches, coastal plains, and deposits located inland from the shoreline. { ′bēch ,mīn·iŋ }

beach plain [GEOL] Embankments of wave-deposited material added to a prograding shoreline. { ′bēch ,plān }

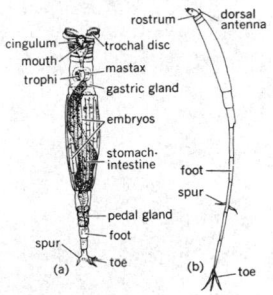

BDELLOIDEA

Bdelloid rotifers, with elongated, segmented bodies. *(a) Rotaria. (b) R. neptunia. (From L. Hyman, The Invertebrates, vol. 3, McGraw-Hill, 1951)*

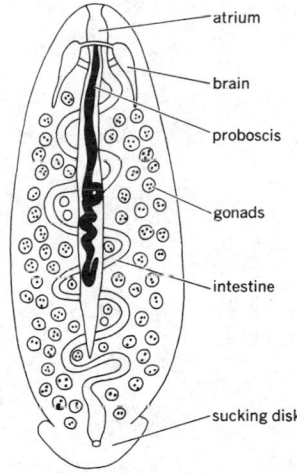

BDELLOMORPHA

Malacobdella grossa. (From W. R. Coe, Biology of the nemerteans of the Atlantic coast of North America, Trans. Conn. Acad. Arts Sci., 35:308, 1943)

B DISPLAY

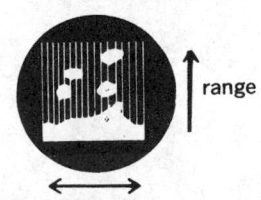

Type B radar display.

beach platform See wave-cut bench. { 'bēch ,plat,fòrm }

beach profile [GEOL] Intersection of a beach's ground surface with a vertical plane perpendicular to the shoreline. { 'bēch 'prō,fīl }

beach ridge [GEOL] A continuous mound of beach material behind the beach that was heaped up by waves or other action. { 'bēch ,rij }

beachrock [PETR] A friable to well-cemented rock made of calcareous skeletal debris that is cemented together by calcium carbonate. { 'bēch,räk }

beach scarp [GEOL] A nearly vertical slope along the beach caused by wave erosion. { 'bēch ,skärp }

beacon [NAV] **1.** A light, group of lights, electronic apparatus, or other device which emits identifying signals related to their positions so that the information so produced can be used by the navigator or pilots of aircraft and ships for guidance orientation or warning. **2.** A structure where such a device is mounted or located. { 'bē·kən }

beaconage [NAV] A system of beacons. { 'bē·kən·ij }

beacon collision avoidance system [NAV] An airborne collision avoidance system that makes use of the air-traffic control radio beacon system (ATCRBS) transponders. Abbreviated BCAS. { 'bē·kən kə'lizh·ən ə'vóid·əns ,sis·təm }

beacon delay [ELECTR] The amount of transponding delay within a beacon, that is, the time between the arrival of a signal and the response of the beacon. { 'bē·kən di'lā }

beacon presentation [ELECTR] The radarscope presentation resulting from radio-frequency waves sent out by a radar beacon. { 'bē·kən ,prē·zən'tā·shən }

beacon skipping [ELECTR] A condition where transponder return pulses from a beacon are missing at the interrogating radar. { 'bē·kən ,skip·iŋ }

beacon stealing [ELECTR] Loss of beacon tracking by one radar due to stronger signals from an interfering radar. { 'bē·kən ,stēl·iŋ }

beacon tracking [ENG] The tracking of a moving object by means of signals emitted from a transmitter or transponder within or attached to the object. { 'bē·kən ,trak·iŋ }

beacon-tracking radar [NAV] Radar equipment used in air-traffic control facilities for beacon tracking. { 'bē·kən ,trak·iŋ ¦rā,där }

bead [COMPUT SCI] A small subroutine. [DES ENG] A projecting rim or band. [ELECTROMAG] A glass, ceramic, or plastic insulator through which passes the inner conductor of a coaxial transmission line and by means of which the inner conductor is supported in a position coaxial with the outer conductor. [MET] **1.** The drop of precious metal obtained by cupellation in fire assaying. **2.** See weld bead. { bēd }

bead and butt [BUILD] Framed work in which the panel is flush with the framing and has a bead run on two edges in the direction of the grain; the ends are left plain. Also known as bead butt; bead butt work. { ¦bēd ən 'bət }

bead-and-flush panel See beadflush panel. { ¦bēd ən 'fləsh ,pan·əl }

bead and quirk See quirk bead. { ¦bēd ən 'kwərk }

bead and reel [BUILD] A semiround convex molding decorated with a pattern of disks alternating with round or elongated beads. Also known as reel and bead. { ¦bēd ən 'rēl }

bead butt See bead and butt. { 'bēd ,bət }

bead, butt, and square [BUILD] Framed work similar to bead and butt but having the panels flush on the beaded face only, and showing square reveals on the other. { ¦bēd ,bət ən 'skwär }

bead butt work See bead and butt. { 'bēd ,bət ,wərk }

beaded lake See paternoster lake. { 'bēd·əd ,lāk }

beaded molding [BUILD] A molding or cornice bearing a cast plaster string of beads. { 'bēd·əd ¦mōl·diŋ }

beaded transmission line [ELECTROMAG] Line using beads to support the inner conductor in coaxial transmission lines. { 'bēd·əd tranz'mish·ən ,līn }

beaded tube end [MECH ENG] The exposed portion of a rolled tube which is rounded back against the sheet in which the tube is rolled. { 'bēd·əd 'tüb ,end }

beadflush panel [BUILD] A panel which is flush with the surrounding framing and finished with a flush bead on all edges of the panel. Also known as bead-and-flush panel. { 'bēd ,fləsh ,pan·əl }

beading [BUILD] Collectively, the bead moldings used in ornamenting a given surface. [MET] The placing of a bead

on a piece of sheet metal for either decorative or strengthening purposes. { 'bēd·iŋ }

beading plane [DES ENG] A plane having a curved cutting edge for shaping beads in wood. Also known as bead plane. { 'bēd·iŋ ,plān }

bead-jointed [ENG] Of a carpentry joint, having a bead along the edge of one piece to make the joint less conspicuous. { 'bēd ,jóin·təd }

bead molding [BUILD] A small, convex molding of semicircular or greater profile. Also known as baguette. { 'bēd ,mōl·diŋ }

bead plane See beading plane. { 'bēd ,plān }

bead test [ANALY CHEM] In mineral identification, a test in which borax is fused to a transparent bead, by heating in a blowpipe flame, in a small loop formed by platinum wire; when suitable minerals are melted in this bead, characteristic glassy colors are produced in an oxidizing or reducing flame and serve to identify elements. { 'bēd ,test }

bead thermistor [ELEC] A thermistor made by applying the semiconducting material to two wire leads as a viscous droplet, which cements the leads upon firing. { 'bēd thər'mis·tər }

beak [BOT] Any pointed projection, as on some fruits, that resembles a bird bill. [INV ZOO] The tip of the umbo in bivalves. [VERT ZOO] **1.** The bill of a bird or some other animal, such as the turtle. **2.** A projecting jawbone element of certain fishes, such as the sawfish and pike. { bēk }

beaker [NAV ARCH] A shipboard vessel, usually used for storing water. [SCI TECH] A deep, open-mouthed, cylindrical vessel with thin walls, which usually has a projecting lip for pouring. { 'bē·kər }

beaker sampler [PETRO ENG] A small, cylindrical vessel with a tapered top used to collect crude oil samples; it is made of low-sparking metal or glass, the bottom is weighted, and there is a small stoppered opening at the top. { 'bē·kər ,sam·plər }

beakhead [NAV ARCH] A space forward of the forecastle of a ship where latrines for crew members are located. { 'bēk,hed }

beaking joint [BUILD] A joint formed by several heading joints occurring in one continuous line; especially used in connection with the laying of floor planks. { 'bēk·iŋ ,jóint }

beam [CIV ENG] A body, with one dimension large compared with the other dimensions, whose function is to carry lateral loads (perpendicular to the large dimension) and bending movements. [NAV ARCH] The width of a ship at its widest point. [PHYS] A concentrated, nearly unidirectional flow of particles, or a like propagation of electromagnetic or acoustic waves. [TEXT] Spool-shaped holder, 8 to 12 feet (2.4 to 3.7 meters) in length, on which is wrapped yarn that is to be transferred to the warp holder of a loom. { bēm }

beam-and-girder construction [BUILD] A system of floor construction in which the load is distributed by slabs to spaced beams and girders. { ¦bēm ən 'gər·dər kən'strək·shən }

beam-and-slab floor [BUILD] A floor system in which a concrete floor slab is supported by reinforced concrete beams. { ¦bēm ən 'slab ,flòr }

beam angle See beam width. { 'bēm ,aŋ·gəl }

Beaman stadia arc [ENG] An attachment to an alidade consisting of a stadia arc on the outer edge of the visual vertical arc; enables the observer to determine the difference in elevation of the instrument and stadia rod without employing vertical angles. { 'bē·mən 'stād·ē·ə ,ärk }

beam antenna [ELECTROMAG] An antenna that concentrates its radiation into a narrow beam in a definite direction. { 'bēm an'ten·ə }

beam approach beacon system See blind approach beacon system. { 'bēm ə'prōch 'bē·kən ,sis·təm }

beam attack [ORD] An attack directed against the side of an aircraft, a tank, or a ship. { 'bēm ə'tak }

beam attenuator [SPECT] An attachment to the spectrophotometer that reduces reference to beam energy to accommodate undersized chemical samples. { 'bēm ə'ten·yə,wād·ər }

beam-balanced pump [PETRO ENG] An oil well pumping unit having a center-pivoted beam with the sucker rod plunger (pump) at the front end and a counterweight on the rearward extension. { 'bēm ,bal·ənst 'pəmp }

beam bearing plate [CIV ENG] A foundation plate (usually of metal) placed beneath the end of a beam, at its point of support, to distribute the end load at the point. { 'bēm ,blāk·iŋ ,plāt }

beam blocking [BUILD] **1.** Boxing-in or covering a joist, beam, or girder to give the appearance of a larger beam. **2.** Strips of wood used to create a false beam. { 'bēm ,blāk·iŋ }

beam bolster [CIV ENG] A rod which provides support for steel reinforcement in formwork for a reinforced concrete beam. { 'bēm ,bōl·stər }

beam box *See* wall box. { 'bēm ,bäks }

beam bracketing *See* beam-rider guidance. { 'bēm ,brak·əd·iŋ }

beam brick [BUILD] A face brick which is used to bond to a poured-in-place concrete lintel. { 'bēm ,brik }

beam bridge [CIV ENG] A fixed structure consisting of a series of steel or concrete beams placed parallel to traffic and supporting the roadway directly on their top flanges. { 'bēm ,brij }

beam building [MIN ENG] A process of rock bolting in flat-lying deposits where the bolts are installed in bedded rock to bind the strata together to act as a single beam capable of supporting itself, thus stabilizing the overlying rock. { 'bēm ,bil·diŋ }

beam ceiling [ARCH] A ceiling having the beams exposed to view. { 'bēm ,sē·liŋ }

beam-climber guidance *See* beam-rider guidance. { 'bēm ,klīm·ər ,gī·dəns }

beam clip [ENG] A device for attaching a pipe hanger to its associated structural beam when it is undesirable to weld the pipe hanger to supporting structural steelwork. Also known as girder clamp; girder clip. { 'bēm ,klip }

beam column [CIV ENG] A structural member subjected simultaneously to axial load and bending moments produced by lateral forces or eccentricity of the longitudinal load. { 'bēm ,käl·əm }

beam compass [GRAPHICS] A compass for drawing large circles, consisting of a beam with sliding sockets carrying steel or pencil points. { 'bēm ,käm·pəs }

beam-condensing unit [SPECT] An attachment to the spectrophotometer that condenses and remagnifies the beam to provide reduced radiation at the sample. { 'bēm kən'den·siŋ ,yü·nət }

beam coupling [ELECTR] The production of an alternating current in a circuit connected between two electrodes that are close to, or in the path of, a density-modulated electron beam. { 'bēm ,kəp·liŋ }

beam current [ELECTR] The electric current determined by the number and velocity of electrons in an electron beam. { 'bēm ,kər·ənt }

beam-deflection amplifier [MECH ENG] A jet-interaction fluidic device in which the direction of a supply jet is varied by flow from one or more control jets which are oriented at approximately 90° to the supply jet. { 'bēm di'flek·shən 'am·plə,fī·ər }

beam-deflection tube [ELECTR] An electron-beam tube in which the current to an output electrode is controlled by transversely moving the electron beam. { 'bēm di'flek·shən ,tüb }

beam diameter [OPTICS] The distance between points on opposite sides of a light beam at which the power per unit area drops to 1/*e* (0.37) times its maximum value. { 'bēm dī,am·əd·ər }

beam divergence [OPTICS] The angle between points on opposite sides of a light beam at which the irradiance drops to 1/*e* (0.37) times its maximum value. [PHYS] The angular spread in the directions of the components of a beam of particles or radiation. { 'bēm də,vər·jəns }

beam drop [ELECTROMAG] Distortion of the normal rectilinear fan pattern of a detection radar in which a portion of the fan is at a lower elevation than the rest of the fan. { 'bēm ,dräp }

beam edge [PHYS] The locus of positions at which the intensity of a beam of particles or radiation is 10% of that along the axis of the beam. { 'bēm ,ej }

beam efficiency [ELECTROMAG] The fraction of the total radiated energy from an antenna contained in a single beam. { 'bēm i,fish·ən·sē }

beam expander [OPTICS] A combination of optical elements used to increase the diameter of a laser beam or other light beam. { 'bēm ik,span·dər }

beam extractor [NUCLEO] A magnet or electrostatic device for removing charged particles from a circular particle accelerator when they have been accelerated to the desired energy. { 'bēm ik,strak·tər }

beam fill [BUILD] Masonry, brickwork, or cement fill, usually between joists or horizontal beams at their supports; provides increased fire resistance. { 'bēm ,fil }

beam-foil spectroscopy [ATOM PHYS] A method of studying the structure of atoms and ions in which a beam of ions energized in a particle accelerator passes through a thin carbon foil from which the ions emerge with various numbers of electrons removed and in various excited energy levels; the light or Auger electrons emitted in the deexcitation of these levels are then observed by various spectroscopic techniques. Abbreviated BFS. { 'bēm ,fȯil spek'träs·kə·pē }

beam form [CIV ENG] A form which gives the necessary shape, support, and finish to a concrete beam. { 'bēm ,fȯrm }

beam-forming electrode [ELECTR] Electron-beam focusing elements in power tetrodes and cathode-ray tubes. { 'bēm ,fȯrm·iŋ i'lek,trōd }

beamguide [ELECTROMAG] A set of elements arranged and spaced so as to form and conduct a beam of electromagnetic radiation. { 'bēm,gīd }

beam hanger [PETRO ENG] An attachment at the end of a walking beam above a well casing to lift the pump rods or sacked rods. { 'bēm ,haŋ·ər }

beam holding [ELECTR] Use of a diffused beam of electrons to regenerate the charges stored on the screen of a cathode-ray storage tube. { 'bēm ,hōl·diŋ }

beam hole [NUCLEO] A hole through the shield, and usually the reflector, of a nuclear reactor which allows a beam of radiation, especially fast neutrons, to escape for experimental purposes. Also known as glory hole. { 'bēm ,hōl }

beamhouse [CHEM ENG] A place where the initial wet operations of tanning, involving soaking in water and solutions of alkali, are carried out. { 'bēm,hau̇s }

beam-indexing tube [ELECTR] A single-beam color television picture tube in which the color phosphor strips are arranged in groups of red, green, and blue. { 'bēm 'in,dek·siŋ ,tüb }

beam lead [ELECTR] A flat thick-film lead, sometimes of gold, deposited on a semiconductor chip chemically or by evaporation, as a connecting lead for a semiconductor device or integrated circuit. { 'bēm ,lēd }

beam lobe switching [ELECTR] Method of determining the direction of a remote object by comparison of the signals corresponding to two or more successive beam angles, differing slightly from the direction of the object. { 'bēm ,lōb ,swich·iŋ }

beam magnet *See* convergence magnet. { 'bēm ,mag·nət }

beam parametric amplifier [ELECTR] Parametric amplifier that uses a modulated electron beam to provide a variable reactance. { 'bēm ,par·ə'me·trik 'am·plə,fī·ər }

beam pattern *See* directivity pattern. { 'bēm ,pad·ərn }

beam pocket [CIV ENG] **1.** In a vertical structural member, an opening to receive a beam. **2.** An opening in the form for a column or girder where the form for an intersecting beam is framed. { 'bēm ,päk·ət }

beam power tube [ELECTR] An electron-beam tube which uses directed electron beams to provide most of its power-handling capability and in which the control grid and screen grid are essentially aligned. Also known as beam tetrode. { 'bēm ,pau̇·ər ,tüb }

beam pumping unit [PETRO ENG] A device used in sucker rod pumping which has a horizontal member (walking beam) that is worked up and down by a rotating crank to produce reciprocating motion. { 'bēm 'pəm·piŋ ,yü·nət }

beam recording [ELECTR] A method of using an electron beam to write data generated by a computer directly on microfilm. { 'bēm ri'kȯrd·iŋ }

beam resonator [OPTICS] A device which acts to confine a laser beam or other beam of electromagnetic radiation to a given region of space without continuous guidance along the beam. Also known as open resonator. { 'bēm 'rez·ən,ād·ər }

beam rider [AERO ENG] A missile for which the guidance system consists of standard reference signals transmitted in a radar beam which enable the missile to sense its location relative to the beam, correct its course, and thereby stay on the beam. { 'bēm ,rīd·ər }

beam-rider guidance [NAV] A system for guiding aircraft

BEAR

The polar bear (*Thalarctos maritimus*).

BEARING

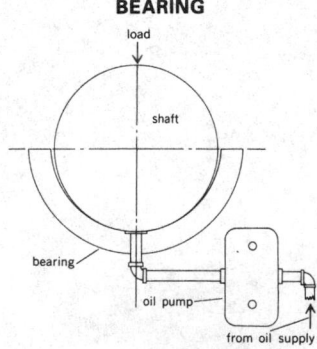

Fluid-film hydrostatic bearing. Hydrostatic oil lift can reduce starting friction drag to less than one-tenth of usual starting drag. (*From W. Staniar, ed., Plant Engineering Handbook, 2d ed., McGraw-Hill, 1959*)

or spacecraft in which a craft follows a radar beam, light beam, or other kind of beam along the desired path. Also known as beam bracketing; beam-climber guidance. { 'bēm ˌrīd·ər 'gī·dəns }

beam riding [AERO ENG] The maneuver of a spacecraft or other vehicle as it follows a beam. { 'bēm ˌrīd·iŋ }

beam sea [NAV] Waves moving in a direction approximately 90° from the heading. { 'bēm ˌsē }

beam splice [CIV ENG] A connection between two lengths of a beam or girder; may be shear or moment connections. { 'bēm ˌsplīs }

beam splitter [OPTICS] A mirror that reflects part of a beam of light falling on it and transmits part. { 'bēm ˌsplid·ər }

beam splitting [ELECTR] Process for increasing accuracy in locating targets by radar; by noting the azimuths at which one radar scan first discloses a target and at which radar data from it ceases, beam splitting calculates the mean azimuth for the target. [OPTICS] The division of a beam of light into two beams by placing a special type of mirror in the path of the beam that reflects part of the light falling on it and transmits part. { 'bēm ˌsplid·iŋ }

beam spread [ENG] The angle of divergence from the central axis of an electromagnetic or acoustic beam as it travels through a material. { 'bēm ˌspred }

Beams servoed rotational method [ENG] A method of measuring the gravitational constant by determining the inertial reaction of a torsional pendulum to the angular acceleration of a rotating table that is required to cancel the attraction of the pendulum to two large masses. { 'bēmz 'sər·vōd rō'tā·shən·əl ˌmeth·əd }

beam steering [ELECTR] Changing the direction of the major lobe of a radiation pattern, usually by switching antenna elements. { 'bēm ˌstir·iŋ }

beam storage [COMPUT SCI] A magnetic storage device that employs electron beams to enter information into, or retrieve information from, storage cells; for example, a cathode-ray-tube storage. { 'bēm ˌstōr·ij }

beam switching [ELECTR] Method of obtaining more accurately the bearing or elevation of an object by comparing the signals received when the beam is in directions differing slightly in bearing or elevation; when these signals are equal, the object lies midway between the beam axes. Also known as lobe switching. { 'bēm ˌswich·iŋ }

beam-switching tube [ELECTR] An electron tube which has a series of electrodes arranged around a central cathode and in which an electron beam is switched from one electrode to another. Also known as cyclophon. { 'bēm ˌswich·iŋ ˌtüb }

beam test [CIV ENG] A test of the flexural strength (modulus of rupture) of concrete from measurements on a standard reinforced concrete beam. { 'bēm ˌtest }

beam tetrode See beam power tube. { 'bēm ˌte‚trōd }

beam well [PETRO ENG] A well pumped by a walking beam. { 'bēm ˌwel }

beam width [ELECTROMAG] The angle, measured in a horizontal plane, between the directions at which the intensity of an electromagnetic beam, such as a radar or radio beam, is one-half its maximum value. Also known as beam angle. { 'bēm ˌwidth }

beam wind [NAV] Nautical term for a crosswind, especially a wind blowing 90° from a ship's heading. { 'bēm ˌwind }

bean [BOT] The common name for various leguminous plants used as food for humans and livestock; important commercial beans are true beans (*Phaseolus*) and California blackeye (*Vigna sinensis*). [ENG] A restriction, such as a nipple, which is placed in a pipe to reduce the rate of fluid flow. { bēn }

bean anthracnose [PL PATH] A fungus disease of the bean caused by *Colletotrichum lindemuthianum*, producing pink to brown lesions on the pod and seed and dark discolorations on the veins on the lower surface of the leaf. { 'bēn an'thrak‚nōs }

bean blight [PL PATH] A bacterial disease of the bean caused by *Xanthomonas phaseoli*, producing water-soaked lesions that become yellowish-brown spots on all plant parts. { bēn ˌblīt }

bean ore [GEOL] A lenticular, pisolitic aggregate of limonite. { 'bēn ˌor }

bear [MIN ENG] To underhole or undermine; to drive in at the top or side of a working. [VERT ZOO] The common name for a few species of mammals in the family Ursidae. { ber }

beard See awn. { bird }

bearded needle [TEXT] A fine steel needle for machine knitting that has a butt at one end and a long, flexible hook at the other that curves back to the shank of the needle. Also known as spring needle. { 'bird·əd 'nēd·əl }

bearding [NAV ARCH] **1.** Cutting a timber so as to form an inclined surface that fits the angle of a ship's side. **2.** The forward edge of a rudder or sternpost. { 'bird·iŋ }

bearding line [NAV ARCH] A line on the side of the sternpost, keel, deadwoods, and stem of a ship that indicates the intersection of these members with the outer face of the frames. { 'bird·iŋ ˌlīn }

Bear Driver See Boötes. { 'ber ˌdrīv·ər }

bearer [CIV ENG] Any horizontal beam, joist, or member which supports a load. [GRAPHICS] **1.** In photoengraving, the metal left on a plate to protect the printing surface while molding. **2.** In composition, one of the type-high slugs locked inside a chase to protect the printing surface. **3.** In printing presses, one of the surface-to-surface ends of cylinders that come in contact. { 'ber·ər }

bearing [CIV ENG] That portion of a beam, truss, or other structural member which rests on the supports. [MECH ENG] A machine part that supports another part which rotates, slides, or oscillates in or on it. [MIN ENG] The direction of a mine drivage, usually given in terms of the horizontal angle turned off a datum direction, such as the true north and south line. [NAV] The horizontal direction from one terrestrial point to another; basically synonymous with azimuth. { 'ber·iŋ }

bearing angle [NAV] Horizontal direction measured from 0° at the reference direction clockwise or counterclockwise through 90° or 180°. { 'ber·iŋ ˌaŋ·gəl }

bearing bar [BUILD] A wrought-iron bar placed on masonry to provide a level support for floor joists. [CIV ENG] A load-carrying bar which supports a grating and which extends in the direction of the grating span. [ENG] See azimuth instrument. { 'ber·iŋ ˌbär }

bearing bronze [MET] A form of bronze usually having a high lead content and antifriction characteristics and used for fabricating bearings. { 'ber·iŋ ˌbränz }

bearing cap [DES ENG] A device designed to fit around a bearing to support or immobilize it. { 'ber·iŋ ˌkap }

bearing capacity [MECH] Load per unit area which can be safely supported by the ground. { 'ber·iŋ kə'pas·əd·ē }

bearing circle [ENG] A ring designed to fit snugly over a compass or compass repeater, and provided with vanes for observing compass bearings. { 'ber·iŋ ˌsər·kəl }

bearing cursor [ENG] Of a radar set, the radial line inscribed on a transparent disk which can be rotated manually about an axis coincident with the center of the plan position indicator; used for bearing determination. Also known as mechanical bearing cursor. { 'ber·iŋ ˌkər·sər }

bearing distance [CIV ENG] The length of a beam between its bearing supports. { 'ber·iŋ ˌdis·təns }

bearing loss [ELEC] Loss of power in a machine caused by friction between the shaft and the bearing. { 'ber·iŋ ˌlos }

bearing partition [BUILD] A partition which supports a vertical load. { 'ber·iŋ pər'tish·ən }

bearing pile [ENG] A vertical post or pile which carries the weight of a foundation, transmitting the load of a structure to the bedrock or subsoil without detrimental settlement. { 'ber·iŋ ˌpīl }

bearing plate [CIV ENG] A flat steel plate used under the end of a wall-bearing beam to distribute the load over a broader area. { 'ber·iŋ ˌplāt }

bearing pressure [MECH] Load on a bearing surface divided by its area. Also known as bearing stress. { 'ber·iŋ ˌpresh·ər }

bearing resolution [ELECTR] Minimum angular separation in a horizontal plane between two targets at the same range that will allow an operator to obtain data on either target. { 'ber·iŋ ˌrez·ə‚lü·shən }

bearing strain [MECH] The deformation of bearing parts subjected to a load. { 'ber·iŋ ˌstrān }

bearing strength [MECH] The maximum load that a column, wall, footing, or joint will sustain at failure, divided by the effective bearing area. { 'ber·iŋ ˌstreŋkth }

bearing stress *See* bearing pressure.   { 'ber·iŋ ˌstres }

bearing test [ENG] A test of the bearing capacities of pile foundations, such as a field loading test of an individual pile; a laboratory test of soil samples for bearing capacities.   { 'ber·iŋ ˌtest }

bearing wall [CIV ENG] A wall capable of supporting an imposed load. Also known as structural wall.   { 'ber·iŋ ˌwȯl }

bear trap gate [CIV ENG] A type of crest gate with an upstream leaf and a downstream leaf which rest in a horizontal position, one leaf overlapping the other, when the gate is lowered.   { 'ber ˌtrap ˌgāt }

beat [PHYS] The periodic variation in amplitude of a wave that is the superposition of two simple harmonic waves of different frequencies.   { bēt }

beat Cepheid [ASTRON] A dwarf Cepheid that displays two or more nearly identical pulsation periods, resulting in periodic amplitude fluctuations in its light curve.   { 'bēt 'sef·ē·əd }

beater [AGR] **1.** A device for chopping or pulverizing unwanted parts of crops such as cornstalks or potato vines. **2.** The part of a thresher that strikes the grains. [ENG] **1.** A tool for packing in material to fill a blasthole containing a charge of powder. **2.** A laborer who shovels or dumps asbestos fibers and sprays them with water in order to prepare them for the beating. [MECH ENG] A machine that cuts or beats paper stock. [TEXT] The section of a loom that drives the weft from the shed into the cloth.   { 'bēd·ər }

beater mill *See* hammer mill.   { 'bēd·ər ˌmil }

beat frequency [ELECTR] The frequency of a signal equal to the difference in frequencies of two signals which produce the signal when they are combined in a nonlinear circuit.   { 'bēt ˌfrē·kwən·sē }

beat-frequency oscillator [ELECTR] An oscillator in which a desired signal frequency, such as an audio frequency, is obtained as the beat frequency produced by combining two different signal frequencies, such as two different radio frequencies. Abbreviated BFO. Also known as heterodyne oscillator.   { 'bēt ˌfrē·kwən·sē 'äs·ə·ˌlād·ər }

beating [ENG] A process that reduces asbestos fibers to pulp for making asbestos paper.   { 'bēd·iŋ }

beating-in [ELECTR] Interconnecting two transmitter oscillators and adjusting one until no beat frequency is heard in a connected receiver; the oscillators are then at the same frequency.   { 'bēd·iŋ ˌin }

beat note [ELECTR] The beat frequency whose signal is produced by two signals having waves that are sinusoidal.   { 'bēt ˌnōt }

beat reception *See* heterodyne reception.   { 'bēt ri'sep·shən }

Beattie and Bridgman equation [THERMO] An equation that relates the pressure, volume, and temperature of a real gas to the gas constant.   { 'bēd·ē ən 'brij·mən i'kwā·zhən }

beat-time programming [COMPUT SCI] A type of programming which requires that data be made available to the computer during some ongoing process prior to a particular point in time.   { 'bēt ˌtīm 'prō·ˌgram·iŋ }

beat tone [ENG ACOUS] Musical tone due to beats, produced by the heterodyning of two high-frequency wave trains.   { 'bēt ˌtōn }

Beaufort force [METEOROL] A number denoting the speed (or so-called strength) of the wind according to the Beaufort wind scale. Also known as Beaufort number.   { 'bō·fərt ˌfȯrs }

Beaufort number *See* Beaufort force.   { 'bō·fərt ˌnəm·bar }

Beaufort wind scale [METEOROL] A system of code numbers from 0 to 12 classifying wind speeds into groups from 0–1 mile per hour or 0–1.6 kilometers per hour (Beaufort 0) to those over 75 miles per hour or 121 kilometers per hour (Beaufort 12).   { 'bō·fərt 'wind ˌskāl }

beaumontage [MATER] A material that is made of a mixture of resin, beeswax, and shellac and used to fill small holes or cracks in wood or metal.   { ˌbō·män'tij }

beauty *See* bottom.   { 'byüd·ē }

beauty quark *See* bottom quark.   { 'byüd·ē ˌkwärk }

beaver [VERT ZOO] The common name for two different and unrelated species of rodents, the mountain beaver (*Aplodontia rufa*) and the true or common beaver (*Castor canadensis*).   { 'bē·vər }

beaver cloth *See* melton.   { 'bē·vər ˌklȯth }

beaverite [MINERAL] Pb(Cu,Fe,Al)₃(SO₄)₂(OH)₆ A canary yellow, hexagonal mineral consisting of a basic sulfate of lead, copper, iron, and aluminum.   { 'bē·və·ˌrīt }

beavertail [ELECTROMAG] Fan-shaped radar beam, wide in the horizontal plane and narrow in the vertical plane, which is swept up and down for height finding.   { 'be·vər·ˌtāl }

bebeerine [ORG CHEM] $C_{36}H_{38}N_2O_6$ An alkaloid derived from the bark of the tropical tree *Nectandra rodiaei*; the dextro form is soluble in acetone, the levo form is soluble in benzene and is an antipyretic; the dextro form is also known as chondrodendrin; the levo, as curine.   { bə'bī·ˌrēn }

Béchamp reduction [ORG CHEM] Reduction of nitro groups to amino groups by the use of ferrous salts or iron and dilute acid.   { bā'shän ri'dək·shən }

bêche [MECH ENG] A pneumatic forge hammer having an air-operated ram and an air-compressing cylinder integral with the frame.   { besh }

Beck effect [ELEC] An increase in the light intensity of an arc lamp whose carbon anode has been treated with rare-earth salts when a certain current is exceeded.   { 'bek i'fekt }

Becker and Kornetzki effect [PHYS] A reduction in the internal friction of a ferromagnetic substance when it is subjected to a magnetic field that is large enough to produce magnetic saturation.   { 'bek·ər ən ˌkȯr'nets·kē i'fekt }

Becke test [MINERAL] A microscope test in which indices of refraction are compared for minerals; the Becke line appears to move toward the material of higher refractivity as the tube of the microscope is raised.   { 'bek·ə ˌtest }

beckerite [MINERAL] A brown variety of the fossil resin retinite with a very high oxygen content.   { 'bek·ə·ˌrīt }

Becklin-Neugebauer object [ASTRON] A compact source of infrared radiation in the Orion Nebula, probably a collapsing protostar of large mass. Abbreviated BN object.   { 'bek·lin 'nȯi·gə·ˌbaú·ər ˌäb·jekt }

Beckmann rearrangement [ORG CHEM] An intramolecular change of a ketoxime into its isomeric amide when treated with phosphorus pentachloride.   { 'bek·män rē·ə'rānj·mənt }

Beckmann thermometer [ENG] A sensitive thermometer with an adjustable range so that small differences in temperature can be measured.   { 'bek·män ther'mäm·əd·ər }

Beckwith-Wiedemann syndrome [MED] A congenital, generalized overgrowth syndrome attributed to a relative deficiency of maternally derived genes that is characterized by visceromegaly and predisposition to childhood tumors, especially Wilms' tumor.   { ˌbek,with 'wēd·ə·män ˌsin,drōm }

becquerel [NUCLEO] The International System unit of activity of a radionuclide, equal to the activity of a quantity of a radionuclide having one spontaneous nuclear transition per second. Symbolized Bq.   { ˌbek·ə·ˌrel *or* be'krel }

Becquerel effect [ELEC] The phenomenon of a current flowing between two unequally illuminated electrodes of a certain type when they are immersed in an electrolyte. [GRAPHICS] The sensitization of a silver halide to long-wavelength light by exposure to diffuse illumination. [OPTICS] *See* paramagnetic Faraday effect.   { ˌbek·ə·ˌrel *or* be'krel i'fekt }

becquerelite [MINERAL] CaU₆O₁₉·11H₂O An orthorhombic mineral consisting of a hydrated oxide of uranium; occurs in tabular, elongated, striated, and massive form.   { be'kre·ˌlīt }

Becquerel rays [NUC PHYS] Formerly, radiation emitted by radioactive substances; later renamed alpha, beta, and gamma rays.   { ˌbek·ə·ˌrel *or* be'krel ˌrāz }

bed [CHEM] The ion-exchange resin contained in the column in an ion-exchange system. [CIV ENG] **1.** In masonry and bricklaying, the side of a masonry unit on which the unit lies in the course of the wall; the underside when the unit is placed horizontally. **2.** The layer of mortar on which a masonry unit is set. [GEOL] **1.** The smallest division of a stratified rock series, marked by a well-defined divisional plane from its neighbors above and below. **2.** An ore deposit, parallel to the stratification, constituting a regular member of the series of formations; not an intrusion. [GRAPHICS] The surface of a flatbed printing press on which the chase of composed type is secured for printing. [HYD] The bottom of a channel for the passage of water. [MECH ENG] The part of a machine having precisely machined ways or bearing surfaces which support or align other machine parts.   { bed }

Bedaux plan [IND ENG] A wage incentive plan in which work is standardized into man-minute units called bedaux (B);

BEAVER

The common beaver (*Castor canadensis*).

BEDBUG

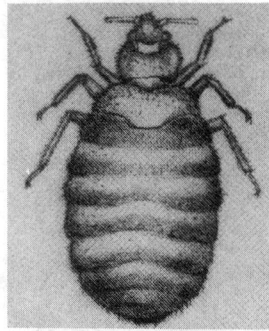

Cimex lectularius, a species of bedbug parasitic on humans.

BEECH

American beech *(Fagus grandifolia).*

60 B per hour is 100% productivity, and earnings are based on work units per length of time. { bə'dō ,plan }

bedbug [INV ZOO] The common name for a number of species of household pests in the insect family Cimicidae that infest bedding, and by biting humans obtain blood for nutrition. { 'bed,bəg }

bed charge [MET] The primary charge of fuel in a cupola to initiate the melting process. { 'bed ,chärj }

bed coke [MET] The primary layer of coke placed in a cupola at a preselected height above the tuyeres for the initial combustion and melting operation. { 'bed ,kōk }

bedded [GEOL] Pertaining to rocks exhibiting depositional layering or bedding formed from consolidated sediments. { 'bed·əd }

bedded chert [PETR] Chert of brittle, close-jointed, rhythmically layered character found over large areas in thick deposits, the usually even-bedded layers separated by partings of dark siliceous shale or by siderite layers. { 'bed·əd ,chərt }

bedded vein [GEOL] A lode occupying the position of a bed that is parallel with the enclosing rock stratification. { 'bed·əd ,vān }

bed detector [PETRO ENG] Apparatus to detect and measure the extent of underground formations that are potential oil and gas reservoirs; methods include induction logs, gamma-ray logs, and sonic logs. { 'bed di'tek·tər }

bedding [CIV ENG] **1.** Mortar, putty, or other substance used to secure a firm and even bearing, such as putty laid in the rabbet of a window frame, or mortar used to lay bricks. **2.** A base which is prepared in soil or concrete for laying masonry or concrete. [GEOL] Condition where planes divide sedimentary rocks of the same or different lithology. { 'bed·iŋ }

bedding cleavage [GEOL] Cleavage parallel to the rock bedding. { 'bed·iŋ ,klēv·ij }

bedding course [CIV ENG] The first layer of mortar at the bottom of masonry. { 'bed·iŋ ,kòrs }

bedding dot [BUILD] A small spot of plaster built out to the face of a finished wall or ceiling; serves as a screed for leveling and plumbing in the application of plaster. { 'bed·iŋ ,dät }

bedding fault [GEOL] A fault whose fault surface is parallel to the bedding plane of the constituent rocks. Also known as bedding-plane fault. { 'bed·iŋ ,fòlt }

bedding fissility [GEOL] Primary foliation parallel to the bedding of sedimentary rocks. { 'bed·iŋ fi'sil·əd·ē }

bedding joint [GEOL] A joint parallel to the rock bedding. { 'bed·iŋ ,jòint }

bedding plane [GEOL] Any of the division planes which separate the individual strata or beds in sedimentary or stratified rock. { 'bed·iŋ ,plān }

bedding-plane fault See bedding fault. { 'bed·iŋ ,plān ,fòlt }

bedding-plane slip See flexural slip. { 'bed·iŋ ,plān ,slip }

bedding schistosity [GEOL] Schistosity that is parallel to the rock bedding. { 'bed·iŋ ,shis'täs·əd·ē }

bedding thrust [GEOL] A thrust fault parallel to bedding. { 'bed·iŋ ,thrəst }

bedding void [GEOL] A void formed between successive batches of lava that are discharged in a single short activity of a volcano, as well as between flows made a long time apart. { 'bed·iŋ ,vòid }

bede [MIN ENG] A miner's pick. { 'bēd }

bedford cord [TEXT] A closely woven, sturdy ribbed fabric with a raised effect; may be made of wool, cotton, silk, acrylic, or polyesters. { 'bed·fərd ,kòrd }

Bedford limestone See spergenite. { 'bed·fərd 'līm,stōn }

bed glazing See back putty. { 'bed ,glā·ziŋ }

bediasite [GEOL] A black to brown tektite found in Texas. { bē'dī·ə,zīt }

bed joint [CIV ENG] **1.** A horizontal layer of mortar on which masonry units are laid. **2.** One of the radial joints in an arch. { 'bed ,jòint }

bed load [GEOL] Particles of sand, gravel, or soil carried by the natural flow of a stream on or immediately above its bed. Also known as bottom load. { 'bed ,lōd }

bed molding [ARCH] A molding of the cornice of an entablature situated beneath the corona and immediately above the frieze. [BUILD] **1.** The lowest member of a band of moldings. **2.** Any molding under a projection, such as between eaves and sidewalls. { 'bed ,mōl·diŋ }

Bedoulian [GEOL] Lower Cretaceous (lower Aptian) geologic time in Switzerland. { bə'dül·ē·ən }

bedrock [GEOL] General term applied to the solid rock underlying soil or any other unconsolidated surficial cover. { 'bed,räk }

Bedsonia [MICROBIO] The psittacosis-lymphogranuloma-trachoma (PLT) group of bacteria belonging to the Chlamydozoaceae; all are obligatory intracellular parasites. { bed'sō·nē·ə }

bedsore See decubitus ulcer. { 'bed,sór }

bedspring array See billboard array. { 'bed,spriŋ ə'rā }

bee [INV ZOO] Any of the membranous-winged insects which compose the superfamily Apoidea in the order Hymenoptera characterized by a hairy body and by sucking and chewing mouthparts. { bē }

beech [BOT] Any of various deciduous trees of the genus *Fagus* in the beech family (Fagaceae) characterized by smooth gray bark, triangular nuts enclosed in burs, and hard wood with a fine grain. { bēch }

beech bark disease [PL PATH] A disease of beech caused by the beech scale (*Cryptococcus fagi*) and a fungus (*Nectria coccinea faginata*) acting together; bark is destroyed, foliage wilts, and the tree eventually dies. { bēch ,bärk di'zēz }

bee dance [INV ZOO] Circling and wagging movements exhibited by worker bees to give other bees in the hive information about the location of a new source of food. { 'bē ,dans }

beef [AGR] The flesh of a bovine animal, such as a cow or steer, used as food. { bēf }

beef stearin See oleostearin. { bēf ,stir·ən }

beegerite [MINERAL] $Pb_6Bi_2S_9$ A light to dark gray mineral consisting of lead bismuth sulfide; usually occurs in granular to dense massive form. { 'be·gə,rīt }

beehive Also known as hive. [AGR] A container that is constructed to house a colony of honeybees. [INV ZOO] A colony of bees. { 'bē,hīv }

Beehive See Praesepe. { 'bē,hīv }

beehive oven [ENG] An arched oven that carbonizes coal into coke by using the heat of combustion of gases that are formed, and of a small part of the coke that is formed, with no recovery of by-products. { 'bē,hīv ,əv·ən }

beekeeping [AGR] The management and maintenance of colonies of honeybees. { 'bē¦kēp·iŋ }

beekite [MINERAL] **1.** A concretionary form of calcite or silica that occurs in small rings on the surface of a fossil shell which has weathered out of its matrix. **2.** White, opaque accretions of silica found on silicified fossils or along joint surfaces as a replacement of organic matter. { 'bē,kīt }

beer [FOOD ENG] A lightly hopped, fermented malt beverage brewed by bottom fermentation. { bir }

beerbachite [PETR] A hornfels with large poikiloblastic crystals of olivine. { 'bir·bə,kīt }

beer drinkers' cardiomyopathy [MED] Congestive heart failure and nonspecific cardiomyopathy presumed due to cobalt added to beer. { 'bir ,driŋ·kərz ¦kärd·ē·ō·mī'äp·ə·thē }

Beer-Lambert-Bouguer law See Bouguer-Lambert-Beer law. { ¦bā·ər ¦läm·bərt bü'ger ,lò }

Beer's law [PHYS CHEM] The law which states that the absorption of light by a solution changes exponentially with the concentration, all else remaining the same. { 'bā·ərz ,lò }

beer stone [FOOD ENG] A mixture of calcium oxalate and organic material that deposits on container surfaces during brewing operations. Also known as beer scale. { 'bir ,stōn }

beeswax [MATER] Yellow to grayish-brown solid wax obtained from bee honeycombs by boiling and straining; used in floor waxes, waxed paper, and textile finishes and in pharmacy. Also known as yellow wax. { 'bēz,waks }

beet [BOT] *Beta vulgaris.* The red or garden beet, a cool-season biennial of the order Caryophyllales grown for its edible, enlarged fleshy root. { bēt }

beetle [ENG] See rammer. [INV ZOO] The common name given to members of the insect order Coleoptera. [MIN ENG] A powerful, cable-hauled propulsion unit, operated under remote control, for moving a train of wagons at the mine surface. { 'bēd·əl }

beetle stone See septarium. { 'bēd·əl ,stōn }

beetling [TEXT] Beating cloth with hammers to force the fibers together and produce a lustrous, linenlike effect. { 'bēt·liŋ }

beet sugar [FOOD ENG] Sugar made from sugarbeets by crystallization of the syrup extracted from the roots. { 'bēt ,shùg·ər }

beet yellows virus [VIROL] The type species of the plant-virus genus *Closterovirus*. Abbreviated BYV. { 'bēt 'yel·ōz ¦vī·rəs }

beet yellows virus group *See* Closterovirus. { 'bēt 'yel·ōz¦vī·rəs ¦grüp }

before the wind [NAV] In a direction approximating that toward which the wind is blowing; applies particularly to the situation where the wind is aft, and the craft is being aided by it. { bə¦fȯr thə 'wind }

Beggiatoa [MICROBIO] A genus of bacteria in the family Beggiatoaceae; filaments are individual, and cells contain sulfur granules when grown on media containing hydrogen sulfide. { bə¦jad·ə·wə }

Beggiatoaceae [MICROBIO] A family of bacteria in the order Cytophagales; cells are in chains in colorless, flexible, motile filaments; microcysts are not known. { bə¦jad·ə·wās·ē¸ē }

Beggiatoales [MICROBIO] Formerly an order of motile, filamentous bacteria in the class Schizomycetes. { bə¸jad·ə'wā¸lēz }

BEGIN [COMPUT SCI] An enclosing statement of ALGOL used to indicate the beginning of a block; any variable in a block enclosed by BEGIN and END is normally local to this block. { bi'gin }

beginning-of-information marker [COMPUT SCI] A section of magnetic tape covered with reflective material that indicates the beginning of the area on which information is to be recorded. { bi'gin·iŋ əv ¸in·fər'mā·shən ¸mär·kər }

Begoniaceae [BOT] A family of dicotyledonous plants in the order Violales characterized by an inferior ovary, unisexual flowers, stipulate leaves, and two to five carpels. { bə¸gō·nē'ās·ē¸ē }

behavior [PSYCH] Any overt activity of an organism. { bi'hāv·yər }

behavioral dynamics [IND ENG] 1. The behavioral operating characteristics of individuals and groups in terms of how these people are conditioned by their working environments. 2. The interactions between individuals or groups in the workplace. { bi'hā·vyə·rəl dī'nam·iks }

behavioral ecology [ECOL] The branch of ecology that focuses on the evolutionary causes of variation in behavior among populations and species. { bi'hāv·yə·rəl ē'käl·ə·jē }

behavioral isolation [BIOL] An isolating mechanism in which two allopatric species do not mate because of differences in courtship behavior. Also known as ethological isolation. { bi'hav·yə·rəl ī·sə'lā·shən }

behavioral psychophysics [PSYCH] A branch of psychology concerned primarily with the measurement of sensory capacities of normal, intact animals. { bi'hāv·yə·rəl ¦sī·kō¦fiz·iks }

behavioral toxicology [MED] The study of behavioral abnormalities induced by exogenous agents such as drugs, chemicals in the general environment, and chemicals encountered in the workplace. { bə¦hāv·yə·rəl ¸täk·sə'käl·ə·jē }

behavior disorder *See* abnormal behavior. { bi'hāv·yər dis¸-ȯrd·ər }

behaviorism [PSYCH] A school of psychology concerned with observable, tangible, and measurable data regarding behavior and human activities, but excluding ideas and emotions as purely subjective phenomena. { bi'hāv·yə¸riz·əm }

behavior therapy [PSYCH] A mode of therapy that focuses on altering observable and quantifiable behavior of an individual by means of systematic manipulation of environmental and behavioral variables that are thought to be functionally related to the individual's behavior. { bi'hāv·yər ¸ther·ə·pē }

Behçet's disease [MED] Chronic disease of young adult males characterized by recurrent painful ulcers of the mouth and genitalia, inflammation of the irises, and joint pains. { 'bā¸chets di¸zēz }

beheaded stream [HYD] A water course whose upper portion, through erosion, has been cut off and captured by another water course. { bi'hed·əd ¦strēm }

behenic acid *See* docosanoic acid.

behenyl alcohol [ORG CHEM] $CH_3(CH_2)_{20}CH_2OH$ A saturated fatty alcohol; colorless, waxy solid with a melting point of 71°C; soluble in ethanol and chloroform; used for synthetic fibers and lubricants. Also known as 1-docosanol. { bə'hen·əl 'al·kə¸hȯl }

Behrens-Fisher problem [STAT] The problem of calculating the probability of drawing two random samples whose means differ by some specified value (which may be zero) from normal populations, when one knows the difference of the means of these populations but not the ratio of their variances. { ¦ber·ənz ¦fish·ər ¸präb·ləm }

beidellite [MINERAL] A clay mineral of the montmorillonite group in which Si^{4+} has been replaced by Al^{3+} and in which there is virtual absence of Mg or Fe replacing Al. { bī'de¸līt }

bei function [MATH] One of the functions that is defined by $ber_n(z) \pm i\ bei_n(z) = J_n(ze^{\pm 3\pi i/4})$, where J_n is the nth Bessel function. { 'bī ¸faŋk·shən }

Beijerinckia [MICROBIO] A genus of bacteria in the family Azotobacteraceae; slightly curved or pear-shaped rods with large, refractile lipoid bodies at the ends of the cell. { bī·zhə'riŋ·kyə }

Beilby layer [MET] An amorphous layer formed on a polished metal surface. { 'bīl·bē ¸lā·ər }

bejel [MED] An infectious nonvenereal treponemal disease occurring principally in children in the Middle East. { 'be·jəl }

Békésy audiometry [ACOUS] A subject-controlled auditory threshold testing procedure. { 'bā¸kā·shē ȯd·ē'äm·ə·trē }

bel [PHYS] A dimensionless unit expressing the ratio of two powers or intensities, or the ratio of a power to a reference power, such that the number of bels is the common logarithm of this ratio. Symbolized b; B. { bel }

belat [METEOROL] A strong land wind from the north or northwest which sometimes blows across the southeastern coast of Arabia and is accompanied by a hazy atmosphere due to sand blown from the interior desert. { bā'lät }

belaying pin [NAV ARCH] A pin-shaped metal rod about which ropes are secured on shipboard. { bi'lā·iŋ ¸pin }

Belemnoidea [PALEON] An order of extinct dibranchiate mollusks in the class Cephalopoda. { ¸bə·ləm'nȯid·ē·ə }

Belfast truss [CIV ENG] A bowstring beam for large spans, having the upper member bent and the lower member horizontal; constructed entirely of timber components. { 'bel¸fast 'trəs }

Belgian block [MATER] 1. A stone block used for paving, having the shape of a truncated pyramid, with a depth of 7–8 inches (18–20 centimeters), a base of 5–6 inches (13–15 centimeters) square, and a face opposite the base that is 1 inch (2.5 centimeters) or less smaller than the base. 2. Any stone block used for paving. { 'bel·jən ¦bläk }

Belgian retort process *See* horizontal retort process. { 'bel·jən ri'tȯrt ¸präs·əs }

B eliminator [ELECTR] Power pack that changes the alternating-current powerline voltage to the direct-current source required by plant circuits of vacuum tubes or semiconductor devices. { 'bē i'lim·ə¸nād·ər }

Belinda [ASTRON] A satellite of Uranus orbiting at a mean distance of 46,760 miles (75,260 kilometers) with a period of 15 hours, and with a diameter of about 42 miles (68 kilometers). { bə'lin·də }

Belinuracea [PALEON] An extinct group of horseshoe crabs; arthropods belonging to the Limulida. { ¸bel·ə·nü'rās·ē·ə }

belite *See* larnite. { 'bē¸līt }

bell [ENG] 1. A hollow metallic cylinder closed at one end and flared at the other; it is used as a fixed-pitch musical instrument or signaling device and is set vibrating by a clapper or tongue which strikes the lip. 2. *See* bell tap. [MET] A conical device that seals the top of a blast furnace. { bel }

belladonna [BOT] *Atropa belladonna.* A perennial poisonous herb that belongs to the family Solanaceae; atropine is produced from the roots and leaves; used as an antispasmodic, as a cardiac and respiratory stimulant, and to check secretions. Also known as deadly nightshade. { ¸bel·ə'dän·ə }

Bellamy drift [NAV] The net drift angle of an aircraft calculated between any two pressure soundings; it is that drift attributable only to the cross-wind component of the wind; the actual drift may be greater or less, but the difference is small for winds normally encountered. { 'bel·ə·mē ¦drift }

bell-and-spigot joint [ENG] A pipe joint in which a pipe ending in a bell-like shape is joined to a pipe ending in a spigotlike shape. { ¦bel ən 'spik·ət ¸jȯint }

bell arch [ARCH] A round arch supported on large corbels, giving rise to a bell-shaped appearance. { 'bel ¸ärch }

Bellatrix [ASTRON] A bluish-white star of stellar magnitude

BELLADONNA

Belladonna (*Atropa belladonna*), flowering branch and isolated flowers (two views).

1.7, spectral classification B2-III, in the constellation Orion; the star γ Orionis.　{ bə'lā·triks }

bell buoy [NAV] A channel buoy with a bell which rings as the buoy swings in the waves, usually marking rocks or shoals. { 'bel ,bȯi }

bell cap [CHEM ENG] A hemispherical or triangular metal casting used on distillation-column trays to force upflowing vapors to bubble through layers of downcoming liquid. { 'bel ,kap }

bell character [COMPUT SCI] A control character that activates a bell, alarm, or other audio device to get someone's attention.　{ 'bel ,kar·ik·tər }

belled caisson [CIV ENG] A type of drilled caisson with a flared bottom.　{ beld 'kā,sän }

Bellerophontacea [PALEON] A superfamily of extinct gastropod mollusks in the order Aspidobranchia. { bə,ler·ə,fän'tās·ē·ə }

bell glass *See* bell jar.　{ 'bel ,glas }

bell hole [MIN ENG] 1. One of the holes or excavations made at the section joints of a pipeline for the purpose of repairs. 2. A conical cavity in a coal mine roof caused by the falling of a large concretion.　{ 'bel ,hōl }

bellingerite [MINERAL] $3Cu(IO_3)_2 \cdot 2H_2O$ A light green triclinic mineral consisting of hydrated copper iodate. { bə'liŋ·ə,rīt }

bellite [MATER] An explosive consisting of five parts of ammonium nitrate to one part of metadinitrobenzene with some potassium nitrate.　{ 'be,līt }

bell jar [ENG] A bell-shaped vessel, usually made of glass, which is used for enclosing a vacuum, holding gases, or covering objects. Also known as bell glass.　{ 'bel jär }

bell-jar testing [ENG] A leak testing method in which a vessel is filled with tracer gas and placed in a vacuum chamber; leaks are evidenced by gas drawn into the vacuum chamber. { 'bel jär ,tes·tiŋ }

bell-joint clamp [ENG] A clamp applied to a bell-and-spigot joint to prevent leakage.　{ 'bel ,jȯint ,klamp }

Bellman's principle of optimality [IND ENG] The principle that an optimal sequence of decisions in a multistage decision process problem has the property that whatever the initial state and decisions are, the remaining decisions must constitute an optimal policy with regard to the state resulting from the first decisions.　{ 'bel·mənz 'prin·sə·pəl əv ,äp·tə'mal·əd·ē }

bell metal [MET] An alloy of copper and tin, containing 15–25% tin but 20–24% for best tonal quality.　{ 'bel ,med·əl }

bell-metal ore *See* stannite.　{ 'bel ,med·əl ,ȯr }

bell mouth [DES ENG] A flared mouth on a pipe opening or other orifice. [ENG] A defect which occurs during metal drilling in which a twist drill produces a hole that is not a perfect circle.　{ 'bel ,maùth }

bell nipple [PETRO ENG] A bell-mouthed nipple inserted into the top of oil well casing to allow easy entry of drilling tools and to protect the top of the casing during drilling.　{ 'bel ,nip·əl }

Bell numbers [MATH] The numbers, B_n, that count the total number of partitions of a set with n elements.　{ 'bel ,nəm·bərz }

bellows [ENG] 1. A mechanism that expands and contracts, or has a rising and falling top, to suck in air through a valve and blow it out through a tube. 2. Any of several types of enclosures which have accordionlike walls, allowing one to vary the volume. *See* aneroid capsule. [OPTICS] An accordionlike component of a camera which forms a passage between the lens and the film and allows one to vary the distance between them.　{ 'bel·ōz }

bellows expansion joint [DES ENG] In a run of piping, a joint formed with a flexible metal bellows which compress or stretch to compensate for linear expansion or contraction of the run of piping.　{ 'bel·ōz ik'span·shən ,jȯint }

bellows gage [ENG] A device for measuring pressure in which the pressure on a bellows, with the end plate attached to a spring, causes a measurable movement of the plate.　{ 'bel·ōz ,gāj }

bellows gas meter [ENG] A device for measuring the total volume of a continuous gas flow stream in which the motion of two bellows, alternately filled with and exhausted of the gas, actuates a register.　{ 'bel·ōz ¦gas ,mēd·ər }

bellows seal [MECH ENG] A boiler seal in the form of a bellows which prevents leakage of air or gas.　{ 'bel·ōz ,sēl }

bells and whistles [COMPUT SCI] Special hardware features that are likely to attract attention but may not be important or even practical.　{ 'belz ən 'wis·əlz }

bell screw *See* bell tap.　{ 'bel ,skrü }

bell-shaped curve [STAT] The curve representing a continuous frequency distribution with a shape having the overall curvature of the vertical cross section of a bell; usually applied to the normal distribution.　{ ¦bel ¦shāpt 'kərv }

Bell's law [NEURO] 1. The law that in the spinal cord the ventral roots are motor and the dorsal roots sensory in function. 2. The law that in a reflex arc the nerve impulse can be conducted in one direction only.　{ 'belz ,lȯ }

bell socket *See* bell tap.　{ 'bel ,säk·ət }

Bell's theorem [QUANT MECH] A theorem which states that any hidden variable that satisifies the condition of locality cannot possibly reproduce all the statistical predictions of quantum mechanics, and which places upper limits, for the predictions of any such theory, on the strength of correlations between measurements of spatially separated objects, whereas quantum mechanics predicts very strong correlations between such measurements.　{ 'belz ,thir·əm }

bell tap [PETRO ENG] A cylindrical fishing tool having an upward-tapered inside surface provided with hardened threads; when slipped over the upper end of lost, cylindrical, downhole drilling equipment and turned, the threaded inside surface of the bell tap cuts into and grips the outside surface of the lost equipment. Also known as bell; bell screw; bell socket; box bill; die; die collar.　{ 'bel ,tap }

bell transformer [ELEC] An iron-core, step-down transformer with a voltage step-down ratio of approximately 6 to 1 or 12 to 1, used in low-current power supplies and frequently in circuits for doorbells, alarm bells, and buzzers.　{ 'bel tranz,fȯr·mər }

bell-type manometer [ENG] A differential pressure gage in which one pressure input is fed into an inverted cuplike container floating in liquid, and the other pressure input presses down upon the top of the container so that its level in the liquid is the measure of differential pressure.　{ 'bel,tīp mə'näm·əd·ər }

bell wire [ELEC] A copper wire, usually solid rather than stranded, and soft-drawn rather than hard-drawn, used in low-current, low-voltage applications.　{ 'bel ,wīr }

belly [ANAT] 1. The abdominal cavity or the abdomen. 2. The most prominent, fleshy, central portion of a muscle. [GRAPHICS] A downward curve in the center of a type line. { 'bel·ē }

Belondiroidea [INV ZOO] A diverse superfamily of nematodes belonging to the order Dorylaimida, consisting of a diverse group whose principal common characteristic is a thick sheath of spiral muscles enclosing the basal swollen portion of the esophagus.　{ bə,län·də'rȯid·ē·ə }

belonephobia [PSYCH] Abnormal fear of sharp-pointed objects.　{ ,be·lə·nə'fōb·ē·ə }

Beloniformes [VERT ZOO] The former ordinal name for a group of fishes now included in the order Atheriniformes. { ,be·lə·nə'fȯr,mēz }

belonite [GEOL] A rod- or club-shaped microscopic embryonic crystal in a glassy rock.　{ 'bel·ə,nīt }

Belostomatidae [INV ZOO] The giant water bugs, a family of hemipteran insects in the subdivision Hydrocorisae.　{ ,bel·ə·stō'mad·ə,dē }

below minimums [METEOROL] Below operational weather limits for aircraft.　{ bi'lō 'min·ə,məmz }

belt [CIV ENG] In brickwork, a projecting row (or rows) of bricks, or an inserted row made of a different kind of brick. [ECOL] 1. Any altitudinal vegetation zone or band from the base to the summit of a mountain. 2. Any benthic vegetation zone or band from sea level to the ocean depths. 3. Any of the concentric vegetation zones around bodies of fresh water. [HYD] A long area or strip of pack ice, with a width of 1 kilometer (0.6 mile) to more than 100 kilometers (60 miles). [MECH ENG] A flexible band used to connect pulleys or to convey materials by transmitting motion and power.　{ belt }

belt conveyor [MECH ENG] A heavy-duty conveyor consisting essentially of a head or drive pulley, a take-up pulley, a level or inclined endless belt made of canvas, rubber, or metal, and carrying and return idlers.　{ belt kən'vā·ər }

BELLED CAISSON

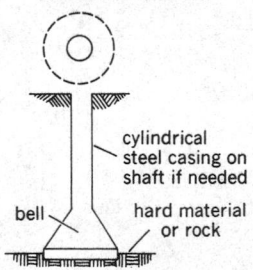

cylindrical steel casing on shaft if needed

bell　　hard material or rock

Belled caisson, showing top view and cross section from side.

belt course See string course. { 'belt ˌkȯrs }

belt dressing [MATER] A material, usually beeswax, applied to leather belts to increase friction between the belt and pulley surface. { 'belt ˌdres·iŋ }

belt drive [MECH ENG] The transmission of power between shafts by means of a belt connecting pulleys on the shafts. { 'belt ˌdrīv }

belted-bias tire See bias-belted tire. { 'bel·təd ˌbī·əs ˌtīr }

belted plain [GEOL] A plain whose surface has been slowly worn down and sculptured into bands or belts of different levels. { 'bel·təd ˌplān }

belteroporic [GEOL] Of crystals in rocks whose growth was determined by the direction of easiest growth. { ˌbel'ter·ə'pȯr·ik }

belt feeder [MECH ENG] A short belt conveyor used to transfer granulated or powdered solids from a storage or supply point to an end-use point; for example, from a bin hopper to a chemical reactor. { 'belt ˌfēd·ər }

belt grinding [MET] Grinding with an abrasive-coated continuous belt. Also known as linishing. { 'belt ˌgrīn·diŋ }

belt guard [MECH ENG] A cover designed to protect a belt as well as the pulleys it connects. { 'belt ˌgärd }

belt highway See beltway. { 'belt 'hī,wā }

belting [MATER] **1.** A sturdy fabric, usually of cotton, used in belts. **2.** A heavy leather, made from hides of cattle, used in power transmission belts. Also known as belting leather. { 'bel·tiŋ }

belting leather See belting. { 'bel·tiŋ ˌleth·ər }

belt of cementation See zone of cementation. { ¦belt əv ˌsi·men'tā·shən }

belt of soil moisture See belt of soil water. { ¦belt əv 'sȯil ˌmȯis·chər }

belt of soil water [GEOL] The upper subdivision of the zone of aeration limited above by the land surface and below by the intermediate belt; this zone contains plant roots and water available for plant growth. Also known as belt of soil moisture; discrete film zone; soil-water belt; soil-water zone; zone of soil water. { ¦belt əv 'sȯil ˌwȯd·ər }

belt printer [COMPUT SCI] A type of impact printer similar to a chain printer in which the characters are carried on a moving belt rather than a chain. { 'belt ˌprint·ər }

belt sander [MECH ENG] A portable sanding tool having a power-driven abrasive-coated continuous belt. { 'belt ˌsan·dər }

belt shifter [MECH ENG] A device with fingerlike projections used to shift a belt from one pulley to another or to replace a belt which has slipped off a pulley. { 'belt ˌshif·tər }

belt slip [MECH ENG] The difference in speed between the driving drum and belt conveyor. { 'belt ˌslip }

belt tightener [MECH ENG] In a belt drive, a device that takes up the slack in a belt that has become stretched and permanently lengthened. { 'belt ˌtīt·nər }

beltway [CIV ENG] A highway that encircles an urban area along its perimeter. Also known as belt highway; ring road. { 'belt,wā }

bemegride [PHARM] $C_8H_{13}NO_2$ Platelets which melt at 127°C and sublime at 100°C (2 mmHg pressure); soluble in water and in acetone; used as an analeptic drug in barbiturate poisoning for humans and as an analeptic drug and central nervous system stimulant in animals. Also known as methetharimide; β,β-methylethylglutarimide. { 'bem·əˌgrīd }

bempa [ORG CHEM] $C_6H_{18}N_3PO$ A white solid soluble in water; used as chemosterilant for insects. Also known as hexamethylphosphorictriamide. { 'bem·pə }

Bénard convection cells [PHYS] A regular array of hexagonal cells which sometimes appear in convection in a layer of liquid heated from below. { bā·när kən'vek·shən ˌselz }

Bence-Jones protein [PATH] An abnormal group of globulins appearing in the serum and urine, usually in association with multiple myeloma and characterized by coagulation at 50–60°C. { ¦bens ¦jōnz ˌprō,tēn }

Bence-Jones proteinuria [MED] The presence of Bence-Jones protein in the urine. { ¦bens ¦jōnz ˌprō·tə'nür·ē·ə }

bench [GEOL] A terrace of level earth or rock that is raised and narrow and that breaks the continuity of a declivity. [MIN ENG] **1.** One of two or more divisions of a coal seam, separated by slate and so forth or simply separated by the process of cutting the coal, one bench or layer being cut before the adjacent one. **2.** A long horizontal ledge of ore in an underground working place. **3.** A ledge in an open-pit mine from which excavation takes place at a constant level. { bench }

bench assembly [ENG] A technique of fitting and joining parts using a bench as a work surface. { 'bench ə'sem·blē }

bench blasting [MIN ENG] A mining system used either underground or in surface pits whereby a thick ore or waste zone is removed by blasting a series of successive horizontal layers called benches. { 'bench 'blas·tiŋ }

bench check [IND ENG] A workshop or servicing bay check which includes the typical check or actual functional test of an item to ascertain what is to be done to return the item to a serviceable condition or ascertain the item's temporary or permanent disposition. { 'bench ˌchek }

bench dog [ENG] A wood or metal peg, placed in a slot or hole at the end of a bench; used to keep a workpiece from slipping. { 'bench ˌdȯg }

bench gravel [GEOL] Gravel beds found on the sides of valleys above the present stream bottoms, representing parts of the bed of the stream when it was at a higher level. { 'bench ˌgrav·əl }

bench height See bank height. { 'bench ˌhīt }

bench hook [ENG] Any device used on a carpenter's bench to keep work from moving toward the rear of the bench. Also known as side hook. { 'bench ˌhuk }

benching [CIV ENG] **1.** Concrete laid on the side slopes of drainage channels where the slopes are interrupted by manholes, and so forth. **2.** Concrete laid on sloping sites as a safeguard against sliding. **3.** Concrete laid along the sides of a pipeline to provide additional support. [MIN ENG] A method of working small quarries or opencast pits in steps or benches. { 'bench·iŋ }

bench lathe [MECH ENG] A small engine or toolroom lathe suitable for attachment to a workbench; bed length usually does not exceed 6 feet (1.8 meters) and workpieces are generally small. { 'bench ˌlāth }

bench lava [GEOL] Semiconsolidated, crusted basaltic lava forming raised platforms and crags about the edges of lava lakes. Also known as bench magma. { 'bench ˌla·və }

bench magma See bench lava. { 'bench ˌmag·mə }

benchmark [ENG] A relatively permanent natural or artificial object bearing a marked point whose elevation above or below an adopted datum—for example, sea level—is known. Abbreviated BM. [IND ENG] A standard of measurement possessing sufficient identifiable characteristics common to the individual units of a population to facilitate economical and efficient comparison of attributes for units selected from a sample. [SCI TECH] A reference value against which a measurement or a series of measurements may be compared. { 'bench,märk }

benchmark index [IND ENG] In manufacturing and mining, an index designed to reflect changes in output occurring between census years. { 'bench,märk 'in,deks }

benchmark job [IND ENG] A job that can be related or compared to other jobs in terms of common characteristics and considered an acceptable gauge for other jobs without the need of direct measurements. { 'bench,märk ˌjäb }

benchmark problem [COMPUT SCI] A problem to be run on computers to evaluate their performances relative to one another. { 'bench,märk 'präb·ləm }

benchmark test [COMPUT SCI] A test of computer software or hardware that is generally run on a number of products to compare their performance. { 'bench,märk ˌtest }

bench photometer [ENG] A device which uses an optical bench with the two light sources to be compared mounted one at each end; the comparison between the two illuminations is made by a device moved along the bench until matching brightnesses appear. { 'bench fə'täm·əd·ər }

bench placer [GEOL] A placer in ancient stream deposits from 50 to 300 feet (15 to 90 meters) above present streams. { 'bench ˌplās·ər }

bench plane [DES ENG] A plane used primarily in benchwork on flat surfaces, such as a block plane or jack plane. { 'bench ˌplān }

bench plot [MIN ENG] A plot of information on a horizontal section of a diagram of a mine; information may include the value of drillhole intercepts, the value of estimated blocks, and geologic contacts. { 'bench ˌplät }

bench sander [MECH ENG] A stationary power sander, usually mounted on a table or stand, which is equipped with a rotating abrasive disk or belt. { 'bench ,san·dər }

bench-scale testing [ENG] Testing of materials, methods, or chemical processes on a small scale, such as on a laboratory worktable. { 'bench ,skāl 'tes·tiŋ }

bench slope *See* bank slope. { 'bench ,slōp }

bench stop [ENG] A bench hook which is used to fasten work in place, often by means of a screw. { 'bench ,stäp }

bench table [BUILD] A projecting course of masonry at the foot of an interior wall or around a column; generally wide enough to form a seat. { 'bench ,tā·bəl }

bench vise [ENG] An ordinary vise fixed to a workbench. { 'bench ,vīs }

benchwork [ENG] Any work performed at a workbench rather than on machines or in the field. { 'bench,wərk }

bend [DES ENG] **1.** The characteristic of an object, such as a machine part, that is curved. **2.** A section of pipe that is curved. **3.** A knot formed by a rope fastened to an object or another rope. [ELECTROMAG] A smooth change in the direction of the longitudinal axis of a waveguide. [GEOL] **1.** A curve or turn occurring in a stream course, bed, or channel which has not yet become a meander. **2.** The land area partly encircled by a bend or meander. [NAV] The departure of a defined navigation course or bearing from a straight line. { bend }

bend allowance [DES ENG] Length of the arc of the neutral axis between the tangent points of a bend in any material. { 'bend ə'laů·əns }

Benday plate [GRAPHICS] A printing plate treated with the Benday process. { 'ben¦dā 'plāt }

Ben Day process *See* Benday process. { ¦ben ¦dā 'prä·səs }

Benday process [GRAPHICS] A process for printing shadings consisting of patterns of lines, dots, stipples, and so on, which involves inking a Benday screen (a rectangle of hardened gelatin with the pattern in relief), printing it on portions of the metal plate on which an outline drawing has been photoprinted, and then etching the metal as a line plate. Also spelled Ben Day process. { 'ben¦dā 'prä·səs }

bender *See* bending machine. { 'ben·dər }

bender element [ELECTR] A combination of two thin strips of different piezoelectric materials bonded together so that when a voltage is applied, one strip increases in length and the other becomes shorter, causing the combination to bend. { 'ben·dər 'el·ə·mənt }

bending [ENG] **1.** The forming of a metal part, by pressure, into a curved or angular shape, or the stretching or flanging of it along a curved path. **2.** The forming of a wooden member to a desired shape by pressure after it has been softened or plasticized by heat and moisture. [HYD] Movement of sea ice up or down resulting from lateral pressure exerted by wind or tide. [MOL BIO] A conformational change characterized by a localized bend or kink in deoxyribonucleic acid due to heterogeneities in local structural composition. [OCEANOGR] The first stage in the formation of pressure ice caused by the action of current, wind, tide, or air temperature changes. { 'ben·diŋ }

bending brake [MECH ENG] A press brake for making sharply angular linear bends in sheet metal. { 'ben·diŋ ,brāk }

bending iron [ENG] A tool used to straighten or to expand flexible pipe, especially lead pipe. { 'ben·diŋ ,ī·ərn }

bending machine [MECH ENG] A machine for bending a metal or wooden part by pressure. Also known as bender. { 'ben·diŋ mə,shēn }

bending moment [MECH] Algebraic sum of all moments located between a cross section and one end of a structural member; a bending moment that bends the beam convex downward is positive, and one that bends it convex upward is negative. { 'ben·diŋ ,mō·mənt }

bending-moment diagram [MECH] A diagram showing the bending moment at every point along the length of a beam plotted as an ordinate. { 'ben·diŋ ,mō·mənt ,dī·ə,gram }

bending schedule [CIV ENG] A chart showing the shapes and dimensions of every reinforcing bar and the number of bars required on a particular job for the construction of a reinforced concrete structure. { 'ben·diŋ ,skej·əl }

bending shackle *See* anchor shackle. { 'ben·diŋ ,shak·əl }

bending slab [NAV ARCH] A slab that is made up of a number of large cast-iron blocks with holes for pins about which frames and other structural members of ships are bent. { 'ben·diŋ ,slab }

bending stress [MECH] An internal tensile or compressive longitudinal stress developed in a beam in response to curvature induced by an external load. { 'ben·diŋ ,stres }

Bendix-Weiss universal joint [MECH ENG] A universal joint that provides for constant angular velocity of the driven shaft by transmitting the torque through a set of four balls lying in the plane that contains the bisector of, and is perpendicular to, the plane of the angle between the shafts. { ¦ben,diks ¦wīs ,yü·nə'vər·səl ,jóint }

bend plane *See* tilt boundary. { 'bend ,plān }

bend radius [DES ENG] The radius corresponding to the curvature of a bent specimen or part, as measured at the inside surface of the bend. { 'bend ,rād·ē·əs }

bends *See* caisson disease. { benz }

bend test [MET] A ductility test in which a specimen is bent through an arc of known radius and angle. { 'bend ,test }

bend wheel [MECH ENG] A wheel used to interrupt and change the normal path of travel of the conveying or driving medium; most generally used to effect a change in direction of conveyor travel from inclined to horizontal or a similar change. { 'bend ,wēl }

Benedicks effect [PHYS] An electromotive force produced in a circuit containing one metal only, but having impurities or internal strains, in the presence of an asymmetrical temperature distribution. { 'ben·ə,diks i'fekt }

Benedict equation of state [PHYS CHEM] An empirical equation relating pressures, temperatures, and volumes for gases and gas mixtures; superseded by the Benedict-Webb-Rubin equation of state. { 'ben·ə,dikt i'kwā·zhən əv 'stāt }

Benedict's solution [ANALY CHEM] A solution of potassium and sodium tartrates, copper sulfate, and sodium carbonate; used to detect reducing sugars. { 'ben·ə,diks sə'lü·shən }

beneficiation [MET] Improving the chemical or physical properties of an ore so that metal can be recovered at a profit. Also known as mineral dressing. { ,ben·ə,fish·ē'ā·shən }

benequinox [ORG CHEM] $C_{13}H_{11}N_3O_2$ A yellow-brown powder that decomposes at 195°C; used as a fungicide for grain seeds and seedlings. { ben'ē·kwə,näks }

bengaline [TEXT] A warp-faced fabric with distinct, fully covered crosswise ribs that are formed by coarse filling. { 'beŋ·gə,lēn }

Benguela Current [OCEANOGR] A strong current flowing northward along the southwestern coast of Africa. { ben'gwel·ə¦kər·ənt }

Benham top [OPTICS] A disk whose surface has black and white portions and which, when rotated at certain speeds and subjected to certain lighting, produces sensations of color. { 'ben·əm ,täp }

benign [MED] Of no danger to life or health. { bə'nīn }

benign lymphoreticulosis *See* cat scratch disease. { bə'nīn ¦lim·fə·rə,tik·yə'lō·səs }

benign myalgic encephalomyelitis *See* neuromyasthenia. { bə'nīn mī'al·jik ,en,sef·ə·lō,mī·ə'līd·əs }

benign tertian malaria *See* vivax malaria. { bə'nīn 'tər·shən mə'ler·ē·ə }

benign tumor [MED] A nonmalignant neoplasm. { bə'nīn 'tü·mər }

Benioff extensometer [ENG] A linear strainmeter for measuring the change in distance between two reference points separated by 60–90 feet (20–30 meters) or more; used to observe earth tides. { 'ben·ē·óf ,ek,sten'säm·əd·ər }

Benioff zone [GEOPHYS] A zone of earthquake hypocenters distributed on well-defined planes that dips from a shallow depth into the earth's mantle to depths as great as 420 miles (700 kilometers). Also known as Benioff-Wadati zone; Wadati-Benioff zone. { 'ben·ē·óf ,zōn }

Benioff-Wadati zone *See* Benioff zone. { ¦ben·ē,óf wə'dä·tē ,zōn }

benitoite [MINERAL] $BaTi(SiO_3)_3$ A blue to violet barium-titanium silicate mineral; at one time it was cut and sold as sapphire. { bə'nēd·ə,wīt }

benjamin gum *See* benzoin. { 'ben·jə·mən ,gəm }

benjaminite [MINERAL] $Pb_2(Cu,Ag)_2Bi_4S_9$ A gray mineral occurring in granular massive form. { 'ben·jə·mə,nīt }

benne oil *See* sesame oil. { 'ben·ē ,óil }

Bennettitales [PALEOBOT] An equivalent name for the Cycadeoidales. { bə,ned·ə'tā,lēz }

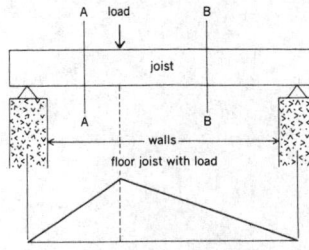

BENDING MOMENT

A   load   B

↓

joist

A   B

walls

floor joist with load

Schematic of bending moment on an end-supported joist with a concentrated load. Bending moment at section A-A is upward reactive force of left wall times distance from wall to A-A. Bending moment at section B-B is upward reactive force of left wall times the distance from wall to B-B plus the moment produced by the downward load acting at its distance from B-B. The resulting bending moment at each cross section is shown on the lower portion of the diagram.

Bennettitatae [PALEOBOT] A class of fossil gymnosperms in the order Cycadeoidales. { be,ned·ə'tā,dē }

ben oil [MATER] Oil from seeds of ben (*Moringa*) trees; used in foods, cosmetics, perfumes, and laxatives and as a lubricant. { 'ben ,oil }

benomyl [ORG CHEM] $C_{14}H_{18}N_4O_3$ Methyl-l-butylcarbamoyl-2-benzimidazole carbamate; a fungicide used to control plant disease. { 'ben·ə,mil }

bensulide [ORG CHEM] $C_{14}H_{24}O_4NPS_3$ An *S-(O,O*-diisopropyl phosphorodithioate) ester of *N-*(2-mercaptoethyl)-benzenesulfonamide; an amber liquid slightly soluble in water; melting point is 34.4°C; used as a preemergent herbicide for annual grasses and for broadleaf weeds in lawns and vegetable and cotton crops. { 'ben·sə,līd }

bent [CIV ENG] A framework support transverse to the length of a structure. { bent }

bent-back occlusion [METEOROL] An occluded front that has reversed its direction of motion as a result of the development of a new cyclone (usually near the point of occlusion) or, less frequently, as the result of the displacement of the old cyclone along the front. Also known as back-bent occlusion. { 'bent ,bak ə'klü·zhən }

bent bar [CIV ENG] A longitudinal reinforcing bar which is bent to pass from one face of a structural member to the other face. { 'bent ,bär }

benthic [OCEANOGR] Of,pertaining to, or living on the bottom or at the greatest depths of a large body of water. Also known as benthonic. { 'ben·thik }

benthiocarb [ORG CHEM] $C_{12}H_{16}NOCl$ An amber liquid with a boiling point of 126–129°C; slightly soluble in water; used as an herbicide to control aquatic weeds in rice crops. { ben'thī·ō,kärb }

benthonic See benthic. { ben'thän·ik }

benthos [ECOL] Bottom-dwelling forms of marine life. Also known as bottom fauna. [OCEANOGR] The floor or deepest part of a sea or ocean. { 'ben,thäs }

bent housing [PETRO ENG] A housing, designed for a positive-displacement downhole turbodrill, having a bend of 1 to 3° to facilitate directional drilling. { 'bent 'haùz·iŋ }

Benton hologram [OPTICS] A type of hologram that can be viewed in white light but lacks parallax in the vertical plane. { 'bent·ən 'häl·ə,gram }

bentonite [GEOL] A clay formed from volcanic ash decomposition and largely composed of montmorillonite and beidellite. Also known as taylorite. { 'bent·ən,īt }

bentonite slurry [MATER] Slurry composed of water and clay powder consisting chiefly of montmorillonite minerals. { 'bent·ən,īt ,slər·ē }

bent-pipe system [COMMUN] A transponder on board a communications satellite that performs no signal processing other than heterodyning (frequency-changing) the uplink frequency bands to those of the downlinks. { ,bent 'pīp ,sis·təm }

bent sub [PETRO ENG] A short cylinder installed in the drill stem between the lowest drill collar and the downhole turbodrill to deflect the turbodrill from vertical in order to drill a directional hole. { 'bent ,səb }

bent-tube boiler [MECH ENG] A water-tube steam boiler in which the tubes terminate in upper and lower steam-and-water drums. Also known as drum-type boiler. { 'bent ,tüb 'bóil·ər }

bentu de soli [METEOROL] An east wind on the coast of Sardinia. { 'ben·tü dā 'sōl·ē }

bentwood [ENG] Wood formed to shape by bending, rather than by carving or machining. { 'bent,wùd }

benzadox [ORG CHEM] $C_6H_5CONHOCH_2COOH$ White crystals with a melting point of 140°C; soluble in water; used as an herbicide to control kochia in sugarbeets. { 'ben·zə,däks }

benzal chloride [ORG CHEM] $C_6H_5CHCl_2$ A colorless liquid that is refractive and fumes in air; boiling point 207°C; used to make benzaldehyde and cinnamic acid. { ,benz·əl 'klòr,īd }

benzaldehyde [ORG CHEM] C_6H_5CHO A colorless, liquid aldehyde, boiling at 170°C and possessing the odor of bitter almonds; used as a flavoring agent and an intermediate in chemical syntheses. { benz'al·də,hīd }

benzaldoxime [ORG CHEM] C_6H_5CHNOH An oxime of benzaldehyde; the antiisomeric form melts at 130°C, the syn form at 34°C; both forms are soluble in ethyl alcohol and ether; used in synthesis of other organic compounds. { ,benz·əl'däk,sēm }

benzalkonium [ORG CHEM] $C_6H_5CH_2N(CH_3)_2R^+$ An organic radical in which R may range from C_8H_{17} to $C_{18}H_{37}$; found in surfactants, as the chloride salt. { ,benz·əl'kòn·ē·əm }

benzalkonium chloride [ORG CHEM] $C_6H_5CH_2(CH_3)_2NRCl$ A yellow-white powder soluble in water; used as a fungicide and bactericide; the R is a mixture of alkyls from C_8H_{17} to $C_{18}H_{37}$. { ,benz·əl'kòn·ē·əm 'klòr,īd }

benzamide [ORG CHEM] $C_6H_5CONH_2$ A compound with melting point 132.5° to 133.5°C; slightly soluble in water, soluble in ethyl alcohol and carbon tetrachloride; used in chemical synthesis. { ben'za,mīd }

benzanilide [ORG CHEM] $C_2H_5CONHC_6H_5$ Leaflet crystals with a melting point of 163°C; soluble in alcohol; used to manufacture dyes and perfumes. { benz'an·ə,līd }

benzanthracene [ORG CHEM] $C_{18}H_{14}$ A weakly carcinogenic material that is isomeric with naphthacene; melting point 162°C; insoluble in water, soluble in benzene. { benz'an·thrə,sēn }

benzanthrone [ORG CHEM] $C_{17}H_{10}O$ A compound with melting point 170°C; insoluble in water; used in dye manufacture. { benz'an,thrōn }

Benzedrine See amphetamine. { 'ben·zə,drēn }

benzene [ORG CHEM] C_6H_6 A colorless, liquid, flammable, aromatic hydrocarbon that boils at 80.1°C and freezes at 5.4–5.5°C; used to manufacture styrene and phenol. Also known as benzol. { 'ben,zēn }

benzenediazonium chloride [ORG CHEM] $C_6H_5N(N)Cl$ An ionic salt soluble in water; used as a dye intermediate. { ,ben,zēn,dī·ə'zōn·ē·əm 'klòr,īd }

benzenephosphorus dichloride [ORG CHEM] $C_6H_5PCl_2$ An irritating, colorless liquid with a boiling point of 224.6°C; soluble in inert organic solvents; used in organic synthesis and oil additives. { 'ben,zēn'fäs·fə·rəs ,dī'klòr,īd }

benzene ring [ORG CHEM] The six-carbon ring structure found in benzene, C_6H_6, and in organic compounds formed from benzene by replacement of one or more hydrogen atoms by other chemical atoms or radicals. { 'ben,zēn ,riŋ }

benzene series [ORG CHEM] A series of carbon-hydrogen compounds based on the benzene ring, with the general formula C_nH_{2n-6}, where *n* is 6 or more; examples are benzene, C_6H_6, toluene, C_7H_8, and xylene, C_8H_{10}. { 'ben,zēn ,sir·ēz }

benzenesulfonate [ORG CHEM] Any salt or ester of benzenesulfonic acid. { ,ben,zēn'səl·fə,nāt }

benzenesulfonic acid [ORG CHEM] $C_6H_5SO_3H$ An organosulfur compound, strongly acidic, water soluble, nonvolatile, and hygroscopic; used in the manufacture of detergents and phenols. { ,ben,zēn,səl'fän·ik 'as·əd }

1,2,4-benzenetricarboxylic acid [ORG CHEM] $C_6H_3(COOH)_3$ Crystals with a melting point of 218–220°C; crystallizes from acetic acid or from dilute alcohol; used as an intermediate in the preparation of adhesives, plasticizers, dyes, inks, and resins. { wən |tü |fòr 'ben,zēn·trī|kär·bäk|sil·ik 'as·əd }

1,2,4-benzenetriol [ORG CHEM] $C_6H_3(OH)_3$ Monoprismatic leaflets with a melting point of 141°C; freely soluble in water, ether, alcohol, and ethylacetate; used in gas analysis. { |wən |tü |fòr 'ben,zēn'trī,ól }

benzenoid [ORG CHEM] Any substance which has the electronic character of benzene. { 'ben·zə,nòid }

benzestrol [PHARM] $C_{20}H_{26}O_2$ A white powder; melting point 162–166°C; soluble in ethanol and acetone, insoluble in water; used as a medicine. { ben'ze,stról }

benzethonium chloride [PHARM] $C_{27}H_{42}ClNO_2$ Crystallizes as thin, hexagonal plates; melting point is 164–166°C; soluble in water, alcohol, chloroform, and acetone; used as a topical antiseptic on humans and animals. { ,ben·zə'thòn·ē·əm 'klòr,īd }

benzhydrol [ORG CHEM] $(C_6H_5)_2CHOH$ Colorless needles; melting point 69°C; slightly soluble in water, very soluble in ethanol and ether; used in preparation of other organic compounds including antihistamines. { benz'hī,dról }

benzidine [ORG CHEM] $NH_2C_6H_4C_6H_4NH_2$ An aromatic amine with a melting point of 128°C; used as an intermediate in syntheses of direct dyes for cotton. { 'ben·zə,dēn }

benzil [ORG CHEM] $C_6H_5COCOC_6H_5$ A yellow powder; melting point 95°C; insoluble in water, soluble in ethanol, ether, and benzene; used in organic synthesis. { 'ben,zil }

benzilic acid [ORG CHEM] $(C_6H_5)_2C(OH)CO_2H$ A white,

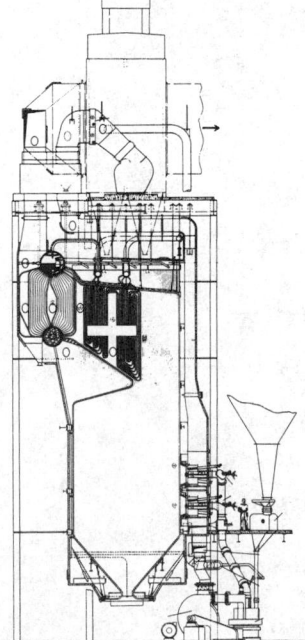

BENT-TUBE BOILER

Diagram of a bent-tube boiler.

crystalline acid, synthesized by heating benzil with alcohol and potassium hydroxide; used in organic synthesis. { 'ben'zil·ik 'as·əd }

benzimidazole [ORG CHEM] $C_7H_6N_2$ Colorless crystals; melting point 170°C; slightly soluble in water, soluble in ethanol; used in organic synthesis. { ‚ben·zə'mid·ə‚zól }

benzin See petroleum benzin. { 'ben·zən }

benzine See petroleum benzin. { 'ben‚zēn }

benzoate [ORG CHEM] A salt or ester of benzoic acid, formed by replacing the acidic hydrogen of the carboxyl group with a metal or organic radical. { 'ben·zə‚wāt }

benzocaine See ethyl-*para*-aminobenzoate. { 'ben·zə‚kān }

benzodiazepine [MED] A group of tranquilizers that are used to combat anxiety and convulsions. { ‚ben·zō‚dī'az·ə‚pēn }

benzodihydropyrone [ORG CHEM] $C_9H_8O_2$ A white to light yellow, oily liquid having a sweet odor; soluble in alcohol, chloroform, and ether; used in perfumery. { 'ben·zō‚dī‚hī·dra'pī‚rōn }

benzoic acid [ORG CHEM] C_6H_5COOH An aromatic carboxylic acid that melts at 122.4°C, boils at 250°C, and is slightly soluble in water and relatively soluble in alcohol and ether; derivatives are valuable in industry, commerce, and medicine. { ben'zō·ik 'as·əd }

benzoic anhydride [ORG CHEM] $(C_6H_5CO)_2O$ An acid anhydride that melts at 42°C, boils at 360°C, and crystallizes in colorless prisms; used in synthesis of a variety of organic chemicals, including some dyes. { ben'zō·ik an'hī‚drīd }

benzoin [MATER] A balsamic resin obtained from trees of the genus *Styrax;* used as an expectorant, as an inhalant in respiratory tract inflammations, and as an antiseptic. Also known as benjamin gum; benzoinam; gum benzoin. [ORG CHEM] $C_{14}H_{12}O_2I$ An optically active compound; white or yellowish crystals, melting point 137°C; soluble in acetone, slightly soluble in water; used in organic synthesis. { 'ben·zə·wən }

α-benzoin oxime [ORG CHEM] $C_6H_5CH(OH)C(NOH)C_6H_5$ Prisms crystallized from benzene; melting point is 151–152°C; soluble in alcohol and in aqueous ammonium hydroxide solution; used in the detection and determination of copper, molybdenum, and tungsten. { 'al·fə 'ben·zə·wən 'äk‚sēm }

benzol See benzene. { 'ben‚zól }

benzol-acetone process [CHEM ENG] A solvent dewaxing process in which a mixture of the solvent and oil containing wax is cooled until the wax solidifies and is then removed by filtration. { 'ben‚zól 'as·ə‚tōn ‚präs·əs }

benzoline See normal benzine. { 'ben·zə‚lēn }

benzomate [ORG CHEM] $C_{18}H_{18}O_5N$ A white solid that melts at 71.5–73°C; used as a wettable powder as a miticide. { 'ben·zə‚māt }

benzonitrile [ORG CHEM] C_6H_5CN A colorless liquid with an almond odor; made by heating benzoic acid with lead thiocyanate and used in the synthesis of organic chemicals. Also known as phenyl cyanide. { ‚ben·zō‚nī·trəl }

benzophenone [ORG CHEM] $C_6H_5COC_6H_6$ A diphenyl ketone, boiling point 305.9°C, occurring in four polymorphic forms (α, β, γ, and δ) each with different melting point; used as a constituent of synthetic perfumes and as a chemical intermediate. Also known as diphenyl ketone; phenyl ketone. { ‚ben·zō·fə'nōn }

benzopyrene [ORG CHEM] $C_{20}H_{12}$ A five-ring aromatic hydrocarbon found in coal tar, in cigarette smoke, and as a product of incomplete combustion; yellow crystals with a melting point of 179°C; soluble in benzene, toluene, and xylene. { ‚ben·zō‚pī‚rēn }

1,2-benzopyrone See coumarin. { ‚wən ‚tü ‚ben·zō‚pī‚rōn }

2,3-benzopyrrole See indole. { ‚tü ‚thrē ‚ben·zō‚pī‚rōl }

5,6-benzoquinoline [ORG CHEM] $C_{13}H_9N$ Crystals which are soluble in dilute acids, alcohol, ether, or benzene; melting point is 93°C; used as a reagent for the determination of cadmium. { ‚fīv ‚siks ‚ben·zō'kwin·ə‚lēn }

benzoquinone See quinone. { ‚ben·zō‚kwə'nōn }

benzoresorcinol [ORG CHEM] $C_{13}H_{10}O_3$ A compound crystallizing as needles from hot-water solution; used in paints and plastics as an ultraviolet light absorber. Also known as resbenzophenone. { ‚ben·zō·ri'sór·sə‚nól }

benzosulfimide See saccharin. { ‚ben·zō'səl·fə‚mīd }

benzothiazole [ORG CHEM] C_6H_4SCHN A thiazole fused to a benzene ring; can be made by ring closure from *o*-amino

thiophenols and acid chlorides; derivatives are important industrial products. { ‚ben·zō'thī·ə‚zól }

4-benzothienyl-*N*-methylcarbamate [ORG CHEM] $C_{10}H_9$- NO_2S A white powder compound with a melting point of 128°C; used as an insecticide for crop insects. { ‚fór ‚ben·zō'thī·ə‚nil ‚en‚meth·əl'kär·bə‚mat }

benzothiofuran See thianaphthene. { ‚ben‚zō‚thī‚ō'fyù‚ran }

1,2,3-benzotriazole [ORG CHEM] $C_6H_5N_3$ A compound with melting point 98.5°C; soluble in ethanol, insoluble in water; derivatives are ultraviolet absorbers; used as a chemical intermediate. { ‚wən ‚tü ‚thrē ‚ben·zō'trī·ə‚zól }

benzotrichloride [ORG CHEM] $C_6H_5CCl_3$ A colorless to yellow liquid that fumes upon exposure to air; has penetrating odor; insoluble in water, soluble in ethanol and ether; used to make dyes. { ‚ben·zō‚trī'klór‚īd }

benzotrifluoride [ORG CHEM] Colorless liquid, boiling point 102.1°C; used for dyes and pharmaceuticals, as solvent and vulcanizing agent, in insecticides. { ‚ben·zō‚trī'flùr‚īd }

benzoyl [ORG CHEM] The radical $C_6H_5ICO^-$ found, for example, in benzoyl chloride. { 'ben·zə·wəl }

benzoylation [ORG CHEM] Introduction of the aryl radical (C_6H_5CO) into a molecule. { ‚ben·zō·ə'lā·shən }

benzoyl chloride [ORG CHEM] C_6H_5COCl Colorless liquid whose vapor induces tears; soluble in ether, decomposes in water; used as an intermediate in chemical synthesis. { 'ben·zə·wəl 'klór‚īd }

benzoyl chloride 2,4,6-trichlorophenylhydrazone [ORG CHEM] $C_6H_5CClN_2HC_6H_2Cl_3$ A white to yellow solid with a melting point of 96.5–98°C; insoluble in water; used as an anthelminthic for citrus. { 'ben·zə·wəl 'klór‚īd ‚tü ‚fór ‚siks ‚trī·klór·ə‚fen·əl'hī·drə‚zōn }

benzoyl peroxide [ORG CHEM] $(C_6H_5CO)_2O_2$ A white, crystalline solid; melting point 103–105°C; explodes when heated above 105°C; slightly soluble in water, soluble in organic solvents; used as a bleaching and drying agent and a polymerization catalyst. { 'ben·zə·wəl pə'räk‚sīd }

benzoylpropethyl [ORG CHEM] $C_{18}H_{17}Cl_2NO_3$ An off-white, crystalline compound with a melting point of 72°C; used as a preemergence herbicide for control of wild oats. { ‚ben·zə·wəl‚prō·pə·thəl }

3,4-benzpyrene [ORG CHEM] $C_{20}H_{12}$ A polycyclic hydrocarbon; a chemical carcinogen that will cause skin cancer in many species when applied in low dosage. { ‚thrē ‚fór ‚benz'pī‚rēn }

benzthiazuron [ORG CHEM] $C_9H_9N_3SO$ A white powder that decomposes at 287°C; slightly soluble in water; used as a preemergent herbicide for sugarbeets and fodder beet crops. { ‚benz‚thī'az·yə‚rän }

benzyl [ORG CHEM] The radical $C_6H_5CH_2^-$ found, for example, in benzyl alcohol, $C_6H_5CH_2OH$. { 'ben·zəl }

benzyl acetate [ORG CHEM] $C_6H_5CH_2OOCCH_3$ A colorless liquid with a flowery odor; used in perfumes and flavorings and as a solvent for plastics and resins, inks, and polishes. Also known as phenylmethyl acetate. { 'ben·zəl 'as·ə‚tāt }

benzylacetone [ORG CHEM] $C_6H_5(CH_2)_2COCH_3$ A liquid with a melting point of 233–234°C; used as an attractant to trap melon flies. { 'ben·zəl'as·ə‚tōn }

benzyl alcohol [ORG CHEM] $C_6H_5CH_2OH$ An alcohol that melts at 15.3°C, boils at 205.8°C, and is soluble in water and readily soluble in alcohol and ether; valued for the esters it forms with acetic, benzoic, and sebacic acids and used in the soap, perfume, and flavor industries. Also known as phenylmethanol. { 'ben·zəl 'al·kə‚hól }

benzylamine [ORG CHEM] $C_6H_5CH_2NH_2$ A liquid that is soluble in water, ethanol, and ether; boils at 185°C (770 mmHg) and at 84°C (24 mmHg); it is toxic; used as a chemical intermediate in dye production. Also known as aminotoluene. { ‚ben·zəl'am‚ēn }

benzyl benzoate [ORG CHEM] $C_6H_5COOCH_2C_6H_5$ An oily, colorless liquid ester; used as an antispasmodic drug and as a scabicide. { 'ben·zəl 'ben·zə‚wāt }

benzyl bromide [ORG CHEM] $C_6H_5CH_2Br$ A toxic, irritating, corrosive clear liquid with a boiling point of 198–199°C; acts as a lacrimator; soluble in alcohol, benzene, and ether; used to make foaming and frothing agents. { 'ben·zəl 'brō‚mīd }

benzyl chloride [ORG CHEM] $C_6H_5CH_2Cl$ A colorless liquid with a pungent odor produced by the chlorination of toluene. { 'ben·zəl 'klór‚īd }

benzyl chloroformate [ORG CHEM] $C_8H_7ClO_2$ An oily

liquid with an acrid odor which causes eyes to tear; boiling point is 103°C (20 mmHg pressure); used to block the amino group in peptide synthesis. { 'ben·zəl ˌklȯr·ə'fȯrˌmāt }

benzyl cinnamate [ORG CHEM] $C_8H_7COOCH_2C_6H_5$ White crystals; melting point 39°C; insoluble in water, soluble in ethanol; used in perfumery. { 'ben·zəl 'sin·əˌmāt }

benzyl cyanide [ORG CHEM] $C_6H_5CH_2CN$ A toxic, colorless liquid; insoluble in water, soluble in alcohol and ethanol; boils at 234°C; used in organic synthesis. { 'ben·zəl 'sī·əˌnīd }

benzyl ether [ORG CHEM] $(C_6H_5CH_2)_2O$ A liquid unstable at room temperature; boiling point 295–298°C; used in perfumes and as a plasticizer for nitrocellulose. Also known as dibenzyl ether. { 'ben·zəl 'ē·thər }

benzyl ethyl ether [ORG CHEM] $C_6H_5CH_2OC_2H_5$ A colorless, oily, combustible liquid with a boiling point of 185°C; used in organic synthesis and as a flavoring. { 'ben·zəl 'eth·əl 'ē·thər }

benzyl fluoride [ORG CHEM] $C_6H_5CH_2F$ A toxic, irritating, colorless liquid with a boiling point of 139.8°C at 753 millimeters of mercury; used in organic synthesis. { 'ben·zəl 'flu̇rˌīd }

benzyl formate [ORG CHEM] $C_6H_5CH_2OOCH$ A colorless liquid with a fruity-spicy odor and a boiling point of 203°C; used in perfumes and as a flavoring. { 'ben·zəl 'fȯrˌmāt }

benzylideneacetone [ORG CHEM] $C_6H_5CH=CHCOCH_3$ A crystalline compound soluble in alcohol, benzene, chloroform, and ether; melting point is 41–45°C; used in perfume manufacture and in organic synthesis. { benˈzil·əˌdēn'asˌə ˌtōn }

benzyl isoeugenol [ORG CHEM] $CH_3CHCHC_6H_3$-$(OCH_3)OCH_2C_6H_5$ A white, crystalline compound with a floral odor; soluble in alcohol and ether; used in perfumery. { 'ben·zəl ˌī·sō'yü·jəˌnȯl }

benzyl mercaptan [ORG CHEM] $C_6H_5CH_2SH$ A colorless liquid with a boiling point of 195°C; soluble in alcohol and carbon disulfide; used as an odorant and for flavoring. { 'ben·zəl mər'kap·tan }

benzyl penicillinic acid [ORG CHEM] $C_{16}H_{18}N_2O_4S$ An amorphous white powder extracted with ether or chloroform from an acidified aqueous solution of benzyl penicillin. { 'ben·zəl ˌpen·ə·sə'lin·ik 'as·əd }

benzyl penicillin potassium [MICROBIO] $C_{16}H_{17}KN_2O_4S$ Moderately hygroscopic crystals; soluble in water; inactivated by acids and alkalies; obtained from fermentation of *Penicillium chrysogenum*; used as an antimicrobial drug in human and animal disease. { 'ben·zəl ˌpen·ə'sil·ən pə'tas·ē·əm }

benzyl penicillin sodium [MICROBIO] $C_{16}H_{17}N_2NaO_4S$ Crystals obtained from a methanol-ethyl acetate acidified extract of fermentation broth of *Penicillium chrysogenum*; used as an antimicrobial in human and animal disease. { 'ben·zəl ˌpen·ə'sil·ən 'sōd·ē·əm }

benzyl propionate [ORG CHEM] $C_2H_5COOCH_2C_6H_5$ A combustible liquid with a sweet odor and a boiling point of 220°C; used in perfumes and for flavoring. { 'ben·zəl 'prō·pē·əˌnāt }

benzyl salicylate [ORG CHEM] $C_{14}H_{12}O_3$ A thick liquid with a slight, pleasant odor; used as a fixer in perfumery and in sunburn preparations. { 'ben·zəl sə'lis·əˌlāt }

benzyne [ORG CHEM] C_6H_4 A chemical species whose structure consists of an aromatic ring in which four carbon atoms are bonded to hydrogen atoms and two adjacent carbon atoms lack substituents; a member of a class of compounds known as arynes. { 'benˌzīn }

Beranek scale [ACOUS] A scale which measures the subjective loudness of a noise; noises are arranged into six arbitrary categories: very quiet, quiet, moderately quiet, noisy, very noisy, and intolerably noisy. { 'bä'ran·ik ˌskāl }

beraunite [MINERAL] $Fe^{2+}Fe^{3+}(PO_4)_3(OH)_5 \cdot 3H_2O$ A reddish-brown to blood red, monoclinic mineral consisting of hydrated basic phosphate of ferric and ferrous iron. { bə'rau̇ˌnīt }

berbamine [ORG CHEM] $C_{37}H_{40}N_2O_6$ An alkaloid; melting point 170°C; slightly soluble in water, soluble in alcohol and ether. { 'bər·bəˌmēn }

Berberidaceae [BOT] A family of dicotyledonous herbs and shrubs in the order Ranunculales characterized by alternate leaves, perfect, well-developed flowers, and a seemingly solitary carpel. { ˌbər·bə·rə'dās·ē·ē }

berberine [ORG CHEM] $C_{20}H_{19}NO_5$ A toxic compound;

melting point 145°C; the anhydrous form is insoluble in water, soluble in alcohol and ether. { 'bär·bəˌrēn }

Berenice's Hair *See* Coma Berenices. { ˌber·ə'nē·səz 'her }

beresorite *See* phoenicochroite. { bə'res·əˌrīt }

ber function [MATH] One of the functions that is defined by $\text{ber}_n(z) \pm i\text{bei}_n(z) = J_n(ze^{\pm 3\pi i/4})$, where J_n is the nth Bessel function. { 'ber ˌfəŋk·shən }

bergamot oil [MATER] A yellow-green essential oil from the rind of bergamot (*Citrus bergamia*) fruit; it is volatile, contains linalyl acetate, limonene, and livalol, and is used in perfumes. { 'bər·gəˌmät ˌȯil }

berg crystal *See* rock crystal. { 'bərg ˌkris·təl }

Bergeron-Findeisen theory [METEOROL] The theoretical explanation that precipitation particles form within a mixed cloud (composed of both ice crystals and liquid water drops) because the equilibrium vapor pressure of water vapor with respect to ice is less than that with respect to liquid water at the same temperature. Also known as ice-crystal theory; Wedener-Bergeron process. { 'berzh·əˌrän ˈfinˌdīz·ən ˌthē·ə·rē }

Bergius process [CHEM ENG] Treatment of carbonaceous matter, such as coal or cellulosic materials, with hydrogen at elevated pressures and temperatures in the presence of a catalyst, to form an oil similar to crude petroleum. Also known as coal hydrogenation. { 'ber·gē·əs ˌpräs·əs }

Bergmann's rule [ECOL] The principle that in a polytypic wide-ranging species of warm-blooded animals the average body size of members of each geographic race varies with the mean environmental temperature. { 'bərg·mənz ˌrül }

Bergman-Turner unit [BIOL] A unit for the standardization of thyroid extract. { ˈbərg·mən ˈtər·nər ˌyü·nət }

bergmehl *See* rock milk. { 'berkˌmel }

bergschrund [HYD] A type of crevice in a glacier; formed when ice and snow break away from a rock face. { 'berk ˌshru̇nt }

Berg's diver method *See* diver method. { 'bərgz 'dīv·ər ˌmeth·əd }

berg till *See* floe till. { 'bərg ˌtil }

bergy-bit *See* growler. { 'bərg·ē ˌbit }

beriberi [MED] A disorder resulting from the deficiency of vitamin B_1 and characterized by neurologic symptoms, cardiovascular abnormalities, edema, and cerebral manifestations. { ˈber·ēˈber·ē }

Bering Sea [GEOGR] A body of water north of the Pacific Ocean, bounded by Siberia, Alaska, and the Aleutian Islands. { 'ber·iŋ 'sē }

Berkefeld filter [MICROBIO] A diatomaceous-earth filter that is used for the sterilization of heat-labile liquids, such as blood serum, enzyme solutions, and antibiotics. { 'berk·əˌfeld ˌfil·tər }

berkelium [CHEM] A radioactive element, symbol Bk, atomic number 97, the eighth member of the actinide series; properties resemble those of the rare-earth cerium. { 'bər·klē·əm }

berkeyite *See* lazulite. { 'bərk·ēˌīt }

berlinite [MINERAL] $Al(PO_4)$ A colorless to gray or pale rose, hexagonal mineral consisting of aluminum orthophosphate; occurs in massive form. { 'bər·ləˌnīt }

Berl saddle [CHEM ENG] A type of column packing used in distillation columns. { 'bərl ˌsad·əl }

berm [CIV ENG] A horizontal ledge cut between the foot and top of an embankment to stabilize the slope by intercepting sliding earth. [GEOL] **1.** A narrow terrace which originates from the interruption of an erosion cycle with rejuvenation of a stream in the mature stage of its development and renewed dissection. **2.** A horizontal portion of a beach or backshore formed by deposit of material as a result of wave action. Also known as backshore terrace; coastal berm. { bərm }

bermanite [MINERAL] $Mn^{2+}Mn_2^{3+}(PO_4)_2(OH)_2 \cdot 4H_2O$ A reddish-brown, orthorhombic mineral consisting of a hydrated basic phosphate of manganese; occurs in crystal aggregates and as lamellar masses. { 'bər·məˌnīt }

berm crest [GEOL] The seaward limit and usually the highest spot on a coastal berm. Also known as berm edge. { 'bərm ˌkrest }

berm edge *See* berm crest. { 'bərm ˌej }

Bermuda grass [BOT] *Cynodon dactylon.* A long-lived perennial in the order Cyperales. { bər'myüd·ə ˌgras }

BERL SADDLE

The shape of the Berl saddle.

Bermuda high [METEOROL] The semipermanent subtropical high of the North Atlantic Ocean, especially when it is located in the western part of that ocean area. { bər'myüd·ə ¦hī }

Bernal chart [CRYSTAL] A chart used to determine the coordinates in reciprocal space of x-ray reflections that produce the spots on an x-ray diffraction photograph of a single crystal. { bər'nal ¸chärt }

bernalite [MINERAL] Fe(OH)₃ An iron hydroxide, yellow-green or dark green in color. { 'bern·ə¸līt }

Bernoulli differential equation See Bernoulli equation. { ber¸nü·lē or 'ber·nü¦yē ¸dif·ə'ren·chəl i'kwā·zhən }

Bernoulli distribution See binomial distribution. { ber¸nü·lē dis·trə'byü·shən }

Bernoulli effect [FL MECH] As a consequence of the Bernoulli theorem, the pressure of a stream of fluid is reduced as its speed of flow is increased. { ber¸nü·lē i'fekt }

Bernoulli equation See Bernoulli theorem. [MATH] A nonlinear first-order differential equation of the form $(dy/dx) + yf(x) = y^n g(x)$, where n is a number different from unity and f and g are given functions. Also known as Bernoulli differential equation. { ber¸nü·lē i'kwā·zhən }

Bernoulli-Euler law [MECH] A law stating that the curvature of a beam is proportional to the bending moment. { ber¸nü·lē ¦òil·ər ¦lò }

Bernoulli experiments See binomial trials. { bər'nü·lē ik¸sper·ə·məns }

Bernoulli law See Bernoulli theorem. { ber¸nü·lē ¸lò }

Bernoulli number [MATH] The numerical value of the coefficient of $x^{2n}/(2n)!$ is the expansion of $xe^x/(e^x-1)$. { ber¸nü·lē ¸nəm·bər }

Bernoulli polynomial [MATH] The nth such polynomial is

$$\sum_{k=0}^{n} \binom{n}{k} B_k Z^{n-k}$$

where $\binom{n}{k}$ is a binomial coefficient, and B_k is a Bernoulli number. { ber¸nü·lē ¸päl·ə'nō·mē·əl }

Bernoulli's lemniscate [MATH] A curve shaped like a figure eight whose equation in rectangular coordinates is expressed as $(x^2 + y^2)^2 = a^2(x^2 - y^2)$. { ber¸nü·lēz lem'nis·kət }

Bernoulli theorem [FL MECH] An expression of the conservation of energy in the steady flow of an incompressible, inviscid fluid; it states that the quantity $(p/\rho) + gz + (v^2/2)$ is constant along any streamline, where p is the fluid pressure, v is the fluid velocity, ρ is the mass density of the fluid, g is the acceleration due to gravity, and z is the vertical height. Also known as Bernoulli equation; Bernoulli law. [STAT] See law of large numbers. { ber¸nü·lē 'thir·əm }

Bernoulli trials See binomial trials. { bər'nül·ē ¸trīlz }

Beroida [INV ZOO] The single order of the class Nuda in the phylum Ctenophora. { bə'rō·ə·də }

Berriasian [GEOL] Part of or the underlying stage of the Valanginian at the base of the Cretaceous. { ¸ber·ē'ā·zhən }

berry [BOT] A usually small, simple, fleshy or pulpy fruit, such as a strawberry, grape, tomato, or banana. { 'ber·ē }

berth deck [NAV ARCH] 1. The deck on a warship where hammocks were formerly swung. 2. A space in which lie the crew's sleeping quarters. { 'bərth ¸dek }

Berthelot equation [PHYS CHEM] A form of the equation of state which relates the temperature, pressure, and volume of a gas with the gas constant. { 'ber·tə·lō i'kwā·zhən }

Berthelot method [THERMO] A method of measuring the latent heat of vaporization of a liquid that involves determining the temperature rise of a water bath that encloses a tube in which a given amount of vapor is condensed. { 'ber·tə¸lō ¸meth·əd }

Berthelot relation [PHYS] A relationship between molecular attraction constants of like and unlike species. { 'ber·tə·lō ri'lā·shən }

Berthelot-Thomsen principle [PHYS CHEM] The principle that of all chemical reactions possible, the one developing the greatest amount of heat will take place, with certain obvious exceptions such as changes of state. { 'ber·tə·lō ¦täm·sən ¸prin·sə·pəl }

berthierite [MINERAL] FeSb₂S₄ A dark steel gray, orthorhombic mineral consisting of iron antimony sulfide. { 'bər·thē·ə¸rīt }

berthollide [CHEM] A compound whose solid phase exhibits a range of composition. { 'bər·thə¸līd }

Berthon dynamometer [ENG] An instrument for measuring the diameters of small objects, consisting of two metal straight-edges inclined at a small angle and rigidly joined together; a scale on one of the straightedges is used to read the diameters of objects inserted between them. { 'bər¸thän ¸dī·nə'mäm·əd·ər }

berthonite See bournonite. { 'bər·thə¸nīt }

Bertrand curve [MATH] One of a pair of curves having the same principal normals. Also known as associate curve; conjugate curve. { 'ber¸tränd ¸kərv }

bertrandite [MINERAL] Be₄Si₂O₇(OH)₂ A colorless or pale-yellow mineral consisting of a beryllium silicate occurring in prismatic crystals; hardness is 6–7 on Mohs scale, and specific gravity is 2.59–2.60. { 'bər·trən¸dīt }

Bertrand lens [OPTICS] An auxiliary lens that can be inserted in the tube of a polarizing microscope to obtain interference figures. { 'ber¸tränd ¸lenz }

Bertrand's postulate [MATH] The proposition that there exists at least one prime number between any integer greater than three and twice the integer minus two. { 'ber¸tränz 'päs·chə·lət }

Bertrand's rule [MICROBIO] The rule stating that in those compounds, and only in those compounds, having cis secondary alcoholic groups containing at least one carbon atom of d configuration which is subtended by a primary alcohol group, or having a methyl-substituted primary alcohol group of d configuration, the d-carbon atom will be dehydrogenated by the vinegar bacteria *Acetobacter suboxydans*, yielding a ketone. { 'ber¸tränz ¸rül }

Beryciformes [VERT ZOO] An order of actinopterygian fishes in the infraclass Teleostei. { bə¸ris·ə'fòr¸mēz }

Berycomorphi [VERT ZOO] An equivalent name for the Beryciformes. { bə¸rik·ō'mòr¸fī }

beryllate [INORG CHEM] 1. BeO₂²⁻ An ion containing beryllium and oxygen. 2. A salt produced by the reaction of a strong alkali such as sodium hydroxide with beryllium oxide. { 'ber·ə¸lāt }

beryllia See beryllium oxide. { bə'ril·ē·ə }

beryllide [INORG CHEM] A chemical combination of beryllium with a metal, such as zirconium or tantalum. { bə'ril·ə¸dē }

berylliosis [MED] Chronic lung inflammation due to inhalation of beryllium oxide dust. { bə¸ril·ē'ō·səs }

beryllium [CHEM] A chemical element, symbol Be, atomic number 4, atomic weight 9.0122. [MET] A rare metal, occurring naturally in combinations, with density about one-third of aluminum; used most commonly in the manufacture of beryllium-copper alloys which find numerous industrial and scientific applications. { bə'ril·ē·əm }

beryllium alloy [MET] Any dilute alloy of base metals containing a few percent of beryllium in a precipitation-hardening system. { bə'ril·ē·əm 'al¸oi }

beryllium bronze See beryllium copper. { bə'ril·ē·əm 'bränz }

beryllium copper [MET] 1. An alloy of copper and beryllium containing not more than about 3% beryllium; used for springs, tools, and plastic molds. Also known as beryllium bronze. 2. An alloy of copper and beryllium specifically for addition to metals in the foundry. { bə'ril·ē·əm 'käp·ər }

beryllium detector [ENG] An instrument designed to detect and analyze for beryllium by gamma-ray activation analysis. Also known as berylometer. { bə'ril·ē·əm di'tek·tər }

beryllium fluoride [INORG CHEM] BeF₂ A hygroscopic, amorphous solid with a melting point of 800°C; soluble in water; used in beryllium metallurgy. { bə'ril·ē·əm ¦flùr¸īd }

beryllium monel [MET] A nickel-copper alloy containing beryllium. { bə'ril·ē·əm mō¸nel }

beryllium nitrate [INORG CHEM] Be(NO₃)₂·3H₂O A compound that forms colorless, deliquescent crystals that are soluble in water; used to introduce beryllium oxide into materials used in incandescent mantles. { bə'ril·ē·əm 'nī¸trāt }

beryllium nitride [INORG CHEM] Be₃N₂ Refractory, white

BERNOULLI'S LEMNISCATE

Curve known as Bernoulli's lemniscate.

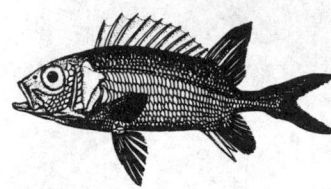

BERYCIFORMES

Squirrelfish (*Holocentrus ascensionis*) of the family Holocentridae. (*After G. B. Goode, Fishery Industries of the United States, 1884*)

crystals with a melting point of 2200±40°C; used in the manufacture of radioactive carbon-14 and in experimental rocket fuels. { bə'ril·ē·əm ˈnīˌtrīd }

beryllium oxide [INORG CHEM] BeO An amorphous white powder, insoluble in water; used to make beryllium salts and as a refractory. Also known as beryllia. { bə'ril·ē·əm 'äkˌsīd }

beryllonite [MINERAL] NaBe(PO₄) A colorless or yellow mineral occurring in short, prismatic or tabular, monoclinic crystals with two good pinacoidal cleavages at right angles; hardness is 5.5–6 on Mohs scale, and specific gravity is 2.85. { bə'ril·əˌnīt }

berylometer See beryllium detector. { ˌber·ə'läm·əd·ər }

Berytidae [INV ZOO] The stilt bugs, a small family of hemipteran insects in the superfamily Pentatomorpha. { bə'rid·əˌdē }

berzelianite [MINERAL] Cu₂Se A silver-white mineral composed of copper selenide and found in igneous rock; specific gravity is 4.03. { ˌbər'zēl·yəˌnīt }

Bessel ellipsoid of 1841 [GEOD] The reference ellipsoid of which the semimajor axis is 6,377,397.2 meters, the semiminor axis is 6,356,079.0 meters, and the flattening or ellipticity equals 1/299.15. Also known as Bessel spheroid of 1841. { 'bes·əl iˈlipˌsoid əv ˌāˌtēnˈfôr·dēˈwən }

Bessel equation [MATH] The differential equation $z^2 f''(z) + zf'(z) + (z^2 - n^2)f(z) = 0$. { 'bes·əl iˈkwā·zhən }

Bessel function [MATH] A solution of the Bessel equation. Also known as cylindrical function. Symbolized $J_n(z)$. { 'bes·əl ˌfəŋk·shən }

Besselian elements [ASTRON] Data on a solar eclipse, giving, for selected times, the coordinates of the axis of the moon's shadow with respect to the fundamental plane, and the radii of umbra and penumbra in that plane; the data allow one to derive local circumstances of the eclipse at any point on the earth's surface. { bə'sel·yən ˈel·ə·mənts }

Besselian star numbers [ASTRON] Constants used in the reduction of a mean position of a star to an apparent position; used to account for short-term variations in precession, nutation, aberration, and parallax. { bə'sel·yən 'stär ˌnəm·bərz }

Besselian year See fictitious year. { bə'sel·yən ˌyir }

Bessel inequality [MATH] The statement that the sum of the squares of the inner product of a vector with the members of an orthonormal set is no larger than the square of the norm of the vector. { 'bes·əl ˌin·ē·kwäl·əd·ē }

Bessel spheroid of 1841 See Bessel ellipsoid of 1841. { 'bes·əl 'sfirˌoid əv ˌāˌtēnˈfôr·dēˈwən }

Bessel transform See Hankel transform. { 'bes·əl 'tranzˌfôrm }

Bessemer converter [MET] A pear-shaped, basic-lined, cylindrical vessel for producing steel by the Bessemer process. { 'bes·ə·mər kənˈvərd·ər }

Bessemer iron [MET] Pig iron with about 1% silicon and a sulfur and a phosphorus content below 0.10; used to make steel by the Bessemer or the acid open-hearth process. Also known as Bessemer pig iron. { 'bes·ə·mər 'ī·ərn }

Bessemer matte [MET] Product of the oxidation of furnace matte; contains nickel, copper, cobalt, precious metals, and about 22% sulfur. { 'bes·ə·mər 'mat }

Bessemer ore [MET] An iron ore containing very little phosphorus, considered suitable for refining by the Bessemer process. { 'bes·ə·mər 'ôr }

Bessemer pig iron See Bessemer iron. { 'bes·ə·mər 'pig ˌī·ərn }

Bessemer process [MET] A steelmaking process in which carbon, silicon, phosphorus, and manganese contained in molten pig iron are oxidized by a strong blast of air. { 'bes·ə·mər 'präs·əs }

Bessemer steel [MET] Steel manufactured by the Bessemer process. { 'bes·ə·mər 'stēl }

Bessey unit [BIOL] A unit for the standardization of phosphatase. { 'bes·ē ˌyü·nət }

Be star [ASTRON] A star of spectral type B in the Draper catalog that has emission lines indicating mass loss and a surrounding gaseous shell. { ˈbēˌē ˌstär }

best commercial practice [ENG] A manufacturing standard for a process vessel which has not been designed according to standard codes, such as the American Society of Mechanical Engineers Boiler Code. { ˌbest kə'mər·shəl 'prak·təs }

best estimate [STAT] A term applied to unbiased estimates which have a minimum variance. { ˌbest 'es·tə·mət }

best fit See goodness of fit. { ˌbest 'fit }

beta [ASTRON] For dust grains ejected from the nucleus of a comet, the ratio of the radiation pressure force to the solar gravitational force. [ELECTR] The current gain of a transistor that is connected as a grounded-emitter amplifier, expressed as the ratio of change in collector current to resulting change in base current, the collector voltage being constant. [NUCLEO] The amount of reactivity of a nuclear reactor corresponding to the delayed neutron fraction. [PL PHYS] The ratio of the ion energy density of a plasma to its magnetic energy diversity, or of the particle pressure to the magnetic-field pressure. [SCI TECH] The second letter of the Greek alphabet; β, B. { 'bād·ə }

beta-absorption gage See beta gage. { 'bād·ə əb'sôrp·shən ˌgāj }

beta blocker [NEURO] An adrenergic blocking agent capable of blocking nerve impulses to special sites (beta receptors) in the cerebellum; reduces the rate of heartbeats and the force of heart contractions. { 'bād·ə ˌbläk·ər }

beta brass [MET] A type of brass containing nearly equal proportions of copper and zinc. { 'bād·ə ˌbras }

beta Canis Majoris stars See beta Cephei stars. { 'bād·ə 'kan·əs mə'jór·əs ˌstärz }

beta carotene [BIOCHEM] C₄₀H₅₆ A carotenoid hydrocarbon pigment found widely in nature, always associated with chlorophylls; converted to vitamin A in the liver of many animals. { 'bād·ə 'kar·əˌtēn }

beta catenin [BIOCHEM] A multifunctional protein that is involved in Wnt signal transduction and plays an essential role in intercellular adhesion. { ˌbād·ə 'kat·ən·in }

beta cell [HISTOL] **1.** Any of the basophilic chromophiles in the anterior lobe of the adenohypophysis. **2.** One of the cells of the islets of Langerhans which produce insulin. { 'bād·ə ˌsel }

beta Centauri [ASTRON] A first-magnitude navigational star in the constellation Centaurus; 200 light-years from the sun; spectral classification B0. Also known as Agena; Hadar. { 'bā·də sen'tó·rē }

beta Cephei stars [ASTRON] A class of pulsating variables lying above the upper main sequence with short periods of 3 1/2–6 hours, spectral classes B0 to B3, and doubly periodic light curves. Also known as beta Canis Majoris stars. { 'bād·ə 'sef·ēˌī ˌstärz }

beta chalcocite See chalcocite. { 'bād·ə 'chal·kəˌsīt }

beta circuit [ELEC] The part of an amplifier circuit that is responsible for the feedback. { 'bād·ə ˌsər·kət }

beta coefficient [STAT] Also known as beta weight. **1.** One of the coefficients in a regression equation. **2.** A moment ratio, especially one used to describe skewness and kurtosis. { 'bād·ə kō·ə'fish·ənt }

beta-cutoff frequency [ELECTR] The frequency at which the current amplification of an amplifier transistor drops to 3 decibels below its value at 1 kilohertz. { 'bād·ə 'kədˌóf ˌfrē·kwən·sē }

betacyanin [BIOCHEM] A group of purple plant pigments found in leaves, flowers, and roots of members of the order Caryophyllales. { ˌbād·əˌsī·ə·nən }

beta decay [NUC PHYS] Radioactive transformation of a nuclide in which the atomic number increases or decreases by unity with no change in mass number; the nucleus emits or absorbs a beta particle (electron or positron). Also known as beta disintegration. { 'bād·ə di'kā }

beta-decay spectrum [NUC PHYS] The distribution in energy or momentum of the beta particles arising from a nuclear disintegration process. { 'bād·ə di'kā ˌspek·trəm }

beta disintegration See beta decay. { 'bād·ə dis,int·ə'grā·shən }

beta distribution [STAT] The probability distribution of a random variable with density function $f(x) = [x^{\alpha-1}(1-x)^{\beta-1}]/B(\alpha,\beta)$, where B represents the beta function, α and β are positive real numbers, and $0 < x < 1$. Also known as Pearson Type I distribution. { 'bād·ə dis·trə'byü·shən }

beta emitter [NUC PHYS] A radionuclide that disintegrates by emission of a negative or positive electron. { 'bād·ə i'mid·ər }

beta factor [PL PHYS] In plasma physics, the ratio of the plasma kinetic pressure to the magnetic pressure. { 'bād·ə ˌfak·tər }

betafite See ellsworthite. { 'bed·əˌfīt }

beta function [MATH] A function of two positive variables, defined by

$$B(m,n) = \int_0^1 x^{m-1}(1-x)^{n-1}dx$$

{ 'bād·ə ˌfəŋk·shən }

beta gage [NUCLEO] A penetration-type thickness gage that measures the absorption of beta rays in the sample. Also known as beta-absorption gage. { 'bād·ə ˌgāj }

beta-gamma survey meter [NUCLEO] An ionization-chamber type of monitor that is sensitive primarily to beta particles and gamma rays. { 'bād·ə ˌgam·ə 'sər,vā ˌmēd·ər }

beta globulin [BIOCHEM] A heterogeneous fraction of serum globulins containing transferrin and various complement components. { 'bād·ə 'gläb·yə·lən }

beta glucanase [FOOD ENG] An enzyme that acts on glucan (a carbohydrate found in barley and oats); used in the brewing industry to hydrolyze glucans which interfere with the clarification of wort and filtration of beer; also used in the animal feed industry. { 'bād·ə 'glü·kə,nās }

beta hemolysis [MICROBIO] A sharply defined, clear, colorless zone of hemolysis surrounding certain streptococci colonies growing on blood agar. { 'bād·ə hə'mäl·ə·səs }

Betaherpesvirinae [VIROL] A subfamily of animal, double-stranded linear deoxyribonucleic acid viruses of the family Herpesviridae, which are enveloped by a lipid bilayer and several glycoproteins. Also known as cytomegalovirus group. { ˌbād·ə'hər·pēz'vī·rə,nē }

betaine [ORG CHEM] $C_5H_{11}O_2N$ An alkaloid; very soluble in water, soluble in ethyl alcohol and methanol; the hydrochloride is used as a source of hydrogen chloride and in medicine. Also known as lycine; oxyneurine. { 'bēd·ə,ēn }

beta interaction See weak interaction. { 'bād·ə in·tər'ak·shən }

beta-lactamase [MICROBIO] A bacterial enzyme that catalyzes the hydrolysis of the lactam ring in some penicillin antibiotics, rendering them ineffective. { 'bād·ə 'lak·tə,mās }

betalain [BIOCHEM] The name for a group of 35 red or yellow compounds found only in plants of the family Caryophyllales, including red beets, red chard, and cactus fruits. { 'bed·ə,lān }

betamethasone [PHARM] A white, crystalline powder with a melting point of 240°C; a corticosteroid hormone. { 'bād·ə'meth·ə,zōn }

betanin [BIOCHEM] An anthocyanin that contains nitrogen and constitutes the principal pigment of garden beets. { 'bēd·ə·nən }

beta oxidation [BIOCHEM] Catabolism of fatty acids in which the fatty acid chain is shortened by successive removal of two carbon fragments from the carboxyl end of the chain. { 'bād·ə äks·ə'dā·shən }

beta particle [NUC PHYS] An electron or positron emitted from a nucleus during beta decay. { 'bād·ə ˌpard·ə·kəl }

beta plane [GEOPHYS] The model, introduced by C.G. Rossby, of the spherical earth as a plane whose rate of rotation (corresponding to the Coriolis parameter) varies linearly with the north-south direction. { 'bād·ə ˌplān }

beta random variable [MATH] A random variable whose probability distribution is a beta distribution. { 'bād·ə ˌran·dəm 'ver·ē·ə·bəl }

beta ray [NUC PHYS] A stream of beta particles. { 'bād·ə ˌrā }

beta-ray spectrometer [SPECT] An instrument used to determine the energy distribution of beta particles and secondary electrons. Also known as beta spectrometer. { 'bād·ə ˌrā spek'träm·əd·ər }

beta rhythm [PHYSIO] An electric current of low voltage from the brain, with a pulse frequency of 13–30 per second, encountered in a person who is aroused and anxious. { 'bād·ə ˌrith·əm }

beta rule See reduction rule. { 'bād·ə ˌrül }

beta software [COMPUT SCI] An application or program that is in development and undergoing testing. Also known as beta version; betaware. { ˌbād·ə 'sóf,wer }

beta spectrometer See beta-ray spectrometer. { 'bād·ə spek'träm·əd·ər }

beta taxonomy [SYST] The level of taxonomic study dealing with the arrangement of species into lower and higher taxa. { 'bād·ə tak'sän·ə·mē }

beta test [COMPUT SCI] The first test of a computer system outside the laboratory, in its actual working environment. { 'bād·ə ˌtest }

beta test site [COMPUT SCI] An organization or company that tests a software or hardware product under actual working conditions and reports the results to the vendor. { 'bād·ə ˌtest ˌsīt }

betatron [NUCLEO] A device for accelerating electrons in an evacuated ring by means of a time-varying magnetic flux encircled by the ring. Also known as rheotron. { 'bād·ə,trän }

betatron oscillations [NUCLEO] Oscillations of particles about an equilibrium orbit in a particle accelerator. { 'bād·ə,trän ˌäs·ə'lā·shənz }

beta version See beta software. { 'bād·ə ˌvər·zhən }

betaware See beta software. { 'bād·ə,wer }

beta weight See beta coefficient. { 'bād·ə ˌwāt }

betaxanthin [BIOCHEM] The name given to any of the yellow pigments found only in plants of the family Caryophyllales; they always occur with betacyanins. { 'bād·ə'zan·thən }

Betelgeuse [ASTRON] An orange-red giant star of stellar magnitude 0.1–1.2, 650 light-years from the sun, spectral classification M2-Iab, in the constellation Orion; the star α Orionis. { 'bed·əl,jüs }

betel nut [BOT] A dried, ripe seed of the palm tree *Areca catechu* in the family Palmae; contains a narcotic. { 'bēd·əl ˌnət }

BET equation See Brunauer-Emmett-Teller equation. { ˌbē,ē,tē i'kwā·zhən }

Bethe-ansatz technique [PHYS] A method for the solution of one-dimensional many-body problems that was first applied to one-dimensional magnets and has been generalized to many-body problems with point interactions. { 'bād·ē 'an,zats tek,nēk }

Bethe-Heitler theory [NUCLEO] A theory for the energy loss of charged particles passing through matter, based on the Dirac equation and the Born approximation for the interaction of the particle with the field of a nucleus. { ˌbet·ə 'hīt·lər ,thē·ə·rē }

Bethell process See full-cell process. { 'beth·əl 'präs·əs }

Bethe-Salpeter equation [PARTIC PHYS] The relativistic analog of the integral form of the two-body Schrödinger equation, the two-particle interaction kernel being the analog of the potential. { 'bāt·ə sal'pād·ər i'kwā·zhən }

Bethe-Slater curve [SOLID STATE] A graph of the exchange energy for the transition elements versus the ratio of the interatomic distance to the radius of the 3*d* shell. { ˌbet·ə 'slād·ər ,kərv }

Bethylidae [INV ZOO] A small family of hymenopteran insects in the superfamily Bethyloidea. { bə'thil·ə,dē }

Bethyloidea [INV ZOO] A superfamily of hymenopteran insects in the suborder Apocrita. { ˌbeth·ə'lói·dē·ə }

betrunked river [GEOL] A river that is shorn of its lower course as a result of submergence of the land margin by the sea. { bē'trəŋkt 'riv·ər }

Betterton-Kroll process [CHEM ENG] A method for obtaining pure bismuth from softened and desilverized lead. { ˌbed·ər·tən ˌkról ,präs·əs }

Betti group See homology group. { 'bāt·tē ,grüp }

Betti number See connectivity number. { 'bāt·tē ,nəm·bər }

Betti reciprocal theorem [MECH] A theorem in the mathematical theory of elasticity which states that if an elastic body is subjected to two systems of surface and body forces, then the work that would be done by the first system acting through the displacements resulting from the second system equals the work that would be done by the second system acting through the displacements resulting from the first system. { 'bāt·tē ri'sip·rə·kəl ,thir·əm }

Betti's method [MECH] A method of finding the solution of the equations of equilibrium of an elastic body whose surface displacements are specified; it uses the fact that the dilatation is a harmonic function to reduce the problem to the Dirichlet problem. { 'bāt·tēz ,meth·əd }

Betts' process [MET] A refining process for electrolytically purifying lead. { 'bets ,präs·əs }

Betula [BOT] The birches, a genus of deciduous trees composing the family Betulaceae. { 'bech·ə·lə }

Betulaceae [BOT] A small family of dicotyledonous plants in the order Fagales characterized by stipulate leaves, seeds without endosperm, and by being monoecious with female flowers mostly in catkins. { ˌbech·ə'lās·ē¸ē }

betula oil *See* methyl salicylate. { 'bech·ə·lə ˌȯil }

betulinic acid [ORG CHEM] $C_{30}H_{48}O_3$ A dibasic acid, slightly soluble in water, ethyl alcohol, and acetone. { ¦bech·ə¦lin·ik 'as·əd }

between decks [NAV ARCH] **1.** The space beneath the main deck of a ship. **2.** A deck beneath the main deck; in particular, a raised deck in a cargo ship's hold. { bə'twēn ˌdeks }

betwixt mountains *See* median mass. { bə'twikst ˌmaủnt·ənz }

Betz cell [HISTOL] Any of the large conical cells composing the major histological feature of the precentral motor cortex in humans. { 'bets ˌsel }

Betz momentum theory [MECH ENG] A theory of windmill performance that considers the deceleration in the air traversing the windmill disk. { 'bets mə'ment·əm ˌthē·ə·rē }

beudantite [MINERAL] $PbFe_3(AsO_4)(SO_4)(OH)_6$ A black, dark green, or brown, hexagonal mineral consisting of a basic sulfate-arsenate of lead and ferric iron; occurs as rhombohedral crystals. { 'byüd·ənˌīt }

Beutler-Fano profile [PHYS] A function that describes a scattering cross section in the vicinity of a resonance in terms of three parameters: the resonance energy, the resonance width, and the profile index, which determines the shape of the resonance. { ˈbȯit·lər 'fan·ō ˌprō¸fīl }

BeV [PHYS] A billion (10^9) electronvolts, a unit used in the United States; the international unit is GeV (gigaelectronvolts), which has the same value. { bev }

bevel [DES ENG] **1.** The angle between one line or surface and another line or surface, or the horizontal, when this angle is not a right angle. **2.** A sloping surface or line. [GRAPHICS] An instrument composed of two rules joined together, opening to any angle, which is used to draw angles and measure and lay off bevels. { 'bev·əl }

beveled closer *See* king closer. { ¦bev·əld 'klō·zər }

bevel gear [MECH ENG] One of a pair of gears used to connect two shafts whose axes intersect. { 'bev·əl ˌgir }

beveling [GEOL] Planing by erosion of the outcropping edges of strata. [MECH ENG] *See* chamfering. { 'bev·ə·liŋ }

Beverage antenna *See* wave antenna. { 'bev·rij an'ten·ə }

beyerite [MINERAL] $(Ca,Pb)Bi_2(CO_3)_2O_2$ A bright yellow to lemon yellow, tetragonal mineral consisting of bismuth and calcium carbonate; occurs as thin plates and compact earthy masses. { 'bī·əˌrīt }

beyond-the-horizon communication *See* scatter propagation. { bə'yänd thə hə'rīz·ən kəˌmyü·nə'kā·shən }

Beyrichacea [PALEON] A superfamily of extinct ostracods in the suborder Beyrichicopina. { ˌbī·rə'kās·ē·ə }

Beyrichicopina [PALEON] A suborder of extinct ostracods in the order Paleocopa. { ˌbī·rə·kə¸kō'pī·nə }

Beyrichiidae [PALEON] A family of extinct ostracods in the superfamily Beyrichacea. { ˌbī·rē'kī·əˌdē }

bezel [DES ENG] **1.** A grooved rim used to hold a transparent glass or plastic window or lens for a meter, tuning dial, or some other indicating device. **2.** A sloping face on a cutting tool. [LAP] The oblique face of a cut gem, between the table and the girdle. { 'bez·əl }

Bézier curve [COMPUT SCI] A curve in a drawing program that is defined mathematically, and whose shape can be altered by dragging either of its two interior determining points with a mouse. [MATH] A simple smooth curve whose shape is determined by a mathematical formula from the locations of four points, the two end points of the curve and two interior points. { 'bāz·yā ˌkərv }

Bezold-Abney phenomenon [PHYSIO] The perception of light at very high intensities as colorless. { 'bāt¸zȯlt 'ab·nē fəˌnäm·əˌnän }

Bézout domain [MATH] An integral domain in which all finitely generated ideals are principal. { ˌbā¸zō dō¸mān }

Bézout's theorem [MATH] The theorem that the product of the degrees of two algebraic plane curves that lack a common component equals the number of their points of intersection, counted to the degree of their multiplicity, including points of intersection at infinity. { 'bā¸zōz ˌthir·əm }

BFL *See* buffered FET logic.

BFO *See* beat-frequency oscillator.

BFS *See* beam-foil spectroscopy.

B girdle [PETR] A circular pattern in petrofabric diagrams that indicates a B axis. { 'bē ˌgərd·əl }

BHA *See* butylated hydroxyanisole.

Bhabha scattering [PARTIC PHYS] Scattering of positrons by electrons. { 'bä·bä 'skad·ər·iŋ }

BHC *See* 1,2,3,4,5,6-hexachlorocyclohexane.

B-H curve [ELECTROMAG] A graphical curve showing the relation between magnetic induction *B* and magnetizing force *H* for a magnetic material. Also known as magnetization curve. { ¦bē¦āch ˌkərv }

B-H meter [ENG] A device used to measure the intrinsic hysteresis loop of a sample of magnetic material. { ¦bē¦āch ˌmēd·ər }

B horizon [GEOL] The zone of accumulation in soil below the A horizon (zone of leaching). Also known as illuvial horizon; subsoil; zone of accumulation; zone of illuviation. { 'bē hə'rīz·ən }

bhp *See* boiler horsepower. *See* brake horsepower.

BHT *See* butylated hydroxytoluene.

Bi *See* abampere; bismuth.

biacetyl *See* diacetyl. { ¦bī·ə'sēd·əl }

Bial's test [PATH] A test for the presence of a pentose in urine, utilizing oracin, hydrochloric acid, and ferric chloride; a green color or green precipitate indicates pentose. { 'byälz ˌtest }

biamperometry [ANALY CHEM] Amperometric titration that uses two polarizing or indicating electrodes to detect the end point of a redox reaction between the substance being titrated and the titrant. { ¦bī¸am·pə'räm·ə·trē }

Bianca [ASTRON] A satellite of Uranus orbiting at a mean distance of 36,760 miles (59,160 kilometers) with a period of 10 hours 27 minutes, and with a diameter of about 27 miles (44 kilometers). { bē'äŋk·ə }

Bianchi classification [RELAT] A classification of possible types of spatially homogeneous space-times. { bē'aŋ·kē ˌklas·ə·fəˌkā·shon }

Bianchi cosmology [ASTRON] A model of the universe which is homogeneous but not necessarily isotropic. { bē'aŋ·kē käz'mäl·ə·jē }

Bianchi identity [MATH] A differential identity satisfied by the Riemann curvature tensor: the antisymmetric first covariant derivative of the Riemann tensor vanishes identically. { 'byäŋ·kē ī'den·əd·ē }

bianchite [MINERAL] $(Fe,Zn)SO_4 \cdot 6H_2O$ A white, monoclinic mineral consisting of iron and zinc sulfate hexahydrate; occurs in crusts of indistinct crystals. { bē'aŋ¸kīt }

bias [ANALY CHEM] A systematic error occurring in a chemical measurement that is inherent in the method itself or caused by some artifact in the system, such as a temperature effect. [ELEC] **1.** A direct-current voltage used on signaling or telegraph relays or electromagnets to secure desired time spacing of transitions from marking to spacing. **2.** The restraint of a relay armature by spring tension to secure a desired time spacing of transitions from marking to spacing. **3.** The effect on teleprinter signals produced by the electrical characteristics of the line and equipment. **4.** The force applied to a relay to hold it in a given position. [ELECTR] **1.** A direct-current voltage applied to a transistor control electrode to establish the desired operating point. **2.** *See* grid bias. [MATER] In a reinforced composite material, the angle made by the reinforcing fibers with the longitudinal direction of the material. [STAT] In estimating the value of a parameter of a probability distribution, the difference between the expected value of the estimator and the true value of the parameter. [TEXT] A direction on fabric that is equivalent to the diagonal of a square of the fabric cut so that the warp and the filling are parallel to the edges. { 'bī·əs }

bias-belted tire [ENG] A motor-vehicle pneumatic tire constructed with a belt of textile cord, steel, or fiber glass around the tire underneath the tread and on top of the ply cords, and laid at an acute angle to the center line of the tread. Also known as belted-bias tire. { 'bī·əs ˌbel·təd 'tīr }

bias cell [ELECTR] A small dry cell used singly or in series to provide the required negative bias for the grid circuit of an electron tube. Also known as grid-bias cell. { 'bī·əs ˌsel }

bias compensation [ENG ACOUS] The application of an outward-directed tension to the pickup arm of a record player to counteract the tendency of the arm to slide toward the center. { 'bī·əs ˌkäm·pənˌsā·shən }

BEVEL GEAR

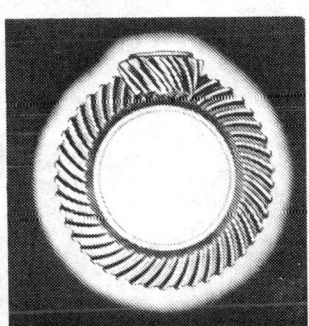

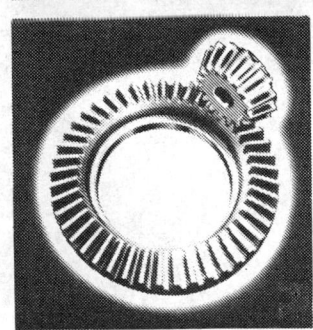

Types of bevel gears. *(Gleason Gear Works)*

bias current [ELECTR] **1.** An alternating electric current above about 40,000 hertz added to the audio current being recorded on magnetic tape to reduce distortion. **2.** An electric current flowing through the base-emitter junction of a transistor and adjusted to set the operating point of the transistor. { 'bī·əs ‚kər·ənt }

bias distortion [COMMUN] *See* bias telegraph distortion. [ELECTR] Distortion resulting from the operation on a nonlinear portion of the characteristic curve of a vacuum tube or other device, due to improper biasing. { 'bī·əs dis'tór·shən }

biased automatic gain control *See* delayed automatic gain control. { 'bī·əst ód·ə'mad·ik 'gān kən‚trōl }

biased sample [STAT] A sample obtained by a procedure that incorporates a systematic error introduced by taking items from a wrong population or by favoring some elements of a population. { ‚bī·əst 'sam·pəl }

biased statistic [STAT] A statistic whose expected value, as obtained from a random sampling, does not equal the parameter or quantity being estimated. { 'bī·əst stə'tis·tik }

bias error [STAT] A measurement error that remains constant in magnitude for all observations; a kind of systematic error. { 'bī·əs ‚er·ər }

bias meter [COMMUN] A meter used in teletypewriter work for measuring signal bias directly in percent; a positive reading indicates a marking signal bias; a negative reading, a spacing signal bias. { 'bī·əs ‚mēd·ər }

bias oscillator [ELECTR] An oscillator used in a magnetic recorder to generate the alternating-current signal that is added to the audio current being recorded on magnetic tape to reduce distortion. { 'bī·əs ‚äs·ə‚lād·ər }

bias-ply tire [ENG] A motor-vehicle pneumatic tire that has crossed layers of ply cord set diagonally to the center line of the tread. { ‚bī·əs ‚plī 'tīr }

bias register [COMPUT SCI] A computer device that stores a number that is added to the memory address each time the computer memory is referenced by the program, thus offsetting the program addresses by a fixed amount. { 'bī·əs ‚rej·ə·stər }

bias resistor [ELECTR] A resistor used in the cathode or grid circuit of an electron tube to provide a voltage drop that serves as the bias. { 'bī·əs ri'sis·tər }

bias telegraph distortion [COMMUN] A distortion that causes telegraph mark-and-space pulses to be lengthened or shortened; often caused by changes in the amplitude of incoming pulses. Also known as bias distortion; spacing bias. { 'bī·əs 'tel·ə‚graf dis'tór·shən }

bias voltage [ELECTR] A voltage applied or developed between two electrodes as a bias. { 'bī·əs ‚vōl·tij }

bias winding [ELEC] A control winding that carries a steady direct current which serves to establish desired operating conditions in a magnetic amplifier or other magnetic device. { 'bī·əs ‚wīn·diŋ }

biaxial crystal [CRYSTAL] A crystal of low symmetry in which the index ellipsoid has three unequal axes. { bī'ak·sē·əl 'krist·əl }

biaxial indicatrix [CRYSTAL] An ellipsoid whose three axes at right angles to each other are proportional to the refractive indices of a biaxial crystal. { bī'ak·sē·əl in'dik·ə‚triks }

biaxial stress [MECH] The condition in which there are three mutually perpendicular principal stresses; two act in the same plane and one is zero. { bī'ak·sē·əl ‚stress }

Biazzi process [CHEM ENG] A continuous-flow process for the nitration of glycerin to nitroglycerin; also used to produce glycol dinitrate and diethylene glycol nitrate. { bē'at·sē ‚präs·əs }

bibb cock *See* bibcock. { 'bib ‚käk }

bibcock [DES ENG] A faucet or stopcock whose nozzle is bent downward. Also spelled bibb cock. { 'bib‚käk }

bibenzyl [ORG CHEM] $C_{14}H_{14}$ A hydrocarbon consisting of two benzene rings attached to ethane. Also known as dibenzyl. { ‚bī'ben·zil }

Bibionidae [INV ZOO] The March flies, a family of orthorrhaphous dipteran insects in the series Nematocera. { ‚bib·ē'än·ə‚dē }

bibliophobia [PSYCH] An abnormal fear of books. { ‚bib·lē·ə'fō·bē·ə }

bicable tramway [MECH ENG] A tramway consisting of two stationary cables on which the wheeled carriages travel, and an endless rope, which propels the carriages. { 'bī‚kā·bəl 'tram‚wā }

bicameral [BIOL] Having two chambers, as the heart of a fish. { bī'kam·ə·rəl }

bicapsular [BIOL] **1.** Having two capsules. **2.** Having a capsule with two locules. { bī'kap·sə·lər }

bicarbonate [INORG CHEM] A salt obtained by the neutralization of one hydrogen in carbonic acid. { bī'kär·bə‚nət }

bicarbonate of soda *See* sodium bicarbonate. { bī'kär·bə·nət əv 'sō·də }

bicarinate [BIOL] Having two keellike projections. { bī 'kar·ə‚nāt }

bicaudal [ZOO] Having two tails. { bī'kód·əl }

bicellular [BIOL] Having two cells. { bī'sel·yə·lər }

bicephalous [ZOO] Having two heads. { bī'sef·ə·ləs }

biceps [ANAT] **1.** A bicipital muscle. **2.** The large muscle of the front of the upper arm that flexes the forearm; biceps brachii. The thigh muscle that flexes the knee joint and extends the hip joint; biceps femoris. { 'bī‚seps }

bichloride of mercury *See* mercuric chloride. { bī'klór‚īd əv 'mər·kyə·rē }

bichromate *See* dichromate. { ‚bī'krō‚māt }

biciliate [BIOL] Having two cilia. { bī'sil·ē‚āt }

bicipital [ANAT] **1.** Pertaining to muscles having two origins. **2.** Pertaining to ribs having double articulation with the vertebrae. [BOT] Having two heads or two supports. { bī'sip·əd·əl }

bicipital groove *See* intertubercular sulcus. { bī'sip·əd·əl ‚grüv }

bicipital tuberosity [ANAT] An eminence on the anterior inner aspect of the neck of the radius; the tendon of the biceps muscle is inserted here. { bī'sip·əd·əl ‚tüb·ə'räs·əd·ē }

BiCMOS technology [ELECTR] An integrated circuit technology that combines bipolar transistors and CMOS devices on the same chip. { ‚bī'sē‚mós tek‚näl·ə·jē }

Bico bi-drop *See* bi-drop. { ‚bēk·ō 'bī‚dräp }

bicollateral bundle [BOT] A vascular bundle in which phloem is located both externally and internally with respect to the xylem, with all tissues lying on the same radius and with the internal phloem lying next to the pith. { ‚bī·kə'lad·ə·rəl 'bənd·əl }

bicompact set *See* compact set. { bī'käm‚pakt ‚set }

bicomponent fiber [TEXT] A fiber manufactured from two different polymers by spinning and joining them in a simultaneous process from one spinneret. Also known as conjugate fiber. { bī·kəm'pō·nənt 'fī·bər }

biconcave lens *See* double-concave lens. { bī'kän‚kāv 'lenz }

biconditional gate *See* equivalence gate. { ‚bī·kən'dish·ən·əl 'gāt }

biconditional operation [MATH] A logic operator on two statements P and Q whose result is true if P and Q are both true or both false, and whose result is false otherwise. Also known as if and only if operation; match. { ‚bī·kən‚dish·ən·əl ‚äp·ə'rā·shən }

biconditional statement [MATH] A statement that one of two propositions is true if and only if the other is true. { ‚bī·kən‚dish·ən·əl 'stāt·mənt }

biconical antenna [ELECTROMAG] An antenna consisting of two metal cones having a common axis with their vertices coinciding or adjacent and with coaxial-cable or waveguide feed to the vertices. { bī'kän·ə·kəl an'ten·ə }

biconnected graph [MATH] A connected graph in which two points must be removed to disconnect the graph. { ‚bī· kə'nek·təd 'graf }

biconstituent fiber [TEXT] A manufactured fiber composed of two or more dissimilar fibers combined before the extrusion process. Also known as matrix fiber. { ‚bī·kən'stich·ə·wənt 'fī·bər }

bicontinuous function *See* homeomorphism. { ‚bī·kən'tin· yə·wəs 'fəŋk·shən }

biconvex lens *See* double-convex lens. { bī'kän‚veks 'lenz }

bicorn [MATH] A plane curve whose equation in cartesian coordinates x and y is $(x^2 + 2ay - a^2)^2 = y^2(a^2 - x^2)$, where a is a constant. { 'bī‚kórn }

bicornuate uterus [ANAT] A uterus with two horn-shaped processes on the superior aspect. { bī'kór·yə‚nāt 'yüd·ə·rəs }

Bicosoecida [INV ZOO] An order of colorless, free-living protozoans, each member having two flagella, in the class Zoomastigophorea. { bī·kō'se·shə·də }

bicostate [BOT] Of a leaf, having two principal longitudinal ribs. { 'bī·kō‚stāt }

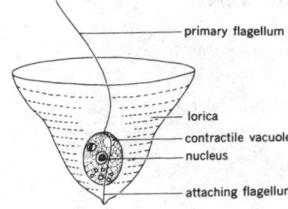

BICOSOECIDA

A bicosoecid, *Codomonas annulata.*

bicuculine [ORG CHEM] $C_{20}H_{17}NO_6$ A convulsant alkaloid found in plants of the family Fumariaceae. { bī'kü·kyə‚lēn }

bicuspid [ANAT] Any of the four double-pointed premolar teeth in humans. [BIOL] Having two points or prominences. { bī'kəs·pəd }

bicycle [MECH ENG] A human-powered land vehicle with two wheels, one behind the other, usually propelled by the action of the rider's feet on the pedals. { 'bī‚sik·əl }

bicyclic [BOT] Having or arranged in two whorls, as in petals. { bī'sī·klik }

bicyclic compound [ORG CHEM] A compound having two rings which share a pair of bridgehead carbon atoms. { bī'sik· lik 'käm‚pau̇nd }

bid [ENG] An estimate of costs for specified construction, equipment, or services proposed to a customer company by one or more supplier or contractor companies. { bid }

bidalotite See anthophyllite. { bə'däl·ə‚tīt }

Bidder's organ [VERT ZOO] A structure in the males of some toad species that may develop into an ovary in older individuals. { 'bid·ərz ‚ȯr·gən }

bidentate [BIOL] Having two teeth or teethlike processes. { bī'den‚tāt }

bidentate ligand [INORG CHEM] A chelating agent having two groups capable of attachment to a metal ion. { bī'den‚tāt 'lig·ənd }

bidirectional [ENG] Being directionally responsive to inputs in opposite directions. { ‚bī·də'rek·shən·əl }

bidirectional antenna [ELECTROMAG] An antenna that radiates or receives most of its energy in only two directions. { ‚bī·də'rek·shən·əl an'ten·ə }

bidirectional clamping circuit [ELECTR] A clamping circuit that functions at the prescribed time irrespective of the polarity of the signal source at the time the pulses used to actuate the clamping action are applied. { ‚bī·də'rek·shən·əl 'klam·piŋ ‚sər·kət }

bidirectional clipping circuit [ELECTR] An electronic circuit that prevents transmission of the portion of an electrical signal that exceeds a prescribed maximum or minimum voltage value. { ‚bī·də'rek·shən·əl 'klip·iŋ ‚sər·kət }

bidirectional counter See forward-backward counter. { ‚bī· də'rek·shən·əl 'kau̇n·tər }

bidirectional data bus [COMPUT SCI] A channel over which data can be transmitted in either direction within a computer system. { ‚bī·də'rek·shən·əl 'dad·ə ‚bəs }

bidirectional microphone [ENG ACOUS] A microphone that responds equally well to sounds reaching it from the front and rear, corresponding to sound incidences of 0 and 180°. { ‚bī· də'rek·shən·əl 'mī·krə‚fōn }

bidirectional parallel port [COMPUT SCI] A parallel port that can transfer data in both directions, and at speeds much greater than a standard parallel port. { ‚bī·də‚rek·shən·əl ‚par· ə‚lel 'pȯrt }

bidirectional printer [COMPUT SCI] A printer in which printing can be done in both a left-to-right and a right-to-left direction. { ‚bī·də'rek·shən·əl 'print·ər }

bidirectional pulse-amplitude modulation See double-polarity pulse-amplitude modulation. { ‚bī·də'rek·shən·əl ‚pəls ‚am·plə‚tüd ‚maj·ə'lā·shən }

bidirectional replication [MOL BIO] A mechanism of replication of deoxyribonucleic acid that involves two replicating forks moving in opposite directions away from the same origin. { ‚bī·də‚rek·shən·əl ‚rep·lə'kā·shən }

bidirectional transducer [ELECTR] A transducer capable of measuring in both positive and negative directions from a reference position. Also known as bilateral transducer. { ‚bī· də'rek·shən·əl tranz'dü·sər }

bidirectional transistor [ELECTR] A transistor that provides switching action in either direction of signal flow through a circuit; widely used in telephone switching circuits. { ‚bī· də'rek·shən·əl tran'zis·tər }

bidirectional triode thyristor [ELECTR] A gate-controlled semiconductor switch designed for alternating-current power control. { ‚bī·də'rek·shən·əl 'trī‚ōd thī'ris·tər }

bi-drop [FL MECH] A device in which two drops of different wetting liquids are juxtaposed inside a tube, resulting in spontaneous motion of the liquid and coating of the inner surface of the tube. Also known as Bico bi-drop. { 'bī‚dräp }

Bieberbach conjecture [MATH] The proposition, proven in 1984, that if a function $f(z)$ is analytic and univalent in the unit disk, and if it has the power series expansion $z + a_2 z^2 + a_3^3 + \cdots$, then, for all n ($n = 2, 3, \ldots$), the absolute value of a_n is equal to or less than n. { 'bē·bə‚bäk kən‚jek·chər }

bieberite [MINERAL] $CoSO_4 \cdot 7H_2O$ A rose red or flesh red, monoclinic mineral consisting of cobalt sulfate heptahydrate; occurs as crusts and stalactites. { 'bē·bə‚rīt }

Biebrich red See scarlet red. { 'bē‚brik 'red }

Biedenharn identity [NUC PHYS] A relationship among the six-j symbols of Wigner. { 'bēd·ən‚härn i'den·ə·dē }

Biela Comet [ASTRON] A comet seen in 1852 at one perihelion passage; presumed to have separated into two bodies. Also known as Comet Biela. { 'bē·lä ‚käm·ət }

Bielids See Andromedids. { 'bē‚lidz }

Bienayme-Chebyshev inequality [STAT] The probability that the magnitude of the difference between the mean of the sample values of a random variable and the mean of the variable is less than st, where s is the standard deviation and t is any number greater than 1, is equal to or greater than $1 - (1/t^2)$. { ‚bē‚nīm·ə chə·bi'shȯf ‚in·i'kwäl·əd·ē }

biennial plant [BOT] A plant that requires two growing seasons to complete its life cycle. { bī'en·ē·əl ‚plant }

Bierbaum scratch hardness test [ENG] A test for the hardness of a solid sample by microscopic measurement of the width of scratch made by a diamond point under preset pressure. { 'bir‚bau̇m ‚skrach 'härd·nəs ‚test }

biface tool [DES ENG] A tool, as an ax, made from a coil flattened on both sides to form a V-shaped cutting edge. { 'bī‚fās 'tül }

bifacial [BOT] Of a leaf, having dissimilar tissues on the upper and lower surfaces. [DES ENG] Of a tool, having both sides alike. { bī'fā·shəl }

bifanged [ANAT] Of a tooth, having two roots. { bī 'faŋgd }

bifenox [ORG CHEM] $C_{14}H_9Cl_2NO_5$ A tan, crystalline compound with a melting point of 84–86°C; insoluble in water; used as a preemergence herbicide for weed control in soybeans, corn, and sorghum, and as a pre- and postemergence herbicide in rice and small greens. { bī'fen‚äks }

bifid [BIOL] Divided into two equal parts by a median cleft. { 'bī‚fid }

Bifidobacterium [MICROBIO] A genus of bacteria in the family Actinomycetaceae; branched, bifurcated, club-shaped or spatulate rods forming smooth microcolonies; metabolism is saccharoclastic. { ‚bī·fə‚dō·bak'tir·ē·əm }

bifilar electromagnetic oscillograph [ELECTROMAG] A writing low-frequency light-beam oscillograph usually using a moving coil with a single U-shaped turn (bifilar type). { bī'fi· lər i‚lek·trō·mag'ned·ik ä'sil·ə‚graf }

bifilar electrometer [ENG] An electrostatic voltmeter in which two conducting quartz fibers, stretched by a small weight or spring, are separated by their attraction in opposite directions toward two plate electrodes carrying the voltage to be measured. { bī'ti·lər i·lek'träm·əd·ər }

bifilar micrometer See filar micrometer. { bī'fi·lər mī'kräm· əd·ər }

bifilar resistor [ELEC] A resistor wound with a wire doubled back on itself to reduce the inductance. { bī'fi·lər ri'zis·tər }

bifilar suspension [ENG] The suspension of a body from two parallel threads, wires, or strips. { bī'fi·lər səs'pen·shən }

bifilar transformer [ELEC] A transformer in which wires for the two windings are wound side by side to give extremely tight coupling. { bī'fi·lər tranz'fȯr·mər }

bifilar winding [ELEC] A winding consisting of two insulated wires, side by side, with currents traveling through them in opposite directions. { bī'fi·lər 'wīn·diŋ }

biflabellate [INV ZOO] The shape of certain insect antennae, characterized by short joints with long, flattened processes on opposite sides. { ‚bī·flə'bel·ət }

biflagellate [BIOL] Having two flagella. { bī'flaj·ə‚lāt }

bifluoride [INORG CHEM] An acid fluoride whose formula has the form MHF_2; an example is sodium bifluoride, $NaHF_2$. { bī'flu̇r‚īd }

bifocal lens [OPTICS] **1.** A lens with two parts having different focal lengths. **2.** In particular, an eyeglass lens having one part that corrects for distant vision and one part for near vision. { bī'fō·kəl 'lenz }

bifoliate [BOT] Two-leaved. { bī'fōl·ē·ət }

biforate [BIOL] Having two perforations. { bī'fə‚rāt }

bifunctional catalyst [CHEM] A catalytic substance that

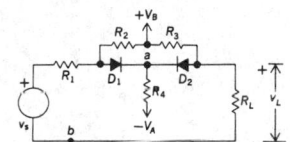

BIDIRECTIONAL CLIPPING CIRCUIT

Circuit diagram of bidirectional clipping obtained by connecting two diodes.

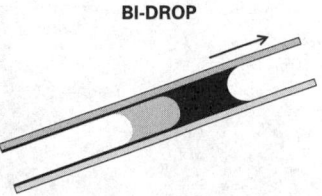

BI-DROP

Self-motion of the Bico bi-drop inside a tube.

possesses two catalytic sites and thus is capable of catalyzing two different types of reactions. Also known as dual-function catalyst. { ¦bī¦fəŋk·shən·əl ′kad·ə‚list }

bifunctional chelating agent [ORG CHEM] A reagent with a molecular structure that contains a strong metal-chelating group and a chemically reactive functional group. { ¦bī¦fəŋk·shən·əl ′kē‚lād·iŋ ‚ā·jənt }

bifunctional vector See shuttle vector. { ¦bī¦fəŋk·shən·əl ′vek·tər }

bifurcated contact [ELEC] A contact having a forked shape such that it can slide over and interlock with an identical mating contact. { ′bī·fər‚kād·əd ′kän‚takt }

bifurcation [MATH] The appearance of qualitatively different solutions to a nonlinear equation as a parameter in the equation is varied. [SCI TECH] **1.** Division into two branches, parts, or aspects. **2.** Point at which division occurs. { bī·fər′kā·shən }

bifurcation buoy [NAV] A red and black horizontally banded buoy used to mark the point at which a channel divides into two branches when proceeding from seaward. { bī·fər′kā·shən ‚bȯi }

bifurcation ratio [HYD] The ratio of number of stream segments of one order to the number of the next higher order. { bī·fər′kā·shən ‚rā·shō }

bifurcation theory [MATH] The study of the local behavior of solutions of a nonlinear equation in the neighborhood of a known solution of the equation; in particular, the study of solutions which appear as a parameter in the equation is varied and which at first approximate the known solution, thus seeming to branch off from it. Also known as branching theory. { ‚bī·fər′kā·shən ‚thē·ə·rē }

big bang theory [ASTRON] A theory of the origin and evolution of the universe which holds that approximately 1.4×10^{10} years ago all the matter in the universe was packed into a small agglomeration of extremely high density and temperature which exploded, sending matter in all directions and giving rise to the expanding universe. Also known as superdense theory. { ¦big ′baŋ ‚thē·ə·rē }

big bud [PL PATH] **1.** A parasitic disease of currants caused by a gall mite (*Eriophyes ribis*) and characterized by abnormal swelling of the buds. **2.** A virus disease of the tomato characterized by swelling of the buds. { ¦big ‚bəd }

big crunch [ASTRON] A singularity at the origin of a black hole into which all the matter and radiation in a closed universe would eventually collapse. Also known as gnab gib. { ¦big ¦krənch }

Big Dipper [ASTRON] A group of stars that is part of the constellation Ursa Major. Also known as Charles' wain. { ¦big ′dip·ər }

bigeminal pulse [PHYSIO] A pulse that is characterized by a double impulse produced by coupled heartbeats, that is, an extra heartbeat occurs just after the normal beat. { bī′jem·ə·nəl ′pəls }

Big Four [ASTRON] A group of large asteroids including Ceres, Pallas, Vesta, and Juno, the first four that were discovered. { ¦big ′fȯr }

bight [GEOL] **1.** A long, gradual bend or recess in the coastline which forms a large, open receding bay. **2.** A bend in a river or mountain range. [OCEANOGR] An indentation in shelf ice, fast ice, or a floe. { bīt }

big inch pipe [PETRO ENG] A pipeline 24 inches (61 centimeters) in diameter which carries oil or gas, usually for great distances. { ‚big ′inch ′pīp }

bigit See bit. { ′bij·ət }

big LEO system [COMMUN] A system of relatively large satellites in low earth orbit (LEO) to provide global mobile handheld telephony and other services. { ‚big ′lē·ō ‚sis·təm }

big M method [COMPUT SCI] A technique for solving linear programming problems in which artificial variables are assigned cost coefficients which are a very large number M, say, $M = 10^{35}$. { ‚big ′em ‚meth·əd }

Bignoniaceae [BOT] A family of dicotyledonous trees or shrubs in the order Scrophulariales characterized by a corolla with mostly five lobes, mature seeds with little or no endosperm and with wings, and opposite or whorled leaves. { ‚big·nō·nē′ās·ē‚ē }

bigraded module [MATH] A collection of modules $E_{s,t}$ indexed by pairs of integers s and t, with each module over a fixed principal ideal domain. { ¦bī‚grād·əd ‚māj·əl }

big vein [PL PATH] A soil-borne virus disease of lettuce characterized by enlargement and yellowing of the leaf veins. { ¦big ‚vān }

bigwoodite [PETR] A medium-grained plutonic rock consisting of microcline, microcline-microperthite, sodic plagioclase, and hornblende, aegirine-augite, or biotite. { big′wu̇‚dīt }

biharmonic function [MATH] A solution to the partial differential equation $\Delta^2 u(x,y,z) = 0$, where Δ is the Laplacian operator; occurs frequently in problems in electrostatics. { ¦bī‚här′män·ik ′fəŋk·shən }

bijection [MATH] A mapping f from a set A onto a set B which is both an injection and a surjection; that is, for every element b of B there is a unique element a of A for which $f(a) = b$. Also known as bijective mapping. { ′bī‚jek·shən }

bijective mapping See bijection. { ‚bī′jek·tiv ′map·iŋ }

bijugate [BOT] Of a pinnate leaf, having two pairs of leaflets. { ′bī·jə‚gāt }

bilabial [ANTHRO] Distance between the highest point on the upper lip and lowest point on the lower lip. { bī′lāb·ē·əl }

bilabiate [BOT] Having two lips, such as certain corollas. { bī′lāb·ē·ət }

bilateral [BIOL] Of or relating to both right and left sides of an area, organ, or organism. [ELECTR] Having a voltage current characteristic curve that is symmetrical with respect to the origin. { bī′lad·ə·rəl }

bilateral amplifier [ELECTR] An amplifier capable of receiving as well as transmitting signals; used primarily in transceivers. { bī′lad·ə·rəl ′am·plə‚fī·ər }

bilateral antenna [ELECTROMAG] An antenna having maximum response in exactly opposite directions, 180° apart, such as a loop. { bī′lad·ə·rəl an′ten·ə }

bilateral circuit [ELEC] Circuit wherein equipment at opposite ends is managed, operated, and maintained by different services. { bī′lad·ə·rəl ′sər·kət }

bilateral cleavage [EMBRYO] The division pattern of a zygote that results in a bilaterally symmetrical embryo. { bī′lad·ə·rəl ′klēv·ij }

bilateral hermaphroditism [ZOO] The presence of an ovary and a testis on each side of the animal body. { bī′lad·ə·rəl hər′maf·rə‚dīd‚iz·əm }

bilateral Laplace transform [MATH] A generalization of the Laplace transform in which the integration is done over the negative real numbers as well as the positive ones. { bī′lad·ə·rəl lə′pläs ′tranz‚fȯrm }

bilateral network [ELEC] A network or circuit in which the magnitude of the current remains the same when the voltage polarity is reversed. { bī′lad·ə·rəl ′net‚wərk }

bilateral observation [ORD] System for determining deviation of impacts or bursts from the target by using two instruments and observers located at a distance from each other. { bī′lad·ə·rəl ‚äb·zər′vā·shən }

bilateral slit [SPECT] A slit for spectrometers and spectrographs that is bounded by two metal strips which can be moved symmetrically, allowing the distance between them to be adjusted with great precision. { ¦bī‚lad·ə·rəl ′slit }

bilateral symmetry [BIOL] Symmetry such that the body can be divided by one median, or sagittal, dorsoventral plane into equivalent right and left halves, each a mirror image of the other. { bī′lad·ə·rəl ′sim·ə·trē }

bilateral tolerance [DES ENG] The amount that the size of a machine part is allowed to vary above or below a basic dimension; for example, 3.650 ± 0.003 centimeters indicates a tolerance of ± 0.003 centimeter. { bī′lad·ə·rəl ′täl·ə·rəns }

bilateral transducer See bidirectional transducer. { bī′lad·ə·rəl tranz′dü·sər }

Bilateria [ZOO] A major division of the animal kingdom embracing all forms with bilateral symmetry. { ‚bī·lə′tir·ē·ə }

bilayer [CHEM] A layer two molecules thick, such as that formed on the surface of the aqueous phase by phospholipids in aqueous solution. { ′bī‚lā·ər }

bile [PHYSIO] An alkaline fluid secreted by the liver and delivered to the duodenum to aid in the emulsification, digestion, and absorption of fats. Also known as gall. { bīl }

bile acid [BIOCHEM] Any of the liver-produced steroid acids, such as taurocholic acid and glycocholic acid, that appear in the bile as sodium salts. { ′bīl ′as·əd }

bile duct [ANAT] Any of the major channels in the liver

through which bile flows toward the hepatic duct. { 'bīl ,dəkt }

bile pigment [BIOCHEM] Either of two colored organic compounds found in bile: bilirubin and biliverdin. { 'bīl 'pig·mənt }

bile salt [BIOCHEM] The sodium salt of glycocholic and taurocholic acids found in bile. { 'bīl ,sȯlt }

bilge [NAV ARCH] **1.** Part of the underwater body of a ship between the flat of the bottom and the straight vertical sides. **2.** Internally, the lowest part of the hull, next to the keelson. { bilj }

bilge block [CIV ENG] A wooden support under the turn of a ship's bilge in dry dock. { 'bilj ,bläk }

bilge board [NAV ARCH] **1.** A wood or metal board that slides in a case like a centerboard, but is built into each bilge of a ship. **2.** See limber board. { 'bilj ,bȯrd }

bilge keel [NAV ARCH] A keel extending outside of the hull and fastened along a ship near the turn of the bilge to reduce rolling. Also known as rolling chock. { 'bilj ,kēl }

bilge keelson [NAV ARCH] A keelson close to the turn of the bilge. { 'bilj ,kēl·sən }

bilge pump [NAV ARCH] A pump to remove water that collects in the bottom of a ship. { 'bilj ,pəmp }

bilge water [NAV ARCH] Water that builds up in a ship's bilges. { 'bilj ,wȯd·ər }

bilharzias See schistosomiasis. { ,bil'härz·ē·əs }

bilharziasis See schistosomiasis. { ,bil,här'ze·ə·səs }

biliary atresia [MED] Failure of the bile ducts to develop in the embryo. { 'bil·ē,er·ē ə'trēzh·ə }

biliary cirrhosis [MED] A progressive inflammatory disease of the liver due to obstruction of bile ducts. { 'bil·ē,er·ē sə'rō·səs }

biliary colic [MED] Severe abdominal pain caused by passage of a gallstone through the bile ducts into the duodenum. { 'bil·ē,er·ē 'käl·ik }

biliary diskinesia [MED] A functional spasticity of the sphincter of Oddi with disturbances in the speed of evacuation of the biliary tract. { 'bil·ē,er·ē ,dis·kə'nēzh·ə }

biliary system [ANAT] The complex of canaliculi, or microscopic bile ducts, that empty into the larger intrahepatic bile ducts. { 'bil·ē,er·ē ,sis·təm }

bilicyanin [BIOCHEM] A blue pigment found in gallstones; an oxidation product of biliverdin or bilirubin. { ,bil·ə'sī·ə·nən }

bilification [PHYSIO] Formation and excretion of bile. { ,bil·ə·fə'kā·shən }

bilinear concomitant [MATH] An expression $B(u,v)$, where u, v are functions of x, satisfying $vL(u) - uL̄(v) = (d/dx)·B(u,v)$, where L, $L̄$ are given adjoint differential equations. { bī'lin·ē·ər kən'käm·ə·tənt }

bilinear expression [MATH] An expression which is linear in each of two variables separately. { bī'lin·ē·ər ik'spresh·ən }

bilinear form [MATH] **1.** A polynomial of the second degree which is homogeneous of the first degree in each of two sets of variables; thus, it is a sum of terms of the form $a_{ij}x_iy_j$, where $x_1, \ldots, x_m$ and $y_1, \ldots, y_n$ are two sets of variables and the a_{ij} are constants. **2.** More generally, a mapping $f(x, y)$ from $E \times F$ into R, where R is a commutative ring and $E \times F$ is the Cartesian product of two modules E and F over R, such that for each x in E the function which takes y into $f(x, y)$ is linear, and for each y in F the function which takes x into $f(x, y)$ is linear. { bī'lin·ē·ər 'fȯrm }

bilinear transformations See Möbius transformations. { bī'lin·ē·ər tranz·fər'mā·shənz }

bilinite [MINERAL] $Fe^{2+}Fe^{3+}_2(SO_4)_4·22H_2O$ A white to yellowish mineral consisting of a hydrated sulfate of divalent and trivalent iron; occurs in radial-fibrous aggregates. { 'bil·ə,nīt }

biliprotein [BIOCHEM] The generic name for the organic compounds in certain algae that are composed of phycobilin and a conjugated protein. { ,bil·ə'prō,tēn }

bilirubin [BIOCHEM] $C_{33}H_{36}N_4O_{66}$ An orange, crystalline pigment occurring in bile; the major metabolic breakdown product of heme. { ,bil·ə'rü·bən }

bilirubinemia [MED] The presence of bilirubin in the blood. { ,bil·ə,rü·bə'nē·mē·ə }

biliverdin [BIOCHEM] $C_{33}H_{34}N_4O_6$ A green, crystalline

pigment occurring in the bile of amphibians, birds, and humans; oxidation product of bilirubin in humans. { ,bil·ə'vərd·ən }

bill [DES ENG] One blade of a pair of scissors. [INV ZOO] A flattened portion of the shell margin of the broad end of an oyster. [NAV ARCH] The point at the end of an anchor fluke. [VERT ZOO] The jaws, together with the horny covering, of a bird. [ZOO] Any jawlike mouthpart. { bil }

billboard array [ELECTROMAG] A broadside antenna array consisting of stacked dipoles spaced one-fourth to three-fourths wavelength apart in front of a large sheet-metal reflector. Also known as bedspring array; mattress array. { 'bil,bȯrd ə'rā }

billet [ENG] In a hydraulic extrusion press, a large cylindrical cake of plastic material placed within the pressing chamber. [MET] A semifinished, short, thick bar of iron or steel in the form of a cylinder or rectangular prism produced from an ingot; limited to 1.5 inches (3.8 centimeters) in width and thickness with a cross-sectional area up to 36 square inches (232 square centimeters). { 'bil·ət }

Billet furnace [MET] A furnace for heating steel in sizes between a bloom and a bar. { 'bil·ət 'fər·nəs }

billet mill [MET] A rolling mill for making billets from ingots. { 'bil·ət ,mil }

Billet split lens [OPTICS] A lens cut into two halves, along the optic axis; used in interferometry. { 'bil·ət 'split ,lenz }

billiard cloth [TEXT] A heavy-weight wool fabric with a smooth, even surface; used on billiard tables. { 'bil·yərd ,klȯth }

Billingsellacea [PALEON] A group of extinct articulate brachiopods in the order Orthida. { ,bil·iŋ·sə'läs·ē·ə }

billion [MATH] **1.** The number 10^9. **2.** In British usage, the number 10^{12}. { 'bil·yən }

billow cloud [METEOROL] Broad, nearly parallel lines of cloud oriented normal to the wind direction, with cloud bases near an inversion surface. Also known as undulatus. { 'bil·ō ,klaůd }

bilobate [BIOL] Divided into two lobes. { 'bī'lō,bāt }

bilobular [BIOL] Having two lobules. { ,bī'läb·yə·lər }

bilocular [BIOL] Having two cells or compartments. { ,bī'läk·yə·lər }

bilophodont [ZOO] Having two transverse ridges, as the molar teeth of certain animals. { bī'läf·ə,dänt }

bimaceral [GEOL] A coal microlithotype that consists of a mixture of two macerals. { bī'mas·ə·rəl }

bimag core See bistable magnetic core. { 'bī,mag ,kȯr }

bimanous [ANAT] Having the distal part of the two forelimbs modified as hands, as in primates. { bī'man·əs }

bimaxillary [ANTHRO] Pertaining to the distance between the lower margins of the sutures of the maxilla and malar bones. [ZOO] Pertaining to the two halves of the maxilla. { ,bī'max·ə,ler·ē }

bimetal [MATER] A laminate of two dissimilar metals, with different coefficients of thermal expansion, bonded together. { ,bī'med·əl }

bimetallic corrosion [MET] Corrosion of two different metals exposed to an electrolyte while in electrical contact. { ,bī·mə'tal·ik kə'rōzh·ən }

bimetallic plate [GRAPHICS] Printing plate in planographic printing composed of two metals; one rejects ink, the other does not. { ,bī·mə'tal·ik ,plāt }

bimetallic strip [ENG] A strip formed of two dissimilar metals welded together; different temperature coefficients of expansion of the metals cause the strip to bend or curl when the temperature changes. { ,bī·mə'tal·ik ,strip }

bimetallic thermometer [ENG] A temperature-measuring instrument in which the differential thermal expansion of thin, dissimilar metals, bonded together into a narrow strip and coiled into the shape of a helix or spiral, is used to actuate a pointer. Also known as differential thermometer. { ,bī·mə'tal·ik thər'mäm·əd·ər }

bimodal distribution [STAT] A probability distribution with two different values that are markedly more frequent than neighboring values. { ,bī·mōd·əl di·strə'byü·shən }

bimolecular [CHEM] Referring to two molecules. { ,bī·mə'lek·yə·lər }

bimolecular reaction [CHEM] A chemical transformation or change involving two molecules. { ,bī·mə'lek·yə·lər rē'ak·shən }

bimorph cell [ELECTR] Two piezoelectric plates cemented together in such a way that an applied voltage causes one to

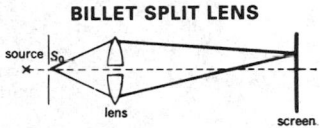

BILLET SPLIT LENS

Billet split-lens interference.

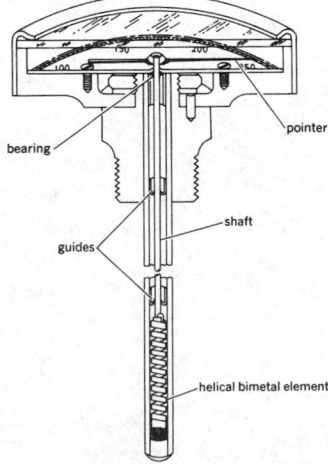

BIMETALLIC THERMOMETER

Cutaway view of bimetallic thermometer. (*Weston Instruments, Division of Daystrom, Inc.*)

expand and the other to contract so that the cell bends in proportion to the applied voltage; conversely, applied pressure generates double the voltage of a single cell; used in phonograph pickups and microphones. { 'bī‚mȯrf ¦sel }

bin [COMPUT SCI] A magnetic-tape memory in which a number of tapes are stored in a single housing. [ENG] An enclosed space, box, or frame for the storage of bulk substance. { bin }

binapacryl [ORG CHEM] $C_{15}H_{18}O_6N_2$ A light tan solid with a melting point of 68–69°C; insoluble in water; used for powdery mildew and for mites on fruits. { bə'nap·ə‚kril }

binary [COMPUT SCI] Possessing a property for which there exists two choices or conditions, one choice excluding the other. [SCI TECH] Composed of or characterized by two parts or elements. { 'bīn·ə·rē }

binary alloy [MET] An alloy composed of two principal metallic components. { 'bīn·ə·rē 'a‚lȯi }

binary arithmetic operation [COMPUT SCI] An arithmetical operation in which the operands are in the form of binary numbers. Also known as binary operation. { 'bīn·ə·rē ‚ar·ith'med·ik äp·ə'rā·shən }

binary cell [COMPUT SCI] An elementary unit of computer storage that can have one or the other of two stable states and can thus store one bit of information. { 'bīn·ə·rē ¦sel }

binary chain [COMPUT SCI] A series of binary circuit elements so arranged that each can change the state of the one following it. { 'bīn·ə·rē ¦chān }

binary chop See binary search. { 'bīn·ə·rē 'chäp }

binary code [COMPUT SCI] A code in which each allowable position has one of two possible states, commonly 0 and 1; the binary number system is one of many binary codes. { 'bīn·ə·rē ¦kōd }

binary coded character [COMPUT SCI] One element of a notation system representing alphanumeric characters such as decimal digits, alphabetic letters, and punctuation marks by a predetermined configuration of consecutive binary digits. { 'bīn·ə·rē ‚kōd·əd 'kar·ik·tər }

binary coded decimal system [COMPUT SCI] A system of number representation in which each digit of a decimal number is represented by a binary number. Abbreviated BCD system. { 'bīn·ə·rē ‚kōd·əd 'des·məl ‚sis·təm }

binary coded decimal-to-decimal converter [COMPUT SCI] A computer circuit which selects one of ten outputs corresponding to a four-bit binary coded decimal input, placing it in the 0 state and the other nine outputs in the 1 state. { 'bīn·ə·rē ‚kōd·əd 'des·məl tə 'des·məl kən'vərd·ər }

binary coded octal system [COMPUT SCI] Octal numbering system in which each octal digit is represented by a three-place binary number. { 'bīn·ə·rē ‚kōd·əd 'äk·təl ‚sis·tem }

binary component [ELECTR] An electronic component that can be in either of two conditions at any given time. Also known as binary device. { 'bīn·ə·rē kəm'pō·nənt }

binary compound [CHEM] A compound that has two elements; it may contain two or more atoms; examples are KCl and $AlCl_3$. { 'bīn·ə·rē 'käm‚paùnd }

binary conversion [COMPUT SCI] Converting a number written in binary notation to a number system with another base, such as decimal, octal, or hexadecimal. { 'bīn·ə·rē kən'vər·zhən }

binary counter See binary scaler. { 'bīn·ə·rē 'kaùnt·ər }

binary decision [COMPUT SCI] A decision between only two alternatives. { 'bīn·ə·rē di'sizh·ən }

binary device See binary component. { 'bīn·ə·rē di'vīs }

binary digit See bit. { 'bīn·ə·rē 'dij·ət }

binary dump [COMPUT SCI] The operation of copying the contents of a computer memory in binary form onto an external storage device. { 'bīn·ə·rē ¦dəmp }

binary encoder [ELECTR] An encoder that changes angular, linear, or other forms of input data into binary coded output characters. { 'bīn·ə·rē en'kōd·ər }

binary encounter approximation [ATOM PHYS] An approximation for predicting the probability that an incident proton will eject an inner shell electron from an atom; it uses a semiclassical treatment of momentum transfer from the incident proton to the ejected electron. { 'bīn·ə·rē in'kaùnt·ər ə‚präk·sə'mā·shən }

binary explosive [MATER] A high explosive composed of a mixture of two high explosives, to secure an explosive which

is superior to its components in regard to sensitivity, fragmentation, blast, or loadability. { 'bīn·ə·rē ik'splō·siv }

binary field [COMPUT SCI] A field that contains data in the form of binary numbers. { 'bīn·ə·rē 'fēld }

binary file [COMPUT SCI] A computer program in machine language that can be directly executed by the computer. { 'bīn·ə·rē 'fīl }

binary fission [BIOL] A method of asexual reproduction accomplished by the splitting of a parent cell into two equal, or nearly equal, parts, each of which grows to parental size and form. { 'bīn·ə·rē 'fish·ən }

binary granite [PETR] **1.** A granite made up of quartz and feldspar. **2.** A granite containing muscovite mica and biotite. { 'bīn·ə·rē 'gran·ət }

binary image [COMPUT SCI] A representation in a computer storage device of each of the holes in a punch card or paper tape (for example, by indicating the places where there are holes with a 1 and the places where there are no holes with a 0), to be differentiated from the characters represented by the combinations of holes. { 'bīn·ə·rē 'im·ij }

binary incremental representation [COMPUT SCI] A type of incremental representation in which the value of change in a variable is represented by one binary digit which is set equal to 1 if there is an increase in the variable and to 0 if there is a decrease. { 'bīn·ə·rē ‚iŋ·krə'men·təl ‚rep·ri‚zen'tā·shən }

binary large object [COMPUT SCI] In a database management system, a file-storage system used most often for multimedia files (large files). Abbreviated BLOB. { 'bīn·ə·rē ¦lärj 'äb‚jekt }

binary loader [COMPUT SCI] A computer program which transfers to main memory an exact image of the binary pattern of a program held in a storage or input device. { 'bīn·ə·rē ¦lōd·ər }

binary logic [ELECTR] An assembly of digital logic elements which operate with two distinct states. { 'bīn·ə·rē 'läj·ik }

binary magnetic core [SOLID STATE] A ferromagnetic core that can be made to take either of two stable magnetic states. { 'bīn·ə·rē mag'ned·ik 'kȯr }

binary notation See binary number system. { 'bīn·ə·rē nō'tā·shən }

binary number [MATH] A number expressed in the binary number system of positional notation. { 'bīn·ə·rē 'nəm·bər }

binary number system [MATH] A representation for numbers using only the digits 0 and 1 in which successive digits are interpreted as coefficients of successive powers of the base 2. Also known as binary notation; binary system; dyadic number system. { 'bīn·ə·rē 'nəm·bər ‚sis·təm }

binary numeral [MATH] One of the two digits 0 and 1 used in writing a number in binary notation. { 'bī‚ner·ē 'nüm·rəl }

binary operation [COMPUT SCI] See binary arithmetic operation. [MATH] A rule for combining two elements of a set to obtain a third element of that set, for example, addition and multiplication. { 'bīn·ə·rē äp·ə'rā·shən }

binary optics [OPTICS] A technology that uses etching technology to produce optical elements with computer-generated microscopic surface relief patterns having two or more levels. { 'bī‚ner·ē 'äp·tiks }

binary phase-shift keying [COMMUN] Keying of binary data or Morse code dots and dashes by ±90° phase deviation of the carrier. Abbreviated BPSK. { 'bīn·ə·rē 'fāz ‚shift 'kē·iŋ }

binary picture [GRAPHICS] A picture that has only two shades of brightness, white and black. { 'bīn·ə·rē 'pik·chər }

binary point [COMPUT SCI] The character, or the location of an implied symbol, that separates the integral part of a numerical expression from its fractional part in binary notation. { 'bīn·ə·rē 'pȯint }

binary pulsar [ASTRON] A pulsar which forms one component of a binary star. { 'bīn·ə·rē 'pəl‚sär }

binary quantic [MATH] A quantic that contains two variables. { ‚bīn·ə·rē 'kwän·tik }

binary row [COMPUT SCI] Pertaining to the binary representation of data on punched cards in which adjacent positions in a row correspond to adjacent bits of the data. { 'bīn·ə·rē ¦rō }

binary scaler [ELECTR] A scaler that produces one output pulse for every two input pulses. Also known as binary counter; scale-of-two circuit. { 'bīn·ə·rē 'skā·lər }

binary search [COMPUT SCI] A dichotomizing search in which the set of items to be searched is divided at each step into two equal, or nearly equal, parts.　Also known as binary chop.　{ 'bīn·ə·rē 'sərch }

binary separation [CHEM ENG]　Separation by distillation or solvent extraction of a fully miscible liquid mixture of two chemical compounds.　{ 'bīn·ə·rē sep·ə'rā·shən }

binary sequence [MATH]　A sequence, every element of which is 0 or 1.　{ ,bīn·ə·rē 'sē·kwəns }

binary signal [ELECTR]　A voltage or current which carries information by varying between two possible values, corresponding to 0 and 1 in the binary system.　{ 'bīn·ə·rē 'sig·nəl }

binary star [ASTRON]　A pair of stars located sufficiently near each other in space to be connected by the bond of mutual gravitational attraction, compelling them to describe an orbit around their common center of gravity.　Also known as binary system.　{ 'bīn·ə·rē 'stär }

binary system [ASTRON]　See binary star. [ENG]　Any system containing two principal components. [MATH]　See binary number system.　{ 'bīn·ə·rē 'sis·təm }

binary-to-decimal conversion [MATH]　The process of converting a number written in binary notation to the equivalent number written in ordinary decimal notation.　{ 'bīn·ə·rē tə 'des·məl kən'vər·zhən }

binary tree [MATH]　A rooted tree in which each vertex has a maximum of two successors.　{ 'bīn·ə·rē 'trē }

binary weapon [ORD]　A weapon that operates by storing less-than-lethal chemicals in separate compartments of a projectile; the wall between the compartments breaks after the munition is fired, allowing the chemicals to combine in a lethal mixture.　{ 'bīn·ə·rē 'wep·ən }

binary word [COMPUT SCI]　A group of bits which occupies one storage address and is treated by the computer as a unit.　{ 'bīn·ə·rē 'wərd }

binate [BOT]　Growing in pairs.　{ 'bī,nāt }

binaural [ACOUS]　Pertaining to sound that reaches the listener over two paths, to give the effect of auditory perspective.　{ bī'nȯr·əl }

binaural hearing [PHYSIO]　The perception of sound by stimulation of two ears.　{ bī'nȯr·əl 'hir·iŋ }

binaural intensity effect [ACOUS]　The relationship wherein, if sound of the same frequency and phase is incident at both ears, the angle between the apparent direction of the sound and the median plane of the line joining the ears is proportional to the logarithm of the ratio of the intensities of sound received at the left and right ears.　{ bī'nȯr·əl in'ten·səd·ē i'fekt }

binaural phase effect [ACOUS]　A displacement in the apparent direction of a sound that results when a difference in phase is introduced between otherwise identical sound signals applied to the two ears; the angular displacement from the median plane is proportional to the phase difference.　{ bī'nȯr·əl 'fāz i'fekt }

binaural sound [ACOUS]　The sound resulting from a reproduction system which has two channels, each fed into a different earphone or loudspeaker, so that a listener hears sounds coming from their original directions (with reference to the separated microphones used in recording the original sounds).　{ bī'nȯr·əl 'saúnd }

binche lace [TEXT]　A durable, firm bobbin lace which has a mesh background interspersed with appliqued simple snowflake motifs.　{ 'banch ,lās }

b-incomplete curve [RELAT]　A curve space of finite length, where length is defined by a generalized affine parameter, that has an end point at a space-time singularity.　{ ¦bē ¦in·kəm,plēt 'kərv }

binder [MATER]　**1.** A resin or other cementlike material used to hold particles together and provide mechanical strength or to ensure uniform consistency, solidification, or adhesion to a surface coating; typical binders are resin, glue, gum, and casein. **2.** See binding agent.　{ 'bīn·dər }

binder course [CIV ENG]　Coarse aggregate with a bituminous binder between the foundation course and the wearing course of a pavement.　{ 'bīn·dər ,kȯrs }

binderless briquetting [ENG]　The briquetting of coal by the application of pressure without the addition of a binder.　{ 'bīn·dər·ləs bri'ked·iŋ }

bindheimite [MINERAL]　$Pb_2Sb_2O_6(O,OH)$　A hydrous lead antimonate mineral produced from natural oxidation of jamesonite; found in Nevada.　{ 'bint,hī,mīt }

B indicator See B scope.　{ ¦bē ¦in·də,kād·ər }

binding agent [MATER]　A liquid component of paint that solidifies as it dries and thereby serves to bind the pigment particles and develop adhesion to a surface.　Also known as binder.　{ 'bīn·diŋ ,ā·jənt }

binding coal See caking coal.　{ 'bīn·diŋ ,kōl }

binding energy [PHYS]　Abbreviated BE.　Also known as total binding energy (TBE).　**1.** The net energy required to remove a particle from a system.　**2.** The net energy required to decompose a system into its constituent particles.　{ 'bīn·diŋ ,en·ər·jē }

binding fraction [NUC PHYS]　The ratio of the binding energy of a nucleus to the atomic mass number.　{ 'bīnd·iŋ ,frak·shən }

binding post [ELEC]　A manually turned screw terminal used for making electrical connections.　{ 'bīn·diŋ ,pōst }

binding site See active site.　{ 'bīn·diŋ ,sīt }

binding time [COMPUT SCI]　**1.** The instant when a symbolic expression in a computer program is reduced to a form which is directly interpretable by the hardware.　**2.** The instant when a variable is assigned its data type, such as integer or string.　{ 'bīn·diŋ ,tīm }

bind-seize See freeze.　{ ¦bīnd ¦sēz }

Binet age [PSYCH]　An individual's mental age as determined by the Binet-Simon intelligence scale.　{ bē'nā ,āj }

Binet's formula [PSYCH]　The premise that children under 9 years of age whose mental development is retarded by 2 years are probably mentally deficient, and children of 9 years or older retarded by 3 years are definitely deficient.　{ bē'nāz 'fȯrm·yə·lə }

Binet-Simon intelligence scale [PSYCH]　Test for determining the relative mental development of children between 3 and 12 years of age; results are expressed as an intelligence quotient, the ratio of mental to chronological age.　{ bē'nā sē'mōn in'tel·ə·jəns ,skāl }

binge eating disorder [PSYCH]　A condition that is characterized by recurrent episodes of excessive eating but, unlike bulimia nervosa, no extreme weight control behaviors (purging, laxatives, fasting) are present; persons with the disorder have chaotic eating patterns and frequently overeat as well as binge.　{ ¦binj'ēd·iŋ dis,ȯrd·ər }

Bingham number [FL MECH]　A dimensionless number used to study the flow of Bingham plastics.　{ 'biŋ·əm ,nəm·bər }

Bingham plastic [FL MECH]　A non-Newtonian fluid exhibiting a yield stress which must be exceeded before flow starts; thereafter the rate of shear versus shear stress curve is linear.　{ 'biŋ·əm ,plas·tik }

bing ore [GEOL]　The purest lead ore, with the largest crystals of galena.　{ 'biŋ ,ȯr }

binistor [ELECTR]　A silicon *npn* tetrode that serves as a bistable negative-resistance device.　{ ,bī'nis·tər }

binnacle [NAV ARCH]　A stand, case, or box for a ship's magnetic compass, which may contain nonmagnetic gear such as a lamp.　{ 'bin·ə·kəl }

binocular [BIOL]　**1.** Of, pertaining to, or used by both eyes.　**2.** Of a type of visual perception which provides depth-of-field focus due to angular difference between the two retinal images. [OPTICS]　Any optical instrument designed for use with both eyes to give enhanced views of distant objects, whose distinguishing performance feature is the depth perception obtainable.　{ bī'näk·yə·lər }

binocular accommodation [PHYSIO]　Automatic lens adjustment by both eyes simultaneously for focusing on distant objects.　{ bī'näk·yə·lər ə,käm·ə'dā·shən }

binocular microscope [OPTICS]　A microscope having two oculars, allowing the use of both eyes at once.　{ bī'näk·yə·lər 'mī·krə,skōp }

binode [ELECTR]　An electron tube with two anodes and one cathode used as a full-wave rectifier.　Also known as double diode.　{ 'bī,nōd }

binomen [SYST]　A binomial name assigned to species, as *Canis familiaris* for the dog.　{ bī'nō·mən }

binomial [MATH]　A polynomial with only two terms.　{ bī'nō·mē·əl }

binomial array See Pascal's triangle.　{ bī'nō·mē·əl ə'rā }

binomial array antenna [ELECTROMAG]　Directional antenna array for reducing minor lobes and providing maximum response in two opposite directions.　{ bī'nō·mē·əl ə'rā an'ten·ə }

binomial coefficient [MATH]　A coefficient in the expansion

BINOCULAR

Prism binocular. (*Bausch and Lomb Optical Co.*)

of $(x + y)^n$, where n is a positive integer; the $(k + 1)$st coefficient is equal to the number of ways of choosing k objects out of n without regard for order. Symbolized $\binom{n}{k}$; $_nC_k$; $C(n,k)$; C_k^n. { bī'nō·mē·əl kō·ə'fish·ənt }

binomial differential [MATH] A differential of the form $x^p(a + bx^q)^r dx$, where p, q, r are integers. { bī'nō·mē·əl ‚dif·ə'ren·chəl }

binomial distribution [STAT] The distribution of a binomial random variable; the distribution (n,p) is given by $P(B = r) = \binom{n}{r} p^r q^{n-r}$, $p + q = 1$. Also known as Bernoulli distribution. { bī'nō·mē·əl ‚dis·trə'byü·shən }

binomial equation [MATH] An equation having the form $x^n - a = 0$. { bī'nō·mē·əl i'kwā·zhən }

binomial expansion *See* binomial series. { bī'nō·mē·əl iks'pan·shən }

binomial law [MATH] The probability of an event occurring r times in n Bernoulli trials is equal to $\binom{n}{r}p^r(1 - p)^{n-r}$, where p is the probability of the event. { bī'nō·mē·əl ‚lò }

binomial nomenclature [SYST] The Linnean system of classification requiring the designation of a binomen, the genus and species name, for every species of plant and animal. { bī'nō·mē·əl ‚nō·mən'klā·chər }

binomial probability paper [STAT] Graph paper designed to aid in the analysis of data from a binomial population, that is, data in the form of proportions or as percentages; both axes are marked so that the graduations are square roots of the variable. { bī'nō·mē·əl ‚prä·bə'bil·əd·ē ‚pā·pər }

binomial random variable [STAT] A random variable, parametrized by a positive integer n and a number p in the closed interval between 0 and 1, whose range is the set $\{0, 1, \ldots, n\}$ and whose value is the number of successes in n independent binomial trials when p is the probability of success in a single trial. { bī'nō·mē·əl ‚ran·dəm 'ver·ē·ə·bəl }

binomial series [MATH] The expansion of $(x + y)^n$ when n is neither a positive integer nor zero. Also known as binomial expansion. { bī'nō·mē·əl 'sir·ēz }

binomial surd [MATH] A sum of two roots of rational numbers, at least one of which is an irrational number. { bī'nō·mē·əl 'sərd }

binomial theorem [MATH] The rule for expanding $(x + y)^n$. { bī'nō·mē·əl 'thir·əm }

binomial trials [STAT] A sequence of trials, each trial offering that a certain result may or may not happen. Also known as Bernoulli experiments; Bernoulli trials. { bī'nō·mē·əl 'trīlz }

binomial trials model [STAT] A product model in which each factor has two simple events with probabilities p and $q = 1 - p$. { bī'nō·mē·əl 'trīlz ‚mäd·əl }

binormal [MATH] A vector on a curve at a point so that, together with the positive tangent and principal normal, it forms a system of right-handed rectangular cartesian axes. { bī'nòr·məl }

binormal indicatrix [MATH] For a space curve, all the end points of those radii of a unit sphere that are parallel to the positive directions of the binormals of the curve. Also known as spherical indicatrix of the binormal. { bī'nòr·məl in'dik·ə‚triks }

binuclear [CYTOL] Having two nuclei. { bī'nü·klē·ər }

binucleolate [CYTOL] Having two nucleoli. { bī·nü'klē·ə‚lāt }

bioacoustics [BIOL] The study of the relation between living organisms and sound. { ‚bī·ō·ə'kü·stiks }

bioactivity [BIOL] The effect that a substance has on a living organism or tissue after interaction. { ‚bī·ō·ak'tiv·əd·ē }

bioarcheology [ARCHEO] A discipline in which the concepts of human biology are integrated with anthropological archeology. { ‚bī·ō‚är·kē'äl·ə·jē }

bioartificial organs [MED] Devices, used for both short-term and long-term organ replacement, that are designed and manufactured for membrane biocompatibility, diffusion limitations, device retrieval in the event of failure, and mechanical stability. { ‚bī·ō‚ard·ə‚fish·əl 'òr·gənz }

bioassay [ANALY CHEM] A method for quantitatively determining the concentration of a substance by its effect on the growth of a suitable animal, plant, or microorganism under controlled conditions. { ‚bī·ō'as‚ā }

bioastronautics [BIOL] The study of biological, behavioral, and medical problems pertaining to astronautics. { ‚bī·ō‚as·trə'nòd·iks }

bioautography [ANALY CHEM] A bioassay based upon the ability of some compounds (for example, vitamin B_{12}) to enhance the growth of some organisms or compounds and to repress the growth of others; used to assay certain antibiotics. { ‚bī·ō‚ó'täg·rə·fē }

bioavailability [PHYSIO] The extent and rate at which a substance, such as a drug, is absorbed into a living system or is made available at the site of physiological activity. { ‚bī·ō·ə‚vāl·ə'bil·əd·ē }

biobubble [ECOL] A model concept of the ecosphere in which all living things are considered as particles held together by nonliving forces. { 'bī·ō‚bəb·əl }

biocalorimetry [BIOPHYS] The measurement of the energetics of biological processes such as biochemical reactions, association of ligands to biological macromolecules, folding of proteins into their native conformations, phase transitions in biomembranes, and enzymatic reactions, among others. { ‚bī·ō‚kal·ə'rim·ə·trē }

biocatalyst [BIOCHEM] A biochemical catalyst, especially an enzyme. { ‚bī·ō'kad·əl·ist }

biocenology [ECOL] The study of natural communities and of interactions among the members of these communities. { ‚bī·ō·sə'näl·ə·jē }

biocenose *See* biotic community. { ‚bī·ō'sē‚nōs }

bioceramic [MATER] Biocompatible or osteoinductive (stimulating bone growth) ceramic material, such as hydroxyapatite or some other type of calcium phosphate ceramic, used for reconstructive bone surgery and dental implants. { ‚bī·ō·sə'ram·ik }

biochemical [BIOCHEM] **1.** A chemical that produces an effect on living organisms after making contact. **2.** Referring to chemical substances found in or having an effect on living organisms. { ‚bī·ō'kem·i·kəl }

biochemical deposit [GEOL] A precipitated deposit formed directly or indirectly from vital activities of organisms, such as bacterial iron ore and limestone. { ‚bī·ō'kem·ə·kəl di'päz·ət }

biochemical engineering [BIOCHEM] The application of chemical engineering principles to conceive, design, develop, operate, or utilize processes and products based on biological and biochemical phenomena; this field is included in a wide range of industries, such as health care, agriculture, food, enzymes, chemicals, waste treatment, and energy. { ‚bī·ō'kem·i·kəl ‚en·jə'nir·iŋ }

biochemical fuel cell [ELEC] An electrochemical power generator in which the fuel source is bioorganic matter; air is the oxidant at the cathode, and microorganisms catalyze the oxidation of the bioorganic matter at the anode. { ‚bī·ō'kem·ə·kəl 'fyül ‚sel }

biochemical oxygen demand [MICROBIO] The amount of dissolved oxygen required to meet the metabolic needs of aerobic microorganisms in water rich in organic matter, such as sewage. Abbreviated BOD. Also known as biological oxygen demand. { ‚bī·ō'kem·ə·kəl 'äk·sə·jən di'mand }

biochemical oxygen demand test [MICROBIO] A standard laboratory procedure for measuring biochemical oxygen demand; standard measurement is made for 5 days at 20°C. Abbreviated BOD test. { ‚bī·ō'kem·ə·kəl 'äk·sə·jən di'mand ‚test }

biochemical pharmacology [MED] The study of the effects of chemicals on biochemical reactions in living systems and the effects of these systems on the chemicals, that is, their metabolism. { ‚bī·ō‚kem·ə·kəl ‚fär·mə'käl·ə·jē }

biochemical profile [IND ENG] Data recorded by both electromyographic and biomechanical means during the performance of a task to evaluate changes in the functional capacity of a worker resulting from modifications in human-equipment interfaces. { ‚bī·ō'kem·ə·kəl 'prō‚fīl }

biochemical rock [PETR] A type of sedimentary rock primarily comprising deposits resulting directly or indirectly from processes and activities of living organisms. { ‚bī·ō'kem·i·kəl 'räk }

biochemistry [CHEM] The study of chemical substances occurring in living organisms and the reactions and methods for identifying these substances. { ‚bī·ō'kem·ə·strē }

biochemorphology [BIOCHEM] The science dealing with the chemical structure of foods and drugs and their reactions on living organisms. { ‚bī·ō‚ke·mòr‚fäl·ə·jē }

biochip [ELECTR] An experimental type of integrated circuit whose basic components are organic molecules. { 'bī·ō‚chip }

biochore [ECOL] A group of similar biotopes. { 'bī·ō ‚kȯr }

biochrome [BIOCHEM] Any naturally occurring plant or animal pigment. { 'bī·ō‚krōm }

biochron [PALEON] A fossil of relatively short range of time. { 'bī·ō‚krän }

biochronology [GEOL] The relative age dating of rock units based on their fossil content. { ‚bī·ō·krə'näl·ə·jē }

biociation [ECOL] A subdivision of a biome distinguished by the predominant animal species. { bī‚äs·ē'ā·shən }

biocide See pesticide. { 'bī·ə‚sīd }

bioclastic rock [PETR] Rock formed from material broken or arranged by animals, humans, or sometimes plants; a rock composed of broken calcareous remains of organisms. { 'bī·ō‚klas·tik 'räk }

bioclimatic law [ECOL] The law which states that phenological events are altered by about 4 days for each 5° change of latitude northward or longitude eastward; events are accelerated in spring and retreat in autumn. { ‚bī·ō‚klī'mad·ik 'lȯ }

bioclimatograph [ECOL] A climatograph showing the relation between climatic conditions and some living organisms. { ‚bī·ō‚klī'mad·ə‚graf }

bioclimatology [ECOL] The study of the effects of the natural environment on living organisms. { ‚bī·ō‚klī·mə'täl·ə·jē }

biocoenosis [ECOL] A group of organisms that live closely together and form a natural ecologic unit. { ‚bī·ō·sə'nō·səs }

biocompatibility [PHYSIO] The condition of being compatible with living tissue by virtue of a lack of toxicity or ability to cause immunological rejection. { ‚bī·ō·kəm‚pad·ə'bil·əd·ē }

biocontainment See biological containment. { ‚bī·ō·kən'tān·mənt }

biocontrol See biological control. { ‚bī·ō·kən'trōl }

biocontrol system [CONT SYS] A mechanical system that is controlled by biological signals, for example, a prosthesis controlled by muscle activity. { ‚bī·ō·kən'trōl ‚sis·təm }

bioconversion [BIOPHYS] The process of converting biomass to a source of usable energy. { ‚bī·ō·kən'vər·zhən }

biocorrosion See biological corrosion. { ‚bī·ō·kə'rō·zhən }

biocycle [ECOL] A group of similar biotopes composing a major division of the biosphere; there are three biocycles: terrestrial, marine, and fresh-water. { 'bī·ō‚sī·kəl }

biocytin [BIOCHEM] $C_{16}H_{28}N_4O_4S$ Crystals with a melting point of 241–243°C; obtained from dilute methanol or acetone solutions; characterized by its utilization by *Lactobacillus casei* and *L. delbrückii* LD5 as a biotin source, and by its unavailability as a biotin source to *L. arabinosus*. Also known as biotin complex of yeast. { ‚bī·ō'sīt·ən }

biocytinase [BIOCHEM] An enzyme present in the blood and liver which hydrolyzes biocytin into biotin and lysine. { ‚bī·ō'sīt·ən‚ās }

biodegradability [MATER] The characteristic of a substance that can be broken down by microorganisms. { ‚bī·ō·di‚grād·ə'bil·əd·ē }

biodegradation [ECOL] The destruction of organic compounds by microorganisms. { ‚bī·ō‚deg·rə'dā·shən }

biodeterioration [MATER] Decay of wood or other material caused by fungi, bacteria, insects, or marine boring organisms. { ‚bī·ō·di‚tir·ē·ər'ā·shən }

biodistribution kinetics [BIOL] A mathematical description of the in vivo distribution of a radionuclide present in various organs as a function of time following its administration. { ‚bī·ō‚dis·trə‚byü·shən ki'ned·iks }

biodiversity [ECOL] The range of living organisms (such as plant and animal species) in an environment during a specific time period. { ‚bī·ō·di'vər·sə·dē }

biodynamic [AGR] Of or pertaining to a system of organic farming. [BIOPHYS] Of or pertaining to the dynamic relation between an organism and its environment. { ‚bī·ō·dī'nam·ik }

biodynamics [BIOPHYS] The study of the effects of dynamic processes (motion, acceleration, weightlessness, and so on) on living organisms. { ‚bī·ō·dī'nam·iks }

bioelectric current [PHYSIO] A self-propagating electric current generated on the surface of nerve and muscle cells by potential differences across excitable cell membranes. { ‚bī·ō·i'lek·trik 'kər·ənt }

bioelectricity [PHYSIO] The generation by and flow of an electric current in living tissue. { ‚bī·ō·i‚lek'tris·əd·ē }

bioelectric model [PHYSIO] A conceptual model for the study of animal electricity in terms of physical principles. { ‚bī·ō·i'lek·trik 'mäd·əl }

bioelectrochemistry [PHYSIO] The study of the control of biological growth and repair processes by electrical stimulation. { ‚bī·ō·i‚lek·trō'kem·ə·strē }

bioelectronics [BIOPHYS] **1.** The application of electronic theories and techniques to the problems of biology. **2.** The use of biotechnology in electronic devices such as biosensors, molecular electronics, and neuronal interfaces; more speculatively, the use of proteins in constructing circuits. { ‚bī·ō‚i‚lek'trän·iks }

bioenergetics [BIOCHEM] The branch of biology dealing with energy transformations in living organisms. { ‚bī·ō‚en·ər'jed·iks }

bioengineer See genetic engineer. { ‚bī·ō‚en·jə'nir }

bioengineering [ENG] The application of engineering knowledge to the fields of medicine and biology. { ‚bī·ō‚en·jə'nir·iŋ }

bioerosion [OCEANOGR] The process by which animals, through drilling, grazing, and burrowing, erode hard substances such as rocks and coral reefs. { ‚bī·ō·i'rōzh·ən }

bioethics [BIOL] A discipline concerned with the application of ethics to biological problems, especially in the field of medicine. { ‚bī·ō'eth·iks }

biofacies [GEOL] **1.** A rock unit differing in biologic aspect from laterally equivalent biotic groups. **2.** Lateral variation in the biologic aspect of a stratigraphic unit. { 'bī·ō‚fā·shēz }

biofeedback [PSYCH] The use of instrumentation to provide immediate visual or auditory information to an individual about changes in one or more of the individual's physiologic processes not ordinarily perceived, such as brainwave activity, muscle tension, or blood pressure, in order to manipulate the process by conscious mental control. { ‚bī·ō'fēd‚bak }

biofilm [MICROBIO] A microbial (bacterial, fungal, algal) community, enveloped by the extracellular biopolymer which these microbial cells produce, that adheres to the interface of a liquid and a surface. { 'bī·ō‚film }

biofilter [ENG] An emission control device that uses microorganisms to destroy volatile organic compounds and hazardous air pollutants. { 'bī·ō‚fil·tər }

bioflavonoid [BIOCHEM] A group of compounds obtained from the rinds of citrus fruits and involved with the homeostasis of the walls of small blood vessels; in guinea pigs a marked reduction of bioflavonoids results in increased fragility and permeability of the capillaries; used to decrease permeability and fragility in capillaries in certain conditions. Also known as citrus flavonoid compound; vitamin P complex. { ‚bī·ō'flav·ə‚nȯid }

biofog [METEOROL] A type of steam fog caused by contact between extremely cold air and the warm, moist air surrounding human or animal bodies or generated by human activity. { 'bī·ō‚fäg }

biogalvanic battery [MED] A battery that is implanted in the body and depends on interaction of metal electrodes with oxygen and fluids of the body to generate sufficient power for a heart pacemaker or other implanted medical device. { ‚bī·ō·gal'van·ik 'bad·ə·rē }

biogas [MATER] A mixture of methane and carbon dioxide generated from the bacterial decomposition of animal and vegetable wastes. { 'bī·ō‚gas }

biogenesis [BIOL] Development of a living organism from a similar living organism. { ‚bī·ō'jen·ə·səs }

biogenetic law See recapitulation theory. { ‚bī·ō·jə'ned·ik 'lȯ }

biogenetics See genetic engineering. { ‚bī·ō·jə'ned·iks }

biogenic [BIOL] **1.** Essential to the maintenance of life. **2.** Produced by actions of living organisms. { ‚bī·ō'jen·ik }

biogenic amine [BIOCHEM] Any of a group of organic compounds that contain one or more amine groups ($-NH_2$) and have a possible role in brain functioning, including catecholamines and indoles. { ‚bī·ō'jen·ik 'a‚mēn }

biogenic amine hypothesis [PSYCH] The hypothesis that abnormalities in the physiology and metabolism of biogenic amines are involved in the development of certain psychiatric illnesses. { ‚bī·ō'jen·ik 'a‚mēn hī‚päth·ə·səs }

biogenic chert [PETR] Chert derived from the tests of pelagic silica-secreting organisms, particularly diatoms and radiolarians. { ‚bī·ō'jen·ik 'chərt }

biogenic mineral [MINERAL] A mineral in sediments or sedimentary rock which represents the hard parts of dead organisms. { ¦bī·ō¦jen·ik 'min·rəl }

biogenic reef [GEOL] A mass consisting of the hard parts of organisms, or of a biogenically constructed frame enclosing detrital particles, in a body of water; most biogenic reefs are made of corals or associated organisms. { ¦bī·ō¦jen·ik 'rēf }

biogenic sediment [GEOL] A deposit resulting from the physiological activities of organisms. { ¦bī·ō¦jen·ik 'sed·ə·mənt }

biogeochemical cycle [GEOCHEM] The chemical interactions that exist between the atmosphere, hydrosphere, lithosphere, and biosphere. { ¦bī·ō¸jē·ō¦kem·ə·kəl 'sīkəl }

biogeochemical prospecting [GEOCHEM] A prospecting technique for subsurface ore deposits based on interpretation of the growth of certain plants which reflect subsoil concentrations of some elements. { ¸bī·ō¸jē·ō¦kem·ə·kəl 'präs¸pek·tiŋ }

biogeochemistry [GEOCHEM] A branch of geochemistry that is concerned with biologic materials and their relation to earth chemicals in an area. { ¸bī·ō¸jē·ō¦kem·ə·strē }

biogeographic realm [ECOL] Any of the divisions of the landmasses of the world according to their distinctive floras and faunas. { ¸bī·ō¸jē·ə¸graf·ik 'relm }

biogeography [ECOL] The science concerned with the geographical distribution of animal and plant life. { ¦bī·ō·jē¦äg·rə·fē }

biogeosphere [ECOL] The region of the earth extending from the surface of the upper crust to the maximum depth at which organic life exists. { ¸bī·ō¦jē·ə¸sfir }

biohazard [BIOL] Any biological agent or condition that presents a hazard to life. { 'bī·ō¸haz·ərd }

bioherm [GEOL] A circumscribed mass of rock exclusively or mainly constructed by marine sedimentary organisms such as corals, algae, and stromatoporoids. Also known as organic mound. { 'bī·ō¸hərm }

biohermal limestone [PETR] Reefs or reeflike mounds of carbonate that accumulated much in the same fashion as modern reefs and atolls of the Pacific Ocean. { ¦bī·ō¦hər·məl 'līm¸stōn }

biohermite [PETR] Limestone formed of debris from a bioherm. { ¦bī·ō¦hər¸mīt }

biohydrology [ECOL] Study of the interactions between water, plants, and animals, including the effects of water on biota as well as the physical and chemical changes in water or its environment produced by biota. { ¦bī·ō¸hī'dräl·ə·jē }

bioinformatics [COMPUT SCI] The use of computers to study biological systems. { ¸bī·ō¸in·fər'mad·iks }

bioinorganic chemistry [BIOCHEM] The application of the principles of inorganic chemistry to problems of biology and biochemistry. Also known as inorganic biochemistry; metallobiochemistry. { ¸bī·ō¸in·ór¦gan·ik 'kem·ə·strē }

bioinstrumentation [ENG] The use of instruments attached to animals and man to record biological parameters such as breathing rate, pulse rate, body temperature, or oxygen in the blood. { ¦bī·ō¸in·strə·mən'tā·shən }

bioleaching [BIOCHEM] The dissolution of metals from their mineral source by naturally occurring microorganisms. Also known as biooxidation. { ¦bī·ō'lēch·iŋ }

biolite [GEOL] A concretion formed of concentric layers through the action of living organisms. [PETR] See biolith. { 'bī·ō¸līt }

biolith [PETR] A rock formed from or by organic material. Also known as biolite. { 'bī·ō¸lith }

biolithite [PETR] An inclusive category for all organic limestone. { ¸bī·ō'lith¸īt }

biological [BIOL] Of or pertaining to life or living organisms. [IMMUNOL] A biological product used to induce immunity to various infectious diseases or noxious substances of biological origin. { ¦bī·ə¦läj·ə·kəl }

biological agent [ORD] Any of the viruses, microorganisms, and toxic substances derived from living organisms and used as offensive weapons to produce death or disease in humans, animals, and growing plants. { ¦bī·ə¦läj·ə·kəl 'ā·jənt }

biological anthropology See physical anthropology. { ¸bī·ə¦läj·i·kəl ¸an·thrə'päl·ə·jē }

biological balance [ECOL] Dynamic equilibrium that exists among members of a stable natural community. { ¦bī·ə¦läj·ə·kəl 'bal·əns }

biological clock [PHYSIO] Any physiologic factor that functions in regulating body rhythms. { ¦bī·ə¦läj·ə·kəl 'kläk }

biological constraint [PSYCH] In learning theory, the observation that certain behaviors are more easily learned by some organisms than by others. { ¦bī·ə¦läj·ə·kəl kən'strānt }

biological containment [GEN] A technique by which the genetic constitution of an organism is altered in order to minimize its ability to grow outside the laboratory. Also known as biocontainment. { ¦bī·ə¦läj·ə·kəl kən'tān·mənt }

biological control [ECOL] Natural or applied regulation of populations of pest organisms, especially insects, through the role or use of natural enemies. Also known as biocontrol. { ¦bī·ə¦läj·ə·kəl kən'trōl }

biological corrosion [MET] Deterioration of metals as a result of the metabolic activity of microorganisms. Also known as biocorrosion. { ¦bī·ə¦läj·ə·kəl kə'rō·zhən }

biological equilibrium [BIOPHYS] A state of body balance for an actively moving animal, when internal and external forces are in equilibrium. { ¦bī·ə¦läj·ə·kəl ¸ē·kwə'lib·rē·əm }

biological half-life [PHYSIO] The time required by the body to eliminate half of the amount of an administered substance through normal channels of elimination. { ¦bī·ə¦läj·ə·kəl 'haf ¸līf }

biological indicator [BIOL] An organism that can be used to determine the concentration of a chemical in the environment. { ¸bī·ə¦läj·ə·kəl 'in·də¸kād·ər }

biological invasion [ECOL] The process by which species (or genetically distinct populations), with no historical record in an area, breach biogeographic barriers and extend their range. { ¸bi·ə¦läj·i·kəl in'vā·zhən }

biological magnification [ECOL] The increasing concentration of toxins from pesticides, herbicides, and various types of waste in living organisms that accompanies cycling of nutrients through the trophic levels of food webs. { ¸bī·ə¦läj·ə·kəl ¸mag·nə·fə'kā·shən }

biological oceanography [OCEANOGR] The study of the flora and fauna of oceans in relation to the marine environment. { ¦bī·ə¦läj·ə·kəl ¸ō·shə'näg·rə·fē }

biological oil-spill control [ECOL] The use of cultures of microorganisms capable of living on oil as a means of degrading an oil slick biologically. { ¦bī·ə¦läj·ə·kəl 'óil ¸spil kən'trōl }

biological oxidation [BIOCHEM] Energy-producing reactions in living cells involving the transfer of hydrogen atoms or electrons from one molecule to another. { ¦bī·ə¦läj·ə·kəl ¸äk·sə'dā·shən }

biological oxygen demand See biochemical oxygen demand. { ¦bī·ə¦läj·ə·kəl 'äk·sə·jən di'mand }

biological productivity [ECOL] The quantity of organic matter or its equivalent in dry matter, carbon, or energy content which is accumulated during a given period of time. { ¦bī·ə¦läj·ə·kəl prə¸dək'tiv·əd·ē }

biological shield [MICROBIO] A structure designed to prevent the migration of living organisms from one part of a system to another; used on sterilized space probes. [NUCLEO] A radiation-absorbing shield used to protect personnel from the effects of nuclear particles or radiation in the vicinity of a nuclear reactor. { ¦bī·ə¦läj·ə·kəl 'shēld }

biological specificity [BIOL] The principle that defines the orderly patterns of metabolic and developmental reactions giving rise to the unique characteristics of the individual and of its species. { ¦bī·ə¦läj·ə·kəl spes·ə'fis·əd·ē }

biological standardization [PHARM] The standardization of drugs or biological products that cannot be chemically analyzed by studying the drugs' pharmacologic action on animals. { ¦bī·ə¦läj·ə·kəl ¸stan·dərd·ə'zā·shən }

biological value [BIOCHEM] A measurement of the efficiency of the protein content in a food for the maintenance and growth of the body tissues of an individual. { ¦bī·ə¦läj·ə·kəl 'val·yü }

biological warfare [ORD] Abbreviated BW. **1.** Employment of living microorganisms, toxic biological products, and plant growth regulators to produce death or injury in humans, animals, or plants. **2.** Defense against such action. { ¦bī·ə¦läj·ə·kəl 'wór¸fer }

biologic artifact [ORG CHEM] An organic compound with a chemical structure that demonstrates the compound's derivation from living matter. { ¦bī·ə¦läj·ik 'ard·ə¸fakt }

biologic weathering See organic weathering. { ¦bī·ə¦läj·ik 'weth·ə·riŋ }

biology [SCI TECH] A division of the natural sciences concerned with the study of life and living organisms. { bī'äl·ə·jē }

bioluminescence [BIOL] The emission of visible light by living organisms. { ¦bī·ō¸lü·mə'nes·əns }

biolysis [BIOL] **1.** Death and the following tissue disintegration. **2.** Decomposition of organic materials, such as sewage, by living organisms. { bī'äl·ə·səs }

biomagnetism [BIOPHYS] The production of a magnetic field by a living organism. { ¦bī·ō'mag·nə¸tiz·əm }

biomarkers [GEOL] Complex organic compounds found in oil, bitumen, rocks, and sediments that are linked with and distinctive of a particular source (such as algae, bacteria, or vascular plants); they are useful dating indicators in stratigraphy and molecular paleontology. Also known as chemical fossils; molecular fossils. { 'bī·ō¸mär·kərz }

biomass [ECOL] The dry weight of living matter, including stored food, present in a species population and expressed in terms of a given area or volume of the habitat. { 'bī·ō¸mas }

biomaterial [MED] A natural or synthetic nondrug material that is compatible with living tissue and is suitable for surgical implanting; it can be used to enhance, treat, or replace organs, tissues, and functions in a living organism. { ¦bī·ō·mə'tir·ē·əl }

biomathematics [BIOPHYS] Mathematical methods applied to the study of living organisms. { ¸bī·ō¸math·ə'mad·iks }

biome [ECOL] A complex biotic community covering a large geographic area and characterized by the distinctive life-forms of important climax species. { 'bī¸ōm }

biomechanics [BIOPHYS] The study of the mechanics of living things. { ¦bī·ō·mə'kan·iks }

biomedical chemical engineering [MED] The application of chemical engineering principles to the solution of medical problems due to physiological impairment. { ¸bī·ō¦med·i·kəl ¦kem·i·kəl ¸en jə'nir·iŋ }

biomedical engineering [ENG] The application of engineering technology to the solution of medical problems; examples are the development of prostheses such as artificial valves for the heart, various types of sensors for the blind, and automated artificial limbs. { ¸bī·ō¦med·ə·kəl ¸en·jə'nir·iŋ }

biomedicine [MED] The science concerned with the study of the environment required for astronauts in space vehicles. { ¸bī·ō¦med·ə·sən }

biomere [ECOL] A biostratigraphic unit bounded by abrupt nonevolutionary changes in the dominant elements of a single phylum. { 'bī·ō¸mir }

biometeorology [BIOL] The study of the relationship between living organisms and atmospheric phenomena. { ¦bī·ō ¸mēd·ē·ə'räl·ə·jē }

biometer [BIOL] An instrument which is used to measure minute amounts of carbon dioxide given off by the functioning tissue of an organism. { bī'äm·ə·tər }

biometric device [COMPUT SCI] A device that identifies persons seeking access to a computing system by determining their physical characteristics through fingerprints, voice recognition, retina patterns, pictures, weight, or other means. { ¸bī·ō¦me·trik di¦vīs }

biometrician [STAT] A person skilled in biometry. Also known as biometricist. { bī¸äm·ə'trish·ən }

biometricist See biometrician. { ¸bī·ō'me·trə¸sist }

biometrics [STAT] The use of statistics to analyze observations of biological phenomena. { ¸bī·ō'me·triks }

biometry [STAT] The use of statistics to calculate the average length of time that a human being lives. { bī'äm·ə·trē }

biomicrite [PETR] A limestone resembling biosparite except that the microcrystalline calcite matrix exceeds calcite cement. { bī'ō'mī¸krīt }

biomicrosparite [PETR] **1.** Biomicrite in which the micrite groundmass has recrystallized to microspar. **2.** Microsparite containing fossil fragments or fossils. { ¸bī·ō¸mī·krō'spär¸īt }

biomicrudite [PETR] Biomicrite with fossil fragments or fossils greater than 1 millimeter in diameter. { ¸bī·ō'mī·krə¸dīt }

biomimetic catalyst [ORG CHEM] A synthetic compound that can simulate the mode of action of a natural enzyme by catalyzing a reaction at ambient conditions. { ¦bī·ō·mə'med·ik 'kad·ə¸list }

biomimetics [BIOCHEM] A branch of science in which synthetic systems are developed by using information obtained from biological systems. { ¦bī·ō·mə¦med·iks }

biomineralization [PHYSIO] A mineralization process carried out within a living organism, such as formation of bone in vertebrates. { ¸bī·ō¸min·rəl·ə'zā·shən }

biomining [MICROBIO] The use of microorganisms to recover metals of value, such as gold, silver, and copper, from sulfide minerals. { 'bī·ō¸mīn·iŋ }

biomolecular [BIOCHEM] Pertaining to organic molecules occurring in living organisms, especially macromolecules. { ¸bī·ō·mə'le·kyə·lər }

biomorph [GRAPHICS] An abstract form (painted, drawn, or sculpted) whose shape resembles that of a living organism. { 'bī·ō¸mòrf }

bion [ECOL] An independent, individual organism. { 'bī¸än }

bionavigation [VERT ZOO] The ability of animals such as birds to find their way back to their roost, even if the landmarks on the outward-bound trip were effectively concealed from them. { ¸bī·ō¸nav·ə'gā·shən }

bionics [ENG] The study of systems, particularly electronic systems, which function after the manner of living systems. { bī'än·iks }

bionomics See ecology. { ¸bī·ō'näm·iks }

biooxidation See bioleaching. { ¸bī·ō¸äk·sə'dā·shən }

biopak [ENG] A container for housing a living organism in a habitable environment and for recording biological functions during space flight. { 'bī·ō¸pak }

biopelite See black shale. { bī'äp·ə¸līt }

biopelmicrite [PETR] A limestone similar to biopelsparite but with a microcrystalline matrix that exceeds calcite cement. { bī·ə'pel·mə¸krīt }

biopelsparite [PETR] A limestone similar to biosparite but with the ratio of fossils and fossil fragments to pellets between 3:1 and 1:3. { bī·ə'pel·spə¸rīt }

biophage See macroconsumer. { 'bī·ō¸fāj }

biophagous [ZOO] Feeding on living organisms. { bī'äf·ə·gəs }

biopharmaceutics [PHARM] The study of the relationships between physical and chemical properties, dosage, and administration of a drug and its activity in humans and animals. { ¦bī·ō¸färm·ə'süd·iks }

biopharming [MED] The application of genetic engineering on living organisms to induce or increase the production of pharmacologically active substances. { 'bī·ō¸färm·iŋ }

biophile [BIOCHEM] Any element concentrated or found in the bodies of living organisms and organic matter; examples are carbon, nitrogen, and oxygen. { 'bī·ə¸fīl }

biophysics [SCI TECH] The hybrid science involving the application of physical principles and methods to study and explain the structures of living organisms and the mechanics of life processes. { ¦bī·ō¦fiz·iks }

biopolymer [BIOCHEM] A biological macromolecule such as a protein or nucleic acid. { ¦bī·ō'päl·ə·mər }

biopotency [BIOCHEM] Capacity of a chemical substance, as a hormone, to function in a biological system. { ¦bī·ō'pōt·ən·sē }

biopotential [PHYSIO] Voltage difference measured between points in living cells, tissues, and organisms. { ¦bī·ō·pə'ten·chəl }

bioprocess [BIOCHEM] **1.** A technique used to produce a commercial substance (such as alcohol) by a biological process (that is, via microbial fermentation). **2.** A technique used to prepare a biological material (usually genetically engineered) for commercial use. { ¦bī·ō¦prä·səs }

bioprospecting [PHARM] The search for new pharmaceutical (and sometimes nutritional or agricultural) products from natural sources, such as plants, microorganisms, and sometimes animals. { ¸bī·ō'prä·spek·tiŋ }

biopsy [PATH] The removal and examination of tissues, cells, or fluids from the living body for the purposes of diagnosis. { 'bī¸äp·sē }

biopyribole [MINERAL] **1.** A collective term for the rock-forming minerals pyroxene, amphibole, and mica. **2.** A chemically diverse but structurally related group of minerals that constitute substantial fractions of both the earth's crust and upper mantle; they exhibit single-chain, double-chain, triple-chain, and sheet silicate structures. { ¸bī·ō'pir·ə¸bōl }

bioreactor [BIOCHEM] A vessel that is used for the fermentation and production of living organisms, such as bacteria or yeast. { 'bī·ō·rē,ak·tər }

biorefinery [BIOCHEM] A large, integrated processing facility that produces chemicals and biochemicals from plant matter, wood waste, and waste paper. { 'bī·ō·ri¦fīn·rē }

bioregion [ECOL] A region with borders that are naturally defined by topographic systems (such as mountains, rivers, and oceans) and ecological systems (such as deserts, rainforests, and tundras). { 'bī·ō,rē·jən }

bioregionalism [ECOL] An environmentalist movement to make political boundaries coincide with bioregions. { ,bī·ō'rē·jən·əl,iz·əm }

bioremediation [ECOL] The use of a biological process (via plants or microorganisms) to clean up a polluted environmental area (such as an oil spill). { ,bī·ō·ri,mē·dē'ā·shən }

biorheology [BIOPHYS] The study of the deformation and flow of biological fluids, such as blood, mucus, and synovial fluid. { ¦bī·ō·rē¦äl·ə·jē }

biorhythm [PHYSIO] A biologically inherent cyclic variation or recurrence of an event or state, such as a sleep cycle or circadian rhythm. { 'bī·ō,rith·əm }

BIOS See basic input/output system.

biosafety [BIOL] The establishment and maintenance of safe conditions in a biological research laboratory to ensure that pathogenic microbes are contained (and not released to workers or the environment). { ¦bī·ō¦sāf·tē }

biosatellite [AERO ENG] An artificial satellite designed to contain and support humans, animals, or other living material in a reasonably normal manner for a period of time and to return safely to earth. { ¦bī·ō'sad·əl,īt }

bioscience [BIOL] The study of the nature, behavior, and uses of living organisms as applied to biology. { ¦bī·ō¦sī·əns }

biosensor [ANALY CHEM] An analytical device that converts the concentration of an analyte in an appropriate sample into an electrical signal by means of a biologically derived sensing element intimately connected to, or integrated into, a transducer. { ¦bī·ō¦sen·sər }

bioseparation [BIOCHEM] The recovery of a product from solutions of cells and media; the process used must avoid harsh conditions that could damage the product. { ,bī·ō,sep·ə'rā·shən }

bioseries [EVOL] A historical sequence produced by the changes in a single hereditary character. { ¦bī·ō,sir·ēz }

biosocial [ZOO] Pertaining to the interplay of biological and social influences. { ,bī·ō'sō·shəl }

biosolid [CIV ENG] A recyclable, primarily organic solid material produced by wastewater treatment processes. { ¦bī·ō,säl·əd }

biosonar [PHYSIO] A guidance system in certain animals, such as bats, utilizing the reflection of sounds that they produce as they move about. { ¦bī·ō¦sō,när }

biosparite [PETR] A limestone made up of less than 25% oolites and less than 25% intraclasts, with the ratio by volume of fossils and fragments to pellets being more than 3:1 and the calcite cement content being greater than the microcrystalline calcite content. { bī'äs·pə,rīt }

biospeleology [BIOL] The study of cave-dwelling organisms. { ,bī·ō,spē·lē'äl·ə·jē }

biosphere [ECOL] The life zone of the earth, including the lower part of the atmosphere, the hydrosphere, soil, and the lithosphere to a depth of about 1.2 miles (2 kilometers). { 'bī·ə,sfir }

biostabilizer [CIV ENG] A component in mechanized composting systems; consists of a drum in which moistened solid waste is comminuted and tumbled for about 5 days until the aeration and biodegradation turns the waste into a fine dark compost. { ,bī·ō'stāb·əl,īz·ər }

biostasy [ECOL] Maximum development of organisms when, during tectonic repose, residual soils form extensively on the land and calcium carbonate deposition is widespread in the sea. { bī'äs·tə·sē }

biostratigraphic unit [GEOL] A stratum or body of strata that is defined and identified by one or more distinctive fossil species or genera without regard to lithologic or other physical features or relations. { ¦bī·ō,strad·ə'graf·ik 'yü·nət }

biostratigraphy [PALEON] A part of paleontology concerned with the study of the conditions and deposition order of sedimentary rocks. { ¦bī·ō·strə'tig·rə·fē }

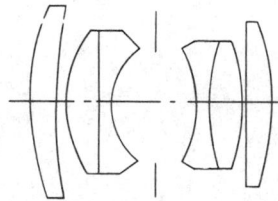

BIOTAR LENS

Biotar camera lens.

BIOTIN

Structural formula of biotin.

biostatistics [STAT] The use of statistics to obtain information from biological data. { ,bī·ō·stə'tis·tiks }

biostromal limestone [GEOL] Biogenic carbonate accumulations that are laterally uniform in thickness, in contrast to the moundlike nature of bioherms. { ¦bī·ə¦strō·məl 'līm,stōn }

biostrome [GEOL] A bedded structure or layer (bioclastic stratum) composed of calcite and dolomitized calcarenitic fossil fragments distributed over the sea bottom as fine lentils, independent of or in association with bioherms or other areas of organic growth. [INV ZOO] A flat-bedded, fossil, reeflike structure. { 'bī·ə,strōm }

biosynthesis [BIOCHEM] Production, by synthesis or degradation, of a chemical compound by a living organism. { ,bī·ō'sin·thə·səs }

biot [ELEC] See abampere. [OPTICS] A unit of rotational strength in substances exhibiting circular dichroism, equal to 10^{-40} times the corresponding centimeter-gram-second unit. { 'bī·ät }

biota [BIOL] **1.** Animal and plant life characterizing a given region. **2.** Flora and fauna, collectively. { bī'ōd·ə }

biotar lens [OPTICS] A modern camera lens which is a modified Gauss objective with a large aperture and a field of about 24°. { 'bī·ō,tär ¦lenz }

biotechnical robot [CONT SYS] A robot that requires the presence of a human operator in order to function. { ¦bī·ō¦tek·nə·kəl 'rō,bät }

biotechnology [GEN] The use of advanced genetic techniques to construct novel microbial, plant, and animal strains or obtain site-directed mutants to improve the quantity or quality of products or obtain other desired phenotypes. { ¦bī·ō·tek'näl·ə·jē }

biotelemetry [ENG] The use of telemetry techniques, especially radio waves, to study behavior and physiology of living things. { ¦bī·ō·tə'lem·ə·trē }

Biot-Fourier equation [THERMO] An equation for heat conduction which states that the rate of change of temperature at any point divided by the thermal diffusivity equals the Laplacian of the temperature. { ¦byō¦für·yā i'kwā·zhən }

biotherapy [MED] Treatment of disease with biologicals, that is, materials produced by living organisms. { ¦bī·ō'ther·ə·pē }

biotic [BIOL] **1.** Of or pertaining to life and living organisms. **2.** Induced by the actions of living organisms. { bī'äd·ik }

biotic community [ECOL] An aggregation of organisms characterized by a distinctive combination of both animal and plant species in a particular habitat. Also known as biocenose. { bī'äd·ik kə'myün·əd·ē }

biotic district [ECOL] A subdivision of a biotic province. { bī'äd·ik 'dis·trikt }

biotic environment [ECOL] That environment comprising living organisms, which interact with each other and their abiotic environment. { bī'äd·ik in'vī·ərn·mənt }

biotic isolation [ECOL] The occurrence of organisms in isolation from others of their species. { bī'äd·ik ,ī·sə'lā·shən }

biotic potential [ECOL] The maximum possible growth rate of living things under ideal conditions. { bī'äd·ik pə'ten·chəl }

biotic province [ECOL] A community, according to some systems of classification, occupying an area where similarity of climate, physiography, and soils leads to the recurrence of similar combinations of organisms. { bī'äd·ik 'präv·əns }

biotin [BIOCHEM] $C_{10}H_{16}N_2O_3S$ A colorless, crystalline vitamin of the vitamin B complex occurring widely in nature, mainly in bound form. { 'bī·ə·tən }

biotin carboxylase [BIOCHEM] An enzyme which condenses bicarbonate with biotin to form carboxybiotin. { 'bī·ə·tən kär'bäk·sə,lās }

biotin complex of yeast See biocytin. { 'bī·ə·tən 'käm,pleks əv 'yēst }

biotite [MINERAL] A black, brown, or dark green, abundant and widely distributed species of rock-forming mineral in the mica group; its chemical composition is variable: $K_2[Fe(II),Mg]_{6-4}[Fe(III),Al,Ti]_{0-2}(Si_{6-5},Al_{2-3})O_{20-22}(OH,F)_{4-2}$. Also known as black mica; iron mica; magnesia mica; magnesium-iron mica. { 'bī·ə,tīt }

biotite schist [PETR] A schist composed of biotite. { 'bī·ə,tīt 'shist }

Biot number [FL MECH] A dimensionless group, used in the study of mass transfer between a fluid and a solid, which gives the ratio of the mass-transfer rate at the interface to the mass-transfer rate in the interior of a solid wall of specified thickness. { 'byō ,nəm·bər }

biotope [ECOL] An area of uniform environmental conditions and biota. { 'bī·ə,tōp }

biotransformation [BIOCHEM] The series of chemical reactions that occur in a compound, especially a drug, as a result of enzymatic or metabolic activities by a living organism. { ¦bī·ō,tranz·fər'mā·shən }

biotron [ENG] A test chamber used for biological research within which the environmental conditions can be completely controlled, thus allowing observations of the effect of variations in environment on living organisms. { 'bī·ə,trän }

Biot-Savart law [ELECTROMAG] A law that gives the intensity of the magnetic field due to a wire carrying a constant electric current. { byō sə'vär 'lò }

Biot's law [OPTICS] The law that an optically active substance rotates plane-polarized light through an angle inversely proportional in its wavelength. { 'byōz ,lò }

bioturbation [GEOL] The disruption of marine sedimentary structures by the activities of benthic organisms. { ¦bī·ō·tər'bā·shən }

biotype [GEN] A group of organisms having the same genotype. { 'bī·ə,tīp }

bioultrasonics [ACOUS] The study of the interaction of sound at frequencies above about 20,000 hertz with living systems. { ¦bī·ō·əl·trə'sän·iks }

biozone [PALEON] The range of a single taxonomic entity in geologic time as reflected by its occurrence in fossiliferous rocks. { 'bī·ō,zōn }

bipack [GRAPHICS] A photographic film composed of two sensitive emulsions; each is sensitive to a different color when exposed simultaneously. { 'bī,pak }

biparasitic [ECOL] Parasitic upon or in a parasite. { ¦bī,par·ə'sid·ik }

biparietal [ANTHRO] Distance between the most distant opposite points of the parietal bone. { ¦bī·pə'rī·əd·əl }

biparous [BOT] Bearing branches on dichotomous axes. [VERT ZOO] Bringing forth two young at a birth. { 'bī·pə·rəs }

bipartite cubic [MATH] The points satisfying the equation $y^2 = x(x - a)(x - b)$. { bī'pär,tīt 'kyü·bik }

bipartite graph [MATH] A linear graph (network) in which the nodes can be partitioned into two groups G_1 and G_2 such that for every arc (i,j) node i is in G_1 and node j in G_2. { bī'pär,tīt 'graf }

bipartite uterus [ANAT] A uterus divided into two parts almost to the base. { bī'pär,tīt 'yüd·ə·rəs }

bipectinate [INV ZOO] Of the antennae of certain moths, having two margins with comblike teeth. [ZOO] Branching like a feather on both sides of a main shaft. { bī'pek·tə,nāt }

biped [VERT ZOO] **1.** A two-footed animal. **2.** Any two legs of a quadruped. { 'bī,ped }

bipedal [BIOL] Having two feet. { bī'ped·əl }

bipedal dinosaur [PALEON] A dinosaur having two long, stout hindlimbs for walking and two relatively short forelimbs. { bī'ped·əl 'dīn·ə,sòr }

bipeltate [BOT] Having two shield-shaped parts. [ZOO] Having a shell or other covering resembling a double shield. { bī'pel,tāt }

bipenniform [ANAT] Of the arrangement of muscle fibers, resembling a feather barbed on both sides. { bī'pen·ə,fòrm }

biphasic [BOT] Possessing both a sporophyte and a gametophyte generation in the life cycle. { bī'fāz·ik }

biphenyl [ORG CHEM] $C_{12}H_{10}$ A white or slightly yellow crystalline hydrocarbon, melting point 70.0°C, boiling point 255.9°C, and density 1.9896, which gives plates or monoclinic prismatic crystals; used as a heat-transfer medium and as a raw material for chlorinated diphenyls. Also known as diphenyl; phenylbenzene. { bī'fen·əl }

***para*-biphenylamine** [ORG CHEM] $C_{12}H_{11}N$ Leaflets with a melting point of 53°C; readily soluble in hot water, alcohol, and chloroform; used in the detection of sulfates and also as a carcinogen in cancer research. { ¦par·ə ¦bī·fə'nil·ə,mēn }

biphyletic [EVOL] Descended in two branches from a common ancestry. { ¦bī·fī'led·ik }

Biphyllidae [INV ZOO] The false skin beetles, a family of coleopteran insects in the superfamily Cucujoidea. { ¦bī'fil·ə,dē }

bipinnaria [INV ZOO] The complex, bilaterally symmetrical, free-swimming larval stage of most asteroid echinoderms. { ¦bī·pi'ner·ē·ə }

bipinnate [BOT] Pertaining to a leaf that is pinnate for both its primary and secondary divisions. { bī'pin,āt }

biplane [AERO ENG] An aircraft with two wings fixed at different levels, especially one above and one below the fuselage. { 'bī,plān }

biplate [OPTICS] **1.** Two plates of glass cemented together with a small angle between them, for producing a double image of a slit in interference experiments. **2.** Two half-wave plates of doubly refracting material, each cut parallel to its optical axis and cemented together with axes perpendicular; used to detect optical polarization. Also known as Bravais biplate. { 'bī,plāt }

bipolar [SCI TECH] **1.** Having two poles. **2.** Capable of assuming positive or negative values, such as an electric charge, or pertaining to a quantity with this property, such as a bipolar transistor. { bī'pō·lər }

bipolar amplifier [ELECTR] An amplifier capable of supplying a pair of output signals corresponding to the positive or negative polarity of the input signal. { bī'pō·lər 'am·plə,fī·ər }

bipolar circuit [ELECTR] A logic circuit in which zeros and ones are treated in a symmetric or bipolar manner, rather than by the presence or absence of a signal; for example, a balanced arrangement in a square-loop-ferrite magnetic circuit. { bī'pō·lər 'sər·kət }

bipolar coordinate system [MATH] **1.** A two-dimensional coordinate system defined by the family of circles that pass through two common points, and the family of circles that cut the circles of the first family at right angles. **2.** A three-dimensional coordinate system in which two of the coordinates depend on the x and y coordinates in the same manner as in a two-dimensional bipolar coordinate system and are independent of the z coordinate, while the third coordinate is proportional to the z coordinate. { ¦bī,pō·lər kō'òrd·ən·ət ,sis·təm }

bipolar disorder [PSYCH] A major affective disorder in which there are episodes of both mania and depression. Also known as manic-depressive illness. { bī'pō·lər dis'òrd·ər }

bipolar electrode [ELEC] Electrode, without metallic connection with the current supply, one face of which acts as anode surface and the opposite face as a cathode surface when an electric current is passed through a cell. { bī'pō·lər i'lek,trōd }

bipolar flagellation [MICROBIO] The presence of flagella at both poles in certain bacteria. { bī'pō·lər ,flaj·ə'lā·shən }

bipolar format [COMPUT SCI] A method of representing binary data in which 0 bits have zero voltage and each 1 bit has a polarity opposite that of the preceding 1 bit. { bī'pō·lər 'fòr,mat }

Bipolarina [INV ZOO] A suborder of protozoan parasites in the order Myxosporida. { ,bī·pō·lə'rī·nə }

bipolar integrated circuit [ELECTR] An integrated circuit in which the principal element is the bipolar junction transistor. { bī'pō·lər 'in·tə,grād·əd 'sər·kət }

bipolar junction transistor [ELECTR] A bipolar transistor that is composed entirely of one type of semiconductor, silicon. Abbreviated BJT. Also known as silicon homojunction. { ¦bī,pōl·ər jəŋk·shən tran'zis·tər }

bipolar magnetic driving unit [ENG ACOUS] Headphone or loudspeaker unit having two magnetic poles acting directly on a flexible iron diaphragm. { bī'pō·lər mag'ned·ik 'driv·iŋ ,yü·nət }

bipolar memory [COMPUT SCI] A computer memory employing integrated-circuit bipolar junction transistors as bistable memory cells. { bī'pō·lər 'mem·rē }

bipolar nebula [ASTRON] A nebula consisting of two relatively symmetrical bright lobes with a star between them. { bī'pōl·ər 'neb·yə·lə }

bipolar power supply [ELEC] A high-precision, regulated, direct-current power supply that can be set to provide any desired voltage between positive and negative design limits, with a smooth transition from one polarity to the other. { bī'pō·lər 'paù·ər sə'plī }

bipolar signal [COMMUN] A signal in which different logical states are represented by electrical voltages of opposite polarity. { bī'pō·lər 'sig·nəl }

bipolar spin device See magnetic switch. { ¦bī,pō·lər 'spin di,vīs }

BIPEDAL DINOSAUR

Restored bipedal carnivorous saurischian *Allosaurus* (about 40 feet or 12 meters long), Late Jurassic. North America.

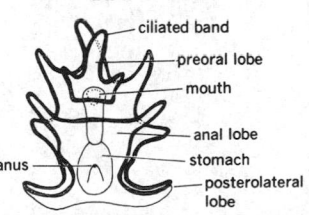

BIPINNARIA

ciliated band — preoral lobe — mouth — anal lobe — stomach — posterolateral lobe — anus

Bipinnaria larva showing component parts.

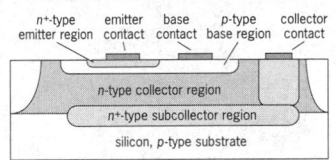

BIPOLAR INTEGRATED CIRCUIT

n^+-type emitter region — emitter contact — base contact — p-type base region — collector contact — n-type collector region — n^+-type subcollector region — silicon, p-type substrate

Isolated n^+ pn bipolar junction transistor for integrated-circuit operation.

bipolar spin switch *See* magnetic switch. { ¦bī¦pō·lər 'spin ¦swich }

bipolar transistor [ELECTR] A transistor that uses both positive and negative charge carriers. { bī'pō·lər tranz'is·tər }

bipotential [BIOL] Having the potential to develop in either of two mutually exclusive directions. { ¦bī·pə'ten·chəl }

bipotential electrostatic lens [ELECTR] An electron lens in which image and object space are field-free, but at different potentials; examples are the lenses formed between apertures of cylinders at different potentials. Also known as immersion electrostatic lens. { ¦bī·pə'ten·chəl i‚lek·trə'stad·ik 'lenz }

bipotentiality [BIOL] **1.** Capacity to function either as male or female. **2.** Hermaphroditism. { ¦bī·pə‚ten·chē'al·əd·ē }

biprism [OPTICS] A prism with apex angle only a little less than 180°, which produces a double image of a point source, giving rise to interference fringes on a nearby screen. { 'bī‚priz·əm }

biprism interference [OPTICS] Light interference fringes seen on a screen near a biprism. { 'bī‚priz·əm‚int·ər'fir·əns }

bipropellant [MATER] A rocket propellant consisting of two unmixed or uncombined chemicals (fuel and oxidizer) fed to the combustion chamber separately. { ¦bī·prə'pel·ənt }

2,2′-bipyridine *See* 2,2′-dipyridyl. { ¦tü ¦tü·prīm ¦bī'pir·ə‚dēn }

bipyramid [CRYSTAL] A crystal having the form of two pyramids that meet at a plane of symmetry. Also known as dipyramid. { ¦bī¦pir·ə‚mid }

biquadratic [MATH] Any fourth-degree algebraic expression. Also known as quartic. { ¦bī·kwə'drad·ik }

biquadratic equation *See* quartic equation. { ¦bī·kwə'drad·ik i'kwā·zhən }

biquartic filter [ELECTR] An active filter that uses operational amplifiers in combination with resistors and capacitors to provide infinite values of Q and simple adjustments for bandpass and center frequency. { ¦bī¦kwórd·ik 'fil·tər }

biquartz [OPTICS] A device consisting of two adjoining pieces of quartz of equal thickness that rotate the plane of polarization of light in opposite directions; used with a Nicol prism or other analyzer to increase the accuracy of the latter in determining the properties of polarized light. { 'bī‚kwórts }

biquinary abacus [MATH] An abacus in which the frame is divided into two parts by a bar which separates each wire into two- and five-counter segments. { bī'kwin·ə·rē 'ab·ə·kəs }

biquinary notation [MATH] A mixed-base notation system in which the first of each pair of digits counts 0 or 1 unit of five, and the second counts 0, 1, 2, 3, or 4 units. Also known as biquinary number system. { bī'kwin·ə·rē nō'tā·shən }

biquinary number system *See* biquinary notation. { bī'kwin·ə·rē 'nəm·bər ‚sis·təm }

biradial symmetry [BIOL] Symmetry both radial and bilateral. Also known as disymmetry. { bī'rād·ē·əl 'sim·ə·trē }

biradical [CHEM] A chemical species having two independent odd-electron sites. { bī'rad·ə·kəl }

biramous [BIOL] Having two branches, such as an arthropod appendage. { bī'rā·məs }

birch [BOT] The common name for all deciduous trees of the genus *Betula* that compose the family Betulaceae in the order Fagales. { bərch }

birch tar oil [MATER] A toxic liquid mixture of guaiacol, phenols, cresol, xylenol, and creosol; derived from distillation of birch tar and by dry distillation of the wood of *Betula alba*; used as a disinfectant, in leather dressing, and in medicine. { 'bərch ‚tär ‚oil }

bird [VERT ZOO] Any of the warm-blooded vertebrates which make up the class Aves. { bərd }

birdcaged wire [ENG] Wire rope whose strands have been distorted into the shape of a birdcage by a sudden release of a load during a hoisting operation. { 'bərd‚kājd ‚wīr }

bird-hipped dinosaur [PALEON] Any member of the order Ornithischia, distinguished by the birdlike arrangement of their hipbones. { 'bərd ‚hipt 'dīn·ə‚sór }

bird louse [INV ZOO] The common name for any insect of the order Mallophaga. Also known as biting louse. { 'bərd ‚laús }

birdnesting [ORD] Clumping together of chaff dipoles after they have been dispensed from an aircraft. { 'bərd‚nest·iŋ }

bird of prey [VERT ZOO] Any of various carnivorous birds of the orders Falconiformes and Strigiformes which feed on meat taken by hunting. { 'bərd əv 'prā }

birdseye [MATER] A small localized area in wood in which the fibers are indented and otherwise contorted to form few to many circular or elliptical figures on the tangential surface. [TEXT] A fabric woven on a Dobby loom that is characterized by diamond-shaped spots. { 'bərdz‚ī }

bird's-eye rot [PL PATH] A fungus disease of the grape caused by *Elsinoe ampelina* and characterized by small, dark, sunken spots with light centers on the fruit. { 'bərdz 'ī ‚rät }

bird's-eye spot [PL PATH] **1.** A plant disease characterized by dark, round spots surrounded by lighter tissue. **2.** A fungus disease of tea leaves caused by *Orcospora theae*. **3.** A leaf spot of the hevea rubber tree caused by the fungus *Helminthosporium heveae*. { 'bərdz 'ī ‚spät }

bird's-foot delta [GEOL] A delta with long, projecting distributary channels that branch outward like the toes or claws of a bird. { 'bərdz ‚füt 'del·tə }

birectangular [MATH] Property of a geometrical object that has two right angles. { ¦bī·rek'taŋ·gyə·lər }

birefringence [OPTICS] **1.** Splitting of a light beam into two components, which travel at different velocities, by a material. **2.** For a light beam that has been split into two components by a material, the difference in the indices of refraction of the components within the material. Also known as double refraction. { ‚bī·ri'frin·jəns }

birefringent filter [OPTICS] A filter consisting of alternate layers of polarizing films and plates cut from a birefringent crystal; transmits light in a series of sharp, widely spaced wavelength bands. Also known as Lyot filter; monochromatic filter. { ‚bī·ri'frin·jənt 'fil·tər }

birefringent plate [OPTICS] A piece of birefringent optical material with parallel plane surfaces. { ‚bī·ri'frin·jənt 'plāt }

Birge-Mieck rule [ATOM PHYS] The rule that the product of the equilibrium vibrational frequency and the square of the internuclear distance is a constant for various electronic states of a diatomic molecule. { 'bir·gə 'mēk ‚rül }

Birge-Sponer extrapolation [SPECT] A method of calculating the dissociation limit of a diatomic molecule when the convergence limit cannot be observed directly, based on the assumption that vibrational energy levels converge to a limit for a finite value of the vibrational quantum number. { ¦bir·gə 'spōn·ər ik‚strap·ə'lā·shən }

birimose [BOT] Opening by two slits, as an anther. { bī 'rī‚mōs }

Birkeland-Eyde process [CHEM ENG] An arc process of nitrogen fixation in which air passes through an alternating-current arc flattened by a magnetic field to form about 1% nitric oxide. { ¦bərk·lənd ¦ī·də 'präs·əs }

Birkhoff's theorem [RELAT] A theorem which states that if a space-time containing matter or energy satisfies Einstein's equations of general relativity and is centrally symmetric, then it is necessarily static and under a coordinate transformation it becomes identical to the Schwarzschild solution. { 'bərk‚hófs ‚thir·əm }

Birkhoff-von Neumann theorem [MATH] The theorem that a matrix is doubly stochastic if and only if it is a convex combination of permutation matrices. { 'bər‚hóf fón 'nói·män ‚thir·əm }

Birmingham wire gage [DES ENG] A system of standard sizes of brass wire, telegraph wire, steel tubing, seamless tubing, sheet spring steel, strip steel, and steel plates, bands, and hoops. Abbreviated BWG. { 'bər·miŋ·əm 'wīr ‚gāj }

birnessite [MINERAL] A manganese oxide mineral often found as a primary constituent of manganese nodules or crusts. { 'bər'nes·īt }

birotulate [INV ZOO] A sponge spicule characterized by two wheel-shaped ends. { bī'räch·ə‚lāt }

birth [BIOL] The emergence of a new individual from the body of its parent. { bərth }

birth canal [ANAT] The channel in mammals through which the fetus is expelled during parturition; consists of the cervix, vagina, and vulva. { 'bərth kə'nal }

birth control [MED] Limitation of the number of children born by preventing or reducing the frequency of impregnation. { bərth kən'trōl }

birth-death process [IND ENG] A simple queuing model in which units to be served arrive (birth) and depart (death) in a completely random manner. [STAT] A method for describing the size of a population in which the population increases or decreases by one unit or remains constant over short time periods. { 'bərth ¦deth ¦prä,səs }

birth defect *See* congenital anomaly. { 'bərth di'fekt }

birthmark [MED] Any abnormal cellular or vascular benign nevus that is present at birth or that appears sometime later. { 'bərth,märk }

birth process [STAT] A stochastic process that defines a population whose members may have offspring; usually applied to the case where the population increases by one. { 'bərth ,prä,ses }

birth rate [BIOL] The ratio between the number of live births and a specified number of people in a population over a given period of time. { 'bərth ,rāt }

bis- [CHEM] A prefix indicating doubled or twice. { bis }

2,2-bis(para-chlorophenyl)-1,1-dichloroethane [ORG CHEM] $C_{14}H_{10}Cl_4$ A colorless, crystalline compound with a melting point of 109–111°C; insoluble in water; used as an insecticide on fruits and vegetables. Also known as DDD; TDE. { ¦tü ¦tü ¦bis 'par·ə ¦klōr·ə'fen·əl ¦wən ¦wən di ¦klōr·ō'e,thän }

bischofite [MINERAL] $MgCl_2 \cdot 6H_2O$ A colorless to white, monoclinic mineral consisting of magnesium chloride hexahydrate. { 'bish·ə,fīt }

biscuit [ENG ACOUS] *See* preform. [MATER] **1.** A clay object that has been fired once prior to glazing. **2.** Pottery that is unglazed in its final form. [MET] An upset blank for drop forging. { 'bis·kət }

biscuit cutter [MIN ENG] A short (6–8 inches or 15–20 centimeters) core barrel that is sharpened at the bottom and forced into the rocks by the jars. { 'bis·kət ,kəd·ər }

bise [METEOROL] A cold, dry wind which blows from a northerly direction in the winter over the mountainous districts of southern Europe. Also spelled bize. { bēz }

bisection algorithm [MATH] A procedure for determining the root of a function to any desired accuracy by repeatedly dividing a test interval in half and then determining in which half the value of the function changes sign. { bī'sek·shən 'al·gə,rith·əm }

bisector [MATH] The ray dividing an angle into two equal angles. { ,bī'sek·tər }

bisectrix [CRYSTAL] A line that is the bisector of the angle between the optic axes of a biaxial crystal. { ,bī'sek,triks }

biserial [BIOL] Arranged in two rows or series. { ,bī'sir·ē·əl }

biserial correlation coefficient [STAT] A measure of the relationship between two qualities, one of which is a measurable random variable and the other a variable which is dichotomous, classified according to the presence or absence of an attribute; not a product moment correlation coefficient. { ¦bī¦sir·ē·əl ,kär·ə'lā·shən ,kō·ə,fish·ənt }

biserrate [BIOL] **1.** Having serrated serrations. **2.** Serrate on both sides. { bī'ser,āt }

bisexual [BIOL] Of or relating to two sexes. [PSYCH] **1.** Possessing mental and behavioral characteristics of both sexes. **2.** Having sexual desires for members of both sexes. { ,bī'sek·shə·wəl }

Bishop's ring [METEOROL] A faint, broad, reddish-brown corona occasionally seen in dust clouds, especially those which result from violent volcanic eruptions. { 'bish·əps 'riŋ }

bisilicate [MET] A type of slag whose silicate degree is 2. [MINERAL] *See* metasilicate. { ,bī'sil·ə·kət }

bismanol [MET] A magnetic alloy of bismuth and manganese. { 'biz·mə,nól }

bismite [MINERAL] Bi_2O_3 A monoclinic mineral composed of bismuth trioxide; native bismuth ore, occurring as a yellow earth. Also known as bismuth ocher. { 'biz,mīt }

bismuth [CHEM] A metallic element, symbol Bi, of atomic number 83 and atomic weight 208.980. [MINERAL] The brittle, rhombohedral mineral form of the native element bismuth. { 'biz·məth }

bismuth alloy [MET] A group of low-melting alloys (many below 100°C) of bismuth combined with lead, tin, and cadmium; used in automatic sprinklers, special solders, safety plugs in compressed-gas cylinders, automatic shutoffs for water-heating systems, castings, and type metal. { 'biz·məth 'al,ói }

bismuthate [INORG CHEM] A compound of bismuth in which the bismuth has a valence of +5; an example is sodium bismuthate, $NaBiO_3$. { 'biz·mə,thāt }

bismuth blende *See* eulytite. { 'biz·məth ¦blend }

bismuth carbonate *See* bismuth subcarbonate. { 'bis·məth 'kär·bə,nāt }

bismuth chloride [INORG CHEM] $BiCl_3$ A deliquescent material that melts at 230–232°C and decomposes in water to form the oxychloride; used to make bismuth salts. Also known as bismuth trichloride. { 'biz·məth 'klōr,īd }

bismuth chromate [INORG CHEM] $Bi_2O_3 \cdot Cr_2O_3$ An orange-red powder, soluble in alkalies and acids; used as a pigment. { 'biz·məth 'krō,māt }

bismuth citrate [ORG CHEM] $BiC_6H_5O_7$ A salt of citric acid that forms white crystals, insoluble in water; used as an astringent. { 'biz·məth 'sī,trāt }

bismuth germinate detector [NUCLEO] A high-efficiency, low-resolution detector of gamma rays that uses bismuth germinate ($Bi_4Ge_3O_{12}$), an intrinsic scintillator, whose large gamma-ray absorption coefficient makes possible a reduction in detector size. { 'biz·məth ¦jer·mə,nāt di,tek·tər }

bismuth glance *See* bismuthinite. { 'biz·məth ¦glans }

bismuth hydroxide [INORG CHEM] $Bi(OH)_3$ A water-insoluble, white powder; precipitated by hydroxyl ion from bismuth salt solutions. { 'biz·məth hī'dräk,sīd }

bismuthinite [MINERAL] Bi_2S_3 A mineral consisting of bismuth trisulfide, which has an orthorhombic structure and is usually found in fibrous or leafy masses that are lead gray with a yellowish tarnish and a metallic luster. Also known as bismuth glance. { 'biz·məth·ə,nīt }

bismuth iodide [INORG CHEM] BiI_3 A bismuth halide that sublimes in grayish-black hexagonal crystals melting at 408°C, insoluble in water; used in analytical chemistry. { 'biz·məth 'ī·ə,dīd }

bismuth nitrate [INORG CHEM] $Bi(NO_3)_3 \cdot 5H_2O$ White, triclinic crystals that decompose in water; used as an astringent and antiseptic. { 'biz·məth 'nī,trāt }

bismuth ocher *See* bismite. { 'biz·məth 'ō·kər }

bismuth oleate [ORG CHEM] $Bi(C_{17}H_{33}COO)_3$ A salt of oleic acid obtained as yellow granules; used in medicines to treat skin diseases. { 'biz·məth 'ō·lē,āt }

bismuth oxide *See* bismuth trioxide. { 'biz·məth 'äk,sīd }

bismuth oxycarbonate *See* bismuth subcarbonate. { 'biz·məth ,äk·sē'kär·bə,nāt }

bismuth oxychloride [INORG CHEM] $BiOCl$ A white powder; insoluble in water, soluble in acid; a toxic material if ingested; used in pigments and cosmetics. { 'biz·məth ,äk·sē'klōr,īd }

bismuth phenate [ORG CHEM] $C_6H_5O \cdot Bi(OH)_2$ An odorless, tasteless, gray-white powder; used in medicine. { 'biz·məth 'fen,āt }

bismuth potassium tartrate *See* potassium bismuth tartrate. { 'biz·məth pe'tas·ē·əm 'tär,trāt }

bismuth pyrogallate [ORG CHEM] $Bi(OH)C_6H_3(OH)O_2$ An odorless, tasteless, yellowish-green, amorphous powder; used in medicine as intestinal antiseptic and dusting powder. { 'biz·məth ¦pī·rō'gal,āt }

bismuth spar *See* bismutite. { 'biz·məth 'spär }

bismuth subcarbonate [INORG CHEM] $(BiO)_2CO_3$ or $Bi_2O_3 \cdot CO_2 \cdot \frac{1}{2}H_2O$ A white powder; dissolves in hydrochloric or nitric acid, insoluble in alcohol and water; used as opacifier in x-ray diagnosis, in ceramic glass, and in enamel fluxes. { 'biz·məth səb'kär·bə,nāt }

bismuth subgallate [ORG CHEM] $C_6H_2(OH)_3COOBi(OH)_2$ A yellow powder; dissolves in dilute alkali solutions, but is insoluble in water, ether, and alcohol; used in medicine. { 'biz·məth ,səb'gal,āt }

bismuth subnitrate [INORG CHEM] $4BiNO_3(OH)_2 \cdot BiO(OH)$ A white, hygroscopic powder; used in bismuth salts, perfumes, cosmetics, ceramic enamels, pharmaceuticals, and analytical chemistry. { 'biz·məth ,səb'ni,trāt }

bismuth subsalicylate [INORG CHEM] $Bi(C_7H_5O)_3Bi_2O_3$ A white powder that is insoluble in ethanol and water; used in medicine and as a fungicide for tobacco crops. { 'biz·məth ,səb·sə'lis·ə,lāt }

bismuth telluride [INORG CHEM] Bi_2Te_3 Gray, hexagonal platelets with a melting point of 573°C; used for semiconductors, thermoelectric cooling, and power generation applications. { 'biz·məth 'tel·yə,rīd }

bismuth trichloride *See* bismuth chloride.

bismuth trioxide [INORG CHEM] Bi_2O_3 A yellow powder; melting point 820°C; insoluble in water, dissolves in acid;

used to make enamels and to color ceramics. Also known as bismuth oxide; bismuth yellow. { 'biz·məth trī'äk,sīd }

bismuth yellow *See* bismuth trioxide. { 'biz·məth 'yel·ō }

bismutite [MINERAL] $(BiO)_2CO_3$ A dull-white, yellowish, or gray, earthy, amorphous mineral consisting of basic bismuth carbonate. Also known as bismuth spar. { 'biz·məd,īt }

bismutotantalite [MINERAL] $Bi(Ta,Nb)O_4$ A pitch black, orthorhombic mineral consisting of an oxide of bismuth and tantalum and occurring in crystals. { ,biz·məd·ə'tan·tə,līt }

bison [VERT ZOO] The common name for two species of the family Bovidae in the order Artiodactyla; the wisent or European bison (*Bison bonasus*), and the American species (*Bison bison*). { 'bīs·ən }

bisphenoid [CRYSTAL] A form apparently consisting of two sphenoids placed together symmetrically. { bī'sfē,noid }

bisphenol A [ORG CHEM] $(CH_3)_2C(C_6H_5OH)_2$ Brown crystals that are insoluble in water; used in the production of phenolic and epoxy resins. { bī'sfēn·ól 'ā }

bisporangiate [BOT] Having two different types of sporangia. { ,bī·spə'ran·jē,āt }

bispore [BOT] In certain red algae, an asexual spore that is produced in pairs. { 'bī,spór }

bistable [SCI TECH] Capable of assuming either of two stable states. { ,bī'stā·bəl }

bistable circuit [ELECTR] A circuit with two stable states such that the transition between the states cannot be accomplished by self-triggering. { ,bī'stā·bəl ,sar·kət }

bistable magnetic core [ELECTR] A magnetic core that can be in either of two possible states of magnetization. Also known as bimag core. { ,bī'stā·bəl mag'ned·ik 'kór }

bistable multivibrator [ELECTR] A multivibrator in which either of the two active devices may remain conducting, with the other nonconducting, until the application of an external pulse. Also known as Eccles-Jordan circuit; Eccles-Jordan multivibrator; flip-flop circuit; trigger circuit. { ,bī'stā·bəl məl·ti'vī,brād·ər }

bistable optical device [OPTICS] A device which can be in either of two stable states of optical transmission for a single value of the input light intensity.

bistable system [CHEM] A chemical system with two relatively stable states which permits an oscillation between domination by one of these states to domination by the other. { ,bī'stā·bəl 'sis·təm }

bistable unit [ENG] A physical element that can be made to assume either of two stable states; a binary cell is an example. { ,bī'stā·bəl 'yü·nət }

bistatic radar [ENG] Radar system in which the receiver is some distance from the transmitter, with separate antennas for each. { 'bī,stad·ik 'rā,där }

bistatic reflectivity [OPTICS] The characteristic of a reflector which reflects energy along a line, or lines, different from or in addition to that of the incident ray. { 'bī,stad·ik rē,flek'·tiv·əd·ē }

bisulfate [INORG CHEM] A compound that has the HSO_4^- radical; derived from sulfuric acid. { bī'səl,fāt }

bisynchronous transmission [COMMUN] A set of procedures for handling synchronous transmission of data and, in particular, for handling a block of data, called a message format, that is transmitted in a single operation. { bī'siŋ·krə·nəs tranz·'mish·ən }

bit [COMPUT SCI] **1.** A unit of information content equal to one binary decision or the designation of one of two possible and equally likely values or states of anything used to store or convey information. **2.** A dimensionless unit of storage capacity specifying that the capacity of a storage device is expressed by the logarithm to the base 2 of the number of possible states of the device. [DES ENG] **1.** A machine part for drilling or boring. **2.** The cutting plate of a plane. **3.** The blade of a cutting tool such as an ax. **4.** A removable tooth of a saw. **5.** Any cutting device which is attached to or part of a drill rod or drill string to bore or penetrate rocks. [MATH] In a pure binary numeration system, either of the digits 0 or 1. Also known as bigit; binary digit. [MET] In soldering, the portion of the iron that transfers either heat or solder to the joint involved. { bit }

bitangent *See* double tangent. { bī'tan·jənt }

bitartrate [ORG CHEM] A salt with the radical $HC_4H_4O_6^-$. Also known as acid tartrate. { bī'tär,trāt }

bit blank [DES ENG] A steel bit in which diamonds or other cutting media may be inset by hand peening or attached by a mechanical process such as casting, sintering, or brazing. Also known as bit shank; blank; blank bit; shank. { 'bit ,blaŋk }

bit block transfer [COMPUT SCI] In computer graphics, a hardware function that moves a rectangular block of bits from the main memory to the display memory at high speed. Abbreviated bitblt. { 'bit ,bläk 'tranz·fər }

bitblt *See* bit block transfer.

bit breaker [DES ENG] A heavy plate that fits in a rotary table for holding the drill bit while it is being inserted or broken out of the drill stem. { 'bit ,bräk·ər }

bit buffer unit [COMMUN] A unit that terminates bit-serial communications lines coming from and going to technical control. { 'bit 'bəf·ər ,yü·nət }

bit cone *See* roller cone bit. { 'bit ,kōn }

bit count appendage [COMPUT SCI] One of the two-byte elements replacing the parity bit stripped off each byte transferred from main storage to disk volume (the other element is the cyclic check); these two elements are appended to the block during the write operation; on a subsequent read operation these elements are calculated and compared to the appended elements for accuracy. { 'bit ,kaúnt ə'pen·dij }

bit density [COMPUT SCI] Number of bits which can be placed, per unit length, area, or volume, on a storage medium; for example, bits per inch of magnetic tape. Also known as record density. { 'bit 'den·səd·ē }

bit depth [COMPUT SCI] In a digital file, the number of colors for an image; calculated as 2 to the power of the bit depth; for example, a bit depth of 8 supports up to 256 colors, and a bit depth of 24 supports up to 16 million colors. { 'bit ,depth }

bit drag [DES ENG] A rotary-drilling bit that has serrated teeth. Also known as drag bit. { 'bit ,drag }

bite [BIOL] **1.** To seize with the teeth. **2.** Closure of the lower teeth against the upper teeth. [ENG] In glazing, the length of overlap of the inner edge of a frame over the edge of the glass. [GRAPHICS] In photoengraving, the various stages of etching accomplished through the action of acid. [MED] Skin injury produced by an animal's teeth or the mouthparts of an insect. { bīt }

bitegmic [BOT] Having two integuments, especially in reference to ovules. { bī'teg·mik }

biternate [BOT] Of a ternate leaf, having each division ternate. { bī'tər,nāt }

bitewing [MED] A dental x-ray film having a central wing on which the teeth can close to hold it in place. { 'bīt,wiŋ }

bit flipping *See* bit manipulation. { 'bit ,flip·iŋ }

bithionol [ORG CHEM] A halogenated form of bisphenol used as an ingredient in germicidal soaps and as a medicine in the treatment of clonorchiases. { bī'thī·ə,nól }

biting angle [ORD] Smallest angle of impact at which a projectile will penetrate or pierce armor. { 'bīd·iŋ ,aŋ·gəl }

biting-in [GRAPHICS] In etching, the action of dilute nitric acid or other mordant upon the bare steel or copper plate. { 'bīd·iŋ 'in }

biting louse *See* bird louse. { 'bīd·iŋ ,laús }

bit location [COMPUT SCI] Storage position on a record capable of storing one bit. { 'bit lō'kā·shən }

bit manipulation [COMPUT SCI] Changing bits from one state to the other, usually to influence the operation of a computer program. Also known as bit flipping. { 'bit mə,nip·yə'lā,shən }

bit-mapped font [COMPUT SCI] A font that is specified by a complete set of dot patterns for each character and symbol. { 'bit ,mapt 'fänt }

bit-mapped graphics *See* raster graphics. { 'bit ,mapt 'graf·iks }

bit mapping [COMPUT SCI] The assignment of each location in a computer's storage to a physical location on an electronic display. { 'bit 'map·iŋ }

bit matrix [ENG] The material, usually powdered and fused tungsten carbide, into which diamonds are set in the manufacture of diamond bits. { 'bit ,mā·triks }

bit-oriented protocol [COMMUN] A communications protocol in which individual bits within a byte are used as control codes. { 'bit ,ór·ē,ent·əd 'prōd·ə,kól }

Bitot's spots [MED] The silver-gray, shiny, triangular spots on the cornea characteristic of xerosis conjunctivae. { 'bē·tōs ,späts }

BISON

North American bison (*Bison bison*), with enormous forequarter development and hump behind the head.

BISTABLE MULTIVIBRATOR

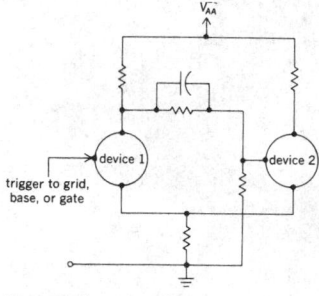

Circuit diagram of an unsymmetrical bistable multivibrator.

bit parallel [COMMUN] Simultaneous transmission of character-forming bits over parallel paths. { 'bit 'par·ə,lel }

bit pattern [COMPUT SCI] A combination of binary digits arranged in a sequence. { 'bit ,pad·ərn }

bit per second [COMMUN] A unit specifying the instantaneous speed at which a device or channel transmits data. Abbreviated bps. { 'bit pər 'sek·ənd }

bit position [COMPUT SCI] The position of a binary digit in a word, generally numbered from the least significant bit. { 'bit pa'zish·ən }

bit rate [COMMUN] Quantity, per unit time, of binary digits (or pulses representing them) which will pass a given point on a communications line or channel in a continuous stream. { 'bit ,rāt }

bitrochanteric width [IND ENG] A measurement corresponding to hip breadth that is used in seating design. { ,bī·trə,kan'ter·ik 'width }

bit serial [COMMUN] Sequential transmission of character-forming bits. { 'bit 'sir·ē·əl }

bit shank *See* bit blank. { 'bit ,shaŋk }

bit-sliced microprocessor [COMPUT SCI] A microprocessor in which the major logic of the central processor is partitioned into a set of large-scale-integration circuits, as opposed to being placed on a single chip. { 'bit ,slīst ,mī·kro'präs·əs·ər }

bit stream [COMPUT SCI] **1.** A consecutive line of bits transmitted over a circuit in a transmission method in which character separation is accomplished by the terminal equipment. **2.** A binary signal without regard to grouping by character. { 'bit ,strēm }

bit-stream generator [COMMUN] An algorithmic procedure for producing an unending sequence of binary digits to implement a stream. { 'bit ,strēm 'jen·ə,rād·ər }

bit string [COMPUT SCI] A set of consecutive binary digits representing data in coded form, in which the significance of each bit is determined by its position in the sequence and its relation to the other bits. { 'bit ,striŋ }

bit stuffing [COMMUN] The insertion of extra bits in a transmitted message in order to fill a frame to a fixed size or to break up a pattern of bits that could be mistaken for control codes. { 'bit ,stəf·iŋ }

bit sub [PETRO ENG] A sub designed to be inserted between the drill collar and the bit. { 'bit ,səb }

bit synchronization [COMMUN] Element of a message header used to synchronize all of the bits and characters that follow. { 'bit ,siŋ·krə·nə'zā·shən }

bitt [NAV ARCH] **1.** A short metal or wood post on the deck of a ship used to secure mooring or other lines; usually in a pair. **2.** To secure a line about a ship's bitt. { bit }

bitter end [NAV ARCH] The end of a line or cable, especially the inboard end. { 'bid·ər 'end }

bitter lake [HYD] A lake rich in alkaline carbonates and sulfates. { 'bid·ər 'lāk }

Bitter magnet [ELECTROMAG] A normal-conductor air-cored magnet consisting of stacked circular copper plates with a number of holes through which cooling water passes. { 'bid·ər 'mag·nət }

bittern [CHEM ENG] Concentrated sea water or brine containing the bromides and magnesium and calcium salts left in solution after sodium chloride has been removed by crystallization. [VERT ZOO] Any of various herons of the genus *Botaurus* characterized by streaked and speckled plumage. { 'bid·ərn }

bitter orange oil [MATER] An essential oil from the rind or peel of the orange (*Citrus vulgaris*); insoluble in water; used as a flavoring and in some perfumes. Also known as orange oil. { 'bid·ər 'är·inj ,óil }

Bitter pattern [SOLID STATE] A pattern produced when a drop of a colloidal suspension of ferromagnetic particles is placed on the surface of a ferromagnetic crystal; the particles collect along domain boundaries at the surface. { 'bid·ər ,pad·ərn }

bitter pit [PL PATH] A disease of uncertain etiology affecting apple, pear, and quince; spots of dead brown tissue appear in the flesh of the fruit and discolored depressions are seen on its surface. Also known as Baldwin spot; stippen. { 'bid·ər ,pit }

bitter rot [PL PATH] A fungus disease of apples, grapes, and other fruit caused by *Glomerella cingulata*. { 'bid·ər ,rät }

bit test [COMPUT SCI] A check by a computer program to determine the status of a particular bit. { 'bit ,test }

Bittner milk factor *See* milk factor. { 'bit·nər 'milk ,fak·tər }

bitudobe [MATER] A sun-baked brick composed of adobe and a binder of emulsified asphalt. { 'bich·yü,dō·bē }

bitumastic [MATER] A combination of asphalt and filler used mainly to coat metals to protect them against corrosion and weathering. { ,bi·tyü'mas·tik }

bitumen [MATER] Naturally occurring or pyrolytically obtained substances of dark to black color consisting almost entirely of carbon and hydrogen, with very little oxygen, nitrogen, or sulfur. { bī'tü·mən }

bitumenite *See* torbanite. { bī'tü·mə,nīt }

bituminization *See* coalification. { bī,tü·mə·nə'zā·shən }

bituminous [MATER] **1.** Containing much organic, or at least carbonaceous, matter, mostly in the form of tarry hydrocarbons which are usually described as bitumen. **2.** Similar to bitumen. **3.** Giving off volatile bituminous substances on heating, as in bituminous coal. [MINERAL] Of a mineral, having the odor of bitumen. { bī'tü·mə·nəs }

bituminous cement [MATER] A bituminous material suitable for use as a binder, having cementing qualities which are dependent mainly on its bituminous character. { bī'tü·mə·nəs si'ment }

bituminous coal [GEOL] A dark brown to black coal that is high in carbonaceous matter and has 15–50% volatile matter. Also known as soft coal. { bī'tü·mə·nəs 'kōl }

bituminous coating [MATER] A coating made principally of bituminous material and used as a surfacing for roads and as a water-repellent barrier in buildings. { bī'tü·mə·nəs 'kōd·iŋ }

bituminous concrete [MATER] A concrete made with bituminous material as a binder for sand and gravel. { bī'tü·mə·nəs kän'krēt }

bituminous distributor [MECH ENG] A tank truck having a perforated spray bar and used for pumping hot bituminous material onto the surface of a road or driveway. { bī'tüm·ə·nəs dis'trib·yəd·ər }

bituminous grout [MATER] A mixture of bitumen and an aggregate, such as sand, that liquefies when heated and cures in air and is used as a sealant for joints or cracks. { bī'tüm·ə·nəs 'graut }

bituminous lignite [GEOL] A brittle, lustrous bituminous coal. Also known as pitch coal. { bī'tü·mə·nəs 'lig,nīt }

bituminous paint [MATER] Paint with a high proportion of bitumen; usually dark in color. { bī'tü·mə·nəs 'pānt }

bituminous rock *See* asphalt rock. { bī'tü·mə·nəs 'räk }

bituminous sand [GEOL] Sand containing bituminous-like material, such as the tar sands at Athabasca, Canada, from which oil is extracted commercially. { bī'tü·mə·nəs 'sand }

bituminous sandstone [PETR] A sandstone containing bituminous matter. { bī'tü·mə·nəs 'sand,stōn }

bituminous shale [PETR] A shale containing bituminous material. { bī'tü·mə·nəs 'shāl }

bituminous wood [GEOL] A variety of brown coal having the fibrous structure of wood. Also known as board coal; wood coal; woody lignite; xyloid coal; xyloid lignite. { bī'tü·mə·nəs 'wud }

bit zone [COMPUT SCI] **1.** One of the two left-most bits in a commonly used system in which six bits are used for each character; related to overpunch. **2.** Any bit in a group of bit positions that are used to indicate a specific class of items; for example, numbers, letters, special signs, and commands. { 'bit ,zōn }

biunique correspondence [MATH] A correspondence that is one to one in both directions. { bī'yü,nēk ,kär·ə'spän·dəns }

biuret [ORG CHEM] $NH_2CONHCONH_2$ Colorless needles that are soluble in hot water and decompose at 190°C; a condensation product of urea. { ,bī·ya'ret }

bivalent [CHEM] Possessing a valence of two. { bī'vā·lənt }

bivalent antibody [IMMUNOL] An antibody possessing two antibody combining sites. { bī'vāl·ənt 'ant·i,bäd·ē }

bivalent chromosome [CYTOL] The structure formed following synapsis of a pair of homologous chromosomes from the zygotene stage of meiosis up to the beginning of anaphase. { bī'vā·lənt 'krō·mə,sōm }

bivalve [INV ZOO] The common name for a number of

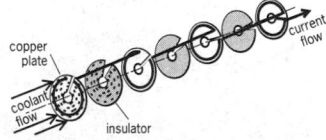

diverse, bilaterally symmetrical animals, including mollusks, ostracod crustaceans, and brachiopods, having a soft body enclosed in a calcareous two-part shell. { 'bī,valv }

Bivalvia [INV ZOO] A large class of the phylum Mollusca containing the clams, oysters, and other bivalves. { bī'val· vē·ə }

bivane [ENG] A double-jointed vane which measures vertical as well as horizontal wind direction. { 'bī,vān }

bivariate distribution [STAT] The joint distribution of a pair of variates for continuous or discontinuous data. { bī'ver·ē· ət ,dis·trə'byü·shən }

biventer [ANAT] A muscle having two bellies. { bī'ven· tər }

bivittate [ZOO] Having a pair of longitudinal stripes. { bī'vi,tāt }

bivium [INV ZOO] The pair of starfish rays that extend on either side of the madreporite. { 'bī·vē·əm }

bivoltine [INV ZOO] **1.** Having two broods in a season, used especially of silkworms. **2.** Of insects, producing two generations a year. { bī'vōl,tēn }

bivouac [ORD] An encampment set up for a short period of time by using tents. { 'biv·ə,wak }

bixbyite [MINERAL] $(Mn,Fe)_2O_3$ A manganese-iron oxide mineral; black cubic crystals found in cavities in rhyolite. Also known as partridgeite; sitaparite. { 'biks·bē,īt }

bixin [ORG CHEM] $C_{25}H_{30}O_4$ A carotenoid acid occurring in the seeds of *Bixa orellano;* used as a fat and food coloring agent. { 'bik·sən }

bize *See* bise. { bēz }

BJT *See* bipolar junction transistor.

Bk *See* berkelium.

BL *See* base line.

Blaauw mechanism [ASTRON] An explanation for the disruption of a binary system as being due to the decrease in the gravitational binding force when a shell of gas ejected by the primary component overtakes the secondary. { 'blö ,mek· ə,niz·əm }

black [CHEM] Fine particles of impure carbon that are made by the incomplete burning of carbon compounds, such as natural gas, naphthas, acetylene, bones, ivory, and vegetables. [COMMUN] *See* black signal. [OPTICS] Quality of an object which uniformly absorbs large percentages of light of all visible wavelengths. { blak }

black acids [MATER] Sulfonates in the sludge formed during treatment of petroleum products with sulfuric acid; soluble in water but insoluble in naphtha, benzene, and carbon tetrachloride. { 'blak 'as·ədz }

black alkali [GEOL] A deposit of sodium carbonate that has formed on or near the surface in arid to semiarid areas. { 'blak 'al·kə,lī }

black amber *See* jet coal. { 'blak 'am·bər }

black-and-white groups *See* Shubnikov groups. { 'blak ən 'wīt 'grüps }

black-and-white television *See* monochrome television. { 'blak ən 'wīt 'tel·ə,vizh·ən }

black annealing [MET] A type of box annealing used to impart a black color to the metal surface; first process in tin plating. { 'blak ə'nēl·iŋ }

black ash [MATER] A carbon product made by furnace heating of black liquor from papermaking processes. { 'blak 'ash }

black balsam *See* Peru balsam. { 'blak 'ból·səm }

blackband [GEOL] An earthy carbonate of iron that is present with coal beds. { 'blak,band }

black band disease [INV ZOO] A coral reef disease that is characterized by a thick black band of tissue that advances rapidly across infected corals, leaving empty coral skeletons behind. { 'blak,band di,zēz }

blackberry [BOT] Any of the upright or trailing shrubs of the genus *Rubus* in the order Rosales; an edible berry is produced by the plant. { 'blak,ber·ē }

blackbird [VERT ZOO] Any bird species in the family Icteridae, of which the males are predominantly or totally black. { 'blak,bərd }

black blight [PL PATH] Any of several diseases of tropical plants caused by superficial sooty molds. { 'blak ,blīt }

blackboard [MATER] A panel, usually black but sometimes colored, for writing on with chalk. Also known as chalkboard. { 'blak,bórd }

blackbody [THERMO] An ideal body which would absorb all incident radiation and reflect none. Also known as hohlraum; ideal radiator. { 'blak,bäd·ē }

blackbody radiation [THERMO] The emission of radiant energy which would take place from a blackbody at a fixed temperature; it takes place at a rate expressed by the Stefan-Boltzmann law, with a spectral energy distribution described by Planck's equation. { 'blak,bäd·ē ,rā·dē'ā·shən }

blackbody temperature [THERMO] The temperature of a blackbody that emits the same amount of heat radiation per unit area as a given object; measured by a total radiation pyrometer. Also known as brightness temperature. { 'blak,bäd·ē ,tem· prə·chər }

black box [ENG] Any component, usually electronic and having known input and output, that can be readily inserted into or removed from a specific place in a larger system without knowledge of the component's detailed internal structure. { 'blak ,bäks }

black-bulb thermometer [ENG] A thermometer whose sensitive element has been made to approximate a blackbody by covering it with lampblack. { 'blak ,bəlb thər'mäm·əd·ər }

black buran *See* karaburan. { 'blak bü'rän }

black canker *See* ink disease. { 'blak 'kaŋ·kər }

black carbon counter [NUCLEO] The original type of radiation counter used in radiocarbon dating, in which the sample, whose carbon has first been converted to carbon black, is mounted on the inside of a steel cylinder which is inserted into a sensitive Geiger counter. Also known as Libby counter. { 'blak 'kär·bən ,kaunt·ər }

black chaff [PL PATH] A bacterial disease of wheat caused by *Xanthomonas translucens undulosa* and characterized by dark, longitudinal stripes on the chaff. { 'blak 'chaf }

black coal *See* natural coke. { 'blak 'kōl }

black cobalt *See* asbolite. { 'blak 'kō·bólt }

black copper [MET] The more or less impure metallic copper (70–99% copper) produced in blast furnaces when running on oxide ores or roasted sulfide material. { 'blak 'käp·ər }

black coral [INV ZOO] The common name for antipatharian cnidarians having black, horny axial skeletons. { 'blak 'kär· əl }

black cotton soil *See* regur. { 'blak 'kat·ən 'sóil }

black cyanide *See* calcium cyanide. { 'blak 'sī·ə,nīd }

blackdamp [MIN ENG] A nonexplosive mixture of carbon dioxide with other gases, especially with 85–90% nitrogen, which is heavier than air and cannot support flame or life. Also known as chokedamp. { 'blak,damp }

black death *See* plague. { 'blak 'deth }

black diamond *See* carbonado. { 'blak 'dī·mənd }

black disease [VET MED] Necrotic hepatitis of sheep, resulting from infection with *Clostridium novyi* type B, with the necessary conditions for the growth of the clostridia provided by the damaged liver tissue produced by the fluke *Fasciola hepatica.* { 'blak di'zēz }

black drop [ASTRON] As seen through a telescope, an apparent dark elongation of the image of Venus or Mercury when the planets' images are at the sun's limb. { 'blak 'dräp }

black durain [GEOL] A durain that has high hydrogen content and volatile matter, many microspores, and some vitrain fragments. { 'blak 'dù,rān }

black dwarf *See* brown dwarf. { 'blak 'dwórf }

black end [PL PATH] **1.** A disease of the pear marked by blackening of the epidermis and flesh in the region of the calyx; believed to be a result of a disturbed water relation. **2.** A fungus disease of the banana caused by several species, especially *Gloesporium musarum,* characterized by discoloration of the stem of the fruit. { 'blak ,end }

blacker-than-black level [COMMUN] In television, a level of greater instantaneous amplitude than the black level, used for synchronization and control signals. { 'blak·ər thən 'blak ,lev·əl }

blackeye bean *See* cowpea. { 'blak,ī 'bēn }

blackfire [PL PATH] A bacterial disease of tobacco caused by *Pseudomonas angulata* and characterized by angular leaf spots which gradually darken and may fall out, leaving ragged holes. { 'blak,fīr }

black frost [HYD] A dry freeze with respect to its effects upon vegetation, that is, the internal freezing of vegetation unaccompanied by the protective formation of hoarfrost. Also known as hard frost. { 'blak 'fróst }

black granite *See* diorite. { 'blak 'gran·ət }

BLACKBERRY

Thorny, biennial stem of blackberry shrub.

black grease [MATER] Grease which has black coloration due to use of asphalt, either naturally occurring or from residues used in the manufacture of the grease. { 'blak 'grēs }

blackhead *See* comedo. { 'blak,hed }

blackhead disease [PL PATH] **1.** A parasitic disease of the banana caused by eelworms of the family Tylenchidae. **2.** A rot disease of the banana rootstock that is caused by the fungus *Thielaviopsis paradoxa*. { 'blak,hed di'zēz }

blackheart malleable iron *See* whiteheart malleable iron. { 'blak,härt 'mal·yə·bəl 'ī·ərn }

black hole [COMPUT SCI] *See* stale link. [RELAT] A region of space-time from which nothing can escape, according to classical physics; quantum corrections indicate a black hole radiates particles with a temperature inversely proportional to the mass and directly proportional to Planck's constant. { 'blak 'hōl }

black ice [HYD] A type of ice forming on lake or salt water; compact, and dark in appearance because of its transparency. { 'blak ,īs }

blacking [MET] Carbonaceous material, such as powdered graphite, used to coat the inner surfaces of a dry-sand mold to improve the separation and finish of a casting. { 'blak·iŋ }

black iron oxide *See* ferrous oxide. { 'blak 'ī·ərn 'äk,sīd }

black kernel [PL PATH] A fungus disease of rice caused by *Curvularia lunata* and characterized by dark discoloration of the kernels. { 'blak 'kərn·əl }

black knot [PL PATH] A fungus disease of certain fruit and nut trees that is characterized by black excrescences on the branches; destructive to plum, cherry, gooseberry, filbert, and hazel. { 'blak 'nät }

black lead *See* graphite. { 'blak 'led }

blackleg [VET MED] An acute, usually fatal bacterial disease of cattle, and occasionally of sheep, goats, and swine, caused by *Clostridium chauvoei*. { 'blak,leg }

black level [ELECTR] The level of the television picture signal corresponding to the maximum limit of black peaks. { 'blak ,lev·əl }

black light [OPTICS] Invisible light, such as ultraviolet rays which fall on fluorescent materials and cause them to emit visible light. { 'blak ,līt }

black lignite [GEOL] A lignite with a fixed carbon content of 35–60% and a total carbon content of 73.6–76.2% that contains between 6300 and 8300 Btu per pound; higher in rank than brown lignite. Also known as lignite A. { 'blak 'lig,nīt }

black line [PL PATH] A disease of walnuts, especially of English varieties grafted to black walnuts, characterized by a black line of dead tissue at the graft union, eventually leading to death of the tree. { 'blak ,līn }

black liquor [MATER] **1.** The liquid material remaining from pulpwood cooking in the soda or sulfate papermaking process. **2.** *See* iron acetate liquor. { 'blak 'lik·ər }

blacklung *See* anthracosis. { 'blak,ləŋ }

black measles [MED] *See* hemorrhagic measles. [PL PATH] A disease of California grapevines of uncertain etiology; characterized by black spotting of the skin of the fruit, browning and dropping of the leaves, and dying back of the canes from the tip. { 'blak 'mē·zəlz }

black membrane [CELL MOL] An artificial planar membrane that forms over a hole in the partition between two aqueous compartments and is optically black when viewed in incident light; used to study the permeability of bilayer membranes and the mobility of bilayer components. { 'blak 'mem,brān }

black mica *See* biotite. { 'blak 'mī·kə }

black mold [MYCOL] Any dark fungus belonging to the order Mucorales. [PL PATH] A fungus disease of rose grafts and onion bulbs marked by black appearance due to the mold. { 'blak 'mōld }

black mud [GEOL] A mud formed where there is poor circulation or weak tides, such as in lagoons, sounds, or bays; the color is due to iron sulfides and organic matter. { 'blak 'məd }

blacknose [PL PATH] A physiological disease of the date, with the distal end of the fruit becoming dark, cracking, and shriveling. { 'blak,nōz }

black ocher *See* wad. { 'blak 'ō·kər }

black oil [MATER] Low-grade, black petroleum oil used to lubricate slow-moving or rough-surfaced machinery where high-grade lubricants are impractical or too expensive. { 'blak 'öil }

black opal [MINERAL] A variety of gem-quality opal displaying internal reflections against a dark background. { 'blak 'ō·pəl }

blackout [COMMUN] *See* radio blackout. [ELEC] The shutting off of power in an electrical power transmission system, either deliberately or through failure of the system. { 'blak,aut }

black patch [PL PATH] A fungus disease of red clover characterized by simultaneous blackening of plants in groups. { 'blak ,pach }

black peak [COMMUN] A peak excursion of the television picture signal in the black direction. { 'blak ,pēk }

black pepper [FOOD ENG] A spice; the dried unripe berries of *Piper nigrum*, a vine of the pepper family (Piperaceae) in the order Piperales. { 'blak 'pep·ər }

black pepper oil [MATER] An essential oil obtained from the fruit of the black pepper plant; colorless to slightly greenish with a pepper odor and mild taste. { 'blak 'pep·ər ,öil }

black pod [PL PATH] A pod rot of cacao caused by the fungus *Phytophthora faveri*. { 'blak ,päd }

black point [PL PATH] A disease of wheat and other cereals characterized by blackening of the embryo ends of the grains. { 'blak ,point }

black powder [MATER] A low explosive consisting of an intimate mixture of potassium nitrate or sodium nitrate, charcoal, and sulfur. { 'blak 'paud·ər }

black print [GRAPHICS] The film having the black component in a four-color separation process, or the comparable plate in the four-color printing process. { 'blak 'print }

black ring [PL PATH] **1.** A virus disease of cabbage and other members of the family Cruciferae characterized by dark necrotic and often sunken rings on the surface of the leaf. **2.** A virus disease of the tomato characterized in the early stage by small black rings on young leaves. { 'blak ,riŋ }

black root [PL PATH] Any plant disease characterized by black discolorations of the roots. { 'blak ,rüt }

black root rot [PL PATH] **1.** Any of several plant diseases characterized by dark lesions of the root. **2.** A fungus disease of the apple caused by *Xylaria mali*. **3.** A fungus disease of tobacco and other plants caused by *Thielaviopsis basicola*. { 'blak 'rüt ,rät }

black rot [PL PATH] Any fungal or bacterial disease of plants characterized by dark brown discoloration and decay of a plant part. { 'blak ,rät }

black sand [GEOL] Heavy, dark, sandlike minerals found on beaches and in stream beds; usually magnetite and ilmenite and sometimes gold, platinum, and monazite are present. { 'blak 'sand }

black scope [ELECTR] Cathode-ray tube operating at the threshold of luminescence when no video signals are being applied. { 'blak 'skōp }

black scour [VET MED] Hemorrhagic enteritis of sheep, swine, and cattle, usually associated with a heavy worm burden but sometimes caused by bacterial infection. { 'blak ,skaur }

Black Sea [GEOGR] A large inland sea, area 163,400 square miles (423,000 square kilometers), bounded on the north and east by the Commonwealth of Independent States (former U.S.S.R.) on the south and southwest by Turkey, and on the west by Bulgaria and Rumania. { 'blak 'sē }

black shale [PETR] Very thinly bedded shale rich in sulfides such as pyrite and organic material deposited under barred basin conditions so that there was an anaerobic accumulation. Also known as biopelite. { 'blak 'shāl }

black shank [PL PATH] A black rot disease of tobacco caused by *Phytophthora parasitica* var. *nicotianae*. { 'blak ,shaŋk }

black signal [COMMUN] Signal at any point in a facsimile system produced by the scanning of a maximum density area of the subject copy. Also known as black; picture black. { 'blak ,sig·nəl }

black silver *See* stephanite. { 'blak 'sil·vər }

black smoke [ENG] A smoke that has many particulates in it from inefficient combustion; comes from burning fossil fuel, either coal or oil. { 'blak 'smōk }

black smoker *See* hydrothermal vent. { 'blak 'smōk·ər }

black snow [HYD] Snow that falls through a particulate-laden atmosphere. { 'blak 'snō }

black spot [PL PATH] Any bacterial or fungal disease of

BLACK PEPPER

Berries on a branch of *Piper nigrum*.

plants characterized by black spots on a plant part. { 'blak ,spät }

black stem [PL PATH] Any of several fungal diseases of plants characterized by blackening of the stem. { 'blak ,stem }

black storm *See* karaburan. { ¦blak 'storm }

blackstrap molasses [FOOD ENG] Syrupy mother liquor left after all economical sucrose has been removed from cane- or beet-sugar juice; used as raw material for butanol, acetone, citric acid, and ethyl alcohol manufacture. { 'blak,strap mə'la·səs }

black stripe *See* black thread. { 'blak,strīp }

black-surface enclosure [THERMO] An enclosure for which the interior surfaces of the walls possess the radiation characteristics of a blackbody. { 'blak ,sər·fəs in'klozh·ər }

black-surface field [ELECTR] A layer of p^+ material which is applied to the back surface of a solar cell to reduce hole-electron recombinations there and thereby increase the cell's efficiency. { 'blak ,sər·fəs ,fēld }

black tellurium *See* nagyagite. { 'blak ta'lùr·ē·əm }

black thread [PL PATH] A fungus disease affecting the para rubber tree at the point where the tree is tapped, caused by *Phytophthora meadii* and characterized by black stripes extending through the exposed bast into the cambium or wood. Also known as black stripe; stripe canker. { 'blak ,thred }

black tip [PL PATH] Any of several plant diseases characterized by dark necrotic areas at the tip of the seed or fruit. { 'blak ,tip }

blacktongue [VET MED] A niacin-deficiency disease of dogs, with black discoloration of the tongue. { 'blak,təŋ }

blacktop [MATER] A black bituminous material that is used to pave roadways; it is spread over a layer of crushed rocks and packed down into a level surface; it may be spread over small areas of roadways in need of repair. { 'blak,täp }

blacktop paver [MECH ENG] A construction vehicle that spreads a specified thickness of bituminous mixture over a prepared surface. { 'blak,täp ,pāv·ər }

black transmission [COMMUN] The amplitude-modulated transmission of facsimile signals in which the maximum signal amplitude corresponds to the greatest copy density or darkest shade. { 'blak tranz'mish·ən }

black vomit [MED] Dark vomited matter, consisting of digested blood and gastric contents. { ¦blak 'väm·ət }

blackwater [MED] Any disease that is characterized by dark-colored urine. { 'blak,wód·ər }

blackwater fever [MED] A complication of falciparum malaria, characterized by intravascular hemolysis, hemoglobinuria, tachycardia, high fever, and poor prognosis. { 'blak,wód·ər 'fev·ər }

bladder [ANAT] Any saclike structure in humans and animals, such as a swimbladder or urinary bladder, that contains a gas or functions as a receptacle for fluid. [GEOL] *See* vesicle. { 'blad·ər }

bladder cell [INV ZOO] Any of the large vacuolated cells in the outer layers of the tunic in some tunicates. { 'blad·ər ,sel }

bladder press [MECH ENG] A machine which simultaneously molds and cures (vulcanizes) a pneumatic tire. { 'blad·ər ,pres }

blade [BOT] The broad, flat portion of a leaf. Also known as lamina. [ELEC] A flat moving conductor in a switch. [ENG] **1.** A broad, flat arm of a fan, turbine, or propeller. **2.** The broad, flat surface of a bulldozer or snowplow by which the material is moved. **3.** The part of a cutting tool, such as a saw, that cuts. [VERT ZOO] A single plate of baleen. { blād }

bladed-surface aerator [CIV ENG] A bladed, rotating component of a water treatment plant; used to infuse air into the water. { 'blad·əd ,sər·fəs 'er,ād·ər }

blade loading [AERO ENG] A rotor's thrust in a rotary-wing aircraft divided by the total area of the rotor blades. { 'blād ,lōd·iŋ }

Blagden's law [PHYS CHEM] The law that the lowering of a solution's freezing point is proportional to the amount of dissolved substance. { 'blag·dənz ,lo }

Blaine formation [GEOL] A Permian red bed formation containing red shale and gypsum beds of marine origin in Oklahoma, Texas, and Kansas. { 'blān fór'mā·shən }

blairmorite [PETR] A porphyritic extrusive rock consisting mainly of analcite phenocrysts in a groundmass of sanidine,

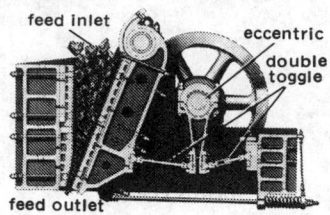

BLAKE JAW CRUSHER

feed inlet
eccentric
double toggle
feed outlet

Drawing of a Blake jaw crusher. *(Allis Chalmers Co.)*

analcite, and alkalic pyroxene, with accessory sphene, melanite, and nepheline. { 'bler·mə,rīt }

blakeite [MINERAL] A deep reddish-brown to deep brown mineral consisting of anhydrous ferric tellurite; occurs in massive form, as microcrystalline crusts. { 'blā,kīt }

Blake jaw crusher [MECH ENG] A crusher with one fixed jaw plate and one pivoted at the top so as to give the greatest movement on the smallest lump. { 'blāk 'jó ,krəsh·ər }

Blake number [FL MECH] A dimensionless number used in the study of beds of particles. { 'blāk ,nəm·bər }

Blancan [GEOL] Upper Pliocene or lowermost Pleistocene geologic time. { 'blaŋ·kən }

blanc fixe [INORG CHEM] $BaSO_4$ A commercial name for barium sulfate, with some use in pure form in the paint, paper, and pigment industries as a pigment extender. { ,bläŋk 'fēks }

blanching [FOOD ENG] A hot-water or steam direct-scalding treatment of raw foodstuffs of particulate type to inactivate enzymes which otherwise might cause quality deterioration, particularly of flavor, during processing and storage. { 'blan·chiŋ }

Blanc rule [ORG CHEM] The rule that glutaric and succinic acids yield cyclic anhydrides on pyrolysis, while adipic and pimelic acids yield cyclic ketones; there are certain exceptions. { 'bläŋk ,rül }

blandel *See* apostilb. { blan'del }

blank [ANALY CHEM] In a chemical analysis, the measured value that is obtained in the absence of a specified component of a sample and that reflects contamination from sources external to the component; it is deducted from the value obtained when the test is performed with the specified component present. Also known as analytical blank. [DES ENG] *See* bit blank. [ELECTR] To cut off the electron beam of a television picture tube, camera tube, or cathode-ray oscilloscope tube during the process of retrace by applying a rectangular pulse voltage to the grid or cathode during each retrace interval. Also known as beam blank. [ENG] **1.** The result of the final cutting operation on a natural crystal. **2.** *See* blind. [MET] **1.** A semifinished piece of metal to be stamped or forged into a tool or implement. **2.** A semifinished, pressed, compacted mass of powdered metal. **3.** Metal sheet prepared for a forming operation. [ORD] Ammunition which contains no projectile but which does contain a charge of low explosive, such as black powder, to produce a noise. { blaŋk }

blank bit *See* bit blank. { 'blaŋk ,bit }

blank carburizing [MET] A simulated carburizing procedure carried out without a carburizing medium. Also known as pseudocarburizing. { 'blaŋk 'kär·bə,rīz·iŋ }

blank casing [PETRO ENG] A casing that has no perforations. { 'blaŋk ,kās·iŋ }

blank cell [COMPUT SCI] A cell of a spreadsheet that contains no text or numeric values, and for which no formatting is specified other than the global formats of the spreadsheet. { 'blaŋk ,sel }

blank character [COMPUT SCI] A character, either printed or appearing as a blank, used to denote a blank space among printed characters. Also known as space character. { 'blaŋk 'kar·ik·tər }

blanket [GRAPHICS] In offset lithography, a rubber sheet covering the cylinder of an offset press that transfers the image from the plate to the paper. [MIN ENG] A textile material used in ore treatment plants for catching coarse free gold and sometimes associated minerals, for example, pyrite. [NUCLEO] A layer of fertile uranium-238 or thorium-232 material placed around or within the core of a nuclear reactor to breed new fuel. { 'blaŋ·kət }

blanket cloth [TEXT] **1.** Heavyweight reversible fabric commonly woven with jacquard figures of one color on a ground of another color; filling is thick and softspun. **2.** Heavyweight overcoating fabric with soft, raised finish. { 'blaŋ·kət ,klóth }

blanket cylinder [GRAPHICS] The cylinder on which the blanket is mounted in offset lithography. { 'blaŋ·kət ¦sil·ən·dər }

blanket deposit [GEOL] A flat deposit of ore; its length and width are relatively great compared with its thickness. { 'blaŋ·kət di'päz·ət }

blanket gas [CHEM ENG] A gas phase introduced into a vessel above a liquid phase to prevent contamination of the liquid, reduce hazard of detonation, or to exert pressure on the liquid. Also known as cushion gas. { 'blaŋ·kət ,gas }

blanketing [COMMUN] Interference due to a nearby transmitter whose signals are so strong that they override other signals over a wide band of frequencies. [MIN ENG] The material caught on the blankets that are used in concentrating gold-bearing sands or slimes. { 'blaŋ·kəd·iŋ }

blanket insulation [MATER] Insulation in the form of a rolled sheet sometimes having a vapor-barrier treated paper backing. { 'blaŋ·kət ‚in·sə'lā·shən }

blanket sand [GEOL] A relatively thin body of sand or sandstone covering a large area. Also known as sheet sand. { 'blaŋ·kət ‚sand }

blank flange [DES ENG] A solid disk used to close off or seal a companion flange. { 'blaŋk 'flanj }

blank form See blank medium. { 'blaŋk ‚fȯrm }

blank holder [MET] A tool to prevent the edge of a sheet metal blank from wrinkling during deep-drawing operations. { 'blaŋk ‚hōl·dər }

blankholder slide [MECH ENG] The outer slide of a double-action power press; it is usually operated by toggles or cams. { 'blaŋk‚hōl·dər ‚slīd }

blanking [ENG] **1.** The closing off of flow through a liquid-containing process pipe by the insertion of solid disks at joints or unions; used during maintenance and repair work as a safety precaution. Also known as blinding. **2.** Cutting of plastic or metal sheets into shapes by striking with a punch. Also known as die cutting. { 'blaŋk·iŋ }

blanking circuit [ELECTR] A circuit preventing the transmission of brightness variations during the horizontal and vertical retrace intervals in television scanning. { 'blaŋk·iŋ ‚sər·kət }

blanking level [ELECTR] The level that separates picture information from synchronizing information in a composite television picture signal; coincides with the level of the base of the synchronizing pulses. Also known as pedestal; pedestal level. { 'blaŋk·iŋ ‚lev·əl }

blanking pulse [ELECTR] A positive or negative square-wave pulse used to switch off a part of a television or radar set electronically for a predetermined length of time. { 'blaŋk·iŋ ‚pəls }

blanking signal [ELECTR] The signal rendering the return trace invisible on the picture tube of a television receiver. { 'blaŋk·iŋ ‚sig·nəl }

blanking time [ELECTR] The length of time that the electron beam of a cathode-ray tube is shut off. { 'blaŋk·iŋ ‚tīm }

blank medium [COMPUT SCI] An empty position on the medium concerned, such as a column without holes on a punch tape, used to indicate a blank character. Also known as blank form. { 'blaŋk 'mēd·ē·əm }

blank nitriding [MET] Simulation of nitriding without the introduction of nitrogen; achieved by use of an inert material or by application of a coating to the piece. { 'blaŋk 'nī‚trīd·iŋ }

blank tape [COMPUT SCI] A portion of a paper tape having sprocket holes only, to indicate a blank character. { 'blaŋk 'tāp }

blank tape halting problem [COMPUT SCI] The problem of finding an algorithm that, for any Turing machine, decides whether the machine eventually stops if it started on an empty tape; it has been proved that no such algorithm exists. { 'blaŋk 'tāp 'hȯl·tiŋ ‚präb·ləm }

Blaschke's theorem [MATH] The theorem that a bounded closed convex plane set of width 1 contains a circle of radius 1/3. { 'bläsh·kəz ‚thir·əm }

Blasius theorem [AERO ENG] A theorem that provides formulas for finding the force and moment on the airfoil profiler. { 'blä·zē·əs ‚thir·əm }

blast [COMPUT SCI] To release internal or external memory areas from the control of a computer program in the course of dynamic storage allocation, making these areas available for reallocation to other programs. [ENG] The setting off of a heavy explosive charge. [PHYS] **1.** The brief and rapid movement of air or other fluid away from a center of outward pressure, as in an explosion. **2.** The characteristic instantaneous rise in pressure, followed by a sudden decrease, that results from this movement, differentiated from less rapid pressure changes. { blast }

blast area [ORD] **1.** Area affected by the blast of an explosion. **2.** Scorched area of ground around the muzzle of a gun, caused by repeated blasts. { 'blast ‚er·ē·ə }

blast burner [ENG] A burner in which a controlled burst of air or oxygen under pressure is supplied to the illuminating gas used. Also known as blast lamp. { 'blast ‚bȯr·nər }

blast cell [HISTOL] An undifferentiated precursor of a human blood cell in the reticuloendothelial tissue. { 'blast ‚sel }

blast chamber [AERO ENG] A combustion chamber, especially in a gas-turbine, jet, or rocket engine. { 'blast ‚chām·bər }

blast cleaning [ENG] Any cleaning process in which an abrasive is directed at high velocity toward the surface being cleaned, for example, sand blasting. { 'blast ‚klēn·iŋ }

blast deflector [AERO ENG] A device used to divert the exhaust of a rocket fired from a vertical position. { 'blast di'flek·tər }

blast ditching [CIV ENG] The use of explosives to aid in ditch excavation, such as for laying pipelines. { 'blast ‚dich·iŋ }

blast effect [PHYS] Violent air movements and pressure changes and the destruction or damage resulting therefrom, generally caused by an explosion on or above the surface of the earth. { 'blast i'fekt }

blastema [EMBRYO] **1.** A mass of undifferentiated protoplasm capable of growth and differentiation. **2.** See anlage. { bla'stēma }

blaster [ENG] A device for detonating an explosive charge; usually consists of a machine by which an operator, by pressing downward or otherwise moving a handle of the device, may generate a powerful transient electric current which is transmitted to an electric blasting cap. Also known as blasting machine. { 'blas·tər }

blast freezer [ENG] An upright freezer in which very cold air circulated by blowers is used for rapid freezing of food. { 'blast ‚frē·zər }

blast furnace [MET] A tall, cylindrical smelting furnace for reducing iron ore to pig iron; the blast of air blown through solid fuel increases the combustion rate. { 'blast ‚fər·nəs }

blast-furnace coke [MATER] Coke which supplies carbon monoxide to reduce the ore in a blast furnace and supplies heat to melt the iron. { 'blast ‚fər·nəs ‚kōk }

blast-furnace gas [MATER] The gas product from iron ore smelting when hot air passes over coke in blast ovens; contains carbon dioxide, carbon monoxide, hydrogen, and nitrogen and is used as fuel gas. { 'blast ‚fər·nəs ‚gas }

blast gate [MET] A sliding plate in a cupola blast pipe to channel airflow in proper proportion. { 'blast ‚gāt }

blast heater [MECH ENG] A heater that has a set of heat-transfer coils through which air is forced by a fan operating at a relatively high velocity. { 'blast ‚hēd·ər }

blasthole [ENG] **1.** A hole that takes a heavy charge of explosive. **2.** The hole through which water enters in the bottom of a pump stock. { 'blast‚hōl }

blasthole drilling [ENG] Drilling to produce a series of holes for placement of blasting charges. { 'blast‚hōl ‚dril·iŋ }

blastic deformation [GEOL] Rock deformation involving recrystallization in which space lattices are destroyed or replaced. { 'blas·tik ‚dē‚fȯr'mā·shən }

blasticidin-S [ORG CHEM] A compound with a melting point of 235–236°C; soluble in water; used as a fungicide for rice crops. { ‚blas'tis·ə·dən 'es }

blasting [ENG] **1.** Cleaning materials by a blast of air that blows small abrasive particles against the surface. **2.** The act of detonating an explosive. [GEOL] Abrasion caused by movement of fine particles against a stationary fragment. { 'blas·tiŋ }

blasting agent [MATER] A compound or mixture, such as ammonium nitrate or black powder, that detonates as a result of heat or shock; used in mining, blasting, pyrotechnics, and propellants. { 'blas·tiŋ ‚ā·jənt }

blasting barrel [MIN ENG] A piece of iron pipe, usually about one-half inch in diameter, used to provide a smooth passageway through the stemming for the miner's squib; it is recovered after each blast and used until destroyed. { 'blas·tiŋ ‚bar·əl }

blasting cap [ENG] A copper shell closed at one end and containing a charge of detonating compound, which is ignited by electric current or the spark of a fuse; used for detonating high explosives. { 'blas·tiŋ ‚kap }

blasting fuse [ENG] A core of gunpowder in the center of jute, yarn, and so on for igniting an explosive charge in a shothole. { 'blas·tiŋ ,fyüz }

blasting gelatin [MATER] A plastic dynamite that contains 5–10% nitrocellulose added to nitroglycerin; used principally in submarine work. { 'blas·tiŋ jel·ət·ən }

blasting machine *See* blaster. { 'blas·tiŋ ma'shēn }

blasting mat [ENG] A heavy, flexible, tear-resistant covering that is spread over the surface during blasting to contain earth fragments. { 'blast·iŋ ,mat }

blasting powder [MATER] A powder containing less nitrate, and in its place more charcoal than black powder; composition is 65–75% sodium or nitrate, potassium nitrate, 10–15% sulfur, and 15–20% charcoal. { 'blas·tiŋ ,paùd·ər }

blast lamp *See* blast burner; blowtorch. { 'blast ,lamp }

blasto- [EMBRYO] A germ or bud, with reference to early embryonic stages of development. [PETR] A prefix indicating the presence in a rock of residual structures somewhat modified by metamorphism. { 'blas·tō }

Blastobacter [MICROBIO] A genus of budding bacteria; cells are rod-, wedge-, or club-shaped, and several attach at the narrow end to form a rosette; reproduce by budding at the rounded end. { 'blas·tō|bak·tər }

Blastobasidae [INV ZOO] A family of lepidopteran insects in the superfamily Tineoidea. { 'blas·tō|bas·ə,dē }

blastocarpous [BOT] Germinating in the pericarp. { |blas·tō|kär·pəs }

blastochyle [EMBRYO] The fluid filling the blastocoele. { 'blas·tə,kīl }

Blastocladiales [MYCOL] An order of aquatic fungi in the class Phycomycetes. { |blas·tō|klā·dē'ā·lēz }

blastocoele [EMBRYO] The cavity of a blastula. Also known as segmentation cavity. { 'blas·tə,sēl }

blastocone [EMBRYO] An incomplete blastomere. { 'blas·tō,kōn }

blastocyst [EMBRYO] A modified blastula characteristic of placental mammals. { 'blas·tə,sist }

blastocyte [EMBRYO] An embryonic cell that is undifferentiated. [INV ZOO] An undifferentiated cell capable of replacing damaged tissue in certain lower animals. { 'blas·tə,sīt }

blastoderm [EMBRYO] The blastodisk of a fully formed vertebrate blastula. { 'blas·tə,dərm }

blastodisk [EMBRYO] The embryo-forming, protoplasmic disk on the surface of a yolk-filled egg, such as in reptiles, birds, and some fish. { 'blas·tə,disk }

blast-off [AERO ENG] The takeoff of a rocket or missile. { 'blast,òf }

blastogranitic rock [PETR] A metamorphic granitic rock which still has parts of the original granitic texture. { |blas·tō·grə'nid·ik 'räk }

Blastoidea [PALEON] A class of extinct pelmatozoan echinoderms in the subphylum Crinozoa. { bla'stòid·ē·ə }

blastokinesis [INV ZOO] Movement of the embryo into the yolk in some insect eggs. { ,blas·tō,kə'nē·səs }

blastoma [MED] **1.** A tumor whose parenchymal cells have certain embryonal characteristics. **2.** A true tumor. { ,bla'stōm·ə }

blastomere [EMBRYO] A cell of a blastula. { 'blas·tə ,mir }

blastomycosis [MED] A term for two infectious, yeastlike fungus diseases of humans: North American blastomycosis, caused by *Blastomyces dermatitidis,* and South American (paracoccidioidomycosis) caused by *Blastomyces brasiliensis.* { ,blas·tə,mī'kō·səs }

blastomylonite [PETR] Rock which has recrystallized after granulation. { ,blas·tə'mī·lə,nīt }

blastopelitic [PETR] Descriptive of the structure of metamorphosed argillaceous rocks. { |blas·tō·pə'lid·ik }

blastophitic [PETR] A metamorphic rock which once contained lath-shaped crystals partly or wholly enclosed in augite and in which part of the original texture remains. { |blas·tō'fid·ik }

blastophore [CYTOL] The cytoplasm that is detached from a spermatid during its transformation to a spermatozoon. [INV ZOO] An amorphous core of cytoplasm connecting cells of the male morula of developing germ cells in oligochaetes. { 'blas·tə,fòr }

blastopore [EMBRYO] The opening of the archenteron. { 'blas·tə,pòr }

blastoporphyritic [PETR] Applied to the textures of metamorphic rocks that are derived from porphyritic rocks; the porphyritic character still remains as a relict feature. { |blas·tō|pòr·fə|rid·ik }

blastopsammite [GEOL] A relict fragment of sandstone that is contained in a metamorphosed conglomerate. { bla'stäp·sə,mīt }

blastopsephitic [GEOL] Descriptive of the structure of metamorphosed conglomerate or breccia. { bla|stäp·sə|fid·ik }

blastospore [MYCOL] A fungal resting spore that arises by budding. { 'blas·tə,spór }

blastostyle [INV ZOO] A zooid on certain hydroids that lacks a mouth and tentacles and functions to produce medusoid buds. { 'blas·tə,stīl }

blastotomy [EMBRYO] Separation of cleavage cells during early embryogenesis. { |blas·'täd·ə·mē }

blastozooid [INV ZOO] A zooid produced by budding. { ,blas·tə'zō,òid }

blast pressure [PHYS] The impact pressure of the air set in motion by an explosion. { 'blast ,presh·ər }

blast roasting [MIN ENG] The roasting of finely divided ores by means of a blast to maintain internal combustion within the charge. Also known as roast sintering. { 'blast ,rō·stiŋ }

blastula [EMBRYO] A hollow sphere of cells characteristic of the early metazoan embryo. { 'blas·chə·lə }

blastulation [EMBRYO] Formation of a blastula from a solid ball of cleaving cells. { ,blas·chə'lā·shən }

blast wall [ENG] A heavy wall used to isolate buildings or areas which contain highly combustible or explosive materials or to protect a building or area from blast damage when exposed to explosions. { 'blast ,wòl }

blast wave [PHYS] An air wave set in motion by an explosion. { 'blast ,wāv }

Blattabacterium [MICROBIO] A genus of the tribe Wolbachiae; straight or slightly curved rod-shaped cells; symbiotic in cockroaches. { ,blad·ə·bak'tir·ē·əm }

Blattidae [INV ZOO] The cockroaches, a family of insects in the order Orthoptera. { 'blad·ə·dē }

blazar [ASTRON] A type of quasar whose light exhibits strong optical polarization and large variability. { 'blā,zär }

blaze-of-grating technique [OPTICS] A technique whereby the ruled grooves of a diffraction grating are given a controlled shape so that they reflect as much as 80% of the incoming light into one particular order for a given wavelength. { ,blāz əv 'grād·iŋ tek'nēk }

BLC *See* boundary-layer control.

bleaching [GRAPHICS] An afterprocess in the production of direct positive photographs, in which an oxidizing solution dissolves the negative silver. [OPTICS] A decrease in the optical absorption of a medium, produced by radiation or by external forces. [TEXT] A process in which natural coloring matter is removed from a fiber, yarn, or fabric to make it white. { 'blēch·iŋ }

bleaching agent [CHEM] An oxidizing or reducing chemical such as sodium hypochlorite, sulfur dioxide, sodium acid sulfite, or hydrogen peroxide. [FOOD ENG] A chemical, such as an aromatic acyl peroxide or monoperoxyphthalic acid, used to bleach flour, fats, oils, and other edibles. { 'blēch·iŋ ,ā·jənt }

bleaching assistant [MATER] A material added to textile bleach baths to cause better penetration by the bleach; includes pine oils, borax, and sulfonated oils. { 'blēch·iŋ a'sis·tənt }

bleaching clay [MATER] Absorbent clay contacted with petroleum and vegetable and other oils to make decolorized products. { 'blēch·iŋ ,klā }

bleaching powder [MATER] A mixture of calcium hydroxide, calcium chloride, and calcium hypochlorite that is used as a bleaching agent. Also known as chloride of lime; chlorinated lime. { 'blēch·iŋ ,paùd·ər }

bleach liquor [MATER] A water solution of calcium hypochlorite used as a laundry and textile bleach, germicide, and deodorant. { 'blēch ,lik·ər }

bleach spot [GEOL] A green or yellow area in red rocks formed by reduction of ferric oxide around an organic particle. Also known as deoxidation sphere. { 'blēch ,spät }

Blears effect [ENG] The dependence of the signal from an ionization gage on the geometry of the system being measured when an organic vapor is present in the vacuum; the effect can

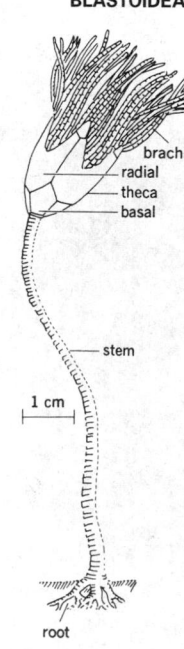

BLASTOIDEA

brachiole
radial
theca
basal

stem

1 cm

root

A blastoid, *Pentremites.*

falsify measurement results by up to an order of magnitude. { 'blirz i̯fekt }

bleb [MATER] A small bubble in a material that has solidified, such as glass. [PETR] A small, usually spherical inclusion in a rock mass. [MED] A localized collection of fluid, as serum or blood, in the epidermis. { bleb }

bleed [CHEM] Diffusion of coloring matter from a substance. [COMPUT SCI] In optical character recognition, the flow of ink in printed characters beyond the limits specified for their recognition by a character reader. [ENG] To let a fluid, such as air or liquid oxygen, escape under controlled conditions from a pipe, tank, or the like through a valve or outlet. [GRAPHICS] The extension of a photograph or other artwork to the very edge of the printed page. [MED] To exude blood from a wound. { blēd }

bleeder [ELECTR] A high resistance connected across the dc output of a high-voltage power supply which serves to discharge the filter capacitors after the power supply has been turned off, and to provide a stabilizing load. [ENG] A connection located at a low place in an air line or a gasoline container so that, by means of a small valve, the condensed water or other liquid can be drained or bled off from the line or container without discharging the air or gas. [MED] **1.** A person subject to frequent hemorrhages, as a hemophiliac. **2.** A blood vessel from which there is persistent uncontrolled bleeding. **3.** A blood vessel which has escaped closure by cautery or ligature during a surgical procedure. [MET] An incomplete casting that results from some molten metal draining out of the mold cavity after pouring has ceased. { 'blēd·ər }

bleeder cloth [MATER] A layer of material placed over a composite material during the curing process to allow escape of excess gas and resin. { 'blēd·ər ‚klȯth }

bleeder current [ELEC] Current drawn continuously from a voltage source to lessen the effect of load changes or to provide a voltage drop across a resistor. { 'blēd·ər ‚kər·ənt }

bleeder resistor [ELEC] A resistor connected across a power pack or other voltage source to improve voltage regulation by drawing a fixed current value continuously; also used to dissipate the charge remaining in filter capacitors when equipment is turned off. { 'blēd·ər ri'zis·tər }

bleeder turbine [MECH ENG] A multistage turbine where steam is extracted (bled) at pressures intermediate between throttle and exhaust, for process or feedwater heating purposes. { 'blēd·ər ‚tər·bən }

bleeding [CHEM ENG] The undesirable movement of certain components of a plastic material to the surface of a finished article. Also known as migration. [ENG] Natural separation of a liquid from a liquid-solid or semisolid mixture; for example, separation of oil from a stored lubricating grease, or water from freshly poured concrete. Also known as bleedout. [MATER] **1.** The outward penetration of a coloring agent from a substrate through the surface coat of paint. **2.** The movement of grout through a pavement from below a road surfacing material to the outer surface. [TEXT] Referring to a fabric in which the dye is not fast and therefore comes out when the fabric is wet. { 'blēd·iŋ }

bleeding canker [PL PATH] A fungus disease of hardwoods caused by *Phytophthora cactorum* and characterized by cankers which exude a reddish ooze on the trunk and branches. { 'blēd·iŋ ‚kaŋ·kər }

bleeding cycle [MECH ENG] A steam cycle in which steam is drawn from the turbine at one or more stages and used to heat the feedwater. Also known as regenerative cycle. { 'blēd·iŋ ‚sī·kəl }

bleeding disease [PL PATH] A disease of the coconut palm caused by *Ceratostomella paradoxa*, characterized by a rust-colored exudation from cracks on the stem. { 'blēd·iŋ di‚zēz }

bleeding time [PHYSIO] The time required for bleeding to stop after a small puncture wound. { 'blēd·iŋ ‚tīm }

bleed line [GRAPHICS] A line-width change usually due to overexposure or overdeveloping in photography. { 'blēd ‚līn }

bleedout *See* bleeding. { 'blēd‚aut }

bleed-through [GRAPHICS] Of records printed on both sides, the obtrusive show-through of printed matter from one side to the other. [MATER] Discoloration in the surface layer of wood veneer construction that is caused by oozing of glue through the face veneers. Also known as strike-through. { 'blēd‚thrü }

bleed valve [ENG] A small-flow valve connected to a fluid process vessel or line for the purpose of bleeding off small quantities of contained fluid. { 'blēd ‚valv }

blend [MATER] A mixture so combined as to render the parts indistinguishable from one another. { blend }

blende *See* sphalerite. { blend }

blended data [ENG] Q point that is the combination of scan data and track data to form a vector. { 'blen·dəd 'dad·ə }

blended fuel oil [MATER] A mixture of petroleum residual and distillate fuel oils. { 'blen·dəd 'fyül ‚ȯil }

blended unconformity [GEOL] An unconformity that is not sharp because the original erosion surface was covered by a thick residual soil that graded downward into the underlying rock. { 'blen·dəd ‚ən·kən'fȯr·məd·ē }

blended whiskey [FOOD ENG] Whiskey containing at least 20% by volume of 100 proof straight whiskey and mixed with other whiskey or neutral spirits, the mixture being not less than 80 proof at the time of bottling. { 'blen·dəd 'wis·kē }

blending [MATER] The process of mixing two or more substances having different properties to obtain a final product having characteristics different from those of the starting materials. [TEXT] Producing uniform yarn and fabric by bringing together two or more distinctive types of fibers of different lengths. { 'blen·diŋ }

blending inheritance [GEN] Inheritance in which the character of the offspring is a blend of those in the parents; a common feature for quantitative characters, such as stature, determined by large numbers of genes and affected by environmental variation. { 'blen·diŋ in'her·ə·təns }

blending naphtha [MATER] A petroleum distillate used to thin or dilute heavy petroleum stocks (for example, lubricating oil) to simplify handling and processing. { 'blen·diŋ ‚naf·thə }

blending problem [IND ENG] A linear programming problem in which it is required to find the least costly mix of ingredients which yields the desired product characteristics. { 'blen·diŋ ‚präb·ləm }

blending stock [CHEM ENG] Any substance used for compounding gasoline, including natural gasoline, catalytically reformed products, and additives. Also known as blendstock. { 'blen·diŋ ‚stäk }

blending value [ENG] Measure of the ability of an added component (for example, tetraethyllead, isooctane, and aromatics) to affect the octane rating of a base gasoline stock. { 'blen·diŋ ‚val·yü }

blendstock *See* blending stock. { 'blend‚stäk }

blend stop [BUILD] A thin wood strip fastened to the exterior vertical edge of the pulley stile or jamb to hold the sash in position. { 'blend ‚stäp }

Blenniidae [VERT ZOO] The blennies, a family of carnivorous marine fishes in the suborder Blennioidei. { ble'nī·ə‚dē }

Blennioidei [VERT ZOO] A large suborder of small marine fishes in the order Perciformes that live principally in coral and rock reefs. { ‚ble·nē'ȯi·dē‚ī }

blennorrhagia [MED] Excessive discharge of mucus. Also known as blennorrhea. { ble·nə'rā·jē·ə }

blennorrhea *See* blennorrhagia. { ‚ble·nə'rē·ə }

Blephariceridae [INV ZOO] A family of dipteran insects in the suborder Orthorrhapha. { ‚blef·ə·ri'ser·ə‚dē }

blepharism [MED] Spasm of the eyelids causing rapid, repetitive involuntary winking. { 'blef·ə‚riz·əm }

blepharitis [MED] Inflammation of the eyelids. { ‚blef·ə'rīd·əs }

blepharoconjunctivitis [MED] Inflammation of the eyelids and the conjunctiva. { ‚blef·ə·rō·kən‚jəŋk·tə'vīd·əs }

blepharoplast [MICROBIO] A cytoplasmic granule bearing a bacterial flagellum. { 'blef·ə·rō‚plast }

blepharoplasty [MED] Any plastic surgical operation on the eyelid. Also known as tarsoplasty. { 'blef·ə·rə‚plas·dē }

blepharospasm [MED] Spasmodic winking due to spasms of the orbicular muscle of the eyelid. { 'blef·ə·rō‚spaz·əm }

BL Herculis stars [ASTRON] W Virginis stars of relatively low luminosity and mass. { 'be‚el 'hər·kyə·ləs ‚stärz }

blight [PL PATH] Any plant disease or injury that results in general withering and death of the plant without rotting. { blīt }

blight canker [PL PATH] A phase of fire blight characterized by cankers. { 'blīt ‚kaŋ·kər }

blimp [AERO ENG] A name originally applied to nonrigid,

BLIMP

Examples of nonrigid airships. *(Goodyear Aircraft Corp.)*

pressure-type airships, usually of small size; now applied to airships with volumes of approximately 1,500,000 cubic feet (42,000 cubic meters). { blimp }

blind [ENG] A solid disk inserted at a pipe joint or union to prevent the flow of fluids through the pipe; used during maintenance and repair work as a safety precaution. Also known as blank. [GEOL] Referring to a mineral deposit with no surface outcrop. { blīnd }

blind approach beacon system [NAV] A pulse-type, ground-based navigation beacon used for runway approach at airports, which sends out signals that produce range and runway position information on the L-scan cathode-ray indicator of an aircraft making an instrument approach. Also known as beam approach beacon system (British usage). Abbreviated babs. { ¦blīnd ə'prōch 'bē·kən ˌsis·təm }

blind bombing See bombing through overcast. { ¦blīnd 'bäm·iŋ }

blind borer [MIN ENG] A shaft-boring machine that does not require a pilot hole. { ¦blīnd 'bȯr·ər }

blind coal See natural coke. { ¦blīnd ¦kōl }

blind controller system [CONT SYS] A process control arrangement that separates the in-plant measuring points (for example, pressure, temperature, and flow rate) and control points (for example, a valve actuator) from the recorder or indicator at the central control panel. { ¦blīnd kən'trōl·ər ˌsis·təm }

blind drainage See closed drainage. { ¦blīnd ¦drā·nij }

blind drift [MIN ENG] In a mine, a horizontal passage not yet connected with the other workings. { ¦blīnd 'drift }

blind drilling [ENG] Drilling in which the drilling fluid is not returned to the surface. { ¦blīnd 'dril·iŋ }

blind embossing [GRAPHICS] A design that is stamped on a material to give the effect of a bas-relief without ink or gold leaf. { ¦blīnd em'bȯs·iŋ }

blind flange [DES ENG] A flange used to close the end of a pipe. { ¦blīnd 'flanj }

blind floor See subfloor. { ¦blīnd 'flȯr }

blind hole [DES ENG] A hole which does not pass completely through a workpiece. [ENG] A type of borehole that does not have the drilling mud or other circulating medium carry the cuttings to the surface. { ¦blīnd 'hōl }

blind image [GRAPHICS] **1.** A lithographic image that can no longer receive ink. { ¦blīnd 'im·ij }

blinding [ENG] **1.** A thin layer of lean concrete, fine gravel, or sand that is applied to a surface to smooth over voids in order to provide a cleaner, drier, or more durable finish. **2.** A layer of small rock chips applied over the surface of a freshly tarred road. **3.** See blanking. [MIN ENG] Interference with the functioning of a screen mesh by a matting of fine materials during screening. Also known as blocking; plugging. { 'blīn·diŋ }

blind joint [ENG] A joint which is not visible from any angle. { ¦blīnd 'jȯint }

blind landing [AERO ENG] Landing an aircraft solely by the use of instruments because of poor visibility. { ¦blīnd 'lan·diŋ }

blindness [MED] **1.** Loss or absence of the ability to perceive visual images. **2.** The condition of a person having less than 1/10 (20/200 on the Snellen test) normal vision. { 'blīnd·nəs }

blind nipple [MECH ENG] A short piece of piping or tubing having one end closed off; commonly used in boiler construction. { ¦blīnd 'nip·əl }

blind passage [MICROBIO] Transfer of some material from an inoculated animal or cell culture that does not exhibit evidence of infection, to a fresh animal or cell culture. { ¦blīnd 'pas·ij }

blind ram [PETRO ENG] A ram on a blowout preventer that acts as the closing element on an open hole; its ends do not fit around the drill pipe, but seal each other and shut off the space below completely. { ¦blīnd 'ram }

blind riser [MET] An internal riser that does not extend to the outer surface of a mold. { ¦blīnd 'rī·zər }

blind rollers [OCEANOGR] Long, high swells which have increased in height, almost to the breaking point, as they pass over shoals or run in shoaling water. Also known as blind seas. { ¦blīnd 'rō·lərz }

blind sample [ANALY CHEM] In chemical analysis, a selected sample whose composition is unknown except to the

person submitting it; used to test the validity of the measurement process. { ¦blīnd 'samp·əl }

blind seas See blind rollers. { ¦blīnd 'sēz }

blind seed [PL PATH] A fungus disease of forage grasses caused by *Phealea temulenta*, resulting in abortion of the seed. { 'blīnd ˌsēd }

blind spot [ENG] An area on a filter screen where no filtering occurs. Also known as dead area. [NEURO] A place on the retina of the eye that is insensitive to light, where the optic nerve passes through the eyeball's inner surface. { 'blīnd ˌspät }

blindstory [ARCH] A floor level that has no exterior windows. { 'blīnd ˌstór·ē }

blind trial See double-blind technique. { ¦blīnd 'trīl }

blind valley [GEOL] A valley that has been made by a spring from an underground channel which emerged to form a surface stream, and that is enclosed at the head of the stream by steep walls. { ¦blīnd 'val·ē }

blind zone [COMMUN] Area from which echoes cannot be received; generally, an area shielded from the transmitter by some natural obstruction and therefore from which there can be no return. { 'blīnd ˌzōn }

B line See index register. { 'bē ˌlīn }

blink [MECH] A unit of time equal to 10^{-5} day or to 0.864 second. [METEOROL] A brightening of the base of a cloud layer, caused by the reflection of light from a snow- or ice-covered surface. { bliŋk }

blink comparator [OPTICS] An optical instrument used to alternately view two pictures in the same visual field in rapid succession, to detect small differences in similar images. { 'bliŋk kəm'par·əd·ər }

blinker light [COMMUN] A light that can be flashed on and off to send a coded message, especially from ship to ship. { 'bliŋ·kər ˌlīt }

blinking [COMMUN] Method of providing information in pulse systems by modifying the signal at its source so that signal presentation on the display scope alternately appears and disappears; in loran, this indicates that a station is malfunctioning. [ELECTR] Electronic-countermeasures technique employed by two aircraft separated by a short distance and within the same azimuth resolution so as to appear as one target to a tracking radar; the two aircraft alternately spot-jam, causing the radar system to oscillate from one place to another, making an accurate solution of a fire control problem impossible. [NAV] Regular shifting right and left or alternate appearance and disappearance of a loran signal to indicate that the signals of a pair of stations are out of synchronization. { 'bliŋ·kiŋ }

blink microscope [OPTICS] A blink comparator which magnifies the compared pictures. { 'bliŋk ¦mī·krə,skōp }

blip [ELECTR] **1.** The display of a received pulse on the screen of a cathode-ray tube. Also known as pip. **2.** An ideal infrared radiation detector that detects with unit quantum efficiency all of the radiation in the signal for which the detector was designed, and responds only to the background radiation noise that comes from the field of view of the detector. [GRAPHICS] A mark, spot, or line on a medium such as microfilm that is optically sensed and is used for counting or timing. { blip }

blip-scan ratio [ELECTR] The ratio of the number of times a target appears on a radarscope to the number of times it could have been seen. { 'blip ˌskan 'rā·shō }

blister [ENG] A raised area on the surface of a metallic or plastic object caused by the pressure of gases developed while the surface was in a partly molten state, or by diffusion of high-pressure gases from an inner surface. [GEOL] A domelike protuberance caused by the buckling of the cooling crust of a molten lava before the flowing mass has stopped. [GRAPHICS] A damaged area on a photographic material where the emulsion has separated from the base. [MATER] A roughly circular or elliptic unbonded area between plies of a laminated material; usually caused by trapped moisture. Also known as steam blow. [MED] A local swelling of the skin resulting from the accumulation of serous fluid between the epidermis and true skin. [MIN ENG] A protrusion, more or less circular in plan, extending downward into a coal seam. [NUCLEO] A protuberance that sometimes develops on the surface of a nuclear-reactor fuel element during use, generally because of entrapped gases. { 'blis·tər }

blister blight [PL PATH] **1.** A fungus disease of the tea plant caused by *Exobasidium vexans* and characterized by blisterlike lesions on the leaves. **2.** A rust disease of Scotch pine caused by *Cronartium asclepiadeum* and characterized by blisterlike lesions on the twigs. { 'blis·tər ,blīt }

blister canker [PL PATH] A fungus disease of the apple tree caused by *Nummularia discreta* and characterized by rough, black cankers on the trunk and large branches. Also known as apple pox. { 'blis·tər ,kaŋ·kər }

blister copper [MET] Copper having 96–99% purity and a blistered appearance and formed by forcing air through molten copper matte. { 'blis·tər ,käp·ər }

blister furnace [MET] A furnace for smelting ore to produce blister copper. { 'blis·tər ,fər·nəs }

blister gas [MATER] Any of several war gases, such as lewisite, which cause burning, inflammation, or tissue destruction internally or externally. { 'blis·tər ,gas }

blister hypothesis [GEOL] A theory of the formation of compressional mountains by a process in which radiogenic heat expands and melts a portion of the earth's crust and subcrust, causing a domed regional uplift (blister) on a foundation of molten material that has no permanent strength. { 'blis·tər hī'päth·ə·səs }

blistering [ENG] The appearance of enclosed or broken macroscopic cavities in a body or in a glaze or other coating during firing. { 'blis·tə·riŋ }

blister rust [PL PATH] Any of several diseases of pines caused by rust fungi of the genus *Cronartium;* the sapwood and inner bark are affected and blisters are produced externally. { 'blis·tər ,rəst }

blister spot [PL PATH] A bacterial disease of the apple caused by *Pseudomonas papulans* and characterized by dark-brown blisters on the fruit and cankers on the branches. { 'blis·tər ,spät }

blister steel [MET] Raw steel, made from wrought iron by cementation followed by slow cooling, in which blisters are formed by gas attempting to escape from the metal. { 'blis·tər ,stēl }

blizzard [METEOROL] A severe weather condition characterized by low temperatures and by strong winds bearing a great amount of snow (mostly fine, dry snow picked up from the ground). { 'bliz·ərd }

BL Lacertae objects [ASTRON] A class of extragalactic sources of extremely intense, highly variable electromagnetic radiation which are related to quasars but have a featureless optical spectrum, and display strong optical polarization and a radio spectrum that increases in intensity at shorter wavelengths. { ¦bē¦el lə'ser·tē ,äb·jiks }

bloat [VET MED] Distension of the rumen in cattle and other ruminants due to excessive gas formation following heavy fermentation of legumes eaten wet. { blōt }

bloatware *See* fatware. { 'blōt,wer }

blob [METEOROL] In radar, oscilloscope evidence of a fairly small-scale temperature and moisture inhomogeneity produced by turbulence within the atmosphere. { bläb }

BLOB *See* binary large object. { bläb or ¦bē¦el¦ō'bē }

Bloch equations [SOLID STATE] Approximate equations for the rate of change of magnetization of a solid in a magnetic field due to spin relaxation and gyroscopic precession. { 'bläk i'kwā·zhənz }

Bloch function [SOLID STATE] A wave function for an electron in a periodic lattice, of the form $u(\mathbf{r}) \exp[i\mathbf{k}\cdot\mathbf{r}]$ where $u(\mathbf{r})$ has the periodicity of the lattice. { 'bläk ,fəŋk·shən }

Bloch theorem [QUANT MECH] The theorem that the lowest state of a quantum-mechanical system without a magnetic field can carry no current. [SOLID STATE] The theorem that, in a periodic structure, every electronic wave function can be represented by a Bloch function. { 'bläk ,thir·əm }

Bloch wall [SOLID STATE] A transition layer, with a finite thickness of a few hundred lattice constants, between adjacent ferromagnetic domains. Also known as domain wall. { 'bläk ,wól }

block [COMPUT SCI] **1.** A group of information units (such as records, words, characters, or digits) that are transported or considered as a single unit by virtue of their being stored in successive storage locations; for example, a group of logical records constituting a physical record. **2.** The section of a computer memory or storage device that stores such a group of information units. Also known as storage block. **3.** To combine two or more information units into a single unit. **4.** A contiguous group of text characters that is marked for moving,

copying, saving, deletion, or some other word-processing operation. [DES ENG] **1.** A metal or wood case enclosing one or more pulleys; has a hook with which it can be attached to an object. **2.** *See* cylinder block. [MIN ENG] **1.** A division of a mine, usually bounded by workings but sometimes by survey lines or other arbitrary limits. **2.** In quarrying, a large portion of rock that is removed from the quarry as a solid mass for further processing at a mill. [PETRO ENG] The subdivision of a sea area for the licensing of oil and gas exploration and production rights. [STAT] In experimental design, a homogeneous aggregation of items under observation, such as a group of contiguous plots of land or all animals in a litter. { bläk }

block and block *See* chockablock. { ¦bläk ən ¦bläk }

block and fall *See* block and tackle. { ¦bläk ən 'fól }

block and tackle [MECH ENG] Combination of a rope or other flexible material and independently rotating frictionless pulleys. Also known as block and fall. { ,bläk ən 'tak·əl }

block body [COMPUT SCI] A list of statements that follows the block head in a computer program with block structure. { 'bläk ,bäd·ē }

block brake [MECH ENG] A brake which consists of a block or shoe of wood bearing upon an iron or steel wheel. { 'bläk ,bräk }

block brazing [MET] The process of joining metals by applying heated blocks to the joint and using a nonferrous filler metal with a melting point above 800°F (427°C). { 'bläk ,brāz·iŋ }

block caving [MIN ENG] A method of caving where a block, 150–250 feet (46–77 meters) on a side and several hundred feet high, is induced to cave in after it is undercut; the broken ore is drawn off at a bell-shaped draw point. { 'bläk ,kāv·iŋ }

block chaining *See* chained block encryption. { 'bläk ,chān·iŋ }

block check character [COMMUN] A character that is added to a block of data to check its accuracy, and consists of parity bits each of which is set by observing a specified set of bits in the block. { 'bläk ,chek ,kar·ik·tər }

block cipher [COMMUN] A cipher that transforms a string of input bits of fixed length into a string of output bits of fixed length. { 'bläk ,sī·fər }

block clay *See* mélange. { 'bläk ,klā }

block coal [MIN ENG] **1.** A bituminous coal that breaks into large lumps or cubical blocks. **2.** Coal that passes over 5-, 6-, and 8-inch (127-, 152-, and 203-millimeter) block screens; used in smelting iron. { 'bläk ,kōl }

block code [COMMUN] An error-correcting code generated by an encoder that produces a fixed-length code word with each incoming fixed-length message block. { 'bläk ,kōd }

block coefficient [NAV ARCH] The ratio of the volume of water displaced by a ship to the product of the ship's length, breadth, and draft. { 'bläk ,kō·ə'fish·ənt }

block copolymer [ORG CHEM] A copolymer in which the like monomer units occur in relatively long alternate sequences on a chain. Also known as block polymer. { 'bläk kō'päl·ə·mər }

block correction [MAP] A corrected reproduction of a small area of a nautical chart that is pasted to the chart for which it is issued. { 'bläk kə'rek·shən }

block data [COMPUT SCI] A statement in FORTRAN which declares that the program following is a data specification subprogram. { 'bläk ,dad·ə }

block diagram [ENG] A diagram in which the essential units of any system are drawn in the form of rectangles or blocks and their relation to each other is indicated by appropriate connecting lines. { 'bläk ,dī·ə,gram }

blocked F-format data set *See* FB data set. { ¦bläkt ¦ef'fór,-mat 'dad·ə ,set }

blocked impedance [ELEC] The impedance at the input of a transducer when the impedance of the output system is made infinite, as by blocking or clamping the mechanical system. { 'bläkt im'pēd·əns }

blocked impurity band detector [ELECTR] A detector of long-wavelength infrared radiation consisting of a heavily doped extrinsic photoconductor on which an undoped intrinsic layer is grown epitaxially to prevent dark current from flowing in the impurity band. { 'bläkt im'pyùr·əd·ē ¦band di,tek·tər }

blocked operation [CHEM ENG] The use of a single chemical or refinery process unit alternately in more than one operation; for example, a catalytic reactor will first produce a

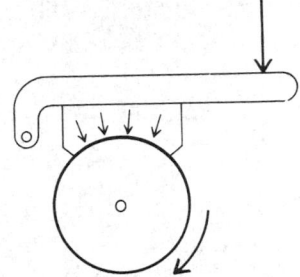

BLOCK BRAKE

Single-block brake. Block is fixed to operating lever, where force is applied in direction of top arrow.

BLOCK CAVING

Block caving technique in underground mining showing the block on the left undercut and caved in; broken ore is drawn off through bell-shaped draw points.

chemical product and then will be blocked from the main process stream during catalyst regeneration. { 'bläkt äp·ə'rā·shən }

blocked-out ore *See* developed reserves. { 'bläkt ¦aút ¦ȯr }

blocked process [COMPUT SCI] A program that is running on a computer but is temporarily prevented from making progress because it requires some resource (such as a printer or user input) that is not immediately available. { 'bläkt 'prä¸ses }

blocked reading frame *See* closed reading frame. { 'bläkt 'rēd·iŋ ¸frām }

blocked resistance [ENG ACOUS] Resistance of an audio-frequency transducer when its moving elements are blocked so they cannot move; represents the resistance due only to electrical losses. { 'bläkt ri'zis·təns }

blocked-up [GRAPHICS] Referring to an area of a negative that is empty of detail; usually caused by overexposure or overdevelopment. { 'bläkt 'əp }

block encryption [COMMUN] The use of a block cipher, usually employing the data encryption standard (DES), in which each 64-bit block of data is enciphered or deciphered separately, and every bit in a given output block depends on every bit in its respective input block and on every bit in the key, but on no other bits. Also known as electronic codebook mode (ECB). { 'bläk en'krip·shən }

blocker-type forging [ENG] A type of forging for designs involving the use of large radii and draft angles, smooth contours, and generous allowances. { 'bläk·ər ¸tīp 'fȯr·jiŋ }

blockette [COMPUT SCI] A subdivision of a group of consecutive machine words transferred as a unit, particularly with reference to input and output. { blä'ket }

block faulting [GEOL] A type of faulting in which fault blocks are displaced at different orientations and elevations. { 'bläk ¸fȯl·tiŋ }

block glide [GEOL] A translational landslide in which the slide mass moves outward and downward as an intact unit. { 'bläk ¸glīd }

block grease [MATER] High-melting-point grease that can be handled as a block or stick at normal temperatures; used for journal bearings. Also known as brick grease. { 'bläk ¸grēs }

block head [COMPUT SCI] A list of declarations at the beginning of a computer program with block structure. { 'bläk ¦hed }

block hole [ENG] A small hole drilled into a rock or boulder into which an anchor bolt or a small charge or explosive may be placed; used in quarries for breaking large blocks of stone or boulders. { 'bläk ¸hōl }

blockhouse [ENG] **1.** A reinforced concrete structure, often built underground or half-underground, and sometimes dome-shaped, to provide protection against blast, heat, or explosion during rocket launchings or related activities, and usually housing electronic equipment used in launching the rocket. **2.** The activity that goes on in such a structure. { 'bläk¸haús }

block identifier [COMPUT SCI] A means of identifying an area of storage in FORTRAN so that this area may be shared by a program and its subprograms. { 'bläk ī'den·tə¸fī·ər }

block ignore character [COMPUT SCI] A character associated with a block which indicates the presence of errors in the block. { 'bläk ig'nȯr ¸kar·ik·tər }

blocking [AGR] The practice of grouping together all experimental units (such as plots of ground or animals) that make up a replication in an agricultural experiment. [CHEM] Undesired adhesion of granular particles; often occurs with damp powders or plastic pellets in storage bins or during movement through conduits. [COMPUT SCI] Combining two or more computer records into one block. [ELECTR] **1.** Applying a high negative bias to the grid of an electron tube to reduce its anode current to zero. **2.** Overloading a receiver by an unwanted signal so that the automatic gain control reduces the response to a desired signal. **3.** Distortion occurring in a resistance-capacitance-coupled electron tube amplifier stage when grid current flows in the following tube. [ENG] Undesired adhesion between layers of plastic materials in contact during storage or use. [HISTOL] **1.** The process of embedding tissue in a solid medium, such as paraffin. **2.** A histochemical process in which a portion of a molecule is treated to prevent it from reacting with some other agent. [MET] **1.** A preliminary hot-forging operation which imparts an approximate shape to the rough stock. **2.** Reducing the oxygen content

of the bath in an open-hearth furnace. [METEOROL] Large-scale obstruction of the normal west-to-east progress of migratory cyclones and anticyclones. [MIN ENG] *See* blinding. [PSYCH] A sudden obstruction or interruption in spontaneous flow of thinking or speaking, perceived as an absence or deprivation of thought. [SOLID STATE] The hindering of motion of dislocations in a solid substance by small particles of a second substance included in the solid; results in hardening of the substance. [STAT] The grouping of sample data into subgroups with similar characteristics. { 'bläk·iŋ }

blocking and wedging [MIN ENG] A method of holding mine timber sets in place; blocks of wood are set on the caps directly over the post supports and have a grain of block parallel with the top of the cap; wedges are driven tightly between the blocks and the roof. { ¦bläk·iŋ ən 'wej·iŋ }

blocking antibody [IMMUNOL] Antibody that inhibits the reaction between antigen and other antibodies or sensitized T lymphocytes. { ¦bläk·iŋ ¦ant·i¸bäd·e }

blocking anticyclone *See* blocking high. { ¦bläk·iŋ ¸an·ti'sī¸klōn }

blocking capacitor *See* coupling capacitor. { 'bläk·iŋ kə'pas·əd·ər }

blocking factor [COMPUT SCI] The largest possible number of records of a given size that can be contained within a single block. { 'bläk·iŋ ¸fak·tər }

blocking group [ORG CHEM] In peptide synthesis, a group that is reacted with a free amino or carboxyl group on an amino acid to prevent its taking part in subsequent formation of peptide bonds. { 'bläk·iŋ ¸grüp }

blocking high [METEOROL] Any high (or anticyclone) that remains nearly stationary or moves slowly compared to the west-to-east motion upstream from its location, so that it effectively blocks the movement of migratory cyclones across its latitudes. Also known as blocking anticyclone. { ¦bläk·iŋ 'hī }

blocking layer *See* depletion layer. { 'bläk·iŋ ¸lā·ər }

blocking oscillator [ELECTR] A relaxation oscillator that generates a short-time-duration pulse by using a single transistor or electron tube and associated circuitry. Also known as squegger; squegging oscillator. { 'bläk·iŋ 'äs·ə¸lād·ər }

blocking oscillator driver [ELECTR] Circuit which develops a square pulse used to drive the modulator tubes, and which usually contains a line-controlled blocking oscillator that shapes the pulse into the square wave. { 'bläk·iŋ 'äs·ə¸lād·ər 'drī·vər }

block input [COMPUT SCI] **1.** A block of computer words considered as a unit and intended or destined to be transferred from an internal storage medium to an external destination. **2.** *See* output area. { 'bläk 'in¸pút }

block lava [GEOL] Lava flows which occur as a tumultuous assemblage of angular blocks. Also known as aa lava. { 'bläk ¸läv·ə }

block length [COMPUT SCI] The total number of records, words, or characters contained in one block. { 'bläk ¸leŋkth }

block loading [COMPUT SCI] A program loading technique in which the control sections of a program or program segment are loaded into contiguous positions in main memory. { 'bläk ¸lōd·iŋ }

block mark [COMPUT SCI] A special character that indicates the end of a block. { 'bläk ¸märk }

block mountain [GEOL] A mountain formed by the combined processes of uplifting, faulting, and tilting. Also known as fault-block mountain. { 'bläk ¸maún·tən }

block move *See* cut and paste. { ¦bläk 'müv }

block multiplexor channel [COMPUT SCI] A transmission channel in a computer system that can simultaneously transmit blocks of data from several high-speed input/output devices by interleaving the data. { 'bläk ¦məlt·i¸plek·sər ¸chan·əl }

block operation [COMPUT SCI] An editing or formatting procedure that is carried out on a selected block of text in a word-processing document. { 'bläk ¸äp·ə¸rā·shən }

block parity [COMMUN] An error-checking technique involving the comparison of a transmitted block check character with one calculated by the receiving device. { 'bläk 'par·əd·ē }

block plane [DES ENG] A small type of hand plane, designed for cutting across the grain of the wood and for planing end grains. { 'bläk ¸plān }

block polymer *See* block copolymer. { 'bläk 'päl·ə·mər }

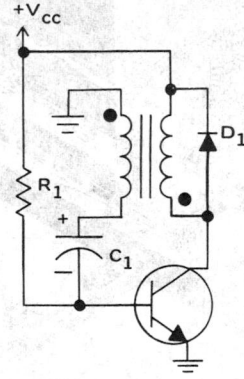

BLOCKING OSCILLATOR

$+V_{CC}$

R_1 D_1 C_1

Circuit diagram of a typical free-running blocking oscillator.

block printing [GRAPHICS] The earliest form of printing, involving the cutting of crude pictures and lettering on blocks of wood. { 'bläk ˌprint·iŋ }

block protection [COMPUT SCI] An instruction in a word-processing or page-layout program that prevents a soft page break from being inserted in a specified block of text, ensuring against a bad page break. { 'bläk prə'tek·shən }

block protector [ELEC] Rectangular piece of carbon, bakelite with a metal insert, or porcelain with a carbon insert which, in combination with each other, make one element of a protector; they form a gap which will break down and provide a path to ground for excessive voltages. { 'bläk prə'tek·tər }

block section [CIV ENG] In a railroad system, a specific length of track that is controlled by stop signals. { 'bläk ˌsek·shən }

block sequence [MET] A procedure by which a continuous multiple-pass joint is completed by alternating the deposition of intermittent cross-sectional buildup and intervening lengths of weld metal. { 'bläk ˌsē·kwəns }

block signal system [CONT SYS] An automatic railroad traffic control system in which the track is sectionalized into electrical circuits to detect the presence of trains, engines, or cars. { 'bläk 'sig·nəl ˌsis·təm }

block sort [COMPUT SCI] A method of sorting a file, usually with punched card sorters, in which the file is first sorted according to the value of the digit in the highest digit position of the key; the resulting collections of records can next be sorted independently in smaller operations, and the separate sections then joined. { 'bläk ˌsȯrt }

block standby [COMPUT SCI] Locations always set aside in storage for communication with buffers in order to make more efficient use of such buffers. { 'bläk ˌstand·bī }

block structure [COMPUT SCI] In computer programming, a conceptual tool used to group sequences of statements into single compound statements and to allow the programmer explicit control over the scope of the program variables. { 'bläk ˌstrək·chər }

block system [CIV ENG] A railroad system for controlling train movements by using signals between block posts, that is, the structures that contain the instruments indicating the positions of trains, conditions within block sections, and control levers for signals and other functions. [MIN ENG] A system of pillars in which a series of entries, rooms, and crosscuts are driven to divide the coal into blocks of about equal size which are then extracted on retreat. { 'bläk ˌsis·təm }

block transfer [COMPUT SCI] The movement of data in blocks instead of by individual records. { 'bläk ˌtrans·fər }

blocky iceberg [OCEANOGR] An iceberg with steep, precipitous side and with a horizontal or nearly horizontal upper surface. { ¦bläk·ē 'īs·ˌbərg }

blödite See bloedite. { 'blō·ˌdīt }

bloedite [MINERAL] $MgSO_4 \cdot Na_2SO_4 \cdot 4H_2O$ A white or colorless monoclinic mineral consisting of magnesium sodium sulfate. Also spelled blödite. Also known as astrakanite; astrochanite. { 'blō·ˌdīt }

blomstrandine See priorite. { ˌblȯm'stran·ˌdēn }

blondel See apostilb. { 'blȯn·del }

Blondel-Rey law [OPTICS] A law utilized to determine the apparent point brilliance of a flashing light. { blȯn·del rā ˌlȯ }

blood [HISTOL] A fluid connective tissue consisting of the plasma and cells that circulate in the blood vessels. { bləd }

blood agar [MICROBIO] A nutrient microbiologic culture medium enriched with whole blood and used to detect hemolytic strains of bacteria. { 'bləd ˌäg·ər }

blood bank [ENG] A place for storing whole blood or plasma under refrigeration. { 'bləd ˌbaŋk }

blood blister [MED] A blister that is filled with blood. { 'bləd ˌblis·tər }

blood-brain barrier [NEUROSCI] A barrier to the entry of substances from the blood into brain tissue; believed to be formed primarily by the endothelial cells of the brain vasculature. { ¦bləd ¦brān 'bar·ē·ər }

blood cell [HISTOL] An erythrocyte or a leukocyte. { 'bləd ˌsel }

blood chimerism [GEN] Having red blood cells of two different genotypes stemming from two different fertilized eggs or individuals. { 'bləd ka'mir·iz·əm }

blood count [PATH] Determination of the number of white and red blood cells in a definite volume of blood. { 'bləd ˌkau̇nt }

blood crisis [MED] The sudden appearance of large numbers of nucleated erythrocytes in the circulating blood. { 'bləd ˌkrī·səs }

blood disease [PL PATH] A bacterial disease affecting the vascular tissues of the banana in the Celebes; believed to be caused by *Xanthomonas celebensis* and characterized by blighting of the leaves and reddish-brown rot of the fruit. { 'bləd di·ˌzēz }

blood dyscrasia [MED] Obsolete term. Any abnormal condition of the formed elements of blood or of the constituents required for clotting. { 'bləd də'skrazh·ə }

blood gas [ORD] War gas which, when absorbed into the body, primarily by breathing, affects body functions by interfering with transfer of oxygen from the lungs via the blood to tissues. { 'bləd ˌgas }

blood group [IMMUNOL] An immunologically distinct, genetically determined class of human erythrocyte antigens, identified as A, B, AB, and O. Also known as blood type. { 'bləd ˌgrüp }

blood island [EMBRYO] One of the areas in the yolk sac of vertebrate embryos allocated to the production of the first blood cells. { 'bləd ˌī·lənd }

bloodline [BIOL] A line of direct ancestors, especially in a pedigree. { 'bləd·ˌlīne }

blood-plate hemolysis [MICROBIO] Destruction of red blood cells in a blood agar medium by a bacterial toxin. { 'bləd ˌplāt hə'mäl·ə·səs }

blood platelet See thrombocyte. { 'bləd ˌplāt·lət }

blood poisoning See septicemia. { 'bləd ˌpȯiz·ən·iŋ }

blood pressure [PHYSIO] Pressure exerted by blood on the walls of the blood vessels. { 'bləd ˌpresh·ər }

blood rain [METEOROL] Rain of a reddish color caused by dust particles containing iron oxide that were picked up by the raindrops during descent. { 'bləd ˌrān }

bloodstone [MINERAL] **1.** A form of deep green chalcedony flecked with red jasper. Also known as heliotrope; oriental jasper. **2.** See hematite. { 'bləd ˌstōn }

bloodstream [PHYSIO] The flow of blood in its circulation through the body. { 'bləd ˌstrēm }

blood sugar [BIOCHEM] The carbohydrate, principally glucose, of the blood. { 'bləd ˌshu̇g·ər }

blood test [PATH] **1.** A serologic test for syphilis. **2.** A blood count. **3.** A test for detection of blood, usually one based on the peroxidase activity of blood, such as the benzidine test or guaiac test. { 'bləd ˌtest }

blood type See blood group. { 'bləd ˌtīp }

blood typing [IMMUNOL] Determination of an individual's blood group. { 'bləd ˌtīp·iŋ }

blood vessel [ANAT] A tubular channel for blood transport. { 'bləd ˌves·əl }

bloocy line [PETRO ENG] The discharge pipe in an air drilling operation that conducts away from the rig the air or gas being used for circulation. { 'blü·ē ˌlīn }

bloom [BOT] **1.** An individual flower. Also known as blossom. **2.** To yield blossoms. **3.** The waxy coating that appears as a powder on certain fruits, such as plums, and leaves, such as cabbage. [ECOL] A colored area on the surface of bodies of water caused by heavy planktonic growth. [ENG] **1.** Fluorescence in lubricating oils or a cloudy surface on varnished or enameled surfaces. **2.** To apply an antireflection coating to glass. [GEOL] See blossom. [GRAPHICS] A milky or foggy defect that may appear on the surface of a varnished painting; caused by moisture. [MATER] Crystals formed on the surface of treated wood by exudation and evaporation of the solvent in preservative solutions. [MET] **1.** A semifinished bar of metal formed from an ingot and having a rectangular cross section exceeding 36 square inches (232 square centimeters). **2.** To hammer or roll metal in order to make its surface bright. **3.** See slag. [MINERAL] See efflorescence. [OPTICS] Color of oil in reflected light, differing from its color in transmitted light. Also known as fluorescence. { blüm }

bloomer [MET] A furnace used to shape wrought iron directly from ore. { 'blü·mər }

blooming [ELECTR] **1.** Defocusing of television picture areas where excessive brightness results in enlargement of spot size and halation of the fluorescent screen. **2.** An increase in radarscope spot size due to an increase in signal intensity.

[MATER] The migration of sulfur or other substances to the surface of a sample of rubber, causing discoloration. { 'blüm·iŋ }

blooming mill [MET] A rolling mill for making blooms from ingots. Also known as cogging mill. { 'blü·miŋ ‚mil }

blossom [BOT] See bloom. [GEOL] The oxidized or decomposed outcrop of a vein or coal bed. Also known as bloom. { 'bläs·əm }

blossom-end rot [PL PATH] **1.** Any rot disease of fruit that originates at the blossom end. **2.** A physiological disease of tomato believed to be caused by extreme fluctuations in available moisture; characterized by shallow leathery depressions with a water-soaked appearance around the tip end of the fruit. { 'bläs·əm ‚end ‚rät }

blotch print [TEXT] Cloth that is printed with both the ground color and the design. { 'bläch ‚print }

blotter [ENG] A disk of compressible material used between a grinding wheel and its flanges to avoid concentrated stress. { 'bläd·ər }

blotter model [PETRO ENG] An analysis device in which the analogous movement of copper ammonium or zinc ammonium ions in blotting paper or gelatin indicates oil-well injection-fluid movement through porous underground reservoirs. { 'bläd·ər ‚mäd·əl }

blotter press [CHEM ENG] A plate-and-frame filter in which the filter medium is blotting paper. { 'bläd·ər ‚press }

blotting [CELL MOL] The transfer of electrophoretically separated polypeptides onto a solid support medium, such as nitrocellulose paper or a nylon membrane. { 'bläd·iŋ }

blotting paper [MATER] An unsized paper used to absorb excess ink from penned letters or signatures; also used for other applications where a soft, spongy paper is required. { 'bläd·iŋ ‚pā·pər }

blow [COMPUT SCI] To write data or code into a programmable read-only memory chip by melting the fuse links corresponding to bits that are to be zero. [ELEC] Opening of a circuit because of excess current, particularly when the current is heavy and a melting or breakdown point is reached. { blō }

blowback [CHEM ENG] **1.** A continuous stream of liquid or gas bled through air lines from instruments and to the process line being monitored; prevents process fluid from backing up and contacting the instrument. **2.** Reverse flow of fluid through a filter medium to remove caked solids. Also known as backwash. [GRAPHICS] To enlarge, or make an enlargement of, an image. [MECH ENG] See blowdown. { 'blō ‚bak }

blowback gun [ORD] An automatic gun utilizing the pressure of the propellant gases to force the bolt to the rear, independently of the barrel which does not move relative to the receiver. { 'blō‚bak ‚gən }

blowball [BOT] A fluffy seed ball, as of the dandelion. { 'blō‚ból }

blowby [MECH ENG] **1.** Leaking of fluid between a cylinder and its piston during operation. { 'blō‚bī }

blowcase [CHEM ENG] A cylindrical or spherical corrosion- and pressure-resistant container from which acid is forced by compressed air to the agitator; used in manufacture of acids but largely superseded by centrifugal pumps. Also known as acid blowcase; acid egg. [PETRO ENG] **1.** A pumping device used to transfer crude oil and water mixtures without agitation in order to prevent emulsions. **2.** See drip. { 'blō‚käs }

blowdown [CHEM ENG] Removal of liquids or solids from a process vessel or storage vessel or a line by the use of pressure. [MECH ENG] The difference between the pressure at which the safety valve opens and the closing pressure. Also known as blowback. [METEOROL] A wind storm that causes trees or structures to be blown down. { 'blō‚daůn }

blowdown line [CHEM ENG] A large conduit to receive and confine fluids forced by pressure from process vessels. { 'blō‚daůn ‚līn }

blowdown stack [CHEM ENG] A vertical stack or chimney into which the contents of a chemical or petroleum process unit are emptied in case of an operational emergency. { 'blō‚daůn ‚stak }

blowdown tunnel [AERO ENG] A wind tunnel in which stored compressed gas is allowed to expand through a test section to provide a stream of gas or air for model testing. { 'blō‚daůn ‚tən·əl }

blowdown turbine [AERO ENG] A turbine attached to a reciprocating engine which receives exhaust gases separately from each cylinder, utilizing the kinetic energy of the gases. { 'blō‚daůn ‚tər·bən }

blower [MECH ENG] A fan which operates where the resistance to gas flow is predominantly downstream of the fan. { 'blō·ər }

blowhole [GEOL] A longitudinal tunnel opening in a sea cliff, on the upland side away from shore; columns of sea spray are thrown up through the opening, usually during storms. [MET] A pocket of air or gas formed in a metal during solidification. [VERT ZOO] The nostril on top of the head of cetacean mammals. { 'blō‚hōl }

blow in [MET] To put a blast furnace into operation. [PETRO ENG] Of an oil well, to begin sending forth oil or gas. { 'blō ‚in }

blowing [CHEM ENG] The introduction of compressed air near the bottom of a tank or other container in order to agitate the liquid therein. [ENG] See blow molding. { 'blō·iŋ }

blowing agent [MATER] A chemical added to plastics and rubbers that generates inert gases on heating, causing the resin to assume a cellular structure. Also known as foaming agent. { 'blō·iŋ ‚ā·jənt }

blowing boundary-layer control [AERO ENG] A technique that is used in addition to purely geometric means to control boundary-layer flow; it consists of reenergizing the retarded flow in the boundary layer by supplying high-velocity flow through slots or jets on the surface of the body. { 'blō·iŋ 'baůn·drē ‚lā·ər kən'trōl }

blowing cave [GEOL] A cave with an alternating air movement. Also known as breathing cave. { ¦blō·iŋ ¦kāv }

blowing dust [METEOROL] Dust picked up locally from the surface of the earth and blown about in clouds or sheets. { ¦blō·iŋ ¦dəst }

blowing fan See forcing fan. { 'blō·iŋ ‚fan }

blowing pressure [ENG] Pressure of the air or other gases used to inflate the parison in blow molding. { 'blō·iŋ ‚presh·ər }

blowing sand [METEOROL] Sand picked up from the surface of the earth by the wind and blown about in clouds or sheets. { ¦blō·iŋ ¦sand }

blowing snow [METEOROL] Snow lifted from the surface of the earth by the wind to a height of 6 feet (1.8 meters) or more (higher than drifting snow) and blown about in such quantities that horizontal visibility is restricted. { ¦blō·iŋ ¦snō }

blowing spray [METEOROL] Spray lifted from the sea surface by the wind and blown about in such quantities that horizontal visibility is restricted. { ¦blō·iŋ ¦sprā }

blowing still [CHEM ENG] A still or process column in which blown or oxidized asphalt is made. { ¦blō·iŋ ¦stil }

blowing-up furnace [MET] A furnace used for sintering ore and for the volatilization of lead and zinc. { 'blō·iŋ ‚əp ‚fər·nəs }

blow-lifting gripper [CONT SYS] A robot component that uses compressed air to lift objects. { 'blō ¦lift·iŋ ‚grip·ər }

blow molding [ENG] A method of fabricating hollow plastic objects, such as bottles, by forcing a parison into a mold cavity and shaping by internal air pressure. Also known as blowing. { 'blō ‚mōl·diŋ }

blown asphalt [MATER] Asphalt which is treated or heated with air or steam within a blowing still at relatively high temperature. { ¦blōn ¦as‚fólt }

blown film [MATER] Film that has been produced by extruding a tube of plastic material which is then expanded in diameter and reduced in thickness by internal air pressure, and then slit along one side. { ¦blōn 'film }

blown foam [MATER] A cellular plastic consisting of an expanded matrix resembling natural foam. { ¦blōn 'fōm }

blown-fuse indicator [ELEC] A neon warning light connected across a fuse so that it lights when the fuse is blown. { ¦blōn ¦fyüz 'in·də‚kād·ər }

blown glass [ENG] Glassware formed by blowing air into a ball of liquefied glass until it reaches the desired shape. { ¦blōn 'glas }

blown oil [MATER] A vegetable or animal oil that has been agitated and partially oxidized by a current of warm air or oxygen, including castor, whale, fish, rape, and linseed oils; used in paints, lubricants, and plasticizers. { ¦blōn 'óil }

blown tubing [ENG] A flexible thermoplastic film tube made by applying pressure inside a molten extruded plastic tube to expand it prior to cooling and winding flat onto rolls. { ¦blōn ′tü·biŋ }

blowoff [AERO ENG] The action of applying an explosive force and separating a package section away from the remaining part of a rocket vehicle or reentry body, usually to retrieve an instrument or to obtain a record made during early flight. { ′blō₁óf }

blowoff valves [MECH ENG] Valves in boiler piping which facilitate removal of solid matter present in the boiler water. { ′blō₁óf ₁valvz }

blowout [ELEC] The melting of an electric fuse because of excessive current. [ELECTROMAG] The extinguishing of an electric arc by deflection in a magnetic field. Also known as magnetic blowout. [ENG] **1.** The bursting of a container (such as a tube pipe, pneumatic tire, or dam) by the pressure of the contained fluid. **2.** The rupture left by such bursting. **3.** The abrupt escape of air from the working chamber of a pneumatic caisson. [GEOL] Any of the various trough-, saucer-, or cuplike hollows formed by wind erosion on a dune or other sand deposit. [HYD] A bubbling spring which bursts from the ground behind a river levee when water at flood stage is forced under the levee through pervious layers of sand or silt. Also known as sand boil. [PETRO ENG] A sudden, unplanned escape of oil or gas from a well during drilling. { ′blō₁aút }

blowout coil [ELECTROMAG] A coil that produces a magnetic field in an electrical switching device for the purpose of lengthening and extinguishing an electric arc formed as the contacts of the switching device part to interrupt the current. { ′blō₁aút ₁kóil }

blowout dune See parabolic dune. { ′blō₁aút ₁dün }

blowout magnet [ELECTROMAG] An electromagnet or permanent magnet used to deflect and extinguish the arc formed when a high-current circuit breaker or switch is opened. { ′blō₁aút ₁mag·nət }

blowout preventer [PETRO ENG] Any one of several types of valves used on the wellhead to prevent the loss of pressure either in the annular space between drill pipe and casing or in the open hole during drilling completion operations. { ′blō ₁aút pri₁ven·tər }

blowpipe [BIOL] A small tube, tapering to a straight or slightly curved tip, used in anatomy and zoology to reveal or clean a cavity. [ENG] **1.** A long, straight tube, used in glass blowing, on which molten glass is gathered and worked. **2.** A small, tapered, and frequently curved tube that leads a jet, usually of air, into a flame to concentrate and direct it; used in flame tests in analytical chemistry and in brazing and soldering of fine work. **3.** See blowtorch. { ′blō₁pīp }

blowpipe reaction analysis [ANALY CHEM] A method of analysis in which a blowpipe is used to heat and decompose a compound or mineral; a characteristic color appears in the flame or a colored crust appears on charcoal. { ′blō₁pīp rē′ak·shən ə′nal·ə·səs }

blowpit See blowtank. { ′blō₁pit }

blow pressure [ENG] Air pressure required for plastics blow molding. { ′blō ₁presh·ər }

blow rate [ENG] The speed of the cycle at which air or an inert gas is applied intermittently during the forming procedure of blow molding. { ′blō ₁rāt }

blowtank [CHEM ENG] A tank or pit, used in papermaking, into which the contents of a digester are blown upon completion of a cook. Also known as blowpit. { ′blō₁taŋk }

blowtorch [ENG] A small, portable blast burner which operates either by having air or oxygen and gaseous fuel delivered through tubes or by having a fuel tank which is pressured by a hand pump. Also known as blast lamp; blowpipe. { ′blō ₁tórch }

blow up See abend. { ′blō ₁əp }

blowup [CIV ENG] The localized buckling or breaking of a rigid pavement caused by excess pressure along its length. [GRAPHICS] An enlargement of a photographic print, or of a detail of the print. { ′blō₁əp }

blowup ratio [ENG] **1.** In blow molding of plastics, the ratio of the diameter of the mold cavity to the diameter of the parison. **2.** In blown tubing, the ratio of the diameter of the finished product to the diameter of the die. { ′blō₁əp ₁rā·shō }

blubber [INV ZOO] A large sea nettle or medusa. [VERT ZOO] A thick insulating layer of fat beneath the skin of whales and other marine mammals. { ′bləb·ər }

blubber oil See whale oil. { ′bləb·ər ₁óil }

blue [OPTICS] The hue evoked in an average observer by monochromatic radiation having a wavelength in the approximate range from 455 to 492 nanometers; however, the same sensation can be produced in a variety of other ways. { blü }

blue annealing [MET] Softening metal sheets by heating in an open furnace and cooling in air; bluish oxide forms on the metal surface. { ¦blü ə′nēl·iŋ }

blue asbestos See crocidolite. { ¦blü as′bes·təs }

blue baby [MED] An infant with congenital cyanosis due to cardiac or pulmonary defect, causing shunting of unoxygenated blood into the systemic circulation. { ¦blü ₁bā·bē }

blue band [ASTRON] A dark band which appears around the polar caps of Mars as they shrink during the spring and early summer. [GEOL] **1.** A layer of bubble-free, dense ice found in a glacier. **2.** A bluish clay found as a thin, persistent bed near the base of No. 6 coal everywhere in the Illinois-Indiana coal basin. { ¦blü ′band }

blueberry [BOT] Any of several species of plants in the genus *Vaccinium* of the order Ericales; the fruit, a berry, occurs in clusters on the plant. { ′blü₁ber·ē }

blue brittleness [MET] Loss of ductility noted for some steels when heated to 400–600°F (204–316°C), the blue heat range. { ¦blü ′brid·əl·nəs }

blue cap [MIN ENG] The characteristic blue halo, or tip, of the flame of a safety lamp when firedamp is present in the air. { ¦blü ₁kap }

blue comb [VET MED] A disease of domestic fowl and other birds that resembles Bright's disease in humans; may be a viral disease or caused by excessive salt intake. { ¦blü ₁kōm }

blue copper ore See azurite. { ¦blü ₁käp·ər ′ór }

blue edge [ASTRON] The curve on the Hertzsprung-Russell diagram given by the maximum temperature, as a function of luminosity, at which a star of specified composition is unstable against small-amplitude pulsations. { ¦blü ′ej }

bluefish [VERT ZOO] *Pomatomus saltatrix*. A predatory fish in the order Perciformes. Also known as skipjack. { ¦blü ₁fish }

blue flash See green flash. { ¦blü ₁flash }

blue gas [MATER] A gas consisting chiefly of carbon monoxide and hydrogen, formed by the action of steam upon hot coke; used mainly as a source of hydrogen and in synthesis of other chemical compounds. Also known as blue water gas. { ¦blü ₁gas }

blue glow [ELECTR] A glow normally seen in electron tubes containing mercury vapor, due to ionization of the mercury molecules. [MET] Luminescence emitted by certain metallic oxides when heated. { ¦blü ₁glō }

bluegrass [BOT] The common name for several species of perennial pasture and lawn grasses in the genus *Poa* of the order Cyperales. { ′blü₁gras }

blue-green algae See cyanobacteria. { ¦blü¦grēn ′al·jē }

blue-green algal virus See cyanophage. { ¦blü¦grēn ₁al·gəl ′vī·rəs }

blue-green flame See green flash. { ¦blü¦grēn ′flām }

blue ground [GEOL] **1.** The decomposed peridotite or kimberlite that carries the diamonds in the South African mines. **2.** Strata of the coal measures, consisting principally of beds of hard clay or shale. { ¦blü ₁graúnd }

blue haze [ASTRON] A condition of the Martian atmosphere that sometimes causes it to be opaque to radiation near the blue end of the visible spectrum. { ¦blü ′hāz }

blue ice [HYD] Pure ice in the form of large, single crystals that is blue owing to the scattering of light by the ice molecules; the purer the ice, the deeper the blue. { ¦blü ′īs }

blueing See bluing. { ′blü·iŋ }

blue iron earth See vivianite. { ¦blü ¦ī·ərn ′ərth }

blue laser [OPTICS] A laser that emits bluish-purple light efficiently at room temperature from a semiconductor diode based on multiple quantum wells of III–V nitrides such as indium gallium nitride. { ¦blü ¦lā·zər }

blue lead See galena. { ¦blü ₁led }

blueline [GRAPHICS] A blue image or outline printed on paper or plastic sheeting and used as a guide for drafting, stripping, or layout; it does not reproduce when the finished work is photographed. { ′blü₁līn }

blue magnetism [GEOPHYS] The magnetism displayed by the south-seeking end of a freely suspended magnet; this is the magnetism of the earth's north magnetic pole. { ¦blü ′mag·nə₁tiz·əm }

BLOWOUT COIL

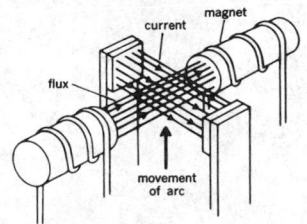

The relation of direction of current, magnetic flux, and movement of arc in a blowout coil.

BLUEBERRY

A cluster of blueberries.

BLUEFISH

The bluefish (*Pomatomus saltatrix*).

blue malachite See azurite. { ¦blü 'mal·ə,kīt }

blue metal [GEOL] The common fine-grained blue-gray mudstone which is part of many of the coal beds of England. { 'blü ,med·əl }

blue mold [MYCOL] Any fungus of the genus *Penicillium*. { 'blü ,mōld }

blue mud [GEOL] A combination of terrigenous and deep-sea sediments having a bluish gray color due to the presence of organic matter and finely divided iron sulfides. { 'blü ,məd }

blue nevus [MED] A nevus composed of spindle-shaped pigmented melanocytes in the middle and lower two-thirds of the dermis. { 'blü 'nē·vəs }

blue ocher See vivianite. { ¦blü 'ō·kər }

blueprint [GRAPHICS] **1.** A contact print, with white lines on a blue background, of a drawing; made on linen or on ferroprussiate paper and developed in water or a special solution. **2.** A photoprint used in offset lithography or photoengraving for use in checking positions of image elements. { 'blü,print }

blueprint machine [GRAPHICS] A machine which exposes and develops blueprint paper. { 'blü,print mə'shēn }

blue rot [PL PATH] A fungus disease of conifers caused by members of the genus *Ceratostomella*, characterized by blue discoloration of the wood. Also known as bluing. { 'blü ,rät }

blueschist facies [PETR] High-pressure, low-temperature metamorphism associated with subduction zones which produces a broad mineral association including glaucophane, actinolite, jadeite, aegirine, lawsonite, and pumpellyite. { 'blü,shist 'fā,shēz }

blue shift [ASTRON] A displacement of lines in the spectrum of a celestial object toward shorter wavelengths, indicating motion of the object toward the observer. { 'blü ,shift }

blue-sky scale See Linke scale. { ¦blü¦skī 'skāl }

blue spar See lazulite. { 'blü ,spär }

blue stain [PL PATH] A bluish stain of the sapwood of many trees that is caused by several fungi, especially of the genera *Fusarium*, *Ceratostomella*, and *Penicillium*. { 'blü ,stān }

blue star [ASTRON] A star of spectral type O, B, A, or F according to the Draper catalog. { 'blü ,stär }

bluestem grass [BOT] The common name for several species of tall, perennial grasses in the genus *Andropogon* of the order Cyperales. { 'blü,stem ¦gras }

bluestone [MINERAL] See chalcanthite. [PETR] **1.** A sandstone that is highly argillaceous and of even texture and bedding. **2.** The commercial name for a feldspathic sandstone that is dark bluish gray; it is easily split into thin slabs and used as flagstone. { 'blü,stōn }

blue straggler star [ASTRON] A member of a star cluster that lies above the turnoff point of the cluster's Hertzsprung-Russell diagram, and lies near the main sequence. { ¦blü 'strag·lər ,stär }

blue tetrazolium [ORG CHEM] $C_{40}H_{32}Cl_2N_8O_2$ Lemon yellow crystals that decompose at 242–245°C; soluble in chloroform, ethanol, and methanol; used in seed germination research, as a stain for molds and bacteria, and in histochemical studies. { ¦blü te·trə'zōl·ē·əm }

bluetongue [VET MED] An arthropod-borne disease of ruminant species that is caused by a ribonucleic acid-containing virus in the genus *Orbivirus*, family Reoviridae; acute infection evokes high fever, excessive salivation, nasal discharge, hyperemia (buccal and nasal mucosa, skin, coronet band), and erosions and ulcerations of mucosal surfaces in the mouth. { 'blü,təŋ }

blue vitriol [MATER] A hydrous solution of copper sulfate that is applied to the surface of a metal for layout purposes. [MINERAL] See chalcanthite. { 'blü 'vit·rē,ól }

blue water gas See blue gas. { 'blü 'wód·ər ,gas }

bluff [GEOGR] **1.** A steep, high bank. **2.** A broad-faced cliff. { bləf }

bluff body [AERO ENG] A body having a broad, flattened front, as in some reentry vehicles. { ¦bləf 'bäd·ē }

bluff-bowed ship [NAV ARCH] A ship that has a full broad bow. { ¦bləf ¦baud 'ship }

bluing [MET] Also spelled blueing. **1.** Formation of a bluish oxide film on polished steel; improves appearance and provides some corrosion resistance. **2.** Heating of formed springs to reduce internal stress. **3.** A blue oxide film formed on the polished surface of a metal due to extremely high temperatures. [PL PATH] See blue rot. { 'blü·iŋ }

Blumeriella jaapii [MYCOL] Previously known as *Coccomyces hiemalis*; a fungal plant pathogen that causes cherry leaf spot. { ,blü·mer·ē,el·ə 'jäp·ē,ē }

blunger [ENG] **1.** A large spatula-shaped wooden implement used to mix clay with water. **2.** A vat, containing a rotating shaft with fixed knives, for mixing clay and water into slip. { 'blən·jər }

blunging [ENG] The mixing or suspending of ceramic material in liquid by agitation, to form slip. { 'blən·jiŋ }

blunt dissection [MED] In surgery, the exposure of structures or separation of tissues without cutting. { ¦blənt di'sek·shən }

blunt file [DES ENG] A file whose edges are parallel. { ¦blənt ¦fīl }

blunting [DES ENG] Slightly rounding a cutting edge to reduce the probability of edge chipping. { 'blən·tiŋ }

blurring [MATH] An operation that decreases the value of the membership function of a fuzzy set if it is greater than 0.5, and increases it if it is less than 0.5. { 'blər·iŋ }

B lymphocyte See B cell. { ¦bē 'lim·fə,sīt }

BM See barrels per month; benchmark.

BMD See ballistic missile defense.

B meson [PARTIC PHYS] **1.** An elementary particle with strong nuclear interactions, baryon number $B = 0$, spin $J = 0$, positive parity, negative charge parity, isotopic spin $I = 1$, and mass 1234 MeV, that decays into an omega meson and a pion. **2.** A meson consisting of a combination of a bottom (*b*) quark with an ordinary up (*u*) or down (*d*) quark, having a mass of approximately 5.2 GeV. { ¦bē 'mä,zän }

BMI See body mass index.

B mode [ACOUS] A form of ultrasonic medical tomography in which a two-dimensional picture is formed by scanning the line-of-sight propagation path and monitoring the position and direction of the path. { 'bē ,mōd }

BMR See basal metabolic rate.

BMT See basic motion-time study.

BMV See brome mosaic virus.

BMX bicycle [MECH ENG] A small, extremely strong, type of bicycle, having generally 20-inch (500-millimeter) wheels, large-cleat (knobbly) tires, upright but not high-rise handlebars, and a seat positioned more towards the rear wheel than on a conventional bicycle, and used for stunt riding and tricks. { ,bē,em,eks 'bī,sik·əl }

BNC connector [ELEC] A small device for connecting coaxial cables, used frequently in low-power, radio-frequency and test applications. Abbreviation for bayonet Neil-Concelman connector. { ,bē,en'sē kə,nek·tər }

BNF See Backus-Naur form.

BNOA See β-naphthoxyacetic acid.

BN object See Becklin-Neugebauer object. { ¦bē'en ,äb·jəkt }

boa [VERT ZOO] Any large, nonvenomous snake of the family Boidae in the order Squamata. { 'bō·ə }

boar See wild boar. { bór }

board [MATER] A piece of lumber whose dimensions are less than 2 inches (5 centimeters) thick and between 4 and 12 inches (10 and 30 centimeters) wide. { bórd }

board coal See bituminous wood. { 'bórd ,kōl }

board drop hammer [MECH ENG] A type of drop hammer in which the ram is attached to wooden boards which slide between two rollers; after the ram falls freely on the forging, it is raised by friction between the rotating rollers. Also known as board hammer. { 'bórd 'dräp ,ham·ər }

board-foot [ENG] Unit of volume in measuring lumber; equals 144 cubic inches (2360 cubic centimeters), or the volume of a board 1 foot square and 1 inch thick. Abbreviated bd-ft. { ¦bórd'fút }

board hammer See board drop hammer. { 'bórd ,ham·ər }

boarding [ENG] **1.** A batch of boards. **2.** Covering with boards. { 'bor·diŋ }

board measure [ENG] Measurement of lumber in board-feet. Abbreviated bm. { 'bórd ,mezh·ər }

Board of Trade unit See kilowatt-hour. { ¦bórd əv 'trād ,yü·nət }

boart See bort. { bórt }

boast [ENG] **1.** To shape stone or curve furniture roughly in preparation for finer work later on. **2.** To finish the face of a building stone by cutting a series of parallel grooves. { bōst }

boaster *See* boasting chisel.   { 'bō·stər }

Boas' test   [PATH]   A test that uses an alcoholic solution of resorcinol and sucrose to determine the presence of free hydrochloric acid in gastric juices.   { 'bō,as ,test }

boasting chisel   [DES ENG]   A broad chisel used in boasting stone. Also known as boaster.   { 'bōs·tiŋ ,chiz·əl }

boat   [CHEM]   A platinum or ceramic vessel for holding a substance for analysis by combustion.   [NAV ARCH]   A small watercraft.   { bōt }

boat boom   [NAV ARCH]   A spar swung out from the side of a vessel, to which boats can be securely attached. Also known as boat spar; riding boom.   { 'bōt ,büm }

boat conformation   [ORG CHEM]   A boat-shaped conformation in space which can be assumed by cyclohexane or similar compounds; a relatively unstable form.   { 'bōt ,kän·fər'mā·shən }

boat deck   [NAV ARCH]   The deck on a ship that usually is used to stow lifeboats.   { 'bōt ,dek }

boat fall   [NAV ARCH]   One of a pair of tackles used in lowering or hoisting a boat from or to the davits of a ship.   { 'bōt ,fȯl }

boat hook   [NAV ARCH]   A long rod with a knob on one end and a metal point and hook at the other; used to push or pull other boats, logs, or objects from or to the side of a boat or to engage lines, rings, or buoys.   { 'bōt ,hùk }

boat plug   [NAV ARCH]   A threaded plug that stops up the drainage hole of a boat near the keel and can be removed when the boat is dry-docked to drain out bilge water.   { 'bōt ,pləg }

boat spar *See* boat boom.   { 'bōt ,spär }

boat spike   [DES ENG]   A long, square spike used in construction with heavy timbers. Also known as barge spike.   { 'bōt ,spīk }

boattail   [AERO ENG]   Of an elongated body such as a rocket, the rear portion having decreasing cross-sectional area.   { 'bōt,tāl }

bob   [MET]   A feeding device for providing molten metal to a casting during solidification to prevent shrinkage.   { bäb }

Bobasatranidae   [PALEON]   A family of extinct palaeonisciform fishes in the suborder Platysomoidei.   { bə,bas·ə'tran·ə,dē }

bobbin   [ELECTROMAG]   An insulated spool serving as a support for a coil.   [TEXT]   A cylinder with projecting edges at one or both ends and a hole along the axis, used for winding twisted strands of textile fiber, thread, or yarn.   { 'bäb·ən }

bobbin core   [ELECTROMAG]   A magnetic core having a form or bobbin on which the ferromagnetic tape is wrapped for support of the tape.   { 'bäb·ən ,kȯr }

bobbinet   [TEXT]   A machine-made net fabric with a hexagonal mesh; made of cotton, silk, rayon, or nylon and used in the manufacture of garments or as a base for embroidered or appliqued laces.   { 'bäb·ə,net }

bobbing   [ELECTR]   Fluctuation of the strength of a radar echo, or its indication on a radarscope, due to alternate interference and reinforcement of returning reflected waves.   { 'bäb·iŋ }

bobbin lace   [TEXT]   A lace made by hand from thread wound on bobbins by following a pattern denoted by pins, usually worked over a fabric pillow. Also known as pillow lace.   { 'bäb·ən ,lās }

Bobeck effect   [SOLID STATE]   The contraction of magnetic strip domains in a thin magnetic film to cylindrical domains called magnetic bubbles.   { 'bäb·ək i,fekt }

bobierrite   [MINERAL] $Mg_3(PO_4)_2 \cdot 8H_2O$ A transparent, colorless or white, monoclinic mineral consisting of octahydrated magnesium phosphate.   { 'bō·bē·ə,rīt }

Bobillier's law   [MECH]   The law that, in general plane rigid motion, when a and b are the respective centers of curvature of points A and B, the angle between Aa and the tangent to the centrode of rotation (pole tangent) and the angle between Bb and a line from the centrode to the intersection of AB and ab (collineation axis) are equal and opposite.   { bō'bil·yāz ,lȯ }

bobtail curtain antenna   [ELECTROMAG]   A bidirectional, vertically polarized, phased-array antenna that has two horizontal sections, each 0.5 electrical wavelength long, that connect three vertical sections, each 0.25 electrical wavelength long.   { 'bäb,tāl 'kərt·ən an,ten·ə }

Bochner integral   [MATH]   The Bochner integral of a function, f, with suitable properties, from a measurable set, A, to a Barach space, B, is the limit of the integrals over A of a sequence of simple functions, s_n, from A to B such that the limit of the integral over A of the norm of $f - s_n$ approaches zero.   { 'bäk·nər int·i·grəl }

bock beer   [FOOD ENG]   A dark, heavy, rich beer.   { 'bäk 'bir }

BOD *See* biochemical oxygen demand.

Bode diagram   [ELECTR]   A diagram in which the phase shift or the gain of an amplifier, a servomechanism, or other device is plotted against frequency to show frequency response; logarithmic scales are customarily used for gain and frequency.   { 'bōd ,dī·ə,gram }

bodenite   [MINERAL]   A metallic, steel-gray mineral consisting of cobalt, nickel, iron, arsenic, and bismuth; occurs in granular to fibrous masses.   { 'bōd·ən,īt }

Bodenstein number   [FL MECH]   A dimensionless group used in the study of diffusion in reactors.   { 'bō·dən,stīn ,nəm·bər }

Bode's law   [ASTRON]   An empirical law giving mean distances of planets to the sun by the formula $a = 0.4 + 0.3 \times 2^n$, where a is in astronomical units and n equals $-\infty$ for Mercury, 0 for Venus, 1 for Earth, and so on; the asteroids are included as planets. Also known as Titius-Bode law.   { 'bōdz ,lȯ }

bodily tide *See* earth tide.   { 'bäd·əl·ē 'tīd }

Bodonidae   [INV ZOO]   A family of protozoans in the order Kinetoplastida characterized by two unequally long flagella, one of them trailing.   { bə'dän·ə,dē }

BOD test *See* biochemical oxygen demand test.   { 'bē¦ō'dē ,test }

body   [AERO ENG]   **1.** The main part or main central portion of an airplane, airship, rocket, or the like; a fuselage or hull. **2.** Any fabrication, structure, or other material form, especially one aerodynamically or ballistically designed; for example, an airfoil is a body designed to produce an aerodynamic reaction.   [GEOGR]   A separate entity or mass of water, such as an ocean or a lake.   [GEOL]   An ore body, or pocket of mineral deposit.   [MATER]   The consistency or viscosity of fluid materials, such as lubricating oils, paints, and cosmetics.   [MECH ENG]   The part of a drill which runs from the outer corners of the cutting lips to the shank or neck.   { 'bäd·ē }

body angle   [AERO ENG]   The angle which the longitudinal axis of the airframe makes with some selected line.   { 'bäd·e ,aŋ·gəl }

body axis   [AERO ENG]   Any one of a system of mutually perpendicular reference axes fixed in an aircraft or a similar body and moving with it.   { 'bäd·ē ,ak·səs }

body burden   [NUCLEO]   The amount of radioactive material present in the body of a human or animal.   { 'bäd·ē ,bərd·ən }

body capacitance   [ELEC]   Capacitance existing between the human hand or body and a circuit.   { 'bäd·ē kə'pas·ə·təns }

body cavity   [ANAT]   The peritoneal, pleural, or pericardial cavities, or the cavity of the tunica vaginalis testis.   { 'bäd·ē ,kav·əd·ē }

body-centered lattice   [CRYSTAL]   A space lattice in which the point at the intersection of the body diagonals is identical to the points at the corners of the unit cell.   { 'bäd·ē ,sen·tərd 'lad·əs }

body centrode   [MECH]   The path traced by the instantaneous center of a rotating body relative to the body.   { 'bäd·ē 'sen,trōd }

body cone   [MECH]   The cone in a rigid body that is swept out by the body's instantaneous axis during Poinsot motion. Also known as polhode cone.   { 'bäd·ē ,kōn }

body force   [MECH]   An external force, such as gravity, which acts on all parts of a body.   { 'bäd·ē ,fȯrs }

body-load aggregate   [IND ENG]   A biomechanical unit that comprises the combined weight of the load being manipulated and the body segments involved in the task.   { 'bäd·ē 'lōd 'a·grə·gət }

body mass index   [MED]   An estimation of the amount of fat stored in adipose tissue that can be calculated by dividing the body weight in kilograms by the square of the height in meters. Abbreviated BMI.   { 'bäd·ē 'mas ,in·deks }

body motion   [IND ENG]   Motion of parts of a human body requiring a change of posture or weight distribution.   { 'bäd·ē ,mō·shən }

body of revolution   [MATH]   A symmetrical body having the form described by rotating a plane curve about an axis in its plane.   { 'bäd·ē əv rev·ə'lü·shən }

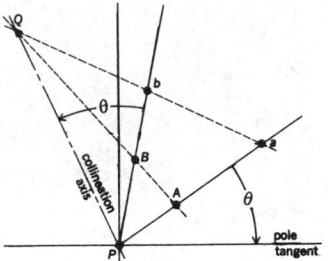

BOBILLIER'S LAW

Application of Bobillier's law of four-bar linkage consisting of fixed member *ab*, cranks *aA* and *bB*, and connecting rod *AB*.

BODONIDAE

Bodo, a genus in the family Bodonidae, showing the two unequally long flagella.

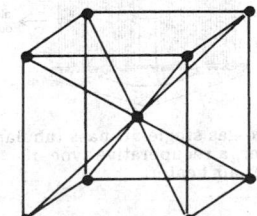

BODY-CENTERED LATTICE

Drawing of a body-centered cubic lattice.

body rhythm [PHYSIO] Any bodily process having some degree of regular periodicity. { 'bäd·ē ,rith·əm }

body-righting reflex [PHYSIO] A postural reflex, initiated by the asymmetric stimulation of the body surface by the weight of the body, so that the head tends to assume a horizontal position. { 'bäd·ē ,rīd·iŋ 'rē,fleks }

body rotation [CONT SYS] An axis of motion of a pick-and-place robot. { 'bäd·e rō,tā·shən }

body wave [GEOPHYS] A seismic wave that travels within the earth, as distinguished from one that travels along the surface. { 'bäd·ē ,wāv }

boehmite [MINERAL] AlO(OH) Gray, brown, or red orthorhombic mineral that is a major constituent of some bauxites. { 'bā,mīt }

boehm lamellae [GEOL] Lines or bands with dusty inclusions that are subparallel to the basal plane of quartz. { ¦bämlə 'mel·ē }

Boersch effect [ELECTR] The deviation of the energy distribution of electrons emitted from a cathode from a Maxwellian distribution, due to broadening of the distribution by a space-charge region in front of the cathode. { 'bersh i,fekt }

Boettger's test [ANALY CHEM] A test for the presence of saccharides, utilizing the reduction of bismuth subnitrate to metallic bismuth, a precipitate. { 'bet,gərz ,test }

bog [ECOL] A plant community that develops and grows in areas with permanently waterlogged peat substrates. Also known as moor; quagmire. { bäg }

bogen structure [GEOL] The structure of vitric tuffs composed largely of shards of glass. { 'bō·gən ,strək·chər }

bogey See bogie. { 'bō·gē }

bog harrow [AGR] A type of disk harrow with extra-large notched disks. { 'bäg ,har·ō }

boghead cannel shale [GEOL] A coaly shale that contains much waxy or fatty algae. { 'bäg,hed ¦kan·əl ,shāl }

boghead coal [GEOL] Bituminous or subbituminous coal containing a large proportion of algal remains and volatile matter; similar to cannel coal in appearance and combustion. { 'bäg,hed ,kōl }

bogie Also spelled bogey; bogy. [AERO ENG] A type of landing-gear unit consisting of two sets of wheels in tandem with a central strut. [ENG] **1.** A supporting and aligning wheel or roller on the inside of an endless track. **2.** A low truck or cart of solid build. **3.** A truck or axle to which wheels are fixed, which supports a railroad car, the leading end of a locomotive, or the end of a vehicle (such as a gun carriage) and which is allowed to swivel under it. **4.** A railroad car or locomotive supported by a bogie. [MECH ENG] The drive-wheel assembly and supporting frame comprising the four rear wheels of a six-wheel truck, mounted so that they can self-adjust to sharp curves and irregularities in the road. [MIN ENG] A small truck or trolley upon which a bucket is carried from the shaft to the spoil bank. { 'bō·gē }

bog iron ore [MINERAL] A soft, spongy, porous deposit of impure hydrous iron oxides formed in bogs, marshes, swamps, peat mosses, and shallow lakes by precipitation from iron-bearing waters and by the oxidation action of algae, iron bacteria, or the atmosphere. Also known as lake ore; limnite; marsh ore; meadow ore; morass ore; swamp ore. { 'bäg ¦ī·ərn ,ór }

bog manganese See wad. { 'bäg 'maŋ·gə,nēs }

bog-mine ore See bog ore. { 'bäg ,mīn ,ór }

bog moss [ECOL] Moss of the genus Sphagnum occurring as the characteristic vegetation of bogs. { 'bäg ,mós }

bog ore [MINERAL] A poorly stratified accumulation of earthy metallic mineral substances, consisting mainly of oxides, that are formed in bogs, marshes, swamps, and other low-lying moist places. Also known as bog-mine ore. { 'bäg ,ór }

bogy See bogie. { 'bō,gē }

Bohemian glass See hard glass. { bō'hem·ē·ən ¦glas }

Bohemian ruby See rose quartz. { bō'hem·ē·ən ¦rü·bē }

Bohemian topaz See citrine. { bō'hem·ē·ən ¦tō,paz }

Bohr atom [ATOM PHYS] An atomic model having the structure postulated in the Bohr theory. { 'bór ,ad·əm }

Bohr-Breit-Wigner theory See Breit-Wigner theory. { ¦bór ¦brīt 'vig·nər ,thē·ə·rē }

Bohr effect [BIOCHEM] The effect of carbon dioxide and pH on the oxygen equilibrium of hemoglobin; increase in carbon dioxide prevents an increase in the release of oxygen from oxyhemoglobin. { 'bór i'fekt }

Bohr frequency condition [ATOM PHYS] The law that the frequency of the radiation emitted or absorbed during the transition of an atomic system between two stationary states equals the difference in the energies of the states divided by Planck's constant. { ¦bór 'frē·kwən·sē kən,dish·ən }

bohrium [CHEM] A synthetic chemical element, symbolized Bh, atomic number 107; the fifteenth transuranium element. { 'bór·ē·əm }

Bohr magneton [ATOM PHYS] The amount $he/4\pi mc$ of magnetic moment, where h is Planck's constant, e and m are the charge and mass of the electron, and c is the speed of light. { 'bór 'mag·nə,tän }

Bohr orbit [ATOM PHYS] One of the electron paths about the nucleus in Bohr's model of the hydrogen atom. { 'bór ,ór·bət }

Bohr radius [ATOM PHYS] The radius of the ground-state orbit of the hydrogen atom in the Bohr theory. { 'bór ,rād·ē·əs }

Bohr's correspondence principle See correspondence principle. { 'bórz kär·ə'spän·dəns ,prin·sə·pəl }

Bohr-Sommerfeld theory [ATOM PHYS] A modification of the Bohr theory in which elliptical as well as circular orbits are allowed. { ¦bór ¦zó·mər,felt ,thē·ə·rē }

Bohr theory [ATOM PHYS] A theory of atomic structure postulating an electron moving in one of certain discrete circular orbits about a nucleus with emission or absorption of electromagnetic radiation necessarily accompanied by transitions of the electron between the allowed orbits. { 'bór ,thē·ə·rē }

Bohr-van Leeuwen theorem [QUANT MECH] The theorem that magnetism is inexplicable in classical physics and is a quantum phenomenon. { ¦bór van'lā·vən ,thir·əm }

Bohr-Wheeler theory of fission [NUC PHYS] A theory accounting for the stability of a nucleus against fission by treating it as a droplet of incompressible and uniformly charged liquid endowed with surface tension. { ¦bór ¦wēl·ər 'thē·ə·rē əv 'fish·ən }

Boidae [VERT ZOO] The boas, a family of nonvenomous reptiles of the order Squamata, having teeth on both jaws and hindlimb rudiments. { 'bō·ə,dē }

boil See furuncle. { bóil }

boil-away test See weathering test. { 'bóil ə¦wā ,test }

boil disease [VET MED] A protozoan disease of fish caused by Myxobolus pfeiffer that forms large tumorous masses in the muscles and connective tissues, finally causing death. { 'bóil di,zēz }

boiled oil [MATER] A drying oil in which metallic driers are added during the cooking; the reaction releases water, resulting in a boiling of the batch. { ¦bóild 'óil }

boiler [MECH ENG] A water heater for generating steam. { 'bóil·ər }

boiler air heater [MECH ENG] A component of a steam-generating unit that transfers heat from the products of combustion after they have passed through the steam-generating and superheating sections to combustion air, which recycles heat to the furnace. { 'bóil·ər 'er ,hēd·ər }

boiler casing [MECH ENG] The gas-tight structure surrounding the component parts of a steam generator. { 'bóil·ər ,kās·iŋ }

boiler circulation [MECH ENG] Circulation of water and steam in a boiler, which is required to prevent overheating of the heat-absorbing surfaces; may be provided naturally by gravitational forces, mechanically by pumps, or by a combination of both methods. { 'bóil·ər sər·kyə'lā·shən }

boiler cleaning [ENG] A mechanical or chemical process for removal of grease, scale, and other deposits from steam boiler surfaces. { 'bóil·ər ,klēn·iŋ }

boiler code [MECH ENG] A code, established by professional societies and administrative units, which contains the basic rules for the safe design, construction, and materials for steam-generating units, such as the American Society of Mechanical Engineers code. { 'bóil·ər ,kōd }

boiler compound [CHEM] Any chemical used to treat boiler water to prevent corrosion, the fouling of heat-absorbing surfaces, foaming, and the contamination of steam. { 'bóil·ər ,kam,paúnd }

boiler controls [MECH ENG] Either manual or automatic devices which maintain desired boiler operating conditions with respect to variables such as feedwater flow, firing rate, and steam temperature. { 'bóil·ər kən'trōlz }

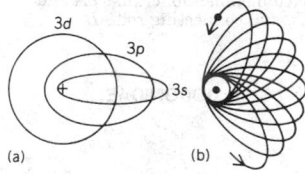

BOHR-SOMMERFELD THEORY

Possible elliptical orbits, according to the Bohr-Sommerfeld theory. (a) The three permitted orbits for $n = 3$. (b) Precession of the 3s orbit caused by the relativistic variation of mass.

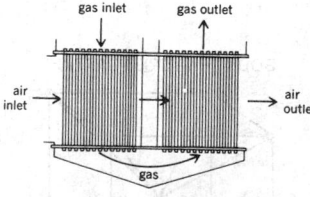

BOILER AIR HEATER

A two-gas single-air-pass tubular heater, a recuperative type of boiler air heater.

boiler draft [MECH ENG] The difference between atmospheric pressure and some lower pressure existing in the furnace or gas passages of a steam-generating unit. { 'bȯil·ər ‚draft }

boiler economizer [MECH ENG] A component of a steam-generating unit that transfers heat from the products of combustion after they have passed through the steam-generating and superheating sections to the feedwater, which it receives from the boiler feed pump and delivers to the steam-generating section of the boiler. { 'bȯil·ər i'kän·ə‚miz·ər }

boiler efficiency [MECH ENG] The ratio of heat absorbed in steam to the heat supplied in fuel, usually measured in percent. { 'bȯil·ər i'fish·ən‚sē }

boiler feedwater [MECH ENG] Water supplied to a steam-generating unit. { 'bȯil·ər 'fēd‚wȯd·ər }

boiler feedwater regulation [MECH ENG] Addition of water to the steam-generating unit at a rate commensurate with the removal of steam from the unit. { 'bȯil·ər 'fēd‚wȯd·ər reg·yə'lā·shən }

boiler fuel [MATER] Natural gas, residual oil, and various solid fuels used to heat a boiler, usually the most economical fuel available. { 'bȯil·ər ‚fyül }

boiler furnace [MECH ENG] An enclosed space provided for the combustion of fuel to generate steam in a boiler. Also known as steam-generating furnace. { 'bȯil·ər ‚fər·nəs }

boiler heat balance [MECH ENG] A means of accounting for the thermal energy entering a steam-generating system in terms of its ultimate useful heat absorption or thermal loss. { 'bȯil·ər 'hēt ‚bal·əns }

boiler horsepower [MECH ENG] A measurement of water evaporation rate; 1 boiler horsepower equals the evaporation per hour of 34 1/2 pounds (15.7 kilograms) of water at 212°F (100°C) into steam at 212°F. Abbreviated bhp. { 'bȯil·ər 'hȯrs‚pau̇·ər }

boiler hydrostatic test [MECH ENG] A procedure that employs water under pressure, in a new boiler before use or in old equipment after major alterations and repairs, to test the boiler's ability to withstand about 1 1/2 times the design pressure. { 'bȯil·ər hī·drə'stad·ik 'test }

boiler layup [MECH ENG] A significant length of time during which a boiler is inoperative in order to allow for repairs or preventive maintenance. { 'bȯil·ər 'lā·əp }

boiler plate [COMPUT SCI] A commonly used expression or phrase that is stored in memory and can be copied into a word-processing document as needed. [GEOL] A fairly smooth surface on a cliff, consisting of flush or overlapping slabs of rock, having little or no foothold. [HYD] A crusty, frozen surface of snow. [MET] Flat-rolled steel, usually 1/4 to 1/2 inch (6 to 13 millimeters) thick; used mainly for covering ships and making boilers and tanks. Also known as boiler steel. { 'bȯil·ər ‚plāt }

boiler-plate model [AERO ENG] A metal copy of a flight vehicle, the structure or components of which are heavier than the flight model. { 'bȯil·ər ‚plāt ‚mäd·əl }

boiler scale [CHEM] Deposits from silica and other contaminants in boiler water that form on the internal surfaces of heat-absorbing components, increase metal temperatures, and result in eventual failure of the pressure parts because of overheating. Also known as scale. { 'bȯil·ər ‚skāl }

boiler setting [MECH ENG] The supporting steel and gastight enclosure for a steam generator. { 'bȯil·ər ‚sed·iŋ }

boiler steel See boiler plate. { 'bȯil·ər ‚stēl }

boiler storage [MECH ENG] A steam-generating unit that, when out of service, may be stored wet (filled with water) or dry (filled with protective gas). { 'bȯil·ər ‚stȯr·ij }

boiler superheater [MECH ENG] A boiler component, consisting of tubular elements, in which heat is added to high-pressure steam to increase its temperature and enthalpy. { 'bȯil·ər ‚sü·pər‚hēd·ər }

boiler trim [MECH ENG] Piping or tubing close to or attached to a boiler for connecting controls, gages, or other instrumentation. { 'bȯil·ər ‚trim }

boiler tube [MECH ENG] One of the tubes in a boiler that carry water (water-tube boiler) to be heated by the high-temperature gaseous products of combustion or that carry combustion products (fire-tube boiler) to heat the boiler water that surrounds them. { 'bȯil·ər ‚tüb }

boiler walls [MECH ENG] The refractory walls of the boiler furnace, usually cooled by circulating water and capable of withstanding high temperatures and pressures. { 'bȯil·ər ‚wȯlz }

boiler water [MECH ENG] Water in the steam-generating section of a boiler unit. { 'bȯil·ər ‚wȯd·ər }

boiling [ASTRON] The telescopic appearance of the limbs of the sun and planets when the earth's atmosphere is turbulent, characterized by a constant rippling motion and lack of a clearly defined edge. [PHYS CHEM] The transition of a substance from the liquid to the gaseous phase, taking place at a single temperature in pure substances and over a range of temperatures in mixtures. { 'bȯil·iŋ }

boiling point [PHYS CHEM] Abbreviated bp. **1.** The temperature at which the transition from the liquid to the gaseous phase occurs in a pure substance at fixed pressure. **2.** See bubble point. { 'bȯil·iŋ ‚pȯint }

boiling-point elevation [CHEM] The raising of the normal boiling point of a pure liquid compound by the presence of a dissolved substance, the elevation being in direct relation to the dissolved substance's molecular weight. { 'bȯil·iŋ ‚pȯint el·ə'vā·shən }

boiling range [CHEM] The temperature range of a laboratory distillation of an oil from start until evaporation is complete. { 'bȯil·iŋ ‚rānj }

boiling spring [HYD] **1.** A spring which emits water at a high temperature or at boiling point. **2.** A spring located at the head of an interior valley and rising from the bottom of a residual clay basin. **3.** A rapidly flowing spring that develops strong vertical eddies. { 'bȯil·iŋ ‚spriŋ }

boiling-water reactor [NUCLEO] A nuclear reactor in which the coolant is water, maintained at such a pressure as to allow it to boil and form steam. Abbreviated BWR. { 'bȯil·iŋ ‚wȯd·ər rē'ak·tər }

boil-off [TEXT] Removal of impurities from fabric by boiling the material in a solution. [THERMO] The vaporization of a liquid, such as liquid oxygen or liquid hydrogen, as its temperature reaches its boiling point under conditions of exposure, as in the tank of a rocket being readied for launch. { 'bȯil‚ȯf }

boil-off assistant [MATER] A material used to facilitate the removal of natural glue from silk during boiling in soap solution (degumming); examples are sulfonated oils, pine oils, and solvents and other wetting agents. { 'bȯil‚ȯf ə'sis·tənt }

boil smut [PL PATH] A fungus disease of corn caused by Ustilago maydis, characterized by galls containing black spores. { 'bȯil ‚smət }

bojite [PETR] **1.** A gabbro with primary hornblende substituting for augite. **2.** Hornblende diorite. { 'bō‚jīt }

boldface [GRAPHICS] A type that has thick heavy lines so that it has a conspicuous black appearance. { 'bōld‚fās }

bole [FOR] The main stem of a tree of substantial diameter; capable of yielding timber, veneer logs, and large poles. [GEOL] Any of various red, yellow, or brown earthy clays consisting chiefly of hydrous aluminum silicates. Also known as bolus; terra miraculosa. [GRAPHICS] The foundation laid for gold leaf. { 'bōl }

boleite [MINERAL] A deep Prussian blue, tetragonal mineral consisting of a hydroxide-chloride of lead, copper, and silver. { bō'lā‚īt }

boletic acid See fumaric acid. { bə'led·ik 'as·əd }

bolide [ASTRON] A brilliant meteor, especially one which explodes; a detonating fireball meteor. { 'bō‚līd }

boll [BOT] A pod or capsule (pericarp), as of cotton and flax. { 'bōl }

bollard [CIV ENG] A heavy post on a dock or ship used in mooring ships. { 'bäl·ərd }

boll rot [PL PATH] A fungus rot of cotton bolls caused by Glomerella gossypii and Xanthomonas malvacearum. { 'bōl ‚rät }

bollseye See sodium cacodylate. { 'bōlz‚ī }

boll weevil [INV ZOO] A beetle, Anthonomus grandis, of the order Coleoptera; larvae destroy cotton plants and are the most important pests in agriculture. { 'bōl ‚wē·vəl }

boll-weevil hanger [PETRO ENG] A screw-on connector used to connect and seal the lower end of a length of oil-well casing to the next smaller casing string. { 'bōl ‚wē·vəl ‚haŋ·ər }

bolograph [ENG] Any graphical record made by a bolometer; in particular, a graph formed by directing a pencil of light reflected from the galvanometer of the bolometer at a moving photographic film. { 'bōl·ə‚graf }

bolometer [ENG] An instrument that measures the energy of electromagnetic radiation in certain wavelength regions by utilizing the change in resistance of a thin conductor caused by the heating effect of the radiation. Also known as thermal detector. { bə'läm·əd·ər }

bolometric correction [ASTRON] The difference between the bolometric and visual magnitude. { ¦bō·lə¦me·trik kə'rek·shən }

bolometric magnitude [ASTRON] The magnitude of a celestial object, as calculated from the total amount of radiation received from the object at all wavelengths. { ¦bō·lə¦me·trik 'mag·nə,tüd }

bolometric neutrino detection [NUCLEO] A technique for detecting low-energy neutrinos (less than 1 MeV) by measuring small temperature changes or the production of phonons (sound waves) generated in ultracold materials by the coherent scattering of the neutrinos off electrons or their coherent elastic scattering off nuclei. { ¦bō·lə¦me·trik nü'trē·no di,tek·shən }

bolson [GEOL] In the southwestern United States, a basin or valley having no outlet. { bōl,sän }

bolster [ENG] A plate for maintaining a fixed space between stacked heat exchangers or heat-exchanger shells. [MET] A block of steel to which drop-forging dies are attached. { 'bōl·stər }

bolster plate [MECH ENG] A plate fixed on the bed of a power press to locate and support the die assembly. { 'bōl·stər ,plāt }

bolt [DES ENG] A rod, usually of metal, with a square, round, or hexagonal head at one end and a screw thread on the other, used to fasten objects together. [FOR] A short section of tree trunk. [MATER] In veneer production, a short log of a length suitable for peeling on a lathe. [MIN ENG] See bolthole. [ORD] The sliding part in a breechloading weapon that pushes a cartridge into position and holds it there as the gun is fired. [TEXT] The entire length of cloth from a loom. { bōlt }

bolt blank [DES ENG] A threadless bolt with a head that can be threaded for specific applications. Also known as screw blank. { 'bōlt ,blaŋk }

bolted joint [ENG] The assembly of two or more parts by a threaded bolt and nut or by a screw that passes through one member and threads into another. { ¦bōl·təd 'jȯint }

bolted rail crossing [CIV ENG] A crossing whose running surfaces are made of rolled rail and whose parts are joined with bolts. { ¦bōl·təd ,rāl 'krȯs·iŋ }

bolt face [ORD] That portion of a gun bolt that abuts the base of the cartridge case. { 'bōlt ,fās }

bolthole [MIN ENG] A short, narrow opening made to connect the main working with the airhead or ventilating drift of a coal mine. Also known as bolt. { 'bōlt,hōl }

bolting [ENG] A fastening system using screw-threaded devices such as nuts, bolts, or studs. [FOOD ENG] The process of refining or purifying, especially of sifting flour or meal through a sieve. [MIN ENG] The use of vibrating sieves to separate particles of different sizes. { 'bōl·tiŋ }

bolting cloth [MATER] A sieve cloth made of wire, hair, or silk or other thread used to remove lumps from flour or to make screen prints or needlework. { 'bōl·tiŋ ,klȯth }

bolt sleeve [DES ENG] A tube designed to surround a bolt in a concrete wall to prevent the concrete from adhering to the bolt. { 'bōlt ,slēv }

boltwoodite [MINERAL] $K_2(UO_2)_2(SiO_3)_2(OH)_2 \cdot 5H_2O$ Yellow mineral consisting of hydrous potassium uranyl silicate. { 'bōlt·wə,dīt }

Boltzmann constant [STAT MECH] The ratio of the universal gas constant to the Avogadro number. { 'bōlts·mən ,kän·stənt }

Boltzmann distribution [STAT MECH] A function giving the probability that a molecule of a gas in thermal equilibrium will have generalized position and momentum coordinates within given infinitesimal ranges of values, assuming that the molecules obey classical mechanics. { 'bōlts·mən dis·trə'byü·shən }

Boltzmann engine [THERMO] An ideal thermodynamic engine that utilizes blackbody radiation; used to derive the Stefan-Boltzmann law. { 'bōlts·mən ,en·jən }

Boltzmann entropy hypothesis [STAT MECH] The hypothesis that the entropy of a system in a given state is directly proportional to the logarithm of the probability of finding it in that state. { 'bōlts·mən 'en·trə·pē hī'päth·ə·səs }

Boltzmann factor [STAT MECH] The factor $\exp(-E/kT)$ that appears in the expression giving the probability for atoms to have an excitation energy E when at temperature T, where k is the Boltzmann constant. { 'bōlts,mən ,fak·tər }

Boltzmann H theorem [STAT MECH] The theorem that the entropy of a system never decreases; Boltzmann proved this for a classical gas of colliding particles. Also known as H theorem of Boltzmann. { 'bōlts·mən 'āch ,thir·əm }

Boltzmann statistics See Maxwell-Boltzmann statistics. { 'bōlts·mən stə'tis·tiks }

Boltzmann transport equation [STAT MECH] An equation used to study the nonequilibrium behavior of a collection of particles; it states that the rate of change of a function which specifies the probability of finding a particle in a unit volume of phase space is equal to the sum of terms arising from external forces, diffusion of particles, and collisions of the particles. Also known as Maxwell-Boltzmann equation. { 'bōlts·mən 'tranz,pȯrt i'kwā·zhən }

Boltzmann-Vlasov equations [PL PHYS] The equations that govern a high-temperature plasma in which the collisional mean free path is much larger than all the characteristic lengths of the system. { 'bōlts·mən 'vlä,sȯf i'kwā·zhənz }

bolus [GEOL] See bole. [PHARM] A pill of large size. [PHYSIO] The mass of food prepared by the mouth for swallowing. { 'bō·ləs }

bolus alba See kaolin. { 'bō·ləs 'äl·bə }

Bolyai geometry See Lobachevski geometry. { 'bōl·yī jē'äm·ə·trē }

Bolzano's theorem [MATH] The theorem that a single-valued, real-valued, continuous function of a real variable is equal to zero at some point in an interval if its values at the end points of the interval have opposite sign. { ,bōl'tsän·ōz ,thir·əm }

Bolzano-Weierstrass property [MATH] The property of a topological space, each of whose infinite subsets has at least one accumulation point. { bōl'tsän·ō 'vī·ər,shträs ,präp·ərd·ē }

Bolzano-Weierstrass theorem [MATH] The theorem that every bounded, infinite set in finite dimensional Euclidean space has a cluster point. { ,bōl'tsän·ō 'vī·ər,shträs ,thir·əm }

bomb [COMPUT SCI] See abend. [GEOL] Any large (greater than 64 millimeters) pyroclast ejected while viscous. [ORD] An explosive or other lethal agent, together with its container or holder, which is planted or thrown by hand, dropped from an aircraft, or projected by some other slow-speed device (such as a mortar) and used to destroy, damage, injure, or kill. { bäm }

Bombacaceae [BOT] A family of dicotyledonous tropical trees in the order Malvales with dry or fleshy fruit usually having woolly seeds. { ,bäm·bə'kās·ē,ē }

bomb adapter-booster [ORD] A device consisting of an adapter and an auxiliary explosive charge, used in an explosive train to detonate a bomb. { ¦bäm ə'dap·tər ,büs·tər }

bombard [NUCLEO] To direct a stream of particles or photons against a target. [ORD] To carry out a sustained attack upon a city, fort, or the like with bombs, projectiles, rockets, or other explosive missiles. { bäm'bärd }

bombardment [ELECTR] The use of induction heating to heat electrodes of electron tubes to drive out gases during evacuation. { bäm'bärd·mənt }

bombardment aircraft See bomber. { bam'bärd·mənt ¦er,kraft }

Bombay blood group system [IMMUNOL] A system comprising an immunologically distinct, genetically determined group of human erythrocytes characterized by the lack of A, B, or H antigens. { bäm'bā 'bləd ,grüp ,sis·təm }

bombazine [TEXT] A fine English fabric with a plain or twill weave, often with silk warp and worsted filling. { ¦bam·bə¦zēn }

bomb ballistics [MECH] The special branch of ballistics concerned with bombs dropped from aircraft. { 'bäm bə'lis·tiks }

bomb bay [AERO ENG] The compartment or bay in the fuselage of a bomber where the bombs are carried for release. { ¦bäm ¦bā }

bomb calorimeter [ENG] A calorimeter designed with a strong-walled container constructed of a corrosion-resistant alloy, called the bomb, immersed in about 2.5 liters of water in a metal container; the sample, usually an organic compound,

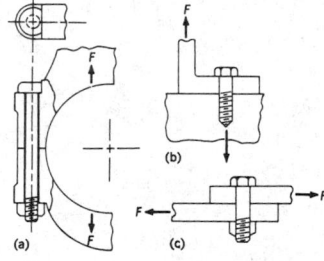

BOLTED JOINT

Forms of bolted joint (*F is force*). (a) Direct tension. (b) Eccentric loading. (c) Single shear.

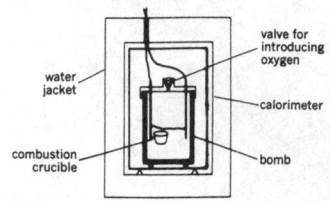

BOMB CALORIMETER

Schematic diagram of bomb calorimeter.

is ignited by electricity, and the heat generated is measured. { 'bäm kȧl·ə'rim·əd·ər }

bomb carpet [ORD] The fall of bombs, or the bombfall pattern produced, in carpet bombing; the area struck by carpet bombing. { 'bäm ,kär·pət }

bomb casing [ORD] The principal container, usually metal, for the main charge of a bomb. { 'bäm ,kās·iŋ }

bomb complete round [ORD] A complete aerial bomb, including all of the components such as arming wires and fuses, necessary to attach the bomb to a release mechanism and to make the bomb function after release. { 'bäm kəm'plēt 'rȧund }

bomb dispenser [ORD] An aerodynamically shaped unit that is externally mounted but not permanently fixed on high-speed aircraft to carry and eject bombs. { 'bäm dis'pen·sər }

bomb ejection cartridge [ORD] An explosive item used to eject a bomb from a bomb cluster or bomb station. { 'bäm i'jek·shən ,kär·trij }

bomber [AERO ENG] An airplane specifically designed to carry and drop bombs. Also known as bombardment aircraft. { 'bäm·ər }

bomb fin [ORD] A fin attached to a bomb in order to afford directional stability. { 'bäm ,fin }

bombiccite See hartite. { 'bäm'bē,chīt }

Bombidae [INV ZOO] A family of relatively large, hairy, black and yellow bumblebees in the hymenopteran superfamily Apoidea. { 'bäm·bə,dē }

bombing [ORD] The action of dropping bombs from an aircraft with the purpose of hitting a target. { 'bäm·iŋ }

bombing table [ORD] A table giving the bombsight settings required for dropping a particular type of bomb at various speeds and altitudes. { 'bäm·iŋ ,tā·bəl }

bombing through overcast [ORD] Blind bombing through clouds, using radar, infrared equipment, or other electronic equipment for guidance. Abbreviated BTO. Also known as blind bombing. { 'bäm·iŋ thrü 'ō·vər,kast }

bomb load [ORD] **1.** The weight or number of bombs carried by an aircraft. **2.** The bomb or bombs carried. { 'bäm ,lōd }

bombproof [ENG] Referring to shelter, building, or other installation resistant or impervious to the effects of bomb explosions. { 'bäm,prüf }

bomb rack [ORD] A mechanical device, fitted to an airplane in the bomb bay or suspended underneath the plane, that releases and arms bombs at the bombardier's or pilot's command. { 'bäm ,rak }

bomb reconnaissance [ORD] Reconnoitering to determine the presence of an unexploded missile, ascertaining its nature, applying all practicable protective measures for the protection of personnel, installations, and equipment, and finally reporting essential information to the authority directing explosive ordnance disposal operations. { 'bäm ri'kän·ə·səns }

bomb-release line [ORD] An imaginary line around a target area at which a bomber, traveling toward it at a constant speed and altitude, releases its first bomb so that it and others will strike the target area. { 'bäm rə'lēs ,līn }

bomb-release point [ORD] The point in flight on a bomb run at which a bombing airplace releases its bomb load. { 'bäm rə'lēs ,pȯint }

bomb run [ORD] The flight course of a bombing airplane just before the release of bombs. { 'bäm ,rən }

bomb sag [GEOL] Depressed and deranged laminae mainly found in beds of fine-grained ash or tuff around an included volcanic bomb or block which fell on and became buried in the deposit. { 'bäm ,sag }

bomb shelter [CIV ENG] A bomb-proof structure for protection of people. { 'bäm ,shel·tər }

bombsight [ORD] A device which determines, or enables a bombardier to determine, the point in space at which a bomb must be released from an aircraft in order to hit a target. { 'bäm ,sīt }

bomb test [ENG] A leak-testing technique in which the vessel to be tested is immersed in a pressurized fluid which will be driven through any leaks present. { 'bäm ,test }

Bombycidae [INV ZOO] A family of lepidopteran insects of the superorder Heteroneura that includes only the silkworms. { 'bäm'bis·ə,dē }

bombykol [BIOL] The first pheromone to be characterized chemically; it is an unsaturated straight-chain alcohol secreted in microgram amounts by females of the silkworm moth (*Bombyx mori*) and is capable of attracting male silkworm moths at large distances. { 'bäm·bə,kȯl }

Bombyliidae [INV ZOO] The bee flies, a family of dipteran insects in the suborder Orthorrhapha. { ,bäm·bə'lī·ə,dē }

Bombyx [INV ZOO] The type genus of Bombycidae. { 'bäm,biks }

Bombyx mori [INV ZOO] The species name of the commercial silkworm. { 'bäm,biks 'mȯr·ē }

bond [CHEM] The strong attractive force that holds together atoms in molecules and crystalline salts. Also known as chemical bond. [CIV ENG] A piece of building material that serves to unite or bond, such as an arrangement of masonry units. [ELEC] The connection made by bonding electrically. [ENG] **1.** A wire rope that fixes loads to a crane hook. **2.** Adhesion between cement or concrete and masonry or reinforcement. [MET] **1.** Material added to molding sand to impart bond strength. **2.** Junction of the base metal and filler metal, or the base metal beads, in a welded joint. { bänd }

Bond albedo [OPTICS] The fraction of the total incident light that is reflected by a spherical body. { 'bänd al,bē·dō }

Bond and Wang theory [MECH ENG] A theory of crushing and grinding from which the energy, in horsepower-hours, required to crush a short ton of material is derived. { 'bänd ən 'waŋ ,thē·ə·rē }

bond angle [PHYS CHEM] The angle between bonds sharing a common atom. Also known as valence angle. { 'bänd ,aŋ·gəl }

bond blister [MET] An unbonded area at the interface between the coating and the metal core of a cladded surface. { 'bänd ,blis·tər }

bond clay [MATER] A type of clay with high plasticity and high dry strength used to bond nonplastic materials; may be refractory. { 'bänd ,klā }

bond coat [MATER] **1.** A coat of bonding agent or plaster applied to a surface to provide a bond for succeeding coats of plaster. **2.** A coat of primer applied to a surface to act as a sealer or to ensure adhesion of paint to the surface. { 'bänd ,kōt }

bond course [BUILD] A course of headers to bond the facing masonry to the backing masonry. { 'bänd ,kȯrs }

bond dissociation energy [PHYS CHEM] The change in enthalpy that occurs with the homolytic cleavage of a chemical bond under conditions of standard state. { ,bänd di,sō·sē'ā·shən 'en·ər·je }

bond distance [PHYS CHEM] The distance separating the two nuclei of two atoms bonded to each other in a molecule. Also known as bond length. { 'bänd ,dis·təns }

bonded coating [MATER] A finishing or protecting layer of any compound affixed to a surface. { 'bän·dəd ,kōd·iŋ }

bonded NR diode [ELECTR] An n^+ junction semiconductor device in which the negative resistance arises from a combination of avalanche breakdown and conductivity modulation which is due to the current flow through the junction. { 'bän·dəd ,en'är 'dī,ōd }

bonded-phase chromatography [ANALY CHEM] A type of high-pressure liquid chromatography which employs a stable, chemically bonded stationary phase. { 'bän·dəd ,fāz ,krō·mə'täg·rə·fē }

bonded strain gage [ENG] A strain gage in which the resistance element is a fine wire, usually in zigzag form, embedded in an insulating backing material, such as impregnated paper or plastic, which is cemented to the pressure-sensing element. { 'bän·dəd 'strān ,gāj }

bonded transducer [ENG] A transducer which employs a bonded strain gage for sensing pressure. { 'bän·dəd tranz'dü·sər }

bond energy [PHYS CHEM] **1.** The average value of specific bond dissociation energies that have been measured from different molecules of a given type. **2.** See average bond dissociation energy. { 'bänd ,en·ər·jē }

bonder See bondstone. { 'bän·dər }

bonderize [MET] To coat steel with a solution of phosphates for corrosion protection. { 'bän·də,rīz }

BOMBIDAE

A drawing of a hairy, black and yellow bumblebee.

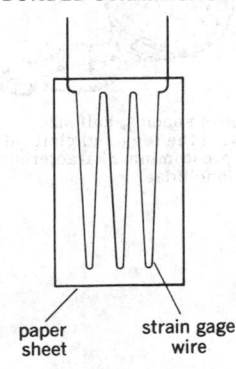

BONDED STRAIN GAGE

paper sheet   strain gage wire

Elements of a bonded strain gage.

BOND HEADER

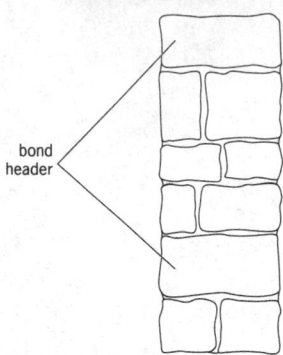

bond header

A masonry section showing bond headers.

BONE

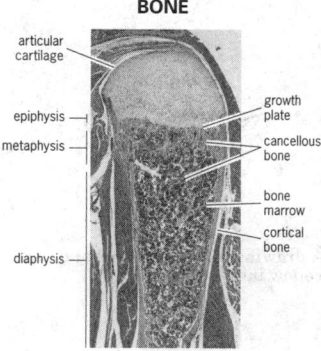

articular cartilage

growth plate

epiphysis

cancellous bone

metaphysis

bone marrow

cortical bone

diaphysis

Longitudinal section through the femur of a 7-day-old mouse.

BONELLIDAE

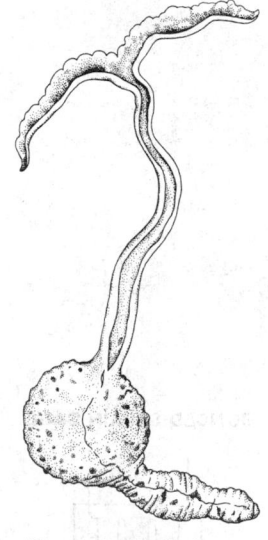

Bonellia species, half size, showing the long and cleft (at the top) prostomium characteristic of the Bonellidae.

bond header [BUILD] In masonry, a stone that extends the full thickness of the wall. Also known as throughstone. { 'bänd ,hed·ər }

bond hybridization [CHEM] The linear combination of two or more simple atomic orbitals. { ¦bänd ,hī·brəd·ə'zā·shən }

bonding [CHEM] The joining together of atoms to form molecules or crystalline salts. [ELEC] The use of low-resistance material to connect electrically a chassis, metal shield cans, cable shielding braid, and other supposedly equipotential points to eliminate undesirable electrical interaction resulting from high-impedance paths between them. [ENG] **1.** The fastening together of two components of a device by means of adhesives, as in anchoring the copper foil of printed wiring to an insulating baseboard. **2.** *See* cladding. [PSYCH] The formation of an emotional attachment between two people whose identities are significantly affected by their mutual interactions. [TEXT] The joining of two fabrics, usually a face fabric and a lining fabric. { 'bän·diŋ }

bonding agent [MATER] Any substance that fixes one material to another. { 'bän·diŋ ,ā·jənt }

bonding electron [PHYS CHEM] An electron whose orbit spans the entire molecule and so assists in holding it together. { 'bän·diŋ i'lek,trän }

bonding orbital [PHYS CHEM] A molecular orbital formed by a bonding electron whose energy decreases as the nuclei are brought closer together, resulting in a net attraction and chemical bonding. { 'bän·diŋ 'ȯr·bəd·əl }

bonding pad [ELECTR] A metallized area on the surface of a semiconductor device, to which connections can be made. { 'bän·diŋ ,pad }

bonding strength [MECH] Structural effectiveness of adhesives, welds, solders, glues, or of the chemical bond formed between the metallic and ceramic components of a cermet, when subjected to stress loading, for example, shear, tension, or compression. { 'bän·diŋ ,streŋkth }

bonding wire [ELEC] Wire used to connect metal objects so they have the same potential (usually ground potential). { 'bän·diŋ ,wīr }

bond length *See* bond distance. { 'bänd ,leŋkth }

bond-line formula [ORG CHEM] A representation of a molecule in which bonds are represented by lines, carbon atoms are represented by line ends and intersections, and atoms other than hydrogen and carbon are represented by their elemental symbols, as is hydrogen when it is bonded to an atom other than hydrogen or carbon. Also known as carbon-skeleton formula; line-segment formula. { 'bänd ,līn ,fȯr·myə·lə }

bond migration [CHEM] The movement of a bond to a different position within the same molecular entity. { 'bänd mī,grā·shən }

bond moment [PHYS CHEM] The degree of polarity of a chemical bond as calculated from the value of the force of the response of the bond when the bond is subjected to an electric field. { 'bänd ,mō·mənt }

Bond number [FL MECH] A dimensionless number used in the study of atomization and the study of bubbles and drops, equal to $(\rho - \rho')L^2 g/\sigma$, where ρ is the density of a bubble or drop, ρ' is the density of the surrounding medium, L is a characteristic dimension, g is the acceleration of gravity, and σ is the surface tension of the bubble or drop. { 'bänd ,nəm·bər }

bond orientational order [PHYS] An ordering of atoms or molecules in an intermediate state of condensed matter in which the atoms or molecules are distributed at random, as in a fluid or glass, but the condensed matter is orientationally anisotropic on a macroscopic scale. { ¦bänd ,ȯr·ē·ən¦tā·shən·əl 'ȯrd·ər }

bond paper [MATER] A paper used for writing paper, business forms, and typewriter paper; the less expensive bond papers are made from wood sulfite pulps; rag-content bonds contain 25, 50, 75, or 100% of pulp made from rags, and offer greater permanence and strength. { 'bänd ,pā·pər }

Bond's law [MECH ENG] A statement that relates the work required for the crushing of solid materials (for example, rocks and ore) to the product size and surface area and the lengths of cracks formed. Also known as Bond's third theory. { 'bänz 'lȯ }

Bond's third theory *See* Bond's law. { 'bänz ,thərd 'thē·ə·rē }

bondstone [BUILD] A stone joining the coping above a gable to the wall. [CIV ENG] A masonry stone set with its longest dimension perpendicular to the wall face to bind the wall together. Also known as bonder. { 'bänd,stōn }

bond strength [CHEM] The strength with which a chemical bond holds two atoms together; conventionally measured in terms of the amount of energy, in kilocalories per mole, required to break the bond. [ENG] The amount of adhesion between bonded surfaces measured in terms of the stress required to separate a layer of material from the base to which it is bonded. { 'bänd ,streŋkth }

bond timber [BUILD] A section of wood built horizontally into a brick or stone wall in order to strengthen it or to hold it together during construction. { 'bänd ,tim·bər }

bone [ANAT] One of the parts constituting a vertebrate skeleton. [HISTOL] A hard connective tissue that forms the major portion of the vertebrate skeleton. { bōn }

Bone Age [ARCHEO] A prehistoric period of human culture characterized by the use of implements made of bone and antler. { 'bōn ,āj }

bone ash [CHEM] A white ash consisting primarily of tribasic calcium phosphate obtained by burning bones in air; used in cleaning jewelry and in some pottery. { 'bōn ,ash }

bone bed [GEOL] Several thin strata or layers with many fragments of fossil bones, scales, teeth, and also organic remains. { 'bōn ,bed }

bone black [MATER] A black substance made by carbonizing crushed, defatted bones in closed vessels; used as a paint and varnish pigment, as a decolorizing absorbent in clarifying shellac, in cementation, and in gas masks. Also known as animal black; bone char. { 'bōn ,blak }

bone char *See* bone black. { 'bōn ,chär }

bone chert [PETR] A weathered residual chert that appears chalky and porous with a white color but may be stained red or other colors. { 'bōn ,chərt }

bone coal [GEOL] Argillaceous coal or carbonaceous shale that is found in coal seams. { 'bōn ,kōl }

bone conduction [BIOPHYS] Transmission of sound vibrations to the internal ear via the bones of the skull. { 'bōn kən'dək·shən }

Bonellidae [INV ZOO] A family of wormlike animals belonging to the order Echiuroinea. { bō'nel·ə,dē }

bone marrow [HISTOL] A vascular modified connective tissue occurring in the long bones and certain flat bones of vertebrates. { 'bōn ,mar·ō }

bone meal [MATER] A substance made by grinding animal bones; steamed meal, made from pressure-steamed bones, is used as a fertilizer; raw meal is used in animal feed. { 'bōn ,mēl }

bone oil [MATER] Dark brown oil with a disagreeable odor, derived by destructive distillation of bones or other animal substances; used as an alcohol denaturant, an insecticide, or a source of pyrrole and for organic preparations. Also known as animal oil; Dippel's oil; hartshorn oil; Jeppel's oil. { 'bōn ,oil }

bone seeker [NUCLEO] A radioisotope that tends to accumulate in the bones when it is introduced into the body; an example is strontium-90, which behaves chemically like calcium. { 'bōn ,sēk·ər }

Böning effect [ELEC] The displacement of associated ions that have been bound to capturing ions in fine channels in a dielectric medium when an electric field is applied. { 'bən·iŋ i,fekt }

boninite [PETR] An andesitic rock that contains much glass and abundant phenocrysts of bronzite and less of olivine and augite. { 'bän·ə,nīt }

bonnaz [TEXT] A machine used for embroidery in which the fabric is held tightly and the needle moves freely to outline a design with the chain stitch. { bə'naz }

Bonne projection [MAP] A type of conical map projection; meridians are plotted as curves and the parallels are spaced along them at true distances. { 'bȯn prə'jek·shən }

Bononian [GEOL] Upper Jurassic (lower Portlandian) geologic time. { bə'nōn·ē·ən }

bonsai [BOT] The production of a mature, very dwarfed tree in a relatively small container. { bōn'sī }

bonus [PETRO ENG] Payment by a lessee of an oil- or gas-production royalty to the landowner at a rate greater than the customary one-eighth of the value of the oil or gas withdrawn. { 'bō·nəs }

bony fish [VERT ZOO] The name applied to all members of the class Osteichthyes. { 'bō·nē ,fish }

bony labyrinth [ANAT] The system of canals within the otic bones of vertebrates that houses the membranous labyrinth of the inner ear. { 'bō·nē 'lab·ə,rinth }

Boo *See* Boötes.

booby trap [ORD] An explosive charge such as a mine, grenade, demolition block, shell, or bulk explosive fitted with a detonator and a firing device, and usually concealed and set to explode when an unsuspecting person touches off its firing mechanism by stepping upon, lifting, or moving a harmless-looking object. { 'büb·ē ,trap }

Boodleaceae [BOT] A family of green marine algae in the order Siphonocladales. { bō,äd·lē'ās·ē,ē }

book *See* mica book. { bůk }

Book A *See* DVD-read-only. { ¦bůk 'ā }

Book B *See* DVD-video. { ¦bůk 'bē }

book capacitor [ELEC] A trimmer capacitor consisting of two plates which are hinged at one end; capacitance is varied by changing the angle between them. { 'bůk kə'pas·əd·ər }

Book D *See* DVD-write once. { bůk 'dē }

Book E *See* DVD-rewritable. { ¦bůk 'ē }

book form drawing [GRAPHICS] A surface portrayal requiring two or more sheets, each identified with the same drawing number and a sequential page number. { 'bůk ¦fòrm ,drô·iŋ }

book gill [INV ZOO] A type of gill in king crabs consisting of folds of membranous tissue arranged like the leaves of a book. { 'bůk ,gil }

bookkeeping operation [COMPUT SCI] A computer operation which does not directly contribute to the result, that is, arithmetical, logical, and transfer operations used in modifying the address section of other instructions in counting cycles and in rearranging data. Also known as red-tape operation. { 'bůk,kēp·iŋ äp·ə'rā·shən }

book louse [INV ZOO] A common name for a number of insects belonging to the order Psocoptera; important pests in herbaria, museums, and libraries. { 'bůk ,laůs }

book lung [INV ZOO] A saccular respiratory organ in many arachnids consisting of numerous membranous folds arranged like the pages of a book. { 'bůk ,ləŋ }

bookmark [COMPUT SCI] **1.** Any method of halting the processing of a transaction and holding it, as far as it has been completed, until processing resumes. **2.** A code that is inserted at a particular place in a document or that is associated with a particular document so that the user can easily return to the specified insertion point or document. **3.** A Web page location (URL) which is saved by a user for quick reference. { 'bůk,märk }

book mold [MET] A mold made in two halves hinged together like a book. { 'bůk ,mōld }

book structure [GEOL] A rock structure of numerous parallel sheets of slate alternating with quartz. { 'bůk ,strək·chər }

boolean [COMPUT SCI] A scalar declaration in ALGOL defining variables similar to FORTRAN's logical variables. { 'bü·lē·ən }

Boolean algebra [MATH] An algebraic system with two binary operations and one unary operation important in representing a two-valued logic. { 'bü·lē·ən 'al·jə·brə }

Boolean calculus [MATH] Boolean algebra modified to include the element of time. { 'bü·lē·ən 'kal·kyə·ləs }

Boolean data type *See* logical data type. { 'bü·lē·ən 'dad·ə ,tīp }

Boolean determinant [MATH] A function defined on Boolean matrices which depends on the elements of the matrix in a manner analogous to the manner in which an ordinary determinant depends on the elements of an ordinary matrix, with the operation of multiplication replaced by intersection and the operation of addition replaced by union. { ¦bül·ē·ən di'tər·mə·nənt }

Boolean function [MATH] A function $f(x,y,\ldots,z)$ assembled by the application of the operations AND, OR, NOT on the variables x, y, . . . , z and elements whose common domain is a Boolean algebra. { 'bü·lē·ən 'fəŋk·shən }

Boolean matrix [MATH] A rectangular array of elements each of which is a member of a Boolean algebra. { ¦bül·ē·ən 'mā,triks }

Boolean operation table [MATH] A table which indicates, for a particular operation on a Boolean algebra, the values that result for all possible combination of values of the operands; used particularly with Boolean algebras of two elements which may be interpreted as "true" and "false." { ¦bül·ē·ən ,äp·ə'rā·shən ,tā·bəl }

Boolean operator [MATH] A logic operator that is one of the operators AND, OR, or NOT, or can be expressed as a combination of these three operators. { ¦bül·ē·ən 'äp·ə,rād·ər }

Boolean ring [MATH] A commutative ring with the property that for every element a of the ring, $a \times a$ and $a + a = 0$; it can be shown to be equivalent to a Boolean algebra. { ¦bül·ē·ən 'riŋ }

Boolean search [COMPUT SCI] A search for selected information, that is, information satisfying conditions that can be expressed by AND, OR, and NOT functions. { 'bü·lē·ən 'sərch }

boom [COMMUN] A movable mechanical support, usually in a television or motion picture studio, to suspend a microphone within range of the performers but above the field of view of the camera. [ENG] **1.** A row of joined floating timbers that extend across a river or enclose an area of water for the purpose of keeping saw logs together. **2.** A temporary floating barrier launched on a body of water to contain material, for example, an oil spill. **3.** A structure consisting of joined floating logs placed in a stream to retard the flow. [MECH ENG] A movable steel arm installed on certain types of cranes or derricks to support hoisting lines that must carry loads. [NAV ARCH] **1.** A spar attached to a mast or kingpost of a ship carrying cargo-hoisting gear. **2.** A spar upon which the lower side of a sail is bent. { büm }

boom cat [MECH ENG] A tractor supporting a boom and used in laying pipe. { 'büm ,kat }

boom crutch [NAV ARCH] A movable prop for supporting the free end of the boom of a ship when it is not being used. { 'büm ,krəch }

boom dog [MECH ENG] A ratchet device installed on a crane to prevent the boom of the crane from being lowered but permitting it to be raised. Also known as boom ratchet. { 'büm ,dòg }

boomer [ENG] A device used to tighten chains on pipe or other equipment loaded on a truck to make the cargo secure. [MIN ENG] In placer mining, an automatic gate in a dam that holds the water until the reservoir is filled, then opens automatically and allows the escape of such a volume that the soil and upper gravel of the placer are washed away. { 'büm·ər }

boomerang sediment corer [ENG] A device, designed for nighttime recovery of a sediment core, which automatically returns to the surface after taking the sample. { 'bü·mə,raŋ 'sed·ə·mənt ,kôr·ər }

boom ratchet *See* boom dog. { 'büm ,rach·ət }

boom stop [MECH ENG] A steel projection on a crane that will be struck by the boom if it is raised or lowered too great a distance. { 'büm ,stäp }

Boöpidae [INV ZOO] A family of lice in the order Mallophaga, parasitic on Australian marsupials. { bō'äp·ə,dē }

Boord synthesis [CHEM ENG] A method of producing alpha olefins by the reduction of alpha bromo ethers with zinc. { 'bòrd ,sin·thə·səs }

boorga *See* burga. { 'bůr·gə }

boost [AERO ENG] **1.** An auxiliary means of propulsion such as by a booster. **2.** To supercharge. **3.** To launch or push along during a portion of a flight. **4.** *See* boost pressure. [ELECTR] To augment in relative intensity, as to boost the bass response in an audio system. [ENG] To bring about a more potent explosion of the main charge of an explosive by using an additional charge to set it off. { büst }

boost charge [ELEC] Partial charge of a storage battery, usually at a high current rate for a short period. { 'büst ,chärj }

booster [AERO ENG] *See* booster engine. *See* booster rocket. *See* launch vehicle. [ELEC] A small generator inserted in series or parallel with a larger generator to maintain normal voltage output under heavy loads. [ELECTR] **1.** A separate radio-frequency amplifier connected between an antenna and a television receiver to amplify weak signals. **2.** A radio-frequency amplifier that amplifies and rebroadcasts a received television or communication radio carrier frequency for reception by the general public. [IMMUNOL] The dose of an immunizing agent given to stimulate the effects of a previous dose of the same agent. [MECH ENG] A compressor that is used as the first stage in a cascade refrigerating system.

[ORD] An assembly of metal parts and explosive charge provided to augment the explosive component of a fuse, to cause detonation of the main explosive charge of the munition. { 'büs·tər }

booster battery [ELECTR] A battery which increases the sensitivity of a crystal detector by maintaining a certain voltage across it and thereby adjusting conditions to increase the response to a given input. { 'büs·tər ¸bad·ə·rē }

booster brake [MECH ENG] An auxiliary air chamber, operated from the intake manifold vacuum, and connected to the regular brake pedal, so that less pedal pressure is required for braking. { 'büs·tər ¸brāk }

booster ejector [MECH ENG] A nozzle-shaped apparatus from which a high-velocity jet of steam is discharged to produce a continuous-flow vacuum for process equipment. { 'büs·tər ē'jek·tər }

booster engine [AERO ENG] An engine, especially a booster rocket, that adds its thrust to the thrust of the sustainer engine. Also known as booster. { 'büs·tər ¸en·jən }

booster fan [MECH ENG] A fan used to increase either the total pressure or the volume of flow. { 'büs·tər ¸fan }

booster pump [MECH ENG] A machine used to increase pressure in a water or compressed-air pipe. { 'büs·tər ¸pəmp }

booster response See anamnestic response. { 'büs·tər ri'späns }

booster rocket [AERO ENG] Also known as booster. **1.** A rocket motor, either solid- or liquid-fueled, that assists the normal propulsive system or sustainer engine of a rocket or aeronautical vehicle in some phase of its flight. **2.** A rocket used to set a vehicle in motion before another engine takes over. { 'büs·tər ¸räk·ət }

booster stations [ENG] Booster pumps or compressors located at intervals along a liquid-products or gas pipeline to boost the pressure of the flowing fluid to keep it moving toward its destination. { 'büs·tər ¸stā·shənz }

booster voltage [ELECTR] The additional voltage supplied by the damper tube to the horizontal output, horizontal oscillator, and vertical output tubes of a television receiver to give greater sawtooth sweep output. { 'büs·tər ¸vōl·tij }

booster well [ORD] A hollow space, in the main explosive charge of an item of ammunition, into which the booster fits. { 'büs·tər ¸wel }

boost-glide vehicle [AERO ENG] An air vehicle capable of aerodynamic lift which is projected to an extreme altitude by reaction propulsion and then coasts down with little or no propulsion, gliding to increase its range when it reenters the sensible atmosphere. { 'büst ¦glīd 've·ə·kəl }

boost pressure [AERO ENG] Manifold pressure greater than the ambient at atmospheric pressure, obtained by supercharging. Also known as boost. { 'büst ¸presh·ər }

boot [COMPUT SCI] To load the operating system into a computer after it has been switched on; usually applied to small computers. [ELEC] A protective covering over any portion of a cable, wire, or connector. [MIN ENG] **1.** A projecting portion of a reinforced concrete beam acting as a corbel to support the facing material, such as brick or stone. **2.** The lower end of a bucket elevator. [PETRO ENG] See surge column. { 'büt }

Boot See Boötes. { 'büt }

boot button See bootstrap button. { 'büt ¸bət·ən }

Boötes [ASTRON] A constellation which lies south and east of Ursa Major; the star Arcturus is a member of the group. Abbreviated Boo; Boot. Also known as Bear Driver. { bō'ō¸tēz }

boothite [MINERAL] CuSO₄·7H₂O A blue, monoclinic mineral consisting of copper sulfate heptahydrate; usually occurs in massive or fibrous form. { 'bü¸thīt }

α Boötis See Arcturus. { ¦al·fə bō'ō·təs }

bootjack [ENG] A fishing tool used in drilling wells. { 'büt¸jak }

bootleg [MIN ENG] A hole, shaped somewhat like the leg of a boot, caused by a blast that has failed to shatter the rock properly. { 'büt¸leg }

boot record [COMPUT SCI] A special area of a floppy diskette or hard drive which is used by the computer during system startup. { 'büt ¸rek·ərd }

bootstrap [COMPUT SCI] The procedures for making a computer or a program function through its own actions. [ENG]

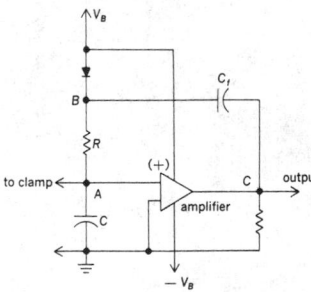

BOOTSTRAP SAWTOOTH GENERATOR

Circuit diagram of bootstrap sawtooth generator.

A technique or device designed to bring itself into a desired state by means of its own action. { 'büt¸strap }

bootstrap button [COMPUT SCI] The first button pressed when a computer is turned on, causing the operating system to be loaded into memory. Also known as boot button; initial program load button; IPL button. { 'büt¸strap ¸bət·ən }

bootstrap circuit [ELECTR] A single-stage amplifier in which the output load is connected between the negative end of the anode supply and the cathode, while signal voltage is applied between grid and cathode; a change in grid voltage changes the input signal voltage with respect to ground by an amount equal to the output signal voltage. { 'büt¸strap ¸sər·kət }

bootstrap driver [ELECTR] Electronic circuit used to produce a square pulse to drive the modulator tube; the duration of the square pulse is determined by a pulse-forming line. { 'büt¸strap ¸drīv·ər }

bootstrap instructor technique [COMPUT SCI] A technique permitting a system to bring itself into an operational state by means of its own action. Also known as bootstrap technique. { 'büt¸strap in'strək·tər tek'nēk }

bootstrap integrator [ELECTR] A bootstrap sawtooth generator in which an integrating amplifier is used in the circuit. Also known as Miller generator. { 'büt¸strap 'in·tə¸grād·ər }

bootstrap loader [COMPUT SCI] A very short program loading routine, used for loading other loaders in a computer; often implemented in a read-only memory. { 'büt¸strap 'lōd·ər }

bootstrap memory [COMPUT SCI] A device that provides for the automatic input of new programs without erasing the basic instructions in the computer. { 'büt¸strap 'mem·rē }

bootstrapping [ELECTR] A technique for lifting a generator circuit above ground by a voltage value derived from its own output signal. { 'büt¸strap·iŋ }

bootstrap process [AERO ENG] A self-generating or self-sustaining process; specifically, the operation of liquid-propellant rocket engines in which, during main-stage operation, the gas generator is fed by the main propellants pumped by the turbopump, and the turbopump in turn is driven by hot gases from the gas generator system. { 'büt¸strap ¸präs·əs }

bootstrap program See loading program. { 'büt¸strap ¸prō·grəm }

bootstrap sawtooth generator [ELECTR] A circuit capable of generating a highly linear positive sawtooth waveform through the use of bootstrapping. { 'büt¸strap ¦sò¸tüth ¦jen·ə¸rād·ər }

bootstrap scheme [PARTIC PHYS] A theory of elementary particles in which the existence of each particle contributes to forces between it and other particles; these forces lead to bound systems which are the particles themselves. { 'büt¸strap ¸skēm }

bootstrap technique See bootstrap instructor technique. { 'büt¸strap tek'nēk }

boot virus [COMPUT SCI] A virus that infects the boot records on floppy diskettes and hard drives and is designed to self-replicate from one disk to another. { 'büt ¸vī·rəs }

Bopyridae [INV ZOO] A family of epicaridean isopods in the tribe Bopyrina known to parasitize decapod crustaceans. { bō'pī·rə¸dē }

Bopyrina [INV ZOO] A tribe of dioecious isopods in the suborder Epicaridea. { bō·pī'rə·nə }

Bopyroidea [INV ZOO] An equivalent name for Epicaridea. { ¸bō·pi'ròid·ē·ə }

bora [METEOROL] A fall wind whose source is so cold that when the air reaches the lowlands or coast the dynamic warming is insufficient to raise the air temperature to the normal level for the region; hence it appears as a cold wind. { 'bòr·ə }

boracic acid See boric acid. { bə'ras·ik 'as·əd }

boracite [MINERAL] Mg₃B₇O₁₃Cl A white, yellow, green, or blue orthorhombic borate mineral occurring in crystals which appear isometric in external form; it is strongly pyroelectric, has a hardness of 7 on Mohs scale, and a specific gravity of 2.9. { 'bòr·ə¸sīt }

bora fog [METEOROL] A dense fog caused when the bora lifts a spray of small drops from the surface of the sea. { 'bòr·ə ¸fäg }

Boraginaceae [BOT] A family of flowering plants in the order Lamiales comprising mainly herbs and some tropical trees. { bə¦ra·jə'nās·ē¸ē }

Boralf [GEOL] A suborder of the soil order Alfisol, dull brown or yellowish brown in color; occurs in cool or cold regions, chiefly at high latitudes or high altitudes. { 'bȯr‚alf }

borane [INORG CHEM] **1.** A class of binary compounds of boron and hydrogen; boranes are used as fuels. Also known as boron hydride. **2.** A substance which may be considered a derivative of a boron-hydrogen compound, such as BCl_3 and $B_{10}H_{12}I_2$. { 'bȯ‚rān }

borate [CHEM] **1.** A generic term referring to salts or esters of boric acid. **2.** Related to boric oxide, B_2O_3, or commonly to only the salts of orthoboric acid, H_3BO_3. { 'bȯ‚rāt }

borate mineral [MINERAL] Any of the large and complex group of naturally occurring crystalline solids in which boron occurs in chemical combination with oxygen. { 'bȯ‚rāt 'min‚rəl }

boration See hydroboration. { bȯ'rā‚shən }

borax [MINERAL] $Na_2B_4O_7\cdot10H_2O$ A white, yellow, blue, green, or gray borate mineral that is an ore of boron and occurs as an efflorescence or in monoclinic crystals; when pure it is used as a cleaning agent, antiseptic, and flux. Also known as diborate; pyroborate; sodium (1:2) borate; sodium tetraborate; tincal. { 'bȯ‚raks }

borax glass [MATER] A glassy, transparent solid formed by fusing borax. { 'bȯ‚raks ‚glas }

borazole [INORG CHEM] $B_3N_3H_6$ A colorless liquid boiling at 53°C; with water it hydrolyzes to form boron hydrides; the borazole molecule is the inorganic analog of the benzene molecule. { 'bȯr‚ə‚zōl }

borazon [INORG CHEM] A form of boron nitride with a zinc blende structure produced by subjecting the ordinary form to high pressure and temperature. { 'bȯr‚ə‚zän }

Borda mouthpiece [FL MECH] A reentrant tube in a hydraulic reservoir, whose contraction coefficient (the ratio of the cross section of the issuing jet of liquid to that of the opening) can be calculated more simply than for other discharge openings. { 'bȯr‚də 'mau̇th‚pēs }

Bordeaux mixture [MATER] A fungicide made from a mixture of lime, copper sulfate, and water. { bȯr'dō ‚miks‚chər }

bordered [BOT] Having a margin with a distinctive color or texture; used especially of a leaf. { 'bȯrd‚ərd }

bordered pit [BOT] A wood-cell pit having the secondary cell wall arched over the cavity of the pit. { 'bȯrd‚ərd 'pit }

border facies [GEOL] The outer portion of an igneous intrusion which differs in composition and texture from the main body. { 'bȯrd‚ər ‚fā‚shēz }

bordering [MATH] For a determinant, the procedure of adding a column and a row, which usually have unity as a common element and all other elements equal to zero. { 'bȯrd‚ər‚iŋ }

borderland [GEOL] One of the crystalline, continental landmasses postulated to have existed on the exterior (oceanward) side of geosynclines. { 'bȯrd‚ər‚land }

borderland slope [GEOL] A declivity which indicates the inner margin of the borderland of a continent. { 'bȯrd‚ər‚land ‚slōp }

borderline mental retardation [PSYCH] Below-normal intellectual functioning associated with an intelligence quotient of about 68–85. { 'bȯrd‚ər‚līn ‚ment‚əl ‚rē‚tär'dā‚shən }

borderline psychosis [PSYCH] The psychiatric diagnosis of an individual whose symptoms are severe but are not clearly psychotic or neurotic. { 'bȯrd‚ər‚līn sī'kō‚səs }

borderline syndrome [PSYCH] An impairment of ego function noted in individuals who have been angry most of their lives and who cannot relate meaningfully to or love other people, and who in consequence develop nonpsychotic but deviant mechanisms, such as withdrawal while expressing angry concern, or showing passive compliance at the price of real involvement, to maintain a precarious mental equilibrium. { 'bȯrd‚ər‚līn 'sin‚drōm }

border-punched card See edge-notched card. { 'bȯrd‚ər ‚pəncht ‚kärd }

Bordetella [MICROBIO] A genus of gram-negative, aerobic bacteria of uncertain affiliation; minute coccobacilli, parasitic and pathogenic in the respiratory tract of mammals. { ‚bȯr‚də'tel‚ə }

Bordetella avium [MICROBIO] A nonsporulating, gram-negative coccobacillus that causes respiratory infections in birds. { ‚bȯr‚də‚tel‚ə 'ā‚vē‚əm }

Bordetella bronchiseptica [MICROBIO] An aerobic, gram-negative bacterium that is a pathogen in many domestic and wild mammals, including horses, swine, dogs, and rodents, and may cause a variety of respiratory diseases in them. { ‚bȯr‚də‚tel‚ə ‚braŋ‚ki'sep‚ti‚kə }

Bordini effect [MET] A phenomenon that takes place when metals having a close-packed crystal structure are subjected to an oscillating stress; there is a peak in the internal friction at a particular frequency of the stress. { bȯr'dēn‚ē i'fekt }

bore [DES ENG] Inside diameter of a pipe or tube. [MECH ENG] **1.** The diameter of a piston-cylinder mechanism as found in reciprocating engines, pumps, and compressors. **2.** To penetrate or pierce with a rotary tool. **3.** To machine a workpiece to increase the size of an existing hole in it. [MIN ENG] **1.** A tunnel under construction. **2.** To cut or drill a hole for blasting, water infusion, exploration, or water or firedamp drainage. [OCEANOGR] **1.** A high, breaking wave of water, advancing rapidly up an estuary. Also known as eager; mascaret; tidal bore. **2.** A submarine sand ridge, in very shallow water, whose crest may rise to intertidal level. [ORD] The interior of a gun barrel or tube. { bȯr }

boreal [ECOL] Of or relating to northern geographic regions. { 'bȯr‚ē‚əl }

boreal forest See taiga. { 'bȯr‚ē‚əl 'fär‚əst }

Boreal life zone [ECOL] The zone comprising the climate and biotic communities between the Arctic and Transitional zones. { 'bȯr‚ē‚əl 'līf ‚zōn }

bore axis [ORD] The longitudinal center line of a gunbore. { 'bȯr ‚ak‚səs }

bore diameter [ORD] The interior diameter or caliber of a gun or launching tube. { 'bȯr dī'am‚əd‚ər }

bore evacuation [ORD] Clearing of the bore of propellant gases and extraneous material after firing. { 'bȯr i‚vak‚yə'wā‚shən }

bore evacuator [ORD] A device placed on the tube of artillery weapons to force propellant gases to flow outward through the muzzle end of the bore from a cylindrical chamber around the gun tube; the chamber encloses a series of jets around the cannon tube through which the gases flow. { 'bȯr i'vak‚yə‚wād‚ər }

bore expansion [ORD] Permanent enlargement of the bore of a gun due to interior pressure; it does not include enlargement due to bore erosion. { 'bȯr ig'span‚shən }

borehole See drill hole. { 'bȯr‚hōl }

borehole bit See noncoring bit. { 'bȯr‚hōl ‚bit }

borehole logging [ENG] The technique of investigating and recording the character of the formation penetrated by a drill hole in mineral exploration and exploitation work. Also known as drill-hole logging. { 'bȯr‚hōl ‚läg‚iŋ }

borehole mining [PETRO ENG] Extraction of minerals as liquid or gas from the earth's crust by means of boreholes and suction pumps. { 'bȯr‚hōl ‚mīn‚iŋ }

borehole survey [ENG] **1.** Also known as drill-hole survey. Determining the course of and the target point reached by a borehole, using an azimuth-and-dip recording apparatus small enough to be lowered into a borehole. **2.** The record of the information thereby obtained. { 'bȯr‚hōl ‚sər‚vā }

Borel measurable function [MATH] **1.** A real-valued function such that the inverse image of the set of real numbers greater than any given real number is a Borel set. **2.** More generally, a function to a topological space such that the inverse image of any open set is a Borel set. { bȯ‚rel ‚mezh‚rə‚bəl 'fəŋk‚shən }

Borel measure [MATH] A measure defined on the class of all Borel sets of a topological space such that the measure of any compact set is finite. { bə'rel ‚mezh‚ər }

Borel set [MATH] A member of the smallest σ-algebra containing the compact subsets of a topological space. { bȯ'rel ‚set }

Borel sigma algebra [MATH] The smallest sigma algebra containing the compact subsets of a topological space. { bȯ‚rel ‚sig‚mə 'al‚jə‚brə }

bore premature [ORD] The premature explosion of a projectile occurring within the bore. { 'bȯr ‚prē‚mə'chu̇r }

borer [INV ZOO] Any insect or other invertebrate that burrows into wood, rock, or other substances. [MECH ENG] An apparatus used to bore openings into the earth up to about 8 feet (2.4 meters) in diameter. { 'bȯr‚ər }

bore riding pin [ORD] A safety pin which is held in place in the fuse while the projectile or missile is within the gun barrel or launching tube and then ejected from the fuse by centrifugal force or spring action beyond the muzzle. { 'bȯr ‚rīd·iŋ ‚pin }

borescope [ENG] A straight-tube telescope using a mirror or prism, used to visually inspect a cylindrical cavity, such as the cannon bore of artillery weapons for defects of manufacture and erosion caused by firing. { 'bȯr‚skōp }

boresight [ORD] A telescopic device utilizing a boresighting reticle to align the axis of a weapon with a target; it is attached to the exterior of a weapon barrel or tube by a mount. { 'bȯr‚sīt }

boresight camera [OPTICS] A camera mounted in the optical axis of a tracking radar to photograph rockets being tracked or known, fixed targets while in camera range and thus to provide a correction for the alignment of the radar. { 'bȯr‚sīt ‚kam·rə }

boresighting [ENG] Initial alignment of a directional microwave or radar antenna system by using an optical procedure or a fixed target at a known location. [ORD] The process by which the axis of a gun bore and the line of sight of a gunsight are made parallel or are made to converge on a point. { 'bȯr‚sīd·iŋ }

Borhyaenidae [VERT ZOO] A family of carnivorous mammals in the superfamily Borhyaenoidea. { ‚bȯr·ē'ēn·ə‚dē }

Borhyaenoidea [VERT ZOO] A superfamily of carnivorous mammals in the order Marsupialia. { ‚bȯr·ē·ə'nȯid·ē·ə }

boric acid [INORG CHEM] H_3BO_3 An acid derived from boric oxide in the form of white, triclinic crystals, melting at 185°C, soluble in water. Also known as boracic acid; orthoboric acid. { ¦bȯr·ik 'as·əd }

boric acid ester [ORG CHEM] Any compound readily hydrolyzed to yield boric acid and the respective alcohol; for example, trimethyl borate hydrolyzes to boric acid and methyl alcohol. { ¦bȯr·ik 'as·əd 'es·tər }

borickite [MINERAL] $CaFe_5(PO_4)_2(OH)_{11}\cdot 3H_2O$ Reddish-brown, isotropic mineral consisting of a hydrated basic phosphate of calcium and iron; occurs in compact reniform masses. { 'bȯr·ə‚kīt }

boric oxide [INORG CHEM] B_2O_3 A trioxide of boron obtained as rhombic crystals melting at 460°C; used as an intermediate in the production of boron halides and metallic borides and as a thermal neutron absorber in nuclear engineering. Also known as boron oxide. { ¦bȯr·ik 'äk‚sīd }

boride [INORG CHEM] A binary compound of boron and a metal formed by heating a mixture of the two elements. { 'bȯr‚īd }

boring bar [MECH ENG] A rigid tool holder used to machine internal surfaces. { 'bȯr·iŋ ‚bär }

boring log *See* drill log. { 'bȯr·iŋ ‚läg }

boring machine [MECH ENG] A machine tool designed to machine internal work such as cylinders, holes in castings, and dies; types are horizontal, vertical, jig, and single. { 'bȯr·iŋ mə'shēn }

boring mill [MECH ENG] A boring machine tool used particularly for large workpieces; types are horizontal and vertical. { 'bȯr·iŋ ‚mil }

boring sponge [INV ZOO] Marine sponge of the family Clionidae represented by species which excavate galleries in mollusks, shells, corals, limestone, and other calcareous matter. { 'bȯr·iŋ ‚spənj }

Born approximation [QUANT MECH] A method used for the computation of cross sections in scattering problems; the interactions are treated as perturbations of free-particle systems. { 'bȯrn ə·präk·sə'mā·shən }

borneol [ORG CHEM] $C_{10}H_{17}OH$ White lumps with camphor odor; insoluble in water, soluble in alcohol; melting point 203°C; used in perfumes, medicine, and chemical synthesis. { 'bȯr·nē‚ȯl }

Born equation [PHYS CHEM] An equation for determining the free energy of solvation of an ion in terms of the Avogadro number, the ionic valency, the ion's electronic charge, the dielectric constant of the electrolytic, and the ionic radius. { 'bȯrn i'kwä·zhən }

Born-Haber cycle [SOLID STATE] A sequence of chemical and physical processes by means of which the cohesive energy of an ionic crystal can be deduced from experimental quantities; it leads from an initial state in which a crystal is at zero pressure and 0 K to a final state which is an infinitely dilute gas of its constituent ions, also at zero pressure and 0 K. { ¦bȯrn 'hā·bər ‚sī·kəl }

bornhardt [GEOL] A large dome-shaped granite-gneiss outcrop having the characteristics of an inselberg. { 'bȯrn‚härt }

bornite [MINERAL] Cu_5FeS_4 A primary mineral in many copper ore deposits; specific gravity 5.07; the metallic and brassy color of a fresh surface rapidly tarnishes upon exposure to air to an iridescent purple. { 'bȯr‚nīt }

Born-Madelung model [SOLID STATE] A classical theory of cohesive energy, lattice spacing, and compressibility of ionic crystals. { ¦bȯrn 'mäd·əl·əŋ 'mäd·əl }

Born-Mayer equation [SOLID STATE] An equation for the cohesive energy of an ionic crystal which is deduced by assuming that this energy is the sum of terms arising from the Coulomb interaction and a repulsive interaction between nearest neighbors. { ¦bȯrn 'mī·ər i'kwä·zhən }

Born-Oppenheimer approximation [PHYS CHEM] The approximation, used in the Born-Oppenheimer method, that the electronic wave functions and energy levels at any instant depend only on the positions of the nuclei at that instant and not on the motions of the nuclei. Also known as adiabatic approximation. { ¦bȯrn 'äp·ən‚hī·mər ə‚präk·sə‚mā·shən }

Born-Oppenheimer method [PHYS CHEM] A method for calculating the force constants between atoms by assuming that the electron motion is so fast compared with the nuclear motions that the electrons follow the motions of the nuclei adiabatically. { ¦bȯrn 'äp·ən‚hīm·ər ‚meth·əd }

Born-von Kármán theory [SOLID STATE] A theory of specific heat which considers an acoustical spectrum for the vibrations of a system of point particles distributed like the atoms in a crystal lattice. { ¦bȯrn fən'kär‚män ‚thē·ə·rē }

bornyl acetate [ORG CHEM] $C_{10}H_{17}OOCCH_3$ A colorless liquid that forms crystals at 10°C; has characteristic piny-camphoraceous odor; used in perfumes and for flavoring. { 'bȯrn·əl 'as·ə‚tāt }

bornyl isovalerate [ORG CHEM] $C_{10}H_{17}OOC_5H_9$ An aromatic fluid with a boiling point of 255–260°C; soluble in alcohol and ether; used in medicine and as a flavoring. { 'bȯrn·əl ¦ī·sō'val·ə‚rāt }

boroarsenate [MINERAL] One of a group of borate minerals containing arsenic; cahnite is an example. { ¦bȯr·ō'är·sə‚nät }

borolanite [PETR] A hypabyssal rock that is essentially orthoclase and melanite with subordinate nepheline, biotite, and pyroxene. { bə'räl·ə‚nīt }

Boroll [GEOL] A suborder of the soil order Mollisol, characterized by a mean annual soil temperature of less than 8°C and by never being dry for 60 consecutive days during the 90-day period following the summer solstice. { 'bȯ‚rȯl }

boron [CHEM] A chemical element, symbol B, atomic number 5, atomic weight 10.811; it has three valence electrons and is nonmetallic. { 'bȯ‚rän }

boron-10 [NUC PHYS] A nonradioactive isotope of boron with a mass number of 10; it is a good absorber for slow neutrons, simultaneously emitting high-energy alpha particles, and is used as a radiation shield in Geiger counters. { 'bȯ‚rän 'ten }

boron alloy [MET] Alloy of boron and iron used to increase the high-temperature strength characteristics of alloy steel. { 'bȯ‚rän 'al‚ȯi }

boronatrocalcite *See* ulexite. { ¦bȯr·ō‚na·trō'kal‚sīt }

boron carbide [ORG CHEM] Any compound of boron and carbon, especially B_4C (used as an abrasive, alloying agent, and neutron absorber). { 'bȯ‚rän 'kär‚bīd }

boron chamber [NUCLEO] An ionization chamber that is lined with boron or boron compounds or filled with a gaseous boron compound. { 'bȯ‚rän ‚chäm·bər }

boron counter tube [NUCLEO] A counter tube filled with boron fluoride or having electrodes coated with boron or boron compounds; used for detecting slow neutrons. { 'bȯ‚rän ‚kaȯnt·ər ‚tüb }

boron fiber [CHEM] Fiber produced by vapor-deposition methods; used in various composite materials to impart a balance of strength and stiffness. Also known as boron filament. { 'bȯ‚rän ‚fī·bər }

boron filament *See* boron fiber. { 'bȯ‚rän ‚fil·ə·mənt }

boron fluoride [INORG CHEM] BF_3 A colorless pungent gas in a dry atmosphere; used in industry as an acidic catalyst for polymerizations, esterifications, and alkylations. Also known as boron trifluoride. { 'bȯ‚rän 'flȯr‚īd }

boron fuel [MATER] The boron compounds alkyl decaborane, diborane, and pentaborane; used for ramjet engines and afterburners. { 'bȯ,rän ,fyül }

boron hydride See borane. { 'bȯ,rän 'hī,drīd }

boron nitride [INORG CHEM] BN A binary compound of boron and nitrogen, especially a white, fluffy powder with high chemical and thermal stability and high electrical resistance. { 'bȯ,rän 'nī,trīd }

boron nitride fiber [INORG CHEM] Inorganic, high-strength fiber, made of boron nitride, that is resistant to chemicals and electricity but susceptible to oxidation above 1600°F (870°C); used in composite structures for yarns, fibers, and woven products. { 'bȯ,rän 'nī,trīd 'fī·bər }

boron oxide See boric oxide. { 'bȯ,rän 'äk,sīd }

boron polymer [ORG CHEM] Macromolecules formed by polymerization of compounds containing, for example, boron-nitrogen, boron-phosphorus, or boron-arsenic bonds. { 'bȯ,rän 'päl·ə·mər }

boron steel [MET] Alloy steel with a small amount (as little as 0.0005) of boron added to increase hardenability; can be used to replace other alloys in short supply. { 'bȯ,rän 'stēl }

boron thermopile [NUCLEO] A thermopile in which alternate thermocouple junctions are coated with boron; exposure to a flux of slow neutrons generates heat in these junctions, producing an output voltage proportional to neutron flux. { 'bȯ,rän 'thər·mə,pīl }

boron trichloride [INORG CHEM] BCl₃ A colorless liquid used as a catalyst and in refining of aluminum, magnesium, zinc, and copper. { 'bȯ,rän trī'klȯr,īd }

boron triethoxide See ethyl borate. { 'bȯ,rän ,trī·ə'thäk,sīd }

boron triethyl See triethylborane. { 'bȯ,rän trī'eth·əl }

boron trifluoride See boron fluoride. { 'bȯ,rän trī'flȯr,īd }

boron trifluoride counter [NUCLEO] A proportional counter filled with boron trifluoride gas, designed to detect and count neutrons. { 'bȯ,rän trī'flȯr,īd ,kaȯnt·ər }

boron trifluoride etherate [ORG CHEM] C₄H₁₀BF₃O A fuming liquid hydrolyzed by air immediately; boiling point is 125.7°C; used as a catalyst in reactions involving condensation, dehydration, polymerization, alkylation, and acetylation. { 'bȯ,rän trī'flȯr,īd 'ē·thə,rāt }

borosilicate [MINERAL] A salt of boric and silicic acids which occurs in the natural minerals tourmaline, datolite, and dumortierite. { 'bȯr ō'sil·i·kət }

borosilicate glass [MATER] A type of glass containing at least 5% boric oxide; used in glassware that resists heat. { 'bȯrō'sil·ə·kət 'glas }

Borrelia [MICROBIO] A genus of bacteria in the family Spirochactaceae; helical cells with uneven coils and parallel fibrils coiled around the cell body for locomotion; many species cause relapsing fever in humans. { bə'rel·ē·ə }

Borrelia anserina [MICROBIO] A motile, helical bacterial pathogen propagated by ticks of the genus *Argas* that causes borreliosis in geese, ducks, turkeys, pheasants, chickens, and other birds. { bə,rel·ē·ə an'ser·ə·nə }

Borrelia burgdorferi [MICROBIO] A gram-negative, helically shaped bacterium that is the causative agent of Lyme disease. { bə'rēl·yə ,bərg'dȯr·fə·rē }

Borrmann effect [PHYS] The irregular transmission of x-rays when a single crystal of high perfection is placed in a monochromatic x-ray beam in a reflecting position. { 'bȯr,-män i'fekt }

borrow [CIV ENG] Earth material such as sand and gravel that is taken from one location to be used as fill at another. [MATH] An arithmetically negative unit; it occurs in direct subtraction by raising the low-order digit of the minuend by one unit of the next-higher-order digit; for example, when subtracting 67 from 92, a tens digit is borrowed from the 9, to raise the 2 to a factor of 12; the 7 of 67 is then subtracted from the 12 to yield 5 as the units digit of the difference; the 6 is then subtracted from 8, or 9 − 1, yielding 2 as the tens digit of the difference. { 'bä·rō }

borrow pit [CIV ENG] An excavation dug to provide material (borrow) for fill elsewhere. { 'bä·rō ,pit }

BORSCHT [COMMUN] An interface circuit between ordinary telephone lines carrying analog voice signals and digital time-division multiplex facilities, which digitizes voice signals, assigns them time slots, and then multiplexes them. Acronym for battery, overvoltage, ringing, supervision, coding, hybrid and test access. { 'bȯrsht }

bort [MINERAL] Imperfectly crystallized diamond material unsuitable for gems because of its shape, size, or color and because of flaws or inclusions; used for abrasive and cutting purposes. Also spelled boart. { bȯrt }

bort bit See diamond bit. { 'bȯrt ,pit }

Boryhaenid [PALEON] A carnivorous marsupial from the Miocene Epoch that resembled the wolf. { ,bȯr·ē'han·əd }

Bosanquet's law [ELECTROMAG] The statement that, in analogy to Ohm's law for the resistance of an electric circuit, in a magnetic circuit the ratio of the magnetomotive force to the magnetic flux is a constant known as the reluctance. { 'bō·zən,kets ,lȯ }

Bosch fuel injection pump [MECH ENG] A pump in the fuel injection system of an internal combustion engine, whose pump plunger and barrel are a very close lapped fit to minimize leakage. { 'bȯsh 'fyül in'jek·shən ,pəmp }

Bosch metering system [MECH ENG] A system having a helical groove in the plunger which covers or uncovers openings in the barrel of the pump; most usually applied in diesel engine fuel-injection systems. { 'bȯsh 'mēd·ə·riŋ ,sis·təm }

Bose distribution See Bose-Einstein distribution. { 'bōz dis·trə'byü·shən }

Bose-Einstein condensate [CRYO] The state of matter of a gas of bosonic particles below a critical temperature such that a large number of particles occupy the ground state of the system. { 'bōz 'īn,stīn 'kan·dən,sāt }

Bose-Einstein condensation [CRYO] A phase transition that occurs when a gas of bosonic particles is cooled below a critical temperature very close to absolute zero, in which a large number of the particles come to occupy the ground state of the system and form a coherent matter wave. { 'bōz 'īn,stīn kän·den'sā·shən }

Bose-Einstein distribution [STAT MECH] For an assembly of independent bosons, such as photons or helium atoms of mass number 4, a function that specifies the number of particles in each of the allowed energy states. Also known as Bose distribution. { 'bōz 'īn,stīn dis·trə'byü·shən }

Bose-Einstein statistics [STAT MECH] The statistical mechanics of a system of indistinguishable particles for which there is no restriction on the number of particles that may exist in the same state simultaneously. Also known as Einstein-Bose statistics { 'bōz 'īn,stīn stə'tis·tiks }

Bose gas [STAT MECH] An assemblage of noninteracting or weakly interacting bosons. { 'bōz ,gas }

bosh [MET] **1.** Tapering lower portion of a blast furnace, from the blast holes of the hearth up to the maximum internal diameter at the bottom of the stack. **2.** Quartz deposited on the furnace lining during the smelting of copper ore. { bäsh }

bosing [ARCHEO] A technique used to locate buried ditches or depressions in chalk subsoil; the soil surface is struck with a special board or tool, and the presence of a depression is then indicated by a characteristic sound. { 'bōz·iŋ }

boson [STAT MECH] A particle that obeys Bose-Einstein statistics; includes photons, pi mesons, and all nuclei having an even number of particles and all particles with integer spin. { 'bō,sän }

bosporus [GEOGR] A strait connecting two seas or a lake and a sea. { 'bäs·pə·rəs }

bosque See temperate and cold scrub. { 'bäsk, 'bä·skā }

boss [DES ENG] Protuberance on a cast metal or plastic part to add strength, facilitate assembly, provide for fastenings, or so forth. [GEOL] A large, irregular mass of crystalline igneous rock that formed some distance below the surface but is now exposed by denudation. [NAV ARCH] See bossing. { bȯs }

bossing [NAV ARCH] **1.** A faired structural extension of the hull covering the propeller shaft where it emerges from the main hull and housing the main shaft bearing. **2.** The hub of a screw propeller. Also known as boss. { 'bȯs·iŋ }

bostonite [PETR] A rock with coarse trachytic texture formed almost wholly of albite and microcline and with accessory pyroxene. { 'bȯs·tə,nīt }

Boston ridge [BUILD] A method of applying shingles to the ridge of a house by which the shingles alternate in overlap from one side of the ridge to the other. { 'bȯs·tən ,rij }

Bostrichidae [INV ZOO] The powder-post beetles, a family of coleopteran insects in the superfamily Bostrichoidea. { bä'strik·ə,dē }

Bostrichoidea [INV ZOO] A superfamily of beetles in the coleopteran suborder Polyphaga. { ,bä·strə'kȯid·ē·ə }

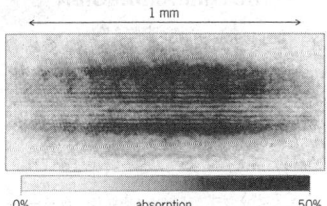

BOSE-EINSTEIN CONDENSATION

1 mm

0%   absorption   50%

Interference pattern of two expanding Bose-Einstein condensates, demonstrating their coherence. Interference fringes have a spacing of 15 micrometers. (*After D. S. Durfee and W. S. Ketterle, Experimental studies of Bose-Einstein condensation, Opt. Express, 2:299-313, Optical Society of America, 1998*)

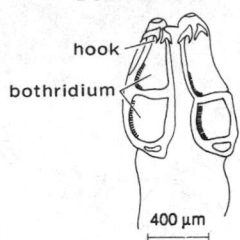

BOTHRIDIUM

hook

bothridium

400 μm

A drawing of the scolex of *Acanthobothrium* species showing the hooks on the bothridium.

BOTHRIOCIDAROIDA

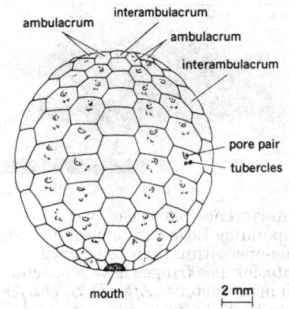

ambulacrum
interambulacrum
ambulacrum
interambulacrum
pore pair
tubercles
mouth
2 mm

A reconstruction of the test of *Bothriocidaris*.

botallackite [MINERAL] $Cu_2(OH)_3Cl\cdot3H_2O$ A pale bluish-green to green, orthorhombic mineral consisting of a basic copper chloride; occurs as crusts of crystals. { bə'tal·ə‚kīt }

botanical *See* crude drug. { bə'tan·ə·kəl }

botanical garden [BOT] An institution for the culture of plants collected chiefly for scientific and educational purposes. { bə'tan·ə·kəl 'gär·dən }

botany [BIOL] A branch of the biological sciences which embraces the study of plants and plant life. [TEXT] *See* merino. { 'bät·ən·ē }

bothridium [INV ZOO] A muscular holdfast organ, often with hooks, on the scolex of tetraphyllidean tapeworms. { bä'thrid·ē·əm }

Bothriocephaloidea [INV ZOO] The equivalent name for the Pseudophyllidea. { ‚bä·thrē·ō‚sef·ə'lóid·ē·ə }

Bothriocidaroida [PALEON] An order of extinct echinoderms in the subclass Perischoechinoidea in which the ambulacra consist of two columns of plates, the interambulacra of one column, and the madreporite is placed radially. { ‚bä·thrē·ō‚sik·ə'róid·ē·ə }

bothrium [INV ZOO] A suction groove on the scolex of pseudophyllidean tapeworms. { 'bäth·rē·əm }

botryogen [MINERAL] $MgFe(SO_4)_2(OH)\cdot7H_2O$ Orange-red, monoclinic mineral consisting of a hydrated basic sulfate of magnesium and trivalent iron. { 'bä·trē·ə‚jen }

botryoid [GEOL] **1.** A mineral formation shaped like a bunch of grapes. **2.** Specifically, such a formation of calcium carbonate occurring in a cave. Also known as clusterite. { 'bä·trē‚óid }

botryoidal [SCI TECH] Of formations and structures, shaped like a bunch of grapes. { ‚bä·trē‚óid·əl }

botryomycosis [VET MED] A chronic infectious bacterial disease of horses caused by *Staphylococcus aureus* and characterized by localized fibromatous tumors. { ‚bä·trē‚mī'kō·səs }

Botrytis disease [PL PATH] Any of various fungus diseases of plants caused by fungi of the genus *Botrytis*; characterized by soft rotting. { bō'trīd·əs di‚zēz }

bottle [ENG] A container made from pipe or plate with drawn, forged, or spun end closures, and used for storing or transporting gas. { 'bäd·əl }

bottle centrifuge [ENG] A centrifuge in which the mixture to be separated is poured into small bottles or test tubes; they are then placed in a rotor assembly which is spun rapidly. { 'bäd·əl 'sen·trə‚fyüj }

bottled gas [MATER] Butane, propane, or butane-propane mixtures liquefied and bottled under pressure for use as a domestic cooking or heating fuel. Also known as bugas. { ‚bäd·əld 'gas }

bottle graft [BOT] A plant graft in which the scion is a detached branch and is protected from wilting by keeping the base of the branch in a bottle of water until union with the stock. { 'bäd·əl ‚graft }

bottleneck [PETRO ENG] A section of reduced diameter in a drill pipe that is caused by excessive longitudinal strain or a combination of such strain and irregular swaying of the mechanism. { 'bäd·əl‚nek }

bottleneck analysis [COMPUT SCI] A detailed study of the manner in which elements of a computer system are related to find out where bottlenecks arise, so that the system's performance can be improved. { 'bäd·əl‚nek ə‚nal·ə·səs }

bottleneck assignment problem [IND ENG] A linear programming problem in which it is required to assign machines to jobs (or vice versa) so that the efficiency of the least efficient operation is maximized. { 'bäd·əl‚nek ə'sīn·mənt ‚präb·ləm }

bottle test [PETRO ENG] An analytical procedure in which a chemical is added to samples of a water-oil emulsion to determine the quantity of chemical needed to separate the emulsion into oil and water fractions. { 'bäd·əl ‚test }

bottle thermometer [ENG] A thermoelectric thermometer used for measuring air temperature; the name is derived from the fact that the reference thermocouple is placed in an insulated bottle. { 'bäd·əl thər'mäm·əd·ər }

bottom [COMPUT SCI] The termination of a file. [GEOL] **1.** The bed of a body of running or still water. **2.** *See* root. [PARTIC PHYS] The new quantum number associated with the bottom quark. Also known as beauty. { 'bäd·əm }

bottom blow [ENG] A type of plastics blow molding machine in which air is injected into the parison from the bottom of the mold. { 'bäd·əm ‚blō }

bottom break [BOT] A branch that arises from the base of a plant stem. [MIN ENG] The break or crack that separates a block of stone from a quarry floor. { 'bäd·əm ‚brāk }

bottom chord [CIV ENG] Any of the bottom series of truss members parallel to the roadway of a bridge. { 'bäd·əm ‚kórd }

bottom dead center [MECH ENG] The position of the crank of a vertical reciprocating engine, compressor, or pump when the piston is at the end of its downstroke. Abbreviated BDC. { 'bäd·əm ‚ded 'sen·tər }

bottom dump [ENG] A construction wagon with movable gates in the bottom to allow vertical discharge of its contents. { 'bäd·əm ‚dəmp }

bottomed hole [ENG] A completed borehole, or a borehole in which drilling operations have been discontinued. { 'bäd·əmd 'hōl }

bottom fauna *See* benthos. { 'bäd·əm ‚fón·ə }

bottom fermentation [FOOD ENG] A slow alcoholic fermentation during which yeast cells accumulate at the bottom of the fermenting liquid; occurs during fermentation of lager beer and wines of low alcoholic content. { 'bäd·əm fər·mən'tā·shən }

bottom flow [ENG] A molding apparatus that forms hollow plastic articles by injecting the blowing air at the bottom of the mold. [HYD] A density current that is denser than any section of the surrounding water and that flows along the bottom of the body of water. Also known as underflow. { 'bäd·əm ‚flō }

bottom hold-down [PETRO ENG] A device located at the lower end of a bottom-hole pump that anchors it in a well. { 'bäd·əm 'hōl ‚daún }

bottom hole [PETRO ENG] The deepest portion of an oil well. { 'bäd·əm ‚hōl }

bottom-hole assembly [PETRO ENG] That portion of the drilling assembly that extends beyond the drill pipe. { 'bäd·əm ‚hōl ə‚sem·blē }

bottom-hole cash [PETRO ENG] Cash which is contributed by mineral-rights lessees adjacent to a drilling lessee and which is payable when the well reaches a specified depth, regardless of whether or not the completed well is a producer. { 'bäd·əm ‚hōl ‚kash }

bottom-hole choke [PETRO ENG] A device containing a restricted opening that is placed in the lower end of the tubing to control fluid flow rate. { 'bäd·əm ‚hōl ‚chōk }

bottom-hole packer [PETRO ENG] An anchored-in-place seal used to provide liquid-proof packing in the annular space between the outside of the oil-producing tubing and the inside of the drill casing. { 'bäd·əm ‚hōl ‚pak·ər }

bottom-hole plug [PETRO ENG] A bridge plug or a cement plug installed near the bottom of the drill hole to shut off a depleted, water-producing, or unproductive zone. { 'bäd·əm ‚hōl ‚pləg }

bottom-hole pressure [PETRO ENG] **1.** Gas-drive pressure recorded at the bottom of an oil-well shaft; used to analyze oil-reservoir performance and evaluate the performance of downhole equipment. **2.** The pressure in a well measured at a point opposite the producing formation. { 'bäd·əm ‚hōl ‚presh·ər }

bottom-hole pressure bomb [PETRO ENG] A bomb for measuring and recording the pressure in a well at a point opposite the producing formation. { 'bäd·əm ‚hōl 'presh·ər ‚bäm }

bottom-hole pump [PETRO ENG] A pump installed near the bottom of a well to lift well fluids; may be a rod pump, high-pressure liquid pump, or a centrifugal pump. { 'bäd·əm ‚hōl ‚pəmp }

bottom-hole samples [PETRO ENG] Fluid samples from gas-condensate-well reservoirs; used to study the state of the hydrocarbon system under reservoir conditions and to estimate total hydrocarbons in place. { 'bäd·əm ‚hōl ‚sam·pəlz }

bottom-hole separator [PETRO ENG] A device designed to separate crude oil and natural gas at the bottom of the well in order to increase the volumetric efficiency of the pumping equipment. { 'bäd·əm ‚hōl 'sep·ə‚rād·ər }

bottom-hole temperature [PETRO ENG] A temperature measurement made at a depth in a well equal to the midpoint of the producing zone. { 'bäd·əm ‚hōl ‚tem·prə·chər }

bottom ice *See* anchor ice. { 'bäd·əm ‚īs }

bottoming drill [DES ENG] A flat-ended twist drill designed to convert a cone at the bottom of a drilled hole into a cylinder. { 'bäd·əm·iŋ ‚dril }

bottomland [GEOL] A lowland formed by alluvial deposit about a lake basin or a stream. { 'bäd·əm‚land }

bottom load *See* bed load. { 'bäd·əm ˌlōd }

bottom loading pressure [PETRO ENG] The pressure that is exerted on the bottom hull of a column-stabilized, semisubmersible drilling rig. { 'bäd·əm 'lōd·iŋ ˌpresh·ər }

bottom moraine *See* ground moraine. { 'bäd·əm mə'rān }

bottomonium [PARTIC PHYS] A meson, such as the upsilon particle, that is made up of the bottom quark *b* and its antiquark *b*. { ˌbäd·ə'mō·nē·əm }

bottom pillar [MIN ENG] A large block of solid coal left unworked around the shaft. { 'bäd·əm ˌpil·ər }

bottom quark [PARTIC PHYS] A quark with a mass of about 4.7 GeV, electric charge of −⅓, zero isotopic spin, strangeness and charm, and a new quantum number associated with it. Also known as *b* quark; beauty quark. Symbolized *b*. { 'bäd·əm ˌkwärk }

bottom rot [PL PATH] **1.** A fungus disease of lettuce, caused by *Pellicularia filamentosa,* that spreads from the base upward. **2.** A fungus disease of tree trunks caused by pore fungi. { 'bäd·əm ˌrät }

bottoms [CHEM ENG] Residual fractions that remain at the bottom of a fractionating tower following distillation of the lighter components. [PETRO ENG] *See* basic sediment and water. { 'bäd·əmz }

bottom sampler [ENG] Any instrument used to obtain a sample from the bottom of a body of water. { 'bäd·əm ˌsam·plər }

bottom sediment *See* basic sediment and water. { 'bäd·əm 'sed·ə·mənt }

bottomset beds [GEOL] Horizontal or gently inclined layers of finer material carried out and deposited on the bottom of a lake or sea in front of a delta. { 'bäd·əmˌset ˌbedz }

bottom settlings *See* basic sediment and water. { 'bäd·əm ˌset·liŋz }

bottom steam [CHEM] Steam piped into the bottom of the still during oil distillation. { 'bäd·əm ˌstēm }

bottom tap [DES ENG] A tap with a chamfer 1 to 1½ threads in length. [MET] A hole in the bottom of a furnace for draining out the slag. { 'bäd·əm ˌtap }

bottom terrace [GEOL] A landform deposited by streams with moderate or small bottom loads of coarse sand and gravel, and characterized by a broad, sloping surface in the direction of flow and a steep escarpment facing downstream. { 'bäd·əm ˌter·əs }

bottom time [OCEANOGR] In an operation involving diving by personnel, the total time of a dive measured from the moment the diver passes below the water surface until the beginning of the ascent. { 'bäd·əm ˌtīm }

bottom-up analysis [COMPUT SCI] A reductive method of syntactic analysis which attempts to reduce a string to a root symbol. { ¦bäd·əm¦əp¦ ə'nal·ə·səs }

bottom water [HYD] Water lying beneath oil or gas in productive formations. [OCEANOGR] The water mass at the deepest part of a water column in the ocean. { 'bäd·əm ˌwȯd·ər }

bottom wiper plug [PETRO ENG] A device, installed in the cementing head of a well, run down the casing ahead of the cement to remove mud from the casing walls so as to prevent contamination of the cement. { 'bäd·əm 'wī·pər ˌpləg }

botulin [MICROBIO] The neurogenic toxin which is produced by *Clostridium botulinum* and *C. parabotulinum* and causes botulism. Also known as botulinus toxin. { 'bäch·ə·lən }

botulinus [MICROBIO] A bacterium that causes botulism. { 'bäch·ə'lī·nəs }

botulinus toxin *See* botulin. { 'bäch·ə'lī·nəs 'täk·sən }

botulism [MED] Food poisoning due to intoxication by the exotoxin of *Clostridium botulinum* and *C. parabotulinum.* { 'bäch·ə,liz·əm }

boturon [ORG CHEM] $C_{12}H_{13}N_2OCl$ A white solid with a melting point of 145–146°C; used as pre- and postemergence herbicide in cereals, orchards, and vineyards. Also known as butyron. { 'bäch·ə,rän }

bouclé [TEXT] **1.** Yarn made with loose loops. **2.** A fabric made from bouclé yarn. { bü'klā }

boudin [GEOL] One of a series of sausage-shaped segments found in a boudinage. { bü'dan }

boudinage [GEOL] A structure in which beds set in a softer matrix are divided by cross fractures into segments resembling pillows. { ˌbüd·ən¦äzh }

bough [BOT] A main branch on a tree. { baů }

bougie decimale [OPTICS] Formerly, a unit of luminous intensity equal to 0.96 international standard candle. { 'bü,zhē des·ə'mäl }

Bouguer correction *See* Bouguer reduction. { bü'ger kə'rek·shən }

Bouguer gravity anomaly [GEOPHYS] A value that corrects the observed gravity for latitude and elevation variations, as in the free-air gravity anomaly, plus the mass of material above some datum (usually sea level) within the earth and topography. { bü'ger 'grav·əd·ē ə'näm·ə·lē }

Bouguer-Lambert-Beer law [ANALY CHEM] The intensity of a beam of monochromatic radiation in an absorbing medium decreases exponentially with penetration distance. Also known as Beer-Lambert-Bouguer law; Lambert-Beer law. { bü'ger ¦läm·bert ¦ber ˌlȯ }

Bouguer-Lambert law [ANALY CHEM] The law that the change in intensity of light transmitted through an absorbing substance is related exponentially to the thickness of the absorbing medium and a constant which depends on the sample and the wavelength of the light. Also known as Lambert's law. { bü'ger ¦läm·bərt ˌlȯ }

Bouguer reduction [GEOL] A correction made in gravity work to take account of the station's altitude and the rock between the station and sea level. Also known as Bouguer correction. { bü'ger ri'dək·shən }

Bouguer's halo [METEOROL] A faint, white circular arc of light of about 39° radius around the antisolar point. Also known as Ulloa's ring. { bü'gerz 'hä,lō }

Bouin's solution [MATER] A picric acid-acetic acid-formaldehyde fixative and preserving fluid for contractile forms. { bü'anz sə'lu·shən }

boulangerite [MINERAL] $Pb_5Sb_4S_{11}$ A bluish-lead-gray, monoclinic mineral consisting of lead antimony sulfide. { bü'lan·jə,rīt }

boulder [GEOL] A worn rock with a diameter exceeding 256 millimeters. Also spelled bowlder. { 'bōl·dər }

boulder barricade [GEOL] An accumulation of large boulders that is visible along a coast between low and half tide. { 'bōl·dər ,bar·ə,kād }

boulder belt [GEOL] A long, narrow accumulation of boulders elongately transverse to the direction of glacier movement. { 'bōl·dər ,belt }

boulder buster [ENG] A heavy, pyramidal- or conical-point steel tool which may be attached to the bottom end of a string of drill rods and used to break, by impact, a boulder encountered in a borehole. Also known as boulder cracker. { 'bōl·dər ,bəs·tər }

boulder clay *See* till. { 'bōl·dər ,klā }

boulder cracker *See* boulder buster. { 'bōl·dər ,krak·ər }

boulder pavement [GEOL] A surface of till with boulders; the till has been abraded to flatness by glacier movement. { 'bōl·dər ,pāv·mənt }

boulder train [GEOL] Glacial boulders derived from one locality and arranged in a right-angled line or lines leading off in the direction in which the drift agency operated. { 'bōl·dər ,trān }

boule [CRYSTAL] A pure crystal, such as silicon, having the atomic structure of a single crystal, formed synthetically by rotating a small seed crystal while pulling it slowly out of molten material in a special furnace. { bül }

bounce cast [GEOL] A short ridge underneath a stratum fading out gradually in both directions. { 'baůns ,kast }

bounce dive [OCEANOGR] A dive performed rapidly with an extremely short bottom time in order to keep decompression time to a minimum. { 'baůns ,dīv }

bounced message [COMPUT SCI] An electronic mail message that is returned to sender because attempts to deliver it have been unsuccessful. { ,baůnst 'mes·ij }

bounce table [MECH ENG] A testing device which subjects devices and components to impacts such as might be encountered in accidental dropping. { 'baůns ,tā·bəl }

bouncing putty [MATER] A silicone polymer in a soft elastic mass; the material's elasticity will increase as the applied force increases. { ¦baůns·iŋ 'pəd·ē }

boundary [ELECTR] An interface between *p*- and *n*-type semiconductor materials, at which donor and acceptor concentrations are equal. [GEOL] A line between areas occupied by rocks or formations of different type and age. [MATH]

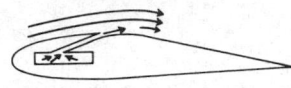

BOUNDARY-LAYER CONTROL

In this airfoil, retarded flow in the boundary layer is reenergized by supplying high-velocity flow through a slot in the surface.

See frontier. [SCI TECH] A line or area which determines inclusion in a system. { 'baún·drē }

boundary condition [MATH] A requirement to be met by a solution to a set of differential equations on a specified set of values of the independent variables. { 'baún·drē kən'dish·ən }

boundary friction [MECH] Friction between surfaces that are neither completely dry nor completely separated by a lubricant. { 'baún·drē ‚frik·shən }

boundary layer [METEOROL] The lower portion of the atmosphere, extending to a height of approximately 1.2 miles (2 kilometers). { 'baún·drē ‚lā·ər }

boundary-layer control [FL MECH] Control over the development of a boundary layer by reduction of surface roughness and choice of surface contours. Abbreviated BLC. { 'baún·drē ‚lā·ər kən'trōl }

boundary-layer flow [FL MECH] The flow of that portion of a viscous fluid which is in the neighborhood of a body in contact with the fluid and in motion relative to the fluid. { 'baún·drē ‚lā·ər ‚flō }

boundary-layer photocell *See* photovoltaic cell. { 'baún·drē ‚lā·ər 'fō·dō‚sel }

boundary-layer separation [FL MECH] That point where the boundary layer no longer continues to follow the contour of the boundary because the residual momentum of the fluid (left after overcoming viscous forces) may be insufficient to allow the flow to proceed into regions of increasing pressure. Also known as flow separation. { 'baún·drē ‚lā·ər sep·ə'rā·shən }

boundary-layer theory *See* film theory. { 'baún·drē ‚lā·ər ‚thē·ə·rē }

boundary line [MAP] A line of demarcation along which two areas meet. [PHYS CHEM] On a phase diagram, the line along which any two phase areas adjoin in a binary system, or the line along which any two liquidus surfaces intersect in a ternary system. { 'baún·drē ‚līn }

boundary lubrication [ENG] A lubricating condition that is a combination of solid-to-solid surface contact and liquid-film shear. { 'baún·drē ‚lü·brə'kā·shən }

boundary map [MAP] A map constructed for the purpose of delineating a boundary line and adjacent territory. { 'baún·drē ‚map }

boundary marker [NAV] A radio transmitter operating at 75 megahertz and installed near the approach end of landing runway (3.9 nautical miles ± 1000 feet, or 7123 ± 305 meters) and approximately on the localizer course line. { 'baún·drē ‚mär·kər }

boundary monument [ENG] A material object placed on or near a boundary line to preserve and identify the location of the boundary line on the ground. { 'baún·drē ‚män·yə·mənt }

boundary of a set *See* frontier. { 'baún·drē əv ə 'set }

boundary pillar [MIN ENG] A pillar left in mines between adjoining properties. { 'baún·drē ‚pil·ər }

boundary point [MATH] In a topological space, a point of a set with the property that every neighborhood of the point contains points of both the set and its complement. { 'baún·drē ‚póint }

boundary survey [ENG] A survey made to establish or to reestablish a boundary line on the ground or to obtain data for constructing a map or plat showing a boundary line. { 'baún·drē ‚sər·vā }

boundary value component *See* perfectly mobile component. { 'baún·drē ‚val·yü kəm‚pō·nənt }

boundary value problem [MATH] A problem, such as the Dirichlet or Neumann problem, which involves finding the solution of a differential equation or system of differential equations which meets certain specified requirements, usually connected with physical conditions, for certain values of the independent variable. { 'baún·drē ‚val·yü ‚präb·ləm }

boundary wave [GEOPHYS] A seismic wave that propagates along a free surface or an interface between defined layers. { 'baún·drē ‚wāv }

boundary wavelength *See* quantum limit. { ‚baún·drē 'wāv‚leŋkth }

bound barrel [ORD] A barrel that is touching parts of the stock in such a manner that expansion due to heat from firing causes the barrel to bind and bend, resulting in inaccurate fire. { ‚baúnd 'bar·əl }

bound charge [ELEC] Electric charge which is confined to

atoms or molecules, in contrast to free charge, such as metallic conduction electrons, which is not. Also known as polarization charge. { ‚baúnd 'chärj }

bounded difference [MATH] For two fuzzy sets A and B, with membership functions m_A and m_B, the fuzzy set whose membership function $m_{A \ominus B}$ has the value $m_A(x) - m_B(x)$ for every element x for which $m_A(x) \geq m_B(x)$, and has the value 0 for every element x for which $m_A(x) \leq m_B(x)$. { ‚baúnd·əd 'dif·rəns }

bounded function [MATH] **1.** A function whose image is a bounded set. **2.** A function of a metric space to itself which moves each point no more than some constant distance. { ‚baúnd·əd 'fəŋk·shən }

bounded growth [MATH] The property of a function f defined on the positive real numbers which requires that there exist numbers M and a such that the absolute value of $f(t)$ is less than Ma^t for all positive values of t. { ‚baúnd·əd 'grōth }

bounded linear transformation [MATH] A linear transformation T for which there is some positive number A such that the norm of $T(x)$ is equal to or less than A times the norm of x for each x. { ‚baúnd·əd 'lin·ē·ər tranz·fər'mā·shən }

bounded product [MATH] For two fuzzy sets A and B, with membership functions m_A and m_B, the fuzzy set whose membership function $m_{A \odot B}$ has the value $m_A(x) + m_B(x) - 1$ for every element x for which $m_A(x) + m_B(x) \geq 1$, and has the value 0 for every element x for which $m_A(x) + m_B(x) \leq 1$. { ‚baúnd·əd 'präd·əkt }

bounded sequence [MATH] A sequence whose members form a bounded set. { ‚baúnd·əd 'sē·kwəns }

bounded set [MATH] **1.** A collection of numbers whose absolute values are all smaller than some constant. **2.** A set of points, the distance between any two of which is smaller than some constant. { ‚baúnd·əd 'set }

bounded sum [MATH] For two fuzzy sets A and B, with membership functions m_A and m_B, the fuzzy set whose membership function $m_{A \oplus B}$ has the value $m_A(x) + m_B(x)$ for every element x for which $m_A(x) + m_B(x) \leq 1$, and has the value 1 for every element x for which $m_A(x) + m_B(x) \geq 1$. { ‚baúnd·əd 'səm }

bounded variation [MATH] A real-valued function is of bounded variation on an interval if its total variation there is bounded. { ‚baúnd·əd ver·ē'ā·shən }

bound electron [ATOM PHYS] An electron whose wave function is negligible except in the vicinity of an atom. { ‚baúnd i'lek‚trän }

bound glue state *See* glueball. { ‚baúnd 'glü ‚stāt }

bounding mine [ORD] Type of antipersonnel mine usually buried just below the surface of the ground; it has a small charge which throws the case up into the air; this case explodes at a height of 3 or 4 feet (1 or 1.2 meters), throwing shrapnel or fragments in all directions. { ‚baúnd·iŋ 'mīn }

bound level [NUC PHYS] An energy level in a nucleus so close to the ground state that it can only decay by gamma emission. { ‚baúnd ‚lev·əl }

bound particle [PHYS] A particle which is confined to some finite region. { ‚baúnd 'pärd·i·kəl }

bounds register [COMPUT SCI] A device which stores the upper and lower bounds on addresses in the memory of a given computer program in a time-sharing system. { 'baúnz ‚rej·ə·stər }

bound symbol [COMMUN] A contextual symbol either not preceded or not followed, or neither preceded nor followed, by space. { ‚baúnd 'sim·bəl }

bound symbol sequence [COMMUN] A symbol sequence either not preceded or not followed, or neither preceded nor followed, by space. { ‚baúnd ‚sim·bəl 'sē·kwəns }

bound variable [MATH] In logic, a variable that occurs within the scope of a quantifier, and cannot be replaced by a constant. { ‚baúnd 'ver·ē·ə·bəl }

bound vector [MECH] A vector whose line of application and point of application are both prescribed, in addition to its direction. { ‚baúnd 'vek·tər }

bound water [CHEM] Water that is a portion of a system such as tissues or soil and does not form ice crystals until the material's temperature is lowered to about −20°C. { ‚baúnd 'wòd·ər }

bourbon [FOOD ENG] A whiskey distilled from a corn mash containing at least 51% corn, with malt and rye composing the remaining ingredients, and aged in containers made of charred oak. { 'bùr·bən }

bourdon [ACOUS] In a carillon, the bell with the lowest tone. { 'bùrd·ən }

Bourdon pressure gage [ENG] A mechanical pressure-measuring instrument employing as its sensing element a curved or twisted metal tube, flattened in cross section and closed. Also known as Bourdon tube. { 'bùr·dən 'presh·ər ¸gāj }

Bourdon tube See Bourdon pressure gage. { 'bùr·dən 'tüb }

Bourges process [GRAPHICS] A process used in art and color preparation for fake color work; transparent films in a variety of colors and densities are used as overlays for the black and white art work. { 'bùrzh 'präs·əs }

bourne [HYD] A small intermittent stream in a dry valley. { bùrn }

Bourne shell [COMPUT SCI] The original Unix shell. { 'bùrn ¸shel }

Bournesville's disease See tuberous sclerosis. { 'bùrn¸vēlz di¸zēz }

bournonite [MINERAL] PbCuSbS₃ Steel-gray to black orthorhombic crystals; mined as an ore of copper, lead, and antimony. Also known as berthonite; cogwheel ore. { 'bùr·nə¸nit }

bourrelet [ORD] The cylindrical surface of a projectile on which the projectile bears while in the bore of the weapon. { ¦bùr·ə¦lā }

Boussinesq approximation [FL MECH] The assumption (frequently used in the theory of convection) that the fluid is incompressible except insofar as the thermal expansion produces a buoyancy, represented by a term $g\alpha T$, where g is the acceleration of gravity, α is the coefficient of thermal expansion, and T is the perturbation temperature. { 'bü·si¸nesk ə¸präk· sə'mā·shən }

Boussinesq equation [ENG] A relation used to calculate the influence of a concentrated load on the backfill behind a retaining wall. { 'bü·si¸nesk i'kwā·shən }

Boussinesq number [FL MECH] A dimensionless number used to study wave behavior in open channels. { 'bü·si¸nesk ¸nəm·bər }

Boussinesq's problem [MECH] The problem of determining the stresses and strains in an infinite elastic body, initially occupying all the space on one side of an infinite plane, and indented by a rigid punch having the form of a surface of revolution with axis of revolution perpendicular to the plane. Also known as Cerruti's problem. { 'bü·si¸nesks ¸präb·ləm }

boussingaultite [MINERAL] $(NH_4)_2Mg(SO_4)_2 \cdot 6H_2O$ A colorless to yellowish-pink, monoclinic mineral consisting of a hydrated sulfate of ammonium and magnesium; usually occurs in massive form, as crusts or stalactites. { ¸büs· ən'gól¸tīt }

bouton [NEURO] A club-shaped enlargement at the end of a nerve fiber. Also known as end bulb. { bü'tōn }

boutonneuse fever See fièvre boutonneuse. { 'büt·ən¸üz ¸fē·vər }

Bouvealt-Blanc method [ORG CHEM] A laboratory method for preparing alcohols by reduction of esters utilizing sodium dissolved in alcohol. { ¦bü¸vō 'blän ¸meth·əd }

Bovidae [VERT ZOO] A family of pecoran ruminants in the superfamily Bovoidea containing the true antelopes, sheep, and goats. { 'bō·və¸dē }

bovine [VERT ZOO] **1.** Any member of the genus *Bos*. **2.** Resembling or pertaining to a cow or ox. { 'bō¸vīn }

bovine mastitis [VET MED] Inflammation of the udder of a cow; may result from injury or bacterial infection. { 'bō¸vīn ma·'stīd·əs }

bovine staggers [VET MED] A disease of cattle in southern Africa caused by eating the poisonous herb *Matricaria nigellae-folia* and characterized by staggering, emaciation, and finally paralysis. { 'bō¸vīn 'stag·ərz }

Bovoidea [VERT ZOO] A superfamily of pecoran ruminants in the order Artiodactyla comprising the pronghorns and bovids. { bō'vóid·ē·ə }

bow [AERO ENG] The forward section of an aircraft. [ARCH] A part of a building shaped as an arc or a polygon and projecting from a straight wall. [MATER] The distortion of lumber in which there is a deviation from a straight line in a direction perpendicular to the flat face. [NAV ARCH] The forward part of a ship. { bù }

bow compass [GRAPHICS] A small compass whose legs are connected by a spring shaped like a bow, rather than by a joint. { 'bù ¸käm·pəs }

Bowden cable [MECH ENG] A wire made of spring steel which is enclosed in a helical casing and used to transmit longitudinal motions over distances, particularly around corners. { 'bōd·ən ¸kā·bəl }

Bowditch curve See Lissajous figure. { 'bù·dich ¸kərv }

bow divider [GRAPHICS] A small divider whose legs are connected by a bow-shaped spring, rather than by a joint. { ¦bō di'vīd·ər }

bowel [ANAT] The intestine. { bùl }

Bowen reaction series [MINERAL] A series of minerals wherein any early-formed phase will react with the melt later in the differentiation to yield a new mineral further in the series. { 'bō·ən rē'ak·shən ¸sir·ēz }

Bowen's disease [MED] **1.** Intraepithelial squamous-cell carcinoma of the skin, forming distinctive plaques. **2.** A similar carcinoma occurring in mucous membranes. { 'bō·ənz di¸zēz }

Bower-Barff process [MET] A method of coating iron or steel with magnetic oxide, such as Fe_3O_4, in order to minimize atmospheric corrosion. { 'bù·ər ¦bärf ¸präs·əs }

bowfin [VERT ZOO] *Amia calva.* A fish recognized as the only living species of the family Amiidae. Also known as dogfish; grindle; mudfish. { 'bō¸fin }

bow flare slamming [NAV ARCH] An extreme pitch-and-heave motion of a ship in which the keel emerges from the water but the forefoot does not. { 'bù ¸fler ¸slam·iŋ }

bow gun [ORD] Gun mounted at the front of a ship or armored vehicle, especially a semifixed, forward-firing gun in tanks. { 'bù ¸gən }

Bowie formula [GEOPHYS] A correction used for calculation of the local gravity anomaly on earth. { 'bō·ē ¸fòrm·yə·lə }

bowk See hoppit. { bōk }

bowl classifier [CHEM ENG] A shallow bowl with a concave bottom so that a liquid-solid suspension can be fed to the center; coarse particles fall to the bottom, where they are raked to a central discharge point, and liquid and fine particles overflow the edges and are collected. { ¦bōl ¸klas·ə¸fī·ər }

bowl crater [ASTRON] A type of lunar crater whose interior cross section is a smooth curve, with no flat floor. { 'bōl ¸krād·ər }

bowlder See boulder. { 'bōl·dər }

bow light [NAV ARCH] A white light near the bow of a ship at anchor or under way. { 'bù ¸līt }

bowline [NAV ARCH] **1.** A rope attached to the vertical edge of a square sail near its midpoint, and used to keep the sail's weather edge taut forward when the vessel is close-hauled. **2.** A knot forming a loop that does not slip under tension, used particularly for mooring and hauling. { 'bù¸līn }

bowlingite See saponite. { 'bō·liŋ¸gīt }

bowl mill See bowl-mill pulverizer. { 'bōl ¸mil }

bowl-mill pulverizer [MECH ENG] A type of pulverizer which directly feeds a coal-fired furnace, in which springs press pivoted stationary rolls against a rotating bowl grinding ring, crushing the coal between them. Also known as a bowl mill. { 'bōl ¸mil ¸pəl·və¸rīz·ər }

bowl scraper [MECH ENG] A towed steel bowl hung within a fabricated steel frame, running on four or two wheels; transports soil, in addition to spreading and leveling it. { 'bōl ¸skrāp·ər }

Bowman's capsule [ANAT] A two-layered membranous sac surrounding the glomerulus and constituting the closed end of a nephron in the kidneys of all higher vertebrates. { ¦bō·mənz 'kap·səl }

bow-on [ORD] Facing the firer; a bow-on target presents its narrower dimension exactly toward the gun firing at it; when an enemy tank is headed exactly at the gun firing at it, the tank is a bow-on target. { 'bù¸ón }

bowshock [ASTROPHYS] The shock wave set up by the inter-action of the supersonic solar wind with a planet's magnetic field. { 'bù¸shäk }

Bow's notation [MECH] A graphical method of representing coplanar forces and stresses, using alphabetical letters, in the solution of stresses or in determining the resultant of a system of concurrent forces. { 'bōz nō'tā·shən }

bowsprit [NAV ARCH] A large spar that projects forward from the forward end of a sailing ship, used to carry sails and support the masts by stays. { 'bù¸sprit }

BOWFIN

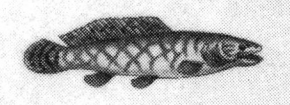

Bowfin (*Amia calva*), which may attain 2½ feet (76 centimeters) and is found only in some parts of the United States.

bowstring beam [CIV ENG] A steel, concrete, or timber beam or girder shaped in the form of a bow and string; the string resists the horizontal forces caused by loads on the arch. { 'bō‚striŋ ‚bēm }

bowtie antenna [ELECTROMAG] An antenna that consists of two triangular pieces of stiff wire or two triangular flat metal plates, arranged in the configuration of a bowtie, with the feed point at the gap between the apexes of the triangles. { 'bō‚tī an‚ten·ə }

bow wave [FL MECH] A shock wave occurring in front of a body, such as an airfoil, or apparently attached to the forward tip of the body. { 'bau̇ ‚wāv }

box [DES ENG] *See* boxing. [ENG] A protective covering or housing. { bäks }

box annealing [MET] Slow heating of metal sheets in a closed metal box to prevent oxidation, followed by cooling; usually limited to iron-base alloys. { 'bäks ə'nēl·iŋ }

box barrage [ORD] In antiaircraft artillery fire, an antiaircraft barrage delivered in a box-shape pattern by guns surrounding a defended target. { 'bäks bə'räzh }

box beam *See* box girder. { 'bäks ‚bēm }

box bill *See* bell tap. { 'bäks ‚bil }

boxboard [MATER] Paperboard used for making cardboard boxes. { 'bäks‚bȯrd }

box caisson [CIV ENG] A floating steel or concrete box with an open top which will be filled and sunk at a foundation site in a river or seaway. Also known as American caisson; stranded caisson. { 'bäks 'kā‚sän }

box camera [OPTICS] A camera that consists of a box, an arrangement for loading and winding film, a simple lens of fixed focus, and a simple shutter with a speed of about $1/30$ second. { 'bäks ‚kam·rə }

box canyon [GEOGR] A canyon with steep rock sides and a zigzag course, that is usually closed upstream. { 'bäks ‚kan·yən }

boxcar [COMMUN] One of a series of long signal-wave pulses which are separated by very short intervals of time. [ENG] A railroad car with a flat roof and vertical sides, usually with sliding doors, which carries freight that needs to be protected from weather and theft. { 'bäks‚kär }

boxcar circuit [ELECTR] A circuit used in radar for sampling voltage waveforms and storing the latest value sampled; the term is derived from the flat, steplike segments of the output voltage waveform. { 'bäks‚kär ‚sər·kət }

boxcar function [MATH] A function whose value is zero except for a finite interval of its argument, for which it has a constant nonzero value. { 'bäks‚kär ‚fəŋk·shən }

box-coking test [ENG] A laboratory test which forecasts the quality of coke producible in commercial practice; uses a specially designed sheet-steel box containing about 60 pounds (27 kilograms) of coal in a commercial coke oven. { 'bäks ‚kōk·iŋ ‚test }

box flue [ARCHEO] A component of a Roman heating system made of four tiles joined at the edges and open at the top and the bottom; designed to conduct hot air up the walls to escape at the eaves. { 'bäks ‚flü }

box fold [GEOL] A fold in which the broad, flat top of an anticline or the broad, flat bottom of a syncline is bordered by steeply dipping limbs. { 'bäks ‚fōld }

box girder [CIV ENG] A hollow girder or beam with a square or rectangular cross section. Also known as box beam. { 'bäks ‚gər·dər }

box-girder bridge [CIV ENG] A fixed bridge consisting of steel girders fabricated by welding four plates into a box section. { 'bäks ‚gər·dər ‚brij }

box header boiler [MECH ENG] A horizontal boiler with a front header and rear inclined rectangular header connected by tubes. { 'bäks ‚hed·ər ‚bȯil·ər }

Box Hole [GEOL] A meteorite crater in central Australia, 575 feet (175 meters) in diameter. { 'bäks ‚hōl }

boxing [DES ENG] The threaded nut for the screw of a mounted auger drill. Also known as box. [ENG] A method of securing shafts solely by slabs and wooden pegs. [MET] Continuing a fillet weld around a corner. Also known as end turning. { 'bäks·iŋ }

boxing shutter [BUILD] A window shutter which can be folded into a boxlike enclosure or recess at the side of the window frame. { 'bäks·iŋ ‚shəd·ər }

boxing the compass [NAV] In nautical practice, calling out

in sequence the names of the points (and sometimes the half and quarter points) indicated by the compass of a maneuvering vessel. { 'bäks·iŋ t͟hə 'käm‚pəs }

box loom [TEXT] A loom that uses several shuttles to weave fabric with more than one type of filling yarn or with a regular filling pattern. { 'bäks ‚lüm }

box piles [CIV ENG] Pile foundations made by welding together two sections of steel sheet piling or combinations of beams, channels, and plates. { 'bäks ‚pīlz }

boxplot [IND ENG] In quality control, a graph summarizing the distribution, central value, and variability of a set of data values; used to identify problems (or potential problems) that affect the quality of processes and products. { 'bäks‚plät }

boxwork [GEOL] Limonite and other minerals which formed at one time as blades or plates along cleavage or fracture planes, after which the intervening material dissolved, leaving the intersecting blades or plates as a network. { 'bäks‚wərk }

box wrench [ENG] A closed-end wrench designed to fit a variety of sizes and shapes of bolt heads and nuts. { 'bäks ‚rench }

Boyle's law [PHYS] The law that the product of the volume of a gas times its pressure is a constant at fixed temperature. Also known as Mariotte's law. { 'bȯilz ‚ló }

Boyle's temperature [THERMO] For a given gas, the temperature at which the virial coefficient B in the equation of state $Pv = RT[1 + (B/v) + (C/v^2) + \cdots]$ vanishes. { 'bȯilz 'tem·prə·chər }

Boys camera [OPTICS] A type of camera used for the observation of lightning flashes. { 'bȯiz ‚kam·rə }

Boys' method [OPTICS] A method of measuring the refractive index of a lens, in which the curvatures of the lens surfaces are determined by positioning a light source so that reflection from a surface gives an image coincident with the object; these curvatures and the focal length are used to calculate the refractive index. { 'bȯiz ‚meth·əd }

bp *See* boiling point.

bpd *See* barrels per day.

B phase *See* liquid B. { 'bē ‚fāz }

bpm *See* barrels per month.

B power supply *See* B supply. { 'bē ‚pau̇·ər ‚sə·plī }

bps *See* bit per second.

BPSK *See* binary phase-shift keying.

Bq *See* becquerel.

b quark *See* bottom quark. { 'bē ‚kwärk }

Br *See* bromine.

BRA *See* β-resorcylic acid.

brace [DES ENG] A cranklike device used for turning a bit. [ENG] A diagonally placed structural member that withstands tension and compression, and often stiffens a structure against wind. { brās }

brace and bit [DES ENG] A small hand tool to which is attached a metal- or wood-boring bit. { 'brās ən 'bit }

braced framing [CIV ENG] Framing a building with post and braces for stiffness. { 'brāst 'frām·iŋ }

braced-rib arch [CIV ENG] A type of steel arch, usually used in bridge construction, which has a system of diagonal bracing. { 'brāst 'rib ‚ärch }

brace head [ENG] A cross handle attached at the top of a column of drill rods by means of which the rods and attached bit are turned after each drop in chop-and-wash operations while sinking a borehole through overburden. Also known as brace key. { 'brās ‚hed }

brace key *See* brace head. { 'brās ‚kē }

brace pile *See* batter pile. { 'brās ‚pīl }

brace root *See* prop root. { 'brās ‚rüt }

brachial [ZOO] Of or relating to an arm or armlike process. { 'brā·kē·əl }

brachial artery [ANAT] An artery which originates at the axillary artery and branches into the radial and ulnar arteries; it distributes blood to the various muscles of the arm, the shaft of the humerus, the elbow joint, the forearm, and the hand. { 'brā·kē·əl 'ärd·ə·rē }

brachial cavity [INV ZOO] The anterior cavity which is located inside the valves of brachiopods and into which the brachia are withdrawn. { 'brā·kē·əl 'kav·əd·ē }

brachial plexus [NEURO] A plexus of nerves located in the neck and axilla and composed of the anterior rami of the lower four cervical and first thoracic nerves. { 'brā·kē·əl 'plek·səs }

BRACED-RIB ARCH

The Bayonne Bridge, a braced-rib arch, across the Kill Van Kull in New Jersey. (*Port of New York Authority*)

Brachiata [INV ZOO] A phylum of deuterostomous, sedentary bottom-dwelling marine animals that live encased in tubes. { ˌbra·kē'ad·ə }

brachiate [BOT] Possessing widely divergent branches. [ZOO] Having arms. { 'bra·kē‚āt }

brachiating motion [CONT SYS] A type of robotic motion that employs legs or other equipment to help the manipulator move in its working environment. { 'brā·kē'ād·iŋ 'mō·shən }

brachiating robot [CONT SYS] A robot that is capable of moving over the surface of an object. { 'brā·kē'ād·iŋ 'rō‚bät }

brachiolaria [INV ZOO] A transitional larva in the development of certain starfishes that is distinguished by three anterior processes homologous with those of the adult. { ˌbra·kē·ō'la·re·ə }

Brachiopoda [INV ZOO] A phylum of solitary, marine, bivalved coelomate animals. { ˌbra·kē'ä·pə·də }

brachioradialis [ANAT] The muscle of the arm that flexes the elbow joint; origin is the lateral supracondylar ridge of the humerus, and insertion is the lower end of the radius. { ˌbrā·kē·ō‚rā·dē'äl·əs }

Brachiosaurus [PALEON] A herbivorous sauropod dinosaur, 90 feet (27 meters) long and weighing 85–110 tons, from the Late Jurassic that had a very long neck. { ˌbrā·kē·ə'sór·əs }

brachistochrone [MECH] The curve along which a smooth-sliding particle, under the influence of gravity alone, will fall from one point to another in the minimum time. { brə'kis·tə‚krōn }

brachium [ANAT] The upper arm or forelimb, from the shoulder to the elbow. [INV ZOO] **1.** A ray of a crinoid. **2.** A tentacle of a cephalopod. **3.** Either of the paired appendages constituting the lophophore of a brachiopod. { 'brā·kē·əm }

Brachyarchus [MICROBIO] A genus of bacteria of uncertain affiliation; rod-shaped cells bent in a bowlike configuration and usually arranged in groups of two, four, or more cells. { ˌbra·kē'är·kəs }

brachyaxis [CRYSTAL] The shorter lateral axis, usually the *a* axis, of an orthorhombic or triclinic crystal. Also known as brachydiagonal. { ˌbra·kē'ak·səs }

brachyblast [BOT] A short shoot often bearing clusters of leaves. { 'bra·kə‚blast }

brachycephalic [ANTHRO] Being short- or broad-headed, that is, having a cephalic index of over 80. { ˌbrak·i·sə'fal·ik }

brachycephaly [ANTHRO] The state of being brachycephalic. { ˌbrak·i'sef·ə‚lē }

brachycerebral [ANTHRO] Having a round or short brain. { ˌbrak·i·sə'rē·brəl }

brachycranial [ANTHRO] Being short- or broad-skulled, that is, with a cephalic index of over 80. { ˌbrak·i'krā·nē·əl }

brachydactylia [MED] Abnormal shortening of fingers or toes. { ˌbrak·i·dak'til·ē·ə }

brachydiagonal *See* brachyaxis. { ˌbrak·i·dī'ag·ə·nəl }

brachydont [ANAT] Of teeth, having short crowns, well-developed roots, and narrow root canals, characteristic of humans. { 'brak·ə‚dänt }

brachyfacial [ANTHRO] Having a short or broad face. { ˌbrak·ə'fāsh·əl }

Brachygnatha [INV ZOO] A subsection of brachyuran crustaceans to which most of the crabs are assigned. { bra'kig·nə·thə }

brachypinacoid [GEOL] A pinacoid parallel to the vertical and the shorter lateral axis. { ˌbrak·i'pin·ə‚kóid }

Brachypsectridae [INV ZOO] A family of coleopteran insects in the superfamily Cantharoidea represented by a single species. { ˌbra·kip'sek·trə‚dē }

Brachypteraciidae [VERT ZOO] The ground rollers, a family of colorful Madagascan birds in the order Coraciiformes. { brə‚kip·ter·ə'sī·ə‚dē }

brachypterous [INV ZOO] Having rudimentary or abnormally small wings, referring to certain insects. { brə'kip·tə·rəs }

brachysclereid [BOT] A sclereid that is more or less isodiametric and is found in certain fruits and in the pith, cortex, and bark of many stems. Also known as stone cell. { ˌbrak·i'sklir·ē·əd }

brachyskelic [ANTHRO] Having legs short in proportion to trunk length, with a skelic index of 75–80. { ˌbrak·i'skel·ik }

brachysm [BOT] Plant dwarfing in which there is shortening of the internodes only. { 'bra‚kiz·əm }

brachysyncline [GEOL] A broad, short syncline. { ˌbrak·i'sin‚klīn }

brachytherapy [MED] Radiation treatment using a solid or enclosed radioisotopic source on the surface of the body or at a short distance from the area to be treated. { ˌbrak·i'ther·ə·pē }

Brachythoraci [PALEON] An order of the joint-neckfishes, now extinct. { ˌbrak·i'thór·ə‚sī }

Brachyura [INV ZOO] The section of the crustacean order Decapoda containing the true crabs. { ˌbra·kē'yùr·ə }

bracing [ENG] The act or process of strengthening or making rigid. { 'brās·iŋ }

brackebuschite [MINERAL] $Pb_4MnFe(VO_4)_4·2H_2O$ Dark brown to black, monoclinic mineral consisting of a hydrated vanadate of lead, manganese, and iron. { 'bra·kə‚bù‚shīt }

bracket [BUILD] A vertical board to support the tread of a stair. [CIV ENG] A projecting support. [ORD] **1.** The distance between two strikes or series of strikes, one of which is over the target and the other short of it, or one of which is to the right and the other to the left of the target. **2.** A group of shots (or bombs) which fall both over and short of the target. { 'brak·ət }

bracket fungus [MYCOL] A basidiomycete characterized by shelflike sporophores, sometimes seen on tree trunks. { 'brak·ət ‚fəŋ·gəs }

bracketing method [ORD] A method of adjusting artillery and mortar fire in which a bracket is established by obtaining an over and a short strike, with respect to the observer, then successively splitting the bracket in half until a target hit is obtained or the smallest practicable range change has been made. { 'brak·əd·iŋ ‚meth·əd }

bracketing salvo [ORD] A series of shots in which the number of shots going over the target equals the number falling short of it. { 'brak·əd·iŋ ‚sal·vō }

Brackett series [SPECT] A series of lines in the infrared spectrum of atomic hydrogen whose wave numbers are given by $R_H[(1/16) − (1/n^2)]$, where R_H is the Rydberg constant for hydrogen and n is any integer greater than 4. { 'brak·ət ‚sir·ēz }

brackish [HYD] **1.** Of water, having salinity values ranging from approximately 0.50 to 17.00 parts per thousand. **2.** Of water, having less salt than sea water, but undrinkable. { 'brak·ish }

Braconidae [INV ZOO] The braconid wasps, a family of hymenopteran insects in the superfamily Ichneumonoidea. { brə'kän·ə‚dē }

bract [BOT] A modified leaf associated with plant reproductive structures. { brakt }

bracteolate [BOT] Having bracteoles. { brak'tē·ə·lət }

bracteole [BOT] A small bract, especially if on the floral axis. Also known as bractlet. { 'brak·tē‚ōl }

bractlet *See* bracteole. { 'brak·lət }

brad [DES ENG] A small finishing nail whose body either is of uniform thickness or is tapered. { brad }

bradding [ENG] A distortion of a bit tooth caused by the application of excessive weight, causing the tooth to become dull so that its softer inner portion caves over the harder case area. { 'brad·iŋ }

Bradfordian [GEOL] Uppermost Devonian geologic time. { ˌbrad'fórd·ē·ən }

Bradley aberration [ASTRON] Stellar aberration with a maximum of 20.5 seconds of arc; can be used to compute an approximate velocity for light. { 'brad·lē ab·ə'rā·shən }

bradleyite [MINERAL] $Na_3Mg(PO_4)(CO_3)$ A light gray mineral consisting of a phosphate-carbonate of sodium and magnesium; occurs as fine-grained masses. { 'brad·lē‚īt }

bradyauxesis [BIOL] Allometric growth in which a part lags behind the body as a whole in development. { ˌbra·dē·óg'zē·səs }

bradycardia [MED] Slow heart rate. { ˌbrad·i'kärd·ē·ə }

bradydiastole [MED] Prolongation of diastole beyond normal limits. { ˌbra·dē·dī'as·tə‚lē }

bradyesthesia [MED] Retardation in the transmission of sensory impressions. { ˌbra·dē·es'thēzh·yə }

bradykinesia [MED] Extreme slowness in movement. { ˌbra·dē·kə'nēzh·yə }

bradykinetic syndrome [MED] A neurologic condition characterized by a generalized slowness of motor activity. { ˌbra·dē·ki'ned·ik 'sin‚drōm }

BRACHYGNATHA

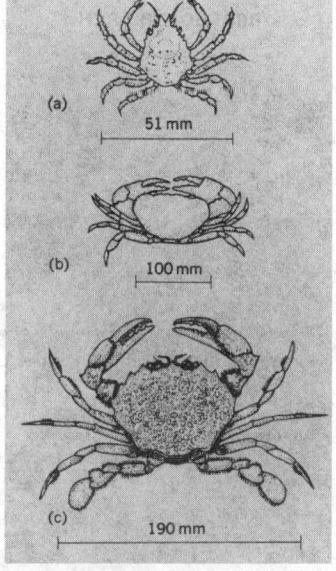

(a) 51 mm

(b) 100 mm

(c) 190 mm

Representative crabs of the Brachygnatha: *(a)* spider crab, *Mithrax acuticornis*; *(b)* freshwater crab, *Epilobocera sinuatifrons*; *(c)* swimming crab, *Ovalipes oceallatus*.

bradykinin [BIOCHEM] $C_{50}H_{73}N_{15}O_{11}$ A polypeptide kinin; forms an amorphous precipitate in glacial acetic acid; released from plasma precursors by plasmin; acts as a vasodilator. Also known as callideic I; kallidin I. { ¦brād·i'kī·nən }

Bradyodonti [PALEON] An order of Paleozoic cartilaginous fishes (Chondrichthyes), presumably derived from primitive sharks. { ¦brā·dē·ō'dän,tī }

bradypnea [MED] Abnormal slowness of respiration. { ¦brād·ē'nē·ə }

Bradypodidae [VERT ZOO] A family of mammals in the order Edentata comprising the true sloths. { ¦brād·i'pä·də,dē }

bradytely [EVOL] Evolutionary change that is either arrested or occurring at a very slow rate over long geologic periods. { 'brād·ə,te·lē }

Bragg angle [SOLID STATE] One of the characteristic angles at which x-rays reflect specularly from planes of atoms in a crystal. { 'brag ,aŋ·gəl }

Bragg cell See acoustooptic modulator. { 'brag ,sel }

Bragg curve [ATOM PHYS] **1.** A curve showing the average number of ions per unit distance along a beam of initially monoenergetic ionizing particles, usually alpha particles, passing through a gas. Also known as Bragg ionization curve. **2.** A curve showing the average specific ionization of an ionizing particle of a particular kind as a function of its kinetic energy, velocity, or residual range. { 'brag ,kərv }

Bragg diffraction See Bragg scattering. { 'brag di'frak·shən }

Bragg effect [OPTICS] A phenomenon observed in the recording of white-light holograms in which a photographic film irradiated from both sides displays, upon development, blackening in a series of film planes. Also known as color-filter effect. { 'brag i,fekt }

Bragg ionization curve See Bragg curve. { 'brag ī·ə·nə'zā·shən ,kərv }

braggite [MINERAL] PtS A steel-gray platinum sulfide mineral with tetragonal crystals. { 'bra,gīt }

Bragg-Kleeman rule See Bragg rule. { 'brag 'klā·mən ,rül }

Bragg-Pierce law [PHYS] A relationship for determining an element's atomic absorption coefficient for x-rays when the atomic number of the element and the wavelength of the x-rays are known. { 'brag ¦pirs ,lȯ }

Bragg reflection See Bragg scattering. { 'brag ri'flek·shən }

Bragg rule [ATOM PHYS] An empirical rule according to which the mass stopping power of an element for alpha particles is inversely proportional to the square root of the atomic weight. Also known as Bragg-Kleeman rule. { 'brag ,rül }

Bragg scattering [SOLID STATE] Scattering of x-rays or neutrons by the regularly spaced atoms in a crystal, for which constructive interference occurs only at definite angles called Bragg angles. Also known as Bragg diffraction; Bragg reflection. { 'brag ,skad·ə·riŋ }

Bragg's equation See Bragg's law. { 'bragz i'kwā·zhən }

Bragg's law [SOLID STATE] A statement of the conditions under which a crystal will reflect a beam of x-rays with maximum intensity. Also known as Bragg's equation; Bravais' law. { 'bragz ,lȯ }

Bragg spectrometer [ENG] An instrument for x-ray analysis of crystal structure and measuring wavelengths of x-rays and gamma rays, in which a homogeneous beam of x-rays is directed on the known face of a crystal and the reflected beam is detected in a suitably placed ionization chamber. Also known as crystal spectrometer; crystal-diffraction spectrometer; ionization spectrometer. { 'brag spek'träm·əd·ər }

braid [MATH] A braid of order n consists of two parallel lines, sets of n points on each of the lines with a one-to-one correspondence between them, and n nonintersecting space curves, each of which connects one of the n points on one of the parallel lines with the corresponding point on the other; the space curves are configured so that no curve turns back on itself, in the sense that its projection on the plane of the parallel lines lies between the parallel lines and intersects any line parallel to them no more than once, and any two such projections intersect at most a finite number of times. { brād }

braided stream [HYD] A stream flowing in several channels that divide and reunite. { 'brād·əd ,strēm }

braided wire [ELEC] A tube of fine wires woven around a conductor or cable for shielding purposes or used alone in flattened form as a grounding strap. { 'brād·əd ,wīr }

braiding [ENG] Weaving fibers into a hollow cylindrical shape. { 'brād·iŋ }

Braille [COMMUN] A system of written communication for the blind in which letters are represented by raised dots over which the trained blind person moves the fingertips. { brāl }

Brailsford-Morquio syndrome See Morquio's syndrome. { 'brālz·fərd 'mȯr·ke·ō ,sin,drōm }

brain [ANAT] The portion of the vertebrate central nervous system enclosed in the skull. [ZOO] The enlarged anterior portion of the central nervous system in most bilaterally symmetrical animals. { brān }

braincase See cranium. { 'brān,kās }

brain coral [INV ZOO] A reef-building coral resembling the human cerebrum in appearance. { 'brān ,kär·əl }

brain hormone [INV ZOO] A neurohormone secreted by the insect brain that regulates the release of ecdysone from the prothoracic glands. { 'brān ,hȯr,mōn }

brainstem [ANAT] The portion of the brain remaining after the cerebral hemispheres and cerebellum have been removed. { 'brān,stem }

brainstorming [IND ENG] A procedure used to find a solution for a problem by collecting all the ideas, without regard for feasibility, which occur from a group of people meeting together. { 'brān ,stȯrm·iŋ }

brain wave [PHYSIO] A rhythmic fluctuation of voltage between parts of the brain, ranging from about 1 to 60 hertz and 10 to 100 microvolts. { 'brān ,wāv }

brake [MECH ENG] A machine element for applying friction to a moving surface to slow it (and often, the containing vehicle or device) down or bring it to rest. { brāk }

brake band [MECH ENG] The contracting element of the band brake. { 'brāk ,band }

brake block [MECH ENG] A portion of the band brake lining, shaped to conform to the curvature of the band and attached to it with countersunk screws. { 'brāk ,bläk }

brake drum [MECH ENG] A rotating cylinder attached to a rotating part of machinery, which the brake band or brake shoe presses against. { 'brāk ,drəm }

brake fluid [MATER] The liquid in the cylinder of an automotive brake which, under the action of a piston, is forced through tubing into cylinders at each car wheel, moving a pair of pistons outward so that the brake shoes are thrust against the revolving brake drums. { 'brāk ,flü·əd }

brake horsepower [MECH ENG] The power developed by an engine as measured by the force applied to a friction brake or by an absorption dynamometer applied to the shaft or flywheel. Abbreviated bhp. { ¦brāk 'hȯrs,pau·ər }

brake line [MECH ENG] One of the pipes or hoses that connect the master cylinder and the wheel cylinders in a hydraulic brake system. { 'brāk ,līn }

brake lining [MECH ENG] A covering, riveted or molded to the brake shoe or brake band, which presses against the rotating brake drum; made of either fabric or molded asbestos material. { 'brāk ,lin·iŋ }

brake mean-effective pressure [MECH ENG] Applied to reciprocating piston machinery, the average pressure on the piston during the power stroke, derived from the measurement of brake power output. { 'brāk ¦mēn i'fek·tiv 'presh·ər }

brake shoe [MECH ENG] The renewable friction element of a shoe brake. Also known as shoe. { 'brāk ,shü }

brake thermal efficiency [MECH ENG] The ratio of brake power output to power input. { 'brāk 'thər·məl ə'fish·ən·sē }

braking effects [PHYS CHEM] The electrophoretic effect and the asymmetry effect, which together control the speed with which ions drift in a strong electrolyte. { 'brāk·iŋ i,feks }

braking ellipses [AERO ENG] A series of ellipses, decreasing in size due to aerodynamic drag, followed by a spacecraft in entering a planetary atmosphere. { 'bra·kiŋ i'lip,sēz }

braking rocket See retrorocket. { 'bra·kiŋ ,räk·ət }

brale [MET] A conical diamond indenter with an angle of 120° used in the Rockwell hardness test. { brāl }

bramble [BOT] **1.** A plant of the genus *Rubus*. **2.** A rough, prickly vine or shrub. { 'bram·bəl }

brammalite [MINERAL] A mica-type clay mineral that is different from illite because it has soda instead of potash; it is the sodium analog of illite. Also known as sodium illite. { 'bram·ə,līt }

branch [BOT] A shoot or secondary stem on the trunk or a limb of a tree. [COMPUT SCI] **1.** Any one of a number of

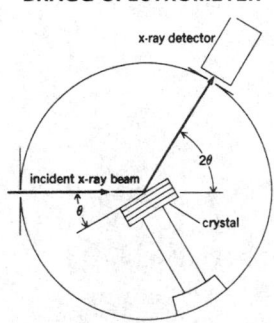

BRAGG SPECTROMETER

Schematic of Bragg spectrometer. θ is angle between incident beam and crystallographic planes; 2θ is angle between incident and diffracted beams.

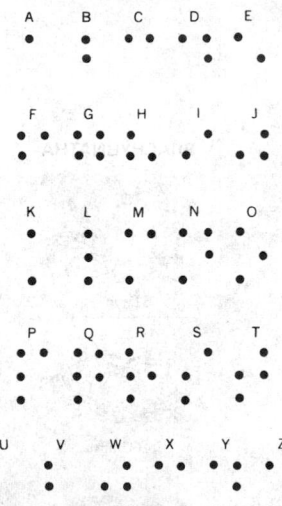

BRAILLE

Sketch showing Braille alphabet. Combinations of dots (up to six per letter) represent letters. The trained blind person runs the fingers over the raised dots.

instruction sequences in a program to which computer control is passed, depending upon the status of one or more variables. **2.** *See* jump. [ELEC] A portion of a network consisting of one or more two-terminal elements in series. Also known as arm. [ENG] In a piping system, a pipe that originates in or discharges into another pipe. Also known as branch line. [HYD] A small stream that merges into another, generally bigger, stream. [MATH] **1.** A complex function which is analytic in some domain and which takes on one of the values of a multiple-valued function in that domain. **2.** A section of a curve that is separated from other sections of the curve by discontinuities, singular points, or other special points such as maxima and minima. [NUC PHYS] A product resulting from one mode of decay of a radioactive nuclide that has two or more modes of decay. [ORG CHEM] *See* side chain. [SCI TECH] An area of study representing an independent offshoot of a related basic discipline. { branch }

branch-and-bound technique [IND ENG] A technique in nonlinear programming in which all sets of feasible solutions are divided into subsets, and those having bounds inferior to others are rejected. { ¦branch ən ¦bau̇nd tek'nēk }

branch circuit [ELEC] A portion of a wiring system in the interior of a structure that extends from a final overload protective device to a plug receptable or a load such as a lighting fixture, motor, or heater. { branch ¦sər·kət }

branch-circuit distribution center [COMMUN] Distribution center at which branch circuits are supplied. { branch ¦sər·kət dis·trə'byü·shən ¦sen·tər }

branch cut [MATH] A line or curve of singular points used in defining a branch of a multiple-valued complex function. { branch ¦kət }

branch cutout [ELEC] The holder for a fuse that protects a branch circuit in an interior wiring system. { 'branch 'kəd,au̇t }

branched acinous gland [ANAT] A multicellular structure with saclike glandular portions connected to the surface of the containing organ or structure by a common duct. { ¦brancht 'as·ə·nəs ,gland }

branched chain *See* side chain. { 'brancht 'chān }

branched-chain ketoaciduria *See* maple syrup urine disease. { ¦brancht ¦chān ,kēd·ō·as·ə'du̇r·ē·ə }

branched polymer [ORG CHEM] A polymer chain having branch points that connect three or more chain segments; examples include graft copolymers, star polymers, comb polymers, and dendritic polymers. { ¦brancht 'päl·ə·mər }

branched tubular gland [ANAT] A multicellular structure with tube-shaped glandular portions connected to the surface of the containing organ or structure by a common secreting duct. { ¦brancht ¦tüb·yə·lər 'gland }

branch gain *See* branch transmittance. { 'branch ,gān }

branchia *See* gill. { 'braŋ·kē·ə }

branchial [ZOO] Of or pertaining to gills. { 'braŋ·kē·əl }

branchial arch [VERT ZOO] One of the series of paired arches on the sides of the pharynx which support the gills in fishes and amphibians. { 'braŋ·kē·əl 'ärch }

branchial basket [ZOO] A cartilaginous structure that supports the gills in protochordates and certain lower vertebrates such as cyclostomes. { 'braŋ·kē·əl 'bas·kət }

branchial cleft [EMBRYO] A rudimentary groove in the neck region of air-breathing vertebrate embryos. [VERT ZOO] One of the openings between the branchial arches in fishes and amphibians. { 'braŋ·kē·əl 'kleft }

branchial heart [INV ZOO] A muscular enlarged portion of a vein of a cephalopod that contracts and forces the blood into the gills. { 'braŋ·kē·əl 'härt }

branchial plume [INV ZOO] An accessory respiratory organ that extends out under the mantle in certain Gastropoda. { 'braŋ·kē·əl 'plüm }

branchial pouch [ZOO] In cyclostomes and some sharks, one of the respiratory cavities occurring in the branchial clefts. { 'braŋ·kē·əl 'pau̇ch }

branchial sac [INV ZOO] In tunicates, the dilated pharyngeal portion of the alimentary canal that has vascular walls pierced with clefts and serves as a gill. { 'braŋ·kē·əl 'sak }

branchial segment [EMBRYO] Any of the paired pharyngeal segments indicating the visceral arches and clefts posterior to and including the third pair in air-breathing vertebrate embryos. { 'braŋ·kē·əl 'seg·mənt }

branchiate [VERT ZOO] Having gills. { 'braŋ·kē,āt }

branching [COMPUT SCI] The selection, under control of a computer program, of one of two or more branches. [NUC PHYS] The occurrence of two or more modes by which a radionuclide can undergo radioactive decay. Also known as multiple decay; multiple disintegration. { 'branch·iŋ }

branching adaptation *See* divergent adaptation. { 'branch·iŋ ,ad,ap'tā·shən }

branching bay *See* estuary. { 'branch·iŋ 'bā }

branching diagram [MATH] In bifurcation theory, a graph in which a parameter characterizing solutions of a nonlinear equation is plotted against a parameter that appears in the equation itself. { 'branch·iŋ ,dī·ə,gram }

branching fraction [NUC PHYS] That fraction of the total number of atoms involved which follows a particular branch of the disintegration scheme; usually expressed as a percentage. { 'branch·iŋ 'frak·shən }

branching process [STAT] A stochastic process in which the members of a population may have offspring and the lines of descent branch out as the new members are born. { 'branch·iŋ 'präs·əs }

branching ratio [NUC PHYS] The ratio of the number of parent atoms or particles decaying by one mode to the number decaying by another mode; the ratio of two specified branching fractions. { 'branch·iŋ 'rā·shō }

branching theory *See* bifurcation theory. { 'branch·iŋ ,thē·ə·rē }

branch instruction [COMPUT SCI] An instruction that makes the computer choose between alternative subprograms, depending on the conditions determined by the computer during the execution of the program. { 'branch in'strək·shən }

branchiocranium [VERT ZOO] The division of the fish skull constituting the mandibular and hyal regions and the branchial arches. { ¦braŋ·kē·ō'krā·nē·əm }

branchiomere [EMBRYO] An embryonic metamere that will differentiate into a visceral arch and cleft; a branchial segment. { 'braŋ·kē·ə,mir }

branchiomeric musculature [VERT ZOO] Those muscles derived from branchial segments in vertebrates. { ¦braŋ·kē·ə¦mer·ik 'məs·kyə·lə,chər }

Branchiopoda [INV ZOO] A subclass of crustaceans containing small or moderate-sized animals commonly called fairy shrimps, clam shrimps, and water fleas. { ,braŋ·kē'äp·ə·də }

branchiostegite [INV ZOO] A gill cover and chamber in certain malacostracan crustaceans, formed by lateral expansion of the carapace. { ,braŋ·kē'äs·tə,jīt }

Branchiostoma [ZOO] A genus of lancelets formerly designated as amphioxus. { ,braŋ·kē'äs·tə·mə }

Branchiotremata [INV ZOO] The hemichordates, a branch of the subphylum Oligomera. { ,braŋ·kē·ə'trem·əd·ə }

branchite *See* hartite. { 'bran,chīt }

Branchiura [INV ZOO] The fish lice, a subclass of fish ectoparasites in the class Crustacea. { ,braŋ·kē'yu̇r·ə }

branch joint [ELEC] Joint used for connecting a branch conductor or cable, where the latter continues beyond the branch. { 'branch ,jȯint }

branch line [CIV ENG] A secondary line in a railroad system that connects to the main line. [ENG] *See* branch. { 'branch ,līn }

branch point [COMPUT SCI] A point in a computer program at which there is a branch instruction. [ELEC] A terminal in an electrical network that is common to more than two elements or parts of elements of the network. Also known as junction point; node. [MATH] **1.** A point at which two or more sheets of a Riemann surface join together. **2.** In bifurcation theory, a value of a parameter in a nonlinear equation at which solutions branch off from the basic solution. { 'branch ,pȯint }

branch prediction [COMPUT SCI] A method whereby a processor guesses the outcome of a branch instruction so that it can prepare in advance to carry out the instructions that follow the predicted outcome. { 'branch prə,dik·shən }

branch sewer [CIV ENG] A part of a sewer system that is larger in diameter than the lateral sewer system; receives sewage from both house connections and lateral sewers. { ¦branch ¦sü·ər }

branch transmittance [CONT SYS] The amplification of current or voltage in a branch of an electrical network; used in the representation of such a network by a signal-flow graph. Also known as branch gain. { 'branch trans'mit·əns }

brandtite [MINERAL] $Ca_2Mn(AsO_4)_2 \cdot 2H_2O$ A colorless to

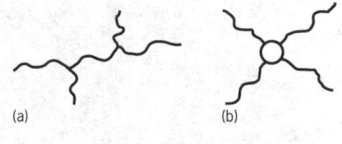

BRANCHED POLYMER

(a)   (b)

(c)   (d)

Examples of branched polymers. (*a*) Branched polymer (if arms are of composition similar to backbone) or graft polymer (if compositions are different). (*b*) Star polymer. (*c*) Comb polymer. (*d*) Dendritic polymer.

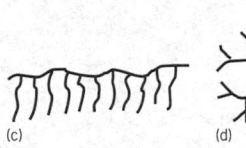

BRANCHED TUBULAR GLAND

common duct

gland

A simple branched tubular gland.

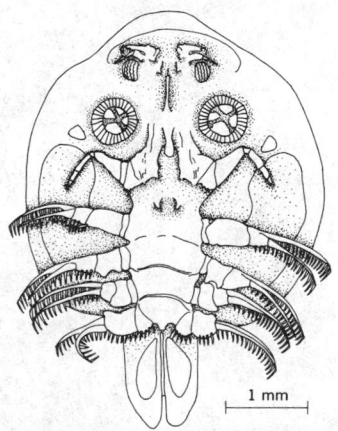

BRANCHIURA

1 mm

Argulus japonicus, male fish louse.

white, monoclinic mineral consisting of a hydrated arsenate of calcium and manganese. { 'brant,īt }

brandy [CHEM ENG] A potable alcoholic beverage distilled from wine or fermented fruit juice, usually after the aging of the wine in wooden casks; cognac is a brandy distilled from wines made from grapes from the Cognac region of France. { 'bran·dē }

Branhamella [MICROBIO] A genus of bacteria in the family Neisseriaceae; cocci occur in pairs with flattened adjacent sides; parasites of mammalian mucous membranes. { ,bran·ə'mel·ə }

Branley-Lenard effect [ELECTR] The strong ionization of air and other gases by ultraviolet radiation with wavelengths in the range 120–150 nanometers. { 'bran·lē 'len·ərd i,fekt }

brannerite [MINERAL] A complex, black, opaque titanite of uranium and other elements in which the weight of uranium exceeds the weight of titanium; monoclinic and possibly $(U,Ca,Fe,Y,Th)_3Ti_5O_6$. { 'bran·ə,rīt }

Brans-Dicke theory [RELAT] A theory of gravitation in which the gravitational field is described by the tensor field of general relativity and by a new scalar field, which is determined by the distribution of mass-energy in the universe and replaces the gravitational constant. { 'bränz ¦dik ,thē·ə·rē }

brashness [MATER] Pertaining to wood that possesses relatively low shock resistance and therefore fails abruptly when broken during bending. { 'brash·nəs }

brass [GEOL] A British term for sulfides of iron (pyrites) in coal. Also known as brasses. [MET] A copper-zinc alloy of varying proportions but typically containing 67% copper and 33% zinc. { bras }

brass chills See metal fume fever. { 'bras ,chilz }

brasses See brass. { 'bras·əz }

brass founder's ague See metal fume fever. { 'bras ,faún·dərz ,āg }

Brassica [BOT] A large genus of herbs in the family Cruciferae of the order Capparales, including cabbage, watercress, and sweet alyssum. { 'bras·ə·kə }

Brassicaceae [BOT] An equivalent name for the Cruciferae. { ,bras·ə'kās·ē,ē }

brassin [BIOCHEM] Any of a class of plant hormones characterized as long-chain fatty-acid esters; brassins act to induce both cell elongation and cell division in leaves and stems. { 'bra·sən }

brass ore See aurichalcite. { 'bras,ór }

Brathinidae [INV ZOO] The grass root beetles, a small family of coleopteran insects in the superfamily Staphylinoidea. { brə'thī·nə,dē }

brattice [MIN ENG] A temporary board or cloth partition in any mine passage to confine the air and force it into the working places. Also spelled brattish; brettice; brettis. { 'brad·əs }

brattice cloth [MIN ENG] Fire-resistant canvas or duck used to erect a brattice. { 'brad·əs ,klóth }

brattish See brattice. { 'brad·ish }

Braulidae [INV ZOO] The bee lice, a family of cyclorrhaphous dipteran insects in the section Pupipara. { 'braúl·ə,dē }

braunite [MINERAL] $3Mn_2O_3 \cdot MnSiO_3$ Brittle mineral that forms tetragonal crystals; commonly found as steel-gray or brown-black masses in the United States, Europe, and South America; it is an ore of manganese. { 'braú,nīt }

Braun tube See cathode-ray tube. { 'braún ,tüb }

Bravais biplate See biplate. { brə¦vä 'bī,plāt }

Bravais indices [CRYSTAL] A modification of the Miller indices; frequently used for hexagonal and trigonal crystalline systems; they refer to four axes, the c axis and three others at 120° angles in the basal plane. { brə'vä 'in·də,sēz }

Bravais lattice [CRYSTAL] One of the 14 possible arrangements of lattice points in space such that the arrangement of points about any chosen point is identical with that about any other point. { brə'vä 'lad·əs }

Bravais' law See Bragg's law. { brə'väz ,ló }

bra vector [QUANT MECH] A vector describing the state of a dynamic system in Hilbert space; the dual of a ket vector. { 'brä ,vek·tər }

brave west winds [METEOROL] A nautical term for the strong and rather persistent westerly winds over the oceans in temperate latitudes, between 40° and 50°S. { ¦brāv ¦west 'winz }

bravoite [MINERAL] $(Ni,Fe)S_2$ A yellow sulfide ore of nickel containing iron. { 'brä,vó,īt }

Braxton-Hicks contractions [MED] Intermittent painless contractions of the uterus that occur throughout pregnancy. { 'brak·stən 'hiks kən,trak·shənz }

brayer [GRAPHICS] A soft rubber roller attached to a wooden or metal handle; used to ink blocks, stones, or printing plates. { 'brā·ər }

Brayton cycle [THERMO] A thermodynamic cycle consisting of two constant-pressure processes interspersed with two constant-entropy processes. Also known as complete-expansion diesel cycle; Joule cycle. { 'brāt·ən ,sī·kəl }

braze [MET] To solder metals by melting a nonferrous filler metal, such as brass or brazing alloy (hard solder), with a melting point lower than that of the base metals, at the point of contact. Also known as hard-solder. { brāz }

brazed joint [MET] The joining of two or more metallic components by brazing or braze welding. { ¦brāzd 'jóint }

brazed shank tool [MECH ENG] A metal cutting tool made of a material different from the shank to which it is brazed. { ¦brāzd 'shaŋk ,tül }

braze welding [MET] A method of welding in which coalescence is produced by heating above 800°F (427°C) and by using a nonferrous filler metal having a melting point below that of the base metals; in distinction to brazing, capillary attraction does not distribute the filler metal in the joint. { 'brāz ,wel·diŋ }

Brazil Current [OCEANOGR] The warm ocean current that flows southward along the Brazilian coast below Natal; the western boundary current in the South Atlantic Ocean. { brə'zil ,kər·ənt }

brazilianite [MINERAL] $NaAl_3(PO_4)_2(OH)_4$ A chartreuse yellow to pale yellow, monoclinic mineral consisting of a basic phosphate of sodium and aluminum. { brə'zil·yə,nīt }

Brazil nut [BOT] Bertholletia excelsa. A large broad-leafed evergreen tree of the order Lecythedales; an edible seed is produced by the tree fruit. { brə'zil ,nət }

Brazil wax See carnauba wax. { brə'zil ,waks }

brazing alloy [MET] 1. An alloy used as a filler metal for brazing; copper alloys and nickel alloys are used for brazing of steels. 2. Solder that does not melt below red heat. Formerly known as hard solder. { 'brāz·iŋ ,al,oi }

brazing brass See spelter solder. { 'brāz·iŋ ,bras }

brazing metal [MET] A nonferrous metal to be added to a joint in braze welding; can be a pure metal such as copper, zinc, or nickel, or a brazing alloy. { 'brāz·iŋ ,med·əl }

brazing sheet [MET] 1. Brazing filler metal in sheet form. 2. Flat-rolled metal clad with brazing filler metal on one or both sides. { 'brāz·iŋ ,shēt }

breached anticline [GEOL] An anticline that has been more deeply eroded in the center. Also known as scalped anticline. { ¦brēcht 'an·ti,klīn }

breached cone [GEOL] A cinder cone in which lava has broken through the sides and broken material has been carried away. { ¦brēcht ,kōn }

breaching [MECH ENG] The space between the end of the tubing and the jacket of a hot-water or steam boiler. { 'brēch·iŋ }

breadboard [ELECTR] A printed circuit board designed so that the user can mount and wire whatever circuitry is desired. { 'bred,bórd }

breadboarding [ELECTR] Assembling an electronic circuit in the most convenient manner, without regard for final locations of components, to prove the feasibility of the circuit and to facilitate changes when necessary. { 'bred,bórd·iŋ }

breadboard model [ENG] Uncased assembly of an instrument or other piece of equipment, such as a radio set, having its parts laid out on a flat surface and connected together to permit a check or demonstration of its operation. { 'bred,bórd ,mäd·əl }

breadcrust [GEOL] A surficial structure resembling a crust of bread, as the concretions formed by evaporation of salt water. { 'bred,krəst }

breadcrust bomb [GEOL] A volcanic bomb with a cracked exterior. { 'bred,krəst ,bäm }

breadfruit [BOT] Artocarpus altilis. An Indo-Malaysian tree, a species of the mulberry family (Moraceae). The tree produces a multiple fruit which is edible. { 'bred,früt }

bread mold [MYCOL] Any fungus belonging to the family Mucoraceae in the order Mucorales. { 'bred ,mōld }

breadth-height index [ANTHRO] Ratio of the maximum

breadth of the skull to its maximum height multiplied by 100. { 'bredth ˌhīt 'inˌdeks }

break [COMPUT SCI] **1.** To interrupt processing by a computer, usually by depressing a key. **2.** A place in a file of records where one or more of the values in the records change. [ELEC] **1.** A fault in a circuit. **2.** The minimum distance in a circuit-opening device between the stationary and movable contacts when these contacts are in the open position. [ELECTR] A reflected radar pulse which appears on a radarscope as a line perpendicular to the base line. [GEOGR] A significant variation of topography, such as a deep valley. [GEOL] *See* knickpoint. [METEOROL] **1.** A sudden change in the weather; usually applied to the end of an extended period of unusually hot, cold, wet, or dry weather. **2.** A hole or gap in a layer of clouds. [MIN ENG] **1.** A plane of discontinuity in the coal seam such as a slip, fracture, or cleat; the surfaces are in contact or slightly separated. **2.** A fracture or crack in the roof beds as a result of mining operations. { brāk }

breakage and reunion [CYTOL] The classical model of crossing over by means of physical breakage and crossways reunion of completed chromatids during the process of meiosis. { 'brāk·ij ən rē'yün·yən }

breakaway [FL MECH] Boundary-layer separation in which the boundary layer does not become reattached to the surface. { 'brāk·əˌwā }

breakaway phenomenon *See* breakoff phenomenon. { 'brāk·əˌwā fə'näm·əˌnän }

breakaway wrist [CONT SYS] A robotic wrist that has a safety feature that guarantees its protection from damage if too much force is exerted on the wrist or end effector. { 'brāk·əˌwā ˌrist }

break-before-make contact [ELEC] One of a pair of contacts that interrupt one circuit before establishing another. { 'brāk bəˌför 'māk ˌkänˌtakt }

breakbone fever *See* dengue. { 'brākˌbōn ˌfē·vər }

break-bulk cargo [IND ENG] Miscellaneous goods packed in boxes, bales, crates, cases, bags, cartons, barrels, or drums; may also include lumber, motor vehicles, pipe, steel, and machinery. { 'brāk ˌbəlk 'kär·gō }

break contact [ELEC] The contact of a switching device which opens a circuit upon the operation of the device. { 'brāk ˌkänˌtakt }

breakdown [ELEC] A large, usually abrupt rise in electric current in the presence of a small increase in voltage; can occur in a confined gas between two electrodes, a gas tube, the atmosphere (as lightning), an electrical insulator, and a reverse-biased semiconductor diode. Also known as electrical breakdown. [MFT] The initial process of rolling and drawing, or a series of such processes, which reduce a casting or extruded shape before its final reduction to desired size. [PETRO ENG] The amount of pressure required at the wellhead to rupture a formation during fracture treatment. { 'brākˌdau̇n }

breakdown diode [ELEC] A semiconductor diode in which the reverse-voltage breakdown mechanism is based either on the Zener effect or the avalanche effect. { 'brākˌdau̇n 'dīˌōd }

breakdown impedance [ELECTR] Of a semiconductor, the small-signal impedance at a specified direct current in the breakdown region. { 'brākˌdau̇n im'pēd·əns }

breakdown law [STAT] The law that if the event E is broken down into the exclusive events $E_1, E_2, \ldots$ so that E is the event E_1 or E_2 or $\ldots$, then if F is any event, the probability of F is the sum of the products of the probabilities of E_i and the conditional probability of F given E_i. { 'brākˌdau̇n ˌló }

breakdown potential *See* breakdown voltage. { 'brākˌdau̇n pə'ten·shəl }

breakdown region [ELECTR] Of a semiconductor diode, the entire region of the volt-ampere characteristic beyond the initiation of breakdown for increasing magnitude of bias. { 'brāk ˌdau̇n ˌrē·jən }

breakdown torque [ELEC] The maximum torque that a motor can develop at its rated applied voltage and frequency without an abrupt drop in speed. { 'brākˌdau̇n ˌtörk }

breakdown voltage [ELEC] **1.** The voltage measured at a specified current in the electrical breakdown region of a semiconductor diode. Also known as Zener voltage. **2.** The voltage at which an electrical breakdown occurs in a dielectric. **3.** The voltage at which an electrical breakdown occurs in a gas. Also known as breakdown potential; sparking potential; sparking voltage. { 'brākˌdau̇n ˌvól·tij }

breaker [MIN ENG] **1.** In anthracite mining, the structure in which the coal is broken, sized, and cleaned for market. Also known as coalbreaker. **2.** One of a row of drill holes above the mining holes in a tunnel face. [OCEANOGR] A wave breaking on a shore, over a reef, or other mass in a body of water. { 'brā·kər }

breaker-and-a-half [ELEC] A substation switching arrangement that involves two buses between which three breaker bays are installed. { 'brā·kər ən ə 'haf }

breaker-and-a-third [ELEC] A substation switching arrangement having four breakers and three connections per bay. { 'brā·kər ən ə 'thərd }

breaker cam [MECH ENG] A rotating, engine-driven device in the ignition system of an internal combustion engine which causes the breaker points to open, leading to a rapid fall in the primary current. { 'brā·kər ˌkam }

breaker depth [OCEANOGR] The still-water depth measured at the point where a wave breaks. Also known as breaking depth. { 'brā·kər ˌdepth }

breaker plate [ENG] In plastics die forming, a perforated plate at the end of an extruder head; often used to support a screen to keep foreign particles out of the die. { 'brā·kər ˌplāt }

breaker points [ELEC] Low-voltage contacts used to interrupt the current in the primary circuit of a gasoline engine's ignition system. { 'brā·kər ˌpóints }

breaker terrace [GEOL] A type of shore found in lakes in glacial drift; the terrace is formed from stones deposited by waves. { 'brā·kər ˌter·əs }

break-even [NUCLEO] The point at which the energy generated by a controlled nuclear fusion reaction equals the energy required to maintain the reaction. { brā'kē·vən }

break-even analysis [IND ENG] Determination of the break-even point. { brā'kē·vən ə'nal·ə·səs }

break-even point [IND ENG] The point at which a company neither makes a profit nor suffers a loss from the operations of the business, and at which total costs are equal to total sales volume. { brā'kē·vən ˌpóint }

break for color [GRAPHICS] In artwork and composition, the separation of parts for printing in different colors. { 'brāk fər ˌkäl·ər }

break frequency [CONT SYS] The frequency at which a graph of the logarithm of the amplitude of the frequency response versus the logarithm of the frequency has an abrupt change in slope. Also known as corner frequency; knee frequency. { 'brāk ˌfrē·kwən·sē }

break-in device [ELECTR] A device in a radiotelegraph communication system allowing an operator to receive signals in intervals between his own transmission signals. { 'brākˌin di'vīs }

breaking circulation [PETRO ENG] Beginning operation of the mud pump in order to restore circulation of the mud column. { 'brāk·iŋ ˌsər·kyəˌlā·shən }

breaking depth *See* breaker depth. { 'brāk·iŋ ˌdepth }

breaking down [PETRO ENG] The procedure of unscrewing the drill stem into single joints and placing them on the pipe rack either after a well has been completed or when pipe size is being changed. { 'brāk·iŋ 'dau̇n }

breaking-down rolls [MET] A rolling mill unit used for breakdown operations. { 'brāk·iŋ ˌdau̇n ˌrōlz }

breaking-drop theory [GEOPHYS] A theory of thunderstorm charge separation based upon the suggested occurrence of the Lenard effect in thunderclouds, that is, the separation of electric charge due to the breakup of water drops. { 'brāk·iŋ 'dräp ˌthē·ə·rē }

breaking load [MECH] The stress which, when steadily applied to a structural member, is just sufficient to break or rupture it. Also known as ultimate load. { 'brāk·iŋ ˌlōd }

breaking pin device [ENG] A device designed to relieve pressure resulting from inlet static pressure by the fracture of a loaded part of a pin. { 'brāk·iŋ ˌpin di'vīs }

breaking radius [MATER] The limiting radius of curvature to which wood or plywood can be bent without breaking. { 'brāk·iŋ ˌrād·ē·əs }

breaking strength [MECH] The ability of a material to resist breaking or rupture from a tension force. { 'brāk·iŋ ˌstreŋkth }

breaking stress [MECH] The stress required to fracture a material whether by compression, tension, or shear. { 'brāk· iŋ ‚stres }

break-in operation [COMMUN] A method of radio communication in which it is possible for the receiving operator to interrupt or break into the transmission. { 'brā‚kin ‚äp·ə‚rā· shən }

break key [COMPUT SCI] A key on a computer keyboard whose depression causes processing to be interrupted. { 'brāk ‚kē }

breakoff phenomenon [AERO ENG] The feeling which sometimes occurs during high-altitude flight of being totally separated and detached from the earth and human society. Also known as breakaway phenomenon. { 'brā‚kóf fə‚näm· ə‚nän }

breakout [ELEC] A joint at which one or more conductors are brought out from a multiconductor cable. [ENG] Failure or collapse of a borehole wall due to stress anisotropy. [MIN ENG] To pull drill rods or casing from a borehole and unscrew them at points where they are joined by threaded couplings to form lengths that can be stacked in the drill tripod or derrick. { 'brā‚kaut }

breakout block [PETRO ENG] A heavy plate fitting in a rotary table that holds a drill bit being unscrewed from a drill collar. { 'brā‚kaut ‚bläk }

breakout box [ELECTR] A device connected to a multiconductor cable that provides terminal connections to test the signals in a transmission. { 'brāk‚aut ‚bäks }

breakoutput [COMPUT SCI] An ALGOL procedure which causes all bytes in a device buffer to be sent to the device rather than wait until the buffer is full. { 'brā‚kaut‚put }

breakout schedule [IND ENG] A schedule for a construction job site, generally in the form of a bar chart, that communicates detailed day-to-day activities to all working levels on the project. { 'brāk‚aut ‚skej·əl }

breakout tongs [MIN ENG] A device used to begin unscrewing one section of pipe from another section. { 'brā‚kaut ‚täŋz }

breakover [ELECTR] In a silicon controlled rectifier or related device, a transition into forward conduction caused by the application of an excessively high anode voltage. [MIN ENG] See conversion. { 'brā‚kō·vər }

breakover voltage [ELECTR] The positive anode voltage at which a silicon controlled rectifier switches into the conductive state with gate circuit open. { 'brā‚kō·vər ‚vól·tij }

break period [COMMUN] Of a dial telephone, the time interval during which the circuit contacts are open. { 'brāk ‚pir· ē·əd }

breakpoint [CHEM ENG] See breakthrough. [COMPUT SCI] A point in a program where an instruction, instruction digit, or other condition enables a programmer to interrupt the run by external intervention or by a monitor routine. [IND ENG] In a time study, the end of an element in a work cycle and the point at which a reading is made. Also known as end point; reading point. { 'brāk‚póint }

breakpoint switch [COMPUT SCI] A manually operated switch which controls conditional operation at breakpoints, used primarily in debugging. { 'brāk‚póint ‚swich }

breakpoint symbol [COMPUT SCI] A symbol which may be optionally included in an instruction, as an indication, tag, or flag, to designate it as a breakpoint. { 'brāk‚póint ‚sim·bəl }

break rolls [FOOD ENG] A pair of corrugated or fluted cylinders which rotate toward each other at different speeds and which are used in wheat milling to break down the kernels. { 'brāk ‚rōlz }

breaks in overcast [METEOROL] In United States weather observing practice, a condition wherein the cloud cover is more than 0.9 but less than 1.0. { 'brāks in 'ō·vər‚kast }

breakthrough [CHEM ENG] **1.** A localized break in a filter cake or precoat that permits fluid to pass through without being filtered. Also known as breakpoint. **2.** In an ion-exchange system, the first appearance of unadsorbed ions of the type which deplete the activity of the resin bed; this indicates that the bed must be regenerated. [COMPUT SCI] An interruption in the intended character stroke in optical character recognition. [MIN ENG] A passage cut through the pillar to allow the ventilating current to pass from one room to another; larger than a doghole. Also known as room crosscut. { 'brāk‚thrü }

breakthrough sweep efficiency [PETRO ENG] The completeness with which an oil-field waterflood sweeps through a reservoir area; related to the critical water saturation (point beyond which oil will not be pushed ahead of the water flow). { 'brāk‚thrü 'swēp i'fish·ən·sē }

break thrust [GEOL] A thrust fault cutting across one limb of a fold. { 'brāk ‚thrəst }

breakup [HYD] The spring melting of snow, ice, and frozen ground; specifically, the destruction of the ice cover on rivers during the spring thaw. { 'brāk‚əp }

breakwater [CIV ENG] A wall built into the sea to protect a shore area, harbor, anchorage, or basin from the action of waves. { 'brāk‚wód·ər }

breast [ANAT] The human mammary gland. [MIN ENG] **1.** In coal mines, a chamber driven in the seam from the gangway, for the extraction of coal. **2.** See face. { brest }

breastbeam [NAV ARCH] A beam at the edge of the quarterdeck or forecastle. { 'brest‚bēm }

breast boards [CIV ENG] Timber planks used to support the tunnel face when excavation is in loose soil. { 'brest ‚bórdz }

breast drill [DES ENG] A small, portable hand drill customarily used by handsetters to drill the holes in bit blanks in which diamonds are to be set; it includes a plate that is pressed against the worker's breast. { 'brest ‚dril }

breast hole [MET] A hole for raking cinders out of a smelting cupola. { 'brest ‚hōl }

breasting dolphin [CIV ENG] A pile or other structure against which a moored ship rests. { 'brest·iŋ ‚däl·fən }

breast wall [CIV ENG] A low wall built to retain the face of a natural bank of earth. { 'brest ‚wól }

breastwork [ORD] Earthwork, constructed wholly or partly above the surface of the ground, which gives protection for defenders in a standing position, firing over the crest. { 'brest‚wərk }

breather pipe [MECH ENG] A pipe that opens into a container for ventilation, as in a crankcase or oil tank. Also known as crankcase breather. { 'brē·thər ‚pīp }

breath-hold diving [ENG] A form of diving without the use of any artificial breathing mixtures. { 'breth ‚hōld ‚div·iŋ }

breathing [ENG] **1.** Opening and closing of a plastics mold in order to let gases escape during molding. Also known as degassing. **2.** Movement of gas, vapors, or air in and out of a storage-tank vent line as a result of liquid expansions and contractions induced by temperature changes. [MATER] Permeability of plastic sheeting to air, bubbles, voids, or trapped gas globules. [PHYSIO] Inhaling and exhaling. { 'brēth· iŋ }

breathing apparatus [ENG] An appliance that enables a person to function in irrespirable or poisonous gases or fluids; contains a supply of oxygen and a regenerator which removes the carbon dioxide exhaled. { 'brēth·iŋ ap·ə'rad·əs }

breathing bag [ENG] A component of a semiclosed-circuit breathing apparatus that mixes the gases to provide low breathing resistance. { 'brēth·iŋ ‚bag }

breathing cave See blowing cave. { 'brēth·iŋ ‚kāv }

breathing line [CIV ENG] A level of 5 feet (1.5 meters) above the floor; suggested temperatures for various occupancies of rooms and other chambers are usually given at this level. { 'brēth·iŋ ‚līn }

breccia [PETR] A rock made up of very angular coarse fragments; may be sedimentary or may be formed by grinding or crushing along faults. { 'brech·ə }

breccia dike [GEOL] A dike formed of breccia injected into the country rock. { 'brech·ə ‚dīk }

breccia marble [PETR] Any marble containing angular fragments. { 'brech·ə ‚mär·bəl }

breccia pipe See pipe. { 'brech·ə ‚pīp }

breech [ORD] The rear part of the bore of a gun, especially the opening that permits the projectile to be inserted at the rear of the bore. { brēch }

breechblock [ORD] A movable steel block in the mechanism of a breech-loading gun that seals the breech opening of the barrel during firing. { 'brēch‚bläk }

breech bolt [ORD] A mechanism which opens and closes the breech in a carbine, machine gun, rifle, and the like; designed to push a cartridge into the chamber by sliding action. { 'brēch ‚bōlt }

breeches buoy [NAV ARCH] A device for carrying people from a stranded ship to shore or between ships. { 'brich· əz ‚bói }

breech flash [ORD] Flames and gas flash occurring at the breech of a weapon. { 'brēch ˌflash }

breeching [MECH ENG] A duct through which the products of combustion are transported from the furnace to the stack; usually applied in steam boilers. { 'brē·chiŋ }

breech interlock [ORD] A safety device used with weapons that are loaded or rammed automatically or in weapons in which the position of the breechblock cannot be readily seen by the loader; prevents the loading or ramming of a round when the breechblock is not fully open. { 'brēch ˌin·tərˌläk }

breech loading [ORD] A method of loading a weapon in which the ammunition is inserted at the rear of the bore. { 'brēch ˌlōd·iŋ }

breech mechanism [ORD] The assembly at the rear of a gun which receives the round of ammunition, inserts it in the chamber, fires the round by detonating the primer, and extracts the empty case. { 'brēch ˌmek·ə·nizˌəm }

breech preponderance [ORD] Unbalance of the tipping parts of a weapon when the weight of the breech exerts a greater moment about the trunnions than does the weight of the muzzle end. { 'brēch prə'pän·drəns }

breech pressure [ORD] In interior ballistics, the pressure from the propellant gases acting against the inner face of the breechblock. { 'brēch ˌpresh·ər }

breed [AGR] A group of animals that have a common origin and possess characteristics that are not common to other individuals of the same species. { brēd }

breeder reactor [NUCLEO] A nuclear reactor that produces more fissionable material that it consumes. { 'brēd·ər rē'ak·tər }

breeding [AGR] The application of genetic principles to the improvement of farm animals and of cultivated plants. [GEN] Controlled mating and selection, or hybridization of plants and animals in order to modify the species with respect to one or more phenotypic traits. [NUCLEO] The production of nuclear fuel, by absorption of neutrons in a nuclear reactor, at a rate exceeding that at which fuel is being consumed. { 'brēd·iŋ }

breeding factor See breeding ratio. { 'brēd·iŋ ˌfak·tər }

breeding gain [NUCLEO] The excess of fissionable atoms produced per fissionable atom consumed in a breeder reactor. { 'brēd·iŋ ˌgān }

breeding ratio [NUCLEO] The ratio of the number of fissionable atoms produced in a breeder reactor to the number of fissionable atoms consumed in the reactor; breeding gain is the breeding ratio minus 1. Also known as breeding factor. { 'brēd·iŋ ˌrā·shō }

breeze [METEOROL] **1.** A light, gentle, moderate, fresh wind. **2.** In the Beaufort scale, a wind speed ranging from 4 to 31 miles (6.4 to 49.6 kilometers) per hour. { brēz }

B register See index register. { 'bē ˌrej·ə·stər }

bregma [ANAT] The point at which the coronal and sagittal sutures of the skull meet. { 'breg·mə }

Breguet range equation [AERO ENG] An equation for the range of an aircraft stating that the range is equal to $K(PE/SFC)(L/D)$ ln (TOW/LW), where PE is propeller efficiency, SFC is specific fuel consumption, L/D is lift-to-drag ratio, TOW is takeoff weight, LW is landing weight, and K is a constant (equal to 375 if the range is expressed in miles and the specific fuel consumption is expressed in pounds per horsepower-hour). { bre·gā ˌränj iˌkwā·zhən }

breithauptite [MINERAL] NiSb A light copper red mineral consisting of nickel antimonide; commonly occurs in association with silver minerals. { 'brītˌhaupˌtīt }

Breit-Wigner formula [NUC PHYS] A formula which relates the cross section of a particular nuclear reaction with the energy of the incident particle, when the energy is near that required to form a discrete resonance level of the component nucleus. { ˌbrīt 'vig·nər ˌfȯr·myə·lə }

Breit-Wigner theory [NUC PHYS] A theory of nuclear reactions from which the Breit-Wigner formula is derived. Also known as Bohr-Breit-Wigner theory. { ˌbrīt 'vig·nər ˌthē·ə·rē }

bremsstrahlung [ELECTROMAG] Radiation that is emitted by an electron accelerated in its collision with the nucleus of an atom. { 'bremˌshträ·ləŋ }

Brennan monorail car [MECH ENG] A type of car balanced on a single rail so that when the car starts to tip, a force automatically applied at the axle end is converted gyroscopically into a strong righting moment which forces the car back into a position of lateral equilibrium. { ˌbren·ən 'män·əˌrāl ˌkär }

brennschluss [AERO ENG] **1.** The cessation of burning in a rocket, resulting from consumption of the propellants, from deliberate shutoff, or from other cause. **2.** The time at which this cessation occurs. { 'brenˌshlùs }

Brentidae [INV ZOO] The straight-snouted weevils, a family of coleopteran insects in the superfamily Curculionoidea. { 'bren·təˌdē }

Bretonian orogeny [GEOL] Post-Devonian diastrophism that is found in Nova Scotia. { bre'tōn·ē·ən ȯ'räj·ə·nē }

Bretonian strata [GEOL] Upper Cambrian strata in Cape Breton, Nova Scotia. { bre'tōn·ē·ən 'strad·ə }

bretonne lace [TEXT] A type of lace having a net ground with designs embroidered in heavy thread, often colored. { brə'tän 'lās }

brettice See brattice. { bred·əs }

brettis See brattice. { bred·əs }

breunnerite [MINERAL] $(Mg,Fe,Mn)CO_3$ A carbonate mineral consisting of an isomorphous system of the metallic components. { 'bróin·əˌrīt }

Brevibacteriaceae [MICROBIO] Formerly a family of gram-positive, rod-shaped, schizomycetous bacteria in the order Eubacteriales. { ˌbrev·əˌbak·tir·ē'ās·ē·ē }

Brevibacterium [MICROBIO] A genus of short, unbranched, rod-shaped bacteria in the coryneform group. { ˌbrev·əˌbak'tir·ē·əm }

brevitoxin [BIOCHEM] One of several ichthyotoxins produced by the dinoflagellate Ptychodiscus brevis. { ˌbrev·ə'täk·sən }

brevity code [COMMUN] Code which has as its sole purpose the shortening of messages rather than the concealment of their content. { 'brev·əd·ē ˌkōd }

Brewer anaerobic jar [MICROBIO] A glass container in which petri dish cultures are stacked and maintained under anaerobic conditions. { 'brü·ər an·ə'rō·bik 'jär }

brewers' yeast [FOOD ENG] Dried yeast cells recovered as a by-product of the brewing of beer and used as a natural source of vitamin B and protein. { ˌbrü·ərz 'yēst }

brewing [FOOD ENG] The process of making beer, ale, and other malt beverages by boiling mashed malt to produce a wort, flavoring the wort with hops, fermenting this mixture with yeast, and drawing off the fermented wort for bottling. { 'brü·iŋ }

brewster [OPTICS] A unit of stress optical coefficient of a material; it is equal to the stress optical coefficient of a material in which a stress of 1 bar produces a relative retardation between the components of a linearly polarized light beam of 1 angstrom when the light passes through a thickness of 1 millimeter in a direction perpendicular to the stress. Abbreviated B. { 'brü·stər }

Brewster fringes [OPTICS] Interference fringes observed when white light is viewed through two plane parallel plates of nearly equal thickness. { 'brü·stər ˌfrin·jəz }

brewsterite [MINERAL] $Sr(Al_2Si_6O_{18})\cdot5H_2O$ A member of the zeolite family of minerals; crystallizes in the monoclinic system and usually contains some calcium. { 'brü·stəˌrīt }

Brewster point [OPTICS] A neutral point located 15 to 20° directly below the sun. { 'brü·stər ˌpȯint }

Brewster process [CHEM ENG] Concentration of dilute acetic acid by use of an extraction solvent (for example, isopropyl ether), followed by distillation. { 'brü·stər ˌpräs·əs }

Brewster's angle [OPTICS] The angle of incidence of light reflected from a dielectric surface at which the reflectivity for light whose electrical vector is in the plane of incidence becomes zero; given by Brewster's law. Also known as polarizing angle. { ˌbrü·stərz ˌaŋ·gəl }

Brewster's law [OPTICS] The law that the index of refraction for a material is equal to the tangent of the polarizing angle for the material. { ˌbrü·stərz ˌlȯ }

Brewster stereoscope [OPTICS] A type of stereoscope that uses prisms to enable the eyes to form a fused image of two pictures whose separation is greater than the interocular distance. { ˌbrü·stər 'ster·ē·əˌskōp }

Brewster window [OPTICS] A special glass window used at opposite ends of some gas lasers to transmit one polarization of the laser output beam without loss. { 'brü·stər ˌwinˌdō }

Brianchon's theorem [MATH] The theorem that if a hexagon circumscribes a conic section, the three lines joining three

BRIDGED-T NETWORK

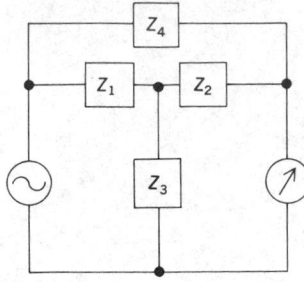

Schematic circuit of bridged-T network.

BRIDGE PIER

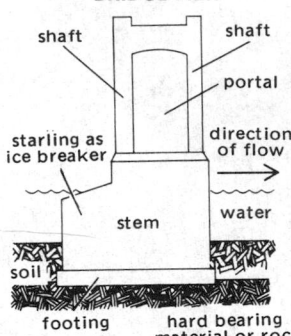

Large bridge pier—vertical shafts with heavy portal and base.

BRIDGMAN ANVIL

Configuration of Bridgman anvil, a basic type of static high-pressure equipment.

pairs of opposite vertices are concurrent (or are parallel). { ¦brē·ən¦känz ¦thir·əm }

brick [MATER] A building material usually made from clay, molded as a rectangular block, and baked or burned in a kiln. { brik }

brickfielder [METEOROL] A hot, dry, dusty north wind blowing from the interior across the southern coast of Australia. { 'brik₁fēl·dər }

brick grease See block grease. { 'brik ₁grēs }

bricking curb [MIN ENG] A curb set in a circular shaft to support the brick walling. { 'brik·iŋ ₁kərb }

brick molding [BUILD] A wooden molding applied to the gap between the frame of a door or window and the masonry into which the frame has been set. { 'brik ₁mōld·iŋ }

brick seat [BUILD] A ledge on a footing or a wall for supporting a course of masonry. { 'brik ₁sēt }

brick walling [MIN ENG] A permanent support for circular shafts by walling or casing. { 'brik 'wȯl·iŋ }

bridge [CIV ENG] A structure erected to span natural or artificial obstacles, such as rivers, highways, or railroads, and supporting a footpath or roadway for pedestrian, highway, or railroad traffic. [COMMUN] A device that joins two networks of the same type. [ELEC] **1.** An electrical instrument having four or more branches, by means of which one or more of the electrical constants of an unknown component may be measured. **2.** An electrical shunt path. [MATH] A line whose removal disconnects a component of a graph. Also known as isthmus. [MIN ENG] A piece of timber held above the cap of a set by blocks and used to facilitate the driving of spiling in soft or running ground. [NAV ARCH] An elevated structure extending across or over the weather deck of a vessel, containing stations for control and visual communications. [ORG CHEM] A connection between two different parts of a molecule consisting of a valence bond, an atom, or an unbranched chain of atoms. [PETRO ENG] **1.** An obstruction in a borehole resulting from the wall caving or the presence of a large boulder. **2.** A device installed in a borehole either permanently or temporarily to retain cement or other material. { brij }

bridge abutment [CIV ENG] The end foundation upon which the bridge superstructure rests. { 'brij ə₁bət·mənt }

bridge bearing [CIV ENG] The support at a bridge pier carrying the weight of the bridge; may be fixed or seated on expansion rollers. { 'brij ₁ber·iŋ }

bridge cable [CIV ENG] Cable from which a roadway or truss is suspended in a suspension bridge; may be of pencil-thick wires laid parallel or strands of wire wound spirally. { 'brij ₁ka·bəl }

bridge circuit [ELEC] An electrical network consisting basically of four impedances connected in series to form a rectangle, with one pair of diagonally opposite corners connected to an input device and the other pair to an output device. { 'brij ₁sər·kət }

bridge crane [MECH ENG] A hoisting machine in which the hoisting apparatus is carried by a bridgelike structure spanning the area over which the crane operates. { 'brij ₁krān }

bridge deck [NAV ARCH] A partial deck above the main deck on merchant ships, usually near the middle of the ship. { 'brij ₁dek }

bridged intermediate See bridged ion. { 'brijd in·tər'mēd·ē·ət }

bridged ion [ORG CHEM] A reactive intermediate in which an atom from one of the reactants is bonded partially to each of two carbon atoms of a reactant containing a double carbon-to-carbon bond. Also known as bridged intermediate; cyclic ion. { 'brijd 'ī·ən }

bridged tap [ELEC] Portion of a cable pair connected to a circuit which is not a part of the useful path. { 'brijd 'tap }

bridged-T network [ELEC] A T network with a fourth branch connected between an input and an output terminal and across two branches of the network. { ¦brijd ¦tē 'net₁wərk }

bridge foundation [CIV ENG] The piers and abutments of a bridge, on which the superstructure rests. { 'brij faún'dā·shən }

bridge graft [BOT] A plant graft in which each of several scions is grafted in two positions on the stock, one above and the other below an injury. { 'brij ₁graft }

bridge house [NAV ARCH] A structure above the main deck near the middle of a ship whose top forms the bridge deck. { 'brij ₁haús }

bridge hybrid See hybrid junction. { 'brij 'hī·brəd }

bridge limiter [ELECTR] A device employed in analog computers to keep the value of a variable within specified limits. { 'brij ₁lim·əd·ər }

bridge magnetic amplifier [ELECTR] A magnetic amplifier in which each of the gate windings is connected in series with an arm of a bridge rectifier; the rectifiers provide self-saturation and direct-current output. { ¦brij mag¦ned·ik 'am·plə₁fī·ər }

bridge oscillator [ELECTR] An oscillator using a balanced bridge circuit as the feedback network. { 'brij äs·ə'lād·ər }

bridge pier [CIV ENG] The main support for a bridge, upon which the bridge superstructure rests; constructed of masonry, steel, timber, or concrete founded on firm ground below river mud. { 'brij ₁pir }

bridge plug [PETRO ENG] A downhole tool used to isolate a lower zone while an upper section is being tested or cemented; consists of slips, a plug mandrel, and a sealing element that is run and set in casing. { 'brij ₁pləg }

bridge rectifier [ELECTR] A full-wave rectifier with four elements connected as a bridge circuit with direct voltage obtained from one pair of opposite junctions when alternating voltage is applied to the other pair. { 'brij ₁rek·tə₁fī·ər }

bridge trolley [MECH ENG] Either of the wheeled attachments at the ends of the bridge of an overhead traveling crane, permitting the bridge to move backward and forward on elevated tracks. { 'brij ₁träl·ē }

bridge vibration [MECH] Mechanical vibration of a bridge superstructure due to natural and human-produced excitations. { 'brij vī'brā·shən }

bridgewall [MECH ENG] A wall in a furnace over which the products of combustion flow. { 'brij₁wȯl }

bridgeware [COMPUT SCI] Software or hardware that translates programs or converts data from one format to another. { 'brij₁wer }

bridging [ELEC] **1.** Connecting one electric circuit in parallel with another. **2.** The action of a selector switch whose movable contact is wide enough to touch two adjacent contacts so that the circuit is not broken during contact transfer. [MATH] The operation of carrying in addition or multiplication. [MET] **1.** Formation of arched cavities in a powder compact. **2.** Jamming of the charge in a blast or a cupola furnace due to adherence of fine ore particles to the inner walls. **3.** Formation of solidified metal over the top of the charge in a mold or crucible. [MIN ENG] The obstruction of the receiving opening in a material-crushing device by two or more pieces wedged together, each of which could easily pass through. { 'brij·iŋ }

bridging amplifier [ELECTR] Amplifier with an input impedance sufficiently high so that its input may be bridged across a circuit without substantially affecting the signal level of the circuit across which it is bridged. { 'brij·iŋ ₁am·plə₁fī·ər }

bridging connection [ELECTR] Parallel connection by means of which some of the signal energy in a circuit may be withdrawn frequently, with imperceptible effect on the normal operation of the circuit. { 'brij·iŋ kə₁nek·shən }

bridging contacts [ELEC] A contact form in which the moving contact touches two stationary contacts simultaneously during transfer. { 'brij·iŋ ₁kän₁taks }

bridging ligand [ORG CHEM] A ligand in which an atom or molecular species which is able to exist independently is simultaneously bonded to two or more metal atoms. { 'brij·iŋ ₁līg·ənd }

bridging loss [ELECT] Loss resulting from bridging an impedance across a transmission system; quantitatively, the ratio of the signal power delivered to that part of the system following the bridging point, and measured before the bridging, to the signal power delivered to the same part after the bridging. { 'brij·iŋ ₁lȯs }

bridging material [MATER] A fibrous, flaky, or granular substance added to a cement slurry or drilling fluid to seal a formation in which lost circulation has occurred. Also known as lost-circulation material. { 'brij·iŋ mə₁tir·ē·əl }

Bridgman anvil [PHYS] A device for producing high static pressures using two large massive opposed pistons bearing on a small thin sample confined by a gasket material. { 'brij·mən 'an·vəl }

Bridgman effect [SOLID STATE] The phenomenon that when an electric current passes through an anisotropic crystal, there is an absorption or liberation of heat due to the nonuniformity in current distribution. { 'brij·mən i'fekt }

Bridgman relation [SOLID STATE] $P = QT\Sigma$ in a metal or semiconductor, where P is the Ettingshausen coefficient, Q the Nernst-Ettingshausen coefficient, T the temperature, and Σ the thermal conductivity in a transverse magnetic field. { 'brij·mən ri'lā·shən }

Bridgman sampler [MIN ENG] A mechanical device that automatically selects two samples as the ore passes through. { 'brij·mən ‚sam·plər }

Bridgman technique [SOLID STATE] A method of growing single crystals in which a vertical cylinder that tapers conically to a point at the bottom and contains the substance to be crystallized in molten form is slowly lowered into a cold zone, resulting in crystallization beginning at the tip. { 'brij·mən tek'nēk }

bridle [ENG] A pumping unit cable that is looped over the horse head and then connected to the carrier bar; supports the polished-rod clamp. { 'brīd·əl }

bridled-cup anemometer [ENG] A combination cup anemometer and pressure-plate anemometer, consisting of an array of cups about a vertical axis of rotation, the free rotation of which is restricted by a spring arrangement; by adjustment of the force constant of the spring, an angular displacement can be obtained which is proportional to wind velocity. { ‚brīd·əl ‚kəp an·ə'mäm·əd·ər }

bridled pressure plate [METEOROL] An instrument for measuring air velocity in which the pressure on a plate exposed to the wind is balanced by the force of a spring, and the deflection of the plate is measured by an inductance-type transducer. { ‚brīd·əld 'presh·ər ‚plāt }

briefing See pilot briefing. { 'brē·fiŋ }

brig [PHYS] A unit to express the ratio of two quantities, as a logarithm to the base 10; that is, a ratio of 10^x is equal to x brig; it is analogous to the bel, but the latter is restricted to power ratios. Also known as dex. { brig }

Briggs equalizer [ENG] A breathing device consisting of head harness, mouthpiece, nose clip, corrugated breathing tube, an equalizing device, 120 feet (37 meters) of reinforced air tubes, and a strainer and spike. { ‚brigz 'ē·kwə‚līz·ər }

Briggsian logarithm See common logarithm. { ‚brigz·ē·ən 'läg·ə‚rith·əm }

Briggs' logarithm See common logarithm. { ‚brigz 'log·ə‚rith·əm }

Briggs pipe thread See American standard pipe thread. { ‚brigz 'pīp ‚thred }

Briggs stretcher carriage [MIN ENG] A stretcher used as an ambulance trolley in transporting casualties from underground workings. { ‚brigz 'strech·ər ‚ka·rij }

bright [MATER] Referring to lubricating oils that are clear, or free from moisture. [OPTICS] Attribute of an area that appears to emit a large amount of light. { brīt }

bright annealing [MET] Heating and cooling a metal in an inert atmosphere to inhibit oxidation; surface remains relatively bright. { 'brīt ə'nēl·iŋ }

bright band [METEOROL] The enhanced echo of snow as it melts to rain, as displayed on a range-height indicator scope. { 'brīt ‚band }

bright-banded coal See bright coal. { 'brīt ‚ban·dəd ‚kōl }

bright coal [GEOL] A jet-black, pitchlike type of banded coal that is more compact than dull coal and breaks with a shell-shaped fracture; microscopic examination shows a consistency of more than 5% anthraxyllon and less than 20% opaque matter. Also known as bright-banded coal; brights. { 'brīt ‚kōl }

bright diffuse nebula [ASTRON] A nebula which is illuminated by the action of embedded or nearby stars. { 'brīt də‚fyüs 'neb·yə·lə }

bright dipping [MET] Immersing metal into an acid solution to give a bright, clean surface. { 'brīt ‚dip·iŋ }

brightener [MET] Any of the agents which are employed in small concentrations in the electrolytic bath for electroplating metal to yield smoother or brighter coatings. { 'brīt·nər }

bright-field [OPTICS] Having a brightly lighted background. { 'brīt ‚fēld }

bright-field microscope [CELL MOL] A type of light microscope that produces a dark image against a brighter background; commonly used for the visualization of stained cells. { 'brīt ‚fēld 'mī·krə‚skōp }

bright-line spectrum [SPECT] An emission spectrum made up of bright lines on a dark background. { 'brīt ‚līn 'spek·trəm }

brightness [OPTICS] **1.** The characteristic of light that gives a visual sensation of more or less light. **2.** See luminance. { 'brīt·nəs }

brightness control [ELECTR] A control that varies the luminance of the fluorescent screen of a cathode-ray tube, for a given input signal, by changing the grid bias of the tube and hence the beam current. Also known as brilliance control; intensity control. { 'brīt·nəs kən'trōl }

brightness level See adaptation luminance. { 'brīt·nəs ‚lev·əl }

brightness temperature See blackbody temperature. { 'brīt·nəs ‚tem·prə·chər }

bright plating [MET] An electroplating process resulting in a smooth, lustrous surface without polishing. { 'brīt ‚plād·iŋ }

bright points [ASTRON] Relatively small regions on the sun, distributed uniformly over the solar disk, from which there is increased x-ray and ultraviolet emission, having lifetimes on the order of 8 hours. { 'brīt ‚póins }

bright rim structures [ASTRON] Bright edges exhibited by many diffuse-emission nebulae, usually on the side facing the exciting star. { 'brīt ‚rim ‚strək·chərz }

brights See bright coal. { 'brīts }

Bright's disease [MED] Any of several kidney diseases attended by glomerulonephritis. { 'brīts di‚zēz }

bright segment [GEOPHYS] A faintly glowing band which appears above the horizon after sunset or before sunrise. Also known as crepuscular arch; twilight arch. { 'brīt ‚seg·mənt }

bright stars catalog [ASTRON] A catalog of stars brighter than 6.5 magnitude, giving positions, motions, parallaxes, and spectral classes. { 'brīt ‚stärz 'kad·ə‚läg }

bright stock [MATER] High-viscosity refined and dewaxed lubricating oils used in the compounding of motor oils. { 'brīt ‚stäk }

bril [OPTICS] A unit of subjective luminance; 100 brils is the luminance level that corresponds to a luminance of 1 millilambert, and a doubling of luminance level corresponds to an increase of 1 bril. { bril }

brilliance [ELECTR] **1.** The degree of brightness and clarity of the display of a cathode-ray tube. **2.** The degree to which the higher audio frequencies of an input sound are reproduced by a radio receiver, by a public address amplifier, or by a sound-recording playback system. { 'bril·yəns }

brilliance control See brightness control. { 'bril·yəns kən'trōl }

brilliant cut [LAP] A style of cutting gemstones in which there are 58 facets, all triangular or kite-shaped, except for a large facet at the top of the stone and a very small facet at the lowest point. { 'bril·yənt ‚kət }

brilliancy [LAP] A characteristic of gems that depends on the refractive index, transparency, polish, and proportions of the cut stone. { 'bril·yən·sē }

Brillouin function [SOLID STATE] A function of x with index (or parameter) n that appears in the quantum-mechanical theories of paramagnetism and ferromagnetism and is expressed as $[(2n + 1)/2n]\coth[(2n + 1)x/2n] - (1/2n)\coth(x/2n)$. { brēy·wan 'fəŋk·shən }

Brillouin scattering [SOLID STATE] Light scattering by acoustic phonons. { brēy·wan 'skad·ər·iŋ }

Brillouin zone [SOLID STATE] A fundamental region of wave vectors in the theory of the propagation of waves through a crystal lattice; any wave vector outside this region is equivalent to some vector inside it. { brēy·wan ‚zōn }

Brill's disease [MED] A mild recurrence of typhus some years after the initial infection. { ‚brilz di‚zēz }

brimstone [MINERAL] A common or commercial name for native sulfur. { 'brim‚stōn }

brine [MATER] A liquid used in a refrigeration system, usually an aqueous solution of calcium chloride or sodium chloride, which is cooled by contact with the evaporator surface and then goes to the space to be refrigerated. [OCEANOGR] Sea water containing a higher concentration of dissolved salt than that of the ordinary ocean. { brīn }

brine cooler [MECH ENG] The unit for cooling brine in a refrigeration system; the brine usually flows through tubes or pipes surrounded by evaporating refrigerant. { 'brīn ‚kül·ər }

Brinell number [ENG] A hardness rating obtained from the

BRILLOUIN ZONE

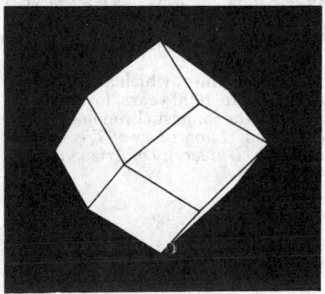

Brillouin zone for the body-centered cubic lattice.

Brinell test; expressed in kilograms per square millimeter. { brə'nel ˌnəm·bər }

Brinell test [ENG] A test to determine the hardness of a material, in which a steel ball 1 centimeter in diameter is pressed into the material with a standard force (usually 3000 kilograms); the spherical surface area of indentation is measured and divided into the load; the results are expressed as Brinell number. { brə'nel ˌtest }

brine spring [HYD] A salt-water spring. { 'brīn ˌspriŋ }

Brinkmann number [FL MECH] A dimensionless number used to study viscous flow. { 'briŋk·män ˌnəm·bər }

briolette [LAP] An oval or pear-shaped gemstone with its surface cut in triangular, or sometimes rectangular, facets. { ˌbrē·ə¦let }

briquet [MATER] A block of some compressed substance, such as coal dust, metal powder, or sawdust, used as a fuel. Also spelled briquette. { bri'ket }

briquette *See* briquet. { bri'ket }

briquetting [ENG] **1.** The process of binding together pulverized minerals, such as coal dust, into briquets under pressure, often with the aid of a binder, such as asphalt. **2.** A process or method of mounting mineral ore, rock, or metal fragments in an embedding or casting material, such as natural or artificial resins, waxes, metals, or alloys, to facilitate handling during grinding, polishing, and microscopic examination. { bri'-ked·iŋ }

brisa [METEOROL] **1.** A northeast wind which blows on the coast of South America or an east wind which blows on Puerto Rico during the trade wind season. **2.** The northeast monsoon in the Philippines. Also spelled briza. { 'brē·sə }

brisa carabinera *See* carabine. { 'brē·sə ˌkär·ə·bi'ner·ə }

brisance [ORD] The ability of an explosive to shatter the medium which confines it; the shattering effect of the explosive. { brə'zäns }

brisance index [ENG] The ratio of an explosive's power to shatter a weight of graded sand as compared to the weight of sand shattered by TNT. { brə'zäns ˌin‚deks }

brise carabinée *See* carabine. { ˌbrēz kä·rä·bē'nä }

Brisingidae [INV ZOO] A family of deep-water echinoderms with as many as 44 arms, belonging to the order Forcipulatida. { brə'sin·jə‚dē }

brisote [METEOROL] The northeast trade wind when it is blowing stronger than usual on Cuba. { brē'sȯ·tä }

bristle [BIOL] A short stiff hair or hairlike structure on an animal or plant. { 'bris·əl }

bristlecone pine [BOT] A small slow-growing evergreen tree of the genus *Pinus* that grows at high altitudes in the western United States, having dense branches with rust-brown bark and short needles in bunches of five and thorn-tipped cone scales. The two types are *P. longaeva*, which lives longer than any other tree (over 4000 years), and *P. monophylla*, the single-leaf pinyon. { ˌbris·əl‚kōn 'pīn }

Bristol board [MATER] Cardboard with a surface smooth enough for painting or writing, usually at least 0.006 inch (0.15 millimeter) thick. { 'brist·əl ˌbȯrd }

britannia cell [MIN ENG] In mineral processing, a pneumatic flotation cell 7 to 9 feet (2.1 to 2.7 meters) deep. { bri'tan·yə ˌsel }

britannia metal [MET] A silver-white tin alloy, similar to pewter, containing about 7% antimony, 2% copper, and often some zinc and bismuth; used in domestic utensils. { bri'tan·yə ˌmed·əl }

britholite [MINERAL] $(Na,Ce,Ca)_5(OH)[(P,Si)O_4]_3$ A rare-earth phosphate found in carbonatites in Kola Peninsula, former Soviet Union. { 'brith·ə‚līt }

British absolute system of units [PHYS] A measurement system based on the foot, the second, and the pound mass; force unit is the poundal. Also known as foot-pound-second system of units (fps system of units). { 'brid·ish 'ab·sə‚lüt ˌsis·təm əv 'yü·nəts }

British antilewisite *See* dimercaprol. { 'brid·ish an·tē'lü·wə‚sīt }

British engineering system of units *See* British gravitational system of units. { 'brid·ish en·jə'nir·iŋ ˌsis·təm əv 'yü·nəts }

British gravitational system of units [PHYS] A measurement system based on the foot, the second, and the slug mass; 1 slug weighs 32.174 pounds at sea level and 45° latitude, and equals 14.594 kilograms. Also known as British engineering

system of units; engineer's system of units. { 'brid·ish grav·ə'tā·shən·əl 'sis·təm əv 'yü·nəts }

British imperial pound [MECH] The British standard of mass, of which a standard is preserved by the government. { 'brid·ish im'pir·ē·əl 'paund }

British thermal unit [THERMO] Abbreviated Btu. **1.** A unit of heat energy equal to the heat needed to raise the temperature of 1 pound of air-free water from 60° to 61°F at a constant pressure of 1 standard atmosphere; it is found experimentally to be equal to 1054.5 joules. Also known as sixty degrees Fahrenheit British thermal unit ($Btu_{60/61}$). **2.** A unit of heat energy that is equal to 1/180 of the heat needed to raise 1 pound of air-free water from 32°F (0°C) to 212°F (100°C) at a constant pressure of 1 standard atmosphere; it is found experimentally to be equal to 1055.79 joules. Also known as mean British thermal unit (Btu_{mean}). **3.** A unit of heat energy whose magnitude is such that 1 British thermal unit per pound equals 2326 joules per kilogram; it is equal to exactly 1055.05585262 joules. Also known as international table British thermal unit (Btu_{IT}). { 'brid·ish 'thər·məl ˌyü·nət }

brittle fracture [MET] A break in a brittle piece of metal which failed because stress exceeded cohesion. { ¦brid·əl 'frak·chər }

brittle mica [MINERAL] Hydrous sodium, calcium, magnesium, and aluminum silicates; a group of more or less related minerals that resemble true micas but cleave to brittle flakes and contain calcium as the essential constituent. { ¦brid·əl 'mī·kə }

brittleness [MECH] That property of a material manifested by fracture without appreciable prior plastic deformation. { 'brid·əl·nəs }

brittle silver ore *See* stephanite. { ¦brid·əl 'sil·vər ˌȯr }

brittle star [INV ZOO] The common name for all members of the echinoderm class Ophiuroidea. { ¦brid·əl 'stär }

brittle temperature [THERMO] The temperature point below which a material, especially metal, is brittle; that is, the critical normal stress for fracture is reached before the critical shear stress for plastic deformation. { 'brid·əl ˌtem·prə·chər }

Brix degree [CHEM ENG] A unit of the Brix scale. { 'briks də‚grē }

Brix scale [CHEM ENG] A hydrometer scale for sugar solutions indicating the percentage by weight of sugar in the solution at a specified temperature. { 'briks ˌskāl }

briza *See* brisa. { 'brē·zə }

broach [MECH ENG] A multiple-tooth, barlike cutting tool; the teeth are shaped to give a desired surface or contour, and cutting results from each tooth projecting farther than the preceding one. { brōch }

broaching [ENG] **1.** The restoration of the diameter of a borehole by reaming. **2.** The breaking down of the walls between two contiguous drill holes. [MECH ENG] The machine-shaping of metal or plastic by pushing or pulling a broach across a surface or through an existing hole in a workpiece. { 'brōch·iŋ }

broaching bit *See* reaming bit. { 'brōch·iŋ bit }

broadband [COMMUN] A band with a wide range of frequencies. { 'brȯd‚band }

broadband amplifier [ELECTR] An amplifier having essentially flat response over a wide range of frequencies. { 'brȯd‚band ¦am·plə‚fī·ər }

broadband antenna [ELECTROMAG] An antenna that functions satisfactorily over a wide range of frequencies, such as for all 12 very-high-frequency television channels. { 'brȯd‚band an'ten·ə }

broadband channel [COMMUN] A data transmission channel that can handle frequencies higher than the normal voice-grade line limit of 3 to 4 kilohertz; it can carry many voice or data channels simultaneously or can be used for high-speed single-channel data transmission. { 'brȯd‚band ¦chan·əl }

broadband klystron [ELECTR] Klystron having three or more resonant cavities that are externally loaded and stagger-tuned to broaden the bandwidth. { 'brȯd‚band ¦klī‚strän }

broadband path [COMMUN] A path having a bandwidth of 20 kilohertz or greater. { 'brȯd‚band ˌpath }

broad beam [PHYS] In measurements of the attenuation of a beam of ionizing radiation, a beam in which much of the scattered radiation reaches the detector, along with the unscattered radiation. { 'brȯd ˌbēm }

BRISTLECONE PINE

Bristlecone pines, which attain ages of more than 4000 years, form the basis for the longest chronology yet developed. (*Laboratory of Tree-Ring Research, University of Arizona*)

broadcast [COMMUN] A television, radio, or data transmission intended for public reception. { 'bród,kast }

broadcast band [COMMUN] The band of frequencies extending from 535 to 1605 kilohertz, corresponding to assigned radio carrier frequencies that increase in multiples of 10 kHz between 540 and 1600 kHz for the United States. Also known as standard broadcast band. { 'bród,kast ,band }

broadcaster [AGR] A machine that utilizes a rotating fanlike distributor for sowing grain, grass, and clover seed or for spreading fertilizer. { 'bród,kas·tər }

broadcast message [COMMUN] A message that is sent to all users of a computer network when they log on to the network. { 'bród,kast 'mes·ij }

broadcast station [COMMUN] A television or radio station used for transmitting programs to the general public. Also known as station. { 'bród,kast ,stā·shən }

broadcast transmitter [ELECTR] A transmitter designed for use in a commercial amplitude-modulation, frequency-modulation, or television broadcast channel. { 'bród,kast tranz'-mid·ər }

broadcloth [TEXT] A closely woven ribbed fabric having the rib running in the direction of the weft. { 'bród,klóth }

broadening of spectral lines [SPECT] A widening of spectral lines by collision or pressure broadening, or possibly by Doppler effect. { 'bród·ən·iŋ əv ¦spek·trəl 'līn }

broad goods [TEXT] Woven fabrics that are over 18 inches (450 millimeters) wide. { 'bród ,gúdz }

broadleaf tree [BOT] Any deciduous or evergreen tree having broad, flat leaves. { 'bród,lēf ,trē }

broad on the beam [NAV] Bearing 090° (broad on the starboard beam) or 270° relative (broad on the port beam); if the bearings are approximate, the expression "on the beam" or "abeam" is used. { ¦bród ón ¦thə 'bēm }

broad on the bow [NAV] Bearing 045° relative (broad on the starboard bow) or 315° relative (broad on the port bow); if the bearings are approximate, the expression "on the bow" should be used. { ¦bród ón ¦thə 'baú }

broad on the quarter [NAV] Bearing 135° relative (broad on the starboard quarter) or 225° relative (broad on the port quarter); if the bearings are approximate, the expression "on the quarter" is used. { ¦bród ón ¦thə 'kwórd·ər }

broad-sense heritability [GEN] The degree to which individual phenotypes are determined by their genotypes; expressed as the ratio of the total genetic variance to the total phenotypic variance. { ¦bród ,sens ,her·ə·tə'bil·ə·dē }

broadside [ELECTROMAG] Perpendicular to an axis or plane. { 'bród,sīd }

broadside array [ELECTROMAG] An antenna array whose direction of maximum radiation is perpendicular to the line or plane of the array. { 'bród,sīd ə'rā }

broadside on [NAV] Orientation of a ship in which the length of the ship is perpendicular to the direction of the wind or of ocean currents. { 'bród,sīd 'ón }

broadside-on position [ELECTROMAG] The position of a point which lies on a line through the center of a magnet, perpendicular to the magnetic axis. Also known as Gauss B position. { ¦bród,sīd 'ón pə,zish·ən }

broad-spectrum antibiotic [MICROBIO] An antibiotic that is effective against both gram-negative and gram-positive bacterial species. { ¦bród ¦spek·trəm ,ant·i·bī'äd·ik }

broad tuning [ELECTR] Poor selectivity in a radio receiver, causing reception of two or more stations at a single setting of the tuning dial. { 'bród ¦tün·iŋ }

brocade [TEXT] Fabric made in a jacquard weave, usually with raised designs, and having a luxurious appearance; made of silk, polyester, or blends. { brō'kād }

Broca galvanometer [ELECTROMAG] A type of astatic galvanometer in which a current-carrying coil encloses consequent poles at the centers of two parallel magnets with opposite polarities. { ¦brō·kə ,gal·və'näm·əd·ər }

Broca's aphasia See nonfluent aphasia. { ¦brō·kəz ə'fā·zhə }

Broca's area [NEURO] In the human brain, an area in the inferior left frontal lobe—one of several areas believed to activate the fibers of the precentral gyrus concerned with movements necessary for speech production; injury to this area generally results in nonfluent aphasia, with effortful articulation, loss of syntax, but relatively well-preserved auditory comprehension. { 'brō·kəz ,er·ē·ə }

brocatelle [TEXT] A heavy cross-ribbed fabric that resembles brocade but has jacquard figures in high relief. { ¦bräk·ə¦tel }

broccoli [BOT] *Brassica oleracea* var. *italica*. A biennial crucifer of the order Capparales which is grown for its edible stalks and buds. { 'brak·ə·lē }

brochanite See brochantite. { brō'shän,īt }

brochanthite See brochantite. { brō'shän,thīt }

brochantite [MINERAL] $Cu_4(SO_4)(OH)_6$ A monoclinic copper mineral, emerald to dark green, commonly found with copper sulfide deposits; a minor copper ore. Also known as brochanite; brochanthite; warringtonite. { brō'shän,tīt }

Brocken bow See anticorona. { ¦bräk·ən ,bō }

Brocken specter [METEOROL] The illusory appearance of a gigantic figure (actually, the observer's shadow projected on cloud surfaces), observed on the Brocken peak in the Hartz Mountains of Saxony, but visible from other mountaintops under suitable conditions. { ¦bräk·ən ,spek·tər }

Brodmann's area 4 See motor area. { 'bräd·mənz ,er·ē·ə 'fór }

Brodmann's area 17 See visual projection area. { 'bräd·mənz ,er·ē·ə sev·ən'tēn }

Brodmann's areas [PHYSIO] Numbered regions of the cerebral cortex used to identify cortical functions. { 'bräd·mənz ,er·ē·əz }

broeboe [METEOROL] A strong, dry east wind in the southwestern part of the island of Celebes. { ¦brō·ə¦bō·ə }

Broenner's acid See Brönner's acid. { 'bren·ərz 'as·əd }

broken [METEOROL] Descriptive of a sky cover of from 0.6 to 0.9 (expressed to the nearest tenth). { 'brō·kən }

broken-back transit [OPTICS] A type of transit telescope in which the light path is broken by insertion of a right-angled prism at the intersection of the optical and rotational axes. Also known as prism transit. { 'brō·kən ,bak 'tranz·ət }

broken belt [OCEANOGR] The transition zone between open water and consolidated ice. { ¦brō·kən 'belt }

broken bracket See angle bracket. { ¦brō·kən ¦brak·ət }

broken-color work See antiquing. { ¦brō·kən ¦kəl·ər ,wərk }

broken ground See loose ground. { ¦brō·kən 'graúnd }

broken line [MATH] A line which is composed of a series of line segments lying end to end, and which does not form a continuous line. { 'brō·kən ,līn }

broken stone See crushed stone. { ¦brō·kən 'stōn }

broken stream [HYD] A stream that repeatedly disappears and reappears, such as occurs in an arid region. { 'brō·kən 'strēm }

broken water [OCEANOGR] Water having a surface covered with ripples or eddies, and usually surrounded by calm water. { ¦brō·kən 'wód·ər }

broken wind See heaves. { ¦brō·kən 'wind }

bromacetone [ORG CHEM] $CH_2BrCOCH_3$ A colorless liquid which is a powerful irritant and lacrimator; used as tear gas and to make other chemicals. { ¦bróm'as·ə,tōn }

bromacil [ORG CHEM] 5-Bromo-3-*sec*-butyl-6-methyluracil, a soil sterilant; general at high dosage and selective at low. { 'brom·ə,sil }

bromadiolone [ORG CHEM] $C_{30}H_{23}BrO_4$ A rodenticide. { ¦brō·mə'dī·ə,lōn }

bromate [CHEM] **1.** BrO_3^- A negative ion derived from bromic acid, $HBrO_3$ **2.** A salt of bromic acid. **3.** $C_9H_9ClO_3$ A light brown solid with a melting point of 118–119°C; used as a herbicide to control weeds in crops such as flax, cereals, and legumes. { 'brō,māt }

bromatium [ECOL] A swollen hyphal tip on fungi growing in ants nests that is eaten by the ants. { brō'māsh·əm }

bromcresol green See bromocresol green. { brōm'krē,sól 'grēn }

bromcresol purple See bromocresol purple. { brōm'krē,sól 'pər·pəl }

bromegrass [BOT] The common name for a number of forage grasses of the genus *Bromus* in the order Cyperales. { 'brōm,gras }

bromelain [BIOCHEM] An enzyme that digests protein and clots milk; prepared by precipitation by acetone from pineapple juice; used to tenderize meat, to chill-proof beer, and to make protein hydrolysates. Also spelled bromelin. { 'brō·mə ,lān }

Bromeliaceae [BOT] The single family of the flowering plant order Bromeliales. { brō,mel·ē'ās·ē,ē }

BROCA'S AREA

frontal lobe · *parietal lobe* · *temporal lobe* · *Broca's area*

Lateral surface view of the cerebral hemisphere showing Broca's area. (*After C. R. Noback, The Human Nervous System, 4th ed., McGraw-Hill, 1991*)

BROKEN-BACK TRANSIT

Broken-back transit at the Sternberg Astronomical Institute, Moscow.

Bromeliales [BOT] An order of monocotyledonous plants in the subclass Commelinidae, including terrestrial xerophytes and some epiphytes. { brō,mel·ē′ā·lēz }

bromelin *See* bromelain. { 'brō·mə·lən }

bromellite [MINERAL] BeO A white hexagonal mineral consisting of beryllium oxide; it is harder than zincite. { brō'me,līt }

brome mosaic virus [VIROL] The type species of the genus *Bromovirus*. Abbreviated BMV. { ¦brōm mō′zā·ik ,vī·rəs }

brome mosaic virus group *See* Bromovirus. { ,brōm mō,zā·ik 'vī·rəs ,grüp }

bromethalin [ORG CHEM] $C_{14}H_7Br_3F_3N_3O_4$ A rodenticide. { ,brō·mə′thal·ən }

bromeosin *See* eosin. { ,brōm′ē·ə·sən }

bromhidrosiphobia [PSYCH] Abnormal fear of offensive body odors. { ,brōm,hī·drə·sə′fō·bē·ə }

bromic acid [INORG CHEM] $HBrO_3$ A liquid, colorless to slightly yellow; boils with decomposition at 100°C; used in dyes and as a chemical intermediate. { 'brō·mik 'as·əd }

bromide [CHEM] A compound derived from hydrobromic acid, HBr, with the bromine atom in the 1-oxidation state. { 'brō,mīd }

bromide paper [GRAPHICS] A photographic paper with high sensitivity, used for short exposures; it is coated with an emulsion of silver bromide. { 'brō,mīd ,pā·pər }

brominating agent [CHEM] A compound capable of introducing bromine into a molecule; examples are phosphorus tribromide, bromine chloride, and aluminum tribromide. { 'brō·mə,nād·iŋ ,ā·jənt }

bromination [CHEM] The process of introducing bromine into a molecule. { brō·mə′nā·shən }

bromine [CHEM] A chemical element, symbol Br, atomic number 35, atomic weight 79.904; used to make dibromide ethylene and in organic synthesis and plastics. { 'brō,mēn }

bromine number [ANALY CHEM] The amount of bromine absorbed by a fatty oil; indicates the purity of the oil and degree of unsaturation. { 'brō,mēn ,nəm·bər }

bromine test [CHEM ENG] A laboratory test in which the unsaturated hydrocarbons present in a crude oil are determined by mixing a sample with bromine; the lower the rate of bromine absorption, the more paraffinic the test sample. { 'brō,mēn ,test }

bromine trifluoride [CHEM] BrF_3 A liquid with a boiling point of 135°C. { 'brō,mēn ,trī′flu̇r,īd }

bromine value [CHEM ENG] An expression representing the number of centigrams of bromine absorbed by 1 gram of oil under test conditions; an indication of the degree of unsaturation of a given oil. { 'brō,mēn ,val·yü }

bromine water [CHEM] An aqueous saturated solution of bromine used as a reagent wherever a dilute solution of bromine is needed. { 'brō,mēn ,wȯd·ər }

bromism [MED] A disease state produced by prolonged usage or overdosage of bromide compounds. { 'brō,miz·əm }

bromlite [MINERAL] $BaCa(CO_3)_3$ An orthorhombic mineral composed of a carbonate of barium and calcium. Also known as alstonite. { 'brōm,līt }

bromo- [CHEM] A prefix that indicates the presence of bromine in a molecule. { 'brō·mō }

***N*-bromoacetamide** [ORG CHEM] $CH_3CONHBr$ Needlelike crystals with a melting point of 102–105°C; soluble in warm water and cold ether; used as a brominating agent and in the oxidation of primary and secondary alcohols. { ¦en ¦brō·mō·ə′sed·ə,mīd }

***para*-bromoacetanilide** [ORG CHEM] C_8H_8BrNO Crystals with a melting point of 168°C; soluble in benzene, chloroform, and ethyl acetate; insoluble in cold water; used as an analgesic and antipyretic. { ¦par·ə ¦brō·mō,a·səd′an·əl,īd }

bromoacetone [ORG CHEM] $BrCH_2COCH_3$ A colorless liquid used as a lacrimatory agent. { ¦brō·mō′as·ə,tōn }

bromo acid *See* eosin. { 'brō·mō 'as·əd }

bromoalkane [ORG CHEM] An aliphatic hydrocarbon with bromine bonded to it. { ¦brō·mō′al,kān }

***para*-bromoaniline** [ORG CHEM] $BrC_6H_4NH_2$ Rhombic crystals with a melting point of 66–66.5°C; soluble in alcohol and in ether; used in the preparation of azo dyes and dihydroquinazolines. { ¦par·ə ¦brō·mō′an·ə·lēn }

***para*-bromoanisole** [ORG CHEM] C_7H_7BrO Crystals which melt at 9–10°C; used in disinfectants. { ¦par·ə ¦brō·mō′an·ə,sōl }

bromobenzene [ORG CHEM] C_6H_5Br A heavy, colorless liquid with a pleasant odor; used as a solvent, in motor fuels and top-cylinder compounds, and to make other chemicals. { ¦brō·mō′ben,zēn }

***para*-bromobenzyl bromide** [ORG CHEM] $BrC_6H_4CH_2Br$ Crystals with an aromatic odor and a melting point of 61°C; soluble in cold and hot alcohol, water, and ether; used to identify aromatic carboxylic acids. { ¦par·ə ,brō·mo′benz·əl 'brō,mīd }

bromobenzylcyanide [ORG CHEM] $C_6H_5CHBrCN$ A light yellow oily compound used as a tear gas for training and for riot control. Abbreviated BBC. { ¦brō·mō¦benz·əl′sī·ə,nīd }

bromochloromethane [ORG CHEM] $BrCH_2Cl$ A clear, colorless liquid with a boiling point of 67°C; volatile, soluble in organic solvents, with a chloroform-like odor; used in fire extinguishers. { ¦brō·mō¦klȯr·ō′me,thān }

bromochloroprene [ORG CHEM] $CHCl=CHCH_2Br$ A compound used as a nematicide and soil fumigant. { ,brō·mō′klȯr·ə,prēn }

bromocresol green [ORG CHEM] Tetrabromo-*m*-cresol sulfonphthalein, a gray powder soluble in water or alcohol; used as an indicator between pH 4.5 (yellow) and 5.5 (blue). Also known as bromcresol green. { ¦brō·mō′krē,sȯl 'grēn }

bromocresol purple [ORG CHEM] Dibromo-*o*-cresol sulfonphthalein, a yellow powder soluble in water; used as an indicator between pH 5.2 (yellow) and 6.8 (purple). Also known as bromcresol purple. { ¦brō·mō′krē,sȯl ,pər·pəl }

bromocriptine [ORG CHEM] $C_{32}H_{40}BrN_5O_5$ A polypeptide alkaloid that is a derivative of the ergotoxin group of ergot alkaloids and is a dopamine receptor agonist. { ,brō·mō′krip,tēn }

bromocyclen [ORG CHEM] $C_8H_5BrCl_6$ A compound used as an insecticide for wheat crops. { ,brō·mō′sī·klən }

5-bromodeoxyuridine [BIOCHEM] $C_9H_{11}O_5NBr$ A thymidine analog that can be incorporated into deoxyribonucleic acid during its replication; induces chromosomal breakage in regions rich in heterochromatin. Abbreviated BUDR. { ¦fīv ¦brō·mō·dē,äk·sē′rur·ə,dēn }

bromodiethylacetylurea [PHARM] $C(C_2H_5)_2BrCONHCO-NH_2$ A white, crystalline powder with a melting point of 116–117°C; soluble in chloroform, ether, and alcohol; used in medicine. { ¦brō·mō,dī′eth·əl·ə,sēd·əl′y u̇·rē·ə }

bromofenoxim [ORG CHEM] $C_{13}H_7N_3O_6Br_2$ A cream-colored powder with melting point 196–197°C; slightly soluble in water; used as herbicide to control weeds in cereal crops. { ¦brō·mō·fə′näk·səm }

bromoform [ORG CHEM] $CHBr_3$ A colorless liquid, slightly soluble in water; used in the separation of minerals. { ,brō·mə′fȯrm }

1-bromonaphthalene [ORG CHEM] $C_{10}H_7Br$ An oily liquid that is slightly soluble in water and miscible with chloroform, benzene, ether, and alcohol; used in the determination of index of refraction of crystals and for refractometric fat determination. { ¦wən ¦brō·mō′naf·thə,lēn }

bromonium ion [ORG CHEM] A halonium ion in which the halogen is bromine; occurs as a bridged structure. { brə′mōn·ē·əm 'ī·ən }

1-bromooctane [ORG CHEM] $CH_3(CH_2)_6CH_2Br$ Colorless liquid that is miscible with ether and alcohol; boiling point is 198–200°C; used in organic synthesis. { ¦wən ¦brō·mō′äk,tān }

***para*-bromophenacyl bromide** [ORG CHEM] $C_8H_6Br_2O$ Crystals with a melting point of 109–110°C; soluble in warm alcohol; used in the identification of carboxylic acids and as a protecting reagent for acids and phenols. { ¦par·ə ¦brō·mō·fə′nas·əl 'brō,mīd }

***para*-bromophenylhydrazine** [ORG CHEM] $C_6H_7BrN_2$ Needlelike crystals with a melting point of 108–109°C; soluble in benzene, ether, chloroform, and alcohol; used in the preparation of indoleacetic acid derivatives and in the study of transosazonation of sugar phenylosazones. { ¦par·ə ¦brō·mō·fen·əl′hī·drə,zēn }

bromophos [ORG CHEM] $C_8H_8SPBrCl_2O_3$ A yellow, crystalline compound with a melting point of 54°C; used as an insecticide and miticide for livestock, household insects, flies, and lice. { 'brō·mə,fäs }

bromopicrin [ORG CHEM] CBr_3NO_2 Prismatic crystals with a melting point of 103°C; soluble in alcohol, benzene, and ether; used for military poison gas.   Also known as nitro-bromoform.   { ˌbrō·mōˈpik·rən }

N-bromosuccinimide [ORG CHEM] $C_4H_4BrNO_2$ Orthorhombic bisphenoidal crystals with a melting point of 173–175°C; used in the bromination of olefins.   { ˌen ˈbrō·mōˌsəkˈsinˌəˌmīd }

bromotrifluoroethylene [ORG CHEM] $BrFC:CF_2$ A colorless gas with a freezing point of −168°C and a boiling point of −58°C; soluble in chloroform; used as a refrigerant, in hardening of metals, and as a low-toxicity fire extinguisher. Abbreviated BFE.   { ˌbrō·mōˌtrīˌflürˈōˈethˌəˌlēn }

bromotrifluoromethane [ORG CHEM] $CBrF_3$ Fluorine compound that has a molecular weight of 148.93, melting point −180°C, boiling point −59°C; used as a fire-extinguishing agent.   { ˌbrō·mōˌtrīˌflürˈōˈmeˌthān }

bromouracil [BIOCHEM] $C_4H_3N_2O_2Br$ 5-Bromouracil, an analog of thymine that can react with deoxyribonucleic acid to produce a polymer with increased susceptibility to mutation.   { ˌbrō·mōˈyůr·əˌsəl }

Bromoviridae [VIROL] A family of ribonucleic acid (RNA)-containing plant viruses that are characterized by nonenveloped virions with three segments of linear, positive-sense, single-stranded RNA. It includes the genera *Bromovirus, Cucumovirus, Ilarvirus, Alfamovirus,* and *Oleavirus.*   { ˌbrō·məˈvirˌəˌdī }

Bromovirus [VIROL] A genus of plant viruses in the family Bromoviridae characterized by icosahedral particles with genomes consisting of four species of single-stranded ribonucleic acid separately encapsidated.   Also known as Brome mosaic virus group.   { ˈbrō·məˌvī·rəs }

α-bromo-*meta*-xylene [ORG CHEM] $CH_3C_6H_4CH_2Br$ A liquid that is a powerful lacrimator; soluble in alcohol and ether; used in organic synthesis and chemical warfare.   { ˌalˈfə ˈbrō·mō ˈmedˈə ˈzīˌlēn }

bromoxynil [ORG CHEM] $C_7H_3OBr_2N$ A colorless solid with a melting point of 194–195°C; slightly soluble in water; used as a herbicide in wheat, barley, oats, rye, and seeded turf.   { brōˈmäkˈsəˌnil }

bromoxynil octanoate [ORG CHEM] $C_{15}H_{17}Br_2NO_2$ A pale brown liquid, insoluble in water; melting point is 45–46°C; used to control broadleaf weeds.   { ˌbrōˈmäkˈsəˌnil ˌäkˈtanˌəˌwāt }

bromthymol blue [ORG CHEM] An acid-base indicator in the pH range 6.0 to 7.6; color change is yellow to blue.   { ˌbrōmˈthīˌmȯl ˈblü }

bromuration [HISTOL] A process in which a tissue section is treated with a solution of bromine or a bromine compound.   { ˌbräm·yəˈrā·shən }

Bromwich contour [MATH] A path of integration in the complex plane running from $c - i\infty$ to $c + i\infty$, where c is a real, positive number chosen so that the path lies to the right of all singularities of the analytic function under consideration.   { ˈbrämˌwich ˌkänˌtůr }

bromyrite [MINERAL] AgBr A secondary ore of silver that occurs in the oxidized zone of silver deposits; exists in crusts and coatings resembling a wax.   { ˈbrō·məˌrīt }

bronchial adenoma [MED] A low-grade malignant or potentially malignant tumor of bronchi.   { ˈbräŋ·kē·əl ˌadˌənˈōˌmə }

bronchial asthma [MED] Asthma usually due to hypersensitivity to an inhaled or ingested allergen.   { ˈbräŋ·kē·əl ˈazˌmə }

bronchial tree [ANAT] The arborization of the bronchi of the lung, considered as a structural and functional unit.   { ˈbräŋ·kē·əl ˈtrē }

bronchiectasis [MED] Dilation of the bronchi and bronchioles following a chronic inflammatory process or an infection attended by pus formation.   { ˌbräŋ·kēˈek·təˌsəs }

bronchiolar carcinoma [MED] Adenocarcinoma of the lung characterized by mucus-producing cells which spread over the alveoli.   { ˌbräŋ·kēˈō·lər ˌkärsˈənˈōˌmə }

bronchiole [ANAT] A small, thin-walled branch of a bronchus, usually terminating in alveoli.   { ˈbräŋ·kēˌōl }

bronchiolitis [MED] Inflammation of the bronchioles.   { ˌbräŋ·kēˈōˈlīdˌəs }

bronchiolitis obliterans [MED] Inflammation of the bronchioles with the formation of an exudate and fibrous tissue that obliterate the lumen.   { ˌbräŋ·kēˈōˈlīdˌəs ōˈblidˌəˌranz }

bronchitis [MED] An inflammation of the bronchial tubes.   { bräŋˈkīdˌəs }

bronchoconstriction [PHYSIO] Narrowing of the air passages in bronchi and bronchioles.   { ˌbräŋ·kō·kənˌstrikˈshən }

bronchoconstrictor [PHARM] Any agent that causes a narrowing of the air passages in bronchi and bronchioles.   { ˌbräŋ·kō·kənˌstrikˌtər }

bronchodilation [PHYSIO] Widening of the air passages in bronchi and bronchioles.   { ˌbräŋ·kōˌdīˈlāˌshən }

bronchodilator [MED] An instrument used to increase the caliber of the pulmonary air passages.   [PHARM] Any agent that causes a widening of the air passages in bronchi and bronchioles.   { ˌbräŋ·kōˌdīˈlādˌər }

bronchoesophagology [MED] The specialty concerned with endoscopic examination through the mouth of the esophagus and tracheobronchial tree.   { ˌbräŋ·kō·ēˌsäfˌəˈgälˌəˌjē }

bronchofiberscope [MED] A fiber-optic endoscope adapted for viewing the trachea and bronchi.   { ˌbräŋ·kəˈfīˌbərˌskōp }

bronchogram [MED] Radiography of the bronchial tree made after the introduction of a radiopaque substance.   { ˈbräŋ·kəˌgram }

bronchography [MED] Roentgenographic visualization of the bronchial tree following injection of a radiopaque material.   { ˌbräŋˈkägˌrəˌfē }

bronchopneumonia [MED] Inflammation of the lungs which has spread from infected bronchi.   Also known as lobular pneumonia.   { ˌbräŋ·kō·nəˈmōˌnyə }

bronchorrhea [MED] Excessive discharge of mucus from the bronchial mucous membranes.   { ˌbräŋ·kəˈrēˌə }

bronchoscope [MED] An instrument for the visual examination of the interior of the bronchi.   { ˈbräŋ·kəˌskōp }

bronchospasm [MED] Temporary narrowing of the bronchi due to violent, involuntary contraction of the smooth muscle of the bronchi.   { ˈbräŋ·kōˌspazˌəm }

bronchospirometry [MED] The determination of various aspects of the functional capacity of a single lung or lung segment.   { ˌbräŋ·kō·spəˈrämˌəˌtrē }

bronchus [ANAT] Either of the two primary branches of the trachea or any of the bronchi's pulmonary branches having cartilage in their walls.   { ˈbräŋˌkəs }

Brönner's acid [ORG CHEM] $C_{10}H_6(NH_2)SO_3H$ A colorless, water-soluble naphthylamine sulfonic acid that forms needle crystals; used in dyes.   Also spelled Broenner's acid.   { ˈbrenˌərz ˈasˌəd }

Brønsted acid [CHEM] A chemical species which can act as a source of protons.   Also known as proton acid; protonic acid.   { ˈbrənˌsteth *or* ˈbrenˌsted ˈasˌəd }

Brønsted base See base.   { ˈbrənˌsteth ˌbās }

Brønsted-Lowry theory [CHEM] A theory that all acid-base reactions consist simply of the transfer of a proton from one base to another.   Also known as Brønsted theory.   { ˈbrənˌsteth ˌlauˌrē ˌthēˌəˌrē }

Brønsted theory See Brønsted-Lowry theory.   { ˈbrənˌsteth ˌthēˌəˌrē }

brontides [GEOPHYS] Low, rumbling, thunderlike sounds of short duration, most frequently heard in active seismic regions and believed to be of seismic origin.   { ˈbränˌtīdz }

brontophobia [PSYCH] An abnormal fear of thunder.   { ˌbränˌtəˈfōˌbēˌə }

Brontosaurus See Apatosaurus.   { ˌbränˌtəˈsȯrˌəs }

Brontotheriidae [PALEON] The single family of the extinct mammalian superfamily Brontotherioidea.   { ˌbränˌtōˈthəˈrīˌəˌdē }

Brontotherioidea [PALEON] The titanotheres, a superfamily of large, extinct perissodactyl mammals in the suborder Hippomorpha.   { ˌbränˌtōˌtheˈrēˈȯidˌēˌə }

bronze [MET] An alloy of copper and tin in varying proportions; other elements such as zinc, nickel, and lead may be added.   { bränz }

bronze mica See phlogopite.   { ˌbränz ˈmīˌkə }

bronzing [GRAPHICS] A process in which bronze powder is applied to a surface that has been printed with a sizing ink while the ink is still wet in order to produce a metallic sheen.   { ˈbranˌziŋ }

bronzing liquid [MATER] A solvent, gloss oil, or varnish containing a bronze powder; used to produce bronze-colored finishes.   { ˈbranˌziŋ ˌlikˌwəd }

bronzite [MINERAL] (Mg,Fe)(SiO$_3$) An orthopyroxene mineral that forms metallic green orthorhombic crystals; a form of the enstatite-hypersthene series. { 'brän,zīt }

bronzitfels See bronzitite. { 'brän·zət,felz }

bronzitite [PETR] A pyroxenite that is composed almost entirely of bronzite. Also known as bronzitfels. { 'brän·zə,tīt }

brood [BOT] Heavily infested by insects. [ZOO] **1.** The young of animals. **2.** To incubate eggs or cover the young for warmth. **3.** An animal kept for breeding. { brüd }

brood capsule [INV ZOO] A secondary scolex-containing cyst constituting the infective agent of a tapeworm. { 'brüd ,kap·səl }

brood parasitism [ECOL] A type of social parasitism among birds characterized by a bird of one species laying and abandoning its eggs in the nest of a bird of another species. { 'brüd ,par·ə·sə,tiz·əm }

brood pouch [VERT ZOO] A pouch of an animal body where eggs or embryos undergo certain stages of development. { 'brüd ,pauch }

brookite [MINERAL] TiO₂ A brown, reddish, or black orthorhombic mineral; it is trimorphous with rutile and anatase, has hardness of 5.5–6 on Mohs scale, and a specific gravity of 3.87–4.08. Also known as pyromelane. { 'brü,kīt }

Brooks variable inductometer [ELEC] An inductometer providing a nearly linear scale and consisting of two movable coils, side by side in a plane, sandwiched between two pairs of fixed coils. { brüks 'ver·ē·ə·bəl ,in,dək'täm·əd·ər }

brooming [CIV ENG] A method of finishing uniform concrete surfaces, such as the tops of pavement slabs or floor slabs, by dragging a broom over the surface to produce a grooved texture. { 'brü·miŋ }

broomy flow [FL MECH] A swirling flow of a fluid in a pipe after passing through a constricted section or after a sudden change of direction. { 'brü·mē ,flō }

Brotulidae [VERT ZOO] A family of benthic teleosts in the order Perciformes. { brō'tü·lə,dē }

Brouwer's theorem [MATH] A fixed-point theorem stating that for any continuous mapping *f* of the solid *n*-sphere into itself there is a point *x* such that $f(x) = x$. { 'brau·ərz ,thir·əm }

brown acid [CHEM ENG] Oil-soluble petroleum sulfonate found in sludge following sulfuric acid treatment of petroleum products. { ¦braun ¦as·əd }

brown algae [BOT] The common name for members of the Phaeophyta. { ¦braun ¦al·jē }

Brown and Sharpe gage See American wire gage. { ¦braun ən ¦shärp ,gāj }

brown bast [PL PATH] A physiological disease of the para rubber tree characterized by grayish- or greenish-brown discolorations of the inner bark near the tapping cut. { ¦braun ¦bast }

brown blight [PL PATH] A virus disease of lettuce characterized by spots and streaks on the leaves, reduction in leaf size, and gradual browning of the foliage, beginning at the base. { ¦braun ¦blīt }

brown blotch [PL PATH] **1.** A bacterial disease of mushrooms caused by *Pseudomonas tolaasi* and characterized by brown blotchy discolorations. **2.** A fungus disease of the pear characterized by brown blotches on the fruit. { ¦braun ¦bläch }

brown canker [PL PATH] A fungus disease of roses caused by *Cryptosporella embrina* and characterized by lesions that are initially purple and gradually become buff. { ¦braun ¦kaŋ·kər }

brown clay See red clay. { ¦braun ¦klā }

brown clay ironstone [GEOL] Limonite in the form of concrete masses, often in concretionary nodules. { ¦braun ¦klā ¦ī·ərn,stōn }

brown coal See lignite. { ¦braun ¦kōl }

brown coat [MATER] Mortar with about 1 to 11/2 bushels (35 to 53 liters) of hair per 200 pounds (91 kilograms) quicklime; used to make a brown coat of plaster, which is covered with a finish coat, and often covers a scratch coat. { ¦braun ¦kōt }

brown dwarf [ASTRON] A starlike body whose mass is too small (less than about 8% that of the sun) to sustain nuclear reactions in its core. Also known as black dwarf; failed star; infrared dwarf; lilliputian star; substellar object; super-Jupiter. { ¦braun ¦dwórf }

brown fat cell [HISTOL] A moderately large, generally

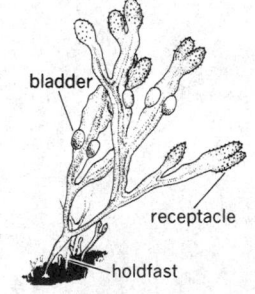

BROWN ALGAE

labels: bladder, receptacle, holdfast

Fucus, a brown alga. (From H. J. Fuller and O. Tippo, College Botany, rev. ed., Holt, 1954)

spherical cell in adipose tissue that has small fat droplets scattered in the cytoplasm. { braun 'fat ,sel }

brown felt blight [PL PATH] A fungus disease of conifers caused by several Ascomycetes, especially *Herpotrichia nigra* and *Neopeckia coulteri;* a dense felty growth of brown or black mycelia forms on the branches. { braun ,felt 'blīt }

brown hematite See limonite. { braun 'hem·ə,tīt }

Brownian movement [STAT MECH] Random movements of small particles suspended in a fluid, caused by the statistical pressure fluctuations over the particle. { braun·ē·ən ,müv·mənt }

brown induration [MED] A pathologic condition marked by acute pulmonary congestion and edema with leakage of blood into the alveoli. { braun in·də'rā·shən }

browning [PL PATH] Any plant disorder or disease marked by brown discoloration of a part. Also known as stem break. { 'brau·niŋ }

Browning automatic rifle [ORD] A gas-operated weapon whose cyclical rate of fire is adjustable but which is capable of firing 200–350 rounds per minute. { 'brau·niŋ ód·ə'mad·ik 'rī·fəl }

Browning machine gun [ORD] Any of certain caliber-.30 and caliber-.50 machine guns designed by, or modified from designs by, John M. Browning. { 'brau·niŋ mə'shēn ,gən }

brown iron ore See limonite. { braun 'ī·ərn ,ór }

brown lead oxide See lead dioxide. { braun 'led 'äk,sīd }

brown leaf rust [PL PATH] A fungus disease of rye caused by *Puccinia dispersa.* { braun 'lēf ,rəst }

brown lignite [GEOL] A type of lignite with a fixed carbon content ranging from 30 to 55% and total carbon from 65 to 73.6; contains 6300 Btu per pound (14.65 megajoules per kilogram). Also known as lignite B. { braun 'lig,nīt }

brownline [GRAPHICS] A print with a white background and brown lines made by contact-printing a negative on sensitized paper. { 'braun,līn }

brown lung disease See byssinosis. { ,braun 'ləŋ di,zēz }

brown mica See phlogopite. { braun 'mī·kə }

brownout [ELEC] **1.** A restriction of electrical power usage during a power shortage, especially for advertising and display purposes. **2.** An extinguishing of some of the lights in a city as a defensive measure against enemy bombardment. { 'braun,aut }

brown patch [PL PATH] A fungus disease of grasses in golf greens and lawns caused by various soil-inhabiting species, typically producing brown circular areas surrounded by a band of grayish-black mycelia. { braun ,pach }

brown petroleum [MATER] A solid or semisolid product formed by air acting on asphalt. { braun pə'trō·lē·əm }

brownprint [GRAPHICS] A print with a brown background and white lines made by contact-printing a negative on a sensitized paper. { 'braun,print }

brown-ring test [ANALY CHEM] A common qualitative test for the nitrate ion; a brown ring forms at the juncture of a dilute ferrous sulfate solution layered on top of concentrated sulfuric acid if the upper layer contains nitrate ion. { 'braun ,riŋ ,test }

brown root [PL PATH] A fungus disease of numerous tropical plants, such as coconut and rubber, caused by *Hymenochaete noxia* and characterized by defoliation and by incrustation of the roots with earth and stones held together by brown mycelia. { braun ,rüt }

brown root rot [PL PATH] **1.** A fungus disease of plants of the pea, cucumber, and potato families caused by *Thielavia basicola* and characterized by blackish discoloration and decay of the roots and stem base. **2.** A disease of tobacco and other plants comparable to the fungus disease but believed to be caused by nematodes. { braun 'rüt ,rät }

brown rot [PL PATH] Any fungus or bacterial plant disease characterized by browning and tissue decay. { braun 'rät }

brown seaweed [BOT] A common name for the larger algae of the division Phaeophyta. { braun 'sē,wēd }

brown smoke [ENG] Smoke with less particulates than black smoke; comes from burning fossil fuel, usually fuel oil. { braun 'smōk }

brown snow [METEOROL] Snow intermixed with dust particles. { braun 'snō }

brown soil [GEOL] Any of a zonal group of soils, with a brown surface horizon which grades into a lighter-colored soil and then into a layer of carbonate accumulation. { 'braun ˌsȯil }

brown spar [GEOL] Any light-colored crystalline carbonate that contains iron, such as ankerite or dolomite, and is therefore brown. { 'braun ˌspär }

brown spot [PL PATH] Any fungus disease of plants, especially Indian corn, characterized by brown leaf spots. { 'braun ˌspät }

brown stem rot [PL PATH] A fungus disease of soybeans caused by *Cephalosporium gregatum* in which there is brownish internal stem rot followed by discoloration and withering of leaves. { 'braun 'stem ˌrät }

brownstone [PETR] Ferruginous sandstone with its grains coated with iron oxide. { 'braunˌstōn }

brown stringy rot [PL PATH] A disease of conifers caused by the Indian point fungus; rusty or brown fibrous streaks appear in the heartwood. { 'braun 'striŋ·ē ˌrät }

browse [BIOL] **1.** Twigs, shoots, and leaves eaten by livestock and other grazing animals. **2.** To feed on this vegetation. [COMPUT SCI] Any method for looking at information stored in a computer in a random manner, usually using a cathode-ray tube or other electronic display. [GRAPHICS] Rapid examination by means of a reader screen of the individual frames of a multi-image microform; used in searching for or processing a specific image. { 'brauz }

browse mode [COMPUT SCI] A mode of operation in which data in a document or database are conveniently displayed for rapid, on-screen review. { 'brauz ˌmōd }

browser [COMPUT SCI] An interactive program (client) that requests, retrieves, and displays pages from the World Wide Web. { 'brauz·ər }

broxyquinoline [ORG CHEM] $C_9H_5Br_2NO$ Crystals with a melting point of 196°C; soluble in acetic acid, chloroform, benzene, and alcohol; used as a reagent for copper, iron, and other metals. { ˌbräk·si'kwin·ə,lēn }

brubru [METEOROL] A squall in Indonesia. { 'brü,brü }

Brucella [MICROBIO] A genus of gram-negative, aerobic bacteria of uncertain affiliation; single, nonmotile coccobacilli or short rods, all of which are parasites and pathogens of mammals. { brü'sel·ə }

Brucellaceae [MICROBIO] Formerly a family of small, coccoid to rod-shaped, gram-negative bacteria in the order Eubacteriales. { ˌbrü·sə'lās·ē,ē }

brucellergen [BIOCHEM] A nucleoprotein fraction of brucellae used in skin tests to detect the presence of *Brucella* infections. { ˌbrü'sel·ər,jen }

brucellergen test [IMMUNOL] A diagnostic skin test for detection of *Brucella* infections. { ˌbrü'sel·ər·jen ˌtest }

brucellosis [MED] An infectious bacterial disease of humans caused by *Brucella* species acquired by contact with diseased animals. Also known as Malta fever; Mediterranean fever; undulant fever. [VET MED] *See* contagious abortion. { ˌbrü·sə'lō·səs }

Bruchidae [INV ZOO] The pea and bean weevils, a family of coleopteran insects in the superfamily Chrysomeloidea. { 'brü·kə,dē }

Bruch's membrane [ANAT] The membrane of the retina that separates the pigmented layer of the retina from the choroid coat of the eye. { 'brüks ˌmem,brān }

brucine [ORG CHEM] $C_{23}H_{26}N_2O_4$ A poisonous alkaloid from the seeds of plant species such as *Nux vomica;* used in alcohol as a denaturant. { 'brü,sīn }

brucite [MINERAL] $Mg(OH)_2$ A hexagonal mineral; native magnesium hydroxide that appears gray and occurs in serpentines and impure limestones; hardness is 2.5 on Mohs scale, and specific gravity is 2.38-2.40. { 'brü,sīt }

Brücke-Abney phenomenon [PHYSIO] The inability to distinguish colors other than blue-violet, green, and red at very low light levels. { 'brük·ə 'ab·nē fə,näm·ə,nän }

Brückner cycle [CLIMATOL] An alternation of relatively cool-damp and warm-dry periods, forming an apparent cycle of about 35 years. { 'brük·nər ,sī·kəl }

brugnatellite [MINERAL] $Mg_6Fe(OH)_{13}CO_3·4H_2O$ A flesh pink to yellowish- or brownish-white, hexagonal mineral consisting of a hydrated carbonate-hydroxide of magnesium and ferric iron; occurs in massive form. { ˌbrü·nyə'te,līt }

bruma [METEOROL] A haze that appears in the afternoons on the coast of Chile when sea air is transported inland. { 'brü·mə }

Brunauer-Emmett-Teller equation [PHYS CHEM] An extension of the Langmuir isotherm equation in the study of sorption; used for surface area determinations by computing the monolayer area. Abbreviated BET equation. { 'brü,naur 'em·ət ˌtel·ər i'kwā·zhən }

Brunner's glands [ANAT] Simple, branched, tubular mucus-secreting glands in the submucosa of the duodenum in mammals. Also known as duodenal glands; glands of Brunner. { 'brən·ərz ,glanz }

Brunt-Douglas isallobaric wind *See* isallobaric wind. { 'brənt ˌdȯg·ləs ˌī,sa·lə,bär·ik 'wind }

Brunton *See* Brunton compass. { 'brənt·ən }

Brunton compass [ENG] A compact field compass, with sights and reflector attached, used for geological mapping and surveying. Also known as Brunton; Brunton pocket transit. { 'brənt·ən ,käm·pəs }

Brunton pocket transit *See* Brunton compass. { 'brənt·ən ,päk·ət 'tran·zət }

brüscha [METEOROL] A northwest wind in the Bergell Valley, Switzerland. { 'brüsh·ə }

brush *See* tropical scrub. [ELEC] A conductive metal or carbon block used to make sliding electrical contact with a moving part. { brəsh }

brush border [CYTOL] A superficial protoplasic modification in the form of filiform processes or microvilli; present on certain absorptive cells in the intestinal epithelium and the proximal convolutions of nephrons. { 'brəsh ,bȯr·dər }

brush discharge [ELEC] A luminous electric discharge that starts from a conductor when its potential exceeds a certain value but remains too low for the formation of an actual spark. { 'brəsh ,dis,chärj }

brush encoder [ELECTR] An encoder in which brushes that make contact with conductive segments on a rotating or linearly moving surface convert positional information to digitally encoded data. { 'brəsh en'kōd·ər }

brush fire [FOR] A fire involving growth that is heavier than grass but less than full tree size. { 'brəsh ,fīr }

brush holder [ELEC] A structure in which a brush can slide in a direction perpendicular to the moving surface of a motor, generator, or other device. { 'brəsh ,hōl·dər }

brush hopper [IND ENG] A rotating brush that wipes quantities of eyelets, rivets, and other small special parts past shaped openings in a chute. { 'brəsh ,häp·ər }

brushing [TEXT] A process for finishing fabrics by rubbing them against a rough surface, such as a wire-covered cylinder, usually to produce a thick nap. Also known as napping. { 'brəsh·iŋ }

brushing shot [MIN ENG] A charge fired in the air of a mine to blow out obnoxious gases or to start an air current. { 'brəsh·iŋ ,shät }

brushite [MINERAL] $CaHPO_4·2H_2O$ A nearly colorless mineral that is a constituent of rock phosphates that crystallizes in slender or massive crystals. { 'brə,shīt }

brush lag [ELEC] The distance that the brushes on a motor are displaced in a direction opposite to the motor's rotation in order to overcome the effect of armature reaction. { 'brəsh ,lag }

brush lead [ELEC] The distance that the brushes on a generator are displaced in the direction of the motor's rotation in order to overcome the effect of armature reaction. { 'brəsh ,lēd }

brush machine [FOOD ENG] A machine used in rice processing to rub the grains smooth; consists of a vertical cylindrical frame covered with soft leather strips revolving at high speed within a cylinder of wire screening. { 'brəsh mə'shēn }

brush plating [MET] Electroplating in which the anode with the solution is in the form of a brush or pad; used for plating equipment too large to be immersed. { 'brəsh ,plād·iŋ }

brush rake [MECH ENG] A device with heavy-duty tines that is fixed to the front of a tractor or other prime mover for use in land clearing. { 'brəsh ,rāk }

brush rocker [ELEC] A yoke to which the brush holders in an electrical machine are attached, and which can be moved to adjust the positions of the brushes. Also known as brush rocker ring. { 'brəsh ,rä·kər }

brush rocker ring *See* brush rocker. { 'brəsh ,rä·kər ,riŋ }

brush-shifting motor [ENG] A category of alternating-current motor in which the brush contacts shift to modify operating speed and power factor. { 'brəsh ,shif·tiŋ ,mōd·ər }

brush station [COMPUT SCI] A location in a device where

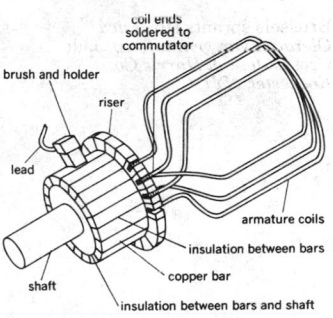

BRUSH

coil ends soldered to commutator

brush and holder

riser

lead

armature coils

insulation between bars

copper bar

shaft

insulation between bars and shaft

Commutator and brush assembly, showing stationary brush that makes sliding contact with rotating commutator.

BRUSSELLS SPROUTS

Brussels sprouts (*Brassica Oleracea* var. *gemmifera*), Jade Cross. (*Joseph Harris Co., Rochester, NY*)

BUBBLE CAP

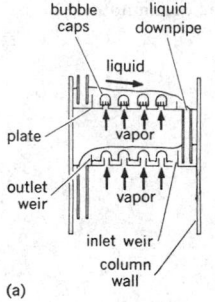

(a)

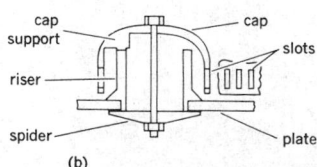

(b)

Schematic of *(a)* cross section through bubble-cap-plate column and *(b)* typical bubble cap.

the holes in a punched card are sensed by brushes sweeping electrical contacts. { 'brəsh ,stā·shən }

brussels sprouts [BOT] *Brassica oleracea* var. *gemmifera*. A biennial crucifer of the order Capparales cultivated for its small, edible, headlike buds. { |brəs·əlz 'spraúts }

brute force attack [COMPUT SCI] An attempt to gain unauthorized access to a computing system by generating and trying all possible passwords. { |brüt |fòrs ə'tak }

brute-force filter [ELEC] Type of powerpack filter depending on large values of capacitance and inductance to smooth out pulsations rather than on resonant effects of tuned filters. { |brüt ,fòrs 'fil·tər }

brute-force technique [COMPUT SCI] Any method that relies chiefly on the advanced processing capabilities of a large computer to accomplish a task. { |brüt ,fòrs tek'nēk }

brute supply [ELEC] A type of power supply that is completely unregulated, employing no circuitry to maintain output voltage constant with changing input line or load variations. { |brüt sə'plī }

Bruton's disease [IMMUNOL] A hereditary type of agammaglobulinemia that is a sex-linked recessive disorder characterized by a deficiency of all types of immunoglobulins, reflecting a failure of the entire humoral antibody marrow system; the thymus may be normal, but the lymph nodes and spleen lack lymph cell follicles. { 'brüt·ənz di,zēz }

Bruxellian [GEOL] Lower middle Eocene geologic time. { brü'sel·yən }

bruxism [MED] A clenching and grinding of the teeth that occurs unconsciously while the individual is awake or sleeping. { 'brək,siz·əm }

Bryales [BOT] An order of the subclass Bryidae; consists of mosses which often grow in disturbed places. { brī'ā·lēz }

Bryatae See Bryopsida. { 'brī·ə,tē }

Bryidae [BOT] A subclass of the class Bryopsida; includes most genera of the true mosses. { 'brī·ə,dē }

bryology [BOT] The study of bryophytes. { brī'äl·ə·jē }

Bryophyta [BOT] A small phylum of the plant kingdom, including mosses, liverworts, and hornworts, characterized by the lack of true roots, stems, and leaves. { brī'ä·fə·də }

Bryopsida [BOT] The mosses, a class of small green plants in the phylum Bryophyta. Also known as Musci. { brī'äp·sə·də }

Bryopsidaceae [BOT] A family of green algae in the order Siphonales. { brī,äp·sə'dās·ē,ē }

Bryopsidales See Siphonales. { brī,äp·sə'dā,lēz }

bryostatin-1 [PHARM] A polyketide isolated from the bryozoan *Bugula neritina* that has both anticancer and immune modulating activity. { |brī·ə,stat·ən 'wən }

Bryoxiphiales [BOT] An order of the class Bryopsida in the subclass Bryidae; consists of a single genus and species, *Bryoxiphium norvegicum*, the sword moss. { brī,äk·sə·fē'ā·lēz }

Bryozoa [INV ZOO] The moss animals, a major phylum of sessile aquatic invertebrates occurring in colonies with hardened exoskeleton. { ,brī·ə'zō·ə }

BS&W See basic sediment and water.

B scan See B scope. { 'bē ,skan }

B scope [ELECTR] A cathode-ray scope on which signals appear as spots, with bearing angle as the horizontal coordinate and range as the vertical coordinate. Also known as B indicator; B scan. { 'bē ,skōp }

BSD See barrels per stream day.

B size [ENG] **1.** One of a series of sizes to which trimmed paper and board are manufactured; for size *BN*, with *N* equal to any integer from 0 to 10, the length of the shorter side is $2^{-N/2}$ meters, and the length of the longer side is $2^{(1-N)/2}$ meters, with both lengths rounded off to the nearest millimeter. **2.** Of a sheet of paper, the dimensions 11 inches by 17 inches (279 millimeters by 432 millimeters). { 'bē ,sīz }

BSPL See band sound pressure level.

b-spline [COMPUT SCI] A curve that is generated by a computer-graphics program, guided by a mathematical formula which ensures that it will be continuous with other such curves; it is mathematically more complex but easier to blend than a Bézier curve. { 'bē,splīn }

BSR See bulk shielding reactor.

B stage [ORG CHEM] An intermediate stage in a thermosetting resin reaction in which the plastic softens but does not fuse when heated, and swells but does not dissolve in contact with certain liquids. { 'bē ,stāj }

B star See B-type star. { 'bē ,stär }

B station [NAV] In loran, the designation applied to one transmitting station of a pair, the signal of which always occurs more than half a repetition period after the succeeding signal and less than half a repetition period before the preceding signal from the other station of the pair, designated an A station. { 'bē ,stā·shən }

B store See index register. { 'bē ,stòr }

B supply [ELECTR] Anode high voltage and screen-grid power source in vacuum tube circuits. Also known as B power supply. { 'bē sə'plī }

BT See bathythermograph.

B Tauri [ASTRON] A daytime meteor shower that occurs at the end of June and has its radiant near the star. { 'bē 'tòr·ē }

B tectonite [PETR] Tectonite with a fabric dominated by linear elements indicating an axial direction rather than a slip surface. { |bē 'tek·tə,nīt }

BTO See bombing through overcast.

B trace [ELECTR] In loran the second trace of an oscilloscope which corresponds to the signal from the B station. { 'bē ,trās }

B-tree See balanced-tree. { 'bē ,trē }

B+-tree [COMPUT SCI] A version of the balanced-tree that maintains a hierarchy of indexes while linking the data sequentially. { |bē 'pləs ,trē }

Btu See British thermal unit.

B-type star [ASTRON] A type in a classification based on stellar spectral characteristics; has strong HeI absorption. Also known as B star. { 'bē ,tīp ,stär }

bu See bushel.

bubble [COMPUT SCI] A circle that represents data in a data flow diagram. [METEOROL] *See* bubble high. [PHYS] **1.** A small, approximately spherical body of fluid within another fluid or solid. **2.** A thin, approximately spherical film of liquid inflated with air or other gas. [SOLID STATE] *See* magnetic bubble. { 'bəb·əl }

bubble cap [CHEM ENG] A metal cap covering a hole in the plate within a distillation tower; designed to permit vapors to rise from below the plate, pass through the cap, and make contact with liquid on the plate. { 'bəb·əl ,kap }

bubble-cap plate [CHEM ENG] One of the devices in large-diameter fractional distillation columns that are designed to produce a bubbling action to exchange the vapor bubbles flowing up the column. { 'bəb·əl ,kap ,plāt }

bubble-cap tray See bubble tray. { 'bəb·əl ,kap ,trā }

bubble cavitation [FL MECH] **1.** Formation of vapor- or gas-filled cavities in liquids by mechanical forces. **2.** The formation of vapor-filled cavities in the interior of liquids in motion when the pressure is reduced without change in ambient temperature. { 'bəb·əl kav·ə'tā·shən }

bubble chamber [NUCLEO] A chamber in which the movements and interactions of charged particles can be observed as visible tracks in a superheated liquid, the tracks being gas bubbles that form along the paths of the moving particles. { 'bəb·əl ,chām·bər }

bubble chart See data flow diagram. { 'bəb·əl ,chärt }

bubble high [METEOROL] A small high, complete with anticyclonic circulation, of the order of 50 to 300 miles (80 to 480 kilometers) across, often induced by precipitation and vertical currents associated with thunderstorms. Also known as bubble. { 'bəb·əl ,hī }

bubble horizon [NAV] An artificial horizon parallel to the celestial horizon, established by means of a bubble level on a bubble sextant. { 'bəb·əl hə'rīz·ən }

bubble-jet printer [GRAPHICS] A nonimpact printer that uses heating elements to shoot ink from a matrix of tiny nozzles. { ,bəb·əl jet 'print·ər }

bubble memory [COMPUT SCI] A computer memory in which the presence or absence of a magnetic bubble in a localized region of a thin magnetic film designates a 1 or 0; storage capacity can be well over 1 megabit per cubic inch. Also known as magnetic bubble memory. { 'bəb·əl ,mem·rē }

bubble mold cooling [ENG] In plastics injection molding, cooling by means of a continuous liquid stream flowing into a cavity equipped with an outlet at the end opposite the inlet. { 'bəb·əl ,mōld ,kü·liŋ }

bubble point [PHYS CHEM] In a solution of two or more components, the temperature at which the first bubbles of gas appear. Also known as boiling point. { 'bəb·əl ,pòint }

bubble-point reservoir *See* dissolved-gas-drive reservoir. { 'bəb·əl ˌpȯint 'rez·əvˌwär }

bubble pulse [GEOPHYS] An extraneous effect during a seismic survey caused by a bubble formed by a seismic charge, explosion, or spark fired in a body of water. { 'bəb·əl ˌpəls }

bubble raft [SOLID STATE] A visual demonstration for the structure of dislocations in metal lattices, showing slip propagation; it consists of many identical bubbles floating on a liquid surface in something like a crystalline array. { 'bəb·əl ˌraft }

bubble sextant [NAV] A sextant with a bubble or spirit level to indicate the horizontal. { 'bəb·əl 'sek·stənt }

bubble sort [COMPUT SCI] A procedure for sorting a set of items that begins by sequencing the first and second items, then the second and third, and so on, until the end of the set is reached, and then repeats this process until all items are correctly sequenced. { 'bəb·əl ˌsȯrt }

bubble test [ENG] Measurement of the largest opening in the mesh of a filter screen; determined by the pressure needed to force air or gas through the screen while it is submerged in a liquid. { 'bəb·əl ˌtest }

bubble tower [CHEM ENG] A plate tower used in distillation, with plates containing bubble caps. { 'bəb·əl ˌtau̇·ər }

bubble train [GEOL] A string or strings of vesicles in lava, indicating the path of rising gas escaping a flow of lava. { 'bəb·əl ˌtrān }

bubble tray [CHEM ENG] A perforated, circular plate placed within a distillation tower at specific places to collect the fractions of petroleum produced in fractional distillation. Also known as bubble-cap tray. { 'bəb·əl ˌtrā }

bubble-tray column [CHEM ENG] A fractionating column whose plates are formed from bubble caps. { 'bəb·əl ˌtrā ˌkäl·əm }

bubble tube [ENG] The glass tube in a spirit level containing the liquid and bubble. { 'bəb·əl ˌtüb }

bubble wall fragment [GEOL] A glassy volcanic shard revealing part of a vesicle surface which may be curved or flat. { 'bəb·əl ˌwȯl ˌfrag·mənt }

bubo [MED] An inflammatory enlargement of lymph nodes, usually of the groin or axilla; commonly associated with chancroid, lymphogranuloma venereum, and plague. { 'bü,bō }

bubonic plague *See* plague. { bü'bän·ik 'plāg }

bubulum oil *See* neatsfoot oil. { bə'bü·ləm ˌȯil }

bucaramangite [MINERAL] A pale yellow variety of retinite that looks like amber but is insoluble in alcohol. { ˌbyü·kə·rə'man,gīt }

buccal cavity [ANAT] The space anterior to the teeth and gums in the mouths of all vertebrates having lips and cheeks. Also known as vestibule. { 'bək·əl ˌkav·əd·ē }

buccal gland [ANAT] Any of the mucous glands in the membrane lining the cheeks of mammals, except aquatic forms. { 'bək·əl ˌgland }

Buccinacea [INV ZOO] A superfamily of gastropod mollusks in the order Prosobranchia. { ˌbək·sin'ās·ē·ə }

Buccinidae [INV ZOO] A family of marine gastropod mollusks in the order Neogastropoda containing the whelks in the genus *Buccinum*. { bək'sin·ə,dē }

Bucconidae [VERT ZOO] The puffbirds, a family of neotropical birds in the order Piciformes. { bə'kän·ə,dē }

Bucerotidae [VERT ZOO] The hornbills, a family of Old World tropical birds in the order Coraciiformes. { ˌbyü·sə'räd·ə,dē }

Bucherer reaction [ORG CHEM] A method of preparation of polynuclear primary aromatic amines; for example, α-naphthylamine is obtained by heating β-naphthol in an autoclave with a solution of ammonia and ammonium sulfite. { 'bük·ər·ər rē'ak·shən }

Buchholz protective device [ELEC] A protective relay which is attached to an oil-filled tank containing a transformer and which is activated either by gas produced by faults or by oil surges produced by explosive faults in the transformer. Also known as gas bubble protective device. { 'bük,hȯls prə'tek·tiv di'vīs }

buchite [PETR] A partially vitrified inclusion of sandstone in basalt. { 'bü,kīt }

buchonite [PETR] An extrusive rock formed of labradorite, titanaugite, and titaniferous hornblende, with nepheline and sodic sanidine and accessory biotite, apatite, and opaque oxides. { 'bü·kə,nīt }

buck [BUILD] The frame into which the finished door fits.

[MIN ENG] **1.** To break up or pulverize ore samples. **2.** A large quartz reef in which there is little or no accessory minerals such as gold. Also known as buck quartz; bull quartz. [VERT ZOO] A male deer. { bək }

bucket [BOT] *See* calyx. [COMPUT SCI] A name usually reserved for a storage cell in which data may be accumulated. [ENG] **1.** A cup on the rim of a Pelton wheel against which water impinges. **2.** A reversed curve at the toe of a spillway to deflect the water horizontally and reduce erosiveness. **3.** A container on a lift pump or chain pump. **4.** A container on some bulk-handling equipment, such as a bucket elevator, bucket dredge, or bucket conveyor. **5.** A water outlet in a turbine. *See* calyx. { 'bək·ət }

bucket brigade device [ELECTR] A semiconductor device in which majority carriers store charges that represent information, and minority carriers transfer charges from point to point in sequence. Abbreviated BBD. { 'bək·ət bri'gād di'vīs }

bucket carrier *See* bucket conveyor. { 'bək·ət ˌkar·ē·ər }

bucket conveyor [MECH ENG] A continuous bulk conveyor constructed of a series of buckets attached to one or two strands of chain or in some instances to a belt. Also called bucket carrier. { 'bək·ət kən'vā·ər }

bucket dredge [MECH ENG] A floating mechanical excavator equipped with a bucket elevator. { 'bək·ət ˌdrej }

bucket drill [MIN ENG] An auger stem drill in which the drill bit is replaced by a bit incorporating a steel cylinder to confine the cutting. { 'bək·ət ˌdril }

bucket dumper *See* lander. { 'bək·ət ˌdəm·pər }

bucket elevator [MECH ENG] A bucket conveyor operating on a steep incline or vertical path. Also known as elevating conveyor. { 'bək·ət 'el·əˌvād·ər }

bucket excavator [MECH ENG] An elevating scraper, that is, one that does the work of a conventional scraper but has a bucket elevator mounted in front of the bowl. { 'bək·ət 'ek·skəˌvād·ər }

bucket ladder *See* bucket-ladder dredge. { 'bək·ət ˌlad·ər }

bucket-ladder dredge [MECH ENG] A dredge whose digging mechanism consists of a ladderlike truss on the periphery of which is attached an endless chain riding on sprocket wheels and carrying attached buckets. Also known as bucket ladder; bucket-line dredge; ladder-bucket dredge; ladder dredge. { 'bək·ət ˌlad·ər ˌdrej }

bucket-ladder excavator *See* trench excavator. { 'bək·ət ˌlad·ər 'ek·skə·vād·ər }

bucket-line dredge *See* bucket-ladder dredge. { 'bək·ət ˌlin ˌdrej }

bucket loader [MECH ENG] A form of portable, self-feeding, inclined bucket elevator for loading bulk materials into cars, trucks, or other conveyors. { 'bək·ət ˌlōd·ər }

bucket temperature [ENG] The surface temperature of ocean water as measured by a bucket thermometer. { 'bək·ət 'tem·prə·chər }

bucket thermometer [ENG] A thermometer mounted in a bucket and used to measure the temperature of water drawn into the bucket from the surface of the ocean. { 'bək·ət thər'mäm·əd·ər }

bucket-wheel excavator [MECH ENG] A continuous digging machine used extensively in large-scale stripping and mining. Abbreviated BWE. Also known as rotary excavator. { 'bək·ət ˌwēl 'ek·skəˌvād·ər }

buckeye [BOT] The common name for deciduous trees composing the genus *Aesculus* in the order Sapindales; leaves are opposite and palmately compound, and the seed is large with a firm outer coat. { 'bək,ī }

bucking [MIN ENG] A hand process for crushing ore. { 'bək·iŋ }

bucking coil [ELECTROMAG] A coil connected and positioned in such a way that its magnetic field opposes the magnetic field of another coil; for example, the hum-bucking coil of an excited-field loudspeaker. { 'bək·iŋ ˌkȯil }

Buckingham's equations [MECH ENG] Equations which give the durability of gears and the dynamic loads to which they are subjected in terms of their dimensions, hardness, surface endurance, and composition. { 'bək·iŋ·əmz i'kwā·zhənz }

Buckingham's π theorem [PHYS] The theorem that if there are n physical quantities, $x_1, x_2, \ldots, x_n$, which can be expressed in terms of m fundamental quantities and if there exists one and only one mathematical expression connecting them which

BUCKEYE

Leaf of horse chestnut (*Aesculus hippocastanum*).

remains formally true no matter how the units of the fundamental quantities are changed, namely $\phi(x_1, x_2, \ldots, x_n) = 0$, then the relation ϕ can be expressed by a relation of the form $F(\pi_1, \pi_2, \ldots, \pi_{n-m}) = 0$, where the π's are $n - m$ independent dimensionless products of $x_1, x_2, \ldots, x_n$. Also known as pi theorem. { 'bək·iŋ ·əmz 'pī ,thir·əm }

bucking transformer [ELEC] A transformer whose voltage opposes that of a second transformer. { 'bək·iŋ tranz'fòr·mər }

bucking voltage [ELEC] A voltage having a polarity opposite to that of another voltage against which it acts. { 'bək·iŋ ,vōl·tij }

bucklandite See allanite. { 'bək·lən,dīt }

buckle [GRAPHICS] Curvature in a film that has been stored at an improper humidity. [MET] An up-and-down wrinkle on the surface of a metal bar or sheet. { 'bək·əl }

buckle fold [GEOL] A double flexure of rock beds formed by compression acting in the plane of the folded beds. { 'bək·əl ,fōld }

buckle plate [CIV ENG] A steel floor plate which is slightly arched to increase rigidity. { 'bək·əl ,plāt }

Buckley gage [ENG] A device that measures very low gas pressures by sensing the amount of ionization produced in the gas by a predetermined electric current. { 'bək·lē ,gāj }

buckling [ENG] Wrinkling or warping of fibers in a composite material. [MECH] Bending of a sheet, plate, or column supporting a compressive load. [NUCLEO] The size-shape factor that appears in the general nuclear reactor equation and is a measure of the curvature of the neutron density distribution in the reactor. { 'bək·liŋ }

buckling stress [MECH] Force exerted by the crippling load. { 'bək·liŋ ,stres }

buckminsterfullerene [CHEM] C_{60} The most abundant and most stable of the fullerenes, containing 60 carbon atoms in a highly spherical arrangement; named in honor of R. Buckminster Fuller, a practitioner of geodesic dome architecture. Also known as buckyball. { ¦bək,min·stər'fùl·ə,rēn }

buckstay [MECH ENG] A structural support for a furnace wall. { 'bək,stā }

buckwheat [AGR] A herbaceous and erect annual belonging to the Polygonaceae family; its dry seed or grain is used as a source of food and animal feed. { 'bək,wēt }

buckwheat coal [GEOL] An anthracite coal that passes through 9/16-inch (14-millimeter) holes and over 5/16-inch (8-millimeter) holes in a screen. { 'bək,wēt ,kōl }

buckwheat no. 3 See barley coal. { 'bək,wēt ,nəm·bər 'thrē }

buckyball See buckminsterfullerene. { 'bək·ē,bòl }

Bucky diaphragm See Potter-Bucky grid. { 'bək·ē ¦dī·ə,fram }

bud [BOT] An embryonic shoot containing the growing stem tip surrounded by young leaves or flowers or both and frequently enclosed by bud scales. { bəd }

Budan's theorem [MATH] The theorem that the number of roots of an nth-degree polynomial lying in an open interval equals the difference in the number of sign changes induced by n differentiations at the two ends of the interval. { 'bü,dänz ,thir·əm }

budbreak [BOT] Initiation of growth from a bud. { 'bəd,brāk }

budding [BIOL] A form of asexual reproduction in which a new individual arises as an outgrowth of an older individual. Also known as gemmation. [BOT] A method of vegetative propagation in which a single bud is grafted laterally onto a stock. [VIROL] A form of virus release from the cell in which replication has occurred, common to all enveloped animal viruses; the cell membrane closes around the virus and the particle exits from the cell. { 'bəd·iŋ }

budding bacteria [MICROBIO] Bacteria that reproduce by budding. { 'bəd·iŋ bak'tir·ē·ə }

buddle [MIN ENG] A device for concentrating ore that uses a circular arrangement from which the finely divided ore is delivered in water from a central point, the heavier particles sinking and the lighter particles overflowing. { 'bəd·əl }

budget year [METEOROL] The 1-year period beginning with the start of the accumulation season at the firn line of a glacier or ice cap and extending through the following summer's ablation season. { 'bəj·ət ,yir }

bud grafting [BOT] Grafting a plant by budding. { 'bəd ,graf·tiŋ }

budling [BOT] The shoot that develops from the bud which was the scion of a bud graft. { 'bəd·liŋ }

BUDR See 5-bromodeoxyuridine.

bud rot [PL PATH] Any plant disease or symptom involving bud decay. { 'bəd ,rät }

bud scale [BOT] One of the modified leaves enclosing and protecting buds in perennial plants. { 'bəd ,skāl }

bud scale scar [BOT] A characteristic marking left on a stem when a bud falls off. { 'bəd ,skāl ,skär }

Buerger precession method [CRYSTAL] The recording on film of a single level of the reciprocal lattice of an individual crystal, by means of x-ray diffraction, for the purpose of determining unit cell dimensions and space groups. { 'bər·gər prē'sesh·ən ,meth·əd }

Buerger's disease See thromboangitis obliterans. { 'bər·gərz di,zēz }

buetschliite [MINERAL] $K_6Ca_2(CO_3)_5 \cdot 6H_2O$ A mineral that is probably hexagonal and consists of a hydrated carbonate of potassium and calcium. { 'bùch·lē,īt }

buffalo [VERT ZOO] The common name for several species of artiodactyl mammals in the family Bovidae, including the water buffalo and bison. { 'bəf·ə,lō }

buffer [CHEM] A solution selected or prepared to minimize changes in hydrogen ion concentration which would otherwise occur as a result of a chemical reaction. Also known as buffer solution. [COMPUT SCI] See buffer storage. [ECOL] An animal that is introduced to serve as food for other animals to reduce the losses of more desirable animals. [ELEC] An electric circuit or component that prevents undesirable electrical interaction between two circuits or components. [ELECTR] **1.** An isolating circuit in an electronic computer used to prevent the action of a driven circuit from affecting the corresponding driving circuit. **2.** See buffer amplifier. [ENG] A device, apparatus, or piece of material designed to reduce mechanical shock due to impact. [MIN ENG] **1.** Blasted material piled against or near a rock face to improve fragmentation and reduce scattering of rock from the next blast. **2.** A movable metal plate set in place in a tunnel excavation to limit the amount of rock scattered during blasting. { 'bəf·ər }

buffer amplifier [ELECTR] An amplifier used after an oscillator or other critical stage to isolate it from the effects of load impedance variations in subsequent stages. Also known as buffer; buffer stage. { ¦bəf·ər 'am·plə,fī·ər }

buffer cap [ORD] A light cap of ductile metal placed over, and in contact with, an armor-piercing cap in some designs of armor-piercing ammunition. { 'bəf·ər ,kap }

buffer capacitor [ELECTR] A capacitor connected across the secondary of a vibrator transformer or between the anode and cathode of a cold-cathode rectifier tube to suppress voltage surges that might otherwise damage other parts in the circuit. { 'bəf·ər kə'pas·əd·ər }

buffer capacity [CHEM] The relative ability of a buffer solution to resist pH change upon addition of an acid or a base. { 'bəf·ər kə'pas·əd·ē }

buffered computer [COMPUT SCI] A computer having a temporary storage device to compensate for differences in transmission speeds. { 'bəf·ərd kəm'pyüd·ər }

buffered device [COMPUT SCI] A piece of peripheral equipment, such as a printer, that is equipped with a buffer storage so that it can accept information more rapidly than it can process it. { 'bəf·ərd di'vīs }

buffered FET logic [ELECTR] A logic gate configuration used with gallium-arsenide field-effect transistors operating in the depletion mode, in which the level shifting required to make the input and output voltage levels compatible is achieved with Schottky barrier diodes. Abbreviated BFL. { 'bəf·ərd ¦ef,ē,tē 'läj·ik }

buffered I/O channel [COMPUT SCI] A storage device located between input/output (I/O) channels and main storage control to free the channels for use by other operations. { 'bəf·ərd ¦ī,ō ,chan·əl }

buffered terminal [COMPUT SCI] A computer terminal which contains storage equipment so that the rate at which it sends or receives data over its line does not need to agree exactly with the rate at which the data are entered or printed. { 'bəf·ərd 'tər·mən·əl }

buffer element [ELEC] A low-impedance inverting driver circuit. { 'bəf·ər ,el·ə·mənt }

BUCKWHEAT

Buckwheat: (a) Mature plant. (b) Seed.

BUFFALO

The Indian or water buffalo (Bubalus bubalis).

buffering agent [FOOD ENG] A chemical, such as lactic, citric, or acetic acid or the sodium salts of various acids, added to processed food to adjust and regulate its pH. { 'bəf·ə·riŋ ˌa·jənt }

buffer pooling [COMPUT SCI] A technique for receiving data in an input/output control system in which a number of buffers are available to the system; when a record is produced, a buffer is taken from the pool, used to hold the data, and returned to the pool after data transmission. { 'bəf·ər ˌpül·iŋ }

buffer solution *See* buffer. { 'bəf·ər sə'lü·shən }

buffer stage *See* buffer amplifier. { 'bəf·ər ˌstāj }

buffer storage [COMPUT SCI] A synchronizing element used between two different forms of storage in a computer; computation continues while transfers take place between buffer storage and the secondary or internal storage. Also known as buffer. { 'bəf·ər ˌstor·ij }

buffer zone [COMPUT SCI] An area of main memory set aside for temporary storage. { 'bəf·ər ˌzōn }

buffeting [AERO ENG] **1.** The beating of an aerodynamic structure or surfaces by unsteady flow, gusts, and so forth. **2.** The irregular shaking or oscillation of a vehicle component owing to turbulent air or separated flow. { 'bəf·əd·iŋ }

buffeting flutter [FL MECH] A phenomenon that can occur when one wire or cable in a strong wind is in the wake of another, whereby flow separated from the front wire induces the rear wire into oscillations with large amplitude. Also known as wake-induced galloping; wire-induced flutter. { 'bəf·əd·iŋ ˌfləd·ər }

buffeting Mach number [AERO ENG] The free-stream Mach number of an aircraft when the local Mach number over the tops of the wings approaches unity. { 'bəf·əd·iŋ 'mäk ˌnəm·bər }

buffing [ENG] The smoothing and brightening of a surface by an abrasive compound pressed against it by a soft wheel or belt. { 'bəf·iŋ }

buffing wheel [DES ENG] A flexible wheel with a surface of fine abrasive particles for buffing operations. { 'bəf·iŋ ˌwēl }

Buffon's problem [STAT] The problem of calculating the probability that a needle of specified length, dropped at random on a plane ruled with a series of straight lines a specified distance apart, will intersect one of the lines. { bü'fonz ˌpräb·ləm }

Bufonidae [VERT ZOO] A family of toothless frogs in the suborder Procoela including the true toads (*Bufo*). { byü'fän·ə,dē }

bufotenine [ORG CHEM] $C_{12}H_{16}N_2O$ An active pressor agent found in the skin of the common toad; a toxic alkaloid with epinephrinelike biological activity. { ˌbyü·fə'te,nēn }

bug [COMPUT SCI] A defect in a program code or in designing a routine or a computer. [ELECTR] **1.** A semiautomatic code-sending telegraph key in which movement of a lever to one side produces a series of correctly spaced dots and movement to the other side produces a single dash. **2.** An electronic listening device, generally concealed, used for commercial or military espionage. [ENG] **1.** A defect or imperfection present in a piece of equipment. **2.** *See* bullet. [INV ZOO] Any insect in the order Hemiptera. { bəg }

bugas *See* bottled gas. { 'byü,gas }

bug dust [MIN ENG] The fine coal or other material resulting from a boring or cutting of a drill, a mining machine, or even a pick. { 'bəg ˌdəst }

buggy [ENG] *See* concrete buggy. [MIN ENG] A four-wheeled steel car used for hauling coal to and from chutes. { 'bəg·ē }

bughole *See* vug. { 'bəg,hōl }

buhrstone [PETR] A silicified fossiliferous limestone with abundant cavities previously occupied by fossil shells. Also known as millstone. { 'bər,stōn }

buhrstone mill [MECH ENG] A mill for grinding or pulverizing grain in which a flat siliceous rock (buhrstone), generally of cellular quartz, rotates against a stationary stone of the same material. { 'bər,stōn ,mil }

build [ELECTR] To increase in received signal strength. { bild }

builder [MATER] An additive used in a detergent to improve the cleaning efficiency of the surfactant, principally by inactivating water-hardness ions. [MET] A fire-clay brick cull used for bottom construction in kilns or for boxing brick during burning. { 'bil·dər }

building [CIV ENG] A fixed structure for human occupancy and use. { 'bil·diŋ }

building block [MATER] A hollow block, either of burned clay or concrete, used in constructing the walls of buildings, and often faced with brick or stone. { 'bil·diŋ ,bläk }

building-block approach [IND ENG] A technique for development of a set of standard data by creating fixed groups or modules of work elements that may be added together to obtain time values for elements and entire operations. { 'bil·diŋ ,bläk ə,prōch }

building code [CIV ENG] Local building laws to promote safe practices in the design and construction of a building. { 'bil·diŋ ,kōd }

building dock [CIV ENG] A type of graving dock or basin, usually built of concrete, in which ships are constructed and then floated out through a caisson gate after flooding the dock. { 'bil·diŋ ,däk }

building envelope [CIV ENG] The interior, enclosed space of a building. { 'bil·diŋ 'en·və,lōp }

building footprint *See* footprint. { 'bil·diŋ ,fut,print }

building line [CIV ENG] A designated line beyond which a building cannot extend. { 'bil·diŋ ,līn }

building-out circuit [ELEC] Short section of transmission line, or a network which is shunted across a transmission line, for the purpose of impedance matching. { 'bil·diŋ ˌaut 'sər·kət }

building-out network [ELEC] Network designed to be connected to a based network so that the combination will simulate the sending-end impedance, neglecting dissipation, of a line having a termination other than that for which the basic network was designed. { 'bil·diŋ ˌaut ,net,wərk }

building-out section [ELEC] Short section of transmission line, either open or short-circuited at the far end, shunted across another transmission line for use on an impedance-matching transformer. { 'bil·diŋ ˌaut ,sek·shən }

building paper [MATER] Heavy, waterproof paper used to cover sheathing and subfloors to prevent passage of air and water. { 'bil·diŋ ,pā·pər }

buildup [ECOL] A significant increase in a natural population, usually as a result of progressive changes in ecological relations. [MET] Deposition of excess metal, either by electroplating or spraying, on worn or undersized machine components to restore required dimensions. [SCI TECH] **1.** A gradual accumulation of material. **2.** An increased value of pressure that results from a series of small increases. { 'bil,dəp }

buildup curve [PETRO ENG] Graph of bottom-hole pressure buildup versus shut-in time for a gas or oil well. { 'bil,dəp ,kərv }

buildup index *See* fire-danger meter. { 'bil,dəp ,in,deks }

buildup pressure [PETRO ENG] The increase in bottom-hole pressure up to an equilibrium value in a shut-in oil or gas well. { 'bil,dəp ,presh·ər }

buildup test [PETRO ENG] A test for determining the effective drainage radius of a wellbore in addition to the presence of permeability barriers or other production deterrents; the well is shut for a specified period, and a bottom-hole pressure bomb is run to record pressure. { 'bil,dəp ,test }

built detergent [MATER] A detergent formulated with a builder. { 'bilt di'tər·jənt }

built-in antenna [ELECTROMAG] An antenna that is located inside the cabinet of a radio or television receiver. { 'bilt,in an'ten·ə }

built-in beam *See* fixed-end beam. { 'bilt,in 'bēm }

built-in check [COMPUT SCI] A hardware device which controls the accuracy of data either moved or stored within the computer system. { 'bilt,in 'chek }

built-in function [COMPUT SCI] A function that is available through a simple reference and specification of arguments in a given higher-level programming language. Also known as built-in procedure; intrinsic procedure; standard function. { 'bilt,in 'fəŋk·shən }

built-in pointing device [COMPUT SCI] A trackball or pointing stick that is built into the case of a portable computer and used to move an on-screen pointer. { 'bilt,in 'point·iŋ di,vīs }

built-in procedure *See* built-in function. { 'bilt,in prə'sēj·ər }

built terrace *See* alluvial terrace. { 'bilt ,ter·əs }

BUHRSTONE MILL

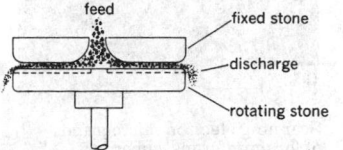

feed
fixed stone
discharge
rotating stone

Buhrstone mill. Material is fed at center of fixed stone and moves toward outer edge of the stones where finely ground product is discharged.

built-up beam [ENG] A structural steel member that is fabricated by welding or riveting rather than being rolled. { 'bilt,əp 'bēm }

built-up edge [ENG] Chip material adhering to the tool face adjacent to a cutting edge during cutting. { 'bilt,əp 'ej }

built-up fraction [GRAPHICS] In typesetting, a fraction set with at least two blocks of type, with each successive block to the right of the preceding one; each block contains one numeral, and the slash is between two blocks. { 'bilt,əp 'frak·shən }

built-up gun [ORD] Gun tube assembled by the shrinkage method; consists of two or more concentric cylinders shrunk one over another; stronger than a simple cylinder of the same dimensions. { 'bilt,əp 'gən }

built-up mica [MATER] Large, laminated plates of mica made by bonding thin splittings of natural mica with shellac, glyptol, or some other suitable adhesive. { 'bilt,əp 'mī·kə }

built-up roof [BUILD] A roof constructed of several layers of felt and asphalt. { 'bilt,əp 'rüf }

built-up roofing [MATER] A seamless piece of flexible, waterproofed roofing material consisting of plies of felt mopped with asphalt or pitch. { 'bilt,əp 'rüf·iŋ }

bulb [BOT] A short, subterranean stem with many overlapping fleshy leaf bases or scales, such as in the onion and tulip. [ELEC] See envelope. { bəlb }

bulb angle [DES ENG] A steel angle iron enlarged to a bulbous thickening at one end. { 'bəlb ,aŋ·gəl }

bulbar paralysis [MED] A clinical syndrome due to involvement of the nuclei of the last four or five cranial nerves, characterized principally by paralysis or weakness of the muscles which control swallowing, talking, movement of the tongue and lips, and sometimes respiratory paralysis. { 'bəl·bər pə'ral·ə·səs }

bulb glacier [HYD] A glacier formed at the foot of a mountain and out into an open slope; the glacier ends spread out into an ice fan. { 'bəlb ,glā·shər }

bulbil [BOT] A secondary bulb usually produced on the aerial part of a plant. { 'bəl·bəl }

bulbocavernosus See bulbospongiosus. { ¦bəl·bō,ka·vər'nō·səs }

bulb of percussion [ARCHEO] A cone-shaped bulge on a fractured flint surface that was made by a blow striking at an angle. { 'bəlb əv pər'kə·shən }

bulb of the penis [ANAT] The expanded proximal portion of the corpus spongiosum of the penis. { ¦bəlb əv thə 'pē·nəs }

bulbospongiosus [ANAT] A muscle encircling the bulb and adjacent proximal parts of the penis in the male, and encircling the orifice of the vagina and covering the lateral parts of the vestibular bulbs in the female. Also known as bulbocavernosus. { ¦bəl·bō,spən·jē'ō·səs }

bulbourethral gland [ANAT] Either of a pair of compound tubular glands, anterior to the prostate gland in males, which discharge into the urethra. Also known as Cowper's gland. { ¦bəl·bō·yú'rēth·rəl ,gland }

bulbous bow [NAV ARCH] A large, elliptical or cylindrical swelling of the lower underwater hull form protruding forward of the bow designed to cancel wave-making at high speeds. { 'bəl·bəl 'baú }

bulge forming [ENG] A process by which contours are formed on the sides of tubular workpieces by exerting pressure inside the tube to force expansion into a die clamped around the exterior. { 'bəlj ,fȯrm·iŋ }

bulgur [FOOD ENG] A lightly milled wheat product that has been boiled, dehydrated, and cracked. { 'bəl·gər }

bulimia [MED] Excessive, insatiable appetite, seen in psychotic states; a symptom of diabetes mellitus and of certain cerebral lesions. Also known as hyperphagia. { bə'lēm·ē·ə }

Buliminacea [INV ZOO] A superfamily of benthic, marine foraminiferans in the suborder Rotaliina. { ¦byü,lim·ə'nās·ē·ə }

bulk [GRAPHICS] In book printing, the thickness of paper, in terms of the number of pages per inch for a given basis weight. { bəlk }

bulk acoustic wave [ACOUS] An acoustic wave that travels through a piezoelectric material, as in a quartz delay line. Also known as volume acoustic wave. { ¦bəlk ə'kü·stik 'wāv }

bulk-acoustic-wave delay line [ELECTR] A delay line in which the delay is determined by the distance traveled by a bulk acoustic wave between input and output transducers mounted on a piezoelectric block. { ¦bəlk ə'kü·stik ¦wāv di'lā ,līn }

bulk cargo [IND ENG] Cargo which is loaded into a ship's hold without being boxed, bagged, or hand stowed, or is transported in large tank spaces. { 'bəlk 'kär,gō }

bulk carrier [NAV ARCH] A vessel designed to transport dry or liquid bulk cargo. { 'bəlk 'kar·ē·ər }

bulk density [ENG] The mass of powdered or granulated solid material per unit of volume. { 'bəlk 'den·səd·ē }

bulk diode [ELECTR] A semiconductor microwave diode that uses the bulk effect, such as Gunn diodes and diodes operating in limited space-charge-accumulation modes. { 'bəlk 'dī,ōd }

bulked-down See solid-piled. { 'bəlkt 'daȯn }

bulked yarn [TEXT] A yarn whose individual fibers have been treated so that they do not follow a linear path; air spaces are developed within the yarn, leading to increased loft. { 'bəlkt 'yärn }

bulk effect [ELECTR] An effect that occurs within the entire bulk of a semiconductor material rather than in a localized region or junction. { 'bəlk i'fekt }

bulk-effect device [ELECTR] A semiconductor device that depends on a bulk effect, as in Gunn and avalanche devices. { 'bəlk i'fekt di'vīs }

bulk eraser [ELECTROMAG] A device used to erase an entire reel of recorded magnetic tape at once without running it through a recorder. { 'bəlk i'rā·sər }

bulk factor [ENG] The ratio of the volume of loose powdered or granulated solids to the volume of an equal weight of the material after consolidation into a voidless solid. { 'bəlk ,fak·tər }

bulk flotation [MIN ENG] The rising of a mineralized froth, of more than one mineral, in a single operation. { 'bəlk flō'tā·shən }

bulk flow See convection. { 'bəlk 'flō }

bulk-handling machine [MECH ENG] Any of a diversified group of materials-handling machines designed for handling unpackaged, divided materials. { 'bəlk ,hand·liŋ mə'shēn }

bulkhead [AERO ENG] A wall, partition, or such in a rocket, spacecraft, airplane fuselage, or similar structure, at right angles to the longitudinal axis of the structure and serving to strengthen, divide, or help give shape to the structure. [MIN ENG] A tight-seal partition of wood, rock, and mud or concrete in mines that serves to protect against gas, fire, and water. [NAV ARCH] An upright partition separating compartments in a ship. { 'bəlk,hed }

bulkhead deck [NAV ARCH] The highest continuous deck of a ship to which all transverse watertight bulkheads extend. { 'bəlk,hed ,dek }

bulkhead line [CIV ENG] The farthest offshore line to which a structure may be constructed without interfering with navigation. { 'bəlk,hed ,līn }

bulkhead wharf [CIV ENG] A bulkhead that may be used as a wharf by addition of mooring appurtenances, paving, and cargo-handling facilities. { 'bəlk,hed ,wȯrf }

bulking [MATER] The difference in volume of a given mass of sand or other fine material in moist and dry conditions. { 'bəl·kiŋ }

bulking power [TEXT] The relative ability of a textile fiber to increase the finished material's volume. { 'bəl·kiŋ ,paú·ər }

bulking value [CHEM ENG] The relative ability of a pigment or other substance to increase the volume of paint. { 'bəl·kiŋ ,val·yü }

bulk insulation [ENG] A type of insulation that retards the flow of heat by the interposition of many air spaces and, in most cases, by opacity to radiant heat. { 'bəlk in·sə'lā·shən }

bulk lifetime [SOLID STATE] The average time that elapses between the formation and recombination of minority charge carriers in the bulk material of a semiconductor. { 'bəlk 'līf,tīm }

bulk material [IND ENG] Material purchased in uniform lots and in quantity for distribution as required for a project. { 'bəlk mə'tir·ē·əl }

bulk memory [COMPUT SCI] A high-capacity memory used in connection with a computer for bulk storage of large quantities of data. { 'bəlk 'mem·rē }

bulk micromachining [ENG] A set of processes that enable the three-dimensional sculpting of single-crystal silicon to make

---

BULIMINACEA

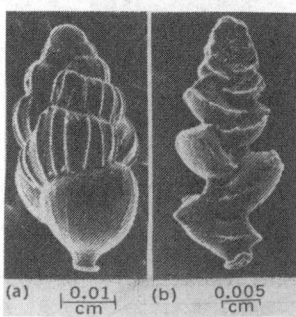

(a) 0.01 cm   (b) 0.005 cm

Scanning electron micrographs of foraminiferans, suborder Rotaliina, superfamily buliminacea, (a) Eouvigerina, from Upper Cretaceous of Texas. (b) Uvigerina, from Miocene of California. (R. B. MacAdam, Chevron Oil Field Research Co.)

small structures that serve as components of microsensors. { ¦bəlk ¦mī·krō·mə'shēn·iŋ }

bulk mining [MIN ENG] Mining in which large quantities of low-grade ore are taken without attempt to segregate the high-grade portions. { ¦bəlk 'mīn·iŋ }

bulk modulus See bulk modulus of elasticity. { ¦bəlk 'mäj·ə·ləs }

bulk modulus of elasticity [MECH] The ratio of the compressive or tensile force applied to a substance per unit surface area to the change in volume of the substance per unit volume. Also known as bulk modulus; compression modulus; hydrostatic modulus; modulus of compression; modulus of volume elasticity. { ¦bəlk 'mäj·,ləs əv i,las'tis·əd·ē }

bulk molding compound [MATER] A mixture of resin, inert fillers, reinforcements, and other additives which forms a puttylike rope, sheet, or preformed shape; used as a premix in composite manufacture. { 'bəlk 'mōld·iŋ ,käm,pau̇nd }

bulk photoconductor [ELECTR] A photoconductor having high power-handling capability and other unique properties that depend on the semiconductor and doping materials used. { ¦bəlk ¦fō·dō·kən,dək·tər }

bulk plant [PETRO ENG] A wholesale receiving and distributing facility for petroleum products; includes storage tanks, warehouses, railroad sidings, truck loading racks, and related elements. Also known as bulk terminal. { 'bəlk ,plant }

bulk resistor [ELECTR] An integrated-circuit resistor in which the *n*-type epitaxial layer of a semiconducting substrate is used as a noncritical high-value resistor; the spacing between the attached terminals and the sheet resistivity of the material together determine the resistance value. { 'bəlk ri'zis·tər }

bulk rheology [MECH] The branch of rheology wherein study of the behavior of matter neglects effects due to the surface of a system. { 'bəlk rē'äl·ə·jē }

bulk sample See gross sample. { 'bəlk ,sam·pəl }

bulk sampling [ANALY CHEM] The taking of samples in arbitrary, irregular units rather than discrete units of uniform size for chemical analysis. { 'bəlk ,sam·pliŋ }

bulk shielding reactor [NUCLEO] The prototype swimming-pool reactor located at Oak Ridge, Tennessee; it uses heterogeneous enriched fuel and provides a combination of high thermal-neutron flux, ready accessibility, and versatility. Abbreviated BSR. { 'bəlk ,shēld·iŋ rē'ak·tər }

bulk solid [MATER] An assembly of solid particles that is large enough for the statistical mean of any property to be independent of the number of particles. Also known as particulate solid; powder. { 'bəlk 'säl·əd }

bulk storage See backing storage. { 'bəlk 'stȯr·ij }

bulk strain [MECH] The ratio of the change in the volume of a body that occurs when the body is placed under pressure, to the original volume of the body. { 'bəlk ,strān }

bulk strength [MECH] The strength per unit volume of a solid. { 'bəlk 'streŋkth }

bulk terminal See bulk plant. { 'bəlk 'tər·mən·əl }

bulk transport [MECH ENG] Conveying, hoisting, or elevating systems for movement of solids such as grain, sand, gravel, coal, or wood chips. { 'bəlk 'tranz,pȯrt }

Bull See Taurus. { bəl }

bullate [BIOL] Appearing blistered or puckered, especially of certain leaves. { 'bə,lāt }

bull bit [MIN ENG] A flat drill bit. { 'bəl ,bit }

bull block [MET] Power-driven machine for drawing wire through a die. { 'bəl ,bläk }

bulldozer [MECH ENG] A wheeled or crawler tractor equipped with a reinforced, curved steel plate mounted in front, perpendicular to the ground, for pushing excavated materials. [MET] A machine for bending, forging, and punching narrow plates and bars, in which a ram is pushed along a horizontal path by a pair of cranks that are linked to two bullwheels with eccentric pins. { 'bu̇l,dōz·ər }

bullet [ENG] **1.** A conical-nosed cylindrical weight, attached to a wire rope or line, either notched or seated to engage and attach itself to the upper end of a wire line core barrel or other retrievable or retractable device that has been placed in a borehole. Also known as bug; go-devil; overshot. **2.** A scraper with self-adjusting spring blades, inserted in a pipeline and carried forward by the fluid pressure, clearing away accumulations or debris from the walls of a pipe. Also known as go-devil. **3.** A bullet-shaped weight or small explosive charge

dropped to explode a charge of nitroglycerin placed in a borehole. Also known as go-devil. **4.** An electric lamp covered by a conical metal case, usually at the end of a flexible metal shaft. **5.** See torpedo. [GRAPHICS] **1.** A hollow hemispherical shell, made of iron and filled with pitch, which holds small objects during the execution of artistic designs in metal. **2.** A circle or other graphic character, about the height of a lowercase letter, used to set off items in a list. [MATER] A small, lustrous, nearly spherical industrial diamond. [ORD] The projectile fired, or intended to be fired, from a small arm. { 'bu̇l·ət }

bullet drop [MECH] The vertical drop of a bullet. { 'bu̇l·ət ,dräp }

bullet group [ORD] The bullet holes in a target made by fire from a given small arm from one position and caused by ballistic variations or inaccurate aim. { 'bu̇l·ət ,grüp }

bulletin board [COMPUT SCI] A collection of information that is stored in a computer system and can be accessed either by a specified group of people or the general public, usually by dialing a number on the public telephone system. { 'bu̇l·ət·ən ,bȯrd }

bulletin board system [COMPUT SCI] A computer system that enables its users, usually members of a particular interest group, to leave messages and to share information and software. Abbreviated BBS. { 'bu̇l·ət·ən ,bȯrd ,sis·təm }

bullet jacket [ORD] A metal shell surrounding a metal core, the combination composing a bullet for small arms; the jacket is either made of or coated with a relatively soft metal which engages the rifling in the bore, causing rotation of the bullet. { 'bu̇l·ət ,jak·ət }

bullet nose [MATH] A plane curve whose equation in cartesian coordinates x and y is $(a^2/x^2) - (b^2/y^2) = 1$, where a and b are constants. { 'bu̇l·ət ,nōz }

bullet perforator [PETRO ENG] A device in the form of a tubular gun that can be lowered to a specified depth in a well and made to fire bullets through the casing and cement, forming holes which permit formation fluids to enter the wellbore. { 'bu̇l·ət 'pər·fə,rād·ər }

bulletproof armor [ORD] Light, homogeneous, hard armor plate, having a Brinell hardness of 400 to 475, resistant to small-caliber armor-piercing bullets. { 'bu̇l·ət,prüf 'är·mər }

bullet splash [ORD] Dispersion of finely divided or melted metal produced upon impact of a projectile with armor plate or other hard objects. { 'bu̇l·ət ,splash }

bull-eye squall [METEOROL] A squall forming in fair weather, characteristic of the ocean off the coast of South Africa; it is named for the peculiar appearance of the small, isolated cloud marking the top of the invisible vortex of the storm. { 'bu̇l ,ī ,skwȯl }

bullfrog See barney. { 'bu̇l,fräg }

bull gear [DES ENG] A bull wheel with gear teeth. { 'bu̇l ,gir }

bullhorn [COMMUN] A portable loudspeaker, generally with built-in amplifier and microphone, used for voice messages to crowds or from one ship to another at sea. { 'bu̇l,hȯrn }

bulliform [BOT] Type of plant cell involved in tissue contraction or water storage, or of uncertain function. { 'bu̇l·ə,fȯrm }

bulliform cell [BOT] One of the large, highly vacuolated cells occurring in the epidermis of grass leaves. Also known as motor cell. { 'bu̇l·ə,fȯrm ,sel }

bulling bar [ENG] A bar for ramming clay into cracks containing blasting charges which are about to be exploded. { 'bu̇l·iŋ ,bär }

bullion [MET] Gold or silver in bulk in the shape of bars or ingots. [MIN ENG] A concretion found in some types of coal, composed of carbonate or silica stained by brown humic derivatives; often well-preserved plant structures form the nuclei. { 'bu̇l·yən }

bull ladle [MET] A ladle used in foundry operations for carrying molten metal. { 'bu̇l ,lād·əl }

bull nose [BUILD] A rounded external angle, as one used at window returns and doorframes. [MET] A ladle used in foundry operations for carrying molten metal. { 'bu̇l ,nōz }

bull-nose bit See wedge bit. { 'bu̇l ,nōz ,bit }

bull-nose plane [DES ENG] A small rabbet plane used to smooth or shape joints or other places that cannot be reached by larger planes. { 'bu̇l ,nōz ,plān }

bullous emphysema [MED] Acute overinflation of the

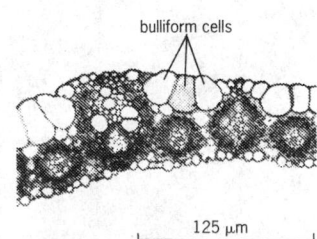

BULLIFORM CELL

bulliform cells

125 μm

Transection of leaf of *Andropogon* showing large bulliform cells in the upper epidermis.

lungs due to extreme efforts to inhale to overcome a bronchial obstruction. { 'bùl·əs ˌem·fə'zē·mə }

Bullpup [ORD] A U.S. Navy air-to-surface radio-guided missile used against enemy tanks, bridges, and other small targets. { 'bùl‚pəp }

bull's-eye rot [PL PATH] A fungus disease of apples caused by either *Neofabraea malicorticis* or *Gloeosporium perennans* and characterized by spots resembling eyes on the fruit. { 'bùlz ‚ī ‚rät }

bull shaker [MIN ENG] A shaking chute where large coal from the dump is cleaned by hand. { 'bùl ‚shā·kər }

bullvalene [ORG CHEM] 1. A compound, molecular formula $C_{10}H_{10}$, that does not have a permanent structure, but has more than 1,200,000 equivalent structures. 2. A fluxional compound. { 'bùl·və‚lēn }

bull wheel [MECH ENG] 1. The main wheel or gear of a machine, which is usually the largest and strongest. 2. A cylinder which has a rope wound about it for lifting or hauling. 3. A wheel attached to the base of a derrick boom which swings the derrick in a vertical plane. { 'bùl ‚wēl }

bulwark [NAV ARCH] The part of a ship's side that extends above the main deck to protect it against heavy weather. { 'bùl‚wərk }

bulwark plating [NAV ARCH] Light plating that extends above a ship's hull plating and acts as a bulwark. { 'bùl‚wərk ‚plād·iŋ }

Bulygen number [THERMO] A dimensionless number used in the study of heat transfer during evaporation. { 'bül·ə·jən ‚nəm·bər }

bumblebee [INV ZOO] The common name for several large, hairy social bees of the genus *Bombus* in the family Apidae. { 'bəm·bəl‚bē }

bumboat [NAV ARCH] A small boat on which a hoist has been installed for operating dredge lines and handling anchors. { 'bəm‚bōt }

bummer [FOR] A low truck with two wheels for carrying logs, or a tracked cart for dragging them. [MIN ENG] The person who runs conveyors in a quarry or mine. { 'bəm·ər }

bump Cepheid [ASTRON] A Cepheid variable star with a period of 5 – 15 days that displays a prominent secondary maximum (bump) in its light and velocity curves. { 'bəmp 'sef·ē·əd }

bump contact [ELECTR] A large-area contact used for alloying directly to the substrate of a transistor for mounting or interconnecting purposes. { 'bəmp ‚kän‚takt }

bump-down [PETRO ENG] In an oil well pumping operation, the pump's hitting the bottom in the downstroke because the rod between the pumping jack and the pump seat is too long. { 'bəmp‚daùn }

bumper [ENG] 1. A metal bar attached to one or both ends of a powered transportation vehicle, especially an automobile, to prevent damage to the body. 2. In a drilling operation, the supporting stay between the main foundation sill and the engine block. 3. In drilling, a fishing tool for loosening jammed cable tools. [MET] A vibrating machine for ramming and consolidating sand into a mold. { 'bəm·pər }

bumper jar [PETRO ENG] An expansion joint installed on the drill stem of a floating offshore drilling rig, permitting vertical movement of the upper section while the lower section of the tool remains stationary. { 'bəm·pər ‚jär }

bumper sub [PETRO ENG] A motion compensator that maintains constant weight on the drilling bit during vertical motion of a floating offshore drilling rig and also provides a jarring action to remove heavy objects in the borehole. { 'bəm·pər ‚səb }

bumpiness [AERO ENG] An atmospheric condition causing aircraft to experience sudden vertical jolts. { 'bəm·pē·nəs }

bumping [AERO ENG] *See* chugging. [CHEM] Uneven boiling of a liquid caused by irregular rapid escape of large bubbles of highly volatile components as the liquid mixture is heated. [MECH ENG] *See* chugging. [MET] Forming a dish in metal by many repeated blows. { 'bəm·piŋ }

bumps [MIN ENG] Sudden, violent expulsion of coal from one or more pillars, accompanied by loud reports and earth tremors. { 'bəmps }

bunched pair [ELEC] Group of pairs tied together or otherwise associated for identification. { 'bəncht 'per }

buncher *See* buncher resonator. { 'bən·chər }

buncher resonator [ELECTR] The first or input cavity resonator in a velocity-modulated tube, next to the cathode; here the faster electrons catch up with the slower ones to produce bunches of electrons. Also known as buncher; input resonator. { 'bən·chər ‚rez·ən‚äd·ər }

bunching [ELECTR] The flow of electrons from cathode to anode of a velocity-modulated tube as a succession of electron groups rather than as a continuous stream. { 'bən·chiŋ }

bunching voltage [ELECTR] Radio-frequency voltage between the grids of the buncher resonator in a velocity-modulated tube such as a klystron; generally, the term implies the peak value of this oscillating voltage. { 'bən·chiŋ ‚vōl·tij }

bunch-map analysis [STAT] A graphic technique in confluence analysis; all subsets of regression coefficients in a complete set are drawn on standard diagrams, and the representation of any set of regression coefficients produces a "bunch" of lines; allows the observer to determine the effect of introducing a new variate on a set of variates. { ‚bənch ‚map ə'nal·ə·səs }

bunchy top [PL PATH] Any viral disease of plants in which there is a shortening of the internodes with crowding of leaves and shoots at the stem apex. { 'bən·chē ‚täp }

bund [CIV ENG] An embankment or embanked thoroughfare along a body of water; the term is used particularly for such structures in the Far East. { bənd }

bundle [MATH] A triple (E, p, B), where E and B are topological spaces and p is a continuous map of E onto B; intuitively E is the collection of inverse images under p of points from B glued together by the topology of X. { 'bən·dəl }

bundle branch [ANAT] Either of the components of the atrioventricular bundle passing to the right and left ventricles of the heart. { 'bənd·əl ‚branch }

bundled program [COMPUT SCI] A computer program written, maintained, and updated by the computer manufacturer, and included in the price of the hardware. { ‚bənd·əld 'prō·grəm }

bundle of His [ANAT] A small region of heart muscle located in the right auricle and specialized to relay contraction impulses from the right auricle to the ventricles. Also known as atrioventricular bundle. { 'bənd·əl əv ‚äch‚ī'es }

bundle of planes *See* sheaf of planes. { ‚bənd·əl əv 'plānz }

bundle scar [BOT] A mark within a leaf scar that shows the point of an abscised vascular bundle. { 'bənd·əl ‚skär }

bundle sheath [BOT] A sheath around a vascular bundle that consists of a layer of parenchyma. { 'bənd·əl ‚shēth }

bundling [COMMUN] The provision of a combination of services, such as cable television and telephone service, over a single communications system. { 'bən·dliŋ }

bundling machine [MECH ENG] A device that automatically accumulates cans, cartons, or glass containers for semiautomatic or automatic loading or for shipping cartons by assembling the packages into units of predetermined count and pattern which are then machine-wrapped in paper, film paperboard, or corrugated board. { 'bənd·liŋ mə'shēn }

bund wall [ENG] A retaining wall designed to contain the contents of a tank or a storage vessel in the event of a rupture or other emergency. { 'bənd ‚wòl }

α-bungarotoxin [NEURO] A neurotoxin found in snake venom which blocks neuromuscular transmission by binding with acetylcholine receptors on motor end plates. { ‚al·fə ‚bəŋ·gə·rə'täk·sən }

Buniakowski's inequality *See* Cauchy-Schwarz inequality. { ‚bùn·yə'kòf·skēz ‚in·i'kwäl·əd·ē }

bunion [MED] A swelling of a bursa of the foot, especially of the metatarsophalangeal joint of the great toe; associated with thickening of the adjacent skin and a forcing of the great toe into adduction. { 'bən·yən }

bunker [CIV ENG] A bin, often elevated, that is divided into compartments for storing material such as coal or sand. [MECH ENG] A space in a refrigerator designed to hold a cooling element. [ORD] A fortified structure for the protection of personnel, a defended gun position, or a defensive position. { 'bəŋ·kər }

bunker C fuel oil [MATER] A special grade of bunker fuel oil. Also known as Navy Heavy; No. 6 fuel. { ‚bəŋ·kər 'sē ‚fyül ‚òil }

bunker fuel oil [MATER] A heavy residual petroleum oil used as fuel by ships, industry, and large-scale heating and power-production installations. { ‚bəŋ·kər ‚fyül ‚òil }

bunkering [ENG] Storage of solid or liquid fuel in containers from which the fuel can be continuously or intermittently withdrawn to feed a furnace, internal combustion engine, or fuel tank, for example, coal bunkering and fuel-oil bunkering. { 'bən·kər·iŋ }

Bunn chart [CRYSTAL] A chart for classifying x-ray diffraction powder photographs of substances whose crystals have tetragonal or hexagonal symmetry. { bən ‚chärt }

bunny suit [ENG] Protective clothing worn by an individual who works in a clean room to prevent contamination of equipment and materials. { 'bən·ē ‚süt }

bunodont [ANAT] Having tubercles or rounded cusps on the molar teeth, as in humans. { 'byü·nə‚dänt }

bunolophodont [VERT ZOO] **1.** Of teeth, having the outer cusps in the form of blunt cones and the inner cusps as transverse ridges. **2.** Having such teeth, as in tapirs. { ‚byü·nə'läf·ə‚dänt }

Bunonematoidea [INV ZOO] A superfamily of nematodes in the order Rhabditida, characterized by asymmetrical bodies in both the labia and the distribution of sensory organs. { ‚bü·nō‚nem·ə'tȯid·ē·ə }

bunoselenodont [VERT ZOO] **1.** Of teeth, having the inner cusps in the form of blunt cones and the outer ones as longitudinal crescents. **2.** Having such teeth, as in the extinct titanotheres. { ‚byü·nō·sə'lē·nə‚dänt }

Bunsen burner [ENG] A type of gas burner with an adjustable air supply. { 'bən·sən 'bər·nər }

Bunsen disk [OPTICS] The screen generally used in a grease-spot photometer, with a circular translucent spot at the center. { 'bən·sən ‚disk }

Bunsen ice calorimeter [ENG] Apparatus to gage heat released during the melting of a compound by measuring the increase in volume of the surrounding ice-water solution caused by the melting of the ice. Also known as ice calorimeter. { 'bən·sən 'īs kal·ə'rim·əd·ər }

bunsenite [MINERAL] NiO A pistachio-green mineral consisting of nickel monoxide and occurring as octahedral crystals. { 'bən·sə‚nīt }

Bunsen-Kirchhoff law [SPECT] The law that every element has a characteristic emission spectrum of bright lines and absorption spectrum of dark lines. { ‚bən·sən 'kir‚kȯf ‚lȯ }

bunt [PL PATH] A fungus disease of wheat caused by two *Tilletia* species and characterized by grain replacement with fishy-smelling smut spores. { bənt }

buntal [TEXT] A fine, white fiber derived from the leaves of the talipot palm. Also known as buri. { bùn'täl }

Bunter [GEOL] Lower Triassic geologic time. Also known as Buntsandstein. { 'bùn·tər }

bunton [MIN ENG] A steel or timber element in the lining of a rectangular shaft. { 'bənt·ən }

Buntsandstein *See* Bunter. { 'bùnt·sən‚shtīn }

Bunyaviridae [VIROL] A family of enveloped spherical viruses whose lipid envelopes contain at least one virus-specific glycopeptide; members develop in the cytoplasm and mature by budding. { ‚bən·yə'vī·rə‚dē }

Bunyavirus [VIROL] A genus of viruses in the family Bunyaviridae; contains a minimum of 87 species which exhibit some degree of antigenic relationship. { 'bən·yə‚vī·rəs }

buoy [ENG] An anchored or moored floating object, other than a lightship, intended as an aid to navigation, to attach or suspend measuring instruments, or to mark the position of something beneath the water. { 'bȯi }

buoyage [NAV] A system of buoys; for example, one in which the buoys are assigned shape, color, and number distinction in accordance with location relative to the nearest obstruction is called a cardinal buoyage system. { 'bȯi·ij }

buoyancy [FL MECH] The resultant vertical force exerted on a body by a static fluid in which it is submerged or floating. { 'bȯi·ən·sē }

buoyancy parameter [FL MECH] The Grashof number divided by the square of the Reynolds number. { 'bȯi·ən·sē pə'ram·əd·ər }

buoyancy pontoons [PETRO ENG] Pontoons that buoy up offshore pipelines during the welding together of sections over bodies of water, after which the pontoons are removed and the pipeline is allowed to sink into position on the bottom. { 'bȯi·ən·sē pän'tünz }

buoyancy tank [NAV ARCH] An airtight tank near one of the ends of a small boat to keep it from sinking if it capsizes or fills with water. { 'bȯi·ən·sē ‚taŋk }

buoyancy-type density transmitter [ENG] An instrument which records the specific gravity of a flowing stream of a liquid or gas, using the principle of hydrostatic weighing. { 'bȯi·ən·sē ‚tīp 'den·səd·ē tranz'mid·ər }

buoyant density [PHYS] A technique that uses the sedimentation equilibrium in a density gradient to characterize a solute. { 'bȯi·ənt 'den·səd·ē }

buoyant force [FL MECH] The force exerted vertically upward by a fluid on a body wholly or partly immersed in it; its magnitude is equal to the weight of the fluid displaced by the body. { 'bȯi·ənt 'fȯrs }

buoy sensor [ENG ACOUS] A hydrophone used as a sensor in buoy projects; some hydrophone arrays are designed for telemetering. { 'bȯi ‚sen·sər }

Buprestidae [INV ZOO] The metallic wood-boring beetles, the large, single family of the coleopteran superfamily Buprestoidea. { byü'pres·tə‚dē }

Buprestoidea [INV ZOO] A superfamily of coleopteran insects in the suborder Polyphaga including many serious pests of fruit trees. { ‚byü·pres'tȯid·ē·ə }

Burali-Forti paradox [MATH] The order-type of the set of all ordinals is the largest ordinal, but that ordinal plus one is larger. { bù'räl·ē 'fȯr·tē 'par·ə‚däks }

buran [METEOROL] A violent northeast storm of south Russia and central Siberia, similar to the blizzard. { bü'rän }

burble [FL MECH] **1.** A separation or breakdown of the laminar flow past a body. **2.** The eddying or turbulent flow resulting from this occurrence. { 'bər·bəl }

burble angle *See* burble point. { 'bər·bəl ‚aŋ·gəl }

burble point [FL MECH] A point reached in an increasing angle of attack at which burble begins. Also known as burble angle. { 'bər·bəl ‚pȯint }

burden [ELEC] The amount of power drawn from the circuit connecting the secondary terminals of an instrument transformer, usually expressed in volt-amperes. [ENG] **1.** The distance from a drill hole to the more or less vertical surface of rock that has already been exposed by blasting or excavating. **2.** The volume of the rock to be removed by blasting in a drill hole. [GEOL] All types of rock or earthy materials overlying bedrock. [MET] **1.** The material which is melted in a direct arc furnace. **2.** In an iron blast furnace, the ratio of iron and flux to coke and other fuels in the charge. { 'bərd·ən }

Burdigalian [GEOL] Upper lower Miocene geologic time. { ‚bərd·i'gäl·yən }

buret [CHEM] A graduated glass tube used to deliver variable volumes of liquid; usually equipped with a stopcock to control the liquid flow. { byü'ret }

burga [METEOROL] A storm of wind and sleet in Alaska. Also spelled boorga. { 'bùr·gə }

Burgers vector [CRYSTAL] A translation vector of a crystal lattice representing the displacement of the material to create a dislocation. { 'bər·gərz ‚vek·tər }

Burgess Shale [GEOL] A fossil deposit in the Canadian Rockies, British Columbia, consisting of a diverse fauna that accumulated in a clay and silt sequence during the Cambrian. { 'bər·jəs 'shäl }

burglar alarm [ENG] An alarm in which interruption of electric current to a relay, caused, for example, by the breaking of a metallic tape placed at an entrance to a building, deenergizes the relay and causes the relay contacts to operate the alarm indicator. Also known as intrusion alarm. { 'bər·glər ə‚lärm }

Burhinidae [VERT ZOO] The thick-knees or stone curlews, a family of the avian order Charadriiformes. { byü'rin·ə‚dē }

buri *See* buntal. { bü'rē }

burial ground [NUCLEO] A place for burying unwanted radioactive objects to prevent escape of their radiations, the earth acting as a shield. Also known as graveyard. { 'ber·ē·əl ‚graùnd }

burial metamorphism [GEOL] A kind of regional metamorphism which affects sediments and interbedded volcanic rocks in a geosyncline without the factors of orogenesis or magmatic intrusions. { 'ber·ē·əl med·ə'mȯr‚fiz·əm }

buried hill [GEOL] A hill of resistant older rock over which later sediments are deposited. { 'ber·ēd 'hil }

buried placer [GEOL] Old deposit of a placer which has been buried beneath lava flows or other strata. { 'ber·ēd 'plās·ər }

buried river [GEOL] A river bed which has become buried beneath streams of alluvial drifts or basalt. { 'ber·ēd 'riv·ər }

BUPRESTIDAE

An example of Buprestidae. *(From T. I. Storer and R. L. Usinger, General Zoology, 3d ed., McGraw-Hill, 1957)*

BURGERS VECTOR

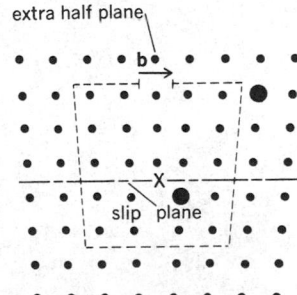

An arrangement of atoms around an edge dislocation. The dashed lines represent Burgers closure circuits which define Burgers vector b. Dislocation is perpendicular to the paper at X. The large dots represent impurity atoms which may be present.

buried set-point method [CONT SYS] A procedure for guiding a robot manipulator along a template, in which low-gain servomechanisms apply a force along the edge of the template, while the manipulator's tool is parallel to, and buried below, the template surface. { 'ber·ēd 'set,point ,meth·əd }

buried soil See paleosol. { 'ber·ēd 'soil }

burin [GRAPHICS] An etching tool used for engraving on metal or wood. { 'bər·ən }

burkeite [MINERAL] $Na_6(CO_3)(SO_4)_2$ A white to pale buff or gray mineral consisting of a carbonate-sulfate of sodium. { 'bər,kīt }

Burkett's lymphoma [MED] A malignant lymphoma of children, typically involving the retroperitoneal area and the mandible, but sparing the peripheral lymph nodes, bone marrow, and spleen. { 'bər·kəts lim'fō·mə }

burl [BOT] A hard, woody outgrowth on a tree, usually resulting from the entwined growth of a cluster of adventitious buds. [MATER] In lumber or veneer, a localized severe distortion of the grain that is generally rounded in outline. { bərl }

burlap [TEXT] A coarse cloth woven from jute fiber. Also known as gunny; hessian. { 'bər,lap }

burling [TEXT] Removing loops, knots, and vegetable matter in the finishing section of a woolen or worsted mill. { 'bər·liŋ }

Burmanniaceae [BOT] A family of monocotyledonous plants in the order Orchidales characterized by regular flowers, three or six stamens opposite the petals, and ovarian nectaries. { bər,man·ē'ās·ē,ē }

burn [ENG] To consume fuel. [MED] An injury to tissues caused by heat, chemicals, electricity, or irradiation effects. { bərn }

burnable poison [NUCLEO] A neutron absorber that is incorporated in the fuel or fuel cladding of a nuclear reactor and gradually burns up under neutron irradiation. Also known as burnout poison. { 'bərn·ə·bəl 'póiz·ən }

burn cut See parallel cut. { 'bərn ,kət }

burned sand [MET] The dissipated claying portion of a casting sand resulting from the heat of the metal. { 'bərnd 'sand }

burner [CHEM ENG] A furnace where sulfur or sulfide ore are burned to produce sulfur dioxide and other gases. [ENG] 1. The part of a fluid-burning device at which the flame is produced. 2. Any burning device used to soften old paint to aid in its removal. 3. A worker who operates a kiln which burns brick or tile. 4. A worker who alters the properties of a mineral substance by burning. 5. A worker who uses a flame-cutting torch to cut metals. [MECH ENG] A unit of a steam boiler which mixes and directs the flow of fuel and air so as to ensure rapid ignition and complete combustion. { 'bər·nər }

burner fuel oil [MATER] Any of the petroleum distillate and residual oils used to heat homes and buildings. { 'bər·nər 'fyül ,oil }

burner windbox [ENG] A chamber surrounding a burner, under positive air pressure, for proper distribution and discharge of secondary air. { 'bər·nər 'wind,bäks }

burnettize [ENG] To saturate fabric or wood with a solution of zinc chloride under pressure to keep it from decaying. { bər'ned,īz }

Burnett's syndrome See milk-alkali syndrome. { bər'nets ,sin,drōm }

burn-in [ELECTR] Operation of electronic components before they are applied in order to stabilize their characteristics and reveal defects. [ENG] See freeze. [GRAPHICS] A method for giving specified areas of the image extra exposure while protecting the rest of the image. { 'bərn ,in }

burning [ENG] The firing of clay products placed in a kiln. [MET] 1. Permanent damage to a metal caused by heating beyond temperature limits of the treatment. 2. Deep pitting of a metal caused by excessive pickling. { 'bər·niŋ }

burning constant [ORD] In interior ballistics, a figure expressing the relative rate of burning of a propellant; this constant has a different value for each type of propellant. Symbolized B. { 'bər·niŋ ,kän·stənt }

burning glass [OPTICS] A converging lens used to produce intense heat by converging the rays of the sun on a small area. { 'bər·niŋ ,glas }

burning index See fire-danger meter. { 'bər·niŋ ,in,deks }

burning line [PETRO ENG] A pipeline used to convey refinery fuel gas, as distinguished from gas intended for subsequent processing. { 'bər·niŋ ,līn }

burning oil [MATER] An oil such as kerosine or mineral seal oil suitable for burning. { 'bər·niŋ ,oil }

burning point [ENG] The lowest temperature at which a volatile oil in an open vessel will continue to burn when ignited by a flame held close to its surface; used to test safety of kerosine and other illuminating oils. { 'bər·niŋ ,point }

burning quality [ENG] Rated performance for a burning oil as determined by specified ASTM (American Society for Testing and Materials) tests. { 'bər·niŋ ,kwal· əd·ē }

burning-quality index [ENG] Prediction of burning performance of furnace and heater oils; derived from ASTM (American Society for Testing and Materials) distillation, API (American Petroleum Institute) gravity, paraffinicity, and volatility. { 'bər·niŋ ,kwäl·əd·ē ,in,deks }

burning rate [MATER] The tendency and rate of materials to burn at given temperatures, in contrast to melting or disintegrating. { 'bər·niŋ ,rāt }

burning-rate constant [AERO ENG] A constant, related to initial grain temperature, used in calculating the burning rate of a rocket propellant grain. { 'bər·niŋ ,rāt ,kän·stənt }

burning train See igniter train. { 'bər·niŋ ,trān }

burning velocity [CHEM] The normal velocity of the region of combustion reaction (reaction zone) relative to nonturbulent unburned gas, in the combustion of a flammable mixture. { 'bər·niŋ və'läs·əd·ē }

burnish [ENG] To polish or make shiny. [MET] To develop a smooth, lustrous surface finish by tumbling with steel balls or rubbing with a hard metal pad. { 'bər·nish }

burnisher [ENG] A tool with a hard, smooth rounded edge or surface; used for finishing the edges of scraper blades, for smoothing or polishing plastic or metal surfaces, or for other applications requiring manipulation by rubbing. { 'bər·nə·shər }

burn off [METEOROL] With reference to fog or low stratus cloud layers, to dissipate by daytime heating from the sun. { 'bərn 'óf }

burnout [AERO ENG] 1. An act or instance of fuel or oxidant depletion or of depletion of both at once. 2. The time at which this depletion occurs. 3. The point on a rocket trajectory at which this depletion occurs. [ELEC] Failure of a device due to excessive heat produced by excessive current. [ENG] An instance of a device or a part overheating so as to result in destruction or damage. [GRAPHICS] A degree of exposure of a diazo-coated material that renders the film incapable of producing density when developed because the photosensitive diazo component has been destroyed. [NUCLEO] 1. To receive the greatest amount of radiation permissible during a given time. 2. The point at which the heat flux across a surface causes film-blanketing of the surface, resulting in a drop in the film heat-transfer coefficient, overheating, and possible surface failure. { 'bərn,aut }

burnout density [GRAPHICS] The density exhibited by a diazo material after it has been completely degraded by actinic light. { 'bərn,aut ,den·səd·ē }

burnout poison See burnable poison. { 'bərn,aut ,póiz·ən }

burnout velocity [AERO ENG] The velocity of a rocket at the time when depletion of the fuel or oxidant occurs. Also known as all-burnt velocity; burnt velocity. { 'bərn,aut və'läs·əd·ē }

burn shoe [PETRO ENG] A special rotary shoe used in fishing operations for milling away tubular metal components that are causing stuck pipe. { 'bərn ,shü }

Burnside boring machine [MECH ENG] A machine for boring in all types of ground with the feature of controlling water immediately if it is tapped. { 'bərn,sīd 'bor·iŋ mə'shən }

Burnside-Frobenius theorem [MATH] Pertaining to a group of permutations on a finite set, the theorem that the sum over all the permutations, g, of the number of fixed points of g is equal to the product of the number of distinct orbits with respect to the group and the number of permutations in the group. { 'bərn,sīd frō'bē·nē·əs ,thir·əm }

burnt deposit [MET] A dark, powdery deposit obtained by excessive current density in electroplating. { 'bərnt di'päz·ət }

burn-through See jammer finder. { 'bərn,thrü }

burnt lime See calcium oxide. { 'bərnt 'līm }

burnt-on sand [MET] A mixture of sand and metal on the surface of a casting due to metal penetration of the sand mold. { ′bərnt ‚ȯn ‚sand }

burnt plate oil [MATER] A material used to thin etching ink too heavy in body; made by boiling linseed oil and, at a certain temperature, igniting it. { ′bərnt ‚plāt ‚ȯil }

burnt shale [MIN ENG] Carbonaceous shale which has remained for a long period in a colliery tip and undergone spontaneous combustion and converted into a coppery slag material. Also known as oxidized shale. { ′bərnt ′shāl }

burnt velocity See burnout velocity. { ′bərnt və′läs·əd·ē }

burnup [NUCLEO] A measure of nuclear-reactor fuel consumption, expressed either as the percentage of fuel atoms that have undergone fission or as the amount of energy produced per unit weight of fuel. { ′bər‚nəp }

burr [BOT] **1.** A rough or prickly envelope on a fruit. **2.** A fruit so characterized. [MET] A thin, ragged fin left on the edge of a piece of metal by a cutting or punching tool. { bər }

burr ball See lake ball. { ′bər ‚bȯl }

burr mill [FOOD ENG] A mill for grinding crops, in which two ribbed plates or disks rub or crush the material between them. { ′bər ‚mil }

burro [VERT ZOO] A small donkey used as a pack animal. { ′bu̇r·ō }

burrow [MIN ENG] A refuse heap at a coal mine. { ′bər·ō }

bursa [ANAT] A simple sac or cavity with smooth walls containing a clear, slightly sticky fluid and interposed between two moving surfaces of the body to reduce friction. { ′bər·sə }

bursa of Fabricius [VERT ZOO] A thymus-like organ in the form of a diverticulum at the lower end of the alimentary canal in birds. { ′bər·sə əv fə′brēsh·əs }

Burseraceae [BOT] A family of dicotyledonous plants in the order Sapindales characterized by an ovary of two to five cells, prominent resin ducts in the bark and wood, and an intrastaminal disk. { ‚bər·sə′rās·ē‚ē }

bursicle [BOT] A purse or pouchlike receptacle. { ‚bər·sə·kəl }

bursitis [MED] Inflammation of a bursa. { ‚bər′sīd·əs }

burst [COMMUN] **1.** A sudden increase in the strength of a signal being received from beyond line-of-sight range. **2.** A group of bits of characters that are transmitted together as a unit. **3.** A group of errors that occur together in a communication and alter its content. **4.** See color burst. [COMPUT SCI] **1.** To separate a continuous roll of paper into stacks of individual sheets by means of a burster. **2.** The transfer of a collection of records in a storage device, leaving an interval in which data for other requirements can be obtained from or entered into the device. **3.** A sequence of signals regarded as a unit in data transmission. [ELECTR] **1.** An exceptionally large electric pulse in the circuit of an ionization chamber due to the simultaneous arrival of several ionizing particles. **2.** A radar term for a single pulse of radio energy. [ORD] **1.** Continuous fire from an automatic weapon, as from an aircraft machine gun, sometimes described as a long or short burst. **2.** The explosion of a projectile, bomb, or similar munition. { bərst }

burst amplifier [ELECTR] An amplifier stage in a color television receiver that is keyed into conduction and amplification by a horizontal pulse at the instant of each arrival of the color burst. Also known as chroma band-pass amplifier. { ′bərst ‚am·plə‚fī·ər }

burst center [ORD] A point about which the bursts of several projectiles from rounds fired under like conditions are evenly distributed. { ′bərst ‚sen·tər }

burst disk [AERO ENG] A diaphragm designed to burst at a predetermined pressure differential; sometimes used as a valve, for example, in a liquid-propellant line in a rocket. Also known as rupture disk. { ′bərst ‚disk }

Burstein effect [SPECT] The shift of the absorption edge in the spectrum of a semiconductor to higher energies at high carrier densities in the semiconductor. { ′bər‚stīn i‚fekt }

burster [ASTRON] A celestial source of radiation, such as x-rays or gamma rays, that is very intense for brief periods of time and whose nature has not yet been established. [BOT] An abnormally double flower having the calyx split or fragmented. [COMPUT SCI] An off-line device in a computer system used to separate the continuous roll of paper produced as output from a printer into individual sheets, generally along perforations in the roll. [ORD] An explosive element used

in chemical ammunition such as bombs, mines, and shells to open the container and disperse the contents. { ′bər·stər }

bursting charge [ORD] The main explosive charge in a mine, bomb, projectile, or the like that breaks the casing and produces fragmentation or demolition. { ′bər·stiŋ ‚chärj }

bursting layer [ORD] Layer of hard material used in the roofs of dugouts or cave shelters; it sets off projectiles fused for short delay or immediate detonation before they can enter deeply enough to cause great destruction. { ′bər·stiŋ ‚lā·ər }

bursting strength [MECH] A measure of the ability of a material to withstand pressure without rupture; it is the hydraulic pressure required to burst a vessel of given thickness. { ′bər·stiŋ ‚streŋkth }

burst mode [COMPUT SCI] A method of transferring data between a peripheral unit and a control processing unit in a computer system in which the peripheral unit sends the central processor a signal to receive data until the peripheral unit signals that the transfer is completed. { ′bərst ‚mōd }

burst pedestal [COMMUN] Rectangular pulselike television signal which may be part of the color burst; the amplitude of the color burst pedestal is measured from the alternating-current axis of the sine-wave portion to the horizontal pedestal. { ′bərst ‚ped·ə·stəl }

burst pressure [MECH] The maximum inside pressure that a process vessel can safely withstand. { ′bərst ‚presh·ər }

burst range [ORD] Horizontal distance from the piece to the point of burst. { ′bərst ‚rānj }

burst separator [ELECTR] The circuit in a color television receiver that separates the color burst from the composite video signal. { ′bərst sep·ə′rād·ər }

burst slug detector [NUCLEO] A radiation detector used for detecting small leaks in a fuel element of a nuclear reactor by measuring the radiation from short-lived fission products that escape into the coolant. { ′bərst ′sləg di′tek·tər }

burst transmission [COMMUN] A radio transmission in which messages stored for a given time are sent at from 10 to more than 100 times the normal rate, recorded when received, and then slowed down to the normal rate for the user. { ′bərst tranz′mish·ən }

burst wave [FL MECH] Wave of compressed air caused by a bursting projectile or bomb; a detonation wave; it may produce extensive local damage. { ′bərst ‚wāv }

burton [MECH ENG] A small hoisting tackle with two blocks, usually a single block and a double block, with a hook block in the running part of the rope. { ′bərt·ən }

bus [AERO ENG] A spacecraft or missile that is designed to carry one or more separable devices, such as probes or warheads. [COMPUT SCI] The circuitry and wiring connecting the various components of a computer through which data is transmitted; for example, in a personal computer the system bus interconnects the CPU, memory, and input/output devices. [ELEC] **1.** A set of two or more electric conductors that serve as common connections between load circuits and each of the polarities (in direct-current systems) or phases (in alternating-current systems) of the source of electric power. **2.** See busbar. [ELECTR] One or more conductors in a computer along which information is transmitted from any of several sources to any of several destinations. [ENG] A motor vehicle for carrying a large number of passengers. { bəs }

bus architecture [COMPUT SCI] A structure for handling data transmission in a computer system or network, in which components are all linked to a common bus. { ′bəs ′är·kə‚tek·chər }

busbar [ELEC] A heavy, rigid metallic conductor, usually uninsulated, used to carry a large current or to make a common connection between several circuits. Also known as bus. { ′bəs‚bär }

bus cable [ELECTR] An electrical conductor that can be attached to a bus to extend it outside the computer housing or join it to another bus within the same computer. { ′bəs ‚kā·bəl }

Busch lemniscate [METEOROL] The locus in the sky, or on a diagrammatic representation thereof, of all points at which the plane of polarization of diffuse sky radiation is inclined 45° to the vertical; a polarization isocline. Also known as neutral line. { ′bu̇sh lem′nis·kət }

bus cycle [COMPUT SCI] A single transaction between the main memory and the CPU. { ′bəs ‚sī·kəl }

bus duct [ELEC] An enclosed metal unit containing copper or aluminum busbars for distribution of large amounts of power between components of the distribution system. { 'bəs ,däkt }

bus extender [ELECTR] A printed circuit board that can be joined to a bus to increase its capacity. { 'bəs ik,sten·dər }

bushbaby [VERT ZOO] Any of six species of African arboreal primates in two genera (*Galago* and *Euoticus*) of the family Lorisidae. Also known as galago; night ape. { 'bùsh,bā·bē }

bushel [MECH] Abbreviated bu. **1.** A unit of volume (dry measure) used in the United States, equal to 2150.42 cubic inches or approximately 35.239 liters. **2.** A unit of volume (liquid and dry measure) used in Britain, equal to 2219.36 cubic inches or 8 imperial gallons (approximately 36.369 liters). { 'bùsh·əl }

bush fallowing [AGR] A type of subsistence agriculture in which land is cultivated for a period of time and then left uncultivated for several years so that its fertility will be restored. { 'bùsh ,fal·ə·wiŋ }

BUSH HAMMER

An example of a bush hammer.

bush hammer [MECH ENG] A hand-held or power-driven hammer that has a serrated face containing pyramid-shaped points and is used to dress a concrete or stone surface. { 'bùsh ,ham·ər }

bushing *See* nipple. [ELEC] *See* sleeve. [MECH ENG] A removable piece of soft metal or graphite-filled sintered metal, usually in the form of a bearing, that lines a support for a shaft. { 'bùsh·iŋ }

Bushveld Complex [GEOL] In South Africa, an enormous layered intrusion, containing over half the world's platinum, chromium, vanadium, and refractory minerals. { 'bùsh,veld 'käm,pleks }

bus mouse [COMPUT SCI] A mouse that is plugged into a printed circuit board inserted into the computer's bus. { 'bəs ,maùs }

bus network [COMMUN] A communications network whose components are joined together by a single cable. { 'bəs 'net,wərk }

bus reactor [ELEC] An air-core inductor connected between two buses or two sections of the same bus in order to limit the effects of voltage transients on either bus. { 'bəs re'ak·tər }

buster [MET] A pair of dies used in press forging for barreling, for flattening a hot metal billet, or for loosening scale on hot, ferrous forging billets. [MIN ENG] An expanding wedge used to break down coal or rock. { 'bəs·tər }

bustite *See* aubrite. { 'bəs,tīt }

busulfan [PHARM] $CH_3SO_2O(CH_2)_4OSO_2CH_3$ Crystals with a melting point of 114–118°C; soluble in acetone at 25°C; used as an antineoplastic drug. { ,byü'səl·fən }

busway [ELEC] A prefabricated assembly of standard lengths of busbars rigidly supported by solid insulation and enclosed in a sheet-metal housing. { 'bəs,wā }

busy test [COMMUN] A test, in telephony, made to find out whether certain facilities which may be desired, such as a subscriber line or trunk, are available for use. { 'biz·ē ,test }

busy tone [COMMUN] Interrupted low tone returned to the subscriber as an indication that the party's line is busy. { 'biz·ē ,tōn }

1,3-butadiene [ORG CHEM] C_4H_6 A colorless gas, boiling point −4.41°C, a major product of the petrochemical industry; used in the manufacture of synthetic rubber, latex paints, and nylon. { 'wən ,thrē ,byüd·ə'dī·ēn }

butadiene dimer [ORG CHEM] C_8H_{12} The third ingredient in ethylene-propylene-terpolymer (EPT) synthetic rubbers; isomers include 3-methyl-1,4,6-heptatriene, vinylcyclohexene, and cyclooctadiene. { ,byüd·ə'dī·ēn 'dī·mər }

butadiene rubber *See* polybutadiene. { ,byüd·ə'dī·ēn 'rəb·ər }

BUTTERFLY

Adult of the marsh (meadow) fritillary.

butadiene-styrene rubber [MATER] A synthetic rubber that is formed by copolymerization of butadiene and styrene. { ,byüd·ə'dī·ēn 'stī,rēn 'rəb·ər }

Butamer process [CHEM ENG] A method of isomerizing normal butane into isobutane in the presence of hydrogen and a solid, noble-metal catalyst; used to prepare raw material in a gasoline alkylation process. { 'byüd·ə·mər ,präs·əs }

butane [ORG CHEM] C_4H_{10} An alkane of which there are two isomers, *n* and isobutane; occurs in natural gas and is produced by cracking petroleum. { 'byü,tān }

butane dehydrogenation [CHEM ENG] A process to remove hydrogen from butane to produce butene or butadiene. { 'byü,tān dē,hī·drə·jə'nā·shən }

2,3-butanediol [ORG CHEM] $CH_3CHOHCHOHCH_3$ A

major fermentation product of several species of bacteria. { ,tü ,thrē ,byüd·ə·nēd·ē,ól }

butane vapor-phase isomerization [CHEM ENG] A process to isomerize normal butane into isobutane in the presence of aluminum chloride catalyst and hydrogen chloride promoter. { 'byü,tān 'vā·pər ,fāz ī,säm·ə·rə'zā·shən }

butanol [ORG CHEM] Any one of four isomeric alcohols having the formula C_4H_9OH; colorless, toxic liquids soluble in most organic liquids. Also known as butyl alcohol. { 'byüt·ən,ól }

butanol fermentation [MICROBIO] Butanol production as a result of the fermentation of corn and molasses by the anaerobic bacterium *Clostridium acetobutylicum*. { 'byüt·ən,ól ,fər·mən'tā·shən }

butazolidine *See* phenylbutazone. { ,byüd·ə'zäl·ə,dēn }

butcher linen [TEXT] A coarse, originally homespun linen that is now imitated in many synthetic fiber fabrics. { 'bùch·ər ,lin·ən }

butene-1 [ORG CHEM] $CH_3CH_2CHCH_2$ A colorless, highly flammable gas; insoluble in water, soluble in organic solvents; used to produce polybutenes, butadiene aldehydes, and other organic derivatives. { 'byü,tēn 'wən }

butene-2 [ORG CHEM] $CH_3CHCHCH_3$ A colorless, highly flammable gas, used to make butadiene and in the synthesis of four- and five-carbon organic molecules; the cis form, boiling point 3.7°C, is insoluble in water, soluble in organic solvents, and is also known as high-boiling butene-2; the trans form, boiling point 0.88°C, is insoluble in water, soluble in most organic solvents, and is also known as low-boiling butene-2. { 'byü,tēn 'tü }

Butler finish *See* satin finish. { 'bət·lər ,fin·ish }

butlerite [MINERAL] $Fe(SO_4)(OH)\cdot2H_2O$ A deep orange, monoclinic mineral consisting of a hydrated basic ferric sulfate. Also known as parabutlerite. { 'bət·lə,rīt }

Butler oscillator [ELEC] Oscillator in which a piezoelectric crystal is connected between the cathode of two tubes, one functioning as a cathode follower, and the other as a grounded-grid amplifier. { 'bət·lər 'äs·ə,lād·ər }

Butomaceae [BOT] A family of monocotyledonous plants in the order Alismatales characterized by secretory canals, linear leaves, and a straight embryo. { ,byüd·ə'mās·ē,ē }

butopyronoxyl [ORG CHEM] $C_{12}H_{18}O_4$ A yellow to amber liquid with a boiling point of 256–260°C; miscible with ether, glacial acetic acid, alcohol, and chloroform; used as an insect repellent for skin and clothing. { ,byüd·ə,pī·rə'näk·səl }

2-butoxyethanol [ORG CHEM] $HOCH_2CH_2OC_4H_9$ A liquid with a boiling point of 171–172°C; soluble in most organic solvents and water; used in dry cleaning as a solvent for nitrocellulose, albumin, resins, oil, and grease. { ,tü ,byü,täk·sē'eth·ə,nól }

butt [BOT] The portion of a plant from which the roots extend, for example, the base of a tree trunk. [BUILD] The bottom or cover edge of a shingle. [DES ENG] The enlarged and squared-off end of a connecting rod or similar link in a machine. [MIN ENG] Coal exposed at right angles to the face and, in contrast to the face, generally having a rough surface. [ORD] Rear end of the stock of a rifle or other small arm. { bət }

butt cable *See* hand cable. { 'bət ,kā·bəl }

butt contact [ELEC] A hemispherically shaped contact designed to mate against a similarly shaped contact. { 'bət ,kän,takt }

butte [GEOGR] A detached hill or ridge which rises abruptly. { byüt }

butter [FOOD ENG] A fatty food product made from milk or cream or both and consisting of a solid emulsion of fat globules, air, and water. { 'bəd·ər }

butterfat [FOOD ENG] A mixture of glycerides derived from fatty acids; the natural fat of milk and butter. { 'bəd·ər,fat }

butter finish [MET] The semilustrous surface produced with a mildly abrasive wheel. { 'bəd·ər ,fin·ish }

butterfly [INV ZOO] Any insect belonging to the lepidopteran suborder Rhopalocera, characterized by a slender body, broad colorful wings, and club-shaped antennae. [MATER] A color imperfection in a lime-putty finish caused by lumps in the lime that were not broken up during mixing. { 'bəd·ər,flī }

butterfly bomb [ORD] A small fragmentation or antipersonnel bomb equipped with two folding wings which rotate and arm the fuse as the bomb descends; designed to be dropped in clusters, they are frequently fitted with antidisturbance or delay fuses. { 'bəd·ər₁flī ₁bäm }

butterfly capacitor [ELEC] A variable capacitor having stator and rotor plates shaped like butterfly wings, with the stator plates having an outer ring to provide an inductance so that both capacitance and inductance can be varied, thereby giving a wide tuning range. { 'bəd·ər₁flī kə'pas·əd·ər }

butterfly damper See butterfly valve. { 'bəd·ər₁flī ₁dam·pər }

butterfly effect [PHYS] In a chaotic system, the ability of miniscule changes in initial conditions (such as the flap of a butterfly's wings) to have far-reaching, large-scale effects on the development of the system (such as the course of weather a continent away). { 'bəd·ər₁flī i₁fekt }

butterfly network [COMPUT SCI] A scheme that connects the units of a multiprocessing system and needs n stages to connect 2^n processors; at each stage a switch is thrown, depending on a particular bit in the addresses of the processors being connected. { 'bəd·ər₁flī ₁net₁wərk }

butterfly nut See wing nut. { 'bəd·ər₁flī ₁nət }

butterfly valve [ENG] A valve that utilizes a turnable disk element to regulate flow in a pipe or duct system, such as a hydraulic turbine or a ventilating system. Also known as butterfly damper. { 'bəd·ər₁flī ₁valv }

buttering [MET] Coating the faces of a weld joint prior to welding to prevent cross contamination of the weld metal and base metal. { 'bəd·ə·riŋ }

buttermilk [FOOD ENG] A fermentation food product prepared by inoculating sweet milk with cultures of *Streptococcus lactis* and *Leuconostoc citrovorum*. { 'bəd·ər₁milk }

butter rock See halotrichite. { 'bəd·ər ₁räk }

butterscotch [FOOD ENG] A confection formed by heating sugar and butter together until a light brown color develops. { 'bəd·ər₁skäch }

Butterworth filter [ELECTR] An electric filter whose pass band (graph of transmission versus frequency) has a maximally flat shape. { 'bəd·ər₁wərth 'fil·tər }

Butterworth head [MECH ENG] A mechanical hose head with revolving nozzles; used to wash down shipboard storage tanks. { 'bəd·ər₁wərth ₁hed }

butt fusion [ENG] The joining of two pieces of plastic or metal pipes or sheets by heating the ends until they are molten and then pressing them together to form a homogeneous bond. { 'bət ₁fyü·zhən }

butt gage [ENG] A tool used to mark the outline for the hinges on a door. { 'bət ₁gāj }

buttgenbachite [MINERAL] $Cu_{19}(NO_3)_2Cl_4(OH)_{32}·3H_2O$ An azure blue, hexagonal mineral consisting of a hydrated basic chloride-sulfate-nitrate of copper. { 'bət·gən₁ba₁kīt }

butt joint [ELEC] A connection formed by placing the ends of two conductors together and joining them by welding, brazing, or soldering. [ELECTROMAG] A connection giving physical contact between the ends of two waveguides to maintain electrical continuity. [ENG] A joint in which the parts to be joined are fastened end to end or edge to edge with one or more cover plates (or other strengthening) generally used to accomplish the joining. { 'bət ₁jóint }

buttock [MIN ENG] A corner formed by two coal faces more or less at right angles, such as the end of a working face. { 'bəd·ək }

buttock lines [ENG] The lines of intersection of the surface of an aircraft or its float, or of the hull of a ship, with its longitudinal vertical planes. Also known as buttocks. { 'bəd·ək ₁līnz }

buttocks [ANAT] The two fleshy parts of the body posterior to the hip joints. [ENG] See buttock lines. [NAV ARCH] The convex part of the stern end of a ship above the water line. Also known as counter. { 'bəd·əks }

butt-off [MET] To supplement ramming in the production of castings by either manual or pneumatic jolting. { 'bət ₁óf }

button [COMPUT SCI] A small circle or rectangle on a graphical user interface, such that moving the pointer to it and clicking the mouse initiates some action. [ELECTR] **1.** A small, round piece of metal alloyed to the base wafer of an alloy-junction transistor. Also known as dot. **2.** The container that holds the carbon granules of a carbon microphone. Also known as carbon button. [MET] **1.** Mass of metal remaining in a crucible after fusion has been completed. **2.** That part of a weld

which tears out in the destructive testing of spot-, seam-, or projection-welded specimens. { 'bət·ən }

button balance [MIN ENG] A small, very delicate balance used for weighing assay buttons. { 'bət·ən ₁bal·əns }

button bit [DES ENG] A drilling bit made with button-shaped tungsten carbide inserts. { 'bət·ən ₁bit }

button die [DES ENG] A mating member, usually replaceable, for a piercing punch. Also known as die bushing. { 'bət·ən ₁dī }

buttonhead [DES ENG] A screw, bolt, or rivet with a hemispherical head. { 'bət·ən₁hed }

buttonhook contact [ELEC] A curved, hooklike contact often used on feed-through terminals of headers to facilitate soldering or unsoldering of leads. { 'bət·ən₁hùk 'kän₁takt }

buttress [ARCH] An upright projection that supports or resists lateral forces in a building. [BOT] A ridge of wood developed in the angle between a lateral root and the butt of a tree. [CIV ENG] A pier constructed at right angles to a restraining wall on the side opposite to the restrained material; increases the strength and thrust resistance of the wall. [PALEON] A ridge on the inner surface of a pelecypod valve which acts as a support for part of the hinge. { 'bə·trəs }

buttress dam [CIV ENG] A concrete dam constructed as a series of buttresses. { 'bə·trəs ₁dam }

buttress sands [GEOL] Sandstone bodies deposited above an unconformity; the upper portion rests upon the surface of the unconformity. { 'bə·trəs ₁sanz }

buttress thread [DES ENG] A screw thread whose forward face is perpendicular to the screw axis and whose back face is at an angle to the axis, so that the thread is both efficient in transmitting power and strong. { 'bə·trəs ₁thred }

buttress-thread casing [PETRO ENG] A drill casing in which the ends of the sections are buttressed together and held in place with a short threaded outer sleeve; used where greater than normal clearance, strength, and leak resistance are needed. { 'bə·trəs ₁thred 'kās·iŋ }

butt rot [PL PATH] A fungus decay of the base of a tree trunk caused by polypores (such as *Fomea* species). { 'bət ₁rät }

butt weld [MET] A weld that joins the ends of two pieces of metal of similar cross section without overlapping. { 'bət ₁weld }

butyl [ORG CHEM] Any of the four variations of the hydrocarbon radical C_4H_9: $CH_3CH_2CH_2CH_2-$, $(CH_3)_2CHCH_2-$, $CH_3CH_2CHCH_3-$, and $(CH_3)_3C-$. { 'byüd·əl }

butyl acetate [ORG CHEM] $CH_3COOC_4H_9$ A colorless liquid slightly soluble in water; used as a solvent. { 'byüd·əl 'as·ə₁tāt }

butyl acetoacetate [ORG CHEM] $C_8H_{14}O_3$ A colorless liquid with a boiling point of 213.9°C; soluble in alcohol and ether; used for synthesis of dyestuffs and pharmaceuticals. { 'byüd·əl ₁as·ə·tō'as·ə₁tāt }

butyl acrylate [ORG CHEM] $CH_2CHCOOC_4H_9$ A colorless liquid that is nearly insoluble in water and polymerizes readily upon heating; used as an intermediate for organic synthesis, polymers, and copolymers. { 'byüd·əl 'ak·rə₁lāt }

butyl alcohol See butanol. { 'byüd·əl 'al·kə₁hól }

***n*-butylamine** [ORG CHEM] $C_4H_9NH_2$ A colorless, flammable liquid; miscible with water and ethanol; used as an intermediate in organic synthesis and to make insecticides, emulsifying agents, and pharmaceuticals. { ₁en 'byüd·əl·ə₁mēn }

***sec*-butylamine** [ORG CHEM] $CH_3CHNH_2C_2H_5$ A flammable, colorless liquid; boils in the range 63–68°C; may be used as an intermediate in organic synthesis. { ₁sek 'byüd·əl·ə₁mēn }

***tert*-butylamine** [ORG CHEM] $(CH_3)_3CNH_2$ A flammable liquid; boiling range 63–68°C; may be used in organic synthesis as an intermediate. { ₁tərt 'byüd·əl·ə₁mēn }

***n*-butyl-*para*-aminobenzoate** [PHARM] $H_2NC_6H_4COO-C_4H_9$ A white, crystalline powder with a melting point of 57–59°C; soluble in dilute acids, alcohol, chloroform, ether, and fatty oils; used as a local anesthetic, in burn treatment and ointments, and in suntan preparations to absorb ultraviolet light. { ₁en 'byüd·əl 'par·ə ₁am·ə·nō'ben·zə₁wāt }

butylate [ORG CHEM] $C_{11}H_{23}NOS$ A colorless liquid used as an herbicide for preplant control of weeds in corn. { 'byüd·əl₁āt }

butylated hydroxyanisole [ORG CHEM] $(CH_3)_3CC_6H_3O-H(OCH_3)$ An antioxidant consisting chiefly of a mixture of

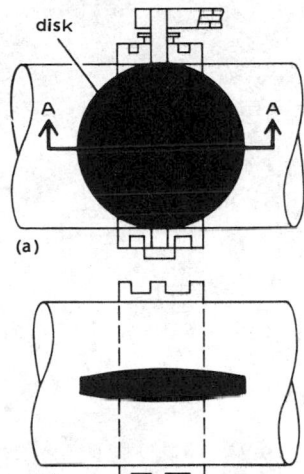

BUTTERFLY VALVE

Butterfly-type valve for penstock is typically 25 feet (7.6 meters) in diameter. (*a*) Elevation. (*b*) Section A-A.

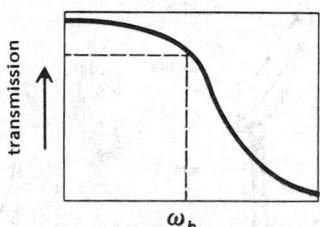

BUTTERWORTH FILTER

Plot of the passband of the Butterworth filter; ω_b is the point where the transmission function falls below the passband tolerance.

2- and 3-*tert*-butyl-4-hydroxyanisole and used to control rancidity of lard and animal fats in foods. Abbreviated BHA. { 'byüd·əl,ād·əd hī,dräk·sē'an·ə,sól }

butylated hydroxytoluene [ORG CHEM] $[(CH_3)_3C]_2$-$C_6H_2(CH_3)OH$ Crystals with a melting point of 72°C; soluble in toluene, methanol, and ethanol; used as an antioxidant in foods, in petroleum products, and for synthetic rubbers. Abbreviated BHT. { 'byüd·əl,ād·əd hī,dräk·sē'täl·yə,wēn }

butylbenzene [ORG CHEM] $C_6H_5C_4H_9$ A colorless liquid used as a raw material for organic synthesis, especially for insecticides; forms are normal (1-phenylbutane), secondary (2-phenylbutane), and tertiary (2-methyl-2-phenylpropane). { 'byüd·əl,ben,zēn }

N-sec-butyl-4-*tert*-butyl-2,6-dinitroaniline [ORG CHEM] $C_{14}H_{21}N_3O_4$ Orange crystals with a melting point of 60–61°C; solubility in water is 1.0 part per million at 24°C; used as a preemergence herbicide. { 'en ¦sek 'byüd·əl ¦for ¦tərt 'byüd·əl ¦tü ¦siks ,dī,nī·trō'an·ə,lēn }

butyl carbinol [ORG CHEM] $(CH_3)_3CCH_2OH$ Colorless crystals that melt at 52°C; slightly soluble in water. { 'byüd·əl 'kär·bə,nól }

butyl chloride [ORG CHEM] C_4H_9Cl A colorless liquid used as an alkylating agent in organic synthesis, as a solvent, and as an anthelminthic; forms are normal (1-chlorobutane), secondary, and iso or tertiary. { 'byüd·əl 'klór,īd }

tert-butyl chloroacetate [ORG CHEM] $ClCH_2COOC(CH_3)_3$ A liquid with a boiling point of 155°C; hydrolyzes to *tert*-butyl alcohol and chloroacetic acid; used in glycidic ester condensation. { ¦tərt 'byüd·əl ,klór·ō'as·ə,tāt }

butyl citrate [ORG CHEM] $C_3H_5O(COOC_4H_9)_3$ A colorless, odorless, nonvolatile liquid, almost insoluble in water; used as a plasticizer, solvent for cellulose nitrate, and antifoam agent. { 'byüd·əl 'sī,trāt }

butyl diglycol carbonate [ORG CHEM] $(C_4H_9OCO_2CH_2$-$CH_2)_2O$ A colorless, combustible liquid with a boiling range of 164–166°C; used as a plasticizer and solvent and in pharmaceuticals and lubricants manufacture. { 'byüd·əl dī'glī,kól 'kär·bə·nāt }

butylene [ORG CHEM] Any of three isomeric alkene hydrocarbons with the formula C_4H_8; all are flammable and easily liquefied gases. { 'byüd·ə,lēn }

1,3-butylene glycol [ORG CHEM] $HOCH_2CH_2CH(OH)CH_3$ A viscous, colorless, hygroscopic liquid; soluble in water and alcohol; used as a solvent, food additive, and flavoring, and for plasticizers and polyurethanes. { ¦wən ¦thrē 'byüd·ə,lēn 'glī,kól }

1,4-butylene glycol [ORG CHEM] $HOCH_2CH_2CH_2CH_2OH$ A colorless, combustible, oily liquid with a boiling point of 230°C; soluble in alcohol; used as a solvent and humectant, and in plastics and pharmaceuticals manufacture. { ¦wən ¦fór 'byüd·ə,lēn 'glī,kól }

1,2-butylene oxide [ORG CHEM] $H_2COCHCH_2CH_3$ A colorless, water-soluble liquid with a boiling point of 63°C; used as an intermediate for various polymers. { ¦wən ¦tü 'byüd·ə,lēn 'äk,sīd }

butyl ether [ORG CHEM] $C_8H_{18}O$ A colorless liquid, boiling at 142°C, and almost insoluble in water; used as an extracting agent, as a medium for Grignard and other reactions, and for purifying other solvents. { 'byüd·əl 'ē·thər }

butyl formate [ORG CHEM] $HCOOC_4H_9$ An ester of formic acid and butyl alcohol. { 'byüd·əl 'fór,māt }

tert-butylhydroperoxide [ORG CHEM] $(CH_3)_3COOH$ A liquid soluble in organic solvents; used as a catalyst in polymerization reactions, to introduce the peroxy group into organic molecules. { ¦tərt 'byüd·əl,hī·drō·pə'räk,sīd }

butyl lactate [ORG CHEM] $CH_3CHOHCOOC_4H_9$ A stable liquid, water-white and nontoxic, miscible with many solvents; used as a solvent for resins and gums, in lacquers and varnishes, and as a chemical intermediate. { 'byüd·əl 'lak,tāt }

butyl mercaptan [ORG CHEM] C_4H_9SH A colorless, odorous liquid, a component of skunk secretion; used commercially as a gas-odorizing agent. { 'byüd·əl mər'kap·tan }

butyl oleate [ORG CHEM] $C_{22}H_{42}O_2$ A butyl ester of oleic acid; used as a plasticizer. { 'byüd·əl 'ō·lē,āt }

para-tert-butylphenol [ORG CHEM] $(CH_3)_3CC_6H_4OH$ Needlelike crystals with a melting point of 98°C; soluble in alcohol and ether; used as an intermediate in production of varnish and lacquer resins, an additive in motor oil, and an

ingredient in deemulsifiers in oil fields. { ¦par·ə ¦tərt 'byüd·əl'fē,nól }

butyl propionate [ORG CHEM] $C_2H_5COOC_4H_9$ A colorless aromatic liquid; used in fruit essences. { 'byüd·əl 'prō·pē·ə,nāt }

butyl rubber [MATER] A synthetic rubber made by the polymerization of isoprene and isobutylene. { 'byüd·əl ,rəb·ər }

butyl stearate [ORG CHEM] $C_{17}H_{35}COOC_4H_9$ A liquid that solidifies at approximately 19°C; mixes with vegetable oils and is soluble in alcohol and ethers but insoluble in water; used as a lubricant, in polishes, as a plasticizer, and as a dye solvent. { 'byüd·əl 'stir,āt }

butynedial [ORG CHEM] $HOCH_2C:CCH_2OH$ White crystals with a melting point of 58°C; soluble in water, aqueous acids, alcohol, and acetone; used as a corrosion inhibitor, defoliant, electroplating brightener, and polymerization accelerator. { ,byüd·ə'nēd·ē·əl }

butyraldehyde [ORG CHEM] $CH_3(CH_2)_2CHO$ A colorless liquid boiling at 75.7°C; soluble in ether and alcohol, insoluble in water; derived from the oxo process. { ¦byüd·ər'al·də,hīd }

butyrate [ORG CHEM] An ester or salt of butyric acid containing the $C_4H_7O_2$ radical. { 'byüd·ə,rāt }

butyric acid [ORG CHEM] $CH_3CH_2CH_2COOH$ A colorless, combustible liquid with boiling point 163.5°C (757 mmHg); soluble in water, alcohol, and ether; used in synthesis of flavors, in pharmaceuticals, and in emulsifying agents. { byü'tir·ik 'as·əd }

butyric anhydride [ORG CHEM] $C_8H_{14}O_3$ A colorless liquid that decomposes in water to form butyric acid; exists in two isomeric forms. { byü'tir·ik an'hī,drīd }

butyric fermentation [BIOCHEM] Fermentation in which butyric acid is produced by certain anaerobic bacteria acting on organic substances, such as butter; occurs in putrefaction and in digestion in herbivorous mammals. { byü'tir·ik fər·mən'tā·shən }

butyrinase [BIOCHEM] An enzyme that hydrolyzes butyrin, found in the blood serum. { 'byüd·ə·rə,nās }

Butyrivibrio [MICROBIO] A genus of gram-negative, strictly anaerobic bacteria of uncertain affiliation; motile curved rods occur singly, in chains, or in filaments; ferment glucose to produce butyrate. { ¦byüd·ə·rə'vī·brē,ō }

butyrolactone [ORG CHEM] $C_4H_6O_2$ A liquid, the anhydride of butyric acid; used as a solvent in the manufacture of plastics. { ¦byüd·ə·rō'lak,tōn }

butyronitrile [ORG CHEM] $CH_3(CH_2)_2CN$ A toxic, colorless liquid with a boiling point of 116–117.7°C; soluble in alcohol and ether; used in industrial, chemical, and pharmaceutical products, and in poultry medicines. { 'byüd·ə,rän'ī,tril }

Buxbaumiales [BOT] An order of very small, atypical mosses (Bryopsida) composed of three genera and found on soil, rock, and rotten wood. { ,bəks,bóm·ē'ā·lēz }

buy-back crude [PETRO ENG] Oil in which the government of the production territory has a right to a share if it is a stockholder in the oil-producing company. Also known as participation crude. { 'bī ,bak ,krüd }

Buys-Ballot's law [METEOROL] A law describing the relationship of the horizontal wind direction in the atmosphere to the pressure distribution: if one stands with one's back to the wind, the pressure to the left is lower than to the right in the Northern Hemisphere; in the Southern Hemisphere the relation is reversed. Also known as baric wind law. { 'bīz bə'läts ,lò }

buzz [AERO ENG] Sustained oscillation of an aerodynamic control surface caused by intermittent flow separation on the surface, or by a motion of shock waves across the surface, or by a combination of flow separation and shock-wave motion on the surface. [CONT SYS] *See* dither. [ELECTR] The condition of a combinatorial circuit with feedback that has undergone a transition, caused by the inputs, from an unstable state to a new state that is also unstable. [FL MECH] In supersonic diffuser aerodynamics, a nonsteady shock motion and airflow associated with the shock system ahead of the inlet. { bəz }

buzzer [ELECTROMAG] Electromagnetic device with an armature that vibrates rapidly, making a buzzing sound. { 'bəz·ər }

B virus [VIROL] An animal virus belonging to subgroup A of the herpesvirus group. { 'bē ,vī·rəs }

BUXBAUMIALES

capsule

(b)

seta

(a) (c) (d)

Morphological features of Buxbaumiales. (a) Sporophyte of *Buxbaumia aphylla (from W. H. Welch, Mosses of Indiana, Indiana Department of Conservation, 1957). (b) diphyscium foliosum, habit sketch, (c) leaves, and (d) perichaetial leaf (from H. S. Conard, How to Know the Mosses and Liverworts, William C. Brown, 1956).*

(b,v,r,k,λ)-design *See* balanced incomplete block design. { ¦bē ¦vē ¦är ¦kā 'lam·də di‚zīn }

BW *See* biological warfare.

BWE *See* bucket-wheel excavator.

BWG *See* Birmingham wire gage.

BWO *See* backward-wave oscillator.

BWR *See* boiling-water reactor.

BX cable [ELEC] Insulated wires in flexible metal tubing used for bringing electric power to electronic equipment. { ¦bē¦eks ¦kā·bəl }

byerite [GEOL] Bituminous coal that does not crack in fire and melts and enlarges upon heating. { 'bī·ə‚rīt }

Byers process [MET] The main process for manufacturing wrought iron; pig iron is melted in a cupola, desulfurized in a ladle, and refined in a Bessemer converter. { 'bī·ərz ‚präs·əs }

byon [GEOL] Gem-bearing gravel, particularly that with brownish-yellow clay in which corundum, rubies, sapphires, and so forth occur. { 'bī‚än }

bypass [CIV ENG] A road which carries traffic around a congested district or temporary obstruction. [COMMUN] The use of alternative systems, such as satellite and microwave, to transmit data and voice signals, avoiding use of the communication lines of the local telephone company. [ELEC] A shunt path around some element or elements of a circuit. [ENG] An alternating, usually smaller, diversionary flow path in a fluid dynamic system to avoid some device, fixture, or obstruction. { 'bī‚pas }

bypass capacitor [ELEC] A capacitor connected to provide a low-impedance path for radio-frequency or audio-frequency currents around a circuit element. Also known as bypass condenser. { 'bī‚pas kə'pas·əd·ər }

bypass channel [CIV ENG] **1.** A channel built to carry excess water from a stream. Also known as flood relief channel; floodway. **2.** A channel constructed to divert water from a main channel. { 'bī‚pas ‚chan·əl }

bypass condenser *See* bypass capacitor. { 'bī‚pas kən'den·sər }

bypass filter [ELECTR] Filter which provides a low-attenuation path around some other equipment, such as a carrier frequency filter used to bypass a physical telephone repeater station. { 'bī‚pas ‚fil·tər }

bypass valve [ENG] A valve that opens to direct fluid elsewhere when a pressure limit is exceeded. { 'bī‚pas ‚valv }

by-product [ENG] A product from a manufacturing process that is not considered the principal material. { 'bī‚präd·əkt }

Byrrhidae [INV ZOO] The pill beetles, the single family of the coleopteran insect superfamily Byrrhoidea. { 'bir·ə‚dē }

Byrrhoidea [INV ZOO] A superfamily of coleopteran insects in the superorder Polyphaga. { bə'rȯid·ē·ə }

bysmalith [GEOL] A body of igneous rock that is more or less vertical and cylindrical; it crosscuts adjacent sediments. { 'biz·mə‚lith }

byssinosis [MED] A pneumoconiosis caused by the inhalation of cotton dust. Also known as brown lung disease. { ¦bīs·ə'nō·səs }

byte [COMPUT SCI] A sequence of adjacent binary digits operated upon as a unit in a computer and usually shorter than a word. { bīt }

byte addressable computer [COMPUT SCI] A computer in which each byte of memory can be addressed independently of the others. { ¦bīt ə¦dres·ə·bəl kəm'pyüd·ər }

bytecode [COMPUT SCI] Compiled Java programs that can be transferred across a network and executed by the Java virtual machine. { 'bīt‚kōd }

byte mode [COMPUT SCI] A method of transferring data between a peripheral unit and a central processor in which one byte is transferred at a time. { 'bīt ‚mōd }

byte multiplexor channel [COMPUT SCI] A transmission channel in a computer system that can transmit data simultaneously from several devices and only one byte at a time. { 'bīt 'məlt·i‚plek·sər ‚chan·əl }

byte-oriented protocol [COMPUT SCI] A communications protocol in which full bytes are used as control codes. Also known as character-oriented protocol. { ¦bīt ‚ȯr·ē‚ent·əd 'prōd·ə‚kȯl }

by the head *See* down by the head. { ‚bī thə 'hed }

by the stern *See* down by the stern. { ‚bī thə 'stərn }

bytownite [MINERAL] A plagioclase feldspar with a composition ranging from $Ab_{30}An_{70}$ to $Ab_{10}An_{90}$, where Ab = $NaAlSi_3O_8$ and An = $CaAl_2Si_2O_8$; occurs in basic and ultrabasic igneous rock. { 'bī·taů‚nīt }

Byturidae [INV ZOO] The raspberry fruitworms, a small family of coleopteran insects in the superfamily Cucujoidea. { bī'tùr·ə‚dē }

BYV *See* beet yellows virus.

c *See* calorie; centi-; charmed quark; curie.

C [CHEM] *See* carbon. [COMPUT SCI] A programming language designed to implement the Unix operating system. [ELEC] *See* capacitance; capacitor; coulomb.

C++ [COMPUT SCI] An object-oriented language that was created as an extension to the C language. { 'sē,pləs,pləs }

C² *See* command and control. { 'sē 'tü }

C³ *See* command, control, and communications. { 'sē 'thrē }

Ca *See* calcium.

⁴⁵Ca *See* calcium-45.

CA *See* conflict alert.

caatinga [ECOL] A sparse, stunted forest in areas of little rainfall in northeastern Brazil; trees are leafless in the dry season. { kä'tiŋ·gə }

Ca²⁺ ATPase *See* calcium pump. { ¦sē,ā¦tü¦pləs ¦ā¦tē'pās *or* ¦kal·sē·əm ¦tü¦pləs ¦ā¦te'pās }

cab [ENG] In a locomotive, truck, tractor, or hoisting apparatus, a compartment for the operator. { kab }

caballing [OCEANOGR] The mixing of two water masses of identical in situ densities but different in situ temperatures and salinities, such that the resulting mixture is denser than its components and therefore sinks. { kə'bal·iŋ }

cabane [AERO ENG] The arrangement of struts used on early types of airplanes to brace the wings. { kə'ban }

Cabannes' factor [ANALY CHEM] An equational factor to correct for the depolarization effect of the horizontal components of scattered light during the determination of molecular weight by optical methods. { kə'bänz ,fak·tər }

Cabannes-Hofmann effect [GRAPHICS] A dependence of photographic film on exposure time whereby a film exposed at low intensity for a long time blackens more rapidly during development than if the same light exposure is provided at higher intensity for a short time. { kə'ban 'hof·mən i,fekt }

cabbage [BOT] *Brassica oleracea* var. *capitata*. A biennial crucifer of the order Capparales grown for its head of edible leaves. { 'kab·ij }

cabbage yellows [PL PATH] A fungus disease of cabbage caused by *Fusarium conglutinans* and characterized by yellowing and dwarfing. { 'kab·ij ,yel·ōz }

cabble [MET] To break up into pieces preparatory to the processes of fagoting, fusing, and rolling into bars, as is done with charcoal iron. { 'ka·bəl }

cabezon [VERT ZOO] *Scorpaenichthys marmoratus*. A fish that is the largest of the sculpin species, weighing as much as 25 pounds (11.3 kilograms) and reaching a length of 30 inches (76 centimeters). { 'kab·ə,zōn, *or* ,ka·bə'zón }

Cabibbo theory [PARTIC PHYS] A theory describing baryon beta-decay processes, according to which the amplitude for such processes is given by $G\{\cos\Theta[V(\Delta s = 0) + A(\Delta s = 0)] + \sin\Theta[V(\Delta s = +1) + A(\Delta s = +1)]\}$, where Θ is the Cabibbo angle, Δs is the change in strangeness for the baryon, G is a universal beta-decay amplitude, and V and A are vector and axial vector amplitudes, respectively; it is experimentally determined that $\sin\Theta \approx 0.25$, so that $\cos\Theta = 0.97$. { ka'bi·bō ,thē·ə·rē }

cabinet file [DES ENG] A coarse-toothed file with flat and convex faces used for woodworking. { 'kab·ə·nət ,fīl }

cabinet hardware [DES ENG] Parts for the final trim of a cabinet, such as fastening hinges, drawer pulls, and knobs. { 'kab·ə·nət ,härd,wer }

cabinet saw [DES ENG] A short saw, one edge used for ripping, the other for crosscutting. { 'kab·ə·nət ,só }

cabinet scraper [DES ENG] A steel tool with a contoured edge used to remove irregularities on a wood surface. { 'kab·ə·nət ,skrāp·ər }

cable [ARCH] A convex molding within one of the vertical grooves of a column or pilaster. [DES ENG] A stranded, ropelike assembly of wire or fiber. [ELEC] Strands of insulated electrical conductors laid together, usually around a central core, and surrounded by a heavy insulation. [OCEANOGR] *See* cable length. { 'kā·bəl }

cable-and-trunk schematic [ELEC] A drawing which shows, in block form, the interconnection between all major electric circuits in an office. { ¦kā·bəl ən 'trəŋk skə'mad·ik }

cable armor [ELEC] One or more layers of extra-strength material, such as steel wire or tape, to reinforce the usual lead wall in cable construction. { 'kā·bəl ,är·mər }

cable bend [NAV ARCH] **1.** A small rope used to tie the end of a cable into a loop to secure an anchor. **2.** A knot used to attach a cable to an anchor. { 'kā·bəl ,bend }

cable bridge [ELEC] A rubber tube that encloses cables running over a floor or other surface. { 'kā·bəl ,brij }

cable buoy [ENG] A buoy used to mark one end of a submarine underwater cable during time of installation or repair. { 'kā·bəl ,bói }

cable chain *See* chain cable. { 'kā·bəl ,chān }

cable code *See* Morse cable code. { 'kā·bəl ,kōd }

cable complement [ELEC] Group of wire pairs in a cable having some common distinguishing characteristic. { 'kā·bəl ,käm·plə·mənt }

cable conveyor [MECH ENG] A powered conveyor in which a trolley runs on a flexible, torque-transmitting cable that has helical threads. { 'kā·bəl kən'vā·ər }

cable delay [COMPUT SCI] The time required for one bit of data to go through a cable, about 1.5 nanoseconds per foot of cable. { 'kā·bəl di'lā }

cable drilling [ENG] Rock drilling in which the rock is penetrated by percussion, at the bottom of the hole, of a bit suspended from a wire line and given motion by a beam pivoted at the center. { 'kā·bəl ,dril·iŋ }

cable duct [ENG] A pipe, either earthenware or concrete, through which prestressing wires or electric cable are pulled. { 'kā·bəl ,dəkt }

cable fill [ELEC] Ratio of the number of wire pairs in use to the total number of pairs in a cable. { 'kā·bəl ,fil }

cable lacquer [MATER] Black, colored, or clear lacquer made from synthetic resins, having a high dielectric strength and being resistant to oils and heat; used to give a tough, flexible coating. { 'kā·bəl ,la·kər }

cable-laid [DES ENG] Consisting of three ropes with a left-hand twist, each rope having three twisted strands. { 'kā·bəl ,lād }

cable layer [NAV ARCH] A type of ship fitted with huge reels of submarine telephone cable and devices for paying out the cable to the sea bottom. { 'kā·bəl ,lā·ər }

cable length [OCEANOGR] A unit of distance, originally equal to the length of a ship's anchor cable, now variously considered to be 600 feet (183 meters), 608 feet (185.3 meters, one-tenth of a British nautical mile), or 720 feet or 120 fathoms (219.5 meters). Also known as cable. { 'kā·bəl ,lengkth }

cableman [ENG] A person who installs, repairs, or otherwise works with cables. { 'kā·bəl·mən }

cable matcher *See* gender changer. { 'kā·bəl ,mach·ər }

CABBAGE

Cabbage (*Brassica olergacea* var. *capitata*), cultivar Golden Acre 84. (*Joseph Harris Co., Inc., Rochester, NY*)

cable messenger [ELEC] Stranded group of wires supported above the ground at intervals by poles or other structures and employed to furnish, within these intervals, frequent points of support for conductors or cables. { 'kā·bəl ,mes·ən·jər }

cable modem [ELEC] A device that converts the signals used in a computer to signals that can be transmitted over cable television networks, and vice versa. { ¦kā·bəl ¦mō,dem }

cable noise [ELECTR] Electrical noise that is picked up by the conductors in a cable. { 'kā·bəl ,nȯiz }

cable paper [MATER] A paper used to insulate electrical cables. { 'kā·bəl ,pā·pər }

cable railway [MECH ENG] An inclined track on which rail cars travel, with the cars fixed to an endless steel-wire rope at equal spaces; the rope is driven by a stationary engine. { 'kā·bəl ¦rāl,wā }

cable release [ENG] A wire plunger to actuate the shutter of a camera, thus avoiding undesirable camera movement. { 'kā·bəl ri'lēs }

cable run [ELEC] Path occupied by a cable on cable racks or other support from one termination to another. { 'kā·bəl ,rən }

cable running list [ELEC] Drawing showing the code of cable, terminations, circuit names, and numbering of cables appearing in an office. { 'kā·bəl ¦rən·iŋ ,list }

cable shield [ELEC] A metallic layer applied over insulation covering a cable, composed of woven or braided wires, foil wrap, or metal tube, which acts to prevent electromagnetic or electrostatic interference from affecting conductors within. { 'kā·bəl ,shēld }

cable skidding [FOR] The use of a chain or cable choker to move tree lengths or tree segments for a short distance over unimproved terrain. { 'kā·bəl ,skid·iŋ }

cable-stayed bridge [CIV ENG] A modification of the cantilever bridge consisting of girders or trusses cantilevered both ways from a central tower and supported by inclined cables attached to the tower at top or sometimes at several levels. { 'kā·bəl ,stād ,brij }

cable stopper [NAV ARCH] A device, such as a short chain fastened to a pad on the deck and attached to the anchor chain, to take the strain off the anchor windlass when a ship is at anchor. { 'kā·bəl ,stäp·ər }

cable system [MIN ENG] A drilling system involving a heavy string of tools suspended from a flexible cable. { 'kā·bəl ,sis·təm }

cable-system drill See churn drill. { 'kā·bəl ¦sis·təm ,dril }

cable television [COMMUN] A television program distribution system in which signals from all local stations and usually a number of distant stations are picked up by one or more high-gain antennas, amplified on individual channels, then fed directly to individual receivers of subscribers by overhead or underground coaxial cable. Also known as community antenna television (CATV). { 'kā·bəl 'tel·ə,vizh·ən }

cabletext [COMMUN] Any videotex service that uses coaxial cable. { 'kā·bəl,tekst }

cable tier [NAV ARCH] **1.** The part of a ship used to store spare rigging and cables. **2.** See chain locker. { 'kā·bəl ,tir }

cable-tool drilling [ENG] A drilling procedure in which a sharply pointed bit attached to a cable is repeatedly picked up and dropped on the bottom of the hole. { 'kā·bəl ¦tül ,dril·iŋ }

cable tools [MIN ENG] The bits and other bottom-hole tools and equipment used to drill boreholes by percussive action, using a cable, instead of rods, to connect the drilling bit with the machine on the surface. { 'kā·bəl ,tülz }

cable trough [ELEC] An enclosed channel, usually beneath a floor, that provides a path for cables. { 'kā·bəl ,trȯf }

cable twist [SCI TECH] A yarn or rope construction in which each successive twist is in the opposite direction to the preceding twist. { 'kā·bəl ,twist }

cable vault [CIV ENG] A manhole containing electrical cables. [ELEC] Vault in which the outside plant cables are spliced to the tipping cables. { 'kā·bəl ,vȯlt }

cableway [MECH ENG] A transporting system consisting of a cable extended between two or more points on which cars are propelled to transport bulk materials for construction operations. [MIN ENG] A cable system of material handling in which carriers are supported by a cable and not detached from the operating span. { 'kā·bəl,wā }

cableway carriage [MECH ENG] A trolley that runs on main load cables stretched between two or more towers. { 'kā·bəl,wā 'kar·ij }

cabochon [LAP] One of two basic types of cutting styles used to fashion gemstones; the gem is cut in convex form, highly polished, but not faceted. { 'kab·ə,shän }

caboose [ENG] A car on a freight train, often the last car, usually for use by the train crew. { kə'büs }

Cabot's ring [CYTOL] A ringlike body in immature erythrocytes that may represent the remains of the nuclear membrane. { 'kab·əts 'riŋ }

CABRA numbers [MET] Copper and Brass Research Association number designations for various wrought copper and copper alloy grades. { 'kab·rə ,nəm·bərz }

cabretta [MATER] Tanned sheepskin leather. { kə'bred·ə }

cab signal [ENG] A signal in a locomotive that informs the engine operator about conditions affecting train movement. { 'kab ,sig·nəl }

cacao [BOT] *Theobroma cacao.* A small tropical tree of the order Theales that bears capsular fruits which are a source of cocoa powder and chocolate. Also known as chocolate tree. { kə'kaü }

cacao butter See cocoa butter. { kə'kaü ,bəd·ər }

Caccioppoli-Banach principle See Banach's fixed-point theorem. { ,kä·chē'äp·ə·lē 'bä,näk ,prin·sə·pəl }

cache [COMPUT SCI] A small, fast storage buffer integrated in the central processing unit of some large computers. { kash }

cachexia [MED] Weight loss, weakness, and wasting of the body encountered in certain diseases or in terminal illnesses. { ka'kek·sē·ə }

cacimbo [METEOROL] Local name in Angola for the wet fogs and drizzles noted with onshore winds from the Benguela Current. { kä'sim,bō }

cacodyl [ORG CHEM] $(CH_3)_2As^-$ A radical found in, for example, cacodylic acid, $(CH_3)_2AsOOH$. { 'kak·ə,dil }

cacodylate [ORG CHEM] Any salt of cacodylic acid. { ,kak·ə'di,lāt }

cacodylic acid [ORG CHEM] $(CH_3)_2AsOOH$ Colorless crystals that melt at 200°C; soluble in alcohol and water; used as a herbicide. { ¦kak·ə¦dil·ik 'as·əd }

cacomistle [VERT ZOO] *Bassariscus astutus.* A raccoonlike mammal that inhabits the southern and southwestern United States; distinguished by a bushy black-and-white ringed tail. Also known as civet cat; ringtail. { 'kak·ə,mis·əl }

caconym [SYST] A taxonomic name that is linguistically unacceptable. { 'kak·ə,nim }

cacotheline [ORG CHEM] $C_{20}H_{22}N_2O_5(NO_2)_2$ An azoic compound used as a metal indicator in chelometric titrations. { kə'käth·ə,lēn }

cacoxenite [MINERAL] $Fe_4(PO_4)_3(OH)_3 \cdot 12H_2O$ Yellow or brownish mineral consisting of a hydrous basic iron phosphate occurring in radiated tufts. { kə'käk·sə,nīt }

Cactaceae [BOT] The cactus family of the order Caryophyllales; represented by the American stem succulents, which are mostly spiny with reduced leaves. { kak'tās·ē,ē }

cactus [BOT] The common name for any member of the family Cactaceae, a group characterized by a fleshy habit, spines and bristles, and large, brightly colored, solitary flowers. { 'kak·təs }

cactus alkaloid See anhalonium alkaloid. { 'kak·təs 'al·kə,lȯid }

CAD See computer-aided design. { kad }

cadalene [ORG CHEM] $C_{15}H_{18}$ A colorless liquid which boils at 291–292°C (720 mmHg; 95,990 pascals) and which is a substituted naphthalene. { 'kad·əl,ēn }

cadang-cadang [PL PATH] An infectious virus disease of the coconut palm characterized by yellow-bronzing of the leaves. { kä,däŋ 'kä,däŋ }

cadastral survey [CIV ENG] A survey made to establish property lines. { kə'das·trəl }

cadaver [MED] A dead animal or human body to be studied by dissection. { kə'dav·ər }

cadaverine [BIOCHEM] $C_5H_{14}N_2$ A nontoxic, organic base produced as a result of the decarboxylation of lysine by the action of putrefactive bacteria on flesh. { kə'dav·ə,rēn }

CADD See computer-aided design and drafting. { cad }

caddis fly [INV ZOO] The common name for all members of the insect order Trichoptera; adults are mothlike and the immature stages are aquatic. { 'kad·əs ,flī }

caddy [COMPUT SCI] In certain types of disk drives, a plastic

CACTUS

Mule cactus with closeup of flower, maximum height of 6 feet (1.8 meters).

CADDIS FLY

Adult caddis fly.

tray in which a CD-ROM disk is placed before loading. { 'kad·ē }

cade oil [MATER] A brown, viscous essential oil that has a tar odor and is slightly soluble in water; it is derived by dry distillation of the wood of the European juniper (*Juniperus oxycedrus*); used in antiseptic soaps, perfumes, and pharmaceuticals. Also known as juniper tar oil. { 'kād ,ȯil }

CADF *See* commutated antenna direction finder.

cadherin [CELL MOL] Any of a family of calcium-dependent cell adhesion glycoproteins that play a fundamental role in tissue differentiation and structure. { kad'hir·ən }

cadinene [ORG CHEM] $C_{15}H_{24}$ A colorless liquid that boils at 274.5°C, and is a terpene derived from cubeb oil, cade oil, juniper berry oil, and other essential oils. { 'kad·ən,ēn }

cadmium [CHEM] A chemical element, symbol Cd, atomic number 48, atomic weight 112.40. [MET] A tin-white, malleable, ductile metal capable of high polish; principal use is in the plating of iron and steel, and to a much less extent of copper, brass, and other alloys, to protect them from corrosion and improve solderability and surface conductivity, and as a control absorber and shield in nuclear reactors. { 'kad·mē·əm }

cadmium acetate [ORG CHEM] $Cd(OOCCH_3)_2 \cdot 3H_2O$ A compound that forms colorless monoclinic crystals, soluble in water and in alcohol; used for chemical testing for sulfides, selenides, and tellurides and for producing iridescent effects on porcelain. { 'kad·mē·əm 'as·ə,tāt }

cadmium blende *See* greenockite. { 'kad·mē·əm ,blend }

cadmium bromate [INORG CHEM] $Cd(BrO_3)_2$ Colorless powder, soluble in water; used as an analytical reagent. { 'kad·mē·əm 'brō,māt }

cadmium bromide [INORG CHEM] $CdBr_2$ A compound produced as a yellow crystalline powder, soluble in water and alcohol; used in photography, process engraving, and lithography. { 'kad·mē·əm 'brō,mīd }

cadmium carbonate [INORG CHEM] $CdCO_3$ A white crystalline powder, insoluble in water, soluble in acids and potassium cyanide; used as a starting compound for other cadmium salts. { 'kad·mē·əm 'kär·bə,nāt }

cadmium cell [ELEC] A standard cell used as a voltage reference; at 20°C its voltage is 1.0186 volts. { 'kad·mē·əm ,sel }

cadmium chlorate [INORG CHEM] $CdClO_3$ White crystals, soluble in water; a highly toxic material. { 'kad·mē·əm 'klȯr,āt }

cadmium chloride [INORG CHEM] $CdCl_2$ A halide in the form of colorless crystals, soluble in water, methanol, and ethanol; used in photography, in dyeing and calico printing, and as a solution to precipitate sulfides. { 'kad·mē·əm 'klȯr,īd }

cadmium cutoff [NUCLEO] The neutron energy, approximately 0.3 electronvolt, below which cadmium has a high neutron absorption cross section but above which this cross section falls off sharply. { 'kad·mē·əm 'kət,ȯf }

cadmium difference [NUCLEO] The difference between the response of an uncovered neutron detector to a neutron beam and the response of the same detector under identical conditions when it is covered with a thin layer of cadmium. { 'kad·mē·əm 'dif·rəns }

cadmium fluoride [INORG CHEM] CdF_2 A crystalline compound with a melting point of 1110°C; soluble in water and acids; used for electronic and optical applications as a starting material for laser crystals. { 'kad·mē·əm 'flu̇r,īd }

cadmium hydroxide [INORG CHEM] $Cd(OH)_2$ A white powder, soluble in dilute acids; used to prepare negative electrodes for cadmium-nickel storage batteries. { 'kad·mē·əm hī'dräk,sīd }

cadmium iodide [INORG CHEM] CdI_2 A cadmium halide that forms lustrous, white, hexagonal scales, consisting of two water-soluble allotropes; used in photography, in process engraving, and formerly as an antiseptic. { 'kad·mē·əm 'ī·ə,dīd }

cadmium lamp [ELEC] A lamp containing cadmium vapor; wavelength (6438.4696 international angstroms, or 643.84696 nanometers) of light emitted is a standard of length. { 'kad·mē·əm ,lamp }

cadmium metallurgy [MET] The extraction of cadmium from zinc ores, or from complex ores as a by-product of zinc, lead, and copper smelting. { 'kad·mē·əm 'med·əl,ər·jē }

cadmium neutron [NUCLEO] A neutron whose energy lies below the cadmium cutoff. { 'kad·mē·əm 'nü,trän }

cadmium-nickel storage cell *See* nickel-cadmium battery. { 'kad·mē·əm ¦nik·əl 'stȯr·ij ,sel }

cadmium nitrate [CHEM] $Cd(NO_3)_2 \cdot 4H_2O$ White, hygroscopic crystals, soluble in water, alcohol, and liquid ammonia; used to give a reddish-yellow luster to glass and porcelain ware. { 'kad·mē·əm 'nī,trāt }

cadmium ocher *See* greenockite. { 'kad·mē·əm 'ō·kər }

cadmium oxide [INORG CHEM] CdO In the cubic form, a brown, amorphous powder, insoluble in water, soluble in acids and ammonia salts; used for cadmium plating baths and in the manufacture of paint pigments. { 'kad·mē·əm 'äk,sīd }

cadmium potassium iodide *See* potassium tetraiodocadmate. { 'kad·mē·əm pə'tas·ē·əm 'ī·ə,dīd }

cadmium ratio [NUCLEO] The ratio of the response of an uncovered neutron detector to that of the same detector under identical conditions when it is covered with cadmium of a specified thickness. { 'kad·mē·əm 'rā·shō }

cadmium red [MATER] A pigment composed of a mixture of cadmium sulfide, cadmium selenide, and barite; used as a red pigment. { 'kad·mē·əm 'red }

cadmium selenide cell [ELECTR] A photoconductive cell that uses cadmium selenide as the semiconductor material and has a fast response time and high sensitivity to longer wavelengths of light. { 'kad·mē·əm 'sel·ə,nīd ,sel }

cadmium silver oxide cell [ELEC] An alkaline-electrolyte cell that may be used without recharging in primary batteries or that may be recharged for secondary-battery use. { 'kad·mē·əm 'sil·vər 'äk,sīd ,sel }

cadmium sulfate [INORG CHEM] $CdSO_4$ A compound that forms colorless, efflorescent crystals, soluble in water; used as an antiseptic and astringent, in the treatment of syphilis, gonorrhea, and rheumatism, and as a detector of hydrogen sulfide and fumaric acid. { 'kad·mē·əm 'səl,fāt }

cadmium sulfide [INORG CHEM] CdS A compound with two forms: orange, insoluble in water, used as a pigment, and also known as orange cadmium; light yellow, hexagonal crystals, insoluble in water, and also known as cadmium yellow. { 'kad·mē·əm 'səl,fīd }

cadmium sulfide cell [ELECTR] A photoconductive cell in which a small wafer of cadmium sulfide provides an extremely high dark-light resistance ratio. { 'kad·mē·əm 'səl,fīd ,sel }

cadmium telluride [INORG CHEM] $CdTe$ Brownish-black, cubic crystals with a melting point of 1090°C; soluble, with decomposition, in nitric acid; used for semiconductors. { 'kad·mē·əm 'tel·yə,rīd }

cadmium telluride detector [ELECTR] A photoconductive cell capable of operating continuously at ambient temperatures up to 750°F (400°C); used in solar cells and infrared, nuclear-radiation, and gamma-ray detectors. { 'kad·mē·əm 'tel·yə,rīd di'tek·tər }

cadmium tungstate [INORG CHEM] $CdWO_4$ White or yellow crystals or powder; soluble in ammonium hydroxide and alkali cyanides; used in fluorescent paint, x-ray screens, and scintillation counters. { 'kad·mē·əm 'təŋ,stāt }

cadmium yellow *See* cadmium sulfide. { 'kad·mē·əm 'yel·ō }

caducicorn [VERT ZOO] Having deciduous horns, as certain deer. { kə'dü·sə,kȯrn }

caducous [BOT] Lasting on a plant only a short time before falling off. { 'kad·ə·kəs }

cadwaladerite [MINERAL] $Al(OH)_2Cl \cdot 4H_2O$ A mineral consisting of a hydrous basic aluminum chloride. { kad'wäl·ə·də,rīt }

Cae *See* Caelum.

caecilian [VERT ZOO] The common name for members of the amphibian order Apoda. { sē'sil·yən }

caecum *See* cecum. { 'sē·kəm }

Caelum [ASTRON] A southern constellation, right ascension 5 hours, declination 40°S. Abbreviated Cae. Also known as Chisel. { 'sē·ləm }

Caenolestidae [PALEON] A family of extinct insectivorous mammals in the order Marsupialia. { ,sē·nə'les·tə,de }

Caenolestoidea [VERT ZOO] A superfamily of marsupial mammals represented by the single living family Caenolestidae. { ,sē·nə·le'stȯid·ē·ə }

caenostylic [VERT ZOO] Having the first two visceral arches attached to the cranium and functioning in food intake; a condition found in sharks, amphibians, and chimaeras. { ¦sē·nə¦stīl·ik }

Caesalpinoidea [BOT] A subfamily of dicotyledonous plants in the legume family, Leguminosae. { ¸sē,zal·pə'nóid·ē·ə }

caffeic acid [ORG CHEM] $C_9H_8O_4$ A yellow crystalline acid that melts at 223–225°C with decomposition; soluble in water and alcohol. { ka'fē·ik 'as·əd }

caffeine [ORG CHEM] $C_8H_{10}O_2N_4 \cdot H_2O$ An alkaloid found in a large number of plants, such as tea, coffee, cola, and mate. { kaf,ēn }

caffetannin [MATER] One of the hydrolyzable tannins obtained from coffee berries and other plant products. { ¦kaf·ə¦tan·ən }

cage [CRYSTAL] A void occurring in a crystal structure capable of trapping one or more foreign atoms. [MECH ENG] A frame for maintaining uniform separation between the balls or rollers in a bearing. Also known as separator. [MIN ENG] The car which carries personnel and materials in a mine hoist. [PETRO ENG] A component in a sucker rod pump that contains the valve ball and maintains it at a correct operating distance from the valve seats. [PHYS CHEM] An aggregate of molecules in the condensed phase that surrounds fragments formed by thermal or photochemical dissociation or pairs of molecules in a solution that have collided without reacting. { kāj }

cage antenna [ELECTROMAG] Broad-band dipole antenna in which each pole consists of a cage of wires whose overall shape resembles that of a cylinder or a cone. { kāj an'ten·ə }

cage compound See clathrate. { ¦kāj ¦käm,paùnd }

cage effect [PHYS CHEM] A phenomenon involving the dissociation of molecules unable to move apart rapidly because of the presence of other molecules, with the result that the dissociation products may recombine. { kāj i,fekt }

cage guides [MIN ENG] Directive apparatus used to guide the cages in the mine shaft and to prevent their swinging or colliding with each other. { kāj ,gīdz }

cage hydrocarbon [ORG CHEM] A compound composed of only carbon and hydrogen atoms that contains three or more rings arranged topologically so as to enclose a volume of space; in general, the space within a cage hydrocarbon is too small to accommodate even a proton. { ¦kāj ,hī·drə'kär·bən }

cage mill [MECH ENG] Pulverizer used to disintegrate clay, press cake, asbestos, packing-house by-products, and various tough, gummy, high-moisture-content or low-melting-point materials. { kāj ,mil }

cage shoes [MIN ENG] Fittings attached on the side of a cage by bolts so that they engage the rigid guides in a shaft. { kāj ,shüz }

caging [NAV] The process of orienting and mechanically locking the spin axis of a gyroscope to an internal reference position. { kāj·iŋ }

cahnite [MINERAL] $Ca_2B(OH)_4(AsO_4)$ A tetragonal borate mineral occurring in white, sphenoidal crystals. { kä,nīt }

CAI See computer-assisted instruction.

Cailletet and Mathias law [PHYS CHEM] The law that describes the relationship between the mean density of a liquid and its saturated vapor at that temperature as being a linear function of the temperature. { kī·ə'tā ən mə'thī·əs ,lò }

caiman [VERT ZOO] Any of five species of reptiles of the genus *Caiman* in the family Alligatoridae, differing from alligators principally in having ventral armor and a sharper snout. { kā·mən }

cainophobia [PSYCH] Abnormal fear of newness. { ,kān·ə'fō·bē·ə }

Cainotheriidae [PALEON] The single family of the extinct artiodactyl superfamily Cainotherioidea. { ,kān·ə·thə'rī·ə,dē }

Cainotherioidea [PALEON] A superfamily of extinct, rabbit-sized tylopod ruminants in the mammalian order Artiodactyla. { ,kān·ə·ther·ē'óid·ē·ə }

Cainozoic See Cenozoic. { ,kān·ə'zō·ik }

cairn [ENG] An artificial mound of rocks, stones, or masonry, usually conical or pyramidal, whose purpose is to designate or to aid in identifying a point of surveying or of cadastral importance. { kern }

cairngorm See smoky quartz. { 'kern,górm }

caisson [CIV ENG] **1.** A watertight, cylindrical or rectangular chamber used in underwater construction to protect workers

from water pressure and soil collapse. **2.** A float used to raise a sunken vessel. **3.** See dry-dock caisson. [ORD] A two-wheeled, horse-drawn vehicle used for transporting ammunition and other military equipment. { 'kā,sän }

caisson disease [MED] A condition resulting from a rapid change in atmospheric pressure from high to normal, causing nitrogen bubbles to form in the blood and body tissues. Also known as bends; compressed-air illness. { 'kā,sän di,zēz }

caisson foundation [CIV ENG] A shaft of concrete placed under a building column or wall and extending down to hardpan or rock. Also known as pier foundation. { 'kā,sän foùn'dā·shən }

cajeput oil [MATER] An essential oil with a blue-green color and a camphorlike odor; chief constituents are eucalyptol, pinene, and alpha terpineol; distilled from leaves and twigs of species of East Indian trees of the genus *Melaleuca*; used in perfumes and medicines. Also spelled cajuput oil. { 'kaj·ə·pùt ,oil }

cajeputol See eucalyptol. { 'kaj·ə·pə,tól }

cajuput oil See cajeput oil. { 'kaj·ə·pùt ,oil }

caju rains [METEOROL] In northeastern Brazil, light showers that occur during the month of October. { ¦kä¦zhü ,rānz }

cake [MIN ENG] **1.** Solidified drill sludge. **2.** That portion of a drilling mud adhering to the walls of a borehole. **3.** To form into a mass, as when ore sinters together in roasting, or coal cakes in coking. { kāk }

cake of gold [MET] Gold formed into a compact mass (though not melted) by distillation of the mercury from a mercury-gold amalgam. Also known as sponge gold. { ¦kāk əv ,góld }

caking [ENG] Changing of a powder into a solid mass by heat, pressure, or water. { 'kāk·iŋ }

caking coal [GEOL] A type of coal which agglomerates and softens upon heating; after volatile material has been expelled at high temperature, a hard, gray cellular mass of coke remains. Also known as binding coal. { 'kāk·iŋ ,kōl }

cal See calorie.

Cal See kilocalorie.

CAL [COMPUT SCI] A higher-level language, developed especially for time-sharing purposes, in which a user at a remote console typewriter is directly connected to the computer and can work out problems on-line with considerable help from the computer. Derived from Conversational Algebraic Language. { kal }

calabarine See physostigmine. { kə'lab·ə,rēn }

Calabar swelling [MED] Edematous, painful, subcutaneous swelling occurring in the body of natives of Calabar and of other parts of West Africa, probably due to an allergic reaction to *Loa loa* infection. { 'kal·ə,bär ,swel·iŋ }

Calabi conjecture [MATH] If the volume of a certain type of surface, defined in a higher dimensional space in terms of complex numbers, is known, then a particular kind of metric can be defined on it; the conjecture was subsequently proved to be correct. { kə'lä·bē kən,jek·chər }

Calabrian [GEOL] Lower Pleistocene geologic time. { kə'läb·rē·ən }

calaite See turquoise. { kə'lā,īt }

calamanthia oil See marjoram oil. { kal·ə'man·thē·ə ,oil }

calamine [MET] An alloy composed of zinc, lead, and tin. See hemimorphite. See smithsonite. [PHARM] A powder mixture of zinc oxide and ferric oxide, used in skin lotions and ointments. { 'kal·ə,mīn }

Calamitales [PALEOBOT] An extinct group of reedlike plants of the subphylum Sphenopsida characterized by horizontal rhizomes and tall, upright, grooved, articulated stems. { kə,lam·ə'tā·lēz }

calamus oil [MATER] A yellow essential oil that is slightly soluble in water; composed mainly of eugenol and 2,4,5-trimethoxy-1-propenyl benzene; derived by steam distillation of the roots of calamus (*Acorus calamus*); used in perfumery and medicine. { 'kal·ə·məs ,oil }

calandria [CHEM ENG] One of the tubes through which the heating fluid circulates in an evaporator. [NUCLEO] In a CANDU nuclear reactor, a horizontal, cylindrical tank filled with heavy water and penetrated by a lattice of fuel channels with pressurized tubes containing the fuel and circulating coolant. { kə'lan,drē·ə }

calandria evaporator See short-tube vertical evaporator. { kə'lan,drē·ə i'vap·ə,rād·ər }

CAFFEINE

Structural formula of caffeine.

CAIMAN

The broad-nosed caiman (*Caiman latirostris*).

CALAMITALES

Scouring rush (*Calamites*) restored with whorls of branches at each joint of articulated stems. (*From R. C. Hussey, Historical Geology, 2d ed., McGraw-Hill, 1947*)

Calanoida [INV ZOO] A suborder of the crustacean order Copepoda, including the larger and more abundant of the pelagic species. { ‚kal·ə'nȯid·ə }

Calappidae [INV ZOO] The box crabs, a family of reptantian decapods in the subsection Oxystomata of the section Brachyura. { kə'lap·ə‚dē }

calathiform [BIOL] Cup-shaped, being almost hemispherical. { kə'lath·ə‚förm }

calaverite [MINERAL] AuTe$_2$ A yellowish or tin-white, monoclinic mineral commonly containing gold telluride and minor amounts of silver. { kə'lav·ə‚rīt }

calc-alkalic series [PETR] Series of igneous rocks in which the weight percentage of silica is 55–61. { ‚kalk ‚al'kal·ik ‚sir·ēz }

calcaneal exostosis See heel spur. { kal'kā·nē·əl ‚ek·sə'stō·səs }

calcaneocuboid ligament [ANAT] The ligament that joins the calcaneus and the cuboid bones. { kal‚kan·ē·ō‚kyü‚bȯid ‚lig·ə·mənt }

calcaneum [ANAT] **1.** A bony projection of the metatarsus in birds. **2.** See calcaneus. { kal'kan·ē·əm }

calcaneus [ANAT] A bone of the tarsus, forming the heel bone in humans. Also known as calcaneum. { kal'kan·ē·əs }

calcar [ZOO] A spur or spurlike process, especially on an appendage or digit. { 'kal‚kär }

calcarate [BOT] Having spurs. { 'kal·kə‚rāt }

Calcarea [INV ZOO] A class of the phylum Porifera, including sponges with a skeleton composed of calcium carbonate spicules. { kal'kar·ē·ə }

calcarenite [PETR] A type of limestone or dolomite composed of coral or shell sand or of sand formed by erosion of older limestones, with particle size ranging from 1/16 to 2 millimeters. { kal·kə'rē‚nīt }

calcareous [SCI TECH] Resembling, containing, or composed of calcium carbonate. { kal'ker·ē·əs }

calcareous algae [BOT] Algae that grow on limestone or in soil impregnated with lime. { kal'ker·ē·əs 'al‚jē }

calcareous crust See caliche. { kal'ker·ē·əs 'krəst }

calcareous duricrust See caliche. { kal'ker·ē·əs 'dur‚i‚krəst }

calcareous ooze [GEOL] A fine grained pelagic sediment containing undissolved sand- or silt-sized calcareous skeletal remains of small marine organisms mixed with amorphous clay-sized material. { kal'ker·ē·əs 'üz }

calcareous schist [PETR] A coarse-grained metamorphic rock derived from impure calcareous sediment. { kal'ker·ē·əs 'shist }

calcareous sinter See tufa. { kal'ker·ē·əs 'sin·tər }

calcareous soil [GEOL] A soil containing accumulations of calcium and magnesium carbonate. { kal'ker·ē·əs 'sȯil }

calcareous tufa See tufa. { kal'ker·ē·əs 'tü·fə }

calcarine fissure See calcarine sulcus. { 'kal·kə‚rēn 'fish·ər }

calcarine sulcus [ANAT] A sulcus on the medial aspect of the occipital lobe of the cerebrum, between the lingual gyrus and the cuneus. Also known as calcarine fissure. { 'kal·kə‚rēn 'səl·kəs }

Calcaronea [INV ZOO] A subclass of sponges in the class Calcarea in which the larva are amphiblastulae. { kal·kə'rō·nē·ə }

calcemia See hypercalcemia. { kal'sē·mē·ə }

calcic [SCI TECH] Derived from or containing calcium. { 'kal·sik }

calciclastic [PETR] Pertaining to calcium carbonate–containing rock eroded from a preexisting source, transported some distance, and then redeposited; for example, calciclastic limestone. { ‚kal·sə'klas·tik }

calcicole [BOT] Requiring soil rich in calcium carbonate for optimum growth. { 'kal·sə‚kōl }

calcicosis [MED] A form of pneumoconiosis caused by the inhalation of marble (calcium carbonate) dust. { ‚kal·sə'kō·səs }

calciferol [BIOCHEM] A synthetic form of vitamin D that is prepared by ultraviolet irradiation of ergosterol, a vitamin D precursor found in plants. { kal'sif·ə‚rȯl }

calciferous [BIOL] Containing or producing calcium or calcium carbonate. { kal'sif·ə·rəs }

calciferous gland [INV ZOO] One of a series of glands that secrete calcium carbonate into the esophagus of certain oligochaetes. { kal'sif·ə·rəs ‚gland }

calcification [GEOCHEM] Any process of soil formation in which the soil colloids are saturated to a high degree with exchangeable calcium, thus rendering them relatively immobile and nearly neutral in reaction. [PHYSIO] The deposit of calcareous matter within the tissues of the body. { ‚kal·sə·fə'kā·shən }

calcifuge [ECOL] A plant that grows in an acid medium that is poor in calcareous matter. { 'kal·sə‚fyüj }

calcilutite [PETR] **1.** A dolomite or limestone formed of calcareous rock flour that is typically nonsiliceous. **2.** A rock of calcium carbonate formed of grains or crystals with average diameter less than $^1/_{16}$ millimeter. { ‚kal·sə'lü‚tīt }

calcimeter [ENG] An instrument for estimating the amount of lime in soils. { kal'sim·əd·ər }

calcimine [MATER] A thin paint, white or colored, made of pigment, glue, and water. Also known as kalsomine. { 'kal·sə‚mīn }

calcination [CHEM ENG] A process in which a material is heated to a temperature below its melting point to effect a thermal decomposition or a phase transition other than melting. { ‚kal·sə'nā·shən }

calcine [ENG] **1.** To heat to a high temperature without fusing, as to heat unformed ceramic materials in a kiln, or to heat ores, precipitates, concentrates, or residues so that hydrates, carbonates, or other compounds are decomposed and the volatile material is expelled. **2.** To heat under oxidizing conditions. [MATER] Product of calcining or roasting. { 'kal‚sīn }

Calcinea [INV ZOO] A subclass of sponges in the class Calcarea in which the larvae are parenchymulae. { kal'sin·ē·ə }

calcined clay [MATER] Clay that has been heated to drive out volatile materials; a natural abrasive. { 'kal‚sīnd 'klā }

calcined coke [MATER] Coke that has been heated to expel volatile material. { 'kal‚sīnd ‚kōk }

calcined gypsum See plaster of paris. { 'kal‚sīnd 'jip·səm }

calcined limestone [MATER] Limestone that has been heated in a vertical-shaft kiln to drive off carbon dioxide. { 'kal‚sīnd 'līm‚stōn }

calcined soda See soda ash. { 'kal‚sīnd 'sō·də }

calcining furnace [ENG] A heating device, such as a vertical-shaft kiln, that raises the temperature (but not to the melting point) of a substance such as limestone to make lime. Also known as calciner. { 'kal‚sin·iŋ ‚fər·nəs }

calcinosis [MED] Deposition of calcium salts in the skin, subcutaneous tissue, or other part of the body in certain pathologic conditions. { ‚kal·sə'nō·səs }

calciocarnotite See tyuyamunite. { ‚kal·sē·ō'kär·nə‚tīt }

calcioferrite [MINERAL] Ca$_2$Fe$_2$(PO$_4$)OH·7H$_2$O A yellow or green mineral consisting of a hydrous basic calcium iron phosphate and occurring in nodular masses. { 'kal·sē·ō'fe‚rīt }

calciovolborthite [MINERAL] CaCu(VO$_4$)(OH) Green, yellow, or gray mineral consisting of a basic vanadate of calcium and copper. Also known as tangeite. { ‚kal·sē·ō 'vȯl‚bȯr‚thīt }

calciphylaxis [IMMUNOL] A sudden local calcification in tissues in response to induced hypersensitivity following systemic sensitization by a calcifying factor. { ‚kal·sə·fə'lak·səs }

calcirudite [PETR] Dolomite or limestone formed of worn or broken pieces of coral or shells or of limestone fragments coarser than sand; the interstices are filled with sand, calcite, or mud, the whole bound together with a calcareous cement. { kal'sir·ə‚dīt }

calcite [MINERAL] CaCO$_3$ One of the commonest minerals, the principal constituent of limestone; hexagonal-rhombohedral crystal structure, dimorphous with aragonite. Also known as calcspar. { 'kal‚sīt }

calcite compensation depth [GEOL] The depth in the ocean (about 5000 meters) below which solution of calcium carbonate occurs at a faster rate than its deposition. Abbreviated CCD. { 'kal‚sīt käm·pən'sā·shən ‚depth }

calcite dolomite [PETR] A carbonate rock with a composition of 10–50% calcite and 90–50% dolomite. { 'kal‚sīt 'dol·ə‚mīt }

calcitonin [BIOCHEM] A polypeptide, calcium-regulating

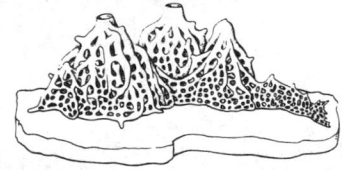

CALCINEA

Clathrina clathrus, a calcinean sponge.

hormone produced by the ultimobranchial bodies in vertebrates. Also known as thyrocalcitonin. { kal·sə'tō·nən }

calcitriol [BIOCHEM] 1,25-dihydroxycholecalciferol, the form of vitamin D that is involved in intestinal absorption of Ca^{2+} and Ca^{2+} resorption by bone. { ˌkal·sə'trī·ȯl }

calcium [CHEM] A chemical element, symbol Ca, atomic number 20, atomic weight 40.08; used in metallurgy as an alloying agent for aluminum-bearing metal, as an aid in removing bismuth from lead, and as a deoxidizer in steel manufacture, and also used as a cathode coating in some types of photo tubes. { 'kal·sē·əm }

calcium acetate [ORG CHEM] $Ca(C_2H_3O_2)_2$ A compound that crystallizes as colorless needles that are soluble in water; formerly used as an important source of acetone and acetic acid; now used as a mordant and as a stabilizer of plastics. { 'kal·sē·əm 'as·ə,tāt }

calcium acrylate [ORG CHEM] $(CH_2CHCOO)_2Ca$ Free-flowing, water-soluble white powder used for soil stabilization, oil-well sealing, and ion exchange and as a binder for clay products and foundry molds. { 'kal·sē·əm 'ak·rə,lāt }

calcium arsenate [INORG CHEM] $Ca_3(AsO_4)_2$ An arsenic compound used as an insecticide to control cotton pests. { 'kal·sē·əm 'ärs·ən,āt }

calcium arsenite [INORG CHEM] $Ca_3(AsO_3)_2$ White granules that are soluble in water; used as an insecticide. { 'kal·sē·əm 'ärs·ən,īt }

calcium bisulfite [INORG CHEM] $Ca(HSO_3)_2$ A white powder, used as an antiseptic and in the sulfite pulping process. { 'kal·sē·əm bī'səl,fīt }

calcium-bomb process [MET] A former process to produce pellets of pure titanium metal by mixing titanium chloride with calcium in a steel bomb and heating. { 'kal·sē·əm ,bäm ,präs·əs }

calcium bromide [INORG CHEM] $CaBr_2$ A deliquescent salt in the form of colorless hexagonal crystals that are soluble in water and absolute alcohol. { 'kal·sē·əm 'brō,mīd }

calcium carbide [INORG CHEM] CaC_2 An alkaline earth carbide obtained in the pure form as transparent crystals that decompose in water; used to make acetylene gas. { 'kal·sē·əm 'kär,bīd }

calcium carbonate [INORG CHEM] $CaCO_3$ White rhombohedrons or a white powder; occurs naturally as calcite; used in paint manufacture, as a dentifrice, as an anticaking medium for table salt, and in manufacture of rubber tires. { 'kal·sē·əm 'kär·bə,nāt }

calcium chlorate [INORG CHEM] $Ca(ClO_3)_2·2H_2O$ White monoclinic crystals, decomposed by heating. { 'kal·sē·əm 'klȯr,āt }

calcium chloride [INORG CHEM] $CaCl_2$ A colorless, deliquescent powder that is soluble in water and ethanol; used as an antifreeze and as an antidust agent. { 'kal·sē·əm 'klȯr,īd }

calcium chromate [INORG CHEM] $CaCrO_4·2H_2O$ Yellow, monoclinic crystals that are slightly soluble in water; used to make other pigments. { 'kal·sē·əm 'krō,māt }

calcium cyanamide [INORG CHEM] $CaCN_2$ In pure form, colorless rhombohedral crystals, the commercial form being a gray material containing 55–70% $CaCN_2$; used as a fertilizer, weed killer, and defoliant. { 'kal·sē·əm sī'an·ə,mīd }

calcium cyanide [INORG CHEM] $Ca(CN)_2$ In pure form, a white powder that gives off hydrogen cyanide in air at normal humidity; prepared commercially in impure black or gray flakes; used as an insecticide and rodenticide. Also known as black cyanide. { 'kal·sē·əm 'sī·ə,nīd }

calcium cyclamate [ORG CHEM] $C_{12}H_{24}O_6N_2S_2Ca_2·H_2O$ White crystals with a very sweet taste, soluble in water; has been used as a low-calorie sweetening agent. { 'kal·sē·əm 'sī·klə,māt }

calcium dihydrogen phosphate See calcium phosphate. { 'kal·sē·əm dī'hī·drə·jən 'fäs,fāt }

calcium fluoride [INORG CHEM] CaF_2 Colorless, cubic crystals that are slightly soluble in water and soluble in ammonium salt solutions; used in etching glass and preparing hydrofluoric acid. { 'kal·sē·əm 'flùr,īd }

calcium-45 [NUC PHYS] A radioisotope of calcium having a mass number of 45, often used as a radioactive tracer in studying calcium metabolism in humans and other organisms; half-life is 165 days. Designated ^{45}Ca. { 'kal·sē·əm ,fȯrd·ē'fīv }

calcium gluconate [ORG CHEM] $Ca(C_6H_{11}O_7)_2·H_2O$ White powder that loses water at 120°C; soluble in hot water

but less soluble in cold water, insoluble in acetic acid and alcohol; used in medicine, as a foaming agent, and as a buffer in foods. { 'kal·sē·əm 'glü·kə,nāt }

calcium hardness [CHEM] Presence of calcium ions in water, from dissolved carbonates and bicarbonates; treated in boiler water by introducing sodium phosphate. { 'kal·sē·əm ,härd·nəs }

calcium hydride [INORG CHEM] CaH_2 In pure form, white crystals that are insoluble in water; used in the production of chromium, titanium, and zirconium in the Hydromet process. { 'kal·sē·əm 'hī,drīd }

calcium hydrogen phosphate See calcium phosphate. { ˌkal·sē·əm ˌhī·drə·jən 'fäs,fāt }

calcium hydroxide [INORG CHEM] $Ca(OH)_2$ White crystals, slightly soluble in water; used in cement, mortar, and manufacture of calcium salts. Also known as hydrated lime. { 'kal·sē·əm hī'dräk,sīd }

calcium hypochlorite [INORG CHEM] $Ca(OCl)_2·4H_2O$ A white powder, used as a bleaching agent and disinfectant for swimming pools. { 'kal·sē·əm hī·pō'klȯr,īt }

calcium iodide [INORG CHEM] CaI_2 A yellow, hygroscopic powder that is very soluble in water; used in photography. { 'kal·sē·əm 'ī·ə,dīd }

calcium iodobehenate [ORG CHEM] $Ca(OOCC_{21}H_{42}I)_2$ A yellowish powder that is soluble in warm chloroform; used in feed additives. { 'kal·sē·əm ˌī·ə·dō'be·ə,nāt }

calcium lactate [ORG CHEM] $Ca(C_3H_5O_3)_2·5H_2O$ A salt of lactic acid in the form of white crystals that are soluble in water; used in calcium therapy and as a blood coagulant. { 'kal·sē·əm 'lak,tāt }

calcium metabolism [BIOCHEM] Biochemical and physiological processes involved in maintaining the concentration of calcium in plasma at a constant level and providing a sufficient supply of calcium for bone mineralization. { 'kal·sē·əm mə'tab·ə,liz·əm }

calcium naphthenate [ORG CHEM] Calcium derivative of cycloparaffin hydrocarbon (generally cyclopentane or cyclohexane base) that is a light, sticky, water-insoluble mass; used as a hardening agent in plastic compounds, in waterproofing, adhesives, wood fillers, and varnishes. { 'kal·se·əm 'naf·thə,nāt }

calcium nitrate [INORG CHEM] $Ca(NO_3)_2·4H_2O$ Colorless, monoclinic crystals that are soluble in water; the anhydrous salt is very deliquescent; used as a fertilizer and in explosives. Also known as nitrocalcite. { 'kal·se·əm 'nī,trāt }

calcium orthoarsenate [ORG CHEM] $Ca_3(AsO_4)_2$ A white powder, insoluble in water; used as a preemergence insecticide and herbicide for turf. { 'kal·se·əm ˌȯr·thō'ärs·ən,āt }

calcium oxalate [ORG CHEM] $CaC_2O_4·H_2O$ A salt of oxalic acid in the form of white crystals that are insoluble in water. { 'kal·se·əm 'äk·sə,lāt }

calcium oxide [INORG CHEM] CaO A caustic white solid sparingly soluble in water; the commercial form is prepared by roasting calcium carbonate limestone in kilns until all the carbon dioxide is driven off; used as a refractory, in pulp and paper manufacture, and as a flux in manufacture of steel. Also known as burnt lime; calx; caustic lime. { 'kal·se·əm 'äk,sīd }

calcium pantothenate [ORG] $(C_9H_{16}NO_5)_2Ca$ White slightly hygroscopic powder; soluble in water, insoluble in chloroform and ether; melts at 170–172°C; found in either the dextro or levo form or as a racemate; used in nutrition and in animal feed. { 'kal·se·əm pan·tə'the,nāt }

calcium peroxide [INORG CHEM] CaO_2 A cream-colored powder that decomposes in water; used as an antiseptic and a detergent. { 'kal·se·əm pə'räk,sīd }

calcium phosphate [INORG CHEM] **1.** Any phosphate of calcium. **2.** Any of the following three calcium orthophosphates, all of which are white or colorless in pure form: $Ca(H_2PO_4)_2$ is used as a fertilizer, as a plastics stabilizer, and in baking powder, and is also known as acid calcium phosphate, calcium dihydrogen phosphate, monobasic calcium phosphate, monocalcium phosphate; $CaHPO_4$ is used in pharmaceuticals, animal feeds, and toothpastes, and is also known as calcium hydrogen phosphate, dibasic calcium phosphate, dicalcium orthophosphate, dicalcium phosphate; $Ca_3(PO_4)_2$ is used as a fertilizer, and is also known as tribasic calcium phosphate, tricalcium phosphate. { 'kal·se·əm 'fäs,fāt }

calcium plumbate [INORG CHEM] $Ca(PbO_3)_2$ Orange crystals that are insoluble in cold water but decompose in hot water; used as an oxidizer in the manufacture of glass and matches. { 'kal·se·əm 'pləm,bāt }

calcium plumbite [INORG CHEM] $CaPbO_2$ Colorless crystals that are slightly soluble in water. { 'kal·se·əm 'pləm,bīt }

calcium pump [CELL MOL] An enzyme that uses the energy generated by the hydrolysis of adenosine triphosphate (ATP) to move calcium ions out of the cytoplasm or, in the case of muscle cells, from the cytoplasm to the sarcoplasmic reticulum. Also known as Ca^{2+} ATPase. { 'cal·se·əm ,pəmp }

calcium pyrophosphate [INORG CHEM] $Ca_2P_2O_7$ White, abrasive powder, used in dentifrice polishes, in metal polishes, and as a food supplement. { 'kal·se·əm ,pī·rō'fäs,fāt }

calcium resinate [ORG CHEM] Yellowish white, amorphous powder that is soluble in acid, insoluble in water; made by boiling rosin with calcium hydroxide and filtering, or by fusion of melted rosin with hydrated lime; used for waterproofing, leather tanning, and the manufacture of paint driers and enamels. Also known as limed rosin. { 'kal·se·əm 'rez·ən,āt }

calcium reversal lines [SPECT] Narrow calcium emission lines that appear as bright lines in the center of broad calcium absorption bands in the spectra of certain stars. { 'kal·se·əm ri'vər·səl ,līnz }

calcium silicate [INORG CHEM] Any of three silicates of calcium: tricalcium silicate, Ca_3SiO_5; dicalcium silicate, Ca_2SiO_4; calcium metasilicate, $CaSiO_3$. { 'kal·se·əm 'sil·ə,kāt }

calcium star [ASTRON] A term sometimes used to denote a star of spectral class F, which has prominent absorption bands of calcium. { 'kal·se·əm ,stär }

calcium stearate [ORG CHEM] $Ca(C_{18}H_{35}O_2)_2$ A metallic soap produced as a white powder that is insoluble in water but slightly soluble in petroleum, benzene, and toluene. { 'kal·se·əm 'stir,āt }

calcium sulfate [INORG CHEM] **1.** $CaSO_4$ A white crystalline salt, insoluble in water; used in Keene's cement, in pigments, as a paper filler, and as a drying agent. **2.** Either of two hydrated forms of the salt: the dihydrate, $CaSO_4 \cdot 2H_2O$, and the hemihydrate, $CaSO_4 \cdot 1/2H_2O$. { 'kal·se·əm 'səl,fāt }

calcium sulfide [INORG CHEM] CaS In pure form, white cubic crystals, slightly soluble in water; used as a base for luminescent materials. Also known as hepar calcies; sulfurated lime. { 'kal·se·əm 'səl,fīd }

calcium sulfite [INORG CHEM] $CaSO_3 \cdot 2H_2O$ A white powder that is soluble in dilute sulfurous acid; may be dehydrated at 150°C to the anhydrous salt; used in the sulfite process for the manufacture of wood pulp. { 'kal·se·əm 'səl,fīt }

calcium tungstate [INORG CHEM] $CaWO_4$ White, tetragonal crystals, slightly soluble in water; used in manufacture of luminous paints. Also known as artificial scheelite; calcium wolframate. { 'kal·se·əm 'təŋ,stāt }

calcium wolframate See calcium tungstate. { 'kal·se·əm 'wùl·frə,māt }

calclacite [MINERAL] $CaCl_2Ca(C_2H_3O_2) \cdot 10H_2O$ A white mineral consisting of a hydrated chloride-acetate of calcium; occurs as hairlike efflorescences. { 'kal·klə,sīt }

Calclamnidae [PALEON] A family of Paleozoic echinoderms of the order Dendrochirotida. { kal'klam·nə,dē }

calclithite [PETR] Limestone with 50% or more fragments of older limestone that was redeposited after being eroded from the land. { 'kal·klə,thīt }

calcrete [GEOL] A conglomerate of surficial gravel and sand cemented by calcium carbonate. { 'kal'krēt }

calc-silicate [GEOL] Referring to a metamorphic rock consisting mainly of calcite and calcium-bearing silicates. { 'kalk 'sil·ə·kət }

calc-silicate hornfels [PETR] A metamorphic rock with a fine grain of calcium silicate minerals. { ,kalk 'sil·ə,kāt 'hórn,felz }

calc-silicate marble [PETR] Marble having conspicuous calcium silicate or magnesium silicate minerals. { ,kalk 'sil·ə,kāt 'mär·bəl }

calcspar See calcite. { 'kalk,spär }

calcsparite See sparry calcite. { kalk'spä,rīt }

calculated address See generated address. { 'kal·kyə,lād·əd 'ad,res }

calculated altitude See computed altitude. { 'kal·kyə,lād·əd 'al·tə,tüd }

calculating machine See calculator. { 'kal·kyə,lād·iŋ mə'shēn }

calculation-based molecular modeling [PHYS CHEM] The use of computers, together with theoretical chemistry and mathematical expressions, to describe the structure of molecules and predict the most favorable conformation of a molecule or to calculate the energy of interaction between two molecules. { ,kal·kyə'lā·shən ¦bäst mə'lek·yə·lər 'mäd·əl·iŋ }

calculator [COMPUT SCI] A device that performs logic and arithmetic digital operations based on numerical data which are entered by pressing numerical and control keys. Also known as calculating machine. { 'kal·kyə,lād·ər }

calculus [ANAT] A small and cuplike structure. [MATH] The branch of mathematics dealing with differentiation and integration and related topics. [PATH] An abnormal, solid concretion of minerals and salts formed around organic materials and found chiefly in ducts, hollow organs, and cysts. { 'kal·kyə·ləs }

calculus of enlargement See calculus of finite differences. { 'kal·kyə·ləs əv in'lärj·mənt }

calculus of finite differences [MATH] A method of interpolation that makes use of formal relations between difference operators which are, in turn, defined in terms of the values of a function on a set of equally spaced points. Also known as calculus of enlargement. { 'kal·kyə·ləs əv 'fī,nīt 'dif·rən·səs }

calculus of residues [MATH] The application of the Cauchy residue theorem and related theorems to compute the residues of a meromorphic function at simple poles, evaluate contour integrals, expand meromorphic functions in series, and carry out related calculations. { 'kal·kyə·ləs əv 'rez·ə,düz }

calculus of tensors [MATH] The branch of mathematics dealing with the differentiation of tensors. { 'kal·kyə·ləs əv 'ten·sərs }

calculus of variations [MATH] The study of problems concerning maximizing or minimizing a given definite integral relative to the dependent variables of the integrand function. { 'kal·kyə·ləs əv ,ver·ē'ā·shənz }

calculus of vectors [MATH] That branch of calculus concerned with differentiation and integration of vector-valued functions. { 'kal·kyə·ləs əv 'vek·tərz }

caldera [GEOL] A large collapse depression at a volcano summit that is typically circular to slightly elongate in shape, with dimensions many times greater than any included vent. It ranges from a few miles to 37 miles (60 kilometers) in diameter. It may resemble a volcanic crater in form, but differs in that it is a collapse rather than a constructional feature. { kal'der·ə }

Caldwell catalog [ASTRON] A catalog of star clusters, nebulae, and galaxies for the use of amateur observers, whose objects are easy to locate with a small telescope but are not included in Messier's Catalog. { 'kól,dwel ¦kad·ə,läg }

Caldwell number [ASTRON] A number by which star clusters, nebulae, and galaxies are listed in the Caldwell Catalog, arranged in order of declination; for example, the Hyades are C41. { 'kól,dwel ¦nəm·bər }

Caledonian orogeny [GEOL] Deformation of the crust of the earth by a series of diastrophic movements beginning perhaps in Early Ordovician and continuing through Silurian, extending from Great Britain through Scandinavia. { ¦kal·ə¦dōn·ē·ən ó'räj·ə·nē }

Caledonides [GEOL] A mountain system formed in Late Silurian to Early Devonian time in Scotland, Ireland, and Scandinavia. { ,kal·ə'dä,nīdz }

caledonite [MINERAL] $Cu_2Pb_5(SO_4)_3CO_3(OH)_6$ A mineral occurring as green, orthorhombic crystals composed of basic copper lead sulfate; found in copper-lead deposits. { ,kal·ə'dä,nīt }

calefaction [ENG] **1.** Warming. **2.** The condition of being warmed. { ¦kal·ə¦fak·shən }

calendar [ASTRON] A system for everyday use in which time is divided into days and longer periods, such as weeks, months, and years, and a definite order for these periods and a correspondence between them are established. { 'kal·ən·dər }

calendar clock [HOROL] A clock that shows day, date, month, and phases of the moon, in addition to time of day. { 'kal·ən·dər ,kläk }

calendar day [ASTRON] The period from midnight to midnight; it is 24 hours of mean solar time in length and coincides with the civil day. { 'kal·ən·dər ,dā }

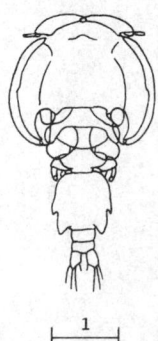

CALENDER

plastic mass

to conditioning
equipment

A four-roll calender.

CALIGOIDA

Pandarus satyrus Dana, male, a
caligoidan.

calendar month [ASTRON] The month of the calendar, varying from 28 to 31 days in length. { 'kal·ən·dər ,mənth }

calendar year [ASTRON] The year in the Gregorian calendar, common years having 365 days and leap years 366. Also known as civil year. { 'kal·ən·dər ,yir }

calender [ENG] **1.** To pass a material between rollers or plates to thin it into sheets or to make it smooth and glossy. **2.** The machine which performs this operation. { 'kal·ən·dər }

calendered paper [MATER] Paper that has passed through the calenders of a paper machine. { 'kal·ən·dərd 'pā·pər }

calendering [TEXT] Mechanical finishing process to produce a hard, shiny fabric. { 'kal·ən·dər·iŋ }

calf [OCEANOGR] *See* calved ice. [VERT ZOO] The young of the domestic cow, elephant, rhinoceros, hippopotamus, moose, whale, and others. { kaf }

calf circumference [ANTHRO] An average of three measurements taken at maximum horizontal distance around the left calf as the subject stands with weight distributed evenly on both feet. { ¦kaf sər'kəm·frəns }

calf diphtheria [VET MED] Disease of calves caused by species of *Sphaerophorus* which affects the mouth and pharynx and develops into pneumonia and septicemia, resulting in death. { ¦kaf dif'thir·ē·ə }

Caliban [ASTRON] A small satellite of Uranus in a retrograde orbit with a mean distance of 4,475,000 miles (7,200,000 kilometers), eccentricity of 0.081, and sidereal period of 1.59 years. { 'kal·ə,ban }

caliber [ORD] The diameter of a projectile or the diameter of the bore of a gun or launching tube; for example, a caliber .22 cartridge has a diameter of approximately 0.22 inch (5.6 millimeters). { 'kal·ə·bər }

calibrant [ANALY CHEM] In chemical analysis, a substance used to calibrate the response of a measurement system to the analyte. { 'kal·ə·brənt }

calibrate [SCI TECH] To determine, by measurement or comparison with a standard, the correct value of each scale reading on a meter or other device, or the correct value for each setting of a control knob. { 'kal·ə,brāt }

calibrated airspeed [AERO ENG] The airspeed as read from a differential-pressure airspeed indicator which has been corrected for instrument and installation errors; equal to true airspeed for standard sea-level conditions. { 'kal·ə,brād·əd 'er,spēd }

calibrated altitude [NAV] Indicated altitude corrected for instrument and installation errors. { 'kal·ə,brād·əd 'al·tə,tüd }

calibrating tank [ENG] A tank having known capacity used to check the volumetric accuracy of liquid delivery by positive-displacement meters. Also known as meter-proving tank. { 'kal·ə,brād·iŋ ,taŋk }

calibration curve [ENG] A plot of calibration data, giving the correct value for each indicated reading of a meter or control dial. { 'kal·ə,brā·shən ,kərv }

calibration fire [ORD] Preparatory fire to determine the separate corrections to be applied to the individual guns or launchers of a battery in order to cause all the weapons to hit the same point, or bursts or impacts to assume a desired pattern. { 'kal·ə,brā·shən ,fīr }

calibration markers [ENG] On a radar display, electronically generated marks which provide numerical values for the navigational parameters such as bearing, distance, height, or time. { 'kal·ə,brā·shən ,mär·kərz }

calibration point [ORD] A point at which calibration fire is directed. { 'kal·ə,brā·shən ,pȯint }

calibration radio beacon [NAV] A special radio beacon operated primarily for calibrating shipboard radio direction-finders; it transmits either continuously during scheduled hours or upon request. { 'kal·ə,brā·shən 'rād·ē·ō ,bē·kən }

calibration reference [ANALY CHEM] Any of the standards of various types that indicate whether an analytical instrument or procedure is working within prescribed limits; examples are test solutions used with pH meters, and solutions with known concentrations (standard solutions) used with spectrophotometers. { 'kal·ə,brā·shən ,ref·rəns }

caliche [GEOL] **1.** Conglomerate of gravel, rock, soil, or alluvium cemented with sodium salts in Chilean and Peruvian nitrate deposits; contains sodium nitrate, potassium nitrate, sodium iodate, sodium chloride, sodium sulfate, and sodium borate. **2.** A thin layer of clayey soil capping auriferous veins

(Peruvian usage). **3.** Whitish clay in the selvage of veins (Chilean usage). **4.** A recently discovered mineral vein. **5.** A secondary accumulation of opaque, reddish brown to buff or white calcareous material occurring in layers on or near the surface of stony soils in arid and semiarid regions of the southwestern United States; called hardpan, calcareous duricrust, and kanker in different geographic regions. Also known as calcareous crust; croute calcaire; nari; sabach; tepetate. { kə'lē·chē }

Caliciaceae [BOT] A family of lichens in the order Caliciales in which the disk of the apothecium is borne on a short stalk. { kə,lē·sē'ās·ē,ē,ē }

Caliciales [BOT] An order of lichens in the class Ascolichenes characterized by an unusual apothecium. { kə,lē·sē'ā,lēz }

Caliciviridae [VIROL] A family of nonenveloped ribonucleic acid viruses with characteristic hollow surface structures that resemble cups. { kə,lē·sē'vī·rə,dē }

calico [TEXT] Any plain-weave or inexpensive figured cotton cloth. { 'kal·ə,kō }

California Current [OCEANOGR] The ocean current flowing southward along the western coast of the United States to northern Baja California. { ¦kal·ə¦fȯr·nyə }

California fog [METEOROL] Fog peculiar to the coast of California and its coastal valleys; off the coast, winds displace warm surface water, causing colder water to rise from beneath, resulting in the formation of fog; in the coastal valleys, fog is formed when moist air blown inland during the afternoon is cooled by radiation during the night. { ¦kal·ə¦fȯr·nyə 'fäg }

California method [HYD] A form of frequency analysis which employs the return period, a parameter that measures the average time period between the occurrence of a quantity in hydrology and that of an equal or greater quantity, as the plotting position. { ¦kal·ə¦fȯr·nyə 'meth·əd }

California polymerization [CHEM ENG] A polymerization process for converting C_3-C_4 olefins to motor fuel by utilizing a catalyst of phosphoric acid on quartz chips. { ¦kal·ə¦fȯr·nyə pə,lim·ə·rə'zā·shən }

California sampler [MIN ENG] A drive sampler equipped with a piston that can be retracted mechanically to any desired point within the barrel of the sampler. { ¦kal·ə¦fȯr·nyə 'sam·plər }

California-type dredge [MIN ENG] A single-lift dredge in which closely spaced buckets deliver to a trommel; oversize rocks are piled behind the dredge by a conveyor or stacker; the undersize are washed on gold-saving tables on the deck, and tailings discharge astern through sluices. { ¦kal·ə¦fȯr·nyə ,tīp 'drej }

californite [MINERAL] $Ca_{10}Al_4(Mg,Fe)_2Si_9O_{34}(OH,F)_4$ A variety of vesuvianite resembling jade; it is dark-, yellowish-, olive-, or grass-green and occurs in translucent to opaque compact or massive form. Also known as American jade. { ,kal·ə'fȯr,nīt }

californium [CHEM] A chemical element, symbol Cf, atomic number 98; all isotopes are radioactive. { ,kal·ə'fȯr·nē·əm }

Caligidae [INV ZOO] A family of fish ectoparasites belonging to the crustacean suborder Caligoida. { kə'lij·ə,dē }

Caligoida [INV ZOO] A suborder of the crustacean order Copepoda, including only fish ectoparasites and characterized by a sucking mouth with styliform mandibles. { kal·ə'gȯid·ə }

calina [METEOROL] A haze prevalent in Spain during the summer, when the air becomes filled with dust swept up from the dry ground by strong winds. { kə'lēn·ə }

caliper [DES ENG] An instrument with two legs or jaws that can be adjusted for measuring linear dimensions, thickness, or diameter. { 'kal·ə·pər }

caliper gage [DES ENG] An instrument, such as a micrometer, of fixed size for calipering. { 'kal·ə·pər ,gāj }

caliper log [MIN ENG] A graphic record showing the diameter of a drilled hole at each depth; measurements are obtained by drawing a caliper upward through the hole and recording the diameter on quadrile paper. { 'kal·ə·pər ,läg }

caliper splint [MED] A splint designed for the leg, consisting of two metal rods from a posterior-thigh band or a padded ischial ring to a metal plate attached to the sole of the shoe at the instep. { 'kal·ə·pər ,splint }

calite [MET] A heat-resistant alloy of iron, nickel, and aluminum which resists oxidation up to 2200°F (1204°C) and is practically noncorrodible under ordinary conditions of exposure. { 'kā,līt }

calixarene [ORG CHEM] A cyclic structure containing the group $(-Ar-CH_2-)_n$, where Ar represents an aryl group. { kə'lik·sə,rēn }

calk *See* caulk. { kòk }

call [COMPUT SCI] **1.** To transfer control to a specified closed subroutine. **2.** A statement in a computer program that references a closed subroutine or program. { kòl }

call announcer [ELECTR] Device for receiving pulses from an automatic telephone office and audibly reproducing the corresponding number in words, so that it may be heard by a manual operator. { 'kòl ə'naùn·sər }

Callao painter *See* painter. { kə'yaù ,pān·tər }

call by location [COMPUT SCI] A method of transferring arguments from a calling program to a subprogram in which the referencing program provides to the subprogram the memory location at which the value of the argument can be found, rather than the value itself. Also known as call by reference. { 'kòl bī ,lō'kā·shən }

call by name [COMPUT SCI] A method of transferring arguments from a calling program to a subprogram in which the actual expression is passed to the subprogram. { 'kòl bī 'nām }

call by reference *See* call by location. { 'kòl bī 'ref·rəns }

call by value [COMPUT SCI] A method of transferring arguments from a calling program to a subprogram in which the subprogram is provided with the values of the argument and on path leads back to the referencing program. { 'kòl bī 'val·yü }

call circuit [ELEC] Communications circuit between switching points used by traffic forces for transmitting switching instructions. { 'kòl ,sər·kət }

called routine [COMPUT SCI] A subroutine that is accessed by a call or branch instruction in a computer program. { 'kòld rü,tēn }

Callendar and Barnes' continuous-flow calorimeter [ENG] A calorimeter in which the heat to be measured is absorbed by water flowing through a tube at a constant rate, and the quantity of heat is determined by the rate of flow and the temperature difference between water at ends of the tube. { 'kal·ən·dər ən 'bärnz kən'tin·yə·wəs ,flō kal·ə'rim·əd·ər }

Callendar's compensated air thermometer [ENG] A type of constant-pressure gas thermometer in which errors resulting from temperature differences between the thermometer bulb and the connecting tubes and manometer used to maintain constant pressure are eliminated by the configuration of the connecting tubes. { 'kal·ən·dərz 'käm·pən,sād·əd 'er thər,mäm·əd·ər }

Callendar's equation [THERMO] **1.** An equation of state for steam whose temperature is well above the boiling point at the existing pressure, but is less than the critical temperature: $(V - b) = (RT/p) - (a/T^n)$, where V is the volume, R is the gas constant, T is the temperature, p is the pressure, n equals 10/3, and a and b are constants. **2.** A very accurate equation relating temperature and resistance of platinum, according to which the temperature is the sum of a linear function of the resistance of platinum and a small correction term, which is a quadratic function of temperature. { 'kal·ən·dərz i'kwā·zhən }

Callendar's thermometer *See* platinum resistance thermometer. { 'kal·ən·dərz thər'mäm·əd·ər }

callenia *See* stromatolite. { kə'lēn·yə }

call fire [ORD] Fire delivered on a specific target in response to a request from the supported unit. { 'kòl ,fīr }

call forwarding [COMMUN] A telephone service that allows a telephone to be programmed to automatically transfer incoming calls to a designated number. { 'kòl 'fòr·wərd·iŋ }

Callichthyidae [VERT ZOO] A family of tropical catfishes in the suborder Siluroidei. { ,ka,lik'thī·ə,dē }

callideic I *See* bradykinin. { ,kal·i,dē·ik 'wən }

Callier's Q factor [GRAPHICS] A relationship between the diffuse transmission density of a photographic negative, which measures the total light transmitted, and the specular density, which measures only the light passing directly through. { 'kal,yāz 'kyü ,fak·tər }

calligraphy [GRAPHICS] Elegant writing or penmanship as an art and as a profession. { kə'lig·rə·fē }

call in [COMPUT SCI] To transfer control of a digital computer, temporarily, from a main routine to a subroutine that is inserted in the sequence of calculating operations, to fulfill an ancillary purpose. { 'kòl ,in }

call indicator [ELECTR] Device for receiving pulses from an automatic switching system and displaying the corresponding called number before an operator at a manual switchboard. { 'kòl 'in·də,kād·ər }

calling device [ELECTR] Apparatus which generates the pulses required for establishing connections in an automatic telephone switching system. { 'kòl·iŋ di'vīs }

calling program [COMPUT SCI] A computer program that initiates a call to another program. { 'kòl·iŋ ,prō·grəm }

calling routine [COMPUT SCI] A subroutine that initiates a call to another subroutine. { 'kòl·iŋ rü,tēn }

calling sequence [COMPUT SCI] A specific set of instructions to set up and call a given subroutine, make available the data required by it, and tell the computer where to return after the subroutine is executed. { 'kòl·iŋ ,sē·kwəns }

calling song [INV ZOO] A high-intensity insect sound which may play a role in habitat selection among certain species. { 'kòl·iŋ ,sòŋ }

Callionymoidei [VERT ZOO] A suborder of fishes in the order Perciformes, including two families of colorful marine bottom fishes known as dragonets. { kə'län·ē,mòid·ē,ī }

Callipallenidae [INV ZOO] A family of marine arthropods in the subphylum Pycnogonida lacking palpi and having chelifores and 10-jointed ovigers. { ,kal·ə·pə'len·ə,dē }

Calliphoridae [INV ZOO] The blow flies, a family of myodarian cyclorrhaphous dipteran insects in the subsection Calypteratae. { ,kal·ə'fòr·ə,dē }

Callipic cycle [ASTRON] Four Metonic cycles, or 76 years. { kə'lip·ik ,sī·kəl }

Callisto [ASTRON] A satellite of Jupiter orbiting at a mean distance of 1,884,000 kilometers. Also known as Jupiter IV. { kə'lis,tō }

Callithricidae [VERT ZOO] The marmosets, a family of South American mammals in the order Primates. { kal·ə'thris·ə,dē }

Callitrichales [BOT] An order of flowering plants, division Magnoliophyta (Angiospermae), in the subclass Asteridae of the class Magnoliopsida (dicotyledons); consists of three small families with about 50 species, most of which are aquatics or small herbs of wet places. { kə,lit·rə'kā·lēz }

call letters [COMMUN] Identifying letters, sometimes including numerals, assigned to radio and television stations by the Federal Communications Commission and other regulatory authorities throughout the world. Also known as call sign. { 'kòl ,led·ərz }

call number [COMPUT SCI] In computer operations, a set of characters identifying a subroutine, and containing information concerning parameters to be inserted in the subroutine, or information to be used in generating the subroutine, or information related to the operands. { 'kòl ,nəm·bər }

Callon's rule [MIN ENG] The rule stating that when a pillar is left in an inclined seam for support in a shaft or a structure on the surface, a greater width should be left on the rise side of the shaft or structure than on the dip side. { 'kal·ənz ,rül }

Callorhinchidae [VERT ZOO] A family of ratfishes of the chondrichthyan order Chimaeriformes. { kal·ə'riŋk·ə,dē }

callose [BIOCHEM] A carbohydrate component of plant cell walls; associated with sieve plates where calluses are formed. [BIOL] Having hardened protuberances, as on the skin or on leaves and stems. { 'ka,lōs }

Callovian [GEOL] A stage in uppermost Middle or lowermost Upper Jurassic which marks a return to clayey sedimentation. { kə'lōv·ē·ən }

Callow flotation cell [MIN ENG] Nonmechanical apparatus for separation of floatable solid gangue from pulverized ore, with the mixture suspended in liquid and aerated by air bubbles coming up through a porous medium, so that the lighter gangue floats away from the heavier ore. { 'kal·ō flō'tā·shən ,sel }

Callow process [MIN ENG] A flotation process in which agitation is provided by the forcing of air into the pulp through the canvas-covered bottom of the cell. { 'kal·ō ,präs·əs }

Callow screen [MIN ENG] A continuous belt system formed of fine screen wire that is used to separate fine solids from coarse ones. { 'kal·ō ,skrēn }

call setup time [COMMUN] The period of time between the lifting of a handset to make a telephone call and the start of voice or data transmission. { 'kòl 'sed·əp ,tīm }

call sign *See* call letters. { 'kòl ,sīn }

call up [COMPUT SCI] To retrieve data from computer memory, especially for display and user interaction. { 'kòl ,əp }

callus [BOT] **1.** A thickened callose deposit on sieve plates. **2.** A hard tissue that forms over a damaged plant surface. [MED] Hard, thick area on the surface of the skin. { 'kal·əs }

calm [METEOROL] The absence of apparent motion of the air; in the Beaufort wind scale, smoke is observed to rise vertically, or the surface of the sea is smooth and mirrorlike; in U.S. weather observing practice, the wind has a speed under 1 mile per hour or 1 knot (1.6 kilometers per hour). { käm }

calmagite [ORG CHEM] $C_{17}H_{14}N_2O_5S$ A compound crystallizing from acetone as red crystals that are soluble in water; used as an indicator in the titration of calcium or magnesium with EDTA. { 'kal·mə‚jīt }

calm belt [METEOROL] A belt of latitude in which the winds are generally light and variable; the principal calm belts are the horse latitudes (the calms of Cancer and of Capricorn) and the doldrums. { 'käm ‚belt }

calmodulin [BIOCHEM] A calcium-modulated protein consisting of a single polypeptide with 148 amino acids and a molecular weight of 16,700, found in all eukaryotes. { kal'-mäj·ə·lən }

calms of Cancer [METEOROL] One of the two light, variable winds and calms which occur in the centers of the subtropical high-pressure belts over the oceans; their usual position is about latitude 30°N, the horse latitudes. { ¦kämz əv 'kan·sər }

calms of Capricorn [METEOROL] One of the two light, variable winds and calms which occur in the centers of the subtropical high-pressure belts over the oceans; their usual position is about latitude 30°S, the horse latitudes. { ¦kämz əv 'kap·ri‚körn }

Calobryales [BOT] An order of liverworts; characterized by prostrate, simple or branched, leafless stems and erect, leafy branches of a radial organization. { ‚kal·ō·brī'ā·lēz }

calomel [MINERAL] Hg_2Cl_2 A colorless, white, grayish, yellowish, or brown secondary, sectile, tetragonal mineral; used as a cathartic, insecticide, and fungicide. Also known as calomelene; calomelite; horn quicksilver; mercurial horn ore. { 'kal·ə·məl }

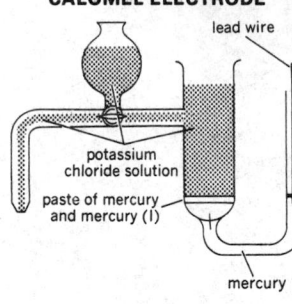

CALOMEL ELECTRODE

lead wire

potassium chloride solution

paste of mercury and mercury (I)

mercury

Diagram of calomel electrode.

calomel electrode [PHYS CHEM] A reference electrode of known potential consisting of mercury, mercury chloride (calomel), and potassium chloride solution; used to measure pH and electromotive force. Also known as calomel half-cell; calomel reference electrode. { 'kal·ə·məl i'lek‚trōd }

calomelene See calomel. { kə'läm·ə‚lēn }

calomel half-cell See calomel electrode. { 'kal·ə·məl 'haf ‚sel }

calomelite See calomel. { ‚kal·ə'me‚līt }

calomel reference electrode See calomel electrode. { 'kal·ə·məl 'ref·rəns i'lek‚trōd }

caloreceptor [PHYSIO] A cutaneous sense organ that is stimulated by heat. { ¦kal·ō·ri'sep·tər }

calorescence [PHYS] The production of visible light by infrared radiation; the transformation is indirect, the light being produced by heat and not by any direct change of wavelength. { ‚kal·ə'res·əns }

calorie [THERMO] Abbreviated cal; often designated c. **1.** A unit of heat energy, equal to 4.1868 joules. Also known as International Table calorie (IT calorie). **2.** A unit of energy, equal to the heat required to raise the temperature of 1 gram of water from 14.5° to 15.5°C at a constant pressure of 1 standard atmosphere; equal to 4.1855 ± 0.0005 joules. Also known as fifteen-degrees calorie; gram-calorie (g-cal); small calorie. **3.** A unit of heat energy equal to 4.184 joules; used in thermochemistry. Also known as thermochemical calorie. { 'kal·ə·rē }

calorific value [ENG] Quantity of heat liberated on the complete combustion of a unit weight or unit volume of fuel. { ¦kal·ə¦rif·ik 'val·yü }

calorifier [ENG] A device that heats fluids by circulating them over heating coils. { kə'lór·ə‚fī·ər }

calorimeter [ENG] An apparatus for measuring heat quantities generated in or emitted by materials in processes such as chemical reactions, changes of state, or formation of solutions. [NUCLEO] A device used for measuring the total energy of both neutral and charged particles generated in a high-energy particle collision, by means of total absorption. { ‚kal·ə'rim·əd·ər }

calorimetric-respirometric ratio [PHYSIO] The ratio expressing that the oxidation of all carbon compounds produces approximately the same amount of heat per amount of oxygen.

Abbreviated CR ratio. { kə¦lór·ə¦me·trik ri¦spī·rə¦me·trik 'rā·shō }

calorimetric test [ENG] The use of a calorimeter to determine the thermochemical characteristics of propellants and explosives; properties normally determined are heat of combustion, heat of explosion, heat of formation, and heat of reaction. { kə¦lór·ə¦me·trik 'test }

calorimetric titration See thermometric titration. { kə¦lór·ə¦me·trik tī'trā·shən }

calorimetry [ENG] The measurement of the quantity of heat involved in various processes, such as chemical reactions, changes of state, and formations of solutions, or in the determination of the heat capacities of substances; fundamental unit of measurement is the joule or the calorie (4.184 joules). { kal·ə'rim·ə·trē }

Caloris Basin [ASTRON] A large depression on Mercury, about 1300 kilometers in diameter. { kə'lór·əs 'bā·sən }

calorizator [FOOD ENG] A device for heating beet juice; used in beet-sugar factories. { kə'lór·ə‚zād·ər }

calorize [MET] To treat by a process by which a coating of aluminum and aluminum-iron alloys is produced on iron and steel (and, less commonly, brass, copper, or nickel), which coating protects the metal from burning in temperatures up to 1800°F (982°C). { 'kal·ə‚rīz }

calorized steel See aluminized steel. { 'kal·ə‚rīzd ‚stēl }

calotte [BIOL] A cap or caplike structure. [INV ZOO] **1.** The four-celled polar cap in Dicyemidae. **2.** A ciliated, retractile disc in certain bryozoan larva. **3.** A dark-colored anterior area in certain nematomorphs. { kə'lät }

calotype [GRAPHICS] An obsolete method of photography in which paper is treated with silver iodide, silver nitrate, and acetic and gallic acids; after exposure the paper is developed in a solution of silver nitrate and gallic acid. { 'kal·ə‚tīp }

calpain [CELL MOL] A calcium-dependent cysteine protease in the cytoplasm that is central to most processes in cell biology (plasma membrane–associated signaling events; cell proliferation, differentiation, activation, and communication; and programmed cell death). Its overactivation has been observed in muscular dystrophy, Alzheimer's disease, AIDS, cataract formation, multiple sclerosis, and arthritis. { 'kal‚pān }

calpastatin [CELL MOL] A protein found in all cells that is both the specific inhibitor and substrate of calpains. { ‚kal·pə'stat·ən }

calsequestrin [CELL MOL] An acidic protein, with a molecular weight of 44,000, which binds calcium inside the sarcoplasmic reticulum. { ¦kal·sə¦kwes·trən }

calthrop [INV ZOO] A sponge spicule having four axes in which the rays are equal or almost equal in length. { 'kal‚thrəp }

calutron [NUCLEO] An electromagnetic apparatus for separating isotopes of uranium and other elements according to their masses, using the principle of the mass spectrograph. { 'kal·yə‚trän }

calvarium [ANAT] A skull lacking facial parts and the lower jaw. { kal'ver·ē·əm }

calved ice [OCEANOGR] A piece of ice floating in a body of water after breaking off from a mass of land ice or an iceberg. Also known as calf. { ‚kavd 'īs }

Calvé's disease See osteochondrosis. { kal'väz di‚zēz }

Calvin-Benson cycle See Calvin cycle. { ¦kal·vən 'ben·sən ‚sī·kəl }

Calvin cycle [CELL MOL] A metabolic process during photosynthesis that uses light indirectly to convert carbon dioxide to sugar in the stroma of chloroplasts. Also known as Calvin-Benson cycle; carbon fixation cycle. { 'kal·vən ‚sī·kəl }

calving [GEOL] The breaking off of a mass of ice from its parent glacier, iceberg, or ice shelf. Also known as ice calving. [VERT ZOO] Giving birth to a calf. { 'kav·iŋ }

calx See calcium oxide. { kalks }

calycanthemy [PL PATH] Abnormal development of calyx structures into petals or petaloid structures. { ‚kal·ə'kan·thə·mē }

Calycerales [BOT] An order of flowering plants, division Magnoliophyta (Angiospermae), in the subclass Asteridae of the class Magnoliopsida (dicotyledons); consists of a single family with about 60 species native to tropical America. { 'kal·ə·sə'rā·lēz }

calyculate [BOT] Having bracts that imitate a second, external calyx. { kə'lik·yə·lət }

calymma [INV ZOO] The outer, vacuolated protoplasmic layer of certain radiolarians. { kə'lim·ə }

Calymmatobacterium [MICROBIO] A genus of gram-negative, usually encapsulated, pleomorphic rods of uncertain affiliation; the single species causes granuloma inguinale in humans. { kə¦lim·əd·ō,bak'tir·ē·əm }

Calymnidae [INV ZOO] A family of echinoderms in the order Holasteroida characterized by an ovoid test with a marginal fasciole. { kə'lim·nə,dē }

Calypso [ASTRON] A small, irregularly shaped satellite of Saturn that librates about the leading Lagrangian point of Tethys's orbit. { kə'lip·sō }

calypter [INV ZOO] A scalelike or lobelike structure above the haltere of certain two-winged flies. Also known as alula; squama. { kə'lip·tər }

Calyptoblastea [INV ZOO] A suborder of cnidarians in the order Hydroida, including the hydroids with protective cups around the hydranths and gonozooids. { kə¦lip·tō¦blas·tē·ə }

calyptra [BOT] **1.** A membranous cap or hoodlike covering, especially the remains of the archegonium over the capsule of a moss. **2.** Tissue surrounding the archegonium of a liverwort. **3.** Root cap. { kə'lip·trə }

Calypteratae [INV ZOO] A subsection of dipteran insects in the suborder Cyclorrhapha characterized by calypters associated with the wings. { kə'lip·tə·rə,tē }

calyptrate [BOT] Having a calyptra. { kə'lip,trāt }

calyptrogen [BOT] The specialized cell layer from which a root cap originates. { kə'lip·trə·jən }

Calyssozoa [INV ZOO] The single class of the bryozoan subphylum Entoprocta. { kə,lis·ə'zō·ə }

calyx [BOT] The outermost whorl of a flower; composed of sepals. [ENG] A steel tube that is a guide rod and is also used to catch cuttings from a drill rod. Also known as bucket; sludge barrel; sludge bucket. [INV ZOO] A cup-shaped structure to which the arms are attached in crinoids. [MED] **1.** A cuplike structure. **2.** In the kidney, a collecting structure extending from the renal pelvis. { 'kā,liks }

calyx drill [ENG] A rotary core drill with hardened steel shot for cutting rock. Also known as shot drill. { 'kā,liks ,dril }

calyx tube [BOT] A tube formation resulting from fusion of the lateral edges of a group of sepals. { 'kā·liks ,tüb }

Calzecchi-Onesti effect [ELEC] A change in the conductivity of a loosely aggregated metallic powder caused by an applied electric field. { ,kält'se·kē ,ō'nes·tē i'fekt }

cam [MECH ENG] A plate or cylinder which communicates motion to a follower by means of its edge or a groove cut in its surface. { kam }

Cam *See* Camelopardalis.

CAM *See* cell adhesion molecule; computer-aided manufacturing; crassulacean acid metabolism. { ¦sē¦ā'em *or* kam }

cam acceleration [MECH ENG] The acceleration of the cam follower. { 'kam ak·sel·ə'rā·shən }

Camacolaimoidea [INV ZOO] A superfamily of nematodes consisting of a single family, the Camacolaimidae; they occur in marine or brackish-water environments. { ,kam·ə·kō·lə'mòid·ē·ə }

Camallanida [INV ZOO] An order of phasmid nematodes in the subclass Spiruria, including parasites of domestic animals. { kam·ə'lan·ə·də }

Camallanoidea [INV ZOO] A superfamily of parasitic nematodes in the subclass Spiruria. { kə,mal·ə'nòid·ē·ə }

camanchaca *See* garúa. { kä·män'chä·kə }

Camarasaurus [PALEON] A herbivorous sauropod dinosaur, 60 feet (18 meters) long and weighing 20 tons, from the Late Jurassic Period that had a very long neck and tail. { ,ka·mə·rə'sòr·əs }

Camarodonta [INV ZOO] An order of Euechinoidea proposed by R. Jackson and abandoned in 1957. { kam·ə·rə'dän·tə }

camarodont dentition [INV ZOO] In echinoderms, keeled teeth meeting the epiphyses so that the foramen magnum of the jaw is closed. { 'kam·ə·rə,dänt den'tish·ən }

Cambaridae [INV ZOO] A family of crayfishes belonging to the section Macrura in the crustacean order Decapoda. { kam'bär·ə,dē }

Cambarinae [INV ZOO] A subfamily of crayfishes in the family Astacidae, including all North American species east of the Rocky Mountains. { kam'bär·ə,nē }

camber [AERO ENG] The rise of the curve of an airfoil section, usually expressed as the ratio of the departure of the curve from a straight line joining the extremities of the curve to the length of this straight line. [DES ENG] Deviation from a straight line; the term is applied to a convex, edgewise sweep or curve, or to the increase in diameter at the center of rolled materials. [GEOL] **1.** A terminal, convex shoulder of the continental shelf. **2.** A structural feature that is caused by plastic clay beneath a bed flowing toward a valley so that the bed sags downward and seems to be draped over the sides of the valley. *See* round of beam. { 'kam·bər }

camber angle [MECH ENG] The inclination from the vertical of the steerable wheels of an automobile. { 'kam·bər ,aŋ·gəl }

camber arch [ARCH] An arch with a slightly curved interior and horizontal exterior. { 'kam·bər ,arch }

camber-keeled [NAV ARCH] Property of a ship whose keel is somewhat convex. { 'kam·bər ,kēld }

cambium [BOT] A layer of cells between the phloem and xylem of most vascular plants that is responsible for secondary growth and for generating new cells. { 'kam·bē·əm }

Cambrian [GEOL] The lowest geologic system that contains abundant fossils of animals, and the first (earliest) geologic period of the Paleozoic era from 570 to 500 million years ago. { 'kam·brē·ən }

cambric [TEXT] **1.** Tightly woven cotton with one side calendered to resemble linen. **2.** Linen with a sheen. { 'kam,brik }

camcorder [ELECTR] A one-piece hand-held television camera with built-in videocassette recorder, microphone, and battery pack, utilizing a charge-coupled device array as its light-sensitive element. { 'kam,córd·ər }

cam cutter [MECH ENG] A semiautomatic or automatic machine that produces the cam contour by swinging the work as it revolves; uses a master cam in contact with a roller. { 'kam ,kəd·ər }

cam dwell [DES ENG] That part of a cam surface between the opening and closing acceleration sections. { 'kam ,dwel }

camel [VERT ZOO] The common name for two species of artiodactyl mammals, the bactrian camel (*Camelus bactrianus*) and the dromedary camel (*C. dromedarius*), in the family Camelidae. { 'kam·əl }

camel hair [TEXT] A fine, natural textile fiber obtained from the bactrian camel. { 'kam·əl ,hcr }

Camelidae [VERT ZOO] A family of tylopod ruminants in the superfamily Cameloidea of the order Artiodactyla, including four species of camels and llamas. { ka'mel·ə,dē }

Cameloidea [VERT ZOO] A superfamily of tylopod ruminants in the order Artiodiodactyla. { kam·ə'lòid·ē·ə }

Cameloopardalis [ASTRON] Latin name for the Giraffe constellation of the northern hemisphere. Abbreviated Cam; Caml. Also known as Camelopardus; Giraffe. { ka,mel·ə'pärd·əl·əs }

Camelopardus *See* Camelopardalis. { ka,mel·ə'pär·dəs }

cam engine [MECH ENG] A piston engine in which a cam-and-roller mechanism seems to convert reciprocating motion into rotary motion. { 'kam ,en·jən }

cameo [LAP] A type of carved gemstone in which the background is cut away to leave the subject in relief. { 'kam·ē·ō }

camera [ELECTR] *See* television camera. [OPTICS] A light-tight enclosure containing an aperture (usually provided with an optical lens or system of lenses) through which the light from an object passes and forms an image, often on a light-sensitive material, inside. { 'kam·rə }

camera base [GRAPHICS] A bed or support below a camera that provides an area for placement of documents for filming. { 'kam·rə ,bās }

camera cable [ELEC] Cable or group of wires that carries the picture from the television camera to the control room. { 'kam·rə ,kā·bəl }

camera card [GRAPHICS] An aperture card that holds unexposed and unprocessed film that will be exposed and processed without being removed from the card. { 'kam·rə ,kärd }

camera chain [COMMUN] A television camera, associated amplifiers, a monitor, and the cable needed to bring the camera output signal to the control room. { 'kam·rə ,chān }

camera head [GRAPHICS] That component in a microfilming device that contains the film, the film-advance mechanism, and the lens. { 'kam·rə ,hed }

camera lucida [OPTICS] An instrument having a peculiarly

CAMARODONT DENTITION

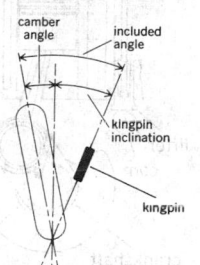

Jaw and tooth arrangement.

CAMBER ANGLE

Front left wheel of an automobile viewed from driver's position. Angles are exaggerated.

CAMCORDER

Television camcorder. (*Thomson Multimedia*)

shaped prism or a system of mirrors, and often a microscope, which causes a virtual image of an object to be produced on a plane surface, enabling the image's outline to be traced. { ¦kam·rə 'lü·səd·ə }

camera obscura [OPTICS] A primitive camera in which the real image of an object can be observed or traced on the wall of the enclosure opposite the aperture, rather than being recorded photographically. { ¦kam·rə əb'skyur·ə }

camera-ready [GRAPHICS] A layout prepared for the offset printer; contains the actual material to be reproduced. { ¦kam·rə 'red·ē }

camera station [MAP] In aerial photography, the point in space occupied by the camera lens at the moment of exposure. Also known as air station. { 'kam·rə ,stā·shən }

camera study *See* memomotion study. { 'kam·rə ,stəd·ē }

Camerata [PALEON] A subclass of extinct stalked echinoderms of the class Crinoidea. { ,kam·ə'räd·ə }

camera tube [ELECTR] An electron-beam tube used in a television camera to convert an optical image into a corresponding charge-density electric image and to scan the resulting electric image in a predetermined sequence to provide an equivalent electric signal. Also known as pickup tube; television camera tube. { 'kam·rə ,tüb }

cam follower [MECH ENG] The output link of a cam mechanism. { 'kam ,fäl·ə·wər }

Camilla [ASTRON] An asteroid with a diameter of about 220 kilometers, mean distance from the sun of 3.49 astronomical units, and C-type surface composition. { kə'mil·ə }

Caml *See* Camelopardalis. { 'kam·əl }

cam mechanism [MECH ENG] A mechanical linkage whose purpose is to produce, by means of a contoured cam surface, a prescribed motion of the output link. { 'kam ,mek·ə,niz·əm }

Cammett table [MIN ENG] A side-moving table for concentrating ore. { 'kam·ət ,tā·bəl }

cam nose [MECH ENG] The high point of a cam, which in a reciprocating engine holds valves open or closed. { 'kam ,nōz }

camouflage [ORD] The method of concealing things or people from the enemy by trying to make them appear to be a section of the natural background. { 'kam·ə,fläzh }

cAMP *See* cyclic adenylic acid.

Campanian [GEOL] European stage of Upper Cretaceous. { kam'pan·ē·ən }

Campanulaceae [BOT] A family of dicotyledonous plants in the order Campanulales characterized by a style without an indusium but with well-developed collecting hairs below the stigmas, and by a well-developed latex system. { kam,pan·yə'lās·ē,ē }

Campanulales [BOT] An order of dicotyledonous plants in the subclass Asteridae distinguished by a chiefly herbaceous habit, alternate leaves, and inferior ovary. { kam,pan·yə'lā,lēz }

campanulate [BOT] Bell-shaped; applied particularly to the corolla. { kam'pan·yə·lət }

cam pawl [MECH ENG] A pawl which prevents a wheel from turning in one direction by a wedging action, while permitting it to rotate in the other direction. { 'kam ,pól }

Campbell bridge [ELEC] **1.** A bridge designed for comparison of mutual inductances. **2.** A circuit for measuring frequencies by adjusting a mutual inductance, until the current across a detector is zero. { 'kam·əl ,brij }

Campbell process [MET] An open-hearth steel manufacturing process in which ore and pig iron are used as raw materials in a tilting furnace. { 'kam·əl ,präs·əs }

Campbell's formula [ELECTROMAG] A formula which relates the propagation constant of a loaded transmission line to the propagation constant and characteristic impedance of an unloaded line and the impedance of each loading coil. { 'kam·əlz ¦fór·myə·lə }

Campbell-Stokes recorder [ENG] A sunshine recorder in which the time scale is supplied by the motion of the sun and which has a spherical lens that burns an image of the sun upon a specially prepared card. { ¦kam·əl ¦stōks ri'kórd·ər }

camp ceiling [BUILD] A ceiling that is flat in the center portion and sloping at the sides. { 'kamp ,sē·liŋ }

campestrian [ECOL] Of or pertaining to the northern Great Plains area. { kam'pes·trē·ən }

camphane [ORG CHEM] $C_{10}H_{18}$ An alicyclic hydrocarbon;

white crystals, soluble in alcohol, with a melting point of 158–159°C. { 'kam,fan }

camphene [ORG CHEM] $C_{10}H_{16}$ A bicyclic terpene used as raw material in the synthesis of insecticides such as toxaphene and camphor. { 'kam,fēn }

camphor [ORG CHEM] $C_{10}H_{16}O$ A bicyclic saturated terpene ketone that exists in optically active dextro and levo forms and as a racemate; the dextro form is obtained from the wood and bark of the camphor tree, the levo form is found in some essential oils, and the inactive form is obtained from an Asiatic chrysanthemum or made synthetically from certain terpenes. { 'kam·fər }

camphoric acid [PHARM] $C_{10}H_{16}O_4$ A compound crystallizing in leaflets from water and in monoclinic prisms from alcohol; melting point is 186–188°C; used as a central respiratory stimulant. { kam'fór·ik 'as·əd }

camphor oil [MATER] An essential oil obtained by steam distillation from the wood of the camphor tree (*Cinnamomum camphora*); used in the manufacture of camphor and safrole. { 'kam·fər ,óil }

d-**camphorsulfonic acid** [ORG CHEM] $C_{10}H_{16}O_4S$ A compound crystallizing as prisms from ethyl acetate or glacial acetic acid; slightly soluble in glacial acetic acid and in ethyl acetate; used in the resolution of optically active isomers. Also known as Reychler's acid. { ¦dē ¦kam·fər,səl'fän·ik 'as·əd }

camphor tree [BOT] *Cinnamomum camphora*. A plant of the laurel family (Lauraceae) in the order Magnoliales from which camphor is extracted. { 'kam·fər ,trē }

Camp-Meidell condition [STAT] For determining the distribution of a set of numbers, the guideline stating that if the distribution has only one mode, if the mode is the same as the arithmetic mean, and if the frequencies decline continuously on both sides of the mode, then more than $1 - (1/2.25t^2)$ of any distribution will fall within the closed range $\bar{X} \pm t\sigma$, where t = number of items in a set, $\bar{X}$ = average, and σ = standard deviation. { ¦kamp ¦mī'del kən,dish·ən }

Campodeidae [INV ZOO] A family of primarily wingless insects in the order Diplura which are most numerous in the Temperate Zone of the Northern Hemisphere. { kam·pə'dē·ə,dē }

campodeiform [INV ZOO] Elongate, flattened, and narrowed posteriorly. { kam'pō·dē·ə,fórm }

camp-on system [COMMUN] A circuit control feature whereby a user attempting to establish a telephone call and encountering a busy station will hold the connection for a preset time, to the exclusion of other callers, in case the original conversation should terminate. { 'kamp ¦ón ,sis·təm }

campos [ECOL] The savanna of South America. { 'käm ,pōs }

cam profile [DES ENG] The shape of the contoured cam surface by means of which motion is communicated to the follower. Also known as pitch line. { 'kam ,prō,fīl }

camptonite [PETR] A lamprophyre containing pyroxene, sodic hornblende, and olivine as dark constituents and labradorite as the light constituent; sodic orthoclase may be present. { 'kam·tə,nīt }

camptothecin [PHARM] An alkaloid belonging to the family of drugs called topoisomerase inhibitors which is used as an anticancer drug; it is the first known naturally produced DNA topoisomerase inhibitor and was originally isolated from the Chinese tree *Camptothecin acuminate*. { ,kam·tō'thē·sən }

Campylobacter [MICROBIO] A genus of bacteria in the family Spirillaceae; spirally curved rods that are motile by means of a polar flagellum at one or both poles. { ,kam·pə·lə'bak·tər }

Campylobacter enteritis [MED] A water-borne gastroenteritis caused by *Campylobacter jejune*. { kam¦pī·lə,bak·tər ,en·tə'rīd·əs }

Campylobacter jejune [MICROBIO] A microaerophilic pathogen associated with raw meats and unpasteurized milk; ingestion of a small amount can cause diarrhea, cramps, and nausea. { kam¦pī·lə,bak·tər jə'jü·nē }

campylotropous [BOT] Having the ovule symmetrical but half inverted, with the micropyle and funiculus at right angles to each other. { ¦kam·pə¦lä·trə·pəs }

camshaft [MECH ENG] A rotating shaft to which a cam is attached. { 'kam,shaft }

can [DES ENG] A cylindrical metal vessel or container, usually with an open top or a removable cover. [NUCLEO] *See* jacket. { kan }

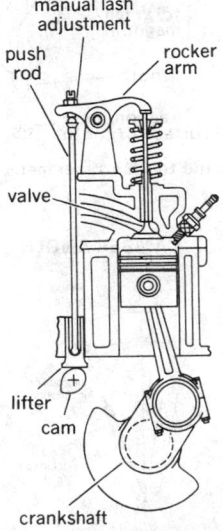

CAM MECHANISM

manual lash adjustment
push rod
rocker arm
valve
lifter
cam
crankshaft

Cam mechanism for opening and closing valves in automotive engine. *(Texaco, Inc.)*

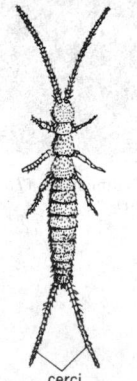

CAMPODEIDAE

cerci

Campodea folsomi, in the family Campodeidae.

Canaceidae [INV ZOO] The seashore flies, a family of myodarian cyclorrhaphous dipteran insects in the subsection Acalypteratae. { ‚kan·ə'sē·ə‚dē }

Canada balsam [MATER] A transparent balsam useful for cementing together lenses and other optical elements because its index of refraction is in the same range as that of glass. { ‚kan·ə·də 'bȯl·səm }

Canadian life zone [ECOL] The zone comprising the climate and biotic communities of the portion of the Boreal life zone exclusive of the Hudsonian and Arctic-Alpine zones. { kə'nād·ē·ən 'līf ‚zōn }

Canadian Shield See Laurentian Shield. { kə'nād·ē·ən 'shēld }

canadol [MATER] A light petroleum naphtha fraction having a specific gravity slightly higher than petroleum ether; used as a solvent and an anesthetic. { 'kan·ə‚dȯl }

canal [BIOL] A tubular duct or passage in bone or soft tissues. [CIV ENG] An artificial open waterway used for transportation, waterpower, or irrigation. [DES ENG] A groove on the underside of a corona. [GEOGR] A long, narrow arm of the sea extending far inland, between islands, or between islands and the mainland. [NUCLEO] A water-filled trench or conduit associated with a nuclear reactor, used for removing and sometimes storing radioactive objects taken from the reactor; the water acts as a shield against radiation. { kə'nal }

canal boat [NAV ARCH] A long, narrow boat used on canals, with nearly vertical bow and stern giving it a large cargo capacity. { kə'nal ‚bōt }

canal cell [BOT] One of the row of cells that make up the axial row within the neck of an archegonium. { kə'nal ‚sel }

canaliculate [BIOL] Having small channels, canals, or grooves. { ‚kan·əl'ik·yə‚lāt }

canaliculus [HISTOL] 1. One of the minute channels in bone radiating from a Haversian canal and connecting lacunae with each other and with the canal. 2. A passage between the cells of the cell cords in the liver. { ‚kan·əl'ik·yə·ləs }

canalization [ENG] Any system of distribution canals or conduits for water, gas, electricity, or steam. [EVOL] The effect of natural selection on development to produce pathways that are insensitive to minor genetic or environmental variation; results in the phenotypic norm of the species. [MED] Surgical method of wound drainage without tubes by forming channels. [PHYSIO] The formation of new channels in tissues, such as the formation of new blood vessels through a thrombus. { ‚kan·əl·ə'zā·shən }

canalized character [GEN] A trait whose variability is restricted within narrow boundaries even when individual members of the species are subjected to enhanced environmental pressures or mutations. { ‚kan·ə‚līzd 'kar·ik·tər }

canal of Schlemm [ANAT] An irregular channel at the junction of the sclera and cornea in the eye that drains aqueous humor from the anterior chamber. { kə'nal əv 'shlem }

canal ray [ATOM PHYS] The name given in early gaseous discharge experiments to the particles passing through a hole or canal in the cathode; the ray comprises positive ions of the gas being used in the discharge. { kə'nal ‚rā }

canal surface [MATH] The envelope of a family of spheres of equal radii whose centers are on a given space curve. { kə'nal ‚sər·fəs }

canal valve [ANAT] The semilunar valve in the right atrium of the heart between the orifice of the inferior vena cava and the right atrioventricular orifice. Also known as eustachian valve. { kə'nal ‚valv }

cananga oil See ilang-ilang oil. { kə'naŋ·gə ‚oil }

canard [AERO ENG] 1. An aerodynamic vehicle in which horizontal surfaces used for trim and control are forward of the wing or main lifting surface. 2. The horizontal trim and control surfaces in such an arrangement. { kə'närd }

Canary Current [OCEANOGR] The prevailing southward flow of water along the northwestern coast of Africa. { kə'ner·ē ‚kar·ənt }

canary-pox virus [VIROL] An avian poxvirus that causes canary pox, a disease closely related to fowl pox. { kə'ner·ē ‚päks ‚vī·rəs }

Canastotan [GEOL] Lower Upper Silurian geologic time. { kə'nas·tə·tən }

canavanine [BIOCHEM] $C_5H_{12}O_3N_4$ An amino acid found in the jack bean. { kə'nav·ə‚nēn }

can buoy [NAV] Floating cylindrical unlighted buoy with a flat top, constructed of metal; depending on position, it may be painted black with or without an odd number on it, or painted with horizontal stripes. { 'kan ‚bȯi }

Canc See Cancer.

cancellate [BIOL] Lattice-shaped. Also known as clathrate. { 'kan·sə‚lāt }

cancellation circuit [ELECTR] A circuit used in providing moving-target indication on a plan position indicator scope; cancels constant-amplitude fixed-target pulses by subtraction of successive pulse trains. { kan·sə'lā·shən ‚sər·kət }

cancellation law [MATH] A rule which allows formal division by common factors in equal products, even in systems which have no division, as integral domains; $ab = ac$ implies that $b = c$. { kan·sə'lā·shən ‚lȯ }

cancellous [BIOL] Having a reticular or spongy structure. { kan'sel·əs }

cancellous bone [HISTOL] A form of bone near the ends of long bones having a cancellous matrix composed of rods, plates, or tubes; spaces are filled with marrow. { kan'sel·əs ‚bōn }

cancer [MED] Any malignant neoplasm, including carcinoma and sarcoma. { 'kan·sər }

Cancer [ASTRON] A constellation with right ascension 9 hours, declination 20°N. Abbreviated Canc. Also known as Crab. { 'kan·sər }

cancer eye [VET MED] A malignant epithelioma of the eye of cattle, common in regions of intense sunlight. { 'kan·sər ‚ī }

cancerphobia [PSYCH] An abnormal fear of acquiring cancer. { ‚kan·sər'fō·bē·ə }

cancrinite [MINERAL] $Na_3CaAl_3Si_3O_{12}CO_3(OH)_2$ A feldspathoid tectosilicate occurring in hexagonal crystals in nepheline syenites, usually in compact or disseminated masses. { 'kaŋ·krə‚nīt }

cancroid [MED] A squamous-cell carcinoma. { 'kaŋ‚krȯid }

candela [OPTICS] A unit of luminous intensity, defined as 1/60 of the luminous intensity per square centimeter of a blackbody radiator operating at the temperature of freezing platinum. Formerly known as candle. Also known as new candle. { kan'del·ə }

candelilla wax [MATER] A wax obtained from the wax-coated stems of candelilla shrubs, especially *Euphorbia antisyphilitica;* used for varnishes and furniture and shoe polishes. { ‚kan·də'lē·ə ‚waks }

Candida [MYCOL] A genus of yeastlike, pathogenic imperfect fungi that produce very small mycelia. { 'kan·də·də }

Candida utilis [MICROBIO] An asexual yeast species used industrially in the production of single-cell protein for food and fodder. { 'kan·də·də 'yü·də·lis }

candidiasis [MED] A fungus infection of the skin, lungs, mucous membranes, and viscera of humans caused by a species of *Candida,* usually *C. albicans.* Also known as moniliasis. { ‚kan·də'dī·ə·səs }

candite See ceylonite. { 'kan‚dīt }

candle See candela. { 'kan·dəl }

Candlemas crack See Candlemas Eve winds. { 'kan·dəl·məs ‚krak }

Candlemas Eve winds [METEOROL] Heavy winds often occurring in Great Britain in February or March (Candlemas is February 2). Also known as Candlemas crack. { 'kan·dəl·məs ‚ēv 'winz }

candlenut oil See lumbang oil. { 'kan·dəl‚nət ‚oil }

candlepower [OPTICS] Luminous intensity expressed in candelas. Abbreviated cp. { 'kan·dəl‚pau̇·ər }

CANDU reactor [NUCLEO] A type of nuclear reactor that uses natural uranium as fuel, hot pressurized heavy water as coolant, and cool unpressurized heavy water as moderator; utilizes pressurized tubes to contain the fuel and circulating coolant. { 'kan‚dü rē‚ak·tər }

cane [BOT] 1. A hollow, usually slender, jointed stem, such as in sugarcane or the bamboo grasses. 2. A stem growing directly from the base of the plant, as in most Rosaceae, such as blackberry and roses. { kān }

cane blight [PL PATH] A fungus disease affecting the canes of several bush fruits, such as currants and raspberries; caused by several species of fungi. { 'kān ‚blīt }

CANDIDA UTILIS

Image of *Candida utilis.*

CANNACEAE

Flowers of a cultivated *Canna* of the order Zingiberales. (*Photograph by Luoma Photos, from National Audubon Society*)

cane molasses [FOOD ENG] The heavy, residual syrup after the crystallization of cane syrup. { 'kān mə'las·əs }

canescent [BOT] Having a grayish epidermal covering of short hairs. { kə'nes·ənt }

cane sugar [ORG CHEM] Sucrose derived from sugarcane. { 'kān ,shug·ər }

Canes Venatici [ASTRON] A northern constellation with right ascension 13 hours, declination 40°N, between Ursa Major and Boötes. Abbreviated CVn. Also known as Hunting Dogs. { 'kä,nēz ·və'nad·ə,sē }

Canes Venatici I cloud [ASTRON] A relatively nearby, loosely clustered group of galaxies consisting chiefly of late-type spirals and irregular galaxies, with recession velocities near 220 miles (350 kilometers) per second. { 'kā,nēz və'nad·ə,sē 'wən ,klaud }

canfieldite [MINERAL] Ag_8SnS_6 A black mineral of the argyrodite series consisting of silver thiostannate, with a specific gravity of 6.28; found in Germany and Bolivia. { 'kan,fēl,dīt }

can hoisting system [MIN ENG] A hoisting method used in shallow lead and zinc mines in which cans are loaded below and hoisted to the surface where they are capsized and the load discharged; the operation is controlled at the top of the shaft. { 'kan ,hois·tiŋ ,sis·təm }

Canidae [VERT ZOO] A family of carnivorous mammals in the superfamily Canoidea, including dogs and their allies. { 'kan·ə,dē }

canine [ANAT] A conical tooth, such as one located between the lateral incisor and first premolar in humans and many other mammals. Also known as cuspid. [VERT ZOO] Pertaining or related to dogs or to the family Canidae. { 'kā,nīn }

canine distemper [VET MED] A pantropic virus disease occurring among animals of the family Canidae. { 'kā,nīn dis'tem·pər }

Canis [VERT ZOO] The type genus of the dog family (Canidae), including dogs, wolves, and jackals. { 'kā·nəs }

Canis Major [ASTRON] A constellation with right ascension 7 hours, declination 20°S. Abbreviated CMa. Also known as Greater Dog. { ,kā·nəs 'mā·jər }

α Canis Majoris See Sirius. { 'al·fə ,kā·nəs 'mā·jər·is }

ε Canis Majoris See Adhara. { 'ed·ə ,kā·nəs 'mā·jər·is }

Canis Minor [ASTRON] A constellation with right ascension 8 hours, declination 5°N. Abbreviated CMi. Also known as Lesser Dog. { ,kā·nəs 'mī·nər }

α Canis Minoris See Procyon. { 'al·fə ,kā·nəs 'mī·nər·is }

canister [MECH ENG] See charcoal canister. [ORD] A special short-range, antipersonnel projectile designed to be fired from rifled guns. { 'kan·ə'stər }

canker [PL PATH] An area of necrosis on a woody stem resulting in shrinkage and cracking followed by the formation of callus, ultimately killing the stem. [VET MED] A localized chronic inflammation of the ear in cats, dogs, foxes, ferrets, and others caused by the mite *Otodectes cynotis*. { 'kaŋ·kər }

canker sore [MED] Small ulceration of the mucous membrane of the mouth, sometimes caused by a food allergy. { 'kaŋ·kər ,sor }

canker stain [PL PATH] A fungus disease of plane trees caused by *Endoconidiophora fimbriata platani* and characterized by bluish-black or reddish-brown discolorations beneath blackened cankers on the trunk and sometimes the branches. { 'kaŋ·kər ,stān }

cankerworm [INV ZOO] Any of several lepidopteran insect larvae in the family Geometridae which cause severe plant damage by feeding on buds and foliage. { 'kaŋ·kər,wərm }

Cannabaceae [BOT] A family of dicotyledonous herbs in the order Urticales, including Indian hemp (*Cannabis sativa*) and characterized by erect anthers, two styles or style branches, and the lack of milky juice. { kan·ə'bās·ē,ē }

cannabidiol [ORG CHEM] $C_{21}H_{28}(OH)_2$ A constituent of cannabis which, upon isomerization to a tetrahydrocannabinol, has some of the physiologic activity of marijuana. { ¦kan·ə·bə'dī,ol }

cannabinoid [ORG CHEM] Any one of the various chemical constituents of cannabis (marijuana), that is, the isomeric tetrahydrocannabinols, cannabinol, and cannabidiol. { kə'nab·ə,noid }

cannabinol [ORG CHEM] $C_{21}H_{26}O_2$ A physiologically inactive phenol formed by spontaneous dehydrogenation of tetrahydrocannabinol from cannabis. { 'kan·ə·bə,nol }

Cannabis [BOT] A genus of tall annual herbs in the family Cannabaceae having erect stems, leaves with three to seven elongate leaflets, and pistillate flowers in spikes along the stem. { 'kan·ə·bəs }

cannabiscetin See myricetin. { ¦kan·ə'bis·ə,tēn }

cannabism [MED] Poisoning resulting from excessive or habitual use of cannabis. { 'kan·ə,biz·əm }

Cannaceae [BOT] A family of monocotyledonous plants in the order Zingiberales characterized by one functional stamen, a single functional pollen sac in the stamen, mucilage canals in the stem, and numerous ovules in each of the one to three locules. { kə'nās·ē,ē }

canned cycle [COMPUT SCI] Any set of operations, either software or hardware, that is activated by a single command. { 'kand 'sī·kəl }

canned motor [MECH ENG] A motor enclosed within a casing along with the driven element (that is, a pump) so that the motor bearings are lubricated by the same liquid that is being pumped. { ¦kand 'mōd·ər }

canned program [COMPUT SCI] A program which has been written to solve a particular problem, is available to users of a computer system, and is usually fixed in form and capable of little or no modification. { ¦kand 'prō·grəm }

canned pump [MECH ENG] A watertight pump that can operate under water. { ¦kand 'pəmp }

cannel coal [GEOL] A fine-textured, highly volatile bituminous coal distinguished by a greasy luster and blocky, conchoidal fracture; burns with a steady luminous flame. Also known as cannelite. { 'kan·əl ,kōl }

cannelite See cannel coal. { 'kan·əl,īt }

canneloid [GEOL] **1.** Coal that resembles cannel coal. **2.** Coal intermediate between bituminous and cannel. **3.** Durain laminae in banded coal. **4.** Cannel coal of anthracite or semianthracite rank. { 'kan·əl,oid }

cannel shale [GEOL] A black shale formed by the accumulation of an aquatic ooze rich in bituminous organic matter in association with inorganic materials such as silt and clay. { 'kan·əl ,shāl }

cannelure [ORD] **1.** A groove in a bullet which contains a lubricant, or into which the cartridge is crimped. **2.** A groove in a cartridge case providing a purchase for the extractor. **3.** A ringlike groove for locking the jacket of an armor-piercing bullet to the core. **4.** A ringlike groove in the rotating band of a gun projectile to lessen the resistance offered to the gun rifling and to prevent fringing grooves. { 'kan·əl·ur }

cannibalize [ENG] To remove parts from one piece of equipment and use them to replace like, defective parts in a similar piece of equipment in order to keep the latter operational. { 'kan·ə·bə,līz }

canning [FOOD ENG] Packing and preserving of food in cans or jars subjected to sterilizing temperatures. [NUCLEO] Placing a jacket around a slug of uranium before inserting the slug in a nuclear reactor. { 'kan·iŋ }

Cannizzaro reaction [ORG CHEM] The reaction in which aldehydes that do not have a hydrogen attached to the carbon adjacent to the carbonyl group, upon encountering strong alkali, readily form an alcohol and an acid salt. { kän·it'sär·ō rē'ak·shən }

cannon [ORD] A complete assembly which consists of a tube and a breech mechanism with a firing mechanism or base cap and which is a component of a gun, howitzer, or mortar; may include muzzle appendages; the term is generally limited to calibers greater than 1 inch (2.5 centimeters). { 'kan·ən }

cannonball [ORD] A missile that is spheroidal in shape and fired from a cannon. { 'kan·ən,bol }

cannon cradle [ORD] A portion of a cannon assembly which is designed to support the cannon and allows the cannon to recoil and counter-recoil. { 'kan·ən ,krād·əl }

cannon pinion [HOROL] A tube that holds the minute hand at one end and meshes with the minute wheel of a timepiece on the other end. { 'kan·ən ,pin·yən }

cannon primer [ORD] Primer used with separate-loading ammunition. { 'kan·ən ,prī·mər }

cannula [MED] A small tube that can be inserted into a body cavity, duct, or vessel. { 'kan·yə·lə }

cannular combustion chambers [AERO ENG] The separate combustion chambers in an aircraft gas turbine. Also known as can-type combustors. { 'kan·yə·lər kəm'bəs·chən ,chām·bərz }

Canoidea [VERT ZOO] A superfamily belonging to the mammalian order Carnivora, including all dogs and doglike species such as seals, bears, and weasels. { kə'nóid·ē·ə }

canola oil [FOOD ENG] An edible vegetable oil derived from rapeseed that is low in saturated fatty acids (less than 7%), high in monosaturated fatty acids (60%), and high in polyunsaturated fatty acids (30%). { kə'nōl·ə ˌóil }

canonical [SCI TECH] Relating to the simplest or most significant form of a general function, equation, statement, rule, or expression. { kə'nän·ə·kəl }

canonical change [ASTRON] A periodic change in one of the components of the orbit of a celestial object. { kə'nän·ə·kəl 'chānj }

canonical coordinates [MATH] Any set of generalized coordinates of a system together with their conjugate momenta. { kə'nän·ə·kəl kō'órd·ən·əts }

canonical correlation [STAT] The maximum correlation between linear functions of two sets of random variables when specific restrictions are imposed upon the coefficients of the linear functions of the two sets. { kə'nän·ə·kəl ˌkór·ə'lā·shən }

canonical distribution [STAT MECH] The density of members of the canonical ensemble in phase space. { kə'nän·ə·kəl ˌdis·trə'byü·shən }

canonical ensemble [STAT MECH] A hypothetical collection of systems of particles used to describe an actual individual system which is in thermal contact with a heat reservoir but is not allowed to exchange particles with its environment. { kə'nän·ə·kəl än'säm·bəl }

canonical equations of motion See Hamilton's equations of motion. { kə'nän·ə·kəl i'kwā·zhənz əv 'mō·shən }

canonical form [CONT SYS] A specific type of dynamical system representation in which the associated matrices possess specific row-column structures. [ORG CHEM] **1.** A resonance structure for a cyclic compound in which the bonds do not intersect. **2.** See contributing structure. [SCI TECH] A particularly clean and simple representation which usually follows or satisfies a general rule. { kə'nän·ə·kəl ˌfórm }

canonically conjugate variables [MECH] A generalized coordinate and its conjugate momentum. { kə'nän·ə·klē ˌkan·jə·gət 'ver·ē·ə·bəlz }

canonical matrix [MATH] A member of an equivalence class of matrices that has a particularly simple form, where the equivalence classes are determined by one of the relations defining equivalent, similar, or congruent matrices. { kə'nän·ə·kəl 'mā·triks }

canonical momentum See conjugate momentum. { kə'nän·ə·kəl mə'ment·əm }

canonical schema [COMPUT SCI] A model that represents the structure and interrelationships of data within a data base. { kə'nän·ə·kəl 'skē·mə }

canonical sequence [MOL BIO] An archetypical nucleotide or amino acid sequence to which all variants are compared. { kə'nän·ə·kəl 'sē·kwəns }

canonical structure See contributing structure. { kə'nän·ə·kəl 'strək·chər }

canonical time unit [ASTRON] For geocentric orbits, the time required by a hypothetical satellite to move one radian in a circular orbit of the earth's equatorial radius, that is, 13.447052 minutes. { kə'nän·ə·kəl 'tīm ˌyü·nət }

canonical transformation [MATH] Any function which has a standard form, depending on the context. [MECH] A transformation which occurs among the coordinates and momenta describing the state of a classical dynamical system and which leaves the form of Hamilton's equations of motion unchanged. Also known as contact transformation. { kə'nän·ə·kəl ˌtranz·fər'mā·shən }

Canopus [ASTRON] A star that is 180 light-years from the sun; spectral classification F0Ia. Also known as α Carinae. { kə'nō·pəs }

canopy [AERO ENG] **1.** The umbrellalike part of a parachute which acts as its main supporting surface. **2.** The overhead, transparent enclosure of an aircraft cockpit. [FOR] The uppermost branching and spreading layer of a forest. { 'kan·ə·pē }

cant [ORD] The leaning or tilt of an object; especially, the sidewise tilt of a gun. { kant }

cantala [BOT] A fiber produced from agave (*Agave cantala*)

leaves; used to make twine. Also known as Cebu maguey; maguey; Manila maguey. { kan'täl·ə }

cantaloupe [BOT] The fruit (pepo) of *Cucumis malo*, a small, distinctly netted, round to oval muskmelon of the family Cucurbitaceae in the order Violales. { 'kant·əl,ōp }

cant body [NAV ARCH] The part of a ship's body where the frames are set at an angle to the keel to form the stern and bow. { 'kant ,bäd·ē }

canted leg [PETRO ENG] A type of jackup rig design in which the legs can be slanted outward to increase support against lateral stresses when the unit is placed on the sea floor. { 'kant·əd 'leg }

Canterbury northwester [METEOROL] A strong northwest foehn descending the New Zealand Alps onto the Canterbury plains of South Island, New Zealand. { ˌkan·tə,ber·ē ˌnórth'wes·tər }

cant file [DES ENG] A fine-tapered file with a triangular cross section, used for sharpening saw teeth. { 'kant ,fīl }

canthariasis [MED] Infection or disease caused by coleopteran insects or their larvae. { kan·thə'rī·ə·səs }

Cantharidae [INV ZOO] The soldier beetles, a family of coleopteran insects in the superfamily Cantharoidea. { kan 'thar·ə,dē }

cantharides camphor See cantharidin. { kan'thar·ə,dēz 'kam·fər }

cantharidin [ORG CHEM] $C_{10}H_{12}O_4$ Colorless crystals that melt at 218°C; slightly soluble in acetone, chloroform, alcohol, and water; used in veterinary medicine. Also known as cantharides camphor. { kan'thar·ə·dən }

Cantharoidea [INV ZOO] A superfamily of coleopteran insects in the suborder Polyphaga. { kan·thə'róid·ē·ə }

cant hook [DES ENG] A lever with a hooklike attachment at one end, used in lumbering. { 'kant ,hùk }

canthus [ANAT] Either of the two angles formed by the junction of the eyelids, designated outer or lateral, and inner or medial. { 'kan·thəs }

cantilever [ENG] A beam or member securely fixed at one end and hanging free at the other end. [ENG] In particular, in an atomic force microscope a very small beam that has a tip attached to its free end; the deflection of the beam is used to measure the force acting on the tip. { 'kant·əl,ē·vər }

cantilever arch [ARCH] An arch supported by flat projections on opposing walls. { 'kant·əl,ē·vər ,àrch }

cantilever bridge [CIV ENG] A fixed bridge consisting of two spans projecting toward each other and joined at their ends by a suspended simple span. { 'kant·əl,ē·vər ,brij }

cantilever footing [CIV ENG] A footing used to carry a load from two columns, with one column and one end of the footing placed against a building line or exterior wall. { 'kant·əl,ē·vər 'fùd·iŋ }

cantilever retaining wall [CIV ENG] A type of wall formed of three cantilever beams: stem, toe projection, and heel projection. { 'kant·əl,ē·vər ri'tān·iŋ wól }

cantilever spring [MECH ENG] A flat spring supported at one end and holding a load at or near the other end. { 'kant·əl,ē·vər ,spriŋ }

cantilever vibration [MECH] Transverse oscillatory motion of a body fixed at one end. { 'kant·əl,ē·vər vī'brā·shən }

canting [MECH] Displacing the free end of a beam which is fixed at one end by subjecting it to a sideways force which is just short of that required to cause fracture. { 'kant·iŋ }

canting strip See water table. { 'kant·iŋ ,strip }

canton crepe [TEXT] Thick, slightly ribbed crepe, originally made of silk in Canton, China. { 'kan'tän ,krāp }

Cantor diagonal process [MATH] A technique of proving statements about infinite sequences, each of whose terms is an infinite sequence by operation on the *n*th term of the *n*th sequence for each *n*; used to prove the uncountability of the real numbers. { 'kän·tór dī'ag·ən·əl ,präs·əs }

Cantor function [MATH] A real-valued nondecreasing continuous function defined on the closed interval [0,1] which maps the Cantor ternary set onto the interval [0,1]. { 'kän·tór ,fəŋk·shən }

Cantor's axiom [MATH] The postulate that there exists a one-to-one correspondence between the points of a line extending indefinitely in both directions and the set of real numbers. { 'kan·tərz 'ak·sē·əm }

Cantor ternary set [MATH] A perfect, uncountable, totally

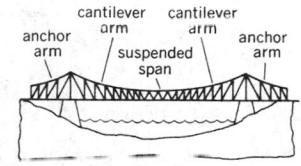

CANTILEVER BRIDGE

Drawing of a cantilever bridge.

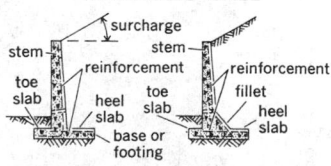

CANTILEVER RETAINING WALL

Cantilever walls, showing the parts.

disconnected subset of the real numbers having Lebesgue measure zero; it consists of all numbers between 0 and 1 (inclusive) with ternary representations containing no ones. { 'kän·tor 'tər·nə·rē ,set }

Cantor theorem [MATH] A theorem that there is no one-to-one correspondence between a set and the collection of its subsets. { 'kan·tòr 'thir·əm }

cant strip [BUILD] **1.** A strip placed along the angle between a wall and a roof so that the roofing will not bend sharply. **2.** A strip placed under the edge of the lowest row of tiles on a roof to give them the same slope as the other tiles. { 'kant ,strip }

can-type combustors *See* cannular combustion chambers. { ¦kan ,tīp kəm'bəs·tərz }

canvas [TEXT] A firm, closely woven fabric of plain weave made principally from hemp, but also from flax, jute, cotton, or a blend of fibers. { 'kan·vəs }

canvas duck [TEXT] A lightweight cotton or linen cloth; the term is occasionally used for heavier canvases as well. { 'kan·vəs 'dək }

canyon [GEOGR] A chasm, gorge, or ravine cut in the surface of the earth by running water; the sides are steep and form cliffs. { 'kan·yən }

canyon bench [GEOL] A steplike level of hard strata in the walls of deep valleys in regions of horizontal strata. { 'kan·yən ,bench }

canyon fill [GEOL] Loose, unconsolidated material which fills a canyon to a depth of 50 feet (15 meters) or more during periods between great floods. { 'kan·yən ,fil }

canyon wind [METEOROL] **1.** Also known as gorge wind. **2.** The mountain wind of a canyon; that is, the nighttime down-canyon flow of air caused by cooling at the canyon walls. **3.** Any wind modified by being forced to flow through a canyon or gorge; its speed may be increased as a jet-effect wind, and its direction is rigidly controlled. { 'kan·yən ¦wind }

caoutchouc [MATER] Formerly, crude rubber which had been cured over a fire into a solid, dark mass for shipment. { kaú'chük }

cap [ENG] A detonating or blasting cap. [GEN] In many eukaryotic messenger ribonucleic acids, the structure at the 5′ end consisting of 7′-methyl-guanosine-pppX, where X is the first nucleotide encoded in the deoxyribonucleic acid; it is added posttranscriptionally. [MATH] The symbol ∩, which indicates the intersection of two sets. [MIN ENG] **1.** A piece of timber placed on top of a prop or post in a mine. **2.** The horizontal section of a set of timber that is used as a support in a mine roadway. { kap }

Cap *See* Capricornus.

capability [COMPUT SCI] A permission that is given to a user of a computing system in advance to access a particular object in the system in a particular way, and that the user can later present to a reference monitor as a prevalidated ticket to gain access. { ,kāp·ə'bil·ə·dē }

capability list [COMPUT SCI] A row of an access matrix that contains the access rights of a given user to various files and other resources of a computer system. { ,kā·pə'bil·əd·ē ,list }

capacitance [ELEC] The ratio of the charge on one of the conductors of a capacitor (there being an equal and opposite charge on the other conductor) to the potential difference between the conductors. Symbolized C. Formerly known as capacity. [ENG] In a closed feedwater heater, the volume of water required for proper operation of the drain control valve. { kə'pas·ə·təns }

capacitance altimeter [ENG] An absolute altimeter which determines height of an aircraft aboveground by measuring the variations in capacitance between two conductors on the aircraft when the ground is near enough to act as a third conductor. { kə'pas·ə·təns al'tim·əd·ər }

capacitance box [ELEC] An assembly of capacitors and switches which permits adjustment of the capacitance existing at the terminals in nominally uniform steps, from a minimum value near zero to the maximum which exists when all the capacitors are connected in parallel. { kə'pas·ə·təns ,bäks }

capacitance bridge [ELEC] A bridge for comparing two capacitances, such as a Schering bridge. { kə'pas·ə·təns ,brij }

capacitance hat [ELECTROMAG] A network of wires that is placed at the top of an antenna either to increase its bandwidth or to lower its resonant frequency. { kə'pas·əd·əns ,hat }

capacitance level indicator [ENG] A level indicator in which the material being monitored serves as the dielectric of a capacitor formed by a metal tank and an insulated electrode mounted vertically in the tank. { kə'pas·ə·təns ¦lev·əl 'in·də,kād·ər }

capacitance meter [ENG] An instrument used to measure capacitance values of capacitors or of circuits containing capacitance. { kə'pas·ə·təns ,mēd·ər }

capacitance-operated intrusion detector [ENG] A boundary alarm system in which the approach of an intruder to an antenna wire encircling the protected area a few feet above ground changes the antenna-ground capacitance and sets off the alarm. { kə'pas·ə·təns ¦áp·ə,rād·əd in'trü·zhən di'tek·tər }

capacitance probe [PETRO ENG] A sensing device that determines the dielectric constants of the oil and water components of an oil-water emulsion. { kə'pas·əd·əns ,prōb }

capacitance relay [ELECTR] An electronic relay that responds to a small change in capacitance, such as that created by bringing a hand near a pickup wire or plate. { kə'pas·ə·təns 'rē,lā }

capacitance standard *See* standard capacitor. { kə'pas·ə·təns ,stan·dərd }

capacitation [PHYSIO] The process of physiological change occurring in mammalian spermatozoa during passage through the female reproductive tract that enables them to penetrate the egg membrane. { kə,pas·ə'tā·shən }

capacitive coupling [ELEC] Use of a capacitor to transfer energy from one circuit to another. { kə'pas·ə·təns ,kəp·liŋ }

capacitive diaphragm [ELECTROMAG] A resonant window used in a waveguide to provide the equivalent of capacitive reactance at the frequency being transmitted. { kə'pas·əd·iv 'dī·ə,fram }

capacitive-discharge ignition [ELECTR] An automotive ignition system in which energy is stored in a capacitor and discharged across the gap of a spark plug through a step-up pulse transformer and distributor each time a silicon controlled rectifier is triggered. { kə'pas·əd·iv ¦dis,chärj ig¦nish·ən }

capacitive-discharge pilot light [ELECTR] An electronic ignition system, operating off an alternating-current power line or battery power supply, that produces a spark for lighting a gas flame. { kə'pas·əd·iv ¦dis,chärj 'pī·lət ,līt }

capacitive divider [ELEC] Two or more capacitors placed in series across a source, making available a portion of the source voltage across each capacitor; the voltage across each capacitor will be inversely proportional to its capacitance. { kə'pas·əd·iv di'vīd·ər }

capacitive electrometer [ENG] An instrument for measuring small voltages; the voltage is applied to the plates of a capacitor when they are close together, then the voltage source is removed and the plates are separated, increasing the potential difference between them to a measurable value. Also known as condensing electrometer. { kə'pas·əd·iv ,lek'träm·əd·ər }

capacitive feedback [ELECTR] Process of returning part of the energy in the plate (or output) circuit of a vacuum tube (or other device) to the grid (or input) circuit by means of a capacitance common to both circuits. { kə'pas·əd·iv 'fēd,bak }

capacitive load [ELECTROMAG] A load in which the capacitive reactance exceeds the inductive reactance; the load draws a leading current. { kə'pas·əd·iv ¦lōd }

capacitive loading [ELECTROMAG] **1.** Raising the resonant frequency of an antenna by connecting a fixed capacitor or capacitors in series with it. **2.** Lowering the resonant frequency of an antenna by installing a capacitance hat. { kə'pas·əd·iv 'lōd·iŋ }

capacitive post [ELECTROMAG] Metal post or screw extending across a waveguide at right angles to the E field, to provide capacitive susceptance in parallel with the waveguide for tuning or matching purposes. { kə'pas·əd·iv ,pōst }

capacitive pressure transducer [ENG] A measurement device in which variations in pressure upon a capacitive element proportionally change the element's capacitive rating and thus the strength of the measured electric signal from the device. { kə'pas·əd·iv 'presh·ər tranz,dü·sər }

capacitive reactance [ELECTROMAG] Reactance due to the capacitance of a capacitor or circuit, equal to the inverse of the product of the capacitance and the angular frequency. { kə'pas·əd·iv rē'ak·təns }

capacitive tuning [ELECTR] Tuning involving use of a variable capacitor. { kə¦pas·əd·iv 'tün·iŋ }

capacitive window [ELECTROMAG] Conducting diaphragm extending into a waveguide from one or both sidewalls, producing the effect of a capacitive susceptance in parallel with the waveguide. { kə¦pas·əd·iv 'win·dō }

capacitor [ELEC] A device which consists essentially of two conductors (such as parallel metal plates) insulated from each other by a dielectric and which introduces capacitance into a circuit, stores electrical energy, blocks the flow of direct current, and permits the flow of alternating current to a degree dependent on the capacitor's capacitance and the current frequency. Symbolized C. Also known as condenser; electric condenser. { kə'pas·əd·ər }

capacitor antenna [ELECTROMAG] Antenna consisting of two conductors or systems of conductors, the essential characteristic of which is its capacitance. Also known as condenser antenna. { kə'pas·əd·ər an'ten·ə }

capacitor bank [ELEC] A number of capacitors connected in series or in parallel. { kə'pas·əd·ər ,baŋk }

capacitor box [ELECTR] A box-shaped structure in which a capacitor is submerged in a heat-absorbing medium, usually water. Also known as condenser box. { kə'pas·əd·ər ,bäks }

capacitor color code [ELEC] A method of marking the value on a capacitor by means of dots or bands of colors as specified in the Electronic Industry Association color code. { kə'pas·əd·ər 'kəl·ər ,kōd }

capacitor hydrophone [ENG ACOUS] A capacitor microphone that responds to waterborne sound waves. { kə'pas·əd·ər 'hī·drə,fōn }

capacitor-input filter [ELECTR] A power-supply filter in which a shunt capacitor is the first element after the rectifier. { kə'pas·əd·ər ¦in,pút ,fil·tər }

capacitor loudspeaker See electrostatic loudspeaker. { kə'pas·əd·ər 'laúd,spēk·ər }

capacitor microphone [ENG ACOUS] A microphone consisting essentially of a flexible metal diaphragm and a rigid metal plate that together form a two-plate air capacitor; sound waves set the diaphragm in vibration, producing capacitance variations that are converted into audio-frequency signals by a suitable amplifier circuit. Also known as condenser microphone; electrostatic microphone. { kə'pas·əd·ər 'mī·krə,fōn }

capacitor motor [ELEC] **1.** A single-phase induction motor having a main winding connected directly to a source of alternating-current power and an auxiliary winding connected in series with a capacitor to the source of ac power. **2.** See capacitor-start motor. { kə'pas·əd·ər ,mōd·ər }

capacitor pickup [ENG ACOUS] A phonograph pickup in which movements of the stylus in a record groove cause variations in the capacitance of the pickup. { kə'pas·əd·ər 'pik·əp }

capacitor-resistor unit See rescap. { kə'pas·əd·ər ri'zis·tər ,yü·nət }

capacitor-start motor [ELEC] A capacitor motor in which the capacitor is in the circuit only during the starting period; the capacitor and its auxiliary winding are disconnected automatically by a centrifugal switch or other device when the motor reaches a predetermined speed. Also known as capacitor motor. { kə'pas·əd·ər 'stärt ,mōd·ər }

capacitor start-run motor See permanent-split capacitor motor. { kə'pas·əd·ər ¦stärt ¦rən ,mōd·ər }

capacity [ANALY CHEM] In chromatography, a measurement used in ion-exchange systems to express the adsorption ability of the ion-exchange materials. [COMPUT SCI] See storage capacity. [ELEC] See capacitance. [SCI TECH] Volume, especially in reference to merchandise or containers thereof. { kə'pas·əd·ē }

capacity cell [ELEC] **1.** Capacitance-type device used to measure the dielectric constants of gases, liquids, or solids. **2.** Capacitance-type device used to monitor certain composition changes in flowing streams. { kə'pas·əd·ē ,sel }

capacity correction [ENG] The correction applied to a mercury barometer with a nonadjustable cistern in order to compensate for the change in the level of the cistern as the atmospheric pressure changes. { kə'pas·əd·ē kə'rek·shən }

capacity factor [IND ENG] The ratio of average actual use to the available capacity of an apparatus or industrial plant to store, process, treat, manufacture, or produce. { kə'pas·əd·ē ,fak·tər }

capacity of the wind [GEOL] The total weight of airborne particles (soil and rock) of given size, shape, and specific gravity, which can be carried in 1 cubic mile (4.17 cubic kilometers) of wind blowing at a given speed. { kə'pas·əd·ē əv thə ¦wind }

capacity-rate product [COMMUN] The product of the capacity of a data-storage device in gigabytes and the data rate in megabits per second. { kə'pas·ə·dē ,rāt ,präd·əkt }

cap-binding protein [MOL BIO] A protein that specifically recognizes the methylated cap of eukaryotic messenger ribonucleic acid (mRNA) and is essential in the regulation of mRNA translation. { ¦kap ¦bīnd·iŋ 'prō,tēn }

cap cloud [METEOROL] An approximately stationary cloud, or standing cloud, on or hovering above an isolated mountain peak; formed by the cooling and condensation of humid air forced up over the peak. Also known as cloud cap. { 'kap ,klaúd }

cap crimper [ENG] A tool resembling a pliers that is used to press the open end of a blasting cap onto the safety fuse before placing the cap in the primer. { 'kap ,krim·pər }

cape [GEOGR] A prominent point of land jutting into a body of water. Also known as head; headland; mull; naze; ness; point; promontory. { kāp }

cape chisel [DES ENG] A chisel that tapers to a flat, narrow cutting end; used to cut flat grooves. { 'kāp ,chiz·əl }

cape doctor [METEOROL] The strong southeast wind which blows on the South African coast. { 'kāp ,däk·tər }

cape foot [MECH] A unit of length equal to 1.033 feet or to 0.3148584 meter. { 'kāp ,fút }

Cape Horn Current [OCEANOGR] That part of the west wind drift flowing eastward in the immediate vicinity of Cape Horn, and then curving northeastward to continue as the Falkland Current. { ,kāp ,hórn 'kər·ənt }

Capella [ASTRON] A star that is 45 light-years from the sun; spectral classification G0IIIp. Also known as α Aurigae. { kə'pel·ə }

Capell fan [MIN ENG] A centrifugal type of mine shaft fan. { kə'pel ,fan }

capers [FOOD ENG] The buds and berries of the caper plant (*Cappuris spinosa*) pickled and used as a condiment. { 'kā·pərz }

Capgras syndrome [PSYCH] A delusional misidentification syndrome commonly seen in schizophrenia that causes the individual to replace a familiar person (usually the spouse) with an impostor with the same or similar physical appearance. { 'käp·grəz ,sin,drōm }

capillarity [FL MECH] The action by which the surface of a liquid where it contacts a solid is elevated or depressed, because of the relative attraction of the molecules of the liquid for each other and for those of the solid. Also known as capillary action. { ,kap·ə'lar·əd·ē }

capillarity correction [ENG] As applied to a mercury barometer, that part of the instrument correction which is required by the shape of the meniscus of the mercury. { ,kap·ə'lar·əd·ē kə,rek·shən }

capillaroscope [MED] A microscope used for diagnostic examination of the cutaneous capillaries, as in the nail beds and conjunctiva. { ,kap·ə'lar·ə,skōp }

capillary [ANAT] The smallest vessel of both the circulatory and lymphatic systems; the walls are composed of a single cell layer. [GEOL] A fissure or a crack in a formation which provides a route for flow of water or hydrocarbons. { 'kap·ə,ler·ē }

capillary action See capillarity. { 'kap·ə,ler·ē 'ak·shən }

capillary angioma See hemangioma. { 'kap·ə,ler·ē ,an·jē'ō·mə }

capillary attraction [FL MECH] The force of adhesion existing between a solid and a liquid in capillarity. { 'kap·ə,ler·ē ə'trak·shən }

capillary bed [ANAT] The capillaries, collectively, of a given area or organ. { 'kap·ə,ler·ē ,bed }

capillary collector [ENG] An instrument for collecting liquid water from the atmosphere; the collecting head is fabricated of a porous material having a pore size of the order of 30 micrometers; the pressure difference across the water-air interface prevents air from entering the capillary system while allowing free flow of water. { 'kap·ə,ler·ē kə'lek·tər }

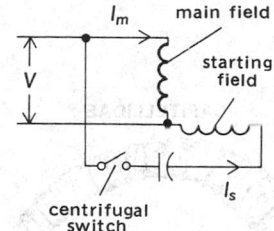

CAPACITOR MOTOR

I_m main field

V starting field

centrifugal switch I_s

Winding connections of the starting winding connected to the supply through a capacitor. I_m = main winding; I_s = starting winding; V = common voltage.

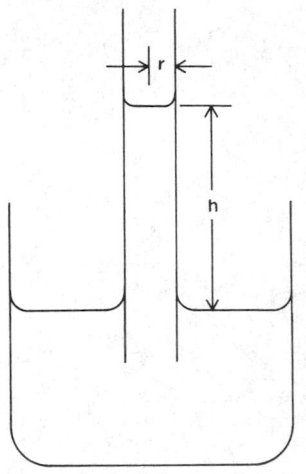

CAPILLARY TUBE

Rise of liquid to a height *h* in a capillary tube whose radius is *r*.

CAPITELLIDAE

Left lateral view of *Capitomastus*, a genus of Capitellidae.

capillary column [ANALY CHEM] One of the long, narrow (100 meters by 0.2–0.5 millimeter or 330 feet by 0.008–0.02 inch) columns used for capillary gas chromatography. Also known as open tubular column. { 'kap·ə,ler·ē ,käl·əm }

capillary condensation [PHYS CHEM] Condensation of an adsorbed vapor within the pores of the adsorbate. { 'kap·ə,ler·ē ,kän,den'sā·shən }

capillary control [PETRO ENG] Discarded theory of reservoir oil flow to a well hole through capillary pores; it attributed flow resistance to gas bubbles within the capillaries. { 'kap·ə,ler·ē kən'trōl }

capillary curve [FL MECH] The curve along which the surface of a liquid intersects a vertical plane perpendicular to a vertical glass plane surface. { 'kap·ə,ler·ē ,kərv }

capillary depression [FL MECH] The depression of the meniscus of a liquid contained in a tube where the liquid does not wet the walls of the container, as in a mercury barometer; the meniscus has a convex shape, resulting in a depression. { 'kap·ə,ler·ē di'presh·ən }

capillary drying [ENG] Progressive removal of moisture from a porous solid mass by surface evaporation followed by capillary movement of more moisture to the drying surface from the moist inner region, until the surface and core stabilize at the same moisture concentration. { 'kap·ə,ler·ē 'drī·iŋ }

capillary ejecta *See* Pele's hair. { 'kap·ə,ler·ē i'jek·tə }

capillary electrochromatography [ANALY CHEM] A separation technique in which analytes are transported through a small-diameter packed column by electroosmosis (electrically induced flow of the mobile phase) by applying a high potential (5–30 kilovolts) across the column. { ¦kap·ə·ler·ē i,lek·trō,krō·mə'täg·rə·fē }

capillary electrometer [ENG] An electrometer designed to measure a small potential difference between mercury and an electrolytic solution in a capillary tube by measuring the effect of this potential difference on the surface tension between the liquids. Also known as Lippmann electrometer. { 'kap·ə,ler·ē i,lek'träm·əd·ər }

capillary electrophoresis [ANALY CHEM] A technique for separating substances from a fluid substrate; the sample is placed in a capillary tube which is then subjected to a high-voltage current that separates its chemical constituents. { ¦kap·ə·ler·ē i,lek·trō·fə'rē·səs }

capillary equilibrium method [PETRO ENG] Test method to predict oil and gas flow through an oil reservoir core by dethrottling the flow to hold capillary flow in equilibrium between oil and gas within the reservoir. { 'kap·ə,ler·ē ,ē·kwə'lib·rē·əm ,meth·əd }

capillary fitting [ENG] A pipe fitting having a socket-type end so that when the fitting is soldered to a pipe end, the solder flows by capillarity along the annular space between the pipe exterior and the socket within it, forming a tight fit. { 'kap·ə,ler·ē ,fid·iŋ }

capillary fringe [HYD] The lower subdivision of the zone of aeration that overlies the zone of saturation and in which the pressure of water in the interstices is lower than atmospheric. { 'kap·ə,ler·ē ,frinj }

capillary gas chromatography [ANALY CHEM] A highly efficient type of gas chromatography in which the gaseous sample passes through capillary tubes with internal diameters between 0.2 and 0.5 millimeter and lengths up to 100 meters, and adsorption takes place on a medium that is spread on the inner walls of these tubes. { 'kap·ə,ler·ē ¦gas krō·mə'täg·rə·fē }

capillary gel electrophoresis [ANALY CHEM] A form of capillary electrophoresis in which a polyacrylamide gel (or other polymeric material) is placed inside the capillary and separation is based on size and charge; often used to separate oligonucleotides and proteins. { ¦kap·ə·ler·ē ¦jel i,lek·trō·fə'rē·səs }

capillary migration [HYD] Movement of water produced by the force of molecular attraction between rock material and the water. { 'kap·ə,ler·ē mī'grā·shən }

capillary number [FL MECH] A dimensionless number associated with a liquid that compares the intensity of liquid viscosity and surface tension, equal to $\mu V/\sigma$, where μ is the viscosity, σ is the surface tension, and V is a fluid velocity such as the deposition velocity on a solid that is drawn out of the liquid. { 'kap·ə·ler·ē ,nəm·bər }

capillary pressure [FL MECH] **1.** The difference of pressure across the interface of two immiscible fluid phases. **2.** The pressure or adhesive force exerted by water in an enclosed space as a result of surface tension. [PHYSIO] Pressure exerted by blood against capillary walls. { 'kap·ə,ler·ē ,presh·ər }

capillary pyrites *See* millerite. { 'kap·ə,ler·ē 'pī,rīts }

capillary ripple *See* capillary wave. { 'kap·ə,ler·ē ,rip·əl }

capillary rise [FL MECH] The rise of a liquid in a capillary tube times the radius of the tube. { 'kap·ə,ler·ē ,rīz }

capillary tube [ENG] A tube sufficiently fine so that capillary attraction of a liquid into the tube is significant. { 'kap·ə,ler·ē ,tüb }

capillary viscometer [ENG] A long, narrow tube that is used to measure the laminar flow of fluids. { 'kap·ə,ler·ē vis'käm·əd·ər }

capillary water [HYD] Soil water held by capillarity as a continuous film around soil particles and in interstices between particles above the phreactic line. { 'kap·ə,ler·ē ,wȯd·ər }

capillary wave [FL MECH] **1.** A wave occurring at the interface between two fluids, such as the interface between air and water on oceans and lakes, in which the principal restoring force is controlled by surface tension. **2.** A water wave of less than 1.7 centimeters. Also known as capillary ripple; ripple. { 'kap·ə,ler·ē ,wāv }

capillary zone electrophoresis [ANALY CHEM] A type of capillary electrophoresis in which the capillary is filled with a homogenous buffer, and compounds are separated on the basis of their relative charge and size. { ¦kap·ə·ler·ē ¦zōn i,lek·trō,fə'rē·səs }

capillitium [MYCOL] A network of threadlike tubes or filaments in which spores are embedded within sporangia of certain fungi, such as the slime molds. { ,kap·ə'lish·ē·əm }

capital [ARCH] The topmost part of a column. { 'kap·ət·əl }

capital amount factor [IND ENG] Any of 20 common compound interest formulas used to calculate the equivalent uniform annual cost of all cash flows. { 'kap·ət·əl ə'maunt ,fak·tər }

capital budgeting [IND ENG] Planning the most effective use of resources to obtain the highest possible level of sustained profits. { 'kap·ət·əl 'bəj·əd·iŋ }

capital expenditure [IND ENG] Money spent for long-term additions or improvements and charged to a capital assets account. { 'kap·ət·əl ik'spen·di·chər }

capital ship [NAV ARCH] A surface warship classified as major, for example, a battleship or aircraft carrier. { 'kap·ət·əl 'ship }

capitate [BIOL] Enlarged and swollen at the tip. [BOT] Forming a head, as certain flowers of the Compositae. { 'kap·ə,tāt }

capitellate [BOT] **1.** Having a small knoblike termination. **2.** Grouped to form a capitulum. { ¦kap·ə¦te,lāt }

Capitellidae [INV ZOO] A family of mud-swallowing annelid worms, sometimes called bloodworms, belonging to the Sedentaria. { ,kap·ə'te·lə,dē }

capitellum [ANAT] A small head or rounded process of a bone. { ,kap·ə'te·ləm }

Capitonidae [VERT ZOO] The barbets, a family of pantropical birds in the order Piciformes. { ,kap·ə'tä·nə,dē }

capitulum [BIOL] A rounded, knoblike, usually terminal proturberance on a structure. [BOT] One of the rounded cells on the manubrium in the antheridia of lichens belonging to the Caliciales. { kə'pich·ə·ləm }

cap lamp [MIN ENG] The lamp a miner wears on his safety hat or cap for illumination. { 'kap ,lamp }

Capnocytophaga [MICROBIO] A genus of bacteria comprising fusiform, fermentative, gram-negative rods which require carbon dioxide for growth and show gliding motility. { ,kap·nō,sī'täf·ə·gə }

capon [AGR] A castrated male chicken. { 'kā,pän }

Caponidae [INV ZOO] A family of arachnid arthropods in the order Araneida characterized by having tracheae instead of book lungs. { kə'pä·nə,dē }

Capparaceae [BOT] A family of dicotyledonous herbs, shrubs, and trees in the order Capparales characterized by parietal placentation; hypogynous, mostly regular flowers; four to many stamens; and simple to trifoliate or palmately compound leaves. { ,kap·ə'rās·ē,ē }

Capparales [BOT] An order of dicotyledonous plants in the subclass Dilleniidae. { ,kap·ə'rā,lēz }

capped column [HYD] A form of ice crystal consisting of

a hexagonal column with plate or stellar crystals (so-called caps) at its ends and sometimes at intermediate positions; the caps are perpendicular to the column. { 'kapt 'käl·əm }

capped fuse [ENG] A length of safety fuse with the cap or detonator crimped on before it is taken to the place of use. { 'kapt 'fyüz }

capped steel [MET] Partially deoxidized steel cast in an open-top mold, which is capped to solidify the top metal and enforce internal pressure, resulting in a surface condition similar to that of rimming steel. { 'kapt 'stēl }

cappelenite [MINERAL] (Ba,Ca,Na)(Y,La)₆B₆Si₁₃(O,OH)₂₇ A greenish-brown hexagonal mineral consisting of a rare yttrium-barium borosilicate occurring in crystals. { 'kap·lə‚nīt }

cap piece [MIN ENG] A piece of wood fitted over a straight post or timber to provide more bearing surface. { 'kap ‚pēs }

capping [ENG] Preparation of a capped fuse. [GEOL] **1.** Consolidated barren rock overlying a mineral or ore deposit. **2.** *See* gossan. [MIN ENG] The attachment at the end of a winding rope. [MOL BIO] Addition of a methylated cap to eukaryotic messenger ribonucleic acid molecules. [PETRO ENG] **1.** The process of sealing or covering a borehole such as an oil or gas well. **2.** The material or device used to seal or cover a borehole. { 'kap·iŋ }

caprate [ORG CHEM] Any of the salts of capric acid, containing the group C₉H₁₉COO⁻. { 'ka‚prāt }

Caprellidae [INV ZOO] The skeleton shrimps, a family of slender, cylindrical amphipod crustaceans in the suborder Caprellidea. { kə'prel·ə‚dē }

Caprellidea [INV ZOO] A suborder of marine and brackish-water animals of the crustacean order Amphipoda. { ‚kap·rə'lid·ē·ə }

capric acid [ORG CHEM] CH₃(CH₂)₈COOH A fatty acid found in oils and animal fats. { 'ka‚prik 'as·əd }

capric anhydride [ORG CHEM] (CH₃(CH₂)₈CO)₂O White crystals that are insoluble in water; used as a chemical intermediate. { 'ka‚prik an'hī‚drīd }

Capricornus [ASTRON] A constellation with right ascension 21 hours, declination 20°S. Abbreviated Cap. Also known as Sea Goat. { ‚kap·rə'kòr·nəs }

Caprifoliaceae [BOT] A family of dicotyledonous, mostly woody plants in the order Dipsacales, including elderberry and honeysuckle; characterized by distinct filaments and anthers, typically five stamens and five corolla lobes, more than one ovule per locule, and well-developed endosperm. { ‚kap·rə‚fōl·ē'ās·e‚ē }

Caprimulgidae [VERT ZOO] A family of birds in the order Caprimulgiformes, including the nightjars, or goatsuckers. { ‚kap·rə'məl·jə‚dē }

Caprimulgiformes [VERT ZOO] An order of nocturnal and crepuscular birds, including nightjars, potoos, and frog-mouths. { ‚kap·rə‚məl·jə'fòr‚mēz }

Capripoxvirus [VIROL] A genus of viruses belonging to the subfamily Chordopoxvirinae; natural hosts are ungulates; these viruses can be mechanically transmitted by arthropods. { 'kap·ri‚päks‚vī·rəs }

capristor *See* rescap. { ka'pris·tər }

caproamide [ORG CHEM] CH₃(CH₂)₄CONH₂ An amide, melting point 100–101°C; used as a chemical intermediate. { ‚ka·prō'am‚īd }

cap rock [GEOL] **1.** An overlying, generally impervious layer or stratum of rock that overlies an oil- or gas-bearing rock. **2.** Barren vein matter, or a pinch in a vein, supposed to overlie ore. **3.** A hard layer of rock, usually sandstone, a short distance above a coal seam. **4.** An impervious body of anhydrite and gypsum in a salt dome. { 'kap ‚räk }

caproic acid [ORG CHEM] CH₃(CH₂)₄COOH A colorless liquid fatty acid found in oils and animal fats; used in synthesizing pharmaceuticals and flavors. { kə'prō·ik 'as·əd }

caproic anhydride [ORG CHEM] [CH₃(CH₂)₄COO]₂ White crystals that are insoluble in water, melting point −40.6°C, boiling point 241–243°C. { kə'prō·ik an'hī‚drīd }

caprolactam [ORG CHEM] (CH₂)₅NH·CO White flakes, melting point 68–69°C, made from cyclohexanone; used to make synthetic fiber, particularly nylon-6. { ‚ka·prō'lak·təm }

ε-caprolactone [ORG CHEM] CH₂(CH₂)₄NHCO White crystals, used to make synthetic fibers, plastics, films, coatings, and plasticizers; its vapors or fine crystals are respiratory irritants. { ¦ā·də ‚ka·prō'lak‚tōn }

caprylamide [ORG CHEM] CH₃(CH₂)₆CONH₂ An amine, melting point 105–110°C; decomposes above 200°C; used as a chemical intermediate. { kə'pril·ə‚mīd }

capryl compounds [ORG CHEM] A misnomer for octyl compounds; that is, the term octyl halide is preferred for caprylic halides, and octanoic acid for caprylic acid. { 'ka‚prəl ‚käm‚paùnz }

1-caprylene *See* 1-octene. { ¦wən 'kap·rə‚lēn }

caprylic acid [ORG CHEM] C₈H₁₆O₂ A liquid fatty acid occurring in butter, coconut oil, and other fats and oils. { kə'pril·ik 'as·əd }

caprylic anhydride [ORG CHEM] [CH₃(CH₂)₆CO]₂O A white solid that melts at −1°C; used as a chemical intermediate. { kə'pril·ik an'hī‚drīd }

capsaicin [ORG CHEM] C₁₈H₂₇O₃N A toxic material extracted from capsicum. { kap'sā·ə·sən }

Capsaloidea [INV ZOO] A superfamily of ectoparasitic trematodes in the subclass Monogenea characterized by a sucker-shaped holdfast with anchors and hooks. { ‚kap·sə'lòid·ē·ə }

capsanthin [BIOCHEM] C₄₀H₅₈IO₃ Carmine-red carotenoid pigment occurring in paprika. { kap'san·thən }

cap screw [DES ENG] A screw which passes through a clear hole in the part to be joined, screws into a threaded hole in the other part, and has a head which holds the parts together. { 'kap ‚skrü }

capsicum [BOT] The fruit of a plant of the genus *Capsicum*, especially *C. frutescens*, cultivated in southern India and the tropics; a strong irritant to mucous membranes and eyes. { 'kap·sə·kəm }

capsid [INV ZOO] The name applied to all members of the family Miridae. [VIROL] In a virus, the protein shell surrounding the nucleic acid and its associated protein core. Also known as protein coat. { 'kap·səd }

capsomere [VIROL] An individual protein subunit of a capsid. { 'kap·sə‚mir }

capstan [ENG] A shaft which pulls magnetic tape through a machine at constant speed. [NAV ARCH] A rotating vertical spindle-mounted drum on which cable is wound for raising an anchor or other heavy weight. { 'kap·stən }

capstan nut [DES ENG] A nut whose edge has several holes, in one of which a bar can be inserted for turning it. { 'kap·stən ‚nət }

capstan screw [DES ENG] A screw whose head has several radial holes, in one of which a bar can be inserted for turning it. { 'kap·stən ‚skrü }

capstone [ARCH] A stone placed at the top of a stone arch. { 'kap‚stōn }

capsular ligament [ANAT] A saclike ligament surrounding the articular cavity of a freely movable joint and attached to the bones. { 'kap·sə·lər 'lig·ə·mənt }

capsulate [BIOL] Enclosed in a capsule. { 'kap·sə‚lāt }

capsule [AERO ENG] A small, sealed, pressurized cabin with an internal environment that will support human or animal life during extremely high-altitude flight, space flight, or escape. [ANAT] A membranous structure enclosing a body part or organ. [BOT] A closed structure bearing seeds or spores; it is dehiscent at maturity. [ENG] A boxlike component or unit, often sealed. [MICROBIO] A thick, mucous envelope, composed of polypeptide or carbohydrate, surrounding certain microorganisms. [PHARM] A soluble shell in which drugs are enclosed for oral administration. { 'kap·səl }

captan [ORG CHEM] C₉H₈O₂NSCl₃ A buff to white solid with a melting point of 175°C; used as a fungicide for diseases of fruits, vegetables, and flowers. { 'kap‚tan }

captive balloon [AERO ENG] A moored balloon, usually held by steel cables. { 'kap·tiv bə'lün }

captive fastener [DES ENG] A screw-type fastener that does not drop out after it has been unscrewed. { 'kap·tiv 'fas·ən·ər }

captive test [ENG] A hold-down test of a propulsion subsystem, rocket engine, or motor. { 'kap·tiv 'test }

Captorhinomorpha [PALEON] An extinct subclass of primitive lizardlike reptiles in the order Cotylosauria. { ‚kap·tə‚rī·nə¦mòr·fə }

capture [AERO ENG] The process in which a missile is taken under control by the guidance system. [ASTROPHYS] Of a central force field, as of a planet, to overcome by gravitational force the velocity of a passing body and bring the body under

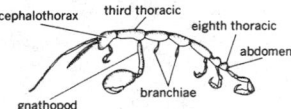

CAPRELLIDEA

Caprella grandimana, a caprellid.

CAPRICORNUS

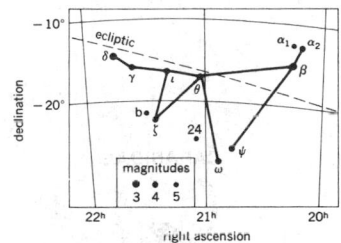

Line pattern of constellation Capricornus. Grid lines represent coordinates of the sky. Apparent brightness, or magnitude, of stars is shown by size of dots, which are graded by appropriate numbers as indicated.

CAPSALOIDEA

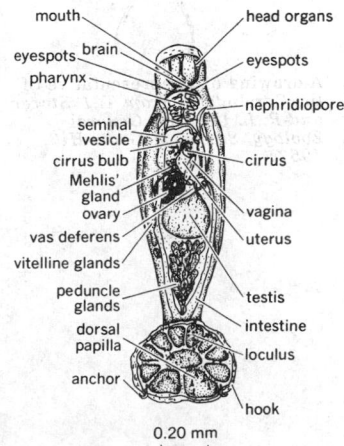

Heterocotyle acetobactis, from the spotted eagle ray, ventral view.

the control of the central force field, in some cases absorbing its mass. [GEOCHEM] In a crystal structure, the substitution of a trace element for a lower-valence common element. [HYD] The natural diversion of the headwaters of one stream into the channel of another stream having greater erosional activity and flowing at a lower level. Also known as piracy; river capture; river piracy; robbery; stream capture; stream piracy; stream robbery. [PHYS] A process in which an atomic or nuclear system acquires an additional particle; for example, the capture of electrons by positive ions, or capture of neutrons by nuclei. { 'kap·chər }

capture area [ENG ACOUS] The effective area of the receiving surface of a hydrophone, or the available power of the acoustic energy divided by its equivalent plane-wave intensity. { 'kap·chər ,er·ē·ə }

capture cross section [NUC PHYS] The cross section that is effective for radiative capture. { 'kap·chər ¦kròs 'sek·shən }

capture effect [ELECTR] The effect wherein a strong frequency-modulation signal in an FM receiver completely suppresses a weaker signal on the same or nearly the same frequency. { 'kap·chər i'fekt }

capture gamma rays [NUC PHYS] The gamma rays emitted in radiative capture. { ¦kap·chər 'gam·ə ¦rāz }

capture ratio [COMMUN] A measure of the ability of a frequency-modulation tuner to reject the weaker of two stations that are on the same frequency; the lower the ratio of desired to undesired signals, the better the performance of the tuner. { 'kap·chər ,rā·shō }

capturing [ENG] The use of a torquer to restrain the spin axis of a gyro to a specified position relative to the spin reference axis. { 'kap·chə·riŋ }

capybara [VERT ZOO] *Hydrochoerus capybara.* An aquatic rodent (largest rodent in existence) found in South America and characterized by partly webbed feet, no tail, and coarse hair. { ¦kap·ə·'bar·ə }

Cap Z protein [BIOCHEM] A microfilament capping protein located in the Z line of skeletal muscle, made up of two subunits that selectively bind to the positive ends of actin filaments, stabilizing them and preventing depolymerization. { ¦kap ¦zē 'prō,tēn }

car *See* automobile. { kär }

CAR *See* computer-assisted retrieval. { kär }

Carabidae [INV ZOO] The ground beetles, a family of predatory coleopteran insects in the suborder Adephaga. { kə'rab·ə,dē }

carabine [METEOROL] In France and Spain, a sudden and violent wind. Also known as brisa carabinera; brise carabinée. { kär·ə·bēn *or* kär·ə·bē·nä }

caraboid larva [INV ZOO] The morphologically distinct larva of certain beetles, characterized by a narrow, elongate body with long legs, a head that occasionally bears a single ocellus on each side, and three-segmented antennae with a well-developed sensorium. { 'kar·ə,bóid 'lar·və }

Caracarinae [VERT ZOO] The caracaras, a subfamily of carrion-feeding birds in the order Falconiformes. { ,karə'kar·ə,nē }

caracolite [MINERAL] A rare, colorless mineral occurring as crystalline incrustations, and consisting of a sulfate and chloride of sodium and lead. { ,kar·ə'kō,līt }

Caradocian [GEOL] Lower Upper Ordovician geologic time. { kar·ə'dō·shən }

caramel [MATER] A dark-brown mass formed by heating sugar in the presence of ammonium salts. { 'kar·ə·məl }

caramelize [FOOD ENG] To convert sugar or the sugar content of a food into a caramel or a caramellike substance. { 'kär·mə,līz }

caranda wax [MATER] A wax similar to carnauba wax; obtained from the tropical palm caranda (*Copernicia australis*). { ¦ka·rən¦dä ,waks }

Carangidae [VERT ZOO] A family of perciform fishes in the suborder Percoidei, including jacks, scads, and pompanos. { kə'ran·jə,dē }

carangiform motion [VERT ZOO] A type of fish locomotion in which the fish moves its head slightly but builds considerable amplitude of motion toward the tail. { kə'ran·jə,fòrm ,mō·shən }

carapace [GEOL] The upper normal limb of a fold having an almost horizontal axial plane. [INV ZOO] A dorsolateral, chitinous case covering the cephalothorax of many arthropods.

CARABIDAE

A drawing of a representative of the Carabidae. (*From T. I. Storer and R. L. Usinger, General Zoology, 3d ed., McGraw-Hill, 1957*)

[VERT ZOO] The bony, dorsal part of a turtle shell. { 'kar·ə,pās }

Carapidae [VERT ZOO] The pearlfishes, a family of sinuous, marine shore fishes in the order Gadiformes that live as commensals in the body cavity of holothurians. { kə'rap·ə,dē }

carat [LAP] A unit of weight of gemstones, equal to 200 milligrams. Also known as metric carat. { 'kar·ət }

carat count [LAP] The number of near-equal-size diamonds having a total weight of 1 carat. { 'kar·ət ,kaúnt }

carate *See* pinta. { kə'räd·ē }

Carathéodory outer measure [MATH] A positive, countably subadditive set function defined on the class of all subsets of a given set; used for defining measures. { ,kär·ə¦tā·ə'dòr·ē ¦aúd·ər 'mezh·ər }

Carathéodory's principle [THERMO] An expression of the second law of thermodynamics which says that in the neighborhood of any equilibrium state of a system, there are states which are not accessible by a reversible or irreversible adiabatic process. Also known as principle of inaccessibility. { ,kär·ə,tā·ə'dòr·ēz 'prin·sə·pəl }

Carathéodory theorem [MATH] The theorem that each point of the convex span of a set in an *n*-dimensional Euclidean space is a convex linear combination of points in that set. { ,kär·ə,tā·ə'dòr·ē ,thir·əm }

caraway [BOT] *Carum carvi.* A white-flowered perennial herb of the family Umbelliferae; the fruit is used as a spice and flavoring agent. { 'kar·ə,wā }

caraway oil [MATER] A pale-yellow to colorless liquid that is slightly soluble in water; distilled from the dried fruit of the caraway plant (*Carum carvi*), the main ingredients being carvone and *d*-limonene; used in flavors, medicines, soaps, and perfumes. Also known as caraway seed oil; carui oil. { 'kar·ə,wā ,òil }

caraway seed oil *See* caraway oil. { 'kar·ə,wā ,sēd ,òil }

carbachol [PHARM] $C_6H_{15}ClN_2O_2$ Hygroscopic, hard, prismatic crystals with a melting point of 200–203°C; soluble in water, methanol, and alcohol; used as a cholinergic drug in humans and parasympathomimetic drug in larger animals. { 'kär·bə,kòl }

carbamate [ORG CHEM] An ester of carbamic acid. { 'kär·bə,māt }

carbamazepine [PHARM] $C_{15}H_{12}N_2O$ An anticonvulsant that is used in the treatment of epilepsy, and is also useful in the treatment of periods of mania associated with bipolar depression. { ,kär·bə'maz·ə,pēn }

carbamic acid [BIOCHEM] NH_2COOH An amino acid known for its salts, such as urea and carbamide. { kar'bam·ik 'as·əd }

carbamide *See* urea. { 'kär·bə,mīd }

carbamino [BIOCHEM] A compound formed by the combination of carbon dioxide with a free amino group in an amino acid or a protein. { kär'bam·ə,nō }

carbamoyl [ORG CHEM] The radical NH_2CO, formed from carbamic acid. { kär'bam·ə,wil }

carbamyl phosphate [BIOCHEM] $NH_2COPO_4H_2$ The ester formed from reaction of phosphoric acid and carbamyl acid. { 'kär·bə,mil 'fäs,fāt }

carbanilide [ORG CHEM] $(NHC_6H_5)CO(NHC_6H_5)$ Colorless crystals that are very slightly soluble in water, and dissolve in ether and alcohol; used in organic synthesis. { kär·bə'nil,īd }

carbanion [CHEM] One of the charged fragments which arise on heterolytic cleavage of a covalent bond involving carbon; the fragment carries an unshared pair of electrons and bears a negative charge. { ¦kärb'an,ī·ən }

carbaryl [ORG CHEM] $C_{12}H_{11}NO_2$ A colorless, crystalline compound with a melting point of 142°C; used as an insecticide for crops, forests, lawns, poultry, and pets. { 'kär·bə,ril }

carbazide *See* carbodihydrazide. { 'kär·bə,zīd }

carbazole [ORG CHEM] One of a group of organic heterocyclic compounds containing a dibenzopyrrole system. Also known as 9-azafluorene. { 'kär·bə,zōl }

carbene [ORG CHEM] A compound of carbon which exhibits two valences to a carbon atom; the two valence electrons are distributed in the same valence; an example is CH_2. { 'kär,bēn }

carbenium ion [ORG CHEM] A cation in which the charged atom is carbon; for example, R_2C^+, where R is an organic group. { kär'bē·nē·əm ,ī·ən }

carbenoid species [ORG CHEM] A species that is not a free carbene but has the characteristics of a carbene when participating in a chemical reaction. { 'kär·bə‚nȯid ‚spē·shēz }

carbide [INORG CHEM] A binary compound of carbon with an element more electropositive than carbon; carbon-hydrogen compounds are excluded. [MATER] A cemented or compacted mixture of powdered carbides of heavy metals forming a hard material used in metal-cutting tools. Also known as cemented carbide. { 'kär‚bīd }

carbide lamp [MIN ENG] A lamp that is charged with calcium carbide and water to form acetylene, which it burns. { 'kär‚bīd ‚lamp }

carbide miner [MIN ENG] An automated coal mining machine; the unit is a continuous miner controlled from outside the coal seam. { 'kär‚bīd ‚mī·nər }

carbide nuclear fuel [NUCLEO] A nuclear reactor fuel which is mixed with carbon compounds and a metal to give structural strength and oxidation resistance. { 'kär‚bīd ‚nü·klē·ər 'fyül }

carbide tool [DES ENG] A cutting tool made of tungsten, titanium, or tantalum carbides, having high heat and wear resistance. { 'kär‚bīd ‚tül }

carbine [ORD] A rifle of short length and light weight. { 'kär‚bēn }

carbinol [ORG CHEM] **1.** A primary alcohol with general formula RCH$_2$OH. **2.** The radical CH$_2$OH of primary alcohols. **3.** An alcohol derived from methanol. { 'kär·bə‚nȯl }

carbinyl See methyl. { 'kär·bə‚nil }

carbocation [ORG CHEM] A positively charged ion whose charge resides, at least in part, on a carbon atom or group of carbon atoms. { ¦kär·bō¦kat‚ī·ən }

carbocyclic compound [ORG CHEM] A compound with a homocyclic ring in which all the ring atoms are carbon, for example, benzene. { ¦kär·bō¦si·klik 'kam‚pau̇nd }

carbodihydrazide [ORG CHEM] CO(NHNH$_2$)$_2$ Colorless crystals that melt at 154°C; very soluble in alcohol and water; used in photographic chemicals. { ¦kär·bō‚dī'hī·drə‚zīd }

carbodiimide [ORG CHEM] **1.** HN=C=NH An unstable tautomer of cyanamide. **2.** Any compound with the general formula RN=C=NR which is a formal derivative of carbodiimide. { ¦kär·bō'dī·ə‚mīd }

carbofuran [ORG CHEM] C$_{12}$H$_{15}$NO$_3$ A white solid with a melting point of 150–152°C; soluble in water; used as an insecticide, miticide, and nematicide in many crops. { ‚kär·bō'fyu̇r‚än }

carbohumin See ulmin. { ‚kär·bō'hyü·mən }

carbohydrase [BIOCHEM] Any enzyme that catalyzes the hydrolysis of disaccharides and more complex carbohydrates. { ‚kär·bō'hī‚drās }

carbohydrate [BIOCHEM] Any of the group of organic compounds composed of carbon, hydrogen, and oxygen, including sugars, starches, and celluloses. { ‚kär·bō'hī‚drāt }

carbohydrate gum [ORG CHEM] A polysaccharide which produces a gel of a viscous solution when it is dispersed in water at low concentrations; examples are agar, guar gum, xanthan gum, gum arabic, and sodium carboxymethyl cellulose. { ‚kär·bō'hī‚drāt 'gəm }

carbohydrate metabolism [BIOCHEM] The sum of the biochemical and physiological processes involved in the breakdown and synthesis of simple sugars, oligosaccharides, and polysaccharides and in the transport of sugar across cell membranes. { ‚kär·bō'hī‚drāt me'tab·ə‚liz·əm }

carbolfuchsin [MATER] A solution of fuchsin, phenol, alcohol, and water; used as a stain in the identification of bacteria. { ‚kär·bäl'fyük·sən }

carbolic acid See phenol. { kär'bäl·ik 'as·əd }

carboloy [MET] A hard alloy, containing carbon, tungsten, and cobalt, used for cutting tools and as an abrasive. { 'kär·bə‚lȯi }

carbometer [ENG] An instrument for measuring the carbon content of steel by measuring magnetic properties of the steel in a known magnetic field. { kär'bäm·əd·ər }

carbomycin [MICROBIO] C$_{42}$H$_{67}$O$_{16}$N Colorless, crystalline antibiotic produced by *Streptomyces halstedii;* principally active against gram-positive bacteria. { ‚kär·bō'mīs·ən }

carbomycin B [MICROBIO] C$_{42}$H$_{67}$O$_{15}$N A colorless, crystalline antibiotic differing from carbomycin only in having one less oxygen atom in its molecule. { ‚kär·bō'mīs·ən 'bē }

carbon [CHEM] A nonmetallic chemical element, symbol C,

atomic number 6, atomic weight 12.01115; occurs freely as diamond, graphite, and coal. { 'kär·bən }

carbon-12 [NUC PHYS] A stable isotope of carbon with mass number of 12, forming about 98.9% of natural carbon; used as the basis of the newer scale of atomic masses, having an atomic mass of exactly 12u (relative nuclidic mass unit) by definition. { 'kär·bən 'twelv }

carbon-13 [NUC PHYS] A heavy isotope of carbon having a mass number of 13. { 'kär·bən 'thər‚tēn }

carbon-14 [NUC PHYS] A naturally occurring radioisotope of carbon having a mass number of 14 and half-life of 5780 years; used in radiocarbon dating and in the elucidation of the metabolic path of carbon in photosynthesis. Also known as radiocarbon. { 'kär·bən 'fȯr‚tēn }

carbonaceous [SCI TECH] Relating to or composed of carbon. { kär·bə'nā·shəs }

carbonaceous chondrite [GEOL] A chondritic meteorite that contains a relatively large amount of carbon and has a resulting dark color. Also known as carbonaceous meteorite. { ‚kär·bə'nā·shəs 'kän‚drīt }

carbonaceous meteorite See carbonaceous chondrite. { kär·bə'nā·shəs 'mēd·ē·ə‚rīt }

carbonaceous rock [PETR] Rock with carbonaceous material included. { kär·bə'nā·shəs 'räk }

carbonaceous sandstone [PETR] Sandstone rich in carbon. { kär·bə'nā·shəs 'san‚stōn }

carbonaceous shale [GEOL] Shale rich in carbon. { kär·bə'nā·shəs 'shāl }

carbonado [MINERAL] A dark-colored, fine-grained diamond aggregate; valuable for toughness and absence of cleavage planes. Also known as black diamond; carbon diamond. { kär·bə'nä·dō }

carbon arc [ELEC] An electric arc between two electrodes, at least one of which is made of carbon; used in welding and high-intensity lamps, such as in searchlights and photography lamps. { ¦kär·bən 'ärk }

carbon-arc brazing [MET] Brazing base metals by heating with an electric arc between a carbon electrode and the workpiece. { ¦kär·bən ¦ärk 'brāz·iŋ }

carbon-arc cutting [MET] An arc cutting process which generates heat in order to melt a base metal with a carbon electrode to eventually produce a cut. { ¦kär·bən ¦ärk 'kəd·iŋ }

carbon-arc lamp [ELEC] An arc lamp in which an electric current flows between two electrodes of pure carbon, with incandescence at one or both electrodes and some light from the luminescence of the arc. { ¦kär·bən ¦ärk 'lamp }

carbon-arc welding [MET] Welding by maintaining an electric arc between a nonconsumable carbon electrode and the work. { ¦kär·bən ¦ärk 'wel·diŋ }

carbonate [CHEM] **1.** An ester or salt of carbonic acid. **2.** A compound containing the carbonate (CO$_3^{2-}$) ion. **3.** Containing carbonates. { 'kär·bə‚nət }

carbonate cycle [GEOCHEM] The biogeochemical carbonate pathways, involving the conversion of carbonate to CO$_2$ and HCO$_3$, the solution and deposition of carbonate, and the metabolism and regeneration of it in biological systems. { 'kär·bə‚nət ‚sī·kəl }

carbonate mineral [MINERAL] A mineral containing considerable amounts of carbonates. { 'kär·bə‚nət 'min·rəl }

carbonate reservoir [GEOL] An underground oil or gas trap formed in reefs, clastic limestones, chemical limestones, or dolomite. { 'kär·bə‚nət 'rez·əv‚wär }

carbonate rock [PETR] A rock composed principally of carbonates, especially if at least 50% by weight. { 'kär·bə‚nət 'räk }

carbonate spring [HYD] A type of spring containing dissolved carbon dioxide gas. { 'kär·bə‚nət 'spriŋ }

carbonation [CHEM] Conversion to a carbonate. [CHEM ENG] The process by which a fluid, especially a beverage, is impregnated with carbon dioxide. [GEOCHEM] A process of chemical weathering whereby minerals that contain soda, lime, potash, or basic oxides are changed to carbonates by the carbonic acid in air or water. { ‚kär·bə'nā·shən }

carbonatite [PETR] **1.** Intrusive carbonate rock associated with alkaline igneous intrusive activity. **2.** A sedimentary rock that is composed of at least 80% calcium or magnesium. { kär'bän·ə‚tīt }

carbon bit [DES ENG] A diamond bit in which the cutting medium is inset carbon. { ¦kär·bən ¦bit }

carbon black [CHEM] **1.** An amorphous form of carbon produced commercially by thermal or oxidative decomposition of hydrocarbons and used principally in rubber goods, pigments, and printer's ink. **2.** *See* gas black. { ¦kär·bən ¦blak }

carbon brush [ELEC] A rod made of carbon that bears against a commutator, collector ring, or slip ring to provide passage for the electric current from a dynamo through an outside circuit or for an external current through a motor. { ¦kär·bən ¦brəsh }

carbon burning [NUC PHYS] The synthesis of nuclei in stars through reactions involving the fusion of two carbon-12 nuclei at temperatures of about 5×10^8 K. { ¦kär·bən ¦bərn·iŋ }

carbon burning rate [CHEM ENG] The weight of carbon burned per unit time from the catalytic-cracking catalyst in the regenerator. { ¦kär·bən 'bərn·iŋ ‚rāt }

carbon button *See* button. { ¦kär·bən ¦bət·ən }

carbon canister *See* charcoal canister. { ¦kär·bən 'kan·ə·stər }

carbon cycle [GEOCHEM] The cycle of carbon in the biosphere, in which plants convert carbon dioxide to organic compounds that are consumed by plants and animals, and the carbon is returned to the biosphere in the form of inorganic compounds by processes of respiration and decay. [NUC PHYS] *See* carbon-nitrogen cycle. { 'kär·bən ‚sī·kəl }

carbon-detonation supernova model [ASTRON] A model for a supernova in a star of 4 to 9 solar masses through the explosive ignition of carbon in a high-density, electron-degenerate core by the formation and propagation of a detonation wave. { ¦kär·bən ‚det·ən'ā·shən ‚sü·pər'nō·və ‚mäd·əl }

carbon diamond *See* carbonado. { ¦kär·bən 'dī·mənd }

carbon dioxide [INORG CHEM] CO_2 A colorless, odorless, tasteless gas about 1.5 times as dense as air. { ¦kär·bən dī'äk‚sīd }

carbon dioxide absorption tube [ANALY CHEM] An absorbent-packed tube used to capture the carbon dioxide formed during the microdetermination of carbon-hydrogen by the Pragl combustion procedure. { ¦kär·bən dī'äk‚sīd əb'sórp·shən ‚tüb }

carbon dioxide fire extinguisher [CHEM ENG] A type of chemical fire extinguisher in which the extinguishing agent is liquid carbon dioxide, stored under 800–900 pounds per square inch (5.5–6.2 megapascals) at normal room temperature. { ¦kär·bən dī'äk‚sīd 'fīr ik‚stiŋ·gwish·ər }

carbon dioxide gas laser [PHYS] A powerful, continuously operating laser in the infrared that can emit several hundred watts of power at a wavelength of 10.6 micrometers. { ¦kär·bən dī'äk‚sīd 'gas ‚lā·zər }

carbon dioxide indicator [MIN ENG] A detector of carbon dioxide in mines based on the gas's absorption by potassium hydroxide. { ¦kär·bən dī'äk‚sīd 'in·də‚kād·ər }

carbon dioxide process [MET] A casting process in which the molding material is a mixture of sand and 1.5–6% liquid silicate as a binder, and the mixture is packed around the pattern and hardened by blowing carbon dioxide gas through it. { ¦kär·bən dī'äk‚sīd ‚präs·əs }

carbon disulfide [ORG CHEM] CS_2 A sulfide, used as a solvent for oils, fats, and rubbers and in paint removers. { ¦kär·bən dī'səl‚fīd }

carbon electrode [MET] A nonfiller-metal electrode consisting of carbon or a graphite rod; sometimes contains copper powder for increased electrical conductivity; used in carbon arc welding. { ¦kär·bən i'lek‚trōd }

carbon equivalent [MET] An empirical relationship of the total carbon (TC), silicon (Si) and phosphorus (P) content of gray iron: CE = %TC + 0.3(%Si + %P). { ¦kär·bən i'kwiv·ə‚lənt }

carbon fiber [MATER] **1.** Commercial material made by pyrolyzing any spun, felted, or woven raw material to a char at temperatures from 700 to 1800°C. **2.** A filamentary form of carbon, usually with a diameter in the 6–10-micrometer range. { 'kär·bən ‚fī·bər }

carbon film [ANALY CHEM] Carbon deposited by evaporation onto a specimen to protect and prepare it for electron microscopy. { ¦kär·bən 'film }

carbon-film hygrometer element [ELEC] An electrical hygrometer element constructed of a plastic strip coated with a film of carbon black dispersed in a hygroscopic binder; variations in atmospheric moisture content vary the volume of the binder and thus change the resistance of the carbon coating. { 'kär·bən ‚film hī'gräm·əd·ər ‚el·ə·mənt }

carbon-film resistor [ELEC] A resistor made by depositing a thin carbon film on a ceramic form. { 'kär·bən ‚film ri'zis·tər }

carbon fixation [CELL MOL] During photosynthesis, the process by which plants convert carbon dioxide from the air into organic molecules. { 'kär·bən fik‚sā·shən }

carbon fixation cycle *See* Calvin cycle. { ¦kär·bən fik'sā·shən ‚sī·kəl }

carbon-14 dating [NUCLEO] Determining the approximate age of organic material associated with archeological or fossil artifacts by measuring the rate of radiation of the carbon-14 isotope. Also known as radioactive carbon dating; radiocarbon dating. { ¦kär·bən ‚fór‚tēn 'dād·iŋ }

carbon-hydrogen analyzer [ANALY CHEM] A device used in the quantitative analysis of the carbon and hydrogen content of organic compounds. { ¦kär·bən ¦hī·drə·jən 'an·ə‚līz·ər }

carbon hydrophone [ENG ACOUS] A carbon microphone that responds to waterborne sound waves. { ¦kär·bən 'hī·drə‚fōn }

carbonic acid [INORG CHEM] H_2CO_3 The acid formed by combination of carbon dioxide and water. { kär'bän·ik 'as·əd }

carbonic anhydrase [BIOCHEM] An enzyme which aids carbon dioxide transport and release by catalyzing the synthesis, and the dehydration, of carbonic acid from, and to, carbon dioxide and water. { kär'bän·ik an'hī‚drās }

Carboniferous [GEOL] A division of late Paleozoic rocks and geologic time including the Mississippian and Pennsylvanian periods. { ‚kär·bə'nif·ə·rəs }

carbonification *See* coalification. { kär‚bän·ə·fə'kā·shən }

carbon isotope ratio [GEOL] Ratio of carbon-12 to either of the less common isotopes, carbon-13 or carbon-14, or the reciprocal of one of these ratios; if not specified, the ratio refers to carbon-12/carbon-13. Also known as carbon ratio. { ¦kär·bən 'is·ə‚tōp ‚rā·shō }

carbonite *See* natural coke. { 'kär·bə‚nīt }

carbonitrided steel [MET] Steel which is produced by carbonitriding. { ‚kär·bə'nī‚trīd·əd 'stēl }

carbonitriding [MET] A case-hardening process that maintains carbon and alloy steels in a hot gaseous atmosphere from which they absorb carbon and nitrogen simultaneously, producing a hard, wear-resistant surface. { ‚kär·bə'nī‚trīd·iŋ }

carbonium ion [ORG CHEM] A carbocation which has a positively charged carbon with a coordination number greater than 3. { kär'bōn·ē·əm ‚ī·ən }

carbonization [CHEM] The conversion of a carbon-containing substance to carbon or a carbon residue as the destructive distillation of coal by heat in the absence of air, yielding a solid residue with a higher percentage of carbon than the original coal; carried on for the production of coke and of fuel gas. [GEOCHEM] **1.** In the coalification process, the accumulation of residual carbon by changes in organic material and their decomposition products. **2.** Deposition of a thin film of carbon by slow decay of organic matter underwater. **3.** A process of converting a carbonaceous material to carbon by removal of other components. { ‚kär·bə·nə'zā·shən }

carbon knock [MECH ENG] Premature ignition resulting in knocking or pinging in an internal combustion engine caused when the accumulation of carbon produces overheating in the cylinder. { 'kär·bən ‚näk }

carbon lamp [ELEC] An arc lamp with carbon electrodes. { 'kär·bən ‚lamp }

carbonless copy paper [MATER] A sheet of paper, used to make duplicate copies of written or printed material, whose back is coated with a layer of microcapsules that contain a dye in colorless form in a hydrocarbon solvent; writing or printing pressure breaks the capsules and releases the dye, which reacts with a clay or phenolic resin coating on top of a second paper sheet, located directly below the first, to produce visible color. { 'kär·bən·ləs 'käp·ē ‚pā·pər }

carbon log [PETRO ENG] A record of hydrocarbon pressure in a well as revealed by determination of the carbon content. { 'kär·bən ‚läg }

CARBONIFEROUS

CENOZOIC	QUATERNARY	
	TERTIARY	
MESOZOIC	CRETACEOUS	
	JURASSIC	
	TRIASSIC	
PALEOZOIC	PERMIAN	
	CARBONIFEROUS	PENNSYLVANIAN
		MISSISSIPPIAN
	DEVONIAN	
	SILURIAN	
	ORDOVICIAN	
	CAMBRIAN	
PRECAMBRIAN		

Position of the Carboniferous and its relationship to the eras and periods of geologic time.

carbon microphone [ENG ACOUS] A microphone in which a flexible diaphragm moves in response to sound waves and applies a varying pressure to a container filled with carbon granules, causing the resistance of the microphone to vary correspondingly. { ¦kär·bən 'mī·krə₁fōn }

carbon molecular sieve [CHEM] A molecular sieve that utilizes a special type of activated carbon for the adsorbent. { ¦kär·bən mə¦lek·yə·lər 'siv }

carbon monoxide [INORG CHEM] CO A colorless, odorless gas resulting from the incomplete oxidation of carbon; found, for example, in mines and automobile exhaust; poisonous to animals. { ¦kär·bən mə'näk₁sīd }

carbon monoxide laser [OPTICS] A molecular gas laser in which the active laser molecule is carbon monoxide, and the strongest wavelengths are 4.9 to 5.7 micrometers. Also known as CO laser. { ¦kär·bən mə'näk₁sīd 'lā·zər }

carbonmonoxyhemoglobin [BIOCHEM] A stable combination of carbon monoxide and hemoglobin formed in the blood when carbon monoxide is inhaled. Also known as carbonylhemoglobin; carboxyhemoglobin. { ¦kär·bən·mə¦näk·sē'hē·mə₁glō·bən }

carbon nanotubes [CHEM] Cylindrical molecules (sealed at both ends with a convex arrangement of atoms) composed of carbon with a diameter of around 1 nanometer and lengths up to a few micrometers. Single-walled nanotubes may be conducting or semiconducting, depending on the diameter and chirality of the tube. Multiwall nanotubes containing coaxial shells of the elemental single-wall nanotubes are also possible. { ₁kär·bən 'nan·ō₁tübz }

carbon-nitrogen cycle [NUC PHYS] A series of thermonuclear reactions, with release of energy, which presumably occurs in stars that are more massive than the sun; the net accomplishment is the synthesis of four hydrogen atoms into a helium atom, the emission of two positrons and much energy, and restoration of a carbon-12 atom with which the cycle began. Also known as carbon cycle; nitrogen cycle. { ¦kär·bən ¦nī·trə·jən ₁sī·kəl }

carbon-nitrogen-oxygen bicycle [NUC PHYS] The pair of carbon-nitrogen-oxygen cycles formed from the original carbon-nitrogen cycle and an alternative cycle that results when proton capture by a nitrogen-15 nucleus results in the formation of an oxygen-16 nucleus. { ¦kär·bən ¦nī·trə·jən ¦äk·sə·jən 'bī₁sī·kəl }

carbon-nitrogen-oxygen cycles [NUC PHYS] A group of nuclear reactions involving the interaction of protons with carbon, nitrogen, and oxygen nuclei; completion of any of the cycles results in the consumption of four protons and the production of a helium-4 nucleus, two positrons, two neutrinos, and energy. Abbreviated CNO cycles. { ¦kär·bən ¦nī·trə·jən ¦äk·sə·jən ₁sī·kəlz }

carbon-nitrogen-phosphorus ratio [OCEANOGR] The relatively constant relationship between the concentrations of carbon (C), nitrogen (N), and phosphorus (P) in plankton, and N and P in sea water, owing to removal of the elements by the organisms in the same proportions in which the elements occur and their return upon decomposition of the dead organisms. { ¦kär·bən ¦nī·trə·jən ¦fäs·fə·rəs ₁rā·shō }

carbon number [ANALY CHEM] The number of carbon atoms in a material under analysis; plotted against chromatographic retention volume for compound identification. { 'kär·bən ₁nəm·bər }

carbon paper [MATER] A paper, coated with dark waxy pigment, used to make duplicate copies while typewriting or handwriting; a sheet of carbon paper is sandwiched between two paper sheets, so that the impression made on the top sheet causes the carbon paper to transfer a pigmented impression onto the bottom sheet. { 'kär·bən ₁pā·pər }

carbon pile [ELEC] A variable resistor consisting of a stack of carbon disks mounted between a fixed metal plate and a movable one that serve as the terminals of the resistor; the resistance value is reduced by applying pressure to the movable plate. { 'kär·bən ₁pīl }

carbon-pile pressure transducer [ENG] A measurement device in which variations in pressure upon a conductive carbon core proportionally change the core's electrical resistance, and thus the strength of the measured electric signal from the device. { 'kär·bən ₁pīl 'presh·ər tranz₁dü·sər }

carbon pool [GEOCHEM] A reservoir with the capacity to store and release carbon, such as soil, terrestrial vegetation, the ocean, and the atmosphere. { 'kär·bən ₁pül }

carbon potential [MET] Determination of the extent to which an environment containing active carbon can affect the carbon content of a steel. { 'kär·bən pə'ten·chəl }

carbon ratio [GEOL] **1.** The ratio of fixed carbon to fixed carbon plus volatile hydrocarbons in a coal. **2.** See carbon isotope ratio. { 'kär·bən ₁rā·shō }

carbon-ratio theory [GEOL] The theory that the gravity of oil in any area is inversely proportional to the carbon ratio of the coal. { 'kär·bən ₁rā·shō ₁thē·ə·rē }

carbon refractory [MATER] Carbon, generally in the form of graphite, used as a refractory in such equipment as crucibles and stopper nozzles for steel casting. { 'kär·bən ri'frak·trē }

carbon replication [ANALY CHEM] A faithful carbon-film, mold of a specimen surface (for example, powders, bones, or crystals) which is thin enough to be studied by electron microscopy. { 'kär·bən ₁rep·lə'kā·shən }

carbon residue [CHEM ENG] The quantity of carbon produced from a lubricating oil heated in a closed container under standard conditions. { 'kär·bən 'rez·ə₁dü }

carbon-residue test [CHEM ENG] A destructive-distillation method for estimation of carbon residues in fuels and lubricating oils. Also known as Conradson carbon test. { 'kär·bən 'rez·ə₁dü ₁test }

carbon resistance thermometer [ENG] A highly sensitive resistance thermometer for measuring temperatures in the range 0.05–20 K; capable of measuring temperature changes of the order 10^{-5} degree. { 'kär·bən ri¦zis·təns thər₁mäm·əd·ər }

carbon resistor [ELECTR] A resistor consisting of carbon particles mixed with a binder, molded into a cylindrical shape, and baked; terminal leads are attached to opposite ends. Also known as composition resistor. { 'kär·bən ri'zis·tər }

carbon restoration [MET] Carburizing of the surface layer of a material to achieve the original level of carbon content which had been depleted during processing. { 'kär·bən res·tə'rā·shən }

carbon sequence [ASTRON] Wolf-Rayet stars in which carbon emission bands dominate the spectrum. { 'kär·bən ₁sē·kwəns }

carbon sequestration [GEOCHEM] The uptake and storage of atmospheric carbon in, for example, soil and vegetation. { ₁kär·bən ₁sē·kwes'trā·shən }

carbon sink [GEOCHEM] A reservoir that absorbs or takes up atmospheric carbon; for example, a forest or an ocean. { 'kär·bən ₁siŋk }

carbon-skeleton formula See bond-line formula. { ₁kär·bən ¦skel·ə·tən ₁för·myə·lə }

carbon star [ASTRON] Any of a class of stars with an apparently high abundance ratio of carbon to hydrogen; a majority of these are low-temperature red giants of the C class. { 'kär·bən ₁star }

carbon steel [MET] Steel containing carbon, to about 2%, as the principal alloying element. { 'kär·bən 'stēl }

carbon suboxide [INORG CHEM] C_3O_2 A colorless lacrimatory gas having an unpleasant odor with a boiling point of $-6.8°C$. { 'kär·bən səb'äk₁sīd }

carbon tetrachloride [ORG CHEM] CCl_4 Colorless dense liquid, specific gravity 1.595, slightly soluble in water; used as a dry-cleaning agent. { 'kär·bən te·trə'klór₁īd }

carbon tetrafluoride [ORG CHEM] CF_4 A colorless gas with a boiling point of $-126°C$; used as a refrigerant. Also known as tetrafluoromethane. { 'kär·bən te·trə'flür₁īd }

carbon transducer [ENG] A transducer consisting of carbon granules in contact with a fixed electrode and a movable electrode, so that motion of the movable electrode varies the resistance of the granules. { 'kär·bən tranz'dü·sər }

carbonyl [ORG CHEM] A functional group found in organic compounds in which a carbon atom is doubly bonded to an oxygen atom ($-CO-$). Also known as carbonyl group. { 'kär·bə₁nil }

carbonylation [CHEM] Introduction of a carbonyl radical into a molecule. { kär₁bän·əl'ā·shən }

carbonyl bromide [ORG CHEM] $COBr_2$ A poisonous liquid boiling at 187.83°C; may be used by the military as a toxic suffocant. { 'kär·bə₁nil 'brō₁mīd }

carbonyl compound [ORG CHEM] A compound containing the carbonyl group (CO). { 'kär·bə₁nil ₁käm₁paùnd }

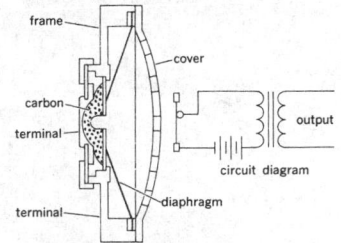

CARBON MICROPHONE

Sectional view and electrical circuit diagram of a carbon microphone.

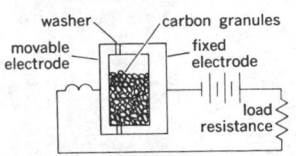

CARBON TRANSDUCER

Sectional view and electrical circuit diagram of a carbon transducer.

***N,N'*-carbonyldiimidazole** [ORG CHEM] $C_7H_6N_4O$ Crystals with a melting point of 115.5–116°C; hydrolyzed by water very quickly; used in the synthesis of peptides. { ¦en ¦en,prīm 'kar·bə,nil,dī·i'mid·ə,zōl }

carbonyl fluoride [ORG CHEM] COF_2 A colorless gas that is soluble in water; used in organic synthesis. { 'kär·bə,nil 'flur,īd }

carbonyl group *See* carbonyl. { 'kär·bə,nil ,grüp }

carbonylhemoglobin *See* carbonmonoxyhemoglobin. { 'kär·bə,nil¦hēm·ə'glō·bən }

carbonyl process [MET] **1.** A process in powder metallurgy for the production of iron, nickel, and iron-nickel alloy powders for magnetic applications. **2.** A process used in putting a metallic coating on molybdenum tungsten and other metals. { 'kär·bə,nil ,präs·əs }

carbophenothion [ORG CHEM] $C_{11}H_{16}ClO_2PS_3$ An amber liquid used to control pests on fruits, nuts, vegetables, and fiber crops. { ¦kär·bō¦fēn·ō'thī,än }

carboplatin [MED] A platinum coordination compound and anticancer agent used for advanced gynecologic malignancies, especially ovarian tumors, and for head and neck cancers and lung cancers. { ¦kär·bō'plat·ən }

carborane [ORG CHEM] **1.** Any of a class of compounds containing boron, carbon, and hydrogen. **2.** $B_{10}C_2H_{12}$ A specific member of the class. { 'kär·bə,rān }

carborundum [MATER] A manufactured crystalline material (silicon carbide), prepared by fusing coke and sand in an electric furnace; used as an abrasive in the grinding of low-tensile-strength materials, and as a semiconductor with a maximum operating temperature of 1300°C, to rectify and detect radio waves. { ,kär·bə'rən·dəm }

carbosand [MATER] Sand treated with an organic solution and then roasted; used as a spray to disperse oil slicks. { 'kär·bō,sand }

carboxin [ORG CHEM] $C_{12}H_{13}NO_2S$ An off-white solid with a melting point of 91.5–92.5°C; used to treat seeds of barley, oats, wheat, corn, and cotton for fungus diseases. Also known as DCMO. { 'kär'bäk·sən }

carboxy group [ORG CHEM] −COOH The functional group of carboxylic acid. Also known as carboxyl group. { kär'bäk·sē ,grüp }

carboxyhemoglobin *See* carbonmonoxyhemoglobin. { kär'¦bäk·sē¦hē·mə,glō·bən }

carboxylase [BIOCHEM] Any enzyme that catalyzes a carboxylation or decarboxylation reaction. { kär'bäk·sə,lās }

carboxylate anion [ORG CHEM] An anion with the general formula $(RCO_2)^-$, which is formed when the hydrogen attached to the carboxyl group of a carboxylic acid is removed. { kär'bäk·sə,lāt 'an,ī·ən }

carboxylation [ORG CHEM] Addition of a carboxyl group into a molecule. { kär,bäk·sə'lā·shən }

carboxyl group *See* carboxy group. { kär'bäk·səl ,grüp }

carboxylic [CHEM] Having chemical properties resembling those of carboxylic acid. { ¦kär,bäk¦sil·ik }

carboxylic acid [ORG CHEM] Any of a family of organic acids characterized by the presence of one or more carboxyl groups. { ¦kär,bäk¦sil·ik 'as·əd }

carboxyl terminal [BIOCHEM] The end of a polypeptide chain with a free carboxyl group. { kär'bäk·səl 'tər·mən·əl }

carboxymethylcellulose [ORG CHEM] An acid ether derivative of cellulose used as a sodium salt; a white, odorless, bulky solid used as a stabilizer and emulsifier; negatively charged resin used in ion-exchange chromatography as a cation exchanger. Also known as cellulose gum. { kär,bäk·sē¦meth·əl 'sel·yə,lōs }

carboxypeptidase [BIOCHEM] Any enzyme that catalyzes the hydrolysis of a peptide at the end containing the free carboxyl group. { kär,bäk·sē'pep·tə,dās }

carbro process [GRAPHICS] A photographic method of making carbon prints in which tanning of pigmented gelatin occurs in a special bleach bath, and the gelatin yields prints made on bromide paper. { 'kär,brō ,präs·əs }

carbuncle [MED] A bacterial infection of subcutaneous tissue caused by *Staphylococcus aureus;* multiple sinuses are created by tissue destruction. { 'kär,bəŋ·kəl }

carbureted water gas [MATER] Water gas that has been enriched by hydrocarbon gases of high fuel value. { 'kär·bə,red·əd 'wòd·ər ,gas }

carburetion [CHEM ENG] The process of enriching a gas by adding volatile carbon compounds, such as hydrocarbons, to it, as in the manufacture of carbureted water gas. [MECH ENG] The process of mixing fuel with air in a carburetor. { ,kär·bə'rā·shən }

carburetor [CHEM ENG] An apparatus for vaporizing, cracking, and enriching oils in the manufacture of carbureted water gas. [MECH ENG] A device that makes and controls the proportions and quantity of fuel-air mixture fed to a spark-ignition internal combustion engine. { 'kär·bə,red·ər }

carburetor icing [MECH ENG] The formation of ice in an engine carburetor as a consequence of expansive cooling and evaporation of gasoline. { 'kär·bə,red·ər ,ī·siŋ }

carburize [MET] To surface-harden steel by converting the outer layer of low-carbon steel to high-carbon steel by heating the steel above the transformation range in contact with a carbonaceous material. { 'kär·bə,rīz }

carbyne [CHEM] Elemental carbon in a triply bonded form. { 'kär,bīn }

carcenet [METEOROL] A very cold and violent gorge wind in the eastern Pyrenees (upper Aude valley). { kärs'nā }

carcerand [ORG CHEM] A macrocyclic compound capable of including organic guest molecules. { 'kär·sə·rənd }

Carcharhinidae [VERT ZOO] A large family of sharks belonging to the charcharinid group of galeoids, including the tiger sharks and blue sharks. { ,kär·kə'rīn·ə,dē }

Carchariidae [VERT ZOO] A family of shallow-water predatory sharks belonging to the isurid group of galeoids. { ,kär·kə'rī·ə,dē }

carcharodont [VERT ZOO] Possessing sharp, flat, triangular teeth with serrated margins, like those of the human-eating sharks. { kär'kar·ə,dänt }

carcinoembryonic antigen [IMMUNOL] A glycoprotein found in tissues of the fetal gut during the first two trimesters of pregnancy and in the peripheral blood of individuals with some forms of cancer, such as digestive-system or breast cancer. { ¦kärs·ən·ō,em·brē¦än·ik 'ant·i·jən }

carcinogen [MED] Any agent that incites development of a carcinoma or any other sort of malignancy. { kär'sin·ə·jən }

carcinogenesis [CELL MOL] The processes of tumor development. { ,kärs·ən·ō'jen·ə·səs }

carcinoid [MED] A potentially malignant tumor of the argentaffin cells of the stomach and intestine. { 'kärs·ən,ȯid }

carcinoid syndrome [MED] A complex of symptoms arising from the metastasis of a carcinoid tumor to the liver. { 'kärs·ən,ȯid 'sin,drōm }

carcinoma [MED] A malignant epithelial tumor. { ,kärs·ən'ō·mə }

carcinoma in situ [MED] A malignant tumor in the premetastatic stage, when cells are at the site of origin. { ,kärs·ən'ō·mə in 'si·chü }

carcinomatosis [MED] Metastasis of a primary carcinoma to many sites throughout the body. { ,kärs·ən,äm·ə'tō·səs }

Carcinus maenas [INV ZOO] A decapod crustacean commonly found on the coasts of northwest Europe and the northeast United States that feeds on invertebrates such as mollusks, polychaete worms, and other crustaceans, and periodically sheds its exoskeleton in order to grow. Also known as shore crab. { ¦kär·sən·əs 'mī·nəs }

carcinotron *See* backward-wave oscillator. { 'kärs·ən·ə,trän }

card [ELECTR] A printed circuit board or other arrangement of miniaturized components that can be plugged into a computer or peripheral device. { kärd }

cardamom *See* cardamon. { 'kärd·ə·məm }

cardamon [BOT] *Elettaria cardamomum.* A perennial herbaceous plant in the family Zingiberaceae; the seed of the plant is used as a spice. Also spelled cardamom. { 'kärd·ə·mən }

cardamon oil [MATER] A pale-yellow, combustible essential oil; insoluble in water, soluble in alcohol and ether; distilled from cardamon seeds, the chief known ingredients being terpinene, borneol, dipentene, limonene, and eucalyptol; used in flavoring and medicine. { 'kärd·ə·mən ,ȯil }

Cardan joint *See* Hooke's joint. { 'kär,dan ,jȯint }

Cardan motion [MECH ENG] The straight-line path followed by a moving centrode in a four-bar centrode linkage. { 'kär,dan 'mō·shən }

Cardan shaft [MECH ENG] A shaft with a universal joint at its end to accommodate a varying shaft angle. { 'kär,dan ,shaft }

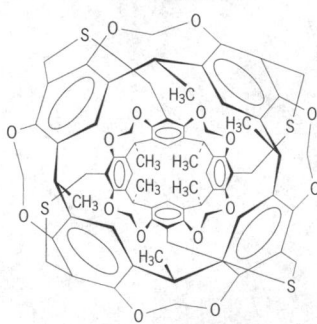

CARCERAND

Encapsulation of a molecular guest by a carcerand.

Cardan's suspension [DES ENG] An arrangement of rings in which a heavy body is mounted so that the body is fixed at one point; generally used in a gyroscope. { 'kär‚danz səs'pen·shən }

cardboard [MATER] A good quality of chemical pulp or rag pasteboard made by combining two or more webs of paper, either with or without paste, while still wet; used for signs, printed material, and high-quality boxes. { 'kärd‚bȯrd }

card cage [ELECTR] A rack built into a computer to hold printed circuit boards and allow them to be installed or removed easily. { 'kärd ‚kāj }

card dialer [COMMUN] A telephone in which a number can be dialed automatically and almost instantly by inserting a coded card for that number in a slot on the dialer. { 'kärd ‚dī·lər }

card-edge connector [ELEC] A connector that mates with printed-wiring leads running to the edge of a printed circuit board on one or both sides. Also known as edgeboard connector. { 'kärd ‚ej kə'nek·tər }

card holder [ELECTR] A U-shaped slot designed to hold the edge of a printed circuit board securely in a card cage. { 'kärd ‚hōl·dər }

cardia [ANAT] **1.** The orifice where the esophagus enters the stomach. **2.** The large, blind diverticulum of the stomach adjoining the orifice. [INV ZOO] Anterior enlargement of the ventriculus in some insects. { 'kärd·ē·ə }

cardiac [ANAT] **1.** Of, pertaining to, or situated near the heart. **2.** Of or pertaining to the cardia of the stomach. { 'kärd·ē‚ak }

cardiac arrest [MED] Cessation of the heartbeat. { 'kärd· ē‚ak ə'rest }

cardiac arrhythmia [MED] Any disturbance or irregularity of the heartbeat. { ‚kärd·ē‚ak ā'rith·mē·ə }

cardiac cirrhosis [MED] Progressive fibrosis of the liver due to prolonged venous blood retention as a result of prolonged and severe heart failure. { 'kärd·ē‚ak sə'rō·səs }

cardiac cycle [PHYSIO] The sequence of events in the heart between the start of one contraction and the start of the next. { 'kärd·ē‚ak ‚sī·kəl }

cardiac edema [MED] Accumulation of fluids throughout the body, as a function of cardiac failure. { 'kärd·ē‚ak i'dē·mə }

cardiac electrophysiology [PHYSIO] The science that is concerned with the mechanism, spread, and interpretation of the electric currents which arise within heart muscle tissue and initiate each heart muscle contraction. { 'kärd·ē‚ak ‚lek·trō‚fiz·ē'äl·ə·jē }

cardiac failure [MED] A complex of symptoms resulting from failure of the heart to pump sufficient quantities of blood. Also known as heart failure. { 'kärd·ē‚ak 'fāl·yər }

cardiac gland [ANAT] Any of the mucus-secreting, compound tubular structures near the esophagus or in the cardia of the stomach of vertebrates; capable of secreting digestive enzymes. { 'kärd·ē‚ak ‚gland }

cardiac glycoside [BIOCHEM] A class of naturally occurring glycosides that exhibit the ability to strengthen the contraction of heart muscles. { ‚kärd·ē‚ak 'glī·kə‚sīd }

cardiac input [PHYSIO] The amount of venous blood returned to the heart during a specified period of time. { 'kärd· ē‚ak 'in‚pùt }

cardiac loop [EMBRYO] The embryonic heart formed by bending and twisting of the cardiac tube. { 'kärd·ē‚ak 'lüp }

cardiac massage [MED] Rhythmic compression of the heart by a physician or other person in the effort to maintain effective circulation following heart failure. { 'kärd·ē‚ak mə'säzh }

cardiac murmur [MED] Any adventitious sound heard in the region of the heart. Also known as heart murmur. { 'kärd· ē‚ak ‚mər·mər }

cardiac muscle [HISTOL] The principal tissue of the vertebrate heart; composed of a syncytium of striated muscle fibers. { 'kärd·ē‚ak 'məs·əl }

cardiac output [PHYSIO] The total blood flow from the heart during a specified period of time. { 'kärd·ē‚ak 'aùt‚pùt }

cardiac pacemaker See pacemaker. { 'kärd·ē‚ak 'pā‚smā· kər }

cardiac plexus [NEURO] A network of visceral nerves situated at the base of the heart; contains both sympathetic and vagal nerve fibers. { 'kärd·ē‚ak 'plek·səs }

cardiac sphincter [ANAT] The muscular ring at the orifice between the esophagus and stomach. { 'kärd·ē‚ak 'sfiŋk·tər }

cardiac tamponade [MED] Cardiac compression caused by an accumulation of fluid within the pericardium. { 'kärd·ē‚ak ‚tam·pə‚nād }

cardiac valve [ANAT] Any of the structures located within the orifices of the heart that maintain unidirectional blood flow. { 'kärd·ē‚ak 'valv }

cardiectomy [MED] Excision of the cardiac end of the stomach. { ‚kärd·ē'ek·tə·mē }

cardinal heading [NAV] A heading in the direction of any of the cardinal points of the compass. { 'kärd·nəl 'hed·iŋ }

cardinal measurement See interval measurement. { 'kärd· nəl 'mezh·ər·mənt }

cardinal number [MATH] The number of members of a set; usually taken as a particular well-ordered set representative of the class of all sets which are in one-to-one correspondence with one another. { 'kärd·nəl 'nəm·bər }

cardinal point [GEOD] Any of the four principal directions: north, east, south, or west of a compass. [OPTICS] Any one of six points in an optical system, namely, the two principal points, two nodal points, and two focal points. Also known as Gauss point. { 'kärd·nəl ‚pȯint }

cardinal point effect [ELECTR] The increased intensity of a line or group of returns on the radarscope occurring when the radar beam is perpendicular to the rectangular surface of a line or group of similarly aligned features in the ground pattern. { 'kärd·nəl ‚pȯint i'fekt }

cardinal system [NAV] In marine operations, a buoyage system in which the buoys are assigned distinctive shape, color, and number in accordance with location relative to the nearest obstruction and pertinent to the cardinal points. { 'kärd·nəl ‚sis·təm }

cardinal teeth [INV ZOO] Ridges and grooves on the inner surfaces of both valves of a bivalve mollusk near the anterior end of the hinge. { 'kärd·nəl 'tēth }

cardinal vein [EMBRYO] Any of four veins in the vertebrate embryo which run along each side of the vertebral column; the paired veins on each side discharge blood to the heart through the duct of Cuvier. { 'kärd·nəl ‚vān }

cardinal winds [METEOROL] Winds from the four cardinal points of the compass, that is, north, east, south, and west winds. { 'kärd·nəl ‚winz }

carding [TEXT] Straightening or smoothing of raw fibers in a parallel fashion, with a carding machine. { 'kärd·iŋ }

carding machine [TEXT] A machine for disentangling fibers before they are spun. { 'kärd·iŋ mə'shēn }

Cardiobacterium [MICROBIO] A genus of gram-negative, rod-shaped bacteria of uncertain affiliation; facultative anaerobes that ferment fructose, glucose, mannose, sorbitol, and sucrose; the single species causes endocarditis in humans. { ‚kärd·ē·ō‚bak'tir·ē·əm }

cardioblast [INV ZOO] Any of certain early embryonic cells in insects from which the heart develops. { 'kärd·ē·ə‚blast }

cardiocirculatory See cardiovascular. { ‚kärd·ē·ō'sər·kyə· lə‚tȯr·ē }

cardiodynamics [PHYSIO] The dynamics of the heart's action in pumping blood. { ‚kärd·ē·ō·dī'nam·iks }

cardiogenic plate [EMBRYO] An area of splanchnic mesoderm in the early mammalian embryo from which the heart develops. { ‚kärd·ē·ō‚jen·ik 'plāt }

cardiogenic shock [MED] Shock due to inadequate arterial blood flow following left ventricular failure or pulmonary embolism. { ‚kärd·ē·ō‚jen·ik 'shäk }

cardiography [MED] Analysis of heart movements in the cardiac cycle by means of electronic instruments, especially by tracings. { ‚kärd·ē'äg·rə·fē }

cardioid [MATH] A heart-shaped curve generated by a point of a circle that rolls without slipping on a fixed circle of the same diameter. { 'kärd·ē‚ȯid }

cardioid condenser [OPTICS] A substage condenser that cuts off the direct light and allows only the light diffracted or dispersed from the object to enter the microscope; used in dark-field microscopes. { 'kärd·ē‚ȯid kən'den·sər }

cardioid microphone [ENG ACOUS] A microphone having a heart-shaped, or cardioid, response pattern, so it has nearly uniform response for a range of about 180° in one direction and minimum response in the opposite direction. { 'kärd·ē‚ȯid 'mī·krə‚fōn }

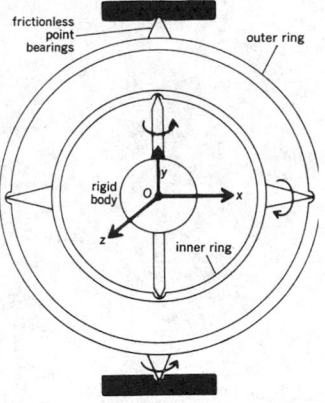

CARDAN'S SUSPENSION

Cardan's suspension. Point *O* of the rigid body is fixed in space while the body is free to rotate about any axis in space under no external moments.

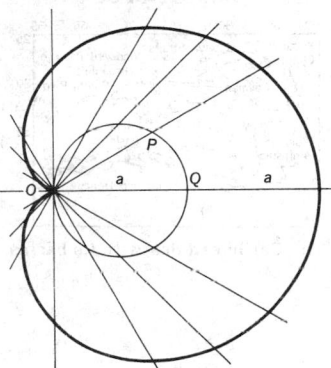

CARDIOID

Point-construction of a cardioid. The fixed circle is *OPQ* which has a diameter *a*. The cardioid is constructed by laying off, along every secant *OP* passing through the fixed point *O*, the distance *a* in both directions from *P*.

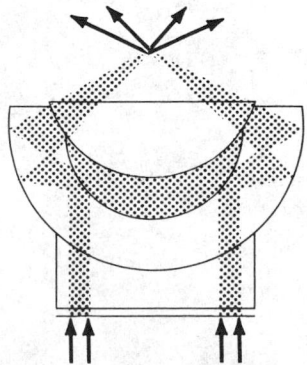

CARDIOID CONDENSER

Cardioid condenser, a dark-field device. Shaded meniscus area is air space; unshaded areas are portions of condenser through which passes the light, indicated by arrows and shaded paths. *(Photographic Service Department, Kodak Research Laboratory)*

cardioid pattern [ENG] Heart-shaped pattern obtained as the response or radiation characteristic of certain directional antennas, or as the response characteristic of certain types of microphones. { 'kärd·ē₁ȯid ,pad·ərn }

cardiolipin [BIOCHEM] A complex phospholipid found in the ether alcohol extract of powdered beef heart; mixed with leutin and cholesterol, it functions as the antigen in the Wassermann complement-fixation test for syphilis. Also known as diphosphatidyl glycerol. { ,kärd·ē·ō'lip·ən }

cardiology [MED] The study of the heart. { ,kärd·ē'äl·ə·jē }

cardiomyopathy See myocardiopathy. { ¦kärd·ē·ō₁mī'äp·ə·thē }

cardiopathy [MED] Any disorder or disease of the heart. { ,kärd·ē'äp·ə·thē }

cardiophobia [PSYCH] An abnormal fear of heart disease. { ,kärd·ē·ō'fō·bē·ə }

cardiopulmonary [PHYSIO] Pertaining to the heart and lungs. { ¦kärd·ē·ō'pu̇l·mə,ner·ē }

cardiopulmonary bypass [MED] A surgical technique for avoiding circulation of the blood through the heart and lungs by directing the flow from the entrance of the right atrium, conditioning it by mechanical means, and reentering it at the aorta. { ¦kärd·ē·ō'pu̇l·mə,ner·ē 'bī·pas }

cardiopulmonary resuscitation [MED] The simultaneous forced ventilation of the lungs and squeezing of the heart ventricles to sustain the flow of oxygenated blood throughout the system; often applied in cases of cardiac arrest. Abbreviated CPR. { ¦kärd·ē·ō'pu̇l·mə,ner·ē ri,səs·ə'tā·shən }

cardiorrhaphy [MED] Suturing of the heart muscle. { ,kärd·ē'ȯr·ə·fē }

cardioscope [MED] 1. An instrument for the examination or visualization of the interior of the cardiac chambers. 2. An instrument which, by means of a cathode-ray oscillograph, projects an electrocardiographic record on a luminous screen. { 'kärd·ē·ə,skōp }

cardiospasm [MED] Failure of the cardiac sphincter to relax; associated with spasm of the cardiac portion of the stomach and dilation of the esophagus. { 'kärd·ē·ō,spaz·əm }

cardiotachometer [MED] An electronic amplifier that times and records pulse rates of the heart. { ¦kärd·ē·ō·tə'käm·əd·ər }

cardiotomy [MED] Dissection or incision of the heart or the cardia of the stomach. { ,kärd·ē'äd·ə·mē }

cardiotonic drug [PHARM] Any agent, such as digitalis, that increases cardiac muscle tonus. { ¦kärd·ē·ō¦tän·ik 'drəg }

cardiovascular [PHYSIO] Pertaining to the heart and circulatory system. Also known as cardiocirculatory. { ¦kärd·ē·ō'vas·kyə·lər }

cardiovascular system [ANAT] Those structures, including the heart and blood vessels, which provide channels for the flow of blood. { ¦kärd·ē·ō'vas·kyə·lər ,sis·təm }

cardiovascular toxicity [MED] The adverse effects on the heart or blood systems which result from exposure to toxic chemicals. { ¦kärd·ē·ō¦vas·kyə·lər tak'sis·əd·ē }

Cardiovirus [VIROL] A genus of viruses of the family Picornaviridae; consists of strains of encephalomyocarditis virus and mouse encephalomyelitis. { 'kär·dē·ō,vī·rəs }

carditis [MED] Inflammation of the heart tissues. { kär'dīd·əs }

card key access [ENG] A physical security system in which doors are unlocked by placing a badge that contains magnetically coded information in proximity to a reading device; some systems also require the typing of this information on a keyboard. { 'kärd ,kē 'ak,ses }

card reader [ELECTR] A device that reads magnetic stripes on the back of an identification card. { 'kärd ,rēd·ər }

card slot [ELECTR] A groove where a printed circuit board fits into a card cage or backplane. { 'kärd ,slät }

car dump [MECH ENG] Any one of several devices for unloading industrial or railroad cars by rotating or tilting the car. { 'kär ,dəmp }

δ-3-carene [ORG CHEM] $C_{10}H_{16}$ A clear, colorless, combustible terpene liquid, stable to about 250°C; used as a solvent and in chemical synthesis. { ¦del·tə ¦thrē 'ka,rēn }

caret [GRAPHICS] A printed or written character having the form of an inverted V. { 'kar·ət }

Carettochelyidae [VERT ZOO] A family of reptiles in the order Chelonia containing only one species, the New Guinea

CARIBBEAN SEA

The Caribbean Sea with its basins.

plateless turtle (*Carettochelys insculpta*). { kə¦red·ō·kə'lī·ə,dē }

car ferry [NAV ARCH] A ferry that is specially designed to carry railroad cars, trucks, or passenger cars. { 'kär ,fer·ē }

car float [NAV ARCH] A barge with railroad tracks on its deck, used to carry railroad cars in harbors or inland waterways. { 'kär ,flōt }

car-following theory [ENG] A mathematical model of the interactions between motor vehicles in terms of relative speed, absolute speed, and separation. { 'kär,fäl·ə·wiŋ ,thē·ə·rē }

cargo boom [MECH ENG] A long spar extending from the mast of a derrick to support or guide objects lifted or suspended. { 'kär·gō ,büm }

cargo liner See cargo ship. { 'kär·gō ,līn·ər }

cargo mill [IND ENG] A sawmill equipped with docks so the product can be loaded directly onto ships. { 'kär·gō ,mil }

cargo ship [NAV ARCH] A power-driven ship employed exclusively in commercial transportation of commodities on the ocean and large inland bodies of water. Also known as cargo liner. { 'kär·gō ,ship }

cargo winch [MECH ENG] A motor-driven hoisting machine for cargo having a drum around which a chain or rope winds as the load is lifted. { 'kär·gō ,winch }

Cariamidae [VERT ZOO] The long-legged cariamas, a family of birds in the order Gruiformes. { ,kär·ē'am·ə,dē }

Caribbean Current [OCEANOGR] A water current flowing westward through the Caribbean Sea. { kar·ə'bē·ən 'kər·ənt }

Caribbean Sea [GEOGR] One of the largest and deepest enclosed basins in the world, surrounded by Central and South America and the West Indian island chains. { kar·ə'bē·ən 'sē }

Caribosireninae [VERT ZOO] A subfamily of trichechiform sirenean mammals in the family Dugongidae. { ,kar·ə,bäs·ə'ren·ə,nē }

Caridea [INV ZOO] A large section of decapod crustaceans in the suborder Natantia including many diverse forms of shrimps and prawns. { kə'rid·ē·ə }

caries [MED] 1. Bone decay. 2. Tooth decay. Also known as dental caries. { 'kar·ēz }

carillon [ENG] A musical instrument played from a keyboard with two or more full chromatic octaves of fine bells shaped for homogeneity of timbre. { 'kär·ə,län }

carina [BIOL] A ridge or a keel-shaped anatomical structure. See keel. { kə'rī·nə }

Carina [ASTRON] A constellation, right ascension 9 hours, declination 60°S. Abbreviated Car. Also known as Keel. { kə'rī·nə }

α Carinae See Canopus. { ¦al·fə kə'rī,nē }

Carina Nebula [ASTRON] A gaseous nebula near the star η Carinae in the Milky Way. { kə'ri·nə 'neb·yə·lə }

carinate [BIOL] Having a ridge or keel, as the breastbone of certain birds. { 'kar·ə,nāt }

carination [ARCHEO] A sharp break in the curve of the profile of a container, resulting in a projecting angle. { ,kar·ə'nā·shən }

Carinomidae [INV ZOO] A monogeneric family of littoral ribbonlike worms in the order Palaeonemertini. { ,kär·ə'näm·ə,dē }

Carinthian furnace [MET] A zinc distillation furnace with small, vertical retorts. [MIN ENG] A small reverberatory furnace with an inclined hearth, in which lead ore is treated by roasting and reaction, wood being the usual fuel. { kə'rin·thē·ən 'fər·nəs }

cariostatic [PHYSIO] Acting to halt bone or tooth decay. { ¦kar·ē·ə¦stad·ik }

Carius method [ANALY CHEM] A procedure used to analyze organic compounds for sulfur, halogens, and phosphorus that involves heating the sample with fuming nitric acid in a sealed tube. { 'kär·ē·əs ,meth·əd }

carling [NAV ARCH] A short girder or timber running fore and aft, used to support or stiffen a ship's deck, or to frame an opening in the deck where beams have been cut. { 'kär·liŋ }

Carlsbad law [CRYSTAL] A feldspar twin law in which the twinning axis is the *c* axis, the operation is rotation of 180°, and the contact surface is parallel to the side pinacoid. { 'kärlz,bad ,lȯ }

Carlsbad turn [CRYSTAL] A twin crystal in the monoclinic system with the vertical axis as the turning axis. { 'kärlz,bad ,tərn }

Carme [ASTRON] A small satellite of Jupiter with a diameter of about 19 miles (31 kilometers), orbiting with retrograde motion at a mean distance of 1.4×10^7 miles (2.3×10^7 kilometers). Also known as Jupiter XI. { 'kär·mä }

carminic acid [ORG CHEM] $C_{22}H_{20}O_{13}$ A glucosidal hydroxyanthrapurin that is derived from cochineal; a red crystalline dye used as a stain for biological materials. Also known as cochinilin. { kär'min·ik 'as·əd }

carminite [MINERAL] $PbFe_2(AsO_4I)_2(OH)_2$ A carmine to tile-red mineral consisting of a basic arsenate of lead and iron. { 'kär·mə,nīt }

carnallite [MINERAL] $KMgCl_3 \cdot 6H_2O$ A milky-white or reddish mineral that crystallizes in the orthorhombic system and occurs in deliquescent masses; it is valuable as an ore of potassium. { 'kärn·əl,īt }

carnassial [ANAT] Of or pertaining to molar or premolar teeth specialized for cutting and shearing. { kär'nas·ē·əl }

carnation ringspot virus [VIROL] The type species of the plant-virus genus *Dianthovirus*; symptoms of infection include leaf mottling and ringspotting, plant stunting and distortion, and flower distortion. Abbreviated CRSV. { kär'nā·shən 'riŋ,spät ,vī·rəs }

carnauba wax [MATER] The hardest natural wax, having a melting point of 85°C, exuded from the leaves of the carnauba palm (*Carnauba cerifera*); used for insulating purposes and in making candles, shoe polish, high-luster wax, varnishes, phonograph records, and surface coating of automobiles. Also known as Brazil wax. { kär'nó·bə 'waks }

carnaubic acid [ORG CHEM] $C_{24}H_{48}O_2$ An acid found in carnauba wax and beef kidney. { kär'nó·bik 'as·əd }

carnegieite [MINERAL] $NaAlSiO_4$ An artificial mineral similar to feldspar; it is triclinic at low temperatures, isometric at elevated temperatures. { 'kär·nə·gē,īt }

Carnian [GEOL] Lower Upper Triassic geologic time. Also spelled Karnian. { 'kärn·ē·ən }

carnitine [BIOCHEM] $C_7H_{15}NO_3$ α-Amino-β-hydroxybutyric acid trimethylbetaine; a constituent of striated muscle and liver, identical with vitamin B_T. { 'kär·nə,tēn }

Carnivora [VERT ZOO] A large order of placental mammals, including dogs, bears, and cats, that is primarily adapted for predation as evidenced by dentition and jaw articulation. { kär'niv·ə·rə }

carnivorous [BIOL] Eating flesh or, as in plants, subsisting on nutrients obtained from animal protoplasm. { kär'niv·ə·rəs }

carnivorous plant See insectivorous plant. { kär'niv·ə·rəs 'plant }

Carnosauria [PALEON] A group of large, predacious saurischian dinosaurs in the suborder Theropoda having short necks and large heads. { ,kär·nə'sór·ē·ə }

carnosine [BIOCHEM] $C_9H_{14}N_4O_3$ A colorless, crystalline dipeptide occurring in the muscle tissue of vertebrates. { 'kär·nə,sēn }

Carnot-Clausius equation [THERMO] For any system executing a closed cycle of reversible changes, the integral over the cycle of the infinitesimal amount of heat transferred to the system divided by its temperature equals 0. Also known as Clausius theorem. { kär',nöt 'klóz·ē·əs i,kwä·zhən }

Carnot cycle [THERMO] A hypothetical cycle consisting of four reversible processes in succession: an isothermal expansion and heat addition, an isentropic expansion, an isothermal compression and heat rejection process, and an isentropic compression. { kär'nō ,sī·kəl }

Carnot efficiency [THERMO] The efficiency of a Carnot engine receiving heat at a temperature absolute T_1 and giving it up at a lower temperature absolute T_2; equal to $(T_1 - T_2)/T_1$. { kär'nō i'fish·ən·sē }

Carnot engine [MECH ENG] An ideal, frictionless engine which operates in a Carnot cycle. { kär'nō 'en·jən }

carnotite [MINERAL] $K(UO_2)_2(VO_4)_2 \cdot nH_2O$ A canary-yellow, fine-grained hydrous vanadate of potassium and uranium having monoclinic microcrystals; an ore of radium and uranium. { 'kär·nə,tīt }

Carnot number [THERMO] A property of two heat sinks, equal to the Carnot efficiency of an engine operating between them. { kär'nō ,nəm·bər }

Carnot's reagent [CHEM] A solution of sodium bismuth thiosulfate in alcohol used for determining potassium. { kär'noz rē'ā·jənt }

Carnot's theorem [THERMO] **1.** The theorem that all Carnot engines operating between two given temperatures have the same efficiency, and no cyclic heat engine operating between two given temperatures is more efficient than a Carnot engine. **2.** The theorem that any system has two properties, the thermodynamic temperature T and the entropy S, such that the amount of heat exchanged in an infinitesimal reversible process is given by $dQ = TdS$; the thermodynamic temperature is a strictly increasing function of the empirical temperature measured on an arbitrary scale. { kär'noz 'thir·əm }

Carnoy's solution [MATER] A tissue fixative composed of mercuric salts and acetic acids; used where penetration of hard objects such as seeds is required. { 'kär,nóiz sə'lü·shən }

caroba See carob wood. { kə'rō·bə }

carob wood [MATER] The wood from a large Brazilian tree, *Jacaranda copaia*. Also known as caroba; jacaranda. { 'kar·əb ,wúd }

Carolina Bays [GEOGR] Shallow, marshy, often ovate depressions on the coastal plain of the mideastern and southeastern United States of unknown origin. { ,kar·ə'lī·nə 'bāz }

Carolinian life zone [ECOL] A zone comprising the climate and biotic communities of the oak savannas of eastern North America. { ,kar·ə,lin·ē·ən 'līf ,zōn }

Caro's acid [INORG CHEM] H_2SO_5 A white solid melting at about 45°C, formed during the acid hydrolysis of peroxydisulfates. { 'kä·roz 'as·əd }

carotenase [BIOCHEM] An enzyme that effects the hydrolysis of carotenoid compounds, used in bleaching of flour. { kə'rät·ən,ās }

carotene [BIOCHEM] $C_{40}H_{56}$ Any of several red, crystalline, carotenoid hydrocarbon pigments occurring widely in nature, convertible in the animal body to vitamin A, and characterized by preferential solubility in petroleum ether. Also known as carotin. { 'kar·ə,tēn }

carotenemia [MED] The presence of carotene in the blood; may cause yellowing of the skin. { ,ka'rä·tə'nēm·ē·ə }

carotenoid [BIOCHEM] A class of labile, easily oxidizable, yellow, orange, red, or purple pigments that are widely distributed in plants and animals and are preferentially soluble in fats and fat solvents. { kə'rät·ən,óid }

carotenol See xanthophyll. { kə'rät·ən,ól }

carotid artery [ANAT] Either of the two principal arteries on both sides of the neck that supply blood to the head and neck. Also known as common carotid artery. { kə'räd·əd 'ärd·ə·rē }

carotid body [ANAT] Either of two chemoreceptors sensitive to changes in blood chemistry which lie near the bifurcations of the carotid arteries. Also known as glomus caroticum. { kə'räd·əd ,bäd·ē }

carotid ganglion [NEURO] A group of nerve cell bodies associated with each carotid artery. { kə'räd·əd 'gaŋ·glē,än }

carotid sinus [ANAT] An enlargement at the bifurcation of each carotid artery that is supplied with sensory nerve endings and plays a role in reflex control of blood pressure. { kə'räd·əd 'sī·nəs }

carotin See carotene. { 'kar·ə,tin }

carotol [BIOCHEM] $C_{15}H_{25}OH$ A sesquiterpenoid alcohol in carrots. { 'kar·ə,tól }

carousel [MECH ENG] A rotating transport system that transfers and presents workpieces for loading and unloading by a robot or other machine. { ,kar·ə'sel }

carp [VERT ZOO] The common name for a number of freshwater, cypriniform fishes in the family Cyprinidae, characterized by soft fins, pharyngeal teeth, and a suckerlike mouth. { kärp }

carpal tunnel syndrome [MED] A condition caused by compression of the median nerve in the passage between the wrist and carpal bones; characterized by nocturnal pain, numbness, and tingling in the hand. { 'kär·pəl ,tən·əl 'sin,dróm }

carpel [BOT] The basic specialized leaf of the female reproductive structure in angiosperms; a megasporophyll. { 'kär·pəl }

carpenter's level [DES ENG] A bar, usually of aluminum or wood, containing a spirit level. { 'kär·pən·tərz ,lev·əl }

carpenter stopper [NAV ARCH] A device, generally made from high-carbon steel, that allows a temporary grip on a wire rope up to its breaking strength without injury or slippage. { 'kär·pən·tər ,stäp·ər }

carpet-bombing [ORD] The laying down of bombs in a

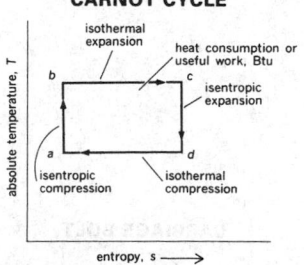

CARNOT CYCLE

Carnot cycle, comprising the four processes *b-c*, *c-d*, *d-a*, and *a-b*.

creeping pattern to cover the area as with a carpet. { 'kär·pət ,bäm·iŋ }

carpholite [MINERAL] MnAl₂Si₂O₆(OH)₄ A straw-yellow fibrous mineral consisting of a hydrous aluminum manganese silicate occurring in tufts; specific gravity is 2.93. { 'kär·fə,līt }

carphosiderite [MINERAL] A yellow mineral consisting of a basic hydrous iron sulfate occurring in masses and crusts. { ,kär·fō'sīd·ə,rīt }

car pincher [MIN ENG] A worker in a mine who uses a pinch bar to position cars under loading chutes. { 'kär ,pinch·ər }

carpincho [MATER] Processed capybara skin, noted for its elastic properties. { kär'pin·chō }

Carpinus betulus *See* European hornbeam. { kär,pēn·əs 'bech·yə·ləs }

Carpinus caroliniana *See* American hornbeam. { kär,pē·nəs ,ka·rə,lin·ē'än·ə }

carpogonium [BOT] The basal, egg-bearing portion of the female reproductive organ in some thallophytes, especially red algae. { ,kär·pə'gō·nē·əm }

Carpoidea [PALEON] Former designation for a class of extinct homalozoan echinoderms. { kär'pòid·ē·ə }

carpoids [PALEON] An assemblage of three classes of enigmatic, rare Paleozoic echinoderms formerly grouped together as the class Carpoidea. { 'kär,pòidz }

carpology [BOT] The study of the morphology of fruit and seeds. { kär'päl·ə·jē }

carpophagous [ZOO] Feeding on fruits. { kär'pa·fə·gəs }

carpophore [BOT] The portion of a flower receptacle that extends between and attaches to the carpels. [MYCOL] The stalk of a fruiting body in fungi. { 'kär·pə,fòr }

carpophyte [BOT] A thallophyte that forms a sporocarp following fertilization. { 'kär·pə,fīt }

carposporangium [BOT] In red algae, a sporangium that forms the cystocarp and contains carpospores. { ,kär·pō·spə'ran·jē·əm }

carpospore [BOT] In red algae, a diploid spore produced terminally by a gonimoblast, giving rise to the diploid tetrasporic plant. { 'kär·pə,spòr }

carpus [ANAT] **1.** The wrist in humans or the corresponding part in other vertebrates. **2.** The eight bones of the human wrist. [INV ZOO] The fifth segment from the base of a generalized crustacean appendage. { 'kär·pəs }

carrageen [BOT] *Chondrus crispus.* A cartilaginous red algae harvested in the northern Atlantic as a source of carrageenan. Also known as Irish moss; pearl moss. { 'kar·ə,gēn }

carrageenan [ORG CHEM] A polysaccharide derived from the red seaweed (Rhodophyceae) and used chiefly as an emulsifying, gelling, and stabilizing agent and as a viscosity builder in foods, cosmetics, and pharmaceuticals. Also spelled carrageenin. { ,kar·ē·ə'gē·nən }

carrageenin *See* carrageenan. { ,kar·ə'gē·nən }

Carrara marble [PETR] All marble quarried near Carrara, Italy, having a prevailing white to bluish color, or white with blue veins. { kə'rä·rə 'mär·bəl }

car retarder [ENG] A device located along the track to reduce or control the velocity of railroad or mine cars. { 'kär ri'tärd·ər }

carriage [ENG] **1.** A device that moves in a predetermined path in a machine and carries some other part, such as a recorder head. **2.** A mechanism designed to hold a paper in the active portion of a printing or typing device, for example, a typewriter carriage. [GRAPHICS] The component in a unitized microform reader or reproduction device that holds the microform. [MECH ENG] A structure on an industrial truck or stacker that supports forks or other attached equipment and travels vertically within the mast. [ORD] Mobile or fixed support for a cannon; sometimes includes the elevating and traversing mechanisms. { 'kar·ij }

carriage bolt [DES ENG] A round-head type of bolt with a square neck, used with a nut as a through bolt. { 'kar·ij ,bōlt }

carriage return [COMPUT SCI] The operation that causes the next character to be printed at the extreme left margin, and usually advances to the next line at the same time. { 'kar·ij ri'tərn }

carriage stop [MECH ENG] A device added to the outer way of a lathe bed for accurately spacing grooves, turning multiple

diameters and lengths, and cutting off pieces of specified thickness. { 'kar·ij ,stäp }

carriage tape *See* control tape. { 'kar·ij ,tāp }

carrier [CHEM] A substance that, when associated with a trace of another substance, will carry the trace with it through a chemical or physical process. [COMMUN] **1.** The radio wave produced by a transmitter when there is no modulating signal, or any other wave, recurring series of pulses, or direct current capable of being modulated. Also known as carrier wave; signal carrier. **2.** A wave generated locally at a receiver that, when combined with the sidebands of a suppressed-carrier transmission in a suitable detector, produces the modulating wave. **3.** *See* carrier system. [GEN] An individual who is heterozygous for a recessive gene. [IMMUNOL] A protein to which a hapten becomes attached, thereby rendering the hapten immunogenic. [MECH ENG] Any machine for transporting materials or people. [MED] A person who harbors and eliminates an infectious agent and so transmits it to others, but who may not show signs of the disease. [NAV ARCH] *See* aircraft carrier. { 'kar·ē·ər }

carrier amplifier [ELECTR] A direct-current amplifier in which the dc input signal is filtered by a low-pass filter, then used to modulate a carrier so it can be amplified conventionally as an alternating-current signal; the amplified dc output is obtained by rectifying and filtering the rectified carrier signal. { 'kar·ē·ər ,am·plə,fī·ər }

carrier amplitude regulation [COMMUN] Change in amplitude of the carrier wave in an amplitude-modulated transmitter when modulation is applied under conditions of symmetrical modulation. { 'kar·ē·ər 'am·plə,tüd reg·yə'lā·shən }

carrier beat [COMMUN] An undesirable heterodyne of facsimile signals, each synchronous with a different stable reference oscillator, causing a pattern in received copy. { 'kar·ē·ər ,bēt }

carrier channel [COMMUN] The equipment and lines that make up a complete carrier-current circuit between two or more points. { 'kar·ē·ər ,chan·əl }

carrier chrominance signal *See* chrominance signal. { 'kar·ē·ər 'krō·mə·nəns ,sig·nəl }

carrier-controlled approach system [NAV] An aircraft-carrier radar system providing information by which aircraft approaches may be directed by radio. { 'kar·ē·ər kən'trōld ə'prōch ,sis·təm }

carrier culture [VIROL] A cell culture exhibiting a persistent infection; only a small fraction of the cell population is infected, but the viruses released when the infected cells die infect a small number of other cells. { 'kar·ē·ər ,kəl·chər }

carrier current [COMMUN] A higher-frequency alternating current superimposed on ordinary telephone, telegraph, and power-line frequencies for communication and control purposes. { 'kar·ē·ər ,kər·ənt }

carrier density [SOLID STATE] The density of electrons and holes in a semiconductor. { 'kar·ē·ər ,den·səd·ē }

carrier detect [COMPUT SCI] A signal sent by a modem to a computer or a terminal to indicate that it is receiving a character. { 'kar·ē·ər di,tekt }

carrier frequency [COMMUN] The frequency generated by an unmodulated radio, radar, carrier communication, or other transmitter, or the average frequency of the emitted wave when modulated by a symmetrical signal. Also known as center frequency; resting frequency. { 'kar·ē·ər ,frē·kwən·sē }

carrier gas [ANALY CHEM] In gas chromatography, a gas used as an eluant for extracting the sample from the column as the gas passes through. Also known as eluant gas. [MET] The gas used in thermal spraying which transmits powder from the feeder to the spray gun. { 'kar·ē·ər ,gas }

carrier isolating choke coil [ELECTROMAG] Inductor inserted in series with a line on which carrier energy is applied to impede the flow of carrier energy beyond that point. { 'kar·ē·ər ,ī·sə,lād·iŋ 'chōk ,kòil }

carrier leak [COMMUN] Carrier remaining after carrier suppression in a suppressed-carrier transmission system. { 'kar·ē·ər ,lēk }

carrier level [COMMUN] The strength or level of an unmodulated carrier signal at a particular point in a radio system, expressed in decibels in relation to some reference level. { 'kar·ē·ər ,lev·əl }

carrier line [ELEC] Any transmission line used for multiple-channel carrier communication. { 'kar·ē·ər ,līn }

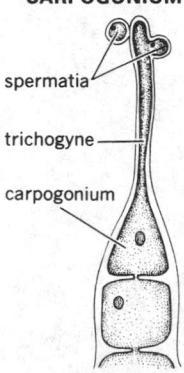

CARPOGONIUM

spermatia

trichogyne

carpogonium

Filament with terminal carpogonium (containing an egg) and bearing an elongated trichogyne. *(From H. J. Fuller and O. Tippo, College Botany, rev. ed., Holt, 1954)*

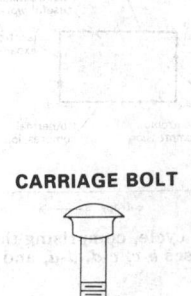

CARRIAGE BOLT

A round-head carriage bolt.

carrier loading [ELECTROMAG] The addition of lumped inductances to the cable section of a transmission line specifically designed for carrier transmission; it serves to minimize impedance mismatch between cable and open wire and to reduce the cable attenuation. { 'kar·ē·ər ¦lōd·iŋ }

carrier mobility [SOLID STATE] The average drift velocity of carriers per unit electric field in a homogeneous semiconductor; the mobility of electrons is usually different from that of holes. { 'kar·ē·ər mō'bil·əd·ē }

carrier molecule [IMMUNOL] An immunogenic molecule such as a foreign protein to which a hapten is coupled, thus enabling the hapten to induce an immune response. { 'kar·ē·ər 'mäl·ə,kyül }

carrier noise [COMMUN] Noise produced by undesired variation of a radio-frequency signal in the absence of any intended modulation. Also known as residual modulation. { 'kar·ē·ər ,nȯiz }

carrier pipe [ENG] Pipe used to carry or conduct fluids, as contrasted with an exterior protective or casing pipe. { 'kar·ē·ər ,pīp }

carrier power output rating [COMMUN] Power available at the output terminals of a transmitter when the output terminals are connected to the normal-load circuit or to a circuit equivalent thereto. { 'kar·ē·ər ¦pau̇·ər 'au̇t,pu̇t ,rād·iŋ }

carrier protein [CELL MOL] A membrane protein that transports a specific solute across the cell membrane by binding to the solute on one side of the membrane and then releasing it on the other. { 'kar·ē·ər ,prō,tēn }

carrier repeater [ELECTR] Equipment designed to raise carrier signal levels to such a value that they may traverse a succeeding line section at such amplitude as to preserve an adequate signal-to-noise ratio; while the heart of a repeater is the amplifier, necessary adjuncts are filters, equalizers, level controls, and so on, depending upon the operating methods. { 'kar·ē·ər ri'pēd·ər }

carrier rig [PETRO ENG] A self-propelled unit designed for servicing oil and gas wells; components include masts, hoists, engines, and other auxiliary equipment. { 'kar·ē·ər ,rig }

carrier rocket [AERO ENG] A rocket vehicle used to carry something, as the carrier rocket of the first artificial earth satellite. { 'kar·ē·ər ,räk·ət }

carrier sense multiple access with collision detection See CSMA/CD. { 'kar·ē·ər ¦sens ¦məl·tə·pəl 'ak,ses with kə'lizh·ən di,tek·shən }

carrier shift [COMMUN] **1.** Transmission of radio-teletypewriter messages by shifting the carrier frequency in one direction for a marking signal and in the opposite direction for a spacing signal. **2.** Condition resulting from imperfect modulation whereby the positive and negative excursions of the envelope pattern are unequal, thus effecting a change in the power associated with the carrier. { 'kar·ē·ər ,shift }

carrier signaling [COMMUN] Method by which busy signals, ringing, or dial signaling relays are operated by the transmission of a carrier-frequency tone. { 'kar·ē·ər ,sig·nəl·iŋ }

carrier suppression [COMMUN] **1.** Suppression of the carrier frequency after conventional modulation at the transmitter, with reinsertion of the carrier at the receiving end before demodulation. **2.** Suppression of the carrier when there is no modulation signal to be transmitted; used on ships to reduce interference between transmitters. { 'kar·ē·ər sə'presh·ən }

carrier swing [COMMUN] The total deviation of a frequency-modulated or phase-modulated wave from the lowest instantaneous frequency to the highest instantaneous frequency. { 'kar·ē·ər ,swiŋ }

carrier system [COMMUN] A system permitting a number of simultaneous, independent communications over the same circuit. Also known as carrier. { 'kar·ē·ər ,sis·təm }

carrier telegraphy [COMMUN] Telegraphy in which a single-frequency carrier wave is modulated by the transmitting apparatus for transmission over wire lines. { 'kar·ē·ər tə'leg·rə·fē }

carrier telephony [COMMUN] Telephony in which a single-frequency carrier wave is modulated by a voice-frequency signal for transmission over wire lines. { 'kar·ē·ər tə'lef·ə·nē }

carrier terminal [ELECTR] Apparatus at one end of a carrier transmission system, whereby the processes of modulation, demodulation, filtering, amplification, and associated functions are effected. { 'kar·ē·ər ¦tərm·ən·əl }

carrier-to-noise ratio [COMMUN] The ratio of the magnitude of the carrier to that of the noise after specified band limiting and before any nonlinear process such as amplitude limiting and detection. { 'kar·ē·ər tə 'nȯiz ,rā·shō }

carrier transfer filters [ELECTR] Filters arranged as a carrier-frequency crossover or bridge between two transmission circuits. { 'kar·ē·ər ¦tranz·fər ,fil·tərz }

carrier transmission [COMMUN] Transmission in which the transmitted electric wave is a wave resulting from the modulation of a single-frequency wave by a modulating wave. { 'kar·ē·ər tranz'mish·ən }

carrier wave See carrier. { 'kar·ē·ər ,wāv }

Carrington rotation number [ASTRON] A method of numbering rotations of the sun based on a mean rotation period of sunspots of 27.2753 days, and starting with rotation number 1 on November 9, 1853. { 'kar·iŋ·tən rō'tā·shən ,nəm·bər }

Carrion's disease [MED] A bacterial infection of humans endemic in the Andes which is caused by *Bartonella bacilliformis* and attacks red blood cells and blood-forming organs. Also known as bartonellosis. { 'kar·ē¦ȯnz di,zēz }

carron oil [MATER] A mixture comprising equal volumes of linseed oil and lime water which is used for the relief of burns. Also known as lime liniment. { 'kar·ən ,ȯil }

carrot [BOT] *Daucus carota*. A biennial umbellifer of the order Umbellales with a yellow or orange-red edible root. { 'kar·ət }

carrot oil [MATER] A light-yellow oil distilled from the seeds of the carrot (*Daucus carota*), and containing carotene, pinene, palmitic acid, limonene, and butyric or isobutyric acid; used as a flavoring. { 'kar·ət ,ȯil }

carrousel [IND ENG] In an assembly-line operation, a conveyor that moves objects in a complete circuit on a horizontal plane. { ka·rə'sel }

carry [MATH] An arithmetic operation that occurs in the course of addition when the sum of the digits in a given position equals or exceeds the base of the number system; a multiple m of the base is subtracted from this sum so that the remainder is less than the base, and the number m is then added to the next-higher-order digit. { 'kar·ē }

carry-complete signal [COMPUT SCI] A signal generated by a digital parallel adder, indicating that all carries from an adding operation have been generated and propagated, and that the addition operation is completed. { ¦kar·ē kəm¦plēt ,sig·nəl }

carry flag [COMPUT SCI] A flip-flop circuit which indicates overflow in arithmetic operations. { 'kar·ē ,flag }

carrying capacity [ECOL] The maximum population size that the environment can support without deterioration. [ELEC] The maximum amount of current or power that can be safely handled by a wire or other component. { 'kar·ō·iŋ kə'pas·əd·ē }

carry lookahead [COMPUT SCI] A circuit which allows low-order carries to ripple through all the way to the highest-order bit to output a completed sum. { 'kar·ē 'lu̇k·ə,hed }

carry-over [CHEM ENG] Unwanted liquid or solid material carried by the overhead effluent from a fractionating column, absorber, or reaction vessel. [HYD] The portion of the stream flow during any month or year derived from precipitation in previous months or years. { 'kar·ē ,ō·vər }

carry-save adder [COMPUT SCI] A device for the rapid addition of three operands; consists of a sequence of full adders, in which one of the operands is entered in the carry inputs, and the carry outputs, instead of feeding the carry inputs of the following full adders, form a second output word which is then added to the ordinary output in a two-operand adder to form the final sum. { ¦kar·ē 'sāv 'ad·ər }

carry signal [COMPUT SCI] A signal produced in a computer when the sum of two digits in the same column equals or exceeds the base of the number system in use or when the difference between two digits is less than zero. { 'kar·ē ,sig·nəl }

carry time [COMPUT SCI] The time needed to transfer all carry digits to the next higher column. { 'kar·ē ,tīm }

car shaker [MECH ENG] A device consisting of a heavy yoke on an open-top car's sides that actively vibrates and rapidly discharges a load, such as coal, gravel, or sand, when an unbalanced pulley attached to the yoke is rotated fast. { 'kär ,shāk·ər }

carsickness [MED] Motion sickness resulting from acceleratory movements of a train or automobile. { 'kär,sik·nəs }

car stop [ENG] An appliance used to arrest the movement of a mine or railroad car. { 'kär ˌstäp }

Carter chart [ELECTROMAG] An Argand diagram of the complex reflection coefficient of a waveguide junction on which are drawn lines of constant magnitude and phase of the impedance. { 'kärd·ər ˌchärt }

Carterinacea [INV ZOO] A monogeneric superfamily of marine, benthic foraminiferans in the suborder Rotaliina characterized by a test with monocrystal calcite spicules in a granular groundmass. { ˌkärd·ər·ə'näs·ē·ə }

Carter's theorem [RELAT] Theorem proving that the only stationary, charged black hole solutions to the equations of general relativity are the Kerr-Newman solutions. { 'kärd·ərz ˌthir·əm }

Cartesian anthropometry [ANTHRO] The measurement of the living human head in three dimensions. { kär,tē·zhən ˌan·thrə'päm·ə·trē }

Cartesian axis [MATH] One of a set of mutually perpendicular lines which all pass through a single point, used to define a Cartesian coordinate system; the value of one of the coordinates on the axis is equal to the directed distance from the intersection of axes, while the values of the other coordinates vanish. { kär'tē·zhən 'ak·səs }

Cartesian-coordinate robot [CONT SYS] A robot having orthogonal, sliding joints and supported by a nonrotary base as the axis. { kär'tē·zhən kō̇'ȯrd·ən·ət 'rō,bät }

Cartesian coordinates [MATH] **1.** The set of numbers which locate a point in space with respect to a collection of mutually perpendicular axes. **2.** *See* rectangular coordinates. { kär'tē·zhən kō'ȯrd·nəts }

Cartesian coordinate system [MATH] A coordinate system in *n* dimensions where *n* is any integer made by using *n* number axes which intersect each other at right angles at an origin, enabling any point within that rectangular space to be identified by the distances from the *n* lines. Also known as rectangular Cartesian coordinate system. { kär'tē·zhən kō'ȯrd·nət ˌsis·təm }

Cartesian diver manostat [ENG] Preset, on-off-control manometer arrangement by which a specified low pressure (high vacuum) is maintained via the rise or submergence of a marginally buoyant float within a liquid mercury reservoir. { kär'tē·zhan ˈdīv·ər ˈman·ə,stat }

Cartesian geometry *See* analytic geometry. { kär'tē·zhan jē'äm·ə·trē }

Cartesian oval [MATH] A plane curve consisting of all points P such that $aFP + bF'P = c$, where F and F' are fixed points and a, b, and c are constants which are not necessarily positive. { kär'tē·zhən 'ō·vəl }

Cartesian plane [MATH] A plane whose points are specified by Cartesian coordinates. { kär'tēzh·ən 'plän }

Cartesian product [MATH] In reference to the product of P and Q, the set $P \times Q$ of all pairs (p,q), where p belongs to P and q belongs to Q. { kär'tē·zhan 'präd·əkt }

Cartesian surface [MATH] A surface obtained by rotating the curve $n_0(x^2 + y^2)^{1/2} \pm n_1[(x - a)^2 + y^2]^{1/2} = c$ about the x axis. { kär'tē·zhan 'sər·fəs }

Cartesian tensor [MATH] The aggregate of the functions of position in a tensor field in an *n*-dimensional Cartesian coordinate system. { kär'tē·zhan 'ten·sər }

cartilage [HISTOL] A specialized connective tissue which is bluish, translucent, and hard but yielding. { 'kärd·əl·ij }

cartilage bone [HISTOL] Bone formed by ossification of cartilage. { 'kärd·əl·ij 'bōn }

cartilaginous fish [VERT ZOO] The common name for all members of the class Chondrichthyes. { ˈkärd·əlˈaj·ə·nas 'fish }

cartogram [MAP] A type of single-factor or topical map that is often diagrammatic to show traffic flow, movement of people or goods, or value by area, where areas of the political subdivisions are distorted so that their size is proportional to their monetary value. { 'kärd·ə,gram }

cartographer [GRAPHICS] An individual who makes charts or maps. { 'kär'täg·rə·fər }

cartographic satellite [AERO ENG] An applications satellite that is used to prepare maps of the earth's surface and of the culture on it. { ˈkärd·əˈgraf·ik 'sad·əl,īt }

cartography [GRAPHICS] The making of maps and charts for the purpose of visualizing spatial distributions over various areas of the earth. { kär'täg·rə·fē }

cartology [GRAPHICS] A method of coal-seam correlation which involves the mapping and drawing of both vertical and horizontal sections. { kär'täl·ə·jē }

cartoon [GRAPHICS] **1.** Animated drawings in a motion picture format. **2.** A drawing on paper that is used as a model for a final work. { kär'tün }

cartouche [GRAPHICS] A border or scroll that is decorative and executed with a pen or brush. { kar'tüsh }

cartridge [COMPUT SCI] A self-contained module that contains disks, magnetic tape, or integrated circuits for storing data. [ENG] A cylindrical, waterproof, paper shell filled with high explosive and closed at both ends; used in blasting. [ENG ACOUS] *See* phonograph pickup; tape cartridge. [GRAPHICS] A container designed to hold microforms to be inserted into a reader, a reader-printer, or a retrieval device; when used for roll microfilm, it is a single-core device. [NUCLEO] *See* jacket. [ORD] **1.** An assemblage of the components required to function a weapon once. **2.** Ammunition for a gun which contains in a unit assembly all components required to function the gun once, and which is loaded into the gun in one operation. { 'kär·trij }

cartridge-actuated initiator [AERO ENG] An item designed to provide gas pressure for activating various aircraft components such as canopy removers, thrusters, and catapults. { 'kär·trij 'ak·chə,wād·əd in'ish·ē,ād·ər }

cartridge belt [ORD] An ammunition belt with loops or pockets for carrying cartridges or clips of cartridges. { 'kär·trij ,belt }

cartridge brass [MET] An alloy containing 70% copper and 30% zinc; uses include cartridge cases, automotive radiator cores and tanks, lighting fixtures, eyelets, rivets, springs, screws, and plumbing products. { 'kär·trij ,bras }

cartridge case [ORD] An item which is designed to hold an ammunition primer and propellant and to which a projectile may be affixed; its profile and size conform to the chamber of the weapon in which the round is fired. { 'kär·trij ,kās }

cartridge clip [ORD] A metallic device used to hold cartridges for ease of loading into a rifle or a revolver. Also known as clip. { 'kär·trij ,klip }

cartridge disk [COMPUT SCI] A type of disk storage device consisting of a single disk encased in a compact container which can be inserted in and removed from the disk drive unit; used extensively with minicomputer systems. { 'kär·trij ,disk }

cartridge filter [ENG] A filter for the clarification of process liquids containing small amounts of solids; turgid liquid flows between thin metal disks, assembled in a vertical stack, to openings in a central shaft supporting the disks, and solids are trapped between the disks. { 'kär·trij ,fil·tər }

cartridge font [COMPUT SCI] A font for a computer printer that is stored on a read-only memory chip within a cartridge (a module that is inserted in a slot in the printer). { 'kär·trīj ,fänt }

cartridge fuse [ELEC] A type of electric fuse in which the fusible element is connected between metal ferrules at either end of an insulating tube. { 'kär·trij ,fyüz }

cartridge lamp [ELEC] A pilot or dial lamp that has a tubular glass envelope with metal-ferrule terminals at each end. { 'kär·trij ,lamp }

cartridge magazine [ORD] A metallic spring-loaded container of rectangular cross section designed to hold cartridges and to be inserted into the magazine well of a pistol, rifle, or machine gun; it is an integral part of a gun, capable of being removed from the magazine well, reloaded, and reused. { 'kär·trij ,mag·ə,zēn }

cartridge receiver [ORD] An item integral to a carbine, rifle, or shotgun, designed to take the charge from a cartridge magazine or clip or a single cartridge or shell and hold it until chambered. { 'kär·trij ri'sē·vər }

cartridge silk *See* powder silk. { 'kär·trij ,silk }

cartridge starter [MECH ENG] An explosive device which, when placed in an engine and detonated, moves a piston, thereby starting the engine. { 'kär·trij ,stärd·ər }

cartridge tape drive [COMPUT SCI] A tape drive which will automatically thread the tape on the takeup reels without human assistance. Formerly known as hypertape drive. { 'kär·trij ,tāp ,drīv }

car tunnel kiln [ENG] A long kiln with the fire located near

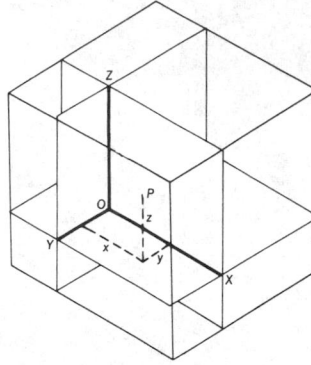

CARTESIAN COORDINATE SYSTEM

Cartesian coordinate system showing the coordinates x, y, and z of point P from the origin O.

the midpoint; ceramic ware is fired by loading it onto cars which are pushed through the kiln. { kär 'tən·əl ˌkil }

Cartwheel [ASTRON] A ring galaxy found in the southern hemisphere. { 'kärtˌwēl }

carui oil *See* caraway oil. { 'kär·ə·wē ˌoil }

caruncle [ANAT] Any normal or abnormal fleshy outgrowth, such as the comb and wattles of fowl or the mass in the inner canthus of the eye. [BOT] A fleshy outgrowth developed from the seed coat near the hilum in some seeds, such as the castor bean. { 'kaˌrəŋ·kəl }

carvacrol [ORG CHEM] $(CH_3)_2CHC_6H_3(CH_3)OH$ A colorless liquid, boiling at 237°C; used in perfumes, flavorings, and fungicides. { 'kär·vəˌkról }

Carvallo paradox [OPTICS] The absurdity that since light is composed from infinitely long wave trains of various frequencies, a spectrograph should show the spectrum of a source both before and after it is illuminated. { kär'väl·ō ˌpar·ə̇ˌdäks }

carved-out payment [PETRO ENG] A proportionate royalty payment based on proceeds from oil or gas production from leased property that has been assigned (carved) out of total payments for the leased property. { ¦kärvˌdaut 'pā·mənt }

carvol *See* carvone. { 'kärˌvól }

carvone [ORG CHEM] $C_{10}H_{14}O$ A liquid ketone that boils at 231°C; soluble in water and alcohol; it is optically active and occurs naturally in both dextro and levo forms; used in flavorings and perfumery. Also known as carvol. { 'kärˌvōn }

Carya ovata *See* shagbark hickory. { ¦kär·ē·ə ō'vä·tə }

caryinite [MINERAL] $(Ca,Pb,Na)_5(Mn,Mg)_4(AsO_4)_5$ A mineral consisting chiefly of a calcium manganese arsenate. { 'kar·ē·əˌnīt }

Caryophanaceae [MICROBIO] Formerly a family of large, gram-negative bacteria belonging to the order Caryophanales and having disklike cells arranged in chains. { ¦kar·ē·ō·fə'nās·ēˌē }

Caryophanales [MICROBIO] Formerly an order of bacteria in the class Schizomycetes occurring as trichomes which produce short structures that function as reproductive units. { kar·ē·äf·ə'nā·lēz }

Caryophanon [MICROBIO] A genus of gram-positive, large, rod-shaped or filamentous bacteria of uncertain affiliation. { ˌkar·ē'äf·əˌnän }

Caryophyllaceae [BOT] A family of dicotyledonous plants in the order Caryophyllales differing from the other families in lacking betalains. { ¦kar·ē·o·fə'lās·ēˌē }

Caryophyllales [BOT] An order of dicotyledonous plants in the subclass Caryophyllidae characterized by free-central or basal placentation. { kar·ē·ō·fə'lā·lēz }

caryophyllene [ORG CHEM] $C_{15}H_{24}$ A liquid sesquiterpene that is found in some essential oils, particularly clove oil. { ˌkar·ē·ō'fiˌlēn }

Caryophyllidae [BOT] A relatively small subclass of plants in the class Magnoliopsida characterized by trinucleate pollen, ovules with two integuments, and a multilayered nucellus. { kar·ē·ō'fil·əˌdē }

caryophyllin [ORG CHEM] $C_{30}H_{48}O_3$ A ketone, soluble in alcohol, extracted from oil of cloves. { ˌkar·ē·ō'fil·ən }

caryophyllus oil *See* oil of cloves. { ˌkar·ē·ō'fil·əs ˌoil }

caryopsis [BOT] A small, dry, indehiscent fruit having a single seed with such a thin, closely adherent pericarp that a single body, a grain, is formed. { ˌkar·ē'äp·səs }

Cas *See* Cassiopeia.

casaba melon [BOT] *Cucumis melo.* A winter muskmelon with a yellow rind and sweet flesh belonging to the family Cucurbitaceae of the order Violales. { kə̇ˌsäb·ə ˌmel·ən }

Casale process [CHEM ENG] A process that employs promoted iron oxide catalyst for synthesis of ammonia from nitrogen and hydrogen. { kə̇ˌsäl·ē 'präs·əs }

cascade [COMPUT SCI] A series of actions that take place in the course of data processing, each triggered by the previous action in the series. [ELEC] An electric-power circuit arrangement in which circuit breakers of reduced interrupting ratings are used in the branches, the circuit breakers being assisted in their protection function by other circuit breakers which operate almost instantaneously. Also known as backup arrangement. [ELECTR] *See* avalanche. [ENG] An arrangement of separation devices, such as isotope separators, connected in series so that they multiply the effect of each individual device. [GEOL] A landform structure formed by gravity collapse, consisting of a bed that buckles into a series of folds as it slides down the flanks of an anticline. [HYD] A small waterfall or series of falls descending over rocks. [MOL BIO] A molecular system that is capable of self-propagation or amplification. [PHYS] The emission of a series of photons by a quantum system, such as an atomic nucleus or a laser, in an excited state, accompanying transitions of the system to successively lower excited states, until the system reaches the ground state. { ka'skād }

cascade amplifier [ELECTR] A vacuum-tube amplifier containing two or more stages arranged in the conventional series manner. Also known as multistage amplifier. { ka'skād ˌam·pləˌfī·ər }

cascade-amplifier klystron [ELECTR] A klystron having three resonant cavities to provide increased power amplification and output; the extra resonator, located between the input and output resonators, is excited by the bunched beam emerging from the first resonator gap and produces further bunching of the beam. { ka'skād ˌam·pləˌfī·ər 'klīˌsträn }

cascade compensation [CONT SYS] Compensation in which the compensator is placed in series with the forward transfer function. Also known as series compensation; tandem compensation. { ka'skād käm·pən'sā·shən }

cascade connection [ELECTR] A series connection of amplifier stages, networks, or tuning circuits in which the output of one feeds the input of the next. Also known as tandem connection. { ka'skād kə'nek·shən }

cascade control [CONT SYS] An automatic control system in which various control units are linked in sequence, each control unit regulating the operation of the next control unit in line. { ka'skād kən·trōl }

cascade converter [ELEC] A rotary converter that is powered from the secondary of an induction motor that is connected to the same shaft. { ka'skad kənˌvərd·ər }

cascade cooler [CHEM ENG] Fluid-cooling device through which the fluid flows in a series of horizontal tubes, one above the other; cooling water from a trough drips over each tube, then to a drain. Also known as serpentine cooler; trickle cooler. { ka'skād ˌkü·lər }

cascaded [ENG] Of a series of elements or devices, arranged so that the output of one feeds directly into the input of another, as a series of dynodes or a series of airfoils. { ka'skād·əd }

cascaded carry [COMPUT SCI] A carry process in which the addition of two numerals results in a sum numeral and a carry numeral that are in turn added together, this process being repeated until no new carries are generated. { ka'skād·əd'karˌē }

cascaded feedback canceler [ELECTR] Sophisticated moving-target-indicator canceler which provides clutter and chaff rejection. Also known as velocity shaped canceler. { ka'skād·əd 'fēdˌbak ˌkan·slər }

cascade gamma emission [NUC PHYS] The emission by a nucleus of two or more gamma rays in succession. { ka'skād 'gam·ə i'mish·ən }

cascade hyperon *See* xi hyperon. { ka'skād ˌhī·pə̇ˌrän }

cascade image tube [ELECTR] An image tube having a number of sections stacked together, the output image of one section serving as the input for the next section; used for light detection at very low levels. { ka'skād 'im·ij ˌtüb }

cascade impactor [ENG] A low-speed impaction device for use in sampling both solid and liquid atmospheric suspensoids; consists of four pairs of jets (each of progressively smaller size) and sampling plates working in series and designed so that each plate collects particles of one size range. { ka'skād im'pak·tər }

cascade junction [ELECTR] Two *pn* semiconductor junctions in tandem such that the condition of the first governs that of the second. { ka'skād 'jəŋk·shən }

cascade limiter [ELECTR] A limiter circuit that uses two vacuum tubes in series to give improved limiter operation for both weak and strong signals in a frequency-modulation receiver. Also known as double limiter. { ka'skād 'lim·əd·ər }

cascade liquefaction [CRYO] A method of liquefying gases in which a gas with a high critical temperature is liquefied by increasing its pressure; evaporation of this liquid cools a second liquid so that it can also be liquefied by compression, and so on. { ka'skād lik·wə'fak·shən }

cascade mixer-settler [CHEM ENG] Series of liquid-holding

CASCADE IMAGE TUBE

Two-stage electrostatically focused cascade-type image tube. (*U.S. Army Engineer Development Corps*)

vessels with stirrers, each connected to an unstirred vessel in which solids or heavy immiscible liquids settle out of suspension; light liquid moves through the mixer-settler units, counterflowing to heavy material, in such a manner that fresh liquid contacts treated heavy material, and spent (used) liquid contacts fresh (untreated) heavy material.   { ka'skād ¦mik·sər ¦set·lər }

cascade mixing   [ELEC]   A mechanism for ion-beam mixing of a film and a substrate in which the recoil of an atom from a collision with an incident ion initiates a series of secondary collisions among the film and substrate atoms, leading to transfer of atoms from the substrate into the film as well as from the film into the substrate.   { ka'skād ,mik·siŋ }

cascade molecule   See dendrimer.   { ka¦skād 'mäl·ə,kyül }

cascade networks   [ELEC]   Two networks in tandem such that the output of the first feeds the input of the second.   { ka'skād 'net,wərks }

cascade noise   [ELECTR]   The noise in a communications receiver after an input signal has been subjected to two tandem stages of amplification.   { ka'skād 'nȯiz }

cascade particle   See xi hyperon.   See xi-minus particle.   { ka'skād 'pärd·ə·kəl }

cascade pulverizer   [MECH ENG]   A form of tumbling pulverizer that uses large lumps to do the pulverizing.   { ka'skād 'pəl·və,rīz·ər }

cascade regulation   [MOL BIO]   **1.** In prokaryotes, a form of genetic regulation in which one operon codes for the production of an internal inducer that turns on one or more operons.   **2.** In eukaryotes, a multistep model of genetic regulation involving mechanisms that interface with messenger ribonucleic acid formation, transport, and translation.   { ka'skād ,reg·yə'lā·shən }

cascade sequence   [MET]   Combined longitudinal and buildup sequence in which weld beads are deposited in overlapping layers.   { ka'skād ,sē·kwəns }

cascade shower   [PARTIC PHYS]   A cosmic-ray shower of electrons, positrons, and gamma rays which grows by pair production and bremsstrahlung events.   { ka'skād ,shau̇·ər }

cascade system   [MECH ENG]   A combination of two or more refrigeration systems connected in series to produce extremely low temperatures, with the evaporator of one machine used to cool the condenser of another.   { ka'skād ,sis·təm }

cascade transformer   [ELEC]   A source of high voltage that is made up of a collection of step-up transformers; secondary windings are in series, and primary windings, except the first, are supplied from a pair of taps on the secondary winding of the preceding transformer.   { ka'skād tranz'fȯr·mər }

cascade tray   [CHEM ENG]   A fractionating apparatus that consists of a series of parallel troughs arranged in stairstep fashion.   { ka'skād ,trā }

cascade unit   See radiation length.   { ka'skād ,yü·nət }

Cascadian orogeny   [GEOL]   Post-Tertiary deformation of the crust of the earth in western North America.   { ka'skād·ē·ən ȯ'räj·ə·nē }

cascading   [ELEC]   An effect in which a failure of an electrical power system causes this system to draw excessive amounts of power from power systems which are interconnected with it, causing them to fail, and these systems cause adjacent systems to fail in a similar manner, and so forth.   [MECH ENG]   An effect in ball-mill rotating devices when the upper level of crushing bodies breaks clear and falls to the top of the crop load.   { ka'skād·iŋ }

cascading drain   [MECH ENG]   A flow of water into the closed shell of a feedwater heater from a water source maintained at a higher pressure.   { ka'skād·iŋ 'drān }

cascading glacier   [HYD]   A glacier broken by numerous crevasses because of passing over a steep irregular bed, giving the appearance of a cascading stream.   { ka'skād·iŋ 'glā·shər }

cascading menu   [COMPUT SCI]   A menu that appears next to a pull-down menu as the result of selecting a choice on the latter.   { ka,skād·iŋ 'men·yü }

cascading windows   [COMPUT SCI]   Two or more windows displayed so that they overlap but their title bars are still visible.   { ka,skād·iŋ 'win,dōz }

cascarilla oil   [MATER]   An essential oil derived from cascarilla bark (*Croton eluteria*), containing cascarillic acid, $C_{11}H_{20}O_2$; used as a flavoring in foods, medicine, and tobacco.   { ,kas·kə'ril·ə ,ȯil }

cascode amplifier   [ELECTR]   An amplifier consisting of a grounded-emitter input stage that drives a grounded-base output

stage; advantages include high gain and low noise; widely used in television tuners.   { 'ka,skōd 'am·plə,fī·ər }

case   [COMPUT SCI]   **1.** In computers, a set of data to be used by a particular program.   **2.** The metal box that houses a computer's circuit boards, disk drives, and power supply. Also known as system unit.   [ENG]   An item designed to hold a specific item in a fixed position by virtue of conforming dimensions or attachments; the item which it contains is complete in itself for removal and use outside the container.   [GRAPHICS]   The cover of a hardbound book.   [MET]   Outer layer of a ferrous alloy which has been made harder than the core by case hardening.   [MIN ENG]   A small fissure admitting water into the mine workings.   [PETRO ENG]   To line a borehole with steel tubing, such as casing or pipe.   { kās }

CASE   See computer-aided software engineering.   { kās }

caseation necrosis   [PATH]   Tissue death involving loss of cellular integrity with the consequent conversion to a cheeselike substance; typical in tuberculosis.   { ,kas·ē'ā·shən nə'krō·səs }

case bay   [BUILD]   A division of a roof or floor, consisting of two principal rafters and the joists between them.   { 'kās ,bā }

cased glass   [MATER]   Glass composed of two or more layers of different colors.   { 'kāst ,glas }

cased hole   [PETRO ENG]   A wellbore which has been lined with casing.   { 'kāst 'hōl }

case I firing   See case I pointing.   { ¦kās 'wən ,fīr·iŋ }

case II firing   See case II pointing.   { ¦kās 'tü ,fīr·iŋ }

case III firing   See case III pointing.   { ¦kās 'thrē ,fīr·iŋ }

case hardening   [GEOL]   Formation of a mineral coating on the surface of porous rock by evaporation of a mineral-bearing solution.   [MATER]   A condition of stress and set in dry lumber characterized by compressive stress in the outer layers and tensile stress in the center or core.   [MET]   Process of carburizing low-carbon steel or other ferrous alloy for making the outer layer (case) harder than the core.   { 'kās ,härd·ən·iŋ }

casein   [ORG CHEM]   The protein of milk; a white solid soluble in acids.   { 'ka,sēn }

casein-formaldehyde   [ORG CHEM]   A modified natural polymer.   { 'ka,sēn fȯr'mal·də,hīd }

casein glue   [MATER]   A produce of dried curds of milk, lime, and other chemical ingredients, mixed cold; used for both plywood and assembly work.   { 'ka,sēn ,glü }

casein paint   [MATER]   A paint with casein substituted for linseed oil.   { 'ka,sēn ¦pānt }

casein plastic   [MATER]   A plastic made with casein and used for buttons, beads, knitting needles, and novelties; thin sheets and rods of casein plastic are cured (hardened) in formaldehyde baths.   { 'ka,sēn ¦plas·tik }

casemate   [ORD]   A bombproof structure used as a powder magazine, gun emplacement, or the like.   { 'kās,māt }

casement window   [BUILD]   A window hinged on the side that opens to the outside.   { 'kās·mənt 'win·dō }

caseous lymphadenitis   [VET MED]   A chronic bacterial disease of sheep and goats caused by *Corynebacterium pseudotuberculosis*, characterized by caseation of the lymph glands and sometimes the lungs, liver, spleen, and kidneys.   { 'kā·shəs lim,fad·ən'ī·dəs }

case I pointing   [ORD]   Direct pointing, laying, or firing; gun pointing, in which direction and elevation are set with sight or telescope pointed at the target.   Also known as case I firing.   { ¦kās 'wən ,pȯint·iŋ }

case II pointing   [ORD]   Combined direct and indirect pointing, laying, or firing; gun pointing, in which direction is set with a sight or telescope pointed at the target; the elevation is set with an elevation quadrant, range quadrant, or range disk. Also known as case II firing.   { ¦kās 'tü ,pȯint·iŋ }

case III pointing   [ORD]   Indirect pointing, laying, or firing; gun pointing, in which direction is set with an azimuth circle or with a sight or telescope pointed at an aiming point other than the target; the elevation is set with an elevation quadrant, range quadrant, or range disk.   Also known as case III firing.   { ¦kās 'thrē ,pȯint·iŋ }

case primer   [ORD]   Primer intended to be assembled into the cartridge case of case ammunition.   { 'kās ,prī·mər }

case-sensitive language   [COMPUT SCI]   A programming language in which upper-case letters are distinguished from lower-case letters.   { 'kās ,sens·ə·tiv 'laŋ·gwij }

case shot   [ORD]   A shell loaded with shrapnel.   { 'kās ,shät }

case structure [COMPUT SCI] A group of program statements in which a condition is tested and, according to the results of the test, one of at least three specific groups of program statements is executed, after which the program returns to the original location. { 'kās ,strək·chər }

cashew [BOT] *Anacardium occidentale.* An evergreen tree of the order Sapindales grown for its kidney-shaped edible nuts and resinous oil. { 'kash·ü }

cashew gum [MATER] A gum obtained from the bark of the cashew tree; hard, yellowish-brown substance used for inks, insecticides, pharmaceuticals, varnishes, and bookbinders' gum. Also known as anacardium gum. { 'kash·ü ,gəm }

cashew nutshell oil [MATER] A toxic, irritating oil obtained from the shell of the cashew nut; contains about 90% anacardic acid; used for varnishes, plasticizers, insecticides, coloring materials, and preservatives. { 'kash·ü 'nət,shel ,ȯil }

cashmere [TEXT] A natural textile fiber obtained from the Cashmere goat of the Himalayan region of China and India. { 'kazh,mir }

Casimir-du Pré theory [SOLID STATE] A theory of spin-lattice relaxation which treats the lattice and spin systems as distinct thermodynamic systems in thermal contact with one another. { 'kaz·ə,mir dyü'prā ,thē·ə·rē }

Casimir effect [QUANT MECH] An attractive force between two parallel, conducting plates in empty space that arises from zero-point quantum fluctuations of the vacuum electromagnetic field and is proportional to $1/d^4$, where d is the plate separation. { 'kaz·ə,mir i,fekt }

casing [BUILD] A finishing member around the opening of a door or window. [DES ENG] The outer portion of a tire assembly consisting of fabric or cord to which rubber is vulcanized. [MECH ENG] A fire-resistant covering used to protect part or all of a steam generating unit. [PETRO ENG] A special steel tubing welded or screwed together and lowered into a borehole to prevent entry of loose rock, gas, or liquid into the borehole or to prevent loss of circulation liquid into porous, cavernous, or crevassed ground. { 'kā,siŋ }

casing burst pressure [PETRO ENG] The pressure that will cause a casing wall to fail when applied to the casing string. { 'kā,siŋ 'bərst ,presh·ər }

casing cementing [PETRO ENG] Filling of the area between a casing and the borehole with cement in order to prevent fluid migration between permeable zones and to support the casing. { 'kā,siŋ si'men·tiŋ }

casing centralizer [PETRO ENG] A device secured around the casing during drilling to hold the pipe in a center position, permitting a uniform cement sheath to be constructed around the pipe. Also known as centralizer. { 'kā,siŋ 'sen·trə,līz·ər }

casing cutter [PETRO ENG] Heavy cylindrical apparatus bearing knives, and run on a string of tubing or drill pipe into a well, to perform cutting on the inner walls of a pipe through rotation, thus freeing a pipe section. { 'kā,siŋ ,kəd·ər }

casing hanger *See* hanger. { 'kā,siŋ ,haŋ·ər }

casinghead [PETRO ENG] A fitting at the head of an oil or gas well that allows the pumping operation to take place, as well as the separation of oil and gas. { 'kā,siŋ,hed }

casinghead gas [MATER] The natural gas that is emitted from the mouth or opening of an oil well. { 'kā,siŋ,hed ,gas }

casinghead gasoline [MATER] Liquid hydrocarbon product removed from casinghead gas by absorption, compression, or refrigeration. { 'kā,siŋ,hed ,gas·ə'lēn }

casinghead tank [PETRO ENG] Storage tank for natural gasoline or other liquids with vapor pressures between 4 and 40 pounds per square inch gage (28 and 280 kilopascals, gage); intermediate between a general-purpose tank and a compressed-gas tank. { 'kā,siŋ,hed ,taŋk }

casing joint [PETRO ENG] Joint or union that connects two lengths of pipe used to form an oil-well casing. { 'kā,siŋ ,jȯint }

casing log [PETRO ENG] Recorded data of a down-hole inspection of an oil or gas well made to determine some characteristic of the formations penetrated by the drill hole; types of logs include resistivity, induction, radioactivity, geologic, temperature, and acoustic. { 'kā,siŋ ,läg }

casing nail [DES ENG] A nail about half a gage thinner than a common wire nail of the same length. { 'kā,siŋ ,nāl }

casing pack [PETRO ENG] A nonsolidifying mud placed in the well ahead of the cement after the casing has been set, so that the casing can be cut above the cemented section and removed when necessary. { 'kā·siŋ ,pak }

casing point [PETRO ENG] The location in a well at which the casing is set; usually corresponds to the depth at which the casing shoe has been positioned. { 'kā·siŋ ,pȯint }

casing pressure [PETRO ENG] Gas pressure which is built up between the casing and tubing in a well. { 'kā,siŋ ,presh·ər }

casing roller [PETRO ENG] A tool, consisting of a mandrel with a series of eccentric roll surfaces fitted with rollers, for restoring distorted casing to normal diameter and shape. { 'kā·siŋ ,rō·lər }

casing seat [PETRO ENG] The position of the end of a casing string where it is cemented in a well; usually a casing shoe is positioned at the bottom of the casing at this point. { 'kā·siŋ ,sēt }

casing shoe [ENG] A ring with a cutting edge on the bottom of a well casing. { 'kā,siŋ ,shü }

casing spear [PETRO ENG] An instrument used for recovering casing which has accidentally fallen into the well. { 'kā·siŋ ,spir }

casing string [PETRO ENG] The total of all lengths of casing that have been run in a well. { 'kā·siŋ ,striŋ }

casing swage [PETRO ENG] A solid cylindrical tool for making an opening in a collapsed casing and restoring it to its original shape. { 'kā·siŋ ,swāj }

casing tester [PETRO ENG] A closely fitting, rubber-flanged bucket or a similar tool let down in a well to determine the location of a leak in the casing. { 'kā·siŋ ,tes·tər }

cask *See* coffin. { kask }

cask buoy [NAV] A buoy in the shape of a cask. { 'kask ,bȯi }

casket *See* coffin. { 'kas·kət }

Casparian strip [BOT] A thin band of suberin- or lignin-like deposition in the radial and transverse walls of certain plant cells during the primary development phase of the endodermis. { ka'spar·ē·ən ,strip }

caspase [CELL MOL] Any of a family of intracellular proteins that mediate apoptosis. { 'kas,pās }

Cassadagan [GEOL] Middle Upper Devonian geologic time, above Chemungian. { kə'sad·ə·gən }

cassava [BOT] *Manihot esculenta.* A shrubby perennial plant grown for its starchy, edible tuberous roots. Also known as manihot; manioc. { kə'sav·ə }

Cassegrain antenna [ELECTROMAG] A microwave antenna in which the feed radiator is mounted at or near the surface of the main reflector and aimed at a mirror at the focus; energy from the feed first illuminates the mirror, then spreads outward to illuminate the main reflector. { kas·gran an'ten·ə }

Cassegrain focus [OPTICS] The principal focus of a Cassegrain telescope, located just behind the primary mirror. { kas·gran 'fō·kəs }

Cassegrain-Newtonian telescope *See* Newtonian-Cassegrain telescope. { kas·gran nü'tōn·ē·ən 'tel·ə,skōp }

Cassegrain telescope [OPTICS] A reflecting telescope in which a small hyperboloidal mirror reflects the convergent beam from the paraboloidal primary mirror through a hole in the primary mirror to an eyepiece in back of the primary mirror. { kas·gran 'tel·ə,skōp }

Cassel green *See* barium manganate. { 'kas·əl ,grēn }

Casselian *See* Chattian. { ka'sel·yən }

cassette [ENG] A light-tight container designed to hold photographic film or plates. [ENG ACOUS] A small, compact container that holds a magnetic tape and can be readily inserted into a matching tape recorder for recording or playback; the tape passes from one hub within the container to the other hub. [GRAPHICS] A double-core container designed to hold processed roll microfilm for inserting into a reader, reader-printer, or other retrieval device. [MYCOL] In yeast, any of the sites lying in tandem that contain nucleotide sequences that can be substituted for one another. { kə'set }

cassette cartridge system [COMPUT SCI] An input system often used in minicomputers; its low cost and ease in mounting often offset its slow access time. { kə'set ,kär·trij ,sis·təm }

cassette memory [COMPUT SCI] A removable magnetic tape cassette that stores computer programs and data. { kə'set 'mem·rē }

cassia oil [MATER] An essential oil extracted from the bark of Chinese cinnamon (*Cinnamomum cassia*), and containing cinnamaldehyde. Also known as Chinese oil of cinnamon. { 'kash·ə ,ȯil }

Cassiar orogeny [GEOL] Orogenic episode in the Canadian Cordillera during late Paleozoic time. { 'kas·ē·ər ȯ'räj·ə·nē }

Cassidulinacea [INV ZOO] A superfamily of marine, benthic foraminiferans in the suborder Rotaliina, characterized by a test of granular calcite with monolamellar septa. { ,kas·ə,dü·lə'nās·ē·ə }

Cassiduloida [INV ZOO] An order of exocyclic Euechinoidea possessing five similar ambulacra which form petal-shaped areas (phyllodes) around the mouth. { ,kas·ə·də'lȯid·ē·ə }

cassidyite [MINERAL] $Ca_2(Ni,Mg)(PO_4)_2 \cdot 2H_2O$ A mineral found in meteorites. { kə'sid·ē,īt }

Cassinian oval See oval of Cassini. { kə'sin·ē·ən 'ō·vəl }

Cassini projection [MAP] A map projection, formerly used for topographic and cadastral mapping, in which scale is preserved along the central meridian and great circle arcs by plotting as rectangular coordinates on a plane the lengths of arcs along a central meridian and along a great circle perpendicular to that meridian. { kə'sē·nē prə'jek·shən }

Cassini's division [ASTRON] The gap, 2500 miles (4000 kilometers) wide, that separates ring A from ring B of the planet Saturn. { kə'sē·nēz di'vizh·ən }

Cassiopeia [ASTRON] A constellation with right ascension 1 hour, declination 60°N. Abbreviated Cas. { ,kas·ē·ə'pē·ə }

Cassiopeia A [ASTRON] One of the strongest discrete radio sources, located in the constellation Cassiopeia, associated with patches of filamentary nebulosity which are probably remnants of a supernova. { ,kas·ē·ə'pē·ə 'ā }

cassiterite [MINERAL] SnO_2 A yellow, black, or brown mineral that crystallizes in the tetragonal system in prisms terminated by dipyramids; the most important ore of tin. Also known as tin stone. { kə'sid·ə,rīt }

cassowary [VERT ZOO] Any of three species of large, heavy, flightless birds composing the family Casuariidae in the order Casuariiformes. { 'kas·ə,wer·ē }

cast [ENG] **1.** To form a liquid or plastic substance into a fixed shape by letting it cool in the mold. **2.** Any object which is formed by placing a castable substance in a mold or form and allowing it to solidify. Also known as casting. [MED] **1.** A rigid dressing used to immobilize a part of the body. **2.** See strabismus. [NAV] **1.** To turn a ship in its own water. **2.** To turn a ship to a desired direction without gaining either headway or sternway. **3.** To take a sounding with the lead. [OPTICS] A change in a color because of the adding of a different hue. [PALEON] A fossil reproduction of a natural object formed by infiltration of a mold of the object by waterborne minerals. [PHYSIO] A mass of fibrous material or exudate having the form of the body cavity in which it has been molded; classified from its source, such as bronchial, renal, or tracheal. { kast }

castable [MATER] A refractory aggregate mixed with a bonding agent such as aluminous hydraulic cement which, with addition of water, will develop structural strength and set in a mold. { 'ka·stə·bəl }

Castaing-Slodzian mass analyzer See direct-imaging mass analyzer. { ¦kas·taŋ ¦slō·zhən ,mas 'an·ə,līz·ər }

castaneous [BIOL] Chestnut-colored. { ka'stān·ē·əs }

cast coated paper [MATER] A paper with a high-gloss enamel finish that has been produced by drying coated paper under pressure from a polished cylinder. { 'kast ,kōd·əd 'pā·pər }

caste [INV ZOO] One of the levels of mature social insects in a colony that carry out a specific function; examples are workers and soldiers. { kast }

castellanus [METEOROL] A cloud species with at least a fraction of its upper part presenting some vertically developed cumuliform protuberances (some of which are more tall than wide) which give the cloud a crenellated or turreted appearance. Previously known as castellatus. { ,kas·tə'län·əs }

castellated bit [DES ENG] **1.** A long-tooth, sawtooth bit. **2.** A diamond-set coring bit with a few large diamonds or hard metal cutting points set in the face of each of several upstanding prongs separated from each other by deep waterways. Also known as padded bit. { 'kas·tə,lād·əd 'bit }

castellated nut [DES ENG] A type of hexagonal nut with a cylindrical portion above through which slots are cut so that a cotter pin or safety wire can hold it in place. { 'kas·tə,lād·əd 'nət }

castellatus See castellanus. { ,kas·tə,läd·əs }

caster [ENG] **1.** The inclination of the kingpin or its equivalent in automotive steering, which is positive if the kingpin inclines forward, negative if it inclines backward, and zero if it is vertical as viewed along the axis of the front wheels. **2.** A wheel which is free to swivel about an axis at right angles to the axis of the wheel, used to support trucks, machinery, or furniture. { 'kas·tər }

cast-film extrusion See chill-roll extrusion. { ¦kast ¦film ik'strü·zhən }

Castigliano's principle See Castigliano's theorem. { ,kas·til'yä·nōz ,prin·sə·pəl }

Castigliano's theorem [MECH] The theorem that the component in a given direction of the deflection of the point of application of an external force on an elastic body is equal to the partial derivative of the work of deformation with respect to the component of the force in that direction. Also known as Castigliano's principle. { ,kas·til'yä·nōz ,thir·əm }

Castile soap [MATER] A white, odorless, hard soap made from sodium hydroxide and olive oil. { ka'stēl 'sōp }

casting See cast. { 'kast·iŋ }

casting alloy [MET] An alloy which cannot be forged or rolled and can be shaped only as a casting. { 'kast·iŋ ,a,lȯi }

casting area [ENG] In plastics injection molding, the moldable area of a thermoplastic material for a given thickness and under given conditions of molding. { 'kast·iŋ ,er·ē·ə }

casting copper [MET] Copper used for making foundry castings; obtained from copper ores, and inferior to electrolytic copper. { 'kast·iŋ ,käp·ər }

casting jet [ARCHEO] A plug of metal that fits exactly into the opening of a mold, having been knocked out when the object was completed. { 'kast·iŋ ,jet }

casting ladle [MET] A refractory-lined steel ladle used to transport molten metal from the furnace to a mold. { 'kast·iŋ ,lād·əl }

casting-out nines [MATH] A method of checking the correctness of elementary arithmetical operations, based on the fact that an integer yields the same remainder as the sum of its decimal digits, when divided by 9. { ¦kast·iŋ ¦aut 'nīnz }

casting plaster [MATER] A white plaster used for castings and carvings. { 'kast·iŋ ,plas·tər }

castings See fecal pellets. { 'kast·iŋz }

casting shrinkage [MET] **1.** Total reduction in volume of a casting due to partial reductions at each stage of solidification. **2.** Reduction in volume at each stage of solidification of a casting. { 'kast·iŋ ,shriŋ·kij }

casting slip [MATER] A slurry of clay and additives mixed in water with deflocculating agents and used for casting in molds. { 'kast·iŋ ,slip }

casting strain [MECH] Any strain that results from the cooling of a casting, causing casting stress. { 'kast·iŋ ,strān }

casting stress [MECH] Any stress that develops in a casting due to geometry and casting shrinkage. { 'kast·iŋ ,stres }

casting wheel [MET] A large turntable with molds mounted on the outer edge; used primarily in the base-metal industries for cast ingots, anodes, and so on. { 'kast·iŋ ,wēl }

cast iron [MET] Any carbon-iron alloy cast to shape and containing 1.8–4.5% carbon, that is, in excess of the solubility in austenite at the eutectic temperature. Abbreviated C.I. { ¦kast 'ī·ərn }

cast-iron front [ARCH] A style of architecture characterized by large window areas and cast-iron columns and spandrels. { ¦kast 'ī·ərn 'frənt }

Castle's intrinsic factor See intrinsic factor. { 'kas·əlz in'trin·zik 'fak·tər }

cast loading See melt loading. { 'kast ,lōd·iŋ }

Castner cell [CHEM ENG] A type of mercury cell used in the commercial production of chlorine and sodium. { 'kast·nər ,sel }

Castner process [CHEM ENG] A process used industrially to make high-test sodium cyanide by reacting sodium, glowed charcoal, and dry ammonia gas to form sodamide, which is converted to cyanamide immediately; the cyanamide is converted to cyanide with charcoal. { 'kast·nər ,präs·əs }

Castniidae [INV ZOO] The castniids; large diurnal, butterflylike moths composing the single, small family of the lepidopteran superfamily Castnioidea. { ,kast'nī·ə,dē }

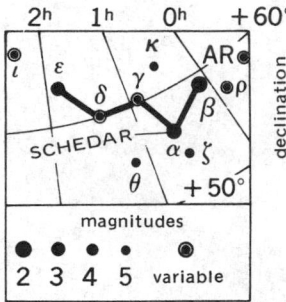

Castnioidea [INV ZOO] A superfamily of neotropical and Indo-Australian lepidopteran insects in the suborder Heteroneura. { ‚kast·nī'óid·ē·ə }

Castor [ASTRON] A multiple star of spectral classification A0 in the constellation Gemini; the star α Geminorum. { 'kas·tər }

castor bean [BOT] The seed of the castor oil plant (*Ricinus communis*), a coarse, erect annual herb in the spurge family (Euphorbiaceae) of the order Geraniales. { 'kas·tər ‚bēn }

castoreum gland [VERT ZOO] A preputial scent gland in the beaver. { ka'stór·ē·əm ‚gland }

castorite [MINERAL] A transparent variety of petalite occurring in crystals. { 'kas·tə‚rīt }

castor machine oil [MATER] Petroleum-base lubricating oil thickened with an aluminum-base soap, such as aluminum oleate. { 'kas·tər mə'shēn ‚óil }

castor oil [MATER] A colorless or greenish nondrying oil extracted from the castor bean; used as a cathartic, in soap, and after processing as a lubricant, and as a leather preservative. Also known as ricinus oil. { 'kas·tər ‚óil }

castor oil acid *See* ricinoleic acid. { 'kas·tər ‚óil 'as·əd }

castration [MED] Removing, or inhibiting the function or development of, the ovaries or testes. { ka'strā·shən }

castration anxiety [PSYCH] Anxiety due to the fear of loss of the genitals or injury to them. { ka'strā·shən aŋ'zī·ə·dē }

cast setting *See* mechanical setting. { 'kast ‚sed·iŋ }

cast steel [MET] Steel shaped by casting. { 'kast ‚stēl }

cast stone [MATER] Building stone molded from concrete so that it resembles natural stone. { 'kast ‚stōn }

cast structure [MET] The microstructure of a casting. { 'kast ‚strək·chər }

cast-weld [MET] Joining parts by pouring molten metal over them in a mold. { 'kast ‚weld }

casual carrier [MED] A person who carries an infectious microorganism but never manifests the disease. { 'kazh·ə·wəl 'kar·ē·ər }

Casuariidae [VERT ZOO] The cassowaries, a family of flightless birds in the order Casuariiformes lacking head and neck feathers and having bony casques on the head. { ‚kazh·ə‚wa'rē·ə‚dē }

Casuariiformes [VERT ZOO] An order of large, flightless, ostrichlike birds of Australia and New Guinea. { ‚kazh·ə‚wa·rē·ə'fór‚mēz }

Casuarinaceae [BOT] The single, monogeneric family of the plant order Casuarinales characterized by reduced flowers and green twigs bearing whorls of scalelike, reduced leaves. { ‚kazh·ə‚wa·rə'nās·ē‚ē }

Casuarinales [BOT] A monofamilial order of dicotyledonous plants in the subclass Hamamelidae. { ‚kazh·ə‚wa·rə'nā·lēz }

cat [NAV ARCH] **1.** A sturdy tackle used for bringing an anchor up to a ship's cathead. **2.** To hoist an anchor up to the cathead of a ship. [VERT ZOO] The common name for all members of the carnivoran mammalian family Felidae, especially breeds of the domestic species, *Felis domestica*. { kat }

CAT *See* clear-air turbulence; computerized tomography.

catabiosis [PHYSIO] Degenerative changes accompanying cellular senescence. { ‚kad·ə‚bī'ō·səs }

catabolism [BIOCHEM] That part of metabolism concerned with the breakdown of large protoplasmic molecules and tissues, often with the liberation of energy. { kə'tab·ə‚liz·əm }

catabolite [BIOCHEM] Any product of catabolism. { kə'tab·ə‚līt }

catabolite repression [BIOCHEM] An intracellular regulatory mechanism in bacteria whereby glucose, or any other carbon source that is an intermediate in catabolism, prevents formation of inducible enzymes. { kə'tab·ə‚līt ri'presh·ən }

catachosis [GEOL] Fracturing or crushing of rock during metamorphism. { ‚kad·ə'kō·səs }

cataclasis [GEOL] Deformation of rock by fracture and rotation of aggregates or mineral grains. { ‚kad·ə'klā·səs }

cataclasite *See* cataclastic rock. { ‚kad·ə'klā‚sīt }

cataclastic metamorphism [PETR] Local metamorphism restricted to a region of faults and overthrusts involving purely mechanical forces resulting in cataclasis. { ‚kad·ə‚klas·tik ‚med·ə'mòr‚fiz·əm }

cataclastic rock [PETR] Rock containing angular fragments formed by cataclasis. Also known as cataclasite. { ‚kad·ə‚klas·tik 'räk }

cataclastic structure *See* mortar structure. { ‚kad·ə‚klas·tik 'strək·chər }

cataclysmic variable [ASTRON] **1.** A star showing a sudden increase in the magnitude of light, followed by a slow fading of light; examples are novae and supernovae. Also known as explosive variable. **2.** In particular, a short-period binary star, one of whose components is a white dwarf star, capable of irregularly timed but recurrent outbursts of brightness by 2 to 10,000. { ‚kad·ə‚kliz·mik 'ver·ē·ə·bəl }

cata-condensed polycyclic [ORG CHEM] An aromatic compound in which no more than two rings have a single carbon atom in common. { ‚kad·ə·kən'denst ‚päl·i'sī·klik }

catadioptric [OPTICS] Involving both reflection and refraction of light. { ‚kad·ə‚dī'äp·trik }

catadromous [VERT ZOO] Pertaining to fishes which live in fresh water and migrate to spawn in salt water. { kə'ta·drə·məs }

catagelophobia [PSYCH] An abnormal fear of ridicule. { ‚kad·ə‚jel·ə‚fō·bē·ə }

Catalan conjecture [MATH] The conjecture that the only pair of consecutive positive integers that are powers of smaller integers is the pair (8,9). { 'kä·tə‚län kən‚jek·chər }

Catalan forge [MET] A furnace or forge for making wrought iron from ore, in which the ore loaded at front and charcoal at rear are covered with fine ore and charcoal dust moistened with water. { 'kad·ə‚lan ‚fòrj }

Catalan numbers [MATH] The numbers, c_n, which count the ways to insert parentheses in a string of n terms so that their product may be unambiguously carried out by multiplying two quantities at a time. { 'kat·əl·ən ‚nəm·bərz }

catalase [BIOCHEM] An enzyme that catalyzes the decomposition of hydrogen peroxide into molecular oxygen and water. { 'kad·əl‚ās }

catalectrotonus [PHYSIO] The negative electric potential during the passage of a current on the surface of a nerve or muscle in the region of the cathode. { ‚kad·əl·ek·'träd·ən·əs }

catalepsy [PSYCH] Suspended animation with loss of voluntary motion associated with hysteria and the schizophrenic reactions in humans, and with organic nervous system disease in animals. { 'kad·əl‚ep·sē }

catalog [COMPUT SCI] **1.** All the indexes to data sets or files in a system. **2.** The index to all other indexes; the master index. **3.** To add an entry to an index or to build an entire new index. **4.** A list of items in a data storage device, usually arranged so that a particular kind of information can be located easily. { 'kad·əl‚äg }

cataloged procedure [COMPUT SCI] A group of control cards (job control language statements) that has been placed in a cataloged data set. { 'kad·əl‚ägd prə'sē·jər }

catalog number [ASTRON] The designation of a star composed of the name of a particular star catalog and the number of the star as listed there. { 'kad·əl‚äg ‚nəm·bər }

catalog-order device [ELECTR] A logic circuit element that is readily obtainable from a manufacturer, and can be combined with other such elements to provide a wide variety of logic circuits. { 'kad·əl‚äg ‚ór·dər di'vīs }

catalysis [CHEM] A phenomenon in which a relatively small amount of substance augments the rate of a chemical reaction without itself being consumed. { kə'tal·ə·səs }

catalyst [CHEM] Substance that alters the velocity of a chemical reaction and may be recovered essentially unaltered in form and amount at the end of the reaction. { 'kad·əl·əst }

catalyst carrier [CHEM] A neutral material used to support a catalyst, such as activated carbon, diatomaceous earth, or activated alumina. { 'kad·əl·əst ‚kar·ē·ər }

catalyst selectivity [CHEM] **1.** The relative activity of a catalyst in reference to a particular compound in a mixture. **2.** The relative rate of a single reactant in competing reactions. { 'kad·əl·əst sə‚lek'tiv·əd·ē }

catalyst stripping [CHEM ENG] Introduction of steam to remove hydrocarbons retained on the catalyst; the steam is introduced where the spent catalyst leaves the reactor. { 'kad·əl·əst ‚strip·iŋ }

catalytic activity [CHEM ENG] The ratio of the space velocity of a catalyst being tested, to the space velocity required for a standard catalyst to give the same conversion as the catalyst under test. { ‚kad·əl‚id·ik ak'tiv·əd·ē }

CASTOR BEAN

Ricinus communis. the castor bean plant. Plant grows 3–40 feet (0.9–12 meters high).

CASUARINACEAE

Casuarina equisetifolia. (Photograph by W. D. Brush)

catalytically cracked gasoline [MATER] Gasoline produced in a refinery reactor equipped with catalytic cracking equipment. { ¦kad·əl¦id·ik·lē ¦krakt gas·ə'lēn }

catalytic antibody [IMMUNOL] A large protein that is naturally produced by the immune system and has the capability of catalyzing a chemical reaction similarly to enzymes. Also known as abzyme. { ¦kad·ə‚lid·ik 'an·ti‚bäd·ē }

catalytic converter [CHEM ENG] A device that is fitted to the exhaust system of an automotive vehicle and contains a catalyst capable of converting potentially polluting exhaust gases into harmless or less harmful products. { ¦kad·əl¦id·ik kən'vərd·ər }

catalytic cracker See catalytic cracking unit. { ¦kad·əl¦id·ik 'krak·ər }

catalytic cracking [CHEM ENG] Conversion of high-boiling hydrocarbons into lower-boiling types by a catalyst. { ¦kad·əl¦id·ik 'krak·iŋ }

catalytic cracking unit [CHEM ENG] A unit in a petroleum refinery in which a catalyst is used to carry out cracking of hydrocarbons. Also known as catalytic cracker. { ¦kad·əl¦id·ik 'krak·iŋ ‚yü·nət }

catalytic hydrogenation [CHEM ENG] Hydrogenating by means of catalysts such as nickel or palladium. { ¦kad·əl¦id·ik ‚hī·drə·jə'nā·shən }

catalytic polymerization [CHEM ENG] Polymerization of monomers to form high-molecular-weight molecules in the presence of catalysts. { ¦kad·əl¦id·ik pə‚lim·ə·rə'zā·shən }

catalytic reforming [CHEM ENG] Rearranging of hydrocarbon molecules in a gasoline boiling-range feedstock to form hydrocarbons having a higher antiknock quality. Abbreviated CR. { ¦kad·əl¦id·ik rē'fȯr·miŋ }

catalytic site See active site. { ¦kad·əl¦id·ik ‚sīt }

catamaran [NAV ARCH] 1. A sailing or powered boat having two rather slender hulls joined by a deck or other structure. 2. A rectangular raft resting on two parallel cylindrical floats. { ¦kad·ə·mə‚ran }

catamount See puma. { 'kad·ə·maȯnt }

cat-and-mouse engine [MECH ENG] A type of rotary engine, typified by the Tschudi engine, which is an analog of the reciprocating piston engine, except that the pistons travel in a circular motion. Also known as scissor engine. { ¦kat ən 'maȯs ‚en·jən }

cataphoresis See electrophoresis. { ‚kad·ə·fə'rē·səs }

catapleiite [MINERAL] $(Na_2,Ca)ZrSi_3O_9·2H_2O$ A yellow or yellowish-brown mineral crystallizing in the hexagonal system, consisting of a hydrous silicate of sodium, calcium, and zirconium, and occurring in thin tabular crystals; hardness is 6 on Mohs scale, and specific gravity is 2.8. { ‚kad·ə'plī‚īt }

cataplexy [MED] 1. Sudden loss of muscle tone provoked by exaggerated emotion, such as excessive anger or laughter, often associated with a profound desire for sleep. 2. Prostration by the sudden onset of disease. 3. Hypnotic sleep. { 'kad·ə‚plek·sē }

Catapochrotidae [INV ZOO] A monospecific family of coleopteran insects in the superfamily Cucujoidea. { ‚kad·ə·pə'kräd·ə‚dē }

catapult [AERO ENG] 1. A power-actuated machine or device for hurling an object at high speed, for example, a device which launches aircraft from a ship deck. 2. A device, usually explosive, for ejecting a person from an aircraft. [ORD] A mechanical device for hurling grenades or bombs. { 'kad·ə‚pəlt }

cataract [HYD] A waterfall of considerable volume with the vertical fall concentrated in one sheer drop. [MED] An opacity in the crystalline lens or the lens capsule of the eye. { 'kad·ə‚rakt }

cataracting [MECH ENG] A motion of the crushed bodies in a ball mill in which some, leaving the top of the crop load, fall with impact to the toe of the load. { 'kad·ə‚rak·tiŋ }

catarobic [ECOL] Pertaining to a body of water characterized by the slow decomposition of organic matter, and oxygen utilization which is insufficient to prevent the activity of aerobic organisms. { ¦kad·ə‚rō·bik }

catarrh [MED] An old term for an inflammation of mucous membranes, particularly of the respiratory tract. { kə'tär }

catarrhal conjunctivitis [MED] A usually acute inflammation of the conjunctiva with smarting of the eyes, heaviness of the lids, photophobia, and excessive mucous or mucopurulent secretion, caused by a variety of contagious organisms, but sometimes becoming chronic as a sequela of the acute form or because of irritation from polluted atmosphere or allergic factors. Also known as pinkeye. { kə'tär·əl kən‚jəŋk·ti'vīd·əs }

catarrhal jaundice See infectious hepatitis. { kə'tär·əl 'jȯn·dəs }

catastrophe theory [MATH] A theory of mathematical structure in which smooth continuous inputs lead to discontinuous responses. { kə'tas·trə·fē ‚thē·ə·rē }

catastrophic error [COMPUT SCI] A situation in which so many errors are detected in a computer program that its compilation or execution is automatically terminated. { ‚kad·ə'straf·ik 'er·ər }

catastrophic failure [ENG] 1. A sudden failure without warning, as opposed to degradation failure. 2. A failure whose occurrence can prevent the satisfactory performance of an entire assembly or system. { ‚kad·ə'straf·ik 'fāl·yər }

catastrophism [GEOL] The theory that most features in the earth were produced by the occurrence of sudden, short-lived, worldwide events. [PALEON] The theory that the differences between fossils in successive stratigraphic horizons resulted from a general catastrophe followed by creation of the different organisms found in the next-younger beds. { kə'tas·trə‚fiz·əm }

catatonia [PSYCH] A type of schizophrenic reaction in which the individual remains speechless and motionless, assumes fixed postures, and lacks the will and resists attempts to activate speech and movement. Also known as catatonic schizophrenia. { ‚kad·ə'tōn·ē·ə }

catatonic schizophrenia See catatonia. { ‚kad·ə¦tän·ik ‚skit·sə'frē·nē·ə }

catazone [GEOL] The deepest zone of rock metamorphism where high temperatures and pressures prevail. { 'kad·ə‚zōn }

cat back [NAV ARCH] A piece of rope sometimes attached to the hook of a cat block to assist in hooking the ring of the anchor. Also known as backrope. { 'kat ‚bak }

cat block [NAV ARCH] A heavy block with a hook, used in hoisting a ship's anchor up to the cathead. { 'kat ‚bläk }

catch [DES ENG] A device used for fastening a door or gate and usually operated manually from only one side, for example, a latch. { kach }

catch basin [CIV ENG] 1. A basin at the point where a street gutter empties into a sewer, built to catch matter that would not easily pass through the sewer. 2. A well or reservoir into which surface water may drain off. { 'kach ‚bā·sən }

catch crop [AGR] A rapidly growing plant that can be intercropped between rows of the main crop; often used as a green manure. { 'kach ‚kräp }

catcher [ELECTR] Electrode in a velocity-modulated vacuum tube on which the spaced electron groups induce a signal; the output of the tube is taken from this element. { 'kach·ər }

catching diode [ELECTR] Diode connected to act as a short circuit when its anode becomes positive; the diode then prevents the voltage of a circuit terminal from rising above the diode cathode voltage. { 'kach·iŋ ‚dī‚ōd }

catching sample [PETRO ENG] During drilling operations, a sample of a cutting obtained from the drilling fluid as it emerges from the well bore or the bailer of a cable tool. { 'kach·iŋ ‚sam·pəl }

catching up [GRAPHICS] The taking of ink or scumming by the nonimage portions of a press plate. { ¦kach·iŋ ‚əp }

catchlights [GRAPHICS] Reflections of light sources in a subject's eyes or eyeglasses. { 'kach‚līts }

catchment area [GEOGR] The rural-urban outskirts of a particular city. [HYD] See drainage basin. { 'kach·mənt ‚er·ē·ə }

catchment glacier See snowdrift glacier. { 'kach·mənt ‚glā·shər }

catch pit [MIN ENG] In mineral processing, a sump in a mill to which the floor slopes gently; spillage gravitates or is hosed to this area. { 'kach ‚pit }

catchwater [CIV ENG] A ditch for catching water on sloping land. { 'kach‚wȯd·ər }

cat cracker [CHEM ENG] A refinery unit where catalytic cracking is done. { 'kat ‚krak·ər }

catechol [ORG CHEM] One of a group of three isomeric dihydroxy benzenes in which the two hydroxyl groups are ortho to each other. Also known as catechin; pyrocatechol; pyrocatechuic acid. { ˈkad·ə‚kȯl }

catecholamine [BIOCHEM] Any one of a group of sympathomimetic amines containing a catechol moiety, including especially epinephrine, norepinephrine (levarterenol), and dopamine. { ‚kad·əˈkäl·ə‚mēn }

categorical data [STAT] Data separable into categories that are mutually exclusive, for example, age groups. { ‚kad·əˈgȯr·i·kəl ˈdad·ə }

categorization [COMPUT SCI] Process of separating multiple addressed messages to form individual messages for singular addresses. { ‚kad·ə·gə·rəˈzā·shən }

category [MATH] A class of objects together with a set of morphisms for each pair of objects and a law of composition for morphisms; sets and functions form an important category, as do groups and homomorphisms. [SYST] In a hierarchical classification system, the level at which a particular group is ranked. { ˈkad·ə‚gȯr·ē }

catena [COMPUT SCI] A series of data items that appears in a chained list. [GEOL] A group of soils derived from uniform or similar parent material which nonetheless show variations in type because of differences in topography or drainage. { kəˈtē·nə }

catenane [ORG CHEM] A supramolecular species consisting of mechanically interlocked macrocyclic rings. { ˈkat·ən‚ān }

catenary [MATER] In fiberglass-reinforced plastics, a measure of the sag in an assemblage of a number of strands, which have a minimal amount of twist, at the midpoint of a specified length. [MATH] The curve obtained by suspending a uniform chain by its two ends; the graph of the hyperbolic cosine function. Also known as alysoid; chainette. { ˈkat·ə‚ner·ē }

catenary suspension [ENG] Holding a flexible wire or chain aloft by its end points; the wire or chain takes the shape of a catenary. { ˈkat·ə‚ner·ē səsˈpen·shən }

catenate [COMPUT SCI] To arrange a collection of items in a chained list or catena. { ˈkat·ən‚āt }

catenation [CHEM] Formation of a chain structure by the bonding of atoms of the same element, for example, carbon in the hydrocarbons. { ‚kat·ənˈā·shən }

catenin [BIOCHEM] Any of a family of 80–102-kilodalton proteins that are thought to have a major role in regulation of cell-to-cell adhesion, which is related to their interaction with E-cadherin and the actin cytoskeleton. { ˈkat·ə·nən }

catenoid [MATH] The surface of revolution obtained by rotating a catenary about a horizontal axis. { ˈkat·ən‚ȯid }

catenulate [BIOL] Having a chainlike form. { kəˈten·yə‚lāt }

Catenulida [INV ZOO] An order of threadlike, colorless fresh-water rhabdocoeles with a simple pharynx and a single, median protonephridium. { kə‚ten·yəˈlī·də }

caterer problem [MATH] A linear programming problem in which it is required to find the optimal policy for a caterer who must choose between buying new napkins and sending them to either a fast or a slow laundry service. { ˈkād·ə·rər ‚präb·ləm }

caterpillar [INV ZOO] **1.** The wormlike larval stage of a butterfly or moth. **2.** The larva of certain insects, such as scorpion flies and sawflies. [MECH ENG] A vehicle, such as a tractor or army tank, which runs on two endless belts, one on each side, consisting of flat treads and kept in motion by toothed driving wheels. { ˈkad·ər‚pil·ər }

caterpillar chain [DES ENG] A short, endless chain on which dogs (grippers) or teeth are arranged to mesh with a conveyor. { ˈkad·ər‚pil·ər ‚chān }

caterpillar fungus See Cordyceps sinensis. { ˈkat·ər‚pil·ər ‚fəŋ·gəs }

caterpillar gate [CIV ENG] A steel gate carried on crawler tracks that is used to control water flow through a spillway. { ˈkad·ər‚pil·ər ‚gāt }

catfish [VERT ZOO] The common name for a number of fishes which constitute the suborder Siluroidei in the order Cypriniformes, all of which have barbels around the mouth. { ˈkat‚fish }

catforming [CHEM ENG] A naphtha-reforming process with a catalyst of platinum-silica-alumina which results in very high hydrogen purity. { ˈkat‚fȯr·miŋ }

catgut [MATER] A thin cord made from the submucosa of sheep and other animal intestine; used for sutures and ligatures, for strings of musical instruments, and for tennis racket strings. Also known as gut. { ˈkat‚gət }

Catharanthus roseus [BOT] The Madagascar periwinkle plant from which the anticancer compounds vinblastine and vincristine are derived. { ‚kath·ə‚ran·thəs ˈrō·zē·əs }

catharsis [PSYCH] Release of tension by releasing deep-seated emotions or reliving a traumatic experience. { kəˈthär·səs }

cathartic [PHARM] Any drug, such as castor oil, mineral oil, or a laxative, that causes defecation. { kəˈthär·dik }

Cathartidae [VERT ZOO] The New World vultures, a family of large, diurnal predatory birds in the order Falconiformes that lack a voice and have slightly webbed feet. { kəˈthär·də‚dē }

cathead [NAV ARCH] A projection on a ship's bow which supports tackle for hoisting an anchor, and to which the anchor is secured. { ˈkat‚hed }

cathead line See catline. { ˈkat‚hed ‚līn }

cathedral glass [MATER] Unpolished translucent sheet glass. { kəˈthē·drəl ‚glas }

cathepsin [BIOCHEM] Any of several proteolytic enzymes occurring in animal tissue that hydrolyze high-molecular-weight proteins to proteoses and peptones. { kəˈthep·sən }

catheter [MED] A hollow, tubular device for insertion into a cavity, duct, or vessel to permit injection or withdrawal of fluids or to establish patency of the passageway. { ˈkath·ə·dər }

catheterization [MED] Insertion or use of a catheter. { ‚kath·əd·ə·rəˈzā·shən }

cathetometer [ENG] An instrument for measuring small differences in height, for example, between two columns of mercury. { ‚kath·əˈtäm·əd·ər }

cathexis [PSYCH] In psychoanalytic theory, the investing of libidinal energy in an activity, object, or person. { kəˈthek·səs }

cathode [ELEC] The terminal at which current leaves a primary cell or storage battery; it is negative with respect to the device, and positive with respect to the external circuit. [ELECTR] **1.** The primary source of electrons in an electron tube; in directly heated tubes the filament is the cathode, and in indirectly heated tubes a coated metal cathode surrounds a heater. Designated K. Also known as negative electrode. **2.** The terminal of a semiconductor diode that is negative with respect to the other terminal when the diode is biased in the forward direction. [PHYS CHEM] The electrode at which reduction takes place in an electrochemical cell, that is, a cell through which electrons are being forced. { ˈkath‚ōd }

cathode bias [ELECTR] Bias obtained by placing a resistor in the common cathode return circuit, between cathode and ground; flow of electrode currents through this resistor produces a voltage drop that serves to make the control grid negative with respect to the cathode. { ˈkath‚ōd ‚bī·əs }

cathode cleaning [MET] Electrolytically removing soil from work connected to the cathode. { ˈkath‚ōd ‚klēn·iŋ }

cathode copper [MET] Copper deposited at the cathode during electrolytic refining; it is melted and marketed as electrolytic copper. { ˈkath‚ōd ‚käp·ər }

cathode corrosion [MET] **1.** Corrosion of the cathode of an electrochemical circuit, usually caused by production of alkaline reaction products. **2.** Corrosion of the cathodic member of a galvanic couple. { ˈkath‚ōd kəˈrō·zhən }

cathode-coupled amplifier [ELECTR] A cascade amplifier in which the coupling between two stages is provided by a common cathode resistor. { ‚kath‚ōd ‚kəp·əld ˈam·plə‚fī·ər }

cathode coupling [ELECTR] Use of an input or output element in the cathode circuit for coupling energy to another stage. { ˈkath‚ōd ‚kəp·liŋ }

cathode crater [ELECTR] A depression formed in the surface of a cathode by sputtering. { ˈkath‚ōd ‚krād·ər }

cathode dark space [ELECTR] The relatively nonluminous region between the cathode glow and the negative flow in a glow-discharge cold-cathode tube. Also known as Crookes dark space; Hittorf dark space. { ˈkath‚ōd ˈdärk ‚spās }

cathode disintegration [ELECTR] The destruction of the active area of a cathode by positive-ion bombardment. { ˈkath‚ōd dis‚int·əˈgrā·shən }

cathode drop [ELECTR] The voltage between the arc stream and the cathode of a glow-discharge tube. Also known as cathode fall. { ˈkath‚ōd ‚dräp }

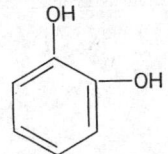

CATECHOL

Structural formula of catechol.

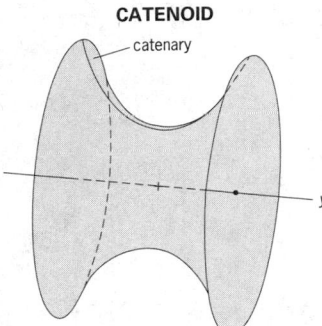

CATENOID

The axis of revolution is the y axis.

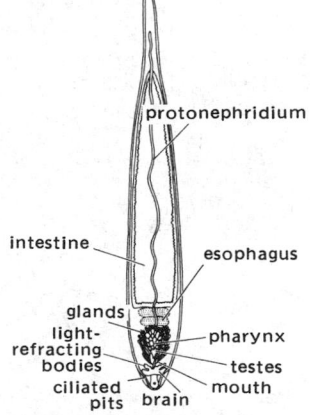

CATENULIDA

Stenostomum grande, the commonest and most widely distributed of all fresh-water rhabdocoeles.

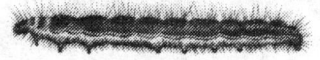

CATERPILLAR

Forest tent caterpillar (*Malacosoma distria*).

cathode efficiency [CHEM ENG] The proportion of current used for completion of a given process at the cathode. { 'kath₁ōd i₁fish·ən·sē }

cathode emission [ELECTR] A process whereby electrons are emitted from the cathode structure. { 'kath₁ōd i'mish·ən }

cathode fall *See* cathode drop. { 'kath₁ōd ₁fȯl }

cathode follower [ELECTR] A vacuum-tube circuit in which the input signal is applied between the control grid and ground, and the load is connected between the cathode and ground. Also known as grounded-anode amplifier; grounded-plate amplifier. { 'kath₁ōd ₁fäl·ə·wər }

cathode glow [ELECTR] The luminous glow that covers all or part of the cathode in a glow-discharge cold-cathode tube. { 'kath₁ōd ₁glō }

cathode interface capacitance [ELECTR] A capacitance which, when connected in parallel with an appropriate resistance, forms an impedance approximately equal to the cathode interface impedance. Also known as layer capacitance. { ¦kath₁ōd ¦in·tər₁fās kə'pas·əd·əns }

cathode interface impedance [ELECTR] The impedance between the cathode base and coating in an electron tube, due to a high-resistivity layer or a poor mechanical bond. Also known as layer impedance. { ¦kath₁ōd ¦in·tər₁fās im'pēd·əns }

cathode keying [ELECTR] Transmitter keying by means of a key in the cathode lead of the keyed vacuum-tube stage, opening the direct-current circuits for the grid and anode simultaneously. { 'kath₁ōd ₁kē·iŋ }

cathode layers [ELECTR] One or more faint layers next to, and on the anode side of, the Aston dark space in a glow-discharge tube. { 'kath₁ōd ₁lā·ərz }

cathode modulation [ELECTR] Amplitude modulation accomplished by applying the modulating voltage to the cathode circuit of an electron tube in which the carrier is present. { 'kath₁ōd ₁mäj·ə'lā·shən }

cathode ray [ELECTR] A stream of electrons, such as that emitted by a heated filament in a tube, or that emitted by the cathode of a gas-discharge tube when the cathode is bombarded by positive ions. { 'kath₁ōd ¦rā }

cathode-ray oscillograph [ELECTR] A cathode-ray oscilloscope in which a photographic or other permanent record is produced by the electron beam of the cathode-ray tube. { 'kath₁ōd ¦rā ä'sil·ə₁graf }

cathode-ray oscilloscope [ELECTR] A test instrument that uses a cathode-ray tube to make visible on a fluorescent screen the instantaneous values and waveforms of electrical quantities that are rapidly varying as a function of time or another quantity. Abbreviated CRO. Also known as oscilloscope; scope. { 'kath₁ōd ¦rā ä'sil·ə₁skōp }

cathode-ray output [COMPUT SCI] A cathode-ray tube used in a computer to display output information in graphic form or by character representation. { 'kath₁ōd ¦rā 'aut₁put }

cathode-ray storage tube [ELECTR] A storage tube in which the information is written by means of a cathode-ray beam. { 'kath₁ōd ¦rā 'stȯr·ij ₁tüb }

cathode-ray tube [ELECTR] An electron tube in which a beam of electrons can be focused to a small area and varied in position and intensity on a surface. Abbreviated CRT. Originally known as Braun tube; also known as electron-ray tube. { 'kath₁ōd ¦rā ₁tüb }

cathode-ray tuning indicator [ELECTR] A small cathode-ray tube having a fluorescent pattern whose size varies with the voltage applied to the grid; used in radio receivers to indicate accuracy of tuning and as a modulation indicator in some tape recorders. Also known as electric eye; electron-ray indicator; magic eye; tuning eye. { 'kath₁ōd ¦rā 'tün·iŋ in·də'kād·ər }

cathode-ray voltmeter [ELEC] An instrument consisting of a cathode-ray tube of known sensitivity, whose deflection can be used to measure voltages. { 'kath₁ōd ¦rā 'vōlt₁mēd·ər }

cathode resistor [ELECTR] A resistor used in the cathode circuit of a vacuum tube, having a resistance value such that the voltage drop across it due to tube current provides the correct negative grid bias for the tube. { 'kath₁ōd ri'zis·tər }

cathode spot [ELECTR] The small cathode area from which an arc appears to originate in a discharge tube. { 'kath₁ōd ₁spät }

cathode sputtering *See* sputtering. { 'kath₁ōd 'spəd·ə·riŋ }

cathodic coating [MET] Material forming a continuous film on a base metal, deposited by mechanical coating or electroplating. { kə'thäd·ik ₁kōd·iŋ }

cathodic disbonding [MET] The destruction of adhesion between a cathodic coating and its substrate by products of a cathodic reduction reaction. { kə'thōd·ik ₁dis'bänd·iŋ }

cathodic inhibitor [CHEM ENG] A compound, such as calcium bicarbonate or sodium phosphate, which is deposited on a metal surface in a thin film that operates at the cathodes to provide physical protection over the entire surface against corrosive attack in a conducting medium. { kə'thäd·ik in'hib·əd·ər }

cathodic polarization [PHYS CHEM] Portion of electric cell polarization occurring at the cathode. { kə'thäd·ik ₁pō·lə·rə'zā·shən }

cathodic protection [MET] Protecting a metal from electrochemical corrosion by using it as the cathode of a cell with a sacrificial anode. Also known as electrolytic protection. { kə'thäd·ik prə'tek·shən }

cathodoluminescence [ELECTR] Luminescence produced when high-velocity electrons bombard a metal in vacuum, thus vaporizing small amounts of the metal in an excited state, which amounts emit radiation characteristic of the metal. Also known as electronoluminescence. { ¦kath·ə₁dō₁lüm·ə'nes·əns }

cathodophosphorescence [ELECTR] Phosphorescence produced when high-velocity electrons bombard a metal in a vacuum. { ¦kath·ə₁dō₁fas·fə'res·əns }

catholyte [CHEM] Electrolyte adjacent to the cathode in an electrolytic cell. { 'kath·ə₁līt }

cation [CHEM] A positively charged atom or group of atoms, or a radical which moves to the negative pole (cathode) during electrolysis. { 'kat₁ī·ən }

cation analysis [ANALY CHEM] Qualitative analysis for cations in aqueous solution. { 'kat₁ī·ən ə'nal·ə·səs }

cation exchange [CHEM] A chemical reaction in which hydrated cations of a solid are exchanged, equivalent for equivalent, for cations of like charge in solution. { 'kat₁ī·ən iks'chānj }

cation exchange resin [ORG CHEM] A highly polymerized synthetic organic compound consisting of a large, nondiffusible anion and a simple, diffusible cation, which later can be exchanged for a cation in the medium in which the resin is placed. { 'kat₁ī·ən iks'chānj 'rez·ən }

cationic detergent [CHEM] A member of a group of detergents that have molecules containing a quaternary ammonium salt cation with a group of 12 to 24 carbon atoms attached to the nitrogen atom in the cation; an example is alkyltrimethyl ammonium bromide. { ₁kad·ē'än·ik di'tər·jənt }

cationic hetero atom [CHEM] A positively charged atom, other than carbon, in an otherwise carbon atomic chain or ring. { ₁kad·ē'än·ik 'hed·ə·rō 'ad·əm }

cationic polymerization [ORG CHEM] A type of polymerization in which Lewis acids act as catalysts. { ₁kad·ē'än·ik pə₁lim·ə·rə'zā·shən }

cationic reagent [CHEM] A surface-active agent with active positive ions used for ore beneficiation (flotation via flocculation); an example of a cationic reagent is cetyl trimethyl ammonium bromide. { ₁kad·ē'än·ik rē'ā·jənt }

cationtrophy [CHEM] The breaking off of an ion, such as a hydrogen ion or metal ion, from a molecule so that a negative ion remains in equilibrium. { ₁kad·ē'än·trə·fē }

catkin [BOT] An indeterminate type of inflorescence that resembles a scaly spike and sometimes is pendant. { 'kat·kən }

catline [PETRO ENG] In oil well drilling, a heavy line which is used for hoisting. Also known as cathead line. { 'kat₁līn }

catoptric light [OPTICS] Light reflected from a mirror, for example, light from a filament, concentrated into a parallel beam by means of a reflector. { kə'täp·trik 'līt }

catoptrite [MINERAL] An iron black to jet black, monoclinic mineral consisting of a silicoantimonate of aluminum and divalent manganese. Also spelled katoptrite. { kə'täp₁trīt }

Catostomidae [VERT ZOO] The suckers, a family of cypriniform fishes in the suborder Cyprinoidei. { ₁kad·ə'stäm·ə₁dē }

CAT scan [MED] An image of a sectional view of a portion of the body made by computerized tomography. { 'kat ₁skan }

CAT scanner [MED] An instrument consisting of integrated x-ray and computing equipment and used for computerized tomography. { 'kat ₁skan·ər }

cat scratch disease [MED] A benign systematic illness in humans characterized by malaise, fever, and a granulomatous lymphadenitis; the causative organism has not been identified. Also known as benign lymphoreticulosis. { 'kat ¦skrach di₍zēz }

cat's eye [LAP] Any of several gems cut in convex form, principally from crysoberyl, exhibiting opalescent reflections. Also known as cymophane. { 'kats ₍ī }

catshaft [PETRO ENG] An axle fitted with a revolving spool (cathead) at either end which crosses through the draw works of an oil well mechanism. { 'kat₍shaft }

cat's paw [METEOROL] A puff of wind; a light breeze affecting a small area, as one that causes patches of ripples on the surface of water. { 'kats ₍pó }

CATT *See* controlled avalanche transit-time triode. { kat }

Cattell infant intelligence scale [PSYCH] A modification of the revised Stanford-Binet test adapted for children from 2 to 30 months of age. { kə'tel 'in·fənt in'tel·ə·jəns ₍skāl }

cattle [AGR] Domesticated bovine animals, including cows, steers, and bulls, raised and bred on a ranch or farm. { 'kad·əl }

CATV *See* cable television.

catwalk [ENG] A narrow, raised platform or pathway used for passage to otherwise inaccessible areas, such as a raised walkway on a ship permitting fore and aft passage when the main deck is awash, a walkway on the roof of a freight car, or a walkway along a vehicular bridge. { 'kat₍wòk }

catwhisker [ELECTR] A sharply pointed, flexible wire used to make contact with the surface of a semiconductor crystal at a point that provides rectification. { 'kat₍wis·kər }

Cauchy boundary conditions [MATH] The conditions imposed on a surface in euclidean space which are to be satisfied by a solution to a partial differential equation. { kō·shē 'baùn·drē kən₍dish·ənz }

Cauchy condensation test [MATH] A monotone decreasing series of positive terms Σa_n converges or diverges as does $\Sigma p^n a_{p^n}$ for any positive integer p. { kō·shē ₍kän·den'sā·shən ₍test }

Cauchy data [RELAT] The Cauchy data for a hyperbolic partial differential equation consist of the value of the field and its time derivative on some spacelike surface. { kō·shē ₍dad·ə }

Cauchy dispersion formula [OPTICS] A semiempirical formula for the index of refraction n of a medium as a function of wavelength λ, according to which $n = A + (B/\lambda^2)$, where A and B are constants. { kō·shē dis'pər·zhən ₍fòr·myə·lə }

Cauchy distribution [STAT] A distribution function having the form $M/[\pi M^2 + (x - a)^2]$, where x is the variable and M and a are constants. Also known as Cauchy frequency distribution. { kō·shē dis·trə'byü·shən }

Cauchy formula [MATH] An expression for the value of an analytic function f at a point z in terms of a line integral

$$ f(z) = \frac{1}{2\pi i} \int_C \frac{f(\zeta)}{\zeta - z}\, d\zeta $$

where C is a simple closed curve containing z. Also known as Cauchy integral formula. { kō·shē ₍fòr·myə·lə }

Cauchy frequency distribution *See* Cauchy distribution. { kō·shē 'frē·kwən·sē dis·trə'byü·shən }

Cauchy-Hadamard theorem [MATH] The theorem that the radius of convergence of a Taylor series in the complex variable z is the reciprocal of the limit superior, as n approaches infinity, of the nth root of the absolute value of the coefficient of z^n. { kō·shē 'had·ə·mär ₍thir·əm }

Cauchy horizon [RELAT] Boundary of the region that can be predicted by Cauchy data set on a spacelike surface (partial Cauchy surface). { kō·shē hə₍rīz·ən }

Cauchy inequality [MATH] The square of the sum of the products of two variables for a range of values is less than or equal to the product of the sums of the squares of these two variables for the same range of values. { kō·shē ₍in·i'kwäl·əd·ē }

Cauchy integral formula *See* Cauchy formula. { kō·shē ₍in·tə·grəl ¦fòr·mya·lə }

Cauchy integral test *See* Cauchy's test for convergence. { kō·shē 'in·tə·grəl ₍test }

Cauchy integral theorem [MATH] The theorem that if γ is a closed path in a region R satisfying certain topological properties, then the integral around γ of any function analytic in R is zero. { kō·shē 'in·tə·grəl ₍thir·əm }

Cauchy mean [MATH] The Cauchy mean-value theorem for the ratio of two continuous functions. { kō·shē ₍mēn }

Cauchy mean-value theorem [MATH] The theorem that if f and g are functions satisfying certain conditions on an interval $[a,b]$, then there is a point x in the interval at which the ratio of derivatives $f'(x)/g'(x)$ equals the ratio of the net change in f, $f(b) - f(a)$, to that of g. { kō·shē ¦mēn ¦val·yü ₍thir·əm }

Cauchy net [MATH] A net whose members are elements of a topological vector space and which satisfies the condition that for any neighborhood of the origin of the space there is an element a of the directed system that indexes the net such that if b and c are also members of this directed system and $b \geq a$ and $c \geq a$, then $x_b - x_c$ is in this nieghborhood. { kō·shē ₍net }

Cauchy number [FL MECH] A dimensionless number used in the study of compressible flow, equal to the density of a fluid times the square of its velocity divided by its bulk modulus. Also known as Hooke number. { kō·shē ₍nəm·bər }

Cauchy principal value [MATH] Also known as principal value. **1.** The Cauchy principal value of

$$ \int_{-\infty}^{\infty} f(x)dx \quad \text{is} \quad \lim_{s \to \infty} \int_{-s}^{s} f(x)dx $$

provided the limit exists. **2.** If a function f is bounded on an interval (a,b) except in the neighborhood of a point c, the Cauchy principal value of

$$ \int_a^b f(x)dx \quad \text{is} \quad \lim_{\Delta \to 0} \left[\int_a^{c-\delta} f(x)dx + \int_{c+\delta}^b f(x)dx \right] $$

provided the limit exists. { kō·shē ¦prin·sə·pəl ¦val·yü }

Cauchy problem [MATH] The problem of determining the solution of a system of partial differential equation of order m from the prescribed values of the solution and of its derivatives of order less than m on a given surface. { kō·shē ₍präb·ləm }

Cauchy product [MATH] A method of multiplying two absolutely convergent series to obtain a series which converges absolutely to the product of the limits of the original series:

$$ \left(\sum_{n=0}^{\infty} a_n\right)\left(\sum_{n=0}^{\infty} b_n\right) = \sum_{n=0}^{\infty} c_n \quad \text{where} \quad c_n = \sum_{k=0}^{n} a_k b_{n-k} $$

{ kō·shē ₍prad·əkt }

Cauchy radical test [MATH] A test for convergence of series of positive terms: if the nth root of the nth term is less than some number less than unity, the series converges; if it remains equal to or greater than unity, the series diverges. { kō·shē 'rad·i·kəl ₍test }

Cauchy random variable [MATH] A random variable that has a Cauchy distribution. { kō·shē ₍ran·dəm 'ver·ē·ə·bəl }

Cauchy ratio test [MATH] A series of nonnegative terms converges if the limit, as n approaches infinity, of the ratio of the $(n + 1)$st to nth term is smaller than 1, and diverges if it is greater than 1; the test fails if this limit is 1. Also known as ratio test. { kō·shē 'rā·shō ₍test }

Cauchy relations [SOLID STATE] A set of six relations between the compliance constants of a solid which should be satisfied provided the forces between atoms in the solid depend only on the distances between them and act along the lines joining them, and provided that each atom is a center of symmetry in the lattice. { kō·shē ri'lā·shənz }

Cauchy residue theorem [MATH] The theorem expressing a line integral around a closed curve of a function which is analytic in a simply connected domain containing the curve, except at a finite number of poles interior to the curve, as a sum of residues of the function at these poles. { kō·shē 'rez·ə₍dü ₍thir·əm }

Cauchy-Riemann equations [MATH] A pair of partial differential equations that is satisfied by the real and imaginary parts of a complex function $f(z)$ if and only if the function is analytic: $\partial u/\partial x = \partial v/\partial y$ and $\partial u/\partial y = -\partial v/\partial x$, where $f(z) = u + iv$ and $z = x + iy$. { kō·shē 'rē₍män i'kwā·zhənz }

Cauchy-Schwarz inequality [MATH] The square of the inner product of two vectors does not exceed the product of the squares of their norms. Also known as Buniakowski's

CAULERPACEAE

Caulerpa showing the stolonlike branches, rhizoidal branches, and erect featherlike frond typical of the Caulerpaceae.

CAULOBACTERACEAE

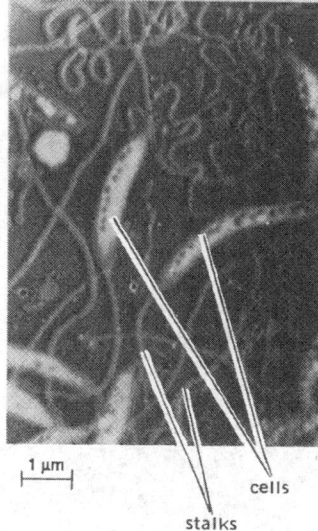

Electron micrograph of *Caulobacter*. (*Courtesy of J. Wachsman*)

inequality; Schwarz' inequality. { kō·shē 'shwȯrts in·i'kwäl·əd·ē }

Cauchy sequence [MATH] A sequence with the property that the difference between any two terms is arbitrarily small provided they are both sufficiently far out in the sequence; more precisely stated: a sequence $\{a_n\}$ such that for every $\epsilon > 0$ there is an integer N with the property that if n and m are both greater than N, then $|a_n - a_m| < \epsilon$. Also known as fundamental sequence; regular sequence. { kō·shē 'sē·kwəns }

Cauchy's mean-value theorem See second mean-value theorem. { kō·shēz ‚mēn ‚val·yü 'thir·əm }

Cauchy's test for convergence [MATH] **1.** A series is absolutely convergent if the limit as n approaches infinity of its nth term raised to the $1/n$ power is less than unity. **2.** A series a_n is convergent if there exists a monotonically decreasing function f such that $f(n) = a_n$ for n greater than some fixed number N, and if the integral of $f(x)dx$ from N to ∞ converges. Also known as Cauchy integral test; Maclaurin-Cauchy test. { kō·shēz ‚test fər kən'vər·jəns }

Cauchy surface [RELAT] A surface S in a space-time M is a (global) Cauchy surface if every nonspacelike curve in M intersects S exactly once; that is, the Cauchy development of S equals M. { kō·shē ‚sər·fəs }

Cauchy transcendental equation [MATH] An equation whose roots are characteristic values of a certain type of Sturm-Liouville problem: $\tan \sigma\pi = (k + K)/(\sigma^2 - kK)$, where k and K are given, and σ is to be determined. { kō·shē ‚trans‚en¦dent·əl i'kwā·zhən }

cauda equina [NEURO] The roots of the sacral and coccygeal nerves, collectively; so called because of their resemblance to a horse's tail. { 'kau̇d·ə i'kwīn·ə }

caudal [ZOO] Toward, belonging to, or pertaining to the tail or posterior end. { 'kȯd·əl }

caudal artery [VERT ZOO] The extension of the dorsal aorta in the tail of a vertebrate. { 'kȯd·əl 'ärd·ə·rē }

caudal vertebra [ANAT] Any of the small bones of the vertebral column that support the tail in vertebrates; in humans, three to five are fused to form the coccyx. { 'kȯd·əl 'vər·tə·brə }

Caudata [VERT ZOO] An equivalent name for Urodela. { kau̇·dad·ə }

caudate [ZOO] **1.** Having a tail or taillike appendage. **2.** Any member of the Caudata. { 'kȯ‚dāt }

caudate lobe [ANAT] The tailed lobe of the liver that separates the right extremity of the transverse fissure from the commencement of the fissure for the inferior vena cava. { 'kȯ‚dāt 'lōb }

caudate nucleus [ANAT] An elongated arched gray mass which projects into and forms part of the lateral wall of the lateral ventricle. { 'kȯ‚dāt 'nü·klē·əs }

caudex [BOT] The main axis of a plant, including stem and roots. { 'kȯ‚deks }

caudicle [BOT] A slender appendage attaching pollen masses to the stigma in orchids. { 'kȯd·ə·kəl }

Cauer filter See elliptic-integral filter. { 'kau̇·ər ‚fil·tər }

Cauer form [ELEC] A continued fraction expansion of the impedance used in the network synthesis for a driving point function resulting in a ladder network. { 'kau̇·ər ‚fȯrm }

caul [ENG] A sheet of metal or other material that is heated and used to equalize pressure during fabricating plywood, shaping surface veneer, and hot-pressing composite materials. { kȯl }

cauldron subsidence [GEOL] **1.** A structure formed by the lowering along a steep ring fracture of a more or less cylindrical block, usually 1 to 10 miles (1.6 to 16 kilometers) in diameter, into a magma chamber. **2.** The process of forming such a structure. { 'kȯl·drən səb'sī·dəns }

Caulerpaceae [BOT] A family of green algae in the order Siphonales. { ‚kȯ·lər'pās·ē‚ē }

Caulerpales See Siphonales. { ‚kȯ·lər'pa‚lēz }

caulescent [BOT] Having an aboveground stem. { kȯ'les·ənt }

cauliflorous [BOT] Producing flowers on the older branches or main stem. { ¦kȯl·ə¦flȯr·əs }

cauliflory [BOT] Of flowers, growth on the main stem of limbs of a tree. { kȯl·ə‚flȯr·ē }

cauliflower [BOT] *Brassica oleracea* var. *botrytis*. A biennial crucifer of the order Capparales grown for its edible white head or curd, which is a tight mass of flower stalks. { 'kȯl·ə‚flau̇·ər }

cauliflower disease [PL PATH] **1.** A disease of the strawberry plant caused by the eelworm and manifested as clustered, puckered, and malformed leaves. **2.** A bacterial disease of the strawberry and some other plants caused by *Corynebacterium fascians*. { 'kȯl·ə‚flau̇·ər di‚zēz }

cauline [BOT] Belonging to or arising from the stem, particularly if on the upper portion. { 'kȯ‚līn }

caulk Also spelled calk. [ENG] To make a seam or point airtight, watertight, or steamtight by driving in caulking compound, dry pack, lead wool, or other material. [MATER] Material used to caulk seams. { kȯk }

caulking compound [MATER] A heavy paste, such as a synthetic, containing a polysulfide rubber and lead peroxide curing agent, or a natural product such as oakum, used for caulking. { 'kȯk·iŋ ‚käm‚pau̇nd }

caulking iron [DES ENG] A tool for applying caulking to a seam. { 'kȯk·iŋ ‚ī·ərn }

Caulobacter [MICROBIO] A genus of prosthecate bacteria; cells are rod-shaped, fusiform, or vibrioid and stalked, and reproduction is by binary fission of cells. { ‚kȯl·ō'bak·tər }

Caulobacteraceae [MICROBIO] Formerly a family of aquatic, stalked, gram-negative bacteria in the order Pseudomonadales. { ¦kȯl·ə‚bak·tə'rās·ē‚ē }

caulocarpic [BOT] Having stems that bear flowers and fruit every year. { ‚kȯl·ō'kär·pik }

Caulococcus [MICROBIO] A genus of bacteria of uncertain affiliation; coccoid cells may be connected by threads; reproduces by budding. { ‚kȯl·ō'käk·əs }

caulome [BOT] The stem structure or stem axis of a plant as a whole. { 'kȯ‚lōm }

causal boundary [RELAT] A boundary attached to a space-time that depends only on the causal structure; it does not distinguish between boundary points at finite distances (singularities) or those at infinity. Also known as C boundary. { ¦kȯz·əl 'bau̇n·drē }

causal curve [RELAT] A curve in space-time that is nowhere spacelike. { ¦kȯz·əl 'kərv }

causal future [RELAT] The causal future relative to a set of points S in a space-time M is the set of points in M which can be reached from S by future-directed timelike or null curves. { ¦kȯz·əl 'fyü·chər }

causalgia [MED] A sensation of burning pain, especially of the palms and soles, which may be of psychic or organic origin. { kȯ'zal·jē·ə }

causality [MECH] In classical mechanics, the principle that the specification of the dynamical variables of a system at a given time, and of the external forces acting on the system, completely determines the values of dynamical variables at later times. Also known as determinism. [PHYS] **1.** The principle that an event cannot precede its cause; in a relativistic theory, an event cannot have an effect outside its future light cone. **2.** In relativistic quantum field theory, the principle that the field operators at different space-time points commute (for boson fields; anticommute in the case of fermion fields) if the separation of the points is spacelike. [QUANT MECH] The principle that the specification of the dynamical state of a system at a given time, and of the interaction of the system with its environment, determines the dynamical state of the system at later times, from which a probability distribution for the observation of any dynamical variable may be determined. Also known as determinism. [SCI TECH] The existence of regularities which control natural phenomena. { kȯ'zal·əd·ē }

causality condition [RELAT] The condition of a space-time requiring there be no closed nonspacelike curves. { kȯ'zal·əd·ē kən‚dish·ən }

causally simple [RELAT] A set of points U in a space-time is said to be causally simple if the causal past and causal future of every compact subset of U is closed in U. { ¦kȯz·əl·ē 'sim·pəl }

causal past [RELAT] The causal past relative to a set of points S in a space-time M is the set of points in M which can be reached from S by past-directed timelike or null curves. { ¦kȯz·əl 'past }

causal system [CONT SYS] A system whose response to an input does not depend on values of the input at later times. Also known as nonanticipatory system; physical system. { 'kȯ‚zəl ‚sis·təm }

caustic [CHEM] **1.** Burning or corrosive. **2.** A hydroxide of a light metal. [OPTICS] A curve or surface which is tangent to the rays of an initially parallel beam after reflection or refraction in an optical system. [PHYS] A curve or surface which is tangent to adjacent orthogonals to waves that have been reflected or refracted from a curved surface. { 'kȯ·stik }

caustic alcohol See sodium ethylate. { 'kȯ·stik 'al·kə‚hȯl }

caustic barley See sabadilla. { 'kȯ·stik 'bär·lē }

caustic cracking See caustic embrittlement. { 'kȯ·stik 'krak·iŋ }

caustic dip [MET] Immersion of metal into a caustic solution such as sodium hydroxide. { 'kȯ·stik 'dip }

caustic embrittlement [MET] Intercrystalline cracking of steel caused by exposure to caustic solutions above 70°C while under tensile stress; once common in riveted boilers. Also known as caustic cracking. { 'kȯ·stik im'brid·əl·mənt }

causticity [CHEM] The property of being caustic. { kȯ'stis·əd·ē }

caustic flooding See alkaline flooding. { 'kȯ·stik 'fləd·iŋ }

causticization [CHEM ENG] A process for converting an alkaline carbonate into lime. { ‚kȯs·tə·sə'zā·shən }

caustic lime See calcium oxide. { 'kȯ·stik'līm }

caustic potash See potassium hydroxide. { 'kȯ·stik 'päd‚ash }

caustic soda [INORG CHEM] See sodium hydroxide. [MATER] Sodium hydroxide that contains 76–78% sodium oxide; the most important of the commercial caustic materials, used in chemical manufacture, petroleum refining, and pulp and paper manufacture. { 'kȯ·stik 'sōd·ə }

caustic treater [CHEM ENG] A vessel containing a strong alkali through which solutions are passed for removal of undesirable substances, for example, sulfides, mercaptans, or acids. { 'kȯ·stik ‚trēd·ər }

caustic wash [CHEM] **1.** Treating a product with a solution of caustic soda to remove impurities. **2.** The solution itself. { 'kȯ·stik 'wäsh }

caustobiolith [GEOL] Combustible organic rock formed by direct accumulation of plant materials; includes coal peat. { ‚kȯ‚stō'bī·ə‚lith }

cauterization [MED] Use of a device or chemical agent to coagulate or destroy tissue. { ‚kȯd·ə·rə'zā·shən }

cautery [MED] Any agent or device used to coagulate or destroy tissue by means of heat, cold, electric current, or caustic chemicals. { 'kȯd·ə·rē }

caution area [NAV] An area in which there is a visible hazard to flight or navigation, such as unusual aircraft maneuvers, high density of aircraft, or parachute drops. { 'kȯ·shən ‚er·ē·ə }

cautionary characteristic [NAV] In marine operations, a unique characteristic of light signifying a particular caution, for example, a quick flashing characteristic indicates a sharp turn in a channel. { 'kȯ·shə‚ner·e ‚kar·ık·tə'ris·tik }

cautious control [CONT SYS] A control law for a stochastic adaptive control system which hedges and uses lower gain when the estimates are uncertain. { 'kȯ·shəs kən'trōl }

cavaburd [METEOROL] Shetland Islands term for a thick fall of snow. Also spelled kavaburd. { 'ka·və·bərd }

Cavalieri's theorem [MATH] The theorem that two solids have the same volume if their altitudes are equal and all plane sections parallel to their bases and at equal distances from their bases are equal. { ‚kav·ə'lyer·ēz ‚thir·əm }

cavaliers [METEOROL] The local term, in the vicinity of Montpelier, France, for the days near the end of March or the beginning of April when the mistral is usually strongest. { kä'väl·yā }

cavalry twill See tricotine. { 'kav·əl·rē ‚twil }

cave [ENG] A pit or tunnel under a glass furnace for collecting ashes or raking the fire. [GEOL] A natural, hollow chamber or series of chambers and galleries beneath the earth's surface, or in the side of a mountain or hill, with an opening to the surface. [MIN ENG] **1.** Fragmented rock materials, derived from the sidewalls of a borehole, that obstruct the hole or hinder drilling progress. Also known as cavings. **2.** The partial or complete failure of borehole sidewalls or mine workings. Also known as cave-in. [NUCLEO] A heavily shielded compartment in which highly radioactive material can be handled, generally by remote control. Also known as hot cell. { kāv }

cave breccia [GEOL] Sharp fragments of limestone debris deposited on the floor of a cave. { ‚kāv 'brech·ə }

cave formation See speleothem. { 'kāv fȯr'mā·shən }

cave-in See cave. { 'kāv‚in }

Cavellinidae [PALEON] A family of Paleozoic ostracodes in the suborder Platycopa. { ‚kav·ə'lin·ə‚dē }

Cavendish balance [ENG] An instrument for determining the constant of gravitation, in which one measures the displacement of two small spheres of mass m, which are connected by a light rod suspended in the middle by a thin wire, caused by bringing two large spheres of mass M near them. { 'kav·ən‚dish 'bal·əns }

caveolae [CYTOL] Tiny indentations in the cell surface membrane which trap fluids during the process of micropinocytosis. { kə'vē·ə·lē }

cave pearl [GEOL] A small, smooth, rounded concretion of calcite or aragonite, formed by concentric precipitation about a nucleus and usually found in limestone caves. { 'kāv 'pərl }

caver [METEOROL] A gentle breeze in the Hebrides, west of Scotland. Also spelled kaver. { 'kāv·ər }

cavern [GEOL] An underground chamber or series of chambers of indefinite extent carved out by rock springs in limestone. { 'kav·ərn }

cavernicolous [BIOL] Inhabiting caverns. { ‚kav·ər‚nik·ə·ləs }

cavernous [GEOL] **1.** Having many caverns or cavities. **2.** Producing caverns. **3.** Of or pertaining to a cavern, that is, suggesting vastness. { 'kav·ər·nəs }

cavernous sinus [ANAT] Either of a pair of venous sinuses of the dura mater located on the side of the body of the sphenoid bone. { 'kav·ər·nəs 'sī·nəs }

Caviidae [VERT ZOO] A family of large, hystricomorph rodents distinguished by a reduced number of toes and a rudimentary tail. { kə'vī·ə‚dē }

caving [MIN ENG] A mining procedure, used when the surface is expendable, in which the ore body is undercut and allowed to fall, breaking into small pieces that are recovered by passages (drifts) driven for that purpose; sublevel caving, block caving, and top slicing are examples. [PETRO ENG] Collapsing of the walls of a wellbore. Also known as sloughing. { 'kāv·iŋ }

caving ground [MIN ENG] Rock formation that will not stand in the walls of an underground opening without support such as that offered by cementation, casing, or timber. { 'kāv·iŋ ‚graúnd }

cavings See slough. { 'kāv·iŋz }

CA virus See croup-associated virus. { sē'ā ‚vī·rəs }

cavitation [CHEM] Emulsification produced by disruption of a liquid into a liquid-gas two-phase system, when the hydrodynamic pressure of the liquid is reduced to the vapor pressure. [ENG] Pitting of a solid surface such as metal or concrete. [FL MECH] Formation of gas- or vapor-filled cavities within liquids by mechanical forces; broadly includes bubble formation when water is brought to a boil and effervescence of carbonated drinks; specifically, the formation of vapor-filled cavities in the interior or on the solid boundaries of vaporized liquids in motion where the pressure is reduced to a critical value without a change in ambient temperature. [PATH] The formation of one or more cavities in an organ or tissue, especially as the result of disease. { ‚kav·ə'tā·shən }

cavitation corrosion See cavitation erosion. { ‚kav·ə'tā·shən kə'rō·zhən }

cavitation damage See cavitation erosion. { ‚kav·ə'tā·shən ‚dam·ij }

cavitation erosion [MET] Attack of metal surfaces caused by the collapse of cavitation bubbles on the surface of the liquid and characterized by pitting. Also known as cavitation corrosion; cavitation damage. { ‚kav·ə'tā·shən i'rō·zhən }

cavitation noise [ACOUS] Noise resulting from the formation of vapor- or gas-filled cavities in liquids by mechanical forces, as occurs near a propeller. { ‚kav·ə'tā·shən ‚nȯiz }

cavitation number [FL MECH] The excess of the local static pressure head over the vapor pressure head divided by the velocity head. { ‚kav·ə'tā·shən ‚nəm·bər }

cavitation resistance inducer [MECH ENG] In liquid flows through rotating machinery, an axial flow pump with high-solidity blades that is used in front of a main pump in order to increase the inlet head and thereby prevent cavitation in the downstream impeller. { ‚kav·ə‚tā·shən ri'sis·təns in‚dü·sər }

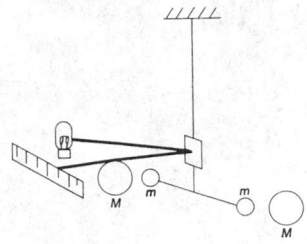

CAVENDISH BALANCE

The torsion balance used by H. Cavendish to determine the gravitational constant G.

caviton [PL PHYS] A region of a plasma having reduced mass density and enhanced wave energy density. { 'kav·ə,tän }

cavity [BIOL] A hole or hollow space in an organ, tissue, or other body part. [ELECTROMAG] See cavity resonator. { 'kav·əd·ē }

cavity charge See shaped charge. { 'kav·əd·ē ,chärj }

cavity coupling [ELECTROMAG] The extraction of electromagnetic energy from a resonant cavity, either waveguide or coaxial, using loops, probes, or apertures. { 'kav·əd·ē ,kəp·liŋ }

cavity filter [ELECTROMAG] A microwave filter that uses quarter-wavelength-coupled cavities inserted in waveguides or coaxial lines to provide band-pass or other response characteristics at frequencies in the gigahertz range. { 'kav·əd·ē ,fil·tər }

cavity frequency meter [ENG] A device that employs a cavity resonator to measure microwave frequencies. { 'kav·əd·ē 'frē·kwən·sē ,mēd·ər }

cavity impedance [ELECTR] The impedance of the cavity of a microwave tube which appears across the gap between the cathode and the anode. { 'kav·əd·ē im'pēd·əns }

cavity magnetron [ELECTR] A magnetron having a number of resonant cavities forming the anode; used as a microwave oscillator. { 'kav·əd·ē 'mag·nə,trän }

cavity oscillator [ELECTR] An ultra-high-frequency oscillator whose frequency is controlled by a cavity resonator. { 'kav·əd·ē 'äs·ə,lād·ər }

cavity radiator [THERMO] A heated enclosure with a small opening which allows some radiation to escape or enter; the escaping radiation approximates that of a blackbody. { 'kav·əd·ē 'rād·ē,ād·ər }

cavity resonance [ELECTROMAG] The resonant oscillation of the electromagnetic field in a cavity. [ENG ACOUS] The natural resonant vibration of a loudspeaker baffle; if in the audio range, it is evident as unpleasant emphasis of sounds at that frequency. { 'kav·əd·ē 'rez·ən·əns }

cavity resonator [ELECTROMAG] A space totally enclosed by a metallic conductor and excited in such a way that it becomes a source of electromagnetic oscillations. Also known as cavity; microwave cavity; microwave resonance cavity; resonant cavity; resonant chamber; resonant element; rhumbatron; tuned cavity; waveguide resonator. { 'kav·əd·ē 'rez·ən,ād·ər }

cavity ringdown laser absorption spectroscopy [SPECT] A direct absorption technique used for measuring short-lived species and for trace-gas analysis in which the rate of decay of light, injected with a pulsed laser and trapped in a cavity formed by two highly reflective mirrors, is measured, allowing the calculation of the amount of light absorbed by the sample. Abbreviated CRLAS. { 'kav·əd·ē 'riŋ,daùn 'lā·zər əb'sörp·shən ,spek'träs·kə·pē }

cavity tuning [ELECTROMAG] Use of an adjustable cavity resonator as a tuned circuit in an oscillator or amplifier, with tuning usually achieved by moving a metal plunger in or out of the cavity to change the volume, and hence the resonant frequency of the cavity. { 'kav·əd·ē ,tün·iŋ }

cavity-type diode amplifier See diode amplifier. { 'kav·əd·ē ,tīp 'dī,ōd ,am·plə,fī·ər }

cavity wall [BUILD] A wall constructed in two separate thicknesses with an air space between; provides thermal insulation. Also known as hollow wall. { 'kav·əd·ē ,wól }

CAVU [AERO ENG] An operational term commonly used in aviation, which designates a condition wherein the ceiling is more than 10,000 feet (3048 meters) and the visibility is more than 10 miles (16 kilometers). Derived from ceiling and visibility unlimited.

cavy [VERT ZOO] Any of the rodents composing the family Caviidae, which includes the guinea pig, rock cavies, mountain cavies, capybara, salt desert cavy, and mara. { 'kā·vē }

CAW See channel address word.

c axis [CRYSTAL] A vertically oriented crystal axis, usually the principal axis; the unique symmetry axis in tetragonal and hexagonal crystals. [GEOL] The reference axis perpendicular to the plane of movement of rock or mineral strata. [MECH ENG] The angle that specifies the rotation of a machine tool about the z axis. { 'sē ,ak·səs }

cay [GEOL] **1.** A flat coral island. **2.** A flat mound of sand built up on a reef slightly above high tide. **3.** A small, low coastal islet or emergent reef composed largely of sand or coral. { kā }

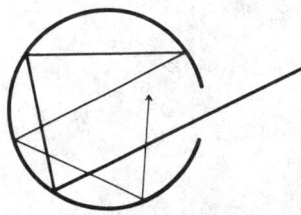

CAVITY RADIATOR

A drawing of a cavity radiator showing the almost complete absorption of entering radiation energy because of the multiple reflections it encounters.

Cayley algebra [MATH] The nonassociative division algebra consisting of pairs of quaternions; it may be identified with an eight-dimensional vector space over the real numbers. { 'kā·lē ,al·jə·brə }

Cayley-Hamilton theorem [MATH] The theorem that a linear transformation or matrix is a root of its own characteristic polynomial. Also known as Hamilton-Cayley theorem. { ¦kāl·ē ¦ham·əl·tən ,thir·əm }

Cayley-Klein parameters [MATH] A set of four complex numbers used to describe the orientation of a rigid body in space, or equivalently, the rotation which produces that orientation, starting from some reference orientation. { ¦kāl·ē ¦klīn pə,ram·əd·ərz }

Cayley numbers [MATH] The members of a Cayley algebra. Also known as octonions. { 'kāl·ē ,nəm·bərz }

Cayley's sextic [MATH] A plane curve with the equation $r = 4a \cos^3 (\theta/3)$, where r and θ are radial and angular polar coordinates and a is a constant. { 'kā,lēz 'sek·stik }

Cayley's theorem [MATH] A theorem that any group G is isomorphic to a subgroup of the group of permutations on G. { 'kā,lēz ,thir·əm }

cay sandstone [GEOL] Firmly cemented or friable coral sand formed near the base of coral reef cays. { ¦kā 'san,stōn }

Caytoniales [PALEOBOT] An order of Mesozoic plants. { ,kā·tän·ē'ā,lēz }

Cayugan [GEOL] Upper Silurian geologic time. { kī'yü·gən }

Cazenovian [GEOL] Lower Middle Devonian geologic time. { kaz·ə'nōv·ē·ən }

C band [COMMUN] A band of radio frequencies extending from 4 to 8 gigahertz. { 'sē ,band }

C-band fixed satellite service [COMMUN] Satellite communication at frequencies in and near the C band, with the uplink frequency in a band from 5.85 to 7.075 gigahertz and the downlink frequency in bands from 3.4 to 4.2 gigahertz and 4.5 to 4.8 gigahertz. { 'sē ,band ¦fikst ¦sad·ə,līt ,sər·vəs }

C-band waveguide [ELECTROMAG] A rectangular waveguide, with dimensions 3.48 by 1.58 centimeters, which is used to excite only the dominant mode (TE_{01}) for wavelengths in the range 3.7–5.1 centimeters. { 'sē ,band 'wāv,gīd }

C battery [ELEC] The battery that supplies the steady bias voltage required by the control-grid electrodes of electron tubes in battery-operated equipment. Also known as grid battery. { 'sē ,bad·ə·rē }

CBC See cipher block chaining.

C bias See grid bias. { 'sē ,bī·əs }

C boundary See causal boundary. { 'sē ,baùn·drē }

CBP See CREB-binding protein.

CBX See computerized branch exchange.

Cc See cirrocumulus cloud.

CCD See calcite compensation depth; charge-coupled device.

C cells [HISTOL] Calcitonin-secreting cells located in the thyroid gland in mammals and in the ultimobranchial body in lower animals. { 'sē ,selz }

C chart [IND ENG] A quality-control chart showing number of defects in subgroups of constant size; gives information concerning quality level, its variability, and evidence of assignable causes of variation. { 'sē ,chärt }

CCIS See common-channel interoffice signaling.

CCIT 2 code [COMMUN] A printing-telegraph code in which each character is represented by five binary digits. Also known as international telegraph alphabet; International Telegraphic Consultative Committee code 2. { ,sē,sē,ī,tē 'tü ,kōd }

C core [ELECTROMAG] A spirally wound magnetic core that is formed to a desired rectangular shape before being cut into two C-shaped pieces and placed around a transformer or magnetic amplifier coil. { 'sē ,kór }

CCR process See cyclic catalytic reforming process. { ,sē,sē'är ,präs·əs }

CCR5 [MED] Belonging to the seven-transmembrane chemokine receptor family, the major cofactor for primary macrophage-tropic human immunodeficiency virus-1 strains.

CCTV See closed-circuit television.

CCU See communications control unit.

CCW See channel command word.

Cd See cadmium.

CD See circular dichroism.

CDC gene See cell-division-cycle gene. { ¦sē¦dē'sē jēn }

CD1 [IMMUNOL] Glycoproteins that are expressed on the surface of immature thymocytes, Langerhans cells, and certain B cells, similar in structure to class I major histocompatibility complex molecules.

CD4 [IMMUNOL] A T-cell signaling and/or co-stimulatory monomeric transmembrane glycoprotein involved in major histocompatibility complex II adhesion.

CD-4 sound *See* compatible discrete four-channel sound. { ¦sē¦de ¦fȯr ¦saúnd }

C damage [ORD] Damage of a degree or nature that prohibits an aircraft from completing its mission. { 'sē ¦dam·ij }

CD galaxy [ASTRON] A supergiant elliptical galaxy with an extended envelope, the largest known type of galaxy. { ¦sē¦dē 'gal·ik·sē }

C display [ELECTR] In radar, a rectangular display in which targets appear as blips with bearing indicated by the horizontal coordinate, and angles of elevation by the vertical coordinate. { 'sē di'splā }

Cdk *See* cyclin-dependent kinase.

CDM *See* code-division multiplex.

CDMA *See* code-division multiple access.

cDNA *See* complementary deoxyribonucleic acid.

CD-R [COMMUN] A compact-disk format that allows users to record audio or other digital data in such a way that the recording is permanent (nonerasable) and may be read indefinitely. Derived from compact-disk recordable. Also known as compact-disk write-once (CD-WO).

CD-ROM *See* compact-disk read-only memory. { ¦sē¦dē 'räm }

CD-RW [COMMUN] A compact-disk format that allows audio or other digital data to be written, read, erased, and rewritten. Derived from compact-disk rewritable. Also known as compact-disk erasable.

CD-WO *See* CD-R.

Ce *See* cerium.

Cebidae [VERT ZOO] The New World monkeys, a family of primates in the suborder Anthropoidea including the capuchins and howler monkeys. { 'seb·ə,dē }

Cebochoeridae [PALEON] A family of extinct palaeodont artiodactyls in the superfamily Entelodontoidae. { ¦seb·ə,kō'er·ə,dē }

cebollite [MINERAL] $H_2Ca_4Al_2Si_3O_{16}$ A greenish to white mineral consisting of hydrous calcium aluminum silicate occurring in fibrous aggregates; hardness is 5 on Mohs scale, and specific gravity is 3. { 'seb·ə,līt }

Cebrionidae [INV ZOO] The robust click beetles, a family of cosmopolitan coleopteran insects in the superfamily Elateroidea. { ¦seb·rē'än·ə,dē }

cecidium [PL PATH] Plant gall produced either by insects in ovipositing or by fungi as a consequence of infection. { sə'sid·ē·əm }

cecilite [PETR] A basaltic rock having few phenocrysts and consisting of at least 50% leucite with augite, melilite, nepheline, olivine, anorthite, magnetite, and apatite. { 'ses·əl,īt }

Cecropidae [INV ZOO] A family of crustaceans in the suborder Caligoida which are external parasites on fish. { sə'kräp·ə,dē }

cecum [ANAT] The blind end of a cavity, duct, or tube, especially the sac at the beginning of the large intestine. Also spelled caecum. { 'sē·kəm }

cedar [BOT] The common name for a large number of evergreen trees in the order Pinales having fragrant, durable wood. { 'sē·dər }

cedarwood oil [MATER] An essential alcohol-soluble oil obtained from the heartwood of various cedars, the chief components being cedrol and cedrene. { 'sē·dər,wúd ,ȯil }

cedricite [MINERAL] A variety of lamproite composed principally of diopside, leucite, and phlogopite and usually containing crystals of serpentine. { 'sed·rə,sīt }

ceiba *See* kapok. { 'sā·bə }

ceiling [BUILD] The covering made of plaster, boards, or other material that constitutes the overhead surface in a room. [MATH] The smallest integer that is equal to or greater than a given real number a; symbolized $\lceil a \rceil$. [METEOROL] In the United States, the height ascribed to the lowest layer of clouds or of obscuring phenomena when it is reported as broken, overcast, or obscuration and not classified as thin or partial. { 'sē·liŋ }

ceiling and visibility unlimited *See* CAVU. { 'sē·liŋ ən ,viz·ə'bil·əd·ē ən'lim·əd·əd }

ceiling balloon [AERO ENG] A small balloon used to determine the height of the cloud base; the height is computed from the ascent velocity of the balloon and the time required for its disappearance into the cloud. { 'sē·liŋ bə'lün }

ceiling classification [METEOROL] In aviation weather observations, a description or explanation of the manner in which the height of the ceiling is determined. { 'sē·liŋ ,klas·ə·fə'kā·shən }

ceiling effect [PSYCH] In testing, the actual limitation on a person's test score as the maximum score is approached or the limit on the performance of some task. { 'sēl·iŋ i,fekt }

ceiling light [ENG] A type of cloud-height indicator which uses a searchlight to project vertically a narrow beam of light onto a cloud base. Also known as ceiling projector. { 'sē·liŋ ,līt }

ceiling projector *See* ceiling light. { 'sē·liŋ prə'jek·tər }

ceiling temperature [ORG CHEM] For addition (chain) polymerization, the temperature at which the propagation and depropagation rates are equal, that is, the net rate of polymer formation is zero. Above the ceiling temperture, depolymerization, an unzipping reaction to reform monomer, occurs. { 'sēl·iŋ ,tem·prə·chər }

ceilometer [ENG] An automatic-recording cloud-height indicator. { sē'läm·əd·ər }

celadonite [MINERAL] A soft, green variety of mica having high iron content and containing silicates of magnesium and potassium. { 'sel·ə·də,nīt }

Celastraceae [BOT] A family of dicotyledonous plants in the order Celastrales characterized by erect and basal ovules, a flower disk that surrounds the ovary at the base, and opposite or sometimes alternate leaves. { ,sel·ə'strās·ē,ē }

Celastrales [BOT] An order of dicotyledonous plants in the subclass Rosidae marked by simple leaves and regular flowers. { ,sel·ə'strā·lēz }

celerity *See* phase velocity. { sə'ler·əd·ē }

celery [BOT] *Apium graveolens* var. *dulce*. A biennial umbellifer of the order Umbellales with edible petioles or leaf stalks. { 'sel·rē }

celery seed oil [MATER] A colorless liquid extracted from celery seeds, containing selinene and apiol. { 'sel·rē ,sēd ,ȯil }

celestial body [ASTRON] Any aggregation of matter in space constituting a unit for astronomical study, as the sun, moon, a planet, comet, star, or nebula. Also known as heavenly body. { sə'les·chəl 'bäd·ē }

celestial coordinates [ASTRON] Any set of coordinates, such as zenithal distance, altitude, celestial latitude, celestial longitude, local hour angle, azimuth and declination, used to define a point on the celestial sphere. { sə'les·chəl kō'órd·nəts }

celestial equator [ASTRON] The primary great circle of the celestial sphere in the equatorial system, everywhere 90° from the celestial poles, the intersection of the extended plane of the equator and the celestial sphere. Also known as equinoctial. { sə'les·chəl i'kwäd·ər }

celestial equator system of coordinates *See* equatorial system. { sə'les·chəl i'kwäd·ər ¦sis·təm əv kō'órd·nəts }

celestial fix [NAV] A position established by means of observation on one or more celestial bodies. { sə'les·chəl 'fiks }

celestial geodesy [GEOD] The branch of geodesy which utilizes observations of near celestial bodies and earth satellites to determine the size and shape of the earth. { sə'les·chəl jē'äd·ə·sē }

celestial globe [ASTRON] A small globe representing the celestial sphere, on which the apparent positions of the stars are located. Also known as star globe. { sə'les·chəl 'glōb }

celestial guidance [NAV] The process of directing movements of an aircraft or spacecraft, especially in the selection of a flight path, by reference to celestial bodies. { sə'les·chəl 'gīd·əns }

celestial horizon [ASTRON] That great circle of the celestial sphere which is formed by the intersection of the celestial sphere and a plane through the center of the earth and is perpendicular to the zenith-nadir line. Also known as rational horizon. { sə'les·chəl hə'rīz·ən }

celestial-inertial guidance [NAV] The process of directing the movements of an aircraft or spacecraft by an inertial guidance system which receives correction inputs from observations of celestial bodies. Also known as automatic celestial navigation. { sə'les·chəl i'nər·shəl 'gīd·əns }

CEDAR

Leaf arrangement for the Atlas cedar (*Cedrus atlantica*).

CELERY

Celery (*Apium graveolens*). (*Joseph Harris Co., Rochester, NY*)

celestial latitude [ASTRON] Angular distance north or south of the ecliptic; the arc of a circle of latitude between the ecliptic and a point on the celestial sphere, measured northward or southward from the ecliptic through 90°, and labeled N or S to indicate the direction of measurement. Also known as ecliptic latitude. { sə'les·chəl 'lad·ə‚tüd }

celestial line of position [NAV] A line of position determined by observation of a celestial body. { sə'les·chəl 'līn əv pə'zish·ən }

celestial longitude [ASTRON] Angular distance east of the vernal equinox, along the ecliptic; the arc of the ecliptic or the angle at the ecliptic pole between the circle of latitude of the vernal equinox and the circle of latitude of a point on the celestial sphere, measured eastward from the circle of latitude of the vernal equinox, through 360°. Also known as ecliptic longitude. { sə'les·chəl 'län·jə‚tüd }

celestial mechanics [ASTROPHYS] The calculation of motions of celestial bodies under the action of their mutual gravitational attractions. Also known as gravitational astronomy. { sə'les·chəl mə'kan·iks }

celestial meridian [ASTRON] A great circle on the celestial sphere, passing through the two celestial poles and the observer's zenith. { sə'les·chəl mə'rid·ē·ən }

celestial navigation See astronavigation. { sə'les·chəl nav·ə'gā·shən }

celestial observation [NAV] **1.** Also known as sight. **2.** The measurement of the altitude, and sometimes the azimuth, of a celestial body for a line of position. **3.** The data obtained by such observation. { sə'les·chəl ‚äb·sər'vā·shən }

celestial parallel See parallel of declination. { sə'les·chəl 'par·ə‚lel }

celestial pole [ASTRON] Either of the two points of intersection of the celestial sphere and the extended axis of the earth, labeled N or S to indicate the north celestial pole or the south celestial pole. { sə'les·chəl 'pōl }

celestial reference system [ASTRON] A system for specifying the locations and times of astronomical objects and events. { si¦les·chəl 'ref·rəns ‚sis·təm }

celestial sphere [ASTRON] An imaginary sphere of indefinitely large radius, which is described about an assumed center, and upon which positions of celestial bodies are projected along radii passing through the bodies. { sə'les·chəl 'sfir }

celestine See celestite. { 'sel·ə‚stēn }

celestite [MINERAL] SrSO₄ A colorless or sky-blue mineral occurring in orthorhombic, tabular crystals and in compact forms; fracture is uneven and luster is vitreous; principal ore of strontium. Also known as celestine. { 'sel·ə‚stīt }

celiac [ANAT] Of, in, or pertaining to the abdominal cavity. { 'sēl·ē‚ak }

celiac syndrome [MED] A complex of symptoms produced by intestinal malabsorption of fat and marked by bulky, loose, foul-smelling stools, high in fatty acid content. Also known as idiopathic steatorrhea. { 'sēl·e‚ak ‚sin‚drōm }

cell [BIOL] The microscopic functional and structural unit of all living organisms, consisting of a nucleus, cytoplasm, and a limiting membrane. [COMPUT SCI] **1.** An elementary unit of data storage. **2.** In a spreadsheet, the intersection of a row and a column. [ELEC] A single unit of a battery. [IND ENG] A manufacturing unit consisting of a group of work stations and their interconnecting materials-transport mechanisms and storage buffers. [MATH] **1.** The homeomorphic image of the unit ball. **2.** One of the (n − 1)-dimensional polytopes that enclose a given n-dimensional polytope. [MIN ENG] A compartment in a flotation machine. [NUCLEO] One of a set of elementary regions in a heterogeneous reactor, all of which have the same geometrical form and the same neutron characteristics. [PHYS CHEM] A cup, jar, or vessel containing electrolyte solutions and metal electrodes to produce an electric current (conductiometric or potentiometric) or for electrolysis (electrolytic). { sel }

cell address [COMPUT SCI] A combination of a letter and a number that specifies the column and row in which a cell is located on a spreadsheet. { 'sel ə‚dres }

cell adhesion molecule [CELL MOL] A class of membrane proteins comprising the outer surfaces of cell membranes in the developing nervous system that is thought to be intimately involved in guiding development during embryonic life. Abbreviated CAM. { ¦sel ad‚hē·zhən 'mäl·ə·kyül }

cellar [COMPUT SCI] See push-down storage. [PETRO ENG] An excavation in the ground for providing additional height between the rig floor and the wellhead to accommodate various well components and provide a place for collecting drainage water and other fluids for subsequent disposal. { 'sel·ər }

cellar deck [PETRO ENG] The lower of the two decks of a double-decked semisubmersible drilling rig. { 'sel·ər ‚dek }

cell-associated virus [VIROL] Virus particles that remain attached to or within the host cell after replication. { ¦sel ə¦sō·shē‚ād·əd 'vī·rəs }

cell body [NEUROSCI] The part of a nerve cell (neuron) containing the nucleus and several cytoplasmic structures such as mitochondria and neurofibrils. { 'sel ¦bäd·ē }

cell complex [MATH] A topological space which is the last term of a finite sequence of spaces, each obtained from the previous by sewing on a cell along its boundary. { 'sel ‚käm‚pleks }

cell constancy [BIOL] The condition in which the entire body, or a part thereof, consists of a fixed number of cells that is the same for all adults of the species. { 'sel 'kän·stən·sē }

cell constant [PHYS CHEM] The ratio of distance between conductance-titration electrodes to the area of the electrodes, measured from the determined resistance of a solution of known specific conductance. { 'sel ‚kän·stənt }

cell culture See tissue culture. { 'sel ‚kəl·chər }

cell cycle [CELL MOL] In eukaryotic cells, the cycle of events consisting of cell division, including mitosis and cytokinesis, and interphase. { 'sel ‚sī·kəl }

cell determination [PHYSIO] The process by which multipotential cells become committed to a particular development pathway. { ¦sel ‚di‚tər·mə¦nā·shən }

cell differentiation [CELL MOL] The series of events involved in the development of a specialized cell having specific structural, functional, and biochemical properties. { 'sel dif·ə‚ren·chē'ā·shən }

cell division [CELL MOL] The process by which living cells multiply; may be mitotic or amitotic. { 'sel di'vizh·ən }

cell-division-cycle gene [CELL MOL] A gene that regulates the cell cycle. Also known as CDC gene. { 'sel də‚vizh·ən ¦sī·kəl jēn }

cell fractionation [CELL MOL] A laboratory technique that uses differential centrifugation to separate the different components of the cell, resulting in nuclear, mitochondrial, microsomal, and soluble fractions. { ¦sel ‚frak·shə'nā·shən }

cell-free extract [CELL MOL] A fluid obtained by breaking open cells; contains most of the soluble molecules of a cell. { 'sel ‚frē 'ek‚strakt }

cell frequency [STAT] The number of observations of specified conditional constraints on one or more variables; used mainly in the analysis of data obtained by performing actual counts. { 'sel ‚frē·kwən·se }

cell inclusion [CELL MOL] A small, nonliving intracellular particle, usually representing a form of stored food, not immediately vital to life processes. { 'sel in'klü·zhan }

cell junction [CELL MOL] A specialized site on a cell at which it is attached to another cell or to the extracellular matrix. { 'sel ¦jəŋk·shen }

cell lineage [EMBRYO] The developmental history of individual blastomeres from their first cleavage division to their ultimate differentiation into cells of tissues and organs. { 'sel 'lin·yəj }

cell-lineage mutant [CELL MOL] Any mutation that affects the division of cells or the fates of their progeny. { ¦sel ‚lin·ē·əj 'myüt·ənt }

cell-mediated immunity [IMMUNOL] Immune responses produced by the activities of T cells rather than by immunoglobulins. { ¦sel ¦mē·dē‚ād·əd i'myü·nəd·ē }

cell membrane [CELL MOL] A thin layer of protoplasm, consisting mainly of lipids and proteins, which is present on the surface of all cells. Also known as plasmalemma; plasma membrane. { 'sel ¦mem‚brān }

cell movement [CELL MOL] **1.** Intracellular movement of cellular components. **2.** The movement of a cell relative to its environment. { 'sel ‚müv·mənt }

cellobiase [BIOCHEM] An enzyme that participates in the hydrolysis of cellobiose into glucose. { ‚sel·ō'bī‚ās }

cellobiose [ORG CHEM] $C_{12}H_{22}O_{11}$ A disaccharide which does not occur freely in nature or as a glucoside; a unit of cellulose and lichenin; crystallizes as minute water-soluble crystals from alcohol. Also known as cellose. { ‚sel·ō′bī‚ōs }

celloidin [MATER] A concentrated solution of pyroxylin used principally in microscopy for embedding specimens or for section-cutting. { se′lóid·ən }

cellophane [MATER] A thin, transparent sheeting of regenerated cellulose; it is moisture-proof, and sometimes dyed, and used chiefly as food wrapping or as bags for dialysis. { ′sel·ə‚fān }

cellose See cellobiose. { ′se‚lōs }

cellosolve [ORG CHEM] $C_2H_5OCH_2CH_2OH$ An important industrial chemical used in varnish removers, in cleaning solutions, and as a solvent for paints, varnishes, and plastics. Also known as 2-ethoxyethanol. { ′sel·ə‚sälv }

cell pathology [PATH] Abnormalities of the events taking place within cells. { ′sel pə′thäl·ə·jē }

cell permeability [CELL MOL] The permitting or activating of the passage of substances into, out of, or through cells. { ′sel pər·mē·ə′bil·əd·ē }

cell plate [CELL MOL] A membrane-bound disk formed during cytokinesis in plant cells which eventually becomes the middle lamella of the wall formed between daughter cells. { ′sel ‚plāt }

cell pointer [COMPUT SCI] A rectangular highlight that indicates the active cell in a spreadsheet program. { ′sel ‚póint·ər }

cell protection [COMPUT SCI] A format applied to a cell or range of cells in a spreadsheet, or to the entire spreadsheet, that prevents the contents of the cells in question from being altered. { ′sel prə‚tek·shən }

cell recognition [CELL MOL] The mutual recognition of cells, as expressed by specific cellular adhesion, due to a specific complementary interaction between molecules on adjacent cell surfaces. { ′sel ‚rek·əg‚nish·ən }

cell reference [COMPUT SCI] The address of a cell that contains a value that is needed to solve a formula in a spreadsheet program. { ′sel ‚ref·rəns }

cell sap [CELL MOL] The liquid content of a plant cell vacuole. { ′sel ‚sap }

cells of Paneth [HISTOL] Coarsely granular secretory cells found in the crypts of Lieberkühn in the small intestine. Also known as Paneth cells. { ‚sclz əv ′pän·ət }

cell-surface differentiation [CELL MOL] The specialization or modification of the cell surface. { ′sel ‚sər·fəs dif·ə‚ren·chē′ā·shən }

cell-surface ionization [PHYSIO] The presence of a negative charge on the surface of all living cells suspended in aqueous salt solutions at neutral pH. { ′sel ‚sər·fəs ‚ī·ən·ə′zā·shən }

cell theory [BIOL] **1.** A principle that describes the cell as the fundamental unit of all living organisms. **2.** A principle that describes the properties of an organism as the sum of the properties of its component cells. { ′sel ‚thē·ə·rē }

cell-type tube [ELECTR] Gas-filled radio-frequency switching tube which operates in an external resonant circuit; a tuning mechanism may be incorporated in either the external resonant circuit or the tube. { ′sel ‚tīp ‚tüb }

cellular [BIOL] Characterized by, consisting of, or pertaining to cells. [PETR] Pertaining to igneous rock having a porous texture, usually with the cavities larger than pore size and smaller than caverns. { ′sel·yə·lər }

cellular affinity [BIOL] The phenomenon of selective adhesiveness observed among the cells of certain sponges, slime molds, and vertebrates. { ′sel·yə·lər ə′fin·əd·ē }

cellular automaton [COMPUT SCI] A theoretical model of a parallel computer which is subject to various restrictions to make practicable the formal investigation of its computing powers. [MATH] A mathematical construction consisting of a system of entities, called cells, whose temporal evolution is governed by a collection of rules, so that its behavior over time may appear highly complex or chaotic. { ′sel·yə·lər ó′täm·ə·tän }

cellular chain [COMPUT SCI] A chain which is not allowed to cross a cell boundary. { ′sel·yə·lər ′chän }

cellular cofferdam [CIV ENG] A cofferdam consisting of interlocking steel-sheet piling driven as a series of interconnecting cells; cells may be of circular type or of straight-wall

diaphragm type; space between lines of pilings is filled with sand. { ′sel·yə·lər ′kóf·ər‚dam }

cellular convection [METEOROL] An organized, convective, fluid motion characterized by the presence of distinct convection cells or convective units, usually with upward motion (away from the heat source) in the central portions of the cell, and sinking or downward flow in the cell's outer regions. { ′sel·yə·lər kən′vek·shən }

cellular glass [MATER] Sheets or blocks of thermal insulating material for walls and roofs made from pulverized glass that is heated with a gas-forming chemical at the flow temperature of glass. Also known as cellulated glass; foamed glass. { ′sel·yə·lər ′glas }

cellular horn See multicellular horn. { ′sel·yə·lər ′hórn }

cellular immunity [IMMUNOL] Immune responses carried out by active cells rather than by antibodies. { ′sel·yə·lər i′myü·nəd·ē }

cellular immunology [IMMUNOL] The study of the cells of the lymphoid organs, which are the main agents of immune reactions in all vertebrates. { ′sel·yə·lər ‚im·yə′näl·ə·jē }

cellular infiltration [MED] **1.** Passage of cells into tissues in the course of inflammation. **2.** Migration of or invasion by cells of neoplasms. { ′sel·yə·lər in·fil′trā·shən }

cellular manufacturing [IND ENG] A type of manufacturing in which equipment is organized into groups or cells according to function and intermachine relationships. { ′sel·yə·lər ‚man·ə′fak·chər·iŋ }

cellular mobile radio [COMMUN] A system that serves portable and mobile radio receivers in which the service area is subdivided into multiple cells or zones, and unique radio channel frequencies are assigned to each cell. { ′sel·yə·lər ′mō·bəl ′rād·ē·ō }

cellular multilist [COMPUT SCI] A type of multilist organization composed of cellular chains. { ′sel·yə·lər ′məl·ti‚list }

cellular plastic [MATER] A type of plastic with apparent density decreased substantially by the presence of numerous cells disposed throughout its mass. { ′sel·yə·lər ′plas·tik }

cellular rubber See rubber sponge. { ′sel·yə·lər ′rəb·ər }

cellular slime molds [BIOL] A group of funguslike protozoa that form slimy aggregations on decaying organic matter; they differ from true slime molds in that the pseudoplasmodium is made of a group of separate cells rather than an amebalike mass. { ′sel·yə·lər ′slīm ‚mōlz }

cellular soil See polygonal ground. { ′sel·yə·lər ′sóil }

cellular splitting [COMPUT SCI] A method of adding records to a file in which the records are grouped into cells and each cell is divided into two when it becomes full. { ′sel·yə·lər ′splid·iŋ }

cellular striation [ENG] Stratum of cells inside a cellular-plastic object that differs noticeably from the cell structure of the remainder of the material. { ′sel·yə·lər strī′ā·shən }

cellular transformation See transformation. { ′sel·yə·lər ‚tranz·fər′mā·shən }

cellulase [BIOCHEM] Any of a group of extracellular enzymes, produced by various fungi, bacteria, insects, and other lower animals, that hydrolyze cellulose. { ′sel·yə‚lās }

cellulated glass See cellular glass. { ′sel·yə‚lād·əd ‚glas }

cellulitis [MED] Inflammation of connective tissue, especially the loose subcutaneous tissue. { ‚sel·yə′līd·əs }

cellulolytic [BIOL] Having the ability to hydrolyze cellulose; applied to certain bacteria and protozoans. { ‚sel·yə‚lid·ik }

Cellulomonas [MICROBIO] A genus of gram-positive, irregular rods in the coryneform group of bacteria; metabolism is respiratory; most strains produce acid from glucose, and cellulose is attacked by all strains. { ‚sel·yə′läm·ə·nəs }

cellulose [BIOCHEM] $(C_6H_{10}O_5)_n$ The main polysaccharide in living plants, forming the skeletal structure of the plant cell wall; a polymer of β-D-glucose units linked together, with the elimination of water, to form chains of 2000–4000 units. { ′sel·yə‚lōs }

α-cellulose See alpha cellulose. { ‚al·fə ′sel·yə‚lōs }

cellulose acetate [ORG CHEM] An acetic acid ester of cellulose; a tough, flexible, slow-burning, and long-lasting thermoplastic material used as the base for magnetic tape and movie film, in acetate rayon, as a plastic film in food packaging, in lacquers, and for molded receiver cabinets. { ′sel·yə‚lōs ′as·ə‚tāt }

cellulose acetate butyrate [ORG CHEM] An ester of cellulose formed by the action of a mixture of acetic acid and butyric

CELLOBIOSE

Structural formula of cellobiose.

CELLULAR COFFERDAM

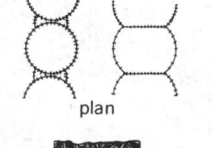

plan

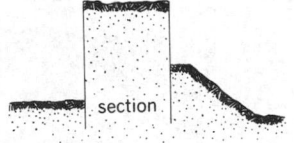

section

Steel-sheet pile cellular cofferdam.

acid and their anhydrides on purified cellulose; has high impact resistance, clarity, and weatherability; used in making plastic film, lacquer, lenses, and outdoor signs. { 'sel·yə‚lōs 'as·ə‚tāt 'byüd·ə‚rāt }

cellulose acetate rayon [TEXT] The spun product of the acetic ester of cellulose. { 'sel·yə‚lōs 'as·ə‚tāt 'rā‚än }

cellulose diacetate [ORG CHEM] The ester formed by esterification of two hydroxyl groups of a cellulose molecule with acetic acid. { 'sel·yə‚lōs dī'as·ə‚tāt }

cellulose ester [ORG CHEM] Cellulose in which the free hydroxyl groups have been replaced wholly or in part by acidic groups. { 'sel·yə‚lōs 'es·tər }

cellulose ether [ORG CHEM] The product of the partial or complete etherification of the hydroxyl groups in a cellulose molecule. { 'sel·yə‚lōs 'ē·thər }

cellulose fiber [ORG CHEM] Any fiber based on esters or ethers of cellulose. { 'sel·yə‚lōs 'fī·bər }

cellulose gum See carboxymethyl cellulose. { 'sel·yə‚lōs 'gəm }

cellulose methyl ether See methylcellulose. { 'sel·yə‚lōs 'meth·əl ‚ē·thər }

cellulose nitrate [ORG CHEM] Any of several esters of nitric acid, produced by treating cotton or some other form of cellulose with a mixture of nitric and sulfuric acids; used as explosive and propellant. Also known as nitrocellulose; nitrocotton. { 'sel·yə‚lōs 'nī‚trāt }

cellulose propionate [ORG CHEM] An ester of cellulose and propionic acid. { 'sel·yə‚lōs 'prō·pē·ə‚nāt }

cellulose triacetate [ORG CHEM] A cellulose resin formed by the complete esterification of the cellulose by acetic acid; used as a base in protective coatings. { 'sel·yə‚lōs trī'as·ə‚tāt }

cellulose xanthate [ORG CHEM] A compound formed by reaction of soda cellulose (prepared by treating cellulose with strong sodium hydroxide solution) with carbon disulfide. { 'sel·yə‚lōs 'zan‚thāt }

cellulosic [ORG CHEM] Any of the derivatives of cellulose, such as cellulose acetate. { ‚sel·yə'lō·sik }

cellulosic resin [ORG CHEM] Any resin based on cellulose compounds such as esters and ethers. { 'sel·yə‚lōs 'rez·ən }

cell wall [CELL MOL] A semirigid, permeable structure that is composed of cellulose, lignin, or other substances and that envelops most plant cells. { ¦sel ¦wol }

celo [MECH] A unit of acceleration equal to the acceleration of a body whose velocity changes uniformly by 1 foot (0.3048 meter) per second in 1 second. { 'se·lō }

Celor lens system [OPTICS] An anastigmatic lens system consisting of two air-spaced achromatic doublet lenses, one on each side of the stop. Also known as Gauss lens system; Gauss objective lens. { 'se·lor 'lenz ‚sis·təm }

CELP coder See code-excited linear predictive coder. { ¦sē¦ē¦el'pē ‚kōd·ər or 'selp ‚kōd·ər }

celsian [MINERAL] $BaAl_2Si_2O_8$ Colorless, monoclinic mineral consisting of barium feldspar. { 'sel·sē‚an }

Celsius degree [THERMO] Unit of temperature interval or difference equal to the kelvin. { 'sel·sē·əs di'grē }

Celsius temperature scale [THERMO] Temperature scale in which the temperature Θ_c in degrees Celsius (°C) is related to the temperature T_k in kelvins by the formula $\Theta_c = T_k - 273.15$; the freezing point of water at standard atmospheric pressure is very nearly 0°C and the corresponding boiling point is very nearly 100°C. Formerly known as centigrade temperature scale. { 'sel·se·əs 'tem·prə·chər ‚skāl }

Celyphidae [INV ZOO] A family of myodarian cyclorrhaphous dipteran insects in the subsection Acalypteratae. { se'lif·ə‚dē }

cement [GEOL] Any chemically precipitated material, such as carbonates, gypsum, and barite, occurring in the interstices of clastic rocks. [HISTOL] Calcified tissue which fastens the roots of teeth to the alveolus. Also known as cementum. [INV ZOO] Any of the various adhesive secretions, produced by certain invertebrates, that harden on exposure to air or water and are used to bind objects. [MATER] **1.** A dry powder made from silica, alumina, lime, iron oxide, and magnesia which hardens when mixed with water; used as an ingredient in concrete. **2.** An adhesive for the assembling of surfaces which are not in close contact. { si'ment }

cementation [CHEM] The setting of a plastic material. [ENG] **1.** Plugging a cavity or drill hole with cement. Also known as dental work. **2.** Consolidation of loose sediments

or sand by injection of a chemical agent or binder. [GEOL] The precipitation of a binding material around minerals or grains in rocks. [MET] **1.** High-temperature impregnation of a metal surface with another material. **2.** Conversion of wrought iron into steel by packing layers of bars in charcoal sealed with clay and heating to 1000°C for 7–10 days. { ‚sē‚men'tā·shən }

cementation factor [PETRO ENG] Mathematical expression in oil-reservoir analysis for the degree to which precipitated minerals have bound together the grains (for example, of sand). { ‚sē‚men'tā·shən ‚fak·tər }

cementation sinking [MIN ENG] A technique of shaft sinking through strata containing water by injecting liquid cement or chemicals into the ground. { ‚sē‚men'tā·shən ‚siŋ·kiŋ }

cement bond [PETRO ENG] The adhesion of a well casing to the cement used for attaching it to the formation. { si'ment ¦bänd }

cement-bond survey [PETRO ENG] A sonic logging procedure for recording the condition of the cement used in the annulus to bond the casing and the formation. { si'ment ¦bänd 'sər‚vā }

cement brick [MATER] A type of brick made from a mixture of cement and sand, molded under pressure and cured under steam at 200°F (93°C); used as backing brick and where there is no danger of attack from acid or alkaline conditions. { si'ment ‚brik }

cement carrier [NAV ARCH] A special ship designed to carry bulk cement; usually fitted with an air-mixing system to allow pumping dry cement powder through pipes to unload or load the ship. { si'ment ‚kar·ē·ər }

cement channeling [PETRO ENG] A failure during cementing of the casing to the formation whereby the cement slurry does not rise uniformly, leaving open spaces and thus preventing a strong bond. { si'ment ‚chan·əl·iŋ }

cement copper [MET] A precipitate of copper from copper sulfate solution by the addition of iron. { si'ment ‚käp·ər }

cemented carbide See carbide. { si'men·təd 'kär‚bīd }

cemented lens See compound lens. { si'men·təd 'lenz }

cement gland [INV ZOO] A structure in many invertebrates that produces cement. { si'ment ‚gland }

cement gravel [GEOL] Gravel consolidated by clay, silica, calcite, or other binding material. { si'ment ‚grav·əl }

cement gun [MECH ENG] **1.** A machine for mixing, wetting, and applying refractory mortars to hot furnace walls. Also known as cement injector. **2.** A mechanical device for the application of cement or mortar to the walls or roofs of mine openings or building walls. { si'ment ‚gən }

cementing basket [PETRO ENG] A collapsible metal cone designed to fit against a wellbore wall and prevent passage of cement. Also known as metal-petal basket. { si'ment·iŋ ‚bas·kət }

cement injector See cement gun. { si'ment in'jek·tər }

cementite [MET] Fe_3C A hard, brittle, crystalline compound occurring as lamellae or plates in steel. Also known as iron carbide. { si'men‚tīt }

cementitious material [MATER] Any of various building materials which may be mixed with a liquid, such as water, to form a plastic paste, and to which an aggregate may be added; includes cements, limes, and mortar. { ¦sē‚men¦tish·əs mə'tir·ē·əl }

cement kiln [ENG] A kiln used to fire cement to less than complete melting. { si'ment ‚kil }

cement-lined casing [PETRO ENG] Steel-, oil-, or gas-well casing pipe with internal lining of special cement; used to withstand severe corrosive conditions. { si'ment ‚līnd 'kās·iŋ }

cement log [PETRO ENG] Gamma-ray measurement and logging of the height and condition of cement surrounding downhole oil-well casing. { si'ment ‚läg }

cement mill [MECH ENG] A mill for grinding rock to a powder for cement. { si'ment ‚mil }

cement mortar [MATER] A mixture of approximately four parts of sand to one part of portland cement with a small amount of lime and enough water to make it plastic. { si'ment 'mord·ər }

cement paint [MATER] A mixture based on portland cement, with filler, accelerator, and water repellent added, that is combined with water and applied to masonry, concrete, or brickwork; provides a waterproof coating. { si'ment ‚pānt }

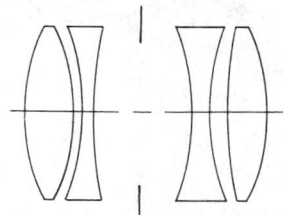

CELOR LENS SYSTEM

Celor lens system, an anastigmatic photographic objective.

cement paste [MATER] A mixture of cement and water, hardened or unhardened. { si'ment ,pāst }

cement plaster [MATER] A gypsum plaster used in mortar mixtures for plastering interior surfaces. { si'ment ,plas·tər }

cement pump [MECH ENG] A piston device used to move concrete through pipes. { si'ment ,pəmp }

cement rock [PETR] An argillaceous limestone containing lime, silica, and alumina in variable proportions and usually some magnesia; used in the manufacture of natural hydraulic cement. { si'ment ,räk }

cement-sand process [MET] A sand casting process in which portland cement is the binder for sand; typical mixtures have 11% portland cement and 89% silica sand. { si'ment 'sand ,präs·əs }

cement silo [ENG] A silo used to store dry, bulk cement. { si'ment 'sī,lō }

cement temper [MATER] Use of portland cement as an additive in a lime plaster preparation to increase strength and durability. { sə'ment ,tem·pər }

cementum [HISTOL] *See* cement. [MED] A tissue closely resembling bone which covers the root of a tooth. { si'men·təm }

cement valve [MECH ENG] A ball-, flapper-, or clack-type valve placed at the bottom of a string of casing, through which cement is pumped, so that when pumping ceases, the valve closes and prevents return of cement into the casing. { si'ment ,valv }

Cen *See* Centaurus.

Cenomanian [GEOL] Lower Upper Cretaceous geologic time. { ¦sen·ə¦mān·ē·ən }

cenote *See* pothole. { sə'nōd·ē }

Cenozoic [GEOL] The youngest of the eras, or major subdivisions of geologic time, extending from the end of the Mesozoic Era to the present, or Recent. Also known as Cainozoic. { ¦sen·ə¦zō·ik }

censored data [STAT] Observations collected by determining in advance whether to record only a specified number of the smallest or largest values, or of the remaining values in a sample of a particular size. { ¦sen·sərd 'dad·ə }

census [STAT] A complete counting of a population, as opposed to a partial counting or sampling. { 'sen·səs }

cent [ACOUS] The interval between two sounds whose basic frequency ratio is the twelve-hundredth root of 2; the interval, in cents, between any two frequencies is 1200 times the logarithm to the base 2 of the frequency ratio. [NUCLEO] A unit of nuclear reactivity equal to one-hundredth of a dollar. { sent }

cental *See* hundredweight. { 'sent·əl }

centare *See* centiare. { 'sen,tär }

Centaurus [ASTRON] A constellation with right ascension 13 hours, declination 50°S. Abbreviated Cen. { sen'tór·əs }

Centaurus A [ASTRON] A strong, discrete radio source in the constellation Centaurus, associated with the peculiar galaxy NGC 5128. { sen'tór·əs 'ā }

Centaurus cluster [ASTRON] A large cluster of galaxies that shows a composite structure, with a concentration of galaxies having recession velocities of about 3000 kilometers (1900 miles) per second, and a weaker concentration having recession velocities of about 4500 kilometers (2800 miles) per second. { sen'tór·əs ,kləs·tər }

Centaurus X-3 [ASTROPHYS] A source of x-rays that pulses with a period of 4.8 seconds and is eclipsed every 2.1 days; believed to be a binary star, one of whose members is a rotating neutron star. Abbreviated Cen X-3. { sen'tór·əs ,eks 'thrē }

center [IND ENG] A manufacturing unit containing a number of interconnected cells. [MATH] **1.** The point that is equidistant from all the points on a circle or sphere. **2.** The point (if it exists) about which a curve (such as a circle, ellipse, or hyperbola) is symmetrical. **3.** The point (if it exists) about which a surface (such as a sphere, ellipsoid, or hyperboloid) is symmetrical. **4.** For a regular polygon, the center of its circumscribed circle. **5.** The subgroup consisting of all elements that commute with all other elements in a given group. **6.** The subring consisting of all elements a such that $ax = xa$ for all x in a given ring. [OPTICS] To adjust the components of an optical system so that their centers of curvature lie on a common optical axis. Also known as square-on. [STAT] For a distribution, the expected value of any random variable which has the distribution. { 'sen·tər }

center-bearing swing bridge [CIV ENG] A type of swing bridge that has a single large bearing on a pier, called the pivot pier, in the waterway. { 'sen·tər ,ber·iŋ 'swiŋ ,brij }

centerboard [NAV ARCH] A metal or wooden slab in a casing along the centerline of a sailboat which may be lowered to increase the boat's resistance to lateral motion, and raised when the boat is in shallow water or is beached. { 'sen·tər,bórd }

center-coupled loop [ELECTR] Coupling loop in the center of one of the resonant cavities of a multicavity magnetron. { 'sen·tər ,kup·əld 'lüp }

center drill [ENG] A two-fluted tool consisting of a twist drill with a 60° countersink; used to drill countersink center holes in a workpiece to be mounted between centers for turning or grinding. { 'sen·tər ,dril }

centered lattice [CRYSTAL] A crystal lattice in which the axes have been chosen according to the rules for the crystal system, and in which there are lattice points at the centers of certain planes as well as at the corners. { 'sen·tərd 'lad·əs }

center-feed tape [COMPUT SCI] Punched tape in which the centers of the sprocket holes are in line with the centers of the holes carrying the data or message. { 'sen·tər ,fēd 'tāp }

center-fire [ORD] Pertaining to a cartridge with primer in the center of the head, or base. { 'sen·tər ,fīr }

center frequency *See* carrier frequency. { 'sen·tər 'frē·kwən·sē }

center gage [DES ENG] A gage used to check angles; for example, the angles of cutting tool points or screw threads, or the angular position of cutting tools. { 'sen·tər ,gāj }

center-gated mold [ENG] A plastics injection mold with the filling orifice interconnected to the nozzle and the center of the cavity area. { 'sen·tər ,gād·əd 'mōld }

centering [CIV ENG] A curved, temporary support for an arch or dome during a casting or laying operations. { 'sen·tə·riŋ }

centering control [ELECTR] One of the two controls used for positioning the image on the screen of a cathode-ray tube; either the horizontal centering control or the vertical centering control. { 'sen·tə·riŋ kən'trōl }

centering machine [MECH ENG] A machine for drilling and countersinking work to be turned on a lathe. { 'sen·tə·riŋ ma'shēn }

center jump [METEOROL] The formation of a second low-pressure center within an already well-developed low-pressure center; the latter diminishes in magnitude as the center of activity shifts or appears to jump to the new center. { 'sen·tər ¦jəmp }

center-justify [GRAPHICS] To center text uniformly between the left and right margins. Also known as flush center. { ¦sen·tər ¦jəs·tə·fī }

centerless grinder [MECH ENG] A cylindrical metal-grinding machine that carries the work on a support or blade between two abrasive wheels. { 'sen·tər·ləs 'grin·dər }

center line [COMPUT SCI] *See* stroke center line. [ENG] A line that represents an axis of symmetry on a plane figure such as a plan for a structure or a machine. { 'sen·tər ,līn }

center loading [ELECTROMAG] Alteration of the resonant frequency of a transmitting antenna by inserting an inductance or capacitance about halfway between the feed point and the end of the antenna. { 'sen·tər 'lōd·iŋ }

center of action [METEOROL] A semipermanent high or low atmospheric pressure system at the surface of the earth; fluctuations in the intensity, position, orientation, shape, or size of such a center are associated with widespread weather changes. { 'sen·tər əv 'ak·shən }

center of area [MATH] For a plane figure, the center of mass of a thin uniform plate having the same boundaries as the plane figure. Also known as center of figure; centroid. { 'sen·tər əv 'er·ē·ə }

center of attraction [MECH] A point toward which a force on a body or particle (such as gravitational or electrostatic force) is always directed; the magnitude of the force depends only on the distance of the body or particle from this point. { 'sen·tər əv ə'trak·shən }

center of buoyancy [MECH] The point through which acts the resultant force exerted on a body by a static fluid in which it is submerged or floating; located at the centroid of displaced volume. { 'sen·tər əv 'bói·ən·sē }

center of burst [ORD] A point in the air about which the

CENOZOIC

CENOZOIC	QUATERNARY	
	TERTIARY	
MESOZOIC	CRETACEOUS	
	JURASSIC	
	TRIASSIC	
PALEOZOIC	PERMIAN	
	CARBONIFEROUS	PENNSYLVANIAN
		MISSISSIPPIAN
	DEVONIAN	
	SILURIAN	
	ORDOVICIAN	
	CAMBRIAN	
PRECAMBRIAN		

Chart showing the relationship of the Cenozoic to other eras.

bursts of several antiaircraft projectiles from rounds fired under like conditions are evenly distributed. { 'sen·tər əv 'bərst }

center of curvature [MATH] At a given point on a curve, the center of the osculating circle of the curve at that point. { 'sen·tər əv 'kər·və·chər }

center of falls *See* pressure-fall center. { 'sen·tər əv 'fȯlz }

center of figure *See* center of area. *See* center of volume. { 'sen·tər əv 'fig·yər }

center of flotation [PETRO ENG] In offshore oil operations, the geometric center of the water plane at which the mobile drilling rig floats and around which the rig rotates when it is subject to an external force without a change in displacement. { 'sen·tər əv flō'tā·shən }

center of force [MECH] The point toward or from which a central force acts. { 'sen·tər əv 'fȯrs }

center of geodesic curvature [MATH] For a curve on a surface at a given point, the center of curvature of the orthogonal projection of the curve onto a plane tangent to the surface at the point. { ¦sen·tər əv jē·ə'des·ik ¦kərv·ə·chər }

center of gravity [MECH] A fixed point in a material body through which the resultant force of gravitational attraction acts. { 'sen·tər əv 'grav·əd·ē }

center of inertia *See* center of mass. { 'sen·tər əv i'nər·shə }

center of inversion [CRYSTAL] A point in a crystal lattice such that the lattice is left invariant by an inversion in the point. [MATH] The point O with respect to which an inversion is defined, so that every point P is mapped by the inversion into a point Q that is collinear with O and P. { 'sen·tər əv in'vər·zhən }

center of lift [AERO ENG] The mean of all the centers of pressure on an airfoil. { 'sen·tər əv 'lift }

center of mass [MECH] That point of a material body or system of bodies which moves as though the system's total mass existed at the point and all external forces were applied at the point. Also known as center of inertia; centroid. { 'sen·tər əv 'mas }

center-of-mass coordinate system [MECH] A reference frame which moves with the velocity of the center of mass, so that the center of mass is at rest in this system, and the total momentum of the system is zero. Also known as center of momentum coordinate system. { 'sen·tər əv 'mas kō'ȯrd·nət ˌsis·təm }

center-of-momentum coordinate system *See* center-of-mass coordinate system. { 'sen·tər əv mə'men·təm kō'ȯrd·nət ˌsis·təm }

center of normal curvature [MATH] For a given point on a surface and for a given direction, the normal section of the surface through the given point and in the given direction. { 'sen·tər əv 'nȯrm·əl 'kər·və·chər }

center of oscillation [MECH] Point in a physical pendulum, on the line through the point of suspension and the center of mass, which moves as if all the mass of the pendulum were concentrated there. { 'sen·tər əv ˌäs·ə'lā·shən }

center of percussion [MECH] If a rigid body, free to move in a plane, is struck a blow at a point O, and the line of force is perpendicular to the line from O to the center of mass, then the initial motion of the body is a rotation about the center of percussion relative to O; it can be shown to coincide with the center of oscillation relative to O. { 'sen·tər əv pər'kəsh·ən }

center of perspective [MATH] The point specified by Desargues' theorem, at which lines passing through corresponding vertices of two triangles are concurrent. { 'sen·tər əv pər'spek·tiv }

center of pressure [AERO ENG] The point in the chord of an airfoil section which is at the intersection of the chord (prolonged if necessary) and the line of action of the combined air forces (resultant air force). [FL MECH] For a body immersed in a fluid, the point through which the resultant of the forces on the surface of the body due to hydrostatic pressure acts. { 'sen·tər əv 'presh·ər }

center-of-pressure coefficient [AERO ENG] The ratio of the distance of a center of pressure from the leading edge of an airfoil to its chord length. { 'sen·tər əv 'presh·ər ˌkō·ə'fish·ənt }

center of principal curvature [MATH] For a given point on a surface, the center of normal curvature at the point in one of the two principal directions. { 'sen·tər əv 'prin·sə·pəl ˌkər·və·chər }

center of projection [MATH] The fixed point in a central projection. { ¦sen·tər əv prə'jek·shən }

center of rises *See* pressure-rise center. { 'sen·tər əv 'rīz·əz }

center of similitude [MATH] **1.** A point of intersection of lines that join the ends of parallel radii of coplanar circles. **2.** *See* homothetic center. { 'sen·tər əv si'mil·ə‚tüd }

center of spherical curvature [MATH] The center of the osculating sphere at a specified point on a space curve. { ¦sen·tər əv ¦sfer·ə·kəl 'kər·və·chər }

center of suspension [MECH] The intersection of the axis of rotation of a pendulum with a plane perpendicular to the axis that passes through the center of mass. { 'sen·tər əv sə'spen·shən }

center of symmetry [SCI TECH] A point in an object through which any straight line encounters exactly similar points on opposite sides. Also known as symmetry center. { 'sen·tər əv 'sim·ə·trē }

center of thrust *See* thrust axis. { 'sen·tər əv 'thrəst }

center of twist [MECH] A point on a line parallel to the axis of a beam through which any transverse force must be applied to avoid twisting of the section. Also known as shear center. { 'sen·tər əv 'twist }

center of volume [MATH] For a three-dimensional figure, the center of mass of a homogeneous solid having the same boundaries as the figures. Also known as center of figure; centroid. { 'sen·tər əv 'väl·yəm }

center plug [DES ENG] A small diamond-set circular plug, designed to be inserted into the annular opening in a core bit, thus converting it to a noncoring bit. { 'sen·tər ‚pləg }

center punch [DES ENG] A tool similar to a prick punch but having the point ground to an angle of about 90°; used to enlarge prick-punch marks or holes. { 'sen·tər ‚pənch }

center square [DES ENG] A straight edge with a sliding square; used to locate the center of a circle. { 'sen·tər ‚skwer }

center staff [HOROL] The arbor that holds the minute hand and cannon pinion of a timepiece. { 'sen·tər ‚staf }

center tap [ELEC] A terminal at the electrical midpoint of a resistor, coil, or other device. Abbreviated CT. { 'sen·tər ‚tap }

center wheel [HOROL] A wheel that drives the wheels leading to the escapement, and that has a pinion post on which the minute hand is mounted. { 'sen·tər ‚wēl }

centesis [MED] Surgical puncture or perforation, as of a tumor or membrane. { ‚sen'tē·səs }

centi- [SCI TECH] A prefix representing 10^{-2}, which is 0.01 or one-hundredth. Abbreviated c. { 'sen·tē *or* 'sent·ə }

centiare [MECH] Unit of area equal to 1 square meter. Also spelled centare. { 'sen·tē‚är }

centibar [MECH] A unit of pressure equal to 0.01 bar or to 1000 pascals. { 'sent·ə‚bär }

centigrade heat unit [THERMO] A unit of heat energy, equal to 0.01 of the quantity of heat needed to raise 1 pound of air-free water from 0 to 100°C at a constant pressure of 1 standard atmosphere; equal to 1900.44 joules. Symbolized CHU; (more correctly) CHU$_{mean}$. { 'sent·ə‚grād 'hēt ‚yü·nət }

centigrade temperature scale *See* Celsius temperature scale. { 'sent·ə‚grād 'tem·prə·chər ‚skāl }

centigram [MECH] Unit of mass equal to 0.01 gram or 10^{-5} kilogram. Abbreviated cg. { 'sent·ə‚gram }

centihg *See* centimeter of mercury. { 'sen‚tig *or* 'sent·ē‚āch'jē }

centiliter [MECH] A unit of volume equal to 0.01 liter or to 10^{-5} cubic meter. { 'sent·ə‚lēd·ər }

centimeter [MECH] A unit of length equal to 0.01 meter. Abbreviated cm. { 'sent·ə‚mēd·ər }

centimeter-candle *See* phot. { 'sent·ə‚mēd·ər 'kand·əl }

centimeter-gram-second system [PHYS] An absolute system of metric units in which the centimeter, gram mass, and the second are the basic units. Abbreviated cgs system. { ¦sent·ə‚mēd·ər ¦gram 'sek·ənd ‚sis·təm }

21-centimeter line [SPECT] A radio-frequency spectral line of neutral atomic hydrogen at a wavelength of approximately 21 centimeters and a frequency of approximately 1420 megahertz, that results from hyperfine transitions between states in which the spins of the electron and proton are parallel and antiparallel. { ¦twen·tē¦wən 'sen·tə‚mēd·ər ‚līn }

centimeter of mercury [MECH] A unit of pressure equal to the pressure that would support a column of mercury 1 centimeter high, having a density of 13.5951 grams per cubic centimeter, when the acceleration of gravity is equal to its standard value (980.665 centimeters per second per second); it is equal to 1333.22387415 pascals; it differs from the dekatorr by less than 1 part in 7,000,000. Abbreviated cmHg. Also known as centihg. { 'sent·ə,mēd·ər əv 'mər·kyə·rē }

centimetric waves [COMMUN] Microwaves having wavelengths between 1 and 10 centimeters, corresponding to frequencies between 3 and 30 gigahertz. { ¦sent·ə¦me·trik 'wāvz }

centimorgan [GEN] A unit of genetic map distance, equal to the distance along a chromosome that gives a recombination frequency of 1%. { 'sent·ə,mórg·ən }

centipede [INV ZOO] The common name for an arthropod of the class Chilopoda. { 'sent·ə,pēd }

centipoise [FL MECH] A unit of viscosity which is equal to 0.01 poise. Abbreviated cp. { 'sent·ə,póiz }

centistoke [FL MECH] A cgs unit of kinematic viscosity in customary use, equal to the kinematic viscosity of a fluid having a dynamic viscosity of 1 centipoise and a density of 1 gram per cubic centimeter. Abbreviated cs. { 'sent·ə,stōk }

centner See hundredweight. { 'sent·nər }

centrad [MATH] A unit of plane angle equal to 0.01 radian or to about 0.573 degree. { 'sent,rad }

central angle [MATH] In a circle, an angle whose sides are radii of the circle. { 'sen·trəl 'aŋ·gəl }

central apnea [MED] A pause in breathing lasting more than 10 seconds that is caused by a failure of commands from the brain. Also known as central sleep apnea. { ¦sen·trəl 'ap·nē·ə }

central apparatus [CYTOL] The centrosome or centrosomes together with the surrounding cytoplasm. Also known as cytocentrum. { 'sen·trəl ,ap·ə'rad·əs }

central-battery system [COMMUN] A telephone or telegraph system which obtains all the energy for signaling (and for speaking, in the case of the telephone) from a single battery of secondary cells located at the main exchange. { ¦sen·trəl 'bad·ə·rē ,sis·təm }

central breaker [MIN ENG] A breaker where the coal from several mines in a district is prepared. { 'sen·trəl 'brā·kər }

central canal [ANAT] The small canal running through the center of the spinal cord from the conus medullaris to the lower part of the fourth ventricle; represents the embryonic neural tube. { 'sen·trəl kə'nal }

central condensation [ASTRON] The bright, central portion of the coma of a comet, containing one or more nuclei. { 'sen·trəl ,kän·dən'sā·shən }

central conic [MATH] A conic that has a center, namely, a circle, ellipse, or hyperbola. { 'sen·trəl 'kän·ik }

central control [AERO ENG] The place, facility, or activity at which the whole action incident to a test launch and flight is coordinated and controlled, from the make-ready at the launch site and on the range, to the end of the rocket flight downrange. [ORD] Fire control of weapons by a central location, not by the individual gunner; especially used in antiaircraft batteries. [SYS ENG] Control exercised over an extensive and complicated system from a single center. { 'sen·trəl kən'trōl }

central deafness [MED] Deafness that results from some injury or failure to function in the central nervous system. { 'sen·trəl 'def·nəs }

central difference [MATH] One of a series of quantities obtained from a function whose values are known at a series of equally spaced points by repeatedly applying the central difference operator to these values; used in interpolation or numerical calculation and integration of functions. { 'sen·trəl 'dif·rəns }

central difference operator [MATH] A difference operator, denoted ∂, defined by the equation $\partial f(x) = f(x + h/2) - f(x - h/2)$, where h is a constant denoting the difference between successive points of interpolation or calculation. { 'sen·trəl ¦dif·rəns 'äp·ə,rād·ər }

central dogma [GEN] The concept, subject to several exceptions, that genetic information is coded in self-replicating deoxyribonucleic acid and undergoes unidirectional transfer to messenger ribonucleic acids in transcription that act as templates for protein synthesis in translation. { 'sen·trəl 'dòg·mə }

central eclipse [ASTRON] An eclipse in which the eclipsing

body passes centrally (midpoints in line) over the body eclipsed. { 'sen·trəl i'klips }

Centrales [BOT] An order of diatoms (Bacillariophyceae) in which the form is often circular and the markings on the valves are radial. { sen'trā·lēz }

central field approximation [PHYS] The approximation that the electrons in an atom or the nucleons in a nucleus move in the potential of a central force which is the same for all the particles. { 'sen·trəl ,fēld ə,präk·sə'mā·shən }

central force [MECH] A force whose line of action is always directed toward a fixed point; the force may attract or repel. { 'sen·trəl 'fórs }

central gear [MECH ENG] The gear on the central axis of a planetary gear train, about which a pinion rotates. Also known as sun gear. { 'sen·trəl 'gir }

central heating [CIV ENG] The use of a single steam or hot-water heating plant to serve a group of buildings, facilities, or even a complete community through a system of distribution pipes. { 'sen·trəl 'hēd·iŋ }

centralized configuration See star network. { 'sen·trə,līzd kən,fig·yə'rā·shən }

centralized data base [COMPUT SCI] A data base at a single physical location, usually employed in conjunction with centralized data processing. { 'sen·trə,līzd 'dad·ə ,bās }

centralized data processing [COMPUT SCI] The processing of all the data concerned with a given activity at one place, usually with fixed equipment within one building. { 'sen·trə,līzd 'dad·ə ,präs,əs·iŋ }

centralized traffic control [CIV ENG] Control of train movements by signal indications given by a train director at a central control point. Abbreviated CTC. { 'sen·trə,līzd 'traf·ik kən'trōl }

centralizer [MATH] The subgroup consisting of all elements which commute with a given element of a group. [PETRO ENG] See casing centralizer. { 'sen·trə,līz·ər }

central-limit theorem [STAT] The theorem that the distribution of sample means taken from a large population approaches a normal (Gaussian) curve. { ¦sen·trəl ¦lim·ət ,thir·əm }

central mean operator [MATH] A difference operator, denoted μ, defined by the equation $\mu f(x) = [f(x + h/2) + f(x - h/2)]/2$, where h is a constant denoting the difference between successive points of interpolation or calculation. Also known as averaging operator. { ¦sen·trəl ¦mēn 'äp·ə,rad·ər }

central meridian [ASTRON] The meridian of a planet that crosses the center of the visible face of the planet at a given instant. { 'sen·trəl mə'rid·ē·ən }

central-meridian transit [ASTRON] The passage of an object on the surface of a planet across the central meridian. { 'sen·trəl mə'rid·ē·ən 'tranz·ət }

central mix concrete [MATER] A concrete prepared at a concrete mixing plant and transported to the building site. { 'sen·trəl ,miks 'kän,krēt }

central nervous system [NEURO] The division of the vertebrate nervous system comprising the brain and spinal cord. { 'sen·trəl 'nər·vəs ,sis·təm }

central office [COMMUN] A switching unit, installed in a telephone system serving the general public, having the necessary equipment and operating arrangements for terminating and interconnecting lines and trunks. Also known as telephone central office. { 'sen·trəl 'ó·fəs }

central office line See subscriber line. { 'sen·trəl ¦ó·fəs ,līn }

central orbit [MECH] The path followed by a body moving under the action of a central force. { 'sen·trəl 'ór·bət }

central paralysis [MED] Paralysis due to a lesion of the brain or spinal cord. { 'sen·trəl pa'ral·ə·səs }

central peak [ASTRON] A mountain located at the center of the floor of a lunar crater. { 'sen·trəl 'pēk }

central placentation [BOT] Having the ovules located in the center of the ovary. { 'sen·trəl ,pla·sən'tā·shən }

central plane [MATH] For a fixed ruling of a ruled surface, the plane tangent to the surface at the central point of the ruling. { 'sen·trəl 'plān }

central pocket loop [ANAT] A whorl type of fingerprint pattern having two deltas and at least one ridge that make a complete circuit. { ¦sen·trəl 'päk·ət ,lüp }

central point [MATH] For a fixed ruling L on a ruled surface, the limiting position, as a variable ruling L' approaches L, of

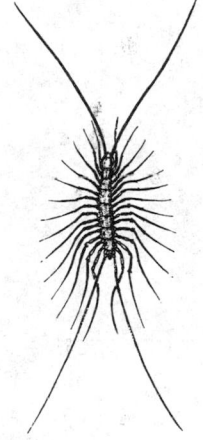

CENTIPEDE

Scutigera coleoptrata, the house centipede; body length about 25 millimeters. *(From R. E. Snodgrass, A Textbook of Arthropod Anatomy, copyright 1952 by Cornell University Press; used by permission)*

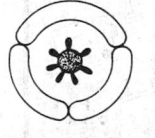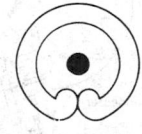

CENTRAL PLACENTATION

Free central placentation, ovules shown in black.

CENTRAL POCKET LOOP

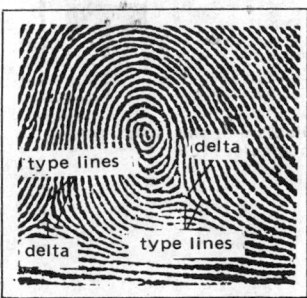

Central pocket loop, a type of whorl pattern. *(Federal Bureau of Investigation)*

the foot on *L* of the common perpendicular to *L* and *L'*. { 'sen·trəl 'pȯint }

central pressure [METEOROL] At any given instant, the atmospheric pressure at the center of a high or low; the highest pressure in a high, the lowest pressure in a low. { 'sen·trəl 'presh·ər }

central processing unit [COMPUT SCI] The part of a computer containing the circuits required to interpret and execute the instructions. Abbreviated CPU. { 'sen·trəl 'präs,əs·iŋ ,yü·nət }

central-processing-unit time [COMPUT SCI] The time actually required to process a set of instructions in the logic unit of a computer. { 'sen·trəl 'präs,es·iŋ ,yü·nət ,tīm }

central projection [MATH] A mapping of a configuration into a plane that associates with any point of the configuration the intersection with the plane of the line passing through the point and a fixed point. { 'sen·trəl prə'jek·shən }

central quadric [MATH] A quadric surface that has a center, namely, a sphere, ellipsoid, or hyperboloid. { 'sen·trəl 'kwä·drik }

central sulcus [ANAT] A groove situated about the middle of the lateral surface of the cerebral hemisphere, separating the frontal from the parietal lobe. { 'sen·trəl 'səl·kəs }

central terminal [COMPUT SCI] A communication device which queues tellers' requests for processing and which channels answers to the consoles originating the transactions. { 'sen·trəl 'tər·mən·əl }

central valley *See* rift valley. { 'sen·trəl 'val·ē }

central water [OCEANOGR] Upper water mass associated with the central region of oceanic gyre. { 'sen·trəl 'wȯd·ər }

Centrarchidae [VERT ZOO] A family of fishes in the order Perciformes, including the fresh-water or black basses and several sunfishes. { ,sen'trär·kə,dē }

centric [ANAT] Having all teeth of both jaws meet normally with perfect distribution of forces in the dental arch. { 'sen·trik }

centrifugal [MECH] Acting or moving in a direction away from the axis of rotation or the center of a circle along which a body is moving. { ,sen'trif·i·gəl }

centrifugal atomizer [MECH ENG] Device that atomizes liquids with a spinning disk; liquid is fed onto the center of the disk, and the whirling motion (3000 to 50,000 revolutions per minute) forces the liquid outward in thin sheets to cause atomization. { ,sen'trif·i·gəl 'ad·ə,mīz·ər }

centrifugal barrier [MECH] A steep rise, located around the center of force, in the effective potential governing the radial motion of a particle of nonvanishing angular momentum in a central force field, which results from the centrifugal force and prevents the particle from reaching the center of force, or causes its Schrödinger wave function to vanish there in a quantum-mechanical system. { ,sen'trif·i·gəl 'bar·ē·ər }

centrifugal brake [MECH ENG] A safety device on a hoist drum that applies the brake if the drum speed is greater than a set limit. { ,sen'trif·i·gəl 'brāk }

centrifugal casting [ENG] A method for casting metals or forming thermoplastic resins in which the molten material solidifies in and conforms to the shape of the inner surface of a heated, rapidly rotating container. { ,sen'trif·i·gəl 'kast·iŋ }

centrifugal clarification [MECH ENG] The removal of solids from a liquid by centrifugal action which decreases the settling time of the particles from hours to minutes. { ,sen'trif·i·gəl ,klar·i·fə'kā·shən }

centrifugal classification [MECH ENG] A type of centrifugal clarification purposely designed to settle out only the large particles (rather than all particles) in a liquid by reducing the centrifuging time. { ,sen'trif·i·gəl ,klas·ə·fə'kā·shən }

centrifugal classifier [MECH ENG] A machine that separates particles into size groups by centrifugal force. { ,sen'trif·i·gəl 'klas·ə,fī·ər }

centrifugal clutch [MECH ENG] A clutch operated by centrifugal force from the speed of rotation of a shaft, as when heavy expanding friction shoes act on the internal surface of a rim clutch, or a flyball-type mechanism is used to activate clutching surfaces on cones and disks. { ,sen'trif·i·gəl 'kləch }

centrifugal collector [MECH ENG] Device used to separate particulate matter of 0.1–1000 micrometers from an airstream; some types are simple cyclones, high-efficiency cyclones, and impellers. { ,sen'trif·i·gəl kə'lek·tər }

centrifugal compressor [MECH ENG] A machine in which

a gas or vapor is compressed by radial acceleration in an impeller with a surrounding casing, and can be arranged multistage for high ratios of compression. { ,sen'trif·i·gəl kəm'pres·ər }

centrifugal cutout [ELEC] A switch that is opened by centrifugal force and is usually closed by a spring when the centrifugal force is reduced. { sen'trif·ə·gəl 'kəd,aut }

centrifugal discharge elevator [MECH ENG] A high-speed bucket elevator from which free-flowing materials are discharged by centrifugal force at the top of the loop. { ,sen'trif·i·gəl 'dis,charj ,el·ə,vād·ər }

centrifugal distortion [PHYS] Tendency of a molecule to stretch slightly as its speed of rotation increases. { ,sen'trif·i·gəl di'stȯr·shən }

centrifugal drainage pattern *See* radial drainage pattern. { ,sen'trif·i·gəl 'drān·ij ,pad·ərn }

centrifugal extractor [CHEM ENG] A device for separating components of a liquid solution, consisting of a series of perforated concentric rings in a cylindrical drum that rotates at 2000–5000 revolutions per minute around a cylindrical shaft; liquids enter and leave through the shaft; they flow radially and concurrently in the rotating drum. { ,sen'trif·i·gəl ik'strak·tər }

centrifugal fan [MECH ENG] A machine for moving a gas, such as air, by accelerating it radially outward in an impeller to a surrounding casing, generally of scroll shape. { 'sen'trif·i·gəl 'fan }

centrifugal filter [ENG] An adaptation of the centrifugal settler; centrifugal action of a spinning container segregates heavy and light materials but heavy materials escape through nozzles as a thick slurry. { ,sen'trif·i·gəl 'fil·tər }

centrifugal filtration [MECH ENG] The removal of a liquid from a slurry by introducing the slurry into a rapidly rotating basket, where the solids are retained on a porous screen and the liquid is forced out of the cake by the centrifugal action. { ,sen'trif·i·gəl fil'trā·shən }

centrifugal force [MECH] **1.** An outward pseudo-force, in a reference frame that is rotating with respect to an inertial reference frame, which is equal and opposite to the centripetal force that must act on a particle stationary in the rotating frame. **2.** The reaction force to a centripetal force. { ,sen'trif·i·gəl 'fȯrs }

centrifugal governor [MECH ENG] A governor whose flyweights respond to centrifugal force to sense speed. { ,sen'trif·i·gəl 'gəv·ə·nər }

centrifugal molecular still [CHEM ENG] A device used for molecular distillation; material is fed to the center of a hot, rapidly rotating cone housed in a chamber at a high vacuum; centrifugal force spreads the material rapidly over the hot surface, where the evaporable material goes off as a vapor to the condenser. { ,sen'trif·i·gəl mə'lek·yə·lər 'stil }

centrifugal moment [MECH] The product of the magnitude of centrifugal force acting on a body and the distance to the center of rotation. { ,sen'trif·i·gəl 'mō·mənt }

centrifugal pump [MECH ENG] A machine for moving a liquid, such as water, by accelerating it radially outward in an impeller to a surrounding volute casing. { ,sen'trif·i·gəl 'pəmp }

centrifugal sedimentation [CHEM ENG] Removing solids from liquids by causing particles to settle through the liquid radially toward or away from the center of rotation (depending on the solid-liquid relative densities) by use of a centrifuge. { ,sen'trif·i·gəl ,sed·ə·mən'tā·shən }

centrifugal separation [MECH ENG] The separation of two immiscible liquids in a centrifuge within a much shorter period of time than could be accomplished solely by gravity. [NUCLEO] Separation of isotopes by spinning a mixture in gas or vapor form at high speed. { ,sen'trif·i·gəl ,sep·ə'rā·shən }

centrifugal settler [CHEM ENG] Spinning container that separates solid particles from liquids; centrifugal force causes suspended solids to move toward or away from the center of rotation, thus concentrating them in one area for removal. { ,sen'trif·i·gəl 'set·lər }

centrifugal stretching [PHYS] Stretching of the bonds of a rotating molecule caused by centrifugal force, resulting in an increase in the molecule's moment of inertia and a modification of its energy levels. { ,sen'trif·i·gəl 'strech·iŋ }

centrifugal switch [MECH ENG] A switch opened or closed by centrifugal force; used on some induction motors to open

CENTRIFUGAL
MOLECULAR STILL

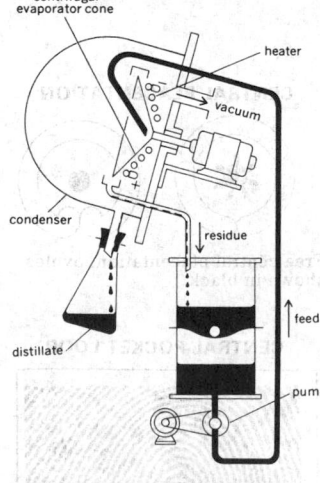

centrifugal evaporator cone

heater

vacuum

condenser

residue

distillate

feed

pump

Diagram of the centrifugal molecular still.

the starting winding when the motor has almost reached synchronous speed. { ˌsen'trif·i·gəl 'swich }

centrifugal tachometer [MECH ENG] An instrument which measures the instantaneous angular speed of a shaft by measuring the centrifugal force on a mass rotating with it. { ˌsen'trif·i·gəl tə'käm·əd·ər }

centrifugation potentials [PHYS CHEM] Electric potential differences between points at different distances from the axis of rotation of a colloidal solution that is being rapidly rotated in a centrifuge. { sen,trif·ə'gā·shən pə,ten·chəlz }

centrifuge [MECH ENG] **1.** A rotating device for separating liquids of different specific gravities or for separating suspended colloidal particles, such as clay particles in an aqueous suspension, according to particle-size fractions by centrifugal force. **2.** A large motor-driven apparatus with a long arm, at the end of which human and animal subjects or equipment can be revolved and rotated at various speeds to simulate the prolonged accelerations encountered in rockets and spacecraft. { 'sen·trə,fyüj }

centrifuge microscope [OPTICS] An instrument which permits magnification and observation of living cells being centrifuged; image of the material magnified by the objective which rotates near the periphery of the centrifuge head is brought to the axis of rotation where it is observed in a stationary ocular. { 'sen·trə,fyüj 'mī·krə,skōp }

centrifuge refining [CHEM ENG] The use of centrifuges for liquids processing, such as separation of solids or immiscible droplets from liquid carriers, or for liquid-liquid solvent extraction. { 'sen·trə,fyüj ri'fīn·iŋ }

centrifuge tube [ANALY CHEM] Calibrated, tube-shaped glass container used with laboratory centrifuges for volumetric analysis of separable (solid-liquid or immiscible liquid) samples. { 'sen·trə,fyüj ,tüb }

centrilobular emphysema [MED] A disorder marked by pulmonary inflation, primarily affecting the respiratory bronchioles and usually more severe in the upper lobes. { ¦sen·trə'lä·byə·lər ,em·fə'sē·mə }

centriole [CYTOL] A complex cellular organelle forming the center of the centrosome in most cells; usually found near the nucleus in interphase cells and at the spindle poles during mitosis. { 'sen·trē,ōl }

centripetal [MECH] Acting or moving in a direction toward the axis of rotation or the center of a circle along which a body is moving. { ˌsen'trip·əd·əl }

centripetal acceleration [MECH] The radial component of the acceleration of a particle or object moving around a circle, which can be shown to be directed toward the center of the circle. Also known as radial acceleration. { ˌsen'trip·əd·əl ik,sel·ə'rā·shən }

centripetal force [MECH] The radial force required to keep a particle or object moving in a circular path, which can be shown to be directed toward the center of the circle. { ˌsen'trip·əd·əl 'fòrs }

centrobaric [MECH] **1.** Pertaining to the center of gravity, or to some method of locating it. **2.** Possessing a center of gravity. { ˌsen·trō¦bar·ik }

centroclinal [GEOL] Referring to geologic strata dipping toward a common center, as in a structural basin. { ¦sen·trō¦klīn·əl }

centrode [MECH] The path traced by the instantaneous center of a plane figure when it undergoes plane motion. { 'sen,trōd }

Centrohelida [INV ZOO] An order of protozoans in the subclass Heliozoia lacking a central capsule and having axopodia or filopodia, and siliceous scales and spines. { ¦sen·trō'hel·ə·də }

centroid See center of area; center of mass; center of volume. { 'sen,tròid }

centroid of asymptotes [CONT SYS] The intersection of asymptotes in a root-locus diagram. { 'sen,tròid əv 'as·əm,tōd·ēz }

centroids of areas and lines [MATH] Points positioned identically with the centers of gravity of corresponding thin homogeneous plates or thin homogeneous wires; involved in the analysis of certain problems of mechanics such as the phenomenon of bending. { 'sen,tròidz əv 'er·ē·əz ən 'līnz }

centrolecithal ovum [CYTOL] An egg cell having the yolk centrally located; occurs in arthropods. { ¦sen·trō'les·ə·thəl 'ō·vəm }

Centrolenidae [VERT ZOO] A family of arboreal frogs in the suborder Procoela characterized by green bones. { ˌsen·trə'len·ə,dē }

centromere [CYTOL] A specialized chromomere to which the spindle fibers are attached during mitosis. Also known as kinetochore; kinomere; primary constriction. { 'sen·trə,mir }

centromere distance [GEN] The distance of a gene from a centromere, measured in terms of recombination frequency. { 'sen·trə,mir ,dis·təns }

centromere effect [GEN] The reduced level of genetic recombination shown by genetic loci close to the centromere. { 'sen·trə,mir i,fekt }

centromere shift [GEN] A type of chromosomal defect in which the centromere changes position during chromosomal rearrangement in the G1 phase of the cell cycle. { 'sen·trə,mir ,shift }

Centronellidina [PALEON] A suborder of extinct articulate brachiopods in the order Terebratulida. { ¦sen·trō·nə'lid·ən·ə }

centrosome [CYTOL] A spherical hyaline region of the cytoplasm surrounding the centriole in many cells; plays a dynamic part in mitosis as the focus of the spindle pole. { 'sen·trə,sōm }

centrosome cycle [CELL MOL] Duplication of the centrosome during interphase (S phase) of the animal cell cycle followed by separation of the resulting centrioles and associated microtubules at the beginning of mitosis to form the poles of the mitotic spindle. Following mitosis, each daughter cell has a new centrosome in association with its chromosomes. { 'sen·trə,sōm ,sī·kəl }

Centrospermae [BOT] An equivalent name for the Caryophyllales. { ˌsen·trō'spər,mē }

Centrospermales [BOT] An equivalent name for the Caryophyllales. { ˌsen·trō·spər'mā·lēz }

centrosphere [CYTOL] The differentiated layer of cytoplasm immediately surrounding the centriole. [GEOL] The central core of the earth. Also known as the barysphere. { 'sen·trə,sfir }

centrosymmetry [PHYS] Property of a body or system which is unchanged under space inversion through a specified point. { ¦sen·trō'sim·ə·trē }

centrum [ANAT] The main body of a vertebra. [BOT] The central space in hollow-stemmed plants. { 'sen·trəm }

century date [HOROL] The number of days that have elapsed in the century, that is, since January 1, 1900. { 'sen·chə·rē ,dāt }

Cen X-3 See Centaurus X-3.

Cep See Cepheus.

CEPHA See ethephon. { 'sef·ə }

cephaeline [ORG CHEM] $C_{14}H_{19}O_2N$ An alkaloid, slightly soluble in water, extracted from the root of ipecac; used as an emetic. { sə'fā·ə,lēn }

cephalalgia [MED] Headache or head pain. { ,sef·ə'lal·jə }

Cephalaspida [PALEON] An equivalent name for the Osteostraci. { ,sef·ə'las·pə·də }

Cephalaspidomorphi [VERT ZOO] An equivalent name for Monorhina. { ,sef·ə¦las·pə·də'mòr·fī }

cephalic [ZOO] Of or pertaining to the head or anterior end. { sə'fal·ik }

cephalic index [ANTHRO] The ratio of maximum breadth to maximum length of the head multiplied by 100. { sə'fal·ik 'in,deks }

cephalic module [ANTHRO] A measure of absolute head size derived by averaging the length, breadth, and auricular height of the head. { sə'fal·ik 'mäj·ül }

cephalic vein [ANAT] A superficial vein located on the lateral side of the arm which drains blood from the radial side of the hand and forearm into the axillary vein. { sə'fal·ik 'vān }

cephalin [BIOCHEM] Any of several acidic phosphatides whose composition is similar to that of lecithin but having ethanolamine, serine, and inositol instead of choline; found in many living tissues, especially nervous tissue of the brain. { 'sef·ə·lən }

Cephalina [INV ZOO] A suborder of protozoans in the order Eugregarinida that are parasites of certain invertebrates. { ,sef·ə'lī·nə }

cephalization [ZOO] Anterior specialization resulting in the

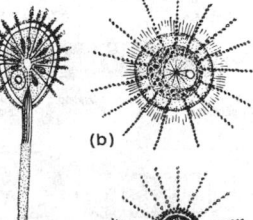

CENTROHELIDA

Some Centrohelida species.
(a) Actinolophus pedunculatus, sessile on Bryozoa.
(b) Heterophrys myriopoda.
(c) Pompholyxophrys punicea.
(From R. P. Hall, Protozoology, Prentice-Hall, 1953)

CEPHALOCARIDA

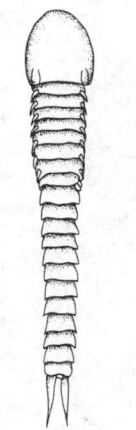

Hutchinsoniella macracantha.

CERAMBYCIDAE

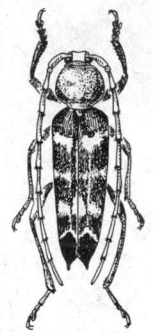

A longhorn beetle. *(From T. I. Storer and R. L. Usinger, General Zoology, 3d ed., McGraw-Hill, 1957)*

CERAMIC CAPACITOR

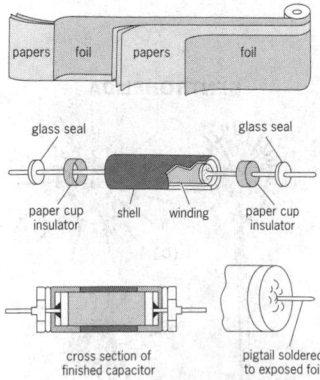

Ceramic capacitor constructed in chip form, showing cutaway of finished chip.

concentration of sensory and neural organs in the head. { ˌsefˈə·lə'zā·shən }

Cephalobaenida [INV ZOO] An order of the arthropod class Pentastomida composed of primitive forms with six-legged larvae. { ˌsef·ə·lō'bēn·ə·də }

Cephaloboidea [INV ZOO] A superfamily of free-living nematodes in the order Rhabditida distinguished by cephalic elaborations or ornamentations. { 'sef·ə·lə'bȯid·ē·ə }

Cephalocarida [INV ZOO] A subclass of Crustacea erected to include the primitive crustacean *Hutchinsoniella macracantha.* { ˌsef·ə·lō'kar·ə·də }

Cephalochordata [VERT ZOO] A subphylum of the Chordata comprising the lancelets, including *Branchiostoma.* { ˌsef·ə·lō·kȯr'däd·ə }

cephalodium [BOT] A small wart-like growth containing nitrogen-fixing cyanobacteria that is found on or in the thallus of some lichens with photobionts. { ˌsef·ə'lō·dē·əm }

Cephaloidae [INV ZOO] The false longhorn beetles, a small family of coleopteran insects in the superfamily Tenebrionoidea. { ˌsef·ə'lȯid·ē }

cephalomere [INV ZOO] One of the somites that make up the head of an arthropod. { sə'fal·ə,mir }

cephalometry [ANTHRO] The science of measuring the head, especially for determining the characteristics of a particular race, sex, or somatotype. { ˌsef·ə'läm·ə·trē }

cephalont [INV ZOO] A sporozoan just prior to spore formation. { 'sef·ə,länt }

cephalopelvic disproportion [MED] A condition in which the fetus is unable to pass safely through the pelvis during labor because of pelvic contraction, an unfavorable fetal position, or a large fetal head in relation to pelvic size. Abbreviated CPD. { ˌsef·ə·lō¦pel'vik ˌdis·prə'pȯr·shən }

Cephalopoda [INV ZOO] Exclusively marine animals constituting the most advanced class of the Mollusca, including squids, octopuses, and *Nautilus.* { ˌsef·ə'läp·ə·də }

cephalosporin [MICROBIO] Any of a group of antibiotics produced by strains of the imperfect fungus *Cephalosporium.* { ˌsef·ə·lə'spȯr·ən }

cephalothin [MICROBIO] An antibiotic derived from the fungus *Cephalosporium,* resembling penicillin units in structure and activity, and effective against many gram-positive cocci that are resistant to penicillin. { 'sef·ə·lə·thən }

cephalothorax [INV ZOO] The body division comprising the united head and thorax of arachnids and higher crustaceans. { ¦sef·ə·lə'thȯr,aks }

Cephalothrididae [INV ZOO] A family of ribbonlike worms in the order Palaeonemertini. { ˌsef·ə·lō'thrī·də,dē }

cephalotrichous flagellation [CYTOL] Insertion of flagella in polar tufts. { ˌsef·ə·lō'trī·kəs ˌflaj·ə'lā·shən }

Cepheid [ASTRON] One of a subgroup of periodic variable stars whose brightness does not remain constant with time and whose period of variation is a function of intrinsic mean brightness. { 'sē·fē·əd }

Cepheus [ASTRON] A constellation with right ascension 22 hours, declination 70°N. Abbreviated Cep. { 'sē·fē·əs }

Cephidae [INV ZOO] The stem sawflies, composing the single family of the hymenopteran superfamily Cephoidea. { 'sef·ə,dē }

Cephoidea [INV ZOO] A superfamily of hymenopteran insects in the suborder Symphyta. { sa'fȯid·ē·ə }

cepstrum [ACOUS] The Fourier transform of the logarithm of a speech power spectrum; used to separate vocal tract information from pitch excitation in voiced speech. { 'sep·trəm }

cepstrum vocoder [ENG ACOUS] A digital device for reproducing speech in which samples of the cepstrum of speech, together with pitch information, are transmitted to the receiver, and are then converted into an impulse response that is convolved with an impulse train generated from the pitch information. { 'sep·trəm 'vō¦kōd·ər }

Ceractinomorpha [INV ZOO] A subclass of sponges belonging to the class Demospongiae. { sə¦rak·tə·nə'mȯr·fə }

ceramagnet [ELECTROMAG] A ferrimagnet composed of the hard magnetic material $BaO·6Fe_2O_3$. { 'se·rə,mag·nət }

ceramal *See* cermet. { sə'ram·əl }

Cerambycidae [INV ZOO] The longhorn beetles, a family of coleopteran insects in the superfamily Chrysomeloidea. { se·rəm'bī·sə,dē }

cerambycoid larva [INV ZOO] A beetle larva that is morphologically similar to a caraboid larva except for the former's absence of legs. { sə'ram·bə,kȯid 'lar·və }

ceramet *See* cermet. { sə'ram·ət }

ceramic [MATER] **1.** Inorganic, nonmetallic materials processed or used at high temperature, generally including oxides, nitrides, borides, carbides, silicides, and sulfides. Intermetallic compounds such as aluminides and beryllides are also considered ceramics, as are phosphides, antimonides, and arsenides. **2.** Consisting of such a product. { sə'ram·ik }

ceramic aggregate [MATER] **1.** Portland cement concrete containing lumps of ceramic material. **2.** Concrete made with porous clay to reduce its weight. { sə'ram·ik 'ag·rə·gət }

ceramic amplifier [ELECTR] An amplifier that utilizes the piezoelectric properties of semiconductors such as silicon. { sə'ram·ik 'am·plə,fī·ər }

ceramic-based microcircuit [ELECTR] A microminiature circuit printed on a ceramic substrate. { sə'ram·ik,bāst 'mī·krō,sər·kət }

ceramic capacitor [ELEC] A capacitor whose dielectric is a ceramic material such as steatite or barium titanate, the composition of which can be varied to give a wide range of temperature coefficients. { sə'ram·ik kə'pas·əd·ər }

ceramic cartridge [ENG ACOUS] A device containing a piezoelectric ceramic element, used in phonograph pickups and microphones. { sə'ram·ik 'kär·trij }

ceramic coating [MET] A nonmetallic, inorganic coating made of sprayed aluminum oxide or of zirconium oxide, or a cemented coating of an intermetallic compound, such as aluminum disilicide, of essentially crystalline nature, applied as a protective film on metal to protect against temperatures above 1100°C. { sə'ram·ik 'kōd·iŋ }

ceramic earphones *See* crystal headphones. { sə'ram·ik 'ir,fōnz }

ceramic fiber [MATER] A small-dimension filament or thread composed of a ceramic material, usually alumina and silica, used in lightweight units for electrical, thermal, and sound insulation, filtration at high temperatures, packing, and reinforcing other ceramic materials. { sə'ram·ik 'fī·bər }

ceramic filter [ELECTR] A type of mechanical filter that uses a series of resonant ceramic disks to obtain a band-pass response. { sə'ram·ik 'fil·tər }

ceramic glaze [ENG] A glossy finish on a clay body obtained by spraying with metallic oxides, chemicals, and clays and firing at high temperature. { sə'ram·ik 'glāz }

ceramicite [PETR] A porcelained pyrometamorphic rock composed of basic plagioclase and cordierite with a small amount of hypersthene and a groundmass of glass. { sə'ram·ə,sīt }

ceramic magnet [ELECTROMAG] A permanent magnet made from pressed and sintered mixtures of ceramic and magnetic powders. Also known as ferromagnetic ceramic. { sə'ram·ik 'mag·nət }

ceramic microphone [ENG ACOUS] A microphone using a ceramic cartridge. { sə'ram·ik 'mī·krə,fōn }

ceramic mold casting [MET] A precision casting process using a ceramic body fired to high temperature as the mold, and carbon, low-alloy, or stainless steel as the casting. { sə'ram·ik 'mōld ,kast·iŋ }

ceramic pickup [ENG ACOUS] A phonograph pickup using a ceramic cartridge. { sə'ram·ik 'pik·əp }

ceramic radiant [ENG] A baked-clay component of a gas heating unit which radiates heat when incandescent from the gas flame. { sə'ram·ik 'rād·ē·ənt }

ceramic reactor [NUCLEO] A nuclear reactor in which the fuel and moderator assemblies are made from high-temperature-resistant ceramic materials such as metal oxides, carbides, or nitrides. { sə'ram·ik rē'ak·tər }

ceramic rod flame spraying [MET] A method of flame spraying in which the ceramic rod is fed into a gun that utilizes an oxyfuel gas flame to atomize and airblast the rod material to the substrate. { sə'ram·ik 'räd ,flām ,sprā·iŋ }

ceramics [ENG] The art and science of making ceramic products. { sə'ram·iks }

ceramic tile [MATER] A burned-clay product composed of a clay body with a decorative surface glaze; used principally for decorative and sanitary effects. { sə'ram·ik 'tīl }

ceramic tool [DES ENG] A cutting tool made from metallic oxides. { sə'ram·ik ,tül }

ceramic transducer *See* electrostriction transducer. { sə'ram·ik tranz'dü·sər }

ceramic tube [ELECTR] An electron tube having a ceramic envelope capable of withstanding operating temperatures over 500°C, as required during reentry of guided missiles. { sə'ram·ik 'tüb }

ceramic veneer [MATER] A burned clay non-load-bearing unit used in masonry construction. { sə'ram·ik və'nir }

ceramide [ORG CHEM] Any of a group of amides formed by linking a fatty acid to sphingosine. { 'ser·ə,mīd }

Ceramonematoidea [INV ZOO] A superfamily of marine nematodes in the order Desmodorida characterized by distinctive cuticular ornamentation, giving the appearance of a body covered with rings of crested tilelike plates. { ,ser·ə·mō,nem·ə'tóid·ē·ə }

ceramoplastic [MATER] A high-temperature insulating material made by bonding synthetic mica with glass. { sə¦ram·ō¦plas·tik }

Ceramoporidae [PALEON] A family of extinct, marine bryozoans in the order Cystoporata. { sə,ram·ə'pór·ə,dē }

Cerapachyinae [INV ZOO] A subfamily of predacious ants in the family Formicidae, including the army ant. { ,ser·ə·pə'kī·ə,nē }

Ceraphronidae [INV ZOO] A superfamily of hymenopteran insects in the superfamily Proctotrupoidea. { ,ser·ə'frän·ə,dē }

cerargyrite [MINERAL] AgCl A colorless to pearl-gray mineral; crystallizes in the isometric system, but crystals, usually cubic, are rare; a secondary mineral that is an ore of silver. Also known as chlorargyrite; horn silver. { sa'rär·jə,rīt }

cerata [INV ZOO] Respiratory papillae of the mantle in certain nudibranchs. { sa'räd·ə }

cerate [ORG CHEM] A metallic salt or soap made from lard. [PHARM] An unguent made of wax, resin, or spermaceti mixed with oil, lard, and medicinal ingredients; used for external application. { 'sir,āt }

ceratine [INV ZOO] A hornlike material secreted by some anthozoans. { 'ser·ə,tēn }

Ceratiomyxaceae [MYCOL] The single family of the fungal order Ceratiomyxales. { sə,räsh·ē·ō,mik'sās·ē,ē }

Ceratiomyxales [MYCOL] An order of myxomycetous fungi in the subclass Ceratiomyxomycetidae. { sə,rash·ē·ō,mik 'sā·lēz }

Ceratiomyxomycetidae [MYCOL] A subclass of fungi belonging to the Myxomycetes. { sə,räsh·ē·ō,mik·sō,mī'sed·ə,dē }

ceratite [PALEON] A fossil ammonoid of the genus *Ceratites* distinguished by a type of suture in which the lobes are further divided into subordinate crenulations while the saddles are not divided and are smoothly rounded. { 'ser·ə,tīt }

ceratitic [PALEON] Pertaining to a ceratite. { ,ser·ə'tid·ik }

Ceratodontidae [PALEON] A family of Mesozoic lungfishes in the order Dipteriformes. { ,ser·ə·tō'dän·tə,dē }

Ceratomorpha [VERT ZOO] A suborder of the mammalian order Perissodactyla including the tapiroids and the rhinoceratoids. { ,ser·ə·tō'mór·fə }

Ceratophyllaceae [BOT] A family of rootless, free-floating dicotyledons in the order Nymphaeales characterized by unisexual flowers and whorled, cleft leaves with slender segments. { ,ser·ə·tō·fə'lās·ē,ē }

Ceratopsia [PALEON] The horned dinosaurs, a suborder of Upper Cretaceous reptiles in the order Ornithischia. { ,ser·ə'täp·sē·ə }

Ceratosaurus [PALEON] A carnivorous therapod dinosaur, 20 feet (6 meters) long, from the Late Jurassic Period that had strong hindlimbs, short and weak forelimbs (with four-fingered hands), and massive jaws lined with enormous teeth. { sə,rad·ə'sór·əs }

ceraunograph [ENG] An instrument that detects radio waves generated by lightning discharges and records their occurrence. { sə'rón·ə,graf }

cercaria [INV ZOO] The larval generation which terminates development of a digenetic trematode in the intermediate host. { sər'kar·ē·ə }

cerci [ZOO] A pair of posterior abdominal appendages with delicate hairs found on many insects, including cockroaches and crickets, that are very sensitive to the air currents generated by a moving object. { 'sər·sē }

Cercopidae [INV ZOO] A family of homopteran insects belonging to the series Auchenorrhyncha. { sar'käp·ə,dē }

Cercopithecidae [VERT ZOO] The Old World monkeys, a family of primates in the suborder Anthropoidea. { ¦sər·kō·pə'thē·sə,dē }

cercopod [INV ZOO] **1.** Either of two filamentous projections on the posterior end of notostracan crustaceans. **2.** *See* cercus. { 'sər·kə,päd }

Cercospora [MYCOL] A genus of imperfect fungi having dark, elongate, multiseptate spores. { sər'käs·pə·rə }

Cercospora leaf spot [PL PATH] Any of several fungus diseases of plants caused by *Cercospora* species and characterized by areas of discoloration on the leaves. { sər'käs·pə·rə 'lēf ,spät }

cercus [INV ZOO] Either of a pair of segmented sensory appendages on the last abdominal segment of many insects and certain other anthropods. Also known as cercopod. { 'sər·kəs }

cere [VERT ZOO] A soft, swollen mass of tissue at the base of the upper mandible through which the nostrils open in certain birds, such as parrots and birds of prey. { sir }

cerea flexibilitas [MED] The flexibility often present in catatonia in which the person's arm or leg remains in the position in which it is placed. { 'ser·ē·ə ,flek·sə'bil·ə,täs }

cereal [BOT] Any member of the grass family (Graminae) which produces edible, starchy grains usable as food by humans and livestock. Also known as grain. { 'sir·ē·əl }

cereal binder [MATER] A binding material derived from flour; used for core mixtures in a casting process. { 'sir·ē·əl ,bīn·dər }

cerebellar ataxia [MED] Incoordination of muscles due to disease of the cerebellum. { ,ser·ə'bel·ər ā'tak·sē·ə }

cerebellum [NEURO] The part of the vertebrate brain lying below the cerebrum and above the pons, consisting of three lobes and concerned with muscular coordination and the maintenance of equilibrium. { ,ser·ə'bel·əm }

cerebral arteriosclerosis [MED] Hardening of the arteries of the brain, sometimes resulting in an organic mental disorder that may be primarily neurologic, primarily psychologic, or a combination of both. Also known as multiple infarct dementia. { sə'rē·brəl ür,tir ē ō·sklə'rō·səs }

cerebral cortex [NEURO] The superficial layer of the cerebral hemispheres, composed of gray matter and concerned with coordination of higher nervous activity. { sə'rē·brəl 'kór,teks }

cerebral hemisphere [NEURO] Either of the two lateral halves of the cerebrum. { sə'rē·brəl 'hem·ə,sfir }

cerebral localization [NEURO] Designation of a specific region of the brain as the area controlling a specific physiologic function or as the site of a lesion. { sə'rē·brəl lō·kə·lə'zā·shən }

cerebral palsy [MED] Any nonprogressive motor disorder in humans caused by brain damage incurred during fetal development. { sə'rē·brəl 'pól·zē }

cerebral peduncle [NEURO] One of two large bands of white matter (containing descending axons of upper motor neurons) which emerge from the underside of the cerebral hemispheres and approach each other as they enter the rostral border of the pons. { sə'rē·brəl pi'dəŋ·kəl }

cerebration [PSYCH] The act or process of using the mind; thinking. { ser·ə'brā·shən }

cerebrose *See* galactose. { 'ser·ə,brōs }

cerebroside [BIOCHEM] Any of a complex group of glycosides found in nerve tissue, consisting of a hexose, a nitrogenous base, and a fatty acid. Also known as galactolipid. { 'ser·ə·brō,sīd }

cerebroside lipoidosis *See* Gaucher's disease. { 'ser·ə·brō,sīd ,li'pói'dō·səs }

cerebrospinal axis [ANAT] The axis of the body composed of the brain and spinal cord. { sə¦rē·brō'spīn·əl 'ak·səs }

cerebrospinal fluid [PHYSIO] A clear liquid that fills the ventricles of the brain and the spaces between the arachnoid mater and pia mater. { sə¦rē·brō'spīn·əl 'flü·əd }

cerebrospinal meningitis [MED] Inflammation of the meninges of the brain and spinal cord. { sə'rē·brō'spīn·əl ,men·ən'jīd·əs }

cerebrotonia [PSYCH] A temperament pattern proposed by W. H. Sheldon as characteristic of an ectomorph, who is an

CERCOSPORA

15 μm

The dark multiseptate spores of *Cercospora*.

individual with an intellectualized, solitudinous, and inhibited approach to life. { sə¦re·brō'tōn·ē·ə }

cerebrovascular accident [MED] A symptom complex resulting from cerebral hemorrhage, embolism, or thrombosis of the cerebral vessels, characterized by sudden loss of consciousness. { sa¦rē·brō'vas·kyə·lər 'ak·sə·dənt }

cerebrum [NEURO] The enlarged anterior or upper part of the vertebrate brain consisting of two lateral hemispheres. { sə'rē·brəm }

Cerelasmidae [INV ZOO] A family of Psamminida with a soft test composed principally of organic cement; xenophyae, when present, are not systematically arranged. { ‚ser·ə'las·mə‚dē }

cerelose See glucose. { 'sir·ə‚lōs }

Cerenkov counter [NUCLEO] An apparatus for detecting high-energy charged particles by observation of the Cerenkov radiation produced. { chə'reŋ·kəf ‚kaün·tər }

Cerenkov radiation [ELECTROMAG] Light emitted by a high-speed charged particle when the particle passes through a transparent, nonconducting material at a speed greater than the speed of light in the material. { chə'reŋ·kəf rād·ē'ā·shən }

Cerenkov rebatron radiator [ELECTR] Device in which a tightly bunched, velocity-modulated electron beam is passed through a hole in a dielectric; the reaction between the higher velocity of the electrons passing through the hole and the slower velocity of the electromagnetic energy passing through the dielectric results in radiation at some frequency higher than the frequency of modulation of the electron beam. { chə'reŋ·kəf ¦rē·bə‚trän ¦rād·ē‚ād·ər }

Ceres [ASTRON] The largest asteroid, with a diameter of about 960 kilometers, mean distance from the sun of 2.766 astronomical units, and C-type surface composition. { 'sir‚ēz }

ceresin [MATER] **1.** A hydrocarbon wax refined from veins of wax shale known as ozocerite; used in manufacture of candles, shoe polish, electrical insulation, and floor waxes because of its great compatibility with other substances. Also known as ceresine; ozocerite. **2.** A mixture of paraffin wax and beeswax, or a mixture of ozocerite and paraffin. { 'se·rə·sən }

ceresine See ceresin. { 'se·rə‚sēn }

ceria See ceric oxide. { 'ser·ē·ə }

Ceriantharia [INV ZOO] An order of the Zoantharia distinguished by the elongate form of the anemone-like body. { ‚ser·ē·ən'thar·ē·ə }

Ceriantipatharia [INV ZOO] A subclass proposed by some authorities to include the anthozoan orders Antipatharia and Ceriantharia. { ‚sir·ē‚an·tə·pə'thar·ē·ə }

ceric oxide [INORG CHEM] CeO₂ A pale-yellow to white powder; soluble in sulfuric acid, insoluble in dilute acid and water; used in ceramics and as a polish for optical glass. Also known as ceria; cerium dioxide; cerium oxide. { 'sir·ik 'äk‚sīd }

ceric sulfate [INORG CHEM] Ce(SO₄)₂·4H₂O Yellow needles forming a basic salt with excess water; used in waterproofing, mildew-proofing, and in dyeing and printing textiles. { 'sir·ik 'səl‚fāt }

cerine See allanite. { 'sir‚ēn }

cerinic acid See cerotic acid. { sə'rēn·ik 'as·əd }

cerite [MINERAL] (Ca,Fe)Ce₃Si₃O₁₂·H₂O A brown rare-earth hydrous silicate of cerium and other metals found in gneiss; hardness is 5.5 on Mohs scale, and specific gravity is 4.86. { 'sir‚īt }

Cerithiacea [INV ZOO] A superfamily of gastropod mollusks in the order Prosobranchia. { ‚ser·ə‚thī'ās·ē·ə }

cerium [CHEM] A chemical element, symbol Ce, atomic number 58, atomic weight 140.12; a rare-earth metal, used as a getter in the metal industry, as an opacifier and polisher in the glass industry, in Welsbach gas mantles, in cored carbon arcs, and as a liquid-liquid extraction agent to remove fission products from spent uranium fuel. { 'sir·ē·əm }

cerium-140 [NUC PHYS] An isotope of cerium with atomic mass number of 140, 88.48% of the known amount of the naturally occurring element. { 'sir·ē·əm ‚wən'fórd·ē }

cerium-142 [NUC PHYS] A radioactive isotope of cerium with atomic mass number of 142; emits α-particles and has a half-life of 5 × 10¹⁵ years. { 'sir·ē·əm ‚wən‚fórd·ē'tü }

cerium-144 [NUC PHYS] A radioactive isotope of the element cerium with atomic mass number of 144; a beta emitter with a half-life of 285 days. { 'sir·ē·əm ‚wən‚fórd·ē'fór }

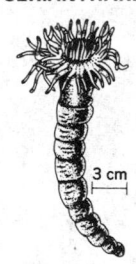

CERIANTHARIA

Cerianthus solitarius.

3 cm

cerium dioxide See ceric oxide. { 'sir·ē·əm dī'äk‚sīd }

cerium fluoride [INORG CHEM] CeF₃ White hexagonal crystals, melting point 1460°C; used in arc carbons to increase the brilliance of carbon-arc lamps. { 'sir·ē·əm 'flúr‚īd }

cerium oxide See ceric oxide. { 'sir·ē·əm 'äk‚sīd }

cerium stearate [ORG CHEM] Ce(C₁₈H₃₅O₂)₂ White, waxy, inert powder, melting point 100–110°C; used in waterproofing compounds. { 'sir·ē·əm 'stir‚āt }

Cermak-Spirek furnace [ENG] An automatic reverberatory furnace of rectangular form divided into two sections by a wall; used for roasting zinc and quicksilver ores. { 'sər‚mak ‚spir·ek ‚fər·nəs }

cermet [MATER] Any of a group of composite materials made by mixing, pressing, and sintering metal with ceramic; examples are silicon-silicon carbide and chromium-alumina carbide. Also known as ceramal; ceramet; metal ceramic. { 'sər‚met }

cermet nuclear fuel [NUCLEO] A nuclear reactor fuel mixed with a heat-resistant ceramic and a metal to give it both refractory and damage-resistant properties. { 'sər‚met 'nü·klē·ər 'fyül }

cermet resistor [ELEC] A metal-glaze resistor, consisting of a mixture of finely powdered precious metals and insulating materials fired onto a ceramic substrate. { 'sər‚met ri'zis·tər }

cernuous [BOT] Drooping or inclining. { 'sərn·yə·wəs }

cerography [GRAPHICS] Painting in which wax is used as a binder for the pigments. { sə'räg·rə·fē }

cerolite [MINERAL] A mixture of serpentine and stevensite occurring in yellow or greenish waxlike masses. { 'sir·ə‚līt }

Cerophytidae [INV ZOO] A small family of coleopteran insects in the superfamily Elateroidea. { ‚ser·ə'fīd·ə‚dē }

cerotic acid [ORG CHEM] CH₃(CH₂)₂₄COOH A fatty acid derived from carnauba wax or beeswax; melts at 87.7°C. Also known as cerinic acid; hexacosanoic acid. { sə'räd·ik 'as·əd }

Cerruti's problem See Boussinesq's problem. { se'rü·dēz ‚präb·ləm }

cers [METEOROL] A term for the mistral in Catalonia, Narbonne, and parts of Provence (southern France and northeastern Spain). { sərz }

certainty equivalence control [CONT SYS] An optimal control law for a stochastic adaptive control system which is obtained by solving the control problem in the case of known parameters and substituting the known parameters with their estimates. { 'sərt·ən·tē i'kwiv·ə·ləns kən'trōl }

certation [BOT] Competition in growth rate between pollen tubes of different genotypes resulting in unequal chances of accomplishing fertilization. { sər'tā·shən }

certificate [COMMUN] A data record containing an identification, a digital signature from a third party who is believed to be trustworthy, attesting to the authenticity of the identity, and an encryption key which provides a basis for two unknown entities to establish a shared encryption. { sər'tif·i·kət }

certified color See food color. { 'sərd·ə‚fīd 'kəl·ər }

certified reference material [ANALY CHEM] A reference material, one or more of whose property values are certified by a technically valid procedure, for which a certificate or other documentation has been issued by an appropriate certifying agency. { 'sərd·ə‚fīd 'ref·rəns ‚mə'tir·ē·əl }

ceruloplasmin [BIOCHEM] The copper-binding serum protein in human blood. { sə¦rül·ō¦plaz·mən }

cerumen [PHYSIO] The waxy secretion of the ceruminous glands of the external ear. Also known as earwax. { sə'rü·mən }

ceruminous gland [ANAT] A modified sweat gland in the external ear that produces earwax. { sə'rü·mə·nəs ‚gland }

cerussite [MINERAL] PbCO₃ A yellow or white member of the aragonite group occurring in orthorhombic crystals; produced by the action of carbon dioxide on lead ore. { sə'rəs‚īt }

cervantite [MINERAL] Sb₂O₄ A white or yellow secondary mineral crystallizing in the orthorhombic system and formed by oxidation of antimony sulfide. { sər'van‚tīt }

cervical [ANAT] Of or relating to the neck, a necklike part, or the cervix of an organ. { 'sər·və·kəl }

cervical canal [ANAT] Canal of the cervix of the uterus. { 'sər·və·kəl kə'nal }

cervical dysplasia [MED] Abnormal growth of the epithelial cells lining the cervix. Also known as cervical intraepithelial neoplasia. { ‚sər·və·kəl di'splāzh·yə }

cervical flexure [EMBRYO] A ventrally concave flexure of the embryonic brain occurring at the junction of hindbrain and spinal cord. { 'sər·və·kəl 'flek·shər }

cervical ganglion [NEURO] Any ganglion of the sympathetic nervous system located in the neck. { 'sər·və·kəl 'gaŋ·glē,än }

cervical plexus [NEURO] A plexus in the neck formed by the anterior branches of the upper four cervical nerves. { 'sər·və·kəl 'plek·səs }

cervical sinus [EMBRYO] A triangular depression caudal to the hyoid arch containing the posterior visceral arches and grooves. { 'sər·və·kəl 'sī·nəs }

cervical vertebra [ANAT] Any of the bones in the neck region of the vertebral column; the transverse process has a characteristic perforation by a transverse foramen. { 'sər·və·kəl 'vərd·ə·brə }

cervicitis [MED] Inflammation of the cervix uteri. { ,sər·və'sīd·əs }

Cervidae [VERT ZOO] A family of pecoran ruminants in the superfamily Cervoidea, characterized by solid, deciduous antlers; includes deer and elk. { 'sər·və,dē }

cervix [ANAT] A constricted or necklike portion of a structure. { 'sər·viks }

Cervoidea [VERT ZOO] A superfamily of tylopod ruminants in infraorder Pecora, including deer, giraffes, and related species. { sər'vóid·ē·ə }

ceryl alcohol [ORG CHEM] $C_{26}H_{53}OH$ An alcohol derived from Chinese wax, melting at 79°C and insoluble in water. { 'sir·əl 'al·kə,hól }

cesarean section [MED] Delivery of the fetus through an abdominal incision. { sə'zer·ē·ən 'sek·shən }

Cesáro equation [MATH] An equation which relates the arc length along a plane curve and the radius of curvature. { chā'zä·rō i,kwā·zhən }

cesarolite [MINERAL] $H_2PbMn_3O_8$ A steel-gray mineral consisting of a hydrous lead manganate occurring in spongy masses. { ,chäz·ə'rō,līt }

Cesáro summation [MATH] A method of attaching sums to certain divergent sequences and series by taking averages of the first n terms and passing to the limit. { chā'zä·rō sə'mā·shən }

cesium [CHEM] A chemical element, symbol Cs, atomic number 55, atomic weight 132.905. { 'sē·zē·əm }

cesium-134 [NUC PHYS] An isotope of cesium, atomic mass number of 134; emits negative beta particles and has a half-life of 2.19 years; used in photoelectric cells and in ion propulsion systems under development. { 'sē·zē·əm ,wən,thərd·ē'fór }

cesium-137 [NUC PHYS] An isotope of cesium with atomic mass number of 137; emits negative beta particles and has a half-life of 30 years; offers promise as an encapsulated radiation source for therapeutic and other purposes. Also known as radiocesium. { 'sē·zē·əm ,wən,thərd·ē'sev·ən }

cesium-antimonide photocathode [ELECTR] A photocathode obtained by exposing a thin layer of antimony to cesium vapor at elevated temperatures; has a maximum sensitivity in the blue and ultraviolet regions of the spectrum. { 'sē·zē·əm 'an·tə·mə,nīd ,fōd·ō'kath,ōd }

cesium-beam atomic clock [NUCLEO] An instrument, used as the primary standard of frequency and time, in which a microwave oscillator, which generates radiation in a microwave cavity, is maintained at a frequency such that a hyperfine transition is induced in cesium atoms in a beam passing through the cavity. Also known as cesium-beam atomic oscillator. { 'sē·zē·əm ,bēm ə'täm·ik 'kläk }

cesium-beam atomic oscillator See cesium-beam atomic clock. { 'sē·zē·əm ,bēm ə'täm·ik 'äs·ə,lād·ər }

cesium-beam sputter source [ELECTR] A source of negative ions in which a beam of positive cesium ions, accelerated through a potential difference on 20–30 kilovolts, sputters the cesium-coated inner surface of a hollow cone fabricated from or containing the element whose negative ion is required, and an appreciable fraction of the negative ions leaving the surface are extracted from the rear hole of the sputter cone. { 'sē·zē·əm ,bēm 'spəd·ər ,sórs }

cesium-beam tube See cesium electron tube. { 'sē·zē·əm ,bēm ,tüb }

cesium bromide [INORG CHEM] CsBr A colorless, crystalline powder with a melting point of 636°C; soluble in water; used in medicine, for infrared spectroscopy, and in scintillation counters. { 'sē·zē·əm 'brō,mīd }

cesium carbonate [INORG CHEM] Cs_2CO_3 A white,

hygroscopic, crystalline powder; soluble in water; used in specialty glasses. { 'sē·zē·əm 'kär·bə,nāt }

cesium chloride [INORG CHEM] CsCl Colorless cuboid crystals, melting point 646°C; used in filaments of radio tubes to increase sensitivity, in photoelectric cells, and for photosensitive deposit on cathodes. { 'sē·zē·əm 'klór,īd }

cesium electron tube [ELECTR] An electronic device used as an atomic clock, producing electromagnetic energy that is accurate and stable in frequency. Also known as cesium beam tube. { 'sē·zē·əm i'lek,trän ,tüb }

cesium fluoride [INORG CHEM] CsF Toxic, irritating, deliquescent crystals with a melting point of 682°C; soluble in water and methanol; used in medicine, mineral water, and brewing. { 'sē·zē·əm 'flúr,īd }

cesium fountain [ATOM PHYS] A device for performing highly accurate frequency measurements, in which cesium atoms are cooled and trapped by pairs of counterpropagating tuned laser beams and are thrown upward by shifting the frequency of the vertical lasers; they are further cooled and placed in one of the hyperfine states, and then pass through a microwave field region before falling into a detection region. { |sēz·ē·əm 'faúnt·ən }

cesium hollow cathode [ELECTR] A cathode in which cesium is heated at the bottom of a cylinder serving as the cathode of an electron tube, to give current densities that can be as high as 800 amperes per square centimeter. { 'sē·zē·əm |häl·ō 'ka,thōd }

cesium hydroxide [INORG CHEM] CsOH Colorless or yellow, fused crystalline mass with a melting point of 272.3°C; soluble in water; used as electrolyte in alkaline storage batteries at subzero temperatures. { 'sē·zē·əm ,hī'dräk,sīd }

cesium iodide [INORG CHEM] CsI A colorless, deliquescent, crystalline powder with a melting point of 621°C; soluble in water and alcohol; crystals used for infrared spectroscopy. { 'sē·zē·əm 'ī·ə,dīd }

cesium-ion engine [AERO ENG] An ion engine that uses a stream of cesium ions to produce a thrust for space travel. { 'sē·zē·əm |ī·ən 'en·jən }

cesium magnetometer [ENG] A magnetometer that uses a cesium atomic-beam resonator as a frequency standard in a circuit that detects very small variations in magnetic fields. { 'sē·zē·əm ,mag·nə'täm·əd·ər }

cesium perchlorate [INORG CHEM] $CsClO_4$ A crystalline solid with a melting point of 250°C; soluble in water; used in optics and for specialty glasses. { 'sē·zē·əm pər'klór,āt }

cesium phototube [ELECTR] A phototube having a cesium-coated cathode; maximum sensitivity in the infrared portion of the spectrum. { 'sē·zē·əm 'fōd·ō,tüb }

cesium sulfate [INORG CHEM] Cs_2SO_4 Colorless crystals with a melting point of 1010°C; soluble in water; used for brewing and in mineral waters. { 'sē·zē·əm 'səl,fāt }

cesium thermionic converter [ELECTR] A thermionic diode in which cesium vapor is stored between the plates to neutralize space charge and to lower the work function of the emitter. { 'sē·zē·əm thər·mē'än·ik kən'vərd·ər }

cesium-vapor lamp [ELECTR] A lamp in which light is produced by the passage of current between two electrodes in ionized cesium vapor. { 'sē·zē·əm |vā·pər ,lamp }

cesium-vapor Penning source [ELECTR] A conventional Penning source modified for negative-ion generation through the introduction or a third, sputter cathode, made from or containing the element of interest, which is the source of negative ions, and through the introduction of cesium vapor into the arc chamber. { 'sē·zē·əm |vā·pər 'pen·iŋ ,sórs }

cesium-vapor rectifier [ELECTR] A gas tube in which cesium vapor serves as the conducting gas and a condensed monatomic layer of cesium serves as the cathode coating. { 'sē·zē·əm |vā·pər |rek·tə,fī·ər }

cespitose [BOT] **1.** Tufted; growing in tufts, as grass. **2.** Having short stems forming a dense turf. { 'ses·pə,tōs }

cesspit See cesspool. { 'ses,pit }

cesspool [CIV ENG] An underground tank for raw sewage collection; used where there is no sewage system. Also known as cesspit. { 'ses,pül }

Cestida [INV ZOO] An order of ribbon-shaped ctenophores having a very short tentacular axis and an elongated pharyngeal axis. { 'ses·tə·də }

Cestoda [INV ZOO] A subclass of tapeworms including most

CERVICAL VERTEBRA

spinous process — cranial articular process

— caudal articular process

— transverse foramen

body — transverse process

Cervical vertebra of a dog. (*From M. E. Miller, G. C. Christensen, and H. E. Evans, Anatomy of the Dog, Saunders, 1964*)

CESTIDA

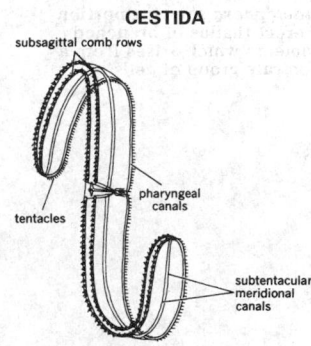

subsagittal comb rows

pharyngeal canals

tentacles

subtentacular meridional canals

Velamen species.

members of the class Cestoidea; all are endoparasites of vertebrates. { se'stō·də }

Cestodaria [INV ZOO] A small subclass of worms belonging to the class Cestoidea; all are endoparasites of primitive fishes. { ˌses·tə'dar·ē·ə }

Cestoidea [INV ZOO] The tapeworms, endoparasites composing a class of the phylum Platyhelminthes. { se'stói·dē·ə }

Cet See Cetus.

Cetacea [VERT ZOO] An order of aquatic mammals, including the whales, dolphins, and porpoises. { sē'tā·shə }

cetane See n-hexadecane. { 'sē,tān }

cetane index [CHEM ENG] An empirical method for finding the cetane number of a fuel based on API gravity and the mid boiling point. { 'sē,tān ˌin,deks }

cetane number [CHEM ENG] The percentage by volume of cetane (cetane number 100) in a blend with α-methylnaphthalene (cetane number 0); indicates the ability of a fuel to ignite quickly after being injected into the cylinder of an engine. { 'sē,tān ˌnəm·bər }

cetane-number improver [CHEM] A chemical which has the effect of increasing a diesel fuel's cetane number; examples are nitrates, nitroalkanes, nitrocarbonates, and peroxides. { 'sē,tān ˌnəm·bər im'prüv·ər }

cetin [ORG CHEM] $C_{15}H_{31}COOC_{16}H_{33}$ A white, crystalline, waxy substance with a melting point of 50°C; soluble in alcohol and ether; used as a base for ointments and emulsions and in the manufacture of soaps and candles. { 'sēt·ən }

cetology [VERT ZOO] The study of whales. { sē'täl·ə·jē }

Cetomimiformes [VERT ZOO] An order of rare oceanic, deepwater, soft-rayed fishes that are structurally diverse. { ˌsēd·ə,mim·ə'fór,mēz }

Cetomimoidei [VERT ZOO] The whalefishes, a suborder of the Cetomimiformes, including bioluminescent, deep-sea species. { ˌsēd·ə·mə'mói·dē,ī }

Cetorhinidae [VERT ZOO] The basking sharks, a family of large, galeoid elasmobranchs of the isurid line. { ˌsēd·ə'rīn·ə,dē }

cetrimonium bromide [ORG CHEM] $CH_3(CH_2)_{15}N(CH_3)_3Br$ Crystals with a melting point of 237–243°C; soluble in alcohol, water, and sparingly in acetone; used as a cationic detergent, antiseptic, and precipitant for nucleic acids and mucopolysaccharides. { ˌse·trə'mōn·ē·əm 'brō,mīd }

Cetus [ASTRON] A constellation with right ascension 2 hours, declination 10°S. Abbreviated Cet. Also known as Whale. { 'sēd·əs }

cetyl [ORG CHEM] The radical represented as $C_{16}H_{33}-$. { 'sēd·əl }

cetyl alcohol [ORG CHEM] $C_{15}H_{33}OH$ A colorless wax, insoluble in water although a solution in kerosine forms an insoluble film on water. { 'sēd·əl 'al·kə,hól }

cetyl vinyl ether [ORG CHEM] $C_{16}H_{33}OCO:CH_2$ A colorless liquid with a boiling point of 142°C; may be copolymerized with unsaturated monomers to make internally plasticized resins. { 'sēd·əl 'vīn·əl 'ē·thər }

Ceva's theorem [MATH] The theorem that if three concurrent straight lines pass through the vertices A, B, and C of a triangle and intersect the opposite sides, produced if necessary, at D, E, and F, then the product $AF·BD·CE$ of the lengths of three alternate segments equals the product $FB·DC·EA$ of the other three. { 'chā·vəz ˌthir·əm }

cevedilla See sabadilla. { ˌsev·ə'dil·ə }

cevian [MATH] A straight line that passes through a vertex of a triangle or tetrahedron and intersects the opposite side or face. { 'chāv·ē·ən }

ceylonite [MINERAL] A dark-green, brown, or black iron-bearing variety of spinel. Also known as candite; pleonaste; zeylanite. { sə'lä,nīt }

Cf See californium.

C factor [MAP] An empirical value in aerial photography which expresses the vertical measuring capability of a given stereoscopic system; the factor is the ratio of the flight height to the smallest contour interval accurately plottable; in planning for aerial photography the C factor is used to determine the flight height required for a specified contour interval, camera, and instrument system. Also known as altitude contour. { 'sē ˌfak·tər }

CFC See chlorofluorocarbon.

CFI See computational flow imaging.

CFIA See component-failure-impact analysis.

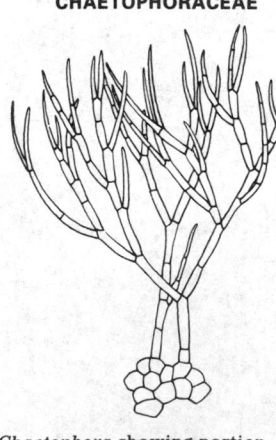

CHAETOPHORACEAE

Chaetophora showing portion of erect thallus of branched filaments which arises from a prostrate group of cells.

C figure See C index. { 'sē ˌfig·yər }

cfs See cusec.

cg See centigram.

CGI See common gateway interface.

CGI script [COMPUT SCI] A program, written in a language such as Perl, that is used for creating interactive Web pages; for example, it allows a Web server to process a request from a user, communicate with a database, and reply to the user by creating a Web page. { 'sē'jē'ī ,skript }

CGM See computer graphics metafile.

cgs system See centimeter-gram-second system. { 'sē'gē'es 'sis·təm }

Cha See Chamaeleon.

chabazite [MINERAL] $CaAl_2Si_4O_{12}·6H_2O$ A white to yellow or red member of the zeolite group occurring in glassy rhombohedral crystals; hardness is 4–5 on Mohs scale, and specific gravity is 2.08–2.16. { 'kab·ə,zīt }

chad [COMPUT SCI] The piece of material removed when forming a hole or notch in a punched tape or punched card. Also known as chip. [NUCLEO] **1.** A unit of neutron flux equal to 1 neutron per square centimeter per second. **2.** A unit of neutron flux equal to 10^{12} neutrons per square centimeter per second. { chad }

chadacryst See xenocryst. { 'kad·ə,krist }

chadless tape [COMPUT SCI] Paper tape in which the perforations for code characters are made by an incomplete circular cut, with the resulting flap of material folded aside. { 'chad·ləs 'tāp }

chad tape [COMPUT SCI] Paper tape in which perforations for code characters are completely punched out. { 'chad ,tāp }

Chadwick-Goldhaber effect See photodisintegration. { 'chad·wik ,gōlt'häb·ər i'fekt }

chaeta See seta. { 'kēd·ə }

Chaetetidae [PALEON] A family of Paleozoic corals of the order Tabulata. { ˌkē'tē·də,dē }

Chaetodontidae [VERT ZOO] The butterflyfishes, a family of perciform fishes in the suborder Percoidei. { ˌkēd·ō'dän·tə,dē }

Chaetognatha [INV ZOO] A phylum of abundant planktonic arrowworms. { ˌkē'täg·nə·thə }

Chaetonotoidea [INV ZOO] An order of the class Gastrotricha characterized by two adhesive tubes connected with the distinctive paired, posterior tail forks. { ˌkē,tän·ō'tóid·ē·ə }

Chaetophoraceae [BOT] A family of algae in the order Ulotrichales characterized as branched filaments which taper toward the apices, sometimes bearing terminal setae. { ˌkēd·ō·fə'rās·ē,ē }

Chaetopteridea [INV ZOO] A family of spioniform polychaete annelids belonging to the Sedentaria. { ˌkē,täp·tə'rīd·ē·ə }

chaff [AGR] Seed coverings and small stem pieces which are separated from grass and grain seeds in threshing or processing. [ORD] Thin, flat pieces of metal foil, plain or backed, designed to act as a countermeasure against enemy radar when released into the atmosphere. { chaf }

chafing corrosion See fretting corrosion. { 'chāf·iŋ kə'rō·zhən }

chafing fatigue [MET] Fatigue induced by corrosion damage between metal surfaces in close contact under pressure. { 'chāf·iŋ fə,tēg }

Chagas' disease [MED] An acute and chronic protozoan disease of humans caused by the hemoflagellate *Trypanosoma* (*Schizotrypanum*) *cruzi*. Also known as South American trypanosomiasis. { 'shäg·əs di,zēz }

chain [CHEM] A structure in which similar atoms are linked by bonds. [CIV ENG] See engineer's chain; Gunter's chain. [COMMUN] A network of radio, television, radar, navigation, or other similar stations connected by special telephone lines, coaxial cables, or radio relay links so all can operate as a group for broadcast purposes, communication purposes, or determination of position. [COMPUT SCI] **1.** A series of data or other items linked together in some way. **2.** A sequence of binary digits used to construct a code. [DES ENG] **1.** A flexible series of metal links or rings fitted into one another; used for supporting, restraining, dragging, or lifting objects or transmitting power. **2.** A mesh of rods or plates connected together, used to convey objects or transmit power. [GEOL] A series of interconnected or related natural features, such as lakes, islands, or seamounts, arranged in a longitudinal sequence. [MATH] *See* linearly ordered set. { 'chān }

chain balance [ANALY CHEM] An analytical balance with one end of a fine gold chain suspended from the beam and the other fastened to a device which moves over a graduated vernier scale. { 'chān ,bal·əns }

chain belt [DES ENG] Belt of flat links to transmit power. { 'chān ,belt }

chain block [MECH ENG] A tackle which uses an endless chain rather than a rope, often operated from an overhead track to lift heavy weights especially in workshops. Also known as chain fall; chain hoist. { 'chān ,bläk }

chain bond [CIV ENG] A masonry bond formed with a chain or bar. { 'chān ,bänd }

chain cable [NAV ARCH] The chain to which an anchor is attached. Also known as cable chain. { 'chān ,kā·bəl }

chain coal cutter [MIN ENG] A cutter which makes a groove in the coal by an endless chain moving around a flat plate called a jib. { 'chān 'kōl ,kəd·ər }

chain code [COMPUT SCI] A binary code consisting of a cyclic sequence of some or all of the possible binary words at a given length such that each word is derived from the previous one by moving the binary digits one position to the left, dropping the leading bit, and inserting a new bit at the end, in such a way that no word recurs before the cycle is complete. { 'chān ,kōd }

chain command [COMPUT SCI] Any input/output command in a sequence of input/output commands such as WRITE, READ, SENSE. { 'chān kə'mand }

chain complex [MATH] A sequence $\{C_n\}$, $-\infty < n < \infty$, of Abelian groups together with a sequence of boundary homomorphisms d_n: $C_n \rightarrow C_{n-1}$ such that $d_{n-1} \circ d_n = 0$ for each n. { 'chān ,käm,pleks }

chain conveyor [MECH ENG] A machine for moving materials that carries the product on one or two endless linked chains with crossbars; allows smaller parts to be added as the work passes. { 'chān kən'vā·ər }

chain course [CIV ENG] A course of stone held together by iron cramps. { 'chān ,kórs }

chain data flag [COMPUT SCI] A value of 1 given to a specific bit of a channel command word, commonly used with scatter read or scatter write operations. { 'chān 'dad·ə ,flag }

chain decay *See* series disintegration. { 'chān di'kā }

chain disintegration *See* series disintegration. { 'chān dis,int·ə·'grā·shən }

chain drive [MECH ENG] A flexible device for power transmission, hoisting, or conveying, consisting of an endless chain whose links mesh with toothed wheels fastened to the driving and driven shafts. { 'chān ,drīv }

chained block encryption [COMMUN] The use of a block cipher in which the bits of a given output block depend not only on the bits in the corresponding input block and in the key, but also on any or all prior data bits, either inputted to or produced during the enciphering or deciphering process. Also known as block chaining. { |chānd 'bläk in'krip·shən }

chained list [COMPUT SCI] A collection of data items arranged in a sequence so that each item contains an address giving the location of the next item in a computer storage device. Also known as linked list. { |chānd 'list }

chained records [COMPUT SCI] A file of records arranged according to the chaining method. { |chānd 'rek·ərdz }

chainette [INV ZOO] In some cestodes, a longitudinal row of similar spines, usually with lateral winglike expansions, found near the base of the tentacles. [MATH] *See* catenary. { chā'net }

chain fall *See* chain block. { 'chān ,fól }

chain fission yield [NUC PHYS] The sum of the independent fission yields for all isobars of a particular mass number. { |chān 'fish·ən ,yēld }

chain-float liquid-level gage [ENG] Float device to measure the level of liquid in a vessel; the float, suspended from a counterweighted chain draped over a toothed sprocket, rises or falls with the liquid level, and the chain movement turns the sprocket to position a calibrated depth-indicator. { 'chān |flōt 'lik·wəd |lev·əl ,gāj }

chain gear [MECH ENG] A gear that transmits motion from one wheel to another by means of a chain. { 'chān ,gir }

chain grate stoker [MECH ENG] A wide, endless chain used to feed, carry, and burn a noncoking coal in a furnace, control

the air for combustion, and discharge the ash. { 'chān ,grāt ,stōk·ər }

chain hoist *See* chain block. { 'chān ,hóist }

chain homomorphism [MATH] A sequence of homomorphisms f_n: $C_n \rightarrow D_n$ between the groups of two chain complexes such that $f_{n-1} d_n = \bar{d}_n f_n$ where d_n and $\bar{d}_n$ are the boundary homomorphisms of $\{C_n\}$ and $\{D_n\}$ respectively. { 'chān ,hō·mō'mór,fiz·əm }

chain index [STAT] An index number derived by relating the value at any given period to the value in the previous period rather than to a fixed base. { 'chān ,in,deks }

chaining [CIV ENG] In land surveying, measuring distance by means of a chain or tape. [COMPUT SCI] A method of storing records which are not necessarily contiguous, in which the records are arranged in a sequence and each record contains means to identify its successor. { 'chān·iŋ }

chaining search [COMPUT SCI] A method of searching for a data item in a chained list in which an initial key is used to obtain the location of either the item sought or another item in the list, and the search then progresses through the chain until the required item is obtained or the chain is completed. { 'chān·iŋ ,sərch }

chain intermittent fillet welding [MET] **1.** The forming of two lines of intermittent fillet welds in a T joint or lap joint so that the increments of welding in one line are approximately opposite those in the other line. **2.** The forming of two lines of equal-length fillet welds concurrently on opposite sides of the perpendicular member of a T joint at intermittent intervals. { 'chān in·tər'mit·ənt 'fil·ət ,wel·diŋ }

chain isomerism [ORG CHEM] A type of molecular isomerism seen in carbon compounds; as the number of carbon atoms in the molecule increases, the linkage between the atoms may be a straight chain or branched chains producing isomers that differ from each other by possessing different carbon skeletons. { 'chān ,ī'säm·ə,riz·əm }

chain lightning [GEOPHYS] A rare form of lightning in a long zigzag or apparently broken line. { 'chān ,līt·niŋ }

chain locker [NAV ARCH] A compartment in the lower part of a ship toward the bow for stowing an anchor chain. { 'chān ,läk·ər }

chain loom [TEXT] A knitting machine used to make flat-knit fabric. { 'chān ,lüm }

chain of simplices [MATH] A member of the free Abelian group generated by the simplices of a given dimension of a simplicial complex. { 'chān əv 'sim·plə,sēz }

chain pipe [NAV ARCH] A heavy steel pipe through which the anchor chain passes from the windlass to the chain locker. { 'chān ,pīp }

chain plate [NAV ARCH] One of the metal plates bolted to the side of a ship, to which lines supporting the masts are fastened. { 'chān ,plāt }

chain pointer [COMPUT SCI] The part of a data item in a chained list that gives the address of the next data item. { 'chān 'póint·ər }

chain printer [COMPUT SCI] A high-speed printer in which the type slugs are carried by the links of a revolving chain. { 'chān ,print·ər }

chain printing [COMPUT SCI] The printing of a group of linked files by placing commands at the end of each file that direct the program to continue printing the next one. { 'chān 'print·iŋ }

chain pump [MECH ENG] A pump containing an endless chain that is fitted at intervals with disks and moves through a pipe and raises sludge. { 'chān ,pəmp }

chain radar beacon [COMMUN] A beacon with a fast recovery time to permit simultaneous interrogation and tracking of the beacon by a number of radars. { 'chān 'rā,där ,bē·kən }

chain radar system [ENG] A number of radar stations located at various sites on a missile range to enable complete radar coverage during a missile flight; the stations are linked by data and communication lines for target acquisition, target positioning, or data-recording purposes. { 'chān 'rā,där ,sis·təm }

chain reaction [CHEM] A chemical reaction in which many molecules undergo chemical reaction after one molecule becomes activated. [NUCLEO] *See* nuclear chain reaction. { 'chān rē'ak·shən }

chain riveting [ENG] Riveting consisting of rivets one behind the other in rows along the seam. { 'chān ,riv·əd·iŋ }

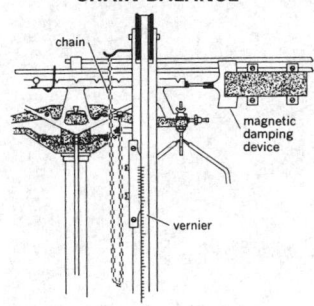

CHAIN BALANCE

Chain balance with notched beam, magnetic damper, and vernier scale.

chain rule [MATH] A rule for differentiating a composition of functions: $(d/dx)f(g(x)) = f'(g(x)) \cdot g'(x)$. { 'chān ,rül }

chain saw [MECH ENG] A gasoline-powered saw for felling and bucking timber, operated by one person; has cutting teeth inserted in a sprocket chain that moves rapidly around the edge of an oval-shaped blade. { 'chān ,só }

chain scission [ORG CHEM] The cleavage of polymer chains, as in natural rubber as a result of heating. { 'chān ,sizh·ən }

chain stopper [NAV ARCH] A short chain with a hook attached to a ship's forecastle, used to secure the anchor chain when the anchor is raised or lowered, thus relieving the windlass from strain, or to quickly release the anchor. { 'chān ,stäp·ər }

chain structure [SOLID STATE] A crystalline structure in which forces between atoms in one direction are greater than those in other directions, so that the atoms are concentrated in chains. { 'chān ,strək·chər }

chain tongs [DES ENG] A tool for turning pipe, using a chain to encircle and grasp the pipe. { 'chān ,täŋz }

chain transfer [ORG CHEM] The abstraction of an atom from another molecule (initiator, monomer, polymer, or solvent) by the radical end of a growing (addition) polymer, which simultaneously terminates the polymer chain and creates a new radical capable of chain polymerization; also occurs in cationic polymerization. { 'chān ,tranz·fər }

chain vise [DES ENG] A vise in which the work is encircled and held tightly by a chain. { 'chān ,vīs }

chainwall [MIN ENG] A coal mining technique in which the mine roof is supported by coal pillars, between which the coal is mined away. { 'chān ,wól }

chain weave [TEXT] Compact, heavyweight cloth weave made with plied yarn in both directions; its pattern is two-up and two-down broken twill, with two ends of right hand and two of left hand repeated on four threads each way. { 'chān ,wēv }

chair conformation See chair form. { 'cher ,kän·fər,mā·shən }

chair form [PHYS CHEM] A particular nonplanar conformation of a cyclic molecule with more than five atoms in the ring; for example, in the chair form of cyclohexane, the hydrogens are staggered and directed perpendicularly to the mean plane of the carbons (axial conformation, *a*) or equatorially to the center of the mean plane (equatorial conformation, *e*). Also known as chair conformation. { 'cher ,fòrm }

chairs See folding boards. { cherz }

chalaza [BOT] The region at the base of the nucellus of an ovule; gives rise to the integuments. [CYTOL] One of the paired, spiral, albuminous bands in a bird's egg that attach the yolk to the shell lining membrane at the ends of the egg. { kə'lāz·ə }

chalazion [MED] A small tumor of the eyelid formed by retention of tarsal gland secretions. Also known as a Meibomian cyst. { kə'lāz·ē·ən }

chalazogamy [BOT] A process of fertilization in which the pollen tube passes through the chalaza to reach the embryo sac. { ,kal·ə'zäg·ə·mē }

chalazoidite See mud ball. { 'kal·ə,zói,dīt }

chalcanthite [MINERAL] $CuSO_4 \cdot 5H_2O$ A blue to bluish-green mineral which occurs in triclinic crystals or in massive fibrous veins or stalactites. Also known as bluestone; blue vitriol. { kal'kan,thīt }

chalcedony [MINERAL] A cryptocrystalline variety of quartz; occurs as crusts with a rounded, mammillary, or botryoidal surface and as a major constituent of nodular and bedded cherts; varieties include carnelian and bloodstone. { kal'sed·ən·ē }

chalcedonyx [MINERAL] A mineral consisting of onyx with alternating gray and white bands; valued as a semiprecious stone. { ,kal·sə'dän·iks }

Chalcididae [INV ZOO] The chalcids, a family of hymenopteran insects in the superfamily Chalcidoidea. { kal'sid·ə,dē }

Chalcidoidea [INV ZOO] A superfamily of hymenopteran insects in the suborder Apocrita, including primarily insect parasites. { ,kal·sə'dóid·ē·ə }

chalcoalumite [MINERAL] $CuAl_4(SO_4)(OH)_{12} \cdot 3H_2O$ A turquoise-green to pale-blue mineral consisting of a hydrous basic sulfate of copper and aluminum. { ¦kal·kō'al·ə,mīt }

chalcocite [MINERAL] Cu_2S A fine-grained, massive mineral with a metallic luster which tarnishes to dull black on exposure; crystallizes in the orthorhombic system, the crystals

being rare and small usually with hexagonal outline as a result of twinning; hardness is 2.5–3 on Mohs scale, and specific gravity is 5.5–5.8. Also known as beta chalcocite; chalcosine; copper glance; redruthite; vitreous copper. { 'kal·kə,sīt }

chalcocyanite [MINERAL] $CuSO_4$ A white mineral consisting of copper sulfate. Also known as hydrocyanite. { ,kal·kə'sī·ə,nīt }

chalcogen [INORG CHEM] Any of the elements that form group 16 of the periodic table; included are oxygen, sulfur, selenium, tellurium, and polonium. { 'kal·kə·jən }

chalcogenide [INORG CHEM] A binary compound containing a chalcogen and a more electropositive element or radical. { 'kal·kə·jə,nīd }

chalcogenide glass [MATER] A type of glass containing large amounts of one of the chalcogens tellurium, selenium, or sulfur; used in glass switches. { 'kal·kə·jə,nīd 'glas }

chalcography [GRAPHICS] The art of copper engraving. { kal'käg·rə·fē }

chalcolite See torbernite. { 'kal·kə,līt }

chalcomenite [MINERAL] $CuSeO_3 \cdot 2H_2O$ A blue mineral consisting of copper selenite occurring in crystals. { ,kal·kə'mē,nīt }

chalcophanite [MINERAL] $(Zn,Mn,Fe)Mn_2O_5 \cdot nH_2O$ A black mineral with metallic luster consisting of hydrous manganese and zinc oxide. { kal'käf·ə,nīt }

chalcophile [GEOL] Having an affinity for sulfur and therefore massing in greatest concentration in the sulfide phase of a molten mass. { 'kal·kə,fīl }

chalcophyllite [MINERAL] $Cu_{18}Al_2(AsO_4)_3(OH)_{27} \cdot 33H_2O$ A green mineral consisting of basic arsenate and sulfate of copper and aluminum occurring in tabular crystals or foliated masses. Also known as copper mica. { ,kal·kō'fi,līt }

chalcopyrite [MINERAL] $CuFeS_2$ A major ore mineral of copper; crystallizes in the tetragonal crystal system, but crystals are generally small with diphenoidal faces resembling the tetrahedron; usually massive with a metallic luster and brass-yellow color; hardness is 3.5–4 on Mohs scale, and specific gravity is 4.1–4.3. Also known as copper pyrite; yellow pyrite. { ,kal·kō'pī,rīt }

chalcopyrrohite [MINERAL] $CuFe_4S_5$ A sulfide mineral occurring in meteorites. { ,kal·kō'pī·rə,nīt }

chalcosiderite [MINERAL] $Cu(Fe,Al)_6(PO_4)_4(OH)_8 \cdot 4H_2O$ A green mineral, isomorphous with turquoise, consisting of a hydrous basic phosphate of copper, iron, and aluminum. { ¦kal·kō'sīd·ə,rīt }

chalcosine See chalcocite. { 'kal·kə,sēn }

chalcostibite [MINERAL] $CuSbS_2$ A lead-gray mineral consisting of antimony copper sulfide. { ,kal·kō'sti,bīt }

chalcotrichite [MINERAL] A capillary variety of cuprite occurring in long needlelike crystals. Also known as hair copper; plush copper ore. { ,kal·kō'tri,kīt }

chaldron [MECH] **1.** A unit of volume in common use in the United Kingdom, equal to 36 bushels, or 288 gallons, or approximately 1.30927 cubic meters. **2.** A unit of volume, formerly used for measuring solid substances in the United States, equal to 36 bushels, or approximately 1.26861 cubic meters. { 'chòl·drən }

chalice cell See goblet cell. { 'chal·əs ,sel }

chalicosis [MED] A pulmonary affection caused by inhalation of stone dust. { ,kal·ə'kō·səs }

Chalicotheriidae [PALEON] A family of extinct perissodactyl mammals in the superfamily Chalicotherioidea. { ,kal·ə,kō·thə'rī·ə,dē }

Chalicotherioidea [PALEON] A superfamily of extinct, specialized perissodactyls having claws rather than hooves. { ¦kal·ə,kō·thi·rē'óid·ē·ə }

chalk [MATER] Artificially prepared pure calcium carbonate; used as the basis for pastels. Also known as whiting. [PETR] A variety of limestone formed from pelagic organisms; it is very fine-grained, porous, and friable; white or very light-colored, it consists almost entirely of calcite. { chók }

chalkboard See blackboard. { 'chók,bòrd }

chalking [CHEM] **1.** Treating with chalk. **2.** Forming a powder which is easily rubbed off. [GRAPHICS] In printing, rubbing off pigment when the ink dried improperly because it was absorbed too rapidly into the paper. [MET] Defect of coated metals in which a layer of powder forms between the coating and the base metal. { 'chók·iŋ }

CHAIR FORM

a a e a
e e a
e a
e a
a a a
a

The chair form conformation of cyclohexane; *a* represents hydrogens that are staggered and perpendicular to the mean plane of the carbon; *e* represents the hydrogens that are equatorially directed to the center of the mean plane of the carbons.

CHALCIDIDAE

A member of the Chalcididae. (*From T. I. Storer and R. L. Usinger, General Zoology, 3d ed. McGraw-Hill, 1957*)

chalkstone [PATH] A gouty deposit, usually of sodium urate, in the hands or feet. { 'chȯk,stōn }

challenge [COMMUN] To cause an interrogator to transmit a signal which puts a transponder into operation. [IMMUNOL] Administration of an antigen to ascertain state of immunity. { 'chal·ənj }

challenger *See* interrogator. { 'chal·ən·jər }

challenge-response [COMPUT SCI] A method of identifying and authenticating persons seeking access to a computing system; each user is issued a device resembling a pocket calculator and is given a different problem to solve (the challenge), to which the calculator provides part of the answer, each time the person seeks authentication. { 'chal·ənj ri'späns }

challenging signal *See* interrogation. { 'chal·ən·jiŋ ,sig·nəl }

challiho [METEOROL] Strong southerly winds which blow for about 40 days in spring in some parts of India; sometimes the winds are violent, causing blinding dust storms. { 'chäl·ə,hō }

challis [TEXT] A soft fabric with a slight nap, woven in a plain or twill weave; may be cotton, wool, or a synthetic fiber. { 'shal·ē }

chalmersite *See* cubanite. { 'chä·mər,zīt }

chalones [BIOCHEM] Substances thought to be molecules of the protein-polypeptide class that are produced as part of the growth-control systems of tissues; known to inhibit cell division by acting on several phases in the mitotic cycle. { 'ka,lōnz }

chalybite *See* siderite. { 'kal·ə,bīt }

Chamaeleon [ASTRON] A constellation, right ascension 11 hours, declination 80°S. Abbreviated Cha. Also spelled Chameleon. { kə'mēl·yən }

Chamaeleontidae [VERT ZOO] The chameleons, a family of reptiles in the suborder Sauria. { kə,mēl·ē'än·tə,dē }

Chamaemyidae [INV ZOO] The aphid flies, a family of myodarian cyclorrhaphous dipteran insects of the subsection Acalypteratae. { ,kam·ə'mī·ə,dē }

chamaephyte [ECOL] Any perennial plant whose winter buds are within 10 inches (25 centimeters) of the soil surface. { 'kam·ə,fīt }

chamaerrhine [ANTHRO] Having a short broad nose with a nasal index of 51–57.9 for a skull, and 85–99.9 for a person. { 'kam·ə,rīn }

Chamaesiphonales [BOT] An order of blue-green algae of the class Cyanophyceae; reproduce by cell division, colony fragmentation, and endospores. { ,kam·ē,sī·fə'nā·lēz }

chamber [CIV ENG] The space in a canal lock between the upper and lower gates. [GRAPHICS] A sleeve or channel of a transparent film jacket. [MIN ENG] **1.** The working place of a miner. **2.** A body of ore with definite boundaries apparently filling a preexisting cavern. [ORD] The part of the gun in which the charge is placed: in a revolver, the hole in the cylinder; in a cannon, the space between the obturator or breechblock and the forcing cone. { 'chām·bər }

chamber acid [INORG CHEM] Sulfuric acid made by the obsolete chamber process. { 'chām·bər 'as·əd }

chamber capacity *See* chamber volume. { 'chām·bər kə'pas·əd·ē }

chambered pith [BOT] A form of pith in which the parenchyma collapses or is torn during development, leaving the sclerenchyma plates to alternate with hollow zones. { ,chām·bərd 'pith }

chambering [MIN ENG] Increasing the size of a drill hole in a quarry by firing a succession of small charges, until the hole can take a proper explosive charge to bring down the face of the quarry. [ORD] That phase of small-arms operation that deals with the placing of the round into the chamber after the round has been fed into the weapon by the feeding device. { 'chām·bə·riŋ }

chamber kiln [ENG] A kiln consisting of a series of adjacent chambers in a ring or oval through which the fire moves, taking several days to make a circuit; waste gas from the fire preheats ware in chambers toward which the fire is moving, while combustion air is preheated by ware in chambers already fired. { 'chām·bər ,kil }

chamber pressure [AERO ENG] The pressure of gases within the combustion chamber of a rocket engine. { 'chām·bər ,presh·ər }

chamber process [CHEM ENG] An obsolete method of manufacturing sulfuric acid in which sulfur dioxide, air, and steam are reacted in a lead chamber with oxides of nitrogen as the catalyst. { 'chām·bər ,präs·əs }

Chambersielloidea [INV ZOO] A superfamily of nematodes in the order Rhabdita characterized by highly unusual elaborations of the oral opening in the form of six mandibles. { chām,bər·sē·ə'lȯid·ē·ə }

chamber test [ENG] A fire test developed specifically for floor coverings that measures the speed and distance of the spread of flames under specified conditions. { 'chām·bər ,test }

chamber volume [AERO ENG] The volume of the rocket combustion chamber, including the convergent portion of the nozzle up to the throat. Also known as chamber capacity. { 'chām·bər ,väl·yəm }

chambray [TEXT] A cotton fabric similar to gingham; it is plain woven, lightweight, and without a pattern, and has a dyed or unbleached warp and white filling. { 'sham,brā }

chamecephalic [ANTHRO] Having a flattened backward-slanting head with a length-height index of 70 or less. { ,kam·ə·sa'fal·ik }

chamecranial [ANTHRO] Having a flat low skull with a length-height index of less than 70. { ,kam·ə'krān·ē·əl }

chameleon [VERT ZOO] The common name for about 80 species of small to medium-size lizards composing the family Chamaeleontidae. { kə'mēl·yən }

Chameleon *See* Chamaeleon. { kə'mel·yən }

chameprosopic [ANTHRO] Having a broad low face with a facial index of 90 or less. { ,kam·ə·prə'sō·pik }

chamfer [ENG] To bevel a sharp edge on a machined part. { 'cham·fər }

chamfer angle [DES ENG] The angle that a beveled surface makes with one of the original surfaces. { 'cham·fər ,aŋ·gəl }

chamfering [MECH ENG] Machining operations to produce a beveled edge. Also known as beveling. { 'cham·fə·riŋ }

chamfer plane [DES ENG] A plane for chamfering edges of woodwork. { 'cham·fər ,plān }

chamois [VERT ZOO] *Rupicapra rupicapra.* A goatlike mammal included in the tribe Rupicaprini of the family Bovidae. { 'sham·ē }

chamosite [MINERAL] A greenish-gray or black mineral consisting of silicate belonging to the chlorite group and having monoclinic crystals; found in many oolitic iron ores. { 'sham·ə,zīt }

Champlainian [GEOL] Middle Ordovician geologic time. { ,sham'plān·ē·ən }

champsosaur [PALEON] A large crocodile-like reptile that lived in freshwater ponds and swamps 55–65 million years ago. { 'champ·sə,sȯr }

chance cause [ANALY CHEM] A cause for variability in a measurement process that occurs randomly and unpredictably and for unknown reasons. { ,chans 'kȯz }

chance-constrained programming [COMPUT SCI] Type of nonlinear programming wherein the deterministic constraints are replaced by their probabilistic counterparts. { ,chans kən'strānd 'prō,gram·iŋ }

chance process [MIN ENG] A method for separating clean coal from slate and other impurities in a mixture of sand and water. { ,chans ,präs·əs }

chance variable *See* random variable. { ,chans 'ver·ē·ə·bəl }

chancre [MED] **1.** A lesion or ulcer at the site of primary inoculation by an infecting organism. **2.** The initial lesion of syphilis. { 'shaŋ·kər }

chancroid [MED] A lesion of the genitalia, usually of venereal origin, caused by *Hemophilus ducreyi.* Also known as soft chancre. { 'chaŋ,krȯid }

Chandler motion *See* polar wandering. { 'chand·lər ,mō·shən }

Chandler period [GEOPHYS] The period of the Chandler wobble. { 'chand·lər ,pir·ē·əd }

Chandler wobble [GEOPHYS] A movement in the earth's axis of rotation, the period of motion being about 14 months. Also known as Eulerian nutation. { 'chand·lər ,wäb·əl }

Chandrasekhar limit [ASTROPHYS] A limiting mass of about 1.44 solar masses above which a white dwarf cannot exist in a stable configuration. { ,chən·drə'shā,kär ,lim·ət }

Chandrasekhar-Schönberg limit [ASTROPHYS] A mass limit for the isothermal, helium core of a main-sequence star above which the star must rapidly increase in radius and evolve away from the main sequence. Also known as Schönberg-Chandrasekhar limit. { ,chən·drə'shā,kär 'shərn,bərg ,lim·ət }

CHAMELEON

Flap-necked chameleon (*Chamaeleo dilepis*); the neck flap is raised.

CHAMOIS

The chamois (*Rupicapra rupicapra*).

CHANIDAE

The milkfish (*Chanos chanos*), length up to 5 feet (1.5 meters).

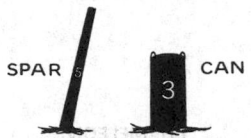

CHANNEL BUOY

SPAR   CAN
     3

Black buoy with odd number indicates channel lies to starboard, proceeding from seaward.

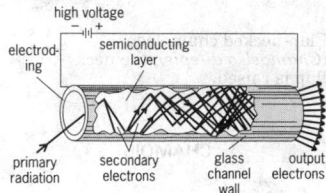

CHANNEL ELECTRON MULTIPLIER

high voltage
− +
electrod- ing
semiconducting layer
primary radiation
secondary electrons
glass channel wall
output electrons

Cutaway view of a straight, single-channel electron multiplier, showing the cascade of secondary electrons resulting from the initial, primary radiation event, which produces an output charge pulse. (*After J. L. Wiza, Microchannel plate detectors, Nucl. Instrum. Meth., 162: 587–601, 1979*)

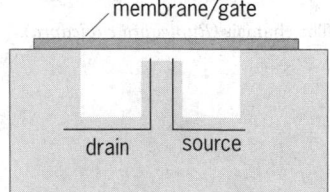

CHANNEL FET MICROPHONE

membrane/gate

drain   source

Cross section of a channel FET microphone.

chandui See chanduy. { 'chän·dwē }

chanduy [METEOROL] A cool, descending wind at Guayaquil, Ecuador, which blows during the dry season (July to November). Also spelled chandui. { 'chän·dwē }

change chart [METEOROL] A chart indicating the amount and direction of change of some meteorological element during a specified time interval; for example, a height-change chart or pressure-change chart. Also known as tendency chart. { 'chānj ‚chärt }

change dump [COMPUT SCI] A type of dump in which only those locations in a computer memory whose contents have changed since some previous event are copied. { 'chānj ‚dəmp }

changed memory routine [COMPUT SCI] A selective memory dump routine in which only those words that have been changed in the course of running a program are printed. { ¦chānjd 'mem·rē rü‚tēn }

change file [COMPUT SCI] A transaction file that is used to update a master file. { 'chānj ‚fīl }

change gear [MECH ENG] A gear used to change the speed of a driven shaft while the speed of the driving remains constant. { 'chānj ‚gir }

change of control [COMPUT SCI] **1.** A break in a series of records at which processing of the records may be interrupted and some predetermined action taken. **2.** See jump. { 'chānj əv kən'trōl }

change of tide [OCEANOGR] A reversal of the direction of motion (rising or falling) of a tide, or in the set of a tidal current. Also known as turn of the tide. { 'chānj əv ‚tīd }

changeover switch [ELEC] A means of moving a circuit from one set of connections to another. { 'chān‚jō·vər ‚swich }

change record [COMPUT SCI] A record that is used to alter information in a corresponding master record. Also known as amendment record; transaction record. { 'chānj ‚rek·ərd }

change tape [COMPUT SCI] A paper tape or magnetic tape carrying information that is to be used to update filed information; the latter is often on a master tape. Also known as transaction tape. { 'chānj ‚tāp }

changing bag [ENG] An enclosure of lightproof material used for operations such as loading of film holders in daylight. { 'chānj·iŋ ‚bag }

Chanidae [VERT ZOO] A monospecific family of teleost fishes in the order Gonorynchiformes which contain the milkfish (*Chanos chanos*), distinguished by the lack of teeth. { 'kan·ə‚dē }

channel [CHEM ENG] In percolation filtration, a portion of the clay bed where there is a preponderance of flow. [CIV ENG] A natural or artificial waterway connecting two bodies of water or containing moving water. [COMMUN] **1.** A band of radio frequencies allocated for a particular purpose; a standard broadcasting channel is 10 kilohertz wide, a television channel 6 megahertz wide. **2.** A path through which electrical transmission of information takes place. [COMPUT SCI] **1.** A path along which digital or other information may flow in a computer. **2.** The section of a storage medium that is accessible to a given reading station in a computer, such as a path parallel to the edge of a magnetic tape or drum or a path in a delay-line memory. **3.** One of the longitudinal rows of intelligence holes punched along the length of paper tape. Also known as level. **4.** A device or portion of a computer that controls and stores data and transfers information between the computer and peripheral equipment. [ELECTR] **1.** A path for a signal, as an audio amplifier may have several input channels. **2.** The main current path between the source and drain electrodes in a field-effect transistor or other semiconductor device. [ENG] The forming of cavities in a gear lubricant at low temperatures because of congealing. [HYD] The deeper portion of a waterway carrying the main current. [NAV] Navigable portion of a body of water. [NUCLEO] A passage for fuel slugs or heat-transfer fluid in a reactor. [PETRO ENG] In a drilling operation, a cavity appearing behind the casing because of a defect in the cement. { 'chan·əl }

channel adapter [COMPUT SCI] Equipment that allows devices operating at different rates of speed to be connected and data to be transferred at the slower data rate. { 'chan·əl ə‚dap·tər }

channel address word [COMPUT SCI] A four-byte code containing the protection key and the main storage address of the first channel command word at the start of an input/output operation. Abbreviated CAW. { 'chan·əl 'ad‚res ‚wərd }

channel-attached device [COMPUT SCI] Equipment that is directly connected to a computer by a channel. { 'chan·əl ə¦tacht di‚vīs }

channel bank [ELECTR] Part of a carrier-multiplex terminal that performs the first step of modulation of the transmitting voice frequencies into a higher-frequency band, and the final step in the demodulation of the received higher-frequency band into the received voice frequencies. { 'chan·əl ‚baŋk }

channel black See gas black. { 'chan·əl ‚blak }

channel buoy [NAV] In marine operations, a buoy marking a channel. { 'chan·əl ‚bói }

channel capacity [COMMUN] The maximum number of bits or other information elements that can be handled in a particular channel per unit time. { 'chan·əl kə'pas·əd·ē }

channel command [COMPUT SCI] The step, equivalent to a program instruction, required to tell an input/output channel what operation is to be performed, and where the data are or should be located. { 'chan·əl kə'mand }

channel command word [COMPUT SCI] A code specifying an operation, one or more flags, a count, and a storage location. Abbreviated CCW. { 'chan·əl kə'mand ‚wərd }

channel configuration [COMPUT SCI] The types, number, and logical relationships of devices connected to a given computer channel. { 'chan·əl kən‚fig·yə‚rā·shən }

channel control [HYD] A condition whereby the stage of a stream is controlled only by discharge and the general configuration of the stream channel, that is, the contours of its bed, banks, and floodplains. { 'chan·əl kən'trōl }

channel control command [COMPUT SCI] An order to a control unit to perform a nondata input/output operation. { 'chan·əl kən'trōl kə'mand }

channel design [COMPUT SCI] The type of channel, characterized by the tasks it can perform, available to a computer. { 'chan·əl di'zīn }

channel director [COMPUT SCI] A unit in some very large computers that controls the functioning of several channels. { 'chan·əl di‚rek·tər }

channel effect [ELECTR] A leakage current flowing over a surface path between the collector and emitter in some types of transistors. { 'chan·əl i'fekt }

channel electron multiplier [ELECTR] A single-particle detector which consists of a hollow glass or ceramic tube with a semiconducting inner surface; it responds to one or more primary particle impact events at its entrance by producing, in a cascade multiplication process, a charge pulse of typically 10^4–10^8 electrons. { ¦chan·əl i¦lek‚trän 'məl·tə‚plī·ər }

channel-end condition [COMPUT SCI] A signal indicating that the use of an input/output channel is no longer required. { 'chan·əl ‚end kən'dish·ən }

channeler See channeling machine. { 'chan·əl·ər }

channel FET microphone [ENG ACOUS] A microphone in which a membrane is used as the gate to a field-effect transistor (FET) located just below it, and motion of the membrane modulates the current between the source and drain of the transistor. { ¦chan·əl ¦fet 'mī·krə‚fōn or ¦ef‚e¦tē }

channel fill [GEOL] Accumulations of sand and detritus in a stream channel where the transporting capacity of the water is insufficient to remove the material as rapidly as it is delivered. { 'chan·əl ‚fil }

channel frequency See stream frequency. { 'chan·əl ‚frē·kwən‚sē }

channel gradient ratio See stream gradient ratio. { 'chan·əl 'grād·ē·ənt ‚rā·shō }

channeling [ANALY CHEM] In chromatography, furrows or breaks in an ion-exchange bed which permit a solution to run through without having contact with active groups elsewhere in the bed. [COMMUN] A type of multiplex transmission in which the separation between communication channels is accomplished through the use of carriers or subcarriers. [NUCLEO] The transmission of extra particles through a medium in a nuclear reactor due to the presence of voids in the medium. [PETRO ENG] A condition occurring in a water-drive reservoir in which oil is bypassed because of erratic or uncontrolled water encroachment; can be aggravated by excessive production rates. [PHYS] The steering of energetic charged particles by the atomic rows or atomic planes of a crystalline solid. { 'chan·əl·iŋ }

channeling machine [MECH ENG] An electrically powered machine that operates by a chipping action of three to five chisels while traveling back and forth on a track; used for primary separation from the rock ledge in marble, limestone, and soft sandstone quarries. Also known as channeler. { 'chan·əl·iŋ mə'shēn }

channeling radiation [PHYS] The radiation emitted by energetic charged particles that pass through a solid. { 'chan·əl·iŋ ,rād·ē'ā·shən }

channel iron [DES ENG] A metal strip or beam with a U-shape. { 'chan·əl ,ī·ərn }

channelization [COMMUN] The division of a single wide-band (high-capacity) communications channel into many relatively narrow-band (lower-capacity) channels. { ,chan·əl·ə'zā·shən }

channelizing [COMMUN] The process of subdividing a wide-band transmission facility so as to handle a number of different circuits requiring comparatively narrow bandwidths. { 'chan·əl,īz·iŋ }

channel-lag deposit [GEOL] Coarse residual material left as accumulations in the channel in the normal processes of the stream. { 'chan·əl ,lag di,päz·ət }

channel light [NAV] In marine operations, a light marking a channel. { 'chan·əl ,līt }

channel marker [NAV] In marine operations, a marker, such as a light or buoy, indicating a channel. { 'chan·əl ,mär·kər }

channel mask [COMPUT SCI] A portion of a program status word indicating which channels may interrupt the task by their completion signals. { 'chan·əl ,mask }

channel miles [COMMUN] The summation, in miles, of the electrical path of individual channels between two points; these points may be connected by wire or radio, or a combination of both. { 'chan·əl ,mīlz }

channel morphology See river morphology. { 'chan·əl ,mòr·fäl·ə·jē }

channel-mouth bar [GEOL] A bar formed where moving water enters a body of still water, due to decreased velocity. { 'chan·əl ,maùth ,bär }

channel net [HYD] Stream channel pattern within a drainage basin. { 'chan·əl ,net }

channel order See stream order. { 'chan·əl ,òrd·ər }

channel pattern [HYD] The configuration of a limited reach of a river channel as seen in plan view from an airplane. { 'chan·əl ,pad·ərn }

channel plate multiplier See microchannel plate. { 'chan·əl ¦plat ¦məl·tə,plī·ər }

channel process [CHEM ENG] A carbon-black process in which iron channel beams are used as depositing surfaces for carbon black. { 'chan·əl ,präs·əs }

channel program [COMPUT SCI] The set of steps, called channel commands, by means of which an input/output channel is controlled. { 'chan·əl ,prō·grəm }

channel protein [CELL MOL] Protein forming an aqueous pore spanning the lipid bilayer of the cell membrane which when open allows certain solutes to traverse the membrane. { 'chan·el ,prō,tēn }

channel read-backward command [COMPUT SCI] A command to transfer data from tape device to main storage while the tape is moving backward. { 'chan·əl 'rēd ¦bak·wərd kə,mand }

channel read command [COMPUT SCI] A command to transfer data from an input/output device to main storage. { 'chan·əl 'rēd kə'mand }

channel reliability [COMMUN] The percent of time a channel was available for use in a specific direction during a specified period of time. { 'chan·əl ri,lī·ə'bil·əd·ē }

channel roughness [GEOL] A measure of the resistivity offered by the material constituting stream channel margins to the flow of water. { 'chan·əl ,rəf·nəs }

channel sample See groove sample. { 'chan·əl ,sam·pəl }

channel sand [GEOL] A sandstone or sand deposited in a stream bed or other channel eroded into the underlying bed. { 'chan·əl ,sand }

channel segment See stream segment. { 'chan·əl ,seg·mənt }

channel selector [ELEC] A control used to tune in the desired channel in a radio or television receiver. { 'chan·əl si'lek·tər }

channel sense command [COMPUT SCI] A command commonly used to denote an unusual condition existing in an input/output device and requesting more information. { 'chan·əl 'sens kə'mand }

channel shifter [ELECTR] Radiotelephone carrier circuit that shifts one or two voice-frequency channels from normal channels to higher voice-frequency channels to reduce cross talk between channels; the channels are shifted back by a similar circuit at the receiving end. { 'chan·əl ,shif·tər }

channel skip [COMPUT SCI] A control character that causes a printer to skip down to a specified line on a page or to the top of the next page. { 'chan·əl ,skip }

channel spacing [COMMUN] The difference in frequency between successive radio or television channels. { 'chan·əl ,spās·iŋ }

channel spin [NUC PHYS] The vector sum of the spins of the particles involved in a nuclear reaction, either before or after the reaction takes place. { 'chan·əl ,spin }

channel splay See floodplain splay. { 'chan·əl ,splā }

channel status table [COMPUT SCI] A table that is set up by an executive program to show the status of the various channels that connect the central processing unit with peripheral units, enabling the program to control input/output operations. { ¦chan·əl ¦stad·əs ,tā·bəl }

channel status word [COMPUT SCI] A storage register containing the status information of the input/output operation which caused an interrupt. Abbreviated CSW. { ¦chan·əl ¦stad·əs ,wərd }

channel steamer [NAV ARCH] A special type of ship which operates across the English Channel, carrying passengers and cars. { 'chan·əl ,stē·mər }

channel synchronizer [ELECTR] An electronic device providing the proper interface between the central processing unit and the peripheral devices. { 'chan·əl 'siŋ·krə,nīz·ər }

channel-to-channel adapter [COMPUT SCI] A device which provides two computer systems with interchannel communications. { ¦chan·əl tə ¦chan·əl ə'dap·tər }

channel width [GEOL] The distance across a stream or channel as measured from bank to bank near bankful stage. [NUC PHYS] The part of the total energy width of a nuclear energy level that corresponds to a particular mode of decay. { 'chan·əl ,width }

channel wing [AERO ENG] A wing that is trough-shaped so as to surround partially a propeller to get increased lift at low speeds from the slipstream. { 'chan·əl ,wiŋ }

channel write command [COMPUT SCI] A command which transfers data from main storage to an input/output device. { ¦chan·əl 'wrīt kə'mand }

Channidae [VERT ZOO] The snakeheads, a family of freshwater perciform fishes in the suborder Anabantoidei. { 'kan·ə,dē }

chantilly [TEXT] Bobbin lace with fine hexagonal mesh ground; the pattern is outlined in heavy thread with a design that usually contains scrolls and florals. { shan'til·ē }

Chaoboridae [INV ZOO] The phantom midges, a family of dipteran insects in the suborder Orthorrhapha. { ,kā·ə'bòr·ə,dē }

chaos See chaotic behavior. { 'kā,äs }

chaotropic [BIOCHEM] Having the ability to destabilize hydrogen bonding and hydrophobic interactions. { ,kā·ə'träp·ik }

chaotic advection [FL MECH] Fluid motion in which the trajectories of particles initially quite close diverge rapidly, even though the flow carrying the particles may be simple. { kā'äd·ik ad'vek·shən }

chaotic behavior [MECH] The behavior of a system whose final state depends so sensitively on the system's precise initial state that the behavior is in effect unpredictable and cannot be distinguished from a random process, even though it is strictly determinate in a mathematical sense. Also known as chaos. { kā'äd·ik bi'hā·vyər }

chaparral [ECOL] A vegetation formation characterized by woody plants of low stature, impenetrable because of tough, rigid, interlacing branches, which have simple, waxy, evergreen, thick leaves. { ¦shap·ə¦ral }

chaplet [MET] Metal support used to space and hold the core in position within a sand mold. { 'chap·lət }

Chaplygin-Kármán-Tsien relation [FL MECH] The relation that in the case of isentropic flow of an ideal gas with negligible viscosity and thermal conductivity, the sum of the pressure and a constant times the reciprocal of the density of the fluid is constant along a streamline; a useful, although physically impossible, approximation. { chə'plē·gən ˌkärˌmän 'tsyen ri'lä·shən }

Chapman-Enskog approximations [STAT MECH] Approximations to a solution of the Boltzmann transport equation in the Chapman-Enskog theory. { ¦chap·mən 'enˌskȯg əˌprák·sə'mā·shənz }

Chapman-Enskog solution [STAT MECH] The solution of the Boltzmann transport equation according to the Chapman-Enskog theory. { ¦chap·mən 'enˌskȯg sə'lü·shən }

Chapman-Enskog theory [STAT MECH] A method of solving the Boltzmann transport equation by successive approximations, essentially in powers of the mean free path. Also known as Enskog theory. { ¦chap·mən 'enˌskȯg ˌthē·ə·rē }

Chapman equation [GEOPHYS] A theoretical relation describing the distribution of electron density with height in the upper atmosphere. [STAT MECH] The relationship that the viscosity of a gas equals $(0.499)mv/[\sqrt{2} \pi\sigma^2(1 + C/T)]$, where m is the mass of a molecule, v its average speed, σ its collision diameter, C the Sutherland constant, and T the absolute temperature (Kelvin scale). { 'chap·mən iˌkwā·zhən }

chapmanite [MINERAL] $Fe_2Sb(SiO_4)_2(OH)$ A mineral consisting of a silicate of iron and antimony. { 'chap·məˌnīt }

Chapman-Jouguet plane [MECH] A hypothetical, infinite plane, behind the initial shock front, in which it is variously assumed that reaction (and energy release) has effectively been completed, that reaction product gases have reached thermodynamic equilibrium, and that reaction gases, streaming backward out of the detonation, have reached such a condition that a forward-moving sound wave located at this precise plane would remain a fixed distance behind the initial shock. { ¦chap·mən zhü¦gwā ˌplān }

Chapman region [GEOPHYS] A hypothetical region in the upper atmosphere in which the distribution of electron density with height can be described by Chapman's theoretical equation. { 'chap·mən ˌrē·jən }

Characeae [BOT] The single family of the order Charales. { kə'rās·ēˌē }

Characidae [VERT ZOO] The characins, the single family of the suborder Characoidei. { kə'ras·əˌdē }

Characoidei [VERT ZOO] A suborder of the order Cypriniformes including fresh-water fishes with toothed jaws and an adipose fin. { ˌkar·ə'kȯid·ēˌī }

character [COMPUT SCI] **1.** An elementary mark used to represent data, usually in the form of a graphic spatial arrangement of connected or adjacent strokes, such as a letter or a digit. **2.** A small collection of adjacent bits used to represent a piece of data, addressed and handled as a unit, often corresponding to a digit or letter. [GEOPHYS] A distinctive aspect of a seismic event, for example, the waveform. [PSYCH] The sum of a person's relatively fixed personality traits and habitual modes of response. { 'kar·ik·tər }

character-addressable computer [COMPUT SCI] A computer that processes data as single characters, and is therefore able to handle words of varying length. { 'kar·ik·tər ə¦dres·ə·bəl kəm'pyüd·ər }

character adjustment [COMPUT SCI] An address modification affecting a specific number of characters of the address part of the instruction. { 'kar·ik·tər ə'jəs·mənt }

character analysis [PSYCH] Psychoanalysis focused on an individual's character defenses. { 'kar·ik·tər əˌnal·ə·səs }

character boundary [COMPUT SCI] In character recognition, a real or imaginary rectangle which serves as the delimiter between consecutive characters or successive lines on a source document. { 'kar·ik·tər ˌbaún·drē }

character cell [COMPUT SCI] A matrix of dots that is used to form a single character on a printer or display screen. { 'kar·ik·tər ˌsel }

character code [COMMUN] A bit pattern assigned to a particular character in a coded character set. { 'kar·ik·tər ˌkȯd }

character convergence [ECOL] A process whereby two relatively evolved species interact so that one converges toward the other with respect to one or more traits. { 'kar·ik·tər kənˌvər·jəns }

character data type [COMPUT SCI] A scalar data type which provides an internal representation of printable characters. { 'kar·ik·tər 'dad·ə ˌtīp }

character defense [PSYCH] Any character trait that serves an unconscious defensive purpose. { 'kar·ik·tər diˌfens }

character density [COMPUT SCI] The number of characters recorded per unit of length or area. Also known as record density. { 'kar·ik·tər ˌden·səd·ē }

character disorder [PSYCH] A pattern of behavior and emotional response that is socially disapproved or unacceptable, with little evidence of anxiety or other symptoms seen in neuroses. { 'kar·ik·tər dis'ȯrd·ər }

character displacement [ECOL] An outcome of competition in which two species living in the same area have evolved differences in morphology or other characteristics that lessen competition for food resources. { 'kar·ik·tər dis'plās·mənt }

character display terminal [COMPUT SCI] A console that can display only alphanumeric characters, and cannot show arbitrary lines or curves. { 'kar·ik·tər di'splā ˌtərm·ə·nəl }

character emitter [COMPUT SCI] In character recognition, an electromechanical device which conveys a specimen character in the form of a time pulse or group of pulses. { 'kar·ik·tər i'mid·ər }

character fill [COMPUT SCI] To fill one or more locations in a computer storage device by repeated insertion of some particular character, usually blanks or zeros. { 'kar·ik·tər ˌfil }

character generator [COMPUT SCI] A hard-wired subroutine which will display alphanumeric characters on a screen. { 'kar·ik·tər ˌjen·əˌrād·ər }

character graphics [COMPUT SCI] A collection of special symbols that can be strung together like letters of the alphabet to generate graphics. { 'kar·ik·tər ˌgraf·iks }

character group [MATH] The set of all continuous homomorphisms of a topological group onto the group of all complex numbers with unit norm. { 'kar·ik·tər ˌgrüp }

characteristic [ELECTR] A graph showing how the voltage or current between two terminals of an electronic device varies with the voltage or current between two other terminals. [MATH] **1.** That part of the logarithm of a number which is the integral (the whole number) to the left of the decimal point in the logarithm. **2.** For a family of surfaces that depend continuously on a parameter, the limiting curve of intersection of two members of the family as the two values of the parameter determining them approach a common value. **3.** For a ring or field, the smallest possible integer whose product with any element of the ring or field equals zero, provided that such an integer exists; otherwise the characteristic is zero. { ˌkar·ik·tə'ris·tik }

characteristic acoustic impedance [ACOUS] The product of the density and the speed of sound in a medium; it is analogous to the characteristic impedance of an infinitely long transmission line. Also known as intrinsic impedance. { ˌkar·ik·tə'ris·tik ə'kü·stik im'pēd·əns }

characteristic chamber length [AERO ENG] The length of a straight, cylindrical tube having the same volume as that of the chamber of a rocket engine if the chamber had no converging section. { ˌkar·ik·tə'ris·tik 'chām·bər ˌleŋkth }

characteristic cone [MATH] A conelike region important in the study of initial value problems in partial differential equations. { ˌkar·ik·tə'ris·tik 'kōn }

characteristic curve [GRAPHICS] In photography, a graph that shows how increases in exposure increase the density of the film. [MATH] **1.** One of a pair of conjugate curves in a surface with the property that the directions of the tangents through any point of the curve are the characteristic directions of the surface. **2.** A curve plotted on graph paper to show the relation between two changing values. **3.** A characteristic curve of a one-parameter family of surfaces is the limit of the curve of intersection of two neighboring surfaces of the family as those surfaces approach coincidence. { ˌkar·ik·tə'ris·tik 'kərv }

characteristic directions [MATH] For a point P on a surface S, the pair of conjugate directions which are symmetric with respect to the directions of the lines of curvature on S through P. { ˌkar·ik·tə'ris·tik də'rek·shənz }

characteristic distortion [COMMUN] **1.** Displacement of signal transitions resulting from the persistence of transients caused by preceding transitions. **2.** In teletypewriter transmission systems, repetitive displacement or disruption peculiar to specific portions of a teletypewriter signal; the two types of characteristic distortions are line and equipment. { ˌkar·ik·tə'ris·tik də'stȯr·shən }

characteristic equation [MATH] **1.** Any equation which has a solution, subject to specified boundary conditions, only when a parameter occurring in it has certain values. **2.** Specifically, the equation $A\mathbf{u} = \lambda\mathbf{u}$, which can have a solution only when the parameter λ has certain values, where A can be a square matrix which multiplies the vector $\mathbf{u}$, or a linear differential or integral operator which operates on the function $\mathbf{u}$, or in general, any linear operator operating on the vector $\mathbf{u}$ in a finite or infinite dimensional vector space.   Also known as eigenvalue equation.   **3.** An equation which sets the characteristic polynomial of a given linear transformation on a finite dimensional vector space, or of its matrix representation, equal to zero. [PHYS] An equation relating a set of variables, such as pressure, volume, and temperature, whose values determine a substance's physical condition. [PL PHYS] An equation whose solutions give the frequencies and modes of those perturbations of a hydromagnetic system which decay or grow exponentially in time, and indicate regions of stability of such a system. { ‚kar·ik·tə'ris·tik i'kwä·zhən }

characteristic exhaust velocity [AERO ENG] Of a rocket engine, a descriptive parameter, related to effective exhaust velocity and thrust coefficient.   Also known as characteristic velocity. { ‚kar·ik·tə'ris·tik ig'zóst və'läs·əd·ē }

characteristic form [MATH] A means of classifying partial differential equations. { ‚kar·ik·tə'ris·tik 'fórm }

characteristic frequency [COMMUN] Frequency which can be easily identified and measured in a given emission. { ‚kar·ik·tə'ris·tik 'frē·kwən·sē }

characteristic function [MATH] **1.** The function χ_A defined for any subset A of a set by setting $\chi_A(x) = 1$ if x is in A and $\chi_A(x) = 0$ if x is not in A.   Also known as indicator function. **2.** See eigenfunction. [PHYS] A function, such as the point characteristic function or the principal function, which is the integral of some property of an optical or mechanical system over time or over the path followed by the system, and whose value for a path actually followed by a system is a maximum or a minimum with respect to nearby paths with the same end points. [STAT] A function that uniquely defines a probability distribution; it is equal to $\sqrt{2\pi}$ times the Fourier transform of the frequency function of the distribution. { ‚kar·ik·tə'ris·tik 'fəŋk·shən }

characteristic impedance [COMMUN] The impedance that, when connected to the output terminals of a transmission line of any length, makes the line appear to be infinitely long, for there are then no standing waves on the line, and the ratio of voltage to current is the same for each point on the line.   Also known as surge impedance. { ‚kar·ik·tə'ris·tik im'pēd·əns }

characteristic length [MECH] A convenient reference length (usually constant) of a given configuration, such as overall length of an aircraft, the maximum diameter or radius of a body of revolution, or a chord or span of a lifting surface. { ‚kar·ik·tə'ris·tik 'leŋkth }

characteristic loss spectroscopy [SPECT] A branch of electron spectroscopy in which a solid surface is bombarded with monochromatic electrons, and backscattered particles which have lost an amount of energy equal to the core-level binding energy are detected.   Abbreviated CLS. { ‚kar·ik·tə'ris·tik 'lós ‚spek'träs·kə·pē }

characteristic manifold [MATH] **1.** A surface used to study the problem of existence of solutions to partial differential equations. **2.** The linear set of eigenvectors corresponding to a given eigenvalue of a linear transformation. { ‚kar·ik·tə'ris·tik 'man·ə‚fóld }

characteristic number See eigenvalue. { ‚kar·ik·tə'ris·tik 'nəm·bər }

characteristic overflow [COMPUT SCI] An error condition encountered when the characteristic of a floating point number exceeds the limit imposed by the hardware manufacturer. { ‚kar·ik·tə'ris·tik 'ō·vər‚flō }

characteristic point [MATH] The characteristic point of a one-parameter family of surfaces corresponding to the value u_0 of the parameter is the limit of the point of intersection of the surfaces corresponding to the values u_0, u_1, and u_2 of the parameter as u_1 and u_2 approach u_0 independently. { ‚kar·ik·tə'ris·tik 'póint }

characteristic polynomial [MATH] The polynomial whose roots are the eigenvalues of a given linear transformation on a finite dimensional vector space. { ‚kar·ik·tə'ris·tik ‚päl·ə'nō·mē·əl }

characteristic radiation [ATOM PHYS] Radiation originating in an atom following removal of an electron, whose wavelength depends only on the element concerned and the energy levels involved. { ‚kar·ik·tə'ris·tik ‚räd·ē'ā·shən }

characteristic ray [MATH] For a differential equation, an integral curve which generates all the others. { ‚kar·ik·tə'ris·tik 'rā }

characteristic root See eigenvalue. { ‚kar·ik·tə'ris·tik 'rüt }

characteristic temperature See Debye temperature. { ‚kar·ik·tə'ris·tik 'tem·prə·chər }

characteristic underflow [COMPUT SCI] An error condition encountered when the characteristic of a floating point number is smaller than the smallest limit imposed by the hardware manufacturer. { ‚kar·ik·tə'ris·tik 'ən·dər‚flō }

characteristic value See eigenvalue. { ‚kar·ik·tə'ris·tik 'val·yü }

characteristic vector See eigenvector. { ‚kar·ik·tə'ris·tik 'vek·tər }

characteristic velocity See characteristic exhaust velocity. { ‚kar·ik·tə'ris·tik və'läs·əd·ē }

characteristic x-rays [ATOM PHYS] Electromagnetic radiation emitted as a result of rearrangements of the electrons in the inner shells of atoms; the spectrum consists of lines whose wavelengths depend only on the element concerned and the energy levels involved. { ‚kar·ik·tə'ris·tik 'eks‚rāz }

characterization factor [CHEM ENG] A number which expresses the variations in physical properties with change in character of the paraffinic stock; ranges from 12.5 for paraffinic stocks to 10.0 for the highly aromatic stocks.   Also known as Watson factor. { ‚kar·ik·tə·rə'zā·shən 'fak·tər }

character mode [COMPUT SCI] A mode of computer operation in which only text is displayed. { 'kar·ik·tər ‚mōd }

character neurosis [PSYCH] In psychoanalysis, any longstanding neurotic syndrome manifested first in childhood. { 'kar·ik·tər nü‚rō·səs }

character of the bottom [NAV] In marine operations, the type of material of which the bottom is composed; a sounding lead is sometimes armed with tallow or similar substance so that a sample of the bottom is obtained when a sounding is made; the sample is often helpful in determining position.   Also known as nature of the bottom. { 'kar·ik·tər əv thə 'bäd·əm }

character-oriented computer [COMPUT SCI] A computer in which the locations of individual characters, rather than words, can be addressed. { ‚kar·ik·tər ‚ór·ē‚en·təd kəm‚pyüd·ər }

character-oriented protocol See byte-oriented protocol. { 'kar·ik·tər ‚ór·ē‚ent·əd 'prōd·ə‚kól }

character outline [COMPUT SCI] The graphic pattern formed by the stroke edges of a printed or handwritten character in character recognition. { 'kar·ik·tər 'aùt‚līn }

character pitch [GRAPHICS] The number of characters printed in 1 inch (25.4 millimeters) by a printer or typewriter. { 'kar·ik·tər ‚pich }

character printer See serial printer. { 'kar·ik·tər ‚prin·tər }

character progression [ECOL] The geographic gradation of expression of specific characters over the range of distribution of a race or species. { 'kar·ik·tər prə‚gresh·ən }

character reader [COMPUT SCI] In character recognition, any device capable of locating, identifying, and translating into machine code the handwritten or printed data appearing on a source document. { 'kar·ik·tər ‚rēd·ər }

character recognition [COMPUT SCI] The technology of using a machine to sense and encode into a machine language the characters which are originally written or printed by human beings. { 'kar·ik·tər ‚rek·ig'nish·ən }

character set [COMMUN] A set of unique representations called characters, for example, the 26 letters of the English alphabet, the Boolean 0 and 1, the set of signals in Morse code, and the 128 characters of the USASCII. { 'kar·ik·tər ‚set }

character skew [COMPUT SCI] In character recognition, an improper appearance of a character to be recognized, in which it appears in a tilted condition with respect to a real or imaginary horizontal base line. { 'kar·ik·tər ‚skyü }

character stasis [GEN] Long-term constancy in a phenotypic character within a lineage. { 'kar·ik·tər ‚stā·səs }

character string [COMPUT SCI] A sequence of characters in a computer memory or other storage device.   Also known as alphabetic string. { 'kar·ik·tər 'striŋ }

character string constant [COMPUT SCI] An arbitrary combination of letters, digits, and other symbols which, in the processing of nonnumeric data involving character strings, performs a function analogous to that of a numeric constant in the processing of numeric data. { 'kar·ik·tər ,striŋ ,kän·stənt }

character stroke *See* stroke. { 'kar·ik·tər ,strōk }

character style [COMPUT SCI] In character recognition, a distinctive construction that is common to all members of a particular character set. { 'kar·ik·tər ,stīl }

character terminal [COMPUT SCI] A screen that can display only text. { 'kar·ik·tər ,tər·mə·nəl }

character-writing tube [ELECTR] A cathode-ray tube that forms alphanumeric and symbolic characters on its screen for viewing or recording purposes. { 'kar·ik·tər ,rīd·iŋ ,tüb }

Charadrii [VERT ZOO] The shore birds, a suborder of the order Charadriiformes. { kə'rad·rē,ī }

Charadriidae [VERT ZOO] The plovers, a family of birds in the superfamily Charadrioidea. { kə·rə'drī·ə,dē }

Charadriiformes [VERT ZOO] An order of cosmopolitan birds, most of which live near water. { kə,rad·rē·ə'fȯr,mēz }

Charadrioidea [VERT ZOO] A superfamily of the suborder Charadrii, including plovers, sandpipers, and phalaropes. { kə,rad·rē'ȯid·ē·ə }

Charales [BOT] Green algae composing the single order of the class Charophyceae. { kə'rā·lēz }

charcoal [MATER] Also known as char. **1.** A porous solid product containing 85–98% carbon and produced by heating carbonaceous materials such as cellulose, wood, or peat at 500–600°C in the absence of air. **2.** The residue obtained from the carbonization of a noncoking coal, such as subbituminous coal, lignite, or anthracite. **3.** *See* low-temperature coke. { 'chär,kōl }

charcoal canister [MECH ENG] In an evaporative control system, a container filled with activated charcoal that traps gasoline vapors emitted by the fuel system. Also known as canister; carbon canister. { 'chär,kōl 'kan·ə'stər }

charcoal rot [PL PATH] A fungus disease of potato, corn, and other plants caused by *Macrophomina phaseoli;* tissues of the root and lower stem are destroyed and blackened. { 'chär,kōl ,rät }

charcoal test [CHEM ENG] A determination of the natural gasoline content of natural gas by adsorbing the gasoline on activated charcoal and then recovering it by distillation. { 'chär,kōl ,test }

Chareae [BOT] A tribe of green algae belonging to the family Characeae. { 'kar·ē,ē }

charge [ELEC] **1.** A basic property of elementary particles of matter; the charge of an object may be a positive or negative number or zero; only integral multiples of the proton charge occur, and the charge of a body is the algebraic sum of the charges of its constituents; the value of the charge may be inferred from the Coulomb force between charged objects. Also known as electric charge quantity of electricity. **2.** To convert electrical energy to chemical energy in a secondary battery. **3.** To feed electrical energy to a capacitor or other device that can store it. [ENG] **1.** A unit of an explosive, either by itself or contained in a bomb, projectile, mine, or the like, or used as the propellant for a bullet or projectile. **2.** To load a borehole with an explosive. **3.** The material or part to be heated by induction or dielectric heating. **4.** The measurement or weight of material, either liquid, preformed, or powder, used to load a mold at one time during one cycle in the manufacture of plastics or metal. [MECH ENG] **1.** In refrigeration, the quantity of refrigerant contained in a system. **2.** To introduce the refrigerant into a refrigeration system. [MET] Material introduced into a furnace for melting. [NUCLEO] The fissionable material or fuel placed in a reactor to produce a chain reaction. { chärj }

charge carrier [SOLID STATE] A mobile conduction electron or mobile hole in a semiconductor. Also known as carrier. { 'chärj ,kar·ē·ər }

charge collector [ELEC] The structure within a battery electrode that provides a path for the electric current to or from the active material. Also known as current collector. { 'chärj kə,lek·tər }

charge conjugation conservation [PARTIC PHYS] The principle that the laws of motion are left unchanged by the charge conjugation operation; it is violated by the weak interactions,

but no other violations have as yet been established. { 'chärj ,kän·jə¦gā·shən ,kän·sər'vā·shən }

charge conjugation operation [PARTIC PHYS] The operation of changing every particle into its antiparticle. { 'chärj ,kän·jə¦gā·shən ,äp·ə'rā·shən }

charge conjugation parity *See* charge parity. { 'chärj ,kän·jə¦gā·shən 'par·əd·ē }

charge conservation *See* conservation of charge. { 'chärj ,kän·sər'vā·shən }

charge-coupled device [ELECTR] A semiconductor device wherein minority charge is stored in a spatially defined depletion region (potential well) at the surface of a semiconductor and is moved about the surface by transferring this charge to similar adjacent wells. Abbreviated CCD. { 'chärj ¦kəp·əld di'vīs }

charge-coupled image sensor [ELECTR] A device in which charges are introduced when light from a scene is focused on the surface of the device; image points are accessed sequentially to produce a television-type output signal. Also known as solid-state image sensor. { 'chärj ¦kəp·əld 'im·ij ,sen·sər }

charge-coupled memory [COMPUT SCI] A computer memory that uses a large number of charge-coupled devices for data storage and retrieval. { 'chärj ¦kəp·əld 'mem·rē }

charge coupling [COMPUT SCI] Transfer of all electric charges within a semiconductor storage element to a similar, nearby element by means of voltage manipulations. { 'chärj ,kəp·liŋ }

charged-current interaction [PARTIC PHYS] A weak interaction in which the charges of the interacting fermions are changed; easily observed processes such as beta decay are of this type. { ¦chärjd ¦kər·ənt in·tər'ak·shən }

charge-delocalized ion [ORG CHEM] A charged species in which the charge is distributed over more than one atom. { 'chärj dē'lōk·əl,īzd 'ī·ən }

charge density [ELEC] The charge per unit area on a surface or per unit volume in space. { 'chärj ,den·səd·ē }

charge-density wave [SOLID STATE] The ground state of a metal in which the conduction-electron charge density is sinusoidally modulated in space. { 'chärj ,den·səd·ē ,wāv }

charged particle [PARTIC PHYS] A particle whose charge is not zero; the charge of a particle is added to its designation as a superscript, with particles of charge +1 and −1 (in terms of the charge of the proton) denoted by + and − respectively; for example, π^+, Σ^-. { 'chärjd 'pärd·ə·kəl }

charged species [CHEM] A chemical entity in which the overall total of electrons is unequal to the overall total of protons. { 'chärjd 'spē·shēz }

charge establishment [ORD] Process of establishing the correct weight of a propelling charge to produce the prescribed muzzle velocity with the prescribed projectile in a particular weapon; performed at a proving ground. { 'chärj i'stab·lish·mənt }

charge exchange [PHYS] The transfer of electric charge from one particle to another during a collision between the two particles. { 'chärj iks,chānj }

charge-exchange source [ELECTR] A source of negative ions, generally negative helium ions, in which positive ions generated in a duoplasmatron are directed through a donor canal, usually containing lithium vapor, where they pick up sequentially two electrons to form negative ions. { 'chärj iks,chānj ,sȯrs }

charge independence [NUC PHYS] The principle that the nuclear (strong) force between a neutron and a proton is identical to the force between two protons or two neutrons in the same orbital and spin state. [PARTIC PHYS] As a generalization of the nuclear physics definition, the principle that the strong interactions of particles are unchanged if a particle is replaced by another particle of the same isotopic spin multiplet. { ¦chärj in·də'pen·dəns }

charge-injection device [ELECTR] A charge-transfer device used as an image sensor in which the image points are accessed by reference to their horizontal and vertical coordinates. Abbreviated CID. { 'chärj in,jek·shən di'vīs }

charge invariance [NUC PHYS] The principle that interactions between nucleons are left unchanged by rotations in isotopic spin space. { 'chärj in'ver·ē·əns }

charge-localized ion [ORG CHEM] A charged species in which the charge is centered on a single atom. { 'chärj ,lō·kə,līzd 'ī·ən }

charge-mass ratio [ELEC] The ratio of the electric charge of a particle to its mass. { ˌchärj ˌmas 'rā·shō }

charge multiplet *See* isospin multiplet. { ¦chärj 'məl·tə·plət }

charge neutrality [PL PHYS] The near equality in the density of positive and negative charges throughout a volume, which is characteristic of a plasma. [SOLID STATE] The condition in which electrons and holes are present in equal numbers in a semiconductor. { ¦chärj nü'tral·əd·ē }

charge parity [PARTIC PHYS] The eigenvalue of the charge conjugation operation; it exists only for a system which goes into itself under this operation. Also known as charge conjugation parity. { 'chärj ˌpar·əd·ē }

charge population [CHEM] The net electric charge on a specified atom in a molecule that, while it cannot be observed physically, can be determined by a prescribed definition. { 'chärj ˌpäp·yə,lā·shən }

charge quantization [ELEC] The principle that the electric charge of an object must equal an integral multiple of a universal basic charge. { 'chärj ˌkwan·tə'zā·shən }

charger *See* battery charger. { 'chär·jər }

charger-eliminator [ELEC] A battery charger with a low-noise, low-impedance output which can either charge a storage battery or supply a dc load directly, without a storage battery in parallel. { 'chär·jər ə'lim·ə,nad·ər }

charger-reader [NUCLEO] An auxiliary device used to charge and read small, portable ionization chambers. { 'chär·jər ¦rēd·ər }

charge-state process [SOLID STATE] A process involving the motion of preexisting crystal defects in a solid, following a change in the charges of the defects. { 'chärj ˌstāt ˌpräs·əs }

charge-storage transistor [ELECTR] A transistor in which the collector-base junction will charge when forward bias is applied with the base at a high level and the collector at a low level. { 'chärj ˌstȯr·ij tranz'is·tər }

charge-storage tube [ELECTR] A storage tube in which information is retained on a surface in the form of electric charges. { 'chärj ˌstȯr·ij ˌtüb }

charge-storage varactor [ELECTR] A varactor that uses semiconductor techniques to achieve power outputs above 50 watts at ultra-high and microwave frequencies. { 'chärj ˌstȯr·ij və'rak·tər }

charge transfer [PHYS CHEM] The process in which an ion takes an electron from a neutral atom, with a resultant transfer of charge. { 'chärj ˌtranz·fər }

charge-transfer complexes [CHEM] Compounds in which electrons move between molecules. { 'chärj ˌtranz·fər 'kam·plek·səs }

charge-transfer device [ELECTR] A semiconductor device that depends upon movements of stored charges between predetermined locations, as in charge-coupled and charge-injection devices. { 'chärj ˌtranz·fər di'vīs }

charge-weight ratio [ORD] The ratio of the weight of a charge, especially an explosive charge, to the total weight of the complete bomb or projectile that contains the charge; the term is not used in connection with propellants. { 'chärj ¦wāt ˌrā·shō }

charging current [ELEC] The current that flows into a capacitor when a voltage is first applied. { 'chär·jiŋ ˌkər·ənt }

charging line [PETRO ENG] A pipeline for transporting fresh charging stock of crude oil, gas, oil, and such to a still. { 'chär·jiŋ ˌlīn }

charging pump [CHEM ENG] Pump that provides pressurized fluid flow for the input of another unit, such as to a triplex pump that requires positive pressure. { 'chär·jiŋ ˌpəmp }

charging stock [PETRO ENG] A product introduced into a still; may be any product recovered through previous distillation such as gas oil or fuel oil, or any product selected for further distillation or refining. { 'chär·jiŋ ˌstäk }

Charles' law [PHYS] The law that at constant pressure the volume of a fixed mass or quantity of gas varies directly with the absolute temperature; a close approximation. Also known as Gay-Lussac's first law. { 'chärlz 'lo }

Charles' Wain *See* Big Dipper. { 'chärlz 'wān }

Charlier polynomials [MATH] Families of polynomials which are orthogonal with respect to Poisson distributions. { shär¦lyā ˌpäl·ə'nō·mē·əlz }

Charlton white *See* lithopone. { 'chärl·tən 'wīt }

charm [PARTIC PHYS] A quantum number which has been proposed to account for an apparent lack of symmetry in the behavior of hadrons relative to that of leptons, to explain why certain reactions of elementary particles do not occur, and to account for the longevity of the J-1 and J-2 particles. { chärm }

Charmat process [FOOD ENG] A bulk process for making champagne in which the wine undergoes secondary fermentation in a glass-lined vat instead of in a bottle. { shär'mä ˌpräs·əs }

charmed particle [PARTIC PHYS] A particle whose total charm is not equal to zero. { ¦chärmd 'pärd·ə·kəl }

charmed quark [PARTIC PHYS] A quark with an electric charge of $+2/3$, baryon number of $1/3$, zero strangeness, and charm of $+1$. Symbolized c. { ¦chärmd 'kwärk }

charmonium [PARTIC PHYS] A meson, such as the J/ψ particle, that is made up of the charmed quark c and its antiparticle $\bar{c}$. { chär'mō·nē·əm }

Charmouthian [GEOL] Middle Lower Jurassic geologic time. { chär'maůth·ē·ən }

charnockite [PETR] Any of various faintly foliated, nearly massive varieties of quartzofeldspathic rocks containing hypersthene. { 'chär·nə,kīt }

charnockite series [GEOL] A series of plutonic rocks compositionally similar to the granitic rock series but characterized by the presence of orthopyroxene. { 'chär·nə,kīt ˌsir·ēz }

Charon [ASTRON] The only known satellite of Pluto, with an orbital period of 6.387 days, distance from Pluto of approximately 19,600 kilometers, and diameter of approximately 1250 kilometers. { 'ka·rən }

Charophyceae [BOT] A class of green algae in the division Chlorophyta. { ˌkar·ə'fīs·ē,ē }

Charophyta [BOT] A group of aquatic plants, ranging in size from a few inches to several feet in height, that live entirely submerged in water. { kə'räf·əd·ə }

Charpak-Massonet current distribution system [PARTIC PHYS] An electronic data readout method used in spark chambers to locate a single spark, as determined by observing how the spark current divides between the two available paths to the ground. { ¦chär,päk ˌmas·ō'nā 'kər·ənt dis·trə'byü·shən ˌsis·təm }

Charpit's method [MATH] A method for finding a complete integral of the general first-order partial differential equation in two independent variables; it involves solving a set of five ordinary differential equations. { 'chär,pits ˌmeth·əd }

Charpy test [MET] An impact test to determine the ductility of a metal; a freely swinging pendulum is allowed to strike and break a notched specimen that has been laid loosely on a support; the work done by the pendulum is obtained by comparing the position of the pendulum before release with the position to which the pendulum swings after breaking the specimen. { 'shär·pē ˌtest }

charring ablator [MATER] An ablation material characterized by the formation of a carbonaceous layer at the heated surface which impedes heat flow into the material by its insulating and reradiating characteristics. { 'chär·iŋ ə'blād·ər }

chart [MAP] **1.** A map, generally designed for navigation or other particular purposes, in which essential map information is combined with various other data critical to the intended use. **2.** To prepare a chart or to engage in a charting operation. [MATH] An n-chart is a pair (U,h), where U is an open set of a topological space and h is a homeomorphism of U onto an open subset of n-dimensional Euclidean space. [SCI TECH] A form, such as a graph, table, or diagram, which gives information about some variable quantity. { chärt }

chartaceous [BOT] Resembling paper. { chär'tā·shəs }

chart catalog [NAV] A list or enumeration of navigational charts, sometimes with index charts indicating the extent of coverage of the various navigational charts. { 'chart ˌkad·əl,äg }

chart comparison unit [ENG] A device that permits simultaneous viewing of a radar plan position indicator display and a navigation chart so that one appears superimposed on the other. Also known as autoradar plot. { 'chart kəm'par·ə·sən ˌyü·nət }

chart convergence [MAP] Convergence of the meridians as shown on a chart. { 'chart kən'vər·jəns }

chart datum *See* datum plane. { 'chärt ˌdad·əm }

chart desk [ENG] A flat surface on which charts are spread out, usually with storage space for charts and other navigating equipment below the plotting surface. { 'chärt ˌdesk }

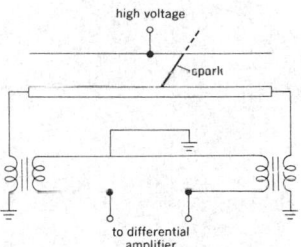

CHARPAK-MASSONET CURRENT DISTRIBUTION SYSTEM

Circuit diagram of an electronic determination of spark position in a narrow gap by the Charpak-Massonet current distribution system.

charted depth [OCEANOGR] The vertical distance from the tidal datum to the bottom. { 'chär·təd 'depth }

charted visibility [NAV] The extreme distance, shown in numbers on a chart, at which a navigational light can be seen. { 'chär·təd viz·ə'bil·əd·ē }

chart house [NAV ARCH] A room, usually adjacent to or on the bridge, where charts and other navigational equipment are stored, and where navigational computations, plots, and so on may be made. Also known as chart room. { 'chärt ‚haùs }

chartlet [NAV] A small chart covering a large geographical area; used principally to show errors in locations for loran A and Decca chains; such charts appear at the beginning of each table and show the correction in the geographical position that should be applied. { 'chärt·lət }

chartometer [MAP] An instrument used for measurement of distances on charts or maps. { ‚chär'täm·əd·ər }

chart projection [MAP] A map projection used for a chart. { 'chärt prə'jek·shən }

chart reading [NAV] Interpretation of the symbols, lines, abbreviations, and terms appearing on charts. { 'chärt ‚rēd·iŋ }

chart recorder [ENG] A recorder in which a dependent variable is plotted against an independent variable by an ink-filled pen moving on plain paper, a heated stylus on heat-sensitive paper, a light beam or electron beam on photosensitive paper, or an electrode on electrosensitive paper. The plot may be linear or curvilinear on a strip chart recorder, or polar on a circular chart recorder. { 'chärt ri'körd·ər }

chartreusin [MICROBIO] $C_{18}H_{18}O_{18}$ Crystalline, greenish-yellow antibiotic produced by a strain of *Streptomyces chartreusis;* active against gram-positive microorganisms, acid-fast bacilli, and phage of *Staphylococcus pyogenes.* { shär'trü·zən }

chart room *See* chart house. { 'chärt ‚rüm }

chart symbol [NAV] A character, letter, or similar graphical representation used on a chart to indicate some object, characteristic, or such. { 'chärt ‚sim·bəl }

chart table [ENG] A flat surface on which charts are spread out, particularly one without storage space below the plotting surface, as in aircraft and VPR (virtual PPI reflectoscope) equipment. { 'chärt ‚tā·bəl }

Charybdis *See* Galofaro. { kə'rib·dəs }

chase [BUILD] A vertical passage for ducts, pipes, or wires in a building. [DES ENG] A series of cuts, each having a path that follows the path of the cut before it; an example is a screw thread. [ENG] **1.** The main body of the mold which contains the molding cavity or cavities. **2.** The enclosure used to shrink-fit parts of a mold cavity in place to prevent spreading or distortion, or to enclose an assembly of two or more parts of a split-cavity block. **3.** To straighten and clean threads on screws or pipes. [GRAPHICS] A rectangular metal frame in which type and plates are locked for letterpress printing. [ORD] The exposed part of a gun (artillery) in front on the trunnion band or cradle. { chās }

chase mortise [DES ENG] A mortise with a sloping edge from bottom to surface so that a tenon can be inserted when the outside clearance is small. { 'chās ‚mörd·əs }

chase pilot [AERO ENG] A pilot who flies an escort airplane and advises another pilot who is making a check, training, or research flight in another craft. { 'chās ‚pī·lət }

chaser [AERO ENG] The vehicle that maneuvers in order to effect a rendezvous with an orbiting object. [ENG] A thread-cutting tool with many teeth. { 'chās·ər }

chase ring [MECH ENG] In hobbing, the ring which restrains the blank from spreading during hob sinking. { 'chās ‚riŋ }

chasing tool [DES ENG] A hammer or chisel used to decorate metal surfaces. { 'chās·iŋ ‚tül }

chasmogamy [BOT] The production of a hermaphroditic floral type that opens at anthesis and may be visited by an insect vector, providing a means of cross-pollination. { kaz'mäg·ə·mē }

chasmophyte [ECOL] A plant that grows in rock crevices. { 'kaz·mə‚fīt }

chassignite [GEOL] An achondritic stony meteorite composed chiefly of olivine (95); resembles dunite. { 'shas·ən‚yīt }

chassis [ENG] **1.** A frame on which the body of an automobile or airplane is mounted. **2.** A frame for mounting the working parts of a radio or other electronic device. { 'chas·ē }

chassis ground [ELEC] A connection made to the metal chassis on which the components of a circuit are mounted, to serve as a common return path to the power source. { 'chas·ē ‚graùnd }

chassis lubricant [MATER] A lubricating grease of consistency to be applied with a grease gun through fittings on autos and farm and industrial equipment. { 'chas·ē 'lü·bri·kənt }

chassis punch [DES ENG] A hand tool used to make round or square holes in sheet metal. { 'chas·ē ‚pənch }

chat mode [COMPUT SCI] A communications option that allows two or more computers to conduct a conversation by typing in turn. { 'chat ‚mōd }

chatoyant [MINERAL] Of a mineral or gemstone, having a changeable luster or color marked by a band of light, resembling the eye of a cat in this respect. { shə'tói·ənt }

chat room [COMPUT SCI] A Web site or server space on the Internet where live keyboard conversations (usually organized around a specific topic) with other people occur. { 'chat ‚rüm }

chatter [ELEC] Prolonged undesirable opening and closing of electric contacts, as on a relay. Also known as contact chatter. [ENG] An irregular alternating motion of the parts of a relief valve due to the application of pressure where contact is made between the valve disk and the seat. [ENG ACOUS] Vibration of a disk-recorder cutting stylus in a direction other than that in which it is driven. { 'chad·ər }

chattering [CONT SYS] A mode of operation of a relay-type control system in which the relay switches back and forth infinitely fast. { 'chad·ə·riŋ }

chatter mark [GEOL] A scar on the surface of bedrock made by the abrasive action of drift carried at the base of a glacier. [MATER] One of a series of marks made in a crosswise direction on a material; caused by vibration during rolling, extrusion, cutting, or drawing. { 'chad·ər ‚märk }

Chattian [GEOL] Upper Oligocene geologic time. Also known as Casselian. { 'chad·ē·ən }

Chattock gage [ENG] A form of micromanometer in which observation of the interface between two immiscible liquids is used to determine when the pressure to be measured has been balanced by the pressure head resulting from tilting of the entire apparatus. { 'chad·ək ‚gāj }

Chauffard-Still disease *See* Still's disease. { shō'fär 'stil di‚zēz }

chaulmoogra oil [MATER] Any of several fixed oils extracted from seeds of trees in the family Flacourtiaceae; widely used at one time to treat leprosy and other diseases. { chôl'mü·grə ‚oil }

Chautauquan [GEOL] Upper Devonian geologic time, below Bradfordian. { shə'täk·wən }

chavicol [ORG CHEM] $C_3H_5C_6H_4OH$ A colorless phenol that is liquid at room temperature; boils at 230°C; soluble in alcohol and water; found in many essential oils. { 'chav·ə‚kól }

Chazyan [GEOL] Middle Ordovician geologic time. { 'chaz·ē·ən }

chE *See* cholinesterase.

Cheadle's disease *See* infantile scurvy. { chēd·əlz diz'ēz }

Chebyshev approximation *See* min-max technique.

Chebyshev filter [ELECTR] A filter in which the transmission frequency curve has an equal-ripple shape, with very small peaks and valleys. { 'cheb·ə·shəf ‚fil·tər }

Chebyshev polynomials [MATH] A family of orthogonal polynomials which solve Chebyshev's differential equation. { 'cheb·ə·shəf ‚päl·i'nō·mē·əlz }

Chebyshev's differential equation [MATH] A special case of Gauss' hypergeometric second-order differential equation: $(1 - x^2)f''(x) - xf'(x) + n^2f(x) = 0.$ { 'cheb·ə·shəfs dif·ə'ren·chəl i'kwā·zhən }

Chebyshev's inequality [STAT] Given a nonnegative random variable $f(x)$, and $k > 0$, the probability that $f(x) \geq k$ is less than or equal to the expected value of f divided by k. { 'cheb·ə·shəfs ‚in·i'kwäl·əd·ē }

check [COMPUT SCI] A test which is necessary to detect a mistake in computer programming or a computer malfunction. [ENG] A device attached to something in order to limit the movement, such as a door check. [MATER] A lengthwise crack in a board. [MET] A minute crack occurring in steel that has been cooled too quickly. { chek }

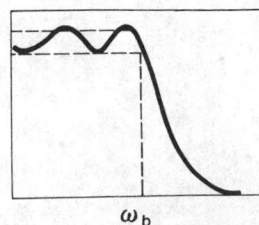

CHEBYSHEV FILTER

Equal ripple shape characteristic in the transmission band of the Chebyshev filter. Frequency ω_b is lower limit of transition region from pass band to stop band.

check bit [COMPUT SCI] A binary check digit. { 'chek ‚bit }

check box [COMPUT SCI] In a graphical user interface, a small box on which an x or check mark appears when the option indicated next to the box is turned on, and disappears when the option is turned off. { 'chek ‚bäks }

check character [COMPUT SCI] A redundant character used to perform a check. { 'chek ‚kar·ik·tər }

check cracks See checking. { 'chek ‚kraks }

check cross [GEN] The crossing of an unknown genotype with a phenotypically similar individual of known genotype. { 'chek ‚krós }

check dam [CIV ENG] A low, fixed structure, constructed of timber, loose rock, masonry, or concrete, to control water flow in an erodable channel or irrigation canal. { 'chek ‚dam }

check digit [COMPUT SCI] A redundant digit used to perform a check. { 'chek ‚dij·ət }

checkerboard regenerator [ENG] An open-checkerwork arrangement of firebrick in a high-temperature chamber that absorbs heat during a batch processing cycle, then releases it to preheat fresh combustion air during the down cycle; used, for example, in the steel industry with open-hearth and heat-treating furnaces. { 'chek·ər‚bórd ri'jen·ə‚rād·ər }

checker plate [ENG] A type of slip-resistant floor plate with a distinctive raised pattern that is used for walkways and platforms. { 'chek·ər ‚plāt }

checkers [ENG] Open brickwork in a checkerboard regenerator allowing for the passage of hot, spent gases. { 'chek·ərz }

check fillet [BUILD] A curb set into a roof to divert or control the flow of rainwater. { 'chek ‚fil·ət }

check flight [AERO ENG] **1.** A flight made to check or test the performance of an aircraft, rocket, or spacecraft, or a piece of its equipment, or to obtain measurements or other data on performance. **2.** A familiarization flight in an aircraft, or a flight in which the pilot or the aircrew are tested for proficiency. { 'chek ‚flīt }

check indicator [COMPUT SCI] A console device, usually a light, informing the operator that an error has occurred. { 'chek ‚in·də‚kād·ər }

check indicator instruction [COMPUT SCI] A computer instruction which directs that a signal device is turned on to call the operator's attention to the fact that there is some discrepancy in the instruction now in use. { 'chek ‚in·də‚kād·ər in'strək·shən }

checking [MATER] Fine, shallow cracks appearing on the surface of a material or in a film of a surface coating. Also known as check cracks. [MET] Temporarily reducing the volume or temperature of the air blast in a blast furnace. { 'chek·iŋ }

checking program [COMPUT SCI] A computer program which detects and determines the nature of errors in other programs, particularly those that involve incorrect coding or punching of wrong characters. Also known as checking routine. { 'chek·iŋ ‚prō·grəm }

checking routine See checking program. { 'chek·iŋ rü'tēn }

check ligament [ANAT] A thickening of the orbital fascia running from the insertion of the lateral rectus muscle to the medial orbital wall (medial check ligament) or from the insertion of the lateral rectus muscle to the lateral orbital wall (lateral check ligament). { 'chek ‚lig·ə·mənt }

checkline [NAV ARCH] A heavy line passed through a chock and fastened to a bitt of a ship that is coming alongside a wharf in order to slow the ship down or to keep it from moving away from the wharf. { 'chek‚līn }

check number [COMPUT SCI] A number denoting a specific type of hardware malfunction. { 'chek ‚nəm·bər }

check observation [METEOROL] An aviation weather observation taken primarily for aviation radio broadcast purposes; usually abbreviated to include just those elements of a record observation that have an important affect on aircraft operations. { 'chek ‚äb·zər'vā·shən }

checkout [COMPUT SCI] A collection of routines that are built into a compiler to test and debug programs. [ENG] A sequence of actions to test or examine a thing as to its readiness for incorporation into a new phase of use or as to the performance of its intended function. { 'chek‚aut }

checkout compiler [COMPUT SCI] A special compiler designed specifically to test and debug programs by using checkout routines. { 'chek‚aut kəm‚pī·lər }

checkpoint [CELL MOL] A point in the eukaryotic cell cycle at which the cycle may continue if specific conditions are present or will stop if conditions are not right. [COMPUT SCI] That place in a routine at which the entire state of the computer (memory, registers, and so on) is written out on auxiliary storage (tape, disk, cards) from which it may be read back into the computer if the program is to be restarted later. [NAV] Geographical location on land or water above which the position of an aircraft in flight may be determined by observation or by electronic means. { 'chek‚póint }

checkpoint/restart [COMPUT SCI] The procedures for resuming a processing run after it has been halted either accidentally or deliberately. { 'chek‚póint 'rē‚stärt }

check problem See check routine. { 'chek ‚präb·ləm }

check protect symbol [COMPUT SCI] A character, usually an asterisk, that is printed in place of leading zeros in a number, such as a dollar amount on a check. { 'chek prə'tekt ‚sim·bəl }

check rail [BUILD] A rail, thicker than the window, that spans the opening between the top and bottom sash; usually beveled and rabbeted. See guardrail. { 'chek ‚rāl }

check register [COMPUT SCI] A register in which transferred data are temporarily stored so that they may be compared with a second transfer of the same data, to verify the accuracy of the transfer. { 'chek ‚rej·ə·stər }

check routine [COMPUT SCI] A routine or problem designed primarily to indicate whether a fault exists in a computer, without giving detailed information on the location of the fault. Also known as check problem; test program; test routine. { 'chek rü'tēn }

check row [COMPUT SCI] A row (or one of two or more rows) on a paper tape which contains the cumulated sum of existing rows, column by column, resulting in either 1 or 0 by column, thus verifying that all rows have been properly read. { 'chek ‚rō }

check sample See control sample. { 'chek ‚sam·pəl }

check screen See oversize control screen. { 'chek ‚skrēn }

check standard [ANALY CHEM] In physical calibration, an artifact that is measured at specified intervals. { 'chek ‚stan·dərd }

check stop [BUILD] A narrow length of wood or metal that is installed to hold a sliding element in place, such as the lower part of a sash of a double-hung window. { 'chek ‚stäp }

check study [IND ENG] A review of a job or operation in part or in its entirety to evaluate the validity of a standard time. { 'chek ‚stəd·ē }

check sum [COMPUT SCI] A sum of digits or numbers used in a summation check. { 'chek ‚səm }

check symbol [COMPUT SCI] One or more digits generated by performing an arithmetic check or summation check on a data item which are then attached to the item and copied along with it through various stages of processing, allowing the check to be repeated to verify the accuracy of the copying processes. { 'chek ‚sim·bəl }

check valve [MECH ENG] A device for automatically limiting flow in a piping system to a single direction. Also known as nonreturn valve. { 'chek ‚valv }

check word [COMPUT SCI] A computer word, containing data from a block of records, that is joined to the block and serves as a check symbol during transfers of the block between different locations. { 'chek ‚wərd }

Chediak-Higashi anomaly [PATH] Deeply staining, coarse, peroxidase-positive granules in the cytoplasm of neutrophils and eosinophils in certain disease states. { 'ched·ē·ak ‚hi'gä‚shē ə‚näm·ə·lē }

cheek [ANAT] The wall of the mouth in humans and other mammals. [MET] Portion of a three-part flask between the cope and the drag. [ZOO] The lateral side of the head in submammalian vertebrates and in invertebrates. { chēk }

cheek pouch [VERT ZOO] A saclike dilation of the cheeks in certain animals, such as rodents, in which food is held. { 'chēk ‚pauch }

cheese [FOOD ENG] A food produced from milk that has been clotted by acid or rennet to form a curd which is cut, shaped, pressed, and salted or brined. [TEXT] Tube of spun yarn to be put on a warp beam for weaving. { chēz }

cheese antenna [ELECTROMAG] An antenna having a parabolic reflector between two metal plates, dimensioned to permit propagation of more than one mode in the desired direction of polarization. { 'chēz an'ten·ə }

CHEETAH

The cheetah (*Acinonyx jubatus*).

CHEILOSTOMATA

|← 1 mm →|

Anascan cheilostome, showing the surface of encrusting sheetlike colony of *Membranipora*, Miocene-Recent.

cheesebox still [CHEM ENG] One of the first types of vertical cylindrical stills designed with a vapor dome. { 'chēz,-bäks ,stil }

cheese cement [MATER] A glue made from cheese or milk curd. { 'chēz si'ment }

cheesecloth [TEXT] A very loosely woven, thin, lightweight cotton fabric; used in unfinished (gray) form for covering tobacco plants and for making tea bags and wiping cloths; the finished form is used for curtains, bedspreads, and such. { 'chēz,klóth }

cheese head [DES ENG] A raised cylindrical head on a screw or bolt. { 'chēz ,hed }

cheetah [VERT ZOO] *Acinonyx jubatus*. A doglike carnivoran mammal belonging to the cat family, having nonretractile claws and long legs. { 'chēd·ə }

cheiloplasty [MED] Any plastic operation upon the lip. { 'kī·lō,plas·tē }

cheilosis [MED] Cracking at the corners of the mouth and scaling of the lips, usually associated with riboflavin deficiency. { kī'lō·səs }

Cheilostomata [INV ZOO] An order of ectoproct bryozoans in the class Gymnolaemata possessing delicate erect or encrusting colonies composed of loosely grouped zooecia. { ,kī·lə'stō·məd·ə }

cheimaphobia [PSYCH] An abnormal fear of cold. { kī·mə'fō·bē·ə }

Cheiracanthidae [PALEON] A family of extinct acanthodian fishes in the order Acanthodiformes. { ,kī·rə'kan·thə,dē }

cheiromegaly [MED] Enlargement of one or both hands that is not attributable to disease of the hypophysis. Also spelled chiromegaly. { ,kī·rə'meg·ə·lē }

cheiroplasty [MED] Any plastic operation performed on the hand. Also spelled chiroplasty. { 'kī·rə,plas·tē }

chela [INV ZOO] **1.** A claw or pincer on the limbs of certain crustaceans and arachnids. **2.** A sponge spicule with talonlike terminal processes. { 'kē·lə }

chelate [INV ZOO] Pertaining to an appendage with a pincerlike organ or claw. [ORG CHEM] A molecular structure in which a heterocyclic ring can be formed by the unshared electrons of neighboring atoms. { 'kē,lāt }

chelate laser [OPTICS] A liquid laser that uses a rare-earth chelate (a metalloorganic compound), with initial excitation taking place within the organic part of the liquid molecule and then transferring to the metallic ions to give lasing action. Also known as rare-earth chelate laser. { 'kē,lāt ,lā·zər }

chelating agent [ORG CHEM] An organic compound in which atoms form more than one coordinate bond with metals in solution. { 'ke,lād·iŋ ,ā·jənt }

chelating resin [ORG CHEM] Any of the ion-exchange resins with unusually high selectivity for specific cations; for example, phenol-formaldehyde resin with 8-quinolinol replacing part of the phenol, particularly selective for copper, nickel, cobalt, and iron(III). { 'ke,lād·iŋ 'rez·ən }

chelation [ORG CHEM] A chemical process involving formation of a heterocyclic ring compound which contains at least one metal cation or hydrogen ion in the ring. { kē'lā·shən }

chelerythrine [ORG CHEM] $C_{21}H_{17}O_4H$ A poisonous, crystalline alkaloid, slightly soluble in alcohol; it is derived from the seeds of the herb celandine (*Chelidonium majus*) and has narcotic properties. { ,kel·ə'rī,thrēn }

cheletropic reaction [PHYS CHEM] A chemical reaction involving the elimination of a molecule in which two sigma bonds terminating at a single atom are made or broken. { ,kel·ə'trä·pik rē'ak·shən }

chelicera [INV ZOO] Either appendage of the first pair in arachnids, usually modified for seizing, crushing, or piercing. { kə'lis·ə·rə }

Chelicerata [INV ZOO] A subphylum of the phylum Arthropoda; chelicerae are characteristically modified as pincers. { kə,lis·ə'räd·ə }

Chelidae [VERT ZOO] The side-necked turtles, a family of reptiles in the suborder Pleurodira. { 'kel·ə,dē }

chelidonic acid [ORG CHEM] $C_7H_4O_6$ A pyran isolated from the perennial herb celandine (*Chelidonium majus*). { ¦kel·ə¦dän·ik 'as·əd }

chelifore [INV ZOO] Either of the first pair of appendages on the cephalic segment of pycnogonids. { 'kel·ə,fór }

cheliform [INV ZOO] Having a forcepslike organ formed by

a movable joint closing against an adjacent segment; referring especially to a crab's claw. { 'kel·ə,fórm }

cheliped [INV ZOO] Either of the paired appendages bearing chelae in decapod crustaceans. { 'kel·ə,ped }

chellin *See* khellin. { 'kel·ən }

chelometry [ANALY CHEM] Analytical technique involving the formation of 1:1 soluble chelates when a metal ion is titrated with aminopolycarboxylate and polyamine reagents; a form of complexiometric titration. { ke'läm·ə·trē }

Chelonariidae [INV ZOO] A family of coleopteran insects in the superfamily Dryopoidea. { ke,län·ə'rī·ə,dē }

Chelonethida [INV ZOO] An equivalent name for the Pseudoscorpionida. { ,kel·ə'neth·ə·də }

Chelonia [VERT ZOO] An order of the Reptilia, subclass Anapsida, including the turtles, terrapins, and tortoises. { ke'lōn·ē·ə }

Cheloniidae [VERT ZOO] A family of reptiles in the order Chelonia including the hawksbill, loggerhead, and green sea turtles. { ,kel·ə'nī·ə,dē }

Cheluridae [INV ZOO] A family of amphipod crustaceans in the suborder Gammaridea. { kə'lür·ə,dē }

Chelydridae [VERT ZOO] The snapping turtles, a small family of reptiles in the order Chelonia. { kə'lid·rə,dē }

chemical [CHEM] **1.** Related to the science of chemistry. **2.** A substance characterized by definite molecular composition. { 'kem·i·kəl }

chemical affinity *See* affinity. { 'kem·i·kəl ə'fin·əd·ē }

chemical agent [MATER] A solid, liquid, or gas employed in three principal categories: war gases, smokes, and incendiaries; through its chemical properties produces lethal, injurious, or irritant effects on humans, makes a screening or colored smoke, or acts as a fire starter. { 'kem·i·kəl 'ā·jənt }

chemical ammunition [ORD] Any ammunition, such as bombs, projectiles, bullets, or flares, containing a chemical agent, such as war gases, smokes, and incendiaries. { 'kem·i·kəl ,am·yə'nish·ən }

chemical bomb [ORD] A bomb with a chemical agent for the main charge. { 'kem·i·kəl ¦bämb }

chemical bond *See* bond. { 'kem·i·kəl ¦bänd }

chemical burn [MED] Tissue destruction caused by caustic agents, irritant gases, or other chemical agents. { 'kem·i·kəl ,bərn }

chemical carcinogenesis [MED] The chemical-induced cancerous transformation of normal cells via a multistep process in which the genetic code of the cells is altered and then the altered cells are promoted to form tumors. { ¦kem·i·kəl ,kär·sə·nə'gen·ə·səs }

chemical-cartridge respirator [MIN ENG] An air purification device worn by miners that removes small quantities of toxic gases or vapors from the inspired air; the cartridge contains chemicals which operate by processes of oxidation, absorption, or chemical reaction. { 'kem·i·kəl 'kär·trij 'res·pə,rād·ər }

chemical cellulose *See* alpha cellulose. { 'kem·i·kəl 'sel·yə,lōs }

chemical compound *See* compound. { 'kem·i·kəl 'käm ,paúnd }

chemical conversion *See* conversion. { 'kem·i·kəl kən'vər·zhən }

chemical conversion coating [MET] A protective or decorative coating formed on the surface of a metal as the result of chemical reaction of the metal with a selected environment. { 'kem·i·kəl kən'vər·zhən ,kōd·iŋ }

chemical crystallography [CRYSTAL] The geometric description, and study, of the internal arrangement of atoms in crystals formed from chemical compounds. { 'kem·i·kəl kris·tə'läg·rə·fē }

chemical cutoff [PETRO ENG] A technique for making a smooth cut in a steel pipe in a well by applying jets of a corrosive solution against the wall of the pipe. { 'kem·i·kəl 'kəd,óf }

chemical dating [ANALY CHEM] The determination of the relative or absolute age of minerals and of ancient objects and materials by measurement of their chemical compositions. { 'kem·i·kəl 'dād·iŋ }

chemical denudation [GEOL] Wasting of the land surface by water transport of soluble materials into the sea. { 'kem·i·kəl ,dē·nü'dā·shən }

chemical deposition [CHEM] Precipitation of a metal from a solution of a salt by introducing another metal. { 'kem·i·kəl ˌdep·ə'zish·ən }

chemical dosimeter [NUCLEO] A dosimeter in which the accumulated radiation-exposure dose is indicated by color changes accompanying chemical reactions induced by the radiation. { 'kem·i·kəl dō'sim·əd·ər }

chemical dynamics [PHYS CHEM] A branch of physical chemistry that seeks to explain time-dependent phenomena, such as energy transfer and chemical reactions, in terms of the detailed motion of the nuclei and electrons that constitute the system. { 'kem·ə·kəl dī'nam·iks }

chemical ecology [ECOL] The study of ecological interactions mediated by the chemicals that organisms produce. { ˌkem·i·kəl ē'käl·ə·jē }

chemical element *See* element. { 'kem·i·kəl 'el·ə·mənt }

chemical energy [PHYS CHEM] Energy of a chemical compound which, by the law of conservation of energy, must undergo a change equal and opposite to the change of heat energy in a reaction; the rearrangement of the atoms in reacting compounds to produce new compounds causes a change in chemical energy. { 'kem·i·kəl 'en·ər·jē }

chemical engineering [ENG] That branch of engineering serving those industries that chemically convert basic raw materials into a variety of products, and dealing with the design and operation of plants and equipment to perform such work; all products are formed in chemical processes involving chemical reactions carried out under a wide range of conditions and frequently accompanied by changes in physical state or form. { 'kem·i·kəl ˌen·jə'nir·iŋ }

chemical equilibrium [CHEM] A condition in which a chemical reaction is occurring at equal rates in its forward and reverse directions, so that the concentrations of the reacting substances do not change with time. Also known as equilibrium. { 'kem·i·kəl ˌē·kwə'lib·rē·əm }

chemical etching [MET] Formation of characteristic surface features when a polished metal surface is etched by suitable reagents. { 'kem·i·kəl 'ech·iŋ }

chemical exchange process [CHEM] A method of separating isotopes of the lighter elements by the repetition of a process of chemical change which involves exchange of the isotopes. { 'kem·i·kəl iks'chānj ˌpräs·əs }

chemical film dielectric [ELEC] An extremely thin layer of material on one or both electrodes of an electrolytic capacitor, which conducts electricity in only one direction and thereby constitutes the insulating element of the capacitor. { 'kem·i·kəl ˌfilm ˌdī·ə'lek·trik }

chemical fire extinguisher [CHEM ENG] Any of three types of fire extinguishers (vaporizing liquid, carbon dioxide, and dry chemical) which expel chemicals in solid, liquid, or gaseous form to blanket or smother a fire. { 'kem·i·kəl 'fīr ik'stiŋ·gwish·ər }

chemical flux [CHEM] In a chemical reaction, the amount of a given substance per unit volume transformed per unit time. Also known as chemiflux. { 'kem·ə·kəl 'fləks }

chemical flux cutting [MET] An oxygen cutting process in which metals are cut by using flux. { 'kem·i·kəl 'fləks ˌkəd·iŋ }

chemical focus *See* actinic focus. { 'kem·i·kəl 'fō·kəs }

chemical fog [GRAPHICS] Background density on a film that occurs either before or during development, due to a chemical reaction. { 'kem·i·kəl 'fäg }

chemical force microscope [ENG] A modification of the atomic force microscope in which an organic monolayer on the probe tip that terminates with specific chemical functional groups is sensitive to specific molecular interactions between these groups and those on the sample surface. { ˌkem·ə·kəl ˌfórs 'mī·krə,skōp }

chemical formula [CHEM] A notation utilizing chemical symbols and numbers to indicate the chemical composition of a pure substance; examples are CH_4 for methane and HCl for hydrogen chloride. { 'kem·i·kəl 'fór·myə·lə }

chemical fossils *See* biomarkers. { ˌkem·i·kəl 'fäs·əlz }

chemical grenade [ORD] General term for any hand grenade or rifle grenade charged with a chemical agent. { 'kem·i·kəl grə'nād }

chemical hygrometer *See* absorption hygrometer. { 'kem·i·kəl hī'gräm·əd·ər }

chemical indicator [ANALY CHEM] **1.** A substance whose physical appearance is altered at or near the end point of a

chemical titration. **2.** A substance whose color varies as the concentration of hydrogen ions in the solution to which it is added varies. Also known as indicator. { 'kem·i·kəl 'in·də,kād·ər }

chemical inhibitor [CHEM] A substance capable of stopping or retarding a chemical reaction. { 'kem·i·kəl in'hib·əd·ər }

chemical ion pump [CHEM ENG] A vacuum pump whose pumping action is based on evaporation of a metal whose vapor then reacts with the chemically active molecules in the gas to be evacuated. { 'kem·i·kəl 'ī·ən ˌpəmp }

chemical kinetics [PHYS CHEM] That branch of physical chemistry concerned with the mechanisms and rates of chemical reactions. Also known as reaction kinetics. { 'kem·i·kəl kə'ned·iks }

chemical laser *See* chemically pumped laser. { 'kem·i·kəl 'lā·zər }

chemically foamed plastic [MATER] A foamed plastic having its cellular structure produced by gases generated from chemical reaction of the components. { 'kem·ik·lē ˌfōmd 'plas·tik }

chemically pumped laser [OPTICS] A laser in which pumping is achieved by using a chemical action rather than electrical energy to produce the required pulses of light. Also known as chemical laser. { 'kem·ik·lē ˌpəmt 'lā·zər }

chemically pure [CHEM] Without impurities detectable by analysis. Abbreviated cp. { 'kem·ik·lē 'pyúr }

chemically sensitive field-effect transistor [ELECTR] A field-effect transistor in which the ordinary gate electrode is replaced by a chemically sensitive membrane so that the gain of the transistor depends on the concentration of chemical substances. { 'kem·ik·lē ˌsen·səd·iv 'fēld i¦fekt tran,zis·tər }

chemical machining [MET] Making of metal parts to specified dimensions by removing surface metal with chemicals (acids or alkalies). Also known as chemical milling. { 'kem·i·kəl mə'shēn·iŋ }

chemical meningitis [MED] Meningeal inflammation brought about by foreign irritants such as alcohol, detergents, chemotherapeutic agents, or contrast agents used in some radiologic imaging procedures. { ˌkem·ə·kəl ˌmen·in'jīd·əs }

chemical metallurgy [MET] The science and technology of extracting metals from ores and refining them. Also known as process metallurgy. { 'kem·i·kəl 'med·əl·ər·jē }

chemical microscopy [ANALY CHEM] Application of the microscope to the solution of chemical problems. { 'kem·i·kəl mī'kräs·kə·pē }

chemical milling *See* chemical machining. { 'kem·i·kəl 'mil·iŋ }

chemical operations *See* chemical warfare. { 'kem·i·kəl ˌäp·ə'rā·shənz }

chemical pathology [PATH] The study of disease by using chemical methods. { 'kem·i·kəl pə'thäl·ə·jē }

chemical polarity [PHYS CHEM] Tendency of a molecule, or compound, to be attracted or repelled by electrical charges because of an asymmetrical arrangement of atoms around the nucleus. { 'kem·i·kəl pə'lar·əd·ē }

chemical polishing [MET] Smoothing and brightening the surface of a metal by treatment with a chemical agent. { 'kem·i·kəl 'päl·ish·iŋ }

chemical porcelain [MATER] High-purity, nonporous grade of porcelain used to make laboratory analysis utensils, such as crucibles, retorts, and spatulas. { 'kem·i·kəl 'pórs·lən }

chemical potential [PHYS CHEM] In a thermodynamic system of several constituents, the rate of change of the Gibbs function of the system with respect to the change in the number of moles of a particular constituent. { 'kem·i·kəl pə'ten·chəl }

chemical precipitates [GEOL] A sediment formed from precipitated materials as distinguished from detrital particles that have been transported and deposited. { 'kem·i·kəl pri'sip·ə,tāts }

chemical pressurization [AERO ENG] The pressurization of propellant tanks in a rocket by means of high-pressure gases developed by the combustion of a fuel and oxidizer or by the decomposition of a substance. { 'kem·i·kəl ˌpresh·ə·rə'zā·shən }

chemical process industry [CHEM ENG] An industry in which the raw materials undergo chemical conversion during their processing into finished products, as well as (or instead of) the physical conversions common to industry in general;

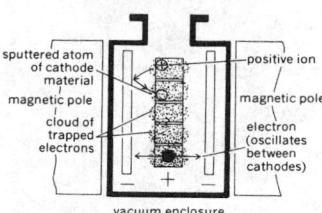

CHEMICAL-ION PUMP

The principal features of an ion pump.

includes the traditional chemical, petroleum, and petrochemical industries. { ¦kem·i·kəl 'präs·əs ¸in·də·strē }

chemical pulp [MATER] Wood pulp made by separating the fibers of wood chips by the action of alkalies or acids. { ¦kem·i·kəl 'pəlp }

chemical pulping [CHEM ENG] Separation of wood fiber for paper pulp by chemical treatment of wood chips to dissolve the lignin that cements the fibers together. { ¦kem·i·kəl 'pəlp·iŋ }

chemical pump [PETRO ENG] Skid-mounted pumping unit used to feed chemicals into the power oil (used to operate bottom-hole pumps in oil wells) to reduce corrosion in the system and to assist in water removal when the power oil and well-produced oil reach the ground-level wash tank. { ¦kem·i·kəl ¦pəmp }

chemical purity See purity. { ¦kem·ə·kə 'pyür·ə·dē }

chemical reaction [CHEM] A change in which a substance (or substances) is changed into one or more new substances; there is only a minute change, Δm, in the mass of the system, given by $\Delta E = \Delta mc^2$, where ΔE is the energy emitted or absorbed and c is the speed of light. { ¦kem·i·kəl rē'ak·shən }

chemical reactivity [CHEM] The tendency of two or more chemicals to react to form one or more products differing from the reactants. { ¦kem·i·kəl rē¸ak'tiv·əd·ē }

chemical reactor [CHEM ENG] Vessel, tube, pipe, or other container within which a chemical reaction is made to take place; may be batch or continuous, open or packed, and can use thermal, catalytic, or irradiation actuation. { ¦kem·i·kəl rē'ak·tər }

chemical relaxation [CHEM] The readjustment of a chemical system to a new equilibrium after the equilibrium of a chemical reaction is disturbed by a sudden change, particularly in an external parameter such as pressure or temperature. { ¦kem·ə·kəl ¸rē¸lak'sā·shən }

chemical remanent magnetization [GEOPHYS] Permanent magnetization of rocks acquired when a magnetic material, such as hematite, is grown at low temperature through the oxidation of some other iron mineral, such as magnetite or goethite; the growing mineral becomes magnetized in the direction of any field which is present. Abbreviated CRM. { ¦kem·i·kəl 'rem·ə·nənt ¸mag·nət·ə'zā·shən }

chemical reservoir [GEOL] An underground oil or gas trap formed in limestones or dolomites deposited in quiescent geologic environments. { ¦kem·i·kəl 'rez·əv¸wär }

chemical resistance [MATER] Ability of solid materials to resist damage by chemical reactivity or solvent action. { ¦kem·i·kəl ri'zis·təns }

chemical rock [PETR] A type of sedimentary rock comprising material deposited directly by precipitation from solution or colloidal suspension and frequently possessing a crystalline texture. { ¦kem·i·kəl 'räk }

chemical sense [NEURO] A process of the nervous system for reception of and response to chemical stimulation by excitation of specialized receptors. { ¦kem·i·kəl 'sens }

chemical shift [PHYS CHEM] Shift in a nuclear magnetic-resonance spectrum resulting from diamagnetic shielding of the nuclei by the surrounding electrons. { ¦kem·i·kəl 'shift }

chemical shim [NUCLEO] A chemical, usually boric acid, that is placed in the coolant system of a nuclear reactor to serve as a neutron absorber and that compensates for fuel burnup during normal operation. { ¦kem·i·kəl 'shim }

chemical shutdown [NUCLEO] Addition of a dissolved poison to the coolant of a nuclear reactor to achieve shutdown. { ¦kem·i·kəl 'shət¸daùn }

chemical similitude [CHEM ENG] A procedure used to ensure satisfactory operation of a full-scale chemical process by comparison with pilot plant data. { ¦kem·i·kəl sə'mil·ə¸tüd }

chemical species See species. { ¦kem·i·kəl 'spē¸shēz }

chemical spray [ORD] Aerial release, or device for aerial release, of liquid war gas for casualty effect, or of liquid smoke for aerial smoke screens. { ¦kem·i·kəl 'sprā }

chemical sterilization [ENG] The use of bactericidal chemicals to sterilize solutions, air, or solid surfaces. { ¦kem·i·kəl ¸ster·ə·lə'zā·shən }

chemical stoneware [MATER] Clay pottery material that resists acids and alkalies; used for ball mills, pipes, laboratory sinks and utensils, and so on. { ¦kem·i·kəl 'stōn¸wer }

chemical symbol [CHEM] A notation for one of the chemical elements, consisting of letters; for example Ne, O, C, and

Na represent neon, oxygen, carbon, and sodium. { ¦kem·i·kəl 'sim·bəl }

chemical synthesis [CHEM] The formation of one chemical compound from another. { ¦kem·i·kəl 'sin·thə·səs }

chemical tanker [NAV ARCH] Ship designed with tanks of stainless steel, or of other materials, capable of containing chemicals. { ¦kem·i·kəl 'taŋ·kər }

chemical thermodynamics [PHYS CHEM] The application of thermodynamic principles to problems of chemical interest. { ¦kem·i·kəl ¸thər·mō·də'nam·iks }

chemical thermometer [ENG] A filled-system temperature-measurement device in which gas or liquid enclosed within the device responds to heat by a volume change (rising or falling of mercury column) or by a pressure change (opening or closing of spiral coil). { ¦kem·i·kəl thər'mäm·əd·ər }

chemical tracer [NUCLEO] A tracer having chemical properties similar to those of the substance with which it is mixed. { ¦kem·i·kəl 'trā·sər }

chemical vapor deposition [SOLID STATE] The growth of thin solid films on a crystalline substrate as the result of thermochemical vapor-phase reactions. Abbreviated CVD. { ¦kem·i·kəl 'vā·pər ¸dep·ə'zish·ən }

chemical warfare [ORD] Originally, the employment of poison gases as antipersonnel agents; later expanded to include flame and incendiary warfare, smoke for screening or signaling purposes, and microorganisms (bacteria and their toxins, rickettsia, viruses) for the production of casualties or destruction of crops. Also known as chemical operations. { ¦kem·i·kəl 'wór¸fer }

chemical weathering [GEOCHEM] A weathering process whereby rocks and minerals are transformed into new, fairly stable chemical combinations by such chemical reactions as hydrolysis, oxidation, ion exchange, and solution. Also known as decay; decomposition. { ¦kem·i·kəl 'weth·ə·riŋ }

chemiclearance [CHEM] The use of chemical analysis to establish the safe use of a substance. { ¦kem·i¸klir·əns }

chemiflux See chemical flux. { ¦kem·ə¸fləks }

chemi-ionization [CHEM] Ionization that occurs as a result of the collison of a particle with a neutral species, usually excited, such as a metastable atom. { ¸kem·ē¸ī·ə·nə'zā·shən }

chemiluminescence [PHYS CHEM] Emission of light as a result of a chemical reaction without an apparent change in temperature. { ¸kem·i¸lüm·ə'nes·əns }

chemimechanical pulp [MATER] Plant material treated by the sulfite, soda, or sulfate process for papermaking. { ¦kem·i·mə¦kan·i·kəl 'pəlp }

chemionics [CHEM] The chemistry of molecular components and devices that operate on photons, electrons, and ions. { ¸kem·ē'än·iks }

chemiosmosis [CHEM] A chemical reaction occurring through an intervening semipermeable membrane. Also known as chemosmosis. { ¦kem·ē¸äs¦mō·səs }

chemiosmotic coupling [BIOCHEM] The mechanism by which adenosinediphosphate is phosphorylated to adenosinetriphosphate in mitochondria and chloroplasts. { ¦kem·ē¸äs¦mäd·ik 'kəp·liŋ }

chemisorption [PHYS CHEM] A chemical adsorption process in which weak chemical bonds are formed between gas or liquid molecules and a solid surface. { ¦kem·i¸sórp·shən }

chemist [CHEM] A scientist specializing in chemistry. { 'kem·əst }

chemistry [SCI TECH] The scientific study of the properties, composition, and structure of matter, the changes in structure and composition of matter, and accompanying energy changes. { 'kem·ə·strē }

chemoautotroph [MICROBIO] Any of a number of autotrophic bacteria and protozoans which do not carry out photosynthesis. { ¸kē·mō¸ód·ə'träf·ik }

chemocline [HYD] The transition in a meromictic lake between the mixolimnion layer (at the top) and the monimolimnion layer (at the bottom). { 'kē·mə¸klīn }

chemodectoma [MED] A benign tumor of the carotid body. { ¦kē·mō¸dek'tō·mə }

chemodifferentiation [EMBRYO] The process of cellular differentiation at the molecular level by which embryonic cells become specialized as tissues and organs. { ¸kē·mō¸dif·ə¸ren·chē'ā·shən }

chemoheterotroph [BIOL] An organism that derives energy

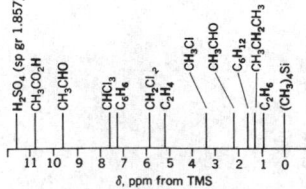

CHEMICAL SHIFT

Chemical shifts for representative compounds. Decreasing values of δ correspond to increasing magnetic field in a constant-frequency spectrometer. The scale calibration is obtained from the resonance signal of a small amount of tetramethylsilane (TMS) placed in the sample tube to provide a zero reference point.

and carbon from the oxidation of preformed organic compounds. { ¦kē·mō'hed·ə·rə,träf }

chemokine [CELL MOL] A small (7–14 kilodaltons of soluble protein) chemoattractant cytokine produced by cells and tissues at the beginning of an immune system response to infection, allergen, injury, and so forth that controls the nature and magnitude of immune cell infiltration and inflammation at the affected site. { 'kē·mə,kīn }

chemometrics [ANALY CHEM] The use of statistics and mathematics for experimental design and analysis of chemical data. { ¦kē·mō'me·triks }

chemonite [MATER] A wood preservative consisting of a water solution of small percentages of copper hydroxide, arsenic trioxide, ammonia, and acetic acid. { 'kem·ə,nīt }

chemoorganotroph [BIOL] An organism that requires an organic source of carbon and metabolic energy. { ¦kē·mō,ór'-gan·ə,träf }

chemoprevention [MED] Prevention of illness through pharmaceutical means; for example, use of drugs to arrest or reverse development of premalignant neoplasia. { ¦kē·mō·pri·¦ven·shən }

chemoprophylaxis [MED] Use of drugs to prevent the development of infectious diseases. { ¦kē·mō,prō·fə'lak·səs }

chemoreception [PHYSIO] Reception of a chemical stimulus by an organism. { ¦kē·mō·ri'sep·shən }

chemoreceptor [PHYSIO] Any sense organ that responds to chemical stimuli. { ¦kē·mō·ri'sep·tər }

chemoselectivity [ORG CHEM] The preferential reaction of a chemical reagent with one functional group in the presence of other similar functional groups; for example, a chemoselective reducing agent might reduce an aldehyde but not a ketone. { ¦kē·mō,si·lek'tiv·əd·ē }

chemosis [MED] An eye disorder characterized by swelling of the mucous membrane that covers the eyeball and lines the inner surface of the eyelids. { ke'mō·səs }

chemosmosis See chemiosmosis. { ¦kem,äs'mō·səs }

chemosphere [METEOROL] The vaguely defined region of the upper atmosphere in which photochemical reactions take place; generally considered to include the stratosphere (or the top thereof) and the mesosphere, and sometimes the lower part of the thermosphere. { 'kē·mō,sfir }

chemostat [MICROBIO] An apparatus, and a principle, for the continuous culture of bacterial populations in a steady state. { 'kē·mə,stat }

chemostratigraphy [GEOCHEM] The correlation and dating of marine sediments and sedimentary rocks through the use of trace-element concentrations, molecular fossils, and certain isotopic ratios that can be measured on components of the rocks. { ¦kē·mō·strə'tig·rə·fē }

chemosurgery [MED] Surgical removal of diseased or unwanted tissue by the application of chemicals { ¦kē·mō¦sərj·ə·rē }

chemosynthesis [BIOCHEM] The synthesis of organic compounds from carbon dioxide by microorganisms using energy derived from chemical reactions. { ¦kē·mō'sin·thə·səs }

chemotaxin [BIOCHEM] A chemical that promotes movement of a cell or microorganism in the process of chemotaxis. { ¦kē·mō,tak·sən }

chemotaxis [BIOL] The orientation or movement of a motile organism with reference to a chemical agent. { ¦kē·mō'tak·səs }

chemotaxonomy [BOT] The classification of plants based on natural products. { ¦kē·mō,tak'sän·ə·mē }

chemotherapeutic [PHARM] Any agent used for chemotherapy. { ¦kē·mō,ther·ə,pyüd·ik }

chemotherapeutic index [PHARM] The relationship between toxicity of a compound for the body and the toxicity for parasites. { ¦kē·mō,ther·ə,pyüd·ik 'in,deks }

chemotherapy [MED] Administering chemical substances for treatment of disease, especially cancer and diseases caused by parasites. { ¦kē·mō'ther·ə·pē }

chemotropism [BIOL] Orientation response of a sessile organism with reference to chemical stimuli. { ¦kē·mō 'trō,piz·əm }

chemotypes [BOT] Plants of the same species that are chemically different but otherwise indistinguishable. { kē·mə,tīps }

Chemungian [GEOL] Middle Upper Devonian geologic time, below Cassodagan. { ke'mən·jē·ən }

chemurgy [CHEM ENG] A branch of chemistry concerned with the profitable utilization of organic raw materials, especially agricultural products, for nonfood purposes such as for paints and varnishes. { 'ke·mər·jē }

chenevixite [MINERAL] $Cu_2Fe_2(AsO_4)_2(OH)_4·H_2O$ A dark-green to greenish-yellow mineral consisting of a hydrous copper iron arsenate occurring in masses. { ¦shen¦ə¦vik,sīt }

chenic acid See chenodeoxycholic acid. { 'kēn·ik 'as·əd }

chenier [GEOL] A continuous ridge of beach material built upon swampy deposits; often supports trees, such as pines or evergreen oaks. { 'shen·yā }

chenille [TEXT] **1.** A wool, cotton, silk, rayon, or synthetic yarn with pile protrudng all around. **2.** A fabric woven from chenille yarn. **3.** A fabric with a pile made by weaving a cloth with warp threads around soft filling threads and then cutting. { shə'nēl }

chenodeoxycholic acid [BIOCHEM] $C_{24}H_{40}O_4$ A constituent of bile; needlelike crystals with a melting point of 119°C; soluble in alcohol, methanol, and acetic acid; used on an experimental basis to prevent and dissolve gallstones. Also known as anthropodesoxycholic acid; chenic acid; gallodesoxycholic acid. { ¦kē·nō,dē,äk·sē¦kō·lik 'as·əd }

Chenopodiaceae [BOT] A family of dicotyledonous plants in the order Caryophyllales having reduced, mostly greenish flowers. { ¦kē·nə,pō·dē'ās·ē,ē }

chenopodium oil [MATER] An alcohol-soluble, colorless to yellow oil, derived from the herb *Chenopodium ambrosioides;* chief constituents are *para*-cymeme, *l*-limonene, and ascaridole; used in medicine. Also known as American wormseed oil; goosefoot oil. { ¦kē·nə'pōd·ē·əm ,oil }

chergui [METEOROL] An eastern or southeastern desert wind in Morocco (North Africa), especially in the north; it is persistent, very dry and dusty, hot in summer, cold in winter. { 'chər·gwē }

Chermidae [INV ZOO] A small family of minute homopteran insects in the superfamily Aphidoidea. { 'kər·mə·dē }

Cherminae [INV ZOO] A subfamily of homopteran insects in the family Chermidae; all forms have a beak and an open digestive tract. { 'kər·mə·nē }

Chernozem [GEOL] One of the major groups of zonal soils, developed typically in temperate to cool, subhumid climate; the Chernozem soils in modern classification include Borolls, Ustolls, Udolls, and Xerolls. Also spelled Tchernozem. { ¦chər·nəz¦yóm }

cherophobia [PSYCH] An abnormal fear of happiness. { ¦kər·ə'fō·bē·ə }

cherry [BOT] **1.** Any trees or shrub of the genus *Prunus* in the order Rosales. **2.** The simple, fleshy, edible drupe or stone fruit of the plant. { 'cher·ē }

Cherry-Burrell process [FOOD ENG] A butter manufacturing process in which high fat cream is pasteurized, cooled, and collected in vats, where color and salt are added and the fat content reduced to 80; it is pumped continuously through a chiller-agitator and a texturator, from which it emerges as butter in a continuous ribbon. { ¦cher·ē ¦bər·əl ,präs·əs }

cherry leaf spot [PL PATH] A fungus disease of the cherry caused by *Coccomyces hiemalis;* spotting and chlorosis of the leaves occurs, with consequent retardation of tree and fruit development. { 'cher·ē ,lēf ,spät }

cherry picker [AERO ENG] A crane used to remove the aerospace capsule containing astronauts from the top of the rocket in the event of a malfunction. [MECH ENG] Any of several small traveling cranes, especially one used to hoist a passenger on the end of a boom. [MIN ENG] A small hoist used to facilitate car changing near the loader in a mine tunnel. { 'cher·ē ,pik·ər }

chersophyte [ECOL] A plant that grows in dry waste lands. { 'kərz·ə,fīt }

chert [PETR] A hard, dense, sedimentary rock composed of fine-grained silica, characterized by a semivitreous to dull luster and a splintery to conchoidal fracture; commonly gray, black, reddish brown, or green. Also known as hornstone; phthanite. { chərt }

chertification [GEOL] A process of replacement by silica in limestone in the form of fine-grained quartz or chalcedony. { ¦chərd·ə·fə'ka·shən }

chessylite See azurite. { 'shes·ə,līt }

chest breadth [ANTHRO] The measurement across the chest at nipple level. { 'chest ,bredth }

CHERT

Bedded chert, Monterery formation (Miocene), California. (*Photograph by M. N. Broamlette, USGS*)

CHESTNUT

Twig, leaf, and bud of American chestnut (*Castanea dentata*).

CHEVALIER LENS

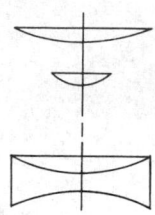

Chevalier magnifying lens.

CHEVROTAIN

Indian chevrotain, distinguished by white spots on coat.

chest circumference [ANTHRO] The horizontal circumference taken just above the nipples during a period of quiet breathing. { 'chest sər'kəm·frəns }

chest depth [ANTHRO] A measurement of the chest taken front to back from the sternum to the spinal groove. { 'chest ,depth }

Chesterian [GEOL] Upper Mississippian geologic time. { che'stir·ē·ən }

chestnut [BOT] The common name for several species of large, deciduous trees of the genus *Castanea* in the order Fagales, which bear sweet, edible nuts. { 'ches,nət }

chestnut blight [PL PATH] A fungus disease of the chestnut caused by *Endothia parasitica*, which attacks the bark and cambium, causing cankers that girdle the stem and kill the plant. Also known as chestnut canker. { 'ches,nət ,blīt }

chestnut canker See chestnut blight. { 'ches,nət 'kaŋ·kər }

chestnut coal [GEOL] Anthracite coal small enough to pass through a round mesh of 1 5/8 inches (3.1 centimeters) but too large to pass through a round mesh of 13/16 inches (1.7 centimeters). { 'ches,nət ,kōl }

Chestnut soil [GEOL] One of the major groups of zonal soils, developed typically in temperate to cool, subhumid to semiarid climate; the Chestnut soils in modern classification include Ustolls, Borolls, and Xerolls. { 'ches,nət ,sȯil }

Chevalier lens [OPTICS] A type of magnifying lens composed of an achromatic negative lens combined with a distant collecting front lens; a magnifying power up to 10X with an object distance up to 3 inches (7.62 centimeters) can be obtained. { shə'val·yā ,lenz }

cheviot [TEXT] **1.** A twill fabric similar to serge but having a slightly rough, napped surface; originally made of wool from sheep of the Cheviot Hills between England and Scotland. **2.** Striped or checked cotton shirting made of coarse yarns, woven to resemble woolen cheviot. { 'shev·ē·ət }

chevkinite [MINERAL] $(Fe,Ca)(Ce,La)_2(Si,Ti)_2O_8$ A mineral consisting of silicotitanate of iron, calcium, and rare-earth elements. { 'chef·kə,nīt }

Chevrel phase [SOLID STATE] One of a series of ternary molybdenum chalcogenide compounds with unusual superconducting properties and the general formula $M_xMo_6X_8$, where M represents any one of a large number of metallic elements, x has values between 1 and 4, and X is a chalcogen (sulfur, selenium, or tellurium). { she'vrel ,fāz }

chevron [VERT ZOO] The bone forming the hemal arch of a caudal vertebra. { 'shev·rən }

chevron fold [GEOL] An accordionlike fold with limbs of equal length. { 'shev·rən ,fōld }

chevrotain [VERT ZOO] The common name for four species of mammals constituting the family Tragulidae in the order Artiodactyla. Also known as mouse deer. { 'shev·rə,tān }

Cheyne-Stokes respiration [MED] Breathing characterized by periods of hyperpnea alternating with periods of apnea; rhythmic waxing and waning of respiration; occurs most commonly in older patients with heart failure and cerebrovascular disease. { 'chān·ē 'stōks res·pə'rā·shən }

Chézy formula [FL MECH] For the velocity V of open-channel flow which is steady and uniform, $V = \sqrt{8g/f} \cdot \sqrt{mS}$, where f is the Darcy-Weisbach friction coefficient, m the hydraulic radius, S the energy dissipation per unit length, and g the acceleration of gravity. { 'shā·zē ,fȯr·myə·lə }

chiasma [ANAT] A cross-shaped point of intersection of two parts, especially of the optic nerves. [CYTOL] The point of junction and fusion between paired chromatids or chromosomes, first seen during diplotene of meiosis. { kī'az·mə }

Chiasmodontidae [VERT ZOO] A family of deep-sea fishes in the order Perciformes. { kī,az·mə'dän·tə,dē }

chiastolite [MINERAL] A variety of andalusite whose crystals have a cross-shaped appearance in cross section due to the arrangement of carbonaceous impurities. Also known as macle. { kī'as·tə,līt }

chibli See ghibli. { 'chib·lē }

Chicago boom [MECH ENG] A hoisting device that is supported on the structure being erected. { shə'kä·gō ,büm }

Chicago caisson [CIV ENG] A cofferdam about 4 feet (1.2 meters) in diameter lined with planks and sunk in medium-stiff clays to hard ground for pier foundations. Also known as open-well caisson. { shə'kä·gō 'kā,sän }

chichili See chili. { chi'chil·ē }

chicken [VERT ZOO] *Galus galus.* The common domestic fowl belonging to the order Galliformes. { 'chik·ən }

chickenpox [MED] A mild, highly infectious viral disease of humans caused by a herpesvirus and characterized by vesicular rash. Also known as varicella. { 'chik·ən,päks }

chick unit [BIOL] A unit for the standardization of pantothenic acid. { 'chick ,yü·nət }

chicle [MATER] A gummy exudate obtained from the bark of *Achras zapota*, an evergreen tree belonging to the sapodilla family (Sapotaceae); used as the principal ingredient of chewing gum. { 'chik·əl }

chicory [BOT] *Cichorium intybus.* A perennial herb of the order Campanulales grown for its edible green leaves. { 'chik·ə·rē }

Chideruan [GEOL] Uppermost Permian geologic time. { chi'der·ə·wən }

chief cell [HISTOL] **1.** A parenchymal, secretory cell of the parathyroid gland. **2.** A cell in the lumen of the gastric fundic glands. { ¦chēf 'sel }

chief ray [OPTICS] A ray in a pencil that passes through the intersection of the axis of an optical system with the plane of the aperture stop. { ¦chēf 'rā }

chi element [GEN] Any of the special sites in bacterial deoxyribonucleic acid near which enhancement of genetic recombination occurs. { 'kī ,el·ə·mənt }

chiffon [TEXT] A very thin transparent fabric in plain weave. { shi'fän }

chigger [INV ZOO] The common name for bloodsucking larval mites of the Trombiculidae which parasitize vertebrates. { 'chig·ər }

chilarium [INV ZOO] One of a pair of processes between the bases of the fourth pair of walking legs in the king crab. { kī'lar·ē·əm }

child [COMPUT SCI] **1.** An element that follows a given element in a data structure. **2.** In object-oriented programming, a subclass. { chīld }

childhood aphasia [PSYCH] Loss or impairment of the use of language in a child. { 'chīld,hůd ə'fā·zhə }

childhood disintegrative disorder [PSYCH] A condition occurring in 3- and 4-year-old children that is characterized by unequivocally normal development in the first several years of life, followed by a marked developmental regression (a child who previously had been speaking in sentences becomes totally mute), and various autistic features develop. Also known as Heller's syndrome. { ¦chīld,hůd dis'in·tə,grād·iv dis,ȯrd·ər }

Child-Langmuir equation See Child's law.

Child-Langmuir-Schottky equation See Child's law. { ¦chīld ¦laŋ·myür 'shät,kē i'kwā·zhən }

child process [COMPUT SCI] One of the subsidiary processes that branches out from the root task in the fork-join model of programming on parallel machines. { ¦chīld ,präs·es }

childrenite [MINERAL] $(Fe,Mn)AlPO_4(OH)_2 \cdot H_2O$ A pale-yellowish to dark-brown orthorhombic mineral consisting of a hydrous basic iron aluminum phosphate occurring as translucent crystals; it is isomorphous with eosphorite; hardness is 4.5-5 on Mohs scale, and specific gravity is 3.18–3.24. { 'chil·drə,nīt }

Child's law [ELECTR] A law stating that the current in a thermionic diode varies directly with the three-halves power of anode voltage and inversely with the square of the distance between the electrodes, provided the operating conditions are such that the current is limited only by the space charge. Also known as Child-Langmuir equation; Child-Langmuir-Schottky equation; Langmuir-Child equation. { 'chīldz ,lȯ }

Chile mill [MECH ENG] A crushing mill having vertical rollers running in a circular enclosure with a stone or iron base or die. Also known as edge runner. { 'chil·ē ,mil }

Chile niter See Chile saltpeter. { ¦chil·ē 'nīd·ər }

Chile saltpeter [MINERAL] Also known as Chile niter. **1.** Soda niter found in large quantities in caliche in arid regions of northern Chile. **2.** Deposits of sodium nitrate. { ¦chil·ē ,sȯlt'pēd·ər }

chili [METEOROL] A warm, dry, descending wind in Tunisia, resembling the sirocco; in southern Algeria it is called chichili. { 'chil·ē }

chill [MET] **1.** A metal plate inserted in the surface of a sand mold or placed in the mold cavity to rapidly cool and solidify the casting, producing a hard surface. **2.** White or mottled iron occurring on the surface of a rapidly cooled gray iron casting. { chil }

chill-block melt spinning [MET] A rapid quenching process in which a jet of molten metal is directed onto a cold moving surface, such as a spinning disk, where the jet is shaped and solidified; quench rates are 1000 to 1,000,000 K per second. { 'chil ,bläk 'melt ,spin·iŋ }

chilled contact [PETR] The finer-grained portion of an igneous rock found near its contact with older rock. { 'child 'kän,takt }

chilled iron [MET] Cast iron made in iron- or steel-faced molds so the surface of the casting cools rapidly, retaining most of the carbon and becoming white and hard. { 'child 'ī·ərn }

chilled roll [MET] A roll consisting of an outer hard layer of white (chilled) iron with a middle transitional layer of mottled iron and a core of full gray iron. { 'child 'rōl }

chilled shot [MET] Lead shots containing 3–6% antimony. { 'child 'shät }

chiller [CHEM ENG] Oil-refining apparatus in which the temperature of paraffin distillates is lowered preparatory to filtering out the solid wax components. { 'chil·ər }

chilling [MET] Rapidly removing the heat from a casting. { 'chil·iŋ }

chill roll [ENG] A cored roll used in chill-roll extrusion of plastics. { 'chil ,rōl }

chill-roll extrusion [ENG] Method of extruding plastic film in which the film is cooled while being drawn around two or more highly polished chill rolls, inside of which there is cooling water. Also known as cast-film extrusion. { 'chil ,rōl ek'strü·zhən }

chill wind factor [METEOROL] An arbitrary index, developed by the Canadian Army, to correlate the performance of equipment and personnel in an Arctic winter; it is equal to the sum of the wind speed in miles per hour and the negative of the Fahrenheit temperature; the term is not to be confused with wind chill. { 'chil ,wind ,fak·tər }

Chilobolbinidae [PALEON] A family of extinct ostracods in the superfamily Hollinacea showing dimorphism of the velar structure. { ,kī·lə,bäl'bīn·ə,dē }

Chilopoda [INV ZOO] The centipedes, a class of the Myriapoda that is exclusively carnivorous and predatory. { kī'läp·ə·də }

Chimaeridae [VERT ZOO] A family of the order Chimaeriformes. { kī'mir·ə,dē }

Chimaeriformes [VERT ZOO] The single order of the chondrichthyan subclass Holocephali comprising the ratfishes, marine bottom-feeders of the Atlantic and Pacific oceans. { kī·mir·ə'fòr,mēz }

chimera [BIOL] An organism or a part made up of tissues or cells exhibiting chimerism. { kī'mir·ə }

chimeric deoxyribonucleic acid [GEN] A recombinant deoxyribonucleic acid (DNA) molecule that contains sequences from more than one organism. [MOL BIO] A deoxyribonucleic acid (DNA) molecule that has resulted from recombination or from the splicing together of DNA from two sources. { kī'mir·ik ,dē,äk·se,rī·bō,nü'klē·ik 'as·əd }

chimerism [BIOL] The admixture of cell populations from more than one zygote. { 'kī·mə,riz·əm }

chi meson [PARTIC PHYS] A meson resonance of mass 958 MeV/c^2, designated χ_0, which has 0 isospin and charge, negative parity, positive G parity, and spin probably equal to 0. Also known as eta-prime meson (η'). Also denoted η'_A (958). { ,kī 'mä,zän }

chimney [BUILD] A vertical, hollow structure of masonry, steel, or concrete, built to convey gaseous products of combustion from a building. [ELECTR] A pipelike enclosure that is placed over a heat sink to improve natural upward convection of heat and thereby increase the dissipating ability of the sink. [GEOL] See pipe; spouting horn. { 'chim,nē }

chimney apron [BUILD] A flashing made of a nonferrous metal, such as copper, that is built into the masonry of the chimney and the roofing material at the place where the roof is penetrated by the chimney. { 'chim·nē ,ā·prən }

chimney bar [BUILD] A wrought-iron or steel lintel which is supported by the sidewalls and carries the masonry above the fireplace opening. Also known as turning bar. { 'chim,nē ,bär }

chimney cap [ARCH] A cornice forming the uppermost portion of a chimney. [CIV ENG] A rotary device fitted to a chimney and moved by the wind so that the chimney is turned away from the wind to permit the escape of smoke while rain or snow is prevented from entering the chimney. { 'chim·nē ,kap }

chimney cloud [METEOROL] A cumulus cloud in the tropics that has much greater vertical than horizontal extent. { 'chim,nē ,klaùd }

chimney core [MECH ENG] The inner section of a double-walled chimney which is separated from the outer section by an air space. { 'chim,nē ,kòr }

chimney effect [FL MECH] The tendency of air or gas in a vertical passage to rise when it is heated because its density is lower than that of the surrounding air or gas. { 'chim,nē i'fekt }

chimney rock [GEOL] **1.** A chimney-shaped remnant of a rock cliff whose sides have been cut into and carried away by waves and the gravel beach. **2.** A rock column rising above its surroundings. [MATER] A porous phosphate rock used principally in chimney construction. { 'chim,nē ,räk }

chimopelagic [ECOL] Pertaining to, belonging to, or being marine organisms living at great depths throughout most of the year; during the winter they move to the surface. { ,kī·mō·pə'laj·ik }

chimpanzee [VERT ZOO] Either of two species of Primates of the genus *Pan* indigenous to central-west Africa. { ,chim,pan'zē }

chin [ANAT] The lower part of the face, at or near the symphysis of the lower jaw. { chin }

china clay [MATER] A high-grade white kaolin composed principally of the mineral kaolinite, and often occurring as a lenticular-shaped body; used in the manufacture of ceramics, paper, rubber, catalysts, and ink. { 'chī·nə ,klā }

China grass See ramie. { 'chī·nə ,gras }

chinaldine See quinaldine. { ki'näl,dēn }

China oil See Peru balsam. { 'chī·nə ,òil }

China syndrome See melt-through. { 'chī·nə ,sin,drōm }

China wood oil See tung oil. { 'chī·nə ,wùd ,òil }

chin breadth [ANTHRO] Contact measurement of the maximum width of the chin, taken between the points of intersection of the mandible and the menton. { chin ,bredth }

chinchilla [VERT ZOO] The common name for two species of rodents in the genus *Chinchilla* belonging to the family Chinchillidae. { chin'chil·ə }

chinchilla cloth [TEXT] A wool or other fabric that has a long nap or is tufted or nubbed. { chin'chil·ə ,klòth }

Chinchillidae [VERT ZOO] A family of rodents comprising the chinchillas and viscachas. { chin'chil·ə,dē }

chine [FOOD ENG] A part of an animal carcass backbone, with parts adjoining it, cut for cooking. [NAV ARCH] **1.** A part of a ship's waterway rising above the deck and hollowed out on the inboard edge to make a path for water. **2.** That point where the sides and bottom meet on a flat- or V-bottomed boat. **3.** A longitudinal member that lies along the bilge where the sides and bottom of the boat meet. { chīn }

Chinese bean oil See soybean oil. { chī'nēz 'bēn ,òil }

Chinese ink See india ink. { chī'nēz 'iŋk }

Chinese oil See Peru balsam. { chī'nēz 'òil }

Chinese oil of cinnamon See cassia oil. { chī'nēz ,òil əv 'sin·ə·mən }

Chinese remainder theorem [MATH] The theorem that if the integers $m_1, m_2, \ldots, m_n$ are relatively prime in pairs and if $b_1, b_2, \ldots, b_n$ are integers, then there exists an integer that is congruent to b_i modulo m_i for $i = 1, 2, \ldots, n$. { chī,nēz ri'mān·dər ,thir·əm }

Chinese vermilion See mercuric sulfide. { chī'nēz vər 'mil·yən }

Chinese wax [MATER] A white or yellowish crystalline wax formed on certain trees by the secretions of a scale insect, especially *Ceroplastes ceriferus*. { chī'nēz 'waks }

Chinese white [CHEM] A term used in the paint industry for zinc oxide and kaolin used as a white pigment. Also known as zinc white. { chī'nēz 'wīt }

chinic acid See quinic acid. { 'kin·ik 'as·əd }

chin-neck projection [ANTHRO] Measurement from the tip of the thyroid cartilage to the midpoint of the menton. { 'chin 'nek prə,jek·shən }

chino [TEXT] A twilled cotton fabric made of combed two-ply yarn. { 'chē·nō }

chinoidine See quinoidine. { ki'nòi,dēn }

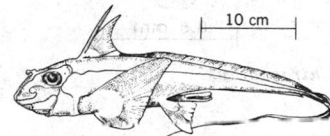

CHIMAERIFORMES

10 cm

Modern chimaeriform of genus *Chimaera*. (*From H. B. Bigelow and W. C. Schroeder, Fishes of the Western North Atlantic, pt. 2, Sears Foundation for Marine Research, Yale University, 1953*)

CHIPMUNK

The eastern chipmunk (*Tamias striatus*)

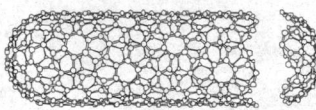

CHIRAL NANOTUBE

Model of a chiral nanotube.

CHIROGNATHIDAE

0.5 mm

Chirognathus.

chinone See quinone. { kin'ōn }

chinook [METEOROL] The foehn on the eastern side of the Rocky Mountains. { shə'nůk }

chinook arch [METEOROL] A foehn cloud formation appearing as a bank of clouds over the Rocky Mountains, generally a flat layer of altostratus, heralding the approach of a chinook. { shə'nůk 'ärch }

chinook salmon [VERT ZOO] *Oncorhynchus tshawytscha.* The Pacific's largest salmon, possibly exceeding 46 kilograms (100 pounds) at maturity, often spawns in tributaries located a considerable distance from the ocean. Also known as king salmon. { shə¦nů 'sa·mən }

chin projection [ANTHRO] Measurement of the distance from the gonion to the most forward point on the vertical midline of the menton. { 'chin prə'jek·shən }

chintz [TEXT] A glazed cotton fabric often printed with figures, birds, and florals. { chins }

chiolite [MINERAL] Na₅Al₃F₁₄ A snow white mineral resembling cryolite. Also known as arksutite. { 'kī·ə,līt }

Chionididae [VERT ZOO] The white sheathbills, a family of birds in the order Charadriiformes. { ,kī·ə'nid·ə,dē }

chionophile [ECOL] Having a preference for snow. { kī'än·ə,fīl }

chionophobia [PSYCH] An abnormal fear of snow. { ,kī·än·ə'fō·bē·ə }

chip [COMPUT SCI] See chad. [ELECTR] **1.** The shaped and processed semiconductor die that is mounted on a substrate to form a transistor, diode, or other semiconductor device. **2.** An integrated microcircuit performing a significant number of functions and constituting a subsystem. Also known as microchip. [GRAPHICS] A section of microfilm smaller in size than a microfiche. [MET] A section of metal that is removed as a workpiece is being machined. { chip }

chip blower [MED] A dental instrument used to blow drilling debris from a tooth cavity that is being prepared for filling. { 'chip ,blō·ər }

chipboard [MATER] A low-density paper board made from mixed waste paper and used where strength and quality are needed. { 'chip,bȯrd }

chip breaker [DES ENG] An irregularity or channel cut into the face of a lathe tool behind the cutting edge to cause removed stock to break into small chips or curls. { 'chip ,brāk·ər }

chip cap [DES ENG] A plate or cap on the upper part of the cutting iron of a carpenter's plane designed to give the tool rigidity and also to break up the wood shavings. { 'chip ,kap }

chip capacitor [ELECTR] A single-layer or multilayer monolithic capacitor constructed in chip form, with metallized terminations to facilitate direct bonding on hybrid integrated circuits. { 'chip kə'pas·əd·ər }

chip card See smart card. { 'chip ,kärd }

chip circuit See large-scale integrated circuit. { 'chip ,sər·kət }

chip log [ENG] A line, marked at intervals (commonly 50 feet or 15 meters), that is paid out over the stern of a moving ship and is pulled out by a drag (the chip), to determine the ship's speed. { 'chip ,läg }

chipmunk [VERT ZOO] The common name for 18 species of rodents belonging to the tribe Marmotini in the family Sciuridae. { 'chip,məŋk }

chipper [ENG] A tool such as a chipping hammer used for chipping. [MECH ENG] A machine with revolving knives for reducing large pieces of wood to chips. { 'chip·ər }

chipping [MET] Removing seams, surface defects, or other excess fragments from semifinished metal products by using a manual or pneumatic chisel or a continuous machine. { 'chip·iŋ }

chipping floor [ARCHEO] A workshop area characterized by the presence of debris remaining from the manufacture of chipped stone tools. { 'chip·iŋ ,flȯr }

chipping hammer [ENG] A hand or pneumatic hammer with chisel-shaped or pointed faces used to remove rust and scale from metal surfaces. { 'chip·iŋ ,ham·ər }

chip resistor [ELECTR] A thick-film resistor constructed in chip form, with metallized terminations to facilitate direct bonding on hybrid integrated circuits. { 'chip ri'zis·tər }

chip sampling [MIN ENG] Taking small pieces of ore or coal from the width of an ore face exposure; may be done at random or along a line. { 'chip ,sam·pliŋ }

chipset [COMPUT SCI] A number of integrated circuits, packaged as one unit, which perform one or more related functions. { 'chip,set }

chiral carbon atom See asymmetric carbon atom. { ¦kī·rəl ¦kär·bən 'ad·əm }

chiral center [ORG CHEM] An atom in a molecule that is attached to four different groups. { 'kī·rəl 'sen·tər }

chirality [CHEM] The handedness of an asymmetric molecule. [PARTIC PHYS] The characteristic of particles of spin ¹/₂ ℏ that are allowed to have only one spin state with respect to an axis of quantization parallel to the particle's momentum; if the particle's spin is always parallel to its momentum, it has positive chirality; antiparallel, negative chirality. [PHYS] The characteristic of an object that cannot be superimposed upon its mirror image. { kī'ral·əd·ē }

chiral molecules [CHEM] Molecules which are not superposable with their mirror images. { 'kī·rəl 'mäl·ə,kyülz }

chiral nanotube [PHYS CHEM] A carbon nanotube formed from a graphite sheet that is rolled up so that the succession of hexagons of carbon atoms on a particular cylinder makes an angle with the axis of the nanotube. { ,kī·rəl 'nan·ō,tüb }

chiral symmetry group [PARTIC PHYS] A group of symmetry transformations that act differently on the left- and right-handed parts of fermion fields. { 'kī·rəl 'sim·ə·trē ,grüp }

chiral twinning See optical twinning. { 'kī·rəl 'twin·iŋ }

Chireix antenna [ELECTROMAG] A phased array composed of two or more coplanar square loops, connected in series. Also known as Chireix-Mesny antenna. { ki'rāks an,ten·ə }

Chireix-Mesny antenna See Chireix antenna. { ki'rāks ,mez,nē an,ten·ə }

Chiridotidae [INV ZOO] A family of holothurians in the order Apodida having six-spoked, wheel-shaped spicules. { ,kī·rə'däd·ə,dē }

Chirodidae [PALEON] A family of extinct chondrostean fishes in the suborder Platysomoidei. { ,kī'räd·ə,dē }

Chirognathidae [PALEON] A family of conodonts in the suborder Neurodontiformes. { ,kī·rəg'näth·ə,dē }

chiromegaly See cheiromegaly. { ,kī·rə'meg·ə·lē }

Chiron [ASTRON] An object circling the sun in an eccentric orbit which takes it from inside the orbit of Saturn out to near the orbit of Uranus, and which has a period of 50.7 years. { 'kī,rän }

chiroplasty See cheiroplasty. { 'kī·rə,plas·tē }

chiropodist [MED] One who treats minor ailments of the feet. Also known as podiatrist. { kə'räp·əd·əst }

chiropractic [MED] A system of therapeutics based upon the theory that disease is caused by abnormal function of the nervous system; attempts to restore normal function are made through manipulation and treatment of the structures of the body, especially those of the spinal column. { 'kī·rə,prak·tik }

chiropractor [MED] One who practices the chiropractic arts. { 'kī·rə,prak·tər }

Chiroptera [VERT ZOO] The bats, an order of mammals having the front limbs modified as wings. { kī'räp·tə·rə }

chiropterophilous [BIOL] Pollinated by bats. { kī¦räf·ə·ləs }

chiropterygium [VERT ZOO] A typical vertebrate limb, thought to have evolved from a finlike appendage. { ¦kī·räp·tə¦rij·ē·əm }

Chirotheuthidae [INV ZOO] A family of mollusks comprising several deep-sea species of squids. { ,kī·rō'thyüd·ə,dē }

chirp [COMMUN] **1.** An undesirable variation in the frequency of a continuous-wave carrier when it is keyed. **2.** The sound heard in a code receiver when the transmitted carrier frequency is increased linearly for the duration of a pulse code. { chərp }

chirplet [MATH] A wavelet whose instantaneous frequency drifts upward or downward at a fixed rate throughout its duration. { 'chərp·lət }

chirp modulation [COMMUN] A type of modulation in which the frequency of each of a series of pulses is linearly varied in a systematic way. { ¦chərp ,mäj·ə'lā·shən }

chirp radar [ENG] Radar in which a swept-frequency signal is transmitted, received from a target, then compressed in time to give a narrow pulse called the chirp signal. { 'chərp ,rā,där }

chisel [AGR] A strong, heavy tool with curved points used for tilling; drawn by a tractor, it stirs the soil at an appreciable depth without turning it. [DES ENG] A tool for working the surface of various materials, consisting of a metal bar with a sharp edge at one end and often driven by a mallet. { 'chiz·əl }

Chisel *See* Caelum. { 'chiz·əl }

chisel bit *See* chopping bit. { 'chiz·əl ,bit }

chisel bond [ENG] A thermocompression bond in which a contact wire is attached to a contact pad on a semiconductor chip by applying pressure with a chisel-shaped tool. { 'chiz·əl ,bänd }

chisel-edge angle [DES ENG] The angle included between the chisel edge and the cutting edge, as seen from the end of the drill. Also known as web angle. { 'chiz·əl ,ej 'aŋ·gəl }

chisel-tooth saw [DES ENG] A circular saw with chisel-shaped cutting edges. { 'chiz·əl 'tüth ,sȯ }

chi-square distribution [STAT] The distribution of the sum of the squares of a set of variables, each of which has a normal distribution and is expressed in standardized units. { 'kī ¦skwer dis·trə'byü·shən }

chi-square statistic [STAT] A statistic which is distributed approximately in the form of a chi-square distribution; used in goodness-of-fit. { 'kī ¦skwār stə,tis·tik }

chi-square test [STAT] A generalization, and an extension, of a test for significant differences between a binomial population and a multinomial population, wherein each observation may fall into one of several classes and furnishes a comparison among several samples instead of just two. { 'kī ¦skwer 'test }

chitin [BIOCHEM] A white or colorless amorphous polysaccharide that forms a base for the hard outer integuments of crustaceans, insects, and other invertebrates. { 'kīt·ən }

chitinase [BIOCHEM] An externally secreted digestive enzyme produced by certain microorganisms and invertebrates that hydrolyzes chitin. { 'kīt·ən,ās }

chitinivorous bacterium [MICROBIO] Any bacterium which secretes chitinase and can digest chitin; organisms extract chitin from lobster exoskeletons, causing an infection called soft-shell disease. { ,kit·ən'iv·ə·rəs bak'tir·ē·əm }

Chitinozoa [PALEON] An extinct group of unicellular microfossils of the kingdom Protista. { ¦kīt·ən·ə¦zō·ə }

chiton [INV ZOO] The common name for over 600 extant species of mollusks which are members of the class Polyplacophora. { 'kīt·ən }

Chitral fever *See* phlebotomus fever. { 'chi·trəl ,fe·vər }

chiviatite [MINERAL] Pb$_2$Bi$_2$S$_{11}$ A lead-gray mineral consisting of a lead bismuth sulfide occurring in foliated masses. { ,chiv·ē'ä,tīt }

Chladni's figures [MECH] Figures produced by sprinkling sand or similar material on a horizontal plate and then vibrating the plate while holding it rigid at its center or along its periphery; indicate the nodal lines of vibration. { 'klad,nēz ,fig·yərz }

chlamydeous [BOT] **1.** Pertaining to the floral envelope. **2.** Having a perianth. { klə'mid·ē·əs }

Chlamydia [MICROBIO] The single genus of the family Chlamydiaceae. { klə'mid·ē·ə }

Chlamydiaceae [MICROBIO] The single family of the order Chlamydiales; characterized by a developmental cycle from a small elementary body to a larger initial body which divides, with daughter cells becoming elementary bodies. { klə,mid·ē'ās·ē,ē }

Chlamydiales [MICROBIO] An order of coccoid rickettsias; gram-negative, obligate, intracellular parasites of vertebrates. { klə,mid·ē'ā·lēz }

Chlamydobacteriaceae [MICROBIO] Formerly a family of gram-negative bacteria in the order Chlamydobacteriales possessing trichomes in which false branching may occur. { ¦klam·ə,dō,bak·tir·ē'ās·ē,ē }

Chlamydobacteriales [MICROBIO] Formerly an order comprising colorless, gram-negative, algae-like bacteria of the class Schizomycetes. { ¦klam·ə,dō,bak·tir·ē'ā·lēz }

Chlamydomonadidae [INV ZOO] A family of colorless, flagellated protozoans in the order Volvocida considered to be close relatives of protozoans that possess chloroplasts. { ¦klam·ə,dō·mə'näd·ə,dē }

Chlamydoselachidae [VERT ZOO] The frilled sharks, a family of rare deep-water forms having a combination of hybodont and modern shark characteristics. { ¦klam·ə,dō·se'lak·ə,dē }

chlamydospore [MYCOL] A thick-walled, unicellar resting spore developed from vegetative hyphae in almost all parasitic fungi. { 'klam·ə,spȯr }

Chlamydozoaceae [MICROBIO] A family of small, gram-negative, coccoid bacteria in the order Rickettsiales; members

are obligate intracytoplasmic parasites, or saprophytes. { ¦klam·ə,dō·zō'ās·ē,ē }

chloanthite [MINERAL] NiAs$_{2-3}$ A white or gray mineral with metallic luster forming crystals in the isometric system; it is isomorphous with nickel-skutterudite. { klō'an,thīt }

chloasma [MED] Patchy tan, brown, and black hyperpigmentation, especially on the brow and cheek; of unknown cause, but may be due to the action of sunshine upon perfume or to endocrinopathy. { klō'az·mə }

chloflurecol methyl ester [ORG CHEM] C$_{15}$H$_{11}$ClO$_3$ A white, crystalline compound with a melting point of 152°C; slight solubility in water; used as a growth regulator for grass and weeds. { klō'flür·ə,kȯl 'meth·əl 'es·tər }

Chloracea [MICROBIO] The green sulfur bacteria, a family of photosynthetic bacteria in the suborder Rhodobacteriineae. { klȯr'ās·ē·ə }

chloracne [MED] An acnelike eruption caused by chlorinated hydrocarbons. { klȯr'ak·nē }

chloragogen [INV ZOO] Of or pertaining to certain specialized cells forming the outer layer of the alimentary tract in earthworms and other annelids. { ¦klȯr·ə¦gō·jən }

chloral [ORG CHEM] CCl$_3$CHO A colorless, oily liquid soluble in water; used industrially to prepare DDT; a hypnotic. Also known as trichloroacetic aldehyde; trichloroethanal. { 'klȯr·əl }

chloralase [ORG CHEM] C$_8$H$_{11}$Cl$_3$O$_6$ Colorless, water-soluble crystals, melting at 185°C; made by heating chloral with dextrose; used as a hypnotic. { 'klȯr·ə,lās }

chloral hydrate [ORG CHEM] CCl$_3$CH(OH)$_2$ Colorless, deliquescent needles with slightly bitter caustic taste, soluble in water; a hypnotic. Also known as crystalline chloral; hydrated chloral. { 'klȯr·əl 'hī,drāt }

chloralkali [CHEM ENG] Either of the products of the industrial electrolysis of sodium chloride, that is, sodium hydroxide or chlorine. { klȯr'al·kə,lī }

chloralkali process [CHEM ENG] An industrial chemical process based on the electrolysis of sodium chloride for the production of sodium hydroxide and chlorine. { ,klȯr'al·kə,lī ,prä·səs }

chloralkane [ORG CHEM] Chlorinated aliphatic hydrocarbon of the methane series (C$_n$H$_{2n+2}$). { klȯr'al,kān }

chloralosane *See* chloralose. { klȯr·ə'lō,sān }

chloralose [ORG CHEM] C$_8$H$_{11}$O$_6$Cl$_3$ A crystalline compound with a melting point of 178°C; used as a repellent for birds. Also known as glucochloralose. { 'klȯr·ə,lōs }

α-chloralose [ORG CHEM] C$_8$H$_{11}$O$_6$Cl$_3$ Needlelike crystals with a melting point of 87°C; soluble in glacial acetic acid and ether; used on seed grains as a bird repellent and as a hypnotic for animals. Also known as chloralosane; glucochloral. { 'al·fə ¦klȯr·ə,lōs }

chloraluminite [MINERAL] AlCl$_3$·6H$_2$O A mineral consisting of hydrous aluminum chloride. { ¦klȯr·ə¦lüm·ə,nīt }

chloramine T [ORG CHEM] CH$_3$C$_6$H$_4$SO$_2$NClNa·3H$_2$O A white, crystalline powder that decomposes slowly in air, freeing chlorine; used as an antiseptic, a germicide, and an oxidizing agent and chlorinating agent. { 'klȯr·ə,mēn 'tē }

chloramphenicol [MICROBIO] C$_{11}$H$_{12}$O$_2$N$_2$Cl$_2$ A colorless, crystalline, broad-spectrum antibiotic produced by *Streptomyces venezuelae*; industrial production is by chemical synthesis. Also known as chloromycetin. { ,klȯr,am'fen·ə ,kȯl }

Chlorangiaceae [BOT] A primitive family of colonial green algae belonging to the Tetrasporales in which the cells are directly attached to each other. { ,klȯr,an·jē'ās·ē,ē }

chloranil [ORG CHEM] C$_6$Cl$_4$O$_2$ Yellow leaflets melting at 290°C; soluble in organic solvents; made from phenol by treatment with potassium chloride and hydrochloric acid; used as an agricultural fungicide and as an oxidizing agent in the manufacture of dyes. { klȯr'an·əl }

chloranilic acid [ORG CHEM] C$_6$H$_2$Cl$_2$O$_4$ A relatively strong dibasic acid whose crystals are red and melt between 283 and 284°C; used in spectrophotometry. { klȯr·ə'nil·ik 'as·əd }

chloranthy [BOT] A reverting of normally colored floral leaves or bracts to green foliage leaves. { 'klȯr,an·thē }

chlorapatite [MINERAL] Ca$_5$(PO$_4$)$_3$Cl An apatite mineral containing chlorine. { klȯr'ap·ə,tīt }

chlorargyrite *See* cerargyrite. { klȯr'ar·jə,rīt }

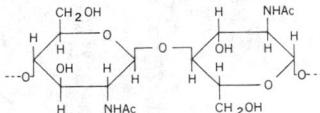

CHITIN

β-*N* acetyl-D-glucosamine unit of chitin.

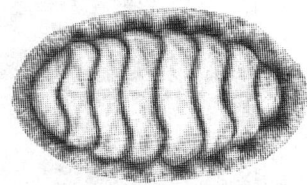

CHITON

Mossy chiton with characteristic shell of eight plates and surrounding girdle bearing spicules on upper surface.

chlorastrolite [MINERAL] A mottled, green variety of pumpellyite occurring as grains or small nodules of a stellate structure in basic igneous rock in the Lake Superior region; used as a semiprecious stone. { ¦klȯr'as·trə¸līt }

chlorate [INORG CHEM] ClO_3^- **1.** A negative ion derived from chloric acid. **2.** A salt of chloric acid. { 'klȯr¸āt }

chlorate candle [MATER] A mixture of solid chemical compounds which, when ignited, liberates free oxygen. { 'klȯr¸āt ¸kand·əl }

chlorate explosive [MATER] A type of explosive with a potassium chlorate base; a substitute for black powder in which potassium chlorate is used in place of potassium nitrate. Also known as chlorate powder. { 'klȯr¸āt ik'splō·siv }

chlorate powder See chlorate explosive. { 'klȯr¸āt ¸paùd·ər }

chlorbenside [ORG CHEM] $C_{13}H_{10}SCl_2$ White crystals with a melting point of 72°C; used as a miticide for spider mites on fruit trees and ornamentals. { klȯr'ben¸sīd }

chlorbromuron [ORG CHEM] $C_9H_{10}ONBrCl$ A white solid with a melting point of 94–96°C; used as a pre- and postemergence herbicide for annual grass and for broadleaf weeds on crops, soybeans, and Irish potatoes. { ¸klȯr·brə'myü·rən }

chlorcyclizine hydrochloride [PHARM] $ClC_6H_4CH(C_6H_5)$-$C_4H_8N_2CH_3$·HCl A white, crystalline solid with a melting point of 222–227°C; soluble in water, chloroform, and alcohol; used as an antihistamine. { klȯr'sī·klə¸zēn hī·drə 'klȯr¸īd }

chlordan See chlordane. { 'klȯr¸dan }

chlordane [ORG CHEM] $C_{10}H_6Cl_8$ A volatile liquid insecticide; a chlorinated hexahydromethanoindene. Also spelled chlordan. { 'klȯr¸dān }

chlordiazepoxide hydrochloride [PHARM] $C_{16}H_{14}ON_3Cl$ A white crystalline conpound, soluble in water; the hydrochloride salt is used as a tranquilizer. { ¦klȯr·dī¸az·ə'pak¸sīd ¸hī·drə'klȯr¸īd }

chlordimeform [ORG CHEM] $C_{10}H_{13}ClN_2$ A tan-colored solid, melting point 35°C; used as a miticide and insecticide for fruits, vegetables, and cotton. { ¸klȯr'dī·mə¸fȯrm }

chlorenchyma [BOT] Chlorophyll-containing tissue in parts of higher plants, as in leaves. { klȯr'eŋ·kə·mə }

chlorendic acid [ORG CHEM] $C_9H_4Cl_6O_4$ White, fine crystals used in fire-resistant polyester resins and as an intermediate for dyes, fungicides, and insecticides. { klȯr'en·dik 'as·əd }

chlorendic anhydride [ORG CHEM] $C_9H_2Cl_6O_3$ White, fine crystals used in fire-resistant polyester resins, in hardening epoxy resins, and as a chemical intermediate. { klȯr'en·dik an'hī¸drīd }

chlorfenethol [ORG CHEM] $C_{14}H_{12}Cl_2O$ A colorless, crystalline compound with a melting point of 69.5–70°C; insoluble in water; used for control of mites in ornamentals and shrub trees. { ¸klȯr'fen·ə¸thȯl }

chlorfenpropmethyl [ORG CHEM] $C_{10}H_{10}OCl_2$ A colorless to brown liquid used as a postemergence herbicide of wild oats, cereals, fodder beets, sugarbeets, and peas. { ¦klȯr·fən¸präp'meth·əl }

chlorfensulfide [ORG CHEM] $C_{12}H_6Cl_4N_2S$ A yellow, crystalline compound with a melting point of 123.5–124°C; used as a miticide for citrus. { ¸klȯr·fən'səl¸fīd }

chlorfenvinphos [ORG CHEM] $C_{12}H_{14}Cl_3O_4P$ An amber liquid with a boiling point of 168–170°C; used as an insecticide for ticks, flies, lice, and mites on cattle. { ¸klȯr·fən'vin¸fäs }

chlorhydrin See chlorohydrin. { klȯr'hī·drən }

chloric acid [INORG CHEM] $HClO_3$ A compound that exists only in solution and as chlorate salts; breaks down at 40°C. { 'klȯr·ik 'as·əd }

chloride [CHEM] **1.** A compound which is derived from hydrochloric acid and contains the chlorine atom in the −1 oxidation state. **2.** In general, any binary compound containing chloride. { 'klȯr¸īd }

chloride benzilate See lachesne. { 'klȯr¸īd 'ben·zə¸lāt }

chloride of lime See bleaching powder. { 'klȯr¸īd əv 'līm }

chloride paper [MATER] A paper made with an emulsion of silver chloride; usually used in photography as contact paper or very-slow-speed enlarging paper. { 'klȯr¸īd ¸pā·pər }

chloride shift [PHYSIO] The reversible exchange of chloride and bicarbonate ions between erythrocytes and plasma to effect transport of carbon dioxide and maintain ionic equilibrium during respiration. { 'klȯr¸īd ¸shift }

chloridization [CHEM] See chlorination. [MET] Treatment of mineral ores with hydrochloric acid or chlorine to form the chloride of the main metal present. { ¸klȯr·ə·də'zā·shən }

chlorimide See dichloramine. { 'klȯr·ə¸mīd }

chlorin [BIOCHEM] A saturated porphyrin for which one double bond at a single pyrrole ring has been reduced. { 'klȯr·ən }

chlorinated lime See bleaching powder. { 'klȯr·ə¸nād·əd 'līm }

chlorinated paraffin [ORG CHEM] One of a group of chlorine derivatives of paraffin compounds. { 'klȯr·ə¸nād·əd 'par·ə·fən }

chlorinated rubber [MATER] A nonrubbery, incombustible rubber derivative produced by the action of chlorine on rubber in solution; used in corrosion-resistant paints and varnishes, and in inks and adhesives. { 'klȯr·ə¸nād·əd 'rəb·ər }

chlorinated wool [TEXT] Wool treated with chlorine chemicals to minimize shrinkage and enhance dye penetration. { 'klȯr·ə¸nād·əd 'wùl }

chlorination [CHEM] **1.** Introduction of chlorine into a compound. Also known as chloridization. **2.** Water sterilization by chlorine gas. [TEXT] A process in which wool is treated with a solution of hypochlorite and an acid or similar mixture to reduce the tendency of the fiber to shrink by matting. { ¸klȯr·ə'nā·shən }

chlorinator [CHEM ENG] The apparatus used in chlorinating. { 'klȯr·ə¸nād·ər }

chlorine [CHEM] A chemical element, symbol Cl, atomic number 17, atomic weight 35.453; used in manufacture of solvents, insecticides, and many non-chlorine-containing compounds, and to bleach paper and pulp. { 'klȯr¸ēn }

chlorine-36 [NUCLEO] A radioactive isotope of chlorine with atomic mass number of 36; a beta emitter with a half-life of 3×10^5 years. { 'klȯr¸ēn ¸thərd-ē'siks }

chlorine dioxide [INORG CHEM] ClO_2 A green gas used to bleach cellulose and to treat water. { 'klȯr¸ēn dī'äk¸sīd }

chlorine log See chlorinolog. { 'klȯr¸ēn 'läg }

chlorine survey [PETRO ENG] A radioactivity-logging survey made inside the casing for measuring the relative chlorine content of the formation. { 'klȯr¸ēn ¸sər¸vā }

chlorine war gas [ORD] Chlorine gas packaged to be released against enemy troops; greenish yellow, toxic, and gaseous at normal temperatures and pressures. { 'klȯr¸ēn 'wȯr ¸gas }

chlorine water [CHEM] A clear, yellowish liquid used as a deodorizer, antiseptic, and disinfectant. { 'klȯr¸ēn ¸wȯd·ər }

chlorinity [OCEANOGR] A measure of the chloride and other halogen content, by mass, of sea water. { klə'rin·əd·ē }

chlorinolog [PETRO ENG] A record of the presence and concentration of chlorine in oil reservoirs, prepared as a method of locating salt-water strata. Also known as chlorine log. { klə'rin·ə¸läg }

chlorite [INORG CHEM] A salt of chlorous acid. [MINERAL] Any of a group of greenish, platyhydrous monoclinic silicates of aluminum, ferrous iron, and magnesium which are closely associated with and resemble the micas. { 'klȯr¸īt }

chlorite schist [PETR] A metamorphic rock whose composition is dominated by members of the chlorite group. { 'klȯr¸īt ¸shist }

chlorite-sericite schist [PETR] A low-grade, fine-grained variety of mica schist without biotite. { 'klȯr¸īt 'ser·ə¸sīt ¸shist }

chloritization [CHEM] The introduction of, production of, replacement by, or conversion into chlorite. { ¸klȯr·əd·ə'zā·shən }

chloritoid [MINERAL] $FeAl_4Si_2O_{10}(OH)_4$ A micaceous mineral related to the brittle mica group; has both monoclinic and triclinic modifications, a gray to green color, and weakly pleochroic crystals. { 'klȯr·ə¸toid }

chloritoid schist [PETR] A variety of mica schist whose composition is dominated by chloritoid. { 'klȯr·ə¸toid ¸shist }

chlormanganokalite [MINERAL] K_4MnCl_6 A wine yellow to lemon or canary yellow, hexagonal mineral consisting of potassium and manganese chloride; occurs as rhombohedrons. { ¸klȯr¦maŋ·gə¸nō'kā¸līt }

chlormephos [ORG CHEM] $C_5H_{12}O_2S_2ClP$ A liquid used as an insecticide for soil. { 'klȯr·mə¸fäs }

chloro- [ORG CHEM] A prefix describing an organic compound which contains chlorine atoms substituted for hydrogen. { 'klȯr·ō }

chloroacetic acid [ORG CHEM] $ClCH_2COOH$ White or colorless, deliquescent crystals that are soluble in water, ether, chloroform, benzene, and alcohol; used as an herbicide and in the manufacture of dyes and other organic molecules. { ¦klȯr·ə¦sēd·ik ˈas·əd }

chloroacetic anhydride [ORG CHEM] $C_4H_4Cl_2O_3$ Crystals with a melting point of 46°C; soluble in chloroform and ether; used in the preparation of cellulose chloracetates and in the *N*-acetylation of amino acids in alkaline solution. { ¦klȯr·ə¦sēd·ik anˈhīˌdrīd }

chloroacetone [ORG CHEM] CH_3COCH_2Cl Pungent, colorless liquid used as military tear gas and in organic synthesis. { klȯrˈas·əˌtōn }

chloroacetonitrile [ORG CHEM] $ClCH_2CN$ A colorless liquid with a pungent odor; soluble in hydrocarbons and alcohols; used as a fumigant. { ¦klȯr·ō¦as·ə·ˈtän·ə·trəl }

chloroacetophenone [ORG CHEM] $C_6H_5COCH_2Cl$ Rhombic crystals melting at 59°C; an intermediate in organic synthesis. { ¦klȯr·ō¦as·əˈtä·fəˌnōn }

chloroacrolein [ORG CHEM] $H_2C{:}ClCHO$ A colorless liquid with a boiling point of 29–31°C; used as a tear gas. { ¦klȯr·ō·əˈkrō·lē·ən }

Chlorobacteriaceae [MICROBIO] The equivalent name for Chlorobiaceae. { ¦klȯr·ōˌbak·tirˈē·ˈāsˌēˌē }

chlorobenzaldehyde [ORG CHEM] C_6H_4CHOCl A colorless to yellowish liquid (ortho form) or powder (para form) with a boiling range of 209–215°C; soluble in alcohol, ether, and acetone; used in dye manufacture. { ¦klȯr·ōˈben·zalˌdəˌhīd }

chlorobenzene [ORG CHEM] C_6H_5Cl A colorless, mobile, volatile liquid with an almondlike odor; used to produce phenol, DDT, and aniline. { ¦klȯr·ōˈbenˌzēn }

chlorobenzilate [ORG CHEM] $C_{16}H_{14}Cl_2O_3$ A yellow-brown, viscous liquid with a melting point of 35–37°C; used as a miticide in agriculture and horticulture. { ¦klȯr·ōˈbenzəˌlāt }

***para*-chlorobenzoic acid** [ORG CHEM] ClC_6H_4COOH A white powder with a melting point of 238°C; soluble in methanol, absolute alcohol, and ether; used in the manufacture of dyes, fungicides, and pharamaceuticals. { ¦par·ə ¦klȯr·ōˌben'zō·ik ˈas·əd }

chlorobenzoyl chloride [ORG CHEM] ClC_6H_4COCl A colorless liquid with a boiling range of 227–239°C; soluble in alcohol, acetone, and water; used in dye and pharmaceuticals manufacture. { ¦klȯr·ōˈbenˌzȯil ˈklȯrˌīd }

chlorobenzyl chloride [ORG CHEM] $ClC_6H_4CH_2Cl$ A colorless liquid with a boiling range of 216–222°C; soluble in acetone, alcohol, and ether; used in the manufacture of organic chemicals. { ¦klȯr·ōˈben·zil ˈklȯrˌīd }

Chlorobiaceae [MICROBIO] A family of bacteria in the suborder Chlorobiineae; cells are nonmotile and contain bacteriochlorophylls *c*, *d*, or *e* in chlorobium vesicles attached to the cytoplasmic membrane. { ¦klȯr·ōˌbīˈāsˌēˌē }

Chlorobiineae [MICROBIO] The green sulfur bacteria, a suborder of the order Rhodospirillales; contains the families Chlorobiaceae and Chloroflexaceae. { ¦klȯr·ōˌbīˈi·əˌnē }

Chlorobium [MICROBIO] A genus of bacteria in the family Chlorobiaceae; cells are ovoid, rod- or vibrio-shaped, and nonmotile, do not have gas vacuoles, contain bacteriochlorophyll *c* or *d*, and are free-living. { klȯrˈō·bē·əm }

chlorobium chlorophyll [BIOCHEM] $C_{51}H_{67}O_4N_4Mg$ Either of two spectral forms of chlorophyll occurs as esters of farnesol in certain (*Chlorobium*) photosynthetic bacteria. Also known as bacterioviridin. { klȯrˈō·bē·əm ˈklȯr·əˌfil }

chlorobromide paper [GRAPHICS] A paper with an emulsion composed of silver bromide and silver chloride; used in photography for fast-speed contact paper, and medium-speed enlarging paper. { ¦klȯr·ōˈbrōˌmīd ˌpā·pər }

chlorobutadiene *See* chloroprene. { ¦klȯr·ōˌbyüd·əˈdīˌēn }

chlorobutanol [ORG CHEM] $Cl_3CC(CH_3)_2OH$ Colorless to white crystals with a melting point of 78°C; soluble in alcohol, glycerol, ether, and chloroform; used as a plasticizer and a preservative for biological solutions. { ¦klȯr·ōˈbyüt·ənˌȯl }

chlorocalcite [MINERAL] $KCaCl_3$ A white mineral consisting of a chloride of potassium and calcium. Also known as hydrophilite. { ¦klȯr·ōˈkalˌsīt }

chlorocarbon [ORG CHEM] A compound of chlorine and carbon only, such as carbon tetrachloride, CCl_4. { ¦klȯr·ōˈkärˌbän }

chlorochromic anhydride *See* chromyl chloride. { ¦klȯr·ōˌkrō·mik anˈhīˌdrīd }

Chlorococcales [BOT] A large, highly diverse order of unicellular or colonial, mostly fresh-water green algae in the class Chlorophyceae. { ¦klȯr·ō·käˈkāˌlēz }

chlorocruorin [BIOCHEM] A green metalloprotein respiratory pigment found in body fluids or tissues of certain sessile marine annelids. { ¦klȯr·ōˈkrȯr·ən }

Chlorodendrineae [BOT] A suborder of colonial green algae in the order Volvocales comprising some genera with individuals capable of detachment and motility. { ¦klȯr·ōˌden'drin·ē·ē }

1,1,1-chlorodifluoroethane [ORG CHEM] CH_3CClF_2 A colorless gas with a boiling point of −130.8°C; used as a refrigerant, solvent, and aerosol propellant. { ¦wən ¦wən ¦wən ¦klȯr·ōˌdīˌflur·ōˈethˌän }

chlorodifluoromethane [ORG CHEM] $CHClF_2$ A colorless gas with a boiling point of −40.8°C and freezing point of −160°C; used as an aerosol propellant and refrigerant. { ¦klȯr·ōˌdīˌflur·ōˈmethˌän }

1-chloro-2,4-dinitrobenzene [ORG CHEM] $C_6H_3ClN_2O_4$ Yellow crystals with a melting point of 52–54°C; soluble in hot alcohol, ether, and benzene; used as a reagent in the determination of pyridine compounds such as nicotinic acid, and nicotinamide. { ¦wən ¦klȯr·ō ¦tü ¦fȯr dīˌnī·trōˈbenˌzēn }

chloroethane *See* ethyl chloride. { ¦klȯr·ōˈethˌän }

chloroethene *See* vinyl chloride. { ¦klȯr·ōˈethˌēn }

chloroethyl alcohol *See* ethylene chlorohydrin. { ¦klȯr·ōˈeth·əl ˈal·kəˌhȯl }

Chloroflexaceae [MICROBIO] A family of phototrophic bacteria in the suborder Chlorobiineae; cells possess chlorobium vesicles and bacteriochlorophyll *a* and *c*, are filamentous, and show gliding motility; capable of anaerobic, phototrophic growth or aerobic chemotrophic growth. { ¦klȯr·ōˈfleks'asˌēˌē }

Chloroflexus [MICROBIO] The single genus of the family Chloroflexaceae. { ¦klȯr·ōˈflekˌsēz }

chlorofluorocarbon [ORG CHEM] A compound consisting of chlorine, fluorine, and carbon; has the potential to destroy ozone in the stratosphere. Abbreviated CFC. Also known as fluorochlorocarbon (FCC). { ¦klȯr·əˌflur·əˈkär·bən }

chlorofluoromethane [ORG CHEM] A compound consisting of chlorine, fluorine, and carbon, has the potential to destroy ozone in the stratosphere. Also known as fluorochlorocarbon (FCC). Abbreviated CFC. { ¦klȯr·əˌflur·əˈmethˌän }

chloroform [ORG CHEM] $CHCl_3$ A colorless, sweet-smelling, nonflammable liquid; used at one time as an anesthetic. Also known as trichloromethane. { ˈklȯr·əˌfȯrm }

chlorogenic acid [BIOCHEM] $C_{16}H_{18}O_9$ An important factor in plant metabolism; isolated from green coffee beans; the hemihydrate crystallizes in needlelike crystals from water. { ¦klȯr·əˌjen·ik ˈas·əd }

chloroguanide hydrochloride [PHARM] $C_{11}H_{16}N_5Cl$ A very effective suppressive drug in low doses, against the three kinds of malaria. { ¦klȯr·əˈgwänˌīd ˌhī·drəˈklȯrˌīd }

chlorohydrin [ORG CHEM] Any of the compounds derived from a group of glycols or polyhydroxy alcohols by chlorine substitution for part of the hydroxyl groups. Also spelled chlorhydrin. { ¦klȯr·əˈhī·drən }

chlorohydrocarbon [ORG CHEM] A carbon- and hydrogen-containing compound with chlorine substituted for some hydrogen in the molecule. { ¦klȯr·ōˈhī·drəˌkär·bən }

chlorohydroquinone [ORG CHEM] $ClC_6H_3(OH)_2$ White to light tan crystals with a melting point of 100°C; soluble in water and alcohol; used as a photographic developer and bactericide and for dyestuffs. { ¦klȯr·ōˌhī·drəˈkwinˈȯn }

5-chloro-8-hydroxyquinoline [ORG CHEM] C_9H_6ClNO Crystals with a melting point of 130°C; used as a fungicide and bactericide. { ¦fīv ¦klȯr·ō ¦āt hīˌdräk·sēˈkwin·əˌlēn }

chloroma [MED] A focal tumorous proliferation of granulocytes, with or without the blood findings of granulocytic leukemia; the sectioned surfaces of the mass are green. { kləˈrō·mə }

chloromagnesite [MINERAL] $MgCl_2$ A mineral consisting of anhydrous magnesium chloride, found on the volcano Vesuvius. { ¦klȯr·ōˈmagˌnəˌsīt }

chloromethane [ORG CHEM] CH_3Cl A colorless, noncorrosive, liquefiable gas which condenses to a colorless liquid; used as a refrigerant, and as a catalyst carrier in manufacture of butyl rubber. Also known as methyl chloride. { ¦klȯr·ōˈmethˌän }

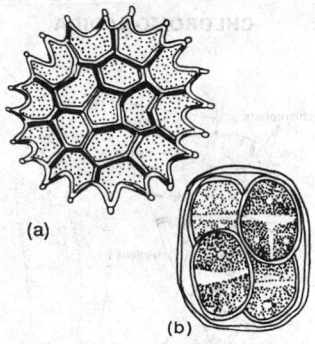

CHLOROCOCCALES

(a)

(b)

Representative colonial forms of chlorococcales. (*a*) *Pediastrum boryanum*. (*b*) *Oöcystis borgei*. (*From G. M. Smith, Fresh-water Algae of the United States, 2d ed., McGraw-Hill, 1950*)

CHLOROFORM

$$\begin{array}{c} Cl \\ | \\ Cl-C-H \\ | \\ Cl \end{array}$$

Structural formula of chloroform.

CHLOROGENIC ACID

Structural formula of chlorogenic acid.

CHLOROMONADIDA

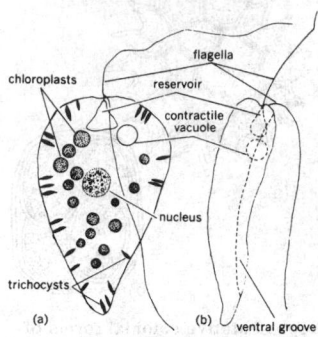

Gonyostomum semen. (a) Dorsal view. *(b)* Side view.

chloromethapyrilene citrate *See* chlorothen citrate. { ¦klȯr·ō‚meth·ə¹pī·rə‚lēn ¹sī‚trāt }

Chloromonadida [INV ZOO] An order of flattened, grass-green or colorless, flagellated protozoans of the class Phytamastigophorea. { ¦klȯr·ō·mə¹näd·ə·də }

Chloromonadina [INV ZOO] The equivalent name for Chloromonadida. { ¦klȯr·ō·mə¹näd·ə·nə }

Chloromonadophyceae [BOT] A group of algae considered by some to be a class of the division Chrysophyta. { ¦klȯr·ō·mə‚näd·ə¹fīs·ē‚ē }

Chloromonadophyta [BOT] A division of algae in the plant kingdom considered by some to be a class, Chloromondophyceae. { ¦klȯr·ō‚mä·nə¹dä·fə·də }

chloromycetin *See* chloramphenicol. { ‚klȯr·ō‚mī¹sēt·ən }

1-chloronaphthalene [ORG CHEM] C₁₀H₇Cl An oily liquid used as an immersion medium in the microscopic determination of refractive index of crystals and as a solvent for oils, fats, and DDT. { ¦wən ‚klȯr·ō¹naf·thə‚lēn }

chloronium ion [ORG CHEM] A halonium ion in which the halogen is chlorine; sometimes occurs as a bridged form. { klə¹rōn·ē·əm ¹ī·ən }

chloropal *See* nontronite. { ¹klȯr·ə‚pal }

***para*-chlorophenol** [PHARM] C₆H₅ClO Crystals with a typical phenol odor and a melting point of 43.2–43.7°C; soluble in alcohol, ether, glycerol, and chloroform; used as a topical antiseptic. { ¦par·ə ¦klȯr·ə¹fen‚ȯl }

chlorophenol red [ORG CHEM] C₁₉H₁₂Cl₂O₅S A dye that is used as an acid-base indicator; yellow in acid solution, red in basic solution. Also spelled chlorphenol red. { ¦klȯr·ə¹fen·ȯl ¹red }

chlorophoenicite [MINERAL] (Mn,An)₅(AsO₄)(OH)₇ Gray-green monoclinic mineral consisting of a basic arsenate of manganese and zinc occurring in crystals. { ‚klȯr·ō¹fēn·ə‚sīt }

Chlorophyceae [BOT] A class of microscopic or macroscopic green algae, division Chlorophyta, composed of fresh- or salt-water, unicellular or multicellular, colonial, filamentous or sheetlike forms. { ‚klȯr·ō¹fīs·ē‚ē }

chlorophyll [BIOCHEM] The generic name for any of several oil-soluble green tetrapyrrole plant pigments which function as photoreceptors of light energy for photosynthesis. { ¹klȯr·ə‚fil }

chlorophyll a [BIOCHEM] C₅₅H₇₂O₅N₄Mg A magnesium chelate of dihydroporphyrin that is esterified with phytol and has a cyclopentanone ring; occurs in all higher plants and algae. { ¹klȯr·ə‚fil ¹ā }

chlorophyllase [BIOCHEM] An enzyme that splits or hydrolyzes chlorophyll. { ¹klȯr·ə·fə‚lās }

chlorophyll b [BIOCHEM] C₅₅H₇₀O₆N₄Mg An ester similar to chlorophyll *a* but with a −CHO substituted for a −CH₃; occurs in small amounts in all green plants and algae. { ¹klȯr·ə‚fil ¹bē }

Chlorophyta [BOT] The green algae, a highly diversified plant division characterized by chloroplasts, having chlorophyll *a* and *b* as the predominating pigments. { klō¹räf·ə·də }

chloropicrin [INORG CHEM] CCl₃NO₂ A colorless liquid with a sweet odor whose vapor is very irritating to the lungs and causes vomiting, coughing, and crying; used as a soil fumigant. Also known as nitrochloroform; trichloronitromethane. { ‚klȯr·ō¹pik·rən }

Chloropidae [INV ZOO] The chloropid flies, a family of myodarian cyclorrhaphous dipteran insects in the subsection Acalypteratae. { klō¹räp·ə‚dē }

chloroplast [BOT] A type of cell plastid occurring in the green parts of plants, containing chlorophyll pigments, and functioning in photosynthesis and protein synthesis. { ¹klȯr·ə‚plast }

chloroplast deoxyribonucleic acid [BIOCHEM] The circular deoxyribonucleic acid duplex, generally 40 to 80 copies, contained within a chloroplast. Abbreviated ctDNA. Also known as chloroplast genome. { ¦klȯr·ə‚plast dē‚äk·sē‚rī·bō·nü¦klē·ik ¹as·əd }

chloroplast genome *See* chloroplast deoxyribonucleic acid. { ¹klȯr·ə‚plast ¹jē‚nōm }

chloroplatinate [INORG CHEM] **1.** A double salt of platinic chloride and another chloride. **2.** A salt of chloroplatinic acid. Also known as platinochloride. { ‚klȯr·ō¹plat·ən‚āt }

chloroplatinic acid [INORG CHEM] H₂PtCl₆ An acid obtained as red-brown deliquescent crystals; used in chemical

analysis. Also known as platinic chloride. { ‚klȯr·ə·plə¹tin·ik ¹as·əd }

chloroprene [ORG CHEM] C₄H₅Cl A colorless liquid which polymerizes to chloroprene resin. Also known as chlorobutadiene. { ¹klȯr·ə‚prēn }

chloroprene resin [ORG CHEM] A polymer of chloroprene used to form materials resembling natural rubber. { ¹klȯr·ə‚prēn ¹rez·ən }

chloropropane [ORG CHEM] Propane molecules with chlorine substituted in various amounts for the hydrogen atoms. { ¦klȯr·ō¹prō‚pān }

3-chloro-1,2-propanediol [ORG CHEM] ClCH₂CH(OH)-CH₂OH A sweetish-tasting liquid that has a tendency to turn a straw color; soluble in ether, alcohol, and water; used to manufacture dye intermediates and to lower the freezing point of dynamite. { ¦thrē ‚klȯr·ō ¦wən ¦tü ¹prō‚pān¹dī‚ȯl }

chloropropene [CHEM] Propene molecules with chlorine substituted for some hydrogen atoms. { ¦klȯr·ə¹prō‚pēn }

β-chloropropionitrile [ORG CHEM] ClCH₂CH₂CN A liquid with an acrid odor; miscible with various organic solvents such as ethanol, ether, and acetone; used in polymer synthesis and in the synthesis of pharmaceuticals. { ¦bād·ə ‚klȯr·ō¦prō·pē·ō¹nī‚trəl }

Chloropseudomonas [MICROBIO] An invalid genus of bacteria; originally described as a member of the family Chlorobiaceae, with motile rod-shaped cells, without gas vacuoles, containing bacteriochlorophyll *c*, and capable of photoheterotrophic growth; cultures now known to be symbiotic of one of a number of typical Chlorobiaceae and a chemoorganotrophic bacterium capable of reducing elemental sulfur to sulfide. { ‚klȯr·ō‚süd·ə¹mō·nəs }

chloropsia [MED] A defect of vision in which all objects appear green. { klō¹räp·sē·ə }

6-chloropurine [PHARM] C₅H₃ClN₄ A compound crystallizing as blunt needles from water; crystals decompose at 175–177°C; soluble in dimethylformamide and ether; used as an antineoplastic drug. { ¦siks ‚klȯr·ō¹pyu̇r‚ēn }

chlorosis [MED] A form of macrocytic anemia in young females characterized by marked reduction in hemoglobin and a greenish skin color. [PL PATH] A disease condition of green plants seen as yellowing of green parts of the plant. { klə¹rō·səs }

chlorosity [OCEANOGR] The chlorine and bromide content of one liter of sea water; equals the chlorinity of the sample times its density at 20°C. { klə¹räs·əd·ē }

***N*-chlorosuccinimide** [ORG CHEM] C₄H₄ClNO₂ Orthorhombic crystals with the smell of chlorine; melting point is 150–151°C; soluble in water, benzene, and alcohol; used as a chlorinating agent. { ¦en ‚klȯr·ō¦sək¹sin·ə‚mīd }

chlorosulfonic acid [INORG CHEM] ClSO₂OH A fuming liquid that decomposes in water to sulfuric acid and hydrochloric acid; used in pharmaceuticals, pesticides, and dyes, and as a chemical intermediate. { ¦klȯr·ō¦səl¹fän·ik ¹as·əd }

chlorothalonil [ORG CHEM] C₈Cl₄N₂ Colorless crystals with a melting point of 250–251°C; used as a fungicide for crops, turf, and ornamental flowers. { ‚klȯr·ə¹thal·ə·nəl }

chlorothen citrate [PHARM] C₁₄H₁₈ClN₃S·C₆H₈O₇ A white, crystalline powder with a melting range of 112–116°C; used in medicine. Also known as chloromethapyrilene citrate. { ¹klȯr·ə·thən ¹sī‚trāt }

chlorothionite [MINERAL] K₂Cu(SO₄)Cl₂ Bright-blue secondary mineral consisting of potassium copper sulfate chloride, found on the volcano Vesuvius. { ‚klȯr·ə¹thī·ə‚nīt }

chlorothymol [ORG CHEM] CH₃C₆H₂(OH)(C₃H₇)Cl White crystals melting at 59–61°C; soluble in benzene alcohol, insoluble in water; used as a bactericide. { ‚klȯr·ə¹thī‚mȯl }

chlorotic streak [PL PATH] A systemic virus disease of sugarcane characterized by yellow or white streaks on the foliage. { klə¹räd·ik ¹strēk }

***ortho*-chlorotoluene** [ORG CHEM] CH₃C₆H₄Cl A liquid with a boiling point of 158.97°C; soluble in alcohol, chloroform, benzene, and ether; used in organic synthesis, as a solvent, and as an intermediate in dyestuff manufacture. { ¦ȯr·thō ‚klȯr·ō¹täl·yə‚wēn }

chlorotrifluoroethylene polymer [ORG CHEM] A colorless, noninflammable, heat-resistant resin, soluble in most organic solvents, and with a high impact strength; can be made into transparent filling and thin sheets; used for chemical piping, fittings, and insulation for wire and cables, and in electronic components. Also known as fluorothene; polytrifluorochloro-ethylene resin. { ¦klȯr·ō·trī¦flu̇r·ō'eth·əl,en 'pal·ə·mər }

chlorotrifluoromethane [ORG CHEM] $CClF_3$ A colorless gas having a boiling point of $-81.4°C$ and a freezing point of $-181°C$; used as a dielectric and aerospace clinical, refrigerant, and aerosol propellant, and for metals hardening and pharmaceuticals manufacture. { ¦klȯr·ō·trī¦flu̇r·ō'me,thān }

chloroxine [ORG CHEM] $C_9H_5Cl_2NO$ Crystals with a melting point of $179-180°C$; soluble in benzene and in sodium and potassium hydroxides; used as an analytical reagent. { klə'räk·sən }

chloroxiphite [MINERAL] $Pb_3CuCl_2(OH)_2O_2$ A dull-olive or pistachio-green mineral consisting of a basic chloride of lead and copper, found in the Mendip Hills of England. { klə'räk·sə,fīt }

4-chloro-3,5-xylenol [ORG CHEM] $ClC_6H_2(CH_3)_2OH$ Crystals with a melting point of $115.5°C$; soluble in water, 95% alcohol, benzene, terpenes, ether, and alkali hydroxides; used as an antiseptic and germicide and to stop mildew; used in humans as a topical and urinary antiseptic and as a topical antiseptic in animals. { ¦fȯr ¦klȯr·ō ¦thrē ¦fīv 'zī·lə,nȯl }

chlorphenol red See chlorophenol red. { klȯr'fē,nȯl 'red }

chlorpromazine [PHARM] $C_{17}H_{19}ClN_2S$ A gray-white, crystalline compound used as a sedative and in preventing or relieving nausea and vomiting. { klȯr'prō·mə,zēn }

chlorpropamide [PHARM] $C_3H_7NHCONHSO_2C_6H_4Cl$ A crystalline compound with a melting point of $127-129°C$; soluble in alcohol; used in the treatment of diabetes. { klȯr'prō·prə,mīd }

chlortetracycline [MICROBIO] $C_{22}H_{23}O_8N_2Cl$ Yellow, crystalline, broad-spectrum antibiotic produced by a strain of *Streptomyces aureofaciens*. { ¦klȯr,te·trə'sī,klēn }

chlorthiamid [ORG CHEM] $C_7H_5Cl_2NS$ An off-white, crystalline compound with a melting point of $151-152°C$; used as a herbicide for selective weed control in industrial sites. { klȯr'thī·ə,mid }

chlorzoxazone [PHARM] $C_7H_4ClNO_2$ Crystals with a melting point of $191-191.5°C$; soluble in aqueous solutions of alkali hydroxides and in ammonia; used as a skeletal muscle relaxant. { klȯr'zäks·ə,zōn }

choana [ANAT] A funnel shaped opening, especially the posterior nares. [INV ZOO] A protoplasmic collar surrounding the basal ends of the flagella in certain flagellates and in the choanocytes of sponges. { 'kō·ə·nə }

choanate fish [PALEON] Any of the lobefins composing the subclass Crossopterygii { 'kō·ə,nāt ,fish }

Choanichthyes [VERT ZOO] An equivalent name for the Sarcopterygii. { ,kō·ə'nik·thē,ēz }

choanocyte [INV ZOO] Any of the choanate, flagellate cells lining the cavities of a sponge. Also known as collar cell. { kō'an·ə,sīt }

Choanoflagellida [INV ZOO] An order of single-celled or colonial, colorless flagellates in the class Zoomastigophorea; distinguished by a thin protoplasmic collar at the anterior end. { ¦kō·ə·nō·flə'jel·ə·də }

Choanolaimoidea [INV ZOO] A superfamily of marine nematodes in the order Chromadoria distinguished by a complex stoma in two parts. { ¦kō·ə·nō·lə'mȯid·ē·ə }

choanosome [INV ZOO] The inner layer of a sponge; composed of choanocytes. { kō'an·ə,sōm }

chock [MIN ENG] A square pillar for supporting the roof in a mine, constructed of prop timber laid up in alternate cross layers, in log-cabin style, the center being filled with waste. [NAV ARCH] **1.** An open or closed metal fitting through which ropes, wires, or cables are passed. **2.** A block or wedge for supporting a boat that is being repaired. { chäk }

chockablock [NAV ARCH] **1.** Also known as block and block. **2.** Brought close together, especially in reference to two blocks of tackle. **3.** Hoisted all the way up. { 'chäk·ə,bläk }

chocolate [FOOD ENG] A dark, bitter, or sweet product that is manufactured from chocolate liquor (or nibs), sugar, and cocoa butter. { 'chak·lət }

chocolate gale See chocolatero. { 'chäk·lət 'gāl }

chocolate liquor [FOOD ENG] In chocolate manufacture, the liquid coming from the dried cocoa nibs during the grinding process. { 'chäk·lət 'lik·ər }

chocolatero [METEOROL] A moderate norther in the Gulf region of Mexico. Also known as chocolate gale. { ,chō·kō·lä'ter·ō }

chocolate spot [PL PATH] A fungus disease of legumes caused by species of *Botrytis* and characterized by brown spots on leaves and stems, with withering of shoots. { 'chäk·lət ,spät }

chocolate tree See cacao. { 'chäk·lət ,trē }

Choeropotamidae [PALEON] A family of extinct palaeodont artiodactyls in the superfamily Entelodontoidae. { ,kir·ə·pə'täm·ə,dē }

choke [ELEC] An inductance used in a circuit to present a high impedance to frequencies above a specified frequency range without appreciably limiting the flow of direct current. Also known as choke coil. [ELECTROMAG] A groove or other discontinuity in a waveguide surface so shaped and dimensioned as to impede the passage of guided waves within a limited frequency range. [MECH ENG] To increase the fuel feed to an internal combustion engine through the action of a choke valve. See choke valve. [ORD] A narrowing toward the muzzle in the bore of gun, hence the choked bore; often applied to shotguns. [PETRO ENG] A removable nipple inserted in a flow line to control oil or gas flow. { chōk }

choke coil See choke. { 'chōk ,kȯil }

choke coupling [ELECTROMAG] Coupling between two parts of a waveguide system that are not in direct mechanical contact with each other. { 'chōk ,kəp·liŋ }

choke crushing [MIN ENG] A recrushing of fine ore. { 'chōk ,krəsh·iŋ }

chokedamp See blackdamp. { 'chōk,damp }

choked disk See papilledema. { 'chōkt 'disk }

choked flow [FL MECH] Flow in a duct or passage such that the flow upstream of a certain critical section cannot be increased by a reduction of downstream pressure. { 'chōkt 'flō }

choked neck [DES ENG] Container neck which has a narrowed or constricted opening. { 'chōkt 'nek }

choke filter See choke input filter. { 'chōk ,fil·tər }

choke flange [ELECTROMAG] A waveguide flange having in its mating surface a slot (choke) so shaped and dimensioned as to restrict leakage of microwave energy within a limited frequency range. { 'chōk ,flanj }

choke input filter [ELEC] A power-supply filter in which the first filter element is a series choke. Also known as choke filter. { 'chōk 'in,pu̇t ,fil·tər }

choke joint [ELECTROMAG] A connection between two waveguides that uses two mating choke flanges to provide effective electrical continuity without metallic continuity at the inner walls of the waveguide. { 'chōk ,jȯint }

choke piston [ELECTROMAG] A piston in which there is no metallic contact with the walls of the waveguide at the edges of the reflecting surface; the short circuit for high-frequency currents is achieved by a choke system. Also known as noncontacting piston; noncontacting plunger. { 'chōk ,pis·tən }

choke ring [ORD] Metal ring used in the reaction chambers of certain recoilless weapons to control gas escape; the same function is carried out by throat rings, throat blocks, and restricting plugs in other types of recoilless weapons. { 'chōk ,riŋ }

choke valve [MECH ENG] A valve which supplies the higher suction necessary to give the excess fuel feed required for starting a cold internal combustion engine. Also known as choke. { 'chōk ,valv }

choking [FL MECH] The condition prevailing in compressible fluid flow when the upper limit of mass flow is reached, or when the speed of sound is reached in a duct. { 'chōk·iŋ }

choking gas [ORD] Casualty gas which causes irritation and inflammation of the bronchial tubes and lungs; for example, phosgene. { 'chōk·iŋ ,gas }

choking Mach number [FL MECH] The Mach number at some reference point in a duct or passage (for example, at the inlet) at which the flow in the passage becomes choked. { 'chōk·iŋ 'mäk ,nəm·bər }

cholagogic [PHYSIO] Inducing the flow of bile. { ,käl·ə¦gäj·ik }

cholagogue [PHARM] Any agent that causes an increased flow of bile into the intestine. { 'kō·lə,gäg }

cholaic acid See taurocholic acid. { kō'lā·ik 'as·əd }

cholane [BIOCHEM] $C_{24}H_{42}$ A tetracyclic hydrocarbon

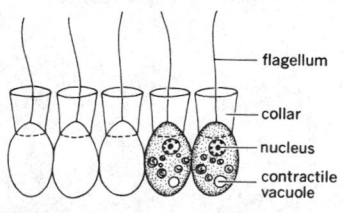

CHOANOFLAGELLIDA

— flagellum
— collar
— nucleus
— contractile vacuole

A linear colony of the choanoflagellate *Desmarella moniliformis*.

which may be considered as the parent substance of sterols, hormones, bile acids, and digitalis aglycons. { 'kō,lān }

cholangiogram [MED] The x-ray film produced by means of cholangiography. { kə'lan·jē·ə,gram }

cholangiography [MED] Roentgenography of the bile ducts. { kə,lan·jē'äg·rə·fē }

cholangiolitis [MED] Inflammation of the bile capillaries. { kə,lan·jē·ə'līd·əs }

cholangioma [MED] Adenocarcinoma of the bile ducts. { kə,lan·jē'ō·mə }

cholangitis [MED] Inflammation of the bile ducts. { ,kō·lən'jīd·as }

cholate [BIOCHEM] Any salt of cholic acid. { 'kō,lāt }

cholecalciferol [PHARM] $C_{27}H_{44}O$ Colorless crystals with a melting range of 84–88°C; soluble in alcohol, chloroform, and fatty oils; derived from the vitamin D_3 of tuna liver oil and used as an antirachitic vitamin. Also known as vitamin D_3. { ,kō·lə,kal'sif·ə,ról }

cholecystectomy [MED] Surgical removal of the gallbladder and cystic duct. { ,kō·lə,sis'tek·tə·mē }

cholecystitis [MED] Inflammation of the gallbladder. { ,kō·lə,sis'tīd·əs }

cholecystography [MED] Radiography of the gallbladder following injection or ingestion of a radiopaque substance excreted in bile. Also known as Graham-Cole test. { ,kō·lə,si'stäg·rə·fē }

cholecystokinin [BIOCHEM] A hormone produced by the mucosa of the upper intestine which stimulates contraction of the gallbladder. { ,kō·lə,sis·tə'kī·nən }

cholecystostomy [MED] The establishment of an opening into the gallbladder, usually for external drainage of its contents. { ,kō·lə,si'stä·stə·mē }

choledochoduodenal junction [ANAT] The point where the common bile duct enters the duodenum. { ¦kō·lə¦däk·ə,dü'wäd·ən·əl 'jəηk·shən }

choledocholithiasis [MED] The presence of calculi in the common bile duct. { ,kō·lə,däk·ə,li'thī·ə·səs }

choledochostomy [MED] The draining of the common bile duct through the abdominal wall. { ,kō·lə,dä'kä·stə·mē }

choleglobin [BIOCHEM] Combined native protein (globin) and open-ring iron-porphyrin, which is bile pigment hemoglobin; a precursor of biliverdin. { ¦kō·lə¦glō·bən }

cholelithiasis [MED] The production of or the condition associated with gallstones in the gallbladder or bile ducts. { ,kō·lə,li'thī·ə·səs }

cholera [MED] **1.** An acute, infectious bacterial disease of humans caused by *Vibrio comma;* characterized by diarrhea, delirium, stupor, and coma. **2.** Any condition characterized by profuse vomiting and diarrhea. { 'käl·ə·rə }

cholera [MED] An acute, severe gastroenteritis. { 'käl·ə·rə }

cholera vibrio [MICROBIO] *Vibrio comma,* the bacterium that causes cholera. { 'käl·ə·rə 'vib·rē,ō }

cholerophobia [PSYCH] Abnormal fear of cholera. { ,käl·ə·rə'fō·bē·ə }

cholesteatoma [MED] An epidermal inclusion cyst of the middle ear, or mastoid bone, sometimes in the external ear canal, brain, or spinal cord. Also known as pearly tumor. { kə,les·tē·ə'tō·mə }

cholesteric material [PHYS CHEM] A liquid crystal material in which the elongated molecules are parallel to each other within the plane of a layer, but the direction of orientation is twisted slightly from layer to layer to form a helix through the layers. { kə'les·tə·rik mə'tir·ē·əl }

cholesteric phase [PHYS CHEM] A form of the nematic phase of a liquid crystal in which the molecules are spiral. { kə'les·tə·rik ,fāz }

cholesterol [BIOCHEM] $C_{27}H_{46}O$ A sterol produced by all vertebrate cells, particularly in the liver, skin, and intestine, and found most abundantly in nerve tissue. { kə'les·tə,ról }

cholic acid [BIOCHEM] $C_{24}H_{40}O_5$ An unconjugated, crystalline bile acid. { 'kō·lik 'as·əd }

choline [BIOCHEM] $C_5H_{15}O_2N$ A basic hygroscopic substance constituting a vitamin of the B complex; used by most animals as a precursor of acetylcholine and a source of methyl groups. { 'kō,lēn }

choline acetyltransferase [BIOCHEM] An enzyme that transfers the acetyl group to choline in the synthesis of acetylcholine from acetyl coenzyme A and choline. { ¦kō,lēn ə,sed·əl'tranz·fə,rās }

cholinergic [PHYSIO] Liberating, activated by, or resembling the physiologic action of acetylcholine. { ¦kō·lə¦nər·jik }

cholinergic nerve [NUERO] Any nerve, such as autonomic preganglionic nerves and somatic motor nerves, that releases a cholinergic substance at its terminal points. { ¦kō·lə¦nər·jik 'nərv }

cholinesterase [BIOCHEM] An enzyme found in blood and in various other tissues that catalyzes hydrolysis of choline esters, including acetylcholine. Abbreviated chE. { 'kō·lə'nes·tə,rās }

choline succinate dichloride dihydrate *See* succinylcholine chloride. { 'kō,lēn 'sək·sə,nāt di'klór,īd di'hī,drāt }

choluria [MED] The presence of bile in the urine. { kō'lú·rē·ə }

cholytaurine *See* taurocholic acid. { ,käl·ə'tó,rēn }

Chondrichthyes [VERT ZOO] A class of vertebrates comprising the cartilaginous, jawed fishes characterized by the absence of true bone. { kän'drik,thē,ēz }

chondrification [PHYSIO] Formation of or conversion into cartilage. { ,kän·drə·fə'kā·shən }

chondrin [BIOCHEM] A horny gelatinous protein substance obtainable from the collagen component of cartilage. { 'kän·drən }

chondrioid [MICROBIO] A cell organelle in bacteria that is functionally equivalent to the mitochondrion of eukaryotes. { 'kän·drē,óid }

chondriome [CYTOL] Referring collectively to the chondriosomes (mitochondria) of a cell as a functional unit. { 'kän·drē,ōm }

chondriosome [CYTOL] Any of a class of self-perpetuating lipoprotein complexes in the form of grains, rods, or threads in the cytoplasm of most cells; thought to function in cellular metabolism and secretion. { 'kän·drē·ə,sōm }

chondrite [GEOL] A stony meteorite containing chondrules. { 'kän,drīt }

chondroblast [HISTOL] A cell that produces cartilage. { 'kän·drō,blast }

Chondrobrachii [VERT ZOO] The equivalent name for Ateleopoidei. { 'kän·drō'brä·kē,ī }

chondroclast [HISTOL] A cell that absorbs cartilage. { 'kän·drō,klast }

chondrocranium [ANAT] The part of the adult cranium derived from the cartilaginous cranium. [EMBRYO] The cartilaginous, embryonic cranium of vertebrates. { ¦kän·drō'krā·nē·əm }

chondrocyte [HISTOL] A cartilage cell. { 'kän·drō,sīt }

chondrodendrin *See* bebeerine. { ¦kän·drō¦den·drən }

chondrodite [MINERAL] $Mg_5(SiO_4)_2(F,OH)_2$ A monoclinic mineral of the humite group; has a resinous luster, yellow-red in color, and occurs in contact-metamorphosed dolomites. { 'kän·drō,dīt }

chondrodysplasia *See* enchondromatosis. { ¦kän·drō·də'splā·zhə }

chondrodystrophy fetalis *See* achondroplasia. { ¦kän·drō¦dis·trə·fē fə'tal·əs }

chondrogenesis [EMBRYO] The development of cartilage. { ¦kän·drō·jə'nē·səs }

chondroitin [BIOCHEM] A nitrogenous polysaccharide occurring in cartilage in the form of condroitinsulfuric acid. { kän'drō·ə·tən }

chondrology [ANAT] The anatomical study of cartilage. { kän'dräl·ə·jē }

chondroma [MED] A benign tumor of bone, cartilage, or other tissue which simulates the structure of cartilage in its growth. { kän'drō·mə }

chondromalacia [MED] Softening of a cartilage. { ¦kän·drō·mə'lā·shə }

chondromucoid [BIOCHEM] A mucoid found in cartilage; a glycoprotein in which chondroitinsulfuric acid is the prosthetic group. { ¦kän·drō¦myü,kóid }

Chondromyces [MICROBIO] A genus of bacteria in the family Polyangiaceae; sporangia are stalked, and vegetative cells are short rods or spheres. { ,kän·drō'mī,sēz }

chondrophone [INV ZOO] In bivalve mollusks, a structure

or cavity supporting the internal hinge cartilage. { 'kän·drō,fōn }

Chondrophora [INV ZOO] A suborder of polymorphic, colonial, free-floating cnidarians of the class Hydrozoa. { kän'drä·fə·rə }

chondroprotein [BIOCHEM] A protein (glycoprotein) occurring normally in cartilage. { ¦kän·drō'prō,tēn }

chondrosarcoma [MED] A malignant tumor of cartilage. { ¦kän·drō·sär'kō·mə }

chondroskeleton [ANAT] **1.** The parts of the bony skeleton which are formed from cartilage. **2.** Cartilaginous parts of a skeleton. [VERT ZOO] A cartilaginous skeleton, as in Chondrostei. { ¦kän·drō'skel·ə·tən }

Chondrostei [PALEON] The most archaic infraclass of the subclass Actinopterygii, or rayfin fishes. { kän'dräs·tē,ī }

Chondrosteidae [PALEON] A family of extinct actinopterygian fishes in the order Acipenseriformes. { ¦kän·drə'stē·ə,dē }

chondrule [GEOL] A spherically shaped body consisting chiefly of pyroxene or olivine minerals embedded in the matrix of certain stony meteorites. { 'kän,drül }

Chonetidina [PALEON] A suborder of extinct articulate brachiopods in the order Strophomenida. { ¦kän·ə·tə·dī·nə }

Chonotrichida [INV ZOO] A small order of vase shaped ciliates in the subclass Holotrichia; commonly found as ectocommensals on marine crustaceans. { ¦kän·ə'trik·ə·də }

chopper [ENG] Any knife, axe, or mechanical device for chopping or cutting an object into segments. [PHYS] A device for interrupting an electric current, beam of light, or beam of infrared radiation at regular intervals, to permit amplification of the associated electrical quantity or signal by an alternating-current amplifier; also used to interrupt a continuous stream of neutrons to measure velocity. { 'chäp·ər }

chopper amplifier [ELECTR] A carrier amplifier in which the direct-current input is filtered by a low-pass filter, then converted into a square-wave alternating-current signal by either one or two choppers. { 'chäp·ər 'am·plə,fī·ər }

chopper-stabilized amplifier [ELECTR] A direct-current amplifier in which a direct-coupled amplifier is in parallel with a chopper amplifier. { ¦chäp·ər ¦stā·bə,līzd 'am·plə,fī·ər }

chopper transistor [ELECTR] A bipolar or field-effect transistor operated as a repetitive "on/off" switch to produce square-wave modulation of an input signal. { 'chäp·ər tran'zis·tər }

chopping [ELECTR] The removal, by electronic means, of one or both extremities of a wave at a predetermined level. [PHYS] The act of interrupting an electric current, beam of light, beam of infrared radiation, or stream of neutrons at regular intervals. { 'chäp·iŋ }

chopping bit [MECH ENG] A steel bit with a chisel-shaped cutting edge, attached to a string of drill rods to break up, by impact, boulders, hardpan, and a lost core in a drill hole. Also known as chisel bit. { 'chäp·iŋ ,bit }

choppy sea [OCEANOGR] In popular usage, short, rough, irregular wave motion on a sea surface. { 'chäp·ē ¦sē }

chop-type feeder [MECH ENG] Device for semicontinuous feed of solid materials to a process unit, with intermittent opening and closing of a hopper gate (bottom closure) by a control arm actuated by an eccentric cam. { 'chäp,tīp ,fēd·ər }

Choquet theorem [MATH] Let K be a compact convex set in a locally convex Hausdorff real vector space and assume that either (1) the set of extreme points of K is closed or (2) K is metrizable; then for every point x in K there is at least one Radon probability measure m on X, concentrated on the set of extreme points of K, such that x is the centroid of m. { shō'kā ,thir·əm }

chord [ACOUS] A combination of two or more tones. [AERO ENG] **1.** A straight line intersecting or touching an airfoil profile at two points. **2.** Specifically, that part of such a line between two points of intersection. [ARCH] The span of an arch. [CIV ENG] The top or bottom, generally horizontal member of a truss. [MATH] A line segment which intersects a curve or surface only at the endpoints of the segment. { kòrd }

chordal thickness [DES ENG] The tangential thickness of a tooth on a circular gear, as measured along a chord of the pitch circle. { ¦kòrd·əl 'thik·nəs }

chordamesoderm [EMBRYO] The portion of the mesoderm in the chordate embryo from which the notochord and related structures arise, and which induces formation of ectodermal neural structures. { ¦kòrd·ə'mes·ə,dərm }

Chordata [ZOO] The highest phylum in the animal kingdom, characterized by a notochord, nerve cord, and gill slits; includes the urochordates, lancelets, and vertebrates. { kòr'däd·ə }

chord length [AERO ENG] The length of the chord of an airfoil section between the extremities of the section. { 'kòrd ,leŋkth }

Chordodidae [INV ZOO] A family of worms in the order Gordioidea distinguished by a rough cuticle containing thickenings called areoles. { kòr'däd·ə,dē }

chordoma [MED] A rarely malignant tumor derived from persistent remnants of the notochord. { kòr'dō·mə }

Chordopoxvirinae [VIROL] A subfamily of vertebrate deoxyribonucleic acid viruses of the Poxviridae family whose members replicate within the cytoplasm. { ¦kòr·dō,päk·sə'vī·rə,nē }

chordotomy [MED] Surgical division of a spinal nerve tract to relieve severe intractable pain. { kòr'däd·ə·mē }

chorea [MED] A nervous disorder seen as part of a syndrome following an organic dysfunction or an infection and characterized by irregular, involuntary movements of the body, especially of the face and extremities. { kə'rē·ə }

choreiform syndrome [MED] A complex of symptoms representing a form or component of minimal brain dysfunction in children, manifested by twitching movements of the face, trunk, and extremities. { kə'rē·ə,fòrm ,sin,drōm }

chorioadenoma [MED] A tumor intermediate in malignancy between a hydatidiform mole and choriocarcinoma. { ¦kòr·ē·ō,ad·ən'ō·mə }

chorioallantois [EMBRYO] A vascular fetal membrane that is formed by the close association or fusion of the chorion and allantois. { ¦kòr·ē·ō·ə'lant·ə·wəs }

choriocarcinoma [MED] A highly malignant tumor derived from chorionic tissue; found most commonly in the uterus and testis. Also known as chorioepithelioma. { ¦kòr·ē·ō,kärs·ən'ō·mə }

chorioepithelioma See choriocarcinoma. { ¦kòr·ē·ō,ep·ə,thē·lē'ō mbmə }

chorion [EMBRYO] The outermost of the extraembryonic membranes of amniotes, enclosing the embryo and all of its other membranes. { 'kòr·ē·än }

chorionic adenoma [MED] A benign tumor of the placenta. { ¦kòr·ē'än·ik ad·ən'ō·mə }

chorionic gonadotropin See human chorionic gonadotropin. { ¦kòr·ē'än·ik gō,nad·ə'trō·pən }

chorionic villus sampling [MED] A technique in which samples of chorionic villi are taken from the placenta for the purpose of genetic testing; usually performed at the end of the second month of pregnancy. { ¦kòr·ē,an·ik 'vil·əs ,sam·pliŋ }

chorionitis See scleroderma. { ¦kòr·ē·ə'nīd·əs }

chorioretinal [ANAT] Pertaining to the choroid and retina of the eye. { ¦kòr·ē·ə'ret·ən·əl }

chorioretinitis [MED] Inflammation of the choroid and retina of the eye. { ¦kòr·ē·ō,ret·ən'īd·əs }

choripetalous See polypetalous. { ¦kòr·ə'ped·əl·əs }

chorisepalous See polysepalous. { ¦kòr·ə'sep·əl·əs }

chorisis [BOT] Separation of a leaf or floral part into two or more parts during development. { 'kòr·ə·səs }

chorismite [PETR] A mixed rock whose fabric is macropolyschematic and which consists of petrologically dissimilar materials of varied origins. { kə'riz,mīt }

Choristida [INV ZOO] An order of sponges in the class Demospongiae in which at least some of the megascleres are tetraxons. { kə'ris·tə·də }

Choristodera [PALEON] A suborder of extinct reptiles of the order Eosuchia composed of a single genus, Champsosaurus. { ¦kòr·ə'städ·ə·rə }

C horizon [GEOL] The portion of the parent material in soils which has been penetrated with roots. { 'sē hə'rīz·ən }

chorography [MAP] All of the methods used to map a region or district. { kə'räg·rə·fē }

choroid [ANAT] The highly vascular layer of the vertebrate eye, lying between the sclera and the retina. { 'kòr,òid }

choroiditis [MED] Inflammation of the choroid. { ,kòr,òi'dīd·əs }

choroid plexus [ANAT] Any of the highly vascular, folded processes that project into the third, fourth, and lateral ventricles of the brain. { ¦kòr,òid 'plek·səs }

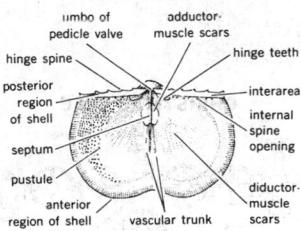

CHONETIDINA

Neochonetes showing features of the pedicle valve.

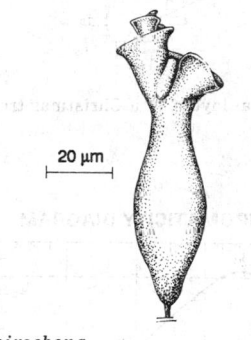

CHONOTRICHIDA

20 µm

Spirochona.

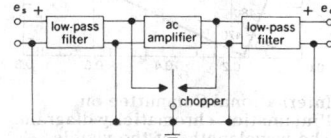

CHOPPER AMPLIFIER

Diagram of chopper amplifier.

chorology [ECOL] The study of how organisms are distributed geographically. { kə'räl·ə·jē }

choropleth [MAP] A map showing the distribution of a phenomenon, usually using various colors; color gradations are correlated to the density per unit area of the phenomenon. { 'klór·ə‚pleth }

chosen position *See* assumed position. { 'chō·zən pə'zish·ən }

chrematophobia [PSYCH] An abnormal fear of money. { krə‚mad·ə'fō·bē·ə }

Christiansen effect [ANALY CHEM] Monochromatic transparency effect when finely powdered substances, such as glass or quartz, are immersed in a liquid having the same refractive index. { 'kris·chən·sən i'fekt }

Christiansen filter [OPTICS] A type of color filter, a solid-in-liquid suspension, which scatters all incident energy except that of a narrow frequency range out of the direct beam. Also known as band-pass filter. { 'kris·chən·sən 'fil·tər }

Christmas disease [MED] A hereditary, sex-linked, hemophilia-like disease involving failure of the clotting mechanism due to a deficiency of Christmas factor. Also known as Factor IX deficiency. { 'kris·məs di‚zēz }

Christmas factor [BIOCHEM] A soluble protein blood factor involved in blood coagulation. Also known as factor IX; plasma thromboplastin component (PTC). { 'kris·məs ‚fak·tər }

Christmas tree [PETRO ENG] An assembly of valves, tees, crosses, and other fittings at the wellhead, used to control oil or gas production and to give access to the well tubing. { 'kris·məs ‚trē }

Christmas tree model [ASTRON] An explanation of superluminal motion wherein randomly flashing lights represent the superluminal features. { 'kris·məs ‚trē ‚mäd·əl }

Christoffel symbols [MATH] Symbols that represent particular functions of the coefficients and their first-order derivatives of a quadratic form. Also known as three-index symbols. { 'kris·tóf·əl ‚sim·bəlz }

christophite *See* marmatite. { 'kris·tə‚fīt }

chroma [OPTICS] **1.** The dimension of the Munsell system of color that corresponds most closely to saturation, which is the degree of vividness of a hue. Also known as Munsell chroma. **2.** *See* color saturation. { 'krō·mə }

chroma band-pass amplifier *See* burst amplifier. { 'krō·mə 'band ‚pas 'am·plə‚fī·ər }

chroma control [ELECTR] The control that adjusts the amplitude of the carrier chrominance signal fed to the chrominance demodulators in a color television receiver, so as to change the saturation or vividness of the hues in the color picture. Also known as color control; color-saturation control. { 'krō·mə kən'trōl }

chromadizing [MET] Treating the surface of aluminum or aluminum alloys with chromic acid to improve paint adhesion. { 'krō·mə‚dīz·iŋ }

Chromadoria [INV ZOO] A subclass of nematode worms in the class Adenophorea. { ‚krō·mə'dór·ē·ə }

Chromadorida [INV ZOO] An order of principally aquatic nematode worms in the subclass Chromadoria. { ‚krō·mə'dór·ə·də }

Chromadoridae [INV ZOO] A family of soil and fresh-water, free-living nematodes in the superfamily Chromadoroidea; generally associated with algal substances. { ‚krō·mə'dór·ə‚dē }

Chromadoroidea [INV ZOO] A superfamily of small to moderate-sized, free-living nematodes with spiral, transversely ellipsoidal amphids and a striated cuticle. { ‚krō·mə·də'róid·ē·ə }

chromaffin [BIOL] Staining with chromium salts. { 'krō·mə·fən }

chromaffin body *See* paraganglion. { 'krō·mə·fən ‚bäd·ē }

chromaffin cell [HISTOL] Any cell of the suprarenal organs in lower vertebrates, of the adrenal medulla in mammals, of the paraganglia, or of the carotid bodies that stains with chromium salts. { 'krō·mə·fən ‚sel }

chromaffin system [PHYSIO] The endocrine organs and tissues of the body that secrete epinephrine; characterized by an affinity for chromium salts. { 'krō·mə·fən ‚sis·təm }

chroma oscillator [ELECTR] A crystal oscillator used in color television receivers to generate a 3.579545-megahertz signal for comparison with the incoming 3.579545-megahertz chrominance subcarrier signal being transmitted. Also known

as chrominance-subcarrier oscillator; color oscillator; color-subcarrier oscillator. { 'krō·mə 'äs·ə‚lād·ər }

chromascope [OPTICS] An instrument used to determine the optical effects of color. { 'krō·mə‚skōp }

chromate [INORG CHEM] **1.** CrO_4^{2-} **2.** An ion derived from the unstable acid H_2CrO_4. **3.** A salt or ester of chromic acid. [MINERAL] A mineral characterized by the cation CrO_4^{2-}. { 'krō‚māt }

chromate treatment [MET] Treatment of metal with a solution of a hexavalent chromium compound to produce a protective coating of metal chromate. { 'krō‚māt ‚trēt·mənt }

Chromatiaceae [MICROBIO] A family of bacteria in the suborder Rhodospirillineae; motile cells have polar flagella, photosynthetic membranes are continuous with the cytoplasmic membrane, all except one species are anaerobic, and bacteriochlorophyll *a* or *b* is present. { ‚krō·mad·ē'as·ē‚ī }

chromatic [OPTICS] Relating to color. { krō'mad·ik }

chromatic aberration [ELECTR] An electron-gun defect causing enlargement and blurring of the spot on the screen of a cathode-ray tube, because electrons leave the cathode with different initial velocities and are deflected differently by the electron lenses and deflection coils. [OPTICS] An optical lens defect causing color fringes, because the lens material brings different colors of light to focus at different points. Also known as color aberration. { krō'mad·ik ab·ə'rā·shən }

chromatic adaptation [PHYSIO] A decrease in sensitivity to a color stimulus with prolonged exposure. [PSYCH] Modification in the perceived hue or saturation of a light stimulus resulting from prior viewing of a light of different hue or saturation. { krō'mad·ik ‚ad‚ap'tā·shən }

chromatic diagram *See* chromaticity diagram. { krō'mad·ik 'dī·ə‚gram }

chromatic difference of magnification [OPTICS] Variation in the size of the image produced by an optical system with the wavelength (or, equivalently, color) of light. Also known as lateral chromatic aberration. { krō'mad·ik 'dif·rəns əv ‚mag·nə·fə'kā·shən }

chromaticity [OPTICS] The color quality of light that can be defined by its chromaticity coordinates; depends only on hue and saturation of a color, and not on its luminance (brightness). { ‚krō·mə'tis·əd·ē }

chromaticity coordinates [OPTICS] The fractional amounts of the x, y, and z primary colors, specified by the International Committee on Illumination, in a color sample; more precisely, $x = X/(X + Y + Z)$, $y = Y/(X + Y + Z)$, $z = Z/(X + Y + Z)$, where X, Y, and Z are the integrals over wavelength λ of the product of the amount of light emerging from the sample per unit wavelength, and the tristimulus values, $\bar{x}(\lambda)$, $\bar{y}(\lambda)$, and $\bar{z}(\lambda)$, respectively. { ‚krō·mə'tis·əd·ē kō'órd·ən‚āts }

chromaticity diagram [OPTICS] A triangular graph for specifying colors, whose ordinate is the y chromaticity coordinate and whose abscissa is the x chromaticity coordinate; the apexes of the triangle represent primary colors. Also known as chromatic diagram. { ‚krō·mə'tis·əd·ē dī·ə‚gram }

chromatic mineral [MINERAL] A mineral with color. { krō'mad·ik ‚min·rəl }

chromatic number [MATH] For a specified surface, the smallest number n such that for any decomposition of the surface into regions the regions can be colored with n colors in such a way that no two adjacent regions have the same color. { krō'mad·ik 'nəm·bər }

chromatic parallax [OPTICS] A type of optical parallax that arises from the dependence of the position of the focal plane on the wavelength of light. { krō'mad·ik 'par·ə‚laks }

chromatic resolving power [OPTICS] The difference between two equally strong spectral lines that can barely be separated by a spectroscopic instrument, divided into the average wavelength of these two lines; for prisms and gratings Rayleigh's criteria are used, and the term is defined as the width of the emergent beam times the angular dispersion. { krō'mad·ik rə'zälv·iŋ ‚pau̇·ər }

chromatics [OPTICS] **1.** The branch of optics concerned with the properties of colors. **2.** The part of colorimetry concerned with hue and saturation. { krō'mad·iks }

chromatic sensitivity [OPTICS] The smallest change in wavelength of light that produces a change in hue which is just large enough to be detected by human vision. { krō'mad·ik sen·sə'tiv·əd·ē }

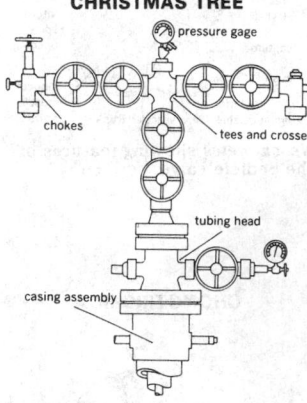

CHRISTMAS TREE

pressure gage

chokes

tees and crosses

tubing head

casing assembly

Typical layout of a Christmas tree.

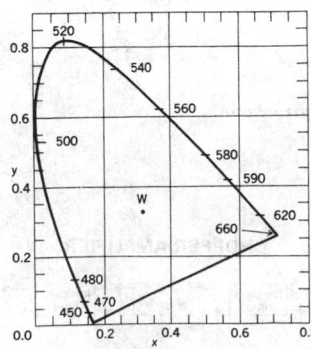

CHROMATICITY DIAGRAM

International Committee on Illumination chromaticity diagram. The wavelengths of the visible spectrum in units of 10^{-9} meter are indicated along the curve. W represents a white composed of equal amounts of the three primaries.

chromatic valence [PHYSIO] A relative measure of the hue-producing effectiveness of a chromatic stimulus. { krō'mad·ik 'val·əns }

chromatic vision [PHYSIO] Vision pertaining to the color sense, that is, the perception and evaluation of the colors of the spectrum. { krō'mad·ik 'vizh·ən }

chromatid [CYTOL] **1.** One of the pair of strands formed by longitudinal splitting of a chromosome which are joined by a single centromere in somatic cells during mitosis. **2.** One of a tetrad of strands formed by longitudinal splitting of paired chromosomes during diplotene of meiosis. { 'krō·mə·təd }

chromatin [BIOCHEM] The deoxyribonucleoprotein complex forming the major portion of the nuclear material and of the chromosomes. { 'krō·mə·tən }

chromatin diminution [GEN] Elimination during development of one or more deoxyribonucleic acid sequences from the genome of somatic cells. { 'krō·mə·ten ˌdim·yə'nü·shən }

chromating [MET] Performing a chromate treatment. { 'krō͵mād·iŋ }

Chromatium [MICROBIO] A genus of bacteria in the family Chromatiaceae; cells are ovoid to rod-shaped, are motile, do not have gas vacuoles, and contain bacteriochlorophyll *a* on vesicular photosynthetic membranes. { krō'māsh·ē·əm }

chromatogram [ANALY CHEM] The pattern formed by zones of separated pigments and of colorless substance in chromatographic procedures. { krō'mad·ə͵gram }

chromatograph [ANALY CHEM] To employ chromatography to separate substances. { krō'mad·ə͵graf }

chromatographic adsorption [ANALY CHEM] Preferential adsorption of chemical compounds (gases or liquids) in an ascending molecular-weight sequence onto a solid adsorbent material, such as activated carbon, alumina, or silica gel; used for analysis and separation of chemical mixtures. { krō'mad·ə͵graf·ik ad'sȯrp·shən }

chromatographic bed [ANALY CHEM] Any of the different configurations in which the stationary phase is contained. { kro'mad·ə͵graf·ik 'bed }

chromatography [ANALY CHEM] A method of separating and analyzing mixtures of chemical substances by chromatographic adsorption. { ˌkrō·mə'täg·rə·fē }

chromatophobia [PSYCH] An abnormal fear of colors. { krō͵mad·ə'fō·bē·ə }

chromatophore [CELL MOL] A type of pigment cell found in the integument and certain deeper tissues of lower animals that contains color granules capable of being dispersed and concentrated. { krō'mad·ə͵fȯr }

chromatophorotrophin [INV ZOO] Any crustacean neurohormone which controls the movement of pigment granules within chromatophores. { krō͵mad·ə͵fȯr·ə'trō·fən }

chromatoplasm [BOT] The peripheral protoplasm in blue-green algae containing chlorophyll, accessory pigments, and stored materials. { krō'mad·ə͵plaz·əm }

chromatopsia [MED] A disorder of visual sensation in which color impressions are disturbed or arise subjectively, with objects appearing as unnaturally colored or colorless objects as colored; may be caused by a disturbance of the optic centers, psychic disturbance, or drugs. { ˌkrō·mə'täp·sē·ə }

chromatoscope [OPTICS] An instrument in which light beams are used to mix color stimuli. { krō'mad·ə͵skōp }

chromatosis [MED] A pathologic process or pigmentary disease in which there is a deposit of coloring matter in a normally unpigmented site, or an excessive deposit in a normally pigmented area. { ˌkrō·mə'tō·səs }

chromatron [ELECTR] A single-gun color picture tube having color phosphors deposited on the screen in strips instead of dots. Also known as Lawrence tube. { 'krō·mə'trän }

chrome alum [INORG CHEM] $KCr(SO_4)_2 \cdot 12H_2O$ An alum obtained as purple crystals and used as a mordant, in tanning, and in photography in the fixing bath. Also known as potassium chromium sulfate. { 'krōm 'al·əm }

chrome brick *See* chrome refractory. { 'krōm ͵brik }

chrome diopside [MINERAL] A bright green variety of diopside containing a small amount of Cr_2O_3. { 'krōm dī'äp͵sīd }

chrome dye [CHEM] One of a class of acid dyes used on wool with a chromium compound as mordant. { 'krōm ͵dī }

chrome green *See* chromic oxide. { 'krōm ͵grēn }

chrome iron ore *See* chromite. { 'krōm 'ī·ərn ͵ȯr }

chromel [METEOROL] An alloy containing 90% nickel and 10% chromium, used to form the chromel-alumel thermocouple. { 'krō͵mel }

chrome leather [MATER] A leather tanned with chromium salts and used in making shoe uppers. { 'krōm 'leth·ər }

chrome plating [MET] A thin plate of chromium deposited by electrolysis on a corrodible metal, giving a bright, metallic surface which is highly resistant to tarnish; used to coat automobile trimming, bathroom fixtures, and many household and other articles. Also known as chromium coating; chromium plating. { 'krōm 'plād·iŋ }

chrome red [CHEM] **1.** A pigment containing basic lead chromate. **2.** Any of several mordant acid dyes. { 'krōm 'red }

chrome refractory [MATER] A ceramic material made from chrome ore and used to line steel furnaces. Also known as chrome brick. { 'krōm ri'frak·trē }

chrome spinel *See* picotite. { 'krōm spə'nel }

chrome steel *See* chromium steel. { 'krōm 'stēl }

chromesthesia [PSYCH] A form of synesthesia in which nonvisual stimuli produce the experience of color sensations. { ˌkrō·mə'sthē·zhə }

chrome tanning [CHEM ENG] Tanning treatment of animal skin with chromium salts. { 'krōm 'tan·iŋ }

chrome-vanadium steel *See* chromium-vanadium steel. { 'krōm və'nād·ē·əm 'stēl }

chrome yellow [CHEM] **1.** A yellow pigment composed of normal lead chromate, $PbCrO_4$, or other lead compounds. **2.** Any of several mordant acid dyes. { 'krōm 'yel·ō }

chromic acid [INORG CHEM] H_2CrO_4 The hydrate of CrO_3; exists only as salts or in solution. { 'krō·mik 'as·əd }

chromic chloride [INORG CHEM] $CrCl_3$ Crystals that are pinkish violet shimmering plates, almost insoluble in water, but easily soluble in presence of minute traces of chromous chloride; used in calico printing, as a mordant for cotton and silk. { 'krō·mik 'klȯr͵īd }

chromic fluoride [INORG CHEM] $CrF_3 \cdot 4H_2O$ Crystals that are green, soluble in water; used in dyeing cottons. { 'krō·mik 'flu̇r͵īd }

chromic hydroxide [INORG CHEM] $Cr(OH)_3 \cdot 2H_2O$ Gray-green, gelatinous precipitate formed when a base is added to a chromic salt; the precipitate dries to a bluish, amorphous powder; prepared as an intermediate in the manufacture of other soluble chromium salts. { 'krō·mik hī'dräk͵sīd }

chromic nitrate [INORG CHEM] $Cr(NO_3)_3 \cdot 9H_2O$ Purple, rhombic crystals that are soluble in water; used as a mordant in textile dyeing. { 'krō·mik 'nī͵trāt }

chromic oxide [INORG CHEM] Cr_2O_3 A dark green, amorphous powder, forming hexagonal crystals on heating that are insoluble in water or acids; used as a pigment to color glass and ceramic ware and as a catalyst. Also known as chrome green. { 'krō·mik 'äk͵sīd }

chrominance [OPTICS] The difference between any color and a specified reference color of equal brightness; in color television, this reference color is white having coordinates $x = 0.310$ and $y = 0.316$ on the chromaticity diagram. { 'krō·mə·nəns }

chrominance carrier *See* chrominance subcarrier. { 'krō·mə·nəns ͵kar·ē·ər }

chrominance-carrier reference [COMMUN] A continuous signal having the same frequency as the chrominance subcarrier in a color television system and having fixed phase with respect to the color burst; this signal is the reference with which the phase of a chrominance signal is compared for the purpose of modulation or demodulation. Also known as chrominance-subcarrier reference; color-carrier reference; color-subcarrier reference. { 'krō·mə·nəns ͵kar·ē·ər ͵ref·rəns }

chrominance channel [COMMUN] Any path that is intended to carry the chrominance signal in a color television system. { 'krō·mə·nəns ͵chan·əl }

chrominance demodulator [ELECTR] A demodulator used in a color television receiver for deriving the I and Q components of the chrominance signal from the chrominance signal and the chrominance-subcarrier frequency. Also known as chrominance-subcarrier demodulator. { 'krō·mə·nəns dē'mäj·ə͵lād·ər }

chrominance frequency [COMMUN] The frequency of the chrominance subcarrier, equal to 3.579545 megahertz. { 'krō·mə·nəns ͵frē·kwən·sē }

chrominance gain control [ELECTR] Variable resistors in

red, green, and blue matrix channels that individually adjust primary signal levels in color television. { 'krō·mə·nəns 'gän kən'trōl }

chrominance modulator [ELECTR] A modulator used in a color television transmitter to generate the chrominance signal from the video-frequency chrominance components and the chrominance subcarrier. Also known as chrominance-subcarrier modulator. { 'krō·mə·nəns 'mäj·ə‚lād·ər }

chrominance signal [COMMUN] One of the two components, called the I signal and Q signal, that add together to produce the total chrominance signal in a color television system. Also known as carrier chrominance signal. { 'krō·mə·nəns ‚sig·nəl }

chrominance subcarrier [COMMUN] The 3.579545-megahertz carrier whose modulation sidebands are added to the monochrome signal to convey color information in a color television receiver. Also known as chrominance carrier; color carrier; color subcarrier; subcarrier. { 'krō·mə·nəns səb'kar·ē·ər }

chrominance-subcarrier demodulator See chrominance demodulator. { 'krō·mə·nəns səb'kar·ē·ər dē'mäj·ə‚lād·ər }

chrominance-subcarrier modulator See chrominance modulator. { 'krō·mə·nəns səb'kar·ē·ər 'mäj·ə‚lād·ər }

chrominance-subcarrier oscillator See chroma oscillator. { 'krō·mə·nəns səb'kar·ē·ər 'äs·ə‚lād·ər }

chrominance-subcarrier reference See chrominance-carrier reference. { 'krō·mə·nəns səb'kar·ē·ər 'ref·rəns }

chrominance video signal [ELECTR] Voltage output from the red, green, or blue section of a color television camera or receiver matrix. { 'krō·mə·nəns 'vid·ē·ō ‚sig·nəl }

chromite [MINERAL] FeCr$_2$O$_4$ A mineral of the spinel group; crystals and pure form are rare, and it usually is massive; the only important ore mineral of chromium. Also known as chrome iron ore. { 'krō‚mīt }

chromium [CHEM] A metallic chemical element, symbol Cr, atomic number 24, atomic weight 51.996. [MET] A blue-white, hard, brittle metal used in chrome plating, in chromizing, and in many alloys. { 'krō·mē·əm }

chromium-51 [NUC PHYS] A radioactive isotope with atomic mass 51 made by neutron bombardment of chromium; radiates gamma rays. { 'krō·mē·əm ‚fif-tē'wən }

chromium carbide [INORG CHEM] Cr$_3$C$_2$ Orthorhombic crystals with a melting point of 1890°C; resistant to oxidation, acids, and alkalies; used for hot-extrusion dies, in spray-coating materials, and as a component for pumps and valves. { 'krō·mē·əm 'kär‚bīd }

chromium chloride [INORG CHEM] A group of compounds of chromium and chloride; chromium may be in the +2, +3, or +6 oxidation state. { 'krō·mē·əm 'klór‚īd }

chromium coating See chrome plating. { 'krō·mē·əm 'kōd·iŋ }

chromium dioxide [INORG CHEM] CrO$_2$ Black, acicular crystals; a semiconducting material with strong magnetic properties used in recording tapes. { 'krō·mē·əm dī'äk‚sīd }

chromium dioxide tape [ELECTR] A magnetic recording tape developed primarily to improve quality and brilliance of reproduction when used in cassettes operated at 1$\frac{7}{8}$ inches per second (4.76 centimeters per second); requires special recorders that provide high bias. { 'krō·mē·əm dī'äk‚sīd 'tāp }

chromium-gold metallizing [ELECTR] A metal film used on a silicon or silicon oxide surface in semiconductor devices because it is not susceptible to purple plague deterioration; a layer of chromium is applied first for adherence to silicon, then a layer of chromium-gold mixture, and finally a layer of gold to which bonding contacts can be applied. { ‚krō·mē·əm ‚gōld 'med·əl·īz·iŋ }

chromium-iron alloy [MET] Any of several acid- and corrosion-resistant alloys containing chromium and iron. { ‚krō·mē·əm ‚ī·ərn 'al‚ói }

chromium molybdenum steel [MET] Cast steel containing up to 1% carbon, 0.7–1.1% chromium, and 0.2–0.4% molybdenum; characterized by high strength and ductility. { ‚krō·mē·əm mə‚lib·də·nəm 'stēl }

chromium-nickel alloy [MET] Any of several alloys containing chromium and nickel in various proportions together with small amounts of other metals. { ‚krō·mē·əm ‚nik·əl 'al‚ói }

chromium oxide [INORG CHEM] A compound of chromium and oxygen; chromium may be in the +2, +3, or +6 oxidation state. { 'krō·mē·əm 'äk‚sīd }

chromium oxychloride See chromyl chloride. { 'krō·mē·əm äk·sē'klór‚īd }

chromium plating See chrome plating. { 'krō·mē·əm 'plād·iŋ }

chromium stearate [ORG CHEM] Cr(C$_{18}$H$_{35}$O$_2$)$_3$ A dark-green powder, melting at 95–100°C; used in greases, ceramics, and plastics. { 'krō·mē·əm 'stir‚āt }

chromium steel [MET] Hard, wear-resistant steel containing chromium as the predominating alloying element. Also known as chrome steel. { 'krō·mē·əm 'stēl }

chromium-vanadium steel [MET] Any of several strong, hard alloy steels containing 0.15–0.25% vanadium, 0.50–1% chromium, and 0.45–0.55% carbon. Also known as chrome-vanadium steel. { ‚krō·mē·əm və‚nād·ē·əm 'stēl }

chromizing [MET] Surface-alloying of metals in which an alloy is formed by diffusion of chromium into the base metal. { 'krō‚miz·iŋ }

Chromobacterium [MICROBIO] A genus of gram-negative, aerobic or facultatively anaerobic, motile, rod-shaped bacteria of uncertain affiliation; they produce violet colonies and violacein, a violet pigment with antibiotic properties. { ‚krō·mō‚bak'tir·ē·əm }

chromoblastomycosis [MED] A granulomatous skin disease caused by any of several fungi, usually *Hormodendrum pedrosoi,* and characterized by warty nodules which may ulcerate. Also known as chromomycosis. { ‚krō·mō‚blas·tō·mī'kō·səs }

chromocenter [CYTOL] An irregular, densely staining mass of heterochromatin in the chromosomes, with six armlike extensions of euchromatin, in the salivary glands of *Drosophila.* { 'krō·mō‚sen·tər }

chromocratic See melanocratic. { ‚krō·mə'krad·ik }

chromocyte [HISTOL] A pigmented cell. { 'krō·mə‚sīt }

chromodynamics [PARTIC PHYS] A theory of the interaction between quarks carrying color in which the quarks exchange gluons in a manner analogous to the exchange of photons between charged particles in electrodynamics. { ‚krō·mō·dī'nam·iks }

chromogen [BIOCHEM] A pigment precursor. [MICROBIO] A microorganism capable of producing color under suitable conditions. { 'krō·mə‚jen }

chromogenesis [BIOCHEM] Production of colored substances as a result of metabolic activity; characteristic of certain bacteria and fungi. { ‚krō·mō'jen·ə·səs }

chromolipid See lipochrome. { ‚krō·mō'lip·id }

chromolithography [GRAPHICS] Lithographic printing with several colors, requiring a stone for each color. { ‚krō·mō·li'thäg·rə·fē }

chromomere [CYTOL] Any of the linearly arranged chromatin granules in leptotene and pachytene chromosomes and in polytene nuclei. { 'krō·mō‚mir }

chromometer See colorimeter. { krə'mäm·əd·ər }

chromomycin [MICROBIO] Any of five components of an antibiotic complex produced by a strain of *Streptomyces griseus;* components are designated A$_1$ to A$_5$, of which A$_3$ (C$_{51}$H$_{72}$O$_{32}$) is biologically active. { ‚krō·mō'mī·sən }

chromomycosis See chromoblastomycosis. { ‚krō·mō·mī'kō·səs }

chromonema [CYTOL] The coiled core of a chromatid; it is thought to contain the genes. { ‚krō·mō'nē·mə }

chromophile [BIOL] Staining readily. { 'krō·mō‚fīl }

chromophobe [BIOL] Not readily absorbing a stain. { 'krō·mə‚fōb }

chromophore [CHEM] An arrangement of atoms that gives rise to color in many organic substances. { 'krō·mə‚fòr }

Chromophycota [BOT] A division of the plant kingdom comprising nine classes of algae ranging in size and complexity from unicellular flagellates to gigantic kelps; distinguished by the presence (in almost all) of chlorophyll *c* to complement chlorophyll *a*, and usually having brownish or yellowish chloroplasts. Also known as Chromophyta. { ‚krō·mō'fī·kəd·ə }

chromophyll [BIOCHEM] Any plant pigment. { 'krō·mə‚fil }

Chromophyta See Chromophycota. { krō'mäf·əd·ə }

chromoplasm [BOT] The pigmented, peripheral protoplasm of blue-green algae cells; contains chlorophyll, carotenoids, and phycobilins. { 'krō·mō‚plaz·əm }

chromoplast [CYTOL] Any colored cell plastid, excluding chloroplasts. { 'krō·mō‚plast }

chromoprotein [BIOCHEM] Any protein, such as hemoglobin, with a metal-containing pigment. { ¦krō·mō'prō‚tēn }

chromoradiometer [ENG] A radiation meter that uses a substance whose color changes with x-ray dosage. { ¦krō·mō·‚rād·ē'äm·əd·ər }

chromoscope [OPTICS] An instrument for analyzing color values and intensities. { 'krō·mə‚skōp }

chromosomal hybrid sterility [GEN] Sterility caused by inability of homologous chromosomes to pair during meiosis due to a chromosome aberration. { ‚krō·mə'sō·məl 'hī·brəd stə'ril·əd·ē }

chromosomal mosaic [CYTOL] An individual showing at least two cell lines with different karyotypes. { ¦krō·mə‚sō·məl mō'zā·ik }

chromosome [GEN] A linear (usually) or circular structure containing deoxyribonucleic acid (DNA) complexed with histone and nonhistone proteins, a centromere, and a telomere at each end, if linear. Chromosomes are seen in animals, plants, and other eukaryotes during mitotic and meiotic cell divisions. The single DNA molecule in each chromosome carries a unique complement of linearly arranged genes. { 'krō·mə‚sōm }

chromosome aberration [GEN] Modification of the normal chromosome complement due to deletion, duplication, or rearrangement of genetic material. { 'krō·mə‚sōm ab·ə'rā·shən }

chromosome arm [CYTOL] One of the two main segments of the chromosome that are separated by the centromere. { 'krō·mə‚sōm 'ärm }

chromosome banding [GEN] **1.** Unique patterns of cross striations seen on polytene chromosomes, as in *Drosophila*. **2.** Unique pattern of stain intensities along each chromosome in mammals and some other species, using any one of various staining procedures; examples are C-, Q-, R-, and G-banding. { 'krō·mə‚sōm 'ban·diŋ }

chromosome breakage syndrome [MED] Any of a number of human genetic disorders characterized by increased frequencies of broken and rearranged chromosomes. { 'krō·mə‚sōm 'brā·kij ‚sin‚drōm }

chromosome complement [GEN] The species-specific, normal diploid set of chromosomes in somatic cells. { 'krō·mə‚sōm 'käm·plə‚mənt }

chromosome condensation [CYTOL] The process whereby chromosomes become shorter and thicker during prophase as a consequence of coiling and supercoiling of chromatic strands. { 'krō·mə‚sōm ‚kän·dən'sā·shən }

chromosome congression *See* congression. { ¦krō·mə‚sōm kən'gresh·ən }

chromosome diminution [EMBRYO] During embryogenesis, the elimination of certain chromosomes from cells that form somatic tissues. Also known as chromosome elimination. { ¦krō·mə‚sōm ‚dim·ə'nyü·shən }

chromosome elimination *See* chromosome diminution. { 'krō·mə‚sōm i‚lim·ə'nā·shən }

chromosome instability syndrome [GEN] Disorders due to defective DNA repair; in humans, these are marked by chromosome changes, increased risk of cancer, and other phenotypic changes. { 'krō·mə‚sōm ‚in·stə'bil·əd·ē ‚sin‚drōm }

chromosome loss [CYTOL] Failure of a chromosome to become incorporated into a daughter nucleus at cell division. { 'krō·mə‚sōm 'lòs }

chromosome map *See* genetic map. { 'krō·mə‚sōm ‚map }

chromosome painting [GEN] Rendering a specific chromosome or chromosome segment distinguishable by deoxyribonucleic acid hybridization with a pool of many fluorescence-labeled DNA fragments derived from the full length of a chromosome or segment. { 'krō·mə‚sōm ‚pānt·iŋ }

chromosome puff [CYTOL] Chromatic material accumulating at a restricted site on a chromosome; thought to reflect functional activity of the gene at that site during differentiation. { 'krō·mə‚sōm ‚pəf }

chromosome transfer [GEN] The transfer of isolated metaphase chromosomes into cultured mammalian cells. { ¦krō·mə‚sōm ‚tranz·fər }

chromosome walking [GEN] Sequential isolation of overlapping molecular clones in order to span large intervals on the chromosome. { 'krō·mə‚sōm ‚wók·iŋ }

chromosphere [ASTRON] A transparent, tenuous layer of gas that rests on the photosphere in the atmosphere of the sun. { 'kró·mə‚sfir }

chromospheric network [ASTRON] A large-scale cellular pattern into which the motion of gas in the chromosphere is ordered by magnetic folds, and which is visible in spectroheliograms taken at the Hα (hydrogen alpha) line at a wavelength of about 656 nanometers and in other spectral regions. { ¦krō·mə‚sfir·ik 'net‚wərk }

chromotropic acid [ORG CHEM] $C_{10}H_8O_8S_2$ White, needlelike crystals that are soluble in water; used as an analytical reagent and azo dye intermediate. { ¦krō·mə‚träp·ik 'as·əd }

chromyl chloride [INORG CHEM] CrO_2Cl_2 A dark-red, toxic, fuming liquid that boils at 116°C; reacts with water to form chromic acid; used to make dyes and chromium complexes. Also known as chlorochromic anhydride; chromium oxychloride. { 'krō·məl 'klòr‚īd }

chron [GEOL] **1.** The time unit equivalent to the stratigraphic unit, subseries, and geologic name of a division of geologic time. **2.** The geochronological equivalent of chronozone. { krän }

chronaxie [PHYSIO] The time interval required to excite a tissue by an electric current of twice the galvanic threshold. { 'krä‚nak·sē }

chronic [MED] Long-continued; of long duration. { 'krän·ik }

chronic alcoholism [MED] Excessive consumption of alcohol over a prolonged period of time. { 'krän·ik 'al·kə‚hó‚liz·əm }

chronic appendicitis [MED] Inflammation of the vermiform appendix characterized by recurring attacks of right-sided abdominal pain over an extended period of time. { 'krän·ik ə‚pen·də'sīd·əs }

chronic carrier [MED] A person who harbors and transmits an infectious agent for an indefinite period. { 'krän·ik 'kar·ē·ər }

chronic catarrhal enteritis [MED] Inflammation of the intestinal tract associated with vascular congestion, mucosal edema, and increased outpouring of mucus in the intestinal lumen. { 'krän·ik kə'tär·əl en·tə'rīd·əs }

chronic glomerulonephritis [MED] Diffuse inflammation of the glomeruli over a prolonged period of time; characterized by progressive fibrosis and associated with hypertension and uremia. { 'krän·ik glə‚mer·ə·lō·nə'frīd·əs }

chronic granulomatosis [MED] A disorder of the phagocytes in which they ingest bacteria normally but fail to kill them; disease is usually fatal due to overwhelming bacterial infection. { 'krän·ik ‚gran·yə‚lō·mə'tō·səs }

chronic hepatitis [MED] A syndrome that is defined clinically by evidence of liver disease for at least six consecutive months. { ‚krän·ik ‚hep·ə'tīd·əs }

chronic hyperplastic perihepatitis *See* polyserositis. { 'krän·ik hī·pər'plas·tik ‚per·ē‚hep·ə'tīd·əs }

chronic infectious arthritis *See* rheumatoid arthritis. { 'krän·ik in'fek·shəs är'thrīd·əs }

chronic leukemia [MED] A leukemia in which the life expectancy is prolonged; leukemias are classified as to acute or chronic, and according to the predominant cell type; the life expectancy is highly variable depending on the latter. { 'krän·ik lü'kēm·e·ə }

chronic myeloid leukemia [MED] A form of leukemia in which immature granulocytes are predominant and the life expectancy is 1–20 years or more. { 'krän·ik 'mī·ə‚loid lü'kem·ē·ə }

chronic rhinitis [MED] Inflammation of the nasal mucous membrane due to repeated attacks of acute rhinitis; associated with membrane hypertrophy and later atrophy. { 'krän·ik rī'nīd·əs }

chronistor [ELECTR] A subminiature elapsed-time indicator that uses electroplating principles to totalize operating time of equipment up to several thousand hours. { krə'nis·tər }

chronoamperometry [ANALY CHEM] Electroanalysis by measuring at a working electrode the rate of change of current versus time during a titration; the potential is controlled. { ¦krän·ō‚am·pə'räm·ə‚trē }

chronocline [PALEON] A cline shown by successive morphological changes in the members of a related group, such as a species, in successive fossiliferous strata. { 'krän·ō‚klīn }

chronocoulometry [ANALY CHEM] The study of electrode surface properties, such as surface area. { ‚krä·nō‚kü'läm·ə‚trē }

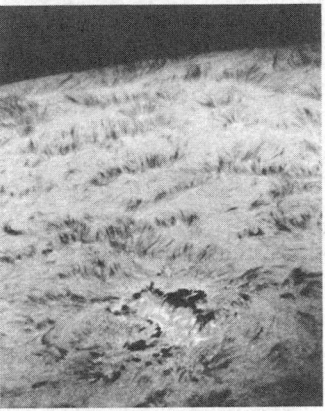

CHROMOSPHERIC NETWORK

Hydrogen-alpha filtergram showing details of the chromospheric network. (*R. Dunn, National Solar Observatory, Sunspot, New Mexico*)

CHROMOTROPIC ACID

Structural formula.

chronocyclegraph [IND ENG] A device used in micromotion studies to record a complete work cycle by taking still pictures with long exposures, the motion paths being traced by small electric lamps fastened to the worker's hands or fingers; time is obtained by interrupting the light circuits with a controlled frequency which produces dots on the film. { 'krän·ō'sī·klə,graf }

chronograph [ENG] An instrument used to register the time of an event or graphically record time intervals such as the duration of an event. { 'krän·ə,graf }

chronolith See time-stratigraphic unit. { 'krän·ə,lith }

chronolithologic unit See time-stratigraphic unit. { ¦krän·ə¦lith·ə'läj·ik 'yü·nət }

chronological future [RELAT] The chronological future relative to a set of points S in a space-time M is the set of points in M which can be reached from S by future-directed timelike curves. { ¦krän·ə¦läj·ə·kəl 'fyü·chər }

chronological past [RELAT] The chronological past relative to a set of points S in a space-time M is the set of points in M which can be reached from S by past-directed timelike curves. { ¦krän·ə¦läj·ə·kəl 'past }

chronology [SCI TECH] The arrangement of data in order of time of appearance. { krə'näl·ə·jē }

chronometer [HOROL] 1. Any extremely accurate watch. 2. A large, strongly built timepiece that beats half seconds and is especially designed for precise timekeeping on ships at sea. { krə'näm·əd·ər }

chronometer time [HOROL] The hour of the day as indicated by a chronometer, generally set approximately to Greenwich mean time; unless the chronometer has a 24-hour dial, chronometer time is usually expressed on a 12-hour cycle and labeled A.M. or P.M. { krə'näm·əd·ər ‚tīm }

chronometric data [ENG] Data in which the desired quantity is the time of occurrence of an event or the time interval between two or more events. { 'krän·ə‚me·trik 'dad·ə }

chronometric encoder [ELECTR] An encoder that uses an electronic counter to time or count electrical events and deliver in digital form a number equivalent to the input magnitude. { 'krän·ə‚me·trik en'kōd·ər }

chronometric radiosonde [ENG] A radiosonde whose carrier wave is switched on and off in such a manner that the interval of time between the transmission of signals is a function of the magnitude of the meteorological elements being measured. { ¦krän·ə¦me·trik 'rād·ē·ō‚sänd }

chronometric tachometer [ENG] A tachometer which repeatedly counts the revolutions during a fixed interval of time and presents the average speed during the last timed interval. { ¦krän·ə¦me·trik tə'käm·əd·ər }

chronometry [HOROL] 1. The science of measuring time. 2. The measurement of time by a chronometer. { krə'näm·ə·trē }

chronon [PHYS] A hypothetical quantum of time, given approximately by the time taken to traverse the classical electron radius, on the order of 10^{-23} second. { 'krän·ən }

chronopher [ELECTR] Instrument for emitting standard time signal impulses from a standard clock or timing device. { 'krän·ə·fər }

chronopotentiometry [ANALY CHEM] Electroanalysis based on the measurement at a working electrode of the rate of change in potential versus time; the current is controlled. { ¦krän·ō·pə‚ten·chē'äm·ə·trē }

chronoscope [HOROL] An electronic instrument used for measuring extremely short intervals of time, such as the time of passage of a rifle bullet between two points. { 'krän·ə,skōp }

chronostratic unit See time-stratigraphic unit. { ¦krän·ə¦strad·ik 'yü·nət }

chronostratigraphic unit See time-stratigraphic unit. { ¦krän·ə¦strad·ə'graf·ik 'yü·nət }

chronostratigraphic zone See chronozone. { ‚krän·ə‚strad·ə'graf·ik 'zōn }

chronostratigraphy [GEOL] A division of stratigraphy that uses age determination and time sequence of rock strata to develop an interpretation of the earth's geologic history. { ‚krän·ə·strə'tig·rə·fē }

chronotherapy [PHARM] Pharmacological treatment timed to the biological rhythms of the person or organism being treated in order to enhance the effect of the drugs used or to reduce undesirable side effects. { ¦krä·nō'ther·ə·pē }

chronothermometer [ENG] A thermometer consisting of a clock mechanism whose speed is a function of temperature; automatically calculates the mean temperature. { ¦krän·ō·thər'mäm·əd·ər }

chronotron [ELECTR] A device that measures millimicrosecond time intervals between pulses on a transmission line to determine the time between the events which initiated the pulses. { 'krän·ə,trän }

chronozone [GEOL] 1. A formal time-stratigraphic unit used to specify strata equivalent in time span to a zone in another type of classification, for example, a biostratigraphic zone. Also known as chronostratigraphic zone. 2. The smallest subdivision of chronostratigraphic units, below stage, composed of rocks formed during a chron of geologic time. { 'krän·ə,zōn }

Chroococcales [BOT] An order of blue-green algae (Cyanophyceae) that reproduce by cell division and colony fragmentation only. { ‚krō·ō·kə'kā·lēz }

Chryomyidae [INV ZOO] A family of myodarian cyclorrhaphous dipteran insects in the subsection Acalypteratae. { ‚krī·ō'mī·ə,dē }

chrysazin See 1,8-dihydroxyanthraquinone. { 'krī·sə·sən }

chrysene [ORG CHEM] $C_{18}H_{12}$ An organic, polynuclear hydrocarbon which when pure gives a bluish fluorescence; a component of short afterglow or luminescent paint. { 'krī‚sēn }

Chrysididae [INV ZOO] The cuckoo wasps, a family of hymenopteran insects in the superfamily Bethyloidea having brilliant metallic blue and green bodies. { krə'sid·ə,dē }

chrysoberyl [MINERAL] $BeAl_2O_4$ A pale green, yellow, or brown mineral that crystallizes in the orthorhombic system and is found most commonly in pegmatite dikes; used as a gem. Also known as chrysopal; gold beryl. { 'kris·ə,ber·əl }

chrysocarpous [BOT] Bearing yellow fruits. { kris·ə¦kär·pəs }

Chrysochloridae [PALEON] The golden moles, a family of extinct lipotyphlan mammals in the order Insectivora. { ¦kris·ə'klȯr·ə,dē }

chrysocolla [MINERAL] $CuSiO_3 \cdot 2H_2O$ A silicate mineral ordinarily occurring in impure cryptocrystalline crusts and masses with conchoidal fracture; a minor ore of copper; luster is vitreous, and color is normally emerald green to greenish-blue. { ‚kris·ə'käl·ə }

chrysoidine [ORG CHEM] $C_6H_5NNC_6H_3(NH_2)_2 \cdot HCl$ Large, black crystals or a red-brown powder that melts at 117°C; soluble in water and alcohol; used as an orange dye for silk and cotton. { kri'sȯ·ə,dēn }

chrysolite [MINERAL] 1. A gem characterized by light-yellowish-green hues, especially the gem varieties of olivine, but also including beryl, topaz, and spinel. 2. A variety of olivine having a magnesium to magnesium-iron ratio of 0.90–0.70. { 'kris·ə,līt }

Chrysomelidae [INV ZOO] The leaf beetles, a family of coleopteran insects in the superfamily Chrysomeloidea. { ‚kris·ə'mel·ə,dē }

Chrysomeloidea [INV ZOO] A superfamily of coleopteran insects in the suborder Polyphaga. { ¦kris·ə·mə'lȯid·ē·ə }

Chrysomonadida [INV ZOO] An order of yellow to brown, flagellated colonial protozoans of the class Phytamastigophorea. { ¦kris·ə·mə'näd·ə·də }

Chrysomonadina [INV ZOO] The equivalent name for the Chrysomonadida. { ¦kris·ə,män·ə'dī·nə }

chrysopal See chrysoberyl. { kri'sō·pəl }

Chrysopetalidae [INV ZOO] A small family of scale-bearing polychaete worms belonging to the Errantia. { ‚kris·ō·pə'tal·ə,dē }

chrysophanic acid [ORG CHEM] $C_{15}H_{10}O_4$ Yellow leaves that melt at 196°C; soluble in ether, chloroform, and hot alcohol; extracted from senna leaves and rhubarb root; used in medicine as a mild laxative. { ¦kris·ō'fan·ik 'as·əd }

Chrysophyceae [BOT] Golden-brown algae making up a class of fresh- and salt-water unicellular forms in the division Chrysophyta. { ‚kris·ō'fīs·ē,ē }

Chrysophyta [BOT] The golden-brown algae, a division of plants with a predominance of carotene and xanthophyll pigments in addition to chlorophyll. { krə'säf·ə·də }

chrysoprase [MINERAL] An apple-green variety of chalcedony that contains nickel; used as a gem. Also known as green chalcedony. { 'kris·ə,prāz }

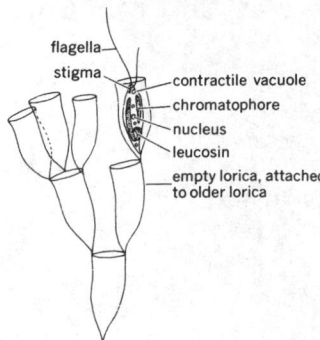

chrysotherapy [MED] The use of gold compounds in the treatment of disease. { 'kris·ō'ther·ə·pē }

chrysotile [MINERAL] $Mg_3Si_2O_5(OH)_4$ A fibrous form of serpentine that constitutes one type of asbestos. { 'kris·ō,tīl }

CH star [ASTRON] A type of metal-poor carbon star that shows especially strong CH, CN, and C_2 bands in its spectra as well as enhanced bands due to the s-process elements; found in the halo of the Milky Way Galaxy. { 'sē'äch ,stär }

Chthamalidae [INV ZOO] A small family of barnacles in the suborder Thoracica. { thə'mal·ə,dē }

CHU See centigrade heat unit.

CHU$_{mean}$ See centigrade heat unit.

chubasco [METEOROL] A severe thunderstorm with vivid lightning and violent squalls coming from the land on the west coast of Nicaragua and Costa Rica in Central America. { chü'bä,skō }

Chubb [GEOL] A meteorite crater in Ungava, Quebec, Canada. { chəb }

chuck [DES ENG] A device for holding a component of an instrument rigid, usually by means of adjustable jaws or set screws, such as the workpiece in a metalworking or woodworking machine, or the stylus or needle of a phonograph pickup. [MET] A small bar between flask bars to secure the sand in the upper box (cope) of a flask. { chək }

chucking [MECH ENG] The grasping of an outsize workpiece in a chuck or jawed device in a lathe. { 'chək·iŋ }

chucking lug [MET] A projection forged or cast onto a piece of metal that functions as a location marker when the work is being machined. { 'chək·iŋ ,ləg }

chucking machine [MECH ENG] A lathe or grinder in which the outsize workpiece is grasped in a chuck or jawed device. { 'chək·iŋ mə'shēn }

chuffing See chugging. { 'chəf·iŋ }

Chugaev reaction [ORG CHEM] The thermal decomposition of methyl esters of xanthates to yield olefins without rearrangement. { chü'gä·əv rē,ak·shən }

chugging [AERO ENG] Also known as bumping; chuffing. **1.** A form of combustion instability in a rocket engine, characterized by a pulsing operation at a fairly low frequency, sometimes defined as occurring between particular frequency limits. **2.** The noise that is made in this kind of combustion. [NUCLEO] An instability in a water-moderated reactor in which the formation of steam bubbles in the core and their subsequent collapse cause oscillations in the reactivity. { 'chəg·iŋ }

chum salmon [INV ZOO] *Oncorhynchus keta.* The salmon with the broadest geographic distribution, from the south coast of Korea to the central coast of California. Also known as dog salmon. { 'chəm ,sa·mən }

churada [METEOROL] A severe rain squall in the Mariana Islands (western Pacific Ocean) during the northeast monsoon; these squalls occur from November to April or May, but especially from January through March. { chü'rä də }

churchite See weinschenkite. { 'chər,chīt }

Church-Rosser theorem [MATH] If for a lambda expression there is a terminating reduction sequence yielding a reduced form *B*, then the leftmost reduction sequence will yield a reduced form that is equivalent to *B* up to renaming. { 'chərch 'rós·ər 'thir·əm }

Church's thesis [MATH] The claim that a function is computable in the intuitive sense if and only if it is computable by a Turing machine. Also known as Turing's thesis. { 'chərch·əz 'thē·səs }

churn drill [MECH ENG] Portable drilling equipment, with drilling performed by a heavy string of tools tipped with a blunt-edge chisel bit suspended from a flexible cable, to which a reciprocating motion is imparted by its suspension from an oscillating beam or sheave, causing the bit to be raised and dropped. Also known as American system drill; cable-system drill. { 'chərn ,dril }

churn hole See pothole. { 'chərn ,hōl }

churning [FOOD ENG] A mechanical mixing process used to separate the fat phase from a fat-water system; universally used in the manufacture of butter. { 'chərn·iŋ }

churn shot drill [MECH ENG] A boring rig with both churn and shot drillings. { 'chərn ,shät ,dril }

chute [ENG] A conduit for conveying free-flowing materials at high velocity to lower levels. [HYD] A short channel across a narrow land area which bypasses a bend in a river; formed by the river's breaking through the land. { shüt }

chute blades [COMPUT SCI] Thin metal bands which form channels to the various pockets of a sorter. { 'shüt ,blādz }

chute conveyor See jigging conveyor. { 'shüt kən'vā·ər }

chute spillway [CIV ENG] A spillway in which the water flow passes over a crest into a sloping, lined, open channel; used for earth and rock-fill dams. { 'shüt 'spil,wā }

chute system [MIN ENG] A method of mining by which ore is broken from the surface downward into chutes and is removed through passageways below. Also known as glory-hole system; milling system. { 'shüt ,sis·təm }

Chworinov rule [MET] The postulation that total freezing time for a casting is a function of the ratio of volume to surface area. { 'shvór·ə·nóv ,rül }

chyle [PHYSIO] Lymph containing emulsified fat, present in the lacteals of the intestine during digestion of ingested fats. { kīl }

chylomicron [BIOCHEM] One of the extremely small lipid droplets, consisting chiefly of triglycerides, found in blood after ingestion of fat. { ,kīl·ə'mī,krän }

chylophyllous [BOT] Having succulent or fleshy leaves. { ,kīl·ō,fil·əs }

chylothorax [MED] An accumulation of chyle in the pleural cavity. { ,kīl·ō,thór,aks }

chylurla [MED] The presence of chyle or lymph in the urine, usually caused by a fistulous communication between the urinary and lymphatic tracts or by lymphatic obstruction. { kī'lùr·ē·ə }

chyme [PHYSIO] The semifluid, partially digested food mass that is expelled into the duodenum by the stomach. { kīm }

chymopapain [BIOCHEM] Any one of several proteolytic enzymes obtained from papaya. { ,kī·mō·pə'pī·ən }

chymosin See rennin. { 'kī·mə·sən }

chymotrypsin [BIOCHEM] A proteinase in the pancreatic juice that clots milk and hydrolyzes casein and gelatin. { ,kī·mə'trip·sən }

chymotrypsinogen [BIOCHEM] An inactive proteolytic enzyme of pancreatic juice; converted to the active form, chymotrypsin, by trypsin. { ,kī·mō,trip'sin·ə·jən }

Chytridiales [MYCOL] An order of mainly aquatic fungi of the class Phycomycetes having a saclike to rhizoidal thallus and zoospores with a single posterior flagellum. { kī,trid·ē'ā·lēz }

Chytridiomycetes [MYCOL] A class of true fungi. { kī,trid·ē·ō,mī'sēd·ēz }

Ci See cirrus cloud; curie.

CI See color index; cropping index; temperature-humidity index.

C.I. See cast iron.

C^{3}I See command, control, communications, and intelligence. { 'sē 'thrē 'ī }

cibarium [INV ZOO] In insects, the space anterior to the mouth cavity in which food is chewed. { sə'bar·ē·əm }

cibophobia [PSYCH] An abnormal aversion to food. { ,sib·ə'fō·bē·ə }

Cicadellidae [INV ZOO] Large family of homopteran insects belonging to the series Auchenorrhyncha; includes leaf hoppers. { ,sik·ə'del·ə,dē }

Cicadidae [INV ZOO] A family of large homopteran insects belonging to the series Auchenorrhyncha; includes the cicadas. { sə'kad·ə,dē }

cicatrix [BIOL] A scarlike mark, usually caused by previous attachment of a part or organ. [MED] The connective-tissue scar formed at the site of a healing wound. { 'sik·ə,triks }

cicero [GRAPHICS] In Europe, a printer's unit of type measurement equal to 12 didot points or 4.51 millimeters (0.178 inch). { 'sis·ə·ro or 'chē·cha·rō }

Cichlidae [VERT ZOO] The cichlids, a family of perciform fishes in the suborder Percoidei. { 'sik·lə,dē }

Cicindelidae [INV ZOO] The tiger beetles, a family of coleopteran insects in the suborder Adephaga. { ,si·sən'del·ə,dē }

Ciconiidae [VERT ZOO] The tree storks, a family of wading birds in the order Ciconiiformes. { ,si·kə'nī·ə,dē }

Ciconiiformes [VERT ZOO] An order of predominantly long-legged, long-necked birds, including herons, storks, ibises, spoonbills, and their relatives. { sə,kōn·ē·ə'fór,mēz }

CID See charge-injection device.

Cidaroida [INV ZOO] An order of echinoderms in the subclass Perischoechinoidea in which the ambulacra comprise two columns of simple plates. { ,sid·ə'rói·də }

CICINDELIDAE

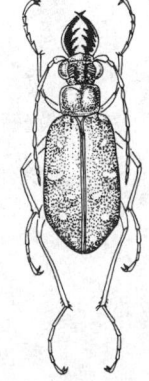

A tiger beetle. *(From T. I. Storer and R. L. Usinger, General Zoology, 3d ed., McGraw-Hill, 1957)*

CID disease *See* combined immunological deficiency disease. { ¦sē¦ī¦dē di¸zēz }

cider [FOOD ENG] Juice extracted from apples or similar fruits that has not been subjected to processing. { 'sīd·ər }

Cifax [COMMUN] Enciphered facsimile communication in which the output of a keyed pulse generator is mixed with the output of the facsimile converter. { 'sī¸faks }

cigarette burning [CHEM] In rocket propellants, black powder, gasless delay elements, and pyrotechnic candles, the type of burning induced in a solid grain by permitting burning on one end only, so that the burning progresses in the direction of the longitudinal axis. { 'sig·ə¸ret ¦bərn·iŋ }

ciguatoxin [BIOCHEM] A toxin produced by the benthic dinoflagellate *Gambierdiscus toxicus*. { ¦sēg·wə¦täk·sən }

Ciidae [INV ZOO] The minute, tree-fungus beetles, a family of coleopteran insects in the superfamily Cucujoidea. { 'sī·ə¸dē }

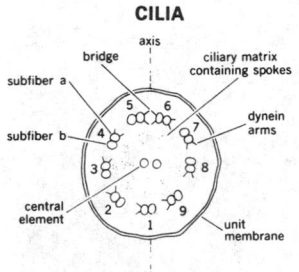

CILIA

Cross section of a cilium with the peripheral filaments numbered. *(From P. Satir, Studies on Cilia, II: Examination of the distal region of the ciliary shaft and the role of the filaments in motility, J. Cell Biol, 26:805–834, 1965)*

cilia [ANAT] Eyelashes. [CYTOL] Relatively short, centriole-based, hairlike processes on certain anatomical cells and motile organisms. { 'sil·ē·ə }

ciliary body [ANAT] A ring of tissue lying just anterior to the retinal margin of the eye. { ¦sil·ē¸er·ē ¦bäd·ē }

ciliary movement [BIOL] A type of cellular locomotion accomplished by the rhythmical beat of cilia. { ¦sil·ē¸er·ē 'müv·mənt }

ciliary muscle [ANAT] The smooth muscle of the ciliary body. { ¦sil·ē¸er·ē 'məs·əl }

ciliary process [ANAT] Circularly arranged choroid folds continuous with the iris in front. { ¦sil·ē¸er·ē 'präs·əs }

Ciliatea [INV ZOO] The single class of the protozoan subphylum Ciliophora. { ¸sil·ē'ad·ē·ə }

ciliated epithelium [HISTOL] Epithelium composed of cells bearing cilia on their free surfaces. { 'sil·ē¸ād·əd ep·ə'thēl·ē·əm }

ciliolate [BIOL] Ciliated to a very minute degree. { 'sil·ē·ə¸lāt }

Ciliophora [INV ZOO] The ciliated protozoans, a homogeneous subphylum of the Protozoa distinguished principally by a mouth, ciliation, and infraciliature. { ¸sil·ē'äf·ə·rə }

CIM *See* computer input from microfilm; computer-integrated manufacturing.

Cimbicidae [INV ZOO] The cimbicid sawflies, a family of hymenopteran insects in the superfamily Tenthredinoidea. { ¸sim'bis·ə¸dē }

Cimex [INV ZOO] The type genus of Cimicidae, including bedbugs and related forms. { 'sī¸meks }

Cimicidae [INV ZOO] The bat, bed, and bird bugs, a family of flattened, wingless, parasitic hemipteran insects in the superfamily Cimicimorpha. { sī'mis·ə¸dē }

Cimicimorpha [INV ZOO] A superfamily, or group according to some authorities, of hemipteran insects in the subdivision Geocorisae. { ¸sī·mə·sə'mȯr·fə }

Cimicoidea [INV ZOO] A superfamily of the Cimicimorpha in some systems of classification. { ¸sī·mə'kȯid·ē·ə }

ciminite [PETR] An extrusive rock consisting essentially of olivine with sanidine and pyroxene and basic plagioclase. { 'chīm·ə¸nīt }

Cimmeria [PALEON] In the Jurassic, a narrow continent that extended east-west at the southern margin of Eurasia. The name comes from the Crimean peninsula of Russia, where there is well-displayed evidence of an intra-Jurassic orogenic disturbance, indicative of continental collision. { sə'mer·ē·ə }

cimolite [MINERAL] $2Al_2O_3 \cdot 9SiO_3 \cdot 6H_2O$ A white, grayish, or reddish mineral consisting of hydrous aluminum silicate occurring in soft, claylike masses. { 'sim·ə¸līt }

cinching [COMPUT SCI] Creases produced in magnetic tape when the supply reel is wound at low tension and suddenly stopped during playback. [GRAPHICS] The tightening of successive loops of film on a roll. { 'sin·chiŋ }

cinch mark [GRAPHICS] A defect on roll film caused by abrasion when the film is tightened by being pulled. { 'sinch ¸märk }

cincholepidine *See* lepidine. { ¸sin·kə'lep·ə¸dēn }

cinchona [BOT] The dried, alkaloid-containing bark of trees of the genus *Cinchona*. { siŋ'kō·nə }

cinchonamine [ORG CHEM] $C_{19}H_{24}N_2O$ A yellow, crystalline, water-insoluble alkaloid that melts at 184°C; derived from the bark of *Remijia purdieana*, a member of the madder family of shrubs. { siŋ'kän·ə¸mēn }

cinchonine [ORG CHEM] $C_{19}H_{22}N_2O$ A colorless, crystalline alkaloid that melts at about 245°C; extracted from cinchona bark, it is used as a substitute for quinine and as a spot reagent for bismuth. { 'siŋ·kə¸nēn }

Cincinnatian [GEOL] Upper Ordovician geologic time. { sin·sə'nad·ē·ən }

Cinclidae [VERT ZOO] The dippers, a family of insect-eating songbirds in the order Passeriformes. { 'siŋ·klə¸dē }

cinclides [INV ZOO] Pores in the body wall of some sea anemones for the release of water and stinging cells. { siŋ'klī¸dēz }

cinclis [INV ZOO] Singular of cinclides. { 'siŋ·kləs }

cinder [GEOL] Fine-grained pyroclastic material ranging in diameter from 0.16 to 1.28 inch (4 to 32 millimeters). [MATER] Slag from a metal furnace. [MET] Scale cast off in forging metal. { 'sin·dər }

cinder block [MATER] A hollow block made of cinder concrete. [MET] A block which closes the front of a blast furnace, containing the cinder notch. { 'sin·dər ¸bläk }

cinder coal *See* natural coke. { 'sin·dər ¸kōl }

cinder concrete [MATER] A concrete containing cinders as the aggregate. { 'sin·dər 'kän¸krēt }

cinder cone [GEOL] A conical elevation formed by the accumulation of volcanic debris around a vent. { 'sin·dər ¸kōn }

cinder notch [MET] An opening in a blast furnace that allows molten slag to flow out. { 'sin·dər ¸näch }

cinder pig [MET] Pig iron produced from a mixture of slag in the furnace and crude metal or ore. { 'sin·dər ¸pig }

cinders [MATER] Incombustible residue from a burning process; in particular, small pieces of clinker from the burning of soft coal. { 'sin·dərz }

C index [GEOPHYS] A subjectively obtained daily index of geomagnetic activity, in which each day's record is evaluated on the basis of 0 for quiet, 1 for moderately disturbed, and 2 for very disturbed. Also known as C figure; magnetic character figure. { 'sē ¸in¸deks }

C indicator *See* C scope. { 'sē ¸in·də¸kād·ər }

cinefluorography [GRAPHICS] The motion picture recording of fluoroscopic images. { ¦sin·ə·flə'räg·rə·fē }

cinema *See* motion picture. { 'sin·ə·mə }

cinematography [GRAPHICS] Motion picture photography. { ¸sin·ə·mə'täg·rə·fē }

cinemicrography [GRAPHICS] The photography of objects formed by a microscope, using a motion picture camera. { ¸sin·ə·mī'kräg·rə·fē }

cineol *See* eucalyptol. { 'sin·ē¸ȯl }

cine-oriented image [GRAPHICS] A microfilm image whose top and bottom are perpendicular to the outer, long edge of the film. { ¸sīn ¦ȯr·ē¸ent·əd 'im·ij }

cineplasty *See* kineplasty. { 'sin·ə¸plas·tē }

cineradiography [GRAPHICS] A version of flash radiography in which a succession of flashes is used to form a moving picture of an object. { ¸sin·ə¸rād·ē'äg·rə·fē }

cinerary urn [ARCHEO] A vessel used for burial of the cremated remains of a person. { 'sin·ə¸rer·ē 'ərn }

cinereous [BIOL] **1.** Ashen in color. **2.** Having the inert and powdery quality of ashes. { sə'nir·ē·əs }

cinetheodolite [ENG] A surveying theodolite in which 35-millimeter motion picture cameras with lenses of 60- to 240-inch (1.5- to 6.1-meter) focal length are substituted for the surveyor's eye and telescope; used for precise time-correlated observation of distant airplanes, missiles, and artificial satellites. { ¸sin·ə·thē'äd·ə¸līt }

Cingulata [VERT ZOO] A group of xenarthran mammals in the order Edentata, including the armadillos. { ¸siŋ·gyə'läd·ə }

cingulate [BIOL] Having a girdle of bands or markings. { 'siŋ·gyə·lət }

cingulum [ANAT] The ridge around the base of the crown of a tooth. [BOT] The part of a plant between stem and root. [INV ZOO] **1.** Any girdlelike structure. **2.** A band of color or a raised line on certain bivalve shells. **3.** The outer zone of cilia on discs of certain rotifers. **4.** The clitellum in annelids. [NEURO] The tract of association nerve fibers in the brain, connecting the callosal and the hippocampal convolutions. { 'siŋ·gyə·ləm }

cinnabar [MINERAL] HgS A vermilion-red mineral that crystallizes in the hexagonal system, although crystals are rare, and commonly occurs in fine, granular, massive form; the only important ore of mercury.　Also known as cinnabarite; vermilion.　{ 'sin·ə,bär }

cinnabarite See cinnabar.　{ ,sin·ə'bä,rīt }

cinnamate [ORG CHEM] A salt of cinnamic acid, containing the functional group $C_9H_7O_2-$.　{ 'sin·ə,māt }

cinnamic acid [ORG CHEM] $C_6H_5CHCHCOOH$ Colorless, monoclinic acid; forms scales, slightly soluble in water; found in natural balsams.　{ sə'nam·ik 'as·əd }

cinnamic alcohol [ORG CHEM] $C_6H_5CH:CHCH_2OH$ White needles that congeal upon heating and are soluble in alcohol; used in perfumery.　{ sə'nam·ik 'al·kə,hól }

cinnamic aldehyde [ORG CHEM] $C_6H_5CH:CHCHO$ A yellow oil with a cinnamon odor, sweet taste, and a boiling point of 248°C; used in flavors and perfumes.　{ sə'nam·ik 'al·də,hīd }

cinnamon [BOT] *Cinnamomum zeylanicum.* An evergreen shrub of the laurel family (Lauraceae) in the order Magnoliales; a spice is made from the bark.　{ 'sin·ə·mən }

cinnamoyl chloride [ORG CHEM] $C_6H_5CHCHCOCl$ Yellow crystals that melt at 35°C, and decompose in water; used as a chemical intermediate.　{ 'sin·ə,móil 'klór,īd }

cipher [COMMUN] A transposition or substitution code for transmitting secret messages.　{ 'sī·fər }

cipher block chaining [COMMUN] A technique for block chaining in which each block of ciphertext is produced by adding, through the EXCLUSIVE OR operation, the previous block of ciphertext to the current block of plaintext.　Abbreviated CBC.　{ 'sī·fər ,bläk ,chān·iŋ }

cipher feedback [COMMUN] An implementation of ciphertext autokey cipher in which the leftmost *n* bits of the data encryption standard (DES) output are added by the EXCLUSIVE OR operation to *N* bits of plaintext to produce *N* bits of ciphertext (where *N* is the number of bits enciphered at one time), and these *N* bits of ciphertext are fed back into the algorithm by first shifting the current DES input *N* bits to the left, and then appending the *N* bits of ciphertext to the right-hand side of the shifted input to produce a new DES input used for the next iteration of the algorithm.　{ 'sī·fər 'fēd,bak }

cipher machine [COMMUN] Mechanical or electrical apparatus for enciphering and deciphering.　{ 'sī·fər mə'shēn }

ciphertext [COMMUN] A message which has been transformed by a cipher so that it can be read only by those privy to the secrets of the cipher.　{ 'sī·fər,tekst }

ciphertext autokey cipher [COMMUN] A stream cipher in which the cryptographic bit stream generated at a given time is determined by the ciphertext generated at earlier times.　{ 'sī·fər,tekst 'ȯd·ō,kē ,sī·fər }

ciphony [COMMUN] A technique by which security is accomplished by converting speech into a series of on-off pulses and mixing these with the pulses supplied by a key generator; to recover the original speech, the identical key must be subtracted and the resultant on-off pulses reconverted into the original speech pattern; unauthorized listeners are unable to reconstruct the plain text unless they have an identical key generator and the daily key setting.　{ 'sī·fə·nē }

ciphony equipment [ELECTR] Any equipment attached to a radio transmitter, radio receiver, or telephone for scrambling or unscrambling voice messages.　{ 'sī·fə·nē jə,kwip·mənt }

Cipolletti weir [CIV ENG] Trapezoidal weir in which the sides of the notch slope are one horizontal to four vertical; used to measure water flow in open channels, especially streams and rivers.　{ chip·ə'led·ē 'wer }

CIPW classification [PETR] A designation for the Norm system of classifying igneous rocks; from the initial letters of the names of those who devised it: Cross, Iddings, Pirsson, and Washington.　{ ,sē'ī,pē'dəb·əl·yü ,klas·ə·fə'kā·shən }

Cir See Circinus.

circadian rhythm [PHYSIO] A self-sustained cycle of physiological changes that occurs over an approximately 24-hour cycle, generally synchronized to light-dark cycles in an organism's environment.　{ sər'kād·ē·ən 'rith·əm }

circinate [BIOL] Having the form of a flat coil with the apex at the center.　{ 'sərs·ən,āt }

circinate vernation [BOT] Uncoiling of new leaves from the base toward the apex, as in ferns.　{ 'sərs·ən,āt vər'nā·shən }

Circinus [ASTRON] A constellation, right ascension 15 hours, declination 60°S.　Abbreviated Cir.　Also known as Compasses.　{ 'sərs·ən·əs }

circle [MATH] **1.** The set of all points in the plane at a given distance from a fixed point.　**2.** A unit of angular measure, equal to one complete revolution, that is, to 2π radians or 360°.　Also known as turn.　{ 'sər·kəl }

circle diagram [ELEC] A diagram which gives a graphical solution of equations for a transmission line, giving the input impedance of the line as a function of load impedance and electrical length of the line.　{ 'sər·kəl 'dī·ə,gram }

circle-dot mode [ELECTR] Mode of cathode-ray storage of binary digits in which one kind of digit is represented by a small circle of excitation of the screen, and the other kind by a similar circle with a concentric dot.　{ 'sər·kəl 'dät ,mōd }

circle graph See pie chart.　{ 'sər·kəl ,graf }

circle haul [MIN ENG] A haulage system in strip mining; empty units enter the mine over one lateral and leave, loaded, over the lateral nearest the tipple.　{ 'sər·kəl ,hȯl }

circle of confusion [OPTICS] The blurred circular image of a point object which is formed by a camera lens, even with the best focusing.　Also known as circle of least confusion.　{ 'sər·kəl əv kən'fyü·zhən }

circle of convergence [MATH] The region in which a power series possesses a limit.　{ 'sər·kəl əv kən'vər·jəns }

circle of curvature [MATH] The circle tangent to a curve on the concave side and having the same curvature at the point of tangency as does the curve.　{ 'sər·kəl əv 'kər·və·chər }

circle of declination See hour circle.　{ 'sər·kəl əv ,dek·lə'nā·shən }

circle of equal altitude [GEOD] A circle on the surface of the earth, on every point of which the altitude of a given celestial body is the same at a given instant; the pole of this circle is the geographical position of the body, and the great-circle distance from this pole to the circle is the zenith distance of the body.　{ 'sər·kəl əv 'ē·kwəl 'al·tə,tüd }

circle of equal declination See parallel of declination.　{ 'sər·kəl əv 'ē·kwəl ,dek·lə'na·shən }

circle of equal probability [AERO ENG] A measure of the accuracy with which a rocket or missile can be guided; the radius of the circle at a specific distance in which 50% of the reliable shots land.　Also known as circle of probable error; circular error probable.　{ 'sər·kəl əv 'ē·kwəl präb·ə'bil·əd·ē }

circle of illumination [GEOL] The edge of the sunlit hemisphere, which forms a circular boundary separating the earth into a light half and a dark half.　{ 'sər·kəl əv ə,lü·mə'nā·shən }

circle of inertia See inertial circle.　{ 'sər·kəl əv i'nər·shə }

circle of inversion [MATH] A circle with respect to which two specified curves are inverse curves.　{ 'sər·kəl əv in'vər·zhən }

circle of latitude [ASTRON] A great circle of the celestial sphere passing through the ecliptic poles, and hence perpendicular to the plane of the ecliptic.　Also known as parallel of latitude.　[GEOD] A meridian of the terrestrial sphere along which latitude is measured.　{ 'sər·kəl əv 'lad·ə,tüd }

circle of least confusion See circle of confusion.　{ 'sər·kəl əv 'lēst kən'fyü·zhən }

circle of longitude [ASTRON] A circle of the celestial sphere, parallel to the ecliptic.　[GEOD] See parallel.　{ 'sər·kəl əv 'län·jə,tüd }

circle of perpetual apparition [ASTRON] That circle of the celestial sphere, centered on the polar axis and having a polar distance from the elevated pole approximately equal to the latitude of the observer, within which celestial bodies do not set.　{ 'sər·kəl əv pər'pech·ə·wəl ap·ə'rish·ən }

circle of perpetual occultation [ASTRON] That circle of the celestial sphere, centered on the polar axis and having a polar distance from the depressed pole approximately equal to the latitude of the observer, within which celestial bodies do not rise.　{ 'sər·kəl əv pər'pech·ə·wəl ,äk·əl'tā·shən }

circle of position [NAV] A circular line of position, used most frequently with reference to the circle of equal altitude surrounding the geographical position of a celestial body, but also used to refer to the lines of position produced by the distance-measuring equipment system.　{ 'sər·kəl əv pə'zish·ən }

circle of probable error See circle of equal probability.　{ 'sər·kəl əv 'präb·ə·bəl 'er·ər }

circle of right ascension See hour circle.　{ 'sər·kəl əv 'rīt ə'sen·shən }

circle of uncertainty [NAV] A circle having as its center a position and as its radius the probable error of the position; the percent of the probable error must be specified error; it is a circle within which a craft is considered to be located. { 'sər‧kəl əv ən'sərt‧ən‧tē }

circle of Willis [ANAT] A ring of arteries at the base of the cerebrum. { 'sər‧kəl əv 'wil‧əs }

circle shear [MECH ENG] A shearing machine that cuts circular disks from a metal sheet rolling between the cutting wheels. { 'sər‧kəl ‚shēr }

circle sheet [NAV] A chart with curves enabling a graphical solution of a three-point problem rather than using a three-arm protractor. { 'sər‧kəl ‚shēt }

circling approach area [NAV] The area in which aircraft circle to land under visual conditions after completing an instrument approach. { 'sər‧kliŋ ə'prōch ‚er‧ē‧ə }

circuit [ELEC] See electric circuit. [ELECTROMAG] A complete wire, radio, or carrier communications channel. [MATH] See cycle. { 'sər‧kət }

circuital field See rotational field. { sə¦kyü‧əd‧əl 'fēld }

circuit analyzer See volt-ohm-milliammeter. { 'sər‧kət ‚an‧ə‚līz‧ər }

circuit board See printed circuit board. { 'sər‧kət ‚bórd }

circuit breaker [ELEC] An electromagnetic device that opens a circuit automatically when the current exceeds a predetermined value. { 'sər‧kət ‚brāk‧ər }

circuit capacity [COMMUN] Number of communications channels which can be handled by a given circuit at the same time. { 'sər‧kət kə'pas‧əd‧ē }

circuit conditioning [ELECTR] Test, analysis, engineering, and installation actions to upgrade a communications circuit to meet an operational requirement; includes the reduction of noise, the equalization of phase and level stability and frequency response, and the correction of impedance discontinuities, but does not include normal maintenance and repair activities. { 'sər‧kət kən'dish‧ə‧niŋ }

circuit design [ELEC] The art of specifying the components and interconnections of an electrical network. { 'sər‧kət də'zīn }

circuit diagram [ELEC] A drawing, using standardized symbols, of the arrangement and interconnections of the conductors and components of an electrical or electronic device or installation. Also known as schematic circuit diagram; wiring diagram. { 'sər‧kət ‚dī‧ə‚gram }

circuit efficiency [ELECTR] Of an electron tube, the power delivered to a load at the output terminals of the output circuit at a desired frequency divided by the power delivered by the electron stream to the output circuit at that frequency. { 'sər‧kət i'fish‧ən‧sē }

circuit element See component. { 'sər‧kət ‚el‧ə‧mənt }

circuit grade [COMMUN] A circuit rating defining the ability to carry information; grades include telegraph, voice, and broad-band. { 'sər‧kət ‚grād }

circuit interrupter [ELEC] A device in a circuit breaker to remove energy from an arc in order to extinguish it. { 'sər‧kət ‚in‧tə‚rəp‧tər }

circuit loading [ELEC] Power drawn from a circuit by an electric measuring instrument, which may alter appreciably the quantity being measured. { 'sər‧kət ‚lōd‧iŋ }

circuit noise [COMMUN] In telephone practice, the noise which is brought to the receiver electrically from a telephone system, excluding noise picked up acoustically by telephone transmitters. { 'sər‧kət ‚nóiz }

circuit noise level [COMMUN] Ratio of the circuit noise at that point to some arbitrary amount of circuit noise chosen as a reference; usually expressed in decibels above reference noise, signifying the reading of a circuit noise meter, or in adjusted decibels, signifying circuit noise meter reading adjusted to represent interfering effect under specified conditions. { 'sər‧kət ‚nóiz ‚lev‧əl }

circuit protection [ELECTR] Provision for automatically preventing excess or dangerous temperatures in a conductor and limiting the amount of energy liberated when an electrical failure occurs. { 'sər‧kət prə'tek‧shən }

circuit reliability [COMMUN] The percent of time a circuit was available to the user during a specified period of time. { 'sər‧kət ri‚lī‧ə'bil‧əd‧ē }

circuitron [ELECTR] Combination of active and passive components mounted in a single envelope like that used for

tubes, to serve as one or more complete operating stages. { 'sər‧kyə‚trän }

circuitry [ELEC] The complete combination of circuits used in an electrical or electronic system or piece of equipment. { 'sər‧kə‚trē }

circuit shift See cyclic shift. { 'sər‧kət ‚shift }

circuit switching [COMMUN] 1. The method of providing communication service through a switching facility, either from local users or from other switching facilities. 2. A method of transmitting messages through a communications network in which a path from the sender to the receiver of fixed bandwidth or speed is set up for the entire duration of a communication or call. { 'sər‧kət ‚swich‧iŋ }

circuit testing [ELEC] The testing of electric circuits to determine and locate an open circuit, or a short circuit or leakage. { 'sər‧kət ‚tes‧tiŋ }

circuit theory [ELEC] The mathematical analysis of conditions and relationships in an electric circuit. Also known as electric circuit theory. { 'sər‧kət ‚thē‧ə‧rē }

circulant determinant [MATH] A determinant in which the elements of each row are the same as those of the previous row moved one place to the right, with the last element put first. { 'sər‧kyə‧lənt də'tər‧mə‧nənt }

circulant matrix [MATH] A matrix in which the elements of each row are those of the previous row moved one place to the right. { 'sər‧kyə‧lənt 'mā‚triks }

circular accelerator See circular particle accelerator. { 'sər‧kyə‧lər ak'sel‧ə‚rād‧ər }

circular antenna [ELECTROMAG] A folded dipole that is bent into a circle, so the transmission line and the abutting folded ends are at opposite ends of a diameter. { 'sər‧kyə‧lər an 'ten‧ə }

circular argument [MATH] An argument that is not valid because it uses the theorem to be proved or a consequence of that theorem that is not proven. { 'sər‧kyə‧lər 'är‧gyə‧mənt }

circular arc See arc. { 'sər‧kyə‧lər 'ärk }

circular behavior [PSYCH] Behavior that stimulates similar behavior in another individual or group. { 'sər‧kyə‧lər bə'hāv‧yər }

circular birefringence [OPTICS] The phenomenon in which an optically active substance transmits right circularly polarized light with a different velocity from left circularly polarized light. { 'sər‧kyə‧lər ‚bī‧rə'frin‧jəns }

circular buffering [COMPUT SCI] A technique for receiving data in an input-output control system which uses a single buffer that appears to be organized in a circle, with data wrapping around it. { 'sər‧kyə‧lər 'bəf‧ə‧riŋ }

circular burner [ENG] A fuel burner having a round opening. { 'sər‧kyə‧lər 'bərn‧ər }

circular channel [ENG] Continuous-length opening with circular cross section through which liquid or gas can be made to flow. { 'sər‧kyə‧lər 'chan‧əl }

circular-chart recorder [ENG] Graphic pen-and-ink recorder where measured values are drawn onto a rotating circular chart by the backward and forward movement of a pivoted pen actuated by the input signal (such as temperature, pressure, flow, or force) from an instrument transmitter. { 'sər‧kyə‧lər ‚chärt ri'kórd‧ər }

circular chromatography See radial chromatography. { 'sər‧kyə‧lər ‚krō‧mə'täg‧rə‧fē }

circular coal See eye coal. { 'sər‧kyə‧lər ‚kōl }

circular coil [ELECTROMAG] In eddy-current nondestructive tests, a type of test coil which surrounds an object. { 'sər‧kyə‧lər ‚kóil }

circular collider [NUCLEO] A type of colliding-beam accelerator in which both beams are stored in large circular rings of magnets and are brought into collision repeatedly at several interaction points. { 'sər‧kyə‧lər kə'līd‧ər }

circular cone [MATH] A cone whose base is a circle. { 'sər‧kyə‧lər 'kōn }

circular conical surface [MATH] The lateral surface of a right circular cone. { ¦sər‧kyə‧lər ¦kän‧ə‧kəl 'sər‧fəs }

circular current [ELEC] An electric current moving in a circular path. { 'sər‧kyə‧lər 'kər‧ənt }

circular cutter [MECH ENG] A rotating blade with a square or knife edge used to slit or shear metal. { 'sər‧kyə‧lər 'kəd‧ər }

circular cylinder [MATH] A solid bounded by two parallel planes and a cylindrical surface whose intersections with planes

CIRCUIT BREAKER

Bulk oil circuit breaker for 138-kilovolt application.

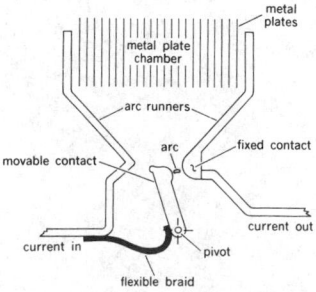

CIRCUIT INTERRUPTER

metal plates
metal plate chamber
arc runners
arc
movable contact
fixed contact
current out
current in
pivot
flexible braid

Cross section of interrupter for a typical medium-voltage circuit breaker.

perpendicular to the straight lines forming the surface are circles. { ¦sər·kyə·lər 'sil·ən·dər }

circular deoxyribonucleic acid [BIOCHEM] A single- or double-stranded ring of deoxyribonucleic acid found in certain bacteriophages and in human wart virus. Also known as ring deoxyribonucleic acid. [MOL BIO] A deoxyribonucleic acid molecule that has no free 5' or 3' ends; characteristic of prokaryotes but also found in mitochondria, chloroplasts, and some viral genomes. { ¦sər·kyə·lər dē¦äk·sē¦rī¦bō¦nü¦klē·ik 'as·əd }

circular dichroism [OPTICS] A change from planar to elliptic polarization when an initially plane-polarized light wave traverses an optically active medium. Abbreviated CD. { ¦sər·kyə·lər 'dī·krō͵iz·əm }

circular electric wave [ELECTROMAG] A transverse electric wave for which the lines of electric force form concentric circles. { ¦sər·kyə·lər i¦lek·trik 'wāv }

circular error [ORD] **1.** A bombing error measured by the radial distance of a point of bomb impact, or mean point of impact, from the center of the target, excluding gross errors. **2.** With an airburst atomic bomb, the bombing error measured from the point on the ground immediately below the bomb burst to the desired ground zero. { ¦sər·kyə·lər 'er·ər }

circular-error average [ORD] The bombing error in a given bombing attack, expressed as the average radial distance of the bomb impacts, or mean points of impact, from the center of the target. { ¦sər·kyə·lər ¦er·ər 'av·rəj }

circular error probable *See* circle of equal probability. { ¦sər·kyə·lər ¦er·ər 'präb·ə·bəl }

circular file [COMPUT SCI] An organized collection of records, generally with a high turnover, in which new records are inserted by replacing the oldest records. { ¦sər·kyə·lər 'fīl }

circular flow method [FL MECH] A method to determine viscosities of Newtonian fluids by measuring the torque from viscous drag of sample material between a closely spaced rotating plate-stationary cone assembly. { ¦sər·kyə·lər ͵flō 'meth·əd }

circular form tool [DES ENG] A round or disk-shaped tool with the cutting edge on the periphery. { ¦sər·kyə·lər ͵fȯrm ͵tül }

circular functions *See* trigonometric functions. { ¦sər·kyə·lər 'fəŋk·shənz }

circular helix [MATH] A curve that lies on a right circular cylinder and intersects all the elements of the cylinder at the same angle. { ¦sər·kyə·lər 'hē͵liks }

circular horn [ELECTROMAG] A circular-waveguide section that flares outward into the shape of a horn, to serve as a feed for a microwave reflector or lens. { ¦sər·kyə·lər 'hȯrn }

circular inch [MECH] The area of a circle 1 inch (25.4 millimeters) in diameter. { ¦sər·kyə·lər 'inch }

circular magnetic wave [ELECTROMAG] A transverse magnetic wave for which the lines of magnetic force form concentric circles. { ¦sər·kyə·lər mag¦ned·ik 'wāv }

circular magnetostriction *See* Wiedemann effect. { ¦sər·kyə·lər mag¦ned·ə͵strik·shən }

circular mil [MECH] A unit equal to the area of a circle whose diameter is 1 mil (0.001 inch); used chiefly in specifying cross-sectional areas of round conductors. Abbreviated cir mil. { ¦sər·kyə·lər 'mil }

circular motion [MECH] **1.** Motion of a particle in a circular path. **2.** Motion of a rigid body in which all its particles move in circles about a common axis, fixed with respect to the body, with a common angular velocity. { ¦sər·kyə·lər 'mō·shən }

circular nomograph [MATH] A chart with concentric circular scales for three variables, laid out so that any straight line passes through values of the variables satisfying a given equation. { ¦sər·kyə·lər 'nō·mə͵graf }

circular orbit [ASTRON] An orbit comprising a complete constant-altitude revolution around the earth. { ¦sər·kyə·lər 'ȯr·bət }

circular paper chromatography [ANALY CHEM] A paper chromatographic technique in which migration from a spot in the sheet takes place in 360° so that zones separate as a series of concentric rings. { ¦sər·kyə·lər 'pā·pər ͵krō·mə'täg·rə·fē }

circular particle accelerator [NUCLEO] A particle accelerator which utilizes a magnetic field to bend charged-particle orbits and confine the extent of particle motion. Also known as circular accelerator. { ¦sər·kyə·lər 'pärd·ə·kəl ak'sel·ə͵rād·ər }

circular permutation [MATH] An arrangement of objects around a circle. { ¦sər·kyə·lər ͵pər·myə'tā·shən }

circular pitch [DES ENG] The linear measure in inches along the pitch circle of a gear between corresponding points of adjacent teeth. { ¦sər·kyə·lər 'pich }

circular plane [DES ENG] A plane that can be adjusted for convex or concave surfaces. { ¦sər·kyə·lər 'plān }

circular point [MATH] A point on a surface at which the normal curvature is the same in all directions. { ¦sər·kyə·lər 'pȯint }

circular point at infinity [MATH] In projective geometry, one of two points at which every circle intersects the ideal line. { ¦sər·kyə·lər ¦pȯint at in'fin·əd·ē }

circular polarization [PHYS] Attribute of a transverse wave (either of electromagnetic radiation, or in an elastic medium) whose electric or displacement vector is of constant amplitude and, at a fixed point in space, rotates in a plane perpendicular to the propagation direction with constant angular velocity. { ¦sər·kyə·lər ͵pō·lə·rə'zā·shən }

circular polarized loop vee [ELECTROMAG] Airborne communications antenna with an omnidirectional radiation pattern to provide optimum near-horizon communications coverage. { ¦sər·kyə·lər ¦pō·lə͵rīzd 'lüp ͵vē }

circular polling [COMMUN] A form of polling in which each terminal is interrogated exactly once in every pass, regardless of its level of activity. { ¦sər·kyə·lər 'pōl·iŋ }

circular reference [COMPUT SCI] A situation created by a programming error in which two or more entities each refer to the other so that the execution of the program is carried on endlessly with no resolution. { ¦sər·kyə·lər 'ref·rəns }

circular saw [MECH ENG] Any of several power tools for cutting wood or metal, having a thin steel disk with a toothed edge that rotates on a spindle. { ¦sər·kyə·lər 'sȯ }

circular scanning [ENG] Radar scanning in which the direction of maximum radiation describes a right circular cone. { ¦sər·kyə·lər 'skan·iŋ }

circular screen [GRAPHICS] A circular halftone screen designed to permit adjustments for correct angles for halftone color photography without disturbing the copy. { ¦sər·kyə·lər 'skrēn }

circular segment [MATH] Portion of circle cut off from the main body of the circle by a straight line (chord) through the circle. { ¦sər·kyə·lər 'seg·mənt }

circular shaft [MIN ENG] A shaft excavated in a round shape. { ¦sər·kyə·lər 'shaft }

circular shift *See* cyclic shift. { ¦sər·kyə·lər 'shift }

circular slide rule [MATH] A slide rule in a circular form whose advantages over a straight slide rule are its precision, because it is equivalent to a straight slide rule many times longer than the circular slide rule's diameter, and ease of multiplication, because the scale is continuous. { ¦sər·kyə·lər 'slīd ͵rül }

circular spike [ENG] A metal timber connector fitted with a circular series of sharp teeth that dig into the wood, preventing lateral motion, as a bolt is tightened through the wood and the spike. { ¦sər·kyə·lər 'spīk }

circular sweep generation [ELECTR] The use of electronic circuits to provide voltage or current which causes an electron beam in a device such as a cathode-ray tube to move in a circular deflection path at constant speed. { ¦sər·kyə·lər 'swēp ͵jen·ə͵rā·shən }

circular velocity [MECH] At any specific distance from the primary, the orbital velocity required to maintain a constant-radius orbit. { ¦sər·kyə·lər və'läs·əd·ē }

circular vortex [METEOROL] An atmospheric flow in parallel planes in which streamlines and other isopleths are concentric circles about a common axis; an atmospheric model of easterly and westerly winds is a circular vortex about the earth's polar axis. { ¦sər·kyə·lər 'vȯr͵teks }

circular wait *See* mutual deadlock. { ¦sər·kyə·lər 'wāt }

circular waveguide [ELECTROMAG] A waveguide whose cross-sectional area is circular. { ¦sər·kyə·lər 'wāv͵gīd }

circular word [MATH] A sequence of elements arranged clockwise around a circle. { ¦sər·kyə·lər 'wərd }

circulate-and-weight method [PETRO ENG] During drilling operations, a method of controlling well pressure in which circulation is begun immediately and mud weight is increased gradually on a predetermined schedule. Also known as concurrent method. { ¦sər·kyə͵lāt ən ¦wāt ͵meth·əd }

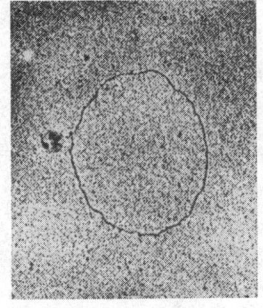

Electron micrograph of circular deoxyribonucleic acid extracted from the human wart virus. *(Courtesy of E. A. C. Follett)*

CIRCULAR NOMOGRAPH

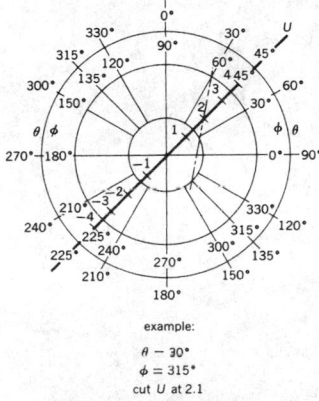

example:

$\theta = 30°$
$\phi = 315°$
cut U at 2.1

Circular nomograph which results from a trigonometric equation expressed in the form of a determinant.

CIRCULAR SAW

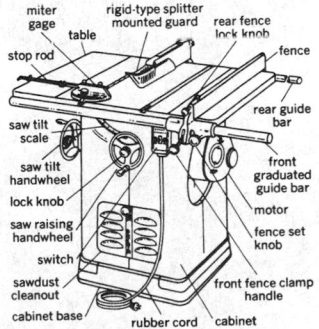

Bench circular saw with tilting arbor is used for parting or slotting and can make cuts as long as working space permits. *(Delta)*

CIRCULAR SPIKE

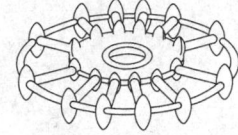

An example of a circular spike.

circulated gas-oil ratio [PETRO ENG] The volume (cubic feet) of gas introduced into a well during gas-lift operations in comparison with the volume (barrel) of oil that is lifted. { 'sər·kyə,lād·əd ¦gas ¦öil 'rā·shō }

circulating current [ASTRON] A current that circulated in an abnormal direction in the atmosphere of the southern hemisphere of Jupiter between 1919 and 1934; its presence was indicated by the behavior of dark spots in the region. { 'sər·kyə,lād·iŋ 'kər·ənt }

circulating electromagnetic wave [ELECTROMAG] An electromagnetic wave whose equiphase surfaces are half-planes originating at a common axis. { ¦sər·kyə,lād·iŋ i,lek·trō·mag¦ned·ik 'wāv }

circulating fluid [ENG] A fluid pumped into a borehole through the drill stem, the flow of which cools the bit and transports the cuttings out of the borehole. { 'sər·kyə,lād·iŋ 'flü·əd }

circulating memory [ELECTR] A digital computer device that uses a delay line to store information in the form of a pattern of pulses in a train; the output pulses are detected electrically, amplified, reshaped, and reinserted in the delay line at the beginning. Also known as delay-line memory; delay-line storage; circulating storage. { 'sər·kyə,lād·iŋ 'mem·rē }

circulating pump [CHEM ENG] Pump used to circulate process liquid out of and back into a process system, as in the circulation of distillation column bottoms through an external heater, or the circulation of storage tank bottoms to mix tank contents. { 'sər·kyə,lād·iŋ 'pəmp }

circulating reactor [NUCLEO] A nuclear reactor in which the fissionable material circulates through the core in fluid form or as small particles suspended in a fluid. { 'sər·kyə,lād·iŋ rē'ak·tər }

circulating register [COMPUT SCI] A shift register in which data move out of one end and reenter the other end, as in a closed loop. { 'sər·kyə,lād·iŋ 'rej·ə·stər }

circulating scrap [MET] At steelworks and foundries, scrap arising during the manufacture of finished iron and steel or of castings. { 'sər·kyə,lād·iŋ 'skrap }

circulating storage See circulating memory. { 'sər·kyə,lād·iŋ 'stór·ij }

circulating system [CHEM ENG] Fluid system in which the process fluid is taken from and pumped back into the system, as in the circulation of distillation column bottoms through an external heater. { 'sər·kyə,lād·iŋ 'sis·təm }

circulation [FL MECH] The flow or motion of fluid in or through a given area or volume. [MATH] For the circulation of a vector field around a closed path, the line integral of the field vector around the path. [METEOROL] For an air mass, in the line integral of the tangential component of the velocity field about a closed curve. [OCEANOGR] A water current flow occurring within a large area, usually in a closed circular pattern. [PHYSIO] The movement of blood through defined channels and tissue spaces; movement is through a closed circuit in vertebrates and certain invertebrates. { ,sər·kyə·'lā·shən }

circulation area [BUILD] The area required for human traffic in a building, including permanent corridors, stairways, elevators, escalators, and lobbies. { ,sər·kyə·'lā·shən ,er·ē·ə }

circulation flux [METEOROL] Flux due to mean atmospheric motion as opposed to eddy flux; the dominant flux in low latitudes. { ,sər·kyə·'lā·shən ,fləks }

circulation index [METEOROL] A measure of the magnitude of one of several aspects of large-scale atmospheric circulation patterns; indices most frequently measured represent the strength of the zonal (east-west) or meridional (north-south) components of the wind, at the surface or at upper levels, usually averaged spatially and often averaged in time. { ,sər·kyə·'lā·shən 'in,deks }

circulation map See traffic-circulation map. { ,sər·kyə·'lā·shən ,map }

circulation pattern [METEOROL] The general geometric configuration of atmospheric circulation usually applied, in synoptic meteorology, to the large-scale features of synoptic charts and mean charts. { ,sər·kyə·'lā·shən ,pad·ərn }

circulator [ELECTROMAG] A waveguide component having a number of terminals so arranged that energy entering one

terminal is transmitted to the next adjacent terminal in a particular direction. Also known as microwave circulator. { ,sər·kyə·'lād·ər }

circulatory system [ANAT] The vessels and organs composing the lymphatic and cardiovascular systems. { 'sər·kyə·lə,tór·ē ,sis·təm }

circulin [MICROBIO] Any of a group of peptide antibiotics produced by *Bacillus circulans* which are related to polymixin and are active against both gram-negative and gram-positive bacteria. { 'sər·kyə·lən }

circulus [BIOL] Any of various ringlike structures, such as the vascular circle of Willis or the concentric ridges on fish scales. { 'sər·kyə·ləs }

circumaural cushion [ACOUS] An earphone cushion that completely surrounds the auricle. { ,sər·kəm,ór·əl 'kúsh·ən }

circumboreal distribution [ECOL] The distribution of a Northern Hemisphere organism whose habitat includes North American, European, and Asian stations. { ¦sər·kəm¦bór·ē·əl ,dis·trə'byü·shən }

circumcenter [MATH] For a triangle or a regular polygon, the center of the circle that is circumscribed about the triangle or polygon. { ¦sər·kəm¦sen·tər }

circumcircle [MATH] A circle that passes through all the vertices of a given polygon, if such a circle exists. { 'sər·kəm,sər·kəl }

circumcision [MED] Surgical excision of the foreskin. { ,sər·kəm'sizh·ən }

circumduction [ANAT] Movement of the distal end of a body part in the form of an arc; performed at ball-and-socket and saddle joints. { ,sər·kəm'dək·shən }

circumference [MATH] **1.** The length of a circle. **2.** For a sphere, the length of any great circle on the sphere. { sər'kəm·fə·rəns }

circumferentor [ENG] A horizontal compass used in surveying that has arms diametrically placed with vertical slit sights in them. { sər'kəm·fə,ren·tər }

circumflex artery [ANAT] Any artery that follows a curving or winding course. { 'sər·kəm,fleks 'ärd·ə·rē }

circumhorizontal arc [OPTICS] A halo phenomenon consisting of a colored arc, red on its upper margin; it extends for about 90° parallel to the horizon and lies about 46° below the sun. { ¦sər·kəm,här·ə'zänt·əl 'ärk }

circumlunar [ASTRON] Around the moon; generally applied to trajectories. { ¦sər·kəm'lü·nər }

circummeridian altitude See exmeridian altitude. { ¦sər·kəm·mə'rid·ē·ən 'al·tə,tüd }

circumnutation [BOT] The bending or turning of a growing stem tip that occurs as a result of unequal rates of growth along the stem. { ¦sər·kəm·nü'tā·shən }

circum-Pacific province See Pacific suite. { ¦sər·kəm·pə'sif·ik 'prä·vəns }

circumpharyngeal connective [INV ZOO] One of a pair of nerve strands passing around the esophagus in annelids and anthropods, connecting the brain and subesophageal ganglia. { ¦sər·kəm·fə'rin·jē·əl kə'nek·tiv }

circumpolar [ASTRON] Revolving about the elevated pole without setting. [GEOGR] Located around one of the polar regions of earth. { ¦sər·kəm'pō·lər }

circumpolar star [ASTRON] A star with its polar distance approximately equal to or less than the latitude of the observer. { ¦sər·kəm'pō·lər 'stär }

circumpolar westerlies See westerlies. { ¦sər·kəm'pō·lər 'wes·tər,lēz }

circumpolar whirl See polar vortex. { ¦sər·kəm'pō·lər 'wərl }

circumradius [MATH] The radius of a circle that is circumscribed about a polygon. { ¦sər·kəm'rād·ē·əs }

circumscissile [BOT] Dehiscing along the line of a circumference, as exhibited by a pyxidium. { ¦sər·kəm'sis·əl }

circumscribed [MATH] **1.** A closed curve (or surface) is circumscribed about a polygon (or polyhedron) if every vertex of the polygon (or polyhedron) is incident upon the curve (or surface) and the polygon (or polyhedron) is contained in the curve (or surface). **2.** A polygon (or polyhedron) is circumscribed about a closed curve (or surface) if every side of the polygon (or face of the polyhedron) is tangent to the curve (or surface) and the curve (or surface) is contained within the polygon (or polyhedron). { 'sər·kəm,skrībd }

circumstellar disk [ASTRON] A flattened cloud of gas or small particles that undergoes approximately circular motion about a star, and in which the material velocity is determined primarily by the balance of gravity and centrifugal force. { ¦sər·kəm¦stel·ər 'disk }

circumvallate papilla See vallate papilla. { ¦sər·kəm¦va,lāt pa'pil·ə }

circumzenithal arc [OPTICS] A brilliant rainbow-colored arc of about a quarter of a circle with its center at the zenith and about 46° above the sun, produced by refraction and dispersion of the sun's light striking the top of prismatic ice crystals in the atmosphere, and usually lasting only a few minutes. { ¦sər·kəm¦zē·nə·thəl 'ärk }

ciré [TEXT] Any fabric treated with wax, heat, and pressure to produce a glossy appearance. { sə'rā }

cir mil See circular mil.

Cirolanidae [INV ZOO] A family of isopod crustaceans in the suborder Flabellifera composed of actively swimming predators and scavengers with biting mouthparts. { ¦sir·ə'lan·ə,dē }

cirque [GEOL] A steep elliptic to elongated enclave high on mountains in calcareous districts, usually forming the blunt end of a valley. Also known as corrie; cwm. { sərk }

cirque lake [HYD] A small body of water occupying a cirque. { 'sərk ,lāk }

Cirratulidae [INV ZOO] A family of fringe worms belonging to the Sedentaria which are important detritus feeders in coastal waters. { ¦sir·ə'tül·ə,dē }

cirrhosis [MED] A progressive, inflammatory disease of the liver characterized by a real or apparent increase in the proportion of hepatic connective tissue. { sə'rō·səs }

cirriform [METEOROL] Descriptive of clouds composed of small particles, mostly ice crystals, which are fairly widely dispersed, usually resulting in relative transparency and whiteness and often producing halo phenomena not observed with other cloud forms. [ZOO] Having the form of a cirrus; generally applied to a prolonged, slender process. { 'sir·ə,fòrm }

Cirripedia [INV ZOO] A subclass of the Crustacea, including the barnacles and goose barnacles; individuals are free-swimming in the larval stages but permanently fixed in the adult stage. { ¦sir·ə'pēd·ē·ə }

cirrocumulus cloud [METEOROL] A principal cloud type, appearing as a thin, white path of cloud without shadows, composed of very small elements in the form of grains, ripples, and so on. Abbreviated Cc. { ¦sir·ō'kyü·myə·ləs ¦klaùd }

Cirromorpha [INV ZOO] A suborder of cephalopod mollusks in the order Octopoda. { ¦sir·ō¦mòr·fə }

cirrostratus cloud [METEOROL] A principal cloud type, appearing as a whitish veil, usually fibrous but sometimes smooth, which may totally cover the sky and often produces halo phenomena, either partial or complete. Abbreviated Cs. { ¦sir·ō'strad·əs ¦klaùd }

cirrus [INV ZOO] **1.** The conical locomotor structure composed of fused cilia in hypotrich protozoans. **2.** Any of the jointed thoracic appendages of barnacles. **3.** Any hairlike tuft on insect appendages. **4.** The male copulatory organ in some mollusks and trematodes. [VERT ZOO] Any of the tactile barbels of certain fishes. [ZOO] A tendrillike animal appendage. { 'sir·əs }

cirrus cloud [METEOROL] A principal cloud type composed of detached cirriform elements in the form of white, delicate filaments, of white (or mostly white) patches, or narrow bands. Abbreviated Ci. { 'sir·əs ¦klaùd }

cirrus sac [INV ZOO] A pouch or channel containing the copulatory organ (cirrus) in certain invertebrates. { 'sir·əs ,sak }

cis [ORG CHEM] A descriptive term indicating a form of isomerism in which atoms are located on the same side of an asymmetric molecule. { sis }

cis-active [MOL BIO] Describing a genetic element (such as a promoter or other regulatory locus) that promotes or suppresses two unrelated targets (such as genes) on the same chromosome as a result of their relative positions on the chromosome. { ¦sis 'ak·tiv }

CISC See complex instruction set computer. { sisk }

cisco [VERT ZOO] Any of several North American freshwater whitefishes of the genus *Coregonus*. { 'sis,kō }

cisele [TEXT] A velvet fabric on which the pattern is formed by contrast between cut and uncut pile loops. { ¦sēz·ə¦lā }

cislunar [ASTRON] Of or pertaining to phenomena, projects, or activity in the space between the earth and moon, or between the earth and the moon's orbit. { ¦sis'lü·nər }

cisplatin [MED] A transition-metal complex that is used in the treatment of cancer, with particular effectiveness in the treatment of testicular and ovarian cancers; second- and third-generation variants of this drug have been developed to mitigate undesirable side effects. { ¦sis'plat·ən }

cissoid [MATH] A plane curve consisting of all points which lie on a variable line passing through a fixed point, and whose distance from the fixed point is equal to the distance between the intersections of the line with two given curves. { 'sis,òid }

cissoid of Diocles [MATH] The cissoid of a circle and a tangent line with respect to a fixed point on the circumference of the circle diametrically opposite the point of tangency. { 'si,sòid əv 'dī·ə·klēz }

cistern [ANAT] A closed, fluid-filled sac or vesicle, such as the subarachnoid spaces or the vesicles comprising the dictyosomes of a Golgi apparatus. [CIV ENG] A tank for storing water or other liquid. [GEOL] A hollow that holds water. { 'sis·tərn }

cistern barometer [ENG] A pressure-measuring device in which pressure is read by the liquid rise in a vertical, closed-top tube as a result of system pressure on a liquid reservoir (cistern) into which the bottom, open end of the tube is immersed. { 'sis·tərn bə'ram·əd·ər }

cis-trans isomerism [ORG CHEM] A type of geometrical isomerism found in alkenic systems in which it is possible for each of the doubly bonded carbons to carry two different atoms or groups; two similar atoms or groups may be on the same side (cis) or on opposite sides (trans) of a plane bisecting the alkenic carbons and perpendicular to the plane of the alkenic system. { 'si¦stranz ī'säm·ə,riz·əm }

cistron [MOL BIO] The genetic unit (deoxyribonucleic acid fragment) that codes for a particular polypeptide; mutants do not complement each other within a cistron. Also known as structural gene. { 'sis,trän }

Citheroniinae [INV ZOO] Subfamily of lepidopteran insects in the family Saturniidae, including the regal moth and the imperial moth. { ,sith·ə·rō'nī·ə,nē }

citizens' band [COMMUN] A frequency band allocated for citizens' radio service (462.550–467.425, 72–76, or 26.965–27.405 megahertz). Also known as citizens' waveband. { 'sit·ə·zənz ,band }

citizens' radio service [COMMUN] A radio communication service intended for private or personal radio communication, including radio signaling and control of objects by radio. { 'sit·ə·zənz 'rād·ē·ō ,sər·vəs }

citraconic acid [ORG CHEM] $C_5H_6O_4$ A dicarboxylic acid; hygroscopic crystals that melt at 91°C; derived from citric acid by heating. { ,si·trə¦kän·ik 'as·əd }

citral [ORG CHEM] $C_{10}H_{16}O$ A pale-yellow liquid that in commerce is a mixture of two isomeric forms, alpha and beta; insoluble in water, soluble in glycerin or benzyl benzoate; used in perfumery and as an intermediate to form other compounds. Also known as geranial; geranialdehyde. { 'si,tral }

citramalase [BIOCHEM] An enzyme that is involved in the fermentation of glutamate by *Clostridium tetanomorphum*; catalyzes the breakdown of citramalic acid to acetate and pyruvate. { ,si·trə'ma,lās }

citrate [BIOCHEM] A salt or ester of citric acid. { 'si,trāt }

citrate test [MICROBIO] A differential cultural test to identify genera within the bacterial family Enterobacteriaceae that are able to utilize sodium citrate as a sole source of carbon. { 'si,trāt ,test }

citric acid [BIOCHEM] $C_6H_8O_7 \cdot H_2O$ A colorless crystalline or white powdery organic, tricarboxylic acid occurring in plants, especially citrus fruits, and used as a flavoring agent, as an antioxidant in foods, and as a sequestering agent; the commercially produced form melts at 153°C. { 'si,trik 'as·əd }

citric acid cycle See Krebs cycle. { 'si,trik 'as·əd 'sī·kəl }

citriculture [BOT] The cultivation of citrus fruits. { 'si·trə,kəl·chər }

citrine [MINERAL] An important variety of crystalline quartz, yellow to brown in color and transparent. Also known as Bohemian topaz; false topaz; quartz topaz; topaz quartz; yellow quartz. { 'si,trēn }

Citrobacter [MICROBIO] A genus of bacteria in the family Enterobacteriaceae; motile rods that utilize citrate as the only carbon source. { ,si·trō'bak·tər }

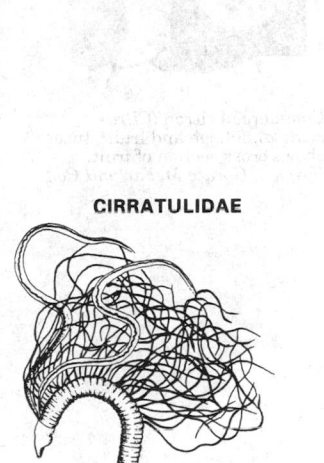

CIRRATULIDAE

Chaetozone in left lateral view.

CIS-TRANS ISOMERISM

cis-2-Butylene

trans-2-Butylene

Cis-trans isomerism among olefins.

CITRON

Commercial citron (*Citrus medica*), foliage and fruit. Inset shows cross section of fruit. (*From J. Horace McFarland Co.*)

CLADOSELACHII

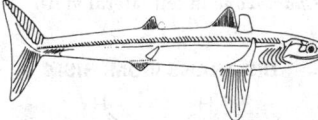

Cladoselache, a Late Devonian sharklike fish.

citron [BOT] *Citrus medica.* A shrubby, evergreen citrus tree in the order Sapindales cultivated for its edible, large, lemonlike fruit. { 'si·trən }

citronella [BOT] *Cymbopogon nardus.* A tropical grass; the source of citronella oil. { ˌsi·trə'nel·ə }

citronella oil [MATER] A yellowish oil distilled from the leaves of either of two grasses, *Cymbopogon nardus* or *C. winterianus*; used as an insect repellent. Also known as Java citronella oil. { ˌsi·trə'nel·ə 'öil }

citronellal hydrate *See* hydroxycitronellal. { ˌsi·trə'nel·əl 'hī,drāt }

citronellol [ORG CHEM] $C_{10}H_{19}OH$ A liquid derived from citronella oil; soluble in alcohol; used in perfumery. { ˌsi·trə'nel,öl }

citrulline [BIOCHEM] $C_6H_{13}O_3N_3$ An amino acid formed in the synthesis of arginine from ornithine. { 'si·trə,lēn }

citrus anthracnose [PL PATH] A fungus disease of citrus plants caused by *Colletotrichum gloeosporioides* and characterized by tip blight, stains on the leaves, and spots, stains, or rot on the fruit. { 'si·trəs ˌan'thrak,nōs }

citrus blast [PL PATH] A bacterial disease of citrus trees caused by *Pseudomonas syringae* and marked by drying and browning of foliage and twigs and black pitting of the fruit. { 'si·trəs ˌblast }

citrus canker [PL PATH] A bacterial disease of citrus plants caused by *Xanthomonas citri* and producing lesions on twigs, foliage, and fruit. { 'si·trəs ˌkaŋ·kər }

citrus flavonoid compound *See* bioflavonoid. { 'si·trəs 'flav·ə,nóid ˌkäm,pau̇nd }

citrus fruit [BOT] Any of the edible fruits having a pulpy endocarp and a firm exocarp that are produced by plants of the genus *Citrus* and related genera. { 'si·trəs ˌfrüt }

citrus gummosis [PL PATH] A disease of citrus trees caused by the fungus *Phytophthora citrophthora*, characterized by the formation of narrow cracks in the bark which exude a pale yellow gum; infection is favored by excessive moisture. { ˌsi·trəs gə'mō·səs }

citrus scab [PL PATH] A fungus disease of citrus plants caused by *Sphaceloma rosarum*, producing scablike lesions on all plant parts. { 'si·trəs ˌskab }

city plan [MAP] A large-scale, comprehensive map of a city delineating streets, important buildings, and other urban elements; relief is shown when important. Also known as town plan. { 'sid·ē ˌplan }

civet [PHYSIO] A fatty substance secreted by the civet gland; used as a fixative in perfumes. [VERT ZOO] Any of 18 species of catlike, nocturnal carnivores assigned to the family Viverridae, having a long head, pointed muzzle, and short limbs with nonretractile claws. { 'siv·ət }

civet cat *See* cacomistle. { 'siv·ət ˌkat }

civet gland [VERT ZOO] A large anal scent gland in civet cats that secretes civet. { 'siv·ət ˌgland }

civetone [BIOCHEM] $C_{17}H_{30}O$ 9-Cycloheptadecen-1-one, a macrocyclic ketone component of civet used in perfumes because of its pleasant odor and lasting quality; believed to function as a sex attractant among civet cats. { 'siv·ə,tōn }

civil airway [NAV] An airway designated for air commerce. { 'siv·əl 'er,wā }

civil day [ASTRON] A mean solar day beginning at midnight instead of at noon; may be based on either apparent solar time or mean solar time. { 'siv·əl ˌdā }

civil engineering [ENG] The planning, design, construction, and maintenance of fixed structures and ground facilities for industry, transportation, use and control of water, or occupancy. { 'siv·əl en·jə'nir·iŋ }

civil time [ASTRON] Solar time in a day (civil day) that begins at midnight; may be either apparent solar time or mean solar time. { 'si·vəl ˌtīm }

civil twilight [ASTRON] The interval of incomplete darkness between sunrise (or sunset) and the time when the center of the sun's disk is 6° below the horizon. { 'si·vəl ˌtwī,līt }

civil year *See* calendar year. { 'si·vəl ˌyir }

Cl *See* chlorine.

cladding [COMMUN] A plastic or glass sheath that is fused to and surrounds the core of an optical fiber. [ENG] Process of covering one material with another and bonding them together under high pressure and temperature. Also known as bonding. [NUCLEO] An outer jacket, usually metallic, for a nuclear fuel element; prevents corrosion of fuel and release of fission products into the coolant. { 'klad·iŋ }

clade [SYST] A taxonomic group containing a common ancestor and its descendants. { klād }

cladism [SYST] A theory of taxonomy by which organisms are grouped and ranked on the basis of the most recent phylogenetic branching point. { 'kla,diz·əm }

Cladistia [VERT ZOO] The equivalent name for Polypteriformes. { kla'dis·tē·ə }

clad metal [MET] A metal overlaid on one or both sides with a different metal. { 'klad ˌmed·əl }

Cladocera [INV ZOO] An order of small, fresh-water branchiopod crustaceans, commonly known as water fleas, characterized by a transparent bivalve shell. { kla'däs·ə·rə }

Cladocopa [INV ZOO] A suborder of the order Myodocopida including marine animals having a carapace that lacks a permanent aperture when the two valves are closed. { kla'däk·ə·pə }

Cladocopina [INV ZOO] The equivalent name for Cladocopa. { ˌklad·ə'käp·ə·nə }

cladode *See* cladophyll. { 'kla,dōd }

cladodont [PALEON] Pertaining to sharks of the most primitive evolutionary level. { 'klad·ə,dänt }

cladogenesis [EVOL] Evolution associated with altered habit and habitat, usually in isolated species populations. { ˌklad·ə'jen·ə·səs }

cladogenic adaptation *See* divergent adaptation. { ˌkladə'jen·ik ˌad,ap'tā·shən }

cladogram [EVOL] A dendritic diagram which shows the evolution and descent of a group of organisms. { 'klad·ə,gram }

Cladoniaceae [BOT] A family of lichens in the order Lecanorales, including the reindeer mosses and cup lichens, in which the main thallus is hollow. { klaˌdō·nē'as·ē,ē }

Cladophorales [BOT] An order of coarse, wiry, filamentous, branched and unbranched algae in the class Chlorophyceae. { klaˌdäf·ə'rā·lēz }

cladophyll [BOT] A branch arising from the axil of a true leaf and resembling a foliage leaf. Also known as cladode. { 'klad·ə,fil }

cladoptosis [BOT] The annual abscission of twigs or branches instead of leaves. { ˌkla,däp'tō·səs }

Cladoselachii [PALEON] An order of extinct elasmobranch fishes including the oldest and most primitive of sharks. { ˌklad·ō·sə'lāk·ē,ī }

cladus [BOT] A branch of a ramose spicule. { 'klā·dəs }

Claibornian [GEOL] Middle Eocene geologic time. { ˌkler'bór·n·ē·ən }

claim *See* mining claim. { klām }

Clairaut's formula [GEOD] An approximate formula for gravity at the earth's surface, assuming that the earth is an ellipsoid; states that the gravity is equal to $g_e [1+(^5/_2 m' - f) \sin^2 \theta]$, where θ is the latitude, g_e is the gravity at the equator, m' is the ratio of centrifugal acceleration to gravity at the equatorial surface, and f is the earth's flattening, equal to $(a - b)/a$, where a is the semimajor axis and b is the semiminor axis. { 'kler·óz ˌfòr·myə·lə }

clairite *See* enargite. { 'kle,rīt }

clairvoyance [PSYCH] A form of extrasensory perception in which there is a receiver and an extant event of which he has knowledge that has not been conveyed to him through his sensory channels. { kler'vói·əns }

Claisen condensation [ORG CHEM] **1.** Condensation, in the presence of sodium ethoxide, of esters or of esters and ketones to form β-dicarbonyl compounds. **2.** Condensation of arylaldehydes and acylphenones with esters or ketones in the presence of sodium ethoxide to yield unsaturated esters. Also known as Claisen reaction. { 'klās·ən känd·ən'sā·shən }

Claisen flask [CHEM] A glass flask with a U-shaped neck, used for distillation. { 'klās·ən ˌflask }

Claisen reaction *See* Claisen condensation. { 'klās·ən ri'äk·shən }

Claisen rearrangement [ORG CHEM] A thermally induced sigmatrophic shift in which an allyl phenyl ether is rearranged to yield an *ortho*-allylphenol. { 'klā·sən ˌrē·ə'rānj·mənt }

Claisen-Schmidt condensation [ORG CHEM] A reaction employed for preparation of unsaturated aldehydes and ketones by condensation of aromatic aldehydes with aliphatic aldehydes or ketones in the presence of sodium hydroxide. { ¦klās·ən ¦shmit känd·on'sā·shən }

clam [INV ZOO] The common name for a number of species of bivalve mollusks, many of which are important as food. { klam }

Clambidae [INV ZOO] The minute beetles, a family of coleopteran insects in the superfamily Dascilloidea. { 'klam bə¸dē }

clammy [BIOL] Moist and sticky, as the skin or a stem. { 'klam·ē }

clamp [DES ENG] A tool for binding or pressing two or more parts together, by holding them firmly in their relative positions. [ELECTR] *See* clamping circuit. { klamp }

clamp connection [MYCOL] In the Basidiomycetes, a lateral connection formed between two adjoining cells of a filament and arching over the septum between them and permitting a type of pseudosexual activity. { 'klamp kə¸nek·shən }

clamper *See* direct-current restorer. { 'klamp·ər }

clamping [ELECTR] The introduction of a reference level that has some desired relation to a pulsed waveform, as at the negative or positive peaks. Also known as direct-current reinsertion; direct-current restoration. { 'klamp·iŋ }

clamping circuit [ELECTR] A circuit that reestablishes the direct-current level of a waveform; used in the dc restorer stage of a television receiver to restore the dc component to the video signal after its loss in capacitance-coupled alternating-current amplifiers, to reestablish the average light value of the reproduced image. Also known as clamp. { 'klamp·iŋ ¸sər·kət }

clamping coupling [MECH ENG] A coupling with a split cylindrical element which clamps the shaft ends together by direct compression, through bolts or rings, and by the wedge action of conical sections; not considered a permanent part of the shaft. { 'klamp·iŋ ¸kəp·liŋ }

clamping diode [ELECTR] A diode used to clamp a voltage at some point in a circuit. { 'klamp·iŋ ¸dī¸ōd }

clamping gripper [CONT SYS] A robot element that uses two-link movements, parallel-jaw movements, and combination movements to grasp and handle objects. { 'klamp·iŋ 'grip·ər }

clamping plate [ENG] A plate on a mold which attaches the mold to a machine. { 'klamp·iŋ ¸plāt }

clamping pressure [ENG] In injection and transfer-molding of plastics, the pressure applied to keep the mold closed in opposition to the fluid pressure of the molding material. { 'klamp·iŋ ¸presh·ər }

clamp-on [COMMUN] A method of holding a call for a line that is in use and of signaling when it becomes free. { 'klamp¸ón }

clamp-on ammeter *See* snap-on ammeter. { 'klamp¸ón 'a¸mēd·ər }

clamp screw [DES ENG] A screw that holds a part by forcing it against another part. { 'klamp ¸skrü }

clamp-screw sextant [ENG] A marine sextant having a clamp screw for controlling the position of the tangent screw. { 'klamp ¸skrü ¸seks·tənt }

clamshell bucket [MECH ENG] A two-sided bucket used in a type of excavator to dig in a vertical direction; the bucket is dropped while its leaves are open and digs as they close. Also known as clamshell grab. { 'klam¸shel ¸bək·ət }

clamshell grab *See* clamshell bucket. { 'klam¸shel ¸grab }

clamshell snapper [MECH ENG] A marine sediment sampler consisting of snapper jaws and a footlike projection which, upon striking the bottom, causes a spring mechanism to close the jaws, thus trapping a sediment sample. { 'klam¸shel ¸snap·ər }

clam worm [INV ZOO] The common name for a number of species of dorsoventrally flattened annelid worms composing the large family Nereidae in the class Polychaeta; all have a distinct head, with numerous appendages. { 'klam ¸wərm }

clan [ECOL] A very small community, perhaps a few square yards in area, in climax formation, and dominated by one species. [PETR] A category of igneous rocks defined in terms of similarities in mineralogical or chemical composition. { klan }

clapboard [MATER] A board, thicker at one edge than the other, used to cover exterior walls. { 'kla¸bərd }

Clapeyron-Clausius equation *See* Clausius-Clapeyron equation. { kla·pā·rōn ¦klōz·ē·əs i¸kwā·zhən }

Clapeyron equation *See* Clausius-Clapeyron equation. { kla·pā·rōn i'kwā·zhən }

Clapeyron's theorem [MECH] The theorem that the strain energy of a deformed body is equal to one-half the sum over three perpendicular directions of the displacement component times the corresponding force component, including deforming loads and body forces, but not the six constraining forces required to hold the body in equilibrium. { kla·pā·rōnz ¸thir·əm }

clapper [ELEC] A hinged or pivoted relay armature. { 'klap·ər }

clapper box [MECH ENG] A hinged device that permits a reciprocating cutting tool (as in a planer or shaper) to clear the work on the return stroke. { 'klap·ər ¸bäks }

Clapp oscillator [ELECTR] A series-tuned Colpitts oscillator, having low drift. { 'klap ¸äs·ə¸läd·ər }

clarain [GEOL] A coal lithotype appearing as stratifications parallel to the bedding plane and usually having a silky luster and scattered or diffuse reflection. Also known as clarite. { 'kla¸rān }

Clarendonian [GEOL] Lower Pliocene or upper Miocene geologic time. { ¸kla·rən'dōn·ē·ən }

clarification [CHEM ENG] The removal of small amounts (usually less than 0.2%) of fine particulate solids from liquids (such as drinking water) by methods such as gravity sedimentation, centrifugal sedimentation, filtration, and magnetic separation. { ¸klar·ə·fə'kā·shən }

clarified oil [MATER] The heavy oil which is taken from the bottom of a fractionator in a catalytic cracking process and from which residual catalyst has been removed. { 'klar·ə¸fīd 'óil }

clarifier [ENG] A device for filtering a liquid. { 'klar·ə¸fī·ər }

clarifying agent *See* fining. { 'klar·ə¸fī·iŋ ¸ā·jənt }

clarifying centrifuge [MECH ENG] A device that clears liquid of foreign matter by centrifugation. { 'klar·ə¸fī·iŋ 'sen·trə¸fyüj }

clarifying filter [ENG] Any filter, such as a sand filter or a cartridge filter, used to purify liquids with a low solid-liquid ratio; in some instances color may be removed as well. { 'klar·ə¸fī·iŋ ¸fil·tər }

Clariidae [VERT ZOO] A family of Asian and African catfishes in the suborder Siluroidei. { kla'rī·ə¸dē }

clarinite [MINERAL] A heterogeneous, generally translucent material making up the major micropetrological ingredient of clarain. { 'klar·ə¸nīt }

clarite *See* clarain. { 'kla¸rīt }

clarity [CHEM ENG] Measure of the amount of opaque suspended solids in a liquid, determined by visual or optical methods. { 'klar·əd·ē }

Clark cell [ELEC] An early form of standard cell, having 1.433 volts at 15°C, now largely replaced by the Weston standard cell as a voltage standard. { 'klärk ¸sel }

Clark degree *See* English degree. { 'klärk də¸grē }

clarke [GEOCHEM] A unit of the average abundance of an element in the earth's crust, expressed as a percentage. Also known as crustal abundance. { klärk }

Clarkecarididae [PALEON] A family of extinct crustaceans in the order Anaspidacea. { ¸klär·kə'rid·ə¸dē }

Clarke ellipsoid of 1866 [GEOD] The reference ellipsoid adopted by the U.S. Coast and Geodetic Survey of 1880 for charting North America. { 'klärk ə'lip¸sóid əv ā'tēn'sik·stē'siks }

clarkeite [MINERAL] $(Na,Ca,Pb)_2U_2(O,OH)_7$ A dark reddish-brown or dark brown mineral consisting of a hydrous or hydrated uranium oxide. { 'klär¸kīt }

Clarke's soap solution [MATER] Reagent used in standard APHA (American Public Health Association) method to estimate hardness in water; consists of powdered castile soap in 80% ethyl alcohol solution. { 'klärks 'sōp sə'lü·shən }

Clark process [CHEM ENG] Softening of water by adding alkaline solutions of calcium hydroxide so that the acid carbonates are converted to normal carbonates. { 'klärk ¸präs·əs }

clarodurain [GEOL] A transitional lithotype of coal composed of vitrinite and other macerals, principally micrinite and exinite. { ¦kla·rō'dü¸rān }

clarofusain [GEOL] A transitional lithotype of coal composed of fusinite and vitrinite and other macerals. { ¦kla·rō'fyü¸zān }

clarovitrain [GEOL] A transitional lithotype of coal rock

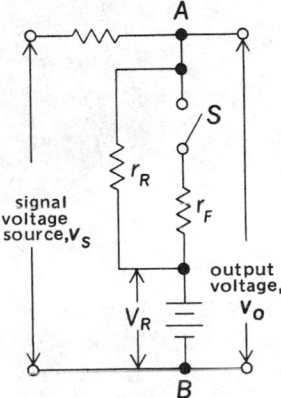

CLAMPING CIRCUIT

Circuit diagram of the elements of a clamping circuit.

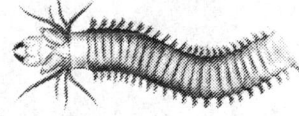

CLAM WORM

Anterior end of a clam worm.

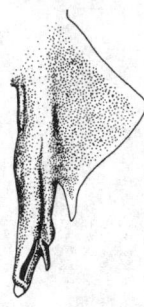

CLASPER

The clasper of the dogfish
(*Squalus*).

composed primarily of the maceral vitrinite, with lesser amounts of other macerals. { ˌklaˈrōˈviˌtrān }

clasp [DES ENG] A releasable catch which holds two or more objects together. { klasp }

clasper [VERT ZOO] A modified pelvic fin of male elasmobranchs and holocephalians used for the transmission of sperm. { 'klasp·ər }

clasp lock [DES ENG] A spring lock with a self-locking feature. { 'klasp ˌläk }

clasp nut [DES ENG] A split nut that clasps a screw when closed around it. { 'klasp ˌnət }

class [COMPUT SCI] In object-oriented programming, a description of the structure and operations of an object. A new class is defined by stating how it differs from an existing class. The new (more specific) class is said to inherit from the original (general) class and is referred to as a subclass of the original class. The original class is referred to as the superclass of the new class. [MATH] **1.** A set that consists of all the sets having a specified property. **2.** The class of a plane curve is the largest number of tangents that can be drawn to the curve from any point in the plane that is not on the curve. [STAT] A collection of adjacent values of a random variable. [SYST] A taxonomic category ranking above the order and below the phylum or division. { klas }

class A amplifier [ELECTR] **1.** An amplifier in which the grid bias and alternating grid voltages are such that anode current in a specific tube flows at all times. **2.** A transistor amplifier in which each transistor is in its active region for the entire signal cycle. { ˌklas 'ā 'am·plə,fī·ər }

class AB amplifier [ELECTR] **1.** An amplifier in which the grid bias and alternating grid voltages are such that anode current in a specific tube flows for appreciably more than half but less than the entire electric cycle. **2.** A transistor amplifier whose operation is class A for small signals and class B for large signals. { ˌklas ˈāˌbē 'am·plə,fī·ər }

class A modulator [ELECTR] A class A amplifier used to supply the necessary signal power to modulate a carrier. { ˌklas 'ā 'mäj·ə,lād·ər }

class A push-pull sound track [ENG ACOUS] Two single photographic sound tracks side by side, the transmission of one being 180° out of phase with the transmission of the other; both positive and negative halves of the sound wave are linearly recorded on each of the two tracks. { ˌklas 'ā ˌpushˌpul 'saun ,trak }

class B amplifier [ELECTR] **1.** An amplifier in which the grid bias is approximately equal to the cutoff value, so that anode current is approximately zero when no exciting grid voltage is applied, and flows for approximately half of each cycle when an alternating grid voltage is applied. **2.** A transistor amplifier in which each transistor is in its active region for approximately half the signal cycle. { ˌklas 'bē 'am·plə,fī·ər }

class B auxiliary power [ELEC] Standby power plant to cover extended outages (days) of primary power. { ˌklas 'bē óg'zil·yə·rē 'paur }

class B modulator [ELECTR] A class B amplifier used to supply the necessary signal power to modulate a carrier; usually connected in push-pull. { ˌklas 'bē 'mäj·ə,lād·ər }

class B push-pull sound track [ENG ACOUS] Two photographic sound tracks side by side, one of which carries the positive half of the signal only, and the other the negative half; during the inoperative half-cycle, each track transmits little or no light. { ˌklas 'bē ˌpushˌpul 'saun ,trak }

class C^n [MATH] The class of all functions that are continuous on a given domain and have continuous derivatives of all orders up to and including the *n*th. { ˌklas 'sē 'en }

class C amplifier [ELECTR] **1.** An amplifier in which the bias on the control element is appreciably greater than the cutoff valve, so that the output current in each device is zero when no alternating control signal is applied, and flows for appreciably less than half of each cycle when an alternating control signal is applied. **2.** A transistor amplifier in which each transistor is in its active region for significantly less than half the signal cycle. { ˌklas 'sē 'am·plə,fī·ər }

class C auxiliary power [ELEC] Quick start (10-60 seconds) power unit to cover short-term outages (hours) of primary power. { ˌklas 'sē óg'zil·yə·rē 'paur }

class D amplifier [ELECTR] A power amplifier that employs a pair of transistors that are connected in push-pull and driven to act as a switch, and a series-tuned output filter, which allows only the fundamental-frequency component of the resultant square wave to reach the load. { ˌklas 'dē 'am·plə,fī·ər }

class D auxiliary power [ELEC] Uninterruptible (no-break) power unit using stored energy to provide continuous power within specified voltage and frequency tolerances. { ˌklas 'dē óg'zil·yə·rē 'paur }

class E amplifier [ELECTR] A power amplifier that employs a single transistor driven to act as a switch, and an output filter selected to bring the drain voltage to zero at the instant the transistor is switched on. { ˌklas 'ē 'am·plə,fī·ər }

class F amplifier [ELECTR] A power amplifier that employs a single transistor and a multiple-resonance output circuit. { ˌklas 'ef 'am·plə,fī·ər }

class formula [MATH] A formula which states that the order of a finite group G is equal to the sum, over a set of representatives x_i of the distinct conjugacy classes of G, of the index of the normalizer of x_i in G. { 'klas ˌfòr·myə·lə }

class frequency [STAT] The frequency with which a random variable assumes the values included in a given class interval. { 'klas ˌfrē·kwən·sē }

classical approximation [QUANT MECH] The approximation that Planck's constant may be considered infinitely small; the laws of quantum mechanics must then reduce to those of classical mechanics. { 'klas·ə·kəl ə,präk·sə'mā·shən }

classical attenuation [ACOUS] Sound absorption through mechanisms that do not involve molecular relaxation, namely, shear viscosity, heat conduction, heat radiation, and diffusion. Also known as thermoviscous attenuation. { ˌklas·i·kəl ə,ten·yə'wā·shən }

classical canonical matrix [MATH] A form to which any matrix can be reduced by a collineatory transformation, with zeros except for a sequence of Jordan matrices siutated along the principal diagonal. { 'klas·ə·kəl kə'nän·ə·kəl 'mā·triks }

classical conductivity theory [STAT MECH] A theory which treats the system of electrons in a metal as a gas and uses the Boltzmann transport equation to calculate conductivity. Also known as Lorentz conductivity theory. { 'klas·ə·kəl ,kän·dək'tiv·əd·ē 'thē·ə·rē }

classical electron radius [ELECTROMAG] The quantity expressed as e^2/m_ec^2, where e is the electron's charge in electrostatic units, m_e its mass, and c the speed of light; equal to approximately 2.82×10^{-13} centimeter. { 'klas·ə·kəl i'lek,trän 'rād·ē·əs }

classical field theory [PHYS] The study of distributions of energy, matter, and other physical quantities under circumstances where their discrete nature is unimportant, and they may be regarded as (in general, complex) continuous functions of position. Also known as c-number theory; continuum mechanics; continuum physics. { 'klas·ə·kəl 'fēld ,thē·ə·rē }

classical Kuiper Belt object [ASTRON] A member of the Kuiper belt that has a near-circular orbit, almost in the ecliptic plane, and has a period of revolution which is outside the resonances with Neptune's period. { ˌklas·i·kəl 'kī·pər ,belt ˌäb·jekt }

classical mechanics [MECH] Mechanics based on Newton's laws of motion. { 'klas·ə·kəl mə'kan·iks }

classical pathway [IMMUNOL] The pathway by which antigen-antibody complex activates the complement system. { ˌklas·i·kəl 'path,wā }

classical physics [PHYS] The branch of physics that is based on the assumption of Newtonian mechanics and excludes relativity and quantum mechanics. { 'klas·ə·kəl 'fiz·iks }

classical T Tauri star [ASTRON] A T Tauri star that exhibits strong emission lines in its optical spectrum, emits a strong stellar wind, and accretes material from a circumstellar disk. { ˌklas·ə·kəl ˌtē ˌtòr·ē 'stär }

classical wave equation See wave equation. { 'klas·ə·kəl 'wāv i'kwā·zhən }

classic botulism [MED] Botulism typically due to ingestion of preformed toxin. { ˌklas·ik 'bäch·ə,liz·əm }

classic epidemic typhus [MED] An epidemic disease caused by *Rickettsia prowazeki* var. *prowazekii*, and characterized by violent headache, a rash, neurological symptoms, and high fever. Also known as epidemic typhus. { 'klas·ik ,ep·ə'dem·ik 'tī,fəs }

classification [ENG] **1.** Sorting out or categorizing of particles or objects by established criteria, such as size, function, or color. **2.** Stratification of a mixture of various-sized particles (that is, sand and gravel), with the larger particles migrating to the bottom. [IND ENG] *See* grading. [ORD] Placing of military documents in special groups for safeguarding defense information. [SYST] A systematic arrangement of plants and animals into categories based on a definite plan, considering evolutionary, physiologic, cytogenetic, and other relationships. { ‚klas·ə·fə'kā·shən }

classification societies [NAV ARCH] Private organizations which issue rules for the construction, equipment, and maintenance of merchant ships. { ‚klas·ə·fə'kā·shən sə'sī·ə,dēz }

classification track [CIV ENG] A railroad track used to separate cars from a train according to destination. { ‚klas·ə·fə'kā·shən ‚trak }

classification yard [CIV ENG] A railroad yard for separating trains according to car destination. { ‚klas·ə·fə'kā·shən ‚yärd }

classifier [MECH ENG] Any apparatus for separating mixtures of materials into their constituents according to size and density. { 'klas·ə‚fī·ər }

classify [SCI TECH] To sort into groups that have common properties. { 'klas·ə·fī }

class interval [STAT] One of several convenient intervals into which the values of the variate of a frequency distribution may be grouped. { 'klas 'int·ər·vəl }

class limits [STAT] The lower and upper limits of a class interval. { 'klas ‚lim·its }

class mark [STAT] The mid-value of a class interval, or the integral value nearest the midpoint of the interval. { 'klas ‚märk }

class NP problems [COMPUT SCI] Problems that cannot necessarily be solved in polynomial time on a sequential computer but can be solved in polynomial time on a nondeterministic computer which, roughly speaking, guesses in turn each of $2N$ possible values of some N-bit quantity. { 'klas ‚en‚pē ‚präb·ləmz }

classons [PARTIC PHYS] Massless bosons which are quanta of the two classical fields, gravitational and electromagnetic. { 'kla‚sänz }

class P problems [COMPUT SCI] Problems that can be solved in polynomial time on a conventional sequential computer. { 'klas 'pē ‚präb·ləmz }

class S modulator [ELECTR] A modulator that is based on pulse-width modulation with a switching frequency several times the highest output frequency, and in which the pulse-width modulated signal is boosted to the desired power level by switching amplifiers, after which the desired audio output is obtained by a low-pass filter. { ‚klas 'es 'mäj·ə‚lād·ər }

class switch [IMMUNOL] A switch of B-lymphocyte expression from one antibody class to another. { 'klas ‚swich }

class II-associated invariant chain peptide [IMMUNOL] A residual fragment of the invariant chain that is essential for proper loading of exogenous peptide on the class II major histocompatibility complex. { ‚klas ‚tü ə‚sō·se‚ād·əd in‚ver·ē·ənt ‚chān 'pep‚tīd }

clast [GEOL] An individual grain, fragment, or constituent of detrital sediment or sedimentary rock produced by physical breakdown of a larger mass. { klast }

clastation *See* weathering. { kla'stā·shən }

clastic [GEOL] Rock or sediment composed of clasts which have been transported from their place of origin, as sandstone and shale. { 'klas·tik }

clastic dike [GEOL] A tabular-shaped sedimentary dike composed of clastic material and transecting the bedding of a sedimentary formation; represents invasion by extraneous material along a crack of the containing formation. { 'klas·tik 'dīk }

clastic pipe [GEOL] A cylindrical body of clastic material having an irregular columnar or pillarlike shape, standing approximately vertically through enclosing formations (usually limestone), and measuring a few centimeters to 50 meters (165 feet) in diameter and 1 to 60 meters (3 to 200 feet) in height. { 'klas·tik 'pīp }

clastic ratio [GEOL] The ratio of the percentage of clastic rocks to that of nonclastic rocks in a geologic section. Also known as detrital ratio. { 'klas·tik 'rā·shō }

clastic reservoir [GEOL] An underground oil or gas trap formed in clastic limestone. { 'klas·tik 'rez·əv‚wär }

clastic sediment [GEOL] Deposits of clastic materials transported by mechanical agents. Also known as mechanical sediment. { 'klas·tik 'sed·ə·mənt }

clastic wedge [GEOL] The sediments of the exogeosyncline, derived from the tectonic landmasses of the adjoining orthogeosyncline. { 'klas·tik 'wej }

clastogenesis [MOL BIO] The loss, addition, or rearrangement of chromosomes. { ‚klas·tə'jen·ə·səs }

clathrate [BIOL] *See* cancellate. [CHEM] An inclusion compound in which the guest species is enclosed on all sides by the species forming the crystal lattice. Also known as cage compound; inclusion compound. [GEOCHEM] *See* gas hydrate. [PETR] Pertaining to a condition, chiefly in leucite rock, in which clear leucite crystals are surrounded by tangential leucite crystals to give the rock an appearance of a net or a section of sponge. Also known as enclosure compound. { 'klath‚rāt }

clathrate hydrate *See* gas hydrate. { 'klath‚rāt 'hī‚drāt }

clathrin [CELL MOL] A protein that forms a lattice-shaped coating, through the assembly of subunits called triskelions, on the cytosolic side of membrane regions called coated pits during the initial stages of receptor-mediated endocytosis. Invagination of the pit results in a clathrin-coated vesicle. { 'klath·rən }

clathrin-coated pit [CELL MOL] A partially invaginated membrane structure (bud or pit) involved in receptor-mediated endocytosis consisting of a cluster of receptor proteins attached on the cytosolic side by means of adaptin molecules to the protein clathrin, which forms a lattice-shaped coating. Complete invagination of the pit and release from the membrane results in a clathrin-coated vesicle containing cargo molecules. { 'klath·rən ‚kōd·əd 'pit }

clathrin-coated vesicle *See* clathrin-coated pit. { 'klath·rən ‚kōd·əd 'ves·ə·kəl }

Clathrinida [INV ZOO] A monofamilial order of sponges in the subclass Calcinea having an asconoid structure and lacking a true dermal membrane or cortex. { kla'thrin·ə·də }

Clathrinidae [INV ZOO] The single family of the order Clathrinida. { kla'thrin·ə‚dē }

clathrochelate [INORG CHEM] A type of coordination compound containing a metal ion both coordinately saturated and encapsulated by a single ligand. { ‚klath·rō'kē‚lāt }

Clathrochloris [MICROBIO] A genus of bacteria in the family Chlorobiaceae; cells are spherical to ovoid and arranged in chains united in trellis-like aggregates, are nonmotile, contain gas vacuoles, and are free-living. { ‚klath·rō'klór·əs }

Claude process [CHEM ENG] A process of ammonia synthesis which uses high operating pressures and a train of converters. [CRYO] A method of liquefying air or other gases in stages, in which the gas is cooled by doing work in an expansion engine and then undergoing the Joule-Thomson effect as it passes through an expansion valve. { 'klōd 'präs·əs }

claudetite [MINERAL] As_2O_3 A mineral containing arsenic that is dimorphous with arsenolite; crystallizes in the monoclinic system. { 'klōd·ə‚tīt }

clause [COMPUT SCI] A part of a statement in the COBOL language which may describe the structure of an elementary item, give initial values to items in independent and group work areas, or redefine data previously defined by another clause. { klōz }

clausius [THERMO] A unit of entropy equal to the increase in entropy associated with the absorption of 1000 international table calories of heat at a temperature of 1 K, or to 4186.8 joules per kelvin. { 'klóz·ē·əs }

Clausius-Clapeyron equation [THERMO] An equation governing phase transitions of a substance, $dp/dT = \Delta H/(T\Delta V)$, in which p is the pressure, T is the temperature at which the phase transition occurs, ΔH is the change in heat content (enthalpy), and ΔV is the change in volume during the transition. Also known as Clapeyron-Clausius equation; Clapeyron equation. { 'klóz·ē·əs kla·pā‚rōn i‚kwā·zhən }

Clausius-Dickel column *See* thermogravitational column. { 'klóz·ē·əs 'dik·əl 'käl·əm }

Clausius equation [THERMO] An equation of state in reference to gases which applies a correction to the van der Waals equation: $\{P + (n^2a/[T(V + c)^2])\}(V - nb) = nRT$, where P is the pressure, T the temperature, V the volume of the gas, n the number of moles in the gas, R the gas constant, a depends only on temperature, b is a constant, and c is a function of a and b. { 'klóz·ē·əs i'kwā·zhən }

Clausius inequality [THERMO] The principle that for any system executing a cyclical process, the integral over the cycle of the infinitesimal amount of heat transferred to the system divided by its temperature is equal to or less than zero. Also known as Clausius theorem; inequality of Clausius. { 'klȯz·ē·əs in·i'kwäl·əd·ē }

Clausius law [THERMO] The law that an ideal gas's specific heat at constant volume does not depend on the temperature. { 'klȯz·ē·əs ,lȯ }

Clausius-Mosotti equation [ELEC] An expression for the polarizability γ of an individual molecule in a medium which has the relative dielectric constant ϵ and has N molecules per unit volume: $\gamma = (3/4\pi N) [(\epsilon - 1)/(\epsilon + 2)]$ (Gaussian units). { 'klȯz·ē·əs mə'zäd·ē i'kwä·zhən }

Clausius-Mosotti-Lorentz-Lorenz equation [ELECTRO-MAG] The equation that results from replacing the real relative dielectric constant in the Clausius-Mosotti equation, or the real index of refraction in the Lorentz-Lorenz equation, with its complex counterpart. { 'klȯz·ē·əs mə'zäd·ē 'lȯ·rens'lȯ·rens i'kwä·zhən }

Clausius number [THERMO] A dimensionless number used in the study of heat conduction in forced fluid flow, equal to $V^3L\rho/k\Delta T$, where V is the fluid velocity, ρ is its density, L is a characteristic dimension, k is the thermal conductivity, and ΔT is the temperature difference. { 'klȯz·ē·əs ,nəm·bər }

Clausius range [STAT MECH] The condition in which the mean free path of molecules in a gas is much smaller than the dimensions of the container. { 'klȯz·ē·əs 'rānj }

Clausius' statement [THERMO] A formulation of the second law of thermodynamics, stating it is not possible that, at the end of a cycle of changes, heat has been transferred from a colder to a hotter body without producing some other effect. { 'klȯz·ē·əs 'stāt·mənt }

Clausius theorem *See* Clausius inequality. { 'klȯz·ē·əs 'thir·əm }

Clausius virial theorem [STAT MECH] The theorem that in a system of particles whose positions and velocities are bounded, the total kinetic energy of the system averaged over a long period of time equals the virial of the system. Also known as virial theorem. { 'klȯz·ē·əs 'vir·ē·əl 'thir·əm }

Claus method [CHEM ENG] Industrial method of obtaining sulfur by a partial oxidation of gaseous hydrogen sulfide in the air to give water and sulfur. { 'klaus ,meth·əd }

clausthalite [MINERAL] PbSe A mineral consisting of lead selenide and resembling galena; specific gravity is 7.6–8.8. { 'klaus·tə,līt }

claustrophobia [PSYCH] An abnormal fear of confined spaces. { ,klȯs·trə'fō·bē·ə }

claustrum [ANAT] A thin layer of gray matter in each cerebral hemisphere between the lenticular nucleus and the island of Reil. { 'klȯ,strəm }

clava [BIOL] A club-shaped structure, as the tip on the antennae of certain insects or the fruiting body of certain fungi. { 'klā·və }

clavate [BIOL] Club-shaped. Also known as claviform. { 'klā,vāt }

Clavatoraceae [PALEOBOT] A group of middle Mesozoic algae belonging to the Charophyta. { ,klav·əd·ə'rās·ē,ē }

Clavaxinellida [INV ZOO] An order of sponges in the class Demospongiae; members have monaxonid megascleres arranged in radial or plumose tracts. { kla¦vak·sə'nel·ə·də }

clavicle [ANAT] A bone in the pectoral girdle of vertebrates with articulation occurring at the sternum and scapula. { 'klav·ə·kal }

claviculate [ANAT] Having a clavicle. { kla'vik·yə·lət }

claviform *See* clavate. { 'klav·ə,fȯrm }

clavus [INV ZOO] Any of several rounded or fingerlike processes, such as the club of an insect antenna or the pointed anal portion of the hemelytron in hemipteran insects. { 'klāv·əs }

claw [ANAT] A sharp, slender, curved nail on the toe of an animal, such as a bird. [DES ENG] A fork for removing nails or spikes. [INV ZOO] A sharp-curved process on the tip of the limb of an insect. { klȯ }

claw bar *See* ripping bar. { 'klȯ ,bär }

claw clutch [MECH ENG] A clutch consisting of claws that interlock when pushed together. { 'klȯ ,kləch }

claw coupling [MECH ENG] A loose coupling having projections or claws cast on each face which engage in corresponding notches in the opposite faces; used in situations in which shafts require instant connection. { 'klȯ ,kəp·liŋ }

claw hammer [DES ENG] A woodworking hammer with a flat working surface and a claw to pull nails. { 'klȯ ,ham·ər }

clay [GEOL] **1.** A natural, earthy, fine-grained material which develops plasticity when mixed with a limited amount of water; composed primarily of silica, alumina, and water, often with iron, alkalies, and alkaline earths. **2.** The fraction of an earthy material containing the smallest particles, that is, finer than 3 micrometers. [MATER] A special grade of absorbent clay used as a filtering medium in refineries for removing solids or colorizing matter from lubricating oils. { klā }

clay atmometer [ENG] An atmometer consisting of a porous porcelain container connected to a calibrated reservoir filled with distilled water; evaporation is determined by the depletion of water. { klā at'mäm·əd·ər }

Clay Belt [GEOL] A lowland area bordering on the western and southern portions of Hudson and James bays in Canada, composed of clays and silts recently deposited in large glacial lakes during the withdrawal of the continental glaciers. { 'klā ,belt }

clay bit [ENG] **1.** A bit designed for use on a clay barrel. **2.** *See* mud auger. { 'klā ,bit }

clay brick [MATER] Brick made from diverse types of clays and used for normal constructional purposes. { ¦klā ¦brik }

Clayden effect *See* dark lightning. { 'klād·ən i'fekt }

clay digger [MECH ENG] A power-driven, hand-held spade for digging hard soil or soft rock. { 'klā ,dig·ər }

clay gall [GEOL] A dry, curled clay shaving derived from dried, cracked mud and embedded and flattened in a sand stratum. { 'klā ,gȯl }

clay ironstone [PETR] **1.** A clayey rock containing large quantities of iron oxide, usually limonite. **2.** A clayey-looking stone occurring among carboniferous and other rocks; contains 20;–30% iron. { 'klā 'ī·ərn,stōn }

clay loam [GEOL] Soil containing 27–40% clay, 20–45% sand, and the remaining portion silt. { ¦klā 'lōm }

clay marl [GEOL] A chalky clay, whitish with a smooth texture. { ¦klā 'märl }

clay mineral [MINERAL] One of a group of finely crystalline, hydrous silicates with a two-or three-layer crystal structure; the major components of clay materials; the most common minerals belong to the kaolinite, montmorillonite, attapulgite, and illite groups. { ¦klā 'min·rəl }

claypan [GEOL] A stratum of compact, stiff, relatively impervious noncemented clay; can be worked into a soft, plastic mass if immersed in water. { 'klā,pan }

clay plug [GEOL] Sediment, with a great deal of organic muck, deposited in a cutoff river meander. { 'klā ,pləg }

clay press [ENG] A press used to remove excess water from a pottery-clay slurry. { 'klā ,pres }

clay refining [CHEM ENG] A treating process for vaporized gasoline or other light petroleum product; the material is passed through a bed of granular clay, and certain olefins are polymerized to gums and absorbed by the clay. { 'klā rə'fīn·iŋ }

clay regeneration [CHEM ENG] Cleaning coarse-grained absorbent clays for reuse in percolation processes by deoiling them with naphtha, steaming out excess naphtha, and roasting in a stream of air to remove carbonaceous matter. { 'klā ri·jen·ə'rā·shən }

clay shale [GEOL] **1.** Shale composed wholly or chiefly of clayey material which becomes clay again on weathering. **2.** Consolidated sediment composed of up to 10% sand and having a silt to clay ratio of less than 1:2. { ¦klā ¦shāl }

clay slip [MATER] A slurry of clay and water used in glazing pottery. { ¦klā ¦slip }

clay soil [GEOL] A fine-grained inorganic soil which forms hard lumps when dry and becomes sticky when wet. { ¦klā ¦sȯil }

claystone [GEOL] Indurated clay, consisting predominantly of fine material of which a major proportion is clay mineral. { 'klā,stōn }

clay vein [GEOL] A body of clay which is similar to an ore vein in form and fills a crevice in a coal seam. Also known as dirt slip. { 'klā ,vān }

clay wash [MATER] A light oil such as naphtha or kerosine used to clean fuller's earth after it has been used in a filter. { 'klā ,wäsh }

clay worsted [TEXT] Worsted made in a weave of three-up and three-down right-handed twill. { ¦klā 'wus·təd }

clean and certify [COMPUT SCI] To prepare a magnetic tape for a computer system by running it through a machine that cleans it, writes a data test pattern on it, and checks it for errors. { 'klēn ən 'sərd·ə,fī }

clean bomb [ORD] A nuclear bomb that produces relatively little radioactive fallout. { ¦klēn ¦bäm }

clean compile [COMPUT SCI] Conversion of a computer program from source to object language with no detection of significant errors by the compiler; logic errors not identified by the compiler may exist. { 'klēn kəm'pīl }

cleaning eye *See* cleanout. { 'klēn·iŋ ,ī }

cleaning lane [ENG] A space that is located between adjacent rows of tubes in a heat exchanger and allows passage of a cleaning device. { 'klēn·iŋ ,lān }

cleaning turbine [MECH ENG] A tool for cleaning the interior surfaces of heat exchangers and boiler tubes; consists of a drive motor, a flexible drive cable or hose, and a head that is an arrangement of blades, modified drill bits, or brushes. { 'klēn·iŋ ,tər·bən }

cleanout [ENG] A pipe fitting containing a removable plug that provides access for inspection or cleaning of the pipe run. Also known as access eye; cleaning eye. { 'klēn,aút }

cleanout auger *See* cleanout jet auger. { 'klēn,aút ,óg·ər }

cleanout door [ENG] An opening in the side of a tank usually at ground level and covered by a plate to provide access for removal of sediments from the bottom of the tank. { 'klēn,-aút ,dór }

cleanout jet auger [ENG] An auger equipped with water-jet orifices designed to clean out collected material inside a driven pipe or casing before taking soil samples from strata below the bottom of the casing. Also known as cleanout auger. { 'klēn,aút 'jet ,óg·ər }

clean room [ENG] A room in which elaborate precautions are employed to reduce dust particles and other contaminants in the air, as required for assembly of delicate equipment. { 'klēn ,rüm }

cleanser [MATER] Any material used to remove dirt, soil, and impurities from surfaces of all kinds. { 'klen·zər }

clean ship [NAV ARCH] A tanker used to carry refined light-petroleum products instead of crude oil or heavy fuel oils. { ¦klēn ¦ship }

clean track [ENG ACOUS] A sound track having no leakage from other tracks. { 'klēn ¦trak }

cleanup [AERO ENG] Improving the external shape and smoothness of an aircraft to reduce its drag. [ELECTR] Gradual disappearance of gases from an electron tube during operation, due to absorption by getter material or the tube structure. [ENG] The time required for a leak-testing system to reduce its signal output to 37% of the signal transmitted at the instant when tracer gases enter the system. [MIN ENG] **1.** The collecting of all the valuable product of a given period of operation in a stamp mill or in a hydraulic or placer mine. **2.** The valuable material resulting from a cleanup. { 'klē,nəp }

clean weapon [ORD] A nuclear warhead in which the amount of residual radioactivity is comparable to that of a normal weapon which has the same energy yield. { ¦klēn ¦wep·ən }

clear [COMPUT SCI] **1.** To restore a storage device, memory device, or binary stage to a prescribed state, usually that denoting zero. Also known as reset. **2.** A function key on calculators, to delete an entire problem or just the last keyboard entry. [METEOROL] **1.** After United States weather observing practice, the state of the sky when it is cloudless or when the sky cover is less than 0.1 (to the nearest tenth). **2.** To change from a stormy or cloudy weather condition to one of no precipitation and decreased cloudiness. [NAV] In marine navigation, to leave or pass safely, as to clear port or clear a shoal. [ORD] **1.** To give a person a security clearance. **2.** To operate a gun so as to unload it or make certain no ammunition remains; to free a gun of stoppages. { klir }

clear-air turbulence [METEOROL] A meteorological phenomenon occurring in the upper troposphere and lower stratosphere, in which high-speed aircraft are subject to violent updrafts and downdrafts. Abbreviated CAT. { ¦klir 'er 'tərb·yə·ləns }

clearance [ENG] Unobstructed space required for occasional removal of parts of equipment. [MECH ENG] **1.** In a piston-and-cylinder mechanism, the space at the end of the cylinder when the piston is at dead-center position toward the end of the cylinder. **2.** The ratio of the volume of this space to the piston displacement during a stroke. [MIN ENG] The space between the top or side of a car and the roof or wall. [NAV] **1.** The clear space between a vessel and an object such as a navigation light, hazard to navigation, or another vessel. **2.** A specific message from air-traffic control to a pilot of an aircraft allowing him to proceed in accordance with the flight plan which the pilot had filed, or with some modification of the original plan. **3.** In the instrument landing system, the difference in the depth of modulation which is required to produce a full-scale deflection of the course deviation indicator needle in any flight sector outside the on-course sectors. [ORD] Elevation of a gun at such an angle that a projectile will not strike an obstacle between the muzzle and the target. [PETRO ENG] The annular space between down-hole drill-string equipment, such as bits, core barrels, and casing, and the walls of the borehole with the down-hole equipment centered in the hole. { 'klir·əns }

clearance angle [MECH ENG] The angle between a plane containing the end surface of a cutting tool and a plane passing through the cutting edge in the direction of cutting motion. { 'klir·əns ,aŋ·gəl }

clearance sector [NAV] In instrument-landing-system localizers, that section which extends from the front to the back course of sectors and within which the deviation indicator provides the required off-course indications. { 'klir·əns ,sek·tər }

clearance test [PATH] The use of a substance such as urea or creatinine, or an injected foreign substance such as inulin to measure renal excretory activity; the ratio of the amount of these excreted substances in two 1-hour periods contrasted with the level of these substances in the blood is calculated in terms of the amount of blood cleared of these substances in a given unit time. { 'klir·əns ,test }

clearance volume [MECH ENG] The volume remaining between piston and cylinder when the piston is at top dead center. { 'klir·əns ,väl·yəm }

clear area [COMPUT SCI] In optical character recognition, any area designated to be kept free of printing or any other extraneous markings. { 'klir ,er·ē·ə }

clear band [COMPUT SCI] In character recognition, a continuous horizontal strip of blank paper which must be obtained between consecutive code lines on a source document. { 'klir ,band }

clear base [GRAPHICS] A colorless material of low known density that can be coated with a photographic emulsion. { 'klir 'bās }

clear-cell carcinoma *See* renal-cell carcinoma. { ¦klir ¦sel ,kärs·ən'ō·mə }

clear channel [COMMUN] A standard broadcast channel in which the dominant station or stations render service over wide areas; stations are cleared of objectionable interference within their primary service areas and over all or a substantial portion of their secondary service areas. { ¦klir 'chan·əl }

clear-cutting [FOR] Felling and removing all trees in a forest area. { 'klir ,kəd·iŋ }

clear-face worsted [TEXT] Closely woven worsted fabric made of twisted yarns and with the nap removed so that the weave is visible. { 'klir ,fās 'wús·təd }

clear gasoline [MATER] Gasoline which is free from antiknock additives. { 'klir ,gas·ə'lēn }

clear ice [HYD] Generally, a layer or mass of ice which is relatively transparent because of its homogeneous structure and small number and size of air pockets. { ¦klir ¦īs }

clearing [GRAPHICS] **1.** The process of removing silver halides from developed films by using a fixer. **2.** The process of exposing a vesicular film to ultraviolet light after the development step in order to decompose any remaining diazonium salts. { 'klir·iŋ }

clear-line image [GRAPHICS] A microfilm image that reproduces the dark lines of the original as clear or transparent lines, yielding a negativelike reproduction. { 'klir ¦līn 'im·ij }

clear octane [ENG] The octane number of a particular gasoline before it has been blended with antiknock additives. { 'klir 'äk,tān }

clear text [COMMUN] Text or language which conveys an intelligible meaning in the language in which it is written with no hidden meaning. { 'klir ,tekst }

clear-voice override [COMMUN] The ability of a speech

scrambler to receive a clear message even when the scrambler is set for scrambler operation. { ¦klir ¦vȯis 'ō·və¸rīd }

cleat [CIV ENG] A strip of wood, metal, or other material fastened across something to serve as a batten or to provide strength or support. [DES ENG] A fitting having two horizontally projecting horns around which a rope may be made fast. [GEOL] Vertical breakage planes found in coal. Also spelled cleet. { klēt }

cleat spar See ankerite. { 'klēt ¸spär }

cleavage [CRYSTAL] Splitting, or the tendency to split, along planes determined by crystal structure and always parallel to a possible face. [EMBRYO] The subdivision of activated eggs into blastomeres. [GEOL] Splitting, or the tendency to split, along parallel, closely positioned planes in rock. { 'klēv·ij }

cleavage banding [GEOL] A compositional banding, usually formed from incompetent material such as argillaceous rocks, that is parallel to the cleavage rather than the bedding. { 'klēv·ij ¸band·iŋ }

cleavage crystal [CRYSTAL] A crystal fragment bounded by cleavage faces giving it a regular form. { 'klēv·ij ¸kris·təl }

cleavage fracture [CRYSTAL] **1.** Manner of breaking a crystalline substance along the cleavage plane. **2.** The appearance of such a broken surface. { 'klēv·ij ¸frak·chər }

cleavage nucleus [EMBRYO] The nucleus of a zygote formed by fusion of male and female pronuclei. { 'klēv·ij ¸nü·klē·əs }

cleavage plane [CRYSTAL] Plane along which a crystalline substance may be split. { 'klēv·ij ¸plān }

cleavelandite [MINERAL] A white, lamellar variety of albite that is almost pure NaAlSi$_3$O$_8$ and has a tabular habit, with individuals often showing mosaic developments and tending to occur in fan-shaped aggregates. { 'klēv·lən¸dīt }

Clebsch-Gordan coefficient See vector coupling coefficient. { ¦klepsh ¦gȯrd·ən ¸kō·ə'fish·ənt }

cleet See cleat. { klēt }

cleft grafting [BOT] A top-grafting method in which the scion is inserted into a cleft cut into the top of the stock. { 'kleft ¸graft·iŋ }

cleft lip See harelip. { ¦kleft 'lip }

cleft palate [MED] A birth defect resulting from incomplete closure of the palate during embryogenesis. { ¦kleft 'pal·ət }

cleft weld [MET] A weld in which a V-shaped projection on one piece is joined to a V-shaped groove in the other. { ¦kleft ¦weld }

cleidocranial dysostosis [MED] A congenital defect in which there is deficient formation of bone in the skull and clavicle. { ¦klī·dō¦krān·ē·əl ¦dis·ä'stō·səs }

cleistocarp See cleistothecium. { 'klī·stə¸kärp }

cleistocarpous [BOT] Of mosses, having the capsule opening irregularly without an operculum. [MYCOL] Forming or having cleistothecia. { ¦klī·stə¦kär·pəs }

cleistogamy [BOT] The production of small closed flowers that are self-pollinating and contain numerous seeds. { ¸klī'stäg·ə·mē }

cleistothecium [MYCOL] A closed sporebearing structure in Ascomycetes; asci and spores are freed of the fruiting body by decay or desiccation. Also known as cleistocarp. { ¦klī·stə'thē·sē·əm }

cleithrophobia [PSYCH] An abnormal fear of being locked in. { ¸klī·thrə'fō·bē·ə }

cleithrum [VERT ZOO] A bone external and adjacent to the clavicle in certain fishes, stegocephalians, and primitive reptiles. { 'klī·thrəm }

clerestory [ARCH] The upward extension of enclosed space achieved by bringing a windowed wall up to interrupt the slope of the roof. { 'klir¸stȯr·ē }

Cleridae [INV ZOO] The checkered beetles, a family of coleopteran insects in the superfamily Cleroidea. { 'kler·ə¸dē }

Cleroidea [INV ZOO] A superfamily of coleopteran insects in the suborder Polyphaga. { klə'rȯid·ē·ə }

Cleveland open-cup tester [ANALY CHEM] A laboratory apparatus used to determine flash point and fire point of petroleum products. { 'klev·lənd ¸ō·pən 'kəp ¸test·ər }

clevis [DES ENG] A U-shaped metal fitting with holes in the open ends to receive a bolt or pin; used for attaching or suspending parts. [MIN ENG] A spring hook or snap hook which, in coal mining, is used to attach the bucket to the hoisting rope. Also known as clivvy. { 'klev·əs }

clevis pin [DES ENG] A fastener with a head at one end, used to join the ends of a clevis. { 'klev·əs ¸pin }

cliachite [MINERAL] A group of brownish, colloidal aluminum hydroxides that constitutes most bauxite. { 'klī·ə¸kīt }

click [COMMUN] A short-duration electric disturbance, such as that sometimes produced by a code-sending key or a switch. [COMPUT SCI] To select an object when the pointer is touching it by pressing and quickly releasing a button on a mouse. [ENG ACOUS] A perforation in a sound track which produces a clicking sound when passed over the projector sound head. { klik }

click filter [ELECTR] A capacitor connected across a switch, relay, or key to lengthen the decay time from the closed to the open condition when the device is opened or closed. { 'klik ¸fil·tər }

click track [ENG ACOUS] A sound track containing a series of clicks, which may be spaced regularly (uniform click track) or irregularly (variable click track). { 'klik ¸trak }

client [COMPUT SCI] A hardware or software entity that requests shared services from a server. { 'klī·ənt }

client-based application [COMPUT SCI] An application that runs on a work station or personal computer in a network and is not available to others in the network. { 'klī·ənt ¸bāst ¸ap·lə¦kā·shən }

client-server system [COMPUT SCI] A computing system composed of two logical parts: a server, which provides information or services, and a client, which requests them. On a network, for example, users can access server resources from their personal computers using client software. { ¦klī·ənt 'sər·vər ¸sis·təm }

cliff [GEOGR] A high, steep, perpendicular or overhanging face of a rock; a precipice. { klif }

cliff of displacement See fault scarp. { 'klif əv dis'plā·smənt }

Cliftonian [GEOL] Middle Middle Silurian geologic time. { klif'tän·ē·ən }

climacophobia [PSYCH] An abnormal fear of staircases. { ¸klī·mə·kə'fō·bē·ə }

climacteric See menopause. { klī'mak·tə·rik }

climagram See climatic diagram. { 'klī·mə¸gram }

climagraph See climatic diagram. { 'klī·mə¸graf }

climate [CLIMATOL] The long-term manifestations of weather. { 'klī·mət }

climate change [METEOROL] Any change in global temperatures and precipitation over time due to natural variability or to human activity. { 'klī·mət ¸chānj }

climate control [CLIMATOL] **1.** Schemes for artificially altering or controlling the climate of a region. **2.** See air conditioning. { 'klī·mət kən'trōl }

climate model [CLIMATOL] A mathematical representation of the earth's climate system capable of simulating its behavior under present and altered conditions. { 'klī·mət ¸mäd·əl }

climate therapy See climatotherapy. { 'klīm·ət ¸ther·ə·pē }

climatic classification [CLIMATOL] The division of the earth's climates into a system of contiguous regions, each one of which is defined by relative homogeneity of the climate elements. { klī'mad·ik ¸klas·ə·fə'kā·shən }

climatic climax [ECOL] A climax community viewed, by some authorities, as controlled by climate. { klī'mad·ik 'klī¸maks }

climatic controls [CLIMATOL] The relatively permanent factors which govern the general nature of the climate of a portion of the earth, including solar radiation, distribution of land and water masses, elevation and large-scale topography, and ocean currents. { klī'mad·ik kən'trōlz }

climatic cycle [CLIMATOL] A long-period oscillation of climate which recurs with some regularity, but which is not strictly periodic. Also known as climatic oscillation. { klī'mad·ik 'sī·kəl }

climatic diagram [CLIMATOL] A graphic presentation of climatic data; generally limited to a plot of the simultaneous variations of two climatic elements, usually through an annual cycle. Also known as climagram; climagraph; climatograph; climogram; climograph. { klī'mad·ik 'dī·ə¸gram }

climatic divide [CLIMATOL] A boundary between regions having different types of climate. { klī'mad·ik də'vīd }

climatic factor [CLIMATOL] Climatic control, but regarded as including more local influences; thus city smoke and the

---

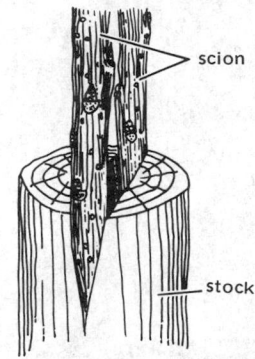

CLEFT GRAFTING

scion

stock

The components of a cleft graft. *(From H. J. Fuller and Z. B. Carothers, The Plant World, 4th ed., Holt, Rinehart, and Winston, 1963)*

CLEVIS

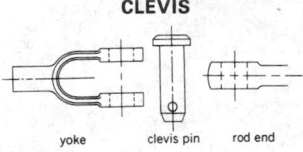

yoke   clevis pin   rod end

A clevis (yoke), with clevis pin and rod end.

CLIENT-SERVER SYSTEM

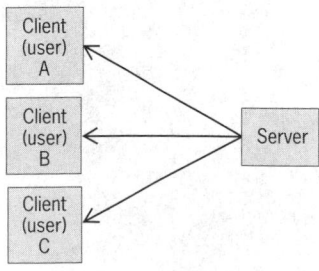

Client (user) A

Client (user) B

Client (user) C

Server

Clients request services of the server independently but use the same interface.

extent of the builtup metropolitan area are climatic factors, but not climatic controls.   { klī'mad·ik 'fak·tər }

climatic forecast [CLIMATOL]   A forecast of the future climate of a region; that is, a forecast of general weather conditions to be expected over a period of years.   { klī'mad·ik 'fȯr‚kast }

climatic optimum [CLIMATOL]   The period in history (about 5000–2500 B.C.) during which temperatures were warmer than at present in nearly all parts of the world.   { klī'mad·ik 'äp·tə·məm }

climatic oscillation   See climatic cycle.   { klī'mad·ik ‚äs·ə'lā·shən }

climatic prediction [METEOROL]   The description of the future state of the climate, that is, the average or expected atmospheric and earth-surface conditions, for example, temperature, precipitation, humidity, winds, and their range of variability. Seasonal and interannual climate predictions, made many months in advance, provide useful information for planners and policy makers.   { klī'mad·ik prə'dik·shən }

climatic province [CLIMATOL]   A region of the earth's surface characterized by an essentially homogeneous climate.   { klī'mad·ik 'prä·vəns }

climatic snow line [METEOROL]   The altitude above which a flat surface (fully exposed to sun, wind, and precipitation) would experience a net accumulation of snow over an extended period of time; below this altitude, ablation would predominate.   { klī'mad·ik 'snō ‚līn }

climatic zone [CLIMATOL]   A belt of the earth's surface within which the climate is generally homogeneous in some respect; an elemental region of a simple climatic classification.   { klī'mad·ik ‚zōn }

Climatiidae [PALEON]   A family of archaic tooth-bearing fishes in the suborder Climatioidei.   { ‚klī·mə'tī·ə‚dē }

Climatiiformes [PALEON]   An order of extinct fishes in the class Acanthodii having two dorsal fins and large plates on the head and ventral shoulder.   { ‚klī·mə‚tī·ə'fȯr‚mēz }

Climatioidei [PALEON]   A suborder of extinct fishes in the order Climatiiformes.   { ‚klī·mə‚tī'ȯid·ē‚ī }

climatochronology [GEOL]   The absolute age dating of recent geologic events by using the oxygen isotope ratios in ice, shells, and so on.   { klī‚mad·ō·krə'näl·ə·jē }

climatograph   See climatic diagram.   { klī'mad·ə‚graf }

climatography [CLIMATOL]   A quantitative description of climate, particularly with reference to the tables and charts which show the characteristic values of climatic elements at a station or over an area.   { ‚klī·mə'täg·rə·fē }

climatological forecast [METEOROL]   A weather forecast based upon the climate of a region instead of upon the dynamic implications of current weather, with consideration given to such synoptic weather features as cyclones and anticyclones, fronts, and the jet stream.   { ‚klī·məd·əl'äj·ə·kəl 'fȯr‚kast }

climatological station elevation [CLIMATOL]   The elevation above mean sea level chosen as the reference datum level for all climatological records of atmospheric pressure in a given locality.   { ‚klī·məd·əl'äj·ə·kəl 'stā·shən ‚el·ə'vā·shən }

climatological station pressure [CLIMATOL]   The atmospheric pressure computed for the level of the climatological station elevation, used to give all climatic records a common reference; it may or may not be the same as station pressure.   { ‚klī·məd·əl'äj·ə·kəl 'stā·shən ‚presh·ər }

climatological substation [CLIMATOL]   A weather-observing station operated (by an unpaid volunteer) for the purpose of recording climatological observations.   { ‚klī·məd·əl'äj·ə·kəl 'səb‚stā·shən }

climatology [METEOROL]   That branch of meteorology concerned with the mean physical state of the atmosphere together with its statistical variations in both space and time as reflected in the weather behavior over a period of many years.   { ‚klī·mə'täl·ə·jē }

climatopathology [MED]   The study of disease in relation to the effects of the natural environment.   { ‚klī·mə·tō·pə'thäl·ə·jē }

climatophysiology [PHYSIO]   The study of the interaction of the natural environment with physiologic factors.   { ‚klī·mə·tō‚fiz·ē'äl·ə·jē }

climatotherapy [MED]   Placing a person in a suitable climate to treat a certain disease. Also known as climate therapy; climotherapy.   { ‚klī·mə·tō'ther·ə·pē }

climax [ECOL]   A mature, relatively stable community in an area, which community will undergo no further change under the prevailing climate; represents the culmination of ecological succession.   { 'klī‚maks }

climax community [ECOL]   The final stage in ecological succession in which a relatively constant environment is reached and species composition no longer changes in a directional fashion, but fluctuates about some mean, or average, community composition.   { 'klī‚maks kə‚myü·nə·dē }

climax plant formation [ECOL]   A mature, stable plant population in a climax community.   { 'klī‚maks 'plant fȯr'mā·shən }

climb [AERO ENG]   The gain in altitude of an aircraft.   { klīm }

climb cutting [MET]   A milling technique in which the teeth of a cutting tool advance into the work in the same direction as the feed. Also known as climb milling; down cutting; down milling.   { 'klīm ‚kəd·iŋ }

climbing bog [ECOL]   An elevated boggy area on a swamp margin, usually occurring where there is a short summer and considerable rainfall.   { 'klīm·iŋ 'bäg }

climbing crane [MECH ENG]   A crane used on top of a high-rise construction that ascends with the building as work progresses.   { 'klīm·iŋ 'krān }

climbing dune [GEOL]   A dune that develops on the windward side of mountains or hills.   { 'klīm·iŋ 'dün }

climbing irons [DES ENG]   Spikes attached to a steel framework worn on shoes to climb wooden utility poles and trees.   { 'klīm·iŋ ‚ī·ərnz }

climbing stem [BOT]   A long, slender stem that climbs up a support or along the tops of other plants by using spines, adventitious roots, or tendrils for attachment.   { 'klīm·iŋ 'stem }

climb milling   See climb cutting.   { 'klīm ‚mil·iŋ }

climogram   See climatic diagram.   { 'klī·mə‚gram }

climograph   See climatic diagram.   { 'klī·mə‚graf }

climotherapy   See climatotherapy.   { 'klīm·ə‚ther·ə·pē }

cline [BIOL]   A graded series of morphological or physiological characters exhibited by a natural group (as a species) of related organisms, generally along a line of environmental or geographic transition.   { klīn }

clingage [PETRO ENG]   Oil adhering to the wall of a measuring or prover tank after the tank has been drained.   { 'kliŋ·ij }

clinical chemistry [PATH]   The science involving chemical analysis of body tissues to diagnose disease.   { 'klin·ə·kəl 'kem·ə·strē }

clinical genetics [GEN]   The study of the role of genetic factors in human disease susceptibility by observation of patients and their families.   { 'klin·ə·kəl jə'ned·iks }

clinical microbiology [MED]   The adaptation of microbiological techniques to the study of the etiological agents of infectious disease.   { ‚klin·ə·kəl ‚mī·krō·bī'äl·ə·jē }

clinical pathology [PATH]   A medical specialty encompassing the diagnostic study of disease by means of laboratory tests of material from the living patient.   { 'klin·ə·kəl pə'thäl·ə·jē }

clinical pharmacology [PHARM]   The study and evaluation of the effects of drugs in humans.   { 'klin·ə·kəl ‚fär·mə'käl·ə·jē }

clinical sports medicine [MED]   Sports medicine that focuses on the prevention and treatment of athletic injuries and the design of exercise and nutrition programs for maintaining peak physical performance.   { ‚klin·ə·kəl 'spȯrts ‚med·ə·sən }

clinical teleconferencing [MED]   The use of teleconferencing in medical diagnosis and treatment, allowing rural healthcare facilities to perform diagnosis and treatment that would otherwise be available only in metropolitan areas.

clinical thermometer [ENG]   A thermometer used to accurately determine the temperature of the human body; the most common type is a mercury-in-glass thermometer, in which the mercury expands from a bulb into a capillary tube past a constriction that prevents the mercury from receding back into the bulb, so that the thermometer registers the maximum temperature attained.   { 'klin·ə·kəl thər'mäm·əd·ər }

clinical trial [MED]   A research study used to find better ways to treat individuals with a specific disease, patients are evaluated after being administered a new treatment or drug.   { 'klin·i·kəl 'trīl }

clinker [GEOL]   Burnt or vitrified stony material, as ejected by a volcano or formed in a furnace. [MATER]   An overburned brick.   { 'kliŋ·kər }

clinker building [DES ENG] A method of building ships and boilers in which the edge of the wooden planks or steel plates used for the outside covering overlap the edge of the plank or plate next to it; clinched nails fasten the planks together, and rivets fasten the steel plates. { 'kliŋ·kər ,bil·diŋ }

clinoamphibole [MINERAL] A group of amphiboles which crystallize in the monoclinic system. { ¦klī·nō¦am·fə,bōl }

clinoaxis [CRYSTAL] The inclined lateral axis that makes an oblique angle with the vertical axis in the monoclinic system. Also known as clinodiagonal. { ¦klī·nō¦ak·səs }

clinochlore [MINERAL] $(Mg,Fe,Al)_3(Si,Al)_2O_5(OH)_4$ Green mineral of the chlorite group, occurring in monoclinic crystals, in folia or scales, or massive. { 'klī·nə,klȯr }

clinoclase [MINERAL] $Cu_3(AsO_4)(OH)_3$ A dark-green mineral consisting of basic copper arsenate occurring in translucent prismatic crystals or massive. Also known as clinoclasite. { 'klī·nə,klās }

clinoclasite See clinoclase. { ¦klī·nə¦klā,sīt }

clinodiagonal See clinoaxis. { ¦klī·nō·dī'ag·ən·əl }

clinoenstatite [MINERAL] $Mg_2(Si_2O_6)$ A monoclinic pyroxene consisting principally of magnesium silicate; occurs frequently in stony meteorites, but is rare in terrestrial environments. { ¦klī·nō'enz·tə,tīt }

clinoferrosilite [MINERAL] $Fe_2(Si_2O_6)$ A monoclinic pyroxene consisting of iron silicate. { ¦klī·nō,fe·rō'sī,līt }

clinoform [GEOL] A subaqueous landform, such as the continental slope of the ocean or the foreset bed of a delta. { 'klī·nə,fȯrm }

clinograph [ENG] A type of directional surveying instrument that records photographically the direction and magnitude of deviations from the vertical of a borehole, well, or shaft; the information is obtained by the instrument in one trip into and out of the well. { 'klī·nə,graf }

clinographic projection [GRAPHICS] A method of representing objects, especially crystals, in which each point P of the object to be represented is projected onto the foot of a perpendicular from P to a plane which is located so that no place surface of the object is represented by a line. { ¦klī·nə¦graf·ik prə'jek·shən }

clinohedral class [CRYSTAL] A rare class of crystals in the monoclinic system having a plane of symmetry but no axis of symmetry. Also known as domatic class. { ¦klī·nō¦hē·drəl ,klas }

clinohedrite [MINERAL] $CaZnSiO_3(OH)_2$ A colorless, white, or purplish monoclinic mineral consisting of a calcium zinc silicate occurring in crystals; hardness is 5.5 on Mohs scale, and specific gravity is 3.33. { ¦klī·nō¦hē,drīt }

clinohumite [MINERAL] $Mg_9(SiO_4)_4(F,OH_2)$ A monoclinic mineral of the humite group. { ¦klī·nō'hyü,mīt }

clinometer [ENG] **1.** A hand-held surveying device for measuring vertical angles; consists of a sighting tube surmounted by a graduated vertical arc with an attached level bubble; used in meteorology to measure cloud height at night, in conjunction with a ceiling light, and in ordnance for boresighting. Also known as Abney level. **2.** A device for measuring the amount of roll aboard ship. { klə'näm·əd·ər }

clinopinacoid [CRYSTAL] A form of monoclinic crystal whose faces are parallel to the inclined and vertical axes. { ¦klī·nə'pin·ə,kȯid }

clinoptilolite [MINERAL] $(Na,K,Ca)_{2-3}Al_3(Al,Si)_2Si_{13}O_{36}·12H_2O$ A zeolite mineral that is considered to be a potassium-rich variety of heulandite. { 'klin·əp'til·ə,līt }

clinopyroxene [MINERAL] The general term for any of those pyroxenes that crystallize in the monoclinic system; on occasion, these pyroxenes have large amounts of calcium with or without aluminum and the alkalies. Also known as monopyroxene clinoaugite. { ¦klī·nə·pə'räk,sēn }

clinozoisite [MINERAL] $Ca_2Al_3(SiO_4)_3(OH)$ A grayish-white, pink, or green monoclinic mineral of the epidote group. { ¦klī·nə'zō·i,sīt }

clint [GEOL] A hard or flinty rock, such as a projecting rock or ledge. { klint }

Clintonian [GEOL] Lower Middle Silurian geologic time. { klin'tōn·ē·ən }

clintonite [MINERAL] $Ca(Mg,Al)_3(Al,Si)O_{10}(OH)_2$ A reddish-brown, copper-red, or yellowish monoclinic mineral of the brittle mica group occurring in crystals or foliated masses. Also known as seybertite; xanthophyllite. { 'klint·ən,īt }

Clionidae [INV ZOO] The boring sponges, a family of marine sponges in the class Demospongiae. { klī'än·ə,dē }

clip [DES ENG] A device that fastens by gripping, clasping, or hooking one part to another. [ORD] See cartridge clip. { klip }

clip and shave [MET] Dual forging operation in which one cutter removes the flash and then another cutter shaves and sizes the piece. { ¦klip ən 'shāv }

clip art [COMPUT SCI] A collection of graphic images that are stored on a computer disk for use in desktop publishing, word processing, and presentation graphics programs. { 'klip ,ärt }

clipboard [COMPUT SCI] An area in memory or a file where cut or copied material is held temporarily before being inserted elsewhere in the same document or in another document. { 'klip,bȯrd }

clip bond [CIV ENG] A bond in which the inner edge of face brick is cut off so that bricks laid diagonal to a wall can be joined to those laid parallel to it. { 'klip ,bänd }

clip-dot fabric [TEXT] A fabric decorated with small woven spots of extra warp or filling yarn-the floating threads between the spots being clipped or sheared in finishing. Also known as clip-spot fabric. { 'klip ,dät ,fab·rik }

clip lead [ELEC] A short piece of flexible wire with an alligator clip or similar temporary connector at one or both ends. { 'klip ,lēd }

clipper See limiter. { 'klip·ər }

Clipper Chip [COMPUT SCI] A chip proposed by the United States government to be used in all devices that might use encryption, such as computers and communications devices, for which the government would have at least some access or control over the decryption key for purposes of surveillance. { 'klip·ər ,chip }

clipper diode [ELECTR] A bidirectional breakdown diode that clips signal voltage peaks of either polarity when they exceed a predetermined amplitude. { 'klip·ər ,dī,ōd }

clipper-limiter [ELECTR] A device whose output is a function of the instantaneous input amplitude for a range of values lying between two predetermined limits but is approximately constant, at another level, for input values above the range. { ¦klip·ər ¦lim·əd·ər }

clipping [COMMUN] The perceptible mutilation of signals or speech syllables during transmission, often due to limiting. [COMPUT SCI] See scissoring. [ELECTR] See limiting. { 'klip·iŋ }

clipping circuit See limiter. { 'klip·iŋ ,sər·kət }

clipping edge [MET] Area of a forging where flash is removed. { 'klip·iŋ ,ej }

clipping level [ELECTR] The level at which a clipping circuit is adjusted; for example, the magnitude of the clipped wave shape. { 'klip·iŋ ,lev·əl }

clip-spot fabric See clip-dot fabric. { 'klip ,spät ,fab,rik }

clique [MATH] In a graph, a complete subgraph of that graph. { klēk }

clisere [ECOL] The succession of ecological communities, especially climax formations, as a consequence of intense climatic changes. { klī,sir }

CLIST [COMPUT SCI] A file containing a series of commands that are processed in the order given when the file is entered. Acronym for command list. { 'sē,list }

clitellum [INV ZOO] The thickened, glandular, saddlelike portion of the body wall of some annelid worms. { klə'tel·əm }

clitoris [ANAT] The homolog of the penis in females, located in the anterior portion of the vulva. { 'klid·ə·rəs }

clivvy See clevis. { 'kliv·ē }

clo [ENG] The amount of insulation which will maintain normal skin temperature of the human body when heat production is 50 kilogram-calories per meter squared per hour, air temperature is 70°F (21°C), and the air is still. { klō }

cloaca [INV ZOO] The chamber which functions as a respiratory, excretory, and reproductive duct in certain invertebrates. [VERT ZOO] The chamber which receives the discharges of the intestine, urinary tract, and reproductive canals in monotremes, amphibians, birds, reptiles, and many fish. { klō'ā·kə }

cloacal bladder [VERT ZOO] A diverticulum of the cloacal wall in monotremes, amphibians, and some fish, into which urine is forced from the cloaca. { klō'ā·kəl 'blad·ər }

cloacal gland [VERT ZOO] Any of the sweat glands in the

CLINOGRAPH

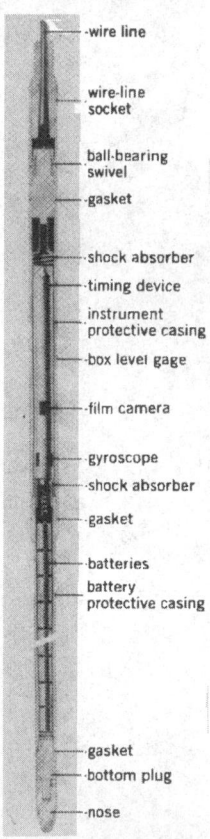

-wire line

wire-line socket

ball-bearing swivel

-gasket

-shock absorber

-timing device

instrument protective casing

-box level gage

-film camera

-gyroscope

-shock absorber

-gasket

-batteries

battery protective casing

-gasket

-bottom plug

-nose

Vertical section through a clinograph. (*Sperry-Sun Well Surveying Co.*)

CLINOMETER

A typical clinometer. (*Keuffel and Esser Co.*)

cloaca of lower vertebrates, as snakes or amphibians. { klō'ā·kəl 'gland }

clobber [COMPUT SCI] To write new data and thereby erase good data in a file, or to otherwise damage the file so that it becomes useless. { 'kläb·ər }

clock [ELECTR] A source of accurately timed pulses, used for synchronization in a digital computer or as a time base in a transmission system. [HOROL] A device for indicating the passage of time, usually containing a means for producing a regularly recurring action. Also known as timing circuit. { kläk }

Clock *See* Horologium. { kläk }

clock control system [CONT SYS] A system in which a timing device is used to generate the control function. Also known as time-controlled system. { 'kläk kən'trōl ,sis·təm }

clock-doubled [COMPUT SCI] Describing a microprocessor that operates at twice the clock speed of the bus or motherboard to which it is attached. { 'kläk ¦dəb·əld }

clock drive [ENG] The mechanism that causes an equatorial telescope to revolve about its polar axis so that it keeps the same star in its field of view. { 'kläk ,drīv }

clocked flip-flop [ELECTR] A flip-flop circuit that is set and reset at specific times by adding clock pulses to the input so that the circuit is triggered only if both trigger and clock pulses are present simultaneously. { 'kläkt 'flip,fläp }

clocked logic [ELECTR] A logic circuit in which the switching action is controlled by repetitive pulses from a clock. { ¦kläkt ¦läj·ik }

clock frequency [ELECTR] The master frequency of the periodic pulses that schedule the operation of a digital computer. Also known as clock rate; clock speed. { 'kläk ,frē·kwən·sē }

clock gene [GEN] Any of a number of genes that interact to determine the duration of development and lifespan. { 'kläk ,jēn }

clock method [ORD] Method of calling artillery shots by reference to the figures on an imaginary clock dial assumed to have the target at its center; thus a shot directly above the target is at 12 o'clock. { 'kläk ,meth·əd }

clock motor *See* timing motor. { 'kläk ,mōd·ər }

clock oscillator [ELECTR] An oscillator that controls an electronic clock. { 'kläk 'äs·ə,lād·ər }

clock paradox [RELAT] The apparent contradiction between the principle of relativity, which asserts the equivalence of different observers, and the prediction, also part of the theory of relativity, that the clock of an observer who passes back and forth will be slower than the clock of an observer at rest. Also known as twin paradox. { 'kläk ,par·ə,däks }

clock pulses [COMPUT SCI] Electronic pulses which are emitted periodically, usually by a crystal device, to synchronize the operation of circuits in a computer. Also known as clock signals. { 'kläk ,pəl·səz }

clock rate *See* clock frequency. { 'kläk ,rāt }

clock signals *See* clock pulses. { 'kläk ,sig·nəlz }

clock speed *See* clock frequency. { 'kläk ,spēd }

clock star [ASTRON] Any star that is used to measure time; always a bright star, whose right ascension is well known. { 'kläk ,stär }

clock time *See* internal cycle time. { 'kläk ,tīm }

clock track [COMPUT SCI] A track on a magnetic recording medium that generates clock pulses for the synchronization of read and write operations. { 'kläk ,trak }

clock-tripled [COMPUT SCI] Describing a microprocessor that operates at three times the clock speed of the bus or motherboard to which it is attached. { 'kläk ¦trip·əld }

clock watch [HOROL] A watch that strikes the hours. { 'kläk ,wäch }

clockwork [HOROL] A timing mechanism. { 'kläk,wərk }

clod [AGR] A compact mass of soil, ranging from about 0.2 to 10 inches (0.5 to 25 centimeters) in size, which is produced by plowing and digging of excessively wet or dry soil. { kläd }

clog snow [HYD] A skiing term for wet, sticky, new snow. { 'kläg ,snō }

cloister vault [ARCH] A vault resembling a pyramid with outward-curving sides. { 'klȯi·stər ,vȯlt }

cloithoid *See* Cornu's spiral. { 'klȯi,thȯid }

clonal selection theory [IMMUNOL] Theory to explain the specificity of the adaptive immune response according to which there is a large pool of lymphocytes, each having genetically predetermined specificity for only one of a vast array of possible antigens. Upon encountering an antigen, the lymphocytes sensitive to it reproduce much more rapidly than the others, thus leading to a buildup of antigen-specific cells large enough to mount the response. { ¦klōn·əl si'lek·shən ,thē·ə·rē }

clone [BIOL] All individuals, considered collectively, produced asexually or by parthenogenesis from a single individual. [COMPUT SCI] A hardware or software product that closely resembles another product created by a different manufacturer or developer, in operation, appearance, or both. [GEN] **1.** An organism whose diploid nuclear genome was derived from a somatic cell of another organism of the same species using biotechnology. **2.** A copy of a genetically engineered DNA sequence. { klōn }

cloning vector [CELL MOL] A carrier, such as a bacterial plasmid or bacteriophage, used to insert a genetic sequence, such as a deoxyribonucleic acid fragment or a complete gene, into a host cell such that the foreign genetic material is capable of replication. { 'klōn·iŋ ,vek·tər }

clonorchiasis [MED] A parasitic infection of humans and other fish-eating mammals which is caused by the trematode *Opisthorchis* (*Clonorchis*) *sinensis*, which is usually found in the bile ducts. { ,klōn·ȯr'kī·ə·səs }

Clonothrix [MICROBIO] A genus of sheathed bacteria; cells are attached and encrusted with iron and manganese oxides, and filaments are tapered. { 'klän·ə,thriks }

clonus [PHYSIO] Irregular, alternating muscular contractions and relaxations. { 'klō·nəs }

close [COMPUT SCI] To make a file unavailable to a computer program which previously had access to it. [METEOROL] Colloquially, descriptive of oppressively still, warm, moist air, frequently applied to indoor conditions. { klōs }

close-control radar [ENG] Ground radar used with radio to position an aircraft over a target that is normally difficult to locate or is invisible to the pilot. { ¦klōs kən'trōl 'rā,där }

close-coupled pump [MECH ENG] Pump with built-in electric motor (sometimes a steam turbine), with the motor drive and pump impeller on the same shaft. { ¦klōs ¦kəp·əld 'pəmp }

close coupling [ELEC] **1.** The coupling obtained when the primary and secondary windings of a radio-frequency or intermediate-frequency transformer are close together. **2.** A degree of coupling that is greater than critical coupling. Also known as tight coupling. { ¦klōs 'kəp·liŋ }

close-coupling method [ATOM PHYS] A method of approximating the wave function describing a process in which an initially free electron impinges on an isolated atom; in this method, the unknown solution of the Schrödinger equation is expanded in terms of the known wave functions of the target atom. { ,klōs 'kəp·liŋ ,meth·əd }

closed aerodrome [NAV] An aerodrome at which ceiling and visibility are below prescribed minimum for ordinary landings and takeoffs. { ¦klōzd 'er·ə,drōm }

closed architecture [COMPUT SCI] A computer architecture whose detailed, technical specifications are available only to those authorized by the manufacturer. { ¦klōzd 'ärk·ə,tek·chər }

closed association [ARCHEO] The relationship between two or more objects that are found together and that can be proved to have been deposited together. { 'klōzd ə,sō·sē'ā·shən }

closed ball [MATH] In a metric space, a closed set about a point *x* which consists of all points that are equal to or less than a fixed distance from *x*. { ¦klōzd 'bȯl }

closed-belt conveyor [MECH ENG] Solids-conveying device with zipperlike teeth that mesh to form a closed tube wrapped snugly around the conveyed material; used with fragile materials. { ¦klōzd ¦belt kən'vā·ər }

closed-box system [ELECTR] A loudspeaker system in which the woofer is mounted in a sealed box. { ¦klōzd 'bäks ,sis·təm }

closed braid [MATH] A modification of a braid in which plane curves are added that connect each of the *n* points on one of the parallel lines specified in the definition of the braid to one of the *n* points on the other in such a way that no two of these curves intersect or terminate at the same point, and the parallel lines themselves are deleted. { ¦klōzd 'brād }

closed-bus system [COMPUT SCI] A computer that lacks receptacles for expansion boards and is difficult to upgrade. { ¦klōd 'bəs ,sis·təm }

CLOSED-BOX SYSTEM

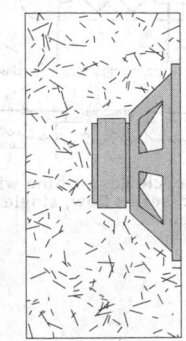

The woofer is mounted in a filled closed box.

closed-caption television [COMMUN] A method of captioning or subtitling television programs by coding captions as a vertical-interval data signal that is decoded at the receiver and superimposed on the normal television picture. { ¦klōzd ¦kap·shən 'tel·ə‚vizh·ən }

closed-cell foam [MATER] A cellular plastic in which there is a predominance of noninterconnecting cells. { ¦klōzd ¦sel 'fōm }

closed circuit [COMMUN] Program source that is not broadcast for general consumption but is fed to remote monitoring units. [ELEC] A complete path for current. { ¦klōzd 'sər·kət }

closed-circuit communications system [COMMUN] A communications systems which is entirely self-contained, and does not exchange intelligence with other facilities and systems. { ¦klōzd ¦sər·kət kə‚myü·nə'kā·shənz ‚sis·təm }

closed-circuit grinder [MIN ENG] A grinder connected to a size classifier (cyclone or screen) to return oversized particles to the grinding operation in closed-circuit pulverizing. { ¦klōzd ¦sər·kət 'grīn·dər }

closed-circuit pulverizing [MIN ENG] A process used in ore dressing in which the material discharged from the pulverizer is passed through an external classifier where the finished product is removed and oversized particles are returned to the pulverizer. { ¦klōzd ¦sər·kət 'pəl·və‚rīz·iŋ }

closed-circuit signaling [COMMUN] Signaling in which current flows in the idle condition, and a signal is initiated by increasing or decreasing the current. { ¦klōzd ¦sər·kət 'sig·nə·liŋ }

closed-circuit telegraph system [COMMUN] Telegraph system in which, when no station is transmitting, the circuit is closed and current flows through the circuit. { ¦klōzd ¦sər·kət 'tel·ə‚graf ‚sis·təm }

closed-circuit television [COMMUN] Any application of television that does not involve broadcasting for public viewing; the programs can be seen only on specified receivers connected to the television camera by circuits, which include microwave relays and coaxial cables. Abbreviated CCTV. { ¦klōzd ¦sər·kət 'tel·ə‚vizh·ən }

closed circular region [MATH] The union of the interior of a circle with the circle itself. { ¦klōzd ¦sər·kyə·lər 'rē·jən }

closed-coil armature [ELEC] The configuration of an armature in which the connection of all the coils forms a closed circuit. { ¦klōzd ¦koil 'är·mə·chər }

closed covering [MATH] A closed covering of a set S in a topological space is a collection of closed sets whose union contains S. { ¦klōzd 'kəv·ər·iŋ }

closed curve [MATH] A curve that has no end points. { ¦klōzd 'kərv }

closed cycle [THERMO] A thermodynamic cycle in which the thermodynamic fluid does not enter or leave the system, but is used over and over again. { ‚klōzd 'sī·kəl }

closed-cycle fuel cell [ELEC] A fuel cell in which the reactants are regenerated by an auxiliary process, such as electrolysis. { ¦klōzd ¦sī·kəl 'fyül ‚sel }

closed-cycle reactor [NUCLEO] A nuclear reactor in which the primary coolant flows to a heat exchanger and then recirculates through the core in a completely closed circuit. { ¦klōzd ¦sī·kəl rē'ak·tər }

closed-cycle turbine [MECH ENG] A gas turbine in which essentially all the working medium is continuously recycled, and heat is transferred through the walls of a closed heater to the cycle. { ¦klōzd ¦sī·kəl 'tər‚bīn }

closed die [MET] A forming or forging die in which the flow of metal is restricted to the cavity of the die set. { ‚klōzd 'dī }

closed dipath [MATH] A directed path whose initial and final vertices are the same. { ¦klōzd 'dī‚path }

closed disk [MATH] A circle and its interior. Also known as disk. { ¦klōzd 'disk }

closed drainage [HYD] Drainage in which the surface flow of water collects in sinks or lakes having no surface outlet. Also known as blind drainage. { ¦klōzd 'drā·nij }

closed ecological system [AERO ENG] A system used in spacecraft that provides for the maintenance of life in an isolated living chamber through complete reutilization of the material available, in particular, by means of a cycle wherein exhaled carbon dioxide, urine, and other waste matter are converted chemically or by photosynthesis into oxygen, water, and food. [ECOL] A community into which a new species cannot enter

due to crowding and competition. { ¦klōzd ek·ə'läj·ə·kəl ‚sis·təm }

closed file [COMPUT SCI] A file that cannot be accessed for reading or writing. { ¦klōzd 'fīl }

closed fireroom system [MECH ENG] A fireroom system in which combustion air is supplied via forced draft resulting from positive air pressure in the fireroom. { ¦klōzd 'fīr‚rüm ‚sis·təm }

closed fold [GEOL] A fold whose limbs have been compressed until they are parallel, and whose structure contour lines form a closed loop. Also known as tight fold. { ¦klōzd ¦fōld }

closed frame [MIN ENG] A mine support frame that is completely closed; especially in inclined shafts, it is used to protect all sides from rock pressure. { ¦klōzd ¦frām }

closed graph theorem [MATH] If T is a linear transformation on Banach space X to Banach space Y whose domain $D(T)$ is closed and whose graph, that is, the set of pairs (x,Tx) for x in $D(T)$, is closed in $X \times Y$, then T is bounded (and hence continuous). { ¦klōzd ¦graf 'thir·əm }

closed half plane [MATH] A half plane that includes the line that bounds it. { ¦klōzd ¦haf 'plān }

closed half space [MATH] A half space that includes the plane that bounds it. { ¦klōzd ¦half 'spās }

closed high [METEOROL] A high that may be completely encircled by an isobar or contour line. { ¦klōzd 'hī }

closed intervals [MATH] A closed interval of real numbers, denoted by $[a,b]$, consists of all numbers equal to or greater than a and equal to or less than b. { ¦klōzd 'in·tər·vəlz }

closed lake [HYD] A lake that does not have a surface effluent and that loses water by evaporation or by seepage. { ¦klōzd 'lāk }

closed linear manifold [MATH] A topologically closed vector subspace of a topological vector space. { ¦klōzd ¦lin·ē·ər 'man·ə‚fōld }

closed linear transformation [MATH] A linear transformation T such that the set of points of the form $[x, T(x)]$ is closed in the Cartesian product $\bar{D} \times \bar{R}$ of the closure of the domain D and the closure of the range R of T. { ¦klōzd ¦lin·ē·ər ‚tranz·fər'mā·shən }

closed loop [COMPUT SCI] A loop whose execution continues indefinitely in the absence of any external intervention. [CONT SYS] A family of automatic control units linked together with a process to form an endless chain; the effects of control action are constantly measured so that if the controlled quantity departs from the norm, the control units act to bring it back. { ¦klōzd 'lüp }

closed-loop control system See feedback control system. { ¦klōzd ¦lüp kən'trōl ‚sis·təm }

closed-loop telemetry system [ENG] **1.** A telemetry system which is also used as the display portion of a remote-control system. **2.** A system used to check out test vehicle or telemetry performance without radiation of radio-frequency energy. { ¦klōzd ¦lüp tə'lem·ə·trē ‚sis·təm }

closed-loop voltage gain [ELECTR] The voltage gain of an amplifier with feedback. { ¦klōzd ¦lüp 'vōl·tij ‚gān }

closed low [METEOROL] A low that may be completely encircled by an isobar or contour line, that is, an isobar or contour line of any value, not necessarily restricted to those arbitrarily chosen for the analysis of the chart. { ¦klōzd 'lō }

closed magnetic circuit [ELECTROMAG] A complete circulating path for magnetic flux around a core of ferromagnetic material. { ¦klōzd mag'ned·ik 'sər·kət }

closed map [MATH] A function between two topological spaces which sends each closed set of one into a closed set of the other. { ¦klōzd 'map }

closed-mapping theorem [MATH] The theorem that a linear, surjective mapping between two Banach spaces is continuous if and only if it is closed. { ¦klōzd 'map·iŋ ‚thir·əm }

closed n-cell [MATH] A set that is homeomorphic with the set of points in n-dimensional Euclidean space ($n = 1, 2, \ldots$) whose distance from the origin is equal to or less than unity. { ¦klōzd 'en ‚sel }

closed nozzle [MECH ENG] A fuel nozzle having a built-in valve interposed between the fuel supply and combustion chamber. { ¦klōzd 'näz·əl }

closed operator [MATH] A linear transformation f whose domain A is contained in a normed vector space X satisfying the condition that if $\lim x_n = x$ for a sequence x_n in A, and $\lim f(x_n) = y$, then x is in A and $f(x) = y$. { ¦klōzd 'äp·ə‚rād·ər }

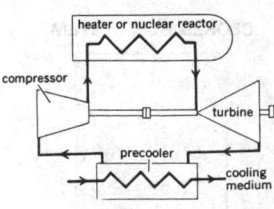

CLOSED-CYCLE TURBINE

A closed-cycle gas turbine with precooler, series flow, single-shaft power plant.

closed orthonormal set *See* complete orthonormal set. { ¦klōzd ¦ȯr·thō¦nȯr·məl 'set }

closed pair [MECH] A pair of bodies that are subject to constraints which prevent any relative motion between them. { ¦klōzd 'per }

closed pass [MET] A metal-rolling operation in which the top roll has a collar that fits a groove on the bottom roll, allowing a flash-free shape to be formed in the rolled metal. { ¦klōzd 'pas }

closed path [MATH] In a graph, a path whose initial and final vertices are the same. { ¦klōzd 'path }

closed polygonal region [MATH] The union of the interior of a polygon with the polygon itself. { ¦klōzd pə¦lig·ən·əl 'rē·jən }

closed pyramidal surface [MATH] A surface generated by a line passing through a fixed point and moving along a polygon in a plane not containing that point. { ¦klōzd ¦pir·ə¦mid·əl 'sər·fəs }

closed reading frame [MOL BIO] A reading frame containing terminator codons that prevent the translation of subsequent nucleotides into protein. Also known as blocked reading frame. { ¦klōzd 'rēd·iŋ ˌfrām }

closed rectangular region [MATH] The union of the interior of a rectangle with the rectangle itself. { ¦klōzd rek¦taŋ·gyə·lər 'rē·jən }

closed reduction [MED] Reduction of fractures or dislocations by manipulation without surgical intervention. { ¦klōzd ri'dək·shən }

closed region [MATH] The closure of an open, connected set. { ¦klōzd 'rē·jən }

closed respiratory gas system [ENG] A self-contained system within a sealed cabin, capsule, or spacecraft that will provide adequate oxygen for breathing, maintain adequate cabin pressure, and absorb the exhaled carbon dioxide and water vapor. { ¦klōzd 'res·prə¸tȯr·ē 'gas ˌsis·təm }

closed rotative gas lift [PETRO ENG] Oil-well control system in which high-pressure compressor gas is injected into a well to force oil fluids from the reservoir, with spent lift gas recompressed for reinjection. { ¦klōzd 'rō¸tad·iv 'gas ˌlift }

closed sea [OCEANOGR] **1.** That part of the ocean enclosed by headlands, within narrow straits, or within other landforms. **2.** That part of the ocean within the territorial jurisdiction of a country. { ¦klōzd 'sē }

closed set [MATH] A set of points which contains all its cluster points. Also known as topologically closed set. { ¦klōzd 'set }

closed shell [PHYS] An atomic or nuclear shell containing the maximum number of electrons or nucleons allowed by the Pauli exclusion principle. { ¦klōzd 'shel }

closed shop [COMPUT SCI] A data-processing center so organized that only professional programmers and operators have access to the center to meet the needs of users. [IND ENG] An establishment permitting only union members to be employed. { ¦klōzd 'shäp }

closed steam [ENG] Steam that flows through a heating coil or annulus so that there is no direct contact between the steam and the material being heated. { ¦klōzd 'stēm }

closed subroutine [COMPUT SCI] A subroutine that can be stored outside the main routine and can be connected to it by linkages at one or more locations. { ¦klōzd 'səb·rü¸tēn }

closed surface [MATH] A surface that has no bounding curve. { ¦klōzd 'sər·fəs }

closed system [ENG] A system for water handling that does not permit air to enter. [THERMO] A system which is isolated so that it cannot exchange matter or energy with its surroundings and can therefore attain a state of thermodynamic equilibrium. Also known as isolated system. { ¦klōzd 'sis·təm }

closed trapped surface [RELAT] A compact spacelike two-surface in space-time such that outgoing null rays perpendicular to the surface are not expanding. { ¦klōzd 'trapt 'sər·fəs }

closed triangular region [MATH] The union of the interior of a triangle with the triangle itself. { ¦klōzd trī¦aŋ·gyə·lər 'rē·jən }

closed universe [ASTRON] A cosmological model in which the volume of the universe is finite and in which the expansion of the universe will slow to a halt billions of years in the future, and the universe will then contract, becoming progressively denser, until it ends in a fireball similar to the big bang. { ¦klōzd 'yü·nə¸vərs }

closefile [COMPUT SCI] A procedure call in time sharing which enables an ALGOL program to close a file no longer required. { 'klōz¸fīl }

close-grained [MATER] Consisting of fine, closely spaced particles, crystals, or other elements. { 'klōs ¸grānd }

close in [PETRO ENG] **1.** To temporarily shut a well that still has the capacity to produce oil or gas. **2.** To close the blowout preventers of a well to control the entry of water, gas, oil, or other formation fluids into the wellbore during a drilling operation. { 'klōz 'in }

close-joints cleavage *See* slip cleavage. { 'klōs 'jȯins 'klē· vij }

close nipple [ENG] A short length of pipe that is completely threaded. { 'klōs 'nip·əl }

close-off rating [MECH ENG] **1.** The maximum allowable pressure drop to which a valve can be subjected at commercial shutoff. **2.** The maximum allowable pressure drop between the outlet of a three-way valve and either of the two inlets, or between the inlet and either of the two outlets. { 'klōz ¸ȯf ¸rād·iŋ }

close-out file [COMPUT SCI] A file created at the end of a processing cycle, usually encompassing a specified period of time. { 'klōz ¸aůt ¸fīl }

close-packed crystal [CRYSTAL] A crystal structure in which the lattice points are centers of spheres of equal radius arranged so that the volume of the interstices between the spheres is minimal. { 'klōs ¸pakt 'kris·təl }

closer [CIV ENG] **1.** In masonry work, the last brick or other masonry component that is laid in a horizontal course. Also known as closure. **2.** A stone course that extends from one windowsill to another. { 'klō·zər }

close routine [COMPUT SCI] A computer program that changes the state of a file from open to closed. { 'klōz rü'tēn }

close sand *See* tight sand. { 'klōs ¸sand }

closest approach [ASTRON] **1.** The event that occurs when two planets or other celestial bodies are nearest to each other as they orbit about the sun or other primary. **2.** The place or time of such an event. { 'klō·səst ə'prōch }

close-talking microphone [ENG ACOUS] A microphone designed for use close to the mouth, so noise from more distant points is suppressed. Also known as noise-canceling microphone. { 'klōs ¸tȯk·iŋ 'mī·krō¸fōn }

close-tolerance forging [MET] Forging in which draft angles are on the order of 1–3°, tolerances are less than half of those for commercial designs, and there is little or no allowance for finish. { ¦klōs 'täl·ə·rəns 'fȯr·jiŋ }

close work [MIN ENG] Driving a tunnel or drifting between two coal seams. { 'klōs ¸wərk }

closing line [MECH] The vector required to complete a polygon consisting of a set of vectors whose sum is zero (such as the forces acting on a body in equilibrium). { 'klōz·iŋ ¸līn }

clooing machine [ENG] A machine for manufacturing wire rope by braiding wire into strands, and strands into rope. Also known as stranding machine. { 'klōz·iŋ mə¸shēn }

closing plug [ORD] A plug used to close openings of various components of a round of ammunition, that is, primer or nose of an unfused projectile. { 'klōz·iŋ ¸pləg }

closing pressure [MECH ENG] The amount of static inlet pressure in a safety relief valve when the valve disk has a zero lift above the seat. { 'klōz·iŋ ¸presh·ər }

closing rate [AERO ENG] The speed at which two aircraft or missiles come closer together. { 'klōz·iŋ ¸rāt }

closing ratio [PETRO ENG] During completion operations, the ratio of the pressure in a hole to the pressure needed by the operating piston to close the rams of the blowout preventer. { 'klōz·iŋ ¸rā·shō }

closing-unit pump [PETRO ENG] A pump installed on an accumulator to control operation of the blowout preventers by pumping hydraulic fluid to them under high pressure. { 'klōz· iŋ 'yü·nət ¸pəmp }

Closteroviridae [VIROL] A family of plant viruses characterized by flexuous rod-shaped particles containing one molecule of single-stranded positive-sense ribonucleic acid; includes the genera *Closterovirus* and *Crinivirus*. { ¦klä·stə·rə'vir· ə¸dī }

Closterovirus [VIROL] A genus of plant viruses belonging to the family Closteroviridae that has a wide host range and is transmitted primarily by aphids; beet yellows virus is the type species. { ¦klä·stə·rə'vī·rəs }

Clostridium [MICROBIO] A genus of bacteria in the family Bacillaceae; usually motile rods which form large spores that distend the cell; anaerobic and do not reduce sulfate. { klä′strid·ē·əm }

Clostridium perfringens [MICROBIO] A spore-forming, toxin-producing bacterium that can contaminate meat left at room temperature. The ingested cells release toxin in the digestive tract, resulting in cramps and diarrhea. { klä‚strid·ē·əm pər′frin·jənz }

Clostridium tetani [MED] A spore-forming bacterium that produces a powerful toxin, tetanospasmin, that blocks inhibitory synapses in the central nervous system and thus causes the severe muscle spasms characteristic of tetanus. { klä‚strid·ē·əm ′tet·ən‚ī }

closure [CIV ENG] See closer. [GEOL] The vertical distance between the highest and lowest point on an anticline which is enclosed by contour lines. [MATH] **1.** The union of a set and its cluster points; the smallest closed set containing the set. **2.** Property of a mathematical set such that a specified mathematical operation that is applied to elements of the set produces only elements of the same set { ′klō·zhər }

closure domain [SOLID STATE] A small ferromagnetic domain whose position and orientation ensure that the flux lines between adjacent larger domains close on themselves. Also known as flux-closure domain. { ′klō·zhər dō‚mān }

closure parameter [ASTRON] The ratio of the actual mean mass density of the observable universe to the critical density for a Friedmann universe. { ′klō·zhər pə‚ram·əd·ər }

clot [PHYSIO] A semisolid coagulum of blood or lymph. { klät }

cloth [TEXT] **1.** A sheet of fibers assembled by weaving, knitting, felting, or some other similar process. **2.** A nonfibrous material of similar properties. { klȯth }

cloth beam [TEXT] One of five essential parts of a loom, upon which the newly constructed fabric is wound. { ′klȯth ‚bēm }

cloth count See thread count. { ′klȯth ‚kaunt }

clothing monitor [NUCLEO] An instrument designed for monitoring radioactive contamination on clothing. { ′klō·thin ‚män·əd·ər }

clothoid See Cornu's spiral. { ′klȯth‚ȯid }

cloth print [GRAPHICS] A photo reproduction image on a woven fiber base. { ′klȯth ‚print }

cloth wheel [DES ENG] A polishing wheel made of sections of cloth glued or sewn together. { ′klȯth ‚wēl }

clot retraction time [PATH] The length of time required for the appearance or completion of the contraction or shrinkage of a blood clot, resulting in the extrusion of serum. { ′klät ri′trak·shən ‚tīm }

clotting factor [PHYSIO] Any of several plasma components that are involved in the clotting of blood, such as fibrinogen, prothrombin, and thromboplastin. { ′kläd·in ‚fak·tər }

clotting time [PHYSIO] The length of time required for shed blood to coagulate under standard conditions. Also known as coagulation time. { ′kläd·in ‚tīm }

cloud [METEOROL] Suspensions of minute water droplets or of ice crystals produced by the condensation of water vapor. [NUC PHYS] The nucleons that are in the nucleus of an atom but not in closed shells. [SCI TECH] Any suspension of particulate matter, such as dust or smoke, dense enough to be seen. { klaud }

cloud absorption [GEOPHYS] The absorption of electromagnetic radiation by the waterdrops and water vapor within a cloud. { ′klaud əb′sȯrp·shən }

cloudage See cloud cover. { ′klau·dij }

cloud albedo [GEOPHYS] The fraction of the incident solar radiation reflected to space by clouds, which depends on the drop size, liquid water content, water vapor content, and thickness of the cloud, and the sun's elevation. The smaller the drops and the greater the liquid water content, the greater the cloud albedo. { ′klaud al′bēd·ō }

cloud attenuation [ELECTROMAG] The attenuation of microwave radiation by clouds (for the centimeter-wavelength band, clouds produce Rayleigh scattering); due largely to scattering, rather than absorption, for both ice and water clouds. { ′klaud ə‚ten·yə′wā·shən }

cloud band [METEOROL] A broad band of clouds, about 10 to 100 or more miles (16 to 160 kilometers) wide, and varying

in length from a few tens of miles to hundreds of miles. { ′klaud ‚band }

cloud bank [METEOROL] A fairly well-defined mass of cloud observed at a distance; covers an appreciable portion of the horizon sky, but does not extend overhead. { ′klaud ‚bank }

cloud banner See banner cloud. { ′klaud ‚ban·ər }

cloud bar [METEOROL] **1.** A heavy bank of clouds that appears on the horizon with the approach of an intense tropical cyclone (hurricane or typhoon); it is the outer edge of the central cloud mass of the storm. **2.** Any long, narrow, unbroken line of cloud, such as a crest cloud or an element of billow cloud. { ′klaud ‚bär }

cloud base [METEOROL] For a given cloud or cloud layer, that lowest level in the atmosphere at which the air contains a perceptible quantity of cloud particles. { ′klaud ‚bās }

cloudburst [METEOROL] In popular terminology, any sudden and heavy fall of rain, usually of the shower type, and with a fall rate equal to or greater than 100 millimeters (3.94 inches) per hour. Also known as rain gush; rain gust. { ′klaud‚bərst }

cloudburst hardness test [MET] A procedure in which a shower of steel balls, dropped from a predetermined height, dulls the surface of a hardened part in proportion to its softness and thus reveals defective areas. { ′klaud‚bərst ′härd·nəs ‚test }

cloudburst treatment [MET] Cold-working the surface of a metal by impingement of an avalanche of metal shot; a form of shot-peening. { ′klaud‚bərst ‚trēt·mənt }

cloud cap See cap cloud. { ′klaud ‚kap }

cloud chamber [NUCLEO] A particle detector in which the path of a charged particle is made visible by the formation of liquid droplets along the trail of ions left by the particle as it passes through the gas of the chamber. Also known as expansion chamber; fog chamber. { ′klaud ‚chām·bər }

cloud classification [METEOROL] **1.** A scheme of distinguishing and grouping clouds according to their appearance and, where possible, to their process of formation. **2.** A scheme of classifying clouds according to their altitudes: high, middle, or low clouds. **3.** A scheme of classifying clouds according to their particulate composition: water clouds, ice-crystal clouds, or mixed clouds. { ′klaud ‚klas·ə·fə′kā·shən }

cloud column [NUCLEO] The column of smoke extending upward from the point of burst of an atomic weapon. { ′klaud ‚käl·əm }

cloud cover [METEOROL] That portion of the sky cover which is attributed to clouds, usually measured in tenths of sky covered. Also known as cloudage; cloudiness. { ′klaud ‚kəv·ər }

cloud crest See crest cloud. { ′klaud ‚krest }

cloud deck [METEOROL] The upper surface of a cloud. { ′klaud ‚dek }

cloud-detection radar [ENG] A type of weather radar designed specifically for the detection of clouds (rather than precipitation). { ′klaud di′tek·shən ‚rā‚där }

cloud discharge [GEOPHYS] A lightning discharge occurring between a positive charge center and a negative charge center, both of which lie in the same cloud. Also known as cloud flash; intracloud discharge. { ′klaud ′dis‚chärj }

cloud droplet [METEOROL] A particle of liquid water from a few micrometers to tens of micrometers in diameter, formed by condensation of atmospheric water vapor and suspended in the atmosphere with other drops to form a cloud. { ′klaud ‚dräp·lət }

cloud-drop sampler [ENG] An instrument for collecting cloud particles, consisting of a sampling plate or cylinder and a shutter, which is so arranged that the sampling surface is exposed to the cloud for a predetermined length of time; the sampling surface is covered with a material which either captures the cloud particles or leaves an impression characteristic of the impinging elements. { ′klaud ‚dräp ′sam·plər }

cloud echo [METEOROL] The radar target signal returned from clouds alone, as detected by cloud detection radars or other very-short-wavelength equipment. { ′klaud ‚ek·ō }

cloud flash See cloud discharge. { ′klaud ‚flash }

cloud forest See temperate rainforest. { ′klaud ‚fär·əst }

cloud formation [METEOROL] **1.** The process by which various types of clouds are formed, generally involving adiabatic

CLOUD

Cumulonimbus clouds. Note the rain showers which appear under some of the clouds. (*Lt. B.H. Wyatt, U.S.N., U.S. Weather Bureau*)

CLOUD CHAMBER

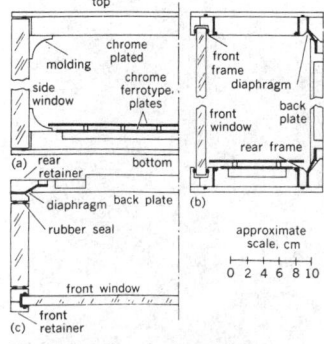

Cloud chamber designed for use in a magnetic field. (*a*) Vertical section parallel to front. (*b*) Vertical section parallel to side. (*c*) Horizontal section. The back plate moves to produce the expansion.

cooling of ascending moist air. **2.** A particular arrangement of clouds in the sky, or a striking development of a particular cloud. { 'klaůd fȯr'mā·shən }

cloud height [METEOROL] The absolute altitude of the base of a cloud. { 'klaůd ˌhīt }

cloud-height indicator [ENG] General term for an instrument which measures the height of cloud bases. { 'klaůd ˌhīt ˌin·dəˌkād·ər }

cloudiness See cloud cover. { 'klaůd·ē·nəs }

cloud-ion chamber [NUCLEO] An instrument combining the functions of a Wilson cloud chamber with those of an ionization chamber. { 'klaůd 'ī͟ˌän ˌchām·bər }

cloud layer [METEOROL] An array of clouds, not necessarily all of the same type, whose bases are at approximately the same level; may be either continuous or composed of detached elements. { 'klaůd ˌlā·ər }

cloud level [METEOROL] **1.** A layer in the atmosphere in which are found certain cloud genera; three levels are usually defined: high, middle, and low. **2.** At a particular time, the layer in the atmosphere bounded by the limits of the bases and tops of an existing cloud form. { 'klaůd ˌlev·əl }

cloud mirror See mirror nephoscope. { 'klaůd ˌmir·ər }

cloud modification [METEOROL] Any process by which the natural course of development of a cloud is altered by artificial means. { 'klaůd ˌmäd·ə·fə'kā·shən }

cloud particle [METEOROL] A particle of water, either a drop of liquid water or an ice crystal, comprising a cloud. { 'klaůd ˌpärd·ə·kəl }

cloud-phase chart [METEOROL] A chart designed to indicate and distinguish supercooled water clouds from ice-crystal clouds. { 'klaůd ˌfāz ˌchärt }

cloud physics [METEOROL] The study of the physical and dynamical processes governing the structure and development of clouds and the release from them of snow, rain, and hail. { 'klaůd ˌfiz·iks }

cloud point [CHEM ENG] The temperature at which paraffin wax or other solid substance begins to separate from a solution of petroleum oil; a cloudy appearance is seen in the oil at this point. { 'klaůd ˌpȯint }

cloud pulse [ELECTR] The output resulting from space charge effects produced by turning the electron beam on or off in a charge-storage tube. { 'klaůd ˌpəls }

cloud seeding [METEOROL] Any technique carried out with the intent of adding to a cloud certain particles that will alter its natural development. { 'klaůd ˌsēd·iŋ }

cloud shield [METEOROL] The principal cloud structure of a typical wave cyclone, that is, the cloud forms found on the cold-air side of the frontal system. { 'klaůd ˌshēld }

cloud street [METEOROL] A line of cumuliform clouds frequently one cumulus element wide, but ranging upward in width so that it is sometimes difficult to differentiate between streets and bands. { 'klaůd ˌstrēt }

cloud symbol [METEOROL] One of a set of specified ideograms that represent the various cloud types of greatest significance or those most commonly observed, and entered on a weather map as part of a station model. { 'klaůd ˌsim·bəl }

cloud system [METEOROL] An array of clouds and precipitation associated with a cyclonic-scale feature of atmospheric circulation, and displaying typical patterns and continuity. Also known as nephsystem. { 'klaůd ˌsis·təm }

cloud test [CHEM ENG] An American Society for Testing and Materials method for determining the cloud point of petroleum oil. { 'klaůd ˌtest }

cloud-to-cloud discharge [GEOPHYS] A lightning discharge occurring between a positive charge center of one cloud and a negative charge center of a second cloud. Also known as intercloud discharge. { 'klaůd tə 'klaůd 'disˌchärj }

cloud-to-ground discharge [GEOPHYS] A lightning discharge occurring between a charge center (usually negative) in the cloud and a center of opposite charge at the ground. Also known as ground discharge. { 'klaůd tə 'graůnd 'disˌchärj }

cloud top [METEOROL] The highest level in the atmosphere at which the air contains a perceptible quantity of cloud particles for a given cloud or cloud layer. { 'klaůd ˌtäp }

cloud track [NUCLEO] The string of minute water droplets that forms along the path of an ionizing particle in the supersaturated vapor of a cloud chamber. { 'klaůd ˌtrak }

cloudy [METEOROL] The character of a day's weather when the average cloudiness, as determined from frequent observations, is more than 0.7 for the 24-hour period. { 'klaůd·ē }

cloudy-crystal-ball model [NUC PHYS] An optical analogy used in explaining scattering of nucleons by nuclei, in which the nucleus is thought of as a sphere of nuclear matter which partially refracts and partially absorbs the incident nucleon (de Broglie) wave. Also known as optical model. { 'klaůd·ē ˌkris·təl ˌbȯl ˌmäd·əl }

cloudy swelling [PATH] A retrogressive change in the cytoplasm of parenchymatous cells, whereby the cell enlarges and its outline becomes irregular, with resultant swelling of the organ. { 'klaůd·ē 'swel·iŋ }

clough [GEOGR] A cleft in a hill; a ravine or narrow valley. { kləf }

clout nail [DES ENG] A nail with a large, thin, flat head used in building. { 'klaůt ˌnāl }

clove [BOT] **1.** The unopened flower bud of a small, conical, symmetrical evergreen tree, *Eugenia caryophyllata*, of the myrtle family (Myrtaceae); the dried buds are used as a pungent, strongly aromatic spice. **2.** A small bulb developed within a larger bulb, as in garlic. { klōv }

clove oil See oil of cloves. { 'klōv ˌȯil }

clover [BOT] **1.** A common name designating the true clovers, sweet clovers, and other members of the Leguminosa. **2.** A herb of the genus *Trifolium*. { 'klō·vər }

cloverleaf [CIV ENG] A highway intersection resembling a clover leaf and designed to allow movement and interchange of traffic without direct crossings and left turns. { 'klō·vərˌlēf }

cloverleaf antenna [ELECTROMAG] Antenna having radiating units shaped like a four-leaf clover. { 'klō·vərˌlēf an ˌten·ə }

cloze procedure [PSYCH] A procedure used in the study of reading processes in which one or more words are deleted from a prose passage and the subject is required to fill in the blanks. { 'klōz prəˌsē·jər }

CLS See characteristic loss spectroscopy.

clubfoot [MED] Congenital malpositioning of a foot such that the forefoot is inverted and rotated with a shortened Achilles tendon. { 'kləbˌfůt }

club fungi [MYCOL] The common name for members of the class Basidiomycetes. { 'kləb ˌfənˌjī }

clubhead fungus See Cordyceps ophioglossoides. { 'kləbˌhed 'fəŋ·gəs }

club moss [BOT] The common name for members of the class Lycopodiatae. { 'kləb ˌmȯs }

clubroot [PL PATH] A disease principally of crucifers, such as cabbage, caused by the slime mold *Plasmodiophora brassicae* in which roots become enlarged and deformed, leading to plant death. { 'kləbˌrüt }

Clupeidae [VERT ZOO] The herrings, a family of fishes in the suborder Clupoidea composing the most primitive group of higher bony fishes. { klü'pē·əˌdē }

Clupeiformes [VERT ZOO] An order of teleost fishes in the subclass Actinopterygii, generally having a silvery, compressed body. { ˌklü·pē·ə'fȯrˌmēz }

clupeine [BIOCHEM] A protamine found in salmon sperm, mainly composed of arginine (74.1%) and small percentages of threonine, serine, proline, alanine, valine, and isoleucine. { 'klü·pēˌēn }

Clupoidea [VERT ZOO] A suborder of fishes in the order Clupeiformes comprising the herrings and anchovies. { ˌklü'pȯidē·ə }

cluse [GEOL] A narrow gorge, trench, or water gap with steep sides that cuts transversely through an otherwise continuous ridge. { klüz }

clusec [MECH ENG] A unit of power used to measure the power of evacuation of a vacuum pump, equal to the power associated with a leak rate of 1 centiliter per second at a pressure of 1 millitorr, or to approximately 1.33322×10^{-6} watt. { 'klüˌsek }

Clusiidae [INV ZOO] A family of myodarian cyclorrhaphous dipteran insects in the subsection Acalypteratae. { ˌklü'sī·əˌdē }

Clusius column [NUCLEO] A device for separating isotopes by thermal diffusion, consisting of a long vertical tube with an electrically heated wire along its axis that produces a temperature gradient, causing the lighter isotope to concentrate around the wire and the heavier isotope to concentrate near the walls. { 'klüz·ē·əs 'käl·əm }

CLOVE

Closed flower buds (cloves) and open buds on a branch of the evergreen tree *Eugenia caryophyllata*.

cluster [ASTRON] *See* star cluster. [COMPUT SCI] **1.** In a clustered file, one of the classes into which records with similar sets of content identifiers are grouped. **2.** A grouping of hardware devices in a distributed processing system. **3.** A group of disk sectors that is treated as a single entity by the operating system [ENG] **1.** A pyrotechnic signal consisting of a group of stars or fireballs. **2.** A grouping of rocket motors fastened together. [ORD] A collection of small bombs held together by an adapter for dropping. { 'kləs·tər }

cluster aggregation [PHYS] A mathematical model of a coagulation process in which a collection of particles all move randomly at once, and two particles, or a particle and a previously formed cluster, stick together whenever they come within a certain fixed distance of each other. { 'kləs·tər ‚ag·rə'gā·shən }

cluster analysis [STAT] A general approach to multivariate problems whose aim is to determine whether the individuals fall into groups or clusters. { 'kləs·tər ə'nal·ə·səs }

cluster bean *See* guar. { 'kləs·tər ‚bēn }

cluster cepheids *See* RR Lyrae stars. { 'kləs·tər 'sef·ē·ədz }

cluster controller [COMPUT SCI] A control unit to which several peripheral devices are assigned. { 'kləs·tər kən‚trōl·ər }

clustered file [COMPUT SCI] A collection of records organized so that items which exhibit similar sets of content identifiers are automatically grouped into common classes. { 'kləs·tərd 'fīl }

cluster expansion [STAT MECH] A virial expansion in which the virial coefficients (of inverse powers of the volume of the gas in question) are obtained from integrals, over positions of a small number of molecules, of functions involving intermolecular potentials. { 'kləs·tər ik'span·shən }

cluster gene [GEN] A gene that codes for a multifunctional enzyme. { 'kləs·tər ‚jēn }

cluster headache [MED] A type of migraine that is characterized by severe unilateral pain in the eye or temple and tends to recur in a series of attacks. { 'kləs·tər ‚hed‚āk }

clustering algorithm [COMPUT SCI] A computer program that attempts to detect and locate the presence of groups of vectors, in a high-dimensional multivariate space, that share some property of similarity. { 'kləs·tə‚riŋ ‚al·gə‚rith·əm }

clusterite *See* botryoid. { 'klə‚stə‚rīt }

cluster mill [MET] A rolling mill in which small-diameter rolls are supported by larger rolls. { 'kləs·tər ‚mil }

cluster point [MATH] A cluster point of a set in a topological space is a point *p* whose neighborhoods all contain at least one point of the set other than *p*. Also known as accumulation point; limit point. { 'kləs·tər ‚point }

cluster radioactivity [NUC PHYS] A process in which a nucleus emits a fragment that is heavier than an alpha particle but lighter than a fission fragment, such as carbon-14 or neon-24. { 'kləs·tər 'rād·ē·ō ak‚tiv·əd·ē }

cluster sampling [STAT] A random sampling plan in which the population is subdivided into groups called clusters so that there is small variability within clusters and large variability between clusters. { 'kləs·tər ‚sam·pliŋ }

cluster variables *See* RR Lyrae stars. { ‚kləs·tər ‚ver·ē·ə·bəlz }

clutch [MECH ENG] A machine element for the connection and disconnection of shafts in equipment drives, especially while running. [VERT ZOO] A nest of eggs or a brood of chicks. { kləch }

clutch point [COMPUT SCI] The moment in time at which the clutch is engaged in a peripheral device such as a card punch. { 'kləch ‚point }

clutter [ELECTROMAG] Unwanted echoes on a radar screen, such as those caused by the ground, sea, rain, stationary objects, chaff, enemy jamming transmissions, and grass. Also known as background return; radar clutter. [MATH] *See* Sperner set. { 'kləd·ər }

clutter gating [ELECTR] A technique which provides switching between moving-target-indicator and normal videos; this results in normal video being displayed in regions with no clutter and moving-target-indicator video being switched in only for the clutter areas. { 'kləd·ər ‚gad·iŋ }

Clypeasteroida [INV ZOO] An order of exocyclic Euechinoidea having a monobasal apical system in which all the genital plates fuse together. { ‚klip·ē‚as·tə'roíd·ə }

clypeus [INV ZOO] An anterior medial plate on the head of

an insect, commonly bearing the labrum on its anterior margin. [MYCOL] A disk of black tissue about the mouth of the perithecia in certain ascomycetes. { 'klip·ē·əs }

clysis [MED] **1.** Administration of an enema. **2.** Subcutaneous or intravenous administration of fluids. { 'klī·səs }

Clythiidae [INV ZOO] The flat-footed flies, a family of cyclorrhaphous dipteran insects in the series Aschiza characterized by a flattened distal end on the hind tarsus. { klə'thī·ə‚dē }

cm *See* centimeter.

Cm *See* curium.

CMa *See* Canis Major.

cmHg *See* centimeter of mercury.

CMi *See* Canis Minor.

CMI *See* computer-managed instruction.

CML *See* current-mode logic.

C mode [ACOUS] A form of acoustic tomography that provides a two-dimensional image display at constant time delay, and presumably at constant distance from the ultrasonic transducer. { 'sē ‚mōd }

CMOS device [ELECTR] A device formed by the combination of a PMOS (*p*-type-channel metal oxide semiconductor device) with an NMOS (*n*-type-channel metal oxide semiconductor device). Derived from complementary metal oxide semiconductor device. { 'se‚mós di'vīs }

CMRR *See* common-mode rejection ratio.

CM Tauri [ASTRON] A supernova observed by the Chinese and Japanese in 1054; remnants are still seen as the Crab Nebula. { ‚sē‚em 'taúr·ē }

CMV *See* cucumber mosaic virus.

CMYK [GRAPHICS] A color model that synthesizes all colors as combinations of cyan, magenta, yellow, and black; it begins with white, and subtracts the appropriate color to yield the desired color. Also known as CMY.

CNC *See* computer numerical control.

C network [ELECTR] Network composed of three impedance branches in series, the free ends being connected to one pair of terminals, and the junction points being connected to another pair of terminals. { 'sē ‚net‚wərk }

C neutron [NUCLEO] A neutron of such energy, up to about 0.3 electronvolt, that it is strongly absorbable in cadmium. { ‚sē 'nü‚trän }

Cnidaria [INV ZOO] A phylum of the Radiata whose members typically bear tentacles and possess intrinsic nematocysts. Also known as Coelenterata. { nī'dar·ē·ə }

cnidoblast [INV ZOO] A cell that produces nematocysts. Also known as cnidocyte; nettle cell; stinging cell. { 'nīd·ə‚blast }

cnidocil [INV ZOO] The trigger on a cnidoblast that activates discharge of the nematocyst when touched. { 'nīd·ə‚sil }

cnidocyte *See* cnidoblast. { 'knīd·ə‚sīt }

cnidophore [INV ZOO] A modified structure bearing nematocysts in certain cnidarians. { 'nīd·ə‚fòr }

Cnidospora [INV ZOO] A subphylum of spore-producing protozoans that are parasites in cells and tissues of invertebrates, fishes, a few amphibians, and turtles. { nī'däs·pə·rə }

CNO cycles *See* carbon-nitrogen-oxygen cycles. { ‚sē‚en‚ō ‚sī·kəlz }

c-number theory *See* classical field theory. { 'sē ‚nəm·bər 'thē·ə·rē }

Co *See* cobalt.

⁶⁰Co *See* cobalt-60.

CoA *See* coenzyme A.

coacervate [CHEM] An aggregate of colloidal droplets bound together by the force of electrostatic attraction. { kō'as·ər‚vāt }

coacervation [CHEM] The separation, by addition of a third component, of an aqueous solution of a macromolecule colloid (polymer) into two liquid phases, one of which is colloid-rich (the coacervate) and the other an aqueous solution of the coacervating agent (the equilibrium liquid). { kō‚as·ər'vā·shən }

coach screw [DES ENG] A large, square-headed, wooden screw used to join heavy timbers. Also known as lag bolt; lag screw. { 'kōch ‚skrü }

coadaptation [EVOL] The selection process that tends to accumulate favorably interacting genes in the gene pool of a population. { ‚kō‚ad·əp'tā·shən }

coagulability test [PATH] Any of several clinical tests of the

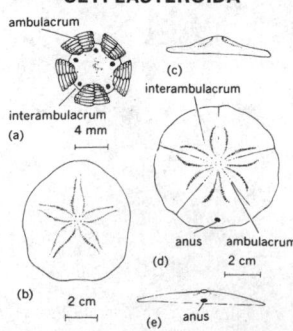

CLYPEASTEROIDA

ambulacrum

(c)
interambulacrum

interambulacrum
(a)
4 mm

anus   ambulacrum
(d)
2 cm

(b)
2 cm

(e)   anus

Typical clypeasteroids. *Laganum,* *(a)* monobasal apical system, *(b)* aboral aspect, and *(c)* posterior aspect. *Arachnoides, (d)* aboral aspect, and *(e)* posterior aspect.

ability of blood to coagulate, such as clot retraction time and quantification, prothrombin time, partial thromboplastin time, and platelet enumeration. { kō‚ag·yə·lə'bil·əd·ē test }

coagulant [CHEM] An agent that causes coagulation. { kō'ag·yə·lənt }

coagulase [BIOCHEM] Any enzyme that causes coagulation of blood plasma. { kō'ag·yə‚lās }

coagulation [CHEM] A separation or precipitation from a dispersed state of suspensoid particles resulting from their growth; may result from prolonged heating, addition of an electrolyte, or from a condensation reaction between solute and solvent; an example is the setting of a gel. *See* agglomeration. { kō‚ag·yə'lā·shən }

coagulation time *See* clotting time. { kō‚ag·yə'lā·shən ‚tīm }

Coahuilan [GEOL] A North American provincial series in Lower Cretaceous geologic time, above the Upper Jurassic and below the Comanchean. { kō·ə'wēl·ən }

coak [DES ENG] **1.** A projection from the end of a piece of wood or timber that is designed to fit into a hole in another piece so that they can be joined to form a continuous unit. **2.** A dowel or hardwood pin that joins overlapping timbers. { kōk }

coal [GEOL] The natural, rocklike, brown to black derivative of forest-type plant material, usually accumulated in peat beds and progressively compressed and indurated until it is finally altered into graphite or graphite-like material. { kōl }

coal auger [MIN ENG] A type of continuous miner which consists of a screw drill of large diameter and which cuts, transports, and loads the coal. { 'kōl ‚óg·ər }

coal ball [GEOL] A subspherical mass containing mineral matter embedded with plant material, found in coal seams and overlying beds of the late Paleozoic. { 'kōl ‚ból }

coal bank [MIN ENG] A seam of coal that is exposed. { 'kōl ‚baŋk }

coal barrier [MIN ENG] A protective pillar composed of coal. { 'kōl ‚bar·ē·ər }

coal bed [GEOL] A seam or stratum of coal parallel to the rock stratification. Also known as coal rake; coal seam. { 'kōl ‚bed }

coal blasting [MIN ENG] Breaking coal with explosives. { 'kōl ‚blast·iŋ }

coalbreaker *See* breaker. { 'kōl‚brāk·ər }

coal breccia [GEOL] Angular fragments of coal within a coal bed. { 'kōl ‚brech·ə }

coal chemicals [MATER] Chemicals obtained as by-products in the primary processing of coal to metallurgical coke; the main source of aromatic compounds used as intermediates in the synthesis of dyes, drugs, antiseptics, and solvents. { 'kōl ‚kem·i·kəlz }

coal clay *See* underclay. { 'kōl ‚klā }

coal cutter [MIN ENG] A power-operated machine which cuts out a thin strip of coal from the bottom of the seam; it draws itself by rope haulage along the coal face. { 'kōl ‚kəd·ər }

coal digger *See* faceman. { 'kōl ‚dig·ər }

coal drill [MIN ENG] Usually, an electric drill of a compact, light design; however, also a light pneumatic drill. { 'kōl ‚dril }

coal dust [MIN ENG] A finely divided coal, sometimes defined as coal that will pass through 100-mesh screens (100 wires to the inch or 40 wires to the centimeter). { 'kōl ‚dəst }

coalesce [SCI TECH] To come together to form a whole. { kō·ə'les }

coalesced copper [MET] A mass, oxygen-free copper made by compacting and sintering cathode copper at high pressure and temperature. { ‚kō·ə'lest 'käp·ər }

coalescence [BOT] The union of plant parts of the same kind such as the united sepals of flowering plants. [MET] The bonding of welded materials into one body. [METEOROL] In cloud physics, merging of two or more water drops into a single larger drop. [PHYS] The uniting by growth in one body, as particles, gas, or a liquid. { ‚kō·ə'les·əns }

coalescence efficiency [METEOROL] The fraction of all collisions which occur between waterdrops of a specified size and which result in actual merging of two drops into a single larger drop. { ‚kō·ə'les·əns i'fish·ən·sē }

coalescence process [METEOROL] The growth of raindrops by the collision and coalescence of cloud drops or small precipitation particles. { ‚kō·ə'les·əns ‚präs·əs }

coalescent [CHEM] Chemical additive used in immiscible liquid-liquid mixtures to cause small droplets of the suspended liquid to unite, preparatory to removal from the carrier liquid. { ‚kō·ə'les·ənt }

coalescent pack [CHEM ENG] High-surface-area packing to consolidate liquid droplets for gravity separation from a second phase (for example, gas or immiscible liquid); packing must be wettable by the droplet phase; Berl saddles, Raschig rings, knitted wire mesh, excelsior, and similar materials are used. { ‚kō·ə'les·ənt 'pak }

coalescer [CHEM ENG] Mechanical process vessel with wettable, high-surface area packing on which liquid droplets consolidate for gravity separation from a second phase (for example, gas or immiscible liquid). { ‚kō·ə'les·ər }

coal face [MIN ENG] The mining face from which coal is extracted. { 'kōl ‚fās }

coalfield [MIN ENG] A region containing coal deposits. { 'kōl‚fēld }

coal gas [MATER] **1.** Flammable gas derived from coal either naturally in place, or by induced methods of industrial plants and underground gasification. **2.** Specifically, fuel gas obtained from carbonization of coal. { 'kōl ‚gas }

coal gasification [CHEM ENG] The conversion of coal, char, or coke to a gaseous product by reaction with air, oxygen, steam, carbon dioxide, or mixtures of these. { 'kōl ‚gas·ə·fə'kā·shən }

coal getter *See* faceman. { 'kōl ‚ged·ər }

coal hydrogenation *See* Bergius process. { 'kōl ‚hī·drə·jə'nā·shən }

coalification [GEOL] Formation of coal from plant material by the processes of diagenesis and metamorphism. Also known as bituminization; carbonification; incarbonization; incoalation. { ‚kōl·ə·fə'kā·shən }

coal-in-oil suspension [MATER] A fluid mixture of pulverized coal dispersed in either fuel oil or coal tar oil, used as a fuel chiefly in large installations. Also known as colloidal fuel. { ‚kōl in ‚óil səs'pen·shən }

coal liquefaction [CHEM ENG] The conversion of coal (with the exception of anthracite) to petroleum-like hydrocarbon liquids, which are used as refinery feedstocks for the manufacture of gasoline, heating oil, diesel fuel, jet fuel, turbine fuel, fuel oil, and petrochemicals. { ‚kōl lik·wə'fak·shən }

Coal Measures [GEOL] The sequence of rocks typically containing coal of the Upper Carboniferous. { 'kōl ‚mezh·ərz }

coal mining [MIN ENG] The technical and mechanical job of removing coal from the earth and preparing it for market. { 'kōl ‚mīn·iŋ }

coal oil [MATER] **1.** Condensed liquid from coal distillation. **2.** An archaic term for kerosine made from petroleum. { 'kōl ‚óil }

coal paleobotany [PALEOBOT] A branch of the paleobotanical sciences concerned with the origin, composition, mode of occurrence, and significance of fossil plant materials that occur in or are associated with coal seams. { 'kōl ‚pā·lē·ō'bät·ən·ē }

coal pebbles [GEOL] Rounded masses of coal occurring in sedimentary rock. { 'kōl ‚peb·əlz }

coal petrology [GEOL] The science that deals with the origin, history, occurrence, structure, chemical composition, and classification of coal. { 'kōl pə'träl·ə·jē }

coal planer [MIN ENG] A type of continuous coal mining machine for longwall mining; consists of a heavy steel plow with cutting knives, with power equipment to drag it back and forth across a coal face. { 'kōl ‚plān·ər }

coal plough [MIN ENG] A device with steel blades which shears off coal and pushes it onto the face conveyor. { 'kōl ‚plaù }

coal rake *See* coal bed. { 'kōl ‚rāk }

Coalsack [ASTRON] An area in one of the brighter regions of the Southern Milky Way which to the naked eye appears entirely devoid of stars and hence dark with respect to the surrounding Milky Way region. { 'kōl‚sak }

coal seam *See* coal bed. { 'kōl ‚sēm }

coal-sensing probe [MIN ENG] A nucleonic instrument to measure the thickness of coal left in the seam floor by means of a gamma-ray backscattering unit. { 'kōl ‚sen·siŋ 'prōb }

coal sizes [MIN ENG] The sizes by which anthracite coal is marketed. { 'kōl ‚sī·zəz }

coal split *See* split. { 'kōl ‚split }

COAL BALL

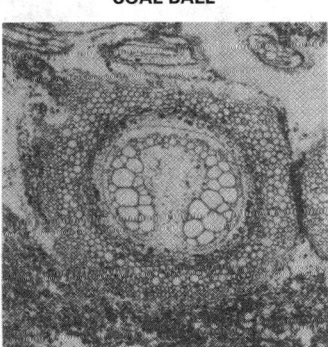

Photomicrograph of the peel made from a coal ball containing the branch of a fern.

coal tar [MATER] A tar obtained from carbonization of coal, usually in coke ovens or retorts, containing several hundred organic chemicals. { 'kōl ,tär }

coal-tar dye [ORG CHEM] Dye made from a coal-tar hydrocarbon or a derivative such as benzene, toluene, xylene, naphthalene, or aniline. { 'kōl ,tär ,dī }

coal-tar enamel [MATER] Coal tar used as a paintlike coating for petroleum-product pipelines; provides protection from both water and cathodic corrosion. { 'kōl ‚tär i,nam·əl }

coal-tar epoxy [MATER] A thermosetting resin produced as a by-product of the carbonization of bituminous coal. { 'kōl ‚tär ə'päk·sē }

coal-tar light oil See light oil. { 'kōl ‚tär 'līt ,òil }

coal-tar pitch [MATER] Dark-brown to black amorphous residue from the redistillation of coal tar; melts at 150°F (66°C); used as a thermoplastic. { 'kōl ‚tär ,pich }

co-altitude See zenith distance. { kō'al·tə,tüd }

coaming [NAV ARCH] A rim placed on a roof or around a hatch, deck, or bulkhead opening to stop water from entering. { 'kōm·iŋ }

coancestry [GEN] The degree of relationship between two parents of a diploid individual. { ‚kō'an,səs·trē }

Coanda effect [FL MECH] The tendency of a gas or liquid coming out of a jet to travel close to the wall contour even if the wall's direction of curvature is away from the jet's axis; a factor in the operation of a fluidic element. { kō'an·də i'fekt }

coarctation [MED] **1.** A compression of the wall of a vessel, narrowing the lumen and reducing the volume (or flow). **2.** A stricture or occlusion resulting from an outside force deforming a vessel. { kō·ärk'tā·shən }

coarse See low-grade. { kórs }

coarse aggregate [MATER] Crushed stone or gravel used in concrete; will not, when dry, pass through a sieve with $1/4$-inch-diameter (6-millimeter) holes. { kórs 'ag·rə·gət }

coarse fragment [GEOL] A rock or mineral fragment in the soil with an equivalent diameter greater than 0.08 inch (2 millimeters). { kórs 'frag·mənt }

coarse-grained [MATER] Having a coarse texture. [PETR] See phaneritic. { kórs ¦grānd }

coarser [MATH] A partition P of a set is coarser than another partition Q of the same set if each member of Q is a subset of a member of P. { kórs·ər }

coarse roll [MIN ENG] A large roll for the preliminary crushing of large pieces of ore, rock, or coal. { kórs 'rōl }

coarse setting [ORD] Preliminary adjustment of a sight in laying a gun, made first on the main scale; then the fine setting is made on the associated scale of smaller graduations. { kórs 'sed·iŋ }

coarse sighting [ORD] Adjustment of the sight of a gun so that a part of the front sight is seen through the notch in the rear sight. { kórs 'sīd·iŋ }

coast [ENG] A memory feature on a radar which, when activated, causes the range and angle systems to continue to move in the same direction and at the same speed as that required to track an original target. [GEOGR] The general region of indefinite width that extends from the sea inland to the first major change in terrain features. { kōst }

coastal berm See berm. { 'kōs·təl 'bərm }

coastal current [OCEANOGR] An offshore current flowing generally parallel to the shoreline with a relatively uniform velocity. { 'kōs·təl 'kər·ənt }

coastal dune [GEOL] A mobile mound of windblown material found along many sea and lake shores. { 'kōs·təl 'dün }

coastal engineering [CIV ENG] A branch of civil engineering pertaining to the study of the action of the seas on shorelines and to the design of structures to protect against this action. { 'kōs·təl en·jə'nir·iŋ }

coastal ice See fast ice. { 'kōs·təl 'īs }

coastal landform [GEOGR] The characteristic features and patterns of land in a coastal zone subject to marine and subaerial processes of erosion and deposition. { 'kōs·təl 'land,fórm }

coastal plain [GEOL] An extensive, low-relief area that is bounded by the sea on one side and by a high-relief province on the landward side. Its geologic province actually extends beyond the shoreline across the continental shelf; it is linked to the stable part of a continent on the trailing edge of a plate. Typically, it has strata that dip gently and uniformly toward the sea. { 'kōs·təl 'plān }

coastal refraction [ELECTROMAG] An apparent change in the direction of travel of a radio wave when it crosses a shoreline obliquely. Also known as land effect. { 'kōs·təl ri'frak·shən }

coastal sediment [GEOL] The mineral and organic deposits of deltas, lagoons, and bays, barrier islands and beaches, and the surf zone. { 'kōs·təl 'sed·ə·mənt }

coast chart [NAV] A nautical chart for use in inshore, coastwise navigation when a course carries a vessel inside outlying reefs and shoals, for use in entering or leaving bays and harbors of considerable size, or for use in navigating larger inland waterways. { kōst ,chärt }

coaster [NAV ARCH] A small merchant ship, about 200 feet (61 meters) long, which operates near coasts, in rivers and estuaries, and on short ocean passages. { 'kō·stər }

coast guard [ORD] A naval force which guards a coast and ensures the order, safety, and effective operation of traffic on the coastal waters. { 'kōst ,gärd }

coast guard cutter [NAV ARCH] A small, armed boat in a coast guard. { 'kōst ,gärd ,kəd·ər }

Coast Guard lines [NAV] Lines established by the U.S. Coast Guard for separating areas of the sea where the inland rules of the road apply, from those areas where the international rules apply. { 'kōst ,gärd ,līnz }

Coast Guard station [NAV] In American usage, any building on the coast used to house personnel and equipment for saving life at sea. Also known as life-saving station. { 'kōst ,gärd ,stā·shən }

coast ice See fast ice. { 'kōst ,īs }

coasting [NAV] Proceeding approximately parallel to a coastline and near enough to be in pilot waters most of the time. { 'kō·stiŋ }

coasting flight [AERO ENG] The flight of a rocket between burnout or thrust cutoff of one stage and ignition of another, or between burnout and summit altitude or maximum horizontal range. { 'kō·stiŋ ,flīt }

coastline [GEOGR] **1.** The line that forms the boundary between the shore and the coast. **2.** The line that forms the boundary between the water and the land. { 'kōst,līn }

coastlining [MAP] The process of obtaining data from which the coastline can be drawn on a chart. { 'kōst,līn·iŋ }

coast pilot [NAV] A book serving as an adjunct to nautical charts, containing important information which cannot be shown conveniently on the charts, and not readily available elsewhere; prepared by the U.S. Coast and Geodetic Survey for coastal waters of continental United States, Hawaii, the Virgin Islands, and Puerto Rico; and by the U.S. Naval Oceanographic Office for foreign waters. Also known as sailing directions. { 'kōst ,pī·lət }

coast piloting [NAV] The directing of the movements of a vessel near a coast by means of terrestrial reference points. { 'kōst ,pī·lət·iŋ }

coast shelf See submerged coastal plain. { 'kōst ,shelf }

coastwise navigation [NAV] Navigation in the vicinity of a coast, in contrast to offshore navigation at a distance from a coast. { 'kōst,wīz ,nav·ə'gā·shən }

coated abrasive [MATER] An abrasive product having the abrasive particles attached to a backing material with glue or a synthetic resin. { 'kōd·əd ə'brā·siv }

coated cathode [ELECTR] A cathode that has been coated with compounds to increase electron emission. { 'kōd·əd 'kath,ōd }

coated electrode [MET] A wire covered with metal oxides and silicates and used as a filler-metal electrode in arc welding. Also known as covered electrode. { 'kōd·əd i'lek,trōd }

coated fabric [TEXT] A fabric that has been coated, covered, or impregnated with substances such as lacquer, varnish, rubber, or polymers. { 'kōd·əd 'fab·rik }

coated filament [ELECTR] A vacuum-tube filament coated with metal oxides to provide increased electron emission. { 'kōd·əd 'fil·ə·mənt }

coated lens [OPTICS] A lens whose surfaces have been coated with a thin, transparent film having an index of refraction that minimizes light loss by reflection. { 'kōd·əd 'lenz }

coated paper [MATER] Paper with a surface coating of clay and other materials to produce a smooth, shiny surface; especially useful for fine, detailed, blur-free reproductions in color or black and white. Also known as enamel paper. { 'kōd·əd 'pā·pər }

coated pit [CYTOL] A cell surface depression that is coated

COASTAL PLAIN

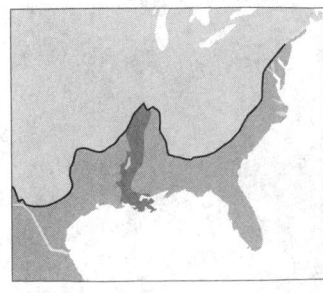

Coastal plains of the United States.

with clathrin on its cytoplasmic surface and functions in receptor-mediated endocytosis. { 'kōd·əd 'pit }

coated vesicle [CELL MOL] An intracellular structure formed by an invagination of the membrane surrounded by a cagelike protein coating. { ¦kōd·əd 'ves·ə·kəl }

coat hanger die [ENG] A plastics-sheet slot die shaped like a coat hanger on the inside. { 'kōt ¦haŋ·ər ‚dī }

coati [VERT ZOO] The common name for three species of carnivorous mammals assigned to the raccoon family (Procyonidae) characterized by their elongated snout, body, and tail. { kə'wäd·ē }

coating [MATER] **1.** Any material that will form a continuous film over a surface. **2.** The film formed by the material. { 'kōd·iŋ }

coating density ratio [MET] In thermal spraying, the ratio of actual density to theoretical density of the coating material used. { 'kōd·iŋ ‚den·səd·ē ‚rā·shō }

coax See coaxial cable. { 'kō‚aks }

coaxial [MECH] Sharing the same axes. [MECH ENG] Mounted on independent concentric shafts. { kō'ak·sē·əl }

coaxial antenna [ELECTROMAG] An antenna consisting of a quarter-wave extension of the inner conductor of a coaxial line and a radiating sleeve that is in effect formed by folding back the outer conductor of the coaxial line for a length of approximately a quarter wavelength. { kō'ak·sē·əl an'ten·ə }

coaxial attenuator [ELECTROMAG] An attenuator that has a coaxial construction and terminations suitable for use with coaxial cable. { kō'ak·sē·əl ə'ten·yə‚wäd·ər }

coaxial bolometer [ELECTR] A bolometer in which the desired square-law detection characteristic is provided by a fine Wollaston wire element that has been thoroughly cleaned before being axially located and soldered in position in its cylinder. { kō'ak·sē·əl bə'läm·əd·ər }

coaxial cable [ELECTROMAG] A transmission line in which one conductor is centered inside and insulated from an outer metal tube that serves as the second conductor. Also known as coax; coaxial line; coaxial transmission line; concentric cable; concentric line; concentric transmission line. { kō'ak·sē·əl 'kā·bəl }

coaxial capacitor See cylindrical capacitor. { kō'ak·sē·əl kə'pas·əd·ər }

coaxial cavity [ELECTROMAG] A cylindrical resonating cavity having a central conductor in contact with its pistons or other reflecting devices. { kō'ak·sē·əl 'kav·əd·ē }

coaxial cavity magnetron [ELECTR] A magnetron which achieves mode separation, high efficiency, stability, and ease of mechanical tuning by coupling a coaxial high Q cavity to a normal set of quarter-wavelength vane cavities. { kō'ak·sē·əl ‚kav·əd·ē 'mag·nə‚trän }

coaxial circles [MATH] Family of circles such that any pair have the same radical axis. { kō'ak·sē·əl 'sər·kəlz }

coaxial connector [ELECTROMAG] An electric connector between a coaxial cable and an equipment circuit, so constructed as to maintain the conductor configuration, through the separable connection, and the characteristic impedance of the coaxial cable. { kō'ak·sē·əl kə'nek·tər }

coaxial-cylinder magnetron [ELECTR] A magnetron in which the cathode and anode consist of coaxial cylinders. { kō'ak·sē·əl ‚sil·ən·dər 'mag·nə‚trän }

coaxial cylinders [MATH] Two cylinders whose cylindrical surfaces consist of the lines that pass through concentric circles in a given plane and are perpendicular to this plane. { kō'ak·sē·əl 'sil·ən·dərz }

coaxial diode [ELECTR] A diode having the same outer diameter and terminations as a coaxial cable, or otherwise designed to be inserted in a coaxial cable. { kō'ak·sē·əl 'dī‚ōd }

coaxial filter [ELECTROMAG] A section of coaxial line having reentrant elements that provide the inductance and capacitance of a filter section. { kō'ak·sē·əl 'fil·tər }

coaxial hybrid [ELECTROMAG] A hybrid junction of coaxial transmission lines. { kō'ak·sē·əl 'hī‚brəd }

coaxial isolator [ELECTROMAG] An isolator used in a coaxial cable to provide a higher loss for energy flow in one direction than in the opposite direction; all types use a permanent magnetic field in combination with ferrite and dielectric materials. { kō'ak·sē·əl 'ī·sə‚läd·ər }

coaxial line See coaxial cable. { kō'ak·sē·əl 'līn }

coaxial-line resonator [ELECTROMAG] A resonator consisting of a length of coaxial line short-circuited at one or both ends. { kō'ak·sē·əl ‚līn 'rez·ən‚ād·ər }

coaxially fed linear array [ELECTROMAG] A beacon antenna having a uniform azimuth pattern. { kō'ak·sē·ə·lē ‚fed 'lin·ē·ər ə'rā }

coaxial machine gun [ORD] A machine gun mounted integrally with the primary gun of a tank. { kō'ak·sē·əl mə'shēn ‚gən }

coaxial planes [MATH] Planes that pass through the same straight line. Also known as collinear planes. { kō'ak·sē·əl 'planz }

coaxial relay [ELECTROMAG] A relay designed for opening or closing a coaxial cable circuit without introducing a mismatch that would cause wave reflections. { kō'ak·sē·əl 'rē‚lā }

coaxial speaker [ENG ACOUS] A loudspeaker system comprising two, or less commonly three, speaker units mounted on substantially the same axis in an integrated mechanical assembly, with an acoustic-radiation-controlling structure. { kō'ak·sē·əl 'spēk·ər }

coaxial stub [ELECTROMAG] A length of nondissipative cylindrical waveguide or coaxial cable branched from the side of a waveguide to produce some desired change in its characteristics. { kō'ak·sē·əl 'stəb }

coaxial switch [ELEC] A switch that changes connections between coaxial cables going to antennas, transmitters, receivers, or other high-frequency devices without introducing impedance mismatch. { kō'ak·sē·əl 'swich }

coaxial transistor [ELECTR] A point-contact transistor in which the emitter and collector are point electrodes making pressure contact at the centers of opposite sides of a thin disk of semiconductor material serving as base. { kō'ak·sē·əl tran'zis·tər }

coaxial transmission line See coaxial cable. { kō'ak·sē·əl tranz'mish·ən ‚līn }

coaxial wavemeter [ENG] A device for measuring frequencies above about 100 megahertz, consisting of a rigid metal cylinder that has an inner conductor along its central axis, and a sliding disk that shorts the inner conductor and the cylinder. { kō'ak·sē·əl 'wāv‚mēd·ər }

coaxing [MET] Improving fatigue strength of a metal by increasing the stress range, beginning just below the fatigue limit. { 'kōks·iŋ }

cob [MIN ENG] To chip away waste material from an ore, using hand hammers. { käb }

cobalamin See vitamin B_{12}. { kə'bol·ə·mən }

cobalt [CHEM] A metallic element, symbol Co, atomic number 27, atomic weight 58.93; used chiefly in alloys. { 'kō‚bolt }

cobalt-60 [NUC PHYS] A radioisotope of cobalt, symbol ^{60}Co, having a mass number of 60; emits gamma rays and has many medical and industrial uses; the most commonly used isotope for encapsulated radiation sources. { 'kō‚bolt 'siks·tē }

cobalt-beam therapy [MED] Therapy involving the use of gamma radiation from a cobalt-60 source mounted in a cobalt bomb. Also known as cobalt therapy. { ¦kō‚bolt ¦bēm 'ther·ə‚pē }

cobalt bloom See erythrite. { 'kō‚bolt ‚blüm }

cobalt blue [CHEM] A green-blue pigment formed of alumina and cobalt oxide. Also known as cobalt ultramarine; king's blue. { ¦kō‚bolt ¦blü }

cobalt bomb [NUCLEO] A quantity of cobalt-60 mounted in a housing with walls having up to 8 inches (20 centimeters) of lead for protection, and with means for removing a lead plug to release a beam of gamma rays for use in cobalt-beam therapy. [ORD] A theoretical atomic or hydrogen bomb encased in cobalt, which cobalt would be transformed into deadly radioactive dust upon detonation. { 'kō‚bolt ‚bäm }

cobalt bromide See cobaltous bromide. { 'kō‚bolt 'brō‚mīd }

cobalt chloride See cobaltous chloride. { 'kō‚bolt 'klor‚īd }

cobalt glance See cobaltite. { 'kō‚bolt 'glans }

cobaltic fluoride See cobalt trifluoride. { kə'bol·tik 'flur‚īd }

cobaltite [MINERAL] CoAsS A silver-white mineral with a metallic luster that crystallizes in the isometric system, resembling crystals of pyrite; it is one of the chief ores of cobalt. Also known as cobalt glance; gray cobalt; white cobalt. { kə'bol‚tīt }

COATI

Common coati (*Nasua nasua*).

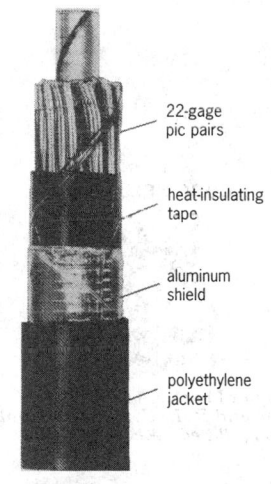

COAXIAL CABLE

22-gage pic pairs

heat-insulating tape

aluminum shield

polyethylene jacket

Cutaway view of coaxial transmission line.

cobalt-molybdate desulfurization [CHEM ENG] A process for desulfurization of petroleum by using cobalt molybdate as a catalyst. { ¦kō¦bȯlt mə¦lib‚dāt dē‚səl·fə·ri'zā·shən }

cobalt nitrate *See* cobaltous nitrate. { 'kō‚bȯlt 'nī‚trāt }

cobaltocalcite [MINERAL] A red, cobalt-bearing variety of calcite. { kə¦bȯl·tō'kal‚sīt }

cobalt ocher *See* asbolite; erythrite. { 'kō‚bȯlt 'ō·kər }

cobaltomenite [MINERAL] $CoSeO_3 \cdot 2H_2O$ A mineral consisting of a hydrous cobalt selenium oxide. { ‚kō‚bȯl'tä·mə‚nīt }

cobaltous acetate [ORG CHEM] $Co(C_2H_3O_2)_2 \cdot 4H_2O$ Reddish-violet, deliquescent crystals; soluble in water, alcohol, and acids; used in paint and varnish driers, for anodizing, and as a feed additive mineral supplement. Also known as cobalt acetate. { kō'bȯl·təs 'as·ə‚tāt }

cobaltous bromide [INORG CHEM] $CoBr_2 \cdot 6H_2O$ Red-violet crystals with a melting point of 47–48°C; soluble in water, alcohol, and ether; used in hygrometers. Also known as cobalt bromide. { kō'bȯl·təs 'brō‚mīd }

cobaltous chloride [INORG CHEM] $CoCl_2$ or $CoCl_2 \cdot 6H_2O$ A compound whose anhydrous form consists of blue crystals and sublimes when heated, and whose hydrated form consists of red crystals and melts at 86.8°C; both forms are used as an absorbent for ammonia in dyes and as a catalyst. Also known as cobalt chloride. { kō'bȯl·təs 'klȯr‚īd }

cobaltous fluorosilicate [INORG CHEM] $CoSiF_6 \cdot H_2O$ A water-soluble, orange-red powder, used in toothpastes. { kō'bȯl·təs ¦flùr·ō'sil·ə‚kāt }

cobaltous nitrate [INORG CHEM] $Co(NO_3)_2 \cdot 6H_2O$ A red crystalline compound with a melting point of 56°C; soluble in organic solvents; used in sympathetic inks, as an additive to soils and animal feeds, and for vitamin preparations and hair dyes. Also known as cobalt nitrate. { kō'bȯl·təs 'nī‚trāt }

cobalt oxide [INORG CHEM] CoO A grayish brown powder that decomposes at 1935°C, insoluble in water; used as a colorant in ceramics and in manufacture of glass. { 'kō‚bȯlt 'äk‚sīd }

cobalt potassium nitrite [INORG CHEM] $K_3Co(NO_2)_6$ A yellow powder which decomposes at the melting point of 200°C; used in medicine and as a yellow pigment. Also known as cobalt yellow; Fischer's salt; potassium cobaltinitrite. { 'kō‚bȯlt pə'tas·ē·əm 'nī‚trīt }

cobalt pyrites *See* linnaeite. { 'kō‚bȯlt 'pī‚rīts }

cobalt sulfate [INORG CHEM] Any compound of either divalent or trivalent cobalt and the sulfate group; anhydrous cobaltous sulfate, $CoSO_4$, contains divalent cobalt, has a melting point of 96.8°C, is soluble in methanol, and is utilized to prepare pigments and cobalt salts; cobaltic sulfate, $Co_2(SO_4)_3 \cdot 18H_2O$, contains trivalent cobalt, is soluble in sulfuric acid, and functions as an oxidizing agent. { 'kō‚bȯlt 'səl‚fāt }

cobalt therapy *See* cobalt-beam therapy. { 'kō‚bȯlt ‚ther·ə·pē }

cobalt trifluoride [INORG CHEM] CoF_3 A brownish powder that reacts with water to form a precipitate of cobaltic hydroxide; used as a fluorinating agent. Also known as cobaltic fluoride. { 'kō‚bȯlt trī'flùr‚īd }

cobalt ultramarine *See* cobalt blue. { 'kō‚bȯlt ‚əl·trə·mə'rēn }

cobalt yellow *See* cobalt potassium nitrite. { 'kō‚bȯlt 'yel·ō }

cobber [MIN ENG] **1.** A device used to reject waste materials from ore concentrates. **2.** A person who breaks fibers from asbestos rocks or chips low-grade material from ore. { 'käb·ər }

cobble [GEOL] A rock fragment larger than a pebble and smaller than a boulder, having a diameter in the range of 64–256 millimeters (2.5–10.1 inches), somewhat rounded or otherwise modified by abrasion in the course of transport. { 'käb·əl }

cobble beach *See* shingle beach. { 'käb·əl ‚bēch }

Cobb's disease [PL PATH] A bacterial disease of sugarcane caused by *Xanthomonas vascularum* and characterized by a slime in the vascular bundles, dwarfing, streaking of leaves, and decay. Also known as sugarcane gummosis. { 'käbz di‚zēz }

Cobitidae [VERT ZOO] The loaches, a family of small fishes, many eel-shaped, in the suborder Cyprinoidei, characterized by barbels around the mouth. { kə'bid·ə‚dē }

Coblentzian [GEOL] Upper Lower Devonian geologic time. { kō'blens·ē·ən }

COBOL [COMPUT SCI] A business data-processing language that can be given to a computer as a series of English statements describing a complete business operation. Derived from common business-oriented language. { 'kō‚bȯl }

coboundary [MATH] An image under the coboundary operator. { kō'baún‚drē }

coboundary operator [MATH] If {C^n} is a sequence of Abelian groups, coboundary operators are homomorphisms {δ^n} such that $\delta^n: C^n \to C^{n+1}$ and $\delta^{n+1} \circ \delta^n = 0$. { kō'baún‚drē 'äp·ə‚rād·ər }

cobra [VERT ZOO] Any of several species of venomous snakes in the reptilian family Elaphidae characterized by a hoodlike expansion of skin on the anterior neck that is supported by a series of ribs. { 'kō·brə }

cobra maneuver [AERO ENG] An aircraft maneuver in which the aircraft pitches up to beyond vertical altitude (reaching an angle of 110–130°) while remaining at almost level flight, and them levels to a regular flight attitude. Also known as Pougachev maneuver. { 'kō·brə mə‚nü·vər }

coca [BOT] *Erythroxylon coca*. A shrub in the family Erythroxylaceae; its leaves are the source of cocaine. { 'kō·kə }

cocaine [PHARM] $C_{17}H_{21}O_4N$ An alkaloid obtained from coca leaves that is used for local anesthesia and as a tonic in digestive and nervous disorders. { kō'kān }

cocarboxylase *See* thiamine pyrophosphate. { ¦kō·kär'bäk·sə‚lās }

cocarcinogen [MED] A noncarcinogenic agent which augments the carcinogenic process. { 'kō·kär'sin·ə·jən }

Coccidia [INV ZOO] A subclass of protozoans in the class Telosporea; typically intracellular parasites of epithelial tissue in vertebrates and invertebrates. { käk'sid·ē·ə }

Coccidioides immitis [MED] A mold primarily found in desert soil that converts into spherules containing endospores when growing within the body and that causes coccidioidomycosis or San Joaquin valley fever. { ‚käk·sid·ē¦ȯi‚dēz i'mīd·əs *or* i'mēd·əs }

coccidioidomycosis [MED] An infectious fungus disease of humans and animals of either a pulmonary or a cutaneous nature; caused by *Coccidioides immitis*. Also known as San Joaquin Valley fever. { käk¦sid·ē¦ȯid·ō·mī'kō·səs }

coccidiosis [MED] The state of or the conditions associated with being infected by coccidia. { käk¦sid·ē¦ō·səs }

coccine [INV ZOO] For protozoa, denoting the sessile state during which reproduction does not occur. { 'käk‚sēn }

Coccinellidae [INV ZOO] The ladybird beetles, a family of coleopteran insects in the superfamily Cucujoidea. { käk·sə'nel·ə‚dē }

coccobacillus [MICROBIO] A short, thick, oval bacillus, midway between the coccus and the bacillus in appearance. { ¦kä·kō·bə'sil·əs }

coccoid [MICROBIO] A spherical bacterial cell. { 'kä‚kȯid }

Coccoidea [INV ZOO] A superfamily of homopteran insects belonging to the Sternorrhyncha; includes scale insects and mealy bugs. { kä'kȯid·ē·ə }

coccolith [BOT] One of the small, interlocking calcite plates covering members of the Coccolithophorida. { 'käk·ə‚lith }

coccolith ooze [GEOL] A fine-grained pelagic sediment containing undissolved sand-or silt-sized particles of coccoliths mixed with amorphous clay-sized material. { 'käk·ə‚lith ‚üz }

Coccolithophora [INV ZOO] An order of phytoflagellates in the protozoan class Phytamastigophorea. { ‚käk·ō·li'thäf·ə·rə }

Coccolithophorida [BOT] A group of unicellular, biflagellate, golden-brown algae characterized by a covering of coccoliths. { ‚käk·ō‚lith·ə'fȯr·ə·də }

Coccomyces hiemalis *See* Blumeriella jaapii. { ‚käk·ō‚mī‚sēz hē'mäl·əs }

Coccomyxaceae [BOT] A family of algae belonging to the Tetrasporales composed of elongate cells which reproduce only by vegetative means. { ‚käk·ō‚mik'sās·ē‚ē }

coccosphere [PALEOBOT] The fossilized remains of a member of Coccolithophorida. { 'käk·ə‚sfir }

Coccosteomorphi [PALEON] An aberrant lineage of the joint-necked fishes. { kä¦kä·stē·ə¦mȯr·fē }

cocculin *See* picrotoxin. { 'käk·yə·lən }

coccus [MICROBIO] A form of eubacteria which are more or less spherical in shape. { 'käk·əs }

coccygeal body [ANAT] A small mass of vascular tissue near the tip of the coccyx. { käk'sij·ē·əl 'bä‚dē }

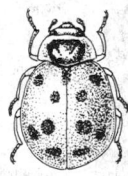

COCCINELLIDAE

A ladybird beetle. *(From T. I. Storer and R. L. Usinger, General Zoology, 3d ed., McGraw-Hill, 1957)*

COCCOSPHERE

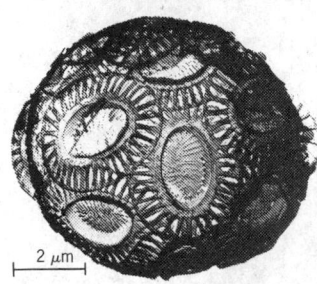

2 μm

Scanning electron micrograph of coccosphere. *(A. McIntyre, Lamont-Doherty Geological Observatory of Columbia University)*

coccygectomy [MED] Surgical excision of the coccyx. { käk·sə'jek·tə·mē }

coccyx [ANAT] The fused vestige of caudal vertebrae forming the last bone of the vertebral column in humans and certain other primates. { 'käk,siks }

cochain complex [MATH] A sequence of Abelian groups C^n, $-\infty < n < \infty$, together with coboundary homomorphisms $\delta^n: C^n \to C^{n+1}$ such that $\delta^{n+1} \circ \delta^n = 0$. { 'kō,chān 'käm,pleks }

cochannel cells [COMMUN] Two cells in a cellular mobile radio system that use the same frequency. { 'kō,chan·əl 'selz }

cochannel interference [COMMUN] Interference caused on one communication channel by a transmitter operating in the same channel. { 'kō,chan·əl ,in·tər'fir·əns }

cochannel interference reduction factor [COMMUN] The ratio of the minimum separation between two cochannel cells without interference to the radius of a cell. { 'kō,chan·əl ,in·tər,fir·əns ri'dək·shən ,fak·tər }

cochineal [CHEM] A red dye made of the dried bodies of the female cochineal insect (*Coccus cacti*), found in Central America and Mexico; used as a biological stain and indicator. { 'käch·ə,nēl }

cochineal solution [ANALY CHEM] An indicator in acid-base titration. { 'käch·ə,nēl sə'lü·shən }

cochinilin See carminic acid. { kō'chin·ə·lən }

cochlea [ANAT] The snail-shaped canal of the mammalian inner ear; it is divided into three channels and contains the essential organs of hearing. { 'kōk·lē·ə }

cochlear duct See scala media. { 'kōk·lē·ər 'dəkt }

Cochleariidae [VERT ZOO] A family of birds in the order Ciconiiformes composed of a single species, the boatbill. { ,kōk·lē·ə'rī·ə,dē }

cochlear implant [NEURO] A sensory prosthesis that restores some hearing to deaf people by electrically stimulating the auditory nerve. { 'kō·klē·ər 'im,plant }

cochlear nerve [NEURO] A sensory branch of the auditory nerve which receives impulses from the organ of Corti. { 'kok·lē·ər 'nərv }

cochlear nucleus [NEURO] One of the two nuclear masses in which the fibers of the cochlear nerve terminate; located ventrad and dorsad to the inferior cerebellar peduncle. { 'kok·lē·ər 'nük·lē·əs }

cochleate [BIOL] Spiral; shaped like a snail shell. { 'kōk·lē,āt }

cochleoid [MATH] A plane curve whose equation in polar coordinates is $r\theta = a \sin \theta$. { 'käk·lē,oid }

Cochliodontidae [PALEON] A family of extinct chondrichthian fishes in the order Bradyodonti. { ,kōk·lē·ō'dän·tə,dē }

Cochran's test [STAT] A test used when one estimated variance appears to be very much larger than the remainder of the estimated variances; based on the ratio of the largest estimate of the variance to the total of all the estimates. { 'käk·rənz ,test }

cocinerite [MINERAL] Cu_4AgS A silver gray mineral consisting of copper and silver sulfide; occurs in massive form. { kō·sə'ne,rīt }

cock [ENG] Any mechanism which starts, stops, or regulates the flow of liquid, such as a valve, faucet, or tap. [VERT ZOO] The adult male of the domestic fowl and of gallinaceous birds. { käk }

Cockcroft-Walton accelerator [NUCLEO] An electrostatic particle accelerator utilizing as a source of high voltage a transformer and an array of rectifiers and condensers, giving voltage multiplication. { 'käk,kroft 'wolt·ən ak'sel·ə,rād·ər }

cockeyed bob [METEOROL] A thunder squall occurring during the summer, on the northwest coast of Australia. { 'käk,īd 'bäb }

cocking lever [ORD] A lever for drawing back or sometimes lowering the striker of an automatic firearm. { 'käk·iŋ ,lev·ər }

cockle [INV ZOO] The common name for a number of species of marine mollusks in the class Bivalvia characterized by a shell having convex radial ribs. { 'käk·əl }

cockle finish [MATER] An irregular surface usually produced on rag bond and ledger paper, obtained by coating the paper with sizing and drying it in loop or festoon fashion in heated air. { 'käk·əl ,fin·ish }

cockpit [AERO ENG] A space in an aircraft or spacecraft where the pilot sits. [NAV ARCH] A sunken area on the deck of a small vessel, near the stern, from which the craft is steered. { 'käk,pit }

cockpit karst See cone karst. { 'käk,pit 'karst }

cockroach See roach. { 'käk,rōch }

cocoa butter [MATER] A brown fat obtained from cacao seeds; melts at 30–35°C; used in the manufacture of chocolate, cosmetics, and pharmaceuticals. Also known as cacao butter; oleum theobromatis; theobroma oil. { 'kō,kō ,bəd·ər }

cocoa nibs [FOOD ENG] In chocolate manufacture, the cleaned broken pieces of the cotyledon of the cacao seed which remain after the roasting, crushing, and winnowing processes. { 'kō,kō ,nibz }

cocodyl oxide [ORG CHEM] $(CH_3)_2AsOAs(CH_3)_2$ A liquid that has an obnoxious odor; slightly soluble in water, soluble in alcohol and ether; boils at 150°C. Also known as alkarsine; bisdimethyl arsenic oxide; dicacodyl oxide. { 'kō·kə·dəl 'äk,sīd }

coconsciousness [PSYCH] A dissociated mental state coexisting with a person's consciousness, but without his awareness, though it is psychodynamically active and may account for various normal and abnormal mental phenomena. { kō'kän·chə·snəs }

coconut [BOT] *Cocos nucifera*. A large palm in the order Arecales grown for its fiber and fruit, a large, ovoid, edible drupe with a fibrous exocarp and a hard, bony endocarp containing fleshy meat (endosperm). { 'kō·kə,nət }

coconut bud rot [PL PATH] A fungus disease of the coconut palm caused by *Phytophthora palmivora* and characterized by destruction of the terminal bud and adjacent leaves. { 'kō·kə,nət 'bəd ,rät }

coconut oil [MATER] A nearly colorless or yellow oil from fresh coconut (*Cocos nucifera*) or from copra (dried coconut); used in foods, in making soap, and as a raw material in fatty-acid production. { 'kō·kə,nət ,oil }

cocoon [INV ZOO] **1.** A protective case formed by the larvae of many insects, in which they pass the pupa stage. **2.** Any of the various protective egg cases formed by invertebrates. { kə'kün }

cocurrent line [OCEANOGR] A line through places having the same tidal current hour. { kō'kər·ənt 'līn }

cocycle [MATH] A chain of simplices whose coboundary is 0. { 'kō,sī·kəl }

cod [VERT ZOO] The common name for fishes of the subfamily Gadidae, especially the Atlantic cod (*Gadus morrhua*). { käd }

codan [ELECTR] A device that silences a receiver except when a modulated carrier signal is being received. { 'kō,dan }

CODAR See correlation, detection, and ranging. { 'kō,där }

Coddington lens [OPTICS] A magnifier consisting of a glass sphere with a deep groove cut around a great circle to serve as a stop. { 'käd·iŋ·tən ,lenz }

Coddington shape factor See shape factor. { 'käd·iŋ·tən 'shāp ,fak·tər }

code [COMMUN] A system of symbols and rules for expressing information, such as the Morse code, Electronic Industries Association color code, and the binary and other machine languages used in digital computers. { kōd }

code area [GRAPHICS] On a microform, a portion of the image area or film frame that is reserved for retrieval coding. { 'kōd ,er·ē·ə }

code beacon [NAV] A beacon that flashes a characteristic signal by which it may be recognized. { 'kōd ,bē·kən }

code book [COMMUN] A book containing a large number of plaintext words, phrases, and sentences and their codetext equivalents. { 'kōd ,bùk }

codec [ELECTR] A device that converts analog signals to digital form for transmission and converts signals traveling in the opposite direction from digital to analog form. Derived from coder-decoder. { 'kō,dek }

codecarboxylase [BIOCHEM] The prosthetic component of the enzyme carboxylase which catalyzes decarboxylation of D-amino acids. Also known as pyridoxal phosphate. { 'kō·də·kär'bäk·sə,lās }

code-check [COMPUT SCI] To remove mistakes from a coded routine or program. { 'kōd ,chek }

code checking time [COMPUT SCI] Time spent checking out a problem on the computer, making sure that the problem is set up correctly and that the code is correct. { 'kōd ,chek·iŋ ,tīm }

codeclination [NAV] In celestial navigation, 90° minus the declination; when the declination and latitude are of the same

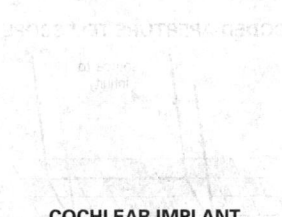

COCHLEAR IMPLANT

Idealized picture of a cochlear implant electrode inserted into the scala tympani.

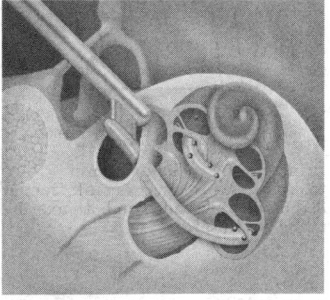

COCKLE

The shell of the rock cockle (*Protothaca laciniata*), covered with radiating ribs crossed with concentric ribs.

COD

Gadua morrhua, a codfish.

name, codeclination is the same as polar distance measured from the elevated pole. { ¦kō¦dek·lə'nā·shən }

code converter [COMPUT SCI] A converter that changes coded information to a different code system. { 'kōd kən'vərd·ər }

coded-aperture telescope [ASTRON] A soft gamma-ray telescope that uses a coded mask to image celestial sources; the position of a gamma-ray source is determined by comparing the observed projection pattern of the mask with all possible projection patterns. { ¦kōd·əd ¦ap·ə·chər 'tel·ə¸skōp }

coded character set [COMPUT SCI] A set of characters together with the code assigned to each character for computer use. { ¦kōd·əd 'kar·ik·tər ¸set }

coded decimal *See* decimal-coded digit. { 'kōd·əd 'des·məl }

code density [GRAPHICS] The number of code elements per unit length that can appear on a microfilm. { 'kōd ¸den·səd·ē }

coded interrogator [COMMUN] An interrogator whose output signal forms the code required to trigger a specific radio or radar beacon. { 'kōd·əd in'ter·ə¸gād·ər }

code-division multiple access [COMMUN] The transmission of messages from a large number of transmitters over a single channel by assigning each transmitter a pseudorandom noise code (typically more than 2000 symbols long for each bit of information) so that the codes are mathematically independent of each other. Abbreviated CDMA. { 'kōd də¦vizh·ən 'məl·tə·pəl 'ak¸ses }

code-division multiplex [COMMUN] Multiplex in which two or more communication links occupy the entire transmission channel simultaneously, with code signal structures designed so a given receiver responds only to its own signals and treats the other signals as noise. Abbreviated CDM. { 'kōd də¦vizh·ən 'məlt¸i¸pleks }

coded mask [ENG] A pattern of tungsten blocks that absorb gamma-ray photons in a gamma-ray telescope, and are arranged so that an astronomical gamma-ray source projects on a position-sensitive detector a pattern that is characteristic of the direction of arrival of the photons. { 'kōd·əd ¸mask }

coded passive reflector antenna [ELECTROMAG] An object intended to reflect Hertzian waves and having variable reflecting properties according to a predetermined code for the purpose of producing an indication on a radar receiver. { 'kōd·əd 'pas·iv ri'flek·tər an¸ten·ə }

coded program [COMPUT SCI] A program expressed in the required code for a computer. { 'kōd·əd 'prō·grəm }

coded stop [COMPUT SCI] A stop instruction built into a computer routine. { 'kōd·əd 'stäp }

code element [COMMUN] One of the separate elements or events constituting a coded message, such as the presence or absence of a pulse, dot, dash, or space. { 'kōd ¸el·ə·mənt }

code error [COMPUT SCI] A surplus or lack of a bit or bits in a machine instruction. { 'kōd ¸er·ər }

code-excited linear predictive coder [COMMUN] A speech coder that uses both short-term and long-term predictors, vector quantization techniques, and an analysis-by-synthesis approach to search for the best combination of coder parameters. Abbreviated CELP coder. { 'kōd i¦sīd·əd ¦lin·ē·ər prə¦dik·tiv 'kōd·ər }

code extension [COMPUT SCI] A method of increasing the number of characters that can be represented by a code by combining characters into groups. { 'kōd ik¸sten·chən }

code group [COMMUN] A combination of letters or numerals or both, assigned to represent one or more words of plain text in a coded message. { 'kōd ¸grüp }

code holes [COMPUT SCI] The informational holes in perforated tape, as opposed to the feed holes or other holes. { 'kōd ¸hōlz }

codehydrogenase I *See* diphosphopyridine nucleotide. { ¦kō·dē'hī·drə·jə¸nās ¦wən }

codehydrogenase II *See* triphosphopyridine nucleotide. { ¦kō·dē'hī·drə·jə¸nās ¦tü }

codeine [PHARM] $C_{18}H_{21}NO_3$ An alkaloid prepared from morphine; used as mild analgesic and cough suppressant. { 'kō¸dēn }

code line [COMPUT SCI] In character recognition, the area reserved for the inscription of the printed or handwritten characters to be recognized. { 'kōd ¸līn }

code medium [GRAPHICS] A reflective or transmissive material used for coding on a microform. { 'kōd ¸mē·dē·əm }

code position [COMPUT SCI] A location in a data-recording

medium at which data may be entered, such as the intersection of a column and a row on a punch card, at which a hole may be punched. { 'kōd pə'zish·ən }

code practice oscillator [ELECTR] An oscillator used with a key and either headphones or a loudspeaker to practice sending and receiving Morse code. { ¦kōd ¦prak·təs 'äs·ə¸lād·ər }

coder [COMMUN] A device that generates a code by generating pulses having varying lengths or spacings, as required for radio beacons and interrogators. Also known as moder; pulse coder; pulse-duration coder. [COMPUT SCI] A person who translates a sequence of computer instructions into codes acceptable to the machine. { 'kōd·ər }

coder-decoder *See* codec. { ¦kōd·ər dē¦kōd·ər }

code reader [COMPUT SCI] A scanning device used for automated identification of a two-dimensional pattern, one part after the other, and generation of either analog or digital signals that correspond to the pattern. Also known as code scanner. { 'kōd ¸rēd·ər }

code ringing [COMMUN] In telephone switching, party-line ringing wherein the number or duration of rings indicates which station is being called. { 'kōd ¸riŋ·iŋ }

code scanner *See* code reader. { 'kōd ¸skan·ər }

code-sending radiosonde [ENG] A radiosonde which transmits the indications of the meteorological sensing elements in the form of a code consisting of combinations of dots and dashes. Also known as code-type radiosonde; contracted code sonde. { 'kōd ¸send·iŋ 'rād·ē·ō¸sänd }

code sensitivity [COMPUT SCI] Property of hardware or software that can handle only data presented in a particular code. { 'kōd ¸sen·sə¸tiv·əd·ē }

code signal [COMMUN] A sequence of discrete conditions or events corresponding to a coded message. { 'kōd ¸sig·nəl }

codetext [COMMUN] A message which has been transformed by a code into a form which can be read only by those privy to the secrets of the code. { 'kōd¸tekst }

code translation [COMMUN] Conversion of a directory code or number into a predetermined code for controlling the selection of an outgoing trunk or line. { 'kōd tranz¸lā·shən }

code transparency [COMPUT SCI] Property of hardware or software that can handle data regardless of what form it is in. { 'kōd tranz¸par·ən¸sē }

code-type radiosonde *See* code-sending radiosonde. { 'kōd ¸tīp 'rād·ē·ō¸sänd }

Codiaceae [BOT] A family of green algae in the order Siphonales having macroscopic thalli composed of aggregates of tubes. { ¸kō·dē'as·ē¸ē }

codimer [ORG CHEM] **1.** A copolymer formed from the polymerization of two dissimilar olefin molecules. **2.** The product of polymerization of isobutylene with one of the two normal butylenes. { ¦kō'dī·mər }

coding [COMPUT SCI] **1.** The process of converting a program design into an accurate, detailed representation of that program in some suitable language. **2.** A list, in computer code, of the successive operations required to carry out a given routine or solve a given problem. { 'kōd·iŋ }

coding delay [NAV] An arbitrary time delay in the transmission of pulse signals; in the loran system it is inserted between transmission of the master and slave signals to prevent zero or small readings, and thus to aid in distinguishing between master and slave station signals. { 'kōd·iŋ di¸lā }

coding disk [COMMUN] Disk with small projections for operating contacts to give a certain predetermined code to a transmission. { 'kōd·iŋ ¸disk }

coding form *See* coding sheet. { 'kōd·iŋ ¸fòrm }

coding line *See* instruction word. { 'kōd·iŋ ¸līn }

coding ratio [BIOCHEM] The number of bases in nucleic acids divided by the number of amino acids whose sequence the bases determine in a particular polypeptide. { 'kōd·iŋ ¸rā·shō }

coding sheet [COMPUT SCI] A sheet of paper printed with a form on which one can conveniently write a coded program. Also known as coding form. { 'kōd·iŋ ¸shēt }

coding strand *See* sense strand. { 'kōd·iŋ ¸strand }

codistor [ELECTR] A multijunction semiconductor device which provides noise rejection and voltage regulation functions. { kō'dis·tər }

cod-liver oil [MATER] A yellow oil, high in vitamin D, extracted from the liver of the Atlantic cod (*Gadus morrhua*);

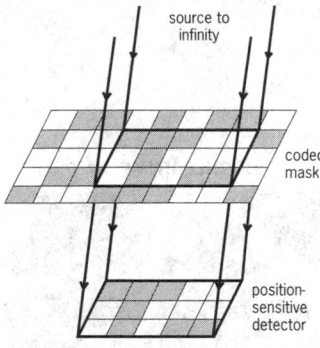

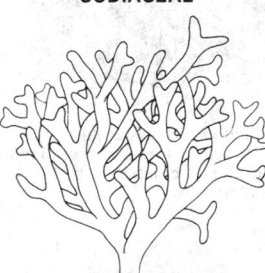

CODIACEAE

A portion of a dichotomously branched thallus of *Codium*.

soluble in alcohol. Also known as banks oil; oleum morrhuae. { 'käd ,liv·ər ,óil }

codominance [GEN] A condition in which each allele of a heterozygous pair expresses itself fully, as in human blood group AB individuals. { kō'däm·ə·nəns }

codon [GEN] The basic unit of the genetic code, comprising sequential, nonoverlapping three-nucleotide sequences in messenger ribonucleic acid, each of which is translated into one amino acid; 61 of the 64 codons code for a specific protein synthesis; the other 3 are stop codons that specify termination of the growing polypeptide or protein chain. { 'kō,dän }

codon family [GEN] A group of codons that code for the same amino acid and differ only in the nucleotide that occupies the third codon position. { 'kō,dän ,fam·lē }

codon fidelity [MOL BIO] The constancy of the genetic coding process as maintained during deoxyribonucleic acid replication and the synthesis of proteins by a series of proofreading reactions that remove errors. { 'kō,dän fi,del·əd·ē }

codon misreading [MOL BIO] The mistranslation of a codon in messenger ribonucleic acid that increases errors in protein synthesis by generation of amino acid substitutions. { 'kō,dän mis,rēd·iŋ }

coefficient [MATH] A factor in a product. { kō·ə'fish·ənt }

coefficient A [NAV] A component of magnetic compass deviation of constant value with compass heading, resulting from mistakes in calculations, compass and pelorus misalignment, and unsymmetrical arrangements of horizontal soft iron. { kō·ə'fish·ənt 'ā }

coefficient B [NAV] A component of magnetic compass deviation, varying with the sine function of the compass heading, resulting from the fore-and-aft component of the craft's permanent magnetic field and induced magnetism in unsymmetrical vertical iron forward or aft of the compass. { kō·ə'fish·ənt 'bē }

coefficient C [NAV] A component of magnetic compass deviation, varying with the cosine function of the compass heading, resulting from the athwartship component of the craft's permanent magnetic field and induced magnetism in unsymmetrical vertical iron port or starboard of the compass. { kō·ə'fish·ənt 'sē }

coefficient D [NAV] A component of magnetic compass deviation, varying with the sine function of twice the compass heading, resulting from induced magnetism in all symmetrical arrangements of the craft's horizontal soft iron. { kō·ə'fish·ənt 'dē }

coefficient E [NAV] A component of magnetic compass deviation, varying with the cosine function of twice the compass heading, resulting from induced magnetism in all unsymmetrical arrangements of the craft's horizontal soft iron. { kō·ə'fish·ənt 'ē }

coefficient J [NAV] A change in magnetic compass deviation, varying with the cosine function of the compass heading for a given value of J, where J is the change of deviation for a heel of 1° on compass heading 000°. { kō·ə'fish·ənt 'jā }

coefficient of absorption See absorption coefficient. { kō·ə'fish·ənt əv əb'sórp·shən }

coefficient of alienation [STAT] A statistic that measures the lack of linear association between two variables; computed by taking the square root of the difference between 1 and the square of the correlation coefficient. { kō·ə'fish·ənt əv ,ā·lē·ə'nā·shən }

coefficient of association [STAT] A statistic used as a measure of the association of data grouped in a 2 × 2 table; the value of the statistic ranges from −1 to +1, with the former indicating perfect negative association and the latter perfect positive association. Usually designated as Q. { kō·ə'fish·ənt əv ə,sō·sē'ā·shən }

coefficient of capacitance [ELEC] One of the coefficients which appears in the linear equations giving the charges on a set of conductors in terms of the potentials of the conductors; a coefficient is equal to the ratio of the charge on a given conductor to the potential of the same conductor when the potentials of all the other conductors are 0. { kō·ə'fish·ənt əv kə'pas·ə·təns }

coefficient of compressibility [MECH] The decrease in volume per unit volume of a substance resulting from a unit increase in pressure; it is the reciprocal of the bulk modulus. { kō·ə'fish·ənt əv kəm,pres·ə'bil·əd·ē }

coefficient of concordance [STAT] A statistic that measures the agreement among sets of rankings by two or more judges. { kō·ə'fish·ənt əv kən'kórd·əns }

coefficient of condensation [STAT MECH] The ratio of the number of molecules condensed on the surface of a solid or liquid in equilibrium with its vapor phase to the total number of vapor molecules striking the surface. { kō·ə'fish·ənt əv ,kän·dən'sā·shən }

coefficient of conductivity See thermal conductivity. { kō·ə'fish·ənt əv ,kän·dək'tiv·əd·ē }

coefficient of contingency [STAT] A measure of the strength of dependence between two statistical variables, based on a contingency table. { kō·ə'fish·ənt əv kən'tin·jən·sē }

coefficient of contraction [FL MECH] The ratio of the minimum cross-sectional area of a jet of liquid discharging from an orifice to the area of the orifice. Also known as contraction coefficient. { kō·ə'fish·ənt əv kən'trak·shən }

coefficient of coupling See coupling constant. { kō·ə'fish·ənt əv 'kəp·liŋ }

coefficient of cubical expansion [THERMO] The increment in volume of a unit volume of solid, liquid, or gas for a rise of temperature of 1° at constant pressure. Also known as coefficient of expansion; coefficient of thermal expansion; coefficient of volumetric expansion; expansion coefficient; expansivity. { kō·ə'fish·ənt əv 'kyüb·ə·kəl ik'span·shən }

coefficient of determination [STAT] A statistic which indicates the strength of fit between two variables implied by a particular value of the sample correlation coefficient r. Designated by r^2. { kō·ə'fish·ənt əv di,tər·mə'nā·shən }

coefficient of discharge See discharge coefficient. { kō·ə'fish·ənt əv 'dis,chärj }

coefficient of eddy diffusion See eddy diffusivity. { kō·ə'fish·ənt əv 'ed·ē də'fyü·zhən }

coefficient of eddy viscosity [FL MECH] The portion of the kinematic viscosity of a turbulent fluid that is associated with its eddy viscosity. Also known as coefficient of turbulence. { kō·ə'fish·ənt əv 'ed·ē vi,skäs·əd·ē }

coefficient of elasticity See modulus of elasticity. { kō·ə'fish·ənt əv i,las'tis·əd·ē }

coefficient of expansion See coefficient of cubical expansion. { kō·ə'fish·ənt əv ik'span·shən }

coefficient of friction [MECH] The ratio of the frictional force between two bodies in contact, parallel to the surface of contact, to the force, normal to the surface of contact, with which the bodies press against each other. Also known as friction coefficient. { kō·ə'fish·ənt əv 'frik·shən }

coefficient of friction of rest See coefficient of static friction. { kō·ə'fish·ənt əv 'frik·shən əv 'rest }

coefficient of induction [ELEC] One of the coefficients which appears in the linear equations giving the charges on a set of conductors in terms of the potentials of the conductors, a coefficient is equal to the ratio of the charge on a given conductor to the potential on another conductor, when the potentials of all the other conductors equal 0. { kō·ə'fish·ənt əv in'dək·shən }

coefficient of kinematic viscosity See kinematic viscosity. { kō·ə'fish·ənt əv ,kin·ə,mad·ik vis'käs·əd·ē }

coefficient of kinetic friction [MECH] The ratio of the frictional force, parallel to the surface of contact, that opposes the motion of a body which is sliding or rolling over another, to the force, normal to the surface of contact, with which the bodies press against each other. { kō·ə'fish·ənt əv kə'ned·ik 'frik·shən }

coefficient of linear expansion [THERMO] The increment of length of a solid in a unit of length for a rise in temperature of 1° at constant pressure. Also known as linear expansivity. { kō·ə'fish·ənt əv 'lin·ē·ər ik'span·shən }

coefficient of multiple correlation [STAT] A measure used as an index of the strength of a relationship between a variable y and a set of one or more variables x_i; computed by deriving the square root of the ratio of the explained variation to the total variation. { kō·ə'fish·ənt əv ,məl·tə·pəl ,kär·ə'lā·shən }

coefficient of nondetermination [STAT] The coefficient of alienation squared; represents that part of the dependent variable's total variation not accounted for by linear association with the independent variable. { kō·ə'fish·ənt əv ,nän·di,tər·mə'nā·shən }

coefficient of performance [THERMO] In a refrigeration

COELACANTH

Living coelacanth, *Latimeria calumnae*; 5 feet (1.5 meters).

cycle, the ratio of the heat energy extracted by the heat engine at the low temperature to the work supplied to operate the cycle; when used as a heating device, it is the ratio of the heat delivered in the high-temperature coils to the work supplied. { ¦kō·ə'fish·ənt əv pər'fȯr·məns }

coefficient of permeability *See* permeability coefficient. { ¦kō·ə'fish·ənt əv ˌpər·mē·ə'bil·əd·ē }

coefficient of potential [ELEC] One of the coefficients which appears in the linear equations giving the potentials of a set of conductors in terms of the charges on the conductors. { ¦kō·ə'fish·ənt əv pə'ten·chəl }

coefficient of reflection *See* reflection coefficient. { ¦kō·ə'fish·ənt əv ri'flek·shən }

coefficient of resistance [FL MECH] The ratio of the loss of head of fluid, issuing from an orifice or passing over a weir, to the remaining head. { ¦kō·ə'fish·ənt əv ri'zis·təns }

coefficient of restitution [MECH] The constant *e*, which is the ratio of the relative velocity of two elastic spheres after direct impact to that before impact; *e* can vary from 0 to 1, with 1 equivalent to an elastic collision and 0 equivalent to a perfectly elastic collision. Also known as restitution coefficient. { ¦kō·ə'fish·ənt əv ˌres·tə'tü·shən }

coefficient of rigidity *See* modulus of elasticity in shear. { ¦kō·ə'fish·ənt əv rə'jid·əd·ē }

coefficient of rolling friction [MECH] The ratio of the frictional force, parallel to the surface of contact, opposing the motion of a body rolling over another, to the force, normal to the surface of contact, with which the bodies press against each other. { ¦kō·ə'fish·ənt əv 'rōl·iŋ 'frik·shən }

coefficient of sliding friction [MECH] The ratio of the frictional force, parallel to the surface of contact, opposing the motion of a body sliding over another, to the force, normal to the surface of contact, with which the bodies press against each other. { ¦kō·ə'fish·ənt əv 'slīd·iŋ 'frik·shən }

coefficient of static friction [MECH] The ratio of the maximum possible frictional force, parallel to the surface of contact, which acts to prevent two bodies in contact, and at rest with respect to each other, from sliding or rolling over each other, to the force, normal to the surface of contact, with which the bodies press against each other. Also known as coefficient of friction of rest. { ¦kō·ə'fish·ənt əv 'stad·ik 'frik·shən }

coefficient of strain [MATH] Multiplier used in transformations to elongate or compress configurations in a direction parallel to an axis. [MECH] For a substance undergoing a one-dimensional strain, the ratio of the distance along the strain axis between two points in the body, to the distance between the same points when the body is undeformed. { ¦kō·ə'fish·ənt əv 'strān }

coefficient of superficial expansion [THERMO] The increment in area of a solid surface per unit of area for a rise in temperature of 1° at constant pressure. Also known as superficial expansivity. { ¦kō·ə'fish·ənt əv ˌsü·pər'fish·əl ik'span·chən }

coefficient of thermal expansion *See* coefficient of cubical expansion. { ¦kō·ə'fish·ənt əv 'thər·məl ik'span·shən }

coefficient of turbulence *See* coefficient of eddy viscosity. { ¦kō·ə'fish·ənt əv 'tər·byə·ləns }

coefficient of variation [STAT] The ratio of the standard deviation of a distribution to its arithmetic mean. { ¦kō·ə'fish·ənt əv ˌver·ē'ā·shən }

coefficient of velocity *See* velocity coefficient. { ¦kō·ə'fish·ənt əv və'läs·əd·ē }

coefficient of viscosity *See* absolute viscosity. { ¦kō·ə'fish·ənt əv vis'käs·əd·ē }

coefficient of volumetric expansion *See* coefficient of cubical expansion. { ¦kō·ə'fish·ənt əv ¦väl·yə¦me·trik ik'span·chən }

coefficients of form [NAV ARCH] Parameters depending on the form of a ship, such as the block coefficient, midship section coefficient, mean length coefficient, water plane coefficient, and displacement coefficient. { ¦kō·ə'fish·əns əv 'fȯrm }

coelacanth [VERT ZOO] Any member of the Coelacanthiformes, an order of lobefin fishes represented by a single living genus, *Latimeria*. { 'sē·lə,kanth }

COELUROSAURIA

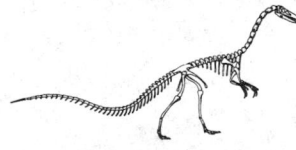

Skeleton of *Coelophysis* (about 8 feet, or 2.5 meters long), a Late Triassic theropod from New Mexico.

Coelacanthidae [PALEON] A family of extinct lobefin fishes in the order Coelacanthiformes. { ˌsē·lə'kan·thə,dē }

Coelacanthiformes [VERT ZOO] An order of lobefin fishes in the subclass Crossopterygii which were common fresh-water animals of the Carboniferous and Permian; one genus, *Latimeria*, exists today. { ¦sē·lə¦kan·thə'fȯr,mēz }

Coelacanthini [VERT ZOO] The equivalent name for Coelacanthiformes. { ˌsē·lə'kan·thə,nī }

Coelenterata *See* Cnidaria. { sə,len·tə'räd·ə }

coelenteron [INV ZOO] The internal cavity of cnidarians. { sə'len·tə,rän }

coeloblastula [EMBRYO] A simple, hollow blastula with a single-layered wall. { ˌsē·lō'blas·chə·lə }

Coelolepida [PALEON] An order of extinct jawless vertebrates (Agnatha) distinguished by skin set with minute, close-fitting scales of dentine, similar to placoid scales of sharks. { ˌsē·lō'lep·ə·də }

coelom [ZOO] The mesodermally lined body cavity of most animals higher on the evolutionary scale than flatworms and nonsegmented roundworms. { 'sē·ləm }

Coelomata [ZOO] The equivalent name for Eucoelomata. { ˌsē·lə'mad·ə }

coelomocyte [INV ZOO] A corpuscle, including amebocytes and eleocytes, in the coelom of certain animals, especially annelids. { 'sē·lō·mə,sīt }

coelomoduct [INV ZOO] Either of a pair of ciliated excretory and reproductive channels passing from the coelom to the exterior in certain invertebrates, including annelids and mollusks. { 'sē·lō·mə,dəkt }

Coelomomycetaceae [MYCOL] A family of entomophilic fungi in the order Blastocladiales which parasitize primarily mosquito larvae. { ¦sē·lō·mə,mī·sə'tās·ē,ē }

coelomostome [INV ZOO] The opening of a coelomoduct into the coelom. { 'sē·lō·mə,stōm }

Coelomycetes [MYCOL] A group set up by some authorities to include the Sphaerioidaceae and the Melanconiales. { ¦sē·lō·mī'sēd,ēz }

Coelopidae [INV ZOO] The seaweed flies, a family of myodarian cyclorrhaphous dipteran insects in the subsection Acalypteratae whose larvae breed on decomposing seaweed. { sə'lō·pə,dē }

coeloplanula [INV ZOO] A hollow planula having a wall of two layers of cells. { ˌsē·lə'plan·yə·lə }

coelostat [ENG] A device consisting of a clockwork-driven mirror that enables a fixed telescope to continuously keep the same region of the sky in its field of view. { 'sē·lə,stat }

Coelurosauria [PALEON] A group of small, lightly built saurischian dinosaurs in the suborder Theropoda having long necks and narrow, pointed skulls. { sə,lur·ə'sȯr·ē·ə }

coelute [ANALY CHEM] In chromatography, two or more chemical compounds that do not separate. { 'kō·ə,lüt }

Coenagrionidae [INV ZOO] A family of zygopteran insects in the order Odonata. { ˌsē,nag·rē'än·ə,dē }

coencytic [MYCOL] Pertaining to filaments or mycelia that lack septa. Also known as nonseptate. { ,kō·in'sī·tik }

coenenchyme [INV ZOO] The mesagloea surrounding and uniting the polyps in compound anthozoans. Also known as coenosarc. { sə'neŋ,kīm }

Coenobitidae [INV ZOO] A family of terrestrial decapod crustaceans belonging to the Anomura. { ˌsē·nə'bid·ə,dē }

coenobium [INV ZOO] A colony of protozoans having a constant size, shape, and cell number, but with undifferentiated cells. { sə'nō·bē·əm }

coenocyte [BIOL] A multinucleate mass of protoplasm formed by repeated nucleus divisions without cell fission. { 'sē·nə,sīt }

Coenomyidae [INV ZOO] A family of orthorrhaphous dipteran insects in the series Brachycera. { ˌsē·nə'mī·ə,dē }

Coenopteridales [PALEOBOT] A heterogeneous group of fernlike fossil plants belonging to the Polypodiophyta. { ˌsē·näp,ter·ə'dā·lēz }

coenosarc [INV ZOO] **1.** The living axial part of a hydroid colony. **2.** *See* coenenchyme. { 'sē·nə,särk }

coenosteum [INV ZOO] The calcareous skeleton of a compound coral or bryozoan colony. { sə'näs·tē·əm }

Coenothecalia [INV ZOO] An order of the class Alcyonaria that forms colonies; lacks spicules but has a skeleton composed of fibrocrystalline argonite. { ˌsē·nō·thə'kāl·ē·ə }

coenotype [BIOL] An organism having the characteristic structure of the group to which it belongs. { 'sē·nə,tīp }

coenurosis [VET MED] An infestation by a coenurus, the metacestode of *Taenia* species; most common in sheep, rabbits, and other herbivores. { ,sē·nyə'rō·səs }

coenzyme [BIOCHEM] The nonprotein portion of an enzyme; a prosthetic group which functions as an acceptor of electrons or functional groups. { kō'en,zīm }

coenzyme I *See* diphosphopyridine nucleotide. { kō'en,zīm 'wən }

coenzyme II *See* triphosphopyridine nucleotide. { kō'en,zīm 'tü }

coenzyme A [BIOCHEM] $C_{21}H_{36}O_{16}N_7P_3S$ A coenzyme in all living cells; required by certain condensing enzymes to act in acetyl or other acyl-group transfer and in fatty-acid metabolism. Abbreviated CoA. { kō'en,zīm 'ā }

coercimeter [ENG] An instrument that measures the magnetic intensity of a natural magnet or electromagnet. { ,kō,ər'sim·əd·ər }

coercion [COMPUT SCI] A method employed by many programming languages to automatically convert one type of data to another. { kō'ər,shən }

coercive force [ELECTROMAG] The magnetic field H which must be applied to a magnetic material in a symmetrical, cyclicly magnetized fashion, to make the magnetic induction B vanish. Also known as magnetic coercive force. { kō'ər·siv 'fòrs }

coercivity [ELECTROMAG] The coercive force of a magnetic material in a hysteresis loop whose maximum induction approximates the saturation induction. { ,kō·ər'siv·əd·ē }

coeruleolactite [MINERAL] $(Ca,Cu)Al_6(PO_4)_4(OH)_8 \cdot 4\text{-}5H_2O$ A milky-white to sky-blue mineral consisting of an aluminum phosphate. { sə,rül·ē·ō'lak,tīt }

coesite [MINERAL] A high-pressure polymorph of SiO_2 formed in nature only under unique physical conditions, requiring pressures of more than 20 kilobars (2 gigapascals); usually found in meteor impact craters. { 'sē,zīt }

coetaneous [SCI TECH] Contemporary. { ,kō·ə'tā·nē·əs }

coevolution [EVOL] An evolutionary pattern based on the interaction among major groups or organisms with an obvious ecological relationship; for example, plant and plant-eater, flower and pollinator. { ,kō,ev·ə'lü·shən }

coextrusion [ENG] Extrusion-forming of plastic or metal products in which two or more compatible feed materials are used in physical admixture through the same extrusion die. { ,kō,ik'strü·zhən }

cofactor [BIOCHEM] A specific substance required for the activity of an enzyme, such as a coenzyme or metal ion. *See* minor. { 'kō,fak·tər }

coffee [BOT] Any of various shrubs or small trees of the genus *Coffea* (family Rubiaceae) cultivated for the seeds (coffee beans) of its fruit; most coffee beans are obtained from the Arabian species, *C. arabica*. { 'kòf·ē }

cofferdam [CIV ENG] A temporary damlike structure constructed around an excavation to exclude water. [NAV ARCH] A void between two bulkheads designed to separate two adjacent liquid-containing compartments. { 'kò·fər,dam }

coffered ceiling [BUILD] An ornamental ceiling constructed of panels that are sunken or recessed. { 'kò·fərd 'sēl·iŋ }

coffin [NUCLEO] A box of heavy shielding material, usually lead, used for transporting radioactive objects and having walls thick enough to attenuate radiation from the contents to an allowable level. Also known as cask; casket. { 'kò·fən }

coffin corner [AERO ENG] The range of Mach numbers between the buffeting Mach number and the stalling Mach number within which an aircraft must be operated. { 'kò·fən ,kòr·nər }

coffinite [MINERAL] $USiO_4$ A black silicate important as a uranium ore; found in sandstone deposits and hydrothermal veins in New Mexico, Utah, and Wyoming. { 'kòf·ə,nīt }

cofinal [MATH] A subset C of a directed set D is cofinal if for each element of D there is a larger element in C. { kō'fīn·əl }

cog [DES ENG] A tooth on the edge of a wheel. [ELEC] A fluctuation in the torque delivered by a motor when it runs at low speed, due to electromechanical effects. Also known as torque ripple. { käg }

cog belt [MECH ENG] A flexible device used for timing and for slip-free power transmission. { 'käg ,belt }

cogeneration [MECH ENG] The simultaneous on-site generation of electric energy and process steam or heat from the same plant. { ,kō,jen·ə'rā·shən }

cogged belt *See* timing belt. { 'kägd ,belt }

cogging [ELECTROMAG] Variations in torque and speed of an electric motor due to variations in magnetic flux as rotor poles move past stator poles. { 'käg·iŋ }

cogging mill *See* blooming mill. { 'käg·iŋ ,mil }

cognac [FOOD ENG] Brandy distilled from grapes grown mostly in the Charente and Charente-Maritime departments of France. { 'kōn,yak }

cognac oil *See* ethyl enanthate. { 'kōn,yak ,òil }

cognate [GEOL] Pertaining to contemporaneous fractures in a system with regard to time of origin and deformational type. { ,käg,nāt }

cognate ejecta [GEOL] Essential or accessory pyroclasts derived from the magmatic materials of a current volcanic eruption. { ,käg,nāt ē'jek·tə }

cognate inclusion *See* autolith. { ,käg,nāt in'klü·zhən }

cognition [PSYCH] The act or process of knowing, including comprehension, judgment, memory, perception, and reasoning. { käg'nish·ən }

cognitive-behavioral therapy [PSYCH] A form of psychotherapy that focuses on changing dysfunctional attitudes into more realistic and positive ones and providing new information-processing skills. { 'käg·nə·tiv bə'hāv·yə·rəl 'ther·ə·pē }

cognitive dissonance [PSYCH] Psychological conflict that results from incongruous beliefs and attitudes held simultaneously. { 'käg·nəd·iv 'dis·ən·əns }

cognitive mapping [PSYCH] A group of mental processes that involve acquisition, coding, storing, manipulation, and recall of spatial information. { 'käg·nəd·iv 'map·iŋ }

cognitive therapy [PSYCH] A method of psychological treatment that emphasizes changing a person's maladaptive processes of thinking, perceptions, and attitudes. { 'käg·nəd·iv 'ther·ə·pē }

COGO [COMPUT SCI] A higher-level computer language oriented toward civil engineering, enabling one to write a program in a technical vocabulary familiar to engineers and feed it to the computer; several versions have been implemented. Derived from coordinated geometry. { 'kō,gō }

cogon [BOT] *Imperate cylindrica*. A grass found in rainforests. Also known as alang-alang. { kō'gōn }

cog railway [CIV ENG] A steep railway that employs a cograil that meshes with a cogwheel on the locomotive to ensure traction. { 'käg 'rāl,wa }

cog region [BIOCHEM] Any group of similar sequences of nucleotides that occurs in deoxyribonucleic acid molecules and may specifically be recognized by endonucleases or other enzymes. { 'käg ,rē·jən }

cogwheel [DES ENG] A wheel with teeth around its edge. { 'käg,wel }

cogwheel ore *See* bournonite. { 'käg,wel ,òr }

cohenite [MINERAL] $(Fe,Ni,Co)_3C$ A tin-white, isometric mineral found in meteorites. { 'kō·ə,nīt }

cohered video [ELECTR] The video detector output signal in a coherent moving-target indicator radar system. { kō'hird 'vid·ē·ō }

coherence [PHYS] **1.** The existence of a correlation between the phases of two or more waves, so that interference effects may be produced between them, or of a correlation between the phases of part of a single wave. **2.** Property of moving in unison, such as is characteristic of the particles in a synchrotron. { kō'hir·əns }

coherence area [OPTICS] A quantitative measure of the spatial coherence of a light beam, equal to the largest cross-sectional area such that light passing through any two pinholes placed in this area will produce interference fringes. { kō'hir·əns ,er·ē·ə }

coherence distance *See* coherence length. { kō'hir·əns ,dis·təns }

coherence length [PHYS] For a beam of particles, the typical length of a wave packet along the beam; the more monochromatic the beam, the greater its coherence length. [SOLID STATE] A measure of the distance through which the effect of any local disturbance is spread out in a superconducting material. Also known as coherence distance. { kō'hir·əns ,leŋkth }

coherence time [PHYS] The average time required for the

COFFEE

Branch of *Coffee arabica*.

COG BELT

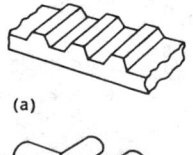

(a)

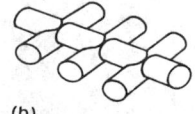

(b)

Cog belts for various uses. *(a)* Flat belt for timing or high-speed power transmission. *(b)* Projections integrally molded with self-lubricating plastic belt for engaging gears from either side and twisting to mesh with misaligned gears.

relative phase of two waves, or the phase of a single wave, to fluctuate appreciably. { kō'hir·əns ˌtīm }

coherent carrier system [NAV] Transponder system in which the interrogating carrier is retransmitted at a definite multiple frequency for comparison. { kō'hir·ənt 'kar·ē·ər ˌsis·təm }

coherent deposit [GEOL] A consolidated sedimentary deposit that is not easily shattered. { kō'hir·ənt di'päz·ət }

coherent detector [ELECTR] A detector used in moving-target indicator radar to give an output-signal amplitude that depends on the phase of the echo signal instead of on its strength, as required for a display that shows only moving targets. { kō'hir·ənt di'tek·tər }

coherent echo [ELECTR] A radar echo whose phase and amplitude at a given range remain relatively constant. { kō'hir·ənt 'ek·ō }

coherent interrupted waves [COMMUN] Interrupted continuous waves occurring in wave trains in which the phase of the waves is maintained through successive wave trains. { kō'hir·ənt in·tə'rəp·təd 'wāvz }

coherent light [OPTICS] Radiant electromagnetic energy of the same, or almost the same, wavelength, and with definite phase relationships between different points in the field. { kō'hir·ənt 'līt }

coherent light communications [COMMUN] Communications using the optical band as a transmission medium by modulating a laser in amplitude or pulse frequency. { kō'hir·ənt 'līt kə,myü·nə'kā·shənz }

coherent moving-target indicator [ENG] A radar system in which the Doppler frequency of the target echo is compared to a local reference frequency generated by a coherent oscillator. { kō'hir·ənt ¦müv·iŋ ¦tär·gət ˌin·də,kād·ər }

coherent noise [ENG] Noise that affects all tracks across a magnetic tape equally and simultaneously. { kō'hir·ənt 'nȯiz }

coherent oscillator [ELECTR] An oscillator used in moving-target indicator radar to serve as a reference by which changes in the radio-frequency phase of successively received pulses may be recognized. Abbreviated coho. { kō'hir·ənt 'äs·ə,lād·ər }

coherent precipitate [PHYS CHEM] A precipitate that is a continuation of the lattice structure of the solvent and has no phase or grain boundary. { kō'hir·ənt prə'sip·ə,tāt }

coherent-pulse radar [ELECTR] A radar in which the radio-frequency oscillations of recurrent pulses bear a constant phase relation to those of a continuous oscillation. { kō'hir·ənt ˌpəls 'rā,där }

coherent pulses [ELECTR] Characterizing pulses in which the phase of the radio-frequency waves is maintained through successive pulses. { kō'hir·ənt 'pəl·səz }

coherent radiation [PHYS] Radiation in which there are definite phase relationships between different points in a cross section of the beam. { kō'hir·ənt ˌrād·ē'ā·shən }

coherent reference [ELECTR] A reference signal, usually of stable frequency, to which other signals are phase-locked to establish coherence throughout a system. { kō'hir·ənt 'ref·rəns }

coherent scattering [PHYS] Scattering in which there is a definite phase relationship between incoming and scattered particles or photons. { kō'hir·ənt 'skad·ə·riŋ }

coherent signal [ELECTR] In a pulsed radar system, a signal having a constant phase; it is mixed with the echo signal, whose phase depends upon the range of the target, in order to detect the phase shift and measure the target's range. { kō'hir·ənt 'sig·nəl }

coherent source [PHYS] A source in which there is a constant phase difference between waves emitted from different parts of the source. { kō'hir·ənt 'sȯrs }

coherent system [NAV] A navigation system in which the signal output is obtained by demodulating the received signal after mixing with a local signal having a fixed phase relation to that of the transmitted signal, to permit use of the information carrier by the phase of the received signal. { kō'hir·ənt 'sis·təm }

coherent transponder [ELECTR] A transponder in which a fixed relation between frequency and phase of input and output signals is maintained. { kō'hir·ənt tranz'pänd·ər }

coherent units [PHYS] A system of units, such as the International System, in which the units of derived quantities are formed as products or quotients of units of the base quantities according to the algebraic relations linking these quantities. { kō'hir·ənt 'yü·nəts }

coherent video [ELECTR] The video signal produced in a moving-target indicator system by combining a radar echo signal with the output of a continuous-wave oscillator; after delay, this signal is detected, amplified, and subtracted from the next pulse train to give a signal representing only moving targets. { kō'hir·ənt 'vid·ē·ō }

coherer [ELEC] A cell containing a granular conductor between two electrodes; the cell becomes highly conducting when it is subjected to an electric field, and conduction can then be stopped only by jarring the granules. { kō'hir·ər }

cohesion [BOT] The union of similar plant parts or organs, as of the petals to form a corolla. [PHYS] The tendency of parts of a body of like composition to hold together, as a result of intermolecular attractive forces. [SCI TECH] The state or process of sticking together. { kō'hē·zhən }

cohesional work [PHYS] The work per unit area required to separate a column of liquid into two parts. { kō'hēzh·ən·əl 'wərk }

cohesionless [GEOL] Referring to a soil having low shear strength when dry, and low cohesion when wet. Also known as frictional; noncohesive. { kō'hē·zhən·ləs }

cohesive end See sticky end. { kō'hē·siv ˌend }

cohesive energy [SOLID STATE] The difference between the energy per atom of a system of free atoms at rest far apart from each other, and the energy of the solid. { kō'hē·siv 'en·ər·jē }

cohesiveness [GEOL] Property of unconsolidated fine-grained sediments by which the particles stick together by surface forces. { kō'hē·siv·nəs }

cohesive soil [GEOL] A sticky soil, such as clay or silt; its shear strength equals about half its unconfined compressive strength. { kō'hē·siv 'sȯil }

cohesive strength [MECH] **1.** Strength corresponding to cohesive forces between atoms. **2.** Hypothetically, the stress causing tensile fracture without plastic deformation. { kō'hē·siv 'streŋkth }

cohesive terminus [MOL BIO] Either of the ends of single-stranded deoxyribonucleic acid that are complementary in the nucleotide sequences and can join, by base pairing, to form circular molecules. { kō'hē·siv 'tər·mə·nəs }

coho See coherent oscillator. { 'kō,hō }

cohomology group [MATH] One of a series of Abelian groups $H^n(K)$ that are used in the study of a simplicial complex K and are closely related to homology groups, being associated with cocycles and coboundaries in the same manner as homology groups are associated with cycles and boundaries. { 'kō·hə'mäl·ə·jē ,grüp }

cohomology theory [MATH] A theory which uses algebraic groups to study the geometric properties of topological spaces; closely related to homology theory. { kō·hō'mäl·ə·jē 'thē·ə·rē }

cohort [STAT] A group of individuals who experience a significant event, such as birth, during the same period of time. { 'kō,hȯrt }

cohort selection [EVOL] A type of natural selection due to interactions among groups of similar ages in a population. { 'kō,hȯrt si,lek·shən }

coho salmon [VERT ZOO] Oncorhynchus kisutch. A species that is widespread across both the Asian and western North American coasts, but is the rarest salmon throughout much of its range. Also known as silver salmon. { 'kō,hō ,sa·mən }

coil [CONT SYS] Any discrete and logical result that can be transmitted as output by a programmable controller. [ELECTROMAG] A number of turns of wire used to introduce inductance into an electric circuit, to produce magnetic flux, or to react mechanically to a changing magnetic flux; in high-frequency circuits a coil may be only a fraction of a turn. Also known as electric coil; inductance; inductance coil; inductor. [SCI TECH] An arrangement of flexible material into a spiral or helix. { kȯil }

coil antenna [ELECTROMAG] An antenna that consists of one or more complete turns of wire. { 'kȯil an'ten·ə }

coil breaks [MET] Creases across a metal strip transverse to the direction of coiling and representing areas of reduced thickness. { 'kȯil ,brāks }

coiled tubular gland [ANAT] A structure having a duct interposed between the surface opening and the coiled glandular portion; an example is a sweat gland. { ¦koild ¦tü·byə·lər 'gland }

coil form [ELECTROMAG] The tubing or spool of insulating material on which a coil is wound. { 'koil ˌform }

coil loading [COMMUN] Loading in which inductors, commonly called loading coils, are inserted in a line at intervals. { 'koil ˌlōd·iŋ }

coil method [GRAPHICS] One of the methods used in terracotta sculpture; the clay is rolled into cylindrical strips about the size of an ordinary pencil and wound up to create the desired shape. { 'koil ˌmeth·əd }

coil neutralization See inductive neutralization. { 'koil nü·trə·lə'zā·shən }

coil serving See serving. { 'koil ˌsərv·iŋ }

coil spring [DES ENG] A helical or spiral spring, such as one of the helical springs used over the front wheels in an automotive suspension. { 'koil ˌspriŋ }

coil weld [MET] A butt weld joining the ends of two metal sheets; forms a continuous strip for coiling. { 'koil ˌweld }

coil winder [ENG] A manual or motor-driven mechanism for winding coils individually or in groups. { 'koil ˌwīn·dər }

coincidence [GEN] A numerical value equal to the number of double crossovers observed, divided by the number expected; a measure of interference. { kō'in·sə·dəns }

coincidence amplifier [ELECTR] An electronic circuit that amplifies only that portion of a signal present when an enabling or controlling signal is simultaneously applied. { kō'in·sə·dəns ˌam·plə·fī·ər }

coincidence boundary [CRYSTAL] A grain boundary separating crystal lattices which are rotated with respect to each other by an angle with a special value, resulting in a periodic grain boundary structure and an extension of a sublattice of the original lattice across the boundary. { kō'in·sə·dəns ˌbaún·drē }

coincidence circuit [ELECTR] A circuit that produces a specified output pulse only when a specified number or combination of two or more input terminals receives pulses within an assigned time interval. Also known as coincidence counter; coincidence gate. { kō'in·sə·dəns ˌsər·kət }

coincidence correction See dead-time correction. { kō'in·sə·dəns kə'rek·shən }

coincidence counter See coincidence circuit. { kō'in·sə·dəns ˌkaúnt·ər }

coincidence counting [NUCLEO] A method of distinguishing particular types of events from background events and of measuring the velocities or directions of particles, by registering the occurrence of counts in two or more particle detectors within a given time interval by means of coincidence circuits. { kō'in·sə·dəns ˌkaúnt·iŋ }

coincidence effect See track adaptation effect. { kō'in·sə·dəns i¸fekt }

coincidence gate See coincidence circuit. { kō'in·sə·dəns ˌgāt }

coincidence magnet [NUCLEO] An electromagnet used in one type of scram mechanism for a nuclear reactor; has three electrically independent coils and releases when any two coils are deenergized. { kō'in·sə·dəns ˌmag·nət }

coincidence rangefinder [OPTICS] An optical rangefinder in which one-eyed viewing through a single eyepiece provides the basis for manipulation of the rangefinder adjustment to cause two images of the target or parts of each, viewed over different paths, to match or coincide. { kō'in·sə·dəns 'rānj¸fīnd·ər }

coincidental evolution [EVOL] The maintenance of sequence homology among nonallelic members of a multigene family within a species. Also known as concerted evolution; horizontal evolution. { ¸kō¸in·sə¦dent·əl ˌev·ə'lü·shən }

coincident-current selection [ELECTR] The selection of a particular magnetic cell, for reading or writing in computer storage, by simultaneously applying two or more currents. { kō¦in·sə·dənt 'kər·ənt si'lek·shən }

co-inducer [BIOCHEM] A molecule that interacts with a repressor to free the operon from restraints on its transcription into messenger ribonucleic acid. { ˌkō·in'dü·sər }

coin gold [MET] Gold of the legal fineness for coins. { 'koin ˌgōld }

coining [MET] **1.** A process of forming metals by squeezing between two dies so as to impress well-defined imprints on both surfaces of the work; usually performed cold. **2.** Final pressing of a sintered compact in powder metallurgy. { 'koin·iŋ }

coinjection molding [ENG] A technique used in polymer processing whereby two or more materials are simultaneously injected into the cavity of a mold. Also known as sandwich molding. { ˌkō·in'jek·shən ˌmōld·iŋ }

coin silver [MET] An alloy of 90% silver, 10% copper; has been used for coining American currency. { 'koin ˌsil·vər }

cointegrate structure [MOL BIO] The circular molecule formed by fusing two replicons, one possessing a transposon, the other lacking it. { kō¦in·tə·grāt 'strək·chər }

coion [ANALY CHEM] Any of the small ions entering a solid ion exchanger and having the same charge as that of the fixed ions. { kō'ī¸än }

coir [MATER] A coarse, brown fiber obtained from the husk of the coconut. { koir }

coisogenic strain [GEN] An animal strain known to differ from the inbred partner strain at a single locus. { ¦kō'i·sə¸jen·ik 'strān }

coitophobia [PSYCH] An abnormal fear of coitus. { ¦kō·əd·ə'fō·bē·ə }

coitus [ZOO] The act of copulation. Also known as intercourse. { 'kō·əd·əs }

coke [MATER] A coherent, cellular, solid residue remaining from the dry (destructive) distillation of a coking coal or of pitch, petroleum, petroleum residues, or other carbonaceous materials; contains carbon as its principal constituent, together with mineral matter and volatile matter. { kōk }

coke breeze [MECH ENG] Undersized coke screenings passing through a screen opening of approximately 5/8 inch (16 millimeters). { 'kōk ¸brēz }

coke coal See natural coke. { 'kōk ¸kōl }

coke drum [CHEM ENG] A vessel in which coke is produced. { 'kōk ¸drəm }

cokeite [GEOL] Naturally occurring coke formed by the action of magma on coal or by natural combustion of coal. { 'kō¸kīt }

coke knocker [MECH ENG] A mechanical device used to break loose coke within a drum or tower. { 'kōk ¸näk·ər }

coke number [CHEM ENG] A number used to report the results of the Ramsbottom carbon residue test. { 'kōk ¸nəm·bər }

coke oven [CHEM ENG] A retort in which coal is converted to coke by carbonization. { 'kōk ¸ōv·ən }

coke-oven gas [MATER] A gas produced during carbonization of coal to form coke. { 'kōk ¸ōv·ən 'gas }

coke-oven regenerator [CHEM ENG] Arrangement of refractory blocks in the flue system of a coke oven to recover waste heat from hot, exiting combustion gases; the blocks, in turn, release heat to warm, incoming fuel gas. { 'kōk ¸ōv·ən ri'jen·ə¸rād·ər }

coker [CHEM ENG] The processing unit in which coking occurs. { 'kōk·ər }

coking [CHEM ENG] **1.** Destructive distillation of coal to make coke **2.** A process for thermally converting the heavy residual bottoms of crude oil entirely to lower-boiling petroleum products and by-product petroleum coke. { 'kok·iŋ }

coking coal [GEOL] A very soft bituminous coal suitable for coking. { 'kok·iŋ ¸kōl }

coking still [CHEM ENG] A still in which coking is done; usually, it is a batch still. { 'kok·iŋ ¸stil }

col [GEOL] A high, sharp-edged pass occurring in a mountain ridge, usually produced by the headward erosion of opposing cirques. [METEOROL] The point of intersection of a trough and a ridge in the pressure pattern of a weather map; it is the point of relatively lowest pressure between two highs and the point of relatively highest pressure between two lows. Also known as neutral point; saddle point. { käl }

Col See Columba.

cola [BOT] *Cola acuminata.* A tree of the sterculia family (Sterculiaceae) cultivated for cola nuts, the seeds of the fruit; extract of cola nuts is used in the manufacture of soft drinks. { 'kō·lə }

CO laser See carbon monoxide laser. { ¦sē¦ō ¸lā·zər }

colatitude [GEOD] Ninety degrees minus the latitude. { kō'lad·ə¸tüd }

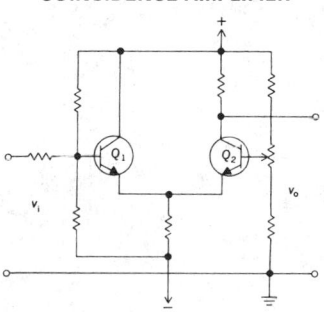

COINCIDENCE AMPLIFIER

Simple coincidence amplifier.

COLA

Branch of *Cola acuminata.*

Colburn analogy [FL MECH] Dimensionless Reynolds equation for fluid-flow resistance modified to be analogous to the Colburn *j* factor heat-transfer equation. { 'kōl·bərn ə'nal·ə·jē }

Colburn j factor equation [THERMO] Dimensionless heat-transfer equation to calculate the natural convection movement of heat from vertical surfaces or horizontal cylinders to fluids (gases or liquids) flowing past these surfaces. { 'kol·bərn 'jā ‚fak·tər i'kwā·zhən }

Colburn method [CHEM ENG] Graphical method, and equations to calculate the theoretical number of plates (trays) needed to separate light and heavy liquids in a distillation column. { 'kōl·bərn ‚meth·əd }

colchicine [ORG CHEM] $C_{22}H_{25}O_6N$ An alkaloid extracted from the stem of the autumn crocus; used experimentally to inhibit spindle formation and delay centromere division, and medicinally in the treatment of gout. { 'käl·chə‚sēn }

colchicum [BOT] Any plant of the genus *Colchicum*, a part of the lily family (mainly the autumn crocus). { 'käl·chi·kəm }

colcothar [INORG CHEM] Red ferric oxide made by heating ferrous sulfate in the air; used as a pigment and as an abrasive in polishing glass. { 'käl·kə‚thär }

cold [ELEC] Pertaining to electrical circuits that are disconnected from voltage supplies and at ground potential; opposed to hot, pertaining to carrying an electrical charge. { kōld }

cold agglutination phenomenon [IMMUNOL] Clumping of human blood group O erythrocytes at 0–4°C, but not at body temperature; occurs in primary atypical pneumonia, trypanosomiasis, and other unidentifiable states. { 'kōld ə‚glüt·ən'ā·shən fə'näm·ə‚nän }

cold agglutinin [IMMUNOL] A nonspecific panagglutinin found in many normal human serums which produce maximum clumping of erythrocytes at 4°C and none at 37°C. { 'kōld ə'glüt·ən·ən }

cold-air drop *See* cold pool. { 'kōld ‚er ‚dräp }

cold-air machine [MECH ENG] A refrigeration system in which air serves as the refrigerant in a cycle of adiabatic compression, cooling to ambient temperature, and adiabatic expansion to refrigeration temperature; the air is customarily reused in a closed superatmospheric pressure system. Also known as dense-air system. { 'kōld ‚er mə‚shēn }

cold-air outbreak *See* polar outbreak. { 'kōld ‚er 'aùt‚brāk }

cold anticyclone *See* cold high. { 'kōld ‚an·tē'sī‚klōn }

cold area [SCI TECH] Laboratory area that is continuously held at a specified degree of coldness to facilitate the conduct of physical and chemical tests. { 'kōld ‚er·ē·ə }

cold bending [MET] The bending of metal rods, especially concrete-reinforcing rods, without heat. { 'kōld ‚bend·iŋ }

cold-blooded [PHYSIO] Having body temperature approximating that of the environment and not internally regulated. { 'kōld ‚bləd·əd }

cold boot [COMPUT SCI] To turn the power on and boot a computer. { ‚kōld 'büt }

cold cathode [ELECTR] A cathode whose operation does not depend on its temperature being above the ambient temperature. { 'kōld 'kath‚ōd }

cold-cathode counter tube [ELECTR] A counter tube having one anode and three sets of 10 cathodes; two sets of cathodes serve as guides that direct the flow discharge to each of the 10 output cathodes in correct sequence in response to driving pulses. { 'kōld 'kath‚ōd 'kaùnt·ər ‚tüb }

cold-cathode discharge *See* glow discharge. { 'kōld 'kath‚ōd 'dis‚chärj }

cold-cathode ionization gage *See* Philips ionization gage. { 'kōld 'kath‚ōd ‚ī·ən·ə'zā·shən ‚gāj }

cold-cathode rectifier [ELECTR] A cold-cathode gas tube in which the electrodes differ greatly in size so electron flow is much greater in one direction than in the other. Also known as gas-filled rectifier. { 'kōld 'kath‚ōd 'rek·tə‚fī·ər }

cold-cathode tube [ELECTR] An electron tube containing a cold cathode, such as a cold-cathode rectifier, mercury-pool rectifier, neon tube, phototube, or voltage regulator. { 'kōld 'kath‚ōd ‚tüb }

cold-chamber die casting [ENG] A die-casting process in which molten metal is ladled either manually or mechanically into a relatively cold cylinder from which it is forced into the die cavity. { 'kōld ‚chām·bər 'dī ‚kast·iŋ }

cold chisel [DES ENG] A chisel specifically designed to cut or chip cold metal; made of specially tempered tool steel machined into various cutting edges. Also known as cold cutter. { 'kōld ‚chiz·əl }

cold color [GRAPHICS] In printing, a color that has bluish tones. { 'kōld ‚kəl·ər }

cold-conductor effect [SOLID STATE] A sudden increase in resistivity, up to seven orders of magnitude, as the temperature increases over a narrow range; observed in certain semiconducting materials, particularly ferroelectric titanate ceramics. { ‚kōld kən'dək·tər i‚fekt }

cold-core cyclone *See* cold low. { ‚kōld ‚kòr 'sī‚klōn }

cold-core high *See* cold high. { ‚kōld ‚kòr 'hī }

cold-core low *See* cold low. { ‚kōld ‚kòr 'lō }

cold cure [CHEM ENG] Vulcanization of rubber at nonelevated temperatures with a solution of a sulfur compound. { 'kōld ‚kyúr }

cold cutter *See* cold chisel. { 'kōld ‚kəd·ər }

cold dark matter [ASTRON] A hypothetical type of dark matter consisting of particles that would have been in thermal equilibrium while traveling at nonrelativistic velocities in the early universe; possibilities include axions, photinos, gravitinos, heavy magnetic monopoles, and weakly interacting massive particles. { 'kōld 'därk ‚mad·ər }

cold desert *See* tundra. { ‚kōld 'dez·ərt }

cold differential test pressure [ENG] The inlet pressure of a pressure-relief valve at which the valve is set to open during testing. { 'kōld ‚dif·ə'ren·chəl 'test ‚presh·ər }

cold dome [METEOROL] A cold air mass, considered as a three-dimensional entity. { 'kōld ‚dōm }

cold drawing [MET] Drawing a tube or wire through a series of successively smaller dies, without the application of heat, to reduce its diameter. [TEXT] Drawing a textile, as nylon, when cold. { 'kōld ‚dró·iŋ }

cold drop *See* cold pool. { 'kōld ‚dräp }

cold emission *See* field emission. { 'kōld i'mish·ən }

cold extrusion [MET] Shaping cold metal by striking a slug in a closed cavity with a punch so that the metal is forced up around the punch. Also known as cold forging; cold pressing; extrusion pressing; impact extrusion. { 'kōld ik'strü·zhən }

cold-finished steel [MET] Steel bars which have been cold-drawn, cold-rolled, centerless-ground, or turned smooth. { 'kōld ‚fin·isht 'stēl }

cold flow [MECH] Creep in polymer plastics. { 'kōld ‚flō }

cold-flow test [AERO ENG] A test of a liquid rocket without firing it to check or verify the integrity of a propulsion subsystem, and to provide for the conditioning and flow of propellants (including tank pressurization, propellant loading, and propellant feeding). { ‚kōld ‚flō ‚test }

cold forging *See* cold extrusion. { 'kōld ‚fòrj·iŋ }

cold forming [MET] Any forging operation performed cold, such as cold extrusion, cold drawing, or coining, which enables close dimensional accuracy to be achieved. { 'kōld ‚fòrm·iŋ }

cold front [METEOROL] Any nonoccluded front, or portion thereof, that moves so that the colder air replaces the warmer air; the leading edge of a relatively cold air mass. { 'kōld ‚frənt }

cold-front-like sea breeze [METEOROL] Sea breeze that forms over the ocean, moves slowly toward the land, and then moves inland quite suddenly. Also known as sea breeze of the second kind. { 'kōld ‚frənt ‚līk 'sē ‚brēz }

cold-front thunderstorm [METEOROL] A thunderstorm attending a cold front. { 'kōld ‚frənt 'thən·dər‚stòrm }

cold galvanizing [MET] Painting iron with a suspension of zinc particles in an organic solvent, so that a zinc coating remains following evaporation of the solvent. { ‚kōld ‚gal·və‚nīz·iŋ }

cold gas approximation [PL PHYS] An approximation according to which the sound speed is much smaller than the Alfvén speed or the gas pressure is much smaller than the magnetic pressure. { 'kōld ‚gas ə‚präk·sə‚mā·shən }

cold glacier [GEOL] A glacier whose base is at a temperature much below 32°F (0°C) and frozen to the bedrock, resulting in insignificant movement and almost no erosion. { ‚kōld 'glā·shər }

cold heading [MET] The cold working of metal in order to increase part or all of the cross-sectional area of the stock. { 'kōld ‚hed·iŋ }

cold hemagglutination [IMMUNOL] A phenomenon caused by the presence of cold agglutinin. { 'kōld ‚hēm·ə‚glüt·ən'ā·shən }

cold high [METEOROL] At a given level in the atmosphere, any high that is generally characterized by colder air near its center than around its periphery. Also known as cold anticyclone; cold-core high. { 'kōld 'hī }

cold-induced vasodilation [MED] A sequence of vasoconstriction followed by vasodilation that acts as a protective mechanism to prevent cold weather injury to the extremities. { 'kōld in,düst ,vā·zō·di'lā·shən }

cold injury [MED] Physical trauma following exposure to very low temperatures. { 'kōld ,in·jə·rē }

cold inspection [MET] The inspection of a forging at room temperature by visible or nondestructive means to detect surface conditions or defects at room temperature. { 'kōld ,in'spek·shən }

cold joint [ENG] A soldered connection which was inadequately heated, with the result that the wire is held in place by rosin flux, not solder. { 'kōld 'jóint }

cold junction [ELECTR] The reference junction of thermocouple wires leading to the measuring instrument; normally at room temperature. { 'kōld 'jəŋk·shən }

cold lap See cold shut. { 'kōld ,lap }

cold light [PHYS] **1.** Light emitted in luminescence. **2.** Visible light which is accompanied by little or no infrared radiation, and therefore has little heating effect. { 'kōld ,līt }

cold lime-soda process [CHEM ENG] A water-softening process in which water is treated with hydrated lime (sometimes in combination with soda ash), which reacts with dissolved calcium and magnesium compounds to form precipitates that can be removed as sludge. { 'kōld |līm |sō·də ,präs·əs }

cold link [COMPUT SCI] A linking of information in two documents in which updating the link requires recopying the information from the source document to the target document. { 'kōld 'liŋk }

cold low [METEOROL] At a given level in the atmosphere, any low that is generally characterized by colder air near its center than around its periphery. Also known as cold-core cyclone; cold-core low. { 'kōld 'lō }

cold molding [ENG] Shaping of an unheated compound in a mold under pressure, followed by heating the article to cure it. { 'kōld ,mōld·iŋ }

cold neutron [SOLID STATE] A very-low-energy neutron in a reactor, used for research into solid-state physics because it has a wavelength of the order of crystal lattice spacings and can therefore be diffracted by crystals. { 'kōld 'nü,trän }

cold nosing See wildcat drilling. { 'kōld ,noz·iŋ }

cold plasma [CHEM ENG] Low-energy ionized gas. { 'kōld 'plaz·mə }

cold plate [MECH ENG] An aluminum or other plate containing internal tubing through which a liquid coolant is forced, to absorb heat transferred to the plate by transistors and other components mounted on it. Also known as liquid-cooled dissipator. { 'kōld ,plāt }

cold pole [CLIMATOL] The location which has the lowest mean annual temperature in its hemisphere. { 'kōld ,pōl }

cold pool [METEOROL] A region of relatively cold air surrounded by warmer air; the term is usually applied to cold air of appreciable vertical extent that has been isolated in lower latitudes as part of the formation of a cutoff low. Also known as cold-air drop; cold drop. { 'kōld ,pül }

cold pressing See cold extrusion. { 'kōld ,pres·iŋ }

cold rolling [MET] Rolling metal at room temperature to reduce thickness or harden the surface; results in a smooth finish and improved resistance to fatigue. { 'kōld ,rōl·iŋ }

cold rubber [MATER] Butadiene-styrene type of synthetic rubber produced by polymerization at about 40°F (4°C), instead of the conventional 120°F (49°C); has improved strength and abrasion resistance. { 'kōld ,rəb·ər }

cold saw [MECH ENG] **1.** Any saw for cutting cold metal, as opposed to a hot saw. **2.** A disk made of soft steel or iron which rotates at a speed such that a point on its edge has a tangential velocity of about 15,000 feet per minute (75 meters per second), and which grinds metal by friction. { 'kōld ,só }

cold-sensitive mutation [GEN] An alteration that causes a gene to be inactive at low temperature. { 'kōld ,sen·səd·iv myü'tā·shən }

cold-setting adhesive [MATER] A synthetic resin that can harden at normal room temperature without the addition of a hardener. { 'kōld ,sed·iŋ əd'hē·ziv }

cold settling [CHEM ENG] A process that removes wax from high-viscosity stocks. { 'kōld ,set·liŋ }

cold-short [MET] Pertaining to lack of ductility in some metals at temperatures below the recrystallization temperature. { 'kōld ,shórt }

cold shot [MET] Intensely hard, globular portions of the surface of an ingot or casting formed by premature solidification upon first contact with the cold sand during pouring. { 'kōld ,shät }

cold shut [MET] **1.** A surface defect of a metal casting in the form of a discontinuity where two streams failed to unite. Also known as cold lap. **2.** Freezing of the top surface of an ingot before the mold is full. { 'kōld ,shət }

cold slug [ENG] The first material to enter an injection mold in plastics manufacturing. { 'kōld ,sləg }

cold-slug well [ENG] The area in a plastic injection mold which receives the cold slug from the sprue opening. { 'kōld ,sləg 'wel }

cold soldering [MET] Soldering of parts without heat. { 'kōld 'säd·ə·riŋ }

cold-spot hygrometer See dew-point hygrometer. { 'kōld ,spät hī'gräm·əd·ər }

cold start [COMPUT SCI] To start running a computer program from the very beginning, without being able to continue the processing that was occurring previously when the system was interrupted. { 'kōld 'stärt }

cold storage [ENG] The storage of perishables at low temperatures produced by refrigeration, usually above freezing, to increase storage life. { 'kōld 'stór·ij }

cold-storage locker plant [ENG] A plant with many rental steel lockers, each with a capacity of about 6 cubic feet (0.17 cubic meter) and generally for food storage by an individual family, placed in refrigerated rooms, at about 0°F (−18°C). { 'kōld 'stór·ij 'läk·ər ,plant }

cold stress [MECH] Forces tending to deform steel, cement, and other materials, resulting from low temperatures. { 'kōld ,stres }

cold stretch [ENG] A pulling operation on extruded plastic filaments in which little or no heat is used; improves tensile properties. { 'kōld ,strech }

cold test [CHEM ENG] A test to determine the temperature at which clouding or coagulation is first visible in a sample of oil, as the temperature of the sample is reduced. { 'kōld ,test }

cold tongue [METEOROL] In synoptic meteorology, a pronounced equatorward extension or protrusion of cold air. { 'kōld ,təŋ }

cold torpor [PHYSIO] Condition of reduced body temperature in poikilotherms. { 'kōld ,tór·pər }

cold trap [MECH ENG] A tube whose walls are cooled with liquid nitrogen or some other liquid to condense vapors passing through it; used with diffusion pumps and to keep vapors from entering a McLeod gage. { 'kōld ,trap }

cold treatment [MET] Subzero cooling, to a temperature of −100°F (−73°C). { 'kōld ,trēt·mənt }

cold trimming [MET] The removal of excess metal from a forging at room temperature by means of a trimming press. { 'kōld ,trim·iŋ }

cold-type composition [GRAPHICS] Any typesetting method which produces copy suitable for offset lithography; copy may be obtained from a typewriter or photocomposition equipment. { 'kōld ,tīp käm·pə'zish·ən }

cold wall [OCEANOGR] The line or surface along which two water masses of significantly different temperature are in contact. { 'kōld ,wól }

cold-water desert [GEOGR] An arid, often foggy region characterized by sparse precipitation because incoming airstreams are cooled over an offshore coastal current and deposit rain over the sea. { 'kōld ,wód·ər 'dez·ərt }

cold-water sphere [OCEANOGR] Those portions of the ocean water having a temperature below 8°C. Also known as oceanic stratosphere. { 'kōld ,wód·ər ,sfir }

cold wave [METEOROL] A rapid fall in temperature within 24 hours to a level requiring substantially increased protection to agriculture, industry, commerce, and social activities. { 'kōld ,wāv }

cold welding [MET] Welding in which a molecular bond is obtained by a cold flow of metal under extremely high pressures, without heat; widely used for sealing transistors and quartz crystal holders. { 'kōld ,weld·iŋ }

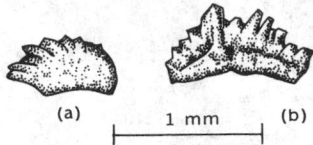

COLEODONTIDAE

(a)   1 mm   (b)

Representatives of Coleodontidae. *(a) Coleodus,* a typical denticulate blade. *(b) Erismodus.*

COLLAGEN

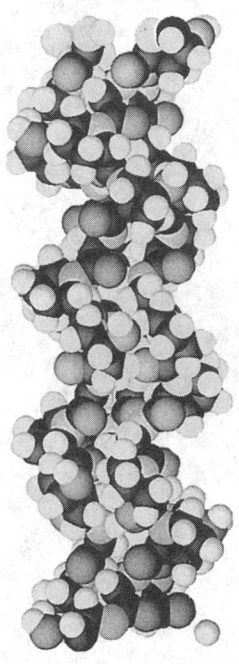

Collagen molecule. Each molecule contains three chains, each coiling into a threefold helix. The chains then coil around one another in a right-handed manner to generate the collagen molecule.

COLLARD

Collard (*Brassica oleracea* var. *acephala*).

cold working [MET] Plastic deformation of a metal below the annealing temperature to cause permanent strain hardening. { 'kōld ˌwərk·iŋ }

Colebrook equation [FL MECH] An empirical equation for the flow of liquids in ducts, relating the friction factor to the Reynolds number and the relative roughness of the duct. { 'kōl,brùk ik'wā·zhən }

Cole-Cole plot [ELEC] For a substance displaying orientation polarization, a graph of the imaginary part versus the real part of the complex relative permittivity that is a circular arc, with its center below the abscissa. { 'kōl 'kōl ,plät }

colectomy [MED] Excision of all or a portion of the colon. { kə'lek·tə·mē }

Cole-Davidson plot [ELEC] For a substance displaying orientation polarization, a graph of the real part versus the imaginary part of the complex relative permittivity that is a skewed arc which approximates a straight line at the high-frequency end and a circular arc at the low-frequency end. { 'kōl 'dā·vəd·sən ,plät }

colemanite [MINERAL] $Ca_2B_6O_{11}\cdot5H_2O$ A colorless or white hydrated borate mineral that crystallizes in the monoclinic system and occurs in massive crystals or as nodules in clay. { 'kōl·mə,nīt }

Coleochaetaceae [BOT] A family of green algae in the suborder Ulotrichineae; all occur as attached, disklike, or parenchymatous thalli. { ˌkō·lē,ō·kē'tās·ē,ē }

Coleodontidae [PALEON] A family of conodonts in the suborder Neurodontiformes. { ˌkō·lē·ō'dän·tə,dē }

Coleoidea [INV ZOO] A subclass of cephalopod mollusks including all cephalopods except *Nautilus,* according to certain systems of classification. { ˌkō·lē'óid·ē·ə }

Coleophoridae [INV ZOO] The case bearers, moths with narrow wings composing a family of lepidopteran insects in the suborder Heteroneura; named for the silk-and-leaf shell carried by larvae. { ˌkō·lē·ō'fór·ə,dē }

coleopter [AERO ENG] An aircraft having an annular (barrel-shaped) wing, the engine and body being mounted within the circle of the wing. { 'kō·lē,äp·tər }

Coleoptera [INV ZOO] The beetles, holometabolous insects making up the largest order of the animal kingdom; general features of the Insecta are found in this group. { ˌkō·lē'äp·tə·rə }

coleoptile [BOT] The first leaf of a monocotyledon seedling. { ˌkō·lē'äp·təl }

coleorhiza [BOT] The sheath surrounding the radicle in monocotyledons. { ˌkō·lē·ə'rīz·ə }

Coleorrhyncha [INV ZOO] A monofamilial group of homopteran insects in which the beak is formed at the anteroventral extremity of the face and the propleura form a shield for the base of the beak. { ˌkō·lē·ə'riŋ·kə }

Coleosporaceae [MYCOL] A family of parasitic fungi in the order Uredinales. { ˌkō·lē·ō·spə'rās·ē,ē }

colic [MED] **1.** Acute paroxysmal abdominal pain usually caused by smooth muscle spasm, obstruction, or twisting. **2.** In early infancy, paroxysms of pain, crying, and irritability caused by swallowing air, overfeeding, intestinal allergy, and emotional factors. { 'käl·ik }

colic artery [ANAT] Any of the three arteries that supply the colon. { 'käl·ik 'ärd·ə·rē }

colicin [MICROBIO] A bacteriocin produced by coliform bacteria, such as *Escherichia coli.* { 'käl·ə·sən }

colidar *See* ladar. { 'käl·ə,där }

coliform bacteria [MICROBIO] Colon bacilli, or forms which resemble or are related to them. { 'käl·ə,fórm bak'tir·ē·ə }

Coliidae [VERT ZOO] The colies or mousebirds, composing the single family of the avian order Coliiformes. { kə'lī·ə,dē }

Coliiformes [VERT ZOO] A monofamilial order of birds distinguished by long tails, short legs, and long toes, all four of which are directed forward. { kə,lī·ə'fór,mēz }

colinearity [MOL BIO] The relationship between the linear sequence of codons in deoxyribonucleic acid and the order of amino acids in the polypeptide product that it specifies. Also spelled collinearity. { ˌkō,lin·ē'ar·əd·ē }

coliphage [VIROL] Any bacteriophage able to infect *Escherichia coli.* { 'käl·ə,fāj }

colistin [MICROBIO] $C_{45}H_{85}O_{10}N_{13}$ A basic polypeptide antibiotic produced by *Bacillus colistinus;* consists of an A and B component, active against a broad spectrum of gram-positive

microorganisms and some gram-negative microorganisms. { kə'lis·tən }

colitis [MED] Inflammation of the large bowel, or colon. { kə'līd·əs }

colk *See* pothole. { kōk }

colla [METEOROL] In the Philippines, a fresh or strong (less than 39–46 miles per hour, or 63–74 kilometers per hour) south to southwest wind, accompanied by heavy rain and severe squall. Also known as colla tempestade. { 'kōl·yə }

collada [METEOROL] A strong wind (35–50 miles per hour, or 56–80 kilometers per hour) in the Gulf of California, blowing from the north or northwest in the upper part, and from the northeast in the lower part of the gulf. { kə'yäd·ə }

collage [GRAPHICS] A composition consisting of paper, cloth, wood, photographs, and so on, pasted together to form a texture or pattern. { kə'läzh }

collagen [BIOCHEM] A fibrous protein found in all multicellular animals, especially in connective tissue. { 'kä·lə·jən }

collagenase [BIOCHEM] Any proteinase that decomposes collagen and gelatin. { 'kä·lə·jə,nās }

collagen disease [MED] Any of various clinical syndromes characterized by widespread alterations of connective tissue, including inflammation and fibrinoid degeneration. { 'kä·lə·jən di,zēz }

collagraph [GRAPHICS] A print made from a plate produced by the collage method. { 'kä·lə,graf }

collapsar [ASTRON] A black hole that forms during the gravitational collapse of a massive star. { kə'lap,sär }

collapse [ENG] Contraction of plastic container walls during cooling; produces permanent indentation. [MATER] The flattening of cells in heartwood during drying or pressure treatment; often characterized by a caved-in or corrugated surface appearance. { kə'laps }

collapse breccia [GEOL] Angular rock fragments derived from the collapse of rock overlying a hollow space. { kə'laps ,brech·ə }

collapse caldera [GEOL] A caldera formed primarily as a result of collapse due to withdrawal of magmatic support. { kə'laps kal'dir·ə }

collapse properties [MECH] Strength and dimensional attributes of piping, tubing, or process vessels, related to the ability to resist collapse from exterior pressure or internal vacuum. { kə'laps ,präp·ərd·ēz }

collapse sink [GEOL] A sinkhole resulting from local collapse of a cavern that has been enlarged by solution and erosion. { kə'laps ,siŋk }

collapse structure [GEOL] A structure resulting from rock slides under the influence of gravity. Also known as gravity-collapse structure. { kə'laps ,strək·chər }

collapsing pressure [MECH] The minimum external pressure which causes a thin-walled body or structure to collapse. { kə'lap·siŋ ,presh·ər }

collar [DES ENG] A ring placed around an object to restrict its motion, hold it in place, or cover an opening. [MIN ENG] The mouth of a mine shaft. [NAV ARCH] **1.** An opening in the end or bight of a rope or cable supporting a mast that goes over the masthead. **2.** A ring or loop of metal, rope, or other material, used to secure a heart or deadeye. **3.** A fitting over a structural part passing through a bulkhead or deck. { 'käl·ər }

collar beam [BUILD] A tie beam in a roof truss connecting the rafters well above the wall plate. { 'käl·ər ,bēm }

collar bearing [MECH ENG] A bearing that resists the axial force of a collar on a rotating shaft. { 'käl·ər ,ber·iŋ }

collar cell *See* choanocyte. { 'käl·ər ,sel }

collard [BOT] *Brassica oleracea* var. *acephala.* A biennial crucifer of the order Capparales grown for its rosette of edible leaves. { 'käl·ərd }

collared hole [ENG] A started hole drilled sufficiently deep to confine the drill bit and prevent slippage of the bit from normal position. { 'käl·ərd ,hōl }

collar locator log [PETRO ENG] Down-hole nuclear-log measurement to locate drill-hole casing collars, usually for precise location of perforating points. { 'käl·ər 'lō,kād·ər ,läg }

collar vortex *See* vortex ring. { 'käl·ər ,vór,teks }

collate [COMPUT SCI] To combine two or more similarly ordered sets of values into one set that may or may not have the same order as the original sets. [GRAPHICS] To assemble in proper sequence all the sheets, signatures, or insertions for a printed piece. { 'kä,lāt }

colla tempestade See colla. { 'kä·lə ,tem·pə'städ·ə }

collateral [ANAT] A side branch of a blood vessel or nerve. { kə'lad·ə·rəl }

collateral bud [BOT] An accessory bud produced beside an axillary bud. { kə¦lad·ə·rəl 'bəd }

collateral bundle [BOT] A vascular bundle in which the phloem and xylem lie on the same radius, with the phloem located toward the periphery of the stem and the xylem toward the center. { kə¦lad·ə·rəl 'bənd·əl }

collateral circulation [PHYSIO] The circulation established for an organ or a part of an organ through the intercommunication of blood vessels when the original direct blood supply is obstructed or abolished. { kə¦lad·ə·rəl ,sər·kyə'lā·shən }

collateral fiber [NEURO] A lateral branch of an axon. { kə¦lad·ə·rəl 'fīb·ər }

collateral ligament [ANAT] Any of various stabilizing ligaments on either side of a hinge joint such as the knee or elbow. { kə¦lad·ə·rəl 'lig·ə·mənt }

collateral respiration [PHYSIO] The passage of air between lobules within the same lobe of a lung, enabling ventilation of a lobule whose branchiole is obstructed. { ke¦lad·ə·rəl ,res·pə'rā·shən }

collateral series [NUC PHYS] A radioactive decay series, initiated by transmutation, that eventually joins into one of the four radioactive decay series encountered in natural radioactivity. { kə¦lad·ə·rəl 'sir,ēz }

collating sequence [COMPUT SCI] The ordering of a set of items such that sets in that assigned order can be collated. { 'kä,lād·iŋ ,sē'kwəns }

collating unit [COMPUT SCI] **1.** An electromechanical device capable of performing singly or simultaneously the merging, sequence-checking, selection, and matching of punched cards. Also known as collator. **2.** A utility program for merging records from two or more files into a single file. { 'kä,lād·iŋ ,yü·nət }

collator See collating unit. { 'kä,lād·ər }

collect [DES ENG] A sleeve or flange that can be tightened about a rotating shaft to halt motion. { kə'lekt }

collecting power [OPTICS] The power of a lens to make parallel rays converge or reduce the divergence of divergent rays. { kə'lek·tiŋ ,pau̇·ər }

collecting tubule [ANAT] One of the ducts conveying urine from the renal tubules (nephrons) to the minor calyces of the renal pelvis. { kə'lek·tiŋ ,tü,byül }

collection trap [ANALY CHEM] Cooled device to collect gas-chromatographic eluent, holding it for subsequent compound-identification analysis. { kə'lek·shən ,trap }

collective [METEOROL] In aviation weather observations, a group of observations transmitted in prescribed order by stations on the same long-line teletypewriter circuit. Also known as sequence. { kə'lek·tiv }

collective bargaining [IND ENG] The negotiation for mutual agreement in the settlement of a labor contract between an employer or his representatives and a labor union or its representatives. { kə'lek·tiv 'bär·gən·iŋ }

collective electron theory [SOLID STATE] A theory of ferromagnetism in which electrons responsible for ferromagnetism are supposed to move more or less freely throughout a crystal, and to align with one another as the result of an exchange interaction. { kə¦lek·tiv i'lek,trän ,thē·ə·re }

collective fire [ORD] Combined fire of various small arms concentrated on a given target or area. { kə'lek·tiv 'fīr }

collective ion accelerator [NUCLEO] A particle accelerator in which charges and currents directly in the beam path are used for both acceleration and focusing. { kə¦lek·tiv 'ī·ən ak'sel·ə,rād·ər }

collective mode [PHYS] A weakly damped and therefore long-lived coherent motion of a large fraction of the particles in a system. { kə'lek·tiv ,mōd }

collective motion [NUC PHYS] Motion of nucleons in a nucleus correlated so that their overall space pattern is essentially constant or undergoes changes which are slow compared to the motions of individual nucleons. { kə'lek·tiv 'mō·shən }

collective paramagnetism [ELECTROMAG] Magnetization of a collection of extremely small ferromagnetic particles, each containing only one magnetic domain, that resembles paramagnetism of a collection of atoms or molecules. Also known as superparamagnetism. { kə'lek·tiv ,par·ə'mag·nə,tiz·əm }

collective transition [NUC PHYS] A nuclear transition from one state of collective motion to another. { kə'lek·tiv tran·z'ish·ən }

collective unconscious [PSYCH] In Jungian theory, a part of the unconscious that theoretically is inherited and common to all people. { kə'lek·tiv ən'kän·shəs }

collector [ELECTR] **1.** A semiconductive region through which a primary flow of charge carriers leaves the base of a transistor; the electrode or terminal connected to this region is also called the collector. **2.** An electrode that collects electrons or ions which have completed their functions within an electron tube; a collector receives electrons after they have done useful work, whereas an anode receives electrons whose useful work is to be done outside the tube. Also known as electron collector. [ENG] A class of instruments employed to determine the electric potential at a point in the atmosphere, and ultimately the atmospheric electric field; all collectors consist of some device for rapidly bringing a conductor to the same potential as the air immediately surrounding it, plus some form of electrometer for measuring the difference in potential between the equilibrated collector and the earth itself; collectors differ widely in their speed of response to atmospheric potential changes. { kə'lek·tər }

collector capacitance [ELECTR] The depletion-layer capacitance associated with the collector junction of a transistor. { kə'lek·tər kə'pas·əd·əns }

collector current [ELECTR] The direct current that passes through the collector of a transistor. { kə'lek·tər ,kər·ənt }

collector cutoff [ELECTR] The reverse saturation current of the collector-base junction. { kə'lek·tər 'kəd,óf }

collector junction [ELECTR] A semiconductor junction located between the base and collector electrodes of a transistor. { kə'lek·tər ,jəŋk·shən }

collector modulation [ELECTR] Amplitude modulation in which the modulator varies the collector voltage of a transistor. { kə'lek·tər ,mäj·ə'lā·shən }

collector plate [ELEC] One of several metal inserts that are sometimes embedded in the lining of an electrolyte cell to make the resistance between the cell lining and the current leads as small as possible. { kə'lek·tər ,plāt }

collector resistance [ELECTR] The back resistance of the collector-base diode of a transistor. { kə'lek·tər ri'zis·təns }

collector ring See slip ring. { kə'lek·tər ,riŋ }

collector voltage [ELECTR] The direct-current voltage, obtained from a power supply, that is applied between the base and collector of a transistor. { kə'lek·tər ,vól·tij }

Collembola [INV ZOO] The springtails, an order of primitive insects in the subclass Apterygota having six abdominal segments. { kə'lem·bə·lə }

collenchyma [BOT] A primary, or early-differentiated, subepidermal supporting tissue in leaf petioles and vein ribs formed before vascular differentiation. { kə'leŋ·kə·mə }

collenchyme [INV ZOO] A loose mesenchyme that fills the space between ectoderm and endoderm in the body wall of many lower invertebrates, such as sponges. { 'kä·lən,kīm }

collenia [PALEOBOT] A convex, slightly arched, or turbinate stromatolite produced by late Precambrian blue-green algae of the genus *Collenia*. { kə'len·ē·ə }

Colles' fracture [MED] A fracture of the radius about 1 inch (2.5 centimeters) above the wrist with dorsal displacement of the distal fragment. { 'kä·lə·səz ,frak·chər }

collet [DES ENG] A split, coned sleeve to hold small, circular tools or work in the nose of a lathe or other type of machine. [ENG] **1.** The glass neck remaining on a bottle after it is taken off the glass-blowing iron. **2.** Pieces of glass, ordinarily discarded, that are added to a batch of glass. Also spelled cullet. [HOROL] A small, friction-tight collar on a balance staff which holds the inner end of a balance spring. [LAP] The small, horizontal face at the bottom of a brilliant-cut gemstone. { käl·ət }

Colletidae [INV ZOO] The colletid bees, a family of hymenopteran insects in the superfamily Apoidea. { kə'led·ə,dē }

colliculus [ANAT] **1.** Any of the four prominences of the corpora quadrigemina. **2.** The anterolateral, apical elevation of the arytenoid cartilages. [NEURO] The elevation where the optic nerve enters the retina. { kə'lik·yə·ləs }

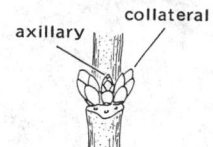

COLLATERAL BUD

Collateral bud in a red maple.

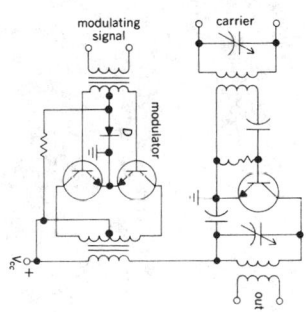

COLLECTOR MODULATION

Circuit diagram of a collector-modulated transistor.

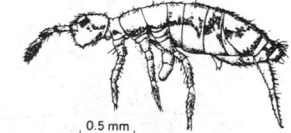

COLLEMBOLA

A collembolan, *Entomobrya cubensis*. (From J. W. Folsom, *Proc. U.S. Nat. Mus., 72(6), plate 6, 1927)*

collider See colliding-beam accelerator. { kə'līd·ər }

2,4,6-collidine [ORG CHEM] $(CH_3)_3C_5H_2N$ A liquid boiling at 170.4°C; slightly soluble in water, soluble in alcohol; used as a chemical intermediate. { ¦tü ¦fór ¦siks 'käl·ə‚dēn }

colliding-beam accelerator [PARTIC PHYS] A particle accelerator in which two beams of high-energy particles are allowed to collide head-on, resulting in high center-of-mass energies. Also known as collider. { kə'līd·iŋ ¦bēm ək'sel·ə‚rād·ər }

colliding-beam source [ELECTR] A device for generating beams of polarized negative hydrogen or deuterium ions, in which polarized negative hydrogen or deuterium atoms are converted to negative ions through charge exchange during collisions with cesium atoms. { kə'līd·iŋ ‚bēm ‚sórs }

colliding-pulse-ring dye laser [OPTICS] A laser consisting of a series of mirrors that form a ring cavity containing an optically pumped saturable gain dye and a saturable absorber dye. { kə'līd·iŋ ¦pəls ‚riŋ 'dī 'lā·zər }

colliery [MIN ENG] A whole coal mining plant; generally the term is used in connection with anthracite mining but sometimes to designate the mine, shops, and preparation plant of a bituminous operation. { 'käl·yə·rē }

colligative properties [PHYS CHEM] Properties dependent on the number of molecules but not their nature. { kə'lig·ə‚div ‚präp·ərd·ēz }

collimate [PHYS] To render parallel to a certain line or direction; paths of electrons in a flooding beam, or paths of various rays of a scanning beam are collimated to cause them to become more nearly parallel as they approach the storage assembly of a storage tube. { 'käl·ə‚māt }

collimated beam [PHYS] A beam of radiation or matter whose rays or particles are nearly parallel so that the beam does not converge or diverge appreciably. { 'käl·ə‚mād·əd 'bēm }

collimating lens [OPTICS] A lens on a collimator used to focus light from a source near one of its focal points into a parallel beam. { 'käl·ə‚mād·iŋ ‚lenz }

collimating sight [ORD] Sight equipped with a collimator, set parallel with the axis of the bore of the gun in horizontal direction, but adjustable in elevation, so that it can be kept focused on an aiming point while the gun is raised or lowered. { 'käl·ə‚mād·iŋ ‚sīt }

collimation error [ASTRON] The amount by which the angle between the optical axis of a transit telescope and its east-west mechanical axis deviates from 90°. [ENG] **1.** Angular error in magnitude and direction between two nominally parallel lines of sight. **2.** Specifically, the angle by which the line of sight of a radar differs from what it should be. { ‚käl·ə'mā·shən ‚er·ər }

collimation tower [ENG] Tower on which a visual and a radio target are mounted to check the electrical axis of an antenna. { ‚käl·ə'mā·shən ‚taú·ər }

collimator [OPTICS] An instrument which produces parallel rays of light. [PHYS] A device for confining the elements of a beam within an assigned solid angle. { 'käl·ə‚mād·ər }

collinear [MATH] Lying on a single straight line. { kə'lin·ē·ər }

collinear array See linear array. { kə'lin·ē·ər ə'rā }

collinear heterodyning [ELECTR] An optical processing system in which the correlation function is developed from an ultrasonic light modulator; the output signal is derived from a reference beam in such a way that the two beams are collinear until they enter the detection aperture; variations in optical path length then modulate the phase of both signal and reference beams simultaneously, and phase differences cancel out in the heterodyning process. { kə'lin·ē·ər 'hed·ə·rə‚dīn·iŋ }

collinearity See colinearity. { kō‚lin·ē'ar·əd·ē }

collinear planes See coaxial planes. { kō'lin·ē·ər 'plānz }

collinear transformation [OPTICS] The mapping of object space into image space produced by an ideal optical image-forming system, in which a unique image point corresponds to each object point and every straight line in the object space has as its corresponding image a unique straight line. { kə'lin·ē·ər ‚tranz·fər'mā·shən }

collinear vectors [MATH] Two vectors, one of which is a non-zero scalar multiple of the other. { kə'lin·ē·ər 'vek·tərz }

collineation [MATH] A mapping which transforms points into points, lines into lines, and planes into planes. Also known as collineatory transformation. { kə‚lin·ē'ā·shən }

collineatory transformation See collineation. { kə'lin·yə‚tór·ē ‚tranz·fər'mā·shən }

collinite [GEOL] The maceral, of collain consistency, of jellified plant material precipitated from solution and hardened; a variety of euvitrinite. { 'käl·ə‚nīt }

Collins helium liquefier [CRYO] A machine which uses the Joule-Thomson effect and work done by helium gas in expansion against a movable piston to liquefy helium. { 'käl·ənz 'hē·lē·əm 'lik·wə‚fī·ər }

collinsite [MINERAL] $Ca_2(Mg,Fe)(PO_4)_2$ A phosphate mineral occurring in concentric layers in phosphoric nodules; found in meteorites. { 'käl·ən‚zīt }

Collins miner [MIN ENG] A type of remote-controlled continuous miner for thin-seam extraction. { 'käl·ənz ‚mīn·ər }

collision [PHYS] An interaction resulting from the close approach of two or more bodies, particles, or systems of particles, and confined to a relatively short time interval during which the motion of at least one of the particles or systems changes abruptly. { kə'lizh·ən }

collision-avoidance radar [ENG] Radar equipment utilized in a collision-avoidance system. { kə'lizh·ən ə'vóid·əns ‚rā‚där }

collision-avoidance system [ENG] Electronic devices and equipment used by a pilot to perform the functions of conflict detection and avoidance. { kə'lizh·ən ə'vóid·əns ‚sis·təm }

collision bearing [NAV] A constant bearing maintained while the distance between two craft is decreasing. { kə'lizh·ən ‚ber·iŋ }

collision blasting [ENG] The blasting out of different sections of rocks against each other. { kə'lizh·ən ‚blast·iŋ }

collision broadening See collision line-broadening. { kə'lizh·ən ‚bród·ən·iŋ }

collision bulkhead [NAV ARCH] A watertight partition in a ship, perpendicular to the fore and aft centerline of the ship, usually near the bow, for keeping out water in the event of a collision. { kə'lizh·ən ‚bəlk‚hed }

collision course [NAV] A course which if followed will bring two craft together. { kə'lizh·ən ‚kórs }

collision-course homing [ORD] Homing in which an offset antenna is used on the missile in conjunction with built-in computers that anticipate the motion of the target and direct the missile ahead of the present position of the target on a converging course that gives a collision in minimum missile travel time. { kə'lizh·ən ¦kórs 'hōm·iŋ }

collision cross section See cross section. { kə'lizh·ən 'krós ‚sek·shən }

collision density [PHYS] The number of collisions of a specified type per unit volume per unit time. { kə'lizh·ən ‚den·səd·ē }

collision detection [COMPUT SCI] A procedure in which a computer network senses a situation where two computer devices attempt to access the network at the same time and blocks the messages, requiring each device to resubmit its message at a randomly selected time. { kə'lizh·ən di‚tek·shən }

collision diameter [PHYS CHEM] The distance between the centers of two molecules taking part in a collision at the time of their closest approach. { kə'lizh·ən dī‚am·əd·ər }

collision efficiency [METEOROL] The fraction of all waterdrops which, initially moving on a collision course with respect to other drops, actually collide (make surface contact) with the other drops. { kə'lizh·ən i'fish·ən·sē }

collision excitation [ATOM PHYS] The excitation of a gas by collisions of moving charged particles. { kə'lizh·ən ‚ek‚sī'tā·shən }

collision frequency [PHYS] The average number of collisions undergone by a particle traveling through a material, such as an electron traveling through a gas, in a unit time. { kə'lizh·ən ‚frē·kwən·sē }

collision ionization [ATOM PHYS] The ionization of atoms or molecules of a gas or vapor by collision with other particles. { kə'lizh·ən ‚ī·ən·ə'zā·shən }

collisionless Boltzmann equation See Vlasov equation. { kə'lizh·ən·ləs 'bōlts‚män i'kwä·zhən }

collisionless plasma [PL PHYS] A plasma in which particles interact through the mutually induced space-charge field, and collisions are assumed to be negligible. { kə'lizh·ən·ləs 'plaz·mə }

collision line-broadening [SPECT] Spreading of a spectral line due to interruption of the radiation process when the radiator collides with another particle. Also known as collision broadening. { kə'lizh·ən 'līn ˌbrȯd·ən·iŋ }

collision matrix See scattering matrix. { kə'lizh·ən ˌmā ˌtriks }

collision of the first kind [PHYS] An inelastic collision in which some of the kinetic energy of translational motion is converted to internal energy of the colliding systems. Also known as endoergic collision. { kə'lizh·ən əv thə 'fərst ˌkīnd }

collision of the second kind [PHYS] An inelastic collision in which some of the internal energy of the colliding systems is converted to kinetic energy of translation. Also known as exoergic collision. { kə'lizh·ən əv thə 'sek·ənd ˌkīnd }

collision parameter [AERO ENG] In orbit computation, the distance between a center of attraction of a central force field and the extension of the velocity vector of a moving object at a great distance from the center. { kə'lizh·ən pə'ram·əd·ər }

collision probability [PHYS] The ratio of the cross section for a given type of collision between two particles to the total cross section for all types of collision between the particles. { kə'lizh·ən ˌpräb·ə,bil·əd·ē }

collision-radiative recombination [ATOM PHYS] The capture of an electron by an ion in a gas, accompanied by the emission of one or more photons. { kə'lizh·ən ˌrād·ē·ād·iv ri,käm·bə'nā·shən }

collision theory [PHYS CHEM] Theory of chemical reaction proposing that the rate of product formation is equal to the number of reactant-molecule collisions multiplied by a factor that corrects for low-energy-level collisions. [QUANT MECH] Theory to describe collisions of simple or complex particles, the derivation of collision cross sections from postulated interactions and the study of properties of collision amplitudes which follow from invariance principles such as conservation of probability and time-reversal invariance. { kə'lizh·ən ,thē· ə·rē }

colloblast [INV ZOO] An adhesive cell on the tentacles of ctenophores. { 'käl·ə,blast }

collodion [ORG CHEM] Cellulose nitrate deposited from a solution of 60% ether and 40% alcohol, used for making fibers and film and in membranes for dialysis. { kə'lōd·ē·ən }

collodion cotton See pyroxylin. { kə'lōd·ē·ən ,kät·ən }

collodion replication [ANALY CHEM] Production of a faithful collodion-film mold of a specimen surface (for example, powders, bones, microorganisms, crystals) which is sufficiently thin to be studied by electron microscopy. [GRAPHICS] A process of image formation utilizing a negative that is a glass plate coated with a collodion containing iodides. { kə'lōd·ē· ən rep·lə'kā·shən }

colloform [GEOL] Pertaining to the rounded, globular texture of mineral formed by colloidal precipitation. { 'käl· ə,fȯrm }

colloid [CHEM] The phase of a colloidal system made up of particles having dimensions of 10–10,000 angstroms (1–1000 nanometers) and which is dispersed in a different phase. { 'käl,ȯid }

colloidal crystal [CHEM] A periodic array of suspended colloidal particles that can arise spontaneously in a monodisperse colloidal system under appropriate conditions. { kə'lȯid·əl 'krist·əl }

colloidal dispersion See colloidal system. { kə'lȯid·əl dis'pər·zhən }

colloidal electrolyte [PHYS CHEM] An electrolyte that yields at least one type of ion in the colloidal size range. { kə'lȯid· əl i'lek·trə,līt }

colloidal fuel See coal-in-oil suspension. { kə'lȯid·əl ,fyül }

colloidal graphite [MATER] Extremely fine flakes of graphite suspended in water, petroleum oil, castor oil, glycerin, or other liquids; used to provide conductive shields on the inside or outside surfaces of electron tubes. { kə'lȯid·əl 'gra,fīt }

colloidal instability [METEOROL] A property attributed to clouds, by which the particles of the cloud tend to aggregate into masses large enough to precipitate. { kə'lȯid·əl in· stə,bil·əd·ē }

colloidal osmotic pressure See oncotic pressure. { kə'lȯid· əl äz'mäd·ik ,presh·ər }

colloidal silver [MATER] Finely divided particles of silver, sometimes used on terminals of electronic components to give a larger surface area for connections. { kə'lȯid·əl 'sil·vər }

colloidal suspension See colloidal system. { kə'lȯid·əl səs'pen·shən }

colloidal system [CHEM] An intimate mixture of two substances, one of which, called the dispersed phase (or colloid), is uniformly distributed in a finely divided state through the second substance, called the dispersion medium (or dispersing medium); the dispersion medium or dispersed phase may be a gas, liquid, or solid. Also known as colloidal dispersion; colloidal suspension. { kə'lȯid·əl 'sis·təm }

colloid chemistry [PHYS CHEM] The scientific study of matter whose size is approximately 10 to 10,000 angstroms (1 to 1000 nanometers), and which exists as a suspension in a continuous medium, especially a liquid, solid, or gaseous substance. { 'käl,ȯid 'kem·ə·strē }

colloider [CIV ENG] A device that removes colloids from sewage. { kə'lȯid·ər }

colloid goiter [MED] A soft, diffuse enlargement of the thyroid gland in which colloid fills the acinar spaces. { 'käl,ȯid 'gȯid·ər }

colloid mill [MECH ENG] A grinding mill for the making of very fine dispersions of liquids or solids by breaking down particles in an emulsion or paste. { 'käl,ȯid ,mil }

collophane [MINERAL] A massive, cryptocrystalline, carbonate-containing variety of apatite and a principal source of phosphates for fertilizers. Also known as collophanite. { 'käl·ə,fān }

collophanite See collophane. { kə'läf·ə,nīt }

Collothecacea [INV ZOO] A monofamilial suborder of mostly sessile rotifers in the order Monogonata; many species are encased in gelatinous tubes. { ,käl·ə·thə'kās·ē·ə }

Collothecidae [INV ZOO] The single family of the Collothecacea. { ,käl·ə'thes·ə,dē }

collotype See photogelatin process. { 'käl·ə,tīp }

colluvium [GEOL] Loose, incoherent deposits at the foot of a slope or cliff, brought there principally by gravity. { kə'lü· vē·əm }

Collyritidae [PALEON] A family of extinct, small, ovoid, exocyclic Euechinoidea with fascioles or a plastron. { ,käl· ə'rid·ə,dē }

Colmol miner [MIN ENG] A continuous miner, in which the coal is completely augered by two banks of cutting arms fitted with picks; the arms rotate in opposite directions to assist in gathering up the cuttings for the central conveyor. { 'käl,mōl ,mīn·ər }

coloboma [MED] A congenital, pathologic, or operative fissure, especially of the eye or eyelid. { ,käl·ə'bō·mə }

colog See cologarithm. { 'kō,läg }

cologarithm [MATH] The cologarithm of a number is the logarithm of the reciprocal of that number. Abbreviated colog. { ,kō'läg·ə,rith·əm }

colon [ANAT] The portion of the human intestine extending from the cecum to the rectum; it is divided into four sections: ascending, transverse, descending, and sigmoid. Also known as large intestine. { 'kō·lən }

colon bacillus See Escherichia coli. { 'kō·lən bə'sil·əs }

colonization [ECOL] The establishment of an immigrant species in a peripherally unsuitable ecological area; occasional gene exchange with the parental population occurs, but generally the colony evolves in relative isolation and in time may form a distinct unit. { ,käl·ə·nə'zā·shən }

colonnade [ARCH] A series of columns placed at regular intervals. { ,käl·ə'nād }

colonoscopy [MED] Visual examination of the inner surface of the colon by means of an endoscope. { ,kō·lə'näs·kə·pē }

colony [BIOL] A localized population of individuals of the same species which are living either attached or separately. [MICROBIO] A cluster of microorganisms growing on the surface of or within a solid medium; usually cultured from a single cell. { 'käl·ə·nē }

colony count [MICROBIO] The number of colonies of bacteria growing on the surface of a solid medium. { 'käl·ə·nē ,kaunt }

colony-stimulating factor [IMMUNOL] A group of lymphocytes that induce the maturation and proliferation of leukocyte, macrophage, and monocyte lines present in bone marrow. { 'käl·ə·nē ,stim·yə,lād·iŋ ,fak·tər }

color [OPTICS] A general term that refers to the wavelength composition of light, with particular reference to its visual appearance. [PARTIC PHYS] A hypothetical quantum number

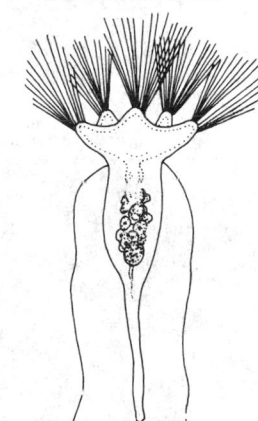

COLLOTHECIDAE

Collotheca species.

carried by quarks, so that each type of quark comes in three varieties which are identical in all measurable qualities but which differ in this additional property; this quantity determines the coupling of quarks to the gluon field. { 'kəl·ər }

color aberration *See* chromatic aberration. { 'kəl·ər ab·ə'rā·shən }

Coloradoan [GEOL] Middle Upper Cretaceous geologic time. { ˌkäl·ə'rad·ə·wən }

coloradoite [MINERAL] HgTe A grayish-black, isometric telluride mineral with a metallic luster; specific gravity is 8.6. { ˌkäl·ə'rad·ə·wīt }

Colorado low [METEOROL] A low which makes its first appearance as a definite center in the vicinity of Colorado on the eastern slopes of the Rocky Mountains; analogous to the Alberta low. { ˌkal·ə'rad·ō ˌlō }

Colorado tick fever [MED] A nonxanthematous acute viral disease of humans occurring in the western United States and transmitted by a bite of the tick *Dermacentor andersoni*; characterized by a short course, intermittent fever, leukopenia, and occasionally meningoencephalitis. Also known as mountain tick fever. { ˌkal·ə'rad·ō 'tik ˌfē·vər }

colorant [FOOD ENG] A substance used to impart color to, or augment color in, a food substance. { 'kəl·ər·ənt }

color attribute [OPTICS] Any of the visual qualities of hue, saturation, or brightness. { 'kəl·ər ˌa·trə,byüt }

color balance [ELECTR] Adjustment of the circuits feeding the three electron guns of a television color picture tube to compensate for differences in light-emitting efficiencies of the three color phosphors on the screen of the tube. { 'kəl·ər ˌbal·əns }

color-bar code [IND ENG] A code that uses one or more different colors of bars in combination with black bars and white spaces, to increase the density of binary coding of data printed on merchandise tags or directly on products for inventory control and other purposes. { 'kəl·ər ˌbär ˌkōd }

color-bar generator [ELECTR] A signal generator that delivers to the input of a color television receiver the signal needed to produce a color-bar test pattern on one or more channels. { 'kəl·ər ˌbär 'jen·ə,rād·ər }

color-bar test pattern [COMMUN] A test pattern of different colors of vertical bars, used to check the performance of a color television receiver. { 'kəl·ər ˌbär 'test ˌpad·ərn }

color-blind [GRAPHICS] Of a photographic emulsion, sensitive only to blue, violet, and ultraviolet light. { 'kəl·ər ˌblīnd }

color blindness [MED] Inability to perceive one or more colors. { 'kəl·ər ˌblīnd·nəs }

color breakup [COMMUN] A transient or dynamic distortion of the color in a color television picture that can originate in videotape equipment, a television camera, or a receiver. { 'kəl·ər ˌbrāk,əp }

color burst [ELECTR] The portion of the composite color television signal consisting of a few cycles of a sine wave of chrominance subcarrier frequency. Also known as burst; reference burst. { 'kəl·ər ˌbərst }

color carrier *See* chrominance subcarrier. { 'kəl·ər ˌkar·ē·ər }

color-carrier reference *See* chrominance-carrier reference. { 'kəl·ər ˌkar·ē·ər ˌref·rəns }

color center [SOLID STATE] A point lattice defect which produces optical absorption bands in an otherwise transparent crystal. { 'kəl·ər ˌsen·tər }

color circle [OPTICS] An arrangement of hues about the circumference of a circle in the order in which they appear in the electromagnetic spectrum, with pairs of complementary colors at opposite ends of diameters. { 'kəl·ər ˌsər·kəl }

color class [MATH] In a given coloring of a graph, the set of vertices which are assigned the same color. { 'kəl·ər ˌklas }

color code [ELEC] A system of colors used to indicate the electrical value of a component or to identify terminals and leads. [ENG] **1.** Any system of colors used for purposes of identification, such as to identify dangerous areas of a factory. **2.** A system of colors used to identify the type of material carried by a pipe; for example, dangerous materials, protective materials, extra valuable materials. { 'kəl·ər ˌkōd }

color coder *See* matrix. { 'kəl·ər ˌkōd·ər }

color-color diagram [ASTRON] A graph whose coordinates are both color indices, showing the distribution of stars or other objects. { 'kəl·ər 'kəl·ər ˌdī·ə,gram }

color comparator [ANALY CHEM] A photoelectric instrument that compares an unknown color with that of a standard

color sample for matching purposes. Also known as photoelectric color comparator. { 'kəl·ər kəm'par·əd·ər }

color control *See* chroma control. { 'kəl·ər kən'trōl }

color contamination [ELECTR] An error in the color rendition of a color television picture that results from incomplete separation of the paths that carry different color components of a picture. { 'kəl·ər kən,tam·ə'nā·shən }

color correction [GRAPHICS] Any method used to improve color rendition; for example, masking, dot etching, reetching, and scanning. [OPTICS] The construction of an optical system so that the image positions of an object are the same for two or more wavelengths, and chromatic aberration is thus minimized. { 'kəl·ər kə'rek·shən }

color decoder *See* matrix. { 'kəl·ər dē'kōd·ər }

color-difference signal [ELECTR] A signal that is added to the monochrome signal in a color television receiver to obtain a signal representative of one of the three tristimulus values needed by the color picture tube. { 'kəl·ər ˌdif·rəns ˌsig·nəl }

color disk [OPTICS] A rotating circular disk having three filter sections to produce the individual red, green, and blue pictures in a field-sequential color television system. { 'kəl·ər ˌdisk }

color Doppler flow image [ACOUS] An acoustic image in which Doppler information is encoded in color, with red and blue representing fluid flow toward and away from the transducer. { ˈkəl·ər ˈdäp·lər ˌflō ˌim·ij }

color Doppler flow imaging scanner [ENG] A device that obtains B-mode images and Doppler blood flow data simultaneously, and superimposes a color Doppler image on the gray-scale B-mode image. { ˈkəl·ər ˈdäp·lər ˈflō ˈim·ij·iŋ ˌskan·ər }

color Doppler optical coherence tomography [OPTICS] An augmentation of optical coherence tomography which uses the interferometric phase information ignored in conventional optical coherence tomography to achieve simultaneous blood flow mapping and spatially resolved imaging. Also known as optical Doppler tomography. { ˈkəl·ər ˈdäp·lər ˈäp·tə·kəl kōˈhēr·əns təˈmä·grə·fē }

colored smoke [MATER] Gaseous products of a distinctive color which forms a colored cloud and may be used for signals or target markers. { ˈkəl·ərd ˈsmōk }

color emissivity *See* monochromatic emissivity. { ˈkəl·ər ˌe·miˈsiv·əd·ē }

color encoder *See* matrix. { 'kəl·ər en'kōd·ər }

color equation [ASTRON] A measure of the color sensitivity and response of a method of observation; photographic, visual, or photoelectric techniques may be employed. [OPTICS] An algebraic equation that expresses a specified color as an additive mixture of primary colors. { 'kəl·ər i'kwā·zhən }

color excess [ASTRON] The difference between the observed color index of a star and the color index corresponding to its spectral type. { 'kəl·ər 'ek,ses }

color facsimile [COMMUN] A facsimile system for transmission of color photographs, in which three separate facsimile transmissions are made from the original color print, using color-separation filters in the optical system of the facsimile transmitter. { 'kəl·ər ˌfak'sim·ə·lē }

colorfast [TEXT] Referring to a fabric that does not fade during normal wear. { 'kəl·ər,fast }

color film [GRAPHICS] A sensitized film used for three-color photography; consists of three emulsions coated one above another and sensitive respectively to blue, green, and red light. { 'kəl·ər ,film }

color filter [OPTICS] An optical element that partially absorbs incident light, consisting of a pane of glass or other partially transparent material, or of films separated by narrow layers; the absorption may be either selective or nonselective with respect to wavelength. Also known as light filter. { 'kəl·ər ,fil·tər }

color-filter effect *See* Bragg effect. { ˈkəl·ər ˈfil·tər iˌfekt }

color force [PARTIC PHYS] The force that acts between quarks to bind them in hadrons and is thought to be the basis of all nuclear forces. { 'kəl·ər ,förs }

color fringing [ELECTR] Spurious chromaticity at boundaries of objects in a television picture. { 'kəl·ər 'frinj·iŋ }

colorimeter [ANALY CHEM] A device for measuring concentration of a known constituent in solution by comparison with colors of a few solutions of known concentration of that constituent. Also known as chromometer. [OPTICS] An instrument that measures color by determining the intensities of the

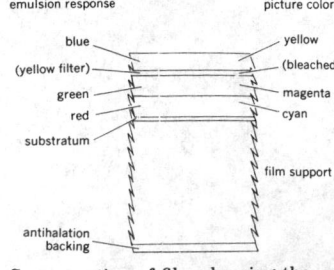

COLOR FILM

emulsion response — picture color
blue — yellow
(yellow filter) — (bleached)
green — magenta
red — cyan
substratum — film support
antihalation backing

Cross section of film showing the various emulsion layers with (left) the emulsion response and (right) the developed color of each layer.

three primary colors that will give that color. { ‚kəl·ə'rim·əd·ər }

colorimetric photometer [OPTICS] A photometer that can measure light intensities in several spectral regions, using color filters placed in the path of the light. { ‚kəl·ə·rə'me·trik fō'täm·əd·ər }

colorimetry [OPTICS] Any technique by which an unknown color is evaluated in terms of standard colors; the technique may be visual, photoelectric, or indirect by means of spectro-photometry; used in chemistry and physics. { ‚kəl·ə'rim·ə·trē }

color index Abbreviated CI. [ASTRON] **1.** Of a star, the numerical difference between the apparent photographic magnitude and the apparent photovisual magnitude. **2.** More generally, the difference in apparent magnitudes between two specified spectral regions. [PATH] The amount of hemoglobin per erythrocyte relative to normal, equal to the percent normal hemoglobin concentration divided by percent normal erythrocyte count. { 'kəl·ər ‚in‚deks }

coloring [MATH] An assignment of colors to the vertices of a graph so that adjacent vertices are assigned different colors. { 'kəl·ər·iŋ }

coloring agent [FOOD ENG] Any substance of natural origin, such as turmeric, annatto, caramel, carmine, and carotine, or a synthetic certified food color added to food to compensate for color changes during processing or to give an appetizing color. { 'kəl·ər·iŋ ‚ā·jənt }

color killer circuit [ELECTR] The circuit in a color television receiver that biases chrominance amplifier tubes to cutoff during reception of monochrome programs. Also known as killer stage. { 'kəl·ər ‚kil·ər ‚sər·kət }

color kinescope See color picture tube. { ¦kəl·ər 'kin·ə‚skōp }

color lake See lake. { 'kəl·ər ‚lāk }

color-magnitude diagram [ASTRON] A graph of the apparent or absolute magnitudes of a group of stars versus their color indices. { 'kəl·ər‚mag·nə‚tüd ‚dī·ə‚gram }

color medium [OPTICS] Any colored, transparent material that is placed in front of a lighting unit to color the light transmitted. { 'kəl·ər ‚mēd·ē·əm }

color model [GRAPHICS] Any method of representing colors for printing or electronic display. { 'kəl·ər ‚mäd·əl }

color oscillator See chroma oscillator. { 'kəl·ər ‚äs·ə‚lād·ər }

color phase [COMMUN] The difference in phase between components (I or Q) of a chrominance signal and the chrominance-carrier reference in a color television receiver. { 'kəl·ər ‚fāz }

color-phase alternation [COMMUN] The periodic changing of the color phase of one or more components of the chrominance subcarrier between two sets of assigned values after every field in a color television system. Abbreviated cpa. { 'kəl·ər ‚fāz ‚ȯl·tər'nā·shən }

color-phase detector [ELECTR] The color television receiver circuit that compares the frequency and phase of the incoming burst signal with those of the locally generated 3.579545-megahertz chroma oscillator and delivers a correction voltage to a reactance tube to ensure that the color portions of the picture will be in exact register with the black-and-white portions on the screen. { 'kəl·ər ‚fāz di'tek·tər }

color photography [GRAPHICS] A film process that uses color film to produce three-color positive pictures. { 'kəl·ər fə'täg·rə·fē }

color picture signal [COMMUN] The electric signal that represents complete color picture information, excluding all synchronizing signals. { 'kəl·ər ‚pik·chər ‚sig·nəl }

color picture tube [ELECTR] A cathode-ray tube having three different colors of phosphors, so that when these are appropriately scanned and excited in a color television receiver, a color picture is obtained. Also known as color kinescope; color television picture tube; tricolor picture tube. { 'kəl·ər ‚pik·chər ‚tüb }

color printing [GRAPHICS] The art and craft of embellishing designs, pictures, and typographic pages with color for a more pleasing effect than obtained in black and white; in addition, pictures more closely represent the original object or painting. { 'kəl·ər ‚print·iŋ }

color purity [ELECTR] Absence of undesired colors in the spot produced on the screen by each beam of a television color picture tube. { 'kəl·ər ‚pyür·əd·ē }

color radiography [GRAPHICS] A radiographic technique in which various intensities are displayed as different colors. { ¦kəl·ər ‚rād·ē'äg·rə·fē }

color rendering [OPTICS] For a light source, the extent of the agreement between the perceived color of a surface illuminated by the source and that of the same surface illuminated by a reference source under specified viewing conditions, measured and expressed in terms of the chromaticity coordinates of the source and the luminance of the source in agreed spectral bands. { ¦kəl·ər 'ren·dər·iŋ }

color saturation [OPTICS] The degree to which a color is mixed with white; high saturation means little white, low saturation means much white. Also known as chroma; saturation. { 'kəl·ər sach·ə'rā·shən }

color-saturation control See chroma control. { 'kəl·ər sach·ə'rā·shən kən'trōl }

color-sensitive [GRAPHICS] Referring to a photographic emulsion which is not color-blind. { 'kəl·ər ‚sen·səd·iv }

color separation [GRAPHICS] The process of preparing a separate drawing, engraving, or negative for each color required in the reproduction of a colored picture. { 'kəl·ər sep·ə'rā·shən }

color signal [COMMUN] Any signal that controls the chromaticity values of a color television picture, such as the color picture signal and the chrominance signal. { 'kəl·ər ‚sig·nəl }

color solid [OPTICS] A three-dimensional diagram which represents the relationship of three attributes of surface color: hue, saturation, and brightness. { 'kəl·ər ‚säl·əd }

color stability [CHEM] Resistance of materials to change in color that can be caused by light or aging, as of petroleum or whiskey. { 'kəl·ər stə'bil·əd·ē }

color standard [ANALY CHEM] Liquid solution of known chemical composition and concentration, hence of known and standardized color, used for optical analysis of samples of unknown strength. { 'kəl·ər ‚stan·dərd }

color stripe [GRAPHICS] A band of color affixed to the top edge of a microform and used for identification and retrieval. { 'kəl·ər ‚strīp }

color SU₃ [PARTIC PHYS] A unitary symmetry based on the equivalence of the three differently colored quarks of a given flavor, which form a fundamental multiplet. { 'kəl·ər ‚es‚yü 'thrē }

color subcarrier See chrominance subcarrier. { 'kəl·ər səb'kar·ē·ər }

color-subcarrier oscillator See chroma oscillator. { 'kəl·ər səb'kar·ē·ər 'ä·sə‚lād·ər }

color-subcarrier reference See chrominance-carrier reference. { 'kəl·ər səb'kar·ē·ər 'ref·rəns }

color sync signal [COMMUN] A signal that is transmitted with each line of a color television broadcast to ensure that the color relationships in the transmitted signal are established and maintained in the receiver. { 'kəl·ər 'siŋk ‚sig·nəl }

color system [OPTICS] Any three-component coordinate system used to represent the attributes of colors. { 'kəl·ər ‚sis·təm }

color television [COMMUN] A television system that reproduces an image approximately in its original colors. { ¦kəl·ər ‚tel·ə‚vizh·ən }

color television picture tube See color picture tube. { ¦kəl·ər ‚tel·ə‚vizh·ən 'pik·chər ‚tüb }

color temperature [STAT MECH] Of a solid surface, that temperature of a blackbody from which the radiant energy has essentially the same spectral distribution as that from the surface. { ¦kəl·ər ¦tem·prə·chər }

color tempering [MET] Reheating of hardened steel and observing color changes to determine quenching temperature and to obtain the desired hardness. { ¦kəl·ər ¦tem·pə·riŋ }

color test [ANALY CHEM] The quantitative analysis of a substance by comparing the intensity of the color produced in a sample by a reagent with a standard color produced similarly in a solution of known strength. { 'kəl·ər ‚test }

color throw [ANALY CHEM] In an ion-exchange process, discoloration of the liquid passing through the bed. { 'kəl·ər ‚thrō }

color-translating microscope [OPTICS] A type of compound microscope that employs three different wavelengths of light to reveal details produced by ultraviolet or other nonvisible radiation. { ¦kəl·ər tranz'lād·iŋ 'mī·krə‚skōp }

color transmission [COMMUN] In television, the transmission of a signal wave which represents both the brightness values and the chromaticity values in the picture. { 'kəl·ər tranz'mish·ən }

color triangle [OPTICS] A triangle on a chromaticity diagram that represents the range of chromaticities that can be obtained as additive mixtures of three prescribed primary colors represented by the corners of the triangle. { 'kəl·ər 'trī,aŋ·gəl }

color vision [PHYSIO] The ability to discriminate light on the basis of wavelength composition. { 'kəl·ər ,vizh·ən }

colossal magnetoresistance [SOLID STATE] A very large magnetoresistance associated with magnetic phase transitions in certain homogeneous materials, particularly a class of rare-earth perovskite manganites. { kə¦läs·əl mag,ned·ō·ri'zis·təns }

Colossendeidae [INV ZOO] A family of deep-water marine arthropods in the subphylum Pycnogonida, having long palpi and lacking chelifores, except in polymerous forms. { ,käl·ə·sen'dā·ə,dē }

colostomy [MED] Surgical formation of an artificial anus by joining the colon to an opening in the anterior abdominal wall. { kə'läs·tə·mē }

colostrum [PHYSIO] The first milk secreted by the mammary gland during the first days following parturition. { kə'las·trəm }

colotomy [MED] Incision of the colon; may be abdominal, lateral, lumbar, or iliac, according to the region of entrance. { kə'läd·ə·mē }

Colpitts oscillator [ELECTR] An oscillator in which a parallel-tuned tank circuit has two voltage-dividing capacitors in series, with their common connection going to the cathode in the electron-tube version and the emitter circuit in the transistor version. { 'kōl,pits ,äs·ə,läd·ər }

colposcope [MED] An instrument for the visual examination of the vagina and cervix; a vaginal speculum. { 'käl·pə,skōp }

colposcopy [MED] Visual examination of cells of the vagina and cervix by means of an endoscope. { käl'päs·kə·pē }

colpotomy [MED] Incision of the vagina. { käl'päd·ə·mē }

colter [AGR] A cutting tool attached to the moldboard plow beam for cutting the sward in advance of the plowshare. Also known as coulter; cutting coulter. { 'kōl·tər }

Colubridae [VERT ZOO] A family of cosmopolitan snakes in the order Squamata. { kə'lü·brə,dē }

Columba [ASTRON] A constellation, right ascension 6 hours, declination 35°S. Abbreviated Col. Also known as Dove. { kə'ləm·bə }

Columbidae [VERT ZOO] A family of birds in the order Columbiformes composed of the pigeons and doves. { kə'ləm·bə,dē }

Columbiformes [VERT ZOO] An order of birds distinguished by a short, pointed bill, imperforate nostrils, and short legs. { kə,ləm·bə'fòr,mēz }

columbite [MINERAL] $(Fe,Mn)(Cb,Ta)_2O_6$ An iron-black mineral with a submetallic luster that crystallizes in the orthorhombic system; the chief ore mineral of niobium (columbium); hardness is 6 on Mohs scale, and specific gravity is 5.4–6.5. Also known as dianite; greenlandite; niobite. { kə'ləm,bīt }

columbium See niobium. { kə'ləm·bē·əm }

columella [ANAT] See stapes. [BIOL] Any part shaped like a column. [BOT] A sterile axial body within the capsules of certain mosses, liverworts, and many fungi. { ,käl·yə'mel·ə }

column [ANALY CHEM] In chromatography, a tube holding the stationary phase through which the mobile phase is passed. [CHEM ENG] See tower. [COMPUT SCI] A vertical arrangement of characters or other expressions, usually referring to a specific print position on a printer. [ENG] A vertical shaft designed to bear axial loads in compression. [GEOL] See geologic column; stalacto-stalagmite. [MATH] See place. [NUCLEO] A hollow cylinder of water and spray thrown up from an underwater burst of an atomic weapon, through which hot, high-pressure gases are vented to the atmosphere; a somewhat similar column of dirt is formed in an underground explosion. Also known as plume. { 'käl·əm }

columnar epithelium [HISTOL] Epithelium distinguished by elongated, columnar, or prismatic cells. { kə'ləm·nər ep·ə'thēl·ē·əm }

columnar ionization [PHYS] Ionization of atoms in a region confined to one or more paths of very small cross-sectional area. { kə'ləm·nər ,ī·ə·nə'zā·shən }

columnar jointing [GEOL] Parallel, prismatic columns that are formed as a result of contraction during cooling in basaltic flow and other extrusive and intrusive rocks. Also known as columnar structure; prismatic jointing; prismatic structure. { kə'ləm·nər 'jóint·iŋ }

columnar resistance [GEOPHYS] The electrical resistance of a column of air 1 centimeter square, extending from the earth's surface to some specified altitude. { kə'ləm·nər ri'zis·təns }

columnar section [GEOL] A vertical strip or scale drawing of the strip taken from a given area or locality showing the sequence of the rock units and their stratigraphic relationship, and indicating the thickness, lithology, age, classification, and fossil content of the rock units. Also known as section. { kə'ləm·nər 'sek·shən }

columnar stem [BOT] An unbranched, cylindrical stem bearing a set of large leaves at its summit, as in palms, or no leaves, as in cacti. { kə'ləm·nər ,stem }

columnar structure [GEOL] See columnar jointing. [MINERAL] Mineral structure consisting of parallel columns of slender prismatic crystals. [PETR] A primary sedimentary structure consisting of columns arranged perpendicular to the bedding. { kə'ləm·nər ,strək·chər }

column bleed [ANALY CHEM] The loss of carrier liquid during gas chromatography due to evaporation into the gas under analysis. { 'käl·əm ,blēd }

column chromatography [ANALY CHEM] Chromatographic technique of two general types: packed columns usually contain either a granular adsorbent or a granular support material coated with a thin layer of high-boiling solvent (partitioning liquid); open-tubular columns contain a thin film of partitioning liquid on the column walls and have an opening so that gas can pass through the center of the column. { 'käl·əm ,krō·mə'täg·rə·fē }

column crane [MECH ENG] A jib crane whose boom pivots about a post attached to a building column. { 'käl·əm ,krān }

column development chromatography [ANALY CHEM] Columnar apparatus for separating or concentrating one or more components from a physical mixture by use of adsorbent packing; as the specimen percolates along the length of the adsorbent, its various components are preferentially held at different rates, effecting a separation. { 'käl·əm də'vel·əp·mənt ,krō·mə'täg·rə·fē }

column drill [MECH ENG] A tunnel rock drill supported by a vertical steel column. { 'käl·əm ,dril }

column formation See trail formation. { 'käl·əm fòr'mā·shən }

column matrix See column vector. { 'käl·əm ,mā·triks }

column operations [MATH] A set of rules for manipulating the columns of a matrix so that the image of the corresponding linear transformation remains unchanged. { 'käl·əm ,äp·ə'rā·shənz }

column order [COMPUT SCI] The storage of a matrix $a(m,n)$ as $a(1,1), a(2,1), \ldots, a(m,1), a(1,2), \ldots$. { 'käl·əm ,òr·dər }

column pipe [MIN ENG] The large cast-iron (or wooden) pipe through which the water is conveyed from the mine pumps to the surface. { 'käl·əm ,pīp }

column printer [COMPUT SCI] A small line printer used with some calculators to provide hard-copy printout of input and output data; typically consists of 20 columns of numerals and a limited number of alphabetic or other identifying characters. { 'käl·əm ,print·ər }

column rank [MATH] The number of linearly independent columns of a matrix; the dimension of the image of the corresponding linear transformation. { 'käl·əm ,raŋk }

column space [MATH] The vector space spanned by the columns of a matrix. { 'käl·əm ,spās }

column splice [CIV ENG] A connection between two lengths of a compression member (column); an erection device rather than a stress-carrying element. { 'käl·əm ,splīs }

column vector [MATH] A matrix consisting of only one column. Also known as column matrix. { 'käl·əm ,vek·tər }

colure [ASTRON] A great circle of the celestial sphere through the celestial poles and either the equinoxes or solstices,

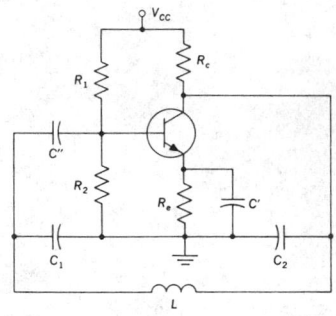

COLPITTS OSCILLATOR

A transistor Colpitts oscillator.

called respectively the equinoctial colure or the solstitial colure. { kə'lūr }

colusite [MINERAL] $Cu_3(As,Sn,V,Fe,Te)S_4$ A bronze-colored mineral consisting of a sulfide of copper and arsenic with vanadium, iron, and telluride substituting for arsenic; usually occurs in massive form. { kə'lü,sīt }

Colydiidae [INV ZOO] The cylindrical bark beetles, a large family of coleopteran insects in the superfamily Cucujoidea. { käl·ə'dī·ə,dē }

colza oil See rape oil. { 'käl·zə ,oil }

Com See Coma Berenices.

COM See computer output on microfilm.

coma [ASTRON] The gaseous envelope that surrounds the nucleus of a comet. Also known as head. [ELECTR] A cathode-ray tube image defect that makes the spot on the screen appear comet-shaped when away from the center of the screen. [MED] Unconsciousness from which the patient cannot be aroused. [OPTICS] A manifestation of errors in an optical system, so that a point has an asymmetrical image (that is, appears as a pear-shaped spot). { 'kō·mə }

Coma Berenices [ASTRON] A constellation, right ascension 13 hours, declination 20°N. Abbreviated Com. Also known as Berenice's Hair. { 'kō·mə ,ber·ə'nī·sēz }

Coma cluster [ASTRON] **1.** A group of over 1000 bright galaxies having a recession velocity of about 4300 miles (6900 kilometers) per second. **2.** An open cluster of about 100 stars at a distance of about 80 parsecs (1.5×10^{15} miles or 2.5×10^{15} kilometers). { 'kō·mə ,kləs·tər }

comagmatic province See petrographic province. { ‖kō·mag-'mad·ik 'prä·vəns }

coma lobe [ELECTROMAG] Side lobe that occurs in the radiation pattern of a microwave antenna when the reflector alone is tilted back and forth to sweep the beam through space because the feed is no longer always at the center of the reflector; used to eliminate the need for a rotary joint in the feed waveguide. { 'kō·mə ,lōb }

Comanchean [GEOL] A North American provincial series in Lower and Upper Cretaceous geologic time, above Coahuilan and below Gulfian. { kə'man·chē·ən }

Comasteridae [INV ZOO] A family of radially symmetrical Crinozoa in the order Comatulida. { ‖kō·mə'ster·ə,dē }

Coma supercluster [ASTRON] A supercluster that is centered on the Coma cluster of galaxies and has several extensions, including the Great Wall. { 'kō·mə 'sü·pər,kləs·tər }

comatic circle [OPTICS] A circle formed in the focal plane by rays from an off-axis point passing through a given zone of a lens that displays coma. { kō‖mad·ik 'sər·kəl }

comatose [MED] In a condition of coma; resembling coma. { 'käm·ə,tōs }

Comatulida [INV ZOO] The feather stars, an order of free-living echinoderms in the subclass Articulata. { ,kō·mə'tül·ə·də }

comb [DES ENG] See drag. [INV ZOO] **1.** A system of hexagonal cells constructed of beeswax by a colony of bees. **2.** A comblike swimming plate in ctenophores. [VERT ZOO] A crest of naked tissue on the head of many male fowl. { kōm }

comb antenna [ELECTROMAG] A broad-band antenna for vertically polarized signals, in which half of a fishbone antenna is erected vertically and fed against ground by a coaxial line. { 'kōm an,ten·ə }

combat analysis [ORD] A theoretical analysis conducted to determine the probable effectiveness of an existing or projected weapons system under combat conditions. { 'käm,bat ə'nal·ə·səs }

combatant ship [NAV ARCH] A ship whose main function is combat with an enemy. { kəm'bat·ənt ,ship }

combat chart [MAP] A chart overprinted with a system of grid lines as an aid in fire control. { 'käm,bat ,chärt }

combat development [ORD] The research, development, and testing of new doctrine and organization, and their integration with new materiel into the army services to obtain the greatest combat effectiveness. { 'käm,bat də'vel·əp·mənt }

combat development field experiment [ORD] A field trial usually under controlled conditions designed to collect data on operations, organizations, or materiel for use in preparing new or modified operational or organizational concepts or qualitative materiel requirements. { 'käm,bat də'vel·əp·mənt 'fēld ,ek'sper·ə·mənt }

combat development study project [ORD] A study directed toward a determination of operational concepts and techniques, new organizations, or qualitative materiel requirements for the army in the field, or a study which contributes to such determination. { 'käm,bat də'vel·əp·mənt 'stəd·ē ,prä·jekt }

combat development test project [ORD] A troop test directed at the determination of the feasibility and desirability of operational or organizational concepts, the suitability of materiel, or a combination of these tasks. { 'käm,bat də'vel·əp·mənt 'test ,prä·jekt }

combat information center [COMMUN] A shipboard location at which tactical information from radar, sonar, and other equipment is received, displayed for rapid analysis, and evaluated. { 'käm,bat in·fər'mā·shən ,sen·tər }

combat loading [ORD] The division of military cargo into categories for loading aboard ships. { 'käm,bat ,lōd·iŋ }

combat serviceable item [ORD] An item of equipment which is complete, ready to perform immediately at its rated capacity, and will remain serviceable under severe operating conditions for a reasonable length of time. { ‖käm,bat ‖sər·və·sə·bəl ,īd·əm }

combat tire [ORD] Pneumatic tire of heavy construction which is designed to operate without air pressure for a limited distance in an emergency. { 'käm,bat ,tīr }

combat vehicle [ORD] A land or amphibious vehicle, with or without armor or armament, designed for specific functions in combat or battle. { 'käm,bat ,vē·ə·kəl }

combat zone [ORD] A region in a theater of operations where fighting takes place, or where space is designated for the operations of friendly combat forces, extending from the front line to a line or boundary designated by the theater commander. { 'käm,bat ,zōn }

combed cotton [TEXT] Cotton yarn that has been cleaned with wire brushes and roller cards after carding to remove short fibers and other impurities. { ‖kōmd 'kät·ən }

comber [OCEANOGR] A deep-water wave of long, curling character with a high, breaking crest pushed forward by a strong wind. { 'kōm·ər }

Combescure transformation [MATH] A one-to-one continuous mapping of one space curve onto another space curve so that tangents to corresponding points are parallel. { 'kōm·bes,kyùr tranz·fər'mā·shən }

comb filter [ELECTR] A wave filter whose frequency spectrum consists of a number of equispaced elements resembling the teeth of a comb. { 'kōm ,fil·tər }

comb growth unit [BIOL] A unit for the standardization of male sex hormones. { 'kōm ,grōth ,yü·nət }

combination [MATH] A selection of one or more of the elements of a given set without regard to order. { ,käm·bə'nā·shən }

combinational circuit [ELECTR] A switching circuit whose outputs are determined only by the concurrent inputs. { ,käm·bə'nā·shən·əl 'sər·kət }

combination buoy [NAV] A buoy in which a light and a sound signal are combined such as a lighted bell buoy, lighted gong buoy, lighted whistle buoy, or lighted horn buoy. { ,käm·bə'nā·shən ,bói }

combination cable [ELEC] A cable having conductors grouped in both quads and pairs. { ,käm·bə'nā·shən ,kā·bəl }

combination chuck [DES ENG] A chuck used in a lathe whose jaws either move independently or simultaneously. { ,käm·bə'nā·shən ‖chək }

combination coefficient [GEOPHYS] A measure of the specific rate of disappearance of small ions in the atmosphere due to either union with neutral Aitken nuclei to form new large ions, or union with large ions of opposite sign to form neutral Aitken nuclei. { ,käm·bə'nā·shən ‖kō·i'fish·ənt }

combination collar [DES ENG] A collar that has left-hand threads at one end and right-hand threads at the other. { ,käm·bə'nā·shən ‖käl·ər }

combination cycle See mixed cycle. { ,käm·bə'nā·shən ‖sik·əl }

combination die [MET] A die having more than one cavity for different castings. { ,käm·bə'nā·shən ‖dī }

combination distributing frame [ELEC] Frame which combines the functions of a main distributing frame and an intermediate distributing frame. { ,käm·bə'nā·shən dis'trib·yəd·iŋ ,frām }

combination-drive reservoir [PETRO ENG] A type of reservoir in which hydrocarbons are swept (displaced) toward the drill hole by injection of water followed by liquefied-petroleum-gas or gas injection. { ¦käm·bə'nā·shən ¦drīv 'res·əv‚wär }

combination fuse [ORD] A fuse combining two different types of fuse mechanisms, especially one combining impact and time mechanisms. { ¦käm·bə'nā·shən ¦fyüz }

combination lock [ENG] A lock that can be opened only when its dial has been set to the proper combination of symbols, in the proper sequence. { ¦käm·bə'nā·shən ¦läk }

combination mill [MET] A rolling mill arranged with continuous rolls for roughing and a guide or looping mill for shaping. { ¦käm·bə'nā·shən ¦mil }

combination plate [GRAPHICS] A photoengraved plate on which halftone and line work have been combined. { ¦käm·bə'nā·shən ¦plāt }

combination pliers [DES ENG] Pliers that can be used either for holding objects or for cutting and bending wire. { ¦käm·bə'nā·shən 'plī·ərz }

combination principle See Ritz's combination principle. { ¦käm·bə'nā·shən ‚prin·sə·pəl }

combination reaction [CHEM] A chemical reaction in which two reactions combine to form a single product. { ¦käm·bə'nā·shən rē‚ak·shən }

combination rig [PETRO ENG] A drilling rig that is fitted with components for both rotary and cable-tool drilling. { ¦käm·bə'nā·shən 'rig }

combination saw [MECH ENG] A saw made in various tooth arrangement combinations suitable for ripping and crosscut mitering. { ¦käm·bə'nā·shən ¦só }

combination spectrum [ASTRON] The composite spectrum characteristic of a symbiotic star. { ¦käm·bə'nā·shən 'spek·trəm }

combination square [DES ENG] A square head and steel rule that when used together have both a 45° and 90° face to allow the testing of the accuracy of two surfaces intended to have these angles. { ¦käm·bə'nā·shən ¦skwer }

combination string [PETRO ENG] A casing string designed to be used at different depths in a given well. { ¦käm·bə'nā·shən 'striŋ }

combination tone [ACOUS] A subjective tone produced by simultaneously sounding two pure tones whose frequencies differ by a large amount. { ¦käm·bə'nā·shən 'tōn }

combination trap [GEOL] Underground reservoir structure closure, deformation, or fault where reservoir rock covers only part of the structure. { ¦käm·bə'nā·shən ¦trap }

combination unit [CHEM ENG] A processing unit that combines more than one process, such as straight-run distillation together with selective cracking. { ¦käm·bə'nā·shən 'yü·nət }

combination vibration [SPECT] A vibration of a polyatomic molecule involving the simultaneous excitation of two or more normal vibrations. { ¦käm·bə'nā·shən vī'brā·shən }

combination wrench [DES ENG] A wrench that is an open-end wrench at one end and a socket wrench at the other. { ¦käm·bə'nā·shən ¦rench }

combinatorial analysis [MATH] **1.** The determination of the number of possible outcomes in ideal games of chance by using formulas for computing numbers of combinations and permutations. **2.** The study of large finite problems. { kəm‚bī·nə'tór·ē·əl ə'nal·ə·səs }

combinatorial chemistry [ORG CHEM] A method for reacting a small number of chemicals to produce simultaneously a very large number of compounds, called libraries, which are screened to identify useful products such as drug candidates. { kəm‚bīn·ə‚tór·ē·əl 'kem·ə·strē }

combinatorial control [CELL MOL] Control of gene expression requiring presence or absence of a particular combination of regulatory proteins. { ¦käm·bə·nə¦tór·ē·əl kən'trōl }

combinatorial proof [MATH] A proof that uses combinatorial reasoning instead of calculation. { ¦käm·bə·nə‚tór·ē·l prüf }

combinatorial theory [MATH] The branch of mathematics which studies the arrangements of elements into sets. { kəm‚bī·nə'tór·ē·əl 'thē·ə·rē }

combinatorial topology [MATH] The study of polyhedrons, simplicial complexes, and generalizations of these. Also known as piecewise linear topology. { kəm‚bī·nə'tór·ē·əl tə'päl·ə·jē }

combinatorics [MATH] Combinatorial topology which studies geometric forms by breaking them into simple geometric figures. { ¦käm·bə·nə'tór·iks }

combined carbon [CHEM] Carbon that is chemically combined within a compound, as contrasted with free or uncombined elemental carbon. { kəm'bīnd 'kär·bən }

combined cyanide [ORG CHEM] The cyanide portion of a complex ion composed of cyanide and a metal. { kəm'bīnd 'sī·ə‚nīd }

combined flexure [MECH] The flexure of a beam under a combination of transverse and longitudinal loads. { kəm'bīnd 'flek·shər }

combined footing [CIV ENG] A footing, either rectangular or trapezoidal, that supports two columns. { kəm'bīnd 'füd·iŋ }

combined head See read/write head. { kəm'bīnd 'hed }

combined immunological deficiency disease [MED] A severe and usually fatal disease in which the individual lacks not only the T (thymus-derived) cells, which are responsible for graft rejection and for defense against viruses, but also the B (bone-marrow-derived) cells, which are responsible for production of globulins and antibodies. Also known as CID disease. { kəm'bīnd ‚im·yə·nə'läj·ə·kəl də'fish·ən·sē di‚zēz }

combined moisture [MIN ENG] Moisture in coal that cannot be removed by ordinary drying. { kəm'bīnd 'mois·chər }

combined sewers [CIV ENG] A drainage system that receives both surface runoff and sewage. { kəm'bīnd 'sü·ərz }

combined stresses [MECH] Bending or twisting stresses in a structural member combined with direct tension or compression. { kəm'bīnd 'stres·əz }

combined water [GEOCHEM] Water attached to soil minerals by means of chemical bonds. { kəm'bīnd 'wòd·ər }

combine-harvester [AGR] A machine for harvesting legumes and grasses; cuts off the plant tops, then beats and cleans the grain while the machine continues to move across the field. { 'käm‚bīn 'här·və·stər }

combiner circuit [ELECTR] The circuit that combines the luminance and chrominance signals with the synchronizing signals in a color television camera chain. { kəm'bīn·ər ‚sər·kət }

combing [BUILD] In roofing, the topmost row of shingles which project above the ridge line. [ENG] **1.** Using a comb or stiff bristle brush to create a pattern by pulling through freshly applied paint. **2.** Scraping or smoothing a soft stone surface. [TEXT] Elimination of short, curly fibers by means of a machine process. { 'kōm·iŋ }

combing machine [TEXT] A machine that combs fibers, separating the longer ones from the shorter. { 'kōm·iŋ mə'shēn }

combining glass [OPTICS] A glass screen designed to reflect display imagery to the viewer, usually at selected wavelengths of light, while being sufficiently transmissive for the viewer to see the scene beyond. { kəm'bīn·iŋ ‚glas }

combining network [COMPUT SCI] A switching system for accessing memory modules in a multiprocessor, in which each switch remembers the memory addresses it has used, and can then satisfy several requests with a single memory access. { kəm'bīn·iŋ 'net‚wərk }

combining-volumes principle [CHEM] The principle that when gases take part in chemical reactions the volumes of the reacting gases and those of the products (if gaseous) are in the ratio of small whole numbers, provided that all measurements are made at the same temperature and pressure. Also known as Gay-Lussac's law of volumes. { kəm¦bīn·iŋ ¦väl·yəmz ‚prin·sə·pəl }

combining weight [CHEM] The weight of an element that chemically combines with 8 grams of oxygen or its equivalent. { kəm'bīn·iŋ ‚wāt }

comb nephoscope [ENG] A direct-vision nephoscope constructed with a comb (a crosspiece containing equispaced vertical rods) attached to the end of a column 8–10 feet (2.4–3 meters) long and supported on a mounting that is free to rotate about its vertical axis; in use, the comb is turned so that the cloud appears to move parallel to the tips of the vertical rods. { ¦kōm ¦nef·ə‚skōp }

combplate [MECH ENG] The toothed portion of the stationary threshold plate that is set into both ends of an escalator or

---

COMBINED FOOTING

trapezoidal footing for unequal loads

◻ load   load ◻

rectangular footing for equal loads

Rectangular and trapezoidal shapes of combined footing.

moving sidewalk and meshes with the grooved surface of the moving steps or treadway. { 'kōm,plāt }

comb polymer [ORG CHEM] A macromolecule in which the main chain has one long branch per repeat unit. { ¦kōm 'päl·ə·mər }

combustible gas [MATER] A gas that burns, including the fuel gases, hydrogen, hydrocarbon, carbon monoxide, or a mixture of these. { kəm'bəs·tə·bəl ¦gas }

combustible loss [ENG] Thermal loss resulting from incomplete combustion of fuel. { kəm'bəs·tə·bəl ¦lös }

combustible shale See tasmanite. { kəm'bəs·tə·bəl ¦shāl }

combustion [CHEM] The burning of gas, liquid, or solid, in which the fuel is oxidized, evolving heat and often light. { kəm'bəs·chən }

combustion chamber [AERO ENG] That part of the rocket engine in which the combustion of propellants takes place at high pressure. Also known as firing chamber. [ENG] Any chamber in which a fuel such as oil, coal, or kerosine is burned to provide heat. [MECH ENG] The space at the head end of an internal combustion engine cylinder where most of the combustion takes place. { kəm'bəs·chən ,chām·bər }

combustion-chamber volume [MECH ENG] The volume of the combustion chamber when the piston is at top dead center. { kəm'bəs·chən ,chām·bər ,väl·yəm }

combustion deposit [ENG] A layer of ash on the heat-exchange surfaces of a combustion chamber, resulting from burning of a fuel. { kəm'bəs·chən də'päz·ət }

combustion efficiency [CHEM] The ratio of heat actually developed in a combustion process to the heat that would be released if the combustion were perfect. { kəm'bəs·chən i'fish·ən·sē }

combustion engine [MECH ENG] An engine that operates by the energy of combustion of a fuel. { kəm'bəs·chən ,en·jən }

combustion engineering [MECH ENG] The design of combustion furnaces for a given performance and thermal efficiency, involving study of the heat liberated in the combustion process, the amount of heat absorbed by heat elements, and heat-transfer rates. { kəm'bəs·chən en·jə'nir·iŋ }

combustion furnace [ANALY CHEM] A heating device used in the analysis of organic compounds for elements. [ENG] A furnace whose source of heat is the energy released in the oxidation of fossil fuel. { kəm'bəs·chən ,fər·nəs }

combustion instability [AERO ENG] Unsteadiness or abnormality in the combustion of fuel, as may occur in a rocket engine. { kəm'bəs·chən ,in·stə'bil·əd·ē }

combustion knock See engine knock. { kəm'bəs·chən ,näk }

combustion nucleus [METEOROL] A condensation nucleus formed as a result of industrial or natural combustion processes. { kəm'bəs·chən ,nü·klē·əs }

combustion rate [CHEM] The rate of burning of any substance. { kəm'bəs·chən ,rāt }

combustion shock [ENG] Shock resulting from abnormal burning of fuel in an internal combustion engine, caused by preignition or fuel-air detonation; or in a diesel engine, the uncontrolled burning of fuel accumulated in the combustion chamber. { kəm'bəs·chən ,shäk }

combustion train [ANALY CHEM] The arrangement of apparatus for elementary organic analysis. { kəm'bəs·chən ,trān }

combustion tube [ANALY CHEM] A glass, silica, or porcelain tube, resistant to high temperatures, that is a component of a combustion train. { kəm'bəs·chən ,tüb }

combustion turbine See gas turbine. { kəm'bəs·chən 'tər,bīn }

combustion wave [CHEM] **1.** A zone of burning propagated through a combustible medium. **2.** The zoned, reacting, gaseous material formed when an explosive mixture is ignited. { kəm'bəs·chən ,wāv }

combustor [MECH ENG] The combustion chamber together with burners, igniters, and injection devices in a gas turbine or jet engine. { kəm'bəs·tər }

come-along [DES ENG] **1.** A device for gripping and effectively shortening a length of cable, wire rope, or chain by means of two jaws which close when one pulls on a ring. **2.** See puller. { 'kəm ə,lȯŋ }

comedo [MED] A collection of sebaceous material and keratin retained in the hair follicle and excretory duct of the sebaceous gland, whose surface is covered with a black dot caused by oxidation of sebum at the follicular orifice. Also known as blackhead. { 'kä·mə,dō }

comedocarcinoma [MED] A type of adenocarcinoma of the breast in which the ducts are filled with cells which, when expressed from the cut surface, resemble comedos. { 'kä·mə·dō,kärs·ən'ō·mə }

comendite [GEOL] A white, sodic rhyolite containing alkalic amphibole or pyroxene. { kə'men,dīt }

comes [ASTRON] The smaller star in a binary system. Also known as companion. { 'kō,mēz }

Comesomatidae [INV ZOO] A family of free-living nematodes in the superfamily Comesomatoidea found as deposit feeders on soft bottom sediments. { ¦kō·mə·sō'mad·ə,dē }

Comesomatoidea [INV ZOO] A superfamily of marine nematodes in the order Chromadorida distinguished by their wide multispiral amphids that make at least two complete turns. { ¦kō·mə·sō·mə'tȯid·ē·ə }

comet [ASTRON] A nebulous celestial body having a fuzzy head surrounding a bright nucleus, one of two major types of bodies moving in closed orbits about the sun; in comparison with the planets, the comets are characterized by their more eccentric orbits and greater range of inclination to the ecliptic. { 'käm·ət }

cometary kilometric radiation [ASTRON] Radio waves detected by space probes encountering Comet Halley, consisting of several different emission patterns ranging from intense, sporadic bursts of broad-band noise to continuously rising and falling tones. { 'käm·ə,ter·ē ¦kil·ə¦me·trik ,rād·ē'ā·shən }

cometary nebula [ASTRON] A fan-shaped reflection nebula that resembles a comet in appearance. { 'käm·ə,ter·ē 'neb·yə·lə }

Comet Biela See Biela Comet. { ¦käm·ət 'byel·ə }

comet family [ASTRON] Those short-period comets whose aphelia correspond closely to Jupiter's orbit. { 'kam·ət ,fam·lē }

comet group [ASTRON] A division of comets on the basis of their period; the short-period group (periods of less than 200 years) contains orbits that show a strong preference for the plan of the solar system; the planes of long-period comets are randomly distributed. { 'käm·ət ,grüp }

Comet Hale-Bopp [ASTRON] A very large comet which was discovered on July 23, 1995, and reached perihelion on April 1, 1997, when its brightness was magnitude −1. { ¦käm·ət ¦hāl 'bäp }

Comet Halley See Halley's Comet. { ¦käm·ət 'ha·lē }

Comet Hyakutake [ASTRON] A comet that passed within about 0.1 astronomical unit of earth in late March 1996. { ¦käm·ət ,hyä·kü'tä·ke }

Comet Kohoutek See Kohoutek's Comet. { ¦käm·ət kə'hō,tek }

Comet Shoemaker-Levy [ASTRON] A comet that was torn apart by an encounter with Jupiter in 1992, and whose pieces collided with Jupiter over a period of a few days around July 22, 1994. { ¦käm·ət ,shü,māk·ər 'lē·vē }

cometabolism [ECOL] A process in which compounds not utilized for growth or energy are transformed to other products by microorganisms. { ,kō·mə'tab·ə,liz·əm }

comfort chart [ENG] A diagram showing curves of relative humidity and effective temperature superimposed upon rectangular coordinates of wet-bulb temperature and dry-bulb temperature. { 'kəm·fərt ,chärt }

comfort control [ENG] Control of temperature, humidity, flow, and composition of air by using heating and air-conditioning systems, ventilators, or other systems to increase the comfort of people in an enclosure. { 'kəm·fərt kən'trōl }

comfort curve [ENG] A line drawn on a graph of air temperature versus some function of humidity (usually wet-bulb temperature or relative humidity) to show the varying conditions under which the average sedentary person feels the same degree of comfort; a curve of constant comfort. { 'kəm·fərt ,kərv }

comfort index See temperature-humidity index. { 'kəm·fərt ,in,deks }

comfort standard See comfort zone. { 'kəm·fərt ,stan·dərd }

comfort temperature [MECH ENG] Any one of the indexes in which air temperatures have been adjusted to represent human comfort or discomfort under prevailing conditions of temperature, humidity, radiation, and wind. { 'kəm·fərt ,tem·prə·chər }

comfort zone [ENG] The ranges of indoor temperature, humidity, and air movement, under which most persons enjoy

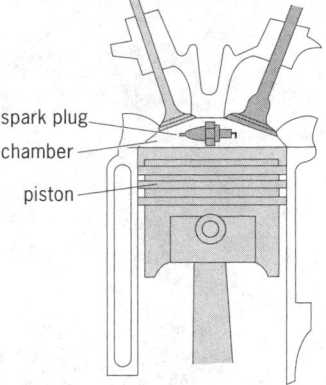
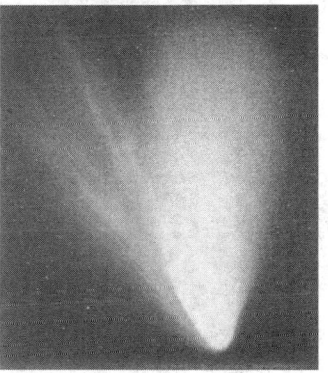
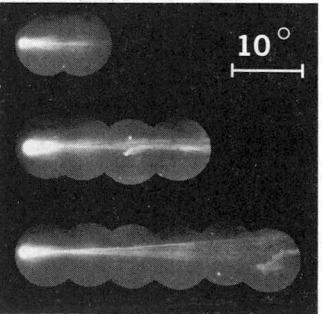

mental and physical well-being. Also known as comfort standard. { 'kəm·fərt ‚zōn }

comic-strip oriented image [GRAPHICS] A microfilm image whose top and bottom are parallel to the outer, long edge of the film. { ¦käm·ik ‚strip ‚ór·ē‚ent·əd 'im·ij }

COMIT [COMPUT SCI] A user-oriented, general-purpose, symbol-manipulation programming language for computers. { 'kō‚mit }

Comleyan [GEOL] Lower Cambrian geologic time. { 'käm·lā·ən }

comma [ACOUS] The difference between the larger and smaller whole tones in the just scale, corresponding to a frequency ratio of 81/80. { 'käm·ə }

command [COMPUT SCI] A signal that initiates a predetermined type of computer operation that is defined by an instruction. [CONT SYS] An independent signal in a feedback control system, from which the dependent signals are controlled in a predetermined manner. { kə'mand }

command and control [SYS ENG] The process of military commanders and civilian managers identifying, prioritizing, and achieving strategic and tactical objectives by exercising authority and direction over human and material resources by utilizing a variety of computer-based and computer-controlled systems, many driven by decision-theoretic methods, tools, and techniques. Abbreviated C². { kə'mand ən kən'trōl }

command button [COMPUT SCI] A small rectangle on a graphical user interface with a command, such as open, close, OK, or print, that is immediately activated upon selection of the button. { kə'mand ‚bət·ən }

command code See operation code. { kə'mand ‚kōd }

command control See command guidance. { kə'mand kən‚trōl }

command, control, and communications [SYS ENG] A version of command and control in which the role of communications equipment is emphasized. Abbreviated C³. { kə'mand kən'trōl ən kə‚myü·ne'kā·shənz }

command, control, communications, and intelligence [SYS ENG] A version of command and control in which the roles of communications equipment and intelligence are emphasized. Abbreviated C³I. { kə'mand kən'trōl kə‚myü·nə'kā·shənz ən in'tel·ə·jəns }

command control program [COMPUT SCI] The interface between a time-sharing computer and its users by means of which they can create, edit, save, delete, and execute their programs. { kə'mand kən‚trōl ‚prō·gram }

command destruct [CONT SYS] A command control system that destroys a flightborne test rocket or a guided missile, actuated by the safety officer whenever the vehicle's performance indicates a safety hazard. { kə'mand di'strəkt }

command-driven program [COMPUT SCI] A computer program that accepts command words and statements typed in by the user. { kə'mand ‚driv·ən 'prō·gram }

command guidance [AERO ENG] The guidance of a missile, rocket, or spacecraft by means of electronic signals sent to receiving devices in the vehicle. [ENG] A type of electronic guidance of guided missiles or other guided aircraft wherein signals or pulses sent out by an operator cause the guided object to fly a directed path. Also known as command control. { kə'mand ‚gīd·əns }

command interpreter [COMPUT SCI] A program that processes commands and other input and output from an active terminal in a time-sharing system. { kə'mand ‚in'tər·prə·tər }

command language [COMPUT SCI] The language of an operating system, through which the users of a data-processing system describe the requirements of their tasks to that system. Also known as job control language. { kə'mand ‚laŋ·gwij }

command level [COMPUT SCI] The ability to control a computer's operating system through the use of commands, normally available only to computer operators. { kə'mand ‚lev·əl }

command line [COMPUT SCI] On a display screen, the space following a prompt (such as $) where a text instruction to a computer or device is typed. { kə'mand ‚līn }

command list See CLIST. { kə'mand ‚list }

command mode [COMPUT SCI] The status of a terminal in a time-sharing environment enabling the programmer to use the command control program. { kə'mand ‚mōd }

command module [AERO ENG] The spacecraft module that carries the crew, the main communication and telemetry equipment, and the reentry capsule during cruising flight. { kə'mand ‚maj·ül }

command net [ORD] The several points that may exercise control over airborne guided missiles. { kə'mand ‚net }

command processor [COMPUT SCI] A computer program that converts a limited number of user commands into the machine commands that direct the operating system. Also known as command shell. { kə'mand 'prä‚ses·ər }

command pulses [ELECTR] The electrical representations of bit values of 1 or 0 which control input/output devices. { kə'mand ‚pəl·səs }

command set [COMMUN] A radio set used to receive or give commands, as between one aircraft and another or between an aircraft and the ground. { kə'mand ‚set }

command shell See command processor. { kə'mand ‚shel }

Commelinaceae [BOT] A family of monocotyledonous plants in the order Commelinales characterized by differentiation of the leaves into a closed sheath and a well-defined, commonly somewhat succulent blade. { ‚kä·mə·lə'nās·ē‚ē }

Commelinales [BOT] An order of monocotyledonous plants in the subclass Commelinidae marked by having differentiated sepals and petals but lacking nectaries and nectar. { ‚kä·mə·lə'nā·lēz }

Commelinidae [BOT] A subclass of flowering plants in the class Liliopsida. { ‚kä·mə'li·nə‚dē }

commensal [ECOL] An organism living in a state of commensalism. { kə'men·səl }

commensalism [ECOL] An interspecific, symbiotic relationship in which two different species are associated, wherein one is benefited and the other neither benefited nor harmed. { kə'men·sə‚liz·əm }

commensurable motions [ASTRON] Mean motions of the planets, or of the satellites of a planet, which satisfy simple arithmetic relationships. { kə'mens·rə·bəl 'mō·shənz }

commensurate orbits [ASTRON] Orbits of two celestial objects about a common center of gravity so that the period of one is a rational fraction of that of the other. { kə'mench·ə·rət 'or·bəts }

comment [COMPUT SCI] An expression identifying or explaining one or more steps in a routine, which has no effect on execution of the routine. { 'käm‚ent }

comment code [COMPUT SCI] One or more characters identifying a comment. { 'käm‚ent ‚kōd }

comment out [COMPUT SCI] To render a statement in a computer program inactive by making it a comment. { 'kä‚ment 'aut }

commercial diesel cycle See mixed cycle. { kə'mər·shəl 'dē·zəl ‚sī·kəl }

commercial harbor [CIV ENG] A harbor in which docks are provided with cargo-handling facilities. { kə'mər·shəl 'här·bər }

commercial lecithin See lecithin. { kə'mər·shəl 'les·ə·thən }

commercial mine [MIN ENG] A coal mine operated to supply purchasers in general, as contrasted with a captive mine. { kə'mər·shəl ¦mīn }

commercial ore [MIN ENG] Mineralized material profitable at prevailing metal prices. { kə'mər·shəl ¦ór }

commercial-type vehicle [ORD] A motor vehicle designed to meet civilian requirements and used by the armed services, without major modification, for routine purposes in connection with the transportation of supplies, personnel, and equipment. { kə'mər·shəl ‚tīp ‚vē·ə·kəl }

commingled yarn [TEXT] A yarn in which two types of materials are intermingled to form a single yarn. { ¦kō‚miŋ·gəld ‚yärn }

commingling [PETRO ENG] Blending of petroleum products with similar properties; usually performed to facilitate transportation in a pipeline. { kō'miŋ·gliŋ }

comminution [MECH ENG] Breaking up or grinding into small fragments. Also known as pulverization. { ‚käm·ə'nü·shən }

comminutor [MECH ENG] A machine that breaks up solids. { 'käm·ə‚nüd·ər }

commissioned [NAV] Pertaining to an aid to navigation which, after it has been constructed, tested, and placed in operation, is declared suitable for use by vessels or crafts. { kə'mish·ənd }

commission ore [MIN ENG] Uranium-bearing ore of 0.10%

U_3O_8 or higher, for which the U.S. Atomic Energy Commission has an established price. { kə'mish·ən ,ȯr }

commissure [BIOL] A joint, seam, or closure line where two structures unite. { 'käm·ə,shu̇r }

commissurotomy [MED] The surgical destruction of a commissure, usually the anterior commissure, particularly in the treatment of certain psychiatric disorders. { ,käm·ə,shu̇'räd·ə·mē }

commitment [CYTOL] The establishment of a unique developmental sequence in a cell that differs from the prior state. { kə'mit·mənt }

committed [ORD] Pertaining to the condition of a fuse when the arming process has reached the point from which arming will continue to completion in spite of the absence of any arming forces. { kə'mid·əd }

common area [COMPUT SCI] An area of storage which two or more routines share. { ¦käm·ən 'er·ē·ə }

common-base connection See grounded-base connection. { ¦käm·ən 'bās kə'nek·shən }

common-base feedback oscillator [ELECTR] A bipolar transistor amplifier with a common-base connection and a positive feedback network between the collector (output) and the emitter (input). { ¦käm·ən 'bās 'fēd,bak ,äs·ə,lād·ər }

common battery [COMMUN] System of current supply where all direct current energy for a unit of a telephone system is supplied by one source in a central office or exchange. { ¦käm·ən 'bäd·ə·rē }

common bile duct [ANAT] The duct formed by the union of the hepatic and cystic ducts. { ¦käm·ən 'bīl ,dəkt }

common bond See American bond. { ¦käm·ən 'bänd }

common branch [ELEC] A branch of an electrical network which is common to two or more meshes. Also known as mutual branch. { ¦käm·ən 'branch }

common brick [MATER] Brick made from natural clay. { ¦käm·ən ¦brik }

common business-oriented language See COBOL. { ¦käm·ən ¦biz·nəs ¦ȯr·ē,ent·əd ,laŋ·gwij }

common carotid artery See carotid artery. { ¦käm·ən kə'räd·əd 'ärd·ə·rē }

common carriage See transmission access. { ¦käm·ən 'kar·ij }

common carrier [IND ENG] A company recognized by an appropriate regulatory agency as having a vested interest in furnishing communications services or in transporting commodities or people. { ¦käm·ən 'kar·ē·ər }

common cause [ANALY CHEM] A cause of variability in a measurement process that is inherent in and common to the process itself. { 'käm·ən 'kȯz }

common-channel interoffice signaling [COMMUN] A method of signaling in a telecommunications switching system in which a network of data communication paths separate from the communications transmission is used for transmitting all signaling information between offices. Abbreviated CCIS. { ¦käm·ən ¦chan·əl ,in·tər,ȯ·fəs 'sig·nəl·iŋ }

common cold [MED] A viral disease of humans most frequently caused by the rhinovirus and accompanied by inflammation of the mucous membranes of the nose, throat, and eyes. { ¦käm·ən 'kōld }

common-collector connection See grounded-collector connection. { ¦käm·ən kə'lek·tər kə'nek·shən }

common control unit [COMPUT SCI] Control unit that is shared by more than one machine. { ¦käm·ən kən'trōl ,yü·nət }

common declaration statement [COMPUT SCI] A nonexecutable statement in FORTRAN which allows specified arrays or variables to be stored in an area available to other programs. { ¦käm·ən ,dek·lə·lə'rā·shən ,stāt·mənt }

common denominator [MATH] Any common multiple of the denominators of a collection of fractions. { ¦käm·ən də'näm·ə,nād·ər }

common difference [MATH] The fixed difference between any term in an arithmetic progression and the preceding term. { ¦käm·ən 'dif·rəns }

common divisor [MATH] For a set of integers, an integer c such that each of the integers in the set is divisible by c. Also known as common factor. { ¦käm·ən di'vīz·ər }

common-drain amplifier [ELECTR] An amplifier using a field-effect transistor so that the input signal is injected between gate and drain, while the output is taken between the source and drain. Also known as source-follower amplifier. { ¦käm·ən 'drān 'am·plə,fī·ər }

common-emitter connection See grounded-emitter connection. { ¦käm·ən i'mid·ər kə'nek·shən }

common establishment See high-water full and change. { ¦käm·ən ə'stab·lish·mənt }

common factor [MATH] See common divisor { ¦käm·ən 'fak·tər }

common feldspar See orthoclase. { ¦käm·ən 'feld,spär }

common fraction [MATH] A fraction whose numerator and denominator are both integers. Also known as simple fraction; vulgar fraction. { ¦käm·ən 'frak·shən }

common-gate amplifier [ELECTR] An amplifier using a field-effect transistor in which the gate is common to both the input circuit and the output circuit. { ¦käm·ən 'gāt 'am·plə,fī·ər }

common gateway interface [COMPUT SCI] A protocol that allows the secure data transfer to and from a server and a network user by means of a program which resides on the server and handles the transaction. For example, if an intranet user sent a request with a Web browser for database information, a CGI program would execute on the server, retrieve the information from the database, format it in HTML, and send it back to the user. Abbreviated CGI. { ¦käm·ən ,gāt,wā 'in·tər,fās }

common hepatic duct See hepatic duct. { ¦käm·ən he'pad·ik 'dəkt }

common iliac artery See iliac artery. { ¦käm·ən ,il·ē,ak 'ärd·ə·rē }

common impedance coupling [ELECTROMAG] The interaction of two circuits by means of an inductance or capacitance in a branch which is common to both circuits. { ¦käm·ən im'ped·əns ,kəp·liŋ }

common-ion effect [CHEM] The lowering of the degree of ionization of a compound when another ionizable compound is added to a solution; the compound added has a common ion with the other compound. { ¦käm·ən ¦ī,än i'fekt }

common item [ORD] Piece of equipment or materiel of supply, used or procured by more than one technical service. { ¦käm·ən ¦ī·dəm }

common joist [BUILD] An ordinary floor beam to which floor boards are attached. { ¦käm·ən ¦jȯist }

common labor [IND ENG] Unskilled workers. { ¦käm·ən ¦lā·bər }

common language [COMPUT SCI] A machine-readable language that is common to a group of computers and associated equipment. { ¦käm·ən ¦laŋ·gwij }

common language family [COMPUT SCI] Various types of business machines, capable of being operated automatically and interchangeably by the same piece of punched paper tape, are said to be of a common language family. { ¦käm·ən ¦laŋ·gwij ¦fam·lē }

common logarithm [MATH] The exponent in the representation of a number as a power of 10. Also known as Briggsian logarithm; Briggs' logarithm. { ¦käm·ən 'läg·ə,rith·əm }

common mica See muscovite. { ¦käm·ən 'mī·kə }

common mode [ELECTR] Having signals that are identical in amplitude and phase at both inputs, as in a differential operational amplifier. { ¦käm·ən ,mōd }

common-mode error [ELECTR] The error voltage that exists at the output terminals of an operational amplifier due to the common-mode voltage at the input. { ¦käm·ən ,mōd 'er·ər }

common-mode gain [ELECTR] The ratio of the output voltage of a differential amplifier to the common-mode input voltage. { ¦käm·ən ,mōd 'gān }

common-mode input capacitance [ELECTR] The equivalent capacitance of both inverting and noninverting inputs of an operational amplifier with respect to ground. { ¦käm·ən ,mōd 'in,pu̇t kə'pas·əd·əns }

common-mode input impedance [ELECTR] The open-loop input impedance of both inverting and noninverting inputs of an operational amplifier with respect to ground. { ¦käm·ən ,mōd 'in,pu̇t im'ped·əns }

common-mode input resistance [ELECTR] The equivalent resistance of both inverting and noninverting inputs of an operational amplifier with respect to ground or reference. { ¦käm·ən ,mōd 'in,pu̇t ri'zis·təns }

common-mode rejection [ELECTR] The ability of an amplifier to cancel a common-mode signal while responding to an out-of-phase signal. Also known as in-phase rejection. { ˈkäm·ən ˌmōd ri'jek·shən }

common-mode rejection ratio [ELECTR] The ratio of the gain of an amplifier for difference signals between the input terminals, to the gain for the average or common-mode signal component. Abbreviated CMRR. { ˈkäm·ən ˌmōd ri'jek·shən 'rā·shō }

common-mode signal [ELECTR] A signal applied equally to both ungrounded inputs of a balanced amplifier stage or other differential device. Also known as in-phase signal. { ˈkäm·ən ˌmōd 'sig·nal }

common-mode voltage [ELECTR] A voltage that appears in common at both input terminals of a device with respect to the output reference (usually ground). { ˈkäm·ən ˌmōd 'vōl·tij }

common multiple [MATH] A quantity (polynomial number) divisible by all quantities in a given set. { ˈkäm·ən 'məl·tə·pəl }

common object request broker [COMPUT SCI] A system that provides interoperability among objects in a heterogeneous, distributed, object-oriented environment in a way that is transparent to the programmer; its design is based on the OMG object model. Abbreviated CORBA. { ˈkäm·ən ˈäb·jekt ri'kwest ˌbrō·kər }

common projectile [ORD] A penetrating type of projectile containing a bursting charge of high explosive, intended to explode after passing through the lighter protective armor of a tank or naval vessel. { ˈkäm·ən prə'jek·təl }

common pyrite *See* pyrite. { ˈkäm·ən 'pī,rīt }

common rafter [BUILD] A rafter which extends from the plate of the roof to the ridge board at right angles to both members, and to which roofing is attached. { ˈkäm·ən 'raf·tər }

common-rail injection [MECH ENG] A type of diesel engine fuel-injection system in which one rail maintains the fuel at a specified pressure while feed lines run from the rail to each fuel injector. { 'käm·ən ˌrāl in'jek·shən }

common return [ELECTR] A return conductor that serves two or more circuits. { ˈkäm·ən ri'tərn }

common salt *See* halite; sodium chloride. { ˈkäm·ən 'sȯlt }

common-source amplifier [ELECTR] An amplifier stage using a field-effect transistor in which the input signal is applied between gate and source and the output signal is taken between drain and source. { ˈkäm·ən ˌsȯrs 'am·plə,fī·ər }

common storage [COMPUT SCI] A section of memory in certain computers reserved for temporary storage of program outputs to be used as input for other programs. { ˈkäm·ən 'stȯr·ij }

common tangent [MATH] A common tangent of two circles is a line that is tangent to both circles. { ˈkäm·ən 'tan·jənt }

common-user channel [COMMUN] Any of the communications channels which are available to all authorized agencies for transmission of command, administrative, and logistic traffic. { ˈkäm·ən ˌyü·zər ˌchan·əl }

common-user circuit [ELEC] A circuit designated to furnish a communications service to a number of users. { ˈkäm·ən ˌyü·zər ˌsər·kət }

common wall [BUILD] A wall that is shared by two dwelling units. { ˈkäm·ən ˈwȯl }

common year [ASTRON] A calendar year of 365 days. { ˈkäm·ən ˈyir }

communicable disease [MED] An infectious disease that can be transmitted from one individual to another either directly by contact or indirectly by fomites and vectors. { kə'myü·nə·kə·bəl di,zēz }

communicating hydrocephaly [MED] A form of hydrocephaly in which there is normal communication between the ventricles and the subarachnoid space of the brain. { kə'myü·nə,kād·iŋ ˌhī·drō'sef·ə·lē }

communicating junction *See* gap junction. { kə'myü·nə,kād·iŋ ˌjəŋk·shən }

communicating word processor [COMPUT SCI] A word processor that can be linked to other word processors to exchange information. { kə'myü·nə,kād·iŋ 'wȯrd ˌprä,ses·ər }

communication [COMMUN] The transmission of intelligence between two or more points over wires or by radio; the terms telecommunication and communication are often used interchangeably, but telecommunication is usually the preferred term when long distances are involved. { kə,myü·nə'kā·shən }

communication band [COMMUN] The band of frequencies effectively occupied by a radio transmitter for the type of transmission and the speed of signaling used. { kə,myü·nə'kā·shən ˌband }

communication bus [COMMUN] A device that transfers control, timing, and data signals between switching processor subsystems; designed to provide physical and electrical isolation, to provide for simple addition of units on an in-service basis, and to provide pluggable connection for efficient factory testing, installation, and maintenance. { kə,myü·nə'kā·shən ˌbəs }

communication cable [COMMUN] A uniform conductive circuit used in the telephone industry to connect customers to their local switching centers and to interconnect local and long-distance switching centers. { kə,myü·nə'kā·shən ˌkā·bəl }

communication channel [COMMUN] The wire or radio channel that serves to convey intelligence between two or more terminals. { kə,myü·nə'kā·shən ˌchan·əl }

communication countermeasure [COMMUN] Any electronic countermeasure against communications, such as jamming. { kə,myü·nə'kā·shən 'kaunt·ər,mezh·ər }

communication disorder [MED] An interference with an individual's ability to comprehend or express ideas, experiences, knowledge, and feelings. { kə,myü·nə'kā·shən ˌdis'ȯrd·ər }

communication engineering [COMMUN] The design, construction, and operation of all types of equipment used for radio, wire, or other types of communication. { kə,myü·nə'kā·shən en·jə'nir·iŋ }

communication link *See* data link. { kə,myü·nə'kā·shən ˌliŋk }

communication protocol [COMPUT SCI] Procedures that enable devices within a computer network to exchange information. Also known as protocol. { kə,myü·nə'kā·shən 'prōd·ə,kȯl }

communication receiver [ELECTR] A receiver designed especially for reception of voice or code messages transmitted by radio communication systems. { kə,myü·nə'kā·shən ri'sē·vər }

communications [ENG] The science and technology by which information is collected from an originating source, transformed into electric currents or fields, transmitted over electrical networks or space to another point, and reconverted into a form suitable for interpretation by a receiver. { kə,myü·nə'kā·shənz }

communications control unit [COMMUN] A device that handles data transmission between components of a communications network, and performs related functions such as multiplexing, message switching, and code conversion. Abbreviated CCU. { kə,myü·nə'kā·shənz kən'trōl ˌyü·nət }

communications intelligence [COMMUN] Technical and intelligence information derived from communications by other than the intended recipients. { kə,myü·nə'kā·shənz in'tel·ə·jəns }

communications language [COMMUN] A language structure complete with conventions, syntax, and character set, used primarily for conveying knowledge of processes between two participants. { kə,myü·nə'kā·shənz ˌlaŋ·gwij }

communications network [COMMUN] Organization of stations capable of intercommunications but not necessarily on the same channel. { kə,myü·nə'kā·shənz ˌnet,wərk }

communications package [COMPUT SCI] A software product that specifies communications protocols for data transmission within a computer network or between a computer and its peripheral equipment. { kə,myü·nə'kā·shənz ˌpak·ij }

communication speed [COMMUN] The rate at which information is transmitted over a communications channel, adjusted for redundancies. { kə,myü·nə'kā·shən ˌspēd }

communications program [COMPUT SCI] A computer program that transmits data to and receives data from local and remote terminals and other computers. { kə,myü·nə'kā·shənz ˌprō·grəm }

communications relay station [COMMUN] Facility for rapidly passing message traffic from one tributary to another by automatic, semiautomatic, or manual means, or by electrically connecting circuits (circuit switching) between two tributaries for direct transmission. { kə‚myü·nə'kā·shənz 'rē‚la ‚stā·shən }

communications satellite [AERO ENG] An orbiting, artificial earth satellite that relays radio, television, and other signals between ground terminal stations thousands of miles apart. Also known as radio relay satellite; relay satellite. { kə‚myü·nə'kā·shənz 'sad·ə‚līt }

communications traffic [COMMUN] All transmitted and received messages. { kə‚myü·nə'kā·shənz ‚traf·ik }

communication system [COMMUN] A telephone, telegraph, teletypewriter, television, data transmission, or other system in which electrical impulses originated at one place are reproduced at a distant point. { kə‚myü·nə'kā·shən ‚sis·təm }

communications zone indicator [ELECTR] Device to indicate whether or not long-distance high-frequency broadcasts are successfully reaching their destinations. { kə‚myü·nə'kā·shənz ‚zōn 'in·də‚kād·ər }

communication theory [COMMUN] The mathematical theory of the communication of information from one point to another. { kə‚myü·nə'kā·shən ‚thē·ə·rē }

community [ECOL] Aggregation of organisms characterized by a distinctive combination of two or more ecologically related species; an example is a deciduous forest. Also known as ecological community. { kə'myü·nə·dē }

community antenna television See cable television. { kə'myü·nə·dē an'ten·ə 'tel·ə‚vizh·ən }

community classification [ECOL] Arrangement of communities into classes with respect to their complexity and extent, their stage of ecological succession, or their primary production. { kə'myü·nə·dē ‚klas·ə·fə'kā·shən }

community dial office [COMMUN] Small dial office with no employees located in the building serving an exchange area. { kə'myü·nə·dē 'dīl ‚óf·əs }

commutated antenna direction finder [NAV] A direction finder utilizing a number of antenna elements and a switching device which connects the elements sequentially to the receiver. Abbreviated CADF. { 'käm·yə‚tād·əd an'ten·ə də'rek·shən ‚find·ər }

commutating capacitor [ELECTR] A capacitor used in gas-tube rectifier circuits to prevent the anode from going highly negative immediately after extinction. { 'kam·yə‚tād·iŋ kə'pas·əd·ər }

commutating pole [ELECTROMAG] One of several small poles between the main poles of a direct-current generator or motor, which serves to neutralize the flux distortion in the neutral plane caused by armature reaction. Also known as compole; interpole. { 'käm·yə‚tād·iŋ ‚pōl }

commutating reactance [ELECTR] An inductive reactance placed in the cathode lead of a three-phase mercury-arc rectifier to ensure that tube current holds over during transfer of conduction from one anode to the next. { 'käm·yə‚tād·iŋ rē'ak·təns }

commutating reactor [ELEC] A reactor found primarily in silicon controlled rectifier (SCR) converters where it is connected in series with a commutation capacitor to form a highly efficient resonant circuit used to cause a current oscillation which turns off (commutates) the conducting SCR. { 'käm·yə‚tād·iŋ rē'ak·tər }

commutating zone [ELECTROMAG] The part of the armature of an electric machine that contains the windings which are short-circuited by the brush on the commutator at a particular instant. { 'käm·yə‚tād·iŋ ‚zōn }

commutation [COMMUN] The sampling of various quantities in a repetitive manner for transmission over a single channel in telemetering. [ELECTR] The transfer of current from one channel to another in a gas tube. [ELECTROMAG] The process of current reversal in the armature windings of a direct-current rotating machine to provide direct current at the brushes. { ‚käm·yə'tā·shən }

commutation rules [QUANT MECH] The specification of the commutators of operators corresponding to the dynamical variables of a system, which are equal to $i\hbar$ times the Poisson brackets of the classical variables to which the operators correspond. { ‚käm·yə'tā·shən ‚rülz }

commutative algebra [MATH] An algebra in which the multiplication operation obeys the commutative law. { ‚käm·yə‚tād·iv 'al·jə·brə }

commutative diagram [MATH] A diagram in which any two mappings between the same pair of sets, formed by composition of mappings represented by arrows in the diagram, are equal. { ‚käm·yə‚tād·iv 'dī·ə‚gram }

commutative group See Abelian group. { ‚käm·yə‚tād·iv ‚grüp }

commutative law [MATH] A rule which requires that the result of a binary operation be independent of order; that is, $ab=ba$. { ‚käm·yə‚tād·iv 'ló }

commutative operation [MATH] A binary operation that obeys a commutative law, such as addition and multiplication on the real or complex numbers. Also known as Abelian operation. { ‚käm·yə‚tād·iv ‚äp·ə'rā·shən }

commutative ring [MATH] A ring in which the multiplication obeys the commutative law. Also known as Abelian ring. { ‚käm·yə‚tād·iv ‚riŋ }

commutator [ELECTROMAG] That part of a direct-current motor or generator which serves the dual function, in combination with brushes, of providing an electrical connection between the rotating armature winding and the stationary terminals, and of permitting reversal of the current in the armature windings. [MATH] The commutator of a and b is the element c of a group such that $bac = ab$. [QUANT MECH] The commutator of a and b is $[a,b] = ab - ba$. { 'käm·yə‚tād·ər }

commutator-controlled welding [MET] Spot or projection welding in which several electrodes in contact with the work simultaneously are operated under the control of an electrical commutating device. Also known as ultraspeed welding. { 'käm·yə‚tād·ər kən‚trōld 'weld·iŋ }

commutator head [ELEC] The butt end of a commutator. { 'käm·yə‚tād·ər ‚hed }

commutator motor [ELEC] An electric motor having a commutator. { 'käm·yə‚tād·ər ‚mōd·ər }

commutator pulse [COMPUT SCI] One of a series of pulses indicating the beginning or end of a signal representing a single binary digit in a computer word. Also known as position pulse (P pulse). { 'käm·yə‚tād·ər ‚pəls }

commutator subgroup [MATH] The subgroup of a given group G consisting of all products of the form $g_1 g_2 \ldots g_n$, where each g_i is the commutator of some pair of elements in G. { 'käm·yə‚tād·ər 'səb‚grüp }

commutator switch [ELEC] A switch, usually rotary and mechanically driven, that performs a set of switching operations in repeated sequential order, such as is required for telemetering many quantities. Also known as sampling switch; scanning switch. { 'käm·yə‚tād·ər ‚swich }

comonomer [CHEM] One of the compounds used to produce a specific polymeric product. { ‚kō'män·ə·mər }

comose [BOT] Having a tuft of soft hairs. { 'kō‚mōs }

comovement effect [ATOM PHYS] The effect on atomic energy levels of the movement of the atomic nucleus, together with the electrons, about their common center of mass. { ‚kō'müv·mənt i‚fekt }

Comoviridae [VIROL] A family of plant viruses that are characterized by icosahedral particles containing a bipartite genome consisting of two single-stranded positive-sense polyadenylated ribonucleic acid molecules; includes the genus Nepovirus. { ‚kō·mə'vir·ə‚dī }

Comovirus [VIROL] A genus of plant viruses belonging to the family Comoviridae; the type species is cowpea mosaic virus. Also known as cowpea mosaic virus group. { 'kō·mə‚vī·rəs }

compact [MET] A briquette made by the compression of metal powder, with or without the addition of nonmetallic constituents. { 'käm‚pakt }

compact disk [COMMUN] A nonmagnetic disk, usually $4\frac{3}{4}$ inches (12 centimeters) in diameter, used for audio or video recording or for data storage; information is recorded using a laser beam to burn microscopic pits into the surface and is accessed by means of a lower-power laser to sense the presence or absence of pits. { 'käm‚pakt 'disk }

compact-disk erasable See CD-RW. { ‚käm‚pak ‚disk i'rās·ə·bəl }

compact-disk read-only memory [COMPUT SCI] A compact disk used for the permanent storage of up to approximately 500 megabytes of data. Abbreviated CD-ROM. { 'käm‚pakt ‚disk ‚rēd ‚ōn·lē 'mem·rē }

compact-disk recordable See CD-R. { ‚käm‚pak ‚disk ri'kórd·ə·bəl }

compact-disk rewritable See CD-RW. { ‚käm‚pak ‚disk ‚rē'rīd·ə·bəl }

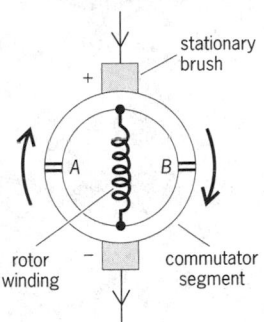

COMMUTATOR

stationary brush
rotor winding
commutator segment

Basic commutator for a dc motor.

compact-disk write-once See CD-R. { ¦käm,pak ,disk ¦rīt 'wəns }

compact H II region [ASTRON] A region of dense ionized hydrogen in interstellar space, not greater than 1 parsec (1.9 × 10^{13} miles or 3.1 × 10^{13} kilometers) in diameter. { 'käm,pakt 'āch 'tü ,rē·jən }

compactification [MATH] For a topological space X, a compact topological space that contains X. { käm'pak·tə·fe,kā·shən }

compacting garbage collection [COMPUT SCI] The physical rearrangement of data cells so that those cells whose contents are no longer useful (garbage) are compressed into a contiguous array. { ,käm'pak·tiŋ 'gär·bij kə'lek·shən }

compaction [COMPUT SCI] A technique for reducing the space required for data storage without losing any information content. Also known as squishing. [ENG] Increasing the dry density of a granular material, particularly soil, by means such as impact or by rolling the surface layers. [GEOL] Process by which soil and sediment mass loses pore space in response to the increasing weight of overlying material. { kəm'pak·shən }

compact-open topology [MATH] A topology on the space of all continuous functions from one topological space into another; a subbase for this topology is given by the sets $W(K,U)$ = $\{f : f(K) \subset U\}$, where K is compact and U is open. { ¦käm,-pakt ¦ō·pən tə'päl·ə·jē }

compact operator [MATH] A linear transformation from one normed vector space to another, with the property that the image of every bounded set has a compact closure. { ¦käm,-pakt 'äp·ə,rād·ər }

compactor [MECH ENG] **1.** Machine designed to consolidate earth and paving materials by kneading, weight, vibration, or impact, to sustain loads greater than those sustained in an uncompacted state. **2.** A machine that compresses solid waste material for convenience in disposal. { kəm'pak·tər }

compact radio source [ASTRON] A source of radio-frequency radiation outside the solar system whose flux at an intermediate radio frequency is dominated by the contribution from a single bright component less than 1 kiloparsec (1.9 × 10^{16} miles or 3.1 × 10^{16} kilometers) in diameter. { 'käm·pakt 'rād·ē·ō ,sórs }

compact set [MATH] A set in a topological space with the property that every open cover has a finite subset which is also a cover. Also known as bicompact set. { 'käm,pakt 'set }

compact space [MATH] A topological space which is a compact set. { ¦käm,pakt 'spās }

compact support [MATH] The property of a function whose support is a compact set. { 'käm,pak sə,pórt }

compactum [MATH] A topological space that is metrizable and compact. { käm'pak·təm }

companded single-sideband system [COMMUN] A long-haul microwave telecommunications system that employs repeaters and single-sideband amplitude modulation and achieves subjective noise improvement by companding to reduce circuit noise between syllables and during pauses in speech. Abbreviated CSSB system. { kəm'pan·dəd ¦siŋ·gəl ¦sīd,band ¦sis·təm }

companding [ELECTR] A process in which compression is followed by expansion; often used for noise reduction in equipment, in which case compression is applied before noise exposure and expansion after exposure. { kəm'pand·iŋ }

compandor [ELECTR] A system for improving the signal-to-noise ratio by compressing the volume range of the signal at a transmitter or recorder by means of a compressor and restoring the normal range at the receiving or reproducing apparatus with an expander. { kəm'pand·ər }

companion See comes. { kəm'pan·yən }

companion body [AERO ENG] A nose cone, last-stage rocket, or other body that orbits along with an earth satellite or follows a space probe. { kəm'pan·yən ,bäd·ē }

companion cell [BOT] A specialized parenchyma cell occurring in close developmental and physiologic association with a sieve-tube member. { kəm'pan·yən ,sel }

companion flange [DES ENG] A pipe flange that can be bolted to a similar flange on another pipe. { kəm'pan·yən ,flanj }

companionway [NAV ARCH] A stairway that runs from one deck of a ship to another. { kəm'pan·yən,wā }

compar [MATER] Generic term for a family of compounded,

modified, and plasticized polyvinyl alcohol resins; used for making full- and solvent-resistant tubing and hose, and printing rolls. { 'käm,pär }

comparable functions [MATH] Two real-valued functions with a common domain of definition such that the values of one of the functions are equal to or greater than the values of the other for all the points in this domain. { 'käm·prə·bəl 'faŋk·shənz }

comparable pair [MATH] A pair of elements, x and y, of a partially ordered set such that either $x \leq y$ or $y \leq x$. { ¦käm·prə·bəl 'per }

comparative embryology [EMBRYO] A branch of embryology that deals with the similarities and differences in the development of animals or plants of different orders. { kəm'par·əd·iv ,em·brē'äl·ə·jē }

comparative experiments [STAT] Experiments conducted to determine statistically whether one procedure is better than another. { kəm'par·əd·iv ik'sper·ə·məns }

comparative genomic hybridization [GEN] A method that uses fluorescence in situ hybridization and comparison of the strength of hybridization signal to determine any differences in copy number of deoxyribonucleic acid sequences anywhere in the nuclear genome. { kəm¦par·əd·iv jə,nō·mik ,hī·brəd·ə'zā·shen }

comparative lifetime [NUC PHYS] The product of the mean life of a nucleus that undergoes beta decay and the probability per unit time that beta decay would occur if the matrix element between the initial and final states of this transition were unity. { kəm'par·əd·iv 'līf,tīm }

comparative pathology [MED] Investigation and comparison of disease in various animals, including humans, to arrive at resemblances and differences which may clarify disease as a phenomenon of nature. { kəm'par·əd·iv pə'thäl·ə·jē }

comparative rabal [ENG] A rabal observation (that is, a radiosonde balloon tracked by theodolite) taken simultaneously with the usual rawin observation (tracking by radar or radio direction-finder), to provide a rough check on the alignment and operating accuracy of the electronic tracking equipment. { kəm'par·əd·iv 'rä,bal }

comparator [ANALY CHEM] An instrument used to determine the concentration of a solution by comparing the intensity of color with a series of standard colors. [COMPUT SCI] A device that compares two transcriptions of the same information to verify the accuracy of transcription, storage, arithmetical operation, or some other process in a computer, and delivers an output signal of some form to indicate whether or not the two sources are equal or in agreement. [CONT SYS] A device which detects the value of the quantity to be controlled by a feedback control system and compares it continuously with the desired value of that quantity. [ENG] A device used to inspect a gaged part for deviation from a specified dimension, by mechanical, electrical, pneumatic, or optical means. { kəm'par·əd·ər }

comparator circuit [ELECTR] An electronic circuit that produces an output voltage or current whenever two input levels simultaneously satisfy predetermined amplitude requirements; may be linear (continuous) or digital (discrete). { kəm'par·əd·ər ,sər·kət }

comparator-densitometer [ANALY CHEM] Device that projects a labeled spectrum onto a screen adjacent to an enlarged image of the spectrum to be analyzed, allowing visual comparison. { ¦käm'par·əd·ər den·sə'täm·əd·ər }

comparator method [THERMO] A method of determining the coefficient of linear expansion of a substance in which one measures the distance that each of two traveling microscopes must be moved in order to remain centered on scratches on a rod-shaped specimen when the temperature of the specimen is raised by a measured amount. { kəm'par·əd·ər ,meth·əd }

comparator probe [COMPUT SCI] A component of a hardware monitor that is used to sense the number of bits that appear in parallel, as in an address register. { kəm'par·əd·ər ,prōb }

comparing brushes [COMPUT SCI] Sets of metallic brushes which verify that all the cards in a gang-punching operation have been properly punched. { kəm'per·iŋ ,brəsh·əz }

comparing control change See control change. { kəm'per·iŋ kən'trōl ,chānj }

comparing unit [ELECTR] An electromechanical device

which compares two groups of timed pulses and signals to establish either identity or nonidentity. { kəm'per·iŋ ,yü·nət }

comparing watch [HOROL] A hack watch, particularly one having an error determined by comparison with a chronometer. { kəm'per·iŋ ,wach }

comparison [COMPUT SCI] A computer operation in which two numbers are compared as to identity, relative magnitude, or sign. { kəm'par·ə·sən }

comparison bridge [ELECTR] A bridge circuit in which any change in the output voltage with respect to a reference voltage creates a corresponding error signal, which, by means of negative feedback, is used to correct the output voltage and thereby restore bridge balance. { kəm'par·ə·sən ,brij }

comparison indicators [COMPUT SCI] Registers, one of which is activated during the comparison of two quantities to indicate whether the first quantity is lower than, equal to, or greater than the second quantity. { kəm'par·ə·sən ,in·də,kād·ərz }

comparison lamp [OPTICS] An incandescent lamp whose luminous intensity is constant (although not necessarily known), and which is compared against other lamps in a photometer. { kəm'par·ə·sən ,lamp }

comparison microscope [OPTICS] 1. An arrangement of two microscopes connected by a special receiving ocular so that the field of one microscope is seen at one side of a dividing line and the field of the other microscope at the opposite side. 2. A projection type of microscope in which the image is compared with a template or known pattern. { kəm'par·ə·sən 'mī·krə,skōp }

comparison property See trichotomy property. { kəm'par·ə·sən ,präp·ərd·ē }

comparison spectrum [SPECT] A line spectrum whose wavelengths are accurately known, and which is matched with another spectrum to determine the wavelengths of the latter. { kəm'par·ə·sən ,spek·trəm }

comparison star [ASTRON] A star of known brightness used as a standard for comparison in determining the magnitude of a nearby celestial object. { kəm'par·ə·sən ,stär }

comparison test [MATH] A simple test for the convergence of an infinite series, according to which a series converges if the absolute values of each of its terms are equal to or less than the corresponding term of a series that is known to converge, and diverges if each of its terms is equal to or greater than the absolute value of the corresponding term of a series that is known to diverge. { kəm'par·ə·sən ,test }

compartment [CYTOL] Any of the membrane-bound organelles within cells. [MIN ENG] A section of a mine shaft separated by framed timbers and planking. { kəm'pärt·mənt }

compartment mill [MECH ENG] A multisection pulverizing device divided by perforated partitions, with preliminary grinding at one end in a short ball-mill operation, and finish grinding at the discharge end in a longer tube-mill operation. { kəm'pärt·mənt ,mil }

compass [ENG] An instrument for indicating a horizontal reference direction relative to the earth. [GRAPHICS] An instrument used for describing arcs or circles with pencil or pen; has two legs hinged together at the top. { 'käm·pəs }

compass adjustment [NAV] The process of neutralizing the magnetic effect a craft exerts on a magnetic compass: permanent magnets and soft iron correctors are arranged about the binnacle so that their effects are nearly equal and opposite to the magnetic material in the craft, thus reducing the deviations and eliminating the sectors of sluggishness and unsteadiness. { 'käm·pəs ə'jəs·mənt }

compass amplitude [NAV] In marine navigation, amplitude relative to compass east or west. { 'käm·pəs ,am·plə,tüd }

compass azimuth [NAV] Azimuth relative to compass north. { 'käm·pəs ,az·ə·məth }

compass bearing [NAV] Direction relative to north as indicated by a compass. { 'käm·pəs ,ber·iŋ }

compass bowl [ENG] That part of a compass in which the compass card is mounted. { 'käm·pəs ,bōl }

compass calibration See swinging ship. { 'käm·pəs ,kal·ə'brā·shən }

compass card [DES ENG] The part of a compass on which the direction graduations are placed, it is usually in the form of a thin disk or annulus graduated in degrees, clockwise from 0° at the reference direction to 360°, and sometimes also in compass points. { 'käm·pəs ,kärd }

compass card axis [DES ENG] The line joining 0° and 180° on a compass card. { 'käm·pəs ,kärd ,ak·səs }

compass compensation [NAV] 1. The process of neutralizing the effects degaussing currents exert on a marine magnetic compass. 2. The process of neutralizing the magnetic effect an aircraft exerts on a magnetic compass. { 'käm·pəs ,käm·pən'sā·shən }

compass correction card [NAV] A card posted in a holder near the magnetic compass, on which there is recorded the difference between the readings of the compass and the correct geomagnetic directions; these errors (deviations) are given for at least the four cardinal points; sometimes the card lists the compass bearings to be flown when it is desired to fly corresponding magnetic headings. { 'käm·pəs kə'rek·shən ,kärd }

compass corrector [NAV] A magnet or piece of iron placed close to a ship's or aircraft's compass to neutralize the effect of the ship's or aircraft's magnetic field. Also known as Flinder's bar. { 'käm·pəs kə'rek·tər }

compass course [NAV] Course relative to compass north. { 'käm·pəs ,körs }

compass declinometer [ENG] An instrument used for magnetic distribution surveys; employs a thin compass needle 6 inches (15 centimeters) long, supported on a sapphire bearing and steel pivot of high quality; peep sights serve for aligning the compass box on an azimuth mark. { 'käm·pəs ,dek·lə'näm·əd·ər }

compass deviation [NAV] The difference between the readings of a compass which is without mechanical defects and is held motionless in space, and the same instrument when it is installed in the same geographic position but is mounted on a ship or aircraft; deviation is a systematic error which is compensated by placing iron bars in places about the compass; residual deviation errors are calibrated and noted on a card so it can be used by the pilot or navigator. { 'käm·pəs ,dēv·ē'ā·shən }

compass direction [NAV] Horizontal direction expressed as angular distance from compass north. { 'käm·pəs də'rek·shən }

compass error [NAV] The angle by which a compass direction differs from the true geographical direction; assuming that there is no lubber line error, it is the algebraic sum of the declination, variation, and motion errors. { 'käm·pəs ,er·ər }

Compasses See Circinus. { 'käm·pə·səz }

compass heading [NAV] Heading relative to compass north. { 'käm·pəs ,hed·iŋ }

compass locator [NAV] A low-power, low-frequency, nondirectional radio beacon installed near an airfield to facilitate instrument approaches. { 'käm·pəs ,lō,kād·ər }

compass meridian [NAV] A line through the north-south points of a magnetic compass, in which the compass card axis lies. { 'käm·pəs mə'rid·ē·ən }

compass motion error [NAV] Errors in compass reading when the craft or vessel is under way; these errors are caused by northerly turning, speed acceleration and deceleration, the liquid in the compass swirling on turns, and vibration caused by engines; these are all random errors, and compensation is not attained in the simple compass. { 'käm·pəs ,mō·shən ,er·ər }

compass north [NAV] The direction of north as indicated by a magnetic compass; the reference direction for measurement of compass directions. { 'käm·pəs 'north }

compass points [GEOD] The 32 divisions of a compass at intervals of 11 1/4, with each division further divided into quarter points. Also known as points of the compass. { 'käm·pəs ,póins }

compass prime vertical [NAV] The vertical circle through the compass east and west points of the horizon. { 'käm·pəs ¦prīm 'vərd·ə·kəl }

compass repeater [NAV] That part of a remote-indicating compass system which repeats at a distance the indications of the master compass. Also known as repeater compass. { 'käm·pəs ri'pēd·ər }

compass roof [BUILD] A roof in which each truss is in the form of an arch. { 'käm·pəs ¦rüf }

compass rose [NAV] A graduated circle, usually marked in degrees, indicating directions north, south, east, and west, and inscribed on a chart; used for the calibration of compasses on crafts. { 'käm·pəs ,rōz }

compass saw [DES ENG] A handsaw which has a handle

with several attachable thin, tapering blades of varying widths, making it suitable for a variety of work, such as cutting circles and curves. { 'käm·pəs ˌsò }

compass track [NAV] The direction of the track relative to compass north. { 'käm·pəs ˌtrak }

compass transmitter [NAV] That part of a remote-indicating compass system which sends the direction indications to the repeaters. { 'käm·pəs ˌtranzˌmid·ər }

compass variation error [NAV] The angular difference between the geographic and magnetic meridians at any place as read on the magnetic compass; variation errors vary in time and place, but they are systematic and can readily be canceled out merely by referring to government-published variation charts; the importance of this error may be judged by this example: the variation in northern California is 20° east while in northern New York it is 10° west, thus a pilot attempting to fly between the two places who failed to correct would be off by 30°. Also known as declination error. { 'käm·pəs verˌē'ā·shən ˌer·ər }

compatibility [COMPUT SCI] The ability of one device to accept data handled by another device without conversion of the data or modification of the code. [IMMUNOL] Ability of two bloods or other tissues to unite and function together. [ORD] In ammunition, the ability of a given material to exist unchanged under certain conditions of temperature and moisture when in the presence of some other specific material. [PHARM] The capacity of two or more ingredients in a medicine to mix without chemical change or loss of therapeutic effectiveness. [SYS ENG] The ability of a new system to serve users of an old system. { kəmˌpad·ə'bil·ə·dē }

compatibility conditions [MECH] A set of six differential relations between the strain components of an elastic solid which must be satisfied in order for these components to correspond to a continuous and single-valued displacement of the solid. { kəmˌpad·ə'bil·əd·ē kənˌdish·ənz }

compatibility mode [COMPUT SCI] A feature of a computer or operating system that enables it to run programs written for another system. { kəmˌpad·ə'bil·əd·ē ˌmōd }

compatibilizer [ORG CHEM] Any polymeric interfacial agent that facilitates formation of uniform blends of normally immiscible polymers with desirable end properties. { kəm'pad·ə·bəˌlīz·ər }

compatible color television system [COMMUN] A color television system that permits the substantially normal monochrome reception of the transmitted color picture signal on a typical unaltered monochrome receiver. { kəmˌpad·ə·bəl 'kəl·ər 'tel·əˌvizh·ən ˌsis·təm }

compatible discrete four-channel sound [ENG ACOUS] A sound system in which a separate channel is maintained from each of the four sets of microphones at the recording studio or other input location to the four sets of loudspeakers that serve as the output of the system. Abbreviated CD-4 sound. { kəmˌpad·ə·bəl dis'krēt ˌfòr ˌchan·əl 'saùnd }

compatible monolithic integrated circuit [ELECTR] Device in which passive components are deposited by thin-film techniques on top of a basic silicon-substrate circuit containing the active components and some passive parts. { kəmˌpad·ə·bəl ˌmän·ə'lith·ik 'in·təˌgrād·əd 'sər·kət }

compatible single-sideband system [COMMUN] A single-sideband system that can be received by an ordinary amplitude-modulation radio receiver without distortion. { kəmˌpad·ə·bəl ˌsiŋ·gəl'sīdˌband ˌsis·təm }

compensated amplifier [ELECTR] A broad-band amplifier in which the frequency range is extended by choice of circuit constants. { 'käm·pənˌsād·əd 'am·pləˌfī·ər }

compensated ionization chamber [NUCLEO] An arrangement of two ionization chambers in parallel, with potentials reversed, used as a radiation null indicator. { 'käm·pənˌsād·əd ˌī·ən·ə'zā·shən ˌchām·bər }

compensated-loop direction finder [ELECTR] A direction finder employing a loop antenna and a second antenna system to compensate for polarization error. { 'käm·pənˌsād·əd ˌlüp də'rek·shən ˌfind·ər }

compensated neutron logging [ENG] Neutron well logging using one source and two detectors; the apparent limestone porosity is calculated by computer from the ratio of the count rate of one detector to that of the other. { 'käm·pənˌsād·əd ˌnü·trän ˌläg·iŋ }

compensated pendulum [DES ENG] A pendulum made of

two materials with different coefficients of expansion so that the distance between the point of suspension and center of oscillation remains nearly constant when the temperature changes. { 'käm·pənˌsād·əd 'pen·jə·ləm }

compensated semiconductor [ELECTR] Semiconductor in which one type of impurity or imperfection (for example, donor) partially cancels the electrical effects on the other type of impurity or imperfection (for example, acceptor). { 'käm·pənˌsād·əd 'sem·i·kənˌdək·tər }

compensated volume control See loudness control. { 'käm·pənˌsād·əd 'väl·yəm kən'trōl }

compensated winding See pole-face winding. { 'käm·pənˌsād·əd 'wīnd·iŋ }

compensating capacitor See balancing capacitor. { 'käm·pənˌsād·iŋ kə'pas·əd·ər }

compensating eyepiece [OPTICS] A type of Huygens eyepiece in which the eye lens is achromatized to compensate for the color errors of the objective. { 'käm·pənˌsād·iŋ 'īˌpēs }

compensating impurity [SOLID STATE] A semiconductor impurity that is of the opposite electrical type to a given impurity, and that reduces the concentration of charge carriers (electrons or holes) that resulted from the given impurity. { 'käm·pənˌsād·iŋ im'pyùr·əd·ē }

compensating leads [ENG] A pair of wires, similar to the working leads of a resistance thermometer or thermocouple, which are run alongside the working leads and are connected in such a way that they balance the effects of temperature changes in the working leads. { 'käm·pənˌsād·iŋ 'lēdz }

compensating network [CONT SYS] A network used in a low-energy-level method for suppression of excessive oscillations in a control system. { 'käm·pənˌsād·iŋ 'netˌwərk }

compensating plate [OPTICS] The first of two plates in a Brace compensator, which covers the entire field of view. { 'käm·pənˌsād·iŋ ˌplāt }

compensation [CONT SYS] Introduction of additional equipment into a control system in order to reshape its root locus so as to improve system performance. Also known as stabilization. [ELECTR] The modification of the amplitude-frequency response of an amplifier to broaden the bandwidth or to make the response more nearly uniform over the existing bandwidth. Also known as frequency compensation. [PSYCH] Counterbalancing a weakness or failure in one area by stressing or substituting a strength or success in another area. { ˌkäm·pən'sā·shən }

compensation balance [HOROL] A balance that compensates for changes caused by temperature variations. { ˌkäm·pən'sā·shən ˌbal·əns }

compensation depth [OCEANOGR] The depth at which the light intensity is just sufficient to bring about a balance between the oxygen produced and that consumed by algae. { ˌkäm·pən'sā·shən ˌdepth }

compensation effect See Jaccarino-Peter effect. { ˌkäm·pən'sā·shən iˌfekt }

compensation pendulum [HOROL] A clock pendulum which has compensating mechanisms that are designed to maintain constant pendulum length in spite of temperature changes. { ˌkäm·pən'sā·shən 'pen·jə·ləm }

compensation point [BOT] The light intensity at which the amount of carbon dioxide released in respiration equals the amount used in photosynthesis, and the amount of oxygen released in photosynthesis equals the amount used in respiration. { ˌkäm·pən'sā·shən ˌpòint }

compensation signals [ENG] In telemetry, signals recorded on a tape, along with the data and in the same track as the data, used during the playback of data to correct electrically the effects of tape-speed errors. { ˌkäm·pən'sā·shən ˌsig·nəlz }

compensator [CONT SYS] A device introduced into a feedback control system to improve performance and achieve stability. Also known as filter. [ELECTR] A component that offsets an error or other undesired effect. [OPTICS] A device, usually consisting of two quartz wedges, for determining the phase difference between the two components of elliptically polarized light. { 'käm·pənˌsād·ər }

compensatory emphysema [MED] Simple, nonobstructive overdistension of lung segments or an entire lung in intrathoracic adaptation to collapse, destruction, or removal of portions of the lung or the opposite lung. { kəm'pen·səˌtòr·ē ˌem·fə'sē·mə }

compensatory hypertrophy [MED] An increase in the size

of an organ following injury or removal of the opposite paired organ or of part of the same organ. { kəmˈpenˌsəˌtȯrˈē hīˈpərˌtrəˌfē }

competence [EMBRYO] The ability of a reacting system to respond to the inductive stimulus during early developmental stages. [GEOL] The ability of the wind to transport solid particles either by rolling, suspension, or saltation (intermittent rolling and suspension); usually expressed in terms of the weight of a single particle. [HYD] The ability of a stream, flowing at a given velocity, to move the largest particles. [MIN ENG] A property of rock strata which possess sufficient strength to span a mine opening without failure. { ˈkämˌpədˈəns }

competent beds [GEOL] Beds or strata capable of withstanding the pressures of folding without flowing or changing in original thickness. { ˈkämˌpədˌənt ˌbedz }

competent cell [CYTOL] A cell that is able to incorporate exogenous deoxyribonucleic acid and undergo genetic transformation. [EMBRYO] A cell that can respond to an inducer during embryonic development. { ˈkämˌpədˌənt ˈsel }

competing equilibria condition [CHEM] The competition for a reactant in a complex chemical system in which several reactions are taking place at the same time. { kəmˈpēdˌiŋ ˌēˌkwəˈlibˌrēˌə kənˌdishˌən }

competition [ECOL] The inter- or intraspecific interaction resulting when several individuals share an environmental necessity. { ˌkämˌpəˈtishˌən }

competitive displacement [ECOL] The inability of a species to successfully live in an area because a second species dominates local resources. { kəmˌpedˌədˌiv diˈsplāsˌmənt }

competitive enzyme inhibition [BIOCHEM] Prevention of an enzymatic process resulting from the reversible interaction of an inhibitor with a free enzyme. { kəmˈpedˌədˌiv ˈenˌzīm ˌinˌəˈbishˌən }

competitive exclusion [ECOL] The result of a competition in which one species is forced out of part of the available habitat by a more efficient species. { kəmˈpedˌədˌiv iksˈklüzhˌən }

competitive-exclusion principle See Gause's principle. { kəmˈpedˌədˌiv iksˈklüzhˌən ˌprinˌsəˌpəl }

competitive inhibition [BIOCHEM] Enzyme inhibition in which the inhibitor competes with the natural substrate for the active site of the enzyme; may be overcome by increasing substrate concentration. { kəmˈpedˌədˌiv ˌinˌəˈbishˌən }

compilation [MAP] The selecting, extracting, and assembling of map detail from various sources (such as existing maps, aerial photographs, and surveys), followed by the production of a new or improved map based on this data. { ˌkämˌpəˈlāshˌən }

compile [COMPUT SCI] To prepare a machine-language program automatically from a program written in a higher programming language, usually generating more than one machine instruction for each symbolic statement. { kəmˈpīl }

compile-and-go [COMPUT SCI] A continuous sequence of steps that combine compilation, loading, and execution of a computer program. { kəmˈpīl ən ˈgō }

compiler [COMPUT SCI] A program to translate a higher programming language into machine language. Also known as compiling routine. { kəmˈpīlˌər }

compiler-level language [COMPUT SCI] A higher-level language normally supplied by the computer manufacturer. { kəmˈpīlˌər ˌlevˌəl ˌlaŋˌgwij }

compiler listing [COMPUT SCI] A report that is produced by a compiler and contains an annotated printout of the source program together with other useful information. { kəmˈpīlˌər ˌlistˌiŋ }

compiler system [COMPUT SCI] The set consisting of a higher-level language, such as FORTRAN, and its compiler which translates the program written in that language into machine-readable instructions. { kəmˈpīlˌər ˌsisˌtəm }

compiler toggle [COMPUT SCI] A piece of information transmitted to a compiler to activate some special feature or otherwise control the way in which the compiler operates. { kəmˈpīlˌər ˌtägˌəl }

compiling routine See compiler. { kəmˈpilˌiŋ rüˌtēn }

compital [BOT] 1. Of the vein of a leaf, intersecting at a wide angle. 2. Of a fern, bearing sori at the intersection of two veins. { ˈkämˌpədˌəl }

complement [IMMUNOL] A heat-sensitive, complex system in fresh human and other sera which, in combination with

antibodies, is important in the host defense mechanism against invading microorganisms. [MATH] 1. The complement of a number A is another number B such that the sum $A + B$ will produce a specified result. 2. For a subset of a set, the collection of all members of the set which are not in the given subset. 3. For a fuzzy set A with membership function m_A, the complement of A is the fuzzy set $\bar{A}$ whose membership function $m_{\bar{A}}$ has the value $1 - m_A(x)$ for every element x. 4. The complement of a simple graph, G, is the graph, $\bar{G}$, with the same vertices as G, in which there is an edge between two vertices if and only if there is no edge between those vertices in G. 5. The complement of an angle A is another angle B such that the sum $A + B$ equals 90°. 6. See radix complement. { ˈkämˌpləˌmənt }

complemental air [PHYSIO] The amount of air that can still be inhaled after a normal inspiration. { ˌkämˌpləˈmentˌəl ˈer }

complementarity [QUANT MECH] The principle that nature has complementary aspects, particle and wave; the two aspects are related by $p = h/\lambda$ and $E = h\nu$, where p and E are the momentum and energy of the particle, λ and ν are the length and frequency of the wave, and h is Planck's constant. { ˌkämˌpləˌmənˈtarˌōdˌē }

complementary [ELECTR] Having pnp and npn or p- and n-channel semiconductor elements on or within the same integrated-circuit substrate or working together in the same functional amplifier state. { ˌkämˌpləˈmenˌtrē }

complementary angle [MATH] One of a pair of angles whose sum is 90°. { ˌkämˌpləˈmenˌtrē ˈaŋˌgəl }

complementary base pairing [MOL BIO] The formation of weak hydrogen bonds between complementary nitrogenous bases (for example, guanine and cytosine) on opposite strands of a double-stranded nucleic acid molecule (such as deoxyribonucleic acid), contributing to the overall stability of the double-stranded structure. { ˌkämˌpləˌmenˌtəˌrē ˈbās ˈperˌiŋ }

complementary colors [OPTICS] Two colors which lie on opposite sides of the white point in the chromaticity diagram so that an additive mixture of the two, in appropriate proportions, can be made to yield an achromatic mixture. { ˌkämˌpləˈmenˌtrē ˈkəlˌərz }

complementary constant-current logic [ELECTR] A type of large-scale integration used in digital integrated circuits and characterized by high density and very fast switching times. Abbreviated CCCL; C³L. { ˌkämˌpləˈmenˌtrē ˌkänˌstənt ˌkəˌrənt ˈläjˌik }

complementary deoxyribonucleic acid [MOL BIO] A deoxyribonucleic acid molecule that is synthesized by reverse transcriptase from a ribonucleic acid template. Abbreviated cDNA. Also known as copy DNA. { ˌkämˌpləˈmenˌtrē ˌdēˌäkˌsēˌrīˌbōˌnüˈklēˌik ˈasˌəd }

complementary deoxyribonucleic acid library [MOL BIO] A collection of complementary deoxyribonucleic acid molecules, representative of all the various messenger ribonucleic acid molecules produced by a specific type of cell of a given species, spliced into a corresponding collection of DNA vectors. { ˌkämˌpləˌmenˌtrē dēˌäkˌsēˌrībˌō,nüˈklēˌik ˈasˌəd }

complementary function [MATH] Any solution of the equation obtained from a given linear differential equation by replacing the inhomogeneous term with zero. { ˌkämˌpləˌmenˌtrē ˈfəŋkˌshən }

complementary genes [GEN] Nonallelic genes for which a wild type allele of one abolishes the phenotypic effect of a mutation of the other. { ˌkämˌpləˈmenˌtrē ˈjēnz }

complementary logic switch [ELECTR] A complementary transistor pair which has a common input and interconnections such that one transistor is on when the other is off, and vice versa. { ˌkämˌpləˈmenˌtrē ˈläjˌik ˌswich }

complementary metal oxide semiconductor device See CMOS device. { ˌkämˌpləˌmenˌtrē ˌmedˌəl ˌäkˌsīd ˈsemˌiˌkənˌdəkˌtər diˈvīs }

complementary minor See minor. { ˌkämˌpləˈmenˌtrē ˈmīˌnər }

complementary operation [MATH] An operation on a Boolean algebra of two elements (labeled "true" and "false") whose result is the negation of a given operation; for example, NAND is complementary to the AND function. { ˌkämˌpləˈmenˌtrē ˌäpˌəˈrāˌshən }

complementary rocks [GEOL] Rocks which are differentiated from the same magma, and whose average composition is the same as the parent magma. { ˌkämˌpləˈmenˌtrē ˈräks }

complementary symmetry [ELECTR] A circuit using both *pnp* and *npn* transistors in a symmetrical arrangement that permits push-pull operation without an input transformer or other form of phase inverter. { ˌkäm·plə'men·trē 'sim·ə·trē }

complementary transistors [ELECTR] Two transistors of opposite conductivity (*pnp* and *npn*) in the same functional unit. { ˌkäm·plə'men·trē tran'zis·tərs }

complementary variables *See* conjugate variables. { ˌkäm·plə'men·trē 'ver·ē·ə·bəlz }

complementary wave [ELECTROMAG] Wave brought into existence at the ends of a coaxial cable, or two-conductor transmission lines, or any discontinuity along the line. { ˌkäm·plə'men·trē 'wāv }

complementary wavelength [OPTICS] The wavelength of light that, when combined with a sample color in suitable proportions, matches a reference standard light. { ˌkäm·plə'men·trē 'wāv,leŋkth }

complementation [MATH] The act of replacing a set by its complement. { ˌkäm·plə·mən'tā·shən }

complementation group [GEN] Different mutations in the same cistron or gene. { ˌkäm·plə·mən'tā·shən ˌgrüp }

complementation law [STAT] The law that the probability of an event *E* is 1 minus the probability of the event not *E*. { ˌkäm·plə·mən'tā·shən ˌlö }

complementation test [GEN] An analytic procedure for determining whether two nonexpressed mutants are in the same cistron or gene. { ˌkäm·plə·mən'tā·shən ˌtest }

complement cascade [IMMUNOL] The sequential activation of complement proteins resulting in lysis of a target cell. { 'käm·plə·mənt kas'kād }

complemented lattice [MATH] A lattice with distinguished elements *a* and *b*, and with the property that corresponding to each point *x* of the lattice, there is a *y* such that the greatest lower bound of *x* and *y* is *a*, and the least upper bound of *x* and *y* is *b*. { 'käm·plə,ment·əd 'lad·əs }

complement fixation [IMMUNOL] The binding of complement to an antigen-antibody complex so that the complement is unavailable for subsequent reaction. { 'käm·plə·mənt ˌfik'sā·shən }

complement-fixation test [IMMUNOL] A diagnostic test to determine the presence of antigen or antibody in the blood by adding complement to the test system; used especially in diagnosing syphilis. { 'käm·plə·mənt ˌfik'sā·shən ˌtest }

complement number system [COMPUT SCI] System of number handling in which the complement of the actual number is operated upon; used in some computers to facilitate arithmetic operations. { 'käm·plə·mənt 'nəm·bər ˌsis·təm }

complete antibody [IMMUNOL] An antibody that can directly agglutinate saline-suspended red blood cells. { kəm'plēt 'ant·i,bäd·ē }

complete bipartite graph [MATH] A graph whose vertices can be partitioned into two sets such that every edge joins a vertex in one set with a vertex in the other, and each vertex in one set is joined to each vertex in the other by exactly one edge. { kəm'plēt bī'pär,tīt ,graf }

complete blood count [PATH] Differential and absolute determinations of the numbers of each type of blood cell in a sample and, by extrapolation, in the general circulation. { kəm'plēt 'bləd ,kaunt }

complete carry [COMPUT SCI] In parallel addition, an arrangement in which the carries that result from the addition of carry digits are allowed to propagate from place to place. { kəm'plēt 'kar·ē }

complete class of decision functions [STAT] A concept in decision theory which states that for a class of decision functions to be complete it must include a uniformly better decision function, which is a decision function that is sometimes better but never worse (according to some criterion) than each decision function not in the class. { kəm'plēt ˌklas əv di'sizh·ən ˌfəŋk·shənz }

complete combustion [CHEM] Combustion in which the entire quantity of oxidizable constituents of a fuel is reacted. { kəm'plēt kəm'bəs·chən }

complete degeneracy [QUANT MECH] The condition in which all the states of interest have the same energy. { kəm'plēt di'jen·ə·rə·sē }

complete electron shell [ATOM PHYS] An inner electron shell of an atom that contains its maximun number of electrons. { kəm'plēt i,lek,trän 'shel }

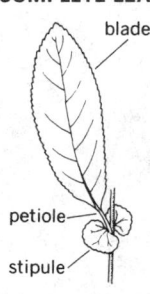

COMPLETE LEAF

blade

petiole

stipule

Drawing of a complete leaf.

complete-expansion diesel cycle *See* Brayton cycle. { kəm'plēt ik'span·shən 'dē·zəl ,si·kəl }

complete flower [BOT] A flower having all four floral parts, that is, having sepals, petals, stamens, and carpels. { kəm'plēt 'flau·ər }

complete four-point *See* four-point. { kəm'plēt 'fòr ,pöint }

complete fusion [MET] Fusion which has occurred over all surfaces of the base metal exposed for welding. { kəm'plēt 'fyü·zhən }

complete graph [MATH] A graph with exactly one edge connecting each pair of distinct vertices and no loops. { kəm'plēt 'graf }

complete induction *See* mathematical induction. { kəm'plēt in'dək·shən }

complete integral [MATH] **1.** A solution of an *n*th order ordinary differential equation which depends on *n* arbitrary constants as well as the independent variable. Also known as complete primitive. **2.** A solution of a first-order partial differential equation with *n* independent variables which depends upon *n* arbitrary parameters as well as the independent variables. { kəm'plēt 'in·tə·grəl }

complete joint penetration [MET] The fusion of weld metal to base metal throughout the entire thickness of the base metal, that is, the deposited weld metal occupies the entire groove. { kəm'plēt 'jöint pen·ə'trā·shən }

complete lattice [MATH] A partially ordered set in which every subset has both a supremum and an infimum. { kəm'plēt 'lad·əs }

complete leaf [BOT] A dicotyledon leaf consisting of three parts: blade, petiole, and a pair of stipules. { kəm'plēt 'lēf }

complete learning method [PSYCH] An experimental procedure in which the subject works with test material until one complete errorless trial is achieved. { kəm'plēt ¦lərn·iŋ ,meth·əd }

complete limit *See* limit superior. { kəm'plēt 'lim·ət }

complete linear topological space [MATH] A topological vector space in which each Cauchy net undergoes Moore-Smith convergence to some point in the space. { kəm'plēt ¦lin·ē·ər ,täp·ə¦läj·ə·kəl 'spās }

complete lubrication [ENG] Lubrication taking place when rubbing surfaces are separated by a fluid film, and frictional losses are due solely to the internal fluid friction in the film. Also known as viscous lubrication. { kəm'plēt ,lüb·rə'kā·shən }

completely additive set function *See* countably additive set function. { kəm'plēt·lē ¦ad·əd·iv 'set ,fəŋk·shən }

completely ordered set *See* linearly ordered set. { kəm'plēt·lē ,órd·ərd 'set }

completely inelastic collision *See* perfectly inelastic collision. { kəm'plēt·lē in·ə'las·tik kə'lizh·ən }

completely normal space [MATH] A topological space with the property that any pair of sets with disjoint closures can be separated by open sets. { kəm'plēt·lē ¦nór·məl 'spās }

completely reducible representation [MATH] A representation of a group as a family of linear operators of a vector space *V* such that *V* is the direct sum of subspaces $V_1, \ldots, V_n$ which are invariant under these operators, but $V_1, \ldots, V_n$ do not have any proper closed subspaces which are also invariant under these operators. Also known as semisimple representation. { kəm'plēt·lē ri¦düs·ə·bəl ,rep·ri·zen'tā·shən }

completely regular space [MATH] A topological space *X* where for every point *x* and neighborhood *U* of *x* there is a continuous function from *X* to [0,1] with *f*(*x*) = 1 and *f*(*y*) = 0, if *y* is not in *U*. { kəm'plēt·lē ¦reg·yə·lər 'spās }

completely separable space *See* perfectly separable space. { kəm'plēt·lē ¦sep·rə·bəl 'spās }

complete matching [MATH] A subset of the edges of a bipartite graph that consists of edges joining each of the vertices in one of the sets of vertices defining the bipartite structure with distinct vertices in the other such set. { kəm'plēt 'mach·iŋ }

complete metric space [MATH] A metric space in which every Cauchy sequence converges to a point of the space. Also known as complete space. { kəm'plēt ¦me·trik 'spās }

complete normed linear space *See* Banach space. { kəm'plēt ¦nórmd ¦lin·ē·ər 'spas }

complete operation [COMPUT SCI] An operation which includes obtaining all operands from storage, performing the operation, returning resulting operands to storage, and obtaining the next instruction. { kəm'plēt äp·ə'rā·shən }

complete order *See* linear order.　{ kəm'plēt 'ȯrd·ər }

complete ordered field [MATH] An ordered field in which every nonempty set that has an upper bound also has a least upper bound.　{ kəm'plēt 'ȯrd·ərd 'fēld }

complete orthonormal set [MATH] A set of mutually orthogonal unit vectors in a (possibly infinite dimensional) vector space which is contained in no larger such set, that is no nonzero vector is perpendicular to all the vectors in the set. Also known as closed orthonormal set.　{ kəm'plēt 'ȯr·thō'nȯr·məl 'set }

complete primitive *See* complete integral.　{ kəm'plēt 'prim·əd·iv }

complete quadrangle [MATH] A plane figure consisting of a quadrangle and its two diagonals. Also known as complete quadrilateral.　{ kəm'plēt 'kwä,draŋ·gəl }

complete quadrilateral *See* complete quadrangle.　{ kəm'plēt ,kwä·drə'lad·ə·rəl }

complete residue system modulo n [MATH] A set of integers that includes one and only one member of each number class modulo *n*.　{ kəm'plēt 'rez·ə·dü 'sis·təm 'mäj·ə,lō 'en }

complete round [ORD] All components of ammunition that are needed to fire a given gun or firearm once.　{ kəm'plēt raůnd }

complete routine [COMPUT SCI] A routine, generally supplied by a computer manufacturer, which does not have to be modified by the user before being applied.　{ kəm'plēt rü'tēn }

complete space *See* complete metric space.　{ kəm'plēt 'spās }

complete system of representations [MATH] A set of representations of a group by matrices (or operators) such that, for any member of the group other than the identity, there is at least one representation for which this member does not correspond to the identity matrix (or the identity operator)　{ kəm'plēt 'sis·təm əv ,rep·ri·zen'tā·shənz }

completing the square [MATH] A method of solving quadratic equations, consisting of moving all terms to the left side of the equation, dividing through by the coefficient of the square term, and adding to both sides a number sufficient to make the left side a perfect square.　{ kəm'plēd·iŋ thə 'skwer }

completion [MATH] For a metric space *X*, a complete metric space obtained from *X* by formally adding limits to Cauchy sequences.　{ kəm'plē·shən }

completion fluid [PETRO ENG] A drilling mud with special properties that permit control of formation pressure and minimize formation damage during well completion.　{ kəm'plē·shən ,flü·əd }

complex [GEOL] An assemblage of rocks that has been folded together, intricately mixed, involved, or otherwise complicated. [MATH] A space which is represented as a union of simplices which intersect only on their faces. [MED] *See* syndrome. [MINERAL] Composed of many ingredients. [PSYCH] A group of associated ideas with strong emotional tones, which have been transferred from the conscious mind into the unconscious and which influence the personality.　{ 'käm,pleks }

complexation *See* complexing.　{ ,käm,plek'sā·shən }

complexation analysis [ANALY CHEM] The determination of the ligand/metal ratio in a coordination complex.　{ ,käm·plek'sā·shən ə,nal·ə·səs }

complexation indicator *See* metal ion indicator.　{ ,käm·plek'sā·shən ,in·də,kād·ər }

complexation reaction [CHEM] A chemical reaction that takes place between a metal ion and a molecular or ionic entity known as a ligand that contains at least one atom with an unshared pair of electrons.　{ ,käm·plek'sā·shən rē,ak·shən }

complex chemical reaction [CHEM] A chemical system in which a number of chemical reactions take place simultaneously, including reversible reactions, consecutive reactions, and concurrent or side reactions.　{ 'käm,pleks 'kem·i·kəl rē'ak·shən }

complex climatology [CLIMATOL] Analysis of the climate of a single space, or comparison of the climates of two or more places, by the relative frequencies of various weather types or groups of such types; a type is defined by the simultaneous occurrence within specified narrow limits of each of several weather elements.　{ 'käm,pleks 'klī·mə'täl·ə·jē }

complex compound [CHEM] Any of a group of chemical compounds in which a part of the molecular bonding is of the coordinate type. Also known as coordination complex.　{ 'käm,pleks 'käm,paůnd }

complex conjugate [MATH] **1.** One of a pair of complex numbers with identical real parts and with imaginary parts differing only in sign. Also known as conjugate. **2.** The matrix whose elements are the complex conjugates of the corresponding elements of a given matrix.　{ 'käm,pleks 'kän·jə·gət }

complex data type [COMPUT SCI] A scalar data type which contains two real fields representing the real and imaginary components of a complex number.　{ 'käm,pleks 'dad·ə ,tīp }

complex declaration statement [COMPUT SCI] A nonexecutable statement in FORTRAN used to specify that the type of identifier appearing in the program is of the form $a + bi$, where i is the square root of -1.　{ 'käm,pleks ,dek·lə'rā·shən ,stāt·mənt }

complex degree of coherence [PHYS] A measure of the coherence between two waves, equal to the cross-correlation function between the normalized amplitude of one wave and the complex conjugate of the normalized amplitude of the other.　{ 'käm,pleks də'grē əv ,kō'hir·əns }

complex dune [GEOL] A dune of varying forms, often very large, and produced by variable, shifting winds and the merging of various dune types.　{ 'käm,pleks 'dün }

complex fold [GEOL] A fold whose axial line is also folded.　{ 'käm,pleks 'fōld }

complex Fourier series [MATH] For a function $f(x)$, the series

$$\sum_{n=-\infty}^{\infty} c_n\, e^{inx}$$

with

$$c_n = \frac{1}{2\pi} \int_{-\pi}^{\pi} f(x)\, e^{-inx}\, dx$$

{ ,käm·pleks 'für·yā ,sir·ēz }

complex fraction [MATH] A fraction whose numerator or denominator is a fraction.　{ 'käm,pleks 'frak·shən }

complex frequency [ENG] A complex number used to characterize exponential and damped sinusoidal motion in the same way that an ordinary frequency characterizes simple harmonic motion; designated by the constant *s* corresponding to a motion whose amplitude is given by Ae^{st}, where *A* is a constant and *t* is time.　{ 'käm,pleks 'frē·kwən·sē }

compleximetric titration *See* complexometric titration.　{ kəm,plek·sə'me·trik tī'trā·shən }

complex impedance *See* electrical impedance; impedance.　{ 'käm,pleks im'pēd·əns }

complexing [CHEM] Formation of a complex compound. Also known as complexation.　{ 'käm,plek·siŋ }

complexing agent [CHEM] A substance capable of forming a complex compound with another material in solution.　{ 'käm,plek·siŋ ,ā·jənt }

complex instruction set computer [COMPUT SCI] A computer in which relatively high-level or complex hardware incorporating microcode is used to implement a relatively large number of instructions. Abbreviated CISC.　{ 'käm,pleks in'strək·shən ,set kəm,pyüd·ər }

complex integer *See* Gaussian integer.　{ 'käm,pleks 'int·ə·jər }

complex ion [CHEM] A complex, electrically charged group of atoms or radical, for example, $Cu(NH_3)_2^{+2}$.　{ 'käm,pleks 'ī,än }

complexity [COMPUT SCI] The number of elementary operations used by a program or algorithm to accomplish a given task.　{ kəm'plek·səd·ē }

complex low [METEOROL] An area of low atmospheric pressure within which more than one low-pressure center is found.　{ 'käm,pleks 'lō }

complex measure [MATH] A function whose domain is a sigma algebra of subsets of a particular set, whose range is in the complex numbers, whose value on the empty set is 0, and whose value on a countable union of pairwise disjoint sets is the sum of its values on each of these sets.　{ 'käm,pleks 'mezh·ər }

complex notation [PHYS] The representation of a physical

COMPLIANT SUBSTRATE

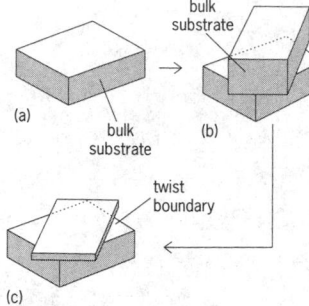

(a) bulk substrate

(b) bulk substrate

twist boundary

(c)

Process flow of forming a semiconductor compliant substrate containing a twist boundary. (*a*) Initial substrate. (*b*) Two substrates bonded at an angle. (*c*) Compliant substrate with a twist boundary, created by removing most of one substrate.

quantity by a complex number whose real component equals the instantaneous value of the physical quantity, a sinusoidally varying quantity thus being represented by a point rotating in a circle centered at the origin of the complex plane with uniform speed. { 'käm,pleks nō'tā·shən }

complex number [MATH] Any number of the form $a + bi$, where a and b are real numbers, and $i^2 = -1$. { 'käm,pleks 'nəm·bər }

complex number system [MATH] The field of complex numbers. { 'käm,pleks 'nəm·bər ,sis·təm }

complexometric titration [ANALY CHEM] A technique of volumetric analysis in which the formation of a colored complex is used to indicate the end point of a titration. Also known as chelatometry. Also spelled compleximetric titration. { kəm'plek·sə¦me·trik ,tī'trā·shən }

complex permeability [ELECTROMAG] A property, designated by μ^*, of a magnetic material, equal to $\mu_0 (L/L_0)$, where L is the complex inductance of an inductance coil in which the magnetic material forms the core when the coil is connected to a sinusoidal voltage source, and L_0 is the vacuum inductance of the coil. { 'käm,pleks ,pər·mē·ə'bil·əd·ē }

complex permittivity [ELEC] A property of a dielectric, equal to $\epsilon_0(C/C_0)$, where C is the complex capacitance of a capacitor in which the dielectric is the insulating material when the capacitor is connected to a sinusoidal voltage source, and C_0 is the vacuum capacitance of the capacitor. { 'käm,pleks ,pər·mə'tiv·əd·ē }

complex plane [MATH] A plane whose points are assigned the real and imaginary parts of complex numbers for coordinates. { 'käm,pleks 'plān }

complex potential [FL MECH] An analytic function in ideal aerodynamics whose real part is the velocity potential and whose imaginary part is the stream function. [NUC PHYS] A generalization of the potential in the Schrödinger equation describing the scattering of a nucleon by a nucleus in the cloudy crystal-ball model. { 'käm,pleks pə'ten·chəl }

complex reflector [ENG] A structure or group of structures having many radar-reflecting surfaces facing in different directions. { 'käm,pleks ri'flek·tər }

complex relative attenuation [ELECTR] The ratio of the peak output voltage, in complex notation, of an electric filter to the output voltage at the frequency being considered. { 'käm,pleks ¦rel·əd·iv ə,ten·yə'wā·shən }

complex salt [INORG CHEM] A class of salts in which there are no detectable quantities of each of the metal ions existing in solution; an example is $K_3Fe(CN)_6$, which in solution has K^+ but no Fe^{3+} because Fe is strongly bound in the complex ion, $Fe(CN)_6^{3-}$. { 'käm,pleks 'sôlt }

complex sphere *See* Riemann sphere. { 'käm,pleks 'sfir }

complex target [ENG] A radar target composed of a number of reflecting surfaces that, in the aggregate, are smaller in all dimensions than the resolution capabilities of the radar. { 'käm,pleks 'tär·gət }

complex tombolo [GEOL] A system resulting when several islands and the mainland are interconnected by a complex series of tombolos. Also known as tombolo cluster; tombolo series. { 'käm,pleks 'täm·bə,lō }

complex tone [ACOUS] A sound wave produced by the combination of simple sinusoidal components of different frequencies. { 'käm,pleks 'tōn }

complex unit [MATH] Any complex number, $x + iy$, whose absolute value, $\sqrt{(x^2 + y^2)}$, equals 1. { 'käm,pleks 'yü·nət }

complex variable [MATH] A variable which assumes complex numbers for values. { 'käm,pleks 'ver·ē·ə·bəl }

complex velocity [FL MECH] In ideal aerodynamic flow, the derivative of the complex potential with respect to $z = x + iy$, where x and y are the chosen coordinates. { 'käm,pleks və'läs·əd·ē }

complex wave [PHYS] A waveform which varies from instant to instant, but can be resolved into a number of sine-wave components, each of a different frequency and probably of a different amplitude. { 'käm,pleks 'wāv }

compliance [MECH] The displacement of a linear mechanical system under a unit force. { kəm'plī·əns }

compliance constant [MECH] Any one of the coefficients of the relations in the generalized Hooke's law used to express strain components as linear functions of the stress components. Also known as elastic constant. { kəm'plī·əns ,kän·stənt }

compliant character [PSYCH] In psychoanalytic theory, traits that include neurotic self-effacement, deference, and inappropriate yielding to another person. { kəm'plī·ənt 'kar·ik·tər }

compliant substrate [ELECTR] A semiconductor substrate into which an artificially formed interface is introduced near the surface which makes the substrate more readily deformable and allows it to support a defect-free semiconductor film of essentially any lattice constant, with dislocations forming in the substrate instead of in the film. Also known as sacrificial compliant substrate. { kəm'plī·ənt 'səb,strāt }

complicate [INV ZOO] Folded lengthwise several times, as applied to insect wings. { 'käm·plə,kāt }

compo board *See* composition board. { 'käm,pō ,bórd }

compole *See* commutating pole. { 'käm,pōl }

component [CHEM] **1.** A part of a mixture. **2.** The smallest number of chemical substances which are able to form all the constituents of a system in whatever proportion they may be present. [ELEC] Any electric device, such as a coil, resistor, capacitor, generator, line, or electron tube, having distinct electrical characteristics and having terminals at which it may be connected to other components to form a circuit. Also known as circuit element; element. [MATH] **1.** In a graph system, a connected subgraph which is not a subgraph of any other connected subgraph. **2.** For a set S, a connected subset of S that is not a subset of any other connected subset of S. **3.** The projection of a vector in a given direction of a coordinate system. [SCI TECH] A constituent part of a system; examples are a vector term which when added to others gives a vector sum, an ingredient of a chemical system, or the mineral portion of a rock. { kəm'pō·nənt }

component bar chart [STAT] A bar chart which shows within each bar the components that make up the bar; each component is represented by a section proportional in size to its representation in the total of each bar. { kəm'pō·nənt 'bär ,chärt }

component distillation [CHEM ENG] A distillation process in which a fraction that cannot normally be separated by distillation is removed by forming an azeotropic mixture. { kəm'pō·nənt dis·tə'lā·shən }

component-failure-impact analysis [SYS ENG] A study that attempts to predict the consequences of failures of the major components of a system. Abbreviated CFIA. { kəm'pō·nənt 'fāl·yər 'im,pakt ə,nal·ə·səs }

component name *See* metavariable. { kəm'pō·nənt ,nām }

component-substances law [CHEM] The law that each substance, singly or in mixture, composing a material exhibits specific properties that are independent of the other substances in that material. { kəm'pō·nənt 'sub·stən·səs ,ló }

component symbol [ELEC] A graphical design used to represent a component in a circuit diagram. { kəm'pō·nənt ,sim·bəl }

component vectors [MATH] Vectors parallel to specified (usually perpendicular) axes whose sum equals a given vector. { kəm'pō·nənt ,vek·tərz }

composing rule *See* composing stick. { kəm'pōz·iŋ ,rül }

composing stick [GRAPHICS] A tool designed for holding type which is being assembled and justified. { kəm'pōz·iŋ ,stik }

Compositae [BOT] The single family of the order Asterales; perhaps the largest family of flowering plants, it contains about 19,000 species. { kəm'päz·ə,tē }

composite [ENG ACOUS] A re-recording consisting of at least two elements. [MATER] A material that results when two or more materials, each having its own, usually different characteristics, are combined, giving useful properties for specific applications. Also known as composite material. { kəm'päz·ət }

composite balance [ELEC] An electric balance made by modifying the Kelvin balance to measure amperage, voltage, or wattage. { kəm'päz·ət 'bal·əns }

composite beam [CIV ENG] A structural member composed of two or more dissimilar materials joined together to act as a unit in which the resulting system is stronger than the sum of its parts. An example in civil structures is the steel-concrete composite beam in which a steel wide-flange shape (I or W shape) is attached to a concrete floor slab. { kəm'päz·ət 'bēm }

COMPOSITE BEAM

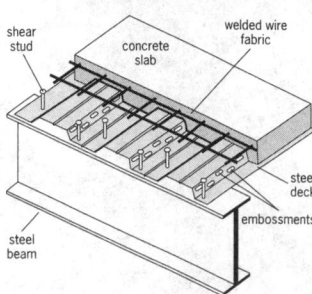

shear stud

concrete slab

welded wire fabric

steel deck

embossments

steel beam

Steel-concrete composite beam in which a steel wide-flange shape (W-shaped deck) is attached to a concrete floor slab.

composite cable [ELEC] Cable in which conductors of different gages or types are combined under one sheath. { kəm'päz·ət 'kā·bəl }

composite circuit [ELECTR] A circuit used simultaneously for voice communication and telegraphy, with frequency-discriminating networks serving to separate the two types of signals. { kəm'päz·ət 'sər·kət }

composite color signal [COMMUN] The color television picture signal plus all blanking and synchronizing signals. Also known as composite picture signal. { kəm'päz·ət 'kəl·ər ,sig·nəl }

composite color sync [COMMUN] The signal comprising all the synchronization signals necessary for proper operation of a color television receiver. { kəm'päz·ət 'kəl·ər ,siŋk }

composite column [CIV ENG] A concrete column having a structural-steel or cast-iron core with a maximum core area of 20. { kəm'päz·ət 'käl·əm }

composite compact [MET] A powder compact composed of more than one layer of different components with each layer retaining its identity. { kəm'päz·ət 'käm,pakt }

composite cone [GEOL] A large volcanic cone constructed of lava and pyroclastic material in alternating layers. { kəm'päz·ət 'kōn }

composite defense [ORD] In antiaircraft artillery, a defense that employs two or more types of fire units which are integrated into a single defense. { kəm'päz·ət də'fens }

composite dike [GEOL] A dike consisting of several intrusions differing in chemical and mineralogical composition. { kəm'päz·ət 'dīk }

composite electrode [MET] A filler-metal electrode composed of more than one metal. { kəm'päz·ət i'lek,trōd }

composite explosive [MATER] A mixture of substances which consume and give off oxygen, together with one or several simple explosives; dynamite is an example. { kəm'päz·ət ik'splō·siv }

composite filter [ELECTR] A filter constructed by linking filters of different kinds in series. { kəm'päz·ət 'fil·tər }

composite flash [GEOPHYS] A lightning discharge which is made up of a series of distinct lightning strokes with all strokes following the same or nearly the same channel, and with successive strokes occurring at intervals of about 0.05 second. Also known as multiple discharge. { kəm'päz·ət 'flash }

composite fold [GEOL] A fold having smaller folds on its limbs. { kəm'päz·ət 'fōld }

composite fuel [MATER] A broad class of solid chemical fuels composed of a fuel and oxidizer and used as propellants in rockets; an example of a fuel is phenol formaldehyde, and an oxidizer is ammonium perchlorate. Also known as composite propellant. { kəm'päz·ət 'fyül }

composite function [MATH] A function of one or more independent variables that are themselves functions of one or more other independent variables. { kəm'päz·ət 'fəŋk·shən }

composite gene [GEN] Any gene arising by recombination between two nonallelic genes and containing portions of both genes. { kəm'päz·ət 'jēn }

composite gneiss [PETR] A banded rock formed by intimate penetration of magma into country rocks. { kəm'päz·ət 'nīs }

composite grain [GEOL] A sedimentary clast formed of two or more original particles. { kəm'päz·ət 'grān }

composite group [MATH] A group that contains normal subgroups other than the identity element and the whole group. { kəm'päz·ət 'grüp }

composite hypothesis [STAT] A hypothesis that specifies a range of values for the distribution of the observed random variables. { kəm'päz·ət hī'päth·ə·səs }

composite I-beam bridge [CIV ENG] A beam bridge in which the concrete roadway is mechanically bonded to the I beams by means of shear connectors. { kəm'päz·ət 'I ,bēm ,brij }

composite joint [MET] A joint connected by welding in conjunction with one or more mechanical means. { kəm'päz·ət 'jóint }

composite macromechanics [ENG] The study of composite material behavior wherein the material is presumed homogeneous and the effects of the constituent materials are detected only as averaged apparent properties of the composite. { kəm'päz·ət ¦mak·rō·mə'kan·iks }

composite map [MIN ENG] A map in which several levels of a mine are shown on a single sheet. { kəm'päz·ət 'map }

composite material See composite. { kəm'päz·ət mə'tir·ē·əl }

composite micromechanics [ENG] The study of composite material behavior wherein the constituent materials are studied on a microscopic scale with specific properties being assigned to each constituent; the interaction of the constituent materials is used to determine the properties of the composite. { kəm'päz·ət ¦mik·rō·mə'kan·iks }

composite nerve [NEURO] A nerve containing both sensory and motor fibers. { kəm'päz·ət 'nərv }

composite number [MATH] Any positive integer which is not prime. Also known as composite quantity. { kəm'päz·ət 'nəm·bər }

composite photograph [GRAPHICS] An assembly of separate photographs, made by several lenses of a multiple-lens camera in simultaneous exposure, into the equivalent of a photograph made with a wide-angle lens. { kəm'päz·ət 'fōd·ə,graf }

composite picture signal See composite color signal. { kəm'päz·ət 'pik·chər ,sig·nəl }

composite pile [CIV ENG] A pile in which the upper and lower portions consist of different types of piles. { kəm'päz·ət 'pīl }

composite plate [MET] A layer of electrodeposited material consisting of at least two different constituents. { kəm'päz·ət 'plāt }

composite profile [MAP] A profile comprising the highest points of a series of profiles that are drawn along several regularly spaced and parallel map lines. { kəm'päz·ət 'prō,fīl }

composite propellant See composite fuel. { kəm'päz·ət prə'pel·ənt }

composite pulse [ELECTR] A pulse composed of a series of overlapping pulses received from the same source over several paths in a pulse navigation system. { kəm'päz·ət 'pəls }

composite quantity See composite number. { kəm'päz·ət 'kwän·əd·ē }

composite sailing [NAV] In marine operations, a modification of great-circle sailing, used when limiting the highest latitude. { kəm'päz·ət 'sāl·iŋ }

composite sample [ANALY CHEM] A sample comprising two or more increments selected to represent the material being analyzed. { kəm'päz·ət 'sam·pəl }

composite sampler [ENG] A hydrometer cylinder equipped with sample cocks at regular intervals along its vertical height; used to take representative (vertical composite) samples of oil from storage tanks. { kəm'päz·ət 'sam·plər }

composite sequence [GEOL] An ideal sequence of cyclic sediments containing all the lithological types in their proper order. { kəm'päz·ət 'sē·kwəns }

composite set [ELECTR] Assembly of apparatus designed to provide one end of a composite circuit. { kəm'päz·ət 'set }

composite sill [GEOL] A sill consisting of several intrusions differing in chemical and mineralogical compositions. { kəm'päz·ət 'sil }

composite steel [MET] Bar steel machined along the entire length which is cast around an insert of tool steel welded to the backing of mild steel; used for shear blades and die parts. { kəm'päz·ət 'stēl }

composite stream [PETRO ENG] A flow of oil and gas or a flow of two or more different hydrocarbons in one stream. { kəm'päz·ət ,strēm }

composite topography [GEOL] A topography whose features have developed in two or more erosion cycles. { kəm'päz·ət tə'päg·rə·fē }

composite track [NAV] A modified great-circle track consisting of an initial great-circle track from the point of departure with its vertex on a limiting parallel of latitude; a parallel-sailing track from this vertex along the limiting parallel to the vertex of a final great-circle track passing through the destination. { kəm'päz·ət 'trak }

composite truss [CIV ENG] A truss having compressive members and tension members. { kəm'päz·ət 'trəs }

composite unconformity [GEOL] An unconformity that has resulted from more than one episode of nondeposition and possible erosion. { kəm'päz·ət ,ən·kən'fór·məd·ē }

composite vein [GEOL] A large fracture zone composed of

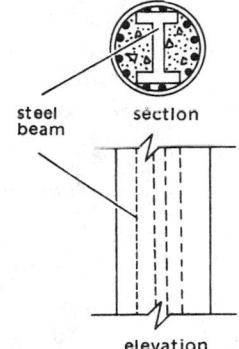

COMPOSITE COLUMN

steel beam   section

elevation

Cross section and elevation of a composite column.

parallel ore-filled fissures and converging diagonals, whose walls and intervening country rock have been replaced to a certain degree. { kəm'päz·ət 'vān }

composite video signal [COMMUN] The video-only portion of the standard color television signal used in the United States, in which red, green, and blue signals are mixed. { kəm'päz·ət 'vid·ē·ō ˌsig·nəl }

composite volcano See stratovolcano. { kəm'päz·ət väl'kā·nō }

composite wave filter [ELECTR] A combination of two or more low-pass, high-pass, band-pass, or band-elimination filters. { kəm'päz·ət 'wāv ˌfil·tər }

composition [CHEM] The elements or compounds making up a material or produced from it by analysis. [GRAPHICS] The act of composing or combining type for printing, either by hand or by machine. [MATH] **1.** The composition of two mappings, f and g, denoted $g \circ f$, where the domain of g includes the range of f, is the mapping which assigns to each element x in the domain of f the element $g(y)$, where $y = f(x)$. **2.** See addition. [MECH] The determination of a force whose effect is the same as that of two or more given forces acting simultaneously; all forces are considered acting at the same point. { ˌkäm·pə'zish·ən }

compositional maturity [GEOL] Concept of a type of maturity in sedimentary rocks in which a sediment approaches the compositional end product to which formative processes drive it. { ˌkäm·pə'zish·ən·əl mə'chùr·əd·ē }

composition board [MATER] A sheet product composed of vegetable fibers mechanically or chemically formed into a pulp which is rolled and pressed. Also known as compo board. { ˌkäm·pə'zish·ən ˌbȯrd }

composition diagram [CHEM ENG] Graphical plots to show the solvent-solute concentration relationships during various stages of extraction operations (leaching, or solid-liquid extraction; and liquid-liquid extraction). { ˌkäm·pə'zish·ən ˌdī·əˌgram }

composition face See composition surface. { ˌkäm·pə'zish·ən ˌfās }

composition metal [MET] A cast copper alloy having a composition of more than 80% copper, with tin, zinc, and lead. { ˌkäm·pə'zish·ən ˌmed·əl }

composition-of-velocities law [MECH] A law relating the velocities of an object in two references frames which are moving relative to each other with a specified velocity. { ˌkäm·pə'zish·ən əv və'läs·əd·ēz ˌlȯ }

composition plane [CRYSTAL] A planar composition surface in a crystal uniting two individuals of a contact twin. { ˌkäm·pə'zish·ənˌplān }

composition resistor See carbon resistor. { ˌkäm·pə'zish·ən ri'zis·tər }

composition series [MATH] A normal series $G_1, G_2, \ldots$, of a group, where each G_i is a proper normal subgroup of G_{i-1} and no further normal subgroups both contain G_i and are contained in G_{i-1}. { ˌkäm·pə'zish·ən 'sir,ēz }

composition surface [CRYSTAL] The surface uniting individuals of a crystal twin; may or may not be planar. Also known as composition face. { ˌkäm·pə'zish·ən ˌsər·fəs }

compositum [MATH] Let E and F be fields, both contained in some field L; the compositum of E and F, denoted EF, is the smallest subfield of L containing E and F. { kəm'päz·əd·əm }

compost [MATER] A mixture of decaying organic matter used to fertilize and condition the soil. { 'kämˌpōst }

compound [CHEM] A substance whose molecules consist of unlike atoms and whose constituents cannot be separated by physical means. Also known as chemical compound. [PETRO ENG] A power transmission mechanism that transfers power from the engines to the pump, drawworks, and other machinery on a drilling rig. { 'kämˌpaùnd }

compound acinous gland [ANAT] A structure with spherical secreting units connected to many ducts that empty into a common duct. { 'kämˌpaùnd 'as·ə·nəs ˌgland }

compound alluvial fan [GEOL] Structure formed by the lateral growth and merger of fans made by neighboring streams. { 'kämˌpaùnd ə'lü·vē·əl ˌfan }

compound angle [ENG] The angle formed by two mitered angles. { 'kämˌpaùnd 'aŋ·gəl }

compound compact [MET] A powder compact made from

a mixture of metals, with each particle retaining its original composition. { 'kämˌpaùnd 'kämˌpakt }

compound cryosar [ELECTR] A cryosar consisting of two normal cryosars with different electrical characteristics in series. { 'kämˌpaùnd 'krī·ōˌsär }

compound curve [MATH] A curve made up of two arcs of differing radii whose centers are on the same side, connected by a common tangent; used to lay out railroad curves because curvature goes from nothing to a maximum gradually, and vice versa. { 'kämˌpaùnd 'kərv }

compound die [MET] A die designed to perform more than one operation on the work with each stroke of the press. { 'kämˌpaùnd 'dī }

compound distribution [STAT] A frequency distribution resulting from the combining of two or more separate distributions of the same general type. { ˌkämˌpaùnd ˌdis·trə'byü·shən }

compound document [COMPUT SCI] A document that contains two or more different data structures, such as text, graphics, and sound. { ˌkämˌpaùnd 'däk·yə·mənt }

compound elastic scattering [NUC PHYS] Scattering in which the final state is the same as the initial state, but there is an intermediate state with the colliding systems amalgamating to form a compound system. { 'kämˌpaùnd i'las·dik 'skad·ə·riŋ }

compound engine [MECH ENG] A multicylinder-type displacement engine, using steam, air, or hot gas, where expansion proceeds successively (sequentially). { 'kämˌpaùnd 'en·jən }

compound event [STAT] **1.** An event whose probability of occurrence depends upon the probability of occurrence of two or more independent events. **2.** An event that consists of two or more events that are not mutually exclusive. { 'kämˌpaùnd i'vent }

compound eye [INV ZOO] An eye typical of crustaceans, insects, centipedes, and horseshoe crabs, constructed of many functionally independent photoreceptor units (ommatidia) separated by pigment cells. { 'kämˌpaùnd 'ī }

compound fault [GEOL] A zone or series of essentially parallel faults, closely spaced. { 'kämˌpaùnd 'fȯlt }

compound field winding [ELEC] A winding composed of shunt and series coils that act either together or against each other. { 'kämˌpaùnd 'fēld ˌwind·iŋ }

compound generator [ELEC] A direct-current generator which has both a series field winding and a shunt field winding, both on the main poles with the shunt field winding on the outside. { 'kämˌpaùnd 'jen·ə'rād·ər }

compound gland [ANAT] A secretory structure with many ducts. { 'kämˌpaùnd 'gland }

compounding [MECH ENG] The series placing of cylinders in an engine (such as steam) for greater ratios of expansion and consequent improved engine economy. { 'kämˌpaùnd·iŋ }

compound layering [BOT] A plant propagation technique in which more than one portion of the same stem is buried. { ˌkämˌpaùnd 'lā·ər·iŋ }

compound leaf [BOT] A type of leaf with the blade divided into two or more separate parts called leaflets. { 'kämˌpaùnd 'lēf }

compound lens [OPTICS] **1.** A combination of two or more lenses in which the second surface of one lens has the same radius as the first surface of the following lens, and the two lenses are cemented together. Also known as cemented lens. **2.** Any optical system consisting of more than one element, even when they are not in contact. { 'kämˌpaùnd 'lenz }

compound lever [MECH ENG] A train of levers in which motion or force is transmitted from the arm of one lever to that of the next. { 'kämˌpaùnd 'lev·ər }

compound magnet [ELEC] A permanent magnet that is constructed from a number of thin magnets having the same shape. { 'kämˌpaùnd 'mag·nət }

compound microscope [OPTICS] A microscope which utilizes two lenses or lens systems; one lens forms an enlarged image of the object, and the second magnifies the image formed by the first. { 'kämˌpaùnd 'mī·krəˌskōp }

compound modulation See multiple modulation. { 'kämˌpaùnd ˌmäj·ə'lā·shən }

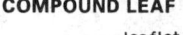

COMPOUND LEAF

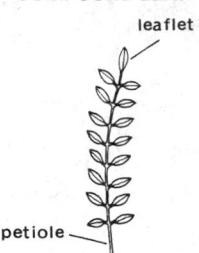

leaflet

petiole

Odd-pinnately compound leaf.

COMPOUND MICROSCOPE

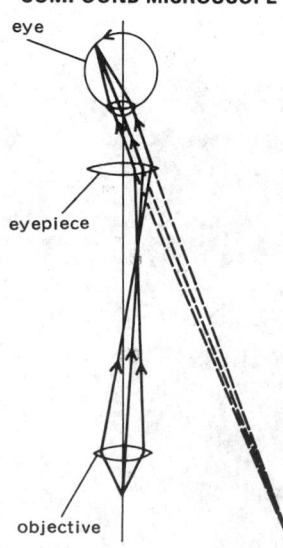

eye

eyepiece

objective

Compound microscope diagram. *(From F. A. Jenkins and H. E. White, Fundamentals of Optics, 3d ed., McGraw-Hill, 1957)*

compound motor [ELEC] A direct-current motor with two separate field windings, one connected in parallel with the armature circuit, the other connected in series with the armature circuit. { 'käm,paùnd 'mōd·ər }

compound nucleus [NUC PHYS] An intermediate state in a nuclear reaction in which the incident particle combines with the target nucleus and its energy is shared among all the nucleons of the system. { 'käm,paùnd 'nü·klē·əs }

compound number [MATH] A quantity which is expressed as the sum of two or more quantities in terms of different units, for example, 3 feet 10 inches, or 2 pounds 5 ounces. { 'käm,paùnd 'nəm·bər }

compound pendulum See pendulum. { 'käm,paùnd 'pen·jə·ləm }

compound pistil [BOT] A pistil composed of two or more united carpels. { 'käm,paùnd 'pis·təl }

compound rest [MECH ENG] A principal component of a lathe consisting of a base and an upper part dovetailed together; the base is graduated in degrees and can be swiveled to any angle; the upper part includes the tool post and tool holder. { 'käm,paùnd 'rest }

compound ripple marks [GEOL] Complex ripple marks of great diversity which originate by simultaneous interference of wave oscillation with current action. { 'käm,paùnd 'rip·əl ,märks }

compound screw [DES ENG] A screw having different or opposite pitches on opposite ends of the shank. { 'käm,paùnd 'skrü }

compound shaft [MIN ENG] A shaft in which the upper stage is often a vertical shaft, while the lower stage, or stages, may be inclined and driven into the deposit. { 'käm,paùnd 'shaft }

compound statement [COMPUT SCI] A single program instruction that contains two or more instructions which could stand alone. { 'käm,paùnd 'stāt·mənt }

compound sugar See oligosaccharide. { 'käm,paùnd 'shùg·ər }

compound tubular-acinous gland [ANAT] A structure in which the secreting units are simple tubes with acinous side chambers and all are connected to a common duct. { 'käm,paùnd 'tüb·yə·lər ¦as·ə·nəs ,gland }

compound tubular gland [ANAT] A structure having branched ducts between the surface opening and the secreting portion. { 'käm,paùnd 'tüb·yə·lər ,gland }

compound twins [CRYSTAL] Individuals of one mineral group united in accordance with two or more different twin laws. { 'käm,paùnd 'twinz }

compound valley glacier [HYD] A glacier composed of several ice streams emanating from different tributary valleys. { 'käm,paùnd 'val·ē ,glā·shər }

compound volcano [GEOL] **1.** A volcano consisting of a complex of two or more cones. **2.** A volcano with an associated volcanic dome. { 'käm,paùnd väl'kā·nō }

compound wave [FL MECH] A plane wave of finite amplitude in which neither the sum of the velocity potential and the component of velocity in the direction of wave motion, nor the difference of these two quantities, is constant. { 'käm,paùnd 'wāv }

compound winding [ELEC] A winding that is a combination of series and shunt winding. { 'käm,paùnd 'wīnd·iŋ }

compregnate [ENG] Compression of materials into a dense, hard substance with the aid of heat. { kəm'preg,nāt }

compressadensity function [MECH] A function used in the acoustic levitation technique to determine either the density or the adiabatic compressibility of a submicroliter droplet suspended in another liquid, if the other property is known. { kəm,pres·ə'den·səd·ē ,fəŋk·shən }

compressed air [MECH] Air whose density is increased by subjecting it to a pressure greater than atmospheric pressure. { kəm'prest 'er }

compressed-air blasting [MIN ENG] A method for breaking down coal by compressed-air power. { kəm'prest ¦er 'blast·iŋ }

compressed-air diving [ENG] Any form of diving in which air is supplied under high pressure to prevent lung collapse. { kəm'prest ¦er 'dīv·iŋ }

compressed-air illness See caisson disease. { kəm'prest ¦er 'il·nəs }

compressed-air loudspeaker [ENG ACOUS] A loudspeaker having an electrically actuated valve that modulates a stream of compressed air. { kəm'prest ¦er 'laùd,spēk·ər }

compressed-air power [MECH ENG] The power delivered by the pressure of compressed air as it expands, utilized in tools such as drills, in hoists, grinders, riveters, diggers, pile drivers, motors, locomotives, and in mine ventilating systems. { kəm'prest ¦er 'paùr·ər }

compressed file See packed file. { kəm,prest 'fīl }

compressed straw slab See strawboard. { kəm'prest ¦strô 'slab }

compressibility [MECH] The property of a substance capable of being reduced in volume by application of pressure; quantitively, the reciprocal of the bulk modulus. { kəm,pres·ə'bil·əd·ē }

compressibility burble [FL MECH] A region of disturbed flow, produced by and rearward of a shock wave. { kəm,pres·ə'bil·əd·ē ,bər·bəl }

compressibility correction [FL MECH] The correction of the calibrated airspeed caused by compressibility error. { kəm,pres·ə'bil·əd·ē kə'rek·shən }

compressibility error [FL MECH] The error in the readings of a differential-pressure-type airspeed indicator due to compression of the air on the forward part of the pitot tube component moving at high speeds. { kəm,pres·ə'bil·əd·ē ,er·ər }

compressibility factor [THERMO] The product of the pressure and the volume of a gas, divided by the product of the temperature of the gas and the gas constant; this factor may be inserted in the ideal gas law to take into account the departure of true gases from ideal gas behavior. Also known as deviation factor; gas-deviation factor; supercompressibility factor. { kəm,pres·ə'bil·əd·ē ,fak·tər }

compressible flow [FL MECH] Flow in which the fluid density varies. { kəm'pres·ə·bəl 'flō }

compressible-flow principle [FL MECH] The principle that when flow velocity is large, it is necessary to consider that the fluid is compressible rather than to assume that it has a constant density. { kəm¦pres·ə·bal ¦flō 'prin·sə·pəl }

compressible fluid flow [CHEM ENG] Gas flow when the pressure drop due to the flow of a gas through a system is large enough, compared with the inlet pressure, to cause a 10% or greater decrease in gas density. { kəm'pres·ə·bəl 'flü·əd ,flō }

compression [COMPUT SCI] See data compression. [ELECTR] **1.** Reduction of the effective gain of a device at one level of signal with respect to the gain at a lower level of signal, so that weak signal components will not be lost in background and strong signals will not overload the system. **2.** See compression ratio. [GEOD] See flattening. [GEOL] A system of forces which tend to decrease the volume or shorten rocks. [MECH] Reduction in the volume of a substance due to pressure; for example in building, the type of stress which causes shortening of the fibers of a wooden member. [MECH ENG] See compression ratio. { kəm'presh·ən }

compressional wave [PHYS] A disturbance traveling in an elastic medium; characterized by changes in volume and by particle motion parallel with the direction of wave movement. Also known as dilatational wave; irrotational wave; pressure wave; P wave. { kəm'presh·ən·əl ,wāv }

compression cable See pressure cable. { kəm'presh·ən ,kā·bəl }

compression coupling [MECH ENG] **1.** A means of connecting two perfectly aligned shafts in which a slotted tapered sleeve is placed over the junction and two flanges are drawn over the sleeve so that they automatically center the shafts and provide sufficient contact pressure to transmit medium loads. **2.** A type of tubing fitting. { kəm'presh·ən ,kəp·liŋ }

compression cup [ENG] A cup from which lubricant is forced to a bearing by compression. { kəm'presh·ən ,kəp }

compression failure [ENG] Buckling or collapse caused by compression, as of a steel or concrete column or of wood fibers. { kəm'presh·ən ,fāl·yər }

compression fitting [ENG] A leak-resistant pipe joint designed with a tight-fitting sleeve that exerts a large inward pressure on the exterior of the pipe. { kəm'presh·ən ,fid·iŋ }

compression gage [ENG] An instrument that measures pressures greater than atmospheric pressure. { kəm'presh·ən ,gāj }

compression ignition [MECH ENG] Ignition produced by

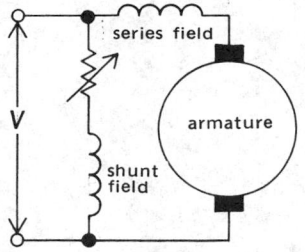

Connection of a compound motor.

One of the distal secreting units of a compound tubular acinous gland.

compression of the air in a cylinder of an internal combustion engine before fuel is admitted. { kəm'presh·ən ig'nish·ən }

compression-ignition engine *See* diesel engine. { kəm'presh·ən ig'nish·ən 'en·jən }

compression machine *See* compressor. { kəm'presh·ən mə'shēn }

compression member [ENG] A beam or other structural member which is subject to compressive stress. { kəm'presh·ən ‚mem·bər }

compression modulus *See* bulk modulus of elasticity. { kəm'presh·ən ‚mäj·ə·ləs }

compression mold [ENG] A mold for plastics which is open when the material is introduced and which shapes the material by heat and by the pressure of closing. { kəm'presh·ən ‚mōld }

compression plant [PETRO ENG] Gas-compression facility used to produce a high-pressure gas stream for injection into reservoir formations to increase oil yield; when the injected gas is that recovered from the well during oil production, the facility is called a gas-cycling plant. { kəm'presh·ən ‚plant }

compression pressure [MECH ENG] That pressure developed in a reciprocating piston engine at the end of the compression stroke without combustion of fuel. { kəm'presh·ən ‚presh·ər }

compression process [CHEM ENG] The recovery of natural gasoline from gas containing a high proportion of hydrocarbons. { kəm'presh·ən ‚prä·səs }

compression ratio [ELECTR] The ratio of the gain of a device at a low power level to the gain at some higher level, usually expressed in decibels. Also known as compression. [MECH ENG] The ratio in internal combustion engines between the volume displaced by the piston plus the clearance space, to the volume of the clearance space. Also known as compression. [MET] Ratio of the volume of loose metal powder to the volume of the compact made from it. { kəm'presh·ən ‚rā·shō }

compression refrigeration [MECH ENG] The cooling of a gaseous refrigerant by first compressing it to liquid form (with resultant heat buildup), cooling the liquid by heat exchange, then releasing pressure to allow the liquid to vaporize (with resultant absorption of latent heat of vaporization and a refrigerative effect). { kəm'presh·ən ri‚frij·ə'rā·shən }

compression release [MECH ENG] Release of compressed gas resulting from incomplete closure of intake or exhaust valves. { kəm'presh·ən ri'lēs }

compression ring [MECH ENG] A ring located at the upper part of a piston to hold the burning fuel charge above the piston in the combustion chamber, thus preventing blowby. { kəm'presh·ən ‚riŋ }

compression spring [ENG] A spring, usually a coil spring, which resists a force tending to compress it. { kəm'presh·ən ‚spriŋ }

compression strength [MECH] Property of a material to resist rupture under compression. { kəm'presh·ən ‚streŋkth }

compression stroke [MECH ENG] The phase of a positive displacement engine or compressor in which the motion of the piston compresses the fluid trapped in the cylinder. { kəm'presh·ən ‚strōk }

compression syndrome *See* crush syndrome. { kəm'presh·ən ‚sin‚drōm }

compression test [ENG] A test to determine compression strength, usually applied to materials of high compression but low tensile strength, in which the specimen is subjected to increasing compressive forces until failure occurs. { kəm'presh·ən ‚test }

compression wave [FL MECH] A wave in a fluid in which a compression is propagated. { kəm'presh·ən ‚wāv }

compression wood [BOT] Dense wood found at the base of some tree trunks and on the undersides of branches. { kəm'presh·ən ‚wu̇d }

compressive intercept receiver [ELECTR] An electromagnetic surveillance receiver that instantaneously analyzes and sorts all signals within a broad radio-frequency spectrum by using pulse compression techniques which perform a complete analysis up to 10,000 times faster than a superheterodyne receiver or spectrum analyzer. { kəm'pres·iv 'in·tər‚sept ri'sē·vər }

compressive member [CIV ENG] A structural member subject to tension. { kəm'pres·iv 'mem·bər }

compressive shrinkage [TEXT] A technique for producing shrink-resistant cotton or linen fabric by lightly dampening the fabric, placing it on a thick blanket, and then pressing it against a heated cylinder, causing it to retract. { kəm'pres·iv 'shriŋk·ij }

compressive strength [MECH] The maximum compressive stress a material can withstand without failure. { kəm'pres·iv 'streŋkth }

compressive stress [MECH] A stress which causes an elastic body to shorten in the direction of the applied force. { kəm'pres·iv 'stres }

compressor [COMPUT SCI] A routine or program that reduces the number of binary digits needed to represent data or information. [ELECTR] The part of a compandor that is used to compress the intensity range of signals at the transmitting or recording end of a circuit. [MECH ENG] A machine used for increasing the pressure of a gas or vapor. Also known as compression machine. { kəm'pres·ər }

compressor blade [MECH ENG] The vane components of a centrifugal or axial-flow, air or gas compressor. { kəm'pres·ər ‚blād }

compressor station [MECH ENG] A permanent facility which increases the pressure on gas to move it in transmission lines or into storage. { kəm'pres·ər ‚stā·shən }

compressor valve [MECH ENG] A valve in a compressor, usually automatic, which operates by pressure difference (less than 5 pounds per square inch or 35 kilopascals) on the two sides of a movable, single-loaded member and which has no mechanical linkage with the moving parts of the compressor mechanism. { kəm'pres·ər ‚valv }

compromise joint [CIV ENG] **1.** A joint bar used for joining rails of different height or section. **2.** A rail that has different joint drillings from that of the same section. { 'käm·prə‚mīz ‚jȯint }

compromise network [ELEC] **1.** Network employed in conjunction with a hybrid coil to balance a subscriber's loop; adjusted for an average loop length or an average subscriber's set, or both, to secure compromise (not precision) isolation between the two directional paths of the hybrid. **2.** Hybrid balancing network which is designed to balance the average of the impedances that may be connected to the switchboard side of a hybrid arrangement of a repeater. { 'käm·prə‚mīz 'net‚wərk }

compromise rail [CIV ENG] A short rail having different sections at the ends to correspond with the rail ends to be joined, thus providing a transition between rails of different sections. { 'käm·prə‚mīz ‚rāl }

compromising emanations [COMMUN] Unintentional data-related or intelligence-bearing signals which, if intercepted and analyzed by any technique, could disclose the classified information transmitted, received, handled, or otherwise processed by equipments. { 'käm·prə‚miz·iŋ ‚em·ə'nā·shənz }

Compton absorption [QUANT MECH] The absorption of an x-ray or gamma-ray photon in Compton scattering, accompanied by the emission of another photon of lower energy. { 'käm·tən əb'sȯrp·shən }

Compton cross section [QUANT MECH] The differential cross section for the elastic scattering of photons by electrons. { 'käm·tən 'krȯs ‚sek·shən }

Compton-Debye effect *See* Compton effect. { ‚käm·tən də'be·ə i'fekt }

Compton effect [QUANT MECH] The increase in wavelength of electromagnetic radiation in the x-ray and gamma-ray region on being scattered by material objects; the scattering is due to the interaction of the photons with electrons that are effectively free. Also known as Compton-Debye effect. { 'käm·tən i'fekt }

Compton electron *See* Compton recoil electron. { 'käm·tən i'lek‚trän }

Compton equation [QUANT MECH] The equation for the change in wavelength $\Delta\lambda$ of radiation scattered by electrons in the Compton effect, $\Delta\lambda = \lambda_c(1 - \cos\theta)$, where λ_c is the Compton wavelength of the electron, and θ is the angle between the directions of incident and scattered radiation. { 'käm·tən i‚kwä·zhən }

Compton-Getting effect [ASTROPHYS] The sidereal diurnal variation of the intensity of cosmic rays which would be expected from the rotation of the galaxy if cosmic radiation originated in extragalactic regions and was isotropic in intergalactic space, and if this radiation was unaffected at entry to and passage through the galaxy. { ˈkäm·tən ˈged·iŋ iˈfekt }

Compton incoherent scattering [NUC PHYS] Scattering of gamma rays by individual nucleons in a nucleus or electrons in an atom when the energy of the gamma rays is large enough so that binding effects may be neglected. { ˈkäm·tən in·kōˈhir·ənt ˈskad·ə·riŋ }

comptonization [ASTRON] The redistribution in the energies of photons in interstellar space that results from their scattering from electrons. { ˌkäm·tə·nəˈzā·shən }

Compton meter [NUCLEO] An ionization chamber having a balance chamber with a uranium source that is adjusted until it balances out normal cosmic radiation; variations in cosmic radiation are then shown on an electrometer. { ˈkäm·tən ˌmēd·ər }

Compton process See Compton scattering. { ˈkäm·tən ˌpräs·əs }

Compton recoil electron [QUANT MECH] An electron set in motion by its interaction with a photon in Compton scattering. Also known as Compton electron. { ˈkäm·tən riˈkȯil iˈlek·trän }

Compton recoil particle [QUANT MECH] Any particle that has acquired its momentum in a scattering process similar to Compton scattering. { ˈkäm·tən riˈkȯil ˌpard·ə·kəl }

Compton rule [PHYS CHEM] An empirical law stating that the heat of fusion of an element times its atomic weight divided by its melting point in degrees Kelvin equals approximately 2. { ˈkäm·tən ˌrül }

Compton scattering [QUANT MECH] The elastic scattering of photons by electrons. Also known as Compton process; gamma-ray scattering. { ˈkäm·tən ˌskad·ə·riŋ }

Compton shift [QUANT MECH] The change in wavelength of scattered radiation due to the Compton effect. { ˈkäm·tən ˌshift }

Compton wavelength [QUANT MECH] A convenient unit of length that is characteristic of an elementary particle, equal to Planck's constant divided by the product of the particle's mass and the speed of light. { ˈkäm·tən ˈwāv·ˌleŋkth }

compulsion [PSYCH] An irresistible, impulsive act performed by an individual against his conscious will and usually arising from an obsession. { kəmˈpəl·shən }

compulsive personality disorder [PSYCH] A personality disorder characterized by a preoccupation with rules, order, organization, efficiency, and detail. { kəmˈpəl·siv ˌpər·səˈnal·əd·ē disˌȯr·dər }

compulsive reaction [PSYCH] Behavior disorder in which a person is uncomfortable with conditions of ambiguity and uncertainty. { kəmˈpəl·siv rēˈak·shən }

compulsory reporting points [NAV] In air operations, geographical points for which an aircraft must report; these points are designated by regulations and can be approved or deleted only by rule-making action. { kəmˈpəl·sə·rē riˈpȯrd·iŋ ˌpȯins }

computable function [MATH] A function whose value can be calculated by some Turing machine in a finite number of steps. Also known as effectively computable function. { kəmˈpyüd·ə·bəl ˈfəŋk·shən }

computation [MATH] **1.** The act or process of calculating. **2.** The result so obtained. { ˌkäm·pyəˈtā·shən }

computational aeroacoustics [ACOUS] The study of the problems of aeroacoustics using computational techniques. { ˌkäm·pyüˈtā·shən·əl ˌer·ō·əˈküs·tiks }

computational chemistry [CHEM] The use of calculations to predict molecular structure, properties, and reactions. { ˌkäm·pyəˈtā·shən·əl ˈkem·ə·strē }

computational flow imaging [FL MECH] A technology for generating digital images of theoretic fluid dynamic phenomena in optical formats that mimic real observations of the corresponding real flow fields. Abbreviated CFI. { ˌkäm·pyə·ˌtā·shən·əl ˈflō ˌim·ij·iŋ }

computational fluid dynamics [FL MECH] A field of study concerned with the use of high-speed digital computers to numerically solve the complete nonlinear partial differential equations governing viscous fluid flows. { ˌkäm·pyəˈtā·shən·əl ˈflü·əd dīˈnam·iks }

computational numerical control See computer numerical control. { ˌkäm·pyəˈtā·shən·əl nüˈmer·ə·kəl kənˈtrōl }

computational statistics [STAT] The conversion of statistical algorithms into computer code that can retrieve useful information from large, complex data sets. Also known as statistical computing. { ˌkäm·pyüˈtā·shən·əl stəˈtis·tiks }

compute-bound program See CPU-bound program. { kəmˈpyüt ˌbau̇nd ˈprō·grəm }

computed altitude [NAV] **1.** In celestial navigation, tabulated altitude interpolated for increments of latitude, declination, or hour angle; if no interpolation is required, the tabulated altitude and computed altitude are identical. **2.** Altitude determined by computation, table, mechanical computer, or graphics, particularly such an altitude of the center of a celestial body measured as an arc on a vertical circle of the celestial sphere from the celestial horizon. Also known as calculated altitude. { kəmˈpyüd·əd ˈal·tə·tüd }

computed azimuth [NAV] In celestial navigation, an azimuth determined by computation, table, mechanical device, or graphics for a given place and time. { kəmˈpyüd·əd ˈaz·ə·məth }

computed azimuth angle [NAV] Azimuth angle determined by computation, table, mechanical device, or graphics for a given place and time. { kəmˈpyüd·əd ˈaz·ə·məth ˌaŋ·gəl }

computed go to [COMPUT SCI] A control procedure in FORTRAN which allows the transfer of control to the *i*th label of a set of *n* labels used as statement numbers in the program. { kəmˈpyüd·əd ˈgō ˌtü }

computed path control [CONT SYS] A control system designed to follow a path calculated to be the optimal one to achieve a desired result. { kəmˈpyüd·əd ˈpath kənˈtrōl }

computed tomography See computerized tomography. { kəmˈpyüd·əd təˈmä·grə·fē }

compute mode [COMPUT SCI] The operation of an analog computer in which input signals are used by the computing units to calculate a solution, in contrast to hold mode and reset mode. { kəmˈpyüt ˌmōd }

computer [COMPUT SCI] A device that receives, processes, and presents data; the two types are analog and digital. Also known as computing machine. { kəmˈpyüd·ər }

computer-aided design [CONT SYS] The use of computers in converting the initial idea for a product into a detailed engineering design. Computer models and graphics replace the sketches and engineering drawings traditionally used to visualize products and communicate design information. Abbreviated CAD. { kəmˈpyüd·ər ˌād·əd dəˈzīn }

computer-aided design and drafting [COMPUT SCI] The carrying out of computer-aided design with a system that has additional features for the drafting function, such as dimensioning and text entry. Abbreviated CADD. { kəmˈpyüd·ər ˌād·əd diˈzīn ən ˈdraft·iŋ }

computer-aided engineering [ENG] The use of computer-based tools to assist in solution of engineering problems. { kəmˈpyüd·ər ˌād·əd ˌen·jəˈnir·iŋ }

computer-aided instruction See computer-assisted instruction. { kəmˈpyüd·ər ˌād·əd inˈstrək·shən }

computer-aided management of instruction See computer-managed instruction. { kəmˈpyüd·ər ˌād·əd ˈman·ij·mənt əv inˈstrək·shən }

computer-aided manufacturing [CONT SYS] The use of computers in converting engineering designs into finished products. Computers assist managers, manufacturing engineers, and production workers by automating many production tasks, such as developing process plans, ordering and tracking materials, and monitoring production schedules, as well as controlling the machines, industrial robots, test equipment, and systems that move and store materials in the factory. Abbreviated CAM. { kəmˈpyüd·ər ˌād·əd ˌman·əˈfak·chə·riŋ }

computer-aided software engineering [COMPUT SCI] The use of software packages to assist in all phases of the development of an information system, including analysis, design, and programming. Abbreviated CASE. { kəmˈpyüd·ər ˌād·əd ˌsȯft·wer en·jəˈnir·iŋ }

computer algebra system See symbolic system. { kəmˈpyüd·ər ˈal·jə·brə ˌsis·təm }

computer analyst [COMPUT SCI] A person who defines a problem, determines exactly what is required in the solution, and defines the outlines of the machine solution; generally, an expert in automatic data processing applications. { kəmˈpyüd·ər ˈan·ə·ˌlist }

computer animation [COMPUT SCI] The use of a computer

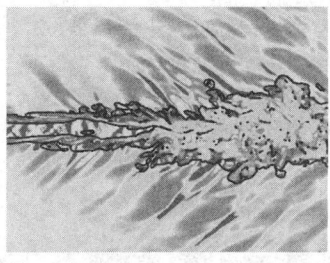

COMPUTATIONAL AEROACOUSTICS

Direct computation of noise radiation from a high-speed turbulent jet. A supersonic jet at Mach number of 1.92 and Reynolds number of 2000 is computed with 2.2×10^7 grid points using high-order compact schemes. (*Courtesy of Jonathan Freund*)

to present, either continuously or in rapid succession, pictures on a cathode-ray tube or other device, graphically representing a time developing system at successive times.   { kəm'pyüd·ər an·ə'mā·shən }

computer architecture [COMPUT SCI] The art and science of assembling logical elements to form a computing device.   { kəm'pyüd·ər 'är·kə͵tek·chər }

computer-assisted instruction [COMPUT SCI] The use of computers to present drills, practice exercises, and tutorial sequences to the student, and sometimes to engage the student in a dialog about the substance of the instruction.   Abbreviated CAI.   Also known as computer-aided instruction; computer-assisted learning (CAL).   { kəm'pyüd·ər ə'sis·təd in'strək·shən }

computer-assisted learning See computer-assisted instruction.   { kəm'pyüd·ər ə'sis·təd 'lərn·iŋ }

computer-assisted retrieval [COMPUT SCI] The use of a computer to locate documents or records stored outside of the computer, on paper or microfilm. Abbreviated CAR.   { kəm'pyüd·ər ə'sis·təd ri'trē·vəl }

computer center See electronic data-processing center.   { kəm'pyüd·ər ͵sen·tər }

computer code [COMPUT SCI] The code representing the operations built into the hardware of a particular computer.   { kəm'pyüd·ər ͵kōd }

computer conferencing See computer networking.   { kəm'pyüd·ər 'kän·frəns·iŋ }

computer control See control. [CONT SYS] Process control in which the process variables are fed into a computer and the output of the computer is used to control the process.   { kəm'pyüd·ər kən'trōl }

computer control counter [COMPUT SCI] Counter which stores the next required address; any counter which furnishes information to the control unit.   { kəm'pyüd·ər kən'trōl ͵kaúnt·ər }

computer-controlled system [CONT SYS] A feedback control system in which a computer operates on both the input signal and the feedback signal to effect control.   { kəm'pyüd·ər kən'trōld ͵sis·təm }

computer control register See program register.   { kəm'pyüd·ər kən'trōl rej·ə·stər }

computer efficiency [COMPUT SCI] **1.** The ratio of actual operating time to scheduled operating time of a computer. **2.** In time-sharing, the ratio of user time to the sum of user time plus system time.   { kəm'pyüd·ər i'fish·ən·sē }

computer forensics [FOREN SCI] The study of evidence from attacks on computer systems in order to learn what has occurred, how to prevent it from recurring, and the extent of the damage.   { kəm'pyüd·ər fə'ren·ziks }

computer graphics [COMPUT SCI] The process of pictorial communication between humans and computers, in which the computer input and output have the form of charts, drawings, or appropriate pictorial representation; such devices as cathode-ray tubes, mechanical plotting boards, curve tracers, coordinate digitizers, and light pens are employed.   { kəm'pyüd·ər 'graf·iks }

computer graphics interface [COMPUT SCI] A standard format for writing graphics drivers. Abbreviated CGI.   { kəm'pyüd·ər 'graf·iks 'in·tər͵fās }

computer graphics metafile [COMPUT SCI] A standard device-independent graphics format that is used to transfer graphics images between computer programs and storage devices. Abbreviated CGM.   { kəm'pyüd·ər 'graf·iks 'med·ə͵fīl }

computer input from microfilm [COMPUT SCI] The technique of reading images on microfilm and transforming them into a form which is understandable to a computer.   Abbreviated CIM.   { kəm'pyüd·ər 'in͵pút frəm 'mī·krə͵film }

computer-integrated manufacturing [IND ENG] A computer-automated system in which individual engineering, production, marketing, and support functions of a manufacturing enterprise are organized; functional areas such as design, analysis, planning, purchasing, cost accounting, inventory control, and distribution are linked through the computer with factory floor functions such as materials handling and management, providing direct control and monitoring of all process operations. Abbreviated CIM.   { kəm'pyüd·ər 'int·ə͵grād·əd ͵man·ə'fak·chər·iŋ }

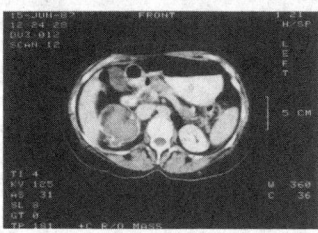

COMPUTERIZED TOMOGRAPHY

Computed tomographic image showing a transverse section through the abdomen. (*American College of Radiology*)

computerized axial tomography See computerized tomography.   { kəm'pyüd·ə͵rīzd 'ak·sē·əl tə'mä·grə·fē }

computerized branch exchange [COMMUN] A computer-controlled telephone switching system that supports such services as conference calling, least-cost routing, direct inward dialing, and automatic reringing of a busy line. Abbreviated CBX.   { kəm'pyüd·ə͵rīzd 'branch iks'chānj }

computerized composition [GRAPHICS] Type composition in which line-end hyphenation and other typographic work has been done by a computer working from unjustified tape.   { kəm'pyüd·ə͵rīzd ͵käm·pə'zish·ən }

computerized tomography [MED] The process of producing a picture showing human body organs in cross section by first electronically detecting the variation in x-ray transmission through the body section at different angles, and then using this information in a digital computer to reconstruct the x-ray absorption of the tissues at an array of points representing the cross section.   Abbreviated CT.   Also known as computed tomography; computerized axial tomography (CAT).   { kəm'pyüd·ə͵rīzd tə'mäg·rə·fē }

computer-limited [COMPUT SCI] Pertaining to a situation in which the time required for computation exceeds the time required to read inputs and write outputs.   { kəm'pyüd·ər ͵lim·əd·əd }

computer literacy [COMPUT SCI] Knowledge and understanding of computers and computer systems and how to apply them to the solution of problems.   { kəm'pyüd·ər 'lit·rə·sē }

computer-managed instruction [COMPUT SCI] The use of computer assistance in testing, diagnosing, prescribing, grading, and record keeping.   Abbreviated CMI.   Also known as computer-aided management of instruction.   { kəm'pyüd·ər 'man·ijd in'strək·shən }

computer memory See memory.   { kəm'pyüd·ər 'mem·rē }

computer modeling [COMPUT SCI] The use of a computer to develop a mathematical model of a complex system or process and to provide conditions for testing it.   { kəm'pyüd·ər 'mäd·əl·iŋ }

computer network [COMPUT SCI] A system of two or more computers that are interconnected by communication channels.   { kəm'pyüd·ər 'net͵wərk }

computer networking [COMMUN] The use of a network of computers and computer terminals by individuals at various locations to interact with each other by entering data into the computer system.   Also known as computer conferencing.   { kəm'pyüd·ər 'net͵wərk·iŋ }

computer numerical control [CONT SYS] A control system in which numerical values corresponding to desired tool or control positions are generated by a computer.   Abbreviated CNC.   Also known as computational numerical control; softwired numerical control; stored-program numerical control.   { kəm'pyüd·ər nü'mer·i·kəl kən'trōl }

computer operation [COMPUT SCI] The electronic action that is required in a computer to give a desired computation.   { kəm'pyüd·ər äp·ə'rā·shən }

computer-oriented language [COMPUT SCI] A low-level programming language developed for use on a particular computer or line of computers produced by a specific manufacturer. Also known as machine-oriented language.   { kəm'pyüd·ər 'òr·ē͵ent·əd 'laŋ·gwij }

computer output on microfilm [COMPUT SCI] The generation of microfilm which displays information developed by a computer. Abbreviated COM.   { kəm'pyüd·ər 'aút͵pút ȯn 'mī·krə͵film }

computer-output typesetting [GRAPHICS] Production of graphic arts quality printout of computer information on photographic paper or film.   { kəm'pyüd·ər 'aút͵pút 'tīp͵sed·iŋ }

computer part programming [CONT SYS] The use of computers to program numerical control systems.   { kəm'pyüd·ər 'pärt 'prō͵gram·iŋ }

computer performance evaluation [COMPUT SCI] The measurement and evaluation of the performance of a computer system, aimed at ensuring that a minimum amount of effort, expense, and waste is incurred in the production of data-processing services, and encompassing such tools as canned programs, source program optimizers, software monitors, hardware monitors, simulation, and bench-mark problems.   Abbreviated CPE.   { kəm'pyüd·ər pər'fȯr·məns i͵val·yə'wā·shən }

computer programming See programming.   { kəm'pyüd·ər 'prō͵gram·iŋ }

computer science [COMPUT SCI] The study of computers and computing, including computer hardware, software, programming, networking, database systems, information technology, interactive systems, and security. { kəm'pyüd·ər 'sī·əns }

computer security [COMPUT SCI] Measures taken to protect computers and their contents from unauthorized use. { kəm'pyüd·ər sə'kyur·əd·ē }

computer storage device See storage device. { kəm'pyüd·ər 'stor·ij di'vīs }

computer system [COMPUT SCI] **1.** A set of related but unconnected components (hardware) of a computer or data-processing system. **2.** A set of hardware parts that are related and connected, and thus form a computer. { kəm'pyüd·ər ˌsis·təm }

computer systems architecture [COMPUT SCI] The discipline that defines the conceptual structure and functional behavior of a computer system, determining the overall organization, the attributes of the component parts, and how these parts are combined. { kəm'pyüd·ər ˌsis·təmz 'ar·kəˌtek·chər }

computer theory [COMPUT SCI] A discipline covering the study of circuitry, logic, microprogramming, compilers, programming languages, file structures, and system architectures. { kəm'pyüd·ər ˌthē·ə·rē }

computer utility [COMPUT SCI] A computer that provides service on a time-sharing basis, generally over telephone lines, to subscribers who have appropriate terminals. { kəm'pyüd·ər yü'til·əd·ē }

computer vision [COMPUT SCI] The use of digital computer techniques to extract, characterize, and interpret information in visual images of a three-dimensional world. { kəm'pyüd·ər 'vizh·ən }

computer word See word. { kəm'pyüd·ər ˌwərd }

computing gunsight [ORD] A sight which includes an electrical or mechanical means for computing the proper angle between the line of sight to the target and the line of departure for the projectile. { kəm'pyüd·iŋ 'gənˌsīt }

computing machine See computer. { kəm'pyüd·iŋ mə'shēn }

computing power [COMPUT SCI] The number of operations that a computer can carry out in 1 second. { kəm'pyüd·iŋ ˌpau̇·ər }

computing unit [COMPUT SCI] The section of a computer that carries out arithmetic, logical, and decision-making operations. { kəm'pyüd·iŋ ˌyü·nət }

Comstock refraction formula [ASTROPHYS] A formula for the apparent angular displacement of an object outside the earth's atmosphere due to refraction, in terms of the barometric pressure, the temperature of the atmosphere, and the observed zenith distance. { 'kämˌstäk ri'frak·shən ˌfor·myə·lə }

conarium See pineal body. { kō'nar·ē·əm }

concatamer [MOL BIO] Two or more identical linear molecules (for example, deoxyribonucleic acid and ribonucleic acid) covalently linked in tandem. { kən'kad·ə·mər }

concatenate [COMPUT SCI] To unite in a sequence, link together, or link to a chain. { kən'kat·ənˌāt }

concatenation [COMPUT SCI] **1.** An operation in which a number of conceptually related components are linked together to form a larger, organizationally similar entity. **2.** In string processing, the synthesis of longer character strings from shorter ones. [ELEC] A method of speed control of induction motors in which the rotors of two wound-rotor motors are mechanically coupled together and the stator of the second motor is supplied with power from the rotor slip rings of the first motor. [ENG ACOUS] The linking together of phonemes to produce meaningful sounds. { kənˌkat·ən'ā·shən }

Concato's disease See polyserositis. { kōn'käd·ōz diz'ēz }

concave [SCI TECH] Having a curved form which bulges inward resembling the interior of a sphere or cylinder or a section of these bodies. { känˌkāv }

concave bit [DES ENG] A type of tungsten carbide drill bit having a concave cutting edge; used for percussive boring. { 'känˌkāv ˌbit }

concave fillet weld [MET] A fillet weld having a concave surface. { 'känˌkāv 'fil·ət ˌweld }

concave function [MATH] A function $f(x)$ is said to be concave over the interval a,b if for any three points x_1, x_2, x_3 such that $a<x_1<x_2<x_3<b$, $f(x_2) \geq L(x_2)$, where $L(x)$ is the equation of the straight line passing through the points $[x_1, f(x_1)]$ and $[x_3, f(x_3)]$. { 'känˌkāv 'fəŋk·shən }

concave grating [SPECT] A reflection grating which both collimates and focuses the light falling upon it, made by spacing straight grooves equally along the chord of a concave spherical or paraboloid mirror surface. Also known as Rowland grating. { 'känˌkāv 'grād·iŋ }

concave polygon [MATH] A polygon at least one of whose angles is greater than 180°. { 'känˌkāv 'päl·əˌgän }

concave polyhedron [MATH] A polyhedron for which there is at least one plane that contains a face of the polyhedron and that is such that parts of the polyhedron are on both sides of the plane. { 'känˌkāv ˌpäl·ə'hē·drən }

concave spherical mirror [OPTICS] A round mirror having a concavely curved surface, in the form of a portion of a sphere. { 'känˌkāv ˌsfer·ə·kəl 'mir·ər }

concavo-convex See convexo-concave. { känˌkāv·ō·kän'veks }

concentrate [CHEM] To increase the amount of a dissolved substance by evaporation. [MIN ENG] **1.** To separate ore or metal from its containing rock or earth. **2.** The clean product recovered in froth flotation or other methods of mineral separation. { 'kän·sənˌtrāt }

concentrated [MATH] A measure (or signed measure) m is concentrated on a measurable set A if any measurable set B with nonzero measure has a nonnull intersection with A. { 'kän·sənˌtrād·əd }

concentrated load [MECH] A force that is negligible because of a small contact area; a beam supported on a girder represents a concentrated load on the girder. { 'kän·sənˌtrād·əd 'lōd }

concentrating table [MIN ENG] A device consisting of a riffled deck to which a reciprocating motion in a horizontal direction is imparted; the material to be separated is fed in a stream of water, the heavy particles collect between the riffles and are conveyed in the direction of the reciprocating motion, while the lighter particles are borne by the water over the riffles to be discharged laterally from the table. { 'kän·sənˌtrād·iŋ ˌtā·bəl }

concentration [CHEM] In solutions, the mass, volume, or number of moles of solute present in proportion to the amount of solvent or total solution. [HYD] The ratio of the area of the sea covered by ice to the total area of sea surface. [MATH] An operation that provides a relatively sharp boundary to a fuzzy set; for a fuzzy set A with membership function m_A, a concentration of A is a fuzzy set whose membership function has the value $[m_A(x)]^\alpha$ for every element x, where α is a fixed number that is greater than 1. [MIN ENG] Separation and accumulation of economic minerals from gangue. { ˌkän·sən'trā·shən }

concentration cell [PHYS CHEM] **1.** Electrochemical cell for potentiometric measurement of ionic concentrations where the electrode potential electromotive force produced is determined as the difference in emf between a known cell (concentration) and the unknown cell. **2.** An electrolytic cell in which the electromotive force is due to a difference in electrolyte concentrations at the anode and the cathode. { ˌkän·sən'trā·shən ˌsel }

concentration-dilution test [PATH] A renal function test to measure the ability of the kidney to concentrate and dilute urine under stress; specific gravity for urine from a normal kidney fluctuates from 1.030 on a restricted fluid intake to 1.003 on a high water intake. { ˌkän·sən'trā·shən də'lü·shən ˌtest }

concentration gradient [CHEM] The graded difference in the concentration of a solute throughout the solvent phase. { ˌkän·sən'trā·shən ˌgrād·ē·ənt }

concentration polarization [PHYS CHEM] That part of the polarization of an electrolytic cell resulting from changes in the electrolyte concentration due to the passage of current through the solution. { ˌkän·sən'trā·shən ˌpō·lə·rə'zā·shən }

concentration potential [CHEM] Tendency for a univalent electrolyte to concentrate in a specific region of a solution. { ˌkän·sən'trā·shən pə'ten·shəl }

concentration ratio [AGR] A measure of a plant's ability to take up a contaminant from soil; it is expressed as the concentration of the element of interest in the dried plant material divided by its concentration in the dried soil. { ˌkän·sən'trā·shən ˌrā·shō }

concentration scale [CHEM] Any of several numerical systems defining the quantitative relation of the components of a mixture; for solutions, concentration is expressed as the mass, volume, or number of moles of solute present in proportion to the amount of solvent or total solution.　{ ˌkän·sən'trā·shən ˌskäl }

concentration time　[HYD]　The time required for water to travel from the most remote portion of a river basin to the basin outlet; it varies with the quantity of flow and channel conditions.　{ ˌkän·sən'trā·shən ˌtīm }

concentrator　[ELECTR]　Buffer switch (analog or digital) which reduces the number of trunks required.　[ENG]　**1.** An apparatus used to concentrate materials.　**2.** A plant where materials are concentrated.　{ 'kän·sənˌtrād·ər }

concentric　[SCI TECH]　Pertaining to the relationship between two different-sized circular, cylindrical, or spherical shapes when the smaller one is exactly centered within the larger one.　{ kən'sen·trik }

concentric bundle　[BOT]　A vascular bundle in which xylem surrounds phloem, or phloem surrounds xylem.　{ kən'sen·trik 'bən·dəl }

concentric cable　See coaxial cable.　{ kən'sen·trik 'kā·bəl }

concentric circles　[MATH]　A family of coplanar circles with the same center.　{ kən'sen·trik 'sər·kəlz }

concentric faults　[GEOL]　Faults that are arranged concentrically.　{ kən'sen·trik 'fȯlts }

concentric fold　[GEOL]　A fold in which the original thickness of the strata is unchanged during deformation.　Also known as parallel fold.　{ kən'sen·trik 'fōld }

concentric fractures　[GEOL]　A system of fractures concentrically arranged about a center.　{ kən'sen·trik 'frak·chərz }

concentric groove　See locked groove.　{ kən'sen·trik 'grüv }

concentric lens　[OPTICS]　A lens whose two spherical surfaces have the same center.　{ kən'sen·trik 'lenz }

concentric line　See coaxial cable.　{ kən'sen·trik 'līn }

concentric locating　[DES ENG]　The process of making the axis of a tooling device coincide with the axis of the workpiece.　{ kən'sen·trik 'lō,kād·iŋ }

concentric orifice plate　[DES ENG]　A fluid-meter orifice plate whose edges have a circular shape and whose center coincides with the center of the pipe.　{ kən'sen·trik 'ȯr·ə·fəs ˌplāt }

concentric reducer　[ENG]　A threaded or butt-welded pipe fitting whose ends are of different sizes but are concentric about a common axis.　{ kən'sen·trik ri'dü·sər }

concentric resonator　[OPTICS]　A beam resonator that consists of a pair of spherical mirrors that have the same axis of rotational symmetry and are positioned so that their centers of curvature coincide on this axis.　{ kən'sen·trik 'rez·ən,ād·ər }

concentric ring structures　[ASTRON]　A formation on the moon's surface consisting of two craters, one inside the other, with approximately the same center.　{ kən'sen·trik 'riŋ ˌstrək·chərz }

concentric slip ring　[ELEC]　A large slip-ring assembly consisting of concentrically arranged insulators and conducting materials.　{ kən'sen·trik 'slip ˌriŋ }

concentric transmission line　See coaxial cable.　{ kən'sen·trik tranz'mish·ən ˌlīn }

concentric tube column　[CHEM ENG]　A carefully insulated distillation apparatus which is capable of very high separating power, and in which the outer vapor-rising annulus of the column is concentric around an inner, bottom-discharging reflux return.　{ kən'sen·trik ˌtüb 'käl·əm }

concentric weathering　See spheroidal weathering.　{ kən'sen·trik 'weth·ə·riŋ }

concentric windings　[ELEC]　Transformer windings in which the low-voltage winding is in the form of a cylinder next to the core, and the high-voltage winding, also cylindrical, surrounds the low-voltage winding.　{ kən'sen·trik 'wīnd·iŋz }

conceptacle　[BOT]　A cavity which is shaped like a flask with a pore opening to the outside, contains reproductive structures, and is bound in a thallus such as in the brown algae.　{ kən'sep·tə·kəl }

concept coordination　[COMPUT SCI]　The basic principles of various punched-card, aspect, and mechanized information retrieval systems in which independently assigned concepts are used to characterize the subject content of documents, and the latter are identified during searching by means of either such assigned concepts or their combination.　{ 'kän,sept kō,ȯrd·ən'ā·shən }

conception　[BIOL]　Fertilization of an ovum by the sperm resulting in the formation of a viable zygote.　[PSYCH]　The mental process of forming ideas, especially abstract ideas.　{ kən'sep·shən }

conceptual modeling　[COMPUT SCI]　Writing a program by means of which a given result will be obtained, although the result is incapable of proof.　Also known as heuristic programming.　{ kən'sep·chə·wəl 'mäd·liŋ }

conceptual schema　[COMPUT SCI]　The logical structure of an entire data base.　{ kən'sep·chə·wəl 'skē·mə }

conceptus　[BIOL]　The product of a conception, including the embryo or fetus and extraembryonic membranes, at any stage of development from fertilization to birth.　{ kən'sep·təs }

concerted evolution　See coincidental evolution.　{ kən'sərd·əd ev·ə'lü·shən }

concerted reaction　[ORG CHEM]　A reaction in which there is a simultaneous occurrence of bond making and bond breaking.　{ kən'sərd·əd rē'ak·shən }

concertina wire　[ORD]　A barbed wire coiled for easier handling and emplacement; when uncoiled, it forms an entanglement for the enemy infantry.　{ ˌkan·sər'tē·nə ˌwīr }

concession lease　[MIN ENG]　A lease form that conveys specified national or state permission to a lessee to explore for or produce minerals (such as oil, gas, or uranium) from specified properties.　{ kən'sesh·ən ˌlēs }

conch　[INV ZOO]　The common name for several species of large, colorful gastropod mollusks of the family Strombidae; the shell is used to make cameos and porcelain.　{ käŋk }

conchiolin　[BIOCHEM]　A nitrogenous substance that is the organic basis of many molluscan shells.　{ käŋ'kī·ə·lən }

conchoid　[MATH]　A plane curve consisting of the locus of both ends of a line segment of constant length on a line which rotates about a fixed point, while the midpoint of the segment remains on a fixed curve which does not contain the fixed point.　{ 'käŋ,kȯid }

conchoidal　[GEOL]　Having a smoothly curved surface; used especially to describe the fracture surface of a mineral or rock.　{ käŋ'kȯid·əl }

conchoid of Nicomedes　[MATH]　The conchoid of a straight line with respect to a fixed point that does not lie on the line.　{ 'käŋ,kȯid əv ,nik·ə'mē·dēz }

Conchorhagae　[INV ZOO]　A suborder of benthonic wormlike animals in the class Kinorhyncha.　{ käŋ'kȯr·ə,gē }

Conchostraca　[INV ZOO]　An order of mussellike crustaceans of moderate size belonging to the subclass Branchiopoda.　{ käŋ'käs·trə·kə }

concomitant　[ANALY CHEM]　Any species in a material undergoing chemical analysis other than the analyte or the solvent in which the sample is dissolved.　{ kən'käm·ə·tənt }

concordance　[GEN]　Similarity in appearance of members of a twin pair with respect to one or more specific traits.　{ kən'kȯrd·əns }

concordant body　[GEOL]　An intrusive igneous body whose contacts are parallel to the bedding of the country rock.　Also known as concordant injection; concordant pluton.　{ kən'kȯrd·ənt 'bäd·ē }

concordant coastline　[GEOL]　A coastline parallel to the land structures which form the margin of an ocean basin.　{ kən'kȯrd·ənt 'kōst,līn }

concordant drainage　See accordant drainage.　{ kən'kȯrd·ənt 'drān·ij }

concordant injection　See concordant body.　{ kən'kȯrd·ənt in'jek·shən }

concordant pluton　See concordant body.　{ kən'kȯrd·ənt 'plü,tän }

concordant segregation　[GEN]　The simultaneous appearance or disappearance of gene markers in hybrid cells undergoing chromosome loss.　{ kən'kȯrd·ənt ,seg·rə'gā·shən }

concrescence　[BIOL]　Convergence and fusion of parts originally separate, as the lips of the blastopore in embryogenesis.　{ kən'krēs·əns }

concrete　[MATER]　A mixture of aggregate, water, and a binder, usually portland cement; hardens to stonelike condition when dry.　{ 'käŋ,krēt }

concrete beam　[CIV ENG]　A structural member of reinforced concrete, placed horizontally over openings to carry loads.　{ 'käŋ,krēt 'bēm }

concrete block　[MATER]　A solid or hollow block of precast concrete.　{ 'käŋ,krēt 'bläk }

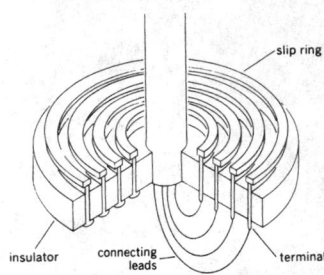

CONCENTRIC SLIP RING

Concentric slip-ring assembly configuration.

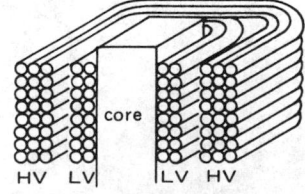

CONCENTRIC WINDINGS

Section through the core and concentric winding of a transformer. LV = low-voltage winding; HV = high-voltage winding.

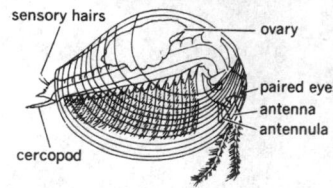

CONCHOSTRACA

Limnadia lenticularis female, lateral aspect.

concrete bridge [CIV ENG] A bridge constructed of pre-stressed or reinforced concrete. { 'kän‚krēt 'brij }

concrete bucket [ENG] A container with movable gates at the bottom that is attached to power cranes or cables to transport concrete. { 'kän‚krēt 'bək·ət }

concrete buggy [ENG] A cart which carries up to 6 cubic feet (0.17 cubic meter) of concrete from the mixer or hopper to the forms. Also known as buggy; concrete cart. { 'kän‚krēt ‚bəg·ē }

concrete caisson sinking [CIV ENG] A shaft-sinking method similar to caisson sinking except that reinforced concrete rings are used and an airtight working chamber is not adopted. { 'kän‚krēt 'kā‚sän ‚siŋk·iŋ }

concrete cart See concrete buggy. { 'kän‚krēt ‚kärt }

concrete chute [ENG] A long metal trough with rounded bottom and open ends used for conveying concrete to a lower elevation. { 'kän‚krēt ‚shüt }

concrete column [CIV ENG] A vertical structural member made of reinforced or unreinforced concrete. { 'kän‚krēt 'käl·əm }

concrete dam [CIV ENG] A dam that is built of concrete. { 'kän‚krēt 'dam }

concrete finish [MATER] The texture or smoothness on the surface of hardened concrete. { 'kän‚krēt ‚fin·ish }

concrete form oil [MATER] Nonviscous, neutral mineral oil used on wooden or metal forms to allow easy removal from set concrete. { 'kän‚krēt ‚fȯrm ‚ȯil }

concrete hardener [MATER] An admixture such as calcium chloride, sodium chloride, or sodium hydroxide that hastens or decreases the hydration rate of cementing material; the concrete takes less time to set and has earlier higher strength. { 'kän‚krēt ‚härd·ən·ər }

concrete masonry [MATER] Building units composed of block, brick, or tile laid by masons. { 'kän‚krēt ‚mās·ən·rē }

concrete mixer [MECH ENG] A machine with a rotating drum in which the components of concrete are mixed. { 'kän‚krēt ‚mik·sər }

concrete nail [DES ENG] A hardened-steel nail that has a flat countersunk head and a tapered point and is used for nailing various materials to concrete or masonry. { 'kän‚krēt 'nāl }

concrete pile [CIV ENG] A reinforced pile made of concrete, either precast and driven into the ground, or cast in place in a hole bored into the ground. { 'kän‚krēt 'pīl }

concrete pipe [CIV ENG] A porous pipe made of concrete and used principally for subsoil drainage; diameters over 15 inches (38 centimeters) are usually reinforced. { 'kän‚krēt 'pīp }

concrete pump [MECH ENG] A device which drives concrete to the placing position through a pipeline of 6-inch (15-centimeter) diameter or more, using a special type of reciprocating pump. { 'kän‚krēt ‚pəmp }

concrete retarder [MATER] A material added to concrete that decreases the hydration rate of cement, thereby increasing the setting time and decreasing the strengthening rate during the early age. { 'kän‚krēt ri'tärd·ər }

concrete slab [CIV ENG] A flat, reinforced-concrete structural member, relatively sizable in length and width, but shallow in depth; used for floors, roofs, and bridge decks. { 'kän‚krēt 'slab }

concrete steel [MET] Steel used in reinforced concrete, which should comply with standard specifications for pre-stressed concrete. { 'kän‚krēt ‚stēl }

concrete thinking [PSYCH] Mental processes characterized by literalness and the tendency to be bound to the most immediate and obvious sense impressions, as well as by a lack of generalization and abstraction. { 'kän‚krēt 'thiŋk·iŋ }

concrete vibrator [MECH ENG] Vibrating device used to achieve proper consolidation of concrete; the three types are internal, surface, and form vibrators. { 'kän‚krēt ‚vī‚brād·ər }

concretion [GEOL] A hard, compact mass of mineral matter in the pores of sedimentary or fragmental volcanic rock; represents a concentration of a minor constituent of the enclosing rock or of cementing material. { kän'krē·shən }

concretionary [GEOL] Tending to grow together, forming concretions. [PATH] A compact mass of inorganic material formed in a body cavity or in tissue. { kən'krē·shə‚ner·ē }

concretioning [GEOL] The process of forming concretions. { kən'krē·shən·iŋ }

concurrency [COMPUT SCI] Referring to two or more tasks of a computer system which are in progress simultaneously. { kən'kər·ən·sē }

concurrent conversion [COMPUT SCI] The transfer of data from one medium to another, such as card to tape, under computer control while programs are being run on the same computer. { kən'kər·ənt kən'vər·zhən }

concurrent engineering [ENG] The simultaneous design of products and related processes, including all product life-cycle aspects such as manufacturing, assembly, test, support, disposal, and recycling. { kən'kər·ənt ‚en·jə'nir·iŋ }

concurrent heating [MET] Application of supplemental heat in metal cutting or welding. { kən'kər·ənt 'hēd·iŋ }

concurrent infection [MED] Two or more forms of an infection existing simultaneously. { kən'kər·ənt in'fek·shən }

concurrent input/output [COMPUT SCI] The simultaneous reading from and writing on different media by a computer. { kən'kər·ənt ‚in‚pu̇t ‚au̇t‚pu̇t }

concurrent line [MAP] A line on a map or chart passing through places having the same current hour. [MATH] One of two or more lines that have a point in common. { kən'kər·ənt ‚līn }

concurrent method See circulate-and-weight method. { kən'kər·ənt ‚meth·əd }

concurrent operations control [COMPUT SCI] The supervisory capability required by a computer to handle more than one program at a time. { kən'kər·ənt äp·ə'rā·shənz kən'trōl }

concurrent plane [MATH] One of three or more planes that have a point in common. { kən'kər·ənt 'plān }

concurrent processing [COMPUT SCI] The conceptually simultaneous execution of more than one sequential program on a computer or network of computers. { kən'kər·ənt 'präs‚əs·iŋ }

concurrent real-time processing [COMPUT SCI] The capability of a computer to process simultaneously several programs, each of which requires responses within a time span related to its particular time frame. { kən'kər·ənt 'rēl ‚tīm ‚präs‚əs·iŋ }

concussion [ENG] Shock waves in the air caused by an explosion underground or at the surface or by a heavy blow directly to the ground surface during excavation, quarrying, or blasting operations. [MED] A state of shock following traumatic injury, especially cerebral trauma, in which there is temporary functional impairment without physical evidence of damage to impaired tissues. { kən'kəsh·ən }

concussion fracture [GEOL] Radiating system of fractures in a shock-metamorphosed rock. { kən'kəsh·ən ‚frak·chər }

concussion fuse [ORD] A bomb fuse designed to function in the air in response to the concussion produced by the explosion of a preceding bomb. { kən'kəsh·ən ‚fyüz }

concussion table [MIN ENG] An inclined table, agitated by a series of shocks and operating like a buddle. Also known as percussion table. { kən'kəsh·ən ‚tā·bəl }

concyclic points [MATH] Points that are located on a common circle. { kən‚sīk·lik 'pȯins }

condensable vapors [CHEM] Gases or vapors which when subjected to appropriately altered conditions of temperature or pressure become liquids. { kən'den·sə·bəl 'vā·pərz }

condensate [MATER] **1.** The liquid product from a condenser. Also known as condensate liquid. **2.** A light hydrocarbon mixture formed as a liquid product in a gas-recycling plant through expansion and cooling of the gas. { 'kän·dən‚sāt }

condensate field [GEOL] A petroleum field developed in predominantly gas-bearing reservoir rocks, but within which condensation of gas to oil commonly occurs with decreases in field pressure. { 'kän·dən‚sāt ‚fēld }

condensate flash [CHEM ENG] Partial evaporation (flash) of hot condensed liquid by a stepwise reduction in system pressure, the hot vapor supplying heat to a cooler evaporator step (stage). { 'kän·dən‚sāt ‚flash }

condensate liquid See condensate. { 'kän·dən‚sāt ‚lik·wəd }

condensate strainer [MECH ENG] A screen used to remove solid particles from the condensate prior to its being pumped back to the boiler. { 'kän·dən‚sāt ‚strān·ər }

condensate well [MECH ENG] A chamber into which condensed vapor falls for convenient accumulation prior to removal. [PETRO ENG] Well that produces a natural gas highly saturated with condensable hydrocarbons heavier than methane and ethane. { 'kän·dən‚sāt ‚wel }

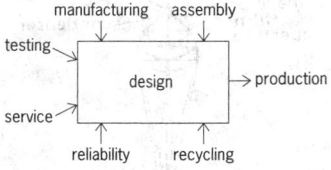

CONCURRENT ENGINEERING

At any given moment, several arrows may be active, and information sometimes flows in both directions along a given arrow.

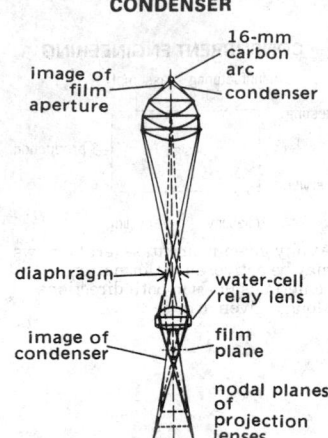

CONDENSER

16-mm carbon arc

image of film aperture

condenser

diaphragm

water-cell relay lens

image of condenser

film plane

nodal planes of projection lenses

A relay condenser system having a water cell incorporated in second stage.

condensation [ACOUS] A measure of the increase in the instantaneous density at a given point owing to a sound wave, namely $(\rho - \rho_0)/\rho_0$, where ρ is the density and ρ_0 is the constant mean density at the point. [CHEM] Transformation from a gas to a liquid. [CRYO] *See* Bose-Einstein condensation. [ELEC] An increase of electric charge on a capacitor conductor. [MECH] An increase in density. [METEOROL] The process by which water vapor becomes a liquid such as dew, fog, or cloud or a solid like snow; condensation in the atmosphere is brought about by either of two processes: cooling of air to its dew point, or addition of enough water vapor to bring the mixture to the point of saturation (that is, the relative humidity is raised to 100). [OPTICS] Focusing or collimation of light. { ¦kän·dən'sā·shən }

condensation cloud [METEOROL] A mist or fog of minute water droplets that temporarily surrounds the fireball following an atomic detonation in a comparatively humid atmosphere. { ¦kän·dən'sā·shən ¦klau̇d }

condensation nucleus [METEOROL] A particle, either liquid or solid, upon which condensation of water vapor begins in the atmosphere. { ¦kän·dən'sā·shən 'nü·klē·əs }

condensation point [MATH] For a set in a topological space, a point whose neighborhoods all contain uncountably many points of the set. { ¦kän·dən'sā·shən ¦pȯint }

condensation polymer [ORG CHEM] A high-molecular-weight compound formed by condensation polymerization. { ¦kän·dən'sā·shən 'päl·ə·mər }

condensation polymerization [ORG CHEM] The stepwise reaction between functional groups of reactants in which a high-molecular-weight polymer is formed only after a large number of steps, for example, the reaction of dicarboxylic acids with diamines to form a polyamide. { ¦kän·dən'sā·shən pə¦lim·ə·rə'zā·shən }

condensation pressure [METEOROL] The pressure at which a parcel of moist air expanded dry adiabatically reaches saturation. Also called adiabatic condensation pressure; adiabatic saturation pressure. { ¦kän·dən'sā·shən 'presh·ər }

condensation reaction [CHEM] One of a class of chemical reactions involving a combination between molecules or between parts of the same molecule. { ¦kän·dən'sā·shən rē'ak·shən }

condensation resin [ORG CHEM] A resin formed by polycondensation. { ¦kän·dən'sā·shən 'rez·ən }

condensation shock wave [FL MECH] A sheet of discontinuity associated with a sudden condensation and fog formation in a field of flow; it occurs, for example, on a wing where a rapid drop in pressure causes the temperature to drop considerably below the dew point. { ¦kän·dən'sā·shən 'shäk ¦wāv }

condensation temperature [ANALY CHEM] In boiling-point determination, the temperature established on the bulb of a thermometer on which a thin moving film of liquid coexists with vapor from which the liquid has condensed, the vapor phase being replenished at the moment of measurement from a boiling-liquid phase. [METEOROL] The temperature at which a parcel of moist air expanded dry adiabatically reaches saturation. Also known as adiabatic condensation temperature; adiabatic saturation temperature. { ¦kän·dən'sā·shən 'tem·prə·chər }

condensation trail [METEOROL] A visible trail of condensed water vapor or ice particles left behind an aircraft, an airfoil, or such, in motion through the air. Also known as contrail; vapor trail. { ¦kän·dən'sā·shən ¦trāl }

condensed matter [PHYS] Matter in the liquid or solid state. { kən¦denst 'mad·ər }

condensed-matter physics [PHYS] The branch of physics concerned with the study of very large numbers of strongly interacting particles, including the study of the solid and liquid states, dense plasmas, liquid crystals, glasses, polymers, and gels. { kən¦denst ¦mad·ər 'fiz·iks }

condensed phase [PHYS CHEM] Either the solid or liquid phase of a material. { kən¦denst ¦fāz }

condensed structural formula [CHEM] A structural representation of a compound that includes all of the atoms present in a molecule or other chemical entity but represents only certain bonds as lines in order to emphasize a structural characteristic. { kən¦denst ¦strək·chər·əl 'fȯr·myə·lə }

condensed system [PHYS CHEM] A chemical system in which the vapor pressure is negligible or in which the pressure maintained on the system is greater than the vapor pressure of any portion. { kən¦denst 'sis·təm }

condensed type [GRAPHICS] A typeface which is narrow or slender. { kən¦denst 'tīp }

condenser [ELEC] *See* capacitor. [MECH ENG] A heat-transfer device that reduces a thermodynamic fluid from its vapor phase to its liquid phase, such as in a vapor-compression refrigeration plant or in a condensing steam power plant. [OPTICS] A system of lenses or mirrors in an optical projection system, which gathers as much of the light from the source as possible and directs it through the projection lens. { kən'den·sər }

condenser antenna *See* capacitor antenna. { kən'den·sər an'ten·ə }

condenser box *See* capacitor box. { kən'den·sər ¦bäks }

condenser bushing [ELEC] An insulation made up of alternate layers of insulating material and metal foil placed between the conductor and outer casing in terminals of transformers and other high-voltage equipment such as switchgears. { kən'den·sər ¦bu̇sh·iŋ }

condenser-discharge anemometer [ENG] A contact anemometer connected to an electrical circuit which is so arranged that the average wind speed is indicated. { kən¦den·sər¦dis¦chärj an·ə'mäm·əd·ər }

condenser ionization chamber [NUCLEO] An ionization chamber which is charged before irradiation, and in which the charge remaining after irradiation indicates the dose received. { kən¦den·sər ¦ī·ə·nə¦zā·shən 'chām·bər }

condenser microphone *See* capacitor microphone. { kən'den·sər 'mī·krə¦fōn }

condenser transducer *See* electrostatic transducer. { kən'den·sər ¦tranz'dü·sər }

condenser tubes [MECH ENG] Metal tubes used in a heat-transfer device, with condenser vapor as the heat source and flowing liquid such as water as the receiver. { kən'den·sər ¦tübs }

condensin [CELL MOL] The class of proteins that form condensin complexes, which drive the coiling of interphase chromosomes, leading to their condensation prior to mitosis. { kən'den·sən }

condensin complex [CELL MOL] One of the protein complexes that drive the coiling of interphase chromosomes, leading to their condensation prior to mitosis. { kən'den·sən ¦käm¦pleks }

condensing electrometer *See* capacitive electrometer. { kən¦dens·iŋ ə¦lek'träm·əd·ər }

condensing engine [MECH ENG] A steam engine in which the steam exhausts from the cylinder to a vacuum space, where the steam is liquefied. { kən¦dens·iŋ ¦en·jən }

condensing flow [FL MECH] The flow and simultaneous condensation (partial or complete) of vapor through a cooled pipe or other closed conduit or container. { kən¦dens·iŋ ¦flō }

condensing gas drive [PETRO ENG] Reservoir-oil displacement by gas where hydrocarbon components of the injected gas condense in the oil that it is displacing. { kən¦dens·iŋ ¦gas ¦drīv }

condensing routine [COMPUT SCI] A routine that converts a program format having one instruction per card to a program format having several instructions per card. { kən¦dens·iŋ rü'tēn }

condition [MATH] The product of the norm of a matrix and of its inverse. [PETRO ENG] To change the properties of a drilling mud by introducing additives. { kən'dish·ən }

conditional [COMPUT SCI] Subject to the result of a comparison made during computation in a computer, or subject to human intervention. { kən'dish·ən·əl }

conditional assembly [COMPUT SCI] A feature of some assemblers which suppresses certain sections of code if stated program conditions are not met at assembly time. { kən'dish·ən·əl ə'sem·blē }

conditional branch *See* conditional jump. { kən'dish·ən·əl 'branch }

conditional breakpoint [COMPUT SCI] A conditional jump that, if a specified switch is set, will cause a computer to stop; the routine may then be continued as coded or a jump may be forced. { kən'dish·ən·əl 'brāk¦pȯint }

conditional convergence [MATH] The property of a series that is convergent but not absolutely convergent. { kən'dish·ən·əl kən'vər·jəns }

conditional distribution [STAT] If W and Z are random variables with discrete values $w_1, w_2, \ldots$, and $z_1, z_2, \ldots$, the conditional distribution of W given $Z = z$ is the distribution which assigns to w_i, $i = 1, 2, \ldots$, the conditional probability of $W = w_i$ given $Z = z$. { kən'dish·ən·əl dis·trə'byü·shən }

conditional expectation [MATH] If X is a random variable on a probability space (Ω, F, P), the conditional expectation of X with respect to a given sub σ-field F' of F is an F'-measurable random variable whose expected value over any set in F' is equal to the expected value of X over this set. [STAT] The expected value of a conditional distribution. { kən'dish·ən·əl ˌek,spek'tā·shən }

conditional expression [COMPUT SCI] A COBOL language expression which is either true or false, depending upon the status of the variables within the expression. { kən'dish·ən·əl ik'spresh·ən }

conditional frequency [STAT] If r and s are possible outcomes of an experiment which is performed n times, the conditional frequency of s given that r has occurred is the ratio of the number of times both r and s have occurred to the number of times r has occurred. { kən'dish·ən·əl 'frē·kwən·sē }

conditional implication See implication. { kən'dish·ən·əl ˌim·plə'kā·shən }

conditional inequality [MATH] An inequality which fails to hold true for some of the values of the variable involved. { kən'dish·ən·əl ˌin·i'kwäl·əd·ē }

conditional instability [METEOROL] The state of a column of air in the atmosphere when its lapse rate of temperature is less than the dry adiabatic lapse rate but greater than the saturation adiabatic lapse rate. { kən'dish·ən·əl ˌin·stə'bil·əd·ē }

conditional jump [COMPUT SCI] A computer instruction that will cause the proper one of two or more addresses to be used in obtaining the next instruction, depending on some property of a numerical expression that may be the result of some previous instruction. Also known as conditional branch; conditional transfer; decision instruction; discrimination; IF statement. { kən'dish·ən·əl 'jəmp }

conditional lethal mutant [GEN] A lethal mutant that expresses characteristics of the wild type when grown under certain conditions, as at a particular temperature, and mutant characteristics under other conditions. { kən'dish·ən·əl 'lē·thəl 'myüt·ənt }

conditionally compact set [MATH] A set whose closure is compact. Also known as relatively compact set. { kən'dish·ən·əl·ē 'käm,pakt ˌset }

conditionally periodic motion [MECH] Motion of a system in which each of the coordinates undergoes simple periodic motion, but the associated frequencies are not all rational fractions of each other so that the complete motion is not simply periodic. { kən'dish·ən·əl·ē ˌpir·ē¦ad·ik ˌmō·shən }

conditionally stable circuit [ELECTR] A circuit which is stable for certain values of input signal and gain, and unstable for other values. { kən'dish·ən·əl·ē ¦sta·bəl ˌsər·kət }

conditional probability [STAT] The probability that a second event will be B if the first event is A, expressed as $P(B/A)$. { kən'dish·ən·əl ˌpräb·ə'bil·əd·ē }

conditional replenishment [COMMUN] A form of differential pulse-code modulation in which the only information transmitted consists of addresses specifying the locations of picture samples in the moving area, and information by which the intensities of moving area picture samples can be reconstructed at the receiver. { kən'dish·ən·əl ri'plen·ish·mənt }

conditional statement [COMPUT SCI] A statement in a computer program that is executed only when a certain condition is satisfied. { kən'dish·ən·əl 'stāt·mənt }

conditional transfer See conditional jump. { kən'dish·ən·əl 'tranz·fər }

condition code [COMPUT SCI] Portion of a program status word indicating the outcome of the most recently executed arithmetic or boolean operation. { kən'dish·ən ˌkōd }

conditioned line [COMPUT SCI] A communications channel, usually a telephone line, that has been adapted for data transmission. { kən'dish·ənd 'līn }

conditioned reflex [PSYCH] Response of an organism to a stimulus which was inadequate to elicit the response until paired for one or more times with an adequate stimulus. { kən'dish·ənd 'rē,fleks }

conditioned stop instruction [COMPUT SCI] A computer instruction which causes the execution of a program to stop if

some given condition exists, such as the specific setting of a switch on a computer console. { kən'dish·ənd 'stäp in'strek·shən }

condition entries [COMPUT SCI] The upper-right-hand portion of a decision table, indicating, for each of the conditions, whether the condition satisfies various criteria listed in the condition stub, or the values of various parameters listed in the condition stub. { kən'dish·ən ˌen,trēz }

conditioning [ELECTR] Equipment modifications or adjustments necessary to match transmission levels and impedances or to provide equalization between facilities. [GRAPHICS] Restoration of microfilm for use after it has been stored for a period of time. [SCI TECH] Subjecting a material or organism to a stipulated treatment or stimulus so that it will respond in a uniform and desired manner to subsequent testing or processing. { kən'dish·ən·iŋ }

condition portion [COMPUT SCI] The upper portion of a decision table, comprising the condition stub and condition entires. { kən'dish·ən ˌpór·shən }

condition stub [COMPUT SCI] The upper-left-hand portion of a decision table, consisting of a single column listing various criteria or parameters which are used to specify the conditions. { kən'dish·ən ˌstəb }

Condon-Shortley-Wigner phase convention [QUANT MECH] Convention relating the phases of states having the same eigenvalue of $J^2 = J_x^2 + J_y^2 + J_z^2$, and different eigenvalues of J_z, where $\mathbf{J}$ is the total angular momentum, according to which the matrix elements of $\mathbf{J}_+ = J_x + iJ_y$ and $\mathbf{J}_- = J_x - iJ_j$ between such states are real. { ¦kän·dən ¦shórt·lē ¦wig·nər 'fās kən,ven·shən }

condor [NAV] A continuous-wave navigation system, similar to benito, that automatically measures bearing and distance from a single ground station; the distance is determined by phase comparison and the bearing by automatic direction finding. [VERT ZOO] Vultur gryphus. A large American vulture having a bare head and neck, dull black plumage, and a white neck ruff. { 'kän,dór }

Condor [ORD] A U.S. Navy air-to-surface missile that uses optoelectronic guidance, developed for use beyond the range of antiaircraft guns which protect heavily defended ground targets; range is about 50 miles (80 kilometers). { 'kän,dór }

conductance [ELEC] The real part of the admittance of a circuit; when the impedance contains no reactance, as in a direct-current circuit, it is the reciprocal of resistance, and is thus a measure of the ability of the circuit to conduct electricity. Also known as electrical conductance. Designated G. [FL MECH] For a component of a vacuum system, the amount of a gas that flows through divided by the pressure difference across the component. [THERMO] See thermal conductance. { kən'dək·təns }

conductance coefficient [PHYS CHEM] The ratio of the equivalent conductance of an electrolyte, at a given concentration of solute, to the limiting equivalent conductance of the electrolyte as the concentration of the electrolyte approaches 0. { kən¦dək·təns ˌkō·ə'fish·ənt }

conductance-variation method [ELEC] A technique for measuring low admittances; measurements in a parallel-resonance circuit with the terminals open-circuited, with the unknown admittance connected, and then with the unknown admittance replaced by a known conductance standard are made; from them the unknown can be calculated. { kən'dək·təns ver·ē'ā·shən ˌmeth·əd }

conducted interference [COMMUN] Interfering signals arriving by direct coupling such as on communications and power lines. { kən'dək·təd ˌin·tər'fir·əns }

conductimetry [CHEM] The scientific study of conductance measurements of solutions; to avoid electrolytic complications, conductance measurements are usually taken with alternating current. { kän·dək'tim·ə·trē }

conducting polymer [MATER] A plastic having high conductivity, approaching that of metals. { kən'dək·tiŋ 'päl·ə·mər }

conduction [ELEC] The passage of electric charge, which can occur by a variety of processes, such as the passage of electrons or ionized atoms. Also known as electrical conduction. [PHYS] Transmission of energy by a medium which does not involve movement of the medium itself. { kən'dək·shən }

conduction aphasia [PSYCH] A form of aphasia featuring

selectively impaired repetition, with relatively fluent speech production and good comprehension, often associated with damage in the white matter underlying the parietal operculum. { kən'dək·shən ə,fā·shə }

conduction band [SOLID STATE] An energy band in which electrons can move freely in a solid, producing net transport of charge. { kən'dək·shən ,band }

conduction cooling [ELECTR] Cooling of electronic components by carrying heat from the device through a thermally conducting material to a large piece of metal with cooling fins. { kən'dək·shən ,kül·iŋ }

conduction current [SOLID STATE] A current due to a flow of conduction electrons through a body. { kən'dək·shən ,kər·ənt }

conduction deafness [MED] Deafness involving an impairment of the mechanism that conducts sound to the sense organ. { kən'dək·shən ,def·nəs }

conduction electron [SOLID STATE] An electron in the conduction band of a solid, where it is free to move under the influence of an electric field. Also known as outer-shell electron; valence electron. { kən'dək·shən i'lek,trän }

conduction field [ELECTROMAG] Energy surrounding a conductor when an electric current is passed through the conductor, which, because of the difference in phase between the electrical field and magnetic field set up in the conductor, cannot be detached from the conductor. { kən'dək·shən ,fēld }

conduction pump [ENG] A pump in which liquid metal or some other conductive liquid is moved through a pipe by sending a current across the liquid and applying a magnetic field at right angles to current flow. { kən'dək·shən ,pəmp }

conductive coating [MATER] A coating used to reduce surface resistance and thus prevent the accumulation of static electric charges. { kən'dək·tiv 'kōd·iŋ }

conductive coupling [ELEC] Electric connection of two electric circuits by their sharing the same resistor. { kən'dək·tiv 'kəp·liŋ }

conductive elastomer [MATER] A rubberlike silicone material in which suspended metal particles conduct electricity. { kən'dək·tiv i'las·tə·mər }

conductive equilibrium See isothermal equilibrium. { kən'dək·tiv ,ē·kwə'lib·rē·əm }

conductive gasket [ELEC] A flexible metallic gasket used to reduce radio-frequency leakage at joints in shielding. { kən'dək·tiv 'gas·kət }

conductive hearing loss [PHYSIO] Failure of sound to be transmitted properly to the receptors in the inner ear so that sounds must be made louder to be heard. { kən,dək·tiv 'hir·iŋ ,lòs }

conductive interference [ELECTR] Interference to electronic equipment that originates in power lines supplying the equipment, and is conducted to the equipment and coupled through the power supply transformer. { kən'dək·tiv ,in·tər'fir·əns }

conductive paste [MED] **1.** A substance applied to the skin to lower its electrical resistance in an area to which electrodes will be applied. **2.** A gel applied to the skin to lower its acoustical impedance and to accommodate an ultrasonic probe; enables ultrasonic energy to penetrate to underlying tissues without severe attenuation at the skin interface. { kən'dək·tiv 'pāst }

conductive rubber [MATER] Rubber that contains suspended carbon or silver spheres; its electrical resistance decreases when it is compressed, making it useful as a contact sensor. { kən'dək·tiv 'rəb·ər }

conductive silver paste [MATER] Silver powder in a suitable vehicle for applying to ceramic or other insulating materials by silk-screening or other methods, then fixing or firing at appropriate temperatures to provide a hard conductive surface or joint. { kən'dək·tiv 'sil·vər ,pāst }

conductivity [ELEC] The ratio of the electric current density to the electric field in a material. Also known as electrical conductivity; specific conductance. [GEOL] See permeability. { ,kän,dək'tiv·əd·ē }

conductivity bridge [ELEC] A modified Kelvin bridge for measuring very low resistances. { ,kän,dək'tiv·əd·ē ,brij }

conductivity cell [ELEC] A glass vessel with two electrodes at a definite distance apart and filled with a solution whose conductivity is to be measured. { ,kän,dək'tiv·əd·ē ,sel }

conductivity current See air-earth conduction current. { ,kän,dək'tiv·əd·ē ,kər·ənt }

conductivity ellipsoid [ELEC] For an anisotropic material, an ellipsoid whose axes are the eigenvectors of the conductivity tensor. { ,kän,dək'tiv·əd·ē i'lip,sòid }

conductivity modulation [ELECTR] Of a semiconductor, the variation of the conductivity of a semiconductor through variation of the charge carrier density. { ,kän,dək'tiv·əd·ē ,mäj·ə'lā·shən }

conductivity modulation transistor [ELECTR] Transistor in which the active properties are derived from minority carrier modulation of the bulk resistivity of the semiconductor. { ,kän,dək'tiv·əd·ē ,mäj·ə'lā·shən tran'zis·tər }

conductivity tensor [ELEC] A tensor which, when multiplied by the electric field vector according to the rules of matrix multiplication, gives the current density vector. { ,kän,dək'tiv·əd·ē ,ten·sər }

conductivity theory [STAT MECH] Theory which treats the system of electrons in a metal as a gas and uses the Boltzmann transport equation to calculate conductivity. { ,kän,dək'tiv·əd·ē ,thē·ə·rē }

conductometer [ENG] An instrument designed to measure thermal conductivity; in particular, one that compares the rates at which different rods transmit heat. { ,kän,dək'täm·əd·ər }

conductometric titration [ANALY CHEM] A titration in which electrical conductance of a solution is measured during the course of the titration. { kən,dək·tə,me,trik tī'trā·shən }

conductor [ELEC] A wire, cable, or other body or medium that is suitable for carrying electric current. Also known as electric conductor. { kən'dək·tər }

conductor pipe [BUILD] A metal pipe through which water is drained from the roof. [PETRO ENG] A short string of large-diameter casing serving primarily to keep the top of a well bore open and to convey upflowing drilling fluid from the well bore to the slush pit. { kən'dək·tər ,pīp }

conductor skin effect See skin effect. { kən'dək·tər ,skin i'fekt }

conduit [ELEC] Solid or flexible metal or other tubing through which insulated electric wires are run. [ENG] Any channel or pipe for conducting the flow of water or other fluid. [GEOL] A water-filled underground passage that is always under hydrostatic pressure. { 'kän·də·wət }

conduplicate [BOT] Folded lengthwise and in half with the upper faces together, applied to leaves and petals in the bud. { kən'düp·lə·kət }

Condylarthra [PALEON] A mammalian order of extinct, primitive, hoofed herbivores with five-toed plantigrade to semidigitigrade feet. { ,kän·də'lär·thrə }

condyle [ANAT] A rounded bone prominence that functions in articulation. [BOT] The antheridium of certain stoneworts. [INV ZOO] A rounded, articular process on arthropod appendages. { 'kän,dīl }

condyloid articulation [ANAT] A joint, such as the wrist, formed by an ovoid surface that fits into an elliptical cavity, permitting all movement except rotation. { 'kän·də,lòid är,tik·yə'lā·shən }

condyloma acuminata [MED] A venereal disease characterized by wartlike growths on the genital organs; thought to be of viral origin. { ,kän·də'lō·mə ə,kyü·mə'näd·ə }

cone [BOT] The ovulate or staminate strobilus of a gymnosperm. [ENG ACOUS] The cone-shaped paper or fiber diaphragm of a loudspeaker. [GEOL] A mountain, hill, or other landform having relatively steep slopes and a pointed top. [HISTOL] A photoceptor of the vertebrate retina that responds differentially to light across the visible spectrum, providing both color vision and visual acuity in bright light. [MATH] A solid bounded by a region enclosed in a closed curve on a plane and a surface formed by the segments joining each point of the closed curve to a point which is not in the plane. [MET] The part of an oxygen gas flame adjacent to the orifice of the tip. [TEXT] A bobbin on which yarn is wound for weaving. { kōn }

cone antenna See conical antenna. { 'kōn an'ten·ə }

cone bearing [MECH ENG] A cone-shaped journal bearing running in a correspondingly tapered sleeve. { 'kōn ,ber·iŋ }

cone-bottom tank [ENG] Liquids-storage tank with downward-pointing conical bottom to facilitate drainage of bottom, as of water or sludge. { 'kōn ,bäd·əm ,taŋk }

CONDUCTOR

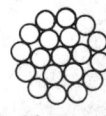

19-strand

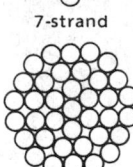

7-strand

37-strand

End views of stranded round electric conductors.

cone brake [MECH ENG] A type of friction brake whose rubbing parts are cone-shaped. { ¦kōn ¦brāk }

cone classifier [MECH ENG] Inverted-cone device for the separation of heavy particulates (such as sand, ore, or other mineral matter) from a liquid stream; feed enters the top of the cone, heavy particles settle to the bottom where they can be withdrawn, and liquid overflows the top edge, carrying the smaller particles or those of lower gravity over the rim; used in the mining and chemical industries. { ¦kōn ¦klas·ə‚fī·ər }

cone clutch [MECH ENG] A clutch which uses the wedging action of mating conical surfaces to transmit friction torque. { ¦kōn ¦kləch }

cone crusher [MECH ENG] A machine that reduces the size of materials such as rock by crushing in the tapered space between a truncated revolving cone and an outer chamber. { ¦kōn ¦krəsh·ər }

cone delta See alluvial cone. { ¦kōn ¦del·tə }

cone dike See cone sheet. { ¦kōn ¦dīk }

cone flow See conical flow. { ¦kōn ‚flō }

conehead rivet [DES ENG] A rivet with a head shaped like a truncated cone. { ¦kōn‚hed ¦riv·ət }

cone-in-cone structure [GEOL] The structure of a concretion characterized by the development of a succession of cones one within another. { ¦kōn in ¦kōn ¦strək·chər }

cone karst [GEOL] A type of karst, typical of tropical regions, characterized by a pattern of steep, convex sides and slightly concave floors. Also known as cockpit karst; Kegel karst. { ¦kōn ‚kärst }

cone key [DES ENG] A taper saddle key placed on a shaft to adapt it to a pulley with a too-large hole. { ¦kōn ‚kē }

cone loudspeaker [ENG ACOUS] A loudspeaker employing a magnetic driving unit that is mechanically coupled to a paper or fiber cone. Also known as cone speaker. { ¦kōn ¦lau̇d‚spēk·ər }

Conelrad [COMMUN] A system for providing official civil defense information and instructions by radio in an emergency without providing radio homing guidance for the enemy. Derived from control of electromagnetic radiation. { ¦kän·əl‚rad }

cone mandrel [DES ENG] A mandrel in which the diameter can be changed by moving conical sleeves. { ¦kōn ‚man·drəl }

Conemaughian [GEOL] Upper Middle Pennsylvanian geologic time. { ¦kän·ə¦mȯg·ē·ən }

cone nozzle [DES ENG] A cone-shaped nozzle that disperses fluid in an atomized mist. { ¦kōn ‚näz·əl }

cone of ambiguity [NAV] The conical volume of airspace above the beacon in which bearing information is unreliable in VOR and Tacan. { ¦kōn əv ‚am·bə¦gyü·ə·dē }

cone of dejection See alluvial cone. { ¦kōn əv di¦jek·shən }

cone of depression [HYD] The depression in the water table around a well defining the area of influence of the well. Also known as cone of influence. { ¦kōn əv di¦presh·ən }

cone of detritus See alluvial cone. { ¦kōn əv di¦trīd·əs }

cone of dispersion [ORD] The pattern in space formed by phenomena from point sources, such as shots on a target from the same gun, that spread out in conical form. { ¦kōn əv dis¦pər·zhən }

cone of escape [GEOPHYS] A hypothetical cone in the exosphere, directed vertically upward, through which an atom or molecule would theoretically be able to pass to outer space without a collision. { ¦kōn əv ə¦skāp }

cone of friction [MECH] A cone in which the resultant force exerted by one flat horizontal surface on another must be located when both surfaces are at rest, as determined by the coefficient of static friction. { ¦kōn əv ¦frik·shən }

cone of influence See cone of depression. { ¦kōn əv ¦in·flü·əns }

cone of nulls [ELECTROMAG] In antenna practice, a conical surface formed by directions of negligible radiation. { ¦kōn əv ¦nəlz }

cone of revolution [MATH] The surface obtained by rotating a line around another line which it intersects, using the intersection point as a pivot. { ¦kōn əv rev·ə¦lü·shən }

cone of silence [NAV] A cone-shaped region, directly over the antenna of a radio-beacon transmitter, in which no signal is detected. { ¦kōn əv ¦sī·ləns }

cone-of-silence marker [NAV] A marker beacon, usually a Z (for zone) marker beacon, located at a radio-range station;

operates at 75 megahertz and produces a light signal accompanied by a 3000-hertz tone. { ¦kōn əv ¦sī·ləns ‚märk·ər }

cone of visibility [AERO ENG] Generally, the right conical space which has its apex at some ground target and within which an aircraft must be located if the pilot is to be able to discern the target while flying at a specified altitude. { ¦kōn əv ‚viz·ə¦bil·əd·ē }

cone pulley See step pulley. { ¦kōn ‚pu̇l·ē }

cone rock bit [MECH ENG] A rotary drill with two hardened knurled cones which cut the rock as they roll. Also known as roller bit. { ¦kōn ¦räk ‚bit }

cone-roof tank [ENG] Liquids-storage tank with flattened conical roof to allow a vapor reservoir at the top for filling operations. { ¦kōn ‚rüf ‚taŋk }

cone settler [MIN ENG] A conical vessel fed centrally with fine ore pulp, in which the apex discharge carries the larger-sized particles, and the peripheral top overflow carries the finer fraction of the solids. { ¦kōn ‚set·lər }

cone sheet [GEOL] An accurate dike forming part of a concentric set that dips inward toward the center of the arc. Also known as cone dike. { ¦kōn ‚shēt }

cone speaker See cone loudspeaker. { ¦kōn ‚spēk·ər }

cone valve [CIV ENG] A divergent valve whose cone-shaped head in a fixed cylinder spreads water around the wide, downstream end of the cone in spillways of dams or hydroelectric facilities. Also known as Howell-Bunger valve. { ¦kōn ‚valv }

Conewangoan [GEOL] Upper Upper Devonian geologic time. { ‚kän·ə¦waŋ·gə·wən }

confectioner's coating [FOOD ENG] A chocolate substitute for coating candy or filling biscuits that is made with a fat other than cocoa butter. { kən¦fek·shən·ərz ‚kōd·iŋ }

conference communications [COMMUN] Communications facilities whereby direct speech conversation may be conducted between three or more locations simultaneously. { ¦kän·frəns kə‚myü·nə¦kā·shənz }

confidence [STAT] The degree of assurance that a specified failure rate is not exceeded. { ¦kän·fə·dəns }

confidence coefficient [STAT] The probability associated with a confidence interval; that is, the probability that the interval contains a given parameter or characteristic. Also known as confidence level. { ¦kän·fə·dəns ‚kō·i¦fish·ənt }

confidence interval [STAT] An interval which has a specified probability of containing a given parameter or characteristic. { ¦kän·fə·dəns ‚in·tər·vəl }

confidence level [IND ENG] The probability in acceptance sampling that the quality of accepted lots manufactured will be better than the rejectable quality level (RQL); 90% level indicates that accepted lots will be better than the RQL 90 times in 100. [STAT] See confidence coefficient. { ¦kän·fə·dəns ‚lev·əl }

confidence limit [STAT] One of the end points of a confidence interval. { ¦kän·fə·dəns ‚lim·ət }

configuration [AERO ENG] A particular type of specific aircraft, rocket, or such, which differs from others of the same model by the arrangement of its components or by the addition or omission of auxiliary equipment; for example, long-range configuration or cargo configuration. [CHEM] The three-dimensional spatial arrangement of atoms in a stable or isolable molecule. [COMPUT SCI] For a computer system, the relationship of hardware elements to each other, and the manner in which they are electronically connected. [ELEC] A group of components interconnected to perform a desired circuit function. [MATH] An arrangement of geometric objects. [MECH] The positions of all the particles in a system. [SYS ENG] A group of machines interconnected and programmed to operate as a system. { kən‚fig·yə¦rā·shən }

configurational free energy [STAT MECH] The free energy of a solid lattice associated with the interaction between neighboring atoms, and with external electric and magnetic fields. { kən‚fig·yə¦rā·shən·əl ¦frē ‚en·ər·jē }

configuration interaction [PHYS CHEM] Interaction between two different possible arrangements of the electrons in an atom (or molecule); the resulting electron distribution, energy levels, and transitions differ from what would occur in the absence of the interaction. { kən‚fig·yə¦rā·shən in·tər'ak·shən }

configuration-interaction method [ATOM PHYS] A method of accounting for the electron correlation in an atom or molecule

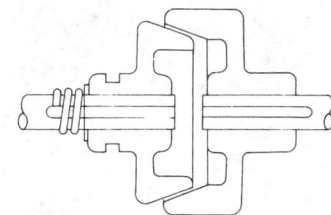

CONE CLUTCH

Cone-type friction clutch.

in which the total many-body wave function is expanded in a basis set and the coefficients in this expansion are determined by minimization of the expectation value of the energy. { kən,fig·yə¦rā·shən ,in·tər'ak·shən ,meth·əd }

confined aquifer *See* artesian aquifer. { kən'fīnd 'ak·wə·fər }

confined charge [ORD] An explosive charge loaded in a resistant container, as opposed to a bare charge. { kən'fīnd 'chärj }

confined explosion [ORD] Explosion occurring in a closed chamber, where the volume is constant. { kən'fīnd ik'splō·zhən }

confined flow [ENG] The flow of any fluid (liquid or gas) through a continuous container (process vessel) or conduit (piping or tubing). { kən'fīnd 'flō }

confinement [ENG] Physical restriction, or degree of such restriction, to passage of detonation wave or reaction zone, for example, that of a resistant container which holds an explosive charge. [PARTIC PHYS] A property of quantum chromodynamics whereby isolated quarks cannot exist, nor can any other isolated particles that carry color charges, such as gluons. [PL PHYS] Restriction of a hot plasma to a given volume as long as possible, by such means as magnetic mirrors and pinch effect. { kən'fīn·mənt }

confining bed [GEOL] An impermeable bed adjacent to an aquifer. { kən'fīn·iŋ ,bed }

confining liquid [CHEM ENG] A liquid seal (most often mercury or sodium sulfate brine) that is displaced during the no-loss transfer of a gas sample from one container to another. { kən'fīn·iŋ ,lik·wəd }

confining pressure [GEOL] An equal, all-sided pressure, such as lithostatic pressure produced by overlying rocks in the crust of the earth. { kən'fīn·iŋ ,presh·ər }

confirmation message [COMPUT SCI] A message that appears on a computer screen asking the user to confirm an action that could have destructive effects, such as loss of data. { ,kän·fər'mā·shən ,mes·ij }

confirmation well [PETRO ENG] The second producing well in a new oil field under development. { ,kän·fər'mā·shən ,wel }

conflict [PSYCH] A mental struggle that arises from the simultaneous operation of opposing impulses, drives, and external or internal demand. { 'kän,flikt }

conflict alert [NAV] A computer program, in use in enroute traffic control centers, which is designed to detect potential encounters between aircraft that are being tracked automatically by air-traffic control radars; in the event of a conflict, a symbol representing the encountering aircraft is flashed on the controller's display. Abbreviated CA. { 'kän,flikt ə'lərt }

confluence [HYD] **1.** A stream formed from the flowing together of two or more streams. **2.** The place where such streams join. { 'kän,flü·əns }

confluent hypergeometric function [MATH] A solution to differential equation $z(d^2w/dz^2) + (\rho - z)(dw/dz) - \alpha w = 0$. { kən'flü·ənt ¦hī·pər,jē·ə¦me,trik 'fəŋk·shən }

confocal conics [MATH] **1.** A system of ellipses and hyperbolas that have the same pair of foci. **2.** A system of parabolas that have the same focus and the same axis of symmetry. { kän'fō·kəl 'kän·iks }

confocal coordinates [MATH] Coordinates of a point in the plane with norm greater than 1 in terms of the system of ellipses and hyperbolas whose foci are at (1,0) and (−1,0). { kän'fō·kəl kō'órd·ən·əts }

confocal laser-scanning fluorescence microscopy [BIOPHYS] A technique that allows three-dimensional microscopic image sectioning of a specimen such as a cell by scanning the object step by step with a laser spot, instead of illuminating it as a whole. From each point the fluorescence is measured and, after analog-to-digital conversion, stored as a matrix in computer memory. { ,kän¦fō·kəl ,lā·zər ,skan·iŋ flə¦res·ənt mī'kräs·kə·pē }

confocal microscope [BIOPHYS] A microscope that creates high-resolution images of very small objects by using a condenser lens to focus the illuminating light from a point source into a very small diffraction-limited spot within the specimen, and an objective lens to focus the light emitted from that spot onto a small pinhole in an opaque screen. { ¦kän,fō·kəl 'mī·krə,skōp }

confocal quadrics [MATH] Quadrics that have the same

principal planes and whose sections by any one of these planes are confocal conics. { kän'fō·kəl 'kwäd·riks }

confocal resonator [ELECTROMAG] A wavemeter for millimeter wavelengths, consisting of two spherical mirrors facing each other; changing the spacing between the mirrors affects propagation of electromagnetic energy between them, permitting direct measurement of free-space wavelength. [OPTICS] A beam resonator that consists of a pair of spherical mirrors which have the same axis of rotational symmetry and are positioned so that their focal points coincide on this axis. { kän'fō·kəl 'rez·ən,ād·ər }

conformability [MET] The ability of a metal that has been cast as a bearing surface to deform plastically to compensate for irregularities in bearing assembly. { kən,fȯr·mə'bil·əd·ē }

conformable [GEOL] **1.** Pertaining to the contact of an intrusive body when it is aligned with the internal structures of the intrusion. **2.** Referring to strata in which layers are formed above one another in an unbroken, parallel order. { kən'fȯr·mə·bəl }

conformable matrices [MATH] Two matrices which can be multiplied together; this is possible if and only if the number of columns in the first matrix equals the number of rows in the second. { kən'fȯr·mə·bəl 'mā·trə,sēz }

conformable optical mask [ELECTR] An optical mask made on a flexible glass substrate so that it can be pulled down under vacuum into intimate contact with the substrate for accurate circuit fabrication. { kən'fȯr·mə·bəl ¦äp·tə·kəl 'mask }

conformal array [ELECTR] A circular, cylindrical, hemispherical, or other shaped array of electronically switched antennas; provides the special radiation patterns required for Tacan, IFF, and other air navigation, radar, and missile control applications. { kən'fȯr·məl ə'rā }

conformal chart [MAP] A chart on a conformal map projection. { kən'fȯr·məl 'chärt }

conformal diagram *See* Penrose diagram. { kən'fȯr·məl 'dī·ə,gram }

conformality [MAP] The retention of angular relationships at each point on a map projection. { ,kän·fər'mal·əd·ē }

conformal mapping [MATH] An angle-preserving analytic function of a complex variable. { kən'fȯr·məl 'map·iŋ }

conformal map projection [MAP] A map projection on which the shape of any small area of the surface mapped is preserved unchanged. Also known as orthomorphic map projection. { kən'fȯr·məl 'map prə¦jek·shən }

conformal optics [OPTICS] The design of optical systems whereby the shape of the outer surfaces is chosen to optimize the interaction with the environment in which the optical system is being used. { kən¦fȯrm·əl 'äp·tiks }

conformal reflection chart [ELECTROMAG] An Argand diagram for plotting the complex reflection coefficient of a waveguide junction and its image, the two being related by a conformal transformation. { kən'fȯr·məl ri'flek·shən ,chärt }

conformation [ORG CHEM] In a molecule, a specific orientation of the atoms that varies from other possible orientations by rotation or rotations about single bonds; generally in mobile equilibrium with other conformations of the same structure. Also known as conformational isomer; conformer. { kän·fər'mā·shən }

conformational analysis [PHYS CHEM] The determination of the arrangement in space of the constituent atoms of a molecule that may rotate about a single bond. { kän·fər'mā·shən·əl ə'nal·ə·səs }

conformational isomer *See* conformation. { kän·fər'mā·shən·əl 'ī·sə·mər }

conformer *See* conformation. { kən'fȯr·mər }

conformity [GEOL] The shared and undisturbed correspondence between adjacent sedimentary strata that have been deposited in orderly sequence with little or no indication of time lapses. { kən'fȯr·məd·ē }

confounding [STAT] Method used in design of factorial experiments in which some information about higher-order interaction is sacrificed so that estimates of main effects in lower-order interactions can be more precise. { kən'faȯnd·iŋ }

confused sea [OCEANOGR] A highly disturbed water surface without a single, well-defined direction of wave travel. { kən'fyüzd 'sē }

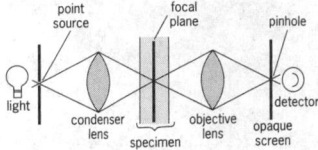

CONFOCAL MICROSCOPE

point source   focal plane   pinhole

light   condenser lens   specimen   objective lens   opaque screen   detector

Schematic diagram of a confocal microscope. (*After R. J. Taylor, Confocal microscopy sheds new light on the dynamics of living cells, J. NIH Res., 1(1):113-115, 1989*)

confusion jamming [ELECTR] An electronic countermeasure technique in which the signal from an enemy tracking radar is amplified and retransmitted with distortion to create a false echo that affects accuracy of target range, azimuth, and velocity data. { kən'fyü·zhən ˌjam·iŋ }

confusion matrix [COMPUT SCI] In pattern recognition, a matrix used to represent errors in assigning classes to observed patterns in which the *ij*th element represents the number of samples from class *i* which were classified as class *j*. { kən'fyü·zhən ˌmā·triks }

confusion reflector [ORD] An electromagnetic-wave reflector dropped from aircraft to create false signals on enemy radarscopes, consisting of strips of aluminum foil or metallized paper, such as chaff or window, cut to lengths that are multiples or submultiples of enemy radar frequencies. Also known as radar confusion reflector. { kən'fyü·zhən ri'flek·tər }

conge [FOOD ENG] A machine in which the ingredients for sweet chocolate are worked to a smooth and viscosity-stable paste. { 'känˌjā }

congelifluction *See* gelifluction. { kən·jel·ə'flək·shən }

congelifraction [GEOL] The splitting or disintegration of rocks as the result of the freezing of the water contained. Also known as frost bursting; frost riving; frost shattering; frost splitting; frost weathering; frost wedging; gelifraction; gelivation. { kən'jel·ə'frak·shən }

congeliturbate [GEOL] Soil or unconsolidated earth which has been moved or disturbed by frost action. { kən·jel·ə'tər·bət }

congeliturbation [GEOL] The churning and stirring of soil as a result of repeated cycles of freezing and thawing; includes frost heaving and surface subsidence during thaws. Also known as cryoturbation; frost churning; frost stirring; geliturbation. { kən·jel·ə·tər'bā·shən }

congener [CHEM] A chemical substance that is related to another substance, such as a derivative of a compound or an element belonging to the same family as another element in the periodic table. { 'kän·jə·nər }

congeneric [SYST] Referring to the species of a given genus. { ˌkän·jə'ner·ik }

congenic [GEN] Describing organisms that differ in genotype at a specified locus. { kən'jen·ik }

congenic strain [GEN] An animal line that differs from its inbred partner strain by only a short chromosomal segment that includes the differential locus (the locus to be studied). { kən'jen·ik 'strān }

congenital [MED] Dating from or existing before birth. { kən'jen·əd·əl }

congenital agammaglobulinemia [MED] A congenital deficiency (in serum) of immunoglobulins, characterized clinically by increased susceptibility to bacterial infections; may be a sex-linked recessive tract, affecting male infants, or sporadic, affecting both sexes, { kən'jen·əd·əl ā¦gam·ə¸gläb·yə·lə'nē·mē·ə }

congenital anomaly [MED] A structural or functional abnormality of the human body that develops before birth. Also known as birth defect. { kən'jen·əd·əl ə'näm·ə·lē }

congenital disease [MED] Any disorder or disease state that is present at birth. { kən'jen·əd·əl di'zēz }

congenital pathology [MED] The study of diseases and defects existing at birth. { kən'jen·əd·əl pə'thäl·ə·jē }

congestin [BIOCHEM] A toxin produced by certain sea anemones. { kən'jes·tən }

congestion [MED] An abnormal accumulation of fluid, usually blood, but occasionally bile or mucus, within the vessels of an organ or part. { kən'jes·chən }

congestive heart failure [MED] A state in which circulatory congestion exists as a result of heart failure. { kən'jes·tiv 'härt ˌfāl·yər }

conging process [FOOD ENG] In making sweet chocolate, working a ground mixture of chocolate liquor, sugar, and cocoa butter with additional cocoa butter in conges. { känj·iŋ ˌpräs·əs }

conglomerate [GEOL] Cemented, rounded fragments of water-worn rock or pebbles, bound by a siliceous or argillaceous substance. *See* racemate. { kən'gläm·ə·rət }

conglomeratic mudstone *See* paraglomerate. { kən¦gläm·ə¦rad·ik 'məd¸stōn }

conglutination [IMMUNOL] The completion of an agglutinating system, or the enhancement of an incomplete one, by the addition of certain substances. [MED] Abnormal union of two contiguous surfaces or bodies. { kənˌglüt·ən'ā·shən }

conglutination phenomenon [IMMUNOL] Clumping of cells or particles, such as red cells or bacteria, when treated with conglutinin in the presence of antibody and nonhemolytic complement. { kənˌglüt·ən'ā·shən fə'näm·ə·nən }

conglutinin [IMMUNOL] A heat-stable substance in bovine and other serums that aids or causes agglomeration or lysis of certain sensitized cells or particles. { kən'glüt·ən·ən }

congo red [ORG CHEM] $C_{32}H_{22}N_6Na_2O_6S_2$ An azo dye, sodium diphenyldiazo-bis-α-naphthylamine sulfonate, used as a biological stain and as an acid-base indicator; it is red in alkaline solution and blue in acid solution. { 'käŋ·gō 'red }

congo red test [PATH] Diagnostic test for amyloidosis, in which congo red is injected intravenously; 30% disappears within 1 hour in normal individuals, but in amyloidosis 40–100% disappears. { 'käŋ·gō 'red ˌtest }

congression [CYTOL] The movement of chromosomes to the spindle equator during mitosis. Also known as chromosome congression. { kən'gresh·ən }

congruence [MATH] **1.** The property of geometric figures that can be made to coincide by a rigid transformation. Also known as superposability. **2.** The property of two integers having the same remainder on division by another integer. { kən'grü·əns }

congruence transformation [MATH] **1.** Also known as transformation. **2.** A mapping which associates with each real quadratic form on a set of coordinates the quadratic form that results when the coordinates are subjected to a linear transformation. **3.** A mapping which associates with each square matrix A the matrix $B = SAT$, where S and T are nonsingular matrices, and T is the transpose of S; if A represents the coefficients of a quadratic form, then this definition is equivalent to definition 1. { kən'grü·əns ˌtranz·fər·mā·shən }

congruent evaporation [MET] In laser deposition, the tendency of the deposited film to have the same composition or stoichiometry as the material that serves as the target of the laser radiation. { kən'grü·ənt i¸vap·ə'rā·shən }

congruent figures [MATH] Two geometric figures (plane or solid), one of which can be made to coincide with the other by a rigid motion in space. { kən¦grü·ənt 'fig·yərz }

congruential generator [COMPUT SCI] A method of generating a sequence of random numbers $x_0, x_1, x_2, \ldots$, in which each member is generated from the previous one by the formula $x_{i+1} \equiv ax_i + b$ modulus m, where a, b, and m are constants. { ˌkän¸grü¦en·chəl 'jen·ə¸rād·ər }

congruent matrices [MATH] Two matrices A and B related by the transformation $B = SAT$, where S and T are nonsingular matrices and T is the transpose of S. { kən'grü·ənt 'mā·trə¸sēz }

congruent melting [GEOL] Melting of a solid substance to a liquid identical in composition. { kən'grü·ənt 'melt·iŋ }

congruent melting point [THERMO] A point on a temperature composition plot of a nonstoichiometric compound at which the one solid phase and one liquid phase are adjacent. { kən'grü·ənt 'melt·iŋ ˌpóint }

congruent numbers [MATH] Two numbers having the same remainder when divided by a given quantity called the modulus. { kən'grü·ənt 'nəm·bərz }

congruent transformation [MET] An isothermal or isobaric phase change in an alloy where the integrity of both phases is maintained throughout the process. { kən'grü·ənt ˌtranz·fər'mā·shən }

Coniacian [GEOL] Lower Senonian geologic time. { ˌkän·ē'ā·shən }

conic [MATH] A curve which may be represented as the intersection of a cone with a plane; the four types of conics are circle, ellipse, parabola, and hyperbola. Also known as conic section. { 'kän·ik }

conical [SCI TECH] Having the shape of or pertaining to a cone. { 'kän·ə·kəl }

conical antenna [ELECTROMAG] A wide-band antenna in which the driven element is conical in shape. Also known as cone antenna. { 'kän·ə·kəl an'ten·ə }

conical ball mill [MECH ENG] A cone-shaped tumbling pulverizer in which the steel balls are classified, with the larger balls at the feed end where larger lumps are crushed, and the smaller balls at the discharge end where the material is finer. { 'kän·ə·kəl 'ból ˌmil }

conical beam [ELECTR] The radar beam produced by conical scanning methods. { 'kän·ə·kəl 'bēm }

conical bearing [MECH ENG] An antifriction bearing employing tapered rollers. { 'kän·ə·kəl 'ber·iŋ }

conical buoy [NAV] A buoy whose part above water is in the shape of a cone. { 'kän·ə·kəl 'bȯi }

conical flow [FL MECH] Steady supersonic flow of a perfect, inviscid gas past a conical solid body in a region of the flow field where the principal physical quantities such as velocity, pressure, and density are constant on rays passing through a fixed point. Also known as cone flow. { 'kän·ə·kəl 'flō }

conical helimagnet [SOLID STATE] A helimagnet in which the directions of atomic magnetic moments all make the same angle (greater than 0° and less than 90°) with a specified axis of the crystal, moments of atoms in successive basal planes are separated by equal azimuthal angles, and all moments have the same magnitude. { 'kän·ə·kəl 'hel·ə,mag·nət }

conical helix [MATH] A curve that lies on a cone and cuts all the elements of the cone at the same angle. { 'kän·ə·kəl 'hē·liks }

conical horn [ACOUS] A horn having a circular cross section and straight sides. { 'kän·ə·kəl 'hȯrn }

conical-horn antenna [ELECTROMAG] A horn antenna having a circular cross section and straight sides. { 'kän·ə·kəl ,hȯrn an'ten·ə }

conical monopole antenna [ELECTROMAG] A variation of a biconical antenna in which the lower cone is replaced by a ground plane and the upper cone is usually bent inward at the top. { 'kän·ə·kəl 'män·ə,pōl an'ten·ə }

conical pendulum [MECH] A weight suspended from a cord or light rod and made to rotate in a horizontal circle about a vertical axis with a constant angular velocity. { 'kän·ə·kəl 'pen·jə·ləm }

conical point *See* inner Lagrangian point. { 'kän·ə·kəl 'pȯint }

conical projection [MATH] A projection which associates with each point P in a plane Q the point p in a second plane q which is collinear with O and P, where O is a fixed point lying outside Q. { 'kän·ə·kəl prə'jek·shən }

conical refiner [MECH ENG] In paper manufacture, a cone-shaped continuous refiner having two sets of bars mounted on the rotating plug and fixed shell for beating unmodified cellulose fibers. { 'kän·ə·kəl ri'fīn·ər }

conical refraction [OPTICS] Phenomenon in which a ray incident on the surface of a biaxial crystal at a certain direction splits into a family of rays which lie along a cone. { 'kän·ə·kəl ri'frak·shən }

conical roll *See* batten roll. { 'kän·ə·kəl ,rōl }

conical scanning [ELECTR] Scanning in radar in which the direction of maximum radiation generates a cone, the vertex angle of which is of the order of the beam width; may be either rotating or nutating, according to whether the direction of polarization rotates or remains unchanged. { 'kän·ə·kəl 'skan·iŋ }

conical surface [MATH] A surface formed by the lines which pass through each of the points of a closed plane curve and a fixed point which is not in the plane of the curve. { 'kän·ə·kəl 'sər·fəs }

conical vault [ARCH] A vault whose inner surface is conical. { 'kän·ə·kəl 'vȯlt }

conic chart [MAP] A chart on a conic projection. { 'kän·ik ,chärt }

conic chart with two standard parallels [MAP] A chart on the conic projection with two standard parallels. Also known as secant conic chart. { 'kän·ik ,chärt with ,tü ¦stan·dərd 'par·ə·lelz }

conichalcite [MINERAL] CaCu(AsO₄)(OH) A grass green to yellowish-green or emerald green, orthorhombic mineral consisting of a basic arsenate of calcium and copper. { ,kän·ə'kal,sīt }

conicoid [MATH] A quadric surface (ellipsoid, paraboloid, or hyperboloid) other than a limiting (degenerate) case of such a surface. { 'kän·ə,kȯid }

Coniconchia [PALEON] A class name proposed for certain extinct organisms thought to have been mollusks; distinguished by a calcareous univalve shell that is open at one end and by lack of a siphon. { ,kän·ə'käŋ·kē·ə }

conic projection [MAP] A map deformation pattern resulting from the transfer of the map to a tangent or intersecting cone. { ¦kän·ik prə'jek·shən }

conic projection with two standard parallels [MAP] A conic map projection in which the surface of a sphere or spheroid, such as the earth, is conceived as developed on a cone which intersects the sphere or spheroid along two standard parallels, the cone being spread out to form a plane; for example, the Lambert conformal projection. Also known as secant conic projection. { ¦kän·ik prə'jek·shən with ,tü ¦stan·dərd 'par·ə·lelz }

conic section *See* conic. { ¦kän·ik 'sek·shən }

Conidae [INV ZOO] A family of marine gastropod mollusks in the order Neogastropoda containing the poisonous cone shells. { 'kän·ə,dē }

conidiophore [MYCOL] A specialized aerial hypha that produces conidia in certain ascomycetes and imperfect fungi. { kə'nid·ē·ə,fȯr }

conidiospore *See* conidium. { kə'nid·ē·ə,spȯr }

conidium [MYCOL] Unicellular, asexual reproductive spore produced externally upon a conidiophore. Also known as conidiospore. { kə'nid·ē·əm }

conifer [BOT] The common name for plants of the order Pinales. { 'kän·ə·fər }

Coniferales [BOT] The equivalent name for Pinales. { kə,nif·ə'rā·lēz }

Coniferophyta [BOT] The equivalent name for Pinicae. { kə,nif·ə'räf·əd·ə }

coniferous forest [ECOL] An area of wooded land predominated by conifers. { kə'nif·ə·rəs 'fär·əst }

coniine [PHARM] C₅H₁₀NC₃H₇ A colorless, oily liquid with a mousy odor and a boiling point of 166°C; soluble in alcohol, ether, and oils; used as a sedative. Also known as propylpiperidine. { 'kō·nē·ən }

coniine hydrobromide [PHARM] C₈H₁₈BrN Prismatic crystals melting at 211°C; soluble in chloroform, water, and alcohol; used as an antispasmodic drug. { 'kō·nē·ən ,hī·drə'brō,mīd }

coning [PETRO ENG] Penetration into the oil column by reservoir water due to uncontrolled production. { kōn·iŋ }

coniscope *See* koniscope. { 'kän·ə,skōp }

conjoint tendon [ANAT] The common tendon of the transverse and internal oblique muscles of the abdomen. { kən'jȯint 'ten·dən }

Conjugales [BOT] An order of fresh-water green algae in the class Chlorophyceae distinguished by the lack of flagellated cells, and conjugation being the method of sexual reproduction. { ,kän·jə'gā·lēz }

conjugase [BIOCHEM] Any of a group of enzymes which catalyze the breakdown of pteroylglutamic acid. { 'kän·jə,gās }

conjugate [GEOL] **1.** Pertaining to fractures in which both sets of veins or joints show the same strike but opposite dip. **2.** Pertaining to any two sets of veins or joints lying perpendicular. [MATH] **1.** An element y of a group related to a given element x by $y = z^{-1}xz$ or $zy = xz$, where z is another element of the group. Also known as transform. **2.** For a quaternion, $x = x_0 + x_1i + x_2j + x_3k$, the quaternion $\bar{x} = x_0 - x_1i - x_2j - x_3k$. **3.** *See* complex conjugate. { 'kän·jə·gət }

conjugate acid-base pair [CHEM] An acid and a base related by the ability of the acid to generate the base by loss of a proton. { 'kän·jə·gət ¦as·əd ¦bās 'per }

conjugate angles [MATH] Two angles whose sum is 360° or 2π radians. Also known as explementary angles. { 'kän·jə·gət 'aŋ·gəlz }

conjugate arcs [MATH] Two arcs of a circle whose sum is the complete circle. { 'kän·jə·gət 'ärks }

conjugate axis [MATH] For a hyperbola whose equation in cartesian coordinates has the standard form $(x^2/a^2) - (y^2/b^2) = 1$, the portion of the y axis from $(0,-b)$ to $(0,b)$. { 'kän·jə·gət 'ak·səs }

conjugate binomial surds *See* conjugate radicals. { 'kän·jə·gət bī'nōm·ē·əl 'sərdz }

conjugate branches [ELEC] Any two branches of an electrical network such that a change in the electromotive force in either does not result in a change in current in the other. Also known as conjugate conductors. { 'kän·jə·gət 'bran·chəz }

conjugate bridge [ELECTR] A bridge in which the detector circuit and the supply circuits are interchanged, as compared with a normal bridge of the given type. { 'kän·jə·gət 'brij }

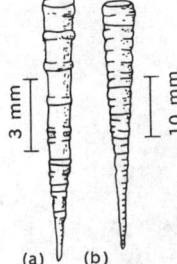

CONICONCHIA

3 mm   10 mm

(a)   (b)

Early Paleozoic *Tentaculites*. *(a) T. gyracanthus* (Eaton), Upper Silurian. *(b) T. scalariforms* Hall, Middle Devonian.

conjugate conductors *See* conjugate branches. { 'kän·jə·gət kən'dək·tərz }

conjugate convex functions [MATH] Two functions $f(x)$ and $g(y)$ are conjugate convex functions if the derivative of $f(x)$ is 0 for $x = 0$ and constantly increasing for $x > 0$, and the derivative of $g(y)$ is the inverse of the derivative of $f(x)$. { 'kän·jə·gət 'kän,veks 'fəŋk·shənz }

conjugate curve [MATH] **1.** A member of one of two families of curves on a surface such that exactly one member of each family passes through each point P on the surface, and the directions of the tangents to these two curves at P are conjugate directions. **2.** *See* Bertrand curve. { 'kän·jə·gət 'kərv }

conjugated diene [ORG CHEM] An acyclic hydrocarbon with a molecular structure containing two carbon-carbon double bonds separated by a single bond. { 'kän·jə,gād·əd 'dī,ēn }

conjugate diameters [MATH] **1.** For a conic section, any pair of straight lines either of which bisects all the chords that are parallel to the other. **2.** For an ellipsoid or hyperboloid, any three lines passing through the point of symmetry of the surface such that the plane containing the conjugate diameters (first definition) of one of the lines also contains the other two lines. { 'kän·jə·gət dī'am·əd·ərz }

conjugate diametral planes [MATH] A pair of diametral planes, each of which is parallel to the chords that define the other. { 'kän·jə·gət ,dī·ə'me·trəl 'plānz }

conjugate directions [MATH] For a point on a surface, a pair of directions, one of which is the direction of a curve on the surface through the point, while the other is the direction of the characteristic of the planes tangent to the surface at points on the curve. { 'kän·jə·gət di'rek·shənz }

conjugate division [MYCOL] Division of dikaryotic cells in certain fungi in which the two haploid nuclei divide independently, each daughter cell receiving one product of each nuclear division. { 'kän·jə·gət də'vizh·ən }

conjugated polyene [ORG CHEM] An acyclic hydrocarbon with a molecular structure containing alternating carbon-carbon double and single bonds. { 'kän·jə,gād·əd 'päl·ē,ēn }

conjugated protein [BIOCHEM] A protein combined with a nonprotein group, other than a salt or a simple protein. { 'kän·jə,gād·əd 'prō,tēn }

conjugate elements [MATH] **1.** Two elements a and b in a group G for which there is an element x in G such that $ax = xb$. **2.** Two elements of a determinant that are interchanged if the rows and columns of the determinant are interchanged. { 'kän·jə·gət 'el·ə·mənts }

conjugate fiber *See* bicomponent fiber. { 'kän·jə·gət 'fī·bər }

conjugate foci *See* conjugate points. { 'kän·jə·gət 'fō,sī }

conjugate hyperbolas [MATH] Two hyperbolas having the same asymptotes with semiaxes interchanged. { 'kän·jə·gət hī'pər·bə·ləz }

conjugate impedances [ELEC] Impedances having resistance components that are equal, and reactance components that are equal in magnitude but opposite in sign. { 'kän·jə·gət im'pēd·ən·səz }

conjugate joint system [GEOL] Two joint sets with a symmetrical pattern arranged about another structural feature or an inferred stress axis. { 'kän·jə·gət ,jóint 'sis·təm }

conjugate lines [MATH] **1.** For a conic section, two lines each of which passes through the intersection of the tangents to the conic at its points of intersection with the other line. **2.** For a quadric surface, two lines each of which intersects the polar line of the other. { 'kän·jə·gət 'līnz }

conjugate momentum [MECH] If q_j ($j = 1,2,\ldots$) are generalized coordinates of a classical dynamical system, and L is its Lagrangian, the momentum conjugate to q_j is $p_j = \partial L/\partial \dot{q}_j$. Also known as canonical momentum; generalized momentum. { 'kän·jə·gət mə'men·təm }

conjugate particles [PARTIC PHYS] A particle and its antiparticle. { 'kän·jə·gət 'pärd·ə·kəlz }

conjugate partition [MATH] If P is a partition, a conjugate partition of P is a partition that is obtained from P by interchanging the rows and columns in its star diagram. { 'kän·jə·gət pär'tish·ən }

conjugate planes [MATH] For a quadric surface, two planes each of which contains the pole of the other. { 'kän·jə·gət 'plānz }

conjugate points [MATH] For a conic section, two points either of which lies on the line that passes through the points of contact of the two tangents drawn to the conic from the other. [OPTICS] Any pair of points such that all rays from one are imaged on the other within the limits of validity of Gaussian optics. Also known as conjugate foci. { 'kän·jə·gət 'póins }

conjugate quaternion [MATH] One of a pair of quaternions that can be expressed as $q = s + ia + jb + kc$ and $\bar{q} = s - (ia + jb + kc)$, where s, a, and c are real numbers and i, j, and k are generators of the quaternions. { ¦kän·ji·gət kwə'tər·nē·ən }

conjugate radicals [MATH] Binomial surds that are of the type $a\sqrt{b} + c\sqrt{d}$ and $a\sqrt{b} - c\sqrt{d}$, where a, b, c, d are rational but $\sqrt{b}$ and $\sqrt{d}$ are not both rational. Also known as conjugate binomial surds. { 'kän·jə·gət 'rad·ə·kəlz }

conjugate roots [MATH] Conjugate complex numbers which are roots of a given equation. { 'kän·jə·gət 'rüts }

conjugate ruled surface [MATH] The ruled surface whose rulings are the lines that are tangent to a given ruled surface at the points of its line of striction and are perpendicular to the rulings of the given ruled surface at these points. { 'kän·jə·gət ¦rüld 'sər·fəs }

conjugate space [MATH] The set of all continuous linear functionals defined on a normed linear space. { 'kän·jə·gət 'spās }

conjugate subgroups [MATH] Two subgroups A and B of a group G for which there exists an element x in G such that B consists of the elements of the form xax^{-1}, where a is in A. { 'kän·jə·gət 'səb,grüps }

conjugate system of curves [MATH] Two one-parameter families of curves on a surface such that a unique curve of each family passes through each point of the surface, and the directions of the tangents to these two curves at any point on the surface are the conjugate directions at that point. { 'kän·jə·gət 'sis·təm əv 'kərvz }

conjugate triangles [MATH] Two triangles in which the poles of the sides of each with respect to a given curve are the vertices of the other. { 'kän·jə·gət 'trī,aŋ·gəlz }

conjugate variables [QUANT MECH] A pair of physical variables describing a quantum-mechanical system such that their commutator is a nonzero constant; either of them, but not both, can be precisely specified at the same time. Also known as complementary variables. { 'kän·jə·gət 'ver·ē·ə·bəlz }

conjugation [BOT] Sexual reproduction by fusion of two protoplasts in certain thallophytes to form a zygote. [INV ZOO] Sexual reproduction by temporary union of cells with exchange of nuclear material between two individuals, principally ciliate protozoans. [MICROBIO] A process involving contact between two bacterial cells during which genetic material is passed from one cell to the other. { ,kän·jə'gā·shən }

conjugon [GEN] Any of a number of different genetic elements in bacterial deoxyribonucleic acid that promote bacterial conjugation and gene transfer. { 'kän·jə,gän }

conjunction [ASTRON] **1.** The situation in which two celestial bodies have either the same celestial longitude or the same sidereal hour angle. **2.** The time at which this conjunction takes place. [MATH] The connection of two statements by the word "and." { kən'jəŋk·shən }

conjunctiva [ANAT] The mucous membrane covering the eyeball and lining the eyelids. { kən'jəŋk·tə·və }

conjunctive matrices [MATH] Two matrices A and B related by the transformation $B = SAT$, where S and T are nonsingular matrices and S is the Hermitian conjugate of I. { kən'jəŋk·tiv 'mā·trə,sēz }

conjunctive search [COMPUT SCI] A search to identify items having all of a certain set of characteristics. { kən'jəŋk·tiv 'sərch }

conjunctive transformation [MATH] The transformation $B = SAT$, where S is the Hermitian conjugate of T, and matrices A and B are equivalent. { kən'jəŋk·tiv ,tranz·fər'mā·shən }

conjunctivitis [MED] Inflammation of the conjunctiva. { kən,jəŋk·tə'vīd·əs }

conn [NAV] To direct or conduct the steering of a vessel; to give orders to the helmsman on steering the ship. { kän }

connarite [MINERAL] A green mineral consisting of hydrous nickel silicate occurring as small crystals or grains. { 'kän·ə,rīt }

connate [GEOL] Referring to materials involved in sedimentary processes that are contemporaneous with surrounding

CONNATE LEAF

Shape of a connate leaf.

materials. [SCI TECH] Born, originated, or produced in a united or fused condition. { kə'nāt }

connate leaf [BOT] A leaf shaped as though the bases of two opposite leaves had fused around the stem. { kə'nāt 'lēf }

connate water [HYD] Water entrapped in the interstices of igneous rocks when the rocks were formed; usually highly mineralized. { kə'nāt 'wòd·ər }

connected graph [MATH] A graph in which each pair of points is connected by a path. { kə'nek·təd 'graf }

connected load [ELEC] The sum of the continuous power ratings of all load-consuming apparatus connected to an electric power distribution system or any part thereof. { kə'nek·təd 'lōd }

connected relation [MATH] A relation such that for any two distinct elements a and b, either (a,b) or (b,a) is a member of the relation. { kə¦nek·təd ri'lā·shən }

connected set [MATH] A set in a topological space which is not the union of two nonempty sets A and B for which both the intersection of the closure of A with B and the intersection of the closure of B with A are empty; intuitively, a set with only one piece. { kə'nek·təd 'set }

connected space [MATH] A topological space which cannot be written as the union of two nonempty disjoint open subsets. { kə'nek·təd 'spās }

connected surface [MATH] A surface between any two points of which there is a continuous path that does not cross the surface's boundary. { kə'nek·təd 'sər·fəs }

connect function [COMPUT SCI] A signal sent over a data line to a selected peripheral device to connect it with the central processing unit. { kə'nekt ,fəŋk·shən }

connecting bar See tombolo. { kə'nekt·iŋ ,bär }

connecting circuit [ELECTR] A functional switching circuit which directly couples other functional circuit units to each other to exchange information as dictated by the momentary needs of the switching system. { kə'nekt·iŋ ,sər·kət }

connecting rod [MECH ENG] Any straight link that transmits motion or power from one linkage to another within a mechanism, especially linear to rotary motion, as in a reciprocating engine or compressor. { kə'nekt·iŋ ,räd }

connection box [COMPUT SCI] A mechanical device for altering electrical connections between various terminals, used to control the operations of a punched-card machine; its function is similar to that of a plug board. { kə'nek·shən ,bäks }

connection gas [PETRO ENG] Gas that is introduced into a well when the mud pump is shut off in order to make a connection. { kə'nek·shən ,gas }

connectionless transmission [COMMUN] Data transmission by packets that include addresses of the source and destination, so that a direct connection between these nodes is unnecessary. { kə¦nek·shən·ləs tranz'mish·ən }

connection-oriented transmission [COMMUN] Data transmission in which a physical path between the source and destination must be established and maintained for the duration of the transmission. { kə¦nek·shən ,òr·ē,ent·əd tranz'mish·ən }

connective tissue [HISTOL] A primary tissue, distinguished by an abundance of fibrillar and nonfibrillar extracellular components. { kə'nek·tiv 'tish·ü }

connectivity number [MATH] **1.** The number of points plus 1 which can be removed from a curve without separating the curve into more than one piece. **2.** The number of closed cuts or cuts joining points of previous cuts (or joining points on the boundary) plus 1 which can be made on a surface without separating the surface. Also known as Betti number. **3.** In general, the n-dimensional connectivity number of a topological space X is the number of infinite cyclic groups whose direct sum with the torsion group $G_n(X)$ forms the homology group $H_n(X)$. { kə¦nek'tiv·əd·ē ,nəm·bər }

connector [COMPUT SCI] In database management, a pointer or link between two data structures. [ELECTR] A switch, or relay group system, which finds the telephone line being called as a result of digits being dialed; it also causes interrupted ringing voltage to be placed on the called line or of returning a busy tone to the calling party if the line is busy. [ENG] **1.** A detachable device for connecting electrical conductors. **2.** A metal part for joining timbers. **3.** A symbol on a flowchart indicating that the flow jumps to a different location on the chart. { kə'nek·tər }

connector block [ELECTR] A device for connecting two cables without using plugs, similar to a barrier strip but larger,

in which wires from one cable are attached to lugs of screws on one side, and wires from the other cable are fastened to corresponding points on the opposite side. { kə'nek·tər ,bläk }

connect time [COMPUT SCI] The time that a user at a terminal is signed on to a computer. { kə'nekt ,tīm }

connellite [MINERAL] $Cu_{19}(SO_4)Cl_4(OH)_{32}·3H_2O$ A deep-blue striated copper mineral; crystals are in the hexagonal system. Also known as footeite. { 'kän·əl,īt }

connexins [CELL MOL] A group of transmembrane proteins that form the intermembrane channels of gap junctions; they are used by inorganic ions and most small organic molecules to pass through cell interiors. { kə'nek·sənz }

connexon [CYTOL] Any of the cylindrical channels associated with gap junctions. { kə'nek,sän }

conning tower [NAV ARCH] **1.** The raised observation post of a submarine, which is in addition usually used as an entrance or exit. **2.** The armored pilothouse of a warship. { 'kän·iŋ ,taú·ər }

connivent [BIOL] Converging so as to meet, but not fused into a single part. { kə'nīv·ənt }

Conoclypidae [PALEON] A family of Cretaceous and Eocene exocyclic Euechinoidea in the order Holectypoida having developed aboral petals, internal partitions, and a high test. { ,kän·ō·klə'pid·ē,ē }

Conocyeminae [PALEON] A subfamily of Mesozoan parasites in the family Dicyemidae. { ,kän·ə,sī'em·ə,nē }

conode See tie line. { 'kō,nōd }

conodont [PALEON] A minute, toothlike microfossil, composed of translucent amber-brown, fibrous or lamellar calcium phosphate; taxonomic identity is controversial. { 'kän·ə,dänt }

Conodontiformes [PALEON] A suborder of conodonts from the Ordovician to the Triassic having a lamellar internal structure. { ,kän·ə,dän·tə'fòr,mēz }

Conodontophoridia [PALEON] The ordinal name for the conodonts. { ¦kän·ə,dän·tə·fə'rid·ē·ə }

conoid [SCI TECH] Shaped somewhat like a cone, but not quite conical. { 'kä,nòid }

conoid of Sturm See astigmatic interval. { ¦kä,nòid əv 'stərm }

Conopidae [INV ZOO] The wasp flies, a family of dipteran insects in the suborder Cyclorrhapha. { kə'näp·ə,dē }

conoplain See pediment. { 'kän·ə,plān }

conoscope [OPTICS] An instrument, essentially a wide-angle microscope, used for study and observation of interference figures and related phenomena of specially cut crystal plates, especially for measuring the axial angle. Also known as hodoscope. { 'kän·ə,skōp }

conotheca [INV ZOO] The thin integument of the phragmocone in certain mollusks. { ¦kō·nə'thē·kə }

Conrad discontinuity [GEOPHYS] A relatively abrupt discontinuity in the velocity of elastic waves in the earth, increasing from 6.1 to 6.4–6.7 kilometers per second; occurs at various depths and marks contact of granitic and basaltic layers. { 'kän·,rad dis,känt·ən'ü·əd·ē }

Conrad machine [MIN ENG] Mechanized pit digger used in checking of alluvial boring; sections of tubing, 5 feet (152 centimeters) long and 2 feet (61 centimeters) in inside diameter, are worked into the ground while spoil is removed by means of a bucket or grab. { 'kän,rad mə'shēn }

Conradson carbon test See carbon-residue test. { 'kän·rəd·sən 'kär·bən ,test }

consanguineous [GEN] Pertaining to two or more individuals that have a common recent ancestor. [GEOL] Of a natural group of sediments or sedimentary rocks, having common or related origin. { ¦kän·saŋ'gwin·ē·əs }

consanguineous ring structures [ASTRON] A formation on the moon's surface consisting of two or more craters that are similar in form and very close to each other. { ¦kän·saŋ¦g-win·ē·əs 'riŋ ,strək·chərz }

consanguinity [GEN] Genetic blood relationship arising from a common ancestor. [PETR] The genetic relationship between igneous rocks in a single petrographic province which are presumably derived from a common parent magma. { ¦kän·saŋ¦gwin·əd·ē }

conscience [PSYCH] The moral, self-critical part of oneself wherein have developed, and reside, standards of behavior and performance and value judgments. { 'kän·chəns }

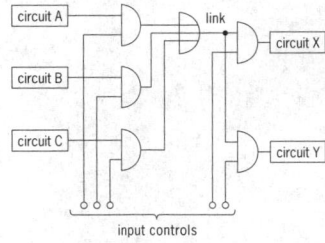

CONNECTING CIRCUIT

Connecting circuit using AND and OR gates.

consciousness [PSYCH] State of being aware of one's own existence, of one's mental states, and of the impressions made upon one's senses. { 'kän·chəs·nəs }

consecutive [MATH] Immediately following one another in a sequence. { kən'sek·yəd·iv }

consecutive angles [MATH] Two angles of a polygon that have a common side. { kən‚sek·yəd·iv 'aŋ·gəlz }

consecutive sides [MATH] Two sides of a polygon that have a common angle. { kən‚sek·yəd·iv 'sīdz }

consensual eye reflex *See* consensual light reflex. { kən¦sench·yə·wəl ¦ī 'rē‚fleks }

consensual light reflex [NEURO] The reaction of both pupils when only one eye is exposed to a change in light intensity. Also known as consensual eye reflex. { kən¦sench·yə·wəl ¦līt 'rē‚fleks }

consensus [SCI TECH] A method of checking or confirming the correctness of an observation or report, based on agreement between different observers. { kən'sen·səs }

consensus sequence [GEN] An average nucleotide sequence; each nucleotide is the most frequent at its position in the sequence. { kən'sen·səs ‚sē·kwəns }

consequence finding program [COMPUT SCI] A computer program that attempts to deduce mathematical consequences from a set of axioms and to select those consequences that will be significant. { 'kän·sə·kwəns ¦fīnd·iŋ ‚prō·grəm }

consequent [GEOL] Of, pertaining to, or characterizing movements of the earth resulting from the external transfer of material in the process of gradation. [MATH] **1.** The second term or denominator of a ratio. **2.** The second of the two statements in an implication. **3.** *See* successor. { 'kän·sə·kwənt }

consequent poles [ELECTROMAG] Pairs of magnetic poles in a magnetized body that are in excess of the usual single pair. { 'kän·sə·kwənt ¦pōlz }

consequent stream [GEOL] A stream whose course is determined by the slope of the land. Also known as superposed stream. { 'kän·sə·kwənt ‚strēm }

consequent valley [GEOL] **1.** A valley whose direction depends on corrugation. **2.** A valley formed by the widening of a trench cut by a consequent stream. { 'kän·sə·kwənt ‚val·ē }

conservation [ECOL] Those measures concerned with the preservation, restoration, beneficiation, maximization, reutilization, substitution, allocation, and integration of natural resources. { ‚kän·sər'vā·shən }

conservation gasoline [MATER] Gasoline that is made from charging stocks consisting of vapors collected from distillation and cracking stills, from storage tanks, and from other points where condensation gasoline vapors may be escaping. { ‚kän·sər'vā·shən ‚gas·ə'lēn }

conservation law [PHYS] A law which states that some physical quantity associated with an isolated system is constant. { ‚kän·sər'vā·shən ‚lo }

conservation of angular momentum [MECH] The principle that, when a physical system is subject only to internal forces that bodies in the system exert on each other, the total angular momentum of the system remains constant, provided that both spin and orbital angular momentum are taken into account. { ‚kän·sər'vā·shən əv 'aŋ·gyə·lər mə'men·təm }

conservation of areas [MECH] A principle governing the motion of a body moving under the action of a central force, according to which a line joining the body with the center of force sweeps out equal areas in equal times. { ‚kän·sər'vā·shən əv 'er·ē·əz }

conservation of charge [ELEC] A law which states that the total charge of an isolated system is constant; no violation of this law has been discovered. Also known as charge conservation. { ‚kän·sər'vā·shən əv 'chärj }

conservation of condensation [FL MECH] The principle that the rapid rise of pressure associated with the spherical wave propagating outward from an explosion must be followed by a region of diminished pressure. { ‚kän·sər'vā·shən əv ‚kän·dən'sā·shən }

conservation of energy [PHYS] The principle that energy cannot be created or destroyed, although it can be changed from one form to another; no violation of this principle has been found. Also known as energy conservation. { ‚kän·sər'vā·shən əv 'en·ər·jē }

conservation of mass [PHYS] The notion that mass can neither be created nor destroyed; it is violated by many microscopic phenomena. { ‚kän·sər'vā·shən əv 'mas }

conservation of matter [PHYS] The notion that matter can be neither created nor destroyed; it is violated by microscopic phenomena. { ‚kän·sər'vā·shən əv 'mad·ər }

conservation of momentum [MECH] The principle that, when a system of masses is subject only to internal forces that masses of the system exert on one another, the total vector momentum of the system is constant; no violation of this principle has been found. Also known as momentum conservation. { ‚kän·sər'vā·shən əv mə'men·təm }

conservation of orbital symmetry *See* Woodward-Hoffmann rule. { ‚kän·sər'vā·shən əv 'or·bəd·əl 'sim·ə·trē }

conservation of parity [QUANT MECH] The law that, if the wave function describing the initial state of a system has even (odd) parity, the wave function describing the final state has even (odd) parity; it is violated by the weak interactions. Also known as parity conservation. { ‚kän·sər'vā·shən əv 'par·əd·ē }

conservation of probability [QUANT MECH] The requirement that the sum of the probabilities of finding a system in each of its possible states is constant. { ‚kän·sər'vā·shən əv ‚präb·ə'bil·əd·ē }

conservation of vorticity [FL MECH] **1.** The principle that the vertical component of the absolute vorticity of each particle in an inviscid, autobarotropic fluid flowing horizontally remains constant. **2.** The hypothesis that the vorticity of fluid particles remains constant during the turbulent mixing of the fluid. { ‚kän·sər'vā·shən əv ‚vor'tis·əd·ē }

conservative concentrations [OCEANOGR] Concentrations such as heat content or salinity occurring in bodies of water that are altered locally, except at the boundaries, by processes of diffusion and advection only. { kən'sər·və·tiv ‚kän·sən'trā·shənz }

conservative force field [MECH] A field of force in which the work done on a particle in moving it from one point to another depends only on the particle's initial and final positions. { kən'sər·və·tiv 'fors ‚fēld }

conservative property [THERMO] A property of a system whose value remains constant during a series of events. { kən'sər·və·tiv 'präp·ərd·ē }

conservative replication [MOL BIO] Replication of a molecule of deoxyribonucleic acid (DNA) such that one DNA molecule would consist of both the original parent strands and the replicated molecule would contain two newly synthesized strands. { kən'sər·vəd·iv ‚rep·lə'kā·shən }

conservative scattering [ELECTROMAG] Scattering of radiation without accompanying absorption. { kən'sər·vəd·iv 'skad·ə·riŋ }

conservative substitution [MOL BIO] Replacement of an amino acid in a polypeptide by one with similar characteristics. { kən'sər·vəd·iv ‚səb·stə'tü·shən }

conservative system [PHYS] A system in which there is no dissipation of energy so that the total energy remains constant with time. { kən'sər·vəd·iv 'sis·təm }

conserved quantity [PHYS] A quantity that remains unchanged with time during the evolution of a dynamical system. { kən'sərvd 'kwän·əd·ē }

conserved sequence [EVOL] A sequence of nucleotides in genetic material or of amino acids in a polypeptide chain that has changed only slightly or not at all during an evolutionary period of time. { kən'sərvd 'sē·kwəns }

conserved vector current [PARTIC PHYS] The hypothesis that the weak hadronic vector current is identical to the conserved isotopic-spin current. Abbreviated CVC. { kən'sərvd 'vek·tər ‚kər·ənt }

consistency [MATER] The degree of solidity or fluidity of a material such as grease, pulp, or slurry. { kən'sis·tən·sē }

consistency condition [MATH] The requirement that a mathematical theory be free from contradiction. { kən'sis·tən·sē kən'dish·ən }

consistency routine [COMPUT SCI] A debugging routine which is used to determine whether the program being checked gives consistent results at specified check points; for example, consistent between runs or with values calculated by other means. { kən'sis·tən·sē rü'tēn }

consistent equations [MATH] Two or more equations that are all satisfied by at least one set of values of the variables. { kən'sis·tənt i'kwā·zhənz }

consistent estimate [STAT] A method of estimation which has the property that the estimate is practically certain to fall very close to a parameter being estimated, provided there are sufficient observations. { kən'sis·tənt 'es·tə·mət }

consociation [ECOL] A climax community of plants which is dominated by a single species. { kən,sō·sē'ā·shən }

Consol [NAV] A radio navigation aid that provides a number of characteristic signal zones which rotate in a time sequence; a bearing may be determined by observation of the instant at which transition occurs from one zone to the following zone. Also known as Sonne. { 'kän,sōl }

Consolan [NAV] A long-range directional navigation system that transmits a slowly rotating keyed radio field pattern; an American version of the German Sonne and British Consol, using two radiators instead of three to minimize night-effect errors. { 'kän,sō,lan }

Consol chart [NAV] A chart showing the lines of position of the Consol navigation system. { 'kän,sōl ,chärt }

console [COMPUT SCI] **1.** The section of a computer that is used to control the machine manually, correct errors, manually revise the contents of storage, and provide communication in other ways between the operator or service engineer and the central processing unit. Also known as master console. **2.** A display terminal together with its keyboard. [ENG] **1.** A main control desk for electronic equipment, as at a radar station, radio or television station, or airport control tower. Also known as control desk. **2.** A large cabinet for a radio or television receiver, standing on the floor rather than on a table. **3.** A grouping of controls, indicators, and similar items contained in a specially designed model cabinet for floor mounting; constitutes an operator's permanent working position. { 'kän,sōl }

console display [COMPUT SCI] The visible representation of information, whether in words, numbers, or drawings, on a console screen connected to a computer. { 'kän,sōl di'splā }

console file adapter [COMPUT SCI] A special input/output device which allows the operator to load reloadable control storage from the system console. { 'kän,sōl 'fīl ə'dap·tər }

console receiver [ELECTR] A television or radio receiver in a console. { 'kän,sōl ri'sēv·ər }

console switch [COMPUT SCI] A switch on a computer console whose setting can be sensed by a computer, so that an instruction in the program can direct the computer to use this setting to determine which of various alternative courses of action should be followed. { 'kän,sōl ,swich }

console typewriter [COMPUT SCI] A typewriter by means of which the computer operator can monitor system and program operations. { 'kän,sōl 'tīp,rīd·ər }

consolidated ice [OCEANOGR] Ice which has been compacted into a solid mass by wind and ocean currents and covers an area of the ocean. { kən'säl·ə,dād·əd 'īs }

consolidation [GEOL] **1.** Processes by which loose, soft, or liquid earth become coherent and firm. **2.** Adjustment of a saturated soil in response to increased load; involves squeezing of water from the pores and a decrease in void ratio. { kən,säl·ə'dā·shən }

consolidation test [MIN ENG] A test in which the specimen is confined laterally in a ring and is compressed between porous plates which are saturated with water. { kən,säl·ə'dā·shən ,test }

Consol station [NAV] A short-base-line directional-antenna system consisting of three low-frequency/medium-frequency vertical antennas in a line on the ground, evenly spaced at a distance of about three times the length of the transmitted continuous wave; coverage is about 240° over long distances, in the form of 22 lobes; the lobes are areas where the signals corresponding to the Morse E and T are heard; these patterns rotate in space and the line of position is determined by counting the numbers of E's and T's that are heard. { 'kän,sōl ,stā·shən }

consolute [CHEM] Of or pertaining to liquids that are perfectly miscible in all proportions under certain conditions. { 'kan·sə,lüt }

consolute temperature [THERMO] The upper temperature of immiscibility for a two-component liquid system. Also known as upper consolute temperature; upper critical solution temperature. { 'kan·sə,lüt 'tem·prə·chər }

consonance [ACOUS] The interval between two tones whose frequencies are in a ratio approximately equal to the

quotient of two whole numbers, each equal to or less than 6, or to such a quotient multiplied or divided by some power of 2. { 'kän·sə·nəns }

consortism See symbiosis. { 'kän,sȯrd,iz·əm }

conspecific [SYST] Referring to individuals or populations of a single species. { ¦kän·spə'sif·ik }

constancy See persistence. { 'kän·stən·sē }

constant [SCI TECH] A value that does not change during a particular process. { 'kän·stənt }

constant-amplitude recording [ENG ACOUS] A sound-recording method in which all frequencies having the same intensity are recorded at the same amplitude. { ¦kän·stənt 'am·plə,tüd ri,kȯrd·iŋ }

constantan [MET] An alloy containing 45% nickel and 55% copper, used to form iron-constantan and copper-constantan thermocouples. { kən'stan·tən }

constant-angle fringes See Haidinger fringes. { ¦kän·stənt 'aŋ·gəl ,frin·jəz }

constant area [COMPUT SCI] A part of storage used for constants. { 'kän·stənt ¦er·ē·ə }

constant-bandwidth analyzer [ACOUS] A tunable sound analyzer which has a fixed pass band that is swept through the frequency range of interest. Also known as constant-bandwidth filter. { ¦kän·stənt 'band,width ,an·ə,līz·ər }

constant-bandwidth filter See constant-bandwidth analyzer. { ¦kän·stənt'band,width 'fil·tər }

constant-conductance network See constant-resistance network. { ¦kän·stənt kən'dək·təns ,net,wərk }

constant-current characteristic [ELECTR] The relation between the voltages of two electrodes in an electron tube when the current to one of them is maintained constant and all other electrode voltages are constant. { ¦kän·stənt 'kər·ənt ,kar·ik·tə'ris·tik }

constant-current dc potentiometer [ELEC] A potentiometer in which the unknown electromotive force is balanced by a constant current times the resistance of a calibrated resistor or slide-wire. Also known as Poggendorff's first method. { ¦kän·stənt 'kər·ənt ¦dē¦sē pə,ten·chē'am·əd·ər }

constant-current electrolysis [CHEM] Electrolysis in which a constant current flows through the cell; used in electro-deposition analysis. { ¦kän·stənt 'kər·ənt i,lek'träl·ə·səs }

constant-current filter [ELECTR] A filter network intended to be connected to a source whose internal impedance is so high it can be assumed as infinite. { ¦kän·stənt 'kər·ənt ¦fil·tər }

constant-current generator [ELECTR] A vacuum-tube circuit, generally containing a pentode, in which the alternating-current anode resistance is so high that anode current remains essentially constant despite variations in load resistance. { ¦kän·stənt 'kər·ənt 'jen·ə,rād·ər }

constant-current modulation [COMMUN] System of amplitude modulation in which output circuits of the signal amplifier and the carrier-wave generator or amplifier are connected via a common coil to a constant-current source. Also known as Heising modulation. { ¦kän·stənt 'kər·ənt ,mäj·ə'lā·shən }

constant-current source [ELECTR] A circuit which produces a specified current, independent of the load resistance or applied voltage. { ¦kän·stənt 'kər·ənt ,sȯrs }

constant-current supply [ELEC] The power supply for repeatered submarine telephone cables; the voltage is varied automatically to maintain a constant current through the use of variable-voltage rectifiers and constant-current regulators at each shore station. { ¦kän·stənt 'kər·ənt sə'plī }

constant-current titration See potentiometric titration. { ¦kän·stənt 'kər·ənt tī'trā·shən }

constant-current transformer [ELEC] A transformer that automatically maintains a constant current in its secondary circuit under varying loads, when supplied from a constant-voltage source. { ¦kän·stənt 'kər·ənt tranz'fȯr·mər }

constant-deviation fringes See Haidinger fringes. { ¦kän·stənt ,dē·vē¦ā·shən ,frin·jəz }

constant-deviation prism [OPTICS] A prism whose deviation is constant and does not depend on the index of refraction or wavelength. { ¦kän·stənt ,dē·vē¦ā·shən 'priz·əm }

constant-deviation spectrometer [SPECT] A spectrometer in which the collimator and telescope are held fixed and the observed wavelength is varied by rotating the prism or diffraction grating. { ¦kän·stənt ,dē·vē¦ā·shən spek'träm·əd·ər }

constant-distance sphere [ENG ACOUS] The relative response of a sonar projector to variations in acoustic intensity, or intensity per unit band, over the surface of a sphere concentric with its center. { ¦kän·stənt 'dis·təns ˌsfir }

constant-effect model [STAT] A model of a test in which the effect of a treatment is the same for all subjects. { ¦kän·stənt i'fekt ˌmäd·əl }

constant element [IND ENG] Under a specified set of conditions, an element for which the standard time allowance should always be the same. { ¦kän·stənt 'el·ə·mənt }

constant-false-alarm rate [ELECTR] Radar system devices used to prevent receiver saturation and overload so as to present clean video information to the display, and to present a constant noise level to an automatic detector. { ¦kän·stənt ˌfȯls ə'lärm ˌrāt }

constant field See stationary field. { ¦kän·stənt 'fēld }

constant-force spring [MECH ENG] A spring which has a constant restoring force, regardless of displacement. { ¦kän·stənt ¦fȯrs ˌspriŋ }

constant function [MATH] A function whose value is the same number for all elements of the function's domain. { ¦kän·stənt ˌfəŋk·shən }

constant-gradient synchrotron [NUCLEO] A synchrotron in which the radial gradient of the magnetic field is constant as a function of angle around the orbit. { ¦kän·stənt ¦grād·ē·ənt 'siŋ·krəˌträn }

constant guidance [CYTOL] The oriented response of isolated tissue cells in culture according to the topography of their substratum. { ¦kän·takt ˌgīd·əns }

constant-head meter [ENG] A flow meter which maintains a constant pressure differential but varies the orifice area with flow, such as a rotameter or piston meter. { ¦kän·stənt ¦hed ˌmēd·ər }

constant-height chart [METEOROL] A synoptic chart for any surface of constant geometric altitude above mean sea level (a constant-height surface), usually containing plotted data and analyses of the distribution of such variables as pressure, wind, temperature, and humidity at that altitude. Also known as constant-level chart; fixed-level chart; isohypsic chart. { ¦kän·stənt ¦hīt ˌchärt }

constant-height surface [METEOROL] A surface of constant geometric or geopotential altitude measured with respect to mean sea level. Also known as constant-level surface; isohypsic surface. { ¦kän·stənt ¦hīt ˌsər·fəs }

constant instruction [COMPUT SCI] A nonexecutable instruction. { ¦kän·stənt in¦strək·shən }

constant-k filter [ELECTR] A filter in which the product of the series and shunt impedances is a constant that is independent of frequency. { ¦kän·stənt ¦kā 'fil·tər }

constant-k lens [ELECTROMAG] A microwave lens that is constructed as a solid dielectric sphere; a plane electromagnetic wave brought to a focus at one point on the sphere emerges from the opposite side of the sphere as a parallel beam. { ¦kän·stənt ¦kā 'lenz }

constant-k network [ELECTR] A ladder network in which the product of the series and shunt impedances is independent of frequency within the operating frequency range. { ¦kän·stənt ¦kā 'net¦wərk }

constant-level balloon [AERO ENG] A balloon designed to float at a constant pressure level. Also known as constant-pressure balloon. { ¦kän·stənt ¦lev·əl bə'lün }

constant-level chart See constant-height chart. { ¦kän·stənt ¦lev·əl 'chärt }

constant-level surface See constant-height surface. { ¦kän·stənt ¦lev·əl 'sər·fəs }

constant-load balance [ENG] An instrument for measuring weight or mass which consists of a single pan (together with a set of weights that can be suspended from a counterpoised beam) that has a constant load (200 grams for the microbalance). { ¦kän·stənt¦lōd 'bal·əns }

constant-load support [ENG] A spring-loaded support designed to maintain a constant and balanced load on a pipe in the event of vertical movement. { ¦kän·stənt ¦lōd sə'pȯrt }

constant-luminance transmission [COMMUN] Type of transmission in which the transmission primaries are a luminance primary and two chrominance primaries. { ¦kän·stənt ¦lü·mə·nəns tranz'mish·ən }

constant of aberration [ASTRON] The maximum aberration of a star observed from the surface of the earth, equal to 20.49 seconds of arc. { ¦kän·stənt əv ab·ə'rā·shən }

constant of gravitation See gravitational constant. { ¦kän·stənt əv grav·ə'tā·shən }

constant of integration [MATH] An arbitrary constant that must be added to an indefinite integral of a function to obtain all the indefinite integrals of that function. Also known as integration constant. { ¦kän·stənt əv ˌin·tə'grā·shən }

constant of motion [MECH] A dynamical variable of a system which remains constant in time. { ¦kän·stənt əv 'mō·shən }

constant of the cone [MAP] The chart convergence factor for a conic projection. { ¦kän·stənt əv thə 'kōn }

constant-potential accelerator [NUCLEO] An accelerator in which constant direct-current voltage is applied to an accelerating tube to produce high-energy ions or electrons. { ¦kän·stənt pə'ten·chəl ik'sel·əˌrād·ər }

constant-potential electrolysis [CHEM] Electrolysis in which a constant voltage is applied to the cell; used in electrodeposition analysis. { ¦kän·stənt pə'ten·chəl iˌlek'träl·ə·səs }

constant-pressure balloon See constant-level balloon. { ¦kän·stənt ¦presh·ər bə'lün }

constant-pressure chart [METEOROL] The synoptic chart for any constant-pressure surface, usually containing plotted data and analyses of the distribution of height of the surface, wind temperature, humidity, and so on. Also known as isobaric chart; isobaric contour chart. { ¦kän·stənt ¦presh·ər ˌchärt }

constant-pressure combustion [MECH ENG] Combustion occurring without a pressure change. { ¦kän·stənt ¦presh·ər kəm'bəs·chən }

constant-pressure gas thermometer [ENG] A thermometer in which the volume occupied by a given mass of gas at a constant pressure is used to determine the temperature. { ¦kän·stənt ¦presh·ər 'gas thərˌmäm·əd·ər }

constant-pressure-pattern flight [NAV] A technique of pressure-pattern flight whereby an aircraft is navigated, in the direction of wind flow, along a height contour line on a constant-pressure surface, thereby assuring a continuous, nearly direct tail wind. { ¦kän·stənt ¦presh·ər ˌpad·ərn ˌflīt }

constant-pressure surface See isobaric surface. { ¦kän·stənt ¦presh·ər ˌsər·fəs }

constant radio code [COMMUN] Code in which all characters are represented by combinations having a fixed ratio of ones to zeros. { ¦kän·stənt 'rād·ē·ō ˌkōd }

constant-resistance dc potentiometer [ELEC] A potentiometer in which the ratio of an unknown and a known potential are set equal to the ratio of two known constant resistances. Also known as Poggendorff's second method. { ¦kän·stənt ri'zis·təns ¦dē¦sē pəˌten·chē'äm·əd·ər }

constant-resistance network [ELECTR] A network having at least one driving-point impedance that is a positive constant. Also known as constant-conductance network. { ¦kän·stənt ri'zis·təns 'net¦wərk }

constant series See displacement series. { ¦kän·stənt 'sir¦ēz }

constant-speed drive [MECH ENG] A mechanism transmitting motion from one shaft to another that does not allow the velocity ratio of the shafts to be varied, or allows it to be varied only in steps. { ¦kän·stənt ¦spēd 'drīv }

constant-speed propeller [AERO ENG] A variable-pitch propeller having a governor which automatically changes the pitch to maintain constant engine speed. { ¦kän·stənt ¦spēd prə'pel·ər }

constant term [MATH] A term that does not contain a variable. Also known as absolute term. { ¦kän·stənt 'tərm }

constant-velocity recording [ENG ACOUS] A sound-recording method in which, for input signals of a given amplitude, the resulting recorded amplitude is inversely proportional to the frequency; the velocity of the cutting stylus is then constant for all input frequencies having that given amplitude. { ¦kän·stənt və'läs·əd·ē riˌkȯrd·iŋ }

constant-velocity universal joint [MECH ENG] A universal joint that transmits constant angular velocity from the driving to the driven shaft, such as the Bendix-Weiss universal joint. { ¦kän·stənt və'läs·əd·ē ˌyü·nəˌvər·səl 'jȯint }

constant-voltage generator [ELEC] An axle generator that is equipped with a regulator which keeps voltage constant. { ¦kän·stənt ¦vōl·tij ¦jen·əˌrād·ər }

constant-voltage transformer [ELEC] A power transformer which will supply a constant voltage to an unvarying

CONSTANT-K NETWORK

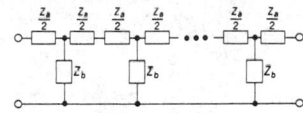

T sections, one form of constant-*k* recurrent ladder network.

CONSTANT-VELOCITY UNIVERSAL JOINT

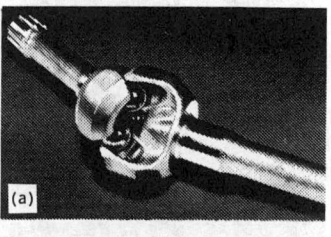

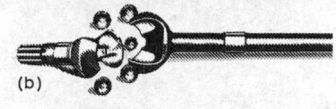

Constant-velocity universal joint. (a) Partially separated cutaway. (b) Disassembled joint showing the arrangement; the right-hand member has been cut away. (Bendix Aviation Corp.)

load, even with changes in the primary voltage. { ¦kän·stənt 'vōl·tij tranz'fôr·mər }

constant-volume gas thermometer See gas thermometer. { ¦kän·stənt 'väl·yəm 'gas thər¦mäm·əd·ər }

Constellariidae [PALEON] A family of extinct, marine bryozoans in the order Cystoporata. { ¦kän·stə·lə'rī·ə‚dē }

constellation [ASTRON] **1.** Any one of the star groups interpreted as forming configurations in the sky; examples are Orion and Leo. **2.** Any one of the definite areas of the sky. { ‚kän·stə'lā·shən }

constipation [MED] The passage of hard, dry stools. { kän·stə'pā·shən }

constituent [SCI TECH] An essential part or component of a system or group: examples are an ingredient of a chemical system, or a component of an alloy. { kən'stich·ə·wənt }

constituent day [ASTRON] The duration of one rotation of the earth on its axis with respect to an astre fictif, that is, a fictitious star representing one of the periodic elements in the tidal forces; approximates the length of a lunar or solar day. { kən'stich·ə·wənt 'dā }

constituent number [OCEANOGR] One of the harmonic elements in a mathematical expression for the tide-producing force, and in corresponding formulas for the tide or tidal current. { kən'stich·ə·wənt 'nəm·bər }

constitutional anthropology [ANTHRO] A branch of physical anthropology concerned with body composition and constitution of an individual. { ‚kän·stə'tü·shən·əl ‚an·thrə'päl·ə·jē }

constitutional isomers [ORG CHEM] Isomers which differ in the manner in which their atoms are linked. Also known as structural isomers. { ‚kän·stə'tü·shən·əl 'ī·sə·mərz }

constitutional unit [CHEM] An atom or group of atoms that is part of a chain in a polymer or oligomer. { ‚kän·stə'tü·shən·əl 'yü·nət }

constitution diagram [MET] Graphical representation of phase-stability relationships in an alloy system as a function of temperature. Also known as phase diagram. { ‚kän·stə'tü·shən 'dī·ə‚gram }

constitutive enzyme [BIOCHEM] An enzyme whose concentration in a cell is constant and is not influenced by substrate concentration. { 'kän·stə‚tüd·iv 'en‚zīm }

constitutive equations [ELECTROMAG] The equations $D = \epsilon E$ and $B = \mu H$, which relate the electric displacement D with the electric field intensity E, and the magnetic induction B with the magnetic field intensity H. { 'kän·stə‚tüd·iv i'kwā·zhənz }

constitutive gene [GEN] A gene that encodes a product required in the maintenance of basic cellular processes or cell architecture. Also known as housekeeping gene. { 'kän·stə‚tüd·iv ‚jēn }

constitutive heterochromatin [GEN] A type of heterochromatin that is always condensed and is often centered on either side of the centromere, and that stains to give a C band. { 'kän·stə‚tüd·iv ¦hed·ə·rō'krō·məd·ən }

constitutive mutation [GEN] A mutation that modifies an operator gene or a regulator gene, resulting in unregulated expression of structural genes that are normally regulated. { 'kän·stə‚tüd·iv myü'tā·shən }

constitutive promoter [MOL BIO] An unregulated promoter segment of deoxyribonucleic acid that allows continuous transcription of its cognate gene. { ¦kän·stə‚tüd·iv prə'mōd·ər }

constitutive property [CHEM] Any physical or chemical property that depends on the constitution or structure of the molecule. { 'kän·stə‚tüd·iv 'präp·ərd·ē }

constitutive secretory pathway [CELL MOL] A secretory pathway found in all cells by which transport vesicles continuously leave the Golgi apparatus and fuse with the plasma membrane, and their contents are exported to the extracellular space or used as components of the plasma membrane. { ¦kän·stə‚tüd·iv ¦sek·rə‚tór·ē 'path‚wā }

constrained mechanism [MECH ENG] A mechanism in which all members move only in prescribed paths. { kən'strānd 'mek·ə‚niz·əm }

constrained optimization problem [MATH] A nonlinear programming problem in which there are constraint functions. { kən'strānd äp·tə·mə'zā·shən ‚präb·ləm }

constraint [ENG] Anything that restricts the transverse contraction which normally occurs in a solid under longitudinal

tension. [MECH] A restriction on the natural degrees of freedom of a system; the number of constraints is the difference between the number of natural degrees of freedom and the number of actual degrees of freedom. [SCI TECH] A condition imposed on a system which limits the freedom of the system; may be physical or mathematical, necessary or incidental. { kən'strānt }

constraint function [MATH] A function defining one of the prescribed conditions in a nonlinear programming problem. { kən'strānt ‚fəŋk·shən }

constraint matrix [COMPUT SCI] The set of equations and inequalities defining the set of admissible solutions in linear programming. { kən'strānt ‚mā·triks }

constraint programming language [COMPUT SCI] A programming language in which constraints (relationships that must hold among a number of variables) are directly usable as programming constructs. { kən¦strānt 'prō‚gram·iŋ ‚laŋ·gwij }

constricting nozzle [MET] A copper nozzle which has a constricting orifice and envelopes the electrode in plasma arc processes. { kən'strik·tiŋ ‚näz·əl }

constriction See stricture. [SCI TECH] Narrowing of a channel or cylindrical member. { kən'strik·shən }

constriction disease [PL PATH] A fungus disease of peach trees caused by a species of *Phomopsis* and marked by death of peripheral structures. { kən'strik·shən di'zēz }

constrictive pericarditis [MED] Inflammation and fibrosis of the pericardium resulting in constriction of the heart and restriction of contraction and blood flow. Also known as Pick's disease. { kən'strik·div per·i‚kär'dīd·əs }

constrictor [AERO ENG] The exit portion of the combustion chamber in some designs of ramjets, where there is a narrowing of the tube at the exhaust. { kən'strik·tər }

constringence See nu value. { kən'strin·jəns }

construction [DES ENG] The number of strands in a wire rope and the number of wires in a strand; expressed as two numbers separated by a multiplication sign. [ENG] **1.** Putting parts together to form an integrated object. **2.** The manner in which something is put together. [MATH] The process of drawing with suitable instruments a geometrical figure satisfying certain specified conditions. [TEXT] A fabric formula, being the number of warp and filling threads per square inch and the weight of the yarns. { kən'strək·shən }

constructional apraxia See optic apraxia. { kən'strək·shən·əl ā'prak·sē·ə }

construction area [BUILD] The area of exterior walls and permanent interior walls and partitions. { kən'strək·shən ‚er·ē·ə }

construction cost [IND ENG] The total costs, direct and indirect, associated with transforming a design plan for material and equipment into a project ready for operation. { kən'strək·shən ‚kóst }

construction engineering [CIV ENG] A specialized branch of civil engineering concerned with the planning, execution, and control of construction operations for projects such as highways, dams, utility lines, and buildings. { kən'strək·shən ‚en·jə'nir·iŋ }

construction equipment [MECH ENG] Heavy power machines which perform specific construction or demolition functions. { kən'strək·shən i'kwip·mənt }

construction joint [CIV ENG] A vertical or horizontal surface in reinforced concrete where concreting was stopped and continued later. { kən'strək·shən ‚jóint }

construction operator [COMPUT SCI] The part of a data structure which is used to construct composite objects from atoms. { kən'strək·shən 'äp·ə‚rād·ər }

construction paper [MATER] A heavy paper made from mechanical pulp and available in a large range of colors, most of which are not lightfast. { kən'strək·shən ‚pā·pər }

construction survey [CIV ENG] A survey that gives locations for construction work. { kən'strək·shən ‚sər‚vā }

construction weight [AERO ENG] The weight of a rocket exclusive of propellant, load, and crew if any. Also known as structural weight. { kən'strək·shən ‚wāt }

construction wrench [DES ENG] An open-end wrench with a long handle; the handle is used to align matching rivet or bolt holes. { kən'strək·shən ‚rench }

constructive interference [PHYS] Phenomenon in which the phases of waves arriving at a specified point over two or more paths of different lengths are such that the square of the resultant amplitude is greater than the sum of the squares of the component amplitudes. { kən'strək·div ˌin·tər'fir·əns }

consumable electrode [MET] A metal electrode that supplies the filler for welding. { kən'süm·ə·bəl i'lek,trōd }

consumable guide electroslag welding [MET] A modification of the wire process of electroslag welding in which the filler metal is supplied by a welding wire held by a stationary metal tube. { kən'süm·ə·bəl 'gīd i'lek·trō,slag 'weld·iŋ }

consumable insert [MET] Filler metal which is located at the root of a welded joint and which becomes part of the final weldment. { kən'süm·ə·bəl 'in·sərt }

consumer [ECOL] A nutritional grouping in the food chain of an ecosystem, composed of heterotrophic organisms, chiefly animals, which ingest other organisms or particulate organic matter. { kən'süm·ər }

consumer's risk [IND ENG] The probability that a lot whose quality equals the poorest quality that a consumer is willing to tolerate in an individual lot will be accepted by a sampling plan. { kən'süm·ərz 'risk }

consummatory behavior [PSYCH] Actions that fulfill a motive and cause appetitive behavior to end. { kən'səm·ə,tór·ē bi'hāv·yər }

consumption See tuberculosis. { kən'səm·shən }

consumptive coagulopathy [MED] Reduction in one or more of the blood elements involved in coagulation as the result of marked blood clotting. { kən'səm·div ˌkō,ag·yə'läp·ə·thē }

consumptive use [HYD] The total annual land water loss in an area, due to evaporation and plant use. { kən'səm·div 'yüs }

contact [ELEC] See electric contact. [ENG] Initial detection of an aircraft, ship, submarine, or other object on a radar scope or other detecting equipment. [FL MECH] The surface between two immiscible fluids contained in a reservoir. [GEOL] The surface between two different kinds of rocks. { 'kän,takt }

contact acid [INORG CHEM] Sulfuric acid produced by the contact process. { 'kän,takt 'as·əd }

contact adsorption [CHEM ENG] Process for removal of minor constituents from fluids by stirring in direct contact with powdered or granulated adsorbents, or by passing the fluid through fixed-position adsorbent beds (activated carbon or ion-exchange resin); used to decolorize petroleum lubricating oils and to remove solvent vapors from air. { 'kän,takt ad'sórp·shən }

contact aerator [CIV ENG] A tank in which sewage that is settled on a bed of stone, cement-asbestos, or other surfaces is treated by aeration with compressed air. { 'kän,takt 'er,ād·ər }

contact anemometer [ENG] An anemometer which actuates an electrical contact at a rate dependent upon the wind speed. Also known as contact-cup anemometer. { 'kän,takt an·ə'mäm·əd·ər }

contact angle See angle of contact. { 'kän,takt ˌaŋ·gəl }

contact arc [ELEC] A spark that occurs immediately after the breaking of an electric contact carrying a current. { 'kän,takt ˌärk }

contact aureole See aureole. { 'kän,takt 'òr·ē,ōl }

contact bed [CIV ENG] A bed of coarse material such as coke, used to purify sewage. { 'kän,takt ˌbed }

contact binary [ASTRON] A binary system at least one of whose components fills its Roche lobe and in which mass exchange is taking place. { 'kän,takt 'bī,ner·ē }

contact block [ELEC] A block of conducting material such as carbon, used in a relay. { 'kän,takt ˌbläk }

contact bounce [ELEC] The uncontrolled making and breaking of contact one or more times, but not continuously, when relay contacts are moved to the closed position. { 'kän,takt ˌbaůns }

contact breccia [PETR] Angular rock fragments resulting from shattering of wall rocks around laccolithic and other igneous masses. { 'kän,takt 'brech·ə }

contact catalysis [CHEM ENG] Process of change in the structure of gas molecules adsorbed onto solid surfaces; the basis of many industrial processes. { 'kän,takt kə'tal·ə·səs }

contact ceiling [BUILD] A ceiling in which the lath and construction are in direct contact, without use of furring or runner channels. { 'kän,takt ˌsēl·iŋ }

contact chart See aeronautical pilotage chart. { 'kän,takt ˌchärt }

contact chatter See chatter. { 'kän,takt ˌchad·ər }

contact clip [ELEC] The clip which the blade of a knife switch is clamped to in the closed condition. { 'kän,takt ˌklip }

contact condenser [MECH ENG] A device in which a vapor, such as steam, is brought into direct contact with a cooling liquid, such as water, and is condensed by giving up its latent heat to the liquid. Also known as direct-contact condenser. { 'kän,takt kən'den·sər }

contact corrosion See crevice corrosion; galvanic corrosion. { 'kän,takt kə'rō·zhən }

contact-cup anemometer See contact anemometer. { 'kän,takt ˌkəp an·ə'mäm·əd·ər }

contact-dependent signaling [CELL MOL] A type of intracellular communication whereby a signal molecule remains bound to the signaling cell surface, rather than being released into the extracellular space, and influences only cells that come into contact with it. { 'kän,tak di,pen·dənt 'sig·nəl·iŋ }

contact dermatitis [MED] An acute or chronic inflammation of the skin resulting from irritation by or sensitization to some substance coming in contact with the skin. { 'kän,takt dər·mə'tīd·əs }

contact drop [ELEC] The voltage drop across the terminals of an electric contact. { 'kän,takt ˌdräp }

contact electricity [ELEC] An electric charge at the surface of contact of two different materials. { 'kän,takt i,lek'tris·əd·ē }

contact electromotive force See contact potential difference. { 'kän,takt i,lek·trə'mōd·iv 'fōrs }

contact filtration [CHEM ENG] A process in which finely divided adsorbent clay is mixed with oil to remove color bodies and to improve the oil's stability. { 'kän,takt fil'trā·shən }

contact flight [NAV] Flight in which visual contact is maintained with the surface of the earth. { 'kän,takt ˌflīt }

contact follow [ELEC] The distance two contacts travel together after just touching. Also known as contact overtravel. { 'kän,takt ˌfäl·ō }

contact force [ELEC] The force exerted by the moving contact of a switch or relay on a stationary contact. { 'kän,takt ˌfórs }

contact fuse See impact fuse. { 'kän,takt ˌfyüz }

contact gear ratio See contact ratio. { 'kän,takt ˌgir ˌrā·shō }

contact grasp [IND ENG] A basic grasp that is used to push an object over a surface, such as using the index finger to push a coin over a flat surface. { 'kän,takt ˌgrasp }

contact head [COMPUT SCI] A read/write head that remains in contact with the recording surface of a hard disk, rather than hovering above it. { 'kän,takt ˌhed }

contact inhibition [CYTOL] Cessation of cell division when cultured cells are in physical contact with each other. { 'kän,takt in·ə'bish·ən }

contact-initiated discharge machining [MECH ENG] An electromachining process in which the discharge is initiated by allowing the tool and workpiece to come into contact, after which the tool is withdrawn and an arc forms. { 'kän,takt ə,nish·ē,ād·əd ˌdis,chärj mə,shēn·iŋ }

contact inspection [ENG] A method by which an ultrasonic search unit scans a test piece in direct contact with a thin layer of couplant for transmission between the search unit and entry surface. { 'kän,takt in'spek·shən }

contact ion pair [ORG CHEM] An ion pair composed of individual ions which retain their stereochemical configuration; no solvent molecules separate the cation and anion. Also known as intimate ion pair. { 'kän,takt 'ī·ən ,per }

contact lens [OPTICS] **1.** A thin lens fitted over the cornea to correct defects of vision. **2.** A similar lens or prism used with a gonioscope in eye examinations. { 'kän,takt ,lenz }

contact log [PETRO ENG] Record of electrical-resistivity data pertaining to strata structures along the depth of a drill hole. { 'kän,takt ,läg }

contact-making meter See instrument-type relay. { 'kän,takt ˌmāk·iŋ ,mēd·ər }

contact-mask read-only memory See last-mask read-only memory. { 'kän,takt ,mask 'rēd ,ōn·lē 'mem·rē }

contact material [MET] A metal having high electrical and thermal conductivity, low contact resistance, minimum sticking or welding tendencies, and high corrosion resistance. { 'kän‚takt mə'tir·ē·əl }

contact metamorphic rock [PETR] A rock formed by the processes of contact metamorphism. { 'kän‚takt ‚med·ə'mȯr·fik 'räk }

contact metamorphism [PETR] Metamorphism that is genetically related to the intrusion or extrusion of magmas and takes place in rocks at or near their contact. { 'kän‚takt ‚med·ə'mȯr·fiz·əm }

contact metasomatism [GEOL] One of the main local processes of thermal metamorphism that is related to intrusion of magmas; takes place in rocks or near their contact with a body of igneous rock. { 'kän‚takt ‚med·ə'sō·mə‚tiz·əm }

contact microphone [ENG ACOUS] A microphone designed to pick up mechanical vibrations directly and convert them into corresponding electric currents or voltages. { 'kän‚takt 'mī·krə‚fōn }

contact mine [ORD] A mine that is fired when struck by the hull of a passing ship. { 'kän‚takt ‚mīn }

contact mineral [MINERAL] A mineral formed by the processes of contact metamorphism. { 'kän‚takt ‚min·rəl }

contact modulation [ELEC] The use of a fast-acting relay, whose contacts make and break at a certain threshold current, to generate square waves from a sine-wave, rectified sine-wave or direct-current source. { 'kän‚takt ‚mäj·ə'lā·shən }

contactor [CHEM ENG] A vessel designed to bring two or more substances into contact. [ELEC] A heavy-duty relay used to control electric power circuits. Also known as electric contactor. { 'kän‚tak·tər }

contactor control system [CONT SYS] A feedback control system in which the control signal is a discontinuous function of the sensed error and may therefore assume one of a limited number of discrete values. { 'kän‚tak·tər kən'trōl ‚sis·təm }

contact overtravel See contact follow. { 'kän‚takt 'ō·vər ‚trav·əl }

contact paper [GRAPHICS] Photographic paper designed to be exposed when directly in contact with a reproducible document. { 'kän‚takt ‚pā·pər }

contact paralysis [CYTOL] The cessation of forward extension of the pseudopods of a cell as a result of its collision with another cell. { 'kän‚takt pə‚ral·ə·səs }

contact piston [ELECTROMAG] A waveguide piston that makes contact with the walls of the waveguide. Also known as contact plunger. { 'kän‚takt ‚pis·tən }

contact plunger See contact piston. { 'kän‚takt ‚plən·jər }

contact point [ELEC] In the ignition system of an internal combustion engine, any of the stationary and movable electrically conducting metal points that open and close to complete or break an electric circuit. { 'kän‚takt ‚pȯint }

contact potential See contact potential difference. { 'kän‚takt pə'ten·chəl }

contact potential difference [ELEC] The potential difference that exists across the space between two electrically connected materials. Also known as contact electromotive force; contact potential; Volta effect. { 'kän‚takt pə'ten·chəl 'dif·rəns }

contact pressure [ELEC] The amount of pressure holding a set of contacts together. { 'kän‚takt ‚presh·ər }

contact-pressure resin See contact resin. { 'kän‚takt ‚presh·ər ‚rez·ən }

contact print [GRAPHICS] A photographic image produced by the exposure of a sensitized emulsion in direct contact with a negative or positive transparency. { 'kän‚takt ‚print }

contact printer [GRAPHICS] **1.** A device which provides a light source and a means for holding the negative and the sensitive material in contact during exposure. **2.** A specialized device for exposing diapositive plates at the same scale as that of the negative. { 'kän‚takt ‚print·ər }

contact process [CHEM ENG] Catalytic manufacture of sulfuric acid from sulfur dioxide and oxygen. { 'kän‚takt ‚präs·əs }

contact protection [ELEC] Any method for suppressing the surge which results when an inductive circuit is suddenly interrupted; the break would otherwise produce arcing at the contacts, leading to their deterioration. { 'kän‚takt prə'tek·shən }

contact ratio [DES ENG] The ratio of the length of the path of contact of two gears to the base pitch, equal to approximately the average number of pairs of teeth in contact. Also known as contact gear ratio. { 'kän‚takt ‚rā·shō }

contact rectifier See metallic rectifier. { 'kän‚takt 'rek·tə ‚fī·ər }

contact resin [MATER] A liquid resin which thickens or polymerizes on heating and, when used for bonding laminates, requires little or no pressure for adherence. Also known as contact-pressure resin. { 'kän‚takt ‚rez·ən }

contact resistance [ELEC] The resistance in ohms between the contacts of a relay, switch, or other device when the contacts are touching each other. { 'kän‚takt ri'zis·təns }

contact screen [GRAPHICS] A photographically made halftone screen on film having a dot structure of graded density, specifically designed for making halftone negatives and positives for photomechanical reproduction; the screen is placed in direct contact with the film or plate to obtain a halftone pattern from a continuous original. { 'kän‚takt ‚skrēn }

contact sensor [ENG] A device that senses mechanical contact and gives out signals when it does so. { 'kän‚takt 'sen·sər }

contact sparking [ELEC] The formation of a spark or arc at the contact points when a circuit is opened while it is carrying a current. { 'kän‚takt ‚spärk·iŋ }

contact thermography [ENG] A method of measuring surface temperature in which a thin layer of luminescent material is spread on the surface of an object and is excited by ultraviolet radiation in a darkened room; the brightness of the coating indicates the surface temperature. { 'kän‚takt thər'mäg·rə· fē }

contact time [ENG] The length of time a substance is held in direct contact with a treating agent. { 'kän‚takt ‚tīm }

contact transformation See canonical transformation. { 'kän‚takt ‚tranz·fər'mā·shən }

contact tube [MET] A device which provides electric current to a continuous electrode in a welding process. { 'kän‚takt ‚tüb }

contact twin [CRYSTAL] Twinned crystals whose members are symmetrically arranged about a twin plane. { 'kän‚takt ‚twin }

contact vein [GEOL] **1.** A variety of fissure vein formed by deposition of minerals in a fault fissure at a rock contact. **2.** A replacement vein formed by mineralized solutions percolating along the more permeable surface areas of the contact. { 'kän‚takt ‚vān }

contact zone See aureole. { 'kän‚takt ‚zōn }

contagion [MED] **1.** The process whereby disease spreads from one person to another, by direct or indirect contact. **2.** The bacterium or virus which transmits disease. { kən'tā· jən }

contagious abortion [VET MED] Brucellosis in cattle caused by *Brucella abortus* and inducing abortion. Also known as Bang's disease; infectious abortion. { kən'tā·jəs ə'bȯr·shən }

contagious disease [MED] An infectious disease communicable by contact with a person suffering from it, with the bodily discharge, or with an object touched by the person. { kən'tā·jəs di'zēz }

contagious distribution [STAT] A probability distribution which is dependent on a parameter that itself has a probability distribution. { kən'tā·jəs dis·trə'byü·shən }

contagious polyarthritis [VET MED] An infectious bacterial disease of mice caused by members of the genus *Bacteroides;* characterized by inflammation and abcess formation in the joints. { kən'tā·jəs ‚päl·ē·är'thrīd·əs }

contagious pustular dermatitis [VET MED] An infectious disease of sheep and goats characterized by vesicles on the skin which are transformed into pustules. { kən'tā·jəs 'pəs· chə·lər ‚dər·mə'tīd·əs }

container [IND ENG] A portable compartment of standard, uniform size, used to hold cargo for air, sea, or ground transport. { kən'tā·nər }

container car [ENG] A railroad car designed specifically to hold containers. { kən'tā·nər ‚kär }

containerization [IND ENG] The practice of placing cargo in large containers such as truck trailers to facilitate loading on and off ships and railroad flat cars. { kən‚tā·nə·rə'zā·shən }

container ship [NAV ARCH] A cargo ship which carries its cargo in weatherproof boxes (usually metal) of standard size, called containers, which need not be opened and are rapidly loaded or unloaded from the ship. { kən'tā·nər ‚ship }

containment [ENG] An enclosed space or facility to contain and prevent the escape of hazardous material. [MOL BIO] Prevention of the replication of the products of recombinant deoxyribonucleic acid technology outside the laboratory. [NUCLEO] **1.** Provision of a gastight enclosure around the highly radioactive components of a nuclear power plant, to contain the radioactivity released by a possible major accident. **2.** The use of remote-control devices (slave apparatus) to remove spent cores from nuclear power plants or, in shielded laboratory hoods, to perform chemical studies of dangerous radioactive materials. { kən'tān·mənt }

containment building [NUCLEO] A steel-reinforced concrete dome built over a nuclear reactor to trap radioactive vapors that might otherwise be released into the enviroment during a nuclear accident. { kən'tān·mənt ‚bil·diŋ }

containment vessel [NUCLEO] A gas-tight shell or other enclosure around a reactor. { kən'tān·mənt ‚ves·əl }

contaminate [SCI TECH] To render unfit or to soil by the introduction of foreign or unwanted material. { kən'tam·ə‚nāt }

contamination [COMPUT SCI] Placement of data at incorrect locations in storage, where it generally overlays valid information or a program code and produces bizarre results. [GEOL] A process in which the chemical composition of a magma changes due to the assimilation of country rocks. [HYD] The addition to water of any substance or property that prevents its use without further treatment. [MICROBIO] The process or act of soiling with bacteria. [NUCLEO] The deposit of radioactive materials, such as fission fragments or radiological warfare agents, on any objective or surface or in the atmosphere. [PSYCH] The fusion of words, resulting in a new word. [SCI TECH] Something that contaminates. { kən‚tam·ə'nā·shən }

contamination monitor [NUCLEO] A radiation counter used to detect radioactive contamination of surface areas or of the atmosphere. { kən‚tam·ə'nā·shən ‚män·əd·ər }

contemporaneous [GEOL] **1.** Formed, existing, or originating at the same time. **2.** Of a rock, developing during formation of the enclosing rock. { kən‚tem·pə'rā·nē·əs }

contemporary carbon [CHEM] The isotopic carbon content of living matter, based on the assumption of a natural proportion of carbon-14. { kən'tem·pə‚rer·ē 'kär·bən }

content See Jordan content. { 'kän‚tent }

content analysis [COMPUT SCI] A method of automatically assigning words that identify the content of information items or search requests in an information retrieval system. { 'kän‚tent ə'nal·ə·səs }

content indicator [COMPUT SCI] Display unit that indicates the content in a computer, and the program or mode being used. { 'kän‚tent ‚in·də‚kād·ər }

contention [COMMUN] A method of operating a multiterminal communication channel in which any station may transmit if the channel is free; if the channel is in use, the queue of contention requests may be maintained in predetermined sequence. [COMPUT SCI] **1.** The condition arising when two or more units attempt to transmit over a time-division-multiplex channel at the same time. **2.** Competition for the same computer resources by two or more devices or programs, such as an attempt by several programs to use the same disk drive simultaneously, or by several users in a multiaccess system to use the system's resources. { kən'ten·chən }

contention resolver [COMPUT SCI] A device that enables a central processing unit, memory, or channel whose attention is being requested over several pathways to give its attention to one pathway and ignore all others. { kən'ten·chən ri'zäl·vər }

contents [COMPUT SCI] The information stored at any address or in any register of a computer. { 'kän‚tens }

context-driven line editor [COMPUT SCI] A line editor in which the user need not know or keep track of line numbers but can call up text by line content; the computer will then search for the indicated pattern. { 'kän‚tekst ‚driv·ən 'līn ‚ed·əd·ər }

context-free grammar [COMPUT SCI] A grammar in which any occurrence of a metavariable may be replaced by one of its alternatives. { 'kän‚tekst ‚frē 'gram·ər }

context-sensitive grammar [COMPUT SCI] A grammar in which the rules are applicable only when a metavariable occurs in a specified context. { 'kän‚tekst ‚sen·səd·iv 'gram·ər }

context-sensitive help [COMPUT SCI] A help screen that provides specific information about the current status or mode of a computer program or instructions for dealing with a particular error condition that has just occurred. { 'kän‚tekst ‚sen·səd·iv 'help }

context switch [COMPUT SCI] The action of a central processing unit that suspends work on one process to work on another. { 'kän‚text ‚swich }

context switching See task switching. { 'kän‚text ‚swich·iŋ }

contextual analysis [COMPUT SCI] A phase of natural language processing, following semantic analysis, whose purpose is to elaborate the semantic representation of what has been made explicit in the utterance with what is implicit from context. { kən'teks·chə·wəl ə'nal·ə·səs }

contextual search [COMPUT SCI] A search for documents or records based upon the data they contain, rather than their file names or key fields. { kən'teks·chə·wəl 'sərch }

contig [GEN] A region of chromosome defined by its hybridization to one or more cloned deoxyribonucleic acid fragments from an overlapping array of clones. [MOL BIO] A group of cloned nucleotide sequences that are contiguous. { kən'tig }

contiguous arc [ASTRON] A crater arc in which successive craters are in contact. { kən'tig·yə·wəs 'ärk }

contiguous chain [ASTRON] A crater chain in which successive craters are in contact. { kən'tig·yə·wəs 'chān }

contiguous craters [ASTRON] A formation on the moon's surface consisting of two craters in contact; the walls at the point of contact are low, broken, or entirely absent. { kən'tig·yə·wəs 'krād·ərz }

contiguous data [COMPUT SCI] Data that are stored in a collection of adjacent locations in a computer memory device. { kən'tig·yə·wəs 'dad·ə }

contiguous functions [MATH] Any pair of hypergeometric functions in which one of the parameters differs by unity and the other two are equal. { kən'tig·yə·wəs 'fəŋk·shənz }

contiguous gene syndrome [GEN] A characteristic complex phenotype produced by deletion of a short chromosome segment, resulting from haplo-insufficiency of several genes in the deleted segment. { kən'tig·yə·wəs 'jēn 'sin‚drōm }

continent [GEOGR] A protuberance of the earth's crustal shell, with an area of several million square miles and sufficient elevation so that much of it is above sea level. { 'känt·ən·ənt }

continental accretion [GEOL] The theory that continents have grown by the addition of new continental material around an original nucleus, mainly through the processes of geosynclinal sedimentation and orogeny. { ‚känt·ən‚ent·əl ə'krē·shən }

continental air [METEOROL] A type of air whose characteristics are developed over a large land area and which therefore has relatively low moisture content. { ‚känt·ən‚ent·əl 'er }

continental anticyclone See continental high. { ‚känt·ən‚ent·əl ‚an·tē'sī‚klōn }

continental borderland [GEOL] The area of the continental margin between the shoreline and the continental slope. { ‚känt·ən‚ent·əl 'bor·dər‚land }

continental climate [CLIMATOL] Climate characteristic of the interior of a landmass of continental size, marked by large annual, daily, and day-to-day temperature ranges, low relative humidity, and a moderate or small irregular rainfall; annual extremes of temperature occur soon after the solstices. { ‚känt·ən‚ent·əl 'klī·mət }

continental code [COMMUN] The code commonly used for manual telegraph communication, consisting of short (dot) and long (dash) symbols, but not the various-length spaces used in the original Morse code. Also known as international Morse code. { ‚känt·ən‚ent·əl 'kōd }

continental crust [GEOL] The basement complex of rock, that is metamorphosed sedimentary and volcanic rock with associated igneous rocks mainly granitic, that underlies the continents and the continental shelves. { ‚känt·ən‚ent·əl 'krəst }

continental deposits [GEOL] Sedimentary deposits laid down within a general land area. { ‚känt·ən‚ent·əl di'päz·əts }

continental displacement See continental drift. { ‚känt·ən‚ent·əl di'splās·mənt }

continental divide [GEOL] A drainage divide of a continent, separating streams that flow in opposite directions; for example,

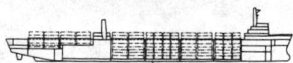

the divide in North America that separates watersheds of the Pacific Ocean from those of the Atlantic Ocean. { ˈkänt·ən¦ent·əl di'vīd }

continental drift [GEOL] The concept of continent formation by the fragmentation and movement of land masses on the surface of the earth. Also known as continental displacement. { ˈkänt·ən¦ent·əl 'drift }

continental geosyncline [GEOL] A geosyncline filled with nonmarine sediments. { ˈkänt·ən¦ent·əl ¦jē·ō'sin‚klīn }

continental glacier [HYD] A sheet of ice covering a large tract of land, such as the ice caps of Greenland and the Antarctic. { ˈkänt·ən¦ent·əl 'glā·shər }

continental growth [GEOL] The processes contributing to growth of continents at the expense of ocean basins. { ˈkänt·ən¦ent·əl 'grōth }

continental heat flow [GEOPHYS] The amount of thermal energy escaping from the earth through the continental crust per unit area and unit time. { ˈkänt·ən¦ent·əl 'hēt ‚flō }

continental high [METEOROL] A general area of high atmospheric pressure which on mean charts of sea-level pressure is seen to overlie a continent during the winter. Also known as continental anticyclone. { ˈkänt·ən¦ent·əl 'hī }

continentality [CLIMATOL] The degree to which a point on the earth's surface is in all respects subject to the influence of a land mass. { ‚känt·ən·en'tal·əd·ē }

continental margin [GEOL] Those provinces between the shoreline and the deep-sea bottom; generally consists of the continental borderland, shelf, slope, and rise. { ˈkänt·ən¦ent·əl 'mär·jən }

continental mass [GEOGR] The continental land rising more or less abruptly from the ocean floor and also the shallow submerged areas surrounding this land. { ˈkänt·ən¦ent·əl 'mas }

continental nucleus [GEOL] A large area of basement rock consisting of basaltic and more mafic oceanic crust and periodotitic mantle from which it is postulated that continents have grown. Also known as continental shield; cratogene; shield. { ˈkänt·ən¦ent·əl 'nü·klē·əs }

continental plate [GEOL] Thick continental crust. { ˈkänt·ən¦ent·əl 'plāt }

continental plateau See tableland. { ˈkänt·ən¦ent·əl plə'tō }

continental platform See continental shelf. { ˈkänt·ən¦ent·əl 'plat‚fȯrm }

continental polar air [METEOROL] Polar air having low surface temperature, low moisture content, and (especially in its source regions) great stability in the lower layers. { ˈkänt·ən¦ent·əl ¦pō·lər 'er }

continental rise [GEOL] A transitional part of the continental margin; a gentle slope with a generally smooth surface, built up by the shedding of sediments from the continental block, and located between the continental slope and the abyssal plain. { ˈkänt·ən¦ent·əl 'rīz }

continental shelf [GEOL] The zone around a continent, that part of the continental margin extending from the shoreline and the continental slope; composes with the continental slope the continental terrace. Also known as continental platform; shelf. { ˈkänt·ən¦ent·əl 'shelf }

continental shield See shield. { ˈkänt·ən¦ent·əl 'shēld }

continental slope [GEOL] The part of the continental margin consisting of the declivity from the edge of the continental shelf extending down to the continental rise. { ˈkänt·ən¦ent·əl 'slōp }

continental terrace [GEOL] The continental shelf and slope together. { ˈkänt·ən¦ent·əl 'ter·əs }

continental tropical air [METEOROL] A type of tropical air produced over subtropical arid regions; it is hot and very dry. { ˈkänt·ən¦ent·əl 'träp·ə·kəl 'er }

continent formation [GEOL] A series of six or seven major episodes, resulting from the buildup of radioactive heat and then the melting or partial melting of the earth's interior; the molten rock melt rises to the surface, differentiating into less primitive lavas; the continent then nucleates, differentiates, and grows from oceanic crust and mantle. { ˈkänt·ən¦ent·əl fər'mā·shən }

contingency interrupt [COMPUT SCI] A processing interruption due to an operator's action or due to an abnormal result from the system or from a program. { kən'tin·jən·sē 'in·tə‚rəpt }

contingency table [STAT] A table for classifying elements

of a population according to two variables, the rows corresponding to one variable and the columns to the other. { kən'tin·jən·sē ‚tā·bəl }

continous-type furnace [MECH ENG] A furnace used for heat treatment of materials, with or without direct firing; pieces are loaded through one door, progress continuously through the furnace, and are discharged from another door. { kən'tin·yə·wəs ‚tīp 'fər·nəs }

continuant [MATH] The determinant of a continuant matrix. { kən'tin·yə·wənt }

continuant matrix [MATH] A square matrix all of whose nonzero elements lie on the principal diagonal or the diagonals immediately above and below the principal diagonal. Also known as triple-diagonal matrix. { kən'tin·yə·wənt 'mā·triks }

continued equality [MATH] An expression in which three or more quantities are set equal by means of two or more equality signs. { kən'tin·yüd i'kwäl·əd·ē }

continued fraction [MATH] The sum of a number and a fraction whose denominator is the sum of a number and a fraction, and so forth; it may have either a finite or an infinite number of terms. { kən'tin·yüd 'frak·shən }

continued-fraction expansion [MATH] **1.** An expansion of a driving-point function about infinity (or zero) in a continued fraction, in which the terms are alternately constants and multiples of the complex frequency (or multiples of the reciprocal of the complex frequency). **2.** A representation of a real number by a continued fraction, in a manner similar to the representation of real numbers by a decimal expansion. { kən'tin·yüd 'frak·shən ik'span·shən }

continued product [MATH] A product of three or more factors, or of an infinite number of factors. { kən'tin·yüd 'präd·əkt }

continue statement [COMPUT SCI] A nonexecutable statement in FORTRAN used principally as a target for transfers, particularly as the last statement in the range of a do statement. { kən'tin·yü ‚stāt·mənt }

continuity [CIV ENG] Joining of structural members to each other, such as floors to beams, and beams to beams and to columns, so they bend together and strengthen each other when loaded. Also known as fixity. [ELEC] Continuous effective contact of all components of an electric circuit to give it high conductance by providing low resistance. [NAV] The ability of a navigational system to let the user navigate without interruption. { ‚känt·ən'ü·əd·ē }

continuity bond [MET] A metallic connection that provides continuous electrical contact between metal structures. { ‚känt·ən'ü·əd·ē ‚bänd }

continuity chart [METEOROL] A chart maintained for weather analysis and forecasting upon which are entered the positions of significant features (pressure centers, fronts, instability lines, through lines, ridge lines) of the regular synoptic charts at regular intervals in the past. { ‚känt·ən'ü·əd·ē ‚chärt }

continuity equation [PHYS] An equation obeyed by any conserved, indestructible quantity such as mass, electric charge, thermal energy, electrical energy, or quantum-mechanical probability, which is essentially a statement that the rate of increase of the quantity in any region equals the total current flowing into the region. Also known as equation of continuity. { ‚känt·ən'ü·əd·ē i'kwā·zhən }

continuity of state [THERMO] Property of a transition between two states of matter, as between gas and liquid, during which there are no abrupt changes in physical properties. { ‚känt·ən'ü·əd·ē əv 'stāt }

continuity test [ELEC] An electrical test used to determine the presence and location of a broken connection. { ‚känt·ən'ü·əd·ē ‚test }

continuous at a point [MATH] A function f is continuous at a point x if, for every sequence $\{x_n\}$ whose limit is x, the sequence $f(x_n)$ converges to $f(x)$; in a general topological space, for every neighborhood W of $f(x)$, there is a neighborhood N of x such that $f^{-1}(W)$ is contained in N. { kən'tin·yə·wəs ad ə 'pȯint }

continuous beam [CIV ENG] **1.** A beam resting upon several supports, which may be in the same horizontal plane. **2.** A beam having several spans in one straight line; generally has at least three supports. { kən'tin·yə·wəs 'bēm }

continuous brake [MECH ENG] A train brake that operates

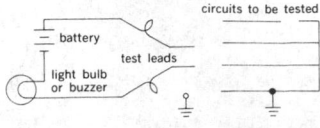

on all cars but is controlled from a single point. { kən¦tin·yə·wəs 'brāk }

continuous bridge [CIV ENG] A fixed bridge supported at three or more points and capable of resisting bending and shearing forces at all sections throughout its length. { kən¦tin·yə·wəs 'brij }

continuous bucket elevator [MECH ENG] A bucket elevator on an endless chain or belt. { kən¦tin·yə·wəs ¦bək·ət 'el·ə‚vād·ər }

continuous bucket excavator [MECH ENG] A bucket excavator with a continuous bucket elevator mounted in front of the bowl. { kən¦tin·yə·wəs ¦bək·ət 'ek·skə‚vād·ər }

continuous carrier [COMMUN] A carrier signal that is transmitted at all times during maintenance of a communications link, whether or not data are being transmitted. { kən¦tin·yə·wəs 'kar·ē·ər }

continuous casting [MET] A technique in which an ingot, billet, tube, or other shape is continuously solidified and withdrawn while it is being poured, so that its length is not determined by mold dimensions. { kən¦tin·yə·wəs 'kast·iŋ }

continuous cell line [CYTOL] A group of morphologically uniform cells that can be propagated in vitro for an indefinite time. { kən¦tin·yə·wəs 'sel ‚līn }

continuous clamp See voltage-amplitude-controlled clamp. { kən¦tin·yə·wəs 'klamp }

continuous coal cutter [MIN ENG] A coal mining machine that cuts the coal face without being withdrawn from the cut. { kən¦tin·yə·wəs 'kōl ‚kəd·ər }

continuous comparator See linear comparator. { kən¦tin·yə·wəs kəm'par·əd·ər }

continuous contact coking [CHEM ENG] A thermal conversion process using the mass-flow lift principle to give continuous coke circulation; oil-wetted particles of coke move downward into the reactor in which cracking, coking, and drying take place; pelleted coke, gas, gasoline, and gas oil are products of the process. { kən¦tin·yə·wəs ¦kän‚takt 'kōk·iŋ }

continuous control [CONT SYS] Automatic control in which the controlled quantity is measured continuously and corrections are a continuous function of the deviation. { kən¦tin·yə·wəs kən'trōl }

continuous countercurrent leaching [CHEM ENG] Process of leaching by the use of continuous equipment in which the solid and liquid are both moved mechanically, and by the use of a series of leach tanks and the countercurrent flow of solvent through the tanks in reverse order to the flow of solid. { kən¦tin·yə·wəs ¦kaúnt·ər‚kər·ənt 'lēch·iŋ }

continuous deformation [MATH] A transformation of an object that magnifies, shrinks, rotates, or translates portions of the object in any manner without tearing. { kən¦tin·yə·wəs ‚dē·fòr'mā·shən }

continuous distillation [CHEM ENG] Separation by boiling of a liquid mixture with different component boiling points; feed is introduced continuously, with continuous removal of overhead vapors and high-boiling bottoms liquids. { kən¦tin·yə·wəs ‚dis·tə'lā·shən }

continuous distribution [STAT] Distribution of a continuous population, which is a class of pairs such that the second member of each pair is a value, and the first member of the pair is a proportion density for that value. { kən¦tin·yə·wəs ‚dis·trə'byü·shən }

continuous dryer [ENG] An apparatus in which drying is accomplished by passing wet material through without interruption. { kən¦tin·yə·wəs 'drī·ər }

continuous-duty rating [ELEC] The rating that defines the load which can be carried for an indefinite time without exceeding a specified temperature rise. { kən¦tin·yə·wəs ‚düd·ē 'rād·iŋ }

continuous dyeing [TEXT] The application of color-producing agents to textiles by impregnating the cloth with dye and then passing it through a series of developing, washing, and drying zones to a final take-up roll. { kən¦tin·yə·wəs 'dī·iŋ }

continuous equilibrium vaporization See equilibrium flash vaporization. { kən¦tin·yə·wəs ‚ē·kwə¦lib·rē·əm vā·pə·rə'zā·shən }

continuous extension [MATH] A continuous function which is equal to another continuous function defined on a smaller domain. { kən¦tin·yə·wəs ik'sten·shən }

continuous filament [TEXT] A long, continuous strand of a manufactured fiber as distinguished from all natural fibers (except raw silk), which are of short staple or length. { kən¦tin·yə·wəs 'fil·ə·mənt }

continuous film scanner [ELECTR] A television film scanner in which the motion picture film moves continuously while being scanned by a flying-spot kinescope. { kən¦tin·yə·wəs 'film ‚skan·ər }

continuous fire [ORD] **1.** Fire conducted at a normal rate without interruption, for application of adjustment corrections or for other causes. **2.** In field artillery, a succession of salvos, the pieces being fired consecutively at the interval designated in the command. { kən¦tin·yə·wəs 'fīr }

continuous-flow conveyor [MECH ENG] A totally enclosed, continuous-belt conveyor pulled transversely through a mass of granular, powdered or small-lump material fed from an overhead hopper. { kən¦tin·yə·wəs ¦flō kən'vā·ər }

continuous flowmeter log [PETRO ENG] A record of surveys made to record changes in the flow pattern of production zones as a function of changes in conditions at the surface, in time, in type of operation, or after stimulation treatments. { kən¦tin·yə·wəs 'flō‚med·ər ‚läg }

continuous footing [CIV ENG] A footing that supports a wall. { kən¦tin·yə·wəs 'fùd·iŋ }

continuous forms [COMPUT SCI] **1.** In character recognition, any batch of source information that exists in reel form, such as tally rolls or cash-register receipts. **2.** Preprinted forms that repeat on each page, with the bottom of one page joined to the top of the next by a perforated attachment, so that they can be fed through a printer. { kən¦tin·yə·wəs 'fòrmz }

continuous function [MATH] A function which is continuous at each point of its domain. Also known as continuous transformation. { kən¦tin·yə wəs 'fəŋk·shən }

continuous furnace [MET] A type of reheating furnace in which the charge introduced at one end moves continuously through the furnace and is discharged at the other end. { kən¦tin·yə·wəs 'fər·nəs }

continuous gas lift [PETRO ENG] Oil production in which reservoir gas pressure (natural or injected) is sufficient to provide a continuous upward flow of oil through the well tubing. { kən¦tin·yə·wəs 'gas ‚lift }

continuous geometry [MATH] A generalization of projective geometry. { kən¦tin·yə·wəs jē'äm·ə·trē }

continuous image [MATH] The image of a set under a continuous function. { kən¦tin·yə·wəs 'im·ij }

continuous industry [IND ENG] An industry in which raw material is subjected to successive operations, turning it into a finished product. { kən¦tin·yə·wəs 'in·dəs·trē }

continuous kiln [ENG] **1.** A long kiln through which ware travels on a moving device, such as a conveyor. **2.** A kiln through which the fire travels progressively. { kən¦tin·yə wəs 'kiln }

continuous leader See dart leader. { kən¦tin·yə·wəs 'lēd·ər }

continuous loading [ELEC] Loading in which the added inductance is distributed uniformly along a line by wrapping magnetic material around each conductor. { kən¦tin·yə·wəs 'lōd·iŋ }

continuously adjustable transformer See variable transformer. { kən¦tin·yə·wəs·lē ə'jəs·tə·bəl tranz'fòr·mər }

continuous mill [MET] A rolling mill in which metal is successively rolled thinner as it passes through a series of synchronized rolls in tandem. { kən¦tin·yə·wəs 'mil }

continuous miner [MIN ENG] Machine designed to remove coal or other soft minerals from the face and to load it into cars or conveyors continuously, without the use of cutting machines, drills, or explosives. { kən¦tin·yə·wəs 'mīn·ər }

continuous mining [MIN ENG] A type of mining in which the continuous miner cuts or rips coal or other soft minerals from the face and loads it in a continuous operation. { kən¦tin·yə·wəs 'mīn·iŋ }

continuous mixer [MECH ENG] A mixer in which materials are introduced, mixed, and discharged in a continuous flow. { kən¦tin·yə·wəs 'mik·sər }

continuous operation [ENG] A process that operates on a continuous flow (materials or time) basis, in contrast to batch, intermittent, or sequenced operations. { kən¦tin·yə·wəs äp·ə'rā·shən }

continuous operator [MATH] A linear transformation of Banach spaces which is continuous with respect to their topologies. { kən¦tin·yə·wəs 'äp·ə‚rād·ər }

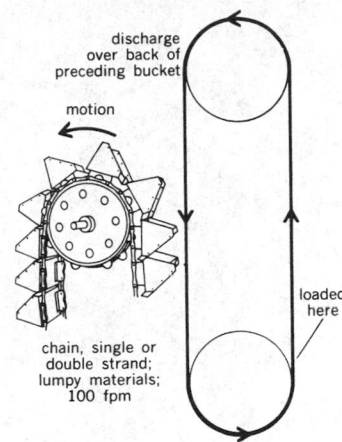

CONTINUOUS BUCKET ELEVATOR

Components of continuous bucket elevator, and typical path of travel.

CONTINUOUS FOOTING

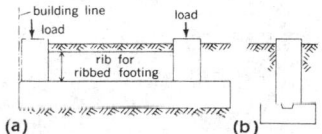

Two types of continuous footings. (*a*) Cantilever and ribbed footings. (*b*) Wall footing.

continuous permafrost zone [GEOL] Regional zone predominantly underlain by permanently frozen subsoil that is not interrupted by pockets of unfrozen ground. { kən¦tin·yə·wəs 'pər·mə‚fròst ‚zōn }

continuous phase [CHEM] The liquid in a disperse system in which solids are suspended or droplets of another liquid are dispersed. Also known as dispersion medium, external phase. [MET] The matrix or background phase of a multiphasic alloy. { kən¦tin·yə·wəs 'fāz }

continuous population [STAT] A population in which a random variable is measuring a continuous characteristic. { kən¦tin·yə·wəs ‚päp·yə'lā·shən }

continuous precipitation [MET] Precipitation that is characteristic of certain alloys, from a supersaturated solid solution, involving a gradual change of the lattice parameter of the matrix with aging time. { kən¦tin·yə·wəs prə‚sip·ə'tā·shən }

continuous production [IND ENG] Manufacture of products, such as chemicals or paper, involving a sequence of processes performed by a series of machines receiving the materials through a closed channel of flow. { kən¦tin·yə·wəs prə'dək·shən }

continuous profiling [GEOL] A method of shooting in seismic exploration in which uniformly placed seismometer stations along a line are shot from holes spaced along the same line so that each hole records seismic ray paths geometrically identical with those from adjacent holes. { kən¦tin·yə·wəs 'prō‚fīl·iŋ }

continuous radiation [ELECTROMAG] Electromagnetic radiation that includes all the wavelengths in some interval. Also known as white radiation. { kən¦tin·yə·wəs ‚rād·ē'ā·shən }

continuous radio beacon [NAV] A single marine radio beacon operating on a frequency without interruption; used specifically with automatic direction finders. { kən¦tin·yə·wəs 'rād·ē·ō ‚bē·kən }

continuous-rail frog [ENG] A metal fitting that holds continuous welded rail sections to railroad ties. { kən¦tin·yə·wəs ‚rāl 'fräg }

continuous rating [ENG] The rating of a component or equipment which defines the substantially constant conditions which can be tolerated for an indefinite time without significant reduction of service life. { kən¦tin·yə·wəs 'rād·iŋ }

continuous reaction series [MINERAL] A branch of Bowen's reaction series comprising the plagioclase mineral group in which reaction of early-formed crystals with water takes place continuously, without abrupt changes in crystal structure. { kən¦tin·yə·wəs rē'ak·shən ‚sir‚ēz }

continuous recorder [ENG] A recorder whose record sheet is a continuous strip or web rather than individual sheets. { kən¦tin·yə·wəs ri'kórd·ər }

continuous sequence [MET] A welding sequence in which each succeeding pass is made longitudinally along the entire length of the joint. { kən¦tin·yə·wəs 'sē·kwəns }

continuous set [STAT] In an infinite number of outcomes of an experiment, those outcomes in which any value in a given interval can occur. { kən¦tin·yə·wəs 'set }

continuous sintering [MET] Sintering process in which materials are moved through the furnace at a fixed rate without interruption. { kən¦tin·yə·wəs 'sin·tə·riŋ }

continuous spectrum [MATH] The portion of the spectrum of a linear operator which is a continuum. [SPECT] A radiation spectrum which is continuously distributed over a frequency region without being broken up into lines or bands. { kən¦tin·yə·wəs 'spek·trəm }

continuous spinning [TEXT] Spinning in which fiber is extruded, coagulated, washed, and wound in a continuous operation on a single machine. { kən¦tin·yə·wəs 'spin·iŋ }

continuous stationery [COMPUT SCI] A continuous ribbon of paper consisting of several hundred or more sheets separated by perforations and folded to form a pack, used to feed a computer printer and generally having sprocket holes along the margin for this purpose. { kən¦tin·yə·wəs 'stā·shə‚ner·ē }

continuous stationery reader [COMPUT SCI] A type of character reader which processes only continuous forms of predefined dimensions. { kən¦tin·yə·wəs 'stā·shə‚ner·ē 'rēd·ər }

continuous still [FOOD ENG] A type of still in which rectification is accomplished, allowing for the collection of several relatively pure fractions of distilled spirits. { kən¦tin·yə·wəs 'stil }

continuous surface [MATH] The range of a continuous function from a plane or a connected region in a plane to three-dimensional Euclidean space. { kən¦tin·yə·wəs 'sər·fəs }

continuous system [CONT SYS] A system whose inputs and outputs are capable of changing at any instant of time. Also known as continuous-time signal system. { kən¦tin·yə·wəs 'sis·təm }

continuous task [IND ENG] A task that requires a continuously changing response by a worker to a continuously changing stimulus. { kən¦tin·yə·wəs 'task }

continuous-time signal system See continuous system. { kən¦tin·yə·wəs ‚tīm 'sig·nəl ‚sis·təm }

continuous titrator [ANALY CHEM] A titrator so equipped that a reservoir refills the buret. { kən¦tin·yə·wəs 'tī‚trād·ər }

continuous tone [GRAPHICS] An image which has not been screened and contains unbroken gradient tones from black to white, and may be either in negative or positive form. { kən¦tin·yə·wəs 'tōn }

continuous-tone squelch [ELECTR] Squelch in which a continuous subaudible tone, generally below 200 hertz, is transmitted by frequency-modulation equipment along with a desired voice signal. { kən¦tin·yə·wəs ‚tōn 'skwelch }

continuous transformation See continuous function. { kən¦tin·yə·wəs tranz·fər'mā·shən }

continuous treatment [PETRO ENG] Introduction of corrosion inhibitors to production fluids at a specified rate so that inhibitor concentration remains constant. { kən¦tin·yə·wəs 'trēt·mənt }

continuous tube process [ENG] Plastics blow-molding process that uses a continuous extrusion of plastic tubing as feed to a series of blow molds as they clamp in sequence. { kən¦tin·yə·wəs ‚tüb ‚präs·əs }

continuous variable [COMPUT SCI] A variable that can take on any of a range of values. { kən¦tin·yə·wəs 'ver·ē·ə·bəl }

continuous vinegar process [FOOD ENG] Continuous, rather than batch, bacterial fermentation of apple cider, wine, other fruit juice, malt, or barley. { kən¦tin·yə·wəs 'vin·ə·gər ‚präs·əs }

continuous wave [ELECTROMAG] A radio or radar wave whose successive sinusoidal oscillations are identical under steady-state conditions. Abbreviated CW. Also known as type A wave. { kən¦tin·yə·wəs 'wāv }

continuous-wave Doppler radar See continuous-wave radar. { kən¦tin·yə·wəs ‚wāv 'däp·lər ‚rā‚där }

continuous-wave gas laser [OPTICS] A laser having a quartz envelope filled with a mixture of helium and neon at low pressure, with Brewster-angle mirrors at opposite ends and an external optical system. { kən¦tin·yə·wəs ¦wāv 'gas ‚lā·zər }

continuous-wave jammer [ELECTR] An electronic jammer that emits a single frequency which gives the appearance of a picket or rail fence on an enemy's radarscope. Also known as rail-fence jammer. { kən¦tin·yə·wəs ¦wāv 'jam·ər }

continuous-wave laser [OPTICS] A laser in which the beam of coherent light is generated continuously, as required for communication and certain other applications. Abbreviated CW laser. { kən¦tin·yə·wəs ¦wāv 'lā·zər }

continuous-wave modulation [COMMUN] Modulation of a continuous wave by modification of its amplitude, frequency, or phase, in contrast to pulse modulation. { kən¦tin·yə·wəs ¦wāv ‚mäj·ə'lā·shən }

continuous-wave radar [ENG] A radar system in which a transmitter sends out a continuous flow of radio energy; the target reradiates a small fraction of this energy to a separate receiving antenna. Also known as continuous-wave Doppler radar. { kən¦tin·yə·wəs ¦wāv 'rā‚där }

continuous-wave tracking system [ELECTR] Tracking system which operates by keeping a continuous radio beam on a target and determining its behavior from changes in the antenna necessary to keep the beam on the target. { kən¦tin·yə·wəs ¦wāv 'trak·iŋ ‚sis·təm }

continuous weld [MET] A weld that is continuous along the entire length of the joint. { kən¦tin·yə·wəs 'weld }

continuous work [IND ENG] A sustained and uninterrupted work activity, for example, exertion of a muscular force. { kən¦tin·yə·wəs 'wərk }

continuous x-rays [ELECTROMAG] The electromagnetic

radiation, having a continuous spectral distribution, that is produced when high-velocity electrons strike a target. { kən'tin·yə·wəs 'eks ,rāz }

continuum [MATH] A compact, connected set. { kən'tin·yə·wəm }

continuum hypothesis [MATH] The conjecture that every infinite subset of the real numbers can be put into one-to-one correspondence with either the set of positive integers or the entire set of real numbers. { kən'tin·yü·əm hī,päth·ə·səs }

continuum mechanics See classical field theory. { kən'tin·yə·wəm mə'kan·iks }

continuum physics See classical field theory. { kən'tin·yə·wəm 'fiz·iks }

contorted [BOT] Twisted; applied to proximate leaves whose margins overlap. { kən'tórd·əd }

contour See contour line. [PHYS] A curve drawn up on a two-dimensional diagram through points which satisfy $f(x,y) = c$, where c is a constant and f is some function, such as the field strength for a transmitter. [SCI TECH] The periphery of a figure or body. { 'kän,túr }

contour analysis [COMPUT SCI] In optical character recognition, a reading technique that employs a roving spot of light which searches out the character's outline by bouncing around its outer edges. { 'kän,túr ə'nal·ə·səs }

contour-change line See height-change line. { 'kän,túr ,chānj ,līn }

contour code [METEOROL] A code in which data on the topography of constant-pressure surfaces are transmitted; a modification of the international analysis code. { 'kän,túr ,kōd }

contour feather [VERT ZOO] Any of the large flight feathers or long tail feathers of a bird. Also known as penna; vane feather. { 'kän,túr ,feth·ər }

contour finder [GRAPHICS] An optical instrument of simple design for use with photographic prints; used to produce contour lines from differences in relief evident in the print. { 'kän,túr ,fīnd·ər }

contour forming [MET] Shaping a sheet of metal onto a shaped die. { 'kän,túr ,fórm·iŋ }

contouring control [COMPUT SCI] The guidance by a computer of a machine tool along a programmed path by interpolating many intermediate points between selected points. { 'kän,túr·iŋ kən'trōl }

contouring temperature recorder [ENG] A device that records data from temperature sensors towed behind a ship and then plots the vertical distribution of isotherms on a continuous basis. { 'kän,túr·iŋ 'tem·prə·chər ri,kórd·ər }

contour integral [MATH] A line integral of a complex function, usually over a simple closed curve. { 'kän,túr ,in·tə·grəl }

contour interval [MAP] The difference in elevation between adjacent contours. { 'kän,túr ,in·tər·vəl }

contourite [OCEANOGR] A marine sediment deposited by swift ocean-bottom currents that generally flow along contours. { 'kän,tú,rīt }

contour line [MAP] A map line representing a contour, that is, connecting points of equal elevation above or below a datum plane, usually mean sea level. Also known as contour; isoheight; isohypse. [METEOROL] A line on a weather map connecting points of equal atmospheric pressure, temperature, or such. { 'kän,túr ,līn }

contour machining [MECH ENG] Machining of an irregular surface. { 'kän,túr mə'shēn·iŋ }

contour map [MAP] A map displaying topographic or structural contour lines. { 'kän,túr ,map }

contour microclimate [CLIMATOL] That portion of the microclimate which is directly attributable to the small-scale variations of ground level. { 'kän,túr ,mī·krō,klī·mət }

contour milling [MET] Milling of an irregular surface. { 'kän,túr ,mil·iŋ }

contour model [COMPUT SCI] A model for describing the run-time execution of programs written in block-structured languages, consisting of a program component, the data component, and the control component. { 'kän,túr ,mäd·əl }

contourograph [ELECTR] Device using a cathode-ray oscilloscope to produce imagery that has a three-dimensional appearance. { ,kän'túr·ə,graf }

contour plan [MIN ENG] A plan showing surface contours or calculated contours of coal seams to be developed. { 'kän,túr ,plan }

contour plowing [AGR] Cultivation of land along lines connecting points of equal elevation, to prevent water erosion. Also known as terracing. { 'kän,túr ,plaú·iŋ }

contour turning [MECH ENG] Making a three-dimensional reproduction of the shape of a template by controlling the cutting tool with a follower that moves over the surface of a template. { 'kän,túr ,tərn·iŋ }

contour value [MAP] A numerical value placed upon a contour line to denote its elevation relative to a given datum, usually mean sea level. { 'kän,túr ,val·yü }

contraception [MED] Prevention of impregnation. { ,kän·trə,sep·shən }

contraceptive [MED] Any mechanical device or chemical agent used to prevent conception. { ,kän·trə,sep·tiv }

contracted code sonde See code-sending radiosonde. { kən'trak·təd ,kōd ,sänd }

contracted curvature tensor [MATH] A symmetric tensor of second order, obtained by summation on two indices of the Riemann curvature tensor which are not antisymmetric. Also known as contracted Riemann-Christoffel tensor; Ricci tensor. { kən'trak·təd 'kər·və·chər ,ten·sər }

contracted pelvis [MED] A pelvis having one or more major diameters reduced in size, interfering with parturition. { kən'trak·təd 'pel·vəs }

contracted Riemann-Christoffel tensor See contracted curvature tensor. { kən'trak·təd ,rē·män kris'tóf·əl ,ten·sər }

contractile [BIOL] Displaying contraction; having the property of contracting. { kən'trak·təl }

contractile ring [CELL MOL] A cytoskeletal structure that forms in animal cells and many unicellular eukaryotes during cell division, contraction of which causes the plasma membrane to pinch inward and the cell to divide. { kən'trak·təl ,riŋ }

contractile vacuole [CYTOL] A tiny, intracellular, membranous bladder that functions in maintaining intra- and extracellular osmotic pressures in equilibrium, as well as excretion of water, such as occurs in protozoans. { kən'trak·təl 'vak·yə·wōl }

contracting stitching [TEXT] Arranging fibers so that they are parallel, and removing those fibers that are shorter than a specified length. { kən,trak·tiŋ 'stich·iŋ }

contraction [GRAPHICS] A microfilm defect in the form of a compressed image that occurs when the film speed is reduced as the document passes through a rotary microfilmer. [MATH] A function f from a metric space to itself for which there is a constant K that is less than 1 such that, for any two elements in the space, a and b, the distance between $f(a)$ and $f(b)$ is less than K times the distance between a and b. [MECH] The action or process of becoming smaller or pressed together, as a gas on cooling. [PHYSIO] Shortening of the fibers of muscle tissue. { kən'trak·shən }

contraction coefficient See coefficient of contraction. { kən'trak·shən ,kō·i'fish·ənt }

contraction crack [ENG] A crack resulting from restriction of metal in a mold while contracting. { kən'trak·shən ,krak }

contraction hypothesis [GEOL] Theory that shrinking of the earth is the cause of compression folding and thrusting. { kən'trak·shən hī'päth·ə·səs }

contraction joint [CIV ENG] A break designed in a structure to allow for drying and temperature shrinkage of concrete, brickwork, or masonry, thereby preventing the formation of cracks. { kən'trak·shən ,jóint }

contraction loss [FL MECH] In fluid flow, the loss in mechanical energy in a stream flowing through a closed duct or pipe when there is a sudden contraction of the cross-sectional area of the passage. { kən'trak·shən ,lós }

contraction rule [MET] A measuring rule having larger divisions than standard measures to allow for shrinkage of a metal casting. Also known as shrinkage rule; shrink rule. { kən'trak·shən ,rül }

contraction semigroup [MATH] A strongly continuous semigroup all of whose elements have norms which are equal to or less than a constant which is, in turn, less than 1. { kən'trak·shən 'sem·i,grüp }

contracture [ARCH] The narrowing of a section of a column. [MED] **1.** Shortening, as of muscle or scar tissue, producing distortion or deformity or abnormal limitation of movement of

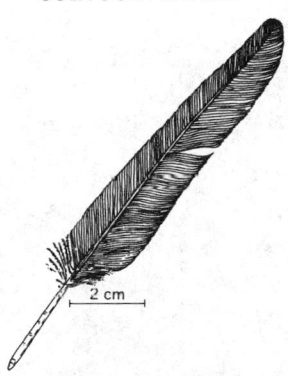

CONTOUR FEATHER

Drawing of a contour feather. *(From J. C. Welty, The Life of Birds, Saunders, 1962)*

a joint. **2.** Retarded relaxation of muscle, as when it is injected with veratrine. { kən'trak·chər }

contracurrent system *See* katoptric system. { ¦kän·trə¦kər·ənt ˌsis·təm }

contraflexure point [CIV ENG] The point in a structure where bending occurs in opposite directions. { ¦kän·trə'flek·shər ˌpȯint }

contrail *See* condensation trail. { 'kän,trāl }

contrail-formation graph [METEOROL] A graph containing the parameters pressure, temperature, and relative humidity for critical values at which condensation trails (contrails) form; used as an aid in forecasting the formation of condensation trails. { 'kän,trāl fȯr'mā·shən ,graf }

contraindication [MED] A symptom, indication, or condition in which a remedy or a method of treatment is inadvisable or improper. { ¦kän·trə,in·də'kā·shən }

contralateral [PHYSIO] Opposite; acting in unison with a similar part on the opposite side of the body. { ¦kän·trə'lad·ə·rəl }

CONTRAN [COMPUT SCI] Computer programming language in which instructions are written at the compiler level, thereby eliminating the need for translation by a compiling routine. { 'kän,tran }

contransformation [GEN] The incorporation of two or more linked genes on the same fragment of foreign deoxyribonucleic acid into a bacterial genome. { ,kän·tranz·fər,mā·shən }

co-transport [CELL MOL] The simultaneous transport of two substrates across a cell membrane, either in the same direction (symport) or in opposite directions (antiport). { ¦kō 'tranz,pȯrt }

contraorbit missile [AERO ENG] A missile that is sent backward along the calculated orbit of an aerospace weapon, satellite, or spacecraft for the purpose of destroying it in a head-on collision with an explosive warhead or through use of a secondary missile. { ¦kän·trə¦ȯr·bət 'mis·əl }

contrapositive [MATH] The contrapositive of the statement "if *p*, then *q*" is the equivalent statement "if not *q*, then not *p*." { ¦kän·trə'päz·əd·iv }

contrapropagating ultrasonic flowmeter [ENG] An instrument for determining the velocity of a fluid flow from the difference between the times required for high-frequency sound to travel between two transducers in opposite directions along a path having a component parallel to the flow. { ¦kän·trə'prä·pə,gād·iŋ 'əl·trə,sän·ik 'flō,mēd·ər }

contrarotating propellers [MECH ENG] A pair of propellers on concentric shafts, turning in opposite directions. { ¦kän·trə'rō,tād·iŋ prə'pel·ərz }

contrarotation [ENG] Rotation in the direction opposite to another rotation. { ¦kän·trə·rō'tā·shən }

contra solem [METEOROL] Characterizing air motion that is counterclockwise in the Northern Hemisphere and clockwise in the Southern Hemisphere; literally, against the sun. { 'kän·trə 'sō,lem }

contrast [COMMUN] The degree of difference in tone between the lightest and darkest areas in a television or facsimile picture. [COMPUT SCI] In optical character recognition, the difference in color, reflectance, or shading between two areas of a surface, for example, a character and its background. { 'kän,trast }

contrast control [ELECTR] A manual control that adjusts the range of brightness between highlights and shadows on the reproduced image in a television receiver. { 'kän,trast kən'trōl }

contrastes [METEOROL] Winds a short distance apart blowing from opposite quadrants, frequent in the spring and fall in the western Mediterranean. { kȯn'tras,tēz }

contrast ratio [ELECTR] The ratio of the maximum to the minimum luminance values in a television picture. { 'kän,trast ,rā·shō }

contrast sensitivity *See* threshold contrast. { 'kän,trast sen·sə'tiv·əd·ē }

contrast sharpening [GRAPHICS] A procedure for increasing local contrasts in an image and reducing blurring by using computer processing of the image to emphasize its high spatial frequencies. { 'kän,trast ,shär·pən·iŋ }

contrast threshold *See* threshold contrast. { 'kän,trast ,thresh,hōld }

contravariant functor [MATH] A functor which reverses the sense of morphisms. { ¦kän·trə¦ver·ē·ənt 'fəŋk·tər }

contravariant index [MATH] A tensor index such that, under a transformation of coordinates, the procedure for obtaining a component of the transformed tensor for which this index has the value p involves taking a sum over q of the product of a component of the original tensor for which the index has the value q times the partial derivative of the pth transformed coordinate with respect to the qth original coordinate; it is written as a superscript. { ¦kän·trə¦ver·ē·ənt 'in,deks }

contravariant tensor [MATH] A tensor with only contravariant indices. { ¦kän·trə¦ver·ē·ənt 'ten·sər }

contravariant vector [MATH] A contravariant tensor of degree 1, such as the tensor whose components are differentials of the coordinates. { ¦kän·trə¦ver·ē·ənt 'vek·tər }

contributing structure [ORG CHEM] A structural formula that is one of a set of formulas, each contributing to the total wave function of a molecule. Also known as canonical form; canonical structure. { kən¦trib·yəd·iŋ 'strək·chər }

contributory *See* tributary. { kən'trib·yə,tȯr·ē }

control [COMPUT SCI] **1.** The section of a digital computer that carries out instructions in proper sequence, interprets each coded instruction, and applies the proper signals to the arithmetic unit and other parts in accordance with this interpretation. **2.** A mathematical check used with some computer operations. [CONT SYS] A means or device to direct and regulate a process or sequence of events. [ELECTR] An input element of a cryotron. [STAT] **1.** A test made to determine the extent of error in experimental observations or measurements. **2.** A procedure carried out to give a standard of comparison in an experiment. **3.** Observations made on subjects which have not undergone treatment, to use in comparison with observations made on subjects which have undergone treatment. { kən'trōl }

control accuracy [CONT SYS] The degree of correspondence between the ultimately controlled variable and the ideal value in a feedback control system. { kən'trōl ,ak·yə·rə·sē }

control agent [CHEM ENG] In process automatic-control work, material or energy within a process system of which the manipulated (controlled) variable is a condition or characteristic. { kən'trōl ,ā·jənt }

control and read-only memory [COMPUT SCI] A read-only memory that also provides storage, sequencing, execution, and translation logic for various microinstructions. Abbreviated CROM. { kən'trōl ən ¦rēd ,ōn·lē 'mem,rē }

control area [NAV] An area over which air-traffic control is exercised by a ground-located traffic-control center. { kən'trōl ,er·ē·ə }

control bit [COMPUT SCI] A bit which marks either the beginning or the end of a character transmitted in asynchronous communication. { kən'trōl ,bit }

control block [COMPUT SCI] A storage area containing (in condensed, formalized form) the information required for the control of a task, function, operation, or quantity of information. { kən'trōl ,bläk }

control board [ELEC] A panel at which one can make circuit changes, as in lighting a theater. [ENG] A panel in which meters and other indicating instruments display the condition of a system, and dials, switches, and other devices are used to modify circuits to control the system. Also known as control panel; panel board. { kən'trōl ,bȯrd }

control break [COMPUT SCI] **1.** A key change which takes place in a control data field, especially in the execution of a report program. **2.** A suspension of computer operation that is accomplished by simultaneously depressing the control key and the break key. { kən'trōl ,brāk }

control character [COMPUT SCI] A character whose occurrence in a particular context initiates, modifies, or stops a control operation in a computer or associated equipment. { kən'trōl ,kar·ik·tər }

control characteristic [ELECTR] **1.** The relation, usually shown by a graph, between critical grid voltage and anode voltage of a gas tube. **2.** The relation between control ampere-turns and output current of a magnetic amplifier. { kən'trōl ,kar·ik·tə'ris·tik }

control chart [IND ENG] A statistical tool used to detect excessive process variability due to specific assignable causes that can be corrected. It serves to determine whether a process is in a state of statistical control, that is, the extent of variation of the output of the process does not exceed that which is expected based on the natural statistical variability of the process. { kən'trōl ,chärt }

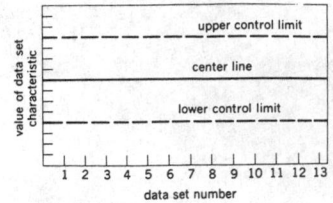

CONTROL CHART

Control chart, consisting of a center line and two sets of control lines.

control circuit [COMPUT SCI] One of the circuits that responds to the instructions in the program for a digital computer. [ELEC] A circuit that controls some function of a machine, device, or piece of equipment. [ELECTR] The circuit that feeds the control winding of a magnetic amplifier. { kən'trōl ˌsər·kət }

control code [COMPUT SCI] A special code that is entered by a user to carry out a particular function, such as the moving or deleting of text in a word-processing program. { kən'trōl ˌkōd }

control column [AERO ENG] A cockpit control lever pivoted or sliding in front of the pilot; controls operation of the elevator and aileron. { kən'trōl ˌkäl·əm }

control computer [COMPUT SCI] A computer which uses inputs from sensor devices and outputs connected to control mechanisms to control physical processes. { kən'trōl kəm'pyüd·ər }

control counter [COMPUT SCI] A counter providing data used to control the execution of a computer program. { kən'trōl ˌkaun·tər }

control data [COMPUT SCI] Data used for identifying, selecting, executing, or modifying another set of data, a routine, a record, or the like. { kən'trōl ˌdad·ə }

control day [METEOROL] One of several days on which the weather is supposed (according to folklore) to provide the key for the weather of a subsequent period. Also known as key day. { kən'trōl ˌdā }

control desk See console. { kən'trōl ˌdesk }

control diagram See flow chart. { kən'trōl ˌdī·ə,gram }

control drive [NUCLEO] The system of control rods which regulate the reaction rate of a nuclear reactor. { kən'trōl ˌdrīv }

control echo [ENG] In an ultrasonic inspection system, consistent reflection from a surface, such as a back reflection, which provides a reference signal. { kən'trōl ˌek·ō }

control electrode [ELECTR] An electrode used to initiate or vary the current between two or more electrodes in an electron tube. { kən'trōl i'lek,trōd }

control element [CONT SYS] The portion of a feedback control system that acts on the process or machine being controlled. [MOL BIO] A site within a gene or operon that acts to control gene expression. { kən'trōl ˌel·ə·mənt }

control feel [AERO ENG] The impression of the stability and control of an aircraft that a pilot receives through the cockpit controls, either from the aerodynamic forces acting on the control surfaces or from forces simulating these aerodynamic forces. { kən'trōl ˌfēl }

control flow graph [COMPUT SCI] A graph describing the logic structure of a software module, in which the nodes represent computational statements or expressions, the edges represent transfer of control between nodes, and each possible execution path of the module has a corresponding path from the entry to the exit node of the graph. { kən'trōl 'flō ˌgraf }

control gene [GEN] A gene that regulates the time and rate at which neighboring structural genes are transcribed in messenger ribonucleic acid. { kən'trōl ˌjēn }

control grid [ELECTR] A grid, ordinarily placed between the cathode and an anode, that serves to control the anode current of an electron tube. { kən'trōl ˌgrid }

control-grid bias [ELECTR] Average direct-current voltage between the control grid and cathode of a vacuum tube. { kən'trōl ˌgrid ˌbī·əs }

control-grid plate transconductance [ELECTR] Ratio of the amplification factor of a vacuum tube to its plate resistance, combining the effects of both into one term. { kən'trōl ˌgrid ˌplāt ˌtranz·kən'dək·təns }

control group [STAT] A sample in which a factor whose effect is being estimated is absent or is held constant, in order to provide a comparison. { kən'trōl ˌgrüp }

control handle See handle. { kən'trōl ˌhand·əl }

control head gap [COMPUT SCI] The distance maintained between the read/write head of a disk drive and the disk surface. { kən'trōl ˌhed ˌgap }

control hierarchy See hierarchical control. { kən'trōl 'hī·ər,är·kē }

control hole See designation punch. { kən'trōl ˌhōl }

control inductor See control winding. { kən'trōl in'dək·tər }

control instructions [COMPUT SCI] Those instructions in a computer program which ensure proper sequencing of instructions so that a programmed task can be performed correctly. { kən'trōl in'strək·shənz }

control joint [CIV ENG] An expansion joint in masonry to allow movement due to expansion and contraction. { kən'trōl ˌjóint }

control key [COMPUT SCI] A special key on a computer keyboard which, when depressed together with another key, generates a different signal than would be produced by the second key alone. { kən'trōl ˌkē }

controllability [AERO ENG] The quality of an aircraft or guided weapon which determines the ease of producing changes in flight direction or in altitude by operation of its controls. [CONT SYS] Property of a system for which, given any initial state and any desired state, there exists a time interval and an input signal which brings the system from the initial state to the desired state during the time interval. { kən,trōl·ə'bil·əd·ē }

controllable-pitch propeller [MECH ENG] An aircraft or ship propeller in which the pitch of the blades can be changed while the propeller is in motion; five types used for aircraft are two-position, variable-pitch, constant-speed, feathering, and reversible-pitch. Abbreviated CP propeller. { kən'trōl·ə·bəl 'pich prə'pel·ər }

control lead [COMPUT SCI] A character or sequence of characters indicating that the information following is a control code and not data. { kən'trōl ˌlēd }

controlled aerodrome [NAV] An aerodrome for which air-traffic control facilities are provided. { kən'trōld 'er·ə,drōm }

controlled airspace [NAV] An airspace of defined dimensions within which air-traffic control is provided. { kən'trōld 'er,spās }

controlled atmosphere [SCI TECH] A specified gas or mixture of gases at a predetermined temperature, and sometimes humidity, in which selected processes take place. { kən'trōld 'at·mə,sfir }

controlled avalanche device [ELECTR] A semiconductor device that has rigidly specified maximum and minimum avalanche voltage characteristics and is able to operate and absorb momentary power surges in this avalanche region indefinitely without damage. { kən'trōld 'av·ə,lanch di'vīs }

controlled avalanche rectifier [ELECTR] A silicon rectifier in which carefully controlled, nondestructive internal avalanche breakdown across the entire junction area protects the junction surface, thereby eliminating local heating that would impair or destroy the reverse blocking ability of the rectifier. { kən'trōld 'av·ə,lanch 'rek·tə,fī·ər }

controlled avalanche transit-time triode [ELECTR] A solid-state microwave device that uses a combination of IMPATT diode and *npn* bipolar transistor technologies; avalanche and drift zones are located between the base and collector regions. Abbreviated CATT. { kən'trōld 'av·ə,lanch 'tranz·ət ,tīm 'trī,ōd }

controlled carrier modulation [COMMUN] System of modulation wherein the carrier is amplitude-modulated by the signal frequencies and, in addition, the carrier is amplitude-modulated according to the envelope of the signal so that the modulation factor remains constant regardless of the amplitude of the signal. Also known as floating carrier modulation; variable carrier modulation. { kən'trōld 'kar·ē·ər ,mäj·ə'lā·shən }

controlled cooling [MET] Process by which an object is cooled from an elevated temperature in a predetermined manner to avoid cracking, internal damage, or hardening, or to produce a desired microstructure. { kən'trōld 'kül·iŋ }

controlled fragment [ORD] One of the pieces produced from warhead casings that have been designed to break up in specific patterns; the fragment takes on its final shape during the detonation of the explosive charge. Also known as fire-formed fragment. { kən'trōld 'frag·mənt }

controlled fusion [NUCLEO] The use of thermonuclear fusion reactions in a controlled manner to generate power. { kən'trōld 'fyü·zhən }

controlled-leakage system [AERO ENG] A system that provides for the maintenance of life in an aircraft or spacecraft cabin by a controlled escape of carbon dioxide and other waste from the cabin, with replenishment provided by stored oxygen and food. { kən'trōld 'lēk·ij ,sis·təm }

controlled medium [CHEM ENG] In process automatic-control work, material within a process system in which a variable (for example, concentration) is controlled. { kən¦trōld 'mēd·ē·əm }

controlled mercury-arc rectifier [ELECTR] A mercury-arc rectifier in which one or more electrodes control the start of the discharge in each cycle and thereby control output current. { kən¦trōld ¦mər·kyə·rē ¦ärk 'rek·tə,fī·ər }

controlled mine [ORD] A mine fitted with firing devices capable of being activated by an electrical system leading to a central control station; may be an underwater or land mine. { kən¦trōld 'mīn }

controlled mosaic [GRAPHICS] An assemblage of aerial photographs in which various adjustments have been made to improve accuracy and correct for distortions. { kən¦trōld mō'zā·ik }

controlled parameter [ENG] In the formulation of an optimization problem, one of the parameters whose values determine the value of the criterion parameter. { kən¦trōld pə'ram·əd·ər }

controlled rectifier [ELECTR] A rectifier that has provisions for regulating output current, such as with thyratrons, ignitrons, or silicon controlled rectifiers. { kən¦trōld 'rek·tə,fī·ər }

controlled thermonuclear reaction [NUCLEO] A fusion reaction generated in a controlled manner for research purposes or for production of useful power. { kən¦trōld ¦thər·mō'nü·klē·ər rē'ak·shən }

controlled thermonuclear reactor [NUCLEO] The heart of a fusion spacecraft propulsion system, based on the thermonuclear reaction of deuterium with a helium-3 isotope to produce helium-4 and protons. Abbreviated CTR. { kən¦trōld ¦thər·mō'nü·klē·ər rē'ak·tər }

controlled time of arrival [NAV] A time of arrival determined in advance, requiring careful planning, accurate navigation, and frequent checking of position and speed. { kən¦trōld 'tīm əv ə'rīv·əl }

controlled variable [CONT SYS] In process automatic-control work, that quantity or condition of a controlled system that is directly measured or controlled. [SCI TECH] The quantity or condition that is measured and controlled. { kən¦trōld 'ver·ē·ə·bəl }

controller *See* automatic controller. { kən'trōl·ər }

controller node [GEN] A genetic unit of regulation consisting of a set of regulators, effectors, and receptors located near the gene that the unit controls. { kən'trō·lər ,nōd }

controller-structure interaction [CONT SYS] Feedback of an active control algorithm in the process of model reduction; this occurs through observation spillover and control spillover. { kən'trōl·ər ,strək·chər in·tər'ak·shən }

control limits [ELECTR] In radar evaluation, upper and lower control limits are established at those performance figures within which it is expected that 95% of quality-control samples will fall when the radar is performing normally. [IND ENG] In statistical quality control, the limits of acceptability placed on control charts; parts outside the limits are defective. { kən'trōl ,lim·əts }

controlling depth [NAV] The least depth in the approach channel to an area, such as a port or anchorage, governing the maximum draft of vessels that can enter. { kən'trōl·iŋ ,depth }

controlling element [GEN] Any of a class of transposable genetic elements that have the capacity to control gene expression at several loci, as well as to render target genes extremely likely to mutate. { kən'trōl·iŋ 'el·ə·mənt }

controlling magnet [ENG] An auxiliary magnet used with a galvanometer to cancel the effect of the earth's magnetic field. { kən'trōl· iŋ ,mag·nət }

controlling obstacle [NAV] In terminal instrument procedures, the highest obstacle relative to a prescribed plane within a specified area; in precision approach procedures where obstacles penetrate the approach surface, the controlling obstacle is the one which results in the requirements for the highest decision height. { kən'trōl·iŋ 'äb·sti·kəl }

control logic [COMPUT SCI] The sequence of steps required to perform a specific function. { kən'trōl ,läj·ik }

control mark *See* tape mark. { kən'trōl ,märk }

control-message display [COMPUT SCI] A device, such as a console typewriter, on which control information, such as information on the progress of a running computer program, is displayed in ordinary language. { kən'trōl ,mes·ij di'splā }

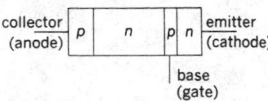

CONTROLLED RECTIFIER

collector (anode) | p | n | p | n | emitter (cathode)

base (gate)

Controlled rectifier diagram.

control module [COMPUT SCI] The set of registers and circuitry required to carry out a specific function. { kən'trōl ,mä·jül }

control-moment gyro [AERO ENG] An internal momentum storage device that applies torques to the attitude-control system through large rotating gyros. { kən'trōl ,mō·mənt ,jī·rō }

control of electromagnetic radiation *See* Conelrad. { kən'trōl əv i¦lek·trō·mag¦ned·ik ,rād·ē'ā·shən }

control operation [COMPUT SCI] Any action that affects data processing but is not directly included, such as managing input/output operations or determining job sequence. { kən'trōl ,äp·ə,rā·shən }

control panel [CIV ENG] *See* panel. [COMPUT SCI] An array of jacks or sockets in which wires (or other elements) may be plugged to control the action of an electromechanical device in a data-processing system such as a printer. Also known as plugboard; wiring board. [ENG] *See* control board. { kən'trōl ,pan·əl }

control plane [AERO ENG] An aircraft from which the movements of another craft are controlled remotely. { kən'trōl ,plān }

control point [COMPUT SCI] **1.** The numerical value of the controlled variable (speed, temperature, and so on) which, under any fixed set of operating conditions, an automatic controller operates to maintain. **2.** One of the hardware locations at which the output of the instruction decoder of the processor activates the input to and output from specific registers as well as operational resources of the system. [MAP] Any station in a horizontal and vertical grid that is identified on a photograph and used for correlating the data shown on that photograph. [NAV] A position marked by a buoy, boat, aircraft, electronic device, conspicuous terrain feature, or other identifiable object which is given a name or number and used as an aid for navigation or control of ships, boats, or aircraft. { kən'trōl ,póint }

control program [COMPUT SCI] A program which carries on input/output operations, loading of programs, detection of errors, communication with the operator, and so forth. { kən'trōl ,prō·grəm }

control punch *See* designation punch. { kən'trōl ,pənch }

control record [COMPUT SCI] A special record added to the end of a file to provide information about the file and the records in it. { kən'trōl ,rek·ərd }

control register [COMPUT SCI] Any one of the registers in a computer used to control the execution of a computer program. { kən'trōl ,rej·ə·stər }

control rocket [AERO ENG] A vernier engine, retrorocket, or other such rocket used to change the attitude of, guide, or make small changes in the speed of a rocket, spacecraft, or the like. { kən'trōl ,räk·ət }

control rod [NUCLEO] Any rod used to control the reactivity of a nuclear reactor; may be a fuel rod or part of the moderator; in a thermal reactor, commonly a neutron absorber. Also known as absorbing rod. { kən'trōl ,räd }

control room [COMMUN] A room from which engineers and production people control and direct a television or radio program or a sound-recording session; the room is adjacent to the main studios and separated from them by large soundproof glass windows. [ENG] A room from which space flights are directed. { kən'trōl ,rüm }

control sample [ANALY CHEM] A material of known composition that is analyzed along with test samples in order to evaluate the accuracy of an analytical procedure. Also known as check sample. { kən'trōl ,sam·pəl }

control section [COMPUT SCI] **1.** The smallest integral subsection of a program, that is, the smallest unit of code that can be separately relocated during loading. **2.** The part of a central processing unit that controls other sections of the unit. { kən'trōl ,sek·shən }

control sequence [COMPUT SCI] The order in which a set of executions are carried to perform a specific function. { kən'trōl ,sē·kwəns }

control signal [COMPUT SCI] A set of pulses used to identify the channels to be followed by transferred data. [CONT SYS] The signal applied to the device that makes corrective changes in a controlled process or machine. { kən'trōl ,sig·nəl }

control spillover [CONT SYS] The excitation by an active control system of modes of motion that have been omitted from the control algorithm in the process of model reduction. { kən'trōl 'spil,ō·vər }

control spring [DES ENG] A spring designed so that its torque cancels that of the instrument of which it is a part, for all deflections of the pointer. { kən'trōl ,spriŋ }

control state [COMPUT SCI] The operating mode of a system which permits it to override its normal sequence of operations. { kən'trōl ,stāt }

control statement [COMPUT SCI] A statement in a computer program that controls program execution, such as a GOTO statement, conditional jump, or a loop. { kən'trōl ,stāt·mənt }

control supervisor [COMPUT SCI] The computer software which controls the processing of the system. { kən'trōl ,sü·pər,vī·zər }

control surface [AERO ENG] **1.** Any movable airfoil used to guide or control an aircraft, guided missile, or the like in the air, including the rudder, elevators, ailerons, spoiler flaps, and trim tabs. **2.** In restricted usage, one of the main control surfaces, such as the rudder, an elevator, or an aileron. { kən'trōl ,sər·fəs }

control switching point [COMMUN] A telephone office which is an important switching center in the routing of long-distance calls in the direct distance dialing system. Abbreviated CSP. { kən'trōl 'swich·iŋ ,point }

control symbol [COMPUT SCI] A symbol which, coded into the machine memory, controls certain steps in the mechanical translation process; since control symbols are not contextual symbols, they appear neither in the input nor in the output. { kən'trōl ,sim·bəl }

control synchro See control transformer. { kən'trōl ,siŋ·krō }

control system [ENG] A system in which one or more outputs are forced to change in a desired manner as time progresses. { kən'trōl ,sis·təm }

control-system feedback [CONT SYS] A signal obtained by comparing the output of a control system with the input, which is used to diminish the difference between them. { kən'trōl ,sis·təm 'fēd,bak }

control systems equipment [COMPUT SCI] Computers which are an integral part of a total facility or larger complex of equipment and have the primary purpose of controlling, monitoring, analyzing, or measuring a process or other equipment. { kən'trōl ,sis·təmz i'kwip·mənt }

control total [COMPUT SCI] The sum of the numbers in a specified record field of a batch of records, determined repetitiously during computer processing so that any discrepancy from the control indicates an error. { kən'trōl ,tōd·əl }

control tower See airport traffic control tower. { kən'trōl ,tau·ər }

control-tower visibility [METEOROL] The visibility that is observed from an airport control tower. { kən'trōl ,tau·ər ,viz·ə'bil·əd·ē }

control track [ENG ACOUS] A supplementary sound track, usually containing tone signals that control the reproduction of the sound track, such as by changing feed levels to loudspeakers in a theater to achieve stereophonic effects. { kən'trōl ,trak }

control transformer [ELEC] A synchro in which the electrical output of the rotor is dependent on both the shaft position and the electric input to the stator. Also known as control synchro. { kən'trōl tranz'fór·mər }

control unit [COMPUT SCI] An electronic device containing data buffers and logical circuitry, situated between the computer channel and the input/output device, and controlling data transfers and such operations as tape rewind. { kən'trōl ,yü·nət }

control unit terminal emulation [COMPUT SCI] A technique that enables a personal computer to imitate a terminal of a main frame. Abbreviated CUT emulation. { kən'trōl ,yü·nət ¦tər·mə·nəl ,em·yə'lā·shən }

control valve [ENG] A valve which controls pressure, volume, or flow direction in a fluid transmission system. { kən'trōl ,valv }

control vane [AERO ENG] A movable vane used for control, especially a movable air vane or jet vane on a rocket used to control flight altitude. { kən'trōl ,vān }

control variable [CONT SYS] One of the input variables of a control system, such as motor torque or the opening of a valve, which can be varied directly by the operator to maximize some measure of performance of the system. { kən'trōl ,ver·ē·ə·bəl }

control winding [ELECTR] A winding used on a magnetic amplifier or saturable reactor to apply control magnetomotive forces to the core. Also known as control inductor. { kən'trōl ,wīnd·iŋ }

control word [COMPUT SCI] A computer word specifying a certain action to be taken. { kən'trōl ,wərd }

control zone [NAV] An airspace of defined dimensions extended upward from the surface to include one or more airports, and within which rules additional to those governing flight in control areas apply for the protection of traffic. Also known as air-traffic control zone. { kən'trōl ,zōn }

contusion [MED] A subcutaneous bruise caused by an injury in which the skin is not broken. { kən'tü·zhən }

Conularida [PALEON] A small group of extinct invertebrates showing a narrow, four-sided, pyramidal-shaped test. { ,kän·əl'ar·ə·də }

Conulata [INV ZOO] A subclass of free-living cnidarians in the class Scyphozoa; individuals are described as tetraramous cones to elongate pyramids having tentacles on the oral margin. { ,kän·əl'äd·ə }

Conulidae [PALEON] A family of Cretaceous exocyclic Euechinoidea characterized by a flattened oral surface. { kə'nü·lə,dē }

conus arteriosus [EMBRYO] The cone-shaped projection from which the pulmonary artery arises on the right ventricle of the heart in man and mammals. { 'kō·nəs är,tir·ē'ō·səs }

convalescence [MED] The period and process of recovery after an illness or injury. { ,kän·və'les·əns }

convalescent carrier [MED] A person who harbors an infectious agent after recovery from a clinical attack of a disease. { ,kän·və'les·ənt 'kar·ē·ər }

convalescent serum [IMMUNOL] The serum of the blood of one or more patients recovering from an infectious disease; used for prophylaxis of the particular infection. { ,kän·və'les·ənt 'sir·əm }

convection [FL MECH] Diffusion in which the fluid as a whole is moving in the direction of diffusion. Also known as bulk flow. [METEOROL] Atmospheric motions that are predominantly vertical, resulting in vertical transport and mixing of atmospheric properties. [OCEANOGR] Movement and mixing of ocean water masses. [PHYS] Transmission of energy or mass by a medium involving movement of the medium itself. { kən'vek·shən }

convectional stability See static stability. { kən'vek·shən·əl stə'bil·əd·ē }

convection cell [GEOPHYS] A concept in plate tectonics that accounts for the lateral or the upward and downward movement of subcrustal mantle material as due to heat variation in the earth. [METEOROL] An atmospheric unit in which organized convective fluid motion occurs. { kən'vek·shən ,sel }

convection coefficient See film coefficient. { kən'vek·shən kō·i'fish·ənt }

convection cooling [ENG] Heat transfer by natural, upward flow of hot air from the device being cooled. { kən'vek·shən ,kül·iŋ }

convection current [ELECTR] The time rate at which the electric charges of an electron stream are transported through a given surface. [GEOPHYS] Mass movement of subcrustal or mantle material as the result of temperature variations. [METEOROL] Any current of air involved in convection; usually, the upward-moving portion of a convection circulation, such as a thermal or the updraft in cumulus clouds. Also known as convective current. { kən'vek·shən ,kər·ənt }

convection modulus [FL MECH] An intrinsic property of a fluid which is important in determining the Nusselt number, equal to the acceleration of gravity times the volume coefficient of thermal expansion divided by the product of the kinematic viscosity and the thermal diffusivity. { kən'vek·shən ,mäj·ə·ləs }

convection oven [ENG] An oven containing a fan that continuously circulates hot air around the food being prepared. { kən'vek·shən ,əv·ən }

convection section [ENG] That portion of the furnace in which tubes receive heat from the flue gases by convection. { kən'vek·shən ,sek·shən }

convection stability See static stability. { kən'vek·shən stə'bil·əd·ē }

convection theory of cyclones [METEOROL] A theory of cyclone development proposing that the upward convection of air (particularly of moist air) due to surface heating can be of

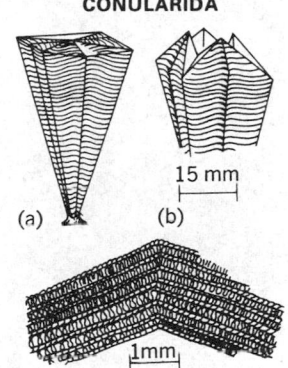

CONULARIDA

15 mm

(a)   (b)

1mm

(c)

Conularida restorations. *(a)* Attachment disk. *(b)* Distal part of same individual with triangular flaps raised. *(c)* Part of exterior showing the ornamentation.

sufficient magnitude and duration that the surface inflow of air will attain appreciable cyclonic rotation. { kən'vek·shən ˌthē·ə·rē əv 'sī,klōnz }

convective activity [METEOROL] Generally, manifestations of convection in the atmosphere, alluding particularly to the development of convective clouds and resulting weather phenomena, such as showers, thunderstorms, squalls, hail, and tornadoes. { kən'vek·div ak'tiv·əd·ē }

convective cloud [METEOROL] A cloud which owes its vertical development, and possibly its origin, to convection. { kən'vek·div 'klaùd }

convective-cloud-height diagram [METEOROL] A graph used as an aid in estimating the altitude of the base of convective clouds; since its basis is the same as that for the dew-point formula, only the surface temperature and dew point need be known to use the diagram. { kən¦vek·div 'klaùd ˌhīt ˌdī·ə,gram }

convective condensation level [METEOROL] On a thermodynamic diagram, the point of intersection of a sounding curve (representing the vertical distribution of temperature in an atmospheric column) with the saturation mixing-ratio line corresponding to the average mixing ratio in the surface layer (that is, approximately the lowest 1500 feet, or 450 meters). { kən'vek·div ˌkän·den'sā,shən ˌlev·əl }

convective current See convection current. { kən'vek·div ˌkər·ənt }

convective discharge [ELECTR] The movement of a visible or invisible stream of charged particles away from a body that has been charged to a sufficiently high voltage. Also known as electric wind; static breeze. { kən'vek·div 'dis,chärj }

convective equilibrium See adiabatic equilibrium. { kən'vek·div ˌē·kwə'lib·rē·əm }

convective instability [METEOROL] The state of an unsaturated layer or column of air in the atmosphere whose wet-bulb potential temperature (or equivalent potential temperature) decreases with elevation. Also known as potential instability. { kən'vek·div in·stə'bil·əd·ē }

convective overturn See overturn. { kən'vek·div 'ō·vər ,tərn }

convective precipitation [METEOROL] Precipitation from convective clouds, generally considered to be synonymous with showers. { kən'vek·div prə,sip·ə'tā·shən }

convective region [METEOROL] An area particularly favorable for the formation of convection in the lower atmosphere, or one characterized by convective activity at a given time. { kən'vek·div ˌrē·jən }

convective zone [ASTROPHYS] A region of instability just below the photosphere of the sun in which part of the heat is carried outward by convective currents. { kən'vek·div ˌzōn }

convector [ENG] A heat-emitting unit for the heating of room air; it has a heating element surrounded by a cabinet-type enclosure with openings below and above for entrance and egress of air. { kən'vek·tər }

convectron [ENG] An instrument for indicating deviation from the vertical which is based on the principle that the convection from a heated wire depends strongly on its inclination; it consists of a Y-shaped tube, each of whose arms contains a wire forming part of a bridge circuit. { kən'vek,trän }

convenience receptacle See outlet. { kən'vēn·yəns ri'sep·tə·kəl }

conventional algorithm [COMMUN] A cryptographic algorithm in which the enciphering and deciphering keys are easily derivable from each other, or are identical, and both must be kept secret. { kən'ven·chən·əl 'al·gə,rith·əm }

conventional bomb [ORD] A nonatomic bomb designed primarily for explosive effect. { kən'ven·chən·əl 'bäm }

conventional completion [PETRO ENG] Well completion in which the tubing is set inside a casing having a minimum diameter of 4.5 inches (11 centimeters). { kən'ven·chən·əl kəm'plē·shən }

conventional current [ELEC] The concept of current as the transfer of positive charge, so that its direction of flow is opposite to that of electrons which are negatively charged. { kən'ven·chən·əl 'kər·ənt }

conventional grouping [COMPUT SCI] When a unit record containing a coding field is used for a single item code and the set of codes of the terms which describe the item, the grouping is conventional; for example, a personnel file in which each individual is represented by a card on which are punched

codes for his or her age, sex, education, and salary. { kən'ven·chən·əl 'grüp·iŋ }

conventional milling [MET] Milling in which the cutter and feed move in opposite directions from the point of contact. { kən'ven·chən·əl 'mil·iŋ }

conventional mining [MIN ENG] The cycle which includes cutting the coal, drilling the shot holes, charging and shooting the holes, loading the broken coal, and installing roof support. Also known as cyclic mining. { kən'ven·chən·əl 'mīn·iŋ }

conventional programming [COMPUT SCI] The use of standard programming languages, as opposed to application development languages, financial planning languages, query languages, and report programs. { kən'ven·chən·əl 'prō,gram·iŋ }

conventional spinning See manual spinning. { kən'ven·chən·əl 'spin·iŋ }

convergence [ANTHRO] Independent development of similarities between unrelated cultures. [EVOL] Development of similarities between animals or plants of different groups resulting from adaptation to similar habitats. [ELECTR] A condition in which the electron beams of a multibeam cathode-ray tube intersect at a specified point, such as at an opening in the shadow mask of a three-gun color television picture tube; both static convergence and dynamic convergence are required. [GEOL] Diminution of the interval between geologic horizons. [HYD] The line of demarcation between turbid river water and clear lake water. [MATH] The property of having a limit for infinite series, sequences, products, and so on. [METEOROL] The increase in wind setup observed beyond that which would take place in an equivalent rectangular basin of uniform depth, caused by changes in platform or depth. [NEURO] The coming together of a group of afferent nerves upon a motoneuron of the ventral horn of the spinal cord. [OCEANOGR] A condition in the ocean in which currents or water masses having different densities, temperatures, or salinities meet; results in the sinking of the colder or more saline water. [PHYS] The intersection of light beams or particles within a small region, or the narrowing of a single beam so that it passes through a small region. { kən'vər·jəns }

convergence circuit [ELECTROMAG] An auxiliary deflection system in a color television receiver which maintains convergence, having separate convergence coils for electromagnetic controls of the positions of the three beams in a convergence yoke around the neck of the kinescope. { kən'vər·jəns ,sər·kət }

convergence coil [ELECTR] One of the coils used to obtain convergence of electron beams in a three-gun color television picture tube. { kən'vər·jəns ,kóil }

convergence constant See convergence factor. { kən'vər·jəns ,kän·stənt }

convergence control [ELECTR] A control used in a color television receiver to adjust the potential on the convergence electrode of the three-gun color picture tube to achieve convergence. { kən'vər·jəns kən'trōl }

convergence electrode [ELECTR] An electrode whose electric field converges two or more electron beams. { kən'vər·jəns i'lek,trōd }

convergence factor [MAP] For a specific chart or map projection, the factor which, when multiplied by the difference of longitude between points on two meridians, will give the convergence of the meridians. Also known as convergence constant. { kən'vər·jəns ,fak·tər }

convergence in measure [MATH] A sequence of functions $f_n(x)$ converges in measure to $f(x)$ if given any $\epsilon > 0$, the measure of the set of points at which $|f_n(x) - f(x)| > \epsilon$ is less than ϵ, provided n is sufficiently large. { kən'vər·jəns in 'mezh·ər }

convergence limit [SPECT] 1. The short-wavelength limit of a set of spectral lines that obey a Rydberg series formula; equivalently, the long-wavelength limit of the continuous spectrum corresponding to ionization from or recombination to a given state. 2. The wavelength at which the difference between successive vibrational bands in a molecular spectrum decreases to 0. { kən'vər·jəns ,lim·ət }

convergence line [METEOROL] Any horizontal line along which horizontal convergence of the airflow is occurring. Also known as asymptote of convergence. { kən'vər·jəns ,līn }

convergence magnet [ELECTR] A magnet assembly whose

magnetic field converges two or more electron beams; used in three-gun color picture tubes. Also known as beam magnet. { kən'vər·jəns ˌmag·nət }

convergence of meridians [MAP] **1.** The angular drawing together of the geographic meridians in passing from the equator to the poles. **2.** The relative difference of directions of meridians at specific points on the meridians; it is equal to the product of the difference of longitude and the convergence factor. { kən'vər·jəns əv mə'rid·ē·ənz }

convergence pressure [PHYS CHEM] The pressure at which the different constant-temperature K (liquid-vapor equilibrium) factors for each member of a two-component system converge to unity. { kən'vər·jəns ˌpresh·ər }

convergence ratio [OPTICS] The ratio of the tangent of the angle between a meridional ray and the optical axis after it passes through an optical system to the tangent of the angle between the ray and the axis before it passes through the system. { kən'vər·jəns ˌrā·shō }

convergence zone [ACOUS] A sound transmission channel produced in sea water by a combination of pressure and temperature changes in the depth range between 2500 and 15,000 feet (750 and 4500 meters); utilized by sonar systems. { kən'vər·jəns ˌzōn }

convergent [MATH] One of the continued fractions that is obtained from a given continued fraction by terminating after a finite number of terms. { kən'vər·jənt }

convergent die [ENG] A die having internal channels which converge. { kən'vər·jənt ˌdī }

convergent-divergent nozzle [DES ENG] A nozzle in which supersonic velocities are attained; has a divergent portion downstream of the contracting section. Also known as supersonic nozzle. { kən'vər·jənt də'vər·jənt ˌnäz·əl }

convergent integral [MATH] An improper integral which has a finite value. { kən'vər·jənt 'in·tə·grəl }

convergent precipitation [METEOROL] A synoptic type of precipitation caused by local updrafts of moist air. { kən'vər·jənt prə,sip·ə'tā·shən }

convergent sequence [MATH] A sequence which has a limit. { kən'vər·jənt 'sē·kwəns }

convergent series [MATH] A series whose sequence of partial sums has a limit. { kən'vər·jənt 'sir,ēz }

convergent zone paths [OCEANOGR] The velocity structure of permanent deep sound channels that produces focusing regions at distant intervals from a shallow source. { kən'vər·jənt 'zōn ,pathz }

converging lens [OPTICS] A lens that has a positive focal length, and therefore causes rays of light parallel to its axis to converge. Also known as positive lens. { kən'vər·jiŋ ,lenz }

converging mirror [OPTICS] A concave mirror that causes rays of light parallel to its axis to converge. Also known as positive mirror. { kən'vərj·iŋ ,mir·ər }

Conversational Algebraic Language See CAL. { kän·vərˌsā·shən·əl al·jəˌbrā·ik 'laŋ·gwij }

conversational compiler [COMPUT SCI] A compiler which immediately checks the validity of each source language statement entered to the computer and informs the user if the next statement can be entered or if a mistake must be corrected. Also known as interpreter. { kän·vər'sā·shən·əl kəm'pīl·ər }

conversational mode [COMMUN] A computer operating mode that permits queries and responses between the computer and human operators at keyboard terminals. { kän·vər'sā·shən·əl ˌmōd }

conversational processing [COMPUT SCI] The operating mode of a computer system which enables a user to have each statement he keys into the system processed immediately. { kän·vər'sā·shən·əl 'präs·əs·iŋ }

conversational time-sharing [COMPUT SCI] The simultaneous utilization of a computer system by multiple users, each user being equipped with a remote terminal with which he communicates with the computer in conversational mode. { kän·vər'sā·shən·əl 'tīm ,sher·iŋ }

converse [MATH] The converse of the statement "if *p*, then *q*" is the statement "if *q*, then *p*." { 'kän,vərs }

conversion See data conversion. [CHEM] Change of a compound from one isomeric form to another. [CHEM ENG] The chemical change from reactants to products in an industrial chemical process. Also known as chemical conversion. [NAV] Determination of the rhumb-line direction of one point from another when the initial great-circle direction is known,

or vice versa, the difference between the two directions being the conversion angle; used in connection with radio bearings, Consol, Consolan, and in great-circle sailing. [NUC PHYS] Nuclear transformation of a fertile substance into a fissile substance. [PETRO ENG] Treatment of a drilling mud to alter its chemical properties. Also known as breakover. [PHYS] Change in a quantity's numerical value as a result of using a different unit of measurement. [PSYCH] A defense mechanism whereby unconscious emotional conflict is transformed into physical disability, the affected part always having symbolic meaning pertinent to the nature of the conflict. { kən'vər·zhən }

conversion angle [NAV] The angle between the rhumb line and the great circle between two points. Also known as arc to chord correction; half-convergency. { kən'vər·zhən ,aŋ·gəl }

conversion coating [MET] A metal-surface coating consisting of a compound of the base metal. { kən'vər·zhən ,kōd·iŋ }

conversion coefficient Also known as conversion fraction; internal conversion coefficient. [NUC PHYS] **1.** The ratio of the number of conversion electrons emitted per unit time to the number of photons emitted per unit time in the de-excitation of a nucleus between two given states. **2.** In older literature, the ratio of the number of conversion electrons emitted per unit time to the number of conversion electrons plus the number of photons emitted per unit time in the de-excitation of a nucleus between two given states. { kən'vər·zhən ,kō·i'fish·ənt }

conversion disorders [PSYCH] Somatoform disorders characterized by physical symptoms referable to the somatosensory nervous system or special sensory organs that cannot be explained on the basis of a medical or neurologic disease and that are caused by psychological factors. Symptoms include paralysis, blindness, ataxia, aphonia, and numbness of the feet. { kən'vərzh·ən dis,ȯrd·ərz }

conversion electron [NUC PHYS] An electron which receives energy directly from a nucleus in an internal conversion process and is thereby expelled from the atom. { kən'vər·zhən i'lek,trän }

conversion equipment [COMPUT SCI] Equipment used for conversion of data from one recording medium to another, as from card to tape. { kən'vər·zhən i'kwip·mənt }

conversion factor [MATH] The numerical factor by which one must multiply (or divide) a quantity that is expressed in terms of a certain unit to express the quantity in terms of another unit. Also known as unit conversion factor. *See* conversion ratio. { kən'vər·zhən ,fak·tər }

conversion fraction *See* conversion coefficient. { kən'vər·zhən ,frak·shən }

conversion gain [ELECTR] **1.** Ratio of the intermediate-frequency output voltage to the input signal voltage of the first detector of a superheterodyne receiver. **2.** Ratio of the available intermediate-frequency power output of a converter or mixer to the available radio-frequency power input. [NUCLEO] The conversion ratio minus one in a nuclear reactor. { kən'vər·zhən ,gān }

conversion length [PHYS] The average distance traveled by an energetic photon in a given medium before it is converted into an electron and a positron through pair production. { kən'vər·zhən ,leŋkth }

conversion polarity [GEN] A gradient in the frequency of gene conversion from one end of a gene to the other. { kən'vər·zhən pə,lar·əd·ē }

conversion program [COMPUT SCI] A set of instructions which allows a program written for one system to be run on a different system. { kən'vər·zhən ,prō·grəm }

conversion rate [COMPUT SCI] The number of complete conversions an analog-to-digital converter can perform per unit time, usually specified in cycles (or conversions) per second. { kən'vər·zhən ,rāt }

conversion ratio [MATH] *See* conversion factor. [NUCLEO] The number of fissionable atoms produced per fissionable atom fissioned in a converter type of nuclear reactor. Also known as conversion factor. { kən'vər·zhən ,rā·shō }

conversion reaction [PSYCH] The form of hysterical neurosis in which the impulse causing anxiety is converted into functional symptoms of the special senses or voluntary nervous system. { kən'vər·zhən rē'ak·shən }

conversion routine [COMPUT SCI] A flexible, self-contained, and generalized program used for data conversion,

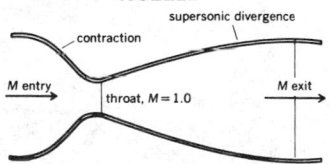

CONVERGENT-DIVERGENT NOZZLE

Diagram of convergent-divergent nozzle. *M* = local Mach number.

which only requires specifications about very few facts in order to be used by a programmer. { kən'vər·zhən rü'tēn }

conversion table [SCI TECH] A list of equivalent values for converting from one set of units to another. { kən'vər·zhən ˌtā·bəl }

conversion time [COMPUT SCI] The time required to read in data from one code into another code. { kən'vər·zhən ˌtīm }

conversive heating [MED] The conversion of some form of energy, especially radio waves, into heat for use in thermotherapy. { kən'vər·siv 'hēd·iŋ }

convert [COMPUT SCI] To transform the representation of data. { kən'vərt }

convertase [BIOCHEM] An enzyme that cleaves inactive protein precursors into smaller biologically active molecules. { 'kän·vər,tās }

converted water See product water. { kən'vərd·əd 'wòd·ər }

converter [COMPUT SCI] A computer unit that changes numerical information from one form to another, as from decimal to binary or vice versa, from fixed-point to floating-point representation, from magnetic tape to disk storage, or from digital to analog signals and vice versa. Also known as data converter. [ELEC] **1.** Any device for changing alternating current to direct current, or direct current to alternating current. **2.** See synchronous converter. [ELECTR] **1.** The section of a superheterodyne radio receiver that converts the desired incoming radio-frequency signal to the intermediate-frequency value; the converter section includes the oscillator and the mixer-first detector. Also known as heterodyne conversion transducer; oscillator-mixer-first detector. **2.** An auxiliary unit used with a television or radio receiver to permit reception of channels or frequencies for which the receiver was not originally designed. **3.** In facsimile, a device that changes the type of modulation delivered by the scanner. **4.** Unit of a radar system in which the mixer of a superheterodyne receiver and usually two stages of intermediate-frequency amplification are located; performs a preamplifying operation. See remodulator. [MET] A type of furnace in which impurities are oxidized out by blowing air through or across a path of molten metal or matte. [NUCLEO] **1.** Also known as nuclear converter. **2.** A nuclear reactor that converts fertile atoms into fuel by neutron capture, using one kind of fuel and producing another. **3.** A nuclear reactor that produces some fissionable fuel, but less than it consumes; the fuel produced may be the same as that consumed or different. { kən'vərd·ər }

converter substation [ELEC] An electric power substation whose main function is the conversion of power from ac to dc, and vice versa. { kən'vərd·ər ˌsəb,stā·shən }

converter tube [ELECTR] An electron tube that combines the mixer and local-oscillator functions of a heterodyne conversion transducer. { kən'vərd·ər ˌtüb }

convertiplane [AERO ENG] A hybrid form of heavier-than-air craft capable, because of one or more horizontal rotors or units acting as rotors, of taking off, hovering, and landing in a fashion similar to a helicopter; and once aloft and moving forward, capable, by means of a mechanical conversion, of flying purely as a fixed-wing aircraft, especially in higher speed ranges. { kən'vərd·ə,plān }

convex [SCI TECH] Having a curved form which bulges outward, resembling the exterior of a sphere or cylinder or a section of these bodies. { 'kän,veks }

convex angle [MATH] A polyhedral angle that lies entirely on one side of each of its faces. { 'kän,veks 'aŋ·gəl }

convex body [MATH] A convex set that has at least one interior point. { 'kän,veks 'bäd·ē }

convex combination [MATH] A linear combination of vectors in which the sum of the coefficients is 1. { 'kän,veks ˌkäm·bə'nā·shən }

convex curve [MATH] A plane curve for which any straight line that crosses the curve crosses it at just two points. { 'kän,veks 'kərv }

convex function [MATH] A function $f(x)$ is considered to be convex over the interval a,b if for any three points x_1, x_2, x_3 such that $a < x_1 < x_2 < x_3 < b$, $f(x_2) \leqslant L(x_2)$, where $L(x)$ is the equation of the straight line passing through the points $[x_1, f(x_1)]$ and $[x_3, f(x_3)]$. { 'kän,veks 'fəŋk·shən }

convex function in the sense of Jensen [MATH] A function $f(x)$ over an interval a, b such that, for any two points x_1 and x_2 satisfying $a < x_1 < x_2 < b$, $f[(x_1 + x_2)/2] \leqslant$ (1/2) $[f(x_1) + f(x_2)]$. { ˈkän,veks ˈfəŋk·shən in t͟hə ˌsens əv 'jen·sən }

convex hull [MATH] The smallest convex set containing a given collection of points in a real linear space. Also known as convex linear hull. { 'kän,veks 'həl }

convex linear combination [MATH] A linear combination in which the scalars are nonnegative real numbers whose sum is 1. { ˈkän,veks ˌlin·ē·ər ˌkäm·bə'nā·shən }

convex linear hull See convex hull. { 'kän,veks 'lin·ē·ər ˌhəl }

convexo-concave [SCI TECH] Having one side convex and the other concave, usually with greater curvature on the convex side. Also known as concavo-convex. { kənˈvek·sō,kän'kāv }

convex polygon [MATH] A polygon all of whose interior angles are less than or equal to 180°. { 'kän,veks 'päl·i,gän }

convex polyhedron [MATH] A polyhedron in the plane which is a convex set, for example, any regular polyhedron. { 'kän,veks ˌpäl·iˌhē·drən }

convex polytope [MATH] A bounded, convex subset of an n-dimensional space enclosed by a finite number of hyperplanes. { ˌkän,veks 'päl·i,tōp }

convex programming [MATH] Nonlinear programming in which both the function to be maximized or minimized and the constraints are appropriately chosen convex or concave functions of the independent variables. { 'kän,veks 'prō ˌgram·iŋ }

convex sequence [MATH] A sequence of numbers, a_1, a_2, ..., such that $a_{i+1} \leqq (1/2)(a_i + a_{i+2})$ for all $i \geqq 1$ (or for all i satisfying $1 \leqq i < n - 2$ if the sequence is a finite sequence with n terms). { 'kän,veks 'sē·kwəns }

convex set [MATH] A set which contains the entire line segment joining any pair of its points. { 'kän,veks 'set }

convex span [MATH] For a set A, the intersection of all convex sets that contain A. { ˌkän,veks 'span }

conveyor [MECH ENG] Any materials-handling machine designed to move individual articles such as solids or free-flowing bulk materials over a horizontal, inclined, declined, or vertical path of travel with continuous motion. { kən'vā·ər }

conveyor belt balance [ENG] A balance used for weighing unpackaged, loose, continuously transported material on a conveyor belt by weighing the load being moved and measuring the belt speed. { kən'vā·ər ˌbelt ˌbal·əns }

convivium [ECOL] A population exhibiting differentiation within the species and isolated geographically, generally a subspecies or ecotype. { kən'viv·ē·əm }

convolute [BIOL] Twisted or rolled together, specifically referring to leaves, mollusk shells, and renal tubules. { 'kän·və,lüt }

convolute bedding [GEOL] The extremely contorted laminae usually confined to a single layer of sediment, resulting from subaqueous slumping. { 'kän·və,lüt ,bed·iŋ }

convolution [ANAT] A fold, twist, or coil of any organ, especially any one of the prominent convex parts of the brain, separated from each other by depressions or sulci. [GEOL] **1.** The process of developing convolute bedding. **2.** A structure resulting from a convolution process, such as a small-scale but intricate fold. [MATH] The convolution of the functions f and g is the function F, defined by

$$F(x) = \int_0^x f(t)g(x - t)\, dt$$

[STAT] A method for finding the distribution of the sum of two or more random variables; computed by direct integration or summation as contrasted with, for example, the method of characteristic functions. { ˌkän·və'lü·shən }

convolutional code [COMMUN] An error-correcting code that processses incoming bits serially rather than in large blocks. { ˌkän·və'lü·shən·əl 'kōd }

convolution family See faltung. { ˌkän·və'lü·shən ,fam·lē }

convolution rule [MATH] The statement that $C(p + q, r)$ is the sum over the index j from $j = 0$ to $j = r$ of the quantity $C(p, j)\, C(q, r - j)$, where, in general, $C(n, r)$ is the number of distinct subsets of r elements in a set of n elements (the binomial coefficient). Also known as Vandermonde's identity. { ˌkän·və'lü·shən ,rül }

convolution theorem [MATH] A theorem stating that, under specified conditions, the integral transform of the convolution of two functions is equal to the product of their integral transforms. { ˌkän·və'lü·shən ˌthir·əm }

convolver [ELECTR] A surface acoustic-wave device in which signal processing is performed by a nonlinear interaction between two waves traveling in opposite directions. Also known as acoustic convolver. { kən'väl·vər }

Convolvulaceae [BOT] A large family of dicotyledonous plants in the order Polemoniales characterized by internal phloem, the presence of chlorophyll, two ovules per carpel, and plicate cotyledons. { kənˌväl·və'lās·ē,ē }

convoy [ORD] An accompanying and protective force on sea or land, such as a fleet or troops. { 'kän,vói }

convulsion [MED] An episode of involuntary, generally violent muscular contractions, rhythmically alternated with periods of relaxation; associated with many systematic and neurological diseases. { kən'vəl·shən }

convulsive disorder [MED] Any pathologic condition in which convulsions are a common symptom and characteristic electroencephalogram patterns are displayed. { kən'vəl·siv dis,ór·dər }

convulsive equivalent See epileptic equivalent. { kən'vəl·siv i'kwiv·ə·lənt }

Conwell-Weisskopf equation [SOLID STATE] An equation for the mobility of electrons in a semiconductor in the presence of donor or acceptor impurities, in terms of the dielectric constant of the medium, the temperature, the concentration of ionized donors (or acceptors), and the average distance between them. { ˌkän,wel ˌvīs,kópf i'kwä·zhən }

cookbook [COMPUT SCI] A document that describes how to install and use a software product or carry out other complex tasks in step-by-step fashion. { 'kúk,búk }

Cooke objective [OPTICS] A three-lens objective consisting of one biconcave lens, the dispersive component, between two biconvex lens, the collective components; used in astronomical cameras. { 'kúk əb'jek·div }

Cooke unit [BIOL] A unit for the standardization of pollen antigenicity. { 'kúk ,yü·nət }

cookie [COMPUT SCI] A data file written to a hard drive by some Web sites; contains information the site can use to track such things as passwords, login, registration or identification, user preferences, online shopping cart information, and lists of pages visited. { 'kúk·ē }

cooking snow See water snow. { 'kúk·iŋ ,snō }

cook-off [ORD] A cartridge fired by environmental heat, because it remained in the chamber of an overheated weapon. { 'kúk,óf }

coolant [MATER] **1.** A cutting fluid for machine operations, which keeps the tool cool to prevent reduction in hardness and resistance to abrasion, and prevents distortion of the work. **2.** A substance, ordinarily fluid, used for cooling any part of a reactor in which heat is generated. **3.** In general, any cooling agent, usually a fluid. { 'kül·ənt }

cooled infrared detector [ELECTR] An infrared detector that must be operated at cryogenic temperatures, such as at the temperature of liquid nitrogen, to obtain the desired infrared sensitivity. { 'küld ,in·frə'red di'tek·tər }

cooled-tube pyrometer [ENG] A thermometer for high-temperature flowing gases that uses a liquid-cooled tube inserted in the flowing gas; gas temperature is deduced from the law of convective heat transfer to the outside of the tube and from measurement of the mass flow rate and temperature rise of the cooling liquid. { 'küld ,tüb pī'räm·əd·ər }

cooler nail [DES ENG] A thin, cement-coated wire nail. { 'kül·ər ,nāl }

cooler ring See cooler-storage ring. { 'kül·ər,riŋ }

cooler-storage ring [NUCLEO] A type of accelerator in which ions are kept circulating in an annular structure, guided by magnetic fields, and the energy spread, angular divergence, and geometrical size of the ion beam are reduced by stochastic, electron, or laser cooling. Also known as cooler ring. { 'kül·ər,stór·ij ,riŋ }

cool flame [CHEM] A faint, luminous phenomenon observed when, for example, a mixture of ether vapor and oxygen is slowly heated; it proceeds by diffusion of reactive molecules which initiate chemical processes as they go. { 'kül ,flām }

Coolidge tube [ELECTROMAG] An x-ray tube in which the needed electrons are produced by a hot cathode. { 'kül·ij ,tüb }

cooling [NUCLEO] Setting aside a highly radioactive material until the radioactivity has diminished to a desired level. { 'kül·iŋ }

cooling channel [ENG] A channel in the body of mold through which a cooling liquid is circulated. { 'kül·iŋ ,chan·əl }

cooling coil [MECH ENG] A coiled arrangement of pipe or tubing for the transfer of heat between two fluids. { 'kül·iŋ ,kóil }

cooling correction [THERMO] A correction that must be employed in calorimetry to allow for heat transfer between a body and its surroundings. Also known as radiation correction. { 'kül·iŋ kə'rek·shən }

cooling curve [THERMO] A curve obtained by plotting time against temperature for a solid-liquid mixture cooling under constant conditions. { 'kül·iŋ ,kərv }

cooling degree day [MECH ENG] A unit for estimating the energy needed for cooling a building; one unit is given for each degree Fahrenheit that the daily mean temperature exceeds 75°F (24°C). { 'kül·iŋ di'grē ,dā }

cooling fin [MECH ENG] The extended element of a heat-transfer device that effectively increases the surface area. { 'kül·iŋ ,fin }

cooling fixture [ENG] A wooden or metal block used to hold the shape or dimensional accuracy of a molding until it cools enough to retain its shape. { 'kül·iŋ ,fiks·chər }

cooling load [MECH ENG] The total amount of heat energy that must be removed from a system by a cooling mechanism in a unit time, equal to the rate at which heat is generated by people, machinery, and processes, plus the net flow of heat into the system not associated with the cooling machinery. { 'kül·iŋ ,lōd }

cooling method [THERMO] A method of determining the specific heat of a liquid in which the times taken by the liquid and an equal volume of water in an identical vessel to cool through the same range of temperature are compared. { 'kül·iŋ ,meth·əd }

cooling pond [CHEM ENG] Outdoor depression into which hot process water is pumped for purposes of cooling by evaporation, convection, and radiation. { 'kül·iŋ ,pänd }

cooling power [MECH ENG] A parameter devised to measure the air's cooling effect upon a human body; it is determined by the amount of heat required by a device to maintain the device at a constant temperature (usually 34°C); the entire system should be made to correspond, as closely as possible, to the external heat exchange mechanism of the human body. { 'kül·iŋ ,paú·ər }

cooling-power anemometer [ENG] Any anemometer operating on the principle that the heat transfer to air from an object at an elevated temperature is a function of airspeed. { 'kül·iŋ ,paúr an·ə'mäm·əd·ər }

cooling process [ENG] Physical operation in which heat is removed from process fluids or solids; may be by evaporation of liquids, expansion of gases, radiation or heat exchange to a cooler fluid stream, and so on. { 'kül·iŋ ,präs·əs }

cooling range [MECH ENG] The difference in temperature between the hot water entering and the cold water leaving a cooling tower. { 'kül·iŋ ,rānj }

cooling stress [MECH] Stress resulting from uneven contraction during cooling of metals and ceramics due to uneven temperature distribution. { 'kül·iŋ ,stres }

cooling table See hotbed. { 'kül·iŋ ,tā·bəl }

cooling tower [ENG] A towerlike device in which atmospheric air circulates and cools warm water, generally by direct contact (evaporation). { 'kül·iŋ ,taú·ər }

coolometer [ENG] An instrument which measures the cooling power of the air, consisting of a metal cylinder electrically heated to maintain a constant temperature; the electrical heating power required is taken as a measure of the air's cooling power. { kü'läm·əd·ər }

cool star [ASTROPHYS] A low-temperature star, generally visible in the infrared range of the electromagnetic spectrum. { ˌkül ,stär }

cool time [MET] The period of time between successive heat times in pulsation and seam welding. { 'kül ,tīm }

Coombs serum [IMMUNOL] An immune serum containing antiglobulin that is used in testing for Rh and other sensitizations. { 'kümz ,sir·əm }

COOKE OBJECTIVE

Cooke objective, a type of wide-field lens objective for astronomical cameras.

COOLING TOWER

Natural-draft cooling towers. (Haman, Inc.)

cooperative multitasking [COMPUT SCI] A method of running more than one program on a computer at a time in which the program currently in control of the processor retains the control until it yields the control to another program voluntarily, which it can do only at certain points in the program. Also known as nonpreemptive multitasking. { kō‚äp·rəd·iv 'məl·tə‚task·iŋ }

cooperative observer [METEOROL] An unpaid observer who maintains a meteorological station for the U.S. National Weather Service. { kō'äp·rəd·iv əb'zər·vər }

cooperative phenomenon [SOLID STATE] A process that involves a simultaneous collective interaction among many atoms or electrons in a crystal, such as ferromagnetism, superconductivity, and order-disorder transformations. { kō'äp·rəd·iv fə'näm·ə‚nän }

cooperative system [ENG] A missile guidance system that requires transmission of information from a remote ground station to a missile in flight, processing of the information by the missile-borne equipment, and retransmission of the processed data to the originating or other remote ground stations, as in azusa and dovap. { kō'äp·rəd·iv ‚sis·təm }

cooperite [MINERAL] (Pt,Pd)S A steel-gray tetragonal mineral of metallic luster consisting of a sulfide of platinum, occurring in irregular grains in igneous rock. { 'kü·pə‚rīt }

Cooper pairs [SOLID STATE] Pairs of bound electrons which occur in a superconducting medium according to the Bardeen-Cooper-Schrieffer theory. { 'kü·pər ‚perz }

coordinate addressing [COMPUT SCI] The use of cartesian coordinates to specify a location, such as the position of a character in an electronic display. { kō'órd·ən·ət 'ad‚res·iŋ }

coordinate axes [MATH] One of a set of lines or curves used to define a coordinate system; the value of one of the coordinates uniquely determines the location of a point on the axis, while the values of the other coordinates vanish on the axis. { kō'órd·ən·ət 'ak‚sēz }

coordinate basis [MATH] A basis for tensors on a manifold induced by a set of local coordinates. { kō'órd·ən·ət 'bā·səs }

coordinate bond See coordinate valence. { kō'órd·ən·ət 'bänd }

coordinate conversion [MAP] Changing the map coordinate values from one system to those of another system. { kō'órd·ən·ət kən'vər·zhən }

coordinate data receiver [ELECTR] A receiver specifically designed to accept the signal of a coordinate data transmitter and reconvert this signal into a form suitable for input to associated equipment such as a plotting board, computer, or radar set. { kō'órd·ən·ət 'dad·ə ri‚sē·vər }

coordinate data transmitter [ELECTR] A transmitter that accepts two or more coordinates, such as those representing a target position, and converts them into a form suitable for transmission. { kō'órd·ən·ət 'dad·ə tranz‚mid·ər }

coordinated-axis control [CONT SYS] Robotic control in which the robot axes reach their end points simultaneously, thus giving the robot's motion a smooth appearance. { kō'órd·ən‚äd·əd 'ak·səs kən‚trōl }

coordinated complex See coordination compound. { kō'órd·ən‚äd·əd 'käm‚pleks }

coordinated geometry See COGO. { kō'órd·ən‚äd·əd jē'äm·ə·trē }

coordinated transpositions [ELEC] Transpositions which are installed in either electric supply or communications circuits or in both, for the purpose of reducing inductive coupling, and which are located effectively with respect to the discontinuities in both the electric supply and communications circuits. { kō'órd·ən‚äd·əd tranz·pə'zish·ənz }

coordinate indexing [COMPUT SCI] An indexing scheme in which equal-rank descriptors are used to describe a document, for information retrieval by a computer or other means. { kō'órd·ən·ət 'in‚deks·iŋ }

coordinate plotter [GRAPHICS] An automated drafting device in which a transverse beam and drafting head are driven over a drawing surface by a computer and tape readers to produce highly precise drawings at high speed. Also known as mechanical plotting board; XY coordinate plotter; XY plotter. { kō'órd·ən·ət 'pläd·ər }

coordinates [MAP] **1.** Linear or angular quantities which designate the position that a point occupies in a given reference frame or system. **2.** A general term to designate the particular kind of reference frame or system, such as plane rectangular

coordinates or spherical coordinates. [MATH] A set of numbers which locate a point in space. { kō'órd·ən·əts }

coordinate storage See matrix storage. { kō'órd·ən·ət 'stór·ij }

coordinate systems [MATH] A rule for designating each point in space by a set of numbers. { kō'órd·ən·ət ‚sis·təmz }

coordinate transformation [MATH] A mathematical or graphic process of obtaining a modified set of coordinates by performing some nonsingular operation on the coordinate axes, such as rotating or translating them. { kō'órd·ən·ət tranz·fər'mā·shən }

coordinate valence [CHEM] A chemical bond between two atoms in which a shared pair of electrons forms the bond and the pair has been supplied by one of the two atoms. Also known as coordinate bond; dative bond. { kō'órd·ən·ət 'vā·ləns }

coordinating holes [DES ENG] Holes in two parts of an assembly which form a single continuous hole when the parts are joined. { kō'órd·ən‚äd·iŋ ‚hōlz }

coordination [ELEC] Design of series-connected circuit breakers whereby breakers with lower current ratings trip before those with higher ratings. { kō‚órd·ən'ā·shən }

coordination chemistry [CHEM] The chemistry of metal ions in their interactions with other molecules or ions. { kō‚órd·ən'ā·shən 'kem·ə·strē }

coordination complex See complex compound. { kō‚órd·ən'ā·shən 'käm‚pleks }

coordination compound [CHEM] A compound with a central atom or ion and a group of ions or molecules surrounding it. Also known as coordinated complex; Werner complex. { kō‚órd·ən'ā·shən ‚käm‚paúnd }

coordination lattice [CRYSTAL] The crystal structure of a coordination compound. { kō‚órd·ən'ā·shən ‚lad·əs }

coordination number [PHYS] The number of nearest neighbors of a point in a space lattice, of an atom or an ion in a solid, or of an anion or cation in a solution. { kō‚órd·ən'ā·shən ‚nəm·bər }

coordination polygon [CHEM] The symmetrical polygonal chemical structure of simple polyatomic aggregates having coordination numbers of 4 or less. { kō‚órd·ən'ā·shən ‚päl·i‚gän }

coordination polyhedron [CHEM] The symmetrical polyhedral chemical structure of relatively simple polyatomic aggregates having coordination numbers of 4 to 8. { kō‚órd·ən'ā·shən ‚päl·i·hē·drən }

coordination polymer [ORG CHEM] Organic addition polymer that is neither free-radical nor simply ionic; prepared by catalysts that combine an organometallic (for example, triethyl aluminum) and a transition metal compound (for example, $TiCl_4$). { kō‚órd·ən'ā·shən ‚päl·ə·mər }

coorongite [GEOL] A boghead coal in the peat stage. { kō'ä·rən‚jīt }

copaiba balsam [MATER] An oleoresin extracted from trees of the genus Copaifera of South America; used as a plasticizer and in medicine. { kō'pī·bə 'bòl·səm }

copal [MATER] Hard, resinous substance exuded from certain trees in the East Indies, South America, and Africa and used in varnish and printing ink. { 'kō·pəl }

cope [MET] The upper portion of a flask, mold, or pattern. { kōp }

cope chisel [DES ENG] A chisel used to cut grooves in metal. { 'kōp ‚chiz·əl }

Copenhagen water See normal water. { ¦kō·pən¦häg·ən ‚wòd·ər }

Copeognatha [INV ZOO] An equivalent name for Psocoptera. { ‚kō·pē'äg·nə·thə }

Copepoda [INV ZOO] An order of Crustacea commonly included in the Entomostraca; contains free-living, parasitic, and symbiotic forms. { kō'pep·ə·də }

Copernican principle [ASTRON] The idea that the earth occupies a typical or unexceptional position in the universe. { kə'pər·nə·kən 'prin·sə·pəl }

Copernican system [ASTRON] The system of planetary motions according to Copernicus, who maintained that the earth revolves about an axis once every day and revolves around the sun once every year while the other planets also move in orbits centered near the sun. { kə'pər·nə·kən ‚sis·təm }

copiapite [MINERAL] **1.** $Fe_5(SO_4)_6(OH)_2 \cdot 20H_2O$ A yellow mineral occurring in granular or scalar aggregates. Also

---

COORDINATION POLYHEDRON

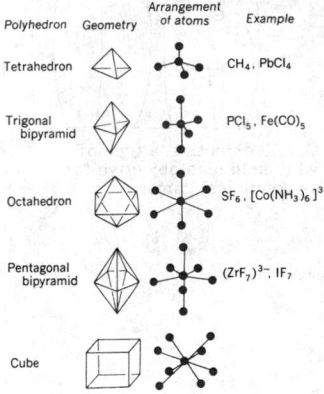

Polyhedron	Geometry	Arrangement of atoms	Example
Tetrahedron			CH_4, $PbCl_4$
Trigonal bipyramid			PCl_5, $Fe(CO)_5$
Octahedron			SF_6, $[Co(NH_3)_6]^{3+}$
Pentagonal bipyramid			$(ZrF_7)^{3-}$, IF_7
Cube			

Types of coordination polyhedrons.

COPEPODA

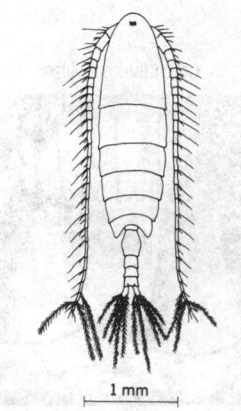

1 mm

Calanus finmarchicus, a calcanoid free-living copepod.

known as ihleite; knoxvillite; yellow copperas. **2.** A group of minerals containing hydrous iron sulfates. { 'kō·pə·ə‚pīt }

coping [BUILD] A covering course on a wall. [MECH ENG] Shaping stone or other nonmetallic substance with a grinding wheel. [MIN ENG] **1.** Process of cutting and trimming the edges of stone slabs. **2.** Process of cutting a stone slab into two pieces. { 'kōp·iŋ }

coping brick [MATER] A brick with a special shape that is used to cap the exposed top of a wall. { 'kōp·iŋ ‚brik }

coping saw [DES ENG] A type of handsaw that has a narrow blade, usually about $1/8$ inch (3 millimeters) wide, held taut by a U-shaped frame equipped with a handle; used for shaping and cutout work. { 'kōp·iŋ ‚sò }

coplanar electrodes [ELECTR] Electrodes mounted in the same plane. { kō'plān·ər i'lek‚trōdz }

coplanar forces [MECH] Forces that act in a single plane; thus the forces are parallel to the plane and their points of application are in the plane. { kō'plān·ər ‚fòrs·əz }

Copodontidae [PALEON] An obscure family of Paleozoic fishes in the order Bradyodonti. { ‚kō·pə'dän·tə‚dē }

copolymer [ORG CHEM] A mixed polymer, the product of polymerization of two or more substances at the same time. { kō'päl·i·mər }

copolymerization [CHEM] A polymerization reaction that forms a copolymer. { ‚kō·pə‚lim·ə·rə'zā·shən }

copper [CHEM] A chemical element, symbol Cu, atomic number 29, atomic weight 63.546. [MET] One of the most important nonferrous metals; a ductile and malleable metal found in various ores and used in industry, engineering, and the arts in both pure and alloyed forms. { 'käp·ər }

copper-64 [NUC PHYS] Radioactive isotope of copper with mass number of 64; derived from pile-irradiation of metallic copper; used as a research aid to study diffusion, corrosion, and friction wear in metals and alloys. { 'käp·ər ‚sik·stē'fòr }

copper acetate See cupric acetate. { 'käp·ər 'as·ə‚tāt }

copper alloy [MET] A solid solution of one or more metals in copper. { 'käp·ər 'al‚òi }

copper amalgam [MET] An alloy of copper and mercury. { 'käp·ər ə'mal·gəm }

copper arsenate [INORG CHEM] $Cu_3(AsO_4)_2 \cdot 4H_2O$ or $Cu_5H_2(AsO_4)_4 \cdot 2H_2O$ Bluish powder, soluble in ammonium hydroxide and dilute acids, insoluble in water and alcohol; used as a fungicide and insecticide. { 'käp·ər 'ärs·ən‚āt }

copper arsenite [INORG CHEM] $CuHAsO_3$ A toxic, light green powder which is soluble in acids and decomposes at the melting point; used as a pigment and insecticide. Also known as copper orthoarsenite; cupric arsenite; Scheele's green. { 'käp·ər 'ärs·ən‚īt }

copperas See ferrous sulfate. { 'käp·ə·rəs }

copper blight [PL PATH] A leaf spot disease of tea caused by the fungus *Guignardia camelliae*. { 'käp·ər ‚blīt }

copper blue See mountain blue. { 'käp·ər ‚blü }

copper brazing [MET] Brazing by using copper as the filler metal. { 'käp·ər ‚brāz·iŋ }

copper bromide See cupric bromide; cuprous bromide. { 'käp·ər 'brō‚mīd }

copper cable [ELEC] A mechanically assembled group of copper wires, used in place of a single, large wire for increased flexibility. { 'käp·ər 'kā·bəl }

copper carbonate [INORG CHEM] $Cu_2(OH)_2CO_3$ A toxic, green powder; decomposes at 200°C and is soluble in acids; used in pigments and pyrotechnics and as a fungicide and feed additive. Also known as artificial malachite; cupric carbonate; mineral green. { 'käp·ər 'kär·bə‚nāt }

copper chloride See cupric chloride; cuprous chloride. { 'käp·ər 'klòr‚īd }

copper chromate See cupric chromate. { 'käp·ər 'krō‚māt }

copper converter [MET] A converter for purifying copper. { 'käp·ər kən'vərd·ər }

copper cyanide See cupric cyanide. { 'käp·ər sī·ə‚nīd }

copper dirt See sour dirt. { 'käp·ər 'dərt }

copper dish gum [CHEM ENG] The milligrams of gum found in 100 milliliters of gasoline when evaporated under controlled conditions in a polished copper dish. { 'käp·ər ‚dish 'gəm }

copper fluoride See cupric fluoride; cuprous fluoride. { 'käp·ər 'flür‚īd }

copper glance See chalcocite. { 'käp·ər 'glans }

copper gluconate [ORG CHEM] $[CH_2OH(CHOH)_4COO]_2Cu$ A light blue, crystalline powder; soluble in water; used in medicine and as a dietary supplement. Also known as cupric gluconate. { 'käp·ər 'glü·kə‚nāt }

copperhead [VERT ZOO] *Agkistrodon contortrix*. A pit viper of the eastern United States; grows to about 3 feet (90 centimeters) in length and is distinguished by its coppery-brown skin with dark transverse blotches. { 'käp·ər‚hed }

copper hydroxide See cupric hydroxide. { 'käp·ər hī'dräk‚sīd }

coppering [ORD] Metal fouling accumulated in the bore of a weapon because of repeated firing; deposited from the rotating bands or jackets of the projectiles. { 'käp·ər·iŋ }

copperite [MINERAL] An important platinum mineral, composed of platinum sulfide. { 'käp·ə‚rīt }

copper loss [ELEC] Power loss in a winding due to current flow through the resistance of the copper conductors. Also known as *IR* loss. { 'käp·ər ‚lós }

copper mica See chalcophyllite. { 'käp·ər 'mī·kə }

copper nickel See niccolite. { 'käp·ər 'nik·əl }

copper nitrite See cupric nitrate. { 'käp·ər 'nī‚trīt }

copper number [ANALY CHEM] The number of milligrams of copper obtained by the reduction of Benedict's or Fehling's solution by 1 gram of carbohydrate. { 'käp·ər ‚nəm·bər }

copper oleate [ORG CHEM] $Cu[OOC(CH_2)_7CH=CH(CH_2)_7CH_3]_2$ A green-blue liquid, used as a fungicide for fruits and vegetables. { 'käp·ər 'ō·lē‚āt }

copperon See cupferron. { 'käp·ə‚rän }

copper ore [GEOL] Rock containing copper minerals. { 'käp·ər ‚òr }

copper orthoarsenite See copper arsenite. { 'käp·ər ‚òr·thō'ärs·ən‚īt }

copper oxide See cupric oxide; cuprous oxide. { 'käp·ər 'äk‚sīd }

copper oxide photovoltaic cell [ELECTR] A photovoltaic cell in which light acting on the surface of contact between layers of copper and cuprous oxide causes a voltage to be produced. { 'käp·ər 'äk‚sīd ‚fōd·ō·vōl'tā·ik 'sel }

copper oxide rectifier [ELECTR] A metallic rectifier in which the rectifying barrier is the junction between metallic copper and cuprous oxide. { 'käp·ər 'äk‚sīd 'rek·tə‚fī·ər }

copper pair See twisted pair. { 'käp·ər ‚per }

copperplate engraving [GRAPHICS] A thin, rigid plate of copper, with the lines of a picture cut, or engraved, into it, used for printing purposes; it is inked over, the ink is removed so that it is retained only in the engraved lines, and the plate is placed in a handpress where ink is transferred from engraved lines to overlying paper. { 'käp·ər‚plāt in'grāv·iŋ }

copper plating [MET] Coating of a substance with copper by an electrolytic process, to minimize corrosion. { 'käp·ər ‚plād·iŋ }

copper powder [MET] A bronzing powder made by saturating nitrous acid with copper and precipitating the latter by the addition of iron. { 'käp·ər ‚paùd·ər }

copper pyrite See chalcopyrite. { 'käp·ər 'pī‚rīt }

copper-8-quinolinolate [ORG CHEM] $C_{18}H_{14}N_2O_2Cu$ A khaki-colored, water-insoluble solid used as a fungicide in fruit-handling equipment. { 'käp·ər ‚āt ‚kwin·ə‚lin·ə‚lāt }

copper resinate [ORG CHEM] Poisonous green powder, soluble in oils and ether, insoluble in water; made by heating rosin oil with copper sulfate, followed by filtering and drying of the resultant solids; used as a metal-paint preservative and insecticide. { 'käp·ər 'rez·ən‚āt }

copper spot [PL PATH] A fungus disease of lawn grasses caused by *Gloeocercospora sorghi* and marked by coppery-red areas. { 'käp·ər ‚spät }

copper steel [MET] Low-carbon steel containing up to 0.25% copper. { 'käp·ər 'stēl }

copper-strip corrosion [ENG] A qualitative method of determining the corrosivity of a petroleum product by observing its effect on a strip of polished copper suspended or placed in the product. Also known as copper strip test. { 'käp·ər ‚strip ki'rō·zhən }

copper-strip test See copper-strip corrosion. { 'käp·ər 'strip ‚test }

copper sulfate See cupric sulfate. { 'käp·ər 'səl‚fāt }

copper sulfide [INORG CHEM] CuS Black, monoclinic or hexagonal crystals that break down at 220°C; used in paints on ship bottoms to prevent fouling. { 'käp·ər 'səl‚fīd }

copper sulfide rectifier [ELECTR] A semiconductor rectifier in which the rectifying barrier is the junction between magnesium and copper sulfide. { 'käp·ər 'səl‚fīd 'rek·tə‚fī·ər }

copper sweetening [CHEM ENG] Those refining processes using cupric chloride to oxidize mercaptans in petroleum. { 'käp·ər ‚swēt·ən·iŋ }

copper uranite *See* torbernite. { 'käp·ər 'yùr·ə‚nīt }

copper vapor laser [OPTICS] A high-power laser that emits intense pulses of very short duration (typically 30 nanoseconds) at a rate of 5000–50,000 pulses per second, at wavelengths of 510.5 nanometers (green) and 578.2 nanometers (yellow). { 'käp·ər ‚vā·pər ‚lās·ər }

copperweld [MET] Copper-covered steel, used as a conductor for high-voltage transmission spans where tensile strength is more important than high conductance. { 'käp·ər‚weld }

copper wire [MET] Wire commonly made from copper by drawing from a hot-rolled rod without annealing; however, the smaller sizes may involve intermediate anneals. { 'käp·ər 'wīr }

coppice [ECOL] A growth of small trees that are repeatedly cut down at short intervals; the new shoots are produced by the old stumps. { 'käp·əs }

coprecipitation [CHEM] Simultaneous precipitation of more than one substance. { ‚kō·prə‚sip·ə'tā·shən }

coproantibody [IMMUNOL] An antibody whose presence in the intestinal tract can be demonstrated by its presence in an extract of the feces. { ‚käp·rō'ant·i‚bäd·ē }

coprocessor [COMPUT SCI] A processing unit that works together with a primary central processing unit to speed a computer's execution of time-consuming operations. { kō'prä‚ses·ər }

coprolalia [MED] Uncontrollable barking or grunting of profane language that is commonly associated with Tourette's syndrome. { ‚käp·rə'lāl·yə }

coprolite [GEOL] Petrified excrement. { 'käp·rə‚līt }

coprophagy [ZOO] Feeding on dung or excrement. { kə'präf·ə·jē }

coprophilous [ECOL] Living in dung. { kə'präf·ə·ləs }

coprophobia [PSYCH] An abnormal fear of fecal matter. { ‚käp·rə'fō·bē·ə }

copulation [ZOO] The sexual union of two individuals, resulting in insemination or deposition of the male gametes in proximity to the female gametes. { ‚käp·yə'lā·shən }

copulatory bursa [INV ZOO] **1.** A sac that receives the sperm during copulation in certain insects. **2.** The caudal expansion of certain male nematodes that functions as a clasper during copulation. { 'käp·yə·lə‚tór·ē 'bər·sə }

copulatory organ [ANAT] An organ employed by certain male animals for insemination. { 'käp·yə·lə‚tór·ē ‚ór·gən }

copulatory spicule *See* spiculum. { 'käp·yə·lə‚tór·ē 'spik·yəl }

copy [COMMUN] **1.** To transcribe Morse code signals into written form. **2.** To reproduce graphical material usually by an electrostatic device. **3.** To reproduce information in a new location and possibly in a different form, leaving the source of the information unchanged. [COMPUT SCI] A string procedure in ALGOL by means of which a new byte string can be generated from an existing byte string. [GRAPHICS] *See* subject copy. { 'käp·ē }

copyboard [GRAPHICS] A frame that holds the original copy in place while it is being photographed. { 'käp·ē‚bórd }

copy deoxyribonucleic acid *See* complementary deoxyribonucleic acid. { ‚käp·ē de‚äk·sē‚rī·bō·nü‚klē·ik 'as·əd }

copy error [MOL BIO] A mutation that occurs during deoxyribonucleic acid replication as a result of an error in base pairing. { 'käp·ē ‚er·ər }

copying program [COMPUT SCI] A system program which copies a data or program file from one peripheral device onto another. { 'käp·ē·iŋ ‚prō·grəm }

copy protection *See* software protection. { 'käp·ē prə‚tek·shən }

coquimbite [MINERAL] Fe$_2$(SO$_4$)$_3$·9H$_2$O A white mineral that crystallizes in the hexagonal system; it is dimorphous with paracoquimbite. { kō'kim‚bīt }

coquina [INV ZOO] A small marine clam of the genus *Donax*. [PETR] A coarse-grained, porous, easily crumbled variety of limestone composed principally of mollusk shell and coral fragments cemented together as rock. { kō'kē·nə }

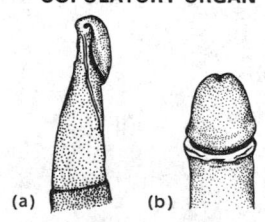

COPULATORY ORGAN

(a)   (b)

Glans penis of (a) bull, (b) man.

coquinoid [PETR] **1.** Of or pertaining to coquina. **2.** Lithified coquina. **3.** An autochthonous deposit of limestone made up of more or less whole mollusk shells. { 'kō·kə‚nóid }

Coraciidae [VERT ZOO] The rollers, a family of Old World birds in the order Coraciiformes. { ‚kór·ə'sī·ə‚dē }

Coraciiformes [VERT ZOO] An order of predominantly tropical and frequently brightly colored birds. { ‚kór·ə‚sī·ə'fór‚mēz }

coracite *See* uraninite. { 'kór·ə‚sīt }

coracoid [ANAT] One of the paired bones on the posterior-ventral aspect of the pectoral girdle in vertebrates. { 'kór·ə‚kóid }

coracoid ligament [ANAT] The transverse ligament of the scapula which crosses over the suprascapular notch. { 'kór·ə‚kóid 'lig·ə·mənt }

coracoid process [ANAT] The beak-shaped process of the scapula. { 'kór·ə‚kóid ‚präs·əs }

coral [INV ZOO] The skeleton of certain solitary and colonial anthozoan cnidarians; composed chiefly of calcium carbonate. { 'kä·rəl }

coral head [GEOL] A small reef patch of coralline material. Also known as coral knoll. { 'kä·rəl ‚hed }

coral knoll *See* coral head. { 'kä·rəl ‚nōl }

Corallanidae [INV ZOO] A family of sometimes parasitic, but often free-living, isopod crustaceans in the suborder Flabellifera. { ‚kä·rə'lan·ə‚dē }

Corallidae [INV ZOO] A family of dimorphic cnidarians in the order Gorgonacea. { kə'ral·ə‚dē }

Corallimorpharia [INV ZOO] An order of solitary sea anemones in the subclass Zoantharia resembling coral in many aspects. { kə‚ral·ə‚mór'far·ē·ə }

Corallinaceae [BOT] A family of red algae, division Rhodophyta, having compact tissue with lime deposits within and between the cell walls. { kə‚ral·ə'nās·ē‚ē }

coralline [INV ZOO] Any animal that resembles coral, such as a bryozoan or hydroid. { 'kär·ə‚lēn }

coralline algae [BOT] Red algae belonging to the family Corallinaceae. { 'kär·ə‚lēn 'al·jē }

corallite [INV ZOO] Skeleton of an individual polyp in a compound coral. { 'kär·ə‚līt }

coralloid [BIOL] Resembling coral, or branching like certain coral. { 'kär·ə‚lóid }

corallum [INV ZOO] Skeleton of a compound coral. { kə'ral·əm }

coral mud [GEOL] Fine-grade deposits of coral fragments formed around coral islands and coasts bordered by coral reefs. { 'kär·əl ‚məd }

coral pinnacle [GEOL] A sharply upward-projecting growth of coral rising from the floor of an atoll lagoon. { 'kär·əl 'pin·ə·kəl }

coral reef [GEOL] A ridge or mass of limestone built up of detrital material deposited around a framework of skeletal remains of mollusks, colonial coral, and massive calcareous algae. { 'kär·əl ‚rēf }

coral-reef lagoon [GEOGR] The central, shallow body of water of an atoll or the water separating a barrier reef from the shore. { 'kär·əl ‚rēf lə'gün }

coral-reef shoreline [GEOL] A shoreline formed by reefs composed of coral polyps. Also known as coral shoreline. { 'kär·əl ‚rēf 'shór‚līn }

coral rock *See* reef limestone. { 'kär·əl ‚räk }

coral sand [GEOL] Coarse-grade deposits of coral fragments formed around coral islands and coasts bordered by coral reefs. { 'kär·əl ‚sand }

coral shoreline *See* coral-reef shoreline. { 'kär·əl 'shór‚līn }

corange line [MAP] A line on a chart joining points of equal tide range. { 'kō'ränj ‚līn }

CORBA *See* common object request broker. { 'kór·bə }

corbel [ARCH] An architectural bracket projecting from within a wall, formed by extensions of the masonry and wood beyond the wall surface, and supporting a burdensome weight above. { 'kór·bəl }

corbel arch [ARCH] A structure resembling an arch, in which successive courses project farther into a gap until they close. { 'kór·bəl ‚ärch }

Corbiculidae [INV ZOO] A family of fresh-water bivalve mollusks in the subclass Eulamellibranchia; an important food in the Orient. { kór·bə'kyül·ə‚dē }

Corbino disk [ELECTROMAG] A variable-resistance device

utilizing the effect of a magnetic field on the flow of carriers from the center to the circumference of a disk made of semiconducting or conducting material. { kȯr'bē·nō ˌdisk }

Corbino effect [ELECTROMAG] The production of an electric current around the circumference of a disk when a magnetic field perpendicular to the disk acts on a radial current in the disk. { kȯr'bēn·ō iˌfekt }

corbinotron [ENG] The combination of a corbino disk, made of high-mobility semiconductor material, and a coil arranged to produce a magnetic field perpendicular to the disk. { kȯr'bē·nəˌträn }

cord [ELEC] A small, very flexible insulated cable. [MATER] **1.** A unit of measure for wood stacked for fuel or pulp; equals 4 × 4 × 8, or 128 cubic feet (approximately 3.6246 cubic meters). **2.** A long, flexible, cylindrical construction of natural or synthetic fibers twisted or woven together. **3.** Strands of material forming the plies in a motor vehicle tire. { kȯrd }

cordage [ENG] Number of cords of lumber per given area. [MATER] Ropes or cords, especially those in the rigging of a ship. { 'kȯrd·ij }

Cordaitaceae [PALEOBOT] A family of fossil plants belonging to the Cordaitales. { ˌkȯr·dāˌī'tās·ēˌē }

Cordaitales [PALEOBOT] An extensive natural grouping of forest trees of the late Paleozoic. { ˌkȯr·dāˌī'tā·lēz }

cordate [BOT] Heart-shaped; generally refers to a leaf base. { 'kȯrˌdāt }

cord circuit [ELEC] Connecting circuit terminating in a plug at one or both ends and used at switchboard positions in establishing telephone connections. { 'kȯrd ˌsər·kət }

Cordelia [ASTRON] A satellite of Uranus orbiting at a mean distance of 30,910 miles (49,750 kilometers) with a period of 8 hours 4 minutes, and with a diameter of about 16 miles (26 kilometers); the inner shepherding satellite for the outermost ring of Uranus. { kȯr'dēl·yə }

cord factor [MICROBIO] A toxic glycolipid found as a surface component of tubercle bacilli that is responsible for virulence and serpentine growth. { 'kȯrd ˌfak·tər }

cord foot [ENG] A stack of wood measuring 16 cubic feet (approximately 0.45307 cubic meter). { 'kȯrd ˌfu̇t }

cordierite [MINERAL] $Mg_2(Al_5Si_5O_{18})$ A blue, orthorhombic magnesium aluminosilicate mineral frequently occurring associated with thermally metamorphosed rocks derived from argillaceous sediments. { 'kȯrd·ē·əˌrīt }

cordillera [GEOGR] A mountain range or group of ranges, including valleys, plains, rivers, lakes, and so on, forming the main mountain axis of a continent. { ˌkȯrd·əl'er·ə }

cordilleran geosyncline [GEOL] The Devonian geosynclinal region of western North America. { ˌkȯrd·əl'er·ən jē·ō'sinˌklīn }

cordite [MATER] A trinitrate cellulose derivative prepared by treating cotton fiber or purified wood pulp with a mixture of nitric and sulfuric acids; an explosive powder. { 'kȯrˌdīt }

cordless switchboard [COMMUN] Manual telephone switchboard which uses manually operated keys to make connections. { 'kȯrd·ləs 'swichˌbȯrd }

cordless telephone [COMMUN] A telephone whose receiver and base are equipped with small antennas and are linked by low-power radio. { 'kȯrd·ləs 'tel·əˌfōn }

cord of ore [MIN ENG] A unit of about 7 tons (6.35 metric tons), but measured by wagonloads and not by weight. { 'kȯrd əv 'ȯr }

cordon [BOT] A plant trained to grow flat against a vertical structure, in a single horizontal shoot or two opposed horizontal shoots. { 'kȯrd·ən }

cordonazo [METEOROL] A southerly wind of hurricane force generated along the western coast of Mexico when a tropical cyclone passes offshore in a northerly direction. { ˌkȯrd·ən'ä·sō }

Cordonnier system See Batten system. { 'kȯrd·ən·yā ˌsis·təm }

cordovan [MATER] Nonporous leather made from split horsehide and tanned with vegetable materials. { 'kȯrd·ə·vən }

cord tire [DES ENG] A pneumatic tire made with cords running parallel to the tread. { 'kȯrd ˌtīr }

Cordulegasteridae [INV ZOO] A family of anisopteran insects in the order Odonata. { ˌkȯrd·yə·lə,ga'ster·əˌdē }

corduroy [TEXT] Cotton, rayon, or other fabric with a cut-pile surface of wales; may be either plain or twill weave. { 'kȯrd·əˌrȯi }

cordwood [MATER] Wood stacked and sold in cords. { 'kȯrdˌwu̇d }

cordwood module [ELECTR] High-density circuit module in which discrete components are mounted between and perpendicular to two small, parallel printed circuit boards to which their terminals are attached. { 'kȯrdˌwu̇d ˌmä·jül }

Cordyceps ophioglossoides [MYCOL] A mushroom that is a parasite on the fruiting bodies of the truffle found in the soil of bamboo, oak, and pine woods that has antitumor properties and is an immune booster. Also known as clubhead fungus. { ˌkȯrd·əˌseps ˌō·fē·ə·glə'sȯiˌdēz }

Cordyceps sinensis [MYCOL] A type of mushroom found on the cold mountain tops and snowy grass marshlands of China that infects insect larvae with spores that germinate before the cocoons are formed; it has been successfully used in clinical trials to treat liver diseases, high cholesterol, and loss of sexual drive. Also known as caterpillar fungus. { ˌkȯrd·əˌseps sī'nen·sis }

cordylite [MINERAL] $(Ce,La)_2Ba(CO_3)_3F_2$ A colorless to wax-yellow mineral consisting of a carbonate and fluoride of cerium, lanthanum, and barium. { 'kȯrd·əlˌīt }

core [ANAT] A fingerprint focal point which is the point on a ridge that is located in the approximate center of the finger impression. [ARCHEO] A piece of stone from which flakes or blades were removed by prehistoric toolmakers; usually it was the by-product of toolmaking but may also have served as an implement. [ATOM PHYS] The electrons in the filled shells of an atom. [ELECTR] See magnetic core. [ELECTROMAG] See magnetic core. [ENG] The inner material of a wall, column, veneered door, or similar structure. [GEOL] **1.** Center of the earth, beginning at a depth of 2900 kilometers. Also known as earth core. **2.** A vertical, cylindrical boring of the earth from which composition and stratification may be determined; in oil or gas well exploration the presence of hydrocarbons or water are items of interest. [GRAPHICS] An unflanged cylindrical reel on which film is wound. [MATER] The center layers of a sheet of plywood. [MET] A specially formed part of a mold used to form internal holes in a casting. [NUC PHYS] The nucleons in the filled shells of a nucleus. [NUCLEO] The active portion of a nuclear reactor, containing the fissionable material. [OCEANOGR] That area within a layer of ocean water where parameters such as temperature, salinity, or velocity reach extreme values. [SCI TECH] The central part of a body or structure. { kȯr }

core analysis [GEOL] The use of core samples taken from the borehole during drilling to give information on strata age, composition, and porosity, and the presence of hydrocarbons or water along the length of the borehole. { 'kȯr ə'nal·ə·səs }

core array [ELECTR] A rectangular grid arrangement of magnetic cores. { 'kȯr ə'rā }

core bank [ELECTR] A stack of core arrays and associated electronics, the stack containing a specific number of core arrays. { 'kȯr ˌbaŋk }

core barrel [DES ENG] A hollow cylinder attached to a specially designed bit; used to obtain a continuous section of the rocks penetrated in drilling. { 'kȯr ˌbar·əl }

core-barrel rod See guide rod. { 'kȯr ˌbar·əl ˌräd }

core binder [MATER] A substance for binding core sand together; an example is core oil. { 'kȯr ˌbīnd·ər }

core bit [DES ENG] The hollow, cylindrical cutting part of a core drill. { 'kȯr ˌbit }

core blower [MET] A machine using compressed air to blow and pack sand into a core box. { 'kȯr ˌblō·ər }

core box [MET] A container for shaping a sand core for a casting. { 'kȯr ˌbäks }

core catcher See split-ring core lifter. { 'kȯrˌkach·ər }

core-catcher case See lifter case. { 'kȯrˌkach·ər ˌkās }

core cutterhead [ENG] The cutting element in a core barrel unit. { 'kȯr ˌkəd·ərˌhed }

cored ammonium nitrate dynamite [MATER] A class of dynamite that has good water resistance and exhibits velocities higher than straight ammonia dynamite; the gelatin core provides for detonation in the complete explosive column. { 'kȯrd ə'mon·ē·əm 'nīˌtrāt 'dī·nəˌmīt }

cored bar [MET] A bar-shaped powder compact which has been heated by electricity to melt the interior. { 'kȯrd ˌbär }

cored electrode [MET] An electrode made of metal with a core of flux or other material. { ¦kȯrd i'lek,trōd }

core drier [MET] A light, skeleton cast-iron or aluminum box, whose internal shape conforms closely to the cope portion of a core for molding, used to support, during baking, a core which cannot be placed on a flat plate. { 'kȯr ,drī·ər }

core drill [MECH ENG] A mechanism designed to rotate and to cause an annular-shaped rock-cutting bit to penetrate rock formations, produce cylindrical cores of the formations penetrated, and lift such cores to the surface, where they may be collected and examined. { 'kȯr ,dril }

cored solder [MET] Soldering wire which has a core consisting of flux. { ¦kȯrd 'säd·ər }

core-dump [COMPUT SCI] To copy the contents of all or part of core storage, usually into an external storage device. { 'kȯr ,dəmp }

core electron [ATOM PHYS] An electron in a filled shell of an atom. { 'kȯr i,lek,trän }

core flow [ENG] A pattern of powder flow occurring in hoppers that is characterized by a central core of flowing powder with the powder near the hopper walls remaining stationary. { 'kȯr ,flō }

core-form transformer [ELECTROMAG] A transformer in which half of the turns of the primary winding and half of those of the secondary are on each of two legs. { ¦kȯr ,fȯrm tranz'fȯr·mər }

core gripper See split-ring core lifter. { 'kȯr ,grip·ər }

core-gripper case See lifter case. { 'kȯr ,grip·ər ,kās }

core-halo galaxy [ASTRON] A radio galaxy characterized by a relatively large region of diffuse radio emission surrounding a central region of more intense emission. { 'kȯr ¦hā·lō ,gal·ik·sē }

core hitch [ELEC] Attachment to a cable core to permit pulling it into a duct without damaging the sheath. { 'kȯr ,hich }

Coreidae [INV ZOO] The squash bugs and leaf-footed bugs, a family of hemipteran insects belonging to the superfamily Coreoidea. { kə'rē·ə,dē }

core image [COMPUT SCI] **1.** A computer program whose storage addresses have been assigned so that it can be loaded directly into main storage for processing. **2.** A visual representation of a computer's main storage. { 'kȯr ,im·ij }

core-image library [COMPUT SCI] A collection of computer programs residing on mass-storage device in ready-to-run form. { 'kȯr ,im·ij ,lī,brer·ē }

core intersection [GEOL] **1.** The point in a borehole where an ore vein or body is encountered as shown by the core. **2.** The width or thickness of the ore body, as shown by the core. Also known as core interval. { 'kȯr ¦in·tər,sek·shən }

core interval See core intersection. { 'kȯr ¦in·tər·vəl }

core iron [MET] A grade of soft iron suitable for cores of chokes, transformers, and relays. { 'kȯr ,ī·ərn }

coreless-type induction heater [ENG] A device in which a charge is heated directly by induction, with no magnetic core material linking the charge. Also known as coreless-type induction furnace. { 'kȯr·ləs ,tīp in'dək·shən ,hēd·ər }

core lifter See split-ring core lifter. { 'kȯr ,lif·tər }

core-lifter case See lifter case. { 'kȯr ,lif·tər ,kās }

core logging [GEOL] The analysis of the strata through which a borehole passes by the taking of core samples at predetermined depth intervals as the well is drilled. { 'kȯr ,läg·iŋ }

core logic [ELECTR] Logic performed in ferrite cores that serve as inputs to diode and transistor circuits. { 'kȯr ,läj·ik }

core loss [ELECTROMAG] The rate of energy conversion into heat in a magnetic material due to the presence of an alternating or pulsating magnetic field. Also known as excitation loss; iron loss. { 'kȯr ,lȯs }

core memory See magnetic core storage. { 'kȯr ,mem·rē }

core memory resident [COMPUT SCI] A control program which is in the main memory of a computer at all times to supervise the processing of the computer. { 'kȯr ,mem·rē ,rez·ə·dənt }

coremium [MYCOL] A small bundle of conidiophores in certain imperfect fungi. { kə'rē·mē·əm }

core molding [MET] A molding process which makes use of assembled cores to construct the mold. { 'kȯr ,mōld·iŋ }

core oil [MATER] An oil compound used with sand to make foundry core. { 'kȯr ,ȯil }

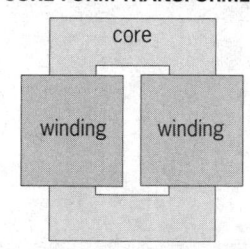

CORE-FORM TRANSFORMER

Arrangement of windings.

core oven [MET] An oven used for baking cores for molding; the walls are constructed of inner and outer layers of sheet metal separated by rock wool or fiber glass insulation, with interlocked joints. { 'kȯr ,əv·ən }

corepressor [MOL BIO] A certain metabolite which, through combination with a repressor apoprotein produced by a regulator gene, can cause the binding of the protein to the operator gene region of a deoxyribonucleic acid chain. { ¦kō·ri,pres·ər }

core print [MET] A projection on a cylindrical casting pattern which supports a core. { 'kȯr ,print }

corer [ENG] An instrument used to obtain cylindrical samples of geological materials or ocean sediments. { 'kȯr·ər }

core rod [MET] The part of a die used to make a hole in a compact. { 'kȯr ,räd }

core rope storage [COMPUT SCI] Direct-access storage consisting of a large number of doughnut-shaped ferrite cores arranged on a common axis, with sense, inhibit, and set wires threaded through or around individual cores in a predetermined manner to provide fixed storage of digital data; each core rope stores one or more complete words, rather than just a single bit. { 'kȯr ,rōp ,stȯr·ij }

core sample [GEOL] A sample of rock, soil, snow, or ice obtained by driving a hollow tube into the undisturbed medium and withdrawing it with its contained sample or core. { 'kȯr ,sam·pəl }

core sand [MATER] Sand used in a core for molding, made from standard molding-sand mixtures or from silica sand, usually with a binder. Also known as foundry core sand. { 'kȯr ,sand }

core set [GRAPHICS] The distortion caused on a photographic film by winding on a spool or core. { 'kȯr ,set }

coresident [COMPUT SCI] A computer program or program module that is stored in a computer memory along with other programs. { kō'rez·ə·dənt }

core spinning [TEXT] A process in which a sheath of nonstretch fibers is applied to fine-denier elastomeric spandex yarns under tension. { 'kȯr ,spin·iŋ }

core-spring case See lifter case. { 'kȯr ,spriŋ ,kās }

core-spun yarn [TEXT] Yarn made by wrapping fibers around an inner fiber. { 'kȯr ,spən ,yärn }

core stack [ELECTR] A number of core arrays, next to one another and treated as a unit. { 'kȯr ,stak }

core state [PHYS] An energy state corresponding to an energy level in a filled shell of an atom or nucleus.

corestone [GEOL] A rounded or broadly rectangular joint block of granite formed as a result of subsurface weathering in a manner similar to a tor but entirely separated from the bedrock. { 'kȯr,stōn }

core storage [COMPUT SCI] **1.** The main memory of a computer. **2.** See magnetic core storage. { 'kȯr ,stȯr·ij }

core test [TEXT] A test for grading wool in which a core sample is withdrawn mechanically from a bag or bale and tested for moisture, ash, vegetable matter, and grease content. { 'kȯr ,test }

core-type induction heater [ELECTROMAG] A device in which a charge is heated by induction, with a magnetic core being used to link the induction coil to the charge. { 'kȯr ,tīp in'dək·shən ,hēd·ər }

core wall See cutoff wall. { 'kȯr ,wȯl }

core wash [MATER] A suspension of fine clay or graphite that is applied to a core in a metal casting to improve that portion's cast surface. { 'kȯr ,wäsh }

core wire [MET] Copper wire having a steel core, often used for antennas. { 'kȯr ,wīr }

core yarn [TEXT] The internal member of a yarn produced by core spinning. { 'kȯr ,yärn }

coriaceous [BIOL] Leathery, applied to leaves and certain insects. { ¦kȯr·ē¦ā·shəs }

coriander [BOT] *Coriandrum sativum.* A strong-scented perennial herb in the order Umbellales; the dried fruit is used as a flavoring. { ¦kȯr·ē'an·dər }

coriander oil [MATER] An essential oil distilled from the fruit of coriander; principal constituents are linalool and pinene; used as a flavoring in gin. { ¦kȯr·ē'an·dər ,ȯil }

coriandrol See linalool. { ¦kȯr·ē'an,drȯl }

Cori ester See glucose-1-phosphate. { 'kȯr·ē ,es·tər }

coring [MET] A variable composition of individual crystals across a casting, due to nonequilibrium growth over a range of temperature; the purest material is near the center. [PETRO ENG] The use of a core barrel (hollow length of tubing) to take samples from the underground formation during the drilling operation; used for core analysis. { 'kȯr·iŋ }

coring reel *See* sand reel. { 'kȯr·iŋ ‚rēl }

Coriolis acceleration [MECH] **1.** An acceleration which, when added to the acceleration of an object relative to a rotating coordinate system and to its centripetal acceleration, gives the acceleration of the object relative to a fixed coordinate system. **2.** A vector which is equal in magnitude and opposite in direction to that of the first definition. { kȯr·ē'ō·ləs ik‚sel·ə'rā·shən }

Coriolis correction [NAV] A correction applied to an assumed position, celestial line of position, or celestial fix or to a computed or observed altitude to allow for apparent acceleration due to the Coriolis force. { kȯr·ē'ō·ləs kə'rek·shən }

Coriolis deflection *See* Coriolis effect. { kȯr·ē'ō·ləs di'flek·shən }

Coriolis effect [MECH] **1.** Also known as Coriolis deflection. **2.** The deflection relative to the earth's surface of any object moving above the earth, caused by the Coriolis force; an object moving horizontally is deflected to the right in the Northern Hemisphere, to the left in the Southern. **3.** The effect of the Coriolis force in any rotating system. [PHYSIO] The physiological effects (nausea, vertigo, dizziness, and so on) felt by a person moving radially in a rotating system, as a rotating space station. { kȯr·ē'ō·ləs i'fekt }

Coriolis force [MECH] A velocity-dependent pseudoforce in a reference frame which is rotating with respect to an inertial reference frame; it is equal and opposite to the product of the mass of the particle on which the force acts and its Coriolis acceleration. { kȯr·ē'ō·ləs ‚fȯrs }

Coriolis operator [SPECT] An operator which gives a large contribution to the energy of an axially symmetric molecule arising from the interaction between vibration and rotation when two vibrations have equal or nearly equal frequencies. { kȯr·ē'ō·ləs ‚äp·ə‚rād·ər }

Coriolis parameter [GEOPHYS] Twice the component of the earth's angular velocity about the local vertical $2\Omega \sin \phi$, where Ω is the angular speed of the earth and ϕ is the latitude; the magnitude of the Coriolis force per unit mass on a horizontally moving fluid parcel is equal to the product of the Coriolis parameter and the speed of the parcel. { kȯr·ē'ō·ləs pə'ram·əd·ər }

Coriolis resonance interactions [SPECT] Perturbation of two vibrations of a polyatomic molecule, having nearly equal frequencies, on each other, due to the energy contribution of the Coriolis operator. { kȯr·ē'ō·ləs 'rez·ən·əns ‚in·tər‚ak·shənz }

Coriolis-type mass flowmeter [ENG] An instrument which determines mass flow rate from the torque on a ribbed disk that is rotated at constant speed when fluid is made to enter at the center of the disk and is accelerated radially. { kȯr·ē'ō·ləs ‚tīp ‚mas 'flō‚med·ər }

corium [ANAT] *See* dermis. [INV ZOO] Middle portion of the forewing of hemipteran insects. { 'kȯr·ē·əm }

Corixidae [INV ZOO] The water boatmen, the single family of the hemipteran superfamily Corixoidea. { kə'rik·sə‚dē }

Corixoidea [INV ZOO] A superfamily of hemipteran insects belonging to the subdivision Hydrocorisae that lack ocelli. { kə‚rik'sȯid·ē·ə }

cork [BOT] A protective layer of cells that replaces the epidermis in older plant stems. { kȯrk }

corkboard [MATER] Board made of compressed cork. { 'kȯrk‚bȯrd }

cork paint [MATER] A paint containing fine cork particles; used on steel parts on ships to prevent sweating. { 'kȯrk ‚pānt }

corkscrew rule [ELECTROMAG] The rule that the direction of the current and that of the resulting magnetic field are related to each other as the forward travel of a corkscrew and the direction in which it is rotated. { 'kȯrk‚skrü ‚rül }

cork tile [MATER] Floor tile made of compressed cork bound with phenolic or other resin binders; used on moisture-free rigid subfloors or on plywood or hardboard. { ‚kȯrk ‚tīl }

Corliss valve [MECH ENG] An oscillating type of valve gear with a trip mechanism for the admission and exhaust of steam to and from an engine cylinder. { 'kȯr·ləs ‚valv }

corm [BOT] A short, erect, fleshy underground stem, usually broader than high and covered with membranous scales. { kȯrm }

cormatose [BOT] Having or producing a corm. { 'kȯr·mə‚tōs }

cormidium [INV ZOO] The assemblage of individuals dangling in clusters from the main stem of pelagic siphonophores. { kȯr'mid·ē·əm }

corn [BOT] *Zea mays.* A grain crop of the grass order Cyperales grown for its edible seeds (technically fruits). [MED] A small, sharply circumscribed, conically shaped deep-seated area of thickened skin composed of the fibrous protein keratin. Also known as heloma. { kȯrn }

Cornaceae [BOT] A family of dicotyledonous plants in the order Cornales characterized by perfect or unisexual flowers, a single ovule in each locule, as many stamens as petals, and opposite leaves. { kȯr'nās·ē‚ē }

Cornales [BOT] An order of dicotyledonous plants in the subclass Rosidae marked by a woody habit, simple leaves, well-developed endosperm, and fleshy fruits. { kȯr'nā·lēz }

cornea [ANAT] The transparent anterior portion of the outer coat of the vertebrate eye covering the iris and the pupil. [INV ZOO] The outer transparent portion of each ommatidium of a compound eye. { 'kȯr·nē·ə }

corneal reflex [MED] Automatic closing of the eyelids as a result of irritation of the cornea. { 'kȯr·nē·əl 'rē‚fleks }

corneite [GEOL] A biotite-hornfels formed during deformation of shale by folding. { 'kȯr·nē‚īt }

corner bead [BUILD] **1.** Any vertical molding used to protect the external angle of the intersecting surfaces. **2.** A strip of formed galvanized iron, sometimes combined with a strip of metal lath, placed on corners to reinforce them before plastering. { 'kȯr·nər ‚bēd }

corner chisel [DES ENG] A chisel with two cutting edges at right angles. { 'kȯr·nər ‚chiz·əl }

corner cut [COMPUT SCI] A corner cut off a punched card at an oblique angle to aid in orientation. [GRAPHICS] A slanted cut at the corner of an aperture card or microfiche that serves to identify the photosensitive side of the film. { 'kȯr·nər ‚kət }

corner effect [ELECTR] The departure of the frequency-response curve of a band-pass filter from a perfect rectangular shape, so that the corners of the rectangle are rounded. [ENG] In ultrasonic testing, reflection of an ultrasonic beam directed perpendicular to the intersection of two surfaces 90° apart. { 'kȯr·nər i'fekt }

corner frequency *See* break frequency. { 'kȯr·nər ‚frē‚kwən·sē }

corner head [BUILD] A metal molding that is built into plaster in corners to prevent plaster from accidentally breaking off. { 'kȯr·nər ‚hed }

cornering tool [DES ENG] A cutting tool with a curved edge, used to round off sharp corners. { 'kȯr·nər·iŋ ‚tül }

cornerite [BUILD] A corner reinforcement for interior plastering. { 'kȯr·nə‚rīt }

corner joint [ENG] An L-shaped joint formed by two members positioned perpendicular to each other. { 'kȯr·nər ‚jȯint }

cornerload test [ENG] A test to determine whether the display of an analytical balance is affected by the load distribution on the weighing pan. { 'kȯr·nər‚lōd ‚test }

corner reflector [ELECTROMAG] An antenna consisting of two conducting surfaces intersecting at an angle that is usually 90°, with a dipole or other antenna located on the bisector of the angle. [OPTICS] A reflector which returns a laser beam in the direction of its source, consisting of perpendicular reflecting surfaces; used to make precise determinations of distances in surveying. { 'kȯr·nər ri'flek·tər }

cornerstone [BUILD] An inscribed stone laid at the corner of a building, usually at a ceremony. { 'kȯr·nər‚stōn }

cornetite [MINERAL] $Cu_3(PO_4)(OH)_3$ A peacock-blue mineral consisting of basic copper phosphate. { 'kȯr·nə‚tīt }

cornice [ARCH] The crowning, overhanging part of an architectural structure. { 'kȯr·nəs }

cornice brake [MECH ENG] A machine used to bend sheet metal into different forms. { 'kȯr·nəs ‚brāk }

cornicle [INV ZOO] Either of two protruding horn-shaped dorsal tubes in aphids which secrete a waxy fluid. { 'kȯr·nə‚kəl }

corniculate [BIOL] Possessing small horns or hornlike processes. { kȯr'nik·yə·lət }

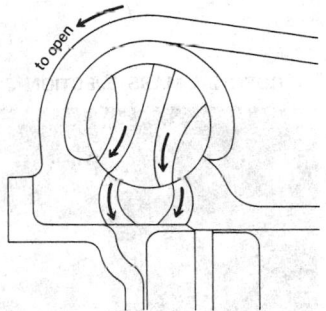

CORLISS VALVE

Corliss steam valve in closed position; arrows indicate path steam will take when the valves are opened.

CORM

A corm, a specialized stem.

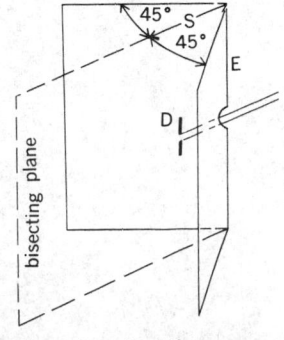

CORNER REFLECTOR

Configuration of a 90° corner reflector. Distance S from the driven radiator D to edge E need not be critically chosen with respect to wavelength; reflector D may lie between 0.25 and 0.7 wavelength.

corniculate cartilage [ANAT] The cartilaginous nodule on the tip of the arytenoid cartilage. { kȯr'nik·yə·lət ‖kärt·lij }

cornification [PHYSIO] Conversion of stratified squamous epithelial cells into a horny layer and into derivatives such as nails, hair, and feathers. { ‚kȯr·nə·fə'kā·shən }

Cornish rolls [MIN ENG] A geared pair of horizontal cylinders, one fixed in a frame and the other held by strong springs; used for grinding. { 'kȯr·nish 'rōlz }

corn oil [MATER] A semidrying, fatty oil of yellowish color, extracted from germs of corn kernels; used mainly as a salad oil, in soft soaps, and in compounded petroleum lubricants. { 'kȯrn ‚ȯil }

corn smut [PL PATH] A fungus disease of corn caused by *Ustilago maydis*. { 'kȯrn ‚smət }

corn snow *See* spring snow. { 'kȯrn ‚snō }

corn sugar *See* dextrose. { 'kȯrn ‚shu̇g·ər }

cornu [ANAT] A horn or hornlike structure. { 'kȯr·nü }

Cornu-Hartmann formula *See* Hartmann dispersion formula. { ‖kȯr·nü 'härt·män ‚fȯr·myə·lə }

Cornu quartz prism [OPTICS] A prism constructed of two 30° quartz prisms, left- and right-handed, used in conjunction with left- and right-handed lenses, so that the rotation of polarization occurring in one half of the optical path is exactly compensated by the reverse rotation in the other; used in a quartz spectrograph. { 'kȯr·nü ‚kwȯrts 'priz·əm }

Cornu's spiral [MATH] A plane curve whose curvature is proportional to its arc length, and whose cartesian coordinates are given in parametric form by the Fresnel integrals. Also known as clothoid; Enler's spiral. { 'kȯr·nüz ‖spī·rəl }

cornwallite [MINERAL] $Cu_5(AsO_4)_2(OH)_4 \cdot H_2O$ A verdigris green to blackish-green mineral consisting of a hydrated basic arsenate of copper; occurs as small botryoidal crusts. { 'kȯrn‚wȯ‚līt }

corn whiskey [FOOD ENG] Whiskey distilled from corn mash containing at least 80% corn, and aged in uncharred oak containers. { 'kȯrn ‚wis·kē }

corolla [BOT] Collectively, the petals of a flower. { kə'räl·ə }

corollate [BOT] Having a corolla. { kə'rä‚lāt }

corolline [BOT] Relating to, resembling, or being borne on a corolla. { 'kȯr·ə‚līn }

coromant cut [MIN ENG] A drill hole pattern; two overlapping holes about $2\frac{1}{4}$ inches (about 5.5 centimeters) in diameter are drilled in the tunnel center and left uncharged; they form a slot roughly 4 by 2 inches (10 by 5 centimeters) to which the easers can break. { 'kȯr·ə·mənt ‚kət }

coromell [METEOROL] A land breeze from the south at La Paz, Mexico, near the mouth of the Gulf of California, prevailing from November to May; it sets in at night and usually persists until 8 or 10 a.m. { kȯr·ə'mel }

corona [ARCH] The overhanging vertical member of a cornice. [ASTRON] *See* solar corona. [BOT] **1.** An appendage or series of fused appendages between the corolla and stamens of some flowers. **2.** The region where stem and root of a seed plant merge. Also known as crown. [ELEC] *See* corona discharge. [GEOL] A mineral zone that is usually radial about another mineral or at the area between two minerals. Also known as kelyphite. [INV ZOO] **1.** The anterior ring of cilia in rotifers. **2.** A sea urchin test. **3.** The calyx and arms of a crinoid. [MET] An area sometimes surrounding the nugget at the faying surfaces of a spot weld which provides a degree of bond strength. [METEOROL] A set of one or more prismatically colored rings of small radii, concentrically surrounding the disk of the sun, moon, or other luminary when veiled by a thin cloud; due to diffraction by numerous waterdrops. [MINERAL] An annular zone of minerals that is disposed either around another mineral or at the contact between two minerals. { kə'rō·nə }

Corona Australis [ASTRON] A constellation, right ascension 19 hours, declination 40°S. Abbreviated CrA. Also known as Southern Crown. { kə'rō·nə ȯs'tral·əs }

Corona Borealis [ASTRON] A constellation, right ascension 16 hours, declination 30°N. Abbreviated CrB. Also known as Northern Crown. { kə'rō·nə bȯr·ē'al·əs }

corona current [ELEC] The current of electricity equivalent to the rate of charge transferred to the air from an object experiencing corona discharge. { kə'rō·nə ‖kər·ənt }

corona discharge [ELEC] A discharge of electricity appearing as a bluish-purple glow on the surface of and adjacent

to a conductor when the voltage gradient exceeds a certain critical value; due to ionization of the surrounding air by the high voltage. Also known as aurora; corona; electric corona. { kə'rō·nə 'dis‚chärj }

coronadite [MINERAL] $Pb(Mn^{2+},Mn^{4+})_8O_{16}$ A black mineral consisting of a lead and manganese oxide, occurring in massive form with fibrous structure; an important constituent of manganese ore. { ‚kȯr·ə'nä‚dīt }

corona failure [ELEC] High-voltage failure initiated by corona discharge at areas of high-voltage stress such as metal inserts or terminals. { kə'rō·nə ‖fāl·yər }

coronagraph [ASTRON] An instrument for photographing the corona and prominences of the sun at times other than at solar eclipse. { kə'rō·nə‚graf }

coronal green line [ASTRON] The strongest emission line in the visible spectrum of the solar corona, located at a wavelength of 530.3 nanometers and resulting from the emission of iron atoms that have lost 13 of their electrons. { 'kȯr·ən·əl grēn ‚līn }

coronal hole [ASTRON] A large-scale, apparently open structure in the solar corona, devoid of any soft x-ray emission and surrounded by diverging boundary structures. { kə'rō·nəl ‚hōl }

coronal mass ejection [ASTRON] A bubble of gas threaded with magnetic field lines, with dimensions of up to hundreds of thousands of miles, that is ejected from the solar corona over the course of several hours and can disrupt the solar wind, resulting in a geomagnetic storm. { kə‚rōn·əl 'mas i‚jek·shən }

coronal suture [ANAT] The union of the frontal with the parietal bones transversely across the vertex of the skull. { kə'rō·nəl 'sü·chər }

corona method [GEOPHYS] A method of estimating drop sizes in clouds by utilizing measurements of the angular radii of the rings of a corona. { kə'rō·nə 'meth·əd }

corona radiata [HISTOL] The layer of cells immediately surrounding a mammalian ovum. { kə'rō·nə ‚rā·dē'äd·ə }

corona resistance [ELEC] Ability of a conductor to resist destruction when a high-voltage electrostatic field ionizes within insulation voids. { kə'rō·nə ri'zis·təns }

coronary artery [ANAT] Either of two arteries arising in the aortic sinuses that supply the heart tissue with blood. { 'kär·ə‚ner·ē 'ärd·ə·rē }

coronary disease [MED] Any condition that reduces the flow of blood to the heart muscles. { 'kär·ə‚ner·ē di'zēz }

coronary failure [MED] Prolonged precordial pain or discomfort without conventional evidence of myocardial infarction; subendocardial ischemia caused by a disparity between coronary blood flow and myocardial needs; this condition is more commonly referred to as coronary artery insufficiency. { 'kär·ə‚ner·ē 'fāl·yər }

coronary occlusion [MED] Complete blockage of a coronary artery. { 'kär·ə‚ner·ē ə'klü·zhən }

coronary sinus [ANAT] A venous sinus opening into the heart's right atrium which drains the cardiac veins. { 'kär·ə‚ner·ē 'sī·nəs }

coronary stenosis [MED] Narrowing of the lumen of a coronary artery. { 'kär·ə‚ner·ē stə'nō·səs }

coronary sulcus [ANAT] A groove in the external surface of the heart separating the atria from the ventricles, containing the trunks of the nutrient vessels of the heart. { 'kär·ə‚ner·ē 'səl·kəs }

coronary thrombosis [MED] Formation of a thrombus in a coronary artery. { 'kär·ə‚ner·ē thräm'bō·səs }

coronary valve [ANAT] A semicircular fold of the endocardium of the right atrium at the orifice of the coronary sinus. { 'kär·ə‚ner·ē 'valv }

coronary vein [ANAT] **1.** Any of the blood vessels that bring blood from the heart and empty into the coronary sinus. **2.** A vein along the lesser curvature of the stomach. { 'kär·ə‚ner·ē 'vān }

corona shield [ELEC] A shield placed about a point of high potential to redistribute electrostatic lines of force. { kə'rō·nə ‚shēld }

corona stabilization [ELEC] The increase in the breakdown voltage of a gas separating two electrodes, where the electric field is very high at one pointed electrode and low at the other, due to the reduction of electric field around the pointed

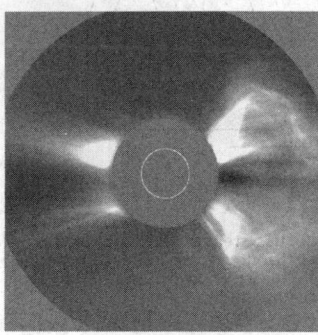

CORONAL MASS EJECTION

Large coronal mass ejection on November 6, 1997, as recorded by the Large-Angle Spectroscopic Coronagraph on the *Solar and Heliospheric Observatory*. The occulting disk is about twice the solar diameter. (*NASA; European Space Agency, the* LASCO *team*)

electrode by corona discharge. { kə'rō·nə ˌstā·bə·lə'zā·shən }

corona start voltage [ELEC] The voltage difference at which corona discharge is initiated in a given system. { kə'rō·nə 'stärt ˌvōl·tij }

Coronatae [INV ZOO] An order of the class Scyphozoa which includes mainly abyssal species having the exumbrella divided into two parts by a coronal furrow. { ˌkȯr·ə'näd·ē }

corona tube [ELEC] A gas-discharge voltage-reference tube employing a corona discharge. { kə'rō·nə ˌtüb }

Coronaviridae [VIROL] A family of vertebrate viruses consisting of the single genus Coronavirus; the prototype, avian infectious virus, has an enveloped spherical form with large spikes and a helical nucleocapsid with single-stranded ribonucleic acid. { kəˌrō·nə'vī·rəˌdē }

coronavirus [VIROL] A major group of animal viruses including avian infectious bronchitis virus and mouse hepatitis virus. { kə'rō·nəˌvī·rəs }

corona voltmeter [ELEC] A voltmeter in which the crest value of a voltage is indicated by the inception of corona at a known electrode spacing. { kə'rō·nə 'vōltˌmēd·ər }

corona wire [GRAPHICS] An electrically charged wire in a laser printer that pulls toner from the drum onto the paper. { kə'rō·nə ˌwīr }

coronet band [ZOO] The area above the hoof containing the germinal cells from which hoof tissue is formed. { 'kär·əˌnet ˌband }

coronizing [MET] Process to electroplate zinc on nickel, thermally treated at 375°C; coating is used on ferrous and copper-based alloys to give resistance to sulfur dioxide, SO₂, and sulfur trioxide, SO₃. { 'kȯr·əˌnīz·iŋ }

coronoid [BIOL] Shaped like a beak. { 'kȯr·əˌnȯid }

coronold fossa [ANAT] A depression in the humerus into which the apex of the coronoid process of the ulna fits in extreme flexion of the forearm. { 'kȯr·əˌnȯid ˌfäs·ə }

coronoid process [ANAT] **1.** A thin, flattened process projecting from the anterior portion of the upper border of the ramus of the mandible, and serving for the insertion of the temporal muscle. **2.** A triangular projection from the upper end of the ulna, forming the lower part of the radial notch. { 'kȯr·əˌnȯid ˌpräs·əs }

coronule [INV ZOO] A peripheral ring of spines on some diatom shells. { 'kȯr·əˌnyül }

Corophiidae [INV ZOO] A family of amphipod crustaceans in the suborder Gammaridea. { kȯr·ə'fī·əˌdē }

corotating [SCI TECH] Rotating about a common axis. { ˌkō'rōˌtād·iŋ }

corotating interaction regions [ASTRON] Regions of enhanced magnetic field bounded by jumps in the solar wind speed that form at distances from the sun greater than 2.5 astronomical units. { ˌkō'rōˌtād·iŋ ˌin·tər'ak·shən ˌrē·jənz }

coroutine [COMPUT SCI] A program module for which the lifetime of a particular activation record is independent of the time when control enters or leaves the module, and in which the activation record maintains a local instruction counter so that, whenever control enters the module, execution begins at the point where it stopped when control last left that particular instance of execution. { 'kō·rüˌtēn }

corpora quadrigemina [ANAT] The inferior and superior colliculi collectively. Also known as quadrigeminal body. { 'kȯr·pə·rə ˌkwäd·rə'jem·ə·nə }

cor pulmonale [MED] Hypertrophy and dilation of the right ventricle that is secondary to obstruction to the pulmonary blood flow and consequent pulmonary hypertension. { 'kȯr ˌpúl·mə'näl·ē }

corpus albicans [HISTOL] The white fibrous scar in an ovary; produced by the involution of the corpus luteum. { 'kȯr·pəs 'al·bəˌkanz }

corpus allatum [INV ZOO] An endocrine structure near the brain of immature arthropods that secretes a juvenile hormone, neotenin. { 'kȯr·pəs ə'lād·əm }

corpus callosum [NEURO] A band of nerve tissue connecting the cerebral hemispheres in humans and higher mammals. { 'kȯr·pəs kə'lō·səm }

corpus cardiacum [INV ZOO] One of a pair of separate or fused bodies of nervous tissue in many insects that lie posterior to the brain and dorsal to the esophagus and that function in the storage and secretion of brain hormone. { 'kȯr·pəs ˌkärd·ē'ak·əm }

corpus cavernosum [ANAT] The cylinder of erectile tissue forming the clitoris in the female and the penis in the male. { 'kȯr·pəs ˌka·vər'nō·səm }

corpus cerebelli [NEURO] The central lobe or zone of the cerebellum; regulates reflex tonus of postural muscles in mammals. { 'kȯr·pəs ˌser·ə'bel·ē }

corpuscle [ANAT] A small, rounded body. [NEURO] An encapsulated sensory-nerve end organ. [OPTICS] A particle of light in the corpuscular theory, corresponding to the photon in the quantum theory. { 'kȯr·pəs·əl }

corpuscular radiation [PHYS] Radiation consisting of subatomic particles, such as electrons, protons, deuterons, and neutrons, as distinguished from electromagnetic radiation. { kȯr'pəs·kyə·lər ˌrād·ē'ā·shən }

corpuscular theory of light [OPTICS] Theory that light consists of a stream of particles; now considered a limiting case of the quantum theory. Also known as Newton's theory of light. { kȯr'pəs·kyə·lər ˌthē·ə·rē əv 'līt }

corpus luteum [HISTOL] The yellow endocrine body formed in the ovary at the site of a ruptured Graafian follicle. { 'kȯr·pəs 'lüd·ē·əm }

corpus striatum [ANAT] The caudate and lenticular nuclei, together with the internal capsule which separates them. { 'kȯr·pəs ˌstrī'ād·əm }

corrasion [GEOL] Mechanical wearing away of rock and soil by the action of solid materials moved along by wind, waves, running water, glaciers, or gravity. Also known as mechanical erosion. { kə'rā·zhən }

corrected altitude [METEOROL] The indicated altitude corrected for temperature deviation from the standard atmosphere. Also known as true altitude. { kə'rek·təd 'al·təˌtüd }

corrected azimuth [ORD] Azimuth of the axis of the bore of a gun firing on a moving target, after allowances have been made for atmospheric, material, and other variable conditions. { kə'rek·təd 'az·ə·məth }

corrected compass course [NAV] Compass course with deviation applied. { kə'rek·təd 'käm·pəs ˌkȯrs }

corrected compass heading [NAV] Compass heading with deviation applied. { kə'rek·təd 'käm·pəs ˌhed·iŋ }

corrected establishment See mean high-water lunitidal interval. { kə'rek·təd i'stab·lish·mənt }

corrected sextant altitude [NAV] Sextant altitude for which index error, height of eye, parallax, refraction, and so on have been applied. Also known as observed altitude. { kə'rek·təd 'sek·stənt ˌal·təˌtüd }

correcting [NAV] The process of applying corrections, particularly that of converting compass to magnetic direction, or compass, magnetic, or gyro to true direction. { kə'rek·tiŋ }

correcting plate See corrector plate. { kə'rek·tiŋ ˌplāt }

correcting wedge [MIN ENG] A deflection wedge used to deflect a crooked borehole back into its intended course. { kə'rek·tiŋ ˌwej }

correction [SCI TECH] A quantity added to a calculated or observed value to obtain the true value. { kə'rek·shən }

correction chamber [ENG] A closable cavity in a weight on an analytical balance; holds material to adjust weight to nominal value. { kə'rek·shən ˌchăm·bər }

correction for attenuation [STAT] A method used to adjust correlation coefficients upward because of errors of measurement when two measured variables are correlated; the errors always serve to lower the correlation coefficient as compared with what it would have been if the measurement of the two variables had been perfectly reliable. { kə'rek·shən fȯr əˌten·yə'wā·shən }

correction of soundings [NAV] In marine operations, the adjustment of soundings for any departure from true depth because of the method of sounding or any fault in the measuring apparatus. { kə'rek·shən əv 'saùndˌiŋz }

correction time [CONT SYS] The time required for the controlled variable to reach and stay within a predetermined band about the control point following any change of the independent variable or operating condition in a control system. Also known as settling time. { kə'rek·shən ˌtīm }

corrective action [CONT SYS] The act of varying the manipulated process variable by the controlling means in order to modify overall process operating conditions. { kə'rek·tiv 'ak·shən }

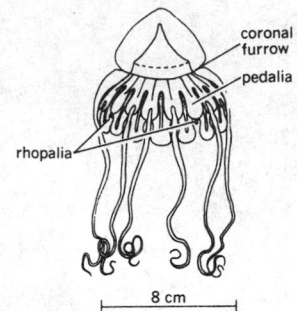

CORONATAE

Periphylla. (From L. H. Hyman, The Invertebrates, vol. 1, McGraw-Hill, 1940)

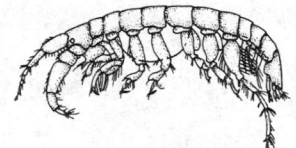

COROPHIIDAE

Lateral view of female *Corophium crassicorne* Bruzelius, a tube-builder. *(From G. O. Sars, An Account of the Crustacea of Norway, vol. 1, 1895)*

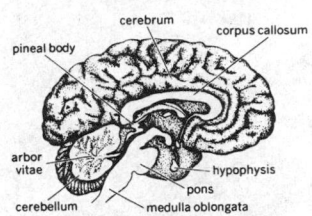

CORPUS CALLOSUM

A sagittal section through the human brain showing corpus callosum.

corrective maintenance [COMPUT SCI] The maintenance performed as required, on an unscheduled basis, by the contractor following equipment failure. Also known as remedial maintenance. [ENG] A procedure of repairing components or equipment as necessary either by on-site repair or by replacing individual elements in order to keep the system in proper operating condition. { kəˈrek·tiv mānt·ən·əns }

corrective network [ELEC] An electric network inserted in a circuit to improve its transmission properties, impedance properties, or both. Also known as shaping circuit; shaping network. { kəˈrek·tiv ˈnet,wərk }

corrective operation See remedial operation. { kəˈrek·tiv äp·əˈrā·shən }

corrective therapy [MED] A program, and the techniques, designed to improve or maintain the health of a patient by improving neuromuscular activities and personal health habits and promoting relaxation by adjustment to stresses. { kəˈrek·tiv ˈther·ə·pē }

corrector [ENG] A magnet, piece of soft iron, or device used in the adjustment or compensation of a magnetic compass. { kəˈrek·tər }

corrector plate [OPTICS] A thin lens or system of lenses used to correct the spherical aberration of a spherical lens or the coma of a parabolic lens; used particularly in telescopes such as the Schmidt telescope. Also known as correcting plate. { kəˈrek·tər ˌplāt }

correed relay [ELEC] Hermetically sealed reed capsule surrounded by a coil winding, used as a switching device with telephone equipment. { ˈkō,rēd ˈrē,lā }

correlated orientation tracking and range See cotar. { ˈkär·əˌlād·əd ˌȯr·ē·ənˈtā·shən ˈtrak·iŋ ən ˈrānj }

correlation [ATOM PHYS] See electron correlation. [GEOL] **1.** The determination of the equivalence or contemporaneity of geologic events in separated areas. **2.** As a step in seismic study, the selecting of corresponding phases, taken from two or more separated seismometer spreads, of seismic events seemingly developing at the same geologic formation boundary. [PHYS] They tendency of two or more systems that independently exhibit simple behavior to show complex and novel behavior together because of their interaction. [STAT] The interdependence or association between two variables that are quantitative or qualitative in nature. { ˌkär·əˈlā·shən }

correlation array See multiplicative acoustic array. { ˌkär·əˈlā·shən əˌrā }

correlation coefficient [STAT] A measurement, which is unchanged by both addition and multiplication of the random variable by positive constants, of the tendency of two random variables X and Y to vary together; it is given by the ratio of the covariance of X and Y to the square root of the product of the variance of X and the variance of Y. { ˌkär·əˈlā·shən ˌkō·iˈfish·ənt }

correlation curve See correlogram. { ˌkär·əˈlā·shən ˌkərv }

correlation detection [ENG] A method of detection of aircraft or space vehicles in which a signal is compared, point to point, with an internally generated reference. Also known as cross-correlation detection. { ˌkär·əˈlā·shən diˈtek·shən }

correlation, detection, and ranging [ORD] A method of detecting submerged submarines with sonobuoys dropped in a characteristic pattern from Navy P-3 aircraft, based on difference in time and phase of signals transmitted to the search plane by the sonobuoys. Abbreviated CODAR. { ˌkär·əˈlā·shən diˈtek·shən ən ˈrän,jiŋ }

correlation direction finder [ENG] Satellite station separated from a radar to receive jamming signals; by correlating the signals received from several such stations, range and azimuth of many jammers may be obtained. { ˌkär·əˈlā·shən dəˈrek·shən ˌfīnd·ər }

correlation distance [COMMUN] In tropospheric scatter propagation, the minimum spatial separation between antennas which will give rise to independent fading of the received signals. { ˌkär·əˈlā·shən ˌdis·təns }

correlation energy [ATOM PHYS] The difference between the experimentally measured energy of a particular energy level of an atom or molecule and the energy calculated in the Hartree-Fock approximation. [SOLID STATE] The modification of the Coulomb energy of a crystal that results from the tendency of electrons to stay apart from each other. { ˌkär·əˈlā·shən ˌen·ər·jē }

correlation ratio [STAT] A measure of the nonlinear relationship between two variables; in a two-way frequency table it may be regarded as the ratio of the variance between arrays to the total variance. { ˌkär·əˈlā·shən ˌrā·shō }

correlation table [STAT] A table designed to categorize paired quantitative data; used to calculate correlation coefficients. { ˌkär·əˈlā·shən ˌtā·bəl }

correlation tracking and triangulation See cotat. { ˌkär·əˈlā·shən ˈtrak·iŋ ən trīˌaŋ·gyəˈlā·shən }

correlation tracking system [ENG] A trajectory-measuring system utilizing correlation techniques where signals derived from the same source are correlated to derive the phase difference between the signals. { ˌkär·əˈlā·shən ˈtrak·iŋ ˌsis·təm }

correlation-type receiver See correlator. { ˌkär·əˈlā·shən ˌtīp riˈsē·vər }

correlation ultrasonic flowmeter [ENG] An instrument for determining the velocity of a fluid flow from the time required for discontinuities in the fluid stream to pass between two pairs of transducers that generate and detect high-frequency sound. { ˌkär·əˈlā·shən əl·trəˈsän·ik ˈflō,mēd·ər }

correlative kinesiology [IND ENG] A field that involves determination of the quantitative relationship between the electrical potential generated by muscular activity and the resultant movement; used in developing a design for a workplace that minimizes fatigue. { kəˈrel·əd·iv kəˌnēz·ē·ˈäl·ə·jē }

correlative rights [PETRO ENG] Legal rights protecting property over a portion of a gas or oil reservoir from excessive or wasteful withdrawal of hydrocarbons by adjoining properties overlying the same reservoir. { kəˈrel·əd·iv ˈrīts }

correlator [ELECTR] A device that detects weak signals in noise by performing an electronic operation approximating the computation of a correlation function. Also known as correlation-type receiver. { ˈkär·əˌlād·ər }

correlogram [MATH] A curve showing the assumed correlation between two mathematical variables. Also known as correlation curve. { kəˈrel·əˌgram }

correspondence See relation. { ˌkär·əˈspän·dəns }

correspondence principle [QUANT MECH] The principle that quantum mechanics has a classical limit in which it is equivalent to classical mechanics. Also known as Bohr's correspondence principle. { ˌkär·əˈspän·dəns ˌprin·sə·pəl }

correspondence printer See letter-quality printer. { ˌkär·əˈspän·dəns ˌprint·ər }

corresponding angles [MATH] For two lines, l_1 and l_2, cut by a transversal t, a pair of angles such that (1) one of the angles has sides l_1 and t while the other has sides l_2 and t; (2) both angles are on the same side of t; and (3) the angles are on the same sides of l_1 and l_2, respectively. { ˈkär·əˌspänd·iŋ ˈaŋ·gəlz }

corresponding points [PHYSIO] Any two retinal areas in the respective eyes so that the area in one eye has an identical direction in the opposite retina. { ˌkär·əˈspänd·iŋ ˈpȯins }

corresponding states [PHYS CHEM] The condition when two or more substances are at the same reduced pressures, the same reduced temperatures, and the same reduced volumes. { ˌkär·əˈspänd·iŋ ˈstāts }

corridor [ECOL] A land bridge that allows free migration of fauna in both directions. { ˈkär·ə·dər }

corrie See cirque. { ˈkȯr·ē }

Corrigan's pulse [MED] A pulse characterized by a rapid, forceful ascent (water-hammer quality) and rapid downstroke or descent (collapsing quality); seen with aortic regurgitation and hyperkinetic circulatory states. { ˈkär·ə·gənz ˈpəls }

corrins [BIOCHEM] Cobalt-containing compounds (porphyrin-like macrocycles) that act in concert with enzymes to catalyze essential reactions in humans. { ˈkō·rinz }

Corrodentia [INV ZOO] The equivalent name for Psocoptera. { ˌkȯr·əˈdench·ə }

corroding lead [MET] Lead that can be corroded to make white lead. { kəˈrōd·iŋ ˌled }

corrosion [GEOCHEM] Chemical erosion by motionless or moving agents. [MET] Gradual destruction of a metal or alloy due to chemical processes such as oxidation or the action of a chemical agent. { kəˈrō·zhən }

corrosion border See corrosion rim. { kəˈrō·zhən ˌbȯrd·ər }

corrosion cell [MET] A condition on a metal surface in which a flow of electric current occurs between the metal surface and an electrolyte with which it is in contact sufficient to cause the metal to degrade. { kəˈrō·zhən ˌsel }

corrosion control See corrosion protection. { kə'rō·zhən kən,trōl }

corrosion coupon See coupon. { kə'rō·zhən ,kü,pän }

corrosion fatigue [MET] Damage to or failure of a metal due to corrosion combined with fluctuating fatigue stresses. { kə'rō·zhən fə'tēg }

corrosion fatigue limit [MET] The maximum stress that a corroded material can withstand for a given number of stress reversals. { kə'rō·zhən fə'tēg 'lim·ət }

corrosion inhibitor [PHYS CHEM] A compound or material deposited as a film on a metal surface that either provides physical protection against corrosive attack or reduces the open-circuit potential difference between local anodes and cathodes and increases the polarization of the former. { kə'rōzh·ən in,hib·əd·ər }

corrosion number See acid number. { kə'rō·zhən ,nəm·bər }

corrosion potential [MET] The measure of corroding surface potential in an electrolyte in relation to a reference electrode while the circuit is open. { kə'rō·zhən pə'ten·chəl }

corrosion protection [MET] The minimization of corrosion by coating with a protective metal, with an oxide or phosphide or similar substance, or with a protective paint, or by rendering the metal passive. Also known as corrosion control. { kə'rō·zhən prə'tek·shən }

corrosion rim [MINERAL] A modification of the outlines of a porphyritic crystal due to the corrosive action of a magma on previously stable minerals. Also known as corrosion border. { kə'rō·zhən ,rim }

corrosion test [MET] Any of various tests to determine the resistance of a metal to chemical attack. { kə'rō·zhən ,test }

corrosive [MATER] A substance that causes corrosion. { kə'rō·siv }

corrosive flux [MET] A soldering flux, usually composed of inorganic salts and acids, which provides oxide removal of the base metal upon application of solder; flux remaining on the base metal is corrosive and should be removed. { kə'rō·siv 'fləks }

corrosiveness [MET] The tendency of a metal to wear away another by chemical attack. { kə'rō·siv·nəs }

corrosive product [CHEM ENG] In petroleum refining, a product that contains a quantity of corrosion-inducing compounds in excess of the limits specified for products classified as sweet. { kə'rō·siv 'präd·əkt }

corrosive sublimate See mercuric chloride. { kə'rō·siv 'səb·lə,māt }

corrugated bar [DES ENG] Steel bar with transverse ridges; used in reinforced concrete. { 'kär·ə,gād·əd 'bär }

corrugated conical-horn antenna [ELECTROMAG] A horn antenna that has a circular cross section and a series of equally spaced ridges protruding from otherwise straight sides. { 'kär·ə,gād·əd ,kän·ə·kəl ,hórn an'ten·ə }

corrugated fastener [DES ENG] A thin corrugated strip of steel that can be hammered into a wood joint to fasten it. { 'kär·ə,gād·əd 'fas·nər }

corrugated lens [OPTICS] A lens having circular sections cut out from the surface to reduce its weight without lowering its focal power. { 'kär·ə,gād·əd 'lenz }

corrugating [DES ENG] Forming straight, parallel, alternate ridges and grooves in sheet metal, cardboard, or other material. { 'kär·ə,gād·iŋ }

corrupt [COMPUT SCI] To destroy or alter information so that it is no longer reliable. { kə'rəpt }

corsite [PETR] A spheroidal variety of gabbro. Also known as miagite; napoleonite. { 'kór,sīt }

cortex [ANAT] The outer portion of an organ or structure, such as of the brain and adrenal glands. [BOT] A primary tissue in roots and stems of vascular plants that extends inward from the epidermis to the phloem. [CYTOL] A peripheral layer in many cells that includes the plasma membrane and associated cytoskeletal and extracellular components. [INV ZOO] The peripheral layer of certain protozoans. { 'kór,teks }

cortical granule [CYTOL] Any of the round to elliptical membrane-bound bodies that occur in the cortex of animal oocytes, contain mucopolysaccharides, and participate in formation of the fertilization membrane. { 'kórd·ə·kəl 'gran·yəl }

cortical stimulator [MED] An electronic instrument used in nerve and mental therapy to deliver an electric shock of prescribed strength by means of a pulsating current. { 'kórd·ə·kəl 'stim·yə,lād·ər }

corticoid See adrenal cortex hormone. { 'kórd·ə,kóid }

corticosteroid [BIOCHEM] **1.** Any steroid hormone secreted by the adrenal cortex of vertebrates. **2.** Any steroid with properties of an adrenal cortex steroid. { ˌkórd·ə,kō'stir,óid }

corticosterone [BIOCHEM] $C_{21}H_{30}O_4$ A steroid hormone produced by the adrenal cortex of vertebrates that stimulates carbohydrate synthesis and protein breakdown and is antagonistic to the action of insulin. { ˌkórd·ə'käs·tə,rōn }

corticotrophic [PHYSIO] Having an effect on the adrenal cortex. { ˌkórd·ə,kō'trä·fik }

corticotropin [BIOCHEM] A hormonal preparation having adrenocorticotropic activity, derived from the adenohypophysis of certain domesticated animals. { ˌkórd·ə,kō'trō·pən }

corticotropin-releasing hormone [BIOCHEM] A substance produced by the hypothalamus that stimulates the pituitary gland to produce adrenocorticotropic hormone (ACTH). Abbreviated CRH. { ˌkórd·ə·kō'trō·pən riˌlēs·iŋ ,hór,mōn }

Corticoviridae [VIROL] A family of nontailed bacterial viruses (bacteriophages) characterized by a nonenveloped icosahedral particle containing a circular double-stranded deoxyribonucleic acid genome. { ˌkórd·i kō'vir·ə,dī }

Corticovirus [VIROL] The only genus of the family Corticoviridae. { ˌkórd·i·kō 'vī·rəs }

cortin unit [BIOL] A unit for the standardization of adrenal cortical hormones. { 'kórt·ən ,yü·nət }

cortisol See hydrocortisone. { 'kórd·ə,sól }

cortisone [BIOCHEM] $C_{21}H_{28}O_5$ A steroid hormone produced by the adrenal cortex of vertebrates that acts principally in carbohydrate metabolism. { 'kórd·ə,sōn }

cortlandite [PETR] A peridotite consisting of large crystals of hornblende with poikilitically included crystals of olivine. Also known as hudsonite. { 'kórt·lən,dīt }

corundum [MINERAL] Al_2O_3 A hard mineral occurring in various colors and crystallizing in the hexagonal system; crystals are usually prismatic or in rounded barrel shapes; gem varieties are ruby and sapphire. { kə'rən·dəm }

corvette [NAV ARCH] **1.** A warship with a continuous deck from fore to stern, usually with no structure above, and usually with only one row of guns. **2.** A very maneuverable escort ship having antisubmarine and antiaircraft guns, depth charges, and detection equipment. { kór'vet }

Corvidae [VERT ZOO] A family of large birds in the order Passeriformes having stout, long beaks; includes the crows, jays, and magpies. { 'kór·və,dē }

Corvus [ASTRON] A constellation, right ascension 12 hours, declination 20°S. Abbreviated Crv. Also known as Crow. { 'kór·vəs }

corvusite [MINERAL] $V_2O_4 \cdot 6V_2O_5 \cdot nH_2O$ A blue-black to brown mineral consisting of a hydrous oxide of vanadium; occurs in massive form. { 'kór·və,sīt }

Corylophidae [INV ZOO] The equivalent name for Orthoperidae. { ˌkór·ə'läf·ə,dē }

corymb [BOT] An inflorescence in which the flower stalks arise at different levels but reach the same height, resulting in a flat-topped cluster. { 'kó,rim }

corymbose [BOT] Resembling or pertaining to a corymb. { kə'rim,bōs }

Corynebacteriaceae [MICROBIO] Formerly a family of nonsporeforming, usually nonmotile rod-shaped bacteria in the order Eubacteriales including animal and plant parasites and pathogens. { ˌkór·ə,nē,bak,tir·ē'ās·ē,ē }

corynebacteriophage [VIROL] Any bacteriophage able to infect *Corynebacterium* species. { ˌkór·ə,nē·bak'tir·ē·ə,fäzh }

Corynebacterium [MICROBIO] A genus of gram-positive, straight or slightly curved rods in the coryneform group of bacteria; club-shaped swellings are common; includes human and animal parasites and pathogens, and plant pathogens. { ˌkór·ə,nē·bak'tir·ē·əm }

Corynebacterium diphtheriae [MICROBIO] A facultatively aerobic, nonmotile species of bacteria that causes diphtheria in humans. Also known as Klebs-Loeffler bacillus. { ˌkór·ə,nē·bak'tir·ē·əm dif'thir·ē,ī }

Coryphaenidae [VERT ZOO] A family of pelagic fishes in the order Perciformes characterized by a blunt nose and deeply forked tail. { ˌkór·ə'fēn·ə,dē }

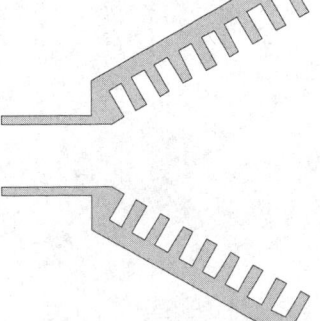

CORRUGATED CONICAL-HORN ANTENNA

Cross section of corrugated conical-horn antenna.

Coryphodontidae [PALEON] The single family of the Coryphodontoidea, an extinct superfamily of mammals. { ‚kȯr·ə·fə'dän·tə‚dē }

Coryphodontoidea [PALEON] A superfamily of extinct mammals in the order Pantodonta. { ‚kȯr·ə·fə‚dän'tȯid·ē·ə }

coryza [MED] Inflammation of the mucous membranes of the nose, usually marked by sneezing and discharge of watery mucous. { kə'rī·zə }

cos See cosine function.

cosalite [MINERAL] $Pb_2Bi_2S_5$ A lead-gray or steel-gray mineral consisting of lead, bismuth, and sulfur; specific gravity is 6.39–6.75. { 'kō·zə‚līt }

cosecant [MATH] The reciprocal of the sine. Denoted csc. { kō'sē‚kant }

cosecant antenna [ELECTROMAG] An antenna that gives a beam whose amplitude varies as the cosecant of the angle of depression below the horizontal; used in navigation radar. { kō'sē‚kant an'ten·ə }

cosecant-squared antenna [ELECTROMAG] An antenna that has a cosecant-squared pattern. { kō'sē‚kant ¦skwerd an'ten·ə }

cosecant-squared pattern [ELECTROMAG] A ground radar-antenna radiation pattern that sends less power to nearby objects than to those farther away in the same sector; the field intensity varies as the square of the cosecant of the elevation angle. { kō'sē‚kant ¦skwerd 'pad·ərn }

coset [MATH] For a subgroup of a group, a set consisting of all elements of the form xh or of all elements of the form hx, where h is an element of the subgroup and x is a fixed element of the group. { 'kō‚set }

cosh See hyperbolic cosine.

cosine emission law [OPTICS] The law that the energy emitted by a radiating surface in any direction is proportional to the cosine of the angle which that direction makes with the normal. { 'kō‚sīn i'mish·ən ‚lȯ }

cosine function [MATH] In a right triangle with an angle θ, the cosine function gives the ratio of adjacent side to hypotenuse; more generally, it is the function which assigns to any real number θ the abscissa of the point on the unit circle obtained by moving from (1,0) counterclockwise θ units along the circle, or clockwise |θ| units if θ is less than 0. Denoted cos. { 'kō‚sīn ‚fəŋk·shən }

cosine pulse [PHYS] A pulse whose amplitude varies during some time interval in proportion to the cosine function over the range from $-\pi/2$ to $\pi/2$, and vanishes outside this time interval. { 'kō‚sīn ‚pəls }

cosine series [MATH] A Fourier series that contains only terms that are even in the independent variable, that is, the constant term and terms involving the cosine function. { 'kō‚sīn ‚sir·ēz }

cosine-squared pulse [PHYS] A pulse whose amplitude varies during some time interval in proportion to the square of the cosine function over the range from $-\pi/2$ to $\pi/2$, and vanishes outside this time interval. { 'kō‚sīn ¦skwerd ‚pəls }

cosine winding [ELECTR] A winding used in the deflection yoke of a cathode-ray tube to prevent changes in focus as the beam is deflected over the entire area of the screen. { 'kō‚sīn ‚wīnd·iŋ }

cosmic [ASTRON] Pertaining to the cosmos, the vast extraterrestrial regions of the universe. { 'käz·mik }

cosmic abundance [ASTRON] The amount of a substance believed to be present in the entire universe, relative to other substances. { 'käz·mik ə'bən·dəns }

cosmic background radiation See cosmic microwave radiation. { 'käz·mik ¦bak‚graund ‚rād·ē‚ā·shən }

cosmic censorship hypothesis [ASTRON] The hypothesis that a system which evolves according to the equations of general relativity from an initial state that does not have singularities or any unusual properties will not develop any space-time singularities that would be visible from large distances. { 'käz·mik 'sen·sər‚ship hī‚päth·ə·səs }

cosmic dust [ASTRON] Fine particles of solid matter forming clouds in interstellar space. { 'käz·mik 'dəst }

cosmic electrodynamics [ASTROPHYS] The science concerned with electromagnetic phenomena in ionized media encountered in interstellar space, in stars, and above the atmosphere. { 'käz·mik i‚lek·trō·də'nam·iks }

cosmic expansion [ASTRON] The recession of all distant galaxies from each other, as manifested in the red shift of their spectral lines. { 'käz·mik ik'span·shən }

cosmic light [ASTRON] The contribution to the brightness of the night sky from all unresolved extragalactic sources. { 'käz·mik 'līt }

cosmic microwave background See cosmic microwave radiation. { 'käz·mik 'mī·krō‚wāv 'bak‚graund }

cosmic microwave radiation [ASTRON] A nearly uniform flux of microwave radiation that is believed to permeate all of space and to have originated in the big bang. Also known as cosmic background radiation; cosmic microwave background; microwave background. { 'käz·mik 'mī·krō‚wāv ‚rād·ē'ā·shən }

cosmic noise [COMMUN] Radio static caused by a phenomenon outside the earth's atmosphere, such as sunspots. { 'käz·mik 'nȯiz }

cosmic radiation See cosmic rays. { 'käz·mik ‚rād·ē'ā·shən }

cosmic radio waves [ASTRON] Radio waves reaching the earth from interstellar or intergalactic sources. { 'käz·mik 'rād·ē·ō ‚wāvz }

cosmic rays [NUC PHYS] Electrons and the nuclei of atoms, largely hydrogen, that impinge upon the earth from all directions of space with nearly the speed of light. Also known as cosmic radiation; primary cosmic rays. { 'käz·mik 'rāz }

cosmic-ray shower [NUC PHYS] The simultaneous appearance of a number of downward-directed ionizing particles, with or without accompanying photons, caused by a single cosmic ray. Also known as air shower; shower. { 'käz·mik ‚rā 'shau·ər }

cosmic-ray telescope [ENG] Any device for detecting and determining the directions of either cosmic-ray primary protons and heavier-element nuclei, or the products produced when these particles interact with the atmosphere. [NUCLEO] An array of counters, sensitive to the direction of the rays detected. { 'käz·mik ‚rā 'tel·ə‚skōp }

cosmic sediment [GEOL] Particles of extraterrestrial origin which are observed as black magnetic spherules in deep-sea sediments. { 'käz·mik 'sed·ə·mənt }

cosmic spherules [GEOCHEM] Solidified, millimeter-sized to microscopic, rounded particles of extraterrestrial materials that melted either during high-velocity entry into the atmosphere or during hypervelocity impact of large meteoroids onto the earth's surface. { ¦käz·mik 'sfe·rülz }

cosmic string [ASTRON] A hypothetical relic of the early universe, postulated to have a diameter of the order of 10^{-35} meter and a linear density of the order of 4×10^{22} kilograms per meter, and to be either infinitely long or in the form of a closed curve. { 'käz·mik 'striŋ }

cosmic year [ASTRON] The period of rotation of the Milky Way Galaxy, about 220 million years. { 'käz·mik 'yir }

Cosmocercidae [INV ZOO] A group of nematodes assigned to the suborder Oxyurina by some authorities and to the suborder Ascaridina by others. { ‚käz·mə'sər·sə‚dē }

Cosmocercoidea [INV ZOO] A superfamily of parasitic nematodes having either three or six lips surrounding a weakly developed stoma. { ‚käz·mō·sər'kȯid·ē·ə }

cosmochemistry [ASTROPHYS] The science of the chemistry of the universe, particularly that beyond earth, concerned primarily with inferences on pre-solar-system events, solar nebular processes, and early planetary processes as deduced from minerals in meteorites and from chemical and isotopic compositions of meteorites and their parts. { ¦käz·mō¦kem·ə·strē }

cosmochlore See ureyite. { 'käz·mə‚klȯr }

cosmogony [ASTROPHYS] Study of the origin and evolution of specific astronomical systems and of the universe as a whole. { käz'mäg·ə·nē }

cosmoid scale [VERT ZOO] A structure in the skin of primitive rhipidistians and dipnoans that is composed of enamel, a dentine layer (cosmine), and laminated bone. { 'käz‚mȯid 'skāl }

cosmological [ASTRON] Relating to the overall structure of the universe. { ¦käz·mə¦läj·ə·kəl }

cosmological constant [RELAT] The multiplicative constant for a term proportional to the metric in Einstein's equation relating the curvature of space to the energy-momentum tensor. { ¦käz·mə¦läj·ə·kəl 'kän·stənt }

cosmological principle [ASTRON] The assumption made in most theories of cosmology that the universe is homogeneous on a large scale. { ¦käz·mə¦läj·ə·kəl 'prin·sə·pəl }

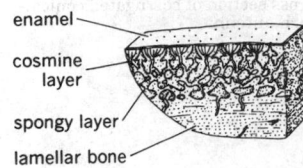

COSMOID SCALE

enamel
cosmine layer
spongy layer
lamellar bone

Cross section through a cosmoid scale. *(From A. S. Romer, The Vertebrate Body, Saunders, 1962)*

cosmological redshift [ASTRON] The red shift that can be ascribed entirely to the general expansion of space-time initiated by the big bang. { ¦käz·mə¦läj·ə·kəl 'red,shift }

cosmological term [RELAT] A term proportional to the metric tensor in Einstein's field equations for special relativity. { ¦käz·mə¦läj·ə·kəl 'tərm }

cosmology [ASTRON] The study of the overall structure of the physical universe. { käz'mäl·ə·jē }

cosmonaut [AERO ENG] An astronaut in the former Soviet Union. { 'käz·mə,nȯt }

cosmopolitan [ECOL] Having a worldwide distribution wherever the habitat is suitable, with reference to the geographical distribution of a taxon. { ¦käz·mə¦päl·ət·ən }

cosolvent [CHEM ENG] During chemical processing, a second solvent added to the original solvent, generally in small concentrations, to form a mixture that has greatly enhanced solvent powers due to synergism. { kō'säl·vənt }

cospectrum [PHYS] **1.** The spectral decomposition of the in-phase components of the covariance of two functions of time. **2.** The real part of the cross spectrum of two functions. { kō'spek·trəm }

Cossidae [INV ZOO] The goat or carpenter moths, a family of heavy-bodied lepidopteran insects in the superfamily Cossoidea having the abdomen extending well beyond the hindwings. { 'käs·ə,dē }

Cosslett process [MET] A process in which iron or steel articles immersed for 3 or 4 hours in a boiling solution, made by mixing iron filings with concentrated phosphoric acid, H_3PO_4 (sufficient to form a paste), and then adding to weak phosphoric acid, become coated with a rust-resisting deposit of basic ferrous phosphate. { 'käs·lət ,präs·əs }

Cossoidea [INV ZOO] A monofamilial superfamily of lepidopteran insects belonging to suborder Heteroneura. { kə'sȯid·ē·ə }

Cossuridae [INV ZOO] A family of fringe worms belonging to the Sedentaria. { kə'syūr·ə,dē }

Cossyphodidae [INV ZOO] The lively ant guest beetles, a small family of coleopteran insects in the superfamily Tenebrionoidea. { ,käs·ə'fä·də,dē }

costa [BIOL] A rib or riblike structure. [BOT] The midrib of a leaf. [INV ZOO] The anterior vein of an insect's wing. { 'käs·tə }

costa bulb [NAV ARCH] A streamlined body of revolution integral with a rudder and directly in line with the propeller, designed to improve propulsion efficiency of ships. { 'käs·tə ,bəlb }

cost accounting [IND ENG] The branch of accounting in which one records, analyzes, and summarizes costs of material, labor, and burden, and compares these actual costs with predetermined budgets and standards. { 'kȯst ə'kaunt·iŋ }

Costaceae [BOT] A family of monocotyledonous plants in the order Zingiberales distinguished by having one functional stamen with two pollen sacs and spirally arranged leaves and bracts. { kȯs'tās·ē,ē }

costal cartilage [ANAT] The cartilage occupying the interval between the ribs and the sternum or adjacent cartilages. { 'käst·əl 'kärd·əl·ij }

costal process [ANAT] An anterior or ventral projection on the lateral part of a cervical vertebra. [EMBRYO] An embryonic rib primordium, the ventrolateral outgrowth of the caudal, denser half of a sclerotome. { 'käst·əl ¦präs·əs }

cost analysis [IND ENG] Analysis of the factors contributing to the costs of operating a business and of the costs which will result from alternative procedures, and of their effects on profits. { 'kȯst ə'nal·ə·səs }

costate [BIOL] Having ribs or ridges. { 'kä,stāt }

cost control See industrial cost control. { 'kȯst kən'trōl }

cost engineering [IND ENG] A branch of industrial engineering concerned with cost estimation, cost control, business planning and management, profitability analysis, and project management, planning, and scheduling. { 'kȯst ,en·jə,nir·iŋ }

Coster-Kronig transition [ATOM PHYS] An Auger transition in which the vacant electron level is filled by an electron from a higher subshell of the same shell. { 'kas·tər 'krō,nig tran,zi·shən }

cost function [SYS ENG] In decision theory, a loss function which does not depend upon the decision rule. { 'kȯst ,fəŋk·shən }

cost-plus contract [ENG] A contract under which a contractor furnishes all material, construction equipment, and labor at actual cost, plus an agreed-upon fee for his services. { 'kȯst 'pləs ,kän,trakt }

cot See cotangent.

cotangent [MATH] The reciprocal of the tangent. Denoted cot; ctn. { kō'tan·jənt }

cotar [ENG] A passive system used for tracking a vehicle in space by determining the line of direction between a remote ground-based receiving antenna and a telemetering transmitter in the missile, using phase-comparison techniques. Derived from correlated orientation tracking and range. { 'kō,tär }

cotat [ENG] A trajectory-measuring system using several antenna base lines, each separated by large distances, to measure direction cosines to an object; then the object's space position is computed by triangulation. Derived from correlation tracking and triangulation. { 'kō,tat }

cotectic [PHYS CHEM] Referring to conditions of pressure, temperature, and composition under which two or more solid phases crystallize at the same time, with no resorption, from a single liquid over a finite range of decreasing temperature. { kō'tek·tik }

cotectic crystallization [PHYS CHEM] Simultaneous crystallization of two or more solid phases from a single liquid over a finite range of falling temperature without resorption. { kō'tek·tik ,krist·əl·ə'zā·shən }

coterminal angles [MATH] Two angles that have the same initial line and the same terminal line and therefore differ by a multiple of 2π radians or 360°. { ¦kō¦tərm·ən·əl 'aŋ·gəlz }

coth See hyperbolic cotangent.

cotidal chart [MAP] A chart of cotidal lines that show approximate locations of high water at hourly intervals as measured from a reference meridian, usually Greenwich. { ,kō'tīd·əl ,chärt }

cotidal hour [ASTRON] The average interval expressed in solar or lunar hours between the moon's passage over the meridian of Greenwich and the following high water at a specified place. { ,kō'tīd·əl 'au·ər }

cotidal line [MAP] A line on a chart passing through all points where high water occurs at the same time. { ,kō'tīd·əl 'līn }

Cotingidae [VERT ZOO] The cotingas, a family of neotropical suboscine birds in the order Passeriformes. { kō'tin·jə,dē }

cotinine [ORG CHEM] The major metabolic product of nicotine which is excreted in the urine; used as a marker for environmental tobacco smoke. { 'kōt·ən,ēn }

co-transport [CELL MOL] The simultaneous transport of two substrates across a cell membrane, either in the same direction (symport) or in opposite directions (antiport). { ,kō'tranz,pȯrt }

cotter [DES ENG] A tapered piece that can be driven in a tapered hole to hold together an assembly of machine or structural parts. { 'käd·ər }

cottered joint [MECH ENG] A joint in which a cotter, usually a flat bar tapered on one side to ensure a tight fit, transmits power by shear on an area at right angles to its length. { 'käd·ərd ,jȯint }

cotter pin [DES ENG] A split pin, inserted into a hole, to hold a nut or cotter securely to a bolt or shaft, or to hold a pair of hinge plates together. { 'käd·ər ,pin }

Cottidae [VERT ZOO] The sculpins, a family of perciform fishes in the suborder Cottoidei. { 'käd·ə,dē }

Cottiformes [VERT ZOO] An order set up in some classification schemes to include the Cottoidei. { ,käd·ə'fȯr,mēz }

Cottoidei [VERT ZOO] The mail-cheeked fishes, a suborder of the order Perciformes characterized by the expanded third infraorbital bone. { kä'tȯid·ē,ī }

cotton [BOT] Any plant of the genus *Gossypium* in the order Malvales; cultivated for the fibers obtained from its encapsulated fruits or bolls. [TEXT] The most economical natural fiber, obtained from plants of the genus *Gossypium*, used in making fabrics, cordage, and padding and for producing artificial fibers and cellulose. { 'kät·ən }

cotton anthracnose [PL PATH] A fungus disease of cotton caused by *Glomerella gossypii* and characterized by reddish-brown to light-colored or necrotic spots. { 'kät·ən an 'thrak,nōs }

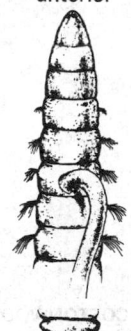

COSSURIDAE

anterior

Cossura, anterior and posterior ends in dorsal view.

COTTER PIN

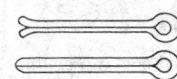

Two common forms of cotter pin.

COTTON EFFECT

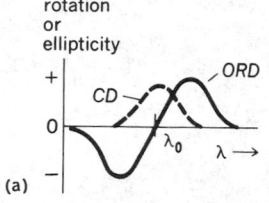

Behavior of optical rotatory dispersion (ORD) and circular dichroism (CD) curves in the vicinity of an absorption band at wavelength λ_0 (idealized). (a) Positive Cotton effect. (b) Negative Cotton effect.

COTTONWOOD

Cottonwood poplar (*Populus deltoides*).

Cotton balance [ENG] A device which employs a current-carrying conductor of special shape to determine the strength of a magnetic field. { 'kät·ən 'bal·əns }

cotton ball *See* ulexite. { 'kät·ən ˌból }

cotton-belt climate [CLIMATOL] A type of warm climate characterized by dry winters and rainy summers; that is, a monsoon climate, in contrast to a Mediterranean climate. { 'kät·ən ˌbelt ˌklī·mət }

Cotton effect [ANALY CHEM] The characteristic wavelength dependence of the optical rotatory dispersion curve or the circular dichroism curve or both in the vicinity of an absorption band. { 'kät·ən i'fekt }

cotton gin [TEXT] A machine that separates cottonseed from the fibers. { 'kät·ən jin }

cottonmouth *See* water moccasin. { 'kät·ənˌmaúth }

Cotton-Mouton birefringence *See* Cotton-Mouton effect. { 'kät·ən ˌmü·ton ˌbī·ri'frin·jəns }

Cotton-Mouton constant [OPTICS] A constant giving the strength of the Cotton-Mouton effect in a liquid; when multiplied by the path length and the square of the magnetic field, it gives the phase difference between the components of light parallel and perpendicular to the field. { 'kät·ən ˌmü·ton ˌkän·stənt }

Cotton-Mouton effect [OPTICS] The double refraction (birefringence) of light in a liquid in a magnetic field at right angles to the direction of light propagation. Also known as Cotton-Mouton birefringence. { 'kät·ən ˌmü·ton i'fekt }

cotton oil [MATER] The yellow, viscous fixed oil, containing principally linoleic acid, pressed from the seeds of various *Gossypium* species; the refined oil is colorless and used in foods and some pharmaceutical preparations. Also known as cottonseed oil; oleum gossypii seminis. { 'kät·ən ˌóil }

cotton root rot [PL PATH] A fungus disease of cotton caused by *Phymatotrichum omnivorum* and marked by bronzing of the foliage followed by sudden wilting and death of the plant. { 'kät·ən 'rüt ˌrät }

cotton rust [PL PATH] A fungus disease of cotton caused by *Puccinia stakmanii* producing low, greenish-yellow or orange elevations on the undersurface of leaves. { 'kät·ən ˌrəst }

cottonseed oil *See* cotton oil. { 'kät·ənˌsēd ˌóil }

cotton staple [TEXT] Natural cotton fiber. { 'kät·ən ˌstā·pəl }

cotton wax [MATER] The wax composing cottonseed coating. { 'kät·ən ˌwaks }

cotton wilt [PL PATH] **1.** A fungus disease of cotton caused by *Fusarium vasinfectum* growing in the water-conducting vessels and characterized by wilt, yellowing, blighting, and death. **2.** A fungus blight of cotton caused by *Verticillium albo-atrum* and characterized by yellow mottling of the foliage. { 'kät·ən ˌwilt }

cottonwood [BOT] Any of several poplar trees (*Populus*) having hairy, encapsulated fruit. { 'kät·ənˌwúd }

cottony rot [PL PATH] A fungus disease of many plants, especially citrus trees, marked by fluffy white growth caused by *Sclerotinia sclerotiorum*, in which there is stem wilt and rot. { 'kät·ən·ē ˌrät }

Cottrell atmosphere [SOLID STATE] A cluster of impurity atoms surrounding a dislocation in a crystal. { kä'trel 'at·məˌsfir }

Cottrell hardening [SOLID STATE] Hardening of a material caused by locking of its dislocations when impurity atoms whose size differs from that of the solvent cluster around them. { 'kä·trəl 'härd·ən·iŋ }

Cottrell precipitator [ENG] A machine for removing dusts and mists from gases, in which the gas passes through a grounded pipe with a fine axial wire at a high negative voltage, and particles are ionized by the corona discharge of the wire and migrate to the pipe. { 'kä·trəl prə'sip·əˌtād·ər }

cotunnite [MINERAL] $PbCl_2$ An alteration product of galena; a soft, white to yellowish mineral that crystallizes in the orthorhombic crystal system. { kə'təˌnīt }

cotyledon [BOT] The first leaf of the embryo of seed plants. { ˌkäd·əl'ēd·ən }

cotylocercous cercaria [INV ZOO] A digenetic trematode larva characterized by a sucker or adhesive gland on the tail. { ˌkäd·əl'äs·ə·rəs ˌsər'kar·ē·ə }

Cotylosauria [PALEON] An order of primitive reptiles in the subclass Anapsida, including the stem reptiles, ancestors of all of the more advanced Reptilia. { ˌkäd·əl·ə'sór·ē·ə }

cotype *See* syntype. { 'kōˌtīp }

coudé focus [OPTICS] Focus achieved with a coudé telescope. { kü'dā ˌfō·kəs }

coudé-Newtonian-Cassegrain telescope [OPTICS] A reflecting telescope designed so that observations can be made at the coudé, Newtonian, or Cassegrain focus. { kü'dā nü'tōn·ē·ən kas·gran 'tel·əˌskōp }

coudé spectrograph [SPECT] A stationary spectrograph that is attached to the tube of a coudé telescope. { kü'dā 'spek·trəˌgraf }

coudé spectroscopy [SPECT] The production and investigation of astronomical spectra using a coudé spectrograph. { kü'dā spek'träs·kə·pē }

coudé telescope [OPTICS] An instrument in which light is reflected along the polar axis to come to focus at a fixed place where it is viewed through a fixed eyepiece or where a spectrograph can be mounted. { kü'dā 'tel·əˌskōp }

Couette flow [FL MECH] Low-speed, steady motion of a viscous fluid between two infinite plates moving parallel to each other. { kü'et ˌflō }

Couette-Taylor flow [FL MECH] The flow of a fluid within the annular space between two concentric cylinders when one or both of the cylinders rotate. { kü'et 'tā·lər ˌflō }

Couette viscometer [ENG] A viscometer in which the liquid whose viscosity is to be measured fills the space between two vertical coaxial cylinders, the inner one suspended by a torsion wire; the outer cylinder is rotated at a constant rate, and the resulting torque on the inner cylinder is measured by the twist of the wire. Also known as rotational viscometer. { kü'et vis'käm·əd·ər }

cougar *See* puma. { 'kü·gər }

cough [MED] A sudden, violent expulsion of air after deep inspiration and closure of the glottis. { kóf }

Couinae [VERT ZOO] The couas, a subfamily of Madagascan birds in the family Cuculidae. { 'kü·əˌnē }

coul *See* coulomb.

coulee [GEOL] **1.** A thick, solidified sheet or stream of lava. **2.** A steep-sided valley or ravine, sometimes with a stream at the bottom. { kü'lā }

coulisse [ENG] A piece of wood that has a groove cut in it to enable another piece of wood to slide in it. Also known as cullis. { kü'lēs }

coulomb [ELEC] A unit of electric charge, defined as the amount of electric charge that crosses a surface in 1 second when a steady current of 1 absolute ampere is flowing across the surface; this is the absolute coulomb and has been the legal standard of quantity of electricity since 1950; the previous standard was the international coulomb, equal to 0.999835 absolute coulomb. Abbreviated coul. Symbolized C. { 'küˌläm }

Coulomb attraction [ELEC] The electrostatic force of attraction exerted by one charged particle on another charged particle of opposite sign. Also known as electrostatic attraction. { 'küˌläm ə'trak·shən }

Coulomb barrier [NUC PHYS] **1.** The Coulomb repulsion which tends to keep positively charged bombarding particles out of the nucleus. **2.** Specifically, the Coulomb potential associated with this force. { 'küˌläm ˌbar·ē·ər }

Coulomb crystal [ATOM PHYS] A crystalline array that is formed from laser-cooled ions stored in an electromagnetic trap and in which the relative positions of the ions are approximately fixed and are determined by the balance between the confining forces of the trap and the Coulomb repulsion of the ions. Also known as ion crystal. [CRYO] A structure formed by electrons trapped at a liquid helium surface at sufficiently high electron densities and low temperatures, in which the electrons occupy the points of a two-dimensional hexagonal lattice. { 'küˌläm ˌkrist·əl }

Coulomb energy [PHYS] The part of the binding energy of a system of particles, such as an atomic nucleus of a solid, which is associated with electrostatic forces between the particles. [PHYS CHEM] The energy associated with the electrostatic interaction between two or more electron distributions in terms of which the actual electron distribution of a covalent bond is described. { 'küˌläm ˌen·ər·jē }

Coulomb excitation [NUC PHYS] Inelastic scattering of a positively charged particle by a nucleus and excitation of the nucleus, caused by the interaction of the nucleus with the rapidly changing electric field of the bombarding particle. { 'kü‚läm ‚ek‚sī'tä·shən }

Coulomb explosion [PHYS] A process in which a molecule moving with high velocity strikes a solid and the electrons that bond the molecule are torn off rapidly in violent collisions with the electrons of the solid; as a result, the molecule is transformed into a cluster of charged atomic constituents that then separate under the influence of their mutual Coulomb repulsion. { 'kü‚läm ik‚splō·zhən }

Coulomb field [ELEC] The electric field created by a stationary charged particle. { 'kü‚läm ‚fēld }

Coulomb force [ELEC] The electrostatic force of attraction or repulsion exerted by one charged particle on another, in accordance with Coulomb's law. { 'kü‚läm ‚fòrs }

Coulomb friction [MECH] Friction occurring between dry surfaces. { 'kü‚läm ‚frik·shən }

Coulomb gage [ELECTROMAG] A gage in which the divergence of the magnetic vector potential is equal to 0. { 'kü‚läm ‚gāj }

Coulomb interactions [ELEC] Interactions of charged particles associated with the Coulomb forces they exert on one another. Also known as electrostatic interactions. { 'kü‚läm in·tər'ak·shənz }

coulombmeter [ENG] An instrument that measures quantity of electricity in coulombs by integrating a stored charge in a circuit which has very high input impedance. { 'kü‚läm‚mēd·ər }

Coulomb potential [ELEC] A scalar point function equal to the work per unit charge done against the Coulomb force in transferring a particle bearing an infinitesimal positive charge from infinity to a point in the field of a specific charge distribution. { kü'läm pə'ten·chəl }

Coulomb repulsion [ELEC] The electrostatic force of repulsion exerted by one charged particle on another charged particle of the same sign. Also known as electrostatic repulsion. { kü'läm ri'pəl·shən }

Coulomb scattering [PHYS] A collision of two charged particles in which the Coulomb force is the dominant interaction. { kü'läm ‚skad·ə·riŋ }

Coulomb's law [ELEC] The law that the attraction or repulsion between two electric charges acts along the line between them, is proportional to the product of their magnitudes, and is inversely proportional to the square of the distance between them. Also known as law of electrostatic attraction. { 'kü'lämz ‚lò }

Coulomb's theorem [ELEC] The proposition that the intensity of an electric field near the surface of a conductor is equal to the surface charge density on the nearby conductor surface divided by the absolute permittivity of the surrounding medium. { 'kü'lämz ‚thir·əm }

coulometer [PHYS CHEM] An electrolytic cell for the precise measurement of electrical quantities or current intensity by quantitative determination of chemical substances produced or consumed. Also known as voltameter. { kü'läm·əd·ər }

coulometric analysis [ANALY CHEM] A technique in which the amount of a substance is determined quantitatively by measuring the total amount of electricity required to deplete a solution of the substance. { ‚kü·lə'me·trik ə'nal·ə·səs }

coulometric titration [ANALY CHEM] The slow electrolytic generation of a soluble species which is capable of reacting quantitatively with the substance sought; some independent property must be observed to establish the equivalence point in the reaction. { ‚kü·lə'me·trik tī'trā·shən }

coulometry [ANALY CHEM] A determination of the amount of an electrolyte released during electrolysis by measuring the number of coulombs used. { kə'läm·ə·trē }

coulostatic analysis [PHYS CHEM] An electrochemical technique involving the application of a very short, large pulse of current to the electrode; the pulse charges the capacitive electrode-solution interface to a new potential, then the circuit is opened, and the return of the working electrode potential to its initial value is monitored; the current necessary to discharge the electrode interface comes from the electrolysis of electroactive species in solution; the change in electrode potential versus time results in a plot, the shape of which is proportional to concentration. { ‚kü·lə'stad·ik ə'nal·ə·səs }

coulter See colter. { 'kōl·tər }

Coulter counter [MICROBIO] An electronic device for counting the number of cells in a liquid culture. { 'kōl·tər ‚kaúnt·ər }

coumachlor [ORG CHEM] $C_{19}H_{15}ClO_4$ A white, crystalline compound with a melting point of 169–171°C; insoluble in water; used as a rodenticide. { 'kü·mə‚klòr }

coumarin [ORG CHEM] $C_9H_6O_2$ The anhydride of o-coumaric acid; a toxic, white, crystalline lactone found in many plants and made synthetically; used in making perfume and soap. Also known as 1,2-benzopyrone. { 'kü·mə·rən }

coumarin glycoside [BIOCHEM] Any of several glycosidic aromatic principles in many plants; contains coumaric acid as the aglycon group. { 'kü·mə·rən 'glī·kə‚sīd }

coumarone [ORG CHEM] C_8H_6O A colorless liquid, boiling point 169°C. { 'kü·mə‚rōn }

coumarone-indene resin [ORG CHEM] A synthetic resin prepared by polymerization of coumarone and indene. { 'kü·mə‚rōn 'in‚dēn ‚rez·ən }

coumatetralyl [ORG CHEM] $C_{19}H_{16}O_3$ A yellow-white, crystalline compound with a melting point of 172–176°C; slightly soluble in water; used as a rodenticide. { ‚kü·mə'te‚trə‚lil }

count [AERO ENG] **1.** To proceed from one point to another in a countdown or plus count, normally by calling a number to signify the point reached. **2.** To proceed in a countdown, for example, T minus 90 and counting. [CHEM] An ionizing event. [DES ENG] The number of openings per linear inch in a wire cloth. [MATH] **1.** To name a set of consecutive positive integers in order of size, usually starting with 1. **2.** To associate consecutive positive integers, starting with 1, with the members of a finite set in order to determine the cardinal number of the set. [NUCLEO] **1.** A single response of the counting system in a radiation counter. **2.** The total number of events indicated by a counter. [TEXT] The number of warp and filling threads per square inch of fabric. { kaúnt }

countability axioms [MATH] Two conditions which are satisfied by a euclidean space and one or the other of which is often assumed in the study of a general topological space; the first states that any point in the topological space has a countable local base, while the second states that the topological space has a countable base. { ‚kaúnt·ə'bil·əd·ē ‚ax·sē·əmz }

countable [MATH] Either finite or denumerable. Also known as enumerable. { 'kaúnt·ə·bəl }

countably additive [MATH] Given a measure m, and a sequence of pairwise disjoint measurable sets, the property that the measure of the union is equal to the sum of the measures of the sets. { 'kaúnt·ə·blē 'ad·əd·iv }

countably additive set function [MATH] A set function with the properties that (1) the union of any finite or countable collection of sets in the range of the function is also in this range, and (2) the value of the function at the union of a finite or countable collection of sets that are in the range of the set function and are pairwise disjoint is equal to the sum of the values at each set in the collection. Also known as completely additive set function. { 'kaúnt·ə·blē 'ad·əd·iv 'set ‚faŋk·shən }

countably compact set [MATH] A set with the property that every cover with countably many open sets contains a finite number of sets which is also a cover. { 'kaúnt·ə·blē ‚käm‚pakt 'set }

countably infinite set See denumerable set. { 'kaúnt·ə·blē ‚in·fə·nət 'set }

countably metacompact space [MATH] A topological space with the property that every open covering F is associated with a point-finite open covering G, such that every element of G is a subset of an element of F. { ‚kaúnt·ə·blē ‚med·ə‚käm‚pakt 'spās }

countably paracompact space [MATH] A topological space with the property that every countable open covering F is associated with a locally finite open covering G, such that every element of G is a subset of an element of F. { ‚kaúnt·ə·blē ‚par·ə‚käm‚pakt 'spās }

countably subaddictive [MATH] A set function m is countably subaddictive if, given any sequence of sets, the measure of the union is less than or equal to the sum of the measures of the sets. { ‚kaúnt·ə·blē səb'ad·əd·iv }

countably subaddictive set function [MATH] A real-valued function defined on a class of sets such that the value of the function on the union of any sequence of sets is equal to or less than the sum of the sequence of the values of the function on the sets. { ¦kaún·tə·blē səb¦ad·əd·iv 'set ¸fəŋk·shən }

count cycle [COMPUT SCI] An increase or decrease of the cycle index by unity or by an arbitrary integer. { 'kaúnt ¸sī·kəl }

countdown [AERO ENG] **1.** The process in the engineering definition, used in leading up to the launch of a large or complicated rocket vehicle, or in leading up to a captive test, a readiness firing, a mock firing, or other firing test. **2.** The act of counting inversely during this process. [COMMUN] The ratio of the number of interrogation pulses not answered by a transponder to the total number received. [ENG] A step-by-step process that culminates in a climatic event, each step being performed in accordance with a schedule marked by a count in inverse numerical order. { 'kaúnt¸daún }

counter [COMPUT SCI] **1.** A register or storage location used to represent the number of occurrences of an event. *See* accumulator. [ELECTR] *See* scaler. [ENG] A complete instrument for detecting, totalizing, and indicating a sequence of events. [NAV ARCH] *See* buttocks. [NUCLEO] *See* radiation counter. { 'kaúnt·ər }

counterbalance *See* counterweight. { ¦kaúnt·ər¦bal·əns }

counterbalanced truck [MECH ENG] An industrial truck configured so that all of its load during a normal transporting operation is external to the polygon formed by the points where the wheels contact the surface. { ¦kaún·tər¦bal·ənst 'trək }

counterbalance system *See* two-step grooving system. { ¦kaúnt·ər¦bal·əns ¸sis·təm }

counterblow hammer [MECH ENG] A forging hammer in which the ram and anvil are driven toward each other by compressed air or steam. { 'kaúnt·ər¸blō ¸ham·ər }

counterbore [DES ENG] A flat-bottom enlargement of the mouth of a cylindrical bore to enlarge a borehole and give it a flat bottom. [ENG] To enlarge a borehole by means of a counterbore. { 'kaúnt·ər¸bór }

counter circuit *See* counting circuit. { 'kaúnt·ər ¸sər·kət }

counter coupling [COMPUT SCI] The technique of combining two or more counters into one counter of larger capacity in electromechanical devices by means of control panel wiring. { 'kaúnt·ər ¸kəp·liŋ }

countercurrent [SCI TECH] A current flowing adjacent to the main current but in the opposite direction. { 'kaúnt·ər¸kər·ənt }

countercurrent cascade [ANALY CHEM] An extraction process involving the introduction of a sample, all at once, into a continuously flowing countercurrent system where both phases are moving in opposite directions and are continuously at equilibrium. { 'kaúnt·ər¸kər·ənt kas'kād }

countercurrent distribution [CHEM ENG] A profile of a compound's concentration in different ratios of two immiscible liquids. { 'kaúnt·ər¸kər·ənt dis·trə'byü·shən }

countercurrent extraction [CHEM ENG] A liquid-liquid extraction process in which the solvent and the process stream in contact with each other flow in opposite directions. Also known as countercurrent separation. { 'kaúnt·ər¸kər·ənt ¸ek'strak·shən }

countercurrent flow [MECH ENG] A sensible heat-transfer system in which the two fluids flow in opposite directions. { 'kaúnt·ər¸kər·ənt 'flō }

countercurrent leaching [CHEM ENG] A process utilizing a series of leach tanks and countercurrent flow of solvent through them in reverse order to the flow of solid. { 'kaúnt·ər¸kər·ənt 'lēch·iŋ }

countercurrent separation *See* countercurrent extraction. { 'kaúnt·ər¸kər·ənt ¸sep·ə'rā·shən }

countercurrent spray dryer [ENG] A dryer in which drying gases flow in a direction opposite to that of the spray. { 'kaúnt·ər¸kər·ənt 'sprā ¸drī·ər }

counter dead time [NUCLEO] The time interval between the start of a counted event and the earliest instant at which a new event can be counted by a radiation counter. { 'kaúnt·ər 'ded ¸tīm }

counter decade *See* decade scaler. { 'kaúnt·ər ¸dek·ād }

counterelectromotive cell [ELEC] Cell of practically no ampere-hour capacity, used to oppose the line voltage. { ¦kaúnt·ər·i¸lek·trō¦mōd·iv 'sel }

counterelectromotive force [ELECTROMAG] The voltage developed in an inductive circuit by a changing current; the polarity of the induced voltage is at each instant opposite that

of the applied voltage. Also known as back electromotive force. { ¦kaúnt·ər·i¸lek·trō¦mōd·iv 'fórs }

counterfire [ORD] **1.** Fire delivered in answer to the fire of an attacker. **2.** Fire intended to destroy or neutralize enemy weapons. { 'kaúnt·ər¸fīr }

counterfloor *See* subfloor. { 'kaún·tər¸flór }

counterflow [ENG] Fluid flow in opposite directions in adjacent parts of an apparatus, as in a heat exchanger. { 'kaúnt·ər¸flō }

counterfort [CIV ENG] A strengthening pier perpendicular and bonded to a retaining wall. { 'kaúnt·ər¸fórt }

counterfort wall [CIV ENG] A type of retaining wall that resembles a cantilever wall but has braces at the back; the toe slab is a cantilever and the main steel is placed horizontally. { 'kaúnt·ər¸fórt ¸wól }

counter-free machine [COMPUT SCI] A sequential machine that cannot count modulo any integer greater than 1. { 'kaúnt·ər ¸frē mə'shēn }

counter/frequency meter [ENG] An instrument that contains a frequency standard and can be used to measure the number of events or the number of cycles of a periodic quantity that occurs in a specified time, or the time between two events. { 'kaúnt·ər 'frē·kwən·sē ¸mēd·ər }

counterglow *See* gegenschein. { 'kaúnt·ər¸glō }

counterimmunoelectrophoresis [IMMUNOL] Immuno-electrophoresis which uses two wells of application, one above the other, along the electrical axis—the anodal well filled with antibody and the cathodal with a negatively charge antigen; electrophoresis results in the antigen and antibody migrating cathodally and anodally, respectively, and a line of precipitation appears where the two meet. { ¦kaúnt·ər¦im·yə·nō·i¸lek·trō·fə'r ē·səs }

counterion [PHYS CHEM] In a solution, an ion with a charge opposite to that of another ion included in the ionic makeup of the solution. { 'kaúnt·ər¸ī¸än }

counterlath [BUILD] **1.** A strip placed between two rafters to support crosswise laths. **2.** A lath placed between a timber and a sheet lath. **3.** A lath nailed at a more or less random spacing between two precisely spaced laths. **4.** A lath put on one side of a partition after the other side has been finished. { 'kaúnt·ər¸lath }

countermeasures [ORD] Devices and techniques intended to impair the operational effectiveness of enemy activity. { 'kaúnt·ər¸mezh·ərz }

countermeasures set [ELECTR] A complete electronic set specifically designed to provide facilities for intercepting and analyzing electromagnetic energy propagated by transmitter and to provide a source of radio-frequency signals which deprive the enemy of effective use of his electronic equipment. { 'kaúnt·ər¸mezh·ərz ¸set }

counterpoise [ELEC] A system of wires or other conductors that is elevated above and insulated from the ground to form a lower system of conductors for an antenna. Also known as antenna counterpoise. [MECH ENG] *See* counterweight. { 'kaúnt·ər¸póiz }

counterpoise gun carriage [ORD] The mechanism that counterbalances the weight of the breechblock of a large gun. { 'kaúnt·ər¸póiz 'gən ¸kar·ij }

counterpoise method *See* substitution weighing. { 'kaún·tər¸póiz ¸meth·əd }

counterpropagating beams [OPTICS] Two light beams that are propagating through a medium in precisely opposite directions. { ¸kaún·tər¸präp·ə¸gäd·iŋ 'bēmz }

counterradiation [GEOPHYS] The downward flux of atmospheric radiation passing through a given level surface, usually taken as the earth's surface. Also known as back radiation. { ¦kaúnt·ər¸rād·ē'ā·shən }

counterrecoil [ORD] Forward movement of a gun returning to firing position after recoil. { ¦kaúnt·ər·ri'kóil }

countershaft [MECH ENG] A secondary shaft that is driven by a main shaft and from which power is supplied to a machine part. { 'kaúnt·ər¸shaft }

countersink [DES ENG] The tapered and relieved cutting portion in a twist drill, situated between the pilot drill and the body. { 'kaúnt·ər¸siŋk }

countersinking [MECH ENG] Drilling operation to form a flaring depression around the rim of a hole. { 'kaúnt·ər¸siŋk·iŋ }

counterstain [BIOL] A second stain applied to a biological

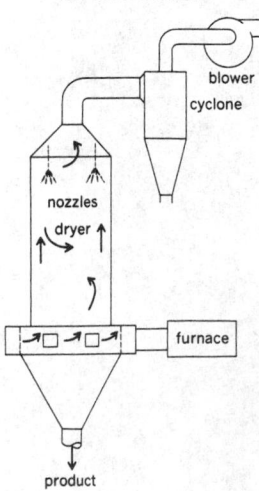

COUNTERCURRENT SPRAY DRYER

blower

cyclone

nozzles

dryer

furnace

product

Liquid feed is atomized from nozzles at the top; arrows show the path that the drying gases take.

specimen to color elements other than those demonstrated by the principal stain. { 'kau̇nt·ər,stān }

counter sun *See* anthelion. { 'kau̇nt·ər,sən }

countersunk bolt [DES ENG] A bolt that has a circular head, a flat top, and a conical bearing surface tapering in from the top; in place, the head is flush-mounted. { ¦kau̇n·tər,səŋk 'bōlt }

counter terms [QUANT MECH] Additional terms added to a Lagrangian in quantum field theory in order to absorb the typical divergences that occur in a perturbation expansion of the theory. { 'kau̇n·tər ,tərmz }

countertransference [PSYCH] The conscious or unconscious emotional reaction of the therapist to the patient, which may interfere with psychotherapy. { 'kau̇nt·ər·tranz'fər·əns }

counter tube [ELECTR] An electron tube having one signal-input electrode and 10 or more output electrodes, with each input pulse serving to transfer conduction sequentially to the next output electrode; beam-switching tubes and cold-cathode counter tubes are examples. [NUCLEO] An electron tube that converts an incident particle or burst of incident radiation into a discrete electric pulse, generally by utilizing the current flow through a gas that is ionized by the radiation; used in radiation counters. Also known as radiation counter tube. { 'kau̇nt·ər ,tüb }

counter voltage [ELEC] The reverse voltage that appears across an inductor when current through the inductor is shut off. { 'kau̇nt·ər ,vōl·tij }

counterweight [MECH ENG] **1.** A device which counterbalances the original load in elevators and skip and mine hoists, going up when the load goes down, so that the engine must only drive against the unbalanced load and overcome friction. **2.** Any weight placed on a mechanism which is out of balance so as to maintain static equilibrium. Also known as counterbalance; counterpoise. { 'kau̇nt·ər,wāt }

counting chamber [MICROBIO] An accurately dimensioned chamber in a microslide which can hold a specific volume of fluid and which is usually ruled into units to facilitate the counting under the microscope of cells, bacteria, or other structures in the fluid. { 'kau̇nt·iŋ ,chām·bər }

counting circuit [ELECTR] A circuit that counts pulses by frequency-dividing techniques, by charging a capacitor in such a way as to produce a voltage proportional to the pulse count, or by other means. Also known as counter circuit. { 'kau̇nt·iŋ ,sər·kət }

counting-down circuit *See* frequency divider. { 'kau̇nt·iŋ ,dau̇n ,sər·kət }

counting glass [TEXT] A magnifying glass for counting the number of threads per inch in textile fabrics. { 'kau̇nt·iŋ ,glas }

counting ionization chamber *See* pulse ionization chamber. { 'kau̇nt·iŋ ,ī·ə·nə'zā·shən ,chām·bər }

counting number [MATH] One of the numbers used in counting objects, either the set of positive integers or the set of positive integers and the number 0. { 'kau̇nt·iŋ ,nəm·bər }

counting rate [PHYS] The average rate of occurrence of events as observed by means of a counting system. { 'kau̇nt·iŋ ,rāt }

counting rate meter [NUCLEO] An instrument that indicates the time rate of occurrence of input pulses to a radiation counter, averaged over a time interval. Also known as rate meter. { 'kau̇nt·iŋ ,rāt ,mēd·ər }

counting rate-voltage characteristic *See* plateau characteristic. { 'kau̇nt·iŋ ,rāt 'vōl·tij ,kar·ik·tə'ris·tik }

country rock [GEOL] **1.** Rock that surrounds and is penetrated by mineral veins. **2.** Rock that surrounds and is invaded by an igneous intrusion. { ¦kən·trē 'räk }

count system *See* numerical system. { 'kau̇nt ,sis·təm }

couplant [ENG] A substance such as water, oil, grease, or paste used to avoid the retarding of sound transmission by air between the transducer and the test piece during ultrasonic examination. { 'kəp·lənt }

couple [CHEM] Joining of two molecules. [ELEC] To connect two circuits so signals are transferred from one to the other. [ELECTR] Two metals placed in contact, as in a thermocouple. [ENG] To connect with a coupling, such as of two belts or two pipes. [MECH] A system of two parallel forces of equal magnitude and opposite sense. { 'kəp·əl }

coupled antenna [ELECTROMAG] An antenna electromagnetically coupled to another. { 'kəp·əld an'ten·ə }

coupled circuits [ELEC] Two or more electric circuits so arranged that energy can transfer electrically or magnetically from one to another. { 'kəp·əld 'sər·kəts }

coupled column [ARCH] One of two columns used as a grouped pair. { 'kəp·əld 'käl·əm }

coupled engine [MECH ENG] A locomotive engine having the driving wheels connected by a rod. { 'kəp·əld 'en·jən }

coupled field vectors [ELECTROMAG] The electric-and magnetic-field vectors, which depend upon each other according to Maxwell's field equations. { 'kəp·əld 'fēld ,vek·tərz }

coupled harmonic oscillators [PHYS] Linear oscillators with an interaction, often also linear or weak. { 'kəp·əld har'män·ik 'äs·ə,lād·ərz }

coupled modes [ACOUS] Modes of acoustic transmission along a duct having a discontinuity, so that the reflected and transmitted waves contain modes other than the incident ones. { 'kəp·əld 'mōdz }

coupled oscillators [ELECTROMAG] A set of alternating-current circuits which interact with each other, for example, through mutual inductances or capacitances. [MECH] A set of particles subject to elastic restoring forces and also to elastic interactions with each other. { 'kəp·əld 'äs·ə,lād·ərz }

coupled reaction [CHEM] A reaction which involves two oxidants with a single reductant, where one reaction taken alone would be thermodynamically unfavorable. { 'kəp·əld rē'ak·shən }

coupled systems [COMPUT SCI] Computer systems that share equipment and can exchange information. [PHYS] Mechanical, electrical, or other systems which are connected in such a way that they interact and exchange energy with each other. { 'kəp·əld 'sis·təmz }

coupled transistors [ELECTR] Transistors connected in series by transformers or resistance-capacitance networks, in much the same manner as electron tubes. { 'kəp·əld tran'zis·tərz }

coupled wave [FL MECH] A surface wave which is being continuously generated by another wave having the same phase velocity. Also known as C wave. { 'kəp·əld 'wāv }

coupler [ELEC] A component used to transfer energy from one circuit to another. [ELECTROMAG] **1.** A passage which joins two cavities or waveguides, allowing them to exchange energy. **2.** A passage which joins the ends of two waveguides, whose cross section changes continuously from that of one to that of the other. [ENG] A device that connects two railroad cars. [GRAPHICS] A substance that can react with the unexposed diazonium salt in a diazo material to produce the visible dye image. [NAV] The portion of a navigation system that receives signals of one type from a sensor and transmits signals of a different type to an actuator. { 'kəp·lər }

coupling [ELEC] **1.** A mutual relation between two circuits that permits energy transfer from one to another, through a wire, resistor, transformer, capacitor, or other device. **2.** A hardware device used to make a temporary connection between two wires. [ENG] **1.** Any device that serves to connect the ends of adjacent parts, as railroad cars. **2.** A metal collar with internal threads used to connect two sections of threaded pipe. [MECH ENG] The mechanical fastening that connects shafts together for power transmission. Also known as shaft coupling. { 'kəp·liŋ }

coupling agent [CHEM] A substance that can react with both reinforcement and matrix components of a composite material to form a binding link at their interface. { 'kəp·liŋ ,ā·jənt }

coupling aperture [ELECTROMAG] An aperture in the wall of a waveguide or cavity resonator, designed to transfer energy to or from an external circuit. Also known as coupling hole; coupling slot. { 'kəp·liŋ ,ap·ə·chər }

coupling capacitor [ELECTR] A capacitor used to block the flow of direct current while allowing alternating or signal current to pass; widely used for joining two circuits or stages. Also known as blocking capacitor; stopping capacitor. { 'kəp·liŋ kə'pas·əd·ər }

coupling coefficient [ELECTR] The ratio of the maximum change in energy of an electron traversing an interaction space to the product of the peak alternating gap voltage and the electronic charge. [PHYS] *See* coupling constant. { 'kəp·liŋ ,kō·i'fish·ənt }

coupling constant [PARTIC PHYS] A measure of the strength of a type of interaction between particles, such as the strong

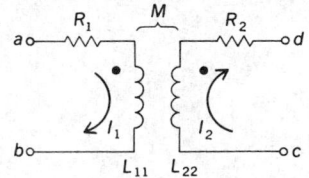

COUPLED CIRCUITS

A pair of coupled circuits.

interaction between mesons and nucleons, and the weak interaction between four fermions; analogous to the electric charge, which is the coupling constant between charged particles and electromagnetic radiation. [PHYS] **1.** A measure of the strength of the coupling between two systems, especially electric circuits; maximum coupling is 1 and no coupling is 0. Also known as coefficient of coupling; coupling coefficient. **2.** A measure of the dependence of one physical quantity on another. { 'kəp·liŋ 'kän·stənt }

coupling hole *See* coupling aperture. { 'kəp·liŋ ,hōl }

coupling loop [ELECTROMAG] A conducting loop projecting into a waveguide or cavity resonator, designed to transfer energy to or from an external circuit. { 'kəp·liŋ ,lüp }

coupling probe [ELECTROMAG] A probe projecting into a waveguide or cavity resonator, designed to transfer energy to or from an external circuit. { 'kəp·liŋ ,prōb }

coupling slot *See* coupling aperture. { 'kəp·liŋ ,slät }

coupon [CHEM ENG] Polished metal strip of specified size and weight used to detect the corrosive action of liquid or gas products or to test the efficiency of corrosion-inhibitor additives. Also known as corrosion coupon. { 'kü,pän }

Courant condition [FL MECH] A condition on numerical hydrodynamics calculations requiring that the time interval employed be no greater than that required for a sound wave to cross a spatial cell. { 'kür,änt kən,dish·ən }

course [CIV ENG] A row of stone, block, or brick of uniform height. [NAV] The intended direction of travel expressed as an angle in the horizontal plane between a reference line (true magnetic north) and the course line (the line connecting the point of origin and the point of destination), usually measured clockwise from the reference line. Also known as desired track. [TEXT] A row of stitches across a knitted fabric; corresponds to the filling in woven fabric. { 'kórs }

course angle [NAV] Course measured from 0° at the reference direction clockwise or counterclockwise through 90° or 180°; it is labeled with the reference direction as a prefix and the direction of measurement from the reference direction as a suffix; for example, course angle S 21° E is 21° east of south, or course 159°. { 'kórs ,aŋ·gəl }

coursed rubble [CIV ENG] Masonry in which rough stones are fitted into approximately level courses. { ¦kórsd 'rəb·əl }

course error [NAV] Angular difference between the course and the course made good. { 'kórs ,er·ər }

course line [NAV] **1.** A line of position plotted on a chart, parallel or substantially parallel to the intended course of a craft, showing whether the craft is to the right or the left of its course. **2.** Any line representing a course. { 'kórs ,līn }

course linearity [NAV] In equisignal radio-range-type beacons, the change in the difference in the depth of modulation of the two signals which produce the course with respect to displacement of the measuring position from the course line but within the course sector. { 'kórs ,lin·ē'ar·əd·ē }

course-line computer [NAV] An airborne computer that accepts bearing and distance information derived from ground facilities and uses this data to compute a course from the aircraft's present position to any other point which the pilot selects, providing only that it be within the coverage of the ground facilities; the steering information is displayed on a right-left indicator, while additional displays such as distance to go are also provided. Also known as arbitrary course computer; bearing distance computer; off-line computer; parallel course computer; rho-theta computer. { 'kórs ,līn kəm'pyüd·ər }

course-line deviation [NAV] The angular or linear difference between the actual track of a vehicle and the intended course line. { 'kórs ,līn ,dēv·ē'ā·shən }

course-line deviation indicator [NAV] An instrument that indicates deviation from a desired course line. { 'kórs ,līn ,dēv·ē'ā·shən ,in·də,kād·ər }

course-line selector [NAV] A device providing means for selecting a course to be followed automatically, usually by means of an electronic system of navigation such as omnirange and distance-measuring equipment. { 'kórs ,līn si'lek·tər }

course made good [NAV] The resultant direction of actual travel of a vehicle, equivalent to its bearing from the point of departure. { 'kórs ¦mād 'güd }

course over the ground [NAV] The direction of the track that a vessel has actually made, measured clockwise from north through 360°. { 'kórs ,ō·vər ‖thə 'graúnd }

course programmer [CONT SYS] An item which initiates and processes signals in a manner to establish a vehicle in which it is installed along one or more projected courses. { 'kórs 'prō,gram·ər }

course recorder [NAV] A device which makes a graphic record of the headings and distances traveled. { 'kórs ri'kórd·ər }

course scalloping [NAV] Irregularities in the field pattern produced by a ground-based beacon caused by terrain features; appears in flight as cyclical variation in the course error. { 'kórs ,skal·əp·iŋ }

course softening [NAV] An intentional decrease in course sensitivity as the navigation aid is approached. { 'kórs ,sóf·ən·iŋ }

courseware [COMPUT SCI] Computer programs designed to be used in computer-aided instruction or computer-managed instruction. { 'kórs,wer }

coursing joint [CIV ENG] A mortar joint connecting two courses of brick or pebble. { 'kórs·iŋ ,jóint }

courtship [ECOL] A sequence of behavioral patterns that eventually may lead to completed mating. { 'kórt,ship }

Couvinian [GEOL] Lower Middle Devonian geologic time. { kü'vin·ē·ən }

covalence [CHEM] The number of covalent bonds which an atom can form. { kō'vā·ləns }

covalent bond [CHEM] A bond in which each atom of a bound pair contributes one electron to form a pair of electrons. Also known as electron pair bond. { kō'vā·lənt 'bänd }

covalent crystal [CRYSTAL] A crystal held together by covalent bonds. Also known as valence crystal. { kō'vā·lənt 'krist·əl }

covalent hydride [INORG CHEM] A compound formed from a nonmetal and hydrogen, for example, H_2S and NH_3. { kō'vā·lənt 'hī,drīd }

covalent radius *See* atomic radius. { kō'vā·lənt 'rād·ē·əs }

covariance [STAT] A measurement of the tendency of two random variables, X and Y, to vary together, given by the expected value of the variable $(X - X[OB])(Y - Y[OB])$, where $X[OB]$ and $Y[OB]$ are the expected values of the variables X and Y respectively. { kō'ver·ē·əns }

covariance analysis [STAT] An extension of the analysis of variance which combines linear regression with analysis of variance; used when members falling into classes have values of more than one variable. { kō'ver·ē·əns ə,nal·ə·səs }

covariant [RELAT] A scalar, vector, or higher-order tensor. { kō'ver·ē·ənt }

covariant components [MATH] Vector or tensor components which, in a transformation from one set of basis vectors to another, transform in the same manner as the basis vectors. { kō'ver·ē·ənt kəm'pō·nəns }

covariant derivative [MATH] For a tensor field at a point P of an affine space, a new tensor field equal to the difference between the derivative of the original field defined in the ordinary manner and the derivative of a field whose value at points close to P are parallel to the value of the original field at P as specified by the affine connection. { kō'ver·ē·ənt də'riv·əd·iv }

covariant equation [PHYS] An equation which has the same form in all inertial frames of reference; that is, its form is unchanged by Lorentz transformations. { kō'ver·ē·ənt i'kwā·zhən }

covariant functor [MATH] A functor which does not change the sense of morphisms. { kō'ver·ē·ənt 'fəŋk·tər }

covariant index [MATH] A tensor index such that, under a transformation of coordinates, the procedure for obtaining a component of the transformed tensor for which this index has value p involves taking a sum over q of the product of a component of the original tensor for which the index has the value q times the partial derivative of the qth original coordinate with respect to the pth transformed coordinate; it is written as a subscript. { kō'ver·ē·ənt 'in,deks }

covariant tensor [MATH] A tensor with only covariant indices. { kō'ver·ē·ənt 'ten·sər }

covariant theory [PHYS] A theory in which the equations have the same form in any inertial reference frame, the frames being related to each other by Lorentz transformations. { kō'ver·ē·ənt 'thē·ə·rē }

covariant vector [MATH] A covariant tensor of degree 1, such as the gradient of a function. { kō'ver·ē·ənt 'vek·tər }

cove [GEOGR] **1.** A small, narrow, sheltered bay, inlet, or creek on a coast. **2.** A deep recess or hollow occurring in a cliff or steep mountainside. { kōv }

covellite [MINERAL] CuS An indigo-blue mineral of metallic luster that crystallizes in the hexagonal system; it is usually massive or occurs in disseminations through other copper minerals and represents an ore of copper. Also known as indigo copper. { kō′ve‚līt }

cover [MATH] **1.** An element, x, of a partially ordered set covers another element y if x is greater than y, and the only elements that are both greater than or equal to y and less than or equal to x are x and y themselves. **2.** See also covering. [MIN ENG] The thickness of rock between the mine workings and the surface. { ′kəv·ər }

coverage [COMMUN] See service area. [GRAPHICS] In microfilming, the portion of a document plane included in the lens field. { ′kəv·rij }

cover crops [AGR] Crops, especially grasses, grown for the express purpose of preventing and protecting a bare soil surface. { ′kəv·ər ‚kräps }

covered electrode See coated electrode. { ′kəv·ərd i′lek‚trōd }

covered smut [PL PATH] A seed-borne smut of certain grain crops caused by *Ustilago hordei* in barley and *U. avenae* in oats. { ′kəv·ərd ′smət }

cover half [MET] The stationary portion of a die. { ′kəv·ər ‚haf }

cover hole [MIN ENG] One of a group of boreholes drilled in advance of mine workings to probe for and detect water-bearing fissures or structures. { ′kəv·ər ‚hōl }

covering [MATH] For a set A, a collection of sets whose union contains A. Also known as cover. { ′kəv·riŋ }

covering power [ENG] The degree to which a coating obscures the underlying material. [MET] The ability of an electroplating bath to produce a coating at a low current density. [OPTICS] The field of view over which a camera lens can produce a sharp image, frequently expressed as an angle. { ′kəv·riŋ ‚paů·ər }

cover plate [ENG] A pane of glass in a welding helmet or goggles which protects the colored lens excluding harmful light rays from damage by weld spatter. { ′kəv·ər ‚plāt }

covers See coversed sine.

coversed sine [MATH] The coversed sine of A is $1 −$ sine A. Denoted covers. Also known as coversine; versed cosine. { ‚kō′vərst ′sīn }

cover sheet See emulsion sheet. { ′kəv·ər ‚shēt }

coversine See coversed sine. { ‚kō‚vər′sīn }

covert [ECOL] A refuge or shelter, such as a coppice, for game animals. [TEXT] A tightly woven woolen twill fabric made by using a single-color yarn for filling threads and yarns in two different shades in the warp. { ′kō·vərt }

covey [VERT ZOO] **1.** A brood of birds. **2.** A small flock of birds of one kind, used typically of partridge and quail. { ′kəv·ē }

covite [PETR] A rock of igneous origin composed of sodic orthoclase, hornblende, sodic pyroxene, nepheline, and accessory sphene, apatite, and opaque oxides. { ′kō‚vīt }

cow [AGR] A domestic bovine of any sex or age. [VERT ZOO] A mature female cattle of the genus *Bos*. { kaů }

Cowdria [MICROBIO] A genus of the tribe Ehrlichieae; coccoid to ellipsoidal, pleomorphic, or rod-shaped cells; intracellular parasites in cytoplasm and vacuoles of vascular endothelium of ruminants. { ′kaů·drē·ə }

Cowell method [AERO ENG] A method of orbit computation using direct step-by-step integration in rectangular coordinates of the total acceleration of the orbiting body. { ′kaů·əl ‚meth·əd }

cowling [AERO ENG] The streamlined metal cover of an aircraft engine. [ENG] A metal cover that houses an engine. { ′kaů·liŋ }

cowoven fabric [TEXT] A fabric woven of two different types of fibers. { ‚kō‚wō·vən ′fab·rik }

cowpea [BOT] *Vigna sinensis*. An annual legume in the order Rosales cultivated for its edible seeds. Also known as black-eye bean. { ′kaů‚pē }

cowpea mosaic virus group See Comovirus. { ′kaů‚pē mō‚zā·ik ′vī·rəs ‚grüp }

Cowper's gland See bulbourethral gland. { ′küp·ərz ‚gland }

cowpox See vaccinia. { ′kaů‚päks }

cowpox virus [VIROL] The causative agent of cowpox in cattle. { ′kaů‚päks ‚vī·rəs }

cowshee See kaus. { ′kaů·shē }

coxa [INV ZOO] The proximal or basal segment of the leg of insects and certain other arthropods which articulates with the body. { ′käk·sə }

coxal cavity [INV ZOO] A cavity in which the coxa of an arthropod limb articulates. { ′käk·səl ′kav·əd·ē }

coxal gland [INV ZOO] One of certain paired glands with ducts opening in the coxal region of arthropods. { ′käk·səl ′gland }

Cox chart [CHEM] A straight-line graph of the logarithm of vapor pressure against a special nonuniform temperature scale; vapor pressure-temperature lines for many substances intersect at a common point on the Cox chart. { ′käks ‚chärt }

Coxiella [MICROBIO] A genus of the tribe Rickettsieae; short rods which grow preferentially in host cell vacuoles. { ‚käk·sē′el·ə }

coxitis [MED] Inflammation of the hip joint. { käk′sīd·əs }

coxopodite [INV ZOO] The basal joint of a crustacean limb. { käk′säp·ə‚dīt }

coxsackie disease [MED] A variety of syndromes resulting from a coxsackievirus infection. { kůk′säk·ē di′zēz }

coxsackievirus [VIROL] A large subgroup of the enteroviruses in the picornavirus group including various human pathogens. { kůk′säk·ē‚vī·rəs }

coyote [VERT ZOO] *Canis latrans*. A small wolf native to western North America but found as far eastward as New York State. Also known as prairie wolf. { ′kī‚ōd·ē }

coyote blasting [MIN ENG] A method of blasting in which large charges are fired in small adits or tunnels driven at the level of the floor, in the face of a quarry or the slope of an open-pit mine. Also known as coyote-hole blasting; gopher hole blasting; heading blasting. { ′kī‚ōd·ē ‚blast·iŋ }

coyote hole See gopher hole. { ′kī‚ōd·ē ‚hōl }

coyote-hole blasting See coyote blasting. { ′kī‚ōd·ē ‚hōl ‚blast·iŋ }

COZI [COMMUN] An ionospheric sounding system for determining propagation characteristics of the ionosphere at various angles at any instant; used to determine how well long-distance, high-frequency broadcasts are reaching their intended destinations. Derived from communications zone indicator. { ‚kō′zī }

cp See candlepower; centipoise; chemically pure.

cpa See color-phase alternation.

CPD See cephalopelvic disproportion.

CPE See computer performance evaluation.

C-peptide [BIOCHEM] A metabolically inactive polypeptide chain that is a by-product of normal insulin production by the beta cells in the pancreas. { ′sē ′pep‚tīd }

CP invariance [PARTIC PHYS] The principle that the laws of physics are left unchanged by a combination of the operations of charge conjugation C and space inversion P; a small violation of this principle has been observed in the decay of neutral K-mesons. { ′sē′pē in′ver·ē·əns }

C₃ plant [BOT] A plant that produces the 3-carbon compound phosphoglyceric acid as the first stage of photosynthesis. { ′sē′thrē ‚plant }

C₄ plant [BOT] A plant that produces the 4-carbon compound oxalocethanoic (oxaloacetic) acid as the first stage of photosynthesis. { ′sē′fōr ‚plant }

cpm See cycle per minute.

CPM See critical path method.

C power supply [ELECTR] A device connected in the circuit between the cathode and grid of a vacuum tube to apply grid bias. { ′sē ′paůr sə‚plī }

CP propeller See controllable-pitch propeller. { ′sē′pē prə′pel·ər }

CPR See cardiopulmonary resuscitation.

cps See hertz.

CPT theorem [PARTIC PHYS] A theorem which states that a Lorentz invariant field theory is invariant to the product of charge conjugation C, space inversion P, and time reversal T. { ′sē′pē′tē ′thir·əm }

CPU See central processing unit.

CPU-bound program [COMPUT SCI] A computer program that involves a large amount of calculation and internal rearrangement of data, so that the speed of execution depends on the speed of the central processing unit (CPU) and memory.

Also known as compute-bound program; cycle-bound program; process-bound program. { ˌsēˌpeˈyü ˈbau̇nd ˌprōˈgrəm }

CPU fan [COMPUT SCI] A fan mounted directly over the integrated-circuit chip containing a computer's central processing unit to prevent overheating. { ˈsēˈpēˈyü ˈfan }

Cr See chromium.

CR See catalytic reforming.

CrA See Corona Australis.

crab [INV ZOO] **1.** The common name for a number of crustaceans in the order Decapoda having five pairs of walking legs, with the first pair modified as chelipeds. **2.** The common name for members of the Merostoma. [NAV] To drift sideways or to leeward, as a ship. { krab }

crabapple [BOT] Any of several trees of the genus *Malus*, order Rosales, cultivated for their small, edible pomes. { ˈkraˌbap·əl }

crabbing [NAV] The horizontal attitude of an aircraft in flight when a crosswind causes its heading to differ from the course. [TEXT] A finishing process that sets warp and weft threads by winding the fabric on a roller under tension and then boiling or steaming. { ˈkrab·iŋ }

crab locomotive [MIN ENG] A type of trolley locomotive equipped with an electric motor, a drum, and haulage cable mounted on a small truck; used to haul mine cars from workings. { ˈkrab ˈlō·kəˈmōd·iv }

Crab Nebula [ASTRON] A gaseous nebula in the constellation Taurus; an amorphous mass which radiates a continuous spectrum involved in a mesh of filaments that radiate a bright-line spectrum. { ˈkrab ˈneb·yə·lə }

Crab pulsar [ASTRON] A pulsar found in the center of the Crab Nebula with a period of about 0.033 second and that emits radiation at all wavelengths from the radio to the x-ray region. { ˈkrab ˈpəlˌsär }

crachin [METEOROL] A period of light rain accompanied by low stratus clouds and poor visibility which frequently occurs in the China Sea between January and April. { kräˈchin }

Cracidae [VERT ZOO] A family of New World tropical upland game birds in the order Galliformes; includes the chachalacas, guans, and curassows. { ˈkra·səˌdē }

crack [CHEM] To break a compound into simpler molecules. [ENG] To open something slightly, for instance, a valve. [SCI TECH] A fissure. { krak }

crack arrester [NAV ARCH] **1.** On a ship, a plate riveted over another plate where the latter has a crack or a stressed area where a crack might begin. **2.** A hole or slot formed at the end of a crack to keep it from spreading, or in a stressed area to prevent a crack from beginning. { ˈkrak əˈres·tər }

cracked [MATER] Applied to those oils produced by the cracking process rather than straight distillation. { krakt }

cracked gasoline [MATER] Gasoline manufactured by heating crude petroleum distillation fractions or residues under pressure, or by heating with or without pressure in the presence of a catalyst, so that heavier hydrocarbons are broken into others, some of which distill in the gasoline range. { ˈkrakt ˈgas·əˌlēn }

cracked residue [CHEM ENG] The residue of fuel resulting from decomposition of hydrocarbons during thermal or catalytic cracking. { ˈkrakt ˈrez·əˌdü }

cracked stem [PL PATH] A boron-deficiency disease of celery characterized by brown mottling of leaves and brittleness and cracking of leaf stalks. { ˈkrakt ˌstem }

cracking [CHEM ENG] A process that is used to reduce the molecular weight of hydrocarbons by breaking the molecular bonds by various thermal, catalytic, or hydrocracking methods. [ENG] Presence of relatively large cracks extending into the interior of a structure, usually produced by overstressing the structural material. { ˈkrak·iŋ }

cracking coil [CHEM ENG] A coil used for cracking heavy petroleum products. { ˈkrak·iŋ ˌkȯil }

cracking still [CHEM ENG] The furnace, reaction chamber, and fractionator for thermal conversion of heavier charging stock to gasoline. { ˈkrak·iŋ ˌstil }

cracovian [MATH] An object which is the same as a matrix except that the product of cracovians *A* and *B* is equal to the matrix product $A'B$, where A' is the transpose of *A*. { krəˈkō·vē·ən }

cradle [CIV ENG] A structure that moves along an inclined track on a riverbank and is equipped with a horizontal deck carrying tracks for transferring railroad cars to and from boats at different water elevations. [ENG] A framework or other resting place for supporting or restraining objects. [ORD] The nonrecoiling structure of a weapon that houses the recoiling parts and rotates to elevate the gun. [TEXT] A device that catches the cards as they fall from a jacquard head. { ˈkrād·əl }

cradle cap [MED] Heavy, greasy crusts on the scalp of an infant; seborrheic dermatitis of infants. { ˈkrād·əl ˌkap }

cradle dump [MIN ENG] A tipple which dumps cars with a rocking motion. { ˈkrād·əl ˌdəmp }

cradle printing [GRAPHICS] Early printing of a crude sort done from movable types. Also known as incunabula printing. { ˈkrād·əl ˌprint·iŋ }

cradle vault See barrel vault. { ˈkrād·əl ˌvȯlt }

crag [GEOL] A steep, rugged point or eminence of rock, as one projecting from the side of a mountain. { krag }

Crambiinae [INV ZOO] The snout moths, a subfamily of lepidopteran insects in the family Pyralididae containing small marshland and grassland forms. { kramˈbī·əˌnē }

Cramér-Rao inequality [STAT] An inequality that is the basis of a method for determining a lower bound to the variance of an estimator of a parameter. { krəˈmä ˈräu̇ˌinˈiˌkwäl·əd·ē }

Cramer's rule [MATH] The method of solving a system of linear equations by means of determinants. { ˈkrä·mərz ˌrül }

cramp [DES ENG] A metal plate with bent ends used to hold blocks together. [MED] **1.** Painful, involuntary contraction of a muscle, such as a leg or foot cramp that may occur in normal individuals at night or in swimming. **2.** Any cramplike pain, as of the intestine, or that accompanying dysmenorrhea. **3.** Spasm of certain muscles, which may be intermittent or constant, from excessive use. { kramp }

crampon [DES ENG] A device for holding heavy objects such as rock or lumber to be lifted by a crane or hoist; shaped like scissors, with points bent inward for grasping the load. Also spelled crampoon. { ˈkramˌpän }

crampoon See crampon. { ˈkramˌpün }

cranberry [BOT] Any of several plants of the genus *Vaccinium*, especially *V. macrocarpon*, in the order Ericales, cultivated for its small, edible berries. { ˈkranˌber·ē }

Cranchiidae [INV ZOO] A family of cephalopod mollusks in the subclass Dibranchia. { ˌkranˈkī·əˌdē }

crandallite [MINERAL] $CaAl_3(PO_4)_2(OH)_5 \cdot H_2O$ A white to light-grayish mineral consisting of a hydrous phosphate of calcium and aluminum occurring in fine, fibrous masses. { ˈkrand·əlˌīt }

crane [MECH ENG] A hoisting machine with a power-operated inclined or horizontal boom and lifting tackle for moving loads vertically and horizontally. [VERT ZOO] The common name for the long-legged wading birds composing the family Gruidae of the order Gruiformes. { krān }

Crane See Grus. { krān }

crane hoist [MECH ENG] A mobile construction machine built principally for lifting loads by means of cables and consisting of an undercarriage on which the unit moves, a cab or house which envelops the main frame and contains the power units and controls, and a movable boom over which the cables run. { ˈkrān ˌhȯist }

crane hook [DES ENG] A hoisting fixture designed to engage a ring or link of a lifting chain, or the pin of a shackle or cable socket. { ˈkrān ˌhu̇k }

crane ship [NAV ARCH] A ship or barge that carries cranes for handling heavy loads. { ˈkrān ˌship }

crane truck [MECH ENG] A crane with a jiblike boom mounted on a truck. Also known as yard crane. { ˈkrān ˌtrək }

Craniacea [INV ZOO] A family of inarticulate branchiopods in the suborder Craniidina. { ˌkrä·nēˈās·ē·ə }

cranial capacity [ANAT] The volume of the cranial cavity. { ˈkran·ē·əl kəˈpas·əd·ē }

cranial flexure [EMBRYO] A flexure of the embryonic brain. { ˈkran·ē·əl ˈflek·shər }

cranial fossa [ANAT] Any of the three depressions in the floor of the interior of the skull. { ˈkran·ē·əl ˈfäs·ə }

cranial index [ANTHRO] The ratio of maximum skull height to maximum skull breadth multiplied by 100. { ˈkran·ē·əl ˈinˌdeks }

cranial nerve [NEURO] Any of the paired nerves which arise in the brainstem of vertebrates and pass to peripheral structures through openings in the skull. { ˈkran·ē·əl ˈnərv }

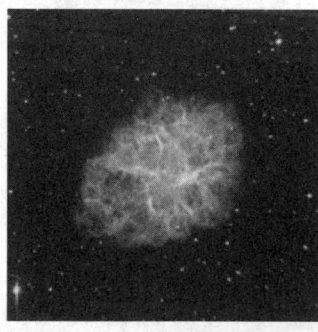

CRAB NEBULA

The Crab Nebula is the 900-year-old remnant of a cataclysmic stellar explosion.

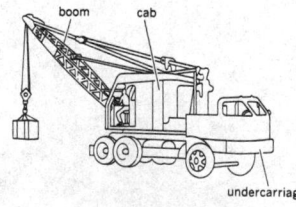

CRANE TRUCK

A truck crane lifting a load. (Link-Belt Co.)

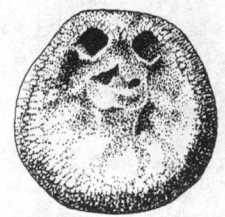

CRANIACEA

Pedicle valve of *Crania*, a genus in Craniacea.

Craniata [VERT ZOO] A major subdivision of the phylum Chordata comprising the vertebrates, from cyclostomes to mammals, distinguished by a cranium. { ‚krā·nē'ad·ə }

craniectomy [MED] Surgical removal of strips or pieces of the cranial bones. { ‚krā·nē·'ek·tə·mē }

Craniidina [INV ZOO] A subdivision of inarticulate branchiopods in the order Acrotretida known to possess a pedicle; all forms are attached by cementation. { ‚krā·ne'id·ən·ə }

craniobuccal pouch [EMBRYO] A diverticulum from the buccal cavity in the embryo from which the anterior lobe of the hypophysis is developed. Also known as Rathke's pouch. { ‚krā·nē·ō'bək·əl 'paúch }

cranioclasis [MED] The operation of breaking the fetal head by means of a cranioclast. { ‚krā·nē·ō'klä·səs }

cranioclast [MED] A heavy forceps for crushing the fetal head. { 'krā·nē·ō‚klast }

craniofacial index [ANTHRO] The ratio of the width of the cranium to the width of the face. { ‚krā·nē·ō'fā·shəl 'in‚deks }

craniofascial dysmorphism [MED] Malformation of the cranium and the low face. { ‚krā·nē·ō‚fā·shəl dis'mor·fiz·əm }

craniometer [ANTHRO] An instrument used to measure a skull. { ‚krā·nē'am·əd·ər }

craniometry [ANTHRO] The science of measuring the skull, especially for determining characteristics of a particular race, sex, or somatotype. { ‚krā·nē'äm·ə·trē }

craniopharyngioma [MED] An epithelial tumor of the craniopharyngeal canal, usually in children. { ‚krā·nē·ō·fə‚rin·jē'ō·mə }

cranioplasty [MED] Surgical correction of defects in the cranial bones, usually by implants of metal, plastic material, or bone. { 'krā·nē·ō‚plas·tē }

cranioscopy [MED] Examination of the human skull. { ‚kra·nē'as·kə·pē }

craniosynostosis [MED] The union of separate cranial bones into a single bone structure. { ‚krā·nē·ō‚sin·ə'stō·səs }

cranium [ANAT] That portion of the skull enclosing the brain. Also known as braincase. { 'krā·nē·əm }

crank [MECH ENG] A link in a mechanical linkage or mechanism that can turn about a center of rotation. { kraŋk }

crank angle [MECH ENG] **1.** The angle between a crank and some reference direction. **2.** Specifically, the angle between the crank of a slider crank mechanism and a line from crankshaft to the piston. { 'kraŋk ‚aŋ·gəl }

crank arm [MECH ENG] The arm of a crankshaft attached to a connecting rod and piston. { 'kraŋk ‚ärm }

crank axle [MECH ENG] **1.** An axle containing a crank. **2.** An axle bent at both ends so that it can accommodate a large body with large wheels. { 'kraŋk ‚ak·səl }

crankcase [MECH ENG] The housing for the crankshaft of an engine, where, in the case of an automobile, oil from hot engine parts is collected and cooled before returning to the engine by a pump. { 'kraŋk‚kās }

crankcase breather See breather pipe. { 'kraŋ·kās ‚brēth·ər }

crankpin [DES ENG] A cylindrical projection on a crank which holds the connecting rod. { 'kraŋk‚pin }

crank press [MECH ENG] A punch press that applies power to the slide by means of a crank. { 'kraŋk ‚pres }

crankshaft [MECH ENG] The shaft about which a crank rotates. { 'kraŋk‚shaft }

crank throw [MECH ENG] **1.** The web or arm of a crank. **2.** The displacement of a crankpin from the crankshaft. { 'kraŋk ‚thrō }

crank web [MECH ENG] The arm of a crank connecting the crankshaft to crankpin, or connecting two adjacent crankpins. { 'kraŋk ‚web }

crannog [ARCHEO] An artificial island constructed from brushwood, stones, peat, and timber, and usually surrounded by a wooden palisade. { 'kran·əg }

crash [COMPUT SCI] **1.** A breakdown, hardware failure, or software problem that renders a computer system inoperative. **2.** See abend. [TEXT] A coarse, rugged fabric woven from linen, cotton, or a combination of both. { krash }

crash bar [ENG] A bar that is installed on a panic exit device located on a door and serves to unlock the door and, sometimes, to activate an alarm. { 'krash ‚bär }

crash locator beacon [COMMUN] An automatic radio beacon carried in aircraft to guide searching forces in the event of a crash. { 'krash 'lō‚kād·ər ‚bē·kən }

craspedon [INV ZOO] A cnidarian medusa stage possessing a velum. { 'kras‚pə‚dän }

craspedote [INV ZOO] Having a velum, used specifically for velate hydroid medusae. { 'kras·pə‚dōt }

Crassulaceae [BOT] A family of dicotyledonous plants in the order Rosales notable for their succulent leaves and resistance to desiccation. { ‚kras·ə'lās·ē‚ē }

crassulacean acid metabolism [BOT] A type of photosynthesis exhibited by many succulent plants in which carbon dioxide is taken up and stored during the night to allow the stomata to remain closed during the daytime, decreasing water loss. Abbreviated CAM. { ‚kras·ə'lā·shən 'as·əd mə'tab·ə‚liz·əm }

crater [GEOL] **1.** A large, bowl-shaped topographic depression with steep sides. **2.** A rimmed structure at the summit of a volcanic cone; the floor is equal to the vent diameter. [MECH ENG] A depression in the face of a cutting tool worn down by chip contact. [MET] A depression at the end of the weld head or under the electrode during welding. { 'krād·ər }

Crater [ASTRON] A constellation, right ascension 11 hours, declination 15°S. Abbreviated Crt. Also known as Cup. { 'krād·ər }

crater arc [ASTRON] A series of lunar craters located along a curved line. { 'krād·ər ‚ärk }

crater chain [ASTRON] A series of lunar craters located along a straight line. { 'krād·ər ‚chān }

crater cone [GEOL] A cone built around a volcanic vent by lava extruded from the vent. { 'krād·ər ‚kōn }

crater cuts [MIN ENG] Cuts with one or more fully charged holes in which blasting is conducted toward the face of the tunnel. { 'krād·ər ‚kəts }

craterization [MED] Surgical excision of part of a bone so that a crater remains. { ‚krād·ə·rə'zā·shən }

crater lake [HYD] A fresh-water lake formed by the accumulation of rain and groundwater in a caldera or crater. { 'krād·ər ‚lāk }

crater lamp [ELECTR] A glow-discharge tube used as a point source of light whose brightness is proportional to the signal current sent through the tube; used for photographic recording of facsimile signals. { 'krād·ər ‚lamp }

craterlet [ASTRON] A very small lunar crater, with diameter less than about 5 miles (8 kilometers), that still has raised walls. { 'krād·ər·lət }

crater pit [ASTRON] A small lunar crater with no raised walls surrounding it. { 'krād·ər ‚pit }

craton [GEOL] A large, stable portion of the continental crust. Cratons are the broad heartlands of continents with subdued topography, encompassing the largest areas of most continents. { 'krā‚tän }

craw [ZOO] **1.** The crop of a bird or insect. **2.** The stomach of a lower animal. { krȯ }

crawler [MECH ENG] **1.** One of a pair of an endless chain of plates driven by sprockets and used instead of wheels by certain power shovels, tractors, bulldozers, drilling machines, and such, as a means of propulsion. **2.** Any machine mounted on such tracks. { 'krȯ·lər }

crawler crane [MECH ENG] A self-propelled crane mounted on two endless tracks that revolve around wheels. { 'krȯ·lər ‚krān }

crawler track See track. { 'krȯl·ər 'trak }

crawler tractor [MECH ENG] A tractor that propels itself on two endless tracks revolving around wheels. { 'krȯ·lər ‚trak·tər }

crawler wheel [MECH ENG] A wheel that drives a continuous metal belt, as on a crawler tractor. { 'krȯ·lər ‚wēl }

crawling [MATER] **1.** Separation and contraction of the glaze on the surface of a ceramic object during drying or firing so that unglazed areas result. **2.** In a film of wet paint, redistribution of the paint so that it is no longer evenly spread, usually due to imperfect bonding of the paint with the surface. { 'krȯl·iŋ }

crawl space [BUILD] **1.** A shallow space in a building which workers can enter to gain access to pipes, wires, and equipment. **2.** A shallow space located below the ground floor of a house and surrounded by the foundation wall. { 'krȯl ‚spās }

crayfish [INV ZOO] The common name for a number of lobsterlike fresh-water decapod crustaceans in the section Astacura. { 'krā‚fish }

crayon [GRAPHICS] A small stick for drawing, usually made

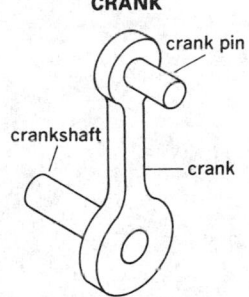

CRANK

A crank for changing radius of rotation.

of a combination of pigments or dyes in a wax or oil medium. { 'krā,än }

crazing [ENG] A network of fine cracks on or under the surface of a material such as enamel, glaze, metal, or plastic. [MET] Development of a network of cracks on a metal surface. { 'krāz·iŋ }

CrB *See* Corona Borealis.

C-reactive protein [IMMUNOL] A plasma protein that is present normally in low concentration, and after trauma or infection in much higher concentration; the biological function is unknown. { ¦sē rē,ak·tiv 'prō,tēn }

cream ice *See* sludge. { 'krēm ,īs }

cream of tartar *See* potassium bitartrate. { 'krēm əv 'tärd·ər }

creatine [BIOCHEM] $C_4H_9O_2N_3$ α-Methylguanidine-acetic acid; a compound present in vertebrate muscle tissue, principally as phosphocreatine. { 'krē·ə,tēn }

creatine kinase [BIOCHEM] An enzyme of vertebrate skeletal and myocardial muscle that catalyzes the transfer of a high-energy phosphate group from phosphocreatine to adenosine diphosphate with the formation of adenosine triphosphate and creatine. { ¦krē·ə,tēn 'kī,nās }

creatinine [BIOCHEM] $C_4H_7ON_3$ A compound present in urine, blood, and muscle that is formed from the dehydration of creatine. { ,krē'at·ən,ēn }

creatinuria [MED] The occurrence of creatine in the urine. { ,krē·ə·tə'nùr·ē·ə }

creation operator [COMPUT SCI] The part of a data structure which allows components to be created. [QUANT MECH] An operator which increases the occupation number of a single state by unity and leaves all the other occupation numbers unchanged. { krē'ā·shən ,äp·ə,rād·ər }

CREB *See* cyclic AMP-responsive element binding protein. { ¦sē¦är¦ē'bē *or* kreb }

CREB-binding protein [MOL BIO] A transcriptional coactivator that binds a phosphorylated domain of cyclic AMP-responsive element binding protein (CREB) to facilitate transcription initiation. Abbreviated CBP. { ¦kreb ¦bīn·diŋ 'prō,tēn }

credence [ELECTROMAG] In radar, a measure of confidence in a target detection, generally proportional to target return amplitude. { 'krēd·əns }

Credé procedure [MED] Instillation of silver nitrate drops into the eyes of a newborn infant to prevent ophthalmia neonatorum. { krə'dā prə'sē·jər }

crednerite [MINERAL] $CuMn_2O_4$ A steel-gray to iron-black foliated mineral consisting of copper, manganese, and oxygen. { 'kred·nə,rīt }

creedite [MINERAL] $Ca_3Al_2(SO_4)(F,OH)_{10}\cdot 2H_2O$ A white or colorless monoclinic mineral consisting of hydrous calcium aluminum fluoride with calcium sulfate, occurring in grains and radiating crystalline masses; hardness is 2 on Mohs scale, and specific gravity is 2.7. { 'krē,dīt }

creek [HYD] A natural stream of water, smaller than a river but larger than a brook. { krēk }

creel [TEXT] A frame for holding the correct number of bobbins or balls of roving in the appropriate position during the winding process. { krēl }

creep [ELECTR] A slow change in a characteristic with time or usage. [ENG] The tendency of wood to move while it is being cut, particularly when being mitered. [GEOL] A slow, imperceptible downward movement of slope-forming rock or soil under sheer stress. [GRAPHICS] A forward movement of the blanket during offset printing. [MECH] A time-dependent strain of solids caused by stress. [MIN ENG] *See* squeeze. { krēp }

creepage [ELEC] The conduction of electricity across the surface of a dielectric. { 'krē·pij }

creep buckling [MECH] Buckling that may occur when a compressive load is maintained on a member over a long period, leading to creep which eventually reduces the member's bending stiffness. { 'krēp ,bək·liŋ }

creeper [ENG] A low platform on small casters that is used for back support and mobility when a person works under a car. [MIN ENG] An endless chain that catches mine car axles on projecting bars. { 'krē·pər }

creep error [ENG] The error that occurs during a mass determination with a digital analytical balance when a value is read, printed, or processed before the display has reached its final position. { 'krēp ,er·ər }

creep-feed grinding *See* creep grinding. { ¦krēp ,fēd 'grīnd·iŋ }

creep grinding [MECH ENG] A grinding operation that uses slow feed rates and produces heavy stock removal. Also known as creep-feed grinding. { 'krēp ,grīnd·iŋ }

creeping barrage [ORD] An artillery barrage that precedes infantry soldiers at a predetermined rate in their attack to protect them and facilitate their advance. { 'krē·piŋ bə'räzh }

creeping disk [INV ZOO] The smooth and adhesive undersurface of the foot or body of a mollusk or of certain other invertebrates, on which the animal creeps. { 'krē·piŋ ,disk }

creeping eruption [MED] A red line of eruption on the skin produced by larva burrowing in the dermis; characterizes the condition of larva migrans. { 'krē·piŋ i'rəp·shən }

creeping flow [FL MECH] Fluid flow in which the velocity of flow is very small. { 'krē·piŋ ,flō }

creep limit [MECH] The maximum stress a given material can withstand in a given time without exceeding a specified quantity of creep. { 'krēp ,lim·ət }

creep recovery [MECH] Strain developed in a period of time after release of load in a creep test. { 'krēp ri'kəv·ə·rē }

creep rupture strength [MECH] The stress which, at a given temperature, will cause a material to rupture in a given time. { 'krēp 'rəp·chər ,streŋkth }

creep strength [MECH] The stress which, at a given temperature, will result in a creep rate of 1% deformation within 100,000 hours. { 'krēp ,streŋkth }

creep test [ENG] Any one of a number of methods of measuring creep, for example, by subjecting a material to a constant stress or deforming it at a constant rate. { 'krēp ,test }

C region [IMMUNOL] The parts of heavy or light chains of immunoglobulin molecules within the same class that have the same amino acid sequence regardless of the molecule. { 'sē ,rē·jən }

cremnophobia [PSYCH] An abnormal fear of steep places or precipices. { ,krem·nə'fō·bē·ə }

cremocarp [BOT] A dry dehiscent fruit consisting of two indehiscent one-seeded mericarps which separate at maturity and remain pendant from the carpophore. { 'krem·ə,kärp }

cremone bolt [DES ENG] A fastening for double doors or casement windows; employs vertical rods that move up and down to engage the top and bottom of the frame. { krə'mōn ,bōlt }

crenate [BIOL] Having a scalloped margin; used specifically for foliar structures, shrunken erythrocytes, and shells of certain mollusks. { 'krē,nāt }

crenation [PHYSIO] A notched appearance of shrunken erythrocytes; seen when they are exposed to the air or to strong saline solutions. { krə'nā·shən }

crenitic [GEOL] Relating to or resulting from the raising of subterranean minerals by the action of spring water. { krə'nid·ik }

Crenothrix [MICROBIO] A genus of sheathed bacteria; cells are nonmotile, sheaths are attached and encrusted with iron and manganese oxides, and filaments may have swollen tips. { 'kren·ə,thriks }

Crenotrichaceae [MICROBIO] Formerly a family of bacteria in the order Chlamydobacteriales having trichomes that are differentiated at the base and tip and attached to a firm substrate. { ¦kren·ə·trə'kās·ē,ē }

crenulate [BIOL] Having a minutely crenate margin. { 'kren·əl,āt }

crenulation cleavage *See* slip cleavage. { ,kren·yə'lā·shən ,klēv·ij }

Creodonta [PALEON] A group formerly recognized as a suborder of the order Carnivora. { ,krē·ə'dän·tə }

creosol [ORG CHEM] $CH_3O(CH_3)C_6H_3OH$ A combination of isomers, derived from coal tar or petroleum; a yellowish liquid with a phenolic odor; used as a disinfectant, in the manufacture of resins, and in flotation of ore. Also known as hydroxymethylbenzene; methyl phenol. { 'krē·ə,sōl }

creosote [MATER] A colorless or yellowish oily liquid containing a mixture of phenolic compounds obtained by distillation of tar; commercial creosote is distilled from coal tar, and pharmaceutical creosote is distilled from wood tar. { 'krē·ə,sōt }

creosote bush [BOT] *Larrea divaricata.* A bronze-green, xerophytic shrub characteristic of all the American warm deserts. { 'krē·ə,sōt ,bùsh }

CRESOL

Structural formulas for the three isomeric methyl phenols, *(a) o-*cresol, *(b) m-*cresol, *(c) p-*cresol.

creosote oil [MATER] A coal tar fraction, boiling between 240 and 270°F (116–132°C); used for producing materials such as creosote and tar acids and used directly as a germicide, insecticide, or pesticide. { 'krē·ə,sōt ,oil }

crepe [TEXT] A silk, polyester, wool, rayon, or other fabric with a crinkled surface obtained by using yarns twisted alternately right and left in the filling. { krāp }

crepe paper [MATER] A lightweight, crinkled paper available in many colors and used for displays, floats, and decorations; it has no strength when wet, and the colors run. { 'krāp ,pā·pər }

Crepe ring See ring C. { 'krāp ,riŋ }

crepitation [MED] A noise produced by the rubbing of fractured ends of bones, by cracking joints, and by pressure upon tissues containing abnormal amounts of air, as in cellular emphysema. { ,krep·ə'tā·shən }

crepuscular [ZOO] Active during the hours of twilight or preceding dawn. { krə'pəs·kyə·lər }

crepuscular arch See bright segment. { krə'pəs·kyə·lər 'ärch }

crepuscular rays [ASTRON] Streaks of light radiating from the sun shortly before and after sunset which shine through breaks in the clouds or through irregular spaces along the horizon. { krə'pəs·kyə·lər 'rāz }

crescent beach [GEOL] A crescent-shaped beach at the head of a bay or the mouth of a stream entering the bay, with the concave side facing the sea. { 'kres·ənt ,bēch }

crescent beam [ENG] A beam bounded by arcs having different centers of curvature, with the central section the largest. { 'kres·ənt ,bēm }

crescentic dune See barchan. { krə'sen·tik 'dün }

crescentic lake See oxbow lake. { krə'sen·tik 'lāk }

crescent phase [ASTRON] A phase of the moon or an inferior planet in which less than half of the visible hemisphere is illuminated. { 'kres·ənt ,fāz }

cresol [ORG CHEM] $CH_3C_6H_4OH$ One of three poisonous, colorless isomeric methyl phenols: o-cresol, m-cresol, p-cresol; used in the production of phenolic resins, tricresyl phosphate, disinfectants, and solvents. { 'krē,sȯl }

cresol red [ORG CHEM] $C_{21}H_{18}O_5S$ A compound derived from o-cresol and used as an acid-base indicator; color change is yellow to red at pH 0.4 to 1.8, or 7.0 to 8.8, depending on preparation. { 'krē,sȯl 'red }

cress [BOT] Any of several prostrate crucifers belonging to the order Capparales and grown for their flavorful leaves; includes watercress (Nasturtium officinale), garden cress (Lepidium sativum), and upland or spring cress (Barbarea verna). { kres }

Cressida [ASTRON] A satellite of Uranus orbiting at a mean distance of 38,380 miles (61,770 kilometers) with a period of 11 hours 9 minutes, and with a diameter of about 41 miles (66 kilometers). { 'kres·əd·ə }

cressing See swaging. { 'kres·iŋ }

crest [DES ENG] The top of a screw thread. [SCI TECH] The highest point of a structure or natural formation, such as the top edge of a dam, the ridge of a roof, the highest point of a gravity wave, or the highest natural projection of a hill or mountain. { krest }

crestal injection See external gas injection. { 'krest·əl in'jek·shən }

crestal plane [GEOL] The plane formed by joining the crests of all beds of an anticline. { 'krest·əl ,plān }

crest clearance [DES ENG] The clearance, in a radial direction, between the crest of the thread of a screw and the root of the thread with which the screw mates. { 'krest ,klir·əns }

crest cloud [METEOROL] A type of standing cloud which forms along a mountain ridge, either on the ridge, or slightly above and leeward of it, and remains in the same position relative to the ridge. Also known as cloud crest. { 'krest ,klaud }

crest factor [PHYS] The ratio of the peak value to the effective value of any periodic quantity such as a sinusoidal alternating current. Also known as amplitude factor; peak factor. { 'krest ,fak·tər }

crest gate [CIV ENG] A gate in the spillway of a dam which functions to maintain or change the water level. { 'krest ,gāt }

crest length [OCEANOGR] The length of a wave measured along its crest. Also known as crest width. { 'krest ,leŋkth }

crest line [GEOL] The line connecting the highest points on the same bed of an anticline in an infinite number of cross sections. { 'krest ,līn }

crest stage [HYD] The highest stage reached at a point along a stream culminating a rise by waters of that stream. { 'krest ,stāj }

crest value See peak value. { 'krest ,val·yü }

crest voltmeter [ELEC] A voltmeter reading the peak value of the voltage applied to its terminals. { 'krest 'vōlt,mēd·ər }

crest width See crest length. { 'krest ,width }

cresylic acid [MATER] **1.** A mixture of phenols containing varying amounts of xylenols, cresols, and other high-boiling fractions. **2.** A crude mixture of the three cresol isomers. { krə'sil·ik 'as·əd }

Cretaceous [GEOL] In geological time, the last period of the Mesozoic Era, preceded by the Jurassic Period and followed by the Tertiary Period; it extended from 144 million years to 65 million years before present. { kri'tā·shəs }

cretin [MED] An individual afflicted with cretinism. { 'krēt·ən or 'kret·ən }

cretinism [MED] A type of dwarfism caused by hypothyroidism and associated with generalized body changes, including mental deficiency. { 'krēt·ən,iz·əm }

cretonne [TEXT] A heavy unglazed fabric that is usually printed, and resembles unglazed chintz. { krə'tän }

crevasse [GEOL] An open, nearly vertical fissure in a glacier or other mass of land ice or the earth, especially after earthquakes. { krə'vas }

crevasse deposit [GEOL] Kame deposited in a crevasse. { krə'vas di'päz·ət }

crevasse hoar [HYD] Ice crystals which form and grow in glacial crevasses and in other cavities where a large cooled space is formed and in which water vapor can accumulate under calm, still conditions. { krə'vas ,hȯr }

crevice [SCI TECH] A deep, narrow opening. { 'krev·əs }

crevice corrosion [MET] Corrosive degradation of metal parts at the crevices left at rolled joints or from other forming procedures; common in stainless steel heat exchangers in contact with chloride-containing fluids or other dissolved corrosives. Also known as contact corrosion. { 'krev·əs kə'rō·zhən }

crew-served [ORD] Of or pertaining to anything served or operated by a crew as distinguished from an individual; for example, a weapon operated by a crew of two or more persons. { 'krü ,sərvd }

CRH See corticotropin-releasing hormone.

criador [METEOROL] The rain-bringing west wind of northern Spain. { krē·ə'dȯr }

crib [CIV ENG] The space between two successive ties along a railway track. [ENG] **1.** Any structure composed of a layer of timber or steel joists laid on the ground, or two layers across each other, to spread a load. **2.** Any structure composed of frames of timber placed horizontally on top of each other to form a wall. [GEOL] See arête. { krib }

crib death See sudden infant death syndrome. { 'krib ,deth }

cribellum [INV ZOO] **1.** A small accessory spinning organ located in front of the ordinary spinning organ in certain spiders. **2.** A chitinous plate perforated with the openings of certain gland ducts in insects. { krə'bel·əm }

Cribrariaceae [BOT] A family of true slime molds in the order Liceales. { krə,brer·ē'ās·ē,ē }

cribriform [BIOL] Perforated, like a sieve. { 'krib·rə,fȯrm }

cribriform fascia [ANAT] The sievelike covering of the fossa ovalis of the thigh. { 'krib·rə,fȯrm 'fā·shə }

cribriform plate [ANAT] **1.** The horizontal plate of the ethmoid bone, part of the floor of the anterior cranial fossa. **2.** The bone lining a dental alveolus. { 'krib·rə,fȯrm ¦plāt }

Cricetidae [VERT ZOO] A family of the order Rodentia including hamsters, voles, and some mice. { krə'sed·ə,dē }

Cricetinae [VERT ZOO] A subfamily of mice in the family Cricetidae. { krə'set·ən,ē }

cricket [BUILD] A device that is used to divert water at the intersections of roofs or at the intersection of a roof and chimney. [INV ZOO] **1.** The common name for members of the insect family Gryllidae. **2.** The common name for any of several related species of orthopteran insects in the families Tettigoniidae, Gryllotalpidae, and Tridactylidae. { 'krik·ət }

cricoid [ANAT] The signet-ring-shaped cartilage forming the

CRETACEOUS

CENOZOIC	QUATERNARY	
	TERTIARY	
MESOZOIC	CRETACEOUS	
	JURASSIC	
	TRIASSIC	
PALEOZOIC	PERMIAN	
	CARBONIFEROUS	PENNSYLVANIAN
		MISSISSIPPIAN
	DEVONIAN	
	SILURIAN	
	ORDOVICIAN	
	CAMBRIAN	
PRECAMBRIAN		

Chart showing relationship of Cretaceous to other geologic periods.

base of the larynx in humans and most other mammals. { 'krī,kóid }

cricondenbar [PHYS CHEM] Maximum pressure at which two phases (for example, liquid and vapor) can coexist. { krə'kän·dən,bär }

cricondentherm [PHYS CHEM] Maximum temperature at which two phases (for example, liquid and vapor) can coexist. { krə'kän·dən,thərm }

Criconematoidea [INV ZOO] A superfamily of plant parasitic nematodes of the order Diplogasterida distinguished by their ectoparasitic habit and males that have atrophied mouthparts and do not feed. { ,krī·kō,nem·ə'tóid·ē·ə }

cri du chat syndrome [MED] An inherited condition characterized by mental subnormality, physical abnormalities, and the emitting of a flat, toneless, catlike cry in infancy. { ,krē dü 'shä ,sin,drōm }

criminal abortion [MED] Illegal interruption of pregnancy. { 'krim·ən·əl ə'bór·shən }

criminalistics *See* forensic science. { ,krim·ən·əl'is·tiks }

crimp [ENG] **1.** To cause something to become wavy, crinkled, or warped, such as lumber. **2.** To pinch or press together, especially a tubular or cylindrical shape, in order to seal or unite. [TEXT] To give curl to synthetic fibers. { krimp }

crimp contact [ELEC] A contact whose back portion is a hollow cylinder that will accept a wire; after a bared wire is inserted, a swaging tool is applied to crimp the contact metal firmly against the wire. Also known as solderless contact. { 'krimp ,kän,takt }

crinal [MECH] A unit of force equal to 0.1 newton. { 'krīn·əl }

crinion-menton [ANTHRO] The measurement from the hairline in the center part of the forehead to the midpoint of the lower edge of the chin. { ¦krin·ē,än ¦men,tän }

crinkling *See* wrinkling. { 'kriŋk·liŋ }

crinoidal limestone [PETR] A rock composed predominantly of crystalline joints of crinoids, with foraminiferans, corals, and mollusks. { krī'nóid·əl 'līm,stōn }

Crinoidea [INV ZOO] A class of radially symmetrical crinozoans in which the adult body is flower-shaped and is either carried on an anchored stem or is free-living. { krə'nóid·ē·ə }

crinoline [TEXT] **1.** A stiff fabric with an open weave that is filled with hard-twist cotton warp and horsehair. **2.** A fabric with a firm starched or permanent resin finish. { 'krin·əl·ən }

Crinozoa [INV ZOO] A subphylum of the Echinodermata comprising radially symmetrical forms that show a partly meridional pattern of growth. { 'krī·nə,zō·ə }

cripple [BUILD] A structural member, such as a stud above a window, that is cut less than full length. { 'krip·əl }

crippled leap-frog test [COMPUT SCI] A variation of the leap-frog test, modified so the computer tests are repeated from a single set of storage locations rather than a changing set of locations. { ¦krip·əld 'lēp ,fräg ,test }

crippled mode [COMPUT SCI] The operation of a computer at reduced capacity when certain parts are not working. { 'krip·əld ,mōd }

crisis [MED] The turning point in the course of a disease. [PSYCH] The psychological events associated with a specific stage of life, as an identity crisis or developmental crisis. { 'krī·səs }

crispation number [PHYS] A dimensionless number used in the study of convection currents, equal to the product of a fluid's dynamic viscosity and its thermal diffusivity, divided by the product of its undisturbed surface tension and a layer thickness. { kri'spā·shən ,nəm·bər }

crisp set [MATH] A conventional set, wherein the degree of membership of any object in the set is either 0 or 1. { 'krisp 'set }

crissum [VERT ZOO] **1.** The region surrounding the cloacal opening in birds. **2.** The vent feathers covering the circumcloacal region. { 'kris·əm }

crista [BIOL] A ridge or crest. [CYTOL] A fold on the inner membrane of a mitochondrion. { 'kris·tə }

cristate [BIOL] Having a crista. { 'kri,stāt }

Cristispira [MICROBIO] A genus of bacteria in the family Spirochaetaceae; helical cells with 3–10 complete turns; they have ovoid inclusion bodies and bundles of axial fibrils; commensals in mollusks. { ,kris·tə'spī·rə }

cristobalite [MINERAL] SiO_2 A silicate mineral that is a high-temperature form of quartz; stable above 1470°C; crystallizes in the tetragonal system at low temperatures and the isometric system at high temperatures. { kri'stō·bə,līt }

crit [NUCLEO] The mass of fissionable material that is critical under a given set of conditions; sometimes applied to the mass of an untamped critical sphere of fissionable material. { krit }

crith [MECH] A unit of mass, used for gases, equal to the mass of 1 liter of hydrogen at standard pressure and temperature; it is found experimentally to equal 8.9885×10^{-5} kilogram. { krith }

critical [NUCLEO] Capable of sustaining a chain reaction at a constant level. { 'krid·ə·kəl }

critical absorption wavelength [SPECT] The wavelength, characteristic of a given electron energy level in an atom of a specified element, at which an absorption discontinuity occurs. { 'krid·ə·kəl əb'sórp·shən 'wāv,leŋkth }

critical altitude [AERO ENG] The maximum altitude at which a supercharger can maintain a pressure in the intake manifold of an engine equal to that existing during normal operation at rated power and speed at sea level without the supercharger. [ORD] The maximum altitude at which the propulsion system of a missile performs satisfactorily. { 'krid·ə·kəl 'al·tə,tüd }

critical angle [ARCH] The angle of pitch of a flight of stairs or a ramp that, if exceeded, is considered uncomfortable and unsafe; it is 50° for stairs and 20° for ramps. [PHYS] An angle associated with total reflection of electromagnetic or acoustic radiation back into a medium from the boundary with another medium in which the radiation has a higher phase velocity; it is the smallest angle with the normal to the boundary at which total reflection occurs. { 'krid·ə·kəl ¦aŋ·gəl }

critical angle of attack [AERO ENG] The angle of attack of an airfoil at which the flow of air about the airfoil changes abruptly so that lift is sharply reduced and drag is sharply increased. Also known as stalling angle of attack. { 'krid·ə·kəl ¦aŋ·gəl əv ə'tak }

critical angle refractometer [OPTICS] A refractometer, such as the Abbe or Pulfrich refractometer, in which the index of refraction of a medium *A* is measured by observing its critical angle with respect to another medium *B* with a known index of refraction, or by measuring the critical angle of *B* with respect to *A*. { 'krid·ə·kəl ¦aŋ·gəl ,rē,frak'täm·əd·ər }

critical anode voltage [ELECTR] The anode voltage at which breakdown occurs in a gas tube. { 'krid·ə·kəl 'a,nōd ,vōl·tij }

critical area *See* picture element. { 'krid·ə·kəl 'er·ē·ə }

critical assembly [NUCLEO] An assembly of sufficient fissionable and moderator material to sustain a fission chain reaction at a low power level. { 'krid·ə·kəl ə'sem·blē }

critical bottom slope [GEOL] The depth distribution in which depth *d* of an ocean increases with latitude ϕ according to an equation of the form $d = d_0 \sin \phi + \text{constant}$. { 'krid·ə·kəl 'bäd·əm ,slōp }

critical care medicine [MED] The treatment of acute, life-threatening disorders, usually in intensive care units. { ,krid·ə·kəl 'ker ,med·ə·sən }

critical compression ratio [MECH ENG] The lowest compression ratio which allows compression ignition of a specific fuel. { 'krid·ə·kəl kəm'presh·ən ,rā·shō }

critical condensation temperature [PHYS CHEM] The temperature at which the sublimand of a sublimed solid recondenses; used to analyze solid mixtures, analogous to liquid distillation. Also known as true condensing point. { 'krid·ə·kəl ,kän·dən'sā·shən ,tem·prə·chər }

critical constant [PHYS CHEM] A characteristic temperature, pressure, and specific volume of a gas above which it cannot be liquefied. { 'krid·ə·kəl 'kän·stənt }

critical cooling rate [MET] The minimum cooling rate that will suppress undesired transformations in a metal. { 'krid·ə·kəl 'kül·iŋ ,rāt }

critical coupling [ELEC] The degree of coupling that provides maximum transfer of signal energy from one radio-frequency resonant circuit to another when both are tuned to the same frequency. Also known as optimum coupling. { 'krid·ə·kəl 'kəp·liŋ }

critical current [SOLID STATE] The current in a superconductive material above which the material is normal and below which the material is superconducting, at a specified temperature and in the absence of external magnetic fields. { 'krid·ə·kəl 'kər·ənt }

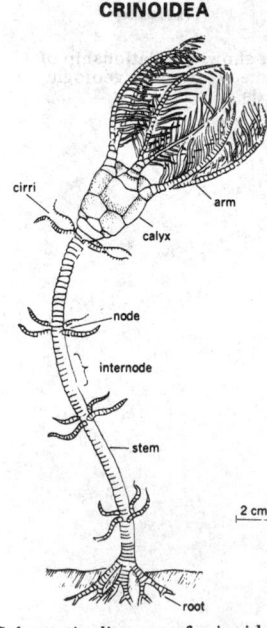

CRINOIDEA

cirri
arm
calyx
node
internode
stem
2 cm
root

Schematic diagram of crinoid.

critical current density [PHYS CHEM] The amount of current per unit area of electrode at which an abrupt change occurs in a variable of an electrolytic process. { 'krid·ə·kəl 'kər·ənt ,den·səd·ē }

critical damping [PHYS] Damping in a linear system on the threshold between oscillatory and exponential behavior. { 'krid·ə·kəl 'dam·piŋ }

critical density [ASTRON] The mass density above which, it is believed, the expansion of the universe will slow down and reverse. [CHEM] The density of a substance exhibited at its critical temperature and critical pressure. [CIV ENG] For a highway, the density of traffic when the volume equals the capacity. [GEOL] That degree of density of a saturated, granular material below which, as it is rapidly deformed, it will decrease in strength and above which it will increase in strength. [THERMO] The density of a substance at the liquid-vapor critical point. { 'krid·ə·kəl 'den·səd·ē }

critical depth [HYD] In a water channel, that depth at which the flow is at its minimum energy with respect to the bottom of the channel. { 'krid·ə·kəl 'depth }

critical elevation [MAP] That elevation which is the high point within the area of a chart. { 'krid·ə·kəl ,el·ə'vā·shən }

critical equation [NUCLEO] Any equation relating parameters of a reactor that must be satisfied for the reactor to be critical. { 'krid·ə·kəl i'kwā·zhən }

critical equatorial velocity [ASTRON] In rotating early-type stars, the velocity at which the centrifugal force at the equator equals the force of gravity there. { 'krid·ə·kəl ,ek·wə'tōr·ē·əl və'läs·əd·ē }

critical experiment [NUCLEO] An experiment in which fissionable material is assembled gradually until the arrangement will support a self-sustaining chain reaction. { 'krid·ə·kəl ik'sper·ə·mənt }

critical exponent [THERMO] A parameter n that characterizes the temperature dependence of a thermodynamic property of a substance near its critical point; the temperature dependence has the form $|T - T_c|^n$, where T is the temperature and T_c is the critical temperature. { 'krid·ə·kəl ik'spō·nənt }

critical facility [NUCLEO] A facility where critical experiments are conducted. { 'krid·ə·kəl fə'sil·əd·ē }

critical field [ELECTR] The smallest theoretical value of steady magnetic flux density that would prevent an electron emitted from the cathode of a magnetron at zero velocity from reaching the anode. Also known as cutoff field. [SOLID STATE] The magnetic field strength below which magnetic flux is excluded from a type I superconductor. Symbolized H_c. { 'krid·ə·kəl 'fēld }

critical flicker frequency [OPTICS] That frequency of an intermittent light source at which the light appears half the time as flickering and half the time as continuous. { 'krid·ə·kəl 'flik·ər ,frē·kwən·sē }

critical flow [FL MECH] The rate of flow of a fluid equivalent to the speed of sound in that fluid. { 'krid·ə·kəl 'flō }

critical flow prover [PETRO ENG] Device used to measure the velocity of gas flow during open-flow testing of gas wells. { 'krid·ə·kəl 'flō ,prü·vər }

critical frequency [ELECTR] See cutoff frequency. [ELECTROMAG] The limiting frequency below which a radio wave will be reflected by an ionospheric layer at vertical incidence at a given time. [GEOPHYS] The minimum frequency of a vertically directed radio wave which will penetrate a particular layer in the ionosphere; for example, all vertical radio waves with frequencies greater than the E-layer critical frequency will pass through the E layer. Also known as penetration frequency. { 'krid·ə·kəl 'frē·kwən·sē }

critical frequency of fusion [NEURO] A sufficiently high flash rate at which the eye fails to detect the flicker of a light; that is, the light pulses seem to fuse to form a steady light indistinguishable from a continuous light that has the same total energy per unit time. { 'krid·ə·kəl 'frē·kwən·sē əv 'fyü·zhən }

critical function [MATH] A function satisfying the Euler equations in the calculus of variations. { 'krid·ə·kəl 'fəŋk·shən }

critical gas saturation See equilibrium gas saturation. { 'krid·ə·kəl 'gas ,sach·ə'rā·shən }

critical grid current [ELECTR] Instantaneous value of grid current when the anode current starts to flow in a gas-filled vacuum tube. { 'krid·ə·kəl 'grid ,kər·ənt }

critical grid voltage [ELECTR] The grid voltage at which anode current starts to flow in a gas tube. Also known as firing point. { 'krid·ə·kəl 'grid ,vōl·tij }

critical gun pull [ORD] The maximum pulling force in pounds developed by the gun feeder mechanism during operation of automatic guns. { 'krid·ə·kəl 'gən ,púl }

critical humidity [CHEM ENG] The humidity of a system's atmosphere above which a crystal of a water-soluble salt will always become damp (absorb moisture from the atmosphere) and below which it will always stay dry (release moisture to the atmosphere). [MET] The atmospheric humidity above which the corrosion rate increases rapidly for a particular metal. { 'krid·ə·kəl yü'mid·əd·ē }

critical isotherm [THERMO] A curve showing the relationship between the pressure and volume of a gas at its critical temperature. { 'krid·ə·kəl 'ī·sə,thərm }

criticality [NUCLEO] The condition in which a nuclear reactor is just self-sustaining. { ,krid·ə'kal·əd·ē }

critical level of escape [GEOPHYS] 1. That level, in the atmosphere, at which a particle moving rapidly upward will have a probability of $1/e$ (e is base of natural logarithm) of colliding with another particle on its way out of the atmosphere. 2. The level at which the horizontal mean free path of an atmospheric particle equals the scale height of the atmosphere. { 'krid·ə·kəl 'lev·əl əv ə'skāp }

critical line See critical locus. { 'krid·ə·kəl 'līn }

critical locus [PHYS CHEM] The line connecting the critical points of a series of liquid-gas phase-boundary loops for multicomponent mixtures plotted on a pressure versus temperature graph. Also known as critical line. { 'krid·ə·kəl 'lō·kəs }

critical Mach number [AERO ENG] The free-stream Mach number at which a local Mach number of 1.0 is attained at any point on the body under consideration. { 'krid·ə·kəl 'mäk ,nəm·bər }

critical magnetic field [SOLID STATE] The field below which a superconductive material is superconducting and above which the material is normal, at a specified temperature and in the absence of current. { 'krid·ə·kəl mag'ned·ik 'fēld }

critical magnetic scattering [SOLID STATE] Intense scattering of low-energy neutrons by a ferromagnetic crystal at temperatures near the Curie point. { 'krid·ə·kəl mag'ned·ik 'skad·ər·iŋ }

critical mass [NUCLEO] The mass of fissionable material of a particular shape that is just sufficient to sustain a nuclear chain reaction. { 'krid·ə·kəl 'mas }

critical micelle concentration [PHYS CHEM] The concentration of a micelle (oriented molecular arrangement of an electrically charged colloidal particle or ion) at which the rate of increase of electrical conductance with increase in concentration levels off or proceeds at a much slower rate. { 'krid·ə·kəl mi'sel ,kän·sən'trā·shən }

critical moisture content [CHEM ENG] The average moisture throughout a solid material being dried, its value being related to drying rate, thickness of material, and the factors that influence the movement of moisture within the solid. { 'krid·ə·kəl 'mòis·chər ,kän·tent }

critical opalescence [OPTICS] Extreme opalescence resulting from strong density fluctuations in a medium near a critical point. { 'krid·ə·kəl ,ōp·ə'les·əns }

critical path method [SYS ENG] A systematic procedure for detailed project planning and control. Abbreviated CPM. { 'krid·ə·kəl 'path ,meth·əd }

critical phenomena [PHYS CHEM] Physical properties of liquids and gases at the critical point (conditions at which two phases are just about to become one); for example, critical pressure is that needed to condense a gas at the critical temperature, and above the critical temperature the gas cannot be liquefied at any pressure. { 'krid·ə·kəl fə'näm·ə·nə }

critical point [MATH] A point at which the first derivative of a function is either 0 or does not exist. [PETRO ENG] A location on the drilling line which is subject to strain when the pipe is run into or pulled out of the drill hole. [PHYS CHEM] 1. The temperature and pressure at which two phases of a substance in equilibrium with each other become identical, forming one phase. 2. The temperature and pressure at which two ordinarily partially miscible liquids are consolute. { 'krid·ə·kəl 'pòint }

critical potential [ATOM PHYS] The energy needed to raise an electron to a higher energy level in an atom (resonance

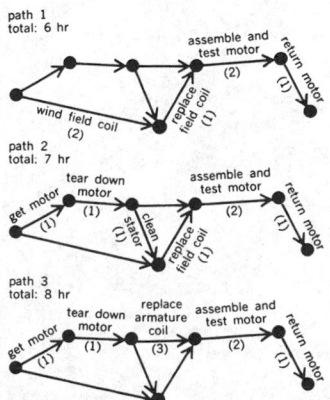

CRITICAL PATH METHOD

The critical path method for motor maintenance job, showing activities involved, duration of the activity in hours (numbers in parentheses), and how the activities dovetail together.

potential) or to remove it from the atom (ionization potential). [ELEC] A potential which results in sudden change in magnitude of the current. { 'krid·ə·kəl pə'ten·chəl }

critical pressure [FL MECH] For a nozzle whose cross section at each point is such that a fluid in isentropic flow just fills it, the pressure at the section of minimum area of the nozzle; if the nozzle is cut off at this point with no diverging section, decrease in the discharge pressure below the critical pressure (at constant admission pressure) does not result in increased flow. [THERMO] The pressure of the liquid-vapor critical point. { 'krid·ə·kəl 'presh·ər }

critical-pressure ratio [FL MECH] The ratio of the critical pressure of a nozzle to the admission pressure of the nozzle (equals 0.53 for gases). { 'krid·ə·kəl 'presh·ər ‚rā·shō }

critical properties [PHYS CHEM] Physical and thermodynamic properties of materials at conditions of critical temperature, pressure, and volume, that is, at the critical point. { 'krid·ə·kəl 'präp·ərd·ēz }

critical range [MET] The temperature range for the reversible change of austenite to ferrite, pearlite, and cementite. { 'krid·ə·kəl 'rānj }

critical ratio [STAT] The ratio of a particular deviation from the mean value to the standard deviation. { 'krid·ə·kəl 'rā·shō }

critical reactor [NUCLEO] A nuclear reactor in which the ratio of moderator to fuel is either subcritical or just critical; used to study the properties of the system and determine critical size. { 'krid·ə·kəl rē'ak·tər }

critical region [GEN] The shortest segment of a chromosome whose gain or loss results in a particular complex phenotype such as Down syndrome. [STAT] In testing hypotheses, the set of sample values leading to rejection of the null hypothesis. { 'krid·ə·kəl 'rē·jən }

critical Reynolds number [FL MECH] The Reynolds number at which there is a transition from laminar to turbulent flow. { 'krid·ə·kəl 'ren·əlz ‚nəm·bər }

critical scattering [PHYS] Intense scattering of some form of radiation by a substance at a temperature near a second-order transition, as in critical opalescence or critical magnetic scattering. { 'krid·ə·kəl 'skad·ər·iŋ }

critical shear stress [SOLID STATE] The shear stress needed to cause slip in a given direction along a given crystallographic plane of a single crystal. { 'krid·ə·kəl 'shir‚stres }

critical size [NUCLEO] A set of physical dimensions for the core and reflector of a nuclear reactor at which a critical chain reaction is maintained. { 'krid·ə·kəl 'sīz }

critical slope [CIV ENG] The maximum angle with the horizontal at which a sloped bank of soil of a given height will remain undeformed without some form of support. [HYD] The channel slope or grade that is equal to the loss of head per foot resulting from flow at a depth which will provide uniform flow at critical depth. { 'krid·ə·kəl 'slōp }

critical solution temperature [PHYS CHEM] The temperature at which a mixture of two liquids, immiscible at ordinary temperatures, ceases to separate into two phases. { 'krid·ə·kəl sə'lü·shən ‚tem·prə·chər }

critical speed [CRYO] *See* critical velocity. [FL MECH] *See* critical velocity. [MECH ENG] The angular speed at which a rotating shaft becomes dynamically unstable with large lateral amplitudes, due to resonance with the natural frequencies of lateral vibration of the shaft. { 'krid·ə·kəl 'spēd }

critical state [PHYS CHEM] Unique condition of pressure, temperature, and composition wherein all properties of coexisting vapor and liquid become identical. { 'krid·ə·kəl 'stāt }

critical strain [MET] The strain at which heating causes rapid growth of large grains in many metals and alloys; phase transformations do not occur. { 'krid·ə·kəl 'strān }

critical table [MATH] A table, usually for a function that varies slowly, which gives only values of the argument near which changes in the value of the function, rounded to the number of decimal places displayed in the table, occur. { 'krid·ə·kəl 'tā·bəl }

critical temperature [AGR] The temperature below which a plant cannot grow. [PHYS CHEM] The temperature of the liquid-vapor critical point, that is, the temperature above which the substance has no liquid-vapor transition. Symbolized T_c. { 'krid·ə·kəl 'tem·prə·chər }

critical value [MATH] The value of the dependent variable at a critical point of a function. [STAT] A number which

causes rejection of the null hypothesis if a given test statistic is this number or more, and acceptance of the null hypothesis if the test statistic is smaller than this number. { 'krid·ə·kəl 'val·yü }

critical velocity [AERO ENG] In rocketry, the speed of sound at the conditions prevailing at the nozzle throat. Also known as throat velocity. [CRYO] The velocity of a superfluid in very narrow channels (on the order of 10^{-5} centimeter), which is nearly constant. Also known as critical speed. [FL MECH] **1.** The speed of flow equal to the local speed of sound. Also known as critical speed. **2.** The speed of fluid flow through a given conduit above which it becomes turbulent. { 'krid·ə·kəl və'läs·əd·ē }

critical vibration [MECH ENG] A vibration that is significant and harmful to a structure. { 'krid·ə·kəl vī'brā·shən }

critical voltage [ELECTR] The highest theoretical value of steady anode voltage, at a given steady magnetic flux density, at which electrons emitted from the cathode of a magnetron at zero velocity would fail to reach the anode. Also known as cutoff voltage. { 'krid·ə·kəl 'vōl·tij }

critical volume [PHYS] The volume occupied by one mole of a substance at the liquid-vapor critical point, that is, at the critical temperature and pressure. { 'krid·ə·kəl 'väl·yəm }

critical wavelength [COMMUN] The free-space wavelength corresponding to the critical frequency. { 'krid·ə·kəl 'wāv‚leŋkth }

critical weight [ENG] In a drilling operation, the weight placed on a bit that will cause the drill string to become resonant with the angular speed at which the rotating shaft is operating. { 'krid·ə·kəl 'wāt }

critical zone [FL MECH] In fluid flow, the area on a graph of the Reynolds number versus friction factor indicating unstable flow (Reynolds number 2000 to 4000) between laminar flow and the transition to turbulent flow. [ORD] Area over which a bombing plane in horizontal-flight or glide bombing must maintain straight flight so that the bombsight can be operated properly and bombs dropped accurately. { 'krid·ə·kəl 'zōn }

crivetz [METEOROL] A wind blowing from the northeast quadrant in Rumania and southern Russia, especially a cold boralike wind from the north-northeast, characteristic of the climate of Rumania. { krə'vets }

CRLAS *See* cavity ringdown laser absorption spectroscopy.

CR law [ELEC] A law which states that when a constant electromotive force is applied to a circuit consisting of a resistor and capacitor connected in series, the time taken for the potential on the plates of the capacitor to rise to any given fraction of its final value depends only on the product of capacitance and resistance. { ‚sē‚är ‚lö }

CRM *See* chemical remanent magnetization.

CRO *See* cathode-ray oscilloscope.

Crocco's equation [FL MECH] A relationship, expressed as $\mathbf{v} \times \boldsymbol{\omega} = -T$ grad S, between vorticity and entropy gradient for the steady flow of an inviscid compressible fluid; $\mathbf{v}$ is the fluid velocity vector, $\boldsymbol{\omega}$ (= curl $\mathbf{v}$) is the vorticity vector, T is the fluid temperature, and S is the entropy per unit mass of the fluid. { 'krä‚kōz i'kwā·zhən }

crochet file [DES ENG] A thin, flat, round-edged file that tapers to a point. { krō'shā ‚fīl }

crocidolite [MINERAL] A lavender-blue, indigo-blue, or leek-green asbestiform variety of riebeckite; occurs in fibrous, massive, and earthy forms. Also known as blue asbestos; krokidolite. { krō'sīd·əl‚īt }

Crockett magnetic separator [MIN ENG] An assembly consisting of a continuous belt submerged in a tank through which ore pulp flows; magnetic solids adhere to the belt, which has a series of flat magnets attached to it, and the solids are dragged clear. { 'kräk·ət mag'ned·ik 'sep·ə‚rād·ər }

crocking [TEXT] Rubbing off of color as a result of improper dyeing, poor penetration, or poor fixation. { 'kräk·iŋ }

crocodile [ELEC] A unit of potential difference or electromotive force, equal to 10^6 volts; used informally at some nuclear physics laboratories. [VERT ZOO] The common name for about 12 species of aquatic reptiles included in the family Crocodylidae. { 'kräk·ə‚dīl }

crocodile clip *See* alligator clip. { 'kräk·ə‚dīl ‚klip }

crocodile shears *See* lever shears. { 'kräk·ə‚dīl ‚shirz }

Crocodilia [VERT ZOO] An order of the class Reptilia which is composed of large, voracious, aquatic species, including the alligators, caimans, crocodiles, and gavials. { 'kräk·ə¦dil· ē·ə }

crocodiling *See* alligatoring. { 'kräk·ə¦dīl·iŋ }

Crocodylidae [VERT ZOO] A family of reptiles in the order Crocodilia including the true crocodiles, false gavial, alligators, and caimans. { ¸kräk·ə'dil·ə¸dē }

Crocodylinae [VERT ZOO] A subfamily of reptiles in the family Crocodylidae containing the crocodiles, *Osteolaemus,* and the false gavial. { ¸kräk·ə'dil·ə¸nē }

crocoisite *See* crocoite. { 'kräk·wə¸zīt }

crocoite [MINERAL] PbCrO₄ A yellow to orange or hyacinth-red secondary mineral occurring as monoclinic, prismatic crystals; it is also massive granular. Also known as crocoisite; red lead ore. { 'kräk·ə¸wīt }

crocus [BOT] A plant of the genus *Crocus*, comprising perennial herbs cultivated for their flowers. [MATER] Finely powdered oxide of iron, of dark red color, used for buffing and polishing. { 'krō·kəs }

crofting [TEXT] The whitening of linen by soaking it in an alkaline solution, then drying it in the sun. { 'krof·tiŋ }

Crohn's disease [MED] Chronic inflammation of the colon and stomach of unknown etiology that involves the full thickness of the intestinal wall, often with bowel narrowing and obstruction of the lumen. It is usually accompanied by granulomas; and abdominal cramps, alteration of bowel function, and diminished food intake are common. { 'krōnz diz¸ēz }

Croixian [GEOL] Upper Cambrian geologic time. { 'krȯi· ən }

CROM *See* control and read-only memory. { 'sē¸räm }

Cro-Magnon man [PALEON] **1.** A race of tall, erect Caucasoid men having large skulls; identified from skeletons found in southern France. **2.** A general term to describe all fossils resembling this race that belong to the upper Paleolithic (35,000–8000 B.C.) in Europe. { krō'mag·nən 'man }

cromfordite *See* phosgenite. { 'kräm·fər¸dīt }

cromolyn sodium [PHARM] C₂₃H₁₄Na₂O₁₁ An inhaled anti-inflammatory agent that prevents the degranulation of mast cells; effective in preventing bronchoconstriction. { 'krō·mə¸lin ¦sōd·ē·əm }

Cromwell Current [OCEANOGR] An eastward-setting subsurface current that extends about 1¹/₂° north and south of the equator, and from about 150°E to 92°W. { 'kräm¸wel ¸kər· ənt }

cron [EVOL] A time unit equal to 10⁶ years; used in reference to evolutionary processes. { krän }

Cronartium ribicola [MYCOL] A heteroecious rust fungus that causes white pine blister rust; it produces pycnia and aecia on pine stems, and uredinia and telia on currants and gooseberries. { krə¦närd·ē·əm ¸rib·i'kō·lə }

croning process [MET] A shell-molding process. { 'krōn· iŋ ¸präs·əs }

cronstedtite [MINERAL] Fe₄²⁺Fe₂³⁺(Fe₂³⁺Si₂)O₁₀(OH)₈ A black to brownish-black mineral consisting of a hydrous iron silicate crystallizing in hexagonal prisms; specific gravity is 3.34–3.35. { 'krän¸sted¸īt }

crooked hole [PETRO ENG] A borehole drilled at an angle, often because of steeply dipping formations; not to be confused with holes deliberately deviated from the vertical to avoid obstacles or to tap otherwise unavailable reservoirs. { 'krük· əd ¦hōl }

crooked-hole country [PETRO ENG] An area where it is difficult to drill completely vertical holes because the formations possess alternating hard and soft strata steeply inclined from the horizontal. { 'krük·əd ¦hōl ¸kən·trē }

Crookes dark space *See* cathode dark space. { ¦krüks 'därk ¸spās }

Crookes glass [MATER] A type of glass that contains cerium and other rare earths and has a high absorption of ultraviolet radiation; used in sunglasses. { 'krüks ¸glas }

crookesite [MINERAL] (Cu,Tl,Ag)₂Se An important selenium mineral occurring in lead-gray masses and having a metallic appearance. { 'krük¸sīt }

Crookes radiometer [PHYS] A radiometer used to demonstrate that radiant energy from the sun can produce motion; a miniature four-vane windmill is mounted in a glass-envelope vacuum tube, with each vane polished on one side and black on the other. { 'krüks ¸rād·ē'äm·əd·ər }

Crookes tube [ELECTR] An early form of low-pressure discharge tube whose cathode was a flat aluminum disk at one end of the tube, and whose anode was a wire at one side of the tube, outside the electron stream; used to study cathode rays. { 'krüks ¸tüb }

crop [AGR] A plant or animal grown for its commercial value. [MET] Defective end portion of an ingot which is removed for scrap before rolling the ingot. [VERT ZOO] A distensible saccular diverticulum near the lower end of the esophagus of birds which serves to hold and soften food before passage into the stomach. { kräp }

crop coal [MIN ENG] Coal of inferior quality found near the surface. { 'kräp ¸kōl }

crop dusting [AGR] Applying fungicides or insecticides in powder form to a crop; usually done from a low-flying aircraft. { 'kräp ¸dəst·iŋ }

crop-flow sensor [AGR] An instrument used in precision agriculture to measure either the volume or the mass of the harvested portion of a crop using a variety of engineering principles, including light interception, radiation absorption, measurement of impact force, and directly weighing the crop. { 'kräp¸flō ¸sen·sər }

crop micrometeorology [AGR] The branch of meteorology that deals with the interaction of crops and their immediate physical environment. { ¦kräp ¸mī·krō¸mēt·ē·ə'räl·ə·jē }

crop out *See* outcrop. { 'kräp ¸aut }

cropping [GRAPHICS] The elimination of unwanted details in a picture by removing portions on the edges of the picture. { 'kräp·iŋ }

cropping index [AGR] The number of crops grown per year on a given land area times 100. Abbreviated CI. { 'kräp· iŋ ¸in¸deks }

crop rotation [AGR] A method of protecting the soil and replenishing its nutrition by planting a succession of different crops on the same land. { 'kräp rō'tā·shən }

cross *See* spider. { krȯs }

Cross *See* Crux. { krȯs }

cross antenna [ELECTROMAG] An array of two or more horizontal antennas connected to a single feed line and arranged in the pattern of a cross. { 'krȯs an¸ten·ə }

cross assembler [COMPUT SCI] An assembly program that allows a computer program written on one type of computer to be used on another type. { 'krȯs ə¸sem·blər }

cross axle [MECH ENG] **1.** A shaft operated by levers at its ends. **2.** An axle with cranks set at 90°. { 'krȯs ¸ak·səl }

crossband [MATER] In plywood comprising three or more plies, a layer of veneer whose grain direction is at right angles to that of the face plies. { 'krȯs¸band }

crossbar [CIV ENG] In a grating, one of the connecting bars which extend across bearing bars, usually perpendicular to them. { 'krȯs¸bär }

crossbar micrometer [ENG] An instrument consisting of two bars mounted perpendicular to each other in the focal plane of a telescope, and inclined to the east-west path of stars by 45°; used to measure differences in right ascension and declination of celestial objects. { 'krȯs¸bär mī'kräm·əd·ər }

crossbar switch [ELEC] A switch having a three-dimensional arrangement of contacts and a magnet system that selects individual contacts according to their coordinates in the matrix. { 'krȯs¸bär ¸swich }

crossbar system [COMMUN] Automatic telephone switching system which is generally characterized by the following features: selecting mechanisms are crossbar switches, common circuits select and test the switching paths and control the operation of the selecting mechanisms, and method of operations is one in which the switching information is received and stored by controlling mechanisms that determine the operations necessary in establishing a telephone connection. { 'krȯs¸bär ¸sis·təm }

crossbeam [BUILD] **1.** Also known as trave. **2.** A horizontal beam. **3.** A beam that runs transversely to the center line of a structure. { 'krȯs¸bēm }

cross bearing [NAV] The determination of the position of a ship or aircraft from the azimuths of two known locations. { 'krȯs ¸ber·iŋ }

cross-bedding [GEOL] The condition of having laminae lying transverse to the main stratification planes of the strata; occurs only in granular sediments. Also known as cross-lamination; cross-stratification. { ¦krȯs 'bed·iŋ }

cross-belt drive [DES ENG] A belt drive having parallel shafts rotating in opposite directions. { 'krȯs ¸belt ¸drīv }

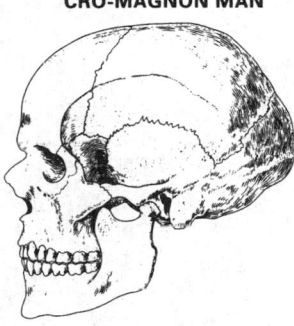

CRO-MAGNON MAN

Skull of a Cro Magnon male. *(From M. F. Ashley Montagu, An Introduction to Physical Anthropology, 2d ed., Charles C. Thomas, 1951)*

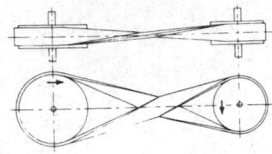

CROSSED BELT

Arrangement of crossed pulley belt showing the directions in which the pulleys rotate.

crossbolt [DES ENG] A lock bolt with two parts which can be moved in opposite directions. { 'krós,bōlt }

cross bond [CIV ENG] A masonry bond in which a course of alternating lengthwise and endwise bricks (Flemish bond) alternates with a course of bricks laid lengthwise. { 'krós ,bänd }

cross box [MECH ENG] A boxlike structure for the connection of circulating tubes to the longitudinal drum of a header-type boiler. { 'krós ,bäks }

cross bracing [BUILD] Boards which are nailed diagonally across studs or other boards so as to impart rigidity to a framework. { 'krós ,brās·iŋ }

crossbreed [BIOL] To propagate new individuals by breeding two distinctive varieties of a species. Also known as outbreed. { 'krós,brēd }

cross-cap [MATH] The self-intersecting surface that results when a Möbius band is deformed so that its boundary is a circle. { 'krós ,kap }

cross-color [ELECTR] In color television, the interference in the receiver chrominance channel caused by cross talk from monochrome signals. { 'krós ,kəl·ər }

cross compiler [COMPUT SCI] A compiler that allows a computer program written on one type of computer to be used on another type. { 'krós kəm,pī·lər }

cross-correlation [STAT] **1.** Correlation between corresponding members of two or more series: if $q_1, \ldots, q_n$ and $r_1, \ldots, r_n$ are two series, correlation between q_i and r_i, or between q_i and r_{i+j} (for fixed j), is a cross correlation. **2.** Correlation between or expectation of the inner product of two series of random variables, where the difference in indices between the corresponding values of the two series is fixed. { 'krós kär·ə'lā·shən }

cross-correlation detection *See* correlation detection. { 'krós kär·ə·lā·shən di'tek·shən }

cross-correlation function [COMMUN] A function, $\phi_{12}(\tau)$, where τ is a time-delay parameter, equal to the limit, as T approaches infinity, of the reciprocal of $2T$ times the integral over t from $-T$ to T of $f_1(t)f_2(t - \tau)$, where f_1 and f_2 are functions of time, such as the input and output of a communication system. { 'krós kär·ə·lā·shən ,fəŋk·shən }

cross-correlator [ELECTR] A correlator in which a locally generated reference signal is multiplied by the incoming signal and the result is smoothed in a low-pass filter to give an approximate computation of the cross-correlation function. Also known as synchronous detector. { ¦krós'kär·ə,läd·ər }

cross-country mill [MET] A rolling mill in which the tables of the mill stands are parallel with a crossover table that connects them; used to produce special forms of bar stock. { ¦krós ¦kən·trē 'mil }

cross-coupling [COMMUN] A measure of the undesired power transferred from one channel to another in a transmission medium. { ¦krós 'kəp·liŋ }

crosscurrent [FL MECH] A current that flows across or opposite to another current. { 'krós,kər·ənt }

crosscurrent extraction [ANALY CHEM] Procedure of batchwise liquid-liquid extraction in a separatory funnel; solvent is added to the sample in the funnel, which is then shaken, and the extract phase is allowed to coalesce, then is drawn off. { 'krós,kər·ənt ik'strak·shən }

cross curve [MATH] A plane curve whose equation in cartesian coordinates x and y is $(a^2/x^2) + (b^2/y^2) = 1$, where a and b are constants. Also known as cruciform curve. { 'krós ,kərv }

crosscut [ENG] A cut made through wood across the grain. [MIN ENG] **1.** A small passageway driven at right angles to the main entry of a mine to connect it with a parallel entry of air course. **2.** A passageway in a mine that cuts across the geological structure. { 'krós,kət }

crosscut file [DES ENG] A file with a rounded edge on one side and a thin edge on the other; used to sharpen straight-sided saw teeth with round gullets. { 'krós,kət ,fīl }

crosscutting relationships [GEOL] Relationships which may occur between two adjacent rock bodies, where the relative age may be determined by observing which rock "cuts" the other, for example, a granitic dike cutting across a sedimentary unit. { 'krós,kəd·iŋ ri'lā·shən,ships }

crosscut saw [DES ENG] A type of saw for cutting across the grain of the wood; designed with about eight teeth per inch. { 'krós,kət ,só }

cross dating [ARCHEO] The matching of patterns of ring widths from one tree with corresponding patterns for the same years from another tree, allowing the accurate assignment of calendar dates to tree rings. { ¦krós 'dād·iŋ }

cross drum boiler [MECH ENG] A sectional header or box header type of boiler in which the axis of the horizontal drum is perpendicular to the axis of the main bank of tubes. { 'krós ,drəm ,bóil·ər }

crossed belt [MECH ENG] A pulley belt arranged so that the sides cross, thereby making the pulleys rotate in opposite directions. { 'króst ¦belt }

crossed cylinder [OPTICS] **1.** A thin lens whose surfaces are portions of circular cylinders whose axes cross at right angles or obliquely. **2.** A weak lens whose effect is equivalent to that of lenses with convex and concave cylindrical surfaces of equal curvature crossed at right angles. { 'króst 'sil·ən·dər }

crossed electrophoresis [IMMUNOL] Immunoelectrophoresis that uses an initial separation along one axis of the plate, after which only a strip of medium along the axis is preserved; new medium containing antisera is poured beside the strip, the plate is turned 90°, and electrophoresis is resumed. { 'króst i,lek·trō·fə'rē·səs }

crossed-field accelerator [AERO ENG] A plasma engine for space travel in which plasma serves as a conductor to carry current across a magnetic field, so that a resultant force is exerted on the plasma. { 'króst ,fēld ik'sel·ə,rād·ər }

crossed-field amplifier [ELECTR] A forward-wave, beam-type microwave amplifier that uses crossed-field interaction to achieve good phase stability, high efficiency, high gain, and wide bandwidth for most of the microwave spectrum. { 'króst ,fēld 'am·plə,fī·ər }

crossed-field backward-wave oscillator [ELECTR] One of several types of backward-wave oscillators that utilize a crossed field, such as the amplitron and carcinotron. { 'króst ,fēld 'bak,wərd ,wāv 'äs·ə,lād·ər }

crossed-field device [ELECTR] Any instrument which uses the motion of electrons in perpendicular electric and magnetic fields to generate microwave radiation, either as an amplifier or oscillator. { 'króst ,fēld di'vīs }

crossed-field multiplier phototube [ELECTR] A multiplier phototube in which repeated secondary emission is obtained from a single active electrode by the combined effects of a strong radio-frequency electric field and a perpendicular direct-current magnetic field. { 'króst ,fēld ,məl·tə,plī·ər 'fōd·ō,tüb }

crossed lens [OPTICS] A lens designed with radii of curvature which give minimum spherical aberration for parallel incident rays. { 'króst 'lenz }

crossed-needle meter [ENG] A device consisting of two pointer-type analog meters inside a single enclosure with pointer movements centered at different positions so that their point of crossing indicates the value of some function of the two readings. { 'króst ¦nēd·əl 'mēd·ər }

crossed paralysis [MED] Paralysis of the arm and leg on one side, associated with contralateral cranial nerve palsies caused by a brainstem lesion involving cranial nerve nuclei and the ipsilateral pyramidal tract. { 'króst pə'ral·ə·səs }

crossed-pointer indicator [NAV] A two-pointer indicator used with an instrument-landing system to indicate the position of an airplane with respect to the glide path. { 'króst ¦point·ər 'in·də,kād·ər }

crossed prisms [OPTICS] A pair of Nicol prisms whose principal planes are perpendicular to each other, so that light passing through one is extinguished by the other. { 'króst 'priz·əmz }

cross effect [PHYS] Any phenomenon in which two or more transport effects are coupled, such as thermal and electrical conductivity, or thermal conductivity and diffusion. { 'krós i,fekt }

cross-eye *See* esotropia. { 'krós ,ī }

cross-fade [ENG ACOUS] In dubbing, the overlapping of two sound tracks, wherein the outgoing track fades out while the incoming track fades in. { 'krós ,fād }

cross fault [GEOL] **1.** A fault whose strike is perpendicular to the general trend of the regional structure. **2.** A minor fault that intersects a major fault. { 'krós ,fólt }

cross-fertilization [BOT] Fertilization between two separate plants. [ZOO] Fertilization between different kinds of individuals. { 'krós ,fərd·əl·ə'zā·shən }

cross fire [COMMUN] Interfering current in one telegraph or signaling channel resulting from telegraph or signaling currents in another channel. { 'krós ˌfīr }

cross-flow [AERO ENG] A flow going across another flow, as a spanwise flow over a wing. { 'krós ˌflō }

cross-flow baffle [ENG] A type of baffle in a shell-and-tube heat exchanger that directs shell-side fluid back and forth or up and down across the tubes. Also known as transverse baffle. { 'krós ˌflō ˌbaf·əl }

cross-flow plane [AERO ENG] A plane at right angles to the free-stream velocity. { 'krós ˌflō ˌplān }

cross flux [ELECTROMAG] A component of magnetic flux perpendicular to that produced by the field magnets in an electrical rotating machine. { 'krós ˌfləks }

cross fold [GEOL] A secondary fold whose axis is perpendicular or oblique to the axis of another fold. Also known as subsequent fold; superimposed fold; transverse fold. { 'krós ˌfōld }

crossfoot [COMPUT SCI] To add numbers in several different ways in a computer, for checking purposes. { 'krós,fút }

cross furring ceiling [BUILD] A ceiling in which furring members are attached perpendicular to the main runners or other structural members. { 'krós ˌfər·iŋ ˌsēl·iŋ }

cross gateway See cross heading. { 'krós 'gāt,wā }

cross hair [ENG] An inscribed line or a strand of hair, wire, silk, or the like used in an optical sight, transit, or similar instrument for accurate sighting. { 'krós ˌher }

crosshatch generator [ELECTR] A signal generator that generates a crosshatch pattern for adjusting color television receiver circuits. { 'krós,hach ˌjen·ə,rād·ər }

crosshaul [MECH ENG] A device for loading objects onto vehicles, consisting of a chain that is hooked on opposite sides of a vehicle, looped under the object, and connected to a power source and that rolls the object onto the vehicle. { 'krós,hól }

crosshead [MECH ENG] A block sliding between guides and containing a wrist pin for the conversion of reciprocating to rotary motion, as in an engine or compressor. [MET] A device generally employed in wire coating which is attached to the discharge end of the extruder cylinder; designed to facilitate extruding material at an angle. [MIN ENG] A runner or guide positioned just above a sinking bucket to restrict excessive swinging. { 'krós,hed }

cross heading [MIN ENG] Mine passage driven for ventilation from the airway to the gangway, or from one breast through the pillar to the adjoining working. Also known as cross gateway; cross hole; headway. { 'krós ˌhed·iŋ }

cross hole See cross heading. { 'krós ˌhōl }

crossing angle [NAV] The angle at which two lines of position intersect. Also known as angle of cut. { 'krós·iŋ ˌaŋ·gəl }

crossing barrier [GEN] Any of the genetically controlled mechanisms that either prevent or significantly reduce the ability of individuals in a population to hybridize with individuals of other populations. { 'krós·iŋ ˌbar·ē·ər }

crossing-over [GEN] The exchange of genetic material between paired homologous chromosomes during meiosis. Also known as crossover. { ˈkrós·iŋ 'ō·vər }

crossing-over map [GEN] A genetic map made by utilizing the frequency of crossing-over as a measure of the relative distances between genes in one linkage group. { ˈkrós·iŋ 'ō·vər ˌmap }

crossing-over value [GEN] The frequency of crossing-over between two linked genes. { ˈkrós·iŋ 'ō·vər ˌval·yü }

crossing plates [CIV ENG] Plates placed between a crossing and the ties to support the crossing and protect the ties. { 'krós·iŋ ˌplāts }

crossing symmetry [PARTIC PHYS] The amplitude for a process that involves creation of a particle with four-momentum P_μ is equal to the amplitude for a process which is the same except it involves destruction of the antiparticle with four-momentum $-P_\mu$. { 'krós·iŋ ˌsim·ə·trē }

cross joint [GEOL] A fracture in igneous rock perpendicular to the lineation caused by flow magma. Also known as transverse joint. { 'krós ˌjoint }

cross-lamination [GEOL] See cross-bedding. [MATER] Construction of a laminated composite material so that some layers are oriented at right angles to the other layers with respect to the grain or the strongest direction in terms of tension. { ˈkrós lam·ə'nā·shən }

crosslap joint [BUILD] A joint in which two wood members cross each other; half the thickness of each is removed so that at the joint the thickness is the same as that of the individual members. { 'krós,lap ˌjoint }

cross-level [ENG] To level at an angle perpendicular to the principal line of sight. { 'krós ˌlev·əl }

crossline screen [GRAPHICS] In halftone photography, a grid that has opaque lines intersecting at right angles, forming transparent squares (screen apertures). Also known as a glass screen. { 'krós,līn ˌskrēn }

crosslink [MOL BIO] A covalent linkage between the complementary strands of deoxyribonucleic acid (DNA) duplex or between bases of a single strand of DNA. [ORG CHEM] The covalent bonds between adjacent polymer chains that lock the chains in place. { 'krós ˌliŋk }

crosslinking [ORG CHEM] The setting up of chemical links between the molecular chains of polymers. { 'krós ˌliŋk·iŋ }

cross-magnetizing effect [ELECTROMAG] The distortion in the flux-density distribution in the air gap of an electric rotating machine caused by armature reaction. { ˈkrós 'mag·nə,tīz·iŋ i'fekt }

crossmarks [GRAPHICS] Register marks used for the exact positioning of images in step-and-repeat, double, or multicolor printing; also used for superimposing overlays onto a base or onto each other. { 'krós,märks }

cross matching [IMMUNOL] Determination of blood compatibility for transfusion by mixing donor cells with recipient serum, and recipient cells with donor serum, and examining for an agglutination reaction. { 'krós ˌmach·iŋ }

cross modulation [COMMUN] A type of interference in which the carrier of a desired signal becomes modulated by the program of an undesired signal on a different carrier frequency; the program of the undesired station is then heard in the background of the desired program. { 'krós ˌmäj·ə'lā·shən }

cross multiplication [MATH] Multiplication of the numerator of each of two fractions by the denominator of the other, as when eliminating fractions from an equation. { 'krós ˌməl·tə·plə'kā·shən }

cross-neutralization [ELECTR] Method of neutralization used in push-pull amplifiers, whereby a portion of the plate-cathode alternating-current voltage of each vacuum tube is applied to the grid-cathode circuit of the other vacuum tube through a neutralizing capacitor. { 'krós ˌnü·trə·lə'zā·shən }

cross office switching time [COMMUN] Time required to connect any input through the switching center to any selected output. { 'krós ˌóf·əs 'swich·iŋ ˌtīm }

Crossopterygii [PALEON] A subclass of the class Osteichthyes comprising the extinct lobefins or choanate fishes and represented by one extant species; distinguished by two separate dorsal fins. { krä,säp·tə'rij·ē,ī }

Crossosomataceae [BOT] A monogeneric family of xerophytic shrubs in the order Dilleniales characterized by perigynous flowers, seeds with thin endosperm, and small, entire leaves. { ˌkrä·sə,sō·mə'tās·ē,ē }

crossover [CIV ENG] **1.** An S-shaped section of railroad track joining two parallel tracks. **2.** A connection between two pipes in the same water supply system or a connection between two water supply systems. [ELEC] A point at which two conductors cross, with appropriate insulation between them to prevent contact. [ELECTR] The plane at which the cross section of a beam of electrons in an electron gun is a minimum. [ENG] The portion of a draw works' drum containing grooves for angle control so the wire rope can cross over to begin a new wrap. Also known as angle-control section. [GEN] See crossing over. { 'krós,ō·vər }

crossover distortion [ELECTR] Amplitude distortion in a class B transistor power amplifier which occurs at low values of current, when input impedance becomes appreciable compared with driver impedance. { 'krós,ō·vər dis'tór·shən }

crossover experiment [MED] An experiment or clinical investigation in which subjects are divided randomly into at least as many groups as there are kinds of treatment to be given, and then the groups are interchanged until every subject has received each treatment. { 'krós,ō·vər ik'sper·ə·mənt }

crossover flange [ENG] Intermediate pipe flange used to connect flanges of different working pressures. { 'krós,ō·vər ˌflanj }

crossover frequency [ENG ACOUS] **1.** The frequency at which a dividing network delivers equal power to the upper

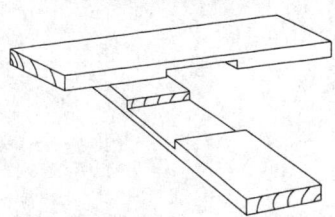

CROSSLAP JOINT

Two wood members connected by a crosslap joint.

and lower frequency channels when both are terminated in specified loads. **2.** *See* transition frequency. { 'krȯs·ō·vər ˌfrē·kwən·sē }

crossover joint [PETRO ENG] A casing length with different threads at either end to permit a change from one thread to another in a casing string. { 'krȯs·ō·vər ˌjȯint }

crossover length [MATH] A length characteristic of a fractal network such that at scales which are small compared with this length the fractal nature of the structure is manifest in its dynamics, whereas at scales which are large compared with this length the dynamics resemble those of a crystalline structure. { 'krȯs·ō·vər ˌleŋkth }

crossover network [ENG ACOUS] A selective network used to divide the audio-frequency output of an amplifier into two or more bands of frequencies. Also known as dividing network; loudspeaker dividing network. { 'krȯs·ō·vər ˌnet·wərk }

crossover spiral *See* lead-over groove. { 'krȯs·ō·vər ˌspī·rəl }

crossover voltage [ELECTR] In a cathode-ray storage tube, the voltage of a secondary writing surface, with respect to cathode voltage, on which the secondary emission is unity. { 'krȯs·ō·vər ˌvōl·tij }

cross-peen hammer [ENG] A hammer with a wedge-shaped surface at one end of the head. { 'krȯs ˌpēn 'ham·ər }

cross-platform computing [COMPUT SCI] The use of very similar user interfaces for versions of programs running on different operating systems and computer architectures. { ˌkrȯs ˌplat·fȯrm kəm'pyüd·iŋ }

cross-pointer indicator [NAV] A flight instrument having two needles which cross in the center when the aircraft is on course; the vertical needle indicates the position of the aircraft with respect to the localizer course, and the horizontal needle serves a similar purpose for the glide slope of an instrument landing system. { 'krȯs ˌpȯint·ər ˌin·də‚kād·ər }

cross-polarization [ELECTROMAG] The component of the electric field vector normal to the desired polarization component. { ˌkrȯs ˌpō·lə·rə'zā·shən }

cross-pollination [BOT] Transfer of pollen from the anthers of one plant to the stigmata of another plant. { ˌkrȯs ˌpä·lə‚nā·shən }

cross product [MATH] **1.** An anticommutative multiplication on the vectors of euclidean three-dimensional space. Also known as vector product. **2.** The product of the two mean terms of a proportion, or the product of the two extreme terms; in the proportion $a/b = c/d$, it is ad or bc. { 'krȯs 'prä·dəkt }

cross ratio [MATH] For four collinear points, A, B, C, and D, the ratio $(AB)(CD)/(AD)(CB)$, or one of the ratios obtained from this quantity by a permutation of A, B, C, and D. { 'krȯs ˌrā·shō }

cross-reacting antibody [IMMUNOL] Antibody that reacts with an antigen that did not stimulate the production of that antibody. { 'krȯs rē‚ak·tiŋ 'ant·i‚bäd·ē }

cross-reacting antigen [IMMUNOL] Antigen that reacts with an antibody whose production was induced by a different antigen. { 'krȯs rē‚ak·tiŋ 'ant·i·jən }

cross-reacting material [BIOCHEM] A protein produced by a mutant gene that is enzymatically inactive but shows serological properties similar to the protein of the wild-type gene. { 'krȯs rē‚ak·tiŋ mə'tir·ē·əl }

cross-reaction [IMMUNOL] Reaction between an antibody and a closely related, but not complementary, antigen. { 'krȯs rē'ak·shən }

cross-referencing program [COMPUT SCI] A computer program used in debugging that produces indexed lists of both the variable names and the statement numbers of the source program. { 'krȯs 'ref·rəns·iŋ ˌprō·grəm }

cross-rolling [MET] **1.** Straightening metal sheets by passing them through rolls at right angles to the principal direction of rolling. **2.** Straightening round bars or tubes by passing the work through parallel to the axes of rolls. { 'krȯs ˌrōl·iŋ }

cross sea [OCEANOGR] A series of waves or swell crossing another wave system at an angle. { ˌkrȯs ˌsē }

cross section [GEOL] **1.** A diagram or drawing that shows the downward projection of surficial geology along a vertical plane, for example, a portion of a stream bed drawn at right angles to the mean direction of the flow of the stream. **2.** An actual exposure or cut which reveals geological features. [GRAPHICS] A diagram or drawing representing a cut at right angles to an axis. [MAP] A horizontal grid system that is laid out on the ground for determining contours, quantities of earthwork, and so on, by means of elevations of the grid points. [MATH] **1.** The intersection of an n-dimensional geometric figure in some euclidean space with a lower dimensional hyperplane. **2.** A right inverse for the projection of a fiber bundle. [PHYS] An area characteristic of a collision reaction between atomic or nuclear particles or systems, such that the number of reactions which occur equals the product of the number of target particles or systems and the number of incident particles or systems which would pass through this area if their velocities were perpendicular to it. Also known as collision cross section. { 'krȯs ˌsek·shən }

cross-sectional study [PSYCH] The study of groups of individuals differing on the basis of specified criteria (for example, age) at the same point in time. { 'krȯs ¦sek·shən·əl 'stəd·ē }

cross section per atom [NUC PHYS] The microscopic cross section for a given nuclear reaction referred to the natural element, even though the reaction involves only one of the natural isotopes. { 'krȯs ˌsek·shən pər 'ad·əm }

cross-section symbols [GRAPHICS] Standardized shadings used to represent various materials in a cross section. { 'krȯs ˌsek·shən ˌsim·bəlz }

cross slide [MECH ENG] A part of a machine tool that allows the tool carriage to move at right angles to the main direction of travel. { 'krȯs ˌslīd }

cross spectrum [PHYS] The complex vector sum of the cospectrum and quadrature spectrum. { ¦krȯs ¦spek·trəm }

cross-staff [NAV] A forerunner of the modern sextant, used for measuring altitudes of celestial bodies, consisting essentially of a wooden rod with one or more perpendicular crosspieces free to slide along the main rod. Also known as forestaff; Jacob's staff. { 'krȯs ˌstaf }

cross-stone *See* harmotome. *See* staurolite. { 'krȯs ˌstōn }

cross-stratification *See* cross-bedding. { 'krȯs ˌstrad·ə·fə'kā·shən }

crosstalk [COMMUN] **1.** The sound heard in a receiver along with a desired program because of cross modulation or other undesired coupling to another communication channel; it is also observed between adjacent pairs in a telephone cable. **2.** Interaction of audio and video signals in a television system, causing video modulation of the audio carrier or audio modulation of the video signal at some point. **3.** Interaction of the chrominance and luminance signals in a color television receiver. [ENG ACOUS] *See* magnetic printing. { 'krȯs ˌtȯk }

crosstalk coupling [COMMUN] The cross coupling between speech communications channels or their component parts. Also known as crosstalk loss. { 'krȯs‚tȯk ˌkəp·liŋ }

crosstalk level [COMMUN] Volume of crosstalk energy, measured in decibels, referred to a reference level. { 'krȯs‚tȯk ˌlev·əl }

crosstalk loss *See* crosstalk coupling. { 'krȯs‚tȯk ˌlȯs }

crosstalk unit [COMMUN] A measure of the coupling between two circuits; the number of crosstalk units is 1 million times the ratio of the current or voltage at the observing point to the current or voltage at the origin of the disturbing signal, the impedances at these points being equal. Abbreviated cu. { 'krȯs‚tȯk ˌyü·nət }

cross-thread [ENG] To screw together two threaded pieces without aligning the threads correctly. { 'krȯs ˌthred }

crosstie [ENG] A timber or metal sill placed transversely under the rails of a railroad, tramway, or mine-car track. { 'krȯs‚tī }

cross-tolerance [MED] Tolerance or resistance to the action of a drug brought about by continued use of another drug of similar pharmacologic action. { ¦krȯs 'täl·ə·rəns }

cross trail error [NAV] The error introduced in a drift observation made by means of an object dropped from an aircraft when the object lands to one side of the track. { 'krȯs ˌtrāl ˌer·ər }

cross turret [MECH ENG] A turret that moves horizontally and at right angles to the lathe guides. { 'krȯs ˌtər·ət }

cross valley *See* transverse valley. { 'krȯs ˌval·ē }

crossvein [MIN ENG] A vein that intersects an older or larger vein. { 'krȯs‚vān }

cross ventilation [ENG] The movement of air from one side of a building or room and out the other side or through a monitor. { 'krȯs ˌvent·əl'ā·shən }

crosswind [METEOROL] A wind which has a component

CROSSTALK

Crosstalk arising in short section of two parallel circuits (def. 1). Conductor closer to disturbing circuit has larger crosstalk voltage than farther conductor; voltage difference is effective at E_2 and W_2 terminals. Energy in disturbing circuit is assumed to flow from east to west.

directed perpendicularly to the course (or heading) of an exposed, moving object. { 'krós,wind }

cross-wire weld [MET] A weld made across wires or bars in order to make wire mesh or other similar products. { 'krós ,wīr ,weld }

Crotalidae [VERT ZOO] A family of proglyphodont venomous snakes in the reptilian suborder Serpentes. { krō'tal·ə,dē }

crotch [SCI TECH] The angular form made by the parting of two branches, legs, or members. { kräch }

crotch height [ANTHRO] The measure of the vertical distance from the crotch of a standing subject to the floor. { 'kräch ,hīt }

crotonaldehyde [ORG CHEM] C_3H_5CHO A colorless liquid boiling at 104°C, soluble in water; vapors are lacrimatory; used as an intermediate in manufacture of *n*-butyl alcohol and quinaldine. Also known as propylene aldehyde. { ¦krōt·ən'al·də,hīd }

crotonic acid [ORG CHEM] C_3H_5COOH An unsaturated acid, with colorless, monoclinic crystals, soluble in water; used in the preparation of synthetic resins, plasticizers, and pharmaceuticals. { krō'tän·ik 'as·əd }

croton oil [MATER] A yellow-brown oil obtained from the seeds of the *Croton tiglium;* used as a purgative and as a substitute for castor oil. { 'krōt·ən ,öil }

croup [MED] Any condition of upper-respiratory pathway obstruction in children, especially acute inflammation of the pharynx, larynx, and trachea, characterized by a hoarse, brassy, and stridulent cough and difficulties in breathing. { krüp }

croup-associated virus [VIROL] A virus belonging to subgroup 2 of the parainfluenza viruses and found in children with croup. Also known as CA virus; laryngotracheobronchitis virus. { ¦krüp ə¦sō·sē,ād·əd 'vī·rəs }

croute calcaire *See* caliche. { ,krüt kal'ker }

Crout reduction [MATH] Modification of the Gauss procedure for numerical solution of simultaneous linear equations; adapted for use on desk calculators and digital computers. { 'kraut ri'dək·shən }

Crova wavelength [STAT MECH] The wavelength in the spectrum of a radiator whose intensity divided by the intensity of the total radiation equals the derivative of the intensity of the wavelength with respect to temperature divided by the derivative of the total intensity with respect to temperature. { 'krō·və 'wāv,leŋkth }

crow [VERT ZOO] The common name for a number of predominantly black birds in the genus *Corvus* comprising the most advanced members of the family Corvidae. { krō }

Crow *See* Corvus. { krō }

crowbar [DES ENG] An iron or steel bar that is usually bent and has a wedge-shaped working end; used as a lever and for prying. [ELEC] A device or action that in effect places a high overload on the actuating element of a circuit breaker or other protective device, thus triggering it. { 'krō,bär }

crowbar voltage protector [ELEC] A separate circuit which monitors the output of a regulated power supply and instantaneously throws a short circuit (or crowbar) across the output terminals of the power supply whenever a preset voltage limit is exceeded. { 'krō,bär 'vōl·tij prə'tek·tər }

Crowe process [MET] In cyanidation, after extraction of the gold, separation of the solution from the ore tailings by filtration or countercurrent decantation and by passage through a vacuum chamber where deaeration occurs. { 'krō ,prä·səs }

crown [ANAT] 1. The top of the skull. 2. The portion of a tooth above the gum. [ARCH] A feature near the top of a terminal, such as the highest point of an arch. [BOT] 1. The topmost part of a plant or plant part. 2. *See* corona. [CIV ENG] 1. Center of a roadway elevated above the sides. 2. In plumbing, that part of a trap where the direction of flow changes from upward to horizontal or downward. [ENG] 1. The part of a drill bit inset with diamonds. 2. The vertex of an arch or arched surface. 3. The top or dome of a furnace or kiln. 4. A high spot forming on a tool joint shoulder as the result of drill pipe wobbling. [LAP] The portion of a faceted gem above the girdle. [MET] That part of the sheet or roll where the thickness or diameter increases from edge to center. [MIN ENG] A horizontal roof member of a timber up to 16 feet (4.9 meters) long and supported at each end by an upright. { kraun }

crown block [PETRO ENG] A wooden or steel beam joined to the tops of derrick posts of an oil well to support pulleys. { 'kraun ,bläk }

crown cell [ELEC] The generic name for alkaline zinc-manganese dioxide dry-cell battery; manganese dioxide-graphite cathode mix is pressed into a steel can onto which a steel cap is spot-welded to contain the amalgamated powdered-zinc anode. { 'kraun ,sel }

crown-cut *See* backsawing. { 'kraun ,kət }

crown ether [ORG CHEM] A macrocyclic polyether whose structure exhibits a conformation with a so-called hole capable of trapping cations by coordination with a lone pair of electrons on the oxygen atoms. { 'kraun ,ē·thər }

crown fire [FOR] A forest fire burning primarily in the tops of trees and shrubs. { 'kraun ,fīr }

crown gall [PL PATH] A bacterial disease of many plants induced by *Bacterium tumefaciens* and marked by abnormal enlargement of the stem near the root crown. { 'kraun ,gól }

crown glass [MATER] A soda-lime glass, typically having 72% SiO_2, 13% CaO, and 15% Na_2O, which is hard and will take a simple polish; highly transparent for visible light. { 'kraun ,glas }

crown grafting [BOT] A method of vegetative propagation whereby a scion 3–6 inches (8–15 centimeters) long is grafted at the root crown, just below ground level. { 'kraun ,graf·tiŋ }

crown platform [PETRO ENG] A platform located at the top of the derrick that provides access to the sheaves of the crown block. { 'kraun ¦plat,form }

crown post [BUILD] Any upright member of a roof truss assembly, such as a king post. { 'kraun ,pōst }

crown rot [PL PATH] Any plant disease or disorder marked by deterioration of the stem at or near ground level. { 'kraun ,rät }

crown rust [PL PATH] A rust disease of oats and certain other grasses caused by varieties of *Puccinia coronata* and marked by light-orange masses of fungi on the leaves. { 'kraun ,rəst }

crown saw [DES ENG] A saw consisting of a hollow cylinder with teeth around its edge; used for cutting round holes. Also known as hole saw. { 'kraun ,só }

crown sheet [MECH ENG] The structural element which forms the top of a furnace in a fire-tube boiler. { 'kraun ,shēt }

crown weir [CIV ENG] The highest point on the internal bottom surface of the crown of a plumbing trap. { 'kraun ,wer }

crown wheel [DES ENG] A gear that is light and crown-shaped. [HOROL] 1. The horizontal escape wheel of a verge escapement timepiece. 2. The wheel in the winding mechanism of a watch that drives the ratchet wheel and is itself driven by the winding-stem pinion. { 'kraun ,wēl }

crow-quill pen [GRAPHICS] An artist's pen made from a crow quill and used to form an extremely fine line. { 'krō ,kwil ,pen }

crow's nest [ENG] An elevated passageway for personnel located at the top of a derrick, refinery, or similar installation. { 'krōz ,nest }

CR ratio *See* calorimetric-respirometric ratio. { ¦sē'är ,rā·shō }

CRSV *See* carnation ringspot virus.

Crt *See* Crater.

CRT *See* cathode-ray tube.

Cru *See* Crux.

cruciate [ANAT] Resembling a cross. { 'krü·shē,āt }

crucible [SCI TECH] A refractory vessel or pot, varying in size from a small laboratory utensil to large industrial equipment for melting or calcining. { 'krü·sə·bəl }

crucible melt extraction [MET] Melt extraction in which the molten metal is contained in a crucible. { 'krü·sə·bəl 'melt ik'strak·shən }

crucible steel *See* drill steel. { 'krü·sə·bəl ,stēl }

Cruciferae [BOT] A large family of dicotyledonous herbs in the order Capparales characterized by parietal placentation; hypogynous, mostly regular flowers; and a two-celled ovary with the ovules attached to the margins of the partition. { krü'sif·ə,rē }

cruciform [SCI TECH] Resembling or arranged like a cross. { 'krü·sə,form }

cruciform core [ELEC] A transformer core in which all windings are on one center leg, and four additional legs arranged in the form of a cross serve as return paths for magnetic flux. { 'krü·sə,form ,kór }

cruciform curve *See* cross curve. { 'krü·sə,form ,kərv }

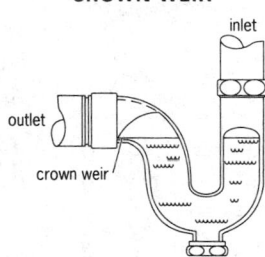

CROWN WEIR

inlet

outlet

crown weir

Diagram of a trap showing the position of a crown weir.

cruciform structure [MOL BIO] A cross-shaped configuration of deoxyribonucleic acid produced by intrastrand base pairing of complementary inverted repeats. { 'krü·sə‚fȯrm ‚strək·chər }

cruciform wing [AERO ENG] An aircraft wing in the shape of a cross. { 'krü·sə‚fȯrm ‚wiŋ }

α Crucis [ASTRON] A double star in the constellation Crux that is 220 light-years from the sun; spectral classification BO.5V. Also known as Acrux. { ‚al·fə 'krü·səs }

β Crucis [ASTRON] A star in the constellation Crux that is 370 light-years from the sun, with magnitude 1.3, spectral classification BO.5IV. Also known as Mimosa. { ‚bā·də 'krü·səs }

crude assay [CHEM ENG] A procedure for determining the general distillation characteristics and other quality information of crude oil. { ‚krüd 'as·ā }

crude desalting [CHEM ENG] The washing of crude oil with water in order to remove materials such as dirt, silt, and water-soluble minerals. { ‚krüd dē'sȯlt·iŋ }

crude drug [PHARM] **1.** A plant or animal drug containing all principles characteristic of the drug. **2.** The dried leaves, bark, or rhizome of a plant containing therapeutically active principles. Also known as botanical; plant extract. { ‚krüd 'drəg }

crude lecithin See lecithin. { ‚krüd 'les·ə·thən }

crude material See raw material. { 'krüd me‚tir·ē·əl }

crude naphtha [MATER] A light distillate made in the fractionation of crude oil. { ‚krüd 'naf·thə }

crude oil [GEOL] A comparatively volatile liquid bitumen composed principally of hydrocarbon, with traces of sulfur, nitrogen, or oxygen compounds; can be removed from the earth in a liquid state. { ‚krüd 'ȯil }

crude ore [MIN ENG] The ore as it leaves the mine in an unconcentrated form. { ‚krüd 'ȯr }

crude scale See scale wax. { ‚krüd 'skāl }

crude still [CHEM ENG] The distillation equipment in which crude oil is separated into various products. { ‚krüd 'stil }

crude yellow scale [MATER] The trade name for a low-grade paraffin wax. { ‚krüd 'yel·ō ‚skāl }

crufomate [ORG CHEM] $C_{12}H_{19}ClNO_3P$ A white, crystalline compound, with a melting point of 61.8°C, which is insoluble in water; used both internally and externally for cattle parasites. { 'krü·fə‚māt }

cruise control [NAV] The act or practice of operating an aircraft so as to achieve the most efficient performance on a given flight under available conditions; it may be instituted to obtain maximum economy, endurance, speed, or range, or maximum efficiency at a predetermined airspeed or power setting. { 'krüz kən'trōl }

cruise missile [AERO ENG] A pilotless airplane that can be launched from a submarine, surface ship, ground vehicle, or another airplane; range can be up to 1500 miles (2400 kilometers), flying at a constant altitude that can be as low as 200 feet (60 meters). { 'krüz ‚mis·əl }

cruiser [NAV ARCH] A type of large warship, but smaller than a battleship, having a displacement of 6000 to 15,000 tons (5442–13,605 metric tons), moderately armed and armored, and capable of any naval duty except combat with battleships. { 'krüz·ər }

cruising altitude [NAV] An indicated pressure altitude maintained in cruising. { 'krüz·iŋ ‚al·tə‚tüd }

cruising radius [NAV] **1.** The maximum distance that an aircraft, starting with full fuel tanks, can cruise under given or specified conditions from its takeoff point before returning with a specified fuel reserve remaining. **2.** The distance a craft can travel at cruising speed without refueling. Also known as cruising range. { 'krüz·iŋ ‚rād·ē·əs }

cruising range See cruising radius. { 'krüz·iŋ ‚rānj }

crumb structure [GEOL] A soil condition in which the particles are crumblike aggregates; suitable for agriculture. { 'krəm ‚strək·chər }

crunode [MATH] A point on a curve through which pass two branches of the curve with different tangents. Also known as node. { 'krü‚nōd }

crura [ANAT] Plural of crus. { 'krür·ə }

crus [ANAT] **1.** The shank of the hindleg, that portion between the femur and the ankle. **2.** Any of various parts of the body resembling a leg or root. { 'krüs }

crush [MET] Casting defect caused by damage to the mold before pouring the metal. [MIN ENG] **1.** A general settlement of the strata above a coal mine due to failure of pillars; generally accompanied by numerous local falls of roof in mine workings. **2.** To reduce ore or quartz by stamps, crushers, or rolls. { krəsh }

crush breccia [GEOL] A breccia formed in place by mechanical fragmentation of rock during movements of the earth's crust. { 'krəsh ‚brech·ə }

crush conglomerate [GEOL] Beds similar to a fault breccia, except that the fragments are rounded by attrition. Also known as tectonic conglomerate. { 'krəsh kən'gläm·ə·rət }

crushed steel [MATER] An abrasive used in the stone, brick, glass, and metal trades, made by heating high-grade crucible steel to white heat, quenching in a bath of cold water, and crushing the fragments. { ‚krəsht 'stēl }

crushed stone [MIN ENG] Irregular fragments of rock crushed or ground to smaller sizes after quarrying. Also known as broken stone. { ‚krəsht 'stōn }

crusher [MECH ENG] A machine for crushing rock and other bulk materials. { 'krəsh·ər }

crush fold [GEOL] A fold of large dimensions that may involve considerable minor folding and faulting such as would produce a mountain chain or an oceanic deep. { 'krəsh ‚fōld }

crush-forming [ENG] Shaping the face of a grinding wheel by forcing a rotating metal roll into it. { 'krəsh ‚fȯr·miŋ }

crushing [MIN ENG] The quantity of ore pulverized or crushed at a single operation in processing. { 'krəsh·iŋ }

crushing mill See stamping mill. { 'krəsh·iŋ ‚mil }

crushing strain [MECH] Compression which causes the failure of a material. { 'krəsh·iŋ ‚strān }

crushing strength [MECH] The compressive stress required to cause a solid to fail by fracture; in essence, it is the resistance of the solid to vertical pressure placed upon it. { 'krəsh·iŋ ‚streŋkth }

crushing test [ENG] A test of the suitability of stone that might be mined for roads or building use. [MET] A test to determine quality of tubing, especially welded tubing, by applying compression parallel to the axis. { 'krəsh·iŋ ‚test }

crush kidney See lower nephron nephrosis. { 'krəsh 'kid·nē }

crush syndrome [MED] A severe, often fatal condition that follows a severe crushing injury, particularly involving large muscle masses, characterized by fluid and blood loss, shock, hematuria, and renal failure. Also known as compression syndrome. { 'krəsh ‚sin‚drōm }

crush zone [GEOL] A zone of fault breccia on fault gouge. { 'krəsh ‚zōn }

crust [GEOL] The outermost solid layer of the earth, mostly consisting of crystalline rock and extending no more than a few miles from the surface to the Mohorovičić discontinuity. Also known as earth crust. [HYD] A hard layer of snow lying on top of a soft layer. { krəst }

Crustacea [INV ZOO] A class of arthropod animals in the subphylum Mandibulata having jointed feet and mandibles, two pairs of antennae, and segmented, chitin-encased bodies. { krə'stā·shə }

crustacyanin [BIOCHEM] A carotenoprotein that determines the color of lobster shells. { ‚krəs·tə'sī·ə·nən }

crustal motion [GEOL] Movement of the earth's crust. { ‚krəst·əl 'mō·shən }

crustal plate See tectonic plate. { 'krəst·əl ‚plāt }

crustecdysone [BIOCHEM] $C_{27}H_{44}O_7$ 20-Hydroxyecdysone, the molting hormone produced by Y organs in crustaceans. { krəs'tek·də‚sōn }

crustose [BOT] Of a lichen, forming a thin crustlike thallus which adheres closely to the substratum of rock, bark, or soil. { 'krəs‚tōs }

crust vegetation [ECOL] Zonal growths of algae, mosses, lichens, or liverworts having variable coverage and a thickness of only a few centimeters. { 'krəst ‚vej·ə'tā·shən }

crutter [MIN ENG] **1.** A worker who drills blasting holes and prepares the blasting charge. **2.** A worker who removes blasted rock. { 'krəd·ər }

Crux [ASTRON] A constellation having four principal bright stars which form the figure of a cross; right ascension 12 hours, declination 60°S. Abbreviated Cru. Also known as Cross; Southern Cross. { krüks }

Crv See Corvus.

cry-, cryo- [SCI TECH] Combining form meaning cold, freezing. { krī, 'krī·ō }

cryalgesia [MED] Pain caused by cold. Also known as crymodynia. { ‚krī·əl'jē·zhə }

cryanesthesia [MED] Loss of sensation of cold. { ‚krī,an·əs'thē·zhə }

cryesthesia [PHYSIO] **1.** The sensation of coldness. **2.** Exceptional sensitivity to low temperatures. { ‚krī·əs'thē·zhə }

crymodynia See cryalgesia. { ‚krī·mə'dīn·ē·ə }

cryoanalgesia [MED] Loss of sensation of pain resulting from the use of a cryoprobe. { ‚krī·ō,an·əl'jē·zhə }

cryoanesthesia [MED] Regional anesthesia produced by localized application of cold. { ‚krī·ō,an·əs'thē·zhə }

cryobiology [BIOL] The use of low-temperature environments in the study of living plants and animals. { ¦krī·ō·bī'äl·ə·jē }

cryobiosis [PHYSIO] A type of cryptobiosis induced by low temperatures. { ¦krī·ō·bī'ō·səs }

cryocautery [MED] A substance or instrument that causes destruction of tissue by freezing. { ‚krī·ō'kȯd·ə·rē }

cryochem process [CHEM ENG] A freeze-drying technique involving conduction heat transfer to the frozen solid held on a metallic surface. { 'krī·ō,kem ‚präs·əs }

cryoconite [GEOL] A dark, powdery dust transported by wind and deposited on the surface of snow or ice; found, however, mainly in cryoconite holes. [MINERAL] A mixture of garnet, sillimanite, zircon, pyroxene, quartz, and various other minerals. { krī'äk·ə,nīt }

cryoconite hole [GEOL] A cylindrical dust well filled with cryoconite; absorbs solar radiation, causing melting of glacier ice around and below it. { krī'äk·ə,nīt ‚hōl }

cryoelectronics [ELECTR] A branch of electronics concerned with the study and application of superconductivity and other low-temperature phenomena to electronic devices and systems. Also known as cryolectronics. { ¦krī·ō·i,lek'trän·iks }

cryoextraction [MED] Removal of a cataract by means of a cryoprobe. { ‚krī·ō·ik'strak·shən }

cryofibrinogen [PATH] An abnormal fibrinogen that precipitates upon cooling but redissolves when warmed to room temperature; rarely found in human plasma. { ‚krī·ō·fī'brin·ə·jən }

cryogen See cryogenic fluid. { 'krī·ə·jən }

cryogenic coil [CRYO] A high-purity coil refrigerated to very low temperatures to reduce effective coil resistivity. { ‚krī·ə'jen·ik 'kȯil }

cryogenic conductor See superconductor. { ‚krī·ə'jen·ik kən'dək·tər }

cryogenic device [CRYO] A device whose operation depends on superconductivity as produced by temperatures near absolute zero. Also known as superconducting device. { ‚krī·ə'jen·ik di'vīs }

cryogenic engineering [ENG] A branch of engineering specializing in technical operations at very low temperatures (about 200 to 400°R, or −160 to −50°C). { ‚krī·ə'jen·ik en·jə'nir·iŋ }

cryogenic film [COMPUT SCI] A storage element using superconducting thin films of lead at liquid-helium temperature. { ‚krī·ə'jen·ik 'film }

cryogenic fluid [CRYO] A liquid which boils at temperatures of less than about 110 K at atmospheric pressure, such as hydrogen, helium, nitrogen, oxygen, air, or methane. Also known as cryogen; cryogenic liquid. { ‚krī·ə'jen·ik 'flü·əd }

cryogenic freezing [FOOD ENG] A freezing technique for preserving shrimp and other foods that are high-priced or have low moisture content by spraying them with liquid nitrogen as they pass through a tunnel on a conveyor belt. { ‚krī·ə'jen·ik 'frēz·iŋ }

cryogenic gyroscope [ENG] A gyroscope in which a spherical rotor of superconducting niobium spins while in levitation at cryogenic temperatures. Also known as superconducting gyroscope. { ‚krī·ə'jen·ik 'jī·rə,skōp }

cryogenic liquid See cryogenic fluid. { ‚krī·ə'jen·ik 'lik·wəd }

cryogenic period [GEOL] A time period in geologic history during which large bodies of ice appeared at or near the poles and climate favored the formation of continental glaciers. { ‚krī·ə'jen·ik 'pir·ē·əd }

cryogenic propellant [MATER] A rocket fuel, oxidizer, or propulsion fluid which is liquid only at very low temperatures. { ‚krī·ə'jen·ik prə'pel·ənt }

cryogenic pump [CRYO] A high-speed vacuum pump that can produce an extremely low vacuum and has a low power consumption; to reduce the pressure, gases are condensed on surfaces within an enclosure at extremely low temperatures, usually attained by using liquid helium or liquid or gaseous hydrogen. Also known as cryopump. { ‚krī·ə'jen·ik 'pəmp }

cryogenics [PHYS] The production and maintenance of very low temperatures, and the study of phenomena at these temperatures. { ‚krī·ə'jen·iks }

cryogenic temperature [CRYO] A temperature within a few degrees of absolute zero. { ‚krī·ə'jen·ik 'tem·prə·chər }

cryogenic transformer [ELECTR] A transformer designed to operate in digital cryogenic circuits, such as a controlled-coupling transformer. { ‚krī·ə'jen·ik tranz'fȯr·mər }

cryoglobulin [PATH] An abnormal protein, usually an immunoglobulin, which precipitates from plasma between 40 and 70°F (4.4 and 21°C). { ‚krī·ō'gläb·yə·lən }

cryohydrate [CHEM] A salt that contains water of crystallization at low temperatures. Also known as cryosel. { ‚krī·ō'hī,drāt }

cryohydric point [PHYS CHEM] The eutectic point of an aqueous salt solution. { ‚krī·ō'hī,drik ‚pȯint }

cryolaccolith See hydrolaccolith. { ‚krī·ō'lak·ə,lith }

cryolectronics See cryoelectronics. { ¦krī·ō·i,lek'trän·iks }

cryolite [MINERAL] Na₃AlF₆ A white or colorless mineral that crystallizes in the monoclinic system but has a pseudocubic aspect; found in masses of waxy luster; hardness is 2.5 on Mohs scale, and specific gravity is 2.95–3.0; used chiefly as a flux in producing aluminum from bauxite and for making salts of sodium and aluminum and porcelaneous glass. Also known as Greenland spar; ice stone. { 'krī·ə,līt }

cryolithionite [MINERAL] Na₃Li₃Al₂F₁₂ A colorless mineral that crystallizes in the isometric system; found in the Ural Mountains. { ‚krī·ō'lith·ē·ə,nīt }

cryology [HYD] The study of ice and snow. [MECH ENG] The study of low-temperature (approximately 200°R, or −160°C) refrigeration. { kı̄'äl·ə·jē }

cryolysis [SCI TECH] Destruction of tissue by very low temperature, about −162°F (−108°C). { krī'äl·ə·səs }

cryomagnetic [CRYO] Pertaining to production of very low temperatures by adiabatic demagnetization of paramagnetic salts. { ¦krī·ō·mag'ned·ik }

cryometer [ENG] A thermometer for measuring low temperatures. { krī'äm·əd·ər }

cryomorphology [GEOL] The branch of geomorphology that treats the processes and topographic features of regions where the ground is permanently frozen. { ¦krī·ō·mȯr'fäl·ə·jē }

cryopedology [GEOL] A branch of geology that deals with the study of intensive frost action and permanently frozen ground. { ¦krī·ō·pə'däl·ə·jē }

cryophilic See cryophilous. { ‚krī·ə'fil·ik }

cryophilous [ECOL] Having a preference for low temperatures. Also known as cryophilic. { krī'äf·ə·ləs }

cryophilous crop [AGR] A crop that will fully flower and seed only after it has experienced low temperatures early in its growth cycle. { krī'äf·ə·ləs 'kräp }

cryophysics [CRYO] Physics as restricted to phenomena occurring at very low temperatures, approaching absolute zero. { ¦krī·ō'fiz·iks }

cryophyte [ECOL] A plant that forms winter buds below the soil surface. { 'krī·ə,fīt }

cryoplanation [GEOL] Land erosion at high latitudes or elevations due to processes of intensive frost action. { ¦krī·ō·plə'nā·shən }

cryoprecipitate [BIOCHEM] The precipitate of a cryoglobulin. { ‚krī·ō·prə'sip·ə,tāt }

cryopreservation [ENG] Preservation of food, biologicals, and other materials at extremely low temperatures. { ¦krī·ō,prez·ər'vā·shən }

cryoprobe [MED] A blunt instrument that can be chilled to a temperature of −162°F (−108°C); used to freeze tissues in cryosurgery. { 'krī·ō,prōb }

cryopump See cryogenic pump. { 'krī·ō,pəmp }

cryoresistive transmission line [ELEC] An electric power transmission line whose conducting cables are cooled to the temperature of liquid nitrogen, 77 K ($-196°C$), resulting in a reduction of the resistance of the conductor by a factor of approximately 10, leading to increased transmission capacity. { ¦krī·ō·ri'zis·tiv tranz'mish·ən ˌlīn }

cryosar [ELECTR] A cryogenic, two-terminal, negative-resistance semiconductor device, consisting essentially of two contacts on a germanium wafer operating in liquid helium. { 'krī·ō,sär }

cryoscope [ENG] A device to determine the freezing point of a liquid. { 'krī·ə,skōp }

cryoscopic constant [ANALY CHEM] Equation constant expressed in degrees per mole of pure solvent; used to calculate the freezing-point-depression effects of a solute. { ¦krī·ə¦skäp·ik 'kän·stənt }

cryoscopy [ANALY CHEM] A phase-equilibrium technique to determine molecular weight and other properties of a solute by dissolving it in a liquid solvent and then ascertaining the solvent's freezing point. { krī'äs·kə,pē }

cryosel *See* cryohydrate. { 'krī·ə,sel }

cryosistor [ELECTR] A cryogenic semiconductor device in which a reverse-biased *pn* junction is used to control the ionization between two ohmic contacts. { ¦krī·ə'zis·tər }

cryosorption pump [MECH ENG] A high-vacuum pump that employs a sorbent such as activated charcoal or synthetic zeolite cooled by nitrogen or some other refrigerant; used to reduce pressure from atmospheric pressure to a few millitorr. { ˌkrī·ə'sȯrp·shən ˌpəmp }

cryosphere [GEOL] That region of the earth in which the surface is perennially frozen. { 'krī·ə,sfir }

cryostat [ENG] An apparatus used to provide low-temperature environments in which operations may be carried out under controlled conditions. { 'krī·ə,stat }

cryostatic pressure [GEOL] Hydrostatic pressure exerted on soil and rocks when soil water freezes. { 'krī·ə,stad·ik 'presh·ər }

cryosurgery [MED] Selective destruction of tissue by freezing, as the use of a liquid nitrogen probe to the brain in parkinsonism. { ¦krī·ō'sərj·ə·rē }

cryotherapy [MED] A form of therapy which consists of local or general use of cold. { ¦krī·ō'ther·ə,pē }

cryotron [ELECTR] A switch that operates at very low temperatures at which its components are superconducting; when current is sent through a control element to produce a magnetic field, a gate element changes from a superconductive zero-resistance state to its normal resistive state. { 'krī·ə,trän }

cryotronics [ELECTR] The branch of electronics that deals with the design, construction, and use of cryogenic devices. { ,krī·ə'trän·iks }

cryoturbation *See* congeliturbation. { ¦krī·ō·tər'bā·shən }

Cryphaeaceae [BOT] A family of mosses in the order Isobryales distinguished by a rough calyptra. { ,krī·fē'ās·ē,ē }

crypt [ANAT] **1.** A follicle or pitlike depression. **2.** A simple glandular cavity. { kript }

cryptanalysis [COMMUN] Steps and operations performed in converting encrypted messages into plain text without previous knowledge of the key employed. { ,krip·tə'nal·ə·səs }

cryptand [ORG CHEM] A macropolycyclic polyazo-polyether, where the three-coordinate nitrogen atoms provide the vertices of a three-dimensional structure. { 'krip,tand }

cryptate [INORG CHEM] The adduct formed between a cryptand and a guest (cation, anion, or neutral species) molecular entity. { 'krip,tāt }

cryptic behavior [ZOO] A behavior pattern that maximizes an organism's ability to conceal itself. { 'krip·tik bə'hāv·yər }

cryptic coloration [ZOO] A phenomenon of protective coloration by which an animal blends with the background through color matching or countershading. { 'krip·tik kəl·ə'rā·shən }

Cryptobiidae [INV ZOO] A family of flagellate protozoans in the order Kinetoplastida including organisms with two flagella, one free and one with an undulating membrane. { ,krip·tō'bī·ə,dē }

cryptobiosis [PHYSIO] A state in which metabolic rate of the organism is reduced to an imperceptible level. { ¦krip·tō·bī'ō·səs }

cryptobiotic [ECOL] Living in concealed or secluded situations. { ¦krip·tō·bī'äd·ik }

Cryptobranchidae [VERT ZOO] The giant salamanders and hellbenders, a family of tailed amphibians in the suborder Cryptobranchoidea. { ¦krip·tə'braŋ·kə,dē }

Cryptobranchoidea [VERT ZOO] A primitive suborder of amphibians in the order Urodela distinguished by external fertilization and aquatic larvae. { ¦krip·tə,braŋ'kȯid·ē·ə }

Cryptocerata [INV ZOO] A division of hemipteran insects in some systems of classification that includes the water bugs (Hydrocorisae). { ¦krip·tō·sə'räd·ə }

Cryptochaetidae [INV ZOO] A family of myodarian cyclorrhaphous dipteran insects in the subsection Acalypteratae. { ¦krip·tə'kēd·ə,dē }

cryptochannel [COMMUN] A complete system of communication that uses electronic encryption and decryption equipment and has two or more radio or wire terminals. { ¦krip·tō'chan·əl }

cryptochromes [CELL MOL] Light-sensitive proteins found in both plants and animals that detect and change conformation in response to blue light; in animals, they play an important role in circadian rhythm. { 'krip·tə,krōmz }

cryptoclastic [GEOL] Composed of extremely fine, almost submicroscopic, broken or fragmental particles. { ¦krip·tə¦klas·tik }

cryptoclimate [ENG] The climate of a confined space, such as inside a house, barn, or greenhouse, or in an artificial or natural cave; a form of microclimate. Also spelled kryptoclimate. { ¦krip·tō'klī·mət }

cryptoclimatology [CLIMATOL] The science of climates of confined spaces (cryptoclimates); basically, a form of microclimatology. Also spelled kryptoclimatology. { ¦krip·tō,klī·mə'täl·ə·jē }

Cryptococcaceae [MYCOL] A family of imperfect fungi in the order Moniliales in some systems of classification; equivalent to the Cryptococcales in other systems. { ,krip·tə·käk'sā·sē,ē }

Cryptococcales [MYCOL] An order of imperfect fungi, in some systems of classification, set up to include the yeasts or yeastlike organisms whose perfect or sexual stage is not known. { ,krip·tə·kä'kā,lēz }

cryptococcal meningitis [MED] Inflammation of the meninges due to yeasts of the genus *Cryptococcus*. { ,krip·tə'käk·əl ,men·ən'jīd·əs }

cryptococcosis [MED] A yeast infection of humans, primarily of the central nervous system, caused by *Cryptococcus neoformans*. Also known as torulosis. { ,krip·tə·kä'kō·səs }

Cryptococcus [MYCOL] A genus of encapsulated pathogenic yeasts in the order Moniliales. { ,krip·tə'käk·əs }

cryptocrystalline [GEOL] Having a crystalline structure but of such a fine grain that individual components are not visible with a magnifying lens. { ¦krip·tō'krist·əl·ən }

Cryptodira [VERT ZOO] A suborder of the reptilian order Chelonia including all turtles in which the cervical spines are uniformly reduced and the head folds directly back into the shell. { ,krip·tə'dī·rə }

cryptogam [BOT] An old term for nonflowering plants. { 'krip·tə,gam }

cryptogram [COMMUN] Information written in code or cipher. { 'krip·tə,gram }

cryptographic algorithm [COMMUN] An unchanging set of rules or steps for enciphering and deciphering messages in a cipher system. { ¦krip·tə¦graf·ik 'al·gə,rith·əm }

cryptographic bitstream [COMMUN] An unending sequence of digits which is combined with ciphertext to produce plaintext or with plaintext to recover ciphertext in a stream cipher system. { ¦krip·tə¦graf·ik 'bit,strēm }

cryptographic key [COMMUN] A sequence of numbers or characters selected by the user of a cipher system to implement a cryptographic algorithm for enciphering and deciphering messages. { ¦krip·tə¦graf·ik 'kē }

cryptography [COMMUN] The science of preparing messages in a form which cannot be read by those not privy to the secrets of the form. { krip'täg·rə·fē }

cryptohalite [MINERAL] $(NH_4)_2SiF_6$ A colorless to white or gray, isometric mineral consisting of ammonium silicon fluoride; occurs in massive and arborescent forms. { ¦krip·tō'ha,līt }

cryptolite *See* monazite. { 'krip·tə,līt }

cryptology [COMMUN] The science of preparing messages in forms which are intended to be unintelligible to those not privy to the secrets of the form, and of deciphering such messages. { krip'täl·ə·jē }

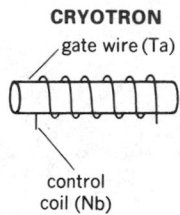

CRYOTRON

gate wire (Ta)

control
coil (Nb)

The wire-wound cryotron operated at temperatures close to absolute zero.

cryptomedusa [INV ZOO] The final stage in the reduction of a hydroid medusa to a rudiment having sex cells within the gonophore. { ¦krip·tō·mə'dü·sə }

cryptomelane [MINERAL] $KMn_8O_{16}·H_2O$ A usually massive mineral, common in manganese ores; contains an oxide of manganese and potassium and crystallizes in the monoclinic system. { ¦krip·tō·mə'lān }

cryptomitosis [INV ZOO] Cell division in certain protozoans in which a modified spindle forms, and chromatin assembles with no apparent chromosome differentiation. { ¦krip·tō·mī'tō·səs }

Cryptomonadida [BIOL] An order of the class Phytamastigophorea considered to be protozoans by biologists and algae by botanists. { ¦krip·tō·mə'näd·ə·də }

Cryptomonadina [BIOL] The equivalent name for Cryptomonadida. { ¦krip·tō·män·ə'dī·nə }

cryptonephridic [INV ZOO] In certain insects, referring to Malpighian tubules independently attached to the hindgut (in contrast to being free). { ¦krip·tō·ne'frid·ik }

cryptopart [COMMUN] One of several portions of a cryptotext; each cryptopart bears a different message indicator. { 'krip·tō¦pärt }

cryptoperthite [MINERAL] A fine-grained, submicroscopic variety of perthite consisting of an intergrowth of potassic and sodic feldspar, detectable only by means of x-rays or with the aid of an electron microscope. { ¦krip·tō'pər¸thīt }

Cryptophagidae [INV ZOO] The silken fungus beetles, a family of coleopteran insects in the superfamily Cucujoidea. { ¸krip·tə'faj·ə¸dē }

Cryptophyceae [BOT] A class of algae of the Pyrrhophyta in some systems of classification; equivalent to the division Cryptophyta. { ¸krip·tə'fīs·ē¸ē }

Cryptophyta [BOT] A division of the algae in some classification schemes; equivalent to the Cryptophyceae. { ¸krip·tə'fīd·ə }

cryptophyte [BOT] A plant that produces buds either underwater or underground on corms, bulbs, or rhizomes. { 'krip·tə¸fīt }

Cryptopidae [INV ZOO] A family of epimorphic centipedes in the order Scolopendromorpha. { krip'täp·ə¸dē }

cryptorchidism *See* cryptorchism. { krip'tór·kə¸diz·əm }

cryptorchism [MED] Failure of the testes to descend into the scrotum from the abdomen or inguinal canals. Also known as cryptorchidism. { krip'tór¸kiz·əm }

Cryptostomata [PALEON] An order of extinct bryozoans in the class Gymnolaemata. { ¸krip·tə'stō·məd·ə }

cryptotext [COMMUN] In cryptology, a text of visible writing which conveys no intelligible meaning in any language, or which apparently conveys an intelligible meaning that is not the real meaning. { 'krip·tō¸tekst }

cryptotope [IMMUNOL] A determinant (or epitope) of an immunological antigen or immunogen which is initially hidden and becomes functional only when the molecule is broken or degraded. { 'krip·tō¸tōp }

cryptoviolin *See* phycobiliviolin. { ¸krip·tə'vī·ə·lən }

cryptovirogenic [VIROL] Possessing the ability to produce infective virus particles after derepression of the viral genome within the cell. { ¦krip·tō¸vī·rə'jen·ik }

cryptovolcanic [GEOL] A small, nearly circular area of highly disturbed strata in which there is no evidence of volcanic materials to confirm the origin as being volcanic. { ¦krip·tō·väl'kan·ik }

cryptoxanthin [BIOCHEM] $C_{40}H_{57}O$ A xanthophyll carotenoid pigment found in plants; convertible to vitamin A by many animal livers. { ¦krip·tō'zan·thən }

cryptozoon [PALEOBOT] A hemispherical or cabbagelike reef-forming fossil algae, probably from the Cambrian and Ordovician. { ¦krip·tō'zō·ən }

crypts of Lieberkühn [ANAT] Simple, tubular glands which arise as evaginations into the mucosa of the small intestine. { 'krips əv 'lē·bər¸kyün }

crystal [CRYSTAL] A homogeneous solid made up of an element, chemical compound or isomorphous mixture throughout which the atoms or molecules are arranged in a regularly repeating pattern. [ELECTR] A natural or synthetic piezoelectric or semiconductor material whose atoms are arranged with some degree of geometric regularity. *See* rock crystal. { 'krist·əl }

crystal activity [ELECTR] A measure of the amplitude of vibration of a piezoelectric crystal plate under specified conditions. { 'krist·əl ak'tiv·əd·ē }

crystal aerugo *See* cupric acetate. { 'krist·əl ē'rü·gō }

crystal-audio receiver [ELECTR] Similar to the crystal-video receiver, except for the path detection bandwidth which is audio rather than video. { ¦krist·əl ¦òd·ē·ō ri'sē·vər }

crystal axis [CRYSTAL] A reference axis used for the vectoral properties of a crystal. { 'krist·əl 'ak·səs }

crystal base [CRYSTAL] The contents of a primitive cell of a crystal. { ¦krist·əl 'bās }

crystal blank [ELECTR] The result of the final cutting operation on a piezoelectric or semiconductor crystal. { 'krist·əl ¸blaŋk }

crystal calibrator [ELECTR] A crystal-controlled oscillator used as a reference standard to check frequencies. { ¦krist·əl 'kal·ə¸brād·ər }

crystal cartridge [ENG ACOUS] A piezoelectric unit used with a stylus in a phonograph pickup to convert disk recordings into audio-frequency signals, or used with a diaphragm in a crystal microphone to convert sound waves into af signals. { 'krist·əl 'kär¸trij }

crystal chemistry [CRYSTAL] The study of the crystalline structure and properties of a mineral or other solid. { 'krist·əl 'kem·ə·strē }

crystal class [CRYSTAL] One of 32 categories of crystals according to the inversions, rotations about an axis, reflections, and combinations of these which leaves the crystal invariant. Also known as symmetry class. { 'krist·əl 'klas }

crystal clock [HOROL] A clock which uses the mechanical resonance of a crystal plate coupled piezoelectrically into an electronic circuit. { 'krist·əl 'kläk }

crystal control [ELECTR] Control of the frequency of an oscillator by means of a quartz crystal unit. { 'krist·əl kən'trōl }

crystal-controlled oscillator [ELECTR] An oscillator whose frequency of operation is controlled by a crystal unit. { ¦krist·əl kən¦trōld 'äs·ə¸lād·ər }

crystal-controlled transmitter [ELECTR] A transmitter whose carrier frequency is directly controlled by the electromechanical characteristics of a quartz crystal unit. { ¦krist·əl kən¦trōld 'tranz¸mid·ər }

crystal counter [NUCLEO] A particle detector in which the sensitive material is a dielectric (nonconducting) crystal mounted between two metallic electrodes. { 'krist·əl 'kaùnt·ər }

crystal current [ELECTR] The actual alternating current flowing through a crystal unit. { 'krist·əl ¸kər·ənt }

crystal cutter [ENG ACOUS] A cutter in which the mechanical displacements of the recording stylus are derived from the deformations of a crystal having piezoelectric properties. { 'krist·əl ¸kəd·ər }

crystal defect [CRYSTAL] Any departure from crystal symmetry caused by free surfaces, disorder, impurities, vacancies and interstitials, dislocations, lattice vibrations, and grain boundaries. Also known as lattice defect. { 'krist·əl 'dē¸fekt }

crystal detector [ELECTR] **1.** A crystal diode, or an equivalent earlier crystal-catwhisker combination, used to rectify a modulated radio-frequency signal to obtain the audio or video signal directly. **2.** A crystal diode used in a microwave receiver to combine an incoming radio-frequency signal with a local oscillator signal to produce an intermediate-frequency signal. { 'krist·əl di'tek·tər }

crystal diffraction [SOLID STATE] Diffraction by a crystal of beams of x-rays, neutrons, or electrons whose wavelengths (or de Broglie wavelengths) are comparable with the interatomic spacing of the crystal. { 'krist·əl di'frak·shən }

crystal-diffraction spectrometer *See* Bragg spectrometer. { 'krist·əl di'frak·shən spek'träm·əd·ər }

crystal diode *See* semiconductor diode. { 'krist·əl 'dī¸ōd }

crystal dynamics *See* lattice dynamics. { ¦krist·əl də'nam·iks }

crystal face [CRYSTAL] One of the outward planar surfaces which define a crystal and reflect its internal structure. Also known as face. { 'krist·əl ¦fās }

crystal field theory [PHYS CHEM] The theory which assumes that the ligands of a coordination compound are the sources of negative charge which perturb the energy levels of the central metal ion and thus subject the metal ion to an electric field

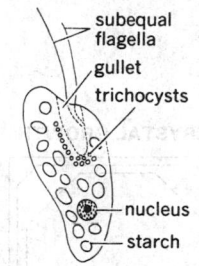

CRYPTOMONADIDA

subequal flagella
gullet
trichocysts
nucleus
starch

Chilomonas paramecium.

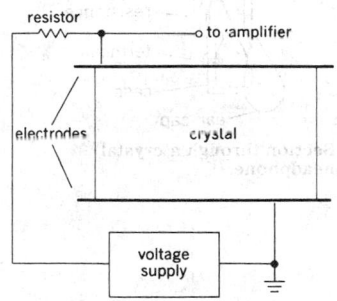

CRYSTAL COUNTER

resistor — to amplifier
electrodes
crystal
voltage supply

Circuit diagram of a crystal counter.

analogous to that within an ionic crystalline lattice. { ¦krist·əl 'fēld ,thē·ə·rē }

crystal filter [ELECTR] A highly selective tuned circuit employing one or more quartz crystals; sometimes used in intermediate-frequency amplifiers of communication receivers to improve the selectivity. { ¦krist·əl 'fil·tər }

crystal form [CRYSTAL] A collection of crystal faces generated by operating on a single face with a subgroup of the symmetry elements of the crystal class. { ¦krist·əl 'form }

crystal glass [MATER] A water-clear lead glass which polishes readily and has a high index of refraction. { ¦krist·əl ,glas }

crystal gliding [CRYSTAL] Slip along a crystal plane due to plastic deformation; often produces crystal twins. Also known as translation gliding. { ¦krist·əl ,glīd·iŋ }

crystal grating [SPECT] A diffraction grating for gamma rays or x-rays which uses the equally spaced lattice planes of a crystal. { ¦krist·əl ,grād·iŋ }

crystal growth [CRYSTAL] The growth of a crystal, which involves diffusion of the molecules of the crystallizing substance to the surface of the crystal, diffusion of these molecules over the crystal surface to special sites on the surface, incorporation of molecules into the surface at these sites, and diffusion of heat away from the surface. { ¦krist·əl ,grōth }

crystal habit [CRYSTAL] The size and shape of the crystals in a crystalline solid. Also known as habit. { ¦krist·əl ,hab·ət }

crystal harmonic generator [ELECTR] A type of crystal-controlled oscillator which produces an output rich in harmonics (overtones or multiples) of its fundamental frequency. { ¦krist·əl har¦män·ik ¦jen·ə,rād·ər }

crystal headphones [ENG ACOUS] Headphones using Rochelle salt or other crystal elements to convert audio-frequency signals into sound waves. Also known as ceramic earphones. { ¦krist·əl 'hed,fōnz }

crystal holder [DES ENG] A housing designed to provide proper support, mechanical protection, and connections for a quartz crystal plate. { ¦krist·əl ,hōl·dər }

crystal hydrophone [ENG ACOUS] A crystal microphone that responds to waterborne sound waves. { ¦krist·əl 'hī·drə,fōn }

crystal indices See Miller indices. { ¦krist·əl 'in·də,sēz }

crystal laser [OPTICS] A laser that uses a pure crystal of ruby or other material for generating a coherent beam of output light. { ¦krist·əl 'lā·zər }

crystal lattice [CRYSTAL] A lattice from which the structure of a crystal may be obtained by associating with every lattice point an assembly of atoms identical in composition, arrangement, and orientation. { ¦krist·əl 'lad·əs }

crystal-lattice filter [ELECTR] A crystal filter that uses two matched pairs of series crystals and a higher-frequency matched pair of shunt or lattice crystals. { ¦krist·əl 'lad·əs ,fil·tər }

crystalliferous bacteria [MICROBIO] Bacteria, especially *Bacillus thuringiensis*, that are characterized by the formation of a protein crystal in the sporangium at the time of spore formation. { ¦kris·tə¦lif·ə·rəs bak'tir·ē·ə }

crystalline [CRYSTAL] Of, pertaining to, resembling, or composed of crystals. { 'kris·tə·lən }

crystalline alumina [MATER] An abrasive which consists of essentially the same mineral as corundum, but whose physical properties such as crystal structure, size, and shape of grain are so controlled as to produce the most desirable abrasives for specific types of grinding. { 'kris·tə·lən ə'lüm·ə·nə }

crystalline anisotropy [SOLID STATE] The tendency of crystals to have different properties in different directions; for example, a ferromagnet will spontaneously magnetize along certain crystallographic axes. { 'kris·tə·lən an·ə'sä·trə·pē }

crystalline chloral See chloral hydrate. { 'kris·tə·lən 'klôr·əl }

crystalline double refraction [OPTICS] The splitting which a wavefront experiences when a wave disturbance propagates through an anisotropic crystal. { 'kris·tə·lən 'dəb·əl ri'frak·shən }

crystalline field [SOLID STATE] The internal electric field in a solid due to localized charges, especially ions, inside. { 'kris·tə·lən 'fēld }

crystalline fracture [MET] A break in a polycrystalline metal, with the fractured surface having a grainy appearance. { 'kris·tə·lən 'frak·chər }

crystalline frost [HYD] Hoarfrost that exhibits a relatively simple macroscopic crystalline structure. { 'kris·tə·lən 'frôst }

crystalline-granular texture [PETR] A primary texture of an igneous rock due to crystallization from a fluid medium. { ¦kris·tə·lən ¦gran·yə·lər 'teks·chər }

crystalline laser [OPTICS] A solid laser in which the lasing material is a pure crystal like ruby or a doped crystal like neodymium-doped ruby or neodymium-doped yttrium aluminum garnet. { 'kris·tə·lən 'lā·zər }

crystalline lens See lens. { 'kris·tə·lən 'lenz }

crystalline polymer [CHEM] A polymer whose sections of adjacent chains are packed in a regular array. { 'kris·tə·lən 'päl·i·mər }

crystalline porosity [GEOL] Porosity in crystalline limestone and dolomite, making possible underground oil reservoirs. { 'kris·tə·lən pə'räs·əd·ē }

crystalline rock [PETR] **1.** Rock made up of minerals in a clearly crystalline state. **2.** Igneous and metamorphic rock, as opposed to sedimentary rock. { 'kris·tə·lən 'räk }

crystallinity [CRYSTAL] The quality or state of being crystalline. [ORG CHEM] The degree to which polymer molecules are oriented into repeating patterns. [PETR] Degree of crystallization exhibited by igneous rock. { ,kris·tə'lin·əd·ē }

crystallin protein [BIOCHEM] Any of a group of stable structural components distributed nonuniformly in the lens of the eye of vertebrates. { 'krist·əl·ən 'prō,tēn }

crystallite [GEOL] A small, rudimentary form of crystal which is of unknown mineralogic composition and which does not polarize light. { 'kris·tə,līt }

crystallization [CRYSTAL] The formation of crystalline substances from solutions or melts. { ,kris·tə·lə'zā·shən }

crystallization differentiation See fractional crystallization. { ,kris·tə·lə'zā·shən ,dif·ə,ren·chē'ā·shən }

crystallizer [CHEM ENG] Process vessel within which dissolved solids in a supersaturated solution are forced out of solution by cooling or evaporation, and then recovered as solid crystals. { 'kris·tə,līz·ər }

crystalloblast [MINERAL] A mineral crystal produced by metamorphic processes. { 'kris·tə·lō,blast }

crystalloblastic series [GEOL] A series of metamorphic minerals ordered according to decreasing formation energy, so crystals of a listed mineral have a tendency to form idioblastic outlines at surfaces of contact with simultaneously developed crystals of all minerals in lower positions. { 'kris·tə·lə'blas·tik 'sir,ēz }

crystalloblastic texture [GEOL] A crystalline texture resulting from metamorphic recrystallization under conditions of high viscosity and directed pressure. { 'kris·tə·lə'blas·tik 'teks·chər }

crystallogram [CRYSTAL] A photograph of the x-ray diffraction pattern of a crystal. { 'kris·tə·lō,gram }

crystallographic axis [CRYSTAL] One of three lines (sometimes four, in the case of a hexagonal crystal), passing through a common point, that are chosen to have definite relation to the symmetry properties of a crystal, and are used as a reference in describing crystal symmetry and structure. { ¦kris·tə·lō¦graf·ik 'ak·səs }

crystallographic texture [MINERAL] A texture of replacement or exsolution mineral deposits, with the distribution and form of the inclusions controlled by the host-mineral crystallography. { ¦kris·tə·lō¦graf·ik 'teks·chər }

crystallography [PHYS] The branch of science that deals with the geometric description of crystals and their internal arrangement. { ,kris·tə'läg·rə·fē }

crystallomagnetic [SOLID STATE] Pertaining to magnetic properties of crystals. { ,kris·tə·lō·mag'ned·ik }

crystallophobia [PSYCH] An abnormal fear of glass. { ,kris·tə·lō'fō·bē·ə }

crystal loudspeaker [ENG ACOUS] A loudspeaker in which movements of the diaphragm are produced by a piezoelectric crystal unit that twists or bends under the influence of the applied audio-frequency signal voltage. Also known as piezoelectric loudspeaker. { ¦krist·əl 'laud,spēk·ər }

crystal microphone [ENG ACOUS] A microphone in which deformation of a piezoelectric bar by the action of sound waves or mechanical vibrations generates the output voltage between the faces of the bar. Also known as piezoelectric microphone. { ¦krist·əl 'mī·krə,fōn }

crystal mixer [ELECTR] A mixer that uses the nonlinear

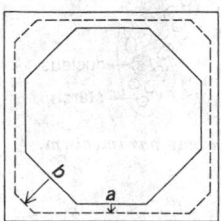
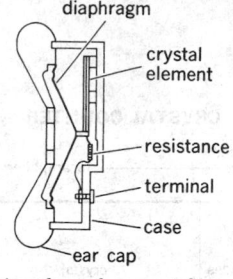

characteristic of a crystal diode to mix two frequencies; widely used in radar receivers to convert the received radar signal to a lower intermediate-frequency value by mixing it with a local oscillator signal. { ¦krist·əl 'mik·sər }

crystal momentum [SOLID STATE] The product of Planck's constant and the wave vector associated with an elementary excitation in a crystal (the magnitude of the wave vector being taken as the reciprocal of the wavelength). { ¦krist·əl mə'men·təm }

crystal monochromator [SPECT] A spectrometer in which a collimated beam of slow neutrons from a reactor is incident on a single crystal of copper, lead, or other element mounted on a divided circle. { krist·əl ¦män·ə'krō¸mād·ər }

crystal operation [ELECTR] Operation using crystal-controlled oscillators. { 'krist·əl 'äp·ə¸rā·shən }

crystal optics [OPTICS] The study of the propagation of light, and associated phenomena, in crystalline solids. { ¦krist·əl 'äp·tiks }

crystal oscillator [ELECTR] An oscillator in which the frequency of the alternating-current output is determined by the mechanical properties of a piezoelectric crystal. Also known as piezoelectric oscillator. { krist·əl 'äs·ə¸lād·ər }

crystal oven [ENG] A temperature-controlled oven in which a crystal unit is operated to stabilize its temperature and thereby minimize frequency drift. { 'krist·əl ¸əv·ən }

crystal phase [MET] A crystal structure formed by an alloy over a certain range of values of the relative proportions of its constituents. { 'krist·əl ¸fāz }

crystal photoeffect [SOLID STATE] An electromotive force induced by illumination of natural cuprite crystals or transparent zinc sulfide, and having a direction dependent on that of the incident light beam. { 'krist·əl 'fōd·ō·i¸fekt }

crystal pickup [ENG ACOUS] A phonograph pickup in which movements of the needle in the record groove cause deformation of a piezoelectric crystal, thereby generating an audio-frequency output voltage between opposite faces of the crystal. Also known as piezoelectric pickup. { ¦krist·əl 'pik¸əp }

crystal plane [CRYSTAL] One of a set of parallel, equally spaced planes in a crystal structure, each of which contains an infinite periodic array of lattice points. { ¦krist·əl 'plān }

crystal plate [ELECTR] A precisely cut slab of quartz crystal that has been lapped to final dimensions, etched to improve stability and efficiency, and coated with metal on its major surfaces for connecting purposes. Also known as quartz plate. { 'krist·əl ¸plāt }

crystal projection [CRYSTAL] Any method of displaying the positions of the poles of a crystal by projecting them on a plane. { ¦krist·əl prə'jek·shən }

crystal pulling [CRYSTAL] A method of crystal growing in which the developing crystal is gradually withdrawn from a melt. { 'krist·əl ¸pul·iŋ }

crystal rectifier See semiconductor diode. { ¦krist·əl 'rek·tə¸fī·ər }

crystal resonator [ELECTR] A precisely cut piezoelectric crystal whose natural frequency of vibration is used to control or stabilize the frequency of an oscillator. Also known as piezoelectric resonator. { ¦krist·əl 'rez·ən¸ād·ər }

crystal sandstone [GEOL] Siliceous sandstone in which deposited silica is precipitated upon the quartz grains in crystalline position. { ¦krist·əl 'sand¸stōn }

crystal set [ELECTR] A radio receiver having a crystal detector stage for demodulation of the received signals, but no amplifier stages. { 'krist·əl ¸set }

crystal settling [GEOL] Sinking of crystals in magma from the liquid in which they formed, by the action of gravity. { ¦krist·əl 'set¸liŋ }

crystal shutter [ELECTROMAG] Mechanical waveguide or coaxial-cable shorting switch that, when closed, prevents undesired radio-frequency energy from reaching and damaging a crystal detector. { ¦krist·əl 'shəd·ər }

crystals of Venus See cupric acetate. { 'krist·əlz əv 've·nəs }

crystal spectrometer See Bragg spectrometer. { 'krist·əl spek'träm·əd·ər }

crystal-stabilized transmitter [ELECTR] A transmitter employing automatic frequency control, in which the reference frequency is that of a crystal oscillator. { ¦krist·əl 'stā·bə¸līzd 'tranz¸mid·ər }

crystal structure [CRYSTAL] The arrangement of atoms or ions in a crystalline solid. { ¦krist·əl 'strək·chər }

crystal symmetry [CRYSTAL] The existence of nontrivial operations, consisting of inversions, rotations around an axis, reflections, and combinations of these, which bring a crystal into a position indistinguishable from its original position. { ¦krist·əl 'sim·ə·trē }

crystal system [CRYSTAL] One of seven categories (cubic, hexagonal, tetragonal, trigonal, orthorhombic, monoclinic, and triclinic) into which a crystal may be classified according to the shape of the unit cell of its Bravais lattice, or according to the dominant symmetry elements of its crystal class. { ¦krist·əl 'sis·təm }

crystal transducer [ELECTR] A transducer in which a piezoelectric crystal serves as the sensing element. { 'krist·əl tran·z'dü·sər }

crystal tuff [GEOL] Consolidated volcanic ash in which crystals and crystal fragments predominate. { ¦krist·əl 'təf }

crystal twin See twin crystal. { 'krist·əl ¸twin }

crystal unit [ELECTR] A complete assembly of one or more quartz plates in a crystal holder. { ¦krist·əl ¦yü·nət }

crystal video receiver [ELECTR] A broad-tuning radar or other microwave receiver consisting only of a crystal detector and a video or audio amplifier. { ¦krist·əl ¦vid·ē·ō ri'sē·vər }

crystal video rectifier [ELECTR] A crystal rectifier transforming a high-frequency signal directly into a video-frequency signal. { ¦krist·əl ¦vid·ē·ō 'rek·tə¸fī·ər }

crystal violet See methyl violet. { ¦krist·əl 'vī·lət }

crystal-vitric tuff [GEOL] Consolidated volcanic ash composed of 50–75% crystal fragments and 25–50% glass fragments. { ¦krist·əl ¦vi·trik 'təf }

crystal whisker [CRYSTAL] A single crystal that has grown in a filamentary form. Also known as whisker. { ¦krist·əl 'wis·kər }

crystogen See cystamine. { 'kris·tə·jən }

crystosphene [HYD] A buried sheet or mass of ice, as in the tundra of northern America, formed by the freezing of rising and spreading springwater beneath alluvial deposits. { 'kris·tə¸sfēn }

cs See centistoke.

Cs See cesium; cirrostratus cloud.

csc See cosecant.

C scan See C scope. { 'sē ¸skan }

csch See hyperbolic cosecant.

C scope [ELECTR] A cathode-ray scope on which signals appear as spots, with bearing angle as the horizontal coordinate and elevation angle as the vertical coordinate. Also known as C indicator; C scan. { 'sē ¸skōp }

C size [ENG] One of a series of sizes to which trimmed paper and board are manufactured; for size CN, with N equal to any integer, the length of the longer side is $2^{3/8 - N/2}$ meters, while the length of the shorter side is $2^{1/8 - N/2}$ meters, with both lengths rounded off to the nearest millimeter. { 'sē ¸sīz }

CSMA/CD [COMPUT SCI] A method of controlling multiaccess computer networks in which each station on the network senses traffic and waits for it to clear before sending a message, and two devices that try to send concurrent messages must both step back and try again. Abbreviation for carrier-sense multiple access with collision detection.

CSP See control switching point.

CSSB system See companded single-sideband system. { ¦sē¸es¸es¦bē ¸sis·təm }

C stage [ORG CHEM] The final stage in a thermosetting resin reaction in which the material is relatively insoluble and infusible; the resin in a fully cured thermoset molding is in this stage. Also known as resite. { 'sē ¸stāj }

CSW See channel status word.

CT See center tap; computerized tomography.

CIM See computer input from microfilm.

CTC See centralized traffic control.

ctDNA See chloroplast deoxyribonucleic acid.

CTD recorder See salinity-temperature-depth recorder. { ¦sē¦t·ē¦dē ri'kórd·ər }

ctenidium [INV ZOO] **1.** The comb- or featherlike respiratory apparatus of certain mollusks. **2.** A row of spines on the head or thorax of some fleas. { tə'nid·ē·əm }

Ctenodrilidae [INV ZOO] A family of fringe worms belonging to the Sedentaria. { ¸ten·ə'dril·ə¸dē }

ctenoid scale [VERT ZOO] A thin, acellular structure composed of bonelike material and characterized by a serrated

CRYSTAL OSCILLATOR

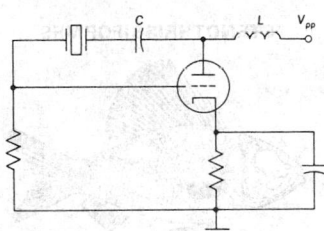

Circuit diagram of Pierce crystal oscillator; C is capacitor, L is inductor, and V_{pp} is plate voltage.

CTENOID SCALE

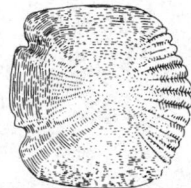

Ctenoid scale from carp.

CTENOPHORA

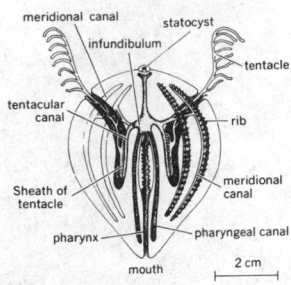

Structure of a cydippid ctenophore.

CTENOTHRISSIFORMES

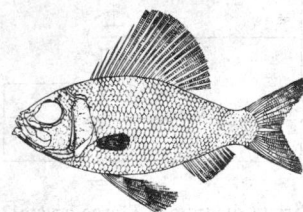

Cretaceous ctenothrissid, *Ctenothrissus radians*; length 10–12 inches (25–30 centimeters).

C-TUBE BOURDON ELEMENT

C-tube bourdon element with mechanism composing an industrial pressure gage. (*Foxboro Co.*)

margin; found in the skin of advanced teleosts. { 'ten,oid ,skāl }

Ctenophora [INV ZOO] The comb jellies, a phylum of marine organisms having eight rows of comblike plates as the main locomotory structure. { tə'näf·ə·rə }

Ctenostomata [INV ZOO] An order of bryozoans in the class Gymnolaemata recognized as inconspicuous, delicate colonies made up of relatively isolated, short, tubular zooecia with chitinous walls. { ,ten·ə'stäm·ə·də }

Ctenostomatida [INV ZOO] The equivalent name for Odontostomatida. { ,ten·ə·stə'mad·ə·də }

Ctenothrissidae [PALEON] A family of extinct teleostean fishes in the order Ctenothrissiformes. { ten·ə'thris·ə,dē }

Ctenothrissiformes [PALEON] A small order of extinct teleostean fishes; important as a group on the evolutionary line leading from the soft-rayed to the spiny-rayed fishes. { ,ten·ə,thris·ə'fòr,mēz }

ctn See cotangent.

CTR See controlled thermonuclear reactor.

C-tube bourdon element [ENG] Hollow tube of flexible (elastic) metal shaped like the arc of a circle; changes in internal gas or liquid pressure flexes the tube to a degree related to the pressure change; used to measure process-stream pressures. { 'sē ,tüb 'bùrd·ən ,el·ə·mənt }

C-type asteroid [ASTRON] A type of asteroid whose surface is very dark and neutral-colored, and probably is of carbonaceous composition similar to primitive carbonaceous chondritic meteorites. { 'sē ,tīp 'as·tə,róid }

C-type virus particle [VIROL] One of a morphologically similar group of enveloped virus particles having a central, spherical ribonucleic acid-containing nucleoid; associated with certain cancers, as sarcomas and leukemias. { 'sē ,tīp 'vī·rəs ,pärd·ə·kəl }

cu See crosstalk unit; cubic.

Cu See copper.

cubane [ORG CHEM] C_8H_8 (polycyclic octane) A cage hydrocarbon with carbons and their bonds at the corners and along the edges, forming a cube. { 'kyü·bān }

cubanite [MINERAL] $CuFe_2S_3$ Bronze-yellow mineral that crystallizes in the orthorhombic system. Also known as chalmersite. { 'kyü·bə,nīt }

cubature [MATH] The numerical integration of a function of two variables. { 'kyüb·ə·chər }

cube [MATH] **1.** Regular polyhedron whose faces are all square. **2.** For a number a, the new number obtained by taking the threefold product of a with itself: $a \times a \times a$. { kyüb }

cubeb [BOT] The dried, nearly ripe fruit (berries) of a climbing vine, *Piper cubeba*, of the pepper family (Piperaceae). { 'kyü,beb }

cubeb oil See oil of cubeb. { 'kyü,beb ,òil }

cube ore See pharmacosiderite. { 'kyüb ,òr }

cube root [MATH] Another number whose cube is the original number. { 'kyüb 'rüt }

cube spar See anhydrite. { 'kyüb ,spär }

cube-surface coil [ELECTROMAG] A system of five equally spaced square coils that produces a region of uniform magnetic field over a large volume which is easily accessible from outside the coils. { 'kyüb ,sər·fəs ,kòil }

cubic [MECH] Denoting a unit of volume, so that if x is a unit of length, a cubic x is the volume of a cube whose sides have length $1x$; for example, a cubic meter, or a meter cubed, is the volume of a cube whose sides have a length of 1 meter. Abbreviated cu. { 'kyü·bik }

cubical antenna [ELECTROMAG] An antenna array, the elements of which are positioned to form a cube. { 'kyü·bə·kəl an'ten·ə }

cubical dilation [MECH] The isotropic part of the strain tensor describing the deformation of an elastic solid, equal to the fractional increase in volume. { 'kyü·bə·kəl di'lā·shən }

cubical expansion [PHYS] The increase in volume of a substance with a change in temperature or pressure. { 'kyü·bə·kəl ik'span·shən }

cubical parabola [MATH] A plane curve whose equation in cartesian coordinates x and y is $y = x^3$. { 'kyüb·ə·kəl pə'rab·ə·lə }

cubic boron nitride [MECH ENG] A synthetic material composed of boron and nitrogen (1:1) that is almost as hard as diamond, used as a superabrasive powder and for cutting and grinding applications. { ¦kyü·bik¦bò·rän 'nī,trīd }

cubic cleavage [CRYSTAL] Isometric crystal cleavage occuring parallel to the faces of a cube. { 'kyü·bik 'klē·vij }

cubic crystal [CRYSTAL] A crystal whose lattice has a unit cell with perpendicular axes of equal length. { 'kyü·bik 'krist·əl }

cubic curve [MATH] A plane curve which has an equation of the form $f(x, y) = 0$, where $f(x, y)$ is a polynomial of degree three in x and y. { 'kyü·bik 'kərv }

cubic determinant [MATH] A mathematical form analogous to an ordinary determinant, with the elements forming a cube instead of a square. { 'kyü·bik di'tər·mə·nənt }

cubic equation [MATH] A polynomial equation with no exponent larger than 3. { 'kyü·bik i'kwā·zhən }

cubic foot per minute [MECH] A unit of volume flow rate, equal to a uniform flow of 1 cubic foot in 1 minute; equal to 1/60 cusec. Abbreviated cfm. { ¦kyü·bik ¦fùt pər 'min·ət }

cubic foot per second See cusec. { ¦kyü·bik ¦fùt pər 'sek·ənd }

cubicle [BUILD] Any small, approximately square room or compartment. [ENG] An enclosure for high-voltage equipment. { 'kyü·bə·kəl }

cubic measure [MECH] A unit or set of units to measure volume. { 'kyü·bik 'mezh·ər }

cubic packing [CRYSTAL] The spacing pattern of uniform solid spheres in a clastic sediment or crystal lattice in which the unit cell is a cube. { 'kyü·bik 'pak·iŋ }

cubic plane [CRYSTAL] A plane that is at right angles to any one of the three crystallographic axes of the cubic system. { 'kyü·bik 'plān }

cubic polynomial [MATH] A polynomial in which all exponents are no greater than 3. { 'kyü·bik ,päl·ə'nō·mē·əl }

cubic quantic [MATH] A quantic of the third degree. { ,kyüb·ik 'kwän·tik }

cubic spline [MATH] One of a collection of cubic polynomials used in interpolating a function whose value is specified at each of a collection of distinct ordered values, X_i ($i=1, \ldots, n$), and whose slope is specified at X_1 and X_n; one cubic polynomial is found for each interval, such that the interpolating system has the prescribed values at each of the X_i, the prescribed slope at X_n and X_n, and a continuous slope at each of the X_i. { 'kyü·bik 'splīn }

cubic surd [MATH] A cube root of a rational number that is itself an irrational number. { 'kyü·bik 'sərd }

cubic system See isometric system. { 'kyü·bik 'sis·təm }

cuboctahedron [MATH] A polyhedron whose faces consist of six equal squares and eight equal equilateral triangles, and which can be formed by cutting the corners off a cube; it is one of the 13 Archimedean solids. Also spelled cubooctahedron. { ,kyü·bik¦bäk·tə'hē·drən }

cuboid [ANAT] The outermost distal tarsal bone in vertebrates. [INV ZOO] Main vein of the wing in many insects, particularly the flies (Diptera). [MATH] See rectangular parallelepiped. [SCI TECH] Nearly cubic in shape. { 'kyü,bóid }

cuboidal epithelium [HISTOL] A single-layered epithelium made up of cubelike cells. { kyü'bóid·əl ,ep·ə'thēl·ē·əm }

Cubomedusae [INV ZOO] An order of cnidarians in the class Scyphozoa distinguished by a cubic umbrella. { ¦kyü·bo·mə'dü·sē }

cubooctahedron See cuboctahedron. { ¦kyü·bo,äk·tə'hē·drən }

Cuccia coupler See electron coupler. { 'kü·chē·ə 'kəp·lər }

cuckoo [VERT ZOO] The common name for about 130 species of primarily arboreal birds in the family Cuculidae; some are social parasites. { 'kù,kü }

cucoloris [GRAPHICS] Material having a cutout that allows light to pass through it and project a form. { ,kük·əl'òr·əs }

Cucujidae [INV ZOO] The flat-back beetles, a family of predatory coleopteran insects in the superfamily Cucujoidea. { kə'kü·yə,dē }

Cucujoidea [INV ZOO] A large superfamily of coleopteran insects in the suborder Polyphaga. { ,kü·kə'yòid·ē·ə }

Cuculidae [VERT ZOO] A family of perching birds in the order Cuculiformes, including the cuckoos and the roadrunner, characterized by long tails, heavy beaks and conspicuous lashes. { kə'kyü·lə,dē }

Cuculiformes [VERT ZOO] An order of birds containing the cuckoos and allies, characterized by the zygodactyl arrangement of the toes. { kə,kyü·lə'fór,mēz }

cucullus [INV ZOO] A transverse flap at the anterior edge of the carapace that hangs over the mouthparts of certain arachnids. { kyü'kəl·əs }

Cucumariidae [INV ZOO] A family of dendrochirotacean holothurian echinoderms in the order Dendrochirotida. { ,kü·kə·mə'rī·ə,dē }

cucumber [BOT] *Cucumis sativus.* An annual cucurbit, in the family Cucurbitaceae grown for its edible, immature fleshy fruit. { 'kyü·kəm·bər }

cucumber mildew [PL PATH] **1.** A downy mildew of cucumbers and melons caused by *Peronoplasmopara cubensis.* **2.** A powdery mildew of cucumbers and melons caused by *Erysiphe cichoracearum.* { 'kyü·kəm·bər 'mil,dü }

cucumber mosaic [PL PATH] A virus disease of cucumbers and related fruits, producing mottling of terminal leaves and fruits and dwarfing of vines. { 'kyu·kəm·bər mō'zā·ik }

cucumber mosaic virus [VIROL] The type species of the genus *Cucumovirus.* Abbreviated CMV. { 'kyü·kəm·bər mō¦zāik 'vī·rəs }

cucumber mosaic virus group See Cucumovirus. { ¦kyü·kəm·bər mō¦zā·ik 'vī·rəs ,grüp }

cucumber tree [FOR] *Magnolia acuminata.* A species that grows in the Appalachian and Ozark mountains and may reach a height of 80 feet (20 meters), rarely 100 feet (30 meters). The ripe fruit is red and resembles a small cucumber in shape. { 'kyü·kəm·bər ,trē }

Cucumovirus [VIROL] A genus of the family Bromoviridae that is characterized by icosahedral particles containing one molecule of one of the four single-stranded ribonucleic acid species; the type species is cucumber mosaic virus. Also known as cucumber mosaic virus group. { 'kyü·kəm,o'vī·rəs }

cucurbit wilt [PL PATH] A bacterial disease of cucumbers and related plants caused by *Erwinia tracheiphila,* characterized by sudden wilting of the plant. { kə'kər·bət ,wilt }

cue circuit [ELECTR] A one-way communication circuit used to convey program control information. { 'kyü ,sər·kət }

cuesta [GEOGR] A gently sloping plain which terminates in a steep slope on one side. { 'kwes·tə }

cul-de-sac [ANAT] Blind pouch or diverticulum. [CIV ENG] A dead-end street with a circular area for turning around. { 'kəl·də,sak }

culdoscope [MED] An instrument used to visualize female pelvic organs, introduced through the vagina or a perforation into the retrouterine pouch. { 'kəl·də,skōp }

culet [LAP] The sharp point of a brilliant-cut stone. { 'kyü·lət }

Culex [INV ZOO] A genus of mosquitoes important as vectors for malaria and several filarial parasites. { 'kyü,leks }

Culicidae [INV ZOO] The mosquitoes, a family of slender, orthorrhaphous dipteran insects in the series Nematocera having long legs and piercing mouthparts. { kyü'lis·ə,dē }

Culicinae [INV ZOO] A subfamily of the dipteran family Culicidae. { kyü'lis·ə,nē }

culinary steam [FOOD ENG] A steam that can be used for food processing by direct injection into the food being prepared. { 'kəl·ə,ner·ē 'stēm }

cull [CHEM ENG] In a plastics molding operation, material remaining in the transfer chamber after the mold has been filled. [SCI TECH] Material rejected for being below standard grade. { 'kəl }

Cullen number [MATH] A number having the form $C_n = (n \cdot 2^n) + 1$ for $n = 0, 1, 2, \ldots$ { 'kəl·ən ,nəm·bər }

cullet See collet. { 'kəl·ət }

cullis See coulisse. { 'kəl·əs }

culm [BOT] **1.** A jointed and usually hollow grass stem. **2.** The solid stem of certain monocotyledons, such as the sedges. [MIN ENG] Fine, refuse coal, screened and separated from larger pieces. { kəlm }

culmen [VERT ZOO] The edge of the upper bill in birds. { 'kəl·mən }

culmination [ASTRON] **1.** The position of a heavenly body

when at highest apparent altitude. **2.** For a heavenly body which is continually above the horizon, the position of lowest apparent altitude. [GEOL] A high point on the axis of a fold. { kəl·mə'nā·shən }

cultellation [ENG] Transferring a surveyed point from a high level (such as on overhang) to a lower level by dropping a marking pin. { kəl·tə'lā·shən }

cultigen [BIOL] A cultivated variety or species of organism for which there is no known wild ancestor. Also known as cultivar. { 'kəl·tə·jən }

cultivar See cultigen. { 'kəl·tə,vär }

cultivate [AGR] To prepare soil for the raising of crops. { 'kəl·tə,vāt }

cultivator [AGR] A farm implement pulled behind a powered machine that is used to break up soil, kill weeds, and create a surface mulch for moisture. { 'kəl·tə,vād·ər }

cultural anthropology [ANTHRO] The division of anthropology dealing with the study of all aspects of culture. { ¦kəl·chə·rəl an·thrə'päl·ə·jē }

cultural ecology [ECOL] The branch of ecology that involves the study of the interaction of human societies with one another and with the natural environment. { ¦kəl·chər·əl ē'käl·ə·jē }

cultural-familial mental retardation [PSYCH] Subnormal general intellectual functioning, usually borderline or mild, presumably on the basis of some degree of environmental deprivation resulting from familial retardation as evidenced by its presence in one parent and one or more siblings. { ¦kəl·chə·rəl fə¦mil·yəl ,ment·əl ,rē,tär'dā·shən }

cultural geography See anthropography. { ¦kəl·chər·əl jē'ag·rə·fē }

cultural psychiatry [PSYCH] A branch of social psychiatry concerned with the mentally ill in relation to their cultural environment. { ¦kəl·chə·rəl sī'kī·ə·trē }

culture [ANTHRO] The complex pattern of behavior that distinguishes a social, ethnic, or religious group. [BIOL] A growth of living cells or microorganisms in a controlled artificial environment. { 'kəl·chər }

culture alteration [CYTOL] A persistent change in the properties of cultured cells, such as altered morphology, virus susceptibility, nutritional requirements, or proliferative capacity. { 'kəl·chər ,ól·tə,rā·shən }

culture community [ECOL] A plant community which is established or modified through human intervention; for example, a fencerow, hedgerow, or windbreak. { 'kəl·chər kə'myü·nəd·ē }

cultured pearl [INV ZOO] A natural pearl grown by means of controlled stimulation of the oyster. { 'kəl·chərd 'pərl }

cultured sour cream [FOOD ENG] A product made by pasteurizing cream containing 18–20% milk fat and inoculating it with bacterial cultures. { ¦kəl·chərd ¦saú·ər 'krēm }

culture medium [MICROBIO] The nutrients and other organic and inorganic materials used for the growth of microorganisms and plant and animal tissue in culture. { ¦kəl·chər ,mēd·ē·əm }

culture-specific syndrome [PSYCH] Any form of disturbed behavior that is specific to a certain cultural system and does not conform to western classification of diseases. { ¦kəl·chər spə¦sif·ik 'sin,drōm }

culvert [ENG] A covered channel or a large-diameter pipe that takes a watercourse below ground level. { 'kəl·vərt }

Cumacea [INV ZOO] An order of the class Crustacea characterized by a well-developed carapace which is fused dorsally with at least the first three thoracic somites and overhangs the sides. { kyü'mās·ē·ə }

cumatophyte [ECOL] A plant that grows under surf conditions. { kyü'mad·ə,fīt }

cumberlandite [PETR] A coarse-grained, ultramafic, ultrabasic rock composed principally of olivine crystals in a ground mass of magnetite and ilmenite with minor plagioclase. { 'kəm·bər·lən,dīt }

cumbraite [PETR] A variety of dacite or rhyodacite containing very calcic plagioclase and pyroxene in a glassy groundmass. { kyüm'brā,īt }

cumec [MECH] A unit of volume flow rate equal to 1 cubic meter per second. { 'kyü,mek }

cumene [ORG CHEM] $C_6H_5CH(CH_3)_2$ A colorless, oily benzenoid hydrocarbon cooling at 152.4°C; used as an additive for high-octane motor fuel. { 'kyü,mēn }

CUCUMBER TREE

Cucumber tree (*Magnolia acuminata*).

CULTIVATOR

Coil-shank field cultivator. (*Allis-Chalmers*)

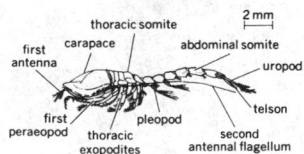

CUMACEA

Typical adult male cumacean.

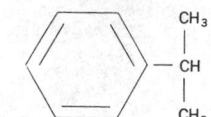

CUMENE

Structural formula of cumene.

cumene hydroperoxide [ORG CHEM] $C_6H_5C(CH_3)_2OOH$ An isopropyl hydroperoxide of cumene; an oily liquid, used to make phenol and acetone. { 'kyü‚mēn ‚hī·drō·pə'räk‚sīd }

cumengite [MINERAL] $Pb_4Cu_4Cl_8(OH)_8·H_2O$ A deep-blue or light-indigo-blue tetragonal mineral consisting of a basic lead-copper chloride occurring in crystals. { kyü'men‚jīt }

cumidine [ORG CHEM] $C_9H_{13}N$ A colorless, water-insoluble liquid, boiling at 225°C. { 'kyü·mə‚dēn }

cumin [BOT] *Cuminum cyminum* An annual herb in the family Umbelliferae; the fruit is valuable for its edible, aromatic seeds. { 'kyü·mən }

cumin oil [MATER] A colorless to yellow liquid with a sharp, spicy taste; soluble in alcohol, ether, and chloroform; used in medicine, flavoring, and perfumes. { 'kyü·mən ‚oil }

cummingtonite [MINERAL] $(Fe,Mg)_7Si_8O_{22}(OH)_2$ A brownish mineral that crystallizes in the monoclinic system; usually occurs as lamellae or fibers in metamorphic rocks. { 'kəm·iŋ·tə‚nīt }

cum sole [GEOPHYS] With the sun; hence anticyclonic or clockwise in the Northern Hemisphere. { ‚kùm 'sōl·ə }

cumulants [STAT] A set of parameters k_h ($h = 1, \ldots r$) of a one-dimensional probability distribution defined by

$$\ln \chi_x(q) = \sum_{h=1}^{r} k_h[(iq)^h/h!] + o(q^r)$$

where $\chi_x(q)$ is the characteristic function of the probability distribution of x. Also known as semi-invariants. { 'kyü·myə·ləns }

cumulate [PETR] Any igneous rock formed by the accumulation of crystals settling out of a magma. { 'kyü·myə‚lāt }

cumulated double bonds [CHEM] Two double bonds on the same carbon atom, as in $>C=C=C<$. { 'kyü·myə‚lād·əd ‚dəb·əl 'bändz }

cumulative compound generator [ELEC] A compound generator in which the series field is connected to aid the shunt field magnetomotive force. { 'kyü·myə·ləd·iv ‚käm‚paúnd 'jen·ə‚rād·ər }

cumulative compound motor [MECH ENG] A motor with operating characteristics between those of the constant-speed (shunt-wound) and the variable-speed (series-wound) types. { 'kyü·myə·ləd·iv ‚käm‚paúnd 'mōd·ər }

cumulative dose [NUCLEO] The total dose resulting from repeated exposures to radiation. { 'kyü·myə·ləd·iv 'dōs }

cumulative double bonds [ORG CHEM] Double bonds joining at least three contiguous carbon atoms in a single structure, for example, $H_2C=C=CH_2$ (allene). Also known as twinned double bonds. { 'kyü·myə·ləd·iv 'dəb·əl 'bänz }

cumulative error [STAT] An error whose magnitude does not approach zero as the number of observations increases. Also known as accumulative error. { 'kyü·myə·ləd·iv 'er·ər }

cumulative excitation [ATOM PHYS] Process by which the atom is raised from one excited state to a higher state by collision, for example, with an electron. { 'kyü·myə·ləd·iv ek·sə'tā·shən }

cumulative frequency distribution [STAT] The frequency with which a variable assumes values less than or equal to some number, obtained by summing the values in a frequency distribution. { 'kyü·myə·ləd·əv 'frē·kwən·sē ‚di·strə‚byü·shən }

cumulative gas [PETRO ENG] Measurement of total gas produced from a reservoir, usually expressed in graphical relationship to total (cumulative) oil produced from the same reservoir. { 'kyü·myə·ləd·iv 'gas }

cumulative ionization [ATOM PHYS] Ionization of an excited atom in the metastable state by means of cumulative excitation. [ELECTR] *See* avalanche. { 'kyü·myə·ləd·iv ‚ī·ən·ə'zā·shən }

cumulative sum chart [IND ENG] A statistical control chart on which the cumulative sum of deviations is plotted over a period of time and which often has a sliding V-shaped mask for comparing the plot with allowable limits. Also known as cusum chart. { 'kyü·myə·ləd·iv 'səm ‚chärt }

cumulative trauma [MED] An injury or work strain that results from the repeated or continuous application of a work stress that would not ordinarily be harmful in single applications or in multiple applications of short duration. { 'kyü·myə·ləd·iv 'traú·mə }

CUMULUS CLOUD

Small cumulus cloud formation.
(*U. S. Weather Bureau*)

cumulene [ORG CHEM] A compound with a molecular structure which contains two or more double bonds in succession. { 'kyü·myə‚lēn }

cumuliform cloud [METEOROL] A fundamental cloud type, showing vertical development in the form of rising mounds, domes, or towers. { 'kyü·myə·lə‚fòrm ‚klaúd }

cumulonimbus calvus cloud [METEOROL] A species of cumulonimbus cloud evolving from cumulus congestus: the protuberances of the upper portion have begun to lose the cumuliform outline; they loom and usually flatten, then transform into a whitish mass with a more or less diffuse outline and vertical striation; cirriform cloud is not present, but the transformation into ice crystals often proceeds with great rapidity. { ‚kyü·myə·lō'nim·bəs 'kal·vəs ‚klaúd }

cumulonimbus capillatus cloud [METEOROL] A species of cumulonimbus cloud characterized by the presence of distinct cirriform parts, frequently in the form of an anvil, a plume, or a vast and more or less disorderly mass of hair, and usually accompanied by a thunderstorm. { ‚kyü·myə·lō'nim·bəs kap·ə'lad·əs ‚klaúd }

cumulonimbus cloud [METEOROL] A principal cloud type, exceptionally dense and vertically developed, occurring either as isolated clouds or as a line or wall of clouds with separated upper portions. { ‚kyü·myə·lō'nim·bəs ‚klaúd }

cumulus [GEOCHEM] The accumulation of minerals which have precipitated from a liquid without having been modified by later crystallization. { 'kyü·myə·ləs }

cumulus cloud [METEOROL] A principal type of cloud in the form of individual, detached elements which are generally dense and possess sharp nonfibrous outlines; these elements develop vertically, appearing as rising mounds, domes, or towers, the upper parts of which often resemble a cauliflower. { 'kyü·myə·ləs ‚klaúd }

cumulus congestus cloud [METEOROL] A strongly sprouting cumulus species with generally sharp outline and sometimes a great vertical development, and with cauliflower or tower aspect. { 'kyü·myə·ləs kən'jes·təs ‚klaúd }

cumulus humilis cloud [METEOROL] A species of cumulus cloud characterized by small vertical development and a generally flattened appearance, vertical growth is usually restricted by the existence of a temperature inversion in the atmosphere, which in turn explains the unusually uniform height of the cloud. Also known as fair-weather cumulus. { 'kyü·myə·ləs 'hyü·mə·ləs ‚klaúd }

cumulus mediocris cloud [METEOROL] A cloud species unique to the species cumulus, of moderate vertical development, the upper protuberances or sproutings being not very marked; there may be a small cauliflower aspect; while this species does not give any precipitation, it frequently develops into cumulus congestus and cumulonimbus. { 'kyü·myə·ləs mē·dē'ō·krəs ‚klaúd }

cumulus oophorus [HISTOL] The layer of gelatinous, follicle cells surrounding the ovum in a Graafian follicle. { 'kyü·myə·ləs ‚ō·ə'fòr·əs }

cuneate [BIOL] Wedge-shaped with the acute angle near the base, as in certain insect wings and the leaves of various plants. { 'kyü·nē‚āt }

cuneiform [ANAT] **1.** Any of three wedge-shaped tarsal bones. **2.** Either of a pair of cartilages lying dorsal to the thyroid cartilage of the larynx. **3.** Wedge-shaped, chiefly referring to skeletal elements. { 'kyü·nē·ə‚fòrm }

cunnus [ANAT] The vulva. { 'kən·əs }

cup [DES ENG] A cylindrical part with only one end open. [ENG] A low spot forming on a tool joint shoulder as a result of wobbling. [MATH] The symbol ∪, which indicates the union of two sets. [MET] Sheet metal part formed during the first deep-drawing operation. { kəp }

Cup *See* Crater. { kəp }

cup-and-ball joint [GEOL] A dish-shaped transverse fracture which divides a basalt column into segments. Also known as ball-and-socket joint. { ‚kəp ən 'bòl ‚jòint }

cup-and-cone fracture *See* cup fracture. { ‚kəp ən 'kōn ‚frak·chər }

cup anemometer [ENG] A rotation anemometer, usually consisting of three or four hemispherical or conical cups mounted with their diametral planes vertical and distributed symmetrically about the axis of rotation; the rate of rotation of the cups, which is a measure of the wind speed, is determined by a counter. { 'kəp an·ə'mäm·əd·ər }

cup barometer [ENG] A barometer in which one end of a graduated glass tube is immersed in a cup, both cup and tube containing mercury. { 'kəp bə'räm·əd·ər }

cup-case thermometer [ENG] Total-immersion type of thermometer with a cup container at the bulb end to hold a specified amount and depth of the material whose temperature is to be measured. { 'kəp ˌkās thər'mäm·əd·ər }

cup core [ELECTROMAG] A core that encloses a coil to provide magnetic shielding; usually has a powdered iron center post through the coil. { 'kəp ˌkȯr }

cup crystal [HYD] A crystal of ice in the form of a hollow hexagonal cup; a common form of depth hoar. { 'kəp ˌkrist·əl }

Cupedidae [INV ZOO] The reticulated beetles, the single family of the coleopteran suborder Archostemata. { kyü'ped·ə,dē }

cupel [MET] A cup made of bone ash or magnesite, used in assaying precious metals. { 'kyü·pel }

cup electrometer [ENG] An electrometer that has a metal cup attached to its plate so that a charged body touching the inside of the cup gives up its entire charge to the instrument. { 'kəp i,lek'träm·əd·ər }

cupellation [MET] **1.** Method using a cupel for assaying precious metals. **2.** Process for refining gold and silver by alloying them with lead and then oxidizing the molten lead to separate the base metal from the precious metal. { ˌkyü·pə'lā·shən }

cupferron [ORG CHEM] $NH_4ONONC_6H_5$ A colorless salt that forms crystals with a melting point of 164°C; its acid solution is a precipitating reagent. Also known as copperon. { 'kəp·fə,rän }

cup fracture [MET] A break in a ductile material under tensile stress in which the surface of failure on one piece has a central flat area with an exterior extended rim. Also known as cup-and-cone fracture. { 'kəp ˌfrak·chər }

cup grease [MATER] A lubricating grease, usually lime base, for many applications. { 'kəp ˌgrēs }

cupola [GEOL] An isolated, upward-projecting body of plutonic rock that lies near a larger body; both bodies are presumed to unite at depth. [MET] A vertical cylindrical furnace for melting gray iron for foundry use; the metal, coke, and flux are put into the top of the furnace onto a bed of coke through which air is blown. Also known as furnace cupola. { 'kyü·pə·lə }

cupola drop [MET] The bed and unmelted charges dropped from a cupola at the end of a heat. { 'kyü·pə·lə ˌdräp }

cup packer [PETRO ENG] A cup-shaped sealing device inserted into the drill stem and lowered into a well to permit pressure testing of the casing and blowout preventers. { 'kəp ˌpak·ər }

cupped pebble [GEOL] A pebble fragment that has become hollow after being subjected to solution. { 'kəpt ˌpeb·əl }

cupping [MET] **1.** First operation of a deep-drawing process. **2.** Fracture of a wire or rod in which one fracture surface is conical and the other concave. { 'kəp·iŋ }

cup product [MATH] A multiplication defined on cohomology classes; it gives cohomology a ring structure. { 'kəp ˌpräd·əkt }

cuprammonium cellulose [TEXT] The cellulose of cotton linters treated with copper sulfate and ammonium. { ˌkyü·prə¦mō·nē·əm 'sel·yə,los }

cuprammonium process [TEXT] A process by which rayon is made from regenerated cellulose in a solution of ammoniacal copper oxide. { ˌkyü·prə¦mō·nē·əm 'präs·əs }

cuprammonium rayon [TEXT] Rayon made from cuprammonium cellulose. { ˌkyü·prə¦mō·nē·əm 'rā,än }

cupreine [ORG CHEM] $C_{19}H_{22}O_2N_2·H_2O$ Colorless, anhydrous crystals with a melting point of 198°C; soluble in chloroform and ether; used in medicine. Also known as hydroxycinchonine. { 'kyü·prē·ēn }

cupreous [SCI TECH] Containing or resembling copper. { 'kyü·prē·əs }

cupric [CHEM] The divalent ion of copper. { 'kyü·prik }

cupric acetate [ORG CHEM] $Cu(C_2H_3O_2)_2·H_2O$ Blue-green crystals, soluble in water; used as a raw material to make paris green. Also known as copper acetate; crystal aerugo; crystals of Venus; verdigris. { 'kyü·prik 'as·ə,tāt }

cupric arsenite See copper arsenite. { 'kyü·prik ärs·ən,īt }

cupric bromide [INORG CHEM] $CuBr_2$ Black prismatic crystals; used in photography as an intensifier and in organic synthesis as a brominating agent. Also known as copper bromide. { 'kyü·prik 'brō,mīd }

cupric carbonate See copper carbonate. { 'kyü·prik 'kär·bə,nāt }

cupric chloride [INORG CHEM] **1.** Also known as copper chloride. **2.** $CuCl_2$ Yellowish-brown, deliquescent powder soluble in water, alcohol, and ammonium chloride. **3.** $CuCl_2·H_2O$ A dihydrate of cupric chloride forming green crystals soluble in water; used as a mordant in dyeing and printing textile fabrics and in the refining of copper, gold, and silver. { 'kyü·prik 'klȯr,īd }

cupric chromate [INORG CHEM] $CuCrO_4$ A yellow liquid, used as a mordant. Also known as copper chromate. { 'kyü·prik 'krō,māt }

cupric cyanide [INORG CHEM] $Cu(CN)_2$ A green powder, insoluble in water; used in electroplating copper on iron. Also known as copper cyanide. { 'kyü·prik 'sī·ə,nīd }

cupric fluoride [INORG CHEM] CuF_2 White crystalline powder used in ceramics and in the preparation of brazing and soldering fluxes. Also known as copper fluoride. { 'kyü·prik 'flur,īd }

cupric gluconate See copper gluconate. { 'kyü·prik 'glü·kə,nāt }

cupric hydroxide [INORG CHEM] $Cu(OH)_2$ Blue macro- or microscopic crystals; used as a mordant and pigment, in manufacture of many copper salts, and for staining paper. Also known as copper hydroxide. { 'kyü·prik hī'dräk,sīd }

cupric nitrate [INORG CHEM] $Cu(NO_3)_2·3H_2O$ Green powder or blue crystals soluble in water; used in electroplating copper on iron. Also known as copper nitrate. { 'kyü·prik 'nī,trāt }

cupric oxide [INORG CHEM] CuO Black, monoclinic crystals, insoluble in water; used in making fibers and ceramics, and in organic and gas analyses. Also known as copper oxide. { 'kyü·prik 'äk,sīd }

cupric sulfate [INORG CHEM] $CuSO_4$ A water-soluble salt used in copper-plating baths; crystallizes as hydrous copper sulfate, which is blue. Also known as copper sulfate. { 'kyü·prik 'səl,fāt }

cuprite [MINERAL] Cu_2O A red mineral that crystallizes in the isometric system and is found in crystals and fine-grained aggregates or is massive; a widespread supergene copper ore. Also known as octahedral copper ore; red copper ore; ruby copper ore. { 'kyü,prīt }

cuprocopiapite [MINERAL] $CuFe_4(SO_4)_6(OH)_2·20H_2O$ A sulfur yellow to orange-yellow, triclinic mineral consisting of a hydrated basic sulfate of copper and iron. { ¦kyü·prō'kō·pē·ə,pīt }

cuprodescloizite See mottramite. { ¦kyü·prō·des'klȯ·ə,zīt }

cupronickel [MET] A copper-base alloy with 10–30% nickel and small amounts of manganese and iron; used in industrial and marine installations as condenser and heat-exchanger tubing. { ¦kyü·prō'nik·əl }

cuprotungstite [MINERAL] $Cu_2(WO_4)(OH)_2$ A green mineral that forms compact masses; soluble in acids; the crystal system is not known. { ¦kyü·prō'təŋ,stīt }

cuprouranite See torbernite. { ¦kyü·prō'yur·ə,nīt }

cuprous bromide [INORG CHEM] Cu_2Br_2 White or gray crystals slightly soluble in cold water. Also known as copper bromide. { 'kyü·prəs 'brō,mīd }

cuprous chloride [INORG CHEM] $CuCl$ or Cu_2Cl_2 Green, tetrahedral crystals, insoluble in water. Also known as copper chloride; resin of copper. { 'kyü·prəs 'klȯr,īd }

cuprous fluoride [INORG CHEM] Cu_2F_2 Red, crystalline powder, melting point 908°C. Also known as copper fluoride. { 'kyü·prəs 'flur,īd }

cuprous oxide [INORG CHEM] Cu_2O An oxide of copper found in nature as cuprite and formed on copper by heat; used chiefly as a pigment and as a fungicide. Also known as copper oxide. { 'kyü·prəs 'äk,sīd }

cupule [BOT] **1.** The cup-shaped involucre characteristic of oaks. **2.** A cup-shaped corolla. **3.** The gemmae cup of the Marchantiales. [INV ZOO] A small sucker on the feet of certain male flies. { 'kyü,pyül }

curare [ORG CHEM] Poisonous extract from the plant *Strychnos toxifera* containing a mixture of alkaloids that produce paralysis of the voluntary muscles by acting on synaptic junctions; used as an adjunct to anesthesia in surgery. { kyü'rä·rē }

curb [CIV ENG] A border of concrete or row of joined stones forming part of a gutter along a street edge. [MIN ENG] A timber frame, circular or square, wedged in a shaft to make a foundation for walling or tubbing, or to support, with or without other timbering, the walls of the shaft. { kərb }

curb roof [ARCH] A roof with a ridge at its center and a parallel lower ridge on both sloping sides. { 'kərb ,rüf }

curb weight [MECH ENG] The weight of a motor vehicle plus fuel and other components or equipment necessary for standard operation; does not include driver weight or payload. { 'kərb ,wāt }

Curculionidae [INV ZOO] The true weevils or snout beetles, a family of coleopteran insects in the superfamily Curculionoidea. { kər,kyü·lē'än·ə,dē }

Curculionoidea [INV ZOO] A superfamily of coleopteran insects in the suborder Polyphaga. { kər,kyü·lē·ə'nȯid·ē·ə }

curculionoid larva [INV ZOO] A kind of beetle larva having a highly reduced and grublike body. { kər'kyü·lē·ə,nȯid 'lär·və }

Curcurbitaceae [BOT] A family of dicotyledonous herbs or herbaceous vines in the order Violales characterized by an inferior ovary, unisexual flowers, one to five stamens but typically three, and a sympetalous corolla. { kər,kər·bə'tās·ē,ē }

Curcurbitales [BOT] The ordinal name assigned to the Curcurbitaceae in some systems of classification. { kər,kər·bə'tā·lēz }

curd [BOT] The edible flower heads of members of the mustard family such as broccoli. [FOOD ENG] **1.** The clotted portion of soured milk or milk treated with an acid or enzyme; used in making cheese. **2.** Any food resembling milk curd. { kərd }

cure [CHEM] To change the properties of a resin material by chemical polycondensation or addition reactions. *See* vulcanization. [ENG] A process by which concrete is kept moist for its first week or month to provide enough water for the cement to harden. Also known as mature. { kyür }

curet [MED] An instrument, shaped like a spoon or scoop, for scraping away tissue. { kyü'ret }

cure time [CHEM ENG] The amount of time required for a rubber compound to reach maximum viscosity or modulus at a given temperature. { 'kyür ,tīm }

curettage [MED] Scraping of the inside of a body cavity or the hollow of an organ with a curet. { ,kyü·rə'täzh }

curie [NUCLEO] A unit of radioactivity, defined as that quantity of any radioactive nuclide which has 3.700×10^{10} disintegrations per second. Abbreviated c; Ci. { 'kyür·ē }

Curle balance [ENG] An instrument for determining the susceptibility of weakly magnetic materials, in which the deflection produced by a strong permanent magnet on a suspended tube containing the specimen is measured. { 'kyür·ē ,bal·əns }

Curie constant [ELECTROMAG] The electric or magnetic susceptibility at some temperature times the difference of the temperature and the Curie temperature, which is a constant at temperatures above the Curie temperature according to the Curie-Weiss law. { 'kyür·ē 'kän·stənt }

Curie point *See* Curie temperature. { 'kyür·ē ,pȯint }

Curie principle [THERMO] The principle that a macroscopic cause never has more elements of symmetry than the effect it produces; for example, a scalar cause cannot produce a vectorial effect. { 'kyür·ē ,prin·sə·pəl }

Curle scale of temperature [THERMO] A temperature scale based on the susceptibility of a paramagnetic substance, assuming that it obeys Curie's law; used at temperatures below about 1 kelvin. { 'kyür·ē ,skāl əv 'tem·prə·chər }

Curie's law [ELECTROMAG] The law that the magnetic susceptibilities of most paramagnetic substances are inversely proportional to their absolute temperatures. { 'kyür,ēz ,lȯ }

Curie temperature [ELECTROMAG] The temperature marking the transition between ferromagnetism and paramagnetism, or between the ferroelectric phase and paraelectric phase. Also known as Curie point. { 'kyür·ē ,tem·prə·chər }

Curie-Weiss law [ELECTROMAG] A relation between magnetic or electric susceptibilities and the absolute temperatures which is followed by ferromagnets, antiferromagnets, nonpolar ferroelectrics, antiferroelectrics, and some paramagnets. { ,kyür·ē ,vīs ,lȯ }

curine *See* bebeerine. { 'kyü,rēn }

CURCULIONIDAE

Snout beetle. *(From T. I. Storer and R. L. Usinger, General Zoology, 3d ed., McGraw-Hill, 1957)*

---

curing [CHEM ENG] A process in which polymers or oligomers are chemically cross-linked to form polymer networks. [CIV ENG] A process for bringing freshly placed concrete to required strength and quality by maintaining the humidity and temperature at specified levels for a given period of time. Also known as seasoning. [SCI TECH] Any one of various processes whereby a product is preserved, perfected, or readied for use. { 'kyür·iŋ }

curing agent *See* hardener. { 'kyür·iŋ ,ā·jənt }

curing temperature [CHEM] That temperature at which a resin or adhesive is subjected to curing. { 'kyür·iŋ ,tem·prə·chər }

curing time [CHEM] The period of time in which a part is subjected to heat or pressure to cure the resin. [ENG] Time interval between the stopping of moving parts during thermoplastics molding and the release of mold pressure. Also known as molding time. { 'kyür·iŋ ,tīm }

curite [MINERAL] $Pb_2U_5O_{17} \cdot 4H_2O$ An orange-red radioactive mineral, occurring in acicular crystals, an alteration product of uraninite. { 'kyü,rīt }

curium [CHEM] An element, symbol Cm, atomic number 96; the isotope of mass 244 is the principal source of this artificially produced element. { 'kyür·ē·əm }

curium-242 [NUC PHYS] An isotope of curium, mass number 242; half-life is 165.5 days for α-particle emission; 7.2×10^6 years for spontaneous fission. { 'kyür·ē·əm ,tü¦fȯrd·ē¦tü }

curium-244 [NUC PHYS] An isotope of curium, mass number 244; half-life is 16.6 years for α-particle emission; 1.4×10^7 years for spontaneous fission; potential use as compact thermoelectric power source. { 'kyür·ē·əm ,tü¦fȯrd·ē¦fȯr }

curl [FOR] A block of timber cut from a crotch for cutting into veneers. [MATER] A defect of paper caused by unequal alteration in the dimensions of the top and underside of the sheet. [MATH] The curl of a vector function is a vector which is formally the cross product of the del operator and the vector. Also known as rotation (rot). { kərl }

curling [MECH ENG] A forming process in which the edge of a sheet-metal part is rolled over to produce a hollow tubular rim. { 'kərl·iŋ }

curling dies [MECH ENG] A set of tools that shape the ends of a piece of work into a form with a circular cross section. { 'kərl·iŋ ,dīz }

curling factor *See* griseofulvin. { 'kərl·iŋ ,fak·tər }

curling machine [MECH ENG] A machine with curling dies; used to curl the ends of cans. { 'kərl·iŋ mə'shēn }

Curling's ulcer [MED] An acute gastric ulcer associated with severe skin burns. { 'kərl·iŋz ,əl·sər }

curly top [PL PATH] A virus disease of sugarbeets and certain other plants that is transmitted by a leafhopper; affected plants are dwarfed and have curled, upturned leaves. { 'kər·lē ,täp }

currant [BOT] A shrubby, deciduous plant of the genus *Ribes* in the order Rosales; the edible fruit, a berry, is borne in clusters on the plant. { 'kər·ənt }

currant leaf spot [PL PATH] **1.** An angular leaf spot of currants caused by the fungus *Cercospora angulata*. **2.** An anthracnose of currants caused by *Pseudopeziza ribis* and characterized by brown or black spots. { 'kər·ənt ,lēf ,spät }

current [ELEC] The net transfer of electric charge per unit time; a specialization of the physics definition. Also known as electric current. [PHYS] **1.** The rate of flow of any conserved, indestructible quantity across a surface per unit time. **2.** *See* current density. { 'kər·ənt }

current algebra [PARTIC PHYS] The application of algebraic relationships among currents derived from approximate symmetries, such as broken SU_3 symmetry, to the study of hadrons. { 'kər·ənt ,al·jə·brə }

current amplification [ELECTR] The ratio of output-signal current to input-signal current for an electron tube, transistor, or magnetic amplifier, the multiplier section of a multiplier phototube, or any other amplifying device; often expressed in decibels by multiplying the common logarithm of the ratio by 20. { 'kər·ənt am·plə·fə'kā·shən }

current amplifier [ELECTR] An amplifier capable of delivering considerably more signal current than is fed in. { 'kər·ənt ,am·plə,fī·ər }

current antinode [ELEC] A point at which current is a maximum along a transmission line, antenna, or other circuit element having standing waves. Also known as current loop. { 'kər·ənt 'an·tə,nōd }

current attenuation [ELECTR] The ratio of input-signal current for a transducer to the current in a specified load impedance connected to the transducer; often expressed in decibels. { 'kər·ənt ə,ten·yə'wā·shən }

current awareness system [COMPUT SCI] A system for notifying users on a periodic basis of the acquisition, by a central file or library, of information (usually literature) which should be of interest to the user. { 'kər·ənt ə'wer·nəs ,sis·təm }

current balance [ELEC] An apparatus with which force is measured between current-carrying conductors, with the purpose of assigning the value of the ampere. Also known as ampere balance. { 'kər·ənt ,bal·əns }

current-bedding [GEOL] Cross-bedding resulting from water or air currents. { 'kər·ənt ,bed·iŋ }

current-carrying capacity [ELEC] The maximum current that can be continuously carried without causing permanent deterioration of electrical or mechanical properties of a device or conductor. { 'kər·ənt ,kar·ē·iŋ kə'pas·əd·ē }

current cell See active cell. { 'kər·ənt 'sel }

current chart [MAP] A map of a water area depicting current speeds and directions by current roses, vectors, or other means. { 'kər·ənt ,chärt }

current collector See charge collector. { 'kər·ənt kə,lek·tər }

current comparator [ELEC] An instrument for determining the ratio of two direct or alternating currents, based on Ampère's laws, in which the two currents are passed through a toroid by two windings of known numbers of turns and the ampere-turn unbalance is measured by a detection winding. { 'kə·rənt kəm,par·əd·ər }

current constants [OCEANOGR] Tidal current relations that remain practically constant for any particular locality. { 'kər·ənt 'kän·stənts }

current-controlled switch [ELECTR] A semiconductor device in which the controlling bias sets the resistance at either a very high or very low value, corresponding to the "off" and "on" conditions of a switch. { 'kər·ənt kən,trōld 'swich }

current curve [OCEANOGR] In marine operations, a graphic representation of the flow of a current, consisting of a rectangular-coordinate graph on which speed is represented by the ordinates and time by the abscissas. { 'kər·ənt ,kərv }

current cycle [OCEANOGR] A complete set of tidal current conditions, as those occurring during a tidal day, lunar month, or Metonic cycle. { 'kər·ənt ,sī·kəl }

current decay [MET] In certain types of welding operations, controlled reduction of the welding impulse over a predetermined time interval to prevent rapid cooling of the weld nugget. { 'kər·ənt di,kā }

current density [ELEC] The current per unit cross-sectional area of a conductor; a specialization of the physics definition. Also known as electric current density. [PHYS] A vector quantity whose component perpendicular to any surface equals the rate of flow of some conserved, indestructible quantity across that surface per unit area per unit time. Also known as current. { 'kər·ənt ,den·səd·ē }

current diagram [OCEANOGR] A graph showing the average speeds of flood and ebb currents throughout the current cycle for a considerable part of a tidal waterway. { 'kər·ənt 'dī·ə,gram }

current difference [OCEANOGR] In marine operations, the difference between the time of slack water or strength of current at a subordinate station and its reference station. { 'kər·ənt ,dif·rəns }

current divider [ELEC] A device used to deliver a desired fraction of a total current to a circuit. { 'kər·ənt di,vīd·ər }

current drain [ELEC] The current taken from a voltage source by a load. Also known as drain. { 'kər·ənt ,drān }

current drift [HYD] A broad, shallow, slow-moving ocean or lake current. { 'kər·ənt ,drift }

current drogue [ENG] A current-measuring assembly consisting of a weighted current cross, sail, or parachute, and an attached surface buoy. { 'kər·ənt ,drōg }

current efficiency [PHYS CHEM] The ratio of the amount of electricity, in coulombs, theoretically required to yield a given quantity of material in an electrochemical process, to the amount actually consumed. { 'kər·ənt i,fish·ən·sē }

current ellipse [OCEANOGR] In marine operations, a graphic representation of a rotary current, in which the speed and direction of the current at various hours of the current cycle are represented by radius vectors; a line connecting the ends of the radius vectors approximates an ellipse. { 'kər·ənt ə,lips }

current-equalizing reactor [ELEC] A reactor that is used to achieve a desired division of current between several circuits operating in parallel. { 'kər·ənt ,ē·kwəˌlīz·iŋ rē'ak·tər }

current feed [ELECTR] Feed to a point where current is a maximum, as at the center of a half-wave antenna. { 'kər·ənt ,fēd }

current feedback [ELECTR] Feedback introduced in series with the input circuit of an amplifier. { 'kər·ənt ,fēd,bak }

current feedback circuit [ELECTR] A circuit used to eliminate effects of amplifier gain instability in an indirect-acting recording instrument, in which the voltage input (error signal) to an amplifier is the difference between the measured quantity and the voltage drop across a resistor. { 'kər·ənt ,fēd,bak ,sər·kət }

current function See Lagrange stream function. { 'kər·ənt ,fəŋk·shən }

current gain [ELECTR] The fraction of the current flowing into the emitter of a transistor which flows through the base region and out the collector. { 'kər·ənt ,gān }

current generator [ELECTR] A two-terminal circuit element whose terminal current is independent of the voltage between its terminals. { 'kər·ənt ,jen·ə,rād·ər }

current hogging [ELECTR] A condition in which the largest fraction of a current passes through one of several parallel logic circuits because it has a lower resistance than the others. { 'kər·ənt ,häg·iŋ }

current hour [OCEANOGR] The average time interval between the moon's transit over the meridian of Greenwich and the time of the following strength of flood current modified by the times of slack water and strength of ebb. { 'kər·ənt ,aü·ər }

current-instruction register See instruction register. { 'kər·ənt in'strək·shən ,rej·ə·stər }

current intensity [ELEC] The magnitude of an electric current. Also known as current strength. { 'kər·ənt in'ten·səd·ē }

current interrupter [ELEC] Mechanism connected into a current-carrying line to periodically interrupt current flow to allow no-current tests of system components. { 'kər·ənt in·tə'rəp·tər }

current limiter [ELECTR] A device that restricts the flow of current to a certain amount, regardless of applied voltage. Also known as demand limiter. { 'kər·ənt ,lim·əd·ər }

current-limiting reactor See series reactor. { 'kər·ənt ,lim·əd·iŋ rē'ak·tər }

current-limiting resistor [ELEC] A resistor inserted in an electric circuit to limit the flow of current to some predetermined value; used chiefly to protect tubes and other components during warm-up. { 'kər·ənt ,lim·əd·iŋ ri'zis·tər }

current line [ENG] In marine operations, a graduated line attached to a current pole, used to measure the speed of a current; as the pole moves away with the current, the speed of the current is determined by the amount of line paid out in a specified time. Also known as log line. { 'kər·ənt ,līn }

current lineation See parting lineation. { 'kər·ənt lin·ē'ā·shən }

current location reference [COMPUT SCI] A symbolic expression, such as a star, which indicates the current location reached by the program; a transfer to * + 2 would bring control to the second statement after the current statement. { 'kər·ənt lō'kā·shən ,ref·rəns }

current loop See current antinode. { 'kər·ənt ,lüp }

current margin [COMMUN] Difference between the steady-state currents flowing through a telegraph receiving instrument corresponding respectively to the two positions of the telegraph transmitter. { 'kər·ənt ,mär·jən }

current mark [GEOL] Any structure formed by direct or indirect action of a water current on a sedimentary surface. { 'kər·ənt ,märk }

current measurement [ELEC] The measurement of the flow of electric current. { 'kər·ənt ,mezh·ər·mənt }

current meter See ammeter. See velocity-type flowmeter. { 'kər·ənt ,mēd·ər }

current mirror [ELECTR] An electronic circuit that generates, at a high-impedance output node, an inflowing or outflowing current that is a scaled replica of an input current

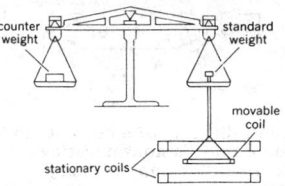

CURRENT BALANCE

Rayleigh current balance.

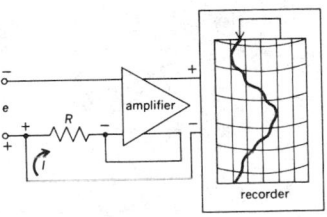

CURRENT FEEDBACK CIRCUIT

Diagram of a current feedback circuit.

CURRENT-MODE LOGIC

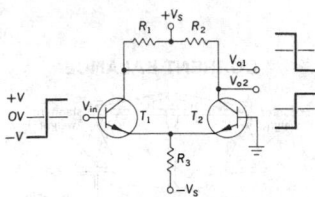

Circuit diagram of a current-mode logic circuit, a nonsaturating circuit, with complementary outputs.

flowing into or out of a low-impedance input node. { 'kər·ənt ,mir·ər }

current-mode filter [ELECTR] An integrated-circuit filter in which the signals are represented by current levels rather than voltage levels. { 'kər·ənt‚mōd ‚fil·tər }

current-mode logic [ELECTR] Integrated-circuit logic in which transistors are paralleled so as to eliminate current hogging. Abbreviated CML. { 'kər·ənt ‚mōd 'läj·ik }

current node [ELEC] A point at which current is zero along a transmission line, antenna, or other circuit element having standing waves. { 'kər·ənt ‚nōd }

current noise [ELECTR] Electrical noise of uncertain origin which is observed in certain resistances when a direct current is present, and which increases with the square of this current. { 'kər·ənt ‚noiz }

current phasor [ELEC] A line referenced to a point, whose length and angle represent the magnitude and phase of a current. { 'kər·ənt ‚fā·zər }

current pole [ENG] A pole used to determine the direction and speed of a current; the direction is determined by the direction of motion of the pole, and the speed by the amount of an attached current line paid out in a specified time. { 'kər·ənt ‚pōl }

current ratio [ELECTROMAG] In a waveguide, the ratio of maximum to minimum current. { 'kər·ənt ‚rā·shō }

current regulator [ELECTR] A device that maintains the output current of a voltage source at a predetermined, essentially constant value despite changes in load impedance. { 'kər·ənt ‚reg·yə‚lād·ər }

current relay [ELEC] A relay that operates at a specified current value rather than at a specified voltage value. { 'kər·ənt ‚rē‚lā }

current ripple [GEOL] A type of ripple mark having a long, gentle slope toward the direction from which the current flows, and a shorter, steeper slope on the lee side. { 'kər·ənt ‚rip·əl }

current rips [OCEANOGR] Small waves formed on the surface of water by the meeting of opposing ocean currents; vertical oscillation, rather than progressive waves, is characteristic of current rips. { 'kər·ənt ‚rips }

current rose [MAP] A graphic presentation of ocean currents for specified areas, utilizing arrows at the cardinal and intercardinal compass points to show the direction toward which the prevailing current flows and the present frequency of set for a given period of time. { 'kər·ənt ‚rōz }

current saturation *See* anode saturation. { 'kər·ənt sach·ə'rā·shən }

current source [ELECTR] An electronic circuit that generates a constant direct current into or out of a high-impedance output node. { 'kər·ənt ‚sórs }

current strength *See* current intensity. { 'kər·ənt ‚streŋkth }

current tables [OCEANOGR] Tables listing predictions of the time and speeds of tidal currents at various places. { 'kər·ənt ‚tā·bəlz }

current tap *See* multiple lamp holder. *See* plug adapter lamp holder. { 'kər·ənt ‚tap }

current transformer [ELEC] An instrument transformer intended to have its primary winding connected in series with a circuit carrying the current to be measured or controlled; the current is measured across the secondary winding. { 'kər·ənt tranz'fór·mər }

current-transformer phase angle [ELEC] Angle between the primary current vector and the secondary current vector reversed; it is conveniently considered as positive when the reversed secondary current vector leads the primary current vector. { 'kər·ənt tranz'fór·mər 'fāz ‚aŋ·gəl }

current-type flowmeter [ENG] A mechanical device to measure liquid velocity in open and closed channels; similar to the vane anemometer (where moving liquid turns a small windmill-type vane), but more rugged. { 'kər·ənt ‚tīp 'flō‚mēd·ər }

current-type telemeter [COMMUN] A telemeter in which the magnitude of a single current is the translating means. { 'kər·ənt ‚tīp tə'lem·əd·ər }

current-voltage dual [ELEC] A circuit which is equivalent to a specified circuit when one replaces quantities with dual quantities; current and voltage impedance and admittance, and meshes and nodes are examples of dual quantities. { 'kər·ənt ‚vōl·tij ‚dül }

curry [FOOD ENG] A mixture of plant spices including turmeric, coriander, cinnamon, cumin, ginger, cardamon, cayenne pepper, cloves, and nutmeg. { 'kər·ē }

cursor [COMPUT SCI] A movable spot of light that appears on the screen of a visual display terminal and can be positioned horizontally and vertically through keyboard controls to instruct the computer at what point a change is to be made. [DES ENG] A clear or amber-colored filter that can be placed over a radar screen and rotated until an etched diameter line on the filter passes through a target echo; the bearing from radar to target can then be read accurately on a stationary 360° scale surrounding the filter. { 'kər·sər }

cursor arrows [COMPUT SCI] Arrows marked on keys of a computer keyboard that control the movement of the cursor. { 'kər·sər ‚ar·ōz }

cursorial [VERT ZOO] Adapted for running. { kər'sór·ē·əl }

curtain [GEOL] **1.** A thin sheet of dripstone that hangs or projects from a cave wall. **2.** A rock formation connecting two adjacent bastions. [NUCLEO] A thin shield, usually cadmium, used in a nuclear reactor to shut off a flow of slow neutrons. { 'kərt·ən }

curtain array [ELECTROMAG] An antenna array consisting of vertical wire elements stretched between two suspension cables. { 'kərt·ən ə'rā }

curtain board [BUILD] A fire-retardant partition applied to a ceiling. { 'kərt·ən ‚bórd }

curtain coating [CHEM ENG] A method in which the substrate to be coated with low-viscosity resins or solutions is passed through, and is perpendicular to, a freely falling liquid curtain. { 'kərt·ən ‚kōd·iŋ }

curtain rhombic antenna [ELECTROMAG] A multiple-wire rhombic antenna having a constant input impedance over a wide frequency range; two or more conductors join at the feed and terminating ends but are spaced apart vertically from 1 to 5 feet (30 to 150 centimeters) at the side poles. { 'kərt·ən 'räm·bik an'ten·ə }

curtain wall [CIV ENG] An external wall that is not load-bearing. { 'kərt·ən ‚wól }

curtate cycloid [MATH] A trochoid in which the distance from the center of the rolling circle to the point describing the curve is less than the radius of the rolling circle. { 'kər‚tāt 'sī‚klóid }

curtate distance [ASTRON] The distance between the earth or the sun and the foot of a perpendicular from a planet or comet to the plane of the earth's orbit. { 'kər‚tāt ‚dis·təns }

Curtius reaction [ORG CHEM] A laboratory method for degrading a carboxylic acid to a primary amine by converting the acid to an acyl azide to give products which can be hydrolyzed to amines. { 'kərd·ē·əs rē‚ak·shən }

curvature [MATH] The reciprocal of the radius of the circle which most nearly approximates a curve at a given point; the rate of change of the unit tangent vector to a curve with respect to arc length of the curve. { 'kər·və·chər }

curvature correction [ASTRON] A correction applied to the mean of a series of observations on a star or planet to take account of the divergence of the apparent path of the star or planet from a straight line. [GEOD] The correction applied in some geodetic work to take account of the divergence of the surface of the earth (spheroid) from a plane. { 'kər·və·chər kə'rek·shən }

curvature effect [ELECTR] Generally, the condition in which the dielectric strength of a liquid or vacuum separating two electrodes is higher for electrodes of smaller radius of curvature. { 'kər·və·chər i'fekt }

curvature of field [OPTICS] Error in the image of a plane object formed on a flat screen by an optical system when the best image lies on a curved surface. Also known as field curvature. { 'kər·və·chər əv 'fēld }

curvature of space [RELAT] **1.** The deviation of a spacelike three-dimensional subspace of curved space-time from euclidean geometry. **2.** The Gaussian curvature of a spacelike three-dimensional subspace of curved space-time. { 'kər·və·chər əv 'spās }

curvature tensor *See* Riemann-Christoffel tensor. { 'kər·və·chər ‚ten·sər }

curve [MATH] The continuous image of the unit interval. { kərv }

curved beam [ENG] A beam bounded by circular arcs. { ˈkərvd ˈbēm }

curved plate [GRAPHICS] In letterpress printing, pieces of type which have been curved to fit the cylinder of a rotary press. { ˈkərvd ˈplāt }

curved space-time [RELAT] A four-dimensional Riemannian space, in which there are no straight lines but only curves, which is a generalization of the Minkowski universe in the general theory of relativity. { ˈkərvd ˈspās ˈtīm }

curved surface [MATH] A surface having no part that is a plane surface. { ˈkərvd ˈsər·fəs }

curve fitting [STAT] The calculation of a curve of some particular character (as a logarithmic curve) that most closely approaches a number of points in a plane. { ˈkərv ˌfid·iŋ }

curve follower [COMPUT SCI] A device in which a photoelectric, capacitive or inductive pick-off guided by a servomechanism reads data in the form of a graph, such as a curve drawn on paper with suitable ink. Also known as graph follower. { ˈkərv ˌfäl·ə·wər }

curve of constant bearing See curve of equal bearing. { ˈkərv əv ˈkän·stənt ˌber·iŋ }

curve of equal bearing [NAV] On a map plot, a curve connecting all points at which the great-circle bearing of a given point is the same. Also known as curve of constant bearing. { ˈkərv əv ˈē·kwəl ˌber·iŋ }

curve of growth [ASTROPHYS] A graph of the equivalent width of an absorption line versus the number of atoms that produce it. { ˈkərv əv ˈgrōth }

curve resistance [MECH] The force opposing the motion of a railway train along a track due to track curvature. { ˈkərv riˈzis·təns }

curves of form [NAV ARCH] Graphs of properties of a vessel's form, such as the displacement and the area of wetted surface, versus the vessel's draft. { ˈkərvz əv ˈfórm }

curve tracer [ENG] An instrument that can produce a display of one voltage or current as a function of another voltage or current, with a third voltage or current as a parameter. { ˈkərv ˌtrā·sər }

curve tracing [MATH] The method of graphing a function by plotting points and analyzing symmetries, derivatives, and so on. { ˈkərv ˌtrās·iŋ }

curvilinear [SCI TECH] Pertaining to curved lines, as in curvilinear coordinates or curvilinear motion. { ˌkər·vəˈlin·ē·ər }

curvilinear coordinates [MATH] Any linear coordinates which are not cartesian coordinates; frequently used curvilinear coordinates are polar coordinates and cylindrical coordinates. { ˌkər·vəˈlin·ē·ər kōˈórd·ən·əts }

curvilinear motion [MECH] Motion along a curved path. { ˌkər·vəˈlin·ē·ər ˈmō·shən }

curvilinear regression [STAT] Regression study of jointly distributed random variables where the function measuring their statistical dependence is analyzed in terms of curvilinear coordinates. Also known as nonlinear regression. { ˌkər·vəˈlin·ē·ər riˈgresh·ən }

curvilinear solid [MATH] A solid whose surfaces are not planes. { ˌkər·vəˈlin·ē·ər ˈsäl·əd }

curvilinear transformation [MATH] A transformation from one coordinate system to another in which the coordinates in the new system are arbitrary twice-differentiable functions of the coordinates in the old system. { ˌkər·vəˈlin·ē·ər tranzˌfərˈmā·shən }

curvilinear trend [STAT] A nonlinear trend which may be expressed as a polynomial or a smooth curve. { ˌkər·vəˈlin·ē·ər ˈtrend }

cuscus oil See vetiver oil. { ˈkəs·kəs ˌóil }

Cuscutaceae [BOT] A family of parasitic dicotyledonous plants in the order Polemoniales which lack internal phloem and chlorophyll, have capsular fruit, and are not rooted to the ground at maturity. { ˌkə·skyüˈtās·ē·ē }

cusec [MECH] A unit of volume flow rate, used primarily to describe pumps, equal to a uniform flow of 1 cubic foot in 1 second. Also known as cubic foot per second (cfs). { ˈkyü·sek }

Cushing's syndrome [MED] A complex of symptoms including facial and truncal obesity, hypertension, edema, and osteoporosis, resulting from oversecretion of adrenocortical hormones. { ˈkush·iŋz ˌsin·drōm }

cushion [PETRO ENG] A volume of water, drilling fluid, or compressed gas injected into the drill pipe or tubing to control both annular and formation pressures. { ˈkush·ən }

cushion effect See Poisson effect. { ˈkəsh·ən iˌfekt }

cushion gas See blanket gas. { ˈkush·ən ˌgas }

cusp [ANAT] 1. A pointed or rounded projection on the masticating surface of a tooth. 2. One of the flaps of a heart valve. [ARCH] A pointed projection or peak created by the intersection of two arcs. [GEOL] One of a series of low, crescent-shaped mounds of beach material separated by smoothly curved, shallow troughs spaced at more or less regular intervals along and generally perpendicular to the beach face. Also known as beach cusp. [GEOPHYS] Any of the funnel-shaped regions in the magnetosphere extending from the front magnetopause to the polar ionosphere, and filled with solar wind plasma. [MATH] A singular point of a curve at which the limits of the tangents of the portions of the curve on either side of the point coincide. Also known as spinode. { kəsp }

cuspate bar [GEOL] A crescentic bar joining with the shore at each end. { ˈkəˌspāt ˈbär }

cuspate ripple mark See linguoid ripple mark. { ˈkəˌspāt ˈrip·əl ˌmärk }

cusp cap [ASTRON] One of the 10 bright areas observed near one of the extremities of the illuminated portion of Venus during the crescent phase. { ˈkəsp ˌkap }

cusped magnetic field [ELECTROMAG] A magnetic field created by adjacent parallel coils that carry current in opposite directions; used in fusion research, to contain a plasma of high-energy deuterium ions. { ˈkəspt magˈned·ik ˈfēld }

cuspid See canine. { ˈkəs·pəd }

cuspidal cubic [MATH] A cubic curve that has one cusp, one point of inflection, and no node. { ˈkəs·pəd·əl ˈkyü·bik }

cuspidal locus [MATH] A curve consisting of the cusps of some family of curves. { ˈkəs·pəd·əl ˈlō·kəs }

cuspidate [BIOL] Having a cusp; terminating in a point. { ˈkəs·pəˌdāt }

cusp of the first kind [MATH] A cusp such that the two portions of the curve adjacent to the cusp lie on opposite sides of the limiting tangent to the curve at the cusp. Also known as simple cusp. { ˈkəsp əv thə ˈfərst ˌkīnd }

cusp of the second kind [MATH] A cusp such that the two portions of the curve adjacent to the cusp lie on the same side of the limiting tangent to the curve at the cusp. { ˈkəsp əv thə ˈsek·ənd kīnd }

custard winds [METEOROL] Cold easterly winds on the northeastern coast of England. { ˈkəs·tərd ˌwinz }

custodial area [BUILD] Area of a building designated for service and custodial personnel; includes rooms, closets, storage, toilets, and lockers. { kəˈstōd·ē·əl ˌer·ē·ə }

custom-designed device [ELECTR] An integrated logic circuit element that is generated by a series of steps resembling photographic development from highly complicated artwork patterns. { ˈkəs·təm dəˈzīnd diˈvīs }

customer substation [ELEC] A distribution substation located on the premises of a larger customer, such as a shopping center, commercial building, or industrial plant. { ˈkəs·tə·mər ˈsəbˌstā·shən }

custom millwork See architectural millwork. { ˈkəs·təm ˈmilˌwərk }

cusum chart See cumulative sum chart. { ˈkyüˌsəm ˌchärt }

cut [BIOCHEM] A double-strand incision in a duplex deoxyribonucleic acid molecule. [CHEM ENG] A fraction obtained by a separation process. [CRYSTAL] A section of a crystal having two parallel major surfaces; cuts are specified by their orientation with respect to the axes of the natural crystal, such as X cut, Y cut, BT cut, and AT cut. [GRAPHICS] A photoengraving used in letterpress printing. [LAP] The style in which a gem is cut, such as brilliant cut, single cut, or rose cut. [MATH] 1. A subset of a given set whose removal from the original set leaves a set that is not connected. 2. See fraction. [MIN ENG] 1. To intersect a vein or a working. 2. To excavate coal. 3. To shear one side of an entry or a crosscut by digging out the coal from floor to roof with a pick. [MOL BIO] A double-strand incision in a duplex deoxyribonucleic acid molecule. [NUCLEO] The fraction that is removed as product or advanced to the next separative element in an isotope separation process. [TEXT] The number of needles per inch in the cylinder or needle bed in a knitting frame. { kət }

cut-and-carry method [MET] A die-fabricating method in which the part remains attached to the strip or is forced back into the strip to be fed through the succeeding stations of a progressive die. { ¦kət ən 'kar·ē ˌmeth·əd }

cut and fill [CIV ENG] Construction of a road, a railway, or a canal which is partly embanked and partly below ground. [GEOL] **1.** Lateral corrosion of one side of a meander accompanied by deposition on the other. **2.** A sedimentary structure consisting of a small filled-in channel. { ¦kət ən 'fil }

cut and paste [COMPUT SCI] An editing function of a word processing system in which a portion of text is marked with a particular character at the beginning and at the end and is then copied to another location within the text. Also known as block move. { ¦kət ən 'pāst }

cut-and-paste transposition [CELL MOL] A form of deoxyribonucleic acid (DNA) transposition in which the transposed segment is cut from the donor DNA and inserted in the receptor DNA location. { ¦kət ən ¦pāst ˌtranz·pə'zish·en }

cutaneous anaphylaxis [IMMUNOL] Hypersensitivity that is marked by an intense skin reaction following parenteral contact with a sensitizing agent. { kyü'tā·nē·əs an·ə·fə'lak·səs }

cutaneous anthrax *See* malignant pustule. { kyü'tā·nē·əs 'an,thraks }

cutaneous appendage [ANAT] Any of the epidermal derivatives, including the nails, hair, sebaceous glands, mammary glands, and sweat glands. { kyü'tā·nē·əs ə'pen·dij }

cutaneous blastomycosis [MED] A form of North American blastomycosis considered by some to be a clinical manifestation of the systemic form. { kyü'tā·nē·əs ¦blas·tō·mī'kō·səs }

cutaneous coccidioidomycosis [MED] A primary skin infection by the fungus *Coccidioides immitis;* a skin infection secondary to a pulmonary lesion. { kyü'tā·nē·əs käk¦sid·ē¦oid·ō·mī'kō·səs }

cutaneous leishmaniasis [MED] A parasitic skin infection by *Leishmania tropica* characterized by deep ulcers of the skin and subcutaneous tissue. { kyü'tā·nē·əs lēsh·mə'nī·ə·səs }

cutaneous mycosis [MED] Any of a group of infections (collectively known as dermatophytoses, ringworms, or tineas) that are caused by keratinophilic fungi (dermatophytes). In general, the infections are limited to the nonliving keratinized layers of skin, hair, and nails, but a variety of pathologic changes can occur depending on the etiologic agent, site of infection, and immune status of the host. { kyü¦tān·ē·əs mī'kō·səs }

cutaneous pain [PHYSIO] A sensation of pain arising from the skin. { kyü'tā·nē·əs 'pān }

cutaneous reaction [MED] **1.** Any change in the outer layers of the skin, as in sunburn or the rash in measles. **2.** Any immediate or delayed immune reaction in the skin resulting from antigen-antibody interaction. { kyü'tā·nē·əs rē'ak·shən }

cutaneous sensation [PHYSIO] Any feeling originating in sensory nerve endings of the skin, including pressure, warmth, cold, and pain. { kyü'tā·nē·əs sen'sā·shən }

cutaway [GRAPHICS] An illustration of an object with a part of its covering or surface removed to show its interior construction or movement. { 'kəd·ə,wā }

cutback [CHEM ENG] Blending of heavier oils with lighter ones to bring the heavier to desired specifications. { 'kət,bak }

cutback asphalt [MATER] Asphalt which has been softened or liquefied by blending with petroleum distillates. { 'kət,bak ,as,fölt }

cutbank [GEOL] The concave bank of a winding stream that is maintained as a steep or even overhanging cliff by the action of water at its base. { 'kət,baŋk }

cut capacity [MATH] For a network whose points have been partitioned into two specified classes, C_1 and C_2, the sum of the capacities of all the segments directed from a point in C_1 to a point in C_2. Also known as cut value. { 'kət kə'pas·əd·ē }

cutch [MATER] Tannin extracted from mangrove bark. { kəch }

cut constraint [SYS ENG] A condition sometimes imposed in an integer programming problem which excludes parts of the feasible solution space without excluding any integer points. { 'kət kən'stränt }

CUT emulation *See* control unit terminal emulation. { 'kət ,em·yə,lā·shən }

Cuterebridae [INV ZOO] The robust botflies, a family of myodarian cyclorrhaphous dipteran insects in the subsection Calypteratae. { kyü·də'reb·rə,dē }

cut form [COMPUT SCI] In optical character recognition, any document form, receipt, or such, of standard dimensions which must be issued a separate read command in order to be recognized. { 'kət ¦förm }

cut glass [MATER] Flint glass ornamented with patterns cut into its surface. { ¦kət ¦glas }

cuticle [ANAT] The horny layer of the nail fold attached to the nail plate at its margin. [BIOL] A noncellular, hardened or membranous secretion of an epithelial sheet, such as the integument of nematodes and annelids, the exoskeleton of arthropods, and the continuous film of cutin on certain plant parts. { 'kyüd·ə·kəl }

cutie pie [NUCLEO] A radiation dose-rate meter having a pistol grip, a plastic cylinder or barrel containing an ionization chamber, and an indicating meter mounted above the grip. { 'kyüd·ē ,pī }

cut in [NAV] To observe and plot lines of position locating an object or craft, particularly by bearings. { 'kət ,in }

cut-in [CONT SYS] A value of temperature or pressure at which a control circuit closes. [ELEC] An electrical device that allows current to flow through an electric circuit. { 'kət ,in }

cutin [BIOCHEM] A mixture of fatty substances characteristically found in epidermal cell walls and in the cuticle of plant leaves and stems. { 'kyüt·ən }

cut-in angle [ELECTR] The phase angle at which a semiconductor diode begins to conduct; it is slightly greater than 0° because the diode requires some forward bias to conduct. { 'kət ,in ,aŋ·gəl }

cutinite [GEOL] A variety of exinite consisting of plant cuticles. { 'kyüt·ən,īt }

cutis *See* dermis. { 'kyüd·əs }

Cutler feed [ELECTROMAG] A resonant cavity that transfers radio-frequency energy from the end of a waveguide to the reflector of a radar spinner assembly. { 'kət·lər ,fēd }

cut methods [SYS ENG] Methods of solving integer programming problems that employ cut constraints derived from the original problem. { 'kət ,meth·əds }

cut nail [DES ENG] A flat, tapered nail sheared from steel plate; it has greater holding power than a wire nail and is generally used for fastening flooring. { 'kət ,nāl }

cutoff [AERO ENG] The shutting off of the propellant flow in a rocket, or the stopping of the combustion of the propellant. [CIV ENG] **1.** A channel constructed to straighten a stream or to bypass large bends, thereby relieving an area normally subjected to flooding or channel erosion. **2.** An impermeable wall, collar, or other structure placed beneath the base or within the abutments of a dam to prevent or reduce losses by seepage along otherwise smooth surfaces or through porous strata. [ELECTR] **1.** The minimum value of bias voltage, for a given combination of supply voltages, that just stops output current in an electron tube, transistor, or other active device. **2.** *See* cutoff frequency. [ENG] **1.** A misfire in a round of shots because of severance of fuse owing to rock shear as adjacent charges explode. **2.** The line on a plastic object formed by the meeting of the two halves of a compression mold. Also known as flash groove; pinch-off. [GEOL] A new, relatively short channel formed when a stream cuts through the neck of an oxbow or horseshoe bend. [MECH ENG] **1.** The shutting off of the working fluid to an engine cylinder. **2.** The time required for this process. [MIN ENG] **1.** A quarryman's term for the direction along which the granite must be channeled, because it will not split. **2.** The number of feet a bit may be used in a particular type of rock (as specified by the drill foreman). **3.** Minimum percentage of mineral in an ore that can be mined profitably. [PHYS] Technique used when the contribution to the value of a physical quantity given by integration over a certain variable is absurd (in particular, when the contribution is infinite); involves cutting off the integral at some limit. { 'kət,öf }

cutoff attenuator [ELECTROMAG] Variable length of waveguide used below its cutoff frequency to introduce variable nondissipative attenuation. { 'kət,öf ə'ten·yə,wād·ər }

cutoff bias [ELECTR] The direct-current bias voltage that must be applied to the grid of an electron tube to stop the flow of anode current. { 'kət,öf ,bī·əs }

cutoff field *See* critical field. { 'kət,öf ,fēld }

cutoff frequency [ELECTR] A frequency at which the attenuation of a device begins to increase sharply, such as the limiting frequency below which a traveling wave in a given mode cannot be maintained in a waveguide, or the frequency above which an electron tube loses efficiency rapidly. Also known as critical frequency; cutoff. { 'kət₁óf ₁frē·kwən·sē }

cutoff high [METEOROL] A warm high which has become displaced out of the basic westerly current, and lies to the north of this current. { 'kət₁óf ₁hī }

cutoff lake *See* oxbow lake. { 'kət₁óf ₁lāk }

cutoff limiting [ELECTR] Limiting the maximum output voltage of a vacuum tube circuit by driving the grid beyond cutoff. { 'kət₁óf ₁lim·əd·iŋ }

cutoff low [METEOROL] A cold low which has become displaced out of the basic westerly current, and lies to the south of this current. { 'kət₁óf ₁lō }

cutoff point [MECH ENG] 1. The point at which there is a transition from spiral flow in the housing of a centrifugal fan to straight-line flow in the connected duct. 2. The point on the stroke of a steam engine where admission of steam is stopped. { 'kət₁óf ₁póint }

cutoff tool [MECH ENG] A tool used on bar-type lathes to separate the finished piece from the bar stock. { 'kət₁óf ₁tül }

cutoff trench [CIV ENG] A trench which is below the foundation base line of a dam or other structure and is filled with an impervious material, such as clay or concrete, to form a watertight barrier. { 'kət₁óf ₁trench }

cutoff valve [MECH ENG] A valve used to stop the flow of steam to the cylinder of a steam engine. { 'kət₁óf ₁valv }

cutoff voltage [ELECTR] 1. The electrode voltage value that reduces the dependent variable of an electron-tube characteristic to a specified low value. 2. *See* critical voltage. { 'kət₁óf ₁vól·tij }

cutoff wall [CIV ENG] A thin, watertight wall of clay or concrete built up from a cutoff trench to reduce seepage. Also known as core wall. { 'kət₁óf ₁wól }

cutoff wavelength [ELECTROMAG] 1. The ratio of the velocity of electromagnetic waves in free space to the cutoff frequency in a uniconductor waveguide. 2. The wavelength corresponding to the cutoff frequency. { 'kət₁óf 'wāv₁leŋkth }

cutoff wheel [MECH ENG] A thin wheel impregnated with an abrasive used for severing or cutting slots in a material or part. { 'kət₁óf ₁wēl }

cut oil [MATER] An oil which has been partially emulsified with water in the presence of air. { 'kət ₁óil }

cut-out [CONT SYS] A value of temperature or pressure at which a control circuit opens. { 'kət ₁aút }

cutout [ELEC] 1. Pairs brought out of a cable and terminated at some place other than at the end of the cable. 2. An electrical device that is used to interrupt the flow of current through any particular apparatus or instrument, either automatically or manually. Also known as electric cutout. [GEOL] *See* horseback. { 'kət₁aút }

cutout angle [ELECTR] The phase angle at which a semiconductor diode ceases to conduct; it is slightly less than 180° because the diode requires some forward bias to conduct. { 'kət₁aút ₁aŋ·gəl }

cutout box [ELEC] A fireproof cabinet or box with one or more hinged doors that contains fuses and switches for various leads in an electrical wiring system. Also known as fuse box. { 'kət₁aút ₁bäks }

cut over [FOR] To cut marketable timber. { 'kət 'ō·vər }

cutover [ENG] 1. To place equipment in active use. 2. The time when testing of equipment is completed and regular usage begins. { 'kət₁ō·vər }

cut plane [GRAPHICS] The intersection of a plane with a three-dimensional object. { 'kət ₁plān }

cut platform *See* wave-cut platform. { 'kət ₁plat₁fórm }

cut point [CHEM ENG] The boiling-temperature division between cuts of a crude oil or base stock. [MATH] A point in a component of a graph whose removal disconnects that component. Also known as articulation point. { 'kət ₁póint }

cutscore [ENG] A knife used in die-cutting processes, designed to cut just partway into the paper or board so that it can be folded. { 'kət₁skór }

cut-set [ELEC] A set of branches of a network such that the cutting of all the branches of the set increases the number of separate parts of the network, but the cutting of all the branches except one does not. { 'kət ₁set }

cut-sheet printer [COMPUT SCI] A printer designed to print on separate sheets of paper. { 'kət ₁shēt ₁print·ər }

cut shot [MIN ENG] A shot designed to bring down coal which has been sheared or opened on one side. { 'kət ₁shät }

cut-signal-branch operation [ELECTR] In systems where radio reception continues without cutting off the carrier, the cut-signal-branch operation technique disables a signal branch in one direction when it is enabled in the other to preclude unwanted signal reflections. { ¦kət ¦sig·nəl ¦branch ₁äp·ə₁rā·shən }

cutter [ENG ACOUS] 1. An electromagnetic or piezoelectric device that converts an electric input to a mechanical output, used to drive the stylus that cuts a wavy groove in the highly polished wax surface of a recording disk. Also known as cutting head; head; phonograph cutter; recording head. 2. *See* cutting tool. [MIN ENG] 1. An operator of a coal-cutting or rock-cutting machine, or a worker engaged in underholing by pick or drill. 2. A joint, usually a dip joint, running in the direction of working; usually in the plural. { 'kəd·ər }

cutter bar [MECH ENG] The bar that supports the cutting tool in a lathe or other machine. { 'kəd·ər ₁bär }

cutter compensation [CONT SYS] The process of taking into account the difference in radius between a cutting tool and a programmed numerical control operation in order to achieve accuracy. { 'kəd·ər ₁käm·pən'sā·shən }

cutterhead [MECH ENG] A device on a machine tool for holding a cutting tool. { 'kəd·ər₁hed }

cutter sweep [MECH ENG] The section that is cut off or eradicated by the milling cutter or grinding wheel in entering or leaving the flute. { 'kəd·ər ₁swēp }

cutting [BOT] A piece of plant stem with one or more nodes, which, when placed under suitable conditions, will produce roots and shoots resulting in a complete plant. { 'kəd·iŋ }

cutting angle [MECH ENG] The angle that the cutting face of a tool makes with the work surface back of the tool. { 'kəd·iŋ ₁aŋ·gəl }

cutting coulter *See* colter. { 'kəd·iŋ ₁kōl·tər }

cutting down [MECH ENG] Removing surface roughness or irregularities from metal by the use of an abrasive. { 'kəd·iŋ 'daún }

cutting drilling [MECH ENG] A rotary drilling method in which drilling occurs through the action of the drill steel rotating while pressed against the rock. { 'kəd·iŋ ₁dril·iŋ }

cutting edge [DES ENG] 1. The point or edge of a diamond or other material set in a drill bit. Also known as cutting point. 2. The edge of a lathe tool in contact with the work during a machining operation. { 'kəd·iŋ 'ej }

cutting fluid [MATER] A fluid flowed over the tool and work in metal cutting to reduce heat generated by friction, lubricate, prevent rust, and flush away chips. { 'kəd·iŋ ₁flü·əd }

cutting head *See* cutter. { 'kəd·iŋ ₁hed }

cutting in [MECH ENG] An undesirable action occurring during loose-drum spooling in which a layer of wire rope spreads apart and forms grooves in which the next layer travels. { 'kəd·iŋ 'in }

cutting machine [MIN ENG] A power-driven apparatus used to undercut or shear the coal to help in its removal from the face. { 'kəd·iŋ mə'shēn }

cutting-off machine [MECH ENG] A machine for cutting off metal bars and shapes; includes the lathe type using single-point cutoff tools, and several types of saws. { 'kəd·iŋ ₁óf mə'shēn }

cutting-off process [METEOROL] A sequence of events by which a warm high or cold low, originally within the westerlies, becomes displaced either poleward (cutoff high) or equatorward (cutoff low) out of the westerly current; this process is evident at very high levels in the atmosphere, and it frequently produces, or is part of the production of, a blocking situation. { 'kəd·iŋ ₁óf ₁präs·əs }

cutting oil [MATER] A type of cutting fluid used in machining metals to lubricate the tool and workpiece, reducing tool wear, increasing cutting speeds, and decreasing power needs; there are two types: active and inactive. { 'kəd·iŋ ₁óil }

cutting pliers [DES ENG] Pliers with cutting blades on the jaws. { 'kəd·iŋ ₁plī·ərz }

cutting point *See* cutting edge. { 'kəd·iŋ ₁póint }

cutting process [MET] A process where metal is severed by the application of a gas or an electric arc. { 'kəd·iŋ ,präs·əs }

cutting ratio [ENG] As applied to metal cutting, the ratio of depth of cut to chip thickness for a given shear angle. { 'kəd·iŋ ,rā·shō }

cutting rule [ENG] A sharp steel rule used in a machine for cutting paper or cardboard. { 'kəd·iŋ ,rül }

cuttings [MIN ENG] Rock fragments broken from the penetrated rock during drilling operations. { kəd·iŋz }

cutting speed [MECH ENG] The speed of relative motion between the tool and workpiece in the main direction of cutting. Also known as feed rate; peripheral speed. { 'kəd·iŋ ,spēd }

cutting stylus [ENG ACOUS] A recording stylus with a sharpened tip that removes material to produce a groove in the recording medium. { 'kəd·iŋ ,stī·ləs }

cutting tip [ENG] The end of the snout of a cutting torch from which gas flows. { 'kəd·iŋ ,tip }

cutting tool [MECH ENG] The part of a machine tool which comes into contact with and removes material from the workpiece by the use of a cutting medium. Also known as cutter. { 'kəd·iŋ ,tül }

cutting torch [ENG] A torch that preheats metal while the surface is rapidly oxidized by a jet of oxygen issuing through the flame from an additional feed line. { 'kəd·iŋ ,tórch }

cuttlefish [INV ZOO] An Old World decapod mollusk of the genus *Sepia;* shells are used to manufacture dentifrices and cosmetics. { 'kəd·əl,fish }

cut value *See* cut capacity. { 'kət ,val·yü }

cutwater [CIV ENG] A sharp-edged structure built around a bridge pier to protect it from the flow of water and material carried by the water. { 'kət,wód·ər }

Cuvieroninae [PALEON] A subfamily of extinct proboscidean mammals in the family Gomphotheriidae. { küv·yə'rän·ə,nē }

C value [MOL BIO] The amount (mass or molecular weight) of deoxyribonucleic acid per haploid cell. { 'sē ,val·yü }

C value paradox [MOL BIO] The observation that the amount of deoxyribonucleic acid in the haploid genome is not related to its evolutionary complexity. { ¦sē ,val·yü 'par·ə,däks }

CVC *See* conserved vector current.

CVD *See* chemical vapor deposition.

CVn *See* Canes Venatici.

CW *See* continuous wave.

C wave *See* coupled wave. { 'sē ,wāv }

CW laser *See* continuous-wave laser. { ¦sē¦dəb·əl,yü 'lā·zər }

cwm *See* cirque. { küm }

cwt *See* hundredweight.

Cyamidae [INV ZOO] The whale lice, a family of amphipod crustaceans in the suborder Caprellidea that bear a resemblance to insect lice. { sī'am·ə,dē }

cyanalcohol *See* cyanohydrin. { ,sī·ən'al·kə,hól }

cyanamide [INORG CHEM] NHCNH An acidic compound that forms colorless needles, melting at 46°C, soluble in water. Also known as urea anhydride. { sī'an·ə,mid }

cyanate [INORG CHEM] A salt or ester of cyanic acid containing the radical CNO. { 'sī·ə,nāt }

cyanazine [ORG CHEM] $C_9H_{13}N_6Cl$ A white solid with a melting point of 166.5–167°C; used as a pre- and postemergence herbicide for corn, sorghum, soybeans, alfalfa, cotton, and wheat. { sī'an·ə,zēn }

cyanic acid [ORG CHEM] HCNO A colorless, poisonous liquid, which polymerizes to cyamelide and fulminic acid. { sī'an·ik 'as·əd }

cyanidation [CHEM] Joining of cyanide to an atom or molecule. [MET] *See* cyanide process. { ,si·ə·nə'dā·shən }

cyanide [INORG CHEM] Any of a group of compounds containing the CN group and derived from hydrogen cyanide, HCN. { 'sī·ə,nīd }

cyanide copper [MET] **1.** An electrolytic solution containing a complex of copper and the cyanide radical. **2.** Copper electrodeposited from the solution. { 'sī·ə,nīd 'käp·ər }

cyanide process [MET] Process of dissolving powdered gold and silver ores in a weak solution of sodium cyanide or potassium cyanide; the precious metals are precipitated from solution by zinc. Also known as cyanidation. { 'sī·ə,nīd ,präs·əs }

cyanide pulp [MET] The mixture resulting from grinding of gold and silver ore, then dissolving out the precious-metal content in a solution of sodium cyanide. { 'sī·ə,nīd ,pəlp }

cyanide slime [MET] Minute particles of precious metals precipitated from cyanide solutions used in extracting the metals from ore. { 'sī·ə,nīd ,slīm }

cyaniding [MET] Introduction of carbon and nitrogen simultaneously into a ferrous alloy by heating while in contact with molten cyanide; usually followed by quenching to produce a hardened case. { 'sī·ə,nīd·iŋ }

cyanine dye [ORG CHEM] $C_{29}H_{35}N_2I$ Green metallic crystals, soluble in water; unstable to light, the dye is used in the photography industry as a chemical sensitizer for film. Also known as iodocyanin; quinoline blue. { 'sī·ə·nən ,dī }

cyanite *See* kyanite. { 'sī·ə,nīt }

cyano- [CHEM] Combining form indicating the radical CN. { 'sī·ə·nō }

cyanoacetamide [ORG CHEM] $C_3H_4N_2O$ Needlelike crystals with a melting point of 119.5°C; soluble in water; used in organic synthesis. Also known as malonamide nitrile. { ¦sī·ə·nō·ə'sed·ə·mīd }

cyanoacetic acid [ORG CHEM] NCCH₂COOH Hygroscopic crystals with a melting point of 66°C; decomposes at 160°C; soluble in ether, water, and alcohol; used in the synthesis of intermediates and in the commercial preparation of barbital. Also known as malonic mononitrile. { ¦sī·ə·nō·ə'sēd·ik 'as·əd }

cyanoacrylate adhesive [MATER] An adhesive having a base of an alkyl 2-cyanoacrylate compound and characterized by excellent polymerizing and bonding properties; used for rubber printing plates, tools, and rubber swimming masks. { ¦sī·ə·nō'ak·rə,lāt ad'hē·ziv }

cyanobacteria [MICROBIO] A group of one-celled to many-celled aquatic organisms. Also known as blue-green algae. { ¦sī·ə·no,bak'tir·ē·ə }

cyanocarbon [ORG CHEM] A derivative of hydrocarbon in which all of the hydrogen atoms are replaced by the CN group. { ¦sī·ə·nō'kär·bən }

cyanochroite [MINERAL] $K_2Cu(SO_4)_2·6H_2O$ A blue mineral consisting of a hydrous sulfate of potassium and copper. { ,sī·ə'nä·krə,wīt }

cyanocobalamin *See* vitamin B_{12}. { ¦sī·ə·nō·kō'bal·ə·mən }

cyano complex [CHEM] A coordination compound containing the CN group. { sī'an·ō 'käm,pleks }

cyanoethylation [ORG CHEM] A chemical reaction involving the addition of acrylonitrile to compounds with a reactive hydrogen. [TEXT] Treating cotton fibers with acrylonitrile and caustic soda to improve dyeing, rot-resistance, and strength characteristics. { ¦sī·ə·nō,e·thə'lā·shən }

2-cyanoethanol *See* ethylene cyanohydrin. { ¦tü ¦sī·ə·nō'eth·ə,nól }

cyanogen [CHEM] A univalent radical, CN. [INORG CHEM] C_2N_2 A colorless, highly toxic gas with a pungent odor; a starting material for the production of complex thiocyanates used as insecticides. Also known as dicyanogen. { sī'an·ə·jən }

cyanogen absorption [ASTROPHYS] Bands in the absorption spectra of stars at wavelengths near 418 nanometers, caused by atmospheric cyanogen; used as a measure of absolute stellar magnitude. { sī'an·ə·jən əb'sórp·shən }

cyanogen bromide [INORG CHEM] CNBr White crystals melting at 52°C, vaporizing at 61.3°C, and having toxic fumes that affect nerve centers; used in the synthesis of organic compounds and as a fumigant. { sī'an·ə·jən 'brō,mīd }

cyanogen chloride [INORG CHEM] ClCN A poisonous, colorless gas or liquid, soluble in water; used in organic synthesis. { sī'an·ə·jən 'klór,īd }

cyanogen fluoride [INORG CHEM] CNF A toxic, colorless gas, used as a tear gas. { sī'an·ə·jən 'flùr,īd }

cyanogen iodide *See* iodine cyanide. { sī'an·ə·jən 'i·ə,dīd }

cyanohydrin [ORG CHEM] A compound containing the radicals CN and OH. Also known as cyanalcohol. { 'sī·ə·nō'hī·drən }

cyanometer [OPTICS] An instrument designed to measure or estimate the degree of blueness of light, as of the sky. { sī·ə'näm·əd·ər }

cyanometry [OPTICS] The study and measurement of the blueness of light. { ,sī·ə'nam·ə·trē }

cyanophage [VIROL] A virus that replicates in blue-green algae. Also known as algophage; blue-green algal virus. { sī'an·ə,fāj }

cyanophilous [BIOL] Having an affinity for blue or green dyes. { ¦sī·ə¦näf·ə·ləs }

cyanophosphos [ORG CHEM] C₁₅H₁₄NO₂PS A white, crystalline solid with a melting point of 83°C; used as an insecticide to control larval pests on rice and vegetables. { ‚sī·ə·nō'fäs‚fōs }

Cyanophyceae [BOT] A class of photosynthetic monerans distinguished by their algalike biology and bacteriumlike cell organization. { ‚sī·ə·nō'fīs·ē‚ē }

cyanophycin [BIOCHEM] A granular protein food reserve in the cells of blue-green algae, especially in the peripheral cytoplasm. { ‚sī·ə·nō'fīs·ən }

Cyanophyta [BOT] An equivalent name for the Cyanophyceae. { ‚sī·ə'näf·ə·də }

cyanoplatinate See platinocyanide. { ¦sī·ə·nō'plat·ən‚āt }

cyanosis [MED] A bluish coloration in the skin and mucous membranes due to deficient levels of oxygen in the blood. { ‚sī·ə'nō·səs }

cyanotrichite [MINERAL] Cu₄Al₂(SO₄)(OH)₁₂·2H₂O A bright-blue or sky-blue mineral consisting of a hydrous basic copper aluminum sulfate. { ‚sī·ə'nä·trə‚kīt }

cyanuric acid [ORG CHEM] HOC(NCOH)₂N·2H₂O Colorless, monoclinic crystals, slightly soluble in water; formed by polymerization of cyanic acid. Also known as pyrolithic acid. { ¦sī·ə¦nůr·ik 'as·əd }

Cyatheaceae [BOT] A family of tropical and pantropical tree ferns distinguished by the location of sori along the veins. { sī·‚ath·ə'ās·ē‚ē }

cyathium [BOT] An inflorescence in which the flowers arise from the base of a cuplike involucre. { sī'ath·ē·əm }

Cyathoceridae [INV ZOO] The equivalent name for the Lepiceridae. { sī‚ath·ə'räs·ə‚dē }

Cyatholaimoidea [INV ZOO] A superfamily of nematodes of the order Chromadorida, distinguished by tightly coiled multispiral amphids located a short distance posterior to the cephalic sensilla. { ‚sī·ə·thō·lə'móid·ē·ə }

Cybele [ASTRON] An asteroid with a diameter of about 167 miles (269 kilometers), mean distance from the sun of 3.423 astronomical units, and C-type surface composition. { 'sib·ə·lē }

cybernation [IND ENG] The use of computers in connection with automation. { sī·bər'nā·shən }

cybernetics [SCI TECH] **1.** The science of control and communication in all of their manifestations within and between machines, animals, and organizations. **2.** Specifically, the interaction between automatic control and living organisms, especially humans and animals. { sī·bər'ned·iks }

cyberspace [COMPUT SCI] The digital realms, including Web sites and virtual worlds. { 'sī·bər‚spās }

cybotaxis [PHYS] A transient molecular orientation in a liquid evidenced by x-ray diffraction effects. { ‚sī·bə'tak·səs }

cybrid [GEN] A hybrid cell produced by fusing a cell nucleus with a cell of the same or a different species whose nucleus has been removed; some are able to proliferate. In plants, an individual produced following fusion of protoplasts from different species with complete elimination of the chromosomes of one of the species. { 'sī·brəd }

Cycadales [BOT] An ancient order of plants in the class Cycadopsida characterized by tuberous or columnar stems that bear a crown of large, usually pinnate leaves. { ‚sī·kə'dā·lēz }

Cycadatae See Cycadopsida. { sī'kad·ə‚dē }

Cycadeoidaceae [PALEOBOT] A family of extinct plants in the order Cycadeoidales characterized by sparsely branched trunks and a terminal crown of leaves. { sī‚kad·ē·óid'ās·ē‚ē }

Cycadeoidales [PALEOBOT] An order of extinct plants that were abundant during the Triassic, Jurassic, and Cretaceous periods. { sī‚kad·ē·óid'ā·lēz }

Cycadicae [BOT] A subdivision of large-leaved gymnosperms with stout stems in the plant division Pinophyta; only a few species are extant. { sī'kad·ə‚sē }

Cycadofilicales [PALEOBOT] The equivalent name for the extinct Pteridospermae. { ¦sī·kə·dō‚fil·ə'kā·lēz }

Cycadophyta [BOT] An equivalent name for Cycadecae elevated to the level of a division. { sī·kə'däf·əd·ə }

Cycadophytae [BOT] An equivalent name for Cycadicae. { sī·kə'däf·ə‚tē }

Cycadopsida [BOT] A class of gymnosperms in the plant subdivision Cycadicae. { sī·kə'däp·sə·də }

cyclamate [ORG CHEM] The calcium or sodium salt of cyclohexylsulfamate, an artificial sweetener. { 'sī·klə‚māt }

cyclane See alicyclic. { 'sī‚klān }

Cyclanthaceae [BOT] The single family of the order Cyclanthales. { ‚sī‚klan'thās·ē‚ē }

Cyclanthales [BOT] An order of monocotyledonous plants composed of herbs; or, seldom, composed of more or less woody plants with leaves that usually have a bifid, expanded blade. { ‚sī‚klan'thā·lēz }

cyclase [BIOCHEM] An enzyme that catalyzes cyclization of a compound. { 'sī‚klās }

cycle [ENG] To run a machine through a single complete operation. [FL MECH] A system of phases through which the working substance passes in an engine, compressor, pump, turbine, power plant, or refrigeration system. [MATH] **1.** A member of the kernel of a boundary homomorphism. **2.** A closed path in a graph that does not pass through any vertex more than once and passes through at least three vertices. Also known as circuit. **3.** See cyclic permutation. [SCI TECH] **1.** One complete sequence of values of an alternating quantity. **2.** A set of operations that is repeated as a unit. [STAT] A periodic movement in a time series. { 'sī·kəl }

cycle annealing [MET] Annealing at a controlled time-temperature cycle to achieve a specific microstructure. { 'sī·kəl ə'nēl·iŋ }

cycle-bound program See CPU-bound program. { 'sī·kəl ¦baůnd 'prō·grəm }

cycle checkout [AERO ENG] The periodic action of carrying out a complete series of operational and calibrational tests on missiles held in alert status. { 'sī·kəl 'chek‚aůt }

cycle count [COMPUT SCI] The operation of keeping track of the number of cycles a computer system goes through during processing time. { 'sī·kəl ‚kaůnt }

cycle criterion [COMPUT SCI] Total number of times a cycle in a computer program is to be repeated. { 'sī·kəl krī'tir·ē·ən }

cycle delay selector [COMPUT SCI] An electromechanical device in a sorter which causes a cycle to be skipped so that the card out of sequence may be directed to a different pocket. { 'sī·kəl di¦lā sə'lek·tər }

cycle gas [PETRO ENG] Gas returned to the reservoir after being compressed to minimize the drop in reservoir pressure. { 'sī·kəl ‚gas }

cyclegraph technique [IND ENG] Recording a brief work cycle by attaching small lights to various parts of a worker and then exposing the work motions on a still-film time plate; motion will appear on the plate as superimposed streaks of light constituting a cyclegraph. { 'sī·klə‚graf ‚tek‚nēk }

cycle index [COMPUT SCI] **1.** The number of times a cycle has been carried out by a computer. **2.** The difference, or its negative, between the number of executions of a cycle which are desired and the number which have actually been carried out. { 'sī·kəl ‚in‚deks }

cycle index counter [COMPUT SCI] A device that counts the number of times a given cycle of instructions in a computer program has been carried out. { 'sī·kəl ‚in‚deks ‚kaůnt·ər }

cycle-matching loran See low-frequency loran. { 'sī·kəl ‚mach·iŋ ‚ló'ran }

cycle of erosion See geomorphic cycle. { 'sī·kəl əv i'rō·zhən }

cycle of sedimentation [GEOL] **1.** Also known as sedimentary cycle. **2.** A series of related processes and conditions appearing repeatedly in the same sequence in a sedimentary deposit. **3.** The sediments deposited from the beginning of one cycle to the beginning of a second cycle of the spread of the sea over a land area, consisting of the original land sediments, followed by those deposited by shallow water, then deep water, and then the reverse process of the receding water. See cyclothem. { 'sī·kəl əv ‚sed·ə·mən'tā·shən }

cycle per minute [PHYS] A unit of frequency of action, equal to 1/60 hertz. Abbreviated cpm. { 'sī·kəl pər 'min·ət }

cycle per second See hertz. { 'sī·kəl pər 'sek·ənd }

cycle plant [CHEM ENG] A plant in which the liquid hydrocarbons are removed from natural gas and then the gas is put back into the earth to maintain pressure in the oil reservoir. { 'sī·kəl ‚plant }

cycle reset [COMPUT SCI] The resetting of a cycle index to its initial or other specified value. { 'sī·kəl 'rē‚set }

cycle skip See skip logging. { 'sī·kəl ‚skip }

CYCADALES

ovulate cones

Sago palm (*Cycas revoluta*). (*New York Botanical Garden*)

cycle stealing [COMPUT SCI] A technique for memory sharing whereby a memory may serve two autonomous masters, commonly a central processing unit and an input-output channel or device controller, and in effect provide service to each simultaneously. { 'sī·kəl ˌstēl·iŋ }

cycle stock [CHEM ENG] The unfinished product taken from a stage of a refinery process and recharged to the process at an earlier stage in the operation. { 'sī·kəl ˌstäk }

cyclethrin [ORG CHEM] $C_{21}H_{28}O_3$ A viscous, brown liquid, soluble in organic solvents; used as an insecticide. { sī'klē·thrən }

cycle time [COMPUT SCI] The shortest time elapsed between one store (or fetch) and the next store (or fetch) in the same memory unit. Also known as memory cycle. [PETRO ENG] In a drilling operation, the time needed for the pump to move the drilling fluid in a bore hole. [SCI TECH] The time required to carry out a sequence of activities repeated in each performance of an operation. { 'sī·kəl ˌtīm }

cycle timer [ELECTR] A timer that opens or closes circuits according to a predetermined schedule. { 'sī·kəl ˌtīm·ər }

cycle timing diagram [COMPUT SCI] A diagram showing the activity that occurs in each clock cycle of a computer during the execution of a machine-language instruction. { 'sī·kəl ˌtīm·iŋ ˌdī·ə,gram }

cyclic [SCI TECH] **1.** Pertaining to some cycle. **2.** Repeating itself in some manner in space or time. { 'sīk·lik }

cyclic adenylic acid [BIOCHEM] $C_{10}H_{12}N_5O_6P$ An isomer of adenylic acid; crystal platelets with a melting point of 219–220°C; a key regulator which acts to control the rate of a number of cellular processes in bacteria, most animals, and some higher plants. Abbreviated cAMP. Also known as adenosine 3',5'-cyclic monophosphate; adenosine 3',5'-cyclic phosphate; adenosine 3',5'-monophosphate; 3',5'-AMP; cyclic AMP. { 'sīk·lik ˌad·ənˌil·ik 'as·əd }

cyclic amide [ORG CHEM] An amide arranged in a ring of carbon atoms. { 'sīk·lik 'a,mīd }

cyclic AMP See cyclic adenylic acid. { 'sīk·lik ˈāˌemˈpē }

cyclic AMP-dependent protein kinase [BIOCHEM] A serine/threonine protein kinase that phosphorylates a variety of substrates and regulates many important processes such as cell growth and differentiation and the flow of ions across the cell membrane. Also known as protein kinase A; PKA. { 'sī·klik ˈāˌemˈpē diˈpen·dənt ˌpro,tēn 'kī,nās }

cyclic AMP-responsive element binding protein [MOL BIO] A deoxyribonucleic acid-binding transcription factor that becomes modified in response to an extracellular signal. Abbreviated CREB. { 'sī·klik ˈāˌemˈpē riˈspän·siv ˈel·ə·mənt ˈbīnd·iŋ 'prō,tēn }

cyclic anhydride [ORG CHEM] A ring compound formed by the removal of water from a compound; an example is phthalic anhydride. { 'sīk·lik an'hī,drīd }

cyclic catalytic reforming process [CHEM ENG] A method for the production of low-Btu reformed gas consisting of the conversion of carbureted water-gas sets by installing a bed of nickel catalyst in the superheater and using the carburetor as a combustion chamber and process steam superheater. Abbreviated CCR process. { 'sīk·lik ˈkäd·əˌlid·ik riˈfor·miŋ ˌpräs·əs }

cyclic chronopotentiometry [ANALY CHEM] An analytic electrochemical method in which instantaneous current reversal is imposed at the working electrode, and its potential is monitored with time. { 'sīk·lik ˌkrän·ō·pəˌten·chē'äm·ə·trē }

cyclic code [COMPUT SCI] A code, such as a binary code, that changes only in one digit when going from one number to the number immediately following, and in that digit by only one unit. { 'sīk·lik 'kōd }

cyclic coil See random coil. { 'sīk·lik 'koil }

cyclic compound [ORG CHEM] A compound that contains a ring of atoms. { 'sīk·lik 'käm,paund }

cyclic coordinate [MECH] A generalized coordinate on which the Lagrangian of a system does not depend explicitly. Also known as ignorable coordinate. { 'sīk·lik kō'órd·ən·ət }

cyclic currents See mesh currents. { 'sīk·lik ˌkər·ənts }

cyclic curve [MATH] **1.** A curve (such as a cycloid, cardioid, or epicycloid) generated by a point of a circle that rolls (without slipping) on a given curve. **2.** The intersection of a quadric surface with a sphere. Also known as spherical cyclic curve. **3.** The stereographic projection of a spherical cyclic curve. Also known as plane cyclic curve. { 'sīk·lik 'kərv }

cyclic element [IND ENG] An element of an operation or process that occurs in each of its cycles. { 'sīk·lik 'el·ə·mənt }

cyclic extension [MATH] A Galois extension whose Galois group is cyclic. { 'sīk·lik ik'sten·chən }

cyclic feeding [COMPUT SCI] In character recognition, a system employed by character readers in which each input document is issued to the document transport in a predetermined and constant period of time. { 'sīk·lik 'fēd·iŋ }

cyclic GMP [BIOCHEM] A 3',5'-cyclic ester of guanosine monophosphate that is involved in vision transduction through its direct effects on Na^+ and Ca^{2+} channels in the plasma membrane of rod cells. { ˈsī·klik ˈjēˌemˈpē }

cyclic graph [MATH] A graph whose vertices correspond to the vertices of a regular polygon and whose edges correspond to the sides of the polygon. { ˈsī·klik 'graf }

cyclic group [MATH] A group that has an element a such that any element in the group can be expressed in the form a^n, where n is an integer. { 'sīk·lik ˌgrüp }

cyclic identity [MATH] The principle that the sum of any component of the Riemann-Christoffel tensor and two other components obtained from it by cyclic permutation of any three indices, while the fourth is held fixed, is zero. { 'sīk·lik ī,den·təd·ē }

cyclic ion See bridged ion. { 'sīk·lik 'ī·ən }

cyclic left module [MATH] A left module over a ring A that has a member x such that any member of the module has the form ax, where a is a member of A. { 'sī·klik ˌleft 'mäj·əl }

cyclic magnetization [ELECTROMAG] A magnetizing force varying between two specific limits long enough so that the magnetic induction has the same value for corresponding points in successive cycles. { 'sīk·lik mag·nə·təˈzā·shən }

cyclic mining See conventional mining. { 'sīk·lik ˌmīn·iŋ }

cyclic nucleotide phosphodiesterase [BIOCHEM] Any of a group of enzymes that degrade cyclic nucleotides. { ˈsīk·lik ˌnü·klē·ə,tīd ˌfäs·fə·dīˈes·tə,rās }

cyclic permeability See normal permeability. { 'sīk·lik ˌpər·mē·əˈbil·əd·ē }

cyclic permutation [MATH] A permutation of an ordered set of symbols which sends the first to the second, the second to the third, . . . , the last to the first. Also known as cycle. { 'sīk·lik pər·myəˈtā·shən }

cyclic polygon [MATH] A polygon whose vertices are located on a common circle. { ˈsī·klik 'päl·iˌgän }

cyclic redundancy check [COMPUT SCI] A block check character in which each bit is calculated by adding the first bit of a specified byte to the second bit of the next byte, and so forth, spiraling through the block. { 'sīk·lik riˈdən·dən·sē ˌchek }

cyclic salt [OCEANOGR] Salt removed from the sea as spray, blown inland, and returned to its source by land drainage. { 'sīk·lik 'solt }

cyclic sedimentation [GEOL] Deposition of various kinds of sediment in a repeated regular sequence. { 'sīk·lik ˌsed·ə·mənˈtā·shən }

cyclic shift [COMPUT SCI] A computer shift in which the digits dropped off at one end of a word are returned at the other end of the word. Also known as circuit shift; circular shift; end-around shift; nonarithmetic shift; ring shift. { 'sīk·lik 'shift }

cyclic storage [COMPUT SCI] A computer storage device, such as a magnetic drum, whose storage medium is arranged in such a way that information can be read into or extracted from individual locations at only certain fixed times in a basic cycle. { 'sīk·lik 'stor·ij }

cyclic testing [ENG] The repeated testing of a device or system at regular intervals to be assured of its reliability. { 'sīk·lik 'test·iŋ }

cyclic train [MECH ENG] A set of gears, such as an epicyclic gear system, in which one or more of the gear axes rotates around a fixed axis. { 'sīk·lik 'trān }

cyclic transfer [COMPUT SCI] The automatic transfer of data from some medium to memory or from memory to some medium until all the data are read. { 'sīk·lik 'tranz·fər }

cyclic twinning [CRYSTAL] Repeated twinning of three or more individuals in accordance with the same twinning law but without parallel twinning axes. { 'sīk·lik 'twin·iŋ }

cyclic voltammetry [PHYS CHEM] An electrochemical technique for studying variable potential at an electrode involving application of a triangular potential sweep, allowing one to sweep back through the potential region just covered. { 'sīk·lik vōl'täm·ə·trē }

cyclin [CELL MOL] Any member of a family of proteins that regulate the cell cycle and whose cellular levels rise steadily until mitosis, then fall abruptly to zero. As cyclins reach a threshold level, they are thought to drive cells into G2 phase and thus toward mitosis. { 'sī·klin }

cyclin-Cdk complex [CELL MOL] A complex of the regulatory protein cyclin with the catalytic protein Cdk that regulates the cell cycle by selectively phosphorylating various protein substrates at different stages of the cycle. { ¦sīk·lin ¦sē¦dē¦kā 'käm,pleks }

cyclin-dependent kinase [CELL MOL] A family of kinases that, once activated by cyclin, regulate the cell cycle by adding phosphate groups to a variety of protein substrates that control processes in the cycle. Abbreviated Cdk. { 'sī·klin di¦pen·dənt 'kī,nās }

cyclin-dependent kinase activating kinase [CELL MOL] A kinase that activates cyclin-dependent kinases via phosphorylation. { 'sī·klin di¦pen·dənt ¦kī,nās ¦ak·tə,vād·iŋ 'kī,nās }

cyclindroid [MATH] **1.** A cylindrical surface generated by the lines perpendicular to a plane that pass through an ellipse in the plane. **2.** A surface that is generated by a straight line that moves so as to intersect two curves and remain parallel to a given plane. { si'klin,dróid }

cycling [CHEM ENG] A series of operations in petroleum refining or natural-gas processing in which the steps are repeated periodically in the same sequence. [CONT SYS] A periodic change of the controlled variable from one value to another in an automatic control system. [PETRO ENG] An operation in which gas flowing from a gas reservoir is passed through a processing plant or separation system, with the gas remaining being returned to the reservoir. Also known as oscillation. { 'sīk·liŋ }

cyclitis [MED] Inflammation of the ciliary body of the eye. { sə'klīd·əs }

cyclitol [ORG CHEM] A cycloalkane that contains one hydroxyl group on each of three or more of the atoms constituting the ring. { 'sī·klə,tól }

cyclization [ORG CHEM] Changing an open-chain hydrocarbon to a closed ring. { ,sī·klə'zā·shən }

cyclized rubber [MATER] A thermoplastic, nonrubbery, tough or hard rubber derivative formed by the action of certain agents, such as sulfonic acid and chlorostannic acid, on rubber; used in paints and adhesives and for insulation. { 'sī,klīzd 'rəb·ər }

cyclizine hydrochloride [PHARM] $(C_6H_5)_2CHC_4H_8N_2CH_3 \cdot HCl$ A white, crystalline powder with a melting point of 285°C; used in medicine. { 'sī·klə,zēn ,hī·drə'klór,īd }

cycloaddition [ORG CHEM] A reaction in which unsaturated molecules combine to form a cyclic compound. { ¦sī·klō·ə'dish·ən }

cycloaliphatic See alicyclic. { ¦sī·klō·al·ə'fad·ik }

cycloalkane See alicyclic. { ¦sī·klō'al,kān }

cycloalkene [ORG CHEM] An unsaturated, monocyclic hydrocarbon having the formula C_nH_{2n-2}. Also known as cycloolefin. { ¦sī·klō'al,kēn }

cycloalkylaryl compound [ORG CHEM] A compound with a multiringed molecular structure containing both aromatic and saturated rings. { ¦sī·klō¦al·kə'lar·əl 'käm,paúnd }

cycloalkyne [ORG CHEM] A cyclic compound containing one or more triple bonds between carbon atoms. { ¦sī·klō'al,kīn }

cycloamylose [BIOCHEM] A member of a group of cyclic oligomers of glucose in which the individual glucose units are connected by 1,4 bonds. Also known as cyclodextrin; Schardinger dextrin. { ¦sī·klō'am·ə,lōs }

cycloate [ORG CHEM] $C_{11}H_{21}NOS$ A yellow liquid with limited solubility in water; boiling point is 145–146°C; used as an herbicide to control weeds in sugarbeets, spinach, and table beets. { 'sī·klə,wāt }

cyclobarbital [PHARM] $C_{12}H_{16}N_2O_3$ A hypnotic and sedative of short duration. { ¦sī·klō'bär·bə,tól }

cyclobutadiene [ORG CHEM] C_4H_4 A cyclic compound containing two alternate double bonds; used in organic synthesis. Also known as butene. { ¦sī·klō,byüd·ə'dī,ēn }

cyclobutane [ORG CHEM] C_4H_8 An alicyclic hydrocarbon, boiling point 11°C; synthesized as a condensable gas; used in

organic synthesis. Also known as tetramethylene. { ¦sī·klō'byü,tān }

cyclobutene [ORG CHEM] C_4H_6 An asymmetrical cyclic hydrocarbon occurring in several isomeric forms. Also known as cyclobutylene. { ¦sī·klō'byü,tēn }

cycloconverter [ELEC] A device that produces an alternating current of constant or precisely controllable frequency from a variable-frequency alternating-current input, with the output frequency usually one-third or less of the input frequency. { ¦sī·klō·kən'vərd·ər }

Cyclocystoidea [PALEON] A class of small, disk-shaped, extinct echinozoans in which the lower surface of the body probably consisted of a suction cup. { ¦sī·klō·si'stóid·ē·ə }

cyclodextrin See cycloamylose. { ¦sī·klō'dek·strən }

cyclodialysis [MED] Detaching the ciliary body from the sclera in order to effect reduction of intraocular tension in certain cases of glaucoma, especially in aphakia. { ¦sī·klō·dī'al·ə·səs }

cyclodiathermy [MED] Destruction, by diathermy, of the ciliary body. { ¦sī·klō'dī·ə,thər·mē }

cyclodiolefin [ORG CHEM] A cycloalkene with two double bonds; sometimes included with alkenes, cycloalkenes, and hydrocarbons containing more than one ethylene bond as olefins in a generic sense. { ¦sī·klō·dī'ō·lə·fən }

cyclododecatriene [ORG CHEM] $C_{12}H_{18}$ One of two cyclic hydrocarbons with three double bonds; the two forms are stereoisomeric; used to make nylon-6 and nylon-12. { ¦sī·klō,dō·dek·ə'trī,ēn }

cycloelimination [ORG CHEM] The reverse of cycloaddition. Also known as cycloreversion. { ,sī·klō·i,lim·ə'nā·shən }

cyclogenesis [METEOROL] Any development or strengthening of cyclonic circulation in the atmosphere. { ¦sī·klō'jen·ə·səs }

cyclograph [ENG] An electronic instrument that produces on a cathode-ray screen a pattern which changes in shape according to core hardness, carbon content, case depth, and other metallurgical properties of a test sample of steel inserted in a sensing coil. { 'sī·klə,graf }

cycloheptaamylose [BIOCHEM] A cycloamylose with seven glucose units in a cyclic array. { ,sī·klə,hep·tə'am·ə,lōs }

cyclohexaamylose [BIOCHEM] A cycloamylose with six glucose units in a cyclic array. { ,sī·klə,hek·sə'am·ə,lōs }

1,3-cyclohexadiene [ORG CHEM] C_6H_8 A partly saturated benzene compound with two double bonds; used in organic synthesis. { ¦wən ¦thrē ¦sī·klō,hek·sə'dī,ēn }

cyclohexane [ORG CHEM] C_6H_{12} A colorless liquid that is a cyclic hydrocarbon synthesized by hydrogenation of benzene; used in organic synthesis. Also known as hexamethylene. { ¦sī·klō'hek,sān }

cyclohexanol [ORG CHEM] $C_6H_{11}OH$ An oily, colorless, hygroscopic liquid with a camphorlike odor and a boiling point of 160.9°C; used in soapmaking, insecticides, dry cleaning, plasticizers, and germicides. Also known as hexahydrophenol. { ¦sī·klō'hek·sə,nól }

cyclohexanone [ORG CHEM] $C_6H_{10}O$ An oily liquid with an odor suggesting peppermint and acetone; soluble in alcohol, ether, and other organic solvents; used as an industrial solvent, in the production of adipic acid, and in the preparation of cyclohexanone resins. { ¦sī·klō'hek·sə,nōn }

cyclohexene [ORG CHEM] C_6H_{10} A compound that occurs in coal tar; a liquid that is used as an alkylation component; used in the manufacture of hexahydrobenzoic acid, adipic acid, and maleic acid. { ¦sī·klō'hek,sēn }

cycloheximide [MICROBIO] $C_{15}H_{23}NO_4$ Colorless crystals with a melting point of 119.5–121°C; soluble in water, in amyl acetate, and in common organic solvents such as ether, acetone, and chloroform; used as an agricultural fungicide. { ¦sī·klō'hek·sə,mīd }

cyclohexylamine [ORG CHEM] $C_6H_{11}NH_2$ A liquid with a strong, fishy, amine odor; miscible with water and common organic solvents; used in organic synthesis and in the manufacture of plasticizers, rubber chemicals, corrosion inhibitors, dyestuffs, dry-cleaning soaps, and emulsifying agents. { ¦sī·klō·hek'sil·ə,mēn }

cycloid [MATH] The curve traced by a point on the circumference of a circle as the circle rolls along a straight line. { 'sī,klóid }

cycloidal gear teeth [DES ENG] Gear teeth whose profile is

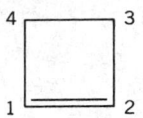

CYCLOBUTANE

Structural formula of cyclobutane.

CYCLOBUTENE

Structural formula of cyclobutene showing numbered carbons.

formed by the trace of a point on a circle rolling without slippage on the outside or inside of the pitch circle of a gear; now used only for clockwork and timer gears. { sī'klȯid·əl 'gir ,tēth }

cycloidal mass spectrometer [SPECT] Small mass spectrometer of limited mass range fitted with a special-type analyzer that generates a cycloidal-path beam of the sample mass. { sī'klȯid·əl 'mas spek'träm·əd·ər }

cycloidal pendulum [MECH] A modification of a simple pendulum in which a weight is suspended from a cord which is slung between two pieces of metal shaped in the form of cycloids; as the bob swings, the cord wraps and unwraps on the cycloids; the pendulum has a period that is independent of the amplitude of the swing. { sī'klȯid·əl 'pen·jə·ləm }

cycloidal propeller [NAV ARCH] A type of vertical-axis propeller; used especially on shallow-draft craft. { sī'klȯid·əl prə'pel·ər }

cycloidal wave [FL MECH] A very steep, symmetrical wave in the form of a cycloid whose crest forms an angle of 120°. { sī'klȯid·əl ,wāv }

cycloid scale [VERT ZOO] A thin, acellular structure which is composed of a bonelike substance and shows annual growth rings; found in the skin of soft-rayed fishes. { 'sī,klȯid ,skāl }

cyclolysis [METEOROL] The weakening or decay of cyclonic circulation in the atmosphere. { sī'kläl·ə·səs }

cyclomatic complexity [COMPUT SCI] A measure of the complexity of a software module, equal to $e - n + 2$, where e is the number of edges in the control flow graph and n is the number of nodes in this graph (that is, the cyclomatic number of the graph plus one). { 'st·klə,mad·ik kəm'plek·səd·ē }

cyclomatic number [MATH] For a graph, the number $e - n + 1$, where e is the number of edges and n is the number of nodes. { ¦sī·klə,mad·ik 'nəm·bər }

cyclomorphosis [ECOL] Cyclic recurrent polymorphism in certain planktonic fauna in response to seasonal temperature or salinity changes. { ,sī·klō'mȯr·fə·səs }

cyclone [CHEM ENG] A static reaction vessel in which fluids under pressure form a vortex. [MECH ENG] Any cone-shaped air-cleaning apparatus operated by centrifugal separation that is used in particle collecting and fine grinding operations. [METEOROL] A low-pressure region of the earth's atmosphere with roundish to elongated-oval ground plan, in-moving air currents, centrally upward air movement, and generally outward movement at various higher elevations in the troposphere. { 'sī,klōn }

cyclone cellar [CIV ENG] An underground shelter, often built in areas frequented by tornadoes. Also known as storm cellar; tornado cellar. { 'sī,klōn ,sel·ər }

cyclone classifier See cyclone separator. { 'sī,klōn ,klas·ə,fī·ər }

cyclone family [METEOROL] A series of wave cyclones occurring in the interval between two successive major outbreaks of polar air, and traveling along the polar front, usually eastward and poleward. { 'sī,klōn ,fam·lē }

cyclone furnace [ENG] A water-cooled, horizontal cylinder in which fuel is fired cyclonically and heat is released at extremely high rates. { 'sī,klōn ,fər·nəs }

cyclone separator [MECH ENG] A funnel-shaped device for removing particles from air or other fluids by centrifugal means; used to remove dust from air or other fluids, steam from water, and water from steam, and in certain applications to separate particles into two or more size classes. Also known as cyclone classifier. { 'sī,klōn 'sep·ə,rād·ər }

cyclone wave [METEOROL] **1.** A disturbance in the lower troposphere, of wavelength 1000–2500 kilometers; cyclone waves are recognized on synoptic charts as migratory high-and low-pressure systems. **2.** A frontal wave at the crest of which there is a center of cyclonic circulation, that is, the frontal wave of a wave cyclone. { 'sī,klōn ,wāv }

cyclonic [GEOPHYS] Having a sense of rotation about the local vertical that is the same as that of the earth's rotation: as viewed from above, counterclockwise in the Northern Hemisphere, clockwise in the Southern Hemisphere, undefined at the Equator. { sī'klän·ik }

cyclonic scale [METEOROL] The scale of the migratory high-and low-pressure systems (or cyclone waves) of the lower troposphere, with wavelengths of 1000–2500 kilometers. Also known as synoptic scale. { sī'klän·ik 'skāl }

cyclonic shear [METEOROL] Horizontal wind shear of such a nature that it contributes to the cyclonic vorticity of the flow; that is, it tends to produce cyclonic rotation of the individual air particles along the line of flow. { sī'klän·ik 'shir }

cyclonite [ORG CHEM] $(CH_2)_3N_3(NO_2)_3$ A white, crystalline explosive, consisting of hexahydro-trinitro-triazine, and having high sensitivity and brisance; mixed with other explosives or substances. Abbreviated RDX. Also known as cyclotrimethylenetrinitramine. { 'sī·klə,nīt }

1,5-cyclooctadiene [ORG CHEM] C_8H_{12} A cyclic hydrocarbon with two double bonds; prepared from butadiene and used to make cyclooctene and cyclooctane, which are intermediates for the production of plastics, fibers, and so on. { ¦wən ¦fīv ¦sī·klō,äk·tə'dī,ēn }

cyclooctane [ORG CHEM] $(CH_2)_8$ A cyclic alkane melting at 9.5°C; used as an intermediate in production of plastics, fibers, adhesives, and coatings. Also known as octomethylene. { ¦sī·klō'äk,tān }

cyclooctatetraene [ORG CHEM] C_8H_8 A cyclic olefin with alternate double bonds; highly reactive; rearranges to styrene. { ¦sī·klō,äk·tə'te·tra,ēn }

cycloolefin See cycloalkene. { ¦sī·klō'ō·lə·fən }

cyclooxygenase [BIOCHEM] An enzyme that catalyzes the conversion of arachidonic acid into prostaglandins. { ,sī·klō'äks·ə·jə,nās }

cycloparaffin See alicyclic. { ¦sī·klō'par·ə·fən }

cyclopean [MATER] Mass concrete with aggregate larger than 6 inches (15 centimeters); used for thick structures such as dams. [PETR] See mosaic. { ¦sī·klə¦pē·ən }

cyclopean stairs [GEOL] The landscape that results in a glacial trough after the ice has melted away, and that consists of an irregular series of rock steps, with steep cliffs on the down-valley side and small lakes in the shallow excavated depressions of the rock steps. { ,sī·klə¦pē·ən 'sterz }

1,3-cyclopentadiene [ORG CHEM] C_5H_6 A colorless liquid boiling at 41.5°C; used to make resins. { ¦wən ¦thrē ¦sī·klō,pen·tə'dī,ēn }

cyclopentadienyl anion [ORG CHEM] C_5H_5- A radical formed from cyclopentadiene. { ¦sī·klō,pen·tə,dī'e·nil 'an,ī·ən }

cyclopentane [ORG CHEM] C_5H_{10} A cyclic hydrocarbon that is a colorless liquid; present in crude petroleum, it is converted during refining to aromatics which improve antiknock and combustion properties of gasoline. { ¦sī·klō'pen,tān }

cyclopentanoid [ORG CHEM] A compound whose key structural unit consists of five carbon atoms arranged in a ring. { ¦sī·klō'pen·tə,nȯid }

cyclopentanol [ORG CHEM] C_5H_9OH A colorless liquid boiling at 139°C; used as a solvent for perfumes and pharmaceuticals. Also known as cyclopentyl alcohol. { ¦sī·klō'pen·tə,nȯl }

cyclopentanone [ORG CHEM] C_5H_8O A saturated monoketone; a colorless liquid boiling at 130°C; used as an intermediate in pharmaceutical preparation. { ¦sī·klō'pen·tə,nōn }

cyclopentene [ORG CHEM] $(CH_2)_3CHCH$ A colorless liquid boiling at 45°C; used as a chemical intermediate in petroleum chemistry. { ¦sī·klō'pen,tēn }

cyclopentenylundecylic acid See hydrocarpic acid. { ¦sī·klō'pen·tə,nil,ən·də'sil·ik 'as·əd }

cyclopentyl alcohol See cyclopentanol. { ¦sī·klō'pent·əl 'al·kə,hȯl }

cyclophane [ORG CHEM] A molecule composed of an aromatic ring (most frequently a benzene ring) and an aliphatic unit which forms a bridge between two (or more) positions of the aromatic ring. { 'sī·klə,fān }

cyclophilin [BIOCHEM] An abundant cytoplasmic protein that catalyzes cis-trans isomerizations; it has a high affinity for the immunosuppressive drug cyclosporin A. { ,sī·kə'fil·ən }

cyclophon See beam-switching tube. { 'sī·klə,fän }

Cyclophoracea [INV ZOO] A superfamily of gastropod mollusks in the order Prosobranchia. { ,sī·klō·fə'rās·ē·ə }

Cyclophoridae [INV ZOO] A family of land snails in the order Pectinibranchia. { ,sī·klō'fȯr·ə,dē }

Cyclophyllidea [INV ZOO] An order of platyhelminthic worms comprising most tapeworms of warm-blooded vertebrates. { ,sī·klō·fə'lid·ē·ə }

cyclopia [MED] A congenital anomaly characterized by fusion of the eye sockets with various degrees of fusion of the eyes, to the occurrence of a single median eye. { sī'klō·pē·ə }

CYCLOID SCALE

Cycloid scale, a primitive vertebrate scale.

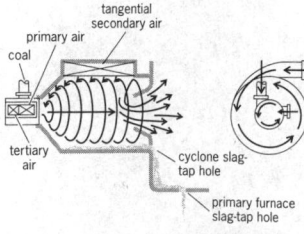

CYCLONE FURNACE

tangential secondary air
primary air
coal
tertiary air
cyclone slag-tap hole
primary furnace slag-tap hole

Schematic diagram of cyclone furnace. (After E. A. Avallone and T. Baumeister III, eds., Marks' Standard Handbook for Mechanical Engineers, 10th ed., McGraw-Hill, 1996)

Cyclopinidae [INV ZOO] A family of copepod crustaceans in the suborder Cyclopoida, section Gnathostoma. { ˌsī·klō'pin·ə,dē }

Cyclopoida [INV ZOO] A suborder of small copepod crustaceans. { ˌsī·klō'póid·ē·ə }

cyclopropane [ORG CHEM] C_3H_6 A colorless gas, insoluble in water; used as an anesthetic. { ¦sī·klō'prō,pān }

cyclopropanoid [ORG CHEM] A compound whose key structural unit consists of three carbon atoms arranged in a ring. { ¦sī·klō'prō·pə,nóid }

Cyclopteridae [VERT ZOO] The lumpfishes and snailfishes, a family of deep-sea forms in the suborder Cottoidei of the order Perciformes. { ˌsī,kläp'ter·ə,dē }

cyclorama [GRAPHICS] A vertical surface, often curved, used to form the background for theatrical settings; an illusion of depth is achieved by even lighting. { ¦sī·klə'räm·ə }

cycloreversion See cycloelimination. { ˌsī·klə·ri'ver·zhən }

Cyclorhagae [INV ZOO] A suborder of benthonic, microscopic marine animals in the class Kinorhyncha of the phylum Aschelminthes. { sī'klór·ə,gē }

Cyclorrhapha [INV ZOO] A suborder of true flies, order Diptera, in which developing adults are always formed in a puparium from which they emerge through a circular opening. { sī'klór·ə·fə }

cycloserine [MICROBIO] $C_3H_6O_2N_2$ Broad-spectrum, crystalline antibiotic produced by several species of *Streptomyces*; useful in the treatment of tuberculosis and urinary-tract infections caused by resistant gram-negative bacteria. { ¦sī·klō'se,rēn }

cyclosilicate [MINERAL] A silicate having the SiO_4 tetrahedra linked to form rings, with a silicon-oxygen ratio of 1:3, such as $Si_3O_9{}^{6-}$ or $Si_6O_{18}{}^{12-}$. Also known as ring silicate. { ¦sī·klō'sil·ə,kāt }

cyclosis [CYTOL] Massive rotational streaming of cytoplasm in certain vacuolated cells, such as the stonewort *Nitella* and *Paramecium*. { sī'klō·səs }

Cyclosporeae [BOT] A class of brown algae, division Phaeophyta, in which there is only a free-living diploid generation. { ¦sī·klō'spór·ē,ē }

cyclosporin A [BIOCHEM] A cyclic peptide produced by some fungi. It inactivates helper T cells, making it useful as an immunosuppressive drug, especially in the prevention of graft rejection in transplantation surgery. { sī·klə,spór·ən 'ā }

Cyclosteroidea [PALEON] A class of Middle Ordovician to Middle Devonian echinoderms in the subphylum Echinozoa. { ¦sī·klo·stə'róid·ē·ə }

Cyclostomata [INV ZOO] An order of bryozoans in the class Stenolaemata. [VERT ZOO] A subclass comprising the simplest and most primitive of living vertebrates characterized by the absence of jaws and the presence of a single median nostril and an uncalcified cartilaginous skeleton. { ¦sī·klō'stō·mə·də }

cyclostrophic flow [FL MECH] A form of gradient flow in which the centripetal acceleration exactly balances the horizontal pressure force. { ¦sī·klō¦strä·fik 'flō }

cyclostrophic wind [METEOROL] The horizontal wind velocity for which the centripetal acceleration exactly balances the horizontal pressure force. { ¦sī·klō¦strä·fik 'wind }

cyclosymmetric function [MATH] A function whose value is unchanged under a cyclic permutation of its variables. { ˌsī·klō·si¦me·trik 'fəŋk·shən }

cyclothem [GEOL] A rock stratigraphic unit associated with unstable shelf of interior basin conditions, in which the sea has repeatedly covered the land. { 'sī·klə,them }

cyclothymia [PSYCH] A disposition marked by alterations of mood between elation and depression out of proportion to apparent external events and stimulated, rather, by internal factors. { ˌsī·klō'thī·mē·ə }

cyclothymic disorder [PSYCH] A mild form of bipolar disorder in which the intensity of the depressive or manic episodes does not reach full criteria. { ˌsī·klō'thī·mik dis'órd·ər }

cyclotol [MATER] High explosive composed of RDX (cyclonite) and TNT. { 'sī·klə,tól }

cyclotomic equation [MATH] An equation which has the form $x^{n-1} + x^{n-2} + \cdots + x + 1 = 0$, where n is a prime number. { ˌsī·klō¦täm·ik i'kwā·zhən }

cyclotomic field [MATH] The extension field of a given field K which is the smallest extension field of K that includes the nth roots of unity for some integer n. { ˌsī·klə¦täm·ik 'fēld }

cyclotomic integer [MATH] A number of the form $a_0 + a_1z + a_2z^2 + \cdots + a_{n-1}z^{n-1}$, where z is a primitive nth root of unity and each a_i is an ordinary integer. { ˌsī·klə,täm·ik 'in·ə·jər }

cyclotomic polynomial [MATH] The nth cyclotomic polynomic is the monic polynomial of degree $\phi(n)$ [where ϕ represents Euler's phi function] whose zeros are the primitive nth roots of unity. { sī·klə,täm·ik 'päl·ə'nō·mē·əl }

cyclotomy [MATH] Theory of dividing the circle into equal parts or constructing regular polygons or, analytically, of finding the nth roots of unity. { sī'kläd·ə·mē }

cyclotrimethylenetrinitramine See cyclonite. { ¦sī·klō,trī¦meth·əl,ēn,trī'nī·trə,mēn }

cyclotron [NUCLEO] An accelerator in which charged particles are successively accelerated by a constant-frequency alternating electric field that is synchronized with movement of the particles on spiral paths in a constant magnetic field normal to their path. Also known as phasotron. { 'sī·klə,trän }

cyclotron cataract See irradiation cataract. { 'sī·klə,trän 'kad·ə,rakt }

cyclotron D See dee. { 'sī·klə,trän 'dē }

cyclotron emission See cyclotron radiation. { 'sī·klə,trän i'mish·ən }

cyclotron frequency [ELECTROMAG] The angular frequency of the motion of a charged particle in a uniform magnetic field in a plane perpendicular to the field. Also known as gyrofrequency. { 'sī·klə,trän 'frē·kwən·sē }

cyclotron-frequency magnetron [ELECTR] A magnetron whose frequency of operation depends on synchronism between the alternating-current electric field and the electrons oscillating in a direction parallel to this field. { 'sī·klə,trän ¦frē·kwən·sē 'mag·nə,trän }

cyclotron magnets [NUCLEO] The magnets which bend charged-particle orbits and confine the extent of particle motion in a cyclotron. { 'sī·klə,trän ,mag·nəts }

cyclotron radiation [ELECTROMAG] The electromagnetic radiation emitted by charged particles as they orbit in a magnetic field, at a speed which is not close to the speed of light. Also known as cyclotron emission. { 'sī·klə,trän ,rād·ē'ā·shən }

cyclotron resonance [PHYS] Resonance absorption of energy from an alternating-current electric field by electrons or ions in a uniform magnetic field when the frequency of the electric field equals the cyclotron frequency, or the cyclotron corresponds to the effective mass of electrons in a solid. Also known as diamagnetic resonance. { 'sī·klə,trän 'rez·ən·əns }

cyclotron-resonance heating [PL PHYS] A modification of magnetic pumping that involves compressing and expanding plasma at a frequency approximating the cyclotron frequency of the ions in the plasma; the goal is temperatures above several million degrees. { 'sī·klə,trän 'rez·ən·əns 'hēd·iŋ }

cyclotron-resonance maser See gyrotron. { 'sī·klə,trän 'rez·ən·əns 'mā·zər }

cyclotron wave [ELECTROMAG] A wave associated with the electron beam of a traveling-wave tube. { 'sī·klə,trän ,wāv }

Cydippida [INV ZOO] An order of the pelagic ctenophores; members retain the cydippid state (resemble the cydippid larva) until the adult stage is reached in development. { sī'dip·ə·də }

Cydippidea [INV ZOO] An order of the Ctenophora having well-developed tentacles. { ˌsī·də'pid·ē·ə }

Cydnidae [INV ZOO] The ground or burrower bugs, a family of hemipteran insects in the superfamily Pentatomorpha. { 'sid·nə,dē }

Cyg See Cygnus.

Cygnus [ASTRON] A conspicuous northern summer constellation; the five major stars are arranged in the form of a cross, but the constellation is represented by a swan with spread wings flying southward; right ascension 21 hours, declination 40°N. Abbreviated Cyg. Also known as Northern Cross; Swan. { 'sig·nəs }

Cygnus A [ASTRON] A strong, discrete radio source in the constellation Cygnus, associated with two spiral galaxies in collision. { 'sig·nəs 'ā }

Cygnus loop [ASTRON] A supernova remnant about 17,000 years old, and 30–40 parsecs across and probably about 770 parsecs distant that emits radio waves and x-rays as well as visible light. Also known as Veil Nebula. { 'sig·nəs ,lüp }

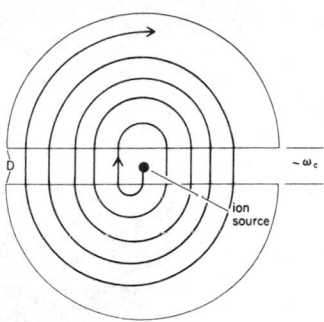

CYCLOTRON

Principle of the cyclotron. The ions are formed in the ion source and are drawn out by one of the D's (electrodes). Arrows show circular orbit path of the ions. Period of orbit equals period $(2\pi/\omega_c)$ of applied D voltage.

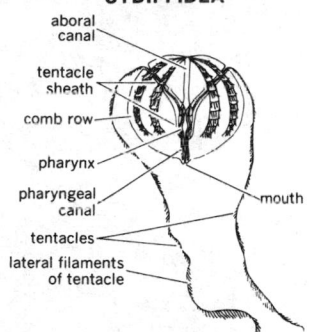

CYDIPPIDEA

Pleurobranchia. (From L. H. Hyman, The Invertebrates, vol. 1, McGraw-Hill, 1940)

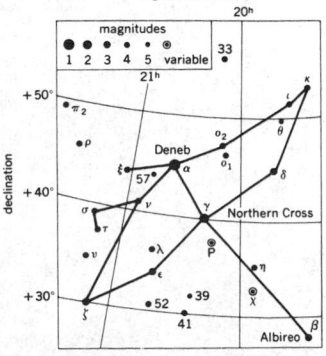

CYGNUS

Line pattern of constellation Cygnus. Grid lines represent coordinates of sky. Apparent brightness, or magnitude, of stars is shown by sizes of dots, which are graded by appropriate numbers as indicated.

Cygnus X-1 [ASTROPHYS] A source of x-rays whose intensity varies in an irregular manner, associated with a weak variable radio source and a ninth-magnitude spectroscopic binary star, designated HDE226868, that consists of a blue supergiant and an invisible companion, which may be a black hole. Abbreviated Cyg X-1. { 'sig·nəs ,eks 'wən }

Cygnus X-3 [ASTROPHYS] A variable source of x-rays, with a period of 4.8 hours, associated with a variable radio source that flared up to enormous levels in September 1972 with no observed increase in x-ray emission. Abbreviated Cyg X-3. { 'sig·nəs ,eks 'thrē }

Cyg X-1 See Cygnus X-1.

Cyg X-3 See Cygnus X-3.

cyhexatin [ORG CHEM] $C_{18}H_{34}OSn$ A whitish solid, insoluble in water; used as a miticide to control plant-feeding mites. { sī'hek·sə·tən }

cylinder [CIV ENG] **1.** A steel tube 10–60 inches (25–152 centimeters) in diameter with a wall at least 1/8 inch (3 millimeters) thick that is driven into bedrock, excavated inside, filled with concrete, and used as a pile foundation. **2.** A domed, closed tank for storing hot water to be drawn off at taps. Also known as storage calorifier. [COMPUT SCI] **1.** The virtual cylinder represented by the tracks of equal radius of a set of disks on a disk drive. **2.** See seek area. [ENG] **1.** A container used to hold and transport compressed gas for various pressurized applications. **2.** The piston chamber in a pump from which the liquid is expelled. [MATH] **1.** A solid bounded by a cylindrical surface and two parallel planes, or the surface of such a solid. **2.** See cylindrical surface. [MECH ENG] See engine cylinder. { 'sil·ən·dər }

cylinder actuator [MECH ENG] A device that converts hydraulic power into useful mechanical work by means of a tight-fitting piston moving in a closed cylinder. { 'sil·ən·dər ,ak·chə,wād·ər }

cylinder block [DES ENG] The metal casting comprising the piston chambers of a multicylinder internal combustion engine. Also known as block; engine block. { 'sil·ən·dər ,bläk }

cylinder bore [DES ENG] The internal diameter of the tube in which the piston of an engine or pump moves. { 'sil·ən·dər ,bȯr }

cylinder function [MATH] Any solution of the Bessel equation, including Bessel functions, Neumann functions, and Hankel functions. { 'sil·ən·dər ,fəŋk·shən }

cylinder gap [GRAPHICS] The space in the cylinders of a printing press where a mechanism for plate (or blanket) clamps and grippers is located. { 'sil·ən·dər ,gap }

cylinder head [MECH ENG] The cap that serves to close the end of the piston chamber of a reciprocating engine, pump, or compressor. { 'sil·ən·dər ,hed }

cylinder liner [MECH ENG] A separate cylindrical sleeve inserted in an engine block which serves as the cylinder. { 'sil·ən·dər ,līn·ər }

cylinder machine [ENG] A paper-making machine consisting of one or a series of rotary cylindrical filters on which wet paper sheets are formed. { 'sil·ən·dər mə'shēn }

cylinder oil [MATER] A viscous lubricating oil for the cylinders and valves of steam engines. { 'sil·ən·dər ,ȯil }

cylinder press [GRAPHICS] A large printing press in which paper is rolled against a flat, reciprocating printing surface by a rotating cylinder. { 'sil·ən·dər ,press }

cylinder stock [MATER] **1.** Residual material in a still after vaporization of lighter petroleum stock. **2.** Compounded or straight oil used for lubricating steam cylinders. { 'sil·ən·dər ,stäk }

cylinder stop [ORD] Device on a pistol that checks the action of the cylinder and acts as a lock to prevent it from moving. { 'sil·ən·dər ,stäp }

cylindrarthrosis [ANAT] A joint characterized by rounded articular surfaces. { ¦sil·ən,drär'thrō·səs }

cylindrical [SCI TECH] Having the shape of a cylinder. { sə'lin·drə·kəl }

cylindrical antenna [ELECTROMAG] An antenna in which hollow cylinders serve as radiating elements. { sə'lin·drə·kəl an'ten·ə }

cylindrical array [ELECTR] An electronic scanning antenna that may consist of several hundred columns of vertical dipoles mounted in cylindrical radomes arranged in a circle. { sə'lin·drə·kəl ə'rā }

cylindrical bow [NAV ARCH] A type of bow used principally on full, slow-speed, bulk-cargo-carrying ships; in general, the bow ends in a large-radius, vertical cylinder. { sə'lin·drə·kəl 'baú }

cylindrical cam [MECH ENG] A cam mechanism in which the cam follower undergoes translational motion parallel to the camshaft as a roller attached to it rolls in a groove in a circular cylinder concentric with the camshaft. { sə'lin·drə·kəl 'kam }

cylindrical capacitor [ELEC] A capacitor made of two concentric metal cylinders of the same length, with dielectric filling the space between the cylinders. Also known as coaxial capacitor. { sə'lin·drə·kəl kə'pas·əd·ər }

cylindrical cavity [ELECTROMAG] A cavity resonator in the shape of a right circular cylinder. { sə'lin·drə·kəl 'kav·əd·ē }

cylindrical-coordinate robot [CONT SYS] A robot in which the degrees of freedom of the manipulator arm are defined chiefly by cylindrical coordinates. { sə'lin·drə·kəl kō¦ȯrd·ən·ət 'rō,bät }

cylindrical coordinates [MATH] A system of curvilinear coordinates in which the position of a point in space is determined by its perpendicular distance from a given line, its distance from a selected reference plane perpendicular to this line, and its angular distance from a selected reference line when projected onto this plane. { sə'lin·drə·kəl ,kō'ȯrd·ən·əts }

cylindrical cutter [DES ENG] Any cutting tool with a cylindrical shape, such as a milling cutter. { sə'lin·drə·kəl 'kəd·ər }

cylindrical-film storage [ELECTR] A computer storage in which each storage element consists of a short length of glass tubing having a thin film of nickel-iron alloy on its outer surface. { sə'lin·drə·kəl 'film ,stȯr·ij }

cylindrical function See Bessel function. { sə'lin·dri·kəl ,fəŋk·shən }

cylindrical grinder [MECH ENG] A machine for doing work on the peripheries or shoulders of workpieces composed of concentric cylindrical or conical shapes, in which a rotating grinding wheel cuts a workpiece rotated from a power headstock and carried past the face of the wheel. { sə'lin·drə·kəl 'grīnd·ər }

cylindrical helix [MATH] A curve lying on a cylinder which intersects the elements of the cylinder at a constant angle. { sə'lin·drə·kəl 'hē,liks }

cylindrical lens [OPTICS] A lens one or both of whose surfaces are a portion of a circular cylinder. { si'lin·drə·kəl 'lenz }

cylindrical map projection [MAP] A map projection produced by projecting the geographic meridians and parallels onto a cylinder which is tangent to (or intersects) the surface of a sphere, and then developing the cylinder into a plane. { sə'lin·drə·kəl 'map prə,jek·shən }

cylindrical pinch See pinch effect. { sə'lin·drə·kəl 'pinch }

cylindrical reflector [ELECTROMAG] A reflector that is a portion of a cylinder; this cylinder is usually parabolic. { sə'lin·drə·kəl ri'flek·tər }

cylindrical surface [MATH] A surface consisting of each of the straight lines which are parallel to a given straight line and pass through a given curve. Also known as cylinder. { sə'lin·drə·kəl 'sər·fəs }

cylindrical wave [ELECTROMAG] A wave whose equiphase surfaces form a family of coaxial cylinders. { sə'lin·drə·kəl 'wāv }

cylindrical winding [ELEC] The current-carrying element of a core-type transformer, consisting of a single coil of one or more layers wound concentrically with the iron core. { sə'lin·drə·kəl 'wīnd·iŋ }

cylindrite [MINERAL] $Pb_3Sn_4Sb_2S_{14}$ A blackish-gray mineral consisting of sulfur, lead, antimony, and tin, occurring in cylindrical forms that separate under pressure into distinct sheets or folia. { sə'lin,drīt }

Cylindrocapsaceae [BOT] A family of green algae in the suborder Ulotrichineae comprising thick-walled, sheathed cells having massive chloroplasts. { sə,lin·drō,kap'sās·ē,ē }

Cylindrocorporoidea [INV ZOO] A superfamily of both free-living and parasitic nematodes of the order Diplogasterida having well-developed lips surrounding the oral opening and lateral lips bearing small amphids. { sə,lin·drə,kȯr·pə'rȯid·ē·ə }

cymbocephalic [ANTHRO] Of a head or skull, having an unusually prolonged receding forehead and a protrudent occiput. { ¦sim·bō·sə'fal·ik }

cyme [BOT] An inflorescence in which each main axis terminates in a single flower; secondary and tertiary axes may also have flowers, but with shorter flower stalks. { sīm }

cymene [ORG CHEM] Any of the isomeric hydrocarbons metacymene, paracymene, and orthocymene; paracymene is a liquid that is colorless, has a pleasant odor, and is made from oil of cumin or oil of wild thyme. { 'sī,mēn }

cymophane See cat's eye. { 'sī·mə,fān }

cymose [BOT] Of, pertaining to, or resembling a cyme. { 'sī,mōs }

Cymothoidae [INV ZOO] A family of isopod crustaceans in the suborder Flabellifera; members are fish parasites with reduced maxillipeds ending in hooks. { ,sī·mə'thói,dē }

cymrite [MINERAL] Ba$_2$Al$_5$Si$_5$O$_{19}$(OH)·3H$_2$O Zeolite mineral consisting of a basic aluminosilicate of barium. { 'kəm,rīt }

Cynipidae [INV ZOO] A family of hymenopteran insects in the superfamily Cynipoidea. { sə'nip·ə,dē }

Cynipoidea [INV ZOO] A superfamily of hymenopteran insects in the suborder Apocrita. { ,sin·ə'póid·ē·ə }

Cynoglossidae [VERT ZOO] The tonguefishes, a family of Asiatic flatfishes in the order Pleuronectiformes. { ,sin·ə'gläs·ə,dē }

cynophobia [PSYCH] An abnormal fear of dogs. { ,sin·ə'fō·bē·ə }

Cyperaceae [BOT] The sedges, a family of monocotyledonous plants in the order Cyperales characterized by spirally arranged flowers on a spike or spikelet; a usually solid, often triangular stem; and three carpels. { ,sip·ə'rās·ē,ē }

Cyperales [BOT] An order of monocotyledonous plants in the subclass Commelinidae with reduced, mostly wind-pollinated or self-pollinated flowers that have a unilocular, two-or three-carpellate ovary bearing a single ovule. { sip·ə'rā·lēz }

Cypheliaceae [BOT] A family of typically crustose lichens with sessile apothecia in the order Caliciales. { sə,fel·ē'ās·ē,ē }

cyphonautes [INV ZOO] The free-swimming bivalve larva of certain bryozoans. { ,sī·fə'nōd·ēz }

Cyphophthalmi [INV ZOO] A family of small, mitelike arachnids in the order Phalangida. { ,sī·fə'thal,mī }

Cypovirus [VIROL] A genus in the family Reoviridae that infects invertebrates; the type species is cytoplasmic polyhedrosis virus. { sip,o'vī·rəs }

Cypraecea [INV ZOO] A superfamily of gastropod mollusks in the order Prosobranchia. { sī'prēsh·ē·ə }

Cypraeidae [INV ZOO] A family of colorful marine snails in the order Pectinibranchia. { sī'prē·ə,dē }

cypress [BOT] The common name for members of the genus *Cupressus* and several related species in the order Pinales. { 'sī·prəs }

Cypridacea [INV ZOO] A superfamily of mostly fresh-water ostracods in the suborder Podocopa. { ,sī·prə'dās·ē·ə }

Cypridinacea [INV ZOO] A superfamily of ostracods in the suborder Myodocopa characterized by a calcified carapace and having a round back with a downward-curving rostrum. { ,sī,prid·ə'nās·ē·ə }

cypridophobia [PSYCH] An abnormal fear of acquiring venereal disease. { sī,prid·ə'fō·bē·ə }

Cyprinidae [VERT ZOO] The largest family of fishes, including minnows and carps in the order Cypriniformes. { sī'prin·ə,dē }

Cypriniformes [VERT ZOO] An order of actinopterygian fishes in the suborder Ostariophysi. { sī,prin·ə'fór,mēz }

Cyprinodontidae [VERT ZOO] The killifishes, a family of actinopterygian fishes in the order Atheriniformes that inhabit ephemeral tropical ponds. { sī,prin·ə'dän·tə,dē }

Cyprinoidei [VERT ZOO] A suborder of primarily freshwater actinopterygian fishes in the order Cypriniformes having toothless jaws, no adipose fin, and faliciform lower pharyngeal bones. { ,sī·prə'nói·dē,ī }

cypris [INV ZOO] An ostracod-like, free-swimming larval stage in the development of Cirripedia. { 'sī·prəs }

Cyrtophorina [INV ZOO] The equivalent name for Gymnostomatida. { ,sərd·ə'fə'rī·nə }

cyrtopia [INV ZOO] A type of crustacean larva (Ostracoda) characterized by an elongation of the first pair of antennae and loss of swimming action in the second pair. { sər'tō·pē·ə }

cyrtosis [PL PATH] A virus disease of cotton characterized by stunting, distortion, and abnormal branching and coloration. { sər'tō·səs }

cyst [MED] A normal or pathologic sac with a distinct wall, containing fluid or other material. { sist }

cystacanth [INV ZOO] The infective larva of the Acanthocephala; lies in the hemocele of the intermediate host. { 'sis·tə,kanth }

cystamine [ORG CHEM] (CH$_2$)$_6$N$_4$ A white, crystalline powder, melting at 280°C; used to make synthetic resins. Also known as aminiform; crystogen; cystamine methenamine; hexamethylene tetramine; urotropin. { 'sis·tə,mēn }

cystamine methenamine See cystamine. { 'sis·tə,mēn mə'then·ə,mēn }

L-cystathionine [BIOCHEM] C$_7$H$_{14}$N$_2$O$_4$S An amino acid formed by condensation of homocysteine with serine, catalyzed by an enzyme transsulfurase; found in high concentration in the brain of primates. { el ,sis·tə'thī·ə,nēn }

cystectomy [MED] **1.** Excision of the gallbladder, or part of the urinary bladder. **2.** Removal of a cyst. **3.** Removal of a piece of the anterior capsule of the lens for the extraction of a cataract. { si'stek·tə·mē }

cysteine [BIOCHEM] C$_3$H$_7$O$_2$NS A crystalline amino acid occurring as a constituent of glutathione and cystine. { 'si,stēn }

cystic disease [MED] A disorder of women, usually at or near menopause, characterized by the development of large cysts in the breast. { 'sis·tik di'zēz }

cystic duct [ANAT] The duct of the gallbladder. { 'sis·tik 'dəkt }

cysticercosis [MED] The infestation in humans by cysticerci of the genus *Taenia*. { ,sis·tə·sər'kō·səs }

cysticercus [INV ZOO] A larva of tapeworms in the order Cyclophyllidea that has a bladder with a single invaginated scolex. { ¦sis·tə'sər·kəs }

cystic fibrosis [MED] A hereditary disease of the pancreas transmitted as an autosomal recessive; involves obstructive lesions, atrophy, and fibrosis of the pancreas and lungs, and the production of mucus of high viscosity. Also known as mucoviscidosis. { ¦sis·tik fī'brō·səs }

cystic fibrosis transmembrane conductance regulator [CELL MOL] A specialized chloride channel that is regulated by cyclic adenosine monophosphate; its disruption has been implicated in cystic fibrosis. { ¦sis·tik fī¦brō·səs tranz¦mem ,brān kən'dək·təns ,reg·yə,lād·ər }

cystic kidney [MED] A kidney with one or more cysts. { 'sis·tik 'kid·nē }

cystine [BIOCHEM] C$_6$H$_{12}$N$_2$S$_2$ A white, crystalline amino acid formed biosynthetically from cysteine. { 'si,stēn }

cystinosis [MED] A congenital metabolic disorder involving sulfur-containing amino acids, usually cystine; characterized by deposits of cystine crystals in the body organs. { ,sis·tə'nō·səs }

cystinuria [MED] The presence in the urine of crystals of cystine together with some lysine, arginine, and ornithine. { ,sis·tə'núr·ē·ə }

cystitis [MED] Inflammation of a fluid-filled organ, especially the urinary bladder. { si'stīd·əs }

Cystobacter [MICROBIO] A genus of bacteria in the family Cystobacteraceae; vegetative cells are tapered, sporangia are sessile, and microcysts are rigid rods. { ¦sis·tə'bak·tər }

Cystobacteraceae [MICROBIO] A family of bacteria in the order Myxobacterales; vegetative cells are tapered, and microcysts are rod-shaped and enclosed in sporangia. { ¦sis·tə,bak·tə'rās·ē,ē }

cystocarp [BOT] A fruiting structure with a special protective envelope, produced after fertilization in red algae. { 'sis·tə,kärp }

cystocele [MED] Herniation of the urinary bladder into the vagina. { 'sis·tə,sēl }

cystocercous cercaria [INV ZOO] A digenetic trematode larva that can withdraw the body into the tail. { ¦sis·tə¦sər· kəs sər'kar·ē,ə }

cystography [MED] Radiography of the urinary bladder after the injection of a contrast medium. { si'stäg·rə·fē }

Cystoidea [PALEON] A class of extinct crinozoans characterized by an ovoid body that was either sessile or attached by a short aboral stem. { si'stóid·ē·ə }

cystolith [BOT] A concretion of calcium carbonate arising

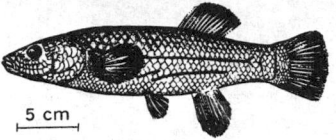

CYPRINODONTIDAE

Striped killifish (*Fundulus majalis*).

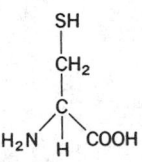

CYSTEINE

Structural formula of cysteine.

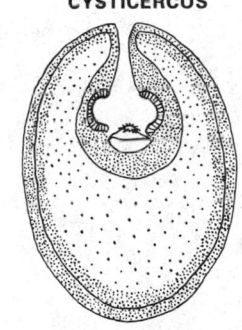

CYSTICERCUS

Cysticercus, the metacestode larva of cyclophyllidean tapeworms.

from the cell walls of modified epidermal cells in some flowering plants. { 'sis·tə‚lith }

cystoma [MED] A cystic mass, especially in or near the ovary. { si'stō·mə }

cystometer [MED] An instrument used to determine pressure in the urinary bladder under standard conditions. { si'stäm·əd·ər }

Cystoporata [PALEON] An order of extinct, marine bryozoans characterized by cystopores and minutopores. { ‚sis·tə'pór·əd·ə }

cystopyelitis [MED] Inflammation of the urinary bladder and the renal pelvis. { ¦sis·tə‚pī·ə'līd·əs }

cystopyelography [MED] Radiography of the urinary bladder, ureter, and renal pelvis after injection of a radiopaque material. { ¦sis·tə‚pī·ə'läg·rə·fē }

cystopyelonephritis [MED] Inflammation of the urinary bladder, renal pelvis, and renal parenchyma. { ¦sis·tə‚pī·ə‚lō·nə'frīd·əs }

cystoscope [MED] An optical instrument for visual examination of the urinary bladder, ureters, and kidneys. { 'sis·tə‚skōp }

cystoureteritis [MED] Inflammation of the urinary bladder and the ureters. { ¦sis·tə·yủ‚rēd·ə'rīd·əs }

Cystoviridae [VIROL] A family of enveloped ribonucleic acid (RNA)-containing bacteriophages characterized by a spherical virion containing three molecules of linear double-stranded RNA. { ‚sis·tə'vir·ə‚dī }

Cystovirus [VIROL] The sole genus of the family Cystoviridae containing the type species *Pseudomonas* Phage phi6. { 'sis·tə‚vī·rəs }

cytac [NAV] A system in which hyperbolic lines of position are determined by measuring the time relationship between two synchronized radio signals; similar to loran but capable of higher accuracy and greater range. { 'sī‚tak }

cytase [BIOCHEM] Any of several enzymes in the seeds of cereals and other plants, which hydrolyze the cell-wall material. { 'sī‚tās }

Cytheracea [INV ZOO] A superfamily of ostracods in the suborder Podocopa comprising principally crawling and digging marine forms. { ‚sith·ə'rās·ē·ə }

Cytherellidae [INV ZOO] The family comprising all living members of the ostracod suborder Platycopa. { ‚sith·ə'rel·ə‚dē }

cytidine [BIOCHEM] $C_9H_{13}N_3O_5$ Cytosine riboside, a nucleoside composed of one molecule each of cytosine and D-ribose. { 'sid·ə‚dēn }

cytidylic acid [BIOCHEM] $C_9H_{14}O_8N_3P$ A nucleotide synthesized from the base cytosine and obtained by hydrolysis of nucleic acid. { ¦sid·ə¦dil·ik 'as·əd }

cytocentrum *See* central apparatus. { ¦sīd·ō'sen·trəm }

cytochalasin [BIOCHEM] One of a series of structurally related fungal metabolic products which, among other effects on biological systems, selectively and reversibly block cytokinesis while not affecting karyokinesis; the molecule with minor variations consists of a benzyl-substituted hydroaromatic isoindolone system, which in turn is fused to a small macrolide-like cyclic ring. { ¦sīd·ō·kə'lā·sən }

cytochemistry [CYTOL] The science concerned with the chemistry of cells and cell components, primarily with the location of chemical constituents and enzymes. { ¦sīd·ō'kem·ə·strē }

cytochrome [BIOCHEM] Any of the complex protein respiratory pigments occurring within plant and animal cells, usually in mitochondria, that function as electron carriers in biological oxidation. { 'sīd·ə‚krōm }

cytochrome a₃ *See* cytochrome oxidase. { 'sīd·ə‚krōm ¦ā səb'thrē }

cytochrome oxidase [BIOCHEM] Any of a family of respiratory pigments that react directly with oxygen in the reduced state. Also known as cytochrome a₃. { 'sīd·ə‚krōm 'äk·sə‚dās }

cytocidal [CYTOL] Causing cell death. { sī'tās·əd·əl }

cytocidal unit [BIOL] A unit for the standardization of adrenal cortical hormones. { sī'tās·əd·əl ‚yü·nət }

cytocrine gland [CYTOL] A cell, especially a melanocyte, that passes its secretion directly to another cell. { 'sīd·ə·krən ‚gland }

cytodiagnosis [PATH] The determination of the nature of an abnormal liquid by the study of cells it contains. { ¦sīd·ō‚dī·ig'nō·səs }

cytoduction [MYCOL] In yeast, the production of cells with mixed cytoplasm but with the nucleus of one or the other parent. { 'sīd·ə‚dək·shən }

cytogamy [CYTOL] Fusion or conjugation of cells. { sī'täg·ə·mē }

cytogenetics [CYTOL] The comparative study of the mechanisms and behavior of chromosomes in populations and taxa, and the effect of chromosomes on inheritance and evolution. { ¦sīd·ō·jə'ned·iks }

cytogenous gland [PHYSIO] A structure that secretes living cells; an example is the testis. { sī'tä·jə·nəs ‚gland }

cytohet [CYTOL] A cell containing two genetically distinct types of a specific organelle. { 'sīd·ə‚het }

cytokine [CELL MOL] Any of a group of peptides that are released by some cells and affect the behavior of other cells, serving as intercellular signals. { 'sīd·ə‚kīn }

cytokine receptor [CELL MOL] A type of cell-surface receptor that binds to cytokines, initiating the Jak-STAT signaling pathway within the cell. { 'sīd·ə‚kīn ri‚sep·tər }

cytokinesis [CYTOL] Division of the cytoplasm following nuclear division. { ¦sīd·ō·kə'nē·səs }

cytokinin [BIOCHEM] Any of a group of plant hormones which elicit certain plant growth and development responses, especially by promoting cell division. { ¦sīd·ō'kī·nən }

cytology [BIOL] A branch of the biological sciences which deals with the structure, behavior, growth, and reproduction of cells and the function and chemistry of cell components. { sī'täl·ə·jē }

cytolysin *See* perforin. { ‚sīd·ə'līs·ən }

cytolysis [PATH] Disintegration or dissolution of cells, usually associated with a pathologic process. { sī'täl·ə·səs }

cytolysosome [CYTOL] An enlarged lysosome that contains organelles such as mitochondria. { ¦sīd·ō'lī·sə‚sōm }

cytomegalic [MED] Of, pertaining to, or characterizing the greatly enlarged cells with enlarged nuclei and inclusion bodies found in tissues in cytomegalic inclusion disease. { ¦sīd·ō·mə¦gal·ik }

cytomegalic inclusion disease [MED] A virus infection primarily of infants characterized by jaundice, liver enlargement, and circulatory disturbances. { ¦sīd·ō·mə¦gäl·ik in'klü·zhən di'zēz }

cytomegalovirus [VIROL] An animal virus belonging to subgroup B of the herpesvirus group; causes cytomegalic inclusion disease and pneumonia. { ¦sīd·ō¦meg·ə·lō'vī·rəs }

cytomegalovirus group *See* Betaherpesvirinae. { ¦sīd·ō¦meg·ə·lō'vī·rəs ‚grüp }

cytomegalovirus infection [MED] A common asymptomatic infection caused by cytomegalovirus, which can produce life-threatening illnesses in the immature fetus and in immunologically deficient subjects. { ¦sīd·ō¦meg·ə·lō‚vī·rəs in'fek·shən }

cytomegalovirus mononucleosis [MED] A self-limited illness such as infectious mononucleosis, the main manifestation of which is fever; it is the only cytomegalovirus illness clearly described in mature, immunologically normal subjects. { ¦sīd·ō¦meg·ə·lō‚vī·rəs ‚män·ə‚nü·klē'ō·səs }

cytomixis [CYTOL] Extrusion of chromatin from one cell into the cytoplasm of an adjoining cell. { ‚sīd·ə'mik·səs }

cytomorphosis [CYTOL] All the structural alterations which cells or successive generations of cells undergo from the earliest undifferentiated stage to their final destruction. { ‚sīd·ə'mór·fə·səs }

cyton [NEURO] The central body of a neuron containing the nucleus and excluding its processes. { 'sī‚tän }

cytopathic effect [CYTOL] A change in the microscopic appearance of cells in a culture after being infected with a virus. { ¦sīd·ə¦path·ik i‚fekt }

cytopathology [PATH] A branch of pathology concerned with abnormalities within cells. { ¦sīd·ō·pə'thäl·ə·jē }

cytopenia [PATH] A blood cell count below normal. { ‚sīd·ō'pē·nē·ə }

Cytophaga [MICROBIO] A genus of bacteria in the family Cytophagaceae; cells are unsheathed, unbranched rods or filaments and are motile; microcysts are not known; decompose agar, cellulose, and chitin. { sī'täf·ə·gə }

Cytophagaceae [MICROBIO] A family of bacteria in the order Cytophagales; cells are rods or filaments, unsheathed cells are motile, filaments are not attached, and carotenoids are present. { ˌsīd·ō·fəˈgäs·ē,ē }

Cytophagales [MICROBIO] An order of gliding bacteria; cells are rods or filaments and motile by gliding, and fruiting bodies are not produced. { ˌsīd·ō·fəˈgä·lēz }

cytopharynx [INV ZOO] A channel connecting the surface with the protoplasm in certain protozoans; functions as a gullet in ciliates. { ˌsīd·ōˈfar·iŋks }

cytophilic antibody [IMMUNOL] A substance capable of combining directly with the receptors of a corresponding antigenic cell. Also known as cytotropic antibody. { ˌsid·əˌfil·ik ˈant·iˌbäd·ē }

cytoplasm [CYTOL] The protoplasm of an animal or plant cell external to the nucleus. { ˈsīd·əˌplaz·əm }

cytoplasmic inheritance [GEN] The control of genetic difference by genes carried in cytoplasmic organelles such as mitochondria or chloroplasts. Also known as extrachromosomal inheritance. { ˌsīd·əˈplaz·mik inˈher·ə·təns }

cytoplasmic male sterility [BOT] The maternally inherited inability of a higher plant to produce viable pollen. { ˈsīd·əˌplaz·mik ˈmäl stəˈril·əd·ē }

cytoplasmic streaming [CYTOL] Intracellular movement involving irreversible deformation of the cytoplasm produced by endogenous forces. { ˌsīd·əˈplaz·mik ˈstrem·iŋ }

cytoplast [CYTOL] The cytoplasmic substance of eukaryotic cells, including a network of proteins forming an internal skeleton and the attached nucleus and organelles. { ˈsīd·əˌplast }

cytopyge [INV ZOO] A fixed point for waste discharge in the body of a protozoan, especially a ciliate. { ˈsīd·əˌpīj }

cytosine [BIOCHEM] $C_4H_5ON_3$ A pyrimidine occurring as a fundamental unit or base of nucleic acids. { ˈsīd·əˌsēn }

cytoskeleton [CYTOL] Protein fibers composing the structural framework of a cell. { ˈsīd·ōˈskel·ə·tən }

cytosol [CYTOL] The fluid portion of the cytoplasm, that is, the cytoplasm exclusive of organelles and membranes. { ˈsīd·əˌsäl or ˈsīd·əˌsōl }

cytosome [CYTOL] The cytoplasm of the cell, as distinct from the nucleus. { ˈsīd·əˌsōm }

cytostasis [CYTOL] Inhibition of the ability of cells to continue growing. { ˌsīd·əˈstä·səs }

cytostatic [CYTOL] Inhibiting cell development. { ˌsīd·ō ˈstad·ik }

cytostome [INV ZOO] The mouth-like opening in many unicellular organisms, particularly Ciliophora. { ˈsīd·əˌstōm }

cytotaxis [PHYSIO] Attraction of motile cells by specific diffusible stimuli emitted by other cells. { ˈsīd·əˈtak·səs }

cytotechnologist [PATH] A person trained to prepare smears of and examine exfoliated cells, referring abnormalities to a physician. { ˌsīd·ōˌtekˈnäl·ə·jəst }

cytotoxic [CYTOL] Pertaining to an agent, such as a drug or virus, that exerts a toxic effect on cells. { ˈsīd·əˈtäk·sik }

cytotoxic T cell [IMMUNOL] A type of T cell which protects against pathogens that invade host cell cytoplasm, where they cannot be bound by antibodies, by recognizing and killing the host cell before the pathogens can proliferate and escape. { ˈsīd·əˌtäk·sik ˈtē ˌsel }

cytotrophoblast [EMBRYO] The inner, cellular layer of a trophoblast, covering the chorion and the chorionic villi during the first half of pregnancy. { ˈsīd·ōˈtrō·fəˌblast }

cytotropic antibody See cytophilic antibody. { ˈsīd·ōˈträ·pik ˈan·təˌbäd·ē }

cytotropism [BIOL] The tendency of individual cells and groups of cells to move toward or away from each other. { sīˈtä·trəˌpiz·əm }

Czapek's agar [MICROBIO] A nutrient culture medium consisting of salt, sugar, water, and agar; used for certain mold cultures. { ˈchä·peks ˌäg·ər }

Czerny-Turner spectrograph [SPECT] A spectrograph used chiefly in laboratory work, which has a plane reflection grating and spherical reflectors for the collimator and camera. { ˈcherˌnēˌtərn·ər ˈspek·trəˌgraf }

Czochralski process [CRYSTAL] A method of producing large single crystals by inserting a small seed crystal of germanium, silicon, or other semiconductor material into a crucible filled with similar molten material, then slowly pulling the seed up from the melt while rotating it. { chəˈkräl·skē ˌpräs·əs }

CYTOSINE

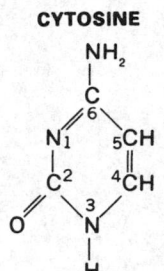

Structural formula of cytosine.

CZERNY-TURNER SPECTROGRAPH

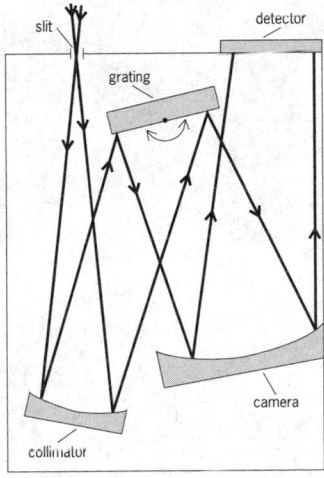

Czerny-Turner plane grating spectrograph.

D

D *See* dee; diopter.

2,4-D *See* 2,4-dichlorophenoxyacetic acid.

DAB *See* digital audio broadcasting.

DABS *See* Mode S. { dabz *or* ¦dē,ā,bē'es }

dac *See* digital-to-analog converter.

dachiardite [MINERAL] $(Na_2Ca)_2(Al_4Si_{20}O_{48})\cdot12H_2O$ A white to colorless mineral in the mordenite group of the zeolite family that crystallizes in the monoclinic system. { ¦däk·ē'är,dīt }

Dacian [GEOL] Lower upper Pliocene geologic time. { 'dā·shən }

dacite [GEOL] Very fine crystalline or glassy rock of volcanic origin, composed chiefly of sodic plagioclase and free silica with subordinate dark-colored minerals. { 'dā,sīt }

dacite glass [GEOL] A natural glass formed by rapid cooling of dacite lava. { 'dā,sīt ,glas }

Dacromycetales [MYCOL] An order of jelly fungi in the subclass Heterobasidiomycetidae having branched basidia with the appearance of a tuning fork. { ¦dak·rə,mī·sə'tā·lēz }

dacryoblennorrhea [MED] Chronic inflammation of the lacrimal sac of the eye accompanied by discharge of mucus. { ¦dak·rə,blen·ə'rē·ə }

dacryocyst *See* lacrimal sac. { 'dak·rə,sist }

dacryocystitis [MED] Inflammation of the lacrimal sac. { ,dak·rə,sis'tīd·əs }

dacryon [ANAT] The point of the face where the frontomaxillary, the maxillolacrimal, and frontolacrimal sutures meet. { 'dak·rē,än }

dactylitic [GEOL] Of a rock texture, characterized by fingerlike projections of a mineral that penetrate another mineral. { dak·tə'lid·ik }

Dactylochirotida [INV ZOO] An order of dendrochirotacean holothurians in which there are 8–30 digitate or digitiform tentacles, which sometimes bifurcate. { ,dak·tə·lō·kə'räd·ə·də }

dactylognathite [INV ZOO] The distal segment of a maxilliped in crustaceans. { ,dak·tə·lō'na,thīt }

dactylography [FOREN SCI] The scientific study of fingerprints as a device for identifyingpeople. { ,dak·tə'läg·rə·fē }

Dactylogyroidea [INV ZOO] A superfamily of trematodes in the subclass Monogenea; all are fish ectoparasites. { ¦dak·tə·lo,ji'rȯid·ē·ə }

dactylopodite [INV ZOO] The distal segment of ambulatory limbs in decapods and of certain limbs in other arthropods. { ,dak·tə'läp·ə,dīt }

dactylopore [INV ZOO] Any of the small openings on the surface of Milleporina through which the bodies of the polyps are extended. { dak'til·ə,pȯr }

Dactylopteridae [VERT ZOO] The flying gurnards, the single family of the perciform suborder Dactylopteroidei. { ¦dak·tə·lō'ter·ə,dē }

Dactylopteroidei [VERT ZOO] A suborder of marine shore fishes in the order Perciformes, characterized by tremendously expansive pectoral fins. { ,dak·tə·lō·tə'rȯid·ē,ī }

Dactyloscopidae [VERT ZOO] The sand stargazers, a family of small tropical and subtropical perciform fishes in the suborder Blennioidei. { ,dak·tə·lō·'skäp·ə,dē }

Dactylosporangium [MICROBIO] A genus of bacteria in the family Actunoplanaceae; fingerlike sporangia are formed in clusters, each containing a single row of three or four motile spores. { ,dak·tə·lō·spə'rän·jē·əm }

dactylosternal [VERT ZOO] Of turtles, having marginal fingerlike processes in joining the plastron to the carapace. { ¦dak·tə·lō'stərn·əl }

dactylozooid [INV ZOO] One of the long defensive polyps of the Milleporina, armed with stinging cells. { ¦dak·tə·lə'zō,ȯid }

dactylus [INV ZOO] The structure of the tarsus of certain insects which follows the first joint; usually consists of one or more joints. { ¦dak·tə·ləs }

dado [ARCH] **1.** The lower portion of an interior wall set off by molding or other decoration. **2.** The portion of a pedestal between surbase and base. **3.** The portion of a wall basement between surbase and base course. { 'dā·dō }

dado head [MECH ENG] A machine consisting of two circular saws with one or more chippers in between; used for cutting flat-bottomed grooves in wood. { 'dā·dō ,hed }

dado joint [BUILD] A joint made by fitting the full thickness of the edge or the end of one board into a corresponding groove in another board. Also known as housed joint. { 'dā·dō jȯint }

dado plane [DES ENG] A narrow plane for cutting flat grooves in woodwork. { 'dā·dō ,plān }

dadur [METEOROL] In India, a wind blowing down the Ganges Valley from the Siwalik hills at Hardwar. { dä'dùr }

daemon [COMPUT SCI] In Unix, a program that runs in the background, such as a server. { 'dē·mən }

DAF *See* decay accelerating factor.

Da Fano bodies [VIROL] Minute basophilic areas of abnormal staining found within cells infected with human herpesvirus 1 or 2. { dä'fän·ō ¦bäd·ēz }

dagger board [NAV ARCH] A device used in a sailing vessel to increase the area of lateral resistance; it is a narrow metal flat or wooden board that slides up and down in the trunk, which extends from inside the hull to the bottom. { 'dag·ər ,bȯrd }

Dagor lens [OPTICS] An anastigmatic lens consisting of two lens systems that are nearly symmetrical with respect to the stop, each system containing three or more lenses. { 'dā,gȯr ,lenz }

daguerreotype [GRAPHICS] A photograph produced on a silver plate or a copper plate coated with silver sensitized by the action of iodine; after exposure of the plate in a camera, a latent image is developed by use of mercury vapor. { də'ger·ə,tīp }

Dahlin's algorithm [CONT SYS] A digital control algorithm in which the requirement of minimum response time used in the deadbeat algorithm is relaxed to reduce ringing in the system response. { 'dä·lənz ,al·gə,rith·əm }

dahoma *See* African greenheart. { də,hō·mə }

daily aberration *See* diurnal aberration. { ¦dā·lē ab·ə'rā·shən }

daily forecast [METEOROL] A forecast for periods of from 12 to 48 hours in advance. { ¦dā·lē 'fȯr,kast }

daily keying element [COMMUN] Part of a specific cipher key that changes at predetermined intervals, usually daily. { ¦dā·lē ,kē·iŋ ,el·ə·mənt }

daily mean [METEOROL] The average value of a meteorological element over a period of 24 hours. { ¦dā·lē ,mēn }

daily rate *See* watch rate. { ¦dā·lē ,rāt }

daily retardation [OCEANOGR] The amount of time by which corresponding tidal phases grow later day by day; averages approximately 50 minutes. { ¦dā·lē ,re,tär'dā·shən }

DACTYLOGYROIDEA

Pseudohaliotrema carbunculus from the pinfish.

DAGOR LENS

Diagram of Dagor lens.

daily variation [GEOPHYS] Oscillation occurring in the earth's magnetic field in a 1-day period. { 'dā·lē ,ver·ē'ā·shən }

dairy [AGR] A farm concerned with the production of milk. [FOOD ENG] An establishment where milk products are made. { 'der·ē }

dairy cattle [AGR] A cattle breed selected, raised, and bred for ability to produce large quantities of milk. { 'der·ē ,kad·əl }

dairy machinery [FOOD ENG] Equipment, such as pasteurizers, used in the production and processing of milk and milk products. { 'der·ē mə'shēn·rē }

dairy product [FOOD ENG] Any food made from milk or milk products. { 'der·ē prä·dəkt }

daisy chain [COMPUT SCI] A means of connecting devices (readers, printers, and so on) to a central processor by party-line input/output buses which join these devices by male and female connectors, the last female connector being shorted by a suitable line termination. { 'dāz·ē ,chān }

daisy wheel printer [COMPUT SCI] A serial printer in which the printing element is a plastic hub that has a large number of flexible radial spokes, each spoke having one or more different raised printing characters; the wheel is rotated as it is moved horizontally step by step under computer control, and stops when a desired character is in a desired print position so a hammer can drive that character against an inked ribbon. { 'dāz·ē ,wēl ,print·ər }

Dakotan [GEOL] Lower Upper Cretaceous geologic time. { də'kot·ən }

dalapon [ORG CHEM] Generic name for 2,2-dichloropropionic acid; a liquid with a boiling point of 185–190°C at 760 mmHg; soluble in water, alcohol, and ether; used as a herbicide. { 'dal·ə,pän }

Dalatiidae [VERT ZOO] The spineless dogfishes, a family of modern sharks belonging to the squaloid group. { ,dal·ə'tī·ə,dē }

d'Alembertian [MATH] A differential operator in four-dimensional space,

$$\frac{\partial^2}{\partial x^2} + \frac{\partial^2}{\partial y^2} + \frac{\partial^2}{\partial z^2} - \frac{1}{c^2}\frac{\partial^2}{\partial t^2}$$

which is used in the study of relativistic mechanics. { ,dal·əm¦bər·shən }

d'Alembert's paradox [FL MECH] The paradox that no forces act on a body moving at constant velocity in a straight line through a large mass of incompressible, inviscid fluid which was initially at rest, or in uniform motion. { ,dal·əm¦bərz 'par·ə,däks }

d'Alembert's principle [MECH] The principle that the resultant of the external forces and the kinetic reaction acting on a body equals zero. { ,dal·əm¦bərz ,prin·sə·pəl }

d'Alembert's solution [PHYS] A general solution to the linearized small-amplitude one-dimensional wave equation, consisting of two traveling waves of arbitrary shape which travel in opposite directions with a constant wave speed and with no change in shape or amplitude. { ,da·ləm'berz sə,ü·shən }

d'Alembert's test for convergence [MATH] A series Σa_n converges if there is an N such that the absolute value of the ratio a_n/a_{n-1} is always less than some fixed number smaller than 1, provided n is at least N, and diverges if the ratio is always greater than 1. Also known as generalized ratio test. { ,dal·əm¦bərz ,test fər kən'vər·jəns }

d'Alembert's wave equation *See* wave equation. { ,dal·əm¦bərz 'wāv i¦kwā·zhən }

Dalitz pair [PARTIC PHYS] The electron and positron resulting from the decay of a neutral pion to these particles and a photon. { 'dä·lits ,per }

Dalitz plot [PARTIC PHYS] Pictorial representation for data on the distribution of certain three-particle configurations that result from elementary-particle decay processes or high-energy nuclear reactions. { 'dä·lits ,plät }

Dallis grass [BOT] The common name for the tall perennial forage grasses composing the genus *Paspalum* in the order Cyperales. { 'da·ləs ,gras }

Dall tube [MECH ENG] Fluid-flow measurement device, similar to a venturi tube, inserted as a section of a fluid-carrying pipe; flow rate is measured by pressure drop across a restricted throat. { 'dol ,tüb }

dalmatian sage oil *See* oil of sage. { dal'mā·shən 'sāj ,oil }

Dalmatian wettability [PETRO ENG] Theory of wettability that some in situ reservoir rocks are partly preferentially oil-wet and partly preferentially water-wet. { dal'mā·shən wed·ə'bil·əd·ē }

dalton *See* atomic mass unit. { 'dol·tən }

Dalton's atomic theory [CHEM] Theory forming the basis of accepted modern atomic theory, according to which matter is made of particles called atoms, reactions must take place between atoms or groups of atoms, and atoms of the same element are all alike but differ from atoms of another element. { 'dol·tənz ə,täm·ik 'thē·ə·rē }

Dalton's law [PHYS] The law that the pressure of a gas mixture is equal to the sum of the partial pressures of the gases composing it. Also known as law of partial pressures. { 'dol·tənz ,lo }

Dalton's temperature scale [THERMO] A scale for measuring temperature such that the absolute temperature T is given in terms of the temperature on the Dalton scale τ by $T = 273.15(373.15/273.15)^{\tau/100}$. { 'dol·tənz 'tem·prə·chər ,skāl }

dam [CIV ENG] **1.** A barrier constructed to obstruct the flow of a watercourse. **2.** A pair of cast-steel plates with interlocking fingers built over an expansion joint in the road surface of a bridge. { dam }

DAMA *See* demand assignment multiple access. { ¦dē¦ā¦em'ā *or* 'däm·ə }

damage [ORD] **1.** An injury short of complete destruction inflicted upon persons, equipment, or installations. **2.** To cause damage. { 'dam·ij }

damage assessment [ORD] The result of examination of combat material, particularly aircraft, ships, and armored vehicles, after a simulated attack, to determine the category in which the damage resulting from the attack would be placed. { 'dam·ij ə,ses·mənt }

damage control [ORD] Procedures for maintaining or restoring integrity, stability, or weapon power of a ship or aircraft. { 'dam·ij kən,trōl }

damaged pack [COMPUT SCI] A disk drive whose use is impaired by physical damage such as a scratch on the recording surface or by a serious software error that renders control information on the disk unreadable. { 'dam·ijd 'pak }

damage potential [ORD] The damage which a projectile or explosive can be expected to do to a specific target. { 'dam·ij pə,ten·chəl }

damage radius [ORD] **1.** The distance at which, in terms of experience or theoretical calculations, certain types of damage can be expected from a specified type of explosive item. **2.** In atomic explosion, the distance from ground zero at which there is a 50% probability that a target element susceptible to the weapon effect considered will be damaged. { 'dam·ij ,rād·ē·əs }

damage tolerance [ENG] The ability of a structure to maintain its load-carrying capability after exposure to a sudden increase in load. { 'dam·ij ,täl·ə·rəns }

damaging stress [MECH] The minimum unit stress for a given material and use that will cause damage to the member and make it unfit for its expected length of service. { 'dam·ə·jiŋ 'stres }

damask [TEXT] A fabric with a satin-weave pattern against a plainwoven background, made on a jacquard loom. { 'dam·əsk }

damkjernite [PETR] A melanocratic dike rock composed of biotite and pyroxene phenocrysts in a groundmass of pyroxene, biotite, and magnetite. { 'dam·kyər,nīt }

Damköhler number I [PHYS] A dimensionless number, equal to the ratio of the time it takes a fluid to flow some characteristic distance, to the time it takes some chemical reaction or other physical process to be completed. Symbolized *Da* I. Also known as Damköhler's ratio. { 'däm,kər·lər ¦nəm·bər ¦wən }

Damköhler number II [PHYS] A measure of the ratio of the rate of a chemical reaction to the rate of molecular diffusion, equal to the square of a characteristic length divided by the product of the diffusivity and the time it takes for a chemical reaction or other physical process to be completed. Symbolized *Da* II. { 'däm,kər·lər ¦nəm·bər ¦tü }

Damköhler number III [PHYS] A measure of the ratio of the heat liberated by a chemical reaction to the bulk transport of heat in a fluid, equal to the time it takes the fluid to travel

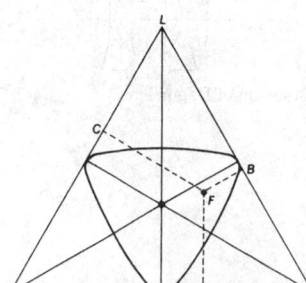

DALITZ PLOT

Configuration of a three-particle system *(abc)* in its barycentric frame is specified by a point F such that the three perpendiculars *FA*, *FB*, and *FC* to the sides of an equilateral triangle *LMN* (of height Q) are equal in magnitude to the kinetic energies T_a, T_b, T_c, where Q denotes the sum. Heavy curve encloses points which correspond to physically allowed configurations.

a characteristic length, divided by the product of the fluid temperature and the time it would take for the chemical reaction to raise this temperature one unit if all the heat liberated by it were immediately absorbed by the fluid. Symbolized *Da* III. { 'däm,kər·lər ¦nəm·bər ¦thrē }

Damköhler number IV [PHYS] A measure of the ratio of the heat liberated by a chemical reaction to the conductive heat transfer, equal to a characteristic length times the heat liberated per unit volume per unit time divided by the product of the thermal conductivity and the temperature. Symbolized *Da* IV. { 'däm,kər·lər ¦nəm·bər ¦fōr }

Damköhler number V *See* Reynolds number. { 'däm,kər·lər ¦nəm·bər ¦fīv }

Damköhler's ratio *See* Damköhler number I. { 'däm,kər·lərz ,rā·shō }

dammar [MATER] **1.** A type of hard resin obtained from evergreen trees of the genus *Agathis*. **2.** A type of soft, clear to yellow East Indian resin derived from several trees of the family Dipterocarpaceae. Also known as gum dammar. { 'dam·ər }

dammar varnish [MATER] Varnish made from East Indian dammar. { 'dam·ər ,vär·nəsh }

damp [ENG] To reduce the fire in a boiler or a furnace by putting a layer of damp coals or ashes on the fire bed. [MIN ENG] A poisonous gas in a coal mine. [PHYS] To gradually diminish the amplitude of a vibration or oscillation. { damp }

damp air [METEOROL] Air that has a high relative humidity. { ¦damp ¦er }

damp course [CIV ENG] A layer of impervious material placed horizontally in a wall to keep out water. { ¦damp ,kórs }

damp down [MET] To stop the blast in a blast furnace by closing the openings in the furnace. { ¦damp 'daún }

damped harmonic motion [PHYS] **1.** Also known as damped oscillation; damped vibration. **2.** The linear motion of a particle subject both to an elastic restoring force proportional to its displacement and to a frictional force in the direction opposite to its motion and proportional to its speed. **3.** A similar variation in a quantity analogous to the displacement of a particle, such as the charge on a capacitor in a simple series circuit containing a resistance. { ¦dampt har'män·ik 'mō·shən }

damped oscillation [PHYS] **1.** Any oscillation in which the amplitude of the oscillating quantity decreases with time. Also known as damped vibration. **2.** *See* damped harmonic motion. { ¦dampt ,äs·ə'lā·shən }

damped regression analysis *See* ridge regression analysis. { ¦dampt ri'gresh·ən ə,nal·ə·səs }

damped vibration *See* damped harmonic motion; damped oscillation. { ¦dampt vī'brā·shən }

damped wave [PHYS] **1.** A wave whose amplitude drops exponentially with distance because of energy losses which are proportional to the square of the amplitude. **2.** A wave in which the amplitudes of successive cycles progressively diminish at the source. { ¦dampt ,wāv }

dampener [ENG] A device for damping spring oscillations after abrupt removal or application of a load. [GRAPHICS] In offset printing, one of the rollers that transfers water to the nonprinting areas of the total surface that is printed. { 'dam·pə·nər }

damper [ELECTR] A diode used in the horizontal deflection circuit of a television receiver to make the sawtooth deflection current decrease smoothly to zero instead of oscillating at zero; the diode conducts each time the polarity is reversed by a current swing below zero. [MECH ENG] A valve or movable plate for regulating the flow of air or the draft in a stove, furnace, or fireplace. { 'dam·pər }

damper loss [ENG] The reduction in rate of flow or of pressure of gas across a damper. { 'dam·pər ,lós }

damper pedal [ENG] A pedal that controls the damping of piano strings. { 'dam·pər ,ped·əl }

damper winding [ELEC] A winding consisting of several conducting bars on the field poles of a synchronous machine, short-circuited by conducting rings or plates at their ends, and used to prevent pulsating variations of the position or magnitude of the magnetic field linking the poles. Also known as amortisseur winding. { 'dam·pər ,wīnd·iŋ }

damp haze [METEOROL] Small water droplets or very hygroscopic particles in the air, reducing the horizontal visibility somewhat, but to not less than $1\frac{1}{4}$ miles (2 kilometers);

similar to a very thin fog, but the droplets or particles are more scattered than in light fog and presumably smaller. { ¦damp 'hāz }

damping [ENG] Reducing or eliminating reverberation in a room by placing sound-absorbing materials on the walls and ceiling. Also known as soundproofing. [PHYS] **1.** The dissipation of energy in motion of any type, especially oscillatory motion and the consequent reduction or decay of the motion. **2.** The extent of such dissipation and decay. { 'dam·piŋ }

damping capacity [MECH] A material's capability in absorbing vibrations. { 'dam·piŋ kə'pas·əd·ē }

damping coefficient *See* damping factor; resistance. { 'dam·piŋ ,kō·i,fish·ənt }

damping constant *See* resistance. { 'dam·piŋ ,kän·stənt }

damping factor [PHYS] **1.** The ratio of the logarithmic decrement of any underdamped harmonic motion to its period. Also known as damping coefficient. **2.** *See* decrement. { 'dam·piŋ ,fak·tər }

damping magnet [ELECTROMAG] A permanent magnet used in conjunction with a disk or other moving conductor to produce a force that opposes motion of the conductor and thereby provides damping. { 'dam·piŋ ,mag·nət }

damping-off [PL PATH] A fungus disease of seedlings and cuttings in which the parasites invade the plant tissues near the ground level, causing wilting and rotting. { ¦dam·piŋ 'óf }

damping ratio [PHYS] The ratio of the actual resistance in damped harmonic motion to that necessary to produce critical damping. Also known as relative damping ratio. { 'dam·piŋ ,rā·shō }

damping resistor [ELEC] **1.** A resistor that is placed across a parallel resonant circuit or in series with a series resonant circuit to decrease the Q factor and thereby eliminate ringing. **2.** A noninductive resistor placed across an analog meter to increase damping. { 'dam·piŋ ri,zis·tər }

damp sheet [MIN ENG] A curtain used in a mine to direct airflow, thus preventing gas accumulation. { ¦damp ¦shēt }

Danaidae [INV ZOO] A family of large tropical butterflies, order Lepidoptera, having the first pair of legs degenerate. { də'nā·ə,dē }

danalite [MINERAL] $(Fe,Mn,Zn)_4Be_3(SiO_4)_3S$ A mineral consisting of a silicate and sulfide of iron and beryllium; it is isomorphous with helvite and genthelvite. { 'dä·nə,līt }

dan buoy [NAV] A floating marker buoy used temporarily, as on fishing grounds and in minesweeping operations. { 'dan ,bói }

danburite [MINERAL] $CaB_2(SiO_4)_2$ An orange-yellow, yellowish-brown, grayish, or colorless transparent to translucent borosilicate mineral with a feldspar structure crystallizing in the orthorhombic system; it resembles topaz and is used as an ornamental stone. { 'dan·bə,rīt }

dance-hall machine [COMPUT SCI] A multiprocessor in which the memory is spread over several modules, and a switch is used to make connections between memory modules and processors, so that several processors can use the memory simultaneously. { 'dans ,hól mə,shēn }

dancing dervish *See* dust whirl. { ¦dan·siŋ 'dər·vish }

dancing devil *See* dust whirl. { ¦dan·siŋ ¦dev·əl }

dancing step *See* balanced step. { ¦dan·siŋ ¦step }

dancing winder *See* balanced step. { ¦dan·siŋ ¦wīn·dər }

Danckwerts model [CHEM ENG] Theory applied to liquid flow across packing in a liquid-gas absorption tower; allows for liquid eddies that bring fresh liquid from the interior of the liquid body to the surface, thus contacting the gas in the column. { 'daŋk·verts ,mäd·əl }

Dandelin sphere [MATH] For a conic that is represented as the intersection of a plane and a circular cone, a sphere that is a tangent to both the plane and the cone. { 'dänd,laŋ ,sfir }

dandruff [MED] Scales of dry sebum formed on the scalp in seborrhea. { 'dan·drəf }

dandy fever *See* dengue. { 'dan·dē ,fē·vər }

dandy roll [MECH ENG] A roll in a Fourdrinier paper-making machine; used to compact the sheet and sometimes to imprint a watermark. { 'dan·dē ,rōl }

Dane particle [VIROL] The causative virus of type B viral hepatitis visualized ultrastructurally in its complete form. { 'dān ,pard·ə·kəl }

danger angle [NAV] The angle between two known points as measured by an observer at a third point (the vessel), which is the safety limit for a vessel approaching a dangerous obstacle such as a reef. { 'dān·jər ‚aŋ·gəl }

danger area [NAV] A specified area within, below, or over which there may exist activities constituting potential danger to aircraft flying over it, or to persons, property, and traffic on land or sea. { 'dān·jər 'er·ē·ə }

danger bearing [NAV] The bearing of any object or obstruction as measured on board a vessel which will put a ship in jeopardy. { 'dān·jər ‚ber·iŋ }

danger buoy [NAV] A buoy marking an isolated danger to navigation. { 'dān·jər ‚bȯi }

danger coefficient [NUCLEO] The change in reactivity per unit mass of a substance resulting from inserting the substance in a particular nuclear reactor. { 'dān·jər ‚kō·i‚fish·ənt }

danger line [NAV] A line on a chart representing a boundary, beyond which some hazard will be encountered. { 'dān·jər ‚līn }

dangerous semicircle [METEOROL] The half of the circular area of a tropical cyclone having the strongest winds and heaviest seas, where a ship tends to be drawn into the path of the storm. { 'dān·jə·rəs 'sem·i‚sər·kəl }

danger sounding [NAV] A minimum sounding chosen for a vessel of specific draft in a given area to indicate the limit of safe navigation. { 'dān·jər ‚saùnd·iŋ }

danger space [ORD] **1.** That portion of the range within which a target of given dimensions could be hit by a projectile with a given angle of fall. **2.** Space around the bursting point of an antiaircraft projectile. { 'dān·jər ‚spās }

dangler [MET] The flexible electrode used in barrel plating. { 'daŋ·glər }

dangling bond [SOLID STATE] A chemical bond associated with an atom in the surface layer of a solid that does not join the atom with a second atom but extends in the direction of the solid's exterior. { ‚daŋ·gliŋ 'bänd }

dangling ELSE [COMPUT SCI] A situation in which it is not clear to which part of a compound conditional statement an ELSE instruction belongs. { ‚daŋ·gliŋ 'els }

Danian [GEOL] Lowermost Paleocene or uppermost Cretaceous geologic time. { 'dān·ē·ən }

Daniell cell [PHYS CHEM] A primary cell with a constant electromotive force of 1.1 volts, having a copper electrode in a copper sulfate solution and a zinc electrode in dilute sulfuric acid or zinc sulfate, the solutions separated by a porous partition or by gravity. { 'dan·yəl ‚sel }

Daniell hygrometer [ENG] An instrument for measuring dew point; dew forms on the surface of a bulb containing ether which is cooled by evaporation into another bulb, the second bulb being cooled by the evaporation of ether on its outer surface. { 'dan·yəl hī'gräm·əd·ər }

Danjon prismatic astrolabe [ENG] A type of astrolabe in which a Wollaston prism just inside the focus of the telescope converts converging beams of light into parallel beams, permitting a great increase in accuracy. { 'dän·yən priz'mad·ik 'as·trə‚lāb }

dannemorite [MINERAL] (Fe,Mn,Mg)$_7$Si$_8$O$_{22}$(OH)$_2$ A yellowish-brown to greenish-gray monoclinic mineral consisting of a columnar or fibrous amphibole. { ‚dan·ə'mȯr‚īt }

dansyl chloride [ORG CHEM] (CH$_3$)$_2$NC$_{10}$H$_6$SO$_2$Cl A reagent for fluorescent labeling of amines, amino acids, proteins, and phenols. { 'dans·əl 'klȯr‚īd }

Danysz reaction [IMMUNOL] A toxin-antitoxin reaction that occurs when an exact equivalence of toxin is added to antitoxin, not in one portion but in successive increments. { 'dä·nish rē'ak·shən }

DAP *See* diallyl phthalate; diaminopimelate.

Daphniphyllales [BOT] An order of dicotyledonous plants in the subclass Hamamelidae, consisting of a single family with one genus, *Daphniphyllum*, containing about 35 species; dioecious trees or shrubs native to eastern Asia and the Malay region, they produce a unique type of alkaloid and often accumulate aluminum and sometimes produce iridoid compounds. { ‚daf·ni·fə'lā·lēz }

daphnite [MINERAL] (MgFe)$_3$(Fe,Al)$_3$(Si,Al)$_4$O$_{10}$(OH)$_8$ A mineral of the chlorite group consisting of a basic aluminosilicate of magnesium, iron, and aluminum. { 'daf‚nīt }

Daphoenidae [PALEON] A family of extinct carnivoran mammals in the superfamily Miacoidea. { də'fēn·ə‚dē }

dapsone *See* 4,4'-sulfonyldianiline. { 'dap‚sōn }

daraf [ELEC] The unit of elastance, equal to the reciprocal of 1 farad. { 'da‚raf }

darapskite [MINERAL] Na$_3$(NO$_3$)(SO$_4$)·H$_2$O A naturally occurring hydrate mineral consisting of a hydrous nitrate and sulfate of sodium. { də'rap‚skīt }

Darboux's monodromy theorem [MATH] The proposition that, if the function $f(z)$ of the complex variable z is analytic in a domain D bounded by a simple closed curve C, and $f(z)$ is continuous in the union of D and C and is injective for z on C, then $f(z)$ is injective for z in D. { 'där·büz ‚män·ə‚drä·mē ‚thir·əm }

darby [ENG] A flat-surfaced tool for smoothing plaster. { 'där·bē }

darcy [PHYS] A unit of permeability, equivalent to the passage of 1 cubic centimeter of fluid of 1 centipoise viscosity flowing in 1 second under a pressure of 1 atmosphere through a porous medium having a cross-sectional area of 1 square centimeter and a length of 1 centimeter. { 'där·sē }

Darcy number 1 [FL MECH] A dimensionless group, equal to four times the Fanning friction factor. Symbolized Da_1. Also known as Darcy-Weisbach coefficient; resistance coefficient 2. { 'där·sē ‚nəm·bər 'wən }

Darcy number 2 [FL MECH] A dimensionless group used in the study of the flow of fluids in porous media, equal to the fluid velocity times the flow path divided by the permeability of the medium. Symbolized Da_2. { 'där·sē ‚nəm·bər ‚tü }

Darcy's law [FL MECH] The law that the rate at which a fluid flows through a permeable substance per unit area is equal to the permeability, which is a property only of the substance through which the fluid is flowing, times the pressure drop per unit length of flow, divided by the viscosity of the fluid. { 'där·sēz ‚lȯ }

Darcy-Weisbach coefficient *See* Darcy number 1. { ‚där·sē 'vīs‚bäk ‚kō·i‚fish·ənt }

Darcy-Weisbach equation [FL MECH] An equation for the loss of head due to friction h_f during turbulent flow of a fluid through a duct of any shape; in the case of a circular pipe, $h_f = f(L/d)(V^2/2g)$, where L and d are the length and diameter of the pipe, V is the fluid velocity, g the acceleration of gravity, and f a dimensionless number called Darcy number 1. { ‚där·sē 'vīs‚bäk i‚kwā·zhən }

dark box [GRAPHICS] A light-proof box used to store certain light-sensitive photographic papers. { 'därk ‚bäks }

dark cloud [ASTRON] A relatively dense, cool cloud of interstellar gas, chiefly molecular, whose dust particles obscure the light of stars behind it. { 'därk ‚klaùd }

dark conduction [ELECTR] Residual conduction in a photosensitive substance that is not illuminated. { ‚därk kən‚dək·shən }

dark current *See* electrode dark current. { 'därk ‚kər·ənt }

dark-current pulse [ELECTR] A phototube dark-current excursion that can be resolved by the system employing the phototube. { 'därk ‚kər·ənt ‚pəls }

dark discharge [ELECTR] An invisible electrical discharge in a gas. { ‚därk 'dis‚chärj }

dark-eclipsing variables [ASTRON] A binary star system, comprising a bright star and an almost dark companion that revolve about each other. { ‚därk ə‚klip·siŋ 'ver·ē·ə·bəlz }

dark-field illumination [OPTICS] A method of microscope illumination in which the illuminating beam is a hollow cone of light formed by an opaque stop at the center of the condenser large enough to prevent direct light from entering the objective; the specimen is placed at the concentration of the light cone, and is seen with light scattered or diffracted by it. { 'därk ‚fēld ə‚lüm·ə'nā·shən }

dark lightning [GRAPHICS] A photographic effect in which lightning gives black photographic streaks instead of white due to multiple exposures caused by successive members of a composite flash. Also known as Clayden effect. { ‚därk ‚līt·niŋ }

dark-line spectrum [SPECT] The absorption spectrum that results when white light passes through a substance, consisting of dark lines against a bright background. { 'därk ‚līn 'spek·trəm }

dark matter [ASTRON] Matter that is postulated to exist to explain the rotational motion of the Milky Way Galaxy and other galaxies, to explain the motions of galaxies in clusters, and, in certain cosmological theories, to achieve the critical

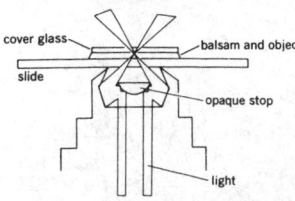

DARK-FIELD ILLUMINATION

cover glass — balsam and object
slide
— opaque stop
— light

Diagram of dark-field illumination. *(American Optical Corp.)*

density of matter in the universe that is just sufficient to close the universe. Also known as missing mass. { 'därk ‚mad·ər }

dark nebula [ASTRON] A cloud of solid particles which absorbs or scatters away radiation directed toward an observer and becomes apparent when silhouetted against a bright nebula or rich star field. Also known as absorption nebula. { 'därk ‚neb·yə·lə }

dark of the moon [ASTRON] **1.** The time period of approximately a week at the time of a new moon, when the light of the moon is absent at night. **2.** Any period in which the light of the moon is obscured. { 'därk əv thə 'mün }

dark plaster [MATER] A plaster made from calcined, unground gypsum. { 'därk 'plas·tər }

dark-red silver ore See pyrargyrite. { 'därk 'red 'sil·vər ‚ȯr }

dark resistance [ELECTR] The resistance of a selenium cell or other photoelectric device in total darkness. { 'därk ri‚zis·təns }

darkroom [GRAPHICS] A light-free room illuminated by a safelight for processing light-sensitive materials. { 'därk ‚rüm }

darkroom filter [OPTICS] An optical component of glass, gelatin, or other material used to alter the radiation emitted by the darkroom light source so that only specific wavelengths are transmitted. { 'därk‚rüm ‚fil·tər }

dark-ruby silver See pyrargyrite. { 'därk ‚rü·bē ‚sil·vər }

dark satellite [AERO ENG] Satellite that gives no information to a friendly ground environment, either because it is controlled or because the radiating equipment is inoperative. { 'därk 'sad·əl‚īt }

dark segment [METEOROL] A bluish-gray band appearing along the horizon opposite the rising or setting sun and lying just below the antitwilight arch. Also known as earth's shadow. { 'därk ‚seg·mənt }

dark space [ELECTR] A region in a glow discharge that produces little or no light. { 'därk ‚spās }

dark spot [ELECTR] A spot on a television receiver tube that results from a spurious signal generated in the television camera tube during rescan, generally from the redistribution of secondary electrons over the mosaic in the tube. { 'därk ‚spät }

dark star [ASTRON] A star that is not visible but is a part of a binary star system; in particular, a star which causes, in an eclipsing variable, a primary eclipse. { 'därk 'stär }

dark-trace tube [ELECTR] A cathode ray tube with a bright face that does not necessarily luminesce, on which signals are displayed as dark traces or dark blips where the potassium chloride screen is hit by the electron beam. Also known as skiatron. { 'därk ‚trās ‚tüb }

Darling shower [METEOROL] A dust storm caused by cyclonic winds in the vicinity of the River Darling in Australia. { 'där·liŋ ‚shau̇·ər }

Darlington amplifier [ELECTR] A current amplifier consisting essentially of two separate transistors and often mounted in a single transistor housing. { 'där·liŋ·tən ‚am·plə‚fī·ər }

Darrier's disease [MED] A genetically determined disease characterized by patches of papules of the horny layer of skin. Also known as keratosis follicularis. { 'där·ē‚āz diz‚ēz }

DARS See direct audio radio service. { ‚dē‚ä'är'es or därz }

d'Arsonval current [ELEC] A current consisting of isolated trains of heavily damped high-frequency oscillations of high voltage and relatively low current, used in diathermy. { 'dars·ən‚vȯl ‚kər·ənt }

d'Arsonval galvanometer [ENG] A galvanometer in which a light coil of wire, suspended from thin copper or gold ribbons, rotates in the field of a permanent magnet when current is carried to it through the ribbons; the position of the coil is indicated by a mirror carried on it, which reflects a light beam onto a fixed scale. Also known as light-beam galvanometer. { 'dars·ən‚vȯl gal·və'näm·əd·ər }

dart [INV ZOO] A small sclerotized structure ejected from the dart sac of certain snails into the body of another individual as a stimulant before copulation. { därt }

dart configuration [AERO ENG] An aerodynamic configuration in which the control surfaces are at the tail of the vehicle. { 'därt kən‚fig·yə'rā·shən }

dart leader [GEOPHYS] The leader which, after the first stroke, initiates each succeeding stroke of a composite flash of lightning. Also known as continuous leader. { 'därt ‚lēd·ər }

dart sac [INV ZOO] A dart-forming pouch associated with the reproductive system of certain snails. { 'därt ‚sak }

darwin [EVOL] A unit of evolutionary rate of change; if some dimension of a part of an animal or plant, or of the whole animal or plant, changes from l_o to l_t over a time of t years according to the formula $l_t = l_o \exp (Et/10^6)$, its evolutionary rate of change is equal to E darwins. { 'där·wən }

Darwin curve [CRYSTAL] A plot of the intensity of diffracted x-rays from a perfect crystal as a function of angle. { 'där·win ‚kərv }

Darwin-Doodson system [GEOPHYS] A method for predicting tides by expressing them as sums of harmonic functions of time. { 'där·wən 'düd·sən ‚sis·təm }

Darwin ellipsoids [ASTRON] Ellipsoidal figures of equilibrium of homogeneous bodies moving about each other in circular orbits, calculated by making certain approximations about their mutual tidal influences. { 'där·wən ə'lip‚sȯidz }

Darwin glass [GEOL] A highly siliceous, vesicular glass shaped in smooth blobs or twisted shreds, found in the Mount Darwin range in western Tasmania. Also known as queenstownite. { 'där·wən ‚glas }

Darwinism [BIOL] The theory of the origin and perpetuation of new species based on natural selection of those offspring best adapted to their environment because of genetic variation and consequent vigor. Also known as Darwin's theory. { 'där·wə‚niz·əm }

Darwin's finch [VERT ZOO] A bird of the subfamily Fringillidae; Darwin studied the variation of these birds and used his data as evidence for his theory of evolution by natural selection. { 'där·winz 'finch }

Darwin's theory See Darwinism. { 'där·winz 'thē·ə·rē }

Darwinulacea [INV ZOO] A small superfamily of nonmarine, parthenogenetic ostracods in the suborder Podocopa. { ‚där‚win·ə'lās·ē·ə }

Darzen's procedure [ORG CHEM] Preparation of alkyl halides by refluxing a molecule of an alcohol with a molecule of thionyl chloride in the presence of a molecule of pyridine. { 'där·zənz prə‚sē·jər }

Darzen's reaction [ORG CHEM] Condensation of aldehydes and ketones with α-haloesters to produce glycidic esters. { 'där·zənz rē‚ak·shən }

Dasyatidae [VERT ZOO] The stingrays, a family of modern sharks in the batoid group having a narrow tail with a single poisonous spine. { ‚da·sā'ad·ə‚dē }

Dascillidae [INV ZOO] The soft-bodied plant beetles, a family of coleopteran insects in the superfamily Dascilloidea. { də'sil·ə‚dē }

Dascilloidea [INV ZOO] Superfamily of coleopteran insects in the suborder Polyphaga. { ‚das·ə'lȯid·ē·ə }

DASD See direct-access storage device. { daz‚dē }

dasheen [BOT] Colocasia esculenta. A plant in the order Arales, grown for its edible corm. { da'shēn }

dashkesanite [MINERAL] $(Na,K)Ca_2(Fe,Mg)_5(Si,Al)_8O_{22}Cl_2$ A monoclinic mineral of the amphibole group consisting of a chloroaluminosilicate of sodium, potassium, iron, and magnesium. { ‚dash·kə'sa‚nīt }

dashpot [MECH ENG] A device used to dampen and control a motion, in which an attached piston is loosely fitted to move slowly in a cylinder containing oil. { 'dash‚pät }

Dasycladaceae [BOT] A family of green algae in the order Dasycladales comprising plants formed of a central stem from which whorls of branches develop. { ‚das·ə·klə'dās·ē‚ē }

Dasycladales [BOT] An order of lime-encrusted marine algae in the division Chlorophyta, characterized by a thallus composed of nonseptate, highly branched tubes. { ‚das·ə·klə'dā·lēz }

dasymeter [PHYS] A thin glass globe used to measure the density of gas by weighing the globe in the gas. { da'sim·əd·ər }

Dasyonygidae [INV ZOO] A family of biting lice, order Mallophaga, that are confined to rodents of the family Procaviidae. { ‚das·ē·ə'nij·ə‚dē }

Dasypodidae [VERT ZOO] The armadillos, a family of edentate mammals in the infraorder Cingulata. { ‚das·ə'päd·ə‚dē }

Dasytidae [INV ZOO] An equivalent name for Melyridae. { də'sid·ə‚dē }

Dasyuridae [VERT ZOO] A family of mammals in the order Marsupialia characterized by five toes on each hindfoot. { das·ē'yu̇r·ə‚dē }

Dasyuroidea [VERT ZOO] A superfamily of marsupial mammals. { ‚das·ē·yə'rȯid·ē·ə }

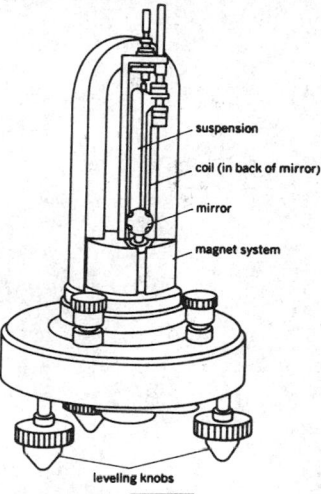

D'ARSONVAL GALVANOMETER

- suspension
- coil (in back of mirror)
- mirror
- magnet system
- leveling knobs

Drawing of d'Arsonval galvanometer. (From D. M. Considine, ed., Process Instruments and Control Handbook, McGraw-Hill, 1957)

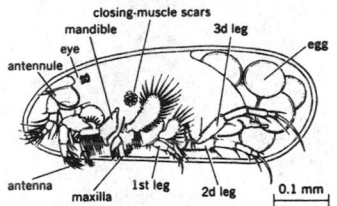

DARWINULACEA

closing-muscle scars, mandible, eye, antennule, 3d leg, egg, antenna, maxilla, 1st leg, 2d leg, 0.1 mm

Darwinula stevensoni.

DAT *See* digital audio tape.

data [COMPUT SCI] **1.** General term for numbers, letters, symbols, and analog quantities that serve as input for computer processing. **2.** Any representations of characters or analog quantities to which meaning, if not information, may be assigned. [SCI TECH] Numerical or qualitative values derived from scientific experiments. { 'dad·ə, 'dād·ə, *or* 'däd·ə }

data acquisition [COMMUN] The phase of data handling that begins with the sensing of variables and ends with a magnetic recording or other record of raw data; may include a complete radio telemetering link. { 'dad·ə ,ak·wə,zish·ən }

data acquisition computer [COMPUT SCI] A computer that is used to acquire and analyze data generated by instruments. { 'dad·ə ,ak·wə,zish·ən kəm'pyüd·ər }

data aggregate [COMPUT SCI] The set of data items within a record. { 'dad·ə ,ag·rə·gət }

data analysis [COMPUT SCI] The evaluation of digital data. { 'dad·ə ə,nal·ə·səs }

data attribute [COMPUT SCI] A characteristic of a block of data, such as the type of representation used or the length in characters. { 'dad·ə ¦a·trə'byüt }

data automation [COMPUT SCI] The use of electronic, electromechanical, or mechanical equipment and associated techniques to automatically record, communicate, and process data and to present the resultant information. { ¦dad·ə öd·ə'mā·shən }

data bank [COMPUT SCI] A complete collection of information such as contained in automated files, a library, or a set of computer disks. { 'dad·ə ,baŋk }

database [COMPUT SCI] A nonredundant collection of interrelated data items that can be shared and used by several different subsystems. { 'dad·ə,bās }

database/data communication [COMPUT SCI] An advanced software product that combines a database management system with data communications procedures. Abbreviated DB/DC. { 'dad·ə,bās 'dad·ə kə,myü·nə'kā·shən }

database machine [COMPUT SCI] A computer that handles the storage and retrieval of data into and out of a database. { 'dad·ə,bās mə,shēn }

database management system [COMPUT SCI] A special data processing system, or part of a data processing system, which aids in the storage, manipulation, reporting, management, and control of data. Abbreviated DBMS. { 'dad·ə,bās 'man·ij·mənt ,sis·təm }

database server [COMPUT SCI] An independently functioning computer in a local-area network that holds and manages the database. { 'dad·ə,bās ,sər·vər }

data break [COMPUT SCI] A facility which permits input/output transfers to occur without disturbing program execution in a computer. { 'dad·ə ,brāk }

data buffering [COMPUT SCI] The temporary collection and storage of data awaiting further processing in physical storage devices, allowing a computer and its peripheral devices to operate at different speeds. { 'dad·ə ,bəf·ə·riŋ }

data bus [ELECTR] An internal channel that carries data between a computer's central processing unit and its random-access memory. { 'dad·ə ,bəs }

data capture [COMPUT SCI] The acquisition of data to be entered into a computer. { 'dad·ə ,kap·chər }

data carrier [COMPUT SCI] A medium on which data can be recorded, and which is usually easily transportable, such as cards, tape, paper, or disks. { 'dad·ə ,kar·ē·ər }

data carrier storage [COMPUT SCI] Any type of storage in which the storage medium is outside the computer, such as tape, cards, or disks, in contrast to inherent storage. { 'dad·ə ,kar·ē·ər ,stór·ij }

data cartridge [COMPUT SCI] A tape cartridge used for nonvolatile and removable data storage in small digital systems. { 'dad·ə ,kar·trij }

data cell drive [COMPUT SCI] A large-capacity storage device consisting of strips of magnetic tape which can be individually transferred to the read-write head. { 'dad·ə ,sel ,drīv }

data center [COMPUT SCI] An organization established primarily to acquire, analyze, process, store, retrieve, and disseminate one or more types of data. { 'dad·ə ,sen·tər }

data chain [COMPUT SCI] Any combination of two or more data elements, data items, data codes, and data abbreviations

in a prescribed sequence to yield meaningful information; for example, "date" consists of data elements year, month, and day. { 'dad·ə ,chān }

data chaining [COMPUT SCI] A technique used in scatter reading or scatter writing in which new storage areas are defined for use as soon as the current data transfer is completed. { 'dad·ə ,chān·iŋ }

data channel [COMPUT SCI] A bidirectional data path between input/output devices and the main memory of a digital computer permitting one or more input/output operations to proceed concurrently with computation. { 'dad·ə ,chan·əl }

data circuit [ELECTR] A telephone facility that allows transmission of digital data pulses with minimum distortion. { 'dad·ə ,sər·kət }

data code [COMPUT SCI] A number, letter, character, symbol, or any combination thereof, used to represent a data item. { 'dad·ə ,kōd }

data collection [COMPUT SCI] The process of sending data to a central point from one or more locations. { 'dad·ə kə,lek·shən }

data communication network [COMPUT SCI] A set of nodes, consisting of computers, terminals, or some type of communication control units in various locations, connected by links consisting of communication channels providing a data path between the nodes. { 'dad·ə kə,myü·nə,kā·shən 'net,wərk }

data communications [COMMUN] The conveying from one location to another by electrical means of information that originates or is recorded in alphabetic, numeric, or pictorial form, or as a signal that represents a measurement; includes telemetering, telegraphy, and facsimile but not voice or television. Also known as data transmission. { 'dad·ə kə,myü·nə'kā·shənz }

data communications processor [COMPUT SCI] A small computer used to control the flow of data between machines and terminals over communications channels. { 'dad·ə kə,myü·nə¦kā·shənz 'präs,es·ər }

data compression [COMPUT SCI] The technique of reducing the number of binary digits required to represent data. { 'dad·ə kəm,presh·ən }

data concentrator [ELECTR] A device, such as a microprocessor, that takes data from several different teletypewriter or other slow-speed lines and feeds them to a single higher-speed line. { 'dad·ə kän·sən,trād·ər }

data conversion [COMPUT SCI] The changing of the representation of data from one form to another, as from binary to decimal, or from one physical recording medium to another, as from card to disk. Also known as conversion. { 'dad·ə kən,vər·zhən }

data conversion line [COMPUT SCI] The channel, electronic or manual, through which data elements are transferred between data banks. { 'dad·ə kən,vər·zhən ,līn }

data converter *See* converter. { 'dad·ə kən,vərd·ər }

data definition [COMPUT SCI] The statements in a computer program that specify the physical attributes of the data to be processed, such as location and quantity of data. { 'dad·ə ,def·ə'nish·ən }

data dependence graph [COMPUT SCI] A chart that represents a program in a data flow language, in which each node is a function and each arc carries a value. { 'dad·ə di,pen·dəns ,graf }

data description language [COMPUT SCI] A programming language used to specify the arrangement of data items within a data base. { 'dad·ə di¦skrip·shən ,laŋ·gwij }

data descriptor [COMPUT SCI] A pointer indicating the memory location of a data item. { 'dad·ə di'skrip·tər }

data dictionary [COMPUT SCI] A catalog which contains the names and structures of all data types. { 'dad·ə ,dik·shə,ner·ē }

data display [COMPUT SCI] Visual presentation of processed data by specially designed electronic or electromechanical devices through interconnection (either on- or off-line) with digital computers or component equipments; although line printers and punch cards may display data, they are not usually categorized as displays but as output equipments. { 'dad·ə di,splā }

data distribution [COMPUT SCI] Data transmission to one or more locations from a central point. { 'dad·ə ,dis·trə,byü·shən }

data division [COMPUT SCI] The section of a program (written in the COBOL language) which describes each data item used for input, output, and storage. { 'dad·ə di,vizh·ən }

data-driven execution [COMPUT SCI] A mode of carrying out a program in a data flow system, in which an instruction is carried out whenever all its input values are present. { 'dad·ə ,driv·ən ,ek·sə'kyü·shən }

data element [COMPUT SCI] A set of data items pertaining to information of one kind, such as months of a year. { 'dad·ə ,el·ə·mənt }

data encryption standard [COMMUN] A cryptographic algorithm of validated strength which is in the public domain and is accepted as a standard. Abbreviated DES. { 'dad·ə en,krip·shən 'stan·dərd }

data entry [COMPUT SCI] The procedures for placing data in a computer system. { 'dad·ə ,en·trē }

data entry program [COMPUT SCI] An application program that receives data from a keyboard or other input device and stores it in a computer system. Also known as input program. { 'dad·ə ¦en·trē ,prō·grəm }

data entry terminal [COMPUT SCI] A portable keyboard and small numeric display designed for interactive communication with a computer. { 'dad·ə ¦en·trē ,tər·mon·əl }

data error [COMPUT SCI] A deviation from correctness in data, usually an error, which occurred prior to processing the data. { 'dad·ə ,er·ər }

data exchange system [COMPUT SCI] A combination of hardware and software designed to accept data from various sources, sort the data according to its destination and priority, carry out any necessary code conversions, and transmit the data to its destination. { 'dad·ə iks¦chānj ,sis·təm }

data expansion [COMPUT SCI] The reproduction in its original form of information that has undergone data compression. { 'dad·ə ik,span·chən }

data field [COMPUT SCI] An area in the main memory of the computer in which a data record is contained. { 'dad·ə ,fēld }

data flow [COMMUN] The route followed by a data message from its origination to its destination, including all the nodes through which it travels. [COMPUT SCI] The transfer of data from an external storage device, through the processing unit and memory, and out to an external storage device. { 'dad·ə ,flō }

data flow analysis [COMPUT SCI] The development of models for the movement of information within an organization, indicating the sources and destinations of information and where and how information is transmitted, processed, and stored. { 'dad·ə ¦flō ə,nal·ə·səs }

data flow diagram [COMPUT SCI] A chart that traces the movement of data in a computer system and shows how the data is to be processed, using circles to represent data. Also known as bubble chart; system flowchart. { 'dad·ə ¦flō ,dī·ə,gram }

data flow language [COMPUT SCI] A programming language used in a data flow system. { 'dad·ə ¦flō ,laŋ·gwij }

data flow system [COMPUT SCI] An alternative to conventional programming languages and architectures which is able to achieve a high degree of parallel computation, in which values rather than value containers are dealt with, and in which all processing is achieved by applying functions to values to produce new values. { 'dad·ə ¦flō ,sis·təm }

data flow technique [COMPUT SCI] A method of computer system design in which diagrams and charts that show how data is to be handled by the system are used to prepare detailed specifications from which actual programs can be written. { 'dad·ə ¦flō tek,nēk }

data formatting [COMPUT SCI] Structuring the presentation of data as numerical or alphabetic and specifying the size and type of each datum. { 'dad·ə fòr'mad·iŋ }

data generator [COMPUT SCI] A specialized word generator in which the programming is designed to test a particular class of device, the pulse parameters and timing are adjustable, and selected words may be repeated, reinserted later in the sequence, omitted, and so forth. { 'dad·ə ,jen·ə,rād·ər }

data glove [GRAPHICS] A device capable of recording hand movements, both the position of the hand and its orientation as well as finger movements; it is capable of simple gesture recognition and general tracking of three-dimensional hand orientation. { 'dad·ə ,gləv }

datagram [COMPUT SCI] A unit of information in the Internet Protocol (IP) containing both data and address information. In TCP/IP networks, datagrams are referred to as packets. { 'dad·ə,gram }

data-handling system [COMPUT SCI] Automatically operated equipment used to interpret data gathered by instrument installations. Also known as data reduction system. { 'dad·ə ,hand·liŋ ,sis·təm }

data independence [COMPUT SCI] Separation of data from processing, either so that changes in the size or format of the data elements require no change in the computer programs processing them or so that these changes can be made automatically by the database management system. { 'dad·ə in·də'pen·dəns }

data-initiated control [COMPUT SCI] The automatic handling of a program dependent only upon the value of input data fed into the computer. { 'dad·ə i,nish·ē,ād·əd kən'trōl }

data-intense application [COMPUT SCI] A program or computer system that handles large quantities of data and extremely repetitive tasks. { 'dad·ə in¦tens ,ap·lə'kā·shən }

data interchange [COMPUT SCI] Switching of data in and out of storage units. { 'dad·ə 'in·tər,chānj }

data item [COMPUT SCI] A single member of a data element. Also known as datum. { 'dad·ə ,ī·dəm }

data level [COMPUT SCI] The rank of a data element in a source language with respect to other elements in the same record. { 'dad·ə ,lev·əl }

data library [COMPUT SCI] A center for the storage of data not in current use by the computer. { 'dad·ə lī,brer·ē }

data line [COMMUN] An individual circuit that transmits data within a communications or computer channel. { 'dad·ə ,līn }

data line monitor [COMMUN] A test instrument that analyzes the signals transmitted over a communications line and provides a visual display or stores the results for further analysis, or both. { ¦dad·ə ,līn 'män·əd·ər }

data link [COMMUN] The physical equipment for automatic transmission and reception of information. Also known as communication link; information link; tie line; tie-link. { 'dad·ə ,liŋk }

data logging [COMPUT SCI] Conversion of electrical impulses from process instruments into digital data to be recorded, stored, and periodically tabulated. { 'dad·ə ,läg·iŋ }

data management [COMPUT SCI] The collection of functions of a control program that provide access to data sets, enforce data storage conventions, and regulate the use of input/output devices. { 'dad·ə ,man·ij·mənt }

data management program [COMPUT SCI] A computer program that keeps track of what is in a computer system and where it is located, and of the various means to store and access the data efficiently. { 'dad·ə ,man·ij·mənt ,prō·grəm }

data manipulation [COMPUT SCI] The standard operations of sorting, merging, input/output, and report generation. { 'dad·ə mə,nip·yə,lā·shən }

data manipulation language [COMPUT SCI] The interface between a data base and an applications program, which is embedded in the language of the applications program and provides the programmer with procedures for accessing data in the data base. { 'dad·ə mə,nip·yə¦lā·shən ,laŋ·gwij }

data mining [COMPUT SCI] **1.** The identification or extraction of relationships and patterns from data using computational algorithms to reduce, model, understand, or analyze data. **2.** The automated process of turning raw data into useful information by which intelligent computer systems sift and sort through data, with little or no help from humans, to look for patterns or to predict trends. { 'dad·ə ,mīn·iŋ }

data module [COMPUT SCI] A sealed disk drive unit that includes mechanical and electronic components for handling data stored on the disk. { 'dad·ə ,mäj·yül }

data move instruction [COMPUT SCI] An instruction in a computer program to transfer data between memory locations and registers or between the central processor and peripheral devices. { 'dad·ə ,müv in'strək·shən }

data name [COMPUT SCI] A symbolic name used to represent an item of data in a source program, in place of the address of the data item. { 'dad·ə ,nām }

data organization [COMPUT SCI] Any one of the data management conventions for physical and spatial arrangement of the physical records of a data set. Also known as data set organization. { 'dad·ə ,òr·gə·nə,zā·shən }

data origination [COMPUT SCI] The process of putting data in a form that can be read by a machine. { 'dad·ə ə‚rij·ə'nā·shən }

data patch panel [COMMUN] A plugboard used to rearrange communications lines and modems by connecting them with double-ended cables, or to attach monitoring devices to analyze circuit signals. { 'dad·ə 'pach ‚pan·əl }

data plotter [COMPUT SCI] A device which plots digital information in a continuous fashion. { 'dad·ə ‚pläd·ər }

data processing [COMPUT SCI] Any operation or combination of operations on data, including everything that happens to data from the time they are observed or collected to the time they are destroyed. Also known as information processing. { 'dad·ə 'präs‚es·iŋ }

data processing center [COMPUT SCI] A computer installation providing data processing service for others, sometimes called customers, on a reimbursable or nonreimbursable basis. { 'dad·ə ¦präs‚es·iŋ ‚sent·ər }

data processing inventory [COMPUT SCI] An identification of all major data processing areas in an agency for the purpose of selecting and focusing upon those in which the use of automatic data processing (ADP) techniques appears to be potentially advantageous, establishing relative priorities and schedules for embarking on ADP studies, and identifying significant relationships among areas to pinpoint possibilities for the integration of systems. { 'dad·ə ¦präs‚es·iŋ ‚in·vən‚tòr·ē }

data processor [COMPUT SCI] **1.** Any device capable of performing operations on data, for instance, a desk calculator, an analog computer, or a digital computer. **2.** Person engaged in processing data. { 'dad·ə 'präs‚es·ər }

data protection [COMPUT SCI] The safeguarding of data against unauthorized access or accidental or deliberate loss or damage. { 'dad·ə prə‚tek·shən }

data purification [COMPUT SCI] The process of removing as many inaccurate or incorrect items as possible from a mass of data before automatic data processing is begun. { 'dad·ə pyür·ə·fə'kā·shən }

data rate [COMMUN] The number of digital bits per second that are recorded or retrieved from a data storage device during the transfer of a large data block. { 'dad·ə ‚rāt }

data record [COMPUT SCI] A collection of data items related in some fashion and usually contiguous in location. { 'dad·ə ‚rek·ərd }

data recorder [COMPUT SCI] A keyboard device for entering data onto magnetic tape. { 'dad·ə ri‚kòr·dər }

data reduction [COMPUT SCI] The transformation of raw data into a more useful form. [STAT] The conversion of all information in a data set into fewer dimensions for a particular purpose, as, for example, a single measure such as a reliability measure. { 'dad·ə ri‚dək·shən }

data reduction system See data-handling system. { ‚dad·ə ri‚dək·shən ‚sis·təm }

data redundancy [COMPUT SCI] The occurrence of values for data elements more than once within a file or database. { 'dad·ə ri‚dən·dən·sē }

data register [COMPUT SCI] A register used in microcomputers to temporarily store data being transmitted to or from a peripheral device. { 'dad·ə ‚rej·ə·stər }

data representation [COMPUT SCI] **1.** The way that the physical properties of a medium are used to represent data. **2.** The manner in which data is expressed symbolically by binary digits in a computer. { 'dad·ə ‚rep·ri·zen'tā·shən }

data retrieval [COMPUT SCI] The searching, selecting, and retrieving of actual data from a personnel file, data bank, or other file. { 'dad·ə ri'trē·vəl }

data rules [COMPUT SCI] Conditions which must be met by data to be processed by a computer program. { 'dad·ə ‚rülz }

data scope [ELECTR] An electronic display that shows the content of the information being transmitted over a communications channel. { 'dad·ə ‚skōp }

data security [COMPUT SCI] The protection of data against the deliberate or accidental access of unauthorized persons. Also known as file security. { 'dad·ə sə‚kyür·əd·ē }

data set [COMPUT SCI] **1.** A named collection of similar and related data records recorded upon some computer-readable medium. **2.** A data file in IBM 360 terminology. { 'dad·ə ‚set }

data set coupler [COMPUT SCI] The interface between a parallel computer input/output bus and the serial input/output of a modem. { 'dad·ə ‚set ‚kəp·lər }

data set label [COMPUT SCI] A data element that describes a data set, and usually includes the name of the data set, its boundaries in physical storage, and certain characteristics of data items within the set. { 'dad·ə ‚set ‚lā·bəl }

data set migration [COMPUT SCI] The process of moving inactive data sets from on-line storage to back up storage in a time-sharing environment. { 'dad·ə ‚set mī‚grā·shən }

data set organization See data organization. { 'dad·ə ‚set ‚òr·gə·nə‚zā·shən }

data sink [COMPUT SCI] A memory or recording device capable of accepting data signals from a data transmission device and storing data for future use. { 'dad·ə ‚siŋk }

data source [COMPUT SCI] A device capable of originating data signals for a data transmission device. { 'dad·ə ‚sòrs }

data stabilization [ELECTR] Stabilization of the display of radar signals with respect to a selected reference, regardless of changes in radar-carrying vehicle attitude, as in azimuth-stabilized plan-position indicator. { 'dad·ə ‚stā·bə·lə‚zā·shən }

data statement [COMPUT SCI] An instruction in a source program that identifies an item of data in the program and specifies its format. { 'dad·ə ‚stāt·mənt }

data station [COMPUT SCI] A remote input/output device which handles a variety of transmissions to and from certain centralized computers. { 'dad·ə ‚stā·shən }

data station control [COMPUT SCI] The supervision of a data station by means of a program resident in the central computer. { 'dad·ə ‚stā·shən kən‚tròl }

data stream [COMMUN] The continuous transmission of data from one location to another. { 'dad·ə ‚strēm }

data striping See disk striping. { 'dad·ə ‚strīp·iŋ }

data structure [COMPUT SCI] A collection of data components that are constructed in a regular and characteristic way. { 'dad·ə ‚strək·chər }

data switch [COMPUT SCI] A manual or automatic device that connects data-processing machines to one another. { 'dad·ə ‚swich }

data system [COMPUT SCI] The means, either manual or automatic, of converting data into action or decision information, including the forms, procedures, and processes which together provide an organized and interrelated means of recording, communicating, processing, and presenting information relative to a definable function or activity. { 'dad·ə ‚sis·təm }

data system interface [COMPUT SCI] **1.** A common aspect of two or more data systems involving the capability of intersystem communications. **2.** A common boundary between automatic data-processing systems or parts of a single system. { 'dad·ə ¦sis·təm ‚in·tər‚fās }

data systems integration [COMPUT SCI] Achievement through systems design of an improved or broader capability by functionally or technically relating two or more data systems, or by incorporating a portion of the functional or technical elements of one data system into another. { 'dad·ə ‚sis·təmz ‚in·tə‚grā·shən }

data system specifications [COMPUT SCI] **1.** The delineation of the objectives which a data system is intended to accomplish. **2.** The data processing requirements underlying that accomplishment; includes a description of the data output, the data files and record content, the volume of data, the processing frequencies, training, and such other facts as may be necessary to provide a full description of the system. { 'dad·ə ‚sis·təm ‚spes·ə·fə‚kā·shən }

data table [COMPUT SCI] An on-screen display of the information in a database management system, presented in columnar format, with field names at the top. { 'dad·ə ‚tā·bəl }

data tablet See electronic tablet. { 'dad·ə ‚tab·lət }

data tracks [COMPUT SCI] Information storage positions on drum storage devices; information is stored on the drum surface in the form of magnetized or nonmagnetized areas. { 'dad·ə ‚traks }

data transcription equipment [COMPUT SCI] Those devices or equipment designed to convey data from its original state to a data processing media. { ¦dad·ə tranz‚krip·shən i‚kwip·mənt }

data transfer [COMPUT SCI] The technique used by the hardware manufacturer to transmit data from computer to storage device or from storage device to computer; usually under specialized program control. { 'dad·ə 'tranz·fər }

data transmission *See* data communications. { 'dad·ə tranz'-mish·ən }

data transmission equipment [COMPUT SCI] The communications equipment used in direct support of data processing equipment. { 'dad·ə tranz'mish·ən i‚kwip·mənt }

data transmission line [ELEC] A system of electrical conductors, such as a coaxial cable or pair of wires, used to send information from one place to another or one part of a system to another. { 'dad·ə tranz'mish·ən ‚līn }

data transmission-utilization measure [COMMUN] The ratio of useful data output to the sum total of data input. { ¦dad·ə tranz'mish·ən yüd·əl·ə'zā·shən ‚mezh·ər }

data type [COMPUT SCI] The manner in which a sequence of bits represents data in a computer program. { 'dad·ə ‚tīp }

data under voice [COMMUN] A telephone digital data service that allows digital signals to travel on the lower portion of the frequency spectrum of existing microwave radio systems; digital channels initially available handled speeds of 2.4, 4.8, 9.6, and 56 kilobits per second. Abbreviated DUV. { 'dad·ə ‚ən·dər 'vȯis }

data unit [COMPUT SCI] A set of digits or characters treated as a whole. { 'dad·ə ‚yü·nət }

data validation [COMPUT SCI] The checking of data for correctness, or the determination of compliance with applicable standards, rules, and conventions. { 'dad·ə val·ə'dā·shən }

data warehouse [COMPUT SCI] **1.** A large specialized database, holding perhaps hundreds of terabytes of data. **2.** A database specifically structured for information access and reporting. { ¦dad·ə 'wer‚haús }

data word [COMPUT SCI] A computer word that is part of the data which the computer is manipulating, in contrast with an instruction word. Also known as information word. { 'dad·ə ‚wərd }

date time group [COMMUN] The date and time, expressed in digits and zone suffix, at which the message was prepared for transmission (expressed as six digits followed by the zone suffix; first pair of digits denoting the date, second pair the hours, third pair the minutes). { ¦dāt ¦tīm ‚grüp }

dating [SCI TECH] The use of methods and techniques to fix dates, assign periods of time, and determine age in archeology, biology, and geology. { 'dād·iŋ }

dative bond *See* coordinate valence. { ¦dād·iv 'bänd }

datolite [MINERAL] CaBSiO₄(OH) A mineral nesosilicate crystallizing in the monoclinic system; luster is vitreous, and crystals are colorless or white with a greenish tinge. { 'dad·əl‚īt }

datum *See* data item. [ENG] **1.** A direction, level, or position from which angles, heights, speeds or distances are conveniently measured. **2.** Any numerical or geometric quantity or value that serves as a base reference for other quantities or values (such as a point, line, or surface in relation to which others are determined). [GEOD] The latitude and longitude of an initial point; the azimuth of a line from this point. [GEOL] The top or bottom of a bed of rock on which structure contours are drawn. { 'dad·əm, 'dād·əm, *or* 'däd·əm }

datum level *See* datum plane. { 'dad·əm ‚lev·əl }

datum line *See* reference line. { 'dad·əm ‚līn }

datum plane [ENG] A permanently established horizontal plane, surface, or level to which soundings, ground elevations, water surface elevations, and tidal data are referred. Also known as chart datum; datum level; reference level; reference plane. { 'dad·əm ‚plān }

datum point [MAP] Any reference point of known or assumed coordinates from which calculation or measurements may be taken. { 'dad·əm ‚pȯint }

Daubentoniidae [VERT ZOO] A family of Madagascan prosimian primates containing a single species, the aye-aye. { ‚dȯ·bən·tō'nī·ə‚dē }

Daubenton's plane [ANTHRO] A plane passing through the opisthion and the orbital points on the skull. { 'dȯ·bən‚tōnz ‚plān }

daubreeite [MINERAL] FeCr₂S₄ A mineral composed of a black chromium iron sulfide; occurs in some meteors. { 'dȯ·brē‚īt }

daughter [NUC PHYS] The immediate product of radioactive decay of an element, such as uranium. Also known as decay product; radioactive decay product. { 'dȯd·ər }

daughter board [COMPUT SCI] A small printed circuit board that is attached to another printed circuit board. { 'dȯd·ər ‚bȯrd }

daughter nucleus [CYTOL] One of the two cell nuclei resulting from a nuclear division. { 'dȯd·ər ‚nü·klē·əs }

Dauphine law [CRYSTAL] A twin law in which the twinned parts are related by a rotation of 180° around the *c* axis. { dȯ‚fēn ‚lȯ }

Davian [GEOL] A subdivision of the Upper Cretaceous in Europe; a limestone formation with abundant hydrocorals, bryozoans, and mollusks in Denmark; marine limestone and nonmarine rocks in southeastern France; and continental formations in the Davian of Spain and Portugal. { 'dä·vē·ən }

Davida [ASTRON] An asteroid with a diameter of about 322 kilometers, mean distance from the sun of 3.18 astronomical units, and C-type surface composition. { 'dä·və·də }

davidite [MINERAL] A black primary pegmatite uranium mineral of the general formula A₆B₁₅(O,OH)₃₆, where A = Fe²⁺, rare earths, uranium, calcium, zirconium, and thorium, and B = titanium, Fe³⁺, vanadium, and chromium. { 'dä·və‚dīt }

Davidson Current [OCEANOGR] A coastal countercurrent of the Pacific Ocean running north, inshore of the California Current, along the western coast of the United States (from northern California to Washington to at least latitude 48°N) during the winter months. { 'dä·vəd·sən ‚kər·ənt }

daviesite [MINERAL] An orthorhombic mineral consisting of a lead oxychloride, occurring in minute crystals. { 'dä·vē‚zīt }

Davis correction [FL MECH] Empirical relation of flow-line diameters used to correct data calculated from the Atherton equation (friction loss in annular passages). { 'dā·vəs kə'rek·shən }

Davis-Gibson color filter [OPTICS] A two-component filter for converting the spectral energy distribution of an incandescent light source to that of white light. { ¦dā·vəs ¦gib·sən 'kəl·ər ‚fil·tər }

Davis magnetic tester [MIN ENG] An instrument used to determine the magnetic contents of ores. { 'dā·vəs mag¦ned·ik 'tes·tər }

davisonite [MINERAL] Ca₃Al(PO₄)₂(OH)₃·H₂O A white mineral consisting of a hydrous basic phosphate of calcium and aluminum. { 'dā·və·sə‚nīt }

Davisson-Calbick formula [ELECTR] A formula which states that the focal length of a simple electrostatic lens consisting of a circular hole in a conducting plate is equal to four times the potential of the plate divided by the difference in the potential gradients on either side of the plate. { ¦da·və·sən 'kal·bik ‚fȯr·myə·lə }

Davisson-Germer experiment [QUANT MECH] The first experiment to demonstrate electron diffraction, in which a beam of electrons was directed at the surface of a nickel crystal, and the distribution of electrons scattered back from the crystal was measured by a Faraday cylinder. { ¦da·və·sən ¦ger·mər ik‚sper·ə·mənt }

Davis wing [AERO ENG] A narrow-chord wing that has comparatively low drag and a stable center of pressure and develops lift at relatively small angles of attack. { 'da·vəs ‚wiŋ }

davit [NAV ARCH] A fixed or movable shipboard crane projecting over the side or over a hatchway, which hoists and lowers boats, anchors, or cargo. { 'dav·ət }

Davy lamp [MIN ENG] An early safety lamp with a mantle of wire gauze around the flame to dissipate the heat from the flame to below the ignition temperature of methane. { 'dā·vē ‚lamp }

Dawes' limit [OPTICS] The resolving power of a telescope, limited by diffraction effects, is 4.5/*a* seconds of arc, where *a* is the aperture in inches. { 'dȯz ‚lim·ət }

dawn [ASTRON] The first appearance of light in the eastern sky before sunrise, or the time of that appearance. Also known as daybreak. { dȯn }

dawn side [ASTRON] That side of a celestial object, such as a planet, which points in the direction of its orbital movement. { dȯn ‚sīd }

Dawsoniales [BOT] An order of mosses comprising rigid plants with erect stems rising from a rhizomelike base. { dȯ‚sō·nē'ā·lēz }

dawsonite [MINERAL] NaAl(OH)₂CO₃ A white, bladed mineral found in certain oil shales that contains large quantities of alumina; specific gravity is 2.40. { 'dȯs·ən‚īt }

day [ASTRON] One of various units of time equal to the period of rotation of the earth with respect to one or another direction in space; specific examples are the mean solar day and the sidereal day. ｛ dā ｝

day beacon [NAV] An unlighted structure serving as an aid to navigation in the daytime. ｛ 'dā ˌbē·kən ｝

daybreak See dawn. ｛ 'dā,brāk ｝

day clock [COMPUT SCI] An internal binary counter, with a resolution usually of a microsecond and a cycle measured in years, providing an accurate measure of elapsed time independent of system activity. ｛ 'dā ˌkläk ｝

day drift [MIN ENG] A mine passageway that has one end at the surface. ｛ 'dā ˌdrift ｝

dayglow [ASTRON] Airglow of the day sky. ｛ 'dā,glō ｝

daylight [ASTRON] Light of the day, from sun and sky. See daylight opening. ｛ 'dā,līt ｝

daylight controls [ENG] Special devices which automatically control the electric power to the lamp, causing the light to operate during hours of darkness and to be extinguished during daylight hours. ｛ 'dā,līt kən'trōlz ｝

daylight glass [MATER] A glass that absorbs red light; used in incandescent lamps to remove excess red emission so that the light spectrum resembles natural daylight. ｛ 'dā,līt ˌglas ｝

daylighting [CIV ENG] To light an area with daylight. ｛ 'dā,līd·iŋ ｝

daylight lamp [ELEC] An incandescent or fluorescent lamp that emits light whose spectral distribution is approximately that of daylight. ｛ 'dā,līt ˌlamp ｝

daylight opening [ENG] The space between two press platens when open. Also known as daylight. ｛ 'dā,līt ˌō·pən·iŋ ｝

daylight saving meridian [ASTRON] The meridian used for reckoning daylight saving time; generally 15° east of the zone of standard meridian. ｛ ¦dā,līt 'sāv·iŋ mə'rid·ē·ən ｝

daylight saving noon [ASTRON] Twelve o'clock daylight saving time, or the instant the mean sun is over the upper branch of the daylight saving meridian; during a war, when daylight saving time may be used throughout the year and called war time, the expression war noon applies. Also known as summer noon. ｛ ¦dā,līt 'sāv·iŋ 'nün ｝

daylight saving time [ASTRON] A variation of zone time, usually 1 hour more advanced than standard time, frequently kept during the summer to make better use of daylight. Also known as summer time. ｛ ¦dā,līt 'sāv·iŋ ˌtīm ｝

daymark [NAV] The daytime-identifying characteristics of an aid to navigation, unique and distinctive to facilitate its recognition; a conspicuous target added to a day beacon or light. ｛ 'dā,märk ｝

day neutral [BOT] Reaching maturity regardless of relative length of light and dark periods. ｛ 'dā ˌnü·trəl ｝

day-neutral response [PHYSIO] A photoperiodic response that is independent or nearly independent of day length. ｛ 'dā ˌnü·trəl ri'späns ｝

Day's multiplexing system [COMMUN] A system of phase-discrimination multiplexing applicable to pairs of channels; used to multiplex the two so-called color components in color television broadcasting. ｛ ¦dāz 'məl·tə,plek·siŋ ˌsis·təm ｝

day's run [NAV] The distance traveled by a vessel in 1 day, usually reckoned from noon to noon. ｛ ¦dāz 'rən ｝

day's work [NAV] In marine operations, the daily routine of the navigation of a vessel at sea, usually consisting principally of the dead reckoning from noon to noon, evening and morning twilight observations, a morning sun observation for a line of position and for checking the compass, and a sun observation at or near noon to obtain a running fix. ｛ ¦dāz 'wərk ｝

day wage [IND ENG] A fixed rate of pay per shift or per daily hours of work, irrespective of the amount of work completed. ｛ 'dā ˌwāj ｝

dazomet [ORG CHEM] $C_5H_{10}N_2S_2$ A white, crystalline compound that decomposes at 100°C; used as a herbicide and nematicide for soil fungi and nematodes, weeds, and soil insects. Also known as tetrahydro-3,5-dimethyl-2H-1,3,5-thiadiazine-6-thione. ｛ 'dā· zə ˌmət ｝

dB See decibel.

dBa See adjusted decibel.

DBCP See dibromochloropropane.

DB/DC See data base/data communication.

dBf See decibels above 1 femtowatt.

db galaxy [ASTRON] A dumbbell-shaped radio galaxy, believed to consist of two elliptical nuclei surrounded by a common extended envelope. ｛ ¦dē¦bē 'gal·ik·sē ｝

dBic [ELECTROMAG] The directive gain of a circularly polarized antenna, expressed as the ratio, in decibels, of the antenna's directivity to that of an isotropic antenna with the same polarization characteristic. Derived from decibels over isotropic. ｛ 'dē,bik ｝

dBk See decibels above 1 kilowatt.

d-block element [CHEM] A transition element occupying the first, second, and third long periods of the periodic table. ｛ 'dē ˌbläk ˌel·ə·mənt ｝

dBm See decibels above 1 milliwatt.

DBMS See database management system.

dBp See decibels above 1 picowatt.

D-brane [PARTIC PHYS] In superstring theory, a point, curve, surface, or higher-dimensional surface to which the end points of strings can be attached; its existence is implied by string duality. ｛ 'dē,brān ｝

dBrn See decibels above reference noise.

DBRT diode See double-barrier resonant tunneling diode. ｛ ¦dē¦bē¦är¦tē 'dī,ōd ｝

DB server [COMPUT SCI] The database portion of a Web server, which serves as a repository of data and content. ｛ ¦dē¦bē ˌsər·vər ｝

DBS system See direct broadcasting satellite system. ｛ ¦dē¦bē 'es ˌsis·təm ｝

dBV See decibels above 1 volt.

dBW See decibels above 1 watt.

dBx See decibels above reference coupling.

dc See direct current.

D cable [ELEC] Two-conductor cable, each conductor having the shape of the letter D, with insulation between the conductors and between the conductors and the sheath. ｛ 'dē ˌkā·bəl ｝

DCB See 1,4-dichlorobutane.

DCC See dicyclohexylcarbodiimide.

dc casting See direct-chill casting. ｛ 'dē,sē ˌkast·iŋ ｝

DCCI See dicyclohexylcarbodiimide.

D center See R center. ｛ 'dē ˌsen·tər ｝

DCFL See direct-coupled FET logic.

DCNA See 2,6-dichloro-4-nitroaniline.

d constant [SOLID STATE] The ratio of the induced strain in a piezoelectric material to the applied electric field that produces this strain. ｛ 'dē ˌkän·stənt ｝

DCPA See dimethyl-2,3,5,6-tetrachloroterephthalate.

DCTL See direct-coupled transistor logic.

dc-to-ac converter See inverter. ｛ ¦dē,sē tü ¦ā,sē kən'vərd·ər ｝

dc-to-ac inverter See inverter. ｛ ¦dē,sē tü ¦ā,sē in'vərd·ər ｝

dc-to-dc converter [ELEC] An electronic circuit which converts one direct-current voltage into another, consisting of an inverter followed by a step-up or step-down transformer and rectifier. ｛ ¦dē,sē tü ¦dē,sē kən'vərd·ər ｝

dcwv See direct-current working volts.

DDA See digital differential analyzer.

DDA value See depth-duration-area value. ｛ ¦dē¦dē'ā ˌval·yü ｝

DDBS See Digital Data Broadcast System.

DDD See 2,2-bis(para-chlorophenyl)-1,1-dichloroethane.

D display [ELECTR] In radar, a C display in which the blips extend vertically to give a rough estimate of distance. ｛ 'dē di,splā ｝

DDR See double data rate.

DDS See digital data service.

DDT [ORG CHEM] Common name for an insecticide; melting point 108.5°C, insoluble in water, very soluble in ethanol and acetone, colorless, and odorless; especially useful against agricultural pests, flies, lice, and mosquitoes. Also known as dichlorodiphenyltrichloroethane.

DDTA See derivative differential thermal analysis.

DDVP See dichlorvos.

DEA See diethanolamine.

deaccentuator [ELECTR] A circuit used in a frequency-modulation receiver to offset the preemphasis of higher audio frequencies introduced at the transmitter. ｛ ¦dē·ak'sen·chə,wād·ər ｝

deacetylation [ORG CHEM] The removal of an acetyl group from a molecule. ｛ ˌdē·ə,sēd·əl'ā·shən ｝

deacidification [CHEM] **1.** Removal of acid. **2.** A process for reducing acidity. ｛ ˌdē·ə,sid·ə·fə'kā·shən ｝

Deacon process [CHEM ENG] A method of chlorine production by passing a hot mixture of gaseous hydrochloric acid with oxygen over a cuprous chloride catalyst. ｛ 'dēk·ən ˌpräs·əs ｝

deactivation [CHEM] **1.** Rendering inactive, as of a catalyst. **2.** Loss of radioactivity. [MET] Chemical removal of the active constituents of a corrosive liquid. [MIN ENG] Treatment of one or more species of mineral particles to reduce floating during froth flotation. { dē̩ak·tə'vā·shən }

deacylation [ORG CHEM] Removal of an acyl group from a compound. { dē̩as·ə'lā·shən }

dead [ELEC] Free from any electric connection to a source of potential difference from electric charge; not having a potential different from that of earth; the term is used only with reference to current-carrying parts which are sometimes alive or charged. [GEOL] In economic geology, designating a region with no economic value. [MIN ENG] An area of subsidence that has totally settled and is not likely to move. { ded }

dead ahead [NAV] In marine operations, a position directly ahead of the ship; a relative bearing of 000°. { ¦ded ə'hed }

dead air [MIN ENG] Air in a mine when it is stagnant or contains carbonic acid. { ¦ded 'er }

dead-air space [BUILD] A sealed air space, such as in a hollow wall. { ¦ded 'er ̩spās }

dead area *See* blind spot. { ¦ded ̩er·ē·ə }

dead arm [PL PATH] A fungus disease caused by *Cryptosporella viticola* in which the main lateral shoots are destroyed; common in grapes. { ¦ded 'ärm }

dead astern [NAV] In marine navigation, a bearing 180° relative; if the bearing is approximate, the term astern should be used. Also known as right astern. { ¦ded ə'stərn }

dead axle [MECH ENG] An axle that carries a wheel but does not drive it. { ¦ded 'ak·səl }

dead band [ELEC] The portion of a potentiometer element that is shortened by a tap; when the wiper traverses this area, there is no change in output. [ENG] The range of values of the measured variable to which an instrument will not effectively respond. Also known as dead zone; neutral zone. { 'ded ̩band }

deadbeat [MECH] Coming to rest without vibration or oscillation, as when the pointer of a meter moves to a new position without overshooting. Also known as deadbeat response. { 'ded̩bēt }

deadbeat algorithm [CONT SYS] A digital control algorithm which attempts to follow set-point changes in minimum time, assuming that the controlled process can be modeled approximately as a first-order plus dead-time system. { 'ded̩bēt 'al·gə̩rith·əm }

deadbeat compass *See* aperiodic compass. { 'ded̩bēt 'käm·pəs }

deadbeat escapement [HOROL] A watch escapement without recoil, having arresting faces of the pallets described by a circular arc whose center is at the pivot point of the anchor; escape-wheel teeth are contoured to give impulses to these pallet faces, over which they slide without recoil. { 'ded̩bēt ə'skāp·mənt }

deadbeat response *See* deadbeat. { 'ded̩bēt ri'späns }

dead block [ENG] A device placed on the ends of railroad passenger cars to absorb the shock of impacts. { 'ded 'bläk }

dead bolt [DES ENG] A lock bolt that is moved directly by the turning of a knob or key, not by spring action. { 'ded 'bōlt }

dead-bright [MET] Of a metal, polished to remove tool marks. { 'ded 'brīt }

dead-burn [MIN ENG] A calcination to produce a dense refractory substance; done at a higher temperature and for a longer time than for normal calcination. { 'ded ̩bərn }

dead-burnt gypsum *See* anhydrous calcium sulfate. { 'ded ̩bərnt 'jip·səm }

dead cave [GEOL] A cave where there is no moisture or no growth of mineral deposits associated with moisture. { ¦ded 'kāv }

dead center [MECH ENG] **1.** A position of a crank in which the turning force applied to it by the connecting rod is zero; occurs when the crank and rod are in a straight line. **2.** A support for the work on a lathe which does not turn with the work. { ¦ded 'sen·tər }

dead-center position [ELEC] Position in which a brush would be placed on the commutator of a direct-current motor or generator if the field flux were not distorted by armature reaction. { ¦ded 'sen·tər pə'zish·ən }

dead code [COMPUT SCI] Statements in a computer program that are not executed, usually as the result of modification of a large program. { ¦ded 'kōd }

dead earth [ELEC] A connection between a line conductor and earth by means of a path of low resistance. { ¦ded 'ərth }

dead end [ACOUS] The end of a sound studio that has the greater sound-absorbing characteristics. [ELEC] The portion of a tapped coil through which no current is flowing at a particular switch position. [SCI TECH] The end of a conduit, passage, power line, or similar system having no exit or continuation. { 'ded 'end }

dead-end effect [ELEC] Absorption of energy by unused portions of a tapped coil. { 'ded ̩end i'fekt }

dead-end switch [ELEC] A switch used to short-circuit unused portions of a tapped coil to prevent dead-end effects. { 'ded ̩end ̩swich }

dead-end tower [CIV ENG] Antenna or transmission line tower designed to withstand unbalanced mechanical pull from all the conductors in one direction together with the wind strain and vertical loads. { 'ded ̩end ̩taù·ər }

deadeye [NAV ARCH] A flat, rounded wooden block usually pierced with three holes to receive the lanyard; used to fasten shrouds and stays. { 'ded̩ī }

dead flat [NAV ARCH] A portion of length in the middle of a ship in which all sections have the same shape. Also known as parallel middle body. { ¦ded 'flat }

dead ground [ELEC] A low-resistance connection between the ground and an electric circuit. { ¦ded 'graùnd }

dead halt *See* drop-dead halt. { ¦ded 'hólt }

deadhead [MET] The portion of a casting that fills up the ingate. [MIN ENG] To begin a new cut without excavating the material from the preceding cut. { 'ded̩hed }

dead letter box [COMMUN] A file for storing undeliverable messages in a data communications system, particularly a message switching system. { ¦ded 'led·ər ̩bäks }

deadlight [NAV ARCH] **1.** A strong plate or a cover with light-obstructing baffles fitted over ventilation ports or windows in stormy weather. **2.** A strong glass or plastic window set in the deck or side of a ship to admit light. { 'ded̩līt }

deadline [PETRO ENG] The end of the drilling line that is not reeled on the hoisting drum of the rotary rig; usually it is anchored to the derrick substructure and does not move as the traveling block is hoisted. { 'ded̩līn }

dead load *See* static load. { ¦ded ̩lōd }

deadlock [COMPUT SCI] A situation in which a task in a multiprogramming system cannot proceed because it is waiting for an event that will never occur. Also known as deadly embrace; interlock; knot. { 'ded̩läk }

deadlocking latch bolt *See* auxiliary dead latch. { 'ded̩läk·iŋ 'lach ̩bōlt }

deadly embrace *See* deadlock. { ¦ded·lē im'brās }

deadly nightshade *See* belladonna. { ¦ded·lē 'nīt̩shād }

deadman [CIV ENG] **1.** A buried plate, wall, or block attached at some distance from and forming an anchorage for a retaining wall. Also known as anchorage; anchor block; anchor wall. **2.** *See* anchor log. { 'ded̩man }

deadman's brake [MECH ENG] An emergency device that automatically is activated to stop a vehicle when the driver removes his or her foot from the pedal. { ¦ded̩manz 'brāk }

deadman's handle [MECH ENG] A handle on a machine designed so that the operator must continuously press on it in order to keep the machine running. { ¦ded̩manz 'han·dəl }

deadman switch [ELEC] An electrical switch that activates some function if it is turned off. { 'ded̩man ̩swich }

dead oil [MATER] An oil, of density greater than water, distilled from tar. { ¦ded 'óil }

dead rail [CIV ENG] One of two rails on a railroad weighing platform that permit an excessive load to leave the platform. { ¦ded ̩rāl }

dead reckoning [NAV] Determination of position of a craft by advancing a previous position to a new one on the basis of assumed distance and direction traveled. { ¦ded 'rek·ən·iŋ }

dead-reckoning analyzer *See* dead-reckoning equipment. { ¦ded 'rek·ən·iŋ 'an·ə̩līz·ər }

dead-reckoning equipment [NAV] A computing device which continuously indicates the dead-reckoning position of a craft, vessel, or vehicle and which may provide a graphical record of the dead-reckoning track. Also known as dead-reckoning analyzer. { ¦ded 'rek·ən·iŋ i̩kwip·mənt }

dead-reckoning plot [NAV] **1.** In marine navigation, the graphic plot of consecutive course lines representing courses steered or planned to be steered, originating from a fix or

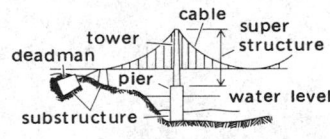

DEADMAN

A drawing of a bridge showing buried block used as a deadman.

running fix position, and suitably labeled as to course, speeds, and times of periodic dead-reckoning positions. **2.** In air navigation, the graphic plot of track and ground speed, suitably labeled. Also known as ground plot. { ¦ded ′rek·ən·iŋ ˌplät }

dead-reckoning position [NAV] A position determined by advancing a known position of the past to the position of the present time by adding one or more vectors representing known or assumed speed, bearing, and wind or current. { ¦ded ′rek·ən·iŋ pəˌzish·ən }

dead-reckoning tracer [NAV] A mechanical device used to produce a continuous plot of all ships within range of a ship's radar. { ¦ded ′rek·ən·iŋ ˌtrā·sər }

dead-reckoning track [NAV] A line representing successive dead-reckoning positions of a craft. Also known as DR track line. { ¦ded ′rek·ən·iŋ ˌtrak }

dead rise [NAV ARCH] The vertical distance between the bottom of a vessel's keel and the intersection of a line drawn along the bottom of a vessel's midship section with a vertical line tangent to the widest part of the underwater body. Also known as rise of bottom; rise of floor. { ¦ded ¦rīz }

dead rising [NAV ARCH] In a wooden ship's profile or sheer plan, a curved fore-and-aft line that passes through the upper end of the floor timbers. { ¦ded ¦rīz·iŋ }

dead-roast [MIN ENG] **1.** A roasting process for driving off sulfur. Also known as sweet roast. **2.** Removing volatiles by roasting within a specified temperature range. { ¦ded ˌrōst }

dead room See anechoic chamber. { ¦ded ˌrüm }

dead sea [HYD] A body of water that has undergone precipitation of its rock salt, gypsum, or other evaporites. { ¦ded ′sē }

Dead Sea [GEOGR] A salt lake between Jordan and Israel. { ¦ded ′sē }

dead sheave [ENG] A grooved wheel on a crown block over which the deadline is fastened. { ¦ded ′shēv }

dead short [ELEC] A short-circuit path that has extremely low resistance. { ¦ded′short }

dead soft steel [MET] **1.** Steel very low in carbon. **2.** Steel annealed until it is very soft. { ¦ded ′sȯft ′stēl }

dead space [ANAT] The space in the trachea, bronchi, and other air passages which contains air that does not reach the alveoli during respiration, the amount of air being about 140 milliliters. Also known as anatomical dead space. [MED] A cavity left after closure of a wound. [PHYSIO] A calculated expression of the anatomical dead space plus whatever degree of overventilation or underperfusion is present; it is alleged to reflect the relationship of ventilation to pulmonary capillary perfusion. Also known as physiological dead space. [THERMO] A space filled with gas whose temperature differs from that of the main body of gas, such as the gas in the capillary tube of a constant-volume gas thermometer. { ¦ded ′spās }

dead spot [COMMUN] A geographic location in which signals from a radio, television, or radar transmitter are received poorly or not at all. [ELECTR] A portion of the tuning range of a receiver in which stations are heard poorly or not at all, due to improper design of tuning circuits. { ¦ded ′spät }

dead stick [AERO ENG] The propeller of an airplane that is not rotating because the engine has stopped. { ¦ded ′stik }

dead-stroke [MECH ENG] Having a recoilless or nearly recoilless stroke. { ¦ded ˌstrōk }

dead-stroke hammer [MECH ENG] A power hammer provided with a spring on the hammer head to reduce recoil. { ¦ded ˌstrōk ′ham·ər }

dead time [CONT SYS] The time interval between a change in the input signal to a process control system and the response to the signal. [ENG] The time interval, after a response to one signal or event, during which a system is unable to respond to another. Also known as insensitive time. { ¦ded ˌtīm }

dead-time compensation [CONT SYS] The modification of a controller to allow for time delays between the input to a control system and the response to the signal. { ¦ded ˌtīm käm·pən′sā·shən }

dead-time correction [ENG] A correction applied to an observed counting rate to allow for the probability of the occurrence of events within the dead time. Also known as coincidence correction. { ¦ded ˌtīm kə′rek·shən }

dead track [CIV ENG] **1.** Railway track that is no longer used. **2.** A section of railway track that is electrically isolated from the track signal circuits. { ¦ded ¦trak }

dead water [OCEANOGR] The mass of eddying water associated with formation of internal waves near the keel of a ship;

forms under a ship of low propulsive power when it negotiates water which has a thin layer of fresher water over a deeper layer of more saline water. { ¦ded ˌwȯd·ər }

deadweight capacity See deadweight tonnage. { ′dedˌwāt kə′pas·əd·ē }

deadweight gage [ENG] An instrument used as a standard for calibrating pressure gages in which known hydraulic pressures are generated by means of freely balanced (dead) weights loaded on a calibrated piston. { ′dedˌwāt ′gāj }

deadweight tonnage [NAV ARCH] The total carrying capacity of a ship expressed in long tons (1 long ton = 2240 pounds); displacement of a fully loaded ship, less the weight of the ship itself. Abbreviated dwt. Also known as deadweight capacity. { ′dedˌwāt ′tən·ij }

dead well [PETRO ENG] A well that is not producing oil or gas, either temporarily or permanently. { ′ded ′wel }

dead wood [PETRO ENG] The ladders, braces, piping, and other internal fixtures in an oil storage tank which reduce the oil capacity of the tank. { ′ded ˌwu̇d }

dead work [MIN ENG] Preparatory work which is for future operations and not directly productive. { ¦ded ¦wərk }

dead wraps [PETRO ENG] The first several turns of wire rope wrapped around the draw works drum and which remain on the drum permanently. { ′ded ′raps }

dead zone See dead band. { ′ded ˌzōn }

dead zone unit [COMPUT SCI] An analog computer device that maintains an output signal at a constant value over a certain range of values of the input signal. { ′ded ˌzōn ˌyü·nət }

DEAE-cellulose See diethylaminoethyl cellulose. { ′dēˌēˌē′äˌē ′sel·yəˌlōs }

deaeration [ENG] Removal of gas or air from a substance, as from feedwater or food. { dēˌer′ā·shən }

deaerator [MECH ENG] A device in which oxygen, carbon dioxide, or other noncondensable gases are removed from boiler feedwater, steam condensate, or a process stream. { dē′erˌād·ər }

deafness [MED] Temporary or permanent impairment or loss of hearing. { ′def·nəs }

deagglomeration [CHEM ENG] Size-reduction process in which loosely adhered clumps (agglomerates) of powders or crystals are broken apart without further disintegration of the powder or crystal particles themselves. { ˌdē·əˌgläm·ə′rā·shən }

deaister See doister. { dē′ās·ter }

deal [DES ENG] **1.** A face on which numbers are registered by means of a pointer. **2.** A disk usually with a series of markings around its border, which can be turned to regulate the operation of a machine or electrical device. { dēl }

dealkalization [CHEM] **1.** Removal of alkali. **2.** Reduction of alkalinity, as in the process of neutralization. { dēˌal·kə·lə′zā·shən }

dealkylate [CHEM] To remove alkyl groups from a compound. { dē′al·kəˌlāt }

deallocation [COMPUT SCI] The release of a portion of computer storage or a peripheral unit from control by a computer program when it is no longerneeded. { dēˌal·ə′kā·shən }

dealuminization [CHEM] Removal of aluminum. { ˌdē·əˌlü·mə·nə′zā·shən }

deamidase [BIOCHEM] An enzyme that catalyzes the removal of an amido group from a compound. { dē′am·əˌdās }

deamidation [ORG CHEM] Removal of the amido group from a molecule. { dēˌam·ə′dā·shən }

deaminase [BIOCHEM] An enzyme that catalyzes the hydrolysis of amino compounds, removing the amino group. { dē′am·əˌnās }

deamination [ORG CHEM] Removal of an amino group from a molecule. { dēˌam·ə′nā·shən }

Dean number [FL MECH] A dimensionless number giving the ratio of the viscous force acting on a fluid flowing in a curved pipe to the centrifugal force; equal to the Reynolds number times the square root of the ratio of the radius of the pipe to its radius of curvature. Symbolized N_D. { ′dēn ˌnəm·bər }

deashing [CHEM] A form of deionization in which inorganic salts are removed from solution by the adsorption of both the anions and cations by ion-exchange resins. { dē′ash·iŋ }

deasphalting [CHEM ENG] The process of removing asphalt from petroleum fractions. { dē′asˌfȯl·tiŋ }

death [MED] Cessation of all life functions; can involve the whole organism, an organ, individual cells, or cell parts. { deth }

death assemblage See thanatocoenosis. { 'deth ə‚sem·blij }

death instinct [PSYCH] In psychoanalytic theory, the unconscious drive which leads the individual toward dissolution and death, and which coexists with the life instinct. { 'deth ‚in‚stiŋkt }

death point [PHYSIO] The limit (as of extremes of temperature) beyond which an organism cannot survive. { 'deth ‚pȯint }

death rate See mortality rate. { 'deth ‚rāt }

debatable time [COMPUT SCI] In the keeping of computer usage statistics, time that cannot be attributed with certainty to any one of various categories of computer use. { di'bād·ə·bəl 'tīm }

debenzylation [ORG CHEM] Removal from a molecule of the benzyl group. { dē‚ben·zə'lā·shən }

de Beurmann-Gougerot disease See sporotrichosis. { dē 'bu̇r·män 'güzh·rō di'zēz }

debittering [FOOD ENG] A process for removing the bitter component from an edible material. { dē'bid·ər·iŋ }

deblocking [COMPUT SCI] Breaking up a block of records into individual records. { dē'bläk·iŋ }

deblooming [CHEM ENG] The process by which the fluorescence, or bloom, is removed from petroleum oils by exposing them in shallow tanks to the sun and atmospheric conditions or by using chemicals. { dē'blüm·iŋ }

Deborah number [MECH] A dimensionless number used in rheology, equal to the relaxation time for some process divided by the time it is observed. Symbolized D. { də'bȯr·ə ‚nəm·bər }

deboss [MET] To press a design into a metal surface. { dē'bȯs }

debossed [GRAPHICS] Having a depressed pattern on the surface of a material. { dē'bȯst }

Debot effect [GRAPHICS] A variation of the Herschel effect, in which red or infrared radiation converts an internal latent image into a surface latent image. { də'bō i'fekt }

debridement [MED] A surgical procedure for removing lacerated, morbid, or contaminated tissue. { dē'brīd·mənt }

debriefing [AERO ENG] The relating of factual information by a flight crew at the termination of a flight, consisting of flight weather encountered, the condition of the aircraft, or facilities along the airways or at the airports. { dē'brēf·iŋ }

debris [GEOL] Large fragments arising from disintegration of rocks and strata. { də'brē }

debris avalanche [GEOL] The sudden and rapid downward movement of incoherent mixtures of rock and soil on deep slopes. { də'brē 'av·ə‚lanch }

debris cone [GEOL] **1.** A mound of fine-grained debris piled atop certain boulders moved by a landslide. **2.** A mound of ice or snow on a glacier covered with a thin layer of debris. { də'brē ‚kōn }

debris dam [CIV ENG] A fixed dam across a stream channel for the retention of sand, gravel, driftwood, or other debris. { də'brē ‚dam }

debris fall [GEOL] A relatively free downward or forward falling of unconsolidated or poorly consolidated earth or rocky debris from a cliff, cave, or arch. { də'brē ‚fȯl }

debris flow [GEOL] A variety of rapid mass movement involving the downslope movement of high-density coarse clast-bearing mudflows, usually on alluvial fans. { də'brē ‚flō }

debris glacier [HYD] A glacier formed from ice fragments that have fallen from a larger and taller glacier. { də'brē ‚glā·shər }

debris line See swash mark. { də'brē ‚līn }

debris slide [GEOL] A type of landslide involving a rapid downward sliding and forward rolling of comparatively dry, unconsolidated earth and rocky debris. { də'brē ‚slīd }

debris slope See talus slope. { də'brē ‚slōp }

de Broglie equation See de Broglie relation. { də'brō‚glē i'kwā·zhən }

de Broglie relation [QUANT MECH] The relation in which the de Broglie wave associated with a free particle of matter, and the electromagnetic wave in a vacuum associated with a photon, has a wavelength equal to Planck's constant divided by the particle's momentum and a frequency equal to the particle's

energy divided by Planck's constant. Also known as de Broglie equation. { də‚brō‚glē ri'lā·shən }

de Broglie's theory [QUANT MECH] The theory that particles of matter have wavelike properties which can give rise to interference effects, and electrons in an atom are associated with standing waves on a Bohr orbit. { də‚brō‚glēz ‚thē·ə·rē }

de Broglie wave [QUANT MECH] The quantum-mechanical wave associated with a particle of matter. Also known as matter wave. { də‚brō‚glē ‚wāv }

de Broglie wavelength [QUANT MECH] The wavelength of the wave associated with a particle as given by the de Broglie relation. { də‚brō‚glē 'wāv‚leŋkth }

debromoaplysiatoxin [BIOCHEM] A bislactone toxin related to aplysiatoxin and produced by the blue-green alga *Lyngbya majuscula*. { dē‚brō·mō·ə'plizh·ə‚täk·sən }

de Brun-van Eckstein rearrangement [ORG CHEM] The isomerization of an aldose or ketose when mixed with aqueous calcium hydroxide to form a mixture of various monosaccharides and unfermented ketoses; used to prepare certain ketoses. { də‚brün van'ek‚shtīn ‚rē·ə‚ranj·mənt }

debubblizer [ENG] A worker who removes bubbles from plastic rods and tubing. { dē‚bə·bə‚līz·ər }

debug [COMPUT SCI] To test for, locate, and remove mistakes from a program or malfunctions from a computer. [ELECTR] To detect and remove secretly installed listening devices popularly known as bugs. [ENG] To eliminate from a newly designed system the components and circuits that cause early failures. { dē'bəg }

debugging routine [COMPUT SCI] A routine to aid programmers in the debugging of their routines; some typical routines are storage printout, tape printout, and drum printout routines. { dē'bəg·iŋ rü‚tēn }

debugging statement [COMPUT SCI] Temporary instructions inserted into a program being tested so as to pinpoint problem areas. { dē'bəg·iŋ ‚stāt·mənt }

debug on-line [COMPUT SCI] **1.** To detect and correct errors in a computer program by using only certain parts of the hardware of a computer, while other routines are being processed simultaneouly. **2.** To detect and correct errors in a program from a console distant from a computer in a multiaccess system. { dē'bəg·iŋ ȯn 'līn }

debunching [ELECTR] A tendency for electrons in a beam to spread out both longitudinally and transversely due to mutual repulsion; the effect is a drawback in velocity modulation tubes. { dē'bənch·iŋ }

deburr [MET] To remove burrs, sharp edges, or fins from metal parts by placing them in a revolving barrel containing abrasives suspended in a liquid. { dē'bər }

debutanization [CHEM ENG] Removal of butane and lighter components in a natural-gasoline plant. { dē‚byüt·ən·ə'zā·shən }

debutanizer [CHEM ENG] The fractionating column in a natural-gasoline plant in which butane and lighter components are removed. { dē'byüt·ən‚īz·ər }

debye [ELEC] A unit of electric dipole moment, equal to 10^{-18} Franklin centimeter. { də'bī }

Debye effect [ELECTROMAG] Selective absorption of electromagnetic waves by a dielectric, due to molecular dipoles. { də'bī i'fekt }

Debye equation [SOLID STATE] The equation for the Debye specific heat, which satisfies the Dulong and Petit law at high temperatures and the Debye T^3 law at low temperatures. { də'bī i'kwā·zhən }

Debye equation for polarization [STAT MECH] The Langevin-Debye formula for the polarization of a dielectric material, relating the total polarization for n molecules to the permanent moment of the specific molecule and its polarizability. { də'bī i‚kwā·zhən fȯr pō·lə·rə'zā·shən }

Debye-Falkenhagen effect [PHYS CHEM] The increase in the conductance of an electrolytic solution when the applied voltage has a very high frequency. { də‚bī 'fäl·kən‚häg·ən i‚fekt }

Debye force See induction force. { də'bī ‚fȯrs }

Debye frequency [SOLID STATE] The maximum allowable frequency in the computation of the Debye specific heat. { də'bī ‚frē·kwən·sē }

Debye-Hückel screening radius See Debye shielding length. { də‚bi 'hik·əl 'skrēn·iŋ ‚rā·dē·əs }

Debye-Hückel theory [PHYS CHEM] A theory of the behavior of strong electrolytes, according to which each ion is surrounded by an ionic atmosphere of charges of the opposite sign whose behavior retards the movement of ions when a current is passed through the medium. { də¦bī 'hik·əl ˌthē·ə·rē }

Debye-Jauncey scattering [SOLID STATE] Incoherent background scattering of x-rays from a crystal in directions between those of the Bragg reflections. { də¦bī 'jón·sē ˌskad·ə·riŋ }

Debye length See Debye shielding length. { də'bī ˌleŋkth }

Debye potentials [ELECTROMAG] Two scalar potentials, designated Π_e and Π_m, in terms of which one can express the electric and magnetic fields resulting from radiation or scattering of electromagnetic waves by a distribution of localized sources in a homogeneous isotropic medium. { də'bī pə'ten·chəlz }

Debye relaxation time [PHYS CHEM] According to the Debye-Hückel theory, the time required for the ionic atmosphere of a charge to reach equilibrium in a current-carrying electrolyte, during which time the motion of the charge is retarded. { də'bī ˌrē,lak'sā·shən ˌtīm }

Debye-Scherrer method [SOLID STATE] An x-ray diffraction method in which the sample, consisting of a powder stuck to a thin fiber or contained in a thin-walled silica tube, is rotated in a monochromatic beam of x-rays, and the diffraction pattern is recorded on a cylindrical film whose axis is parallel to the axis of rotation of the sample. { də¦bī 'sher·ər ˌmeth·əd }

Debye-Sears ultrasonic cell [ACOUS] A process in ultrasonic imaging for which the acoustic wavefronts act as optical gratings to diffract the light on either side of the central spot. { də¦bī 'sirz ¦əl·trə¦sän·ik 'sel }

Debye shielding length [PL PHYS] A characteristic distance in a plasma beyond which the electric field of a charged particle is shielded by particles having charges of the opposite sign. Also known as Debye-Hückel screening radius; Debye length; shielding distance. { də'bī 'shēld·iŋ ˌleŋkth }

Debye specific heat [SOLID STATE] The specific heat of a solid under the assumption that the energy of the lattice arises entirely from acoustic lattice vibration modes which all have the same sound velocity, and that frequencies are cut off at a maximum such that the total number of modes equals the number of degrees of freedom of the solid. { də'bī spə,sif·ik 'hēt }

Debye T³ law [SOLID STATE] The law that the specific heat of a solid at constant volume varies as the cube of the absolute temperature T at temperatures which are small with respect to the Debye temperature. { də'bī ˌtē'kyübd ˌló }

Debye temperature [SOLID STATE] The temperature Θ arising in the computation of the Debye specific heat, defined by $k\Theta = h\nu$, where k is the Boltzmann constant, h is Planck's constant, and ν is the Debye frequency. Also known as characteristic temperature. { də'bī 'tem·prə·chər }

Debye theory [ELEC] The classical theory of the orientation polarization of polar molecules in which the molecules have a single relaxation time, and the plot of the imaginary part of the complex relative permittivity against the real part is a semicircle. { də'bī ˌthē·ə·rē }

Debye-Waller factor [SOLID STATE] A reduction factor for the intensity of coherent (Bragg) scattering of x-rays, neutrons, or electrons by a crystal, arising from thermal motion of the atoms in the lattice. { də¦bī 'väl·ər ˌfak·tər }

deca- [SCI TECH] A prefix denoting 10. { 'dek·ə }

decaborane (14) [INORG CHEM] $B_{10}H_{14}$ A binary compound of boron and hydrogen that is relatively stable at room temperature; melting point 99.5°C, boiling point 213°C. { ¦dek·ə¦bór,ān 'fór¦tēn }

decade [ELEC] A group or assembly of 10 units; for example, a decade counter counts 10 in one column, and a decade box inserts resistance quantities in multiples of powers of 10. [SCI TECH] The interval between any two quantities having the ratio of 10 to 1. { de'kād }

decade box [ELEC] An assembly of precision resistors, coils, or capacitors whose individual values vary in submultiples and multiples of 10; by appropriately setting a 10-position selector switch for each section, the decade box can be set to any desired value within its range. { de'kād ˌbäks }

decade bridge [ELECTR] Electronic apparatus for measurement of unknown values of resistances or capacitances by comparison with known values (bridge); one secondary section

of the oscillator-driven transformer is tapped in decade steps, the other in 10 uniform steps. { de'kād ˌbrij }

decade counter See decade scaler. { de'kād ˌkaúnt·ər }

decade scaler [ELECTR] A scaler that produces one output pulse for every 10 input pulses. Also known as counter decade; decade counter; scale-of-ten circuit. { de'kād ˌskāl·ər }

decagon [MATH] A 10-sided polygon. { 'dek·ə,gän }

decahedron [MATH] A polyhedron that has 10 faces. { ˌdek·ə'hē·drən }

decahydrate [CHEM] A compound that has 10 water molecules. { ˌdek·ə'hī,drāt }

decahydronaphthalene [ORG CHEM] $C_{10}H_{18}$ A liquid hydrocarbon, used in some paints and lacquers as a solvent. { ˌdek·ə,hī·drō'naf·thə,lēn }

decalcification [CHEM] Loss or removal of calcium or calcium compounds from a calcified material such as bone or soil. { dē,kal·sə·fə'kā·shən }

decalescence [MET] Darkening of a metal surface due to isothermal absorption of the latent heat of phase transformation. { ˌde·kə'les·əns }

decaliter [MECH] A unit of volume, equal to 10 liters, or to 0.01 cubic meter. { 'dek·ə,lēd·ər }

decameter [MECH] A unit of length in the metric system equal to 10 meters. { 'dek·ə,mēd·ər }

decametric wave [ELECTROMAG] British term for a radio wave ranging from 10 to 100 meters long. { 'dek·ə,me·trik ˌwāv }

decane [ORG CHEM] $C_{10}H_{22}$ Any of several saturated aliphatic hydrocarbons, especially $CH_3(CH_2)_8CH_3$. { 'dē,kān }

decanning [NUCLEO] Removing the outer container of an enriched uranium fuel rod, in preparation for reprocessing of the fuel. { dē'kan·iŋ }

decanol See decyl alcohol. { 'dek·ə,nól }

decantation [ENG] A method for mechanical dewatering of a wet solid by pouring off the liquid without disturbing underlying sediment or precipitate. { 'dē,kan'tā·shən }

decanter [ENG] Tank or vessel in which solids or immiscible dispersions in a carrier liquid settle or coalesce, with clear upper liquid withdrawn (decanted) as overflow from the top. { də'kant·ər }

decanth larva See lycophore larva. { ˌdē'kanth ˌlär·və }

Decapoda [INV ZOO] **1.** A diverse order of the class Crustacea including the shrimps, lobsters, hermit crabs, and true crabs; all members have a carapace, well-developed gills, and the first three pairs of thoracic appendages specialized as maxillipeds. **2.** An order of dibranchiate cephalopod mollusks containing the squids and cuttle fishes, characterized by eight arms and two long tentacles. { də'kap·əd·ə }

decarbonize [CHEM] To remove carbon by chemical means. { dē'kär·bə,nīz }

decarboxylase [BIOCHEM] An enzyme that hydrolyzes the carboxyl radical, COOH. { ˌdē·kär'bäk·sə,lās }

decarboxylate [ORG CHEM] To remove the carboxyl radical, especially from amino acids and protein. { ˌdē·kär'bäk·sə,lāt }

decarburize [MET] To remove carbon from the surface of a ferrous alloy, particularly steel, by heating in a medium that reacts with carbon. { dē'kär·bə,rīz }

decastere [MECH] A unit of volume, equal to 10 cubic meters. { 'dek·ə,stir }

decating [TEXT] A finishing process in which fabric is sponged or steamed in rollers to set the width and length as well as to achieve smoothness. Also known as decatizing. { də'kād·iŋ }

decatizing See decating. { 'dek·ə,tīz·iŋ }

decavanadate [INORG CHEM] A deep-orange polyvanadate $(V_{10}O_{28}^{6-})$, composed of 10 fused VO_6 octahedra. { ˌdē·kə'van·ə,dāt }

decay [GEOCHEM] See chemical weathering. [MATER] To undergo decomposition. [NUC PHYS] See radioactive decay. [OCEANOGR] In ocean-wave studies, the loss of energy from wind-generated ocean waves after they have ceased to be acted on by the wind; this process is accompanied by an increase in length and a decrease in height of the wave. [PHYS] Gradual reduction in the magnitude of a quantity, as of current, magnetic flux, a stored charge, or phosphorescence. { di'kā }

decay accelerating factor [IMMUNOL] A plasma protein involved with complement regulation on the red blood cell surface; it accelerates the breakdown of C3 and C5 convertases. Abbreviated DAF. { di¦kā ik'sel·ə¦rād·iŋ ¦fak·tər }

decay area [OCEANOGR] The area into which ocean waves travel (as swell) after leaving the generating area. { di'kā ¦er·ē·ə }

decay chain See radioactive series. { di'kā ¦chān }

decay coefficient See decay constant. { di'kā ¦kō·i₁fish·ənt }

decay constant [PHYS] The constant c in the equation $I = I_0 e^{-ct}$, for the time dependence of rate of decay of a radioactive species; here, I is the number of disintegrations per unit time. Also known as decay coefficient; disintegration constant; radioactive decay constant; transformation constant. { di'kā ₁kän·stənt }

decay cooling [NUCLEO] A stage in nuclear fuel reprocessing in which the irradiated fuel elements are stored for a period of time sufficient for the decay of short-lived isotopes. { di'kā ₁kül·iŋ }

decay curve [NUC PHYS] A graph showing how the activity of a radioactive sample varies with time; alternatively, it may show the amount of radioactive material remaining at any time. { di'kā ₁kərv }

decay distance [OCEANOGR] The distance through which ocean waves pass after leaving the generating area. { di'kā ₁dis·təns }

decay family See radioactive series. { di'kā ₁fam·lē }

decay gammas [NUC PHYS] The characteristic gamma rays emitted during the decay of most radioisotopes. { di'kā ₁gam·əz }

decay heat [NUCLEO] Heat produced by the decay of radioactive nuclides. { di'kā ₁hēt }

decay mode [NUC PHYS] A possible type of decay of a radionuclide or elementary particle. { di'kā ₁mōd }

decay of waves [OCEANOGR] The decrease in height and increase in length of waves after leaving a generating area and passing through a calm, or region of lighter winds. { di'kā əv 'wavz }

decay product See daughter. { di'kā ₁prä·dəkt }

decay rate [NUC PHYS] The time rate of disintegration of radioactive material, generally accompanied by emission of particles or gamma radiation. { di'kā ₁rāt }

decay series See radioactive series. { di'kā ₁sir·ēz }

decay theory [PSYCH] A model of forgetting which assumes that memories fade and will gradually be lost if they are not occasionally refreshed. { di'kā ₁thē·ə·rē }

decay time [PHYS] The time taken by a quantity to decay to a stated fraction of its initial value; the fraction is commonly $1/e$. Also known as storage time (deprecated). { di'kā ₁tīm }

Decca [NAV] A hyperbolic navigation system which establishes a line of position from measurement of the phase difference between two continuous-wave signals; the intersection of the two lines of position from two pairs of transmitting stations establishes a navigational fix, or location. { dek·ə }

Decca chain [NAV] A system or combination of three slave radio transmitting stations disposed about a master Decca station. Also known as star chain. { 'dek·ə ₁chān }

Decca chart [NAV] A chart showing Decca lines of position. { 'dek·ə ₁chärt }

Deccan basalt [GEOL] Fine-grained, nonporphyritic, tholeiitic basaltic lava consisting essentially of labradorite, clinopyroxene, and iron ore; found in the Deccan region of southeastern India. Also known as Deccan trap. { 'dek·ən bə'sólt }

Deccan trap See Deccan basalt. { 'dek·ən 'trap }

decelerating electrode [ELECTR] Of an electron-beam tube, an electrode to which a potential is applied to decrease the velocity of the electrons in the beam. { də'sel·ə₁rād·iŋ i'lek₁trōd }

deceleration [MECH] The rate of decrease of speed of a motion. { dē₁sel·ə'rā·shən }

deceleration parachute See drogue. { dē₁sel·ə'rā·shən 'par·ə₁shüt }

deceleration time [COMPUT SCI] For a storage medium, such as magnetic tape that must be physically moved in order for reading or writing to take place, the minimum time that must elapse between the completion of a reading or writing operation and the moment that motion ceases. Also known as stop time. { dē₁sel·ə'rā·shən ₁tīm }

decelerometer [ENG] An instrument that measures the rate at which the speed of a vehicle decreases. { dē₁sel·ə'räm·əd·ər }

deceleron [AERO ENG] A lateral control surface of an airplane that is divided so as to combine the functions of an airbrake and an aileron. { dē'sel·ə₁rän }

December solstice [ASTRON] Winter solstice in the Northern Hemisphere. { di'sem·bər 'säl·stəs }

decentered lens [OPTICS] A lens whose optical center does not coincide with the geometrical center of the rim of the lens; has the effect of a lens combined with a weak prism. { dē'sent·ərd 'lenz }

decentralized data processing [COMPUT SCI] An arrangement comprising a data-processing center for each division or location of a single organization. { dē'sen·trə₁lizd 'dad·ə 'präs₁es·iŋ }

deception [ELECTR] The deliberate radiation, reradiation, alteration, absorption, or reflection of electromagnetic energy in a manner intended to mislead an enemy in the interpretation of information received by his electronic systems. { di'sep·shən }

decerebellate [MED] Lacking the cerebellum, generally by experimental removal. { ¦dē₁ser·ə¦bel·ət }

decerebrate [MED] Lacking the cerebrum either by experimental removal or by disconnection. { dē'ser·ə₁brāt }

decerebrate rigidity [MED] Exaggerated postural tone in the antigravity muscles due to release of vestibular nuclei from cerebral control. { dē'ser·ə₁brāt ri'jid·əd·ē }

dechlorination [CHEM] Removal of chlorine from a substance. { dē₁klór·ə'nā·shən }

deci- [SCI TECH] A prefix indicating 10^{-1}, 0.1, or a tenth. { 'des·ē }

deciare [MECH] A unit of area, equal to 0.1 are or 10 square meters. { 'des·ē₁er }

decibar [MECH] A metric unit of pressure equal to one-tenth bar. { 'des·ə₁bär }

decibel [PHYS] A unit for describing the ratio of two powers or intensities, or the ratio of a power to a reference power; in the measurement of sound intensity, the pressure of the reference sound is usually taken as 2×10^{-4} dyne per square centimeter; equal to one-tenth bel; if P_1 and P_2 are two amounts of power, the first is said to be n decibels greater, where $n = 10 \log_{10}(P_1/P_2)$. Abbreviated dB. { 'des·ə₁bel }

decibel adjusted See adjusted decibel. { 'des·ə₁bel ə'jəs·təd }

decibel loss [COMMUN] Signal attenuation over a transmission path or a conductor expressed in decibels. { 'des·ə₁bel ₁lós }

decibel meter [ENG] An instrument calibrated in logarithmic steps and labeled with decibel units and used for measuring power levels in communication circuits. { 'des·ə₁bel ₁mēd·ər }

decibels above 1 femtowatt [ELEC] A power level equal to 10 times the common logarithm of the ratio of the given power in watts to 1 femtowatt (10^{-15} watt). Abbreviated dBf. { 'des·ə₁bəlz ə¦bəv ¦wən 'fem·tō₁wät }

decibels above 1 kilowatt [ELEC] A measure of power equal to 10 times the common logarithm of the ratio of a given power to 1000 watts. Abbreviated dBk. { 'des·ə₁bəlz ə¦bəv ¦wən 'kil·ə₁wät }

decibels above 1 milliwatt [ELEC] A measure of power equal to 10 times the common logarithm of the ratio of a given power to 0.001 watt; a negative value, such as -2.7 dBm, means decibels below 1 milliwatt. Abbreviated dBm. { 'des·ə₁bəlz ə¦bəv ¦wən 'mil·i₁wät }

decibels above 1 picowatt [ELEC] A measure of power equal to 10 times the common logarithm of the ratio of a given power to 1 picowatt. Abbreviated dBp. { 'des·ə₁bəlz ə¦bəv ¦wən 'pē·kō₁wät }

decibels above 1 volt [ELEC] A measure of voltage equal to 20 times the common logarithm of the ratio of a given voltage to 1 volt. Abbreviated dBV. { 'des·ə₁bəlz ə¦bəv ¦wən 'vōlt }

decibels above 1 watt [ELEC] A measure of power equal to 10 times the common logarithm of the ratio of a given power to 1 watt. Abbreviated dBW. { 'des·ə₁bəlz ə¦bəv ¦wən 'wät }

decibels above reference coupling [ELEC] A measure of the coupling between two circuits, expressed in relation to a reference value of coupling that gives a specified reading on a specified noise-measuring set when a test tone of 90 dBa is impressed on one circuit. Abbreviated dBx. { 'des·ə₁bəlz ə¦bəv 'ref·rəns ₁kəp·liŋ }

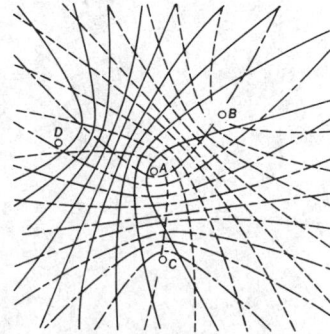

DECCA CHART

Decca station locations and lines of position. *A* is master station and *B*, *C*, and *D* are slave stations. *(From P. C. Sandretto, Electronic Avigation Engineering, International Telephone and Telegraph Corp., 1958)*

decibels above reference noise [ELEC] Units used to show the relationship between the interfering effect of a noise frequency, or band of noise frequencies, and a fixed amount of noise power commonly called reference noise; a 1000-hertz tone having a power level of −90 dBm was selected as the reference noise power; superseded by the adjusted decibel unit. Abbreviated dBrn. { 'des·ə·bəlz ə¦bəv 'ref·rəns ˌnȯiz }

decibels over isotropic *See* dBic. { 'des·ə·bəlz ō·vər ˌī·sə'träp·ik }

decidable predicate [MATH] A predicate for which there exists an algorithm which, for any given value of its independent variables, provides a definite answer as to whether or not it is true. { di'sīd·ə·bəl 'pred·ə·kət }

decidua [MED] The endometrium of pregnancy and associated fetal membranes which are cast off at parturition. { di'sij·ə·wə }

deciduitis [MED] Inflammation of the decidua. { də,sid·yə'wīd·əs }

deciduoma [MED] **1.** A mass of tissue formed in the uterus following pregnancy as the result of hyperplasia of chorionic or decidual cells. **2.** Decidual tissue induced in the uterus, usually by physical trauma. { ,des·i'dwō·mə }

deciduous [BIOL] Falling off or being shed at the end of the growing period or season. { di'sij·ə·wəs }

deciduous teeth [ANAT] Teeth of a young mammal which are shed and replaced by permanent teeth. Also known as milk teeth. { di'sij·ə·wəs 'tēth }

decigram [MECH] A unit of mass, equal to 0.1 gram. { 'des·ə,gram }

decile [STAT] Any of the points which divide the total number of items in a frequency distribution into 10 equal parts. { 'des,īl }

deciliter [MECH] A unit of volume, equal to 0.1 liter, or 10^{-4} cubic meter. { 'des·ə,lēd·ər }

decimal [MATH] A number expressed in the scale of tens. { 'des·məl }

decimal attenuator [ELECTR] System of attenuators arranged so that a voltage or current can be reduced decimally. { 'des·məl ə'ten·yə,wād·ər }

decimal balance [ENG] A balance having one arm 10 times the length of the other, so that heavy objects can be weighed by using light weights. { 'des·məl ,bal·əns }

decimal-binary switch [ELEC] A switch that connects a single input lead to appropriate combinations of four output leads (representing 1, 2, 4, and 8) for each of the decimal-numbered settings of its control knob; thus, for position 7, output leads 1, 2, and 4 would be connected to the input. { ¦des·məl ¦bīn·ə·rē 'swich }

decimal code [COMPUT SCI] A code in which each allowable position has one of 10 possible states; the conventional decimal number system is a decimal code. { ¦des·məl ¦kōd }

decimal-coded digit [COMPUT SCI] One of 10 arbitrarily selected patterns of 1 and 0 used to represent the decimal digits. Also known as coded decimal. { ¦des·məl ¦kōd·əd 'dij·ət }

decimal fraction [MATH] Any number written in the form: an integer followed by a decimal point followed by a (possibly infinite) string of digits. { ¦des·məl 'frak·shən }

decimal number [MATH] A number signifying a decimal fraction by a decimal point to the left of the numerator with the number of figures to the right of the point equal to the power of 10 of the denominator. { ¦des·məl 'nəm·bər }

decimal number system [MATH] A representational system for the real numbers in which place values are read in powers of 10. { ¦des·məl 'nəm·bər ,sis·təm }

decimal place [MATH] Reference to one of the digits following the decimal point in a decimal fraction; the kth decimal place registers units of 10^{-k}. { ¦des·məl 'plās }

decimal point [MATH] A dot written either on or slightly above the line; used to mark the point at which place values change from positive to negative powers of 10 in the decimal number system. { 'des·məl ,pȯint }

decimal processor [COMPUT SCI] A digital computer organized to calculate by decimal arithmetic. { ¦des·məl 'präs,es·ər }

decimal system [MATH] A number system based on the number 10; in theory, each unit is 10 times the next smaller one. { 'des·məl ,sis·təm }

decimal-to-binary conversion [COMPUT SCI] The mathematical process of converting a number written in the scale of 10 into the same number written in the scale of 2. { ¦des·məl tə ¦bin·ə·re kən'vər·zhən }

decimeter [MECH] A metric unit of length equal to one-tenth meter. { 'des·ə,mēd·ər }

decimetric wave [ELECTROMAG] An electromagnetic wave having a wavelength between 0.1 and 1 meter, corresponding to a frequency between 300 and 3000 megahertz. { ¦des·ə¦me·trik 'wāv }

decineper [PHYS] One-tenth of a neper. { 'des·ə,nep·ər }

decinormal [CHEM] Pertaining to a chemical solution that is one-tenth normality in reference to a 1 normal solution. { ¦des·ə'nȯr·məl }

decision [COMPUT SCI] The computer operation of determining if a certain relationship exists between words in storage or registers, and taking alternative courses of action; this is effected by conditional jumps or equivalent techniques. { di'sizh·ən }

decision box [COMPUT SCI] A flow-chart symbol indicating a decision instruction; usually diamond-shaped. { di'sizh·ən ,bäks }

decision calculus [SYS ENG] A guide to the process of decision-making, often outlined in the following steps: analysis of the decision area to discover applicable elements; location or creation of criteria for evaluation; appraisal of the known information pertinent to the applicable elements and correction for bias; isolation of the unknown factors; weighting of the pertinent elements, known and unknown, as to relative importance; and projection of the relative impacts on the objective, and synthesis into a course of action. { di'sizh·ən 'kal·kyə·ləs }

decision element [ELECTR] A circuit that performs a logical operation such as "and," "or," "not," or "except" on one or more binary digits of input information representing "yes" or "no" and that expresses the result in its output. Also known as decision gate. { di'sizh·ən ,el·ə·mənt }

decision gate [ELECTR] *See* decision element. [NAV] In an instrument landing, that point along the path at which the pilot must decide to land or to execute a missed-approach procedure. { di'sizh·ən ,gāt }

decision height [NAV] A height specified in MSL (mean sea level) above the highest runway elevation in the touchdown zone at which a missed approach shall be initiated if the required visual reference has not been established; this term is used only in procedure where an electronic glide slope provides a reference for descent or in ILS (instrument landing systems) or PAR (precision approach radar). Abbreviated d.h. { di'sizh·ən ,hīt }

decision instruction *See* conditional jump. { di'sizh·ən in'strək·shən }

decision-making under uncertainty [STAT] The process of drawing conclusions from limited information or conjecture. { di'sizh·ən ¦māk·iŋ ,ən·dər ən'sərt·ən·tē }

decision mechanism [COMPUT SCI] In character recognition, that component part of a character reader which accepts the finalized version of the input character and makes an assessment as to its most probable identity. { di'sizh·ən ,mek·ə,niz·əm }

decision rule [SYS ENG] In decision theory, the mathematical representation of a physical system which operates upon the observed data to produce a decision. { di'sizh·ən ,rül }

decision support [COMPUT SCI] The process of filtering, optimizing, and organizing mined information to support decision making. { di'sizh·ən sə,pȯrt }

decision support system [COMPUT SCI] A computer-based system that enables management to interrogate the computer system on an ad hoc basis for various kinds of information on the organization and to predict the effect of potential decisions beforehand. Abbreviated DSS. { di'sizh·ən sə'pȯrt ,sis·təm }

decision table [COMPUT SCI] **1.** A table of contingencies to be considered in the definition of a problem, together with the actions to be taken; sometimes used in place of a flow chart for program documentation. **2.** *See* DETAB. { di'sizh·ən ,tā·bəl }

decision theory [SYS ENG] A broad spectrum of concepts and techniques which have been developed to both describe and rationalize the process of decision making, that is, making a choice among several possible alternatives. { di'sizh·ən ,the·ə·rē }

decision tree [IND ENG] Graphic display of the underlying decision process involved in the introduction of a new product by a manufacturer. { di'sizh·ən ˌtrē }

deck [COMPUT SCI] A set of punched cards. [CIV ENG] **1.** A floor, usually of wood, without a roof. **2.** The floor or roadway of a bridge. [ENG] A magnetic-tape transport mechanism. [NAV ARCH] Horizontal or cambered and sloping surfaces on a ship, corresponding to the floors of a building. { dek }

deck beam [NAV ARCH] A strengthening member of the deck that may run the length or width of the deck and is connected at both ends to the ship's frames or bulkheads by girders. { 'dek ˌbēm }

deck bridge [CIV ENG] A bridge that carries the deck on the very top of the superstructure. { 'dek ˌbrij }

deck charge [MIN ENG] A charge that is separated into several smaller components and placed along a quarry borehole. { 'dek ˌchärj }

Deck effect [PARTIC PHYS] The simulation of resonances in multiple-particle production processes in high-energy scattering, wherein the true resonance pertains to a subset of the particles produced. { 'dek iˌfekt }

deck erection [NAV ARCH] A forecastle, bridge, poop, or deckhouse erected on the upper deck of a ship. { 'dek əˌrek·shən }

deck fitting [NAV ARCH] **1.** A fitting where a pipeline penetrates the deck to maintain watertightness. **2.** Any fitting attached to the deck. { 'dek ˌfid·iŋ }

deck height [NAV ARCH] Vertical distance between the molded lines of two adjacent decks. { 'dek ˌhīt }

deckhouse [NAV ARCH] A low building or superstructure, such as a cabin, constructed on the top deck of a ship which may or may not extend to the edges of the deck. { 'dek ˌhaús }

decking [CIV ENG] Surface material on a deck. [ENG] Separating explosive charges containing primers with layers of inert material to prevent passage of concussion. [MIN ENG] Changing tubs on a cage at both ends of a shaft. { 'dek·iŋ }

deckle [ENG] A detachable wood frame fitted around the edges of a papermaking mold. [MATER] In paper manufacturing, the width of the wet sheet as it comes off the wire of a paper machine. { 'dek·əl }

deckle edge [MATER] The unfinished edge of paper having a characteristic appearance as a result of leakage under the frame (deckle) in which the paper is made; handmade paper has a deckle edge on all four sides, machine-made paper only on two sides. { ˌdek·əl 'ej }

deckle rod [ENG] A small rod inserted at each end of the extrusion coating die to adjust the die opening length. { 'dek·əl ˌräd }

deckle strap [ENG] An endless rubber band which runs longitudinally along the wire edges of a paper machine and determines web width. { 'dek·əl ˌstrap }

deck light [NAV ARCH] Any opening in the deck covered with glass; allows daylight to come in below deck. { 'dek ˌlīt }

deck line [NAV ARCH] A line that passes through the intersection of the molded line of the deck beams and the molded line of the frames. { 'dek ˌlīn }

deck loading [MIN ENG] The method of loading deck charges in a quarry borehole. { 'dek ˌlōd·iŋ }

deck log [NAV] A record of significant navigational data and events of a ship's voyage written (usually by the quartermaster) in chronological order. { 'dek ˌläg }

deck machinery [NAV ARCH] Any machinery on the decks of a ship; includes capstans, windlasses, and winches. { 'dek məˌshēn·rē }

deck roof [BUILD] A roof that is nearly flat and without parapet walls. { 'dek ˌrüf }

deck stringer plate [NAV ARCH] The outboard fore-and-aft plate on any deck. { 'dek 'striŋ·ər ˌplāt }

deck switch See gang switch. { 'dek ˌswich }

deck truss [CIV ENG] The frame of a deck. { 'dek ˌtrəs }

deckzelle [INV ZOO] In certain hydroids, one of the supporting or epithelial cells which are usually columnar or cuboidal. { 'dekˌzel }

declaration See declarative statement. { ˌdek·lə'rā·shən }

declarative language [COMPUT SCI] A nonprocedural programming language that allows the programmer to state the task to be accomplished without specifying the procedures needed to carry it out. { diˌklar·əd·iv 'laŋ·gwij }

declarative macroinstruction [COMPUT SCI] An instruction

in an assembly language which directs the compiler to take some action or take note of some condition and which does not generate any instruction in the object program. { diˌklar·əd·iv ˌmak·rō·inˌstrək·shən }

declarative markup language [COMPUT SCI] A system of codes for identifying the subdivisions of a text-processing document, without carrying out the actual formatting. { diˌklar·əd·iv 'mär·kəp ˌlaŋ·gwij }

declarative memory See explicit memory. { diˌklar·əd·iv 'mem·rē }

declarative statement [COMPUT SCI] Any program statement describing the data which will be used or identifying the memory locations which will be required. Also known as declaration. { diˌklar·əd·iv 'stāt·mənt }

declinate [BIOL] Curved toward one side or downward. { 'dek·ləˌnāt }

declination [ASTRON] The angular distance of a celestial object north or south of the celestial equator. [GEOPHYS] The angle between the magnetic and geographical meridians, expressed in degrees and minutes east or west to indicate the direction of magnetic north from true north. Also known as magnetic declination; variation. [NAV] **1.** In a system of polar or spherical coordinates, the angle at the origin between a line to a point and the equatorial plane, measured in a plane perpendicular to the equatorial plane. **2.** The arc between the equator and the point measured on a great circle perpendicular to the equator. { ˌdek·lə'nā·shən }

declination axis [ENG] For an equatorial mounting of a telescope, an axis of rotation that is perpendicular to the polar axis and allows the telescope to be pointed at objects of different declinations. { ˌdek·lə'nā·shən ˌak·səs }

declination circle [ENG] For a telescope with an equatorial mounting, a setting circle attached to the declination axis that shows the declination to which the telescope is pointing. { ˌdek·lə'nā·shən ˌsər·kəl }

declination compass See declinometer. { ˌdek·lə'nā·shən ˌkəm·pəs }

declination difference [NAV] The difference between two declinations, particularly between the declination of a celestial body and the value used as an argument for entering a table. { ˌdek·lə'nā·shən ˌdif·rəns }

declination of grid north See grid declination. { ˌdek·lə'nā·shən əv ˌgrid 'nórth }

declination variometer [ENG] An instrument that measures changes in the declination of the earth's magnetic field, consisting of a permanent bar magnet, usually about 0.4 inch (1 centimeter) long, suspended with a plane mirror from a fine quartz fiber 2–6 inches (5–15 centimeters) in length; a lens focuses to a point a beam of light reflected from the mirror to recording paper mounted on a rotating drum. Also known as D variometer. { ˌdek·lə'nā·shən ˌver·ē'äm·əd·ər }

declining population [ECOL] A population in which old individuals outnumber young individuals. { də'klin·iŋ ˌpäp·yə'lā·shən }

declinometer [ENG] A magnetic instrument similar to a surveyor's compass, but arranged so that the line of sight can be rotated to conform with the needle or to any desired setting on the horizontal circle; used in determining magnetic declination. Also known as declination compass. { ˌdek·lə'näm·əd·ər }

declivity [GEOL] **1.** A slope descending downward from a point of reference. **2.** A downward deviation from the horizontal. { də'kliv·əd·ē }

decode [COMMUN] **1.** To translate coded characters into a more understandable form. **2.** See demodulate. { dē'kōd }

decoder [COMMUN] A device that decodes. [ELECTR] **1.** A matrix of logic elements that selects one or more output channels, depending on the combination of input signals present. **2.** See decoder circuit; matrix; tree. { dē'kōd·ər }

decoder circuit [ELECTR] A circuit that responds to a particular coded signal while rejecting others. Also known as decoder. { dē'kōd·ər ˌsər·kət }

decoding gate [COMPUT SCI] The use of combinatorial logic in circuitry to select a device identified by a binary address code. Also known as recognition gate. { dē'kōd·iŋ ˌgāt }

decoherence [QUANT MECH] The process whereby the quantum-mechanical state of any macroscopic system is rapidly correlated with that of its environment in such a way that no measurement on the system alone (without a simultaneous measurement of the complete state of the environment) can

Declination variometer equipped with Helmholtz coil for calibration. (*U. S. Coast and Geodetic Survey*)

demonstrate any interference between two quantum states of the system. { ¦dē·kō'hir·əns }

decoking [CHEM ENG] Removal of petroleum coke from equipment. { dē'kōk·iŋ }

decollator [COMPUT SCI] A device which separates the sheets of continuous stationery that form the output of a computer printer into separate stacks. { dē'kō‚lād·ər }

décollement [GEOL] Folding or faulting of sedimentary beds by sliding over the underlying rock. { dā'käl·mənt }

decolorize [CHEM ENG] To remove the color from, as from a liquid. { dē'kəl·ə‚rīz }

decolorizer [CHEM ENG] An agent used to decolorize; the removal of color may occur by a chemical reaction or a physical reaction. { dē'kəl·ə‚rīz·ər }

decolorizing carbon [CHEM] Porous or finely divided carbon (activated or bone) with large surface area; used to adsorb colored impurities from liquids, such as lube oils. { dē¦kəl·ə‚rīz·iŋ 'kär·bən }

decometer [ELECTR] An adding-type phasemeter which rotates continuously and adds up the total number of degrees of phase shift between two signals, such as those received from two transmitters in the Decca navigation system. { də'käm·əd·ər }

decomino [MATH] One of the 4655 plane figures that can be formed by joining 10 unit squares along their sides. { ‚dek·ə'mē·nō }

decommissioning [NUCLEO] The process of shutting down a nuclear facility such as a nuclear reactor or reprocessing plant so as to provide adequate protection from radiation exposure and to isolate radioactive contamination from the human environment. { dē·kə'mish·ən·iŋ }

decommutation [ELECTR] The process of recovering a signal from the composite signal previously created by a commutation process. { dē‚käm·yə'tā·shən }

decommutator [ELECTR] The section of a telemetering system that extracts analog data from a time-serial train of samples representing a multiplicity of data sources transmitted over a single radio-frequency link. { dē'käm·yə‚tād·ər }

decompensation [PSYCH] The deterioration of existing defense mechanisms, leading to an exacerbation of pathologic behavior. { dē‚käm·pən'sā·shən }

decomposable process [MATH] A process which can be reduced to several basic events. { dē·kəm'pō·zə·bəl 'präs·əs }

decomposer [ECOL] A heterotrophic organism (including bacteria and fungi) which breaks down the complex compounds of dead protoplasm, absorbs some decomposition products, and releases substances usable by consumers. Also known as microcomposer; microconsumer; reducer. { de·kəm'pō·zər }

decomposition [CHEM] The more or less permanent structural breakdown of a molecule into simpler molecules or atoms. *See* chemical weathering. [MATH] **1.** The expression of a fraction as a sum of partial fractions. **2.** The representation of a set as the union of pairwise disjoint subsets. { dē‚käm·pə'zish·ən }

decomposition potential [PHYS CHEM] The electrode potential at which the electrolysis current begins to increase appreciably. Also known as decomposition voltage. { dē‚käm·pə'zish·ən pə‚ten·chəl }

decomposition voltage *See* decomposition potential. { dē‚käm·pə'zish·ən ‚vōl·tij }

decompound [BOT] Divided or compounded several times, with each division being compound. { dē'käm‚paúnd }

decompression [ENG] Any procedure for the relief of pressure or compression. { dē·kəm'presh·ən }

decompression chamber [ENG] **1.** A steel chamber fitted with auxiliary equipment to raise its air pressure to a value two to six times atmospheric pressure; used to relieve a diver who has decompressed too quickly in ascending. **2.** Such a chamber in which conditions of high atmospheric pressure can be simulated for experimental purposes. { dē·kəm'presh·ən ‚chām·bər }

decompression illness *See* aeroembolism. { dē·kəm'presh·ən ‚il·nəs }

decompression table [ENG] A diving guide that lists ascent rates and breathing mixtures to provide safe pressure reduction to atmospheric pressure after a dive. { dē·kəm'presh·ən ‚tā·bəl }

deconcentrator [ENG] An apparatus for removing dissolved or suspended material from feedwater. { dē'käns·ən‚trād·ər }

decontamination [ENG] The removing of chemical, biological, or radiological contamination from, or the neutralizing of it on, a person, object, or area. { dē·kən‚tam·ə'nā·shən }

decontamination factor [NUCLEO] The ratio of initial specific radioactivity to final specific radioactivity resulting from a separation process. { dē·kən‚tam·ə'nā·shən ‚fak·tər }

decontamination index [NUCLEO] The logarithm of the ratio of initial specific radioactivity to final specific radioactivity resulting from a separation process. { dē·kən‚tam·ə'nā·shən ‚in‚deks }

decoppering agent [ORD] Material, such as finely divided tin, included in a propelling charge, or inserted in the chamber with the propelling charge, for the purpose of removing the coppering from the surface of the bore. { dē'käp·ə·riŋ ‚ā·jənt }

decorative stone [MATER] A stone that serves for architectural decoration, as in mantles or store fronts. { ¦dek·rəd·iv 'stōn }

decorticate [BIOL] Lacking a cortical layer. [TEXT] To remove woody residual matter, such as flax from bast fibers after retting. { dē'kórd·ə‚kāt }

decouple [ENG] **1.** To minimize or eliminate airborne shock waves of a nuclear or other explosion by placing the explosives deep under the ground. **2.** To minimize the seismic effect of an underground explosion by setting it off in the center of an underground cavity. { dē'kəp·əl }

decoupler [IND ENG] A materials handling device designed specifically for cellular manufacturing. { dē'kəp·lər }

decoupling [ELEC] Preventing transfer or feedback of energy from one circuit to another. { dē'kəp·liŋ }

decoupling era [ASTRON] The time about 300,000 years after the big bang when matter and radiation, which had previously been strongly coupled, practically ceased to interact, electrons were able to attach to nuclei and form atoms, and photons could propagate freely. { dē'kəp·liŋ ‚ir·ə }

decoupling filter [ELECTR] One of a number of low-pass filters placed between each of several amplifier stages and a common power supply. { dē'kəp·liŋ ‚fil·tər }

decoupling network [ELEC] Any combination of resistors, coils, and capacitors placed in power supply leads or other leads that are common to two or more circuits, to prevent unwanted interstage coupling. { dē'kəp·liŋ ‚net‚wərk }

decoy [ORD] **1.** An object with reflective characteristics of a target, used in radar deception. **2.** Underwater device which reflects acoustic radiation, used by submarines in the deception of acoustic listening devices and acoustic homing torpedoes. { 'dē‚kói }

decoy target [ORD] A device assembled from prefabricated materials, designed to simulate miscellaneous types of field equipment. { 'dē‚kói ‚tär·gət }

decoy transponder [ELECTR] A transponder that returns a strong signal when triggered directly by a radar pulse, to produce large and misleading target signals on enemy radar screens. { 'dē‚kói tran‚spän·dər }

decreasing function [MATH] A function, f, of a real variable, x, whose value gets smaller as x gets larger; that is, if $x < y$ then $f(x) > f(y)$. Also known as strictly decreasing function. { di'krēs·iŋ ‚fəŋk·shən }

decreasing sequence [MATH] A sequence of real numbers in which each term is less than the preceding term. { di¦krēs·iŋ 'sē·kwəns }

decrement [COMPUT SCI] **1.** A specific part of an instruction word in some binary computers, thus a set of digits. **2.** For a counter, to subtract 1 or some other number from the current value. [HYD] *See* groundwater discharge. [MATH] The quantity by which a variable is decreased. [PHYS] The ratio of the amplitudes of an underdamped harmonic motion during two successive oscillations. Also known as damping factor; numerical decrement. { 'dek·rə·mənt }

decremental arc [ASTRON] A crater arc in which the diameters of the craters decrease from one end of the arc to the other. { 'dek·rə‚ment·əl 'ärk }

decremental chain [ASTRON] A crater chain in which the diameters of the craters decrease from one end of the chain to the other. { 'dek·rə‚ment·əl ‚chān }

decrement field [COMPUT SCI] That part of an instruction word which is used to modify the contents of a storage location or register.   { 'dek·rə·mənt ˌfēld }

decrement gage   [ENG]   A type of molecular gage consisting of a vibrating quartz fiber whose damping is used to determine the viscosity and, thereby, the pressure of a gas.   Also known as quartz-fiber manometer.   { 'de·krə·mənt ˌgāj }

decremeter   [ENG]   An instrument for measuring the logarithmic decrement (damping) of a train of waves.   { 'dek·rə·mēd·ər }

decrepitation   [GEOPHYS]   Breaking up of mineral substances when exposed to heat; usually accompanied by a crackling noise.   { di,krep·ə'tā·shən }

decrypt   [ELECTR]   To convert a cryptogram or series of electronic pulses into plain text by electronic means.   { dē'kript }

Dectra   [NAV]   A radio navigation aid that provides coverage over a specific section of a long ocean route, using equipment and techniques similar to Decca; a master station and a slave station are located at each end of the route; by comparing the phase difference of a master and its slave, guidance is provided to the right or left of the course; comparison of transmission with those of a high-precision quartz clock provides distance information along the course.   { 'dek·trə }

decubitus ulcer   [MED]   An ulcer of the skin and subcutaneous tissues following prolonged lying down, due to pressure on bony protuberances.   Also known as bedsore; pressure ulcer.   { də'kyüb·əd·əs 'əl·sər }

decumbent   [BOT]   Lying down on the ground but with an ascending tip, specifically referring to a stem.   { di'kəm·bənt }

decurrent   [BOT]   Running downward, especially of a leaf base extended past its insertion in the form of a winged expansion.   { di'kər·ənt }

decussate   [BOT]   Of the arrangement of leaves, occurring in alternating pairs at right angles.   [SCI TECH]   To intersect in the form of an X.   { 'dek·ə,sāt }

decussate structure   [GEOL]   A crisscross microstructure of certain minerals; most noticeable in rocks composed predominantly of minerals with a columnar habit.   { 'dek·ə,sāt ˌstrək·chər }

decyl   [ORG CHEM]   An isomeric grouping of univalent radicals, all with formulas $C_{10}H_{21}$, and derived from the decanes by removing one hydrogen.   { 'des·əl }

decyl acetate   [ORG CHEM] $CH_3(CH_2)_9OOCCH_3$ Perfumery liquid with a floral orange-rose aroma.   { 'des·əl 'as·ə,tāt }

decyl alcohol   [ORG CHEM] $C_{10}H_{21}OH$ A colorless oil, boiling at 231°C; used in plasticizers, synthetic lubricants, and detergents.   Also known as decanol.   { 'des·əl 'al·kə,hȯl }

decyl aldehyde   [ORG CHEM] $CH_3(CH_2)_8CHO$ A liquid aldehyde, found in essential oils; used in flavorings and perfumes.   { 'des·əl 'al·də,hīd }

decylene   [ORG CHEM]   Any of a group of isomeric hydrocarbons with formula $C_{10}H_{20}$; the group is part of the ethylene series.   { 'des·ə,lēn }

decyltrichlorosilane   [ORG CHEM] n-$C_{10}H_{21}SiCl_3$ An organochlorosilane that boils at 183°C at 84 mmHg; used in coupling agents or primers to obtain improved bonding between organic polymers and mineral surfaces.   { ¦des·əl·trī'klȯr·ō'sī,lān }

Dedekind cut   [MATH]   A set of rational numbers satisfying certain properties, with which a unique real number may be associated; used to define the real numbers as an extension of the rationals.   { 'dā·də·kint ˌkət }

Dedekind test   [MATH]   If the series

$$\sum_i (b_i - b_{i+1})$$

converges absolutely, the b_i converge to zero, and the series

$$\sum_i a_i$$

has bounded partial sums, then the series

$$\sum_i a_i b_i$$

converges.   { 'dā·də·kint ˌtest }

dedendum   [DES ENG]   The difference between the radius of the pitch circle of a gear and the radius of its root circle.   { də'den·dəm }

dedendum circle   [DES ENG]   A circle tangent to the bottom of the spaces between teeth on a gear wheel.   { də'den·dəm ˌsər·kəl }

dedicated file server   [COMPUT SCI]   A computer that operates solely to provide services to other computers in a particular local-area network and to manage the network operating system.   Also known as dedicated server.   { ˌded·ə,kād·əd 'fīl ˌsər·vər }

dedicated line   [COMPUT SCI]   A permanent communications link that is used solely to transmit information between a computer and a data-processing system.   { ˌded·ə,kād·əd 'līn }

dedicated server   *See* dedicated file server.   { ˌded·ə,kād·əd ˌsər·vər }

dedicated terminal   [COMPUT SCI]   A computer terminal that is permanently connected to a data-processing system by a communications link that is used only to transmit information between the two.   { 'ded·ə,kād·əd 'tərm·ən·əl }

dedifferentiation   [BIOL]   Disintegration of a specialized habit or adaptation.   [CYTOL]   Loss of recognizable specializations that define a differentiated cell.   [PHYSIO]   Return of a specialized cell or structure to a more general or primitive condition.   { dē,dif·ə,ren·chē'ā·shən }

dedolomitization   [GEOL]   Destruction of dolomite to form calcite and periclase, usually by contact metamorphism at low pressures.   { dē,dō·lə,mīd·ə'zā·shən }

deduction   [MATH]   The process of deriving a statement from certain assumed statements by applying the rules of logic.   { di'dək·shən }

dedusting   [MIN ENG]   Cleaning ore, using pneumatic means and screening, to remove dust and other fine impurities.   Also known as aspirating.   { dē'dəst·iŋ }

dee   [NUCLEO]   A hollow accelerating-cyclotron electrode in the shape of the letter D.   Also known as cyclotron D; D.   { dē }

dee line   [NUCLEO]   A structural member that supports the dee of a cyclotron and acts with the dee to form the resonant circuit.   { 'dē ˌlīn }

deemphasis   [ENG ACOUS]   A process for reducing the relative strength of higher audio frequencies before reproduction, to complement and thereby offset the preemphasis that was introduced to help override noise or reduce distortion.   Also known as postemphasis; postequalization.   { dē'em·fə·səs }

deemphasis network   [ENG ACOUS]   An *RC* filter inserted in a system to restore preemphasized signals to their original form.   { dē'em·fə·səs ˌnet,wərk }

deenergize   [ELEC]   To disconnect from the source of power.   { dē'en·ər,jīz }

deep   [OCEANOGR]   An area of great depth in the ocean, representing a depression in the ocean floor.   { dēp }

deep-casting   [OCEANOGR]   Sampling ocean water at great depths by lowering a number of self-sealing bottles, usually made of brass or bronze, on a cable.   { 'dēp ˌkast·iŋ }

deep-draw   [MET]   To form shapes with large depth-diameter ratios in sheet or strip metal by considerable plastic distortion in dies.   { ¦dēp 'drȯ }

deep-draw mold   [ENG]   A mold for plastic material that is long in relation to the thickness of the mold wall.   { ¦dēp ¦drȯ 'mōld }

deep easterlies   *See* equatorial easterlies.   { ¦dēp 'ē·stər·lēz }

deepening   [METEOROL]   A decrease in the central pressure of a pressure system on a constant-height chart, or an analogous decrease in height on a constant-pressure chart.   { 'dēp·ə·niŋ }

deep-etch   [GRAPHICS]   The etching of an offset printing plate so that the printing area becomes slightly recessed, thereby leading to sharper definition and longer life on press.   [MET]   Severe etching of a metal surface to reveal gross features, such as abnormal grain size, segregation, or cracks, at magnifications of 10 diameters or less.   Also known as macroetching.   { ¦dēp 'ech }

deep-etch printing   [GRAPHICS]   Printing with plates produced by deep-etching.   { ¦dēp 'ech ˌprint·iŋ }

deep fascia   [ANAT]   The fibrous tissue between muscles and forming the sheaths of muscles, or investing other deep, definitive structures, as nerves and blood vessels.   { ¦dēp 'fā·shə }

deep fording   [ORD]   The ability of a gun or vehicle, equipped with built-in waterproofing with its suspension in contact with the ground, to negotiate a water obstacle by application of a special waterproofing kit. { ¦dēp 'fȯrd·iŋ }

deep hibernation [PHYSIO] Profound decrease in metabolic rate and physiological function during winter, with a body temperature near 0°C, in certain warm-blooded vertebrates. Also known as hibernation. { 'dēp ¦hī·bər'nā·shən }

deep inelastic collision [NUC PHYS] A nuclear reaction in which two nuclei interact strongly, dissipating sizable amounts of energy and exchanging energy and nucleons, while their surfaces overlap for a brief period corresponding to a partial rotation of the intermediate dinuclear complex. Also known as deep inelastic transfer; incomplete fusion; quasi-fission; relaxed peak process; strongly damped collision. { 'dēp ¦in·ə'las·tik kə'lizh·ən }

deep inelastic transfer See deep inelastic collision. { 'dēp in·ə'las·tik 'tranz·fər }

deep inland sea [GEOGR] A sea adjacent to but in restricted communication with the sea; depth exceeds 660 feet (200 meters). { 'dēp 'in·lənd 'sē }

deep-marine sediments [GEOL] Sedimentary environments occurring in water deeper than 200 meters (660 feet), seaward of the continental shelf break, on the continental slope and the basin. { ¦dēp·mə¦rēn 'sed·ə·mins }

deep mining [MIN ENG] Exploitation of mineral or coal deposits at depths in excess of 3000 feet (900 meters). { 'dēp 'mīn·iŋ }

deep pain [NEUROSCI] Pattern of somesthetic sensation of pain, usually indefinitely localized, originating in the viscera, muscles, and other deep tissues. { 'dēp 'pān }

deep palmar arch [ANAT] The anastomosis between the terminal part of the radial artery and the deep palmar branch of the ulnar artery. Also known as deep volar arch; palmar arch. { 'dēp 'pä·mər ,ärch }

deep-penetration bomb [ORD] A bomb designed to enter deeply into a target before it explodes. { ¦dēp ,pen·ə¦trā·shən 'bäm }

deep-pile fabric [TEXT] Any of various woven or knitted fabrics that simulate fur. { ¦dēp ¦pīl 'fa·brik }

deep scattering layer [OCEANOGR] The stratified populations of organisms which scatter sound in most oceanic waters. { ¦dēp 'skad·ə·riŋ ,lā·ər }

deep-sea basin [GEOL] A depression of the sea floor more or less equidimensional in form and of variable extent. { ¦dēp ¦sē 'bās·ən }

deep-sea channel [GEOL] A trough-shaped valley of low relief beyond the continental rise on the deep-sea floor. Also known as mid-ocean canyon. { ¦dēp ¦sē 'chan·əl }

deep-sea plain [GEOL] A broad, almost level area forming the predominant portion of the ocean floor. { ¦dēp ¦sē 'plān }

deep-seated See plutonic. { ¦dēp 'sēd·əd }

deep-sea trench [GEOL] A long, narrow depression of the deep-sea floor having steep sides and containing the greatest ocean depths; formed by depression, to several kilometers' depth, of the high-velocity crustal layer and the mantle. { ¦dēp ¦sē 'trench }

deep sleep [PSYCH] The third and fourth stage of the sleep cycle, determined by electroencephalographic recording and characterized by slow brain waves. Also known as slow-wave sleep (SWS). { ¦dēp 'slēp }

deep space [ASTRON] Space beyond the gravitational influence of the earth. { ¦dēp 'spās }

Deep Space Network [AERO ENG] A spacecraft network operated by NASA which tracks, commands, and receives telemetry for all types of spacecraft sent to explore deep space, the moon, and solar system planets. Abbreviated DSN. { ¦dēp ¦spās 'net,wərk }

deep-space probe [AERO ENG] A spacecraft designed for exploring space beyond the gravitational and magnetic fields of the earth. { ¦dēp ¦spās 'prōb }

deep-submergence rescue vehicle [NAV ARCH] A small rescue and research submarine; intended primarily to rescue the crew of another submarine to depths of 5000 feet (1500 meters). Abbreviated DSRV. { ¦dēp səb¦mər·jəns 'res·kyü ,vē·ə·kəl }

deep tank [NAV ARCH] A tank that extends from the bottom of a ship up to or higher than the lower deck. { 'dēp ,taŋk }

deep trades See equatorial easterlies. { ¦dēp 'trādz }

deep underwater muon and neutrino detector [ENG] A proposed device for detecting and determining the direction of

extraterrestrial neutrinos passing through a volume of approximately 1 cubic kilometer of ocean water, using an array of several thousand Cerenkov counters suspended in the water to sense the showers of charged particles generated by neutrinos. Abbreviated DUMAND. { ¦dēp ,ən·dər'wȯd·ər 'myü,än ən nü'trē·nō di,tek·tər }

deep volar arch See deep palmar arch. { ¦dēp ¦vō·lər ,ärch }

deep water [OCEANOGR] An ocean area where depth of the water layer is greater than one-half the wave length. { 'dēp 'wȯd·ər }

deep waterline [NAV ARCH] The height of the line on a ship's hull that is reached by the water when the ship is loaded to its maximum safe capacity. { ¦dēp 'wȯd·ər,līn }

deep-water wave [OCEANOGR] A surface wave whose length is less than twice the depth of the water. Also known as short wave. { ¦dēp ,wȯd·ər ,wāv }

deep well [CIV ENG] A well that draws its water from beneath shallow impermeable strata, at depths exceeding 22 feet (6.7 meters). { 'dēp ,wel }

deep-well pump [MECH ENG] A multistage centrifugal pump for lifting water from deep, small-diameter wells; a surface electric motor operates the shaft. Also known as vertical turbine pump. { 'dēp ,wel ,pəmp }

deer [VERT ZOO] The common name for 41 species of even-toed ungulates that compose the family Cervidae in the order Artiodactyla; males have antlers. { dir }

deerhorn antenna [ELECTROMAG] A dipole antenna whose ends are swept back to reduce wind resistance when mounted on an airplane. { 'dir,hȯrn an'ten·ə }

Deerparkian [GEOL] A North American stage of geologic time in the Lower Devonian, above Helderbergian and below Onesquethawan. { dir'pärk·ē·ən }

DEET See diethyltoluamide.

deethanize [CHEM ENG] To separate and remove ethane and sometimes lighter fractions from heavy substances, such as propane, by distillation. { dē'eth·ə,nīz }

deethanizer [CHEM ENG] The equipment used to deethanize. { dē'eth·ə,nīz·ər }

de facto standard [COMPUT SCI] A set of criteria for software, hardware, or communications procedures that is widely accepted because of the dominance of a particular technology over others rather than the action of a recognized standards organization. { dē'fak·tō 'stan·dərd }

default [COMPUT SCI] A value automatically used or an action automatically carried out unless another is specified. { di'fȯlt }

default printer [COMPUT SCI] The printer that is automatically used by a program unless another printer is specifically designated. { di'fȯlt ,print·ər }

defecation [CHEM ENG] Industrial purification, or clarification, of sugar solutions. [PHYSIO] The process by which fecal wastes that reach the lower colon and rectum are evacuated from the body. { ,def·ə'kā·shən }

defect [SCI TECH] An irregularity that spoils the appearance or impairs the usefulness or effectiveness of an object or a material by causing weakness or failure. { 'dē,fekt }

defect chemistry [SOLID STATE] The study of the dynamic properties of crystal defects under particular conditions, such as raising of the temperature or exposure to electromagnetic particle radiation. { 'dē,fekt 'kem·ə·strē }

defect cluster [CRYSTAL] A macroscopic cluster of crystal defects which can arise from attraction among defects. { 'dē,fekt ,kləs·tər }

defect conduction [SOLID STATE] Electric conduction in a semiconductor by holes in the valence band. { 'dē,fekt kən'dək·shən }

defective [SCI TECH] An item or product which has at least one defect. { di'fek·tiv }

defective equation [MATH] An equation that has fewer roots than another equation from which it has been derived. { di'fekt·iv i'kwā·zhən }

defective interfering virus [VIROL] A virus generated at the peak of an infection that can interfere with replication of the normal virus and may modify the outcome of the disease. { di'fek·tiv ,in·tər,fir·iŋ 'vī·rəs }

defective number See deficient number. { di'fek·tiv 'nəm·bər }

DEER

North American elk (*Cervus canadensis*).

defective track [COMPUT SCI] Any circular path on the surface of a magnetic disk which is detected by the system as unable to accept one or more bits of data. { di'fek·tiv 'trak }

defective virus [VIROL] A virus, such as adeno-associated satellite virus, that can grow and reproduce only in the presence of another virus. { di'fek·tiv 'vī·rəs }

defect motion [CRYSTAL] Movement of a point defect from one lattice point to another. { 'dē,fekt 'mō·shən }

defect scattering [SOLID STATE] Scattering of particles or electromagnetic radiation by crystal defects. { di'fekt skad·ər·iŋ }

defect structure [SOLID STATE] A crystal structure in which some atomic positions are occupied by atoms other than those that would be found in a perfect crystal, or are unoccupied. { di'fekt ,strək·chər }

defeminization [PHYSIO] Loss or reduction of feminine attributes, usually caused by ovarian dysfunction or removal. [PSYCH] Psychic process involving a deep and permanent change in the character of a woman, resulting in a giving up of feminine feelings, and the assumption of masculine qualities. { dē,fem·ə·nə'zā·shən }

defender [IND ENG] A machine or facility which is being considered for replacement. { di'fen·dər }

defense information [ORD] Official information which requires protection in the interests of national defense, which is not common knowledge, and which would be of intelligence value to an enemy or potential enemy in the planning or waging of war against a nation. { di'fens ,in·fər'mā·shən }

defense mechanism [PSYCH] Any psychic device, such as rationalization, denial, or repression, for concealing unacceptable feelings or for protecting oneself against unpleasant feelings, memories, or experiences. { di'fens ,mek·ə,niz·əm }

defensive minefield [ORD] A minefield so situated that an adversary's attack will be delayed or repulsed. { di'fen·siv 'mīn,fēld }

deferent [ASTRON] An imaginary circle around the earth, postulated by Ptolemy, in whose circumference a celestial body or its epicycle is supposed to move. { 'def·ə·rənt }

deferment factor *See* discount factor. { di'fər·mənt ,fak·tər }

deferred addressing [COMPUT SCI] A type of indirect addressing in which the address part of an instruction specifies a location containing an address, the latter in turn specifies another location containing an address, and so forth, the number of iterations being controlled by a preset counter. { di'fərd ə'dres·iŋ }

deferred data item [COMPUT SCI] A quantity or attribute that is assigned a value only at the time it is actually processed. { di'fərd 'dad·ə ,īd·əm }

deferred entry [COMPUT SCI] The passing of control of the central processing unit to a subroutine or to an entry point as the result of an asynchronous event. { di'fərd 'en·trē }

deferred mount [COMPUT SCI] Postponement of the placement of a tape on a tape drive until it is actually needed, rather than when the program starts to run. { di'fərd 'maùnt }

deferred processing [COMPUT SCI] The making of computer runs which are postponed until nonpeak periods. { di'fərd 'präs,es·iŋ }

deferrization [CHEM ENG] Removal of iron, for example, from water in an industrial process. { dē,fer·ə'zā·shən }

defervescence *See* lysis. { def·ər'ves·əns }

defibrillation [MED] Stopping a local quivering of muscle fibers, especially of the heart. { dē,fib·rə'lā·shən }

defibrillator [MED] An electronic instrument used for stopping fibrillation during a heart attack by applying controlled electric pulses to the heart muscles. { dē'fib·rə,lād·ər }

deficiency disease [MED] Any disease resulting from a dietary deficiency of minerals, vitamins, or essential nutrients. { də'fish·ən·sē di,zēz }

deficiency index [MATH] For a curve or equation involving two complex variables this is the genus of the Riemann surface associated to the equation. { də'fish·ən·sē ,in,deks }

deficient number [MATH] A positive integer the sum of whose divisors, including 1 but excluding itself, is less than itself. Also known as defective number. { də'fish·ənt 'nəm·bər }

defilade [ORD] An arrangement of fortifications that minimizes the effect of frontal or enfilading fire and plunging or reverse fire. { 'def·ə,läd }

definite-composition law [CHEM] The law that a given chemical compound always contains the same elements in the same fixed proportions by weight. Also known as definite-proportions law. { ¦def·ə·nət ,käm·pə'zish·ən ,lò }

definite network [COMPUT SCI] A sequential network in which no feedback loops exist. { ¦def·ə·nət 'net,wərk }

definite-proportions law *See* definite-composition law. { ¦def·ə·nət prə'pór·shənz ,lò }

definite Riemann integral [MATH] A number associated with a function defined on an interval $[a,b]$ which is

$$\lim_{N\to\infty} \sum_{k=0}^{N-1} f\!\left(a + \frac{k}{N}\right) \cdot \frac{b-a}{N}$$

if f is bounded and continuous; denoted by

$$\int_a^b f(x)dx;$$

if f is a positive function, the definite integral measures the area between the graph of f and the x axis. { ¦def·ə·nət 'rē,män ,in·tə·grəl }

definition [COMMUN] The fidelity with which a television or facsimile receiver forms an image. [ELECTR] The extent to which the fine-line details of a printed circuit correspond to the master drawing. [GRAPHICS] The clarity or sharpness of details in a photographic or microfilm image. [OPTICS] Lens image clarity or discernible detail. { ,def·ə'nish·ən }

definitive host [BIOL] The host in which a parasite reproduces sexually. { də'fin·əd·iv 'hōst }

deflagrating spoon [CHEM] A long-handled spoon used in chemistry to demonstrate deflagration. { 'def·lə,grād·iŋ ,spün }

deflagration [CHEM] A chemical reaction accompanied by vigorous evolution of heat, flame, sparks, or spattering of burning particles. { ,def·lə'grā·shən }

deflashing [ENG] Finishing technique to remove excess material (flash) from a plastic or metal molding. { dē'flash·iŋ }

deflation [GEOL] The sweeping erosive action of the wind over the ground. { di'flā·shən }

deflation basin [GEOL] A topographic depression formed by deflation. { di'flā·shən ,bās·ən }

deflation lake [HYD] A lake in a basin that was formed primarily by wind erosion, especially in arid or semiarid regions. { di'flā·shən ,lāk }

deflected jet fluidic flowmeter *See* fluidic flow sensor. { di¦flek·təd 'jet flü'id·ik 'flō,mēd·ər }

deflecting torque [MECH] An instrument's moment, resulting from the quantity measured, that acts to cause the pointer's deflection. { di'flek·diŋ ,tòrk }

deflection [COMPUT SCI] Encouraging a potential attacker of a computer system to direct the attack elsewhere. [ELECTR] The displacement of an electron beam from its straight-line path by an electrostatic or electromagnetic field. [ENG] **1.** Shape change or reduction in diameter of a conduit, produced without fracturing the material. **2.** Elastic movement or sinking of a loaded structural member, particularly of the mid-span of a beam. [ORD] **1.** Horizontal clockwise angle between the axis of the bore and the line of sighting. **2.** The setting on the scale to compensate for deflection. [PETRO ENG] In oil well drilling, a change in the angle of a well bore. { di'flek·shən }

deflection angle [GEOD] The angle at a point on the earth between the direction of a plumb line (the vertical) and the perpendicular (the normal) to the reference spheroid; this difference seldom exceeds 30 seconds of arc. [ORD] **1.** An angle that measures the departure of a moving object from its directed course. **2.** The angle of a deflection shot in gunnery, measured between the line of sight to the target and the line of sight to the aiming point. { di'flek·shən ,aŋ·gəl }

deflection bit [DES ENG] A long, cone-shaped, noncoring bit used to drill past a deflection wedge in a borehole. { di'flek·shən ,bit }

deflection board [ORD] Instrument used in artillery for figuring azimuth or deflection corrected for wind, drift, and other factors. Also known as gun deflection board. { di'flek·shən ,bórd }

deflection change [ORD] Change in azimuth setting

applying to all guns in a battery when the target moves, or when a shift is made from one target to another; it does not include the deflection difference, which allows for the difference in positions of the various guns firing at the same target. { di'flek·shən ˌchānj }

deflection circuit [ELECTR] A circuit which controls the deflection of an electron beam in a cathode-ray tube. { di'flek·shən ˌsər·kət }

deflection coil [ELECTR] One of the coils in a deflection yoke. { di'flek·shən ˌkȯil }

deflection correction [ORD] A correction that must be applied to the azimuth or shift measured on a firing chart so that the line of fire will pass through the target. { di'flek·shən kəˌrek·shən }

deflection curve [MECH] The curve, generally downward, described by a shot deviating from its true course. { di'flek·shən ˌkərv }

deflection defocusing [ELECTR] Defocusing that becomes greater as deflection is increased in a cathode-ray tube, because the beam hits the screen at a greater slant and the beam spot becomes more elliptical as it approaches the edges of the screen. { di'flek·shən deˌfō·kəs·iŋ }

deflection difference [ORD] The amount that an artillery piece in a battery is traversed toward or away from a given piece. { di'flek·shən ˌdif·rəns }

deflection electrode [ELECTR] An electrode whose potential provides an electric field that deflects an electron beam. Also known as deflection plate. { di'flek·shən iˌlekˌtrōd }

deflection error [ORD] **1.** In artillery, the distance to the right or left of a target between the point aimed at and the burst of the projectile. **2.** In bombing, the distance between the point of impact and the mean point of impact and the center of the target measured at right angles from the line of the aircraft's approach. { di'flek·shən ˌer·ər }

deflection factor [ELECTR] The reciprocal of the deflection sensitivity in a cathode-ray tube. { di'flek·shən ˌfak·tər }

deflection magnetometer [ENG] A magnetometer in which magnetic fields are determined from the angular deflection of a small bar magnet that is pivoted so that it is free to move in a horizontal plane. { di'flek·shən ˌmag·nə'täm·əd·ər }

deflection meter [ENG] A flowmeter that applies the differential pressure generated by a differential-producing primary device across a diaphragm or bellows in such a way as to create a deflection proportional to the differential pressure. { di'flek·shən ˌmēd·ər }

deflection-modulated indicator See amplitude-modulated indicator. { di'flek·shən ˌmäj·əˌlād·əd 'in·dəˌkād·ər }

deflection of the vertical [GEOD] The angle between the direction of gravity, defining astronomical latitude and longitude, and the normal to the reference ellipsoid defining geodetic latitude and longitude. { di'flek·shən əv thə 'vərd·ə·kəl }

deflection plate See deflection electrode. { di'flek·shən ˌplāt }

deflection polarity [ELECTR] Relationship between the direction of a displacement of the cathode beam and the polarity of the applied signal wave. { di'flek·shən pə'lar·əd·ē }

deflection probable error [ORD] The directional error, caused by dispersion, which will be exceeded, as often as not, in an infinite number of rounds fired at a single deflection; it is one-eighth the width of the dispersion pattern at its greatest width; this value is given in the firing tables. { di'flek·shən ˌpräb·ə·bəl 'er·ər }

deflection scale [ORD] Scale on a sight, marked in mils or degrees, for applying corrections in deflection or for moving the piece in direction. { di'flek·shən ˌskāl }

deflection sensitivity [ELECTR] The displacement of the electron beam at the target or screen of a cathode-ray tube per unit of change in the deflection field; usually expressed in inches per volt applied between deflection electrodes or inches per ampere in a deflection coil. { di'flek·shən sen·sə'tiv·əd·ē }

deflection ultrasonic flowmeter [ENG] A flowmeter for determining velocity from the deflection of a high-frequency sound beam directed across the flow. Also known as drift ultrasonic flowmeter. { di'flek·shən ˌəl·trə'sän·ik 'flōˌmēd·ər }

deflection voltage [ELECTR] The voltage applied between a pair of deflection electrodes to produce an electric field. { di'flek·shən ˌvōl·tij }

deflection wedge [DES ENG] A wedge-shaped tool inserted into a borehole to direct the drill bit. { di'flek·shən ˌwej }

deflection yoke [ELECTR] An assembly of one or more electromagnets that is placed around the neck of an electron-beam tube to produce a magnetic field for deflection of one or more electron beams. Also known as scanning yoke; yoke. { di'flek·shən ˌyōk }

deflectometer [ENG] An instrument used for measuring minute deformations in a structure under transverse stress. { ˌdēˌflek'täm·əd·ər }

deflector [ENG] A plate, baffle, or the like that diverts the flow of a forward-moving stream. { di'flek·tər }

deflexed [BIOL] Turned sharply downward. { dē'flekst }

deflocculant [CHEM] An agent that causes deflocculation; examples are sodium carbonate and other basic materials used to deflocculate clay slips. { dē'fläk·yə·lənt }

deflocculate [CHEM ENG] To break up and disperse agglomerates and form a stable colloid. { dē'fläk·yəˌlāt }

defluorination [CHEM] Removal of fluorine. { dēˌflùr·ə'nā·shən }

defluvium [PATH] The pathological loss of a part of an animal or plant, as nails or bark. { dē'flü·vē·əm }

defoaming [CHEM ENG] Reduction or elimination of foam. { dē'fōm·iŋ }

defocus [ENG] To make a beam of x-rays, electrons, light, or other radiation deviate from an accurate focus at the intended viewing or working surface. { dē'fō·kəs }

defocus-dash mode [ELECTR] A mode of cathode-ray tube storage of binary digits in which the writing beam is initially defocused so as to excite a small circular area on the screen; for one kind of binary digit it remains defocused, and for the other kind it is suddenly focused to a concentric dot and drawn out into a dash. { dē'fō·kəs 'dash ˌmōd }

defocus-focus mode [ELECTR] A variation of the defocus-dash mode in which the focused dot is drawn out into a dash. { dē'fō·kəs 'fō·kəs ˌmōd }

defoliant [MATER] A chemical sprayed on plants that causes leaves to fall off prematurely. { dē'fō·lē·ənt }

defoliate [BOT] To remove leaves or cause leaves to fall, especially prematurely. { dē'fō·lēˌāt }

deforestation [FOR] The act or process of removing trees from or clearing aforest. { dēˌfär·ə'stā·shən }

deformation [MATH] A homotopy of the identity map to some other map. [MECH] Any alteration of shape or dimensions of a body caused by stresses, thermal expansion or contraction, chemical or metallurgical transformations, or shrinkage and expansions due to moisture change. { ˌdef·ər'mā·shən }

deformation bands See Lüders' lines. { ˌdef·ər'mā·shən ˌbanz }

deformation curve [MECH] A curve showing the relationship between the stress or load on a structure, structural member, or a specimen and the strain or deformation that results. Also known as stress-strain curve. { ˌdef·ər'mā·shən ˌkərv }

deformation ellipsoid See strain ellipsoid. { ˌdef·ər'mā·shən ə'lipˌsȯid }

deformation energy [NUC PHYS] The energy which must be supplied to an initially spherical nucleus to give it a certain deformation in the Bohr-Wheeler theory. { ˌdef·ər'mā·shən ˌen·ər·jē }

deformation fabric [GEOL] The space orientation of rock elements produced by external stress on the rock. { ˌdefˌərˌmā·shən ˌfab·rik }

deformation lamella [GEOL] A type of slipband in the crystalline grains of a material (particularly quartz) produced by intracrystalline slip during tectonic deformation. { ˌdefˌərˌmā·shən lə'mel·ə }

deformation potential [SOLID STATE] The effective electric potential experienced by free electrons in a semiconductor or metal resulting from a local deformation in the crystal lattice. { ˌdef·ər'mā·shən pə'ten·chəl }

deformation thermometer [ENG] A thermometer with transducing elements which deform with temperature; examples are the bimetallic thermometer and the Bourdon-tube type of thermometer. { ˌdef·er'mā·shən thərˌmäm·əd·ər }

deformed bar [CIV ENG] A steel bar with projections or indentations to increase mechanical bonding; used to reinforce concrete. { dēˌfȯrmd ˌbär }

deformeter [ENG] An instrument used to measure minute

deformations in materials in structural models. { dē'fȯr‚mēd·ər }

defragmentation [COMPUT SCI] A procedure in which portions of files on a computer disk are moved until all parts of each file occupy continuous sectors, resulting in a substantial improvement in disk access times. { ‚dē‚frag·mən'tā·shən }

defragmenter [COMPUT SCI] A program that analyzes storage locations of files on a computer disk and then carries out defragmentation. { ‚dē‚frag'men·tər }

defrost [ENG] To keep free of ice or to remove ice. [THERMO] To thaw out from a frozen state. { dē'frȯst }

defruit [ELECTR] To remove random asynchronous replies from the video input of a display unit in a radar beacon system by comparing the video signals on successive sweeps. { dē'früt }

defuse [ORD] To remove the fuse from a munition. { di'fyüz }

degas [ELECTR] To drive out and exhaust the gases occluded in the internal parts of an electron tube or other gastight apparatus, generally by heating during evacuation. [ENG] To remove gas from a liquid or solid. { dē'gas }

degasifier [MET] An alloy added to molten metal to facilitate the removal of dissolved gases. { dē'gas·ə‚fī·ər }

degasser See getter. { dē'gas·ər }

degassing See breathing. { dē'gas·iŋ }

degauss [ELECTR] To remove, erase, or clear information from a magnetic tape, disk, drum, or core. [ELECTROMAG] To neutralize (demagnetize) a magnetic field of, for example, a ship hull or television tube; a direct current of the correct value is sent through a cable around the ship hull; a current-carrying coil is brought up to and then removed from the television tube. Also known as deperm. { dē'gaùs }

degaussing cable [ELECTROMAG] A single-conductor or multiple-conductor cable used on ships for degaussing. { dē'gaùs·iŋ ‚kā·bəl }

degaussing coil [ELECTROMAG] A plastic-encased coil, about 1 foot (0.3 meter) in diameter, that can be plugged into a 120-volt alternating-current wall outlet and moved slowly toward and away from a color television picture tube to demagnetize adjacent parts. { dē'gaùs·iŋ ‚kȯil }

degaussing control [ELECTROMAG] A control that automatically varies the current in the degaussing cable as a ship changes heading or rolls and pitches. { dē'gaùs·iŋ kən‚trōl }

degeneracy [MATH] The condition in which two characteristic functions of an operator have the same characteristic value. [PHYS] The condition in which two or more modes of a vibrating system have the same frequency; a special case of the mathematics definition. [QUANT MECH] The condition in which two or more stationary states of the same system have the same energy even though their wave functions are not the same; a special case of the mathematics definition. { di'jen·ə·rə·sē }

degeneracy pressure [STAT MECH] The pressure exerted by a degenerate electron or neutron gas. { di'jen·ə·rə·sē ‚presh·ər }

degenerate amplifier [ELECTR] Parametric amplifier with a pump frequency exactly twice the signal frequency, producing an idler frequency equal to that of the signal input; it is considered as a single-frequency device. { di'jen·ə·rət 'am·plə‚fī·ər }

degenerate code [GEN] The observation that more than one triplet sequence of nucleotides (codon) can specify the insertion of the same amino acid into a polypeptide chain. { di'jen·ə·rət 'kōd }

degenerate conduction band [SOLID STATE] A band in which two or more orthogonal quantum states exist that have the same energy, the same spin, and zero mean velocity. { di'jen·ə·rət kən'dək·shən ‚band }

degenerate conic [MATH] A straight line, a pair of straight lines, or a point, which is a limiting form of a conic. { di'jen·ə·rət 'kän·ik }

degenerate electron gas [STAT MECH] An electron gas that is far below its Fermi temperature and is therefore described in first approximation by the Fermi distribution; most of the electrons completely fill the lower energy levels and are unable to take part in physical processes until excited out of these levels. { di'jen·ə·rət i'lek‚trän ‚gas }

degenerate four-wave mixing [OPTICS] A method of achieving optical phase conjugation in which two strong counterpropagating pump beams, having the same frequency, set up a standing wave in a clear material whose index of refraction varies linearly with intensity, thereby providing the conditions in which a third beam, at the same frequency, incident upon the material from any direction, results in a fourth beam which precisely retraces the third one. { di'jen·ə·rə·rət 'fȯr ‚wāv 'mik·siŋ }

degenerate matter [PHYS] Matter that has been stripped of its orbital electrons, so the nuclei are packed close together. { di'jen·ə·rət 'mad·ər }

degenerate semiconductor [SOLID STATE] A semiconductor in which the number of electrons in the conduction band approaches that of a metal. { di'jen·ə·rət 'sem·i·kən‚dək·tər }

degenerate simplex [MATH] A modification of a simplex in which the points $p_0, \ldots, p_n$ on which the simplex is based are linearly dependent. { di'jen·ə·rət 'sim‚pleks }

degeneration [ELECTR] The loss or gain in an amplifier through unintentional negative feedback. [MED] **1.** Deterioration of cellular integrity with no sign of response to injury or disease. **2.** General deterioration of a physical, mental, or moral state. [STAT MECH] A phenomenon which occurs in gases at very low temperatures when the molecular heat drops to less than $^3/_2$ the gas constant. { di‚jen·ə'rā·shən }

degenerative arthritis See degenerative joint disease. { di'jen·ə·rəd·iv är'thrīd·əs }

degenerative disease [MED] General debility and diseases associated with advancing age. { di'jen·ə·rəd·iv di'zēz }

degenerative joint disease [MED] A chronic joint disease characterized pathologically by degeneration of articular cartilage and hypertrophy of bone, clinically by pain on activity which subsides with rest. Also known as degenerative arthritis; hypertrophic arthritis; osteoarthritis; senescent arthritis. { di'jen·ə·rəd·iv 'jȯint di‚zēz }

degenerative recrystallization See degradation recrystallization. { di'jen·ə·rəd·iv rē‚krist·əl·ə'zā·shən }

Degeneriaceae [BOT] A family of dicotyledonous plants in the order Magnoliales characterized by laminar stamens; a solitary, pluriovulate, unsealed carpel; and ruminate endosperm. { ‚dē·jen·ə‚rī'ās·ē‚ē }

deglaciation [HYD] Exposure of an area from beneath a glacier or ice sheet as a result of shrinkage of the ice by melting. { dē‚glās·ē'ā·shən }

deglitcher [ELECTR] A nonlinear filter or other special circuit used to limit the duration of switching transients in digital converters. { dē'glich·ər }

deglutition [PHYSIO] Act of swallowing. { ‚dē‚glü'tish·ən }

degradation [COMPUT SCI] Condition under which a computer operates when some area of memory or some units of peripheral equipment are not available to the user. [GEOL] The wearing down of the land surface by processes of erosion and weathering. [HYD] **1.** Lowering of a stream bed. **2.** Shrinkage or disappearance of permafrost. [ORG CHEM] Conversion of an organic compound to one containing a smaller number of carbon atoms. [PHYS] Loss of energy of a particle, such as a neutron or photon, through a collision. [THERMO] The conversion of energy into forms that are increasingly difficult to convert into work, resulting from the general tendency of entropy to increase. { ‚deg·rə'dā·shən }

degradation failure [ENG] Failure of a device because of a shift in a parameter or characteristic which exceeds some previously specified limit. { ‚deg·rə'dā·shən ‚fāl·yər }

degradation product [PETRO ENG] In petroleum processing, a contaminant or a low-value product formed during a reaction such as cracking, dehydrogenation, or polymerization. { ‚deg·rə'dā·shən ‚präd·əkt }

degradation recrystallization [GEOL] Recrystallization resulting in a decrease in the size of crystals. Also known as degenerative recrystallization; grain diminution. { ‚deg·rə'dā·shən rē‚krist·əl·ə'zā·shən }

degradative plasmid [GEN] A type of plasmid that specifies a set of genes involved in biodegradation of an organic compound. { ‚deg·rə‚dād·iv 'plaz·mid }

degraded illite [MINERAL] Illite with a depleted potassium content because of prolonged leaching. Also known as stripped illite. { dē'grād·əd i'līt }

degrading stream [HYD] A stream actively deepening its

channel or valley and capable of transporting more load than is presently provided. { də'grād·iŋ ,strēm }

degras [MATER] **1.** A semioxidized fat obtained from sheep skins by subjecting them to the action of oxidized fish oil and pressing them; used to dress leather. Also known as moellen. **2.** A mixture of this material with other fatty oils or fats or with wool grease. *See* wool grease. { dā'gräs }

degrease [CHEM ENG] **1.** To remove grease from wool with chemicals. **2.** To remove grease from hides or skins in tanning by tumbling them in solvents. [MET] To remove grease, oil, or fatty material from a metal surface with fumes from a hot solvent. { dē'grēs }

degreaser [ENG] A machine designed to clean grease and foreign matter from mechanical parts and like items, usually metallic, by exposing them to vaporized or liquid solvent solutions confined in a tank or vessel. [MATER] A solvent, such as a polyhalogenated hydrocarbon, that removes fat or oil in many industrial processes. { dē'grēs·ər }

degree [CHEM] Any one of several units for measuring hardness of water, such as the English or Clark degree, the French degree, and the German degree. [FL MECH] One of the units in any of various scales of specific gravity, such as the Baumé scale. [MATH] **1.** A unit for measurement of plane angles, equal to 1/360 of a complete revolution, or 1/90 of a right angle. Symbolized °. **2.** For a term in one variable, the exponent of that variable. **3.** For a term in several variables, the sum of the exponents of its variables. **4.** For a polynomial, the degree of the highest-degree term. **5.** For a differential equation, the greatest power to which the highest-order derivative occurs. **6.** For an algebraic curve defined by the polynomial equation $f(x,y) = 0$, the degree of the polynomial $f(x,y)$. **7.** For a vertex in a graph, the number of arcs which have that vertex as an end point. **8.** For an extension of a field, the dimension of the extension field as a vector space over the original field. [THERMO] One of the units of temperature or temperature difference in any of various temperature scales, such as the Celsius, Fahrenheit, and Kelvin temperature scales (the Kelvin degree is now known as the kelvin). { di'grē }

degree-day [MECH ENG] A measure of the departure of the mean daily temperature from a given standard; one degree-day is recorded for each degree of departure above (or below) the standard during a single day; used to estimate energy requirements for building heating and, to a lesser extent, for cooling. { di'grē ,dā }

degree Engler [FL MECH] A measure of viscosity; the ratio of the time of flow of 200 milliliters of the liquid through a viscometer devised by Engler, to the time for the flow of the same volume of water. { di'grē 'eŋ·glər }

degree of crystallinity [ORG CHEM] In a fairly large sample of a polymer, the fraction that consists of regions showing long-range three-dimensional order. { di'grē əv ,kris·tə'lin·əd·ē }

degree of current rectification [ELECTR] Ratio between the average unidirectional current output and the root mean square value of the alternating current input from which it was derived. { di'grē əv 'kər·ənt ,rek·tə·fə'kā·shən }

degree of curve [CIV ENG] A measure of the curvature of a railway or highway, equal to the angle subtended by a 100-foot (32.8-meter) chord (railway) or by a 100-foot arc (highway). { di'grē əv 'kərv }

degree of degeneracy [MATH] The number of characteristic functions of an operator having the same characteristic value. Also known as order of degeneracy. { di'grē əv di'jen·ə·rə·sē }

degree of enrichment [NUCLEO] The enrichment factor minus 1. { di'grē əv en'rich·mənt }

degree of freedom [MECH] **1.** Any one of the number of ways in which the space configuration of a mechanical system may change. **2.** Of a gyro, the number of orthogonal axes about which the spin axis is free to rotate, the spin axis freedom not being counted; this is not a universal convention; for example, the free gyro is frequently referred to as a three-degree-of-freedom gyro, the spin axis being counted. [PHYS CHEM] Any one of the variables, including pressure, temperature, composition, and specific volume, which must be specified to define the state of a system. [STAT] A number one less than the number of frequencies being tested with a chi-square test. { di'grē əv 'frē·dəm }

degree of polymerization [ORG CHEM] The number of

structural units in the average polymer molecule in a particular sample. Abbreviated D.P. { di'grē əv pə,lim·ə·rə'zā·shən }

degree of voltage rectification [ELECTR] Ratio between the average unidirectional voltage and the root mean square value of the alternating voltage from which it was derived. { di'grē əv 'vōl·tij ,rek·tə·fə'kā·shən }

degrees of frost [METEOROL] In England, the number of degrees Fahrenheit that the temperature falls below the freezing point; thus a day with a minimum temperature of 27°F may be designated as a day of five degrees of frost. { di'grēz əv 'frôst }

degree vector [MATH] The sequence of degrees of the vertices of a simple graph, arranged in nonincreasing order. { di'grē ,vek·tər }

degritting [CHEM ENG] Removal of fine solid particles (grit) from a liquid carrier by gravity separation (settling) or centrifugation. { dē'grid·iŋ }

de Gua's rule [MATH] The rule that if, in a polynomial equation $f(x) = 0$, a group of r consecutive terms is missing, then the equation has at least r imaginary roots if r is even, or the equation has at least $r + 1$ or $r - 1$ imaginary roots if r is odd (depending on whether the terms immediately preceding and following the group have like or unlike signs). { də'gwäz ,rül }

degumming [TEXT] Removing sericin, a gluelike substance, from silk by using a solution of soap, to improve the quality of the silk. { dē'gəm·iŋ }

de Haas-van Alphen effect [SOLID STATE] An effect occurring in many complex metals at low temperatures, consisting of a periodic variation in the diamagnetic susceptibility of conduction electrons with changes in the component of the applied magnetic field at right angles to the principal axis of the crystal. { də¦häs ,van'äl·fən i,fekt }

dehalogenate [ORG CHEM] To remove halogen atoms from a molecule. { dē'hal·ə·jə,nāt }

dehiscence [BOT] Spontaneous bursting open of a mature plant structure, such as fruit, anther, or sporangium, to discharge its contents. [MED] A defect in the boundary of a bony canal or cavity. { də'his·əns }

dehiscent [BOT] Becoming open at maturity to release seeds, as certain fruits. { di'his·ənt }

dehrnite [MINERAL] $(Ca,Na,K)_5(PO_4)_3(OH)$ A colorless to pale green, greenish-white, or gray, hexagonal mineral consisting of a basic phosphate of calcium, sodium, and potassium; occurs as botryoidal crusts and minute hexagonal prisms. { 'der,nīt }

dehumidification [MECH ENG] The process of reducing the moisture in the air; serves to increase the cooling power of air. { ,dē·yü,mid·ə·fə'kā·shən }

dehumidifier [MECH ENG] Equipment designed to reduce the amount of water vapor in the ambient atmosphere. { ,dē·yü'mid·ə,fī·ər }

dehydrase [BIOCHEM] An enzyme which catalyzes the removal of water from a substrate. { dē'hī,drās }

dehydration [CHEM] Removal of water from any substance. [ORG CHEM] An elimination reaction in which a molecule loses both a hydroxyl group (OH) and a hydrogen atom (H) that was bonded to an adjacent carbon. { ,dē·hī'drā·shən }

dehydration tank [CHEM ENG] A tank in which warm air is blown through oil to remove moisture. { ,dē·hī'drā·shən ,taŋk }

dehydrator [CHEM] A substance that removes water from a material; an example is sulfuric acid. [CHEM ENG] Vessel or process system for the removal of liquids from gases or solids by the use of heat, absorbents, or adsorbents. { dē'hī,drād·ər }

dehydroacetic acid [ORG CHEM] $C_8H_8O_4$ Crystals that melt at 108.5°C and are insoluble in water, soluble in acetone; used as a fungicide and bactericide. Abbreviated DHA. { dē¦hī·drō·ə¦sēd·ik 'as·əd }

dehydroascorbic acid [ORG CHEM] $C_6H_6O_6$ A relatively inactive acid resulting from elimination of two hydrogen atoms from ascorbic acid when the latter is oxidized by air or other agents; has potential ascorbic acid activity. { dē¦hī·drō·ə¦skór·bik 'as·əd }

dehydrochlorinase [BIOCHEM] An enzyme that dechlorinates a chlorinated hydrocarbon such as the insecticide DDT; found in some insects that are resistant to DDT. { dē,hī·drō'klór·ə,nās }

dehydrochlorination [BIOCHEM] Removal of hydrogen and

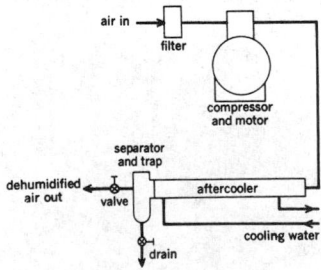

DEHUMIDIFIER

Dehumidifying by compression and aftercooling.

chlorine or hydrogen chloride from a compound. { dē₁hī· drō₁klȯr·ə'nā·shən }

dehydrocholesterol [BIOCHEM] $C_{27}H_{43}OH$ A provitamin of animal origin found in the skin of humans, in milk, and elsewhere, which upon irradiation with ultraviolet rays becomes vitamin D. { dē¦hī·drō·kə'les·tə₁rȯl }

dehydrocholic acid [ORG CHEM] $C_{24}H_{34}O_5$ A white powder melting at 231–240°C, very slightly soluble in water; used as a pharmaceutical intermediate and in medicine. { dē¦hī· drə¦käl·ik 'as·əd }

dehydrocyclization [CHEM ENG] Any process involving both dehydrogenation and cyclization, as in petroleum refining. { dē¦hī·drō₁sīk·lə'zā·shən }

dehydroepiandrosterone [ORG CHEM] $C_{19}H_{28}O_2$ Dimorphous crystals with a melting point of 140–141°C, or leaflet crystals with a melting point of 152–153°C; soluble in alcohol, benzene, and ether; used as an androgen. { dē¦hī·drō·e₁pē· ən'dräs·tə ₁rōn }

dehydrofreezing [FOOD ENG] A dual process for preserving food which involves partial dehydration followed by quick freezing. { dē¦hī·drō'frēz·iŋ }

dehydrogenase [BIOCHEM] An enzyme which removes hydrogen atoms from a substrate and transfers it to an acceptor other than oxygen. { dē'hī·drə·jə₁nās }

dehydrogenation [CHEM] Removal of hydrogen from a compound. { dē¦hī·drə·jə'nā·shən }

dehydrohalogenation [CHEM] Removal of hydrogen and a halogen from a compound. { dē¦hī·drō₁hal·ə·jə'nā·shən }

deicer [AERO ENG] Any device to keep the wings and propeller of an airplane free of ice. [MATER] Any substance used to keep a surface free of ice or to rid it of ice; ethylene glycol is used to deice windshields of automobiles and airplanes. { dē'īs·ər }

deicing [ENG] The removal of ice deposited on any object, especially as applied to aircraft icing, by heating, chemical treatment, and mechanical rupture of the ice deposit. { dē'īs·iŋ }

Deimos [ASTRON] A satellite of Mars orbiting at a mean distance of 14,600 miles (23,500 kilometers). { 'dā₁mȯs }

deinking [CHEM ENG] The process of removing ink from recycled paper so that the fibers can be used again. { dē'iŋk·iŋ }

Deinotheriidae [PALEON] A family of extinct proboscidean mammals in the suborder Deinotherioidea; known only by the genus *Deinotherium*. { ₁dī·nō·thə'rī·ə₁dē }

Deinotherioidea [PALEON] A monofamilial suborder of extinct mammals in the order Proboscidea. { ₁dī·nō₁ther· ē'oid·ē·ə }

deion circuit breaker [ELEC] Circuit breaker built so that the arc that forms when the circuit is broken is magnetically blown into a stack of insulated copper plates, giving the effect of a large number of short arcs in series; each arc becomes almost instantly deionized when the current drops to zero in the alternating current cycle, and the arc cannot reform. { dē'ī₁än 'sər·kət ₁brāk·ər }

deionization [CHEM] An ion-exchange process in which all charged species or ionizable organic and inorganic salts are removed from solution. [ELECTR] The return of an ionized gas to its neutral state after all sources of ionization have been removed, involving diffusion of ions to the container walls and volume recombination of negative and positive ions. { dē₁ī· ən·ə'zā·shən }

deionization potential [ELECTR] The potential at which ionization of the gas in a gas-filled tube ceases and conduction stops. { dē₁ī·ən·ə'zā·shən pə'ten·chəl }

deionization time [ELECTR] The time required for a gas tube to regain its preconduction characteristics after interruption of anode current, so that the grid regains control. Also called recontrol time. { dē₁ī·ən·ə'zā·shən ₁tīm }

Deister phase [GEOL] A subdivision of the late Ammerian phase of the Jurassic period between the Kimmeridgian and lower Portlandian. { 'dī·stər ₁fāz }

dekapoise [FL MECH] A unit of absolute viscosity, equal to 10 poises. { 'dek·ə₁pȯiz }

Delaborne prism [OPTICS] A special compound prism which, when rotated about an axis parallel to the reflecting face and lying in a plane perpendicular to the refracting faces, rotates the image through twice the angle. Also known as Dove prism. { del·ə'bȯrn ₁priz·əm }

delafossite [MINERAL] $CuFeO_2$ A mineral consisting of an oxide of copper and iron. { ₁de·lə'fȯ₁sīt }

Delambre analogies See Gauss formulas. { də'lam·brə ə₁nal· ə·jēz }

delamination [BIOL] The separation of cells into layers. [EMBRYO] Gastrulation in which the endodermal layer splits off from the inner surface of the blastoderm and the space between this layer and the yolk represents the archenteron. [ENG] Separation of a laminate into its constituent layers. { dē₁lam·ə'nā·shən }

de la Rue and Miller's law [ELECTR] The law that in a field between two parallel plates, the sparking potential of a gas is a function of the product of gas pressure and sparking distance only. { del·ə¦rü ən 'mil·ərz ₁lȯ }

de la Tour method [ANALY CHEM] Measurement of critical temperature, involving sealing the sample in a tube and heating it; the temperature at which the meniscus disappears is the critical temperature. { del·ə'tùr ₁meth·əd }

Delaunay orbit element [MECH] In the *n*-body problem, certain functions of variable elements of an ellipse with a fixed focus along which one of the bodies travels; these functions have rates of change satisfying simple equations. { də·lō·nā 'ȯr·bət ₁el·ə·mənt }

de Laval nozzle [AERO ENG] A converging-diverging nozzle used in certain rockets. Also known as Laval nozzle. { də· lä'väl ₁näz·əl }

de Lavaud process [MET] A centrifugal casting process employing water-cooled metal molds, used to produce cast iron pipe, gun barrels, and other cylindrical objects. { də·lä'vō 'präs·əs }

delay [COMMUN] **1.** Time required for a signal to pass through a device or a conducting medium. **2.** Time which elapses between the instant at which any designated point of a transmitted wave passes any two designated points of a transmission circuit; such delay is primarily determined by the constants of the circuit. [IND ENG] Interruption of the normal tempo of an operation; may be avoidable or unavoidable. { di'lā }

delay-action detonator See delay blasting cap. { di'lā ₁ak· shən 'det·ən₁ād·ər }

delay allowance [IND ENG] A percentage of the normal operating time added to the normal time to allow for delays. { di'lā ə₁laủ·əns }

delay blasting cap [ENG] A blasting cap which explodes at a definite time interval after the firing current has been passed by the exploder. Also known as delay-action detonator. { di'lā 'blast·iŋ ₁kap }

delay circuit See time-delay circuit. { di'lā ₁sər·kət }

delay counter [COMPUT SCI] A counter which inserts a time delay in a sequence of events. { di'lā ₁kaủnt·ər }

delay distortion [ELECTR] Phase distortion in which the rate of change of phase shift with frequency of a circuit or system is not constant over the frequency range required for transmission. Also called envelope delay distortion. { di'lā di'stȯr₁shən }

delay Doppler mapping [MAP] Mapping of a planet by illuminating it with a radar beam and measuring the Doppler shift caused by rotation of the planet. { di'lā 'däp·lər ₁map·iŋ }

delayed action bomb [ORD] A bomb with a delay fuse; the delay action may vary from a fraction of a second to several days after impact. { di'lād ₁ak·shən 'bäm }

delayed action mine [ORD] A mine that explodes some time after being activated. { di'lād ₁ak·shən 'mīn }

delayed alpha particle [NUC PHYS] An alpha particle emitted by an excited nucleus that was formed an appreciable time after a beta disintegration process. { di'lād 'al·fə ₁pard·ə·kəl }

delayed automatic gain control [ELECTR] An automatic gain control system that does not operate until the signal exceeds a predetermined magnitude; weaker signals thus receive maximum amplification. Also known as biased automatic gain control; delayed automatic volume control; quiet automatic volume control. { di'lād ₁ȯd·ə¦mad·ik 'gān kən₁trōl }

delayed automatic volume control See delayed automatic gain control. { di'lād ₁ȯd·ə¦mad·ik 'väl·yəm kən₁trōl }

delayed coincidence [NUCLEO] Occurrence of a count in one detector at a short but measurable time later than a count in another detector, the two counts being due to successive events in the same nucleus. { di'lād kō'in·səd·əns }

delayed coking [CHEM ENG] A semicontinuous thermal

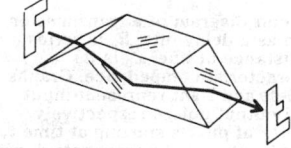

DELABORNE PRISM

Diagram of Delaborne prism.

process for converting heavy petroleum stock to lighter material. { di'lād 'kök·iŋ }

delayed combustion [ENG] Secondary combustion in succeeding gas passes beyond the furnace volume of a boiler. { di'lād kəm'bəs·chən }

delayed critical [NUCLEO] The condition in which a nuclear reactor is critical because of delayed neutrons alone, without requiring the contribution of prompt neutrons. { di'lād 'krid·ə·kəl }

delayed development well See step-out well. { di'lād di'vel·əp·mənt ,wel }

delayed gamma ray [NUCLEO] A gamma ray emitted during radioactive decay of a fission product. { di'lād 'gam·ə ,rā }

delayed hypersensitivity [IMMUNOL] Abnormal reactivity in a sensitized individual beginning several hours after contact with the allergen. { di'lād ,hī·pər,sen·sə'tiv·əd·ē }

delayed neutron [NUC PHYS] A neutron emitted spontaneously from a nucleus as a consequence of excitation left from a preceding radioactive decay event; in particular, a delayed fission neutron. { di'lād 'nü,trän }

delayed neutron fraction [NUC PHYS] The ratio of the mean number of delayed fission neutrons per fission to the mean total number of neutrons (prompt plus delayed) per fission. { di'lād 'nü,trän ,frak·shən }

delayed-opening chaff [ORD] Chaff packages, with parachutes and time fuses, deployed from aircraft or ballistic missiles; intended to make the enemy believe there are flying targets far removed from the dispensing aircraft or missiles. Abbreviated DOC. { di'lād 'ō·pən·iŋ 'chaf }

delayed plan position indicator [ELECTR] A plan position indicator in which initiation of the time base is delayed a fixed time after each transmitted pulse, to give expansion of the range scale for distant targets so that they show more clearly on the screen. { di'lād 'plan pə'zish·ən ,in·də,kād·ər }

delayed proton [NUC PHYS] A proton emitted spontaneously from a nucleus as a consequence of excitation left from a previous radioactive decay event. { di'lād 'prō,tän }

delayed repeater satellite [AERO ENG] Satellite which stores information obtained from a ground terminal at one location, and upon interrogation by a terminal at a different location, transmits the stored message. { di'lād ri'pēd·ər 'sad·əl,īt }

delayed speech [MED] A speech disorder characterized by a complete absence of vocalization or vocalization with no communicative value; speech is considered delayed when it fails to develop by the second year, caused by impaired hearing, severe childhood illness, or emotional disturbance. { di'lād 'spēch }

delayed sweep [ELECTR] A sweep whose beginning is delayed for a definite time after the pulse that initiates the sweep. { di'lād 'swēp }

delayed yield [MET] Time delay between the sudden application of a yield stress and the appearance of yielding. { di'lād 'yēld }

delay element [ORD] A component which provides a specified delay between actuation of the propellant-actuated devices and ignition of the propellant. { di'lā ,el·ə·mənt }

delay equalizer [ELECTR] A corrective network used to make the phase delay or envelope delay of a circuit or system substantially constant over a desired frequency range. { di'lā 'ē·kwə,līz·ər }

delayer [MATER] A substance mixed with solid rocket propellants to decrease the rate of combustion. { di'lā·ər }

delay flip-flop See D flip-flop. { di'lā 'flip,fläp }

delay/frequency distortion [COMMUN] That form of distortion which occurs when the delay of a circuit or system is not constant over the frequency range required for transmissions. { di'lā 'frē·kwən·sē di'stòr·shən }

delay line [ELECTR] A transmission line (as dissipationless as possible), or an electric network approximation of it, which, if terminated in its characteristic impedance, will reproduce at its output a waveform applied to its input terminals with little distortion, but at a time delayed by an amount dependent upon the electrical length of the line. Also known as artificial delay line. { di'lā 'līn }

delay-line memory See circulating memory. { di'lā ,līn 'mem·rē }

delay-line storage See circulating memory. { di'lā ,līn 'stór·ij }

delay multivibrator [ELECTR] A monostable multivibrator that generates an output pulse a predetermined time after it is triggered by an input pulse. { di'lā ,məl·tə'vī,brād·ər }

delay relay [ELEC] A relay having predetermined delay between energization and closing of contacts or between deenergization and dropout. { di'lā 'rē,lā }

delay time [CONT SYS] The amount of time by which the arrival of a signal is retarded after transmission through physical equipment or systems. [ELECTR] The time taken for collector current to start flowing in a transistor that is being turned on from the cutoff condition. [IND ENG] A span of time during which a worker is idle because of factors beyond personal control. { di'lā ,tīm }

delay unit [COMPUT SCI] See transport delay unit. [ELECTR] Unit of a radar system in which pulses may be delayed a controllable amount. { di'lā ,yü·nət }

Delbrück scattering [NUC PHYS] Elastic scattering of gamma rays by a nucleus caused by virtual electron-positron pair production. { 'del·brik ,skad·ə·riŋ }

d electron [ATOM PHYS] An atomic electron that has an orbital angular momentum of 2 in the central field approximation. { 'dē i,lek,trän }

Delepine reaction [ORG CHEM] Slow ammonolysis of alkyl halides in acid to primary amines in the presence of hexamethylenetetramine. { 'del·ə,pīn rē,ak·shən }

deleted representation [COMPUT SCI] In paper tape codes, the superposition of a pattern of holes upon another pattern of holes representing a character, to effectively remove or obliterate the latter. { di'lēd·əd ,rep·rə,zen'tā·shən }

deletion [GEN] Loss of a chromosome segment of any size, down to a part of a single gene. { di'lē·shən }

deletion operator [COMPUT SCI] The part of a data structure which allows components to be deleted. { di'lē·shən ,äp·ə,rād·ər }

deletion record [COMPUT SCI] A record which removes and replaces an existing record when it is added to a file. { di'lē·shən ,rek·ərd }

deliberate fire [ORD] Fire which is conducted at a rate intentionally less than the normal rate of fire for the purpose of applying adjustment corrections between each round or salvo, for tactical reasons, or for conserving ammunition. { də'lib·rət 'fīr }

delignification [CHEM ENG] A chemical process for removing lignin from wood. { dē,lig·nə·fə'kā·shən }

delimiter [COMPUT SCI] A character that separates items of data. { də'lim·əd·ər }

delinquency [PSYCH] **1.** The tendency to commit legal or moral offenses, especially by a minor. **2.** The offense committed. { də'liŋ·kwən·sē }

delinquency proneness [PSYCH] The likelihood of becoming delinquent, measured by estimates of the probability that an individual will become delinquent if exposed to fairly commonplace temptations and opportunities. { də'liŋ·kwən·sē ,prōn·nəs }

deliquescence [BOT] The condition of repeated divisions ending in fine divisions; seen especially in venation and stem branching. [PHYS CHEM] The absorption of atmospheric water vapor by a crystalline solid until the crystal eventually dissolves into a saturated solution. { del·ə'kwes·əns }

delirium [MED] Severely disordered mental state associated with fever, intoxication, head trauma, and other encephalopathies. { di'lir·ē·əm }

delirium tremens [MED] Delirium associated with tremors, insomnia, and other physical and neurological symptoms frequently following chronic alcoholism. { di'lir·ē·əm 'trem·ənz }

deliverability [PETRO ENG] The volume of gas per given unit of time that can be obtained from a well, field, storage reservoir, pipeline, or distribution system. { di,liv·ə·rə'bil·əd·ē }

delivery system [ORD] The means of delivering atomic weapons to the target. { di'liv·ə·rē ,sis·təm }

dell [GEOGR] A small, secluded valley or vale. { del }

dellenite See rhyodacite. { 'del·ə,nīt }

Dellinger fadeout [COMMUN] Type of fadeout that occurs during shortwave reception, believed to be caused by rapid shifting of ionosphere layers during solar eruptions. { 'del·ən·jər 'fād,aut }

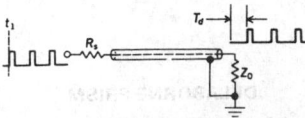

DELAY LINE

Circuit diagram of a transmission line as a delay line. R_s is series resistance of line; Z_0 is its characteristic impedance. Graphs at left and right represent input and output pulses respectively. Series of pulses starting at time t_1 require time T_d to propagate down line.

Delmontian [GEOL] Upper Miocene or lower Pliocene geologic time. { del'män·chən }

delocalized bond [CHEM] A type of molecular bonding in which the electron density of delocalized electrons is regarded as being spread over several atoms or over the whole molecule. Also known as nonlocalized bond. { dē'lō·kə,līzd 'bänd }

delocalized state [QUANT MECH] A state of motion in which a charge carrier is spread over a whole molecule or crystal. { dē'lō·kə,līzd 'stāt }

delomorphous cell See parietal cell. { ¦dē·lō¦mȯr·fəs 'sel }

del operator [MATH] The rule which replaces the function *f* of three variables, *x*, *y*, *z*, by the vector valued function whose components in the *x*, *y*, *z* directions are the respective partial derivatives of *f*. Written ∇f. Also known as nabla. { 'del ,äp·ə,rād·ər }

delorenzite See tanteuxenite. { dē·lə'ren,zīt }

delphidenolon See myricetin. { ¦del·fə¦den·ə,län }

Delphinidae [VERT ZOO] A family of aquatic mammals in the order Cetacea; includes the dolphins. { del'fin·ə,dē }

delphinidin [BIOCHEM] $C_{15}H_{11}O_7Cl$ A purple or brownish-red anthocyanin compound occurring widely in plants. { del'fin·ə·dən }

Delphinus [ASTRON] A northern constellation, right ascension 21 hours, declination 10° north. Also known as Dolphin. [VERT ZOO] A genus of cetacean mammals, including the dolphin. { del'fē·nəs }

Delrac [NAV] An electronic long-range navigation system in which the position of a vessel is ascertained by comparing the phase of signal bursts from different transmission stations. { 'del,rak }

delta [ANAT] A fingerprint focal point which is the point on a ridge at or in front of and nearest the center of the divergence of the type lines. [ELECTR] The difference between a partial-select output of a magnetic cell in a one state and a partial-select output of the same cell in a zero state. [GEOL] An alluvial deposit, usually triangular in shape, at the mouth of a river, stream, or tidal inlet. { 'del·tə }

delta baryon [PARTIC PHYS] **1.** Any excited baryon state belonging to a multiplet having a total isospin of $^3/_2$, a hypercharge of $+1$, positive parity, and an approximate mass of 1232 MeV. Designated Δ(1232). **2.** Any excited baryon state belonging to any multiplet having a total isospin of $^3/_2$ and a hypercharge of $+1$. { 'del·tə 'bar·ē,än }

Delta Cephei [ASTRON] A cepheid variable, from which the name of this type of star is derived; it has a period of 5.3 days. { 'del·tə 'sef·ē,ī }

delta connection [ELEC] A combination of three components connected in series to form a triangle like the Greek letter delta. Also known as mesh connection. { 'del·tə kə'nek·shən }

delta current [ELEC] Electricity going through a delta connection. { 'del·tə ,kər·ənt }

delta E effect [ELECTROMAG] Magnetization of a ferromagnetic substance that is caused by elastic tension. { ¦del·tə 'ē i,fekt }

delta ferrite See delta iron. { 'del·tə 'fe,rīt }

delta function [MATH] A distribution δ such that

$$\int_{-\infty}^{\infty} f(t)\delta(x - t)dt$$

is $f(x)$. Also known as Dirac delta function; Dirac distribution; unit impulse. { 'del·tə ,fəŋk·shən }

delta geosyncline See exogeosyncline. { 'del·tə ,jē·ō'sin,klīn }

delta-gun tube [ELECTR] A color television picture tube in which three electron guns, arranged in a triangle, provide electron beams that fall on phosphor dots on the screen, causing them to emit light in three primary colors; a shadow mask located just behind the screen ensures that each beam excites only dots of one color. { 'del·tə ,gən ,tüb }

deltaic deposits [GEOL] Sedimentary deposits in a delta. { del'tā·ik di'päz·əts }

delta hepatitis [MED] A type of viral hepatitis caused by the delta agent hepatitis D virus, a defective ribonucleic acid virus that requires the helper function of hepatitis B virus for its replication and expression. { ,del·tə ,hep·ə'tīd·əs }

deltahedron [MATH] Any polyhedron whose faces are congruent equilateral triangles. { ,del·tə'hē·drən }

delta iron [MET] The nonmagnetic polymorphic form of iron stable between about 1403°C and the melting point, about 1535°C. Also known as delta ferrite. { 'del·tə ,ī·ərn }

deltaite [MINERAL] A mixture of crandallite and hydroxylapatite. { 'del·tə,īt }

delta matching transformer [ELEC] Impedance device used to match the impedance of an open-wire transmission line to an antenna; the two ends of the transmission line are fanned out so that the impedance of the line gradually increases; the ends of the transmission line are attached to the antenna at points of equal impedance, symmetrically located with respect to the center of the antenna. { 'del·tə ,mach·iŋ tranz,fȯr·mər }

delta modulation [ELECTR] A pulse-modulation technique in which a continuous signal is converted into a binary pulse pattern, for transmission through low-quality channels. { 'del·tə ,mäj·ə'lā·shən }

delta moraine See ice-contact delta. { 'del·tə mə'rān }

delta network [ELEC] A set of three branches connected in series to form a mesh. { ¦del·tə ¦net,wərk }

delta particle [ATOM PHYS] An electron or proton ejected by recoil when a rapidly moving alpha particle or other primary ionizing particle passes through matter. { 'del·tə ,pärd·ə·kəl }

delta plain [GEOL] A plain formed by deposition of silt at the mouth of a stream or by overflow along the lower stream courses. { 'del·tə ,plān }

delta pulse code modulation [ELECTR] A modulation system that converts audio signals into corresponding trains of digital pulses to give greater freedom from interference during transmission over wire or radio channels. { 'del·tə ¦pəls ,kōd ,mäj·ə'lā·shən }

delta region [METEOROL] A region in the atmosphere characterized by difluence. { 'del·tə ,rējən }

delta rhythm [PHYSIO] An electric current generated in slow waves with frequencies of 0.5–3 per second from the forward portion of the brain of normal subjects when asleep. { 'del·tə ,rith·əm }

Delta Scuti stars [ASTRON] A class of pulsating variable stars of spectral type A and with periods of less than 8 hours, relatively small amplitude variations, and masses between 1 and 3 solar masses. { ¦del·tə 'sküd·ē ,stärz }

delta-sigma converter See sigma-delta converter. { ¦del·tə ¦sig·mə kən'vərd·ər }

delta-sigma modulator See sigma-delta modulator. { ¦del·tə ¦sig·mə 'mä·jə,lād·ər }

Deltatheridia [PALEON] An order of mammals that includes the dominant carnivores of the early Cenozoic. { ,del·tə·thə'rid·ē·ə }

delta transformer [ELEC] A three-phase electrical transformer in which the ends of the three windings are connected to form a triangle. { 'del·tə tranz'fȯr·mər }

delta wing [AERO ENG] A triangularly shaped wing of an aircraft. { 'del·tə ,wiŋ }

delta-Y transformation See Y-delta transformation. { 'del·tə ,wī ,tranz·fər'mā·shən }

deltic method [ELECTR] A method of sampling incoming radar, sonar, seismic, speech, or other waveforms along with reference signals, compressing the samples in time, and comparing them by autocorrelation. { 'del·tik ,meth·əd }

deltohedron [CRYSTAL] A polyhedron which has 12 quadrilateral faces, and is the form of a crystal belonging to the cubic system and having hemihedral symmetry. Also known as deltoid dodecahedron; tetragonal tristetrahedron. { ,del·tə'hē·drən }

deltoid [ANAT] The large triangular shoulder muscle; originates on the pectoral girdle and inserts on the humerus. [BIOL] Triangular in shape. [MATH] **1.** The plane curve traced by a point on a circle while the circle rolls along the inside of another circle whose radius is three times as great. **2.** A concave quadrilateral with two pairs of adjacent equal sides. Also known as Steiner's hypocycloid tricuspid. { 'del,tȯid }

deltoid dodecahedron See deltohedron. { 'del,tȯid dō,dek·ə'hē·drən }

deltoid ligament [ANAT] The ligament on the medial side of the ankle joint; the fibers radiate from the medial malleolus to the talus, calcaneus, and navicular bones. { 'del,tȯid 'lig·ə·mənt }

DELTATHERIDIA

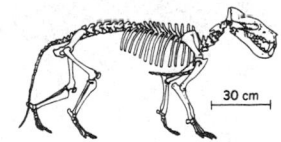

30 cm

The deltatheridian *Hyaenodon*.

delusion [PSYCH] A conviction based on faulty perceptions, feelings, and thinking. { di'lüzh·ən }

deluster [TEXT] To reduce brightness, for example, by adding pigment to the spinning solution of a yarn. { dē'ləs·tər }

delvauxite [MINERAL] A mineral, with the approximate formula $Fe_4(PO_4)_2(OH)_6 \cdot nH_2O$, consisting of a hydrous phosphate of iron. { del'vȯk‚sīt }

demagnetization [ELECTROMAG] **1.** The process of reducing or removing the magnetism of a ferromagnetic material. **2.** The reduction of magnetic induction by the internal field of a magnet. [MIN ENG] Deflocculation in dense-media process using ferrosilicon by passing the fluid through an alternating-current field. { dē‚mag·nəd·ə'zā·shən }

demagnetization coefficient See demagnetizing factor. { dē‚mag·nə·tə'zā·shən ‚kō·ə‚fish·ənt }

demagnetization curve [ELECTROMAG] Graph of magnetic induction B versus magnetic field H in a ferromagnetic material, as the magnetic field is reduced to 0 from its saturation value. { dē‚mag·nə·tə'zā·shən ‚kərv }

demagnetizer [ELECTR] A device for removing undesired magnetism, as from the playback head of a tape recorder or from a recorded reel of magnetic tape that is to be erased. { dē'mag·nə‚tī·zər }

demagnetizing factor [ELECTROMAG] The ratio of the negative of the demagnetizing field to the magnetization of a sample. Also known as demagnetization coefficient. { dē'mag·nə‚tīz·iŋ ‚fak·tər }

demagnetizing field [ELECTROMAG] An additional magnetic field that is produced in a magnetic material subject to an applied magnetic field, due to the magnetic material itself. { dē'mag·nə‚tīz·iŋ ‚fēld }

demal [CHEM] A unit of concentration, equal to the concentration of a solution in which 1 gram-equivalent of solute is dissolved in 1 cubic decimeter of solvent. { 'dem·əl }

demand See demand factor. { də'mand }

demand assignment multiple access [COMMUN] The allocation of bandwidth in a communications system among multiple users based on demand, such as by multiplexing. Abbreviated DAMA. { di¦mand ə‚sīn·mənt ¦məl·tə·pəl 'ak‚ses }

demand-driven execution [COMPUT SCI] A mode of carrying out a program in a data flow system in which no calculation is carried out until its results are demanded as input to another calculation. Also known as lazy evaluation. { də'mand ‚driv·ən ‚ek·sə'kyü·shən }

demanded motions inventory [IND ENG] A list of all motions that are required to perform a specific task, including an exact characterization of each. { də¦man·dəd ¦mō·shənz 'in·vən‚tȯr·ē }

demand factor [ELEC] The ratio of the maximum demand of a building for electric power to the total connected load. Also known as demand. { də'mand ‚fak·tər }

demand limiter See current limiter. { də'mand ‚lim·əd·ər }

demand meter [ENG] Any of several types of instruments used to determine a customer's maximum demand for electric power over an appreciable time interval; generally used for billing industrial users. { də'mand ‚mēd·ər }

demand paging [COMPUT SCI] The characteristic of a virtual memory system which retrieves only that part of a user's program which is required during execution. { də'mand ‚pā·jiŋ }

demand processing [COMPUT SCI] The processing of data by a computer system as soon as it is received, so that it is not necessary to store large amounts of raw data. Also known as immediate processing. { də'mand ‚präs‚es·iŋ }

demand rate [ELEC] The maximum amount of electric power that must be kept available to a customer. { də'mand ‚rāt }

demand reading [COMPUT SCI] A method of carrying out input operations in which blocks of data are transmitted to the central processing unit as needed for processing. { də'mand ‚rēd·iŋ }

demand regulator [ENG] A component of an open-circuit diving system that permits the diver to expel used air directly into the water without rebreathing exhaled carbon dioxide. { də'mand ‚reg·yə‚lād·ər }

demand staging [COMPUT SCI] Moving blocks of data from one storage device to another when programs request them. { də'mand ‚stā·jiŋ }

demand system [ENG] A system in an airplane that automatically dispenses oxygen according to the demand of the flyer's body. { də'mand ‚sis·təm }

demand writing [COMPUT SCI] A method of carrying out output operations in which blocks of data are transmitted from the central processing unit as they are needed by the user. { də'mand ‚rīd·iŋ }

demantoid [MINERAL] A lustrous, green variety of andradite; used as a gem. { də'man‚toid }

demarcation potential See injury potential. { dē‚mär'kā·shən pə‚ten·chəl }

De Marre formula [ORD] A formula expressing the relationship between projectile characteristics and armor-plate penetration capabilities. { də'mär ‚fȯr·myə·lə }

demasking [CHEM] A process by which a masked substance is made capable of undergoing its usual reactions; can be brought about by a displacement reaction involving addition of, for example, another cation that reacts more strongly with the masking ligand and liberates the masked ion. { dē'mask·iŋ }

Dematiaceae [MYCOL] A family of fungi in the order Moniliales; sporophores are not grouped, hyphae are always dark, and the spores are hyaline or dark. { də‚mad·ē'ās·ē‚ē }

Dember effect [ELECTR] Creation of a voltage in a conductor or semiconductor by illumination of one surface. Also known as photodiffusion effect. { däm·bā i'fekt }

Dembowska [ASTRON] Possibly the only moderately large asteroid other than Vesta whose surface composition resembles that of achondritic meteorites; has a diameter of about 190 miles (145 kilometers) and a mean distance from the sun of 2.93 astronomical units. { dem'bȯf·skə }

deme [ECOL] A local population in which the individuals freely interbreed among themselves but not with those of other demes. { dēm }

dementia [PSYCH] Deterioration of intellectual and other mental processes due to organic brain disease. { də'men·chə }

dementia praecox See schizophrenia. { də'men·chə 'prē‚käks }

dementia simplex [PSYCH] A subtype of schizophrenia broadly characterized by a slow, progressive deterioration, often combined with mental deficiency. { də'men·chə 'sim‚pleks }

demersal [BIOL] Living at or near the bottom of the sea. { də'mər·səl }

demethanation See demethanization. { dē‚meth·ə'nā·shən }

demethanator [CHEM ENG] The apparatus in which demethanization is conducted. { dē'meth·ə‚nād·ər }

demethanization [CHEM ENG] The process of distillation in which methane is separated from the heavier components. Also known as demethanation. { dē‚meth·ən·ə'zā·shən }

demethylation [ORG CHEM] Removal of the methyl group from a compound. { de‚meth·ə'lā·shən }

demethylchlortetracycline [MICROBIO] $C_{21}H_{21}O_8N_2Cl$ A broad-spectrum tetracycline antibiotic produced by a mutant strain of *Streptomyces aureofaciens*. { dē¦meth·əl‚klȯr‚te·trə'sī‚klēn }

demeton-S-methyl [ORG CHEM] $C_6H_{15}O_3PS_2$ An oily liquid with a 0.3% solubility in water; used as an insecticide and miticide to control aphids. { 'dem·ə‚tän ¦es 'meth·əl }

demeton-S-methyl sulfoxide [ORG CHEM] $C_6H_{15}O_4PS_2$ A clear, amber liquid; limited solubility in water; used as an insecticide and miticide for pests of vegetable, fruit, and field crops, ornamental flowers, shrubs, and trees. { 'dem·ə‚tän ¦es 'meth·əl səl'fäk‚sīd }

demilune [BIOL] Crescent-shaped. { 'dem·i‚lün }

demineralization [CHEM ENG] Removal of mineral constituents from water. [MED] **1.** Removal or loss of minerals and salts from the body, especially by disease. **2.** In particular, the continual dissolving of tooth mineral that occurs at the surface of teeth as the result of the action of weak acids created by plaque-forming bacteria. { dē‚min·rə·lə'zā·shən }

demister [MECH ENG] A series of ducts in automobiles arranged so that hot, dry air directed from the heat source is forced against the interior of the windscreen or windshield to prevent condensation. { dē'mis·tər }

demister blanket [ENG] A section of knitted wire mesh that is placed below the vapor outlet of a vaporizer or an evaporator to separate entrained liquid droplets from the stream of vapor. Also known as demister pad. { dē'mis·tər ‚blaŋ·kət }

demister pad See demister blanket. { dē'mis·tər ‚pad }

Demjanov rearrangement [ORG CHEM] A structural rearrangement that accompanies treatment of certain primary aliphatic amines with nitrous acid; the amine will undergo a ring contraction or expansion. { dem'yä·nóf rē·ə'ränj·mənt }

DEMO See carboxin.

Demodicidae [INV ZOO] The pore mites, a family of arachnids in the suborder Trombidiformes. { ‚dem·ə'dis·ə‚dē }

demodifier [COMPUT SCI] A data element used to restore part of an instruction which has been modified to its original value. { dē'mäd·ə‚fī·ər }

demodulate [COMMUN] To recover the modulating wave from a modulated carrier. Also known as decode; detect. { dē'mäj·ə‚lāt }

demodulation [COMMUN] The recovery, from a modulated carrier, of a signal having substantially the same characteristics as the original signal. { dē‚mäj·ə'lā·shən }

demodulator See detector. { dē'mäj·ə‚lad·ər }

demographic genetics [BIOL] A branch of population genetics and ecology concerned with genetic differences related to age, population size, genetic alteration in competitive ability, and viability. { ¦dem·ə‚graf·ik jə¦ned·iks }

demography [ECOL] The statistical study of populations with reference to natality, mortality, migratory movements, age, and sex, among other social, ethnic, and economic factors. { də'mäg·rə·fē }

De Moivre's theorem [MATH] The nth power of the quantity $\cos \theta + i \sin \theta$ is $\cos n\theta + i \sin n\theta$ for any integer n. { də'mwäv·rəz ‚thir·əm }

demolition [CIV ENG] The act or process of tearing down a building or other structure. [ORD] Destroying a structure or an area by the use of explosives. { ‚dem·ə'lish·ən }

demolitional measurement [QUANT MECH] A measurement that alters the value of the physical observable being measured. { ‚dem·ə'lish·ən·əl 'mezh·ər·mənt }

demolition block [ORD] An explosive charge, usually in a nonmetallic container, used for demolition purposes. { ‚dem·ə'lish·ən ‚bläk }

demolition bomb [ORD] Former classification for a bomb that explodes after a short penetration, accomplishing damage and destruction by both blast and underground explosion. { ‚dem·ə'lish·ən ‚bäm }

demolition kit [ORD] A group of items of an explosive nature, with the necessary nonexplosive accessories and tools, specially designed containers, and carrying attachments, to enable efficient performance of designated demolition tasks. { ‚dem·ə'lish·ən ‚kit }

demon of Maxwell [THERMO] Hypothetical creature who controls a trapdoor over a microscopic hole in an adiabatic wall between two vessels filled with gas at the same temperature, so as to supposedly decrease the entropy of the gas as a whole and thus violate the second law of thermodynamics. Also known as Maxwell's demon. { 'dē·mən əv 'maks‚wel }

demonophobia [PSYCH] An abnormal fear of devils and demons. { dē‚män·ə'fō·bē·ə }

Demon Star See Algol. { 'dē·mən ‚stär }

De Morgan's rules [MATH] The complement of the union of two sets equals the intersection of their respective complements; the complement of the intersection of two sets equals the union of their complements. { də'mór·gənz ‚rülz }

De Morgan's test [MATH] A series with term u_n, for which $|u_{n+1}/u_n|$ converges to 1, will converge absolutely if there is $c > 0$ such that the limit superior of $n(|u_{n+1}/u_n| - 1)$ equals $-1 - c$. { də'mór·gənz ‚test }

demorphism See weathering. { dē'mór·fiz· əm }

Demospongiae [INV ZOO] A class of the phylum Porifera, including sponges with a skeleton of one-to four-rayed siliceous spicules, or of spongin fibers, or both. { dem·ə'spän·jē‚ē }

demount [COMPUT SCI] To take out a magnetic storage medium from a device that reads or writes on it. { dē'maúnt }

demountable pack [COMPUT SCI] A disk pack that can be taken out and replaced by another. { dē'maúnt·ə·bəl 'pak }

demountable tube [ELECTR] High-power radio tube having a metal envelope with porcelain insulation; can be taken apart for inspection and for renewal of electrodes. { dē'maúnt·ə·bəl 'tüb }

DEMS See Digital Electronic Message Service.

demulsification [CHEM ENG] Prevention or breaking of liquid-liquid emulsions by chemical, mechanical or electrical demulsifiers. { də‚məl·sə·fə'kā·shən }

demulsifier [CHEM ENG] A chemical, mechanical, or electrical system that either breaks liquid-liquid emulsions or prevents them from forming. { dē'məl·sə‚ff·ər }

demultiplexer [ELECTR] A device used to separate two or more signals that were previously combined by a compatible multiplexer and transmitted over a single channel. { dē‚məl·tə‚plek·sər }

demultiplexing [COMMUN] The separation of two or more channels previously multiplexed. { dē'məl·tə‚pleks·iŋ }

demultiplexing circuit [ELECTR] A circuit used to separate the signals that were combined for transmission by multiplex. { dē'məl·tə‚plek·siŋ ‚sər·kət }

demyelinating disease [MED] Any disease associated with the destruction or removal of myelin from nerves. { dē'mī·ə·lə‚nād·iŋ di‚zēz }

demyelination [PATH] Destruction of myelin; loss of myelin from nerve sheaths or nerve tracts. { dē‚mī·ə·lə'nā·shən }

denaturant [CHEM] An inert, bad-tasting, or poisonous chemical substance added to a product such as ethyl alcohol to make it unfit for human consumption. [NUCLEO] A nonfissionable isotope that can be added to fissionable material to make it unsuitable for use in atomic weapons without extensive processing. { dē'nā·chə·rənt }

denaturation map [MOL BIO] A map that shows the positions of denaturation loops of deoxyribonucleic acid and that provides a unique way to distinguish different molecules of deoxyribonucleic acid. { di'nā·chə‚rā·shən ‚map }

denature [CHEM] **1.** To change a protein by heating it or treating it with alkali or acid so that the original properties such as solubility are changed as a result of the protein's molecular structure being changed in some way. **2.** To add a denaturant, such as methyl alcohol, to grain alcohol to make the grain alcohol poisonous and unfit for human consumption. { dē'nā·chər }

denatured alcohol [CHEM] Ethyl alcohol containing a poisonous substance, such as methyl alcohol or benzene, which makes it unfit for human consumption. { dē'nā·chərd 'al·kə‚hól }

dendrimer [ORG CHEM] **1.** A polymer with a well-defined core, an interior, and peripheral surface components constructed in a concentric ordered fashion around the core. **2.** A polymer having a regular branched structure. Also known as dendritic polymer; dendron; starburst polymer. { 'den·drə·mər }

dendrite [NEUROSCI] The part of a neuron that carries the unidirectional nerve impulse toward the cell body. Also known as dendron. [CRYSTAL] A crystal having a treelike structure. { 'den‚drīt }

dendritic [SCI TECH] Having a branching, treelike structure or pattern. { den'drid·ik }

dendritic cell [CELL MOL] A specialized cell of the lymphoid reticuloendothelial system that presents antigens for detection by lymphocytes. { den'drid·ik ¦sel }

dendritic drainage [HYD] Irregular stream branching, with tributaries joining the main stream at all angles. { den'drid·ik 'drān·ij }

dendritic macromolecule [ORG CHEM] A macromolecule whose structure is characterized by a high degree of branching that originates from a single focal point (core). { den'drid·ik ‚mak·rō'mäl·ə‚kyül }

dendritic polymer See dendrimer. { den¦drid·ik 'päl·ə·mər }

dendritic powder [MET] Fine metal particles having a dendritic structure; usually of electrolytic origin. Also known as arborescent powder. { den'drid·ik 'paúd·ər }

dendritic valleys [GEOL] Treelike extensions of the valleys in a region lying upon horizontally bedded rock. { den'drid·ik 'val·ēz }

dendroarcheology [ARCHEO] The science of using tree rings to date wood material from archeological sites or artifacts. { ‚den·drō‚är·kē'äl·ə·jē }

Dendrobatinae [VERT ZOO] A subfamily of anuran amphibians in the family Ranidae, including the colorful poisonous frogs of Central and South America. { ‚den·drō'bat·ən·ē }

DENDROBRANCHIATE GILL

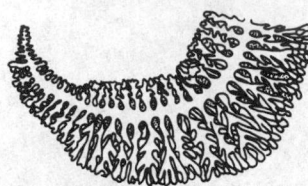

Dendrobranchiate gill of penaeid shrimp. *Benthesicymus.* (*Smithsonian Institution*)

DENDROCERATIDA

Dendritic skeleton of *Dendrilla cactus.*

DENDROIDEA

rhabdosome

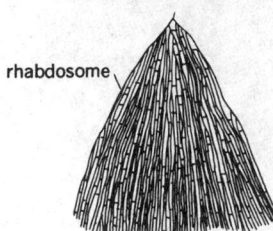

Whole colony of a dendroid graptolite.

dendrobranchiate gill [INV ZOO] A respiratory structure of certain decapod crustaceans, characterized by extensive branching of the two primary series. { ¦den·drō'braŋ·kē‚ät 'gil }

Dendroceratida [INV ZOO] A small order of sponges of the class Demospongiae; members have a skeleton of spongin fibers or lack a skeleton. { ‚den·drō·sə'räd·əd·ə }

dendrochemistry [CHEM] The analysis of the chemical composition of tree rings for naturally occurring or human-manufactured chemicals, especially the mineral elements, to understand the impact of pollution in the air, or surface-water or ground-water supply in ecosystems, or to detect environmental changes over time. { ‚den·drō'kem·i·strē }

Dendrochirotacea [INV ZOO] A subclass of echinoderms in the class Holothuroidea. { ‚den·drō‚kī·rō'täs·ē·ə }

Dendrochirotida [INV ZOO] An order of dendrochirotacean holothurian echinoderms with 10–30 richly branched tentacles. { ‚den·drō‚kī'räd·əd·ə }

dendrochronology [GEOL] The science of measuring time intervals and dating events and environmental changes by reading and dating growth layers of trees as demarcated by the annual rings. { ¦den·drō·krə'näl·ə·jē }

dendroclimatology [METEOROL] The study of the tree-ring record to reconstruct climate history, based on the fact that temperature, precipitation, and other climatic variables affect tree growth. { ‚den·drō‚klī·mə'täl·ə·jē }

Dendrocolaptidae [VERT ZOO] The woodcreepers, a family of passeriform birds belonging to the suboscine group. { ‚den·drō·kə'lap·tə‚dē }

dendroecology [ARCHEO] The science of analyzing changes in ecological processes over time using tree-ring information. [ECOL] The use of tree rings to study changes in ecological processes over time such as defoliation by insect outbreaks; the effects of air, water, and soil pollution on tree growth and forest health; the age, maturity, and successional status of forest stands; and the effects of human disturbances and management on forest vitality. { ‚den·drō·ē'käl·ə·jē }

dendrogeomorphology [GEOGR] The use of tree-ring data to study earth surface processes. Scientists can date when trees were killed (by dating the outer ring of the tree) or bent (by analyzing when dramatic changes in tree growth occurred) by mass movements, such as landslides and snow avalanches. { ‚den·drō‚gē·ō·mòr'fäl·ə·jē }

dendrogram [BIOL] A genealogical tree; the trunk represents the oldest common ancestor, and the branches indicate successively more recent divisions of a lineage for a group. { 'den·drə‚gram }

dendrohydrology [ARCHEO] The science of using tree-ring data to investigate and reconstruct hydrologic properties, such as streamflow and riverflow, runoff, and past lake levels. [HYD] The science of determining hydrologic occurrences by the comparison of tree ring thickness with streamflow or precipitation. Also known as tree-ring hydrology. { ‚den·drō·hī'dräl·ə·jē }

dendroid [BIOL] Branched or treelike in form. { 'den‚drȯid }

Dendroidea [PALEON] An order of extinct sessile, branched colonial animals in the class Graptolithina occurring among typical benthonic fauna. { den'drȯid·ē·ə }

dendrology [FOR] The division of forestry concerned with the classification, identification, and distribution of trees and other woody plants. { den'dräl·ə·jē }

dendrometer [FOR] A device used to measure a tree's height and diameter using principles based on the relation of the sides of similar triangles. { den'dräm·əd·ər }

Dendromurinae [VERT ZOO] The African tree mice and related species, a subfamily of rodents in the family Muridae. { ¦den·drō'myùr·ə‚nē }

dendron *See* dendrimer; dendrite. { 'den‚drän }

dendrophagous [ZOO] Feeding on trees, referring to insects. { den'dräf·ə·gəs }

dendrophysis [MYCOL] A hyphal thread with arboreal branching in certain fungi. { den'dräf·ə·səs }

dendropyrochronology [ARCHEO] The use of tree rings to study and reconstruct the history of wild fires; for example, a low-intensity surface fire will kill a part of the living cambium on the lower portion of the tree trunk which subsequent growth will preserve, leaving a fire scar in the tree-ring record. { ‚den·drō‚pī·rō·krə'näl·ə·jē }

Deneb [ASTRON] A white star of spectral classification A2-Ia in the constellation Cygnus; the star α Cygni. { 'den‚eb }

Denebola [ASTRON] A white star of stellar magnitude 2.2, spectral classification A2, in the constellation Leo; the star β Leonis. { də'neb·ə·lə }

denervate [MED] To interfere with or cut off the nerve supply to a part of the body, or to remove a nerve; may occur by excision, drugs, or a disease process. { dē'nər‚vät }

denervation hypersensitivity [NEUROSCI] Extreme sensitivity of an organ that has recovered from the removal or interruption of its nerve supply. { ‚dē·nər‚vā·shən ‚hīp·ər‚sen·sə'tiv·əde }

dengue [MED] An acute viral disease of humans characterized by fever, rash, prostration, and lymphadenopathy; transmitted by the mosquito *Aedes aegypti.* Also known as breakbone fever; dandy fever. { 'deŋ·gē }

Dengue fever [MED] An infection borne by the *Aedes* female mosquito, and caused by one of four closely related but antigenically distinct Dengue virus serotypes (DEN-1, DEN-2, DEN-3, and DEN-4). It starts abruptly after an incubation period of 2–7 days with high fever, severe headache, myalgia, and rash. It is found throughout the tropical and subtropical zones. Also known as break-bone fever. { 'deŋ·gē ‚fēv·ər }

Dengue hemmorhagic fever [MED] A severe and potentially fatal form of Dengue fever that is characterized by loss of appetite, vomiting, high fever, headache, and abdominal pain; shock and circulatory failure may occur. { ‚deŋ·gē ‚hem·ə‚raj·ik 'fē·vər }

denial [MATH] *See* negation. [PSYCH] An unconscious defense mechanism in which an individual denies herself or himself recognition of an observation in order to avoid pain or anxiety. { di'nī·əl }

denier [TEXT] A unit expressing the mass of a fiber divided by its length, equal to 1 gram for 9000 meters of fiber. { 'den·yər }

denim [TEXT] A sturdy twill-weave cotton fabric having a solid-colored warp and a white filling. { 'den·əm }

Denison sampler [ENG] A soil sampler consisting of a central nonrotating barrel which is forced into the soil as friction is removed by a rotating external barrel; the bottom can be closed to retain the sample during withdrawal. { 'den·ə·sən ‚sam·plər }

Denisyuk hologram [OPTICS] A type of hologram that can be viewed in ordinary white light through use of the depth dimension of the emulsion. { 'den·ə·syùk 'häl·ə‚gram }

denitration [CHEM] Removal of nitrates or nitrogen. Also known as denitrification. { dē‚nī'trā·shən }

denitrification [CHEM] *See* denitration. [MICROBIO] The reduction of nitrate or nitrite to gaseous products such as nitrogen, nitrous oxide, and nitric oxide; brought about by denitrifying bacteria. { dē‚nī·trə·fə'kā·shən }

denitrifying bacteria [MICROBIO] Bacteria that reduce nitrates to nitrites or nitrogen gas; most are found in soil. { dē'nī·trə‚fī·iŋ bak'tir·ē·ə }

denitrogenate [PHYSIO] To remove nitrogen from the body by breathing nitrogen-free gas. { dē'nī·trə·jə‚nät }

denominator [MATH] In a fraction, the term that divides the other term (called the numerator), and is written below the line. { də'näm·ə‚nād·ər }

De Nora cell [CHEM ENG] Mercury-cathode cell used for production of chlorine and caustic soda by electrolysis of sodium chloride brine. { də'nȯr·ə ‚sel }

dense [GRAPHICS] Very opaque because of a concentration of material, as pertaining to a negative or transparency that has been overdeveloped or overexposed. { dens }

dense-air refrigeration cycle *See* reverse Brayton cycle. { ¦dens 'er ri‚frij·ə'rā·shən ‚sī·kəl }

dense-air system *See* cold-air machine. { ¦dens 'er ‚sis·təm }

dense binary code [COMPUT SCI] A code in which all possible states of the binary pattern are used. { 'dens ¦bī·nə·rē 'kōd }

dense connective tissue [HISTOL] A fibrous connective tissue with an abundance of enlarged collagenous fibers which tend to crowd out the cells and ground substance. { ¦dens kə¦nek·tiv 'tish·yü }

dense fibrillar component [CYTOL] A component of the nucleolus that lacks granules and stains more intensely than other nucleolar components. { ¦dens ¦fi·brə·lər kəm¦pōn·ənt }

dense-in-itself set [MATH] A set every point of which is an accumulation point; a set without any isolated points. { 'dens in it'self ˌset }

dense list [COMPUT SCI] A list in which all the cells contain records of the file. { ¦dens ¦list }

dense-media separator [MIN ENG] A device in which a heavy mineral is dispersed in water, causing heavier ores to sink and lighter ores to float. { ¦dens ¦mēd·ē·ə 'sep·ə‚rād·ər }

dense subset [MATH] A subset of a topological space whose closure is the entire space. { ¦dens 'səb‚set }

densification [MATER] In ceramic powder processes, the step where the porous powder shape is converted into a part; the three main processes are sintering, hot pressing, and hot isostatic pressing. { ˌden·si·fə'kā·shən }

densify [ENG] To increase the density of a material such as wood by subjecting it to pressure or impregnating it with another material. { 'den·sə‚fī }

densimeter [ENG] An instrument which measures the density or specific gravity of a liquid, gas, or solid. Also known as densitometer; density gage; density indicator; gravitometer. { den'sim·əd·ər }

densitometer [ENG] **1.** An instrument which measures optical density by measuring the intensity of transmitted or reflected light; used to measure photographic density. **2.** See densimeter. { ˌden·sə'täm·əd·ər }

density [MATER] Closeness of texture or consistency. [MATH] For an increasing sequence of integers, the greatest lower bound of the quantity $F(n)/n$, where $F(n)$ is the number of integers in the sequence (other than zero) equal to or less than n. [MECH] The mass of a given substance per unit volume. [OPTICS] **1.** The degree of opacity of a translucent material. **2.** The common logarithm of opacity. [PHYS] The total amount of a quantity, such as energy, per unit of space. { 'den·səd·ē }

density airspeed [AERO ENG] Calibrated airspeed corrected for pressure altitude and true air temperature. { 'den·səd·ē 'er‚spēd }

density altitude [METEOROL] The altitude, in the standard atmosphere, at which a given density occurs. { 'den·səd·ē 'al·tə‚tüd }

density bombing [ORD] Dropping a given tonnage of bombs onto an area in order to make certain of striking particular targets. { 'den·səd·ē ‚bäm·iŋ }

density bottle See specific-gravity bottle. { 'den·səd·ē ‚bäd·əl }

density channel [METEOROL] A channel used to investigate a density current; for example, in experiments relating to the behavior of cold masses of air in the atmosphere and related frontal structures. { 'den·səd·ē ‚chan·əl }

density correction [AERO ENG] A correction made necessary because the airspeed indicator is calibrated only for standard air pressure; it is applied to equivalent airspeed to obtain true airspeed, or to calibrated airspeed to obtain density airspeed. [ENG] **1.** The part of the temperature correction of a mercury barometer which is necessitated by the variation of the density of mercury with temperature. **2.** The correction, applied to the indications of a pressure-tube anemometer or pressure-plate anemometer, which is necessitated by the variation of air density with temperature. { 'den·səd·ē kə'rek·shən }

density current [METEOROL] Intrusion of a dense air mass beneath a lighter air mass; the usage applies to cold fronts. [OCEANOGR] See turbidity current. { 'den·səd·ē ‚kər·ənt }

density-dependent factor [ECOL] A factor that affects the birth rate or mortality rate of a population in ways varying with the population density. { ¦den·səd·ē ‚di¦pen·dənt ‚fak·tər }

density effect [NUCLEO] The reduction in the stopping power of dense materials for relativistic particles that is caused by the reduction of the effective electric field of the particles by the polarization of adjacent atoms. { 'den·səd·ē i‚fekt }

density error [AERO ENG] The error in the indications of a differential-pressure-type airspeed indicator due to nonstandard atmospheric density. { 'den·səd·ē ‚er·ər }

density function [MATH] A density function for a measure m is a function which gives rise to m when it is integrated with respect to some other specified measure. [STAT] See probability density function.. { 'den·səd·ē ‚fəŋk·shən }

density gage See densimeter. { 'den·səd·ē ‚gāj }

density gradient centrifugation [ANALY CHEM] Separation of particles according to density by employing a gradient of varying densities; at equilibrium each particle settles in the gradient at a point equal to its density. { 'den·səd·ē ¦grād·ē·ənt sen‚trif·ə'gā·shən }

density-independent factor [ECOL] A factor that affects the birth rate or mortality rate of a population in ways that are independent of the population density. { ¦den·səd·ē ‚in·də¦pen·dənt ‚fak·tər }

density indicator See densimeter. { 'den·səd·ē ‚in·də‚kād·ər }

density log [PETRO ENG] Radioactivity logging of reservoir structure densities down an oil-well bore by emission and detection of gamma rays. { 'den·səd·ē ‚läg }

density matrix [QUANT MECH] A matrix ρ_{mn} describing an ensemble of quantum-mechanical systems in a representation based on an orthonormal set of functions ϕ_n; for any operator G with representation G_{mn}, the ensemble average of the expectation value of G is the trace of ρG. { 'den·səd·ē 'mā·triks }

density modulation [ELECTR] Modulation of an electron beam by making the density of the electrons in the beam vary with time. { 'den·səd·ē ‚mäj·ə'lā·shən }

density of states [SOLID STATE] A function of energy E equal to the number of quantum states in the energy range between E and $E + dE$ divided by the product of dE and the volume of the substance. { 'den·səd·ē əv 'stāts }

density packing [COMPUT SCI] In computers, the number of binary digit magnetic pulses stored on tape or drum per linear inch on a single track by a single head. { 'den·səd·ē ‚pak·iŋ }

density ratio [METEOROL] The ratio of the density of the air at a given altitude to the air density at the same altitude in a standard atmosphere. { 'den·səd·ē ‚rā·shō }

density rule [ENG] A grading system for lumber based on the width of annual rings. { 'den·səd·ē ‚rül }

density scale [GRAPHICS] A value for the range density for a photographic material that corresponds to the difference between the maximum density and the minimum density. Also known as net density. { 'den·səd·ē ‚skāl }

density specific impulse [AERO ENG] The product of the specific impulse of a propellant combination and the average specific gravity of the propellants. { 'den·səd·ē spə‚sif·ik 'im‚pəls }

density step tablet [COMMUN] Facsimile test chart consisting of a series of areas; density of the areas increases from a low value to a maximum value in steps. Also known as step tablet. { 'den·səd·ē 'step ‚tab·lət }

density transmitter [ENG] An instrument used to record the density of a flowing stream of liquid by measuring the buoyant force on an air-filled chamber immersed in the stream. { 'den·səd·ē tranz'mid·ər }

density wave [PHYS] A sound wave or other type of material wave which causes the density of the matter through which it passes to alternately rise above and drop below its mean value. { 'den·səd·ē ‚wāv }

density-wave theory [ASTROPHYS] A theory explaining the spiral structure of galaxies by a periodic variation in space in the density of matter which rotates with a fixed angular velocity while the angular velocity of the matter itself varies with distance from the galaxy's center. { 'den·səd·ē ‚wāv ‚thē·ə·rē }

density wedge [GRAPHICS] A band of paper or film showing graduated tones ranging from white to black. { 'den·səd·ē ‚wej }

densofacies See metamorphic facies. { ˌden·sō'fā·shēz }

Densovirus [VIROL] A genus of the animal virus family Parvoviridae whose virion is nonenveloped, with deoxyribonucleic acid single-stranded; replicates autonomously. { ˌden·sō'vī·rəs }

dental [ANAT] Pertaining to the teeth. { 'dent·əl }

dental arch [ANAT] The parabolic curve formed by the cutting edges and masticating surfaces of the teeth. { ¦dent·əl 'ärch }

dental bridge [MED] A prosthetic device used to replace missing teeth. { 'dent·əl ‚brij }

dental calculi [MED] Calcareous deposits of organic and mineral matter on the teeth. Also known as tartar. { ¦dent·əl 'kal·kyə‚lē }

dental caries See caries. { ¦dent·əl 'kar·ēz }

dental coupling [MECH ENG] A type of flexible coupling used to join a steam turbine to a reduction-gear pinion shaft; consists of a short piece of shaft with gear teeth at each end, and mates with internal gears in a flange at the ends of the two shafts to be joined. { 'dent·əl 'kəp·liŋ }

dental epithelium [HISTOL] The cells forming the boundary of the enamel organ. { 'dent·əl ep·ə'thē·lē·əm }

dental follicle *See* dental sac. { |dent·əl 'fäl·ə·kəl }

dental formula [VERT ZOO] An expression of the number and kind of teeth in each half jaw, both upper and lower, of mammals. { |dent·əl 'for·myə·lə }

dental gold [MET] An alloy composed of 5 to 12% silver, 4 to 10% copper, with the balance gold. { 'dent·əl 'gōld }

Dentaliidae [INV ZOO] A family of mollusks in the class Scaphopoda; members have pointed feet. { ,dent·əl'ī·ə,dē }

dental index [ANTHRO] A ratio of the length of the teeth to the distance from the nasion to the basion multiplied by 100, used to determine the relative size of teeth. { |dent·əl 'in,deks }

dental materials science [MED] An interdisciplinary area that applies biology, chemistry, and physics to the development, understanding, and evaluation of materials used in the practice of dentistry; principally involved in restorative dentistry, prosthodontics, pedodontics, and orthodontics. { 'dent·əl mə,tir·ē·əlz 'sī·əns }

dental pad [VERT ZOO] A firm ridge that replaces incisors in the maxilla of cud-chewing herbivores. { 'dent·əl ,pad }

dental papilla [EMBRYO] The mass of connective tissue located inside the enamel organ of a developing tooth, and forming the dentin and dental pulp of the tooth. { 'dent·əl pə'pil·ə }

dental plate [INV ZOO] A flat plate that replaces teeth in certain invertebrates, such as some worms. [VERT ZOO] A flattened plate that represents fused teeth in parrot fishes and related forms. { 'dent·əl ,plāt }

dental pulp [HISTOL] The vascular connective tissue of the roots and pulp cavity of a tooth. { 'dent·əl ,pəlp }

dental ridge [EMBRYO] An elevation of the embryonic jaw that forms a cusp or margin of a tooth. { 'dent·əl ,rij }

dental sac [EMBRYO] The connective tissue that encloses the developing tooth. Also known as dental follicle. { 'dent·əl ,sak }

dental work *See* cementation. { 'dent·əl ,wərk }

dentate [BIOL] **1.** Having teeth. **2.** Having toothlike or conical marginal projections. { 'den,tāt }

dentate fissure *See* hippocampal sulcus. { 'den,tāt 'fish·ər }

dentate nucleus [NEUROSCI] An ovoid mass of nerve cells located in the center of each cerebellar hemisphere, which give rise to fibers found in the superior cerebellar peduncle. { 'den,tāt 'nü·klē·əs }

denticle [ZOO] A small tooth or toothlike projection, as the type of scale of certain elasmobranchs. { 'dent·ə·kəl }

denticulate [ZOO] Having denticles; serrate. { den'tik·yə·lət }

dentifrice [FOOD ENG] A paste, powder, or liquid preparation used to clean teeth. { 'den·tə·fris }

dentigerous [BIOL] Having teeth or toothlike structures. { den'tij·ə·rəs }

dentil [ARCH] One of a series of small rectangular blocks under a cornice. { 'dent·əl }

dentil band [ARCH] A molding or band on a cornice resembling a row of dentils, but with the spaces between dentils filled. { 'dent·əl 'band }

dentin [HISTOL] A bonelike tissue composing the bulk of a vertebrate tooth; consists of 70% inorganic materials and 30% water and organic matter. { 'dent·ən }

dentinoblast [HISTOL] A mesenchymal cell that forms dentin. { den'tēn·ə,blast }

dentinogenesis [PHYSIO] The formation of dentin. { den·|tēn·ə|jen·ə·səs }

dentinogenesis imperfecta [MED] An inherited dental disorder that causes defective formation of dentin. { ,den·tə·nō,jen·ə·səs ,im·pər'fek·tə }

dentinoma [MED] A benign odontogenic tumor made up of dentin. { ,den·tə'nō·mə }

dentistry [MED] A branch of medical science concerned with the prevention, diagnosis, and treatment of diseases of the teeth and adjacent tissues and the restoration of missing dental structures. { 'dent·ə·strē }

dentition [VERT ZOO] The arrangement, type, and number of teeth which are variously located in the oral or in the pharyngeal cavities, or in both, in vertebrates. { den'tish·ən }

denture [MED] A partial or complete prosthetic appliance to replace one or more missing teeth. { 'den·chər }

denudation [GEOL] General wearing away of the land; laying bare of subjacent lands. { ,dē·nü'dā·shən }

denumerable set [MATH] A set which may be put in one-to-one correspondence with the positive integers. Also known as countably infinite set. { də'nüm·rə·bəl 'set }

deodorant [MATER] A substance used to remove, correct, or repress undesirable odors. { dē'ōd·ə·rənt }

deodorized kerosine [MATER] A highly refined petroleum kerosine that has very little odor; used as an illuminant for wick lamps. Also known as refined kerosine. { dē'ōd·ə,rīzd 'ker·ə,sēn }

deodorizing [CHEM ENG] A process for removing odor-creating substances from oil or fat, in which the oil or fat is held at high temperatures and low pressure while steam is blown through. { dē'ōd·ə,rīz·iŋ }

deoil [CHEM ENG] To reduce the amount of liquid oil entrained in solid wax. { dē'oil }

deoperculate [BOT] Of mosses and liverworts, to shed the operculum. { dē·ō'pər·kyə,lāt }

deorbit [AERO ENG] To recover a spacecraft from earth orbit by providing a new orbit which intersects the earth's atmosphere. { dē'or·bət }

deoxidant *See* deoxidizer. { dē'äk·sə·dənt }

deoxidation [CHEM] **1.** The condition of a molecule's being deoxidized. **2.** The process of deoxidizing. { dē,äk·sə'dā·shən }

deoxidation sphere *See* bleach spot. { dē,äk·sə'dā·shən ,sfir }

deoxidize [CHEM] **1.** To remove oxygen by any of several processes. **2.** To reduce from the state of an oxide. [MET] To remove an oxide film from a metal surface. { dē'äk·sə,dīz }

deoxidized copper [MET] Pure copper deoxidized with phosphorus to reduce cuprous oxide and eliminate porosity. { dē'äk·sə,dīzd 'käp·ər }

deoxidizer [CHEM] Any substance which reduces the amount of oxygen in a substance, especially a metal, or reduces oxide compounds. Also known as deoxidant. { dē'äk·sə,dīz·ər }

deoxycholate [BIOCHEM] A salt or ester of deoxycholic acid. { dē|äk·sə'kō,lāt }

deoxycholic acid [BIOCHEM] $C_{24}H_{40}O_4$ One of the unconjugated bile acids; in bile it is largely conjugated with glycine or taurine. { dē|äk·sə'käl·ik 'as·əd }

deoxycorticosterone [BIOCHEM] $C_{21}H_{30}O_3$ A steroid hormone secreted in small amounts by the adrenal cortex. { dē|äk·sē,kord·ə'kä·stə,rōn }

deoxygenation [CHEM] Removal of oxygen from a substance, such as blood or polluted water. { dē,äk·sə·jə'nā·shən }

deoxyribonuclease [BIOCHEM] An enzyme that catalyzes the hydrolysis of deoxyribonucleic acid to nucleotides. Abbreviated DNase. { dē|äk·sē,rī·bō'nü·klē,ās }

deoxyribonucleic acid [BIOCHEM] A linear polymer made up of deoxyribonucleotide repeating units (composed of the sugar 2-deoxyribose, phosphate, and a purine or pyrimidine base) linked by the phosphate group joining the 3′ position of one sugar to the 5′ position of the next; most molecules are double-stranded and antiparallel, resulting in a right-handed helix structure kept together by hydrogen bonds between a purine on one chain and a pyrimidine on another; carrier of genetic information, which is encoded in the sequence of bases; present in chromosomes and chromosomal material of cell organelles such as mitochondria and chloroplasts, and also present in some viruses. Abbreviated DNA. { dē|äk·sē,rī·bō·nü|klē·ik 'as·əd }

deoxyribonucleic acid clone [MOL BIO] A deoxyribonucleic acid segment inserted via a vector into a host cell and replicated along with the vector to form many copies per cell. { dē|äk·sē,rī·bō·nü|klē·ik |as·əd 'klōn }

deoxyribonucleic acid complexity [MOL BIO] A measure of the fraction of nonrepetitive deoxyribonucleic acid that is characteristic of a given sample. { dē|äk·sē,rī·bō·nü|klē·ik |as·əd kəm'plek·səd·ē }

deoxyribonucleic acid-directed ribonucleic acid polymerase [BIOCHEM] An enzyme which transcribes a ribonucleic acid (RNA) molecule complementary to deoxyribonucleic acid

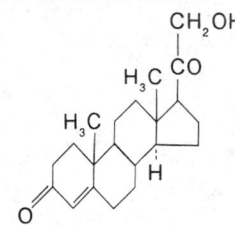
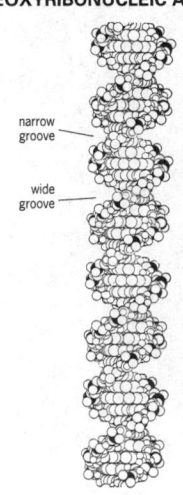

(DNA); required for initiation of DNA replication as well as transcription of RNA. { dē¦äk·sē¸rī·bō·nü¦klē·ik ¦as·əd də'rek·təd ¦rī·bō·nü¦kle·ik ¦as·əd pə'lim·ə¸rās }

deoxyribonucleic acid footprinting [MOL BIO] A method for determining the sequence of deoxyribonucleic acid-binding proteins. { dē¦äk·sē¸rī·bō·nü¦klē·ik ¦as·əd 'fút¸print·iŋ }

deoxyribonucleic acid hybridization [MOL BIO] A technique for selectively binding specific segments of single-stranded deoxyribonucleic acid (DNA) or ribonucleic acid by base pairing to complementary sequences on single-stranded DNA molecules that are trapped on a nitrocellulose filter. { dē¦äk·sē¸rī·bō·nü¦klē·ik ¦as·əd ¸hī·brə·də'zā·shən }

deoxyribonucleic acid lesion [MOL BIO] Deoxyribonucleic acid deformations that may result in gene mutation or changes in chromosome structure. { dē¦ak·sē¸rī·bō·nü¦klē·ik ¦as·əd 'lē·zhən }

deoxyribonucleic acid ligase [BIOCHEM] An enzyme which joins the ends of two deoxyribonucleic acid chains by catalyzing the synthesis of a phosphodiester bond between a 3′-hydroxyl group at the end of one chain and a 5′-phosphate at the end of the other. { dē¦ak·sē¸rī·bō·nü¦klē·ik ¦as·əd 'lī¸gās }

deoxyribonucleic acid polymerase I [BIOCHEM] An enzyme which catalyzes the addition of deoxyribonucleotide residues to the end of a deoxyribonucleic acid (DNA) strand; generally considered to function in the repair of damaged DNA. { dē¦ak·sē¸rī·bō·nü¦klē·ik ¦as·əd pə'lim·ə¸rās ¦wən }

deoxyribonucleic acid polymerase II [BIOCHEM] An enzyme similar in action to DNA polymerase I but with lower activity. { dē¦ak·sē¸rī·bō·nü¦klē·ik ¦as·əd pə'lim·ə¸rās ¦tü }

deoxyribonucleic acid polymerase III [BIOCHEM] An enzyme thought to be the primary enzyme involved in deoxyribonucleic acid replication. { dē¦äk·sē¸rī·bō·nü¦klē·ik ¦as·əd pə'lim·ə¸rās ¦thrē }

deoxyribonucleoprotein [BIOCHEM] A protein containing molecules of deoxyribonucleic acid in close association with protein molecules. { dē¦äk·sē¸rī·bō¸nü·klē·ō 'prō¸tēn }

deoxyribonucleotide [BIOCHEM] A nucleotide that contains deoxyribose and is a constituent of deoxyribonucleic acid. { dē¦äk·sē¸rī·bō'nü·klē·ə¸tīd }

deoxyribose [BIOCHEM] $C_5H_{10}O_4$ A pentose sugar in which the hydrogen replaces the hydroxyl groups of ribose; a major constituent of deoxyribonucleic acid. { dē¦äk·sē¦rī¸bōs }

deoxyribovirus [VIROL] Any virus that contains deoxyribonucleic acid. { dē¦äk·sē¸rī·bō'vī· rəs }

deoxy sugar [BIOCHEM] A substance which has the characteristics of a sugar, but which shows a deviation from the required hydrogen-to-oxygen ratio. { dē¦äk·sē 'shùg·ər }

2,4-DEP See tris[2-(2,4-dichlorophenoxy)ethyl]phosphite.

departure [METEOROL] The amount by which the value of a meteorological element differs from the normal value. [NAV] **1.** The distance between two meridians at any given parallel of latitude, expressed in linear units, usually nautical miles; the distance to the east or west made good by a craft in proceeding from one point to another. **2.** The point at which reckoning of a voyage begins; usually established by bearings of prominent landmarks as the vessel clears a harbor and proceeds to sea; when a person establishes this point, he is said to take departure. Also known as point of departure. **3.** Act of departing or leaving. { di'pär·chər }

departure point [NAV] A navigational checkpoint used by aircraft as a marker for setting course. { di'pär·chər ¸point }

departure track [CIV ENG] A railroad yard track for combining freight cars into outgoing trains. { di'pär·chər ¸trak }

depauperate [BIOL] Inferiority of natural development or size. { dē'pò·pə·rət }

DEPC See diethyl pyrocarbonate.

depegram [METEOROL] On a diagram having entropy and temperature as coordinates, a curve representing the distribution of the dew point as a function of pressure for a given sounding of the atmosphere. { 'dep·ə¸gram }

dependence [MED] Habituation to, abuse of, or addiction to a substance. [STAT] The existence of a relationship between frequencies obtained from two parts of an experiment which does not arise from the direct influence of the result of the first part on the chances of the second part but indirectly from the fact that both parts are subject to influences from a common outside factor. { di'pen·dəns }

dependency [COMPUT SCI] The necessity for a computer to complete work on some job before execution of another can begin. { di'pen·dən·sē }

dependency needs [PSYCH] The vital, originally infantile needs for mothering, love, affection, shelter, protection, security, food, and warmth, which are also present in some degree even in adult life. { di'pen·dən·sē ¸nēdz }

dependent equation [MATH] **1.** An equation is dependent on one or more other equations if it is satisfied by every set of values of the unknowns that satisfy all the other equations. **2.** A set of equations is dependent if any member of the set is dependent on the others. { di'pen·dənt i'kwā·zhən }

dependent events [STAT] Two events such that the occurrence of one affects the probability of the occurrence of the other. { di'pen·dənt i'vens }

dependent segment [COMPUT SCI] In a database management system, a block of data that depends on data at a higher level for its full meaning. { di'pen·dənt 'seg·mənt }

dependent variable [MATH] If y is a function of x, that is, if the function assigns a single value of y to each value of x, then y is the dependent variable. { di'pen·dənt 'ver·ē·ə·bəl }

Dependovirus [VIROL] A genus of the animal-virus family Parvoviridae that is characterized by defective viruses that require a helper virus (usually an adenovirus) for their replication. { di'pen·də¸vī·rəs }

depentanizer [CHEM ENG] A fractionating column for removal of pentane and lighter fractions from a hydrocarbon mixture. { də'pent·ən¸īz·ər }

depeq [METEOROL] Strong winds over Loet Tawar (Sumatra, East Indies) during the southwest monsoon. { də'pek }

depergelation [HYD] The act or process of thawing permafrost. { dē¸pər·jə'lā·shən }

deperm See degauss. { dē'pərm }

depersonalization [PSYCH] Loss of the sense of one's identity or of reality concerning the self. { dē¸pərs·ən·ə·lə'zā·shən }

depersonalization neurosis [PSYCH] Neurosis in which an individual has feelings of unreality and of estrangement from the self, body, or surroundings. { dē¸pərs·ən·əl·ə'zā·shən nù¸rō·səs }

Depertellidae [PALEON] A family of extinct perissodactyl mammals in the superfamily Tapiroidea. { de·pər'tel·ə¸dē }

dephlegmation [CHEM ENG] In a distillation operation, the partial condensation of vapor to form a liquid richer in higher boiling constituents than the original vapor. { dē¸fleg'mā·shən }

dephlegmator [CHEM ENG] An apparatus used in fractional distillation to cool the vapor mixture, thereby condensing higher-boiling fractions. { dē'fleg¸mād·ər }

dephosphorize [MET] Removal of phosphorus from a molten metal such as steel. { dē'fäs·fə¸rīz }

dephosphorylate [BIOCHEM] To remove a phosphate group. { ¸dē·fäs'fór·ə¸lāt }

depilation [ENG] Removal of hair from animal skins in processing leather. { ¸dep·ə'lā·shən }

depilatory [MATER] A chemical that removes hairs from skin. { də'pil·ə¸tór·ē }

depleted material [NUCLEO] Material in which the amount of one or more isotopes of a constituent has been reduced by an isotope separation process or by a nuclear reaction. { də'plēd·əd mə'tir·ē·əl }

depleted uranium [NUCLEO] Uranium having a smaller percentage of uranium-235 than the 0.7% found in natural uranium. { də'plēd·əd yù'rän·ē·əm }

depletion [ECOL] Using a resource, such as water or timber, faster than it is replenished. [ELECTR] Reduction of the charge-carrier density in a semiconductor below the normal value for a given temperature and doping level. [NUCLEO] The percentage reduction in the quantity of fissionable atoms in the fuel assemblies or fuel mixture that occurs during operation of a nuclear reactor. { də'plē·shən }

depletion drive [PETRO ENG] Displacement mechanism (type of drive) to expel hydrocarbons from porous reservoir formations, that is, to remove more hydrocarbon from the reservoir; types of drives are gas or water (natural or injected) and injected liquefied petroleum gas. { də'plē·shən ¸drīv }

depletion layer [ELECTR] An electric double layer formed at the surface of contact between a metal and a semiconductor having different work functions, because the mobile carrier charge density is insufficient to neutralize the fixed charge density of donors and acceptors. Also known as barrier layer (deprecated); blocking layer (deprecated); space-charge layer. { də'plē·shən ,lā·ər }

depletion-layer capacitance *See* barrier capacitance. { di'plē·shən ,lā·ər kə'pas·əd·əns }

depletion-layer rectification [ELECTR] Rectification at the junction between dissimilar materials, such as a *pn* junction or a junction between a metal and a semiconductor. Also known as barrier-layer rectification. { də'plē·shən ,lā·ər ,rek·tə·fə'kā·shən }

depletion-layer transistor [ELECTR] A transistor that relies directly on motion of carriers through depletion layers, such as spacistor. { də'plē·shən ,lā·ər tran'zis·tər }

depletion mode [ELECTR] Operation of a field-effect transistor in which current flows when the gate-source voltage is zero, and is increased or decreased by altering the gate-source voltage. { də'plē·shən ,mōd }

depletion-mode HEMT [ELECTR] A high-electron mobility transistor (HEMT) in which application of negative bias to the gate electrode cuts off the current between source and drain. Abbreviated D-HEMT. { də'plē·shən ,mōd ,āch,ē,em'tē }

depletion region [ELECTR] The portion of the channel in a metal oxide field-effect transistor in which there are no charge carriers. { də'plē·shən ,rē·jən }

depletion-type reservoir [PETRO ENG] Oil reservoir which is initially in (and during depletion remains in) a state of equilibrium between the gas and liquid phases; includes single-phase gas, two-phase bubble-point, and retrograde-gas-condensate (or dew-point) reservoirs. { də'plē·shən ¦tīp 'rez·ə,vwär }

depocenter [GEOL] A site of maximum deposition. { 'dep·ə,sen·tər }

depolarization [ELEC] The removal or prevention of polarization in a substance (for example, through the use of a depolarizer in an electric cell) or of polarization arising from the field due to the charges induced on the surface of a dielectric when an external field is applied. [OPTICS] The resolution of polarized light in an optical depolarizer. { dē,pō·lə·rə'zā·shən }

depolarization factor [ELEC] The ratio of the internal electric field induced by the charges on the surface of a dielectric when an external field is applied to the polarization of the dielectric. { dē,pō·lə·rə'zā·shən ,fak·tər }

depolarizer [PHYS CHEM] A substance added to the electrolyte of a primary cell to prevent excessive buildup of hydrogen bubbles by combining chemically with the hydrogen gas as it forms. Also known as battery depolarizer. { dē'pō·lə,rīz·ər }

depolymerization [ORG CHEM] Decomposition of macromolecular compounds into relatively simple compounds. { ,dē·pə,lim·ə·rə'zā·shən }

deposit [COMPUT SCI] To preserve the contents of a portion of a computer memory by copying it in a backing storage. [GEOL] Consolidated or unconsolidated material that has accumulated by a natural process or agent. [MATER] Any material applied to a base by means of vacuum, electrical, chemical, screening, or vapor methods. [SCI TECH] Any solid matter which is gradually laid down on a surface by a natural process. { də'päz·ət }

deposit attack [MET] Corrosion under or around the edge of a noncontinuous local deposit on a metal surface. { də'päz·ət ə'tak }

deposit dose [NUCLEO] The residual radioactivity deposited on the surface after a nuclear explosion, as by water falling as rain from the base surge of an underwater atomic explosion. { də'päz·ət ,dōs }

deposited carbon resistor [ELECTR] A resistor in which the resistive element is a carbon film pyrolytically deposited on a ceramic substrate. { də'päz·əd·əd 'kär·bən ri'zis·tər }

deposited metal [MET] Molten metal used during a welding operation for the fusion of base metals. { də'päz·əd·əd 'med·əl }

deposit feeder [INV ZOO] Any animal that feeds on the detritus that collects on the substratum at the bottom of water. Also known as detritus feeder. { də'päz·ət ,fēd·ər }

deposit gage [ENG] The general name for instruments used in air pollution studies for determining the amount of material deposited on a given area during a given time. { də'päz·ət ,gāj }

deposition [GEOL] The laying, placing, or throwing down of any material; specifically, the constructive process of accumulation into beds, veins, or irregular masses of any kind of loose, solid rock material by any kind of natural agent. { ,dep·ə'zish·ən }

depositional dip *See* primary dip. { ,dep·ə'zish·ən·əl 'dip }

depositional fabric [PETR] Arrangement of detrital particles settled from suspension or of crystals from a differentiating magma determined by the plane of the surface on which they come to rest. { ,dep·ə'zish·ən·əl 'fab·rik }

depositional sequence [GEOL] A major but informal assemblage of formations or groups and supergroups, bounded by regionally extensive unconformities at both their base and top and extending over broad areas of continental cratons. { ,dep·ə'zish·ən·əl 'sē·kwəns }

depositional strike [GEOL] Sedimentary deposits that are continuous laterally on a gently sloping surface. { ,dep·ə'zish·ən·əl 'strīk }

deposition efficiency [MET] The ratio of the weight of deposited metal to the net weight of the consumed electrodes, exclusive of stubs, in welding. { ,dep·ə'zish·ən ə'fish·ən·sē }

deposition potential [PHYS CHEM] The smallest potential which can produce electrolytic deposition when applied to an electrolytic cell. { ,dep·ə'zish·ən pə'ten·chəl }

deposition rate [MET] The amount of welding material deposited per unit of time, expressed in pounds per hour. { ,dep·ə'zish·ən ,rāt }

deposition sequence [MET] The order of deposition of weld-metal increments. { ,dep·ə'zish·ən ,sē·kwəns }

depot [ORD] **1.** An establishment for storing supplies or for maintaining equipment. **2.** The installation for this establishment. { 'dep·ō }

deprecated usage [SCI TECH] Word usage which is disapproved by experts in the pertinent field because the term in question has misleading connotations; for example, a capacitor has frequently been called a "condenser," but it does not condense anything. { 'dep·rə,kād·əd 'yü·sij }

depreciation [IND ENG] Loss of value due to physical deterioration. { di,prē·shē'ā·shən }

depressed center car [ENG] A flat railroad car having a low center section; used to provide adequate tunnel clearance for oversized loads. { di¦prest 'sent·ər ,kär }

depressed equation [MATH] An equation that results from reducing the number of roots in a given equation with one unknown by dividing the original equation by the difference of the unknown and a root. { di'prest i'kwā·zhən }

depressed fracture [MED] A fracture of the skull in which the fractured part is depressed below the normal level. { di'prest 'frak·chər }

depression [GEOL] **1.** A hollow of any size on a plain surface having no natural outlet for surface drainage. **2.** A structurally low area in the crust of the earth. [METEOROL] An area of low pressure; usually applied to a certain stage in the development of a tropical cyclone, to migratory lows and troughs, and to upper-level lows and troughs that are only weakly developed. Also known as low. [PSYCH] A mood provoked by conscious awareness of an idea or feeling that was previously pushed into the unconscious. { di'presh·ən }

depression angle *See* angle of depression. { di'presh·ən ,aŋ·gəl }

depression spring [HYD] A type of gravity spring that flows onto the land surface because the surface slopes down to the water table. { di'presh·ən 'spriŋ }

depression storage [HYD] Water retained in puddles, ditches, and other depressions in the surface of the ground. { di'presh·ən ,stòr·ij }

depressive neurosis *See* dysthymia. { di¦pres·iv nù'rō·səs }

depressor [ANAT] A muscle that draws a part down. [CHEM ENG] An agent that prevents or retards a chemical reaction or process. { di'pres·ər }

depressor nerve [NEUROSCI] A nerve which, upon stimulation, lowers the blood pressure either in a local part or throughout the body. { di'pres·ər ,nərv }

depropanization [CHEM ENG] In processing of petroleum, the removal of propane and sometimes higher fractions. { dē,prō·pə·nə'zā·shən }

depropanized material [MATER] Material that has undergone distillation to remove lighter components from butanes and heavier material. { dē'prō·pə,nīzd mə'tir·ē·əl }

depropanizer [CHEM ENG] A fractionating column in a gasoline plant for removal of propane and lighter components. { dē'prō·pə‚nīz·ər }

deproteinize [ORG CHEM] To remove protein from a substance. { dē'prō‚tē‚nīz }

depside [ORG CHEM] One of a class of esters that form from the joining of two or more molecules of phenolic carboxylic acid. { dep‚sīd }

depsidone [ORG CHEM] One of a class of compounds that consists of esters such as depsides, but are also cyclic ethers. { dep·sə‚dōn }

depth [NAV ARCH] The vertical distance amidships from the upper surface of the flat plate keel to the underside of the plating of a specified deck at ship's side. [OCEANOGR] The vertical distance from a specified sea level to the sea floor. { depth }

depth bomb [ORD] An explosive item designed to be dropped from an aircraft for use against underwater targets. { 'depth ‚bäm }

depth charge [ORD] A cylindrical or teardrop-shaped container holding a charge of TNT or other explosive, dropped from the deck of a ship, and detonated at a preset depth as an antisubmarine weapon. { 'depth ‚chärj }

depth contour See isobath. { 'depth ‚kän·túr }

depth curve See isobath. { 'depth ‚kərv }

depth dose [NUCLEO] The radiation dose delivered at a particular depth beneath the surface of a body; usually expressed as percent of surface dose or of air dose. { 'depth ‚dōs }

depth-duration-area value [METEOROL] The average depth of precipitation that has occurred within a specified time interval over an area of given size. Abbreviated DDA value. { ¦depth də¦rā·shən 'er·ē·ə ‚val·yü }

depth finder [ENG] A radar or ultrasonic instrument for measuring the depth of the sea. { 'depth ‚fīnd·ər }

depth gage [DES ENG] An instrument or tool for measuring the depth of depression to a thousandth inch. { 'depth ‚gāj }

depth hoar [HYD] A layer of ice crystals formed between the ground and snow cover by sublimation. Also known as sugar snow. { 'depth ‚hór }

depth magnification [OPTICS] The ratio of the distance between two nearby points of the axis on the image side of an optical system to the distance between their conjugate points on the object side. { 'depth ‚mag·nə·fə'kā·shən }

depth marker [ENG] A thin board or other lightweight substance used as a means of identifying the surface of snow or ice which has been covered by a more recent snowfall. { 'depth ‚märk·ər }

depth micrometer [DES ENG] A micrometer used to measure the depths of holes, slots, and distances of shoulders and projections. { 'depth mī'kräm·əd·ər }

depth of compensation [GEOPHYS] That depth at which density differences occurring in the earth's crust are compensated isostatically; calculated to be between 62 and 70–73 miles (100 and 113–117 kilometers). [HYD] The depth in a body of water at which illuminance has diminished to the extent that oxygen production through photosynthesis and oxygen consumption through respiration by plants are equal; it is the lower boundary of the euphotic zone. { 'depth əv ‚käm·pən'sā·shən }

depth of cut [MET] The thickness that is removed as a workpiece is being machined. { ¦depth əv 'kət }

depth of engagement [DES ENG] The depth of contact, in a radial direction, between mating threads. { 'depth əv ‚en'gāj·mənt }

depth of field [OPTICS] The range of distances over which a camera gives satisfactory definition, when its lens is in the best focus for a certain specific distance. { 'depth əv 'fēld }

depth of focus [OPTICS] The range of image distances corresponding to the range of object distances included in depth of field. { ¦depth əv 'fō·kəs }

depth of fusion [MET] The distance that fusion extends from the original surface into the base metal in a welding operation. { 'depth əv 'fyü·zhən }

depth of thread [DES ENG] The distance, in a radial direction, from the crest of a screw thread to the base. { 'depth əv 'thred }

depthometer [PETRO ENG] A device used to measure the depth of a well or the depth to a specific point in the well. { dep'thäm·əd·ər }

depth perception [PHYSIO] Ability to judge spatial relationships. { 'depth pər'sep·shən }

depth sounder [ENG] An instrument for mechanically measuring the depth of the sea beneath a ship. { 'depth ‚saùnd·ər }

depth-type filtration [CHEM ENG] Removal of solids by passing the carrier fluid through a mass-filter medium that provides a tortuous path with many entrapments to catch the solids. { 'depth ‚tīp fil'trā·shən }

depth zone [GEOL] A zone within the earth giving rise to different metamorphic assemblages. [OCEANOGR] Any one of four oceanic environments: the littoral, neritic, bathyal, and abyssal zones. { 'depth ‚zōn }

depurination [BIOCHEM] Detachment of guanine from sugar in a deoxyribonucleic acid molecule. { dē‚pyùr·ə'nā·shən }

de Quervain's disease [MED] Inflammation of tendons and their sheaths at the styloid process of the radius, often causing pain in the inner side of the wrist. { də'ker·vənz diz‚ēz }

dequeue [ENG] To select an item from a queue. { dē'kyü }

derail [ENG] **1.** To cause a railroad car or engine to run off the rails. **2.** A device to guide railway cars or engines off the tracks to avoid collision or other accident. { dē'rāl }

derangement [MATH] A permutation of a finite set of elements that carries no element of the set into itself. { di 'rānj·mənt }

derangement numbers [MATH] The numbers D_n, $n = 1, 2, 3, \ldots$, giving the number of permutations of a set of n elements that carry no element of the set into itself. { di'rānj·mənt ‚nəm·bərz }

derating [ELECTR] The reduction of the rating of a device to improve reliability or to permit operation at high ambient temperatures. { dē'rād·iŋ }

derby [MET] A large, usually cylindrical piece of primary metal, whose weight may exceed 100 pounds (45 kilograms), formed by bomb reduction. { 'dər·bē }

derbylite [MINERAL] $Fe_6Ti_6Sb_2O_{23}$ A black or brown orthorhombic mineral occurring in cinnabar-bearing gravels. { 'dər·bē‚līt }

Derbyshire spar See fluorite. { 'där·bə‚shir ‚spär }

derealization [PSYCH] Loss of the sense of the reality of people or objects in one's environment. { dē‚rē·ə·lə'zā·shən }

derecho See plow wind. { dā'rā·chō }

dereistic [PSYCH] Pertaining to mental activity that is not in accordance with reality, logic, or experience. { ‚dē·rē'ist·ik }

derelict [NAV] Any property abandoned at sea, often of sufficient size as to constitute a menace to navigation; especially an abandoned vessel. { 'der·ə‚likt }

derelict land [ECOL] Land that, because of mining, drilling, or other industrial processes, or by serious neglect, is unsightly and cannot be beneficially utilized without treatment. { 'der·ə‚likt ‚land }

derepression [MICROBIO] Transfer of microbial cells from an enzyme-repressing medium to a nonrepressing medium. [MOL BIO] Increased production of a gene product due to interference with the action of a repressor on the operator portion of the operon. { dē·ri'presh·ən }

derichment [ANALY CHEM] In gravimetric analysis by coprecipitation of salts, a system with λ less than unity, when λ is the logarithmic distribution coefficient expressed by the ratio of the logarithms of the ratios of the initial and final solution concentrations of the two salts. { dē'rich·mənt }

derivation [MATH] **1.** The process of deducing a formula. **2.** A function D on an algebra which satisfies the equation $D(uv) = uD(v) + vD(u)$. { ‚der·ə'vā·shən }

derivative [CHEM] A substance that is made from another substance. [MATH] The slope of a graph $y = f(x)$ at a given point c; more precisely, it is the limit as h approaches zero of $f(c + h) - f(c)$ divided by h. Also known as differential coefficient; rate of change. { də'riv·əd·iv }

derivative action [CONT SYS] Control action in which the speed at which a correction is made depends on how fast the system error is increasing. Also known as derivative compensation; rate action. { də'riv·əd·iv ‚ak·shən }

derivative compensation See derivative action. { də'riv·əd·iv ‚käm·pən'sā·shən }

derivative differential thermal analysis [ANALY CHEM] A method for precise determination in thermograms of slight temperature changes by taking the first derivative of the differential thermal analysis curve (thermogram) which plots time versus differential temperature as measured by a differential thermocouple. Also known as DDTA. { də'riv·əd·iv dif·ə'ren·chəl 'thər·məl ə'nal·ə·səs }

derivative network [CONT SYS] A compensating network whose output is proportional to the sum of the input signal and its derivative. Also known as lead network. { də'riv·əd·iv 'net,wərk }

derivative polarography [ANALY CHEM] Polarography technique in which the rate of change of current with respect to applied potential is measured as a function of the applied potential (di/dE versus E, where i is current and E is applied potential). { də'riv·əd·iv ,pō·lə'räg·rə·fē }

derivative rock *See* sedimentary rock. { də'riv·əd·iv 'räk }

derivative thermometric titration [ANALY CHEM] The use of a special resistance-capacitance network to record first and second derivatives of a thermometric titration curve (temperature versus weight change upon heating) to produce a sharp end-point peak. { də'riv·əd·iv thər·mə'me·trik tī'trā·shən }

derivatized gelatin [MATER] A form of gelatin in which the amino groups have been acylated; used for photographic and microencapsulation applications. { də'riv·ə'tīzd 'jel·ət·ən }

derived curve [MATH] A curve whose ordinate, for each value of the abscissa, is equal to the slope of some given curve. Also known as first derived curve. { də'rīvd 'kərv }

derived fuel [MATER] A fuel produced by some treatment of a raw fuel; for example, coke from coal or gasoline from petroleum. { di'rīvd 'fyül }

derived gust velocity [METEOROL] The maximum velocity of a sharp-edged gust that would produce a given acceleration on a particular airplane flown in level flight at the design cruising speed of the aircraft at a given air density. { də'rīvd 'gəst və,läs·əd·ē }

derived quantity [PHYS] A physical quantity which, in a specified system of measurement, is defined by operations based on other physical quantities. { də'rīvd 'kwän·əd·ē }

derived set [MATH] The set of cluster points of a given set. { də'rīvd 'set }

derived sound system [ENG ACOUS] A four-channel sound system that is artificially synthesized from conventional two-channel stereo sound by an adapter, to provide feeds to four loudspeakers for approximating quadraphonic sound. { də'rīvd 'saúnd ,sis·təm }

derived unit [PHYS] A unit that is formed, in a specified system of measurement, by combining base units and other derived units according to the algebraic relations linking the corresponding quantities. { də'rīvd 'yü·nət }

dermabrasion [MED] The surgical removal of scarred or tatooed skin by mechanical means such as rotating wire brushes or sandpaper. { ¦dər·mə¦brā·zhən }

Dermacentor [INV ZOO] A genus of ticks, important as vectors of disease. { 'dər·mə,sen·tər }

Dermacentor andersoni [INV ZOO] The wood tick, which is the vector of Rocky Mountain spotted fever and tularemia. { 'dər·mə,sen·tər an·dər'sō·nē }

Dermacentor variabilis [INV ZOO] A North American tick which is parasitic primarily on dogs but may attack humans and other mammals. { 'dər·mə,sen·tər ver·ē'ab·ə·ləs }

dermal [ANAT] Pertaining to the dermis. { 'dər·məl }

dermal bone [ANAT] A type of bone that ossifies directly from membrane without a cartilaginous predecessor; occurs only in the skull and shoulder region. Also known as investing bone; membrane bone. { ¦dər·məl 'bōn }

dermal denticle [VERT ZOO] A toothlike scale composed mostly of dentine with a large central pulp cavity, found in the skin of sharks. { ¦dər·məl 'dent·i·kəl }

dermalia [INV ZOO] Dermal microscleres in sponges. { dər'mal·yə }

dermal pore [INV ZOO] One of the minute openings on the surface of poriferans leading to the incurrent canals. { ¦dər·məl 'pōr }

Dermaptera [INV ZOO] An order of small or medium-sized, slender insects having incomplete metamorphosis, chewing mouthparts, short forewings, andcerci. { dər'map·tə·rə }

Dermatemydinae [VERT ZOO] A family of reptiles in the order Chelonia; includes the river turtles. { ,dər·mə·tə'mī·də,nē }

dermatitis [MED] Inflammation of the skin. { ,dər·mə'tīd·əs }

Dermatocarpaceae [BOT] A family of lichens in the order Pyrenulales having an umbilicate or squamulose growth form;

DERMESTIDAE

A skin beetle. *(From T. I. Storer and R. L. Usinger, General Zoology, 3d ed., McGraw-Hill, 1957)*

most members grow on limestone or calcareous soils. { dər¦mad·ō,kär'pās·ē,ē }

dermatocranium [ANAT] Bony parts of the skull derived from ossifications in the dermis of the skin. { dər¦mad·ə'krā·nē·əm }

dermatogen [BOT] The outer layer of primary meristem or the primordial epidermis in embryonic plants. Also known as protoderm. { dər'mad·ə,jen }

dermatoglyphics [ANAT] **1.** The integumentary patterns on the surface of the fingertips, palms, and soles. **2.** The study of these patterns. { dər¦mad·ə¦glif·iks }

dermatograph [MED] A crayonlike or similar instrument, used to mark the skin before surgery to outline positions of organs. { dər'mad·ə,graf }

dermatologist [MED] A physician who specializes in diseases of the skin. { ,dər·mə'täl·ə·jəst }

dermatology [MED] The science of the structure, function, and diseases of the skin. { ,dər·mə'täl·ə·jē }

dermatome [ANAT] An area of skin delimited by the supply of sensory fibers from a single spinal nerve. [EMBRYO] Lateral portion of an embryonic somite from which the dermis will develop. [MED] Instrument for cutting skin for grafting. { 'dər·mə,tōm }

dermatomyositis [MED] An inflammatory reaction of unknown cause involving degenerative changes of skin and muscle. { dər¦mad·ō,mī·ə'sīd·əs }

dermatopathic lymphadenitis *See* lipomelanotic reticulosis. { dər¦mad·ə¦path·ik ,lim¦fad·ən'īd·əs }

dermatopathology [MED] A branch of pathology concerned with diseases of the skin. { dər¦mad·ō·pə'thäl·ə·jē }

dermatopathophobia [PSYCH] An abnormal fear of contracting a skin disease. { dər¦mad·ō¦path·ə'fō·bē·ə }

Dermatophilaceae [MICROBIO] A family of bacteria in the order Actinomycetales; cells produce mycelial filaments or muriform thalli; includes human and mammalian pathogens. { dər¦mad·ō·fə'lās·ē,ē }

Dermatophilus [MICROBIO] A genus of bacteria in the family Dermatophilaceae; mycelial filaments are long, tapering, and branched, and are divided transversely and longitudinally (in two planes) by septa; spherical spores are motile. { dər¦mad·ə'fil·əs }

dermatophyte [MYCOL] A fungus parasitic on skin or its derivatives. { dər'mad·ə,fīt }

dermatophytosis *See* cutaneous mycosis. { dər'mad·ō,fī'tō·səs }

dermatoplast [BOT] In angiosperms, a cell with a cell wall. { dər'mad·ə,plast }

dermatosclerosis *See* scleroderma. { dər¦mad·ō·sklə'rō·səs }

dermatosis [MED] Any skin disease. { ,dər·mə'tō·səs }

Dermestidae [INV ZOO] The skin beetles, a family of coleopteran insects in the superfamily Dermestoidea, including serious pests of stored agricultural grain products. { dər'mes·tə·dē }

Dermestoidea [INV ZOO] A superfamily of coleopteran insects in the suborder Polyphaga. { ,dər·mə'stóid·ē·ə }

dermis [ANAT] The deep layer of the skin, a dense connective tissue richly supplied with blood vessels, nerves, and sensory organs. Also known as corium; cutis. { 'dər·məs }

Dermochelidae [VERT ZOO] A family of reptiles in the order Chelonia composed of a single species, the leatherback turtle. { ,dər·mə'kel·ə,dē }

dermoepidermal junction [HISTOL] The area of separation between the stratum basale of the epidermis and the papillary layer of the dermis. { ¦dər·mō,ep·ə¦dər·məl 'jəŋk·shən }

dermographia [MED] A condition in which the skin is peculiarly susceptible to irritation, characterized by elevations or wheals with surrounding erythematous axon reflex flare, caused by tracing a fingernail or a blunt instrument over the skin. { 'dər·mə'graf·ē·ə }

dermoid cyst [MED] A benign cystic teratoma with skin, skin appendages, and their products as the most prominent components, usually involving the ovary or the skin. { ¦dər,mȯid 'sist }

dermometer [PHYSIO] An instrument used to measure the electrical resistance of the skin. { dər'mäm·əd·ər }

dermonecrotic [MED] Pertaining to or causing necrosis of the skin. { ,dər·mō·nə'kräd·ik }

Dermoptera [VERT ZOO] The flying lemurs, an ancient order of primatelike herbivorous and frugivorous gliding mammals confined to southeastern Asia and eastern India. { dər'map·tə·rə }

Derodontidae [INV ZOO] The tooth-necked fungus beetles, a small family of coleopteran insects in the superfamily Dermestoidea. { ˌder·ə'dän·tə,dē }

derogatory matrix [MATH] A matrix whose order is greater than the order of its reduced characteristic equation. { də'räg·ə,tȯr·ē 'mā·triks }

derosination [CHEM ENG] Removing excess resins from wood by saponification with alkaline aqueous solutions or organic solvents. { dē,räz·ən'ā·shən }

derrick [MECH ENG] A hoisting machine consisting usually of a vertical mast, a slanted boom, and associated tackle; may be operated mechanically or by hand. { 'der·ik }

derrick barge [PETRO ENG] A crane barge used in offshore drilling platform construction and suitable for work in rough seas. { 'der·ik ,bärj }

derrick crane See stiffleg derrick. { 'der·ik ,krān }

derrick post See king post. { 'der·ik ,pōst }

derris [BOT] Any of certain tropical shrubs in the genus *Derris* in the legume family (Leguminosae), having long climbing branches. { 'der·əs }

Derxia [MICROBIO] A genus of bacteria in the family Azotobacteraceae; rod-shaped, pleomorphic, motile cells; older cells contain large refractive bodies. { 'dərk·sē·ə }

Deryagin number [PHYS] A dimensionless group equal to the ratio of the thickness of a film coating a liquid to the capillary length of the liquid. Symbolized *De*. { ˌder·ē'ag·ən ,nəm·bər }

DES See data encryption standard; diethylstilbesterol.

desalination [CHEM ENG] Removal of salt, as from water or soil. Also known as desalting. { dē,sal·ə'nā·shən }

desalinization See desalination. { dē,sal·ə·nə'zā·shən }

desalting [CHEM ENG] **1.** The process of extracting inorganic salts from oil. **2.** See desalination. { dē'sȯl·tiŋ }

desander [ENG] A centrifuge-type device for removing sand from drilling fluid in order to prevent abrasion damage to pumps. { dē'san·dər }

Desarguesian plane [MATH] Any projective plane in which points and lines satisfy Desargues' theorem. Also known as Arguesian plane. { dā·zär,gā·zē·ən 'plān }

Desargues' theorem [MATH] If the three lines passing through corresponding vertices of two triangles are concurrent, then the intersections of the three pairs of corresponding sides lie on a straight line, and conversely. { dā'zärgz ,thir·əm }

desaturated color [OPTICS] A color that is neither a pure spectral color nor a purple formed from a mixture of deep red and violet. { dē,sach·ə,rād·əd 'kəl·ər }

DeSauty's bridge [ELEC] A four-arm bridge used to compare two capacitances; two adjacent arms contain capacitors in series with resistors, while the other two arms contain resistors only. Also known as Wien-DeSauty bridge. { də'sȯd·ēz ,brij }

descaling [ENG] Removing scale, usually oxides, from the surface of a metal or the inner surface of a pipe, boiler, or other object. { dē'skāl·iŋ }

Descartes laws of refraction See Snell laws of refraction. { dā'kärt 'lȯz əv ri'frak·shən }

Descartes ray [OPTICS] A ray of light incident on a sphere of transparent material, such as a water droplet, which after one internal reflection leaves the drop at the smallest possible angle of deviation from the direction of the incident ray; these rays make the primary rainbow. { dā'kärt ,rā }

Descartes' rule of signs [MATH] A polynomial with real coefficients has at most k real positive roots, where k is the number of sign changes in the polynomial. { dā'kärts 'rül əv 'sīnz }

Descemet's membrane [HISTOL] A layer of the cornea between the posterior surface of the stroma and the anterior surface of the endothelium which contains collagen arranged on a crystalline lattice. { des'māz ,mem,brān }

descendant [GEOL] A topographic feature that is formed from the mass beneath an older topographic form, now removed. { di'sen·dənt }

descender [GRAPHICS] The segment of certain letters which extends below the general level of cross-alignment with other letters; for example, the lower part of "g" or "y." { di'sen·dər }

descending [ANAT] Extending or directed downward or caudally, as the descending aorta. [NEUROSCI] In the nervous system, efferent; conducting impulses or progressing down the spinal cord or from central to peripheral. { di'sen·diŋ }

descending branch [MECH] That portion of a trajectory which is between the summit and the point where the trajectory terminates, either by impact or air burst, and along which the projectile falls, with altitude constantly decreasing. Also known as descent trajectory. { di'sen·diŋ 'branch }

descending chain condition [MATH] The condition on a ring that every descending sequence of left ideals (or right ideals) has only a finite number of distinct members. { di'send·iŋ 'chān kən,dish·ən }

descending chromatography [ANALY CHEM] A type of paper chromatography in which the sample-carrying solvent mixture is fed to the top of the developing chamber, being separated as it works downward. { di'sen·diŋ krō·mə'täg·rə·fē }

descending colon [ANAT] The portion of the colon on the left side, extending from the bend below the spleen to the sigmoid flexure. { di'send·iŋ 'kōl·ən }

descending node [AERO ENG] That point at which an earth satellite crosses to the south side of the equatorial plane of its primary. Also known as southbound node. [ASTRON] The point at which a planet, planetoid, or comet crosses the ecliptic from north to south. { di'sen·diŋ 'nōd }

descending sort [COMPUT SCI] The arranging of data records from high to low sequence (9 to 0, and Z to A). { di'send·iŋ 'sȯrt }

descending sequence [MATH] **1.** A sequence of elements in a partially ordered set such that each member of the sequence is equal to or less than the preceding one. **2.** In particular, a sequence of sets such that each member of the sequence is a subset of the preceding one. { di'send·iŋ 'sē·kwəns }

descending vertical angle See angle of depression. { di'sen·diŋ ,vərd·i·kəl 'aŋ·gəl }

descent [AERO ENG] Motion of a craft in which the path is inclined with respect to the horizontal. { di'sent }

descent trajectory See descending branch. { di'sent trə'jek·tə·rē }

describing function [CONT SYS] A function used to represent a nonlinear transfer function by an approximately equivalent linear transfer function; it is the ratio of the phasor representing the fundamental component of the output of the nonlinearity, determined by Fourier analysis, to the phasor representing a sinusoidal input signal. { di'skrīb·iŋ ,fəŋk·shən }

descriptive anatomy [ANAT] Study of the separate and individual portions of the body, with regard to form, size, character, and position. { di'skrip·tiv ə'nad·ə·mē }

descriptive astronomy [ASTRON] Astronomy as presented by graphic and verbal description. { di'skrip·tiv ə'strän·ə·mē }

descriptive botany [BOT] The branch of botany that deals with diagnostic characters or systematic description of plants. { di'skrip·tiv 'bät·ən·ē }

descriptive climatology [CLIMATOL] Climatology as presented by graphic and verbal description, without going into causes and theory. { di'skrip·tiv klī·mə'täl·ə·jē }

descriptive geometry [MATH] The application of graphical methods to the solution of three-dimensional space problems. { di'skrip·tiv jē'äm·ə·trē }

descriptive meteorology [METEOROL] A branch of meteorology which deals with the description of the atmosphere as a whole and its various phenomena, without going into theory. Also known as aerography. { di'skrip·tiv mēd·ē·ə'räl·ə·jē }

descriptive psychiatry [PSYCH] A system of psychiatry based on the study of readily observable external factors. { di'skrip·tiv sī'kī·ə·trē }

descriptive statistics [STAT] Presentation of data in the form of tables and charts or summarization by means of percentiles and standard deviations. { di'skrip·tiv stə'tis·tiks }

descriptor [COMPUT SCI] A word or phrase used to identify a document in a computer-based information storage and retrieval system. { di'skrip·tər }

Desdemona [ASTRON] A satellite of Uranus orbiting at a mean distance of 38,935 miles (62,660 kilometers) with a period of 11 hours 24 minutes, and with a diameter of about 36 miles (58 kilometers). { ˌdez·də'mōn·ə }

deseaming [MET] Removing defects from the surfaces of ingots, blooms, or semifinished products, usually by means of a chipping hammer or an oxy-gas flame. { dē'sēm·iŋ }

desensitization [COMMUN] Reduction in receiver sensitivity due to the presence of a high-level off-channel signal overloading the radio-frequency amplifier or mixer stages, or causing automatic gain control action. [IMMUNOL] Loss or reduction of sensitivity to infection or an allergen accomplished by means of frequent, small doses of the antigen. Also known as hyposensitization. [PSYCH] Relief from or removal of a mental complex. { dē,sen·sə·tə'zā·shən }

desensitizer [GRAPHICS] **1.** A substance whose main ingredient is a gum and is used for rendering the nonimage portions of a lithographic plate nonreceptive to ink. **2.** A chemical used in photography to decrease the color sensitivity of a photographic emulsion. { dē'sen·sə,tīz·ər }

deserialize [COMMUN] To convert a data stream from a serial stream of bits to parallel streams of bits. { dē'sir·ē·ə,līz }

desert [GEOGR] **1.** A wide, open, comparatively barren tract of land with few forms of life and little rainfall. **2.** Any waste, uninhabited tract, such as the vast expanse of ice in Greenland. { 'dez·ərt }

desert climate [CLIMATOL] A climate type which is characterized by insufficient moisture to support appreciable plant life; that is, a climate of extreme aridity. { ¦dez·ərt ¦klī·mət }

desert crust See desert pavement. { ¦dez·ərt ¦krəst }

desert devil See dust whirl. { 'dez·ərt ,dev·əl }

deserticolous [ECOL] Living in a desert. { ¦dez·ər¦tik·ə·ləs }

desertification [ECOL] The creation of desiccated, barren, desertlike conditions due to natural changes in climate or possibly through mismanagement of the semiarid zone. { də,zərd·ə·fə'kā·shən }

desert pavement [GEOL] A mosaic of pebbles and large stones which accumulate as the finer dust and sand particles are blown away by the wind. Also known as desert crust. { ¦dez·ərt 'pāv·mənt }

desert peneplain See pediplain. { ¦dez·ərt 'pen·ə,plān }

desert plain See pediplain. { ¦dez·ərt 'plān }

desert polish [GEOL] A smooth, shining surface imparted to rocks and other hard substances by the action of windblown sand and dust of desert regions. { ¦dez·ərt 'päl·ish }

desert soil [GEOL] In early United States classification systems, a group of zonal soils that have a light-colored surface soil underlain by calcareous material and a hardpan. { ¦dez·ərt 'sȯil }

desert varnish See rock varnish. { ¦dez·ərt 'vär·nish }

desert wind [METEOROL] A wind blowing off the desert, which is very dry and usually dusty, hot in summer but cold in winter, and with a large diurnal range of temperature. { ¦dez·ərt 'wind }

desexualization [PHYSIO] Depriving an organism of sexual characters or power, as by spaying or castration. [PSYCH] Repression of the sexual drive or rechanneling of sexual energy into areas considered by the individual to be more socially acceptable. { dē,seksh·ə·lə'zā·shən }

desiccant See drying agent. { 'des·i·kənt }

desiccation [HYD] The permanent decrease or disappearance of water from a region, caused by a decrease of rainfall, a failure to maintain irrigation, or deforestation or overcropping. [SCI TECH] Thorough removal of water from a substance, often with the use of a desiccant. { ,des·ə'kā·shən }

desiccation breccia [GEOL] Fragments of a mud-cracked layer of sediment deposited with other sediments. { ,des·ə'kā·shən ,brech·ə }

desiccation crack See mud crack. { ,des·ə'kā·shən ,krak }

desiccator [CHEM ENG] A closed vessel, usually made of glass and having an airtight lid, used for drying solid chemicals by means of a desiccant. { 'des·ə,kād·ər }

design [SCI TECH] The act of conceiving and planning the structure and parameter values of a system, device, process, or work of art. { di'zīn }

designated volume [ANALY CHEM] The volume of an item of volumetric glassware as calibrated at a given temperature, frequently 20°C (68°F). { ¦dez·ig,nād·əd 'väl·yəm }

designation [COMPUT SCI] An item of data forming part of a computer record that indicates the type of record and thus determines how it is to be processed. { ,dez·əg'nā·shən }

design climatology [CLIMATOL] The scientific analysis of climatic data for the purpose of improving the design of equipment and structures intended to operate in or withstand extremes of climate. { di'zīn klī·mə'täl·ə·jē }

design engineering [ENG] A branch of engineering concerned with the creation of systems, devices, and processes useful to and sought by society. { di'zīn ,en·jə'nir·iŋ }

design factor [ENG] A safety factor based on the ratio of ultimate load to maximum permissible load that can be safely placed on a structure. { di'zīn ,fak·tər }

design feature [ECOL] An organismal trait that can influence rates of death and reproduction, and hence Darwinian fitness. { di'zīn ,fē·chər }

design flood [CIV ENG] The flood, either observed or synthetic, which is chosen as the basis for the design of a hydraulic structure. { di'zīn ,fləd }

design for environment [SYS ENG] A methodology for the design of products and systems that promotes pollution prevention and resource conservation by including within the design process the systematic consideration of the environmental implications of engineering designs. Abbreviated DFE. { di¦zīn fər in'vī·ərn·mənt }

design gross weight [AERO ENG] The gross weight at take-off that an aircraft, rocket, or such is expected to have, used in design calculations. { di'zīn ¦grōs 'wāt }

design head [CIV ENG] The planned elevation between the free level of a water supply and the point of free discharge or the level of free discharge surface. { di'zīn ,hed }

design heating load [ENG] The space heating needs of a building or an enclosed area expressed in terms of the probable maximum requirement. { di'zīn ,hēd·iŋ ,lōd }

design load [DES ENG] The most stressful combination of weight or other forces a building, structure, or mechanical system or device is designed to sustain. { di'zīn ,lōd }

design-oriented system [COMPUT SCI] A computer system developed primarily to maximize performance of hardware and software, rather than ease of use. { di'zīn ¦ȯr·ē,ent·əd ,sis·təm }

design pressure [CIV ENG] **1.** The force exerted by a body of still water on a dam. **2.** The pressure which the dam can withstand. [DES ENG] The pressure used in the calculation of minimum thickness or design characteristics of a boiler or pressure vessel in recognized code formulas; static head may be added where appropriate for specific parts of the structure. { di'zīn 'presh·ər }

design speed [CIV ENG] The highest continuous safe vehicular speed as governed by the design features of a highway. { di'zīn ,spēd }

design standards [DES ENG] Generally accepted uniform procedures, dimensions, materials, or parts that directly affect the design of a product or facility. { di'zīn ,stan·dərdz }

design storm [CIV ENG] A storm whose magnitude, rate, and intensity do not exceed the design load for a storm drainage system or flood protection project. { di'zīn ,stȯrm }

design stress [DES ENG] A permissible maximum stress to which a machine part or structural member may be subjected, which is large enough to prevent failure in case the loads exceed expected values, or other uncertainties turn out unfavorably. { di'zīn ,stres }

design thickness [DES ENG] The sum of required thickness and corrosion allowance utilized for individual parts of a boiler or pressure vessel. { di'zīn ,thik·nəs }

design water depth [OCEANOGR] **1.** A value based on the sum of the vertical distance from the nominal water level to the ocean bottom and the height of the tides, both astronomical and storm. **2.** The greatest water depth in which an offshore drilling well is able to maintain its operations. { di'zīn 'wȯd·ər ,depth }

design waterline [NAV ARCH] The waterline on a ship when it is floating freely at rest in still water in its normally loaded condition. Abbreviated DWL. { di¦zīn 'wȯd·ər,līn }

desilication [GEOCHEM] Removal of silica, as from rock or a magma. { dē,sil·ə'kā·shən }

desilker [FOOD ENG] A machine consisting of a series of revolving rolls and brushes for removing silk from ears of corn. { dē'silk·ər }

desilter [MECH ENG] Wet, mechanical solids classifier (separator) in which silt particles settle as the carrier liquid is slowly stirred by horizontally revolving rakes; solids are plowed outward and removed at the periphery of the container bowl. { de'sil·tər }

desilting basin [CIV ENG] A space or structure constructed just below a diversion structure of a canal to remove bed, sand, and silt loads. Also known as desilting works. { de'sil·tiŋ ‚ba·sən }

desilting works See desilting basin. { de'sil·tiŋ ‚wərks }

desilverization [MET] The act or process of removing silver; specifically, the process used to remove silver and gold from ore after softening. { de‚sil·vər·ə'za·shən }

desired ground zero [ORD] For a surface burst, the point on the earth's surface where atomic detonation is desired; for an air burst or underground burst, the point on the earth's surface directly below or directly above the desired point of detonation. { də'zīrd ¦graund 'zir·o }

desired track See course. { də'zīrd 'trak }

de Sitter space [RELAT] A constant-curvature, vacuum solution to Einstein's equations of general relativity with cosmological term. { də'sid·ər ‚spas }

desize [TEXT] To remove size or sizing agents from warp yarns prior to weaving to protect them against the abrasive action of loom parts. { de'sīz }

desk calculator [COMPUT SCI] A device that is used to perform arithmetic operations and is small enough to be conveniently placed on a desk. { ¦desk 'kal·kyə‚lād·ər }

desk check See dry run. { 'desk ‚chek }

desktop [COMPUT SCI] In a graphical user interface, a screen on which frequently used software resources are represented by icons. { 'desk‚täp }

desktop accessory software [COMPUT SCI] A set of computer programs providing functions that simulate the office accessories normally found on a desktop, such as a notepad, appointment calendar, and calculator. Also known as desktop application; desktop organizer. { 'desk‚täp ik¦ses·ə·rē 'sof‚wer }

desktop application See desktop accessory software. { ¦desk‚täp ‚ap·lə'ka·shən }

desktop organizer See desktop accessory software. { ¦desk‚täp 'ör·gə‚nīz·ər }

desktop publishing [COMPUT SCI] The use of a personal computer to produce printed output of high quality that is camera-ready for a printing facility. { 'desk‚täp 'pəb·lish·iŋ }

deslimer [MECH ENG] Apparatus, such as a bowl-type centrifuge, used to remove fine, wet particles (slime) from cement rocks and to size pigments and abrasives. { de'slīm·ər }

desma [INV ZOO] A branched, knobby spicule in some Demospongiae. { 'dez·mə }

desmacyte [INV ZOO] A bipolar collencyte found in the cortex of certain sponges. { 'dez·mə‚sīt }

Desmanthos [MICROBIO] A genus of bacteria in the family Pelonemataceae; unbranched, relatively straight filaments with a thickened base which are arranged in bundles partially enclosed in a sheath. { dez'man·thós }

desmetryn [ORG CHEM] $C_9H_{17}N_5S$ A white, crystalline compound with a melting point of 84–86°C; used as a postemergence herbicide for broadleaf and grassy weeds. { dez'me·trən }

desmid [BOT] Any member of a group of microscopic, unicellular green algae of the family Desmidiaceae, having cells of varying shapes but always composed of mirror-image semicells, often demarcated by a median constriction or incision, and a cell wall that has pores, which are frequently ornamented. { 'dez·məd }

Desmidiaceae [BOT] A family of desmids, mostly unicellular algae in the order Conjugales. { dez‚mid·e'as·e‚e }

desmin [BIOCHEM] The muscle protein forming the Z lines in striated muscle. { 'dez·mən }

desmine See stilbite. { 'dez‚men }

desmochore [ECOL] A plant having sticky or barbed disseminules. { 'dez·mə‚kór }

Desmodonta [PALEON] An order of extinct bivalve, burrowing mollusks. { ‚dez·mə'dän·tə }

Desmodontidae [VERT ZOO] A small family of chiropteran mammals comprising the true vampire bats. { ‚dez·mə'dän·tə‚de }

Desmodoroidea [INV ZOO] A superfamily of marine-and brackish-water-inhabiting nematodes with an annulated, usually smooth cuticle. { ‚dez·mə·də'róid·e‚ə }

Des Moinesian [GEOL] Lower Middle Pennsylvanian geologic time. { də'móin·e‚ən }

Desmokontae [BOT] The equivalent name for Desmophyceae. { ‚dez·mə'kän·te }

desmolase [BIOCHEM] Any of a group of enzymes which catalyze rupture of atomic linkages that are not cleaved through hydrolysis, such as the bonds in the carbon chain of D-glucose. { 'dez·mə‚las }

desmoneme [INV ZOO] A nematocyst having a long coiled tube which is extruded and wrapped around the prey. { ‚dez·mə'nem }

desmopelmous [VERT ZOO] A type of bird foot in which the hindtoe cannot be bent independently because planter tendons are united. { ‚dez·mə'pel·mas }

Desmophyceae [BOT] A class of rare, mostly marine algae in the division Pyrrhophyta. { ‚dez·mə'fis·e‚e }

desmoplasia [MED] **1.** The formation and proliferation of connective tissue, frequently in the growth of tumors. **2.** The formation of adhesions. { ‚dez·mə'pla·zhə }

Desmoscolecida [INV ZOO] An order of the class Nematoda. { ‚dez·mə·skə'les·ə·də }

Desmoscolecidae [INV ZOO] A family of nematodes in the superfamily Desmoscolecoidea; individuals resemble annelids in having coarseannulation. { ‚dez·mə·skə'les·ə‚de }

Desmoscolecoidea [INV ZOO] A small superfamily of free-living nematodes characterized by a ringed body, an armored head set, and hemispherical amphids. { ‚dez·mə‚sko·lə'kóid·e·ə }

desmose [INV ZOO] A fibril connecting the centrioles during mitosis in certain protozoans. { 'dez‚mos }

desmosome See adhering junction. { 'dez·mə‚som }

Desmostylia [PALEON] An extinct order of large hippopotamuslike, amphibious, gravigrade, shellfish-eating mammals. { ‚dez·mə'stil·e·ə }

Desmostylidae [PALEON] A family of extinct mammals in the order Desmostylia. { ‚dez·mə'stil·ə‚de }

Desmothoracida [INV ZOO] An order of sessile and free-living protozoans in the subclass Heliozoia having a spherical body with a perforate, chitinous test. { ‚dez·mə·thə'ras·ə·də }

desorption [PHYS CHEM] The process of removing a sorbed substance by the reverse of adsorption or absorption. { de'sórp·shən }

Desor's larva [INV ZOO] An oval, ciliated larva of certain nemertineans in which the gastrula remains inside the egg membrane. { də'zórz ‚lär·va }

despin [AERO ENG] To stop or reduce the rotation of a spacecraft or one of its components. { ‚de'spin }

Despoina [ASTRON] A satellite of Neptune orbiting at a mean distance of 32,500 miles (52,500 kilometers) with a period of 8.0 hours, and with a diameter of about 110 miles (180 kilometers). { des'póin·ə }

despooler [COMPUT SCI] Software that reads computer output information from a buffer and routes it to a printer. { de'spül·ər }

despun antenna [ELECTROMAG] Satellite directional antenna pointed continuously at earth by electrically or mechanically despinning the antenna at the same rate that the satellite is spinning for stabilization. { de'spən an'ten·ə }

desquamation [PHYSIO] Shedding; a peeling and casting off, as of the superficial epithelium, mucous membranes, renal tubules, and the skin. { de·skwə'ma·shən }

destearinate [CHEM ENG] A process of removing from a fatty oil the lower melting point compounds. { de'stir·ə‚nat }

destination [COMPUT SCI] The location (record, file, document, program, device, or disk) to which information is moved or copied. { ‚des·tə'na·shən }

destination address [COMPUT SCI] The location to which a jump instruction passes control in a program. { ‚des·tə'na·shən ə'dres }

destination time [COMPUT SCI] The time involved in a memory access plus the time required for indirect addressing. { ‚des·tə'na·shən ‚tim }

destraction [CHEM ENG] A high-pressure technique for separating high-boiling or nonvolatile material by dissolving it with application of supercritical gases. { di'strak·shən }

destressing [MIN ENG] Relieving stress on the abutments of an excavation by drilling and blasting to loosen peak stress zones. { de'stres·iŋ }

DESMOSTYLIA

Restoration of *Desmostylus* from the Miocene of California.

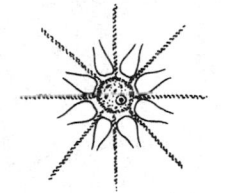

DESMOTHORACIDA

Choanocystis lepidula, a floating species. (*From R. P. Hall, Protozoology, Prentice-Hall, 1953*)

Destriau effect [SOLID STATE] Sustained emission of light by suitable phosphor powders that are embedded in an insulator and subjected only to the action of an alternating electric field. { 'des·trē·aủ i¸fekt }

destroyer [NAV ARCH] A small, fast, lightly armored warship capable of a variety of functions, usually armed with 5-inch (127-millimeter) guns, torpedoes, depth charges, and mines, and sometimes with guided missiles. { di'strói·ər }

destruct [AERO ENG] The deliberate action of destroying a rocket vehicle after it has been launched, but before it has completed its course. { di'strəkt }

destruction operator See annihilation operator. { di'strək·shən 'äp·ə¸rād·ər }

destructive breakdown [ELECTR] Breakdown of the barrier between the gate and channel of a field-effect transistor, causing failure of the transistor. { di'strək·tiv 'brāk¸daủn }

destructive distillation [ORG CHEM] Decomposition of organic compounds by heat without the presence of air. { di'strək·tiv dis·tə'lā·shən }

destructive interference [OPTICS] The interaction of superimposed light from two different sources when the phase relationship is such as to reduce or cancel the resultant intensity to less than the sum of the individual lights. { di'strək·tiv ¸in·tər'fir·əns }

destructive memory See destructive readout memory. { di¦strək·tiv 'mem¸rē }

destructive read [COMPUT SCI] Reading that partially or completely erases the stored information as it is being read. { di'strək·tiv 'rēd }

destructive readout memory [COMPUT SCI] A memory type in which reading the contents of a storage location destroys the contents of that location. Also known as destructive memory. { di'strək·tiv 'rēd¸aủt ¸mem·rē }

destructive testing [ENG] **1.** Intentional operation of equipment until it fails, to reveal design weaknesses. **2.** A method of testing a material that degrades the sample under investigation. { di'strək·tiv 'test·iŋ }

destruct line [AERO ENG] On a rocket test range, a boundary line on each side of the down-range course beyond which a rocket cannot fly without being destroyed under destruct procedures; or a line beyond which the impact point cannot pass. { di'strəkt ¸līn }

destructor [ORD] An explosive device used intentionally by the control center to destroy a missile after launching. Also known as destruct system. { di'strək·tər }

destruct system [ORD] **1.** An explosive device enabling destruction of materials, weapons, or other objects to prevent acquisition by the enemy. **2.** See destructor. { di'strəkt ¸sis·təm }

desulfonation [ORG CHEM] Removal of the sulfonate group from an organic molecule. { dē¸səl·fə'nā·shən }

Desulfotomaculum [MICROBIO] A genus of bacteria in the family Bacillaceae; motile, straight or curved rods with terminal to subterminal spores; anaerobic and reduce sulfate. { dē¸səl·fəd·ə'mak·yə·ləm }

Desulfovibrio [MICROBIO] A genus of gram-negative, strictly anaerobic bacteria of uncertain affiliation; motile, curved rods reduce sulfates and other sulfur compounds to hydrogen sulfide. { dē¸səl·fə'vib·rē¸ō }

desulfurization [CHEM ENG] The removal of sulfur, as from molten metals or petroleum oil. { dē¸səl·fə·rə'zā·shən }

desulfurization unit [CHEM ENG] A unit in petroleum refining for removal of sulfur compounds or sulfur. { dē·səl·fə· rə'zā·shən ¸yü·nət }

desyl [ORG CHEM] The functional group C_6H_5CO-$CH(C_6H_5)$-; may be formed from desoxybenzoin. { 'des·əl }

DET See diethyltoluamide.

DETAB [COMPUT SCI] A programming language based on COBOL in which problems can be specified in the form of decision tables. Acronym for decision table. { 'dē¸tab }

detachable bit [ENG] An all-steel drill bit that can be removed from the drill steel, and can be resharpened. Also known as knock-off bit; rip bit. { di'tach·ə·bəl 'bit }

detachable plugboard [COMPUT SCI] A control panel that can be removed from the computer or other system and exchanged for another without altering the positions of the plugs and cords. Also known as removable plugboard. { di'tach·ə·bəl 'pləg¸bȯrd }

detached binary [ASTRON] A binary system in which the components do not fill their Roche lobes and have little tidal distortion, and in which significant mass exchange is not taking place. { di'tacht 'bī¸ner·ē }

detached core [GEOL] The inner bed or beds of a fold that may become separated or pinched off from the main body of the strata due to extreme folding and compression. { di'tacht 'kȯr }

detached-lever escapement [HOROL] A watch escapement whose regulating device is given an impulse during only a small part of its operating cycle. { di¦tacht ¦lev·ər ə¸skāp·mənt }

detached meristem [BOT] A meristematic region originating from apical meristem but becoming discontinuous with it because of differentiation of intervening tissue. { di'tacht 'mer·ə¸stem }

detached shock wave [FL MECH] A shock wave not in contact with the body which originates it. { di'tacht 'shäk ¸wāv }

detail [GRAPHICS] The extent to which image elements that are close together can be individually distinguished. { 'dē¸tāl }

detail chart [COMPUT SCI] A flow chart representing every single step of a program. { 'dē¸tāl ¸chärt }

detail drawing [GRAPHICS] A large-scale drawing of a small part of a structure or machine. { 'dē¸tāl ¸drȯ·iŋ }

detailed balance [STAT MECH] The hypothesis that when a system is in equilibrium any process occurs with the same frequency as the reverse process. { də'tāld 'bal·əns }

detail file [COMPUT SCI] A file containing current or transient data used to update a master file or processed with the master file to obtain a specific result. Also known as transaction file. { 'dē¸tāl ¸fīl }

detailing See screening. { 'dē¸tāl·iŋ }

detail printing [COMPUT SCI] The printing of information for each card as the card passes through the machine; the function is used to prepare reports that show complete detail about each card; during this listing operation, the machine adds, subtracts, cross-adds, or cross-subtracts, and prints many combinations of totals. { 'dē¸tāl ¸print·iŋ }

det drill See fusion-piercing drill. { 'det ¸dril }

detect See demodulate. { di'tekt }

detection [COMMUN] The recovery of information from an electrical or electromagnetic signal. { di'tek·shən }

detection limit [ANALY CHEM] In chemical analysis, the minimum amount of a particular component that can be determined by a single measurement with a stated confidence level. { di'tek·shən ¸lim·ət }

detectivity [ELECTR] The normalized radiation power required to give a signal from a photoconductor that is equal to the noise. { ¸dē¸tek'tiv·əd·ē }

detector [ELECTR] The stage in a receiver at which demodulation takes place; in a superheterodyne receiver this is called the second detector. Also known as demodulator; envelope detector. [SCI TECH] Apparatus or system used to detect the presence of an object, radiation, chemical compound, or such. { di'tek·tər }

detector balanced bias [ELECTR] Controlling circuit used in radar systems for anticlutter purposes. { di'tek·tər ¦bal·ənst 'bī·əs }

detector bar [CIV ENG] A device that keeps a railroad switch locked while a train is passing over it. { di'tek·tər ¸bär }

detector car [ENG] A railroad car used to detect flaws in rails. { di'tek·tər ¸kär }

detent [MECH ENG] A catch or lever in a mechanism which initiates or locks movement of a part, especially in escapement mechanisms. { 'dē¸tent }

detention basin [CIV ENG] A reservoir without control gates for storing water over brief periods of time until the stream has the capacity for ordinary flow plus released water; used for flood regulation. { di'ten·chən ¸bā·sən }

detergent [MATER] A synthetic cleansing agent resembling soap in the ability to emulsify oil and hold dirt, and containing surfactants which do not precipitate in hard water; may also contain protease enzymes and whitening agents. { di'tər·jənt }

detergent additive [MATER] A substance incorporated in lubricating oils which gives them the property of keeping insoluble material in suspension. { di'tər·jənt ¸ad·ə·tiv }

detergent alkylate See dodecylbenzene. { di'tər·jənt 'al·kə¸lāt }

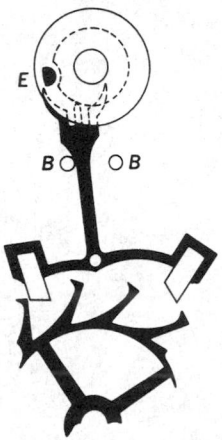

DETACHED-LEVER ESCAPEMENT

Detached-lever escapement, often used in modern watches. Banking pins *B* limit oscillation of anchor and lever; impulse pin *E* causes recoil of the escape wheel to release the pallet. *(From F. J. Britten, Britten's Old Clocks and Watches and Their Makers, 7th ed., Dutton, 1956)*

detergent oil [MATER] A lubricating oil with special sludge-dispersing properties for use in internal combustion engines. { di'tər·jənt ,oil }

deteriorating supplies [ORD] Those items that may reasonably be expected to become unusable within 1 or 2 years, whether used or not. { di'tir·ē·ə,rād·iŋ sə'plīz }

deterioration [ENG] Decline in the quality of equipment or structures over a period of time due to the chemical or physical action of the environment. { di,tir·ē·ə'rā·shən }

determinant [CONT SYS] The product of the partial return differences associated with the nodes of a signal-flow graph. [MATH] A certain real-valued function of the column vectors of a square matrix which is zero if and only if the matrix is singular; used to solve systems of linear equations and to study linear transformations. { də'tər·mə·nənt }

determinant tensor [MATH] A tensor whose components are each equal to the corresponding component of the Levi-Civita tensor density times the square root of the determinant of the metric tensor, and whose contravariant components are each equal to the corresponding component of the Levi-Civita density divided by the square root of the metric tensor. Also known as permutation tensor. { də'tər·mə·nənt 'ten·sər }

determinate [SCI TECH] Bounded by definite limits. { də'tər·mə·nət }

determinate cleavage [EMBRYO] A type of cleavage which separates portions of the zygote with specific and distinct potencies for development as specific parts of the body. { də'tər·mə·nət 'klē·vij }

determinate growth [BOT] Growth in which the axis, or central stem, being limited by the development of the floral reproductive structure, does not grow or lengthen indefinitely. { də'tər·mə·nət 'grōth }

determinate structure [MECH] A structure in which the equations of statics alone are sufficient to determine the stresses and reactions. { də'tər·mə·nət 'strək·chər }

determination [ANALY CHEM] The finding of the value of a chemical or physical property of a compound, such as reaction-rate determination or specific-gravity determination. { də,tər·mə'nā·shən }

determinism See causality. { də'tər·mə,niz·əm }

deterministic algorithm See static algorithm. { də,tər·mə'nis·tik 'al·gə,rith·əm }

deterministic equation [PHYS] An equation that governs the motion of a dynamical system and does not contain terms corresponding to random forces. { də,tər·mə'nis·tik i'kwā·zhən }

deterrence [COMPUT SCI] Making an attack on a computer sufficiently difficult to discourage potential attackers. { di'tər·əns }

detonating agent [MATER] An explosive, such as PETN, contained in the blasting cap or detonator. { 'det·ən,ād·iŋ ,ā·jənt }

detonating fuse [ENG] A device consisting of a core of high explosive within a waterproof textile covering and set off by an electrical blasting cap fired from a distance by means of a fuse line; used in large, deep boreholes. { 'det·ən,ād·iŋ 'fyüz }

detonating net [ORD] Network of detonating cord that is interlaced in a mesh design; used for clearing paths through mine fields by exploding the mines over which the nets are placed and detonated. { 'det·ən,ād·iŋ ,net }

detonating primer [MATER] A primer used to fire high explosives that is exploded by a fuse. { 'det·ən,ād·iŋ 'prīm·ər }

detonating rate [MECH] The velocity at which the explosion wave passes through a cylindrical charge. { 'det·ən,ād·iŋ ,rāt }

detonating relay [ENG] A device used in conjunction with the detonating fuse to avoid short-delay blasting. { 'det·ən,ād·iŋ ,rē,lā }

detonation [CHEM] An exothermic chemical reaction that propagates with such rapidity that the rate of advance of the reaction zone into the unreacted material exceeds the velocity of sound in the unreacted material; that is, the advancing reaction zone is preceded by a shock wave. [MECH ENG] Spontaneous combustion of the compressed charge after passage of the spark in an internal combustion engine; it is accompanied by knock. { ,det·ən'ā·shən }

detonation flame spraying [MET] A flame-spraying method in which the combined mixture of fuel gas, oxygen, and powdered coating liquefies and explodes the material to the workpiece. { ,det·ən'ā·shən 'flām ,sprā·iŋ }

detonation front [ENG] The reaction zone of a detonation. { ,det·ən'ā·shən ,front }

detonation wave [FL MECH] A shock wave that accompanies detonation and has a shock front followed by a region of decreasing pressure in which the reaction occurs. { ,det·ən'ā·shən ,wāv }

detonator [ENG] A device, such as a blasting cap, employing a sensitive primary explosive to detonate a high-explosive charge. { 'det·ən,ād·ər }

detonator safety [ENG] A fuse has detonator safety or is detonator safe when the functioning of the detonator cannot initiate subsequent explosive train components. { 'det·ən,ād·ər ,sāf·tē }

detonics [ENG] The study of detonating and explosives performance. { de'tän·iks }

detorsion [INV ZOO] Untwisting of the 180° visceral twist imposed by embryonic torsion on many gastropod mollusks. [MED] Untwisting of an abnormal torsion, as of a ureter or intestine. { dē'tór·shən }

detoxification [BIOCHEM] The act or process of removing a poison or the toxic properties of a substance in the body. { dē,täk·sə·fə'kā·shən }

detrainment [METEOROL] The transfer of air from an organized air current to the surrounding atmosphere. { dē'trān·mənt }

detrital fan See alluvial fan. { də'trīd·əl 'fan }

detrital minerals [MINERAL] Grains of heavy minerals found in sediment, resulting from mechanical disintegration of the parent rock. { də'trīd·əl 'min·rəlz }

detrital ratio See clastic ratio. { də'trīd·əl 'rā·shō }

detrital remanent magnetization [GEOPHYS] Magnetization acquired by magnetic grains during formation of a sedimentary rock. Abbreviated DRM. { də'trīd·əl 'rem·ə·nənt 'mag·nəd·ə'zā·shən }

detrital reservoir [GEOL] A clastic or detrital-granular reservoir, classified by rock type and other factors such as sediments (quartzose-type, graywacke, or arkose sediments). { də'trīd·əl 'rez·əv,wär }

detrital sediment [GEOL] Accumulations of the organic and inorganic fragmental products of the weathering and erosion of land transported to the place of deposition. { də'trīd·əl 'sed·ə·mənt }

detritivore [ECOL] An organism that consumes dead organic matter. { di'trid·ə,vór }

detritus [ECOL] Dead plants and corpses or cast-off parts of various organisms. [GEOL] Any loose material removed directly from rocks and minerals by mechanical means, such as disintegration or abrasion. { də'trīd·əs }

detritus feeder See deposit feeder. { də'trīd·əs 'fēd·ər }

detritus food web [ECOL] A trophic web that is based on the consumption of dead organic material. { di,trīd·əs 'füd ,web }

detritus tank [CIV ENG] A tank in which heavy suspended matter is removed in sewage treatment. { də'trīd·əs ,taŋk }

detrivorous [BIOL] Referring to an organism that feeds on dead animals or partially decomposed organic matter. { də'triv·ə·rəs }

Detroit rocking furnace [ENG] An indirect arc type of rocking furnace having graphite electrodes entering horizontally from opposite ends. { də'tróit 'räk·iŋ 'fər·nəs }

detune [ELECTR] To change the inductance or capacitance of a tuned circuit so its resonant frequency is different from the incoming signal frequency. { dē'tün }

detuning stub [ELECTROMAG] Quarter-wave stub used to match a coaxial line to a sleeve-stub antenna; the stub detunes the outside of the coaxial feed line while tuning the antenna itself. { dē'tün·iŋ 'stəb }

deutencephalon See epichordal brain. { düt'en'sef·ə,län }

deuteranomaly [MED] A partial deuteranopia. { ,düt·er·ə'näm·ə·lē }

deuteranopia [MED] Defective vision consisting of red-green color confusion, with no marked reduction in the brightness of any color. { ,düd·ər·ə'nō·pē·ə }

deuteration [CHEM] The addition of deuterium to a chemical compound. { ,düd·ər'ā·shən }

deuteric [GEOL] Of or pertaining to alterations in igneous rock during the later stages and as a direct result of consolidation of magma or lava. Also known as epimagmatic; paulopost. { dü'tir·ik }

deuteride [CHEM] A hydride in which the hydrogen is deuterium. { 'düd·ə,rīd }

deuterium [CHEM] The isotope of the element hydrogen with one neutron and one proton in the nucleus; atomic weight 2.0144. Designated D, d, H^2, or ^2H. { dü'tir·ē·əm }

deuterium cycle See proton-proton chain. { dü'tir·ē·əm ,sī·kəl }

deuterium discharge tube [ELECTR] A tube similar to a hydrogen discharge lamp, but with deuterium replacing the hydrogen; source of high-intensity ultraviolet radiation for spectroscopic microanalysis. { dü'tir·ē·əm 'dis,chärj ,tüb }

deuterium oxide See heavy water. { dü'tir·ē·əm 'äk,sīd }

deuterogamy [BOT] Secondary pairing of sexual cells or nuclei replacing direct copulation in many fungi, algae, and higher plants. { ,düd·ə'räg·ə·mē }

Deuteromycetes [MYCOL] The equivalent name for Fungi Imperfecti. { ¦düd·ə·rō,mī'sēd·ēz }

deuteron [NUC PHYS] The nucleus of a deuterium atom, consisting of a neutron and a proton. Designated d. Also known as deuton. { 'düd·ə,rän }

deuteron accelerator [NUCLEO] An accelerator that produces a flux of slow neutrons by bombarding a metal-tritium target with deuterons. { 'düd·ə,rän ak'sel·ə,rād·ər }

deuteron capture [NUC PHYS] The absorption of a deuteron by a nucleus, giving rise to a compound nucleus which subsequently decays. { 'düd·ə,rän ,kap·chər }

Deuterophlebiidae [INV ZOO] A family of dipteran insects in the suborder Cyclorrhapha. { ,düd·ə·rō·flə'bī·ə,dē }

Deuterostomia [ZOO] A division of the animal kingdom which includes the phyla Echinodermata, Chaetognatha, Hemichordata, and Chordata. { ,düd·ə·rō'stō·mē·ə }

deutocerebrum [INV ZOO] The median lobes of the insect brain. { ¦düd·ō'ser·ə·brəm }

deuton See deuteron. { 'dü,tän }

deutoplasm [EMBRYO] The nutritive yolk granules in egg cells. { 'düd·ə,plaz·əm }

DeVecchis process [MIN ENG] A smelting process for pyrites in which the raw material is roasted, concentrated magnetically, and then reduced in a rotary kiln or electric furnace. { də'vek·əs ,präs·əs }

developable surface [MATH] A surface that can be obtained from a plane sheet by deformation, without stretching or shrinking. { di¦vel·əp·ə·bəl 'sər·fəs }

developed blank [MET] A blank requiring little or no trimming after being formed. { də'vel·əpt 'blaŋk }

developed dye [CHEM] A direct azo dye that can be further diazotized by a developer after application to the fiber; it couples with the fiber to form colorfast shades. Also known as diazo dye. { də'vel·əpt 'dī }

developed image [GRAPHICS] A reproducible and visible image that has resulted from development of an exposed sensitized material. { də'vel·əpt 'im·ij }

developed muzzle velocity [ORD] The actual muzzle velocity produced by any gun. { də'vel·əpt 'məz·əl və,läs·əd·ē }

developed ore See developed reserves. { də'vel·əpt 'ȯr }

developed planform [AERO ENG] The plan of an airfoil as drawn with the chord lines at each section rotated about the airfoil axis into a plane parallel to the plane of projection and with the airfoil axis rotated or developed and projected into the plane of projection. { də'vel·əpt 'plan,fȯrm }

developed reserves [MIN ENG] Ore that is exposed on three sides and for which tonnage yield and quality estimates have been made. Also known as assured mineral; blocked-out ore; developed ore; measured ore; ore in sight. { də'vel·əpt ri'zərvz }

developed silver [GRAPHICS] Elemental silver yielded by reduction of a silver salt during the development process. { də'vel·əpt 'sil·vər }

developer [CHEM] An organic compound which interacts on a textile fiber to develop a dye. [GRAPHICS] A chemical solution used to develop exposed photographic materials by reducing silver salts to metallic silver. { də'vel·əp·ər }

developer's toolkit [COMPUT SCI] A collection of program subroutines that are used to help write an application program in a particular programming language or with a particular operating system. { di¦vel·əp·ərz 'tül,kit }

developer streaks [GRAPHICS] Regions of nonuniformities within uniformly exposed regions in a processed image. { də'vel·əp·ər ,strēks }

developing [GRAPHICS] The complete process of producing an image on sensitized material by means of chemical agents. { də'vel·əp·iŋ }

development [ANALY CHEM] In the separation of mixtures by paper chromatography or thin-layer chromatography, the production of colored derivatives of the solutes by spraying the stationary phase with selective reagents in order to establish the location of individual substances. [BIOL] A process of regulated growth and differentiation that results from interaction of the genome with the cytoplasm, the internal cellular environment, and the external environment. [ENG] The exploratory work required to determine the best production techniques to bring a new process or piece of equipment to the production stage. [GEOL] The progression of changes in fossil groups which have succeeded one another during deposition of the strata of the earth. [GRAPHICS] The stage in the processing of an exposed photosensitive material in which the latent image becomes visible. [METEOROL] The process of intensification of an atmospheric disturbance, most commonly applied to cyclones and anticyclones. [MIN ENG] Opening of a coal seam or ore body by sinking shafts or driving levels, as well as installing equipment, for proving ore reserves and exploiting them. { də'vel·əp·mənt }

development age [PSYCH] An index of physical, mental, social, or emotional development stated in age equivalent as determined by standardized methods. { də'vel·əp·mənt ,āj }

developmental alexia [PSYCH] A deficiency in learning to read in the absence of physical defects or mental deficiency. { də,vel·əp'ment·əl ā'lek·sē·ə }

developmental aphasia [PSYCH] A deficiency in learning to speak, in contrast with the child's mental or development age. { də,vel·əp'ment·əl ə'fā·zhə }

developmental control gene [GEN] A gene whose primary function is the regulation of cell fates during development. { di,vel·əp¦men·təl kən'trōl ,jēn }

developmental crisis [PSYCH] A period of childhood stress related to unsuccessful attempts to establish trust, identity, autonomy, or initiative. { də,vel·əp'ment·əl 'krī·səs }

developmental disability [MED] A substantial handicap or impairment originating before the age of 18 that may be expected to continue indefinitely. { də'vel·əp,ment·əl ,dis·ə,bil·əd·ē }

developmental genetics [GEN] A branch of genetics primarily concerned with the manner in which genes control or regulate development. { də'vel·əp,ment·əl jə'ned·iks }

developmental instability [GEN] Variation of development within a genotype due to local fluctuations in internal or external environmental conditions. { di,vel·əp¦men·təl ,in·stə'bil·əd·ē }

developmental noise [GEN] Any uncontrollable variation in phenotype due to random events during development. { di,vel·əp¦men·təl 'nȯiz }

developmental psychology [PSYCH] The branch of psychology that deals with changes in behavior occurring with changes in age. { də,vel·əp'ment·əl sī'käl·ə·jē }

developmental toxicity [MED] Adverse effects on the developing child which result from exposure to toxic chemicals or other toxic substances, can include birth defects, low birth weight, and functional or behavioral weaknesses that show up as the child develops. { di,vel·əp¦ment·əl tak'sis·ə·dē }

development center [GRAPHICS] The portion of an exposed photosensitive material which contains the latent image. { də'vel·əp·mənt ,sen·tər }

development drift [MIN ENG] A tunnel dug in a mine either from the surface or a point underground to get to coal or ore for exploitation or mining purposes. { də'vel·əp·mənt ,drift }

development drilling [MIN ENG] Drilling boreholes to locate, identify, and prove an ore body or coal seam. { də'vel·əp·mənt ,dril·iŋ }

development index [METEOROL] An index used as an aid in forecasting cyclogenesis; the development index I is defined most frequently as the difference in divergence between two well-separated, tropospheric, constant-pressure surfaces. Also known as relative divergence. { də'vel·əp·mənt ,in,deks }

development rock [MIN ENG] Rock containing both barren

and valuable rock, broken during development work. { də'vel·əp·mənt ˌräk }

development system [COMPUT SCI] The computer and software that are used to create a computer program. { di'vel·əp·mənt ˌsis·təm }

development tool [COMPUT SCI] A piece of hardware or software that is used to help design a computer or write a computer program. { di'vel·əp·mənt ˌtül }

development well [PETRO ENG] A well drilled to produce oil or gas from a proven productive area. { də'vel·əp·mənt ˌwel }

Devereaux agitator [MIN ENG] An agitator that utilizes an upthrust propeller to stir pulp; used in leach agitation of minerals. { 'dev·rō 'aj·ə·tād·ər }

devernalization [BOT] Annulment of the vernalization effect. { dē‚vərn·əl·ə'zā·shən }

deviation [ENG] The difference between the actual value of a controlled variable and the desired value corresponding to the set point. [EVOL] Evolutionary differentiation involving interpolation of new stages in the ancestral pattern of morphogenesis. [OPTICS] The angle between the incident ray on an object or optical system and the emergent ray, following reflection, refraction, or diffraction. Also known as angle of deviation. [PETRO ENG] During a drilling operation, the inclination of the borehole from the vertical. [STAT] The difference between any given number in a set and the mean average of those numbers. { ˌdev·ē·ā'shən }

deviation absorption [COMMUN] Distortion in a frequency-modulated receiver due to inadequate bandwidth, inadequate amplitude-modulation rejection, or inadequate discriminator linearity. { ˌdev·ē·ā'shən əb‚sȯrp·shən }

deviation card [NAV] A card having a table of the deviation of an aircraft magnetic compass on various headings; this card is usually mounted on the compass. { ˌdev·ē·ā'shən ˌkärd }

deviation factor *See* compressibility factor. { ˌdev·ē·ā'shən ˌfak·tər }

deviation hole [PETRO ENG] Drilled hole with deviation from true vertical, usually limited by contract to 3–5°; not to be confused with a crooked hole, resulting from carelessness or a steeply dipping formation. { ˌdev·ē·ā'shən ˌhōl }

deviation loss [COMMUN] The transmission loss measured on a transducer when relative response is considered as a function of bearing. { ˌdev·ē·ā'shən ˌlȯs }

deviation ratio [COMMUN] Ratio of the maximum frequency deviation to the maximum modulating frequency of a frequency-modulated system under specified conditions. { ˌdev·ē·ā'shən ˌrā·shō }

deviation sensitivity [NAV] A value expressed as the ratio of the rate of change in course indication to the deviation from the course line. { ˌdev·ē·ā'shən ˌsen·sə'tiv·əd·ē }

deviation survey [PETRO ENG] Measurements made during a drilling operation to determine the angle from which the hole has deviated from the vertical. { ˌdev·ē·ā'shən 'sər‚vā }

deviation table [NAV] A table of the deviation of a magnetic compass on various headings, magnetic or compass; for an aircraft compass, this information is usually placed on a card called a deviation card. Also known as magnetic compass table. { ˌdev·ē·ā'shən ˌtā·bəl }

deviatonic stress [MECH] The portion of the total stress that differs from an isostatic hydrostatic pressure; it is equal to the difference between the total stress and the spherical stress. { ˌdev·ē·ə'tän·ik 'stres }

deviatoric stress [GEOL] A condition in which the stress components operating at a point in a body are not the same in every direction. Also known as differential stress. { ˈdev·ē·ə¦tȯr·ik 'stres }

device [COMPUT SCI] A general-purpose term used, often indiscriminately, to refer to a computer component or the computer itself. [ELECTR] An electronic element that cannot be divided without destroying its stated function; commonly applied to active elements such as transistors and transducers. [ENG] A mechanism, tool, or other piece of equipment designed for specific uses. { di'vīs }

device address [COMPUT SCI] The binary code which corresponds to a unique device, referred to when selecting this specific device. { di'vīs ə'dres }

device assignment [COMPUT SCI] The use of a logical device number used in conjunction with an input/output instruction, and made to refer to a specific device. { di'vīs ə'sīn·mənt }

device cluster [COMPUT SCI] A collection of peripheral devices (usually terminals) that have a common control unit. { di'vīs ˌkləs·tər }

device control character [COMPUT SCI] A special character used to direct a peripheral or communications device to perform a specific function. { di'vīs kən'trōl ˌkar·ik·tər }

device dependence [COMPUT SCI] Property of a computer program that will operate only with specified hardware. { di'vīs de‚pen·dəns }

device driver [COMPUT SCI] A subroutine which handles a complete input/output operation. { di'vīs ˌdrīv·ər }

device-end condition [COMPUT SCI] The completion of an input/output operation, such as the transfer of a complete data block, recognized by the hardware in the absence of a byte count. { di'vīs ˌend kən'dish·ən }

device end pending [COMPUT SCI] A hardware error in which a peripheral device does not respond when addressed by the central processing unit, usually because the device has become inoperative. { di'vīs 'end ˌpend·iŋ }

device flag [COMPUT SCI] A flip-flop output which indicates the ready status of an input/output device. { di'vīs ˌflag }

device independence [COMPUT SCI] Property of a computer program whose successful execution (without recompilation) does not depend on the type of physical unit associated with a given logical unit employed by the program. { di'vīs ˌin·də'pen·dəns }

device-independent colors [COMPUT SCI] Colors produced by printers, monitors, and other output devices that have been modified to conform with a standard method of color description. { di¦vīs ˌin·də‚pen·dənt 'kəl·ərz }

device-name assignment [COMPUT SCI] The designation of a peripheral device by a symbolic name rather than an address. { di'vīs 'nām ə‚sīn·mənt }

device number [COMPUT SCI] The physical or logical number which refers to a specific input/output device. { di'vīs ˌnəm·bər }

device selector [COMPUT SCI] A circuit which gates data-transfer or command pulses to a specific input/output device. { di'vīs si'lek·tər }

devil *See* devil float. { 'dev·əl }

devil float [ENG] A hand float containing nails projecting at each corner and used to roughen the surface of plaster to provide a key for the next coat. Also known as devil; nail coat. { 'dev·əl ˌflōt }

devillite [MINERAL] $Cu_4Ca(SO_4)_2(OH)_6 \cdot 3H_2O$ A dark-green mineral consisting of a hydrous basic sulfate of copper and calcium, occurring in six-sided platy crystals. { də'və‚līt }

devil on two sticks *See* devil's curve. { 'dev·əl ȯn ˌtü 'stiks }

devil's curve [MATH] A plane curve whose equation in Cartesian coordinates x and y is $y^4 - a^2y^2 = x^4 - b^2x^2$, where a and b are constants. Also known as devil on two sticks. { 'dev·əlz 'kərv }

devil's pitchfork [DES ENG] A tool with flexible prongs used in recovery of a bit, underreamer, cutters, or such lost during drilling. { 'de·vəlz 'pich‚fȯrk }

devitrification [CHEM] The process by which the glassy texture of a material is converted into a crystalline texture. { dē‚vi·trə·fə'kā·shən }

devitrified glass [MATER] A glassy material which has been changed from a vitreous to a brittle crystalline state during manufacture. { dē'vi·trə‚fīd 'glas }

devolatilize [CHEM ENG] To remove volatile components from a material. { ˌdē'väl·ə·tə‚līz }

Devonian [GEOL] The fourth period of the Paleozoic Era, covering the geological time span between about 412 and 354 $\times 10^6$ years before present. { di'vō·nē·ən }

De Vries effect [GEOCHEM] A relatively short-term oscillation, on the order of 100 years, in the radiocarbon content of the atmosphere, and the resulting variation in the apparent radiocarbon age of samples. { də'vrēz i'fekt }

devrinol [ORG CHEM] $C_{17}H_{21}O_2N$ A brown solid with a melting point of 68.5–70.5°C; slight solubility in water; used as a herbicide for crops. Also known as 2-(α-naphthoxy)-*N,N*-diethylpropionamide. { 'dev·rə‚nȯl }

dew [HYD] Water condensed onto grass and other objects

DEVONIAN

CENOZOIC	QUATERNARY	
	TERTIARY	
MESOZOIC	CRETACEOUS	
	JURASSIC	
	TRIASSIC	
PALEOZOIC	PERMIAN	
	CARBONIFEROUS	PENNSYLVANIAN
		MISSISSIPPIAN
	DEVONIAN	
	SILURIAN	
	ORDOVICIAN	
	CAMBRIAN	
PRECAMBRIAN		

Chart showing relationship of Devonian to other periods.

near the ground, the temperatures of which have fallen below the dew point of the surface air because of radiational cooling during the night but are still above freezing. { 'dü }

Dewar calorimeter [ENG] **1.** Any calorimeter in which the sample is placed inside a Dewar flask to minimize heat losses. **2.** A calorimeter for determining the mean specific heat capacity of a solid between the boiling point of a cryogenic liquid, such as liquid oxygen, and room temperature, by measuring the amount of the liquid that evaporates when the specimen is dropped into the liquid. { 'dü·ər ,kal·ə'rim·əd·ər }

Dewar flask [PHYS] A vessel having double walls, the space between being evacuated to prevent the transfer of heat and the surfaces facing the vacuum being heat-reflective; used to hold liquid gases and to study low-temperature phenomena. { 'dü·ər ,flask }

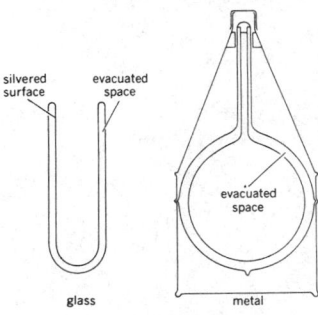

DEWAR FLASK

silvered surface
evacuated space
evacuated space
glass
metal

Typical Dewar containers.

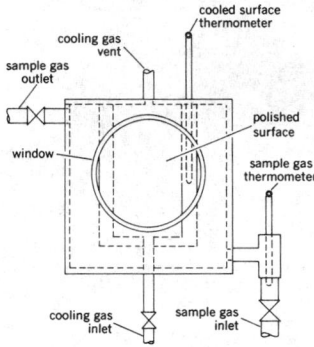

DEW-POINT HYGROMETER

cooled surface thermometer
cooling gas vent
sample gas outlet
polished surface
window
sample gas thermometer
cooling gas inlet
sample gas inlet

Dew-point type of hygrometer.

Dewar structure [ORG CHEM] A structural formula for benzene that contains a bond between opposite atoms. { 'dü·ər ,strək·chər }

dewaterer [MECH ENG] Wet-type mechanical classifier (solids separator) in which solids settle out of the carrier liquid and are concentrated for recovery. { dē'wòd·ər·ər }

dewatering [ENG] **1.** Removal of water from solid material by wet classification, centrifugation, filtration, or similar solid-liquid separation techniques. **2.** Removing or draining water from an enclosure or a structure, such as a riverbed, caisson, or mine shaft, by pumping or evaporation. { dē'wòd·ər·iŋ }

dewaxed oil [MATER] Lubricating oil that has had a portion of the wax removed. { dē'wakst 'òil }

dewaxing [CHEM ENG] Removing wax from a material or object; a process used to separate solid hydrocarbons from petroleum. { dē'waks·iŋ }

dewcap [OPTICS] An open tube attached to the end of a refracting telescope to prevent moisture from condensing on the objective. { 'dü,kap }

dew cell [ENG] An instrument used to determine the dew point, consisting of a pair of spaced, bare electrical wires wound spirally around an insulator and covered with a wicking wetted with a water solution containing an excess of lithium chloride; an electrical potential applied to the wires causes a flow of current through the lithium chloride solution, which raises the temperature of the solution until its vapor pressure is in equilibrium with that of the ambient air. { 'dü ,sel }

dewclaw [VERT ZOO] **1.** A vestigial digit on the foot of a mammal which does not reach the ground. **2.** A claw or hoof terminating such a digit. { 'dü,klò }

dewetting [MET] Flow of solder away from the soldered surface during reheating following initial soldering. { dē'wed·iŋ }

deweylite [MINERAL] A mixture of clinochrysolite and stevensite. Also known as gymnite. { 'dü·ē,līt }

dewindtite [MINERAL] Pb(UO$_2$)$_2$(PO$_4$)$_2$·3H$_2$O A canary-yellow secondary mineral consisting of a hydrous phosphate of lead and uranium. { də'win,tīt }

de Witte relation [GEOPHYS] Graphical plot of the relation between electrical conductivity and distance over which the conductivity is measured through reservoir rock with clay minerals, (the effect is similar to two parallel electrical circuits), the current passing through the conducting clay minerals and the water-filled pores. { də'wit rē'lā·shən }

dewlap [ANAT] A fleshy or fatty fold of skin on the throat of some humans. [BOT] One of a pair of hinges at the joint of a sugarcane leaf blade. [VERT ZOO] A fold of skin hanging from the neck of some reptiles and bovines. { 'dü,lap }

dew point [CHEM] The temperature and pressure at which a gas begins to condense to a liquid. [METEOROL] The temperature at which air becomes saturated when cooled without addition of moisture or change of pressure; any further cooling causes condensation. Also known as dew-point temperature. { 'dü ,pòint }

dew-point boundary [CHEM ENG] On a phase diagram for a gas-condensate reservoir (pressure versus temperature with constant gas-oil ratios), the area along which the gas-oil ratio approaches zero. { 'dü ,pòint ,baún·drē }

dew-point composition [CHEM ENG] The water vapor-air composition at saturation, that is, at the temperature at which water exerts a vapor pressure equal to the partial pressure of water vapor in the air-water mixture. { 'dü ,pòint ,käm·pə'zish·ən }

dew-point curve [CHEM ENG] On a PVT phase diagram, the

line that separates the two-phase (gas-liquid) region from the one-phase (gas) region, and indicates the point at a given gas temperature or pressure at which the first dew or liquid phase occurs. { 'dü ,pòint ,kərv }

dew-point depression [CHEM ENG] Reduction of the liquid-vapor dew point of a gas by removal of a portion of the liquid (such as water) from the gas (such as air). [METEOROL] The number of degrees the dew point is found to be lower than the temperature. { 'dü ,pòint di'presh·ən }

dew-point formula [METEOROL] A formula for the calculation of the approximate height of the lifting condensation level; employed to estimate the height of the base of convective clouds, under suitable atmospheric and topographic conditions. { 'dü ,pòint ,fòr·myə·lə }

dew-point hygrometer [CHEM ENG] An instrument for determining the dew point by measuring the temperature at which vapor being cooled in a silver vessel begins to condense. Also known as cold-spot hygrometer. { 'dü ,pòint hī'gräm·əd·ər }

dew-point pressure [CHEM ENG] The gas pressure at which a system is at its dew point, that is, the conditions of gas temperature and pressure at which the first dew or liquid phase occurs. { 'dü ,pòint ,presh·ər }

dew-point recorder [ENG] An instrument which gives a continuous recording of the dew point; it alternately cools and heats the target and uses a photocell to observe and record the temperature at which the condensate appears and disappears. Also known as mechanized dew-point meter. [PETRO ENG] An instrument used for continuous measurement and recording of the dew point of a gas. { 'dü ,pòint ri'kòrd·ər }

dew-point reservoir [PETRO ENG] A hydrocarbon reservoir in which the temperature lies between the critical temperature and the cricondentherm (maximum temperature and pressure at which two phases can coexist) and in the one-phase region. Also known as retrograde gas-condensate reservoir. { 'dü ,pòint 'rez·əv,wär }

dew-point spread [METEOROL] The difference in degrees between the air temperature and the dew point. { 'dü ,pòint ,spred }

dew-point temperature *See* dew point. { 'dü ,pòint 'tem·prə·chər }

dew retting [MICROBIO] A type of retting process in which the stems of fiber plants are spread out in moist meadows, and the pectin decomposition is accomplished by molds and aerobic bacteria with the formation of CO$_2$ and H$_2$. { 'dü ,red·iŋ }

dex *See* brig. { deks }

Dexaminidae [INV ZOO] A family of amphipod crustaceans in the suborder Gammeridea. { dek'sam·in·ə,dē }

dexterotropic [BIOL] Turning toward the right; applied to cleavage, shell formation, and whorl patterns. { ¦dek·stə·rō¦träp·ik }

dextral drag fold [GEOL] A drag fold in which the trace of a given surface bed is displaced to the right. { 'dek·strəl 'drag ,fōld }

dextral fault [GEOL] A strike-slip fault in which an observer approaching the fault sees the opposite block as having moved to the right. Also known as right-lateral fault; right-lateral slip fault; right-slip fault. { 'dek·strəl 'fòlt }

dextral fold [GEOL] An asymmetric fold in which the long limb appears to be offset to the right to an observer looking along the long limb. { 'dek·strəl 'fōld }

dextran [BIOCHEM] Any of the several polysaccharides, (C$_5$H$_{10}$O$_5$)$_n$, that yield glucose units on hydrolysis. { 'dek ,stran }

dextranase [BIOCHEM] An enzyme that hydrolyzes 1,6-α-glucosidic linkages in dextran. { 'dek·strə,nās }

dextrin [BIOCHEM] A polymer of D-glucose which is intermediate in complexity between starch and maltose. { 'dek·strən }

dextrinization [ORG CHEM] Any process that involves dextrinizing. { ,dek·strə·nə'zā·shən }

dextrinize [ORG CHEM] To convert a starch into dextrins. { 'dek·strə,nīz }

dextro *See* dextrorotatory; dextrorotatory enantiomer. { 'dek,strō }

dextrocardia [MED] The presence of the heart in the right hemithorax, with the cardiac apex directed to the right. { ,dek·strō'kär·dē·ə }

dextromethorphan hydrobromide [PHARM] $C_{18}H_{25}NO \cdot HBr \cdot H_2O$ White crystals or a crystalline powder; soluble in alcohol and chloroform; used in medicine. { |dek·strō·mə|thȯr·fən,hī·drō'brō,mīd }

dextrophobia [PSYCH] An abnormal fear of objects to the right of the body. { ,dek·strə'fō·bē·ə }

dextropimaric acid [ORG CHEM] $C_{19}H_{29}COOH$ A compound found in particular in oleoresins of pine trees. { |dek·strō·pə'mar·ik 'as·əd }

dextrorotatory [OPTICS] Rotating clockwise the plane of polarization of a wave traveling through a medium in a clockwise direction, as seen by an eye observing the light. Abbreviated dextro. { ,dek·strə'rōd·ə,tȯr·ē }

dextrorotatory enantiomer [ORG CHEM] An optically active substance that rotates the plane of plane-polarized light clockwise. Symbolized d. Also known as dextro. { ,dek·strō'rōd·ə,tȯr·ē ə|nan·tē|ō·mər }

dextrorse [BOT] Twining toward the right. { 'dek,strȯrs }

dextrorse curve See right-handed curve. { 'dek,strȯrs ,kərv }

dextrorsum See right-handed curve. { dek'strȯr·səm }

dextrose [BIOCHEM] $C_6H_{12}O_6 \cdot H_2O$ A dextrorotatory monosaccharide obtained as a white, crystalline, odorless, sweet powder, which is soluble in about one part of water; an important intermediate in carbohydrate metabolism; used for nutritional purposes, for the temporary increase of blood volume, and as a diuretic. Also known as corn sugar; grape sugar. { 'dek,strōs }

dextrotopic cleavage [EMBRYO] A clockwise spiral cleavage pattern. { |dek·strə|täp·ik 'klē·vij }

dezincification [CHEM] Removal of zinc. [MET] Corrosion of brass in which both components of the alloy are dissolved and the copper is redeposited as a porous surface residue. { dē,ziŋk·ə·fə'kā·shən }

DFE See design for environment.

D flip-flop [ELECTR] A flip-flop whose output is a function of the input which appeared one pulse earlier. Also known as delay flip-flop. { |dē 'flip,fläp }

DFP See isoflurophate.

D galaxy [ASTRON] A giant galaxy consisting of an elliptically shaped nucleus surrounded by an unusually large envelope. { 'dē ,gal·ik·sē }

D glass [MATER] A type of glass with a high boron content and a precisely controlled dielectric constant. { 'dē ,glas }

DGPS See differential GPS.

DG synchro amplifier [ELECTR] Synchro differential generator driven by servosystem. { |dē|jē |siŋ·krō 'am·plə,fī·ər }

DHA See dehydroacetic acid; dihydroxyacetone.

D-HEMT See depletion-mode HEMT.

D horizon [GEOL] A soil horizon sometimes occurring below a B or C horizon, consisting of unweathered rock. { 'dē hə'rīz·ən }

di- [SCI TECH] Prefix meaning two. { dī }

Di See didymium.

DI See temperature-humidity index.

diabantite [MINERAL] $(Mg,Fe^{2+},Al)_6(Si,Al)_4O_{10}(OH)_8$ Mineral of the chlorite group consisting of a basic silicate of magnesium, iron, and aluminum, occurring in cavities in basic igneous rock. { ,dī·ə'ban,tīt }

diabase [PETR] An intrusive rock consisting principally of labradorite and pyroxene. { 'dī·ə,bās }

diabase amphibolite [PETR] Amphibolite formed by dynamic metamorphism of diabase. { 'dī·ə,bās am'fib·ə,līt }

diabasic [PETR] Denoting igneous rock in which the interstices between the feldspar crystals are filled with discrete crystals or grains of pyroxene. { |dī·ə|bās·ik }

diabatic [THERMO] A thermodynamic change of state of a system in which there is a transfer of heat across the boundaries of the system. Also known as nonadiabatic. { |dī·ə|bad·ik }

diabetes [MED] Any of various abnormal conditions characterized by excessive urinary output, thirst, and hunger; usually refers to diabetes mellitus. { ,dī·ə'bēd·ēz }

diabetes insipidus [MED] A form of diabetes due to a disfunction of the hypothalamus. { ,dī·ə'bēd·ēz in'sip·ə·dəs }

diabetes mellitus [MED] A metabolic disorder arising from a defect in carbohydrate utilization by the body, related to inadequate or abnormal insulin production by the pancreas. { ,dī·ə'bēd·ēz 'mel·ə·dəs }

diabetic acidosis [MED] Metabolic acidosis seen in diabetes mellitus, due to an excess of ketone bodies. { |dī·ə|bed·ik ,as·ə'dō·səs }

diabetic gangrene [MED] A moist form of gangrene occurring in persons with diabetes mellitus, often following a minor injury. { |dī·ə|bed·ik ,gaŋ'grēn }

diabetic glomerulosclerosis See intercapillary glomerulosclerosis. { |dī·ə|bed·ik glə|mer·yə·lō·sklə'rō·səs }

diabetophobia [PSYCH] An abnormal fear of becoming diabetic. { ,dī·ə·bed·ə'fō·bē·ə }

diablastic [PETR] Pertaining to a texture in metamorphic rock that consists of intergrown and interpenetrating rod-shaped components. { dī·ə'blas·tik }

diaboleite [MINERAL] $Pb_2CuCl_2(OH)_4$ A sky-blue mineral consisting of a basic chloride of lead and copper. { |dī·ə·bō'lā,īt }

diac See trigger diode. { 'dī,ak }

diacetate [ORG CHEM] An ester or salt that contains two acetate groups. { dī'as·ə,tāt }

diacetic acid See acetoacetic acid. { dī·ə'sēd·ik 'as·əd }

diacetic ether See ethyl acetoacetate. { dī·ə'sēd,ik 'ē·thər }

diacetin [ORG CHEM] $C_3H_5(OH)(CH_3COO)_2$ A colorless, hygroscopic liquid that is soluble in water, alcohol, ether, and benzene; boiling point 259°C; used as a plasticizer and softening agent and as a solvent. Also known as glyceryl diacetate. { dī'as·əd·ən }

diacetone alcohol [ORG CHEM] $CH_3COCH_2C(CH_3)_2OH$ A colorless liquid used as a solvent for nitrocellulose and resins. { 'dī·as·ə,tōn 'al·kə,hȯl }

diacetyl [ORG CHEM] **1.** $CH_3COCOCH_3$ A yellowish-green liquid with a boiling point of 88°C; has a strong odor that resembles quinone; occurs naturally in bay oil and butter and is produced from methyl ethyl ketone or by a special fermentation of glucose; used as an aroma carrier in food manufacturing. Also known as biacetyl. **2.** A prefix indicating two acetyl groups. { dī'as·əd·əl }

diacetylurea [ORG CHEM] $C_5H_8O_3N_2$ An acyl derivative of urea containing two acetyl groups. { dī|as·əd·əl·yu'rē·ə }

diachronous [GEOL] Of a rock unit, varying in age in different areas or cutting across time planes or biostratigraphic zones. Also known as time-transgressive. { dī'ak·rə·nəs }

diacid [CHEM] An acid that has two acidic hydrogen atoms; an example is oxalic acid. { dī'as·əd }

diaclinal [GEOL] Pertaining to a stream crossing a fold, perpendicular to the strike of the underlying strata it traverses. { |dī·ə|klīn·əl }

Diacodectidae [PALEON] A family of extinct artiodactyl mammals in the suborder Palaeodonta. { ,dī·ə·kə'dek·tə,dē }

diactine [INV ZOO] A type of sponge spicule which develops in two directions from a central point. { dī'ak,tēn }

diactor [ELEC] Direct-acting automatic regulator for control of shunt generator voltage output. { dī'ak·tər }

diacylglycerol [BIOCHEM] A product of the cleavage of membrane-associated phospholipid by phospholipase C that activates isozymes of protein kinase C, an important regulator of cell function. { ,dī·ə·səl'glis·ə,rōl }

diad axis [CRYSTAL] A rotation axis whose multiplicity is equal to 2. { 'dī,ad ,aks·əs }

diadelphous stamen [BOT] A stamen that has its filaments united into two sets. { |dī·ə|del·fəs 'stā·mən }

Diadematacea [INV ZOO] A superorder of Euchinoidea having a rigid or flexible test, perforate tubercles, and branchial slits. { ,dī·ə,dē·mə'tās·ē·ə }

Diadematidae [INV ZOO] A family of large euechinoid echinoderms in the order Diadematoida having crenulate tubercles and long spines. { ,dī·ə·də'mad·ə,dē }

Diadematoida [INV ZOO] An order of echinoderms in the superorder Diadematacea with hollow primary radioles and crenulate tubercles. { ,dī·ə·dē·mə'tȯid·ə }

diadochite [MINERAL] $Fe_2(PO_4)(SO_4)(OH) \cdot 5H_2O$ A brown or yellowish mineral consisting of a basic hydrous ferric phosphate and sulfate. { dī'ad·ə,kīt }

diadochy [CRYSTAL] Replacement or ability to be replaced of one atom or ion by another in a crystal lattice. { dī'ad·ə·kē }

diadromous [BOT] Having venation in the form of fanlike radiations. [VERT ZOO] Of fish, migrating between salt and fresh waters. { dī'ad·rə·məs }

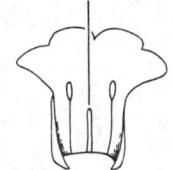

DIADELPHOUS STAMEN

staminode

Cutaway drawing of a flower showing diadelphous stamen.

Diadumenidae [INV ZOO] A family of anthozoans in the order Actiniaria. { ‚dī·ə·dü'men·ə‚dē }

diafocal point [OPTICS] For a ray of light refracted by a lens, a point on the ray which lies on a plane passing through the axis of the lens which is parallel to the ray on the opposite side of the lens. { ¦dī·ə¦fō·kəl 'póint }

diagenesis [GEOL] Chemical and physical changes occurring in sediments during and after their deposition but before consolidation. { ‚dī·ə'jen·ə·səs }

diageotropism [BIOL] Growth orientation of a sessile organism or structure perpendicular to the line of gravity. { ¦dī·ə·jē'ä·trə‚piz·əm }

diagnosis [COMPUT SCI] The process of locating and explaining detectable errors in a computer routine or hardware component. [MED] Identification of a disease from its signs and symptoms. [SYST] In taxonomic study, a statement of the characters that distinguish a taxon from coordinate taxa. { ‚dī·əg'nō·səs }

diagnostic bacteriology [MICROBIO] A branch of medical bacteriology that focuses on the identification of bacteria by their ability to grow on various selective media and by the characteristic appearance of their colonies on test media. { ‚dī·əg‚näs·tik bak‚tir·ē'äl·ə·jē }

diagnostic check See diagnostic routine. { ‚dī·əg'näs·tik 'chek }

diagnostic equation [METEOROL] Any equation governing a system which contains no time derivative and therefore specifies a balance of quantities in space at a moment of time; examples are a hydrostatic equation or a balance equation. { ‚dī·əg'näs·tik i'kwā·zhən }

diagnostic message [COMPUT SCI] A statement produced automatically during some computer processing activity, such as program compilation, that provides information on the status of the computer or its software, particularly errors or potential problems. { ¦dī·əg¦näs·tik 'mes·ij }

diagnostic routine [COMPUT SCI] A routine designed to locate a computer malfunction or a mistake in coding. Also known as diagnostic check; diagnostic subroutine; diagnostic test; error detection routine. { ‚dī·əg'näs·tik rü'tēn }

diagnostics [ENG] Information on what tests a device has failed and how they were failed; used to aid in troubleshooting. { ‚dī·əg'näs·tiks }

diagnostic subroutine See diagnostic routine. { ‚dī·əg'näs·tik 'səb·rü‚tēn }

diagnostic test See diagnostic routine. { ‚dī·əg'näs·tik 'test }

diagnotor [COMPUT SCI] A combination diagnostic and edit routine which questions unusual situations and notes the implied results. { ‚dī·əg'nōd·ər }

diagonal [CIV ENG] A sloping structural member, under compression or tension or both, of a truss or bracing system. [MATH] **1.** The set of points all of whose coordinates are equal to one another in an n-dimensional coordinate system. **2.** A line joining opposite vertices of a polygon with an even number of sides. [OPTICS] A plane mirror or prism face mounted near the eyepiece of a telescope at an angle to the light path, to redirect the light for convenience of observation or to reduce the intensity of the image of the sun so that it can be observed directly. [TEXT] A heavy twilled fabric. { dī'ag·ən·əl }

diagonal bond [CIV ENG] A masonry bond with diagonal headers. { dī'ag·ən·əl 'bänd }

diagonal fault [GEOL] A fault whose strike is diagonal or oblique to the strike of the adjacent strata. Also known as oblique fault. { dī'ag·ən·əl 'fólt }

diagonal horn antenna [ELECTROMAG] Horn antenna in which all cross sections are square and the electric vector is parallel to one of the diagonals; the radiation pattern in the far field has almost perfect circular symmetry. { dī'ag·ən·əl 'hórn an‚ten·ə }

diagonalize [MATH] To convert a square matrix to a diagonal matrix, usually by multiplying it on the left by a second matrix A of the same order, and on the right by the inverse of A. { dī'ag·ən·ə‚līz }

diagonal joint [GEOL] A joint having its strike oblique to the strike of the strata of the sedimentary rock, or to the cleavage plane of the metamorphic rock in which it occurs. Also known as oblique joint. { dī'ag·ən·əl 'jóint }

diagonal Latin square [MATH] A Latin square in which each of the symbols appears exactly once in each diagonal. { dī¦ag·ən·əl ‚lat·ən 'skwer }

diagonally dominant matrix [MATH] A matrix in which the absolute value of each diagonal element is either greater than the sum of the absolute values of the off-diagonal elements of the same row or greater than the sum of the off-diagonal elements in the same column. { dī'ag·ən·əl·ē 'däm·ə·nənt 'mā‚triks }

diagonal matrix [MATH] A matrix whose nonzero entries all lie on the principal diagonal. { dī'ag·ən·əl 'mā·triks }

diagonal pitch [ENG] In rows of staggered rivets, the distance between the center of a rivet in one row to the center of the adjacent rivet in the next row. { dī'ag·ən·əl 'pich }

diagonal pliers [DES ENG] Pliers with cutting jaws at an angle to the handles to permit cutting off wires close to terminals. { dī'ag·ən·əl 'plī·ərz }

diagonal stay [MECH ENG] A diagonal member between the tube sheet and shell in a fire-tube boiler. { dī'ag·ən·əl 'stā }

diagram [COMPUT SCI] A schematic representation of a sequence of subroutines designed to solve a problem; it is a coarser and less symbolic representation than a flow chart, frequently including descriptions in English words. [GRAPHICS] **1.** A line drawing that represents an object or area according to a scale. **2.** A graph which shows the relation between two variables or which plots the occurrence of events or objects as a function of two variables. [MATH] A picture in which sets are represented by symbols and mappings between these sets are represented by arrows. { 'dī·ə‚gram }

diagram factor [MECH ENG] The ratio of the actual mean effective pressure, as determined by an indicator card, to the map of the ideal cycle for a steam engine. { 'dī·ə‚gram ‚fak·tər }

diagram on the plane of the celestial equator See time diagram. { 'dī·ə‚gram ón t͟hə 'plān əv t͟hə sə'les·chəl ē'kwäd·ər }

diagram on the plane of the equinoctial See time diagram. { 'dī·ə‚gram ón t͟hə 'plān əv t͟hə ē·kwə'näk·chəl }

diagravitropism [PL PHYS] A response of plant organs to gravity where growth is horizontal. { ‚dī·ə‚grav·ə'trä‚piz·əm }

diaheliotropism [BOT] Movement of plant leaves which follow the sun such that they remain perpendicular to the sun's rays throughout the day. { ‚dī·ə‚hē·lē·ə'trä‚piz·əm }

diakinesis [CYTOL] The last stage of meiotic prophase, when the chromatids attain maximum contraction and the bivalents move apart and position themselves against the nuclear membrane. { dī·ə·kə'nē·səs }

diakoptics [MATH] A piecewise approach to the solution of large-scale interconnected systems, in which the large system is first broken up into several small pieces or subdivisions, the subdivisions are solved separately, and finally the effect of interconnection is determined and added to each subdivision to yield the complete solution of the system. { ‚dī·ə'käp·tiks }

dial [COMMUN] In automatic telephone switching, a type of calling device which, when wound up and released, generates pulses required for establishing connections. [DES ENG] A separate scale or other device for indicating the value to which a control is set. { dīl }

DIAL See differential absorption lidar. { 'dī‚al }

dial backup [COMMUN] A dial telephone line that can be used in case a point-to-point line fails, so that data transmission can continue. { dīl 'bak‚əp }

dial cable [DES ENG] Braided cord or flexible wire cable used to make a pointer move over a dial when a separate control knob is rotated, or used to couple two shafts together mechanically. { dīl ‚kā·bəl }

dial central office [COMMUN] Telephone or teletypewriter office where necessary automatic equipment is located for connecting two or more users together by wires for communications purposes. { dīl ¦sen·trəl 'óf·əs }

dial cord [DES ENG] A braided cotton, silk, or glass fiber cord used as a dial cable. { 'dīl ‚kórd }

dialdehyde [ORG CHEM] A molecule that has two aldehyde groups, such as dialdehyde starch. { dī'al·də‚hīd }

dialect [COMPUT SCI] A version of a programming language that differs from other versions in some respects but generally resembles them. { 'dī·ə‚lekt }

dial exchange [COMMUN] A telephone exchange area in which all subscribers originate their calls by dialing. { 'dīl iks‚chānj }

dial feed [MECH ENG] A device that rotates workpieces into position successively so they can be acted on by a machine. { 'dīl ˌfēd }

dialifor [ORG CHEM] $C_{14}H_{17}ClNO_4S_2P$ A white, crystalline compound with a melting point of 67–69°C; insoluble in water; used to control pests in citrus fruits, grapes, and pecans. { dī'al·əˌfȯr }

dial indicator [DES ENG] Meter or gage with a calibrated circular face and a pivoted pointer to give readings. { 'dīl ˌin·dəˌkād·ər }

dialing key [COMMUN] Method of dialing in which a set of numerical keys is used to originate dial pulses instead of a dial; generally used in connection with voice-frequency dialing. { 'dī·liŋ ˌkē }

dialing step [ENG] The minimum amount, expressed in units of mass, that can be added or removed on a balance fitted with dial weights. { 'dīl·iŋ ˌstep }

dial jacks [ELEC] Strip of jacks associated with and bridged to a regular out-going trunk jack circuit to provide a connection between the dial cords and the outgoing trunks. { 'dīl ˌjaks }

dial key [ELEC] Key unit of the subscriber's cord circuit used to connect the dial into the line. { 'dīl ˌkē }

dialkyl [ORG CHEM] A molecule that has two alkyl groups. { dī'alˌkəl }

dialkyl amine [ORG CHEM] An amine that has two alkyl groups bonded to the amino nitrogen. { dī'al·kəl 'aˌmēn }

diallage [MINERAL] A green, brown, gray, or bronze-colored clinopyroxene characterized by prominent parting parallel to the front pinacoid a (100). { 'dī·ə·lij }

dial lamp [ELEC] A small lamp used to illuminate a dial. { 'dīl ˌlamp }

dial leg [ELEC] Conductor in a circuit brought out for direct-current dial signaling. { 'dīl ˌleg }

diallyl phthalate [ORG CHEM] $C_6H_4(COOCH_2CH:CH_2)_2$ A colorless, oily liquid with a boiling range of 158–165°C; used as a plasticizer and for polymerization. Abbreviated DAP. { dī'al·əl 'thaˌlāt }

dial office [COMMUN] Central office operating on dial signals. { 'dīl ˌȯf·əs }

dialog [COMPUT SCI] A form of data processing involving an interaction between a computer system and a terminal operator who uses a keyboard and electronic display to enter data which the computer edits and may respond to. { 'dī·əˌläg }

dialog box [COMPUT SCI] On a computer screen, a small window that is used to emphasize the importance of some action or to request an answer to a question. { 'dī·əˌläg ˌbäks }

dial press [MECH ENG] A punch press with dial feed. { 'dīl ˌpres }

dial pulse interpreter [ELECTR] A device that converts the signaling pulses of a dial telephone to a form suitable for data entry to a computer. { 'dīl ˌpəls in'tər·prəd·ər }

dial pulsing See loop pulsing. { 'dīl ˌpəls·iŋ }

dial telephone system [COMMUN] A telephone system in which telephone connections between customers are ordinarily established by electronic and mechanical apparatus, controlled by manipulations of dials operated by calling parties. { 'dīl 'tel·əˌfōn ˌsist·əm }

dial tone [COMMUN] A tone employed in a dial telephone system to indicate that the equipment is ready for dialing operation. { 'dīl ˌtōn }

dial-up [COMMUN] **1.** The service whereby a dial telephone can be used to initiate and effect station-to-station telephone calls. **2.** In computer networks, pertaining to terminals which must dial up to receive service, as contrasted with those hard-wired or permanently connected into the network. { 'dīl ˌəp }

dial-up telephone system [COMMUN] The switched telephone network that is regulated by national governments; operated in the United States by various carriers. { 'dīl ˌəp 'tel·əˌfōn ˌsis·təm }

dialuric acid [ORG CHEM] $C_4H_4N_2O$ An acid that is derived by oxidation of uric acid or by the reduction of alloxan; may be used in organic synthesis. { dī·ə'lur·ik 'as·əd }

dial weight [ENG] A weight piece that acts on the invariable arm of an analytical balance and is added or removed from outside the case by a weight-lifting dialing system. { 'dīl ˌwāt }

dialysis [PHYS CHEM] A process of selective diffusion through a membrane; usually used to separate low-molecular-weight solutes which diffuse through the membrane from the colloidal and high-molecular-weight solutes which do not. { dī'al·ə·səs }

dialyzate [CHEM] The material that does not diffuse through the membrane during dialysis; alternatively, it may be considered the material that has diffused. { dī'al·əˌzāt }

dialyzer [CHEM ENG] **1.** The semipermeable membrane used for dialyzing liquid. **2.** The container used in dialysis; it is separated into compartments by membranes. { 'dī·əˌlīz·ər }

diamagnet [ELECTROMAG] A substance which is diamagnetic, such as the alkali and alkaline earth metals, the halogens, and the noble gases. { ˌdī·ə'magˌnət }

diamagnetic [ELECTROMAG] Having a magnetic permeability less than 1; materials with this property are repelled by a magnet and tend to position themselves at right angles to magnetic lines of force. { ˌdī·ə·mag'ned·ik }

diamagnetic Faraday effect [OPTICS] Faraday effect at frequencies near an absorption line which is split due to the splitting of the upper level only. { ˌdī·ə·magˌned·ik 'far·əˌdā iˌfekt }

diamagnetic resonance See cyclotron resonance. { ˌdī·ə·magˌned·ik 'rez·ən·əns }

diamagnetic susceptibility [ELECTROMAG] The susceptibility of a diamagnetic material, which is always negative and usually on the order of -10^{-5} cm^3/mole. { ˌdī·ə·magˌned·ik səˌsep·tə'bil·əd·ē }

diamagnetism [ELECTROMAG] The property of a material which is repelled by magnets. { ˌdī·ə'magˌnə·tiz·əm }

diamantine [MINERAL] Consisting of or resembling diamond. { ˌdī·ə'manˌtēn }

diameter [MATH] **1.** A line segment which passes through the center of a circle, and whose end points lie on the circle. **2.** The length of such a line. **3.** For a conic, any straight line that passes through the midpoints of all the chords of the conic that are parallel to a given chord. **4.** For a set, the smallest number that is greater than or equal to the distance between every pair of points of the set. { dī'am·əd·ər }

diameter group [MECH ENG] A dimensionless group, used in the study of flow machines such as turbines and pumps, equal to the fourth root of pressure number 2 divided by the square root of the delivery number. { dī'am·əd·ər ˌgrüp }

diameter tape [ENG] A tape for measuring the diameter of trees; when wrapped around the circumference of a tree, it reads the diameter directly. { dī'am·əd·ər ˌtāp }

diametral curve [MATH] A curve that passes through the midpoints of a family of parallel chords of a given curve. { dī'am·ə·trəl 'kərv }

diametral pitch [DES ENG] A gear tooth design factor expressed as the ratio of the number of teeth to the diameter of the pitch circle measured in inches. { dī'am·ə·trəl 'pich }

diametral plane [MATH] **1.** A plane that passes through the center of a sphere. **2.** A plane that passes through the midpoints of a family of parallel chords of a quadric surface that are parallel to a given chord. { dī'am·ə·trəl 'plān }

diametral surface [MATH] A surface that passes through the midpoints of a family of parallel chords of a given surface that are parallel to a given chord. { dī'am·ə·trəl 'sər·fəs }

diamictite [PETR] A calcareous, terrigenous sedimentary rock that is not sorted or poorly sorted and contains particles of many sizes. Also known as mixtite. { dī·ə'mikˌtīt }

diamicton [PETR] A nonlithified diamictite. Also known as symmicton. { dī·ə'mikˌtän }

diamide [ORG CHEM] A molecule that has two amide ($-CONH_2$) groups. { 'dī·əˌmīd }

diamidine [ORG CHEM] A molecule that has two amidine ($-C=NHNH_2$) groups. { dī'am·əˌdēn }

diamine [ORG CHEM] Any compound containing two amino groups. { 'dī·əˌmēn }

diamine oxidase [BIOCHEM] A flavoprotein which catalyzes the aerobic oxidation of amines to the corresponding aldehyde and ammonia. { 'dī·əˌmēn 'äk·səˌdās }

diamino [ORG CHEM] A term used in chemical nomenclature to indicate the presence in a molecule of two amino ($-NH_2$) groups. { dī'am·əˌnō }

3,5-diaminobenzoic acid [ORG CHEM] $C_7H_8N_2O_2$ Monohydrate crystals with a melting point of 228°C; soluble in organic solvents such as alcohol and benzene; used in the detection and determination of nitrites. { ˌthrē ˌfīv dīˌam·əˌnōˌben'zō·ik 'as·əd }

2,7-diaminofluorene [ORG CHEM] $C_{13}H_{12}N_2$ A compound

DIAMAGNETIC FARADAY EFFECT

Diamagnetic Faraday effect which is temperature-independent; ν = frequency of light. Splitting of the absorption line is due to the splitting of the upper level only, and the lower level of the line is not split.

crystallizing as needlelike crystals from water; the melting point is 165°C; soluble in alcohol; used to detect bromide, chloride, nitrate, persulfate, cadmium, zinc, copper, and cobalt. Also known as 2,7-fluorenediamine. { ¦tü ¦sev·ən dī¦am·ə‚nō 'flür‚ēn }

diaminopimelate [BIOCHEM] $C_7H_{14}O_4N_2$ A compound that serves as a component of cell wall mucopeptide in some bacteria and as a source of lysine in all bacteria. Abbreviated DAP. { dī¦am·ə‚nō'pim·ə‚lāt }

diamond [MINERAL] A colorless mineral composed entirely of carbon crystallized in the isometric system as octahedrons, dodecahedrons, and cubes; the hardest substance known; used as a gem and in cutting tools. { 'dī‚mənd }

diamond antenna *See* rhombic antenna. { 'dī‚mənd an 'ten·ə }

diamond anvil [ENG] A brilliant-cut diamond of extremely high quality that is modified to have 16 sides and has the culet cut off to create either a flat tip or a flat surface followed by a bevel of 5–10°. { 'dī·mənd 'an·vəl }

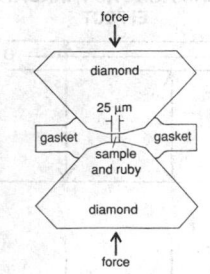

DIAMOND-ANVIL CELL

Configuration of diamond anvils used in a diamond-anvil cell.

diamond-anvil cell [ENG] A device for generating an extremely high pressure in a sample that is sandwiched between two diamond anvils to which forces are applied. { 'dī·mənd ¦an·vəl ‚sel }

diamond bit [DES ENG] A rotary drilling bit crowned with bort-type diamonds, used for rock boring. Also known as bort bit. { 'dī‚mənd ‚bit }

diamond boring [ENG] Boring with a diamond tool. { 'dī·mənd ‚bòr·iŋ }

diamond canker [PL PATH] A virus disease that affects the bark of certain stone-fruit trees, resulting in weakening of the trunk and limbs. { 'dī·mənd ‚kaŋ·kər }

diamond chisel [DES ENG] A chisel having a V-shaped or diamond-shaped cutting edge. { 'dī·mənd ‚chiz·əl }

diamond circuit [ELECTR] A gate circuit that provides isolation between input and output terminals in its off state, by operating transistors in their cutoff region; in the on state the output voltage follows the input voltage as required for gating both analog and digital signals, while the transistors provide current gain to supply output current on demand. { 'dī·mənd ‚sər·kət }

diamond coring [ENG] Obtaining core samples of rock by using a diamond drill. { 'dī·mənd 'kòr·iŋ }

diamond count [DES ENG] The number of diamonds set in a diamond crown bit. { 'dī·mənd ‚kaûnt }

diamond crossing [CIV ENG] An oblique railroad crossing that forms a diamond shape between the tracks. { 'dī·mənd ‚kròs·iŋ }

diamond crown [DES ENG] The cutting bit used in diamond drilling; it consists of a steel shell set with black diamonds on the face and cutting edges. { 'dī·mənd ‚kraûn }

diamond drill [DES ENG] A drilling machine with a hollow, diamond-set bit for boring rock and yielding continuous and columnar rock samples. { 'dī·mənd ‚dril }

Diamond-Hinman radiosonde [ENG] A variable audio-modulated radiosonde used by United States weather services; the carrier signal from the radiosonde is modulated by audio signals determined by the electrical resistance of the humidity- and temperature-transducing elements and by fixed reference resistors; the modulating signals are transmitted in a fixed sequence at predetermined pressure levels by means of a baroswitch. { ¦dī·mənd ¦hin·mən 'rād·ē·ō‚sänd }

diamond indenter [ENG] An instrument that measures hardness by indenting a material with a diamond point. { 'dī·mənd in'den·tər }

diamond matrix [DES ENG] The metal or alloy in which diamonds are set in a drill crown. [GEOL] The rock material in which diamonds are formed. { 'dī·mənd 'mā·triks }

diamond orientation [DES ENG] The set of a diamond in a cutting tool so that the crystal face will be in contact with the material being cut. { 'dī·mənd ‚òr·ē·ən'tā·shən }

diamond-particle bit [DES ENG] A diamond bit set with small fragments of diamonds. { 'dī·mənd¦pärd·ə·kəl ‚bit }

diamond paste [MATER] An abrasive consisting of diamond dust in a viscous material. { 'dī·mənd ‚pāst }

diamond pattern [DES ENG] The arrangement of diamonds set in a diamond crown. { 'dī·mənd ‚pad·ərn }

diamond plate [LAP] A plate with a layer of diamond dust and oil; used for rubbing down gems. { 'dī·mənd ‚plāt }

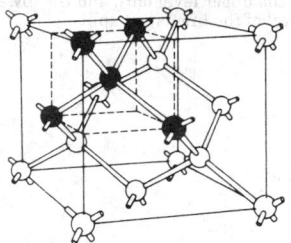

DIAMOND STRUCTURE

Crystal structure of diamond. Large cube is unit cell. Small cube emphasizes arrangement of bonds between atom at center of tetrahedron and nearest neighbors at vertices.

diamond point [DES ENG] A cutting tool with a diamond tip. { 'dī·mənd ‚póint }

diamond-point bit *See* mud auger. { 'dī·mənd ¦póint ‚bit }

diamond-pyramid hardness number [MET] The quotient of the load applied in the diamond-pyramid hardness test divided by the pyramidal area of the impression. { ¦dī·mənd ¦pir·ə·mid 'härd·nəs ‚nəm·bər }

diamond-pyramid hardness test [MET] An indentation hardness test in which a diamond-pyramid indenter, with a 136° angle between opposite faces, is forced under variable loads into the surface of a test specimen. Also known as Vickers hardness test. { ¦dī·mənd ¦pir·ə·mid 'härd·nəs ‚test }

diamond reamer [DES ENG] A diamond-inset pipe behind, and larger than, the drill bit and core barrel that is used for enlarging boreholes. { 'dī·mənd ‚rēm·ər }

diamond-ring effect [ASTRON] A phenomenon observed just before and after the central phase of a total solar eclipse, in which the last Baily's bead glows brightly compared with other visible features, and the solar corona forms a band that is visible on the rest of the lunar edge. { 'dī·mənd 'riŋ i‚fekt }

diamonds [FL MECH] The pattern of shock waves often visible in a rocket exhaust which resembles a series of diamond shapes placed end to end. { 'dī·mənz }

diamond saw [DES ENG] A circular, band, or frame saw inset with diamonds or diamond dust for cutting sections of rock and other brittle substances. { 'dī·mənd ‚sò }

diamond setter [ENG] A person skilled at setting diamonds by hand in a diamond bit or a bit mold. { 'dī·mənd ‚sed·ər }

diamond size [ENG] In the bit-setting and diamond-drilling industries, the number of equal-size diamonds having a total weight of 1 carat; a 10-diamond size means 10 stones weighing 1 carat. { 'dī·mənd ‚sīz }

diamond structure [CRYSTAL] A crystal structure in which each atom is the center of a tetrahedron formed by its nearest neighbors. { ‚dī·mənd ‚strək·chər }

diamond stylus [ENG ACOUS] A stylus having a ground diamond as its point. { 'dī·mənd 'stī·ləs }

diamond tool [DES ENG] **1.** Any tool using a diamond-set bit to drill a borehole. **2.** A diamond shaped to the contour of a single-pointed cutting tool, used for precision machining. { 'dī·mənd ‚tül }

diamond-turned optics [OPTICS] Optical elements that have been machined to a specular finish on metal-working lathes whose precision is so great that these elements can be used in the infrared without further optical working, and small elements with simple shapes can be used even in the visible region. { 'dī·mənd ¦tərnd 'äp·tiks }

diamond washer [MIN ENG] An apparatus for shaking and separating rock gravel containing diamonds, utilizing a vertical series of screens with 8-, 4-, 2-, and 1-millimeter mesh. { 'dī·mənd ‚wäsh·ər }

diamond wheel [DES ENG] A grinding wheel in which synthetic diamond dust is bonded as the abrasive to cut very hard materials such as sintered carbide or quartz. { 'dī·mənd ‚wēl }

diamyl phenol [ORG CHEM] $(C_5H_{11})_2C_6H_3OH$ A straw-colored liquid with a boiling range of 280–295°C; used in synthetic resins, lubricating oil additives, plasticizers, detergents, and fungicides. { dī'am·əl 'fē‚nól }

diamyl sulfide [ORG CHEM] $(C_5H_{11})_2S$ A combustible, yellow liquid with a distillation range of 170–180°C; used as a flotation agent and an odorant. { dī'am·əl 'səl‚fīd }

diandrous [BOT] Having two stamens. { dī'an·drəs }

Dianemaceae [MICROBIO] A family of slime molds in the order Trichales. { ‚dī·ə·nə'mās·ē‚ē }

dianite *See* columbite. { 'dī·ə‚nīt }

Dianthovirus [VIROL] A genus of plant viruses within the family Tombusviridae that are characterized by icosahedral particles containing two single-stranded positive-strand ribonucleic acid molecules; Carnation ringspot virus is the type species. Also known as Carnation ringspot virus group. { dī 'an·thə‚vī·rəs }

Dianulitidae [PALEON] A family of extinct, marine bryozoans in the order Cystoporata. { dī‚an·yə'lid·ə‚dē }

diapause [PHYSIO] A period of spontaneously suspended growth or development in certain insects, mites, crustaceans, and snails. { 'dī·ə‚póz }

diapedesis [MED] Hemorrhage of blood cells, especially erythrocytes, through an intact vessel wall into the tissues. { ‚dī·ə·pə'dē·səs }

Diapensiaceae [BOT] The single family of the Diapensiales, an order of flowering plants. { ‚dī·ə‚pen·sē'ās·ē‚ē }

Diapensiales [BOT] A monofamilial order of dicotyledonous plants in the subclass Dilleniidae comprising certain herbs and dwarf shrubs in temperate and arctic regions of the Northern Hemisphere. { ‚dī·ə‚pen·sē'ā·lēz }

Diaphanocephalidae [INV ZOO] A family of parasitic roundworms belonging to the Strongyloidea; snakes are the principal host. { di¦af·ə·nō·sə'fal·ə‚dē }

Diaphanocephaloidea [INV ZOO] A superfamily of nematodes represented by a single family, Diaphanocephalidae, distinguished by the modification of the stoma into two massive lateral jaws and the absence of a corona radiata or lips. { dī¦af·ə·nō‚sef·ə'lóid·ē‚ pp }

diaphone [NAV] A fog signal device that produces sounds similar to a siren and uses a reciprocating piston actuated by compressed air. { 'dī·ə‚fōn }

diaphorase [BIOCHEM] Mitochondrial flavoprotein enzymes which catalyze the reduction of dyes, such as methylene blue, by reduced pyridine nucleotides such as reduced diphosphopyridine nucleotide. { dī'af·ə‚rās }

diaphorite [MINERAL] PB₂Ag₃Sb₃S₈ A gray-black orthorhombic mineral consisting of sulfide of lead, silver, and antimony, occurring in crystals. Also known as ultrabasite. { dī'af·ə‚rīt }

diaphragm [ANAT] The dome-shaped partition composed of muscle and connective tissue that separates the abdominal and thoracic cavities in mammals. [ELECTROMAG] See iris. [ENG] A thin sheet placed between parallel parts of a member of structural steel to increase its rigidity. [ENG ACOUS] A thin, flexible sheet that can be moved by sound waves, as in a microphone, or can produce sound waves when moved, as in a loudspeaker. [OPTICS] Any opening in an optical system which controls the cross section of a beam of light passing through it, to control light intensity, reduce aberration, or increase depth of focus. Also known as lens stop. [PHYS] **1.** A separating wall or membrane, especially one which transmits some substances and forces but not others. **2.** In general, any opening, sometimes adjustable in size, which is used to control the flow of a substance or radiation. { 'dī·ə‚fram }

diaphragmatic hernia [MED] Protrusion of an abdominal organ through the diaphragm into the thoracic cavity. { ¦di·ə‚frag¦mad·ik 'hər·nē·ə }

diaphragmatic respiration [PHYSIO] Respiration effected primarily by movement of the diaphragm, changing the intrathoracic pressure. { ¦di·ə‚frag¦mad·ik ‚res·pə'rā·shən }

diaphragm cell [CHEM ENG] An electrolytic cell used to produce sodium hydroxide and chlorine from sodium chloride brine; porous diaphragm separates the anode and cathode compartments. { 'dī·ə‚fram ‚sel }

diaphragm compressor [MECH ENG] Device for compression of small volumes of a gas by means of a reciprocally moving diaphragm, in place of pistons or rotors. { 'dī·ə‚fram kəm'pres·ər }

diaphragmed pith [BOT] Pith in which plates or nests of sclerenchyma may be interspersed with the parenchyma. { ‚dī·ə‚framd 'pith }

diaphragm gage [ENG] Pressure- or vacuum-sensing instrument in which pressures act against opposite sides of an enclosed diaphragm that consequently moves in relation to the difference between the two pressures, actuating a mechanical indicator or electric-electronic signal. { 'dī·ə‚fram ‚gāj }

diaphragm horn [ENG ACOUS] A horn that produces sound by means of a diaphragm vibrated by compressed air, steam, or electricity. { 'dī·ə‚fram ‚hórn }

diaphragm jig [MIN ENG] A jig having a flexible diaphragm to pulse water; used in gravity concentration of minerals. { 'dī·ə‚fram ‚jig }

diaphragm meter [ENG] A flow meter which uses the movement of a diaphragm in the measurement of a difference in pressure created by the flow, such as a force-balance-type or a deflection-type meter. { 'dī·ə‚fram ‚mēd·ər }

diaphragm pump [MECH ENG] A metering pump which uses a diaphragm to isolate the operating parts from pumped liquid in a mechanically actuated diaphragm pump, or from hydraulic fluid in a hydraulically actuated diaphragm pump. { 'dī·ə‚fram ‚pəmp }

diaphragm setting [OPTICS] The position of a camera's diaphragm after opening or closing it. { 'dī·ə‚fram ‚sed·iŋ }

diaphragm valve [ENG] A fluid valve in which the open-close element is a flexible diaphragm; used for fluids containing suspended solids, but limited to low-pressure systems. { 'dī·ə‚fram ‚valv }

diaphthoresis See retrograde metamorphism. { dī¦af·thə'rē·səs }

diaphthorite [PETR] Schistose rocks in which minerals have formed by retrograde metamorphism. { dī'af·thə‚rīt }

diaphyseal aclasis See multiple hereditary exostoses. { dī'af·sē·əl 'ak·lə·səs }

diaphysis [ANAT] The shaft of a longbone. { dī'af·ə·səs }

diapir [GEOL] A dome or anticlinal fold in which a mobile plastic core has ruptured the more brittle overlying rock. Also known as diapiric fold; piercement dome; piercing fold. { 'dī·ə‚pir }

diapiric fold See diapir. { ¦dī·ə¦pir·ik 'fōld }

diapophysis [ANAT] The articular portion of a transverse process of a vertebra. { ‚dī·ə'päf·ə·səs }

diapositive See projection slide. { ‚dī·ə'päz·əd·iv }

Diapriidae [INV ZOO] A family of hymenopteran insects in the superfamily Proctotrupoidca. { ‚dī·ə'prī·ə‚dē }

diarch [BOT] Of a plant, having two protoxylem points or groups. { 'dī‚ärch }

diarrhea [MED] The passage of loose or watery stools, usually at more frequent than normal intervals. { ‚dī·ə'rē·ə }

diarsine [ORG CHEM] An arsenic compound containing an As-As bond with the general formula $(R_2As)_2$, where R represents a functional group such as CH_3. { dī'är‚sēn }

diarthrosis [ANAT] A freely moving articulation, characterized by a synovial cavity between the bones. { ¦dī·är'thrō·səs }

diarylamine [ORG CHEM] A molecule that contains an amine group and two aryl groups joined to the amino nitrogen. { ¦dī·ə'ril·ə‚mēn }

diaspore [MINERAL] AlO(OH) A mineral composed of some bauxites occurring in white, lamellar masses; crystallizes in the orthorhombic system. { 'dī·ə‚spór }

diasporometer [OPTICS] Oppositely rotating wedges used in optical rangefinders to obtain deviation of the axis of the image. { ¦dī·ə·spór'äm·əd·ər }

diastase [BIOCHEM] An enzyme that catalyzes the hydrolysis of starch to maltose. Also known as vegetable diastase. { 'dī·ə‚stās }

diastasis [MED] Any simple separation of parts normally joined together, as the separation of an epiphysis from the body of a bone without true fracture, or the dislocation of an amphiarthrosis. [PHYSIO] The final phase of diastole, the phase of slow ventricular filling. { dī'as·tə·səs }

diastem [GEOL] A temporal break between adjacent geologic strata that represents nondeposition or local erosion but not a change in the general regimen of deposition. { 'dī·ə‚stem }

diastema [ANAT] A space between two types of teeth, as between an incisor and premolar. [CYTOL] Modified cytoplasm of the equatorial plane prior to cell division. { ‚dī·ə'stē·mə }

diastereoisomer [ORG CHEM] One of a pair of optical isomers which are not mirror images of each other. Also known as diastereomer. { ‚dī·ə‚ster·ē·ō'ī·sə· mər }

diastereomer See diastereoisomer. { ‚dī·ə¦ster·ē'ō·mər }

diastereotopic ligand [ORG CHEM] A ligand whose replacement or addition gives rise to diastereomers. { ¦dī·ə‚ster·ē·ə'täp·ik 'līg·ənd }

diastole [PHYSIO] The rhythmic relaxation and dilation of a heart chamber, especially a ventricle. { dī'as·tə·lē }

diastolic pressure [PHYSIO] The lowest arterial blood pressure during the cardiac cycle; reflects relaxation and dilation of a heart chamber. { ¦dī·ə¦stäl·ik 'presh·ər }

diastrophism [GEOL] **1.** The general process or combination of processes by which the earth's crust is deformed. **2.** The results of this deforming action. { dī'as·trə‚fiz·əm }

diathermanous [PHYS] Capable of transmitting radiant heat. Also known as diathermic. { ¦dī·ə¦thər·mə·nəs }

diathermic See diathermanous. { ¦dī·ə¦thər·mik }

diathermous envelope [THERMO] A surface enclosing a thermodynamic system in equilibrium that is not an adiabatic envelope; intuitively, this means that heat can flow through the surface. { ¦dī·ə¦thər·məs 'en·və‚lōp }

diathermy [MED] The therapeutic use of high-frequency

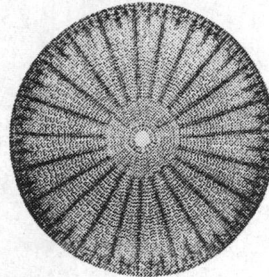

DIATOM

Arachnoidiscus ehrenbergii, a concentric diatom with radial symmetry. (*From H. J. Fuller and O. Tippo, College Botany, rev. ed., Holt, 1954*)

DIATRYMIFORMES

Diatryma steini (reconstruction). (*American Museum of Natural History*)

electric currents to produce localized heat in body tissues. { 'dī·ə,thər·mē }

diathermy interference [COMMUN] Television interference caused by diathermy equipment; produces a herringbone pattern in a dark horizontal band across the picture. { 'dī·ə,thər·mē ,in·tər'fir·əns }

diathermy machine [ELECTR] A radio-frequency oscillator, sometimes followed by rf amplifier stages, used to generate high-frequency currents that produce heat within some part of the body for therapeutic purposes. { 'dī·ə,thər·mē mə,shēn }

diatom [INV ZOO] The common name for algae composing the class Bacillariophyceae; noted for the symmetry and sculpturing of the siliceous cell walls. { 'dī·ə,täm }

diatomaceous earth [GEOL] A yellow, white, or light-gray, siliceous, porous deposit made of the opaline shells of diatoms; used as a filter aid, paint filler, adsorbent, abrasive, and thermal insulator. Also known as kieselguhr; tripolite. { ¦dī·ə·tə¦mā·shəs 'ərth }

diatomaceous ooze [GEOL] A pelagic, siliceous sediment composed of more than 30% diatom tests, up to 40% calcium carbonate, and up to 25% mineral grains. { ¦dī·ə·tə¦mās·shəs 'üz }

diatomic [CHEM] Consisting of two atoms. { ¦dī·ə'täm·ik }

diatomite [GEOL] Dense, chert-like, consolidated diatomaceous earth. { 'dī·ad·ə,mīt }

diatonic scale [ACOUS] A musical scale in which the octave is divided into intervals of two different sizes, five of one and two of the other, with adjustments in tuning systems other than equal temperament. { ¦dī·ə¦tän·ik 'skāl }

diatreme [GEOL] A circular volcanic vent produced by the explosive energy of gas-charged magmas. { 'dī·ə,trēm }

diatropism [BOT] Growth orientation of certain plant organs that is transverse to the line of action of a stimulus. { dī'a·trə,piz·əm }

Diatrymiformes [PALEON] An order of extinct large, flightless birds having massive legs, tiny wings, and large heads and beaks. { dī,a·trə·mə'for,mēz }

diauxic growth [MICROBIO] The diphasic response of a culture of microorganisms based on a phenotypic adaptation to the addition of a second substrate; characterized by a growth phase followed by a lag after which growth is resumed. { dī'ók,-sik 'grōth }

diazepam [PHARM] $C_{16}H_{13}ClN_2O$ A benzodiazepine with a melting point of 125–126°C; used as a minor tranquilizer to relieve muscle spasms, anxiety, and tension. Also known by trade name Valium. { dī'az·ə,pam }

diazine [ORG CHEM] **1.** A hydrocarbon consisting of an unsaturated hexatomic ring of two nitrogen atoms and four carbons. **2.** Suffix indicating a ring compound with two nitrogen atoms. { 'dī·ə,zēn }

diazinon [ORG CHEM] $C_{12}H_{21}N_2O_3PS$ A light amber to dark brown liquid with a boiling point of 83–84°C; used as an insecticide for soil and household pests, and as an insecticide and nematicide for fruits and vegetables. { dī'a·zə,nōn }

diazoalkane [ORG CHEM] A compound with the general formula $R_2C=N_2$ in which two hydrogen atoms of an alkane molecule have been replaced by a diazo group. { dī¦a·zō'al,kān }

diazoamine [ORG CHEM] The grouping $-N=NNH-$. Also known as azimino. { dī¦a·zō'a,mēn }

diazoaminobenzene [ORG CHEM] $C_6H_5NNNHC_6H_5$ Golden yellow scales with a melting point of 96°C; soluble in alcohol, ether, and benzene; used for dyes and insecticides. { dī¦a·zō¦am·ə·nō'ben,zēn }

diazoate [ORG CHEM] A salt with molecular formula of the type $C_6H_5N=NOOM$, where M is a nonvalent metal. { dī'a·zə,wāt }

diazo compound [ORG CHEM] An organic compound containing the radical $-N=N-$. { dī'a·zō 'käm,paúnd }

diazo coupler [GRAPHICS] A substance that combines with the unexposed diazonium salt to form a dye image during development of exposed photosensitive material. { dī'a·zō 'kəp·lər }

diazo dye See developed dye. { dī'a·zō ,dī }

diazo group [ORG CHEM] A functional group with the formula $=N_2$. { dī'a·zō 'grüp }

diazoic acid [ORG CHEM] $C_6H_5N=NOOH$ An isomeric form of phenylnitramine. { 'dī·ə,zō·ik 'as·əd }

diazole [ORG CHEM] A cyclic hydrocarbon with five atoms in the ring, two of which are nitrogen atoms and three are carbon. { 'dī·ə,zōl }

diazo material [GRAPHICS] A sensitized film or paper coated with diazonium salts that produces an image, usually nonreversed, when exposed to light. { dī'a·zō mə,tir·ē·əl }

diazomethane [ORG CHEM] CH_2N_2 A poisonous gas used in organic synthesis to methylate compounds. { dī¦a·zō'me,thān }

diazonium [ORG CHEM] The grouping $=N'''N$. { ,dī·ə'zō·nē·əm }

diazonium salts [ORG CHEM] Compounds of the type R·X·N:N, where R represents an alkyl or aryl group and X represents an anion such as a halide. { ,dī·ə'zō·nē·əm 'sóls }

diazo oxide [ORG CHEM] An organic molecule or a grouping of organic molecules that have a diazo group and an oxygen atom joined to ortho positions of an aromatic nucleus. Also known as diazophenol. { dī'a·zō 'äk,sīd }

diazophenol See diazo oxide. { dī'a·zō'fe,nól }

diazo print [GRAPHICS] A reduction made by the whiteprint process. { dī'a·zō ,print }

diazo process See diazotization. { dī'a·zō ,präs·əs }

diazosulfonate [ORG CHEM] A salt formed from diazosulfonic acid. { dī¦a·zō 'səl·fə,nāt }

diazosulfonic acid [ORG CHEM] $C_6H_5N=NSO_3H$ Any of a group of aromatic acids containing the diazo group bonded to the sulfonic acid group. { dī¦a·zō·səl'fän·ik 'as·əd }

diazotization [ORG CHEM] Reaction between a primary aromatic amine and nitrous acid to give a diazo compound. Also known as diazo process. { dī,az·ət·ə'zā·shən }

diazotroph [MICROBIO] An organism that carries out nitrogen fixation; examples are *Clostridium* and *Azotobacter*. { dī'az·ō,träf }

diazotype [GRAPHICS] A photograph or photocopy produced on a surface by coating with a photosensitive solution containing a diazo compound. { dī'az·ə,tīp }

Dibamidae [VERT ZOO] The flap-legged skinks, a small family of lizards in the suborder Sauria comprising three species confined to southeastern Asia. { dī'bäm·ə,dē }

dibasic [CHEM] **1.** Compounds containing two hydrogens that may be replaced by a monovalent metal or radical. **2.** An alcohol that has two hydroxyl groups, for example, ethylene glycol. { dī'bās·ik }

dibasic acid [CHEM] An acid having two hydrogen atoms capable of replacement by two basic atoms or radicals. { dī'bās·ik 'as·əd }

dibasic calcium phosphate See calcium phosphate. { dī'bās·ik 'kal·sē·əm 'fäs,fāt }

dibasic magnesium citrate [ORG CHEM] $MgHC_6H_5O_7·5H_2O$ A white or yellowish powder soluble in water; used as a dietary supplement or in medicine. { dī'bās·ik mag'nē·zē·əm 'sī,trāt }

dibenzyl See bibenzyl. { dī'ben·zil }

dibenzyl disulfide [ORG CHEM] $C_6H_5CH_2SSCH_2C_6H_5$ A compound crystallizing in leaflets with a melting point of 71–72°C; soluble in hot methanol, benzene, ether, and hot ethanol; used as an antioxidant in compounding of rubber and as an additive to silicone oils. { dī'ben·zil dī'səl,fīd }

dibenzyl ether See benzyl ether. { dī'ben·zil 'ē·thər }

N,N′-dibenzylethylenediamine [PHARM] $C_6H_5CH_2NH-CH_2CH_2NHCH_2C_6H_5$ An oily liquid, soluble in most organic solvents; used in the manufacture of a repository form of penicillin. { ¦en ¦en,prīm dī'ben·zil¦eth·ə,lēn¦dī·ə,mēn }

dibit [COMPUT SCI] A pair of binary digits, used to specify one of four values. { 'dī,bit }

diborane [INORG CHEM] B_2H_6 A colorless, volatile compound that is soluble in ether; boiling point −92.5°C, melting point −165.5°C; can be used to produce pentaborane and decaborane, proposed for use as rocket fuels; also used to synthesize organic boron compounds. { dī'bór,ān }

diborate See borax. { dī'bór,āt }

Dibothriocephalus See Diphyllobothrium. { ¦dī·bə,thrī·ə'sef·ə·ləs }

Dibranchia [INV ZOO] A subclass of the Cephalopoda containing all living cephalopods except *Nautilus*; members possess two gills and, when present, an internal shell. { dī'braŋ·kē·ə }

dibromide [CHEM] Indicating the presence of two bromine atoms in a molecule. { dī'brō,mīd }

dibromo- [CHEM] A prefix indicating two bromine atoms. { dī'brō·mō }

dibromochloropropane [ORG CHEM] $C_3H_5Br_2Cl$ A light yellow liquid with a boiling point of 195°C; used as a nematicide for crops. Abbreviated DBCP. { dī¦brō·mō,klór·ə'prō,pān }

dibromodifluoromethane [ORG CHEM] CF_2Br_2 A colorless, heavy liquid with a boiling point of 24.5°C; soluble in methanol and ether; used in the synthesis of dyes and pharmaceuticals and as a fire-extinguishing agent. { dī¦brō·mō·dī¦flür·ō'me,thän }

dibromomethane *See* methylene bromide. { dī¦brō·mō'me,thän }

2,6-dibromoquinone-4-chlorimide [ORG CHEM] $C_6H_2Br_2$ClNO Yellow prisms, soluble in water; used as a reagent for phenol and phosphatases. { ¦tu ¦siks dī¦brō·mō·kwə'nōn ¦fór 'klór·ə,mīd }

3,5-dibromosalicylaldehyde [ORG CHEM] $Br_2C_6H_2(OH)$CHO Pale yellow crystals with a melting point of 86°C; readily soluble in ether, chloroform, benzene, alcohol, and glacial acetic acid; used as an antibacterial agent. { ¦thrē ¦fīv dī¦brō·mō¦sal·ə·səl'al·də,hīd }

dibucaine [ORG CHEM] $C_{20}H_{29}O_2N_3$ A local anesthetic used both as the base and the hydrochloride salt. { 'dī·byə,kān }

dibutyl [ORG CHEM] Indicating the presence of two butyl groupings bonded through a third atom or group in a molecule. { dī'byüd·əl }

dibutyl amine [ORG CHEM] $C_8H_{19}N$ A colorless, clear liquid with amine aroma; either di-*n*-butylamine, $(C_4H_9)_2$NH, boiling at 160°C, insoluble in water, soluble in hydrocarbon solvents, or di-*sec*-butylamine, $(CH_3CHCH_2CH_3)_2$NH, boiling at 133°C, flammable; used in the manufacture of dyes. { dī'byüd·əl 'a,mēn }

dibutyl maleate [ORG CHEM] $C_4H_9OOCCHCHCOOC_4H_9$ Oily liquid used for copolymers and plasticizers and as a chemical intermediate. { dī'byüd·əl 'mal·ē,āt }

dibutyl oxalate [ORG CHEM] $(COOC_4H_9)_2$ High-boiling, water-white liquid with mild odor, used as a solvent and in organic synthesis. { dī'byüd·əl 'äk·sə,lāt }

dibutyl phthalate [ORG CHEM] $C_{16}H_{22}O_4$ A colorless liquid, used as a plasticizer and insect repellent. { dī'byüd·əl 'tha,lāt }

dibutyl succinate [ORG CHEM] $C_{12}H_{22}O_4$ A colorless liquid, insoluble in water; used as a repellent for cattle flies, cockroaches, and ants around barns. { dī'byüd·əl 'sək·sə,nāt }

dibutyl tartrate [ORG CHEM] $(COOC_4H_9)_2(CHOH)_2$ Liquid used as a solvent and plasticizer for cellulosics and as a lubricant. { dī'byüd·əl 'tär,trāt }

dicalcium [CHEM] A molecule containing two atoms of calcium. { dī'kal·sē·əm }

dicalcium orthophosphate *See* calcium phosphate. { dī'kal·sē·əm ,ór·thō'fäs,fāt }

dicalcium phosphate *See* calcium phosphate. { dī'kal·sē·əm 'fäs,fāt }

di-cap storage [ELECTR] Device capable of holding data in the form of an array of charged capacitors and using diodes for controlling information flow. { 'dī,kap 'stór·ij }

dicarbocyanine [ORG CHEM] **1.** A member of a group of dyes termed the cyanine dyes; the structure consists of two heterocyclic rings joined to the five-carbon chain: =CH—CH=CH—CH=CH—. **2.** A particular dicarbocyanine dye containing two quinoline heterocyclic rings. { dī¦kär·bə¦sī·ə,nēn }

dicarboxylic acid [ORG CHEM] A compound with two carboxyl groups. { dī¦kär·bäk¦sil·ik 'as·əd }

dicarpellate [BOT] Having two carpels. { dī'kär·pə,lāt }

dicaryon *See* dikaryon. { dī'kar·ē,än }

dication [CHEM] A doubly charged cation with the general formula X^{2+}. { dī'kat,ī·ən }

dice *See* die. { dīs }

DICE *See* digital intercontinental conversion equipment.

dicentric [CYTOL] Having two centromeres. { dī'sen·trik }

dicephaly [MED] A severe congenital anomaly in which the infant is born with two distinct heads. { dī'sef·ə·lē }

dicerous [INV ZOO] Having two tentacles or two antennae. { 'dī·sə·rəs }

Dice's life zones [ECOL] Biomes proposed by L.R. Dice based on the concept of the biotic province. { 'dīs·əz 'līf ,zōnz }

dichasium [BOT] A cyme producing two main axes from the primary axis or shoot. { dī'kā·zhē·əm }

Dichelesthiidae [INV ZOO] A family of parasitic copepods in the suborder Caligoida; individuals attach to the gills of various fishes. { dī¦ke·ləs'thī·ə,dē }

dichlamydeous [BOT] Having both calyx and corolla. { dī·klə'mid·ē·əs }

dichlobenil [ORG CHEM] $C_7H_3Cl_2N$ A colorless, crystalline compound with a melting point of 139–145°C; used as a herbicide to control weeds in orchards and nurseries. { dī'klō·bə·nəl }

dichlofenthion [ORG CHEM] $C_{10}H_{13}Cl_2O_3PS$ A white, liquid compound, insoluble in water; used as an insecticide and nematicide for ornamentals, flowers, and lawns. { dī,klō·fən'thī,än }

dichlofluanid [ORG CHEM] $C_9H_{11}Cl_2FN_2O_2S$ A white powder with a melting point of 105–105.6°C; insoluble in water; used as a fungicide for fruits, garden crops, and ornamental flowers. { ,dī·klō·flü'an·əd }

dichlone [ORG CHEM] $C_{10}H_4O_2Cl_2$ A yellow, crystalline compound, used as a fungicide for foliage and as an algicide. { 'dī,klōn }

dichloramine [INORG CHEM] **1.** NH_2Cl_2 An unstable molecule considered to be formed from ammonia by action of chlorine. Also known as chlorimide. **2.** Any chloramine with two chlorine atoms joined to the nitrogen atom. { dī'klór·ə,mēn }

dichloride [CHEM] Any inorganic salt or organic compound that has two chloride atoms in its molecule. { dī'klór,īd }

dichloroacetic acid [ORG CHEM] $CHCl_2COOH$ A strong liquid acid, formed by chlorinating acetic acid; used in organic synthesis. { dī¦klór·ō·ə¦sēd·ik 'as·əd }

dichlorobenzene [ORG CHEM] $C_6H_4Cl_2$ Any of a group of substitution products of benzene and two atoms of chlorine; the three forms are *meta*-dichlorobenzene, colorless liquid boiling at 172°C, soluble in alcohol and ether, insoluble in water, or *ortho*-, colorless liquid boiling at 179°C, used as a solvent and chemical intermediate, or *para*-, volatile white crystals, insoluble in water, soluble in organic solvents, used as a germicide, insecticide, and chemical intermediate. { dī¦klór·ō'ben,zēn }

1,4-dichlorobutane [ORG CHEM] $Cl(CH_2)_4Cl$ A colorless, flammable liquid with a pleasant odor, boiling point 155°C; soluble in organic solvents; used in organic synthesis, including adiponitrile. Abbreviated DCB. { ¦wən ¦fór dī¦klór·ō'byü ,tān }

dichlorodiethylsulfide *See* mustard gas. { dī¦klór·ō·dī¦eth·əl'səl,fīd }

dichlorodifluoromethane [ACOUS] CCl_2F_2 A nontoxic, nonflammable, colorless gas made from carbon tetrachloride; boiling point −30°C; used as a refrigerant and as a propellant in aerosols. { dī¦klór·ō·dī¦flür·ō'me,th aman }

***p,p'*-dichlorodiphenylmethyl carbinol** [ORG CHEM] $(ClC_6H_4)_2C(CH_3)OH$ Crystals with a melting point of 69–69.5°C; soluble in organic solvents; used as an insecticide. Abbreviated DMC. { ¦pē ¦pē,prīm dī¦klór·ō·dī¦fen·əl¦meth·əl 'kär·bə,nól }

dichlorodiphenyltrichloroethane *See* DDT. { dī¦klór·ō·dī¦fen·əl·trī¦klór·ō'e,thän }

***sym*-dichloroethylene** [ORG CHEM] $CHClCHCl$ Colorless, toxic liquid with pleasant aroma, boiling at 59°C; decomposes in light, air, and moisture; soluble in organic solvents, insoluble in water; exists in cis and trans forms; used as solvent, in medicine, and for chemical synthesis. { ¦sim dī¦klór·ō'eth·ə,lēn }

dichloroethyl ether [ORG CHEM] $ClCH_2CH_2OCH_2CH_2Cl$ A colorless liquid insoluble in water, soluble in organic solvents; used as a solvent in paints, varnishes, lacquers, and as a soil fumigant. { dī,klór·ō'eth·əl'ē·thər }

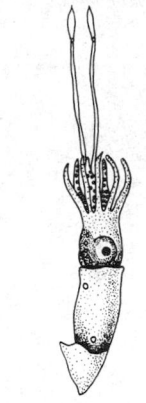

The luminous deep-sea squid *Lycoteuthis diadema*.

dichlorofluoromethane [ORG CHEM] $CHCl_2F$ A colorless, heavy gas with a boiling point of 8.9°C and a freezing point of −135°C; soluble in alcohol and ether; used in fire extinguishers and as a solvent, refrigerant, and aerosol propellant. Also known as fluorocarbon 21; fluorodichloromethane. { dī¦klȯr·ō¦flür·ō'me,thän }

α-dichlorohydrin [ORG CHEM] $CH_2ClCHOHCH_2Cl$ Unstable liquid, the commercial product consisting of a mixture of two isomers; used as a solvent and a chemical intermediate. Abbreviated GDCH. { ¦al·fə dī,klȯr·ō'hī·drən }

dichloromethylsilane See methyldichlorosilane. { ˌdī·klȯr·ō,meth·əl'sī,lān }

2,6-dichloro-4-nitroaniline [ORG CHEM] $C_6H_4Cl_2N_2O_2$ A yellow, crystalline compound that melts at 192–194°C; used as a fungicide for fruits, vegetables, and ornamental flowers. Abbreviated DCNA. { ¦tü ¦siks dī'klȯr·ō ¦fȯr ,nī·trō'an·ə,lēn }

dichloropentane [ORG CHEM] $C_5H_{10}Cl_2$ Mixed dichloro derivatives of normal pentane and isopentane; clear, light-yellow liquid used as solvent, paint and varnish remover, insecticide, and soil fumigant. { dī¦klȯr·ō'pen,tān }

dichlorophen [ORG CHEM] $C_{13}H_{10}Cl_2O_2$ A white, crystalline compound with a melting point of 177–178°C; used as an agricultural fungicide, germicide in soaps, and antihelminthic drug in humans. { dī'klȯr·ə·fən }

2,4-dichlorophenoxyacetic acid [ORG CHEM] $Cl_2C_6H_3OCH_2COOH$ Yellow crystals, melting at 142°C; used as a herbicide and pesticide. Abbreviated 2,4-D. { ¦tü ¦fȯr dī¦klȯr·ō·fə¦näk·sē·ə'sēd·ik 'as·əd }

1,3-dichloro-2-propanol [ACOUS] $ClCH_2CHOH$ CH_2Cl A liquid soluble in water and miscible with alcohol and ether; used as a solvent for nitrocellulose and hard resins, as a binder for watercolors, in the production of photographic lacquer, and in the determination of vitamin A. { ¦wən ¦thrē dī¦klȯr·ō ¦tü 'prō·pə,nȯl }

dichlorotoluene [ORG CHEM] $C_7H_6Cl_2$ A colorless liquid, soluble in organic solvents, insoluble in water; isomers are 2,4-$CH_3C_6H_3Cl_2$, boiling at 200–202°C, and 3,4-$(CH_3C_6H_3Cl_2)$, boiling at 209°C; used as solvent and chemical intermediate. { dī¦klȯr·ō'täl·yə,wēn }

dichlorprop [ORG CHEM] $C_9H_8Cl_2O_3$ A colorless, crystalline solid with a melting point of 117–118°C; used as a herbicide and fumigant for brush control on rangeland and rights-of-way. Abbreviated 2,4-DP. { dī'klȯr,präp }

dichlorvos [ORG CHEM] $C_4H_7O_4Cl_2P$ An amber liquid, used as an insecticide and miticide on public health pests, stored products, and flies on cattle. Abbreviated DDVP. { dī'klȯr,väs }

Dichobunidae [PALEON] A family of extinct artiodactyl mammals in the superfamily Dichobunoidea. { ˌdī·kə'byün·ə,dē }

Dichobunoidea [PALEON] A superfamily of extinct artiodactyl mammals in the suborder Paleodonta composed of small to medium-size forms with tri- to quadritubercular bunodont upper teeth. { ˌdī·kə·byə'nȯid·ē·ə }

dichogamous [BOT] Referring to a type of flower in which the pistils and stamens reach maturity at different times. { dī'käg·ə·məs }

dichogamy [BIOL] Producing mature male and female reproductive structures at different times. { dī'käg·ə·mē }

dichoptous [INV ZOO] Having the margins of the compound eyes separate. { dī'käp·təs }

dichotic listening See dichotic presentation. { dī¦käd·ik 'lis·ən·iŋ }

dichotic presentation [ACOUS] The simultaneous reception of one message through one ear and another message through the other ear. Also known as dichotic listening. { dī¦käd·ik ˌprē·zin'tā·shən }

dichotomic variable [QUANT MECH] A variable with a range consisting of two values; used, for example, to describe a particle with spin $\frac{1}{2}$. { dī¦kə¦täm·ik 'ver·ē·ə·bəl }

dichotomizing search [COMPUT SCI] A procedure for searching an item in a set, in which, at each step, the set is divided into two parts, one part being then discarded if it can be logically shown that the item could not be in that part. { dī'käd·ə,mīz·iŋ ,sərch }

dichotomous venation [BOT] A vascular arrangement in leaves such that the veins are forked, with each vein dividing at intervals into smaller veins of approximately equal size. { dī,käd·ə·məs ve'nā·shən }

dichotomy [ASTRON] The phase of the moon or an inferior planet at which exactly half of its disk is illuminated and the terminator is a straight line. [BIOL] **1.** Divided in two parts. **2.** Repeated branching or forking. [COMPUT SCI] A division into two subordinate classes; for example, all white and all nonwhite, or all zero and all nonzero. { dī'käd·ə·mē }

dichotriaene [INV ZOO] A type of sponge spicule with three rays. { ˌdi·kō'trī,ēn }

dichroic mirror [OPTICS] A glass surface coated with a special metal film that reflects certain colors of light while allowing others to pass through. { dī'krō·ik 'mir·ər }

dichroism [OPTICS] In certain anisotropic materials, the property of having different absorption coefficients for light polarized in different directions. { 'dī·krō,iz·əm }

dichromate [INORG CHEM] A salt of dichromic acid, usually orange or red. Also known as bichromate. { dī'krō,māt }

dichromate treatment [MET] Processing technique involving the formation of a corrosion-resistant film on the surface of a magnesium alloy by boiling the alloy in a sodium dichromate solution. { dī'krō,māt ,trēt·mənt }

dichromatic [BIOL] Having or exhibiting two color phases independently of age or sex. { dī·krə'mad·ik }

dichromatic dye [CHEM] Dye or indicator in which different colors are seen, depending upon the thickness of the solution. { dī·krə'mad·ik 'dī }

dichromatism [MED] Partial color blindness in which vision is apparently based on two primary colors rather than the normal three. { dī'krō·mə,tiz·əm }

dichromic [CHEM] Pertaining to a molecule with two atoms of chromium. { dī'krō·mik }

dichromic acid [INORG CHEM] $H_2Cr_2O_7$ An acid known only in solution, especially in the form of dichromates. { dī'krō·mik 'as·əd }

dicing [ELECTR] Sawing or otherwise machining a semiconductor wafer into small squares, or dice, from which transistors and diodes can be fabricated. { 'dīs·iŋ }

dicing cutter [MECH ENG] A cutting mill for sheet material; sheet is first slit into horizontal strands by blades, then fed against a rotating knife for dicing. { 'dīs·iŋ ,kəd·ər }

Dicke radiometer [ELECTR] A radiometer-type receiver that detects weak signals in noise by modulating or switching the incoming signal before it is processed by conventional receiver circuits. { 'dik·ə ,rād·ē'äm·əd·ər }

Dickinsoniidae [PALEON] A family that comprises extinct flat-bodied, multisegmented coelomates; identified as ediacaran fauna. { ,dik·ən·sə'nī·ə,dē }

dickinsonite [MINERAL] $H_2Na_6(Mn,Fe,Ca,Mg)_{14}(PO_4)_{12}·H_2O$ A green mineral consisting of foliated hydrous acid phosphate, chiefly of manganese, iron, and sodium, and is isostructural with arrojadite; specific gravity is 3.34. { 'dik·ən·sə,nīt }

dickite [MINERAL] $Al_2Si_2O_5(OH)_4$ A mineral of the kaolin group found crystallized in clay in hydrothermal veins; it is polymorphous with kaolinite and nacrite. { 'di,kīt }

Dicksoniaceae [BOT] A family of tree ferns characterized by marginal sori which are terminal on the veins and protected by a bivalved indusium. { ,dik·sə·nē'ās·ē,ē }

Dick test [IMMUNOL] A skin test to determine immunity to scarlet fever; *Streptococcus pyogenes* toxin is injected intracutaneously and produces a reaction if there is no circulating antitoxin. { 'dik ,test }

diclinous [BOT] Having stamens and pistils on different flowers. { dī'klī·nəs }

dicoccous [BOT] Composed of two adherent one-seeded carpels. { dī'käk·əs }

dicotyledon [BOT] Any plant of the class Magnoliopsida, all having two cotyledons. { ,dī,käd·əl'ēd·ən }

Dicotyledoneae [BOT] The equivalent name for Magnoliopsida. { dī,käd·əl·ə'dän·ē,ē }

dicovalent carbon See divalent carbon. { ,dī·kō'vā·lənt 'kär·bən }

Dicranales [BOT] An order of mosses having erect stems, dichotomous branching, and dense foliation. { ,dī·krə'nā·lēz }

dicrotic [MED] Pertaining to a secondary pressure wave in an artery on the descending limb of a main wave during diastole of the heart. { dī'kräd·ik }

dicrotophos [ORG CHEM] $C_8H_{16}O_2P$ The dimethyl phosphate of 3-hydroxy-*N,N*-dimethyl-*cis*-crotonamide; a brown liquid with a boiling point of 400°C; miscible with water; used as an insecticide and miticide for cotton, soybeans, seeds, and ornamental flowers. { dī'kräd·ə‚fäs }

dictionary [COMPUT SCI] A table establishing the correspondence between specific words and their code representations. { 'dik·shə‚ner·ē }

dictionary code [COMPUT SCI] An alphabetical arrangement of English words and terms, associated with their code representations. { 'dik·shə‚ner·ē ‚kōd }

dictionary encoding [COMPUT SCI] A method of data compression in which each word is replaced by a number which is the position of that word in a dictionary. { 'dik·shə‚ner·ē in'kōd·iŋ }

dictionary sort [COMPUT SCI] A sort algorithm that ignores capitalization, punctuation, and spaces, and treats numbers as if they were spelled out alphabetically. { 'dik·shə‚ner·ē ‚sȯrt }

dictyoblastospore [MYCOL] A blastospore with both cross and longitudinal septa. { ‚dik·tē·ō¦blas·tə‚spȯr }

Dictyoceratida [INV ZOO] An order of sponges of the class Demospongiae; includes the bath sponges of commerce. { ‚dik·tē·ō·sə'rad·əd·ə }

Dictyonellidina [PALEON] A suborder of extinct articulate brachiopods. { ‚dik·tē·ō·ne'lid·ən·ə }

dictyonema bed [GEOL] A thin shale bed rich in remains of graptolites of the genus *Dictyonema*. { ‚dik·tē·ə'nē·mə ‚bed }

Dictyonema pavonium [BOT] A common tropical basidiolichen with lobed thalli and the blue-green *Scytonema* as photobiont. { ‚dik·tē·ə‚nē·mə pə'vō·nē·əm }

dictyosome [CYTOL] A stack of two or more cisternae; a component of the Golgi apparatus. { 'dik·tē·ə‚sōm }

Dictyospongiidae [PALEON] A family of extinct sponges in the subclass Amphidiscophora having spicules resembling a one-ended amphidisc (paraclavule). { ‚dik·tē·ō‚spän'jī·ə‚dē }

Dictyosporae [MYCOL] A spore group of the imperfect fungi characterized by multicelled spores with cross and longitudinal septae. { ‚dik·tē·ə'spȯr·ē }

dictyospore [MYCOL] A multicellular spore in certain fungi characterized by longitudinal walls and cross septa. { 'dik·tē·ə‚spȯr }

dictyostele [BOT] A modified siphonostele in which the vascular tissue is dissected into a network of distinct strands; found in certain fern stems. { 'dik·tē·ə‚stēl }

Dictyosteliaceae [MICROBIO] A family of microorganisms belonging to the Acrasiales and characterized by strongly differentiated fructifications. { ¦dik·tē·ō¦stel·ē'ās·ē‚ē }

dicyandiamide [ORG CHEM] NH₂C(NH)(NHCN) White crystals with a melting range of 207–209°C; soluble in water and alcohol; used in fertilizers, explosives, oil well drilling muds, pharmaceuticals, and dyestuffs. Also known as cyanoguanidine. { dī‚sī·ən'dī·ə‚mad }

dicyanide [CHEM] A salt that has two cyanide groups. { dī'sī·ə‚nīd }

dicyanoargentates I See argentocyanides. { dī¦sī·ə·nō‚är·jən'tād·ēz 'wən }

dicycle [MATH] A simple closed dipath. Also known as directed cycle. { 'dī‚sī·kəl }

dicyclohexylamine [ORG CHEM] (C₆H₁₁)₂NH A clear, colorless liquid with a boiling point of 256°C; used for insecticides, corrosion inhibitors, antioxidants, and detergents, and as a plasticizer and catalyst. { dī¦sī·klō‚hek'sil·ə‚mēn }

dicyclohexylcarbodiimide [ORG CHEM] C₁₃H₂₂N₂ Crystals with a melting point of 35–36°C; used in peptide synthesis. Abbreviated DCC; DCCI. { dī¦sī·klō¦hek·səl¦kär·bō'dī·ə‚mad }

Dicyemida [INV ZOO] An order of mesozoans comprising minute, wormlike parasites of the renal organs of cephalopod mollusks. { ‚dī‚sī'em·ə·də }

DIDA See diisodecyl adipate.

didelphic [ANAT] Having a double uterus or genital tract. { dī'del·fik }

Didelphidae [VERT ZOO] The opossums, a family of arboreal mammals in the order Marsupialia. { dī'del·fə‚dē }

dideoxy method [CELL MOL] A method of DNA sequencing utilizing chain-terminating (dideoxy) nucleotides. { ‚dī·dē'äk·sē ‚meth·əd }

didodecyl ether See dilauryl ether. { dī'dō·də·səl 'ē·thər }

Didolodontidae [PALEON] A family consisting of extinct medium-sized herbivores in the order Condylarthra. { ‚dīd·əl·ō'dänt·ə‚dē }

Dido's problem [MATH] The problem of finding the curve, with a given perimeter, that encloses the greatest possible area; the curve is a circle. { 'dē‚dōz ‚präb·ləm }

didot point [GRAPHICS] In Europe, a printer's unit of type measurement equal to 0.0148 inch (0.376 millimeter). { 'dē‚dō ‚point }

DIDP See diisodecyl phthalate.

Didymelales [BOT] An order of dicotyledonous plants in the subclass Hamamelidae, characterized by the primitive nature of the wood, which has vessels with scalariform perforations, and a pistil, which has one carpel; dioecious, evergreen trees restricted to Madagascar. { ‚dī·də·mə'lā·lēz }

Didymiaceae [MICROBIO] A family of slime molds in the order Physarales. { ‚dī·də·mī'ās·ē‚ē }

didymium [CHEM] A mixture of the rare-earth elements praseodymium and neodymium. Abbreviated Di. { dī'dim·ē·əm }

didymolite [MINERAL] Ca₂Al₆Si₉O₂₉ A dark-gray monoclinic mineral consisting of a calcium aluminum silicate, occurring in twinned crystals. { dī'dim·ə‚līt }

Didymosporae [MYCOL] A spore group of the imperfect fungi characterized by two-celled spores. { dī·də·mə'spȯr·ē }

didymous [BIOL] Occurring in pairs. { 'did·ə·məs }

didynamous [BOT] Having four stamens occurring in two pairs, one pair long and the other short. { dī'din·ə·məs }

die [DES ENG] A tool or mold used to impart shapes to, or to form impressions on, materials such as metals and ceramics. [ELECTR] The tiny, sawed or otherwise machined piece of semiconductor material used in the construction of a transistor, diode, or other semiconductor device; plural is dice. [MED] To pass from physical life. See bell tap. { dī }

die adapter [ENG] That part of an extrusion die which holds the die block. { 'dī ə'dap·tər }

dieback [ECOL] A large area of exposed, unprotected swamp or marsh deposits resulting from the salinity of a coastal lagoon. [PL PATH] Of a plant, to die from the top or peripheral parts. { 'dī‚bak }

die blade [ENG] A deformable member attached to a die body which determines the slot opening and is adjusted to produce uniform thickness across plastic film or sheet. { 'dī ‚blād }

die block [ENG] 1. A tool-steel block which is bolted to the bed of a punch press and into which the desired impressions are machined. 2. The part of an extrusion mold die holding the forming bushing and core. { 'dī ‚bläk }

die body [ENG] The stationary part of an extrusion die, used to separate and form material. { 'dī ‚bäd·ē }

die bushing See button die. { 'dī ‚bu̇sh·iŋ }

die casting [ENG] A metal casting process in which molten metal is forced under pressure into a permanent mold; the two types are hot-chamber and cold-chamber. { 'dī ‚kast·iŋ }

die chaser [ENG] One of the cutting parts of a composite die or a die used to cut threads. { 'dī ‚chās·ər }

Dieckman condensation [CHEM ENG] Any condensation of esters of dicarboxylic acids which produce cyclic β-ketoesters. { 'dēk·män ‚kän‚den'sā·shən }

die clearance [ENG] The distance between die members that meet during an operation. { 'dī ‚klir·əns }

die collar See bell tap. { 'dī ‚käl·ər }

die cushion [ENG] A device located in or under a die block or bolster to provide additional pressure or motion for stamping. { 'dī ‚ku̇sh·ən }

die cutting [ENG] See blanking. [GRAPHICS] Cutting special shapes, such as labels, from printed sheets by using sharp steel rules; often done at the same time as the printing. { 'dī ‚kəd·iŋ }

die down [BOT] Normal seasonal death of aboveground parts of herbaceous perennials. { 'dī ‚dau̇n }

die drawing [MET] Reducing the diameter of wire or tubing by pulling it through a die. { 'dī ‚drȯ·iŋ }

die forging [MET] Shaping metal by plastic deformation in a die. { 'dī ‚fȯrj·iŋ }

die forming [MET] Shaping metal by means of a die under pressure. { 'dī ‚fȯr·miŋ }

die gap [ENG] In plastics and metals forming, the distance between the two opposing metal faces forming the opening of a die. { 'dī ‚gap }

Diego blood group [IMMUNOL] A genetically determined, immunologically distinct group of human erythrocyte antigens

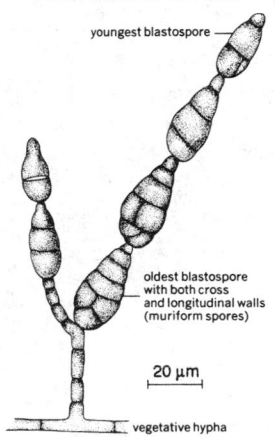

DICTYOBLASTOSPORE

youngest blastospore

oldest blastospore with both cross and longitudinal walls (muriform spores)

20 µm

vegetative hypha

Alternaria tenuis. Sporophore with branched chain of dictyoblastospores. (*After G. Goidanich, 1938*)

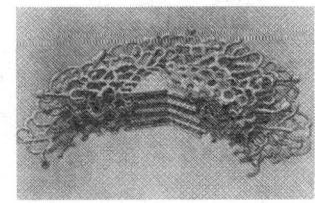

DICTYOSOME

Cutaway view of a model of a portion of a dictyosome composed of six cisternae.

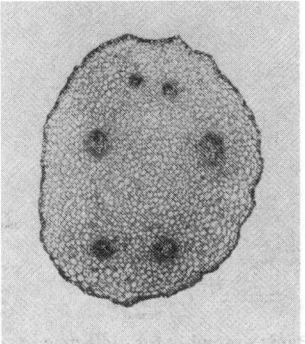

DICTYOSTELE

Dictyostele of *Polypodium*.

recognized by reaction with a specific antibody. { dē'ā·gō 'bləd ‚grüp }

die holder [ENG] A plate or block on which the die block is mounted; it is fastened to the bolster or press bed. { 'dī ‚hōld·ər }

dieing machine [MECH ENG] A vertical press with the slide activated by pull rods attached to the drive mechanism below the bed of the press. { 'dī·iŋ mə'shēn }

die insert [ENG] A removable part or the liner of a die body or punch. { 'dī ‚in·sort }

diel [SCI TECH] Occurring on a 24-hour cycle, as opposed to diurnal (day) or nocturnal (night) occurrences. { 'dī‚el }

dieldrin [ORG CHEM] $C_{12}H_8Cl_6O$ A white, crystalline contact insecticide obtained by oxidation of aldrin; used in mothproofing carpets and other furnishings. { 'dēl·drən }

dielectric *See* dielectric material. { ‚dī·ə'lek·trik }

dielectric absorption [ELEC] The persistence of electric polarization in certain dielectrics after removal of the electric field. *See* dielectric loss. { ‚dī·ə'lek·trik əb'sȯrp·shən }

dielectric amplifier [ELECTR] An amplifier using a ferroelectric capacitor whose capacitance varies with applied voltage so as to give signal amplification. { ‚dī·ə'lek·trik 'am·plə‚fī·ər }

dielectric antenna [ELECTROMAG] An antenna in which a dielectric is the major component used to produce a desired radiation pattern. { ‚dī·ə'lek·trik an'ten·ə }

dielectric breakdown [ELECTR] Breakdown which occurs in an alkali halide crystal at field strengths on the order of 10^6 volts per centimeter. { ‚dī·ə'lek·trik 'brāk‚daȯn }

dielectric circuit [ELEC] Any electric circuit which has capacitors. { ‚di·ə'lek·trik 'sər·kət }

dielectric constant [ELEC] **1.** For an isotropic medium, the ratio of the capacitance of a capacitor filled with a given dielectric to that of the same capacitor having only a vacuum as dielectric. **2.** More generally, $1 + \gamma\chi$, where γ is 4π in Gaussian and cgs electrostatic units or 1 in rationalized mks units, and χ is the electric susceptibility tensor. Also known as relative dielectric constant; relative permittivity; specific inductive capacity (SIC). { ‚dī·ə'lek·trik 'kän·stənt }

dielectric crystal [ELEC] A crystal which is electrically nonconducting. { ‚dī·ə'lek·trik 'krist·əl }

dielectric curing [ENG] A process for curing a thermosetting resin by subjecting it to a high-frequency electric charge. { ‚dī·ə'lek·trik 'kyúr·iŋ }

dielectric current [ELEC] The current flowing at any instant through a surface of a dielectric that is located in a changing electric field. { ‚dī·ə'lek·trik 'kər·ənt }

dielectric displacement *See* electric displacement. { ‚dī·ə'lek·trik di'splās·mənt }

dielectric ellipsoid [ELEC] For an anisotropic medium in which the dielectric constant is a tensor quantity **K,** the locus of points **r** satisfying $\mathbf{r}\cdot\mathbf{K}\cdot\mathbf{r} = 1$. { ‚dī·ə'lek·trik ə'lip‚sȯid }

dielectric fatigue [ELECTR] The property of some dielectrics in which resistance to breakdown decreases after a voltage has been applied for a considerable time. { ‚dī·ə'lek·trik fə'tēg }

dielectric field [ELEC] The average total electric field acting upon a molecule or group of molecules inside a dielectric. Also known as internal dielectric field. { ‚dī·ə'lek·trik 'fēld }

dielectric film [ELEC] A film possessing dielectric properties; used as the central layer of a capacitor. { ‚dī·ə'lek·trik 'film }

dielectric flux density *See* electric displacement. { ‚dī·ə'lek·trik 'fləks ‚den·səd·ē }

dielectric gas [ELEC] A gas having a high dielectric constant, such as sulfur hexafluoride. { ‚dī·ə'lek·trik 'gas }

dielectric heating [ELEC] Heating of a nominally electrical insulating material due to its own electrical (dielectric) losses, when the material is placed in a varying electrostatic field. { ‚dī·ə'lek·trik 'hēd·iŋ }

dielectric hysteresis *See* ferroelectric hysteresis. { ‚dī·ə'lek·trik hi·stə'rē·səs }

dielectric imperfection levels [SOLID STATE] Energy levels that occur in the forbidden zone between the valence and conduction bands of a dielectric crystal, because of imperfections in the crystal. { ‚dī·ə'lek·trik ‚im·pər'fek·shən ‚lev·əlz }

dielectric leakage [ELEC] A very small steady current that flows through a dielectric subject to a steady electric field. { ‚dī·ə'lek·trik 'lēk·ij }

dielectric lens [ELECTROMAG] A lens made of dielectric material so that it refracts radio waves in the same manner that an optical lens refracts light waves; used with microwave antennas. { ‚dī·ə'lek·trik 'lenz }

dielectric-lens antenna [ELECTROMAG] An aperture antenna in which the beam width is determined by the dimensions of a dielectric lens through which the beam passes. { ‚dī·ə'lek·trik ‚lenz an'ten·ə }

dielectric loss [ELECTROMAG] The electric energy that is converted into heat in a dielectric subjected to a varying electric field. Also known as dielectric absorption. { ‚dī·ə'lek·trik 'lȯs }

dielectric loss angle [ELEC] Difference between 90° and the dielectric phase angle. { ‚dī·ə'lek·trik 'lȯs ‚aŋ·gəl }

dielectric loss factor [ELEC] Product of the dielectric constant of a material and the tangent of its dielectric loss angle. { ‚dī·ə'lek·trik 'lȯs ‚fak·tər }

dielectric matching plate [ELECTROMAG] In waveguide technique, a dielectric plate used as an impedance transformer for matching purposes. { ‚dī·ə'lek·trik 'mach·iŋ ‚plāt }

dielectric material [MATER] **1.** Also known as dielectric. **2.** A material which is an electrical insulator or in which an electric field can be sustained with a minimum dissipation of power. **3.** In a more general sense, any material other than a condensed state of a metal. { ‚dī·ə'lək·trik mə‚tir·ē·əl }

dielectric phase angle [ELEC] Angular difference in phase between the sinusoidal alternating potential difference applied to a dielectric and the component of the resulting alternating current having the same period as the potential difference. { ‚dī·ə'lek·trik 'fāz ‚aŋ·gəl }

dielectric polarization *See* polarization. { ‚dī·ə'lek·trik ‚pō·lə·rə'zā·shən }

dielectric power factor [ELEC] Cosine of the dielectric phase angle (or sine of the dielectric loss angle). { ‚dī·ə'lek·trik 'paȯr ‚fak·tər }

dielectric-rod antenna [ELECTROMAG] A surface-wave antenna in which an end-fire radiation pattern is produced by propagation of a surface wave on a tapered dielectric rod. { ‚dī·ə'lek·trik ‚räd an'ten·ə }

dielectric shielding [ELEC] The reduction of an electric field in some region by interposing a dielectric substance, such as polystyrene, glass, or mica. { ‚dī·ə'lek·trik 'shēld·iŋ }

dielectric soak *See* absorption. { ‚dī·ə'lek·trik 'sōk }

dielectric strength [ELEC] The maximum electrical potential gradient that a material can withstand without rupture; usually specified in volts per millimeter of thickness. Also known as electric strength. { ‚dī·ə'lek·trik 'streŋkth }

dielectric susceptibility *See* electric susceptibility. { ‚dī·ə'lek·trik sə‚sep·tə'bil·əd·ē }

dielectric test [ELEC] A test involving application of a voltage higher than the rated value for a specified time, to determine the margin of safety against later failure of insulating materials. { ‚dī·ə'lek·trik 'test }

dielectric vapor detector [ANALY CHEM] Apparatus to measure the change in the dielectric constant of gases or gas mixtures; used as a detector in gas chromatographs to sense changes in carrier gas. { 'dī·ə'lek·trik 'vā·pər di‚tek·tər }

dielectric waveguide [ELEC] A waveguide consisting of a dielectric cylinder surrounded by air. { ‚dī·ə'lek·trik 'wāv‚gīd }

dielectric wedge [ELECTROMAG] A wedge-shaped piece of dielectric used in a waveguide to match its impedance to that of another waveguide. { dī·ə'lek·trik 'wej }

dielectric wire [ELECTROMAG] A dielectric waveguide used to transmit ultra-high-frequency radio waves short distances between parts of a circuit. { dī·ə'lek·trik 'wīr }

dielectronic recombination [ATOM PHYS] The combination of an electron with a positive-ion in a gas, so that the energy released is taken up by two electrons of the resulting atom. { di·ə‚lek'trän·ik ‚rē‚käm·bə'nā·shən }

dielectrophoresis [PHYS CHEM] The ability of an uncharged material to move when subjected to an electric field. { ‚dī·ə‚lek·trō·fə'rē·səs }

die lines [ENG] Lines or markings on the surface of a drawn, formed, or extruded product due to imperfections in the surface of the die. { 'dī ‚līnz }

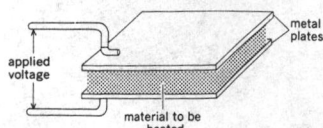

DIELECTRIC HEATING

applied voltage

metal plates

material to be heated

Basic assembly for dielectric heating.

Diels-Alder reaction [ORG CHEM] The 1,4 addition of a conjugated diolefin to a compound, known as a dienophile, containing a double or triple bond; the dienophile may be activated by conjugation with a second double bond or with an electron acceptor. { ¦dēlz ¦äl·dər rē‚ak·shən }

die lubricant [MATER] Any material applied to a die to facilitate movement of the work in the die in certain die-forming operations. { 'dī ¦lü·brə·kənt }

die match [MET] The proper alignment of forging dies in relation to each other in forging equipment. { 'dī ‚mach }

diencephalon [EMBRYO] The posterior division of the embryonic forebrain in vertebrates. { ¦dī·en'sef·ə‚län }

diene [ORG CHEM] One of a class of organic compounds containing two ethylenic linkages (carbon-to-carbon double bonds) in the molecules. Also known as alkadiene; diolefin. { 'dī‚ēn }

diene resin [ORG CHEM] Material containing the diene group of double bonds that may polymerize. { 'dī‚ēn 'rez·ən }

diene value [ORG CHEM] A number that represents the amount of conjugated bonds in a fatty acid or fat. { 'dī‚ēn ‚val·yü }

die nipple [PETRO ENG] A fishing tool similar to a bell tap, except that it is threaded externally instead of internally. { 'dī ‚nip·əl }

dienophile [ORG CHEM] The alkene component of a reaction between an alkene and a diene. { dī'en·ə‚fīl }

die opening [MET] The distance between electrodes in flash or upset welding; it is measured with parts in contact but before the beginning or immediately after completion of the weld cycle. { 'dī ‚ō·pən·iŋ }

die radius [MET] The radius on the exposed edge of a deep-drawing die. { 'dī ‚rād·ē·əs }

die scalping [MET] Drawing wire, tubing, bars, or rods through a sharp-edged die to remove surface layers containing defects. { 'dī ‚skalp·iŋ }

diesel cycle [THERMO] An internal combustion engine cycle in which the heat of compression ignites the fuel. { 'dē·zəl ‚sī·kəl }

diesel electric locomotive [MECH ENG] A locomotive with a diesel engine driving an electric generator which supplies electric power to traction motors for propelling the vehicle. Also known as diesel locomotive. { ¦dē·zəl ə¦lek·trik ‚lō·kə'mōd·iv }

diesel electric power generation [MECH ENG] Electric power generation in which the generator is driven by a diesel engine. { ¦dē·zəl ə¦lek·trik 'pau̇·ər ‚jen·ə‚rā·shən }

diesel engine [MECH ENG] An internal combustion engine operating on a thermodynamic cycle in which the ratio of compression of the air charge is sufficiently high to ignite the fuel subsequently injected into the combustion chamber. Also known as compression-ignition engine. { ¦dē·zəl 'en·jən }

diesel fuel [MATER] Fuel used for internal combustion in diesel engines; usually that fraction of crude oil that distills after kerosine. { 'dē·zəl ‚fyül }

diesel fuel additives [MATER] Compounds added to diesel fuels to improve performance, such as cetane number improvers, metal deactivators, corrosion inhibitors, antioxidants, rust inhibitors, and dispersants. { 'dē·zəl ‚fyül 'ad·əd·ivz }

diesel fuel grades [MATER] Fuels suitable for various classes of engines and service, which must meet specifications for flash point temperature, distillation temperature, cetane number, and concentrations of impurities. { 'dē·zəl ‚fyül ‚grādz }

diesel-fuel water and sediment [MATER] Undesirable constituents of diesel fuel which should not exceed certain limits to ensure clean fuel. { 'dē·zəl ‚fyül 'wȯd·ər ən ‚sed·ə·mənt }

diesel index [CHEM ENG] An empirical expression for the correlation between the aniline number of a diesel fuel and its ignitability. [MECH ENG] Diesel fuel rating based on ignition qualities; high-quality fuel has a high index number. { 'dē·zəl ‚in‚deks }

dieseling [MECH ENG] **1.** Explosions of mixtures of air and lubricating oil in the compression chambers or in other parts of the air system of a compressor. **2.** Continuation of running by a gasoline spark-ignition engine after the ignition is turned off. Also known as run-on. { 'dē·zəl·iŋ }

diesel knock [MECH ENG] A combustion knock caused when the delayed period of ignition is long so that a large quantity of atomized fuel accumulates in the combustion chamber; when combustion occurs, the sudden high pressure resulting from the accumulated fuel causes diesel knock. { 'dē·zəl ‚näk }

diesel locomotive See diesel electric locomotive. { 'dē·zəl ‚lō·kə'mōd·iv }

diesel oil [MATER] Heavy oil residue used as fuel for certain types of diesel engines. { 'dē·zəl ‚ȯil }

diesel rig [MECH ENG] Any diesel engine apparatus or machinery. { 'dē·zəl ‚rig }

diesel squeeze [PETRO ENG] A technique of forcing dry cement mixed with diesel oil through casing openings to repair water-bearing areas without affecting the oil-bearing areas. { 'dē·zəl ‚skwēz }

die set [ENG] A tool or tool holder consisting of a die base for the attachment of a die and a punch plate for the attachment of a punch. { 'dī ‚set }

die shoe [MECH ENG] A block placed beneath the lower part of a die upon which the die holder is mounted; spreads the impact over the die bed, thereby reducing wear. { 'dī ‚shü }

diesinking [ENG] Making a depressed pattern in a die by forming or machining. { 'dī‚siŋk·iŋ }

die slide [MECH ENG] A device in which the lower die of a power press is mounted; it slides in and out of the press for easy access and safety in feeding the parts. { 'dī ‚slīd }

die stamping [GRAPHICS] A procedure in which steel, brass, or bronze dies are used to impress an image on a surface by using a relief counterpart or the reverse thereof. { ‚dī ‚stamp·iŋ }

die steel [MET] Plain carbon steel or alloy steel used in making tools for cutting, machining, shearing, stamping, punching, and chipping. { 'dī ‚stēl }

diester [ORG CHEM] A compound containing two ester groupings. { ‚dī'es·tər }

diesterase [BIOCHEM] An enzyme such as a nuclease which splits the linkages binding individual nucleotides of a nucleic acid. { dī'es·tə‚rās }

diestrus [PHYSIO] The long, quiescent period following ovulation in the estrous cycle in mammals; the stage in which the uterus prepares for the reception of a fertilized ovum. { dī'es·trəs }

die swell ratio [ENG] The ratio of the outer parison diameter (or parison thickness) to the outer diameter of the die (or die gap). { 'dī ‚swel ‚rā·shō }

diet [BIOL] The food or drink regularly consumed. [MED] Food prescribed, regulated, or restricted as to kind and amount, for therapeutic or other purpose. { 'dī·ət }

dietary fiber [FOOD ENG] The plant-cell-wall polysaccharides and lignin in a food or food ingredient that are not broken down by the digestive enzymes of animals and humans. { ¦dī·ə‚ter·ē 'fī·bər }

Dieterici equation of state [THERMO] An empirical equation of state for gases, $pe^{a/RT}(v − b) = RT$, where p is the pressure, T is the absolute temperature, v is the molar volume, R is the gas constant, and a and b are constants characteristic of the substance under consideration. { dē·də'rē·chē i'kwā·zhen əv 'stāt }

Dietert tester [MET] An apparatus for reading Brinell hardness directly from the impression made in the part being tested by means of a depth pin pressed into the depression. { 'dēd·ərt 'tes·tər }

dietetics [MED] The science concerned with applying the principle of nutrition to the feeding of people under various economic conditions or for therapeutic purposes. { ‚dī·ə'ted·iks }

diethanolamine [ORG CHEM] $(HOCH_2CH_2)_2NH$ Colorless, water-soluble, deliquescent crystals, or liquid boiling at 217°C; soluble in alcohol and acetone, insoluble in ether and benzene; used in detergents, as an absorbent of acid gases, and as a chemical intermediate. Also known as DEA. { dī·ə·thə'näl·ə‚mēn }

diether [ORG CHEM] A molecule that has two oxygen atoms with ether bonds. { dī'ē·thər }

1,1-diethoxyethane See acetal. { ¦wən ¦wən ‚dī·ə¦thäk·sē'e‚thän }

diethyl [ORG CHEM] Pertaining to a molecule with two ethyl groups. { dī'eth·əl }

diethyl adipate [ORG CHEM] $C_2H_5OCO(CH_2)_4OCOC_2H_5$

DIELS-ALDER REACTION

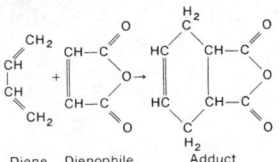

Diels-Alder reaction with structural formulas.

DIESEL ENGINE

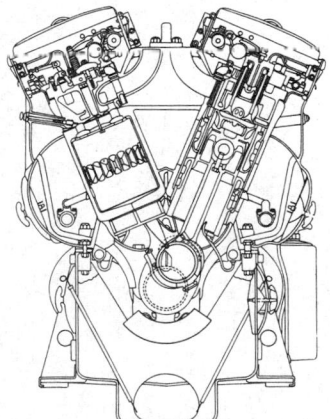

Section through a locomotive diesel engine. *(General Motors, Electromotive Division)*

Water-insoluble, colorless liquid, boiling at 245°C; used as a plasticizer. { dī'eth·əl 'ad·ə‚pāt }

diethylamine [ORG CHEM] $(C_2H_5)_2NH$ Water-soluble, colorless liquid with ammonia aroma, boiling at 56°C; used in rubber chemicals and pharmaceuticals and as a solvent and flotation agent. { ‚dī‚eth·əl'a‚mēn }

diethylaminoethyl cellulose [ORG CHEM] A positively charged resin used in ion-exchange chromatography; an anion exchanger. Also known as DEAE-cellulose. { ‚dī‚eth·əl¦am·ə·nō'eth·əl 'sel·yə‚lōs }

5,5-diethylbarbituric acid See barbital. { ¦fīv ¦fīv ‚dī‚eth·əl'bär·bə'tur·ik 'as·əd }

diethylbenzene [ORG CHEM] $C_6H_4(C_2H_5)_2$ Colorless liquid, boiling at 180–185°C; soluble in organic solvents, insoluble in water; usually a mixture of three isomers, which are 1,2-(or ortho-diethylbenzene), boiling at 183°C, and 1,3- (or meta-), boiling at 181°C, and 1,4- (or para-), boiling at 184°C; used as a solvent. { ‚dī‚eth·əl'ben‚zēn }

diethylcarbamazine [ORG CHEM] $C_{16}H_{29}O_8N_3$ White, water-soluble, hygroscopic crystals, melting at 136°C; used as an anthelminthic. { ‚dī‚eth·əl‚kär'bam·ə‚zēn }

diethyl carbinol [ORG CHEM] $(CH_3CH_2)_2CHOH$ Colorless, alcohol-soluble liquid, boiling at 116°C; slightly soluble in water; used in pharmaceuticals and as a solvent and flotation agent. Also known as sec-n-amyl alcohol. { dī'eth·əl 'kär·bə‚nól }

diethyl carbonate [ORG CHEM] $(C_2H_5)_2CO_3$ Stable, colorless liquid with mild aroma, boiling at 126°C; soluble with most organic solvents; used as a solvent and for chemical synthesis. Also known as ethyl carbonate. { dī'eth·əl 'kär·bə‚nāt }

diethylene glycol [ORG CHEM] $CH_2OHCH_2OCH_2CH_2OH$ Clear, hygroscopic, water-soluble liquid, boiling at 245°C; soluble in many organic solvents; used as a softener, conditioner, lubricant, and solvent, and in antifreezes and cosmetics. { dī'eth·ə‚lēn 'glī‚kól }

diethylene glycol monoethyl ether [ORG CHEM] $C_6H_{14}O_3$ A hygroscopic liquid used as a solvent for cellulose esters and in lacquers, varnishes, and enamels. { dī'eth·ə‚lēn 'glī‚kól ‚män·ō'eth·əl 'ē·thər }

diethylenetriamine [ORG CHEM] $(NH_2C_2H_4)_2NH$ A yellow, hygroscopic liquid with a boiling point of 206.7°C; soluble in water and hydrocarbons; used as a solvent, saponification agent, and fuel component. { dī'eth·ə‚lēn¦trī·ə‚mēn }

diethyl ether [ORG CHEM] $C_4H_{10}O$ A colorless liquid, slightly soluble in water; used as a reagent and solvent. Also known as ethyl ether; ethyl oxide; ethylic ether. { dī'eth·əl'ē·thər }

diethyl maleate [ORG CHEM] $(HCCOOC_2H_5)_2$ Clear, colorless liquid, boiling at 225°C; slightly soluble in water, soluble in most organic solvents; used as a chemical intermediate. { ‚dī'eth·əl 'mal·ē‚āt }

diethyl phosphite [ORG CHEM] $(C_2H_5O)_2HPO$ A colorless liquid with a boiling point of 138°C; soluble in water and common organic solvents; used as a paint solvent, antioxidant, and reducing agent. { dī'eth·əl 'fäs‚fīt }

diethyl phthalate [ORG CHEM] $C_6H_4(CO_2C_2H_5)_2$ Clear, colorless, odorless liquid with bitter taste, boiling at 298°C; soluble in alcohols, ketones, esters, and aromatic hydrocarbons, partly soluble in aliphatic solvents; used as a cellulosic solvent, wetting agent, alcohol denaturant, mosquito repellent, and in perfumes. { dī'eth·əl 'tha‚lāt }

2,2-diethyl-1,3-propanediol [PHARM] $HOCH_2C(C_2H_5)_2CH_2OH$ Crystals with a melting point of 61–61.6°C; used as a skeletal muscle relaxant. { ¦tü ¦tü ‚dī‚eth·əl ¦wən ¦thrē ¦prō‚pān'dī‚ól }

diethyl pyrocarbonate [ORG CHEM] $C_6H_{10}O_5$ A viscous liquid, soluble in alcohols, esters, and ketones; used as a gentle esterifying agent, as a preservative for fruit juices, soft drinks, and wines, and as an inhibitor for ribonuclease. Abbreviated DEPC. { ‚dī‚eth·əl ‚pī·rō'kär·bə‚nāt }

diethylstilbesterol [BIOCHEM] $C_{18}H_{20}O_2$ A white, crystalline, nonsteroid estrogen that is used therapeutically as a substitute for natural estrogenic hormones. Also known as stilbestrol. Abbreviated DES. { ‚dī¦eth·əl·stil'bes·tə‚ról }

diethyl succinate [ORG CHEM] $(CH_2COOC_2H_5)_2$ Water-white liquid with pleasant aroma, boiling at 216°C; soluble in alcohol and ether, slightly soluble in water; used as a chemical intermediate and plasticizer. { ‚dī‚eth·əl 'sək·sə‚nāt }

diethyl sulfate [ORG CHEM] $(C_2H_5)_2SO_4$ A colorless oil with a peppermint odor, and boiling at 208°C; used as an intermediate in organic synthesis. Also known as ethyl sulfate. { ‚dī‚eth·əl 'səl‚fāt }

diethyl sulfide See ethyl sulfide. { ‚dī‚eth·əl 'səl‚fīd }

diethyltoluamide [ORG CHEM] $C_{12}H_{17}ON$ A liquid whose color ranges from off-white to light yellow; used as an insect repellent for people and clothing. Also known as DEET; DET; N,N-diethyl-meta-toluamide. { ‚dī¦eth·əl‚täl·yü'a‚mīd }

dietician [MED] A person trained in dietetics, or the scientific management of meals for individuals or groups. { ‚dī·ə'tish·ən }

Dietl's crisis [MED] Recurrent attacks of radiating pain in the costovertebral angle, accompanied by nausea, vomiting, tachycardia, and hypotension, caused by kinking or twisting of the ureter with intermittent obstructive dilation. { 'dēd·əlz ‚krī·səs }

dietrichite [MINERAL] $(Zn,Fe,Mn)Al_2(SO_4)_4·22H_2O$ Mineral consisting of a hydrous sulfate of aluminum and one or more of the metals zinc, iron, and manganese. { 'dē·tri‚kīt }

dietzeite [MINERAL] $Ca_2(IO_3)_2(CrO_4)$ A dark-golden-yellow iodate mineral commonly in fibrous or columnar form as a component of caliche. { 'dēt·sə‚īt }

die welding [MET] Forge welding in which the weld is completed under pressure between dies. { 'dī ‚weld·iŋ }

diffeomorphic sets [MATH] Sets in Euclidean space such that there is a diffeomorphism between them. { ‚dif·ē·ə‚mòr·fik 'sets }

diffeomorphism [MATH] A bijective function, with domain and range in the same or different Euclidean spaces, such that both the function and its inverse have continuous mixed partial derivatives of all orders in neighborhoods of each point of their respective domains. { ‚dif·ē·ə'mòr·fiz·əm }

difference [MATH] **1.** The result of subtracting one number from another. **2.** The difference between two sets A and B is the set consisting of all elements of A which do not belong to B; denoted $A - B$. { 'dif·rəns }

difference amplifier See differential amplifier. { 'dif·rəns ‚am·plə‚fī·ər }

difference channel [ENG ACOUS] An audio channel that handles the difference between the signals in the left and right channels of a stereophonic sound system. { 'dif·rəns ‚chan·əl }

difference detector [ELECTR] A detector circuit in which the output is a function of the difference between the amplitudes of the two input waveforms. { 'dif·rəns di‚tek·tər }

difference encoding [COMPUT SCI] A method of data compression that takes advantage of a sequence of data that differs little from one value to the next by encoding each value as the difference from the previous value. { 'dif·rəns in‚kōd·iŋ }

difference equation [MATH] An equation expressing a functional relationship of one or more independent variables, one or more functions dependent on these variables, and successive differences of these functions. { 'dif·rəns i'kwā·zhən }

difference in depth modulation [COMMUN] In directive systems employing overlapping lobes with modulated signals, a ratio obtained by subtracting from the percentage of modulation of the larger signal the percentage of modulation of the smaller signal and dividing by 100. { 'dif·rəns 'in ¦depth ‚məj·ə'lā·shən }

difference limen See just-noticeable difference. { 'dif·rəns ‚lī·mən }

difference mapping [COMMUN] A method of coding information in which a sample value is presented as an error term formed by the difference between the sample and the previous sample. { 'dif·rəns ‚map·iŋ }

difference methods [MATH] Versions of the predictor-corrector methods of calculating numerical solutions of differential equations in which the prediction and correction formulas express the value of the solution function in terms of finite differences of a derivative of the function. { 'dif·rəns ‚meth·ədz }

difference number See neutron excess. { 'dif·rəns ‚nəm·bər }

difference of latitude [GEOD] The shorter arc of any meridian between the parallels of two places, expressed in angular measure. { 'dif·rəns əv 'lad·ə‚tüd }

difference of longitude [GEOD] The smaller angle at the

pole or the shorter arc of a parallel between the meridians of two places, expressed in angular measure. { 'dif·rəns əv 'län·jə,tüd }

difference of meridional parts *See* meridional difference. { 'dif·rəns əv me'rid·ē·ən·əl 'pärts }

difference operator [MATH] One of several operators, such as the displacement operator, forward difference operator, or central mean operator, which can be used to conveniently express formulas for interpolation or numerical calculation or integration of functions and can be manipulated as algebraic quantities. { 'dif·rəns ,äp·ə,rād·ər }

difference quotient [MATH] The increment of the value of a function divided by the increment of the independent variable; for the function $y = f(x)$, it is $\Delta y / \Delta x = [f(x + \Delta x) - f(x)]\Delta x$, where Δx and Δy are the increments of x and y. { 'dif·rəns ,kwō·shənt }

difference spectrophotometer *See* absorption spectrophotometer. { 'dif·rəns ,spek·trə·fə'täm·əd·ər }

difference threshold *See* just-noticeable difference. { 'dif·rəns ,thresh,hōld }

difference tone [ACOUS] A combination tone whose frequency equals the difference of the frequencies of the pure tones producing it. { 'dif·rəns ,tōn }

differentiable atlas [MATH] A family of embeddings $h_i : E^n \rightarrow M$ of euclidean space into a topological space M with the property that $h_i^{-1} h_j : E^n \rightarrow E^n$ is a differentiable map for each pair of indices, i, j. { ,dif·ə'ren·chə·bəl 'at·ləs }

differentiable function [MATH] A function which has a derivative at each point of its domain. { ,dif·ə'ren·chə·bəl 'fəŋk·shən }

differentiable manifold [MATH] A topological space with a maximal differentiable atlas; roughly speaking, a smooth surface. { ,dif·ə'ren·chə·həl 'man·ə,fōld }

differential [CONT SYS] The difference between levels for turn-on and turn-off operation in a control system. [MATH] **1.** The differential of a real-valued function $f(x)$, where x is a vector, evaluated at a given vector c, is the linear, real-valued function whose graph is the tangent hyperplane to the graph of $f(x)$ at $x = c$; if x is a real number, the usual notation is $df = f'(c)dx$. **2.** *See* total differential. [MECH ENG] Any arrangement of gears forming an epicyclic train in which the angular speed of one shaft is proportional to the sum or difference of the angular speeds of two other gears which lie on the same axis; allows one shaft to revolve faster than the other, the speed of the main driving member being equal to the algebraic mean of the speeds of the two shafts. Also known as differential gear. { ,dif·ə'ren·chəl }

differential absorption lidar [ENG] A technique for the remote sensing of atmospheric gases, in which lasers transmit pulses of radiation into the atmosphere at two wavelengths, one of which is absorbed by the gas to be measured and one is not, and the difference between the return signals from atmospheric backscattering on the absorbed and nonabsorbed wavelengths is used as a direct measure of the concentration of the absorbing species. Abbreviated DIAL. { ,dif·ə'ren·chəl əb'sórp·shən 'lī,där }

differential aeration cell [PHYS CHEM] An electrolytic cell whose electromotive force derives from a difference in concentration of atmospheric oxygen at one electrode with reference to another electrode of the same material. Also known as oxygen concentration cell. { ,dif·ə'ren·chəl e'rā·shən ,sel }

differential air thermometer [ENG] A device for detecting radiant heat, consisting of a U-tube manometer with a closed bulb at each end, one clear and the other blackened. { ,dif·ə'ren·chəl 'er thər'mäm·əd·ər }

differential amplifier [ELECTR] An amplifier whose output is proportional to the difference between the voltages applied to its two inputs. Also called difference amplifier. { ,dif·ə'ren·chəl 'am·plə,fī·ər }

differential analysis [METEOROL] Synoptic analysis of change charts or of vertical differential charts (such as thickness charts) obtained by the graphical or numerical subtraction of the patterns of some meteorological variable at two times or two levels. { 'dif·ə'ren·chəl ə'nal·ə·səs }

differential analyzer [COMPUT SCI] A mechanical or electromechanical device designed primarily to solve differential equations. { ,dif·ə'ren·chəl 'an·ə,līz·ər }

differential backup [COMPUT SCI] Backup of only files that

have been changed or added since the last backup. { ,dif·ə,ren·chəl 'bak,əp }

differential ballistic wind [ORD] In bombing, a hypothetical wind equal to the difference in velocity between the ballistic wind and the actual wind at release altitude. { ,dif·ə'ren·chəl bə'lis·tik 'wind }

differential blood count *See* differential leukocyte count. { ,dif·ə'ren·chəl 'bləd ,kaùnt }

differential brake [MECH ENG] A brake in which operation depends on a difference between two motions. { ,dif·ə'ren·chəl 'brāk }

differential calculus [MATH] The study of the manner in which the value of a function changes as one changes the value of the independent variable; includes maximum-minimum problems and expansion of functions into Taylor series. { ,dif·ə'ren·chəl 'kal·kyə·ləs }

differential calorimetry [THERMO] Technique for measurement of and comparison (differential) of process heats (reaction, absorption, hydrolysis, and so on) for a specimen and a reference material. { ,dif·ə'ren·chəl ,kal·ə'rim·ə·trē }

differential capacitance [ELECTR] The derivative with respect to voltage of a charge characteristic, such as an alternating charge characteristic or a mean charge characteristic, at a given point on the characteristic. { ,dif·ə'ren·chəl kə'pas·əd·əns }

differential capacitor [ELEC] A two-section variable capacitor having one rotor and two stators so arranged that as capacitance is reduced in one section it is increased in the other. { ,dif·ə'ren·chəl kə'pas·əd·ər }

differential centrifugation [CYTOL] The separation of mixtures such as cellular particles in a medium at various centrifugal forces to separate particles of different density, size, and shape from each other. { ,dif·ə'ren·chəl ,sen·trə·fə'gā·shən }

differential chart [METEOROL] A chart showing the amount and direction of change of a meteorological quantity in time or space. { ,dif·ə'ren·chəl 'chärt }

differential chemical reactor [CHEM ENG] A flow reactor operated at constant temperature and very low concentrations (resulting from very short residence times), with product and reactant concentrations essentially constant at the levels in the feed. { ,dif·ə'ren·chəl 'kem·i·kəl rē'ak·tər }

differential coefficient *See* derivative. { ,dif·ə'ren·chəl ,kō·i'fish·ənt }

differential compaction [GEOL] Compression in sediments, such as sand or limestone, as the weight of overburden causes reduction in pore space and forcing out of water. { ,dif·ə'ren·chəl kəm'pak·shən }

differential comparator [ELECTR] A comparator having at least two high-gain differential-amplifier stages, followed by level-shifting and buffering stages, as required for converting a differential input to single-ended output for digital logic applications. { ,dif·ə'ren·chəl kəm'par·əd·ər }

differential compound motor [ELEC] A direct-current motor whose speed may be made nearly constant or may be adjusted to increase with increasing load. { ,dif·ə'ren·chəl 'käm,paùnd ,mōd·ər }

differential correction [ASTRON] A method for finding from the observed residuals minus the computed residuals $(O - C)$ small corrections which, when applied to the orbital elements or constants, will reduce the deviations from the observed motion to a minimum. { ,dif·ə'ren·chəl kə'rek·shən }

differential cross section [PHYS] The cross section for a collision process resulting in the emission of particles or photons at a specified angle relative to the direction of the incident particles, per unit angle or per unit solid angle. { ,dif·ə'ren·chəl 'krós ,sek·shən }

differential delay [COMMUN] The difference between the maximum and minimum frequency delays occurring across a band. { ,dif·ə'ren·chəl di'lā }

differential diagnosis [MED] Distinguishing between diseases of similar character by comparing their signs and symptoms. [SYST] In taxonomic study, a statement of the characters that distinguish a given taxon from other, specifically mentioned equivalent taxa. { ,dif·ə'ren·chəl ,dī·əg'nō·səs }

differential discriminator [ELECTR] A discriminator that passes only pulses whose amplitudes are between two predetermined values, neither of which is zero. { ,dif·ə'ren·chəl di'skrim·ə,nād·ər }

differential duplex system [ELECTR] System in which the sent currents divide through two mutually inductive sections of a receiving apparatus, connected respectively to the line and to a balancing artificial line in opposite directions, so that there is substantially no net effect on the receiving apparatus; the received currents pass mainly through one section, or through the two sections in the same direction, and operate the apparatus. { ‚dif·ə'ren·chəl 'dü‚pleks ‚sis·təm }

differential ebuliometer [ANALY CHEM] Apparatus for precise and simultaneous measurement of both the boiling temperature of a liquid and the condensation temperature of the vapors of the boiling liquid. { ‚dif·ə'ren·chəl ə‚bü·lē'äm·əd· ər }

differential effects [MECH] The effects upon the elements of the trajectory due to variations from standard conditions. { ‚dif·ə'ren·chəl i'feks }

differential electromagnet [ELEC] An electromagnet having part of its winding opposed to the other part, so that the force exerted by the magnet can be adjusted. { ‚dif·ə'ren·chəl i‚lek·trō'mag·nət }

differential encoding [COMMUN] A method of compressing television signals by transmitting only differences between pixels in neighboring lines and successive frames. { ‚dif·ə‚ren·chəl in'kōd·iŋ }

differential entrapment [PETRO ENG] Controlling oil and gas migration and accumulation by means of selective trapping or gas flushing in interconnecting reservoirs. { ‚dif·ə'ren·chəl en'trap·mənt }

differential equation [MATH] An equation expressing a relationship between functions and their derivatives. { ‚dif·ə'ren·chəl i'kwā·zhən }

differential erosion [GEOL] Rapid erosion of one area of the earth's surface relative to another. { ‚dif·ə'ren·chəl i'rō·zhən }

differential extraction [CHEM ENG] Theoretical limiting case of crosscurrent extraction in a single vessel where feed is continuously extracted with infinitesimal amounts of fresh solvent; true differential extraction cannot be achieved. { ‚dif·ə'ren·chəl ik'strak·shən }

differential fault See scissors fault. { ‚dif·ə'ren·chəl 'fólt }

differential fill-up collar [PETRO ENG] A mechanism installed in the bottom of a well, when casing is set in a drill hole that automatically admits drilling fluids into the casing as required. { ‚dif·ə'ren·chəl 'fil ‚əp ‚käl·ər }

differential flotation [MIN ENG] Separation of a complex ore into two or more mineral components and gangue by flotation. { ‚dif·ə'ren·chəl flō'tā·shən }

differential form [MATH] A homogeneous polynomial in differentials. { ‚dif·ə'ren·chəl 'fórm }

differential frequency circuit [ELEC] A circuit that provides a continuous output frequency equal to the absolute difference between two continuous input frequencies. { ‚dif·ə'ren·chəl ¦frē·kwən·sē ¦sər·kət }

differential frequency meter [ENG] A circuit that converts the absolute frequency difference between two input signals to a linearly proportional direct-current output voltage that can be used to drive a meter, recorder, oscilloscope, or other device. { ‚dif·ə'ren·chəl 'frē·kwən·sē ‚mēd·ər }

differential gain control [ELECTR] Device for altering the gain of a radio receiver according to expected change of signal level, to reduce the amplitude differential between the signals at the output of the receiver. Also known as gain sensitivity control. { ‚dif·ə'ren·chəl 'gān kən‚trōl }

differential galvanometer [ELEC] A galvanometer having a magnetic needle which is free to rotate in the magnetic field produced by currents flowing in opposite directions through two separate identical coils, so that there is no deflection when the currents are equal. { ‚dif·ə'ren·chəl ‚gal·və'näm·əd·ər }

differential game [CONT SYS] A two-sided optimal control problem. [MATH] A game in which the describing equations are differential equations. { ‚dif·ə'ren·chəl 'gām }

differential gap controller [CONT SYS] A two-position (on-off) controller that actuates when the manipulated variable reaches the high or low value of its range (differential gap). { ‚dif·ə'ren·chəl 'gap kən‚trōl·ər }

differential gear See differential. { ‚dif·ə'ren·chəl 'gir }

differential generator [ELEC] A generator whose shunt and series windings are opposed to each other, to limit the maximum current. { ‚dif·ə'ren·chəl 'jen·ə‚rād·ər }

differential geometry [MATH] The study of curves and surfaces using the methods of differential calculus. { ‚dif·ə'ren·chəl jē'äm·ə·trē }

differential GPS [NAV] A technique for improving the accuracy of the Global Positioning System (GPS) in which error corrections are transmitted to users based on measurements of GPS signals by one or more reference receivers situated at known locations. Abbreviated DGPS. { ‚dif·ə'ren·chəl ¦jē¦pē'es }

differential heating [MET] A thermal gradient caused as heating takes place; can result in a distribution of stress in a material. { ‚dif·ə'ren·chəl 'hēd·iŋ }

differential heat of dilution See heat of dilution. { ‚dif·ə'ren·chəl 'hēt əv də'lü·shən }

differential heat of solution [THERMO] The partial derivative of the total heat of solution with respect to the molal concentration of one component of the solution, when the concentration of the other component or components, the pressure, and the temperature are held constant. { ‚dif·ə'ren·chəl 'hēt əv sə'lü·shən }

differential indexing [MECH ENG] A method of subdividing a circle based on the difference between movements of the index plate and index crank of a dividing engine. { ‚dif·ə'ren·chəl 'in‚deks·iŋ }

differential input [ELECTR] Amplifier input circuit that rejects voltages that are the same at both input terminals and amplifies the voltage difference between the two input terminals. { ‚dif·ə'ren·chəl 'in‚pùt }

differential-input capacitance [ELECTR] The capacitance between the inverting and noninverting input terminals of a differential amplifier. { ‚dif·ə¦ren·chəl ¦in‚pùt kə'pas·əd·əns }

differential-input impedance [ELECTR] The impedance between the inverting and noninverting input terminals of a differential amplifier. { ‚dif·ə¦ren·chəl ¦in‚pùt im'ped·əns }

differential-input measurement [ELECTR] A measurement in which the two inputs to a differential amplifier are connected to two points in a circuit under test and the amplifier displays the difference voltage between the points. { ‚dif·ə¦ren·chəl ¦in‚pùt 'mezh·ər·mənt }

differential-input resistance [ELECTR] The resistance between the inverting and noninverting input terminals of a differential amplifier. { ‚dif·ə¦ren·chəl ¦in‚pūt ri'zis·təns }

differential-input voltage [ELECTR] The maximum voltage that can be applied across the input terminals of a differential amplifier without causing damage to the amplifier. { ‚dif·ə¦ren·chəl ¦in‚pùt 'vōl‚tij }

differential instrument [ENG] Galvanometer or other measuring instrument having two circuits or coils, usually identical, through which currents flow in opposite directions; the difference or differential effect of these currents actuates the indicating pointer. { ‚dif·ə'ren·chəl 'in·strə·mənt }

differential interferometer See lateral shear interferometer. { ‚dif·ə'ren·chəl ‚in·tər·fə'räm·əd·ər }

differential ionization chamber [NUCLEO] A two-section ionization chamber in which electrode potentials are such that output current is equal to the difference between the separate ionization currents of the two sections. { ‚dif·ə'ren·chəl ‚ī·ən·ə'zā·shən ‚chäm·bər }

differential keying [ELECTR] Method for obtaining chirp-free break-in keying of continuous wave transmitters by using circuitry that arranges to have the oscillator turn on fast before the keyed amplifier stage can pass any signal, and turn off fast after the keyed amplifier stage has cut off. { ‚dif·ə'ren·chəl 'kē·iŋ }

differential leak detector [ENG] A leak detector consisting of two tubes and a trap which directs the tracer gas from the system into the desired tube. { ‚dif·ə'ren·chəl 'lēk di'tek·tər }

differential leukocyte count [PATH] The percentage of each variety of leukocytes in the blood, usually based on counting 100 leukocytes. Also known as differential blood count. { ‚dif·ə'ren·chəl 'lü·kə‚sīt ‚kaúnt }

differential leveling [ENG] A surveying process in which a horizontal line of sight of known elevation is intercepted by a graduated standard, or rod, held vertically on the point being checked. { ‚dif·ə'ren·chəl 'lev·əl·iŋ }

differentially coherent phase-shift keying See differential phase-shift keying. { ‚dif·ə'ren·chə·lē kō'hir·ənt 'fāz ‚shift ‚kē·iŋ }

differential manometer [ENG] An instrument in which the difference in pressure between two sources is determined from the vertical distance between the surfaces of a liquid in two legs of an erect or inverted U-shaped tube when each of the legs is connected to one of the sources. { ‚dif·ə'ren·chəl mə'näm·əd·ər }

differential microphone *See* double-button microphone. { ‚dif·ə'ren·chəl 'mī·krə‚fōn }

differential-mode gain [ELECTR] The ratio of the output voltage of a differential amplifier to the differential-mode input voltage. { ‚dif·ə¦ren·chəl ¦mōd ‚gān }

differential-mode input [ELECTR] The voltage difference between the two inputs of a differential amplifier. { ‚dif·ə¦ren·chəl ¦mōd ‚in‚pút }

differential-mode signal [ELECTR] A signal that is applied between the two ungrounded terminals of a balanced three-terminal system. { ‚dif·ə¦ren·chəl ¦mōd ‚sig·nəl }

differential modulation [COMMUN] Modulation in which the choice of the significant condition for any signal element is dependent on the choice for the previous signal element. { ‚dif·ə'ren·chəl ‚mäj·ə'lā·shən }

differential motion [MECH ENG] A mechanism in which the follower has two driving elements; the net motion of the follower is the difference between the motions that would result from either driver acting alone. { ‚dif·ə'ren·chəl 'mō·shən }

differential motor [ELEC] A direct-current motor whose shunt and series field windings oppose each other to produce a constant speed. { ‚dif·ə'ren·chəl 'mōd·ər }

differential operational amplifier [ELECTR] An amplifier that has two input terminals, used with additional circuit elements to perform mathematical functions on the difference in voltage between the two input signals. { ‚dif·ə'ren·chəl əp·ə'rā·shən·əl 'am·plə‚fī·ər }

differential operator [MATH] An operator on a space of functions which maps a function *f* into a linear combination of higher-order derivatives of *f*. { ‚dif·ə'ren·chəl 'äp·ə‚rād·ər }

differential output voltage [ELECTR] The difference between the values of two ac voltages, 180° out of phase, present at the output terminals of an amplifier when a differential input voltage is applied to the input terminals of the amplifier. { ‚dif·ə'ren·chəl 'aut‚pút ‚vōl·tij }

differential permeability [ELECTROMAG] The slope of the magnetization curve for a magnetic material. { ‚dif·ə'ren·chəl ‚pər·me·ə'bil·əd·ē }

differential phase [ELECTR] Difference in output phase of a small high-frequency sine-wave signal at two stated levels of a low-frequency signal on which it is superimposed in a video transmission system. { ‚dif·ə'ren·chəl 'fāz }

differential phase-shift keying [COMMUN] Form of phase-shift keying in which the reference phase for a given keying interval is the phase of the signal during the preceding keying interval. Also known as differentially coherent phase-shift keying. { ‚dif·ə'ren·chəl 'fāz ‚shift ‚kē·iŋ }

differential piece-rate system [IND ENG] A wage plan based on a standard task time whereby the worker receives increased or decreased piece rates as his or her production varies from that expected for the standard time. Also known as accelerating incentive. { ‚dif·ə'ren·chəl 'pēs ‚rāt ‚sis·təm }

differential polarography [ANALY CHEM] Technique of polarographic analysis which measures the difference in current flowing between two identical dropping-mercury electrodes at the same potential but in different solutions. { ‚dif·ə'ren·chəl ‚pō·lə'räg·rə·fē }

differential pressure [PHYS] The difference in pressure between two points of a system, such as between the well bottom and wellhead or between the two sides of an orifice. { ‚dif·ə'ren·chəl 'presh·ər }

differential-pressure fuel valve [MECH ENG] A needle or spindle normally closed, with seats at the back side of the valve orifice. { ‚dif·ə'ren·chəl ¦presh·ər 'fyül ‚valv }

differential-pressure gage [ENG] Apparatus to measure pressure differences between two points in a system; it can be a pressured liquid column balanced by a pressured liquid reservoir, a formed metallic pressure element with opposing force, or an electrical-electronic gage (such as strain, thermal-conductivity, or ionization). { ‚dif·ə'ren·chəl 'presh·ər ‚gāj }

differential-pressure pickup [ELEC] An instrument that measures the difference in pressure between two pressure

sources and translates this difference into a change in inductance, resistance, voltage, or some other electrical quality. { ‚dif·ə¦ren·chəl ¦presh·ər ‚pik‚əp }

differential process [CHEM ENG] A process in which a system is caused to move through a bubble point and as a result to form two phases, the minor phase being removed from further contact with the major phase; thus the system continuously changes in quantity and composition. { ‚dif·ə'ren·chəl 'präs·əs }

differential-producing primary device [ENG] An instrument that modifies the flow pattern of a fluid passing through a pipe, duct, or open channel, and thereby produces a difference in pressure between two points, which can then be measured to determine the rate of flow. { ‚dif·ə'ren·chəl prə‚düs·iŋ ¦prī‚mer·ē di'vīs }

differential psychology [PSYCH] The comparative study of the psychological differences between groups of people, for example, men and women. { ‚dif·ə'ren·chəl sī'käl·ə·jē }

differential pulley [MECH ENG] A tackle in which an endless cable passes through a movable lower pulley, which carries the load, and two fixed coaxial upper pulleys having different diameters; yields a high mechanical advantage. { ‚dif·ə'ren·chəl 'púl·ē }

differential pulse-code modulation [COMMUN] A type of pulse-code modulation proposed for television transmission, in which only the differences between the continuous picture elements on the scanning lines are transmitted, enabling the bandwidth of the signal to be reduced. Abbreviated DPCM. { ‚dif·ə'ren·chəl 'pəls ‚kōd ‚mäj·ə'lā·shən }

differential reaction rate [PHYS CHEM] The order of a chemical reaction expressed as a differential equation with respect to time; for example, $dx/dt = k(a - x)$ for first order, $dx/dt = k(a - x)(b - x)$ for second order, and so on, where k is the specific rate constant, a is the concentration of reactant *A*, *b* is the concentration of reactant *B*, and dx/dt is the rate of change in concentration for time *t*. { ‚dif·ə'ren·chəl rē'ak·shən ‚rāt }

differential relay [ELEC] A two-winding relay that operates when the difference between the currents in the two windings reaches a predetermined value. { ‚dif·ə'ren·chəl 'rē‚la }

differential scanning calorimeter [CHEM ENG] An instrument for studying overall chemical reactions by measuring the associated exothermic and endothermic reactions that occur over a specified temperature cycle. { ‚dif·ə'ren·chəl ¦skan·iŋ ‚kal·ə'rim·əd·ər }

differential scanning calorimetry [ANALY CHEM] A method in which a sample and a reference are individually heated (by separately controlled resistance heaters, at a predetermined rate), and enthalpic (heat-generating or -absorbing) processes are detected as differences in electrical energy supplied to either the sample or the reference material to maintain this heating rate. This difference in electrical energy, in milliwatts per second, of the heat flow into or out of the sample is due to the occurrence of a physical or chemical process. { ‚dif·ə¦ren·chəl ¦skan·iŋ ‚kal·ə'rim·ə·trē }

differential scatter [ENG] A technique for the remote sensing of atmospheric particles in which the ackscattering from laser beams at a number of infrared wavelengths is measured and correlated with scattering signatures that are uniquely related to particle composition. Abbreviated DISC. { ‚dif·ə'ren·chəl 'skad·ər }

differential screw [MECH ENG] A type of compound screw which produces a motion equal to the difference in motion between the two component screws. { ‚dif·ə'ren·chəl 'skrü }

differential selection [STAT] A biased selection of a conditioned sample. { ‚dif·ə'ren·chəl si'lek·shən }

differential selsyn [ELEC] Selsyn in which both rotor and stator have similar windings that are spread 120° apart; position of the rotor corresponds to the algebraic sum of the fields produced by the stator and rotor. { ‚dif·ə'ren·chəl 'sel·sən }

differential separation [CHEM ENG] Release of gas (vapor) from liquids by a reduction in pressure that allows the vapor to come out of the solution, so that the vapor can be removed from the system; differs from flash separation, in which the vapor and liquid are kept in contact following pressure reduction. { ‚dif·ə'ren·chəl ‚sep·ə'rā·shən }

differential signal [ELECTR] In a circuit, a signal that is the voltage difference between two nodes, neither of which is at ground potential. Also known as floating signal. { ‚dif·ə'ren·chəl 'sig·nəl }

differential spectrophotometry [SPECT] Spectrophotometric analysis of a sample when a solution of the major component of the sample is placed in the reference cell; the recorded spectrum represents the difference between the sample and the reference cell. { ‚dif·ə'ren·chəl ‚spek·trō·fə'täm·ə·trē }

differential stage [ELECTR] A symmetrical amplifier stage with two inputs balanced against each other so that with no input signal or equal input signals, no output signal exists, while a signal to either input, or an input signal unbalance, produces an output signal proportional to the difference. { ‚dif·ə'ren·chəl 'stāj }

differential steam calorimeter [ENG] An instrument for measuring small specific-heat capacities, such as those of gases, in which the amount of steam condensing on a body containing the substance whose heat capacity is to be measured is compared with the amount condensing on a similar body which is evacuated or contains a substance of known heat capacity. { ‚dif·ə'ren·chəl 'stēm kal·ə'rim·əd·ər }

differential stress *See* deviatoric stress. { ‚dif·ə'ren·chəl 'stres }

differential synchro *See* synchro differential receiver; synchro differential transmitter. { ‚dif·ə'ren·chəl 'siŋ·krō }

differential temperature survey [PETRO ENG] Well-temperature logging method that detects very small temperature anomalies; two thermometers, 6 feet (1.8 meters) apart, record the temperature gradient down the well bore, with small difference changes showing up anomalies. { ‚dif·ə'ren·chəl 'tem·prə·chər ‚sər‚vā }

differential thermal analysis [THERMO] A method of determining the temperature at which thermal reactions occur in a material undergoing continuous heating to elevated temperatures; also involves a determination of the nature and intensity of such reactions. { ‚dif·ə'ren·chəl 'thər·məl ə'nal·ə·səs }

differential thermogravimetric analysis [THERMO] Thermal analysis in which the rate of material weight change upon heating versus temperature is plotted; used to simplify reading of weight-versus-temperature thermogram peaks that occur close together. { ‚dif·ə'ren·chəl ¦thər·mō‚grav·ə¦me·trik ə'nal·ə·səs }

differential thermometer *See* bimetallic thermometer. { ‚dif·ə'ren·chəl thər'mäm·əd·ər }

differential thermometric titration [ANALY CHEM] Thermometric titration in which titrant is added simultaneously to the reaction mixture and to a blank in identically equipped cells. { ‚dif·ə'ren·chəl ¦thər·mə¦me·trik tī'trā·shən }

differential timing [IND ENG] A time-study technique in which the time value of an element of extremely short duration is determined by various calculations involving cycle values that first include and then exclude the element under consideration. { ‚dif·ə'ren·chəl 'tīm·iŋ }

differential topology [MATH] The branch of mathematics dealing with differentiable manifolds. { ‚dif·ə'ren·chəl tə'päl·ə·jē }

differential transducer [ELEC] A transducer that simultaneously senses two separate sources and provides an output proportional to the difference between them. { ‚dif·ə'ren·chəl tranz'dü·sər }

differential transformer [ELEC] A transformer used to join two or more sources of signals to a common transmission line. { ‚dif·ə'ren·chəl tranz'fòr·mər }

differential-transformer transducer [ELEC] A transducer in which movement of the iron core of a transformer varies the output voltage across two series-opposing secondary windings. { ‚dif·ə'ren·chəl tranz¦fòr·mər tranz'dü·sər }

differential voltage gain [ELECTR] Ratio of the change in output signal voltage at either terminal, or in a differential device, to the change in signal voltage applied to either input terminal, all voltages being measured to common reference. { ‚dif·ə'ren·chəl 'vōl·tij ‚gān }

differential voltmeter [ELEC] A voltmeter that measures only the difference between a known voltage and an unknown voltage. { ‚dif·ə'ren·chəl 'vōlt‚mēd·ər }

differential wind [ORD] In bomb ballistics, the vector difference between the wind at the bomb-release altitude and the wind at some other specific lower altitude; differential winds are required for the computation of the differential ballistic wind. { ‚dif·ə'ren·chəl 'wind }

differential winding [ELEC] A winding whose magnetic field opposes that of a nearby winding. { ‚dif·ə'ren·chəl 'wīnd·iŋ }

differential windlass [MECH ENG] A windlass in which the barrel has two sections, each having a different diameter; the rope winds around one section, passes through a pulley (which carries the load), then winds around the other section of the barrel. { ‚dif·ə'ren·chəl 'wind·ləs }

differential wound field [ELEC] Type of motor or generator field having both series and shunt coils that are connected to oppose each other. { ‚dif·ə'ren·chəl ¦waùnd 'fēld }

differentiating circuit [ELEC] A circuit whose output voltage is proportional to the rate of change of the input voltage. Also known as differentiating network. { ‚dif·ə¦ren·che‚ād·iŋ ¦sər·kət }

differentiating network *See* differentiating circuit. { ‚dif·ə¦ren·che‚ād·iŋ 'net‚wərk }

differentiation [MATH] The act of taking a derivative. { ‚dif·ə‚ren·che'ā·shən }

differentiation antigen [IMMUNOL] A cell surface antigen that is expressed only during a specific period of embryological differentiation. { dif·ə‚ren·che¦ā·shən 'ant·i·jən }

differentiator [ELECTR] A device whose output function is proportional to the derivative, or rate of change, of the input function with respect to one or more variables. { ‚dif·ə'ren·che‚ād·ər }

diffluence [FL MECH] A region of fluid flow in which the fluid is diverging from the direction of flow. { 'di‚flü·əns }

diffracted wave [PHYS] A wave whose front has been changed in direction by an obstacle or other nonhomogeneity in a medium, other than by reflection or refraction. { di'frak·təd 'wāv }

diffraction [PHYS] Any redistribution in space of the intensity of waves that results from the presence of an object causing variations of either the amplitude or phase of the waves; found in all types of wave phenomena. { di'frak·shən }

diffractional pulse-height discriminator *See* pulse-height selector. { di'frak·shən·əl 'pəls ‚hīt di'skrim·ə‚nād·ər }

diffraction analysis [PHYS] The study of the atomic structure of solids, liquids, or gases by means of diffraction of x-rays or particles, such as neutrons or electrons. { di'frak·shən ə‚nal·ə·səs }

diffraction grating [SPECT] An optical device consisting of an assembly of narrow slits or grooves which produce a large number of beams that can interfere to produce spectra. Also known as grating. { di'frak·shən ‚grād·iŋ }

diffraction instrument *See* diffractometer. { di'frak·shən ‚in·strə·mənt }

diffraction-limited [OPTICS] Capable of producing images whose separations are as small as the theoretical limit imposed by diffraction effects. { di'frak·shən 'lim·əd·əd }

diffraction loss [PHYS] That part of the reduction in power of a propagating wave or beam that results from diffraction. { di'frak·shən ‚lòs }

diffraction mottle [GRAPHICS] A mottled appearance of a radiograph caused by superposition of diffraction effects. { di'frak·shən ‚mäd·əl }

diffraction pattern [PHYS] Pattern produced on a screen or plate by waves which have undergone diffraction. { di'frak·shən ‚pad·ərn }

diffraction propagation [ELECTROMAG] Propagation of electromagnetic waves around objects or over the horizon by diffraction. { di'frak·shən präp·ə'gā·shən }

diffraction ring [OPTICS] Circular light pattern which appears to surround particles in a microscope field. { di'frak·shən ‚riŋ }

diffraction scattering [PHYS] Elastic scattering that occurs when inelastic processes remove particles from a beam. { di'frak·shən ‚skad·ə·riŋ }

diffraction spectrum [SPECT] Parallel light and dark or colored bands of light produced by diffraction. { di'frak·shən ‚spek·trəm }

diffraction symmetry [CRYSTAL] Any symmetry in a crystal lattice which causes the systematic annihilation of certain beams in x-ray diffraction. { di'frak·shən ‚sim·ə·trē }

diffraction velocimeter [OPTICS] A velocity-measuring instrument that uses a continuous-wave laser to send a beam of coherent light at objects moving at right angles to the beam; the needlelike diffraction lobes reflected by the moving objects sweep past the optical grating in the receiver, thereby generating in a photomultiplier a series of impulses from which velocity can be determined and read out. Also known as optical diffraction velocimeter. { di'frak·shən ‚vel·ə'sim·əd·ər }

diffraction zone [ELECTROMAG] The portion of a radio propagation path which lies outside a line-of-sight path. { ‚di'frak·shən ‚zōn }

diffractometer [PHYS] An instrument used to study the structure of matter by means of the diffraction of x-rays, electrons, neutrons, or other waves. Also known as diffraction instrument. { ‚di‚frak'täm·əd·ər }

diffractometry [CRYSTAL] The science of determining crystal structures by studying the diffraction of beams of x-rays or other waves. { ‚di‚frak'täm·ə·trē }

diffuse aurora [GEOPHYS] A widespread and relatively uniform type of aurora which is easily overlooked from the ground but is prominent in satellite pictures. { də'fyüs ə'rôr·ə }

diffuse-cutting filter [OPTICS] A color filter that gradually changes in absorption with wavelength. { də'fyüs ¦kəd·iŋ ‚fil·tər }

diffused-alloy transistor [ELECTR] A transistor in which the semiconductor wafer is subjected to gaseous diffusion to produce a nonuniform base region, after which alloy junctions are formed in the same manner as for an alloy-junction transistor; it may also have an intrinsic region, to give a *pnip* unit. Also known as drift transistor. { də'fyüzd 'al‚ȯi tran'zis·tər }

diffused-base transistor [ELECTR] A transistor in which a nonuniform base region is produced by gaseous diffusion; the collector-base junction is also formed by gaseous diffusion, while the emitter-base junction is a conventional alloy junction. { də'fyüzd ¦bās tran'zis·tər }

diffused emitter-collector transistor [ELECTR] A transistor in which both the emitter and collector are produced by diffusion. { də'fyüzd i'mid·ər kə'lek·tər tran'zis·tər }

diffused junction [ELECTR] A semiconductor junction that has been formed by the diffusion of an impurity within a semiconductor crystal. { də'fyüzd 'jəŋk·shən }

diffused-junction rectifier [ELECTR] A semiconductor diode in which the *pn* junction is produced by diffusion. { də'fyüzd ¦jəŋk·shən 'rek·tə‚fī·ər }

diffused-junction transistor [ELECTR] A transistor in which the emitter and collector electrodes have been formed by diffusion by an impurity metal into the semiconductor wafer without heating. { də'fyüzd ¦jəŋk·shən tran'zis·tər }

diffused mesa transistor [ELECTR] A diffused-junction transistor in which an *n*-type impurity is diffused into one side of a *p*-type wafer; a second *pn* junction, required for the emitter, is produced by alloying or diffusing a *p*-type impurity into the newly formed *n*-type surface; after contacts have been applied, undesired diffused areas are etched away to create a flat-topped peak called a mesa. { də'fyüzd ¦mā·sə tran'zis·tər }

diffused resistor [ELECTR] An integrated-circuit resistor produced by a diffusion process in a semiconductor substrate. { də'fyüzd ri'zis·tər }

diffuse front [METEOROL] A front across which the characteristics of wind shift and temperature change are weakly defined. { də'fyüs 'frənt }

diffuse galactic light [ASTRON] Starlight that has been scattered or reflected by interstellar dust near the galactic plane. { də'fyüs gə'lak·tik 'līt }

diffuse hypergammaglobulinemia [MED] General increase in serum immunoglobulins due to infection, hepatic disease, collagen diseases, and advanced sarcoidosis. { də'fyüs ¦hī·pər‚gam·ə¦gläb·yə·lə'nēm·ē·ə }

diffuse illumination [OPTICS] Lighting so arranged that the object is illuminated from many directions or sources. { də'fyüs i‚lüm·ə'nā·shən }

diffuse nebula [ASTRON] A type of nebula ranging from huge masses presenting relatively high surface brightness down to faint, milky structures that are detectable only with long exposures and special filters; may contain both dust and gas or may be purely gaseous. { də'fyüs 'neb·yə·lə }

diffuse placenta [EMBRYO] A placenta having villi diffusely scattered over most of the surface of the chorion; found in whales, horses, and other mammals. { də'fyüs plə'sent·ə }

diffuser [ENG] A duct, chamber, or section in which a high-velocity, low-pressure stream of fluid (usually air) is converted into a high-velocity, high-pressure flow. { də'fyüz·er }

diffuse radiation [PHYS] Radiant energy propagating in many different directions through a given small volume of space. { də'fyüs ‚rād·ē'ā·shən }

diffuse reflection [PHYS] Reflection of light, sound, or radio waves from a surface in all directions according to the cosine law. { də'fyüs ri'flek·shən }

diffuse reflection model [PHYS] A model for the behavior of gas molecules striking the surface of a solid body, in which the molecules are absorbed and reemitted with a Maxwellian velocity distribution corresponding to a temperature intermediate between that of the surface and that of the incoming flow of gas. { də'fyüs ri'flek·shən ‚mäd·əl }

diffuse reflector [OPTICS] Any surface whose irregularities are so large compared to the wavelength of the incident radiation that the reflected rays are sent back in a multiplicity of directions. { də'fyüs ri'flek·tər }

diffuse series [SPECT] A series occurring in the spectra of many atoms having one, two, or three electrons in the outer shell, in which the total orbital angular momentum quantum number changes from 2 to 1. { də'fyüs 'sir·ēz }

diffuse skylight *See* diffuse sky radiation. { də'fyüs 'skī‚līt }

diffuse sky radiation [ASTROPHYS] Solar radiation reaching the earth's surface after having been scattered from the direct solar beam by molecules or suspensoids in the atmosphere. Also known as diffuse skylight; skylight; sky radiation. { də'fyüs ¦skī ‚rād·ē'ā·shən }

diffuse sound [ACOUS] Sound that has uniform energy density in a given region so that all directions of energy flux at all parts of the region are equally probable. { də'fyüs 'saùnd }

diffuse spectrum [SPECT] Any spectrum having lines which are very broad even when there is no possibility of line broadening by collisions. { də'fyüs 'spek·trəm }

diffuse transmission [PHYS] Transmission of electromagnetic or acoustic radiation in all directions by a transmitting body. { də'fyüs tranz'mish·ən }

diffuse transmission density [OPTICS] The value of the photographic transmission density obtained when light flux impinges normally on the sample and all the transmitted flux is collected and measured. { də'fyüs tranz'mish·ən ‚den·səd·ē }

diffusing disk *See* diffusion disk. { də'fyüz‚iŋ ‚disk }

diffusing screen [GRAPHICS] A translucent screen used in contact printing to ensure even diffusion of light; the screen is sometimes denser at those points immediately facing the electric lamps, to avoid increased light in those areas. { də'fyüz·iŋ ‚skrēn }

diffusiometer [PHYS] An instrument which measures diffusion in liquids. { də‚fyüz·ē'äm·əd·ər }

diffusion [ACOUS] The degree of variation in the propagation directions of sound waves over the volume of a sound field. [ELECTR] A method of producing a junction by difusing an impurity metal into a semiconductor at a high temperature. [MECH ENG] The conversion of air velocity into static pressure in the diffuser casing of a centrifugal fan, resulting from increases in the radius of the air spin and in area. [METEOROL] The exchange of fluid parcels (and hence the transport of conservative properties) between regions in space, in the apparently random motions of the parcels on a scale too small to be treated by the equations of motion; the diffusion of momentum (viscosity), vorticity, water vapor, heat (conduction), and gaseous components of the atmospheric mixture have been studied extensively. [OPTICS] **1.** The distribution of incident light by reflection. **2.** Transmission of light through a translucent material. [PHYS] **1.** The spontaneous movement and scattering of particles (atoms and molecules), of liquids, gases, and solids. **2.** In particular, the macroscopic motion of the components of a system of fluids that is driven by differences in concentration. [SOLID STATE] **1.** The actual transport of mass, in the form of discrete atoms, through the lattice of a crystalline solid. **2.** The movement of carriers in a semiconductor. { də'fyü·zhən }

diffusion annealing [MET] Heat treatment of metal to promote homogeneity by diffusion of components. { də'fyü·zhən ə‚nēl·iŋ }

diffusion area [PHYS] One-sixth of the mean-square displacement between the appearance and disappearance of a subatomic particle of a given type. { də'fyü·zhən ‚er·ē·ə }

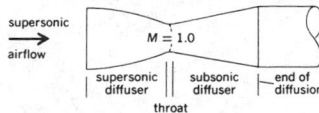

DIFFUSER

Idealized diffuser for decreasing supersonic velocities to subsonic velocities. Velocity equals speed of sound ($M = 1.0$) at throat.

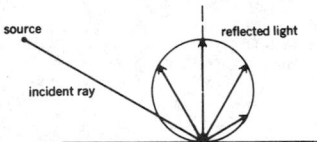

DIFFUSE REFLECTION

Diffuse reflection, such as from a mat surface of microscopic roughness. *(From W. B. Boast, Illumination Engineering, 2d ed., McGraw-Hill, 1953)*

diffusion barrier [CHEM ENG] Porous barrier through which gaseous mixtures are passed for enrichment of the lighter-molecular-weight constituent of the diffusate; used as a many-stage cascade system for the recovery of $^{235}UF_6$ isotopes from a $^{238}UF_6$ stream. { də'fyü·zhən ,bar·ē·ər }

diffusion bonding [MET] A solid-state process for joining metals by using only heat and pressure to achieve atomic bonding. { də'fyü·zhən ,bänd·iŋ }

diffusion brazing [MET] A process which produces bonding of the faying surfaces by heating them to suitable temperatures; the filler metal is diffused with the base metal and approaches the properties of the base metal. Also known as transient liquid phase-bonding. { də'fyü·zhən ,brāz·iŋ }

diffusion capacitance [ELECTR] The rate of change of stored minority-carrier charge with the voltage across a semiconductor junction. { də'fyü·zhən kə'pas·əd·əns }

diffusion cloud chamber [NUCLEO] A cloud chamber in which vapor diffuses from a source near a hot plate and condenses on a cold plate; the resulting layer of supersaturated vapor between the plates is sensitive to the passage of ionizing particles. { də'fyü·zhən 'klaùd ,chām·bər }

diffusion coating [MET] An alloy coating produced by allowing the coating material to diffuse into the base at high temperature. { də'fyü·zhən ,kōd·iŋ }

diffusion coefficient [PHYS] The weight of a material, in grams, diffusing across an area of 1 square centimeter in 1 second in a unit concentration gradient. Also known as diffusivity. { də'fyü·zhən ,kō·i'fish·ənt }

diffusion constant [SOLID STATE] The diffusion current density in a hologeneous semiconductor divided by the charge carrier concentration gradient. { də'fyü·zhən ,kän·stənt }

diffusion current [ANALY CHEM] In polarography with a dropping-mercury electrode, the flow that is controled by the rate of diffusion of the active solution species across the concentration gradient produced by the removal of ions or molecule at the electrode surface. { də'fyü·zhən ,kər·ənt }

diffusion diagram [METEOROL] A diagram for displaying the comparative properties of various diffusion processes, with coordinates of the mean free path or mixing length and mean molecular speed or diffusion velocity, for molecular or eddy diffusion, respectively; each point of the diagram determines diffusivity. { də'fyü·zhən ,dī·ə,gram }

diffusion diameter [STAT MECH] For a gas, the diameter of identical hard spheres that display the same diffusion as that observed for the molecules of the actual gas when their motion is treated classically. { di'fyü·zhən dī,am·əd·ər }

diffusion disk [OPTICS] A piece of transparent material that is marked or embossed, and is used with a camera lens to give the image a hazy softened quality. Also known as diffusing disk. { də'fyü·zhən ,disk }

diffusion equation [PHYS] **1.** An equation for diffusion which states that the rate of change of the density of the diffusing substance, at a fixed point in space, equals the sum of the diffusion coefficient times the Laplacian of the density, the amount of the quantity generated per unit volume per unit time, and the negative of the quantity absorbed per unit volume per unit time. **2.** More generally, any equation which states that the rate of change of some quantity, at a fixed point in space, equals a positive constant times the Laplacian of that quantity. { də'fyü·zhən i'kwā·zhən }

diffusion extraction [FOOD ENG] Extraction of juice by countercurrent flow of hot water through fruit slices. { də'fyü·zhən ik,strak·shən }

diffusion flame [CHEM] A long gas flame that radiates uniformly over its length and precipitates free carbon uniformly. { də'fyü·zhən ,flām }

diffusion gradient [PHYS] The graphed distance of penetration (diffusion) versus concentration of the material (or effect) diffusing through a second material; applies to heat, liquids, solids, or gases. { də'fyü·zhən ,grād·ē·ənt }

diffusion hygrometer [ENG] A hygrometer based upon the diffusion of water vapor through a porous membrane; essentially, it consists of a closed chamber having porous walls and containing a hygroscopic compound, whose absorption of water vapor causes a pressure drop within the chamber that is measured by a manometer. { də'fyü·zhən hī'gräm·əd·ər }

diffusion kernel [NUCLEO] The neutron flux resulting from a point source emitting one neutron per second; it is a function of the distance between the source and the point where the flux is measured. { də'fyü·zhən ,kər·nəl }

diffusion length [PHYS] The average distance traveled by a particle, such as a minority carrier in a semiconductor or a thermal neutron in a nuclear reactor, from the point at which it is formed to the point at which it is absorbed. { də'fyü·zhən ,leŋkth }

diffusion-limited aggregation [PHYS] A mathematical model for particle aggregation processes, such as the growth of a metal deposit on an electrochemical cell, in which particles move according to a random walk process until they arrive at a certain fixed distance from the current aggregate, where they stick to it. { də'fyü·zhən ¦lim·əd·əd ag·rə'gā·shən }

diffusion-limited current density [MET] The density corresponding to the maximum transfer rate that a material can sustain due to diffusion limits. { də'fyü·zhən ¦lim·əd·əd 'kər·ənt ,den·səd·ē }

diffusion number [FL MECH] A dimensionless number used in the study of mass transfer, equal to the diffusivity of a solute through a stationary solution contained in the solid, times a characteristic time, divided by the square of the distance from the midpoint of the solid to the surface. Symbolized β. { də'fyü·zhən ,nəm·bər }

diffusion plant [NUCLEO] A plant which separates isotopes by isotopic diffusion or thermal diffusion. { də'fyü·zhən ,plant }

diffusion potential [PHYS CHEM] A potential difference across the boundary between electrolytic solutions with different compositions. Also known as liquid junction potential. { də'fyü·zhən pə,ten·chəl }

diffusion pump [ENG] A vacuum pump in which a stream of heavy molecules, such as mercury vapor, carries gas molecules out of the volume being evacuated; also used for separating isotopes according to weight, the lighter molecules being pumped preferentially by the vapor stream. { də'fyü·zhən ,pəmp }

diffusion respiration [PHYSIO] Exchange of gases through the cell membrane, between the cells of unicellular or other simple organisms and the environment. { də'fyü·zhən res·pə'rā·shən }

diffusion theory [ELEC] The theory that in semiconductors, where there is a variation of carrier concentration, a motion of the carriers is produced by diffusion in addition to the drift determined by the mobility and the electric field. { də'fyü·zhən ,thē·ə·rē }

diffusion-transfer process [GRAPHICS] Any of several photographic processes for copying documents in which the copy is produced by developing a photographic image, transferring by diffusion the silver salts in undeveloped areas of the receiving paper, and developing the transferred image. { də¦fyü·zhən 'tranz·fər ,präs·əs }

diffusion transistor [ELECTR] A transistor in which current flow is a result of diffusion of carriers, donors, or acceptors, as in a junction transistor. { də'fyü·zhən tran,zis·tər }

diffusion velocity [FL MECH] **1.** The relative mean molecular velocity of a selected gas undergoing diffusion in a gaseous atmosphere, commonly taken as a nitrogen (N_2) atmosphere; a molecular phenomenon that depends upon the gaseous concentration as well as upon the pressure and temperature gradients present. **2.** The velocity or speed with which a turbulent diffusion process proceeds as evidenced by the motion of individual eddies. { də'fyü·zhən və'läs·əd·ē }

diffusion welding [MET] A welding process which utilizes high temperatures and pressures to coalesce the faying surfaces by solid-state bonding; there is no physical movement, visible deformation of the parts involved, or melting. { də'fyü·zhən ,weld·iŋ }

diffusiophoresis [CHEM ENG] A process in a scrubber whereby water vapor moving toward the cold water surface carries particulates with it. { də¦fyü·zē·ō·fə'rē·səs }

diffusive equilibrium [METEOROL] The steady state resulting from the diffusion process, primarily of interest when external forces and sources and sinks exist within the field; in such a state the constituent gases of the atmosphere would be distributed independently of each other, the heavier decreasing more rapidly with height than the lighter; but the presence of turbulent mixing precludes establishment of complete diffusive equilibrium. { də'fyü·ziv ,ē·kwə'lib·rē·əm }

diffusivity [PHYS] *See* diffusion coefficient. [THERMO]

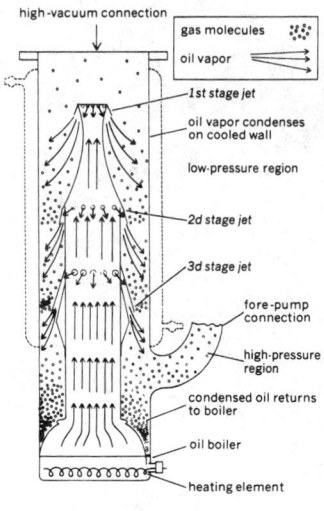

DIFFUSION PUMP

high-vacuum connection
gas molecules
oil vapor
1st stage jet
oil vapor condenses on cooled wall
low-pressure region
2d stage jet
3d stage jet
fore-pump connection
high-pressure region
condensed oil returns to boiler
oil boiler
heating element

Main operating features of diffusion pump.

The quantity of heat passing normally through a unit area per unit time divided by the product of specific heat, density, and temperature gradient. Also known as thermal diffusivity; thermometric conductivity. { dif·yü′ziv·əd·ē }

diffusivity analysis [ANALY CHEM] Analysis of difficult-to-separate materials in solution by diffusion effects, using, for example, dialysis, electrodialysis, interferometry, amperometric titration, polarography, or voltammetry. { dif·yü′ziv·əd·ē ə′nal·ə·səs }

difunctional molecule [ORG CHEM] An organic structure possessing two sites that are highly reactive. { ˌdī′fəŋk·shən·əl ′mäl·ə‚kyül }

digallic acid See tannic acid. { dī′gal·ik ′as·əd }

digamma function [MATH] The derivative of the natural logarithm of the gamma function. { ′dī‚gam·ə ‚fəŋk·shən }

digastric [ANAT] Of a muscle, having a fleshy part at each end and a tendinous part in the middle. { dī′gas·trik }

Digenea [INV ZOO] A group of parasitic flatworms or flukes constituting a subclass or order of the class Trematoda and having two types of generations in the life cycle. { dī′jē·nē·ə }

digenesis [BIOL] Sexual and asexual reproduction in succession. { dī′jen·ə·səs }

digenite [MINERAL] Cu_9S_5 A blue to black mineral consisting of an isometric copper sulfide having a variable deficiency in copper. Also known as alpha chalcocite; blue chalcocite. { ′dī·jə‚nīt }

Di George's syndrome See thymic aplasia. { də′jórj·əz ‚sin‚drōm }

digested sludge [CIV ENG] Sludge or thickened mixture of sewage solids with water that has been decomposed by anaerobic bacteria. { də′jes·təd ′sləj }

digester [CHEM ENG] A vessel used to produce cellulose pulp from wood chips by cooking under pressure. [CIV ENG] A sludge-digestion tank containing a system of hot water or steam pipes for heating the sludge. { də′jes·tər }

digestion [CHEM ENG] **1.** Preferential dissolving of mineral constituents in concentrations of ore. **2.** Liquefaction of organic waste materials by action of microbes. **3.** Separation of fabric from tires by the use of hot sodium hydroxide. **4.** Removing lignin from wood in manufacture of chemical cellulose paper pulp. [CIV ENG] The process of sewage treatment by the anaerobic decomposition of organic matter. [PHYSIO] The process of converting food to an absorbable form by breaking it down to simpler chemical compounds. { də′jes·chən }

digestive efficiency [ECOL] A measure of the amount of ingested chemical energy actually absorbed by an animal. { dī′jes·tiv i′fish·ən·sē }

digestive enzyme [BIOCHEM] Any enzyme that causes or aids in digestion. { də′jes·tiv ′en‚zīm }

digestive gland [PHYSIO] Any structure that secretes digestive enzymes. { də′jes·tiv ‚gland }

digestive system [ANAT] A system of structures in which food substances are digested. { də′jes·tiv ‚sis·təm }

digestive tract [ANAT] The alimentary canal. { də′jes·tiv ‚trakt }

digger [ENG] A tool or apparatus for digging in the ground. [MIN ENG] A person who digs in the ground; usually refers to a coal miner. { ′dig·ər }

digging [ENG] A sudden increase in cutting depth of a cutting tool due to an erratic change in load. { ′dig·iŋ }

digging height See bank height. { ′dig·iŋ ‚hīt }

digging line See inhaul cable. { ′dig·iŋ ‚līn }

diggings [SCI TECH] **1.** Excavated materials. **2.** A place of excavating. { ′dig·iŋz }

digicitrin [BIOCHEM] $C_{21}H_{21}O_{10}$ A flavone compound that is found in foxglove leaves. { ‚dīj·ə′si·trən }

digicom [COMMUN] A wire communication system that transmits speech signals in the form of corresponding trains of pulses and transmits digital information directly from computers, radar, tape readers, teleprinters, and telemetering equipment. { ′dīj·ə‚käm }

digicon [ELECTR] An image tube in which the image produced by electrons from the photocathode is focused directly on a silicon diode array and each incoming photoelectron produces an electrical pulse that is amplified and recorded. { ′dīj·ə‚kän }

digit [COMPUT SCI] In a decimal digital computer, the space reserved for storage of one digit of information. [MATH] A character used to represent one of the nonnegative integers smaller than the base of a system of positional notation. Also known as numeric character. { ′dij·ət }

digit absorbing selector [ELECTR] Dial switch arranged to set up and then fall back on the first one of two digits dialed; it then operates on the next digit dialed. { ′dij·ət əb‚sórb·iŋ si′lek·tər }

digital [COMPUT SCI] Pertaining to data in the form of digits. { ′dij·əd·əl }

digital audio broadcasting [COMMUN] The radio broadcasting of audio signals encoded in digital form. Abbreviated DAB. { ¦dij·əd·əl ′ȯd·ē·ō ′bród‚kast·iŋ }

digital audio tape [COMPUT SCI] A magnetic tape on which sound is recorded and played back in digital form. Abbreviated DAT. { ¦dij·əd·əl ′ȯd·ē·ō ‚tāp }

digital camera [ELECTR] A television camera that breaks up a picture into a fixed number of pixels and converts the light intensity (or the intensities of each of the primary colors) in each pixel to one of a finite set of numbers. { ′dij·əd·əl ′kam·rə }

digital channel [COMMUN] A transmission path that carries only digital signals. { ′dij·əd·əl ′chan·əl }

digital chart [NAV] A navigational chart encoded in a computer-usable format and used, in combination with electronic devices, to produce a computer-generated video display which provides the navigator with an accurate pictorial presentation of the information normally gathered from a paper chart. Also known as electronic chart. { ′dij·əd·əl ′chärt }

digital circuit [ELECTR] A circuit designed to respond at input voltages at one of a finite number of levels and, similarly, to produce output voltages at one of a finite number of levels. { ′dij·əd·əl ′sər·kət }

digital circuit multiplication equipment [COMMUN] Equipment that uses digital compression techniques to increase the capacity of digital satellite and cable links carrying voice, facsimile, and voice-frequency modem traffic. { ‚dij·əd·əl ‚sər·kət ‚məl·tə·plə′kā·shən i‚kwip·mənt }

digital communications [COMMUN] System of telecommunications employing a nominally discontinuous signal that changes in frequency, amplitude, time, or polarity. { ′dij·əd·əl kə‚myü·nə′kā·shənz }

digital comparator [ELECTR] A comparator circuit operating on input signals at discrete levels. Also known as discrete comparator. { ′dij·əd·əl kəm′par·əd·ər }

digital computer [COMPUT SCI] A computer operating on discrete data by performing arithmetic and logic processes on these data. { ′dij·əd·əl kəm′pyüd·ər }

digital control [CONT SYS] The use of digital or discrete technology to maintain conditions in operating systems as close as possible to desired values despite changes in the operating environment. { ′dij·əd·əl kən′trōl }

digital converter [ELECTR] A device that converts voltages to digital form; examples include analog-to-digital converters, pulse-code modulators, encoders, and quantizing encoders. { ′dij·əd·əl kən′vərd·ər }

digital counter [ELECTR] A discrete-state device (one with only a finite number of output conditions) that responds by advancing to its next output condition. { ′dij·əd·əl ′kaunt·ər }

digital data [COMPUT SCI] Data that are electromagnetically stored in the form of discrete digits. { ′dij·əd·əl ′dad·ə }

Digital Data Broadcast System [NAV] A system that will provide information aiding air-traffic control; digital data to aircraft over vortac channels will carry information on the geographic location, elevation, magnetic variation, and related data of the vortac station being received. Abbreviated DDBS. { ‚dij·əd·əl ′dad·ə ′bród‚kast ‚sis·təm }

digital data modulation system [COMMUN] A digital communications system in which the information source consists of a finite number of discrete messages which are coded into a sequence of waveforms or symbols, each one selected from a specified and finite set. { ′dij·əd·əl ′dad·ə ‚mäj·ə′lā·shən ‚sis·təm }

digital data recorder [COMPUT SCI] Electronic device that converts continuous electrical analog signals into number (digital) values and records these values onto a data log via a high-speed typewriter. { ′dij·əd·əl ′dad·ə ri‚kórd·ər }

digital data service [COMMUN] A telephone communication system developed specifically for digital data, using existing local digital lines combined with data-under-voice

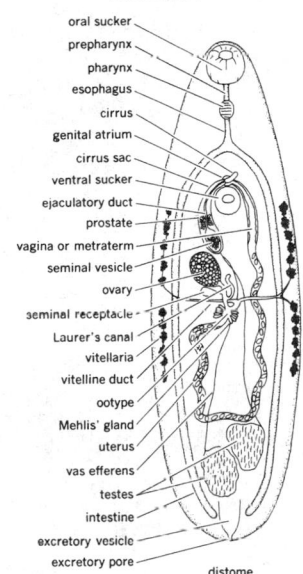

DIGENEA

oral sucker
prepharynx
pharynx
esophagus
cirrus
genital atrium
cirrus sac
ventral sucker
ejaculatory duct
prostate
vagina or metraterm
seminal vesicle
ovary
seminal receptacle
Laurer's canal
vitellaria
vitelline duct
ootype
Mehlis' gland
uterus
vas efferens
testes
intestine
excretory vesicle
excretory pore
distome

Diagram of an adult digenetic trematode. *(From R. M. Cable, An Illustrated Laboratory Manual of Parasitology, Burgess, 1940)*

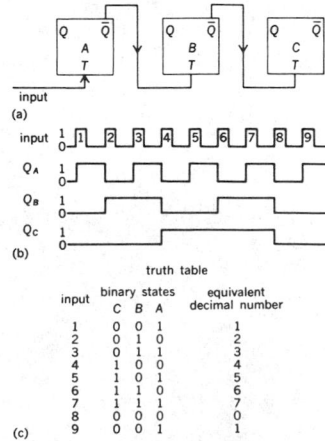

DIGITAL COUNTER

input	binary states			equivalent
	C	B	A	decimal number
1	0	0	1	1
2	0	1	0	2
3	0	1	1	3
4	1	0	0	4
5	1	0	1	5
6	1	1	0	6
7	1	1	1	7
8	0	0	0	0
9	0	0	1	1

An octal counter. T stands for trigger input, $\overline{Q}$ and Q represent output terminals, and A, B, and C identify different flip-flop stages. *(a)* Three successive flip-flop stages. *(b)* Input signal and Q-terminal states of each flip-flop. *(c)* Truth table.

microwave transmission facilities. Abbreviated DDS. { 'dij·əd·əl 'dad·ə ˌsər·vəs }

digital delayer [ENG ACOUS] A device for introducing delay in the audio signal in a sound-reproducing system, which converts the audio signal to digital format and stores it in a digital shift register before converting it back to analog form. { 'dij·əd·əl di'lā·ər }

digital delay generator [ELECTR] A high-precision adjustable time-delay generator in which delays may be selected in increments such as 1, 10, or 100 nanoseconds by means of panel switches and sometimes by remote programming. { 'dij·əd·əl diˌlā jen·ə'rād·ər }

digital differential analyzer [COMPUT SCI] A differential analyzer which uses numbers to represent analog quantities. Abbreviated DDA. { 'dij·əd·əl ˌdif·ə₊ren·chəl 'an·əˌlīz·ər }

digital display [COMPUT SCI] A display in which the result is indicated in directly readable numerals. { 'dij·əd·əl di'splā }

Digital Electronic Message Service [COMMUN] A communication system whose purpose is to provide efficient means for two-way high-speed data communications, transfer of graphic images (facsimile), and teleconferencing between cities and within a city environment. Abbreviated DEMS. { 'dij·əd·əl iˌlek'trän·ik 'mes·ij ˌsər·vəs }

digital filter [ELECTR] An electrical filter that responds to an input which has been quantified, usually as pulses. { 'dij·əd·əl 'fil·tər }

digital flight-control computer [AERO ENG] A device that contains a digital model of the aerodynamic performance of an aircraft under various conditions of flight and uses this information to interpret pilot commands into optimum motions of the aircraft's control surfaces. { 'dij·əd·əl 'flīt kən₊trōl kəmˌpyüd·ər }

digital format [COMPUT SCI] Use of discrete integral numbers in a given base to represent all the quantities that occur in a problem or calculation. { 'dij·əd·əl 'fȯr·mat }

digital frequency meter [ELECTR] A frequency meter in which the value of the frequency being measured is indicated on a digital display. { 'dij·əd·əl 'frē·kwən·sē ˌmēd·ər }

digital incremental plotter [COMPUT SCI] A device for converting digital signals in the output of a computer into graphical form, in which the digital signals control the motion of a plotting pen and of a drum that carries the paper on which the graph is drawn. { 'dij·əd·əl ˌiŋ·krə₊ment·əl 'pläd·ər }

digital integrator [COMPUT SCI] A device for computing definite integrals in which increments in the input variables and output variable are represented by digital signals. { 'dij·əd·əl 'in₊tə₊grād·ər }

digital intercontinental conversion equipment [ELECTR] Equipment which uses pulse-code modulation to convert a 525-line, 60-frame-per-second television signal used in the United States into a 625-line, 50-frame-per-second phase-alternation line signal used in Europe; the 525-line signal is sampled and quantized into a pulse-code modulation signal which is stored in shift registers from which the phase-alternation line signal is read out. Abbreviated DICE. { 'dij·əd·əl ˌin·tər₊känt·ən'ent·əl kən'vər·zhən iˌkwip·mənt }

digitalis [PHARM] The dried leaf of the purple foxglove plant (*Digitalis purpurea*), containing digitoxin and gitoxin; constitutes a powerful cardiac stimulant and diuretic. { dij·ə'tal·əs }

Digitalis [BOT] A genus of herbs in the figwort family, Scrophulariaceae. { dij·ə'tal·əs }

digital log [ENG] A well log that has undergone discrete sampling and recording on a magnetic tape preparatory to use in computerized interpretation and plotting. { 'dij·əd·əl 'läg }

digital loop carrier [COMMUN] A technology for providing 24 telephone circuits on pairs of wires, in which analog input signals are first sampled and digitized, and the binary digital signals from 24 users are then time-multiplexed into a single bit stream. { 'dij·əd·əl 'lüp ˌkar·ē·ər }

digital message entry system [ELECTR] A system that encodes formatted messages in digital form; it enters the encoded digital information into a voice communications transceiver by frequency shift techniques. { ˌdij·əd·əl 'mes·ij 'en₊trē ˌsis·təm }

digital microwave radio [COMMUN] Transmission of voice and data signals in digital form on microwave links, as in the 2-gigahertz common-carrier bands; pulse-code modulation is used. { ˌdij·əd·əl ˌmī·krō₊wāv 'rād·ē·ō }

digital modulation [COMMUN] A method of placing digital traffic on a microwave system without use of modems, by transmitting the information in the form of discrete phase or frequency states determined by the digital signal. { 'dij·əd·əl ˌmäj·ə'lā·shən }

digital monitor [ELECTR] A display unit that accepts digital signals and converts them to analog signals internally in order to illuminate the screen. { 'dij·əd·əl 'män·əd·ər }

Digital Multiplexed Interface [COMPUT SCI] A cost-effective, high-speed interconnection between terminals and host computers in a private branch exchange environment. { 'dij·əd·əl 'məl·tə₊plekst 'in·tər₊fās }

digital multiplier [ELECTR] A multiplier that accepts two numbers in digital form and gives their product in the same digital form, usually by making repeated additions; the multiplying process is simpler if the numbers are in binary form wherein digits are represented by a 0 or 1. { 'dij·əd·əl 'məl·tə₊plī·ər }

digital navigation system [NAV] An electronic digital device which provides an aircraft, spacecraft, or surface vehicle with frequently updated positional information in two or three dimensions in relation to some reference source and translates this information into a form usable either by human operators or directly by vehicle control systems. { 'dij·əd·əl ˌnav·ə'gā·shən ˌsis·təm }

digital object identifier [COMPUT SCI] A system for identifying and exchanging intellectual properties (including, for example, physical objects as well as digital files) in the digital environment. { ˌdij·əd·əl ˌäb₊jekt ī'den·tə₊fī·ər }

digital output [ELECTR] An output signal consisting of a sequence of discrete quantities coded in an appropriate manner for driving a printer or digital display. { 'dij·əd·əl 'au̇t₊pu̇t }

digital phase shifter [ELECTR] Device which provides a signal phase shift by the application of a control pulse; a reversal or phase shift requires a control pulse of opposite polarity. { 'dij·əd·əl 'fāz ˌshif·tər }

digital plotter [ELECTR] A recorder that produces permanent hard copy in the form of a graph from digital input data. { 'dij·əd·əl 'pläd·ər }

digital press [GRAPHICS] A machine used to reproduce documents directly from a digital file created on a desktop computer, through the process of toner-based electrophotography (a technology used by the common laser printer); used for short-run quantities (fewer than 2000), multiple-page documents, and printing that is needed quickly. { ˌdij·əd·əl 'pres }

digital printer [COMPUT SCI] A printer that provides a permanent readable record of binary-coded decimal or other coded data in a digital form that may include some or all alphanumeric characters and special symbols along with numerals. Also known as digital recorder. { 'dij·əd·əl 'print·ər }

digital private automatic branch exchange [COMMUN] A central communications switching system for a local-area network, which employs existing telephone wires in a building for the connection of telephones and computer terminals and systems. { 'dij·əd·əl ˌprīv·ət ˌȯd·ə₊mad·ik 'branch iksˌchānj }

digital proofing [GRAPHICS] Processes used to obtain hardcopy output directly from digital files for purposes of predicting the appearance of printed matter. { ˌdij·əd·əl 'prüf·iŋ }

digital radio [COMMUN] The microwave transmission of digital signals through space or the atmosphere. { ˌdij·əl·əl 'rād·ē·ō }

digital radiography [NUCLEO] The technique of producing images of objects by electronically detecting the arrival of x-ray photons transmitted through the object or emitted from it on various media, such as thin films of rare-earth elements, and converting the sensed analog signals to digital signals representing the intensity of x-ray photons at each position. { 'dij·əd·əl ˌrād·ē'äg·rə·fē }

digital recorder *See* digital printer. { 'dij·əd·əl ri'kȯrd·ər }

digital recording [ELECTR] Magnetic recording in which the information is first coded in a digital form, generally with a binary code that uses two discrete values of residual flux. { 'dij·əd·əl ri'kȯrd·iŋ }

digital representation [COMPUT SCI] The use of discrete impulses or quantities arranged in coded patterns to represent variables or other data in the form of numbers or characters. { 'dij·əd·əl ˌrep·rə₊zen'tā·shən }

digital resolution [COMPUT SCI] The ability of a digital computer to approach a truly correct answer, generally established by the number of places expressed, and the value of the least significant digit in a digitally coded representation. { 'dij·əd·əl ˌrez·ə'lü·shən }

digital set-top box [COMMUN] A device that is attached to a television receiver and can collect, store, and display digitally compressed television signals. { ˌdij·əd·əl 'set,täp ˌbäks }

digital signal analyzer [ELECTR] A signal analyzer in which one or more analog inputs are sampled at regular intervals, converted to digital form, and fed to a memory. { ˈdij·əd·əl 'sig·nəl ˌan·ə,liz·ər }

digital signal processing See signal processing. { ˌdij·əd·əl ˌsig·nəl 'prä·səs·iŋ }

digital signal processing chip [COMPUT SCI] A digital device for executing algorithms for the transformation or extraction of information from signals originally in analog form, such as audio or images. Abbreviated DSP chip. Also known as digital signal processor. { ˌdij·əd·əl ˌsig·nəl 'prä·səs·iŋ ˌchip }

digital signal processor See digital signal processing chip. { ˌdij·əd·əl 'sig·nəl ˌprä,ses·ər }

digital signature [COMMUN] A set of alphabetic or numeric characters used to authenticate a cryptographic message by ensuring that the sender cannot later disavow the message, the receiver cannot forge the message or signature, and the receiver can prove to others that the contents of the message are genuine and originated with the sender. { 'dij·əd·əl 'sig·nə·chər }

digital simulation [COMPUT SCI] The representation of a system in a form acceptable to a digital computer as opposed to an analog computer. { ˌdij·əd·əl ˌsim·yə'lā·shən }

digital speech communications [COMMUN] Transmission of voice in digitized or binary form via landline or radio. { 'dij·əd·əl 'spēch kə,myün·ə,kā·shənz }

digital speech interpolation [COMMUN] In digital speech communications, the use of periods of inactivity or constant signal level to increase the transmission efficiency by insertion of additional signals. Abbreviated DSI. { ˌdij·əd·əl 'spēch ,in·tər·pə,lā·shən }

digital subscriber line [COMMUN] A system that provides subscribers with continuous, uninterrupted connections to the Internet over existing telephone lines, offering a choice of speeds ranging from 32 kilobits per second to more than 50 megabits per second. Abbreviated DSL. { ˌdij·əd·əl səb'skrīb·ər ˌlīn }

digital subtraction angiography [MED] A form of digital radiography that delineates blood vessels by subtracting a digitized tissue background image from an image of tissue injected with an intravascular contrast material with a high content of iodine that attenuates x-rays. Abbreviated DSA. { ˌdij·əd·əl səb'trak·shən ,an·jē'äg·rə·fē }

digital synchronometer [ELECTR] A time comparator that provides a direct-reading digital display of time with high precision by making accurate comparisons between its own digital clock and high-accuracy time transmissions from radio station WWV or a loran C station. { ˌdij·əd·əl ,siŋ·krə'näm·əd·ər }

digital system [COMPUT SCI] Any of the levels of operation for a digital computer, including the wires and mechanical parts, the logical elements, and the functional units for reading, writing, storing, and manipulating information. { 'dij·əd·əl 'sis·təm }

digital telemetering [COMPUT SCI] Conversion of a continuous electrical analog signal into a digital (number system) code prior to transmitting the signal to a receiver (such as digital readout, card punch, or tape). { ˌdij·əd·əl ˌtel·ə,mēd·ər·iŋ }

digital television [COMMUN] Television in which picture information is encoded into digital signals at the transmitter, and decoded at the receiver. { 'dij·əd·əl 'tel·ə,vizh·ən }

digital television converter [ELECTR] A converter used to convert television programs from one system to another, such as for converting 525-line 60-field United States broadcasts to 625-line 50-field European PAL (phase-alternation line) or SECAM (sequential couleur á memoire) standards; the video signal is digitized before conversion. { ˌdij·əd·əl ˌtel·ə,vizh·ən kən'vərd·ər }

Digital Termination System [COMMUN] A microwave communication system that can send and receive information at speeds up to 2.1×10^6 bits per second, and provides intracity digital radio links between a central serving office and subscribers in the Digital Electronic Message Service. Abbreviated DTS. { 'dij·əd·əl ,tər·mə'nā·shən ,sis·təm }

digital-to-analog converter [ELECTR] A converter in which

digital input signals are changed to essentially proportional analog signals. Abbreviated dac. { 'dij·əd·əl tü ˈan·ə,läg kən'vərd·ər }

digital-to-synchro converter [ELECTR] A converter that changes binary-coded decimal or other digital input data to a three-wire synchro output signal representing corresponding angular data. { ˌdij·əd·əl tü ˈsiŋ·krō kən'vərd·ər }

digital transducer [ELECTR] A transducer that measures physical quantities and transmits the information as coded digital signals rather than as continuously varying currents or voltages. { 'dij·əd·əl tranz'dü·sər }

digital versatile disk See DVD. { ˌdij·əd·əl 'vər·səd·əl ,disk }

digital video disk See DVD. { ˌdij·əd·əl 'vid·ē·ō ,disk }

digital voltmeter [ELECTR] A voltmeter in which the unknown voltage is compared with an internally generated analog voltage, the result being indicated in digital form rather than by a pointer moving over a meter scale. { 'dij·əd·əl 'vōlt,mēd·ər }

digital watermark [COMPUT SCI] Invisible or inaudible data (a random pattern of bits or noise) permanently embedded in a graphic, video, or audio file for protecting copyright or authenticating data. { ˌdij·əd·əl 'wód·ər,märk }

digitate [ANAT] Having digits or digitlike processes. { 'dij·ə,tāt }

digitation [GEOL] A secondary recumbent anticline emanating from a larger recumbent anticline. { ˌdij·ə'tā·shən }

digit-coded voice [COMPUT SCI] A limited, spoken vocabulary, each word of which corresponds to a code and which, upon keyed inquiry, can be strung in meaningful sequence and can be outputted as audio response to the inquiry. { 'dij·ət ,kōd·əd 'vois }

digit compression [COMPUT SCI] Any process which increases the number of digits stored at a given location. { 'dij·ət kəm'presh·ən }

digit delay element [ELECTR] A logic element that introduces a delay of one digit period in a series of signals or pulses. { 'dij·ət di'lā ,el·ə·mənt }

digitellum [INV ZOO] A tentacle-like gastric filament in scyphozoans. { ˌdij·ə'tel·əm }

digit emitter [COMPUT SCI] A character emitter limited to the twelfth-row pulses in a punched card. { 'dij·ət i'mid·ər }

digit filter [COMPUT SCI] A device used with punched-card equipment that detects the presence of a designation punch in a specified card column. { 'dij·ət ,fil·tər }

digitigrade [VERT ZOO] Pertaining to animals, such as dogs and cats, which walk on the digits with the posterior part of the foot raised from the ground. { 'dij·ə·də,grād }

digitinervate [BOT] Having straight veins extending from the petiole like fingers. { ˌdij·ə·də'nər,vāt }

digitipinnate [BOT] Having digitate leaves with pinnate leaflets. { ˌdij·ə·də'pin,āt }

digitize [COMPUT SCI] To convert an analog measurement of a quantity into a numerical value. { 'dij·ə,tīz }

digitizer [COMPUT SCI] A large drawing table connected to a computer video display and equipped with a penlike or pucklike instrument whose motions are reproduced on the screen. Also known as digitizer tablet. { 'dij·ə,tīz·ər }

digitizer tablet See digitizer. { 'dij·ə,tīz·ər ,tab·lət }

digitonin [PHARM] $C_{41}H_{64}O_{13}$ A glycoside derived from the purple foxglove plant (Digitalis purpurea); a white powder melting at 255–256°C; used as a medicine for cardiac conditions. { ˌdij·ə'tō·nən }

digitoxigenin [ORG CHEM] $C_{23}H_{34}O_4$ The steroid aglycone formed by removal of three molecules of the sugar digitoxose from digitoxin. { ˌdij·ə,täk·sə'jen·ən }

digitoxin [ORG CHEM] $C_{41}H_{64}O_{13}$ A poisonous steroid glycoside found as the most active principle of digitalis, from the foxglove leaf. { ˌdij·ə'täk·sən }

digit period [ELECTR] The time interval between successive pulses, usually representing binary digits, in a computer or in pulse modulation, determined by the pulse-repetition frequency. Also known as digit time. { 'dij·ət 'pir·ē·əd }

digit place See digit position. { 'dij·ət ,plās }

digit plane [COMPUT SCI] In a computer memory consisting of magnetic cores arranged in a three-dimensional array, a plane containing elements for a particular digit position in various words. { 'dij·ət ,plān }

digit position [MATH] The position of a particular digit in a number that is expressed in positional notation, usually numbered from the lowest significant digit of the number. Also known as digit place. { 'dij·ət pə,zish·ən }

digit pulse [ELECTR] An electrical pulse which induces a magnetizing force in a number of magnetic cores in a computer storage, all corresponding to a particular digit position in a number of different words. { 'dij·ət ,pəls }

digit rearrangement [COMPUT SCI] A method of hashing which consists of selecting and shifting digits of the original key. { 'dij·ət ,rē·ə'ränj·mənt }

digit selector [COMPUT SCI] A device which separates a card column into individual pulses corresponding to punched row positions. { 'dij·ət si,lek·tər }

digit time See digit period. { 'dij·ət ,tīm }

digitus [INV ZOO] In insects, the claw-bearing terminal segment of the tarsus. { 'dij·əd·əs }

diglucoside [BIOCHEM] A compound containing two glucose molecules. { dī'glü·kə,sīd }

diglycerol [ORG CHEM] A compound that is a diester of glycerol. { dī'glis·ə,rȯl }

diglycine See iminodiacetic acid. { dī'glī,sēn }

diglycolic acid [ORG CHEM] $O(CH_2COOH)_2$ A white powder that forms a monohydrate; used in the manufacture of plasticizers, in organic synthesis, and to break emulsions. { dī·glī'käl·ik 'as·əd }

diglycol laurate [ORG CHEM] $C_{11}H_{23}COOC_2H_4OC_2H_4OH$ A light, straw-colored, oily liquid; soluble in methanol, ethanol, toluene, and mineral oil; used in emulsions and as an antifoaming agent. { dī'glī,kȯl 'lȯr,āt }

diglycol stearate [ORG CHEM] $(C_{17}H_{35}COOC_2H_4)_2O$ A white, waxy solid with a melting point of 54–55°C; used as an emulsifying agent, suspending medium for powders in the manufacture of polishes, and thickening agent, and in pharmaceuticals. { dī'glī,kȯl 'stir,āt }

digoxin [ORG CHEM] $C_{41}H_{64}O_{14}$ A crystalline steroid obtained from a foxglove leaf (*Digitalis lanata*); similar to digitalis in pharmacological effects. { dī'gäk·sən }

digram encoding [COMPUT SCI] A method of data compression that relies on the fact that there are unused characters in the alphabet and uses these characters to represent common pairs of characters. { 'dī,gram in,kōd·iŋ }

digraph See directed graph. { 'dī,graf }

dihalide [CHEM] A molecule containing two atoms of halogen combined with a radical or element. { dī'ha,līd }

dihedral [AERO ENG] The upward or downward inclination of an airplane's wing or other supporting surface in respect to the horizontal; in some contexts, the upward inclination only. [MATH] See dihedron. { dī'hē·drəl }

dihedral angle [MATH] The angle between two planes; it is said to be zero if the planes are parallel; if the planes intersect, it is the plane angle between two lines, one in each of the planes, which pass through a point on the line of intersection of the two planes and are perpendicular to it. { dī'hē·drəl ,aŋ·gəl }

dihedral group [MATH] The group of rotations of three-dimensional space that carry a regular polygon into itself. { dī'hē·drəl ,grüp }

dihedral reflector [OPTICS] A corner reflector having two sides meeting at a line. { dī'hē·drəl ri'flek·tər }

dihedron [MATH] A geometric figure formed by two half planes that are bounded by the same straight line. Also known as dihedral. { dī'hē·drən }

diheptal base [ELECTR] A tube base having 14 pins or 14 possible pin positions; used chiefly on television cathode-ray tubes. { dī'hept·əl 'bās }

dihexagonal [CRYSTAL] Of crystals, having a symmetrical form with 12 sides. { dī·hek'sag·ən·əl }

dihexagonal-dipyramidal [CRYSTAL] Characterized by the class of crystals in the hexagonal system in which any section perpendicular to the sixfold axis is dihexagonal. { dī·hek'sag·ən·əl dī·pir·ə'mid·əl }

dihexahedron [CRYSTAL] A type of crystal that has 12 faces, such as a double six-sided pyramid. { dī,hek·sə'hē·drən }

dihexy See dodecane. { dī'hek·sē }

dihydrate [CHEM] A compound with two molecules of water of hydration. { dī'hī,drāt }

dihydrazone [ORG CHEM] A molecule containing two hydrazone radicals. { dī'hī·drə,zōn }

dihydro- [CHEM] A prefix indicating combination with two atoms of hydrogen. { dī¦hī·drō }

dihydrochloride [CHEM] A compound containing two molecules of hydrochloric acid. { dī¦hī·drə'klȯr,īd }

dihydrostreptomycin [MICROBIO] $C_{21}H_{41}O_{12}N_7$ A hydrogenated derivative of streptomycin having the same action as streptomycin. { dī¦hī·drō,strep·tə'mī·sən }

dihydroxy [CHEM] A molecule containing two hydroxyl groups. { dī¦hī¦dräk·sē }

dihydroxyacetone [ORG CHEM] $(HOCH_2)_2CO$ A colorless, crystalline solid with a melting point of 80°C; soluble in water and alcohol; used in medicine, fungicides, plasticizers, and cosmetics. Abbreviated DHA. { dī¦hī¦dräk·sē'as·ə ,tōn }

dihydroxyacetonephosphoric acid [BIOCHEM] $C_3H_7O_6P$ A phosphoric acid ester of dehydroxyacetone, produced as an intermediate substance in the conversion of glycogen to lactic acid during muscular contraction. { dī¦hī¦dräk·sē¦as·ə,tōn·fäs'fȯr·ik 'as·əd }

2,4′-dihydroxyacetophenone [ORG CHEM] $(HO)_2C_6H_3-COCH_3$ Needlelike or leafletlike crystals with a melting point of 145–147°C; soluble in pyridine, warm alcohol, and glacial acetic acid; used as a reagent for the determination of iron. { ¦tü ¦fȯr,prīm ¦dī,hī¦dräk·sē¦as·ə,tä·fə'nōn }

dihydroxy alcohol See glycol. { dī,hī¦dräk·sē 'al·kə,hȯl }

1,8-dihydroxyanthraquinone [ORG CHEM] $C_{14}H_8O_4$ Orange, needlelike crystals that dissolve in glacial acetic acid; used as an intermediate in the commercial preparation of indanthrene and alizarin dyestuffs. Also known as chrysazin. { ¦wən ¦āt ¦dī,hī¦dräk·sē,an·thrə·kwə'nōn }

2,2′-dihydroxy-4,4′-dimethoxybenzophenone [ORG CHEM] $[CH_3OC_6H_3(OH)]_2CO$ Crystals with a melting point of 139–140°C; used in paint and plastics as a light absorber. { ¦tü ¦tü,prīm ¦dī,hī¦dräk·sē ¦fȯr ¦fȯr,prīm ,dī·mə¦thäk·sē¦ben·zō·fə'nōn }

dihydroxymaleic acid [ORG CHEM] $C_4H_4O_6$ Crystals soluble in alcohol; used in the detection of titanium and fluorides. { ¦dī,hī¦dräk·sē,mə'lā·ik 'as·əd }

dihydroxyphenylalanine [BIOCHEM] $C_9H_{11}NO_4$ An amino acid that can be formed by oxidation of tyrosine; it is converted by a series of biochemical transformations, utilizing the enzyme dopa oxidase, to melanins. Also known as dopa. { ¦dī,hī¦dräk·sē,fen·əl'al·ə,nēn }

diiodomethane See methylene iodide. { dī¦ī·ə,dō'me,thān }

3,5-diiodosalicylic acid [ORG CHEM] $C_7H_4I_2O_3$ Crystals with a sweetish, bitter taste and a melting point of 235–236°C; soluble in most organic solvents; used as a source of iodine in foods and a growth promoter in poultry, hog, and cattle feeds. { ¦thrē ¦fīv dī¦ī·ə,dō,sal·ə'sil·ik 'as·əd }

diisobutylene [ORG CHEM] C_8H_{16} Any one of a number of isomers, but most often 2,4,4-trimethylpentene-1 and 2,4,4-trimethylpentene-2; used in alkylation and as a chemical intermediate. { dī¦ī,sō'byüd·əl,ēn }

diisobutyl ketone [ORG CHEM] $(CH_3)_2CHCH_2COCH_2CH-(CH_3)_2$ Stable liquid, boiling at 168°C; soluble in most organic liquids; toxic and flammable; used as a solvent, in lacquers and coatings, and as a chemical intermediate. { dī¦ī,sō'byüd·əl 'kē,tōn }

diisocyanate [ORG CHEM] A compound that contains two NCO (isocyanate) groups; used to produce polyurethane foams, resins, and rubber. { dī¦ī,sō'sī·ə,nāt }

diisodecyl adipate [ORG CHEM] $(C_{10}H_{21}OOC)_2(CH_2)_4$ A light-colored, oily liquid with a boiling range of 239–246°C; used as a primary plasticizer for polymers. Abbreviated DIDA. { dī¦ī,sō'de,səl 'ad·ə,pāt }

diisodecyl phthalate [ORG CHEM] $C_6H_4(COOC_{10}H_{21})_2$ A clear liquid with a boiling point of 250–257°C; used as a plasticizer. Abbreviated DIDP. { dī¦ī,sō'de,səl 'tha,lāt }

diisopropanolamine [ORG CHEM] $(CH_3CHOHCH_2)_2NH$ A white, crystalline solid with a boiling point of 248.7°C; used as an emulsifying agent for polishes, insecticides, and water paints. Abbreviated DIPA. { dī¦ī,sō,prō·pə'näl·ə,mēn }

diisopropyl [ORG CHEM] **1.** A molecule containing two isopropyl groups. **2.** See 2,3-dimethylbutane. { dī¦ī,sō'prō·pəl }

diisopropyl ether See isopropyl ether. { dī¦ī,sō'prō·pəl 'ē,thər }

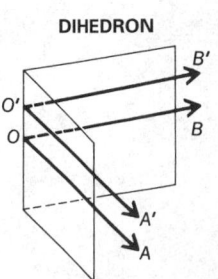

DIHEDRON

Dihedron is formed by the half-planes bounded by the line passing through O and O'. AOB and $A'O'B'$ are dihedral angles.

diisopropyl phosphorofluoridate [PHARM] $[(CH_3)_2CHO]_2FPO$ A colorless, oily liquid that inhibits cholinesterase, prolongs meiosis, and is effective in treating glaucoma. { dī¦ī,sō'prō·pəl ,fäs·fə,rō'flür·ə,dāt }

dikaryon [MYCOL] **1.** Also spelled dicaryon. **2.** A pair of distinct, unfused nuclei in the same cell brought together by union of plus and minus hyphae in certain mycelia. **3.** Mycelium containing dikaryotic cells. { dī'kar·ē,än }

dike [CIV ENG] An embankment constructed on dry ground along a riverbank to prevent overflow of lowlands and to retain floodwater. [GEOL] A tabular body of igneous rock that cuts across adjacent rocks or cuts massive rocks. { dīk }

dike ridge [GEOL] Any small wall-like ridge created by differential erosion. { 'dīk ,rij }

dike set [GEOL] A small group of dikes arranged linearly or parallel to each other. { 'dīk ,set }

dike swarm [GEOL] A large group of parallel, linear, or radially oriented dikes. { 'dīk ,swörm }

diketene [ORG CHEM] $CH_3COCHCO$ A colorless, readily polymerized liquid with pungent aroma; insoluble in water, soluble in organic solvents; used as a chemical intermediate. { dī'kē,tēn }

diketone [ORG CHEM] A molecule containing two ketone carbonyl groups. { dī'kē,ton }

diketopiperazine [ORG CHEM] **1.** $C_4H_6N_2O_2$ A compound formed by dehydration of two molecules of glycine. **2.** Any of the cyclic molecules formed from α-amino acids other than glycine or by partial hydrolysis of protein. { dī¦kē·dō·pi'per·ə,zēn }

diktoma See neuroepithelioma. { dik'tō·mə }

dilactone [ORG CHEM] A molecule that contains two lactone groups. { dī'lak,tōn }

dilatancy [CHEM] The property of a viscous suspension which sets solid under the influence of pressure. [GEOL] Expansion of deformed masses of granular material, such as sand, due to rearrangement of the component grains. { dī'lāt·ən,sē }

dilatant [CHEM] A material with the ability to increase in volume when its shape is changed. { dī'lāt·ənt }

dilatant fluid [FL MECH] A fluid whose apparent viscosity increases simultaneously with an increase in shear rate. Also known as inverted pseudoplastic fluid. { dī¦lāt·ənt 'flü·əd }

dilatation [PHYS] The increase in volume per unit volume of any continuous substance, caused by deformation. { ,dil·ə'tā·shən }

dilatational wave See compressional wave. { ,dil·ə'tā·shən·əl 'wāv }

dilation [MATH] **1.** A transformation which changes the size, and only the size, of a geometric figure. **2.** An operation that provides a relatively flexible boundary to a fuzzy set; for a fuzzy set A with membership function m_A, a dilation of A is a fuzzy set whose membership function has the value $[m_A(x)]^\beta$ for every element x, where β is a fixed number that is greater than 0 and less than 1. [SCI TECH] The act or process of stretching or expanding. { də'lā·shən }

dilatometer [ENG] An instrument for measuring thermal expansion and dilation of liquids or solids. { ,dil·ə'täm·əd·ər }

dilatometry [PHYS] The measurement of changes in the volume of a liquid or dimensions of a solid which occur in phenomena such as allotropic transformations, thermal expansion, compression, creep, or magnetostriction. { ,dil·ə'täm·ə·trē }

dilaton [PARTIC PHYS] A hypothetical elementary particle having zero mass and zero spin, which is introduced in constructing a scale invariant theory involving massive particles. { 'dī·lə,tän }

dilator [PHYSIO] Any muscle, instrument, or drug causing dilation of an organ or part. { dī'lād·ər }

dilauryl ether [ORG CHEM] $(C_{12}H_{25})_2NH$ A liquid with a boiling point of 190–195°C; used for electrical insulators, water repellents, and antistatic agents. Also known as didodecyl ether. { dī'lör·əl 'ē·thər }

dilauryl thiodipropionate [ORG CHEM] $(C_{12}H_{25}OOCCH_2CH_2)_2S$ White flakes with a melting point of 40°C; soluble in most organic solvents; used as an antioxidant, plasticizer, and preservative, and in food wraps and edible fats and oils. { dī'lör·əl ,thī·ō,dī'prō·pē·ə,nāt }

dilepton event [PARTIC PHYS] The inelastic scattering of a neutrino or antineutrino from a nucleus in which there are two leptons among the products of the collision. { dī'lep,tän ə'vent }

dilinoleic acid [ORG CHEM] $C_{34}H_{62}(COOH)_2$ A light yellow, viscous liquid used as an emulsifying agent and shellac substitute. { dī¦lin·ə¦lā·ik 'as·əd }

dill [BOT] *Anethum graveolens.* A small annual or biennial herb in the family Umbelliferae; the aromatic leaves and seeds are used for food flavoring. { dil }

Dilleniaceae [BOT] A family of dicotyledonous trees, woody vines, and shrubs in the order Dilleniales having hypogynous flowers and mostly entire leaves. { di,len·ē'ās·ē,ē }

Dilleniales [BOT] An order of dicotyledonous plants in the subclass Dilleniidae characterized by separate carpels and numerous stamens. { di,len·ē'ā·lēz }

Dilleniidae [BOT] A subclass of plants in the class Magnoliopsida distinguished by being syncarpous, having centrifugal stamens, and usually having bitegmic ovules and binucleate pollen. { dil·ə'nī·ə,dē }

dill oil [MATER] A yellowish essential oil, soluble in propylene glycol, slightly soluble in glycerine, obtained by steam distillation of the dill plant *Anethum graveolens;* chief ingredient is carvone. { 'dil ,öil }

diluent [CHEM] An inert substance added to some other substance or solution so that the volume of the latter substance is increased and its concentration per unit volume is decreased. { 'dil·yə·wənt }

dilute [CHEM] To make less concentrated. { dī'lüt }

dilute phase [CHEM ENG] In liquid-liquid extraction, the liquid phase that is dilute with respect to the material being extracted. { də'lüt ,fāz }

dilution [CHEM] Increasing the proportion of solvent to solute in any solution and thereby decreasing the concentration of the solute per unit volume. [MET] The use of a welding filler metal deposit with a base metal or a previously deposited weld material having a lower alloy content. [OPTICS] Reducing the intensity of a color by adding white. { də'lü·shən }

dilution factor [ELECTROMAG] The energy density of a radiation field divided by the equilibrium value for radiation of the same color temperature. { də'lü·shən ,fak·tər }

dilution gene [GEN] Any modifier gene that acts to reduce the effect of another gene. { də'lü·shən ,jēn }

dilution method [MICROBIO] A technique in which a series of cultures is tested with various concentrations of an antibiotic to determine the minimum inhibiting concentration of antibiotic. { də'lü·shən ,meth·əd }

Dilworth's theorem [MATH] The theorem that, in a finite partially ordered set, the maximum cardinality of an antichain is equal to the minimum number of disjoint chains into which the partially ordered set can be partitioned. { 'dil,wərths ,thir·əm }

DIM See nonthermal decimetric emission.

dimedone See 5,5-dimethyl-1,3-cyclohexanedione. { 'dī·mə,dōn }

dimension [COMPUT SCI] A declarative statement that specifies the width and height of an array of data items. [GRAPHICS] In a mechanical drawing, a labeled measure in a straight line of the breadth, height, or thickness of a part, the angular position of a line, or the location of a detail such as a hole or boss. [MATH] **1.** The number of coordinates required to label the points of a geometrical object. **2.** For a vector space, the number of vectors in any basis of the vector space. **3.** For a simplex, one less than the number of vertices of the simplex. **4.** For a simplicial complex, the largest of the dimensions of the simplices that make up the complex. **5.** The length of one of the sides of a rectangle. **6.** The length of one of the edges of a rectangular parallelepiped. { də'men·chən }

dimensional analysis [PHYS] A technique that involves the study of dimensions of physical quantities, used primarily as a tool for obtaining information about physical systems too complicated for full mathematical solutions to be feasible. { də'men·chən·əl ə'nal·ə·səs }

dimensional constant [PHYS] A physical quantity whose numerical value depends on the units chosen for fundamental quantities but not on the system being considered. { də'men·chən·əl 'kän·stənt }

dimensional formula [PHYS] The expression of a derived quantity as a product of powers of the fundamental quantities. { də'men·chən·əl 'för·myə·lə }

DIMENSIONING

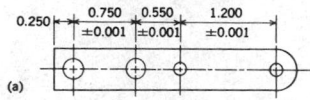

(a)

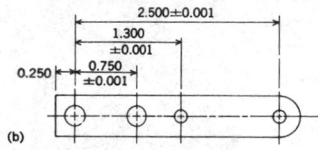

(b)

Two standard methods for marking dimensions on mechanical drawings. *(a)* Point-to-point dimensioning. *(b)* Datum dimensioning.

dimensional regularization [QUANT MECH] A method of extracting a finite piece from an infinite result in quantum field theory based on analytically continuing a typically divergent integral in its number of space-time dimensions. { di'men·chən·əl ˌreg·yəl·ər·ə'zā·shən }

dimensional stability [GRAPHICS] The percentage of change in size of paper under two different conditions of temperature and humidity. [MATER] The ability of a material, such as a textile or plastic, to hold its shape over a period of time and under specific conditions. { də'men·chən·əl stə'bil·əd·ē }

dimension declaration statement [COMPUT SCI] A FORTRAN statement identifying arrays and specifying the number and bounds of the subscripts. { də'men·chən·əl dek·lə'rā·shən ˌstāt·mənt }

dimensioning [GRAPHICS] Assigning of dimensions to a mechanical drawing. { də'men·chən·iŋ }

dimensionless group [PHYS] Any combination of dimensional or dimensionless quantities possessing zero overall dimensions; an example is the Reynolds number. { də'men·chən·ləs 'grüp }

dimensionless number [MATH] A ratio of various physical properties (such as density or heat capacity) and conditions (such as flow rate or weight) of such nature that the resulting number has no defining units of weight, rate, and so on. Also known as nondimensional parameter. { də'men·chən·ləs 'nəm·bər }

dimension line [GRAPHICS] A line on a drawing pointing to another line or part to which the dimensions relate. { də'men·chən ˌlīn }

dimensions [PHYS] The product of powers of fundamental quantities (or of convenient derived quantities) which are used to define a physical quantity; the fundamental quantities are often mass, length, and time. { də'men·chənz }

dimension stone [MATER] Large, sound, relatively flawless blocks of stone used as building stone, monumental stone, paving stone, curbing, and flagging. { də'men·chən ˌstōn }

dimension theory [MATH] The study of abstract notions of dimension, which are topological invariants of a space. { də'men·chən ˌthē·ə·rē }

dimer [CHEM] A molecule that results from a chemical combination of two entities of the same species, for example, the chlorine molecule (Cl_2) or cyanogen (NCCN). { 'dī·mər }

dimercaprol [PHARM] $C_3H_8OS_2$ 2,3-Dimercapto-1-propanol, a colorless, water-soluble oily liquid with a mercaptanlike odor; used as an antidote for arsenic, gold, and mercury poisoning. Also known as British antilewisite (BAL). { 'dī·mər·ka,prōl }

dimeric water [INORG CHEM] Water in which pairs of molecules are joined by hydrogen bonds. { dī'mer·ik 'wȯd·ər }

dimerization [CHEM] A chemical reaction in which two identical molecular entities react to form a single dimer. { ˌdī·mər·ə'zā·shən }

dimerous [BIOL] Composed of two parts. { 'di·mər·əs }

dimetan [ORG CHEM] The generic name for 5,5-dimethyldehydroresorcinol dimethylcarbamate, a synthetic carbamate insecticide. { 'dī·mə,tan }

dimethachlon [ORG CHEM] $C_{10}H_7Cl_2NO_2$ A yellowish, crystalline solid with a melting point of 136.5–138°C; insoluble in water; used as a fungicide. { dī·mə'tha,klän }

dimethicone [PHARM] $CH_3[Si(CH_3)_2O]Si(CH_3)_3$ A colorless oil consisting of dimethylsiloxane polymers; used in ointments and topical drugs. { dī'meth·ə,kōn }

dimethoate [ORG CHEM] $C_5H_{12}NO_3PS_2$ A crystalline compound, soluble in most organic solvents; used as an insecticide. { dī'meth·ə,wāt }

dimethrin [ORG CHEM] $C_{19}H_{28}O_2$ An amber liquid with a boiling point of 175°C; soluble in petroleum hydrocarbons, alcohols, and methylene chloride; used as an insecticide for mosquitoes, body lice, stable flies, and cattle flies. { dī'me·thrən }

dimethyl [ORG CHEM] A compound that has two methyl groups. { dī'meth·əl }

dimethylamine [ORG CHEM] $(CH_3)_2NH$ Flammable gas with ammonia aroma, boiling at 7°C; soluble in water, ether, and alcohol; used as an acid-gas absorbent, solvent, and flotation agent, in pharmaceuticals and electroplating, and in dehairing hides. { ˌdī'meth·əl'am,ēn }

para-dimethylaminobenzalrhodanine [ORG CHEM] $C_{12}H_{12}N_2OS_2$ Deep red, needlelike crystals that decompose at 270°C; soluble in strong acids; used in acetone solution for the detection of ions such as silver, mercury, copper, gold, palladium, and platinum. { ˈpar·ə ˌdī¦meth·əl¦am·ə,nō,ben·zəl'rō·də,nēn }

2-dimethylaminoethanol [ORG CHEM] $(CH_3)_2NCH_2CH_2OH$ A colorless liquid with a boiling point of 134.6°C; used for the synthesis of dyestuffs, pharmaceuticals, and corrosion inhibitors, in medicine, and as an emulsifier. { ¦tü ˌdī¦meth·əl¦am·ə,nō'eth·ə,nȯl }

N,N-dimethylaniline [ORG CHEM] $C_6H_5N(CH_3)_2$ A yellowish liquid slightly soluble in water; used in dyes and solvent and in the manufacture of vanillin. Also known as aniline N,N-dimethyl. { ¦en ¦en ˌdī¦meth·əl'an·ə,lēn }

dimethylbenzene See xylene. { ˌdī¦meth·əl'ben,zēn }

2,3-dimethylbutane [ORG CHEM] $(CH_3)_2CHCH(CH_3)_2$ A colorless liquid with a boiling point of 57.9°C; used as a high-octane fuel. Also known as diisopropyl. { ¦tü ¦thrē ˌdī¦meth·əl'byü,tān }

dimethyl carbate [ORG CHEM] $C_{11}H_{14}O_4$ A colorless liquid with a boiling point of 114–115°C; used as an insect repellent. { ˌdī'meth·əl 'kär,bāt }

5,5-dimethyl-1,3-cyclohexanedione [ORG CHEM] $C_8H_{12}O_2$ Crystals that decompose at 148–150°C; soluble in water and inorganic solvents such as methanol and ethanol; used as a reagent for the identification of aldehydes. Also known as dimedone. { ¦fīv ¦fīv ˌdī¦meth·əl ¦wən ¦thrē ˌsī·klō¦hek,sān 'dī,ōn }

dimethyl diaminophenazine chloride See neutral red. { ˌdī¦meth·əl dī¦am·ə,nō'fen·ə,zēn 'klȯr,īd }

2,2-dimethyl-1,3-dioxolane-4-methanol [ORG CHEM] $C_6H_{12}O_3$ The acetone ketal of glycerin; a liquid miscible with water and many organic solvents; used as a plasticizer and a solvent. { ¦tü ¦tü ˌdī¦meth·əl ¦wən ¦thrē dī¦äk·sə,lān ¦fȯr 'meth·ə,nȯl }

dimethyl ether [ORG CHEM] CH_3OCH_3 A flammable, colorless liquid, boiling at −25°C; soluble in water and alcohol; used as a solvent, extractant, reaction medium, and refrigerant. Also known as methyl ether; wood ether. { ˌdī'meth·əl 'ē·thər }

dimethylethylene See butylene. { ˌdī¦meth·əl'eth·ə,lēn }

N,N-dimethylformamide [ORG CHEM] $HCON(CH_3)_2$ A liquid that boils at 152.8°C; extensively used as a solvent for organic compounds. Abbreviated DMF. { ¦en ¦en ˌdī¦meth·əl'fȯr·mə,mīd }

dimethylglyoxime [ORG CHEM] $(CH_3)_2C_2(NOH)_2$ White, crystalline or powdered solid, used in analytical chemistry as a reagent for nickel. { ˌdī¦meth·əl·glī'äk,sīm }

uns-dimethylhydrazine [ORG CHEM] $(CH_3)_2NNH_2$ A flammable, highly toxic, colorless liquid; used as a component of rocket and jet fuels and as a stabilizer for organic peroxide fuel additives. Abbreviated UDMH. { ¦əns dī¦meth·əl'hī·drə,zēn }

dimethylisopropanolamine [ORG CHEM] $(CH_3)_2NCH_2CH(OH)CH_3$ A colorless liquid with a boiling point of 125.8°C; soluble in water; used in methadone synthesis. { ˌdī¦meth·əl,ī·sə,prō·pə'näl·ə,mēn }

dimethylolurea [ORG CHEM] $CO(NHCH_2OH)_2$ Colorless crystals melting at 126°C, soluble in water; used to increase fire resistance and hardness of wood, and in textiles to prevent wrinkles. Also known as 1,3-bis-hydroxymethylurea; DMU. { ˌdī¦meth·ə·lȯl·yü'rē·ə }

dimethyl phthalate [ORG CHEM] $C_6H_4(COOCH_3)_2$ Odorless, colorless liquid, boiling at 282°C; soluble in organic solvents, slightly soluble in water; used as a plasticizer, in resins, lacquers, and perfumes, and as an insect repellent. { ˌdī¦meth·əl 'tha,lāt }

dimethyl sebacate [ORG CHEM] $[(CH_2)_4COOCH_3]_2$ Clear, colorless liquid, boiling at 294°C; used as a vinyl resin, nitrocellulose solvent, or plasticizer. { ˌdī¦meth·əl 'seb·ə,kāt }

dimethyl sulfate [ORG CHEM] $(CH_3)_2SO_4$ Poisonous, corrosive, colorless liquid, boiling at 188°C; slightly soluble in water, soluble in ether and alcohol; used to methylate amines and phenols. Also known as methyl sulfate. { ˌdī¦meth·əl 'səl,fāt }

dimethyl sulfide See methyl sulfide. { ˌdī¦meth·əl 'səl,fīd }

2,4-dimethylsulfolane [ORG CHEM] $C_6H_{12}O_2S$ A yellow to colorless liquid miscible with lower aromatic hydrocarbons; used as a solvent in liquid-liquid and vapor-liquid extraction processes. { ¦tü ¦fȯr ˌdi¦meth·əl'sȯl·fəˌlān }

dimethyl sulfoxide [ORG CHEM] $(CH_3)_2SO$ A colorless liquid used as a local analgesic and anti-inflammatory agent, as a solvent in industry, and in laboratories as a medium for carrying out chemical reactions. Abbreviated DMSO. { ˌdi¦meth·əl səl'fäkˌsīd }

dimethyl terephthalate [ORG CHEM] $C_6H_4(COOCH_3)_2$ Colorless crystals, melting at 140°C and subliming above 300°; slightly soluble in water, soluble in hot alcohol and ether; used to make polyester fibers and film. Abbreviated DMT. { ˌdi¦meth·əl ¦ter·ə'thaˌlāt }

dimethyl-2,3,5,6-tetrachloroterephthalate [ORG CHEM] $C_{10}H_6Cl_4O_4$ A colorless, crystalline compound with a melting point of 156°C; used as an herbicide for turf, ornamental flowers, and certain vegetables and berries. Abbreviated DCPA. { ˌdi¦meth·əl ¦tü ¦thrē ¦fīv ¦siks ˌte·trə·klȯr·ō,ter·ə'thaˌlāt }

dimictic lake [HYD] A lake which circulates twice a year. { dī'mik·tik 'lāk }

diminution [BOT] Increasing simplification of inflorescences on successive branches. [COMPUT SCI] Limiting the negative effect of an attack on a computer system. { dim·ə'nü·shən }

dimity [TEXT] Sheer fabric with lengthwise cords or checks formed by bunching two, three, or more warp and filling threads together; usually made of combed cotton. { 'dim·əd·ē }

DIMM [COMPUT SCI] A small circuit board that holds semiconductor memory chips with two independent rows of input/output contacts. Derived from dual in-line memory module. { ¦dē¦i¦em'em }

dimmer [ELEC] An electrical or electronic control for varying the intensity of a lamp or other light source. { 'dim·ər }

dimmerfoehn [METEOROL] A rare form of foehn where, during a very strong upper wind from the south, a pressure difference of 12 millibars or more exists between the south and north sides of the Alps; a stormy foehn wind then overleaps the upper valleys in the northern slopes, reaches the ground in the lower parts of the valleys, and enters the foreground as a very strong wind; the foehn wall and the precipitation area extend beyond the crest across the almost calm surface area in the upper valleys. { 'dim·ərˌfān }

dimorphism [CHEM] Having crystallization in two forms with the same chemical composition. [SCI TECH] Existing in two distinct forms, with reference to two members expected to be identical. { dī'mȯrˌfiz·əm }

dimorphite [MINERAL] As_4S_3 An orange-yellow mineral consisting of arsenic sulfide. { dī'mȯrˌfīt }

dimple crystal [CRYO] A periodic pattern of hexagons that forms on a liquid helium surface when electrons, trapped at the surface, and an intense electric charge sheet, are trapped at the surface, and an intense electric field is applied; the hexagons are typically about 2 millimeters across and the interiors of the hexagons are depressions about 10 micrometers deep. { 'dim·pəl ˌkrist·əl }

dimpling [ENG] Forming a conical depression in a metal surface in order to countersink a rivet head. { 'dim·pliŋ }

dimuon event [PARTIC PHYS] An inelastic collision of a neutrino or antineutrino with a nucleus in which there are two muons among the products of the collision. { dī'myüˌän i'vent }

Dimylidae [PALEON] A family of extinct lipotyphlan mammals in the order Insectivora; a side branch in the ancestry of the hedgehogs. { dī'mil·əˌdē }

dina [ELECTR] An airborne radar-jamming transmitter operating in the band from 92 to 210 megahertz with an output of 30 watts, radiating noise in one side band for spot or barrage jamming; the carrier and the other side band are suppressed. { 'dī·nə }

Dinantian [GEOL] Lower Carboniferous geologic time. Also known as Avonian. { di'nan·chən }

Dinarides [GEOGR] A mountain system, east of the Adriatic Sea, in Yugoslavia. { di'nar·əˌdēz }

D indicator See D scope. { 'dē ˌin·dəˌkād·ər }

dineric [PHYS CHEM] **1.** Having two liquid phases. **2.** Pertaining to the interface between two liquids. { dī'ner·ik }

Dines anemometer [ENG] A pressure-tube anemometer in which the pressure head on a weather vane is kept facing into the wind, and the suction head, near the bearing which supports the vane, develops a suction independent of wind direction;

the pressure difference between the heads is proportional to the square of the wind speed and is measured by a float manometer with a linear wind scale. { 'dīnz an·ə'mäm·əd·ər }

dineutron [NUC PHYS] **1.** A hypothetical bound state of two neutrons, which probably does not exist. **2.** A combination of two neutrons which has a transitory existence in certain nuclear reactions. { dī'nü,trän }

dingbat [GRAPHICS] In typography, a special symbol or ornamental character used to illustrate text. Also known as pi symbol. { 'diŋˌbat }

dinghy [NAV ARCH] A boat, less than 20 feet (6 meters) long, propelled by oars or sails, that may be used as a tender to a ship or yacht. { 'diŋ·ē }

dingot [MET] A massive derby, usually a ton or more, produced in a bomb reaction. { 'diŋ·gət }

Dings magnetic separator [MECH ENG] A device which is suspended above a belt conveyor to pull out and separate magnetic material from burden as thick as 40 inches (1 meter) and at belt speeds up to 750 feet (229 meters) per minute. { 'diŋz mag'ned·ik ˌsep·ə,rād·ər }

Dini condition [MATH] A condition for the convergence of a Fourier series of a function f at a number x, namely, that the limits of f at x on the left and right, $f(x-)$ and $f(x+)$, both exist, and that the function given by the absolute value of $[f(x + t) - f(x+) + f(x - t) - f(x-)]/t$ be integrable on some closed interval, $-d \le t \le d$, where d is a positive number. { 'dē·nē kən¦dish·ən }

Dinidoridae [INV ZOO] A family of hemipteran insects in the superfamily Pentatomoidea. { ˌdī·nə'dȯr·ə,dē }

Dini theorem [MATH] The theorem that, if a monotone sequence of continuous real-valued functions converges to a continuous function f on a compact set C, this convergence is uniform; that is, the sequence converges uniformly to f on C. { 'dē·nē ,thir·əm }

dinitramine [ORG CHEM] $C_{11}H_{13}N_3O_4F_3$ A yellow solid with a melting point of 98–99°C; used as a preemergence herbicide for annual grass and broadleaf weeds in cotton and soybeans. { dī'nī·trə,mēn }

dinitrate [CHEM] A molecule that contains two nitrate groups. { dī'nī,trāt }

dinitrite [CHEM] A molecule that has two nitrite groups. { dī'nī,trīt }

2,4-dinitroaniline [ORG CHEM] $(NO_2)_2C_6H_3NH_2$ A compound which crystallizes as yellow needles or greenish-yellow plates, melting at 187.5–188°C, soluble in alcohol; used in the manufacture of azo dyes. { ¦tü ¦fȯr dī¦nī·trō'an·ə,lēn }

2,4-dinitrobenzaldehyde [ORG CHEM] $(NO_2)_2C_6H_3CHO$ Yellow to light brown crystals with a melting point of 72°C; soluble in alcohol, ether, and benzene; used to make Schiff bases. { ¦tü ¦fȯr dī¦nī·tro,ben'zal·də,hīd }

dinitrobenzene [ORG CHEM] Any one of three isomeric substitution products of benzene having the empirical formula $C_6H_4(NO_2)_2$. { dī,nī·trō'ben,zēn }

2,4-dinitrobenzenesulfenyl chloride [ORG CHEM] $(NO_2)_2C_6H_3SCl$ Crystals soluble in glacial acetic acid, with a melting point of 96°C; used as a reagent for separation and identification of naturally occurring indoles. { ¦tü ¦fȯr dī¦nī·trō,ben,zēn'səl·fə,nil 'klȯr,īd }

3,4-dinitrobenzoic acid [ORG CHEM] $C_7H_4N_2O_6$ Crystals with a bitter taste and a melting point of 166°C; used in quantitative sugar analysis. { ¦thrē ¦fȯr dī¦nī·trō·ben'zō·ik 'as·əd }

dinitrogen [CHEM] N_2 The diatomic molecule of nitrogen. { dī'nī·trə·jən }

dinitrogen fixation See nitrogen fixation. { dī'nī·trə·jən fik'sā·shən }

dinitrogen tetroxide See nitrogen dioxide. { dī'nī·trə·jən te'träk,sīd }

dinitrophenol [ORG CHEM] Any one of six isomeric substituent products of benzene having the empirical formula $(NO_2)_2C_6H_3OH$. { dī'nī·trō'fē,nȯl }

2,4-dinitrophenylhydrazine [ORG CHEM] $(NO_2)_2C_6H_3NHNH_2$ A red, crystalline powder with a melting point of approximately 200°C; soluble in dilute inorganic acids; used as a reagent for determination of ketones and aldehydes. { ¦tü ¦fȯr dī¦nī·trō,fen·əl'hī·drə,zēn }

dinitrotoluene [ORG CHEM] Any one of six isomeric substitution products of benzene having the empirical formula $CH_3C_6H_3(NO_2)_2$; they are high explosives formed by nitration of toluene. Abbreviated DNT. { dī¦nī·trō'täl·yə,wēn }

dinking [MECH ENG] Using a sharp, hollow punch for cutting light-gage soft metals or nonmetallic materials. { 'diŋk·iŋ }

Dinocerata [PALEON] An extinct order of large, herbivorous mammals having semigraviportal limbs and hoofed, five-toed feet; often called uintatheres. { ‚dī·nō'ser·ə·də }

Dinoflagellata [INV ZOO] The equivalent name for Dinoflagellida. { ¦dī·nō‚flaj·ə'läd·ə }

Dinoflagellida [INV ZOO] An order of flagellate protozoans in the class Phytamastigophorea; most members have fixed shapes determined by thick covering plates. { ¦dī·nō‚flə'jel·ə·də }

dinokaryon [CYTOL] Nuclear organization peculiar to dinoflagellates and characterized by the absence of a chromosome coiling cycle. { ‚dī·nə'kar·ē‚än }

Dinophilidae [INV ZOO] A family of annelid worms belonging to the Archiannelida. { ‚dī·nō'fil·ə‚dē }

Dinophyceae [BOT] The dinoflagellates, a class of thallophytes in the division Pyrrhophyta. { ‚dī·nō'fīs·ē‚ē }

Dinornithiformes [PALEON] The moas, an order of extinct birds of New Zealand; all had strong legs with four-toed feet. { ‚dīn·ȯr‚nith·ə'fȯr‚mēz }

dinosaur [PALEON] The name, meaning terrible lizard, applied to the fossil bones of certain large, ancient bipedal and quadripedal reptiles placed in the orders Saurischia and Ornithischia. { 'dī·nə‚sȯr }

dinoseb [ORG CHEM] $C_{10}H_{12}O_5N_2$ A reddish-brown liquid with a melting point of 32°C; used as an insecticide and herbicide for numerous crops and in fruit and nut orchards. { 'dī·nə‚seb }

dinoterb acetate [ORG CHEM] $C_{12}H_{14}N_2O_6$ A yellow, crystalline compound with a melting point of 133–134°C; used as a preemergence herbicide for sugarbeets, legumes, and cereals, and as a postemergence herbicide for maize, sorghum, and alfalfa. { 'dī·nə‚tərb 'as·ə‚tāt }

DIN system [GRAPHICS] A system in photography used to find the speed of photographic emulsions; it is stated in terms of the logarithm of the reciprocal of the exposure needed to obtain a density of 0.1 above fog density. { 'din or ‚dē‚ī'en ‚sis·təm }

diocoel [EMBRYO] The cavity of the diencephalon, which becomes the third brain ventricle. { 'dī·ə‚sēl }

dioctahedral [CRYSTAL] Pertaining to a crystal structure in which only two of the three available octahedrally coordinated positions are occupied. [MATH] Having 16 faces. { ‚dī‚äk·tə'hē·drəl }

Dioctophymatida [INV ZOO] An order of parasitic nematode worms in the subclass Enoplia. { dī‚äk·tə·fə'mad·ə·də }

Dioctophymoidea [INV ZOO] An order or superfamily of parasitic nematodes characterized by the peculiar structure of the copulatory bursa of the male. { dī‚äk·tə·fə'mȯid·ē·ə }

dioctyl [ORG CHEM] A compound that has two octyl groups. { dī'äkt·əl }

dioctyl phthalate [ORG CHEM] $(C_8H_{17}OOC)_2C_6H_4$ Pale, viscous liquid, boiling at 384°C; insoluble in water; used as a plasticizer for acrylate, vinyl, and cellulosic resins, and as a miticide in orchards. Abbreviated DOP. { dī¦äkt·əl 'tha‚lāt }

dioctyl phthalate test [ENG] A method used to evaluate air filters to be used in critical air-cleaning applications; a light-scattering technique counts the number of particles of controlled size (0.3 micrometer) entering and emerging from the test filter. Abbreviated DOP test. { dī¦äkt·əl ¦tha‚lāt ‚test }

dioctyl sebacate [ORG CHEM] $(CH_2)_8(COOC_8H_{17})_2$ Water-insoluble, straw-colored liquid, boiling at 248°C; used as a plasticizer for vinyl, cellulosic, and styrene resins. { dī¦äkt·əl 'seb·ə‚kāt }

diode [ELECTR] **1.** A two-electrode electron tube containing an anode and a cathode. **2.** *See* semiconductor diode. { 'dī‚ōd }

diode alternating-current switch *See* trigger diode. { 'dī‚ōd ¦ȯl·tər‚nād·iŋ ¦kər·ənt ‚swich }

diode amplifier [ELECTR] A microwave amplifier using an IMPATT, TRAPATT, or transferred-electron diode in a cavity, with a microwave circulator providing the input/output isolation required for amplification; center frequencies are in the gigahertz range, from about 1 to 100 gigahertz, and power outputs are up to 20 watts continuous-wave or more than 200 watts pulsed, depending on the diode used. { 'dī‚ōd 'am·plə‚fī·ər }

diode bridge [ELECTR] A series-parallel configuration of four diodes, whose output polarity remains unchanged whatever the input polarity. { 'dī‚ōd ‚brij }

diode-capacitor transistor logic [ELECTR] A circuit that uses diodes, capacitors, and transistors to provide logic functions. { ¦dī‚ōd kə¦pas·əd·ər tran'zis·tər ‚läj·ik }

diode characteristic [ELECTR] The composite electrode characteristic of an electron tube when all electrodes except the cathode are connected together. { 'dī‚ōd ‚kar·ik·tə·'ris·tik }

diode clamp *See* diode clamping circuit. { 'dī‚ōd ‚klamp }

diode clamping circuit [ELECTR] A clamping circuit in which a diode provides a very low resistance whenever the potential at a certain point rises above a certain value in some circuits or falls below a certain value in others. Also known as diode clamp. { ¦dī‚ōd 'klamp·iŋ ‚sər·kət }

diode clipping circuit [ELECTR] A clipping circuit in which a diode is used as a switch to perform the clipping action. { ¦dī‚ōd 'klip·iŋ ‚sər·kət }

diode-connected transistor [ELECTR] A bipolar transistor in which two terminals are shorted to give diode action. { 'dī‚ōd kə¦nek·təd tran'zis·tər }

diode demodulator [ELECTR] A demodulator using one or more diodes to provide a rectified output whose average value is proportional to the original modulation. Also known as diode detector. { 'dī‚ōd dē'mäj·ə‚lād·ər }

diode detector *See* diode demodulator. { 'dī‚ōd di'tek·tər }

diode drop *See* diode forward voltage. { 'dī‚ōd ‚dräp }

diode forward voltage [ELECTR] The voltage across a semiconductor diode that is carrying current in the forward direction; it is usually approximately constant over the range of currents commonly used. Also known as diode drop; diode voltage; forward voltage drop. { 'dī‚ōd ¦fȯr·wərd 'vōl·tij }

diode function generator [ELECTR] A function generator that uses the transfer characteristics of resistive networks containing biased diodes; the desired function is approximated by linear segments. { 'dī‚ōd 'feŋk·shən ‚jen·ə‚rād·ər }

diode gate [ELECTR] An AND gate that uses diodes as switching elements. { 'dī‚ōd ‚gāt }

diode laser *See* semiconductor laser. { 'dī‚ōd ‚lāz·ər }

diode limiter [ELECTR] A peak-limiting circuit employing a diode that becomes conductive when signal peaks exceed a predetermined value. { ‚dī‚ōd 'lim·əd·ər }

diode logic [ELECTR] An electronic circuit using current-steering diodes, such that the relations between input and output voltages correspond to AND or OR logic functions. { 'dī‚ōd ‚läj·ik }

diode matrix [ELECTR] A two-dimensional array of diodes used for a variety of purposes such as decoding and read-only memory. { 'dī‚ōd ‚mā·triks }

diode mixer [ELECTR] A mixer that uses a crystal or electron tube diode; it is generally small enough to fit directly into a radio-frequency transmission line. { 'dī‚ōd ‚mik·sər }

diode modulator [ELECTR] A modulator using one or more diodes to combine a modulating signal with a carrier signal; used chiefly for low-level signaling because of inherently poor efficiency. { 'dī‚ōd 'mäj·ə‚lād·ər }

diode pack [ELECTR] Combination of two or more diodes integrated into one solid block. { dī‚ōd ‚pak }

diode peak detector [ELECTR] Diode used in a circuit to indicate when peaks exceed a predetermined value. { 'dī‚ōd 'pēk di‚tek·tər }

diode-pentode [ELECTR] Vacuum tube having a diode and a pentode in the same envelope. { ¦dī‚ōd ¦pen‚tōd }

diode rectifier [ELECTR] A half-wave rectifier of two elements between which current flows in only one direction. { 'dī‚ōd 'rek·tə‚fī·ər }

diode rectifier-amplifier meter [ELECTR] The most widely used vacuum tube voltmeter for measurement of alternating-current voltage; has separate tubes for rectification and direct-current amplification, permitting an optimum design for each. { ¦dī‚ōd ¦rek·tə‚fī·ər 'am·plə‚fī·ər ‚mēd·ər }

diode switch [ELECTR] Diode which is made to act as a switch by the successive application of positive and negative biasing voltages to the anode (relative to the cathode), thereby allowing or preventing, respectively, the passage of other applied waveforms within certain limits of voltage. { 'dī‚ōd ‚swich }

diode theory [ELEC] The theory that in a semiconductor, when the barrier thickness is comparable to or smaller than the mean free path of the carriers, then the carriers cross the barrier

---

DINOPHILIDAE

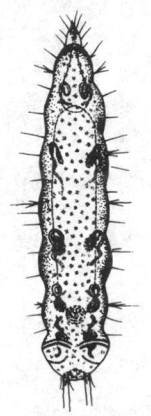

Dinophilus female, dorsal view.

DIODE DEMODULATOR

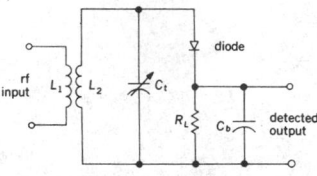

Circuit diagram of diode demodulator.

without being scattered, much as in a vacuum tube diode. { 'dī‚ōd ‚thē·ə·rē }

diode transistor logic [ELECTR] A circuit that uses diodes, transistors, and resistors to provide logic functions. Abbreviated DTL. { ¦dī‚ōd tran'zis·tər ‚läj·ik }

diode-triode [ELECTR] Vacuum tube having a diode and a triode in the same envelope. { ¦dī‚ōd 'trī‚ōd }

diode voltage *See* diode forward voltage. { 'dī‚ōd 'vōl·tij }

diode voltage regulator [ELECTR] A voltage regulator with a Zener diode, making use of its almost constant voltage over a range of currents. Also known as Zener diode voltage regulator. { 'dī‚ōd 'vōl·tij ‚reg·yə‚lād·ər }

diiodide [CHEM] A molecule that contains two iodine atoms bonded to an element or radical. { 'dī·ə‚dīd }

dioecious [BIOL] Having the male and female reproductive organs on different individuals. Also known as dioic. { dī'ē·shəs }

d-i offset system *See* direct-image offset system. { ¦dē¦ī'óf‚set ‚sis·təm }

diogenite [MINERAL] An achondritic stony meteorite composed essentially of iron-rich pyroxene minerals. Also known as rodite. { dī'ä·jə‚nīt }

dioic *See* dioecious. { dī'ō·ik }

diolefin *See* diene. { dī'ō·lə‚fən }

diolefin hydrogenation [CHEM ENG] A fixed-bed catalytic process used to hydrogenate diolefins in C_4 and C_5 fractions to mono-olefin in alkylation feedstocks. { dī'ō·lə‚fən ‚hī·drə·jə'nā·shən }

Diomedeidae [VERT ZOO] The albatrosses, a family of birds in the order Procellariiformes. { ‚dī·ə·mə'dī·ə‚dē }

-dione [ORG CHEM] Suffix indicating the presence of two keto groups. { 'dī‚ōn }

Dione [ASTRON] A satellite of Saturn that orbits at a mean distance of 2.35×10^5 miles (3.78×10^5 kilometers) and has a diameter of about 700 miles (1120 kilometers). { 'dī·ə‚nē }

diophantine analysis [MATH] A means of determining integer solutions for certain algebraic equations. { ¦dī·ə¦fant·ən ə'nal·ə·səs }

diophantine equations [MATH] Equations with more than one independent variable and with integer coefficients for which integer solutions are desired. { ¦dī·ə¦fant·ən i'kwā·zhənz }

Diopsidae [INV ZOO] The stalk-eyed flies, a family of myodarian cyclorrhaphous dipteran insects in the subsection Acalypteratac. { dī'äp·sə‚dē }

diopside [MINERAL] $CaMg(SiO_3)_2$ A white to green monoclinic pyroxene mineral which forms gray to white, short, stubby, prismatic, often equidimensional crystals. Also known as malacolite. { dī'äp‚sīd }

dioptase [MINERAL] $CuSiO_2(OH)_2$ A rare emerald-green mineral that forms hexagonal, hydrous crystals. { dī'äp‚tās }

diopter [OPTICS] A measure of the power of a lens or a prism, equal to the reciprocal of its focal length in meters. Abbreviated D. { dī'äp·tər }

dioptometer [OPTICS] An instrument for determining ocular refraction. { ‚dī‚äp'täm·əd·ər }

dioptric [OPTICS] **1.** Serving in or effecting refraction. **2.** Produced by means of refraction. { dī'äp·trik }

dioptrics [OPTICS] The branch of optics that treats of the refraction of light, especially by the transparent medium of the eye, and by lenses. { dī'äp·triks }

diorchism [ANAT] Having two testes. { dī'ór‚kiz·əm }

diorite [PETR] A phaneritic plutonic rock with granular texture composed largely of plagioclase feldspar with smaller amounts of dark-colored minerals; used occasionally as ornamental and building stone. Also known as black granite. { 'dī·ə‚rīt }

Dioscoreaceae [BOT] A family of monocotyledonous, leafy-stemmed, mostly twining plants in the order Liliales, having an inferior ovary and septal nectaries and lacking tendrils. { ‚dī·ə‚skór·ē'ās·ē‚ē }

1,4-dioxane [ORG CHEM] $C_4H_8O_2$ The cyclic ether of ethylene glycol; it is soluble in water in all proportions and is used as a solvent. { ¦wən ¦fór dī'äk‚sān }

dioxide [CHEM] A compound containing two atoms of oxygen. { dī'äk‚sīd }

dioxin [ORG CHEM] A member of a family of highly toxic chlorinated aromatic hydrocarbons; found in a number of chemical products as lipophilic contaminants. Also known as polychlorinated dibenzo-*para*-dioxin. { dī'äk·sən }

dioxolane [ORG CHEM] $C_3H_6O_2$ A cyclic acetal that is a liquid; used as a solvent and extractant. { dī'äk·sə‚lān }

dioxopurine *See* xanthine. { dī¦äk·sō'pyúr‚ēn }

dioxygen [CHEM] O_2 Molecular oxygen. { dī'äk·si‚jən }

dioxygenase [BIOCHEM] Any of a group of enzymes which catalyze the insertion of both atoms of an oxygen molecule into an organic substrate according to the generalized formula $AH_2 + O_2 \rightarrow A(OH)_2$. { dī'äk·sə·jə‚nās }

dip [ENG] The vertical angle between the sensible horizon and a line to the visible horizon at sea, due to the elevation of the observer and to the convexity of the earth's surface. Also known as dip of horizon. [GEOL] **1.** The angle that a stratum or fault plane makes with the horizontal. Also known as angle of dip; formation dip; true dip. **2.** A pronounced depression in the land surface. { dip }

DIP *See* dual in-line package. { dip }

DIPA *See* diisopropanolamine. { 'dip·ə *or* ¦dē¦ī¦pē¦ā }

dlpath *See* directed path. { 'dī‚path }

dip brazing [MET] Soldering by dipping the work into a hot, molten salt or metal bath and by using a nonferrous metal with a melting point above 800°F (427°C). { 'dip ‚brāz·iŋ }

dip circle *See* inclinometer. { 'dip ‚sər·kəl }

dip coating [ENG] A coating applied to ceramic ware or metal by immersion into a tank of melted nonmetallic material, such as resin or plastic, then chilling the adhering melt. { 'dip ‚kōd·iŋ }

dip correction *See* height-of-eye correction. { 'dip kə'rek·shən }

dip-dye [TEXT] To dye after knitting. { 'dip ‚dī }

dipentene [ORG CHEM] A racemate of limonene. { dī'pen‚tēn }

dipentene glycol *See* terpin hydrate. { dī'pen‚tēn 'glī‚kól }

dipentene hydrochloride *See* terpene hydrochloride. { dī'pen‚tēn ‚hī·drə'klór‚īd }

dipeptidase [BIOCHEM] An enzyme that hydrolyzes a dipeptide. { dī'pep·tə‚dās }

dip fault [GEOL] A type of fault that strikes parallel with the dip of the strata involved. { 'dip ‚fólt }

diphacinone [ORG CHEM] $C_{23}H_{16}O_3$ A yellow powder with a melting point of 145–147°C; used to control rats, mice, and other rodents; acts as an anticoagulant. { də'fas·ə‚nón }

diphase cleaning [MET] Removing soilage from metal surfaces in a cleaning tank incorporating a solvent phase and an aqueous phase. { 'dī‚fāz 'klēn·iŋ }

diphase generator [ELEC] A generator that produces two alternating currents in quadrature. { 'dī‚fāz 'jen·ə‚rād·ər }

diphead [MIN ENG] A passage that follows the inclination of a coal seam. { 'dip‚hed }

diphenamid [ORG CHEM] $C_{16}H_{17}ON$ An off-white, crystalline compound with a melting point of 134–135°C; used as a preemergence herbicide for food crops, fruits, and ornamentals. { dī'fen·ə‚məd }

diphenatrile [ORG CHEM] $C_{14}H_{11}N$ A yellow, crystalline compound with a melting point of 73–73.5°C; used as a preemergence herbicide for turf. { dī'fen·ə‚trəl }

diphenol [ORG CHEM] A compound that has two phenol groups, for example, resorcinol. { dī'fē‚nól }

diphenyl *See* biphenyl. { dī'fen·əl }

diphenylamine [ORG CHEM] $(C_6H_5)_2NH$ Colorless leaflets, sparingly soluble in water; melting point 54°C; used as an additive in propellants to increase the storage life by neutralizing the acid products formed upon decomposition of the nitrocellulose. Also known as phenylaniline. { dī¦fen·əl'am‚ēn }

diphenylaminechloroarsine *See* adamsite. { dī¦fen·əl'am‚ēn‚klór·ō'är‚sēn }

diphenylcarbazide [ORG CHEM] $CO(NHNHC_6H_5)_2$ A white powder, melting point 170°C; used as an indicator, pink for alkalies, colorless for acids. { dī¦fen·əl'kär·bə‚zīd }

diphenyl carbonate [ORG CHEM] $(C_6H_5O)_2CO$ Easily hydrolyzed, white crystals, melting at 78°C; soluble in organic solvents, insoluble in water; used as a solvent, plasticizer, and chemical intermediate. { dī¦fen·əl 'kär·bə‚nāt }

diphenylchloroarsine [ORG CHEM] $(C_6H_5)_2AsCl$ Colorless crystals used during World War I as an antipersonnel device

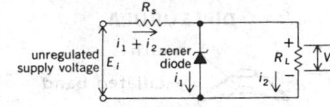

DIODE VOLTAGE REGULATOR

A Zener diode voltage regulator.

to generate a smoke causing sneezing and vomiting. { dī¦fen·əl‚klȯr·ō'är‚sēn }

diphenylene oxide [ORG CHEM] C₁₂H₈O A crystalline solid derived from coal tar; melting point is 87°C; used as an insecticide. { dī¦fen·əl‚ēn 'äk‚sīd }

diphenyl ether See diphenyl oxide. { dī¦fen·əl 'e‚thər }

diphenylethylene See stilbene. { dī¦fen·əl'eth·ə‚lēn }

diphenylguanidine [ORG CHEM] HNC(NHC₆H₅)₂ A white powder, melting at 147°C; used as a rubber accelerator. Also known as DPG; melaniline. { dī¦fen·əl'gwän·ə‚dēn }

diphenyl ketone See benzophenone. { dī¦fen·əl 'kē‚tōn }

diphenylmethane [ORG CHEM] (C₆H₅)₂CH₂ Combustible, colorless crystals melting at 26.5°C; used in perfumery, dyes, and organic synthesis. { dī¦fen·əl'me‚thān }

diphenyl oxide [ORG CHEM] (C₆H₅)₂O A colorless liquid or crystals with a melting point of 27°C and a boiling point of 259°C; soluble in alcohol and ether; used in perfumery, soaps, and resins for laminated electrical insulation. Also known as diphenyl ether; phenyl ether. { dī¦fen·əl 'äk‚sīd }

diphenyl phthalate [ORG CHEM] C₆H₄(COOC₆H₅)₂ White powder, melting at 80°C; soluble in chlorinated hydrocarbons, esters, and ketones, insoluble in water; used as a plasticizer for cellulosic and other resins. { dī¦fen·əl 'tha‚lāt }

diphosgene See trichloromethyl chloroformate. { dī'fäz‚jēn }

diphosphate [CHEM] A salt that has two phosphate groups. { dī'fäs‚fāt }

diphosphatidyl glycerol See cardiolipin. { dī‚fäs'fad·əd·əl 'glis·ə‚rȯl }

diphosphoglyceric acid [ORG CHEM] C₃H₈O₉P₂ An ester of glyceric acid, with two molecules of phosphoric acid, characterized by a high-energy phosphate bond. { dī‚fäs·fə·glə'ser·ik 'as·əd }

diphosphopyridine nucleotide [BIOCHEM] C₂₁H₂₇O₁₄N₇P₂ An organic coenzyme that functions in enzymatic systems concerned with oxidation-reduction reactions. Abbreviated DPN. Also known as codehydrogenase 1; coenzyme 1; nicotinamide adenine dinucleotide (NAD). { dī‚fäs·fə·pir·ə‚dēn 'nü·klē·ə‚tīd }

diphtheria [MED] A communicable bacterial disease of humans caused by the growth of *Corynebacterium diphtheriae* on any mucous membrane, especially of the throat. { dif'thir·ē·ə }

diphtheria antitoxin [IMMUNOL] An antibody produced in animals or in humans after contact with diphtheria toxin or toxoid. { dif ‚thir·ē·ə ‚ant·i'täk·sən }

diphtheritic myocarditis [MED] Inflammation of the cardiac muscle arising from local or generalized diphtheria. { ‚dif·thə'rid·ik ‚mī·ə·kär'dīd·əs }

diphycercal [VERT ZOO] Pertaining to a tail fin, having symmetrical upper and lower parts, and with the vertebral column extending to the tip without upturning. { ¦dī·fə'ser·kəl }

diphyletic [EVOL] Originating from two lines of descent. { dī·fə'led·ik }

Diphyllidea [INV ZOO] A monogeneric order of tapeworms in the subclass Cestoda; all species live in the intestine of elasmobranch fishes. { dī·fə'lid·ē·ə }

Diphyllobothrium [INV ZOO] A genus of tapeworms; including parasites of humans, dogs, and cats. Formerly known as *Dibothriocephalus*. { dī‚fil·ō'bäth·rē·əm }

Diphyllobothrium latum [INV ZOO] A large tapeworm that infects humans, dogs, and cats; causes anemia and disorders of the nervous and digestive systems in humans. { dī‚fil·ō'bäth·rē·əm 'läd·əm }

diphyllous [BOT] Having two leaves. { dī'fil·əs }

diphyodont [ANAT] Having two successive sets of teeth, deciduous followed by permanent, as in humans. { dī'fī·ə‚dänt }

dipicolinic acid [BIOCHEM] C₇H₅O₄N·1¹/₂H₂O A chelating agent composing 5–15% of the dry weight of bacterial spores. { ‚dī‚pik·ə¦lin·ik 'as·əd }

dipicrylamine [ORG CHEM] [(NO₂)₃C₆H₂]₂NH Yellow, prismlike crystals used in the gravimetric determination of potassium. { dī·pə'kril·ə‚mēn }

dip inductor See earth inductor. { 'dip in‚dək·tər }

dip joint [GEOL] A joint that strikes approximately at right angles to the cleavage or bedding of the constituent rock. { 'dip jȯint }

Diplacanthidae [PALEON] A family of extinct acanthodian fishes in the suborder Diplacanthoidei. { ‚dip·lə'kan·thə‚dē }

Diplacanthoidei [PALEON] A suborder of extinct acanthodian fishes in the order Climatiiformes. { ‚dip·lə‚kan'thȯid·ē‚ī }

diplacusis [MED] A difference in the pitch perceptions of the two ears when stimulated by the same sound frequency. { ‚dī·plə'kyü·səs }

Diplasiocoela [VERT ZOO] A suborder of amphibians in the order Anura typically having the eighth vertebra biconcave. { di‚plā·zē·ō'sē·lə }

diplegia [MED] Paralysis of similar parts on the two sides of the body. { dī'plē·jə }

dipleurula [INV ZOO] **1.** A hypothetical bilaterally symmetrical larva postulated to be an ancestral form of echinoderms and chordates. **2.** Any bilaterally symmetrical, ciliated echinoderm larva. { dī'plùr·ə·lə }

diplexer [ELECTR] A coupling system that allows two different transmitters to operate simultaneously or separately from the same antenna. { 'dī‚plek·sər }

diplex operation [COMMUN] Simultaneous transmission or reception of two signals using a specified common element, such as a single antenna or a single carrier. { 'dī‚pleks ‚äp·ə‚rā'shən }

diplex radio transmission [COMMUN] The simultaneous transmission of two signals by using a common carrier wave. { ¦dī‚pleks 'rād·ē·ō tranz‚mish·ən }

diplex reception [ELEC] Simultaneous reception of two signals which have some features in common, such as a single receiving antenna or a single carrier frequency. { 'dī‚pleks ri'sep·shən }

Diplobathrida [PALEON] An order of extinct, camerate crinoids having two circles of plates beneath the radials. { ‚dip·lō'bath·rə·də }

diplobiont [BIOL] An organism characterized by alternating, morphologically dissimilar haploid and diploid generations. { ¦dip·lō¦bī‚änt }

diploblastic [ZOO] Having two germ layers, referring to embryos and certain lower invertebrates. { ¦dip·lō¦blas·tik }

diploblastula [INV ZOO] A two-layered, flagellated larva of certain ceractinomorph sponges. Also known as parenchymella. { ¦dip·lō¦blas·chə·lə }

diplococci [MICROBIO] A pair of micrococci. { ‚dīp·lō'kä·kē *or* 'käk·sī }

Diplodocus [PALEON] Herbivorous sauropod dinosaur, approximately 100 feet (30 meters) long and weighing 12 tons, from the Late Jurassic Period that had a very long neck and tail and a very small body. { di'pläd·ə·kəs }

dip log [GEOL] A log of the dips of formations traversed by boreholes. { 'dip ‚läg }

Diplogasteroidea [INV ZOO] A superfamily of nematodes in the subclass Diplogasteria, having a stoma of variable and often complex shape, a very distinctive esophagus, and a muscular corpus with well-developed valve. { ‚dip·lō‚gas·tə'rȯid·ē·ə }

diploglossate [VERT ZOO] Pertaining to certain lizards, having the ability to retract the end of the tongue into the basal portion. { ¦dip·lō¦glä‚sāt }

diplohaplont [BIOL] An organism characterized by alternating, morphologically similar haploid and diploid generations. { ¦dip·lō¦hap‚länt }

diploid [CRYSTAL] A crystal form in the isometric system having 24 similar quadrilateral faces arranged in pairs. [GEN] Having two complete chromosome pairs in a nucleus (2N). { 'di‚plȯid }

diploidization [GEN] The process by which a tetraploid organism attains the diploid state, involving repeated chromosome loss. { ‚di‚plȯid·ə'zā·shən }

diploid merogony [EMBRYO] Development of a part of an egg in which the nucleus is the normal diploid fusion product of egg and sperm nuclei. { 'di‚plȯid mə'räg·ə·nē }

diplolepidious [BOT] Double-scaled, specifically referring to the peristome of mosses with two rows of scales on the outside and one row on the inner. { ¦dip·lō·lə'pid·ē·əs }

Diplomonadida [INV ZOO] An order of small, colorless protozoans in the class Zoomastigophorea, having a bilaterally symmetrical body with four flagella on each side. { ‚dip·lō‚mə'nad·ə·də }

Diplomystidae [VERT ZOO] A family of catfishes in the suborder Siluroidei confined to the waters of Chile and Argentina. { ‚dip·lō'mis·tə‚dē }

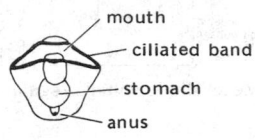

DIPLEURULA

mouth — ciliated band — stomach — anus

Dipleurula larva.

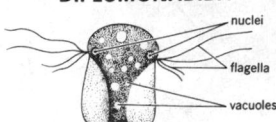

DIPLOMONADIDA

nuclei — flagella — vacuoles

A diplomonad, *Trepomonas rotans*.

diplont [BIOL] An organism with diploid somatic cells and haploid gametes. { 'dip,länt }

diplopia [MED] A disorder characterized by double vision. { də'plō·pē·ə }

Diplopoda [INV ZOO] The millipeds, a class of terrestrial tracheate, oviparous arthropods; each body segment except the first few bears two pairs of walking legs. { də'plä·pə·də }

Diploporita [PALEON] An extinct order of echinoderms in the class Cystoidea in which the thecal canals were associated in pairs. { ,dip·lə'pòr·əd·ə }

Diplorhina [VERT ZOO] The subclass of the class Agnatha that includes the jawless vertebrates with paired nostrils. { ,dip·lə'rī·nə }

diplosome [CYTOL] A double centriole. { 'dip·lə,sōm }

diplospondyly [ANAT] Having two centra in one vertebra. { dip·lə'spän·də·lē }

diplotene [CYTOL] The stage of meiotic prophase during which pairs of nonsister chromatids of each bivalent repel each other and are kept from falling apart by the chiasmata. { 'dip·lō,tēn }

Diplura [INV ZOO] An order of small, primarily wingless insects of worldwide distribution. { də'plùr·ə }

dipmeter [ENG] **1.** An instrument used to measure the direction and angle of dip of geologic formations. **2.** An absorption wavemeter in which bipolar or field-effect transistors replace the electron tubes used in older grid-dip meters. { 'dip,mēd·ər }

dipmeter log [GEOL] A dip log produced by reading of the direction and angle of formation dip as analyzed from impulses from a dipmeter consisting of three electrodes 120° apart in a plane perpendicular to the borehole. { 'dip,mēd·ər ,läg }

dip mold [ENG] A one-piece glassmaking mold with an open top; used to mold patterns. { 'dip ,mōld }

dip needle [ENG] An obsolete type of magnetometer consisting of a magnetized needle that rotates freely in the vertical plane, with an adjustable weight on one side of the pivot. { 'dip ,nēd·əl }

Dipneumonomorphae [INV ZOO] A suborder of the order Araneida comprising the spiders common in the United States, including grass spiders, hunting spiders, and black widows. { dī,nü·mən·ō'mòr·fē }

Dipneusti [VERT ZOO] The equivalent name for Dipnoi. { dip'nü,stī }

Dipnoi [VERT ZOO] The lungfishes, a subclass of the Osteichthyes having lungs that arise from a ventral connection in the gut. { 'dip,nòi }

dipnone [ORG CHEM] $C_{16}H_{14}O$ A liquid ketone, formed by condensation of two acetophenone molecules; used as a plasticizer. { 'dip,nōn }

Dipodidae [VERT ZOO] The Old World jerboas, a family of mammals in the order Rodentia. { də'päd·ə,dē }

dip of horizon See dip. { 'dip əv hə'rīz·ən }

dip oil [MATER] Oil containing about 25% tar acids; used as dip for animals to kill insect parasites. { 'dip ,òil }

dipolar gas [PHYS CHEM] A gas whose molecules have a permanent electric dipole moment. { dī,pōl·ər 'gas }

dipolar ion [CHEM] An ion carrying both a positive and a negative charge. Also known as zwitterion. { dī,pō·lər 'ī,än }

dipolar aprotic solvent [ORG CHEM] A solvent with characteristically high polarity and low reactivity, that is, a solvent having a sizable permanent dipole moment that cannot donate labile hydrogen atoms to form strong hydrogen bonds; examples include acetonitrile, dimethyl sulfoxides, and hexamethylphosphoramide. { dī,pō·lər ā,präd·ik 'säl·vənt }

dipole [ELECTROMAG] Any object or system that is oppositely charged at two points, or poles, such as a magnet or a polar molecule; more precisely, the limit as either charge goes to infinity, the separation distance to zero, while the product remains constant. Also known as doublet; electric doublet. { 'dī,pōl }

dipole anisotropy [ASTRON] A deviation of the equivalent blackbody temperature of the cosmic microwave radiation from its average value which is proportional to the cosine of the angle with some given direction, thus resembling the form of radiation from a dipole antenna. { 'dī,pōl ,an·ə'sä·trə·pē }

dipole antenna [ELECTROMAG] An antenna approximately one-half wavelength long, split at its electrical center for connection to a transmission line whose radiation pattern has a maximum at right angles to the antenna. Also known as doublet antenna; half-wave dipole. { 'dī,pōl an'ten·ə }

dipole-dipole force See orientation force. { ¦dī,pōl ¦dī,pōl ,fòrs }

dipole-dipole interaction [ATOM PHYS] The interaction of two atoms, molecules, or nuclei by means of their electric or magnetic dipole moments. The interaction energy depends on the strength and relative orientation of the two dipoles, as well as on the distance between the centers and the orientation of the radius vector connecting the centers with respect to the dipole vectors. { ¦dī,pōl ¦dī,pōl ,in·tər 'ak·shən }

dipole disk feed [ELECTROMAG] Antenna, consisting of a dipole near a disk, used to reflect energy to the disk. { 'dī,pōl 'disk ,fēd }

dipole moment [ELEC] See electric dipole moment. [ELECTROMAG] See magnetic dipole moment. [PHYS CHEM] The vector sum of the bond moments in a molecule, a measure of the polarity of the molecule. { 'dī,pōl ,mō·mənt }

dipole polarization See orientation polarization. { 'dī,pōl ,pō·lə·rə'zā·shən }

dipole radiation [ELECTROMAG] The electromagnetic radiation generated by an oscillating electric or magnetic dipole. { 'dī,pōl ,rād·ē'ā·shən }

dipole relaxation [ELEC] The process, occupying a certain period of time after a change in the applied electric field, in which the orientation polarization of a substance reaches equilibrium. { 'dī,pōl ,rē,lak'sā·shən }

dipole sound field [ACOUS] A sound field generated by an oscillating dipole source. { 'dī,pōl 'saùnd ,fēld }

dipole transition [ATOM PHYS] A transition of an atom or nucleus from one energy state to another in which dipole radiation is emitted or absorbed. { 'dī,pōl tran'zish·ən }

diporpa larva [INV ZOO] A developmental stage of a monogenean trematode. { di'pòr·pə 'lär·və }

Dippel's oil See bone oil. { 'di·pəlz ,òil }

dipper dredge [MECH ENG] A power shovel resembling a grab crane mounted on a flat-bottom boat for dredging under water. Also known as dipper shovel. { 'dip·ər ,drej }

dipper stick [MECH ENG] A straight shaft connecting the digging bucket of an excavating machine or power shovel with the boom. { 'dip·ər ,stik }

dipper trip [MECH ENG] A device which releases the door of a shovel bucket. { 'dip·ər ,trip }

dipping acid See sulfuric acid. { 'dip·iŋ ,as·əd }

dipping refractometer See immersion refractometer. { 'dip·iŋ ,rē,frak'täm·əd·ər }

dipping sonar [ENG] A sonar transducer that is lowered into the water from a hovering antisubmarine-warfare helicopter and recovered after the search is complete. Also known as dunking sonar. { 'dip·iŋ 'sō,när }

dip plating See immersion plating. { 'dip ,plād·iŋ }

dip pole See magnetic pole. { 'dip ,pōl }

dip reversal See reversal of dip. { 'dip ri'vər·səl }

Diprionidae [INV ZOO] The conifer sawflies, a family of hymenopteran insects in the superfamily Tenthredinoidea. { ,dip·rē'än·ə,dē }

dipropyl [ORG CHEM] A compound containing two propyl groups. { dī'prō·pəl }

dipropylene glycol [ORG CHEM] $(CH_3CHOHCH_2)_2O$ A colorless, slightly viscous liquid with a boiling point of 233°C; soluble in toluene and in water; used as a solvent and for lacquers and printing inks. { dī'prō·pə,lēn 'glī,kòl }

diprotic [CHEM] Pertaining to a chemical structure that has two ionizable hydrogen atoms. { dī'präd·ik }

diprotic acid [CHEM] An acid that has two ionizable hydrogen atoms in each molecule. { di'präd·ik 'as·əd }

Diprotodonta [VERT ZOO] A proposed order of marsupial mammals to include the phalangers, wombats, koalas, and kangaroos. { dī,prōd·ə'dän·tə }

Diprotodontidae [PALEON] A family of extinct marsupial mammals. { dī,prōd·ə'dän·tə,dē }

diproton [NUC PHYS] A hypothetical bound state of two protons, which probably does not exist. { dī'prō,tän }

Dipsacales [BOT] An order of dicotyledonous herbs and shrubs in the subclass Asteridae characterized by an inferior ovary and usually opposite leaves. { ,dip·sə'kā·lēz }

dip slip [GEOL] The component of a fault parallel to the dip of the fault. Also known as normal displacement. { 'dip ,slip }

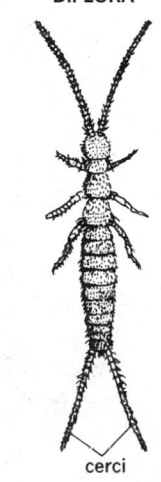

DIPLURA

Campodea folsomi. (From E. O. Essig, College Entomology, Macmillan, 1942)

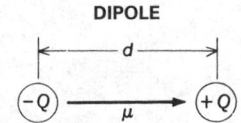

DIPOLE

Electric dipole with moment $\mu = Qd$.

dip slope [GEOL] A slope of the surface of the land determined by and conforming approximately to the dip of the underlying rocks. Also known as back slope; outface. { 'dip ‚slōp }

Dipsocoridae [INV ZOO] A family of hemipteran insects in the superfamily Dipsocoroidea; members are predators on small insects under bark or in rotten wood. { ‚dip·sə'kôr·ə‚dē }

Dipsocoroidea [INV ZOO] A superfamily of minute, ground-inhabiting hemipteran insects belonging to the subdivision Geocorisae. { ‚dip·sə·kə'róid·ē·ə }

dip soldering [MET] A method similar to dip brazing but using a filler metal having a melting point below 800°F (427°C). { 'dip ‚säd·ər·iŋ }

dipsophobia [PSYCH] An abnormal fear of drinking. { ‚dip·sə'fō·bē·ə }

dipstick [ENG] A graduated rod which measures depth when dipped in a liquid, used, for example, to measure the oil in an automobile engine crankcase. { 'dip‚stik }

dipstick microscopy [ENG] A technique for mapping the variation of thickness of a thin liquid film by repeatedly dipping the tip of an atomic force microscope into the film at different locations and calculating its thickness at each location. { 'dip‚stik mī'kräs·kə·pē }

dip stream [HYD] A consequent stream that flows in the direction of the dip of the strata it traverses. { 'dip ‚strēm }

dip-strike symbol [GEOL] A geologic symbol used on maps to show the strike and dip of a planar feature. { 'dip ‚strīk ‚sim·bəl }

DIP switch [COMPUT SCI] A unit with several small rocker-type switches that plugs into a dual in-line package (DIP) on a printed circuit board. { 'dip ‚swich }

Diptera [INV ZOO] The true flies, an order of the class Insecta characterized by possessing only two wings and a pair of balancers. { 'dip·tə·rə }

Dipteriformes [VERT ZOO] The single order of the subclass Dipnoi, the lungfishes. { ‚dip·tə·rə'fôr·mēz }

Dipterocarpaceae [BOT] A family of dicotyledonous plants in the order Theales having mostly stipulate, alternate leaves, a prominently exserted connective, and a calyx that is mostly winged in fruit. { ‚dip·tə·rō‚kär'pās·ē‚ē }

dipterous [BIOL] **1.** Of, related to, or characteristic of Diptera. **2.** Having two wings or winglike structures. { 'dip·tə·rəs }

dipulse [COMMUN] Transmission of a binary code in which the presence of one cycle of a sine-wave tone represents a binary "1" and the absence of one cycle represents a binary "0." { 'dī‚pəls }

dipyramid See bipyramid. { dī'pir·ə‚mid }

dipyre See mizzonite. { 'dī‚pīr }

2,2′-dipyridyl [ORG CHEM] $C_{10}H_8N_2$ A crystalline substance soluble in organic solvents; melting point is 69.7°C; used as a reagent for the determination of iron. Also known as 2,2′-bipyridine. { 'tü ‚tü‚prīm dī'pir·ə·dəl }

diquat [ORG CHEM] $C_{12}H_{12}N_2Br_2$ A yellow water-soluble solid used as a herbicide. { 'dī‚kwät }

Dirac charge [QUANT MECH] A fundamental magnetic charge g such that only integral multiples of g are consistent with quantum mechanics. { di'rak ‚chärj }

Dirac covariants [QUANT MECH] Quantities which behave as a scalar, a pseudoscalar, a vector, an axial vector, or a second-rank tensor under Lorentz transformations, and whose elements consist of basis elements of the Dirac gamma algebra multiplied by the Dirac wave function on the right and its adjoint on the left. { di'rak kō'ver·ē·əns }

Dirac delta function See delta function. { di'rak 'del·tə ‚faŋk·shən }

Dirac distribution See delta function. { də¦rak di·strə'byü·shən }

Dirac electron theory See Dirac theory. { di'rak i'lek‚trän ‚thē·ə·rē }

Dirac equation [QUANT MECH] A relativistic wave equation for an electron in an electromagnetic field, in which the wave function has four components corresponding to four internal states specified by a two-valued spin coordinate and an energy coordinate which can have a positive or negative value. { di'rak i'kwā·zhən }

Dirac fields [QUANT MECH] Operators, arising in the second quantization of the Dirac theory, which correspond to the Dirac wave functions in the original theory. { di'rak ‚fēlz }

Dirac gamma algebra [QUANT MECH] An algebra whose basis consists of 16 linearly independent 4 × 4 matrices constructed from products of the four basic Dirac matrices. { di'rak ¦gam·ə ¦al·jə·brə }

Dirac h See h-bar. { di'rak 'āch }

Dirac hole theory [QUANT MECH] The theory that the continuum of negative energy states that are solutions to the Dirac equation are filled with electrons, and the vacancies in this continuum (holes) are manifested as positrons with energy and momentum that are the negative of those of the state. { di'rak 'hōl ‚thē·ə·rē }

Dirac matrix [QUANT MECH] Any one of four matrices, designated γ_μ (μ = 1, 2, 3, 4), each having four rows and four columns and satisfying $\gamma_\mu\gamma_\nu + \gamma_\nu\gamma_\mu = \delta_{\mu\nu}$, where $\delta_{\mu\nu}$ is the Kronecker delta function, which matrices operate on the four-component wave function in the Dirac equation. Also known as gamma matrix. { di'rak 'mā‚triks }

Dirac moment [QUANT MECH] Magnetic moment of the electron according to the Dirac theory, equal to $e\hbar/2mc$, where e and m are the charge and mass of the positron respectively, $\hbar$ is Planck's constant divided by 2π, and c is the speed of light. { di'rak 'mō·mənt }

Dirac monopole [QUANT MECH] A hypothetical magnetic monopole whose magnetic charge is an integral multiple of $\hbar c/(2e)$, where $\hbar$ is Planck's constant divided by 2π, c is the speed of light, and e is the charge of the electron. { di'rak 'män·ə‚pōl }

Dirac particle [PARTIC PHYS] A particle behaving according to the Dirac theory, which describes the behavior of electrons and muons except for radiative corrections, and is envisaged as describing a central core of a hadron of spin $\frac{1}{2}\hbar$ which remains when the effects of nuclear forces are removed. { di'rak ‚pärd·ə·kəl }

Dirac quantization [QUANT MECH] The condition, arising from conservation of angular momentum, that for any electric charge q and magnetic monopole with magnetic charge m, one has $2qm = n\hbar c$, where n is an integer, $\hbar$ is Planck's constant divided by 2π, and c is the speed of light (gaussian units). { di'rak ‚kwän·tə'zā·shən }

Dirac sea [QUANT MECH] The continuum of negative energy states that are solutions of the Dirac equation, and that are filled with electrons according to Dirac hole theory. { di'rak ‚sē }

Dirac spinor See spinor. { di'rak 'spin·ər }

Dirac theory [QUANT MECH] Theory of the electron based on the Dirac equation, which accounts for its spin angular momentum and gives its magnetic moment and its behavior in an electromagnetic field (except for higher-order corrections). Also known as Dirac electron theory. { di'rak 'thē·ə·rē }

Dirac wave function [QUANT MECH] A function appropriate for describing a spin $\frac{1}{2}$ particle and antiparticle; it is a column matrix with four entries, each of which is a function of the space and time coordinates; the four-components form two first-rank Lorentz spinors. { di'rak 'wāv ‚faŋk·shən }

direct access See random access. { də'rekt 'ak·ses }

direct-access library [COMPUT SCI] A disk-stored set of programs, each of which is directly accessible without sequential search. { də¦rekt ¦ak·ses 'lī‚brer·ē }

direct-access memory See random-access memory. { də¦rekt ¦ak·ses 'mem·rē }

direct-access method [COMPUT SCI] A technique for directly determining the location of data on a disk (track and sector address) from an identifying key in the record. { də¦rekt 'ak‚ses ‚meth·əd }

direct-access storage See random-access memory. { də¦rekt ¦ak·ses 'stôr·ij }

direct-access storage device [COMPUT SCI] Any peripheral storage device, such as a disk or drum, that can be directly addressed by a computer. Abbreviated DASD. { də¦rekt ¦ak‚ses 'stôr·ij di‚vīs }

direct-acting pump [MECH ENG] A displacement reciprocating pump in which the steam or power piston is connected to the pump piston by means of a rod, without crank motion or flywheel. { də¦rekt ¦akt·iŋ 'pəmp }

direct-acting recorder [ENG] A recorder in which the marking device is mechanically connected to or directly operated by the primary detector. { də¦rekt ¦akt·iŋ ri'kórd·ər }

direct address [COMPUT SCI] Any address specifying the location of an operand. { də¦rekt 'a‚dres }

direct-address processing [COMPUT SCI] Any computer

operation during which data are accessed by means of addresses rather than contents. { də¦rekt ¦a‚dres 'präs‚es·iŋ }

direct air cycle [AERO ENG] A thermodynamic propulsion cycle involving a nuclear reactor and gas turbine or ramjet engine, in which air is the working fluid. Also known as direct cycle. { də¦rekt ¦er ‚sī·kəl }

direct allocation [COMPUT SCI] A system in which the storage locations and peripheral units to be assigned to use by a computer program are specified when the program is written, in contrast to dynamic allocation. { də¦rekt ‚al·ə‚kā·shən }

direct-aperture antenna [ELECTROMAG] An antenna whose conductor or dielectric is a surface or solid, such as a horn, mirror, or lens. { də¦rekt ¦ap·ə·chər an'ten·ə }

direct-arc furnace [ENG] A furnace in which a material in a refractory-lined shell is rapidly heated to pour temperature by an electric arc which goes directly from electrodes to the material. { də¦rekt ¦ärk ‚fər·nəs }

direct audio radio service [COMMUN] Radio broadcasting from satellites directly to receivers on the ground. Abbreviated DARS. { də‚rekt ¦ȯd·ē·ō 'rād·ē·ō ‚sər·vəs }

direct-band-gap semiconductor [SOLID STATE] A semiconductor material in which the state of minimum energy in the conduction band and the state of maximum energy in the valence band have the same momentum, so that optical transitions between free electrons and holes are allowed. { də¦rekt 'band ‚gap 'sem·i·kən‚dək·tər }

direct bearing [CIV ENG] A direct vertical support in a structure. { də¦rekt 'ber·iŋ }

direct-bonded bearing [MECH ENG] A bearing formed by pouring molten babbitt metal directly into the bearing housing, allowing it to cool, and then machining the metal to the specified diameter. { də¦rekt ¦bän·dəd 'ber·iŋ }

direct broadcasting satellite system [COMMUN] A television broadcasting system in which program signals are transmitted from ground stations to satellite repeater stations in geostationary orbit, and from there directly to home consumer terminals. Abbreviated DBS. { də¦rekt 'brȯd‚kast·iŋ 'sad·əl‚īt ‚sis·təm }

direct broadcast radio satellite [COMMUN] A satellite in geosynchronous orbit that broadcasts radio programming directly to inexpensive home, car-mounted, and portable radio receivers. { di¦rekt ¦brȯd‚kast 'rād·ē·ō ‚sad·əl‚īt }

direct cell [METEOROL] A closed thermal circulation in a vertical plane in which the rising motion occurs at higher potential temperature than the sinking motion. { də¦rekt 'sel }

direct-chill casting [MET] A continuous ingot-or-billet-casting process in which metal is poured into short molds on a platform and then cooled when the platform is lowered into a water bath. Abbreviated dc casting. Also known as semi-continuous casting. { də¦rekt ¦chil 'kast·iŋ }

direct code [COMPUT SCI] A code in which instructions are written in the basic machine language. { də¦rekt 'kōd }

direct command guidance [ENG] Control of a missile or drone entirely from the launching site by radio or by signals sent over a wire. { də¦rekt kə'mand 'gīd·əns }

direct-connected [MECH ENG] The connection between a driver and a driven part, as a turbine and an electric generator, without intervening speed-changing devices, such as gears. { də¦rekt kə'nek·təd }

direct connect modem [COMMUN] A device that transforms binary signals into electronic pulses (as opposed to sound modulations) that can be carried over a communications channel. { də'rekt kə'nekt 'mō‚dem }

direct-contact condenser See contact condenser. { də¦rekt ¦kän‚takt kən‚den·sər }

direct control [COMPUT SCI] The control of one machine in a data-processing system by another, without human intervention. { də¦rekt kən'trōl }

direct control function See regulatory control function. { də¦rekt kən'trōl ‚fəŋk·shən }

direct cost [IND ENG] The cost in goods and labor to produce a product which would not be spent if the product were not made. { də¦rekt 'kȯst }

direct-coupled [MECH ENG] Joined without intermediate connections. { də¦rekt 'kəp·əld }

direct-coupled amplifier [ELECTR] A direct-current amplifier in which a resistor or a direct connection provides the coupling between stages, so small changes in direct currents can be amplified. { də¦rekt ¦kəp·əld 'am·plə‚fī·ər }

direct-coupled FET logic [ELECTR] A logic gate configuration used with gallium arsenide field-effect transistors operating in the enhancement mode, whose low power consumption and circuit simplicity lead to high packing density and potential use in very large-scale integrated circuits. Abbreviated DCFL. { də¦rekt ¦kəp·əld ¦ef‚e¦tē 'läj·ik }

direct-coupled transistor logic [ELECTR] Integrated-circuit logic using only resistors and transistors, with direct conductive coupling between the transistors; speed can be up to 1 megahertz. Abbreviated DCTL. { də¦rekt ¦kəp·əld tran'zis·tər 'läj·ik }

direct coupling [ELEC] Coupling of two circuits by means of a non-frequency-sensitive device, such as a wire, resistor, or battery, so both direct and alternating current can flow through the coupling path. [MECH ENG] The direct connection of the shaft of a prime mover (such as a motor) to the shaft of a rotating mechanism (such as a pump or compressor). { də¦rekt 'kəp·liŋ }

direct current [ELEC] Electric current which flows in one direction only, as opposed to alternating current. Abbreviated dc. { də¦rekt 'kə·rənt }

direct-current amplifier [ELECTR] An amplifier that is capable of amplifying dc voltages and slowly varying voltages. { də¦rekt ¦kə·rənt 'am·plə‚fī·ər }

direct-current circuit [ELEC] Any combination of dc voltage or current sources, such as generators and batteries, in conjunction with transmission lines, resistors, and power converters such as motors. { də¦rekt ¦kə·rənt 'sər·kət }

direct-current circuit theory [ELEC] An analysis of relationships within a dc circuit. { də¦rekt ¦kə·rənt 'sər·kət ‚thē·ə·rē }

direct-current component [COMMUN] The average value of a signal; in television, it represents the average luminance of the picture being transmitted; in radar, the level from which the transmitted and received pulses rise. { də¦rekt ¦kə·rənt kəm'pō·nənt }

direct-current continuity [ELEC] Property of a circuit in which there is an established pathway for conduction of current from a direct-current source. { də¦rekt ¦kə·rənt ‚känt·ən'ü·əd·ē }

direct-current coupling [ELECTR] That type of coupling in which the zero-frequency term of the Fourier series representing the input signal is transmitted. { də¦rekt ¦kə·rənt 'kəp·liŋ }

direct-current discharge [ELECTR] The passage of a direct current through a gas. { də¦rekt ¦kə·rənt 'dis‚chärj }

direct-current dump [ELECTR] Removal of all direct-current power from a computer system or component intentionally, accidentally, or conditionally; in some types of storage, this results in loss of stored information. { də¦rekt ¦kə·rənt 'dəmp }

direct-current electrode negative [MET] In direct-current arc welding, the arrangement of leads where the surface to be welded is the positive and the electrode is the negative relative to the welding arc. { də¦rekt ¦kə·rənt i‚lek‚trōd 'neg·əd·iv }

direct-current electrode positive [MET] In direct-current arc welding, the arrangement of leads where the surface to be welded is the negative and the electrode is the positive relative to the welding arc. { də¦rekt ¦kə·rənt i‚lek‚trōd 'päz·əd·iv }

direct-current erase [ELECTR] Use of direct current to energize an erasing head of a tape recorder. { də¦rekt ¦kə·rənt ə'rās }

direct-current generator [ELEC] A rotating electric machine that converts mechanical power into dc power. { də¦rekt ¦kə·rənt 'jen·ə‚rād·ər }

direct-current inserter [ELECTR] A television transmitter stage that adds to the video signal a dc component known as the pedestal level. { də¦rekt ¦kə·rənt in'sərd·ər }

direct-current Josephson effect [CRYO] The current flow resulting from the tunneling of electron pairs through a thin insulating barrier between two superconductors in the absence of a voltage drop across the barrier. { di¦rekt ¦kər·ənt 'jō·sef·sən i‚fekt }

direct-current motor [ELEC] An electric rotating machine energized by direct current and used to convert electric energy to mechanical energy. { də¦rekt ¦kə·rənt 'mōd·ər }

direct-current motor control See electronic motor control. { də¦rekt ¦kə·rənt 'mōd·ər kən‚trōl }

direct-current offset [ELECTR] A direct-current level that

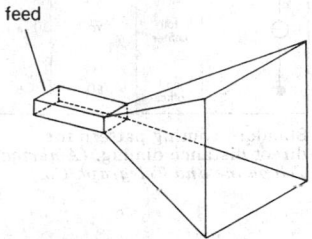

DIRECT-APERTURE ANTENNA

Pyramidal horn, a type of direct-aperture antenna.

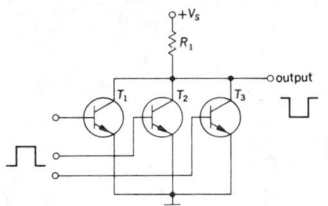

DIRECT-COUPLED TRANSISTOR LOGIC

Direct-coupled transistor logic circuit.

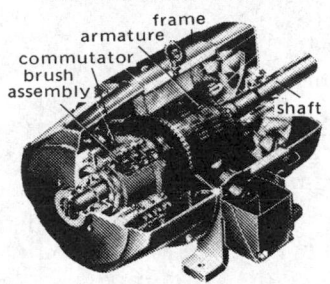

DIRECT-CURRENT MOTOR

Cutaway view of typical direct-current motor. (*General Electric*)

DIRECT DISTANCE DIALING

symbols	names	abbreviation	class number
to other regional centers	regional center	RC	1
	sectional center	SC	2
	primary center	PC	3
	toll center	TC	4
	end office	EO	5

Standard routing pattern for direct distance dialing. (*American Telephone and Telegraph Co.*)

may be added to the input signal of an amplifier or other circuit. { də¦rekt ¦kə·rənt 'óf,set }

direct-current picture transmission [COMMUN] Television transmission in which the signal contains a dc component that represents the average illumination of the entire scene. Also known as direct-current transmission. { də¦rekt ¦kə·rənt 'pik·chər tranz,mish·ən }

direct-current plate resistance [ELECTR] Value or characteristic used in vacuum-tube computations; it is equal to the direct-current plate voltage divided by the direct-current plate current. { də¦rekt ¦kə·rənt 'plāt ri,zis·təns }

direct-current power [ELEC] The power delivered by a dc power system, equal to the line voltage times the load current. { də¦rekt ¦kə·rənt 'pau̇·ər }

direct-current power supply [ELEC] A power supply that provides one or more dc output voltages, such as a dc generator, rectifier-type power supply, converter, or dynamotor. { də¦rekt ¦kə·rənt 'pau̇·ər sə,plī }

direct-current quadruplex system [COMMUN] Direct-current telegraph system which affords simultaneous transmission of two messages in each direction over the same line, achieved by superimposing neutral telegraph upon polar telegraph. { də¦rekt ¦kə·rənt 'kwä·drə,pleks ,sis·təm }

direct-current receiver [ELECTR] A radio receiver designed to operate directly from a 115-volt dc power line. { də¦rekt ¦kə·rənt ri'sēv·ər }

direct-current reinsertion See clamping. { də¦rekt ¦kə·rənt ,rē·in'sər·shən }

direct-current restoration See clamping. { də¦rekt ¦kə·rənt res·tə'rā·shən }

direct-current restorer [ELECTR] A clamp circuit used to establish a dc reference level in a signal without modifying to any important degree the waveform of the signal itself. Also known as clamper; reinserter. { də¦rekt ¦kə·rənt ri'stór·ər }

direct-current signaling [ELEC] A transmission method that uses direct current. { də¦rekt ¦kə·rənt 'sig·nəl·iŋ }

direct-current SQUID [ELECTR] A type of superconducting quantum interference device (SQUID) which contains two Josephson junctions in a superconducting loop; its state is determined from direct-current measurements. { də¦rekt ¦kə·rənt 'skwid }

direct-current tachometer [ELEC] A dc generator operating with negligible load current and with constant field flux provided by a permanent magnet, so its dc output voltage is proportional to speed. { də¦rekt ¦kə·rənt tə'käm·əd·ər }

direct-current telegraphy [COMMUN] Telegraphy in which direct current controlled by the transmitting apparatus is supplied to the line to form the transmitted signal. { də¦rekt ¦kə·rənt tə'leg·rə·fē }

direct-current transducer [ELECTR] A transducer that requires dc excitation and provides a dc output that varies with the parameter being sensed. { də¦rekt ¦kə·rənt tranz'düs·ər }

direct-current transmission See direct-current picture transmission. { də¦rekt ¦kə·rənt tranz'mish·ən }

direct-current vacuum-tube voltmeter [ELECTR] The amplifying and indicating portions of the diode rectifier-amplifier meter, which are usually designed so that the diode rectifier can be disconnected for dc measurements. { də¦rekt ¦kə·rənt ¦vak·yəm ¦tüb 'vōlt,mēd·ər }

direct-current voltage See direct voltage. { də¦rekt ¦kə·rənt 'vōl·tij }

direct-current working volts [ELEC] The maximum continuously applied dc voltage for which a capacitor is rated. Abbreviated dcwV. { də¦rekt ¦kə·rənt 'wərk·iŋ ,vōlts }

direct cycle See direct air cycle. { də¦rekt 'sī·kəl }

direct-cycle reactor [NUCLEO] A nuclear power plant in which the heat-transfer fluid circulates through the reactor and then passes directly to the turbine in a continuous cycle. { də¦rekt 'sī·kəl rē'ak·tər }

direct digital control [CONT SYS] The use of a digital computer generally on a time-sharing or multiplexing basis, for process control in petroleum, chemical, and other industries. { də¦rekt ¦dij·əd·əl kən'trōl }

direct distance dialing [COMMUN] A telephone exchange service that allows a telephone user to dial subscribers outside the local area using a standard routing pattern from the local or end office. { də¦rekt ¦dis·təns 'dīl·iŋ }

direct drive [MECH ENG] A drive in which the driving part is directly connected to the driven part. { də¦rekt 'drīv }

direct-drive approach [NUCLEO] An approach to inertial-confinement fusion in which laser or particle beams are arrayed around the target in a near-uniform pattern and aimed directly at the fuel pellet. { də¦rekt 'drīv ə,prōch }

direct-drive arm [CONT SYS] A robot arm whose joints are directly coupled to high-torque motors. { də¦rekt 'drīv ,ärm }

direct-drive vibration machine [MECH ENG] A vibration machine in which the vibration table is forced to undergo a displacement by a positive linkage driven by a direct attachment to eccentrics or camshafts. { də¦rekt ¦drīv vī'brā·shən mə,shēn }

direct dye [MATER] A group of coal tar dyes that act without mordants, for example, benzidine dyes. Also known as substantive dye. { də¦rekt 'dī }

directed angle [MATH] An angle for which one side is designated as initial, the other as terminal. { də¦rek·təd 'aŋ·gəl }

directed cycle See dicycle. { də¦rek·təd 'sī·kəl }

directed graph [MATH] A graph in which a direction is shown for every arc. Also known as digraph. { də¦rek·təd 'graf }

directed line [MATH] A line on which a positive direction has been specified. { də¦rek·təd 'līn }

directed network [MATH] A directed graph in which each arc is assigned a unique nonnegative integer called its weight. { də¦rek·təd 'net,wərk }

directed number [MATH] A number together with a sign. { də¦rek·təd 'nəm·bər }

directed path [MATH] A sequence of vertices, $v_1, v_2, \ldots v_n$, in a directed graph such that there is an arc from v_i to v_{i+1} for $i = 1, 2, \ldots, n-1$. Also known as dipath. { də¦rek·təd 'path }

directed set [MATH] A partially ordered set with the property that for every pair of elements a,b in the set, there is a third element which is larger than both a and b. Also known as directed system; Moore-Smith set. { də¦rek·təd 'set }

directed system See directed set. { də¦rek·təd 'sis·təm }

direct effect [PHYS CHEM] A chemical effect caused by the direct transfer of energy from ionizing radiation to an atom or molecule in a medium. { də¦rekt i'fekt }

direct electromotive force [ELEC] Unidirectional electromotive force in which the changes in values are either zero or so small that they may be neglected. { də¦rekt i,lek·trō'mōd·iv 'fórs }

direct energy conversion [ENG] Conversion of thermal or chemical energy into electric power by means of direct-power generators. { də¦rekt 'en·ər·jē kən,vər·zhən }

direct-entry terminal [COMPUT SCI] A device from which data are received into a computer immediately, and which edits data at the time of receipt, allowing computer files to be accessed to validate the information entered, and allowing the terminal operator to be notified immediately of any errors. { də¦rekt 'en·trē 'term·ən·əl }

direct-expansion coil [MECH ENG] A finned coil, used in air cooling, inside of which circulates a cold fluid or evaporating refrigerant. Abbreviated DX coil. { də¦rekt ik'span·chən ,kȯil }

direct expert control system [CONT SYS] An expert control system that contains rules that directly associate controller output values with different values of the controller measurements and set points. Also known as rule-based control system. { də¦rekt ¦eks·pərt kən'trōl ,sis·təm }

direct extrusion [ENG] Extrusion by movement of ram and product in the same direction against a die orifice. { də¦rekt ik'strü·zhən }

direct-feedback system [CONT SYS] A system in which electrical feedback is used directly, as in a tachometer. { də¦rekt 'fēd,bak ,sis·təm }

direct fire [ORD] Fire delivered on a target in which the weapon sights are brought directly on the target. { də¦rekt 'fīr }

direct-fire [ENG] To fire a furnace without preheating the air or gas. { də¦rekt ,fīr }

direct-fired evaporator [CHEM ENG] An evaporator in

which the flame and combustion gases are separated from the boiling liquid by a metal wall, or other heating surface. { də¦rekt ¦fīrd i'vap·ə‚rād·ər }

direct-gap semiconductor [SOLID STATE] A semiconductor in which the minimum of the conduction band occurs at the same wave vector as the maximum of the valence band, and recombination radiation consequently occurs with relatively large intensity. { də¦rekt ¦gap 'sem·i·kən‚dək·tər }

direct-geared [MECH ENG] Joined by a gear on the shaft of one machine meshing with a gear on the shaft of another machine. { də'rekt ¦gird }

direct grid bias See grid bias. { də¦rekt ¦grid ‚bī·əs }

direct hierarchy control [COMPUT SCI] A method of manipulating data in a computer storage hierarchy in which data transfer is completely under the control of built-in algorithms and the user or programmer is not concerned with the various storage subsystems. { də¦rekt 'hī·ər‚är·kē kən‚trōl }

direct-image offset system [GRAPHICS] A system for converting letterpress color plates to offset that involves pulling a proof of the letterpress form directly onto a special paper-backed aluminum foil plate, which then receives a slight treatment. Also known as d i offset system. { də¦rekt ¦im·ij 'óf‚set ‚sis·təm }

direct-imaging mass analyzer [ENG] A type of secondary ion mass spectrometer in which secondary ions pass through an electrostatic immersion lens which forms an image that bears a point-to-point relation to the ion's place of origin on the sample surface, and then traverse magnetic sectors which effect mass separation. Also known as Castaing-Slodzian mass analyzer. { də¦rekt ¦im·ij·iŋ ¦mas 'an·ə‚līz·ər }

direct immunofluorescence [IMMUNOL] The use of labeled reactant to reveal the presence of an unlabeled one. { də¦rekt ¦im·yə‚nō·flúr'es·əns }

direct-indicating compass [NAV] A compass in which the dial, scale, or index is carried on the sensing element. { də¦rekt ¦in·də‚kād·iŋ 'käm·pəs }

direct input/output [COMPUT SCI] The transfer of data to and from a computer's main storage by passing it through the central processing unit. { də¦rekt 'in‚pùt 'aùt‚pùt }

direct-insert subroutine [COMPUT SCI] A body of coding or a group of instructions inserted directly into the logic of a program, often in multiple copies, whenever required. { də¦rekt ¦in·sərt 'səb·rü‚tēn }

direct instruction [COMPUT SCI] An instruction containing the address of the operand on which the operation specified in the instruction is to be performed. { də¦rekt in'strək·shən }

direct interelectrode capacitance See interelectrode capacitance. { də¦rekt ‚in·tər·i'lek‚trōd kə'pas·əd·əns }

direct inward dialing [COMMUN] The capability for dialing individual telephone extensions in a large organization directly from outside, without going through a central switchboard. { də¦rekt ¦in·wərd 'dīl·iŋ }

direction [ENG] The position of one point in space relative to another without reference to the distance between them; may be either three-dimensional or two-dimensional, the horizontal being the usual plane of the latter; usually indicated in terms of its angular distance from a reference direction. [GEOL] See trend. { də'rek·shən }

directional antenna [ELECTROMAG] An antenna that radiates or receives radio waves more effectively in some directions than others. { də'rek·shən·əl an'ten·ə }

directional beacon See direction-giving beacon. { də'rek·shən·əl 'bē·kən }

directional beam [ELECTROMAG] A radio or radar wave that is concentrated in a given direction. { də'rek·shən·əl 'bēm }

directional control [ENG] Control of motion about the vertical axis; in an aircraft, usually by the rudder. { də'rek·shən·əl kən'trōl }

directional control valve [ENG] A control valve serving primarily to direct hydraulic fluid to the point of application. { də'rek·shən·əl kən'trōl ‚valv }

directional counter [NUCLEO] A counter that is more sensitive to nuclear radiation from some directions than from others. { də'rek·shən·əl 'kaùnt·ər }

directional coupler [ELECTR] A device that couples a secondary system only to a wave traveling in a particular direction in a primary transmission system, while completely ignoring a wave traveling in the opposite direction. Also known as directive feed. { də'rek·shən·əl 'kəp·lər }

directional derivative [MATH] The rate of change of a function in a given direction; more precisely, if f maps an n-dimensional euclidean space into the real numbers, and $\mathbf{x} = (x_1, \ldots, x_n)$ is a vector in this space, and $\mathbf{u} = (u_1, \ldots, u_n)$ is a unit vector in the space (that is, $u_1^2 + \cdots + u_n^2 = 1$), then the directional derivative of f at $\mathbf{x}$ in the direction of $\mathbf{u}$ is the limit as h approaches zero of $[f(\mathbf{x} + h\mathbf{u}) - f(\mathbf{x})]/h$. { də'rek·shən·əl də'riv·əd·iv }

directional drilling [ENG] A drilling method involving intentional deviation of a wellbore from the vertical. { də'rek·shən·əl 'dril·iŋ }

directional filter [ELECTR] A low-pass, band-pass, or high-pass filter that separates the bands of frequencies used for transmission in opposite directions in a carrier system. Also known as directional separation filter. { də'rek·shən·əl 'fil·tər }

directional gain See directivity index. { də'rek·shən·əl 'gān }

directional gyro [AERO ENG] A flight instrument incorporating a gyro that holds its position in azimuth and thus can be used as a directional reference. Also known as direction indicator. [MECH] A two-degrees-of-freedom gyro with a provision for maintaining its spin axis approximately horizontal. { də'rek·shən·əl 'jī·rō }

directional homing [NAV] Homing in which the bearing is the parameter which is maintained constant. { də'rek·shən·əl 'hōm·iŋ }

directional hydrophone [ENG ACOUS] A hydrophone whose response varies significantly with the direction of sound incidence. { də'rek·shən·əl 'hī·drə‚fōn }

directional log [PETRO ENG] A record of the wellhole drift, from the vertical, and the direction of that drift. { də'rek·shən·əl 'läg }

directional microphone [ENG ACOUS] A microphone whose response varies significantly with the direction of sound incidence. { də'rek·shən·əl 'mī·krə‚fōn }

directional pattern See radiation pattern. { də'rek·shən·əl 'pad·ərn }

directional phase shifter [ELEC] Passive phase shifter in which the phase change for transmission in one direction differs from that for transmission in the opposite direction. { də'rek·shən·əl 'fāz ‚shif·tər }

directional property [MET] Any property of a metal whose magnitude varies with the orientation of the test axis to a specific direction within the metal. { də'rek·shən·əl 'präp·ərd·ē }

directional relay [ELEC] Relay which functions in conformance with the direction of power, voltage, current, pulse, rotation, and so on. { də'rek shən·əl 'rē‚lā }

directional response pattern See directivity pattern. { də'rek·shən·əl ri'späns ‚pad·ərn }

directional separation filter See directional filter. { də'rek·shən·əl sep·ə'rā·shən ‚fil·tər }

directional solidification [MET] Controlled solidification of molten metal in a casting so as to provide feed metal to the solidifying front of the casting. { də'rek·shən·əl sə‚lid·ə·fə'kā·shən }

directional stability [AERO ENG] The property of an aircraft, rocket, or such, enabling it to restore itself from a yawing or side-slipping condition. Also known as weathercock stability. { də'rek·shən·əl stə'bil·əd·ē }

directional structure [GEOL] Any sedimentary structure having directional significance; examples are cross-bedding and ripple marks. Also known as vectorial structure. { də'rek·shən·əl 'strək·chər }

directional well [PETRO ENG] A well drilled at an angle up to 70° from the vertical to avoid obstacles over the reservoir, such as towns, beaches, or bodies of water. { də'rek·shən·əl 'wel }

direction angles [MATH] The three angles which a line in space makes with the positive x, y, and z axes. { də'rek·shən 'aŋ·gəlz }

direction cosine [ENG] In tracking, the cosine of the angle between a baseline and the line connecting the center of the baseline with the target. [MATH] The cosine of one of the direction angles of a line in space. { də'rek·shən 'kō‚sīn }

direction finder See radio direction finder. { də'rek·shən ‚fīnd·ər }

direction-finder-bearing indicator [NAV] That portion of a direction finder on which the bearing (relative, magnetic, true bearing or reciprocal) of the received transmission is indicated. { də¦rek·shən ¦fīnd·ər 'ber·iŋ ,in·də,kād·ər }

direction finder deviation [NAV] The angular difference between a bearing observed by a radio direction finder and the correct bearing, caused by disturbances due to the characteristics of the receiving craft or the station site. { də'rek·shən ,fīnd·ər ,dē·vē,ā·shən }

direction-finding net [NAV] Two or more direction-finding facilities furnishing coordinated information to a designated control-evaluation agency facility for the purpose of providing position or course information to a pilot. { də'rek·shən ,fīnd·iŋ ,net }

direction-giving beacon [NAV] A beacon (light or radio) which, when observed or received by suitable equipment on the craft or vehicle, gives the observer the bearing (usually magnetic) from the beacon; although it operates in a different mode from that of a direction-finding station, the information produced by both types of stations is identical; however, a direction-giving station (which is shore- or ground-based) produces navigational parameters that differ from those of a craft or vehicle-borne direction finder. Also known as directional beacon. { də'rek·shən ,giv·iŋ ,bē·kən }

direction-independent radar [ENG] Doppler radar used in sentry applications. { də'rek·shən ,in·də¦pen·dənt 'rā,där }

direction indicator See directional gyro. { də'rek·shən ,in·də,kād·ər }

direction numbers [MATH] Any three numbers proportional to the direction cosines of a line in space. Also known as direction ratios. { di'rek·shən ,nəm·bərz }

direction of propagation [PHYS] **1.** The normal to a surface of constant phase, in a propagating wave. **2.** The direction of the group velocity. **3.** The direction of time-average energy flow.(In a homogeneous isotropic medium, these three directions coincide.) { də'rek·shən əv ,präp·ə'gā·shən }

direction of relative movement [NAV] The direction of motion relative to a reference point, itself usually in motion. { də'rek·shən əv ¦rel·ə·tiv 'müv·mənt }

direction ratios See direction numbers. { di'rek·shən ,rā·shōz }

direction rectifier [ELECTR] A rectifier that supplies a direct-current voltage whose magnitude and polarity vary with the magnitude and relative polarity of an alternating-current synchro error voltage. { də'rek·shən 'rek·tə,fī·ər }

directive [COMPUT SCI] An instruction in a source program that guides the compiler in making the translation to machine language, and is usually not translated into instructions in the object program. [INV ZOO] Any of the dorsal and ventral paired mesenteries of certain anthozoan cnidarians. { də'rek·tiv }

directive feed See directional coupler. { də'rek·tiv ,fēd }

directive gain [ELECTROMAG] Of an antenna in a given direction, 4π times the ratio of the radiation intensity in that direction to the total power radiated by the antenna. { də'rek·tiv ,gān }

directive therapy [PSYCH] A method of psychiatric treatment by which the therapist, assuming complete understanding of the patient's needs, endeavors to change the patient's attitudes, behavior, or mode of living. { də'rek·tiv 'ther·ə·pē }

directivity [ELECTR] The ability of a logic circuit to ensure that the input signal is not affected by the output signal. [ELECTROMAG] **1.** The value of the directive gain of an antenna in the direction of its maximum value. **2.** The ratio of the power measured at the forward-wave sampling terminals of a directional coupler, with only a forward wave present in the transmission line, to the power measured at the same terminals when the direction of the forward wave in the line is reversed; the ratio is usually expressed in decibels. { də,rek'tiv·əd·ə }

directivity factor [ENG ACOUS] **1.** The ratio of radiated sound intensity at a remote point on the principal axis of a loudspeaker or other transducer, to the average intensity of the sound transmitted through a sphere passing through the remote point and concentric with the transducer; the frequency must be stated. **2.** The ratio of the square of the voltage produced by sound waves arriving parallel to the principal axis of a microphone or other receiving transducer, to the mean square of the voltage that would be produced if sound waves having the same frequency and mean-square pressure were arriving simultaneously from all directions with random phase; the frequency must be stated. { də,rek'tiv·əd·ə ,fak·tər }

directivity index [ENG ACOUS] The directivity factor expressed in decibels; it is 10 times the logarithm to the base 10 of the directivity factor. Also known as directional gain. { də,rek'tiv·əd·ə ,in,deks }

directivity pattern [ENG ACOUS] A graphical or other description of the response of a transducer used for sound emission or reception as a function of the direction of the transmitted or incident sound waves in a specified plane and at a specified frequency. Also known as beam pattern; directional response pattern. { də,rek'tiv·əd·ə ,pad·ərn }

direct keying device [COMPUT SCI] A computer input device which enables direct entry of information by means of a keyboard. { də¦rekt 'kē·iŋ di,vīs }

direct labor [IND ENG] The labor or effort actually producing goods or services. { də¦rekt 'lā·bər }

direct labor standard See standard time. { də¦rekt ¦lā·bər 'stan·dərd }

direct-line drive [PETRO ENG] Waterflood operation involving a network of wells in a direct (straight) line. { də¦rekt ¦līn ,drīv }

directly congruent figures [MATH] Two solid geometric figures, one of which can be made to coincide with the other by a rigid motion in space, without reflection. { də,rek·lē kən ¦grü·ənt 'fig·yərz }

directly heated cathode See filament. { də,rek·lē ¦hēd·əd 'kā,thōd }

direct-map cache [COMPUT SCI] A cache memory that is organized by linking it to locations in random-access memory. { də,rekt ,map 'kash }

direct material [IND ENG] Any raw or semifinished material which will be incorporated into the product. { də¦rekt mə'tir·ē·əl }

direct memory access [COMPUT SCI] The use of special hardware for direct transfer of data to or from memory to minimize the interruptions caused by program-controlled data transfers. Abbreviated dma. { də¦rekt ¦mem·rē 'ak,ses }

direct motion [ASTRON] Eastward, or counterclockwise, motion of a planet or other object as seen from the North Pole (motion in the direction of increasing right ascension). { də¦rekt 'mō·shən }

direct nuclear reaction [NUC PHYS] A nuclear reaction which is completed in the time required for the incident particle to transverse the target nucleus, so that it does not combine with the nucleus as a whole but interacts only with the surface or with some individual constituent. { də¦rekt ¦nü·klē·ər rē'ak·shən }

direct numerical control [COMPUT SCI] The use of a computer to program, service, and log a process such as a machine-tool cutting operation. { də¦rekt nü¦mer·i·kəl kən'trōl }

direct numerical simulation [FL MECH] Simulation of fluid flow that is carried out through numerical integration of the Navier-Stokes equation in a two-stage process: a temporal approximation involving the splitting of the equation into linear and nonlinear parts and the approximation of time derivatives with finite differences, and a spatial approximation involving the replacement of spatial derivatives with finite difference, finite element, spectral, or hybrid numerical approximations. Abbreviated DNS. { də¦rekt nü¦mer·ə·kəl ,sim·yə'lā·shən }

director [ELECTR] Telephone switch which translates the digits dialed into the directing digits actually used to switch the call. [ELECTROMAG] A parasitic element placed a fraction of a wavelength ahead of a dipole receiving antenna to increase the gain of the array in the direction of the major lobe. [ORD] Electromechanical equipment which is used to track a moving target in azimuth and angular height and which, with the addition of other necessary information from an outside source, such as a radar set or a range finder, continuously computes firing data and transmits them to the guns. { də'rek·tər }

director circle [MATH] A circle consisting of the points of intersection of pairs of perpendicular tangents to an ellipse or hyperbola. { di'rek·tər 'sər·kəl }

direct organization [COMPUT SCI] A type of processing in which records within data sets stored on direct-access devices may be fetched directly if their physical locations are known. { də'rekt ór·gə·nə'zā·shən }

director-sight system [ORD] A sighting system used on airborne flexible guns in which the gunner has direct control of the tracking line (line of aim) and indirect control of the gun or guns. { də'rek·tər ,sīt ,sis·təm }

director-type computer [COMPUT SCI] A gunsight computer used in the director-sight system which, in response to the gunner's action of tracking, computes the angle at which a gun must be fired in order to hit a target. { də'rek·tər ˌtīp kəmˌpyüd·ər }

director-type fire control [ORD] The control of gun fire by use of a director-type computer. { də'rek·tər ˌtīp 'fīr kənˌtrōl }

directory [COMPUT SCI] The listing and description of all the fields of the records making up a file. { də'rek·trē }

directory service [COMPUT SCI] **1.** A directory of the names and addresses of all the mail recipients on a particular network, which provides electronic mail addresses. **2.** A provider of online directories of Web sites and search engines. { də'rek·trē ˌsər·vəs }

directory tree [COMPUT SCI] A graphic representation of the hierarchical branching structure in which files are organized in a hard disk or other storage device. { də'rek·trē ˌtrē }

direct outward dialing [COMMUN] A private automatic branch telephone exchange that permits all local stations to dial outside numbers. Abbreviated DOD. { də'rekt ˌaut·wərd 'dīl·iŋ }

direct pickup [COMMUN] The transmission of television images without intermediate photographic or magnetic recording. { də'rekt 'pikˌəp }

direct piezoelectricity [SOLID STATE] Name sometimes given to the piezoelectric effect in which an electric charge is developed on a crystal by the application of mechanical stress. { də'rekt pē¦ā·zō͵i͵lek'tris·əd·ē }

direct plate exposure [GRAPHICS] A method of printing in which printing plates are directly exposed from copy in special cameras, without the use of an intermediate negative. { də'rekt ¦plāt ik'spō·zhər }

direct pointing [ORD] Pointing a piece either in a direction or in a range and direction by means of a sight directed at the target. { də'rekt 'pöint·iŋ }

direct point repeater [ELECTR] Telegraph repeater in which the receiving relay controlled by the signals received over a line repeats corresponding signals directly into another line or lines without the interposition of any other repeating or transmitting apparatus. { də'rekt ¦pöint ri'pēd·ər }

direct positive [GRAPHICS] A photographic positive made by exposure to light and developed without a negative. { də'rekt 'päz·əd·iv }

direct-power generator [ENG] Any device which converts thermal or chemical energy into electric power by methods more direct than the conventional thermal cycle. { də'rekt ¦paù·ər ¦jen·əˌrād·ər }

direct printing [TEXT] Printing by passing the textile through a series of rollers, each printing a different color or different part of the pattern. { də'rekt 'print·iŋ }

direct process [MET] A process in which the metal is produced from the ore in a single step (for example, steel without intermediate pig iron). { də'rekt 'präs·əs }

direct product [MATH] Given a finite family of sets $A_1, \ldots, A_n$, the direct product is the set of all n-tuples $(a_1, \ldots, a_n)$, where a_i belongs to A_i for $i = 1, \ldots, n$. { də'rekt 'präd·əkt }

direct proof [MATH] An argument that establishes the truth of a statement by making direct use of the hypotheses, as opposed to a proof by contradiction. { dəˌrekt 'prüf }

direct proportion [MATH] A statement that the ratio of two variable quantities is equal to a constant. { də'rekt prə'pör·shən }

direct quenching [MET] Rapid cooling of carburized parts directly from the carburizing process. { də'rekt 'kwench·iŋ }

direct-radiator speaker [ENG ACOUS] A loudspeaker in which the radiating element acts directly on the air, without a horn. { də'rekt ¦rād·ēˌād·ər ˌspēk·ər }

direct read after write [COMPUT SCI] The reading of data immediately after the data have been written in order to check for errors in the recoding process. Abbreviated DRAW. { də'rekt ¦rēd ˌaf·tər 'rīt }

direct reading [GRAPHICS] In microfilm technology, that image which is legible in a normal reading position. { də'rekt 'rēd·iŋ }

direct-reading gage [ENG] Gage that records directly (instead of inferentially) measured values, for example, a liquid-level gage pointer actuated by direct linkage with a float. { də'rekt ¦rēd·iŋ 'gāj }

direct realization [ELECTR] An active filter configuration that is derived by systematically replacing the elements of a passive RLC prototype filter (a filter that consists entirely of resistors, inductors, and capacitors) according to some rule. { di¦rekt ˌrē·ə·lə'zā·shən }

direct recording [ENG ACOUS] Recording in which a record is produced immediately, without subsequent processing, in response to received signals. { də'rekt ri'körd·iŋ }

direct-reduction process [MET] Any of several methods for extracting iron ore below the melting point of iron, to produce solid reduced iron that may be converted to steel with little further refining. { dəˌrekt ri'dək·shən ˌpräs·əs }

direct reflection See specular reflection. { dəˌrekt ri'flek·shən }

direct repeat [MOL BIO] Identical or closely related nucleotide sequences present in two or more copies in the same orientation within the same molecule. { di¦rekt ri'pēt }

direct resistance-coupled amplifier [ELECTR] Amplifier in which the collector, drain, or plate of one stage is connected either directly or through a resistor to the base, gate, or control grid of the next stage; used to amplify small changes in direct current. { də¦rekt ri¦zis·təns ˌkəp·əld 'am·plə͵fī·ər }

direct return system [MECH ENG] In a heating or cooling system, a piping arrangement in which the fluid is returned to its origin (boiler or evaporator) by the shortest direct path after it has passed through each heat exchanger. { di¦rekt ri'tərn ˌsis·təm }

directrix [MATH] **1.** A fixed line used in one method of defining a conic; the distance from this line divided by the distance from a fixed point (called the focus) is the same for all points on the conic. **2.** A curve through which a line generating a given ruled surface always passes. { də'rek·triks }

direct route [ELEC] In wire communications, the trunks that connect a pair of switching centers, regardless of the geographical direction the actual trunk facilities may follow. [NAV] The shortest navigational distance between two points on the earth's surface, for example, the great circle. { də'rekt 'rüt }

direct scanning [COMMUN] A scanning method in which the subject is illuminated at all times and only one elemental area of the subject is viewed at a time by the television camera. { də'rekt 'skan·iŋ }

direct scattering theory See scattering theory. { də'rekt 'skad·ə·riŋ ˌthē·ə·rē }

direct screen halftone [GRAPHICS] In color separation, a halftone negative that is produced by direct exposure on an enlarger or by contact through a halftone screen. { də¦rekt ¦skrēn 'hafˌtōn }

direct sequence system [COMMUN] A system for generating spread spectrum transmissions by phase-modulating a sine wave pseudorandomly by an unending string of pseudonoise code symbols, each of duration much smaller than a bit. { də'rekt 'sē·kwəns ˌsis·təm }

direct simulation Monte Carlo [FL MECH] An iterated two-step procedure for simulation of fluid flow, in which the calculation of the outcomes of collisions of paired particles based on intermolecular potentials alternates with the partitioning of the phase space into cells and the random selection of 10–100 particle pairs as candidates for collision, consistent with collision frequency predictions based on kinetic theory. { də¦rekt ˌsim·yə¦lā·shən ¦män·tə kär·lō }

direct solar radiation [ASTROPHYS] That portion of the radiant energy received at the actinometer direct from the sun, as distinguished from diffuse sky radiation, effective terrestrial radiation, or radiation from any other source. { də'rekt ¦sō·lər rād·ē'ā·shən }

direct sound [ACOUS] The portion of the room impulse response consisting of the combination of the true direct sound, which has traveled directly from the sound source to the listener, and the various reflections within the first 20 milliseconds after it. { di¦rekt 'saùnd }

direct stratification See primary stratification. { də'rekt ˌstrad·ə·fə'kā·shən }

direct stroke [ELEC] A lightning stroke that actually strikes some part of a power or communication system. { də'rekt 'strōk }

direct sum [MATH] If each of the sets in a finite direct product of sets has a group structure, this structure may be imposed on the direct product by defining the composition "componentwise"; the resulting group is called the direct sum. { də¦rekt 'səm }

direct support artillery [ORD] Artillery assigned the mission of providing the fire requested by the supported unit. { dəˈrekt səˈpȯrt ärˈtil·ə·rē }

direct symbol recognition [COMPUT SCI] Recognition by sensing the unique geometrical properties of symbols. { dəˈrekt ˈsim·bəl ˌrek·igˌnish·ən }

direct tide [GEOPHYS] A gravitational solar or lunar tide in the ocean or atmosphere which is in phase with the apparent motions of the attracting body, and consequently has its local maxima directly under the tide-producing body, and on the opposite side of the earth. { dəˈrekt ˈtīd }

direct variation [MATH] **1.** A relationship between two variables wherein their ratio remains constant. **2.** An equation or function expressing such a relationship. { dəˈrekt ˌver·ē·ˈā·shən }

direct viewfinder [OPTICS] A viewfinder in which the user views the subject directly through a lens or sight. { dəˈrekt ˈvyüˌfīnd·ər }

direct-view storage tube [ELECTR] A cathode-ray tube in which secondary emission of electrons from a storage grid is used to provide an intensely bright display for long and controllable periods of time. Also known as display storage tube; viewing storage tube. { dəˈrekt ˈvyü ˈstȯr·ij ˌtüb }

direct-vision nephoscope [OPTICS] A type of nephoscope in which the cloud motion is observed by looking directly into the instrument. { dəˈrekt ˈvizh·ən ˈnef·əˌskōp }

direct-vision prism *See* Amici prism. { dəˈrekt ˈvizh·ən ˈpriz·əm }

direct-vision spectroscope [SPECT] A spectroscope that allows the observer to look in the direction of the light source by means of an Amici prism. { dəˈrekt ˈvizh·ən ˈspek·trəˌskōp }

direct voltage [ELEC] A voltage that forces electrons to move through a circuit in the same direction continuously, thereby producing a direct current. Also known as direct-current voltage. { dəˈrekt ˈvōl·tij }

direct wave [COMMUN] A radio wave that is propagated directly through space from transmitter to receiver without being refracted by the ionosphere. { dəˈrekt ˈwāv }

direct-wire circuit [ELEC] Supervised protective signaling circuit usually consisting of one metallic conductor and a ground return and having signal-receiving equipment responsive to either an increase or a decrease in current. { dəˈrekt ˈwīr ˈsər·kət }

direct-writing galvanometer [ENG] A direct-writing recorder in which the stylus or pen is attached to a moving coil positioned in the field of the permanent magnet of a galvanometer. { dəˈrekt ˈwrīd·iŋ ˌgal·və·ˈnäm·əd·ər }

direct-writing recorder [ENG] A recorder in which the permanent record of varying electrical quantities or signals is made on paper, directly by a pen attached to the moving coil of a galvanometer or indirectly by a pen moved by some form of motor under control of the galvanometer. Also known as mechanical oscillograph. { dəˈrekt ˈwrīd·iŋ riˈkȯrd·ər }

Dirichlet conditions [MATH] The requirement that a function be bounded, and have finitely many maxima, minima, and discontinuities on the closed interval $[-\pi, \pi]$. { ˌdē·rē·ˈklä kənˌdish·ənz }

Dirichlet drawer principle *See* pigeonhole principle. { ˌdē·rē·ˈklä ˈdrȯ·ər ˌprin·sə·pəl }

Dirichlet problem [MATH] To determine a solution to Laplace's equation which satisfies certain conditions in a region and on its boundary. { ˌdē·rē·ˈklä ˌpräb·ləm }

Dirichlet series [MATH] A series whose nth term is a complex number divided by n to the zth power. { ˌdē·rē·ˈklä ˌsir·ēz }

Dirichlet test for convergence [MATH] If Σb_n is a series whose sequence of partial sums is bounded, and if $\{a_n\}$ is a monotone decreasing null sequence, then the series

$$\sum_{n=1}^{\infty} a_n b_n$$

converges. { ˌdē·rē·ˈklä ˌtest fər kənˈvər·jəns }

Dirichlet theorem [MATH] The theorem that, if a and b are relatively prime numbers, there are infinitely many prime numbers of the form $a + nb$, where n is an integer. { dē·rē·ˈklä ˌthir·əm }

Dirichlet transform [MATH] For a function $f(x)$, this is the integral of $f(x)\cdot\sin(kx)/x$; its convergence determines the convergence of the Fourier series of $f(x)$. { ˌdē·rē·ˈklä ˌtranzˌfȯrm }

dirigible [AERO ENG] A lighter-than-air craft equipped with means of propelling and steering for controlled flight. { dəˈrij·ə·bəl }

dirt band [GEOL] A dark layer in a glacier representing a former surface, usually a summer surface, where silt and debris accumulated. { ˈdərt ˌband }

dirt bed [GEOL] A buried soil containing partially decayed organic material; sometimes occurs in glacial drift. { ˈdərt ˌbed }

dirt slip *See* clay vein. { ˈdərt ˌslip }

dirty bomb [ORD] An explosive based on nuclear fission that emits many long-lived radioactive isotopes. { ˈdər·dē ˈbäm }

dirty ice [ASTRON] Interstellar ice particles with particles of graphite or other impurities adsorbed on their surfaces. { ˈdər·dē ˈīs }

dirty ship [NAV ARCH] Tanker carrying oil or heavy petroleum products. { ˈdər·dē ˈship }

dirty snowball model [ASTRON] A model of comet structure in which the nucleus of the comet resembles a large dirty snowball. { ˈdər·dē ˈsnōˌbȯl ˌmäd·əl }

disability glare *See* glare. { disˈə·ˈbil·əd·ē ˌglär }

disable [COMPUT SCI] **1.** To prevent some action from being carried out. **2.** To turn off a computer system or a piece of equipment. { disˈā·bəl }

disaccharide [BIOCHEM] Any of the class of compound sugars which yield two monosaccharide units upon hydrolysis. { dīˈsak·əˌrīd }

disappearing carriage [ORD] A movable part for raising a heavy gun above a parapet and lowering it automatically after firing. { ˈdis·əˌpir·iŋ ˌkar·ij }

disappearing filament pyrometer *See* optical pyrometer. { ˈdis·əˌpir·iŋ ˌfil·ə·mənt pīˈräm·əd·ər }

disappearing stair [BUILD] A stair that can be swung up into a ceiling space. { ˈdis·əˌpir·iŋ ˈster }

disappearing target [ORD] Target that is exposed to the firer's view for a short time; for example, bobbing targets or targets raised from target pits for short periods of time. { ˈdis·əˌpir·iŋ ˈtär·gət }

disarm [ORD] To remove the detonating device or fuse of a bomb, mine, or other piece of explosive ordnance, or otherwise render it incapable of exploding in its usual manner. { disˈärm }

disassemble [COMPUT SCI] To translate a program from machine language to assembly language to aid in its understanding. [ENG] To take apart into constituent parts. { ˌdisˈə·sem·bəl }

disassembler [COMPUT SCI] A program that translates machine language into assembly language. { ˌdis·əˈsem·blər }

disaster dump [COMPUT SCI] A listing of the contents of a computer's central processing unit that is created when the computer detects an error that it cannot handle in the course of processing. { diˈzas·tər ˌdəmp }

Disasteridae [PALEON] A family of extinct burrowing, exocyclic Euechinoidea in the order Holasteroida comprising mainly small, ovoid forms without fascioles or a plastron. { ˌdis·əˈster·əˌdē }

disc *See* disk. { disk }

DISC *See* differential scatter. { disk }

discarding petal [ORD] A part of a discarding sabot that is composed of a base and attached pieces, or petals, which surround the core, and are peeled back under centrifugal and aerodynamic forces and discarded just in front of the gun muzzle. { diˈskärd·iŋ ˈped·əl }

Discellaceae [MYCOL] A family of fungi of the order Sphaeropsidales, including saprophytes and some plant pathogens. { ˌdis·əˈlās·ē·ē }

discharge [ELEC] To remove a charge from a battery, capacitor, or other electric-energy storage device. [ELECTR] The passage of electricity through a gas, usually accompanied by a glow, arc, spark, or corona. Also known as electric discharge. [FL MECH] The flow rate of a fluid at a given instant expressed as volume per unit of time. { ˈdisˌchärj }

discharge channel [MECH ENG] The passage in a pressure-relief device through which the fluid is released to the outside of the device. { ˈdisˌchärj ˌchan·əl }

discharge coefficient [FL MECH] In a nozzle or other constriction, the ratio of the mass flow rate at the discharge end of the nozzle to that of an ideal nozzle which expands an identical working fluid from the same initial conditions to the same exit pressure. Also known as coefficient of discharge. { 'dis‚chärj ‚kō·i'fish·ənt }

discharged solids See residue. { ‚dis‚chärjd 'säl·ədz }

discharge head [MECH ENG] Vertical distance between the intake level of a water pump and the level at which it discharges water freely to the atmosphere. { 'dis‚chärj ‚hed }

discharge hydrograph [CIV ENG] A graph showing the discharge or flow of a stream or conduit with respect to time. { 'dis‚chärj ‚hī·drə‚graf }

discharge key [ELEC] Device for switching a capacitor suddenly from a charging circuit to a load through which it can discharge. { 'dis‚chärj ‚kē }

discharge lamp [ELECTR] A lamp in which light is produced by an electric discharge between electrodes in a gas (or vapor) at low or high pressure. Also known as electric-discharge lamp; gas-discharge lamp; vapor lamp. { 'dis‚chärj ‚lamp }

discharge line [ENG] The length of pipe through which drilling mud travels from the mud pump through the standpipe on its way to the borehole. { 'dis‚chärj ‚līn }

discharge liquor [CHEM ENG] Liquid that has passed through a processing operation. Also known as effluent; product. { 'dis‚chärj ‚lik·ər }

discharge printing [GRAPHICS] A method of printing in which an electric discharge is shaped to produce characters. [TEXT] Using bleaching chemicals on a previously dyed fabric to remove the dye and thus imprint a pattern. { 'dis‚chärj ‚print·iŋ }

discharger [ELEC] A silver-impregnated cotton wick encased in a flexible plastic tube with an aluminum mounting lug, used on aircraft to reduce precipitation static. { 'dis‚chärj·ər }

discharge tube [ELECTR] An evacuated enclosure containing a gas at low pressure, through which current can flow when sufficient voltage is applied between metal electrodes in the tube. Also known as electric-discharge tube. [MECH ENG] A tube through which steam and water are released into a boiler drum. { 'dis‚chärj ‚tüb }

discharge-tube leak indicator [ENG] A device which detects the presence of a tracer gas by using a glass tube attached to a high-voltage source; the presence of leaked gas is indicated by the color of the electric discharge. { 'dis‚chärj ‚tüb ‚lēk ‚in·də‚kād·ər }

discharging agent [TEXT] A stripping agent such as sodium hyposulfite which is used to remove dyes from fabric that has been vat-dyed or printed. { 'dis‚chärj·iŋ ‚ā·jənt }

discharging arch [CIV ENG] A support built over, and not touching, a weak structural member, such as a wooden lintel, to carry the main load. Also known as relieving arch. { 'dis‚chärj·iŋ ‚ärch }

dischronation [PSYCH] A disturbance in the consciousness of time. { ‚dis·krə'nā·shən }

discifloral [BOT] Having flowers with enlarged, disklike receptacles. { ‚dis·kə'flȯr·əl }

disciform [BIOL] Disk-shaped. { 'dis·kə‚fȯrm }

Discinacea [INV ZOO] A family of inarticulate brachiopods in the suborder Acrotretidina. { ‚dis·kə'nās·ē·ə }

disclimax [ECOL] A climax community that includes foreign species following a disturbance of the natural climax by humans or domestic animals. Also known as disturbance climax. { dis'klī·maks }

discoaster [BOT] A star-shaped coccolith. { dis'kō·ə‚stər }

discoblastula [EMBRYO] A blastula formed by cleavage of a meroblastic egg; the blastoderm is disk-shaped. { ‚dis·kō'blas·chə·lə }

discocephalous [INV ZOO] Having a sucker on the head. { ‚dis·kō'sef·ə·ləs }

discoctaster [INV ZOO] A type of spicule with eight rays terminating in discs in hexactinellid sponges. { dis'käk·tə‚stər }

discodactylous [VERT ZOO] Having sucking disks on the toes. { ‚dis·kō'dak·tə·ləs }

discodermolide [PHARM] A polyketide isolated from deepwater sponges of the genus *Discodermia* that is a potent antitumor agent which inhibits the proliferation of cancer cells by interfering with the cell's microtubule network. { ‚disk·ə'dər·mə‚līd }

discogastrula [EMBRYO] A gastrula formed from a blastoderm. { ‚dis·kō'gas·trə·lə }

Discoglossidae [VERT ZOO] A family of anuran amphibians in and typical of the suborder Opisthocoela. { ‚dis·kō'gläs·ə‚dē }

discoid [BIOL] **1.** Being flat and circular in form. **2.** Any structure shaped like a disc. { 'dis‚kȯid }

discoidal cleavage [EMBRYO] A type of cleavage producing a disc of cells at the animal pole. { dis'kȯid·əl 'klē·vij }

Discoidiidae [PALEON] A family of extinct conical or globular, exocyclic Euechinoidea in the order Holectypoida distinguished by the rudiments of internal skeletal partitions. { dis‚kȯi'dī·ə‚dē }

Discolichenes [BOT] The equivalent name for Lecanorales. { ‚dis·kō'lī‚kē·nēz }

Discolomidae [INV ZOO] The tropical log beetles, a family of coleopteran insects in the superfamily Cucujoidea. { ‚dis·kō'läm·ə‚dē }

discomfort glare See glare. { dis'kəm·fərt ‚gler }

discomfort index See temperature-humidity index. { dis'kəm·fərt ‚in‚deks }

discomposition [NUCLEO] The process in which an atom is knocked out of its position in a crystal lattice by direct nuclear impact, as by fast neutrons or by fast ions that have been previously knocked out of their lattice positions. { dis'käm·pə‚zish·ən }

discomposition effect [NUCLEO] Changes in physical or chemical properties of a substance caused by discomposition. Also known as Wigner effect. { dis'käm·pə‚zish·ən i‚fekt }

Discomycetes [MYCOL] A group of fungi in the class Ascomycetes in which the surface of the fruiting body is exposed during maturation of the spores. { ‚dis·kō‚mī'sēd·ēz }

discone antenna [ELECTROMAG] A biconical antenna in which one of the cones is spread out to 180° to form a disk; the center conductor of the coaxial line terminates at the center of the disk, and the cable shield terminates at the vertex of the cone. { 'dis‚kōn an'ten·ə }

disconformity [GEOL] Unconformity between parallel beds or strata. { ‚dis·kən'fȯr·məd·ē }

disconnect [ELEC] To open a circuit by removing wires or connections, as distinguished from opening a switch to stop current flow. [ENG] To sever a connection. { ‚dis·kə'nekt }

disconnected set [MATH] A set in a topological space that is the union of two nonempty sets A and B for which both the intersection of the closure of A with B and the intersection of the closure of B with A are empty. { ‚dis·kə'nek·təd 'set }

disconnect fitting [ELEC] An electrical connection that can be disconnected without tools. { ‚dis·kə'nekt ‚fid·iŋ }

disconnecting switch [ELEC] A switch that isolates a circuit or piece of electrical apparatus after interruption of the current. Also known as disconnector. { ‚dis·kə'nek·tiŋ ‚swich }

disconnector See disconnecting switch. { ‚dis·kə'nek·tər }

disconnector release [ELEC] Device which disengages the apparatus used in a telephone connection to restore it to its original condition when not in use. { ‚dis·kə'nek·tər ri'lēs }

discontinuity [ELECTROMAG] An abrupt change in the shape of a waveguide. Also known as waveguide discontinuity. [GEOL] **1.** An interruption in sedimentation. **2.** A surface that separates unrelated groups of rocks. [GEOPHYS] A boundary at which the velocity of seismic waves changes abruptly. [MATH] A point at which a function is not continuous. [MET] The place where the structural nature of a weldment is interfered with because of the materials involved or where the mechanical, physical, or metallurgical aspects are not homogeneous. [PHYS] A break in the continuity of a medium or material at which a reflection of wave energy can occur. { dis‚känt·ən'ü·əd·ē }

discontinuous amplifier [ELECTR] Amplifier in which the input waveform is reproduced on some type of averaging basis. { ‚dis·kən'tin·yə·wəs 'am·plə‚fī·ər }

discontinuous coding sequence [MOL BIO] The coding sequence in deoxyribonucleic acid of eukaryotic split genes

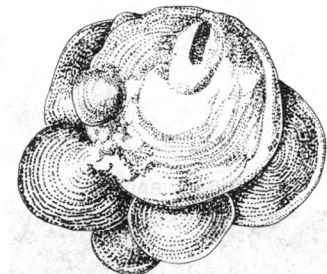

DISCINACEA

A cluster of *Discinisca*. (From R. C. Moore, ed., Treatise on Invertebrate Paleontology, pt. H, Geological Society of America, Inc., and University of Kansas Press, 1965)

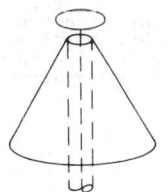

DISCONE ANTENNA

A high-frequency discone antenna.

consisting of exons and introns. { ‚dis·kən'tin·yə·wəs 'kōd·iŋ ‚sē·kwəns }

discontinuous construction [BUILD] A building in which there is no solid connection between the rooms and the building structure or between different sections of the building; the design aims to reduce the transmission of noise. { ‚dis·kən'tin·yə·wəs kən'strək·shən }

discontinuous phase *See* disperse phase. { ‚dis·kən'tin·yə·wəs 'fāz }

discontinuous precipitation [MET] Precipitation principally at and away from the grain boundaries in a supersaturated solid solution; diffraction patterns show two lattice parameters, the solute in solution and the precipitate. { ‚dis·kən'tin·yə·wəs prə‚sip·ə'tā·shən }

discontinuous reaction series [GEOL] The branch of Bowen's reaction series that include olivine, pyroxene, amphibole, and biotite; each change in the series represents an abrupt change in phase. { ‚dis·kən'tin·yə·wəs rē'ak·shən ‚sir·ēz }

discontinuous yielding [MET] The nonuniform plastic deformation of a metal along the length strained in tension. { ‚dis·kən'tin·yə·wəs 'yēld·iŋ }

discopodous [INV ZOO] Having a disk-shaped foot. { di'skäp·ə·dəs }

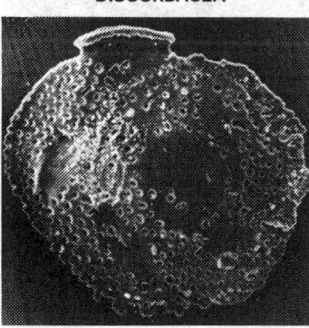

DISCORBACEA

Scanning electron micrograph of *Siphonina* from upper Eocene of Mississippi. (*R. B. MacAdam, Chevron Oil Field Research Co.*)

Discorbacea [INV ZOO] A superfamily of foraminiferan protozoans in the suborder Rotaliina characterized by a radial, perforate, calcite test and a monolamellar septa. { ‚dis·kər'bās·ē·ə }

discord *See* dissonance. { 'di‚skȯrd }

discordance [GEOL] An unconformity characterized by lack of parallelism between strata which touch without fusion. { di'skȯrd·əns }

discordant pluton [GEOL] An intrusive igneous body that cuts across the bedding or foliation of the intruded formations. { di'skȯrd·ənt 'plü‚tän }

DISCOS *See* disturbance compensation system. { 'dis‚kōs }

discount [IND ENG] A reduction from the gross amount, price, or value. { 'dis‚kaunt }

discount factor [PETRO ENG] The ratio of the present worth of one or a series of future payments to the total undiscounted amount of such future payments. Also known as average discount factor; deferment factor; present-worth factor. { 'dis‚kaunt ‚fak·tər }

discovery [MIN ENG] Finding of a valuable mineral deposit. { di'skəv·ə·rē }

discovery claim [MIN ENG] The first claim for the finding of a mineral deposit. { di'skəv·ə·rē ‚klām }

discovery vein [MIN ENG] The vein on which a mining claim is based. { di'skəv·ə·rē ‚vān }

discovery well [PETRO ENG] A successful exploration well. { di'skəv·ə·rē ‚wel }

discrete [SCI TECH] 1. Composed of separate and distinct parts. 2. Having an individually distinct identity. { di'skrēt }

discrete address beacon system *See* Mode S. { di‚skrēt 'ad·res 'bē·kən ‚sis·təm }

discrete comparator *See* digital comparator. { di'skrēt kəm'par·əd·ər }

discrete-film zone *See* belt of soil water. { di'skrēt ‚film ‚zōn }

discrete mathematics *See* finite mathematics. { di‚skrēt ‚math·ə'mat·iks }

discrete Fourier transform [MATH] A generalization of the Fourier transform to finite sets of data; for a function f defined at N data values, 0, 1, 2, . . . , $N - 1$, the discrete Fourier transform is a function, F, also defined on the set (0, 1, 2, . . . , $N - 1$), whose value at n is the sum over the variable r, from 0 through $N - 1$, of the quantity $N^{-1} f(r) \exp(-i2\pi nr/N)$. { di‚skrīt für·yā 'tranz‚fȯrm }

discrete radio source [ASTROPHYS] A source of radio waves coming from a small area of the sky. { di'skrēt 'rād·ē·ō ‚sȯrs }

discrete sampling [ELECTR] Sampling in which the individual samples are of such long duration that the frequency response of the channel is not deteriorated by the sampling process. { di'skrēt 'sam‚pliŋ }

discrete set [MATH] A set with no cluster points. { di'skrēt 'set }

discrete sound system [ENG ACOUS] A quadraphonic sound system in which the four input channels are preserved as four discrete channels during recording and playback processes; sometimes referred to as a 4-4-4 system. { di'skrēt 'saund ‚sis·təm }

discrete spectrum [SPECT] A spectrum in which the component wavelengths constitute a discrete sequence of values rather than a continuum of values. { di'skrēt 'spek·trəm }

discrete system [CONT SYS] A control system in which signals at one or more points may change only at discrete values of time. Also known as discrete-time system. { di'skrēt 'sis·təm }

discrete-time system *See* discrete system. { di'skrēt ‚tīm 'sis·təm }

discrete topology [MATH] For a set A, the set of all subsets of A. { di‚skrēt tə'päl·ə·jē }

discrete transfer function *See* pulsed transfer function. { di¦skrēt 'tranz·fər ‚fəŋk·shən }

discrete variable [MATH] A variable for which the possible values form a discrete set. { di¦skrēt 'ver·ē·ə·bəl }

discrete-word intelligibility [COMMUN] The percent of intelligibility obtained when the speech units under consideration are words, usually presented so as to minimize the contextual relation between them. { di¦skrēt ‚wərd in‚tel·ə·jə'bil·əd·ē }

discretization [MATH] A procedure in the numerical solution of partial differential equations in which the domain of the independent variable is subdivided into cells or elements and the equations are expressed in discrete form at each point by finite difference, finite volume, or finite element methods. { dis‚krēd·ə'zā·shən }

discretization error [MATH] The error in the numerical calculation of an integral that results from using an approximate expression for the true mathematical function to be integrated. { ‚dis·krə·də'zā·shən ‚er·ər }

discriminant [MATH] 1. The quantity $b^2 - 4ac$, where a,b,c are coefficients of a given quadratic polynomial: $ax^2 + bx + c$. 2. More generally, for the polynomial equation $a_0 x^n + a_1 x^{n-1} + \cdots + a_n x_0 = 0$, a_0^{2n-2} times the product of the squares of all the differences of the roots of the equation, taken in pairs. { di'skrim·ə·nənt }

discriminant function [STAT] A linear combination of a set of variables that will classify events or items for which the variables are measured with the smallest possible proportion of misclassifications. { di¦skrim·ə·nənt 'fəŋk·shən }

discrimination [COMMUN] 1. In frequency-modulated systems, the detection or demodulation of the imposed variations in the frequency of the carriers. 2. In a tuned circuit, the degree of rejection of unwanted signals. 3. Of any system or transducer, the difference between the losses at specified frequencies with the system or transducer terminated in specified impedances. [COMPUT SCI] *See* conditional jump. { di‚skrim·ə'nā·shən }

discrimination learning [PSYCH] A learning procedure in which a response is reinforced in the presence of one stimulus but not in the presence of others, enabling the experimental animal to discriminate between the one stimulus and the others. { di‚skrim·ə'nā·shən 'lərn·iŋ }

discriminator [ELECTR] A circuit in which magnitude and polarity of the output voltage depend on how an input signal differs from a standard or from another signal. { di'skrim·ə‚nād·ər }

discriminator transformer [ELECTR] A transformer designed to be used in a stage where frequency-modulated signals are converted directly to audio-frequency signals or in a stage where frequency changes are converted to corresponding voltage changes. { di'skrim·ə‚nād·ər tranz'fȯr·mər }

disdrometer [ENG] Equipment designed to measure and record the size distribution of raindrops as they occur in the atmosphere. { diz'dräm·əd·ər }

disease [MED] An alteration of the dynamic interaction between an individual and his or her environment which is sufficient to be deleterious to the well-being of the individual and produces signs and symptoms. { di'zēz }

disengage [ENG] To break the contact between two objects. { ‚dis·ən'gāj }

dish *See* parabolic reflector. { dish }

dishabituation [PSYCH] Restoration to full strength of a response that has become weakened by habituation. { ,dis·hə,bich·əˈwā·shən }

disharmonic fold [GEOL] A fold in which changes in form or magnitude occur with depth. { dis·härˈmän·ik ˈfōld }

dishing [ENG] In metal-forming or plastics-molding operations, producing a shallow concave surface. { ˈdish·iŋ }

dishpan experiment [METEOROL] A model experiment carried out by differential heating of fluid in a flat, rotating pan; it establishes similarity with the atmosphere and is used to reproduce many important features of the general circulation and, on a smaller scale, atmospheric motion. { ˈdish,pan ik,sper·ə·mənt }

disilane [INORG CHEM] Si₂H₆ A spontaneously flammable compound of silicon and hydrogen; it exists as a liquid at room temperature. { dīˈsi,lān }

disilicate [CHEM] A silicate compound that has two silicon atoms in the molecule. { dīˈsil·ə,kāt }

disilicide [CHEM] A compound that has two silicon atoms joined to a radical or another element. { dīˈsil·ə,sīd }

disinclination [CRYSTAL] A type of crystal imperfection in which one part of the crystal is rotated and therefore displaced relative to the rest of the crystal; observed in liquid crystals and protein coats of viruses. { dis,in·kləˈnā·shən }

disinfectant [MATER] A chemical agent that destroys microorganisms but not bacterial spores. { dis·ənˈfek·tənt }

disintegration [NUC PHYS] Any transformation of a nucleus, whether spontaneous or induced by irradiation, in which particles or photons are emitted. { dis,in·təˈgrā·shən }

disintegration chain See radioactive series. { dis,in·təˈgrā·shən ,chān }

disintegration constant See decay constant. { dis,in·təˈgrā·shən ,kän·stənt }

disintegration energy [NUC PHYS] The energy released, or the negative of the energy absorbed, during a nuclear or particle reaction. Designated Q. Also known as Q value; reaction energy. { dis,in·təˈgrā·shən ,en·ər·jē }

disintegration family See radioactive series. { dis,in·təˈgrā·shən ,fam·lē }

disintegration of measure [MATH] The representation of a measure as an integral of a family of positive measures. { dis,in·təˈgrā·shən əv ˈmezh·ər }

disintegration rate [NUCLEO] **1.** The absolute rate of decay of a radioactive substance, usually expressed in terms of disintegrations per unit of time. **2.** The absolute rate of transformation of a nuclide under bombardment. { dis,in·təˈgrā·shən ,rāt }

disintegration series See radioactive series. { dis,in·təˈgrā·shən ,sir·ēz }

disintegration voltage [ELECTR] The lowest anode voltage at which destructive positive-ion bombardment of the cathode occurs in a hot-cathode gas tube. { dis,in·təˈgrā·shən ,vōl·tij }

disintegrator [MECH ENG] An apparatus used for pulverizing or grinding substances, consisting of two steel cages which rotate in opposite directions. { disˈin·tə,grād·ər }

disjoint sets [MATH] Sets with no elements in common. { disˈjȯint ˈsets }

disjunct endemism [PALEON] A type of regionally restricted distribution of a fossil taxon in which two or more component parts are separated by a major physical barrier and hence not readily explicable in terms of present-day geography. { dis,jəŋkt ˈen·də,miz·əm }

disjunction [CYTOL] Separation of chromatids or homologous chromosomes during anaphase. [MATH] The connection of two statements by the word "or." Also known as alternation. { disˈjəŋk·shən }

disjunctive search [COMPUT SCI] A search to find items that have at least one of a given set of characteristics. { disˈjəŋk·tiv ˈsərch }

disjunctor [MYCOL] A small cellulose body between the conidia of certain fungi, which eventually breaks down and thus frees the conidia. { disˈjəŋk·tər }

disk Also spelled disc. [ASTRON] A relatively thin layer of material distributed in the central plane of a spiral galaxy, in contrast to the nucleus or halo. [BIOL] Any of various rounded and flattened animal and plant structures. [COMPUT SCI] A rotating circular plate having a magnetizable surface on which information may be stored as a pattern of polarized spots on concentric recording tracks. Also known as magnetic

disk. See phonograph record. [MATH] **1.** The region in the plane consisting of all points with norm less than 1 (sometimes less than or equal to 1). **2.** See closed disk.. { disk }

disk-and-doughnut [CHEM ENG] A type of fractionating tower construction of alternating disks and plates that are doughnut-shaped, to provide mixing. { disk ən ˈdō·nət }

disk armature [ELEC] The armature in a motor that has a disk winding or is made up of a metal disk. { ˈdisk ,är·mə·chər }

disk attrition mill See disk mill. { ˈdisk əˈtrish·ən ,mil }

disk brake [MECH ENG] A type of brake in which disks attached to a fixed frame are pressed against disks attached to a rotating axle or against the inner surfaces of a rotating housing. { ˈdisk ˈbrāk }

disk cache [COMPUT SCI] A portion of random-access memory that contains the data most recently read from or written to the disk, allowing rapid access by the central-processing unit. { ˈdisk ,kash }

disk cam [MECH ENG] A disk with a contoured edge which rotates about an axis perpendicular to the disk, communicating motion to the cam follower which remains in contact with the edge of the disk. { ˈdisk ,kam }

disk camera [OPTICS] A camera that uses a disk of color negative film enclosed in a light-tight plastic pack and containing 15 rectangular frames measuring approximately 8 by 10 millimeters. { ˈdisk ,kam·rə }

disk canvas wheel [DES ENG] A polishing wheel made of disks of canvas sewn together with heavy twine or copper wire, and reinforced by steel side plates and side rings with bolts or screws. { ˈdisk ˈkan·vəs ,wēl }

disk capacitor [ELEC] A small, flat, circular capacitor that usually has a ceramic dielectric. { ˈdisk kəˌpas·əd·ər }

disk cartridge [COMPUT SCI] A removable module that contains a single magnetic disk platter which remains attached to the housing when placed into the disk drive. { ˈdisk ,kär·trij }

disk centrifuge [MECH ENG] A centrifuge with a large bowl having a set of disks that separate the liquid into thin layers to create shallow settling chambers. { ˈdisk ˈsen·trə,fyüj }

disk clutch [MECH ENG] A clutch in which torque is transmitted by friction between friction disks with specially prepared friction material riveted to both sides and contact plates keyed to the inner surface of an external hub. { ˈdisk ˈkləch }

disk colorimeter [ANALY CHEM] A device for comparing standard and sample colors by means of rotating color disks. { ˈdisk kəˈlə·rim·əd·ər }

disk coupling [MECH ENG] A flexible coupling in which the connecting member is a flexible disk. { ˈdisk ,kəp·liŋ }

disk crash See head crash. { ˈdisk ,krash }

disk cultivator [AGR] A cultivator consisting of pairs of oppositely inclined disks. { ˈdisk ˈkəl·tə,vād·ər }

disk drive [COMPUT SCI] The physical unit that holds, spins, reads, and writes the magnetic disks. Also known as disk unit. { ˈdisk ,drīv }

disk drive controller [COMPUT SCI] A device that enables a microcomputer to control the functioning of a disk drive. { ˈdisk ˈdrīv kənˈtrō·lər }

disk engine [MECH ENG] A rotating engine in which the piston is a disk. { ˈdisk ,en·jən }

diskette See floppy disk. { diˈsket }

disk file [COMPUT SCI] An organized collection of records held on a magnetic disk. { ˈdisk ,fīl }

disk filter [ENG] A filter in which the substance to be filtered is drawn through membranes stretched on segments of revolving disks by a vacuum inside each disk; the solids left on the membrane are lifted from the tank and discharged. Also known as American filter. { ˈdisk ˈfil·tər }

disk flower [BOT] One of the flowers on the disk of a composite plant. { ˈdisk ,flaů·ər }

disk furrower [AGR] A furrower in which concave disks, at an angle to the direction of motion, are used to cut the soil. { ˈdisk ,fər·ə·wər }

disk grinder [MECH ENG] A grinding machine that employs abrasive disks. { ˈdisk ,grīnd·ər }

disk grinding [MECH ENG] Grinding with the flat side of a rigid, bonded abrasive disk or segmental wheel. { ˈdisk ,grīnd·iŋ }

disk harrow [AGR] A harrow which has two or more opposed gangs of 3–12 disks for cutting clods and trash, destroying weeds, cutting in cover crops, and smoothing and

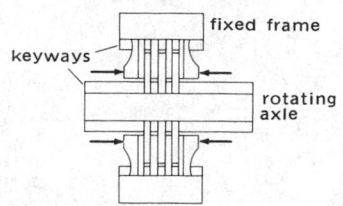

DISK BRAKE

Forces in direction of arrows press disks keyed to axle against disks keyed to frame.

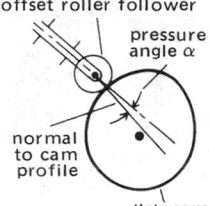

DISK CAM

Diagram of disk cam.

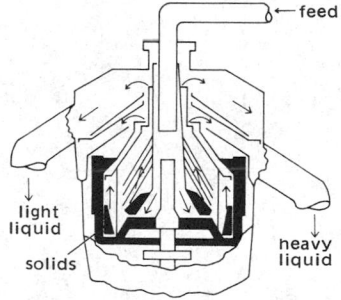

DISK CENTRIFUGE

Cutaway view of disk centrifuge bowl.

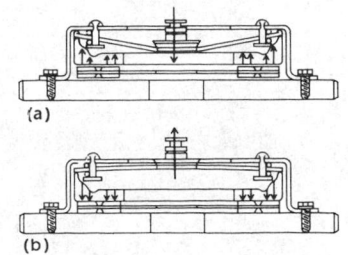

DISK CLUTCH

Disk clutch (a) in disengaged position, (b) in engaged position.

preparing the surface for various farming operations. { 'disk ˌha·rō }

disk leather wheel [DES ENG] A polishing wheel made of leather disks glued together. { ¦disk 'leth·ər ˌwēl }

diskless work station [COMPUT SCI] A computer in a network that has no disk storage of its own. { ¦disk·ləs 'wərk ˌstā·shən }

disk loading [AERO ENG] A measure which expresses the design gross weight of a helicopter as a function of the swept areas of the lifting rotor. { 'disk ˌlōd·iŋ }

disk memory See disk storage. { 'disk ˌmem·rē }

disk meter [ENG] A positive displacement meter to measure flow rate of a fluid; consists of a disk that wobbles or nutates within a chamber so that each time the disk nutates a known volume of fluid passes through the meter. { 'disk ˌmēd·ər }

disk method [MATH] A method of computing the volume of a solid of revolution, by integrating over the volumes of infinitesimal disk-shaped slices bounded by planes perpendicular to the axis of revolution. { 'disk ˌmeth·əd }

disk mill [MECH ENG] Size-reduction apparatus in which grinding of feed solids takes place between two disks, either or both of which rotate. Also known as disk attrition mill. { 'disk ˌmil }

disk operating system [COMPUT SCI] An operating system which uses magnetic disks as its primary on-line storage. Abbreviated DOS. { ¦disk ¦äp·ə‚rād·iŋ ˌsis·təm }

disk pack [COMPUT SCI] A set of magnetic disks that can be removed from a disk drive as a unit. { 'disk ˌpak }

disk plow [AGR] A plow consisting of a number of disk blades attached to one axle or gang bolt; used for rapid, shallow plowing. { 'disk ˌplau̇ }

disk population [ASTRON] The older Population I stars such as the sun. { 'disk ˌpäp·yə'lā·shən }

disk recording [ENG ACOUS] **1.** The process of inscribing suitably transformed acoustical or electrical signals on a phonograph record. **2.** See phonograph record. { ¦disk ri'kȯrd·iŋ }

disk sander [MECH ENG] A machine that uses a circular disk coated with abrasive to smooth or shape surfaces. { 'disk ˌsand·ər }

disk-seal tube [ELECTR] An electron tube having disk-shaped electrodes arranged in closely spaced parallel layers, to give low interelectrode capacitance along with high power output, up to 2500 megahertz. Also known as lighthouse tube; megatron. { 'disk ˌsēl ˌtüb }

disk signal [CIV ENG] Automatic block signal with colored disks that indicate train movements. { ¦disk ¦sig·nəl }

disk spring [MECH ENG] A mechanical spring that consists of a disk or washer supported by one force (distributed by a suitable chuck or holder) at the periphery and by an opposing force on the center or hub of the disk. { 'disk ˌspriŋ }

disk storage [ELECTR] An external computer storage device consisting of one or more disks spaced on a common shaft, and magnetic heads mounted on arms that reach between the disks to read and record information on them. Also known as disk memory; magnetic disk storage. { 'disk ˌstȯr·ij }

disk striping [COMPUT SCI] The distribution of a unit of data over two or more hard disks, enabling the data to be read more quickly. Also known as data striping. { 'disk ˌstrīp·iŋ }

disk telescope [OPTICS] A telescope designed for observations of the brilliant solar disk; examples are the tower telescope and the horizontal fixed telescope. { 'disk ˌtel·ə‚skōp }

disk thermistor [ELECTR] A thermistor which is produced by pressing and sintering an oxide binder mixture into a disk, 0.2–0.6 inch (5–15 millimeters) in diameter and 0.04–0.5 inch (1.0–13 millimeters) thick, coating the major surfaces with conducting material, and attaching leads. { ¦disk thər'mis·tər }

disk unit See disk drive. { 'disk ˌyü·nət }

disk-wall packer [PETRO ENG] A disklike seal between the outside of the well tubing and the inside of the well casing; used to prevent fluid movement from the pressure differential above and below the sealing point. { 'disk ˌwȯl ˌpak·ər }

disk wheel [DES ENG] A wheel in which a solid metal disk, rather than separate spokes, joins the hub to the rim. { 'disk ˌwēl }

dislocation [CRYSTAL] A defect occurring along certain lines in the crystal structure and present as a closed ring or a line anchored at its ends to other dislocations, grain boundaries, the surface, or other structural feature. Also known as line

defect. [GEOL] Relative movement of rock on opposite sides of a fault. Also known as displacement. [MED] Displacement of one or more bones of a joint. { ˌdis·lō'kā·shən }

dislocation breccia See fault breccia. { ˌdis·lō'kā·shən 'brech·ə }

dislocation line [CRYSTAL] A curve running along the center of a dislocation. { ˌdis·lō'kā·shən ‚līn }

dismicrite [GEOL] Fine-grained limestone of obscure origin, resembling micrite but containing sparry calcite bodies. { diz'mī‚krīt }

dismount [ORD] To remove a weapon or piece of equipment from its setting, mount, or carriage. { dis'maunt }

disodium hydrogen phosphate See disodium phosphate. { dī'sōd·ē·əm 'hī·drə·jən 'fäs‚fāt }

disodium methylarsonate [ORG CHEM] $CH_3AsO(ONa)_2$ A colorless, hygroscopic, crystalline solid; soluble in water and methanol; used in pharmaceuticals and as a herbicide. Abbreviated DMA. { dī'sōd·ē·əm ¦meth·əl'ärs·ən‚āt }

disodium phosphate [INORG CHEM] Na_2HPO_4 Transparent crystals, soluble in water; used in the textile processing and other industries to control pH in the range 4–9, as an additive in processed cheese to maintain spreadability, and as a laxative and antacid. Also known as disodium hydrogen phosphate. { dī'sōd·ē·əm 'fäs‚fāt }

disodium tartrate See sodium tartrate. { dī'sōd·ē·əm 'tär ‚trāt }

disomaty [CYTOL] Duplication of chromosomes unaccompanied by nuclear division. { dī'sō·məd·ē }

Disomidae [INV ZOO] A family of spioniform annelid worms belonging to the Sedentaria. { də'säm·ə‚dē }

disophenol [PHARM] $I_2C_6H_2(NO_2)OH$ Light yellow, feathery crystals with a melting point of 157°C; soluble in alcohol; used as an antihelminthic drug in animals. { də'sä·fə‚nȯl }

disorder [CRYSTAL] Departures from regularity in the occupation of lattice sites in a crystal containing more than one element. { dis'ȯrd·ər }

disordered crystalline alloy [SOLID STATE] A mixture of two elements in which the atoms of the mixture are found at more or less random positions on a crystal lattice. { dis'ȯrd·ərd ¦krist·əl·ən 'al‚ȯi }

disorientation [MED] Mental confusion as to one's normal relationship to his or her environment, especially time, place, and people; associated with organic brain disorders. { dis‚ȯr·ē·ən'tā·shən }

dispatching [COMPUT SCI] The control of priorities in a queue of requests in a multiprogramming or multitasking environment. [IND ENG] The selecting and sequencing of tasks to be performed at individual work stations and the assigning of these tasks to the personnel. { dis'pach·iŋ }

dispatching priority [COMPUT SCI] In a multiprogramming or multitasking environment, the priority assigned to an active (non-real time, nonforeground) task. { dis'pach·iŋ prī‚är·əd·ē }

dispenser [ENG] Device that automatically dispenses radar chaff from an aircraft. { də'spen·sər }

dispenser cathode [ELECTR] An electron tube cathode having provisions for continuously replacing evaporated electron-emitting material. { də'spen·sər ‚kath‚ōd }

dispermy [PHYSIO] Entrance of two spermatozoa into an ovum. { 'dī‚spər·mē }

dispersal [CIV ENG] The practice of building or establishing industrial plants, government offices, or the like, in separated areas, to reduce vulnerability to enemy attack. [ORD] The spreading out of equipment, supplies, or personnel, especially for protection against enemy action. { də'spər·səl }

dispersal barrier [ECOL] A physical structure that prevents organisms from crossing into new space. { də'spər·səl ‚bar·ē·ər }

dispersal pattern [GEOCHEM] Distribution pattern of metals in soil, rock, water, or vegetation. { də'spər·səl ‚pad·ərn }

dispersant [MATER] **1.** Also known as dispersing agent. **2.** An additive that can hold finely ground materials in suspension; used as a thinning agent for a slurry. **3.** A material added to a paste, mortar, or concrete to improve the flow properties. { di'spərs·ənt }

disperse [COMPUT SCI] A data-processing operation in which grouped input items are distributed among a larger number of groups in the output. { də'spərs }

dispersed elements [GEOCHEM] Elements which form few or no independent minerals but are present as minor ingredients in minerals of abundant elements. { də'spərst 'el·ə·mənts }

dispersed gas injection [PETRO ENG] Gas-injection pressure maintenance of an oil reservoir in which the injection wells are arranged geometrically to distribute the gas uniformly throughout the oil-productive portions of the reservoir. { də'spərst 'gas in‚jek·shən }

disperse dye [MATER] A very slightly water-soluble, colored material for use on cellulose acetate and other synthetic fibers; color is transferred to the fiber as extremely finely divided particles, resulting in a solution of the dye in the solid fiber. { də'spərs ‚dī }

disperse phase [CHEM] The phase of a disperse system consisting of particles or droplets of one substance distributed through another system. Also known as discontinuous phase; internal phase. { də'spərs ‚fāz }

disperser [MATER] Material added to solid-in-liquid or liquid-in-liquid suspensions to separate the individual suspended particles; used in pigment grinding and dye dispersion. Also known as dispersing agent; emulsifier; emulsifying agent. { də'spər·sər }

disperse system [CHEM] A two-phase system consisting of a dispersion medium and a disperse phase. { də'spərs ‚sis·təm }

dispersible inhibitor [CHEM] An additive that can be dispersed in a liquid with only moderate agitation to retard undesirable chemical action. { di'spər·sə·bəl in'hib·əd·ər }

dispersing agent See dispersant; disperser. { də'spərs·iŋ ‚ā·jənt }

dispersing prism [OPTICS] An optical prism which deviates light of different wavelengths by different amounts and can therefore be used to separate white light into its monochromatic parts. { də'spərs·iŋ ‚priz·əm }

dispersion [AERO ENG] Deviation from a prescribed flight path; specifically, circular dispersion especially as applied to missiles. [ASTRON] The frequency dependence of the retardation of radio waves (such as those emitted by a pulsar) when they pass through an ionized gas. [CHEM] A distribution of finely divided particles in a medium. [COMMUN] The entropy of the output of a communications channel when the input is known. [ELECTROMAG] Scattering of microwave radiation by an obstruction. [MINERAL] In optical mineralogy, the constant optical values at different positions on the spectrum. [PHYS] **1.** The separation of a complex of electromagnetic or sound waves into its various frequency components. **2.** Quantitatively, the rate of change of refractive index with wavelength or frequency at a given wavelength or frequency. **3.** The rate of change of deviation with wavelength or frequency. **4.** In general, any process separating radiation into components having different frequencies, energies, velocities, or other characteristics, such as the sorting of electrons according to velocity in a magnetic field. [STAT] The degree of spread shown by observations in a sample or a population. { də'spər·zhən }

dispersion equation See dispersion formula. { də'spər·zhən i'kwā·zhən }

dispersion error [ORD] Chance variation in a series of shots even though firing conditions are kept as constant as possible. { də'spər·zhən ‚er·ər }

dispersion force [PHYS CHEM] The force of attraction that exists between molecules that have no permanent dipole. { də'spər·zhən ‚fōrs }

dispersion formula [PHYS] Any formula which gives the refractive index as a function of wavelength of electromagnetic radiation. Also known as dispersion equation. { də'spər·zhən ‚fōr·myə·lə }

dispersion fuel [NUCLEO] A fuel mixture consisting of a nuclear fuel dispersed in a nonfissionable matrix. { də'spər·zhən ‚fyül }

dispersion index [STAT] Statistics used to determine the homogeneity of a set of samples. { di'spər·zhən ‚in‚deks }

dispersion ladder [ORD] Table showing the probable distribution of a succession of shots made with the same firing data; specifically, a diagram made up of eight zones, showing the percentage of shots which may be expected to fall within each zone, based on direction (deflection) or range. { də'spər·zhən ‚lad·ər }

dispersion measure [ASTRON] A quantity that describes the dispersion of a radio signal, proportional to the product of the density of interstellar electrons and the distance to the source. { də'spər·zhən ‚mezh·ər }

dispersion medium See continuous phase. { də'spər·zhən ‚mēd·ē·əm }

dispersion mill [MECH ENG] Size-reduction apparatus that disrupts clusters or agglomerates of solids, rather than breaking down individual particles; used for paint pigments, food products, and cosmetics. { də'spər·zhən ‚mil }

dispersion pattern [ORD] The distribution of a series of shots by using coordinate settings as nearly identical as possible. { də'spər·zhən ‚pad·ərn }

dispersion relation [NUC PHYS] A relation between the cross section for a given effect and the de Broglie wavelength of the incident particle, which is similar to a classical dispersion formula. [PHYS] An integral formula relating the real and imaginary parts of some function of frequency or energy, such as a refractive index or scattering amplitude, based on the causality principle and the Cauchy integral formula. [PL PHYS] A relation between the radian frequency and the wave vector of a wave motion or instability in a plasma. { də'spər·zhən ri‚lā·shən }

dispersion strengthening [SOLID STATE] The reduction of plastic deformation of a solid by the presence of a uniform dispersion of another substance which inhibits the motion of plastic dislocations. { də'spər·zhən ‚streŋk·thən·iŋ }

dispersion zone [ORD] The area over which shots scatter when fired with the same sight setting. { də'spər·zhən ‚zōn }

dispersive line [ELECTROMAG] A delay line that delays each frequency a different length of time. { də'spər·siv 'līn }

dispersive medium [ELECTROMAG] A medium in which the phase velocity of an electromagnetic wave is a function of frequency. { də'spər·siv 'mē·dē·əm }

dispersive power [OPTICS] A measure of the power of a medium to separate different colors of light, equal to $(n_2 - n_1)/(n - 1)$, where n_1 and n_2 are the indices of refraction at two specified widely differing wavelengths, and n is the index of refraction for the average of these wavelengths, or for the D line of sodium. { də'spər·siv ‚pau̇·ər }

dispersoid [CHEM] Matter in a form produced by a disperse system. { də'spər‚sȯid }

disphenoid [CRYSTAL] **1.** A crystal form with four similar triangular faces combined in a wedge shape; can be tetragonal or orthorhombic. **2.** A crystal form with eight scalene triangles combined in pairs. { dī'sfē‚nȯid }

displaced ore body [GEOL] An ore body which has been subjected to displacement or disruption after its initial deposition. { dis'plāst 'ȯr ‚bäd·e }

displacement [CHEM] A chemical reaction in which an atom, radical, or molecule displaces and sets free an element of a compound. [COMPUT SCI] The number of character positions or memory locations from some point of reference to a specified character or data item. Also known as offset. [ELEC] See electric displacement. [FL MECH] **1.** The weight of fluid which is displaced by a floating body, equal to the weight of the body and its contents; the displacement of a ship is generally measured in long tons (1 long ton = 2240 pounds). **2.** The volume of fluid which is displaced by a floating body. [GEOL] See dislocation. [MECH] **1.** The linear distance from the initial to the final position of an object moved from one place to another, regardless of the length of path followed. **2.** The distance of an oscillating particle from its equilibrium position. [MECH ENG] The volume swept out in one stroke by a piston moving in a cylinder as for an engine, pump, or compressor. [PSYCH] A defense mechanism in which emotions, ideas, or wishes are transferred from their original object to a more acceptable substitute. { dis'plās·mənt }

displacement angle [ELEC] The change in the phase of an alternator's terminal voltage when a load is applied. { dis'plās·mənt ‚aŋ·gəl }

displacement boat [NAV ARCH] Any ship or boat which travels immersed and operates at relatively lower speeds than craft, such as hydroplanes, which plane on the surface of the water at high speed. { dis'plās·mənt ‚bōt }

displacement chromatography [ANALY CHEM] Variation of column-development or elution chromatography in which

DISPERSING PRISM

Two types of dispersing prisms. (a) Rayleigh prism system. (b) Amici direct-vision system consisting of flint-glass prism and two crown-glass prisms.

the solvent is sorbed more strongly than the sample components; the freed sample migrates down the column, pushed by the solvent. { dis'plās·mənt ‚krō·mə'täg·rə·fē }

displacement collision [NUCLEO] The collision of an energetic particle with an atom in a solid resulting in the atom being moved permanently from its original site. { dis'plās·mənt kə,lizh·ən }

displacement compressor [MECH ENG] A type of compressor that depends on displacement of a volume of air by a piston moving in a cylinder. { dis'plās·mənt kəm,pres·ər }

displacement current [ELECTROMAG] The rate of change of the electric displacement vector, which must be added to the current density to extend Ampère's law to the case of time-varying fields (meter-kilogram-second units). Also known as Maxwell's displacement current. { dis'plās·mənt ‚kə·rənt }

displacement curve [NAV ARCH] A graph of the displacement of a vessel versus its draft; it is a curve of form. { dis'plās·mənt ‚kərv }

displacement efficiency [PETRO ENG] In a gas condensate reservoir, the proportion (by volume) of wet hydrocarbons swept out of pores during dry-gas cycling. { dis'plās·mənt ə'fish·ən·sē }

displacement engine See piston engine. { dis'plās·mənt ‚en·jən }

displacement fluid [MATER] A fluid material, usually drilling mud or salt water, that is pumped into a well after the cement to force the cement out of the casing and into the annulus. { dis'plās·mənt ‚flü·əd }

displacement gyroscope [ENG] A gyroscope that senses, measures, and transmits angular displacement data. { dis'plās·mənt 'jī·rə,skōp }

displacement kernel [NUCLEO] In nuclear reactor theory, a function of two locations that depends only on the distance between the locations, such as the diffusion kernel or slowing-down kernel. { dis'plās·mənt ‚kər·nəl }

displacement law See radioactive displacement law; Wien's displacement law. { dis'plās·mənt ‚lȯ }

displacement length coefficient [NAV ARCH] The displacement, in tons of sea water, of a ship divided by the length over 100 cubed. { dis'plās·mənt ‚leŋkth ‚kō·i'fish·ənt }

displacement loop [MOL BIO] In circular deoxyribonucleic acid (DNA), a small region in which ribonucleic acid is paired with one strand of DNA, effectively displacing the other DNA strand. Also known as D-loop. { dis'plās·mənt ‚lüp }

displacement manometer [ENG] A differential manometer which indicates the pressure difference across a solid or liquid partition which can be displaced against a restoring force. { dis'plās·mənt mə'näm·əd·ər }

displacement meter [ENG] A water meter that measures water flow quantitatively by recording the number of times a vessel of known capacity is filled and emptied. { dis'plās·mənt ‚mēd·ər }

displacement operator [MATH] A difference operator, denoted E, defined by the equation $Ef(x) = f(x + h)$, where h is a constant denoting the difference between successive points of interpolation or calculation. Also known as forward shift operator. { dis'plās·mənt ‚äp·ə,rād·ər }

displacement pump [MECH ENG] A pump that develops its action through the alternate filling and emptying of an enclosed volume as in a piston-cylinder construction. { dis'plās·mənt ‚pəmp }

displacement rate [PETRO ENG] In oil well cementing, the speed at which a given volume of cement slurry or mud is pumped down the borehole. { dis'plās·mənt ‚rāt }

displacement series [CHEM] The elements in decreasing order of their negative potentials. Also known as constant series; electromotive series; Volta series. { dis'plās·mənt ‚sir·ēz }

displacement spike [NUCLEO] A region in a solid in which atoms have been permanently moved from their original locations as the result of energetic particle bombardment. { dis'plās·mənt ‚spīk }

displacer-type meter [ENG] Apparatus to detect liquid level or gas density by measuring the effect of the fluid (gas or liquid) on the buoyancy of a displacer unit immersed within the fluid. { di'splās·ər ‚tīp ‚mēd·ər }

display [ELECTR] **1.** A visible representation of information, in words, numbers, or drawings, as on the cathode-ray tube screen of a radar set, navigation system, or computer console.

2. The device on which the information is projected. Also known as display device. **3.** The image of the information. { di'splā }

display adapter See video display board. { di'splā ə,dap·tər }

display console [COMPUT SCI] A cathode-ray tube or other display unit on which data being processed or stored in a computer can be presented in graphical or character form; sometimes equipped with a light pen with which the user can alter the information displayed. { di'splā ,kän,sōl }

display control [COMPUT SCI] A unit in a computer system consisting of channels and associated control circuitry that connect a number of visual display units with a central processor. { di'splā kən,trōl }

display cycle [COMPUT SCI] In computer graphics, the sequence of operations carried out to display an image. { di,splā 'sī·kəl }

display device See display. { di'splā di,vīs }

display element [COMPUT SCI] In computer graphics, a basic component of a display, such as a circle, line, or dot. { di'splā ,el·ə·mənt }

display entity [COMPUT SCI] In computer graphics, a group of display elements that can be manipulated as a unit. { di'splā ,en·təd·ē }

display frame [COMPUT SCI] In computer graphics, one of a sequence of frames making up a computer-generated animation. { di'splā ,frām }

display information processor [COMPUT SCI] Computer used to generate situation displays in a combat operations center. { di'splā in·fər'mā·shən ,präs,es·ər }

display list [COMPUT SCI] In computer graphics, a set of vectors that form an image stored in vectors graphics format. { di'splā ,list }

display loss See visibility factor. { di'splā ,lȯs }

display packing [COMPUT SCI] An efficient means of transmitting the x and y coordinates of a point packed in a single word to halve the time required to freshen the spot on a cathode-ray tube display. { di'splā ,pak·iŋ }

display power management signaling [COMPUT SCI] Signaling whereby a video adapter can instruct a monitor to reduce its power level to conserve electricity. Abbreviated DPMS. { di'splā 'pau̇·ər ,man·ij·mənt ,sig·nəl·iŋ }

display primary [COMMUN] One of the primary colors produced in a television receiver that, when mixed in proper proportions, serve to produce the other desired colors. Also known as receiver primary. { di'splā 'prī,mer·ē }

display processor [COMPUT SCI] A section of a computer, or a minicomputer which handles the routines required to display an output on a cathode-ray tube. { di'splā ,präs,es·ər }

display screen See video monitor. { di'splā ,skrēn }

display storage tube See direct-view storage tube. { di'splā 'stȯr·ij ,tüb }

display system [COMPUT SCI] The total system, combining hardware and software, needed to achieve a visible representation of information in a data-processing system. { di'splā ,sis·təm }

display terminal [COMPUT SCI] A computer output device in which characters, and sometimes graphic information, appear on the screen of a cathode-ray tube. Also known as display unit; video display terminal (VDT). { di'splā ,tər·mən·əl }

display tube [ELECTR] A cathode-ray tube used to provide a visual display. Also known as visual display unit. { di'splā ,tüb }

display type [GRAPHICS] In composition, type which is larger in size than the regular text type. { di'splā ,tīp }

display unit See display terminal. { di'splā ,yü·nət }

display window [COMMUN] Width of the portion of the frequency spectrum presented on panoramic presentation; expressed in frequency units, usually megahertz. { di'splā ,win,dō }

disposable [ENG] Within a manufacturing system, designed to be discarded after use and replaced by an identical item, such as a filter element. { də'spō·zə·bəl }

disposal area [NAV] On U.S. Coast and Geodetic Survey charts, an area established or approved by the Corps of Engineers for depositing dredged material where existing depths indicate that the intent is not to cause sufficient shoaling to create a danger to surface navigation; soundings and curves are retained. { də'spō·zəl ,er·ē·ə }

disposal field See absorption field. { də'spō·zəl ,fēld }

disposition [COMPUT SCI] The status of a file after it has been closed by a computer program, for example, retained or deleted. { ‚dis·pə'zish·ən }

disproportionation [CHEM] The changing of a substance, usually by simultaneous oxidation and reduction, into two or more dissimilar substances. { ‚dis·prə‚pȯr·shə'nā·shən }

disruptive discharge [ELEC] A sudden and large increase in current through an insulating medium due to complete failure of the medium under electrostatic stress. { dis¦rəp·tiv 'dis‚chärj }

disruptive strength [MET] Failure stress caused by hydrostatic tension. { dis¦rəp·tiv 'strenkth }

dissect [BIOL] To divide, cut, and separate into different parts. { də'sekt }

dissected topography [GEOGR] Physical features marked by erosive cutting. { də'sek·təd tə'päg·rə·fē }

dissecting microscope [OPTICS] Either of two types of optical microscope used to magnify materials undergoing dissection. { də'sek·tiŋ 'mī·krə‚skōp }

dissection [GEOL] Destruction of the continuity of the land surface by erosive cutting of valleys or ravines into a relatively even surface. { də'sek·shən }

dissector tube [ELECTR] Camera tube having a continuous photo cathode on which is formed a photoelectric emission pattern which is scanned by moving its electron-optical image over an aperture. { də'sek·tər ‚tüb }

disseminated necrotizing periarteritis See polyarteritis nodosa. { də'sem·ə‚nād·əd 'nek·rə‚tiz·iŋ ¦per·ē‚ärd·ə'rīd·əs }

disseminule [BIOL] An individual organism or part of an individual adapted for the dispersal of a population of the same organisms. { də'sem·ə‚nyül }

dissepiment [BOT] A partition which divides a fruit or an ovary into chambers. [PALEON] One of the vertically positioned thin plates situated between the septa in extinct corals of the order Rugosa. { də'sep·ə·mənt }

dissimilar terms [MATH] Terms that do not contain the same unknown factors or that do not contain the same powers of these factors. { di¦sim·ə·lər 'tərmz }

dissipation [PHYS] Any loss of energy, generally by conversion into heat; quantitatively, the rate at which this loss occurs. Also known as energy dissipation. { ‚dis·ə'pā·shən }

dissipation coefficient See scattering coefficient. { ‚dis·ə'pā·shən ‚kō·i'fish·ənt }

dissipation constant [GEOPHYS] In atmospheric electricity, a measure of the rate at which a given electrically charged object loses its charge to the surrounding air. { ‚dis·ə'pā·shən ‚kän·stənt }

dissipation factor [ELEC] The inverse of Q, the storage factor. { ‚dis·ə'pā·shən ‚fak·tər }

dissipation function See Rayleigh's dissipation function; viscous dissipation function. { ‚dis·ə'pā·shən ‚fəŋk·shən }

dissipation line [ELECTROMAG] A length of stainless steel or Nichrome wire used as a noninductive terminating impedance for a rhombic transmitting antenna when several kilowatts of power must be dissipated. { ‚dis·ə'pā·shən ‚līn }

dissipation loss [ELEC] A measure of the power loss of a transducer in transmitting signals, expressed as the ratio of its input power to its output power. { ‚dis·ə'pā·shən ‚lȯs }

dissipation trail [FL MECH] A clear rift left behind an aircraft as it flies in a thin cloud layer; the opposite of a condensation trail. Also known as distrail. { ‚dis·ə'pā·shən ‚trāl }

dissipative muffler [ENG] A device which absorbs sound energy as the gas passes through it; a duct lined with sound-absorbing material is the most common type. { ‚dis·ə'pād·iv 'məf·lər }

dissipative tunneling [SOLID STATE] Quantum-mechanical tunneling of individual electrons, rather than pairs, across a thin insulating layer separating two superconducting metals when there is a voltage across this layer, resulting in partial disruption of cooperative motion. { ‚dis·ə'pād·iv 'tən·əl·iŋ }

dissipator See heatsink. { 'dis·ə‚pād·ər }

dissociation [MED] Independent, uncoordinated functioning of the atria and ventricles. [MICROBIO] The appearance of a novel colony type on solid media after one or more subcultures of the microorganism in liquid media. [PHYS CHEM] Separation of a molecule into two or more fragments (atoms, ions, radicals) by collision with a second body or by the absorption of electromagnetic radiation. [PSYCH] The segregation of ideas from their affects or feelings, resulting in independent

functioning of these components of a person's mental processes. { də‚sō·sē'ā·shən }

dissociation constant [PHYS CHEM] A constant whose numerical value depends on the equilibrium between the undissociated and dissociated forms of a molecule; a higher value indicates greater dissociation. { də‚sō·sē'ā·shən ‚kän·stənt }

dissociation energy [PHYS CHEM] The energy required for complete separation of the atoms of a molecule. { də‚sō·sē'ā·shən ‚en·ər·jē }

dissociation limit [SPECT] The wavelength, in a series of vibrational bands in a molecular spectrum, corresponding to the point at which the molecule dissociates into its constituent atoms; it corresponds to the convergence limit. { də‚sō·sē'ā·shən ‚lim·ət }

dissociation pressure [PHYS CHEM] The pressure, for a given temperature, at which a chemical compound dissociates. { də‚sō·sē'ā·shən ‚presh·ər }

dissociation-voltage effect [PHYS CHEM] A change in the dissociation of a weak electrolyte produced by a strong electric field. { də‚sō·sē¦ā·shən 'vōl·tij i‚fekt }

dissociative reaction [PSYCH] A neurotic disorder leading to gross disorganization of the personality. { də¦sō·shəd·iv rē'ak·shən }

dissociative recombination [ATOM PHYS] The combination of an electron with a positive molecular ion in a gas followed by dissociation of the molecule in which the resulting atoms carry off the excess energy. { də¦sō·shəd·iv ‚rē·käm·bə'nā·shən }

dissogeny [ZOO] Having two sexually mature stages, larva and adult, in the life of an individual. { də'sä·jə·nē }

dissolution [CHEM] Dissolving of a material. { ‚dis·ə'lü·shən }

dissolve [CHEM] **1.** To cause to disperse. **2.** To cause to pass into solution. [GRAPHICS] A superimposing of one television or motion picture shot upon another, the emergent shot gradually brightening and the overlapped shot gradually darkening, so that as one scene disappears, another gradually appears. Also known as lap dissolve. { də'zälv }

dissolved air flotation [CHEM ENG] A liquid-solid separation process wherein the main mechanism of suspended-solids removal is the change of apparent specific gravity of those suspended solids in relation to that of the suspending liquid by the attachment of small gas bubbles formed by the release of dissolved gas to the solids. Also known as air flotation. { də'zälvd ‚er flō'tā·shən }

dissolved gas See solution gas. { də'zälvd 'gas }

dissolved-gas drive See internal gas drive. { də'zälvd ‚gas 'drīv }

dissolved-gas-drive reservoir [PETRO ENG] Oil reservoir in which the temperature of the liquid phase is below critical, and the liquid is driven from the reservoir by the expansion of dissolved gas. Also known as a bubble-point reservoir. { də'zälvd ‚gas ‚drīv 'rez·əv‚wär }

dissolved-gas reservoir See solution-gas reservoir. { də'zälvd ‚gas 'rez·əv‚wär }

dissolved load [HYD] Material carried in solution by a stream or river. { di¦zälvd 'lōd }

dissonance [ACOUS] An unpleasant combination of harmonics heard when certain musical tones are played simultaneously. Also known as discord. { 'dis·ə·nəns }

dissymmetrical network See dissymmetrical transducer. { ‚dis·ə'me·trə·kəl 'net‚wərk }

dissymmetrical transducer [ELECTR] A transducer whose input and output image impedances are not equal. Also known as dissymmetrical network. { ‚dis·ə'me·trə·kəl tranz'dü·sər }

dissymmetry [SCI TECH] Lack of symmetry. { di'sim·ə·trē }

dissymmetry coefficient [ANALY CHEM] Ratio of the intensities of scattered light at 45 and 135°, used to correct for destructive interference encountered in light-scattering-photometric analyses of liquid samples. { di'sim·ə·trē ‚kō·i'fish·ənt }

dissymmetry factor [OPTICS] A quantity which expresses the strength of circular dichroism, equal to the difference in the absorption indices for left and right circularly polarized light divided by the absorption index for ordinary light of the same wavelength. Also known as anisotropy factor. { di'sim·ə·trē ‚fak·tər }

Distacodidae [PALEON] A family of conodonts in the suborder Conodontiformes characterized as simple curved cones with deeply excavated attachment scars. { dis·tə'käd·ə,dē }

distal [BIOL] Located away from the point of origin or attachment. { 'dist·əl }

distal convoluted tubule [ANAT] The portion of the nephron in the vertebrate kidney lying between the loop of Henle and the collecting tubules. { 'dist·əl ,kän·və'lüd·əd 'tü·byül }

distance [MATH] **1.** A nonnegative number associated with pairs of geometric objects. **2.** The spatial separation of two points, measured by the length of a hypothetical line joining them. **3.** For two parallel lines, two skew lines, or two parallel planes, the length of a line joining the two objects and perpendicular to both. **4.** For a point and a line or plane, the length of the perpendicular from the point to the line or plane. [MECH] The spatial separation of two points, measured by the length of a hypothetical line joining them. { 'dis·təns }

distance-finding station [NAV] A radio beacon equipped with a synchronized sound signal to provide the pilot or marine with a means of determining distance from the source of the sound, by measuring the difference in the time of reception of the two signals; the sound may be transmitted through air or water and from the same location as the radio signal or a location remote from it. { 'dis·təns ,fīnd·iŋ ,stā·shən }

distance-luminosity relation [ASTRON] The relation in which the light intensity from a star is inversely proportional to the square of its distance. { ¦dis·təns lü·mə'näs·əd·ē ri'lā·shən }

distance mark [ELECTR] A movable point produced on a radar display by a special signal generator, so that when the mark is moved to a target position on the screen the range to the target can be read on the calibrated dial of the signal generator; usually used for gun laying where highly accurate distance is important. { 'dis·təns ,märk }

distance marker [ENG] One of a series of concentric circles, painted or otherwise fixed on the screen of a plan position indicator, from which the distance of a target from the radar antenna can be read directly; used for surveillance and navigation where the relative distances between a number of targets are required simultaneously. Also known as radar range marker; range marker. { 'dis·təns ,märk·ər }

distance marking light [NAV] An approach light indicating distance from the end of a runway, landing strip, or channel. { 'dis·təns ,märk·iŋ ,līt }

distance-measuring equipment [NAV] A radio aid to navigation that provides distance information by measuring total round-trip time of transmission from an airborne interrogator to a ground-based transponder and return. Abbreviated DME. { 'dis·təns ,mezh·ər·iŋ i'kwip·mənt }

distance modulus See modulus of distance. { 'dis·təns ,mäj·ə·ləs }

distance protection [ELEC] Effect of a device operative within a predetermined electrical distance on the protected circuit to cause and maintain an interruption of power in a faulty circuit. { 'dis·təns prə,tek·shən }

distance ratio [MECH ENG] The ratio of the distance moved by the effort or input of a machine in a specified time to the distance moved by the load or output. { 'dis·təns ,rā·shō }

distance reception [COMMUN] Reception of messages from, or communication with, distant radio stations. Abbreviated DX. { 'dis·təns ri'sep·shən }

distance relay [ELEC] Protective relay, the operation of which is a function of the distance between the relay and the point of fault. { 'dis·təns ,rē,lā }

distance resolution [ENG] The minimum radial distance by which targets must be separately distinguishable by a particular radar. Also known as range discrimination; range resolution. { 'dis·təns ,rez·ə,lü·shən }

distance/velocity lag [CONT SYS] The delay caused by the amount of time required to transport material or propagate a signal or condition from one point to another. Also known as transportation lag; transport lag. { ¦dis·təns və'läs·əd·ē ,lag }

distant early-warning line [ORD] Defense line of radar stations at about the 70th parallel on the North American continent. { 'dis·tənt ,ər·lē 'wórn·iŋ ,līn }

distant field [ELECTROMAG] The electromagnetic field at a distance of five wavelengths or more from a transmitter, where the radial electric field becomes negligible. { ¦dis·tənt ¦fēld }

distant signal [CIV ENG] A signal placed at a distance from a block of track to give advance warning when the block is closed. { ¦dis·tənt 'sig·nəl }

distemper [VET MED] Any of several contagious virus diseases of mammals, especially the form occurring in dogs, marked by fever, respiratory inflammation, and destruction of myelinated nerve tissue. { dis'tem·pər }

disthene See kyanite. { 'dis,thēn }

distichous [BIOL] Occurring in two vertical rows. { 'dis·tə·kəs }

distillate [CHEM] The products of distillation formed by condensing vapors. { 'dis·tə,lāt }

distillate fuel [MATER] Any one of the wide variety of fuels obtained from fractions boiling above the temperature at which gasoline comes off in the distillation of petroleum. { 'dis·tə,lāt ,fyül }

distillate fuel oil [MATER] A classification for one of the overhead fractions produced from crude oil in conventional distillation operations. { 'dis·tə,lāt 'fyül ,oil }

distillation [CHEM] The process of producing a gas or vapor from a liquid by heating the liquid in a vessel and collecting and condensing the vapors into liquids. { ,dis·tə'lā·shən }

distillation column [CHEM] A still for fractional distillation. { ,dis·tə'lā·shən ,käl·əm }

distillation curve [CHEM] The graphical plot of temperature versus overhead product (distillate) volume or weight for a distillation operation. { ,dis·tə'lā·shən ,kərv }

distillation loss [CHEM] In a laboratory distillation, the difference between the volume of liquid introduced into the distilling flask and the sum of the residue and condensate received. { ,dis·tə'lā·shən ,lós }

distillation range [CHEM] The difference between the temperature at the initial boiling point and at the end point of a distillation test. { ,dis·tə'lā·shən ,rānj }

distillation test [CHEM ENG] A standardized procedure for finding the initial, intermediate, and final boiling points in the boiling range of petroleum products. { ,dis·tə'lā·shən ,test }

distilled liquor [FOOD ENG] Alcoholic beverages obtained by distilling an alcohol-containing liquid such as wine or fermented fruit juice and then further treating the distillate to obtain a beverage of specific character. Also known as hard liquor. { də'stild 'lik·ər }

distilled mustard gas [ORG CHEM] A delayed-action casualty gas (mustard gas) that has been distilled, or purified, to greatly reduce the odor and thereby increase its difficulty of detection. { də'stild 'məs·tərd ,gas }

distilled water [CHEM] Water that has been freed of dissolved or suspended solids and organisms by distillation. { də'stild 'wód·ər }

distillery [FOOD ENG] The building where distillation of alcoholic beverages occurs. { də'stil·ə·rē }

distilling flask [CHEM] A round-bottomed glass flask that is capable of holding a liquid to be distilled. { də'stil·iŋ ,flask }

distocclusion [MED] Malocclusion of the teeth in which those of the lower jaw are in distal relation to the upper teeth. { ¦dis·tə¦klü·zhən }

distome [INV ZOO] A digenetic trematode characterized by possession of an oral and a ventral sucker. { 'dī,stōm }

distorted water [METEOROL] A multimolecular layer of water, at the boundary between a mass of liquid water and the surrounding vapor, whose structure is not identical with that of bulk water. { di'stórd·əd 'wód·ər }

distortion [ELECTR] Any undesired change in the waveform of an electric signal passing through a circuit or other transmission medium. [ENG] In general, the extent to which a system fails to accurately reproduce the characteristics of an input signal at its output. [ENG ACOUS] Any undesired change in the waveform of a sound wave. [OPTICS] A type of aberration in which there is variation in magnification with the distance from the axis of an optical system, so that images are not geometrically similar to their objects. { di'stór·shən }

distortional wave See S wave. { di'stór·shən·əl 'wāv }

distortion factor [COMMUN] Ratio of the effective value of the residue of a wave after elimination of the fundamental to the effective value of the original wave. { di'stór·shən ,fak·tər }

distortion meter [ENG] An instrument that provides a visual indication of the harmonic content of an audio-frequency wave. { di'stór·shən ,mēd·ər }

distrail See dissipation trail. { 'dis,trāl }

distress frequency [COMMUN] A frequency allotted to distress calls, generally by international agreement; for ships at sea and aircraft over the sea, it is 500 kilohertz. { də'stres ˌfrē·kwən·sē }

distress signal [COMMUN] An international signal used when a ship, aircraft, or other vehicle is threatened by grave and imminent danger and requests immediate assistance; examples are special radiotelegraph and radiotelephone signals or special signal flags or flares. { də'stres ˌsig·nəl }

distributary [HYD] An irregular branch flowing out from a main stream and not returning to it, as in a delta. Also known as distributary channel. { də'strib·yəˌter·ē }

distributary channel See distributary. { də'strib·yəˌter·ē ˌchan·əl }

distributed amplifier [ELECTR] A wide-band amplifier in which tubes are distributed along artificial delay lines made up of coils acting with the input and output capacitances of the tubes. { di'strib·yəd·əd 'am·pləˌfī·ər }

distributed bulletin board [COMPUT SCI] A collection of newsgroups on a wide-area network, whose postings are available to every user. { diˌstrib·yəd·əd 'bul·ət·ən ˌbȯrd }

distributed capacitance [ELEC] Capacitance that exists between the turns in a coil or choke, or between adjacent conductors or circuits, as distinguished from the capacitance concentrated in a capacitor. { di'strib·yəd·əd kə'pas·əd·əns }

distributed circuit [ELECTR] A film circuit whose effective components cannot be easily recognized as discrete. { di'strib·yəd·əd 'sər·kət }

distributed collector [ENG] A component of a solar heating system comprising a series of modular focusing collectors that are interconnected with an absorber pipe network to carry the working fluid to a heat exchanger. { di'strib·yəd·əd kə'lek·tər }

distributed communications [COMMUN] Information transfer beyond the local level that may involve the originating source to transmit information to all communications centers on any one network, and may also cause an interchange of communications among several whole networks. { di'strib·yəd·əd kə'myü·nəˌkā·shənz }

distributed computing [COMPUT SCI] The use of multiple network-connected computers for solving a problem or for information processing. { diˌstrib·yəd·əd kəm'pyüd·iŋ }

distributed constant [ELECTROMAG] A circuit parameter that exists along the entire length of a transmission line. Also known as distributed parameter. { di'strib·yəd·əd 'kän·stənt }

distributed control system [CONT SYS] A collection of modules, each with its own specific function, interconnected tightly to carry out an integrated data acquisition and control application. { di'strib·yəd·əd kən'trōlˌsis·təm }

distributed database [COMPUT SCI] A database maintained in physically separated locations and supported by a computer network so that it is possible to access all parts of the database from various points in the network. { di'strib·yəd·əd 'dad·əˌbās }

distributed-emission photodiode [ELECTR] A broad-band photodiode proposed for detection of modulated laser beams at millimeter wavelengths; incident light falls on a photocathode strip that generates a traveling wave of photocurrent having the same wave velocity as the transmission line which the photodiode feeds. { di'strib·yəd·əd ə'mish·ən ˌfōd·ō·dī·ōd }

distributed fault See fault zone. { di'strib·yəd·əd ˌfȯlt }

distributed free space [COMPUT SCI] Empty spaces in a data layout to allow new data to be inserted at a future time. { di'strib·yəd·əd ˌfrē ˌspās }

distributed inductance [ELECTROMAG] The inductance that exists along the entire length of a conductor, as distinguished from inductance concentrated in a coil. { di'strib·yəd·əd in'dək·təns }

distributed intelligence [COMPUT SCI] The existence of processing capability in terminals and other peripheral devices of a computer system. Also known as distributed logic. { di'strib·yəd·əd in'tel·ə·jəns }

distributed logic See distributed intelligence. { di'strib·yəd·əd 'läj·ik }

distributed logic cluster word processor [COMPUT SCI] A system of word processors each of which can operate independently, although printers are generally shared by a number of terminals. { di'strib·yəd·əd 'läj·ikˌkləs·tər 'wərd ˌpräs·es·ər }

distributed network [COMMUN] A communications network in which there exist alternative routings between the various nodes. [COMPUT SCI] A computer network in which at least some of the processing is done at individual work stations and information is shared by and often stored at the work stations. { di'strib·yəd·əd 'netˌwərk }

distributed numerical control [CONT SYS] The use of central computers to distribute part-classification data to machine tools which themselves are controlled by computers or numerical control tapes. { di'strib·yəd·əd nü'mer·ə·kəl kən'trōl }

distributed parameter See distributed constant. { di'strib·yəd·əd pə'ram·əd·ər }

distributed-parameter system See distributed system. { di'strib·yəd·əd pə'ram·əd·ər ˌsis·təm }

distributed paramp [ELECTR] Paramagnetic amplifier that consists essentially of a transmission line shunted by uniformly spaced, identical varactors; the applied pumping wave excites the varactors in sequence to give the desired traveling-wave effect. { di'strib·yəd·əd 'parˌamp }

distributed processing system [COMPUT SCI] An information processing system consisting of two or more programmable devices, connected so that information can be exchanged. { di'strib·yəd·əd 'präs·es·iŋ ˌsis·təm }

distributed system [COMPUT SCI] A computer system consisting of a collection of autonomous computers linked by a network and equipped with software that enables the computers to coordinate their activities and to share the resources of system hardware, software, and data, so that users perceive a single, integrated computing facility. [CONT SYS] A collection of modules, each with its own specific function, interconnected to carry out integrated data acquisition and control in a critical environment. [SYS ENG] A system whose behavior is governed by partial differential equations, and not merely ordinary differential equations. Also known as distributed-parameter system. { di'strib·yəd·əd 'sis·təm }

distributing frame [ELECTR] Structure for terminating permanent wires of a central office, private branch exchange, or private exchange, and for permitting the easy change of connections between them by means of cross-connecting wires. { di'strib·yəd·iŋ ˌfrām }

distributing point [ORD] Point at which supplies, obtained from the supply point by a division or other unit, are broken down for immediate distribution to subordinate units. { di'strib·yəd·iŋ ˌpoint }

distributing roller [GRAPHICS] A rubber-covered roller whose function is to deliver ink from the supply reservoir to the ink drum of the printing press. { di'strib·yəd·iŋ ˌrōl·ər }

distributing terminal assembly [ELECTR] Frame situated between each pair of selector bays to provide terminal facilities for the selector bank wiring and facilities for cross-connection to trunks running to succeeding switches. { di'strib·yəd·iŋ 'term·ən·əl əˌsem·blē }

distribution [IND ENG] All activities that involve efficient movement of finished products from the end of the production line to the consumer. [MATH] An abstract object which generalizes the idea of function; used in applied mathematics, quantum theory, and probability theory; the delta function is an example. Also known as generalized function. [STAT] For a discrete random variable, a function (or table) which assigns to each possible value of the random variable the probability that this value will occur; for a continuous random variable x, the monotone nondecreasing function which assigns to each real t the probability that x is less than or equal to t. Also known as distribution function; probability distribution; statistical distribution. { ˌdis·trə'byü·shən }

distribution amplifier [ELECTR] A radio-frequency power amplifier used to feed television or radio signals to a number of receivers, as in an apartment house or a hotel. [ENG ACOUS] An audio-frequency power amplifier used to feed a speech or music distribution system and having sufficiently low output impedance so changes in load do not appreciably affect the output voltage. { ˌdis·trə'byü·shən 'am·pləˌfī·ər }

distribution box [CIV ENG] In sanitary engineering, a box in which the flow of effluent from a septic tank is distributed equally into the lines that lead to the absorption field. { ˌdis·trə'byü·shən 'bäks }

distribution cable [ELEC] Cable extending from a feeder cable into a specific area for the purpose of providing service to that area. { ˌdis·trə'byü·shən ˌkā·bəl }

distribution center [ELEC] In an alternating-current power system, the point at which control and routing equipment is installed. { ‚dis·trə'byü·shən ‚sen·tər }

distribution coefficient [OPTICS] One of the tristimulus values of monochromatic radiations having equal power, usually denoted by x, y, z. [PHYS CHEM] The ratio of the amounts of solute dissolved in two immiscible liquids at equilibrium. { ‚dis·trə'byü·shən ‚kō·i'fish·ənt }

distribution control *See* linearity control. { ‚dis·trə'byü·shən kən'trōl }

distribution curve [STAT] The graph of the distribution function of a random variable. { ‚dis·trə'byü·shən 'kərv }

distribution deck [COMPUT SCI] A card file which duplicates all or part of a master card file and used for disseminating or decentralizing. { ‚dis·trə'byü·shən ‚dek }

distribution factor [NUCLEO] A term used to express the modification of the effect of radiation in a biological system attributable to the nonuniform distribution of an internally deposited isotope, such as radium's being concentrated in bones. { ‚dis·trə'byü·shən ‚fak·tər }

distribution frame [COMMUN] A place where a number of cables converge and signals are redistributed among them. { ‚dis·trə'byü·shən ‚frām }

distribution-free method [STAT] Any method of inference that does not depend on the characteristics of the population from which the samples are obtained. { ‚dis·trə'byü·shən ‚frē ‚meth·əd }

distribution function *See* distribution. { ‚dis·trə'byü·shən ‚fəŋk·shən }

distribution graph [HYD] A statistically derived hydrograph for a storm of specified duration, graphically representing the percent of total direct runoff passing a point on a stream, as a function of time; usually presented as a histogram or table of percent runoff within each of successive short time intervals. { ‚dis·trə'byü·shən ‚graf }

distribution law [ANALY CHEM] The law stating that if a substance is dissolved in two immiscible liquids, the ratio of its concentration in each is constant. [STAT MECH] A law which gives a density function specifying the probability of finding a particle in a unit volume of phase space, or the number of particles in each of the states which a particle may occupy, or the number of particles per unit volume of phase space. { ‚dis·trə'byü·shən ‚lō }

distribution ratio [ANALY CHEM] The ratio of the concentrations of a given solute in equal volumes of two immiscible solvents after the mixture has been shaken and equilibrium established. { ‚di·strə'byü·shən ‚rā·shō }

distribution photometer *See* light-distribution photometer. { ‚dis·trə'byü·shən fō'täm·əd·ər }

distribution reservoir [CIV ENG] A service reservoir connected with the conduits of a primary water supply; used to supply water to consumers according to fluctuations in demand over short time periods and serves for local storage in case of emergency. { ‚dis·trə'byü·shən 'rez·əv‚wär }

distribution substation [ELEC] An electric power substation associated with the distribution system and the primary feeders for supply to residential, commercial, and industrial loads. { ‚dis·trə'byü·shən 'səb‚stā·shən }

distribution switchboard [ELEC] Power switchboard used for the distribution of electrical energy at the voltage common for each distribution within a building. { ‚dis·trə'byü·shən 'swich‚bōrd }

distribution system [ELEC] Circuitry involving high-voltage switchgear, step-down transformers, voltage dividers, and related equipment used to receive high-voltage electricity from a primary source and redistribute it at lower voltages. Also known as electric distribution system. { ‚dis·trə'byü·shən ‚sis·təm }

distribution transformer [ELEC] An element of an electric distribution system located near consumers which changes primary distribution voltage to secondary distribution voltage. { ‚dis·trə'byü·shən tranz'fōr·mər }

distributive analysis and synthesis [PSYCH] A method of therapy that entails extensive guided and directed investigation and analysis of an individual's past experience, stressing the individual's assets and liabilities. { di‚strib·yəd·iv ə¦nal·ə·səs ən 'sin·thə·səs }

distributive fault *See* step fault. { di'strib·yəd·iv 'fólt }

distributive lattice [MATH] A lattice in which "greatest

lower bound" obeys a distributive law with respect to "least upper bound," and vice versa. { di'strib·yəd·iv 'lad·əs }

distributive law [MATH] A rule which stipulates how two binary operations on a set shall behave with respect to one another; in particular, if +, ∘ are two such operations then ∘ distributes over + means $a \circ (b + c) = (a \circ b) + (a \circ c)$ for all a,b,c in the set. { di'strib·yəd·iv 'lō }

distributor [ELEC] **1.** Any device which allocates a telegraph line to each of a number of channels, or to each row of holes on a punched tape, in succession. **2.** A rotary switch that directs the high-voltage ignition current in the proper firing sequence to the various cylinders of an internal combustion engine. [ELECTR] The electronic circuitry which acts as an intermediate link between the accumulator and drum storage. [ENG] A device for delivering an exact amount of fuel at the exact time at which it is required. { də'strib·yəd·ər }

distributor gear [MECH ENG] A gear which meshes with the camshaft gear to rotate the distributor shaft. { də'strib·yəd·ər ‚gir }

distributor points [ELEC] Cam-operated contacts, the opening of which triggers the ignition pulse in an internal combustion engine. { də'strib·yəd·ər ‚póins }

district forecast [METEOROL] In U.S. Weather Bureau usage, a general weather forecast for conditions over an established geographical "forecast district." { 'di·strikt 'fór‚kast }

district heating [MECH ENG] The supply of heat, either in the form of steam or hot water, from a central source to a group of buildings. { 'di·strikt 'hēd·iŋ }

disturbance [COMMUN] An undesired interference or noise signal affecting radio, television, or facsimile reception. [CONT SYS] An undesired command signal in a control system. [GEOL] Folding or faulting of rock or a stratum from its original position. [METEOROL] **1.** Any low or cyclone, but usually one that is relatively small in size and effect. **2.** An area where weather, wind, pressure, and so on show signs of the development of cyclonic circulation. **3.** Any deviation in flow or pressure that is associated with a disturbed state of the weather, such as cloudiness and precipitation. **4.** Any individual circulatory system within the primary circulation of the atmosphere. { də'stər·bəns }

disturbance climax *See* disclimax. { də'stər·bəns 'klī‚maks }

disturbance compensation system [AERO ENG] A system applied to navigational satellites to remove the along-the-track component of drag and radiation forces. Abbreviated DISCOS. { də'stər·bəns käm·pən'sā·shən ‚sis·təm }

disturbed-one output [ELECTR] One output of a magnetic cell to which partial-read pulses have been applied since that cell was last selected for writing. { də¦stərbd ¦wən 'aùt‚pùt }

disturbed-sun noise [ASTROPHYS] Noise at times of sunspot or solar flare activity. { də¦stərbd 'sən ‚nóiz }

disubstituted alkene [ORG CHEM] An alkene with the general formula $R_2C=CH_2$ or $RHC=CHR$, where R is any organic group; a carbon atom is bonded directly to each end of the double bond. { dī¦səb·stə‚tüd·əd 'al‚kēn }

disulfate [CHEM] A compound that has two sulfate radicals. { dī'səl‚fāt }

disulfide [CHEM] **1.** A compound that has two sulfur atoms bonded to a radical or element. **2.** One of a group of organosulfur compounds RSSR′ that may be symmetrical (R=R′) or unsymmetrical (R and R′, different). { dī'səl‚fīd }

disulfide bond *See* disulfide bridge. { dī¦səl‚fīd 'bänd }

disulfide bridge [ORG CHEM] A sulfur-to-sulfur bond linking the sulfur atoms of two polypeptide chains. Also known as disulfide bond. { dī¦səl‚fīd 'brij }

disulfide oil [MATER] One of the oils obtained by oxidizing to disulfides the mercaptans extracted from light petroleum distillates; the disulfides separate from the extract as an oily layer. { dī'səl‚fīd ‚oil }

disulfiram [PHARM] $C_{10}H_{20}N_2S_4$ A drug used to treat alcohol abuse that blocks the metabolism of acetaldehyde, the major metabolite of ethanol, causing a rapid buildup of acetaldehyde and a severe physiological syndrome intended to prevent or modify further immediate drinking behavior. Also known as Antabuse. { di'səl·fə‚ram }

disulfonate [CHEM] A molecule that has two sulfonate groups. { dī'səl·fə‚nāt }

disulfonic acid [CHEM] A molecule that has two sulfonic acid groups. { ‚dī'səl'fän·ik 'as·əd }

ditch [CIV ENG] **1.** A small artificial channel cut through earth or rock to carry water for irrigation or drainage. **2.** A long narrow cut made in the earth to bury pipeline, cable, or similar installations. [PETRO ENG] On a drilling rig, a mudflow trench leading from the conductor-pipe outlet. { dich }

ditch check [CIV ENG] A small dam positioned at intervals in a road ditch to prevent erosion. { 'dich ˌchek }

ditcher See trench excavator. { 'dich·ər }

ditching [AERO ENG] A forced landing on water, or the process of making such a landing. [ENG] The digging of ditches, as around storage tanks or process areas to hold liquids in the event of a spill or along the sides of a roadway for drainage. { 'dich·iŋ }

diterpene [ORG CHEM] $C_{20}H_{32}$ **1.** A group of terpenes that have twice as many atoms in the molecule as monoterpenes. **2.** Any derivative of diterpene. { dī'tər‚pēn }

dither [COMMUN] A technique for representing the entire gray scale of a picture by picture elements with only one of two levels ("white" and "black"), in which a multilevel input image signal is compared with a position-dependent set of thresholds, and picture elements are set to "white" only where the image input signal exceeds the threshold. [CONT SYS] A force having a controlled amplitude and frequency, applied continuously to a device driven by a servomotor so that the device is constantly in small-amplitude motion and cannot stick at its null position. Also known as buzz. { 'dith·ər }

dither matrix [COMMUN] A square matrix of threshold values that is repeated as a regular array to provide a threshold pattern for an entire image in the dither method of image representation. { 'dith·ər ˌmā·triks }

dithiocarbamate [ORG CHEM] **1.** A salt of dithiocarbamic acid. **2.** Any other derivative of dithiocarbamic acid. { ¦dī‚thī·ō'kär·bə‚māt }

dithiocarbamic acid [ORG CHEM] NH_2CS_2H A colorless, unstable powder; various metal salts are readily obtained, and used as strong accelerators for rubber. Also known as aminodithioformic acid. { ¦dī‚thī·ō‚kär'bam·ik 'as·əd }

dithioic acid [ORG CHEM] An organic acid in which sulfur atoms have replaced both oxygen atoms of the carboxy group. { ¦dī‚thī¦ō·ik 'as·əd }

dithionate [CHEM] Any salt formed from dithionic acid. { dī'thī·ə‚nāt }

dithionic acid [INORG CHEM] $H_2S_2O_6$ A strong acid formed by the oxidation of sulfurous acid, and known only by its salts and in solution. { ¦dī‚thī'än·ik 'as·əd }

dithiooxamide [ORG CHEM] $NH_2CSCSNH_2$ Red crystals soluble in alcohol; used as a reagent for copper, cobalt, and nickel, and for the determination of osmium. { ¦dī‚thī·ō'äk·sə‚mīd }

1,4-dithiothreitol [ORG CHEM] $C_4H_{10}O_2S_2$ Needlelike crystals soluble in water, ethanol acetone, ethylacetate; used as a protective agent for thiol (SH) groups. { ¦wən ¦fȯr ¦dī‚thī·ō'thrē·ə‚tȯl }

ditokous [VERT ZOO] Producing two eggs or giving birth to two young at one time. { 'did·ə·kəs }

Dittus-Boelter equation [FL MECH] An equation used to calculate the surface coefficient of heat transfer for fluids in turbulent flow inside clean, round pipes. { ¦did·əs ¦bel·tər iˌkwā·zhən }

ditty box [NAV ARCH] A small box with a hinged lid and lock used by a crew member for personal belongings. { 'did·ē ‚bäks }

ditungsten carbide [INORG CHEM] W_2C A gray powder having hardness approaching that of diamond; forms hexagonal crystals with specific gravity 17.2; melting point 2850°C. { 'dī‚təŋ·stən 'kär‚bīd }

diuresis [MED] Increased excretion of urine. { ‚dī·yù'rē·səs }

diuretic [PHARM] Any agent that increases the volume and flow of urine. { ‚dī·yù'red·ik }

diuretic hormone [BIOCHEM] A neurohormone that promotes water loss in insects by increasing the volume of fluid secreted into the Malpighian tubules. { ‚dī·yù'red·ik 'hȯr‚mōn }

diurnal [BIOL] Active during daylight hours. [SCI TECH] Occurring during the daytime. { dī'ərn·əl }

diurnal aberration [ASTRON] Aberration caused by the rotation of the earth; its value varies with the latitude of the observer

and ranges from zero at the poles to 0.31 second of arc. Also known as daily aberration. { dī'ərn·əl ‚ab·ə'rā·shən }

diurnal age See age of diurnal inequality. { dī'ərn·əl 'āj }

diurnal arc [ASTRON] That part of a celestial body's diurnal circle which lies above the horizon of the observer. { dī'ərn·əl 'ärk }

diurnal circle [ASTRON] The apparent daily path of a celestial body, approximating a parallel of declination. { dī'ərn·əl 'sər·kəl }

diurnal inequality [OCEANOGR] The difference between the heights of the two high waters or the two low waters of a lunar day. { dī'ərn·əl ‚in·ə'kwäl·əd·ē }

diurnal migration [BIOL] The daily rhythmic movements of organisms in the sea from deeper water to the surface at the approach of darkness and their return to deeper water before dawn. { dī'ərn·əl mī'grā·shən }

diurnal motion [ASTRON] The apparent daily motion of a celestial body as observed from a rotating body. { dī'ərn·əl 'mō·shən }

diurnal parallax See geocentric parallax. { dī'ərn·əl 'par·ə‚laks }

diurnal range See great diurnal range. { dī'ərn·əl 'rānj }

diurnal tide [OCEANOGR] A tide in which there is only one high water and one low water each lunar day. { dī'ərn·əl 'tīd }

diurnal variation [GEOPHYS] Daily variations of the earth's magnetic field at a given point on the surface, with both solar and lunar periods having their source in the horizontal movements of air in the ionosphere. { dī'ərn·əl ‚ver·ē'ā·shən }

divagation [HYD] Lateral shifting of the course of a stream caused by extensive deposition of alluvium in its bed and frequently accompanied by the development of meanders. { ‚div·ə'gā·shən }

divalent carbon [ORG CHEM] A charged or uncharged carbon atom that has formed only two covalent bonds. Also known as dicovalent carbon. { dī'vā·lənt 'kär·bən }

divalent metal [CHEM] A metal whose atoms are each capable of chemically combining with two atoms of hydrogen. { dī'vā·lənt 'med·əl }

divariant system [THERMO] A system composed of only one phase, so that two variables, such as pressure and temperature, are sufficient to define its thermodynamic state. { dī'ver·ē·ənt 'sis·təm }

divaricate [BIOL] Broadly divergent and spread apart. { dī'var·ə‚kāt }

divaricator [ZOO] A muscle that causes separation of parts, as of brachiopod shells. { dī'var·ə‚kād·ər }

dive [AERO ENG] A rapid descent by an aircraft or missile, nose downward, with or without power or thrust. [ENG] To submerge into an underwater environment so that it may be studied or utilized; includes the use of specialized equipment such as scuba, diving helmets, diving suits, diving bells, and underwater research vessels. [NAV] To submerge a submarine under power. { dīv }

dive bomber [AERO ENG] An aircraft designed to release bombs during a steep dive. { 'dīv ‚bäm·ər }

dive brake [AERO ENG] An air brake designed for operation in a dive; flaps at the following edge of one wing that can be extended into the airstream to increase drag and hold the aircraft to its "never exceed" dive speed in a vertical dive; used on dive bombers and sailplanes. { 'dīv ‚brāk }

divergence [ELECTR] The spreading of a cathode-ray stream due to repulsion of like charges (electrons). [FL MECH] The ratio of the area of any section of fluid emerging from a nozzle to the area of the throat of the nozzle. [MATH] For a vector-valued function, the sum of the diagonal entries of the Jacobian matrix; it is the scalar product of the del operator and the vector. [METEOROL] The two-dimensional horizontal divergence of the velocity field. [NUCLEO] In a nuclear reactor, the condition wherein the number of neutrons produced increases in each succeeding generation. [OCEANOGR] A horizontal flow of water, in different directions, from a common center or zone. [PHYS] The spreading apart of a beam of particles of light. { də'vər·jəns }

divergence loss [GEOPHYS] During geophysical prospecting, the portion of the power lost in transmitting signals that is caused by the spreading of seismic or sound rays by the geometry of the geologic features. { də'vər·jəns ‚lȯs }

divergence speed [AERO ENG] The speed of an aircraft above which no statically stable equilibrium condition exists and the deformation will increase to a point of structural failure. { də'vər·jəns ,spēd }

divergence theorem *See* Gauss' theorem. { də'vər·jəns ,thir·əm }

divergent adaptation [EVOL] Adaptation to different kinds of environment that results in divergence from a common ancestral form. Also known as branching adaptation; cladogenic adaptation. { də'vər·jənt ,ad,ap'tā·shən }

divergent I-beam technique [PHYS] A method of x-ray diffraction analysis in which a divergent beam of x-rays is used to produce Kossel lines. { də'vər·jent 'ī,bēm 'tek'nēk }

divergent die [ENG] A die with the internal channels that lead to the orifice diverging, such as the dies used for manufacture of hollow-body plastic items. { də'vər·jənt 'dī }

divergent integral [MATH] An improper integral which does not have a finite value. { də'vər·jənt 'in·tə·grəl }

divergent nozzle [DES ENG] A nozzle whose cross section becomes larger in the direction of flow. { də'vər·jənt 'näz·əl }

divergent sequence [MATH] A sequence which does not converge. { də'vər·jənt 'sē·kwəns }

divergent series [MATH] An infinite series whose sequence of partial sums does not converge. { də'vər·jənt 'sir·ēz }

divergent transcription [MOL BIO] The initiation of genetic transcription at two promoters that are facing in opposite directions. { də'vər·jənt ,tran'skrip·shən }

diverging duct [DES ENG] Fluid-flow conduit whose internal cross-sectional area increases in the direction of flow. { də'vərj·iŋ ,dəkt }

diverging lens [OPTICS] A lens whose focal length is negative, so that light incident parallel to its axis diverges after passing through it. Also known as negative lens. { də'vərj·iŋ ,lenz }

diverging meniscus lens *See* negative meniscus lens. { də'vərj·iŋ mə'nis·kəs ,lenz }

diverging mirror [OPTICS] A convex mirror that causes rays of light parallel to its axis to diverge. Also known as negative mirror. { də'vərj·iŋ ,mir·ər }

diverging yaw [AERO ENG] In the flight of a projectile, an angle of yaw increasing from the initial yaw, so that the projectile is unstable. { də'vərj·iŋ 'yȯ }

diver method [PHYS CHEM] Measure of the size of suspended solid particles; small glass divers of known density sink to the level where the liquid-suspension density is equal to that of the diver, allowing calculation of particle size. Also known as Berg's diver method. { 'dī·vər ,meth·əd }

diverse vector [NAV] In air operations, an instruction issued by a radar controller to fly a specific course which is not a part of the predetermined radar pattern. Also known as random vector. { də'vərs 'vek·tər }

diversionary missile *See* missile decoy. { də'vər·zhə,ner·ē 'mis·əl }

diversion canal [CIV ENG] An artificial channel for diverting water from one place to another. { də'vər·zhən kə,nal }

diversion chamber [ENG] A chamber designed to direct a stream into a channel or channels. { də'vər·zhən ,chām·bər }

diversion dam [CIV ENG] A fixed dam for diverting stream water away from its course. { də'vər·zhən ,dam }

diversion gate [CIV ENG] A gate which may be closed to divert water from the main conduit or canal to a lateral or some other channel. { də'vər·zhən ,gāt }

diversion tunnel [CIV ENG] An underground passageway used to divert flowing water around a construction site. { də'vər·zhən ,tən·əl }

diversity [COMMUN] Method of signal extraction by which an optimum resultant signal is derived from a combination of, or selection from, a plurality of transmission paths, channels, techniques, or physical arrangements; the system may employ space diversity, polarization diversity, frequency diversity, or any other arrangement by which a choice can be made between signals. { də'vər·səd·ē }

diversity factor [ELEC] Ratio of the sum of the individual maximum demands to total maximum demand, as applied to an electrical distribution system. { də'vər·səd·ē ,fak·tər }

diversity gain [COMMUN] Gain in reception as a result of the use of two or more receiving antennas. { də'vər·səd·ē ,gān }

diversity radar [ENG] A radar that uses two or more transmitters and receivers, each pair operating at a slightly different frequency but sharing a common antenna and video display,

to obtain greater effective range and reduce susceptibility to jamming. { də'vər·səd·ē 'rā,där }

diversity receiver [ELECTR] A radio receiver designed for space or frequency diversity reception. { də'vər·səd·ē ri'sē·vər }

diversity reception [COMMUN] Radio reception in which the effects of fading are minimized by combining two or more sources of signal energy carrying the same modulation. { də'vər·səd·ē ri'sep·shən }

diverter [ELEC] A low resistance which is connected in parallel with the series or commutating pole winding of a direct-current machine and diverts current from it, causing the magnetomotive force produced by the winding to vary. [PETRO ENG] **1.** Equipment installed in a well for controlling a well blowout occurring at a relatively shallow depth. **2.** A system designed to protect a floating rig during a blowout by conducting the flow away from the rig. { də'vərd·ər }

diverter-pole generator [ELEC] Compound wound direct-current generator with the series winding of the diverter pole opposing the flux generated by the shunt wound main pole; provides a close voltage regulation. { də'vərd·ər ,pōl 'jen·ə,rād·ər }

diverter valve *See* air bypass valve. { də'vərd·ər ,valv }

diverticulitis [MED] Inflammation of a diverticulum. { ,dī·vər,tik·yə'līd·əs }

diverticulosis [MED] Presence of many diverticula in the intestine. { ,dī·vər,tik·yə'lō·səs }

diverticulum [MED] An abnormal outpocketing or sac on the wall of a hollow organ. { ,dī·vər'tik·yə·ləm }

diverting agent [PETRO ENG] A viscous gel or suspension of graded solids used during acidizing of an oil reservoir to temporarily block off the most permeable sections of the pay zone to force the acid into less permeable sections. { də'vərd·iŋ ,ā·jənt }

Divesian *See* Oxfordian. { də'vēzh·ən }

divide [GEOGR] A ridge or section of high ground between drainage systems. [MATH] One object (integer, polynomial) divides another if their quotient is an object of the same type. [SCI TECH] A point or line of division. { də'vīd }

divide-and-conquer relation [MATH] A recurrence relation which expresses the value of a number-theoretic function for an argument n in terms of its value for an argument n/b, where b is an integer greater than 1. { di¦vīd ən 'käŋ·kər ri,lā·shən }

divide check [COMPUT SCI] An error signal indicating that an illegal division (such as dividing by zero) was attempted. { də'vīd ,chek }

divided differences [MATH] Quantities which are used in the interpolation or numerical calculation or integration of a function when the function is known at a series of points which are not equally spaced, and which are formed by various operations on the difference between the values of the function at successive points. { də'vīd·əd 'dif·rən·səs }

divided lane [CIV ENG] A highway divided into lanes by a median strip. { də'vīd·əd 'lān }

divided pitch [DES ENG] In a screw with multiple threads, the distance between corresponding points on two adjacent threads measured parallel to the axis. { də'vīd·əd 'pich }

divided slit scan [COMPUT SCI] In optical character recognition, a device consisting of a narrow column of photoelectric cells which scans an input character at given intervals for the purpose of obtaining its horizontal and vertical components. { də'vīd·əd 'slit ,skan }

dividend [MATH] A quantity which is divided by another quantity in the operation of division. { 'div·ə,dend }

divider [DES ENG] A tool like a compass, used in metalworking to lay out circles or arcs and to space holes or other dimensions. { də'vīd·ər }

dividing network *See* crossover network. { də'vīd·iŋ ,net,wərk }

divine proportion *See* golden section. { di,vīn prə'pȯr·shən }

diving bell [ENG] An early diving apparatus constructed in the shape of a box or cylinder without a bottom and connected to a compressed-air hose. { 'dīv·iŋ ,bel }

diving bird [VERT ZOO] Any bird adapted for diving and swimming, including loons, grebes, and divers. { 'dīv·iŋ ,bərd }

diving plane *See* diving rudder. { 'dīv·iŋ ,plān }

diving rudder [NAV ARCH] One of the movable plane surfaces attached to the hull of a submarine that control the vertical

DIVING BIRD

The common loon (*Gavia immer*).

motion of a submarine. Also known as diving plane; hydroplane. { 'dīv·iŋ ,rad·ər }

diving suit [ENG] A waterproof outfit designed for diving, especially one with a helmet connected to a compressed-air hose. { 'dīv·iŋ ,süt }

divining [MIN ENG] An unscientific method for searching for subsurface water or minerals by means of a divining rod. Also known as dowsing. { də'vīn·iŋ }

divining rod [MIN ENG] An unscientific device in the form of a forked rod or tree branch that is supposed to dip when held over water or minerals, depending on the specialty of the operator, or dowser. Also known as dowsing rod; wiggle stick. { də'vīn·iŋ ,räd }

divinyl [ORG CHEM] **1.** A molecule that has two vinyl groups. **2.** See 1,3-butadiene. { dī'vīn·əl }

divinyl acetylene [ORG CHEM] C_6H_6 A linear trimer of acetylene, made by passing acetylene into a hydrochloric acid solution that has metallic catalysts; used as an intermediate in neoprene manufacture. { dī'vīn·əl ə'sed·əl,ēn }

divinylbenzene [ORG CHEM] $C_6H_4(CHCH_2)_2$ Polymerizable, water-white liquid used to make rubbers, drying oils, and ion-exchange resins and other polymers; forms include ortho, meta, and para isomers. Also known as vinylstyrene. { dī'vīn·əl'ben,zēn }

divinyl ether See vinyl ether. { dī'vīn·əl 'ē·thər }

divinyl oxide See vinyl ether. { dī'vīn·əl 'äk,sīd }

division [COMPUT SCI] One of four required parts of a COBOL program, labeled identification, environment, data, and procedure, each with a set of rules governing the contents. [MATH] The inverse operation of multiplication; the number a divided by the number b is the number c such that b multiplied by c is equal to a. { də'vizh·ən }

division algebra [MATH] A hypercomplex system that is also a skew field. { də'vizh·ən ,al·jə·brə }

division algorithm [MATH] The theorem that, for any integer m and any positive integer n, there exist unique integers q and r such that $m = qn + r$ and r is equal to or greater than 0 and less than n. { di'vizh·ən 'al·gə,rith·əm }

division circle [ASTRON] A large circular structure attached to the horizontal axis of a transit circle with accurately calibrated markings; used to determine the inclination of the instrument. { də'vizh·ən ,sər·kəl }

division modulo p [MATH] Division in the finite field with p elements, where p is a prime number. { də'vizh·ən ¦mäj·ə·lō 'pē }

division I of meiosis [CELL MOL] The first nuclear division of meiosis, which results in two daughter cells, each containing either the maternal or paternal chromosome from each homologous pair; it is divided into four stages: prophase I (homologous chromosome pairing, synapsis, and crossing over), metaphase I (interlocked homologous chromosomes line up on the middle of the meiotic spindle), anaphase I (homologous chromosome pairs separate and move to opposite poles), and telephase I (nuclear envelope reforms around each daughter nucleus). { də,vizh·ən ¦wən əv mī'ō·səs }

division II of meiosis [CELL MOL] The second division stage of meiosis, in which the daughter cells resulting from division I divide themselves to produce a total of four gametes, each with a haploid number of cells; it is divided into four stages: prophase II (nuclear envelope breaks down and new meiotic spindle forms), metaphase II (chromosomes recondense and align themselves on a new pair of spindles), anaphase II (separation of sister centromeres and movement of the two sister chromatids to opposite poles), and telophase II (the nuclei begin to reform and the second cell division occurs). { də,vizh·ən ¦tü əv mī'ō·səs }

division plate [MECH ENG] A diaphragm which surrounds the piston rod of a crosshead-type engine and separates the crankcase from the lower portion of the cylinder. { də'vizh·ən ,plāt }

division ring [MATH] **1.** A ring in which the set of nonzero elements form a group under multiplication. **2.** More generally, a nonassociative ring with nonzero elements in which, for any two elements a and b, there are elements x and y such that $ax = b$ and $ya = b$. { di'vizh·ən ,riŋ }

division sign [MATH] **1.** The symbol ÷, used to indicate division. **2.** The diagonal /, used to indicate a fraction. { di'vizh·ən ,sīn }

division subroutine [COMPUT SCI] A built-in program

which achieves division by methods such as repetitive subtraction. { də'vizh·ən 'səb·rü,tēn }

division wall [BUILD] A wall used to create major subdivisions in a building. { də'vizh·ən ,wȯl }

divisor [MATH] **1.** The quantity by which another quantity is divided in the operation of division. **2.** An element b in a commutative ring with identity is a divisor of an element a if there is an element c in the ring such that $a = bc$. { də'vīz·ər }

divisor of zero [MATH] A nonzero element x of a commutative ring such that $xy = 0$ for some nonzero element y of the ring. Also known as zero divisor. { di¦vī·zər əv 'zir·ō }

dixenite [MINERAL] $Mn_5(SiO_3)(AsO_3)(OH)_2$ A black hexagonal mineral consisting of a manganese arsenite and silicate, occurring in scales. { 'dik·sə,nīt }

Dixidae [INV ZOO] A family of orthorrhaphous dipteran insects in the series Nematocera. { 'dik·sə,dē }

dizygotic twins [BIOL] Twins derived from two eggs. Also known as fraternal twins. { ¦dī·zī'gäd·ik 'twinz }

djalmaite See microlite. { 'jal·mə,īt }

djave butter See adjab butter. { 'jä·və ,bəd·ər }

djerfisherite [MINERAL] $K_3CuFe_{12}S_{14}$ A sulfide mineral found only in meteorites. { jər'fish·ə,rīt }

Djulfian [GEOL] Upper upper Permian geologic time. { 'jùl·fē·ən }

D layer [GEOL] The lower mantle of the earth, between a depth of 600 and 1800 miles (1000 and 2900 kilometers). [GEOPHYS] The lowest layer of ionized air above the earth, occurring in the D region only in the daytime hemisphere; reflects frequencies below about 50 kilohertz and partially absorbs higher-frequency waves. { 'dē ,lā·ər }

D line [SPECT] The yellow line that is the first line of the major series of the sodium spectrum; the doublet in the Fraunhofer lines whose almost equal components have wavelengths of 5895.93 and 5889.96 angstroms respectively. { 'dē ,līn }

D-loop See displacement loop. { 'de ,lüp }

DM See adamsite.

dma See direct memory access.

DMA See disodium methylarsonate.

DMB See hydroquinone dimethyl ether.

DMC See p,p'-dichlorodiphenylmethyl carbinol; penicillamine.

DMDT See methoxychlor.

DME See distance-measuring equipment.

D meson [PARTIC PHYS] Collective name for four charmed mesons that form two isotopic spin doublets, have masses of approximately 1865 MeV, and are pseudoscalar particles. { ¦dē 'mā,sän }

DMF See N,N-dimethylformamide.

DMRT 1 gene [GEN] A gene widely required for male sex differentiation, for example in *Drosophila melanogaster* (called dsx), the roundworm *Caenorhabditis elegans* (called mab-3), chickens, and probably humans. { ¦dē¦em¦är¦tē 'wən jēn }

DMSO See dimethyl sulfoxide.

DMT See dimethyl terephthalate.

DMU See dimethylolurea.

DNA See deoxyribonucleic acid.

DNA fingerprint [GEN] Each human individual's virtually unique pattern of deoxyribonucleic acid (DNA) fragment sites produced by restriction enzyme digestion, separated by gel electrophoresis, and hybridized to labeled DNA. { ¦dē¦en¦ā 'fiŋ·gər,print }

DNA fingerprinting See genetic fingerprinting. { ,dē,en·ā 'fiŋ·gər,print·iŋ }

DNA microarray [CELL MOL] A microscopic spot containing identical single-stranded polymeric molecules of deoxyribonucleotides (DNAs), usually oligonucleotides or complementary DNAs, attached to a solid support (such as a membrane, a polymer, or glass) used to simultaneously analyze the expression levels of the corresponding genes. { ¦dē¦en¦ā 'mī·krō·ə,rā }

DNA primase [CELL MOL] An enzyme involved in the initiation of deoxyribonucleic acid (DNA) replication that catalyzes the polymerization of short ribonucleic acid (RNA) primers on the template DNA. { ¦dē¦en¦ā 'prī,mās }

DNase See deoxyribonuclease.

DNA sequencing [CELL MOL] The determination of the sequence of nucleotides in deoxyribonucleic acid (DNA) molecules. { ¦dē¦en¦ā 'sē·kwən,siŋ }

DNA vaccine [IMMUNOL] A type of noninfectious vaccine that directly injects deoxyribonucleic plasmids that express antigens of interest, resulting in foreign protein expression within the cells of the vaccine; however, the vaccine itself is unable to replicate. { ¦de¦en¦ā vak'sēn }

dneprovskite *See* wood tin. { ne'prŏv,skīt }

D nickel [MET] A nickel-manganese (4.75%) alloy of medium strength that resists spark erosion and hence is used for spark-plug electrodes and for some electronic applications. { 'dē ,nik·əl }

DNS *See* direct numerical simulation; domain name system.

Dobbin's reagent [ANALY CHEM] A mercuric chloride-potassium iodide reagent used to test for caustic alkalies in soap. { 'däb·ənz rē'ā·jənt }

dobby loom [TEXT] A weaving loom on which small all-over designs, dots, and stripes can be made in raised woven or self-color effects by means of special attachments. { 'däb·ē ,lüm }

Dobinski's equality [MATH] A formula which expresses a Bell number as the sum of an infinite series. { dō¦bin·skēz ē'kwäl·əd·ē }

Dobrowolsky generator [ELEC] Three-wire, direct-current generator with a balance coil connected across the armature; the coil's midpoint produces the midpoint voltage for the system. { ,dō·brə'väl·skē 'jen·ə,rād·ər }

Dobson prop [MIN ENG] A hydraulic supporting post used in mine tunnel construction. { 'däb·sən ,präp }

Dobson spectrophotometer [SPECT] A photoelectric spectrophotometer used in the determination of the ozone content of the atmosphere; compares the solar energy at two wavelengths in the absorption band of ozone by permitting the radiation of each to fall alternately upon a photocell. { 'däb·sən ,spek·trō·fə'täm·əd·ər }

Dobson support system [MIN ENG] A self-advancing unit consisting of three Dobson props used to support longwall faces. { 'däb·sən sə'pôrt ,sis·təm }

Dobson unit [METEOROL] The unit of measure for atmospheric ozone; one Dobson unit is equal to 2.7×10^{16} ozone molecules per square centimeter, which would be equivalent to a layer of ozone 0.001 centimeter thick, at 1 atmosphere and 0°C. { 'däb·sən ,yü·nət }

DOC *See* delayed-opening chaff.

dock [CIV ENG] **1.** The slip or waterway that is between two piers or cut into the land for the berthing of ships. **2.** A basin or enclosure for reception of vessels, provided with means for controlling the water level. { däk }

docking [AERO ENG] The mechanical coupling of two or more human-made orbiting objects. { däk·iŋ }

docking block [CIV ENG] A timber used to support a ship in dry dock. { däk·iŋ ,bläk }

docking bridge [NAV ARCH] A platform that rises above the deck of a large ship at the stern. { däk·iŋ ,brij }

docking keel [NAV ARCH] A keel placed on some ships parallel to and between the bilge keels and the main keel to support the vessel while in dry dock. { däk·iŋ ,kēl }

docking station [COMPUT SCI] A device that connects a portable computer with peripherals such as an external monitor, keyboard, and so on, allowing a portable computer to function as a desktop computer. { däk·iŋ ,stā·shən }

dock landing ship [NAV ARCH] An amphibious warfare ship which transports and supports landing craft and can quickly unload troops and equipment onto a beach. Designated LSD. { däk 'land·iŋ ,ship }

dockyard [CIV ENG] A yard utilized for ship construction and repair. { 'däk,yärd }

Docodonta [PALEON] A primitive order of Jurassic mammals of North America and England. { ,däk·ə'dän·tə }

docosane [ORG CHEM] $C_{22}H_{46}$ A paraffin hydrocarbon, especially the normal isomer $CH_3(CH_2)_{20}CH_3$. { 'däk·ə,sān }

docosanoic acid [ORG CHEM] $CH_3(CH_2)_{20}CO_2H$ A crystalline fatty acid, melting at 80°C, slightly soluble in water and alcohol, and found in the fats and oils of some seeds such as peanuts. Also known as behenic acid. { ¦dak·ə·sə¦nō·ik 'as·əd }

1-docosanol *See* behenyl alcohol. { ¦wən də'käs·ə,nól }

docosapentanoic acid [ORG CHEM] $C_{21}H_{33}CO_2H$ A pale-yellow liquid, boils at 236°C (5 mmHg), insoluble in water, soluble in ether, and found in fish blubber. { ¦däk·ə·sə,pen·tə¦nō·ik 'as·əd }

doctor [METEOROL] A cooling sea breeze in the tropics. { 'däk·tər }

doctor bar *See* doctor blade. { 'däk·tər ,bär }

doctor blade [ENG] A device for regulating the amount of liquid material on the rollers of a spreader. Also known as doctor bar; doctor knife; doctor roll. { 'däk·tər ,blād }

doctor knife *See* doctor blade. { 'däk·tər ,nīf }

doctor roll [CHEM ENG] **1.** Roller device used to remove accumulated filter cake from rotary filter drums. **2.** *See* doctor blade. { 'däk·tər ,rōl }

doctor solution [CHEM ENG] Sodium plumbite solution used to remove mercaptan sulfur from gasoline and other light petroleum distillates; used in doctor treatment. { 'däk·tər sə'lü·shən }

doctor test [CHEM ENG] A procedure using doctor solution (sodium plumbite) to detect sulfur compounds in light petroleum distillates which react with the sodium plumbite. { 'däk·tər ,test }

doctor treatment [CHEM ENG] Refining process to sweeten (reduce the odor) of gasoline, solvents, and kerosine; sodium plumbite and sulfur convert the odoriferous mercaptans into disulfides. { 'däk·tər ,trēt·mənt }

Doctrine of Signatures [MED] An archaic concept that a medicinal plant was often stamped with some clear indication (signature) of its specific remedial power; for example, plants with yellow sap were said to cure jaundice. { ¦däk·trən əv 'sig·nə·chərz }

document [COMPUT SCI] **1.** Any record, printed or otherwise, that can be read by a human or a machine. **2.** To prepare a written text and charts describing the purpose, nature, usage, and operation of a program or a system of programs. { 'däk·yə·mənt }

document alignment [COMPUT SCI] The phase of the reading process in which a transverse force is applied to a document to line up its reference edge with that of the reading station. { 'däk·yə·mənt ə,līn·mənt }

documentation [COMPUT SCI] The collection, organized and stored, of records that describe the purpose, use, structure, details, and operational requirements of a program, for the purpose of making this information easily accessible to the user. { ,däk·yə·mən'tā·shən }

document comparison utility [COMPUT SCI] A program that compares two documents created by word-processing programs and provides a display of the differences between them. { ,däk·yə·mənt kəm'par·ə·sən yü,til·əd·ē }

document flow [COMPUT SCI] The path taken by documents as they are processed through a record handling system. { 'däk·yə·mənt ,flō }

document handling [COMPUT SCI] In character recognition, the process of loading, feeding, transporting, and unloading a cut-form document that has been submitted for character recognition. { 'däk·yə·mənt ,hand·liŋ }

document image processing [COMPUT SCI] The scanning of paper documents followed by the storage, retrieval, display, and management of the resulting electronic images. Also known as document imaging. { ¦däk·yə·mənt 'im·ij ,prä,ses·iŋ }

document imaging *See* document image processing. { 'däk·yə·mənt ,im·ij·iŋ }

document leading edge [COMPUT SCI] In character recognition, that edge which is the foremost one encountered during the reading process and whose relative position defines the document's direction of travel. { 'däk·yə·mənt ,lēd·iŋ 'ej }

document mark [GRAPHICS] A small mark on each frame of a roll of microfilm, used by the microfilm reader to keep count of the frames. { 'däk·yə·mənt ,märk }

document misregistration [COMPUT SCI] In character recognition, the improper state of appearance of a document, on site in a character reader, with respect to real or imaginary horizontal baselines. { 'däk·yə·mənt ,mis·rej·ə'strā·shən }

document number [COMPUT SCI] The number given to a document by its originators to be used as a means for retrieval; it will follow any one of various systems, such as chronological, subject area, or accession. { 'däk·yə·mənt ,nəm·bər }

document plane [GRAPHICS] In microfilm technology, that surface or area in space at which the document is positioned during exposure. { 'däk·yə·mənt ,plān }

document platen [GRAPHICS] A device for positioning a document in the document plane. { 'däk·yə·mənt ,plat·ən }

document processing [COMPUT SCI] The creation, handling, labeling, and modification of text documents, such as in word processing and in the indexing of documents for retrieval based on their content. { ¦dak·yə·mənt 'prä¸ses·iŋ }

document reader [COMPUT SCI] An optical character reader which reads a limited amount of information (one to five lines) and generally operates from a predetermined format. { 'dak·yə·mənt ¸rēd·ər }

document reference edge [COMPUT SCI] In character recognition, that edge of a source document which provides the basis of all subsequent reading processes, insofar as it indicates the relative position of registration marks, and the impending text. { 'dak·yə·mənt 'ref·rəns ¸ej }

Document Type Definition [COMPUT SCI] In Standard Generalized Markup Language, a file that specifies the tags in a particular document and the relationships among the fields that they represent. Abbreviated DTD. { 'dak·yə·mənt ¸tīp ¸def·ə¸nish·ən }

docuterm [COMPUT SCI] A word or phrase descriptive of the subject matter or concept of an item of information and considered important for later retrieval of information. { 'dak·yə¸tərm }

DOD See direct outward dialing.

dodecagon [MATH] A 12-sided polygon. { dō'dek·ə¸gän }

dodecahedrane [ORG CHEM] $C_{20}H_{20}$ A highly strained saturated hydrocarbon cage structure in the shape of a dodecahedron (12 faces). { dō¸dek·ə'hē·drän }

dodecahedron [MATH] A polyhedron with 12 faces. { dō¸dek·ə'hē¸drən }

dodecahydrate [CHEM] A hydrated compound that has a total of 12 water molecules associated with it. { dō¸dek·ə'hī¸drāt }

dodecamerous [BOT] Having the whorls of floral parts in multiples of 12. { ¦dō·də¦kam·ə·rəs }

dodecane [ORG CHEM] $CH_3(CH_2)_{10}CH_3C_{12}H_{26}$ An oily paraffin compound, a colorless liquid, boiling at 214.5°C, insoluble in water; used as a solvent and in jet fuel research. Also known as dihexy; propylene tetramer; tetrapropylene. { 'dō·də¸kān }

1-dodecene [ORG CHEM] $CH_2CH(CH_2)_9CH_3$ A colorless liquid, boiling at 213°C, insoluble in water; used in flavors, dyes, perfumes, and medicines. { ¦wən 'dō·də¸sēn }

dodecomino [MATH] One of the 63,000 plane figures that can be formed by joining 12 unit squares along their sides. { ¸dō·dek·ə'mē·nō }

dodecyl [ORG CHEM] $C_{12}H_{25}$ A radical derived from dodecane by removing one hydrogen atom; in particular, the normal radical, $CH_3(CH_2)_{10}CH_2-$. { 'dō·də¸sil }

dodecylbenzene [ORG CHEM] Blend of isomeric (mostly monoalkyl) benzenes with saturated side chains averaging 12 carbon atoms; used in the alkyl amyl sulfonate type of detergents. Also known as detergent alkylate. { 'dō·də¸sil'ben¸zēn }

dodecyl sodium sulfate See sodium lauryl sulfate. { 'dō·də¸sil 'sōd·ē·əm 'səl¸fāt }

dodge chain [DES ENG] A chain with detachable bearing blocks between the links. { 'däj ¸chān }

Dodge crusher [MIN ENG] A type of jaw crusher with the movable jaw hinged at the bottom, allowing a highly uniform product to be discharged. { 'däj ¦krəsh·ər }

Dodge pulverizer [MIN ENG] A hexagonal drum-shaped pulverizer that rotates on a horizontal axis and contains steel balls for reducing rock and ore. { 'däj ¦pəl·və¸rīz·ər }

Dodge-Romig tables [IND ENG] Tabular data for acceptance sampling, including lot tolerance and AOQL tables. { ¦däj ¦rō·mig ¸tā·bəlz }

dodging [GRAPHICS] A technique of shading a portion of the photographic paper during exposure so that the tone of certain portions of the print is modified. { 'däj·iŋ }

dodo [ENG] A rectangular groove cut across the grain of a board. [VERT ZOO] *Raphus calcullatus.* A large, flightless, extinct bird of the family Raphidae. { 'dō¸dō }

doe [VERT ZOO] The adult female deer, antelope, goat, rabbit, or any other mammal of which the male is referred to as buck. { dō }

Doebner-Miller synthesis [CHEM ENG] Synthesis of methylquinoline by heating aniline with paraldehyde in the presence of hydrochloric acid. { ¦deb·nər ¦mil·ər 'sin·thə·səs }

doffer [TEXT] **1.** A device that strips material from another part on a textile machine. **2.** A worker who replaces full bobbins or cones with empty ones. { 'däf·ər }

doffing [TEXT] Removing fibers from a cotton carding machine by means of a toothed bar or cylinder. { 'däf·iŋ }

dog [COMPUT SCI] A name for the hexadecimal digit whose decimal equivalent is 13. [DES ENG] **1.** Any of various simple devices for holding, gripping, or fastening, such as a hook, rod, or spike with a ring, claw, or lug at the end. **2.** An iron for supporting logs in a fireplace. **3.** A drag for the wheel of a vehicle. [VERT ZOO] Any of various wild and domestic animals identified as *Canis familiaris* in the family Canidae; all are carnivorous and digitigrade, are adapted to running, and have four toes with nonretractable claws on each foot. { dòg }

dog clutch [DES ENG] A clutch in which projections on one part fit into recesses on the other part. { 'dòg ¸kləch }

dog days [CLIMATOL] The period of greatest heat in summer. { 'dòg ¸dāz }

dogfish See bowfin. { 'dòg¸fish }

dogfish oil See shark liver oil. { 'dòg¸fish ¸òil }

dogger [GEOL] Concretionary masses of calcareous sandstone or ironstone. { 'dòg·ər }

doghole [MIN ENG] A small opening in a mine. { 'dòg ¸hōl }

doghole mine [MIN ENG] A small coal mine employing 15 or less miners. { 'dòg¸hōl ¸mīn }

doghouse [ELECTR] Small enclosure placed at the base of a transmitting antenna tower to house antenna tuning equipment. [PETRO ENG] A small enclosed space on the drilling rig floor used to provide storage for small objects. { 'dòg¸haus }

dog iron [DES ENG] **1.** A short iron bar with ends bent at right angles. **2.** An iron pin that can be inserted in stone or timber in order to lift it. { 'dòg ¸ī·ərn }

dogleg [NAV] That portion of a flight which does not lead directly to the destination or way point, followed to comply with established flight procedures, avoid possible dangers or bad weather areas, or delay time of arrival. [PETRO ENG] Bend or sudden direction change in a wellhole that can cause tubing wear and failure. { 'dòg¸leg }

dogs See folding boards. { 'dògz }

dog screw [DES ENG] A screw with an eccentric head; used to mount a watch in its case. { 'dòg ¸skrü }

Dog Star See Sirius. { 'dòg ¸stär }

dog's tooth [CIV ENG] A masonry string course in which the brick corner projects. { 'dògz ¸tüth }

Doherty amplifier [ELECTR] A linear radio-frequency power amplifier that is divided into two sections whose inputs and outputs are connected by quarter-wave networks; for all values of input signal voltage up to one-half maximum amplitude, section no. 1 delivers all the power to the load; above this level, section no. 2 comes into operation. { 'dō·ərd·ē ¸am·plə¸fī·ər }

doister [METEOROL] In Scotland, a severe storm from the sea. Also known as deaister; dyster. { 'dòis·tər }

Doisy unit [BIOL] A unit for standardization of vitamin K. { 'dòi·zē ¸yü·nət }

dolabriform [BIOL] Shaped like an ax head. { dō'lab·rə¸fòrm }

Dolan equation [PETRO ENG] Empirical equation for reservoir-permeability damage factor by the invasion of drilling mud or other foreign materials. { 'dō·lən i¸kwā·zhən }

doldrums [METEOROL] A nautical term for the equatorial trough, with special reference to the light and variable nature of the winds. Also known as equatorial calms. { 'dōl¸drəmz }

dolerophanite [MINERAL] $Cu_2(SO_4)O$ A brown, monoclinic mineral consisting of a basic copper sulfate, occurring in crystals. { ¸däl·ə'räf·ə¸nīt }

dolichol [BIOCHEM] Any of a group of long-chain unsaturated isoprenoid alcohols containing up to 84 carbon atoms; found free or phosphorylated in membranes of the endoplasmic reticulum and Golgi apparatus. { 'däl·ə¸kòl }

Dolichopodidae [INV ZOO] The long-legged flies, a family of orthorrhaphous dipteran insects in the series Brachycera. { ¸däl·ə·kō'päd·ə¸dē }

Dolichothoraci [PALEON] A group of joint-necked fishes assigned to the Arctolepiformes in which the pectoral appendages are represented solely by large fixed spines. { ¸däl·ə·kō'thòr·ə¸sī }

doline [GEOL] A general term for a closed depression in an area of karst topography that is formed either by solution of

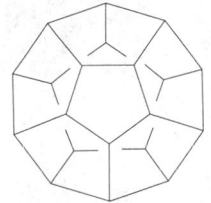

DODECAHEDRANE

Dodecahedrane ring structure.

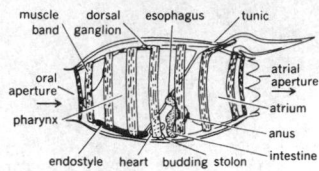

DOLIOLIDA

muscle band, dorsal ganglion, esophagus, tunic, oral aperture, pharynx, atrial aperture, atrium, anus, endostyle, heart, budding stolon, intestine

Doliolum, a doliolid, solitary asexual form.

the surficial limestone or by collapse of underlying caves. { də'lēn }

dolioform [BIOL] Barrel-shaped. { 'dō·lē·ə,fòrm }

doliolaria larva [INV ZOO] A free-swimming larval stage of crinoids and holothurians having an apical tuft and four or five bands of cilia. { ,dō·lē·ə'lar·ē·ə 'lär·və }

Doliolida [INV ZOO] An order of pelagic tunicates in the class Thaliacea; transparent forms, partly or wholly ringed by muscular bands. { ,dō·lē'ä·lə·də }

dollar [NUCLEO] A unit of reactivity, equal to the difference between the reactivities for delayed critical and prompt critical conditions in a given nuclear reactor. { 'däl·ər }

dollar spot [PL PATH] A fungus disease of lawn grasses caused by *Sclerotinia homeocarpa* and characterized by small, round, brownish areas which gradually coalesce. { 'däl·ər ,spät }

dolly [ENG] Any of several types of industrial hand trucks consisting of a low platform or specially shaped carrier mounted on rollers or combinations of fixed and swivel casters; used to carry such things as furniture, milk cans, paper rolls, machinery weighing up to 80 tons, and television cameras short distances. { 'däl·ē }

dolocast [GEOL] The cast or impression of a dolomite crystal. { 'dō·lə,kast }

dolomite [MINERAL] CaMg(CO₃)₂ The carbonate mineral; white or colorless with hexagonal symmetry and a structure similar to that of calcite, but with alternate layers of calcium ions being completely replaced by magnesium. { 'dō·lə,mīt }

dolomite rock See dolomitic limestone. { 'dō·lə,mīt 'räk }

dolomitic limestone [PETR] A limestone whose carbonate fraction contains more than 50% dolomite. Also known as dolomite rock; dolostone. { ¦dō·lə¦mid·ik 'līm,stōn }

dolomitization [GEOL] Conversion of limestone to dolomite rock by replacing a portion of the calcium carbonate with magnesium carbonate. { ,dō·lə·məd·ə'zā·shən }

dolomol See magnesium stearate. { 'dä·lə,mól }

do loop [COMPUT SCI] A FORTRAN iterative technique which enables any number of instructions to be executed repeatedly. { 'dü 'lüp }

dolostone See dolomitic limestone. { 'dō·lə,stōn }

dolphin [CIV ENG] **1.** A group of piles driven close and tied together to provide a fixed mooring in the open sea or a guide for ships coming into a narrow harbor entrance. **2.** A mooring post on a wharf. [VERT ZOO] The common name for about 33 species of cetacean mammals included in the family Delphinidae and characterized by the pronounced beak-shaped mouth. { 'däl·fən }

Dolphin See Delphinus. { 'däl·fən }

domain [COMPUT SCI] **1.** The set of all possible values contained in a particular field for every record of a file. **2.** The protected resources that are surrounded by the security perimeter of a distributed computer system. Also known as enclave; protected subnetwork. **3.** The final two or three letters of an Internet address, which specifies the highest subdivision; in the United States this is the type of organization, such as commercial, educational, or governmental, while outside the United States it is usually a country. [MATH] **1.** For a function, the set of values of the independent variable. **2.** A nonempty open connected set in Euclidean space. Also known as open region; region. **3.** See Abelian field. [SOLID STATE] A region in a solid within which elementary atomic or molecular magnetic or electric moments are uniformly arrayed. { dō'mān }

domain growth [SOLID STATE] A stage in the process of magnetization in which there is a growth of those magnetic domains in a ferromagnet oriented most nearly in the direction of an applied magnetic field. { dō'mān ,grōth }

domain name [COMPUT SCI] An alphanumeric string which identifies a particular computer or a network on the Internet. { dō'mān ,nām }

domain name system [COMPUT SCI] Abbreviated DNS. **1.** A system used on the Internet to map the easily remembered names of host computers (domain names) to their respective Internet Protocol (IP) numbers. **2.** A software database program that converts domain names to Internet Protocol addresses, and vice versa. { dō,mān 'nām ,sis·təm }

domain of dependence [MATH] For an initial-value problem for a partial differential equation, a portion of the range

such that the initial values on this portion determine the solution over the entire range. { ¦dō¦mān əv di'pen·dəns }

domain rotation [SOLID STATE] The stage in the magnetization process in which there is rotation of the direction of magnetization of magnetic domains in a ferromagnet toward the direction of a magnetic applied field and against anisotropy forces. { dō'mān rō'tā·shən }

domain theory [SOLID STATE] A theory of the behavior of ferromagnetic and ferroelectric crystals according to which changes in the bulk magnetization and polarization arise from changes in size and orientation of domains that are each polarized to saturation but which point in different directions. { dō'mān ,thē·ə·rē }

domain-tip memory [COMPUT SCI] A computer memory in which the presence or absence of a magnetic domain in a localized region of a thin magnetic film designates a 1 or 0. Abbreviated DOT memory. Also known as magnetic domain memory. { dō'mān ,tip 'mem·rē }

domain wall See Bloch wall. { dō'mān ,wòl }

domatic class See clinohedral class. { dō'mad·ik 'klas }

domatophobia [PSYCH] An abnormal fear of being in a house. { dō,mad·ə'fō·bē·ə }

dome [ARCH] A hemispherical roof. [ASTRON] A shallow raised structure on the moon's surface with a smooth convex cross section and a diameter anywhere from a few kilometers up to about 80 kilometers (50 miles). [CRYSTAL] An open crystal form consisting of two faces astride a symmetry plane. [ENG] The portion of a cylindrical container used in a filament-winding process that forms an integral end of the container. [ENG ACOUS] An enclosure for a sonar transducer, projector, or hydrophone and associated equipment; designed to have minimum effect on sound waves traveling underwater. [GEOL] A circular or elliptical, almost symmetrical upfold or anticlinal type of structural deformation. A large igneous intrusion whose surface is convex upward. [ORD] The mound of water spray created in air when the shock wave from an underwater detonation of an atomic weapon reaches the surface. { dōm }

Domerian [GEOL] Upper Charmouthian geologic time. { dō'mer·ə·ən }

domestication [BIOL] The adaptation of an animal or plant through breeding in captivity to a life intimately associated with and advantageous to humans. { də,mes·tə'kā·shən }

domestic coke [MATER] Coke for residential heating, which must have as low an ash content and as high a softening temperature (preferably above 2300°F, or 1260°C) for the ash as possible. { də'mes·tik 'kōk }

domestic induction heater [ENG] A cooking utensil heated by current (usually of commercial power line frequency) induced in it by a primary inductor. { də'mes·tik in'dək·shən ,hēd·ər }

domestic public-frequency bands [COMMUN] Radio-frequency bands reserved for public service within the United States. { də'mes·tik ¦pəb·lik 'frē·kwən·sē ,banz }

domestic refrigerator [MECH ENG] A refrigeration system for household use which typically has a compression machine designed for continuous automatic operation and for conservation of the charges of refrigerant and oil, and is usually motor-driven and air-cooled. Also known as refrigerator. { də'mes·tik ri'frij·ə,rād·ər }

domestic satellite [AERO ENG] A satellite in stationary orbit 22,300 miles (35,680 kilometers) above the equator for handling up to 12 separate color television programs, up to 14,000 private-line telephone calls, or an equivalent number of channels for other communication services within the United States. Abbreviated DOMSAT. { də'mes·tik 'sad·əl,īt }

dome theory [MIN ENG] The theory that the movements of strata resulting from underground excavations are limited by a dome whose base is the area of excavation, and that the movements decrease in intensity as they extend upward from the center of the base. { 'dōm ,thē·ə·rē }

domeykite [MINERAL] Cu₃As A tin-white or steel-gray mineral consisting of copper arsenide; specific gravity is 7.2–7.75. { dō'mā,kīt }

dominance [ECOL] The influence that a controlling organism has on numerical composition or internal energy dynamics in a community. [GEN] The expression of a heritable trait in the heterozygote such as to make it phenotypically indistinguishable from the homozygote. { 'däm·ə·nəns }

DOLPHIN

The common dolphin (*Delphinus delphis*)

dominant allele [GEN] The member of a pair of alleles which is phenotypically indistinguishable in both the homozygous and heterozygous condition. { 'däm·ə·nənt ə'lēl }

dominant energy condition [RELAT] The condition used in general relativity theory that all observers see a nonnegative energy density and a nonnegative energy flux. { 'däm·ə·nənt 'en·ər·jē kən,dish·ən }

dominant hemisphere [PHYSIO] The cerebral hemisphere which controls certain motor activities; usually the left hemisphere in right-handed individuals. { 'däm·ə·nənt 'hem·ə,sfir }

dominant mode *See* fundamental mode. { 'däm·ə·nənt 'mōd }

dominant negative mutation [CELL MOL] Mutation resulting in a gene product that can interfere with the function of the normal gene product in heterozygotes. { 'däm·ə·nənt ¦neg·əd·iv myü'tā·shən }

dominant species [ECOL] A species of plant or animal that is particularly abundant or controls a major portion of the energy flow in a community. { 'däm·ə·nənt 'spē,shēz }

dominant strategy [MATH] Relative to a given pure strategy for one player of a game, a second pure strategy for that player that has at least as great a payoff as the given strategy for any pure strategy of the opposing player. { 'däm·ə·nənt 'strad·ə·jē }

dominant wave [ELECTROMAG] The electromagnetic wave that has the lowest cutoff frequency in a given uniconductor waveguide. { 'däm·ə·nənt 'wāv }

dominant wavelength [OPTICS] The single wavelength of light that, when combined in suitable proportions with a reference standard light, matches the color of a given sample. { 'däm·ə·nənt 'wāv,leŋkth }

dominated convergence theorem [MATH] If a sequence $\{f_n\}$ of Lebesgue measurable functions converges almost everywhere to f and if the absolute value of each f_n is dominated by the same integrable function, then f is integrable and $\lim \int f_n \, dm = \int f \, dm$. { 'däm·ə,nād·əd kən'vər·jəns ,thir·əm }

dominating edge set [MATH] A set of edges of a graph such that every edge is either a member of this set or has a vertex in common with a member of this set. { ¦däm·ə,nād·iŋ 'ej ,set }

dominating integral [MATH] An improper integral whose nonnegative, nonincreasing integrand function has the property that its value for all sufficiently large positive integers n is no smaller than the nth term of a given series of positive terms; used in the integral test for convergence. { 'däm·ə,nad·iŋ 'in·tə·grəl }

dominating series [MATH] A series, each term of which is larger than the respective term in some other given series; used in the comparison test for convergence of series. { 'däm·ə,nad·iŋ 'sir·ēz }

dominating vertex set [MATH] A set of vertices in a simple graph such that every vertex of the graph is either a member of this set or is adjacent to a member of this set. Also known as external dominating set. { ¦däm·ə,nād·iŋ 'vər,teks ,set }

doming [MIN ENG] Setting up a domelike region above the open space created by stope excavation. { 'dōm·iŋ }

domino [MATH] The plane figure formed by joining two unit squares along a common side; a rectangle whose length is twice its width. { 'däm·ə,nō }

DOMSAT *See* domestic satellite. { 'däm,sat }

donation [MYCOL] In conjugation, a process involving a nonconjugative plasmid and a conjugative plasmid in which the latter provides the missing conjugative function to the former so that the former may be transferred. { dō'nā·shən }

Donau glaciation [GEOL] A Pleistocene glacial time unit in the Alps region in Europe. { 'dō,naù glā·sē'ā·shən }

Donders reduced eye [OPTICS] An optical model used to simplify calculations of the size and position of images produced by the human eye, consisting of a convex refracting surface that separates air in front from water, with a refractive index of 4/3, behind, and having an anterior focal distance of 15 millimeters and a posterior focal distance of 20 millimeters. { 'dän·dərz ri¦düst 'ī }

dongle [COMPUT SCI] A hardware device that plugs into a computer or printer port and serves as a copy-protection device for certain software, which must verify its presence in order to run properly. Also known as hardware key. { 'däŋ·gəl }

donkey [MIN ENG] *See* barney. [VERT ZOO] A domestic ass (*Equus asinus*); a perissodactyl mammal in the family Equidae. { 'däŋ·kē }

donkey boiler [NAV ARCH] A steam boiler on a ship deck used to supply steam to deck machinery when main boilers are shut down. { 'däŋ·kē ,bòil·ər }

donkey engine [MECH ENG] A small auxiliary engine which is usually portable or semiportable and powered by steam, compressed air, or other means, particularly one used to power a windlass to lift cargo on shipboard or to haul logs. { 'däŋ·kē ,en·jən }

donkey power [PHYS] A unit of power equal to 250 watts; it is approximately 1/3 horsepower. { 'däŋ·kē ,paù·ər }

Donnan distribution coefficient [PHYS CHEM] A coefficient in an expression giving the distribution, on two sides of a boundary between electrolyte solutions in Donnan equilibrium, of ions which can diffuse across the boundary. { 'dän·ən ,dis·trə'byü·shən ,kō·ə,fish·ənt }

Donnan equilibrium [PHYS CHEM] The particular equilibrium set up when two coexisting phases are subject to the restriction that one or more of the ionic components cannot pass from one phase into the other; commonly, this restriction is caused by a membrane which is permeable to the solvent and small ions but impermeable to colloidal ions or charged particles of colloidal size. Also known as Gibbs-Donnan equilibrium. { 'dō·nən ē·kwə'lib·rē·əm }

Donnan potential [PHYS CHEM] The potential difference across a boundary between two electrolytic solutions in Donnan equilibrium. { 'dän·ən pə,ten·chəl }

Donohue equation [THERMO] Equation used to determine the heat-transfer film coefficient for a fluid on the outside of a baffled shell-and-tube heat exchanger. { 'dän·ə·hü i,kwā·zhən }

donor [SOLID STATE] An impurity that is added to a pure semiconductor material to increase the number of free electrons. Also known as donor impurity; electron donor. { 'dō·nər }

donor impurity *See* donor. { 'dō·nər im,pyùr·əd·ē }

donor level [SOLID STATE] An intermediate energy level close to the conduction band in the energy diagram of an extrinsic semiconductor. { 'dō·nər ,lev·əl }

donor splicing site [MOL BIO] The boundary between the left (5′) end of an intron and the right (3′) end of an exon in messenger ribonucleic acid. Also known as left splicing junction. { 'dō·nər ,splīs·iŋ ,sīt }

do nothing instruction *See* NO OP. { 'dü ,nəth·iŋ in,strək·shən }

Donovan body [MED] The causative microorganism of granuloma inguinale, demonstrated in stained mononuclear cells and characterized by one or two opposite polar chromatin masses. { 'dän·ə·vən ,bäd·ē }

donut *See* doughnut. { 'dō·nət }

doodlebug [GEOL] Also known as douser. **1.** Any unscientific device or apparatus, such as a divining rod, used to locate subsurface water, minerals, gas, or oil. **2.** A scientific instrument used for locating minerals. [INV ZOO] The larva of an ant lion. [MECH ENG] **1.** A small tractor. **2.** A motor-driven railcar used for maintenance and repair work. [MIN ENG] The treatment plant or washing unit of a dredge which is mounted on a pontoon and can be floated in an excavation dug by a dragline. [ORD] A small military tank or utility truck. An airborne, magnetic submarine-detecting device. { 'düd·əl,bəg }

door [ENG] A piece of wood, metal, or other firm material pivoted or hinged on one side, sliding along grooves, rolling up and down, revolving, or folding, by means of which an opening into or out of a building, room, or other enclosure is open or closed to passage. { 'dòr }

door check *See* door closer. { 'dòr ,chek }

door closer [DES ENG] A device that makes use of a spring for closing, and a compression chamber from which liquid or air escapes slowly, to close a door at a controlled speed. Also known as door check. In elevators, a device or assembly of devices which closes an open car or hoistway door by the use of gravity or springs. { 'dòr ,klōz·ər }

doorknob capacitor [ELEC] A high-voltage, plastic-encased capacitor resembling a doorknob in size and shape. { 'dòr,näb kə,pas·əd·ər }

doorstop [BUILD] A strip positioned on the doorjamb for the door to close against. { 'dòr,stäp }

DOP *See* dioctyl phthalate.

DONKEY

The donkey, sometimes referred to as the burro.

dopa *See* dihydroxyphenylalanine. { 'dō·pə }

L-dopa *See* levodopa. { ¦el 'dō·pə }

dopamine [BIOCHEM] $C_8H_{11}O_2N$ An intermediate in epinephrine and norepinephrine biosynthesis; the decarboxylation product of dopa. { 'dō·pə,mēn }

dopamine hypothesis [MED] A theory that explains the pathogenesis of schizophrenia and other psychotic states as due to excesses in dopamine activity in various brain areas. { 'dōp·ə,mēn hī,päth·ə·səs }

dopant *See* doping agent. { 'dō·pənt }

dopa oxidase [BIOCHEM] An enzyme that catalyzes the oxidation of dihydroxyphenylalanine to melanin; occurs in the skin. { 'dō·pə 'äk·sə,dās }

dope [ELECTR] *See* doping agent. [MATER] A cellulose ester lacquer used as an adhesive or a coating. { dōp }

doped junction [ELECTR] A junction produced by adding an impurity to the melt during growing of a semiconductor crystal. { ¦dōpt 'jəŋk·shən }

doped solder [MET] Solder having an element added to it to ensure retention of a quality of the base metal on which it is used. { ¦dōpt 'säd·ər }

dope dyeing *See* solution dyeing. { 'dōp ,dī·iŋ }

doping [ELECTR] The addition of impurities to a semiconductor to achieve a desired characteristic, as in producing an *n*-type or *p*-type material. Also known as semiconductor doping. [ENG] Coating the mold or mandrel with a substance which will prevent the molded plywood part from sticking to it and will facilitate removal. { 'dōp·iŋ }

doping agent [ELECTR] An impurity element added to semiconductor materials used in crystal diodes and transistors. Also known as dopant; dope. { 'dōp·iŋ ,ā·jənt }

doping compensation [ELECTR] The addition of donor impurities to a *p*-type semiconductor or of acceptor impurities to an *n*-type semiconductor. { 'dōp·iŋ käm·pən'sā·shən }

Doppler-averaged cross section [PHYS] A cross section averaged over the energy of the incident particles and weighted to take into account the Doppler shifts associated with the thermal motions of the target particles. { ¦däp·lər ¦av·rijd 'krós ,sek·shən }

Doppler broadening [SPECT] Frequency spreading that occurs in single-frequency radiation when the radiating atoms, molecules, or nuclei do not all have the same velocity and may each give rise to a different Doppler shift. { 'däp·lər ,bród·ən·iŋ }

Doppler current meter [ENG] An acoustic current meter in which a collimated ultrasonic signal of known frequency is projected into the water and the reverberation frequency is measured; the difference in frequencies (Doppler shift) is proportional to the speed of water traveling past the meter. { 'däp·lər ,kər·ənt ,mēd·ər }

Doppler effect [PHYS] The change in the observed frequency of an acoustic or electromagnetic wave due to relative motion of source and observer. { 'däp·lər i,fekt }

Doppler-free spectroscopy [SPECT] Any of several techniques which make use of the intensity and monochromatic nature of a laser beam to overcome the Doppler broadening of spectral lines and measure their wavelengths with extremely high accuracy. { 'däp·lər ,frē spek'träs·kə·pē }

Doppler-free two-photon spectroscopy [SPECT] A version of Doppler free spectroscopy in which the wavelength of a transition induced by the simultaneous absorption of two photons is measured by placing a sample in the path of a laser beam reflected on itself, so that the Doppler shifts of the incident and reflected beams cancel. { däp·lər ,frē ¦tü ¦fō,tän spek'träs·kə·pē }

Doppler frequency *See* Doppler shift. { 'däp·lər ,frē·kwən·sē }

Dopplergram [ASTRON] An image of the sun showing line-of-sight (approaching and receding) gas motions, obtained using a spectroheliograph that has been modified to use the Doppler effect. { 'däp·lər,gram }

Doppler-inertial navigation equipment [NAV] Navigation equipment utilizing an inertial navigator and a Doppler navigator in combination so that inherent weaknesses of both systems will be bolstered to increase the reliability of the resultant navigational information. { 'däp·lər ə'nər·shəl ,nav·ə'gā·shən i,kwip·mənt }

dopplerite [GEOL] A naturally occurring gel of humic acids found in peat bags or where an aqueous extract from a low-rank coal can collect. { 'däp·lə,rīt }

Doppler lidar [OPTICS] A type of lidar that measures the frequency shift of the scattered laser pulse; this frequency shift is proportional to the velocity of the scatterer along the propagation path. { 'dop·lər 'lī,där }

Doppler navigation [NAV] Dead reckoning performed automatically by a device which gives a continuous indication of position by integrating the speed and the crab angle of the aircraft as derived from measurement of the Doppler effect of echoes from directed beams of radiant energy transmitted from the craft. { 'däp·lər ,nav·ə'gā·shən }

Doppler radar [ENG] A radar that makes use of the Doppler shift of an echo due to relative motion of target and radar to differentiate between fixed and moving targets and measure target velocities. { 'däp·lər 'rā,där }

Doppler range *See* doran. { 'däp·lər ,rānj }

Doppler shift [PHYS] The amount of the change in the observed frequency of a wave due to Doppler effect, usually expressed in hertz. Also known as Doppler frequency. { 'däp·lər ,shift }

Doppler sonar [ENG] Sonar based on Doppler shift measurement technique. Abbreviated DS. { 'däp·lər 'sō,när }

Doppler spectroscopy [SPECT] A technique for measuring the speed with which an object is moving toward or away from the observer by measuring the amount that light from the object is shifted to a higher or lower frequency by the Doppler effect. { ,däp·lər spek'träs·kə·pē }

Doppler tracking [ENG] Tracking of a target by using Doppler radar. { 'däp·lər ,trak·iŋ }

Doppler ultrasonic flowmeter [ENG] An instrument for determining the velocity of fluid flow from the Doppler shift of high-frequency sound waves reflected from particles or discontinuities in the flowing fluid. { 'däp·lər əl·trə'sän·ik 'flō,mēd·ər }

Doppler velocity and position [NAV] A continuous-wave trajectory-measuring system using the Doppler effect caused by a target moving relative to a ground transmitter and receiving station. Abbreviated Dovap. { 'däp·lər və'lä·səd·ē ən pə'zish·ən }

Doppler VOR [NAV] A ground-based navigational aid operating at very high frequency and using a wide-aperture radiation system to reduce azimuth errors caused by reflection from terrain and other obstacles; makes use of the Doppler principle to solve the problem of ambiguity that arises from the use of a radiation system with apertures that exceed one-half wavelength; the system is so designed that its signals may be received on the equipment used for the narrow-aperture VOR (very-high-frequency omnidirectional radio range). { 'däp·lər ¦vē¦ō'är }

DOP test *See* dioctyl phthalate test. { 'däp ,test }

Doradidae [VERT ZOO] A family of South American catfishes in the suborder Siluroidei. { də'ra·də,dē }

Dorado [ASTRON] A constellation of the southern hemisphere, right ascension 5 hours, declination 65° south. Also known as Swordfish. { də'rä·dō }

30 Doradus *See* Loop Nebula. { ¦thər·dē də'rä·dəs }

doran [ENG] A Doppler ranging system that uses phase comparison of three different modulation frequencies on the carrier wave, such as 0.01, 0.1, and 1 megahertz, to obtain missile range data with high accuracy. Derived from Doppler range. { 'dó,rän }

doraphobia [PSYCH] An abnormal fear of touching animal skin or fur. { ,dór·ə'fō·bē·ə }

dore [MET] Gold and silver bullion remaining in a cupeling furnace after removal of the oxidized lead. { də'rā }

Dorilaidae [INV ZOO] The big-headed flies, a family of cyclorrhaphous dipteran insects in the series Aschiza. { dór·ə'lā·ə,dē }

Dorippidae [INV ZOO] The mask crabs, a family of brachyuran decapods in the subsection Oxystomata. { də'rip·ə,dē }

dormancy [BOT] A state of quiescence during the development of many plants characterized by their inability to grow, though continuing their morphological and physiological activities. { 'dór·mən·sē }

dormant bud *See* latent bud. { 'dór·mənt 'bəd }

dormer window [BUILD] An extension of an attic room through a sloping roof to accommodate a vertical window. { 'dór·mər 'win·dō }

dormouse [VERT ZOO] The common name applied to members of the family Gliridae; they are Old World arboreal rodents intermediate between squirrels and rats. { 'dȯr‚maus }

Dorn effect [PHYS CHEM] A difference in a potential resulting from the motions of particles through water; the potential exists between the particles and the water. { 'dȯrn i‚fekt }

Dorngeholz See thornbush. { 'dȯrn‚gə‚hōlts }

Dorngestrauch See thornbush. { 'dȯrn‚gə‚strauk }

dornveld See thornbush. { 'dȯrn‚felt }

doroid [ELECTROMAG] A coil resembling half a toroid, using a removable core segment to simplify the winding process. { 'dȯ‚rȯid }

Dorr agitator [MECH ENG] A tank used for batch washing of precipitates which cannot be leached satisfactorily in a tank; equipped with a slowly rotating rake at the bottom, which moves settled solids to the center, and an air lift that lifts slurry to the launders. Also known as Dorr thickener. { 'dȯr 'aj‚ə‚tād‚ər }

Dorr classifier [MECH ENG] A horizontal flow classifier consisting of a rectangular tank with a sloping bottom, a rake mechanism for moving sands uphill along the bottom, an inlet for feed, and outlets for sand and slime. { 'dȯr 'klas‚ə‚fī‚ər }

Dorr thickener See Dorr agitator. { 'dȯr 'thik‚ə‚nər }

dorsal [ANAT] Located near or on the back of an animal or one of its parts. { 'dȯr‚səl }

dorsal aorta [ANAT] The portion of the aorta extending from the left ventricle to the first branch. [INV ZOO] The large, dorsal blood vessel in many invertebrates. { 'dȯr‚səl ā'ȯrd‚ə }

dorsal column [NEUROSCI] A column situated dorsally in each lateral half of the spinal cord which receives the terminals of some afferent fibers from the dorsal roots of the spinal nerves. { 'dȯr‚səl 'käl‚əm }

dorsal fin [VERT ZOO] A median longitudinal vertical fin on the dorsal aspect of a fish or other aquatic vertebrate. { 'dȯr‚səl 'fin }

dorsalia [INV ZOO] Paired sensory bristles on the dorsal aspect of the head of gnathostomalids. { 'dȯr'sal‚yə }

dorsal lip [EMBRYO] In an amphibian embryo, the margin or lip of the fold of blastula wall marking the dorsal limit of the blastopore during gastrulation and constituting the primary organizer, is necessary to the development of neural tissue, and forms the originating point of chordamesoderm. { 'dȯr‚səl 'lip }

dorsiferous [BOT] Of ferns, bearing sori on the back of the frond. [ZOO] Bearing the eggs or young on the back. { 'dȯr‚sif‚ə‚rəs }

dorsiflex [ZOO] To flex or cause to flex in a dorsal direction. { 'dȯr‚sə‚fleks }

dorsiflexion sign See Homan's sign. { ‚dȯr‚sə'flek‚shən ‚sīn }

dorsigrade [VERT ZOO] Walking on the back of the toes. { 'dȯr‚sə‚grād }

dorsocaudad [ANAT] To or toward the dorsal surface and caudal end of the body. { ‚dȯr‚sō‚kȯ‚dad }

dorsomedial [ANAT] Located on the back, toward the midline. { ‚dȯr‚sō‚mēd‚ē‚əl }

dorsoposteriad [ANAT] To or toward the dorsal surface and posterior end of the body. { ‚dȯr‚sō‚pō‚stir‚ē‚ad }

dorsum [ANAT] The entire dorsal surface of the animal body. The upper part of the tongue, opposite the velum. { 'dȯr‚səm }

Dorvilleidae [INV ZOO] A family of minute errantian annelids in the superfamily Eunicea. { ‚dȯr‚və'lē‚ə‚dē }

Dorylaimoidea [INV ZOO] An order or superfamily of nematodes inhabiting soil and fresh water. { ‚dȯr‚ə‚lə'mȯid‚ē‚ə }

Dorylinae [INV ZOO] A subfamily of predacious ants in the family Formicidae, including the army ant (*Eciton hamatum*). { dȯ'rī‚lə‚nē }

Dorypteridae [PALEON] A family of Permian palaeonisciform fishes sometimes included in the suborder Platysomoidei. { dȯ‚rip'ter‚ə‚dē }

DOS See disk operating system. { däs }

dosage [GEN] The number of copies of a particular gene. [MED] The prescribed or correct amount of medicine or other therapeutic agent administered to treat a given illness. Also known as dose. [NUCLEO] See absorbed dose. { 'dō‚sij }

dosage compensation [GEN] **1.** A mechanism that equalizes the expression in males and females of genes located on the X chromosome, despite their presence in two doses in the homogametic sex and a single dose in the heterogametic sex. **2.** A mechanism that equalizes the expression of X-linked and autosomal genes by doubling the expression level of X-linked genes in male *Drosophila* and in both male and female mammals with their single active X. { 'dō‚sij ‚käm‚pən‚sā‚shən }

dose [MED] **1.** The measure, expressed in number of roentgens, of a property of x-rays at a particular place; used in radiology. **2.** See dosage. [NUCLEO] See absorbed dose. { dōs }

dose equivalent [NUCLEO] The product of absorbed dose, in rads, and a number of modifying factors due to nonuniform distribution of internally deposited isotopes in radiobiology; the unit is the rem. { ‚dōs ə'kwiv‚ə‚lənt }

dose fractionation [BIOPHYS] The application of a radiation dose in two or more fractions separated by a certain minimal time interval. { ‚dōs ‚frak‚shə‚nā‚shən }

dosemeter See dosimeter. { 'dōs‚mēd‚ər }

dose rate [NUCLEO] The rate at which nuclear radiation is delivered. { 'dōs ‚rāt }

dose-rate meter [NUCLEO] An instrument that measures radiation dose rate. { 'dōs ‚rāt ‚mēd‚ər }

dosimeter [NUCLEO] An instrument that measures the total dose of nuclear radiation received in a given period. Also spelled dosemeter. { dō'sim‚əd‚ər }

dosimetry [NUCLEO] Measurement of the power, energy, irradiance, or radiant exposure of high-energy, ionizing radiation. Also known as radiation dosimetry. { dō'sim‚ə‚trē }

dosing tank [CIV ENG] A holding tank that discharges sewage at a rate required by treatment processes. { 'dōs‚iŋ ‚taŋk }

dot [ELECTR] See button. [GRAPHICS] A subdivision of the printing surface into minute units, formed by a halftone screen and separated by the etching process. { dät }

dot-addressable [COMPUT SCI] The ability of an electronic display or a dot-matrix printer to specify the individual dots that form images of characters. { 'dät ə'dres‚ə‚bəl }

dot angel [ELECTROMAG] An angel that appears on the screens of vertically pointing radars, often on clear cloudless days, as a bright dot; believed to be produced by a vertical column of rising air passing through air layers having different indices of refraction. { 'dät ‚ān‚jəl }

dot character printer See dot matrix printer. { 'dät 'kar‚ik‚tər ‚print‚ər }

dot cycle [COMMUN] In teletypewriter systems, an on-off or mark-space cycle in which both mark and space have the same length as the unit pulse. { 'dät ‚sī‚kəl }

dot etching [GRAPHICS] A technique in correcting the color of a positive or halftone negative employing the chemical reduction of halftone dots. { 'dät ‚ech‚iŋ }

dot generator [ELECTR] A signal generator that produces a dot pattern on the screen of a three-gun color television picture tube, for use in convergence adjustments. { 'dät jen‚ə‚rād‚ər }

dot matrix [COMPUT SCI] An array of dots that forms a character or graphic symbol. { 'dät 'mā‚triks }

dot matrix printer [COMPUT SCI] A type of printer that forms each character as a group of small dots, using a group of wires located in the printing element. Also known as dot character printer. { 'dät 'mā‚triks 'print‚ər }

dot pitch [GRAPHICS] The width of a dot in a dot matrix. { 'dät ‚pich }

dot product See inner product. { 'dät ‚präd‚əkt }

dotriacontane [ORG CHEM] $C_{32}H_{66}$ A paraffin hydrocarbon, in particular, the normal isomer $CH_3(CH_2)_{30}CH_3$, which is crystalline. { dō‚trī‚ə'kän‚tān }

dot-sequential color television [ELECTR] A color television system in which the red, blue, and green primary-color dots are formed in rapid succession along each scanning line. { 'dät sə‚kwen‚chəl ‚kəl‚ər 'tel‚ə‚vizh‚ən }

dots per inch [GRAPHICS] Abbreviated dpi. The number of dots that can be printed, side by side, along a line one inch long; a measure of the resolution of a printing device, printed image, or an image on an electronic display screen such that a greater number of dots per inch represents higher image quality. { ‚däts pər 'inch }

dot system [ELECTR] Manufacturing technique for producing microelectronic circuitry. { 'dät ‚sis‚təm }

dotted swiss [TEXT] A sheer fabric ornamented with slightly raised woven-in dots. { ‚däd‚əd 'swis }

dotter [OPTICS] A worker who uses a centering machine to

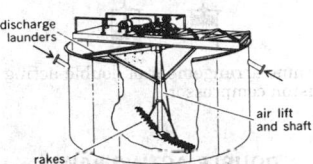

DORR AGITATOR

Dorr agitator for batch washing of precipitates. (*From W. L. Badger and J. T. Banchero, Introduction to Chemical Engineering, McGraw-Hill, 1955*)

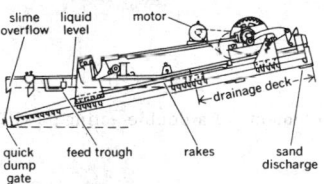

DORR CLASSIFIER

Diagram of Dorr classifier. (*From J. H. Perry, ed., Chemical Engineers' Handbook, 4th ed., McGraw-Hill, 1963*)

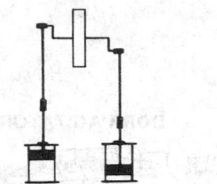

DOUBLE-ACTING COMPRESSOR

Frame arrangement of double-acting piston compressor.

DOUBLE-ACTING PAWL

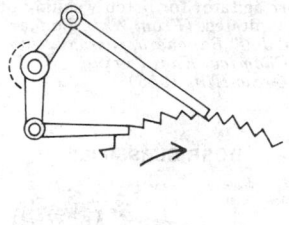

Elements of a double-acting pawl.

locate the optical center and optical axis of a lens, for the guidance of workers who will cut, edge, trim, and mount the lens. Also known as spotter. { 'däd·ər }

double-acting [MECH ENG] Acting in two directions, as with a reciprocating piston in a cylinder with a working chamber at each end. { ¦dəb·əl 'ak·tiŋ }

double-acting compressor [MECH ENG] A reciprocating compressor in which both ends of the piston act in working chambers to compress the fluid. { ¦dəb·əl ¦ak·tiŋ kəm'pres· ər }

double-acting hammer [MET] A forging hammer in which the ram is raised and forced down by a charge of air or steam. { ¦dəb·əl ¦ak·tiŋ 'ham·ər }

double-acting pawl [MECH ENG] A double pawl which can drive in either direction. { ¦dəb·əl ¦ak·tiŋ 'pól }

double action [ORD] Method of fire in a revolver and in old-style rifles and shotguns in which a single pull of the trigger both cocks and fires the weapon. { ¦dəb·əl 'ak·shən }

double-action die [MET] A die designed to perform more than one operation with each stroke of the press. { ¦dəb·əl ¦ak·shən 'dī }

double-action forming [MET] Forming in which more than one shape is imparted by each stroke of the press. { ¦dəb·əl ¦ak·shən 'fór·miŋ }

double-action mechanical press [MECH ENG] A press having two slides which move one within the other in parallel movements. { ¦dəb·əl ¦ak·shən mə¦kan·ə·kəl 'pres }

double aging [MET] Introduction of a primary (stabilizing) and a secondary aging treatment to control the precipitate formed from a supersaturated alloy and to achieve specific properties in the material. { ¦dəb·əl 'āj·iŋ }

double altitudes See equal altitudes. { ¦dəb·əl 'al·tə,tüdz }

double-amplitude-modulation multiplier [ELECTR] A multiplier in which one variable is amplitude-modulated by a carrier, and the modulated signal is again amplitude-modulated by the other variable; the resulting double-modulated signal is applied to a balanced demodulator to obtain the product of the two variables. { ¦dəb·əl ¦am·plə,tüd ¦mäj·ə,lā·shən 'məl· tə,plī·ər }

double angle formula [MATH] An equation that expresses a trigonometric function of twice an angle in terms of trigonometric functions of the angle. { ¦dəb·əl 'aŋ·gəl ,fór·myə·lə }

double arcing [MET] An occurrence in plasma arc welding and cutting where a secondary electric arc displaces the main arc at the outlet of a welding nozzle. { ¦dəb·əl 'ärk·iŋ }

double armature [ELEC] An armature with two separate windings on a single core. { 'dəb·əl 'är·mə·chər }

double-barrier resonant tunneling diode [ELECTR] A variant of the tunnel diode with thin layers of aluminum gallium arsenide and gallium arsenide that have sharp interfaces and have widths comparable to the Schrödinger wavelengths of the electrons, permitting resonant behavior. Abbreviated DBRT diode. { ¦dəb·əl ¦bar·ē·ər ¦rez·ən·ənt ,tən·əl·iŋ 'dī,ōd }

double-base diode See unijunction transistor. { ¦dəb·əl ¦bās 'dī,ōd }

double-base junction diode See unijunction transistor. { ¦dəb·əl ¦bās ¦jəŋk·shən 'dī,ōd }

double-base junction transistor [ELECTR] A tetrode transistor that is essentially a junction triode transistor having two base connections on opposite sides of the central region of the transistor. Also known as tetrode junction transistor. { ¦dəb· əl ¦bās ¦jəŋk·shən tran'zis·tər }

double-base rocket propellant [MATER] A solid rocket propellant using two unstable compounds, such as nitrocellulose and nitroglycerine. { ¦dəb·əl ¦bās ¦räk·ət prə,pel·ənt }

double-beam cathode-ray tube [ELECTR] A cathode-ray tube having two beams and capable of producing two independent traces that may overlap; the beams may be produced by splitting the beam of one gun or by using two guns. { ¦dəb· əl ¦bēm ¦kāth,ōd 'rā ,tüb }

double-beam spectrophotometer [SPECT] An instrument that uses a photoelectric circuit to measure the difference in absorption when two closely related wavelengths of light are passed through the same medium. { ¦dəb·əl ¦bēm spek'trō· fə'täm·əd·ər }

double beta decay [NUC PHYS] A nuclear transformation in which the atomic number changes by 2 and the mass number does not change; either two electrons are emitted or two orbital electrons are captured. { ¦dəb·əl ¦bād·ə di'kā }

double-bevel groove weld [MET] A type of groove weld in which one member has a joint edge beveled on both sides. Also known as double-V groove weld. { ¦dəb·əl ¦bev·əl 'grüv ,weld }

double-blind sample [ANALY CHEM] In chemical analysis, a sample submitted in such a way that neither its composition nor its identification as a check sample is known to the analyst. { ¦dəb·əl ¦blīnd 'sam·pəl }

double-blind technique [STAT] An experimental procedure in which neither the subjects nor the experimenters know the makeup of the test and control group during the actual course of the experiments. Also known as blind trial. { ¦dəb·əl 'blīnd ,tek,nēk }

double block and bleed system [ENG] A valve system configuration in which a full-flow vent valve is installed in a pipeline between two shutoff valves to provide a means of releasing excess pressure between them. { 'dəb·əl ¦bläk ən 'blēd ,sis·təm }

double-block brake [MECH ENG] Two single-block brakes in symmetrical opposition, where the operating force on one lever is the reaction on the other. { ¦dəb·əl ¦bläk 'brāk }

double blossom [PL PATH] A fungus disease of dewberry and blackberry caused by *Fusarium rubi* and characterized by witches'-brooms and enlargement and malformation of the flowers. { ¦dəb·əl 'bläs·əm }

double board [PETRO ENG] The platform from which the derrick operator works, located at a height on the derrick equal to two joined pipe lengths. { ¦dəb·əl 'bórd }

double bond [PHYS CHEM] A type of linkage between atoms in which two pair of electrons are shared equally. { ¦dəb· əl 'bänd }

double-bond isomerism [PHYS CHEM] Isomerism in which two or more substances possess the same elementary composition but differ in having double bonds in different positions. { ¦dəb·əl ¦bänd ī'säm·ə,riz·əm }

double-bond shift [ORG CHEM] In an organic molecular structure, the occurrence when a pair of valence bonds that join a pair of carbons (or other atoms) shifts, via chemical reaction, to a new position, for example, $H_2C=C-C-CH_2$ (butene-1) to $H_2C-C=C-CH_2$ (butene-2). { ¦dəb·əl ¦bänd 'shift }

double bottom [NAV ARCH] A ship bottom with space between the inner and outer plating. { ¦dəb·əl 'bäd·əm }

double-bottom cellular [NAV ARCH] Pertaining to division of the double bottom of a ship into numerous rectangular compartments by transverse and longitudinal framing. { ¦dəb·əl ¦bäd·əm 'sel·yə·lər }

double-bounce calibration [ELECTR] Method of radar calibration which is used to determine the zero set error by using round-trip echoes; the correct range is the difference between the first and second echoes. { ¦dəb·əl ¦baùns kal·ə'brā·shən }

double-break switch [ELEC] Switch which opens the connected circuit at two points. { ¦dəb·əl ¦brāk 'swich }

double bridge See Kelvin bridge. { ¦dəb·əl 'brij }

double brilliant [LAP] A brilliant cut with 40 facets above the girdle and 32 below. { ¦dəb·əl 'bril·yənt }

double-buffered data transfer [COMPUT SCI] The transmission of data into the buffer register and from there into the device register proper. { ¦dəb·əl ¦bəf·ərd'dad·ə ,trans·fər }

double bus-double breaker [ELEC] A substation switching arrangement having two common buses and two breakers per connection. { ¦dəb·əl ¦bəs ¦dəb·əl ,brāk·ər }

double bus-single breaker [ELEC] A substation switching arrangement that involves two common buses and only one breaker per connection. { ¦dəb·əl ¦bəs ¦siŋ·gəl ,brāk·ər }

double-button microphone [ENG ACOUS] A carbon microphone having two carbon-filled buttonlike containers, one on each side of the diaphragm, to give twice the resistance change obtainable with a single button. Also known as differential microphone. { ¦dəb·əl ¦bət·ən 'mī·krə,fōn }

double-channel duplex [COMMUN] A method that provides for simultaneous communication between two stations through use of two radio-frequency channels, one in each direction. { ¦dəb·əl ¦chan·əl 'dü,pleks }

double-channel simplex [COMMUN] A method that provides for nonsimultaneous communication between two stations through use of two radio-frequency channels, one in each direction. { ¦dəb·əl ¦chan·əl 'sim,pleks }

double circulation [PHYSIO] A circulatory system in which blood flows through two separate circuits, as pulmonary and systemic. { ¦dəb·əl ¦sər·kyə'lā·shən }

double-click [COMPUT SCI] To depress and release a mouse button twice in quick succession; often used to initiate an action such as opening a file, and to extend actions that result from a single click. { ¦dəb·əl 'klik }

double cluster [ASTRON] A pair of globular clusters that are physically close to each other, near the northern boundary of the constellation Perseus. { ¦dəb·əl 'kləs·tər }

double Compton scattering [QUANT MECH] A process in which a photon collides with a free electron and two photons are given off. { ¦dəb·əl ¦käm·tən ˌskad·ər·iŋ }

double-concave lens [OPTICS] A lens having surfaces that are adjacent portions of nonintersecting spheres whose centers lie on opposite sides of the plane of the lens. Also known as biconcave lens. { ¦dəb·əl ¦kän,kāv 'lenz }

double-cone bit [DES ENG] A type of roller bit having only two cone-shaped cutting members. { ¦dəb·əl ¦kōn 'bit }

double-convex lens [OPTICS] A lens having surfaces that are adjacent portions of intersecting spheres whose centers lie on opposite sides of the plane of the lens. Also known as biconvex lens. { ¦dəb·əl ¦kän,veks 'lenz }

double-core barrel drill [DES ENG] A core drill consisting of an inner and an outer tube; the inner member can remain stationary while the outer one revolves. { ¦dəb·əl ¦kȯr 'bar·əl ˌdril }

double-coursed [BUILD] Covered with a material such as shingles in such a way that no area is covered with less than two thicknesses. { ¦dəb·əl 'kȯrst }

double-crank press [MECH ENG] A mechanical press with a single wide slide operated by a crankshaft having two crank pins. { ¦dəb·əl ¦kraŋk 'pres }

double cropping [AGR] A form of multiple cropping in which two crops are grown on a field at different times of the year. { ¦dəb·əl 'kräp·iŋ }

double crossover See scissors crossover. { ¦dəb·əl 'krȯs,ō·vər }

double-current cable code [COMMUN] A cable code in which characters are determined by bipolar characters of equal length. { ¦dəb·əl ¦kə·rənt 'kā·bəl ˌkōd }

double-current generator [ELEC] Machine which supplies both direct and alternating current from the same armature winding. { ¦dəb·əl ¦kə·rənt 'jen·ə,rād·ər }

double-current signaling [COMMUN] A system of telegraph signaling that uses both positive and negative currents. { ¦dəb·əl ¦kə·rənt 'sig·nəl·iŋ }

double cusp [MATH] A point on a curve through which two branches of the curve with the same tangent pass, and at which each branch extends in both directions of the tangent. Also known as point of osculation; tacnode. { ¦dəb·əl ¦kəsp }

double-cut file [DES ENG] A file covered with two series of parallel ridges crossing at angles to each other. { ¦dəb·əl ¦kət 'fīl }

double-cut planer [MECH ENG] A planer designed to cut in both the forward and reverse strokes of the table. { ¦dəb·əl ¦kət 'plān·ər }

double-cut saw [DES ENG] A saw with teeth that cut during the forward and return strokes. { ¦dəb·əl ¦kət 'sȯ }

double data rate [COMPUT SCI] A clocking technique that increases the transfer speeds of synchronous memories by using both the leading and trailing edges of the clock signal to transfer data, effectively doubling the transfer rate or bandwidth. { ¦dəb·əl 'dad·ə ˌrāt }

double decomposition [CHEM] The simple exchange of elements of two substances to form two new substances; for example, $CaSO_4 + 2NaCl \rightarrow CaCl_2 + Na_2SO_4$. { ¦dəb·əl dē,käm·pə'zish·ən }

double density [COMPUT SCI] Property of a computer storage medium that holds twice as much data per unit of storage space as the standard; applied particularly to floppy disks. { ¦dəb·əl 'den·səd·ē }

double-diffused transistor [ELECTR] A transistor in which two *pn* junctions are formed in the semiconductor wafer by gaseous diffusion of both *p*-type and *n*-type impurities; an intrinsic region can also be formed. { ¦dəb·əl də'fyüzd tran'zis·tər }

double diffusion [FL MECH] A type of convective transport in fluids that depends on the difference in diffusion rates of at least two density-affecting components. Also, it is necessary

to have an unstable or top-heavy distribution of one component. For example, in oceanography, the two components are heat and dissolved salts, and the unstable component is the slower-diffusing salt. { ¦dəb·əl də'fyü·zhən }

double diode See binode; duodiode. { ¦dəb·əl 'dī,ōd }

double-diode limiter [ELECTR] Type of limiter which is used to remove all positive signals from a combination of positive and negative pulses, or to remove all the negative signals from such a combination of positive and negative pulses. { ¦dəb·əl ¦dī,ōd 'lim·əd·ər }

double distribution [CHEM ENG] The product distribution resulting from counter double-current extraction, a scheme in which each of the two liquid phases is transferred simultaneously and continuously in opposite directions through an interconnected train of contact vessels. { ¦dəb·əl dis·trə'byü·shən }

double-doped transistor [ELECTR] The original grown-junction transistor, formed by successively adding *p*-type and *n*-type impurities to the melt during growing of the crystal. { ¦dəb·əl ¦dōpt tran'zis·tər }

double dot halftone [GRAPHICS] A halftone in which two halftone negatives have been combined into one printing plate, resulting in a greater tonal range than in a conventional halftone. { ¦dəb·əl ¦dät 'haf,tōn }

double-doublet antenna [ELECTROMAG] Two half-wave doublet antennas criss-crossed at their center, one being shorter than the other to give broader frequency coverage. { ¦dəb·əl ¦dəb·lət an'ten·ə }

double drift [NAV] A method of determining the speed and direction of the wind by observing the drift angle on two aircraft headings and deriving the wind information therefrom. { ¦dəb·əl 'drift }

double drill column [MIN ENG] Two drill columns connected by a horizontal bar on which a drill machine can be mounted. Also known as double jack. { ¦dəb·əl 'dril ˌkäl·əm }

double-drum hoist [MECH ENG] A hoisting device consisting of two cable drums which rotate in opposite directions and can be operated separately or together. { ¦dəb·əl ¦drəm 'hoist }

double ebb [OCEANOGR] An ebb current comprising two maxima of velocity that are separated by a smaller ebb velocity. { ¦dəb·əl 'eb }

double electron excitation [ATOM PHYS] An excited state of an atom in which two electrons are excited rather than one. { ¦dəb·əl i¦lek,trän ˌek·sī'tā·shən }

double-ended ferry [NAV ARCH] A ferry vessel with a propeller and rudder at each end and a hull form that is identical at each end, so that the vessel can shuttle back and forth between two terminals without turning. { ¦dəb·əl ¦en·dəd 'fer·ē }

double-ended Q machine [PL PHYS] A Q machine in which the plasma is generated at a hot tungsten plate at one end and the plasma column is reflected at the other end by a second hot tungsten plate. { ¦dəb·əl ¦end·əd 'kyü mə,shēn }

double-entry method [MIN ENG] A mining arrangement involving twin entries in flat or gently dipping coal, so that rooms can be extended from both entryways. { ¦dəb·əl 'en·trē ˌmeth·əd }

double exposure [GRAPHICS] The act of recording two images on top of each other completely or in part. { ¦dəb·əl ik'spō·zhər }

double-exposure holographic interferometry [OPTICS] The study of the interference fringes generated by the superposition of two holograms of the same object, one with the object in an undeformed state and the second after a small deformation. { ¦dəb·əl ik'spō·zhər ¦häl·ə,graf·ik ˌin·tər·fə'räm·ə·trē }

double-face satin [TEXT] Satin cloth made with two warps and one filling to obtain satin effects on both the face and reverse sides. { ¦dəb·əl ¦fās 'sat·ən }

double fertilization [BOT] In most seed plants, fertilization involving fusion between the egg nucleus and one sperm nucleus, and fusion between the other sperm nucleus and the polar nuclei. { ¦dəb·əl ˌfərd·əl·ə'zā·shən }

double floor [BUILD] A floor in which binding joists support the ceiling joists below as well as the floor joists above. { ¦dəb·əl 'flȯr }

DOUBLE-CONCAVE LENS

Double-concave lens, a type of diverging or negative lens.

DOUBLE-CONVEX LENS

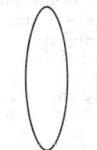

Double-convex lens, a type of collecting or positive lens. (*From F. A. Jenkins and H. E. Whites, Fundamentals of Optics, 3d ed., McGraw-Hill, 1957*)

double frequency shift keying [COMMUN] Multiplex system in which two telegraph signals are combined and transmitted simultaneously by a method of frequency shifting between four radio frequencies. { ¦dəb·əl 'frē·kwən·sē ¦shift 'kē·iŋ }

double group [QUANT MECH] A type of group useful in studying systems of half-integral spin; it is formed by modifying a finite point group by introducing an element which is a rotation through an angle of 2π about an arbitrary axis and which is not the unit element but gives the unit element when applied twice. { ¦dəb·əl 'grüp }

doublehand drilling [ENG] A rock-drilling method performed by two men, one striking the rock with a long-handled sledge hammer while a second holds the drill and twists it between strokes. Also known as double jacking. { 'dəb·əl‚hand 'dril·iŋ }

double headings [MIN ENG] A pair of coal headings driven parallel to each other and positioned side by side about 10–20 yards (9–18 meters) apart. { ¦dəb·əl 'hed·iŋz }

double Hooke's joint [MECH ENG] A universal joint which eliminates the variation in angular displacement and angular velocity between driving and driven shafts, consisting of two Hooke's joints with an intermediate shaft. { 'dəb·əl 'hùks ‚jóint }

double-housing planer [MECH ENG] A planer having two housings to support the cross rail, with two heads on the cross rail and one sidehead on each housing. { ¦dəb·əl 'haùz·iŋ 'plān·ər }

double-hump fission barrier [NUC PHYS] Two separated maxima in a plot of potential energy against nuclear deformation of an actinide nucleus, which inhibit spontaneous fission of the nucleus and give rise to isomeric states in the valley between the two maxima. { ¦dəb·əl ¦həmp 'fish·ən ‚bar·ē·ər }

double-hung [BUILD] Of a window, having top and bottom sashes which are counterweighted or equipped with a spring on each side for easier raising and lowering. { ¦dəb·əl 'həŋ }

double image [ELECTR] A television picture consisting of two overlapping images due to reception of the signal over two paths of different length so that signals arrive at slightly different times. { ¦dəb·əl 'im·ij }

double impeller breaker See impact breaker. { ¦dəb·əl im'pel·ər ‚brāk·ər }

double integral [MATH] The Riemann integral of functions of two variables. { ¦dəb·əl 'in·tə·grəl }

double-integrating gyro [MECH] A single-degree-of-freedom gyro having essentially no restraint of its spin axis about the output axis. { ¦dəb·əl ¦in·tə‚grād·iŋ 'jī·rō }

double jack [DES ENG] A heavy hammer, weighing about 10 pounds (4.5 kilograms), requiring the use of both hands. See double drill column. { ¦dəb·əl 'jak }

double jacking See doublehand drilling. { ¦dəb·əl 'jak·iŋ }

double-J groove weld [MET] A groove weld in which one member has a joint edge in the form of a double J or two half U's, one from each side. Also known as double-U groove weld. { ¦dəb·əl 'jā ‚grüv ‚weld }

double knit [TEXT] Firm, often reversible knitted fabric made by using a double set of needles to produce a single two-layer fabric; weave may be plain or fancy patterns or textures. { ¦dəb·əl 'nit }

double law of the mean See second mean-value theorem. { ¦dəb·əl 'lò əv thə 'mēn }

double layer See electric double layer. { ¦dəb·əl 'lā·ər }

double-length number [COMPUT SCI] A number having twice as many digits as are ordinarily used in a given computer. Also known as double-precision number. { ¦dəb·əl 'leŋkth 'nəm·bər }

double limiter See cascade limiter. { ¦dəb·əl 'lim·əd·ər }

double-list sorting [COMPUT SCI] A method of internal sorting in which the entire unsorted list is first placed in one portion of main memory and sorting action then takes place, creating a sorted list, generally in another area of memory. { ¦dəb·əl ‚list 'sórd·iŋ }

double load [ENG] A charge separated by inert material in a borehole. { ¦dəb·əl 'lōd }

double-loop pattern [FOREN SCI] A whorl type of fingerprint pattern consisting of two separate loop formations and two deltas. { ¦dəb·əl 'lüp 'pad·ərn }

double mast See A frame. { ¦dəb·əl 'mast }

double minimal surface [MATH] A minimal surface that is also a one-sided surface. { ¦dəb·əl ¦min·ə·məl 'sər·fəs }

double minute chromosomes [GEN] Chromatin circles that vary in size from tiny dots (common) to the size of a large chromosome, consisting of multiple copies of a short rearranged DNA segment that has undergone amplification. { ¦dəb·əl mī‚nyüt 'krō·mə‚sōmz }

double mirror [OPTICS] Two plane mirrors inclined at an angle to each other. { ¦dəb·əl 'mir·ər }

double moding [ELECTR] Undesirable shifting of a magnetron from one frequency to another at irregular intervals. { ¦dəb·əl 'mōd·iŋ }

double modulation [COMMUN] A method of modulation in which a subcarrier is first modulated with the desired intelligence, and the modulated subcarrier is then used to modulate a second carrier having a higher frequency. { ¦dəb·əl ‚mäj·ə'lā·shən }

double nickel salt See nickel ammonium sulfate. { ¦dəb·əl ¦nik·əl 'sólt }

double pendulum [MECH] Two masses, one suspended from a fixed point by a weightless string or rod of fixed length, and the other similarly suspended from the first; often the system is constrained to remain in a vertical plane. { ¦dəb·əl 'pen·jə·ləm }

double-pipe exchanger [CHEM ENG] Fluid-fluid heat exchanger made of two concentric pipe sections; one fluid (such as a coolant) flows in the annular space between pipes, and the other fluid (such as hot process stream) flows through the inner pipe. { ¦dəb·əl ‚pīp iks'chān·jər }

double point [MATH] A point on a curve at which a curve crosses or touches itself, or has a cusp; that is, a point at which the curve has two tangents (which may be coincident). { ¦dəb·əl 'póint }

double-polarity pulse-amplitude modulation [COMMUN] Pulse-amplitude modulation employing pulses of positive and negative polarity, the average value being equal to zero. Also known as bidirectional pulse-amplitude modulation. { ¦dəb·əl pə'lar·əd·ē 'pəls ¦am·plə‚tüd ‚mäj·ə'lā·shən }

double-pole double-throw switch [ELEC] A six-terminal switch or relay contact arrangement that simultaneously connects one pair of terminals to either of two other pairs of terminals. Abbreviated dpdt switch. { ¦dəb·əl ¦pōl ¦dəb·əl ¦thrō 'swich }

double-pole single-throw switch [ELEC] A four-terminal switch or relay contact arrangement that simultaneously opens or closes two separate circuits or both sides of the same circuit. Abbreviated dpst switch. { ¦dəb·əl ¦pōl ¦siŋ·gəl ¦thrō 'swich }

double-pole switch [ELEC] A switch that operates simultaneously in two separate electric circuits or in both lines of a single circuit. { ¦dəb·əl ¦pōl 'swich }

double precision [COMPUT SCI] The use of two computer words to represent a double-length number. { ¦dəb·əl prə'sizh·ən }

double-precision hardware [COMPUT SCI] Special arithmetic units in a computer designed to handle double-length numbers, employed in operations in which greater accuracy than normal is desired. { ¦dəb·əl prə'sizh·ən 'härd‚wer }

double-precision number See double-length number. { ¦dəb·əl prə'sizh·ən 'nəm·bər }

double-pulse recording [COMPUT SCI] A technique for recording binary digits in magnetic cells in which each cell consists of two regions that can be magnetized in opposite directions and the value of each bit (0 or 1) is determined by the order in which the regions occur. { ¦dəb·əl 'pəls ri'kórd·iŋ }

double pulsing [NAV] The transmitting of loran signals of two rates by a single station. { ¦dəb·əl 'pəls·iŋ }

double-pulsing station [NAV] A master loran station that controls two slaves and therefore emits pulses at two different rates. { ¦dəb·əl 'pəls·iŋ ‚stā·shən }

double-quantum stimulated-emission device [OPTICS] A laser in which the crystal contains two species of fluorescent ions whose fluorescence frequencies are so related that, when the flash lamp coils produce pumping action, the ions of one species contribute photons to the fluorescence of the other species, which is the active ion. { ¦dəb·əl 'kwänt·əm ¦stim·yə‚läd·əd ē'mish·ən də‚vīs }

double quantum transition [ATOM PHYS] A radiative transition between atomic or molecular states in which two or more photons are simultaneously emitted or absorbed. { ¦dəb·əl 'kwänt·əm tran'sish·ən }

DOUBLE HOOKE'S JOINT

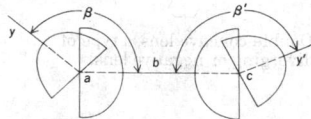

The arrangement in a double Hooke's joint; axes y and y' of the two yokes attached to intermediate shaft b lie respectively in planes containing axes of adjoining shafts (a, y, b in the same plane; b, y', c in the same plane) and the angle β between the driving and intermediate shafts equal angle β' between the intermediate and driven shafts. (*From C. W. Ham, E. J. Crane, and W. L. Rogers, Mechanics of Machinery, McGraw-Hill, 1958*)

DOUBLE-LOOP PATTERN

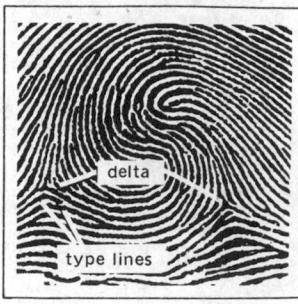

Double-loop pattern in a fingerprint. (*Federal Bureau of Investigation*)

double-quirked bead *See* quirk bead. { ¦dəb·əl ¦kwərkt ′bēd }

doubler [ELEC] *See* frequency doubler; voltage doubler. [MATER] A localized area of extra layers of reinforcement on a section of composite material to provide extra strength to a specified site. { ′dəb·lər }

double-recoil system [ORD] A system in which the gun recoils on the top carriage and the top carriage recoils on the bottom carriage or trail. { ¦dəb·əl ¦rē,kȯil ‚sis·təm }

double refraction *See* birefringence. { ¦dəb·əl ri′frak·shən }

double-replacement reaction [CHEM] A chemical reaction between compounds in which the elements in the reactants recombine to form two different compounds, each of the products having one element from each of the reactants. { ¦dəb·əl ri′plās·mənt rē‚ak·shən }

double-rivet [ENG] To rivet a lap joint with two rows of rivets or a butt joint with four rows. { ¦dəb·əl ′riv·ət }

double roll [GRAPHICS] A technique in which the ink rollers cover the form twice for each printing impression. { ¦dəb·əl ′rōl }

double-roll crusher [MECH ENG] A machine which crushes materials between teeth on two roll surfaces; used mainly for coal. { ¦dəb·əl ′rōl ′krəsh·ər }

double root [MATH] For an algebraic equation, a number a such that the equation can be written in the form $(x-a)^2 p(x) = 0$ where $p(x)$ is a polynomial of which a is not a root. { ¦dəb·əl ′rüt }

double rose [LAP] A gem cut to resemble two roses base to base, with 48 facets. { ¦dəb·əl ′rōz }

double salt [INORG CHEM] **1.** A salt that upon hydrolysis forms two different anions and cations. **1.** A salt that is a molecular combination of two other salts. { ¦dəb·əl ′sȯlt }

double sampling [IND ENG] Inspecting one sample and then deciding whether to accept or reject the lot or to defer action until a second sample is inspected. { ¦dəb·əl ′sam·pliŋ }

double screen [ELECTR] Three-layer cathode-ray tube screen consisting of a two-layer screen with the addition of a second long-persistence coating having a different color and different persistence from the first. { ¦dəb·əl ′skrēn }

double series [MATH] A two-dimensional array of numbers whose sum is the limit of Sm,n, the sum of the terms in the rectangular array formed by the first n terms in each of the first m rows, as m and n increase. { ¦dəb·əl ′sir‚ēz }

double sextant [NAV] A sextant designed to enable the observer to measure simultaneously the left and right horizontal sextant angles of a three-point problem; this sextant differs from a standard sextant in that it comprises a circle with two indices inscribed and two coaxial index mirrors; the horizon glass is silvered top and bottom but is clear in the middle to permit direct observation of the center object ashore. { ¦dəb·əl ′sek·stənt }

double-shield enclosure [ELEC] Type of shielded enclosure or room in which the inner wall is partially isolated electrically from the outer wall. { ¦dəb·əl ‚shēld in′klō·zhər }

double-shot molding [ENG] A means of turning out two-color parts in thermoplastic materials by successive molding operations. { ¦dəb·əl ‚shät ′mōld·iŋ }

double-sideband modulation [COMMUN] Amplitude modulation in which the modulated wave is composed of a carrier, an upper sideband whose frequency is the sum of the carrier and modulation frequencies, and a lower sideband whose frequency is the difference between the carrier and modulation frequencies. Abbreviated DSB. Also known as double-sideband transmitted-carrier modulation (DSB-TC modulation; DSTC modulation). { ¦dəb·əl ¦sīd,band ‚māj·ə′lā·shən }

double-sideband reduced-carrier modulation [COMMUN] A form of amplitude modulation in which both the upper and lower sidebands are transmitted but the power contained in the unmodulated carrier is reduced to a fixed level below that provided to the modulator. Abbreviated DSB-RC modulation. { ‚dəb·əl ¦sīd,band ri‚düst ¦kar·ē·ər ‚mä·jə‚lā·shən }

double-sideband suppressed-carrier modulation [COMMUN] A form of amplitude modulation in which both the upper and lower sidebands are transmitted but the power contained in the unmodulated carrier is reduced to a fixed level below that provided to the modulator. Abbreviated DSB-SC modulation. { ‚dəb·əl ¦sīd,band sə‚prest ¦kar·ē·ər ‚mäj·ə‚lā·shən }

double-sideband transmission [COMMUN] The transmission of a modulated carrier wave accompanied by both of the sidebands resulting from modulation; the upper sideband corresponds to the sum of the carrier and modulation frequencies, whereas the lower sideband corresponds to the difference between the carrier and modulation frequencies. { ¦dəb·əl ¦sīd,band tranz′mish·ən }

double-sideband transmitted-carrier modulation *See* double-sideband modulation. { ¦dəb·əl ¦sīd,band tranz¦mid·əd ¦kar·ē·ər ‚mäj·ə′lā·shən }

double-sided board [ELECTR] A printed wiring board that contains circuitry on both external layers. { ¦dəb·əl ‚sīd·əd ′bȯrd }

double-sided disk [COMPUT SCI] A diskette that can be written on both of its sides. { ¦dəb·əl ¦sīd·əd ′disk }

double-slider coupling *See* slider coupling. { ¦dəb·əl ¦slīd·ər ′kəp·liŋ }

double-slit interference *See* Young's two-slit interference. { ′dəb·əl ‚slit ‚in·tər′fir·əns }

double-solvent refining [CHEM ENG] Petroleum-refining process using two solvents to simultaneously deasphalt and solvent-treat lubricating-oil stocks. { ¦dəb·əl ¦säl·vənt rə′fīn·iŋ }

double square *See* adjustable square. { ¦dəb·əl ′skwer }

double star [ASTRON] A star which appears as a single point of light to the eye but which can be resolved into two points by a telescope. { ¦dəb·əl ′stär }

double-stream amplifier [ELECTR] Microwave traveling-wave amplifier in which amplification occurs through interaction of two electron beams having different average velocities. { ¦dəb·əl ‚strēm ′am·plə‚fī·ər }

double-strike [GRAPHICS] To print a character twice, with only a small displacement between the first and second printings, in order to darken the image. { ′dəb·əl ‚strīk }

double-stub tuner [ELECTROMAG] Impedance-matching device, consisting of two stubs, usually fixed three-eighths of a wavelength apart, in parallel with the main transmission lines. { ¦dəb·əl ‚stəb ′tün·ər }

double-superheterodyne reception [COMMUN] Method of reception in which two frequency converters are employed before final detection. Also known as triple detection. { ¦dəb·əl ‚sü·pər¦het·rə‚dīn ri′sep·shən }

doublet [ATOM PHYS] Two stationary states which have the same orbital and spin angular momentum but which have different total angular momenta, and therefore have slightly different energies due to spin-orbit coupling. [ELECTROMAG] *See* dipole. [FL MECH] A source and a sink separated by an infinitesimal distance, each having an infinitely large strength so that the product of this strength and the separation is finite. [OPTICS] A lens made up of two components, especially an achromat. [PARTIC PHYS] Two elementary particles which have slightly differing masses and the same baryon number, spin, parity, and charge conjugation parity (if self-conjugate), but have different charges. [PHYS CHEM] Two electrons which are shared between two atoms and give rise to a nonpolar valence bond. [SPECT] Two closely separated spectral lines arising from a transition between a single state and a pair of states forming a doublet as described in the atomic physics definition. { ′dəb·lət }

double tangent [MATH] A line which is tangent to a curve at two distinct noncoincident points. Also known as bitangent. Two coincident tangents to branches of a curve at a given point, such as the tangents to a cusp. { ¦dəb·əl ′tan·jənt }

doublet antenna *See* dipole antenna. { ′dəb·lət an′ten·ə }

doublet tempering [MET] A technique for ensuring the stability of the microstructure of a quench-hardened steel by subjecting it to two tempering cycles at approximately the same temperature. { ′dəb·əl ′tem·pər·iŋ }

doublet flow [FL MECH] The motion of a fluid in the vicinity of a doublet; can be superposed with uniform flow to yield flow around a cylinder or a sphere. { ′dəb·lət ‚flō }

double-theodolite observation [ENG] A technique for making winds-aloft observations in which two theodolites located at either end of a base line follow the ascent of a pilot balloon; synchronous measurements of the elevation and azimuth angles of the balloon, taken at periodic intervals, permit computation of the wind vector as a function of height. { ¦dəb·əl thē′äd·əl‚īt äb·zər′vā·shən }

double-throw circuit breaker [ELEC] Circuit breaker by

DOUBLET

Equipotential lines and streamlines for the two-dimensional doublet, as used in fluid mechanics.

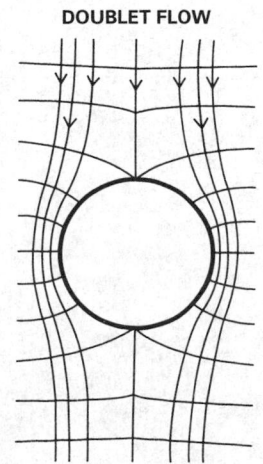

DOUBLET FLOW

Streamlines and equipotential lines for uniform flow about a sphere at rest.

means of which a change in the circuit connections can be obtained by closing either of two sets of contacts. { ¦dəb·əl ‚thrō 'sər·kət ‚brāk·ər }

double-throw switch [ELEC] A switch that connects one set of two or more terminals to either of two other similar sets of terminals. { ¦dəb·əl ‚thrō 'swich }

double tide [OCEANOGR] A high tide comprising two maxima of nearly identical height separated by a relatively small depression, or low tide comprising two minima separated by a relatively small elevation. { ¦dəb·əl 'tīd }

double-track tape recorder [ENG ACOUS] A tape recorder with a recording head that covers half the tape width, so two parallel tracks can be recorded on one tape. Also known as dual-track tape recorder; half-track tape recorder. { ¦dəb·əl ‚trak 'tāp ri‚kȯrd·ər }

double triode [ELECTR] An electron tube having two triodes in the same envelope. Also known as duotriode. { ¦dəb·əl 'trī‚ōd }

doublet trigger [ELECTR] A trigger signal consisting of two pulses spaced a predetermined amount for coding purposes. { 'dəb·lət ‚trig·ər }

double-tuned amplifier [ELECTR] Amplifier of one or more stages in which each stage uses coupled circuits having two frequencies of resonance, to obtain wider bands than those obtainable with single tuning. { ¦dəb·əl ‚tünd 'am·plə‚fī·ər }

double-tuned circuit [ELECTR] A circuit that is resonant to two adjacent frequencies, so that there are two approximately equal values of peak response, with a dip between. { ¦dəb·əl ‚tünd 'sər·kət }

double-tuned detector [ELECTR] A type of frequency-modulation discriminator in which the limiter output transformer has two secondaries, one tuned above the resting frequency and the other tuned an equal amount below. { ¦dəb·əl ‚tünd di'tek·tər }

double-U groove weld See double-J groove weld. { ¦dəb·əl 'yü ‚grüv ‚weld }

double valves [PETRO ENG] Two valves in series used as subsurface traveling or standing valves in wells, the dual arrangement being more reliable than a single valve. { ¦dəb·əl 'valvz }

double-V groove weld See double-bevel groove weld. { ¦dəb·əl 'vē ‚grüv ‚weld }

double-wall cofferdam [CIV ENG] A cofferdam consisting of two lines of steel piles tied to each other, and having the space between filled with sand. { ¦dəb·əl ‚wȯl 'kȯf·ər‚dam }

double-wedge cut [MIN ENG] A drill-hole pattern composed of a shallow wedge within a larger, outer wedge. { ¦dəb·əl ‚wej 'kət }

double weighing [MECH] A method of weighing to allow for differences in lengths of the balance arms, in which object and weights are balanced twice, the second time with their positions interchanged. Also known as Gauss method of weighing. { ¦dəb·əl 'wā·iŋ }

double-welded joint [MET] Any joint that has been welded from both sides. { ¦dəb·əl ‚weld·əd 'jȯint }

double-winding synchronous generator [ELEC] Synchronous generator which has two similar windings, in phase with one another, mounted on the same magnetic structure but not connected electrically, designed to supply power to two independent external circuits. { ¦dəb·əl ‚wīnd·iŋ ¦siŋ·krə·nəs 'jen·ə‚rād·ər }

double word [COMPUT SCI] A unit containing twice as many bits as a word. { ¦dəb·əl 'wərd }

double-word addressing [COMPUT SCI] An addressing mode in computers with short words (less than 16 bits) in which the second of two consecutive instruction words contains the address of a location. { ¦dəb·əl ‚wərd a‚dres·iŋ }

double-work [BOT] In plant propagation, to graft or bud a scion to an intermediate variety that is itself grafted on a stock of still another variety. { ¦dəb·əl 'wərk }

doubling dose [GEN] The radiation dose that would double the rate of spontaneous mutation. { 'dəb·liŋ ‚dōs }

doubling the angle on the bow [NAV] A method of obtaining a running fix by measuring the distance a craft travels while the relative bearing (right or left) of a fixed object doubles; the distance from the object at the time of the second bearing is equal to the run between bearings, neglecting drift. { 'dəb·liŋ ‚thə 'aŋ·gəl ȯn ‚thə 'baů }

DOUGLAS-FIR

Cone on a branch of Douglas-fir
(*Pseudotsuga menziesii*)

doubling time [NUCLEO] The time required for a breeding reactor to double its fuel inventory. { 'dəb·liŋ ‚tīm }

doubly linked ring [COMPUT SCI] A cycle arrangement of data elements in which searches are possible in both directions. { ¦dəb·lē ¦liŋkt 'riŋ }

doubly plunging fold [GEOL] A fold that plunges in opposite directions, either away from or toward a central point. { ¦dəb·lē ‚plənj·iŋ 'fōld }

doubly ruled surface [MATH] A ruled surface that can be generated by either of two distinct moving straight lines; quadric surfaces are the only surfaces of this type. { ¦dəb·lē ¦rüld 'sər·fəs }

doubly stochastic matrix [MATH] A matrix of nonnegative real numbers such that every row sum and every column sum are equal to 1. { ¦dəb·lē stō¦kas·tik 'mā·triks }

doughnut [NUCLEO] Also spelled donut. **1.** The toroidal vacuum chamber in which electrons are accelerated in a betatron or synchrotron. Also known as toroid. **2.** An assembly of enriched fissionable material, often doughnut-shaped, used in a thermal reactor to provide a local increase in fast neutron flux for experimental purposes. [PETRO ENG] A ring of wedges or a threaded, tapered ring that supports a pipe string. { 'dō‚nət }

Douglas-fir [BOT] *Pseudotsuga menziesii*. A large coniferous tree in the order Pinales; cones are characterized by bracts extending beyond the scales. Also known as red fir. { ¦dəg·ləs 'fər }

Douglas-fir oil See pine needle oil. { ¦dəg·ləs ‚fər 'ȯil }

douglasite [MINERAL] $K_2FeCl_4·2H_2O$ Ore from Stassfurt, Germany; a member of the erythrosiderite group; orthorhombic, in the isomorphous series. { 'dəg·lə‚sīt }

Dounce homogenizer [BIOL] An apparatus consisting of a glass tube with a tight-fitting glass pestle used manually to disrupt tissue suspensions to obtain single cells or subcellular fractions. { 'daůns hə'mäj·ə‚nīz·ər }

do-until structure [COMPUT SCI] A set of program statements that is executed once, and may then be executed repeatedly, depending on the results of a test specified in the first statement. { 'dü ən'til ‚strək·chər }

douse [MET] To thrust a hot piece of metal into a liquid during the hardening process. [MIN ENG] To locate and delineate subsurface resources such as water, oil, gas, or minerals. { daůs }

douser See doodlebug. { 'daůs·ər }

Dovap See Doppler velocity and position. { 'dō‚vap }

dove [VERT ZOO] The common name for a number of small birds of the family Columbidae. { dəv }

Dove See Columba. { dəv }

Dove prism See Delaborne prism. { 'dəv ‚priz·əm }

dovetailing [MET] In thermal spraying, roughening the surface by angular cutting prior to the deposit of sprayed material. { 'dəv‚tāl·iŋ }

dovetail joint [DES ENG] A joint consisting of a flaring tenon in a fitting mortise. { 'dəv‚tāl 'jȯint }

dovetail molding [ARCH] A molding in a zig-zag pattern resembling a series of dovetails. { 'dəv‚tāl 'mōld·iŋ }

dovetail saw [DES ENG] A short stiff saw with a thin blade and fine teeth; used for accurate woodwork. { 'dəv‚tāl 'sȯ }

dowel [DES ENG] **1.** A headless, cylindrical pin which is sunk into corresponding holes in adjoining parts, to locate the parts relative to each other or to join them together. Also known as dowel pin. **2.** A round wooden stick from which dowel pins are cut. { 'daůl }

dowel pin See dowel. { 'daůl ‚pin }

dowel plate [DES ENG] A hardened steel plate with drilled holes that is used to fashion dowels by driving pegs through the holes to remove excess wood. { 'daůl ‚plāt }

dowel screw [DES ENG] A dowel with threads at both ends. { 'daůl ‚skrü }

do-while structure [COMPUT SCI] A set of program statements that is executed repeatedly, as long as some condition, specified in the first statement, remains in effect. { 'dü 'wīl ‚strək·chər }

down [ENG] Not in operation. [GEOL] Hillock of sand thrown up along the coast by the sea or the wind. A flat eminence on the top of a hill or mountain. { daůn }

down by the head [NAV ARCH] Having greater draft at the bow than at the stern. Also known as by the head. { ¦daůn bī ‚thə 'hed }

down by the stern [NAV ARCH] Having a greater draft at the stern than at the bow. Also known as by the stern. { ¦daủn bī thə 'stərn }

downcast [MIN ENG] Intake shaft for air in a mine. { 'daủn‚kast }

downcomer [BUILD] *See* downspout. [CHEM ENG] A method of conveying liquid from one tray to the one below in a bubble-tray column. [ENG] In an air-pollution control system, a pipe that conducts gases downward to a device that removes undesirable substances. [MECH ENG] A tube in a boiler waterwall system wherein the fluid flows downward. { 'daủn‚kəm·ər }

down cutting *See* climb cutting. { 'daủn ‚kəd·iŋ }

downcutting [GEOL] Stream erosion in which the cutting is directed in a downward direction. { 'daủn‚kəd·iŋ }

downdip [GEOL] Pertaining to a position parallel to or in the direction of the dip of a stratum or bed. { 'daủn‚dip }

down-Doppler [ACOUS] The sonar situation wherein the target is moving away from the transducer, so that the frequency of the echo is less than the frequency of the reverberations received immediately after the end of the outgoing ping; opposite of up-Doppler. { 'daủn ‚däp·lər }

downdraft [PHYS] A current of air or other gas that travels downward, as during a thunderstorm or in a mine shaft. { 'daủn‚draft }

downdraft carburetor [MECH ENG] A carburetor in which the fuel is fed into a downward current of air. { 'daủn‚draft 'kär·bə‚rād·ər }

down-feed system [MECH ENG] In a heating or cooling system, a piping arrangement in which the fluid is circulated through supply mains that are located above the levels of the units they serve. { 'daủn ‚fēd ‚sis·təm }

downflow [CHEM] In an ion-exchange system, the direction of the flow of the solution being processed. { 'daủn‚flō }

downhand welding *See* flat-position welding. { 'daủn‚hand 'weld·iŋ }

downhole drill [MIN ENG] A hammer or percussive drill in which a reciprocating pneumatic piston is located immediately behind the drill bit and can follow and enter the bit down the hole, for minimizing energy losses. { 'daủn‚hōl 'dril }

downhole equipment *See* drill fittings. { 'daủn‚hōl i¦kwip·mənt }

down-lead *See* lead-in. { 'daủn ‚lēd }

downlink [COMMUN] The radio or optical transmission path downward from a communications satellite to the earth or an aircraft, or from an aircraft to the earth. { 'daủn‚liŋk }

download [COMPUT SCI] To transfer a program or data file from a central computer to a remote computer or to the memory of an intelligent terminal. { 'daủn‚lōd }

down lock [AERO ENG] An airplane mechanism that locks the landing gear in a down position after the gear is lowered. { 'daủn ‚läk }

down milling *See* climb cutting. { 'daủn ‚mil·iŋ }

down quark [PARTIC PHYS] A quark with an electric charge of $-1/3$, baryon number of $1/3$, and 0 strangeness and charm. { 'daủn ‚kwärk }

downrange [AERO ENG] Any area along the flight course of a rocket or missile test range. { 'daủn‚rānj }

downrush [METEOROL] A term sometimes applied to the strong downward-flowing air current that marks the dissipating stages of a thunderstorm. { 'daủn‚rəsh }

Downs cell [CHEM ENG] A brick-lined steel vessel with four graphite anodes projecting upward from the bottom, with cathodes in the form of steel cylinders concentric with the anodes, containing an electrolyte which is 40% sodium chloride (NaCl) and 60% calcium chloride ($CaCl_2$) at 590°C; used to make sodium. { 'daủnz ‚sel }

downslope time [MET] Time necessary for current decrease when using slope control in resistance welding. { 'daủn‚slōp ‚tīm }

downspout [BUILD] A vertical pipe that leads water from a roof drain or gutter down to the ground or a cistern. Also known as downcomer; leader. { 'daủn‚spaủt }

Down's process [CHEM ENG] A method for producing sodium and chlorine from sodium chloride; potassium chloride and fluoride are added to the sodium chloride to reduce the melting point; the fused mixture is electrolyzed, with sodium forming at the cathode and chlorine at the anode. { 'daủnz ‚präs·əs }

Down's syndrome [MED] A syndrome of congenital defects, especially mental retardation, typical facies responsible for the term mongolism, and cytogenetic abnormality consisting of trisomy 21 or its equivalent in the form of an unbalanced translocation. Also known as mongolism; trisomy 21 syndrome. { 'daủnz ‚sin‚drōm }

downstream [CHEM ENG] Portion of a product stream that has already passed through the system; that portion located after a specific process unit. [GEN] Further along in the direction of transcription on one strand of the deoxyribonucleic acid sequence of a gene; for a linked gene, downstream may be in the opposite direction along the chromosome. [HYD] In the direction of flow, as a current or waterway. { 'daủn‚strēm }

downthrow [GEOL] The side of a fault whose relative movement appears to have been downward. { 'daủn‚thrō }

downtime [IND ENG] The lost production time during which a piece of equipment is not operating correctly due to a breakdown, maintenance, necessities, or power failure. { 'daủn‚tīm }

downward compatibility [COMPUT SCI] The ability of an older or smaller computer to accept programs from a newer or larger one. Also known as backward compatibility. { 'daủn·wərd kəm‚pad·ə'bil·əd·ē }

downwarp [GEOL] A segment of the earth's crust that is broadly bent downward. { 'daủn‚wȯrp }

downwash [FL MECH] The downward deflection of air, relative to the direction of motion of an airfoil. { 'daủn‚wäsh }

downwelling *See* sinking. { 'daủn‚wel·iŋ }

downwind [NAV] In the direction toward which the wind is blowing; applies particularly to moving downwind, whether desired or not. { 'daủn‚wind }

downy mildew [PL PATH] A fungus disease of higher plants caused by members of the family Peronosporaceae and characterized by a white, downy growth on the diseased plant parts. { ¦daủn·ē 'mil·dü }

Dow oscillator *See* electron-coupled oscillator. { ¦daủ 'äs·ə‚läd·ər }

dowsing *See* divining. { 'daủz·iŋ }

dowsing rod *See* divining rod. { 'daủz·iŋ ‚räd }

Dowtonian [GEOL] Uppermost Silurian or lowermost Devonian geologic time. { daủ'tōn·ē·ən }

Dowty prop [MIN ENG] A self-contained hydraulic supporting post consisting of two telescoping tubes; the upper (inner) tube contains the oil, pump, yield valve, and other accessories. { 'daủ·dē ‚präp }

doz *See* dozen. { }

dozen [SCI TECH] A group of 12 items. Abbreviated doz. { 'dəz·ən }

D.P. *See* degree of polymerization.

2,4-DP *See* dichlorprop.

DPCM *See* differential pulse-code modulation.

dpdt switch *See* double-pole double-throw switch. { ¦dē¦pē¦d‚ē¦tē ‚swich }

DPE *See* direct plate exposure.

DPG *See* diphenylguanidine.

dpi *See* dots per inch.

DPMS *See* display power management signaling.

DPN *See* diphosphopyridine nucleotide.

dpst switch *See* double-pole single-throw switch. { ¦dē¦pē¦es‚tē ‚swich }

DQ Herculis star *See* intermediate polar. { ¦dē¦kyü 'hər·kyə·ləs ‚stär }

dr *See* dram.

Dra *See* Draco.

Drac *See* Draco.

drachm *See* dram. { dram }

Draco [ASTRON] A long, serpentine constellation that surrounds half of the Little Dipper in the north. Abbreviated Dra; Drac. Also known as Dragon. { 'drā‚kō }

Draconematoidea [INV ZOO] A superfamily of marine nematodes in the order Desmodorida distinguished by a body that, when relaxed, is dorsally and then ventrally arched into a shallow sigmoid shape. { ‚drā‚kō‚nem·ə'tȯid·ē·ə }

draconic month *See* nodical month. { drə‚kän·ik 'mənth }

draconic year *See* eclipse year. { drə‚kän·ik 'yir }

Draconids [ASTRON] Several meteor showers whose radiants lie in the constellation Draco. { drə'kän·ədz }

Draco system [ASTRON] A dwarf elliptical galaxy in the

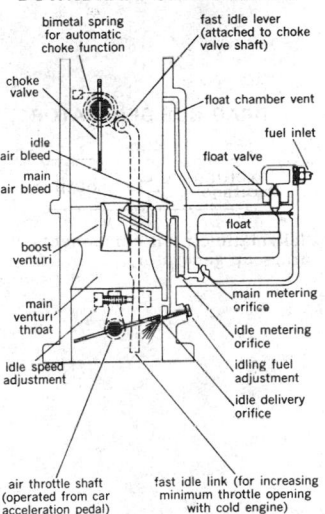

DOWNDRAFT CARBURETOR

bimetal spring for automatic choke function
fast idle lever (attached to choke valve shaft)
choke valve
float chamber vent
idle air bleed
fuel inlet
main air bleed
float valve
float
boost venturi
main venturi throat
main metering orifice
idle metering orifice
idling fuel adjustment
idle speed adjustment
idle delivery orifice
air throttle shaft (operated from car acceleration pedal)
fast idle link (for increasing minimum throttle opening with cold engine)

Idling fuel circuit and choke device for starting and warm-up in typical downdraft carburetor.

Local Group about 250,000 light-years (1.5×10^{18} miles or 2.4×10^{18} kilometers) distant having a diameter of about 3700 light-years (2.2×10^{16} miles or 3.5×10^{16} kilometers) and consisting chiefly of older stars. { 'drā·kō ,sis·təm }

Dracunculoidea [INV ZOO] An order or superfamily of parasitic nematodes characterized by their habitat in host tissues and by the way larvae leave the host through a skin lesion. { drə,kəŋ·kyə'lōid·ē·ə }

Draeger breathing apparatus [MIN ENG] A long-service, self-contained oxygen-breathing apparatus with the oxygen feed governed by the lungs; allows the user to do hard work for up to 5 hours and normal work for 7 hours, and can sustain a resting individual for 18 hours. { 'drāg·ər 'brēth·iŋ ,ap·ə,rad·əs }

Draeger escape apparatus [MIN ENG] A portable, self-contained oxygen-breathing apparatus that is carried on the back of the user; protects against poisonous gases or oxygen shortages for 1 hour. { 'drāg·ər ə'skāp ,ap·ə,rad·əs }

draft Also spelled draught. [CIV ENG] A line of a traverse survey. [ENG] **1.** In molds, the degree of taper on a side wall or the angle of clearance present to facilitate removal of cured or hardened parts from a mold. **2.** The area of a water discharge opening. [FL MECH] **1.** An air current in a confined space, such as that in a cooling tower or chimney. **2.** The difference between atmospheric pressure and some lower pressure in a confined space that causes air to flow, such as exists in the furnace or gas passages of a steam-generating unit or in a chimney. [MET] **1.** The act or process of drawing, with dies. **2.** The work or quantity of work drawn. [NAV ARCH] The vertical distance from the top of the keel plate or bar keel to the load waterline. { draft }

draft differential [FL MECH] The difference in static pressure between two locations of gas flow. { 'draft dif·ə'ren·chəl }

draft gage [ENG] **1.** A modified U-tube manometer used to measure draft of low gas heads, such as draft pressure in a furnace, or small differential pressures, for example, less than 2 inches (5 centimeters) of water. **2.** A hydrostatic depth indicator, installed in the side of a vessel below the light load line, to indicate amount of submergence. { 'draft ,gāj }

draft hood [ENG] A device used to facilitate the escape of combustion products from the combustion chamber of an appliance, to prevent a backdraft in the combustion chamber, and to neutralize the effect of stack action of the chinney or gas vent on the efficient operation of the appliance. { 'draft ,hud }

drafting [GRAPHICS] The making of drawings of objects, structures, or systems that have been visualized by engineers, scientists, and others. [TEXT] **1.** The process of lengthening raw fibers, in the form of slubbing, sliver, or roving, to make the stock look more like yarn. **2.** Plotting directions for weaving on cross-section paper, showing the movement of the threads. { 'draf·tiŋ }

drafting machine *See* parallel motion protractor. { 'draf·tiŋ mə'shēn }

drafting paper [MATER] A fine white or cream-colored paper that is hard-surfaced and has good erasing characteristics. { 'draf·tiŋ ,pā·pər }

draft loss [MECH ENG] A decrease in the static pressure of a gas in a furnace or boiler due to flow resistance. { 'draft ,lòs }

draft marks [NAV ARCH] Numbers at the bow and stern of a ship to indicate the height from the bottom of keel or lowest projection. { 'draft ,märks }

draftsman [ENG] An individual skilled in drafting, especially of machinery and structures. { 'draf·smən }

draft tube [MECH ENG] The piping system for a reaction-type hydraulic turbine that allows the turbine to be set safely above tail water and yet utilize the full head of the site from head race to tail race. { 'draf ,tüb }

drag [COMPUT SCI] To move an object across a screen by moving a pointing device while holding down the control button. [ENG] **1.** A tool fashioned from sheet steel and having a toothed edge along the long dimension; used to level and scratch plaster to produce a key for the next coat of plaster. Also known as comb. **2.** A tool consisting of a steel plate with a finely serrated edge; dragged over the surface to dress stone. [FL MECH] Resistance caused by friction in the direction opposite to that of the motion of the center of gravity of a moving body in a fluid. [MET] **1.** The bottom part of a flask used in casting. **2.** In thermal cutting, the distance deviating from the theoretical vertical line of cutting measured along

the bottom surface of the material. [MIN ENG] Movement of the hanging wall with respect to the foot wall due to the weight of the arch block in an inclined slope. { drag }

drag and drop [COMPUT SCI] A feature whereby operations are performed on objects, such as icons or blocks of text, by dragging them across the screen with a mouse. { ¦drag ən 'dräp }

drag bit *See* bit drag. { 'drag ,bit }

drag-body flowmeter [ENG] Device to meter liquid flow; measures the net force parallel to the direction of flow; the resulting pressure difference is used to solve flow equations. { 'drag ,bäd·ē 'flō,mēd·ər }

drag chain [ENG] **1.** A chain dragged along the ground from a motor vehicle chassis to prevent the accumulation of static electricity. **2.** A chain for coupling rail cars. { 'drag ,chān }

drag-chain conveyor [MECH ENG] A conveyor in which the open links of a chain drag material along the bottom of a hard-faced concrete or cast iron trough. Also known as dragline conveyor. { 'drag ,chān kən'vā·ər }

drag chute *See* drag parachute. { 'drag ,shüt }

drag classifier [MECH ENG] A continuous belt containing transverse rakes, used to separate coarse sand from fine; the belt moves up through an inclined trough, and fast-settling sands are dragged along by the rakes. { 'drag 'klas·ə,fī·ər }

drag coefficient [FL MECH] A characteristic of a body in a flowing inviscous fluid, equal to the ratio of twice the force on the body in the direction of flow to the product of the density of the fluid, the square of the flow velocity, and the effective cross-sectional area of the body. { 'drag ,kō·i'fish·ənt }

drag conveyor *See* flight conveyor. { 'drag kən'vā·ər }

drag-cup generator [ENG] A type of tachometer which uses eddy currents and functions in control systems; it consists of two stationary windings, positioned so as to have zero coupling, and a nonmagnetic metal cup, which is revolved by the source whose speed is to be measured; one of the windings is used for excitation, inducing eddy currents in the rotating cup. Also known as drag-cup tachometer. { 'drag ,kəp 'jen·ə,rād·ər }

drag-cup motor [ELEC] An induction motor having a cup-shaped rotor or conducting material, inside of which is a stationary magnetic core. { 'drag ,kəp 'mōd·ər }

drag-cup tachometer *See* drag-cup generator. { 'drag ,kəp tə'käm·əd·ər }

drag cut [ENG] A drill hole pattern for breaking out rock, in which angled holes are drilled along a floor toward a parting, or on a free face and then broken by other holes drilled into them. { 'drag ,kət }

drag direction [AERO ENG] In stress analysis of a given airfoil, the direction of the relative wind. { 'drag də'rek·shən }

drag factor [CHEM ENG] Ratio of hindered diffusion rate to unhindered rate through a swollen dialysis membrane. Also known as Faxen drag factor; hindrance factor. { 'drag ,fak·tər }

drag fold [GEOL] A minor fold formed in an incompetent bed by movement of a competent bed so as to subject it to couple; the axis is at right angles to the direction in which the beds slip. { 'drag ,fōld }

drag force [PL PHYS] A force on an electrically conducting fluid arising from inelastic collisions of electrons and ions and proportional to the fluid velocity. { 'drag ,fòrs }

dragging of inertial frames [RELAT] A relativistic effect whereby, loosely speaking, a spinning body drags space around with it; this causes a gyroscope in orbit about the earth to display a precession distinct from the geodetic precession. Also known as frame dragging; Lense-Thirring effect. { 'drag·iŋ əv ə¦nər·shəl 'frāmz }

drag-in [MET] Solution carried by the work and handling equipment to another solution. { 'drag ,in }

dragline [MECH ENG] An excavator operated by pulling a bucket on ropes towards the jib from which it is suspended. Also known as dragline excavator. { 'drag,līn }

dragline conveyor *See* drag-chain conveyor. { 'drag,līn kən'vā·ər }

dragline excavator *See* dragline. { 'drag,līn 'eks·kə,vād·ər }

dragline scraper [MECH ENG] A machine with a flat, plow-like blade or partially open bucket pulled on rope for withdrawing piled material, such as stone or coal, from a stockyard to the loading platform; the empty bucket is subsequently returned

---

DRAG-CUP GENERATOR

Schematic circuit components of a drag-cup generator.

to the pile of material by means of a return rope. { 'drag,līn 'skrāp·ər }

drag link [MECH ENG] A four-bar linkage in which both cranks traverse full circles; the fixed member must be the shortest link. { 'drag ,liŋk }

drag mark [GEOL] Long, even mark usually having longitudinal striations produced by current drag of an object across a sedimentary surface. { 'drag ,märk }

Dragon See Draco. { 'drag·ən }

dragonfly [INV ZOO] Any of the insects composing six families of the suborder Anisoptera and having four large, membranous wings and compound eyes that provide keen vision. { 'drag·ən,flī }

drag-out [MET] Solution taken from a bath by the work or handling equipment. { 'drag ,aut }

drag parachute [AERO ENG] Any of various types of parachutes that can be deployed from the rear of an aircraft, especially during landings, to decrease speed and also, under certain flight conditions, to control and stabilize the aircraft. Also known as drag chute. { 'drag 'par·ə,shüt }

drag rope [AERO ENG] A long, heavy rope carried in the basket of a balloon and permitted to hang over the side and drag on the ground in order to lighten the basket. { 'drag ,rōp }

dragsaw [DES ENG] A saw that cuts on the pulling stroke; used in power saws for cutting felled trees. { 'drag,so }

drag-stone mill [MIN ENG] A mill in which a heavy stone is dragged over ore to grind it. { 'drag ,stōn ,mil }

drag technique [MET] An arc-welding method in which the electrode is in contact with the joint being welded without being in short circuit. { 'drag tek,nēk }

drag truss [AERO ENG] A truss that is positioned horizontally between the wing spars; used to stiffen the wing structure and as a resistance for drag forces acting on the airplane wing. { 'drag ,trəs }

drag-type tachometer See eddy-current tachometer. { 'drag ,tīp tə'käm·əd·ər }

drag-weight ratio [AERO ENG] The ratio of the drag of a missile to its total weight. { 'drag ,wāt 'rā·shō }

drag wire [AERO ENG] A part of the truss in an airplane wing and also in the wing support; used to sustain the backward reaction due to the wing's drag. { 'drag ,wīr }

drain [CIV ENG] **1.** A channel which carries off surface water. **2.** A pipe which carries off liquid sewage. [ELEC] See current drain. [ELECTR] The region into which majority carriers flow in a field-effect transistor; it is comparable to the collector of a bipolar transistor and the anode of an electron tube. { drān }

drainage [CIV ENG] Removal of groundwater or surface water, or of water from structures, by gravity or pumping. [HYD] The pattern followed by the waters of an area as they pass or flow off in surface or subsurface streams. [PETRO ENG] The movement of reservoir oil or gas toward a wellbore due to the reduced pressure that results from penetration of the reservoir by the well. { 'drān·ij }

drainage area See drainage basin. { 'drān·ij ,er·ē·ə }

drainage basin [HYD] An area in which surface runoff collects and from which it is carried by a drainage system, as a river and its tributaries. Also known as catchment area; drainage area; feeding ground; gathering ground; hydrographic basin. { 'drān·ij ,bā·sən }

drainage canal [CIV ENG] An artificial canal built to drain water from an area having no natural outlet for precipitation accumulation. { 'drān·ij kə,nal }

drainage density [HYD] Ratio of the total length of all channels in a drainage basin to the basin area. { 'drān·ij ,den·səd·ē }

drainage divide [GEOL] **1.** The border of a drainage basin. **2.** The boundary separating adjacent drainage basins. { 'drān·ij də,vīd }

drainage gallery [CIV ENG] A gallery in a masonry dam parallel to the top of the dam, to intercept seepage from the upstream face and conduct it away from the downstream face. { 'drān·ij ,gal·rē }

drainage lake [HYD] An open lake which loses water via a surface outlet or whose level is essentially controlled by effluent discharge. { 'drān·ij ,lāk }

drainage pattern [HYD] The configuration of a natural or artificial drainage system; stream patterns reflect the topography and rock patterns of the area. { 'drān·ij ,pad·ərn }

drainage point [PETRO ENG] A wellbore that is draining a petroleum or gas reservoir. { 'drān·ij ,point }

drainage ratio [HYD] The ratio expressing runoff compared with precipitation in a specific area for a given time period. { 'drān·ij ,rā·shō }

drainage system [HYD] A surface stream or a body of impounded surface water, together with all other such streams and bodies that are tributary, by which a geographical area is drained. { 'drān·ij ,sis·təm }

drainage well [CIV ENG] A vertical shaft in a masonry dam to intercept seepage before it reaches the downstream side. { 'drān·ij ,wel }

drainage wind See gravity wind. { 'drān·ij ,wind }

drain tile [BUILD] A cylindrical tile with holes in the walls used at the base of a building foundation to carry away groundwater. { 'drān ,tīl }

drain valve [CHEM ENG] A valve used to drain off material that has separated from a fluid or gas stream, or one used to empty a process line, vessel, or storage tank. { 'drān ,valv }

drain wire [ELEC] Metallic conductor frequently used in contact with foil-type signal-cable shielding to provide a low-resistance ground return at any point along the shield. { 'drān ,wīr }

Drake equation [ASTRON] An equation which gives the number of advanced technological civilizations curently active in the Galaxy as the product of the rate at which new stars are born in the Galaxy, the probability (actually a product of probabilities) that any one of these stars will possess the necessary conditions for life to originate and to slowly evolve to a technological civilization, and the average longevity of such civilizations. { 'drāk i,kwā·zhən }

dram [MECH] **1.** A unit of mass, used in the apothecaries' system of mass units, equal to $^1/_8$ apothecaries' ounce or 60 grains or 3.8879346 grams. Also known as apothecaries' dram (dram ap); drachm (British). **2.** A unit of mass, formerly used in the United Kingdom, equal to $^1/_{16}$ ounce (avoirdupois) or approximately 1.77185 grams. Abbreviated dr. { dram }

DRAM See dynamic random-access memory. { 'dē,ram }

dram ap See dram. { 'dram ,ap }

drape forming [ENG] A method of forming thermoplastic sheet in which the sheet is clamped into a movable frame, heated, and draped over high points of a male mold; vacuum is then applied to complete the forming operation. { 'drāp ,fòr·miŋ }

Draper catalog [ASTRON] A nine-volume catalog of stars completed in 1924; it gives positions, magnitudes, and spectral classes of 225,300 stars. { 'drā·pər 'kad·əl,äg }

Draper effect [CHEM ENG] The increase in volume at constant pressure at the start of the reaction of hydrogen and chlorine to form hydrogen chloride; the volume increase is caused by an increase in temperature of the reactants, due to heat released in the reaction. { 'drā·pər i,fekt }

draping [GEOL] Structural concordance of the strata overlying a limestone reef or other hard core to the surface of the reef or core. { 'drāp·iŋ }

draught See draft. { draft }

draught stop See fire stop. { 'draf ,stòp }

draw [ENG] To haul a load. [MET] **1.** A fissure or pocket in a casting formed when the supply of molten metal is inadequate during solidification. **2.** To remove a pattern from a foundry flask. [MIN ENG] **1.** To remove timber supports, allowing overhanging coal to fall down for collection. **2.** To allow ore to run down chutes from stopes, chambers, or ore bins. **3.** To collect broken coal in trucks. **4.** To hoist coal, rock, ore, or other materials to the surface. **5.** The horizontal distance to which creep extends on the surface beyond the stopes. { drò }

DRAW See direct read after write. { drò }

drawability [MET] The ability of a metal to be deep-drawn. { ,drò·ə'bil·əd·ē }

drawbar [ENG] **1.** A bar used to connect a tender to a steam locomotive. **2.** A beam across the rear of a tractor for coupling machines or other loads. **3.** A clay block submerged in a glass-making furnace to define the point at which sheet glass is drawn. { 'drò,bär }

drawbar horsepower [MECH ENG] The horsepower available at the drawbar in the rear of a locomotive or tractor to pull the vehicles behind it. { 'drò,bär 'hòrs,pau·ər }

DRAGONFLY

Winged adult dragonfly.

drawbar pull [MECH ENG] The force with which a locomotive or tractor pulls vehicles on a drawbar behind it. { 'drȯ ˌbär ˌpu̇l }

draw bead [MET] A projection on the surface of a metal sheet to control its flow during drawing. { 'drȯ ˌbēd }

drawbench [MET] A stand on which metal is drawn through dies; used in wire-making, or for drawing of rods and tubing. { 'drȯˌbench }

drawbridge [CIV ENG] Any bridge that can be raised, lowered, or drawn aside to provide clear passage for ships. { 'drȯˌbrij }

drawdown [GRAPHICS] In inkmaking, a procedure for obtaining a rough estimation of a color shade in which a small sample of ink is placed on a piece of paper and spread with a spatula to yield a thin film of ink. [HYD] The magnitude of the change in water surface level in a well, reservoir, or natural body of water resulting from the withdrawal of water. [PETRO ENG] The difference between the static and the flowing bottom-hole pressure. { 'drȯˌdau̇n }

drawdown ratio [ENG] The ratio of die opening thickness to product thickness. { 'drȯˌdau̇n ˌrä·shō }

drawer [ENG] A box or receptacle that slides or rolls on tracks within a cabinet. { 'drȯ·ər }

draw-filing [ENG] Filing by pushing and pulling a file sideways across the work. { 'drȯ ˌfīl·iŋ }

drawhead [MET] A group of rollers through which strip tubing or solid stock is drawn to form angled sections. { 'drȯˌhed }

drawhole [MIN ENG] The aperture in a battery through which coal or ore is drawn. { 'drȯˌhōl }

drawing [CHEM ENG] Removing ceramic ware from a kiln after it has been fired. [GRAPHICS] A surface portrayal of a form or figure in line. [MET] **1.** Pulling a wire or tube through a die to reduce the cross section. **2.** Forcing plastic deformation of metal in a die to form recessed parts. [TEXT] A textile process in which the sliver is prepared for spinning by being held at one end while the other end is pulled to increase the length and decrease the diameter of the ropelike form. { 'drȯ·iŋ }

drawing back [MET] Reheating hardened steel to a temperature below the critical temperature in order to change its hardness. { 'drȯ·iŋ 'bak }

drawing bristol [MATER] A cardboard made of 100% cotton in the higher grades; has good characteristics of permanence, strength, and erasability. { 'drȯ·iŋ 'brist·əl }

drawing cloth [MATER] A linen cloth that is specially treated to be smooth and translucent so that it may be used for ink tracings. { 'drȯ·iŋ ˌklȯth }

drawing compound [MET] A material applied to the work during drawing or pressing operations to eliminate draw marks by preventing direct contact between the work and die. { 'drȯ·iŋ ˌkäm,pau̇nd }

drawing die [MET] A die that forms sheet metal into cuplike, wrinkle-free shapes. { 'drȯ·iŋ ˌdī }

drawing in [TEXT] The process of running the warp threads through the loom harness in order to set up the loom for weaving. { ˌdrȯ·iŋ 'in }

drawing of temper [MET] The process of heating steel to red heat and then letting it cool slowly; opposite of hardening or tempering. { 'drȯ·iŋ əv 'tem·pər }

drawing out [MET] Lengthening of a piece of metal through a heating and hammering process, resulting in a proportional reduction in section area. [TEXT] The action of pulling staple textile fibers lengthwise over each other, producing longer and thinner slivers. { ˌdrȯ·iŋ 'au̇t }

drawing paper [MATER] One of a wide variety of papers used for pen- and-pencil drawing by artists and architects. { 'drȯ·iŋ ˌpā·pər }

drawing program [COMPUT SCI] A graphics program that maintains images in vector graphics format, allowing the user to design and illustrate objects on the display screen. Also known as illustration program. { 'drȯ·iŋ ˌprō·grəm }

drawing timber [MIN ENG] The act of withdrawing timber and other supports from abandoned or worked-out mines. { ˌdrȯ·iŋ 'tim·bər }

drawknife [DES ENG] A woodcutting tool with a long, narrow blade and two handles mounted at right angles to the blade. { 'drȯˌnīf }

draw mark [MET] An impairment of the die or metal surface caused during drawing due to friction or a defect in the die; examples are scoring and die lines. { 'drȯ ˌmärk }

drawn finish [MET] A smooth, bright finish on metal fabrications such as tubing or wire that is obtained by drawing the metal through a die. { 'drȯn 'fin·ish }

drawn glass [MATER] Glass made automatically by drawing the molten material through rollers. { ˌdrȯn 'glas }

draw piece [MET] Any part made by drawing. { 'drȯ ˌpēs }

drawplate [MET] A circular plate having a central hole through which wire is drawn by a punch. Also known as draw ring. { 'drȯˌplāt }

drawpoint [ENG] A steel point used to scratch lines or to pierce holes. { 'drȯˌpȯint }

draw radius [MET] A measure of cutting edge of a die or punch over which the metal is drawn. { 'drȯˌrād·ē·əs }

draw ring *See* drawplate. { 'drȯ ˌriŋ }

drawsheet [GRAPHICS] On a platen press, the uppermost sheet of a tympan to which the guides and fenders are attached. { 'drȯˌshēt }

draw slate [MIN ENG] A soft rock, 2 inches to 2 feet (5 centimeters to 0.6 meter) thick, that lies above a coal seam and falls with the coal as it is excavated, or soon after. { 'drȯ ˌslāt }

draw works [PETRO ENG] An oil-well drilling mechanism used to supply driving power and to lift heavy objects; consists of a countershaft and drum. { 'drȯ ˌwərks }

dream [PSYCH] An involuntary series of visual, auditory, or kinesthetic images, emotions, and thoughts occurring in the mind during sleep or a sleeplike state, which take the form of a sequence of events or of a story, having a feeling of reality but totally lacking a feeling of free will. { drēm }

dredge [ENG] A cylindrical or rectangular device for collecting samples of bottom sediment and benthic fauna. [MECH ENG] A floating excavator used for widening or deepening channels, building canals, constructing levees, raising material from stream or harbor bottoms to be used elsewhere as fill, or mining. { drej }

dredge ship [NAV ARCH] A watercraft used as a dredge. { 'drej ˌship }

dredging [ENG] Removing solid matter from the bottom of a water area. { 'drej·iŋ }

dredging buoy [NAV] A buoy marking the limit of an area where dredging is being performed. { 'drej·iŋ ˌbȯi }

D region [GEOPHYS] The region of ionosphere up to about 60 miles (97 kilometers) above the earth, below the E and F regions, in which the D layer forms. { 'dē ˌrē·jən }

dreikanter [GEOL] A pebble with three facets shaped by sandblasting. { 'drī,kän·tər }

Drepanellacea [PALEON] A monomorphic superfamily of extinct paleocopan ostracods in the suborder Beyrichicopina having a subquadrate carapace, many with a marginal rim. { drə,pan·əl'ās·ē·ə }

Drepanellidae [PALEON] A monomorphic family of extinct ostracodes in the superfamily Drepanellacea. { ˌdre·pə'nel·ə,dē }

Drepanidae [INV ZOO] The hooktips, a small family of lepidopteran insects in the suborder Heteroneura. { dre'pan·ə,dē }

Dresbachian [GEOL] Lower Croixan geologic time. { drez'bäk·ē·ən }

dress [CIV ENG] To smooth the surface of concrete or stone. [ELECTR] The arrangement of connecting wires in a circuit to prevent undesirable coupling and feedback. [MECH ENG] **1.** To shape a tool. **2.** To restore a tool to its original shape and sharpness. [MIN ENG] To sort, grind, clean, and concentrate ore. { dres }

dresser [ENG] Any tool or apparatus used for dressing something. { 'dres·ər }

dressing [AGR] Manure or compost used as a fertilizer. [CIV ENG] The process of smoothing or squaring lumber or stone for use in a building. [ENG] The sharpening, repairing, and replacing of parts, notably drilling bits and tool joints, to ready equipment for reuse. [MED] **1.** Application of various materials for protecting a wound and encouraging healing. **2.** Material so applied. { 'dres·iŋ }

Dressler kiln [MECH ENG] The first successful muffle-type tunnel kiln. { 'dres·lər ˌkil }

Drew number [PHYS CHEM] A dimensionless group used in

the study of diffusion of a solid material A into a stream of vapor initially composed of substance B, equal to

$$\frac{Z_A(M_A - M_B) + M_B}{(Z_A - Y_{AW})(M_B - M_A)} \cdot \ln \frac{M_V}{M_W}$$

where M_A and M_B are the molecular weights of components A and B, M_V and M_W are the molecular weights of the mixture in the vapor and at the wall, and Y_{AW} and Z_A are the mole fractions of A at the wall and in the diffusing stream, respectively. Symbolized N_D. { 'drü ,nəm·bər }

drewite [GEOL] Calcareous ooze composed of impalpable calcareous material. { 'drü,īt }

drex [TEXT] A unit of yarn density (mass per unit length), equal to 1 gram per 10 kilometers of yarn fiber. { dreks }

dribbling [MIN ENG] Fall of debris from the roof of an excavation, usually preceding a heavy fall or cave-in. { 'drib·liŋ }

drier [ENG] A device to remove water. [MATER] **1.** A substance that absorbs water. **2.** A substance that is used to hasten solidification. **3.** Material, such as salts of lead, manganese, and cobalt, which facilitates the oxidation of oils; used in paints and varnishes to speed drying. { 'drī·ər }

drift [ENG] **1.** A gradual deviation from a set adjustment, such as frequency or balance current, or from a direction. **2.** The deviation, or the angle of deviation, of a borehole from the vertical or from its intended course. **3.** To measure the size of a pipe opening by passing a mandrel through it. [GEOL] **1.** Rock material picked up and transported by a glacier and deposited elsewhere. **2.** Detrital material moved and deposited on a beach by waves and currents. [MECH ENG] The water lost in a cooling tower as mist or droplets entrained by the circulating air, not including the evaporative loss. [MIN ENG] A horizontal mine opening which follows a vein or lies within the trend of an ore body. Also known as gallery. [NAV] **1.** The movement of a craft caused by the action of wind or current. **2.** To move gradually from a set position without control. [OCEANOGR] *See* drift current. [SOLID STATE] The movement of current carriers in a semiconductor under the influence of an applied voltage. { drift }

Drift I [ASTRON] A group of stars that tend to move in a stream, traveling in the direction of the constellation Orion; it comprises 60% of the stars whose proper motions are known. { 'drift ¦wən }

Drift II [ASTRON] A group of stars that tend to move in a stream, traveling in the direction of the constellation Scutum; it comprises 40% of the brighter stars. { 'drift ¦tü }

drift angle [NAV] **1.** The horizontal angle between the axis of a ship and the tangent to its path. Also known as drift correction angle. **2.** The angle between the longitudinal axis of an aircraft and its path relative to the ground. { 'drift ,aŋ·gəl }

drift axis [NAV] Of a gyroscope, the axis about which drift occurs; for example, in a directional gyro with the spin axis mounted horizontally, the drift axis is the vertical axis. { 'drift ,ak·səs }

drift bolt [ENG] **1.** A bolt used to force out other bolts or pins. **2.** A metal rod used to secure timbers. { 'drift ,bōlt }

drift bottle [OCEANOGR] A bottle which is released into the sea for studying currents; contains a card, identifying the date and place of release, to be returned by the finder with date and place of recovery. Also known as floater. { 'drift ,bäd·əl }

drift card [OCEANOGR] A card, such as is used in a drift bottle, encased in a buoyant, waterproof envelope and released in the same manner as a drift bottle. { 'drift ,kärd }

drift-corrected amplifier [ELECTR] A type of amplifier that includes circuits designed to reduce gradual changes in output, used in analog computers. { ¦drift kə¦rek·təd 'am·plə,fī·ər }

drift correction angle *See* drift angle. { 'drift kə,rek·shən ,aŋ·gəl }

drift current [OCEANOGR] **1.** A wide, slow-moving ocean current principally caused by winds. Also known as drift; wind drift; wind-driven current. **2.** Current determined from the differences between dead reckoning and a navigational fix. [PL PHYS] A current of free charged particles in perpendicular electric and magnetic fields that results from an average motion of the particles in a direction perpendicular to both fields. { 'drift ,kə·rənt }

drift dam [GEOL] A dam formed by glacial drift in a stream valley. { 'drift ,dam }

drift diameter [PETRO ENG] **1.** The effective width of a hole in a drilling operation. **2.** The minimum diameter of the casing being installed in a wellbore. { 'drift dī'am·əd·ər }

drifter [MECH ENG] A rock drill, similar to but usually larger than a jack hammer, mounted for drilling holes up to $4\frac{1}{2}$ inches (11.4 centimeters) in diameter. [MIN ENG] **1.** A person who excavates mine drifts. **2.** An air-driven rock drill used for excavating mine drifts and crosscuts. { 'drif·tər }

drift error [COMPUT SCI] An error arising in the use of an analog computer due to gradual changes in the output of circuits (such as amplifiers) in the computer. { 'drift ,er·ər }

drift glacier *See* snowdrift glacier. { 'drift ,glā·shər }

drift ice [OCEANOGR] Sea ice that has drifted from its place of formation. { 'drift ,īs }

drift ice foot *See* ramp. { 'drift ,īs ,fůt }

drift indicator [ENG] Device used to record directional logs; records only the amount of drift (deviation from the vertical), and not the direction. [NAV] *See* drift meter. { 'drift ,in·də,kād·ər }

drifting [MIN ENG] Tunneling along the strike of a lode. { 'drif·tiŋ }

drifting mine [ORD] An underwater mine adjusted to float unanchored on or just below the surface of the water. { 'drif·tiŋ ,mīn }

drifting snow [METEOROL] Wind-driven snow raised from the surface of the earth to a height of less than 6 feet (1.8 meters). { 'drif·tiŋ 'snō }

drift lead [NAV] A lead placed on the bottom to indicate movement of a vessel; at anchor the lead line is usually secured to the rail with a little slack and if the ship drags anchor, the line tends forward; also used to indicate when a vessel coming to anchor is dead in the water or when it is moving astern, or to indicate current if a ship is dead in the water. { 'drift ,led }

drift meter [NAV] An instrument for measuring drift angle. Also known as drift indicator; drift sight. { 'drift ,mēd·ər }

drift mining [MIN ENG] Working of shallow veins or beds through drifts or shafts from the surface. { 'drift ,mīn·iŋ }

drift mobility [SOLID STATE] The average drift velocity of carriers per unit electric field in a homogeneous semiconductor. Also known as mobility. { 'drift mō'bil·əd·ē }

drift observation [NAV] Also known as drift sight. **1.** The process of observing drift or leeway. **2.** The value obtained by such an observation. { 'drift ,äb·zər,vā·shən }

driftpin [DES ENG] A round, tapered metal rod that is driven into matching rivet holes of two metal parts for stretching the parts and bringing them into alignment. { 'drift,pin }

drift plug [ENG] A plug that can be driven into a pipe to straighten it or to flare its opening. { 'drift ,pləg }

drift sight *See* drift meter. { 'drift ,sīt }

drift space [ELECTR] A space in an electron tube which is substantially free of externally applied alternating fields and in which repositioning of electrons takes place. { 'drift ,spās }

drift speed [ELEC] Average speed at which electrons or ions progress through a medium. { 'drift ,spēd }

drift station [OCEANOGR] **1.** A scientific station established on the ice of the Arctic Ocean, generally based on an ice flow. **2.** A set of observations made over a period of time from a drifting vessel. { 'drift ,stā·shən }

drift terrace *See* alluvial terrace. { 'drift ,ter·əs }

drift transistor [ELECTR] **1.** A transistor having two plane parallel junctions, with a resistivity gradient in the base region between the junctions to improve the high-frequency response. **2.** *See* diffused-alloy transistor. { 'drift tran,zis·tər }

drift tube [NUCLEO] A tubular electrode placed in the vacuum chamber of a circular accelerator, to which radio-frequency voltage is applied to accelerate the particles. { 'drift ,tüb }

drift ultrasonic flowmeter *See* deflection ultrasonic flowmeter. { ¦drift ¦əl·trə¦sän·ik 'flō,mēd·ər }

drift velocity [SOLID STATE] The average velocity of a carrier that is moving under the influence of an electric field in a semiconductor, conductor, or electron tube. { 'drift və'läs·əd·ē }

drift wave [PL PHYS] An oscillation in a magnetically confined plasma which arises in the presence of density gradients, for example, at the plasma's surface, and which resembles the waves that propagate at the interface of two fluids of different density in a gravity field. { 'drift ,wāv }

Drilidae [INV ZOO] The false firefly beetles, a family of coleopteran insects in the superfamily Cantharoidea. { 'dril·ə‚dē }

drill [ENG] A rotating-end cutting tool for creating or enlarging holes in a solid material. Also known as drill bit. [TEXT] Strong twilled carded cotton cloth. { dril }

drillability [ENG] Fitness for being drilled, denoting ease of penetration. { ‚dril·ə'bil·əd·ē }

drill angle gage See drill grinding gage. { 'dril ‚aŋ·gəl ‚gāj }

drill bit See drill. { 'dril ‚bit }

drill cable [ENG] A cable used to pull up drill rods, casing, and other drilling equipment used in making a borehole. { 'dril ‚kā·bəl }

drill capacity [MECH ENG] The length of drill rod of specified size that the hoist on a diamond or rotary drill can lift or that the brake can hold on a single line. { 'dril kə‚pas·əd·ē }

drill carriage [MECH ENG] A platform or frame on which several rock drills are mounted and which moves along a track, for heavy drilling in large tunnels. Also known as jumbo. { 'dril ‚kar·ij }

drill chuck [DES ENG] A chuck for holding a drill or other cutting tool on a spindle. { 'dril ‚chək }

drill circuit [COMMUN] A telegraph circuit used only to practice sending and receiving. { 'dril ‚sər·kət }

drill collar [DES ENG] A ring which holds a drill bit and gives it radial location with respect to a bearing. { 'dril ‚käl·ər }

drill column [MIN ENG] A steel pipe that can be wedged across an underground opening in a vertical or horizontal position to serve as a base on which to mount a diamond or rock drill. { 'dril ‚käl·əm }

drill cuttings [ENG] Cuttings of rock and other subterranean materials brought to the surface during the drilling of wellholes. { 'dril ‚kəd·iŋz }

drill doctor [MIN ENG] **1.** A person who services drill bits, tools, and steels. **2.** A shop where the mechanic works. { 'dril ‚däk·tər }

drill down [COMPUT SCI] In data mining, viewing data at a greater level of detail; for example, viewing individual sales as opposed to viewing total sales. { ‚dril 'daùn }

drill drift [ENG] A steel wedge used to remove tapered shank tools from spindles, sockets, and sleeves. { 'dril ‚drift }

drilled caisson [CIV ENG] A drilled hole filled with concrete and lined with a cylindrical steel casing if needed. { ‚drild 'kā‚sän }

drilled extrusion ingot [MET] A hollow extrusion ingot made from a solid cast extrusion ingot by drilling. { ‚drild ik'strü·zhən ‚iŋ·gət }

driller [ENG] A person who operates a drilling machine. [MECH ENG] See drilling machine. { 'dril·ər }

driller's method [PETRO ENG] A well-killing method that involves two circulations, one to circulate the formation fluids out of the well and the other to circulate heavier mud through the wellbore. { 'dril·ərz ‚meth·əd }

drill extractor [ENG] A tool for recovering broken drill pieces or a detached drill from a borehole. { 'dril ik‚strak·tər }

drill feed [MECH ENG] The mechanism by which the drill bit is fed into the borehole during drilling. { 'dril ‚fēd }

drill fittings [ENG] All equipment used in a borehole during drilling. Also known as downhole equipment. { 'dril ‚fid·iŋz }

drill floor [ENG] A work area covered with planks around the collar of a borehole at the base of a drill tripod or derrick. { 'dril ‚flòr }

drill footage [ENG] The lineal feet of borehole drilled. { 'dril ‚fùd·ij }

drill gage [DES ENG] A thin, flat metal plate that has accurate holes for many sizes of drills; each hole, identified as to drill size, enables the diameter of a drill to be checked. [ENG] Diameter of a borehole. { 'dril ‚gāj }

drill grinding gage [DES ENG] A tool that checks the angle and length of a twist drill while grinding it. Also known as drill angle gage; drill point gage. { 'dril ‚grīnd·iŋ ‚gāj }

drill hole [ENG] A hole created or enlarged by a drill or auger. Also known as borehole. { 'dril ‚hōl }

drill-hole logging See borehole logging. { 'dril ‚hōl ‚läg·iŋ }

drill-hole pattern [ENG] The number, position, angle, and depth of the shot holes forming the round in the face of a tunnel or sinking pit. { 'dril ‚hōl ‚pad·ərn }

drill-hole survey See borehole survey. { 'dril ‚hōl ‚sər‚vā }

drill-in [MIN ENG] The act or process of setting casting through overburden by using a drill machine. { 'dril ‚in }

drilling [ENG] The creation or enlarging of a hole in a solid material with a drill. { 'dril·iŋ }

drilling break [PETRO ENG] An abrupt increase in the rate of penetration of the drill bit during a well-drilling operation. { 'dril·iŋ ‚brāk }

drilling column [ENG] The column of drill rods, with the drill bit attached to the end. { 'dril·iŋ ‚käl·əm }

drilling fluid See drilling mud. { 'dril·iŋ ‚flü·əd }

drilling machine [MECH ENG] A device, usually motor-driven, fitted with an end cutting tool that is rotated with sufficient power either to create a hole or to enlarge an existing hole in a solid material. Also known as driller. { 'dril·iŋ mə‚shēn }

drilling mud [MATER] A suspension of finely divided heavy material, such as bentonite and barite, pumped through the drill pipe during rotary drilling to seal off porous zones and flush out chippings, and to lubricate and cool the bit. Also known as drilling fluid. { 'dril·iŋ ‚məd }

drilling platform [ENG] The structural base upon which the drill rig and associated equipment is mounted during the drilling operation. { 'dril·iŋ ‚plat‚fòrm }

drilling rate [MECH ENG] The number of lineal feet drilled per unit of time. { 'dril·iŋ ‚rāt }

drilling time [ENG] **1.** The time required in rotary drilling for the bit to penetrate a specified thickness (usually 1 foot) of rock. **2.** The actual time the drill is operating. { 'dril·iŋ ‚tīm }

drilling time log [ENG] Foot-by-foot record of how fast a formation is drilled. { 'dril·iŋ 'tīm ‚läg }

drill jig [MECH ENG] A device fastened to the work in repetition drilling to position and guide the drill. { 'dril ‚jig }

drill jumbo [MIN ENG] A mobile carriage or platform fitted with mechanical arms upon which several drilling machines are mounted. { 'dril ‚jəm·bō }

drill log [ENG] **1.** A record of the events and features of the formations penetrated during boring. Also known as boring log. **2.** A record of all occurrences during drilling that might help in a complete logging of the hole or in determining the cost of the drilling. { 'dril ‚läg }

drill out [ENG] **1.** To complete one or more boreholes. **2.** To penetrate or remove a borehole obstruction. **3.** To locate and delineate the area of a subsurface ore body or of petroleum by a series of boreholes. { ‚dril 'aùt }

drill-over [ENG] The act or process of drilling around a casing lodged in a borehole. { 'dril ‚ō·vər }

drill pipe [MIN ENG] A pipe used for driving a revolving drill bit, used especially in drilling wells; consists of a casing within which tubing is run to conduct oil or gas to ground level; drilling mud flows in the annular space between casing and tubing during the drilling operation. { 'dril ‚pīp }

drill point gage See drill grinding gage. { 'dril ‚pòint ‚gāj }

drill press [MECH ENG] A drilling machine in which a vertical drill moves into the work, which is stationary. { 'dril ‚pres }

drill rod [ENG] The long rod that drives the drill bit in drilling boreholes. { 'dril ‚räd }

drill runner [MIN ENG] A tunnel miner who operates rock drills. { 'dril ‚rən·ər }

drill sleeve [ENG] A tapered, hollow steel shaft designed to fit the tapered shank of a cutting tool to adapt it to the drill press spindle. { 'dril ‚slēv }

drill socket [ENG] An adapter to fit a tapered shank drill to a taper hole that is larger than that in the drill press spindle. { 'dril ‚säk·ət }

drill steel [MET] Steel with at least 0.85% carbon content made by the electric furnace process. Formerly known as crucible steel, when made by the crucible process. { 'dril ‚stēl }

drill stem See bent sub. { 'dril ‚stem }

drill-stem test [PETRO ENG] Bottom-hole pressure information obtained and used to determine formation productivity. { 'dril ‚stem ‚test }

drill string [MECH ENG] The assemblage of drill rods, core barrel, and bit, or of drill rods, drill collars, and bit in a borehole, which is connected to and rotated by the drill collar of the borehole. { 'dril ‚striŋ }

drill up [COMPUT SCI] In data mining, viewing data in less detail; for example, viewing total sales as opposed to individual sales. { ¦dril ¦əp }

drill weave [TEXT] Special fabric sometimes used in filtration; a three-harness, warp-face twill weave, having the two-up and one-down twill effect. { 'dril ‚wēv }

Drilonematoidea [INV ZOO] A superfamily of parasitic nematodes in the subclass Spruria. { ‚drī·lō‚nem·ə'tȯid·ē·ə }

drip [ARCH] *See* hoodmold. [HYD] Condensed or otherwise collected moisture falling from leaves, twigs, and so forth. [MATER] **1.** Oil which comes through the cloth of a paraffin wax press. **2.** Filter drainings too dark to be included in filter stock. [PETRO ENG] A discharge mechanism installed at a low point in a gas transmission line to collect and remove liquid accumulations. Also known as blowcase. { drip }

drip cap [BUILD] A horizontal molding installed over the frame for a door or window to direct water away from the frame. { 'drip ‚kap }

drip-dry [TEXT] Of a fabric, shedding water or moisture rapidly without squeezing, spinning, or wringing. { 'drip ‚drī }

drip edge [BUILD] A metal strip that extends beyond the other parts of the roof and is used to direct rainwater off. { 'drip ‚ej }

drip irrigation [AGR] A method of providing water to plants, almost continuously, through small-diameter tubes and emitters. { 'drip ‚ir·i‚gā·shən }

drip line [NUC PHYS] The boundary, on a chart of the nuclides, beyond which a nucleon (proton or neutron) is no longer bound to the nucleus. { 'drip ‚līn }

dripping drop atomization [HYD] A type of natural gravitational atomization process in which there is periodic emission of drops from the bottom side of a surface to which a liquid is fed continuously, as in dripping of water from leaves. { 'drip·iŋ ¦dräp ‚ad·ə·mə'zā·shən }

dripstone [GEOL] A cave feature, such as a stalagmite, which is formed by precipitation of calcium carbonate or another mineral from dripping water. { 'drip‚stōn }

drive [ELECTR] *See* excitation. [MECH ENG] The means by which a machine is given motion or power (as in steam drive, diesel-electric drive), or by which power is transferred from one part of a machine to another (as in gear drive, belt drive). [MIN ENG] **1.** To excavate in a horizontal or inclined plane. **2.** A horizontal underground tunnel along or parallel to a lode, vein, or ore body. [PSYCH] A strong impetus to behavior or active striving. { drīv }

drive array [COMPUT SCI] A collection of hard disks organized to increase speed and improve reliability, often with the help of data stripping. { 'drīv ə‚rā }

drive bay [COMPUT SCI] A space in the cabinet of a personal computer where disk drives, tape drives, and CD-ROM drives can be installed. Also known as bay. { 'drīv ‚bā }

drive-by-wire [MECH ENG] Electronic throttle control in automobiles. { ¦drīv bī ¦wīr }

drive chuck [MECH ENG] A mechanism at the lower end of a diamond-drill drive rod on the swivel head by means of which the motion of the drive rod can be transmitted to the drill string. { drīv ‚chək }

drive control *See* horizontal drive control. { 'drīv kən‚trōl }

drive fit [DES ENG] A fit in which the larger (male) part is pressed into a smaller (female) part; the assembly must be effected through the application of an external force. { ‚drīv ‚fit }

drivehead [ENG] A cap fitted over the end of a mechanical part to protect it while it is being driven. { 'drīv‚hed }

driveless work station [COMPUT SCI] A computer or terminal in a local area network that does not have its own disk drives and relies on a central mass storage facility for information storage. { 'drīv·ləs 'wərk‚stā·shən }

drive light [COMPUT SCI] A lamp on the front of a disk drive that lights to indicate when the unit is reading or writing data. { 'drīv ‚līt }

driveline [MECH ENG] In an automotive vehicle, the group of parts, including the universal joint and the drive shaft, that connect the transmission with the driving wheels. { 'drīv‚līn }

driven array [ELECTROMAG] An antenna array consisting of a number of driven elements, usually half-wave dipoles, fed in phase or out of phase from a common source. { ¦drīv·ən ə'rā }

driven blocking oscillator *See* monostable blocking oscillator. { ¦drīv·ən ¦bläk·iŋ 'äs·ə‚lād·ər }

driven caisson [CIV ENG] A caisson formed by driving a cylindrical steel shell into the ground with a pile-driving hammer and then placing concrete inside; the shell may be removed when concrete sets. { ¦driv·ən 'kā‚sän }

driven element [ELECTROMAG] An antenna element that is directly connected to the transmission line. { ¦driv·ən 'el·ə‚mənt }

driven gear [MECH ENG] The member of a pair of gears to which motion and power are transmitted by the other. { ¦driv·ən 'gir }

driven snow [METEOROL] Snow which has been moved by wind and collected into snowdrifts. { ¦driv·ən 'snō }

drive pattern [COMMUN] In a facsimile system, undesired pattern of density variations caused by periodic errors in the position of the recording spot. { 'drīv ‚pad·ərn }

drivepipe [ENG] A thick-walled casing pipe that is driven through overburden or into a deep drill hole to prevent caving. { 'drīv‚pīp }

drive pulley [MECH ENG] The pulley that drives a conveyor belt. { 'drīv ‚pül·ē }

drive pulse [ELECTR] An electrical pulse which induces a magnetizing force in an element of a magnetic core storage, reversing the polarity of the core. { 'drīv ‚pəls }

driver [COMPUT SCI] A sequence of program instructions that controls an input/output device such as a tape drive or disk drive. [ELECTR] The amplifier stage preceding the output stage in a receiver or transmitter. [ENG ACOUS] The portion of a horn loudspeaker that converts electrical energy into acoustical energy and feeds the acoustical energy to the small end of the horn. { 'drī·vər }

driver element [ELECTROMAG] Antenna array element that receives power directly from the transmitter. { 'drī·vər ‚el·ə·mənt }

drive rod [ENG] Hollow shaft in the swivel head of a diamond-drill machine through which energy is transmitted from the drill motor to the drill string. Also known as drive spindle. { 'drīv ‚räd }

driver sweep [ELECTR] Sweep triggered only by an incoming signal or trigger. { 'drī·vər ‚swēp }

driver transformer [ELECTR] A transformer in the input circuit of an amplifier, especially in the transmitter. { 'drī·vər tranz'fȯr·mər }

drive sampling [ENG] The act or process of driving a tubular device into soft rock material for obtaining dry samples. { 'drīv ‚sam·pliŋ }

drivescrew [DES ENG] A screw that is driven all the way in, or nearly all the way in, with a hammer. { 'drīv‚skrü }

drive shaft [MECH ENG] A shaft which transmits power from a motor or engine to the rest of a machine. { 'drīv ‚shaft }

drive shoe [DES ENG] A sharp-edged steel sleeve attached to the bottom of a drivepipe or casing to act as a cutting edge and protector. { 'drīv ‚shü }

drive spindle *See* drive rod. { 'drīv ‚spin·dəl }

drive train *See* power train. { 'drīv ‚trān }

drive winding [ELECTR] A coil of wire that is inductively coupled to an element of a magnetic memory. Also known as drive wire. { 'drīv ‚wīn·diŋ }

drive wire *See* drive winding. { 'drīv ‚wīr }

driving clock [ENG] A mechanism for driving an instrument at a required rate. { 'drīv·iŋ ‚kläk }

driving force [CHEM] In a chemical reaction, the formation of products such as an insoluble compound, a gas, a nonelectrolyte, or a weak electrolyte that enable the reaction to go to completion as a metathesis. { 'drīv·iŋ ‚fȯrs }

driving pinion [MECH ENG] The input gear in the differential of an automobile. { 'drīv·iŋ ‚pin·yən }

driving-point function [CONT SYS] A special type of transfer function in which the input and output variables are voltages or currents measured between the same pair of terminals in an electrical network. { 'drīv·iŋ ‚pȯint ‚faŋk·shən }

driving-point impedance [ELECTR] The complex ratio of applied alternating voltage to the resulting alternating current in an electron tube, network, or other transducer. { 'drīv·iŋ ‚pȯint im'pēd·əns }

driving resistance [MECH] The force exerted by soil on a pile being driven into it. { 'drīv·iŋ ri'zis·təns }

driving signal [ELECTR] Television signal that times the scanning at the pickup point. { 'drīv·iŋ ‚sig·nəl }

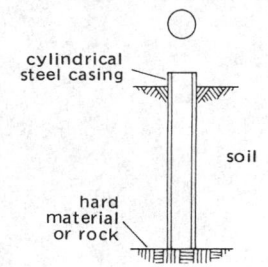

DRIVEN CAISSON

cylindrical steel casing

soil

hard material or rock

Driven caisson, top view and cross-section from the side.

driving wheel [MECH ENG] A wheel that supplies driving power. { 'drīv·iŋ ‚wēl }

drizzle [METEOROL] Very small, numerous, and uniformly dispersed water drops that may appear to float while following air currents; unlike fog droplets, drizzle falls to the ground; it usually falls from low stratus clouds and is frequently accompanied by low visibility and fog. { 'driz·əl }

drizzle drop [METEOROL] A drop of water of diameter 0.2 to 0.5 millimeter falling through the atmosphere; however, all water drops of diameter greater than 0.2 millimeter are frequently termed raindrops, as opposed to cloud drops. { 'driz·əl ‚dräp }

DRM See detrital remanent magnetization.

drogue [AERO ENG] **1.** A small parachute attached to a body for stabilization and deceleration. Also known as deceleration parachute. **2.** A funnel-shaped device at the end of the hose of a tanker aircraft in flight, to receive the probe of another aircraft that will take on fuel. [ENG] **1.** A device, such as a sea anchor, usually shaped like a funnel or cone and dragged or towed behind a boat or seaplane for deceleration, stabilization, or speed control. **2.** A current-measuring assembly consisting of a weighted current cross, sail, or parachute and an attached surface buoy. Also known as drag anchor; sea anchor. { drōg }

Dromadidae [VERT ZOO] A family of the avian order Charadriiformes containing a single species, the crab plover (*Dromas ardeola*). { drō'mad·ə‚dē }

dromedary [VERT ZOO] *Camelus dromedarius* The Arabian camel, distinguished by a single hump. { 'dräm·ə‚der·ē }

Dromiacea [INV ZOO] The dromiid crabs, a subsection of the Brachyura in the crustacean order Decapoda. { ‚drō·mē'ā·shē·ə }

Dromiceidae [VERT ZOO] The emus, a monospecific family of flightless birds in the order Casuariiformes. { ‚drō·mə'sē·ə‚dē }

dromophobia [PSYCH] An abnormal fear of walking about. { ‚drō·mə'fō·bē·ə }

drone [AERO ENG] A pilotless aircraft usually subordinated to the controlling influences of a remotely located command station, but occasionally preprogrammed. [INV ZOO] A haploid male bee or ant; one of the three castes in a colony. { drōn }

drooped ailerons [AERO ENG] Ailerons that are of the hinged trailing-edge type and are so arranged that both the right and left one have a 10 to 15° positive downward deflection with the control column in a neutral position. { ‚drüpt 'ā·lə‚räns }

droop governor [MECH ENG] A governor whose equilibrium speed decreases as the load on the machinery controlled by the governor increases. { 'drüp ‚gə·vər·nər }

drop [FL MECH] The quantity of liquid that coalesces into a single globule; sizes vary according to physical conditions and the properties of the fluid itself. [HYD] The difference in water-surface elevations is measured up-and downstream from a narrowing in the stream. [MET] A casting defect due to the falling of a portion of sand from an overhanging section of the mold. [MINERAL] A funnel-shaped downward intrusion of sedimentary rock into the roof of a coal seam. [PL PATH] A fungus disease of various vegetables caused by *Sclerotinia sclerotiorum* and characterized by wilt and stem rot. { dräp }

drop ball [ENG] A ball, weighing 3000–4000 pounds (1400–1800 kilograms), dropped from a crane through about 20–33 feet (6–10 meters) onto oversize quarry stones left after blasting; this method is used to avoid secondary blasting. { 'dräp ‚bòl }

drop bar [ELEC] Protective device used to ground a high-voltage capacitor when opening a door. [MECH ENG] A bar that guides sheets of paper into a printing or folding machine. { 'dräp ‚bär }

drop black [MATER] Black pigment shaped into droplets. { 'dräp ‚blak }

drop-bottom car [MIN ENG] A mine car designed so that flaps drop open in the bottom to allow the coal to fall out as the car passes over the dump; flaps close as the car leaves. { 'dräp ‚bäd·əm ‚kär }

drop bracket transposition [ELEC] Reversal of the relative positions of two parallel wire conductors while depressing one, so that the crossover is in a vertical plane. { 'dräp ‚brak·ət tranz·pə'zish·ən }

drop-dead halt [COMPUT SCI] A machine halt from which there is no recovery; such a halt may occur through a logical error in programming; examples in which a drop-dead halt could occur are division by zero and transfer to a nonexistent instruction word. Also known as dead halt. { 'dräp ‚ded 'hòlt }

dropfoot [MED] A condition in which the foot drags along the ground and gives the person a characteristic shuffling gait, caused by the failure of the muscle responsible for raising the foot during walking as the leg is swung forward. { 'dräp‚fùt }

drop forging [MET] Plastic deformation of hot metal under a falling weight, such as a drop hammer. { 'dräp ‚fòrj·iŋ }

drop hammer [MECH ENG] See pile hammer. [MET] A hammer used in forging that is raised and then dropped on the metal resting on an anvil or on a die. { 'dräp ‚ham·ər }

drop-in [COMPUT SCI] The accidental appearance of an unwanted bit, digit, or character on a magnetic recording surface or during reading from or writing to a magnetic storage device. { 'dräp ‚in }

drop indicator [COMMUN] Indicator for signaling, consisting of a hinged flap normally held up by a catch; the catch is released by an electromagnet, allowing the flap to drop when a signal is received. { 'dräp 'in·də‚kād·ər }

droplet [MED] A tiny drop of matter consisting of water, mucus, and bacterial products, released through the nasal passages or expectorated. [METEOROL] A water droplet in the atmosphere; there is no defined size limit separating droplets from drops of water, but sometimes a maximum diameter of 0.2 millimeter is the limit for droplets. { 'dräp·lət }

droplet condensation [THERMO] The formation of numerous discrete droplets of liquid on a wall in contact with a vapor, when the wall is cooled below the local vapor saturation temperature and the liquid does not wet the wall. { 'dräp·lət ‚kän·dən'sā·shən }

droplet infection [MED] Infection by contact with airborne droplets of sputum carrying infectious agents. { 'dräp·let in‚fek·shən }

drop log [MIN ENG] A timber which can be dropped across a mine track by remote control to derail cars. { 'dräp ‚läg }

drop model of nucleus See liquid-drop model. { 'dräp ‚mäd·əl əv 'nü·klē·əs }

dropout [COMPUT SCI] The accidental disappearance of a valid bit, digit, or character from a storage medium or during reading from or writing to a storage device. [ELEC] Of a relay, the maximum current, voltage, power, or such, at which it will release from its energized position. [ELECTR] A reduction in output signal level during reproduction of recorded data, sufficient to cause a processing error. [GRAPHICS] A halftone negative, print, or plate on which some of the original image has been removed by masking or opaquing. { 'dräp‚aùt }

dropout current [ELEC] The maximum current at which a relay or other magnetically operated device will release to its deenergized position. { 'dräp‚aùt ‚kə·rənt }

dropout error [ELECTR] Loss of a recorded bit or any other error occurring in recorded magnetic tape due to foreign particles on or in the magnetic coating or to defects in the backing. { 'dräp‚aùt ‚er·ər }

dropout fuse [ELEC] A fuse used on utility line poles which springs open when the fuse metal melts to provide rapid arc extinction, and which drops to an open-circuit position readily distinguishable from the ground. Also known as flip-open cutout fuse. { 'dräp‚aùt ‚fyüz }

dropout voltage [ELEC] The maximum voltage at which a relay or other magnetically operated device will release to its deenergized position. { 'dräp‚aùt ‚vōl·tij }

dropping angle See range angle. { 'dräp·iŋ ‚aŋ·gəl }

dropping fraction [COMPUT SCI] In punched cards, the chance that a given sorting operation will cause a card taken at random to be selected. { 'dräp·iŋ ‚frak·shən }

dropping-mercury electrode [PHYS CHEM] An electrode consisting of a fine-bore capillary tube above which a constant head of mercury is maintained; the mercury emerges from the tip of the capillary at the rate of a few milligrams per second and forms a spherical drop which falls into the solution at the rate of one every 2–10 seconds. { 'dräp·iŋ ‚mər·kyə·rē i'lek·trōd }

DROPOUT FUSE

A dropout fuse rated 50 amperes at 7.8 kilovolts used on utility-line poles. (*General Electric Co.*)

dropping point [CHEM] The temperature at which grease changes from a semisolid to a liquid state under standardized conditions. { 'dräp·iŋ ˌpȯint }

dropping resistor [ELEC] A resistor used in series with a load to decrease the voltage applied to the load. { 'dräp·iŋ ri͵zis·tər }

dropping test [MET] A chemical method for determining thickness of zinc and cadmium plated coatings on metal in which a reagent is dropped on the surface until the basis metal is exposed. { 'dräp·iŋ ˌtest }

drop press *See* punch press. { 'dräp ˌpres }

drop relay [ELEC] Relay activated by incoming ringing current to call an operator's attention to a subscriber's line. { 'dräp 'rē͵lā }

drop repeater [ELECTR] Microwave repeater that is provided with the necessary equipment for local termination of one or more circuits. { 'dräp ri͵pēd·ər }

drop siding [BUILD] Building siding with a shiplap joint. { 'dräp ˌsīd·iŋ }

drop-size distribution [METEOROL] The frequency distribution of drop sizes (diameters, volumes) that is characteristic of a given cloud or rainfall. { 'dräp ˌsīz ˌdis·trə'byü·shən }

dropsonde [ENG] A radiosonde dropped by parachute from a high-flying aircraft to measure weather conditions and report them back to the aircraft. Also known as dropwindsonde; parachute radiosonde. { 'dräp͵sänd }

dropsonde dispenser [ENG] A chamber from which dropsonde instruments are released from weather reconnaissance aircraft; used only for some models of equipment, ejection chambers being used for others. { 'dräp͵sänd də'spen·sər }

dropsonde observation [METEOROL] An evaluation of the significant radio signals received from a descending dropsonde, and usually presented in terms of height, temperature, and dew point at the mandatory and significant pressure levels; it is comparable to a radiosonde observation. { 'dräp͵sänd ˌäb·sər'vā·shən }

drop spillway [CIV ENG] A spillway usually less than 20 feet (6 meters) high having a vertical downstream face, and water drops over the face without touching the face. { 'dräp 'spil͵wā }

dropstone [GEOL] A rock that was carried by a glacier or iceberg, and deposited as the ice melted. { 'dräp͵stōn }

dropsy *See* edema. { 'dräp·sē }

drop tank [AERO ENG] A fuel tank on an airplane that may be jettisoned. { 'dräp ˌtaŋk }

drop theory *See* barrier theory of cyclones. { 'dräp ˌthē·ə·rē }

drop vent [ENG] In a plumbing system, a type of vent that is connected to a drain or vent pipe at a point below the fixture it is serving. { 'dräp ˌvent }

drop weight [FL MECH] The weight of the largest drop that can hang from the end of a tube of given radius. { 'dräp ˌwāt }

drop-weight method [FL MECH] A method of measuring surface tension by measuring the weight of a slowly increasing drop of the liquid hanging from the end of a tube, just before it is detached from the tube. { 'dräp ˌwāt ˌmeth·əd }

dropwindsonde *See* dropsonde. { ˌdräp'wind͵sänd }

drop wire [ELEC] Wire suitable for extending an open wire or cable pair from a pole or cable terminal to a building. { 'dräp ˌwīr }

dropwise condensation [THERMO] Condensation of a vapor on a surface in which the condensate forms into drops. { 'dräp͵wīz ˌkän·dən'sā·shən }

Droseraceae [BOT] A family of dicotyledonous plants in the order Sarraceniales, distinguished by leaves that do not form pitchers, parietal placentation, and several styles. { ˌdrä·sə'rās·ē͵ē }

drosometer [ENG] An instrument used to measure the amount of dew deposited on a given surface. { drō'säm·əd·ər }

Drosophilidae [INV ZOO] The vinegar flies, a family of myodarian cyclorrhaphous dipteran insects in the subsection Acalypteratae, including the fruit fly (*Drosophila melanogaster*). { ˌdrä·sə'fil·ə͵dē }

dross [MET] An impurity, usually an oxide, formed on the surface of a molten metal. { drös }

drossing [MET] A process used in nonferrous pyrometallurgy for removing solid oxide deposits on the surface of a molten metal. { 'dräs·iŋ }

drought [CLIMATOL] A period of abnormally dry weather sufficiently prolonged so that the lack of water causes a serious hydrologic imbalance (such as crop damage, water supply shortage, and so on) in the affected area; in general, the term should be reserved for relatively extensive time periods and areas. { draut }

drowned atoll [GEOL] An atoll which has not reached the water surface. { ¦draund 'a͵tȯl }

drowned coast [GEOL] A shoreline transformed from a hilly land surface to an archipelago of small islands by inundation by the sea. { ¦draund 'kōst }

drowned river mouth *See* estuary. { ¦draund 'riv·ər ˌmauth }

drowned stream [HYD] A stream that has been flooded over by the ocean. Also known as flooded stream. { ¦draund 'strēm }

drowned valley [GEOL] A valley whose lower part has been inundated by the sea due to submergence of the land margin. { ¦draund 'val·ē }

droxtal [HYD] An ice particle measuring 10–20 micrometers in diameter, formed by direct freezing of supercooled water droplets at temperatures below $-30°C$. { 'dräk͵stȯl }

DR track line *See* dead-reckoning track. { ¦dē¦är ¦trak ˌlīn }

Drude equation [OPTICS] An equation which states that the rotation of the plane of polarization of plane-polarized light passing through an optically active substance is inversely proportional to the difference between the square of the wavelength of the light and the square of a constant wavelength. { 'drüd i͵kwā·zhən }

Drude's theory of conduction [SOLID STATE] A theory which treats the electrons in a metal as a gas of classical particles. { 'drüdz ˌthē·ə·rē əv kən'dək·shən }

drug [PHARM] **1.** Any substance used internally or externally as a medicine for the treatment, cure, or prevention of a disease. **2.** A narcotic preparation. { drəg }

drug idiosyncrasy [MED] A peculiarity of constitution that makes an individual respond differently to a drug or treatment than do most people. { ¦drəg ˌid·ī·ō'siŋ·krə·sē }

drug-induced parkinsonism *See* pseudoparkinsonism. { ¦drəg in¦düst 'pärk·ən·sən͵iz·əm }

drug pacemaker [PHARM] A pharmaceutical agent capable of increasing the ventricular rate in a diseased heart. { ¦drəg 'pās͵māk·ər }

drug resistance [MICROBIO] A decreased reactivity of living organisms to the injurious actions of certain drugs and chemicals. { ¦drəg ri'zis·təns }

drug sensitivity gene [MED] A gene that encodes enzymes which catalyze the conversion of a prodrug into active anticancer metabolites. { ¦drəg 'sen·si͵tiv·ə͵tē ˌgēn }

drug tolerance [MED] Condition that may follow repeated ingestion of a drug in so that the effect produced by the original dose no longer occurs. { ¦drəg ˌtä·lə·rəns }

drum [CHEM ENG] Tower or vessel in a refinery into which heated products are conducted so that volatile portions can separate. [DES ENG] **1.** A hollow, cylindrical container. **2.** A metal cylindrical shipping container for liquids having a capacity of 12–110 gallons (45–416 liters). [ELECTR] A computer storage device consisting of a rapidly rotating cylinder with a magnetizable external surface on which data can be read or written by many read/write heads floating a few millionths of an inch off the surface. Also known as drum memory; drum storage; magnetic drum; magnetic drum storage. [MECH ENG] **1.** A horizontal cylinder about which rope or wire rope is wound in a hoisting mechanism. **2.** A hollow or solid cylinder or barrel that acts on, or is acted upon by, an exterior entity, such as the drum in a drum brake. Also known as hoisting drum. { drəm }

drum armature [ELEC] An armature that has a drum winding. { 'drəm ˌärm·ə·chər }

drum brake [MECH ENG] A brake in which two curved shoes fitted with heat- and wear-resistant linings are forced against the surface of a rotating drum. { 'drəm ˌbrāk }

drum cam [MECH ENG] A device consisting of a drum with a contoured surface which communicates motion to a cam follower as the drum rotates around an axis. { 'drəm ˌkam }

drum controller [ELEC] An electric device that has a drum switch for its main switching element; used to govern the way electric power is delivered to a motor. { 'drəm kən͵trō·lər }

drum disk rectifier [ELEC] A mechanical rectifier using synchronous contacts and a copper oxide dry disk. { 'drəm ˌdisk 'rek·tə͵fī·ər }

DRUM BRAKE

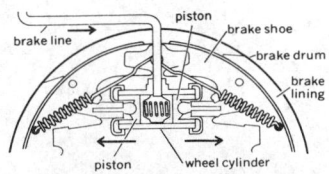

Schematic view of part of hydraulic braking system showing the brake line leading into the brakes which are shown applied. *(Pontiac Motor Division, General Motors Corp.)*

DRUM DRYER

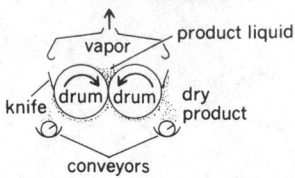

A top-fed double drum dryer.

drum dryer [MECH ENG] A machine for removing water from substances such as milk, in which a thin film of the product is moved over a turning steam-heated drum and a knife scrapes it from the drum after moisture has been removed. { 'drəm ˌdrī·ər }

drum feeder [MECH ENG] A rotating drum with vanes or buckets to lift and carry parts and drop them into various orienting or chute arrangements. Also known as tumbler feeder. { 'drəm ˌfēd·ər }

drum filter [MECH ENG] A cylindrical drum that rotates through thickened ore pulp, extracts liquid by a vacuum, and leaves solids, in the form of a cake, on a permeable membrane on the drum end. Also known as rotary filter; rotary vacuum filter. { 'drəm ˌfil·tər }

drum gate [CIV ENG] A movable crest gate in the form of an arc hinged at the apex and operated by reservoir pressure to open and close a spillway. { 'drəm ˌgāt }

drumlin [GEOL] A hill of glacial drift or bedrock having a half-ellipsoidal streamline form like the inverted bowl of a spoon, with its long axis paralleling the direction of movement of the glacier that fashioned it. { 'drəm·lən }

drumlinoid See rock drumlin. { 'drəm·ləˌnȯid }

drum mark [COMPUT SCI] A character indicating the termination of a record on a magnetic drum. { 'drəm ˌmärk }

drum memory See drum. { ¦drəm 'mem·rē }

drum meter See liquid-sealed meter. { 'drəm ˌmēd·ər }

drummy [MIN ENG] Loose rock or coal, especially in a mine roof, that produces a hollow, weak sound when tapped with a bar. { 'drəm·ē }

drum parity error [COMPUT SCI] Parity error occurring during transfer of information onto or from drums. { ¦drəm 'par·əd·ē ˌer·ər }

drum plotter [ENG] A graphics output device that draws lines with a continuously moving pen on a sheet of paper rolled around a rotating drum that moves the paper in a direction perpendicular to the motion of the pen. { 'drəm ˌpläd·ər }

drum printer [COMPUT SCI] An impact printer in which a complete set of characters for each print position on a line is on a continuously rotating drum behind an inked ribbon, with paper in front of the ribbon; identical characters are printed simultaneously at all required positions on a line, on the fly, by signal-controlled hammers. { 'drəm ˌprint·ər }

drum printing [TEXT] Dyeing yarn by winding it on a drum and applying color in bands of various widths. { 'drəm ˌprint·iŋ }

drum recorder [ELECTR] A facsimile recorder in which the record sheet is mounted on a rotating drum or cylinder. { 'drəm riˌkȯrd·ər }

drum separator [MIN ENG] A cylindrical vessel which rotates slowly and separates run-of-mine coal into clean coal, middlings, and refuse; can be adjusted for different specific gravities. { ¦drəm 'sep·əˌrād·ər }

drum storage See drum. { 'drəm ˌstȯr·ij }

drum switch [ELEC] A switch in which the electrical contacts are made on pins, segments, or surfaces on the periphery of a rotating cylinder or sector, or by the operation of a rotating cam. { 'drəm ˌswich }

drum trap [ENG] In plumbing, a trap in the form of a cylinder with a vertical axis that is fitted with a removable cover plate. { 'drəm ˌtrap }

drum transmitter [ELECTR] A facsimile transmitter in which the subject copy is mounted on a rotating drum or cylinder. { ¦drəm tranz¦mid·ər }

drum-type boiler See bent-tube boiler. { 'drəm ˌtīp ˌbȯil·ər }

drum winding [ELEC] A type of winding in electric machines in which coils are housed in long, narrow gaps either in the outer surface of a cylindrical core or in the inner surface of a core with a cylindrical bore. { 'drəm ˌwīnd·iŋ }

drunk mouse [COMPUT SCI] A mouse whose pointer jumps irrationally, usually as a result of dirt or grease on the rollers. { ¦drəŋk 'maùs }

drupaceous [BOT] Of, pertaining to, or characteristic of a drupe. { drü'pā·shəs }

drupe [BOT] A fruit, such as a cherry, having a thin or leathery exocarp, a fleshy mesocarp, and a single seed with a stony endocarp. Also known as stone fruit. { drüp }

drupelet [BOT] An individual drupe of an aggregate fruit. Also known as grain. { 'drüp·lət }

druse [GEOL] A small cavity in a rock or vein encrusted with aggregates of crystals of the same minerals which commonly constitute the enclosing rock. { drüz }

drusy [GEOL] Of or pertaining to rocks containing numerous druses. { 'drüz·ē }

dry [SCI TECH] Free from or deficient in moisture. { drī }

dry abrasive cutting [MECH ENG] Frictional cutting using a rotary abrasive wheel without the use of a liquid coolant. { ¦drī əˌbrā·siv 'kəd·iŋ }

dry acid [CHEM] Nonaqueous acetic acid used for oil-well reservoir acidizing treatment. { ¦drī 'as·əd }

dry adiabat [METEOROL] A line of constant potential temperature on a thermodynamic diagram. { ¦drī 'ad·ē·əˌbat }

dry adiabatic lapse rate [METEOROL] A special process lapse rate of temperature, defined as the rate of decrease of temperature with height of a parcel of dry air lifted adiabatically through an atmosphere in hydrostatic equilibrium. Also known as adiabatic lapse rate; adiabatic rate. { ¦drī ˌad·ē·ə¦bad·ik 'laps ˌrāt }

dry adiabatic process [METEOROL] An adiabatic process in a system of dry air. { ¦drī ˌad·ē·ə¦bad·ik 'präs·əs }

dry air [METEOROL] Air that contains no water vapor. { ¦drī 'er }

dry ashing [ORG CHEM] The conversion of an organic compound into ash (decomposition) by a burner or in a muffle furnace. { ¦drī 'ash·iŋ }

dry assay [MET] Determination of the amount of a desired constituent in ores, metallurgical residues, and alloys by methods other than those involving liquid means of separation. { ¦drī 'aˌsā }

dry-back boiler See scotch boiler. { ¦drī ˌbak 'bȯil·ər }

dry battery [ELEC] A battery made up of a series, parallel, or series-parallel arrangement of dry cells in a single housing to provide desired voltage and current values. { ¦drī 'bad·ə·rē }

dry bed [CHEM ENG] A configuration of solid adsorption materials, for example molecular sieves or charcoal, used to recover liquid from or purify a gas stream. { ¦drī 'bed }

dry blast cleaning [ENG] Cleaning of metallic surfaces by blasting with abrasive material traveling at a high velocity; abrasive may be accelerated by an air nozzle or a centrifugal wheel. { ¦drī ˌblast 'klēn·iŋ }

dry-bone ore See smithsonite. { ¦drī ˌbōn ˌȯr }

dry box [CHEM] A container or chamber filled with argon, or sometimes dry air or air with no carbon dioxide (CO_2), to provide an inert atmosphere in which manipulation of very reactive chemicals is carried out in the laboratory. { 'drī ˌbäks }

dry-box process [CHEM ENG] The passing of coke-oven or other industrial gases through boxes containing trays of iron oxide coated on wood shavings or other supporting material in order to remove hydrogen sulfide. { 'drī ˌbäks ˌpräs·əs }

dry-bulb temperature [PHYS] The actual air temperature as measured by a dry-bulb thermometer. { ¦drī ˌbəlb 'tem·prə·chər }

dry-bulb thermometer [ENG] An ordinary thermometer, especially one with an unmoistened bulb; not dependent upon atmospheric humidity. { ¦drī ˌbəlb thər'mäm·əd·ər }

dry cargo [IND ENG] Nonliquid cargo, including minerals, grain, boxes, and drums. { ¦drī 'kär·gō }

dry-cargo ship [NAV ARCH] A ship which carries miscellaneous dry cargo, such as boxes, bales, bags, or lumber, which is normally hand-stowed. { ¦drī 'kär·gō ˌship }

dry cell [ELEC] A voltage-generating cell having an immobilized electrolyte. { ¦drī ˌsel }

dry-cell cap light [MIN ENG] A headlamp with a focusing lens lamp and a dry-cell battery unit clipped to the belt; to prevent explosion in a mine, the bulb is ejected automatically in case of its breakage. { ¦drīˌ sel 'kap ˌlīt }

dry-charged battery [ELEC] A storage battery in which the electrolyte is drained from the battery for storage, and which is filled with electrolyte and charged for a few minutes to prepare for use. { ¦drī ˌchärjd 'bad·ə·rē }

dry-chemical fire extinguisher [CHEM ENG] A dry powder, consisting principally of sodium bicarbonate, which is used for extinguishing small fires, especially electrical fires. { ¦drī ˌkem·i·kəl 'fīr ikˌstiŋ·gwə·shər }

dry circuit [ELEC] A relay circuit in which open-circuit voltages are very low and closed-circuit currents extremely small, so there is no arcing to roughen the contacts. { ¦drī ¦sər·kət }

dry-cleaned coal [MIN ENG] Coal that has been mechanically separated from impurities without the use of liquids. { 'drī ¦klēnd 'kōl }

dry cleaning [ENG] To utilize dry-cleaning fluid to remove stains from textile. { 'drī klēn·iŋ }

dry-cleaning fluid [MATER] An organic solvent such as chlorinated hydrocarbons or petroleum naphtha with narrow, carefully selected boiling points; used in dry cleaning. { 'drī ¦klēn·iŋ ‚flü·əd }

dry climate [CLIMATOL] 1. In W. Köppen's climatic classification, the major category which includes steppe climate and desert climate, defined strictly by the amount of annual precipitation as a function of seasonal distribution and of annual temperature. 2. In C. W. Thornwaite's climatic classification, any climate type in which the seasonal water surplus does not counteract seasonal water deficiency, and having a moisture index of less than zero; included are the dry subhumid, semiarid, and arid climates. { 'drī 'klī·mət }

dry coloring [CHEM ENG] A plastics coloring method in which uncolored particles of the plastic material are tumble-blended with selected dyes and pigments. [ENG] A method to color plastics by tumbleblending colorless plastic particles with dyes and pigments. [MATER] A powdered form of pigment. { 'drī ‚kəl·ə·riŋ }

dry contact [ELEC] A contact that does not break or make current. { 'drī 'kän‚takt }

dry cooling tower [MECH ENG] A structure in which water is cooled by circulation through finned tubes, transferring heat to air passing over the fins; there is no loss of water by evaporation because the air does not directly contact the water. { 'drī ‚kül·iŋ ‚tau̇·ər }

dry corrosion [MET] Destruction of a metal or alloy by chemical processes resulting from attack by gases in the atmosphere above the dew point. { 'drī kə'rō·zhən }

dry cough [MED] A cough not accompanied by expectoration. Also known as nonproductive cough; unproductive cough. { 'drī 'kȯf }

dry course [BUILD] An initial roofing course of felt or paper not bedded in tar or asphalt. { 'drī 'kȯrs }

dry criticality [NUCLEO] Reactor criticality achieved without a coolant. { 'drī ‚krid·ə'kal·əd·ē }

dry delta See alluvial fan. { 'drī 'del·tə }

dry-desiccant dehydration [CHEM ENG] Use of silica gel or other solid absorbent to remove liquids from gases, such as water from air, or liquid hydrocarbons from natural gas. { 'drī ¦des·ə·kənt ‚dē·hī'drā·shən }

dry-disk rectifier See metallic rectifier. { 'drī ‚disk 'rek·tə‚fī·ər }

dry distillation [CHEM] A process in which a solid is heated in the absence of liquid to release vapors or liquids from the solid, for example, heating a hydrate to produce the anhydrous salt. { 'drī dis·tə'lā·shən }

dry dock [CIV ENG] A dock providing support for a vessel and a means for removing the water so that the bottom of the vessel can be exposed. { 'drī 'däk }

dry-dock caisson [CIV ENG] The floating gate to a dry dock. Also known as caisson. { 'drī ‚däk 'kā‚sän }

dry-dock iceberg See valley iceberg. { 'drī ‚däk 'īs‚bərg }

dry drilling [MIN ENG] Drilling in which chippings and cuttings are lifted out of a borehole by a current of air or gas. { 'drī 'dril·iŋ }

dry electrolytic capacitor [ELEC] An electrolytic capacitor in which the electrolyte is a paste rather than a liquid; the dielectric is a thin film of gas formed on one of the plates by chemical action. { 'drī i¦lek·trə¦lid·ik kə'pas·əd·ər }

dry farming [AGR] Production of crops in regions having sparse rainfall without the use of irrigation by employing cultivation techniques that conserve soil moisture. { 'drī 'färm·iŋ }

dry firn See polar firn. { 'drī 'fərn }

dry flashover voltage [ELECTR] Voltage at which the air surrounding a clean dry insulator or shell completely breaks down between electrodes. { 'drī 'flash‚ō·vər ‚vōl·tij }

dry fog [METEOROL] A fog that does not moisten exposed surfaces. { 'drī 'fäg }

dry forest [FOR] A type of forest characterized by relatively sparse distributions of pine, juniper, oak, olive, acacia, mesquite, and other drought-resistant species growing in scrub woodland, savanna, or chaparral settings; occurs in the southwestern United States, Mediterranean region, sub-Saharan Africa, and semiarid regions of Mexico, India, and Central and South America. { 'drī 'fär·əst }

dry freeze [HYD] The freezing of the soil and terrestrial objects caused by a reduction of temperature when the adjacent air does not contain sufficient moisture for the formation of hoarfrost on exposed surfaces. { 'drī 'frēz }

dry friction [MECH] Resistance between two dry solid surfaces, that is, surfaces free from contaminating films or fluids. { 'drī 'frik·shən }

dry-fuel rocket [AERO ENG] A rocket that uses a mixture of rapidly burning powders; used especially as a booster rocket. { 'drī 'fyül 'räk·ət }

dry gangrene [MED] Local death of a part caused by arterial obstruction without associated venous obstruction or infection. { 'drī 'gaŋ‚grēn }

dry gas [MATER] A gas that does not contain fractions which may easily condense under normal atmospheric conditions, for example, natural gas with methane and ethane. { 'drī 'gas }

dry grinding [ENG] Reducing particle sizes without a liquid medium. { 'drī 'grīnd·iŋ }

dry haze [METEOROL] Fine dust or salt particles in the air, too small to be individually apparent but in sufficient number to reduce horizontal visibility, and to give the atmosphere a characteristic hazy appearance. { 'drī 'hāz }

dry hole [ENG] A hole driven without the use of water. [PETRO ENG] A well in which no oil or gas is found. { 'drī 'hōl }

dry-hot-rock geothermal system [GEOL] A water-deficient hydrothermal reservoir dominated by the presence of rocks at depths in which large quantities of heat are stored. { 'drī 'hät ‚räk jē·ō¦thər·məl 'sis·təm }

dry ice [INORG CHEM] Carbon dioxide in the solid form, usually made in blocks to be used as a coolant; changes directly to a gas at $-78.5°C$ as heat is absorbed. { 'drī 'īs }

drying [CHEM] 1. An operation in which a liquid, usually water, is removed from a wet solid in equipment termed a dryer. 2. A process of oxidation whereby a liquid such as linseed oil changes into a solid film. { 'drī·iŋ }

drying agent [CHEM] Soluble or insoluble chemical substance that has such a great affinity for water that it will abstract water from a great many fluid materials; soluble chemicals are calcium chloride and glycerol, and insoluble chemicals are bauxite and silica gel. Also known as desiccant. { 'drī·iŋ ‚ā·jənt }

drying oil [MATER] Relatively highly unsaturated oil, such as cottonseed, soybean, and linseed oil, that is easily oxidized and polymerized to form a hard, dry film on exposure to air; used in paints and varnish. { 'drī·iŋ ‚ȯil }

drying oven [ENG] A closed chamber for drying an object by heating at relatively low temperatures. { 'drī·iŋ ‚əv·ən }

Dryinidae [INV ZOO] A family of hymenopteran insects in the superfamily Bethyloidea. { drī'in·ə‚dē }

dry ink [MATER] A finely powdered mixture of resin and pigment that is deposited to form an image in electrophotography. { 'drī 'iŋk }

dry kiln [ENG] A heated room or chamber used to dry and season cut lumber. { 'drī 'kil }

dry limestone process [CHEM ENG] An air-pollution control method in which sulfur oxides are exposed to limestone to convert them to disposable residues. { 'drī 'līm‚stōn ‚präs·əs }

dryline [METEOROL] The boundary separating warm dry air from warm moist air along which thunderstorms and tornadoes may develop. { 'drī‚līn }

dry machining [MECH ENG] Cutting, drilling, and grinding operations in which the use of a cutting fluid (lubricant) has been eliminated. { 'drī mə'shēn·iŋ }

dry measure [MECH] A measure of volume for commodities that are dry. { 'drī ¦mezh·ər }

dry mill [FOOD ENG] A machine for processing corn consisting of a horizontal, revolving conical drum covered with metal projections, in a housing also studded with metal projections and having small perforations through which pass fine particles of hull and germ. [MECH ENG] Grinding device used to powder or pulverize solid materials without an associated liquid. { 'drī ¦mil }

dry mining [MIN ENG] Mining operation in which there is no moisture in the ventilating air. { 'drī 'mīn·iŋ }

dry mortar [MATER] A mortar that is significantly stiffer and has lower water content than standard mortar but contains sufficient water for hydration.　{ ¦drī ˈmȯrd·ər }

dry mounting [GRAPHICS] A method for mounting photographs and other paper materials on cardboard without paste or rubber cement; a light, thin tissue (mounting tissue) is placed over the photograph and heat is applied with slight pressure so that the photo will adhere to the cardboard.　{ 'drī ˌmaùnt·iŋ }

dry offset *See* letterset.　{ ¦drī 'ȯf,set }

dryography *See* waterless offset lithography.　{ ¦drī'äg·rə·fē }

dry oil [MATER] Oil that has been rendered relatively free of water or impurities.　{ 'drī 'ȯil }

Dryomyzidae [INV ZOO] A family of myodarian cyclorrhaphous dipteran insects in the subsection Acalypteratae.　{ ˌdrī·ō'mīz·ə,dē }

Dryopidae [INV ZOO] The long-toed water beetles, a family of coleopteran insects in the superfamily Dryopoidea.　{ drī'äp·ə,dē }

Dryopoidea [INV ZOO] A superfamily of coleopteran insects in the suborder Polyphaga, including the nonpredatory aquatic beetles.　{ drī·ə'pȯid·ē·ə }

dry ore [MIN ENG] An ore of gold or silver which requires added lead and fluxes for treatment.　{ ¦drī 'ȯr }

dry permafrost [GEOL] A loose and crumbly permafrost which contains little or no ice.　{ ¦drī 'pər·mə,frȯst }

dry permeability [ENG] A property of dried bonded sand to permit passage of gases while molten material is poured into a mold.　{ ¦drī ,pər·mē·ə'bil·əd·ē }

dry pint *See* pint.　{ ¦drī 'pīnt }

dry pipe [MECH ENG] A perforated metal pipe above the normal water level in the steam space of a boiler which prevents moisture or extraneous matter from entering steam outlet lines.　{ ¦drī ¦pīp }

dry-pipe system [ENG] A sprinkler system that admits water only when the air it normally contains has been vented; used for systems subjected to freezing temperatures.　{ ¦drī ,pīp ,sis·təm }

dry-pit pump [MECH ENG] A pump operated with the liquid conducted to and from the unit by piping.　{ 'drī ,pit ,pəmp }

dry placer [MIN ENG] A gold-bearing alluvial deposit found in arid regions; it cannot be mined due to lack of water.　{ ¦drī 'plā·sər }

dry plasma etching *See* plasma etching.　{ ¦drī 'plaz·mə }

dry plate [GRAPHICS] A photographic plate that has a sensitized coating of an emulsion of silver halide in gelatin which is dried before exposure to light in the photographic process.　{ 'drī ,plāt }

dry-plate rectifier *See* metallic rectifier.　{ ¦drī ,plāt 'rek·tə,fī·ər }

dry point [ANALY CHEM] The temperature at which the last drop of liquid evaporates from the bottom of the flask.　{ 'drī ,pȯint }

drypoint etching [GRAPHICS] Etching in which a sharp tool (an etching needle) scratches through only the etching ground that is placed on the surface of the copper plate; the plate is then placed in an acid bath, and the chemical action produces a line deep enough to hold ink.　{ 'drī,pȯint 'ech·iŋ }

dry pressing [ENG] Molding clayware by compressing moist clay powder in metal dies.　{ ¦drī 'pres·iŋ }

dry processing [GRAPHICS] Development of a latent image without use of a solution.　{ 'drī 'präs,es·iŋ }

dry pt *See* pint.

dry quicksand [GEOL] An accumulation of alternate layers of firmly compacted sand and loose sand that cannot support heavy loads.　{ ¦drī 'kwik,sand }

dry reed relay [ELEC] Reed-type relay which does not use mercury at the relay contacts.　{ ¦drī ,rēd 'rē,lā }

dry reed switch [ELEC] A switch having contacts mounted on magnetic reeds in a vacuum enclosure, designed for reliable operation in dry circuits.　{ ¦drī ,rēd 'swich }

dry-relief offset [GRAPHICS] Referring to plates made for use on offset-lithographic presses, but printed without the use of dampeners and water; the plates are made photomechanically, and nonprinting areas are etched from 0.008 to 0.015 inch (0.2 to 0.4 millimeter) below the surface. Also known as high etch.　{ ¦drī rə,lēf 'ȯf,set }

dry rot [MICROBIO] A rapid decay of seasoned timber caused by certain fungi which cause the wood to be reduced to a dry, friable texture. [PL PATH] Any of various rot diseases of plants characterized by drying of affected tissues.　{ 'drī ,rät }

DRYSDALE AC POLAR POTENTIOMETER

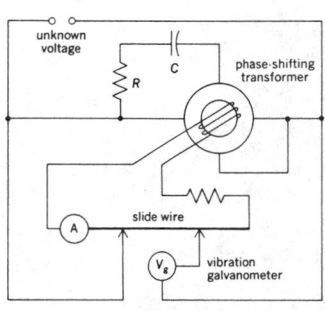

Drysdale ac potentiometer circuit. *(General Electric Co.)*

dry run [COMPUT SCI] A check of the logic and coding of a computer program in which the program's operations are followed from a flow chart and written instructions, and the results of each step are written down, before the program is run on a computer. Also known as desk check. [ENG] Any practice test or session. [ORD] Any simulated firing practice, particularly a dive-bombing approach made without the release of a bomb.　{ 'drī 'rən }

dry sample [MIN ENG] A sample of ore obtained by dry drilling.　{ ¦drī 'sam·pəl }

dry sand [GEOL] **1.** A formation, underlying the production sand, into which oil has leaked due to careless drilling practices. **2.** A nonproductive oil sand.　{ ¦drī ¦sand }

dry sand mold [MET] A mold made of greensand and then dried in an oven to increase its strength.　{ ¦drī ¦sand ,mōld }

Drysdale ac polar potentiometer [ENG] A potentiometer for measuring alternating-current voltages in which the voltage is applied across a slide-wire supplied with current by a phase-shifting transformer; this current is measured by an ammeter and brought into phase with the unknown voltage by adjustment of the transformer rotor, and the unknown voltage is measured by observation of the slide-wire setting for a null indication of a vibration galvanometer.　{ 'drīz,dāl ¦a¦sē ¦pō·lər pə,ten·chē'äm·əd·ər }

dry season [CLIMATOL] In certain types of climate, an annually recurring period of one or more months during which precipitation is at a minimum for the region.　{ 'drī 'sēz·ən }

dry sieving [ENG] Particle-size distribution analysis of powdered solids; the sample is placed on the top sieve screen of a nest (stack), with mesh openings decreasing in size from the top to the bottom of the nest.　{ 'drī 'siv·iŋ }

dry-silver material [GRAPHICS] Photosensitive film or paper that is developed by application of heat.　{ 'drī 'sil·vər mə'tir·ē·əl }

dry sleeve [MECH ENG] A cylinder liner which is not in contact with the coolant.　{ ¦drī ¦slēv }

dry socket [MED] Inflammation of the dental alveolus, especially the inflamed condition following the removal of a tooth. Also known as alveolitis.　{ ¦drī 'säk·ət }

dry spell [CLIMATOL] A period of abnormally dry weather, generally reserved for a less extensive, and therefore less severe, condition than a drought; in the United States, describes a period lasting not less than 2 weeks, during which no measurable precipitation was recorded.　{ 'drī 'spel }

dry spot [CHEM ENG] **1.** An open area of an incomplete surface film on laminated plastic. **2.** A section of laminated glass where the interlayer and glass are not bonded.　{ ¦drī ,spät }

dry start [AERO ENG] The starting up of a liquid-fuel rocket engine without having previously filled the regenerative cooling tubes.　{ 'drī 'stärt }

dry steam [PHYS] Steam with no liquid phase dispersed in it.　{ 'drī 'stēm }

dry-steam drum [MECH ENG] **1.** Pressurized chamber into which steam flows from the steam space of a boiler drum. **2.** That portion of a two-stage furnace that extends forward of the main combustion chamber; fuel is dried and gasified therein, with combustion of gaseous products accomplished in the main chamber; the refractory walls of the Dutch oven are sometimes water-cooled.　{ ¦drī ,stēm 'drəm }

dry-steam energy system [ENG] **1.** A geothermal energy source that produces superheated steam. **2.** A hydrothermal convective system driven by vapor with a temperature in excess of 300°F (150°C).　{ 'drī ,stēm 'en·ər·jē ,sis·təm }

drystone [GEOL] A stalagmite or stalactite formed by dropping water.　{ 'drī,stōn }

dry storage [MECH ENG] Cold storage in which refrigeration is provided by chilled air.　{ 'drī ,stȯr·ij }

dry strength [ENG] The strength of an adhesive joint determined immediately after drying under specified conditions or after a period of conditioning in the standard laboratory atmosphere.　{ 'drī ,streŋkth }

dry string [PETRO ENG] The drill pipe from which drilling mud has been removed as it is extracted from the wellbore.　{ 'drī 'striŋ }

dry tabling [MIN ENG] A process similar to wet tabling, but without the water; used to separate two or more minerals based on specific gravity differences.　{ 'drī ,tāb·liŋ }

dry-tape fuel cell [ELEC] A fuel cell in which the fuel is in the form of a dry tape, coated with fuel, oxidant, and electrolyte, which is fed into the cell at a rate corresponding to the demand for electric energy. { 'drī ,tāp 'fyül ,sel }

dry test meter [ENG] Gas-flow rate meter with two compartments separated by a movable diaphragm which is connected to a series of gears that actuate a dial; when one chamber is full, a valve switches to the other, empty chamber; used to measure household gas-flow rates and to calibrate flow-measurement instruments. { 'drī 'test ,mēd·ər }

dry ticket [IND ENG] Tank inspection form signed by shore and ship inspectors before loading and after discharging the ship. { 'drī ,tik·ət }

dry tongue [METEOROL] In synoptic meteorology, a pronounced protrusion of relatively dry air into a region of higher moisture content. { 'drī ,təŋ }

dry-type transformer [ELECTROMAG] A transformer that is designed to operate in air, without oil cooling. { 'drī,tīp tranz 'fȯr·mər }

dry valley [GEOL] A valley, usually in a chalk or karst type of topography, that has no permanent water course along the valley floor. { 'drī 'val·ē }

dry wall [BUILD] A wall covered with wallboard, in contrast to plaster. [ENG] A wall constructed of rock without cementing material. { 'drī ,wȯl }

dry wash [GEOL] A wash, arroyo, or coulee whose bed lacks water. { 'drī ,wäsh }

dry washer [MIN ENG] A machine for extracting gold mined from dry placers. { 'drī ,wäsh·ər }

dry well [CIV ENG] 1. A well that has been completely drained. 2. An excavated well filled with broken stone and used to receive drainage when the water percolates into the soil. 3. Compartment of a pumping station in which the pumps are housed. [NUCLEO] The first containment tank surrounding a water-cooled nuclear reactor that uses the pressure-suppressing containment system. { 'drī ,wel }

dry whole milk [FOOD ENG] A product made by removing water from whole milk until the dry product contains no more than 4% water and not less than 26% fat. { 'drī ,hōl ,milk }

dry wire drawing [MET] The drawing of dry steel process wire, pretreated by acid cleaning, lime coating, and baking, through a lubricant and wire drawing frame. { 'drī ,wīr ,drō·iŋ }

Drzewiecki theory [MECH ENG] In theoretical investigations of windmill performance, a theory concerning the air forces produced on an element of the blade. { 'dərz·vē·kē ,thē·ə·rē }

DS See Doppler sonar.

DSA See digital subtraction angiography.

DSB See double sideband modulation.

DSB-RC modulation See double-sideband reduced-carrier modulation. { 'dē¦es'bē ¦är'sē ,mäj·ə,lā·shən }

DSB-SC modulation See double-sideband suppressed-carrier modulation. { 'dē¦es'bē ¦es'sē ,mäj·ə,lā·shən }

DSB-TC modulation See double-sideband modulation. { 'dē¦es'bē ¦tē'sē ,mäj·ə,lā·shən }

D scan See D scope. { 'dē ,skan }

D scope [ELECTR] A cathode-ray scope which combines the features of B and C scopes, the signal appearing as a spot with bearing angle as the horizontal coordinate and elevation angle as the vertical coordinate, but with each spot expanded slightly in a vertical direction to give a rough range indication. Also known as D indicator; D scan. { 'dē ,skōp }

DSECT See dummy section. { 'dē'sekt }

D-shell connector [COMPUT SCI] The connector at the end of the cable between a video adapter and a monitor that is plugged into the video adapter. { 'dē,shel kə,nek·tər }

DSI See digital speech interpolation.

DSL See digital subscriber line.

DSN See Deep Space Network.

D sounding See D value. { 'dē ,saund·iŋ }

DSP chip See digital signal processing chip. { 'dē¦es'pē ,chip }

DSRV See deep-submergence rescue vehicle.

DSS See decision support system.

Dst [GEOPHYS] The "storm-time" component of variation of the terrestrial magnetic field, that is, the component which correlates with the interval of time since the onset of a magnetic storm; used as an index of intensity of the ring current.

DSTC modulation See double-sideband modulation. { 'dē¦es ¦tē'sē ,mäj·ə,lā·shən }

DTD See Document Type Definition.

DTL See diode transistor logic.

D₁ trisomy [MED] A syndrome resulting from the presence in triplicate of chromosomes 13–15; manifested in severe congenital anomalies and usually resulting in death in infancy. Also known as trisomy 13–15. { 'dē ,səb,wən 'trī,sō·mē }

DTS See Digital Termination System.

D2 radio source [ASTRON] A radio source consisting of a small, variable nuclear component sometimes coincident with an optical object, and a second, larger component with a much steeper radio spectrum. { 'dē¦tü 'rād·ē·ō ,sȯrs }

D-type symbiotic star [ASTRON] A member of a class of symbiotic stars that show infrared emission indicative of astronomical dust, that is, thermal radiation at average temperatures of typically 1000 K; it generally contains Mira variables and has binary periods longer than 10 years. { 'dē,tīp ,sim·bē,äd·ik 'stär }

dual-actuator hard disk [COMPUT SCI] A hard disk that is equipped with two read/write heads. { 'dül 'ak·chə,wād·ər ¦härd ,disk }

Dualayer distillate process [CHEM ENG] A process for the removal of mercaptan and oxygenated compounds from distillate fuel oils; treatment is with concentrated caustic Dualayer solution and electrical precipitation of the impurities. { 'dü·ə,lā·ər 'dis·təl·ət ,präs·əs }

Dualayer solution [CHEM ENG] A concentrated potassium or sodium hydroxide solution containing a solubilizer; used in the Dualayer distillate process. { 'dü·ə,lā·ər sə'lü·shən }

dual basis [MATH] 1. For a finite-dimensional vector space with basis $x_1, x_2, \ldots, x_n$, the dual basis of the conjugate space is the set of linear functionals $f_1, f_2, \ldots, f_n$ with $f_i(x_i) = 1$ and $f_i(x_j) = 0$ for i not equal to j. 2. For a Banach space with basis $x_1, x_2, \ldots$, the dual basis of the conjugate space is the sequence of continuous linear functionals, $f_1, f_2, \ldots$, defined by $f_i(x_i) = 1$ and $f_i(x_j) = 0$ for i not equal to j, provided that the conjugate space is shrinking. { 'dü·əl 'bā·ses }

dual-bed dehumidifier [MECH ENG] A sorbent dehumidifier with two beds, one bed dehumidifying while the other bed is reactivating, thus providing a continuous flow of air. { 'dü·əl ¦bed ,dē·yü'mid·ə,fī·ər }

dual-channel amplifier [ENG ACOUS] An audio-frequency amplifier having two separate amplifiers for the two channels of a stereophonic sound system, usually operating from a common power supply mounted on the same chassis. { 'dü·əl ¦chan·əl 'am·plə,fī·ər }

dual completion well [PETRO ENG] Single well casing containing two production tubing strings, each in a different zone of the reservoir (one higher, one lower) and each separately controlled. { 'dü·əl kəm¦plē·shən 'wel }

dual control [CONT SYS] An optimal control law for a stochastic adaptive control system that gives a balance between keeping the control errors and the estimation errors small. { 'dü·əl kən'trōl }

dual coordinates [MATH] Point coordinates and plane coordinates are dual in geometry since an equation about one determines an equation about the other. { 'dü·əl kō'ȯrd·ən·əts }

dual-cycle boiling-water reactor [NUCLEO] A boiling-water reactor in which part of the steam used to run the steam turbine is generated in the reactor core and part is generated in an external heat exchanger. Also known as dual-cycle reactor system. { 'dü·əl ¦sī·kəl ¦bȯil·iŋ ¦wȯd·ər rē,ak·tər }

dual-cycle reactor system See dual-cycle boiling-water reactor. { 'dü·əl ¦sī·kəl rē'ak·tər ,sis·təm }

dual diversity receiver [ELECTR] A diversity radio receiver in which the two antennas feed separate radio-frequency systems, with mixing occurring after the converter. { 'dü·əl də'vər·səd·ē ri,sē·vər }

dual-emitter transistor [ELECTR] A passivated *pnp* silicon planar epitaxial transistor having two emitters, for use in low-level choppers. { 'dü·əl i'mid·ər tran,zis·tər }

dual-flow oil burner [MECH ENG] An oil burner with two sets of tangential slots in its atomizer for use at different capacity levels. { 'dü·əl ¦flō 'ȯil ,bər·nər }

dual-fuel engine [MECH ENG] Internal combustion engine that can operate on either of two fuels, such as natural gas or gasoline. { 'dü·əl ¦fyül 'en·jən }

dual-function catalyst See bifunctional catalyst. { 'dül ¦fəŋk·shən 'kad·ə·list }

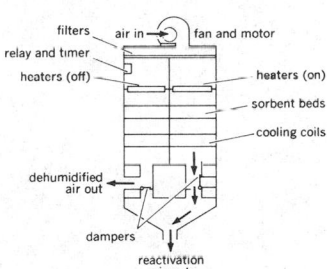

DUAL-BED DEHUMIDIFIER

Dual-bed solid-sorbent dehumidifier. Air is being dehumidified through the left bed at the same time that the right bed is being reactivated.

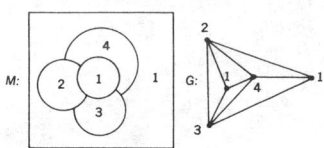

DUAL GRAPH

$M:$ $G:$

A map requiring four colors, indicated by numbers, and its planar graph.

dual graph [MATH] A planar graph corresponding to a planar map obtained by replacing each country with its capital and each common boundary by an arc joining the two countries. { ¦dü·əl 'graf }

dual-gravity valve [CHEM ENG] A float-operated valve that operates on the interface between two immiscible liquids of different specific gravities. { ¦dü·əl 'grav·əd·ē ,valv }

dual group [MATH] The group of all homomorphisms of an Abelian group G into the cyclic group of order n, where n is the smallest integer such that g^n is the identity element of G. { ¦dü·əl 'grüp }

dual-gun cathode-ray tube [ELECTR] A dual-trace oscilloscope in which beams from two electron guns are controlled by separate balanced vertical-deflection plates and also have separate brightness and focus controls. { ¦dü·əl ¦gən ,kath,ōd 'rā ,tüb }

dual in-line package [ELECTR] Microcircuit package with two rows of seven vertical leads that are easily inserted into an etched circuit board. Abbreviated DIP. { ¦dü·əl ¦in ,līn 'pak·ij }

duality principle Also known as principle of duality. [ELEC] The principle that for any theorem in electrical circuit analysis there is a dual theorem in which one replaces quantities with dual quantities; current and voltage, impedance and admittance, and meshes and nodes are examples of dual quantities. [ELECTR] The principle that analogies may be drawn between a transistor circuit and the corresponding vacuum tube circuit. [ELECTROMAG] The principle that one can obtain new solutions of Maxwell's equations from known solutions by replacing **E** with **H**, **H** with $-$**E**, ϵ with μ, and μ with ϵ. [MATH] A principle that if a theorem is true, it remains true if each object and operation is replaced by its dual; important in projective geometry and Boolean algebra. [QUANT MECH] *See* wave-particle duality. { dü'al·əd·ē ,prin·sə·pəl }

duality theorem [MATH] **1.** A theorem which asserts that for a given n-dimensional space, the $(n - p)$ dimensional homology group is isomorphic to a p-dimensional cohomology group for each $p = 0, \ldots, n$, provided certain conditions are met. **2.** Let G be either a compact group or a discrete group, let X be its character group, and let G' be the character group of X; then there is an isomorphism of G onto G' so that the groups G and G' may be identified. **3.** If either of two dual linear-programming problems has a solution, then so does the other. { dü'al·əd·ē ,thir·əm }

dual laser [OPTICS] A gas laser having Brewster windows and concave mirrors at opposite ends, the mirrors having different reflectivities so as to produce two different visible or infrared wavelengths from a helium-neon laser beam. { 'dü·əl 'lā·zər }

dual linear programming [MATH] Linear programming in which the maximum and minimum number are the same number. { 'dü·əl ¦lin·ē·ər 'prō,gram·iŋ }

dual meter [ENG] Meter constructed so that two aspects of an electric circuit may be read simultaneously. { 'dü·əl ¦mēd·ər }

dual-mode control [CONT SYS] A type of control law which consists of two distinct types of operation; in linear systems, these modes usually consist of a linear feedback mode and a bang-bang-type mode. { 'dü·əl ,mōd kən'trōl }

dual modulation [COMMUN] The process of modulating a common carrier wave or subcarrier with two different types of modulation, each conveying separate information. { 'dü·əl ,mäj·ə'lā·shən }

dual network [ELEC] A network which has the same number of terminal pairs as a given network, and whose open-circuit impedance network is the same as the short-circuit admittance matrix of the given network, and vice versa. { 'dü·əl 'net ,wərk }

dual operation [MATH] In projective geometry, an operation that is obtained from a given operation by replacing points with lines, lines with points, the drawing of a line through a point with the marking of a point on a line, and so forth. { ¦dül äp·ə'rā·shən }

dual-purpose gun [ORD] Gun so designed and constructed that effective fire may be delivered against either aerial or surface targets. { ¦dü·əl ¦pər·pəs 'gən }

dual-purpose reactor [NUCLEO] Any nuclear reactor which both acts as a source of heat energy for a power plant and produces fissionable material. { ¦dü·əl ¦pər·pəs rē'ak·tər }

dual radioactive decay [NUC PHYS] Property exhibited by a nucleus which has two or more independent and alternative modes of decay. { ¦dü·əl ,rād·ē·ō¦ak·tiv di'kā }

dual-scanned liquid-crystal display [ELECTR] A passive matrix liquid-crystal display that is improved by being refreshed twice as frequently as standard displays of this type. { ¦dül ,skand ,lik·wəd ¦krist·əl di,splā }

dual-seal tubing joint [PETRO ENG] Tubing connection joint with two sealing surfaces to assure a leak-free connection between sections. { 'dü·əl ,sēl 'tüb·iŋ ,jóint }

dual space [MATH] The vector space consisting of all linear transformations from a given vector space into its scalar field. { 'dü·əl 'spās }

dual-stripe magnetoresistive head [COMPUT SCI] A type of read/write head for hard disks that has separate areas for reading and writing, reduced vulnerability to outside interference, and the ability to pack data densely on disks { ¦dül ¦strīp mag,ned·ō·ri,zis·div 'hed }

dual tensor [MATH] The product of a given tensor, covariant in all its indices, with the contravariant form of the determinant tensor, contracting over the indices of the given tensor. { 'dü·əl 'ten·sər }

dual theorem [MATH] In projective geometry, the theorem that is obtained from a given theorem by replacing points with lines, lines with points, and operations with their dual operations. Also known as reciprocal theorem. { ¦dül 'thir·əm }

dual thrust [AERO ENG] A rocket thrust derived from two propellant grains and using the same propulsion section of a missile. { 'dü·əl 'thrəst }

dual-thrust motor [AERO ENG] A solid-propellant rocket engine built to obtain dual thrust. { 'dü·əl ¦thrəst 'mōd·ər }

dual-tone multifrequency [COMMUN] Signaling method employing set combinations of two specific frequencies used by subscribers and telephone private branch exchange attendants, if their switchboard positions are so equipped, to indicate telephone address digits, precedence ranks, and end of signaling. { 'dü·əl ,tōn ,məl·tē'frē·kwən·sē }

dual-trace amplifier [ELECTR] An oscilloscope amplifier that switches electronically between two signals under observation in the interval between sweeps, so that waveforms of both signals are displayed on the screen. { 'dü·əl ,trās 'am·plə,fī·ər }

dual-trace oscilloscope [ELECTR] An oscilloscope which can compare two waveforms on the face of a single cathode-ray tube, using any one of several methods. { 'dü·əl ,trās ä'sil·ə,skōp }

dual-track tape recorder *See* double-track tape recorder. { 'dü·əl ,trak 'tāp ri,kórd·ər }

dual-use line [COMMUN] Communications link normally used for more than one mode of transmission, such as voice and data. { 'dü·əl ,yüs ,līn }

dual variables [MATH] Mutually dependent variables. { 'dü·əl 'ver·ē·ə·bəlz }

Duane-Hunt law [QUANT MECH] The law that the frequency of x-rays resulting from electrons striking a target cannot exceed eV/h, where e is the charge of the electron, V is the exciting voltage, and h is Planck's constant. { ¦dwān ¦hənt ,ló }

Duane-Hunt limit [QUANT MECH] The upper limit on the frequency of radiation from an x-ray tube given by the Duane-Hunt law. { ¦dwān ¦hənt 'lim·ət }

dub [ENG ACOUS] **1.** To transfer recorded material from one recording to another, with or without the addition of new sounds, background music, or sound effects. **2.** To combine two or more sources of sound into one record. **3.** To add a new sound track or new sounds to a motion picture film, or to a recorded radio or television production. { dəb }

Dubbs cracking [CHEM ENG] A continuous, liquid-phase, thermal cracking process. { ¦dəbz 'krak·iŋ }

dubnium [CHEM] A chemical element, symbolized Db, atomic number 105, a synthetic element; the thirteenth transuranium element. { 'düb·nē·əm }

Duchemin's formula [PHYS] An expression for normal wind pressure per square foot on an inclined surface, $N = F \cdot [(2\sin a)/(1 + \sin^2 a)]$, where $F =$ normal wind force in pounds per square foot on a vertical surface, and $a =$ angle of inclination of inclined surface. { dü·shmanz ,fór·myə·lə }

Duchenne's dystrophy [MED] A sex-linked or autosomal

recessive form of muscular dystrophy, which is progressive with pseudohypertrophy. { dü′shenz ′dis·trə·fē }

duck [ORD] *See* DUKW. [TEXT] Close-woven, heavy fabric made of cotton and used for liquid filtration in the process industries, as well as for making sails, tents, and clothing. [VERT ZOO] The common name for a number of small water-fowl in the family Anatidae, having short legs, a broad, flat bill, and a dorsoventrally flattened body. { dək }

duckbill [MECH ENG] A shaking type of combination loader and conveyor whose loading end is generally shaped like a duck's bill. { ′dək‚bil }

duck-billed dinosaur [PALEON] Any of several herbivorous, bipedal ornithopods having the front of the mouth widened to form a ducklike beak. { ′dək ‚bild ′dīn·ə‚sȯr }

duckbill platypus *See* platypus. { ‚dək‚bil ′plad·ə·pəs }

duckfoot [ENG] In a piping system, a support fitted to the bend of a vertical pipe to permit the direct load of the pipework and fittings to be transferred to the floor, foundation, or associated installations. { ′dək‚fut }

duck's nest [PETRO ENG] A small pit dug in the earth near a wellbore to receive excess quantities of drilling mud. { ′dəks ‚nest }

duck wheat *See* tartary buckwheat. { ′dək ‚wēt }

Ducrey test [IMMUNOL] A skin test to determine past or present infection with *Hemophilus ducreyi*. { dü′krā ‚test }

duct [ANAT] An enclosed tubular channel for conducting a glandular secretion or other body fluid. [COMMUN] An enclosed runway for cables. [GEOPHYS] The space between two air layers, or between an air layer and the earth's surface, in which microwave beams are trapped in ducting. Also known as radio duct; tropospheric duct. [MECH ENG] A fluid flow passage which may range from a few inches in diameter to many feet in rectangular cross section, usually constructed of galvanized steel, aluminum, or copper, through which air flows in a ventilation system or to a compressor, supercharger, or other equipment at speeds ranging to thousands of feet per minute. { dəkt }

ducted fan [MECH ENG] A propeller or multibladed fan inside a coaxial duct or cowling. Also known as ducted propeller; shrouded propeller. { ′dək·təd ′fan }

ducted-fan engine [AERO ENG] An aircraft engine incorporating a fan or propeller enclosed in a duct; especially, a jet engine in which an enclosed fan or propeller is used to ingest ambient air to augment the gases of combustion in the jetstream. { ‚dək·təd ‚fan ′en·jən }

ducted propeller *See* ducted fan. { ‚dək·təd prə′pel·ər }

ducted rocket *See* rocket ramjet. { ‚dək·təd ′räk·ət }

ductile fracture *See* fibrous fracture. { ‚dək·təl ′frak·chər }

ductile iron *See* nodular cast iron. { ‚dək·təl ′ī·ərn }

ductility [MATER] The ability of a material to be plastically deformed by elongation, without fracture. { dək′til·əd·ē }

ducting [GEOPHYS] An atmospheric condition in the troposphere in which temperature inversions cause microwave beams to refract up and down between two air layers, so that microwave signals travel 10 or more times farther than the normal line-of-sight limit. Also known as superrefraction; tropospheric ducting. { ′dək·tiŋ }

ductless gland *See* endocrine gland. { ‚dək·ləs ′gland }

duct of Cuvier [EMBRYO] Either of the paired common cardinal veins in a vertebrate embryo. { ‚dəkt əv küv′yā }

duct of Santorini [ANAT] The dorsal pancreatic duct in a vertebrate embryo; persists in adult life in some species and serves as the pancreatic duct in the adult elasmobranch, pig, and ox. { ‚dəkt əv san·tə′rē·nē }

ductor roller [GRAPHICS] In a lithographic process, a roller for inking and dampening mechanisms on a press which has alternate contacts to a fountain roller and vibrating drum roller. { ′dək·tər ‚rōl·ər }

duct propulsion [AERO ENG] A means of propelling a vehicle by ducting a surrounding fluid through an engine, adding momentum by mechanical or thermal means, and ejecting the fluid to obtain a reactive force. { ‚dəkt prə‚pəl·shən }

ductus arteriosus [EMBRYO] Blood shunt between the pulmonary artery and the aorta of the mammalian embryo. { ′dək·təs ‚är·tir·ē′ō·səs }

ductus deferens *See* vas deferens. { ′dək·təs ′def·ə‚renz }

ductus venosus [EMBRYO] Blood shunt between the left umbilical vein and the right sinus venosus of the heart in the mammalian embryo. { ′dək·təs ve′nō·səs }

dud [ORD] An explosive munition that has failed to explode, although such was intended. { dəd }

Duddell oscillograph [ELECTROMAG] A moving-coil oscillograph; the current to be observed passes through a coil in a magnetic field and a mirror attached to the coil reveals its movement. { dü′del ä′sil·ə‚graf }

Duffy blood group [IMMUNOL] A genetically determined, immunologically distinct group of human erythrocyte antigens defined by their reaction with anti-Fy[a] serum. { ′dəf·ē ′bləd ‚grüp }

Dufour effect [THERMO] Energy flux due to a mass gradient occurring as a coupled effect of irreversible processes. { dü′fȯr i′fekt }

Dufour number [THERMO] A dimensionless number used in studying thermodiffusion, equal to the increase in enthalpy of a unit mass during isothermal mass transfer divided by the enthalpy of a unit mass of mixture. Symbol Du_2. { dü′fȯr ‚nəm·bər }

dufrenite [MINERAL] A blackish-green, fibrous ferric phosphate mineral; commonly massive or in nodules. { dü′frā‚nīt }

dufrenoysite [MINERAL] $Pb_2As_2S_5$ A lead gray to steel gray, monoclinic mineral consisting of lead arsenic sulfide. { ‚dü·frə′nȯi‚zīt }

duftite [MINERAL] $PbCu(AsO_4)(OH)$ Orthorhombic mineral that is composed of a basic arsenate of lead and copper. { ′dəf‚tīt }

Dugongidae [VERT ZOO] A family of aquatic mammals in the order Sirenia comprising two species, the dugong and the sea cow. { dü′gän·jə‚dē }

Dugonginae [VERT ZOO] The dugongs, a subfamily of sirenian mammals in the family Dugongidae characterized by enlarged, sharply deflected premaxillae and the absence of nasal bones. { dü′gän·jə‚nē }

dugout [ORD] Underground shelter built to protect troops, ammunition, and material from gunfire. { ′dəg‚aut }

Duhamel's theorem [MATH] If *f* and *g* are continuous functions, then

$$\lim_{|\Delta_x| \to 0} \sum_{i=1}^{n} f(x_i')g(x_i'')\Delta x_i = \int_a^b f(x)g(x)dx$$

where x_i' and x_i'' are between x_{i-1} and x_i, $i = 1, \ldots, n$, and $|\Delta x| = \max \{x_i - x_{i-1}\}$ for a partition $a = x_0 < x_1 < \cdots < x_n = b$. { dyə′melz ‚thir·əm }

Duhem-Margules equation [THERMO] An equation showing the relationship between the two constituents of a liquid-vapor system and their partial vapor pressures:

$$\frac{d \ln p_A}{d \ln x_A} = \frac{d \ln p_B}{d \ln x_B}$$

where x_A and x_B are the mole fractions of the two constituents, and p_A and p_B are the partial vapor pressures. { dü′em ′mär·gyə·lēz i‚kwā·zhən }

Duhem's equation *See* Gibbs-Duhem equation. { dü′emz i‚kwā·zhən }

Dühring's rule [PHYS CHEM] The rule that a plot of the temperature at which a liquid exerts a particular vapor pressure against the temperature at which a similar reference liquid exerts the same vapor pressure produces a straight or nearly straight line. { ′dir·iŋz ‚rül }

Dukler theory [CHEM ENG] Relationship of velocity and temperature distribution in thin films on vertical walls; used to calculate eddy viscosity and thermal conductivity near the solid boundary. { ′dük·lər ‚thē·ə·rē }

DUKW [ORD] A 2½-ton amphibious truck used by the U.S. Army to transport cargo on land or water; 36 feet (11 meters) long, with a cargo capacity of 5175 pounds (2347 kilograms). Also known as duck. { ‚dək ′dəb·əl‚yü }

dulcitol [ORG CHEM] $C_6H_8(OH)_6$ A sugar with a slightly sweet taste; white, crystalline powder with a melting point of 188.5°C; soluble in hot water; used in medicine and bacteriology. { ′dəl·sə‚tȯl }

dulcose *See* dulcitol. { ′dəl‚kōs }

dull coal [GEOL] A component of banded coal with a grayish color and dull appearance, consisting of small anthraxylon constituents in addition to cuticles and barklike constituents embedded in the attritus. { ′dəl ‚kōl }

DUCK-BILLED DINOSAUR

Restoration of the duck-billed dinosaur *Anatosaurus* (*Trachodon*) from the Late Cretaceous of North America; this dinosaur was 30–40 feet (9–12 meters) long.

dull emitter [ELECTR] An electron tube whose cathode is a filament that does not glow brightly. { ¦dəl ə'mid·ər }

Dulong number *See* Eckert number. { də'lȯŋ ,nəm·bər }

Dulong-Petit law [THERMO] The law that the product of the specific heat per gram and the atomic weight of many solid elements at room temperature has almost the same value, about 6.3 calories (264 joules) per degree Celsius. { də'lȯŋ pə'tē ,lȯ }

Dulong's formula [ENG] A formula giving the gross heating value of coal in terms of the weight fractions of carbon, hydrogen, oxygen, and sulfur from the ultimate analysis. { də'lȯŋz ,fȯr·myə·lə }

dulse [BOT] Any of several species of red algae of the genus *Rhodymenia* found below the intertidal zone in northern latitudes; an important food plant. { dəls }

Dultgen process [GRAPHICS] A halftone intaglio process for color work which makes it possible to vary the size as well as the depth of the dots. { 'dəlt·gən ,präs·əs }

DUMAND *See* deep underwater muon and neutrino detector. { 'dü,mand }

Dumas method [ANALY CHEM] A procedure for the determination of nitrogen in organic substances by combustion of the substance. { 'dü·mä ,meth·əd }

Dumbbell Nebula [ASTRON] A planetary nebula of large apparent diameter and low surface brightness in the constellation Vulpecula, about 220 parsecs (4.2×10^{15} miles or 6.8×10^{15} kilometers) away. { 'dəm,bel 'neb·yə·lə }

dumb iron [ENG] **1.** A rod for opening seams prior to caulking. **2.** A rigid connector between the frame of a motor vehicle and the spring shackle. { dəm ,ī·ərn }

dumb terminal [COMPUT SCI] A computer input/output device that lacks the capability to process or format data, and is thus entirely dependent on the main computer for these activities. { dəm 'term·ən·əl }

dumbwaiter [MECH ENG] An industrial elevator which carries small objects but is not permitted to carry people. { 'dəm,wād·ər }

dumdum [ORD] A bullet that flattens excessively on contact, or one especially designed to flatten excessively. { 'dəm,dəm }

dummy [COMMUN] Telegraphy network simulating a customer's loop for adjusting a telegraph repeater; the dummy side of the repeater is that toward the customer. [COMPUT SCI] An artificial address, instruction, or other unit of information inserted in a digital computer solely to fulfill prescribed conditions (such as word length or block length) without affecting operations. [ENG] Simulating device with no operating features, as a dummy heat coil. [GRAPHICS] A preliminary layout which shows the placement of illustrations and text as they will appear in the final printing. [MET] A cathode that undergoes electroplating at low current densities. [ORD] **1.** A nonexplosive bomb, projectile, or the like, or an object made to appear as one of these. **2.** An object made to appear as an airplane, gun emplacement, or the like from the air. { 'dəm·ē }

dummy antenna [ELECTR] A device that has the impedance characteristic and power-handling capacity of an antenna but does not radiate or receive radio waves; used chiefly for testing a transmitter. Also known as artificial antenna. { ¦dəm·ē an'ten·ə }

dummy argument [COMPUT SCI] The variable appearing in the definition of a macro or function which will be replaced by an address at call time. { ¦dəm·ē 'är·gyə·mənt }

dummy block [MET] A thick disk positioned between the ram and billet in extrusion working to prevent the ram from overheating. { 'dəm·ē ,bläk }

dummy file [COMPUT SCI] A nonexistent file which is treated by a computer program as if it were receiving its output data, when in fact the data are being ignored; used to suppress the creation of files that are needed only occasionally. { 'dəm·ē 'fīl }

dummy instruction [COMPUT SCI] An artificial instruction or address inserted in a list to serve a purpose other than the execution as an instruction. { ¦dəm·ē in'strək·shən }

dummy joint [ENG] A groove cut into the top half of a concrete slab, sometimes packed with filler, to form a line where the slab can crack with only minimum damage. { 'dəm·ē ,jȯint }

dummy load [ELECTR] A dissipative device used at the end of a transmission line or waveguide to convert transmitted energy into heat, so that essentially no energy is radiated outward or reflected back to its source. { 'dəm·ē ,lōd }

dummy message [COMMUN] A message sent for some purpose other than its content, which may consist of dummy groups or may have a meaningless text. { ¦dəm·ē 'mes·ij }

dummy parameter [COMPUT SCI] A parameter whose value has no significance but which is included in an instruction or command to satisfy the requirements of the system. { 'dəm·ē pə'ram·əd·ər }

dummy record [COMPUT SCI] Meaningless information that is stored for some purpose such as fulfillment of a length requirement. { 'dəm·ē 'rek·ərd }

dummy section [COMPUT SCI] The part of an assembly language program in which the arrangement of the data in memory is specified. Abbreviated DSECT. { 'dəm·ē 'sek·shən }

dummy suffix [MATH] A suffix which has no true mathematical significance and is used only to facilitate notation; usually an index which is summed over. { ¦dəm·ē 'səf·iks }

dummy variable [MATH] A variable which has no true mathematical significance and is used only to facilitate notation; usually a variable which is integrated over. { ¦dəm·ē 'ver·ē·ə·bəl }

dumontite [MINERAL] $Pb_2(UO_2)_3(PO_4)_2(OH)_4 \cdot 3H_2O$ Yellow orthorhombic mineral consisting of a hydrated phosphate of uranium and lead, occurring in crystals. { dü'män,tīt }

dumortierite [MINERAL] $Al_8BSi_3O_{19}(OH)$ A pink, green, blue, or violet mineral that crystallizes in the orthorhombic system but commonly occurs in parallel or radiating fibrous aggregates; mined for the manufacture of high-grade porcelain. { dü·mȯr'tir,īt }

dump [COMPUT SCI] To copy the contents of all or part of a storage, usually from an internal storage device into an external storage device. [ELECTR] To withdraw all power from a system or component accidentally or intentionally. [ORD] A temporary storage area, usually in the open, for bombs, ammunition, equipment, or supplies. { dəmp }

dump bailer [ENG] A cylindrical vessel designed to deliver cement or water into a well which otherwise might cave in if fluid was poured from the top. { 'dəmp ,bāl·ər }

dump bucket [MECH ENG] A large bucket with movable discharge gates at the bottom; used to move soil or other construction materials by a crane or cable. { 'dəmp ,bək·ət }

dump car [MECH ENG] Any of several types of narrow-gage rail cars with bodies which can easily be tipped to dump material. { 'dəmp ,kär }

dump check [COMPUT SCI] A computer check that usually consists of adding all the digits during dumping, and verifying the sum when retransferring. { 'dəmp ,chek }

dumping syndrome [MED] An imperfectly understood symptom complex of disagreeable or painful epigastric fullness, nausea, weakness, giddiness, sweating, palpitations, and diarrhea, occurring after meals in patients who have gastric surgery which interferes with the function of the pylorus. { 'dəmp·iŋ ,sin,drōm }

dump power [ELEC] Electric power, generated by any source, which is in excess of the needs of the electric system and which cannot be stored or conserved. { 'dəmp ,paù·ər }

dump routine [COMPUT SCI] A program within a computer's operating system that handles the processing of dumps. { 'dəmp rü,tēn }

dump scow [NAV ARCH] A craft used for transporting rubbish, normally equipped with doors in the bottom for dumping. { 'dəmp ,skaù }

dump tank *See* measuring tank. { 'dəmp ,taŋk }

dump truck [ENG] A motor or hand-propelled truck for hauling and dumping loose materials, equipped with a body that discharges its contents by gravity. { 'dəmp ,trək }

dump valve [ENG] A large valve located at the bottom of a tank or container used in emergency situations to empty the tank quickly; for example, to jettison fuel from an airplane fuel tank. { 'dəmp ,valv }

dumpy level [ENG] A surveyor's level which has the telescope with its level tube rigidly attached to a vertical spindle and is capable only of horizontal rotary movement. { 'dəm·pē 'lev·əl }

dundasite [MINERAL] $PbAl_2(CO_3)_2(OH)_4 \cdot 2H_2O$ A white mineral consisting of a basic lead aluminum carbonate, occurring in spherical aggregates. { 'dən·də,sīt }

DUMPY LEVEL

Engineer's dumpy level. (*W. and L. E. Gurley Co.*)

dune [GEOL] A mound or ridge of unconsolidated granular material, usually of sand size and of durable composition (such as quartz), capable of movement by transfer of individual grains entrained by a moving fluid. { dün }

dune complex [GEOGR] The totality of topographic forms, especially dunes, which comprise the moving landscape. { 'dün ˌkäm,pleks }

dunite [PETR] An ultrabasic rock consisting almost solely of a magnesium-rich olivine with some chromite and picotite; an important source of chromium. { 'dü,nīt }

dunking sonar See dipping sonar. { 'dəŋk·iŋ ˌsō,när }

dunnage [ENG] A configuration of members that forms a structural support for a cooling tower or similar appendage to a building but is not part of the building itself. [IND ENG] **1.** Padding material placed in a container to protect shipped goods from damage. **2.** Loose wood or waste material placed in the ship's hold to protect the cargo from shifting and damage. { 'dən·ij }

dunnite See ammonium picrate. { 'də,nīt }

duo [GRAPHICS] Recording of images on one-half of the film width during one passage of the film, then turning the film end for end and rerunning it to utilize the unused half of the film width. { 'dü,ō }

duodecimal number system [MATH] A representation system for real numbers using 12 as the base. { ˌdü·ə'des·məl 'nəm·bər ˌsis·təm }

duodenal glands See Brunner's glands. { dü¦äd·ən·əl 'glanz }

duodenal ulcer [MED] A peptic ulcer occurring in the wall of the duodenum, the first portion of the small intestine. { dü¦äd·ən·əl 'əl·sər }

duodenum [ANAT] The first section of the small intestine of mammals, extending from the pylorus to the jejunum. { dü¦äd·ən·əm or dü·ə'dē·nəm }

duodiode [ELECTR] An electron tube having two diodes in the same envelope, with either a common cathode or separate cathodes. Also known as double diode. { ˌdü·ō'dī,ōd }

duodiode-pentode [ELECTR] An electron tube having two diodes and a pentode in the same envelope, generally with a common cathode. { ˌdü·ō'dī,ōd 'pen,tōd }

duodiode-triode [ELECTR] An electron tube having two diodes and a triode in the same envelope, generally with a common cathode. { ˌdü·ō'dī,ōd 'trī,ōd }

duolateral coil See honeycomb coil. { ˌdü·ō'lad·ə·rəl 'kȯil }

duoplasmatron [ELECTR] An ion-beam source in which electrons from a hot filament are accelerated sufficiently to ionize a gas by impact; the resulting positive ions are drawn out by high-voltage electrons and focused into a beam by electrostatic lens action. { ˌdü·ō'plaz·mə,trän }

duoprimed word [COMPUT SCI] A computer word containing a representation of the sixth, seventh, eighth, and ninth rows of information from an 80-column card. { ¦dü ō,prīmd 'wərd }

duotone [GRAPHICS] A process in which two halftone cuts, one with a screen angle 30° different from the other, are made from the same black-and-white photograph, and the picture is printed in two tones, usually black and a color such as blue or green. { 'dü·ə,tōn }

duotriode See double triode. { ˌdü·ō'trī,ōd }

duotype [GRAPHICS] A process in which two halftone plates, for letterpress, are produced from a black-and-white original, each plate being etched differently; one plate is etched for detail and printed in a dark color, and the other is etched for a flat effect and printed in a light color. { 'dü·ə,tīp }

Duovac method [MET] Technique for testing for defects in magnetic parts; a moving magnetic field magnetizes the part in many directions, and the part is then sprayed with fluorescent magnetic particles and examined under ultraviolet light so that defects become apparent. { 'dü·ə,vak ,meth·əd }

Dupin's theorem [MATH] The proposition that, given three families of mutually orthogonal surfaces, the line of intersection of any two surfaces of different families is a line of curvature for both the surfaces. { dyü'paz ,thir·əm }

dupion [TEXT] A thick, irregular, double silk fiber obtained from two cocoons nested together. { 'dü·pē,än }

duplet lens system [OPTICS] A system of lenses in which there are two groups of lenses separated by a space, and successive lenses in each group are in contact. { 'düp·lət 'lenz ,sis·təm }

duplex [ENG] Consisting of two parts working together or in a similar fashion. [GRAPHICS] The technique of recording on roll microfilm the front and back image of a document in a single exposure. { 'dü,pleks }

duplex artificial line [ELEC] A balancing network, simulating the impedance of the real line and distant terminal apparatus, which is employed in a duplex circuit for the purpose of making the receiving device unresponsive to outgoing signal currents. { 'dü,pleks ärd·ə,fish·əl 'līn }

duplex cable [ELEC] Two insulated stranded conductors twisted together; they may have a common insulating covering. { 'dü,pleks 'kā·bəl }

duplex channel [COMMUN] A communication channel providing simultaneous transmission in both directions. { 'dü,pleks 'chan·əl }

duplex computer [COMPUT SCI] Two identical computers, either one of which can ensure continuous operation of the system when the other is shut down. { 'dü,pleks kəm'pyüd·ər }

duplex deoxyribonucleic acid [MOL BIO] The deoxyribonucleic acid double helix. { 'dü,pleks dē¦äk·se,rī·bo·nu¦klē·ik 'as·əd }

duplexed system [ENG] A system with two distinct and separate sets of facilities, each of which is capable of assuming the system function while the other assumes a standby status. Also known as redundant system. { 'dü,plekst ,sis·təm }

duplexer [ELECTR] A switching device used in radar to permit alternate use of the same antenna for both transmitting and receiving; other forms of duplexers serve for two-way radio communication using a single antenna at lower frequencies. Also known as duplexing assembly. { 'dü,plek·sər }

duplexing [COMMUN] See duplex operation. [COMPUT SCI] The provision of redundant hardware or excess capacity which can pick up the work load in the event of failure of one part of a computer system. [MET] See duplex process. { 'dü,pleks·iŋ }

duplexing assembly See duplexer. { 'dü,pleks·iŋə,sem·blē }

duplex iron [MET] Cast iron heated in an electric furnace after it has been melted in a cupola. { 'dü,pleks ,ī·ərn }

duplexite [MINERAL] $Ca_4BeAl_2Si_9O_{24}(OH)_2$ A white fibrous mineral consisting of hydrous beryllium calcium aluminosilicate. Also known as bavenite. { 'dü,plek,sīt }

duplex lock [DES ENG] A lock with two independent pin-tumbler cylinders on the same bolt. { 'dü,pleks 'läk }

duplex operation [COMMUN] The operation of associated transmitting and receiving apparatus concurrently, as in ordinary telephones, without manual switching between talking and listening periods. Also known as duplexing; duplex transmission. [ENG] In radar, a condition of operation when two identical and interchangeable equipments are provided, one in an active state and the other immediately available for operation. { 'dü,pleks äp·ə'rā·shən }

duplex paper [GRAPHICS] Photosensitized paper with emulsion on both sides, one side having a smooth finish and the other a rough finish, or having sides of different color. { 'dü,pleks 'pā·pər }

duplex practice See duplex process. { 'dü,pleks ,prak·təs }

duplex prints [TEXT] Designs printed on both face and back of fabrics in two distinct operations, usually to get a woven effect. { 'dü,pleks ,prins }

duplex process [MET] A two-step procedure in which steel is refined by one process (usually the Bessemer process) and finished by another process (usually open-hearth or electric-furnace). Also known as duplexing; duplex practice. { 'dü,pleks ,präs·əs }

duplex pump [MECH ENG] A reciprocating pump with two parallel pumping cylinders. { 'dü,pleks ,pəmp }

duplex tandem compressor [MECH ENG] A compressor having cylinders on two parallel frames connected through a common crankshaft. { 'dü,pleks ¦tan·dəm kəm'pres·ər }

duplex transmission See duplex operation. { 'dü,pleks tranz'mish·ən }

duplex tube [ELECTR] Combination of two vacuum tubes in one envelope. { 'dü,pleks 'tüb }

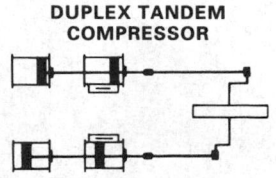

DUPLEX TANDEM COMPRESSOR

Frame arrangement of duplex tandem piston steam compressor.

DUPLEX UTERUS

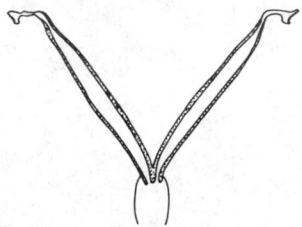

Duplex uterus of the rat. *(From C. K. Weichert, Elements of Chordate Anatomy, 3d ed., McGraw-Hill, 1967)*

duplex uterus [ANAT] A condition in certain primitive mammals, such as rodents and bats, that have two distinct uteri opening separately into the vagina. { ¦dü¦pleks 'yüd·ə·rəs }

duplicate cavity plate [ENG] In plastics molds, the removable plate in which the molding cavities are retained; used in operating where two plates are necessary for insert loading. { ¦düp·lə·kət 'kav·əd·ē ‚plāt }

duplicate film [GRAPHICS] Film generated in a camera containing a dual supply of film. { 'düp·lə·kət 'film }

duplicate measurement [ANALY CHEM] An additional measurement made on the same (identical) sample of material to evaluate the variance in the measurement. { ¦düp·lə·kət 'mezh·ər·mənt }

duplicate record [COMPUT SCI] An unwanted record that has the same key as another record in the same file. { 'düp·lə·kət 'rek·ərd }

duplicate sample [ANALY CHEM] A second sample randomly selected from a material being analyzed in order to evaluate sample variance. { ¦düp·lə·kət 'sam·pəl }

duplicating film [GRAPHICS] A special color film used for making copies of color transparencies to size so that they can be stripped together and color-separated as a single unit. { 'düp·lə‚kād·iŋ ‚film }

duplication check [COMPUT SCI] A check based on the identity in results of two independent performances of the same task. { ‚düp·lə'kā·shən ‚chek }

duplicatus [METEOROL] A cloud variety composed of superposed layers, sheets, or patches, at slightly different levels and sometimes partly merged. { ‚dü·plə'kād·əs }

Dupré equation [THERMO] The work W_{LS} done by adhesion at a gas-solid-liquid interface, expressed in terms of the surface tensions γ of the three phases, is $W_{LS} = \gamma_{GS} + \gamma_{GL} - \gamma_{LS}$. { dü'prā i‚kwā·zhən }

durability [ENG] The quality of equipment, structures, or goods of continuing to be useful after an extended period of time and usage. { ‚dùr·ə'bil·əd·ē }

durable goods [ENG] Products whose usefulness continues for a number of years and that are not consumed or destroyed in a single usage. Also known as durables; hard goods. { ¦dùr·ə·bəl 'gùdz }

durable press [TEXT] A finish, given to textiles by applying synthetic resins and then curing, that imparts lasting crease resistance. Also known as permanent press. { ¦dùr·ə·bəl 'pres }

durable-press resin See permanent-press resin. { ¦dùr·ə·bəl ¦pres 'rez·ən }

durables See durable goods. { 'dùr·ə·bəlz }

durain [GEOL] A hard, granular ingredient of banded coal which occurs in lenticels and shows a close, firm texture. Also known as durite. { 'dü‚rān }

dura mater [ANAT] The fibrous membrane forming the outermost covering of the brain and spinal cord. Also known as endocranium. { 'dùr·ə ‚mā·dər }

durangite [MINERAL] NaAlF(AsO₄) An orange-red, monoclinic mineral consisting of a fluoarsenate of sodium and aluminum; occurs in crystals. { də'ran‚jīt }

Durargid [GEOL] A great soil group constituting a subdivision of the Argids, indicating those soils with a hardpan cemented by silica and called a duripan. { dùr'är·jəd }

duration [MECH] A basic concept of kinetics which is expressed quantitatively by time measured by a clock or comparable mechanism. [OCEANOGR] The interval of time of the rising or falling tide, or the length of time of flood or ebb tidal currents. { də'rā·shən }

duration control [ELECTR] Control for adjusting the time duration of reduced gain in a sensitivity-time control circuit. { də'rā·shən kən‚trōl }

durene [ORG CHEM] C₆H₂(CH₃)₄ Colorless crystals with camphor aroma; boiling point 190°C; soluble in organic solvents, insoluble in water; used as a chemical intermediate. Also known as durol. { 'dü‚rēn }

Durer's conchoid [MATH] A plane curve consisting of points that lie on a variable line passing through points Q and R and are a constant distance a from Q, where Q and R have cartesian coordinates $(q,0)$ and $(0,r)$ and q and r satisfy the equation $q + r = b$, where b is a constant. { 'dùr·ərz 'kän‚kòid }

Durfee square [MATH] The largest square that is filled with

asterisks in the star diagram of a particular partition. { 'dər·fē ‚skwer }

Durham fermentation tube [MICROBIO] A test tube containing lactose or lauryl tryptose and an inverted vial for gas collection; used to test for the presence of coliform bacteria. { 'dùr·əm ‚fər·mən'tā·shən ‚tüb }

duricrust [GEOL] The case-hardened soil crust formed in semiarid climates by precipitation of salts; contains aluminous, ferruginous, siliceous, and calcareous material. { 'dùr·ə‚krəst }

durinite [GEOL] The principal maceral of durain; a heterogeneous material, semiopaque in section (including all parts of plants); micrinite, exinite, cutinite, resinite, collinite, xylinite, suberinite, and fusinite may be present. { 'dùr·ə‚nīt }

duripan [GEOL] A horizon in mineral soil characterized by cementation by silica. { 'dùr·ə‚pan }

durite See durain. { 'dù·rīt }

durol See durene. { 'dù‚rōl }

durometer [ENG] An instrument consisting of a small drill or blunt indenter point under pressure; used to measure hardness of metals and other materials. { də'räm·əd·ər }

durometer hardness [ENG] The hardness of a material as measured by a durometer. { də'räm·əd·ər ‚härd·nəs }

Durville process [MET] A casting process involving the attachment of an inverted mold to the top of a crucible; the metal is melted in the bottom of the crucible, and then the molten metal is decanted into the mold by inverting the entire apparatus. { 'dər·vil ‚präs·əs }

düsenwind [METEOROL] The mountain-gap wind of the Dardanelles; a strong east-northeast wind which blows out of the Dardanelles into the Aegean Sea, penetrating as far as the island of Lemnos, and caused by a ridge of high pressure over the Black Sea. { 'dēz·ən‚vint }

Dushman equation See Richardson-Dushman equation. { 'dùsh·mən i‚kwā·zhən }

dusk [ASTRON] That part of either morning or evening twilight between complete darkness and civil twilight. { dəsk }

dusk side [ASTRON] The side of a planet or other celestial body pointing away from its orbital movement direction. { 'dəsk ‚sīd }

dussertite [MINERAL] BaFe₃(AsO₄)₂(OH)₅ A mineral consisting of a hydrous basic arsenate of barium and iron. { 'dəs·ər‚tīt }

dust [GEOL] Dry solid matter of silt and clay size (less than ¹/₁₆ millimeter). [PHYS] A loose term applied to solid particles predominantly larger than colloidal size and capable of temporary gas suspension. { dəst }

dust and fume monitor [MIN ENG] An instrument designed to measure and record concentrations of dust, fume, and gas in mine environments over an extended period of time. { ¦dəst ən 'fyüm ‚män·əd·ər }

dust avalanche [GEOL] An avalanche of dry, loose snow. { 'dəst ‚av·ə‚lanch }

dust bowl [CLIMATOL] A name given, early in 1935, to the region in the south-central United States afflicted by drought and dust storms, including parts of Colorado, Kansas, New Mexico, Texas, and Oklahoma, and resulting from a long period of deficient rainfall combined with loosening of the soil by destruction of the natural vegetation; dust bowl describes similar regions in other parts of the world. { 'dəst ‚bōl }

dust chamber [ENG] A chamber through which gases pass to permit deposition of solid particles for collection. Also known as ash collector; dust collector. { 'dəst ‚chām·bər }

dust collector See dust chamber. { 'dəst kə‚lek·tər }

dust control system [ENG] System to capture, settle, or inert dusts produced during handling, drying, or other process operations; considered important for safety and health. { 'dəst kən‚trōl ‚sis·təm }

dust core See ferrite core. { 'dəst ‚kòr }

dust counter [ENG] A photoelectric apparatus which measures the size and number of dust particles per unit volume of air. Also known as Kern counter. { 'dəst ‚kaùnt·ər }

dust-counting microscope [ENG] A microscope equipped for quantitative dust sample analysis; magnification is usually 100X. { 'dəst ‚kaùnt·iŋ 'mī·krə‚skōp }

dust devil [METEOROL] A small but vigorous whirlwind, usually of short duration, rendered visible by dust, sand, and debris picked up from the ground; diameters range from about

10 to 100 feet (3 to 30 meters), and average height is about 600 feet (180 meters). { 'dəst ˌdev·əl }

dust-devil effect [GEOPHYS] In atmospheric electricity, rather sudden and short-lived change (positive or negative) of the vertical component of the atmospheric electric field that accompanies passage of a dust devil near an instrument sensitive to the vertical gradient. { 'dəst ˌdev·əl iˌfekt }

dust explosion [ENG] An explosion following the ignition of flammable dust suspended in the air. { 'dəst ik'splō·zhən }

dust extinction [OPTICS] The contribution to total extinction of light made by scattering and absorption by dust particles in the path of a light beam. { 'dəst ik'stiŋk·shən }

dust filter [ENG] A gas-cleaning device using a dry or viscous-coated fiber or fabric for separation of particulate matter. { 'dəst ˌfil·tər }

dust horizon [METEOROL] The top of a dust layer which is confined by a low-level temperature inversion and has the appearance of the horizon when viewed from above, against the sky; the true horizon is usually obscured by the dust layer. { 'dəst həˌrīz·ən }

dusting [MET] Spontaneous disintegration of a material on cooling due to expansion or inversion. { 'dəst·iŋ }

dusting clay [MATER] Finely pulverized clay used as an extender or carrier in insecticide dust formulations. { 'dəst·iŋ ˌklā }

dust separator [ENG] Device or system to remove dust from a flowing stream of gas; includes electrostatic precipitators, wet scrubbers, bag filters, screens, and cyclones. { 'dəst ˌsep·əˌrād·ər }

dust storm [METEOROL] A strong, turbulent wind carrying large clouds of dust. { 'dəst ˌstórm }

dust tail [ASTRON] A comet tail that consists of particles, typically 1 micrometer in diameter and primarily silicate in composition, and is usually curved with a length in the range from 10^6 to 10^7 kilometers. { 'dəst ˌtāl }

dust well [HYD] A pit in an ice surface produced when small, dark particles on the ice are heated by sunshine and sink down into the ice. { 'dəst ˌwel }

dust whirl [METEOROL] A rapidly rotating column of air over a dry and dusty or sandy area, carrying dust, leaves, and other light material picked up from the ground; when well developed, it is known as a dust devil. Also known as dancing dervish; dancing devil; desert devil; sand auger; sand devil. { 'dəst ˌwərl }

Dutch door [BUILD] A door with upper and lower parts that can be opened and closed independently. { ˌdəch 'dór }

Dutch elm disease [PL PATH] A lethal fungus disease of elm trees caused by *Graphium ulmi,* which releases a toxic substance that destroys vascular tissue; transmitted by a bark beetle. { ˌdəch 'elm diˌzēz }

Dutch liquid *See* ethylene chloride. { ˌdəch 'lik·wəd }

dutchman [ENG] A filler piece for closing a gap between two pipes or between a pipe or fitting and a piece of equipment, if the pipe is too short to achieve closure or if the pipe and equipment are not aligned. { 'dəch·mən }

Dutchman's log [ENG] A buoyant object thrown overboard to determine the speed of a vessel; the time required for a known length of the vessel to pass the object is measured, and the speed can then be computed. { ˌdəch·mənz 'läg }

Dutch metal [MET] An alloy of 80% copper and 20% zinc that is ductile, is easily drawn, and takes a high polish; used for low-priced jewelry. { ˌdəch 'med·əl }

Dutch process [CHEM ENG] A process for making white lead; metallic lead is placed in vessels containing a dilute acetic acid, and the vessels are stacked in bark or manure. [FOOD ENG] A chocolate manufacturing process in which cocoa nibs are treated with alkali to neutralize the natural acids present and to enhance color. { ˌdəch ˌpräs·əs }

Dutch roll [AERO ENG] A motion of an airplane which consists of simultaneous oscillations of the bank (or roll) angle, the slideslip angle, and the heading angle, and which, when poorly damped, is annoying to passengers and pilots. { ˌdəch 'rōl }

duty classification of a relay [ELEC] Expression of the frequency with which the relay may be required to operate without exceeding prescribed limitations. { 'düd·ē ˌklas·ə·fəˌkā·shən əv ə 'rēˌlā }

duty cycle [COMMUN] The product of the pulse duration and pulse repetition frequency of a pulse carrier, equal to the time per second that pulse power is applied. Also known as duty factor. [ELECTR] *See* duty ratio. [ENG] **1.** The time intervals devoted to starting, running, stopping, and idling when a device is used for intermittent duty. **2.** The ratio of working time to total time for an intermittently operating device, usually expressed as a percent. Also known as duty factor. [MET] The percentage of time that current flows in equipment over a specific period during electric resistance welding. [NUCLEO] The fraction of time during which a pulsed accelerator beam is on target, usually expressed as a percent. Also known as duty factor. { 'düd·ē ˌsī·kəl }

duty cyclometer [ENG] Test meter which gives direct reading of duty cycle. { 'düd·ē sī'kläm·əd·ər }

duty factor *See* duty cycle. { 'düd·ē ˌfak·tər }

duty of water [HYD] The total volume of irrigation water required to mature a particular type of crop, including consumptive use, evaporation and seepage from ditches and canals, and the water eventually returned to streams by percolation and surface runoff. { 'düd·ē əv 'wód·ər }

duty ratio [ELECTR] In a pulse radar or similar system, the ratio of average to peak pulse power. Also known as duty cycle. { 'düd·ē ˌrā·shō }

DUV *See* data under voice.

duvetyn [TEXT] A twill fabric with a napped velvety surface which obscures the weave. { 'dü·və·tən }

D value [NAV] The difference between pressure altitude and absolute altitude, as determined at a given time in flight, expressed algebraically; the absolute altitude is always minuend. Also known as D sounding. { 'dē ˌval·yü }

D variometer *See* declination variometer. { 'dē ˌver·ē'äm·əd·ər }

DVD [COMMUN] An optical disk that has formats for audio, video, and computer storage applications, and that uses the same basic structure as the compact disk (CD) to store data, but achieves a greater storage capability by using a track pitch less than half that of the CD, pits and lands as little as half as long as the shortest on a CD, and two substrates, bonded together. Derived from digital versatile disk; digital video disk.

DVD-audio [COMMUN] A DVD format for digital storage of audio information. Also known as Book C. { ˌdēˌvēˌdē 'óde·ō }

DVD-RAM *See* DVD-rewritable. { ˌdēˌvēˌdē 'ram }

DVD-read-only [COMMUN] A DVD format in which data written on the disk at the time of its manufacture are permanent, and the disk cannot be written or erased after that. Also known as Book A; DVD-ROM. { ˌdēˌvēˌdē ˌred 'ōn·lē }

DVD-rewritable [COMMUN] A DVD format that allows audio or other digital data to be written, read, erased, and rewritten. Also known as Book E; DVD-RAM. { ˌdēˌvēˌdē rē'rīd ə bəl }

DVD-ROM *See* DVD-read-only. { ˌdēˌvēˌdē 'ram }

DVD-video [COMMUN] A DVD format for digital storage of video information. Also known as Book B. { ˌdēˌvēˌdē 'vid·ē·ō }

DVD-write once [COMMUN] A DVD format that allows users to record audio or other digital data in such a way that the recording is permanent and may be read indefinitely but cannot be erased. Also known as Book D. { ˌdēˌvēˌdē ˌrīt 'wəns }

Dvorak keyboard [ENG] A keyboard whose layout is altered from that of the standard qwerty keyboard to speed up typing; more of the frequently used keys are on the home row. { də'vór·ak ˌkē·bórd }

dwarf [BIOL] Being an atypically small form or variety of something. [MED] An abnormally small individual; especially one whose bodily proportions are altered. { dwórf }

dwarf Cepheids [ASTRON] A class of pulsating variable stars with periods of less than 6 hours and spectral type A or F; similar to δ Scuti stars but sometimes distinguished from them by the slightly larger amplitudes of their light curves. Also known as AI Velorum stars. { 'dwórf 'sef·ē·ədz }

dwarf disease [PL PATH] A virus disease marked by the inhibition of fruit production; common in plum trees. { 'dwórf diˌzēz }

dwarf galaxy [ASTRON] An elliptical galaxy with low mass and low luminosity, having at most a few tens of millions of stars. { 'dwórf 'gal·ik·sē }

DVORAK KEYBOARD

home row

Layout of the Dvorak keyboard.

dwarfism [MED] Underdevelopment of the body due to surgical removal of the pituitary gland or hyposecretion of growth hormone. { 'dwȯr,fiz·əm }

dwarf mouse unit [BIOL] A unit for the standardization of somatotropin. { ¦dwȯrf 'maůs ,yü·nət }

dwarf novae [ASTRON] A class of irregular variable stars which undergo rapid increases in brightness of several magnitudes at semiperiodic intervals, and then decrease more slowly to the normal minimum; they may be divided into U Geminorum stars and Z Camelopardalis stars. { ¦dwȯrf 'nō,vī }

dwarf spheroidal galaxy [ASTRON] One of the smallest and faintest of the dwarf galaxies, with an effective radius of 200–1000 parsecs and an absolute visual magnitude between −8 and −13. { ¦dwȯrf sfir¦ȯid·əl 'gal·ik·sē }

dwarf star [ASTRON] A star that typically has surface temperature of 5730 K, radius of 428,000 miles (690,000 kilometers), mass of 2×10^{33} grams, and luminosity of 4×10^{33} ergs per second. Also known as main sequence star. { ¦dwȯrf 'stär }

dwell [DES ENG] That part of a cam that allows the cam follower to remain at maximum lift for a period of time. [ELEC] The number of degrees through which the distributor cam rotates from the time that the contact points close to the time that they open again. Also known as dwell angle. [ENG] A pause in the application of pressure to a mold. { dwel }

dwell angle See dwell. { 'dwel ,aŋ·gəl }

dwey See dwigh. { dwā }

dwigh [METEOROL] In Newfoundland, a sudden shower or snow storm. Also known as dwey; dwoy. { dwī }

Dwight-Lloyd machine [MIN ENG] A continuous sintering machine in which the feed is moved on articulated plates pulled by chains in conveyor-belt fashion. { ¦dwīt ¦lȯid mə,shēn }

Dwight-Lloyd process [MIN ENG] Blast roasting, with air currents being drawn downward through the ore. { ¦dwīt ¦lȯid ,präs·əs }

DWL See design waterline.

dwoy See dwigh. { dwȯi }

dwt See deadweight tonnage; pennyweight.

Dwyka tillite [GEOL] A glacial Permian deposit that is widespread in South Africa. { də¦vīk·ə 'ti,līt }

DX See distance reception.

DX coil See direct-expansion coil. { ¦dē¦eks ,kȯil }

Dy See dysprosium.

dyad [CELL MOL] Either of the two pair of chromatids produced by separation of a tetrad during the first meiotic division. [MATH] An abstract object which is a pair of vectors **AB** in a given order on which certain operations are defined. { 'dī,ad }

dyadic expansion [MATH] The representation of a number in the binary number system. { dī¦ad·ik ik'span·chən }

dyadic number system See binary number system. { dī¦ad·ik 'nəm·bər ,sis·təm }

dyadic operation [MATH] An operation that has only two operands. { dī'ad·ik ,äp·ə'rā·shən }

dyadic processor [COMPUT SCI] A type of multiprocessor that includes two processors which operate under control of the same copy of the operating system. { dī'ad·ik 'präs,es·ər }

dyadic rational [MATH] A fraction whose denominator is a power of 2. { dī'ad·ik 'rash·ən·əl }

dye [CHEM] A colored substance which imparts more or less permanent color to other materials. Also known as dyestuff. { dī }

dyecrete process [ENG] A process of adding permanent color to concrete with organic dyes. { 'dī,krēt ,präs·əs }

dyeing [CHEM ENG] The application of color-producing agents to material, usually fibrous or film, in order to impart a degree of color permanence demanded by the projected end use. { 'dī·iŋ }

dyeing assistant [CHEM] Material such as sodium sulfate added to a dye bath to control or promote the action of a textile dye. { 'dī·iŋ ə,sis·tənt }

dye laser [OPTICS] A type of tunable laser in which the active material is a dye such as acridine red or esculin, with very large molecules, and laser action takes place between the first excited and ground electronic states, each of which comprises a broad vibrational-rotational continuum. { 'dī ,lā·zər }

dye penetrant [MET] A dye-containing liquid used for detecting cracks or other surface defects in nonmagnetic materials. { ¦dī 'pen·ə·trənt }

dye polymer recording [COMPUT SCI] An optical recording technique in which dyed plastic layers are used as the recording medium. { ¦dī 'päl·ə·mər ri'kȯrd·iŋ }

dye-retarding agent [MATER] Materials that decrease the rate of dye absorption, preventing rapid exhaustion of dye baths. { 'dī ri,tärd·iŋ ,ā·jənt }

dyestuff See dye. { 'dī,stəf }

dye sublimation [GRAPHICS] Printing process in which rolls of film coated with sublimable dyes are transferred to a receiver sheet by a digitally driven thermal printhead; variations in temperature control the amount of dye that sublimates, thus varying the color intensity. { ¦dī ,səb·lə¦mā·shən }

dye toning [GRAPHICS] The process whereby the color of a developing image is altered by changing the color into a mordant and then placing the film in a suitable dye solution. { 'dī ,tōn·iŋ }

dynamic acceleration See dynamic resolution. { dī¦nam·ik ik,sel·ə'rā·shən }

dynamic address translator [COMPUT SCI] A hardware device used in a virtual memory system to automatically identify a virtual address inquiry in terms of segment number, page number within the segment, and position of the record with reference to the beginning of the page. { dī¦nam·ik 'a,dres ,tranz,lād·ər }

dynamic beam forming [ELECTR] A cathode-ray-tube design that ensures that the electron beam will impact a perfectly circular area of the display screen regardless of the location on the screen to which it is directed. { dī¦nam·ik 'bēm ,fȯrm·iŋ }

dynamical friction [PHYS] **1.** The drag force between electrons and ions drifitng with respect to each other. **2.** Sliding friction, in contrast to static friction. { dī¦nam·ə·kəl 'frik·shən }

dynamic algorithm [COMPUT SCI] An algorithm whose operation is, to some extent, unpredictable in advance, generally because it contains logical decisions that are made on the basis of quantities computed during the course of the algorithm. Also known as heuristic algorithm. { dī¦nam·ik 'al·gə,rith·əm }

dynamical halo model [ASTRON] A model for the behavior of cosmic rays in the Galaxy in which the cosmic rays are produced in a thin disk near the central plane and then diffuse through the disk and into an outwardly convecting halo to an outer boundary at a distance of perhaps several kiloparsecs from the central plane. { dī¦nam·ə·kəl 'hā·lō ,mäd·əl }

dynamic allotropy [CHEM] A phenomenon in which the allotropes of an element exist in dynamic equilibrium. { dī¦nam·ik ə'lä·trə·pē }

dynamical parallax [ASTRON] A parallax of binary stars that is computed from the sum of the masses of the binary system. { dī¦nam·ə·kəl 'par·ə,laks }

dynamical similarity [MECH] Two flow fields are dynamically similar if one can be transformed into the other by a change of length and velocity scales. All dimensionless numbers of the flows must be the same. { dī¦nam·ə·kəl sim·ə'lar·əd·ē }

dynamical symmetry [PHYS] A type of symmetry of the Hamiltonian of a physical system that can specify detailed properties of the system. { dī,nam·ə·kəl 'sim·ə·trē }

dynamical system [MATH] An abstraction of the concept of a family of solutions to an ordinary differential equation; namely, an action of the real numbers on a topological space satisfying certain "flow" properties. { dī¦nam·ə·kəl 'sis·təm }

dynamical variable [MECH] One of the quantities used to describe a system in classical mechanics, such as the coordinates of a particle, the components of its velocity, the momentum, or functions of these quantities. { dī¦nam·ə·kəl 'ver·ē·ə·bəl }

dynamic analogies [PHYS] Analogies that make it possible to convert the differential equations for mechanical and acoustical systems to equivalent electrical equations that can be represented by electric networks and solved by circuit theory. { dī¦nam·ik ə'nal·ə·jēz }

dynamic anthropometry [ANTHRO] The study of functional range and pattern of body movements and of the operations that can be performed by the limbs in various positions. { dī¦nam·ik ,an·thrə'päm·ə·trē }

dynamic augment [MECH ENG] Force produced by unbalanced reciprocating parts in a steam locomotive. { dī¦nam·ik 'ȯg‚ment }

dynamic balance [MECH] The condition which exists in a rotating body when the axis about which it is forced to rotate, or to which reference is made, is parallel with a principal axis of inertia; no products of inertia about the center of gravity of the body exist in relation to the selected rotational axis. { dī¦nam·ik 'bal·əns }

dynamic behavior [ENG] A description of how a system or an individual unit functions with respect to time. { dī¦nam·ik bə'hāv·yər }

dynamic boundary condition [FL MECH] The condition that the pressure must be continuous across an internal boundary or free surface in a fluid. { dī¦nam·ik 'baủn·drē kən‚dish·ən }

dynamic braking [MECH] A technique of electric braking in which the retarding force is supplied by the same machine that originally was the driving motor. { dī¦nam·ik 'brāk·iŋ }

dynamic breccia See tectonic breccia. { dī¦nam·ik 'brech·ə }

dynamic capillary pressure [PETRO ENG] Capillary-pressure saturation curves of a core sample determined by the simultaneous steady-state flow of two fluids through the sample; capillarity pressures are determined by the difference in the pressures of the two fluids. { dī¦nam·ik 'kap·ə‚ler·ē ‚presh·ər }

dynamic characteristic See load characteristic. { dī¦nam·ik kar·ik·tə'ris·tik }

dynamic check [ENG] Check used to ascertain the correct performance of some or all components of equipment or a system under dynamic or operating conditions. { dī¦nam·ik 'chek }

dynamic circuit [ELECTR] A metal oxide semiconductor circuit designed to make use of its high input impedance to store charge temporarily at certain nodes of the circuit and thereby increase the speed of the circuit. { dī¦nam·ik 'sər‚kət }

dynamic climatology [CLIMATOL] The climatology of atmospheric dynamics and thermodynamics, that is, a climatological approach to the study and explanation of atmospheric circulation. { dī¦nam·ik ‚klī·mə'täl·ə·jē }

dynamic compressor [MECH ENG] A compressor which uses rotating vanes or impellers to impart velocity and pressure to the fluid. { dī¦nam·ik kəm'pres·ər }

dynamic condenser electrometer [ELEC] A sensitive voltage-measuring instrument in which an object carrying charge resulting from the voltage is moved back and forth in an electrostatic field and the resulting alternating-current signal is observed. { dī¦nam·ik kən¦den·sər i‚lek'träm·əd·ər }

dynamic convergence [ELECTR] The process whereby the locus of the point of convergence of electron beams in a color-television or other multibeam cathode-ray tube is made to fall on a specified surface during scanning. { dī¦nam·ik kən'vər·jəns }

dynamic creep [MECH] Creep resulting from fluctuations in a load or temperature. { dī¦nam·ik 'krēp }

dynamic debugging routine [COMPUT SCI] A debugging routine which operates in conjunction with the program being checked and interacts with it while the program is running. { dī¦nam·ik dē'bəg·iŋ rü‚tēn }

dynamic dump [COMPUT SCI] A dump performed during the execution of a program. { dī¦nam·ik 'dəmp }

dynamic equilibrium Also known as kinetic equilibrium. [MECH] The condition of any mechanical system when the kinetic reaction is regarded as a force, so that the resultant force on the system is zero according to d'Alembert's principle. [PHYS] A condition in which several processes act simultaneously to maintain a system in an overall state that does not change with time. { dī¦nam·ik ē·kwə'lib·rē·əm }

dynamic error [ELECTR] Error in a time-varying signal resulting from inadequate dynamic response of a transducer. { dī¦nam·ik 'er·ər }

dynamic factor [AERO ENG] A ratio formed from the load carried by any airplane part when the airplane is accelerating or subjected to abnormal conditions to the load carried in the conditions of normal flight. { dī¦nam·ik 'fak·tər }

dynamic fluidity [FL MECH] The reciprocal of the dynamic viscosity. { dī¦nam·ik flü'id·əd·ē }

dynamic focusing [ELECTR] The process of varying the focusing electrode voltage for a color picture tube automatically so the electron-beam spots remain in focus as they sweep over the flat surface of the screen. { dī¦nam·ik 'fō·kəs·iŋ }

dynamic forecasting See numerical forecasting. { dī¦nam·ik 'fȯr‚kast·iŋ }

dynamic geomorphology [GEOL] The quantitative analysis of steady-state, self-regulatory geomorphic processes. Also known as analytical geomorphology. { dī¦nam·ik jē·ō·mȯr'fäl·ə·jē }

dynamic height [GEOPHYS] As measured from sea level, the distance above the geoid of points on the same equipotential surface, in terms of linear units measured along a plumb line at a given latitude, generally 45°. [PHYS] The amount of work done when a water particle of unit mass is moved vertically from one level to another. Also known as geodynamic height. { dī¦nam·ik 'hīt }

dynamic-height anomaly [OCEANOGR] The excess of the actual geopotential difference, between two given isobaric surfaces, over the geopotential difference in a homogeneous water column of salinity 35 per mille and temperature 0°C. Also known as anomaly of geopotential difference. { dī¦nam·ik ¦hīt ə'nam·ə·lē }

dynamic holdup [CHEM ENG] Liquid held by a tank or process vessel, with constant introduction of fresh material and counteracting withdrawal of held material to maintain a constant liquid level. { dī¦nam·ik 'hōld‚əp }

dynamic ileus See spastic ileus. { dī¦nam·ik 'il·ē·əs }

dynamic impedance [ELEC] The impedance of a circuit having an inductance and a capacitance in parallel at the frequency at which this impedance has a maximum value. Also known as rejector impedance. { dī¦nam·ik im'ped·əns }

dynamic instability See inertial instability. { dī¦nam·ik ‚in·stə'bil·əd·ē }

dynamicizer [COMPUT SCI] A device that converts a collection of data represented by a spatial arrangement of bits in a computer storage device into a series of signals occurring in time. { dī'nam·ə‚sīz·ər }

dynamic leak test [ENG] A type of leak test in which the vessel to be tested is evacuated and an external tracer gas is applied; an internal leak detector will respond if gas is drawn through any leaks. { dī¦nam·ik 'lēk ‚test }

dynamic level [PETRO ENG] In an oil well, the location of the liquid surface when the well is producing oil. Also known as pumping level. { dī'nam·ik 'lev·əl }

dynamic link [COMPUT SCI] A linking of data in two different programs, whereby modification in either program causes a similar change of the data in the other. { dī¦nam·ik 'liŋk }

dynamic load [AERO ENG] With respect to aircraft, rockets, or spacecraft, a load due to an acceleration of craft, as imposed by gusts, by maneuvering, by landing, by firing rockets, and so on. [CIV ENG] A force exerted by a moving body on a resisting member, usually in a relatively short time interval. Also known as energy load. { dī¦nam·ik 'lōd }

dynamic loudspeaker [ENG ACOUS] A loudspeaker in which the moving diaphragm is attached to a current-carrying voice coil that interacts with a constant magnetic field to give the in-and-out motion required for the production of sound waves. Also known as dynamic speaker; moving-coil loudspeaker. { dī¦nam·ik 'laủd‚spēk·ər }

dynamic memory See dynamic storage. { dī¦nam·ik 'mem·rē }

dynamic memory allocation See dynamic storage allocation. { dī¦nam·ik 'mem·rē al·ə‚kā·shən }

dynamic metamorphism [GEOL] Metamorphism resulting exclusively or largely from rock deformation, principally faulting and folding. Also known as dynamometamorphism. { dī¦nam·ik ‚med·ə'mȯr‚fiz·əm }

dynamic meteorology [METEOROL] The study of atmospheric motions as solutions of the fundamental equations of hydrodynamics or other systems of equations appropriate to special situations, as in the statistical theory of turbulence. { dī¦nam·ik mēd·ē·ə'räl·ə·jē }

dynamic meter [PHYS] The standard unit of dynamic height expressed as 10 square meters per second per second. { dī¦nam·ik 'mēd·ər }

DYNAMIC MICROPHONE

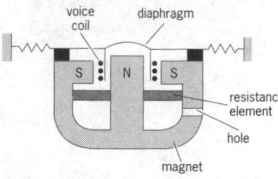

Dynamic microphone of the moving-coil type.

dynamic microphone [ENG ACOUS] A moving-conductor microphone in which the flexible diaphragm is attached to a coil positioned in the fixed magnetic field of a permanent magnet. Also known as moving-coil microphone. { dī¦nam·ik 'mī·krə,fōn }

dynamic model [ENG] A model of an aircraft or other object which has its linear dimensions and its weight and moments of inertia reproduced in scale in proportion to the original. { dī¦nam·ik 'mäd·əl }

dynamic noise suppressor [ENG ACOUS] An audio-frequency filter circuit that automatically adjusts its band-pass limits according to signal level, generally by means of reactance tubes; at low signal levels, when noise becomes more noticeable, the circuit reduces the low-frequency response and sometimes also reduces the high-frequency response. { dī¦nam·ik 'nóiz sə,pres·ər }

dynamic nuclear polarization [NUC PHYS] The creation of assemblies of nuclei whose spin axes are not oriented at random, and which are in a steady state that is not a state of thermal equilibrium. { dī¦nam·ik 'nü·klē·ər ,pō·lə·rə'zā·shən }

dynamic packing [ENG] Any packing that operates on moving surfaces; in functioning, to retain fluid under pressure, they carry the hydraulic load and therefore operate like bearings. { dī¦nam·ik 'pak·iŋ }

dynamic parallax [ASTRON] A value for the parallax of a binary star computed from the observations of the period and angular dimensions of the orbit by assuming a value for the mass of the binary system. Also known as hypothetical parallax. { dī¦nam·ik 'par·ə,laks }

dynamic pickup [ELECTR] A pickup in which the electric output is due to motion of a coil or conductor in a constant magnetic field. Also known as dynamic reproducer; moving-coil pickup. { dī¦nam·ik 'pik,əp }

dynamic plate impedance [ELECTR] Internal resistance to the flow of alternating current between the cathode and plate of a tube. { dī¦nam·ik 'plāt im,pēd·əns }

dynamic plate resistance [ELECTR] Opposition that the plate circuit of a vacuum tube offers to a small increment of plate voltage; it is the ratio of a small change in plate voltage to the resulting change in the plate current, other tube voltages remaining constant. { dī¦nam·ik 'plāt ri,zis·təns }

dynamic positioning [PETRO ENG] A system that generally uses computer-driven propulsion units to maintain a floating offshore drilling rig in position over the well. { dī¦nam·ik pə'zish·ən·iŋ }

dynamic pressure [FL MECH] **1.** The pressure that a moving fluid would have if it were brought to rest by isentropic flow against a pressure gradient. Also known as impact pressure; stagnation pressure; total pressure. **2.** The difference between the quantity in the first definition and the static pressure. { dī¦nam·ik 'presh·ər }

dynamic printout [COMPUT SCI] A printout of data which occurs during the machine run as one of the sequential operations. { dī¦nam·ik 'print,aút }

dynamic problem check [COMPUT SCI] Any dynamic check used to ascertain that the computer solution satisfies the given system of equations in an analog computer operation. { dī¦nam·ik 'präb·ləm ,chek }

dynamic programming [MATH] A mathematical technique, more sophisticated than linear programming, for solving a multidimensional optimization problem, which transforms the problem into a sequence of single-stage problems having only one variable each. { dī¦nam·ik 'prō·grə·miŋ }

dynamic program relocation [COMPUT SCI] The act of moving a partially executed program to another location in main memory, without hindering its ability to finish processing normally. { dī¦nam·ik 'prō·grəm ,rē·lō,kā·shən }

dynamic random-access memory [COMPUT SCI] A read-write random-access memory whose storage cells are based on transistor-capacitor combinations, in which the digital information is represented by charges that are stored on the capacitors and must be repeatedly replenished in order to retain the information. Abbreviated DRAM. { dī¦nam·ik ,ran·dəm 'ak·ses ,mem·rē }

dynamic range [ELECTR] The ratio of the specified maximum signal level capability of a system or component to its noise level; usually expressed in decibels. [GRAPHICS] In scanning an image, the range of gradations of tones in the original from the lightest highlight to the darkest shadow that must be converted by the scanner. { dī¦nam·ik 'rānj }

dynamic regulator [ELECTR] Transmission regulator in which the adjusting mechanism is in self-equilibrium at only one or a few settings and requires control power to maintain it at any other setting. { dī¦nam·ik 'reg·yə,lād·ər }

dynamic relocation [COMPUT SCI] The ability to move computer programs or data from auxiliary memory into main memory at any convenient location. { dī¦nam·ik ,rē·lō'kā·shən }

dynamic reproducer *See* dynamic pickup. { dī¦nam·ik rē·prə'dü·sər }

dynamic resistance [ELEC] A device's electrical resistance when it is in operation. { dī¦nam·ik ri'zis·təns }

dynamic resolution [COMPUT SCI] A feature of some mice whereby the pointer moves a larger distance in proportion to the mouse's actual displacement when the mouse is moved quickly and a smaller distance when it is moved slowly. Also known as automatic acceleration; ballistic tracking; dynamic acceleration; variable acceleration. { dī¦nam·ik ,rez·ə'lü·shən }

dynamic roughness [OCEANOGR] A quantity, designated z_0, dependent on the shape and distribution of the roughness elements of the sea surface, and used in calculations of wind at the surface. Also known as roughness length. { dī¦nam·ik 'rəf·nəs }

dynamics [MECH] That branch of mechanics which deals with the motion of a system of material particles under the influence of forces, especially those which originate outside the system under consideration. { dī¦nam·iks }

dynamic scattering device [OPTICS] A type of numerical display device in which a voltage is applied to a cell containing a nematic liquid crystal of negative dielectric anisotropy; opposing influences from the electric field and electrical conduction produce turbulence, causing the cell to become milk white. { dī¦nam·ik 'skad·ə·riŋ di,vīs }

dynamic sensitivity [ENG] The minimum leak rate which a leak detector is capable of sensing. { dī¦nam·ik sen·sə'tiv·əd·ē }

dynamic sequential control [COMPUT SCI] Method of operation of a digital computer through which it can alter instructions as the computation proceeds, or the sequence in which instructions are executed, or both. { dī¦nam·ik sə¦kwen·chəl kən'trōl }

dynamic shift register [COMPUT SCI] A shift register that stores information by using temporary charge storage techniques. { dī¦nam·ik 'shift ,rej·ə·stər }

dynamic similarity [MECH ENG] A relation between two mechanical systems (often referred to as model and prototype) such that by proportional alterations of the units of length, mass, and time, measured quantities in the one system go identically (or with a constant multiple for each) into those in the other; in particular, this implies constant ratios of forces in the two systems. { dī¦nam·ik ,sim·ə'lar·əd·ē }

dynamic speaker *See* dynamic loudspeaker. { dī¦nam·ik 'spēk·ər }

dynamic stability [MECH] The characteristic of a body, such as an aircraft, rocket, or ship, that causes it, when disturbed from an original state of steady motion in an upright position, to damp the oscillations set up by restoring moments and gradually return to its original state. Also known as stability. { dī¦nam·ik stə'bil·əd·ē }

dynamic stop [COMPUT SCI] A loop in a computer program which is created by a branch instruction in the presence of an error condition, and which signifies the existence of this condition. { dī¦nam·ik 'stäp }

dynamic storage [COMPUT SCI] **1.** Computer storage in which information at a certain position is not always available instantly because it is moving, as in an acoustic delay line or magnetic drum. Also known as dynamic memory. **2.** Computer storage consisting of capacitively charged circuit elements which must be continually refreshed or recharged at regular intervals. { dī¦nam·ik 'stòr·ij }

dynamic storage allocation [COMPUT SCI] A computer system in which memory capacity is made available to a program on the basis of actual, momentary need during program execution, and areas of storage may be reassigned at any time. Also known as dynamic allocation; dynamic memory allocation. { dī¦nam·ik ¦stòr·ij ,al·ə'kā·shən }

dynamic subroutine [COMPUT SCI] Subroutine that

involves parameters, such as decimal point position or item size, from which a relatively coded subroutine is derived by the computer itself. { dī¦nam·ik 'səb·rü͵tēn }

dynamic symmetry [PHYS] A symmetry law related, not to the geometric structure of the constituents of matter, but to the laws which govern the dynamic behavior of these constituents. { dī¦nam·ik 'sim·ə·trē }

dynamic temperature difference [PHYS] The difference between the temperature of a static medium and the surface temperature at the stagnation point of a heat-insulated body immersed in a flowing medium of the same composition. { dī¦nam·ik 'tem·prə·chər ͵dif·rəns }

dynamic test [ENG] A test conducted under active or simulated load. { dī¦nam·ik 'test }

dynamic thickness [OCEANOGR] The vertical separation between two isobaric surfaces in the ocean. { dī¦nam·ik 'thik·nəs }

dynamic time warping [ENG ACOUS] In speech recognition, the operation of compressing or stretching the temporal pattern of speech signals to take speaker variations into account. { dī͵nam·ik 'tīm ͵wȯrp·iŋ }

dynamic topography [MAP] A topographic map indicating the dynamic depth of an isobaric surface. { dī¦nam·ik tə'päg·rə·fē }

dynamic trough [METEOROL] A pressure trough formed on the lee side of a mountain range across which the wind is blowing almost at right angles. Also known as lee trough. { dī¦nam·ik 'trȯf }

dynamic unbalance [MECH ENG] Failure of the rotation axis of a piece of rotating equipment to coincide with one of the principal axes of inertia due to forces in a single axial plane and on opposite sides of the rotation axis, or in different axial planes. { dī¦nam·ik ən'bal·əns }

dynamic vertical See apparent vertical. { dī¦nam·ik 'vərd·ə·kəl }

dynamic viscosity See absolute viscosity. { dī¦nam·ik vis 'käs·əd·ē }

dynamic work [BIOPHYS] Performance of work by a muscle in which one end of the muscle moves with respect to the other, resulting in external movement. [IND ENG] A sustained pattern of work that results in motion around an anatomical joint, for example, a handling or assembly task. { dī¦nam·ik ͵wərk }

dynamite [MATER] A generic term covering a class of nitroglycerin-sensitized mixtures of carbonaceous materials (wood, flour, starch) and oxygen-supplying salts, used as explosives for blasting and mining. { 'dī·nə͵mīt }

Dynamitron accelerator [NUCLEO] A particle accelerator that utilizes a steady high-voltage potential, produced by a large number of sources of constant voltage connected in series, to accelerate charged particles. { dī'nam·ə͵trän ak'sel·ə͵rād·ər }

dynamo See generator. { 'dī·nə͵mō }

dynamo effect [GEOPHYS] A process in the ionosphere in which winds and the resultant movement of ionization in the geomagnetic field give rise to induced current. { 'dī·nə͵mō i͵fekt }

dynamoelectric [PHYS] Pertaining to the conversion of mechanical energy to electric energy, or vice versa. { ¦dī·nə͵mō·i'lek·trik }

dynamoelectric amplifier generator [ELEC] A generator that serves as a power amplifier at low frequencies or direct current; the input signal is applied to the stationary field to change the excitation, and the amplified output is taken from the rotating armature. { ¦dī·nə͵mō·i'lek·trik 'am·plə͵fī·ər jen·ə͵rād·ər }

dynamometamorphism See dynamic metamorphism. { ¦dī·nə͵mō͵med·ə'mȯr͵fiz·əm }

dynamometer [ENG] **1.** An instrument in which current, voltage, or power is measured by the force between a fixed coil and a moving coil. **2.** A special type of electric rotating machine used to measure the output torque or driving torque of rotating machinery by the elastic deformation produced. { ͵dī·nə'mäm·əd·ər }

dynamometer multiplier [ELEC] A multiplier in which a fixed and a moving coil are arranged so that the deflection of the moving coil is proportional to the product of the currents flowing in the coils. { dī·nə'mäm·əd·ər 'məl·tə͵plī·ər }

dynamostatic [ELEC] Pertaining to a machine that uses direct or alternating current to produce static electricity. { ¦dī·nə͵mō'stad·ik }

dynamo theory [GEOPHYS] The hypothesis which explains the regular daily variations in the earth's magnetic field in terms of electrical currents in the lower ionosphere, generated by tidal motions of the ionized air across the earth's magnetic field. { 'dī·nə͵mō ͵thē·ə·rē }

dynamotor [ELEC] A rotating electric machine having two or more windings on a single armature containing a commutator for direct-current operation and slip rings for alternating-current operation; when one type of power is fed in for motor operation, the other type is delivered by generator action. Also known as rotary converter; synchronous inverter. { 'dī·nə͵mō·dər }

dynatron [ELECTR] A screen-grid tube in which secondary emission of electrons from the anode causes the anode current to decrease as anode voltage increases, resulting in a negative resistance characteristic. Also known as negatron. { 'dī·nə͵trän }

dynatron oscillator [ELECTR] An oscillator in which secondary emission of electrons from the anode of a screen-grid tube causes the anode current to decrease as anode voltage is increased, giving the negative resistance characteristic required for oscillation. { 'dī·nə͵trän ͵äs·ə͵lād·ər }

dyne [MECH] The unit of force in the centimeter-gram-second system of units, equal to the force which imparts an acceleration of 1 cm/s² to a 1 gram mass. { dīn }

dyne-centimeter See erg. { dīn 'sen·tə͵mēd·ər }

dyne-cm See erg.

dynein [CELL MOL] A large enzyme complex that hydrolyzes adenosine triphosphate to provide energy to power retrograde [from (+) to (−)] transport along microtubules. { dī'nē·ən }

dynode [ELECTR] An electrode whose primary function is secondary emission of electrons; used in multiplier phototubes and some types of television camera tubes. Also known as electron mirror. { 'dī͵nōd }

dypnone [ORG CHEM] $C_6H_5COCHC(CH_3)C_6H_5$ A light-colored liquid with a boiling point of 246°C at 50 mmHg; used as a plasticizer and perfume base and in light-stable coatings. { 'dip͵nōn }

dysanalyte [MINERAL] A variety of the mineral perovskite in which Nb^{5+} substitutes for Ti^{5+}, and Na^{5+} for Ca^{2+} in the formula $Ca[TiO_3]$. { də'san·əl͵īt }

dysarthria [MED] Impairment of articulation caused by any disorder or lesion affecting the tongue or speech muscles. { di'sär·thrē·ə }

dysarthrosis [MED] Deformity, dislocation, or disease of a joint. A false joint. { ¦dis·är'thrō·səs }

dysautonomia [MED] **1.** Abnormal functioning of the autonomic nervous system. **2.** A congenital syndrome with aberrations in the autonomic nervous system function, including indifference to pain, diminished secretion of tears, poor vasomotor control, motor incoordination, labile cardiovascular reactions, frequent attacks of bronchial pneumonia, and hypersalivation with aspiration and trouble in swallowing. Also known as Riley-Day syndrome. { ¦dis¦ȯd·ə¦näm·ē·ə }

dysbarism [MED] A condition of the body resulting from the existence of a pressure differential between the total ambient pressure and the total pressure of dissolved and free gases within the body tissues, fluids, and cavities; characteristic symptoms include aeroembolism and abdominal gas pains. { 'dis·bə͵riz·əm }

dyschondroplasia See enchondromatosis. { di͵skän·drō 'plā·zhə }

dyscrasite [MINERAL] Ag_2Sb A gray mineral that forms rhombic crystals. { 'dis·krə͵sīt }

dysentery [MED] Inflammation of the intestine characterized by pain, intense diarrhea, and the passage of mucus and blood. { 'dis·ən͵ter·ē }

dysfibrinogenemia [MED] The presence of abnormal fibrinogens in the blood. { ͵dis·fī͵brin·ō·jə'nē·mē·ə }

dysfunction [MED] Impaired or abnormal functioning of a body part. { ͵dis'fəŋk·shən }

dysgammaglobulinemia [MED] A quantitative or qualitative abnormality of serum globulins. { ¦dis͵gam·ə͵gläb·yə·lə'nē·mē·ə }

dysgerminoma [MED] An ovarian tumor composed of large polygonal cells of germ-cell origin, resembling seminoma of the testis, but less malignant. Also known as embryoma of the ovary. { ¦dis·jər·mə'nō·mə }

dyshidrosis [MED] Any disturbance in sweat production or excretion. { ¦dis·hī'drō·səs }

dyshistogenesis [MED] The morphological result of an abnormal organization of cells into tissues. { ¦dis¦his·tə¦jen· ə·səs }

Dysideidae [INV ZOO] A family of sponges in the order Dictyoceratida. { ¦dis·ə¦dē·ə¦dē }

dyskaryosis [PATH] Any abnormality of the nuclei of exfoliated cells, without significant change in cell integrity. { di ¦skar·ē¦ō·səs }

dyskeratosis [MED] **1.** Imperfect keratinization of individual epidermal cells. **2.** Keratinization of corneal epithelium. { di¦sker·ə¦tō·səs }

dyskinesia [MED] **1.** Disordered movements of voluntary or involuntary muscles, particularly those seen in disorders of the extrapyramidal system. **2.** Impaired voluntary movements. { ¦dis·kə¦nē·zhə }

dyslexia [MED] Impairment of the ability to read. { dis ¦lek·sē·ə }

dyslogia [MED] **1.** Difficulty in the expression of ideas by speech. **2.** Impairment of reasoning or the faculty to think logically. { dis¦lō·jē·ə }

dysmenorrhea [MED] Difficult or painful menstruation. { dis¦men·ə¦rē·ə }

dysmnesia [PSYCH] General intellectual impairment. { ¦dis·¦nēzh·ə }

dysmorphophobia [PSYCH] An abnormal fear of deformity. { dis¦mór·fə¦fō·bē·ə }

Dysodonta [PALEON] An order of extinct bivalve mollusks with a nearly toothless hinge and a ligament in grooves or pits. { ¦dis·ə¦dän·tə }

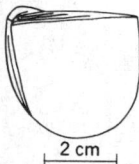

DYSODONTA

2 cm

Dysodont hinge, *Ambonychinia*, Upper Ordovician, Sweden.

Dyson microscope [OPTICS] A type of interference microscope, now obsolete, in which a light ray is split into two parallel beams and then recombined by reflections from surfaces of parallel plates, and one of the beams passes through the object under observation. { ¦dī·sən ¦mī·krə¦skōp }

Dyson notation [ORG CHEM] A notation system for representing organic chemicals developed by G. Malcolm Dyson; the compound is described on a single line, symbols are used for the chemical elements involved as well as for the functional groups and various ring systems; for example, methyl alcohol is C.Q and phenol is B6.Q. { ¦dī·sən nō¦tā·shən }

dysostosis [MED] Defective formation of bone. { ¦dis· ä¦stō·səs }

dyspareunia [MED] The occurrence of pain during sexual intercourse. { ¦dis·pə¦rün·ē·ə }

dyspepsia [MED] Disturbed digestion. { dis¦pep·sē·ə }

dysphagia [MED] Difficulty in swallowing, or inability to swallow, of organic or psychic causation. { dis¦fā·jə }

dysphasia [MED] Partial aphasia due to a brain lesion. { dis¦fā·zhə }

dysphonia [MED] An impairment of the voice. { dis¦fō· nē·ə }

dysphoria [MED] **1.** The condition of not feeling well or of being ill at ease. **2.** Morbid impatience and restlessness, anxiety, or fidgetiness. { dis¦fór·ē·ə }

dysplasia [PATH] Abnormal development or growth, especially of cells. { di¦splā·zhə }

dyspnea [MED] Difficult or labored breathing. { ¦dis· nē·ə }

dysprosium [CHEM] A metallic rare-earth element, symbol Dy, atomic number 66, atomic weight 162.50. { dis¦prō·zē· əm }

dysrhythmia [MED] Disordered rhythm of the brain waves. { dis¦rith·mē·ə }

dyssebacia [MED] Plugging of the sebaceous glands, especially around the nose, mouth, and forehead, with a dry, yellowish material. { ¦di·sə¦bā·shə }

dyssocial behavior [PSYCH] A behavior pattern characteristic of persons who manifest disregard for social codes by following illegal pursuits such as gambling or selling narcotics. { ¦di¦sō·shəl bə¦hāv·yər }

dyssomnia [MED] A group of disorders characterized by difficulty in going to sleep or staying asleep or excessive daytime sleepiness. { di¦säm·nē·ə }

dyster *See* doister. { ¦dī·stər }

dystetic mixture [PHYS CHEM] A mixture of two or more substances that has the highest possible melting point of all mixtures of these substances. { di¦sted·ik ¦miks·chər }

dysthymia [MED] Any childhood condition caused by malfunction of the thymus. [PSYCH] A mood disorder characterized by chronic depression for a period of at least 2 years. Also known as depressive neurosis. { ¦dis¦thī·mē·ə }

dysthymic disorder [PSYCH] Mild chronic depression that often results in impairment of social functioning, family relations, and work performance. { dis¦thī·mik dis¦órd·ər }

dystonia [PHYSIO] Disorder or lack of muscle tonicity. { di¦stōn·ē·ə }

dystophic [BIOL] Pertaining to an environment that does not supply adequate nutrition. { di¦stäf·ik }

dystrophy [MED] **1.** Defective nutrition. **2.** Defective or abnormal development or degeneration. { ¦dis·trə·fē }

dysuria [MED] Painful urination. { dis¦yūr·ē·ə }

Dytiscidae [INV ZOO] The predacious diving beetles, a family of coleopteran insects in the suborder Adephaga. { dī¦tis·ə¦dē }

e [MATH] The base of the natural logarithms; the number defined by the equation

$$\int_1^e \frac{1}{x}\, dx = 1;$$

approximately equal to 2.71828.

E *See* electric-field vector; exa-.

EADI *See* electronic attitude directional indicator.

eager *See* bore. { 'ē·gər }

eagle [VERT ZOO] Any of several large, strong diurnal birds of prey in the family Accipitridae. { 'ē·gəl }

eagle mounting [SPECT] A mounting for a diffraction grating, based on the principle of the Rowland circle, in which the diffracted ray is returned along nearly the same direction as the incident beam. { 'ē·gəl ,maun·tiŋ }

Eagle Nebula [ASTRON] A large emission nebula in the constellation Serpens, about 2500 parsecs away. { 'ē·gəl 'neb·yə·lə }

Eagle's medium [MICROBIO] A tissue-culture medium, developed by H. Eagle, containing vitamins, amino acids, inorganic salts and serous enrichments, and dextrose. { 'ē·gəlz ,mēd·ē·əm }

EAM *See* electric accounting machine.

E and M lead signaling [COMMUN] Communications between a trunk circuit and a separate signaling unit over two leads: an M lead that transmits battery or ground signals to the signaling equipment, and an E lead which receives open or ground signals from the signaling unit. { 'ē ən 'em 'lēd ,sig·nəl·iŋ }

ear [ANAT] The receptor organ that sends both auditory information and space orientation information to the brain in vertebrates. { ir }

eardrum *See* tympanic membrane. { 'ir,drəm }

ear implantation length [ARCH] A measure of the distance from the otobasion superior to the otobasion inferior. { ¦ir ,im,plan'tā·shən ,leŋkth }

earing [MET] Formation of scallops around the top edge of a deep-drawn product due to differences in the directional properties of the metal sheet. { 'ir·iŋ }

earlandite [MINERAL] $Ca_3(C_6H_5O_7)_2 \cdot 4H_2O$ A mineral consisting of a hydrous citrate of calcium; found in sediments in the Weddell Sea. { 'ir·lən,dīt }

ear length [ARCH] A measure of the maximum distance along the anterior-posterior axis of the ear. { 'ir ,leŋkth }

earliest finish time [IND ENG] The earliest time for completion of an activity of a project; for the entire project, it equals the earliest start time of the final event included in the schedule. { ¦ər·lē·əst 'fin·ish ,tīm }

earliest start time [IND ENG] The earliest time at which an activity may begin in the schedule of a project; it equals the earliest time that all predecessor activities can be completed. { ¦ər·lē·əst 'start ,tīm }

earlobe [ANAT] The pendulous, fleshy lower portion of the auricle or external ear. { 'ir,lōb }

early binding [COMPUT SCI] The assignment of data types (such as integer or string) to variables during the compilation of a computer program rather than at run time. { 'ər·lē ¦bīnd·iŋ }

early effect [ELECTR] A change in the base width of a bipolar transistor as a function of base-collector bias voltage. { 'ər·lē i,fekt }

early enzyme [BIOCHEM] Any of the enzymes that are synthesized in a bacterial cell under the direction of an invading bacteriophage. { ¦ər·lē 'en,zīm }

early finish date [IND ENG] The earliest time that an activity can be completed. { ¦ər·lē 'fin·ish ,dāt }

early gene [GEN] Any gene expressed very soon after a growth stimulus that initiates cell proliferation; it is divided into immediate early and delayed early classes. { 'ər·lē ,jēn }

early start date [IND ENG] The earliest time that an activity may be commenced. { ¦ər·lē 'start ,dāt }

early-type spiral [ASTRON] A spiral galaxy with a large nuclear bulge and tightly wound arms. { 'ər·lē ¦tīp 'spī·rəl }

early-type star [ASTROPHYS] A star with relatively high surface temperature, in spectral class O or B. { 'ər·lē ¦tīp 'stär }

early-warning/control and reporting post [ORD] Long-range supplementary radar surveillance data collection and reporting facility for a sector system and a long-range supplementary control facility for offensive, defensive, and return-to-base-type missions. { ¦ər·lē ,wor·niŋ kən'trōl ən ri'pord·iŋ ,pōst }

early-warning radar [ORD] A line of air defense radar units along the perimeter of a defended area to provide the earliest possible warning of approaching aircraft. { ¦ər·lē ,wor·niŋ 'rā,där }

early-warning station *See* aircraft early-warning station. { ¦ər·lē ,wor·niŋ ,stā·shən }

earlywood [BOT] The portion of the annual ring that is formed during the early part of a tree's growing season. { 'ər·lē,wud }

earned value [IND ENG] The budgeted cost of the work performed for a given project. { ¦ərnd 'val·yü }

Earnshaw's theorem [ELEC] The theorem that a charge cannot be held in stable equilibrium by an electrostatic field. { 'ərn,shoz ,thir·əm }

EAROM *See* electrically alterable read-only memory. { 'ē,räm }

earphone [ENG ACOUS] **1.** An electroacoustical transducer, such as a telephone receiver or a headphone, actuated by an electrical system and supplying energy to an acoustical system of the ear, the waveform in the acoustical system being substantially the same as in the electrical system. **2.** A small, lightweight electroacoustic transducer that fits inside the ear, used chiefly with hearing aids. { 'ir,fōn }

earplug [ENG] A device made of a pliable substance which fits into the ear opening; used to protect the ear from excessive noise or from water. { 'ir,pləg }

ear protector [ENG] A device, such as a plug or ear muff, used to protect the human ear from loud noise that may be injurious to hearing, such as that of jet engines. { 'ir prə,tek·tər }

ear rot [PL PATH] Any of several fungus diseases of corn, occurring both in the field and in storage and marked by decay and molding of the ears. { 'ir ,rät }

ear shell *See* abalone. { 'ir ,shel }

earth [ASTRON] The third planet in the solar system, lying between Venus and Mars; sometimes capitalized. [ELEC] *See* ground. [GEOL] **1.** Solid component of the globe, distinct from air and water. **2.** Soil; loose material composed of disintegrated solid matter. { ərth }

earth connection *See* ground. { 'ərth kə,nek·shən }

earth core *See* core. { 'ərth ,kor }

earth crust *See* crust. { 'ərth ,krəst }

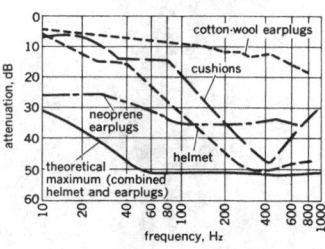

EAR PROTECTOR

Comparison of various ear protectors in the attenuation of sound.

earth current [ELEC] Return, fault, leakage, or stray current passing through the earth from electrical equipment. Also known as ground current. [GEOPHYS] A current flowing through the ground and due to natural causes, such as the earth's magnetic field or auroral activity. Also known as telluric current. { 'ərth ,kə·rənt }

earth-current storm [GEOPHYS] Irregular fluctuations in an earth current in the earth's crust, often associated with electric field strengths as large as several volts per kilometer, and superimposed on the normal diurnal variation of the earth currents. { 'ərth ,kə·rənt ,stȯrm }

earth dam [CIV ENG] A dam having the main section built of earth, sand, or rock, and a core of impervious material such as clay or concrete. { 'ərth ,dam }

earth detector See leakage indicator. { 'ərth di·tek·tər }

earthed system See grounded system. { 'ərtht ,sis·təm }

earth electrode See ground electrode. { 'ərth i,lek,trōd }

earthenware [ENG] Ceramic products of natural clay, fired at 1742–2129°F (950–1165°C), that is slightly porous, opaque, and usually covered with a nonporous glaze. { 'ər·thən,wer }

earth figure [GEOD] The shape of the earth. { 'ərth ,fig·yər }

earthflow [GEOL] A variety of mass movement involving the downslope slippage of soil and weathered rock in a series of subparallel sheets. { 'ərth,flō }

earth hummock [GEOL] A small, dome-shaped uplift of soil caused by the pressure of groundwater. Also known as earth mound. { 'ərth ,həm·ək }

earth inductor [ENG] A type of inclinometer that has a coil which rotates in the earth's field and in which a voltage is induced when the rotation axis does not coincide with the field direction; used to measure the dip angle of the earth's magnetic field. Also known as dip inductor; earth inductor compass; induction inclinometer. { 'ərth in,dək·tər }

earth inductor compass See earth inductor. { 'ərth in'dək·tər ,käm·pəs }

earthing reactor See grounding reactor. { 'ərth·iŋ rē,ak·tər }

earth interior [GEOL] The portion of the earth beneath the crust. { |ərth in|tir·ē·ər }

earth-layer propagation [GEOPHYS] **1.** Propagation of electromagnetic waves through layers of the earth's atmosphere. **2.** Electromagnetic wave propagation through layers below the earth's surface. { 'ərth ,lā·ər ,präp·ə,gā·shən }

earthlight [ASTRON] The illumination of the dark part of the moon's disk, produced by sunlight reflected onto the moon from the earth's surface and atmosphere. Also known as earthshine. { 'ərth,līt }

earth mound See earth hummock. { 'ərth ,maúnd }

earth movements [GEOPHYS] Movements of the earth, comprising revolution about the sun, rotation on the axis, precession of equinoxes, and motion of the surface of the earth relative to the core and mantle. { 'ərth 'müv·məns }

earthmover [MECH ENG] A machine used to excavate, transport, or push earth. { 'ərth,müv·ər }

earth-nut oil See peanut oil. { 'ərth ,nət ,óil }

earth orbit [ASTRON] The elliptical motion of the earth about the sun (eccentricity 0.01675, average radius 9.296×10^7 miles or 1.496×10^8 kilometers) in a sidereal year. { |ərth 'ȯr·bət }

earth oscillations [GEOPHYS] Any rhythmic deformations of the earth as an elastic body; for example, the gravitational attraction of the moon and sun excite the oscillations known as earth tides. { 'ərth ,äs·ə'lā·shənz }

earth pig See aardvark. { 'ərth ,pig }

earth pillar [GEOL] A tall, conical column of earth materials, such as clay or landslide debris, that has been sheltered from erosion by a cap of hard rock. { 'ərth 'pil·ər }

earth pressure [CIV ENG] The pressure which exists between earth materials (such as soil or sediments) and a structure (such as a wall). { 'ərth ,presh·ər }

earthquake [GEOPHYS] A sudden movement of the earth caused by the abrupt release of accumulated strain along a fault in the interior. The released energy passes through the earth as seismic waves (low-frequency sound waves), which cause the shaking. { 'ərth,kwāk }

earthquake-resistant [CIV ENG] Of a structure or building, able to withstand lateral seismic stresses at the base. { 'ərth,kwāk ri,zis·tənt }

earthquake tremor See tremor. { 'ərth,kwāk ,trem·ər }

earthquake zone [GEOL] An area of the earth's crust in which movements, sometimes with associated volcanism, occur. Also known as seismic area. { 'ərth,kwāk ,zōn }

earth radiation See terrestrial radiation. { |ərth ,rād·ē'ā·shən }

Earth Radiation Budget Experiment [METEOROL] A satellite observational program to study the earth's radiation budget. Abbreviated ERBE. { ,ərth ,rād·ē|ā·shən |bəj·ət ik,sper·ə·mənt }

earth rate [ASTRON] The angular velocity or rate of the earth's rotation. { 'ərth ,rāt }

earth-rate correction [NAV] A rate applied to a gyroscope to compensate for the apparent precession of the spin axis caused by the rotation of the earth. { 'ərth ,rāt kə,rek·shən }

earth resources technology satellite [AERO ENG] One of a series of satellites designed primarily to measure the natural resources of the earth; functions include mapping, cataloging water resources, surveying crops and forests, tracing sources of water and air pollution, identifying soil and rock formations, and acquiring oceanographic data. Abbreviated ERTS. { |ərth ri|sȯr·səz tek|näl·ə·je 'sad·əl,īt }

earthrise [ASTRON] The rising of the earth above the horizon of the moon, as viewed from the moon. { 'ərth,rīz }

earth rotation [ASTRON] Motion about the earth's axis that occurs 365.2422 times over a year's period. { |ərth rō|tā·shən }

earth satellite [AERO ENG] An artificial satellite placed into orbit about the earth. [ASTRON] A natural body that revolves about the earth, such as the moon. { 'ərth ,sad·əl,īt }

earth science [SCI TECH] The science that deals with the earth or any part thereof; includes the disciplines of geology, geography, oceanography, and meteorology, among others. { 'ərth ,sī·əns }

earth shadow [METEOROL] Any shadow projecting into a hazy atmosphere from mountain peaks at times of sunrise or sunset. { 'ərth ,shad·ō }

earthshine See earthlight. { 'ərth,shīn }

earth's shadow See dark segment. { |ərths 'shad·ō }

earthstar [MYCOL] A fungus of the genus *Geastrum* that resembles a puffball with a double peridium, the outer layer of which splits into the shape of a star. { 'ərth,stär }

earth station [COMMUN] A facility with a land-based antenna used to transmit and receive information to and from a communications satellite. { 'ərth ,stā·shən }

earth's way [ASTRON] The angle between the direction of the earth's motion and the apparent direction of a star. { 'ərths ,wā }

earth system [GEOPHYS] The atmosphere, oceans, biosphere, cryosphere, and geosphere, together. { 'ərth , sis·təm }

earth thermometer See soil thermometer. { 'ərth thər,mäm·əd·ər }

earth tide [GEOPHYS] The periodic movement of the earth's crust caused by forces of the moon and sun. Also known as bodily tide. { 'ərth ,tīd }

earth tremor See tremor. { 'ərth ,trem·ər }

earth wax See ozocerite. { 'ərth,waks }

earthwork [CIV ENG] **1.** Any operation involving the excavation or construction of earth embankments. **2.** Any construction made of earth. [ORD] A temporary or permanent fortification for attack or defense, made chiefly of earth. { 'ərth,wərk }

earthworm [INV ZOO] The common name for certain terrestrial members of the class Oligochaeta, especially forms belonging to the family Lumbricidae. { 'ərth,wərm }

earthy cobalt See asbolite. { |ərth·ē 'kō,bȯlt }

earthy manganese See wad. { |ərth·ē 'maŋ·gə,nēs }

earwax See cerumen. { 'ir,waks }

earwig [INV ZOO] The common name for members of the insect order Dermaptera. { 'ir,wig }

easel [GRAPHICS] A standing frame, often adjustable, used to hold a painting in process, or to display a chart in meetings. { 'ē·zəl }

easement [CIV ENG] The right held by one person over another person's land for a specific use; rights of tenants are excluded. { 'ēz·mənt }

easement curve [CIV ENG] A curve, as on a highway, whose degree of curvature is varied to provide a gradual transition between a tangent and a simple curve, or between two simple

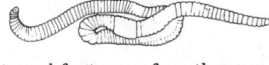
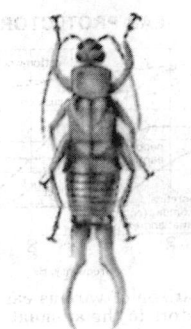

curves which it connects. Also known as transition curve. { 'ēz·mənt 'kərv }

easin [ORG CHEM] $C_{20}H_6O_5I_4Na_2$ The sodium salt of tetraiodofluorescein; a brown powder, insoluble in water; used as a dye and a pH indicator (hydrogen ion) at pH 2.0. Also known as iodoeasin; sodium tetrafluorescein. { 'ē·ə·zən }

east [GEOD] The direction 90° to the right of north. { ēst }

East Africa Coast Current [OCEANOGR] A current that is influenced by the monsoon drifts of the Indian Ocean, flowing southwestward along the Somalia coast in the Northern Hemisphere winter and northeastward in the Northern Hemisphere summer. Also known as Somali Current. { ¦ēst ¦af·rə·kə ¦kōst ,kə·rənt }

East African sleeping sickness See Rhodesian trypanosomiasis. { ¦ēst ¦af·rə·kən 'slēp·iŋ ,sik·nəs }

East Australia Current [OCEANOGR] The current which is formed by part of the South Equatorial Current and flows southward along the eastern coast of Australia. { ¦ēst ò'strāl·yə ,kə·rənt }

Easter-egging [ELECTR] An undirected procedure for checking electronic equipment, which derives its name from the children's activity of searching for hidden eggs at Eastertime. { 'ē·stər ,eg·iŋ }

easterly wave [METEOROL] A long, weak migratory low-pressure trough occurring in the tropics. { 'ēs·tər·lē 'wāv }

eastern equine encephalitis [MED] A mosquito-borne virus infection of horses and mules in the eastern and southern United States caused by a member of arbovirus group A. { ¦ē·stərn ¦ē,kwīn en,sef·ə'līd·əs }

Eastern Hemisphere [GEOGR] The half of the earth lying mostly to the east of the Atlantic Ocean, including Europe, Africa, and Asia. { ¦ē·stərn 'hem·ə,sfir }

Eastern hemlock [FOR] Tsuga canadensis. A type of hemlock that occurs in eastern Canada, the Great Lakes states, and the Appalachians; it grows to a height of about 90 ft (27 m) and has minutely toothed leaves, with some of the smaller ones growing upside down. The bark is a principal domestic source of tannin. { ,ē·stərn 'hem,läk }

East Greenland Current [OCEANOGR] A current setting south along the eastern coast of Greenland and carrying water of low salinity and low temperature. { ¦ēst 'grēn·lənd ,kə·rənt }

East Indian geranium oil See palmarosa oil. { ¦ēst ¦in·dē·ən jə'rān·ē·əm ,oil }

East Indian sandalwood oil See sandalwood oil. { ¦ēst ¦in·dē·ən 'san·dəl,wùd ,oil }

easting [NAV] The distance a craft makes good to the east. { 'ē·stiŋ }

eastonite [MINERAL] $K_2Mg_5AlSi_5Al_3O_{20}(OH)_4$ A mineral consisting of basic silicate of potassium, magnesium, and aluminum; it is an end member of the biotite system. { 'ē·stə,nīt }

east point [GEOD] That intersection of the prime vertical with the horizon which lies to the right of the observer when facing north. { ¦ēst ,point }

east-west effect [ASTRON] The phenomenon due to the fact that a greater number of cosmic-ray particles approach the earth from a westerly direction than from an easterly. { ¦ēst ¦west i,fekt }

easy [COMPUT SCI] A name for the hexadecimal digit whose decimal equivalent is 14. { 'ē·zē }

easy glide [SOLID STATE] A large increase in plastic deformation of a single crystal accompanying a small increase in stress as the result of the passage of many thousands of dislocations through the crystal along a single glide system. { 'ē·zē ,glīd }

Eaton agent [MICROBIO] The name applied to Mycoplasma pneumoniae when it was regarded as a virus. { 'ēt·ən ,ā·jənt }

Eaton agent pneumonia [MED] Pneumonitis in humans, caused by Mycoplasma pneumoniae. Also known as primary atypical pneumonia. { 'ēt·ən ,ā·jənt nə,mōn·yə }

eave [BUILD] The border of a roof overhanging a wall. { ēv }

eaves board [BUILD] A strip nailed along the eaves of a building to raise the end of the bottom course of tile or slate on the roof. { 'ēvz ,bòrd }

eaves molding [BUILD] A cornicelike molding below the eaves of a building. { 'ēvz ,mōl·diŋ }

EBAM See electron-beam memory. { 'ē,bam }

ebb-and-flow structure [GEOL] Rock strata with alternating horizontal and cross-bedded layers, believed to have been produced by ebb and flow of tides. { 'eb ən 'flō ,strək·chər }

ebb current [OCEANOGR] The tidal current associated with the decrease in the height of a tide. { 'eb ,kə·rənt }

ebb tide [OCEANOGR] The portion of the tide cycle between high water and the following low water. Also known as falling tide. { 'eb ,tīd }

EBCDIC See extended binary-coded decimal interchange code. { 'eb·sə,dik }

E beam [PHYS] An intense burst of fast electrons from a small particle accelerator, used to excite a pulsed gas laser in which the gas pressure is too high to permit an electric discharge. { 'ē ,bēm }

Ebenaceae [BOT] A family of dicotyledonous plants in the order Ebenales, in which a latex system is absent and flowers are mostly unisexual with the styles separate, at least distally. { ,eb·ə'nās·ē,ē }

Ebenales [BOT] An order of woody, sympetalous dicotyledonous plants in the subclass Dilleniidae, woody having axile placentation and usually twice as many stamens as corolla lobes. { ,eb·ə'nā·lēz }

E bend [ELECTROMAG] A smooth change in the direction of the axis of a waveguide, throughout which the axis remains in a plane parallel to the direction of polarization. Also known as E-plane bend. { 'ē ,bend }

Eberhard effect [GRAPHICS] The phenomenon due to the fact that the density of a photographic plate, given a uniform exposure through a metal plate with an opening, varies with the size of the opening, when an organic developer is used. { 'ā·bər·härt i,fekt }

Ebert ion counter [ENG] An ion counter of the aspiration condenser type, used for the measurement of the concentration and mobility of small ions in the atmosphere. { 'ā·bərt ī·ən ,kaùnt·ər }

EBIS See electron-beam ion source. { 'ē,bis }

EBM See electron-beam machining.

EBIT See electron-beam ion trap. { 'ē,bit or ¦ē¦bē¦ī'tē }

ebonite See hard rubber. { 'eb·ə,nīt }

ebony [BOT] Any of several African and Asian trees of the genus Diospyros, providing a hard, durable wood. { 'eb·ə nē }

ebracteate [BOT] Without bracts, or much reduced leaves. { ē'brak·tē,āt }

ebracteolate [BOT] Without bracteoles. { ē'brak·tē·ə ,lāt }

Ebriida [INV ZOO] An order of flagellate protozoans in the class Phytamastigophorea characterized by a solid siliceous skeleton. { ē'brī·ə·də }

ebullating-bed reactor [CHEM ENG] A type of fluidized bed in which catalyst particles are held in suspension by the upward movement of the liquid reactant and gas flow. Also known as slurry-bed reactor. { ēb·yə,lād·iŋ ¦bed ,rē,ak·tər }

ebulliometer [PHYS CHEM] The instrument used for ebulliometry. Also known as ebullioscope. { ,ə,bù·lē'äm·əd·ər }

ebulliometry [PHYS CHEM] The precise measurement of the absolute or differential boiling points of solutions. { ,ə,bù·lē'äm·ə·trē }

ebullioscope See ebulliometer. { ə'bù·lē·ə,skōp }

ebullioscopic constant [PHYS CHEM] The ratio of the elevation of the boiling point of a solvent caused by dissolving a solute to the molality of the solution, taken at extremely low concentrations. Also known as molal elevation of the boiling point. { e'bü·lē·ə,skōp·ik ,kän·stənt }

ebullism [PHYSIO] The formation of bubbles, especially of water vapor bubbles in biological fluids, owing to reduced ambient pressure. { 'eb·yə,liz·əm }

ebullition [PHYS] The process or state of a liquid bubbling up or boiling. { ,eb·ə'li·shən }

e-business See electronic commerce. { 'ē,biz·nəs }

ecad [ECOL] A type of plant which is altered by its habitat and possesses nonheritable characteristics. { 'ē,kad }

ECB See block encryption.

EC blank fire See EC smokeless powder. { ¦ē¦sē 'blaŋk ¦fīr }

EC blank powder See EC smokeless powder. { ¦ē¦sē 'blaŋk ¦paùd·ər }

eccentric [SCI TECH] Situated to one side with reference to a center. { ek'sen·trik }

eccentric angle [MATH] 1. For an ellipse having semimajor and semiminor angles of lengths a and b respectively, lying

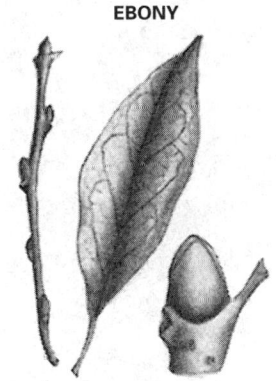

EBONY

Twig, leaf, and bud of persimmon (Diospyros virginiana).

along the x and y axes of a coordinate system respectively, and for a point (x,y) on the ellipse, the angle

$$\text{arc cos}\frac{x}{a} = \text{arc sin}\frac{y}{b}$$

2. For a hyperbola having semitransverse and semiconjugate axes of lengths a and b respectively, lying along the x and y axes of a coordinate system respectively, and for a point (x,y) on the hyperbola, the angle

$$\text{arc sec}\frac{x}{a} = \text{arc tan}\frac{y}{b}$$

{ ek¦sen·trik 'ang·əl }

eccentric anomaly [ASTRON] For a planet in an elliptical orbit, the eccentric angle corresponding to the planet's location. { ek¦sen·trik ə'näm·ə·lē }

eccentric bit [DES ENG] A modified chisel for drilling purposes having one end of the cutting edge extended further from the center of the bit than the other. { ek¦sen·trik 'bit }

eccentric cam [DES ENG] A cylindrical cam with the shaft displaced from the geometric center. { ek¦sen·trik 'kam }

eccentric circles [MATH] **1.** For an ellipse, two circles whose centers are at the center of the ellipse and whose diameters are, respectively, the major and minor axes of the ellipse. **2.** For a hyperbola, two circles whose centers are at the center of symmetry of the hyperbola and whose diameters are, respectively, the transverse and conjugate axes of the hyperbola. { ek¦sen·trik 'sərk·əlz }

eccentric contraction [BIOPHYS] The increase in tension that occurs in a muscle as it lengthens. { ek¦sen·trik kən'trak·shən }

eccentric gear [DES ENG] A gear whose axis deviates from the geometric center. { ek¦sen·trik 'gir }

eccentricity [MATH] The ratio of the distance of a point on a conic from the focus to the distance from the directrix. [MECH] The distance of the geometric center of a revolving body from the axis of rotation. { ek·sən'tris·əd·ē }

eccentric load [ENG] A load imposed on a structural member at some point other than the centroid of the section. { ek¦sen·trik 'lōd }

eccentric orbit [ASTRON] An orbit of a celestial body that deviates markedly from a circle. { ek¦sen·trik 'ȯr·bət }

eccentric reducer [ENG] A threaded or butt-welded fitting for pipes whose ends are not the same size and are eccentric to each other. { ek¦sen·trik ri'düs·ər }

eccentric ring structure [ASTRON] A formation on the moon's surface consisting of two craters, one inside the other, with the inner crater touching the wall of the outer one at one point. { ek¦sen·trik 'riŋ ,strək·chər }

eccentric rotor engine [MECH ENG] A rotary engine, such as the Wankel engine, wherein motion is imparted to a shaft by a rotor eccentric to the shaft. { ek¦sen·trik 'rōd·ər ,en·jən }

eccentric signal [ENG] A survey signal whose position is not in a vertical line with the station it is representing. { ek¦sen·trik 'sig·nəl }

eccentric station [ENG] A survey point over which an instrument is centered and which is not positioned in a vertical line with the station it is representing. { ek¦sen·trik 'stā·shən }

eccentric valve [ENG] A rubber-lined slurry or fluid valve with an eccentric rotary cut-off body to reduce corrosion and wear on mechanical moving valve parts. { ek¦sen·trik 'valv }

ecchymosis [MED] A subcutaneous hemorrhage marked by purple discoloration of the skin. { ¦ek·ə'mō·səs }

Eccles-Jordan circuit *See* bistable multivibrator. { ¦ek·əlz 'jȯrd·ən ,sər·kət }

Eccles-Jordan multivibrator *See* bistable multivibrator. { ¦ek·əlz 'jȯrd·ən ,məl·ti'vī,brād·ər }

eccrine gland [PHYSIO] One of the small sweat glands distributed all over the human body surface; they are tubular coiled merocrine glands that secrete clear aqueous sweat. { 'ek·rən ,gland }

ecdemite [MINERAL] $Pb_6As_2O_7Cl_4$ A greenish-yellow to yellow, tetragonal mineral consisting of an oxychloride of lead and arsenic; occurs as coatings of small tabular crystals and as coarsely foliated masses. { 'ek·də,mīt }

ECDIS *See* electronic chart display and information system. { 'ek,dis or ¦ē¦sē¦dē¦ī'es }

ecdysis [INV ZOO] Molting of the outer cuticular layer of the body, as in insects and crustaceans. { 'ek·də·səs }

ecdysone [BIOCHEM] The molting hormone of insects. { 'ek·də,sōn }

E cell [ELEC] A timing device that converts the current-time integral of an electrical function into an equivalent mass integral (or the converse operation) up to a maximum of several thousand microampere-hours. { 'ē ,sel }

E center *See* R center. { 'ē ,sen·tər }

ecesis [ECOL] Successful naturalization of a plant or animal population in a new environment. { ə'sē·səs }

ECG *See* electrocardiogram.

ecgonine [ORG CHEM] $C_9H_{15}NO_3$ An alkaloid obtained in crystalline form by the hydrolysis of cocaine. { 'ek·gə,nēn }

echelette grating [SPECT] A diffraction grating with coarse groove spacing, designed for the infrared region; has grooves with comparatively flat sides and concentrates most of the radiation by reflection into a small angular coverage. { ¦esh·ə¦let or ¦āsh¦let ,grād·iŋ }

echelle grating [SPECT] A diffraction grating designed for use in high orders and at angles of illumination greater than 45° to obtain high dispersion and resolving power by the use of high orders of interference. { ā'shel ,grād·iŋ }

echelle spectrograph [SPECT] A spectrograph that employs gratings intended to be used in very high orders (greater than 10), and is equipped with a second dispersal element (another grating or a prism) at right angles to the first in order to separate the successive spectral strips from each other. { e,shel 'spek·trə,graf }

echelon [ORD] **1.** A formation of troops; the units are parallel but unaligned, in a steplike manner. **2.** A similar arrangement of planes or ships, as planes flying in a V formation. { 'esh·ə,län }

echelon faults [GEOL] Separate, parallel faults having steplike trends. { 'esh·ə,län ,fȯls }

echelon grating [SPECT] A diffraction grating which consists of about 20 plane-parallel plates about 1 centimeter thick, cut from one sheet, each plate extending beyond the next by about 1 millimeter, and which has a resolving power on the order of 10^6. { 'esh·ə,län ,grād·iŋ }

echelon matrix [MATH] A matrix in which the rows whose terms are all zero are below those with some nonzero terms, the first nonzero term in a row is 1, and this 1 appears to the right of the first nonzero term in any row above it. { 'esh·ə,län ,mā·triks }

Echeneidae [VERT ZOO] The remoras, a family of perciform fishes in the suborder Percoidei. { ,ek·ə'nā·ə,dē }

echidna [VERT ZOO] A spiny anteater; any member of the family Tachyglossidae. { ə'kid·nə }

Echinacea [INV ZOO] A suborder of echinoderms in the order Euechinoidea; individuals have a rigid test, keeled teeth, and branchial slits. { ,ek·ə'nā·shə }

echinate [ZOO] Having a dense covering of spines or bristles. { ə'kī,nāt }

Echinidae [INV ZOO] A family of echinacean echinoderms in the order Echinoida possessing trigeminate or polyporous plates with the pores in a narrow vertical zone. { ə'kī'nə,dē }

Echiniscoidea [INV ZOO] A suborder of tardigrades in the order Heterotardigrada characterized by terminal claws on the legs. { ə,kī'nə'skoid·ē·ə }

echinococcosis [MED] Infestation by the larva (hydatid) of *Echinococcus granulosis* in humans, and in some canines and herbivores. Also known as hydatid disease; hydatidosis. { ə,kī·nə·kä'kō·səs }

Echinococcus [INV ZOO] A genus of tapeworms. { ə,kī·nə'kä·kəs }

echinococcus cyst [INV ZOO] A cyst formed in host tissues by the larva of *Echinococcus granulosus.* { ə,kī·nə'kä·kəs 'sist }

Echinocystitoida [PALEON] An order of extinct echinoderms in the subclass Perischoechinoidea. { ,ek·ə·nō,sis·tə'toid·ə }

Echinodera [INV ZOO] The equivalent name for Kinorhyncha. { ,ek·ə'nä·də·rə }

Echinodermata [INV ZOO] A phylum of exclusively marine coelomate animals distinguished from all others by an internal skeleton composed of calcite plates, and a water-vascular system to serve the needs of locomotion, respiration, nutrition, or perception. { ,ek·ə·nə'dər·məd·ə }

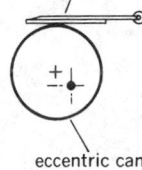

ECCENTRIC CAM

oscillating flat-face follower

eccentric cam

An eccentric cam with a flat-face follower.

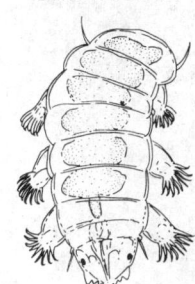

ECHINISCOIDEA

Echiniscoides sigismundi.

Echinoida [INV ZOO] An order of Echinacea with a camarodont lantern, smooth test, and imperforate noncrenulate tubercles. { ‚ek·ə'noid·ə }

Echinoidea [INV ZOO] The sea urchins, a class of Echinozoa having a compact body enclosed in a hard shell, or test, formed by regularly arranged plates which bear movable spines. { ‚ek·ə'noid·e·ə }

Echinometridae [INV ZOO] A family of echinoderms in the order Echinoida, including polyporous types with either an oblong or a sphericaltest. { ‚ek·ə·nō'me·trə‚dē }

echinomycin [MICROBIO] $C_{50}H_{60}O_{12}N_{12}S_2$ A toxic polypeptide antibiotic produced by species of *Streptomyces*. { ‚ek·ə·nō'mīs·ən }

echinopluteus [INV ZOO] The bilaterally symmetrical larva of sea urchins. { ‚ek·ə·nō'plüd·e·əs }

echinopsine [ORG CHEM] $C_{10}H_9O$ An alkaloid obtained from *Echinops* species; crystallizes as needles from benzene solution, melts at 152°C; physiological action is similar to that of brucine and strychnine. { ‚ek·ə'näp‚sēn }

Echinosteliaceae [MYCOL] A family of slime molds in the order Echinosteliales. { ‚ek·ə·nō‚ste·lē'ās·ē‚ē }

Echinosteliales [MYCOL] An order of slime molds in the subclass Myxogastromycetidae. { ‚ek·ə·nō‚ste·lē'ā·lēz }

echinostome cercaria [INV ZOO] A digenetic trematode larva characterized by the large anterior acetabulum and a collar with spines. { ə'kī·nə‚stōm sər'kar·e·ə }

Echinothuriidae [INV ZOO] A family of deep-water echinoderms in the order Echinothurioida in which the large, flexible test collapses into a disk at atmospheric pressure. { ‚ek·ə·nō·thə'rī·ə‚dē }

Echinothurioida [INV ZOO] An order of echinoderms in the superorder Diadematacea with solid or hollow primary radioles, diademoid ambulacral plates, noncrenulate tubercles, and the anus within the apical system. { ‚ek·ə·nō·thü·rē'oid·ə }

Echinozoa [INV ZOO] A subphylum of free-living echinoderms having the body essentially globoid with meridional symmetry and lacking appendages. { ‚ek·ə·nō'zō·ə }

Echiurida [INV ZOO] A small group of wormlike organisms regarded as a separate phylum of the animal kingdom; members have a saclike or sausage-shaped body with an anterior, detachable prostomium. { ‚ek·ē'yùr·ə‚də }

Echiuridae [INV ZOO] A small family of the order Echiuroinea characterized by a flaplike prostomium. { ‚ek·ē'yùr·ə‚dē }

Echiuroidea [INV ZOO] A phylum of schizocoelous animals. { ‚ek·ē·yə'ròid·ē·ə }

Echiuroinea [INV ZOO] An order of the Echiurida. { ‚ek·ē·yə'ròi·nē·ə }

echo [ELECTR] **1.** The signal reflected by a radar target, or the trace produced by this signal on the screen of the cathode-ray tube in a radar receiver. Also known as radar echo; return. **2.** *See* ghost signal. [PHYS] A wave packet that has been reflected or otherwise returned with sufficient delay and magnitude to be perceived as a signal distinct from that directly transmitted. { 'ek·ō }

echo amplitude [ELECTR] In radar, an empirical measure of the strength of a target signal as determined from the appearance of the echo; the amplitude of the echo waveform usually is measured by the deflection of the electron beam from the base line of an amplitude-modulated indicator. { ‚'ek·ō 'am·plə‚tüd }

echo area [ELECTROMAG] In radar, the area of a fictitious perfect reflector of electromagnetic waves that would reflect the same amount of energy back to the radar as the actual target. Also known as radar cross section; target cross section. { 'ek‚ō ‚er·ē·ə }

echo attenuation [ELECTR] The power transmitted at an output terminal of a transmission line, divided by the power reflected back to the same output terminal. { 'ek‚ō ə‚ten·yə'wā·shən }

echo box [ELECTR] A calibrated high-Q resonant cavity that stores part of the transmitted radar pulse power and gradually feeds this energy into the receiving system after completion of the pulse transmission; used to provide an artificial target signal for test and tuning purposes. Also known as phantom target. { 'ek‚ō ‚bäks }

echocardiography [MED] A diagnostic technique for the heart that uses a transducer held against the chest to send high-frequency sound waves which pass harmlessly into the heart;

as they strike structures within the heart, they are reflected back to the transducer and recorded on an oscilloscope. { ‚ek·ō‚kärd·ē·'äg·rə·fē }

echo chamber [ACOUS] A reverberant room or enclosure used in a studio to add echo effects to sounds for radio or television programs. { 'ek·ō ‚chām·bər }

echo check [COMPUT SCI] A method of ascertaining the accuracy of transmission of data in which the transmitted data are returned to the sending end for comparison with original data. Also known as loopback check; loop check; read-back check. { 'ek·ō ‚chek }

echo contour [ELECTR] A trace of equal signal intensity of the radar echo displayed on a range height indicator or plan position indicator scope. { ¦ek·ō 'kän‚tür }

echoencephalograph [MED] An instrument that uses ultrasonic pulses and echo-ranging techniques to give a pictorial representation of intracranial structure. Also known as sonoencephalograph. { ¦ek·ō‚en'sef·lə‚graf }

echo frequency [ELECTR] The number of fluctuations, per unit time, in the power or amplitude of a radar target signal. { 'ek·ō ‚frē·kwən·sē }

echogram [ENG] The graphic presentation of echo soundings recorded as a continuous profile of the sea bottom. [MED] The pictorial display of anatomical structures using pulse-echo techniques. { 'ek·ō‚gram }

echograph [ENG] An instrument used to record an echogram. { 'ek·ō‚graf }

echo intensity [ELECTR] The brightness or brilliance of a radar echo as displayed on an intensity-modulated indicator; echo intensity is, within certain limits, proportional to the voltage of the target signal or to the square root of its power. { ¦ek·ō in'ten·səd·ē }

echolalia [MED] The purposeless, often seemingly involuntary repetition of words spoken by another person; a disorder seen in certain psychotic states and in certain organic brain syndromes. Also known as echophrasia. { ‚ek·ō'lā·lē·ə }

echolic [MED] Producing abortion or accelerating labor. { e'käl·ik }

echolocation [BIOPHYS] An animal's use of sound reflections to localize objects and to orient in the environment. { ‚ek·ō·lō‚kā·shən }

echo matching [ENG] Rotating an antenna to a position in which the pulse indications of an echo-splitting radar arc equal. { 'ek·ō ‚mach·iŋ }

echophrasia *See* echolalia. { ‚ek·ō'frā·zhə }

echoplex technique [COMPUT SCI] A technique for detecting errors in a data communication system with full duplex lines, in which the signal generated when a character is typed on a keyboard is transmitted to a receiver and retransmitted to a display terminal, enabling the operator to check if the character displayed is the same as the character typed. { 'ek·ō‚pleks tek‚nēk }

echo power [ELECTR] The electrical strength, or power, of a radar target signal, normally measured in watts or dBm (decibels referred to 1 milliwatt). { 'ek·ō ‚paù·ər }

echopraxia [PSYCH] Involuntary imitative repetition of the movements of another. { ‚ek·ō'prak·sē·ə }

echo pulse [ELECTR] A pulse of radio energy received at the radar after reflection from a target; that is, the target signal of a pulse radar. { 'ek·ō ‚pəls }

echo ranging [ENG] Active sonar, in which underwater sound equipment generates bursts of ultrasonic sound and picks up echoes reflected from submarines, fish, and other objects within range, to determine both direction and distance to each target. [VERT ZOO] An auditory feedback mechanism in bats, porpoises, seals, and certain other animals whereby reflected ultrasonic sounds are utilized in orientation. { 'ek·ō ‚rānj·iŋ }

echo-ranging sonar [ENG] Active sonar, in which underwater sound equipment generates bursts of ultrasonic sound and picks up echoes reflected from submarines, fish, and other objects within range, to determine both direction and distance to each target. { 'ek·ō ‚rānj·iŋ 'sō‚när }

echo recognition [ENG] Identification of a sonar reflection from a target, as distinct from energy returned by other reflectors. { 'ek·ō ‚rek·ig‚nish·ən }

echo repeater [ENG ACOUS] In sonar calibration and training, an artificial target that returns a synthetic echo by receiving a signal and retransmitting it. { 'ek·ō ri‚pēd·ər }

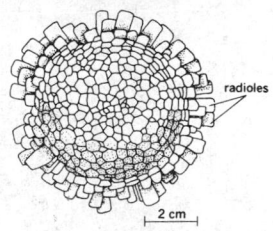

ECHINOIDA

radioles

2 cm

Colobocentrotus atratus, aboral aspect, a Pacific species adapted for life on wave-exposed coral reefs.

ECHINOSTOME CERCARIA

collar with spines

A drawing of an echinostome cercaria. (*From R. M. Cable, An Illustrated Laboratory Manual of Parasitology, Burgess, 1958*)

echo signal *See* target signal. { 'ek·ō ˌsig·nəl }

echosonde [GEOPHYS] An acoustic sounding instrument used to study meteorological disturbances in the lower atmosphere such as wind velocity and turbulence. { 'ek·ō ˌsänd }

echosonogram [ENG] A graphic display obtained with ultrasound pulse-reflection techniques; for example, an echocardiogram. { ˌek·ō 'sän·əˌgram }

echo sounder *See* sonic depth finder. { 'ek·ō ˌsaünd·ər }

echo sounding [ENG] Determination of the depth of water by measuring the time interval between emission of a sonic or ultrasonic signal and the return of its echo from the sea bottom. { 'ek·ō ˌsaünd·iŋ }

echo-splitting radar [ENG] Radar in which the echo is split by special circuits associated with the antenna lobe-switching mechanism, to give two echo indications on the radarscope screen; when the two echo indications are equal in height, the target bearing is read from a calibrated scale. { 'ek·ō ˌsplid·iŋ 'rāˌdär }

echo suppressor [ELECTR] **1.** A circuit that desensitizes radar navigation equipment for a fixed period after the reception of one pulse, for the purpose of rejecting delayed pulses arriving from longer, indirect reflection paths. **2.** A relay or other device used on a transmission line to prevent a reflected wave from returning to the sending end of the line. { 'ek·ō səˌpres·ər }

echo talker [COMPUT SCI] The interference created by the retransmission of a message back to its source while the source is still transmitting. { 'ek·ō ˌtȯk·ər }

echouterograph [MED] An instrument that uses ultrasonic pulses and echo-ranging techniques to give a pictorial representation of the uterus. { ˌek·ō'yüd·ə·rəˌgraf }

echovirus [VIROL] Any member of the Picornaviridae family, genus *Enterovirus;* the name is derived from the group designation enteric cytopathogenic human orphan virus. { 'ek·ōˌvī·rəs }

echylosis [CYTOL] The release of nonparticulate material from the cell through an apparently intact cell membrane. { ē'kī·lə·səs }

eckermannite [MINERAL] Na₃(Mg,Li)₄(Al,Fe)Si₈O₂₂(OH,F)₂ Mineral of the amphibole group containing magnesium, lithium, iron, and fluorine. { 'ek·ərˌmaˌnīt }

Eckert number [PHYS] A dimensionless group used in the study of compressible flow around a body, equal to the square of the fluid velocity far from the body divided by the product of the specific heat of the fluid at constant temperature and the difference between the temperatures of the fluid and the body. Symbolized N^E. Also known as Dulong number. { 'ek·ərt ˌnəm·bər }

Eckert projection [MAP] One of a group of six map projections of the whole earth, each showing the geographic poles represented by parallel straight lines that are one-half the length of the equator. { 'ek·ərt prəˌjek·shən }

ECL *See* emitter-coupled logic.

eclampsia [MED] A disorder occurring during the latter half of pregnancy, characterized by elevated blood pressure, edema, proteinuria, and convulsions or coma. { i'klam·sē·ə }

eclipse [ASTRON] **1.** The reduction in visibility or disappearance of a body by passing into the shadow cast by another body. **2.** The apparent cutting off, wholly or partially, of the light from a luminous body by a dark body coming between it and the observer. Also known as astronomical eclipse. { i'klips }

eclipsed antigen [IMMUNOL] An antigenic determinant of parasitic origin resembling an antigenic determinant of the parasite's host to such a degree that it does not elicit the formation of antibody by the host. { iˌklipst 'ant·i·jen }

eclipsed conformation [PHYS CHEM] A particular arrangement of constituent atoms that may rotate about a single bond in a molecule; for ethane it is such that when viewed along the axis of the carbon-carbon bond the hydrogen atoms of one methyl group are exactly in line with those of the other methyl group. { i'klipst ˌkän·fərˈmā·shən }

eclipse period [VIROL] A phase in the proliferation of viral particles during which the virus cannot be detected in the host cell. { i'klips ˌpir·ē·əd }

eclipse seasons [ASTRON] The two times when the sun is near enough to one of the nodes of the moon's orbit for eclipses to occur; this positioning occurs at nearly opposite times of the

year, and the eclipse seasons vary yearly because of westward regression of the nodes. { i'klips ˌsēz·ənz }

eclipse year [ASTRON] The interval between two successive conjunctions of the sun with the same node of the moon's orbit, equal to 346.62 days. Also known as draconic year; nodical year. { i'klips ˌyir }

eclipsing binary *See* eclipsing variable star. { i'klips·iŋ'bīˌnər·ē }

eclipsing variable star [ASTRON] A binary star whose orbit is such that every time one star passes between the observer and its companion an eclipse results. Also known as eclipsing binary; photometric binary. { i'klips·iŋ ˈver·ē·ə·bəl 'stär }

ecliptic [ASTRON] **1.** The apparent annual path of the sun among the stars; the intersection of the plane of the earth's orbit with the celestial sphere. **2.** The plane of the earth's orbit around the sun. { i'klip·tik }

ecliptic coordinate system [ASTRON] A celestial coordinate system in which the ecliptic is taken as the primary and the great circles perpendicular to it are then taken as secondaries. { i'klip·tik kōˈȯrd·ən·ət ˌsis·təm }

ecliptic diagram [ASTRON] A diagram of the zodiac indicating positions of certain celestial bodies in the ecliptic region. { i'klip·tik 'dī·əˌgram }

ecliptic latitude *See* celestial latitude. { iˌklip·tik 'lad·əˌtüd }

ecliptic limits [ASTRON] The distance of the sun from a node of the moon's orbit such that a solar eclipse cannot occur, or the greatest distance of the moon from a node such that an eclipse of the moon cannot occur. { iˌklip·tik 'lim·əts }

ecliptic longitude *See* celestial longitude. { iˌklip·tik 'länjəˌtüd }

ecliptic pole [ASTRON] On the celestial sphere, either of two points 90° from the ecliptic. { iˌklip·tik 'pōl }

eclogite [PETR] A class of metamorphic rocks distinguished by their composition, consisting essentially of omphacite and pyrope with small amounts of diopside, enstatite, olivine, kyanite, rutile, and rarely, diamond. { 'ek·ləˌjīt }

eclogite facies [PETR] A type of facies composed of eclogite and formed by regional metamorphism at extremely high temperature and pressure. { 'ek·ləˌjīt ˌfā·shēz }

eclosion [INV ZOO] The process of an insect hatching from its egg. { ē'klō·zhən }

ECM *See* electrochemical machining; extracellular matrix.

ecmnesia [MED] A lapse in memory of recent events, with normal memory for earlier events. { ek'nē·zhə }

ecnephias [METEOROL] A squall or thunderstorm in the Mediterranean. { ek·nə'fē·əs }

eco *See* electron-coupled oscillator.

ecocline [ECOL] A genetic gradient of adaptability to an environmental gradient; formed by the merger of ecotypes. { 'ek·ōˌklīn }

ecofallow [AGR] A system for destroying weeds and conserving soil moisture in crop rotation with minimum disturbance of crop residue and soil. { ˌek·ōˌfa·lō }

ecological association [ECOL] A complex of communities, such as an elm-hackberry association, which develops in accord with variations in physiography, soil, and successional history within the major subdivision of a biotic realm. { ek·ə'läj·ə·kəl əˌsō·shēˈā·shən }

ecological climatology [BIOL] A branch of bioclimatology, including the physiological adaptation of plants and animals to their climate, and the geographical distribution of plants and animals in relation to climate. { ek·ə'läj·ə·kəl klī·məˈtäl·ə·jē }

ecological community *See* community. { ek·ə'läj·ə·kəl kəˈmyün·əd·ē }

ecological energetics [ECOL] The study of the flow of energy within an ecological system from the time the energy enters the living system until it is ultimately degraded to heat and irretrievably lost from the system. Also known as production ecology. { ˌek·əˌläj·ə·kəl ˌen·ər'jed·iks }

ecological interaction [ECOL] The relation between species that live together in a community; specifically, the effect an individual of one species may exert on an individual of another species. { ek·ə'läj·ə·kəl in·tərˈak·shən }

ecological modeling [ECOL] The conceptualization and implementation of computer simulations of the behavior of living systems. { ˌek·əˌläj·ə·kəl 'mäd·əl·iŋ }

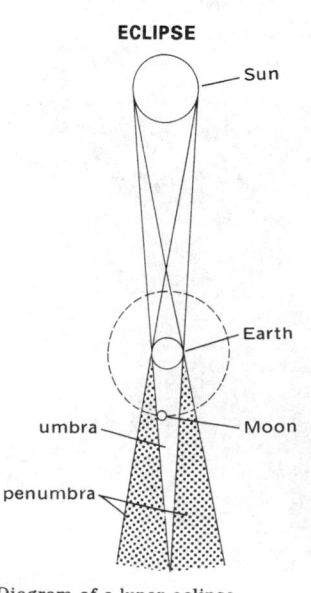

ECLIPSE

Diagram of a lunar eclipse.

ecological physiology [BIOL] The science of the interrelationships between the physiology of organisms and their environment. { ¦ē·kə¦läj·ə·kəl fiz·ē'äl·ə·jē }

ecological pyramid [ECOL] A pyramid-shaped diagram representing quantitatively the numbers of organisms, energy relationships, and biomass of an ecosystem; numbers are high for the lowest trophic levels (plants) and low for the highest trophic level (carnivores). { ek·ə'läj·ə·kəl 'pir·ə·mid }

ecological succession [ECOL] A gradual process incurred by the change in the number of individuals of each species of a community and by establishment of new species populations that may gradually replace the original inhabitants. { ek·ə'läj·ə·kəl sək'sesh·ən }

ecological system See ecosystem. { ek·ə'läj·ə·kəl 'sis·təm }

ecological zoogeography [ECOL] The study of animal distributions in terms of their environments. { ¦ek·ə¦läj·ə·kəl ¦zō·ō·jē'äg·rə·fē }

ecology [BIOL] A study of the interrelationships which exist between organisms and their environment. Also known as bionomics; environmental biology. { ē'käl·ə·jē }

ecomania [PSYCH] A symptom complex characterized by a domineering, haughty, and irritable attitude toward members of one's own family, but an attitude of humility toward those in authority. { ¦ek·ō'mā·nē·ə }

e-commerce See electronic commerce. { 'ē ¦käm·ərs }

econometrics [IND ENG] The application of mathematical and statistical techniques to the estimation of mathematical relationships for testing of economic theories and the solution of economic problems. { ē¦kän·ə¦me·triks }

economic entomology [BIOL] A branch of entomology concerned with the study of economic losses of commercially important animals and plants due to insect predation. [ECOL] The study of insects that have a direct influence on humanity, with an emphasis on pest management. { ¦ek·ə'näm·ik ¦en·tə'mäl·ə·jē }

economic geography [GEOGR] A branch of geography concerned with the relations of physical environment and economic conditions to the manufacture and distribution of commodities. { ¦ek·ə'näm·ik jē'äg·rə·fē }

economic geology [GEOL] 1. Application of geologic knowledge to materials usage and principles of engineering. 2. The study of metallic ore deposits. { ¦ek·ə'näm·ik jē'äl·ə·jē }

economic life [IND ENG] The number of years after which a capital good should be replaced in order to minimize the long-run annual cost of operation, repair, depreciation, and capital. Also known as project life. { ¦ek·ə'näm·ik 'līf }

economic lot size [IND ENG] The number of units of a product or item to be manufactured at each setup or purchased on each order so as to minimize the cost of purchasing or setup, and the cost of holding the average inventory over a given period, usually annual. Also known as project life. { ¦ek·ə'näm·ik 'lät ¦sīz }

economic mineral [MINERAL] Mineral of commercial value. { ¦ek·ə'näm·ik 'min·rəl }

economic mobilization [ORD] The process of preparing for and carrying out such changes in the organization and functioning of the national economy as are necessary to provide the most effective use of resources in a national emergency. { ¦ek·ə'näm·ik ¦mō·bə·lə'zā·shən }

economic order quantity [IND ENG] The number of orders required to fulfill the economic lot size. { ¦ek·ə'näm·ik 'ór·dər ¦kwän·ə·dē }

economic purchase quantity [IND ENG] The economic lot size for a purchased quantity. { ¦ek·ə'näm·ik 'pər·chəs ¦kwän·ə·dē }

economics [IND ENG] A social science that deals with production, distribution, and consumption of commodities, or wealth. { ¦ek·ə'näm·iks or¸ē·kə'näm·iks }

economic tool life [IND ENG] In metal machining, the total time, usually expressed in minutes, during which a given tool performs its required function under the most efficient cutting conditions. { ¦ek·ə'näm·ik 'tül ¸līf }

economic warfare [ORD] The defensive use in peacetime, as well as during a war, of any means by military and civilian agencies to maintain or expand the economic potential for war of a nation and its (probable) allies; and, conversely, the offensive use of any measure in peace or war to diminish or

neutralize the economic potential for war of the (likely) enemy nation and its accomplices. { ¦ek·ə'näm·ik 'wor·fer }

economic war potential [ORD] The segment of the economic capacity of a nation which can be used for purposes of conducting war. { ¦ek·ə'näm·ik 'wor pə¸ten·chəl }

economizer [ENG] A reservoir in a continuous-flow oxygen system in which oxygen exhaled by the user is collected for recirculation in the system. [MECH ENG] A forced-flow, once-through, convection-heat-transfer tube bank in which feedwater is raised in temperature on its way to the evaporating section of a steam boiler, thus lowering flue gas temperature, improving boiler efficiency, and saving fuel. { ē'kän·ə¸miz·ər }

economy [CHEM ENG] In a multiple-effect evaporation system, the total weight of water vaporized in an evaporator per unit weight of the original steam supplied. [COMPUT SCI] The ratio of the number of characters to be coded to the maximum number available with the code; for example, binary-coded decimal using 4 bits provides 16 possible characters but uses only 10 of them. { ē'kän·ə·mē }

ecophene [GEN] The range of phenotypic modifications produced by one genotype within the limits of the habitat under which the genotype is found in nature. { 'ē·kə¸fēn }

ecophenotype [ECOL] A nongenetic phenotypic modification in response to environmental conditions. { ¸ē·kō'phēn·ə¸tīp }

E core [ELECTROMAG] A transformer core made from E-shaped laminations and used in conjunction with I-shaped laminations. { 'ē ¸kor }

E corona [ASTRON] The component of the light seen from the solar corona which consists of radiation emitted from the corona itself, as opposed to scattered light. Also known as emission-line corona. { 'ē kə¸rō·nə }

ecospecies [ECOL] A group of ecotypes capable of interbreeding without loss of fertility or vigor in the offspring. { 'ē·kō¸spē·shēz }

écossais [TEXT] Satin-striped fabric using two colors-one color in lengthwise satin stripes standing out on the face, the other in the underlying plain weave, which may be barred or checked. { ¦ā·kō¦sä }

ecosystem [ECOL] A functional system which includes the organisms of a natural community together with their environment. Derived from ecological system. { 'ek·ō¸sis·təm or 'ē·kō¸sis·təm }

ecosystem mapping [ECOL] The drawing of maps that locate different ecosystems in a geographic area. { 'ek·ō¸sis·təm ¸map·iŋ }

ecotone [ECOL] A zone of intergradation between ecological communities. { 'ek·ə¸tōn }

ecotrine [ECOL] A metabolite produced by one kind of organism and utilized by another. { 'ek·ə¸trēn }

ecotype [ECOL] A subunit, race, or variety of a plant ecospecies that is restricted to one habitat; equivalent to a taxonomic subspecies. { 'ek·ə¸tīp }

ECR See electronic cash register.

ECRIS See electron cyclotron resonance source.

ECR source See electron cyclotron resonance source. { ¦ē¦sē¦är 'sors }

EC smokeless powder [MATER] An explosive powder used chiefly in blank cartridges, but also in some .22-caliber and shotgun ammunition, and formerly in fragmentation grenades. Also known as EC blank fire; EC blank powder. { ¦ē¦sē ¦smōk·ləs 'paud·ər }

ecstasy [MED] A trancelike state with loss of sensory perception and voluntary control. { 'ek·stə¸sē }

ECSW See extended channel status word.

ectasia [MED] Dilation, especially of a hollow organ. { ek'tā·zhə }

Ecterocoelia [INV ZOO] The equivalent name for Protostomia. { ¸ek·tə·rō'sēl·yə }

ectethmoid [ANAT] Either one of the lateral cellular masses of the ethmoid bone. { ek'teth¸moid }

ecthyma [MED] An inflammatory skin disease characterized by large flat pustules that ulcerate and become crusted, and are surrounded by a distinct inflammatory areola. { 'ek·thə·mə }

ectinites [PETR] One of two major groups of metamorphic rocks comprising those formed with no accession or introduction of feldspathic material. { 'ek·tə¸nīts }

ectocardia [MED] An abnormal position of the heart; it may be outside the thoracic cavity or misplaced within the thorax. { ¦ek·tō'kärd·ē·ə }

ectocommensal [ECOL] An organism living on the outer surface of the body of another organism, without affecting its host. { ¦ek·tō·kə'men·səl }

ectocornea [ANAT] The outer layer of the cornea. { ¸ek·tō'kȯr·nē·ə }

ectocyst [INV ZOO] The outer layer of the wall of a zooecium. *See* epicyst. { 'ek·tə¸sist }

ectoderm [EMBRYO] The outer germ layer of an animal embryo. Also known as epiblast. [INV ZOO] The outer layer of a diploblastic animal. { 'ek·tə¸dərm }

ectoenzyme [BIOCHEM] An enzyme which is located on the external surface of a cell. { ¦ek·tō'en¸zīm }

ectogenesis [EMBRYO] Development of an embryo or of embryonic tissue outside the body in an artificial environment. { ¸ek·tō'jen·ə·səs }

ectogony [BOT] The influence of pollination and fertilization on structures outside the embryo and endosperm; effect may be on color, chemical composition, ripening, or abscission. { ek'täg·ə·nē }

ectohumus [GEOL] An accumulation of organic matter on the soil surface with little or no mixing with mineral material. Also known as mor; raw humus. { ¦ek·tō'hyü·məs }

ectomere [EMBRYO] A blastomere that will differentiate into ectoderm. { 'ek·tə¸mir }

ectomesoblast [EMBRYO] An undifferentiated layer of embryonic cells from which arises the epiblast and mesoblast. { ¦ek·tō'me·zō¸blast }

ectomesoderm [EMBRYO] Mesoderm which is derived from ectoderm and is always mesenchymal; a type of primitive connective tissue. { ¦ek·tō'me·zō¸dərm }

ectomorph [PSYCH] A somatotype suggested by W.H. Sheldon to describe a person with a thin physique. { 'ek·tə¸mȯrf }

ectomycorrhizae [ECOL] A type of mycorrhizae composed of a fungus sheath around the outside of root tips, with individual hyphae penetrating between the cortical cells of the root to absorb photosynthates. { ¸ek·tō·mī'kȯr·ə¸zī }

ectoparasite [ECOL] A parasite that lives on the exterior of its host. { ¦ek·tō'par·ə¸sīt }

ectophagous [INV ZOO] The larval stage of a parasitic insect which is in the process of development externally on a host. { ek'täf·ə·gəs }

ectophloic siphonostele [BOT] A type of stele with pith that has the phloem only on the outside of the xylem. { ¸ek·tə'flō·ək sī'fän·ə¸stēl }

ectophyte [ECOL] A plant which lives externally on another organism. { 'ek·tə¸fīt }

ectopia [MED] A congenital or acquired positional abnormality of an organ or other part of the body. { ek'tōp·ē·ə }

ectopic expression [GEN] Phenotypic expression of a gene in a type of cell or tissue in which it is usually inactive. { ek'täp·ik iks'presh·ən }

ectopic pairing [CYTOL] Pairing between nonhomologous segments of the salivary gland chromosomes in *Drosophila*, presumably involving mainly heterochromatic regions. { ek'täp·ik 'per·iŋ }

ectopic pregnancy [MED] Embryonic development outside the uterus, usually within the Fallopian tube. { ek'täp·ik 'preg·nən·sē }

ectoplasm [CYTOL] The outer, gelled zone of the cytoplasmic ground substance in many cells. Also known as ectosarc. { 'ek·tə¸plaz·əm }

Ectoprocta [INV ZOO] A subphylum of colonial bryozoans having eucoelomate visceral cavities and the anus opening outside the circlet of tentacles. { ek·tō'präk·tə }

ectopterygoid [VERT ZOO] A membrane bone located ventrally on the skull, situated behind the palate and extending to the quadrate; found in some fishes and reptiles. { ¦ek·tō'ter·ə¸gȯid }

ectosarc *See* ectoplasm. { 'ek·tə¸särk }

ectosome [INV ZOO] The outer, cortical layer of a sponge. { 'ek·tə¸sōm }

ectostosis [PHYSIO] Formation of bone immediately beneath the perichondrium and surrounding and replacing underlying cartilage. { ¦ek·tə'stō·səs }

ectosymbiont [ECOL] A symbiont that lives on the surface of or is physically separated from its host. { ¦ek·tō'sim·bē¸änt }

ectotherm [PHYSIO] An animal that obtains most of its heat from the environment and therefore has a body temperature very close to that of its environment. { 'ek·tə¸thərm }

Ectothiorhodospira [MICROBIO] A genus of bacteria in the family Chromatiaceae; cells are spiral to slightly bent rods, are motile, contain bacteriochlorophyll *a* on lamellar stock membranes, and produce and deposit sulfur as globules outside the cells. { ¦ek·tō¸thī·ə¸rō'däs·pə·rə }

ectotrophic [BIOL] Obtaining nourishment from outside; applied to certain parasitic fungi that live on and surround the roots of the host plant. { ¦ek·tə'träf·ik }

ectozoa [ECOL] Animals which live externally on other organisms. { ¸ek·tə'zō·ə }

Ectrephidae [INV ZOO] An equivalent name for Ptinidae. { ek'tref·ə¸dē }

ectrodactylia [MED] Congenital absence of any of the fingers or toes or parts of them. { ¸ek·trō·dak'til·ē·ə }

ectromelia [MED] A congenital absence or an anomaly of one or more limbs. { ¸ek·tra'mē·lē·ə }

ectromelia virus [VIROL] A member of subgroup I of the poxvirus group; causes mousepox. { ¸ek·tra'mē·lē·ə 'vī·rəs }

eczema [MED] Any skin disorder characterized by redness, thickening, oozing from blisters or papules, and occasional formation of fissures and crusts. { 'ek·sə·mə }

eczematoid reaction [MED] A dermal and epidermal inflammatory response characterized by erythema, edema, vesiculation, and exudation in the acute stage, and by erythema, edema, thickening of the epidermis, and scaling in the chronic stage. { ek'zē·mə¸tȯid rē¸ak·shən }

ED *See* electronic dummy.

ED₅₀ *See* effective dose 50.

edaphic community [ECOL] A plant community that results from or is influenced by soil factors such as salinity and drainage. { ē'daf·ik kə'myün·əd·ē }

edaphon [BIOL] Flora and fauna in soils. { 'ed·ə¸fän }

Edaphosuria [PALEON] A suborder of extinct, lowland, terrestrial, herbivorous reptiles in the order Pelycosauria. { ¸ed·ə·fō'sȯr·ē·ə }

Eddington limit [ASTROPHYS] A limit on the radiation emitted by a star above which the star becomes unstable. { 'ed·iŋ·tən ¸lim·ət }

Eddington's model [ASTRON] A model of a star in which energy is transported by radiation throughout the star and the ratio of radiation pressure to gas pressure is assumed to be constant. { 'ed·iŋ·tənz ¸mäd·əl }

eddy [FL MECH] A vortexlike motion of a fluid running contrary to the main current. { 'ed·ē }

eddy coefficient *See* exchange coefficient. { 'ed·ē ¸kō·i¸fish·ənt }

eddy conduction *See* eddy heat conduction. { 'ed·ē kən¸dək·shən }

eddy conductivity [THERMO] The exchange coefficient for eddy heat conduction. { 'ed·ē ¸kän¸dək'tiv·əd·ē }

eddy correlation [METEOROL] A method of studying the effects of sea surface on the air above it by measuring simultaneous fluctuations of the horizontal and vertical components of the airflow from the mean. { 'ed·ē ¸kä·rə¸lā·shən }

eddy current [ELECTROMAG] An electric current induced within the body of a conductor when that conductor either moves through a nonuniform magnetic field or is in a region where there is a change in magnetic flux. Also known as Foucault current. { 'ed·ē ¸kə·rənt }

eddy-current brake [MECH ENG] A control device or dynamometer for regulating rotational speed, as of flywheels, in which energy is converted by eddy currents into heat. { 'ed·ē ¸kə·rənt ¸brāk }

eddy-current clutch [MECH ENG] A type of electromagnetic clutch in which torque is transmitted by means of eddy currents induced by a magnetic field set up by a coil carrying direct current in one rotating member. { 'ed·ē ¸kə·rənt ¸kləch }

eddy-current damper [AERO ENG] A device used to damp nutation and other unwanted vibration in spacecraft, based on the principle that eddy currents induced in conducting material by motion relative to magnets tend to counteract that motion. { 'ed·ē ¸kə·rənt ¸dam·pər }

eddy-current heating *See* induction heating. { 'ed·ē ¸kə·rənt ¸hēd·iŋ }

ECTOPHLOIC SIPHONOSTELE

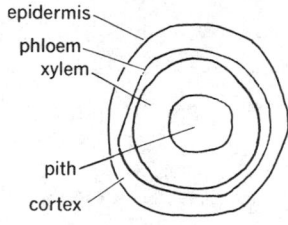

epidermis
phloem
xylem
pith
cortex

Cross section of ectophloic siphonostele.

EDDY CURRENT

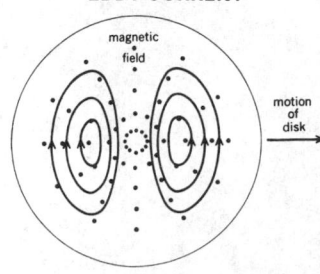

magnetic field

motion of disk

Eddy currents which are induced in a disk moving through a nonuniform magnetic field.

eddy-current loss [ELECTROMAG] Energy loss due to undesired eddy currents circulating in a magnetic core. { 'ed·ē ¦kə·rənt ¸lȯs }

eddy-current sensor [ENG] A proximity sensor which uses an alternating magnetic field to create eddy currents in nearby objects, and then the currents are used to detect the presence of the objects. { 'ed·ē ¸kə·rənt 'sen·sər }

eddy-current tachometer [AERO ENG] A type of tachometer in which a rotating permanent magnet induces currents in a spring-mounted metal cylinder; the resulting torque rotates the cylinder and moves its attached pointer in proportion to the speed of the rotating shaft. Also known as drag-type tachometer. { 'ed·ē ¸kə·rənt tə'käm·əd·ər }

eddy-current test [ELECTROMAG] A nondestructive test in which the change of impedance of a test coil brought close to a conducting specimen indicates the eddy currents induced by the coil, and thereby indicates certain properties or defects of the specimen. { 'ed·ē ¸kə·rənt ¸test }

eddy diffusion [FL MECH] Diffusion which occurs in turbulent flow, by the rapid process of mixing of the swirling eddies of fluid. Also known as turbulent diffusion. { 'ed·ē də¸fyü·zhən }

eddy-diffusion coefficient See eddy diffusivity. { 'ed·ē də¸fyü·zhən ¸kō·i¸fish·ənt }

eddy diffusivity [FL MECH] The exchange coefficient for the diffusion of a conservative property by eddies in a turbulent flow. Also known as coefficient of eddy diffusion; eddy-diffusion coefficient. { 'ed·ē di¸fyü'siv·əd·ē }

eddy flux [FL MECH] The rate of transport (or flux) of fluid properties such as momentum, mass heat, or suspended matter by means of eddies in a turbulent motion; the rate of turbulent exchange. Also known as moisture flux; turbulent flux. { 'ed·ē ¸fləks }

eddy heat conduction [THERMO] The transfer of heat by means of eddies in turbulent flow, treated analogously to molecular conduction. Also known as eddy heat flux; eddy conduction. { 'ed·ē ¸hēt kən'dək·shən }

eddy heat flux See eddy heat conduction. { 'ed·ē ¸hēt ¸fləks }

eddy kinetic energy [FL MECH] The kinetic energy of that component of fluid flow which represents a departure from the average kinetic energy of the fluid, the mode of averaging depending on the particular problem. Also known as turbulence energy. { 'ed·ē kə'ned·ik¸en·ər·jē }

eddy mill See pothole. { 'ed·ē ¸mil }

eddy resistance [FL MECH] Resistance or drag of a ship resulting from eddies that are shed from the hull or appendages of the ship and carry away energy. { 'ed·ē ri'zis·təns }

eddy spectrum [FL MECH] 1. The distribution of frequencies of rotation of eddies in a turbulent flow, or of those eddies having some range of sizes. 2. The distribution of kinetic energy among eddies with various frequencies or sizes. { 'ed·ē 'spek·trəm }

eddy stress See Reynolds stress. { 'ed·ē ¸stres }

eddy velocity [FL MECH] The difference between the mean velocity of fluid flow and the instantaneous velocity at a point. Also known as fluctuation velocity. { 'ed·ē və¸läs·əd·ē }

eddy viscosity [FL MECH] The turbulent transfer of momentum by eddies giving rise to an internal fluid friction, in a manner analogous to the action of molecular viscosity in laminar flow, but taking place on a much larger scale. { 'ed·ē vi'skäs·əd·ē }

eddy viscosity model [FL MECH] A model of the Reynolds stresses in turbulent flow which is based on the idea that turbulent mixing, analogous to molecular mixing, is governed by an effective viscosity (the eddy viscosity) which is not a property of the fluid but a consequence of the local state of turbulence. { 'ed·ē vi'skäs·əd·ē ¸mäd·əl }

Edeleanu process [CHEM ENG] A process for removal of compounds of sulfur from petroleum fractions by an extraction procedure utilizing liquid sulfur dioxide, or liquid sulfur dioxide and benzene. { ə¸del·ē'ä·nü ¸präs·əs }

EDEL room [ENG ACOUS] A control room in a sound-recording studio in which reflective or diffusive surfaces are placed near the loudspeaker and above the mixing console, while the rear wall behind the mixer is made absorptive. Derived from LEDE room (by reverse spelling). { 'ed·əl ¸rüm or ¦ē¦dē¦el ¸rüm }

edema [MED] An excessive accumulation of fluid in the cells, tissue spaces, or body cavities due to a disturbance in the fluid exchange mechanism. Also known as dropsy. { ə'dē·mə }

Edenian [GEOL] Lower Cincinnatian geologic stage in North America, above the Mohawkian and below Maysvillian. { ē'dēn·ē·ən }

Edentata [VERT ZOO] An order of mammals characterized by the absence of teeth or the presence of simple prismatic, unspecialized teeth with no enamel. { ¸ē¸den'tä·də }

edentate [VERT ZOO] 1. Lacking teeth. 2. Any member of the Edentata. { ē'den¸tāt }

edentulous [VERT ZOO] Having no teeth; especially, having lost teeth that were present. { ē'den·chə·ləs }

EDFA See erbium-doped fiber amplifier. { 'ed¸fä or ¦ē¦dē¦ef'ā }

edge [MATH] 1. A line along which two plane faces of a solid intersect. 2. A line segment connecting nodes or vertices in a graph (a geometric representation of the relation among situations). 3. The edge of a half plane is the line that bounds it. Also known as arc. { ej }

edgeboard connector See card-edge connector. { 'ej¸bȯrd kə¸nek·tər }

edge-bridging ligand [ORG CHEM] A ligand that forms a bridge over one edge of the polyhedron of a metal cluster structure. { 'ej ¸brij·iŋ 'lī·gənd }

edge connector [ELECTR] A row of etched lines on the edge of a printed circuit board that is inserted into a slot to establish a connection with another printed circuit board. { 'ej kə¸nek·tər }

edge cover [MATH] A set of edges in a graph such that every vertex of positive degree is the vertex of at least one of the edges in this set. { 'ej ¸kəv·ər }

edge-covering number [MATH] For a graph, the sum of the number of edges in a minimum edge cover and the number of isolated vertices. { 'ej ¸kəv·ər·iŋ ¸nəm·bər }

edge dislocation [CRYSTAL] A dislocation which may be regarded as the result of inserting an extra plane of atoms, terminating along the line of the dislocation. Also known as Taylor-Orowan dislocation. { 'ej ¸dis·lō¸kā·shən }

edge domination number [MATH] For a graph, the smallest possible number of edges in a dominating edge set. { ¦ej ¸däm·ə'nā·shən ¸nəm·bər }

edge effect [ECOL] The influence of adjacent plant communities on the number of animal species present in the direct vicinity. [ELEC] An outward-curving distortion of lines of force near the edges of two parallel metal plates that form a capacitor. { 'ej i¸fekt }

edge excitation [CRYO] An excitation of a droplet of incompressible quantum Hall liquid in which a surface wave propagates along the edge of the droplet in the same direction that electrons drift along the edge, as determined by the direction of the magnetic field. { 'ej ¸ek·sī'tā·shən }

edge focusing [ELECTROMAG] Axial focusing of a stream of ions which occurs when it crosses a fringe magnetic field obliquely; used in mass spectrometers and cyclotrons. { 'ej ¸fō·kəs·iŋ }

edge fog [GRAPHICS] The light fog which appears along the edge of roll film, generally from exposure during loading or unloading. { 'ej ¸fäg }

edge grain [MATER] The grain pattern produced when soft wood is cut so that the tree's annular rings form an angle of more than 45° with the board's surface. { 'ej ¸grān }

edge independence number [MATH] For a graph, the largest possible number of edges in a matching. { ¦ej ¸in·də'pen·dəns ¸nəm·bər }

edge-induced subgraph [MATH] A subgraph whose vertices consist of all the vertices in the original graph that are incident on at least one edge in the subgraph. { ¦ej in¸düst 'səb¸graf }

edge joint [MET] A joint between the edges of welded members which are essentially parallel to each other. { 'ej ¸jȯint }

edge line [MAP] In cartography, a heavy line on a relief map that indicates a sharp change of slope. { 'ej ¸līn }

edge of regression [MATH] The curve swept out by the characteristic point of a one-parameter family of surfaces. { ¦ej əv rē'gresh·ən }

edge printing [GRAPHICS] A preprinted identification along the edge of film. { 'ej ¸print·iŋ }

edger [MET] The part of a forging die which portions out the quantity of metal needed for shaping. { 'ej·ər }

edge runner See Chile mill. { 'ej ¸rən·ər }

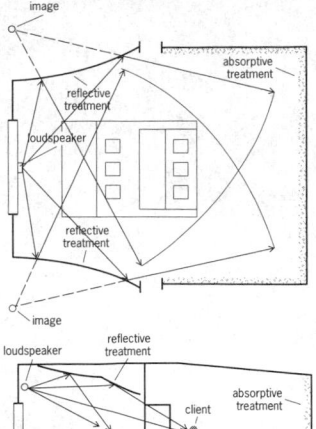

EDEL room with an absorbent rear wall and a live front end as in conventional listening locales.

EDENTATA

The giant anteater (*Myrmecophaga tridactyla*).

EDGE EXCITATION

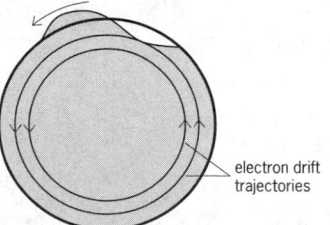

An edge wave on a quantum Hall droplet, which propagates in only one direction.

edge tones [ACOUS] Tones produced when an air jet of sufficient speed is split by a sharp edge. { 'ej ˌtōnz }

edge water [GEOL] In reservoir structures, the subsurface water that surrounds the gas or oil. { 'ej ˌwȯd·ər }

edge wave [OCEANOGR] An ocean wave moving parallel to the coast, with crests normal to the coastline; maximum amplitude is at shore, with amplitude falling off exponentially farther from shore. { 'ej ˌwāv }

edge well [PETRO ENG] A well positioned at the edge of oil or gas accumulations or at the edge of a leased reservoir. { 'ej ˌwel }

Edgeworth-Kuiper Belt See Kuiper Belt. { ¦ej¸wərth 'kī·pər ˌbelt }

edging [MET] Controlling the plate width or the edge shape during rolling operations. { 'ej·iŋ }

Ediacaran fauna [PALEON] The oldest known assemblage of fossil remains of soft-bodied marine animals; first discovered in the Ediacara Hills, Australia. { ˌēd·ē·ə'kar·ən 'fȯn·ə }

edible oil [FOOD ENG] A liquid fat that is capable of being eaten as a food or food accessory. { ¦ed·ə·bəl 'ȯil }

edingtonite [MINERAL] BaAl$_2$Si$_3$O$_{10}$·4H$_2$O Gray zeolite mineral that forms rhombic crystals; sometimes contains large amounts of calcium. { 'ed·iŋ·tə͵nīt }

Edison battery [ELEC] A storage battery composed of cells having nickel and iron in an alkaline solution. Also known as nickel-iron battery. { ¦ed·ə·sən ¦bad·ə·rē }

Edison distribution system [ELEC] Three-wire direct-current distribution system, usually 120 to 240 volts, for combined light and power service from a single set of mains. { ¦ed·ə·sən ˌdis·trə'byü·shən ˌsis·təm }

Edison effect See thermionic emission. { 'ed·ə·sən i͵fekt }

E display [ELECTR] A rectangular radar display in which targets appear as blips with distance indicated by the horizontal coordinate, and elevation by the vertical coordinate. { 'ē di͵splā }

edit [COMPUT SCI] **1.** To modify the form or format of an output or input by inserting or deleting characters such as page numbers or decimal points. **2.** A computer instruction directing that this step be performed. { 'ed·ət }

edit capability [COMPUT SCI] The degree of sophistication available to the programmer to modify his or her statements while in the time-sharing mode. { 'ed·ət ˌkāp·ə͵bil·əd·ē }

edit check [COMPUT SCI] A program instruction or subroutine that tests the validity of input in a data entry program. Also known as edit test. { 'ed·ət ˌchek }

editing See proofreading. { 'ed·əd·iŋ }

edit mask [COMPUT SCI] The receiving word through which a source word is filtered, allowing for the suppression of leading zeroes, the insertion of floating dollar signs and decimal points, and other such formatting. { 'ed·ət ˌmask }

edit mode [COMPUT SCI] A software mode of operation in which previously entered text or data can be modified or replaced. { 'ed·ət ˌmōd }

editor program [COMPUT SCI] A special program by means of which a user can easily perform corrections, insertions, modifications, or deletions in an existing program or data file. { 'ed·ə·tər ˌprō·grəm }

edit test See edit check. { 'ed·ət ˌtest }

EDM See electron discharge machining.

Edman degradation technique [BIOCHEM] In protein analysis, an approach to amino-end-group determination involving the use of a reagent, phenylisothiocyanate, that can be applied to the liberation of a derivative of the amino-terminal residue without hydrolysis of the remainder of the peptide chain. { 'ed·mən ˌdeg·rə'dā·shən tek͵nēk }

EDO RAM See extended data-out random access memory. { ˌā·dō 'ram or ¦ē¦dē¦ō }

EDP See electronic data processing.

EDP center See electronic data-processing center. { ¦ē¦dē'pē ˌsen·tər }

Edrioasteroidea [PALEON] A class of extinct Echinozoa having ambulacral radial areas bordered by tube feet, and the mouth and anus located on the upper side of the theca. { ˌed·rē·ō͵as·tə'rȯid·ē·ə }

Edser and Butler's bands [OPTICS] Dark bands at intervals of equal frequency in a spectrum of light which has passed through a thin plate of transparent material with parallel sides. { 'ed·sər ən 'bət·lərz ˌbanz }

EDTA See ethylenediaminetetraacetic acid.

EDTC See S-ethyl-N,N-dipropylthiocarbamate.

educational age [PSYCH] The average achievement of a pupil or student in school subjects based on average performance for a given chronological age as measured by standard educational tests. { ˌej·ə'kā·shən·əl ˌāj }

educational psychology [PSYCH] A field of psychology that deals with the psychological aspects of teaching and formal learning processes. { ˌej·ə'kā·shən·əl sī͵kä·lə·jē }

eduction [MOL BIO] Loss of host genetic material when the plasmid that had been integrated into the host chromosome exits. { ē'dək·shən }

eductor [ENG] **1.** An ejectorlike device for mixing two fluids. **2.** See ejector. { ē'dək·tər }

eductor pump [MIN ENG] A pump which removes slurried material from a hydraulically disseminated subsurface ore matrix. { ē'dək·tər ˌpəmp }

edulcorate [COMPUT SCI] To eliminate irrelevant data from a data file. { ē'dəl·kə͵rāt }

EDVAC [COMPUT SCI] The first stored program computer, built in 1952. Derived from electron discrete variable automatic compiler. { 'ed͵vak }

Edwardsiella [MICROBIO] A genus of bacteria in the family Enterobacteriaceae; motile rods that produce hydrogen sulfide from TSI agar. { e¦dwärd·zē'el·ə }

Edwards' syndrome See trisomy 18 syndrome. { 'ed·wərdz͵sin͵drōm }

EDXD See energy-dispersive x-ray diffraction.

EEG See electroencephalogram.

eel [VERT ZOO] The common name for a number of unrelated fishes included in the orders Anguilliformes and Cypriniformes; all have an elongate, serpentine body. { ēl }

eel grass See tape grass. { ēl ˌgras }

EELS See electron energy loss spectroscopy.

EEP See electroencephalophone.

EEPROM See electrically erasable programmable read-only memory. { ¦ē'ē͵präm }

EER See equal error rate.

eff See efficiency.

effective address [COMPUT SCI] The address that is obtained by applying any specified indexing or indirect addressing rules to the specified address; the effective address is then used to identify the current operand. { ə¦fek·tiv 'a͵dres }

effective air path [NAV] A straight line on a navigation chart connecting two air positions, commonly used between the air positions of two pressure soundings in order to determine effective true airspeed between the two soundings. { ə¦fek·tiv 'er ˌpath }

effective ampere [ELEC] The amount of alternating current flowing through a resistance that produces heat at the same average rate as 1 ampere of direct current flowing in the same resistance. { ə¦fek·tiv 'am͵pir }

effective angle of attack [AERO ENG] That part of a given angle of attack that lies between the chord of an airfoil and a line representing the resultant velocity of the disturbed airflow. { ə¦fek·tiv ¦aŋ·gəl əv ə'tak }

effective antenna length [ELECTROMAG] Electrical length of an antenna, as distinguished from its physical length. { ə¦fek·tiv an'ten·ə ˌleŋkth }

effective aperture [OPTICS] The diameter of the image of the aperture stop of an optical system, as viewed from the object. { ə¦fek·tiv 'ap·ə·chər }

effective area [CHEM ENG] Absolute or cross-sectional area of process media involved in the process, such as the actual area of filter media through which a fluid passes, or the available surface area of absorbent contacted by a gas or liquid. [ELECTROMAG] Of an antenna in any specified direction, the square of the wavelength multiplied by the power gain (or directive gain) in that direction and divided by 4π (12.57). { ə¦fek·tiv 'er·ē·ə }

effective atmosphere [GEOPHYS] **1.** That part of the atmosphere which effectively influences a particular process or motion, its outer limits varying according to the terms of the process or motion considered. **2.** See optically effective atmosphere. { ə¦fek·tiv 'at·mə͵sfir }

effective bandwidth [ELECTR] The bandwidth of an assumed rectangular band-pass having the same transfer ratio at a reference frequency as a given actual band-pass filter, and passing the same mean-square value of a hypothetical current

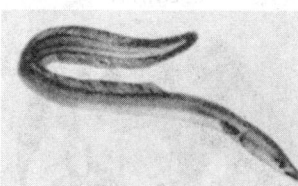

EEL

American eel, showing typical fin structure.

having even distribution of energy throughout that bandwidth. { əˈfek·tiv ˈband·width }

effective capacitance [ELEC] Total capacitance existing between any two given points of an electric circuit. { əˈfek·tiv kəˈpas·əd·əns }

effective center [ENG ACOUS] In a sonar projector, the point where lines coincident with the direction of propagation, as observed at different points some distance from the projector, apparently intersect. Also known as apparent source. { əˈfek·tiv ˈsen·tər }

effective confusion area [ENG] Amount of chaff whose radar cross-sectional area equals the radar cross-sectional area of the particular aircraft at a particular frequency. { əˈfek·tiv kənˈfyü·zhən ˌer·ē·ə }

effective current [ELEC] The value of alternating current that will give the same heating effect as the corresponding value of direct current. Also known as root-mean-square current. { əˈfek·tiv ˈkə·rənt }

effective decline rate [PETRO ENG] The drop in oil or gas production rate over a period of time; equal to unity (1 month or 1 year) divided by the production rate at the beginning of the period. { əˈfek·tiv dəˈklīn ˌrāt }

effective discharge area [DES ENG] A nominal or calculated area of flow through a pressure relief valve for use in flow formulas to determine valve capacity. { əˈfek·tiv ˈdisˌchärj ˌer·ē·ə }

effective dose 50 [PHARM] The amount of a drug required to produce a response in 50% of the subjects to whom the drug is given. Abbreviated ED$_{50}$. Also known as median effective dose. { əˈfek·tiv ˈdōs ˈfifˌtē }

effective earth radius [COMMUN] A radius value used in place of the geometric radius to correct for atmospheric refraction in estimating ranges of antennas when the index of refraction in the atmosphere changes linearly with height; under conditions of standard refraction it is $^4/_3$ the geometric radius. Also known as effective radius of the earth. { əˈfek·tiv ˈərth ˌrād·ē·əs }

effective energy [OPTICS] The energy of a quantum of a beam of monochromatic radiation that is absorbed or scattered by a given medium to the same extent as a given beam of polychromatic radiation. { əˈfek·tiv ˈen·ər·jē }

effective exhaust velocity [AERO ENG] A fictitious exhaust velocity that yields the observed value of jet thrust in calculations. { əˈfek·tiv igˈsȯst vəˌläs·əd·ē }

effective facsimile band [COMMUN] Frequency band of a facsimile signal wave equal in width to that between zero frequency and maximum keying frequency. { əˈfek·tiv fakˈsim·ə·lē ˌband }

effective field intensity [ELECTROMAG] Root-mean-square value of the inverse distance fields at a distance of 1 mile (1.6 kilometers) from the transmitting antenna in all directions in the horizontal plane. { əˈfek·tiv ˈfēld inˌten·səd·ē }

effective firing time [ORD] The period of time during which an aircraft can deliver effective fire at a moving or stationary target. { əˈfek·tiv ˈfīr·iŋ ˌtīm }

effective force See inertial force. { əˈfek·tiv ˈfȯrs }

effective fragment [ORD] In terminal ballistics, a fragment whose mass, velocity, and form, upon impact with the target are such as to enable the fragment to accomplish the desired effect. { əˈfek·tiv ˈfrag·mənt }

effective gun bore line [MECH] The line which a projectile should follow when the muzzle velocity of the antiaircraft gun is vectorially added to the aircraft velocity. { əˈfek·tiv ˈgən ˌbȯr ˌlīn }

effective gust velocity [METEOROL] The vertical component of the velocity of a sharp-edged gust that would produce a given acceleration on a particular airplane flown in level flight at the design cruising speed of the aircraft and at a given air density. { əˈfek·tiv ˈgəst vəˌläs·əd·ē }

effective half-life [NUCLEO] The half-life of a radioisotope in a biological organism, resulting from a combination of radioactive decay and biological elimination. { əˈfek·tiv ˈhafˌlīf }

effective height [ELECTROMAG] The height of the center of radiation of a transmitting antenna above the effective ground level. { əˈfek·tiv ˈhīt }

effective horizon [COMMUN] A horizon whose distance at a given height above sea level is the distance to the horizon of a fictitious earth, having a radius $^4/_3$ times the earth's true

radius; used to estimate ranges of antennas, taking atmospheric refraction into account. { əˈfek·tiv həˈrīz·ən }

effective horsepower [NAV ARCH] The power necessary to overcome the resistance of water to motion of a ship towed at a given speed, measured in horsepower. Abbreviated ehp. Also known as towrope horsepower. { əˈfek·tiv ˈhȯrsˌpau̇·ər }

effective instruction [COMPUT SCI] The computer instruction that results from changing a basic instruction during program modification. Also known as actual instruction. { əˈfek·tiv inˈstrək·shən }

effective isotropic radiated power [COMMUN] A measure of the strength of the signal leaving a satellite antenna in a particular direction, equal to the product of the power supplied to the satellite transmit antenna and its gain in that direction. Abbreviated eirp. { iˌfek·tiv ˌī·səˌträp·ik ˌrād·ē·ād·əd ˈpau̇·ər }

effective launcher line [MECH] The line along which the aircraft rocket would go if it were not affected by gravity. { əˈfek·tiv ˈlȯn·chər ˌlīn }

effective length [NAV ARCH] Mean length of that portion of the ship's hull below the waterline. { əˈfek·tiv ˈleŋkth }

effective lethal phase [GEN] The developmental stage at which a lethal gene generally causes death of the organism carrying it. { əˈfek·tiv ˈlē·thəl ˈfāz }

effectively computable function [MATH] Any function that can be computed on the natural numbers by means of an effective procedure. { əˈfek·tiv·lē kəmˈpyüd·ə·bəl ˈfəŋk·shən }

effectively grounded [ELEC] Grounded through a connection of sufficiently low impedances (inherent or intentionally added) so that fault grounds which may occur cannot build up voltages dangerous to connected personnel or other equipment. { əˈfek·tiv·lē ˈgrau̇nd·əd }

effective magnetic length [ELECTROMAG] The distance between the effective magnetic poles of a magnet. Also known as equivalent magnetic length. { əˈfek·tiv magˈned·ik ˈleŋkth }

effective mass [SOLID STATE] A parameter with the dimensions of mass that is assigned to electrons in a solid; in the presence of an external electromagnetic field the electrons behave in many respects as if they were free, but with a mass equal to this parameter rather than the true mass. { əˈfek·tiv ˈmas }

effective molecular diameter [PHYS CHEM] The general extent of the electron cloud surrounding a gas molecule as calculated in any of several ways. { əˈfek·tiv məˌlek·yə·lər dīˈam·əd·ər }

effective molecular weight [PETRO ENG] Empirical relationship of oil graphed against API (American Petroleum Institute) gravity to give the effective (pseudoaverage) molecular weight of the oil for reservoir calculations. { əˈfek·tiv məˌlek·yə·lər ˈwāt }

effective multiplication factor [NUCLEO] The multiplication factor of an actual reactor, in which there is leakage of neutrons. { əˈfek·tiv ˌməl·tə·pləˈkā·shən ˌfak·tər }

effectiveness level [COMPUT SCI] A measure of the effectiveness of data-processing equipment, equal to the ratio of the operational use time to the total performance period, expressed as a percentage. Also known as average effectiveness level. { əˈfek·tiv·nəs ˌlev·əl }

effective percentage modulation [COMMUN] For a single sinusoidal input component, the ratio of the peak value of the fundamental component of the envelope to the average amplitude of the modulated wave expressed in percent. { əˈfek·tiv pərˈsent·ij ˌmäj·əˈlā·shən }

effective permeability [PHYS CHEM] The observed permeability exhibited by a porous medium to one fluid phase when there is physical interaction between this phase and other fluid phases present. { əˈfek·tiv pərˌmē·əˈbil·əd·ē }

effective pitch [AERO ENG] The distance traveled by an airplane along its flight path for one complete turn of the propeller. { əˈfek·tiv ˈpich }

effective porosity [GEOL] A property of earth containing interconnecting interstices, expressed as a percent of bulk volume occupied by the interstices. { əˈfek·tiv pəˈräs·əd·ē }

effective precipitable water [METEOROL] That part of the precipitable water which, in theory, can actually fall as precipitation. { əˈfek·tiv prəˈsip·əd·ə·bəl ˈwȯd·ər }

effective precipitation [HYD] **1.** The part of precipitation that reaches stream channels as runoff. Also known as effective rainfall. **2.** In irrigation, the portion of the precipitation which remains in the soil and is available for consumptive use. { əˈfek·tiv prəˌsip·əˈtā·shən }

effective pressure *See* effective stress. { əˈfek·tiv ˈpresh·ər }

effective procedure [MATH] A procedure or process determined by a finite list of precise instructions. { iˈfek·div prəˈsē·jər }

effective radiated power [ELECTROMAG] The product of antenna input power and antenna power gain, expressed in kilowatts. Abbreviated ERP. { əˈfek·tiv ˌrād·ē·ˌād·əd ˈpaủ·ər }

effective radiation *See* effective terrestrial radiation. { əˈfek·tiv ˌrād·ē·ˈā·shən }

effective radius [ASTRON] The distance from the center of an external galaxy within which half its luminosity is included. [ORD] The radius of a circle having its center at the center of impact of a nuclear weapon, and within which a specified dosage of radiation is equaled or exceeded within a given period of time. { əˈfek·tiv ˈrād·ē·əs }

effective radius of the earth *See* effective earth radius. { əˈfek·tiv ˈrād·ē·əs əv the ˈərth }

effective rainfall *See* effective precipitation. { əˈfek·tiv ˈrānˌfȯl }

effective rake [MECH ENG] The angular relationship between the plane of the tooth face of the cutter and the line through the tooth point measured in the direction of chip flow. { əˈfek·tiv ˈrāk }

effective range [ORD] Distance at which a weapon may be expected to fire accurately to inflict damage or casualties. { əˈfek·tiv ˈrānj }

effective resistance *See* high-frequency resistance. { əˈfek·tiv riˈzis·təns }

effective snowmelt [HYD] The part of snowmelt that reaches stream channels as runoff. { əˈfek·tiv ˈsnōˌmelt }

effective sound pressure [ACOUS] The root-mean-square value of the instantaneous sound pressure at a point during a complete cycle, expressed in dynes per square centimeter. Also known as root-mean-square sound pressure; sound pressure. { əˈfek·tiv ˈsaủnd ˌpresh·ər }

effective speed [COMPUT SCI] The actual speed that a computer system can sustain over a period of time when the time devoted to various control, error-detection, and other overhead activities is taken into account. { əˈfek·tiv ˈspēd }

effective stress [GEOL] The average normal force per unit area transmitted directly from particle to particle of a rock or soil mass. Also known as effective pressure; intergranular pressure. { əˈfek·tiv ˈstres }

effective surface [ENG] In a heat exchanger, a surface that actively transfers heat. { əˈfek·tiv ˈsər·fəs }

effective temperature [ASTROPHYS] A measure of the temperature of a star, deduced by means of the Stefan-Boltzmann law, from the total energy that is emitted per unit area. [METEOROL] The temperature at which motionless, saturated air would induce, in a sedentary worker wearing ordinary indoor clothing, the same sensation of comfort as that induced by the actual conditions of temperature, humidity, and air movement. [STAT MECH] That temperature which can be inserted in the Boltzmann distribution formula to describe the relative populations of two energy levels which may or may not be in thermal equilibrium. { əˈfek·tiv ˈtem·prə·chər }

effective terrestrial radiation [GEOPHYS] The amount by which outgoing infrared terrestrial radiation of the earth's surface exceeds downcoming infrared counterradiation from the sky. Also known as effective radiation; nocturnal radiation. { əˈfek·tiv təˌres·trē·əl ˌrā·dē·ˈā·shən }

effective thermal resistance [ELECTR] Of a semiconductor device, the effective temperature rise per unit power dissipation of a designated junction above the temperature of a stated external reference point under conditions of thermal equilibrium. Also known as thermal resistance. { əˈfek·tiv ˈthər·məl riˈzis·təns }

effective thrust [AERO ENG] In a rocket motor or engine, the theoretical thrust less the effects of incomplete combustion and friction flow in the nozzle. { əˈfek·tiv ˈthrəst }

effective time [COMPUT SCI] The time during which computer equipment is in actual use and produces useful results. { əˈfek·tiv ˈtīm }

effective transformation group [MATH] A transformation group in which the identity element is the only element to leave all points fixed. { əˈfek·tiv ˌtranz·fərˈmā·shən ˌgrüp }

effective true airspeed [NAV] The effective air distance divided by the elapsed time between two pressure soundings. { əˈfek·tiv ˈtrü ˈerˌspēd }

effective value *See* root-mean-square value. { əˈfek·tiv ˈval·yü }

effector [BIOCHEM] An activator of an allosteric enzyme. [CONT SYS] A motor, solenoid, or hydraulic piston that turns commands to a teleoperator into specific manipulatory actions. [PHYSIO] A structure that is sensitive to a stimulus and causes an organism or part of an organism to react to the stimulus, either positively or negatively. { əˈfek·tər }

effector organ [PHYSIO] Any muscle or gland that mediates overt behavior, that is, movement or secretion. { əˈfek·tər ˌȯr·gən }

effector system [PHYSIO] A system of effector organs in the animal body. { əˌfek·tər ˈsis·təm }

efferent [PHYSIO] Carrying or conducting away, as the duct of an exocrine gland or a nerve. { ˈef·ə·rənt }

effervescence [CHEM] The bubbling of a solution of an element or chemical compound as the result of the emission of gas without the application of heat; for example, the escape of carbon dioxide from carbonated water. { ˌef·ərˈves·əns }

efficiency Abbreviated eff. [CHEM] In an ion-exchange system, a measurement of the effectiveness of a system expressed as the amount of regenerant required to remove a given unit of adsorbed material. [ENG] **1.** Measure of the degree of heat output per unit of fuel when all available oxidizable materials in the fuel have been burned. **2.** Ratio of useful energy provided by a dynamic system to the energy supplied to it during a specific period of operation. [NUCLEO] The probability that a count will be produced in a counter tube by a specified particle or quantum incident. [PHYS] The ratio, usually expressed as a percentage, of the useful power output to the power input of a device. [STAT] **1.** An estimator is more efficient than another if it has a smaller variance. **2.** An experimental design is more efficient than another if the same level of precision can be obtained in less time or with less cost. [THERMO] The ratio of the work done by a heat engine to the heat energy absorbed by it. Also known as thermal efficiency. { əˈfish·ən·sē }

efficiency expert [IND ENG] An individual who analyzes procedures, productivity, and jobs in order to recommend methods for achieving maximum utilization of resources and equipment. { əˈfish·ən·sē ˌek·spərt }

efficient estimator [STAT] A statistical estimator that has minimum variance. { əˈfish·ənt ˈes·təˌmād·ər }

efflorescence [BOT] The period or process of flowering. [CHEM] The property of hydrated crystals to lose water of hydration and crumble when exposed to air. [MATER] A crust of salts, usually white, that forms on the surface of stone, brick, plaster, or mortar because of leaching of free alkalies from adjacent concrete or mortar. [MINERAL] A whitish powder, consisting of one or several minerals produced as an encrustation on the surface of a rock in an arid region. Also known as bloom. { ˌef·ləˈres·əns }

effluent [CHEM ENG] *See* discharge liquor. [CIV ENG] The liquid waste of sewage and industrial processing. [HYD] **1.** Flowing outward or away from. **2.** Liquid which flows away from a containing space or a main waterway. { əˈflü·ənt }

effluent stream [HYD] A stream that is fed by seeping groundwater. { əˈflü·ənt ˈstrēm }

effluent weir [CIV ENG] A dam at the outflow end of a watercourse. { əˈflü·ənt ˈwer }

effluvium [IND ENG] By-products of food and chemical processes, in the form of wastes. { əˈflü·vē·əm }

efflux pump [CELL MOL] An active transport system for the removal of some antibiotics (such as tetracyclines, macrolides, and quinolones) from bacterial cells. { ˈē·fläks ˌpəmp }

effort-controlled cycle [IND ENG] A work cycle which is performed entirely by hand or in which the hand time controls the place. Also known as manually controlled work. { ˈef·ərt kənˌtrōld ˌsī·kəl }

effort rating [IND ENG] Assessing the level of manual effort expended by the operator, based on the observer's concept of normal effort, in order to adjust time-study data. Also known as pace rating; performance rating. { ˈef·ərt ˌrād·iŋ }

effuse [BOT] Expanded; spread out in a definite form. { e'fyüz }

effusion [MED] A pouring out of any fluid into a body cavity or tissue. [PHYS CHEM] The movement of a gas through an opening which is small as compared with the average distance which the gas molecules travel between collisions. [SCI TECH] **1.** The act or process of leaking or pouring out. **2.** Any material that is effused. { e'fyü·zhən }

effusive stage [GEOL] The second cooling stage for volcanic rocks. { e'fyü·siv ˌstāj }

EFL *See* error frequency limit.

e-folding height *See* scale height. { 'ē,fōl·diŋ ,hīt }

e-folding length [PHYS] The distance over which the amplitude of any exponentially varying quantity increases or decreases by a factor of *e* (2.718 …). { 'ē ˌfōld·iŋ ,leŋkth }

e-folding time [PHYS] The time required for the amplitude of an oscillation to increase or decrease by a factor of *e* (2.718...). { 'ē ˌfōld·iŋ ,tīm }

e format [COMPUT SCI] A decimal, normalized form of a floating point number in FORTRAN in which a number such as 18.756 appears as .18756E+02, which stands for .18756 × 10^2. { 'ē ˌfȯr,mat }

EGA *See* evolved gas analysis.

E galaxy *See* elliptical galaxy. { 'ē ˌgal·ik·sē }

Egeria [ASTRON] An asteroid with a diameter of about 139 miles (224 kilometers), mean distance from the sun of 2.58 astronomical units, and G-type (C-like) surface composition. { e'gir·ē·ə }

Egerov's theorem [MATH] If a sequence of measurable functions converges almost everywhere on a set of finite measure to a real-valued function, then given any $\epsilon > 0$ there is a set of measure smaller than ϵ on whose complement the sequence converges uniformly. { 'eg·ə,räfs ˌthir·əm }

Egerton's effusion method [THERMO] A method of determining vapor pressures of solids at high temperatures, in which one measures the mass lost by effusion from a sample placed in a tightly sealed silica pot with a small hole; the pot rests at the bottom of a tube that is evacuated for several hours, and is maintained at a high temperature by a heated block of metal surrounding it. { 'ej·ər·tənz ə'fyü·zhən ˌmeth·əd }

egest [PHYSIO] **1.** To discharge indigestible matter from the digestive tract. **2.** To rid the body of waste. { ē'jest }

egg [CYTOL] **1.** A large, female sex cell enclosed in a porous, calcareous or leathery shell, produced by birds and reptiles. **2.** *See* ovum. { eg }

egg apparatus [BOT] A group of three cells, consisting of the egg and two synergid cells, in the micropylar end of the embryo sac in seed plants. { 'eg ˌap·ə,rad·əs }

egg capsule *See* egg case. { 'eg ,kap·səl }

egg case [INV ZOO] **1.** A protective capsule containing the eggs of certain insects and mollusks. Also known as egg capsule. **2.** A silk pouch in which certain spiders carry their eggs. Also known as egg sac. [VERT ZOO] A soft, gelatinous (amphibians) or strong, horny (skates) envelope containing the egg of certain vertebrates. { 'eg ˌkās }

Eggertz's method [MET] A colorimetric estimation of the carbon content of steel by dissolving the metal in nitric acid and comparing the color with that produced by a similar metal of known carbon content. { 'e·gərts·əz ˌmeth·əd }

Egg Nebula [ASTRON] A reflection nebula consisting of two optical components separated by about 8 arc-seconds, with an infrared source between them. { 'eg ˌneb·yə·lə }

eggplant [BOT] *Solanum melongena.* A plant of the order Polemoniales grown for its edible egg-shaped, fleshy fruit. { 'eg,plant }

egg raft [ZOO] A floating mass of eggs; produced by a variety of aquatic organisms. { 'eg ,raft }

egg sac [ZOO] **1.** The structure containing the eggs of certain microcrustaceans. **2.** *See* egg case. { 'eg ,sak }

eggstone *See* oolite. { 'eg,stōn }

egg tempera [GRAPHICS] A painting process in which the color is bound with egg instead of oil. { 'eg 'tem·pə·rə }

egg tooth [VERT ZOO] A toothlike prominence on the tip of the beak of a bird embryo and the tip of the nose of an oviparous reptile, which is used to break the eggshell. { 'eg ˌtüth }

eglandular [BIOL] Without glands. { ē'glan·dyə·lər }

E glass [MATER] A type of borosilicate glass used to produce glass fibers for reinforced plastics designed for applications

requiring high electrical resistivity. Also known as electric glass. { 'ē ˌglas }

eglestonite [MINERAL] Hg_4Cl_2O Rare mercuric oxide mineral; forms yellow-brown isometric crystals upon exposure to air. { 'eg·əl·stə,nīt }

Egnell's law [METEOROL] The rule stating that above any fixed place the velocity of straight or nearly straight winds in the upper half of the troposphere increases with height at roughly the same rate that the density of the air decreases. { 'eg,nelz ,lȯ }

EGNOS *See* European Geostationary Navigation Overlay System. { 'eg,nōs }

ego [PSYCH] **1.** The self. **2.** The conscious part of the personality that is in contact with reality. { 'ē·gō }

ego analysis [PSYCH] Psychoanalytic analysis of the ways in which the ego resolves or attempts to deal with intrapsychic conflicts, especially in relation to the development of mental mechanisms and the maturation of capacity for rational thought and action. { 'ē,gō ə,nal·ə·səs }

ego-dystonia [PSYCH] A state of mind in which a person's behavior, thoughts, and attitudes are viewed as repugnant or inconsistent with the person's conception of his or her total personality. { 'ē,go dis'tōn·ē·ə }

ego ideal [PSYCH] The part of an individual's personality that is composed of the aims and goals for the self and that usually refers to the conscious or unconscious emulation of significant people with whom the individual has identified. { 'ē,gō ī'dēl }

ego-syntonia [PSYCH] A state of mind in which a person's behavior, thoughts, and attitudes are viewed as acceptable and consistent with the person's conception of his or her total personality. { 'ē,gō sin'tōn·ē·ə }

egress [ASTRON] The departure of the moon from the shadow of the earth in an eclipse, or of a planet from the disk of the sun, or of a satellite (or its shadow) from the disk of the parent planet. { 'ē,gres }

EGT *See* ethylene glycol bis(trichloroacetate).

egueïte [MINERAL] $CaFe_{14}(PO_4)_{10}(OH)_{14} \cdot 21H_2O$ A brownish-yellow mineral consisting of a hydrated basic phosphate of calcium and iron; occurs as small nodules. { e'gwā,īt }

Egyptian asphalt [GEOL] A glance pitch (bituminous mixture similar to asphalt) found in the Arabian Desert. { i'jip·shən 'as,fȯlt }

Egyptian cotton [BOT] Long-staple, high-quality cotton grown in Egypt. { i'jip·shən 'kät·ən }

Egyptian henna *See* henna. { i'jip·shən 'hen·ə }

E-HEMT *See* enhancement-mode HEMT.

EHF *See* extremely high frequency.

ehp *See* effective horsepower.

Ehrenfest's adiabatic law [QUANT MECH] The law that, if the Hamiltonian of a system undergoes an infinitely slow change, and if the system is initially in an eigenstate of the Hamiltonian, then at the end of the change it will be in the eigenstate of the new Hamiltonian that derives from the original state by continuity, provided certain conditions are met. Also known as Ehrenfest's theorem. { 'er·ən,fests ˌad·ē·ə¦bad·ik ¦lȯ }

Ehrenfest's equations [THERMO] Equations which state that for the phase curve $P(T)$ of a second-order phase transition the derivative of pressure P with respect to temperature T is equal to $(C_p^f - C_p^i)/TV(\gamma^f - \gamma^i) = (\gamma^f - \gamma^i)/(K^f - K^i)$, where i and f refer to the two phases, γ is the coefficient of volume expansion, K is the compressibility, C_p is the specific heat at constant pressure, and V is the volume. { 'er·ən,fests i,kwā·zhənz }

Ehrenfest's theorem [QUANT MECH] **1.** The theorem that a quantum-mechanical wave packet obeys the equations of motion of the corresponding classical particle when the position, momentum, and force acting on the particle are replaced by the expectation values of these quantities. **2.** *See* Ehrenfest's adiabatic law. { 'er·ən,fests ,thir·əm }

Ehrenhaft effect [ELECTROMAG] A helical motion of fine particles along the lines of force of a magnetic field during exposure to light, resulting from radiometer effects. { 'er·ən,haft i,fekt }

Ehrlichia [MICROBIO] A genus of the tribe Ehrlichieae; coccoid to ellipsoidal or pleomorphic cells; intracellular parasites in cytoplasm of host leukocytes. { er'lik·ē·ə }

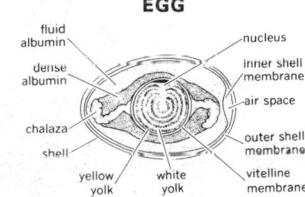

EGG

fluid albumin · nucleus · inner shell membrane · air space · dense albumin · chalaza · outer shell membrane · vitelline membrane · shell · yellow yolk · white yolk

The egg of a bird.

EGGPLANT

Eggplant (*Solanum melongena*), cultivar Black Magic. (*Joseph Harris Co., Rochester, N.Y.*)

Ehrlichieae [MICROBIO] A tribe of the family Rickettsiaceae; spherical and occasionally pleomorphic cells; pathogenic for some mammals, not including humans. { ‚er·lə'kī·ē‚ē }

ehrlichiosis [MED] A tick-borne bacterial infection caused by two distinct *Ehrlichia* species that infect white blood cells; the infection may be asymptomatic, but it also can produce illness ranging from a few mild symptoms to an overwhelming multisystem disease. { är‚lik·ē'ō·səs }

Ehrlich's 606 *See* arsphenamine. { 'er·liks ‚sik·sō'siks }

Ehrlich's reagent [ORG CHEM] $(CH_3)_2NC_6H_4CHO$ Granular or leafletlike crystals that are soluble in many organic solvents; melting point is 74°C; used in the preparation of dyes, as a reagent for arsphenamine, anthranilic acid, antipyrine, indole, and skatole, and as a differentiating agent between true scarlet fever and serum eruptions. { 'er·liks rē‚ā·jənt }

EHSI *See* electronic horizontal-situation indicator.

eht *See* extra-high tension.

E-H T junction [ELECTROMAG] In microwave waveguides, a combination of E- and H-plane T junctions forming a junction at a common point of intersection with the main waveguide. { ‚ē ‚āch 'tē ‚jəŋk·shən }

E-H tuner [ELECTROMAG] Tunable E-H T junction having two arms terminated in adjustable plungers used for impedance transformation. { ‚ē ‚āch 'tün·ər }

ehv *See* extra-high voltage.

eicosane [MATER] A mixture of saturated hydrocarbons mostly straight-chained and averaging 20 carbons in the chain; for this reason, the formula $C_{20}H_{42}$ is given to the technical mixture; used in lubricants and plasticizers. { 'ī·kə‚sān }

eicosanoic acid [ORG CHEM] $CH_3(CH_2)_{18}COOH$ A white, crystalline, saturated fatty acid, melting at 75.4°C; a constituent of butter. Also known as arachic acid; arachidic acid. { ‚ī·kə·sə‚nō·ik 'as·əd }

eicosanoid [BIOCHEM] Any member of a group of naturally occurring substances composed of prostaglandins, thromboxanes, and leukotrienes that are derived from polyunsaturated fatty acids, particularly arachidonic acid, and exhibit various types of biological activity. { ī'käs·ə‚nóid }

eidetic imagery [PSYCH] Imagery in extreme detail; a sort of projection of an image on a mental screen. { ī'ded·ik 'im·ij·rē }

eigenfrequency [PHYS] One of the frequencies at which an oscillatory system can vibrate. { 'ī·gən‚frē·kwən·sē }

eigenfunction [MATH] **1.** Also known as characteristic function. **2.** An eigenvector for a linear operator on a vector space whose vectors are functions. Also known as proper function. **3.** A solution to the Sturm-Liouville partial differential equation. { 'ī·gən‚fəŋk·shən }

eigenfunction expansion [MATH] By using spectral theory for linear operators defined on spaces composed of functions, in certain cases the operator equals an integral or series involving its eigenvectors; this is known as its eigenfunction expansion and is particularly useful in studying linear partial differential equations. { 'ī·gən‚fəŋk·shən ik'span·chən }

eigenmatrix [MATH] Corresponding to a diagonalizable matrix or linear transformation, this is the matrix all of whose entries are 0 save those on the principal diagonal where appear the eigenvalues. { 'ī·gən‚mā·triks }

eigenstate [QUANT MECH] **1.** A dynamical state whose state vector (or wave function) is an eigenvector (or eigenfunction) of an operator corresponding to a specified physical quantity. **2.** *See* energy state. { 'ī·gən‚stāt }

eigenvalue [MATH] The one of the scalars λ such that $T(v) = \lambda v$, where T is a linear operator on a vector space, and v is an eigenvector. Also known as characteristic number; characteristic root; characteristic value; latent root; proper value. { 'ī·gən‚val·yü }

eigenvalue equation *See* characteristic equation. { 'ī·gən‚val·yü i‚kwā·zhən }

eigenvalue problem *See* Sturm-Liouville problem. { 'ī·gən‚val·yü ‚präb·ləm }

eigenvector [MATH] A nonzero vector v whose direction is not changed by a given linear transformation T; that is, $T(v) = \lambda v$ for some scalar λ. Also known as characteristic vector. { 'ī·gən‚vek·tər }

eight curve [MATH] A plane curve whose equation in cartesian coordinates x and y is $x^4 = a^2(x^2 - y^2)$, where a is a constant. Also known as lemniscate of Gerono. { 'āt ‚kərv }

8D technique [METEOROL] A technique for using the radiosonde observation to determine the presence of liquid waterdroplets in supercooled clouds in saturated or nearly saturated layers of air; for each reported level in the sounding, the negative value of eight times the dew-point spread (−8D) is plotted on the pseudoadiabatic chart (or equivalent chart); where the temperature sounding lies to the left of the −8D curve, liquid droplet clouds are considered to be present, and icing is possible on aircraft flying in the cloud layer. Also known as frostpoint technique. { ‚āt 'dē tek‚nēk }

eightfold way [PART PHYS] The classification of hadrons composed of up, down, and strange quarks by SU_3 symmetry; in particular, the eight-dimensional representation of the group SU_3 over its center, which gives rise to unitary octets. { 'āt‚fōld 'wā }

eight-level code [COMMUN] A teletypewriter code that uses eight impulses, in addition to the start and stop impulses, to define a character. { ‚āt ‚lev·əl 'kōd }

eikonal equation [PHYS] An equation for propagation of electromagnetic or acoustic waves in a nonhomogeneous medium; it is valid only when the variation of the properties of the medium is small over the distance of a wavelength. { ī'kōn·əl i‚kwā·zhən }

eikonometer [OPTICS] A scale used to measure sizes of objects viewed through a microscope, usually attached to the eye-piece so that it is seen superimposed on the image. { ī·kə'näm·əd·ər }

Eimeriina [INV ZOO] A suborder of coccidian protozoans in the order Eucoccida in which there is no syzygy and the microgametocytes produce a large number of microgametes. { ‚ī·mə'rī·ə·nə }

E indicator *See* E scope. { 'ē ‚in·də‚kād·ər }

einkanter [GEOL] A stone shaped by windblown sand only upon one facet. { 'īn‚kän·tər }

Einschluss thermometer [ANALY CHEM] All-glass, liquid-filled thermometer, temperature range −201 to +360°C, used for laboratory test work. { 'īn‚shlùs thər‚mäm·əd·ər }

einstein [PHYS] A unit of light energy used in photochemistry, equal to Avogadro's number times the energy of one photon of light of the frequency in question. { 'īn‚stīn }

Einstein-Bohr equation [QUANT MECH] In a system undergoing a transition between two states so that it emits or absorbs radiation, that equation indicating that the radiation frequency equals the difference in energy between the two states divided by Planck's constant. { 'īn‚stīn 'bòr i‚kwā·zhən }

Einstein-Bose statistics *See* Bose-Einstein statistics. { 'īn‚stīn 'bōz stə‚tis·tiks }

Einstein characteristic temperature [SOLID STATE] A temperature, characteristic of a substance, that appears in Einstein's equation for specific heat; it is equal to the product of Planck's constant and the Einstein frequency divided by Boltzmann's constant. { 'īn‚stīn ‚kar·ik·tə‚ris·tik 'tem·prə·chər }

Einstein condensation *See* Bose-Einstein condensation. { 'īn‚stīn ‚kän·dən‚sā·shən }

Einstein-de Haas effect [ELECTROMAG] A freely suspended body consisting of a ferromagnetic material acquires a rotation when its magnetization changes. { 'īn‚stīn də'häs i‚fekt }

Einstein-de Haas method [ELECTROMAG] Method of measuring the gyromagnetic ratio of a ferromagnetic substance; one measures the angular displacement induced in a ferromagnetic cylinder suspended from a torsion fiber when magnetization of the object is reversed, and the magnetization change is measured with a magnetometer. { 'īn‚stīn də'häs ‚meth·əd }

Einstein-de Sitter model [RELAT] A model of the universe in which ordinary euclidean geometry holds good, the distribution of matter extends infinitely at all times, and the universe expands from an infinitely condensed state at such a rate that the density is inversely proportional to the square of the time elapsed since the beginning of the expansion. { 'īn‚stīn də'sid·ər ‚mäd·əl }

Einstein diffusion equation [STAT MECH] An equation which gives the mean square displacement caused by Brownian movement of spherical, colloidal particles in a gas or liquid. { 'īn‚stīn də'fyü·zhən i‚kwā·zhən }

Einstein displacement *See* Einstein shift. { 'īn‚stīn di ‚splās·mənt }

Einstein elevator [RELAT] A windowless elevator freely

falling in its shaft, inside of which conditions resemble interstellar space; used to elucidate the principle of equivalence. { 'īn,stīn 'el·ə,vād·ər }

Einstein equations [STAT MECH] Equations for the density and pressure of a Bose-Einstein gas in terms of power series in a parameter which appears in the Bose-Einstein distribution law. { 'īn,stīn i,kwā·zhənz }

Einstein frequency [SOLID STATE] Single frequency with which each atom vibrates independently of other atoms, in a model of lattice vibrations; equal to the frequency observed in infrared absorption studies. { 'īn,stīn ,frē·kwən·sē }

Einstein frequency condition [SOLID STATE] The assumption that all vibrations of a crystal lattice are harmonic with the same characteristic frequency. { 'īn,stīn 'frē·kwən·se kən,dish·ən }

einsteinium [CHEM] Synthetic radioactive element, symbol Es, atomic number 99; discovered in debris of 1952 hydrogen bomb explosion; now made in cyclotrons. { īn'stīn·ē·əm }

Einstein mass-energy relation [RELAT] The relation in which the energy of a system is equivalent to its mass times the square of the speed of light. { 'īn,stīn 'mas 'en·ər·jē ri,lā·shən }

Einstein number [PL PHYS] A dimensionless number used in magnetofluid dynamics, equal to the ratio of the velocity of a fluid to the speed of light. { 'īn,stīn ,nəm·bər }

Einstein partition function [STAT MECH] The partition function for a solid, based on the Einstein frequency condition. { 'īn,stīn par'tish·ən ,fəŋk·shən }

Einstein photochemical equivalence law [PHYS CHEM] The law that each molecule taking part in a chemical reaction caused by electromagnetic radiation absorbs one photon of the radiation. Also known as Stark-Einstein law. { 'īn,stīn 'fōd·ō,kem·ə·kəl i'kwiv·ə·ləns ,lò }

Einstein photoelectric law [QUANT MECH] The law that the energy of an electron emitted from a system in the photoelectric effect is $h\nu - W$, where h is Planck's constant, ν is the frequency of the incident radiation, and W is the energy needed to remove the electron from the system; if $h\nu$ is less than W, no electrons are emitted. { 'īn,stīn 'fōd·ō·i,lek·trik ,lò }

Einstein-Planck law [QUANT MECH] The law that the energy of a photon is given by Planck's constant times the frequency. [RELAT] The equation of motion of a charged particle in an electromagnetic field, according to which its rate of change of momentum is equal to the Lorentz force, where the magnitude of the momentum is $mv/(1 - v^2/c^2)^{1/2}$, where m and v are the particle's mass and velocity, and c is the speed of light. { 'īn,stīn 'pläŋk ,lò }

Einstein-Podolsky-Rosen experiment [QUANT MECH] A Gedanken experiment which was introduced to argue that quantum mechanics is not a complete theory, involving polarization measurements on two photons emitted in opposite directions in an atomic cascade. Abbreviated EPR experiment. { 'īn,stīn pə'däl·skē 'rōz·ənik,sper·ə·mənt }

Einstein radius [RELAT] The radius of a ring-shaped region through which light is bent by a gravitational lens directly between the light source and the observer. { 'īn,stīn ,rād·ē·əs }

Einstein relation [PHYS] The relation in which the mobility of charges in an ionic solution or semiconductor is equal to the magnitude of the charge times the diffusion coefficient divided by the product of the Boltzmann constant and the absolute temperature. { 'īn,stīn ri,lā·shən }

Einstein-Rosen waves [RELAT] Gravitational waves produced by oscillating ponderable matter, along an infinitely long cylindrical axis, in an exact solution of Einstein's field equations. { 'īn,stīn 'rōz·ən ,wāvz }

Einstein's absorption coefficient [ATOM PHYS] The proportionality constant governing the absorption of electromagnetic radiation by atoms, equal to the number of quanta absorbed per second divided by the product of the energy of the radiation per unit volume per unit wave number and the number of atoms in the ground state. { 'īn,stīnz əb'sórp·shən ,kō·i,fish·ənt }

Einstein's coefficient of spontaneous emission [ATOM PHYS] Proportionality constant governing the rate at which atoms or molecules pass spontaneously from an upper energy state to a lower one by emission of radiation, equal to the number of such transitions per second divided by the number of atoms in the upper state. { 'īn,stīnz ,kō·ə,fish·ənt əv spän,tā·nē·əs ə'mish·ən }

Einstein's coefficient of stimulated emission [ATOM PHYS] Proportionality constant governing the rate at which atoms or molecules pass from an upper energy state to a lower one by stimulated emission of radiation, equal to the number of such transitions per second divided by the product of energy of the radiation inducing the transition per unit volume per unit wave number and the number of atoms in the upper state. { 'īn,stīnz ,kō·ə,fish·ənt əv ,stim·yə,lād·əd ə'mish·ən }

Einstein's equation for specific heat [SOLID STATE] The earliest equation based on quantum mechanics for the specific heat of a solid; uses the assumption that each atom oscillates with the same frequency. { 'īn,stīnz i'kwā·zhən fər spə,si·fik 'hēt }

Einstein's equivalency principle See equivalence principle. { 'īn,stīnz i'kwiv·ə·lən·sē ,prin·sə·pəl }

Einstein's field equations [RELAT] Those equations relevant to the relationship in which the Einstein tensor equals -8π times the energy momentum tensor times the gravitational constant divided by the square of the speed of light. Also known as Einstein's law of gravitation. { 'īn,stīnz 'fēld i,kwā·zhənz }

Einstein shift [RELAT] A shift toward longer wavelengths of spectral lines emitted by atoms in strong gravitational fields. Also known as Einstein displacement. { 'īn,stīn ,shift }

Einstein's law of gravitation See Einstein's field equations. { 'īn,stīnz ,lò əv ,grav·ə'tā·shən }

Einstein space [MATH] A Riemannian space in which the contracted curvature tensor is proportional to the metric tensor. { 'īn,stīn 'spās }

Einstein's principle of relativity [RELAT] The principle that all the laws of physics must assume the same mathematical form in any inertial frame of reference; thus, it is impossible to determine the absolute motion of a system by any means. { 'īn,stīnz 'prin·sə·pəl əv rel·ə'tiv·əd·ē }

Einstein's summation convention [MATH] A notational convenience used in tensor analysis whereupon it is agreed that any term in which an index appears twice will stand for the sum of all such terms as the index assumes all of a preassigned range of values. { 'īn,stīnz sə'mā·shən kən,ven·chən }

Einstein static universe [RELAT] A nonvacuum, globally static solution to Einstein's equations of general relativity with cosmological term. { 'īn,stīn 'stad·ik ,yü·nə,vərs }

Einstein's unified field theories [RELAT] A series of theories attempting to express a general unifying principle underlying electromagnetism and gravity. { 'īn,stīnz ,yü·nə,fīd 'fēld ,thē·ə·rēz }

Einstein tensor [RELAT] The tensor expressed as $E_{\mu\nu} = R_{\mu\nu} - \frac{1}{2}(g_{\mu\nu}R - 2\Lambda)$, where $R_{\mu\nu}$ is the contracted curvature tensor, R is the curvature of space-time, $g_{\mu\nu}$ is the metric tensor, and Λ is the cosmological constant. { 'īn,stīn 'ten·sər }

Einstein universe [RELAT] A model of the universe which is a four-dimensional cylindrical surface in a five-dimensional space. { 'īn,stīn 'yü·nə,vərs }

Einstein viscosity equation [PHYS CHEM] An equation which gives the viscosity of a sol in terms of the volume of dissolved particles divided by the total volume. { 'īn,stīn vis'käs·əd·ē i,kwā·zhən }

Einthoven galvanometer See string galvanometer. { 'īnt,hō·vən ,gal·və'näm·əd·ər }

Einzel lens [ELECTR] An electrostatic lens that consists of three cylindrical tubes through which charged particles pass sequentially, the middle one of which is at a higher potential than the other two. { 'īnt·səl ,lenz }

eirp See effective isotropic radiated power.

Eisenstein irreducibility criterion [MATH] The proposition that a polynomial with integer coefficients is irreducible in the field of rational numbers if there is a prime p that does not divide the coefficient of x^n but divides all the other coefficients, and if p^2 does not divide the coefficient of x^0.

ejaculation [PHYSIO] The act or process of suddenly discharging a fluid from the body; specifically, the ejection of semen during orgasm. { i,jak·yə'lā·shən }

ejaculatory duct [ANAT] The terminal part of the ductus deferens after junction with the duct of a seminal vesicle, embedded in the prostate gland and opening into the urethra. { i'jak·yə·lə,tór·ē 'dəkt }

ejaculatory incompetence [MED] Inability of a male to

reach orgasm and ejaculate during sexual intercourse despite adequacy of erection. { i¦jak·yə·lə₊tór·ē in'käm·pəd·əns }

eject [COMPUT SCI] To move the printing mechanism to the top of the following page, skipping the remainder of the current page. { ē'jekt }

ejecta [GEOL] Material which is discharged by a volcano. [PHYSIO] Excrement. [SCI TECH] Material which is cast out. { ē'jek·tə }

ejection [ENG] The process of removing a molding from a mold impression by mechanical means, by hand, or by compressed air. [ORD] The expelling, by the ejector, of the empty cartridge case from small arms and rapid-fire guns. { ē'jek·shən }

ejection capsule [AERO ENG] In an aircraft or spacecraft, a detachable compartment (serving as a cockpit or cabin) or a payload capsule which may be ejected as a unit and parachuted to the ground. { ē'jek·shən ₊kap·səl }

ejection port [ORD] The opening in the receiver portion of a firearm through which the empty shell cases are thrown from the firearm after firing. { ē'jek·shən ₊pórt }

ejection seat [AERO ENG] Emergency device which expels the pilot safely from a high-speed airplane. { ē'jek·shən 'sēt }

ejector [ENG] **1.** Any of various types of jet pumps used to withdraw fluid materials from a space. Also known as eductor. **2.** A device that ejects the finished casting from a mold. [ORD] A device in the breech mechanism of a gun, rifle, or other firearm which automatically throws out an empty cartridge case, or unfired cartridge, from the breech or receiver. { ē'jek·tər }

ejector condenser [MECH ENG] A type of direct-contact condenser in which vacuum is maintained by high-velocity injection water; condenses steam and discharges water, condensate, and noncondensables to the atmosphere. { ē'jek·tər kən₊den·sər }

ejector half [MET] The movable half of a casting die. { ē'jek·tər ₊haf }

ejector pin [ENG] A pin driven into the rear of a mold cavity to force the finished piece out. Also known as knockout pin. { ē'jek·tər ₊pin }

ejector plate [ENG] The plate backing up the ejector pins and holding the ejector assembly together. { ē'jek·tər ₊plāt }

ejector rod [ENG] A rod that activates the ejector assembly of a mold when it is opened. { ē'jek·tər 'räd }

E-JFET See enhancement-mode junction field-effect transistor.

EKG See electrocardiogram.

Ekman convergence [OCEANOGR] A zone of convergence of warm surface water caused by Ekman transport, creating a marked depression of the ocean's thermocline in the affected area. { 'ek·mən kən'vər·jəns }

Ekman current meter [ENG] A mechanical device for measuring ocean current velocity which incorporates a propeller and a magnetic compass and can be suspended from a moored ship. { 'ek·mən 'kə·rənt ₊mēd·ər }

Ekman dredge [ENG] A special type of dredge for sampling sediment that is fitted with opposable jaws operated by a messenger traveling down a cable to release a spring catch. { 'ek·mən ₊drej }

Ekman layer [METEOROL] The layer of transition between the surface boundary layer of the atmosphere, where the shearing stress is constant, and the free atmosphere, which is treated as an ideal fluid in approximate geostrophic equilibrium. Also known as spiral layer. { 'ek·mən ₊lā·ər }

Ekman spiral [METEOROL] A theoretical representation that a wind blowing steadily over an ocean of unlimited depth and extent and uniform viscosity would cause, in the Northern Hemisphere, the immediate surface water to drift at an angle of 45° to the right of the wind direction, and the water beneath to drift further to the right, and with slower and slower speeds, as one goes to greater depths. { 'ek·mən ₊spī·rəl }

Ekman sucking [FL MECH] A boundary-layer phenomenon in which fluid near the bottom of a spinning vessel is drawn toward the edge of the vessel along the bottom. { 'ek·mən ₊sək·iŋ }

Ekman transport [OCEANOGR] The movement of ocean water caused by wind blowing steadily over the surface; occurs at right angles to the wind direction. { 'ek·mən ₊trans₊pórt }

Ekman water bottle [ENG] A cylindrical tube fitted with plates at both ends and used for deep-water samplings; when

hit by a messenger it turns 180°, closing the plates and capturing the water sample. { 'ek·mən 'wōd·ər ₊bäd·əl }

ekranoplan [AER ENG] See wing-in-ground effect aircraft.

elaboration [COMPUT SCI] A technique, used chiefly in the Ada programming language, of setting up a hierarchy of calculated constants so that the values of one or more of them determine others further down in the hierarchy. { i₊lab·ə'rā·shən }

Elaeagnaceae [BOT] A family of dicotyledonous plants in the order Proteales, noted for peltate leaf scales which often give the leaves a silvery-gray appearance. { ₊el·ē·ag'nās·ē₊ē }

elaidic acid [ORG CHEM] $CH_3(CH_2)_7CH:CH(CH_2)_7COOH$ A transisomer of an unsaturated fatty acid, oleic acid; crystallizes as colorless leaflets, melts at 44°C, boils at 288°C (100 mmHg), insoluble in water, soluble in alcohol and ether; used in chromatography as a reference standard. { ¦el·ə¦id·ik 'as·əd }

elaidinization [ORG CHEM] The process of changing the geometric cis form of an unsaturated fatty acid or a compound related to it into the trans form, resulting in an acid that is more resistant to oxidation. { ə¦lā·ə₊din·ə'zā·shən }

elaidin reaction [ANALY CHEM] A test that differentiates nondrying oils such as olein from semidrying oils and drying oils; nitrous acid converts olein into its solid isomer, while semidrying oils in contact with nitrous acid thicken slowly, and drying oils such as tung oil become hard and resinous. { ə'lā·əd·ən rē₊ak·shən }

elaioplast [HISTOL] An oil-secreting leucoplast. { ə'lī·ə₊plast }

Elaphomycetaceae [MYCOL] A family of underground, saprophytic or mycorrhiza-forming fungi in the order Eurotiales characterized by ascocarps with thick, usually woody walls. { ₊el·ə·fō₊mī·sə'tās·ē₊ē }

Elapidae [VERT ZOO] A family of poisonous reptiles, including cobras, kraits, mambas, and coral snakes; all have a pteroglyph fang arrangement. { ə'lap·ə₊dē }

Elara [ASTRON] A small satellite of Jupiter with a diameter of about 20 miles (32 kilometers), orbiting at a mean distance of 7.29×10^6 miles (11.73×10^6 kilometers). Also known as Jupiter VII. { e'lar·ə }

Elasipodida [INV ZOO] An order of deep-sea aspidochirotacean holothurians in which there are no respiratory trees and bilateral symmetry is often quite conspicuous. { ə₊laz·ə'päd·ə·də }

Elasmidae [INV ZOO] A family of hymenopteran insects in the superfamily Chalcidoidea. { ə'laz·mə₊dē }

Elasmobranchii [VERT ZOO] The sharks and rays, a subclass of the class Chondrichthyes distinguished by separate gill openings, amphistylic or hyostylic jaw suspension, and ampullae of Lorenzini in the head region. { ə₊laz·mə'braŋ·kē₊ī }

Elassomatidae [VERT ZOO] The pygmy sunfishes, a family of the order Perciformes. { ə₊las·ō'mad·ə₊dē }

elastance [ELEC] The reciprocal of capacitance. { i'las·təns }

elastase [BIOCHEM] An enzyme which acts on elastin to change it chemically and render it soluble. { i'la₊stās }

elastic [MECH] Capable of sustaining deformation without permanent loss of size or shape. { i'las·tik }

elastica [MECH] The elastic curve formed by a uniform rod that is originally straight, then is bent in a principal plane by applying forces, and couples only at its ends. { i'las·tə·kə }

elastic aftereffect [MECH] The delay of certain substances in regaining their original shape after being deformed within their elastic limits. Also known as elastic lag. { i'las·tik 'af·tər·i₊fekt }

elastic axis [MECH] The lengthwise line of a beam along which transverse loads must be applied in order to produce bending only, with no torsion of the beam at any section. { i'las·tik 'ak·səs }

elastic bitumen See elaterite. { i'las·tik bī'tü·mən }

elastic body [MECH] A solid body for which the additional deformation produced by an increment of stress completely disappears when the increment is removed. Also known as elastic solid. { i'las·tik 'bäd·ē }

elastic buckling [MECH] An abrupt increase in the lateral deflection of a column at a critical load while the stresses acting on the column are wholly elastic. { i'las·tik 'bək·liŋ }

elastic cartilage [HISTOL] A type of cartilage containing elastic fibers in the matrix. { i'las·tik 'kärt·lij }

elastic center [MECH] That point of a beam in the plane of

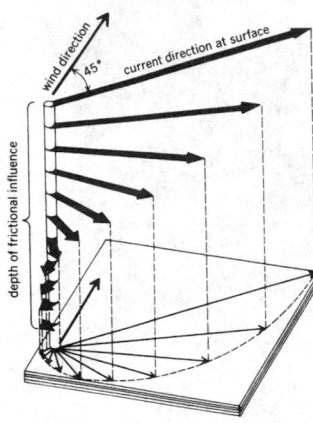

EKMAN SPIRAL

Schematic of pure wind current in deep water, showing decrease in velocity and change of direction at regular intervals of depth (the Ekman spiral).

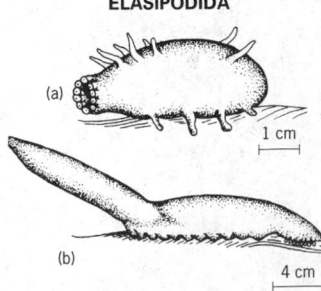

ELASIPODIDA

Two examples of bottom dwelling Elasipodida. (a) *Elpidia*. (b) *Psychropotes*.

the section lying midway between the flexural center and the center of twist in that section. { i'las·tik 'sen·tər }

elastic collision [MECH] A collision in which the sum of the kinetic energies of translation of the participating systems is the same after the collision as before. { i'las·tik kə'lizh·ən }

elastic constant See compliance constant; stiffness constant. { i'las·tik 'kän·stənt }

elastic cross section [PHYS] The cross section for an elastic collision between two particles or systems. { i'las·tik 'krós ‚sek·shən }

elastic curve [MECH] The curved shape of the longitudinal centroidal surface of a beam when the transverse loads acting on it produced wholly elastic stresses. { i'las·tik 'kərv }

elastic deformation [MECH] Reversible alteration of the form or dimensions of a solid body under stress or strain. { i'las·tik ‚dē·fər'mā·shən }

elastic design [CIV ENG] In the design of a structural member, a method of analysis based on a linear stress-strain relationship, with the assumption that the working stresses constitute only a fraction of the elastic limit of the material. { i'las·tik di'zīn }

elastic equilibrium [MECH] The condition of an elastic body in which each volume element of the body is in equilibrium under the combined effect of elastic stresses and externally applied body forces. { i'las·tik ‚ē·kwə'lib·rē·əm }

elastic failure [MECH] Failure of a body to recover its original size and shape after a stress is removed. { i'las·tik 'fāl·yər }

elastic fiber [HISTOL] A homogeneous, fibrillar connective tissue component that is highly refractile and appears yellowish when arranged in bundles. { i'las·tik 'fī·bər }

elastic flow [MECH] Return of a material to its original shape following deformation. { i'las·tik 'flō }

elastic force [MECH] A force arising from the deformation of a solid body which depends only on the body's instantaneous deformation and not on its previous history, and which is conservative. { i'las·tik 'fórs }

elastic hysteresis [MECH] Phenomenon exhibited by some solids in which the deformation of the solid depends not only on the stress applied to the solid but also on the previous history of this stress; analogous to magnetic hysteresis, with magnetic field strength and magnetic induction replaced by stress and strain respectively. { i'las·tik ‚his·tə'rē·səs }

elasticity [MECH] **1.** The property whereby a solid material changes its shape and size under action of opposing forces, but recovers its original configuration when the forces are removed. **2.** The existence of forces which tend to restore to its original position any part of a medium (solid or fluid) which has been displaced. { i‚las'tis·əd·ē }

elasticity modulus See modulus of elasticity. { i‚las'tis·əd·ē ‚mäj·ə·ləs }

elasticity number 1 [FL MECH] A dimensionless number which is a measure of the ratio of elastic forces to inertial forces on a viscoelastic fluid flowing in a pipe, and is equal to the product of the fluid's relaxation time and its dynamic viscosity, divided by the product of the fluid's density and the square of the radius of the pipe. Symbolized N_{Ell}. { i‚las'tis·əd·ē ‚nəm·bər 'wən }

elasticity number 2 [FL MECH] A dimensionless number used in studying the effect of elasticity on a flow process, equal to the fluid's density times its specific heat at constant pressure, divided by the product of its coefficient of bulk expansion and its bulk modulus. Symbolized K_E. { i‚las'tis·əd·ē ‚nəm·bər 'tü }

elastic lag See elastic aftereffect. { i'las·tik 'lag }

elastic limit [MECH] The maximum stress a solid can sustain without undergoing permanent deformation. { i‚las'tis·tik 'lim·ət }

elastic-limit effect [SOLID STATE] A phenomenon in which a material acquires a sharp elastic limit because of the introduction of foreign atoms. { i'las·tik 'lim·ət i‚fekt }

elastic modulus See modulus of elasticity. { i‚las·tik 'mäj·ə·ləs }

elasticoviscosity [FL MECH] That property of a fluid whose rate of deformation under stress is the sum of a part corresponding to a viscous Newtonian fluid and a part obeying Hooke's law. { i'las·tə·kō·vis'käs·əd·ē }

elastic potential energy [MECH] Capacity that a body has

to do work by virtue of its deformation. { i'las·tik pə‚ten·chəl ‚en·ər·jē }

elastic ratio [MECH] The ratio of the elastic limit to the ultimate strength of a solid. { i'las·tik 'rā·shō }

elastic rebound theory [GEOL] A theory which attributes faulting to stresses (in the form of potential energy) which are being built up in the earth and which, at discrete intervals, are suddenly released as elastic energy; at the time of rupture the rocks on either side of the fault spring back to a position of little or no strain. { i'las·tik 'rē‚baund ‚thē·ə·rē }

elastic recovery [MECH] That fraction of a given deformation of a solid which behaves elastically. { i'las·tik ri'kəv·ə·rē }

elastic scattering [MECH] Scattering due to an elastic collision. { i'las·tik 'skad·ə·riŋ }

elastic solid See elastic body. { i'las·tik 'säl·əd }

elastic strain energy [MECH] The work done in deforming a solid within its elastic limit. { i'las·tik 'strān ‚en·ər·jē }

elastic theory [MECH] Theory of the relations between the forces acting on a body and the resulting changes in dimensions. { i'las·tik 'thē·ə·rē }

elastic tissue [HISTOL] A type of connective tissue having a predominance of elastic fibers, bands, or lamellae. { i'las·tik 'tish·ü }

elastic vibration [MECH] Oscillatory motion of a solid body which is sustained by elastic forces and the inertia of the body. { i'las·tik vī'brā·shen }

elastic wave [ACOUS] See acoustic wave. [PHYS] A wave propagated by a medium having inertia and elasticity (the existence of forces which tend to restore any part of a medium to its original position), in which displaced particles transfer momentum to adjoining particles, and are themselves restored to their original position. { i'las·tik 'wāv }

elastin [BIOCHEM] An elastic protein composing the principal component of elastic fibers. { i'las·tən }

elastodynamics [MECH] The study of the mechanical properties of elastic waves. { i‚la·stō·dī'nam·iks }

elastomer [MATER] A polymeric material, such as a synthetic rubber or plastic, which at room temperature can be stretched under low stress to at least twice its original length and, upon immediate release of the stress, will return with force to its approximate original length. { i'las·tə·mər }

elastoplasticity [MECH] State of a substance subjected to a stress greater than its elastic limit but not so great as to cause it to rupture, in which it exhibits both elastic and plastic properties. { i‚las·tō·plə'stis·əd·ē }

elastoresistance [ELEC] The change in a material's electrical resistance as it undergoes a stress within its elastic limit. { i‚las·tō·ri'zis·təns }

elastosis [MED] **1.** Retrogressive change in elastic tissue. **2.** Retrogressive change in cutaneous connective tissue resulting in excessive amounts of material which give the staining reactions for elastin. { i‚la'stō·səs }

elater [BOT] A spiral, filamentous structure that functions in the dispersion of spores in certain plants, such as liverworts and slime molds. { 'el·ə·tər }

Elateridae [INV ZOO] The click beetles, a large family of coleopteran insects in the superfamily Elateroidea; many have light-producing organs. { ‚el·ə'ter·ə‚dē }

elaterite [GEOL] A light-brown to black asphaltic pyrobitumen that is moderately soft and elastic. Also known as elastic bitumen; mineral caoutchouc. { i'lad·ə‚rīt }

Elateroidea [INV ZOO] A superfamily of coleopteran insects in the suborder Polyphaga. { i‚lad·ə'róid·ē·ə }

elaterophore [BOT] A tissue bearing elaters, found in some liverworts. { i'lad·ə·rə‚fór }

E layer [GEOPHYS] A layer of ionized air occurring at altitudes between 60 and 72 miles (100 and 120 kilometers) in the E region of the ionosphere, capable of bending radio waves back to earth. Also known as Heaviside layer; Kennelly-Heaviside layer. { 'ē ‚lā·ər }

elbow [ANAT] The arm joint formed at the junction of the humerus, radius, and ulna. [DES ENG] **1.** A fitting that connects two pipes at an angle, often of 90°. **2.** A sharp corner in a pipe. [ELECTROMAG] In a waveguide, a bend of comparatively short radius, normally 90°, and sometimes for acute angles down to 15°. [GEOGR] A sharp change in direction of a coast line, channel, bank, or so on. { 'el‚bō }

elbow meter [ENG] Pipe elbow used as a liquids flowmeter;

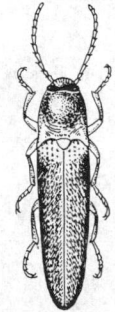

ELATERIDAE

External features of the click beetle. (*From T. I. Storer and R. L. Usinger, General Zoology, 3d ed., McGraw-Hill, 1957*)

flow rate is measured by determining the differential pressure developed between the inner and outer radii of the bend by means of two pressure taps located midway on the bend. { 'el₁bō ₁mēd·ər }

Elbs reaction [ORG CHEM] The formation of anthracene derivatives by dehydration and cyclization of diaryl ketone compounds which have a methyl group or methylene group; heating to an elevated temperature is usually required. { 'elbs rē₁ak·shən }

ELDOR See electron electron double resonance. { 'el₁dȯr or ¦e¦el¦dē¦ō'är }

ELDORA See Electra Doppler Radar. { el'dȯr·ə }

elective culture [MICROBIO] A type of microorganism grown selectively from a mixed culture by culturing in a medium and under conditions selective for only one type of organism. { i¦lek·tiv 'kəl·chər }

electra [NAV] A continuous-wave radio navigation aid that uses special radio beacons to provide a number of equisignal zones (usually 24). { i'lek·trə }

Electra [ASTRON] A small, irregularly shaped satellite of Saturn that librates about the leading Lagrangian point of Dione's orbit. { i'lek·trə }

Electra complex [PSYCH] The female analog of the Oedipus complex, that is, the attraction and attachment of a female child to the father. { i'lek·trə ₁käm₁pleks }

Electra Doppler Radar [METEOROL] An airborne Doppler radar used for detecting and measuring weather phenomena, as well as meteorological research. Abbreviated ELDORA. { i¦lek·trə ¦däp·lər 'rā₁där }

electret [ELEC] A solid dielectric possessing persistent electric polarization, by virtue of a long time constant for decay of a charge instability. { i'lek₁tret }

electret headphone [ENG ACOUS] A headphone consisting of an electret transducer, usually in the form of a push-pull transducer. { i'lek₁tret 'hed₁fōn }

electret microphone [ENG ACOUS] A microphone consisting of an electret transducer in which the foil electret diaphragm is placed next to a perforated, ridged, metal or metal-coated backplate, and output voltage, taken between diaphragm and backplate, is proportional to the displacement of the diaphragm. { i'lek₁tret 'mī·krə₁fōn }

electret transducer [ELECTR] An electroacoustic or electromechanical transducer in which a foil electret, stretched out to form a diaphragm, is placed next to a metal or metal-coated plate, and motion of the diaphragm is converted to voltage between diaphragm and plate, or vice versa. { i'lek₁tret tranz'dü·sər }

electric [ELEC] Containing, producing, arising from, or actuated by electricity; often used interchangeably with electrical. { i'lek·trik }

electrical [ELEC] Related to or associated with electricity, but not containing it or having its properties or characteristics; often used interchangeably with electric. { ə'lek·trə·kəl }

electrical analog [PHYS] An electric circuit whose behavior may be described by the same mathematical equations as some physical system under study. { ə'lek·trə·kəl 'an·ə₁läg }

electrical angle [ELEC] An angle that specifies a particular instant in an alternating-current cycle or expresses the phase difference between two alternating quantities; usually expressed in electrical degrees. { ə'lek·trə·kəl 'aŋ·gəl }

electrical axis [SOLID STATE] The x axis in a quartz crystal; there are three such axes in a crystal, each parallel to one pair of opposite sides of the hexagon; all pass through and are perpendicular to the optical, or z, axis. { ə'lek·trə·kəl 'ak·səs }

electrical blasting cap [ENG] A blasting cap ignited by electric current and not by a spark. { ə'lek·trə·kəl 'blast·iŋ ₁kap }

electrical breakdown See breakdown. { ə'lek·trə·kəl 'brāk₁daún }

electrical calorimeter [ANALY CHEM] Device to measure heat evolved (from fusion or vaporization, for example); measured quantities of heat are added electrically to the sample, and the temperature rise is noted. { ə'lek·trə·kəl kal·ə'rim·əd·ər }

electrical center [ELEC] Point approximately midway between the ends of an inductor or resistor that divides the inductor or resistor into two equal electrical values. { ə'lek·trə·kəl 'sen·tər }

electrical circuit theory See circuit theory. { ə'lek·trə·kəl 'sər·kət ₁thē·ə·rē }

electrical code [ELEC] A systematic body of rules governing the practical application and installation of electrically operated equipment and devices and electric wiring systems. { ə'lek·trə·kəl 'kōd }

electrical conductance See conductance. { ə'lek·trə·kəl kən'dək·təns }

electrical conduction See conduction. { ə'lek·trə·kəl kən'dək·shən }

electrical conductivity See conductivity. { ə'lek·trə·kəl ₁kän₁däk'tiv·əd·ē }

electrical conductivity analyzer [ELEC] Alternating-current, resistance-bridge device used to measure the electrical conductivity of solutions, slurries, or wet solids. { ə'lek·trə·kəl ₁kän₁dək'tiv·əd·ē 'an·ə₁līz·ər }

electrical degree [ELEC] A unit equal to 1/360 cycle of an alternating quantity. { i'lek·trə·kəl də'grē }

electrical discharge machining See electron discharge machining. { i'lek·trə·kəl 'dis₁chärj mə₁shēn·iŋ }

electrical disintegration [MET] Removing excess metal by using an electric spark in air. { i'lek·trə·kəl dis₁in·tə'grā·shən }

electrical distance [ELECTROMAG] The distance between two points, expressed in terms of the duration of travel of an electromagnetic wave in free space between the two points. { i'lek·trə·kəl 'dis·təns }

electrical drainage [ELEC] Diversion of electric currents from subterranean pipes to prevent electrolytic corrosion. { i'lek·trə·kəl 'drān·ij }

electrical engineer [ENG] An engineer whose training includes a degree in electrical engineering from an accredited college or university (or who has comparable knowledge and experience), to prepare him or her for dealing with the generation, transmission, and utilization of electric energy. { i'lek·trə·kəl ₁en·jə'nir }

electrical engineering [ENG] Engineering that deals with practical applications involving current flow through conductors, as in motors and generators. { i'lek·trə·kəl ₁en·jə'nir·iŋ }

electrical equipment [ELEC] Apparatus, appliances, devices, wiring, fixtures, fittings, and material used as a part of or in connection with an electrical installation. { i'lek·trə·kəl i'kwip·mənt }

electrical equivalent [ANALY CHEM] In conductometric analyses of electrolyte solutions, an outside, calibrated current source as compared to (equivalent to) the current passing through the sample under analysis; for example, a Wheatstone-bridge balanced reading. { i'lek·trə·kəl i'kwiv·ə·lənt }

electrical fault See fault. { i'lek·trə·kəl 'fȯlt }

electrical image [ENG] An image that is obtained in the course of borehole logging and is based on electrical rather than optical contrasts. { i¦lek·trə·kəl 'im·ij }

electrical impedance Also known as impedance. [ELEC] **1.** The total opposition that a circuit presents to an alternating current, equal to the complex ratio of the voltage to the current in complex notation. Also known as complex impedance. **2.** The ratio of the maximum voltage in an alternating-current circuit to the maximum current; equal to the magnitude of the quantity in the first definition. { i'lek·trə·kəl im'pēd·əns }

electrical impedance meter [ELEC] An instrument which measures the complex ratio of voltage to current in a given circuit at a given frequency. Also known as impedance meter. { i'lek·trə·kəl im'pēd·əns ₁mēd·ər }

electrical instability [ELEC] A persistent condition of unwanted self-oscillation in an amplifier or other electric circuit. { i'lek·trə·kəl ₁in·stə'bil·əd·ē }

electrical insulating paper See insulating paper. { i'lek·trə·kəl 'in·sə₁lād·iŋ ₁pāp·ər }

electrical insulation See insulation. { i'lek·trə·kəl ₁in·sə'lā·shən }

electrical insulator See insulator. { i'lek·trə·kəl 'in·sə₁lād·ər }

electrical interference See interference. { i'lek·trə·kəl ₁in·tər'fir·əns }

electrical length [ELECTROMAG] The length of a conductor expressed in wavelengths, radians, or degrees. { i'lek·trə·kəl 'leŋkth }

electrical loading See loading. { i'lek·trə·kəl 'lōd·iŋ }

electrical log [ENG] Recorded measurement of the conductivities and resistivities down the length of uncased borehole; gives a complete record of the formations penetrated. { i'lek·trə·kəl 'läg }

electrical logging [ENG] The recording in uncased sections of a borehole of the conductivities and resistivities of the penetrated formations; used for geological correlations of the strata and evaluation of possibly productive horizons. Also known as electrical well logging. { i'lek·trə·kəl 'läg·iŋ }

electrically active fluid [PHYS CHEM] A fluid whose properties are altered by either an electric field (electrorheological fluid) or a magnetic field (ferrofluid). { i'lek·trə·klē ˌak·tiv 'flü·əd }

electrically alterable read-only memory [COMPUT SCI] A read-only memory that can be reprogrammed electrically in the field a limited number of times, after the entire memory is erased by applying an appropriate electric field. Abbreviated EAROM. { i'lek·trə·klē 'ȯl·trə·bəl 'rēd ˌōn·lē 'mem·rē }

electrically connected [ELEC] Connected by means of a conducting path, or through a capacitor, as distinguished from connection merely through electromagnetic induction. { i'lek·trə·klē kə'nek·təd }

electrically erasable programmable read-only memory [COMPUT SCI] An integrated-circuit memory chip that has an internal switch to permit a user to erase the contents of the chip and write new contents into it by means of electrical signals. Abbreviated EEPROM. { i'lek·trə·klē i'rās·ə·bəl prō'gram·ə·bəl 'rēd ˌōn·lē 'mem·rē }

electrically suspended gyro [ENG] A gyroscope in which the main rotating element is suspended by an electromagnetic or an electrostatic field. { i'lek·trə·klē səs'pen·dəd 'jī·rō }

electrical measurement [ELEC] The measurement of any one of the many quantities by which electricity is characterized. { i'lek·trə·kəl 'mezh·ər·mənt }

electrical model [ELEC] A model in the form of a mathematical description or an electrical equivalent circuit that represents the behavior of an electrical device or system. { i'lek·trə·kəl 'mäd·əl }

electrical noise [ELEC] Noise generated by electrical devices, for example, motors, engine ignition, power lines, and so on, and propagated to the receiving antenna direct from the noise source. { i'lek·trə·kəl 'nȯiz }

electrical oil See insulating oil. { i'lek·trə·kəl 'ȯil }

electrical porcelain See insulation porcelain. { i'lek·trə·kəl 'pȯrs·lən }

electrical potential energy [ELEC] Energy possessed by electric charges by virtue of their position in an electrostatic field. { i'lek·trə·kəl pə'ten·chəl'en·ər·jē }

electrical pressure transducer See pressure transducer. { i'lek·trə·kəl 'presh·ər tranz,dü·sər }

electrical properties [ELEC] Properties of a substance which determine its response to an electric field, such as its dielectric constant or conductivity. { i'lek·trə·kəl 'präp·ərd·ēz }

electrical prospecting [ENG] The use of downhole electrical logs to obtain subsurface information for geological analysis. { i'lek·trə·kəl 'präs,pek·tiŋ }

electrical resistance See resistance. { i'lek·trə·kəl ri'zis·təns }

electrical-resistance meter See resistance meter. { i'lek·trə·kəl ri'zis·təns ˌmēd·ər }

electrical-resistance strain gage [ENG] A vibration-measuring device consisting of a grid of fine wire cemented to the vibrating object to measure fluctuating strains. { i'lek·trə·kəl ri'zis·təns 'strān ˌgāj }

electrical-resistance thermometer See resistance thermometer. { i'lek·trə·kəl ri'zis·təns thər'mäm·əd·ər }

electrical resistivity [ELEC] The electrical resistance offered by a material to the flow of current, times the cross-sectional area of current flow and per unit length of current path; the reciprocal of the conductivity. Also known as resistivity; specific resistance. { i'lek·trə·kəl ˌrē·zis'tiv·əd·ē }

electrical resistor See resistor. { i'lek·trə·kəl ri'zis·tər }

electrical resonator See tank circuit. { i'lek·trə·kəl 'rez·ən,ād·ər }

electrical steel [MET] Low carbon-iron alloy containing 0.5–5% silicon, produced in an electric-arc furnace and used for the cores of transformers, alternators, and other iron-core electric machines. { i'lek·trə·kəl 'stēl }

electrical storm [METEOROL] A popular term for a thunderstorm. { i'lek·trə·kəl 'stȯrm }

electrical symbol [ELEC] A simple geometrical symbol used to represent a component of a circuit in a schematic circuit diagram. { i'lek·trə·kəl 'sim·bəl }

electrical synapse [NEUROSCI] An anatomically specialized junction between two nerve cells at which one cell influences the other by means of electrical current from one flowing directly into the other. { i'lek·trə·kəl 'si,naps }

electrical system [ELEC] System of wiring, switches, relays, and other equipment associated with receiving and distributing electricity. { i'lek·trə·kəl ˌsis·təm }

electrical tape See insulating tape. { i'lek·trə·kəl ˌtāp }

electrical thickness [OCEANOGR] The vertical measure between the surface of an ocean current and an isokinetic point having a value of about one-tenth the surface speed. { i'lek·trə·kəl 'thik·nəs }

electrical transcription See transcription. { i'lek·trə·kəl tranz'krip·shən }

electrical unit [ELEC] A standard in terms of which some electrical quantity is evaluated. { i'lek·trə·kəl 'yü·nət }

electrical weighing system [ENG] An instrument which weighs an object by measuring the change in resistance caused by the elastic deformation of a mechanical element loaded with the object. { i'lek·trə·kəl 'wā·iŋ ˌsis·təm }

electrical well logging See electrical logging. { i'lek·trə·kəl 'wel ˌläg·iŋ }

electrical zero [ELEC] A standard reference position from which rotor angles are measured in synchros and other rotating devices. { i'lek·trə·kəl 'zir·ō }

electric anesthesia [MED] Anesthesia produced by electrical means, as with interrupted direct current. { i'lek·trik ˌan·əs'thē·zhə }

electric arc [ELEC] A discharge of electricity through a gas, normally characterized by a voltage drop approximately equal to the ionization potential of the gas. Also known as arc. { i'lek·trik 'ärk }

electric-arc furnace See arc furnace. { i'lek·trik ˌärk 'fər·nəs }

electric-arc heating See arc heating. { i'lek·trik ˌärk 'hēd·iŋ }

electric-arc lamp See arc lamp. { i'lek·trik ˌärk 'lamp }

electric-arc spraying [MET] A thermal spraying process with an electric arc as a heat source and with compressed gas to propel the material. { i'lek·trik ˌärk 'sprā·iŋ }

electric-arc welding [MET] Welding in which the joint is heated to fusion by an electric arc or by a large electric current. Also known as arc welding. { i'lek·trik ˌärk 'weld·iŋ }

electric battery See battery. { i'lek·trik 'bad·ə·rē }

electric boiler [MECH ENG] A steam generator using electric energy, in immersion, resistor, or electrode elements, as the source of heat. { i'lek·trik 'bȯil·ər }

electric brake [MECH ENG] An actuator in which the actuating force is supplied by current flowing through a solenoid, or through an electromagnet which is thereby attracted to disks on the rotating member, actuating the brake shoes; this force is counteracted by the force of a compression spring. Also known as electromagnetic brake. { i'lek·trik 'brāk }

electric bridge See bridge. { i'lek·trik 'brij }

electric burn [MED] A burn caused by electric current. { i'lek·trik 'bərn }

electric calamine See hemimorphite. { i'lek·trik 'kal·ə,mīn }

electric car [MECH ENG] An automotive vehicle that is propelled by one or more electric motors powered by a special rechargeable electric battery rather than by an internal combustion engine. { i'lek·trik 'kär }

electric cell [ELEC] **1.** A single unit of a primary or secondary battery that converts chemical energy into electric energy. **2.** A single unit of a device that converts radiant energy into electric energy, such as a nuclear, solar, or photovoltaic cell. { i'lek·trik 'sel }

electric charge See charge. { i'lek·trik 'chärj }

electric chopper [ELECTROMAG] A chopper in which an electromagnet driven by a source of alternating current sets into vibration a reed carrying a moving contact that alternately touches two fixed contacts in a signal circuit, thus periodically interrupting the signal. { i'lek·trik 'chäp·ər }

electric circuit [ELEC] Also known as circuit. **1.** A path or group of interconnected paths capable of carrying electric

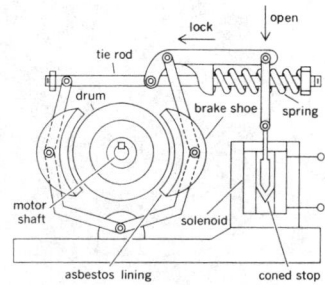

ELECTRIC BRAKE

Actuating force applied to a double-block electric brake.

currents. **2.** An arrangement of one or more complete, closed paths for electron flow. { i¦lek·trik 'sər·kət }

electric circuit theory See circuit theory. { i¦lek·trik 'sər·kət ˌthē·ə·rē }

electric clock [HOROL] **1.** Any clock that is operated by electric power. **2.** Specifically, a clock driven by an alternating-current motor whose current has a definite frequency, controlled at the generator. { i¦lek·trik 'kläk }

electric coil See coil. { i¦lek·trik 'kȯil }

electric comparator [ELEC] A comparator in which movement results in a change in some electrical quantity, which is then amplified by electrical means. { i¦lek·trik kəm'par·əd·ər }

electric condenser See capacitor. { i¦lek·trik kən'den·sər }

electric conductor See conductor. { i¦lek·trik kən'dək·tər }

electric connection [ELEC] A direct wire path for current between two points in a circuit. { i¦lek·trik kə'nek·shən }

electric connector [ELEC] A device that joins electric conductors mechanically and electrically to other conductors and to the terminals of apparatus and equipment. { i¦lek·trik kə'nek·tər }

electric constant [ELEC] The permittivity of empty space, equal to 1 in centimeter-gram-second electrostatic units and to $10^7/4\pi c^2$ farads per meter or, numerically, to 8.854×10^{-12} farad per meter in International System units, where c is the speed of light in meters per second. Symbolized ϵ_0. { i¦lek·trik 'kän·stənt }

electric contact [ELEC] A physical contact that permits current flow between conducting parts. Also known as contact. { i¦lek·trik 'kän,takt }

electric contactor See contactor. { i¦lek·trik 'kän,tak·tər }

electric control [ELEC] The control of a machine or device by switches, relays, or rheostats, as contrasted with electronic control by electron tubes or by devices that do the work of electron tubes. { i¦lek·trik kən'trōl }

electric controller [ELEC] A device that governs in some predetermined manner the electric power delivered to apparatus. { i¦lek·trik kən'trōl·ər }

electric converter See synchronous converter. { i¦lek·trik kən'vərd·ər }

electric corona See corona discharge. { i¦lek·trik kə'rō·nə }

electric coupling [MECH ENG] Magnetic-field coupling between the shafts of a driver and a driven machine. { i¦lek·trik 'kəp·liŋ }

electric current See current. { i¦lek·trik 'kə·rənt }

electric current density See current density. { i¦lek·trik ¦kə·rənt ˌden·səd·ē }

electric current meter See ammeter. { i¦lek·trik ¦kə·rənt ˌmēd·ər }

electric cutout See cutout. { i¦lek·trik 'kəd,aut }

electric delay line [ELECTR] A delay line using properties of lumped or distributed capacitive and inductive elements; can be used for signal storage by recirculating information-carrying wave patterns. { i¦lek·trik di'lā ˌlīn }

electric desalting [CHEM ENG] A process to remove impurities such as inorganic salts from crude oil by settling out in an electrostatic field. { i¦lek·trik dē'sȯlt·iŋ }

electric detonator [ENG] A detonator ignited by a fuse wire which serves to touch off the primer. { i¦lek·trik 'det·ən,ād·ər }

electric dipole [ELEC] A localized distribution of positive and negative electricity, without net charge, whose mean positions of positive and negative charges do not coincide. { i¦lek·trik 'dī,pōl }

electric dipole moment [ELEC] A quantity characteristic of a charge distribution, equal to the vector sum over the electric charges of the product of the charge and the position vector of the charge. { i¦lek·trik 'dī,pōl ˌmō·mənt }

electric dipole transition [ATOM PHYS] A transition of an atom or nucleus from one energy state to another, in which electric dipole radiation is emitted or absorbed. { i¦lek·trik 'dī,pōl tran'zish·ən }

electric discharge See discharge. { i¦lek·trik 'dis,chärj }

electric-discharge lamp See discharge lamp. { i'lek·trik 'dis,chärj ˌlamp }

electric-discharge tube See discharge tube. { i'lek·trik 'dis,chärj ˌtüb }

electric displacement [ELEC] The electric field intensity multiplied by the permittivity. Symbolized D. Also known

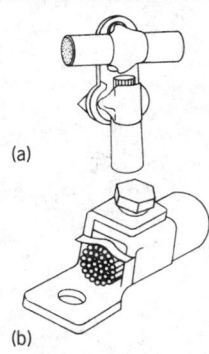

ELECTRIC CONNECTOR

(a)

(b)

Types of electric connectors. (a) T connector. (b) Terminal connector.

as dielectric displacement; dielectric flux density; displacement; electric displacement density; electric flux density; electric induction. { i'lek·trik dis'plās·mənt }

electric displacement density See electric displacement. { i'lek·trik dis'plās·mənt ˌden·səd·ē }

electric distribution system See distribution system. { i'lek·trik ˌdis·trə'byü·shən ˌsis·təm }

electric double layer [PHYS CHEM] A phenomenon found at a solid-liquid interface; it is made up of ions of one charge type which are fixed to the surface of the solid and an equal number of mobile ions of the opposite charge which are distributed through the neighboring region of the liquid; in such a system the movement of liquid causes a displacement of the mobile counterions with respect to the fixed charges on the solid surface. Also known as double layer. { i'lek·trik ¦dəb·əl 'lā·ər }

electric doublet See dipole. { i'lek·trik 'dəb·lət }

electric drive [MECH ENG] A mechanism which transmits motion from one shaft to another and controls the velocity ratio of the shafts by electrical means. { i'lek·trik 'drīv }

electric eel [VERT ZOO] *Electrophorus electricus.* An eellike cypriniform electric fish of the family Gymnotidae. { i'lek·trik 'ēl }

electric energy [ELECTROMAG] **1.** Energy of electric charges by virtue of their position in an electric field. **2.** Energy of electric currents by virtue of their position in a magnetic field. { i'lek·trik 'en·ər·jē }

electric energy measurement [ELEC] The measurement of the integral, with respect to time, of the power in an electric circuit. { i'lek·trik 'en·ər·jē ˌmezh·ər·mənt }

electric energy meter [ELEC] A device which measures the integral, with respect to time, of the power in an electric circuit. { i'lek·trik ¦enər·jē ˌmēd·ər }

electric engine [AERO ENG] A rocket engine in which the propellant is accelerated by some electric device. Also known as electric propulsion system; electric rocket. { i'lek·trik 'en·jən }

electric eye See cathode-ray tuning indicator; photocell; phototube. { i'lek·trik 'ī }

electric fence [ENG] A fence consisting of one or more lengths of wire energized with high-voltage, low-current pulses, and giving a warning shock when touched. { i'lek·trik 'fens }

electric field [ELEC] **1.** One of the fundamental fields in nature, causing a charged body to be attracted to or repelled by other charged bodies; associated with an electromagnetic wave or a changing magnetic field. **2.** Specifically, the electric force per unit test charge. { i'lek·trik 'fēld }

electric-field effect See Stark effect. { i'lek·trik ¦fēld i'fekt }

electric-field intensity See electric-field vector. { i'lek·trik ¦fēld in'ten·səd·ē }

electric-field strength See electric-field vector. { i'lek·trik ¦fēld 'streŋkth }

electric-field vector [ELEC] The force on a stationary positive charge per unit charge at a point in an electric field. Designated E. Also known as electric-field intensity; electric-field strength; electric vector. { i'lek·trik ¦fēld 'vek·tər }

electric filter [ELECTR] **1.** A network that transmits alternating currents of desired frequencies while substantially attenuating all other frequencies. Also known as frequency-selective device. **2.** See filter. { i'lek·trik 'fil·tər }

electric firing mechanism [ORD] Firing mechanism using a firing magneto, battery, or alternating-current power in circuit with an electric primer; one side of the line is connected by an insulated wire to the primer, and the other side is grounded to the frame of the weapon. { i'lek·trik 'fīr·iŋ ˌmek·ə,niz·əm }

electric fish [VERT ZOO] Any of several fishes capable of producing electric discharges from an electric organ. { i'lek·trik 'fish }

electric flowmeter [ELEC] Fluid-flow measurement device relying on an inductance or impedance bridge or on electrical-resistance rod elements to sense flow-rate variations. { i'lek·trik 'flō,mēd·ər }

electric flux [ELEC] **1.** The integral over a surface of the component of the electric displacement perpendicular to the surface; equal to the number of electric lines of force crossing the surface. **2.** The electric lines of force in a region. { i'lek·trik 'fləks }

electric flux density See electric displacement. { i'lek·trik 'fləks ˌden·səd·ē }

electric flux line See electric line of force. { i¦lek·trik 'fləks ˌlīn }

electric forming [ELECTR] The process of applying electric energy to a semiconductor or other device to modify permanently its electrical characteristics. { i¦lek·trik 'fȯr·miŋ }

electric furnace [ENG] A furnace which uses electricity as a source of heat. { i¦lek·trik 'fər·nəs }

electric-furnace steel [MET] Steel produced in an electric furnace. { i¦lek·trik ˌfər·nəs 'stēl }

electric fuse See fuse. { i¦lek·trik 'fyüz }

electric gathering locomotive See gathering motor. { i¦lek·trik ¦gath·ə·riŋ ˌlō·kə'mōd·iv }

electric generator See generator. { i¦lek·trik 'jen·ə·ˌrād·ər }

electric glass See E glass. { i¦lek·trik ˌglas }

electric guitar [ENG ACOUS] A guitar in which a contact microphone placed under the strings picks up the acoustic vibrations for amplification and for reproduction by a loudspeaker. { i¦lek·trik gə'tär }

electric hammer [MECH ENG] An electric-powered hammer; often used for riveting or caulking. { i¦lek·trik 'ham·ər }

electric heating [ENG] Any method of converting electric energy to heat energy by resisting the free flow of electric current. { i¦lek·trik 'hēd·iŋ }

electric hygrometer [ENG] An instrument for indicating by electrical means the humidity of the ambient atmosphere; usually based on the relation between the electric conductance of a film of hygroscopic material and its moisture content. { i¦lek·trik hī'gräm·əd·ər }

electric hysteresis See ferroelectric hysteresis. { i¦lek·trik ˌhis·tə'rē·səs }

electrician [ENG] A skilled worker who installs, repairs, maintains, or operates electric equipment. { i¦lek'trish·ən }

electric ignition [MECH ENG] Ignition of a charge of fuel vapor and air in an internal combustion engine by passing a high-voltage electric current between two electrodes in the combustion chamber. { i¦lek·trik ig'nish·ən }

electric image [ELEC] A fictitious charge used in finding the electric field set up by fixed electric charges in the neighborhood of a conductor; the conductor, with its distribution of induced surface charges, is replaced by one or more of these fictitious charges. Also known as image. { i¦lek·trik 'im·ij }

electric induction See electric displacement. { i¦lek·trik in'dək·shən }

electric instrument [ENG] An electricity-measuring device that indicates, such as an ammeter or voltmeter, in contrast to an electric meter that totalizes or records. { i¦lek·trik 'in·strə·mənt }

electricity [PHYS] Physical phenomenon involving electric charges and their effects when at rest and when in motion. { i¦lek'tris·əd·ē }

electric lamp [ELEC] A lamp in which light is produced by electricity, as the incandescent lamp, arc lamp, glow lamp, mercury-vapor lamp, and fluorescent lamp. { i¦lek·trik 'lamp }

electric line of force [ELEC] An imaginary line drawn so that each segment of the line is parallel to the direction of the electric field or of the electric displacement at that point, and the density of the set of lines is proportional to the electric field or electrical displacement. Also known as electric flux line. { i¦lek·trik ˌlīn əv 'fȯrs }

electric locomotive [MECH ENG] A locomotive operated by electric power picked up from a system of continuous overhead wires, or, sometimes, from a third rail mounted alongside the track. { i¦lek·trik ˌlō·kə'mōd·iv }

electric-magnetic duality [PART PHYS] The property of a physical theory in which the electric and magnetic charges exchange roles when the coupling between electric charges and electric fields, which is actually small, is made to be large, with electric charges becoming fuzzy, heavy, and strongly coupled, while the magnetic charges become pointlike, light, and weakly coupled. { i¦lek·trik mag¦ned·ik dü'al·əd·ē }

electric main See power transmission line. { i¦lek·trik 'mān }

electric meter [ENG] An electricity-measuring device that totalizes with time, such as a watthour meter or ampere-hour meter, in contrast to an electric instrument. { i¦lek·trik 'mēd·ər }

electric moment [ELEC] One of a series of quantities characterizing an electric charge distribution; an l-th moment is given by integrating the product of the charge density, the l-th power

of the distance from the origin, and a spherical harmonic Y^*_{lm} over the charge distribution. { i¦lek·trik 'mō·mənt }

electric monopole [ELEC] A distribution of electric charge which is concentrated at a point or is spherically symmetric. { i¦lek·trik 'män·ə·ˌpōl }

electric motor See motor. { i¦lek·trik 'mōd·ər }

electric multipole [ELECTROMAG] One of a series of types of static or oscillating charge distributions; the multipole of order 1 is a point charge or a spherically symmetric distribution, and the electric and magnetic fields produced by an electric multipole of order 2^n are equivalent to those of two electric multipoles of order 2^{-1} of equal strengths, but opposite sign, separated from each other by a short distance. { i¦lek·trik 'məl·tə·ˌpōl }

electric multipole field [ELECTROMAG] The electric and magnetic fields generated by a static or oscillating electric multipole. { i¦lek·trik 'məl·tə·ˌpōl ˌfēld }

electric network See network. { i¦lek·trik 'net·ˌwərk }

electric octupole moment [ELEC] A quantity characterizing an electric charge distribution; obtained by integrating the product of the charge density, the third power of the distance from the origin, and a spherical harmonic Y^*_{3m} over the charge distribution. { i¦lek·trik 'äk·tə·ˌpōl 'mō·mənt }

electric organ [VERT ZOO] An organ consisting of rows of electroplaques which produce an electric discharge. { i¦lek·trik 'ȯr·gən }

electric outlet See outlet. { i¦lek·trik 'aut·ˌlet }

electric polarizability [ELEC] Induced dipole moment of an atom or molecule in a unit electric field. { i¦lek·trik ˌpō·lə·rī·zə'bil·əd·ē }

electric polarization See polarization. { i¦lek·trik ˌpō·lə·rə'zā·shən }

electric potential [ELEC] The work which must be done against electric forces to bring a unit charge from a reference point to the point in question; the reference point is located at an infinite distance, or, for practical purposes, at the surface of the earth or some other large conductor. Also known as electrostatic potential; potential. Abbreviated V. { i¦lek·trik pə'ten·chəl }

electric power [ELEC] The rate at which electric energy is converted to other forms of energy, equal to the product of the current and the voltage drop. { i¦lek·trik 'pau·ər }

electric power generation [MECH ENG] The large-scale production of electric power for industrial, residential, and rural use, generally in stationary plants designed for that purpose. { i¦lek·trik ˌpau·ər ˌjen·ə'rā·shən }

electric power line See power line. { i¦lek·trik 'pau·ər ˌlīn }

electric power meter [ENG] A device that measures electric power consumed, either at an instant, as in a wattmeter, or averaged over a time interval, as in a demand meter. Also known as power meter. { i¦lek·trik 'pau·ər ˌmēd·ər }

electric power plant [MECH ENG] A power plant that converts a form of raw energy into electricity, for example, a hydro, steam, diesel, or nuclear generating station for stationary or transportation service. { i¦lek·trik 'pau·ər ˌplant }

electric power station [ELEC] A generating station or an electric power substation. { i¦lek·trik 'pau·ər ˌstā·shən }

electric power substation [ELEC] An assembly of equipment in an electric power system through which electric energy is passed for transmission, transformation, distribution, or switching. Also known as substation. { i¦lek·trik ˌpau·ər 'səb·ˌstā·shən }

electric power system [MECH ENG] A complex assemblage of equipment and circuits for generating, transmitting, transforming, and distributing electric energy. { i¦lek·trik ˌpau·ər ˌsis·təm }

electric power transmission [ELEC] Process of transferring electric energy from one point to another in an electric power system. { i¦lek·trik ˌpau·ər tranz·ˌmish·ən }

electric precipitation [CHEM ENG] A process that utilizes an electric field to improve the separation of hydrocarbon reagent dispersions. { i¦lek·trik prə·sip·ə'tā·shən }

electric pressure transducer See pressure transducer. { i¦lek·trik 'presh·ər tranz·ˌdü·sər }

electric probe [PL PHYS] A device used to measure electron temperatures, electron and ion densities, space and wall potentials, and random electron currents in a plasma; consists substantially of one or two small collecting electrodes to which various potentials are applied, with the corresponding collection

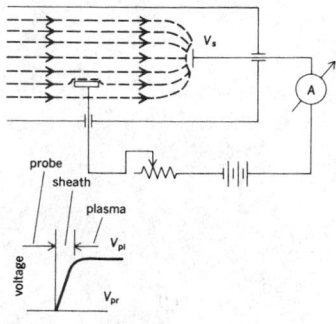

ELECTRIC PROBE

Schematic diagram of a single electric probe.

currents being measured. Also known as electrostatic probe. { i¦lek·trik 'prōb }

electric propulsion [AERO ENG] A general term encompassing all the various types of propulsion in which the propellant consists of electrically charged particles which are accelerated by electric or magnetic fields, or both. { i¦lek·trik prə'pəl·shən }

electric propulsion system *See* electric engine. { i¦lek·trik prə'pəl·shən ‚sis·təm }

electric protective device [ELEC] A particular type of equipment used in electric power systems to detect abnormal conditions and to initiate appropriate corrective action. Also known as protective device. { i¦lek·trik prə'tek·tiv di‚vīs }

electric quadrupole [ELEC] A charge distribution that produces an electric field equivalent to that produced by two electric dipoles whose dipole moments have the same magnitude but point in opposite directions and which are separated from each other by a small distance. { i¦lek·trik 'kwä·drə‚pōl }

electric quadrupole lens [ELECTR] A device for focusing beams of charged particles which has four electrodes with alternately positive and negative polarity; used in electron microscopes and particle accelerators. { i¦lek·trik 'kwä·drə‚pōl ‚lenz }

electric quadrupole moment [ELEC] A quantity characterizing an electric charge distribution, obtained by integrating the product of the charge density, the second power of the distance from the origin, and a spherical harmonic Y^*_{2m} over the charge distribution. { i¦lek·trik 'kwä·drə‚pōl ‚mō·mənt }

electric quadrupole transition [ATOM PHYS] A transition of an atom or molecule from one energy state to another, in which electric quadrupole radiation is emitted or absorbed. { i¦lek·trik 'kwä·drə‚pōl tran'zish·ən }

electric raceway *See* raceway. { i¦lek·trik 'rās‚wā }

electric railroad [MECH ENG] A railroad which has a system of continuous overhead wires or a third rail mounted alongside the track to supply electric power to the locomotive and cars. { i¦lek·trik 'rāl‚rōd }

electric reactor *See* reactor. { i¦lek·trik rē'ak·tər }

electric relay *See* relay. { i¦lek·trik 'rē‚lā }

electric resistance *See* resistance. { i¦lek·trik ri'zis·təns }

electric resistance furnace *See* resistance furnace. { i¦lek·trik ri'zis·təns ‚fər·nəs }

electric rocket *See* electric engine. { i¦lek·trik 'räk·ət }

electric rotating machinery [ELEC] Any form of apparatus which has a rotating member and generates, converts, transforms, or modifies electric power, such as a motor, generator, or synchronous converter. { i¦lek·trik 'rō‚tād·iŋ mə‚shēn·rē }

electric scanning [ELECTR] Scanning in which the required changes in radar beam direction are produced by variations in phase or amplitude of the currents fed to the various elements of the antenna array. { i¦lek·trik 'skan·iŋ }

electric shielding [ELECTROMAG] Any means of avoiding pickup of undesired signals or noise, suppressing radiation of undesired signals, or confining wanted signals to desired paths or regions, such as electrostatic shielding or electromagnetic shielding. Also known as screening; shielding. { i¦lek·trik 'shēld·iŋ }

electric shock [PHYSIO] The sudden pain, convulsion, unconsciousness, or death produced by the passage of electric current through the body. { i¦lek·trik 'shäk }

electric shock tube [PL PHYS] A gas-filled tube used in plasma physics to ionize a gas suddenly; a capacitor bank charged to a high voltage is discharged into the gas at one tube end to ionize and heat the gas, producing a shock wave that may be studied as it travels down the tube. { i¦lek·trik 'shäk ‚tüb }

electric shunt *See* shunt. { i¦lek·trik 'shənt }

electric solenoid *See* solenoid. { i¦lek·trik 'sō·lə‚nȯid }

electric spark *See* spark. { i¦lek·trik 'spärk }

electric spark machining *See* electron discharge machining. { i¦lek·trik ¦spärk mə'shēn·iŋ }

electric stacker [MECH ENG] A stacker whose carriage is raised and lowered by a winch powered by electric storage batteries. { i¦lek·trik 'stak·ər }

electric steel [MET] Steel melted in an electric furnace which permits close control and the addition of alloying elements directly into the furnace. { i¦lek·trik 'stēl }

electric strength *See* dielectric strength. { i¦lek·trik 'streŋkth }

electric surface-recording thermometer [PETRO ENG] Device to measure temperatures during oil-well temperature surveying; has a thermocouple, resistance wire, or thermistor as the temperature-sensitive element. { i¦lek·trik ¦sər·fəs ri¦kȯrd·iŋ thər'mäm·əd·ər }

electric susceptibility [ELEC] A dimensionless parameter measuring the ease of polarization of a dielectric, equal (in meter-kilogram-second units) to the ratio of the polarization to the product of the electric field strength and the vacuum permittivity. Also known as dielectric susceptibility. { i¦lek·trik sə‚sep·tə'bil·əd·ē }

electric switch *See* switch. { i¦lek·trik 'swich }

electric switchboard *See* switchboard. { i¦lek·trik 'swich‚bȯrd }

electric tachometer [ENG] An instrument for measuring rotational speed by measuring the output voltage of a generator driven by the rotating unit. { i¦lek·trik tə'käm·əd·ər }

electric tank *See* electrolytic tank. { i¦lek·trik 'taŋk }

electric telemetering [COMMUN] System to transmit electric impulses from the primary detector to a remote receiving station, with or without wire interconnections. { i¦lek·trik ‚tel·ə'mēd·ə·riŋ }

electric terminal *See* terminal. { i¦lek·trik 'tərm·ən·əl }

electric thermometer [ENG] An instrument that utilizes electrical means to measure temperature, such as a thermocouple or resistance thermometer. { i¦lek·trik thər'mäm·əd·ər }

electric transducer [ELECTR] A transducer in which all of the waves are electric. { i¦lek·trik tranz'dü·sər }

electric transient [ELEC] A temporary component of current and voltage in an electric circuit which has been disturbed. { i¦lek·trik 'tran·zhənt }

electric tuning [ELECTR] Tuning a receiver to a desired station by switching a set of preadjusted trimmer capacitors or coils into the tuning circuits. { i¦lek·trik 'tün·iŋ }

electric twinning [SOLID STATE] A defect occurring in natural quartz crystals, in which adjacent regions of quartz have their electric axes oppositely poled. { i¦lek·trik 'twin·iŋ }

electric typewriter [MECH ENG] A typewriter having an electric motor that provides power for all operations initiated by the touching of the keys. { i¦lek·trik 'tīp‚rīd·ər }

electric vector *See* electric-field vector. { i¦lek·trik 'vek·tər }

electric vehicle [MECH ENG] A ground vehicle propelled by a motor powered by electrical energy from rechargeable batteries or other source onboard the vehicle, or from an external source in, on, or above the roadway; examples include the electrically powered golf cart, automobile, and trolley bus. { i¦lek·trik 'vē·ə·kəl }

electric wave [ELECTROMAG] An electromagnetic wave, especially one whose wavelength is at least a few centimeters. Also known as Hertzian wave. { i¦lek·trik 'wāv }

electric-wave filter *See* filter. { i¦lek·trik ¦wāv 'fil·tər }

electric wind *See* convective discharge. { i¦lek·trik 'wind }

electric wire *See* wire. { i¦lek·trik 'wīr }

electric wiring *See* wiring. { i¦lek·trik 'wīr·iŋ }

electride [INV ZOO] A member of a class of ionic compounds in which the anion is believed to be an electron. { i'lek‚trīd }

electrification [ELEC] **1.** The process of establishing a charge in an object. **2.** The generation, distribution, and utilization of electricity. { i‚lek·trə·fə¦kā·shən }

electrification ice nucleus [METEOROL] An ice nucleus that is formed by the fragmentation of dendritic crystals exposed to an electric field strength of several hundred volts per centimeter; it is a type of fragmentation nucleus. { i‚lek·trə·fə¦kā·shən ‚īs ‚nü·klē·əs }

electrization [ELEC] The electric polarization divided by the permittivity of empty space. { i‚lek·trə'zā·shən }

electroacoustic effect *See* acoustoelectric effect. { i¦lek·trō·ə¦kü·stik i'fekt }

electroacoustics [ENG ACOUS] The conversion of acoustic energy and waves into electric energy and waves, or vice versa. { i¦lek·trō·ə'kü·stiks }

electroacoustic transducer [ENG ACOUS] A transducer that receives waves from an electric system and delivers waves to an acoustic system, or vice versa. Also known as sound transducer. { i¦lek·trō·ə¦kü·stik tranz'dü·sər }

electrobalance [ANALY CHEM] Analytical microbalance utilizing electromagnetic weighing; the sample weight is balanced by the torque produced by current in a coil in a magnetic field, with torque proportional to the current. { i‚lek·trō'-bal·əns }

electrocaloric effect [SOLID STATE] A temperature change in certain crystals caused by alteration of the permanent polarization by application of an external electric field. { i¦lek·trō·kə¦lȯr·ik i'fekt }

electrocapillarity [PHYS] A change in the surface tension of a liquid caused by an electric field at the surface. { i‚lek·trō‚kap·ə'lar·əd·ē }

electrocardiogram [MED] A graphic recording of the electrical manifestations of the heart action as obtained from the body surfaces. Abbreviated ECG; EKG. { i‚lek·trō'kärd·ē·ə‚gram }

electrocardiograph [MED] The instrument used to obtain anelectrocardiogram. { i‚lek·trō'kärd·ē·ə‚graf }

electrocardiography [MED] The medical specialty concerned with the production and interpretation of electrocardiograms. { i¦lek·trō‚kärd·ē'äg·rə·fē }

electrocardiophonograph [MED] An instrument that records graphic traces of heart sounds to fix precisely the times at which valve action occurs and to reveal valvular defects which affect blood flow. { i¦lek·trō¦kärd·ē·ō'fō·nə‚graf }

electrocatalysis [CHEM] Any one of the mechanisms which produce a speeding up of half-cell reactions at electrode surfaces. { i‚lek·trō·kə'tal·ə·səs }

electrocauterization [MED] The application of a direct galvanic current to tissues to cause destruction or coagulation. { i‚lek·trō‚kȯd·ə·rə'zā·shən }

electroceramics [MATER] Ceramic materials having electrical and other properties which make them suitable for use as insulators for power lines and in electrical components. { i‚lek·trō·sə'ram·iks }

electrochemical cell [PHYS CHEM] A combination of two electrodes arranged so that an overall oxidation-reduction reaction produces an electromotive force; includes dry cells, wet cells, standard cells, fuel cells, solid-electrolyte cells, and reserve cells. { i‚lek·trō'kem·ə·kəl 'sel }

electrochemical cleaning [MET] Removing soil by the chemical action caused or sustained by a current of electricity in an electrolyte. Also known as electrolytic cleaning. { i‚lek·trō'kem·ə·kəl 'klēn·iŋ }

electrochemical coating [MET] A coating formed by chemical action on the metal surface and effected by a current of electricity through an electrolyte. { i‚lek·trō'kem·ə·kəl 'kōd·iŋ }

electrochemical corrosion [MET] Corrosion of a metal associated with the flow of electric current in an electrolyte. Also known as electrolytic corrosion. { i‚lek·trō'kem·ə·kəl kə'rō·zhən }

electrochemical effect [PHYS CHEM] Conversion of chemical to electric energy, as in electrochemical cells; or the reverse process, used to produce elemental aluminum, magnesium, and bromine from compounds of these elements. { i‚lek·trō'kem·ə·kəl i'fekt }

electrochemical emf [PHYS CHEM] Electrical force generated by means of chemical action, in manufactured cells (such as dry batteries) or by natural means (galvanic reaction). { i‚lek·trō'kem·ə·kəl ‚ē¦em'ef }

electrochemical equivalent [PHYS CHEM] The weight in grams of a substance produced or consumed by electrolysis with 100% current efficiency during the flow of a quantity of electricity equal to 1 faraday (96,485.34 coulombs). { i‚lek·trō'kem·ə·kəl i'kwiv·ə·lənt }

electrochemical gradient [CELL MOL] The combined effect of a solute's concentration gradient across a membrane and the electric charge gradient across the membrane (the membrane potential), which drives the solute to cross the membrane. { i‚lek·trō‚kem·i·kəl 'grād·ē·ənt }

electrochemical grinding See electrolytic grinding. { i‚lek·trō¦kem·i·kəl 'grīnd·iŋ }

electrochemical machining [MET] Removing excess metal by electrolytic dissolution, effected by the tool acting as the cathode against the workpiece acting as the anode. Abbreviated ECM. Also known as electrolytic machining. { i‚lek·trō'kem·ə·kəl mə'shēn·iŋ }

electrochemical potential [PHYS CHEM] The difference in potential that exists when two dissimilar electrodes are connected through an external conducting circuit and the two electrodes are placed in a conducting solution so that electrochemical reactions occur. { i‚lek·trō'kem·ə·kəl pə'ten·chəl }

electrochemical power generation [ENG] The direct conversion of chemical energy to electric energy, as in a battery or fuel cell. { i‚lek·trō'kem·ə·kəl 'pau̇·ər ‚jen·ə‚rā·shən }

electrochemical process [PHYS CHEM] **1.** A chemical change accompanying the passage of an electric current, especially as used in the preparation of commercially important quantities of certain chemical substances. **2.** The reverse change, in which a chemical reaction is used as the source of energy to produce an electric current, as in a battery. { i‚lek·trō'kem·ə·kəl 'präs·əs }

electrochemical proton gradient [CELL MOL] The combined effect of the electric charge gradient across a membrane and the pH gradient across the membrane (the membrane potential), which drives H⁺ and OH⁻ ions to cross the membrane. { i‚lek·trō‚kem·i·kəl 'prō‚tän ‚grād·ē·ənt }

electrochemical recording [ELECTR] Recording by means of a chemical reaction brought about by the passage of signal-controlled current through the sensitized portion of the record sheet. { i‚lek·trō'kem·ə·kəl ri'kȯrd·iŋ }

electrochemical reduction cell [PHYS CHEM] The cathode component of an electrochemical cell, at which chemical reduction occurs (while at the anode, chemical oxidation occurs). { i‚lek·trō'kem·ə·kəl ri'dək·shən ‚sel }

electrochemical series [PHYS CHEM] A series in which the metals and other substances are listed in the order of their chemical reactivity or electrode potentials, the most reactive at the top and the less reactive at the bottom. Also known as electromotive series. { i‚lek·trō'kem·ə·kəl 'sir·ēz }

electrochemical techniques [PHYS CHEM] The experimental methods developed to study the physical and chemical phenomena associated with electron transfer at the interface of an electrode and solution. { i‚lek·trō'kem·ə·kəl tek'nēks }

electrochemical thermodynamics [THERMO] The application of the laws of thermodynamics to electrochemical systems. { i‚lek·trō'kem·ə·kəl ‚thərm·ō·dī'nam·iks }

electrochemical transducer [ENG] A device which uses a chemical change to measure the input parameter; the output is a varying electrical signal proportional to the measurand. { i‚lek·trō'kem·ə·kəl tranz'dü·sər }

electrochemical valve [ELEC] Electric valve consisting of a metal in contact with a solution or compound, across the boundary of which current flows more readily in one direction than in the other direction, and in which the valve action is accompanied by chemical changes. { i‚lek·trō'kem·ə·kəl 'valv }

electrochemiluminescence [PHYS CHEM] Emission of light produced by an electrochemical reaction. Also known as electrogenerated chemiluminescence. { i‚lek·trō‚kem·ē·ə‚lüm·ə'nes·əns }

electrochemistry [PHYS CHEM] A branch of chemistry dealing with chemical changes accompanying the passage of an electric current; or with the reverse process, in which a chemical reaction is used to produce an electric current. { i¦lek·trō¦kem·ə·strē }

electrochromatography [ANALY CHEM] Type of chromatography that utilizes application of an electric potential to produce an electric differential. Also known as electropherography. { i¦lek·trō‚krō·mə'täg·rə·fē }

electrochromic device [ENG] A self-contained, hermetically sealed, two-electrode electrolytic cell that includes one or more electrochromic materials and an electrolyte. { i‚lek·trə¦krōm·ik di'vīs }

electrochromic display [ELECTR] A solid-state passive display that uses organic or inorganic insulating solids which change color when injected with positive or negative charges. { i¦lek·trō¦krō·mik di'splā }

electrochromic material [MATER] An organic or inorganic substance that can interconvert between two or more colored states upon oxidation or reduction, that is, upon electrolytic loss or gain of electrons. { i‚lek·trə¦krōm·ik mə'tir·ē·əl }

electrocoagulation [MED] The coagulation of tissue by means of a high-frequency electric current. { i¦lek·trō·kō‚ag·yə'lā·shən }

electroconvulsive shock [MED] The technique of eliciting convulsions by applying an electric current through the brain for a brief period. { i¦lek·trō·kən¦vəl·siv 'shäk }

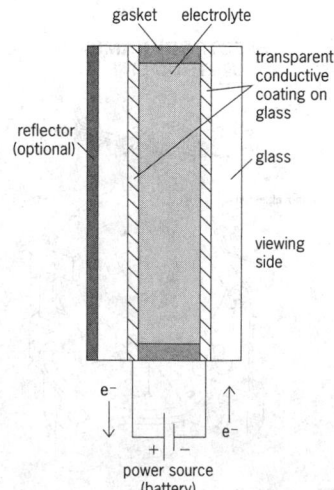

ELECTROCHROMIC DEVICE

gasket · electrolyte · transparent conductive coating on glass · reflector (optional) · glass · viewing side · e⁻ · e⁻ · power source (battery)

The electrochromic materials can be either dissolved in the electrolyte or coated on the transparent electrodes.

electroconvulsive therapy [PSYCH] The technique of eliciting convulsions by applying an electric current through the brain of a human or an experimental animal for a brief period by means of electrodes that are placed on the head; sometimes used as a treatment for severe mental depression. { i¦lek·trō·kən¦vəl·səv 'ther·ə·pē }

electrocorticogram [MED] The record obtained by electrocorticography. { i¦lek·trō'kȯrd·ə·kə‚gram }

electrocorticography [MED] The technique of surveying the electrical activity of the cerebral cortex. { i¦lek·trə‚kȯrd·ə'käg·rə·fē }

electrocratic [CHEM] Referring to the repulsion exhibited by soap films and other colloids in solutions; such repulsion involves a strong osmotic contribution but is largely controlled by electrical forces. { i‚lek·trō'krad·ik }

electrocyclic reaction [PHYS CHEM] The interconversion of a linear π-system containing n π-electrons and a cyclic molecule containing $(n-2)$ π-electrons which is formed by joining the ends of the linear molecule. { i‚lek·trō¦sī·klik rē'ak·shən }

electrode [ELEC] **1.** An electric conductor through which an electric current enters or leaves a medium, whether it be an electrolytic solution, solid, molten mass, gas, or vacuum. **2.** One of the terminals used in dielectric heating or diathermy for applying the high-frequency electric field to the material being heated. { i'lek‚trōd }

electrode admittance [ELECTR] Quotient of dividing the alternating component of the electrode current by the alternating component of the electrode voltage, all other electrode voltages being maintained constant. { i'lek‚trōd ad'mit·əns }

electrodecantation [PHYS CHEM] A modification of electrodialysis in which a cell is divided into three sections by two membranes and electrodes are placed in the end sections; colloidal matter is concentrated at the sides and bottom of the middle section, and the liquid that floats to the top is drawn off. { i¦lek·trō‚dē‚kan'tā·shən }

electrode capacitance [ELECTR] Capacitance between one electrode and all the other electrodes connected together. { i'lek‚trōd kə'pas·əd·əns }

electrode characteristic [ELECTR] Relation between the electrode voltage and the current to an electrode, all other electrode voltages being maintained constant. { i'lek‚trōd ‚kar·ik·tə'ris·tik }

electrode conductance [ELECTR] Quotient of the inphase component of the electrode alternating current by the electrode alternating voltage, all other electrode voltage being maintained constant; this is a variational and not a total conductance. Also known as grid conductance. { i'lek‚trōd kən'dək·təns }

electrode couple [ELEC] The pair of electrodes in an electric cell, between which there is a potential difference. { i'lek‚trōd ‚kə·pəl }

electrode current [ELECTR] Current passing to or from an electrode, through the interelectrode space within a vacuum tube. { i'lek‚trōd ‚kə·rənt }

electrode dark current [ELECTR] The electrode current that flows when there is no radiant flux incident on the photocathode in a phototube or camera tube. Also known as dark current. { i'lek‚trōd ¦därk ‚kə·rənt }

electrode dissipation [ELECTR] Power dissipated in the form of heat by an electrode as a result of electron or ion bombardment. { i'lek‚trōd ‚dis·ə'pā·shən }

electrode drop [ELECTR] Voltage drop in the electrode due to its resistance. { i'lek‚trōd ‚dräp }

electrode efficiency [PHYS CHEM] The ratio of the amount of metal actually deposited in an electrolytic cell to the amount that could theoretically be deposited as a result of electricity passing through the cell. { i'lek‚trōd ə‚fish·ən·sē }

electrode force [MET] The force that occurs between electrodes during seam, spot, and projection welding. Also known as welding force. { i'lek‚trōd ‚fȯrs }

electrode impedance [ELECTR] Reciprocal of the electrode admittance. { i'lek‚trōd im'pēd·əns }

electrode inverse current [ELECTR] Current flowing through an electrode in the direction opposite to that for which the tube is designed. { i'lek‚trōd 'in·vərs ‚kə·rənt }

electrodeless discharge [ELECTR] An electric discharge generated by placing a discharge tube in a strong, high-frequency electromagnetic field. { i¦lek‚trōd·ləs 'dis‚chärj }

electrodeless lamp [ELECTR] A lamp based on an electrodeless discharge. { i¦lek‚trōd·ləs 'lamp }

electrodeposition [MET] Electrolytic process in which a metal is deposited at the cathode from a solution of its ions; includes electroplating and electroforming. Also known as electrolytic deposition. { i¦lek·trō‚dep·ə'zish·ən }

electrodeposition analysis [ANALY CHEM] An electroanalytical technique in which an element is quantitatively deposited on an electrode. { i¦lek·trō‚dep·ə'zish·ən ə'nal·ə·səs }

electrode potential Also known as electrode voltage. [ELECTR] The instantaneous voltage of an electrode with respect to the cathode of an electron tube. [PHYS CHEM] The voltage existing between an electrode and the solution or electrolyte in which it is immersed; usually, electrode potentials are referred to a standard electrode, such as the hydrogen electrode. { i'lek‚trōd pə'ten·chəl }

electrode resistance [ELECTR] Reciprocal of the electrode conductance; this is the effective parallel resistance and is not the real component of the electrode impedance. { i'lek‚trōd ri'zis·təns }

electrodermal response See galvanic skin response. { i‚lek·trə¦dərm·əl ri'späns }

electrodermography [MED] The recording of the electrical resistance of the skin. { i¦lek·trō·dər'mäg·rə·fē }

electrodesiccation [MED] The use of a single terminal electrode with a small sparking distance to destroy lesions or seal off blood vessels. { i¦lek·trō‚des·ə'kā·shən }

electrode skid [MET] Sliding of an electrode over the work surface in spot, seam, or projection welding. { i'lek‚trōd ‚skid }

electrode-type liquid-level meter [ENG] Device that senses liquid level by the effect of the liquid-gas interface on the conductance of an electrode or probe. { i'lek‚trōd ‚tīp ¦lik·wəd ¦lev·əl 'mēd·ər }

electrode voltage See electrode potential. { i'lek‚trōd ‚vōl·tij }

electrodiagnosis [MED] Diagnosis of disease states by recording the spontaneous electrical activity of tissue or organs, or by the response to stimulation of electrically excitable tissue. { i¦lek·trō‚dī·əg'nō·səs }

electrodialysis [PHYS CHEM] Dialysis that is conducted with the aid of an electromotive force applied to electrodes adjacent to both sides of the membrane. { i¦lek·trō·dī'al·ə·səs }

electrodialyzer [PHYS CHEM] An instrument used to conduct electrodialysis. { i‚lek·trō'dī·ə‚līz·ər }

electrodisintegration [NUC PHYS] The breakup of a nucleus into two or more fragments as a result of bombardment by electrons. { i¦lek·trō·dis‚int·ə'grā·shən }

electrodrill [MECH ENG] A drilling machine driven by electric power. { i'lek·trō‚dril }

electrodynamic ammeter [ENG] Instrument which measures the current passing through a fixed coil and a movable coil connected in series by balancing the torque on the movable coil (resulting from the magnetic field of the fixed coil) against that of a spiral spring. { i‚lek·trō·dī'nam·ik 'a‚mēd·ər }

electrodynamic drift [GEOPHYS] Motion of charged particles in the upper atmosphere due to the combined effect of electric and magnetic fields; in the ionospheric F region and above, the drift velocity is perpendicular to both the electric and magnetic fields. { i‚lek·trō·dī'nam·ik 'drift }

electrodynamic instrument [ENG] An instrument that depends for its operation on the reaction between the current in one or more movable coils and the current in one or more fixed coils. Also known as electrodynamometer. { i‚lek·trō·dī'nam·ik 'in·strə·mənt }

electrodynamic loudspeaker [ENG ACOUS] Dynamic loudspeaker in which the magnetic field is produced by an electromagnet, called the field coil, to which a direct current must be furnished. { i‚lek·trō·dī'nam·ik 'laud‚spēk·ər }

electrodynamic machine [ELEC] An electric generator or motor in which the output load current is produced by magnetomotive currents generated in a rotating armature. { i‚lek·trō·dī'nam·ik mə'shēn }

electrodynamics [ELECTROMAG] The study of the relations between electrical, magnetic, and mechanical phenomena. { i‚lek·trō·dī'nam·iks }

electrodynamic shaker See shaker. { i‚lek·trō·dī'nam·ik 'shāk·ər }

electrodynamic wattmeter [ENG] An electrodynamic instrument connected as a wattmeter, with the main current

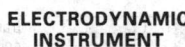
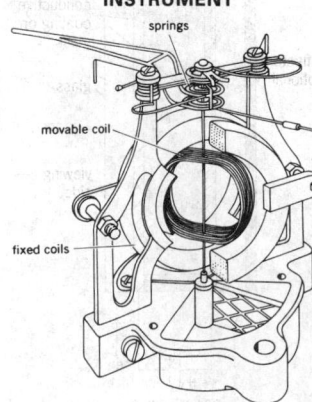

flowing through the fixed coil, and a small current proportional to the voltage flowing through the movable coil.   Also known as moving-coil wattmeter.   { i¦lek·trō·dī'nam·ik 'wät,mēd·ər }

electrodynamometer *See* electrodynamic instrument.   { i¸lek·trō¸dī·nə'mäm·əd·ər }

electroencephalogram [MED] A graphic recording of the electric discharges of the cerebral cortex as detected by electrodes on the surface of the scalp.   Abbreviated EEG.   { i¸lek·trō·en'sef·ə·lə¸gram }

electroencephalograph [MED] An instrument used to make electroencephalograms.   { i¸lek·trō·en'sef·ə·lə¸graf }

electroencephalography [MED] The medical specialty concerned with the production and interpretation of electroencephalograms.   { i¦lek·trō·en¸sef·ə'läg·rə·fē }

electroencephalophone [MED] An instrument that provides an audible presentation of brain waves.   Abbreviated EEP.   { i¸lek·trō·en'sef·ə·lə¸fōn }

electroencephaloscope [MED] An instrument for displaying on a cathode-ray screen the waveforms of voltages generated by various sections of the brain.   { i¦lek·trō·en'sef·ə·lə¸skōp }

electroendosmosis [PHYS] The production of an endosmosis effect by an electrical potential; that is, the use of electricity to cause diffusion of a liquid through an organic membrane.   { i¸lek·trō¦en·däs'mō·səs }

electroerosive machining *See* electron discharge machining.   { i¸lek·trō·ə'rō·siv mə'shēn·iŋ }

electroexplosive [ENG] An initiator or a system in which an electric impulse initiates detonation or deflagration of an explosive.   [MATER] The explosive substance so detonated or deflagrated.   { i¸lek·trō·ik'splō·siv }

electrofiltration [GEOL] Counterprocess during electrical logging of well boreholes, in which mud filtrate forced through the mud cake produces an emf in the mud cake opposite a permeable bed, positive in the direction of filtrate flow.   { i¸lek·trō·fil'trā·shən }

electrofluid [FL MECH] Newtonian (or shear-thinning) fluid whose rheological or flow properties are changed into those of a viscoplastic type by the addition of electric-field modulation.   { i¸lek·trō'flü·əd }

electrofocusing *See* isoelectric focusing.   { i¸lek·trō'fō·kəs·iŋ }

electroformed mold [MET] A mold made by electroplating metal on the reverse pattern on the cavity; molten steel may be then sprayed on the back of the mold to increase its strength.   { i'lek·trə¸fórmd 'mōld }

electroforming [MET] Shaping components by electrodeposition of the metal on a pattern.   { i'lek·trə¸fór·miŋ }

electrogalvanizing [MET] Coating of a metal, especially iron or steel, with zinc by electroplating.   { i¦lek·trō'gal·və¸nīz·iŋ }

electrogasdynamics [PHYS] Conversion of the kinetic energy of a moving gas to electricity, for such applications as high-voltage electric power generation, air-pollution control, and paintspraying.   { i¦lek·trō¸gas·dī'nam·iks }

electrogas flux-cored welding [MET] A modification of the flux-cored welding process in which there is an externally supplied source of gas or gas mixture.   { i¦lek·trō¸gas ¦fləks ¸kórd 'weld·iŋ }

electrogenerated chemiluminescence *See* electrochemiluminescence.   { i¸lek·trō¦jen·ə¸rād·əd ¸kem·ē·¸lüm·ə'nes·əns }

electrogenesis [PHYSIO] The generation of electric current by living tissue.   { i¦lek·trə'jen·ə·səs }

electrogram [ELECTR] A record of an image of an object made by sparking, usually on paper.   [METEOROL] A record, usually automatically produced, which shows the time variations of the atmospheric electric field at a given point.   [PHYSIO] The graphic representation of electric events in living tissues; commonly, an electrocardiogram or electroencephalogram.   { i'lek·trə¸gram }

electrograph [COMMUN] Facsimile transmission equipment.   [ENG] Any plot, graph, or tracing produced by the action of an electric current on prepared sensitized paper (or other chart material) or by means of an electrically controlled stylus or pen.   { i'lek·trə¸graf }

electrographic pencil [ELECTR] A pencil used to make a conductive mark on paper, for detection by a conductive-mark sensing device.   { i'lek·trə¸graf·ik 'pen·səl }

electrographic recording [GRAPHICS] Type of electrography in which the electrostatic image is formed by one or more rows of closely spaced parallel wires to which voltages are applied at appropriate instants to form the desired image charge pattern.   { i'lek·trə¸graf¸ik ri'kórd·iŋ }

electrography [GRAPHICS] The branch of electrostatography in which electrostatic images are formed on an insulating medium without the aid of electromagnetic radiation.   { i¸lek·'träg·rə·fē }

electrogravimetry [ANALY CHEM] Electrodeposition analysis in which the quantities of metals deposited may be determined by weighing a suitable electrode before and after deposition.   { i¸lek·trə·grə'vim·ə·trē }

electrohydraulic [ENG] Operated or effected by a combination of electric and hydraulic mechanisms.   { i¦lek·trō·hī'dról·ik }

electrohydraulic effect [PHYS CHEM] Generation of shock waves and highly reactive species in a liquid as the result of application of very brief but powerful electrical pulses.   { i¦lek·trō·hī¦dról·ik i'fekt }

electrohydrodynamic ionization mass spectroscopy [SPECT] A technique for analysis of nonvolatile molecules in which the nonvolatile material is dissolved in a volatile solvent with a high dielectric constant such as glycerol, and high electric-field gradients at the surface of droplets of the liquid solution induce ion emission.   { i¦lek·trō¦hī·drō·dī'nam·ik ¸ī·ə·nə'zā·shən ¸mas spek'träs·kə·pē }

electroinjection [BIOL] The use of electric-field impulses to introduce foreign deoxyribonucleic acid directly into intact cells.   { i¸lek·trō·in'jek·shən }

electrojet [GEOPHYS] A stream of intense electric current moving in the upper atmosphere around the equator and in polar regions.   { i'lck·trə¸jet }

electrokinetic phenomena [PHYS CHEM] The phenomena associated with movement of charged particles through a continuous medium or with the movement of a continuous medium over a charged surface.   { i¦lek·trō·kə'ned·ik fə'näm·ə·nə }

electrokinetic potential *See* zeta potential.   { i¦lek·trō·kə'ned·ik pə'ten·chəl }

electrokinetics [ELECTROMAG] The study of the motion of electric charges, especially of steady currents in electric circuits, and of the motion of electrified particles in electric or magnetic fields.   { i¦lek·trō·kə'ned·iks }

electrokinetic transducer [ELEC] An instrument which converts dynamic physical forces, such as vibration and sound, into corresponding electric signals by measuring the streaming potential generated by passage of a polar fluid through a permeable refractory-ceramic or fritted-glass member between two chambers.   { i¦lek·trō·kə'ned·ik tranz'dü·ser }

electrokinetograph [ENG] An instrument used to measure ocean current velocities based on their electrical effects in the magnetic field of the earth.   { i¸lek·trō·kə'ned·ə¸graf }

electrokymograph [MED] An instrument that provides a continuous recording of the movements of an internal organ such as the heart, generally by recording the movements or the changes in density of the shadow of the organ as presented on a fluoroscope.   { i¸lek·trō'kī·mə¸graf }

electroless plating [MET] Deposition of a metal coating by immersion of a metal or nonmetal in a suitable bath containing a chemical reducing agent.   { i'lek·trə·ləs'plād·iŋ }

electroluminescence [ELECTR] The emission of light, not due to heating effects alone, resulting from application of an electric field to a material, usually solid.   { i¦lek·trō¸lü·mə'nes·əns }

electroluminescent cell *See* electroluminescent panel.   { i¦lek·trō¸lü·mə'nes·ənt 'sel }

electroluminescent display [ELECTR] A display in which various combinations of electroluminescent segments may be activated by applying voltages to produce any desired numeral or other character.   { i¦lek·trō¸lü·mə'nes·ənt di'splā }

electroluminescent lamp *See* electroluminescent panel.   { i¦lek·trō¸lü·mə'nes·ənt 'lamp }

electroluminescent panel [ELECTR] A surface-area light source employing the principle of electroluminescence; consists of a suitable phosphor placed between sheet-metal electrodes, one of which is essentially transparent, with an alternating current applied between the electrodes.   Also known as electroluminescent cell; electroluminescent lamp; light panel; luminescent cell.   { i¦lek·trō¸lü·mə'nes·ənt 'pan·əl }

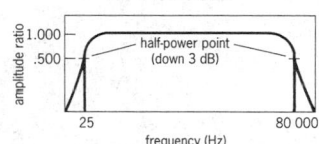

ELECTROKINETIC TRANSDUCER

Typical response curve of unit-cell transducer.

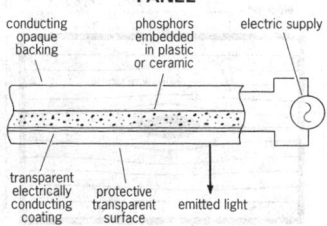

ELECTROLUMINESCENT PANEL

Simplified diagram of electroluminescent panel, not drawn to scale.

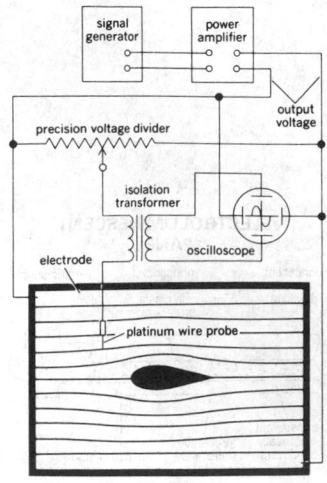

ELECTROLYSIS

Electrolysis of zinc chloride solution.

ELECTROLYTIC TANK

Setup to trace streamlines of nonlift airfoil with an electrolytic tank.

electroluminescent phosphor [MATER] Zinc sulfide powder, with small additions of copper or manganese, which emits light when suspended in an insulator in an intense alternating electric field. Also known as electroluminor. { i¦lek·trō,lü·mə'nes·ənt 'fäs·fər }

electroluminor See electroluminescent phosphor. { i¦lek·trō'lü·mə,nor }

electrolysis [PHYS CHEM] A method by which chemical reactions are carried out by passage of electric current through a solution of an electrolyte or through a molten salt. { i,lek'trä·lə·səs }

electrolyte [PHYS CHEM] A chemical compound which when molten or dissolved in certain solvents, usually water, will conduct an electric current. { i'lek·trə,līt }

electrolyte acid See sulfuric acid. { i'lek·trə,līt 'as·əd }

electrolyte-activated battery [ELEC] A reserve battery in which an aqueous electrolyte is stored in a separate chamber, and a mechanism, which may be operated from a remote location, drives the electrolyte out of the reservoir and into the cells of the battery for activation. { i¦lek·trə,līt ak·tə¦vād·əd 'bad·ə·rē }

electrolyte-MOSFET [ENG] A metal oxide semiconductor field-effect transistor (MOSFET) that is immersed in a solution to determine the concentrations of dissolved redox active species; the bulk part of the work function of the gate electrode of the transistor changes when the sensor membrane is oxidized or reduced. Abbreviated EMOSFET. { i¦lek·trə,līt 'mos ,fet }

electrolytic analysis [ANALY CHEM] Basic electrochemical technique for quantitative analysis of conducting solutions containing oxidizable or reducible material; measurement is based on the weight of material plated out onto the electrode. { i'lek· trə,lid·ik ə'nal·ə·səs }

electrolytic arrester See aluminum-cell arrester. { i'lek· trə,lid·ik ə'res·tər }

electrolytic brightening See electropolishing. { i'lek·trə,lid· ik 'brīt·ən·iŋ }

electrolytic capacitor [ELEC] A capacitor consisting of two electrodes separated by an electrolyte; a dielectric film, usually a thin layer of gas, is formed on the surface of one electrode. Also known as electrolytic condenser. { i'lek·trə,lid·ik kə'pas·əd·ər }

electrolytic cell [PHYS CHEM] A cell consisting of electrodes immersed in an electrolyte solution, for carrying out electrolysis. { i'lek·trə,lid·ik 'sel }

electrolytic cleaning See electrochemical cleaning. { i'lek· trə,lid·ik 'klēn·iŋ }

electrolytic condenser See electrolytic capacitor. { i'lek· trə,lid·ik kən'den·sər }

electrolytic conductance [PHYS CHEM] The transport of electric charges, under electric potential differences, by charged particles (called ions) of atomic or larger size. { i'lek·trə,lid· ik kən'dək·təns }

electrolytic conductivity [PHYS CHEM] The conductivity of a medium in which the transport of electric charges, under electric potential differences, is by particles of atomic or larger size. { i'lek·trə,lid·ik ,kän·dək¦tiv·əd·ē }

electrolytic copper [MET] Metallic copper produced by electrochemical deposition from a copper ion-containing electrolyte. { i'lek·trə,lid·ik 'käp·ər }

electrolytic corrosion See electrochemical corrosion. { i'lek· trə,lid·ik kə'rō·zhən }

electrolytic deposition See electrodeposition. { i'lek·trə,lid· ik dep·ə'zish·ən }

electrolytic development [GRAPHICS] Conversion of a latent image on a photosensitive material into a visible image by means of an electric current. { i¦lek·trə¦lid·ik di'vel·əp· mənt }

electrolytic dissociation [CHEM] The ionization of a compound in a solution. { i'lek·trə,lid·ik di,sō·sē'ā·shən }

electrolytic etching [MET] Engraving the surface of a metal by electrolysis. { i'lek·trə,lid·ik 'ech·iŋ }

electrolytic grinding [MECH ENG] A combined grinding and machining operation in which the abrasive, cathodic grinding wheel is in contact with the anodic workpiece beneath the surface of an electrolyte. Also known as electrochemical grinding. { i'lek·trə,lid·ik 'grīnd·iŋ }

electrolytic interrupter [ELEC] An interrupter that consists of two electrodes in an electrolytic solution; bubbles formed

in the solution continually interrupt the passage of current between the electrodes. { i'lek·trə,lid·ik ,int·ə'rəp·tər }

electrolytic mercaptan process [CHEM ENG] A process in which an aqueous caustic solution is used to extract mercaptans from refinery streams. { i'lek·trə,lid·ik mər'kap·tan ,prä· səs }

electrolytic migration [PHYS CHEM] The motions of ions in a liquid under the action of an electric field. { i¦lek·trə,lid·ik mī'grā·shən }

electrolytic model [PETRO ENG] Laboratory simulation of steady-state fluid flow through porous reservoir mediums; depends on the mobility of ions in absorbent mediums (gelatin or blotter), or through a liquid (potentiometric technique). Also known as gelatin model; oil-field model; potentiometric model. { i'lek·trə,lid·ik 'mäd·əl }

electrolytic photocopying [GRAPHICS] A photocopying process in which an image is projected on a sheet consisting of a paper support, a thin aluminum laminate, and a coating of a white photoconductive substance in contact with an electrolyte, and electrolysis takes place in the exposed areas when a direct current is applied across the electrolyte and the aluminum underlayer. { i'lek·trə,lid·ik 'fōd·ō,käp·ē·iŋ }

electrolytic pickling [MET] Removal of metal by electrolysis using the metal as an electrode in a suitable electrolyte. { i'lek·trə,lid·ik 'pik·liŋ }

electrolytic polarization [PHYS CHEM] The existence of a minimum potential difference necessary to cause a steady current to flow through an electrolytic cell, resulting from the tendency of the products of electrolysis to recombine. { i¦lek· trə,lid·ik pō·lər·ə'zā·shən }

electrolytic polishing See electropolishing. { i'lek·trə,lid·ik 'päl·ish·iŋ }

electrolytic potential [PHYS CHEM] Difference in potential between an electrode and the immediately adjacent electrolyte, expressed in terms of some standard electrode difference. { i'lek·trə,lid·ik pə'ten·chəl }

electrolytic powder [MET] Metal powder produced directly or indirectly by electrodeposition. { i'lek·trə,lid·ik 'paud·ər }

electrolytic process [PHYS CHEM] An electrochemical process involving the principles of electrolysis, especially as relating to the separation and deposition of metals. { i'lek· trə,lid·ik 'präs·əs }

electrolytic protection See cathodic protection. { i'lek· trə,lid·ik prə'tek·shən }

electrolytic recording [ELECTR] Electrochemical recording in which the chemical change is made possible by the presence of an electrolyte. { i'lek·trə,lid·ik ri'kord·iŋ }

electrolytic rectifier [ELEC] A rectifier consisting of metal electrodes in an electrolyte, in which rectification of alternating current is accompanied by electrolytic action; polarizing film formed on one electrode permits current flow in one direction but not the other. { i'lek·trə,lid·ik 'rek·tə,fī·ər }

electrolytic refining See electrorefining. { i'lek·trə,lid·ik rə'fīn·iŋ }

electrolytic rheostat [ELEC] A rheostat that consists of a tank of conducting liquid in which electrodes are placed, and resistance is varied by changing the distance between the electrodes, the depth of immersion of the electrodes, or the resistivity of the solution. Also known as water rheostat. { i'lek· trə,lid·ik 'rē·ə,stat }

electrolytic separation [PHYS CHEM] Separation of isotopes by electrolysis, based on differing rates of discharge at the electrode of ions of different isotopes. { i'lek·trə,lid·ik ,sep·ə'rā·shən }

electrolytic solution [PHYS CHEM] A solution made up of a solvent and an ionically dissociated solute; it will conduct electricity, and ions can be separated from the solution by deposition on an electrically charged electrode. { i'lek·trə,lid· ik sə'lü·shən }

electrolytic strip See humidity strip. { i'lek·trə,lid·ik 'strip }

electrolytic switch [ELEC] A switch having two electrodes projecting into a chamber partly filled with electrolyte, leaving an air bubble of predetermined width; the bubble shifts position and changes the amount of electrolyte in contact with the electrodes when the switch is tilted from true horizontal. { i'lek· trə,lid·ik 'swich }

electrolytic tank [ENG] A tank in which voltages are applied to an enlarged scale model of an electron-tube system or a reduced scale model of an aerodynamic system immersed in a poorly conducting liquid, and equipotential lines between electrodes are traced; used as an aid to electron-tube design or in computing ideal fluid flow; the latter application is based on the fact that the velocity potential in ideal flow and the stream function in planar flow satisfy the same equation, Laplace's equation, as an electrostatic potential. Also known as electric tank; potential flow analyzer. { i'lek·trə,lid·ik 'taŋk }

electrolytic tough pitch [MET] Copper which has been refined electrolytically, containing mostly oxygen as an impurity. { i'lek·trə,lid·ik 'təf 'pich }

electromachining [MECH ENG] The application of electric or ultrasonic energy to a workpiece to effect removal of material. { i¦lek·trō·mə'shēn·iŋ }

electromagnet [ELECTROMAG] A magnet consisting of a coil wound around a soft iron or steel core; the core is strongly magnetized when current flows through the coil, and is almost completely demagnetized when the current is interrupted. { i¦lek·trō'mag·nət }

electromagnetic [PHYS] Pertaining to phenomena in which electricity and magnetism are related. { i¦lek·trō·mag'ned·ik }

electromagnetic amplifying lens [ELECTROMAG] Large numbers of waveguides symmetrically arranged with respect to an excitation medium in order to become excited with equal amplitude and phase to provide a net gain in energy. { i¦lek·trō·mag'ned·ik 'am·plə,fī·iŋ ,lenz }

electromagnetic brake See electric brake. { i¦lek·trō·mag'ned·ik 'brāk }

electromagnetic cathode-ray tube [ELECTR] A cathode-ray tube in which electromagnetic deflection is used on the electron beam. { i¦lek·trō·mag'ned·ik 'ka,thōd 'rā ,tüb }

electromagnetic clutch [MECH ENG] A clutch based on magnetic coupling between conductors, such as a magnetic fluid and powder clutch, an eddy-current clutch, or a hysteresis clutch. { i¦lek·trō·mag'ned·ik 'kləch }

electromagnetic compatibility [ELECTR] The capability of electronic equipment or systems to be operated in the intended electromagnetic environment at design levels of efficiency. { i¦lek·trō·mag'ned·ik kəm,pat·ə'bil·əd·ē }

electromagnetic complex [ELECTROMAG] Electromagnetic configuration of an installation, including all significant radiators of energy. { i¦lek·trō·mag'ned·ik 'käm,pleks }

electromagnetic constant See speed of light. { i¦lek·trō·mag'ned·ik 'kän·stənt }

electromagnetic coupling [ELECTROMAG] Coupling that exists between circuits when they are mutually affected by the same electromagnetic field. { i¦lek·trō·mag'ned·ik 'kəp·liŋ }

electromagnetic crack detector [MET] An instrument that detects cracks in iron or steel objects by applying a strong magnetizing force and measuring the resulting magnetic flux through the object. { i¦lek·trō·mag'ned·ik 'krak di,tek·tər }

electromagnetic current [ELECTR] Motion of charged particles (for example, in the ionosphere) giving rise to electric and magnetic fields. { i¦lek·trō·mag'ned·ik 'kə·rənt }

electromagnetic damping [ELEC] Retardation of motion that results from the reaction between eddy currents in a moving conductor and the magnetic field in which it is moving. { i¦lek·trō·mag'ned·ik 'damp·iŋ }

electromagnetic deflection [ELECTR] Deflection of an electron stream by means of a magnetic field. { i¦lek·trō·mag'ned·ik di'flek·shən }

electromagnetic energy [ELECTROMAG] The energy associated with electric or magnetic fields. { i¦lek·trō·mag'ned·ik 'en·ər·jē }

electromagnetic environment [COMMUN] The radio-frequency fields existing in a given area. { i¦lek·trō·mag'ned·ik en'vī·rən·mənt }

electromagnetic field [ELECTROMAG] An electric or magnetic field, or a combination of the two, as in an electromagnetic wave. { i¦lek·trō·mag'ned·ik 'fēld }

electromagnetic field equations See Maxwell field equations. { i¦lek·trō·mag'ned·ik 'fēld i,kwā·zhənz }

electromagnetic field tensor [ELECTROMAG] An antisymmetric, second-rank Lorentz tensor, whose elements are proportional to the electric and magnetic fields; the Maxwell field equations can be expressed in a simple form in terms of this tensor. { i¦lek·trō·mag'ned·ik 'fēld ,ten·sər }

electrolytic flowmeter [ENG] A flowmeter that offers no obstruction to liquid flow; two coils produce an electromagnetic field in the conductive moving fluid; the current induced in the liquid, detected by two electrodes, is directly proportional to the rate of flow. Also known as electromagnetic meter. { i¦lek·trō·mag'ned·ik 'flō,mēd·ər }

electromagnetic focusing [ELECTR] Focusing the electron beam in a television picture tube by means of a magnetic field parallel to the beam; the field is produced by sending an adjustable value of direct current through a focusing coil mounted on the neck of the tube. { i¦lek·trō·mag'ned·ik 'fō·kəs·iŋ }

electromagnetic horn See horn antenna. { i¦lek·trō·mag'ned·ik 'hórn }

electromagnetic induction [ELECTROMAG] The production of an electromotive force either by motion of a conductor through a magnetic field so as to cut across the magnetic flux or by a change in the magnetic flux that threads a conductor. Also known as induction. { i¦lek·trō·mag'ned·ik in'dək·shən }

electromagnetic inertia [ELECTROMAG] **1.** Characteristic delay of a current in an electric circuit in reaching its maximum value, or in returning to zero, after the source voltage has been removed or applied. **2.** The property of a circuit whereby variation of the current in the circuit gives rise to a voltage in the circuit. { i¦lek·trō·mag'ned·ik i'nər·shə }

electromagnetic interaction [PART PHYS] The interaction of elementary particles that results from the coupling of charge to the electromagnetic field. { i¦lek·trō·mag'ned·ik ,int·ə'rak·shən }

electromagnetic interference [ELEC] Interference, generally at radio frequencies, that is generated inside systems, as contrasted to radio-frequency interference coming from sources outside a system. Abbreviated emi. { i¦lek·trō·mag'ned·ik ,in·tər'fir·əns }

electromagnetic lens [ELECTR] An electron lens in which electron beams are focused by an electromagnetic field. { i¦lek·trō·mag'ned·ik 'lenz }

electromagnetic log [ENG] A log containing an electromagnetic sensing element extended below the hull of the vessel; this device produces a voltage directly proportional to speed through the water. { i¦lek·trō·mag'ned·ik 'läg }

electromagnetic logging [ENG] A method of well logging in which a transmitting coil sets up an alternating electromagnetic field, and a receiver coil, placed in the drill hole above the transmitter coil, measures the secondary electromagnetic field induced by the resulting eddy currents within the formation. Also known as electromagnetic well logging. { i¦lek·trō·mag'ned·ik 'läg·iŋ }

electromagnetic mass [ELECTROMAG] The contribution to the mass of an object from its electric and magnetic field energy. { i¦lek·trō·mag'ned·ik 'mas }

electromagnetic meter See electromagnetic flowmeter. { i¦lek·trō·mag'ned·ik 'mēd·ər }

electromagnetic mirror [ELECTROMAG] Surface or region capable of reflecting radio waves, such as one of the ionized layers in the upper atmosphere. { i¦lek·trō·mag'ned·ik 'mir·ər }

electromagnetic mixing [MET] Mixing of molten alloys by exposing the melt to a strong magnetic field while passing direct current between electrodes at opposite ends of the crucible; stirring action results from interaction of the magnetic field of the current-carrying molten alloy with the external transverse magnetic field. { i¦lek·trō·mag'ned·ik 'mik·siŋ }

electromagnetic moment [ELECTROMAG] The magnetic moment of a current-carrying coil, equal to the product of the current, the number of turns, and the area of the coil. { i¦lek·trō·mag'ned·ik 'mō·mənt }

electromagnetic momentum [ELECTROMAG] The momentum transported by electromagnetic radiation; its volume density equals the Poynting vector divided by the square of the speed of light. { i¦lek·trō·mag'ned·ik mə'men·təm }

electromagnetic noise [ELEC] Noise in a communications system resulting from undesired electromagnetic radiation. Also known as radiation noise. { i¦lek·trō·mag'ned·ik 'nóiz }

electromagnetic oscillograph [ELECTROMAG] An oscillograph in which the recording mechanism is controlled by a moving-coil galvanometer, such as a direct-writing recorder or a light-beam oscillograph. { i¦lek·trō·mag'ned·ik ä'sil·ə,graf }

electromagnetic potential [ELECTROMAG] Collective

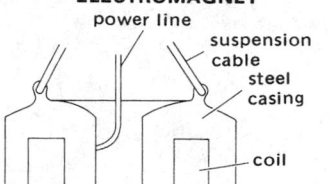

ELECTROMAGNET

power line · suspension cable · steel casing · coil · nonmagnetic manganese steel bumper

Cross section of circular lifting electromagnet.

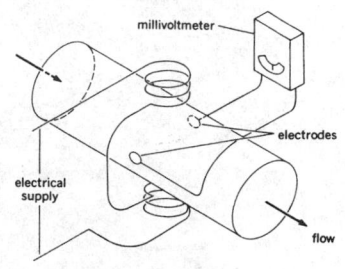

ELECTROMAGNETIC FLOWMETER

millivoltmeter · electrodes · electrical supply · flow

Drawing of electromagnetic flowmeter showing sensing electrodes and other components.

name for a scalar potential, which reduces to the electrostatic potential in a time-independent system, and the vector potential for the magnetic field; the electric and magnetic fields can be written in terms of these potentials. { i¦lek·trō·mag'ned·ik pə'ten·chəl }

electromagnetic properties [ELECTROMAG] The response of materials or equipment to electromagnetic fields, and their ability to produce such fields. { i¦lek·trō·mag'ned·ik 'präp· ərd·ēz }

electromagnetic propulsion [AERO ENG] Motive power for flight vehicles produced by electromagnetic acceleration of a plasma fluid. { i¦lek·trō·mag'ned·ik prə'pəl·shən }

electromagnetic prospecting See electromagnetic surveying. { i¦lek·trō·mag'ned·ik 'prä,spek·tiŋ }

electromagnetic pulse [ELECTROMAG] The pulse of electromagnetic radiation generated by a large thermonuclear explosion. { i¦lek·trō·mag'ned·ik 'pəls }

electromagnetic pump [ELEC] A pump in which a conductive liquid is made to move through a pipe by sending a large current transversely through the liquid; this current reacts with a magnetic field that is at right angles to the pipe and to current flow, to move the current-carrying liquid conductor. { i¦lek· trō·mag'ned·ik 'pəmp }

electromagnetic radiation [ELECTROMAG] Electromagnetic waves and, especially, the associated electromagnetic energy. { i¦lek·trō·mag'ned·ik ,rād·ē'ā·shən }

electromagnetic reconnaissance [ELECTR] Reconnaissance for the purpose of locating and identifying potentially hostile transmitters of electromagnetic radiation, including radar, communication, missile-guidance, and navigation-aid equipment. { i¦lek·trō·mag'ned·ik ri'kän·ə·säns }

electromagnetic relay [ELECTROMAG] A relay in which current flow through a coil produces a magnetic field that results in contact actuation. { i¦lek·trō·mag'ned·ik 'rē,lā }

electromagnetic scattering [PHYS] The process in which energy is removed from a beam of electromagnetic radiation and reemitted without appreciable changes in wavelength. { i¦lek·trō·mag'ned·ik 'skad·ə·riŋ }

electromagnetic separator [ELECTROMAG] Device in which ions of varying mass are separated by a combination of electric and magnetic fields. { i¦lek·trō·mag'ned·ik 'sep· ə,rād·ər }

electromagnetic shielding [ELECTROMAG] Means, similar to electrostatic or magnetostatic shielding, for suppressing changing magnetic fields or electromagnetic radiation at a device. { i¦lek·trō·mag'ned·ik 'shēld·iŋ }

electromagnetic shock wave [ELECTROMAG] Electromagnetic wave of great intensity which results when waves with different intensities propagate with different velocities in a nonlinear optical medium, and faster-traveling waves from a pulse of light catch up with preceding, slower traveling waves. { i¦lek·trō·mag'ned·ik 'shäk ,wāv }

electromagnetic spectrum [ELECTROMAG] The total range of wavelengths or frequencies of electromagnetic radiation, extending from the longest radio waves to the shortest known cosmic rays. { i¦lek·trō·mag'ned·ik 'spek·trəm }

electromagnetic surveying [ENG] Underground surveying carried out by generating electromagnetic waves at the surface of the earth; the waves penetrate the earth and induce currents in conducting ore bodies, thereby generating new waves that are detected by instruments at the surface or by a receiving coil lowered into a borehole. Also known as electromagnetic prospecting. { i¦lek·trō·mag'ned·ik sər'vā·iŋ }

electromagnetic susceptibility [ELECTR] The tolerance of circuits and components to all sources of interfering electromagnetic energy. { i¦lek·trō·mag'ned·ik sə,sep·tə'bil·əd·ē }

electromagnetic system of units [ELECTROMAG] A centimeter-gram-second system of electric and magnetic units in which the unit of current is defined as the current which, if maintained in two straight parallel wires having infinite length and being 1 centimeter apart in vacuum, would produce between these conductors a force of 2 dynes per centimeter of length; other units are derived from this definition by assigning unit coefficients in equations relating electric and magnetic quantities. Also known as electromagnetic units (emu). { i¦lek· trō·mag'ned·ik ¦sis·təm əv 'yü·nəts }

electromagnetic theory of light [ELECTROMAG] Theory according to which light is an electromagnetic wave whose electric and magnetic fields obey Maxwell's equations. { i¦lek·trō·mag'ned·ik ¦thē·ə·rē əv 'līt }

electromagnetic transducer See electromechanical transducer. { i¦lek·trō·mag'ned·ik tranz'dü·sər }

electromagnetic units See electromagnetic system of units. { i¦lek·trō·mag'ned·ik 'yü·nəts }

electromagnetic wave [ELECTROMAG] A disturbance which propagates outward from any electric charge which oscillates or is accelerated; far from the charge it consists of vibrating electric and magnetic fields which move at the speed of light and are at right angles to each other and to the direction of motion. { i¦lek·trō·mag'ned·ik 'wāv }

electromagnetic-wave filter [ELECTROMAG] Any device to transmit electromagnetic waves of desired frequencies while substantially attenuating all other frequencies. { i¦lek·trō· mag'ned·ik ¦wāv ,fil·tər }

electromagnetic well logging See electromagnetic logging. { i¦lek·trō·mag'ned·ik 'wel ,läg·iŋ }

electromagnetism [PHYS] **1.** Branch of physics relating electricity to magnetism. **2.** Magnetism produced by an electric current rather than by a permanent magnet. { i¦lek·trō'-mag·nə,tiz·əm }

electromanometer [ENG] An electronic instrument used for measuring pressure of gases or liquids. { i¦lek·trō·mə'näm· əd·ər }

electromechanical [MECH ENG] Pertaining to a mechanical device, system, or process which is electrostatically or electromagnetically actuated or controlled. { i¦lek·trō·mi'kan·ə· kəl }

electromechanical circuit [ELEC] A circuit containing both electrical and mechanical parameters of consequence in its analysis. { i¦lek·trō·mi'kan·ə·kəl 'sər·kə t }

electromechanical coupling coefficient [SOLID STATE] The ratio of the mutual elastodielectric energy density in a piezoelectric material to the square root of the product of the stored elastic and dielectric energy densities. { i,lek·trō· mi¦kan·ə·kəl 'kəp·liŋ ,kō·ə,fish·ənt }

electromechanical dialer [ELECTR] Telephone dialer which activates one of a set of desired numbers, precoded into it, when the user selects and presses a start button. { i¦lek· trō·mi'kan·ə·kəl 'dī·lər }

electromechanical plotter [COMPUT SCI] An automatic device used in conjunction with a digital computer to produce a graphic or pictorial representation of computer data on hard copy. { i¦lek·trō·mi'kan·ə·kəl 'pläd·ər }

electromechanical recording [ELECTR] Recording by means of a signal-actuated mechanical device, such as a pen arm or mirror attached to the moving coil of a galvanometer. { i¦lek·trō·mi'kan·ə·kəl ri'kórd·iŋ }

electromechanical relay [ELECTROMAG] A protective relay operating on the principle of electromagnetic attraction, as a plunger relay, or of electromagnetic induction. { i¦lek· trō·mi'kan·ə·kəl 'rē,lā }

electromechanical transducer [ELECTR] A transducer for receiving waves from an electric system and delivering waves to a mechanical system, or vice versa. Also known as electromagnetic transducer. { i¦lek·trō·mi'kan·ə·kəl tranz'dü·sər }

electromechanics [MECH ENG] The technology of mechanical devices, systems, or processes which are electrostatically or electromagnetically actuated or controlled. { i¦lek·trō· mi'kan·iks }

electrometallurgy [MET] Industrial recovery and processing of metals by electrical and electrolytic procedures. { i'lek·trō'med·əl·ər·jē }

electrometer [ENG] An instrument for measuring voltage without drawing appreciable current. { i,lek'träm·əd·ər }

electrometer amplifier [ELECTR] A low-noise amplifier having sufficiently low current drift and other characteristics required for measuring currents smaller than 10^{-12} ampere. { i,lek'träm·əd·ər 'am·plə,fī·ər }

electrometer tube [ELECTR] A high-vacuum electron tube having a high input impedance (low control-electrode conductance) to facilitate measurement of extremely small direct currents or voltages. { i,lek'träm·əd·ər ,tüb }

electromigration [ANALY CHEM] A process used to separate isotopes or ionic species by the differences in their ionic mobilities in an electric field. [PHYS CHEM] The movement of ions under the influence of an electrical potential difference. { i¦lek·trō·mī'grā·shən }

electromodulation [SPECT] Modulation spectroscopy in which changes in transmission or reflection spectra induced by a perturbing electric field are measured. { i¦lek·trō,mäj·ə'lā·shən }

electromotance *See* electromotive force. { i¦lek·trō'mōt·əns }

electromotive force [PHYS CHEM] **1.** The difference in electric potential that exists between two dissimilar electrodes immersed in the same electrolyte or otherwise connected by ionic conductors. **2.** The resultant of the relative electrode potential of the two dissimilar electrodes at which electrochemical reactions occur. Abbreviated emf. Also known as electromotance. { i¦lek·trə'mōd·iv 'fōrs }

electromotive series *See* electrochemical series. { i¦lek·trə'mōd·iv 'sir·ēz }

electromyogram [MED] **1.** A graphic recording of the electrical response of a muscle to electrical stimulation. **2.** A graphic recording of eye movements during reading. { i¦lek·trō'mī·ə,gram }

electromyograph [MED] An instrument used for making electromyograms. { i¦lek·trō'mī·ə,graf }

electromyography [MED] A medical specialty concerned with the production and study of electromyograms. { i¦lek·trō·mī'äg·rə·fē }

electron [PHYS] **1.** A stable elementary particle which is the negatively charged constituent of ordinary matter, having a mass of about 9.11×10^{-28} gram (equivalent to 0.511 MeV), a charge of about -1.602×10^{-19} coulomb, and a spin of $^{1}/_{2}$. Also known as negative electron; negatron. **2.** Collective name for the electron, as in the first definition, and the positron. { i¦lek,trän }

electron accelerator [NUCLEO] A device which accelerates electrons to high energies. { i¦lek,trän ak'sel·ə,rād·ər }

electron acceptor [PHYS CHEM] **1.** An atom or part of a molecule joined by a covalent bond to an electron donor. **2.** *See* electrophile. [SOLID STATE] *See* acceptor. { i¦lek,trän ak'sep·tər }

electron-acoustic microscopy [PHYS] A technique for producing images that show variations in an object's thermal and elastic properties; an electron beam generates ultrasonic waves in the specimen which are detected by a piezoelectric transducer whose output controls the brightness of a spot sweeping a cathode-ray tube in synchronism with the electron beam. { i¦lek,trän ə'küs·tik mī'kräs·kə·pē }

electron affinity [ATOM PHYS] The work needed in removing an electron from a negative ion, thus restoring the neutrality of an atom or molecule. { i¦lek,trän ə'fin·əd·ē }

electronarcosis [MED] Profound stupor or unconsciousness produced by passing an electric current through the brain. { i¦lek·trō·när'kō·səs }

electron attachment [ATOM PHYS] The combination of an electron with a neutral atom or molecule to form a negative ion. [NUC PHYS] *See* electron capture. { i¦lek,trän ə'tach·mənt }

electron avalanche *See* avalanche. { i¦lek,trän 'av·ə,lanch }

electron beam [ELECTR] A narrow stream of electrons moving in the same direction, all having about the same velocity. { i¦lek,trän ,bēm }

electron-beam-accessed memory *See* electron-beam memory. { i¦lek,trän ,bēm ak,sest 'mem·rē }

electron-beam channeling [ELECTR] The technique of transporting high-energy, high-current electron beams from an accelerator to a target through a region of high-pressure gas by creating a path through the gas where the gas density may be temporarily reduced; the gas may be ionized; or a current may flow whose magnetic field focuses the electron beam on the target. { i¦lek,trän ,bēm 'chan·əl·iŋ }

electron-beam cutting [MET] A process which uses high-velocity electrons to heat the workpieces to be cut. { i¦lek,trän ,bēm 'kəd·iŋ }

electron-beam drilling [ELECTR] Drilling of tiny holes in a ferrite, semiconductor, or other material by using a sharply focused electron beam to melt and evaporate or sublimate the material in a vacuum. { i¦lek,trän ,bēm 'dril·iŋ }

electron-beam fusion [NUCLEO] The use of intense beams of electrons to implode small pellets of deuterium and tritium so that they reach the temperature and density required for initiating a fusion reaction. { i¦lek,trän ,bēm 'fyü·zhən }

electron-beam generator [ELECTR] Velocity-modulated generator, such as a klystron tube, used to generate extremely high frequencies. { i¦lek,trän ,bēm 'jen·ə,rād·ər }

electron-beam ion source [ELECTR] A source of multiply charged heavy ions which uses an intense electron beam with energies of 5 to 10 kiloelectronvolts to successively ionize injected gas. Abbreviated EBIS. { i¦lek,trän ,bēm 'ī,än ,sòrs }

electron-beam ion trap [ELECTR] A device for producing the highest possible charge states of heavy ions, in which impact ionization or excitation by successive electrons is efficiently achieved by causing the ions to be trapped in a compressed electron beam by the electron beam's space charge. Abbreviated EBIT. { i¦lek,trän ,bēm 'i·ən 'trap }

electron-beam laser [OPTICS] A semiconductor laser in which the electron beam that provides pumping action in a thin plate of cadmium sulfide or other material is swept electrically in two dimensions by a deflection yoke, much as in a cathode-ray tube. { i¦lek,trän ,bēm 'lā·zər }

electron-beam lithography [ELECTR] Lithography in which the radiation-sensitive film or resist is placed in the vacuum chamber of a scanning-beam electron microscope and exposed by an electron beam under digital computer control; after exposure, the film is removed from the vacuum chamber for conventional development and other production processes. { i¦lek,trän ,bēm li'thäg·rə·fē }

electron-beam machining [MET] A machining process in which heat is produced by a focused electron beam at a sufficiently high temperature to volatilize and thereby remove metal in a desired manner; takes place in a vacuum. Abbreviated EBM. { i¦lek,trän ,bēm mə'shēn·iŋ }

electron-beam magnetometer [ENG] A magnetometer that depends on the change in intensity or direction of an electron beam that passes through the magnetic field to be measured. { i¦lek,trän ,bēm mag·nə'täm·əd·ər }

electron-beam melting [MET] A melting process in which an electron beam provides the necessary heat. { i¦lek,trän ,bēm 'melt·iŋ }

electron-beam memory [COMPUT SCI] A memory that uses a high-resolution electron beam to store information on a target in a vacuum tube. Also known as electron-beam-accessed memory (EBAM). { i¦lek,trän ,bēm 'mem·rē }

electron-beam parametric amplifier [ELECTR] A parametric amplifier in which energy is pumped from an electrostatic field into a beam of electrons traveling down the length of the tube, and electron couplers impress the input signal at one end of the tube and translate spiraling electron motion into electric output at the other. { i¦lek,trän ,bēm ,par·ə¦me·trik 'am·plə,fī·ər }

electron-beam pumping [ELECTR] The use of an electron beam to produce excitation for population inversion and lasing action in a semiconductor laser. { i¦lek,trän ,bēm 'pəmp·iŋ }

electron-beam recorder [ELECTR] A recorder in which a moving electron beam is used to record signals or data on photographic or thermoplastic film in a vacuum chamber. { i¦lek,trän ,bēm ri'kòrd·ər }

electron-beam tube [ELECTR] An electron tube whose performance depends on the formation and control of one or more electron beams. { i¦lek,trän ,bēm 'tüb }

electron-beam welding [MET] A technique for joining materials in which highly collimated electron beams are used at a pressure below 10^{-3} mmHg (0.1333 pascal) to produce a highly concentrated heat source; used in outer space. { i¦lek,trän ,bēm 'weld·iŋ }

electron-bombardment-induced conductivity [ELECTR] In a multimode display-storage tube, a process using an electron gun to erase the image on the cathode-ray tube interface. { i¦lek,trän bäm¦bärd·mənt in,düst kän·dək'tiv·əd·ē }

electron bunching *See* bunching. { i¦lek,trän 'bənch·iŋ }

electron capture [ATOM PHYS] The process in which an atom or ion passing through a material medium either loses or gains one or more orbital electrons. [NUC PHYS] A radioactive transformation of nuclide in which a bound electron merges with its nucleus. Also known as electron attachment. { i¦lek,trän 'kap·chər }

electron-capture detector [ANALY CHEM] Extremely sensitive gas chromatography detector that is a modification of the argon ionization detector, with conditions adjusted to favor the formation of negative ions. { i¦lek,trän ,kap·chər di'tek·tər }

ELECTRON ACCELERATOR

A type of electron accelerator, the Stanford Linear Accelerator, installed in underground housing.

electron carrier [CELL MOL] A molecule that accepts electrons from electron donors and donates them to electron acceptors, creating an energy-producing electron transport chain such as occurs in respiration and photosynthesis. { i'lek,trän ,kar·ē·ər }

electron charge [PHYS] The charge carried by an electron, equal to about -1.602×10^{-19} coulomb, or -4.803×10^{-10} statcoulomb. { i'lek,trän ,chärj }

electron cloud [ATOM PHYS] Picture of an electron state in which the charge is thought of as being smeared out, with the resulting charge density distribution corresponding to the probability distribution function associated with the Schrödinger wave function. { i'lek,trän ,klaůd }

electron collector See collector. { i'lek,trän kə,lek·tər }

electron compound [MET] Alloy of two metals in which a progressive change in composition is accompanied by a progression of phases, differing in crystal structure. Also known as Hume-Rothery compound; intermetallic compound. { i'lek,trän ,käm,paůnd }

electron conduction [ELEC] Conduction of electricity resulting from motion of electrons, rather than from ions in a gas or solution, or holes in a solid. [THERMO] The transport of energy in highly ionized matter primarily by electrons of relatively high temperature moving in one direction and electrons of lower temperature moving in the other. { i'lek,trän kən,dək·shən }

electron configuration [ATOM PHYS] The orbital and spin arrangement of an atom's electrons, specifying the quantum numbers of the atom's electrons in a given state. { i'lek,trän kən,fig·yə'rā·shən }

electron cooling [NUCLEO] A method of reducing the energy spread, angular divergence, and geometric size of a charged-particle beam by merging it with a monoenergetic electron beam with which it exchanges energy. { i,lek,trän 'kül·iŋ }

electron correlation [ATOM PHYS] The difference between the actual wave function of an atomic system and the wave function in the Hartree-Fock approximation. Also known as correlation. { i,lek,trän ,kä·rə'ā·shən }

electron-coupled oscillator [ELECTR] An oscillator employing a multigrid tube in which the cathode and two grids operate as an oscillator; the anode-circuit load is coupled to the oscillator through the electron stream. Abbreviated eco. Also known as Dow oscillator. { i'lek,trän ,kəp·əld 'äs·ə,lād·ər }

electron coupler [ELECTR] A microwave amplifier tube in which electron bunching is produced by an electron beam projected parallel to a magnetic field and, at the same time, subjected to a transverse electric field produced by a signal generator. Also known as Cuccia coupler. { i'lek,trän ,kəp·lər }

electron coupling [ELECTR] A method of coupling two circuits inside an electron tube, used principally with multigrid tubes; the electron stream passing between electrodes in one circuit transfers energy to electrodes in the other circuit. Also known as electronic coupling. { i'lek,trän ,kəp·liŋ }

electron cyclotron resonance [PHYS] Resonance absorption of energy from a radio-frequency or microwave-frequency electromagnetic field by electrons in a uniform magnetic field when the frequency of the electromagnetic field equals the cyclotron frequency of the electrons. { i,lek,trän ,sī·klə,trän 'rez·ə·nəns }

electron cyclotron resonance ion source See electron cyclotron resonance source. { i,lek,trän ,sī·klə,trän 'rez·ə·nəns ,ī,än ,sors }

electron cyclotron resonance reactor [ENG] A plasma reactor in which resonant coupling of microwave energy into an electron gas at electron cyclotron resonance accelerates electrons, which in turn ionize and excite the neutral gas, resulting in a low-pressure, almost collisionless plasma. { i,lek,trän 'sī·klə,trän 'rez·ə·nəns rē,ak·tər }

electron cyclotron resonance source [ELECTR] A source of multiply charged heavy ions that uses microwave power to heat electrons to energies of tens of kilovolts in two magnetic mirror confinement chambers in series; ions formed in the first chamber drift into the second chamber, where they become highly charged. Abbreviated ECR source. Also known as electron cyclotron resonance ion source (ECRIS). { i'lek,trän 'sī·klə,trän 'rez·ən·əns ,sors }

electron cyclotron wave [PL PHYS] A wave in a plasma which propagates parallel to the magnetic field produced by currents outside the plasma at frequencies less than that of the electron cyclotron resonance, and which is circularly polarized, rotating in the same sense as electrons in the plasma; responsible for whistlers. Also known as whistler wave. { i'lek,trän 'sī·klə,trän ,wāv }

electron density [PHYS] **1.** The number of electrons in a unit volume. **2.** When quantum-mechanical effects are significant, the total probability of finding an electron in a unit volume. { i'lek,trän 'den·səd·ē }

electron device [ELECTR] A device in which conduction is principally by electrons moving through a vacuum, gas, or semiconductor, as in a crystal diode, electron tube, transistor, or selenium rectifier. { i'lek,trän di'vīs }

electron diffraction [PHYS] The phenomenon associated with the interference processes which occur when electrons are scattered by atoms in crystals to form diffraction patterns. { i'lek,trän di'frak·shən }

electron diffraction analysis [PHYS] Examination of solid surfaces by observing the diffraction of a stream of electrons by the surface. { i'lek,trän di'frak·shən ə,nal·ə·səs }

electron diffraction camera [OPTICS] A camera used to obtain a photographic record of the position and intensity of the diffracted beams produced when a specimen is irradiated by a beam of electrons. { i'lek,trän di'frak·shən ,kam·rə }

electron diffractograph [PHYS] A device, allied to the electron microscope, in which a beam of electrons strikes the sample, showing crystal pattern and other physical attributes on the resulting diffraction pattern; used for chemical analysis, atomic structure determination, and so on. { i'lek,trän di'frak·tə,graf }

electron dipole moment See electron magnetic moment. { i'lek,trän 'dī,pōl ,mō·mənt }

electron discharge machining [MET] A process by which materials that conduct electricity can be removed from a metal by an electric spark; used to form holes of varied shapes in materials of poor machinability. Abbreviated EDM. Also known as electrical discharge machining; electric spark machining; electroerosive machining; electrospark machining. { i'lek,trän 'dis,chärj mə,shēn·iŋ }

electron distribution [PHYS] A function which gives the number of electrons per unit volume of phase space. { i'lek,trän dis·trə'byü·shən }

electron distribution curve [PHYS CHEM] A curve indicating the electron distribution among the different available energy levels of a solid substance. { i'lek,trän dis·trə'byü·shən ,kərv }

electron donor [PHYS CHEM] **1.** An atom or part of a molecule which supplies both electrons of a duplet forming a covalent bond. **2.** See nucleophile. [SOLID STATE] See donor. { i'lek,trän ,dō·nər }

electron-dot formula See Lewis structure. { i,lek,trän ,dät ,fòr·myə·lə }

electron efficiency [ELECTR] The power which an electron stream delivers to the circuit of an oscillator or amplifier at a given frequency, divided by the direct power supplied to the stream. Also known as electronic efficiency. { i'lek,trän ə'fish·ən·sē }

electronegative [ELEC] **1.** Carrying a negative electric charge. **2.** Capable of acting as the negative electrode in an electric cell. [PHYS CHEM] Pertaining to an atom or group of atoms that has a relatively great tendency to attract electrons to itself. { i,lek·trō'neg·əd·iv }

electronegative potential [PHYS CHEM] Potential of an electrode expressed as negative with respect to the hydrogen electrode. { i,lek·trō'neg·əd·iv pə'ten·chəl }

electron-electron double resonance [SPECT] A type of electron paramagnetic resonance (EPR) spectroscopy in which a material is irradiated at two different microwave frequencies, and the changes in the EPR spectrum resulting from sweeping either the second frequency or the magnetic field are monitored through detection at the first frequency. Abbreviated ELDOR. { i,lek,trän i,lek,tran ,dəb·əl 'rez·ən·əns }

electron emission [ELECTR] The liberation of electrons from an electrode into the surrounding space, usually under the influence of heat, light, or a high electric field. { i'lek,trän i'mish·ən }

electron emitter [ELECTR] The electrode from which electrons are emitted. { i'lek,trän i'mid·ər }

electron energy level [ATOM PHYS] A quantum-mechanical concept for energy levels of electrons about the nucleus; electron energies are functions of each particular atomic species. { i'lek,trän 'en·ər·jē ,lev·əl }

electron energy loss spectroscopy [SPECT] A technique for studying atoms, molecules, or solids in which a substance is bombarded with monochromatic energy, and the energies of scattered electrons are measured to determine the distribution of energy loss. Abbreviated EELS. { i'lek,trän 'en·ər·jē ,lòs spek'träs·kə·pē }

electroneutrality principle [PHYS CHEM] The principle that in an electrolytic solution the concentrations of all the ionic species are such that the solution as a whole is neutral. { i¦lek·trō·nü'tral·əd·ē ,prin·sə·pəl }

electron exchanger See redox polymer. { i'lek,trän iks,chān·jər }

electron flow [ELEC] A current produced by the movement of free electrons toward a positive terminal; the direction of electron flow is opposite to that of current. { i'lek,trän ,flō }

electron gas [PHYS] A concentration of electrons whose behavior, is, in first approximation, not governed by forces. { i'lek,trän ,gas }

electron gun [ELECTR] An electrode structure that produces and may control, focus, deflect, and converge one or more electron beams in an electron tube. { i'lek,trän ,gən }

electron-gun density multiplication [ELECTR] Ratio of the average current density at any specified aperture through which the electron stream passes to the average current density at the cathode surface. { i'lek,trän ,gən 'den·səd·ē ,məl·tə·plə'kā·shən }

electron hole See hole. { i'lek,trän ,hōl }

electron-hole droplets [SOLID STATE] A form of electronic excitation observed in germanium and silicon at sufficiently low cryogenic temperatures; it is associated with a liquid-gas phase transition of the charge carriers, and consists of regions of conducting electron-hole Fermi liquid coexisting with regions of insulating exciton gas. { i'lek,trän ¦hōl 'dräp·ləts }

electron-hole recombination [SOLID STATE] The process in which an electron, which has been excited from the valence band to the conduction band of a semiconductor, falls back into an empty state in the valence band, which is known as a hole. { i'lek,trän 'hōl rē,käm·bə'nā·shən }

electron holography [ELECTR] An imaging technique using the wave nature of electrons and light, in which an interference pattern between an object wave and a reference wave is formed using a coherent field-emission electron beam from a sharp tungsten needle, and is recorded on film as a hologram, and the image of the original object is then reconstructed by illuminating a light beam equivalent to the reference wave onto the hologram. { i,lek,trän hō'läg·rə·fē }

electronic [ELECTR] Pertaining to electron devices or to circuits or systems utilizing electron devices, including electron tubes, magnetic amplifiers, transistors, and other devices that do the work of electron tubes. { i,lek'trän·ik }

electronic absorption spectrum [SPECT] Spectrum resulting from absorption of electromagnetic radiation by atoms, ions, and molecules due to excitations of their electrons. { i,lek'trän·ik əb'sòrp·shən ,spek·trəm }

electronically agile radar [ENG] An airborne radar that uses a phased-array antenna which changes radar beam shapes and beam positions at electronic speeds. { i,lek'trän·ik·lē ,a·jəl 'rā,där }

electronic alternating-current voltmeter [ELECTR] A voltmeter consisting of a direct-current milliammeter calibrated in volts and connected to an amplifier-rectifier circuit. { i,lek'trän·ik ¦al·tər,näd·iŋ ¦kə·rənt 'vōlt,mēd·ər }

electronic altimeter See radio altimeter. { i,lek'trän·ik al'tim·əd·ər }

electronic angular momentum [ATOM PHYS] The total angular momentum associated with the orbital motion of the spins of all the electrons of an atom. { i,lek'trän·ik 'aŋ·gyə·lər mə'ment·əm }

electronic attitude directional indicator [NAV] A multicolor cathode-ray-tube display of attitude information (roll and pitch) showing the aircraft's position in relation to the instrument landing system or a very high-frequency omnirange station. Abbreviated EADI. { i,lek'trän·ik 'ad·ə,tüd də'rek·shən·əl 'in·də,kād·ər }

electronic azimuth marker [ELECTR] On an airborne radar plan position indicator (PPI) a bright rotatable radial line used for bearing determination. Also known as azimuth marker. { i,lek'trän·ik 'az·ə·məth ,märk·ər }

electronic band spectrum [SPECT] Bands of spectral lines associated with a change of electronic state of a molecule; each band corresponds to certain vibrational energies in the initial and final states and consists of numerous rotational lines. { i,lek'trän·ik 'band ,spek·trəm }

electronic bearing cursor [ELECTR] Of a marine radar set, the bright rotatable radial line on the plan position indicator used for bearing determination. Also known as electronic bearing marker. { i,lek'trän·ik 'ber·iŋ ,kər·sər }

electronic bearing marker See electronic bearing cursor. { i,lek'trän·ik 'ber·iŋ ,märk·ər }

electronic calculator [ELECTR] A calculator in which integrated circuits perform calculations and show results on a digital display; the displays usually use either seven-segment light-emitting diodes or liquid crystals. { i,lek'trän·ik 'kal·kyə,lād·ər }

electronic camouflage [ELECTR] Use of electronic means, or exploitation of electronic characteristics to reduce, submerge, or eliminate the radar echoing properties of a target. { i,lek'trän·ik 'kam·ə,fläzh }

electronic cash register [ENG] A system for automatically checking out goods from retail food stores, consisting of a device that scans packages and reads symbols imprinted on the label, and a computer that converts the symbol information to tell a cash register the price of the item; the computer can also keep records of sales and inventories. Abbreviated ECR. { i,lek'trän·ik 'kash ,rej·ə·stər }

electronic chart See digital chart. { i,lek'trän·ik 'chärt }

electronic chart display and information system [ENG] A navigation information system with an electronic chart database, as well as navigational and piloting information (typically, vessel-route-monitoring, track-keeping, and track-planning information). Abbreviated ECDIS. { i,lek'trän·ik 'chärt di¦splä ən ,in·fər'mā·shən ,sis·təm }

electronic chart reader [COMPUT SCI] A device which scans curves by a graphical recorder on a continuous paper form and converts them into digital form. { i,lek'trän·ik 'chärt ,rēd·ər }

electronic circuit [ELECTR] An electric circuit in which the equilibrium of electrons in some of the components (such as electron tubes, transistors, or magnetic amplifiers) is upset by means other than an applied voltage. { i,lek'trän·ik 'sər·kət }

electronic clock [HOROL] A clock that uses a ferrite rod and coil to pick up the electromagnetic field of 60-hertz power line wiring in a house; this 60-hertz voltage is amplified by a transistor amplifier and used to control a transistor oscillator that drives a tiny permanent-magnet synchronous clock motor; two mercury cells provide power for the transistors. { i,lek'trän·ik 'kläk }

electronic codebook mode See block encryption. { i,lek'trän·ik 'kōd,bùk ,mōd }

electronic commerce [COMPUT SCI] Business done on the Internet. Also known as e-business; e-commerce. { i·lek¦trän·ik 'kä·mərs }

electronic commutator [ELECTR] An electron-tube or transistor circuit that switches one circuit connection rapidly and successively to many other circuits, without the wear and noise of mechanical switches. { i,lek'trän·ik 'käm·yə,täd·ər }

electronic component [ELECTR] A component which is able to amplify or control voltages or currents without mechanical or other nonelectrical command, or to switch currents or voltages without mechanical switches; examples include electron tubes, transistors, and other solid-state devices. { i,lek'trän·ik kəm'pō·nənt }

electronic composition [GRAPHICS] Typesetting in which characters are generated by electron or laser beams at speeds above about 6000 words per minute. { i,lek'trän·ik ,käm·pə'zish·ən }

electronic computing units [ELECTR] The sensing sections of tabulating equipment which enable the machine to handle the contents of punched cards in a prescribed manner. { i,lek'trän·ik kəm'pyüd·iŋ ,yü·nəts }

electronic confusion area [ELECTROMAG] Amount of

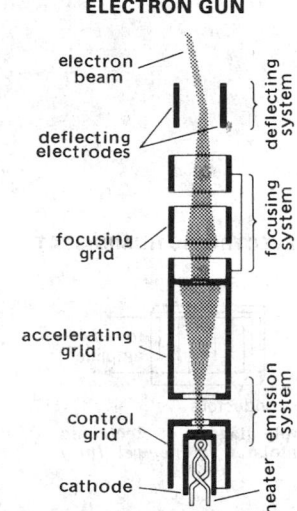

ELECTRON GUN

electron beam

deflecting electrodes

deflecting system

focusing grid

focusing system

accelerating grid

control grid

cathode

heater

emission system

Simplified electron gun employing electrostatic focus and deflection.

space that a target appears to occupy in a radar resolution cell, as it appears to that radar beam. { iˌlek'trän·ik kən'fyü·zhən ˌer·ē·ə }

electronic control [ELECTR] The control of a machine or process by circuits using electron tubes, transistors, magnetic amplifiers, or other devices having comparable functions. { iˌlek'trän·ik kən'trōl }

electronic controller [ELECTR] Electronic device incorporating vacuum tubes or solid-state devices and used to control the action or position of equipment; for example, a valve operator. { iˌlek'trän·ik kən'trōl·ər }

electronic counter [ELECTR] A circuit using electron tubes or equivalent devices for counting electric pulses. Also known as electronic tachometer. { iˌlek'trän·ik 'kaúnt·ər }

electronic countermeasure [ELECTR] An offensive or defensive tactic or device using electronic and reflecting apparatus to reduce the military effectiveness of enemy equipment involving electromagnetic radiation, such as radar, communication, guidance, or other radio-wave devices. Abbreviated ECM. Also known as electromagnetic countermeasure. { iˌlek'trän·ik 'kaúnt·ərˌmezh·ər }

electronic coupling See electron coupling. { iˌlek'trän·ik 'kəp·liŋ }

electronic data processing [COMPUT SCI] Processing data by using equipment that is predominantly electronic in nature, such as an electronic digital computer. Abbreviated EDP. { iˌlek'trän·ik 'dad·ə ˌpräs·əs·iŋ }

electronic data-processing center [COMPUT SCI] The complex formed by the computer, its peripheral equipment, the personnel related to the operation of the center and control functions, and, usually, the office space housing hardware and personnel. Abbreviated EDP center. Also known as computer center. { iˌlek'trän·ik 'dad·ə ˌpräs·əs·iŋ ˌsen·tər }

electronic data-processing management science [COMPUT SCI] The field consisting of a class of management problems capable of being handled by computer programs. { iˌlek'trän·ik 'dad·ə ˌpräs·əs·iŋ 'man·ij·mənt ˌsī·əns }

electronic data-processing system [COMPUT SCI] A system for data processing by means of machines using electronic circuitry at electronic speed, as opposed to electromechanical equipment. { iˌlek'trän·ik 'dad·ə ˌpräs·əs·iŋ ˌsis·təm }

electronic deception [ORD] Radiation or reradiation of electromagnetic waves in a manner intended to mislead the enemy in the interpretation of data received by the enemy's electronic equipment. { iˌlek'trän·ik di'sep·shən }

electronic defense evaluation [ELECTR] A mutual evaluation of radar and aircraft, with the aircraft trying to penetrate the radar's area of coverage in an electronic countermeasure environment. { iˌlek'trän·ik di'fens iˌval·yə'wā·shən }

electronic differential analyzer [COMPUT SCI] A form of analog computer using interconnected electronic integrators to solve differential equations. { iˌlek'trän·ik ˌdif·ə'ren·chəl 'an·əˌliz·ər }

electronic display [ELECTR] An electronic component used to convert electric signals into visual imagery in real time suitable for direct interpretation by a human operator. { iˌlek'trän·ik di'splā }

electronic distance-measuring equipment [NAV] A navigation system consisting of airborne devices that transmit microsecond pulses to special ground beacons, which retransmit the signals to the aircraft; the length of expired time between transmission and reception is measured, converted to kilometers or miles, and presented to the pilot. { iˌlek'trän·ik 'dis·təns ˌmezh·ə·riŋ iˌkwip·mənt }

electronic driftmeter [NAV] An electronic instrument for measuring drift angle; may be an attachment to an airborne radar; is an integral part of a Doppler navigator. { iˌlek'trän·ik 'driftˌmēd·ər }

electronic dummy [ENG ACOUS] A vocal simulator which is a replica of the head and torso of a person, covered with plastisol flesh that simulates the acoustical and mechanical properties of real flesh, and possessing an artificial voice and two artificial ears. Abbreviated ED. { iˌlek'trän·ik 'dəm·ē }

electronic efficiency [ELECTR] Ratio of the power at the desired frequency, delivered by the electron stream to the circuit in an oscillator or amplifier, to the average power supplied to the stream. { iˌlek'trän·ik i'fish·ən·sē }

electronic emission spectrum [SPECT] Spectrum resulting from emission of electromagnetic radiation by atoms, ions, and

molecules following excitations of their electrons. { iˌlek'trän·ik i'mish·ən ˌspek·trəm }

electronic energy curve [PHYS CHEM] A graph of the energy of a diatomic molecule in a given electronic state as a function of the distance between the nuclei of the atoms. { iˌlek'trän·ik 'en·ər·jē ˌkərv }

electronic engineering [ENG] Engineering that deals with practical applications of electronics. { iˌlek'trän·ik ˌen·jə'nir·iŋ }

electronic fix [NAV] A position fix established by means of electronic equipment. { iˌlek'trän·ik 'fiks }

electronic flame safeguard [MECH ENG] An electrode used in a burner system which detects the main burner flame and interrupts fuel flow if the flame is not detected. { iˌlek'trän·ik 'flām 'sāfˌgärd }

electronic flight instrument system [AERO ENG] An aircraft instrumentation system that provides flight and systems information to the flight crew on cathode-ray tubes and flat-panel displays. Also known as glass cockpit. { iˌlek'trän·ik 'flīt ˌin·strə·mənt ˌsis·təm }

electronic fuse [ENG] A fuse, such as the radio proximity fuse, set off by an electronic device incorporated in it. { iˌlek'trän·ik 'fyüz }

electronic heating [ENG] Heating by means of radio-frequency current produced by an electron-tube oscillator or an equivalent radio-frequency power source. Also known as high-frequency heating; radio-frequency heating. { iˌlek'trän·ik 'hēd·iŋ }

electronic horizontal-situation indicator [NAV] An integrated multicolor map display of an airplane's position combined with a color weather radar display, with a scale selected by the pilot, together with information on wind direction and velocity, horizontal situation, and deviation from the planned vertical path. Abbreviated EHSI. { iˌlek'trän·ik ˌhär·ə'zänt·əl ˌsich·ə¦wā·shən ˌin·dəˌkād·ər }

electronic humidistat [ENG] A humidistat in which a change in the relative humidity causes a change in the electrical resistance between two sets of alternate metal conductors mounted on a small flat plate with plastic coating, and this change in resistance is measured by a relay amplifier. { iˌlek'trän·ik hyü'mid·əˌstat }

electronic intelligence [ORD] Electronic systems, apparatus, and operations for obtaining information concerning an enemy's capabilities, intentions, plans, and order of battle. Abbreviated ELINT. { iˌlek'trän·ik in'tel·ə·jəns }

electronic interference [ELECTR] Any electrical or electromagnetic disturbance that causes undesirable response in electronic equipment. { iˌlek'trän·ik ˌint·ər·'fir·əns }

electronic jammer See jammer. { iˌlek'trän·ik 'jam·ər }

electronic jamming See jamming. { iˌlek'trän·ik 'jam·iŋ }

electronic line scanning [ELECTR] Method which provides motion of the scanning spot along the scanning line by electronic means. { iˌlek'trän·ik 'līn ˌskan·iŋ }

electronic listening device [ELECTR] A device used to capture the sound waves of conversation originating in an ostensibly private setting in a form, usually as a magnetic tape recording, which can be used against the target by adverse interests. { iˌlek'trän·ik 'lis·niŋ diˌvīs }

electronic locator See metal detector. { iˌlek'trän·ik 'lōˌkād·ər }

electronic locking [ELECTR] A technique for preventing the operation of a switch until a specific electrical signal (the unlocking signal) is introduced into circuitry associated with the switch; usually, but not necessarily, the unlocking signal is a binary sequence. { iˌlek'trän·ik 'läk·iŋ }

electronic logger See Geiger-Müller probe. { iˌlek'trän·ik 'läg·ər }

electronic magnetic moment [ATOM PHYS] The total magnetic dipole moment associated with the orbital motion of all the electrons of an atom and the electron spins; opposed to nuclear magnetic moment. { iˌlek'trän·ik mag'ned·ik 'mō·mənt }

electronic mail [COMMUN] The electronic transmission of letters, messages, and memos through a communications network. Also known as e-mail. { iˌlek'trän·ik 'māl }

electronic microradiography [ELECTR] Microradiography of very thin specimens in which the emission of electrons from an irradiated object, either the specimen or a lead screen behind it, is used to produce a photographic image of the specimen,

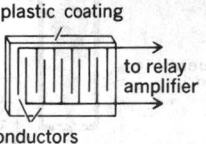

ELECTRONIC HUMIDISTAT

plastic coating

to relay amplifier

conductors

Simple diagram of electronic humidistat. *(Honeywell Inc.)*

which is then enlarged. Also known as e-mail. { i̇,lek'trän·ik 'mī·krō,rād·ē'äg·rə·fē }

electronic motor control [ELECTR] A control circuit used to vary the speed of a direct-current motor operated from an alternating-current power line. Also known as direct-current motor control; motor control. { i̇,lek'trän·ik 'mōd·ər kən,trōl }

electronic multimeter [ELECTR] A multimeter that uses semiconductor or electron-tube circuits to drive a conventional multiscale meter. { i̇,lek'trän·ik 'məl·tē,mēd·ər }

electronic music [ENG ACOUS] Music consisting of tones originating in electronic sound and noise generators used alone or in conjunction with electroacoustic shaping means and sound-recording equipment. { i̇,lek'trän·ik 'myü·zik }

electronic musical instrument [ENG ACOUS] A musical instrument in which an audio signal is produced by a pickup or audio oscillator and amplified electronically to feed a loudspeaker, as in an electric guitar, electronic carillon, electronic organ, or electronic piano. { i̇,lek'trän·ik ¦myü·zə·kəl 'in·strə·mənt }

electronic navigation [NAV] Navigation by means of any electronic device or instrument. { i̇,lek'trän·ik nav·ə'gā·shən }

electronic noise jammer [ELECTR] An electronic jammer which emits a radio-frequency carrier modulated with a white noise signal usually derived from a gas tube; used against enemy radar. { i̇,lek'trän·ik 'nȯiz ,jam·ər }

Electronic Numerical Integrator and Calculator See ENIAC. { i̇,lek'trän·ik nü'mer·ə·kəl 'int·ə,grād·ər ən 'kal·kyə,lād·ər }

electronic organ [ELECTR] A musical instrument which uses electronic circuits to produce music similar to that of a pipe organ. { i̇,lek'trän·ik 'ȯr·gən }

electronic packaging [ENG] The technology of packaging electronic equipment; in current usage it refers to inserting discrete components, integrated circuits, and MSI and LSI chips (usually attached to a lead frame by beam leads) into plates through holes on multilayer circuit boards (also called cards), where they are soldered in place. { i̇,lek'trän·ik 'pak·ij·iŋ }

electronic phase-angle meter [ELECTR] A phasemeter that makes use of electronic devices, such as amplifiers and limiters, that convert the alternating-current voltages being measured to square waves whose spacings are proportional to phase. { i̇,lek'trän·ik 'fāz ,aŋ·gəl ,mēd·ər }

electronic photometer See photoelectric photometer. { i̇,lek'trän·ik fō'täm·əd·ər }

electronic piano [ELECTR] A piano without a sounding board, in which vibrations of each string affect the capacitance of a capacitor microphone and thereby produce audio-frequency signals that are amplified and reproduced by a loudspeaker. { i̇,lek'trän·ik pē'an·ō }

electronic polarization [ELEC] Polarization arising from the displacement of electrons with respect to the nuclei with which they are associated, upon application of an external electric field. { i̇,lek'trän·ik ,pō·lə·rə'zā·shən }

electronic position indicator [NAV] A radio navigation system used in hydrographic surveying which provides circular lines of position. Abbreviated EPI. { i̇,lek'trän·ik pə'zish·ən ,in·də,kād·ər }

electronic power supply See power supply. { i̇,lek'trän·ik 'pau̇·ər sə,plī }

electronic publishing [COMMUN] The provision of information with high editorial and value-added content in electronic form, allowing the user some degree of control and interactivity. { i̇,lek'trän·ik 'pəb·lish·iŋ }

electronic pumping See pumping. { i̇,lek'trän·ik 'pəmp·iŋ }

electronic radiography [ELECTR] Radiography in which the image is detached by direct image converter tubes or by the use of television pickup or electronic scanning, and the resultant signals are amplified and presented for viewing on a kinescope. { i̇,lek'trän·ik rād·ē'äg·ra·fē }

electronic raster scanning See electronic scanning. { i̇,lek'trän·ik 'ras·tər ,skan·iŋ }

electronic reconnaissance [ELECTR] The detection, identification, evaluation, and location of foreign, electromagnetic radiations emanating from other than nuclear detonations or radioactive sources. { i̇,lek'trän·ik ri'kän·ə·səns }

electronic recording [ELECTR] The process of making a graphical record of a varying quantity or signal (or the result of such a process) by electronic means, involving control of

an electron beam by electric or magnetic fields, as in a cathode-ray oscillograph, in contrast to light-beam recording. { i̇,lek'trän·ik ri'kȯrd·iŋ }

electronic robot [CONT SYS] A robot whose motions are powered by a direct-current stepper motor. { i̇,lek'trän·ik 'rō,bät }

electronics [PHYS] Study, control, and application of the conduction of electricity through gases or vacuum or through semiconducting or conducting materials. { i̇,lek'trän·iks }

electronic scanning [ELECTR] Scanning in which an electron beam, controlled by electric or magnetic fields, is swept over the area under examination, in contrast to mechanical or electromechanical scanning. Also known as electronic raster scanning. { i̇,lek'trän·ik 'skan·iŋ }

electronic sculpturing [COMPUT SCI] Procedure for constructing a model of a system by using an analog computer, in which the model is devised at the console by interconnecting components on the basis of analogous configuration with real system elements; then, by adjusting circuit gains and reference voltages, dynamic behavior can be generated that corresponds to the desired response, or is recognizable in the real system. { i̇,lek'trän·ik 'skəlp·chə·riŋ }

electronic security [ELECTR] Protection resulting from all measures designed to deny to unauthorized persons information of value which might be derived from the possession and study of electromagnetic radiations. { i̇,lek'trän·ik sə'kyu̇r·əd·ē }

electronic sky screen equipment [ELECTR] Electronic device that indicates the departure of a missile from a predetermined trajectory. { i̇,lek'trän·ik 'skī ,skrēn i,kwip·mənt }

electronic specific heat [SOLID STATE] Contribution to the specific heat of a metal from the motion of conduction electrons. { i̇,lek'trän·ik spə¦sif·ik 'hēt }

electronic spectrum [SPECT] Spectrum resulting from emission or absorption of electromagnetic radiation during changes in the electron configuration of atoms, ions, or molecules, as opposed to vibrational, rotational, fine-structure, or hyperfine spectra. { i̇,lek'trän·ik 'spek·trəm }

electronic speedometer [ENG] A speedometer in which a transducer sends speed and distance pulses over wires to the speed and mileage indicators, eliminating the need for a mechanical link involving a flexible shaft. { i̇,lek'trän·ik spē'däm·əd·ər }

electronic spreadsheet [COMPUT SCI] A type of computer software for performing mathematical computations on numbers arranged in rows and columns, in which the numbers can depend on the values in other rows and columns, allowing large numbers of calculations to be carried out simultaneously. { i̇,lek'trän·ik 'spred,shēt }

electronic state [QUANT MECH] The physical state of electrons of a system, as specified, for example, by a Schrödinger-Pauli wave function of the positions and spin orientations of all the electrons. { i̇,lek'trän·ik 'stāt }

electronic structure [PHYS] The arrangement of electrons in an atom, molecule, or solid, specified by their wave functions, energy levels, or quantum numbers. { i̇,lek'trän·ik 'strək·chər }

electronic support measures See electronic warfare support measures. { i̇,lek'trän·ik sə'pȯrt ,mezh·ərz }

electronic surge arrester [ELECTR] Device used to switch to ground high-energy surges, thereby reducing transient energy to a level safe for secondary protectors, for example, Zener diodes, silicon rectifiers and so on. { i̇,lek'trän·ik 'sərj ə,res·tər }

electronic switch [ELECTR] **1.** Vacuum tube, crystal diodes, or transistors used as an on and off switching device. **2.** Test instrument used to present two wave shapes on a single gun cathode-ray tube. { i̇,lek'trän·ik 'swich }

electronic switching [COMMUN] Telephone switching using a computer with a storage containing program switching logic, whose output actuates switches that set up telephone connections automatically. [ELECTR] The use of electronic circuits to perform the functions of a high-speed switch. { i̇,lek'trän·ik 'swich·iŋ }

electronic tablet [COMPUT SCI] A data-entry device consisting of stylus, writing surface, and circuitry that produces a pair of digital coordinate values corresponding continuously to the position of the stylus upon the surface. Also known as data tablet. { i̇,lek'trän·ik 'tab·lət }

electronic tachometer *See* electronic counter. { i̯lek'trän·ik tə'käm·əd·ər }

electronic thermometer [ENG] A thermometer in which a sensor, usually a thermistor, is placed on or near the object being measured. { i̯lek'trän·ik thər'mäm·əd·ər }

electronic tonometer *See* tonometer. { i̯lek'trän·ik tō'näm·əd·ər }

electronic tuning [ELECTR] Tuning of a transmitter, receiver, or other tuned equipment by changing a control voltage rather than by adjusting or switching components by hand. { i̯lek'trän·ik 'tün·iŋ }

electronic typewriter [COMPUT SCI] A typewriter whose operation is enhanced through the use of microprocessor technology to provide many of the functions of a word-processing system but which has at most a partial-line visual display. Also known as memory typewriter. { i'lek,trän·ik 'tīp,rīd·ər }

electronic video recording [ELECTR] The recording of black and white or color television visual signals on a reel of photographic film as coded black and white images. Abbreviated EVR. { i̯lek'trän·ik 'vid·ē·ō ri,kòrd·iŋ }

electronic voltage regulator [ELECTR] A device which maintains the direct-current power supply voltage for electronic equipment nearly constant in spite of input alternating-current line voltage variations and output load variations. { i̯lek'trän·ik 'vōl·tij ,reg·yə,lād·ər }

electronic voltmeter [ENG] Voltmeter which uses the rectifying and amplifying properties of electron devices and their associated circuits to secure desired characteristics, such as high-input impedance, wide-frequency range, crest indications, and so on. { i̯lek'trän·ik 'vōlt,mēd·ər }

electronic warfare [ELECTR] Military action involving the use of electromagnetic energy to determine, exploit, reduce, or prevent hostile use of the electromagnetic spectrum, and action which retains friendly use of electromagnetic spectrum. { i̯lek'trän·ik 'wòr,fer }

electronic warfare support measures [ELECTR] That division of electronic warfare involving actions taken to search for, intercept, locate, record, and analyze radiated electromagnetic energy for the purpose of exploiting such radiations in support of military operations. Also known as electronic support measures. { i̯lek'trän·ik 'wòr,fer sə'pòrt ,mezh·ərz }

electronic work function [SOLID STATE] The energy required to raise an electron with the Fermi energy in a solid to the energy level of an electron at rest in vacuum outside the solid. { i̯lek'trän·ik 'wərk ,fəŋk·shən }

electronic writing [ELECTR] The use of electronic circuits and electron devices to reproduce symbols, such as an alphabet, in a prescribed order on an electronic display device for the purpose of transferring information from a source to a viewer of the display device. { i̯lek'trän·ik 'rīd·iŋ }

electron image tube *See* image tube. { i'lek,trän 'im·ij ,tüb }

electron injection [ELECTR] **1.** The emission of electrons from one solid into another. **2.** The process of injecting a beam of electrons with an electron gun into the vacuum chamber of a mass spectrometer, betatron, or other large electron accelerator. { i'lek,trän in'jek·shən }

electron lens [ELECTR] An electric or magnetic field, or a combination thereof, which acts upon an electron beam in a manner analogous to that in which an optical lens acts upon a light beam. Also known as lens. { i'lek,trän 'lenz }

electron lepton number [PART PHYS] The number of electrons and electron-associated neutrinos minus the number of positrons and electron-associated antineutrinos. { i'lek,trän 'lep,tän ,nəm·bər }

electron linear accelerator [NUCLEO] A linear accelerator used to accelerate electrons in a straight line, usually by means of radio-frequency fields which are produced in a loaded waveguide and travel with the electrons. { i'lek,trän ¦lin·ē·ər ak'sel·ə,rād·ər }

electron magnetic moment [ATOM PHYS] The magnetic dipole moment which an electron possesses by virtue of its spin. Also known as electron dipole moment. { i'lek,trän ¦mag·ned·ik 'mō·mənt }

electron mass [PHYS] The mass of an electron, equal to about 9.11×10^{-28} gram, equivalent to 0.511 MeV. Also known as electron rest mass. { i̯lek'trän 'mas }

electron metallography [MET] The study of the microscopic structure of metals employing an electron microscope. { i'lek,trän med·əl'äg·rə·fē }

electron micrograph [GRAPHICS] Photograph of a sample taken with an electron microscope. { i'lek,trän 'mī·krə,graf }

electron microprobe [PHYS] An x-ray machine in which electrons emitted from a hot-filament source are accelerated electrostatically, then focused to an extremely small point on the surface of a specimen by an electromagnetic lens; nondestructive analysis of the specimen can then be made by measuring the backscattered electrons, the specimen current, the resulting x-radiation, or any other resulting phenomenon. Also known as electron probe. { i'lek,trän 'mī·krō,prōb }

electron microscope [ELECTR] A device for forming greatly magnified images of objects by means of electrons, usually focused by electron lenses. { i'lek,trän 'mī·krə,skōp }

electron mirror *See* dynode. { i'lek,trän mir·ər }

electron mobility [SOLID STATE] The drift mobility of electrons in a semiconductor, being the electron velocity divided by the applied electric field. { i'lek,trän mō'bil·əd·ē }

electron multiplicity [ATOM PHYS] In an atom with Russell-Saunders coupling, the quantity $2S + 1$, where S is the total spin quantum number. { i'lek,trän məl·tə'plis·əd·ē }

electron multiplier [ELECTR] An electron-tube structure which produces current amplification; an electron beam containing the desired signal is reflected in turn from the surfaces of each of a series of dynodes, and at each reflection an impinging electron releases two or more secondary electrons, so that the beam builds up in strength. Also known as multiplier. { i'lek,trän 'məl·tə,plī·ər }

electron-multiplier phototube *See* multiplier phototube. { i'lek,trän 'məl·tə,plī·ər 'phōd·ō,tüb }

electron nuclear double resonance [SPECT] A type of electron paramagnetic resonance (EPR) spectroscopy permitting greatly enhanced resolution, in which a material is simultaneously irradiated at one of its EPR frequencies and by a second oscillatory field whose frequency is swept over the range of nuclear frequencies. Abbreviated ENDOR. { i'lek,trän ¦nü·klē·ər ¦dəb·əl 'rez·ən·əns }

electron number [ATOM PHYS] The number of electrons in an ion or atom. { i'lek,trän ,nəm·bər }

electronographic tube [ELECTR] An image tube used in astronomy in which the electron image formed by the tube is recorded directly upon film or plates. { i̯lek'trän·ə¦graf·ik 'tüb }

electronography [ELECTR] The use of image tubes to form intensified electron images of astronomical objects and record them directly on film or plates. { i̯lek·trə'näg·rə·fē }

electronoluminescence *See* cathodoluminescence. { i̯lek¦trän·ə,lü·mə'nes·əns }

electron optics [ELECTR] The study of the motion of free electrons under the influence of electric and magnetic fields. { i'lek,trän 'äp·tiks }

electron orbit [PHYS] The path described by an electron. { i'lek,trän 'òr·bət }

electron pair [PHYS CHEM] A pair of valence electrons which form a nonpolar bond between two neighboring atoms. { i'lek,trän 'per }

electron pair bond *See* covalent bond. { i'lek,trän 'per ,bänd }

electron paramagnetic resonance [PHYS] Magnetic resonance arising from the magnetic moment of unpaired electrons in a paramagnetic substance or in a paramagnetic center in a diamagnetic substance. Abbreviated EPR. Also known as electron spin resonance (ESR); paramagnetic resonance. { i'lek,trän ¦par·ə·mag¦ned·ik 'rez·ən·əns }

electron paramagnetism [PHYS] Paramagnetism in a substance whose atoms or molecules possess a net electronic magnetic moment; arises because of the tendency of a magnetic field to orient the electronic magnetic moments parallel to itself. { i'lek,trän ,par·ə'mag·nə,tiz·əm }

electron-positron pair [PHYS] An electron and a positron produced at the same time in the interaction of a photon with a high-intensity electric field. { i¦lek,trän 'päz·ə,trän ,per }

electron-positron storage ring [NUCLEO] An annular vacuum chamber, enclosed by bending and focusing magnets, in which counterrotating beams of electrons and positrons are stored for several hours and can be made to collide with each other. { i'lek,trän ¦päz·ə,trän 'stòr·ij ,riŋ }

electron probe *See* electron microprobe. { i'lek,trän ,prōb }

electron probe x-ray microanalysis [ANALY CHEM] An analytical technique that uses a narrow electron beam, usually with a diameter less than 1 millimeter, focused on a solid specimen to excite an x-ray spectrum that provides qualitative and quantitative information characteristic of the elements in the sample. Abbreviated EPXMA. { i¦lek,trän ‚prōb ¦eks‚rā ‚mī·krō·ə'nal·ə·səs }

electron radiography [GRAPHICS] A technique for producing a photographic image of an opaque specimen by transmitting electrons through it onto an adjacent photographic film; the electrons are generated in a metal sheet adjacent to the specimen or in the specimen itself by x-rays. { i¦lek,trän ‚rād·ē'ä·grə·fē }

electron radius [PHYS] The classical value r of $2.8179403 \times 10^{-13}$ centimeter for the radius of an electron; obtained by equating mc^2 for the electron to e^2/r, where e and m are the charge and mass of the electron respectively; any classical model for an electron will have approximately this radius. { i'lek,trän 'rād·ē·əs }

electron-ray indicator See cathode-ray tuning indicator. { i'lek,trän ‚rā 'in·də‚kād·ər }

electron-ray tube See cathode-ray tube. { i'lek,trän ‚rā ‚tüb }

electron refraction [ELECTR] The bending of an electron beam passing from one region to another of different electric potential. { i¦lek,trän ri'frak·shən }

electron rest mass See electron mass. { i'lek,trän 'rest ‚mas }

electron ring accelerator [NUCLEO] Proposed particle accelerator in which protons to be accelerated are trapped by the space charge of a ring of relativisitic electrons which is then accelerated. Abbreviated ERA. { i'lek,trän ‚riŋ ak'sel·ə‚rād·ər }

electron shell [ATOM PHYS] **1.** The collection of all the electron states in an atom which have a given principal quantum number. **2.** The collection of all the electron states in an atom which have a given principal quantum number and a given orbital angular momentum quantum number. { i'lek,trän 'shel }

electron spectroscopy [SPECT] The study of the energy spectra of photoelectrons or Auger electrons emitted from a substance upon bombardment by electromagnetic radiation, electrons, or ions; used to investigate atomic, molecular, or solid-state structure, and in chemical analysis. { i'lek,trän spek'träs·kə·pē }

electron spectroscopy for chemical analysis See x-ray photoelectron spectroscopy. { i'lek,trän spek'träs·kə·pē fər 'kem·i·kəl ə'nal·ə·səs }

electron spectrum [SPECT] Visual display, photograph, or graphical plot of the intensity of electrons emitted from a substance bombarded by x-rays or other radiation as a function of the kinetic energy of the electrons. { i'lek,trän 'spek·trəm }

electron spin [QUANT MECH] That property of an electron which gives rise to its angular momentum about an axis within the electron. { i'lek,trän 'spin }

electron spin echo [SOLID STATE] A net magnetization of a material that is sometimes observed at a particular time following the application of two or more short, intense pulses of microwave radiation; in the simplest case, two pulses separated by a time interval t are followed by a net magnetization at time t' after the second pulse. { i‚lek,trän 'spin ‚ek·ō }

electron spin echo envelope modulation [SPECT] **1.** The variation in the intensity of an electron spin echo as the time interval between the two microwave pulses producing the echo is incremented in small steps in the case of a two-pulse echo, or time intervals between suitable pulses are incremented for multiple-pulse echoes. **2.** A type of electron paramagnetic resonance spectroscopy in which this variation is mathematically transformed, using the Fourier transform, to yield the spectrum of nuclear frequencies. Abbreviated ESEEM. { i‚lek,trän ¦spin ‚ek·ō ¦en·və‚lōp ‚mäj·ə'lā·shən }

electron spin density [PHYS] The vector sum of the spin angular momenta of electrons at each point in a substance per unit volume. { i'lek,trän 'spin ‚den·səd·ē }

electron spin resonance See electron paramagnetic resonance. { i'lek,trän 'spin ‚rez·ən·əns }

electron stain [MATER] A substance such as phosphotungstic acid or osmic acid which scatters large numbers of electrons and can therefore be used to stain objects to be examined by an electron microscope. { i'lek,trän 'stān }

electron-stream potential [ELECTR] At any point in an electron stream, the time average of the potential difference between that point and the electron-emitting surface. { i'lek,-trän ‚strēm pə'ten·chəl }

electron-stream transmission efficiency [ELECTR] At an electrode through which the electron stream (beam) passes, the ratio of the average stream current through the electrode to the stream current approaching the electrode. { i'lek,trän ‚strēm tranz'mish·ən ə'fish·ən‚sē }

electron synchrotron [NUCLEO] A circular electron accelerator in which the frequency of the accelerating system is constant, the strength of the magnetic guide field increases, and the electrons move in orbits of nearly constant radius. { i'lek,trän 'siŋ·krə‚trän }

electron telescope [ELECTR] A telescope in which an infrared image of a distant object is focused on the photosensitive cathode of an image converter tube; the resulting electron image is enlarged by electron lenses and made visible by a fluorescent screen. { i'lek,trän 'tel·ə‚skōp }

electron temperature [PL PHYS] The temperature at which ideal gas molecules would have an average kinetic energy equal to that of electrons in a plasma under consideration. { i'lek,trän 'tem·prə·chər }

electron transfer [PHYS] The passage of an electron from one constituent of a system to another. { i'lek,trän 'trans·fər }

electron transition [QUANT MECH] Change of an electron from one state to another, accompanied by emission or absorption of electromagnetic radiation. { i'lek,trän tran'zish·ən }

electron transport system [BIOCHEM] The components of the final sequence of reactions in biological oxidations; composed of a series of oxidizing agents arranged in order of increasing strength and terminating in oxygen. { i'lek,trän 'trans‚pȯrt ‚sis·təm }

electron trap [SOLID STATE] A defect or chemical impurity in a semiconductor or insulator which captures mobile electrons in a special way. { i'lek,trän ‚trap }

electron tube [ELECTR] An electron device in which conduction of electricity is provided by electrons moving through a vacuum or gaseous medium within a gastight envelope. Also known as radio tube; tube; valve (British usage). { i'lek,-trän ‚tüb }

electron-tube amplifier [ELECTR] An amplifier in which electron tubes provide the required increase in signal strength. { i'lek,trän ‚tüb 'am·plə‚fī·ər }

electron-tube generator [ELECTR] A generator in which direct-current energy is converted to radio-frequency energy by an electron tube in an oscillator circuit. { i'lek,trän ‚tüb 'jen·ə‚rād·ər }

electron-tube heater See heater. { i'lek,trän ‚tüb 'hēd·ər }

electron-tube static characteristic [ELECTR] Relation between a pair of variables such as electrode voltage and electrode current with all other voltages maintained constant. { i'lek,trän ‚tüb 'stad·ik kar·ik·tə'ris·tik }

electron tunneling [QUANT MECH] The passage of electrons through a potential barrier which they would not be able to cross according to classical mechanics, such as a thin insulating barrier between two superconductors. { i'lek,trän 'tən·əl·iŋ }

electronuclear breeder See linear accelerator breeder. { i‚lek'trō¦nü·klē·ər 'brēd·ər }

electron vacuum gage [ENG] An instrument used to measure vacuum by the ionization effect that an electron flow (from an incandescent filament to a charged grid) has on gas molecules. { i'lek,trän 'vak‚yüm ‚gāj }

electronvolt [PHYS] A unit of energy which is equal to the energy acquired by an electron when it passes through a potential difference of 1 volt in a vacuum; it is equal to $(1.60217646 \pm 0.00000006) \times 10^{-19}$ volt. Abbreviated eV. { i'lek,trän ‚vōlt }

electron voltaic effect [ELECTR] Sensitivity of photovoltaic cells to electron bombardment. { i¦lek,trän vōl'tā·ik i‚fekt }

electron wave [QUANT MECH] The de Broglie wave or probability amplitude wave of an electron. { i'lek,trän ‚wāv }

electron wave function [QUANT MECH] Function of the spin orientation and position of one or more electrons, specifying the dynamical state of the electrons; the square of the function's modulus gives the probability per unit volume of finding electrons at a given position. { i'lek,trän ‚wāv ‚fəŋk·shən }

electron wavelength [QUANT MECH] The de Broglie wavelength of an electron, given by Planck's constant divided by the momentum. { i'lek,trän 'wāv‚leŋkth }

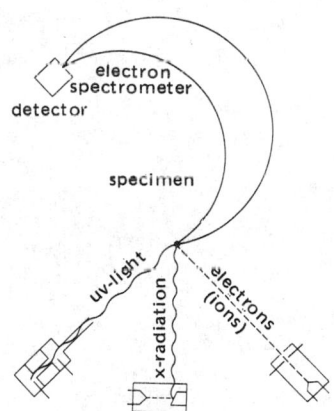

ELECTRON SPECTROSCOPY

Excitation of electron spectra recorded in high-resolution instruments.

ELECTROPHORUS

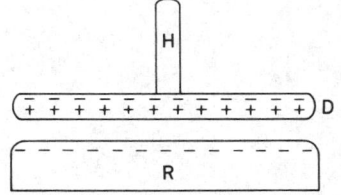

An electrophorus; when the metal plate D with insulating handle H is placed on the rubber plate R, charge is induced as shown.

electrooculogram [PHYSIO] A record of the standing voltage between the front and back of the eye that is correlated with eyeball movement and obtained by electrodes placed on the skin near the eye. { i͵lek·trə·'ok·kyül·ə͵gram }

electrooculography [PHYSIO] The production and study of electrooculograms. { i͵lek·trə͵ok·kyü'läg·rə·fē }

electrooptical birefringence See electrooptical Kerr effect. { i͵lek·trō'äp·tə·kəl bī·ri'frin·jəns }

electrooptical character recognition See optical character recognition. { i͵lek·trō'äp·tə·kəl 'kar·ik·tər ͵rek·ig͵nish·ən }

electrooptical Kerr effect [OPTICS] Birefringence induced by an electric field. Also known as electrooptical birefringence; Kerr effect. { i͵lek·trō'äp·tə·kəl 'kər i͵fekt }

electrooptical modulator [COMMUN] An optical modulator in which a Kerr cell, an electrooptical crystal, or other signal-controlled electrooptical device is used to modulate the amplitude, phase, frequency, or direction of a light beam. { i͵lek·trō'äp·tə·kəl 'mäj·ə͵lād·ər }

electrooptic material [OPTICS] A material in which the indices of refraction are changed by an applied electric field. { i͵lek·trō'äp·tik mə'tir·ē·əl }

electrooptic radar [ENG] Radar system using electrooptic techniques and equipment instead of microwave to perform the acquisition and tracking operation. { i͵lek·trō'äp·tik 'rā͵där }

electrooptics [OPTICS] The study of the influence of an electric field on optical phenomena, as in the electrooptical Kerr effect and the Stark effect. Also known as optoelectronics. { i͵lek·trō'äp·tiks }

electroosmosis [PHYS CHEM] The movement in an electric field of liquid with respect to colloidal particles immobilized in a porous diaphragm or a single capillary tube. { i͵lek·trō·äs'mō·səs }

electroosmotic driver [ELECTR] A type of solion for converting voltage into fluid pressure, which uses depolarizing electrodes sealed in an electrolyte and operates through the streaming potential effect. Also known as micropump. { i͵lek·trō·äz'mäd·ik 'drīv·ər }

electropainting [ENG] Electrolytic deposition of a thin layer of paint on a metal surface which is made an anode. { i'lek·trō͵pānt·iŋ }

electropherography See electrochromatography. { i͵lek·trō·fə'räg·rə·fē }

electrophile [PHYS CHEM] An electron-deficient ion or molecule that takes part in an electrophilic process. { i'lek·trō͵fīl }

electrophilic [PHYS CHEM] **1.** Pertaining to any chemical process in which electrons are acquired from or shared with other molecules or ions. **2.** Referring to an electron-deficient species. { i͵lek·trō'fil·ik }

electrophilic reagent [PHYS CHEM] A reactant which accepts an electron pair from a molecule, with which it forms a covalent bond. { i͵lek·trō͵fil·ik rē'a·jənt }

electrophonic effect [BIOPHYS] The sensation of hearing produced when an alternating current of suitable frequency and magnitude is passed through a person. { i͵lek·trə'fän·ik i'fekt }

electrophoresis [PHYS CHEM] An electrochemical process in which colloidal particles or macromolecules with a net electric charge migrate in a solution under the influence of an electric current. Also known as cataphoresis. { i͵lek·trō·fə'rē·səs }

electrophoretic coating [MET] A surface coating on a metal deposited by electric discharge of particles from a colloidal solution. { i͵lek·trō·fə'red·ik 'kōd·iŋ }

electrophoretic display [OPTICS] A liquid crystal display in which a light-absorbing dye has been added to the liquid to improve both color and luminance contrast. { i͵lek·trō·fə'red·ik di'splā }

electrophoretic effect [PHYS CHEM] Retarding effect on the characteristic motion of an ion in an electrolytic solution subjected to a potential gradient, which results from motion in the opposite direction by the ion atmosphere. { i͵lek·trō·fə'red·ik i'fekt }

electrophoretic mobility [BIOCHEM] A characteristic of living cells in suspension and biological compounds (proteins) in solution to travel in an electric field to the positive or negative electrode, because of the charge on these substances. { i͵lek·trō·fə'red·ik mō'bil·əd·ē }

electrophoretic variants [BIOCHEM] Phenotypically different proteins that are separable into distinct electrophoretic components due to differences in mobilities; an example is erythrocyte acid phosphatase. { i͵lek·trō·fə'red·ik 'ver·ē·əns }

electrophorus [ELEC] A device used to produce electric charges; it consists of a hard-rubber disk, which is negatively charged by rubbing with fur, and a metal plate, held by an insulating handle, which is placed on the disk; the plate is then touched with a grounded conductor, so that negative charge is removed and the plate has net positive charge. { i͵lek'trä·fə·rəs }

electrophotograph [GRAPHICS] An image formed by means of an electrostatic copying system. { i͵lek·trō'fōd·ə͵graf }

electrophotography [GRAPHICS] An electrostatic image-forming process in which light, x-rays, or gamma rays form an electrostatic image on a photoconductive, insulating medium; the charged image areas attract and hold a fine powder called a toner, and the powder image is then transferred to paper or fused there by heat. { i͵lek·trō·fə'täg·rə·fē }

electrophotoluminescence [ELECTR] Emission of light resulting from application of an electric field to a phosphor which is concurrently, or has been previously, excited by other means. { i͵lek·trō͵fōd·ō͵lü·mə'nes·ə ns }

electrophotophoresis [PHYS] Helical motion of small particles suspended in a gas along the direction of an electric field when exposed to a beam of light. { i͵lek·trō͵fōd·ō·fə'rē·səs }

electrophrenic respiration [MED] Artificial respiration in which the nerves that control breathing are stimulated electrically through appropriately placed electrodes. { i͵lek·trə'fren·ik ͵res·pə'rā·shən }

electrophysiology [PHYSIO] The branch of physiology concerned with determining the basic mechanisms by which electric currents are generated within living organisms. { i͵lek·trō͵fiz·ē'äl·ə·jē }

electroplating [MET] Electrodeposition of a metal or alloy from a suitable electrolyte solution; the article to be plated is connected as the cathode in the electrolyte solution; direct current is introduced through the anode which consists of the metal to be deposited. { i'lek·trō͵plād·iŋ }

electroplax [VERT ZOO] One of the structural units of an electric organ of some fishes, composed of thin, flattened plates of modified muscle that appear as two large, waferlike, roughly circular or rectangular surfaces. { i'lek·trō͵plaks }

electropolishing [MET] Smoothing and enhancing the appearance of a metal surface by making it an anode in a suitable electrolyte. Also known as electrolytic brightening; electrolytic polishing. { i͵lek·trō'pä·lə·shiŋ }

electroporation [BIOL] The application of electric pulses to increase the permeability of cell membranes. [CYTOL] The application of electric pulses to animal cells or plant protoplasts to increase membrane permeability. { i͵lek·trō·pə'rā·shən }

electropositive [ELEC] **1.** Carrying a positive electric charge. **2.** Capable of acting as the positive electrode in an electric cell. [PHYS CHEM] Pertaining to elements, ions, or radicals that tend to give up or lose electrons. { i͵lek·trə'päz·əd·iv }

electropositive potential [PHYS CHEM] Potential of an electrode expressed as positive with respect to the hydrogen electrode. { i͵lek·trə͵päz·əd·iv pə'ten·chəl }

electropulse engine [AERO ENG] An engine, for propelling a flight vehicle, that is based on the use of spark discharges through which intense electric and magnetic fields are established for periods ranging from microseconds to a few milliseconds; a resulting electromagnetic force drives the plasma along the leads and away from the spark gap. { i'lek·trō͵pəls ͵en·jən }

electrorefining [CHEM ENG] Petroleum refinery process for light hydrocarbon streams in which an electrostatic field is used to assist in separation of chemical treating agents (acid, caustic, doctor) from the hydrocarbon phase. [MET] Purifying metals by electrolysis using an impure metal as anode from which the pure metal is dissolved and subsequently deposited at the cathode. Also known as electrolytic refining. { i͵lek·trō·ri'fīn·iŋ }

electroreflectance [SPECT] Electromodulation in which reflection spectra are studied. Abbreviated ER. { i͵lek·trō·ri'flek·təns }

electroresistive effect [ELECTR] The change in the resistivity of certain materials with changes in applied voltage. { i¦lek·tro·ri¦zis·tiv i¸fekt }

electroretinogram [MED] A graphic recording of the electric discharges of the retina. Abbreviated ERG. { i¦lek·trō'ret·ən·ə¸gram }

electrorheological fluid [PHYS CHEM] A colloidal suspension of finely divided particles in a carrier liquid, usually an insulating oil, whose rheological properties are changed through an increase in resistance when an electric field is applied. { i¦lek·trō¸rē·ə¦läj·ə·kəl 'flü·əd }

electrorheological material [MATER] A material possessing rheological properties that are controlled by an imposed electric field. { i¸lek·trō¸rē·ə¦läj·ə·kəl mə'tir·ē·əl }

electroscope [ENG] An instrument for detecting an electric charge by means of the mechanical forces exerted between electrically charged bodies. { i'lek·trə¸skōp }

electrosensitive paper [MATER] A conductive paper that darkens when electric current is sent through it. { i¦lek·trō'sen·səd·iv 'pā·pər }

electrosensitive recording [ELECTR] Recording in which the image is produced by passing electric current through the record sheet. { i¦lek·trō'sen·səd·iv ri'kȯrd·iŋ }

electroshock therapy [MED] Treatment of mental patients by passing an electric current of 85–110 volts through the brain. { i'lek·trō¸shäk 'ther·ə·pē }

electroslag welding [MET] A welding process in which consumable electrodes are fed into a joint containing flux; the current melts the flux, and the flux in turn melts the faces of the joint and the electrodes, allowing the weld metal to form a continuously cast ingot between the joint faces. { i'lek¸trō¸slag 'weld¸iŋ }

electrospark machining See electron discharge machining. { i'lek¸trō¸spärk mə'shēn·iŋ }

electrostatic [ELEC] Pertaining to electricity at rest, such as an electric charge on an object. { i¸lek·trə'stad·ik }

electrostatic accelerator [ELECTR] Any instrument which uses an electrostatic field to accelerate charged particles to high velocities in a vacuum. { i¸lek·trə'stad·ik ak¸sel·ə¸rād·ər }

electrostatic actuator See actuator. { i¸lek·trə'stad·ik 'ak·chə¸wād·ər }

electrostatic analyzer [ELECTR] A device which filters an electron beam, permitting only electrons within a very narrow velocity range to pass through. { i¸lek·trə'stad·ik 'an·ə¸līz·ər }

electrostatic atomization [MECH ENG] Atomization in which a liquid jet or film is exposed to an electric field, and forces leading to atomization arise from either free charges on the surface or liquid polarization. { i¸lek·trə'stad·ik ¸ad·ə·mə'zā·shən }

electrostatic attraction See Coulomb attraction. { i¸lek·trə'stad·ik ə'trak·shən }

electrostatic bond [PHYS CHEM] A valence bond in which two atoms are kept together by electrostatic forces caused by transferring one or more electrons from one atom to the other. { i¸lek·trə'stad·ik 'bänd }

electrostatic cathode-ray tube [ELECTR] A cathode-ray tube in which electrostatic deflection is used on the electron beam. { i¸lek·trə'stad·ik ¦kath¸ōd 'rā ¸tüb }

electrostatic coalescence [METEOROL] **1.** The coalescence of cloud drops induced by electrostatic attractions between drops of opposite charges. **2.** The coalescence of two cloud or rain drops induced by polarization effects resulting from an external electric field. { i¸lek·trə'stad·ik kō·ə'les·əns }

electrostatic copier [GRAPHICS] A copier that employs principles of electrostatography. { i¸lek·trə'stad·ik 'käp·ē·ər }

electrostatic copying See electrostatography. { i¸lek·trə'stad·ik 'käp·ē·iŋ }

electrostatic deflection [ELECTR] The deflection of an electron beam by means of an electrostatic field produced by electrodes on opposite sides of the beam; used chiefly in cathode-ray tubes for oscilloscopes. { i¸lek·trə'stad·ik di'flek·shən }

electrostatic detection [ELECTR] The detection and location of any type of solid body, such as a mineral deposit or a mine, by measuring the associated electrostatic field which arises spontaneously or is induced by the detection equipment. { i¸lek·trə'stad·ik di'tek·shən }

electrostatic energy [ELEC] The potential energy which a collection of electric charges possesses by virtue of their positions relative to each other. { i¸lek·trə'stad·ik 'en·ər·jē }

electrostatic error See antenna effect. { i¸lek·trə'stad·ik 'er·ər }

electrostatic field [ELEC] A time-independent electric field, such as that produced by stationary charges. { i¸lek·trə'stad·ik 'fēld }

electrostatic focus [ELECTR] Production of a focused electron beam in a cathode-ray tube by the application of an electric field. { i¸lek·trə'stad·ik 'fō·kəs }

electrostatic force [ELEC] Force on a charged particle due to an electrostatic field, equal to the electric field vector times the charge of the particle. { i¸lek·trə'stad·ik ' fȯrs }

electrostatic force microscopy [ENG] The use of an atomic force microscope to measure electrostatic forces from electric charges on a surface. { i¦lek·trə¸stad·ik ¦fȯrs mī'krä·skə·pē }

electrostatic generator [ELEC] Any machine which produces electric charges by friction or (more commonly) electrostatic induction. { i¸lek·trə'stad·ik 'jen·ə¸rād·ər }

electrostatic gyroscope [ENG] A gyroscope in which a small beryllium ball is electrostatically suspended within an array of six electrodes in a vacuum inside a ceramic envelope. { i¸lek·trə'stad·ik 'jī·rə¸skōp }

electrostatic induction [ELEC] The process of charging an object electrically by bringing it near another charged object, then touching it to ground. Also known as induction. { i¸lek·trə'stad·ik in'dək·shən }

electrostatic instrument [ELEC] A meter that depends for its operation on the forces of attraction and repulsion between electrically charged bodies. { i¸lek·trə'stad·ik 'in·strə·mənt }

electrostatic interactions See Coulomb interactions. { i¸lek·trə'stad·ik int·ə'rak·shənz }

electrostatic ion-cyclotron wave [PL PHYS] A longitudinal wave that propagates in a magnetically confined plasma at a very large angle to the magnetic field, with a frequency somewhat in excess of the ion-cyclotron frequency. { i¦lek·trə¸stad·ik ¦ī·än 'sī·klə¸trän ¸wāv }

electrostatic lens [ELECTR] An arrangement of electrostatic fields which acts upon beams of charged particles similar to the way a glass lens acts on light beams. { i¸lek·trə'stad·ik 'lenz }

electrostatic loudspeaker [ENG ACOUS] A loudspeaker in which the mechanical forces are produced by the action of electrostatic fields; in one type the fields are produced between a thin metal diaphragm and a rigid metal plate. Also known as capacitor loudspeaker. { i¸lek·trə'stad·ik 'laud¸spēk·ər }

electrostatic memory See electrostatic storage. { i'lek·trə¸stad·ik 'mem·rē }

electrostatic microphone See capacitor microphone. { i'lek·trə¸stad·ik 'mī·krə¸fōn }

electrostatic octupole lens [ELECTR] A device for controlling beams of electrons or other charged particles, consisting of eight electrodes arranged in a circular pattern with alternating polarities; commonly used to correct aberrations of quadrupole lens systems. { i'lek·trə¸stad·ik 'äk·tə¸pōl 'lenz }

electrostatic painting [ENG] A painting process that uses the particle-attracting property of electrostatic charges; direct current of about 100,000 volts is applied to a grid of wires through which the paint is sprayed to charge each particle; the metal objects to be sprayed are connected to the opposite terminal of the high-voltage circuit, so that they attract the particles of paint. { i'lek·trə¸stad·ik 'pānt·iŋ }

electrostatic potential See electric potential. { i'lek·trə¸stad·ik pə'ten·chəl }

electrostatic precipitator [ENG] A device which removes dust or other finely divided particles from a gas by charging the particles inductively with an electric field, then attracting them to highly charged collector plates. Also known as precipitator. { i'lek·trə¸stad·ik prə'sip·ə¸tād· ər }

electrostatic printer [GRAPHICS] A line printer in which high-intensity lamps project images of characters onto a sensitized drum to form electrostatic patterns that attract ink powder; the images are then transferred to paper and fused. { i'lek·trə¸stad·ik 'print·ər }

electrostatic probe See electric probe. { i'lek·trə¸stad·ik 'prōb }

electrostatic quadrupole lens [ELECTR] A device for

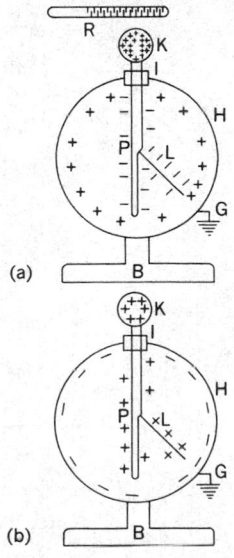

ELECTROSCOPE

Simple gold-leaf electroscope. *(a)* An electroscope being charged by induction by negative charge on hard-rubber rod R. *(b)* Positive charge left on its leaf after induction process is complete. L = gold leaf, P = metal post, I = insulator, K = metal knob, H = metal housing, B = base, R = rubber rod, G = ground.

ELECTROSTATIC QUADRUPOLE LENS

Electrostatic quadrupole lens showing the four electrodes.

focusing beams of electrons or other charged particles, consisting of four electrodes arranged in a circular pattern with alternating polarities. { i'lek·trə,stad·ik ¦kwä·drə,pōl 'lenz }

electrostatic reprography [GRAPHICS] The replication of a document image as a pattern of electric charge on a metal plate; carbon particles attracted to the charge pattern are transferred to and bonded on an accepting surface; for example, multilith mats. { i'lek·trə,stad·ik rə'präg·rə·fē }

electrostatic repulsion See Coulomb repulsion. { i'lek·trə,stad·ik ri'pəl·shən }

electrostatics [ELEC] The study of electric charges at rest, their electric fields, and potentials. { i,lek·trə'stad·iks }

electrostatic scanning [ELECTR] Scanning that involves electrostatic deflection of an electron beam. { i'lek·trə,stad·ik 'skan·iŋ }

electrostatic separation [ENG] Separation of finely pulverized materials by placing them in electrostatic separators. Also known as high-tension separation. { i'lek·trə,stad·ik ,sep·ə'rā·shən }

electrostatic separator [ENG] A separator in which a finely pulverized mixture falls through a powerful electric field between two electrodes; materials having different specific inductive capacitances are deflected by varying amounts and fall into different sorting chutes. { i'lek·trə,stad·ik 'sep·ə,rād·ər }

electrostatic shielding [ELEC] The placing of a grounded metal screen, sheet, or enclosure around a device or between two devices to prevent electric fields from interacting. { i'lek·trə,stad·ik 'shēld·iŋ }

electrostatic storage [ELECTR] A storage in which information is retained as the presence or absence of electrostatic charges at specific spot locations, generally on the screen of a special type of cathode-ray tube known as a storage tube. Also known as electrostatic memory. { i'lek·trə,stad·ik 'stór·ij }

electrostatic storage tube See storage tube. { i'lek·trə,stad·ik 'stór·ij ,tüb }

electrostatic stress [ELEC] An electrostatic field acting on an insulator, which produces polarization in the insulator and causes electrical breakdown if raised beyond a certain intensity. { i'lek·trə,stad·ik 'stres }

electrostatic tape camera [OPTICS] A camera in which images are stored electrostatically on a plastic tape; designed for use in satellites, where the stored image is not damaged by Van Allen or other radiation. { i'lek·trə,stad·ik 'tāp ,kam·rə }

electrostatic transducer [ENG ACOUS] A transducer consisting of a fixed electrode and a movable electrode, charged electrostatically in opposite polarity; motion of the movable electrode changes the capacitance between the electrodes and thereby makes the applied voltage change in proportion to the amplitude of the electrode's motion. Also known as condenser transducer. { i'lek·trə,stad·ik tranz'dü·sər }

electrostatic tweeter [ENG ACOUS] A tweeter loudspeaker in which a flat metal diaphragm is driven directly by a varying high voltage applied between the diaphragm and a fixed metal electrode. { i'lek·trə,stad·ik 'twēd·ər }

electrostatic units [ELEC] A centimeter-gram-second system of electric and magnetic units in which the unit of charge is that charge which exerts a force of 1 dyne on another unit charge when separated from it by a distance of 1 centimeter in vacuum; other units are derived from this definition by assigning unit coefficients in equations relating electric and magnetic quantities. Abbreviated esu. { i'lek·trə,stad·ik 'yü·nəts }

electrostatic valence rule [PHYS CHEM] The postulate that in a stable ionic structure the valence of each anion, with changed sign, equals the sum of the strengths of its electrostatic bonds to the adjacent cations. { i'lek·trə,stad·ik 'vā·ləns ,rül }

electrostatic voltmeter [ENG] A voltmeter in which the voltage to be measured is applied between fixed and movable metal vanes; the resulting electrostatic force deflects the movable vane against the tension of a spring. { i'lek·trə,stad·ik 'vōlt,mēd·ər }

electrostatic wattmeter [ENG] An adaptation of a quadrant electrometer for power measurements in which two quadrants are charged by the voltage drop across a noninductive shunt resistance through which the load current passes, and the line voltage is applied between one of the quadrants and a moving vane. { i'lek·trə,stad·ik 'wät,mēd·ər }

electrostatic wave [PL PHYS] Wave motion of a plasma

whose restoring forces are primarily electrostatic. { i'lek·trə,stad·ik 'wāv }

electrostatography [GRAPHICS] A generic term covering all processes involving the forming and use of electrostatic charged patterns for recording and reproducing images; the field is divided into electrophotography and electrography. Also known as electrostatic copying. { i'lek·trō·stə'täg·rə·fē }

electrostethophone [MED] A stethoscope consisting of a microphone and audio amplifier feeding headphones, used for detection and study of sounds arising within the body. { i'lek·trō'steth·ə,fōn }

electrostriction [MECH] A form of elastic deformation of a dielectric induced by an electric field, associated with those components of strain which are independent of reversal of field direction, in contrast to the piezoelectric effect. Also known as electrostrictive strain. { i¦lek·trō'strik·shən }

electrostriction transducer [ENG ACOUS] A transducer which depends on the production of an elastic strain in certain symmetric crystals when an electric field is applied, or, conversely, which produces a voltage when the crystal is deformed. Also known as ceramic transducer. { i¦lek·trō'strik·shən tranz'dü·sər }

electrostrictive strain See electrostriction. { i¦lek·trō'strik·tiv 'strān }

electrosurgery [MED] The use of electricity to perform surgical procedures, as the use of electricity to simultaneously cut tissue and arrest bleeding. { i¦lek·trō'sərj·ə·rē }

electrosynthesis [CHEM] A reaction in which synthesis occurs as the result of an electric current. { i¦lek·trō'sin·thə·səs }

electrotaxis [BIOL] Movement of an organism in response to stimulation by electric charges. { i¦lek·trō'tak·səs }

electrotherapy [MED] The therapeutic use of electricity. { i¦lek·trō'ther·ə·pē }

electrothermal [PHYS] **1.** Pertaining to both heat and electricity. **2.** In particular, pertaining to conversion of electrical energy into heat energy. { i¦lek·trō'thər·məl }

electrothermal ammeter See thermoammeter. { i¦lek·trō'thər·məl 'a,med·ər }

electrothermal energy conversion [ENG] The direct conversion of electric energy into heat energy, as in an electric heater. { i¦lek·trō'thər·məl 'en·ər·jē kən,vər·zhən }

electrothermal process [ENG] Any process which uses an electric current to generate heat, utilizing resistance, arcs, or induction; used to achieve temperatures higher than can be obtained by combustion methods. { i¦lek·trō'thər·məl 'präs·əs }

electrothermal propulsion [AERO ENG] Propulsion of spacecraft by using an electric arc or other electric heater to bring hydrogen gas or other propellant to the high temperature required for maximum thrust; an arc-jet engine is an example. { i¦lek·trō'thər·məl prə'pəl·shən }

electrothermal recording [ELECTR] Type of electrochemical recording, used in facsimile equipment, wherein the chemical change is produced principally by signal-controlled thermal action. { i¦lek·trō'thər·məl ri'kórd·iŋ }

electrothermal voltmeter [ENG] An electrothermal ammeter employing a series resistor as a multiplier, thus measuring voltage instead of current. { i¦lek·trō'thər·məl 'vōlt,mēd·ər }

electrotinning [MET] Electroplating an object with tin. { i¦lek·trō'tin·iŋ }

electrotonus [PHYSIO] The change of condition in a nerve or a muscle during the passage of a current of electricity. { i,lek'trät·ən·əs }

electrotropism [BIOL] Orientation response of a sessile organism to stimulation by electric charges. { i,lek'trä·trə,piz·əm }

electrotype [GRAPHICS] A duplicate printing surface prepared by making a mold of the type page or halftone plate, then suspending this mold in a bath of copper sulfate and sulfuric acid where, by electrolytic action, a thin shell of copper is deposited on it, and finally pouring molten type metal into this shell to strengthen it for use on the press. { i'lek·trə,tīp }

electrotyping [GRAPHICS] The process of making an electrotrope. { i'lek·trə,tīp·iŋ }

electrovalence [PHYS CHEM] The valence of an atom that has formed an ionic bond. { i¦lek·trō'vā·ləns }

electrovalent bond See ionic bond. { i¦lek·trō¦vā·lənt 'bänd }

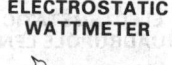
ELECTROSTATIC WATTMETER

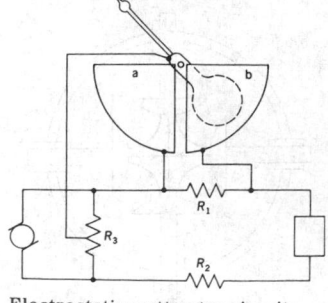

Electrostatic wattmeter circuit diagram showing two quadrants *a* and *b*.

electroviscous effect [FL MECH] Change in a liquid's viscosity induced by a strong electrostatic field. { i¦lek·trō'vis·kəs i'fekt }

electroweak interaction [PART PHYS] The unification of the electromagnetic and weak interactions described by the Weinberg-Salam theory. { i'lek·trō‚wēk ‚in·tər'ak·shən }

electrowinning [MET] Extracting metal from solutions by electrochemical processes. { i¦lek·trō'win·iŋ }

electrum [MET] A naturally occurring alloy of gold and silver. { i'lek·trəm }

Elektra [ASTRON] An asteroid with a diameter of about 113 miles (182 kilometers), mean distance from the sun of 3.117 astronomical units, and C-type surface composition. { i'lek·trə }

Elektrion process [CHEM ENG] A process of condensation and polymerization in which a mixture of a relatively light mineral oil and a fatty oil is subjected to an electric discharge in an atmosphere of hydrogen; the product is a very viscous oil used for blending with lighter lubricating oils. { i'lek·trē‚än ‚präs·əs }

element [CHEM] A substance made up of atoms with the same atomic number; common examples are hydrogen, gold, and iron. Also known as chemical element. [COMPUT SCI] A circuit or device performing some specific elementary data-processing function. [ELEC] **1.** A part of an electron tube, semiconductor device, or antenna array that contributes directly to the electrical performance. **2.** *See* component. [ELECTROMAG] Radiator, active or parasitic, that is a part of an antenna. [IND ENG] A brief, relatively homogeneous part of a work cycle that can be described and identified. [MATH] **1.** In an array such as a matrix or determinant, a quantity identified by the intersection of a given row or column. **2.** In network topology, an edge. **3.** The generatrix of a ruled surface at any one fixed position. **4.** *See* member. { 'el·ə·mənt }

element 110 [CHEM] A synthetic chemical element, atomic number 110; the eighteenth transuranium element. { ¦el·ə·mənt ‚wən'ten }

element 111 [CHEM] A synthetic chemical element, atomic number 111; the nineteenth transuranium element. { ¦el·ə·mənt wən·i'lev·ən }

element 112 [CHEM] A synthetic chemical element, atomic number 112; the twentieth transuranium element. { ¦el·ə·mənt wən'twelv }

elemental area *See* picture element. { ‚el·ə'ment·əl 'er·ē·ə }

elemental motion [IND ENG] In time-and-motion study, a fundamental subdivision of the hand movements in manipulating an object. Also known as basic element; fundamental motion; therblig. { ‚el·ə'ment‚əl ‚mō·shən }

elementary charge [PHYS] An electric charge such that the electric charge of any body is an integral multiple of it, equal to the electron charge. { ‚el·ə'men·trē ‚chärj }

elementary commodity group [IND ENG] The lowest level of goods or services for which consistent values can be determined. Also known as elementary group. { el·ə¦men·trē kə'mad·əd·ē ‚grüp }

elementary event [STAT] A single outcome of an experiment. Also known as simple event. { ‚el·ə¦men·trē i'vent }

elementary excitation [QUANT MECH] The quantum of energy of some vibration or wave, such as a photon, phonon, plasmon, magnon, polaron, or exciton. { ‚el·ə'men·trē ‚ek‚sī'tā·shən }

elementary function [MATH] Any function which can be formed from algebraic functions and the exponential, logarithmic, and trigonometric functions by a finite number of operations consisting of addition, subtraction, multiplication, division, and composition of functions. { ‚el·ə'men·trē 'fəŋk·shən }

elementary group *See* elementary commodity group. { ‚el·ə'men·trē 'grüp }

elementary item [COMPUT SCI] An item considered to have no subordinate item in the COBOL language. { ‚el·ə'men·trē ‚īd·əm }

elementary particle [PART PHYS] A particle which, in the present state of knowledge, cannot be described as compound, and is thus one of the fundamental constituents of all matter. Also known as fundamental particle; particle; subnuclear particle. { ‚el·ə'men·trē 'pärd·i·kəl }

elementary process [PHYS CHEM] In chemical kinetics, the particular events at the atomic or molecular level which make up an overall reaction. { ‚el·ə'men·trē 'präs·əs }

elementary reaction [ORG CHEM] A reaction which involves only a single transition state with no intermediates. Also known as step. { ‚el·ə'men·trē rē'ak·shən }

elementary ring structure [ASTRON] A formation on the moon's surface consisting of a simple wall of uniform cross section enclosing a circular area which has the same elevation as the surrounding surface. Also known as simple ring. { ‚el·ə'men·trē 'riŋ ‚strək·chər }

elementary symmetric functions [MATH] For a set of n variables, a set of n functions, $\sigma_1, \sigma_2, \ldots, \sigma_n$, where σ_k is the sum of all products of k of the n variables. { el·ə'men·trē si¦me·trik 'fəŋk·shənz }

element breakdown [IND ENG] Separation of a work cycle into elemental motions. { 'el·ə·mənt 'brāk‚daùn }

elements [ASTRON] A set of quantities specifying the orbit of a member of the solar system or of a binary star system, used to calculate the body's position at any time. [MECH] The various features of a trajectory such as the angle of departure, maximum ordinate, angle of fall, and so on. [NAV] The specific values of the coordinates used to define the position of an aircraft or vessel. { 'el·ə·mənts }

element time [IND ENG] The time to complete a specific motion element. { 'el·ə·mənt ‚tīm }

elemi [MATER] A soft resin obtained from tropical trees of the family Burseraceae in the Philippines; used as a plasticizer, in cements and printing inks, and for perfumery and waterproofing. { 'el·ə·mē }

eleolite *See* nepheline. { ə'lē·ə‚līt }

eleostearic acid [ORG CHEM] $CH_3(CH_2)_7(CH:CH)_3-(CH_2)_3COOH$ A colorless, water-insoluble, crystalline, unsaturated fatty acid; the glycerol ester is a chief component of tung oil. { ¦el·ē·ō'stir·ik 'as·əd }

elephant [METEOROL] *See* elephanta. [VERT ZOO] The common name for two living species of proboscidean mammals in the family Elephantidae; distinguished by the elongation of the nostrils and upper lip into a sensitive, prehensile proboscis. { 'el·ə·fənt }

elephanta [METEOROL] A strong southeasterly wind on the Malabar coast of southwest India in September and October, at the end of the southwest monsoon, bringing thundersqualls and heavy rain. Also known as elephant; elephanter. { ‚el·ə'fan·tə }

elephant ear [AERO ENG] Thick metal plating that reinforces a missile's skin. { 'el·ə·fənt ‚ir }

elephanter *See* elephanta. { ‚el·ə'fan·tər }

elephant-hide pahoehoe [GEOL] A type of pahoehoe on whose surface are innumerable tummuli, broad swells, and pressure ridges which impart the appearance of elephant hide. { 'el·ə·fənt ‚hīd pa'hō·ē‚hō·ē }

elephantiasis [MED] A parasitic disease of humans caused by the filarial nematode *Wuchereria bancrofti;* characterized by cutaneous and subcutaneous tissue enlargement due to lymphatic obstruction. { ‚el·ə·fan'tī·ə·səs }

Elephantidae [VERT ZOO] A family of mammals in the order Proboscidea containing the modern elephants and extinct mammoths. { ‚el·ə'fan·tə‚dē }

elephant trunks [ASTRON] Long, dark regions that encroach into the bright matter in diffuse nebulae, usually bordered by bright rims. { 'el·ə·fənt ‚trəŋks }

elerwind [METEOROL] A wind of Sun Valley north of Kufstein, in the Tyrol. { 'el·ər‚vint }

elevate [ENG] To increase the angle of elevation of a gun, launcher, optical instrument, or the like. { 'el·ə‚vāt }

elevated flooring *See* raised flooring. { ¦el·ə‚vād·əd 'flór·iŋ }

elevated pole [ASTRON] The celestial pole that appears above the horizon. { 'el·ə‚vād·əd 'pōl }

elevating machine *See* elevator. { 'el·ə‚vād·iŋ mə‚shēn }

elevating mechanism [ORD] Mechanism on a gun carriage or launcher by which the weapon is elevated or depressed. { 'el·ə‚vād·iŋ 'mek·ə‚niz·əm }

elevation [ENG] Vertical distance to a point or object from sea level or some other datum. [GRAPHICS] A graphic projection of a machine or structure on a vertical plane without perspective. [ORD] In antiaircraft artillery, a term sometimes applied to the angular height. { ‚el·ə'vā·shən }

elevation angle [ELECTROMAG] The angle that a radio,

radar, or other such beam makes with the horizontal. [ENG] *See* angle of elevation. { ,el·ə'vā·shən ,aŋ·gəl }

elevation-angle error [ELECTROMAG] In radar, the error in the measurement of the elevation angle of a target resulting from the vertical bending or refraction of radio energy in traveling through the atmosphere. Also known as elevation error. { ,el·ə'vā·shən ,aŋ·gəl ,er·ər }

elevation difference [ORD] The change in elevation which must be applied to a particular gun when firing data are being received from a base piece or other directing point. { ,el·ə'vā·shən ,dif·rəns }

elevation error *See* elevation-angle error. { ,el·ə'vā·shən ,er·ər }

elevation head [FL MECH] The energy per unit mass possessed by a fluid as a result of its height above some reference level. Also known as potential head. { ,el·ə'vā·shən ,hed }

elevation indicator [ORD] A component which presents visually the angle between a fixed reference point and a target in a vertical plane. { ,el·ə'vā·shən 'in·də,kād·ər }

elevation meter [ENG] An instrument that measures the change of elevation of a vehicle. { ,el·ə'vā·shən ,mēd·ər }

elevation of ivory point *See* barometer elevation. { ,el·ə'vā·shən əv 'īv·rē ,point }

elevation prediction correction [ORD] In antiaircraft artillery terminology, the future elevation minus the present elevation. { ,el·ə'vā·shən prə'dik·shən kə,rek·shən }

elevation rate [ORD] Rate of change of present elevation of artillery; it is equal to minus the zenith distance rate of the present line, and may be expressed in mils per second or degrees per second. { ,el·ə'vā·shən ,rāt }

elevation scale [ORD] Scale on a gun carriage that shows the quadrant elevation of the gun. { ,el·ə'vā·shən ,skāl }

elevation stop [ENG] Structural unit in a gun or other equipment that prevents it from being elevated or depressed beyond certain fixed limits. { ,el·ə'vā·shən ,stäp }

elevation table [ORD] Firing table giving a list of ranges, with the corresponding quadrant elevation settings to be applied to a gun. { ,el·ə'vā·shən ,tā·bəl }

elevation tints *See* gradient tints. { ,el·ə'vā·shən ,tins }

elevator [AERO ENG] The hinged rear portion of the longitudinal stabilizing surface or tail plane of an aircraft, used to obtain longitudinal or pitch-control moments. [MECH ENG] Also known as elevating machine. **1.** Vertical, continuous-belt, or chain device with closely spaced buckets, scoops, arms, or trays to lift or elevate powders, granules, or solid objects to a higher level. **2.** Pneumatic device in which air or gas is used to elevate finely powdered materials through a closed conduit. **3.** An enclosed platform or car that moves up and down in a shaft for transporting people or materials. Also known as lift. [PETRO ENG] A clamp gripping a stand or column of casing tubing, drill pipe, or sucker rods so that it can be moved up or down in a borehole being drilled. { 'el·ə,vād·ər }

elevator angle [AERO ENG] The angular displacement of the elevator from its neutral position; it is positive when the trailing edge of the elevator is below the neutral position, and negative when it is above. { 'el·ə,vād·ər ,aŋ·gəl }

elevator bail *See* elevator link. { 'el·ə,vād·ər ,bāl }

elevator dredge [MECH ENG] A dredge which has a chain of buckets, usually flattened across the front and mounted on a nearly vertical ladder; used principally for excavation of sand and gravel beds under bodies of water. { 'el·ə,vād·ər ,drej }

elevator link [PETRO ENG] One of the cylindrical bars that serve to support drilling rig elevators and to attach them to the hook from which the swivel is suspended. Also known as elevator bail. { 'el·ə,vād·ər ,liŋk }

elevator rod [PETRO ENG] A steel block fitted with an opening and latching device to permit insertion of a sucker rod, and provided with two long links to suspend from the elevator clamp when equipping a well with sucker rods for pumping. { 'el·ə,vād·ər ,räd }

eleven punch *See* X punch. { ə'lev·ən ,pənch }

elevon [AERO ENG] The hinged rear portion of an aircraft wing, moved in the same direction on each side of the aircraft to obtain longitudinal control and differentially to obtain lateral control; elevon is a combination of the words elevator and aileron to denote that an elevon combines the functions of aircraft elevators and ailerons. { 'el·ə,vän }

ELF *See* extremely low frequency.

elfinwood *See* krummholz. { 'el·fən,wùd }

Elgin extractor [CHEM ENG] Spray-tower, multistage, counterflow extractor in which the diameter of the base section is expanded to eliminate flow restriction at the light-liquid distribution location. { 'el·jən ik'strak·tər }

eliminant *See* resultant. { i'lim·ə·nənt }

elimination [MATH] A process of deriving from a system of equations a new system with fewer variables, but with precisely the same solutions. { ə,lim·ə'nā·shən }

elimination coefficient [GEN] The frequency with which certain genotypes die prematurely or are hindered during reproduction and are genetically eliminated as a consequence. { ə¦lim·ə'nā·shən ,kō·ə,fish·ənt }

elimination factor [COMPUT SCI] In information retrieval, the ratio obtained in dividing the number of documents that have not been retrieved by the total number of documents in the file. { ə,lim·ə'nā·shən ,fak·tər }

elimination reaction [ORG CHEM] A chemical reaction involving elimination of some portion of a reactant compound, with the production of a second compound. { ə,lim·ə'nā·shən rē,ak·shən }

eliminator [ELECTR] Device that takes the place of batteries, generally consisting of a rectifier operating from alternating current. { ə'lim·ə,nād·ər }

E lines [ELEC] Contour lines of constant electrostatic field strength referred to some reference base. { 'ē ,līnz }

E link [NAV ARCH] A bracket attached to one of the arms of the stand that the compass is on (binnacle); it permits the mounting of a quadrantal error corrector. { 'ē ,liŋk }

ELISA *See* enzyme-linked immunosorbent assay. { ə'līz·ə *or* ¦ē¦el¦ī¦es'ā }

elint *See* electronic intelligence. { 'ē,lint }

elinvar [MET] A nickel-chromium steel alloy containing manganese and tungsten in varying amounts and having a low thermal expansion and almost invariable modulus of elasticity; used for chronometer balances and springs for gages and other instruments. { 'el·in,vär }

elixir [PHARM] A sweetened, aromatic solution, usually hydroalcoholic, sometimes containing soluble medicants; intended for use only as a flavor or vehicle. { ə'lik·sər }

elk [VERT ZOO] *Alces alces.* A mammal (family Cervidae) in Europe and Asia that resembles the North American moose but is smaller; it is the largest living deer. { elk }

elkerite [MATER] A form of bitumen produced during a slow oxidation of petroleum. { 'el·kə,rīt }

ell [BUILD] A wing built perpendicular to the main section of a building. { el }

ellagic acid [ORG CHEM] $C_{14}H_6O_8$ A compound isolated from tannins as yellow crystals that are minimally soluble in hot water. Also known as gallogen. { e'laj·ik 'as·əd }

ellestadite [MINERAL] A pale rose, hexagonal mineral consisting of an apatite-like calcium sulfate-silicate; occurs in granular massive form. { 'el·ə,sta,dīt }

ellipse [MATH] The locus of all points in the plane at which the sum of the distances from a fixed pair of points, the foci, is a given constant. { ə'lips }

ellipsograph [GRAPHICS] A type of compass that draws ellipses. { ə'lip·sə,graf }

ellipsoid [MATH] A surface whose intersection with every plane is an ellipse (or circle). { ə'lip,soid }

ellipsoidal coordinates [MATH] Coordinates in space determined by confocal quadrics. { ə,lip'soid·əl kō'ord·ən·əts }

ellipsoidal floodlight [ELEC] A lighting unit used in theatrical lighting consisting of an ellipsoidal reflector with fixed spacing and a lamp; power requirements are 250–5000 watts and the reflector diameter is 10–24 inches (25–61 centimeters). Also known as scoop. { ə,lip'soid·əl 'fləd,līt }

ellipsoidal harmonics [MATH] Lamé functions that play a role in potential problems on an ellipsoid analogous to that played by spherical harmonics in potential problems on a sphere. { ə¦lip,soid·əl ,här'män·iks }

ellipsoidal lava *See* pillow lava. { ə,lip'soid·əl 'läv·ə }

ellipsoidal of wave normals *See* index ellipsoid. { ə,lip'soid·əl əv ,wāv 'nor·məlz }

ellipsoidal reflector [OPTICS] A concave ellipsoidal surface from which light is specularly reflected; used in a light projector to focus rays from a light source at the near focal point onto the opposite focal point of the ellipse. { ə,lip'soid·əl ri'flek·tər }

ellipsoidal spotlight [ELEC] A lighting unit consisting of a

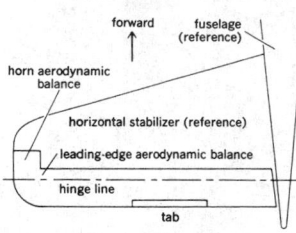

ELEVATOR

forward · fuselage (reference)

horn aerodynamic balance

horizontal stabilizer (reference)

leading-edge aerodynamic balance

hinge line

tab

A typical elevator control surface on the left-hand tail plane of an airplane.

ELK

North American elk (*Cervus canadensis*).

reflector, lamp, single or multiple lens system, and framing device; power requirements are 250–2000 watts. { ə,lip'sȯid· əl 'spät,līt }

ellipsoidal wave functions *See* Lamé wave functions. { ə¦lip-,sȯid·əl 'wāv ,faŋk·shənz }

ellipsoid of revolution [MATH] An ellipsoid generated by rotation of an ellipse about one of its axes. Also known as spheroid. { ə'lip,sȯid əv ,rev·ə'lü·shən }

ellipsoid of wave normals *See* index ellipsoid. { ə'lip,sȯid əv ¦wāv 'nȯr·məlz }

ellipsometer [OPTICS] An instrument for determining the degree of ellipticity of polarized light; used to measure the thickness of very thin transparent films by observing light reflected from the film. { ə,lip'säm·əd·ər }

ellipsometry [OPTICS] A technique for determining the properties of a material from the polarization characteristics of linearly polarized incident light reflected from its surface. { ə,lip'säm·ə·trē }

elliptic [SCI TECH] Oval-shaped. { ə'lip·tik }

elliptical galaxy [ASTRON] A galaxy whose overall shape ranges from a spheroid to an ellipsoid, without any noticeable structural features. Also known as E galaxy; spheroidal galaxy. { ə'lip·tə·kəl 'gal·ik·sē }

elliptical orbit [MECH] The path of a body moving along an ellipse, such as that described by either of two bodies revolving under their mutual gravitational attraction but otherwise undisturbed. { ə'lip·tə·kəl 'ȯr·bət }

elliptical polarization [ELECTROMAG] Polarization of an electromagnetic wave in which the electric field vector at any point in space describes an ellipse in a plane perpendicular to the propagation direction. { ə'lip·tə·kəl ,pō·lə·rə'zā·shən }

elliptical projection [MAP] A map of the surface of the earth formed on an ellipse's interior. { ə'lip·tə·kəl prə'jek·shən }

elliptical ring structure [ASTRON] A lunar crater enclosed by a wall that is elliptical in shape. { ə'lip·tə·kəl 'riŋ ,strək·chər }

elliptical system [ENG] A tracking or navigation system where ellipsoids of position are determined from time or phase summation relative to two or more fixed stations which are the focuses for the ellipsoids. { ə'lip·tə·kəl 'sis·təm }

elliptic arch [ARCH] An arch whose interior curve resembles an ellipse. { ə'lip·tik 'ärch }

elliptic cone [MATH] A cone whose base is an ellipse. { ə¦lip·tik 'kōn }

elliptic conical surface [MATH] A conical surface whose directrix is an ellipse. { ə¦lip·tik 'kän·ə·kəl 'sȯr·fəs }

elliptic coordinates [MATH] The coordinates of a point in the plane determined by confocal ellipses and hyperbolas. { ə'lip·tik kō'ȯrd·ən·əts }

elliptic cylinder [MATH] A cylinder whose directrix is an ellipse. { ə¦lip·tik 'sil·ən·dər }

elliptic differential equation [MATH] A general type of second-order partial differential equation which includes Laplace's equation and has the form

$$\sum_{i,j=1}^{n} A_{ij}\,(\partial^2 u/\partial x_i\,\partial x_j) + \sum_{i=1}^{n} B_i\,(\partial u/\partial x_i) + Cu + F = 0$$

where A_{ij}, B_i, C, and F are suitably differentiable real functions of $x_1, x_2, \ldots, x_n$, and there exists at each point $(x_1, x_2, \ldots, x_n)$ a real linear transformation on the x_i which reduces the quadratic form

$$\sum_{i,j=1}^{n} A_{ij}\, x_i\, x_j$$

to a sum of n squares, all of the same sign. Also known as elliptic partial differential equation. { ə'lip·tik dif·ə¦ren·chəl i'kwā·zhən }

elliptic function [MATH] An inverse function of an elliptic integral; alternatively, a doubly periodic, meromorphic function of a complex variable. { ə'lip·tik 'faŋk·shən }

elliptic gear [MECH ENG] A change gear composed of two elliptically shaped gears, each rotating about one of its focal points. { ə'lip·tik 'gir }

elliptic geometry [MATH] The geometry obtained from euclidean geometry by replacing the parallel line postulate with the postulate that no line may be drawn through a given point, parallel to a given line. Also known as Riemannian geometry. { ə'lip·tik jē'äm·ə·trē }

elliptic integral [MATH] An integral over x whose integrand is a rational function of x and the square root of $p(x)$, where $p(x)$ is a third- or fourth-degree polynomial without multiple roots. { ə'lip·tik 'int·ə·grəl }

elliptic-integral filter [ELECTR] An electronic filter whose gain characteristic has both an equal-ripple shape in the passband and equal minima of attenuation in the stop-band. Also known as Cauer filter. { ə,lip·tik ,int·ə·grəl 'fil·tər }

elliptic integral of the first kind [MATH] Any elliptic integral which is finite for all values of the limits of integration and which approaches a finite limit when one of the limits of integration approaches infinity. { ə¦lip·tik ¦int·ə·grəl əv thə ¦fərst ,kīnd }

elliptic integral of the second kind [MATH] Any elliptic integral which approaches infinity as one of the limits of integration y approaches infinity, or which is infinite for some value of y, but which has no logarithmic singularities in y. { ə¦lip·tik ¦int·ə·grəl əv thə ¦sek·ənd ,kīnd }

elliptic integral of the third kind [MATH] Any elliptic integral which has logarithmic singularities when considered as a function of one of its limits of integration. { ə¦lip·tik ¦int·ə·grəl əv thə ¦thərd ,kīnd }

ellipticity [ASTRON] The difference between the equatorial and polar radii of a planet divided by the mean radius. [ELECTR] *See* axial ratio. [MATH] **1.** Also known as oblateness. **2.** For an ellipse, the difference between the semimajor and semiminor axes of the ellipse, divided by the semimajor axis. **3.** For an oblate spheroid, the difference between the equatorial diameter and the axis of revolution, divided by the equatorial diameter. { ē,lip'tis·əd·ē }

elliptic paraboloid [MATH] A surface which can be so situated that sections parallel to one coordinate plane are parabolas while those parallel to the other plane are ellipses. { ə¦lip·tik pə'rab·ə,lȯid }

elliptic partial differential equation *See* elliptic differential equation. { ə'lip·tik ¦pär·shəl dif·ə¦ren·chəl i'kwā·zhən }

elliptic point [MATH] A point on a surface at which the total curvature is strictly positive. { ə'lip·tik 'pȯint }

elliptic Riemann surface *See* elliptic type. { i¦lip·tik 'rē,män ,sȯr·fəs }

elliptic spring [DES ENG] A spring made of laminated steel plates, arched to resemble an ellipse. { ə'lip·tik 'spriŋ }

elliptic type [MATH] A type of simply connected Riemann surface that can be mapped conformally on the closed complex plane, including the point at infinity. Also known as elliptic Riemann surface. { ə¦lip·tik 'tīp }

elliptic wedge [MATH] The surface generated by a moving straight line that remains parallel to a given plane and intersects both a given straight line and an ellipse whose plane is parallel to the given line but does not contain it. { ə¦lip·tik 'wej }

elliptocytosis [MED] A rare hereditary disease of man characterized by the presence of large numbers of oval or elliptic erythrocytes in the circulating blood. { ə,lip·tə,sī'tō·səs }

ellsworthite [MINERAL] $(Ca,Na,U)_2(Nb,Ta)_2O_6(O,OH)$ A yellow, brown, greenish or black mineral of the pyrochlore group occurring in isometric crystals and consisting of an oxide of niobium, titanium, and uranium. Also known as betafite; hatchettolite. { 'elz·wər,thīt }

elm [BOT] The common name for hardwood trees composing the genus *Ulmus*, characterized by simple, serrate, deciduous leaves. { elm }

Elmidae [INV ZOO] The drive beetles, a small family of coleopteran insects in the superfamily Dryopoidea. { 'el·mə,dē }

El Niño [METEOROL] A warming of the tropical Pacific Ocean that occurs roughly every 4–7 years. { el 'nēn·yō }

El Niño Southern Oscillation [OCEANOGR] **1.** The irregular cyclic swing in atmospheric pressure in the tropical Pacific. **2.** The irregular cyclic swing of warm and cold phases in the tropical Pacific. Abbreviated ENSO. { el 'nēn·yō ¦səth·ərn ,äs·ə'lā·shən }

elongation [ASTRON] The difference between the celestial longitude of the moon or a planet, as measured from the earth, and that of the sun. [COMMUN] The extension of the envelope of a signal due to delayed arrival of multipath components. [MECH] The fractional increase in a material's length due to stress in tension or to thermal expansion. { ē,lȯŋ'gā·shən }

ELM

Twig, leaf, fruit, and terminal and lateral buds of American elm. (*Ulmus americana*).

elongation factor [BIOCHEM] Any of several proteins required for elongation of growing polypeptide chains during protein synthesis. { ē₁loŋ'gā·shən ₁fak·tər }

Elopidae [VERT ZOO] A family of fishes in the order Elopiformes, including the tarpon, ladyfish, and machete. { e'läp·ə₁dē }

Elopiformes [VERT ZOO] A primitive order of actinopterygian fishes characterized by a single dorsal fin composed of soft rays only, cycloid scales, and toothed maxillae. { e₁läp·ə'fōr₁mēz }

elpasolite [MINERAL] K_2NaAlF_6 Mineral composed of sodium potassium aluminum fluoride. { el'pas·ō₁līt }

elpidite [MINERAL] $Na_2ZrSi_6O_{15}\cdot3H_2O$ A white to brickred mineral composed of hydrated sodium zirconium silicate. { 'el·pə₁dīt }

ELR scale See equal listener response scale. { ¦ē¦el¦är ₁skāl }

Elsasser's radiation chart [METEOROL] A radiation chart developed by W. M. Elsasser for the graphical solution of the radiative transfer problems of importance in meteorology: given a radiosonde record of the vertical variation of temperature and water vapor content, one can find with this chart such quantities as the effective terrestrial radiation, net flux of infrared radiation at a cloud base or a cloud top, and radiative cooling rates. { 'el·zə·sərz rād·ē'ā·shən ₁chärt }

ELSE instruction [COMPUT SCI] An instruction in a programming language which tells a program what actions to take if previously specified conditions are not met. { 'els in₁strək·shən }

ELSE rule [COMPUT SCI] A convention in decision tables which spells out which action to take in the case specified conditions are not met. { 'els ₁rül }

Elster-Geitel effect [PHYS] The phenomenon in which a heated conductor acquires a positive or negative electric charge in the presence of a gas, while in a vacuum it always acquires a negative charge. { ¦el·stər ¦gīt·əl i₁fekt }

ELT See emergency locator transmitter.

El Tor vibrio [MICROBIO] Any of the rod-shaped paracholera vibrios; many strains can be agglutinated with anticholera serum. { el ¦tòr 'vib·rē·ō }

eluant [CHEM] A liquid used to extract one material from another, as in chromatography. { 'el·yə·wənt }

eluant gas See carrier gas. { el'yü·ənt ₁gas }

eluate [CHEM] The solution that results from the elution process. { 'el·yə₁wāt }

elusive ulcer See Hunner's ulcer. { i'lü·siv 'əl·sər }

elution [CHEM] The removal of adsorbed species from a porous bed or chromatographic column by means of a stream of liquid or gas. { ē'lü·shən }

elutriation [CHEM ENG] The process of removing substances from a mixture through washing and decanting. [ENG] In a mixture, the separation of finer lighter particles from coarser heavier particles through a slow stream of fluid moving upward so that the lighter particles are carried with it. [GEOL] The washing away of the lighter or finer particles in a soil, especially by the action of raindrops. { ē₁lü·trē'ā·shən }

elutriator [ENG] An apparatus used to separate suspended solid particles according to size by the process of elutriation. { ē'lü·trē₁ad·ər }

eluvial [GEOL] Of, composed of, or relating to eluvium. { ē'lüv·ē·əl }

eluvial placer [GEOL] A placer deposit that is concentrated near the decomposed outcrop of the source. { ē'lüv·ē·əl 'plā·sər }

eluviation [HYD] The process of transporting dissolved or suspended materials in the soil by lateral or downward water flow when rainfall exceeds evaporation. { ē₁lü·ve'ā·shən }

eluvium [GEOL] Disintegrated rock material formed and accumulated in situ or moved by the wind alone. { ē'lü·vē·əm }

elve [METEOROL] A transient luminous event that occurs over a thunderstorm, constituting a broad disk of illumination typically at an altitude of 85–90 kilometers (51–54 miles) with a thickness of about 6 kilometers (4 miles). { elv }

elvegust [METEOROL] A cold descending squall in the upper parts of Norwegian fjords. Also known as sno. { 'el·və₁gəst }

elytron [INV ZOO] **1.** One of the two sclerotized or leathery anterior wings of beetles which serve to cover and protect the membranous hindwings. **2.** A dorsal scale of certain Polychaeta. { 'el·ə₁trän }

em [GRAPHICS] A unit of linear measurement used in printing which is equal to the point size of the type; for example, an em in 6-point type is 6 points wide. { em }

emaciation [MED] A wasted condition of the body; the process of losing flesh so as to become extremely lean. { i₁mā·sē'ā·shən }

emagram [THERMO] A graph of the logarithm of the pressure of a substance versus its temperature, when it is held at constant volume; in meteorological investigations, the potential temperature is often the parameter. { 'em·ə₁gram }

e-mail See electronic mail. { 'ē₁māl }

emanating power [NUCLEO] The fraction of radon atoms, formed in a solid or solution, which escape. { 'em·ə₁nād·iŋ ₁paù·ər }

emanation See radioactive emanation. { ₁em·ə'nā·shən }

emanation security [ELECTR] The protection resulting from all measures designed to deny unauthorized persons information of value which might be derived from unintentional emissions from other than telecommunications systems. { ₁em·ə'nā·shən sə'kyùr·ə·dē }

emanometer [ENG] An instrument for the measurement of the radon content of the atmosphere: radon is removed from a sample of air by condensation or adsorption on a surface, and is then placed in an ionization chamber and its activity determined. { ₁em·ə'näm·əd·ər }

emanometry [NUCLEO] The collective techniques for ionization-chamber determination of the amounts of radioactive gases escaping into the lower atmosphere from the earth's surface; in emanometric measurements, the objective is to count, by typical ionization-chamber methods, all of the ions produced by the alpha particles emitted by the one or more radioactive gases contained in the chamber. { ₁em·ə'näm·ə·trē }

emarginate [BIOL] Having a margin that is notched or slightly forked. { ē'mär·jə₁nāt }

embacle [HYD] The piling up of ice in a stream after a refreeze, and the pile so formed. { em'bak·əl }

EMB agar [MICROBIO] A culture medium containing sugar, eosin, and methylene blue, used in the confirming test for coliform bacteria. { ¦ē¦em¦bē 'äg·ər }

Emballonuridae [VERT ZOO] The sheath-tailed bats, a family of mammals in the order Chiroptera. { em₁bal·ə'nùr·ə₁dē }

embalm [MED] To treat a cadaver with antiseptics and preservatives to prevent decay, before burial or dissection. { em'bäm }

embankment [CIV ENG] **1.** A ridge constructed of earth, stone, or other material to carry a roadway or railroad at a level above that of the surrounding terrain. **2.** A ridge of earth or stone to prevent water from passing beyond desirable limits. Also known as bank. { em'baŋk·mənt }

embarkation [ORD] The loading of troops with their supplies and equipment into ships or aircraft. { ₁em₁bär'kā·shən }

embata [METEOROL] A local onshore southwest wind caused by the reversal of the northeast trade winds in the lee of the Canary Islands. { em'bä·tä }

embatholithic [GEOL] Pertaining to ore deposits associated with a batholith where exposure of the batholith and country rock is about equal. { em₁bath·ə'lith·ik }

embayed [GEOGR] Formed into a bay. [NAV] Pertaining to a vessel in a bay unable to put to sea or to put to sea safely because of wind, current, or sea. { em'bād }

embayed coastal plain [GEOL] A coastal plain that has been partly sunk beneath the sea, thereby forming a bay. { em'bād ¦kōst·əl 'plān }

embayed mountain [GEOL] A mountain that has been depressed enough for sea water to enter the bordering valleys. { em'bād 'maùn₁tən }

embayment [GEOGR] Indentation in a shoreline forming a bay. [GEOL] **1.** Act or process of forming a bay. **2.** A reentrant of sedimentary rock into a crystalline massif. { em'bā·mənt }

Embden-Meyerhof pathway See glycolytic pathway. { ¦em·dən ¦mī·ər₁hòf 'path₁wā }

embed Also spelled imbed. [BIOL] To prepare a specimen for sectioning for microscopic examination by infiltrating with or enclosing in paraffin or other supporting material. [SCI

TECH] **1.** To enclose in a matrix. **2.** To closely surround. { em'bed }

embedability [MET] The ability of a metal to enclose foreign particles within itself. { em‚bed·ə'bil·əd·ē }

embedded command [COMPUT SCI] In word processing, a code inserted in a text document that instructs the printer to change its print attributes. { em¦bed·əd kə'mand }

embedded pointer [COMPUT SCI] A pointer set in a data record instead of in a directory. { em¦bed·əd 'póint·ər }

embedded system [COMPUT SCI] A computer system that cannot be programmed by the user because it is preprogrammed for a specific task and embedded within the equipment which it serves. { em¦bed·əd 'sis·təm }

embedding [MATH] An injective homomorphism between two algebraic systems of the same type. { em¦bed·iŋ }

embedment anchor [NAV ARCH] An anchor that uses a charge of powder or hydrostatic or pneumatic pressure to drive the fluke into the sea floor. { em¦bed·mənt ‚aŋ·kər }

Embiidina [INV ZOO] An equivalent name for Embioptera. { em‚bē·ə'dī·nə }

Embioptera [INV ZOO] An order of silk-spinning, orthopteroid insects resembling the grasshoppers; commonly called the embiids or webspinners. { ‚em·bē¦äp·tə·rə }

Embiotocidae [VERT ZOO] The surfperches, a family of perciform fishes in the suborder Percoidei. { ‚em·bē·ə'täs·ə‚dē }

embolectomy [MED] Surgical removal of an embolus. { ‚em·bə'lek·tə·mē }

embolism [MED] The blocking of a blood vessel by an embolus. { 'em·bə‚liz·əm }

embolite [MINERAL] Ag(Cl,Br) A yellow-green mineral resembling cerargyrite; composed of native silver chloride and silver bromide. { 'em·bə‚līt }

Embolomeri [PALEON] An extinct side branch of slender-bodied, fish-eating aquatic anthracosaurs in which intercentra as well as centra form complete rings. { ‚em·bə'lä·mə‚rī }

embolus [MED] A clot or other mass of particulate matter foreign to the bloodstream which lodges in a blood vessel and causes obstruction. { 'em·bə·ləs }

emboly [EMBRYO] Formation of a gastrula by the process of invagination. { 'em·bə·lē }

embossed plate printer [COMPUT SCI] In character recognition, a data preparation device which accomplishes printing by allowing a raised character behind the paper to push the paper against the printing ribbon in front of the paper. { em¦bäst ¦plāt 'print·ər }

embossing [GRAPHICS] Producing a raised pattern on the surface of paper or wood by means of a die. { em'bäs·iŋ }

embossing die [GRAPHICS] A die used for embossing. { em'bäs·iŋ ‚dī }

embossing stylus [ENG ACOUS] A recording stylus with a rounded tip that forms a groove by displacing material in the recording medium. { em'bäs·iŋ ‚stī·ləs }

embouchure [GEOL] **1.** The mouth of a river. **2.** A river valley widened into a plain. { ¦äm·bə¦shùr }

embrasure [ARCH] **1.** Opening in a wall or parapet, especially one through which a gun is fired. **2.** An opening such as for a door or window with sloping or beveled sides. { em'brā·zhər }

embrechites [PETR] A type of migmatite in which structural features of crystalline shifts are preserved but often partially obliterated by metablastesis. { 'em·brə‚kīts }

Embrithopoda [PALEON] An order established for the unique Oligocene mammal *Arsinoitherium*, a herbivorous animal that resembled the modern rhinoceros. { ‚em·brə'thä·pə·də }

embrittlement [MECH] Reduction or loss of ductility or toughness in a metal or plastic with little change in other mechanical properties. { em'brid·əl·mənt }

embryo [BOT] The young sporophyte of a seed plant. [EMBRYO] **1.** An early stage of development in multicellular organisms. **2.** The product of conception up to the third month of human pregnancy. { 'em·brē·ō }

Embryobionta [BOT] The land plants, a subkingdom of the Plantae characterized by having specialized conducting tissue in the sporophyte (except bryophytes), having multicellular sex organs, and producing an embryo. { ¦em·brē·ō·bī'än·tə }

embryogenesis [EMBRYO] The formation and development of an embryo from an egg. { ‚em·brē·ō'jen·ə·səs }

embryoid [BOT] An embryolike structure originating from somatic cells, such as immature plant embryos, inflorescences, or leaves cultivated in culture. { 'em·brē‚óid }

embryology [BIOL] The study of the development of the organism from the zygote, or fertilized egg. { em·brē'äl·ə·jē }

embryoma of the ovary See dysgerminoma. { ‚em·brē'ō·mə əv the 'ō·və·rē }

embryonal-cell lipoma See liposarcoma. { em'brī·ən·əl ‚sel li'pō·mə }

embryonate [EMBRYO] **1.** To differentiate into a zygote. **2.** Containing an embryo. { 'em·brē·ə'nāt }

embryonated egg culture [VIROL] Embryonated hen's eggs inoculated with animal viruses for the purpose of identification, isolation, titration, or for quantity cultivation in the production of viral vaccines. { 'em·brē·ə‚nād·əd 'eg ‚kəl·chər }

embryonic differentiation [EMBRYO] The process by which specialized and diversified structures arise during embryogenesis. { ‚em·brē¦än·ik ‚dif·ə‚ren·chē'ā·shən }

embryonic inducer [EMBRYO] The acting system in embryos, which contributes to the formation of specialized tissues by controlling the mode of development of the reacting system. { ‚em·brē¦än·ik in'dü·sər }

embryonic induction [EMBRYO] The influence of one cell group (inducer) over a neighboring cell group (induced) during embryogenesis. Also known as induction. { ‚em·brē¦än·ik in'dək·shən }

embryonic stem cell [EMBRYO] Undifferentiated cell derived from the inner cell mass of the blastocyst that can give rise to any of the three embryonic germ layers, and thus can form any cell or tissue type of the body, but cannot give rise to the full spectrum of cells required to complete fetal development. { ‚em·brē¦än·ik 'stem ‚sel }

embryopathy [MED] Any abnormal development of an embryo, either morphological or biochemical. { ‚em·brē¦äp·ə·thē }

Embryophyta [BOT] The equivalent name for Embryobionta. { ‚em·brē'äf·əd·ə }

embryo rescue [GEN] A technique for crossing wild and domestic species of plants in which the wild species is used as the male parent, and the embryos are excised approximately one month after pollination and placed on an artificial medium, where a small fraction survive. { 'em·brē·ō 'res·kyü }

embryo sac [BOT] The female gametophyte of a seed plant, containing the egg, synergids, and polar and antipodal nuclei; fusion of the antipodals and a pollen generative nucleus forms the endosperm. { 'em·brē·ō ‚sak }

embryotomy [MED] Any mutilation of the fetus in the uterus to aid in its removal when natural delivery is impossible. { 'em·brē¦äd·ə·mē }

Emden equation [ASTROPHYS] An equation for stellar structure which arises in a model based on the assumption that the star is a gaseous sphere in adiabatic equilibrium, in which the pressure is proportional to $\rho\gamma$, where ρ is the density and γ is a constant; the equation is $d^2y/dx^2 + (2/x)dy/dx + y^n=0$, where $n=1/(\gamma-1)$, x is proportional to the distance from the center of the sphere, and y is proportional to $\rho^{1/n}$. Also known as Lane-Emden equation. { 'em·dən i‚kwā·zhən }

Emden function [ASTROPHYS] A solution of the Emden equation with the boundary conditions $y=1$ and $dy/dx=0$ at $x=0$. Also known as Lane-Emden function. { 'em·dən ‚fəŋk·shən }

emerald [MINERAL] $Al_2(Be_3Si_6O_{18})$ A brilliant-green to grass-green gem variety of beryl that crystallizes in the hexagonal system; green color is caused by varying amounts of chromium. Also known as smaragd. { 'em·rəld }

emerald cut [LAP] A step cut in which the gem girdle outline is square or rectangular and the steps of elongated facets are parallel to the girdle, with sets on each of the four sides and at the corners. { 'em·rəld ‚kət }

emerged bog [ECOL] A bog which grows vertically above the water table by drawing water up through the mass of plants. { ə¦mərjd 'bäg }

emerged shoreline See shoreline of emergence. { ə¦mərjd 'shór‚līn }

emergence [GEOL] **1.** Dry land which was part of the ocean floor. **2.** The act or process of becoming an emergent land mass. [HYD] See resurgence. { ə'mər·jəns }

emergency alert system [COMMUN] A system of radio,

EMBIOPTERA

Body form of a typical embiid (*Parahagadochir trachelia*). (*From E. S. Ross, Insects Close Up; University of California Press, 1953*).

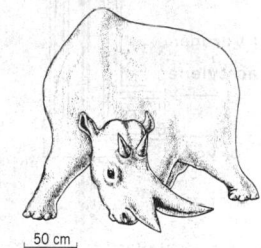

EMBRITHOPODA

50 cm

Arsinoitherium, the early Oligocene embrithopod from Egypt.

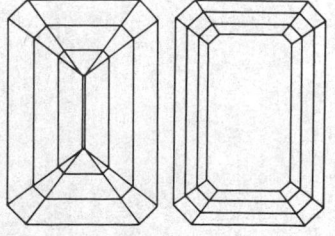

EMERALD CUT

Emerald-cut gemstones. (*Gemological Institute of America*)

television, and cable networks and wire services for communicating with the general public in emergency situations. { ə,mər·jən·sē ə'lərt ,sis·təm }

emergency anchorage [NAV] An anchorage, which may have a limited defense organization, for naval vessels, mobile support units, auxiliaries, or merchant ships. { ə'mər·jən·sē 'aŋ·kə·rij }

emergency brake [MECH ENG] A brake that can be set by hand and, once set, continues to hold until released; used as a parking brake in an automobile. { ə'mər·jən·sē ,brāk }

emergency broadcast system [COMMUN] A system of broadcast stations and interconnecting facilities authorized by the U.S. Federal Communications Commission to operate in a controlled manner during a war, threat of war, state of public peril or disaster, or other national emergency. { ə'mər·jən·sē 'bröd,kast ,sis·təm }

emergency locator transmitter [NAV] A transmitter designed for use in locating downed aircraft or vessels in distress. Abbreviated ELT. { ə'mər·jən·sē 'lō,kād·ər tranz,mid·ər }

emergency medicine [MED] The medical specialty that comprises the immediate decision making and action necessary to prevent death or further disability under emergency conditions. { ə,mər·jən·sē 'med·ə·sən }

emergency position-indicating radio beacon [NAV] In the maritime community, a small hand-held battery-operated transmitter, actuated by water, for use in locating vessels in distress. Abbreviated EPIRB. { ə'mər·jən·sē pə'zish·ən ¦in·də¦kād·iŋ 'rād·ē·ō ,bē·kən }

emergency power supply [ELEC] A source of power that becomes available, usually automatically, when normal power line service fails. { ə'mər·jən·sē 'pau·ər sə,plī }

emergency radio channel [COMMUN] Any radio frequency reserved for emergency use, particularly for distress signals. { ə'mər·jən·sē 'rād·ē·ō ,chan·əl }

emergency receiver [COMMUN] Receiver immediately available in a station for emergency communications and capable of being energized by self-contained or emergency power supply. { ə'mər·jən·sē ri'sē·vər }

emergency recorder plot [NAV] An emergency substitute for a tactical range recorder, consisting of a plastic sheet representing space and time coordinates on which speed and distance data are marked with a grease pencil as they are called off verbally. Abbreviated ERP. { ə'mər·jən·sē ri'kórd·ər ,plät }

emersion [ASTRON] The reappearance of a celestial body after an eclipse or occultation. { ē'mer·zhən }

Emerson wage incentive plan [IND ENG] A plan comprising time wages to $66^2/_3$% of standard performance, empiric bonuses from there to standard performance, ending at 120% time wages, and thereafter a straight-line earning which is 20% above and parallel to basic piece rate. { 'em·ər·sən 'wāj in,sen·tiv ,plan }

emery [MATER] An abrasive which is composed of pulverized, impure corundum; used in polishing and grinding. [MINERAL] A fine, granular, gray-black, impure variety of corundum containing iron oxides, either hematite or magnetite; occurs as masses in limestone and as segregations in igneous rock. { 'em·ə·rē }

emery cake [MATER] Caked, powdered emery in a binding material. { 'em·ə·rē ,kāk }

Emery-Dietz gravity corer [ENG] A tube, with weights attached, which forces sediment samples into its interior as it is dropped on the ocean bottom. { ¦em·ə·rē ¦dēts 'grav·əd·ē ,kór·ər }

emery paper [MATER] An abrasive paper or cloth with an adherent surface layer of emery powder; used for polishing and cleaning metal. { 'em·ə·rē ,pā·pər }

emery rock [PETR] A rock that contains corundum and iron ores. { 'em·ə·rē ,räk }

emery stone [MATER] **1.** A sharpening stone. **2.** A mixture of powdered emery and a binder which can be molded into grinding devices. { 'em·ə·rē ,stōn }

emery wheel [DES ENG] A grinding wheel made of or having a surface of emery powder; used for grinding and polishing. { 'em·ə·rē ,wēl }

emesis [MED] The act of vomiting. { 'em·ə·səs }

emetic [PHARM] Any agent that induces emesis. { i'med·ik }

emetine [PHARM] $C_{29}H_{40}N_2O_4$ Cephaeline methyl ether,

the principal alkaloid of ipecac; a white powder, sparingly soluble in water; it is emetic, diaphoretic, and expectorant, but its chief utility is as an amebicide. { 'em·ə,tēn }

emf See electromotive force.

emi See electromagnetic interference.

emigration [ECOL] The movement of individuals or their disseminules out of a population or population area. { ,em·ə'grā·shən }

emiocytosis [CYTOL] Fusion of intracellular granules with the cell membrane, followed by discharge of the granules outside of the cell; applied chiefly to the mechanism of insulin secretion. Also known as reverse pinocytosis. { ,em·ē,sī'tō·səs }

emissary sky [METEOROL] A sky of cirrus clouds which are either isolated or in small, separated groups; so called because this formation often is one of the first indications of the approach of a cyclonic storm. { 'em·ə,ser·ē ,skī }

emission [ELECTROMAG] Any radiation of energy by means of electromagnetic waves, as from a radio transmitter. [METEOROL] A natural or anthropogenic discharge of particulate, gaseous, or soluble waste material or pollution into the air. { i'mish·ən }

emission characteristics [ELECTR] Relation, usually shown by a graph, between the emission and a factor controlling the emission, such as temperature, voltage, or current of the filament or heater. { i'mish·ən ,kar·ik·tə'ris·tiks }

emission control [METEOROL] A strategy for reducing or preventing atmospheric pollution, such as a catalytic converter used for pollutant removal from automotive exhaust. { i'mish·ən kən,trōl }

emission electron microscope [ELECTR] An electron microscope in which thermionic, photo, secondary, or field electrons emitted from a metal surface are projected on a fluorescent screen, with or without focusing. { i'mish·ən i¦lek,trän 'mī·krə,skōp }

emission flame photometry [ANALY CHEM] A form of flame photometry in which a sample solution to be analyzed is aspirated into a hydrogen-oxygen or acetylene-oxygen flame; the line emission spectrum is formed, and the line or band of interest is isolated with a monochromator and its intensity measured photoelectrically. { i'mish·ən ,flām fō'täm·ə·trē }

emission inventory [ECOL] A quantitative detailed compilation of pollutants emitted into the atmosphere of a given community. { i'mish·ən 'in·vən,tór·ē }

emission-line corona See E corona. { i'mish·ən ¦līn kə'rō·nə }

emission-line galaxy [ASTRON] A galaxy whose spectrum displays narrow, high-excitation emission lines. { i'mish·ən ¦līn 'gal·ik·sē }

emission lines [SPECT] Spectral lines resulting from emission of electromagnetic radiation by atoms, ions, or molecules during changes from excited states to states of lower energy. { i'mish·ən ,līnz }

emission nebula [ASTRON] A type of bright diffuse nebula whose luminosity results from the excitation and ionization of its gas atoms by ultraviolet radiation from a nearby O- or B-type star. { i'mish·ən 'neb·yə·lə }

emission security [ELECTR] That component of communications security which results from all measures taken to protect any unintentional emissions of a telecommunications system from any form of exploitation other than cryptanalysis. { i'mish·ən sə'kyùr·əd·ē }

emission spectrometer [SPECT] A spectrometer that measures percent concentrations of preselected elements in samples of metals and other materials; when the sample is vaporized by an electric spark or arc, the characteristic wavelengths of light emitted by each element are measured with a diffraction grating and an array of photodetectors. { i'mish·ən spek'träm·əd·ər }

emission spectrum [SPECT] Electromagnetic spectrum produced when radiations from any emitting source, excited by any of various forms of energy, are dispersed. { i'mish·ən ,spek·trəm }

emission standard [ENG] The maximum legal quantity of pollutant permitted to be discharged from a single source. { i'mish·ən ,stan·dərd }

emission tomography [MED] A technique which uses the emission of gamma-ray photons from radioactive tracers to

---

EMISSION FLAME PHOTOMETRY

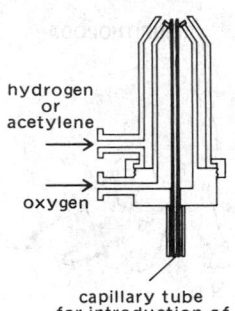

hydrogen or acetylene →

oxygen →

capillary tube for introduction of sample solution

Beckman aspirator-burner used in emission flame photometry.

construct images of the distribution of the tracers in the human body. { i'mish·ən tō'mäg·rə·fē }

emissive power *See* emittance. { i¦mis·iv 'paù·ər }

emissivity [THERMO] The ratio of the radiation emitted by a surface to the radiation emitted by a perfect blackbody radiator at the same temperature. Also known as thermal emissivity. { ¸ē·mə'siv·əd·ē }

emittance [THERMO] The power radiated per unit area of a radiating surface. Also known as emissive power; radiating power. { i'mit·əns }

emitter [COMPUT SCI] A time pulse generator found in some equipment, such as a card punch. [ELECTR] A transistor region from which charge carriers that are minority carriers in the base are injected into the base, thus controlling the current flowing through the collector; corresponds to the cathode of an electron tube. Symbolized E. Also known as emitter region. { i'mid·ər }

emitter barrier [ELECTR] One of the regions in which rectification takes place in a transistor, lying between the emitter region and the base region. { i'mid·ər ¸bar·ē·ər }

emitter bias [ELECTR] A bias voltage applied to the emitter electrode of a transistor. { i'mid·ər ¸bī·əs }

emitter-coupled logic [ELECTR] A form of current-mode logic in which the emitters of two transistors are connected to a single current-carrying resistor in such a way that only one transistor conducts at a time. Abbreviated ECL. { i'mid·ər ¦kəp·əld 'läj·ik }

emitter follower [ELECTR] A grounded-collector transistor amplifier which provides less than unity voltage gain but high input resistance and low output resistance, and which is similar to a cathode follower in its operations. { i'mid·ər ¸fäl·ə·wər }

emitter junction [ELECTR] A transistor junction normally biased in the low-resistance direction to inject minority carriers into a base. { i'mid·ər ¸jəŋk·shən }

emitter pulse [COMPUT SCI] In a punch-card machine, one of a set of pulses associated with a particular row of punch positions on a punch card. { i'mid·ər ¸pəls }

emitter region *See* emitter. { i'mid·ər ¸rē·jən }

emitter resistance [ELECTR] The resistance in series with the emitter lead in an equivalent circuit representing a transistor. { i'mid·ər ri¸zis·təns }

emmetropia [MED] Normal vision. { ¸em·ə'trō·pē·ə }

emmonsite [MINERAL] $Fe_2Te_3O_9 \cdot 2H_2O$ A yellow-green mineral composed of a hydrous oxide of iron and tellurium. { 'em·ən¸zīt }

E mode *See* transverse magnetic mode. { 'ē ¸mōd }

emodin [ORG CHEM] $C_{14}H_4O_2(OH)_3CH_3$ Orange needles crystallizing from alcohol solution, melting point 256–257°C, practically insoluble in water, soluble in alcohol and aqueous alkali hydroxide solutions, occurs as the rhamnoside in plants such as rhubarb root and alder buckthorn; used as a laxative. { 'em·ə·dən }

emollient [PHARM] A softening agent, especially for use on skin and mucous membranes; lanolin is widely used as a base. { ə'mäl·yənt }

EMOSFET *See* electrolyte-MOSFET. { ¦ē'mäs¸fet }

emoticon [COMPUT SCI] A combination of keyboard characters that depicts a sideways face whose expression conveys an emotional response. Also known as smiley. { i'mōd·ə¸kän }

emotion [PSYCH] A strong mental feeling or affect of the consciousness involving visceral and other physiologic changes. { i'mō·shən }

empennage [AERO ENG] The assembly at the rear of an aircraft; it comprises the horizontal and vertical stabilizers. Also known as tail assembly. { ¦am·pə¦näzh }

emphasizer *See* preemphasis network. { 'em·fə¸sīz·ər }

emphysema [MED] A pulmonary disorder characterized by overdistention and destruction of the air spaces in the lungs. { ¸em·fə'sē·mə }

emphysematous chest [MED] The altered contour of the chest seen in pulmonary emphysema, with increased anteroposterior diameter, flaring at the lower rib margins, low position of the diaphragm, and minimal respiratory motion. Also known as barrel chest. { ¸em·fə·sə'mad·əs 'chest }

Empididae [INV ZOO] The dance flies, a family of orthorrhaphous dipteran insects in the series Nematocera. { em'pid·ə¸dē }

empire cloth [MATER] Cotton cloth coated with oxidized oil; used as an electrical insulator. { 'em¸pīr ¸klóth }

empirical [SCI TECH] Based on actual measurement, observation, or experience, rather than on theory. { em'pir·ə·kəl }

empirical curve [MATH] A smooth curve drawn through or close to points representing measured values of two variables on a graph. { em'pir·ə·kəl 'kərv }

empirical formula [CHEM] A chemical formula that indicates the composition of a compound in terms of the relative numbers and kinds of atoms in the simplest ratio; for example, the empirical formula for fluorobenzene is C_6H_5F. { em'pir¸nəmkəl 'fòr¸myəmlə }

empirical probability [STAT] The ratio of the number of times an event has occurred to the total number of trials performed. Also known as a posteriori probability. { em'pir·ə·kəl ¸präb·ə'bil·əd·ē }

empirical rule [SCI TECH] A rule which is derived from measurements or observations, and is not based on any theory. { em'pir·ə·kəl 'rül }

emplacement [GEOL] Intrusion of igneous rock or development of an ore body in older rocks. [ORD] **1.** Prepared position for one or more weapons or pieces of equipment, protecting against hostile fire or bombardment but permitting execution of the mission of the weapons. **2.** Act of fixing a gun in a prepared position from which it may be fired. { em'plās·mənt }

emplectite [MINERAL] $CuBiS_3$ A grayish or white mineral that crystallizes in the orthorhombic system; occurs in masses. { em'plek¸tīt }

employment test [IND ENG] Any of a wide variety of tests to measure intelligence, personality traits, skills, interests, aptitudes, or other characteristics; used to supplement interviews, physical examinations, and background investigations before employment. { em'plói·mənt ¸test }

empodium [INV ZOO] A small peripheral part located between the claws of the tarsi of many insects and arachnids. { em'pōd·ē·əm }

empressite [MINERAL] AgTe An opaque, pale-bronze mineral whose crystal system is unknown. { 'em·prə¸sīt }

empty [ORD] In ammunition nomenclature, indicating that the munition does not contain a payload, but is designed to contain one at the time of final use. { 'em·tē }

empty-cell process [ENG] A wood treatment in which the preservative coats the cells without filling them. { 'em·tē ¸sel 'präs·əs }

empty medium [COMPUT SCI] A material which has been prepared to have data recorded on it by the entry of some preliminary data, such as feed holes punched in a paper tape or header labels written on a magnetic tape; in contrast to a virgin medium. { 'em·tē 'mēd·ē·əm }

empty set [MATH] The set with no elements. { 'em·tē 'set }

empty shell [COMPUT SCI] A room that has been fully prepared for the installation of computer and data-processing equipment. { 'em·tē 'shel }

empyema [MED] The presence of pus in the body cavity, hollow organ, or tissue space; when the term is used without qualification, it generally refers to pus in the pleural space. { ¸em¸pī'ē·mə }

emu [ELECTROMAG] *See* electromagnetic system of units. [VERT ZOO] *Dromiceius novae-hollandiae*. An Australian ratite bird, the second largest living bird, characterized by rudimentary wings and a feathered head and neck without wattles. { 'ē¸myü }

emulation [COMPUT SCI] Imitation of one computer system by another so that the latter functions in exactly the same way and runs the same programs. { ¸em·yə'lā·shən }

emulation mode [COMPUT SCI] A method of operation in which a computer actually executes the instructions of a different (simpler) computer, in contrast to normal mode. { ¸em·yə'lā·shən ¸mōd }

emulator [COMPUT SCI] The microprogram-assisted macroprogram which allows a computer to run programs written for another computer. { 'em·yə¸lād·ər }

emulator circuit [COMPUT SCI] A circuit built into a computer's control section to enable it to process instructions that were written for another computer. { 'em·yə¸lād·ər ¸sər·kət }

emulsan [BIOCHEM] A lipopolysaccharide produced by a strain of *Acinetobacter calcoaceticus*, used to stabilize oil-in-water emulsions. { i'məl·sən }

emulsification [CHEM] The process of dispersing one liquid

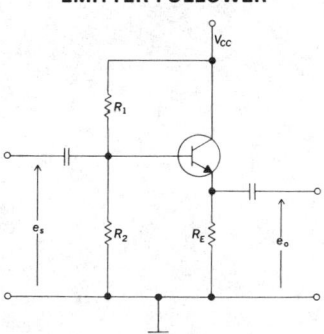

EMITTER FOLLOWER

Circuit diagram of a typical emitter follower.

in a second immiscible liquid; the largest group of emulsifying agents are soaps, detergents, and other compounds, whose basic structure is a paraffin chain terminating in a polar group. { ə,məl·sə·fə'kā·shən }

emulsification test [CHEM ENG] Standard laboratory procedure for evaluating the resistance of insulating oils, turbine oils, and other lubricating oils to emulsification. { ə,məl·sə·fə'kā·shən ,test }

emulsified asphalt [MATER] An emulsion of asphalt cement and water with a small quantity of an emulsifying agent. { ə'məl·sə,fīd 'as,fōlt }

emulsified water [PETRO ENG] Water so thoroughly dispersed in petroleum that special treatment procedures are required to separate it. { ə'məl·sə,fīd 'wȯd·ər }

emulsifier See disperser. { ə'məl·sə,fī·ər }

emulsifying agent See disperser. { ə'məl·sə,fī·iŋ ,ā·jənt }

emulsifying oil See soluble oil. { ə'məl·sə,fī·iŋ ,ȯil }

emulsion [CHEM] A stable dispersion of one liquid in a second immiscible liquid, such as milk (oil dispersed in water). [GRAPHICS] In photography, the photosensitized material on film, plates, and various photographic papers. { ə'məl·shən }

emulsion breaking [CHEM] In an emulsion, the combined sedimentation and coalescence of emulsified drops of the dispersed phase so that they will settle out of the carrier liquid; can be accomplished mechanically (in settlers, cyclones, or centrifuges) with or without the aid of chemical additives to increase the surface tension of the droplets. { ə'məl·shən ,brāk·iŋ }

emulsion cleaner [CHEM ENG] A cleaner composed of organic solvents dispersed in an aqueous solution with the aid of an emulsifying agent. { ə'məl·shən ,klēn·ər }

emulsion jet [NUCLEO] A jetlike formation in a nuclear emulsion caused by an incident particle with very high energy, greater than 100 GeV. { ə'məl·shən ,jet }

emulsion paint [MATER] Paint whose vehicle is an emulsion of a binder (oil, resin, latex, and so on) in water. { ə'məl·shən ,pānt }

emulsion polymerization [ORG CHEM] A polymerization reaction that occurs in one phase of an emulsion. { ə'məl·shən pə,lim·ə·rə'zā·shən }

emulsion speed [GRAPHICS] Sensitivity of a photographic emulsion to light, under standard conditions of exposure and development. { ə'məl·shən ,spēd }

Emydidae [VERT ZOO] A family of aquatic and semiaquatic turtles in the suborder Cryptodira. { e'mid·ə,dē }

en [GRAPHICS] A unit of linear type measure; equal to one-half the point size of the type in question, that is, equal to one-half the width of an em. { en }

enable [COMPUT SCI] **1.** To authorize an activity which would otherwise be suppressed, such as to write on a tape. **2.** To turn on a computer system or a piece of equipment. [ELECTR] To initiate the operation of a device or circuit by applying a trigger signal or pulse. { ə'nā·bəl }

enabled instruction [COMPUT SCI] An instruction in a program in data flow language, all of whose input values are present, so that the instruction may be carried out. { ə'nā·bəld in'strək·shən }

enabling pulse [ELECTR] A pulse that prepares a circuit for some subsequent action. { ə'nāb·liŋ ,pəls }

Enaliornithidae [PALEON] A family of extinct birds assigned to the order Hesperornithiformes, having well-developed teeth found in grooves in the dentary and maxillary bones of the jaws. { e¦nal·ē·ȯr¦nith·ə,dē }

enamel [MATER] A finely ground, resin-containing oil paint that dries relatively harder, smoother, and glossier than ordinary paint. See glaze. { i'nam·əl }

enamel clay [MATER] A ball clay able to float nonplastic enamel slips to make them spray or dip more evenly. { i'nam·əl ,klā }

enameled brick [MATER] Brick with a smooth hard surface acquired from the application of a special wash before burning. { i'nam·əld 'brik }

enameling [ENG] The application of a vitreous glaze to pottery or metal surfaces, followed by fusing in a kiln or furnace. { i'nam·liŋ }

enamel kiln [ENG] A kiln in which enamel colors are fired. { i'nam·əl ,kil }

enamel organ [EMBRYO] The epithelial ingrowth from the dental lamina which covers the dental papilla, furnishes a mold for the shape of a developing tooth, and forms the dental enamel. { i'nam·əl ,ȯr·gən }

enamel oxide [MATER] Any of the mixtures of calcined oxides used to color vitreous enamels used on sheet steel or cast iron. { i'nam·əl 'äk,sīd }

enamel paper See coated paper. { i'nam·əl ,pā·pər }

enamine [ORG CHEM] An amine in which there is a carbon-to-carbon double bond adjacent to the nitrogen, $-C=C-N-$; considered to be the nitrogen analog of an enol. { 'en·ə,mēn }

enantiomer See enantiomorph. { ə¦nan·tē¦ō·mər }

enantiomerically pure [ORG CHEM] Referring to a sample of molecules having the same chirality. IUPAC discourages use of homchiral as a synonym. { ə¦nan·tē·ō¦mer·ə·klē 'pyür }

enantiomeric excess [ORG CHEM] In an asymmetric synthesis, a chemical yield that contains more of the desired enantiomer than other products. { ə¦nan·tē·ō¦mer·ik ek'ses }

enantiomorph [CHEM] One of an isomeric pair of either crystalline forms or chemical compounds whose molecules are nonsuperimposable mirror images. Also known as enantiomer; optical antipode; optical isomer. { ə'nan·tē·ə,mȯrf }

enantiomorphism [CHEM] A phenomenon of mirror-image relationship exhibited by right-handed and left-handed crystals or by the molecular structures of two stereoisomers. { ə¦nan·tē·ə¦mȯr,fiz·əm }

Enantiornithines [VERT ZOO] Opposite birds, so called because their foot bones fuse in the opposite direction of modern birds, from the subclass Sauriuvae. { i,nan·tē¦ȯrn·ə,thēnz }

enantioselective reaction See stereoselective reaction. { ə¦nan·tē·ə·si¦lek·tiv rē'ak·shən }

enantiotopic ligand [ORG CHEM] A ligand whose replacement or addition gives rise to enantiomers. { ə¦nan·tē·ə¦täp·ik 'līg·ənd }

enantiotropy [CHEM] The relation of crystal forms of the same substance in which one form is stable above the transition-point temperature and the other stable below it, so that the forms can change reversibly one into the other. { ə,nan·te'ä·trə·pē }

Enantiozoa [INV ZOO] The equivalent name for Parazoa. { ə¦nan·te·ə'zō·ə }

enargite [MINERAL] A lustrous, grayish-black mineral which is found in orthorhombic crystals but is more commonly columnar, bladed, or massive; hardness is 3 on Mohs scale, specific gravity is 4.44; in some places enargite is a valuable copper ore. Also known as clairite; luzonite. { e'när,jīt }

enarthrosis [ANAT] A freely movable joint that allows a wide range of motion on all planes. Also known as ball-and-socket joint. { ,e,när'thrō·səs }

Encalyptales [BOT] An order of true mosses (subclass Bryidae) characterized by broad papillose leaves and erect capsules covered by very long calyptrae. { en,ka·lip'tā·lēz }

encapsulate [SCI TECH] To surround, encase, or enclose as if in a capsule; for example, the formation of a protective coating around a bacterium, or the enclosure of an item such as an electronic component in plastic. { en'kap·sə,lāt }

encastré beam See fixed-end beam. { än·ka·strā bēm }

encaustic [GRAPHICS] A method of painting in which the pigment is carried in hot wax. { en'kȯs·tik }

Enceladus [ASTRON] A satellite of Saturn orbiting at a mean distance of 153,600 miles (238,000 kilometers). { ,en·se'lä·dùs }

encephalitis [MED] Inflammation of the brain. { en,sef·ə'līd·əs }

encephalitis lethargica [MED] Epidemic encephalitis, probably of viral etiology, characterized by lethargy, ophthalmoplegia, hyperkinesia, and at times residual neurologic disability, particularly parkinsonism with oculogyric crisis. Also known as epidemic encephalitis; sleeping sickness; von Economo's disease. { en,sef·ə'līd·əs lə'thär·jə·kə }

encephalocele [MED] Hernia of the brain through a congenital or traumatic opening in the cranium. { en'sef·ə·lō,sēl }

encephalogram [MED] A roentgenogram of the brain made in encephalography. { en'sef·ə·lə,gram }

encephalography [MED] Roentgenography of the brain following removal of cerebrospinal fluid, by lumbar or cisternal puncture, and its replacement by air or other gas. { en,sef·ə'läg·rə·fē }

encephaloid carcinoma See medullary carcinoma. { en'sef·ə,lȯid ,kärs·ən'ō·mə }

encephalomalacia [MED] **1.** Infarction of the brain. **2.**

Any softening or fragmentation of the brain. { en¦sef·ə·lō·mə'lā·shə }

encephalomyelitis [MED] Inflammation of the brain and spinal cord. { en¦sef·ə·lō,mī·ə'līd·əs }

encephalomyocarditis [MED] An acute febrile RNA virus disease accompanied by pharyngitis, stiff neck, and hyperactive deep reflexes; certain species of wild rats are the reservoir; human infections range from a mild febrile illness to a severe encephalomyelitis. { en¦sef·ə·lō,mī·ə,kär'dīd·əs }

encephalopathy [MED] Any disease of the brain. { en,sef·ə'läp·ə·thē }

Encholaimoidea [INV ZOO] A superfamily of nematodes of the order Dorylaimida, characterized by two circlets of cephalic sense organs on the lips, pouchlike amphids with slitlike openings, a stoma armed with an axial spear, and a body cuticle marked by widely spaced annulations giving a platelike appearance. { ,en·kō·lə'mȯid·ē·ə }

enchondroma [MED] A benign tumor composed of dysplastic cartilage cells, occurring in the metaphysis of cylindric bones, especially of the hands and feet. { ,en,kän'drō·mə }

enchondromatosis [MED] A rare disorder principally involving tubular bones, especially those of the feet and hands, characterized by hamartomatous proliferation of cartilage in the metaphysis, indistinguishable in single lesions from enchondromas. Also known as chondrodysplasia; dyschondroplasia; Ollier's disease. { ,en,kän·drō·mə'tō·səs }

enchymatous [PHYSIO] Of gland cells, distended with secreted material. { en'kim·əd·əs }

encipher [COMMUN] To convert a plain-text message into unintelligible language by means of a cryptosystem. Also known as encrypt. { en'sī·fər }

enciphered facsimile communications [COMMUN] Communications in which security is accomplished by mixing pulses produced by a key generator with the output of the facsimile converter; plain text is recovered by subtracting the identical key at the receiving terminal; unauthorized listeners are unable to reconstruct the plain text unless they have an identical key generator and the daily key setting. { en'sī·fərd fak'sim·ə·lē kə,myün·ə'kā·shənz }

Encke division [ASTRON] A faint line that splits the outer ring of Saturn in two. { 'eŋ·kə də,vizh·ən }

Encke roots [MATH] For any two numbers a_1 and a_2, the numbers $-x_1$ and $-x_2$, where x_1 and x_2 are the roots of the equation $x^2 + a_1 x + a_2 = 0$, with $|x_1| < |x_2|$. { 'eŋ·kə ,rüts }

Encke's Comet [ASTRON] A very faint comet with the shortest period of any known comet, 3.3 years. { 'eŋ·kəz 'käm·ət }

enclave See domain. { 'än,klāv }

enclosed arc lamp [ELEC] An arc lamp in which the arc produced by carbon electrodes is protected from the atmosphere by a translucent enclosure. { in¦klozd 'ärk ,lamp }

enclosure compound See clathrate. { in'klō·zhər ,käm,paůnd }

encode [COMMUN] To express given information by means of a code. [COMPUT SCI] To prepare a routine in machine language for a specific computer. { en'kōd }

encoded abstract [COMPUT SCI] An abstract prepared to be scanned by automatic electronic machines. { en'kōd·əd 'ab,strakt }

encoded question [COMPUT SCI] A question set up and encoded in the form appropriate for operating, programming, or conditioning a searching device. { en'kōd·əd 'kwes·chən }

encoder [COMPUT SCI] In character recognition, that class of printer which is usually designed for the specific purpose of printing a particular type font in predetermined positions on certain size forms. [ELECTR] In an electronic computer, a network or system in which only one input is excited at a time and each input produces a combination of outputs. See matrix. { en'kōd·ər }

encoding strip [COMPUT SCI] In character recognition, the area reserved for the inscription of magnetic-ink characters, as in bank checks. { en'kōd·iŋ 'strip }

encounter [PHYS CHEM] A group of collisions, each of which consists of two molecules that collide without reacting and do not separate immediately because of the cage of surrounding molecules. { en'kaůn·tər }

encrinal limestone [GEOL] A limestone consisting of more than 10% but less than 50% of fossil crinoidal fragments. { en'krīn·əl 'līm,stōn }

encrinite [PALEON] One of certain fossil crinoids, especially of the genus *Encrinus*. { 'eŋ·krə,nīt }

encroachment [MIN ENG] Extraction of coal beyond the boundary that divides one mine area from another. [PETRO ENG] Replacement of oil or gas being withdrawn from a reservoir by groundwater. { in'krōch·mənt }

encrustation [ENG] The buildup of slag or other material inside furnaces and kilns. { en·krə'stā·shən }

encrypt See encipher. { en'kript }

encryption [COMPUT SCI] The coding of a clear text message by a transmitting unit so as to prevent unauthorized eavesdropping along the transmission line; the receiving unit uses the same algorithm as the transmitting unit to decode the incoming message. { en'krip·shən }

Encyrtidae [INV ZOO] A family of hymenopteran insects in the superfamily Chalcidoidea. { en'sərd·ə,dē }

encystment [BIOL] The process of forming or becoming enclosed in a cyst or capsule. { en'sist·mənt }

end See warp. { end }

Endamoeba [INV ZOO] The type genus of the Endamoebidae comprising insect parasites and, in some systems of classification, certain vertebrate parasites. { ¦end·ə'mē·bə }

end-and-end [TEXT] Pertaining to a weave with a minuscule check effect made by alternating white and colored warp yarns. Also known as end-to-end. { ¦end ən ¦end }

endarch [BOT] Formed outward from the center, referring to xylem or its development. { 'en,därk }

end-around carry [COMPUT SCI] A carry from the most significant digit place to the least significant digit place. { ¦end ə¦raůnd 'kar·ē }

end-around shift See cyclic shift. { ¦end ə¦raůnd 'shift }

endarteritis [MED] Inflammation of the lining (tunica intima) of an artery. { ¦end,ärt·ə'rīd·əs }

endarteritis obliterans [MED] Endarteritis, particularly of small arteries, accompanied by degeneration of the intima, leading to occlusion of the blood vessel. Also known as obliterating endarteritis. { ¦end,ärt·ə'rīd·əs ō'blit·ə,ränz }

end-bearing pile [CIV ENG] A bearing pile that is driven down to hard ground so that it carries the full load at its point. Also known as a point-bearing pile. { 'end ,ber·iŋ ,pīl }

end bulb See bouton. { 'end ,bȯlb }

end bulb of Krause See Krause's corpuscle. { 'end ,bȯlb əv 'kraůs }

end cell [ELEC] One of a group of cells in series with a storage battery, which can be switched in to maintain the output voltage of the battery when it is not being charged. { 'end ,sel }

end-cell rectifier [ELECTR] Small trickle charge rectifier used to maintain voltage of the storage battery end cells. { 'end ,sel 'rek·tə,fī·ər }

end construction [CIV ENG] Structural blocks or tiles laid so that the hollow cells run vertically. { 'end kən,strək·shən }

end-construction tile [MATER] A type of structural clay tile designed to receive its principal stress parallel to the axis of the cells. { 'end kən,strək·shən ,tīl }

end correction [ACOUS] A correction that must be made to the assumption that an antinode exists at an open end of a pipe in which air is vibrating, in order to take into account the radiation of sound waves from the pipe. { 'end kə,rek·shən }

end cut See heavy fraction. { 'end ,kət }

end distortion [COMMUN] The displacement of trailing edges of marking pulses transmitted over a teletypewriter circuit relative to the leading edge of the start pulse. { 'end di,stȯr·shən }

end effect [ELECTROMAG] The effect of capacitance at the ends of an antenna; it requires that the actual length of a half-wave antenna be about 5% less than a half wavelength. { 'end i,fekt }

end effector [CONT SYS] The component of a robot that comes into contact with the workpiece and does the actual work on it. Also known as hand { 'end i,fek·tər }

Endeidae [INV ZOO] A family of marine arthropods in the subphylum Pycnogonida. { en'dē·ə,dē }

endellite [MINERAL] $Al_2Sl_2O_5(OH)_4 \cdot 4H_2O$ Term used in the United States for a clay mineral, the more hydrous form of halloysite. Also known as hydrated halloysite; hydrohalloysite; hydrokaolin. { 'en·də,līt }

endemic [MED] Peculiar to a certain region, specifically referring to a disease which occurs more or less constantly in any locality. { en'dem·ik }

endemic goiter [MED] Goiter peculiar to areas that are iodine-poor in food, water, or soil. { en'dem·ik 'gȯid·ər }

endemic rural plague See sylvatic plague. { en'dem·ik ¦rȯr·əl 'plāg }

endemic typhus See murine typhus. { en'dem·ik 'tī·fəs }

endemism [MED] The state or quality of being endemic. { 'en·də,miz·əm }

endergonic [BIOCHEM] Of or pertaining to a biochemical reaction in which the final products possess more free energy than the starting materials; usually associated with anabolism. { ¦en·dər¦gän·ik }

endermic [MED] Acting through the skin by absorption, such as medication applied to the skin. { en'dər·mik }

endexine [BOT] An inner membranous layer of the exosporium. { en'dek,sēn }

end-feed centerless grinding [MECH ENG] Centerless grinding in which the piece is fed through grinding and regulating wheels to an end stop. { 'end ¦fēd ¦sen·tər·ləs 'grīnd·iŋ }

end-fire antenna See end-fire array. { 'end ,fīr an'ten·ə }

end-fire array [ELECTROMAG] A linear array whose direction of maximum radiation is along the axis of the array; it may be either unidirectional or bidirectional; the elements of the array are parallel and in the same plane, as in a fishbone antenna. Also known as end-fire antenna. { 'end ,fīr ə'rā }

end instrument [ELECTR] A pickup used in telemetering to convert a physical quantity to an inductance, resistance, voltage, or other electrical quantity that can be transmitted over wires or by radio. { 'end ,in·strə·mənt }

endite [INV ZOO] **1.** One of the appendages on the inner aspect of an arthropod limb. **2.** A ridgelike chewing surface on the inner part of the pedipalpus or maxilla of many arachnids. { 'en,dīt }

end item [ENG] A final combination of end products, component parts, or materials which is ready for its intended use; for example, ship, tank, mobile machine shop, or aircraft. { 'end ,īd·əm }

end labeling [BIOCHEM] The addition of a radioactively labeled group to one end of a deoxyribonucleic acid strand. { 'end ,lab·əl·iŋ }

end lap [DES ENG] A joint in which two joining members are made to overlap by removal of half the thickness of each. { 'end ,lap }

endless loop [COMPUT SCI] A sequence of instructions in a computer program that is repeated over and over without end, due to a mistake in the programming. { 'end·ləs 'lüp }

endless tangent screw [NAV] That portion of the mechanism of a marine sextant which moves the index arm. { 'end·ləs 'tan·jənt ,skrü }

endless tangent screw sextant [NAV] A marine sextant having an endless tangent screw for controlling the position of the index arm and the vernier or micrometer drum; the index arm may be moved over the entire arc without resetting, by means of the endless tangent screw. { 'end·ləs 'tan·jənt ,skrü 'sek·stənt }

endlichite [MINERAL] A mineral similar to vanadinite, but with the vanadium replaced by arsenic. { 'end·li,kīt }

end loader [MECH ENG] A platform elevator at the rear of a truck. { 'end ,lōd·ər }

end loss [ELECTROMAG] The difference between the actual and the effective lengths of a radiating antenna element. [MET] That portion remaining after designated lengths of bar have been cut into multiples. { 'end ,lȯs }

end mark [COMPUT SCI] A mark which signals the end of a unit of information. { 'end ,märk }

end member [MINERAL] One of the two or more pure chemical compounds that enters into solid solution with other pure chemical compounds to make up a series of minerals of similar crystal structure (that is, an isomorphous, solid-solution series). { 'end ,mem·bər }

end mill [MECH ENG] A machine which has a rotating shank with cutting teeth at the end and spiral blades on the peripheral surface; used for shaping and cutting metal. { 'end ,mil }

end-milled keyway See profiled keyway. { 'end ,mild 'kē,wā }

end moraine [GEOL] An accumulation of drift in the form of a ridge along the border of a valley glacier or ice sheet. { 'end mə,rān }

endo- [ORG CHEM] Prefix that denotes inward-directed valence bonds of a six-membered ring in its boat form. [SCI TECH] Prefix denoting within or inside. { 'en·dō }

endobasion [ANAT] The anteriormost point of the margin of the foramen magnum at the level of its smallest diameter. { ¦en·dō¦bā·sē,än }

endobatholithic [GEOL] Pertaining to ore deposits along projecting portions of a batholith. { ¦en·dō·bath·ə'lid·ik }

endobiotic [ECOL] Referring to an organism living in the cells or tissues of a host. { ¦en·dō·bī'äd·ik }

endobranchiate [ZOO] Animal form with endodermal gills. { ¦en·dō'braŋ·kē,āt }

endocardial fibroelastosis [MED] Fibrous or fibroelastic thickening of the endocardium, of unknown cause. { ¦en·dō¦kärd·ē·əl ¦fī·brō·ə,las'tō·səs }

endocarditis [MED] Inflammation of the endocardium. { ¦en·dō·kär'dīd·əs }

endocardium [ANAT] The membrane lining the heart. { ,en·dō'kärd·ē·əm }

endocarp [BOT] The inner layer of the wall of a fruit or pericarp. { 'en·dō,kärp }

endocast See steinkern. { 'en·dō,kast }

endocervicitis [MED] Inflammation of the mucous membrane of the uterine cervix. { ¦en·dō,sər·və'sīd·əs }

endocervix [ANAT] The glandular mucous membrane of the cervix uteri. { ¦en·dō'sər·viks }

endochondral ossification [PHYSIO] The conversion of cartilage into bone. Also known as intracartilaginous ossification. { ¦en·dō'kän·drəl ,äs·ə·fə'kā·shən }

endocommensal [ECOL] A commensal that lives within the body of its host. { ¦en·dō·kə'men·səl }

endocorpuscular [CYTOL] Located within an erythrocyte. { ¦en·dō·kȯr'pəs·kyə·lər }

endocranium [ANAT] **1.** The inner surface of the cranium. **2.** See dura mater. [INV ZOO] The processes on the inner surface of the head capsule of certain insects. { ¦en·dō'krā·nē·əm }

endocrine gland [PHYSIO] A ductless structure whose secretion (hormone) is passed into adjacent tissue and then to the bloodstream either directly or by way of the lymphatics. Also known as ductless gland. { 'en·də·krən ,gland }

endocrine signaling [PHYSIO] Signaling in which endocrine cells release hormones that act on distant target cells. { 'en·də·krən ,sig·nəl·iŋ }

endocrine system [PHYSIO] The chemical coordinating system in animals, that is, the endocrine glands that produce hormones. { 'en·də·krən ,sis·təm }

endocrine toxicity [MED] Any adverse structural and/or functional changes to the endocrine system which may result from exposure to chemicals; can harm human and animal reproduction and development. { ¦en·də·krən täk'sis·əd·ē }

endocrinology [PHYSIO] The study of the endocrine glands and the hormones that they synthesize and secrete. { ,en·də·krə'näl·ə·jē }

endocuticle [INV ZOO] The inner, elastic layer of an insect cuticle. { ¦en·dō'kyüd·i·kəl }

endocyclic double bond [ORG CHEM] In a molecular structure, a double bond that is part of the ring system. { ¦en·dō'sī·klik ¦dəb·əl 'bänd }

endocyst [INV ZOO] The soft layer consisting of ectoderm and mesoderm, lining the ectocyst of bryozoans. { 'en·də,sist }

endocytic vacuole [CELL MOL] A membrane-bound cellular organelle containing extracellular particles engulfed by the mechanisms of endocytosis. { ¦en·də¦sīd·ik 'vak·yə,wōl }

endocytobiosis [ECOL] Symbiosis in which the symbionts live within host cells. { ,en·dō,sī·tō·bī'ō·səs }

endocytosis [CELL MOL] **1.** An active process in which extracellular materials are introduced into the cytoplasm of cells by either phagocytosis or pinocytosis. **2.** The process by which animal cells internalize large molecules and large collections of fluid. { ¦en·dō·sī'tō·səs }

endodeoxyribonuclease See restriction endonuclease. { ¦en·dō·dē¦äk·sē,rī·bō'nü·klē,ās }

endoderm [EMBRYO] The inner, primary germ layer of an animal embryo; sometimes referred to as the hypoblast. Also known as entoderm; hypoblast. { 'en·dō,dərm }

endodermis [BOT] The innermost tissue of the cortex of most plant roots and certain stems consisting of a single layer of at least partly suberized or cutinized cells; functions to control the movement of water and other substances into and out of the stele. { ¦en·dō¦dər·məs }

endodontics [MED] A branch of dentistry that treats diseases and injuries affecting the root tips or nerves of teeth; root canal is a common procedure. { ‚en·dō'dän·tiks }

endoenzyme [BIOCHEM] An intracellular enzyme, retained and utilized by the secreting cell. { ‚en·dō¦en‚zīm }

endoergic See endothermic. { ¦en·dō¦ər·jik }

endoergic collision See collision of the first kind. { ¦en·dō¦ər·jik kə'lizh·ən }

end-of-arm speed [CONT SYS] The speed at which an end effector arrives at its desired position. { ¦end əv ¦ärm 'spēd }

end-of-block character [COMPUT SCI] A character that indicates the completion of a block of code. { ¦end əv ¦bläk 'kar·ik·tər }

end-of-data mark [COMPUT SCI] A character or word signaling the end of all data held in a particular storage unit. { ¦end əv 'dad·ə ‚märk }

end-of-field mark [COMPUT SCI] A data item signaling the end of a field of data, generally a variable-length field. { ¦end əv 'fēld ‚märk }

end of file [COMPUT SCI] **1.** Termination or point of completion of a quantity of data; end of file marks are used to indicate this point. **2.** Automatic procedures to handle tapes when the end of an input or output tape is reached; a reflective spot, called a record mark, is placed on the physical end of the tape to signal the end. { ¦end əv 'fīl }

end-of-file gap [COMPUT SCI] A gap of precise dimension to indicate the end of a file on tape. Abbreviated EOF gap. { ¦end əv 'fīl ‚gap }

end-of-file indicator [COMPUT SCI] **1.** A device that indicates the end of a file on tape. **2.** See end-of-file mark. { ¦end əv 'fīl 'in·də‚kād·ər }

end-of-file mark [COMPUT SCI] A control character which signifies that the last record of a file has been read. Also known as end-of-file indicator. { ¦end əv 'fīl ‚märk }

end-of-file routine [COMPUT SCI] A program which checks that the contents of a file read into the computer were correctly read; may also start the rewind procedure. { ¦end əv 'fīl rü‚tēn }

end-of-file spot [COMPUT SCI] A reflective piece of tape indicating the end of the tape. { ¦end əv 'fīl ‚spät }

end-of-message [COMMUN] A character or series of characters signifying the end of a message or record, such as a message sent by teletypewriter. { ¦end əv 'mes·ij }

end-of-record gap [COMPUT SCI] A gap of precise dimension (shorter than the end-of-file gap) which indicates the physical end of a record on a magnetic tape. Abbreviated EOR gap. { ¦end əv 're·kərd ‚gap }

end-of-record word [COMPUT SCI] The last word in a record, usually written in a special format that enables identification of the end of the record. { ¦end əv 're·kərd ‚wərd }

end-of-run routine [COMPUT SCI] A routine that carries out various housekeeping operations such as rewinding tapes and printing control totals before a run is completed. { ¦end əv 'rən rü‚tēn }

end-of-tape routine [COMPUT SCI] A program which is brought into play when the end of a tape is reached; may involve a series of validity checks and initiate the tape rewind. { ¦end əv 'tāp rü‚tēn }

end-of-transmission card [COMMUN] Last card of each message; used to signal the end of a transmission and contains the same information as the header card, plus additional data for traffic analysis. { ¦end əv tranz'mish·ən ‚kärd }

end-of-transmission recognition [COMPUT SCI] The capability of a computer to recognize the end of transmission of a data string even if the buffer area is not filled. { ¦end əv tranz'mish·ən rek·ig‚nish·ən }

endogamy [BIOL] Sexual reproduction between organisms which are closely related. [BOT] Pollination of a flower by another flower of the same plant. { en'däg·ə·mē }

endogenetic See endogenic. { ‚en·dō·jə'ned·ik }

endogenic [GEOL] Of or pertaining to a geologic process, or its resulting feature such as a rock, that originated within the earth. Also known as endogenetic; endogenous. { ‚en·dō¦jen·ik }

endogenote [MICROBIO] The genetic complement of the partial zygote formed as a result of gene transfer during the process of recombination in bacteria. { en'däj·ə‚nōt }

endogenous [BIOCHEM] Relating to the metabolism of nitrogenous tissue elements. [GEOL] See endogenic. [MED] Pertaining to diseases resulting from internal causes. [PSYCH] Pertaining to mental disorders caused by hereditary or constitutional factors. { en'däj·ə·nəs }

endogenous pyrogen [BIOCHEM] A fever-inducing substance (protein) produced by cells of the host body, such as leukocytes and macrophages. { en'däj·ə·nəs 'pī·rə·jən }

endogenous variables [MATH] In a mathematical model, the dependent variables; their values are to be determined by the solution of the model equations. { en'däj·ə·nəs 'ver·ē·ə·bəlz }

endogenous virus [GEN] An inactive virus that is integrated into the chromosome of its host cell and can, therefore, exhibit vertical transmission. { en¦däj·ən·əs 'vī‚rəs }

endoglycosidase [BIOCHEM] An enzyme which releases intact glycans from their linkages with amino acids. { ‚en·dō·glī'kō·sə‚dās }

endognath [INV ZOO] The inner and main branch of a crustacean's oral appendage. { 'en·dəg‚nath }

endolecithal [INV ZOO] A type of egg found in turbellarians with yolk granules in the cytoplasm of the egg. Also spelled entolecithal. { en·dō¦les·ə·thəl }

endolithic [ECOL] Living within rocks, as certain algae and coral. { ‚en·dō¦lith·ik }

endolymph [PHYSIO] The lymph fluid found in the membranous labyrinth of the ear. { 'en·də‚limf }

endolymphatic stromomyosis See interstitial endometriosis. { ‚en·də‚lim¦fad·ik ‚strō·mō‚mī'ō·səs }

endomembrane system [CELL MOL] In eukaryotes, the functional continuum of membraneous cell components consisting of the nuclear envelope, endoplastic reticulum, and Golgi apparatus as well as vesicles and other structures derived from these major components. { ‚en·dō'mem‚brān ‚sis·təm }

endomeninx [EMBRYO] The internal part of the meninx primitiva that differentiates into the pia mater and arachnoid membrane. { ‚en·dō¦mē·niŋks }

endomere [EMBRYO] A blastomere that forms endoderm. { 'en·də‚mir }

endometamorphism [GEOL] A phase of contact metamorphism involving changes in an igneous rock due to assimilation of portions of the rocks invaded by its magma. { ¦en·dō‚med·ə'mȯr‚fiz·əm }

endometrioma [MED] Endometriosis in which there is a discrete tumor mass. { ‚en·dō‚mē·trē'ō·mə }

endometriosis [MED] The presence of endometrial tissue in abnormal locations, including the uterine wall, ovaries, or extragenital sites. { ‚en·dō‚mē·trē'ō·səs }

endometritis [MED] Inflammation of the endometrium. { ‚en·dō·mə'trīd·əs }

endometrium [ANAT] The mucous membrane lining the uterus. { ‚en·dō¦mē·trē·əm }

endomitosis [CELL MOL] Division of the chromosomes without dissolution of the nuclear membrane; results in polyploidy or polyteny. { ‚en·dō‚mī'tō·səs }

endomixis [INV ZOO] Periodic division and reorganization of the nucleus in certain ciliated protozoans. { ‚en·dō¦mik·səs }

endomorph [PSYCH] A somatotype suggested by W.H. Sheldon to describe a person with a rounded physique; associated with viscerotonia. { 'en·də‚mȯrf }

endomorphism [MATH] A function from a set with some structure (such as a group, ring, vector space, or topological space) to itself which preserves this structure. { 'en·də'mȯr‚fiz·əm }

Endomycetales [MICROBIO] Former designation for Saccharomycetales. { ‚en·də‚mī·sə'tā·lēz }

Endomycetoideae [MICROBIO] A subfamily of ascosporogenous yeasts in the family Saccharomycetaceae. { ‚en·də‚mī·sə'tȯid·ē‚ē }

Endomychidae [INV ZOO] The handsome fungus beetles, a family of coleopteran insects in the superfamily Cucujoidea. { ‚en·də'mīk·ə‚dē }

endomysium [HISTOL] The connective tissue layer surrounding an individual skeletal muscle fiber. { ‚en·də'miz·ē·əm }

endoneural fibroma See neurofibroma. { ¦en·dō¦nür·əl fī'brō·mə }

endoneurium [HISTOL] Connective tissue fibers surrounding and joining the individual fibers of a nerve trunk. { ‚en·dō'nür·ē·əm }

end-on position [ELECTROMAG] The position of a point which lies on the magnetic axis of a magnet. Also known as Gauss A position. { 'end ¦ȯn pə‚zish·ən }

endonuclease [BIOCHEM] Any of a group of enzymes which degrade deoxyribonucleic acid or ribonucleic acid molecules by attaching nucleotide linkages within the polynucleotide chain. { ¦en·dō'nü·klē‚ās }

endoparasite [ECOL] A parasite that lives inside its host. { ¦en·dō'par·ə‚sīt }

endopeptidase [BIOCHEM] An enzyme that acts upon the centrally located peptide bonds of a protein molecule. { ¦en·dō'pep·tə‚dās }

endoperoxide [BIOCHEM] Any of various intermediates in the biosynthesis of prostaglandins. { ‚en·dō·pə'räk‚sīd }

endophagous [INV ZOO] Of an insect larva, living within and feeding upon the host tissues. { en'däf·ə·gəs }

endophallus [INV ZOO] Inner wall of the phallus of insects. { ¦en·dō'fal·əs }

endophyte [ECOL] A plant that lives within, but is not necessarily parasitic on, another plant. { 'en·də‚fīt }

endoplasm [CELL MOL] The inner, semifluid portion of the cytoplasm. { 'en·də‚plaz·əm }

endoplasmic reticulum [CELL MOL] A vacuolar system of the cytoplasm in differentiated cells that functions in protein synthesis and sequestration. Abbreviated ER. { ¦en·də¦plaz·mik rə'tik·yə·ləm }

endopleurite [INV ZOO] **1.** The portion of a crustacean apodeme which develops from the interepimeral membrane. **2.** One of the laterally located parts on the thorax of an insect which fold inward, extending into the body cavity. { ‚en·də'plür‚īt }

endopodite [INV ZOO] The inner branch of a biramous crustacean appendage. { en'däp·ə‚dīt }

endopolyploid cell [CELL MOL] Any cell whose chromosome number has been increased by endomitosis and for which the degree of ploidy is proportional to the number of times that endomitosis has taken place. { ‚en·dō'päl·ə‚ploid ‚sel }

Endoprocta [INV ZOO] The equivalent name for Entoprocta. { ‚en·də'präk·tə }

endoprosthesis [MED] A prosthesis that is used internally. { ¦en·dō·präs'thē·səs }

endopterygoid [VERT ZOO] A paired dermal bone of the roof of the mouth in fishes. { ¦en·dō'ter·ə‚goid }

Endopterygota [INV ZOO] A division of the insects in the subclass Pterygota, including those orders which undergo a holometabolous metamorphosis. { ¦en·dō‚ter·ə'gäd·ə }

ENDOR See electron nuclear double resonance. { 'en‚dȯr }

endoradiosonde [ENG] A miniature battery-powered radio transmitter encapsulated like a pill, designed to be swallowed for measuring and transmitting physiological data from the gastrointestinal tract. { ¦en·dō'rād·ē·ō‚sänd }

endoreduplication [CELL MOL] Appearance in mitotic cells of certain chromosomes or chromosome sets in the form of multiples. [GEN] Two to twelve or more rounds of replication and chromosome duplication without mitotic cell division, as in the production of polytene chromosomes. { ¦en·dō·rē‚dü·plə'kā·shən }

endoreism See endorheism. { ‚en·dō'rē‚iz·əm }

end organ [NEUROSCI] The expanded termination of a nerve fiber in muscle, skin, mucous membrane, or other structure. { 'end ‚ȯr·gən }

endorheism [HYD] A drainage pattern of a basin or region in which little or none of the surface drainage reaches the ocean. Also spelled endoreism. { ‚en·dō'rē‚iz·əm }

β-endorphin [BIOCHEM] A 31-amino acid peptide fragment of pituitary β-lipotropic hormone having morphinelike activity. { ¦bād·ə en'dȯr·fən }

endorser [COMPUT SCI] A special feature available on most magnetic-ink character-recognition readers that imprints a bank's endorsement on successful document reading. [GRAPHICS] A camera attachment that stamps documents as they are filmed. { en'dȯr·sər }

endosalpingioma See serous cystadenoma. { ¦en·dō·sal‚pin·jē'ō·mə }

endosalpinx [ANAT] The mucous membrane that lines the fallopian tube. { ¦en·dō¦sal‚piŋks }

endosarc [INV ZOO] The inner, relatively fluid part of the protoplasm of certain unicellular organisms. { 'en·də‚särk }

endoscope [MED] An instrument used to visualize the interior of a body cavity or hollow organ. { 'en·də‚skōp }

endosepsis [PL PATH] A fungus disease of figs caused by *Fusarium moniliforme fici*; fruits rot internally. { ‚en·də'sep·səs }

endoskeleton [ZOO] An internal skeleton or supporting framework in an animal. { ¦en·dō'skel·ə·tən }

endosmosis [PHYSIO] The passage of a liquid inward through a cell membrane. { ¦en·dō·äs'mō·səs }

endosome [CELL MOL] A mass of chromatin near the center of a vesicular nucleus. [INV ZOO] The inner layer of certain sponges. { 'en·də‚sōm }

endosperm [BOT] **1.** The nutritive protein material within the embryo sac of seed plants. **2.** Storage tissue in the seeds of gymnosperms. { 'en·də‚spərm }

endosperm nucleus [BOT] The triploid nucleus formed within the embryo sac of most seed plants by fusion of the polar nuclei with one sperm nucleus. { 'en·də‚spərm 'nü·klē·əs }

endospore [BIOL] An asexual spore formed within a cell. { 'en·də‚spȯr }

endosteum [ANAT] The membrane lining of bone marrow cavities. { en'däs·tē·əm }

endostome [BOT] The opening in the inner integument of a bitegmic ovule. { 'en·də‚stōm }

endostyle [INV ZOO] A ciliated groove or pair of grooves in the pharynx of lower chordates. { 'en·də‚stīl }

endosulfan [ORG CHEM] $C_9H_6Cl_6O_3S$ A tan solid that melts between −10 and 100°C; used as an insecticide and miticide on vegetable and forage crops, on ornamental flowers, and in controlling termites and tsetse flies. { ¦en·dō'səl‚fan }

endosymbiont [ECOL] A symbiont that lives within the body of the host without deleterious effect on the host. { ¦en·dō'sim·bē‚änt }

endosymbiont theory [CELL MOL] A theory that the mitochondria of eukaryotes and the chloroplasts of green plants and flagellates originated as free-living prokaryotes that invaded primitive eukaryotic cells and become established as permanent symbionts in the cytoplasm. { ‚en·dō'sim·bē‚änt ‚thē·ə·rē }

endosymbiosis [ECOL] A mutually beneficial relationship in which one organism lives inside the other. { ‚en·dō‚sim·bē'ō·səs }

endosymbiotic infection [VIROL] A virus infection in which virus replication occurs in cells without a cytopathic effect. { ¦en·dō‚sim·bē'äd·ik in'fek·shən }

endotergite [INV ZOO] A dorsal plate to which muscles are attached in the insect skeleton. { ¦en·dō'tər‚jīt }

endotesta [BOT] An inner layer of the testa in various seeds. { 'en·dō‚tes·tə }

endothecium [BOT] The middle of three layers that make up an immature anther; becomes the inner layer of a mature anther. { ‚en·də'thē·shē·əm }

endothelial cell [HISTOL] A type of squamous epithelial cell composing the endothelium. { ‚en·də'thē·lē·əl 'sel }

endotheliochorial placenta [EMBRYO] A type of placenta in which the maternal blood is separated from the chorion by the maternal capillary endothelium; occurs in dogs. { ¦en·də‚thē·lē·ə'kȯr·ē·əl plə'sen·tə }

endothelioma [MED] Any tumor arising from, or resembling, endothelium; usually a benign growth, but occasionally a malignant tumor. { ‚en·dō‚thē·lē'ō·mə }

endothelium [HISTOL] The epithelial layer of cells lining the heart and vessels of the circulatory system. { ‚en·də'thē·lē·əm }

Endotheriidae [PALEON] A family of Cretaceous insectivores from China belonging to the Proteutheria. { ‚en·dō·thə'rī·ə‚dē }

endotherm [PHYS CHEM] In differential thermal analysis, a graph of the temperature difference between a sample compound and a thermally inert reference compound (commonly aluminum oxide) as the substances are simultaneously heated to elevated temperatures at a predetermined rate, and the sample compound undergoes endothermal or exothermal processes. [PHYSIO] An animal that produces enough heat from its own metabolism and employs devices to retard heat loss so that it

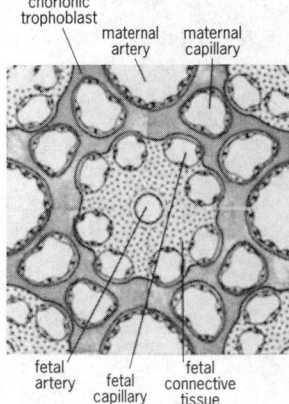

ENDOTHELIOCHORIAL PLACENTA

chorionic trophoblast

maternal artery

maternal capillary

fetal artery

fetal capillary

fetal connective tissue

Section through an endotheliochorial placenta.

is able to keep its body temperature higher than that of its environment. { 'en·də,thərm }

endothermic [NUC PHYS] Petaining to a nuclear or particle reaction in which some of the kinetic energy of the initial particles is converted to mass energy. [PHYS CHEM] Pertaining to a chemical reaction which absorbs heat. Also known as endoergic. { ,en·də'thər·mik }

endothermy [PHYSIO] The utilization of metabolic heat for thermoregulation. { 'en·dō,thər·mē }

Endothyracea [PALEON] A superfamily of extinct benthic marine foraminiferans in the suborder Fusulinina, having a granular or fibrous wall. { ,en·dō·thə'rās·ē·ə }

endotoxin [MICROBIO] A biologically active substance produced by gram-negative bacteria and consisting of lipopolysaccharide, a complex macromolecule containing a polysaccharide covalently linked to a unique lipid structure, termed lipid A. { ,en·dō'täk·sən }

endotracheal [ANAT] Within the trachea. { ,en·dō'trā·kē·əl }

endotrophic [BIOL] Obtaining nourishment from within; applied to certain parasitic fungi that live in the root cortex of the host plant. { 'en·də,trä·fik }

end-plate potential [NEUROSCI] Depolarization of the postsynaptic membrane at the neuromuscular junction, mediated by acetylcholine, in response to action potentials arriving at the endings of presynaptic motor neurons. { 'end ,plāt pə'ten·chəl }

end play [MECH ENG] Axial movement in a shaft-and-bearing assembly resulting from clearances between the components. { 'end ,plā }

end point [ANALY CHEM] That stage in the titration at which an effect, such as a color change, occurs, indicating that a desired point in the titration has been reached. [CHEM ENG] In the distillation analysis of crude petroleum and its products, the highest reading of a thermometer when a specified proportion of the liquid has boiled off. Also known as final boiling point. [COMPUT SCI] In vector graphics, one of the two ends of a line or vector. [CONT SYS] The point at which a robot stops along its path of motion. [IND ENG] *See* breakpoint. [MATH] Either of two values or points that mark the ends of an interval or line segment. { 'end ,póint }

end-point rigidity [CONT SYS] The resistance of a robot to further movement after it has reached its end point. { 'en ,póint ri'jid·əd·ē }

end printing [COMPUT SCI] Printing across one end of a punched card the information punched on the card. { 'end ,print·iŋ }

end product [PHYS] The final product of a chemical or nuclear reaction or process. { 'end ,prä·dəkt }

end-product inhibition [BIOCHEM] In sequential enzyme systems, a control mechanism in which accumulation of final product from a metabolic reaction causes inhibition of product formation. { 'end ,präd·əkt ,in·ə'bish·ən }

end radiation *See* quantum limit. { 'end ,rād·ē,ā·shən }

endrin [ORG CHEM] $C_{12}H_8OCl_6$ Poisonous, white crystals that are insoluble in water; it is used as a pesticide and is a stereoisomer of dieldrin, another pesticide. { 'en·drən }

end section [COMMUN] Additional portion of switchboard added to each end of a large multiple switchboard and used to extend some of the trunks or locals to these end positions to place all jacks within easy reach of the first and last operator. Also known as head section. { 'end ,sek·shən }

end sentinel [COMPUT SCI] A character that indicates the end of a message or record. { 'end ,sent·nəl }

end stone [HOROL] A flat jewel in a timepiece which acts as a bearing for a pivot. { 'end ,stōn }

end stop [MECH ENG] A limit to the movement of a mechanical system or part, usually brought about by valves or shock absorbers. { 'end ,stäp }

end-to-end encryption [COMMUN] Encryption of a message at its point of origination so that it travels in encrypted form all the way to its destination. { ,end·tü,end in'krip·shən }

end turning *See* boxing. { 'end ,tərn·iŋ }

end-vertex [MATH] A vertex of a graph that has exactly one edge incident to it. { 'end ,vər,teks }

endurance [ENG] The time an aircraft, vehicle, or ship can continue operating under given conditions without refueling. { in'dùr·əns }

endurance limit *See* fatigue limit. { in'dùr·əns ,lim·ət }

endurance ratio *See* fatigue ratio. { in'dùr·əns ,rā·shō }

endurance strength *See* fatigue strength. { in'dùr·əns ,streŋkth }

end user [COMPUT SCI] The person for whom the output of a computer is ultimately intended. { 'end ,yüz·ər }

en echelon [GEOL] Referring to an overlapped or staggered arrangement of geologic features. { 'en ,esh·ə,län }

en echelon fault blocks [GEOL] A belt in which the individual fault blocks trend approximately 45° to the trend of the entire fault belt. { 'en ,esh·ə,län 'fólt ,bläks }

enema [MED] A rectal injection of liquid for therapeutic, diagnostic, or nutritive purposes. { 'en·ə·mə }

ene reaction [ORG CHEM] The addition of a compound with a double bond having an allylic hydrogen (ene, such as propene) to a compound with a multiple bond (enophile, such as ethene). { 'ēn rē,ak·shən }

energetics [PHYS] The study of energy and of its transformation from one form to another. { ,en·ər'jed·iks }

energetic solar particles [ASTROPHYS] Electrons and atomic nuclei produced in association with solar flares, with energies mostly in the range 1–100 million electronvolts, but occasionally as high as 15 billion electronvolts. Also known as solar cosmic rays. { ,en·ər'jed·ik ¦sō·lər 'pärd·i·kəlz }

energized [ELEC] Electrically connected to a voltage source. Also known as alive; hot; live. { 'en·ər,jīzd }

energy [PHYS] The capacity for doing work. { 'en·ər·jē }

energy absorption [PHYS] Conversion of mechanical or radiant energy into the internal potential energy or heat energy of a system. { 'en·ər·jē ab,sórp·shən }

energy balance [PHYS] The arithmetic balancing of energy inputs versus outputs for an object, reactor, or other processing system; it is positive if energy is released, and negative if it is absorbed. [PHYSIO] The relation of the amount of utilizable energy taken into the body to that which is employed for internal work, external work, and the growth and repair of tissues. { 'en·ər·jē ,bal·əns }

energy band *See* band. { 'en·ər·jē ,band }

energy-band theory of solids *See* band theory of solids. { 'en·ər·jē ,band ¦thē·ə·rē əv 'säl·ədz }

energy beam [ENG] An intense beam of light, electrons, or other nuclear particles; used to cut, drill, form, weld, or otherwise process metals, ceramics, and other materials. { 'en·ər·jē ,bēm }

energy budget [CLIMATOL] The energy pools, the directions of energy flow, and the rates of energy transformations quantified within a physical or ecological system. { 'en·ər·jē ,bəj·ət }

energy coefficient [OCEANOGR] The ratio between the energy transmitted forward in a wave per unit crest length at a point in shallow water, and the energy transmitted forward in a wave per unit crest length in deep water. { 'en·ər·jē ,kō·i'fish·ənt }

energy conservation *See* conservation of energy. { 'en·ər·jē ,kän·sər'vā·shən }

energy conversion [PHYS] The process of changing energy from one form to another. { 'en·ər·jē kən'vər·zhən }

energy conversion efficiency [MECH ENG] The efficiency with which the energy of the working substance is converted into kinetic energy. { 'en·ər·jē kən'vər·zhən i,fish·ən·sē }

energy density [PHYS] The energy per unit volume of a medium; in the case of an electric or magnetic field, the energy needed to set up the field is thought of as residing in the field. { 'en·ər·jē ,den·səd·ē }

energy diagram *See* energy-level diagram. { 'en·ər·jē ,dī·ə,gram }

energy-dispersive x-ray diffraction [PHYS] A technique in which an energy spectrum is obtained of the x-rays scattered from a polychromatic x-ray beam through a fixed angle by a polycrystalline sample. Abbreviated EDXD. { ¦en·ər·jē di¦spər·səv 'eks,rā di'frak·shən }

energy dissipation *See* dissipation. { 'en·ər·jē dis·ə'pā·shən }

energy efficiency ratio [ELEC] A value that represents the relative electrical efficiency of air conditioners; it is the quotient obtained by dividing Btu-per-hour output by electrical-watts input during cooling. { 'en·ər·jē i'fish·ən·se ,rā·shō }

energy eigenstate *See* energy state. { 'en·ər·jē'ī·gən,stāt }

energy ellipsoid *See* momental ellipsoid. { ¦en·ər·jē i'lip,sóid }

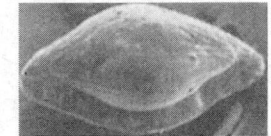

ENDOTHYRACEA

Scanning electron micrograph of *Triticites* from the Pennsylvanian formation of Texas showing the exterior of the fusuline test. (*R. B. MacAdam, Chevron Oil Field Research Co.*)

ENE REACTION

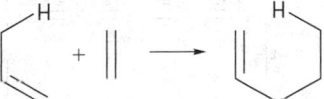

Ene reaction, a pericyclic reaction.

energy flux [PHYS] A vector quantity whose component perpendicular to any surface equals the energy transported across that surface by some medium per unit area per unit time. { 'en·ər·jē ‚fləks }

energy gap [SOLID STATE] A range of forbidden energies in the band theory of solids. { 'en·ər·jē ‚gap }

energy gradient [PHYS] Any change in energy over time or space. { 'en·ər·jē ‚grād·ē·ənt }

energy head [FL MECH] The elevation of the hydraulic grade line at any section of a waterway plus the velocity head of the mean velocity of the water in that section. { 'en·ər·jē ‚hed }

energy integral [MECH] A constant of integration resulting from integration of Newton's second law of motion in the case of a conservative force; equal to the sum of the kinetic energy of the particle and the potential energy of the force acting on it. { 'en·ər·jē 'in·tə·grəl }

energy level [GEOL] The kinetic energy supplied by waves or current action in an aqueous sedimentary environment either at the interface of deposition or several meters above. [QUANT MECH] An allowed energy of a physical system; there may be several allowed states at one level. { 'en·ər·jē ‚lev·əl }

energy-level diagram [QUANT MECH] A diagram in which the energy levels of a quantized system are indicated by distances of horizontal lines from a zero energy level. Also known as energy diagram; level scheme. { 'en·ər·jē ‚lev·əl dī·ə‚gram }

energy load See dynamic load. { 'en·ər·jē ‚lōd }

energy management [AERO ENG] In rocketry, the monitoring of the expenditure of fuel for flight control and navigation. { 'en·ər·jē ‚man·ij·mənt }

energy metabolism [BIOCHEM] The chemical reactions involved in energy transformations within cells. { 'en·ər·jē mə'tab·ə‚liz·əm }

energy momentum tensor [PHYS] A tensor whose 16 elements give the energy density, momentum density, and stresses in a distribution of matter or radiation. { 'en·ər·jē mə'men·təm ‚ten·sər }

energy of a charge [ELEC] Charge energy measured in ergs according to the equation $E = QV$, where Q is the charge and V is the potential in electrostatic units. { 'en·ər·jē əv ə 'chärj }

energy of activation See activation energy. { 'en·ər·jē əv ak·tə'vā·shən }

energy of rotation [PHYS] Kinetic energy of a mass with moment of inertia I rotating with angular velocity ω about the axis, expressed as $E = \frac{1}{2}I\omega^2$. { 'en·ər·jē əv rō'tā·shən }

energy operator [QUANT MECH] The operator corresponding to the energy or Hamiltonian of a classical system. Also known as Hamiltonian operator. { 'en·ər·jē ‚äp·ə‚rād·ər }

energy product curve [ELECTROMAG] Curve obtained by plotting the product of the values of magnetic induction B and demagnetizing force H for each point on the demagnetization curve of a permanent magnet material; usually shown with the demagnetization curve. { 'en·ər·jē 'prä·dəkt ‚kərv }

energy profile [PHYS CHEM] A diagram of the energy changes that take place during a reaction in a chemical system. { 'en·ər·jē ‚prō‚fīl }

energy pyramid [ECOL] An ecological pyramid illustrating the energy flow within an ecosystem. { 'en·ər·jē ‚pir·ə·mid }

energy spectrum [PHYS] Any plot, display, or photographic record of the intensity of some type of radiation as a function of its energy. { 'en·ər·jē ‚spek·trəm }

energy spread [QUANT MECH] The width in energy of a wave packet or metastable state. { 'en·ər·jē ‚spred }

energy state [QUANT MECH] An eigenstate of the energy (Hamiltonian) operator, so that the energy has a definite stationary value. Also known as eigenstate; energy eigenstate; quantum state; stationary state. { 'en·ər·jē ‚stāt }

energy transfer [METEOROL] The transfer of energy of a given form among different scales of motion; for example, kinetic energy may be transferred between the zonal and meridional components of the wind, or between the mean and eddy components of the wind. { 'en·ər·jē ‚trans·fər }

energy-variant sequential detection [COMMUN] Technique of sequential detection consisting of the transmission of a fixed number of pulses of varying energy, received with a single (upper) threshold device. { 'en·ər·jē ‚ver·ē·ənt sə'kwen·chəl di'tek·shən }

energy winds [PHYS] A group of winds which contain the bulk of recoverable kinetic energy for each month. { 'en·ər·jē 'winz }

enfilade [ORD] To rake with gunfire; to fire down the length of a trench or line of troops. { 'en·fə‚lād }

enfleurage [CHEM ENG] Removal of the odoriferous components from flowers by placing them near an odorless mixture of lard and tallow; this mixture absorbs the perfume, which is subsequently extracted. { ‚än‚flü'räzh }

engaged column [CIV ENG] A column partially built into a wall, and not freestanding. { in'gājd 'käl·əm }

Engel-Recklinghausen disease See osteitis fibrosa cystica. { ‚eŋ·gəl 'rek·liŋ‚haúz·ən di‚zēz }

engine [MECH ENG] A machine in which power is applied to do work by the conversion of various forms of energy into mechanical force and motion. { 'en·jən }

engine balance [MECH ENG] Arrangement and construction of moving parts in reciprocating or rotating machines to reduce dynamic forces which may result in undesirable vibrations. { 'en·jən ‚bal·əns }

engine block See cylinder block. { 'en·jən ‚bläk }

engine cooling [MECH ENG] Controlling the temperature of internal combustion engine parts to prevent overheating and to maintain all operating dimensions, clearances, and alignment by a circulating coolant, oil, and a fan. { 'en·jən ‚kúl·iŋ }

engine cycle [THERMO] Any series of thermodynamic phases constituting a cycle for the conversion of heat into work; examples are the Otto cycle, Stirling cycle, and Diesel cycle. { 'en·jən ‚sī·kəl }

engine cylinder [MECH ENG] A cylindrical chamber in an engine in which the energy of the working fluid, in the form of pressure and heat, is converted to mechanical force by performing work on the piston. Also known as cylinder. { 'en·jən ‚sil·ən·dər }

engine displacement [MECH ENG] Volume displaced by each piston moving from bottom dead center to top dead center multiplied by the number of cylinders. { 'en·jən di‚splās·mənt }

engine distillate [MATER] A heavy naphtha-kerosine distillate fuel of low octane number. Also known as tractor fuel. { 'en·jən 'dist·əl·ət }

engine efficiency [MECH ENG] Ratio between the energy supplied to an engine to the energy output of the engine. { 'en·jən i'fish·ən·sē }

engineer [ENG] An individual who specializes in one of the branches of engineering. { ‚en·jə'nir }

engineering [SCI TECH] The science by which the properties of matter and the sources of power in nature are made useful to humans in structures, machines, and products. { ‚en·jə'nir·iŋ }

engineering channel circuit [COMMUN] Auxiliary circuit or channel (radio or wire) for use by operating or maintenance personnel for communications incident to the establishment, operation, maintenance, and control of communications facilities. { ‚en·jə'nir·iŋ 'chan·əl ‚sər·kət }

engineering economy [IND ENG] **1.** Application of engineering or mathematical analysis and synthesis to decision making in economics. **2.** The knowledge and techniques concerned with evaluating the worth of commodities and services relative to their cost. **3.** Analysis of the economics of engineering alternatives. { ‚en·jə'nir·iŋ i'kän·ə·mē }

engineering geology [CIV ENG] The application of education and experience in geology and other geosciences to solve geological problems posed by civil engineering structures. { ‚en·jə'nir·iŋ je'äl·ə·jē }

engineering plastics [ORG CHEM] A class of polymers, based on aromatic backbones, having high strength, stiffness, and toughness together with high thermal and oxidative stability, low creep, and the ability to be processed by standard techniques for thermoplastics; examples include polyacetal, polyamide, polycarbonate, and polysulfone resins. { ‚en·jə'nir·iŋ 'plas·tiks }

engineering time [COMPUT SCI] The nonproductive time of a computer, reserved for maintenance and servicing. { ‚en·jə'nir·iŋ ‚tīm }

engineer's chain [CIV ENG] A surveyor's measuring instrument consisting of 1-foot (30.48-centimeter) steel links joined together by rings, 100 feet (30.5 meters) or 50 feet (15.25 meters) long. Also known as chain. { ‚en·jə'nirz ‚chān }

engineer's scale [GRAPHICS] A rule having a triangular

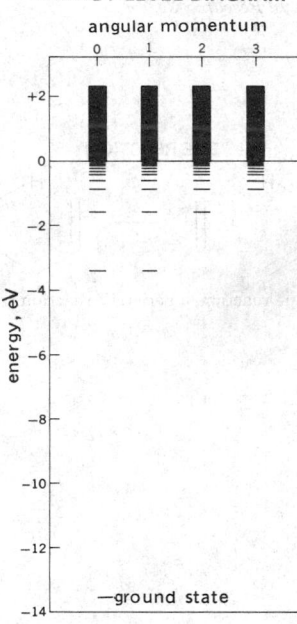

ENERGY-LEVEL DIAGRAM

angular momentum

Energy levels of the hydrogen atom, classified by the orbital angular momentum of the electron, expressed in units of $\hbar$.

cross section and different measurement scales on the two edges of each face. { ‚en·jə'nirz ‚skāl }

engineer's system of units *See* British gravitational system of units. { ‚en·jə'nirz ¦sis·təm əv 'yü·nəts }

engine fuel [MATER] Any of various substances, usually fluid, which provide heat, chemical, or pressure energy for engine operation. { 'en·jən ‚fyül }

engine inlet [MECH ENG] A place of entrance for engine fuel. { 'en·jən ‚in·lət }

engine knock [MECH ENG] In spark ignition engines, the sound and other effects associated with ignition and rapid combustion of the last part of the charge to burn, before the flame front reaches it. Also known as combustion knock. { 'en·jən ‚näk }

engine lathe [MECH ENG] A manually operated lathe equipped with a headstock of the back-geared, cone-driven type or of the geared-head type. { 'en·jən ‚lāth }

engine oil [MATER] Oil used for the bearing lubrication of all types of engines, machines, and shafting and for cylinder lubrication in other than steam engines. { 'en·jən ‚óil }

engine performance [MECH ENG] Relationship between power output, revolutions per minute, fuel or fluid consumption, and ambient conditions in which an engine operates. { 'en·jən pər'fór·məns }

engine revolution counter [NAV ARCH] An instrument for registering the number of revolutions of a propeller shaft of a vessel; this information is useful in estimating a vessel's speed through the water. { 'en·jən rev·ə'lü·shən ‚kaúnt·ər }

engine sludge [ENG] The insoluble products of degradation of lubricating oils and fuels formed during the operation of an internal combustion engine. { 'en·jən ‚sləj }

engine spray [AERO ENG] That part of a pad deluge that is directed at cooling a rocket's engine during launch. { 'en·jən ‚sprā }

engine starter [ELEC] The electric motor in the electrical system of an automobile that cranks the engine for starting. Also known as starter; starting motor. { 'en·jən ‚stärd·ər }

englacial [HYD] Of or pertaining to the inside of a glacier. { en'glā·shəl }

Engler distillation test [CHEM ENG] A standard test for determination of the volatility characteristics of a gasoline by the measurement of the percent of gasoline distilled at various specific temperatures. { 'eŋ·glər dis·tə'lā·shən ‚test }

Engler flask [CHEM ENG] A standardized flask of 100-milliliter volume used in the Engler distillation test. { 'eŋ·glər ‚flask }

Engler viscometer [ENG] An instrument used in the measurement of the degree Engler, a measure of viscosity; the kinematic viscosity v in stokes for this instrument is obtained from the equation $v = 0.00147t - 3.74/t$, where t is the efflux time in seconds. { 'eŋ·glər vi'skäm·əd·ər }

English degree [CHEM] A unit of water hardness, equal to 1 part calcium carbonate to 70,000 parts water; equivalent to 1 grain of calcium carbonate per gallon of water. Also known as Clark degree. { 'iŋ·glish di‚grē }

English garden-wall bond [CIV ENG] A masonry bond in which there are three courses of stretchers to one of headers. { 'iŋ·glish ‚gärd·ən 'wól ‚bänd }

englishite [MINERAL] $K_2Ca_4Al_8(PO_4)_8(OH)_{10} \cdot 9H_2O$ A white mineral composed of hydrous basic phosphate of potassium, calcium, and aluminum. { 'iŋ·gli‚shīt }

English red [MATER] Pigment consisting mostly of red iron oxide. { 'iŋ·glish 'red }

English vermilion [INV ZOO] Bright vermilion pigment of precipitated mercury sulfide; in paints, it tends to darken when exposed to light. { 'iŋ·glish vər'mil·yən }

engobe [MATER] A thin layer of fluid clay applied to a piece of earthenware to support a glaze or enamel or to cover blemishes. { än'gōb }

engram [NEUROSCI] A memory imprint; the alteration that has occurred in nervous tissue as a result of an excitation from a stimulus, which hypothetically accounts for retention of that experience. Also known as memory trace. { 'en‚gram }

Engraulidae [VERT ZOO] The anchovies, a family of herringlike fishes in the suborder Clupoidea. { ‚en'gról·ə‚dē }

engraved-roll coating [GRAPHICS] Pattern coating engraved on a surface by a webbed roll onto which the coating substance is carefully metered. Also known as gravure coating. { in'gravd ‚rōl 'kōd·iŋ }

engraver [GRAPHICS] One who makes engravings, either manually or by machine. { in'grāv·ər }

engraving [GRAPHICS] A photomechanical process in which lines are scribed on line negatives by means of a needle that produces an even, transparent printing line by removal of a thin layer of photographic emulsion. { in'grāv·iŋ }

engrossing [GRAPHICS] The production of decorative designs of letters and illuminations by an artist on citations and diplomas. { in'grōs·iŋ }

engysseismology [GEOPHYS] Seismology dealing with earthquake records made close to the disturbance. { ‚en·jə·sīz'mäl·ə·jē }

enhanceable language [COMPUT SCI] A computer language that has a modest degree of semantic extensibility. { en¦han·sə·bəl 'laŋ·gwij }

enhanced carrier demodulation [COMMUN] Amplitude demodulation system in which a synchronized local carrier of proper phase is fed into the demodulator to reduce demodulation distortion. { en¦hanst 'kar·ē·ər dē‚maj·ə'lā·shən }

enhanced line *See* enhanced spectral line. { en'hanst 'līn }

enhanced recovery [PETRO ENG] Any system that increases the fraction of original oil recovered in place in a reservoir; methods include waterflooding, chemical flooding, gas injection, and thermal methods. { en'hanst ri'kəv·ə·rē }

enhanced small device interface [COMPUT SCI] A standard method of connecting disk and tape drives to computers which allows for the transfer of 1–3 megabytes per second from disk drives holding up to 1 gigabyte of storage. Abbreviated ESDI. { en¦hanst ¦smól di¦vīs 'in·tər‚fās }

enhanced spectral line [SPECT] A spectral line of a very hot source, such as a spark, whose intensity is much greater than that of a line in a flame or arc spectrum. Also known as enhanced line. { en'hanst 'spek·trəl ‚līn }

enhancement [COMPUT SCI] A substantial increase in the capabilities of hardware or software. [ELECTR] An increase in the density of charged carriers in a particular region of a semiconductor. { en'hans·mənt }

enhancement mode [ELECTR] Operation of a field-effect transistor in which no current flows when zero gate voltage is applied, and increasing the gate voltage increases the current. { en'hans·mənt ‚mōd }

enhancement-mode high-electron-mobility transistor [ELECTR] A high-electron-mobility transistor in which application of a positive bias to the gate electrode is required for current to flow between the source and drain electrodes. Abbreviated E-HEMT. { en'hans·mənt ¦mōd 'hī i¦lek‚trän mō¦bil·əd·ē tran'zis·tər }

enhancement-mode junction field-effect transistor [ELECTR] A type of gallium arsenide field-effect transistor in which the gate consists of the junction between the *n*-type gallium arsenide forming the conducting channel and *p*-type material implanted under a metal electrode. Abbreviate E-JFET. { en'hans·mənt ¦mōd 'jəŋk·shən 'fēld i‚fekt tran'zis·tər }

enhancer gene [GEN] Any modifier gene that acts to enhance the action of a nonallelic gene. { en'han·sər ‚jēn }

ENIAC [COMPUT SCI] The first digital computer in the modern sense of the word, built 1942–1945. Derived from Electronic Numerical Integrator and Calculator. { 'ē·nē·ak }

Enicocephalidae [INV ZOO] The gnat bugs, a family of hemipteran insects in the superfamily Enicocephaloidea. { ‚en·ə·kō·sə'fal·ə‚dē }

Enicocephaloidea [INV ZOO] A superfamily of the Hemiptera in the subdivision Geocorisae containing a single family. { ‚en·ə·kō·sef·ə'lóid·ē·ə }

enigmatite [MINERAL] $Na_2Fe_5TiSi_6O_{20}$ A black amphibole mineral occurring in triclinic crystals; specific gravity is 3.14–3.80. Also spelled aenigmatite. { ə'nig·mə‚tīt }

enium ion [ORG CHEM] A cationic portion of an ionic species in which the valence shell of a positively charged nonmetallic atom has two electrons less than normal, and the charged entity has one covalent bond less than the corresponding uncharged species; used as a suffix with the root name. Also known as ylium ion. { 'en·ē·əm ‚ī·ən }

enkephalin [BIOCHEM] A mixture of two polypeptides isolated from the brain; central mode of action is an inhibition of neurotransmitter release. { en'kef·ə·lən }

enlargement [GRAPHICS] Photographic print made in an

enlarger so that the print is bigger than the negative. { en'lärj·mənt }

enlargement loss [FL MECH] Energy loss by friction in a flowing fluid when it moves into a cross-sectional area of sudden enlargement. { en'lärj·mənt ˌlós }

enlarger [OPTICS] An optical projector used to project an enlarged image of a photograph's negative onto photosensitized film or paper. Also known as photoenlarger. { en'lär·jər }

enlarging [GRAPHICS] The process of reproducing an image from a smaller image through a projection process. { en'lärj·iŋ }

enneagon *See* nonagon. { 'en·ē·əˌgän }

enol [ORG CHEM] An organic compound with a hydroxide group adjacent to a double bond; varies with a ketone form in the effect known as enol-keto tautomerism; an example is the compound $CH_3COH=CHCO_2C_2H_5$. { 'ēˌnól }

enolase [BIOCHEM] An enzyme that catalyzes the reversible dehydration of phosphoglyceric acid to phosphopyruvic acid. { 'ē·nəˌlās }

enolate anion [ORG CHEM] The delocalized anion which is left after the removal of a proton from an enol, or of the carbonyl compound in equilibrium with the enol. { 'ē·nəˌlāt 'anˌī·ən }

enol-keto tautomerism [ORG CHEM] The tautomeric migration of a hydrogen atom from an adjacent carbon atom to a carbonyl group of a keto compound to produce the enol form of the compound; the reverse process of hydrogen atom migration also occurs. { ˈē·nól ˈkēd·ō tó'tä·məˌriz·əm }

enology [FOOD ENG] The science of wine and wine production. { ēˈnäl·ə·jē }

enone [BIOCHEM] An alpha-, beta-unsaturated ketone. { 'ē ˌnōn }

enophthalmos [MED] Recession of the eyeball into the orbital cavity. { ˌeˌnäf'thal·məs }

Enopla [INV ZOO] A class or subclass of ribbonlike worms of the phylum Rhynchocoela. { e'näp·ē·ə }

Enoplia [INV ZOO] A subclass of nematodes in the class Adenophorea. { e'näp·lē·ə }

Enoplida [INV ZOO] An order of nematodes in the subclass Enoplia. { e'näp·lə·də }

Enoplidae [INV ZOO] A family of free-living marine nematodes in the superfamily Enoploidea, characterized by a complex arrangement of teeth and mandibles. { e'näp·ləˌdē }

Enoploidea [INV ZOO] A superfamily of small to very large free-living marine nematodes having pocketlike amphids opening to the exterior via slitlike apertures. { e·nə'plóid·ē·ə }

Enoploteuthidae [INV ZOO] A molluscan family of deep-sea squids in the class Cephalopoda. { eˌnäp·lə'tü·thəˌdē }

E notation [COMPUT SCI] A type of scientific notation in which the phrase "times 10 to the power of" is replaced by the letter E; for example, 3.1×10^7 is written 3.1E+7 and 5.1×10^{-9} is written 5.1E−9. { 'ē nōˌtā·shən }

enphytotic [PL PATH] **1.** A disease that occurs regularly among plants of a specific region. **2.** An outbreak of such a disease. { ˌen·fī'täd·ik }

enqueue [ENG] To place a data item in a queue. { en'kyü }

enquiry character [COMPUT SCI] A control character used to request a response from receiving equipment. { in'kwīr·ē ˌkar·ik·tər }

enriched gas [MATER] A motor gasoline containing additives to improve combustion characteristics, as by limiting knock. { in'richt 'gas }

enriched material [NUCLEO] Material in which the amount of one or more isotopes has been increased above that occurring in nature, such as uranium in which the abundance of ^{235}U is increased. { in'richt mə'tir·ē·əl }

enriched reactor [NUCLEO] A nuclear reactor in which the fuel is an enriched material. { in'richt rē'ak·tər }

enriching column [CHEM ENG] The portion of a countercurrent contractor (liquid-liquid extraction or vapor-liquid distillation) above the feed point in which an upward-moving, product-rich stream from the stripping column is further purified by countercurrent contact with a downward-flowing reflux stream from the overhead product-recovery vessel. { in'rich·iŋ ˌkäl·əm }

enrichment [NUCLEO] A process that changes the isotopic ratio in a material; for uranium, for example, the ratio of ^{235}U to ^{238}U may be increased by gaseous diffusion of uranium hexafluoride. { in'rich·mənt }

enrichment culture [MICROBIO] A medium of known composition and specific conditions of incubation which favors the growth of a particular type or species of bacterium. { in'rich·mənt ˌkəl·chər }

enrichment factor [NUCLEO] The ratio of the abundance of a particular isotope in an enriched material to its abundance in the original material. { in'rich·mənt ˌfak·tər }

enrichment medium [MICROBIO] A liquid cultural medium of a given composition which permits preferential emergence of certain organisms that initially may have made up a relatively minute proportion of a mixed inoculum. { in'rich·mənt ˌmē·dē·əm }

enrockment [CIV ENG] A grouping of large stones dropped into water to form a base, such as for supporting a pier. { in'räk·mənt }

enroute chart [NAV] A chart of air routes in specific areas that shows the exact location of electronic aids to navigation, such as radio direction-finder stations, radio and radar beacons, and radio-range stations. Formerly known as radio facility chart. { en'rüt ˌchärt }

enroute turning area [NAV] An area of specified dimensions used by an aircraft when a change of course is necessary; the area extends beyond the primary and secondary obstruction clearance areas and is provided with adequate protection; turning area criteria supplement the basic airways and route segment criteria to protect the aircraft. Also known as turning area. { en'rüt 'tərn·iŋ ˌer·ē·ə }

ensemble [STAT MECH] A collection of systems of particles used to describe an individual system; time averages of quantities describing the individual system are found by averaging over the systems in the ensemble at a fixed time. { än'säm·bəl }

ensialic geosyncline [GEOL] A geosyncline whose geosynclinal prism accumulates on a sialic crust and contains clastics. { en·sē'al·ik jē·ō'sinˌklīn }

ensiform [BIOL] Sword-shaped. { 'en·səˌfórm }

ensiling [AGR] The anaerobic fermentation process used to preserve immature green corn, legumes, grasses, and grain plants; the crop is chopped and packed while at about 70–80% moisture and put into silos or other containers to exclude air. { en'sīl·iŋ }

ensimatic geosyncline [GEOL] A geosyncline whose geosynclinal prism accumulates on a simatic crust and is composed largely of volcanic rock or sediments of volcanic debris. { en·sə'mad·ik jē·ō'sinˌklīn }

Enskog theory *See* Chapman-Enskog theory. { 'enˌskäg ˌthē·ə·rē }

ENSO *See* El Niño Southern Oscillation. { 'enˌsō }

ensonification field [OCEANOGR] The area of the sea floor that is acoustically imaged in the course of a sonar survey. { enˌsän·ə·fə'kā·shən ˌfēld }

ensonify [ACOUS] To fill the ocean or any fluid medium with acoustic radiation, which is then observed and analyzed to study the medium or to locate or image objects within it. Also spelled insonify. { en'sän·i·fī }

enstatite [MINERAL] $MgOSiO_2$ A member of the pyroxene mineral group that crystallizes in the orthorhombic system; usually yellowish gray but becomes green when a little iron is present. { 'en·stəˌtīt }

enstatite chondrite [GEOL] A type of chrondritic meteorite consisting almost entirely of enstatite, with metal inclusions that may be abundant and are usually low in nickel. { 'en·stəˌtīt 'känˌdrīt }

entablature [ARCH] A unit consisting of the architrave, frieze, and cornice of a wall. { en'tab·lə·chər }

Entamoeba [INV ZOO] A genus of parasite amebas in the family Endamoebidae, including some species of the genus *Endamoeba* which are parasites of humans and other vertebrates. { ˌent·ə'mē·bə }

entanglement [ORD] An obstacle, such as barbed wire, that is utilized to stop or hamper the forward movement of troops. { in'taŋ·gəl·mənt }

entasis [ARCH] The slight swelling visible in the profile of a column, used to correct the visual distortion that makes a straight column seem to have a concave profile. { 'en·tə·səs }

Enteletacea [PALEON] A group of extinct articulate brachiopods in the order Orthida. { ‚en·tə·lə′tās·ē·ə }

Entelodontidae [PALEON] A family of extinct palaeodont artiodactyls in the superfamily Entelodontoidea. { ‚en·tə·lə′dän·tə‚dē }

Entelodontoidea [PALEON] A superfamily of extinct piglike mammals in the suborder Palaeodonta having huge skulls and enlarged incisors. { ‚en·tə·lə‚dän′tôid·ē·ə }

enteralgia [MED] Pain in the intestine. { ‚en·tə′ral·jē·ə }

enterectomy [MED] Excision of a part of the intestine. { ‚en·tə′rek·tə·mē }

enteric bacilli [MICROBIO] Microorganisms, especially the gram-negative rods, found in the intestinal tract of humans and animals. { en′ter·ik bə′sil·ī }

enteric cytopathogenic human orphan virus See echovirus. { en′ter·ik ‚sī·dō‚path·ə′jen·ik ‚yü·mən ′ôr·fən ‚vī·rəs }

entering angle [MECH ENG] The angle between the side-cutting edge of a tool and the machined surface of the work; angle is 90° for a tool with 0° side-cutting edge angle effective. { ′ent·ə·riŋ ‚aŋ·gəl }

entering group [ORG CHEM] An atom or group that becomes bonded to the main portion of the substrate during a chemical reaction. { ′ent·ə·riŋ ‚grüp }

enteritis [MED] Inflammation of the intestinal tract. { ‚ent·ə′rīd·əs }

enter key [COMPUT SCI] A key on a computer keyboard that corresponds to the return key on a typewriter and usually signals the computer to act on the information just entered on the keyboard. { ′en·tər ‚kē }

Enterobacter [MICROBIO] A genus of bacteria in the family Enterobacteriaceae; motile rods found in the intestine of humans and other animals; some strains are encapsulated. { ‚ent·ə·rō′bak·tər }

Enterobacteriaceae [MICROBIO] A family of gram-negative, facultatively anaerobic rods; cells are nonsporeforming and may be nonmotile or motile with peritrichous flagella; includes important human and plant pathogens. { ‚ent·ə·rō‚bak·tir·ē′ās·ē‚ē }

enterobiasis [MED] Infestation of the intestinal tract of humans with the nematode *Enterobius vermacularis* (pinworm); characterized by mild enteritis. { ‚ent·ə·rō′bī·ə·səs }

enterococci [MICROBIO] Spherical bacteria in short chains. { ‚en·tə·rə′käk·ē }

enterocoel [ZOO] A coelom that arises by mesodermal outpocketing of the archenteron. { ′ent·ə·rō‚sēl }

Enterocoela [SYST] A section of the animal kingdom that includes the Echinodermata, Chaetognatha, Hemichordata, and Chordata. { ‚ent·ə·rō′sēl·ə }

enterocolitis [MED] Inflammation of the small intestine and colon. { ‚ent·ə·rō·kə′līd·əs }

enterocyte [HISTOL] A cell that lines the intestinal wall. { ′en·tə·rə‚sit }

enterohydrocoel [INV ZOO] In crinoids, an anterior cavity derived from the archenteron. { ‚ent·ə·rō′hī·drə‚sēl }

enterokinase [BIOCHEM] An enzyme which catalyzes the conversion of trypsinogen to trypsin. { ‚ent·ə·rō′kī‚nās }

enterolith [PATH] A concretion formed in the intestine. { ′ent·ə·rō‚lith }

enterolithic [GEOL] Of or pertaining to structures, such as small folds, formed in evaporites due to flowage or hydration. { ‚ent·ə·rə′lith·ik }

enteron [ANAT] The alimentary canal. { ′ent·ə‚rän }

enteropathy [MED] Disease of the intestine. { ‚en·tə′rap·ə‚thē }

Enteropneusta [INV ZOO] The acorn worms or tongue worms, a class of the Hemichordata; free-living solitary animals with no exoskeleton and with numerous gill slits and a straight gut. { ‚ent·ə·rə′nüs·tə }

enteroptosis See visceroptosis. { ‚ent·ə·räp′tō·səs }

enterorrhagia [MED] Intestinal hemorrhage. { ‚ent·ə·rō′rāj·ē·ə }

enterotoxin [MICROBIO] A toxin produced by *Micrococcus pyogenes* var. *aureus* (*Staphylococcus aureus*) which gives rise to symptoms of food poisoning in humans and monkeys. { ‚ent·ə·rō′täk·sən }

enterovirus [VIROL] One of the two subgroups of human picornaviruses; includes the polioviruses, the coxsackieviruses, and the echoviruses. { ‚ent·ə·rō′vī·rəs }

Enterozoa [ZOO] Animals with a digestive tract or cavity; includes all animals except Protozoa, Mesozoa, and Parazoa. { ‚ent·ə·rə′zō·ə }

enthalpimetric analysis [ANALY CHEM] Generic designation for a group of modern thermochemical methodologies such as thermometric enthalpy titrations which rely on monitoring the temperature changes produced in adiabatic calorimeters by heats of reaction occurring in solution; in contradistinction, classical methods of thermoanalysis such as thermogravimetry focus primarily on changes occurring in solid samples in response to externally imposed programmed alterations in temperature. { en‚thal·pə′me·trik ə′nal·ə·səs }

enthalpy [THERMO] The sum of the internal energy of a system plus the product of the system's volume multiplied by the pressure exerted on the system by its surroundings. Also known as heat content; sensible heat; total heat. { en′thal·pē }

enthalpy-entropy chart [THERMO] A graph of the enthalpy of a substance versus its entropy at various values of temperature, pressure, or specific volume; useful in making calculations about a machine or process in which this substance is the working medium. { en‚thal·pē ′en·trə·pē ‚chärt }

enthalpy of reaction [PHYS CHEM] The change in enthalpy accompanying a chemical reaction. { en′thal·pē əv rē′ak·shən }

enthalpy of transition [PHYS CHEM] The change of enthalpy accompanying a phase transition. { en′thal·pē əv tran′zish·ən }

enthalpy of vaporization See heat of vaporization. { en′thal·pē əv ‚vā·pə·rə′zā·shən }

enthalpy-pressure chart See pressure-enthalpy chart. { en‚thal·pē ′presh·ər ‚chärt }

enthalpy titration See thermometric titration. { en′thal·pē tī′trā·shən }

entire [BIOL] Having a continuous, unimpaired margin. { en′tīr }

entire function [MATH] A function of a complex variable which is analytic throughout the entire complex plane. Also known as integral function. { en‚tīr ′fəŋk·shən }

entire ring See integral domain. { en‚tīr ′riŋ }

entire series [MATH] A power series which converges for all values of its variable; a power series with an infinite radius of convergence. { en‚tīr ′sir·ēz }

entire surd [MATH] A surd that does not contain a rational factor or term. { en‚tīr ′sərd }

Entisol [GEOL] An order of soil having few or faint horizons. { ′ent·ə‚sól }

entity See record. { ′ent·ə·tē }

entity type [COMPUT SCI] A particular kind of file in a database, such as an employee, customer, or product file. { ′ent·ə·tē ‚tīp }

Entner-Doudoroff pathway [BIOCHEM] A sequence of reactions for glucose degradation, with the liberation of energy; the distinguishing feature is the formation of 2-keto-3-deoxy-6-phosphogluconate from 6-phosphogluconate and the cleaving of this compound to yield pyruvate and glyceraldehyde-3-phosphate. { ‚ent·nər ‚dō·də‚róf ′path‚wā }

entoblast [EMBRYO] A blastomere that differentiates into endoderm. { ′ent·ə‚blast }

entoderm See endoderm. { ′ent·ə‚dərm }

Entodiniomorphida [INV ZOO] An order of highly evolved ciliated protozoans in the subclass Spirotrichia, characterized by a smooth, firm pellicle and the lack of external ciliature. { ‚ent·ə‚dī·nē·ə′mór·fə·də }

entolecithal See endolecithal. { ‚ent·ə′les·ə·thəl }

entombment [NUCLEO] A method of decommissioning a nuclear facility in which radioactive contamination is made inaccessible by demolition techniques and then the residue is covered with reinforced concrete. { en′tüm·mənt }

Entomoconchacea [PALEON] A superfamily of extinct marine ostracods in the suborder Myodocopa that are without a rostrum above the permanent aperture. { ‚ent·ə‚mō‚kän′kās·ē·ə }

entomogenous [BIOL] Growing on or in an insect body, as certain fungi. { ‚ent·ə′mäj·ə·nəs }

entomology [INV ZOO] A branch of the biological sciences that deals with the study of insects. { ‚ent·ə′mäl·ə·jē }

entomophagous [ZOO] Feeding on insects. { ‚ent·ə′mäf·ə·gəs }

entomophilic fungi [MYCOL] Fungi that parasitize insects. { ‚en·tə·mə‚fil·ik ′fən·jī }

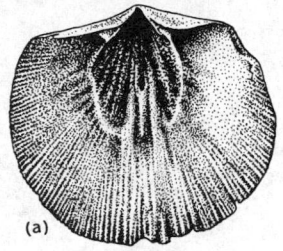

ENTELETACEA

Pionodema, (a) pedicle and (b) brachial valve interiors. (From R. C. Moore, ed., Treatise on Invertebrate Paleontology, pt. H, Geological Society of America and University of Kansas Press, 1965)

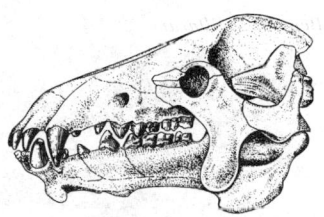

ENTELODONTIDAE

Archaeothesium, Oligocene entelodont, skull length about 24 inches (60 centimeters).

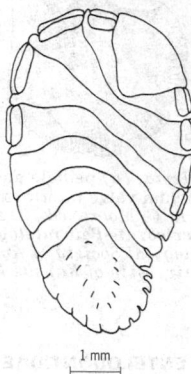

ENTONISCIDAE

Bopyrus squillarum.

entomophilous [ECOL] Pollinated by insects. { ‚ent·ə'mäf·ə·ləs }

entomophobia [PSYCH] An abnormal fear of insects. { ‚ent·ə‚mə'fō·bē·ə }

Entomophthoraceae [MYCOL] The single family of the order Entomophthorales. { ‚ent·ə‚mäf·thə'rās·ē‚; ame }

Entomophthorales [MYCOL] An order of mainly terrestrial fungi in the class Phycomycetes having a hyphal thallus and nonmotile sporangiospores, or conidia. { ‚ent·ə‚mäf·thə'rā·lēz }

Entomostraca [INV ZOO] A group of Crustacea comprising the orders Cephalocarida, Branchiopoda, Ostracoda, Copepoda, Branchiura, and Cirripedia. { ‚ent·ə'mä·strə·kə }

Entoniscidae [INV ZOO] A family of isopod crustaceans in the tribe Bopyrina that are parasitic in the visceral cavity of crabs and porcellanids. { ‚ent·ə'nis·ə‚dē }

entoplastron [VERT ZOO] The anterior median bony plate of the plastron of chelonians. { ‚en·tō‚plas‚trän }

Entoprocta [INV ZOO] A group of bryozoans, sometimes considered to be a subphylum, having a pseudocoelomate visceral cavity and the anus opening inside the circlet of tentacles. { ‚ent·ə'präk·tə }

entrail pahoehoe [GEOL] A type of pahoehoe having a surface that resembles an intertwined mass of entrails. { 'en‚trāl pə'hō·ē‚hō·ē }

entrained fluid [FL MECH] Fluid in the form of mist, fog, or droplets that is carried out of a column or vessel by a rising gas or vapor stream. { en'trānd 'flü·əd }

entrainer [CHEM ENG] An additive that forms an azeotrope with one component of a liquid mixture to aid in otherwise difficult separations by distillation, as in azeotropic distillation. { en'trān·ər }

entrainment [CHEM ENG] A process in which the liquid boils so violently that suspended droplets of liquid are carried in the escaping vapor. [HYD] The pickup and movement of sediment as bed load or in suspension by current flow. [METEOROL] The mixing of environmental air into a preexisting organized air current so that the environmental air becomes part of the current. [OCEANOGR] The transfer of fluid by friction from one water mass to another, usually occurring between currents moving in respect to each other. { en'trān·mənt }

entrance [CIV ENG] The seaward end of a channel, harbor, and so on. [COMPUT SCI] The location of a program or subroutine at which execution is to start. Also known as entry point. [ENG] A place of physical entering, such as a door or passage. [NAV ARCH] The part of a ship's underwater hull which is forward of the amidships. { 'en·trəns }

entrance angle [ENG] In molding, the maximum angle, measured from the center line of the mandrel, at which molten material enters the land area of a die. { 'en·trəns ‚aŋ·gəl }

entrance cable [ELEC] Cable that brings power from an outside power line into a building. { 'en·trəns ‚kā·bəl }

entrance lock [CIV ENG] A lock between the tideway and an enclosed basin made necessary because the levels of the two bodies of water vary; by means of this lock, vessels can pass either way at all states of the tide. Also known as guard lock; tidal lock; tide lock. { 'en·trəns ‚läk }

entrance loss [FL MECH] Energy loss by friction in a flowing fluid when it moves into a cross-sectional area of sudden contraction, as at the entrance of a pipe or a suddenly reduced area of a duct. { 'en·trəns ‚lòs }

entrance pupil [OPTICS] The image of the aperture stop of an optical system formed in the object space by rays emanating from a point on the optical axis in the image space. { 'en·trəns ‚pyü·pəl }

entrance region [METEOROL] The region of confluence at the upwind extremity of a jet stream. { 'en·trəns ‚rē·jən }

entrance slit [SPECT] Narrow slit through which passes the light entering a spectrometer. { 'en·trəns ‚slit }

entrapment [GEOL] The underground trapping of oil or gas reserves by folds, faults, domes, asphaltic seals, unconformities, and such. { en'trap·mənt }

entrenched meander [HYD] A deepened meander of a river which is carried downward further below the valley surface in which the meander originally formed. Also known as inherited meander. { en'trencht mē'an·dər }

entrenched stream [HYD] A stream that flows in a valley or narrow trench cut into a plain or relatively level upland. Also spelled intrenched stream. { en'trencht 'strēm }

entropy [COMMUN] A measure of the absence of information about a situation, or, equivalently, the uncertainty associated with the nature of a situation. [MATH] In a mathematical context, this concept is attached to dynamical systems, transformations between measure spaces, or systems of events with probabilities; it expresses the amount of disorder inherent or produced. [STAT MECH] Measure of the disorder of a system, equal to the Boltzmann constant times the natural logarithm of the number of microscopic states corresponding to the thermodynamic state of the system; this statistical-mechanical definition can be shown to be equivalent to the thermodynamic definition. [THERMO] Function of the state of a thermodynamic system whose change in any differential reversible process is equal to the heat absorbed by the system from its surroundings divided by the absolute temperature of the system. Also known as thermal charge. { 'en·trə·pē }

entropy of activation [PHYS CHEM] The difference in entropy between the activated complex in a chemical reaction and the reactants. { 'en·trə·pē əv‚ak·tə'vā·shən }

entropy of a partition [MATH] If ξ is a finite partition of a probability space, the entropy of ξ is the negative of the sum of the products of the probabilities of elements in ξ with the logarithm of the probability of the element. { 'en·trə·pē əv ə pär'tish·ən }

entropy of a transformation *See* Kolmogorov-Sinai invariant. { 'en·trə·pē əv ə tranz·fər'mā·shən }

entropy of a transformation given a partition [MATH] If T is a measure preserving transformation on a probability space and ξ is a finite partition of the space, the entropy of T given ξ is the limit as $n \rightarrow \infty$ of $1/n$ times the entropy of the partition which is the common refinement of ξ, $T^{-1}\xi$, ..., $T^{-n+1}\xi$. { 'en·trə·pē əv ə tranz·fər'mā·shən 'giv·ən ə pär'tish·ən }

entropy of mixing [PHYS CHEM] After mixing substances, the difference between the entropy of the mixture and the sum of the entropies of the components of the mixture. { 'en·trə·pē əv 'mik·siŋ }

entropy of transition [PHYS CHEM] The heat absorbed or liberated in a phase change divided by the absolute temperature at which the change occurs. { 'en·trə·pē əv tran'zish·ən }

entry [COMPUT SCI] Input data fed during the execution of a program by means of a terminal. { 'en·trē }

entry ballistics [MECH] That branch of ballistics which pertains to the entry of a missile, spacecraft, or other object from outer space into and through an atmosphere. { 'en·trē bə‚lis·tiks }

entry block [COMPUT SCI] The area of main memory reserved for the data which will be introduced at execution time. { 'en·trē ‚bläk }

entry condition [COMPUT SCI] A requirement that must be met before a program or routine can be entered by a computer program. Also known as initial condition. { 'en·trē kən‚dish·ən }

entry corridor [AERO ENG] Depth of the region between two trajectories which define the design limits of a vehicle about to enter a planetary atmosphere, or define the desired landing area (footprint). { 'en·trē ‚kär·ə‚dór }

entry instruction [COMPUT SCI] The first instruction to be executed in a subroutine. { 'en·trē in‚strək·shən }

entry point *See* entrance. { 'en·trē ‚póint }

entry portion [COMPUT SCI] The right-hand portion of a decision table, which comprises the condition entries and action entries, and whose columns are the decision rules. { 'en·trē ‚pór·shən }

entry site [CELL MOL] The ribosome site available for initial binding of transfer ribonucleic acid during genetic translation. { 'en·trē ‚sīt }

entry sorting [COMPUT SCI] A method of internal sorting in which records or blocks of records are placed, one at a time, in a buffer area and then integrated into the sorted list before the next record is placed in the buffer. { 'en·trē ‚sórd·iŋ }

entypy [EMBRYO] The formation of the amnion in certain mammals by the invagination of the embryonic knob into the yolk sac, without the formation of any amniotic folds. { 'ent·ə‚pē }

enucleate [CELL MOL] To remove the nucleus from a cell. [MED] To remove an organ or a tumor in its entirety, as an eye from its socket. { ē'nü·klē‚āt }

enumerable *See* countable.   { ē'nüm·rə·bəl }

enuresis [MED] Urinary incontinence, especially in the absence of organic cause.   { ,en·yə'rē·səs }

envelope [COMMUN] A curve drawn to pass through the peaks of a graph, such as that of a moduated radio-frequency carrier signal. [CELL MOL] The sum of all cell-surface elements that are located outside the plasma membrane. [ENG] The glass or metal housing of an electron tube or the glass housing of an incandescent lamp. [MATH] **1.** The envelope of a one-parameter family of curves is a curve which has a common tangent with each member of the family. **2.** The envelope of a one-parameter family of surfaces is the surface swept out by the characteristic curves of the family. [VIROL] The outer membranous lipoprotein coat of certain viruses. Also known as bulb.   { 'en·və,lōp }

envelope delay [COMMUN] The time required for the envelope of a wave to travel between two points in a system.   { 'en·və,lōp di,lā }

envelope delay distortion *See* delay distortion.   { 'en·və,lōp di,lā di'stòr·shən }

envelope detector *See* detector.   { 'en·və,lōp di,tek·tər }

envelope orography [METEOROL] A method for developing a numerical model for weather forecasting in which it is assumed that mountain passes and valleys are filled mostly with stagnant air, thus increasing the average height of the model mountains and enhancing the blocking effect.   { 'en·və,lōp ò'räg·rə·fē }

envelope soliton [PHYS] A rapidly oscillating wave that propagates with a characteristic constant shape, and can be pictured as cut off by a smoothly modulating envelope.   { 'en·və,lōp 'säl·ə,tän }

envelopmental sound [ACOUS] The portion of the room impulse response consisting of sound that arrives at a listener's location between 20 and 150 milliseconds after the first direct sound, and which has been reflected against walls and ceilings relatively few times.   { in,vel·əp,ment·əl 'saùnd }

envenomation [MATER] The process by which the surface of a plastic close to or in contact with another surface is deteriorated.   { in,ven·ə'mā·shən }

environment [COMPUT SCI] The computer system in which an applications program is running, including the hardware and system software. [ECOL] The sum of all external conditions and influences affecting the development and life of organisms. [ENG] The aggregate of all natural, operational, or other conditions that affect the operation of equipment or components. [PHYS] The aggregate of all the conditions and the influences that determine the behavior of a physical system.   { in'vī·ərn·mənt *or* in'vī·rən·ment }

environmental biology *See* ecology.   { in,vī·ərn,ment·əl bī'äl·ə·jē }

environmental cab [ENG] Operator's compartment in earthmovers equipped with tinted safety glass, soundproofing, air conditioning, and cleaning units.   { in,vī·ərn,ment·əl 'kab }

environmental control [ENG] Modification and control of soil, water, and air environments of humans and other living organisms.   { in,vī·ərn,mənt·əl kən'trōl }

environmental control system [ENG] A system used in a closed area, especially a spacecraft or submarine, to permit life to be sustained; the system provides the occupants with a suitably controlled atmosphere to permit them to live and work in the area.   { in,vī·ərn,mənt·əl kən'trōl ,sis·təm }

environmental engineering [ENG] The technology concerned with the reduction of pollution, contamination, and deterioration of the surroundings in which humans live.   { in,vī·ərn,mənt·əl en·jə'nir·iŋ }

environmental fluid mechanics [FL MECH] The study of the flows of air and water, of the species carried by them (especially pollution), and of their interactions with geological, biological, social, and engineering systems in the vicinity of a planet's surface.   { in,vī·ərn,ment·əl ,flü·əd mi'kan·iks }

environmental impact analysis [IND ENG] Predetermination of the extent of pollution or environmental degradation which will be involved in a mining or processing project.   { in,vī·ərn,mənt·əl 'im,pakt ə,nal·ə·səs }

environmental impact statement [ENG] A report of the potential effect of plans for land use in terms of the environmental, engineering, esthetic, and economic aspects of the proposed objective.   { in,vī·ərn,mənt·əl 'im,pakt ,stāt·mənt }

environmental lapse rate [METEOROL] The rate of decrease of temperature with elevation in the atmosphere. Also known as atmospheric lapse rate.   { in,vī·ərn,mənt·əl 'laps ,rāt }

environmental pathology [MED] A branch of pathology concerned with abiotic environmental agents that influence human health.   { in,vī·ərn,ment·əl pa'thäl·ə·jē }

environmental protection [ENG] The protection of humans and equipment against stresses of climate and other elements of the environment.   { in,vī·ərn,ment·əl prə'tek·shən }

environmental radioactivity [NUCLEO] Radioactivity that originates from natural and anthropogenic sources.   { in,vī·ərn,ment·əl ,rād·ē,ō·ak'tiv·əd·ē }

environmental range [ENG] The range of environment throughout which a system or portion thereof is capable of operation at not less than the specified level of reliability.   { in,vī·ərn,mənt·əl 'rānj }

environmental resistance [ECOL] The effect of physical and biological factors in preventing a species from reproducing at its maximum rate.   { in,vī·ərn,men·təl ri'zis·təns }

environmental stress cracking [MECH] The susceptibility of a material to crack or craze in the presence of surface-active agents or other factors.   { in,vī·ərn,mənt·əl 'stres ,krak·iŋ }

environmental test [ENG] A laboratory test conducted to determine the functional performance of a component or system under conditions that simulate the real environment in which the component or system is expected to operate.   { in,vī·ərn,mənt·əl 'test }

environmental toxicology [MED] A broad field of study encompassing the production, fate, and effects of natural and synthetic pollutants in the environment.   { in,vī·ərn,ment·əl ,täk·sə'käl·ə·jē }

environmental variance [GEN] That portion of the phenotypic variance caused by differences in the environments to which the individuals in a population have been exposed.   { in,vī·ərn,ment·əl 'ver·ē·əns }

environment division [COMPUT SCI] The section of a program written in COBOL which defines the hardware and files to be used by the program.   { in,vī·ərn,mənt di'vizh·ən }

environment of sedimentation [GEOL] A more or less destructive geomorphologic setting in which sediments are deposited as beach environment.   { in,vī·ərn,mənt əv ,sed·ə·men'tā·shən }

environment pointer [COMPUT SCI] **1.** A component of a task descriptor that designates where the instructions and data code for the task are located. **2.** A control component element belonging to the stack model of block structure execution that points to the current environment.   { in,vī·ərn,mənt ,pòint·ər }

environment simulator [ENG] Any machine or artificial device that simulates all or some of the attributes of an environment, such as the solar simulators with artificial suns used in testing spacecraft.   { in,vī·ərn,mənt 'sim·yə,lād·ər }

enzootic [VET MED] **1.** A disease affecting animals in a limited geographic region. **2.** Pertaining to such a disease.   { ,en·zō'äd·ik }

enzyme [BIOCHEM] Any of a group of catalytic proteins that are produced by living cells and that mediate and promote the chemical processes of life without themselves being altered or destroyed.   { 'en,zīm }

enzyme induction [MICROBIO] The process by which a microbial cell synthesizes an enzyme in response to the presence of a substrate or of a substance closely related to a substrate in the medium.   { 'en,zīm in'dək·shən }

enzyme inhibition [BIOCHEM] Prevention of an enzymic process as a result of the interaction of some substance with the enzyme so as to decrease the rate of reaction.   { 'en,zīm ,in·ə·'bish·ən }

enzyme-linked immunosorbent assay [MED] A laboratory technique in which a monoclonal antibody conjugated to an enzyme is used to rapidly detect and quantify the presence of an antigen in a sample. Abbreviated ELISA.   { ,en·zīm ,liŋkt ,im·yə·nə,sòr·bənt 'a,sā }

enzyme-linked receptor [CELL MOL] The type of cell-surface receptor having an intracellular domain that either functions as an enzyme or is enzyme-associated, which is stimulated upon binding of a ligand to the receptor.   { 'en,zīm ,liŋkt ri,sep·tər }

enzyme repression [BIOCHEM] The process by which the rate of synthesis of an enzyme is reduced in the presence of a metabolite, often the end product of a chain of reactions in

ENVELOPE SOLITON

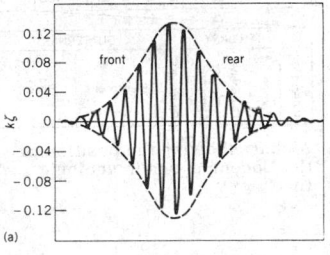

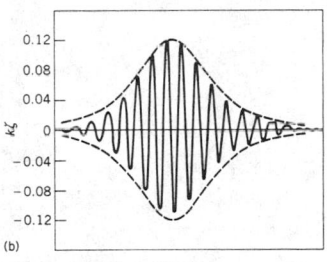

Displacement of the water surface showing the evolution of an envelope soliton at *(a)* 20 feet (6 meters) and *(b)* 98 feet (30 meters) downstream of the maker. The quantity $k\zeta$ measures the displacement in units of wavelength.

which the enzyme in question operates near the beginning. { 'en‚zīm ri'presh·ən }

enzyme unit [BIOCHEM] The amount of an enzyme that will catalyze the transformation of 10^{-6} mole of substrate per minute or, when more than one bond of each substrate is attacked, 10^{-6} of 1 gram equivalent of the group concerned, under specified conditions of temperature, substrate concentration, and pH number. { 'en‚zīm ‚yü·nət }

enzymology [BIOCHEM] A branch of science dealing with the chemical nature, biological activity, and biological significance of enzymes. { ‚en·zə'mäl·ə·jē }

Eocambrian [GEOL] Pertaining to the thick sequences of strata conformably underlying Lower Cambrian fossils. Also known as Infracambrian. { ‚ē·ō'kam·brē·ən }

Eocanthocephala [INV ZOO] An order of the Acanthocephala characterized by the presence of a small number of giant subcuticular nuclei. { ¦ē·ō¦kan·thō'sef·ə·lə }

Eocene [GEOL] The next to the oldest of the five major epochs of the Tertiary period (in the Cenozoic era). { 'ē·ə‚sēn }

Eocrinoidea [PALEON] A class of extinct echinoderms in the subphylum Crinozoa that had biserial brachioles like those of cystoids combined with a theca like that of crinoids. { ‚ē·ō·krə'nóid·ē·ə }

EOD See explosive-ordnance disposal.

EOF gap See end-of-file gap. { ¦ē¦ō'ef ‚gap }

Eogene See Paleogene. { 'ē·ə‚jēn }

Eohippus [PALEON] The earliest, primitive horse, included in the genus *Hyracotherium*; described as a small, four-toed species. { ‚ē·ō'hip·əs }

eolation [GEOL] Any action of wind on the land. { ‚ē·ə'lā·shən }

eolian [METEOROL] Pertaining to the action or the effect of the wind, as in eolian sounds or eolian deposits (of dust). Also spelled aeolian. { ē'ōl·yən }

eolian anemometer [ENG] An anemometer which works on the principle that the pitch of the eolian tones made by air moving past an obstacle is a function of the speed of the air. { ē'ōl·yən an·ə'mäm·əd·ər }

eolian dune [GEOL] A dune resulting from entrainment of grains by the flow of moving air. { ē'ōl·yən 'dün }

eolian erosion [GEOL] Erosion due to the action of wind. { ē'ōl·yən ə'rō·zhən }

eolianite [GEOL] A sedimentary rock consisting of clastic material which has been deposited by wind. { ē'ōl·yə‚nīt }

eolian ripple mark [GEOL] A mark made in sand by the wind. { ē'ōl·yən 'rip·əl ‚märk }

eolian sand [GEOL] Deposits of sand arranged by the wind. { ē'ōl·yən 'sand }

eolian soil [GEOL] A type of soil ranging from sand dunes to loess deposits whose particles are predominantly of silt size. { ē'ōl·yən 'sóil }

eolian sounds [ACOUS] Sounds produced by eddying motions of air in the lee of obstacles, such as wires, twigs, and even the ear itself, when wind blows over those obstacles. { ē'ōl·yən 'saúnz }

eolotropy See anisotropy. { ‚ē·ə'lä·trə·pē }

Eomoropidae [PALEON] A family of extinct perissodactyl mammals in the superfamily Chalicotherioidea. { ‚ē·ō·mə'räp·ə‚dē }

eon [MECH] A unit of time, equal to 10^9 years. { 'ē‚än }

eonothem [GEOL] A chronostratigraphic unit, above erathem, composed of rocks formed during an eon of geologic time. { 'ēn·ə‚them }

EOR See explosive-ordnance reconnaissance.

EORA See explosive-ordnance reconnaissance agent.

EOR gap See end-of-record gap. { ¦ē¦ō'är ‚gap }

Eosentomidae [INV ZOO] A family of primitive wingless insects in the order Protura that possess spiracles and tracheae. { ¦ē·ō‚sen'täm·ə‚dē }

eosin [ORG CHEM] $C_{20}H_8O_5Br_4$ **1.** A red fluorescent dye in the form of triclinic crystals that are insoluble in water; used chiefly in cosmetics and as a toner. Also known as bromeosin; bromo acid; eosine; tetrabromofluorescein. **2.** The red to brown crystalline sodium or potassium salt of this dye; used in organic pigments, as a biological stain, and in pharmaceuticals. { 'ē·ə·sən }

eosinophil [HISTOL] A granular leukocyte having cytoplasmic granules that stain with acid dyes and a nucleus with two lobes connected by a thin thread of chromatin. { ‚ē·ə'sin·ə‚fil }

eosinophil chemotactic factor [IMMUNOL] A peptide released from mast cell granules that stimulates chemotaxis of eosinophils; may be responsible for accumulation of eosinophils at sites of inflammation and allergic reactions. Abbreviated ECF. { ‚ē·ə¦sin·ə·fil ¦kē·mō¦tak·tik 'fak·tər }

eosinophilia [MED] A greater than average number of circulating eosinophils. Also known as acidophilia; oxyphilia. { ‚e·ə‚sin·ə'fil·ē·ə }

eosinophilic erythroblast See normoblast. { ¦ē·ə¦sin·ə¦fil·ik ə'rith·rə‚blast }

eosinophilic granuloma [MED] A disease, principally of childhood, characterized by foci of bone inflammation and granulation containing lipids, mononuclear cells, and eosinophils. { ¦ē·ə¦sin·ə¦fil·ik gran·yə'lō·mə }

eosinophilic pneumonitis See Loeffler's syndrome. { ¦ē·ə¦sin·ə¦fil·ik nü·mə'nīd·əs }

eosphorite [MINERAL] $(Mn,Fe)Al(PO_4)(OH)_2 \cdot H_2O$ A usually rose-pink mineral composed of hydrous aluminum manganese phosphate, found massive or in prismatic crystals. { ē'äs·ə·fə‚rīt }

Eosuchia [PALEON] The oldest, most primitive, and only extinct order of lepidosaurian reptiles. { ‚ē·ō'sü·kē·ə }

eötvös [GEOPHYS] A unit of horizontal gradient of gravitational acceleration, equal to a change in gravitational acceleration of 10^{-9} galileo over a horizontal distance of 1 centimeter. { 'ət·vəsh }

Eötvös constant [PHYS] A constant that appears in an expression for the behavior of the surface tension γ of a liquid as the temperature T drops to a critical temperature T_c at which the surface tension disappears, equal to $\gamma(M/\rho)^{2/3}/(T_c - T)$, where M is the molecular weight and ρ the density of the liquid. { 'ət·vəsh ‚kän·stənt }

Eötvös effect [MECH] An apparent decrease (or increase) in the weight of a body moving from west to east (or east to west) because of its greater (or smaller) centrifugal acceleration. { 'ət·vəsh i‚fekt }

Eötvös experiment [RELAT] An experiment which tests the equality of inertial mass and gravitational mass by balancing on a given body the earth's gravitational attraction against the kinetic reaction arising from the rotation of the earth. { 'ət·vəsh ik‚sper·ə·mənt }

Eötvös number See Bond number. { 'ət·vəsh ‚nəm·bər }

Eötvös rule [THERMO] The rule that the rate of change of molar surface energy with temperature is a constant for all liquids; deviations are encountered in practice. { 'ət·vəsh ‚rül }

Eötvös torsion balance [ENG] An instrument which records the change in the acceleration of gravity over the horizontal distance between the ends of a beam; used to measure density variations of subsurface rocks. { ¦ət·vəsh 'tór·shən ‚bal·əns }

Epacridaceae [BOT] A family of dicotyledonous plants in the order Ericales, distinguished by palmately veined leaves, and stamens equal in number with the corolla lobes. { ‚ep·ə·krə'dās·ē‚ē }

epaulette [INV ZOO] **1.** Any of the branched or knobbed processes on the oral arms of many Scyphozoa. **2.** The first haired scale at the base of the costal vein in Diptera. { ¦ep·ə¦let }

epaxial [BIOL] Above or dorsal to an axis. { e'pak·sē·əl }

epaxial muscle [ANAT] Any of the dorsal trunk muscles of vertebrates. { e¦pak·sē·əl ¦məs·əl }

EPDM See ethylene-propylene terpolymer.

epeiric sea See epicontinental sea. { ə'pīr·ik 'sē }

epeirogeny [GEOL] Movements which affect large tracts of the earth's crust. { ‚e‚pī'räj·ə·nē }

ependyma [HISTOL] The layer of epithelial cells lining the cavities of the brain and spinal cord. Also known as ependymal layer. { e'pen·də·mə }

ependymal layer See ependyma. { e'pen·də·məl ‚lā·ər }

ependymoma [MED] A tumor of the central nervous system whose essential portion consists of cells derived from and resembling ependymal cells. Also known as medulloepithelioma. { e‚pen·də'mō·mə }

Eperythrozoon [MICROBIO] A genus of the family Anaplasmataceae; rings and coccoids occur on erythrocytes and in the plasma of various vertebrates. { ‚ep·ə‚rith·rə'zō·ən }

EOCENE

	PALEOZOIC										MESOZOIC		CENOZOIC

PRECAMBRIAN · CAMBRIAN · ORDOVICIAN · SILURIAN · DEVONIAN · CARBONIFEROUS (Mississippian, Pennsylvanian) · PERMIAN · TRIASSIC · JURASSIC · CRETACEOUS · TERTIARY · QUATERNARY

TERTIARY					QUATERNARY	
Paleocene	Eocene	Oligocene	Miocene	Pliocene	Pleistocene	Recent

A chart showing the position of the Eocene epoch in geologic time.

EOSINOPHIL

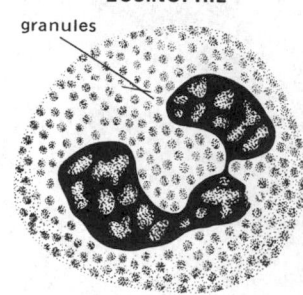

granules

Diagram of an eosinophil showing the typical two-lobed nucleus.

ephapse [NEUROSCI] A false synapse between neighboring neurons where current from one affects the other. { e'faps }

ephaptic transmission [PHYSIO] Electrical transfer of activity to a postephaptic unit by the action current of a preephaptic cell. { e'fap·tik tranz'mish·ən }

Ephedra [BOT] A genus of low, leafless, green-stemmed shrubs belonging to the order Ephedrales; source of the drug ephedrine. { ə'fed·rə }

Ephedrales [BOT] A monogeneric order of gymnosperms in the subdivision Gneticae. { ‚e·fə'drā·lēz }

ephedrine [ORG CHEM] $C_{10}H_{15}NO$ A white, crystalline, water-soluble alkaloid present in several *Ephedra* species and also produced synthetically; a sympathomimetic amine, it is used for its action on the bronchi, blood pressure, blood vessels, and central nervous system. { ə'fed·rən }

ephemeral gully [GEOL] A channel that forms in a cultivated field when precipitation exceeds the rate of soil infiltration. { ə¦fem·ə·rəl ¦gəl·ē }

ephemeral plant [BOT] An annual plant that completes its life cycle in one short moist season; desert plants are examples. { ə'fem·ə·rəl 'plant }

ephemeral stream [HYD] A stream channel which carries water only during and immediately after periods of rainfall or snowmelt. { ə'fem·ə·rəl 'strēm }

Ephemerida [INV ZOO] An equivalent name for the Ephemeroptera. { ‚e·fə'mer·ə·də }

ephemeris [ASTRON] A periodical publication tabulating the predicted positions of celestial bodies at regular intervals, such as daily, and containing other data of interest to astronomers. Also known as astronomical ephemeris. { ə'fem·ə·rəs }

ephemeris day [ASTRON] A unit of time equal to 86,400 ephemeris seconds (International System of Units). { ə'fem·ə·rəs 'dā }

ephemeris second [ASTRON] The fundamental unit of time of the International System of Units from 1960 until 1968, equal to 1/31556925.9747 of the tropical year defined by the mean motion of the sun in longitude at the epoch 1900 January 0 day 12 hours. { ə'fem·ə·rəs 'sek·ənd }

ephemeris time [ASTRON] The uniform measure of time defined by the laws of dynamics and determined in principle from the orbital motions of the planets, specifically the orbital motion of the earth as represented by Newcomb's Tables of the Sun. Abbreviated E.T. { ə'fem·ə·rəs 'tīm }

Ephemeroptera [INV ZOO] The mayflies, an order of exopterygote insects in the subclass Pterygota. { ə‚fem·ə'räp·tə·rə }

ephidrosis *See* hyperhidrosis. { ə'fid·rə·səs }

Ephydridae [INV ZOO] The shore flies, a family of myodarian cyclorrhaphous dipteran insects in the subsection Acalypteratae. { ə'fid·rə‚dē }

ephyra [INV ZOO] A larval, free-swimming medusoid stage of scyphozoans; arises from the scyphistoma by transverse fission. Also known as ephyrula. { 'e·fə·rə }

ephyrula *See* ephyra. { e'fir·ə·lə }

epi- [ORG CHEM] A prefix used in naming compounds to indicate the presence of a bridge or intramolecular connection. [SCI TECH] Prefix denoting upon, beside, near to, over, outer, anterior, prior to, or after. { 'ep·ē }

EPI *See* electronic position indicator.

epiandrum [INV ZOO] The genital orifice of a male arachnid. { ‚ep·ē'an·drəm }

epibasidium [MYCOL] A lengthening of the upper part of each cell of the basidium of various heterobasidiomycetes. { ‚ep·ə·bə'sid·ē·əm }

epibiosis [ECOL] The arrangement in which organisms live on top of each other. { ‚ep·ə·bī'ō·səs }

epibiotic [ECOL] Living, usually parasitically, on the surface of plants or animals; used especially of fungi. { ‚ep·ə·bī'äd·ik }

epiblast *See* ectoderm. { 'ep·ə‚blast }

epiblem [BOT] A tissue that replaces the epidermis in most roots and in stems of submerged aquatic plants. { 'ep·ə‚blem }

epiboly [EMBRYO] The growing or extending of one part, such as the upper hemisphere of a blastula, over and around another part, such as the lower hemisphere, in embryogenesis. { ə'pib·ə·lē }

epibranchial [ANAT] Of or pertaining to the segment below the pharyngobranchial region in a branchial arch. { ‚ep·ə'braŋ·kē·əl }

epicadmium [NUCLEO] Energy above the greatest level at which cadmium shows a large neutron cross section, about 0.3 electronvolt. { ‚ep·ə'kad·mē·əm }

epicalyx [BOT] A ring of fused bracts below the calyx forming a structure that resembles the calyx. { ‚ep·ə'kā·liks }

epicardium [ANAT] The inner, serous portion of the pericardium that is in contact with the heart. [INV ZOO] A tubular prolongation of the branchial sac in certain ascidians which takes part in the process of budding. { ‚ep·ə'kärd·ē·əm }

Epicaridea [INV ZOO] A suborder of the Isopoda whose members are parasitic on various marine crustaceans. { ‚ep·ə·kə'rid·ē·ə }

epicarp [BOT] The outer layer of the pericarp. Also known as exocarp. { 'ep·ə‚kärp }

epicenter [GEOL] A point on the surface of the earth which is directly above the seismic focus of an earthquake and where the earthquake vibrations reach first. [MATH] The center of a circle that generates an epicycloid or hypocycloid. { 'ep·ə‚sen·tər }

epichlorohydrin [ORG CHEM] C_3H_5OCl A colorless, unstable liquid, insoluble in water; used as a solvent for resins. { ‚ep·ə‚klȯr·ə'hī·drən }

epichordal [VERT ZOO] Located upon or above the notochord. { ‚ep·ə'kȯrd·əl }

epichordal brain [EMBRYO] The area of origin of the hindbrain or rhombencephalon, located on the dorsal side of the notochord. Also known as deutencephalon. { ‚ep·ə'kȯrd·əl 'brān }

epiclastic [GEOL] Pertaining to the texture of mechanically deposited sediments consisting of detrital material from preexistent rocks. { ‚ep·ə'klas·tik }

epicnemial [ANAT] Of or pertaining to the anterior portion of the tibia. { ‚ep·ak¦nē·mē·əl }

epicondyle [ANAT] An eminence on the condyle of a bone. { ‚ep·ə'kän‚dīl }

epicondylitis [MED] Infection or inflammation of an epicondyle. { ‚ep·ə‚kän·də'līd·əs }

epicone [INV ZOO] The part anterior to the equatorial groove in a dinoflagellate. { 'ep·ə‚kōn }

epicontinental [GEOL] Located upon a continental plateau or platform. { ‚ep·ə‚kant·ən'ent·əl }

epicontinental sea [OCEANOGR] That portion of the sea lying upon the continental shelf, and the portions which extend into the interior of the continent with similar shallow depths. Also known as epeiric sea; inland sea. { ‚ep·ə‚kant·ən'ent·əl 'sē }

epicotyl [BOT] The embryonic plant stem above the cotyledons. { ‚ep·ə'käd·əl }

epicranium [INV ZOO] The dorsal wall of an insect head. [VERT ZOO] The structures covering the cranium in vertebrates. { ‚ep·ə'krā·nē·əm }

epicuticle [INV ZOO] The outer, waxy layer of an insect cuticle or exoskeleton. { ‚ep·ə'kyüd·i·kəl }

epicycle [MATH] The circle which generates an epicycloid or hypocycloid. { 'ep·ə‚sī·kəl }

epicyclic gear [MECH ENG] A system of gears in which one or more gears travel around the inside or the outside of another gear whose axis is fixed. { ¦ep·ə¦sī·klik 'gir }

epicyclic train [MECH ENG] A combination of epicyclic gears, usually connected by an arm, in which some or all of the gears have a motion compounded of rotation about an axis and a translation or revolution of that axis. { ¦ep·ə¦sī·klik 'trān }

epicycloid [MATH] The curve traced by a point on a circle as it rolls along the outside of a fixed circle. { ‚ep·ə'sī‚klȯid }

epicyst [INV ZOO] The outer layer of a cyst wall in encysted protozoans. Also known as ectocyst. { 'ep·ə‚sist }

epidemic [MED] A sudden increase in the incidence rate of a disease to a value above normal, affecting large numbers of people and spread over a wide area. { ¦ep·ə¦dem·ik }

epidemic diarrhea of the newborn [MED] Contagious, fulminating diarrhea with high mortality, seen in newborns; caused by enteropathogenic strains of *Escherichia coli*, strains of *Staphylococcus*, other bacteria, and possibly viruses. { ¦ep·ə¦dem·ik ‚dī·ə'rē·ə əv thə 'nü‚bȯrn }

epidemic encephalitis *See* encephalitis lethargica. { ¦ep·ə¦dem·ik in‚sef·ə'līd·əs }

EPHEDRA

A species of *Ephedra* growing in Utah. (*Courtesy of Tony Gauba, from National Audubon Society*)

epidemic gastroenteritis [MED] Inflammation of the stomach and intestine, of viral origin; considered to be epidemic when symptoms are manifested by a member of the patient's family within 10 days of the patient's recovery. { ¦ep·ə¦dem·ik ¦gas·trō¸ent·ə'rīd·əs }

epidemic hepatitis *See* infectious hepatitis. { ¦ep·ə¦dem·ik ¸hep·ə'tīd·əs }

epidemic jaundice *See* infectious hepatitis. { ¦ep·ə¦dem·ik 'jȯn·dəs }

epidemic keratoconjunctivitis [MED] Inflammation of the cornea and conjunctiva, caused by a virus; epidemic by nature. { ¦ep·ə¦dem·ik ¸ker·əd·ō·kən¸jəŋk·tə'vīd·əs }

epidemic neuromyasthenia [MED] A prolonged, debilitating disease of the nervous system of adults; characterized by fatigue, headache, muscle pain, paresis, and emotional and mental disturbances; no etiologic agent has been isolated. Also known as acute infective encephalomyelitis; Akureyri disease; benign myalgic encephalomyelitis; epidemic vegetative neuritis; Iceland disease. { ¦ep·ə¦dem·ik ¦nü·rō·mī·əs'thē·nē·ə }

epidemic pleurodynia [MED] An acute epidemic disease of humans, caused by coxsackie B virus; characterized by severe pain in the lower thorax and upper abdomen, and associated with fever and malaise. { ¦ep·ə¦dem·ik ¸plùr·ə'din·ē·ə }

epidemic roseola *See* rubella. { ¦ep·ə¦dem·ik ¸rō·zē'ō·lə }

epidemic typhus *See* classic epidemic typhus. { ¦ep·ə¦dem·ik 'tī·fəs }

epidemic vegetative neuritis *See* epidemic neuromyasthenia. { ¦ep·ə¦dem·ik 'vej·ə¸tād·iv nə'rīd·əs }

epidemiological study [MED] A population study designed to examine associations (commonly, hypothesized causal relations) between personal characteristics and environmental exposures that increase the risk of disease. { ¸ep·ə¸dē·mē·ə¦läj·ə·kəl 'stəd·ē }

epidemiology [MED] The study of the mass aspects of disease. { ¸ep·ə¸dē·mē'äl·ə·jē }

epidermal growth factor [PHYSIO] A polypeptide produced in animals that stimulates and sustains the replication of epidermal cells (of ectodermal or endodermal origin); its human equivalent is urogastrone. { ¸ep·ə¦dərm·əl 'grōth ¸fak·tər }

epidermal ridge [ANAT] Any of the minute corrugations of the skin on the palmar and plantar surfaces of humans and other primates. { ¸ep·ə'dər·məl 'rij }

epidermis [BOT] The outermost layer (sometimes several layers) of cells on the primary plant body. [GISTOL] The outer nonsensitive, nonvascular portion of the skin comprising two strata of cells, the stratum corneum and the stratum germinativum. { ¸ep·ə'dər·məs }

epidermoid carcinoma *See* squamous-cell carcinoma. { ¸ep·ə'dər¸mȯid ¸kärs·ən'ō·mə }

epidermoid cyst [MED] A cyst lined by stratified squamous epithelium without associated cutaneous glands. { ¸ep·ə'dər¸mȯid 'sist }

epidermolysis [MED] The easy separation of various layers of skin, primarily of the epidermis from the corium, observed in certain pathological conditions. { ¸ep·ə·dər'mäl·ə·səs }

epidermolysis bullosa [MED] A congenital skin disease characterized by the development of vesicles and bullae upon slight, or even without, trauma. { ¸ep·ə·dər'mäl·ə·səs bù'lō·sə }

epidiascope [OPTICS] **1.** An optical projection system for forming an enlarged real image of a flat opaque object, in which light is reflected from the object and then from a mirror before being focused by a projection lens. Also known as episcope. **2.** An optical projection system which can easily be altered to project either transparent or opaque objects. { ¸ep·ə'dī·ə¸skōp }

epididymis [ANAT] The convoluted efferent duct lying posterior to the testis and connected to it by the efferent ductules of the testis. { ¸ep·ə'did·ə·məs }

epididymitis [MED] Inflammation of the epididymis. { ¸ep·ə¸did·ə'mīd·əs }

epidiorite [PETR] A dioritic rock formed by alteration of pyroxenic igneous rocks. { ¸ep·ə'dī·ə¸rīt }

epidosite [PETR] A rare metamorphic rock composed of epidote and quartz. { ¸ep·ə'dō¸sīt }

epidote [MINERAL] A pistachio-green to blackish-green calcium aluminum sorosilicate mineral that crystallizes in the monoclinic system; the luster is vitreous, hardness is 61/2 on Mohs scale, and specific gravity is 3.35–3.45. { 'ep·ə¸dōt }

epidote-amphibolite facies [PETR] Metamorphic rocks formed under pressures of 3000–7000 bars and temperatures of 250–450°C with conditions intermediate between those that formed greenschist and amphibolite, or with characteristics intermediate. { ¦ep·ə¸dōt am'fib·ə¸līt ¸fā·shēz }

epidotization [GEOL] The introduction of epidote into, or the formation of epidote from, rocks. { ¸ep·ə¸dōd·ə'zā·shən }

epidural [ANAT] Located on or over the dura mater. { ¦ep·ə¦dür·əl }

epieugeosyncline [GEOL] Deep troughs formed by subsidence which have limited volcanic power and overlie a eugeosyncline. { ¦ep·ē¸yü¸jē·ō'sin¸klīn }

epifauna [ZOO] Benthic fauna that live on a surface, such as the sea floor, other organisms, or objects. { 'ep·ə¸fȯn·ə }

epigaster [EMBRYO] The portion of the intestine in vertebrate embryos which gives rise to the colon. { 'ep·ə¸gas·tər }

epigastric region [ANAT] The upper and middle part of the abdominal surface between the two hypochondriac regions. Also known as epigastrium. { ¦ep·ə¦gas·trik ¸rē·jən }

epigastrium [ANAT] *See* epigastric region. [INV ZOO] The ventral side of mesothorax and metathorax in insects. { ¸ep·ə'gas·trē·əm }

epigean [BOT] Pertaining to a plant or plant part that grows above the ground surface. [ZOO] Living near or on the ground surface, applied especially to insects. { ¦ep·ə¦jē·ən }

epigene [GEOL] **1.** A geologic process originating at or near the earth's surface. **2.** A structure formed at or near the earth's surface. { 'ep·ə¸jēn }

epigenesis [EMBRYO] Development in gradual stages of differentiation. [GEOL] Alteration of the mineral content of rock due to outsideinfluences. { ¸ep·ə'jen·ə·səs }

epigenetic [GEOL] Produced or formed at or near the surface of the earth. { ¦ep·ə·jə¦ned·ik }

epigenetics [GEN] The study of those processes by which genetic information ultimately results in distinctive physical and behavioral characteristics. { ¦ep·ə·jə¦ned·iks }

epigenite [MINERAL] $(Cu,Fe)_5AsS_6$ A steel gray, orthorhombic mineral consisting of copper and iron arsenic sulfide. { ə'pij·ə¸nīt }

epigenotype [GEN] The total developmental system through which the adult form of an organism is realized, comprising the interactions among genes and between genes and the nongenetic environment. { ¸ep·ə'jēn·ə¸tīp }

epigenous [BOT] Developing or growing on a surface, especially of a plant or plant part. { ə'pij·ə·nəs }

epiglottis [ANAT] A flap of elastic cartilage covered by mucous membrane that protects the glottis during swallowing. { ¸ep·ə'gläd·əs }

epigynous [BOT] Having the perianth and stamens attached near the top of the ovary; that is, the ovary is inferior. { ə'pij·ə·nəs }

epigynum [INV ZOO] **1.** The genital pore of female arachnids. **2.** The plate covering this opening. { ə'pij·ə·nəm }

epihydrin alcohol *See* glycidol. { ¦ep·ə¦hī·drən 'al·kə¸hȯl }

epilation [MED] Removal of the hair by the roots by the use of forceps, chemical means, or roentgenotherapy. { ¸ep·ə'lā·shən }

epilemma [HISTOL] The perineurium of very small nerves. { ¸ep·ə'lem·ə }

epilepsy [MED] A condition characterized by the paroxysmal recurrence of transient, uncontrollable episodes of abnormal neurological or mental function, or both. { 'ep·ə¸lep¸sē }

epileptic equivalent [MED] Episodes of sensory or motor phenomena experienced by an epileptic instead of convulsions. Also known as convulsive equivalent. { ¦ep·ə¸lep·tik i'kwiv·ə·lənt }

epilimnion [HYD] A fresh-water zone of relatively warm water in which mixing occurs as a result of wind action and convection currents. { ¸ep·ə'lim·nē¸än }

epimagma [GEOL] A gas-free, vesicular to semisolid magmatic residue of pasty consistency formed by cooling and loss of gas from liquid lava in a lava lake. { ¸ep·ə'mag·mə }

epimagmatic *See* deuteric. { ¸ep·ə·mag'mad·ik }

epimer [ORG CHEM] A type of isomer in which the difference between the two compounds is the relative position of the H

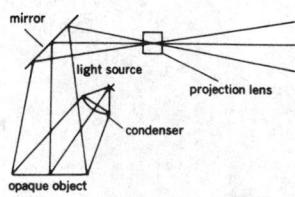

EPIDIASCOPE

mirror

light source

projection lens

condenser

opaque object

Diagram of an epidiascope (def. 1).

EPIGYNOUS

The arrangement of the sepals, petals, and stamens in an epigynous flower.

(hydrogen) group and OH (hydroxyl) group on the last asymmetric C (carbon) atom of the chain, as in the sugars D-glucose and D-mannose. { 'ep·ə·mər }

epimerase [BIOCHEM] A type of enzyme that catalyzes the rearrangement of hydroxyl groups on a substrate. { ə'pim·ə‚rās }

epimere [ANAT] The dorsal muscle plate of the lining of a coelomic cavity. [EMBRYO] The dorsal part of a mesodermal segment in the embryo of chordates. { 'ep·ə‚mir }

epimerization [ORG CHEM] In an optically active compound that contains two or more asymmetric centers, a process in which only one of these centers is altered by some reaction to form an epimer. { ə‚pim·ə·rə'zā·shən }

epimeron [INV ZOO] **1.** The posterior plate of the pleuron in insects. **2.** The portion of a somite between the tergum and the insertion of a limb in arthropods. { ‚ep·ə'mir‚än }

Epimetheus [ASTRON] A satellite of Saturn which orbits at a mean distance of 151,000 kilometers (94,000 miles), near Saturn's rings, in nearly the same orbit as Janus, and has an irregular shape with an average diameter of 120 kilometers (75 miles). { ‚ep·ə'mē‚thē·əs }

epimorphosis [PHYSIO] Regeneration in which cell proliferation precedes differentiation. { ‚ep·ə'mȯr·fə·səs }

epimyocardium [EMBRYO] The layer of the embryonic heart from which both the myocardium and epicardium develop. { ‚ep·ə‚mī·ə'kärd·ē·əm }

epimysium [ANAT] The connective-tissue sheath surrounding a skeletal muscle. { ‚ep·ə'mī·sē·əm }

epinasty [BOT] Growth changes in which the upper surface of a leaf grows, thus bending the leaf downward. { 'ep·ə‚nas·tē }

epinephrine [BIOCHEM] $C_9H_{13}O_3N$ A hormone secreted by the adrenal medulla that acts to increase blood pressure due to stimulation of heart action and constriction of peripheral blood vessels. Also known as adrenaline. { ‚ep·ə'ne·frən }

epineural [ANAT] Arising from or on the outside of a nerve trunk. [INV ZOO] The nervous tissue dorsal to the ventral nerve cord in arthropods. { ‚ep·ə'nur·əl }

epineural canal [INV ZOO] A canal that runs between the radial nerve and the epithelium in echinoids and ophiuroids. { ‚ep·ə'nur·əl kə'nal }

epineurium [ANAT] The connective-tissue sheath of a nerve trunk. { ‚ep·ə'nur·ē·əm }

epipelagic [OCEANOGR] Of or pertaining to the portion of oceanic zone into which enough light penetrates to allow photosynthesis. { ‚ep·ə·pə'laj·ik }

epipelagic zone [OCEANOGR] The region of an ocean extending from the surface to a depth of about 600 feet (200 meters); light penetrates this zone, allowing photosynthesis. { ‚ep·ə·pə'laj·ik 'zōn }

epipetalous [BOT] Having stamens located on the corolla. { ‚ep·ə'ped·əl·əs }

epiphallus [INV ZOO] A sclerite in some orthopterans in the floor of the genital chamber. { ‚ep·ə'fal·əs }

epipharynx [INV ZOO] An organ attached beneath the labrium of many insects. { ‚ep·ə'far·iŋks }

epiphora [MED] An abnormal increase in tearing of one or both eyes. { ‚ep·ə 'fȯr·ə }

epiphragm [BOT] A membrane covering the aperture of the capsule in certain mosses. [INV ZOO] A membranous or calcareous partition that covers the aperture of certain hibernating land snails. { 'ep·ə‚fram }

epiphyll [ECOL] A plant that grows on the surface of leaves. { 'ep·ə‚fil }

epiphyseal arch [EMBRYO] The arched structure in the third ventricle of the embryonic brain, which marks the site of development of the pineal body. { ə‚pif·ə‚sē·əl 'ärch }

epiphyseal plate [ANAT] **1.** The broad, articular surface on each end of a vertebral centrum. **2.** The thin layer of cartilage between the epiphysis and the shaft of a long bone. Also known as metaphysis. { ə‚pif·ə‚sē·əl 'plāt }

epiphysiolysis [MED] The separation of an epiphysis from the shaft of abone. { ə‚pif·ə·sē'äl·ə·səs }

epiphysis [ANAT] **1.** The end portion of a long bone in vertebrates. **2.** See pineal body. { ə'pif·ə·səs }

epiphyte [ECOL] A plant which grows nonparasitically on another plant or on some nonliving structure, such as a building or telephone pole, deriving moisture and nutrients from the air. Also known as aerophyte. { 'ep·ə‚fīt }

epiphytotic [PL PATH] **1.** Any infectious plant disease that occurs sporadically in epidemic proportions. **2.** Of or pertaining to an epidemic plant disease. { ‚ep·ə‚frī'täd·ik }

epiplankton [BIOL] Plankton occurring in the sea from the surface to a depth of about 100 fathoms (180 meters). { ‚ep·ə'plaŋk·tən }

epipleural [ANAT] Arising from a rib. [VERT ZOO] An intramuscular bone arising from and extending between some of the ribs in certain fishes. { ‚ep·ə'plur·əl }

epiploic foramen [ANAT] An aperture of the peritoneal cavity, formed by folds of the peritoneum and located between the liver and the stomach. Also known as foramen of Winslow. { ‚ep·ə‚plō·ik fə'rā·mən }

epipodite [INV ZOO] A branch of the basal joint of the protopodite of thoracic limbs of many arthropods. { ə'pip·ə‚dīt }

epipodium [BOT] The apical portion of an embryonic phyllopodium. [INV ZOO] **1.** A ridge or fold on the lateral edges of each side of the foot of certain gastropod mollusks. **2.** The elevated ring on an ambulacral plate in Echinoidea. { ‚ep·ə'pōd·ē·əm }

Epipolasina [INV ZOO] A suborder of sponges in the order Clavaxinellida having radially arranged monactinal or diactinal megascleres. { ‚ep·ə·pə'laz·ə·nə }

epiproct [INV ZOO] A plate above the anus forming the dorsal part of the tenth or eleventh somite of certain insects. { 'ep·ə‚präkt }

epipubis [VERT ZOO] A single cartilage or bone located in front of the pubis in some vertebrates, particularly in some amphibians. { ‚ep·ə'pyü·bəs }

EPIRB See emergency position-indicating radio beacon. { 'ē‚pərb }

episclera [ANAT] The loose connective tissue lying between the conjunctiva and the sclera. { ‚ep·ə'skler·ə }

episcope See epidiascope. { 'ep·ə‚skōp }

episcotister [OPTICS] A device for reducing the intensity of light by a known fraction, consisting of a rapidly rotating disk with transparent and opaque sectors. { ‚ep·ə·skō'tis·tər }

episepalous [BOT] Having stamens growing on or adnate to the sepals. { ‚ep·ə'sep·ə·ləs }

episiotomy [MED] Medial or lateral incision of the vulva during childbirth, to avoid undue laceration. { ə‚pēz·ē'äd·ə·mē }

episode [GEOL] A distinctive event or series of events in the geologic history of a region or feature. { 'ep·ə‚sōd }

episodic memory [PSYCH] Memory of information about specific past events that involved the self and occurred at a particular time and place. { ‚ep·ə‚säd·ik 'mem·rē }

episome [GEN] A circular genetic element in bacteria, presumably a deoxyribonucleic acid fragment, which is not necessary for survival of the organism and which can be integrated in the bacterial chromosome or remain free. { 'ep·ə‚sōm }

epispadias [MED] A congenital defect of the anterior urethra in which the canal terminates on the dorsum of the penis and posterior to its normal opening. { ‚ep·ə'späd·ē·əs }

episperm See testa. { 'ep·ə‚spərm }

epi spiral [MATH] A plane curve whose equation in polar coordinates (r, θ) is $r \cos n\theta = a$, where a is a constant and n is an integer. { 'ep·ē ‚spī·rəl }

epistasis [GEN] The suppression of the effect of one gene by another. [MED] A checking or stoppage of a hemorrhage or other discharge. [PATH] A scum or film of substance floating on the surface of urine. { ə'pis·tə·səs }

episternum [VERT ZOO] A dermal bone or pair of bones ventral to the sternum of certain fishes and reptiles. { ‚ep·ə'stər·nəm }

epistilbite [MINERAL] $CaAl_2Si_6O_{16}\cdot5H_2O$ A mineral of the zeolite family that contains calcium and aluminosilicate and crystallizes in the monoclinic system; occurs in white prismatic crystals or granular forms. { ‚ep·ə'stil‚bīt }

epistome [INV ZOO] **1.** The area between the mouth and the second antennae in crustaceans. **2.** The plate covering this region. **3.** The area between the labrum and the epicranium in many insects. **4.** A flap covering the mouth of certain bryozoans. **5.** The area just above the labrum in certain dipterans. { 'ep·ə‚stōm }

epitaxial diffused-junction transistor [ELECTR] A junction transistor produced by growing a thin, high-purity layer of semiconductor material on a heavily doped region of the same type. { ‚ep·ə'tak·sē·əl də‚fyüzd ‚jəŋk·shən tran'zis·tər }

epitaxial diffused-mesa transistor [ELECTR] A diffused-mesa transistor in which a thin, high-resistivity epitaxial layer is deposited on the substrate to serve as the collector. { ‚ep·ə'tak·sē·əl də‚fyüzd ‚mā·sə tran'zis·tər }

epitaxial layer [SOLID STATE] A semiconductor layer having the same crystalline orientation as the substrate on which it is grown. { ‚ep·ə'tak·sē·əl ‚lā·ər }

epitaxial transistor [ELECTR] Transistor with one or more epitaxial layers. { ‚ep·ə'tak·sē·əl tran'zis·tər }

epitaxy [CRYSTAL] Growth of one crystal on the surface of another crystal in which the growth of the deposited crystal is oriented by the lattice structure of the substrate. { 'ep·ə‚tak·sē }

epithalamus [ANAT] A division of the vertebrate diencephalon including the habenula, the pineal body, and the posterior commissure. { ‚ep·ə'thal·ə·məs }

epitheca [INV ZOO] **1.** An external, calcareous layer around the basal portion of the theca of many corals. **2.** A protective covering of the epicone. **3.** The outer portion of a diatom frustule. { ‚ep·ə'thē·kə }

epitheliochorial placenta [EMBRYO] A type of placenta in which the maternal epithelium and fetal epithelium are in contact. Also known as villous placenta. { ‚ep·ə‚thē·lē·ō'kôr·ē·əl plə'sen·tə }

epithelioid cell [HISTOL] A macrophage that resembles an epithelial cell. Also known as alveolated cell. { ‚ep·ə'thē·lē‚óid ‚sel }

epithelioma [MED] A tumor derived from epithelium; usually a skin cancer, occasionally cancer of a mucous membrane. { ‚ep·ə‚thē·lē'ō·mə }

epitheliomuscular cell [INV ZOO] An epithelial cell with an elongate base that contains contractile fibrils; common among cnidarians. { ‚ep·ə‚thē·lē·ō'məs·kyə· lər 'sel }

epithelium [HISTOL] A primary animal tissue, distinguished by cells being close together with little intercellular substance; covers free surfaces and lines body cavities and ducts. { ‚ep·ə'thē·lē·əm }

epithema [VERT ZOO] A horny outgrowth on the beak of certain birds. { ‚ep·ə'thē·mə }

epithermal [GEOL] Pertaining to mineral veins and ore deposits formed from warm waters at shallow depth, at temperatures ranging from 50–200°C, and generally at some distance from the magmatic source. { ‚ep·ə'thər·məl }

epithermal deposit [GEOL] Ore deposit formed in and along openings in rocks by deposition at shallow depths from ascending hot solutions. { ‚ep·ə'thər·məldə'päz·ət }

epithermal neutron [NUCLEO] A neutron having an energy in the range immediately above the thermal range, roughly between 0.02 and 100 electronvolts. { ‚ep·ə'thər·məl 'nü‚trän }

epithermal reactor [NUCLEO] A nuclear reactor in which a substantial fraction of fissions is induced by neutrons having more than thermal energy. { ‚ep·ə'thər·məl rē'ak·tər }

epithermal thorium reactor [NUCLEO] A sodium-cooled reactor based on operation with neutrons in the high epithermal energy range; a uranium-thorium fuel mixture is used, with graphite or beryllium as moderator. { ‚ep·ə'thər·məl 'thôr·ē·əm rē'ak·tər }

epitoke [INV ZOO] The posterior portion of marine polychaetes; contains the gonads. { 'ep·ə‚tōk }

epitoky [INV ZOO] In certain polychaetes, development of the posterior sexual part from the anterior sexless part. { 'ep·ə‚täk·ē }

epitope [IMMUNOL] The portion of the antigen molecule that determines its capacity to combine with the specific combining site of its corresponding antibody in an antigen-antibody interaction. { 'ep·ə‚tōp }

epitrichium [EMBRYO] The outer layer of the fetal epidermis of many mammals. { ‚ep·ə'trik·ē·əm }

epitrochlear [ANAT] Of or pertaining to a lymph node that lies above the trochlea of the elbow joint. { ‚ep·ə'trō·klē·ər }

epitrochoid [MATH] A curve traced by a point rigidly attached to a circle at a point other than the center when the circle rolls without slipping on the outside of a fixed circle. { ‚ep·ə'trō‚kóid }

epituberculosis [MED] A massive pulmonary shadow seen in x-ray films in active juvenile tuberculosis, probably caused by bronchial obstruction. { ‚ep·ə·tə‚bər·kyə'lō·s/əs }

epitympanum [ANAT] The attic of the middle ear, or tympanic cavity. { ‚ep·ə'tim·pə·nəm }

epivalve [INV ZOO] **1.** The upper or apical shell of certain dinoflagellates. **2.** The upper shell of a diatom. { 'ep·ə‚valv }

epixylous [ECOL] Growing on wood; used especially of fungi. { ‚ep·ə‚zī·ləs }

epizoic [BIOL] Living on the body of an animal. { ‚ep·ə‚zō·ik }

epizone [GEOL] **1.** The zone of metamorphism characterized by moderate temperature, low hydrostatic pressure, and powerful stress. **2.** The outer depth zone of metamorphic rocks. { 'ep·ə‚zōn }

epizootic [VET MED] **1.** Affecting many animals of one kind in one region simultaneously; widely diffuse and rapidly spreading. **2.** An extensive outbreak of an epizootic disease. { ‚ep·ə·zō‚äd·ik }

epizootiology [VET MED] The study of epizootics. { ‚ep·ə·zō‚äd·ē'äl·ə·jē }

E-plane antenna [ELECTROMAG] An antenna which lies in a plane parallel to the electric field vector of the radiation that it emits. { 'ē ‚plān an‚ten·ə }

E-plane bend See E bend. { 'ē ‚plān ‚bend }

E-plane T junction [ELECTROMAG] Waveguide T junction in which the change in structure occurs in the plane of the electric field. Also known as series T junction. { 'ē ‚plān 'tē ‚jəŋk·shən }

EP lubricant [MATER] A lubricating oil or grease that contains additives to improve ability to adhere to the surfaces of metals under high bearing pressures. Derived from extreme-pressure lubricant. { 'ē‚pē'lü·brə·kənt }

EPMA See electron probe microanalysis.

EPN See O-ethyl-O-para-nitrophenyl phenylphosphonothioate.

epoch [ASTRON] A particular instant for which certain data are valid; for example, star positions in an astronomical catalog, epoch 1950.0. [GEOL] A major subdivision of a period of geologic time. [PHYS] See time. { 'ep·ək }

eponychium [ANAT] The horny layer of the nail fold attached to the nail plate at its margin; represents the remnant of the embryonic condition. [EMBRYO] A horny condition of the epidermis from the second to the eighth month of fetal life, indicating the position of the future nail. { ‚ep·ə'nik·ē·əm }

epoophoron [ANAT] A blind longitudinal duct and 10–15 transverse ductules in the mesosalpinx near the ovary which represent remnants of the reproductive part of the mesonephros in the female; homolog of the head of the epididymis in the male. Also known as parovarium; Rosenmueller's organ. { ‚ep·ō'äf·ə‚rän }

epoxidation [ORG CHEM] Reaction yielding an epoxy compound, such as the conversion of ethylene to ethylene oxide. { e‚päk·sə'dā·shən }

epoxide [ORG CHEM] **1.** A reactive group in which an oxygen atom is joined to each of two carbon atoms which are already bonded. **2.** A three-membered cyclic ether. Also known as oxirane. See ethylene oxide. { e'päk‚sīd }

epoxy- [ORG CHEM] A prefix indicating presence of an epoxide group in a molecule. { ə'päk·sē }

epoxy adhesive [MATER] An adhesive material made of epoxy resin. { ə'päk·sē ad'hē·siv }

1,2-epoxyethane See ethylene oxide. { ‚wən ‚tü ə‚päk·sē'e‚thān }

epoxy resin [ORG CHEM] A polyether resin formed originally by the polymerization of bisphenol A and epichlorohydrin, having high strength, and low shrinkage during curing; used as a coating, adhesive, casting, or foam. { ə'päk·sē 'rez·ən }

Eppley pyrheliometer [ENG] A pyrheliometer of the thermoelectric type; radiation is allowed to fall on two concentric silver rings, the outer covered with magnesium oxide and the inner covered with lampblack; a system of thermocouples (thermopile) is used to measure the temperature difference between the rings; attachments are provided so that measurements of direct and diffuse solar radiation may be obtained. { 'ep·lē ‚pīr‚hē·lē'äm·əd·ər }

EPR See electron paramagnetic resonance.

EPR experiment See Einstein-Podolsky-Rosen experiment. { ‚ē‚pē‚är ik'sper·ə·mənt }

EPROM See erasable programmable read-only memory. { 'ē‚präm }

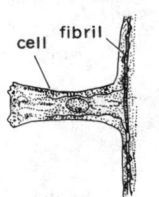

EPITHELIOMUSCULAR CELL

cell   fibril

Microscopic view of an epitheliomuscular cell of *Hydra*.

epsilon chain [MATH] A finite sequence of points such that the distance between any two successive points is less than the positive real number epsilon (ϵ). { 'ep·sə,län ,chān }

Epsilonematoidea [INV ZOO] A superfamily of small (0.5-millimeter) marine nematodes in the order Desmodorida; the body is strongly arched in a sigmoid manner when relaxed. { ,ep·si,län·ə·mə'toid·ē·ə }

epsilon meson [PART PHYS] Neutral, scalar, meson resonance having positive charge conjugation parity and G-parity, a mass of about 730 MeV, and a width of about 600 MeV; decays to two pions. { 'ep·sə,län 'mä,sän }

epsilon neighborhood [MATH] The set of all points in a metric space whose distance from a given point is less than some number; this number is designated ϵ. { 'ep·sə,län 'nā·bər,hůd }

epsilon structure [SOLID STATE] The hexagonal close-packed structure of the ϵ-phase of an electron compound. { 'ep·sə,län ,strək·chər }

epsilon symbols [MATH] The symbols

$$\epsilon^{i_1 i_2 \ldots i_n} \text{ and } \epsilon_{i_1 i_2 \ldots i_n}$$

which are $+1$ if $i_1, i_2, \ldots, i_n$ is an even permutation of 1, 2, $\ldots, n$; -1 if it is an odd permutation; and zero otherwise. { 'ep·sə,län ,sim·bəlz }

epsomite [MINERAL] $MgSO_4 \cdot 7H_2O$ A mineral that occurs in clear, needlelike, orthorhombic crystals; commonly, it is massive or fibrous; luster varies from vitreous to milky, hardness is 2–2.5 on Mohs scale, and specific gravity is 1.68; it has a salty bitter taste and is soluble in water. Also known as epsom salt. { 'ep·sə,mīt }

epsom salt See epsomite. { 'ep·səm ,sȯlt }

Epstein-Barr virus [VIROL] Herpeslike virus particles first identified in cultures of cells from Burkett's malignant lymphoma. { 'ep·stīn ,bär ,vī·rəs }

EPT See ethylene-propylene terpolymer.

epulis [MED] A benign tumorlike lesion of the gingiva. { ə'pyü·ləs }

EPXMA See electron probe x-ray microanalysis.

equal [MATH] Being the same in some sense determined by context. { 'ē·kwəl }

equal altitudes [NAV] In celestial navigation, two altitudes numerically the same; the expression applies particularly to the now obsolescent practice of determining the instant of local apparent noon by observing the altitude of the sun a short time before it reaches the meridian and again at the same altitude after transit; the time of local apparent noon is midway between the times of the two observations; however the second observation must be corrected for the run of the ship which took place between the times of the two readings. Also known as double altitudes. { 'ē·kwəl 'al·tə,tüdz }

equal-area latitude See authalic latitude. { ¦ē·kwəl ¦er·ēə 'lad·ə,tüd }

equal-area map projection [MAP] A map projection having a constant area scale; it is not conformal and is not used for navigation. Also known as authalic map projection; equivalent map projection. { ¦ē·kwəl ¦er·ēə 'map prə,jek·shən }

equal-areas law [ASTRON] The second of Kepler's laws, which states that the line joining a planet and the sun sweeps over equal areas in equal periods of time. { 'ē·kwəl ¦er·ē·əz ,lȯ }

equal-arm balance [MECH] A simple balance in which the distances from the point of support of the balance-arm beam to the two pans at the end of the beam are equal. { ¦ē·kwal ¦ärm 'bal·əns }

equal-energy source [PHYS] Electromagnetic or sound source of energy which emits the same amount of energy for each frequency of the spectrum. { ¦ē·kwal ¦en·ər·jē ,sȯrs }

equal error rate [COMMUN] The error rate of a verification system when the operating threshold for the accept/reject decision is adjusted such that the probability of false acceptance and that of false rejection become equal. Abbreviated EER. { ¦ē·kwəl 'er·ər ,rāt }

equaling file [DES ENG] A slightly bulging double-cut file used in fine toolmaking. { 'ē·kwəl·iŋ ,fīl }

equality [MATH] The state of beingequal. { ē'kwal·əd·ē }

equality gate See equivalence gate. { ē'kwal·əd·ē ,gāt }

equalization [ELECTR] The effect of all frequency-discriminating means employed in transmitting, recording, amplifying, or other signal-handling systems to obtain a desired overall frequency response. Also known as frequency-response equalization. { ,ē·kwə·lə'zā·shən }

equalizer [ELECTR] A network designed to compensate for an undesired amplitude-frequency or phase-frequency response of a system or component; usually a combination of coils, capacitors, and resistors. Also known as equalizing circuit. [MECH ENG] **1.** A bar to which one attaches a vehicle's whiffletrees to make the pull of draft animals equal. Also known as equalizing bar. **2.** A bar which joins a pair of axle springs on a railway locomotive or car for equalization of weight. Also known as equalizing bar. **3.** A device which distributes braking force among independent brakes of an automotive vehicle. Also known as equalizer brake. **4.** A machine which saws wooden stock to equal lengths. [ORD] A device attached to the carriage of those artillery weapons that, when emplaced, rest on two wheels and two trail ends; it is a compensating mechanism to transmit equally the weapon weight and firing shock. { 'ē·kwə,līz·ər }

equalizer brake See equalizer. { 'ē·kwə,līz·ər ,brāk }

equalizing bar See equalizer. { 'ē·kwə,līz·iŋ ,bär }

equalizing circuit See equalizer. { 'ēkwə,līz·iŋ ,sər·kət }

equalizing current [ELEC] Current that circulates between two parallel-connected compound generators to equalize their output. { 'ē·kwə,līz·iŋ ,kər·ənt }

equalizing line [CHEM ENG] A pipe or tubing interconnection between two closed vessels, containers, or process systems to allow pressure equalization. { 'ē·kwə,līz·iŋ ,līn }

equalizing pulses [ELECTR] In television, pulses at twice the line frequency, occurring just before and after the vertical synchronizing pulses, which minimize the effect of line frequency pulses on the interlace. { 'ē·kwə,līz·iŋ ,pəl·səs }

equalizing reservoir [CIV ENG] A reservoir located between a primary water supply and the consumer for the purpose of maintaining equilibrium between different portions of the distribution system. { 'ē·kwə,līz·iŋ 'rez·əv,wär }

equalizing support [ORD] The crossbeam support of an axle-support-type equalizer, on which is mounted the trails and upper carriage; it is connected to the axle by means of a horizontal pintle, which permits a vertical angular difference between the front and rear axis. { 'ē·kwə,līz·iŋ sə,pȯrt }

equal listener response scale [ACOUS] An arbitrary scale of noisiness which measures the average response of a listener to a noise when allowance is made for the apparent increase of intensity of a noise as its frequency increases. Abbreviated ELR scale. { ¦ē·kwəl ¦lis·nər ri'späns ,skāl }

equal loudness contour [ACOUS] A curve on a graph of sound intensity in decibels versus frequency at each point along which sound appears to be equally loud to a listener. Also known as Fletcher-Munson contour. { ¦ē·kwəl 'laůd·nəs ,kän,tůr }

equally likely cases [STAT] All simple events in a trial have the same probability. { ¦ē·kwə·lē ¦līk·lē 'kās·əs }

equally tempered scale [ACOUS] A musical scale formed by dividing the octave into 12 equal intervals and selecting from the resulting notes; thus, the frequency ratio between any two successive notes is exactly $2^{1/12}$ or $2^{1/6}$. Also known as equitempered scale. { ¦ē·kwə·lē ¦tem·pərd 'skāl }

equal ripple [ELECTR] Property of an amplitude or phase characteristic whose local maxima all have the same value, and whose local minima all have the same value, within a specified frequency range. { ¦ē·kwəl 'rip·əl }

equal ripple property [MATH] For any continuous function $f(x)$ on the interval $-1,1$, and for any positive integer n, a property of the polynomial of degree n, which is the best possible approximation to $f(x)$ in the sense that the maximum absolute value of $e_n(x) = f(x) - p_n(x)$ is as small as possible; namely, that $e_n(x)$ assumes its extreme values at least $n + 2$ times, with the consecutive extrema having opposite signs. { ¦ē·kwəl 'rip·əl ,präp·ərd·ē }

equal sets [MATH] Sets with precisely the same elements. { ¦ē·kwəl 'sets }

equals relation See equivalence relation. { 'ē·kwəlz ri,lā·shən }

equal tails test [STAT] A technique for choosing two critical values for use in a two-sided test; it consists of selecting critical values c and d so that the probability of acceptance of the null hypothesis if the test statistic does not exceed c is the same as

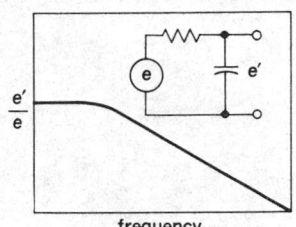

EQUALIZER

Circuit diagram and frequency response characteristics of a type of equalizer that employs a combination of resistance and capacitance; e' = output voltage, e = input voltage.

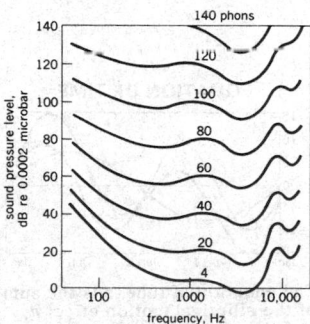

EQUAL LOUDNESS CONTOUR

Equal loudness level contours.

the probability of acceptance of the null hypothesis if the test statistic is not smaller than d. { ¦ē·kwəl ¦tālz ‚test }

equal-zero indicator [COMPUT SCI] A circuit component which is on when the result of an operation is zero. { ¦ē·kwəl ¦zir·ō 'in·də‚kād·ər }

equate [MATH] To state algebraically that two expressions are equal to one another. { ē'kwāt }

equation [CHEM] A symbolic expression that represents in an abbreviated form the laboratory observations of a chemical change; an equation (such as $2H_2 + O_2 \rightarrow 2H_2O$) indicates what reactants are consumed (H_2 and O_2) and what products are formed (H_2O), the correct formula of each reactant and product, and satisfies the law of conservation of atoms in that the symbols for the number of atoms reacting equals the number of atoms in the products. [MATH] A statement that each of two expressions is equal to the other. { i'kwā·zhən }

equation clock [HOROL] A timepiece that indicates the difference between mean solar time and apparent time. { i'kwā·zhən ‚kläk }

equation of continuity See continuity equation. { i'kwā·zhən əv ‚känt·ən'ü·əd·ē }

equation of mixed type [MATH] A partial differential equation which is of hyperbolic, parabolic, or elliptic type in different parts of a region. { i¦kwā·zhən əv ¦mikst 'tīp }

equation of motion [FL MECH] One of a set of hydrodynamical equations representing the application of Newton's second law of motion to a fluid system; the total acceleration on an individual fluid particle is equated to the sum of the forces acting on the particle within the fluid. [MECH] **1.** Equation which specifies the coordinates of particles as functions of time. **2.** A differential equation, or one of several such equations, from which the coordinates of particles as functions of time can be obtained if the initial positions and velocities of the particles are known. [QUANT MECH] A differential equation which enables one to predict the statistical distribution of the results of any measurement upon a system at any time if the initial dynamical state of the system is known. { i'kwā·zhən əv 'mō·shən }

equation of piezotropy [THERMO] An equation obeyed by certain fluids which states that the time rate of change of the fluid's density equals the product of a function of the thermodynamic variables and the time rate of change of the pressure. { i'kwā·zhən əv pē·ə'zä·trə·pē }

equation of state [PHYS CHEM] A mathematical expression which defines the physical state of a homogeneous substance (gas, liquid, or solid) by relating volume to pressure and absolute temperature for a given mass of the material. { i'kwā·zhən əv 'stāt }

equation of the center [ASTRON] The angle between the actual longitude of the moon and the longitude of an imaginary body that moves with constant angular velocity with the same period as the moon. { i'kwā·zhən əv thə 'sen·tər }

equation of time [ASTRON] The addition of a quantity to mean solar time to obtain apparent solar time; formerly, when apparent solar time was in common use, the opposite convention was used; apparent solar time has annual variation as a result of the sun's inclination in the ecliptic and the eccentricity of the earth's elliptical orbit. { i'kwā·zhən əv 'tīm }

equation solver [COMPUT SCI] A machine, usually analog, for solving systems of simultaneous equations, which may be linear, nonlinear, or differential, and for finding roots of polynomials. { i'kwā·zhən ‚sälv·ər }

equator [GEOD] The great circle around the earth, equally distant from the North and South poles, which divides the earth into the Northern and Southern hemispheres; the line from which latitudes are reckoned. { ē'kwād·ər }

equatorial acceleration [ASTROPHYS] A state in which the equatorial atmosphere of a celestial body has a larger absolute angular velocity than the more poleward portions of the atmosphere; exhibited by the sun, Jupiter, and Saturn. { ‚e·kwə'tòr·ē·əl ak‚sel·ə'rā·shən }

equatorial air [METEOROL] The air of the doldrums or the equatorial trough; distinguished somewhat vaguely from the tropical air of the trade-wind zones. { ‚e·kwə'tòr·ē·əl 'er }

equatorial axis [GEOD] The diameter of the earth described between two points on the equator. { ‚e·kwə'tòr·ē·əl 'ak·səs }

equatorial bulge [GEOD] The excess of the earth's equatorial diameter over the polar diameter. { ‚e·kwə'tòr·ē·əl 'bəlj }

equatorial calms See doldrums. { ‚e·kwə'tòr·ē·əl 'kämz }

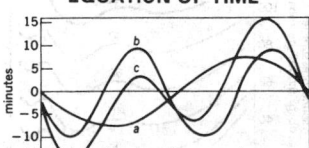

EQUATION OF TIME

The equation of time; c is the sum of the elliptical motion effect a and the inclination effect b.

equatorial chart [MAP] A chart on an equatorial projection. { ‚e·kwə'tòr·ē·əl 'chärt }

equatorial convergence zone See intertropical convergence zone. { ‚e·kwə'tòr·ē·əl kən'vər·jəns ‚zōn }

Equatorial Countercurrent [OCEANOGR] An ocean current flowing eastward (counter to and between the westward-flowing North Equatorial Current and South Equatorial Current) through all the oceans. { ‚e·kwə'tòr·ē·əl 'kaúnt·ər‚kər·ənt }

Equatorial Current See North Equatorial Current. See South Equatorial Current. { ‚e·kwə'tòr·ē·əl 'kə·rənt }

equatorial cylindrical orthomorphic chart See Mercator chart. { ‚e·kwə'tòr·ē·əl sə'lin·drə·kəl ‚òr·thō¦mòr·fik 'chärt }

equatorial dry zone [CLIMATOL] An arid region existing in the equatorial trough; the most famous dry zone is situated a little south of the equator in the central Pacific. Also known as arid zone. { ‚e·kwə'tòr·ē·əl 'drī ‚zōn }

equatorial easterlies [METEOROL] The trade winds in the summer hemisphere when they are very deep, extending at least 5 to 6 miles (8 to 10 kilometers) in altitude, and generally not topped by upper westerlies; if upper westerlies are present, they are too weak and shallow to influence the weather. Also known as deep easterlies; deep trades. { ‚e·kwə'tòr·ē·əl 'ēs·tər‚lēz }

equatorial electrojet [GEOPHYS] A concentration of electric current in the atmosphere found in the magnetic equator. { ‚e·kwə'tòr·ē·əl ə'lek·trə‚jet }

equatorial front See intertropical front. { ‚e·kwə'tòr·ē·əl 'frənt }

equatorial horizontal parallax [ASTRON] The parallax of a member of the solar system measured from positional observations made at the same time at two stations on earth, whose distance apart is the earth's equatorial radius. { ‚e·kwə'tòr·ē·əl ‚här·ə¦zänt·əl 'par·ə‚laks }

equatorial mounting [ENG] The mounting of an equatorial telescope; it has two perpendicular axes, the polar axis (parallel to the earth's axis) that turns on fixed bearings, and the declination axis, supported by the polar axis. { ‚e·kwə'tòr·ē·əl 'maúnt·iŋ }

equatorial orbit [ASTRON] An orbit in the plane of the earth's equator. { ‚e·kwə'tòr·ē·əl 'òr·bət }

equatorial plane [ASTRON] The plane passing through the equator of the earth, or of another celestial body, perpendicular to its axis of rotation and equidistant from its poles. [CELL MOL] The plane in a cell undergoing mitosis that is midway between the centrosomes and perpendicular to the spindle fibers. [MECH] A plane perpendicular to the axis of rotation of a rotating body and equidistant from the intersections of this axis with the body's surface, provided that the body is symmetric about the axis of rotation and is symmetric under reflection through this plane. [OPTICS] See sagittal plane. { ‚e·kwə'tòr·ē·əl 'plān }

equatorial projection [MAP] A map projection centered on the equator. { ‚e·kwə'tòr·ē·əl prə'jek·shən }

equatorial radius [GEOD] The radius assigned to the great circle making up the terrestrial equator; approximately 6,378,139 meters (20,925,653 feet). { ‚e·kwə'tòr·ē·əl 'rād·ē·əs }

equatorial system [ASTRON] A set of celestial coordinates based on the celestial equator as the primary great circle; usually declination and hour angle or sidereal hour angle. Also known as celestial equator system of coordinates; equinoctial system of coordinates. { ‚e·kwə'tòr·ē·əl 'sis·təm }

equatorial telescope [ENG] An astronomical telescope that revolves about an axis parallel to the earth's axis and automatically keeps a star on which it has been fixed in its field of view. { ‚e·kwə'tòr·ē·əl 'tel·ə‚skōp }

equatorial tide [OCEANOGR] **1.** A lunar fortnightly tide. **2.** A tidal component with a period of 328 hours. { ‚e·kwə'tòr·ē·əl 'tīd }

equatorial trough [METEOROL] The quasicontinuous belt of low pressure lying between the subtropical high-pressure belts of the Northern and Southern hemispheres. Also known as meteorological equator. { ‚e·kwə'tòr·ē·əl 'tròf }

Equatorial Undercurrent [OCEANOGR] **1.** A subsurface current flowing from west to east in the Indian Ocean near the 450-foot (150-meter) depth at the equator during the time of the Northeast Monsoon. **2.** A permanent subsurface current in the equatorial region of the Atlantic and Pacific oceans. { ‚e·kwə'tòr·ē·əl 'ən·dər‚kə·rənt }

equatorial vortex [METEOROL] A closed cyclonic circulation with the equatorial trough. { ¦e·kwə'tȯr·ē·əl 'vȯrˌteks }

equatorial wave [METEOROL] A wavelike disturbance of the equatorial easterlies that extends across the equatorial trough. { ¦e·kwə'tȯr·ē·əl 'wāv }

equatorial westerlies [METEOROL] The westerly winds occasionally found in the equatorial trough and separated from the mid-latitude westerlies by the broad belt of easterly trade winds. { ¦e·kwə'tȯr·ē·əl 'wes·tər·lēz }

equiangular polygon [MATH] A polygon all of whose interior angles are equal. { ¦ē·kwē¦aŋ·gyə·lər 'päl·əˌgän }

equiangular spiral See logarithmic spiral. { ¦ē·kwē¦aŋ·gyə·lər 'spī·rəl }

equiangular spiral antenna [ELECTROMAG] A frequency-independent broad-band antenna, cut from sheet metal, that radiates a very broad, circularly polarized beam on both sides of its surface; this bidirectional radiation pattern is its chief limitation. { ¦ē·kwē¦aŋ·gyə·lər ¦spī·rəl an'ten·ə }

equicontinuous at a point [MATH] A family of functions is equicontinuous at a point x if for any $\epsilon > 0$ there is a $\delta > 0$ such that, whenever $|x-y| < \delta$, $|f(x)-f(y)| < \epsilon$ for every function $f(x)$ in the family. { ¦ē·kwē·kən'tin·yə·was at ə 'pȯint }

equicontinuous family of functions [MATH] A family of functions with the property that given any $\epsilon > 0$ there is a $\delta > 0$ such that whenever $|x - y| < \delta$, $|f(x) - f(y)| < \epsilon$ for every function $f(x)$ in the family. Also known as uniformly equicontinuous family of functions. { ¦ē·kwē·kən'tin·yə·wəs 'fam·lē əv 'fəŋk·shənz }

Equidae [VERT ZOO] A family of perissodactyl mammals in the superfamily Equoidea, including the horses, zebras, and donkeys. { 'ek·wəˌdē }

equidecomposable [MATH] The property of two plane or space regions, either of which can be disassembled into finite number of pieces and reassembled to form the other one. { ¦ek·wēˌdē·kəm'pōz·ə·bəl }

equidensity technique [ANALY CHEM] Interference microscopy technique utilizing the Sabattier effect in photographic emulsions; the equidensities (lines of equal density in a photographic emulsion) are produced by exactly superimposing a positive and a negative of the same interferogram, and making a copy; used to measure photographic film emulsion density. { ¦ē·kwə¦den·səd·ē ˌtekˌnēk }

equidistant [MATH] Being the same distance from some given object. { ¦ē·kwə¦dis·tənt }

equidistant system [MATH] A system of parametric curves on a surface obtained by setting surface coordinates u and v equal to various constants, where the coordinates are chosen so that an element of length ds on the surface is given by $ds^2 = du^2 + F\,dudv + dv^2$, where F is a function of u and v. { ¦ē·kwə¦dis·tənt 'sis·təm }

equigeopotential surface See geopotential surface. { ¦ē·kwə·jē·ō·pə'ten·chəl 'sər·fəs }

equigranular [PETR] Pertaining to the texture of rocks whose essential minerals are all of the same order of size. { ¦ē·kwə'gran·yə·lər }

equilateral arch [ARCH] An arch described by two circular curves intersecting at the peak of the arch, each curve having a chord equal to the span. { ¦ē·kwə'lad·ə·rəl 'ärch }

equilateral polygon [MATH] A polygon all of whose sides are the same length. { ¦ē·kwə'lad·ə·rəl 'päl·əˌgän }

equilateral polyhedron [MATH] A polyhedron all of whose faces are identical. { ¦ē·kwə'lad·ə·rəl ¦päl·ē'hē·drən }

equilibrant [MECH] A single force which cancels the vector sum of a given system of forces acting on a rigid body and whose torque cancels the sum of the torques of the system. { i'kwil·ə·brənt }

equilibration [PSYCH] As described by Jean Piaget, the operation together of the two processes of assimilating new information and accommodating to this information. { iˌkwil·ə'brā·shən }

equilibrator [ORD] Force-producing mechanism designed to provide a moment about the trunnions of a gun cradle or launcher which is equal and opposite to that caused by the unbalanced weight of the tipping parts, thus making it easier to elevate the weapon. { i'kwil·əˌbrād·ər }

equilibristat [ENG] A device for measuring the deviation from equilibrium of a railroad car as it goes around a curve. { ˌē·kwə'lib·rəˌstat }

equilibrium [CHEM] See chemical equilibrium. [MECH] Condition in which a particle, or all the constituent particles of a body, are at rest or in unaccelerated motion in an inertial reference frame. Also known as static equilibrium. [PHYS] Condition in which no change occurs in the state of a system as long as its surroundings are unaltered. [STAT MECH] Condition in which the distribution function of a system is time-independent. { ˌē·kwə'lib·rē·əm }

equilibrium brightness [ELECTR] Viewing screen brightness occurring when a display storage tube is in a fully written condition. { ˌēkwə'lib·rē·əm 'brīt·nəs }

equilibrium constant [CHEM] A constant at a given temperature such that when a reversible chemical reaction $cC + bB = gG + hH$ has reached equilibrium, the value of this constant K^0 is equal to

$$\frac{a_G^g\,a_H^h}{a_C^c\,a_B^b}$$

where a_G, a_H, a_C, and a_B represent chemical activities of the species G, H, C, and B at equilibrium. { ˌē·kwə'lib·rē·əm ˌkän·stənt }

equilibrium diagram [PHYS CHEM] A phase diagram of the equilibrium relationship between temperature, pressure, and composition in any system. { ˌē·kwə'lib·rē·əm 'dī·əˌgram }

equilibrium dialysis [ANALY CHEM] A technique used to determine the degree of ion bonding by protein; the protein solution, placed in a bag impermeable to protein but permeable to small ions, is immersed in a solution containing the diffusible ion whose binding is being studied; after equilibration of the ion across the membrane, the concentration of ion in the protein-free solution is determined; the concentration of ion in the protein solution is determined by subtraction; if binding has occurred, the concentration of ion in the protein solution must be greater. { ˌē·kwə'lib·rē·əm dī'al·ə·səs }

equilibrium distillation See equilibrium flash vaporization. { ˌē·kwə'lib·rē·əm ˌdis·tə¦lā·shən }

equilibrium film [PHYS CHEM] A liquid film that is stable or metastable at a certain thickness with respect to small changes in the thickness. { ˌē·kwə'lib·rē·əm ˌfilm }

equilibrium flash vaporization [CHEM ENG] Process in which a continuous liquid-mixture feed stream is partly vaporized in a column or vessel, with continuous withdrawal of vapor and liquid portions, the vapor and liquid in equilibrium. Also known as continuous equilibrium vaporization; equilibrium distillation; flash distillation; simple continuous distillation. { ˌē·kwə'lib·rē·əm 'flash ˌva·pə·rə¦zā·shən }

equilibrium gas saturation [PETRO ENG] Condition of zero relative permeability of a nonwetting/wetting phase system in a reservoir; relation to the nonwetting phase (for example, oil) to the wetting phase (for example, water) when the nonwetting-phase saturation is so small that relatively few pores contain it. Also known as critical gas saturation. { ˌē·kwə'lib·rē·əm 'gas ˌsach·ə'rā·shən }

equilibrium line [HYD] The level on a glacier where the net balance equals zero and accumulation equals ablation. { ˌē·kwə'lib·rē·əm ˌlīn }

equilibrium moisture content [PHYS CHEM] The moisture content in a hydroscopic material that is being dried by contact with air at constant temperature and humidity when a definite, fixed (equilibrium) moisture content in the solid is reached. { ˌē·kwə'lib·rē·əm 'mȯis·chər ˌkän·tent }

equilibrium orbit [NUCLEO] A path, such as a circular path in a synchrotron, or a point moving through space, such as in a linear accelerator, about which the particles in a particle accelerator oscillate, experiencing an effective restoring force toward the path or point. Also known as stable orbit. { ˌē·kwə'lib·rē·əm ˌȯr·bət }

equilibrium population [EVOL] A population in which the gene frequencies have reached an equilibrium between mutation pressure and selection pressure. { ˌē·kwə'lib·rē·əm ˌpäp·yə¦lā·shən }

equilibrium potential [PHYS CHEM] A point in which forward and reverse reaction rates are equal in an electrolytic solution, thereby establishing the potential of an electrode. { ˌē·kwə'lib·rē·əm pə'ten·chəl }

equilibrium prism [PHYS CHEM] Three-dimensional (solid) diagram for multicomponent mixtures to show the effects of composition changes on some key property, such as freezing point. { ¸ē·kwə'lib·rē·əm ¸priz·əm }

equilibrium profile *See* profile of equilibrium. { ¸ē·kwə'lib·rē·əm 'prō¸fīl }

equilibrium ratio [PHYS CHEM] **1.** In any system, relation of the proportions of the various components (gas, liquid) at equilibrium conditions. **2.** *See* equilibrium vaporization ratio. { ¸ē·kwə'lib·rē·əm ¸rā·shō }

equilibrium solar tide [GEOPHYS] The form of the atmosphere which is determined solely by gravitational forces in the absence of any rotation of the earth relative to the sun. { ¸ē·kwə'lib·rē·əm ¸sō·lər 'tīd }

equilibrium solubility [PHYS CHEM] The maximum solubility of one material in another (for example, water in hydrocarbons) for specified conditions of temperature and pressure. { ¸ē·kwə'lib·rē·əm ¸säl·yə'bil·əd·ē }

equilibrium spheroid [GEOPHYS] The shape that the earth would attain if it were entirely covered by a tideless ocean of constant depth. { ¸ē·kwə'lib·rē·əm 'sfir¸oid }

equilibrium state [IND ENG] A state in which the numbers of customers or items waiting in a queue varies in such a way that the mean and distribution remain constant over a long period. { ¸ē·kwə'lib·rē·əm ¸stāt }

equilibrium still [ANALY CHEM] Recirculating distillation apparatus (no product withdrawal) used to determine vapor-liquid equilibria data. { ¸ē·kwə'lib·rē·əm ¸stil }

equilibrium theory [OCEANOGR] An ocean water model which assumes instantaneous response of water bodies to the tide-producing forces of the moon and sun to form an equilibrium surface, and disregards the effects due to friction, inertia, and irregular distribution of land masses. { ¸ē·kwə'lib·rē·əm ¸thē·ə·rē }

equilibrium tide [OCEANOGR] The hypothetical tide due to the tide-producing forces of celestial bodies, particularly the sun and moon. { ¸ē·kwə'lib·rē·əm ¸tīd }

equilibrium vaporization ratio [PHYS CHEM] In a liquid-vapor equilibrium mixture, the ratio of the mole fraction of a component in the vapor phase (y) to the mole fraction of the same component in the liquid phase (x), or $y/x = K$ (the K factor). Also known as equilibrium ratio. { ¸ē·kwə'lib·rē·əm ¸vā·pə·rə'zā·shən ¸rā·shō }

equilibrium vapor pressure [PHYS] The vapor pressure of a system in which two or more phases of water coexist in equilibrium. { ¸ē·kwə'lib·rē·əm 'vā·pər ¸presh·ər }

equine [VERT ZOO] **1.** Resembling a horse. **2.** Of or related to the Equidae. { 'ē¸kwīn }

equine encephalitis [MED] A disease of equines and humans caused by one of three viral strains: eastern, western, and Venezuelan equine viruses. Also known as equine encephalomyelitis. { 'ē¸kwīn en¸sef·ə'līd·əs }

equine encephalomyelitis *See* equine encephalitis. { 'ē¸kwīn en¸sef·ə·lō¸mī·ə'līd·əs }

equinoctial *See* celestial equator. { ¸ē·kwə'näk·shəl }

equinoctial colure [ASTRON] The great circle of the celestial sphere through the celestial poles and the equinoxes; the hour circle of the vernal equinox. { ¸ē·kwə'näk·shəl kə'lúr }

equinoctial point *See* equinox. { ¸ē·kwə'näk·shəl 'póint }

equinoctial rains [METEOROL] Rainy seasons which occur regularly at or shortly after the equinoxes in many places within a few degrees of the equator. { ¸ē·kwə'näk·shəl 'rānz }

equinoctial storm [METEOROL] In semipopular belief, a violent storm of wind and rain which is supposed, both in the United States and in Britain, to occur at or near the time of the equinox. Also known as line gale; line storm. { ¸ē·kwə'näk·shəl 'stórm }

equinoctial system of coordinates *See* equatorial system. { ¸ē·kwə'näk·shəl ¦sis·təm əv kō'órd·ən·əts }

equinoctial tide [OCEANOGR] A tide occurring near an equinox. { ¸ē·kwə'näk·shəl 'tīd }

equinox [ASTRON] **1.** Either of the two points of intersection of the ecliptic and the celestial equator, occupied by the sun when its declination is 0°. Also known as equinoctial point. **2.** That instant when the sun occupies one of the equinoctial points. { 'ē·kwə¸näks }

equinumerable sets *See* equivalent sets. { ¸ek·wə¦nüm·rə·bəl 'sets }

equiparte [METEOROL] In Mexico, heavy cold rains during October to January, which last for several days. Also known as equipatos. { ¦e·kwē¦pär'tā }

equipartition [CHEM] **1.** The condition in a gas where under equal pressure the molecules of the gas maintain the same average distance between each other. **2.** The equal distribution of a compound between two solvents. **3.** The distribution of the atoms in an orderly fashion, such as in a crystal. { ¦e·kwə·pär'tish·ən }

equipartition law [STAT MECH] In a classical ideal gas, the average kinetic energy per molecule associated with any degree of freedom which occurs as a quadratic term in the expression for the mechanical energy, is equal to half of Boltzmann's constant times the absolute temperature. { ¦e·kwə·pär'tish·ən ¸lò }

equipatos *See* equiparte. { ¦e·kwē¦pä·tós }

equiphase wave surface [PHYS] Any surface in a wave over which the field vectors at the same instant are in the same phase or 180° out of phase. { 'e·kwə¸fāz 'wāv ¸sər·fəs }

equiphase zone [GEOPHYS] That region in space where the difference in phase of two radio signals is indistinguishable. { 'e·kwə¸fāz ¸zōn }

equipment [ENG] One or more assemblies capable of performing a complete function. { ə'kwip·mənt }

equipment augmentation [COMPUT SCI] **1.** Procuring additional automatic data-processing equipment capability to accommodate increased work load within an established data system. **2.** Obtaining additional sites or locations. { ə'kwip·mənt ¸óg·mən'tā·shən }

equipment chain [ENG] Group of equipments that are functionally in series; the failure of one or more of the equipments results in loss of the function. { ə'kwip·mənt ¸chān }

equipment characteristic distortion [COMMUN] Teletypewriter transmission repetitive display or disruption peculiar to specific portions of a signal, normally caused by maladjusted or dirty contacts of the sending or receiving equipment. { ə'kwip·mənt ¸kar·ik·tə¸ris·tik di'stór·shən }

equipment compatibility [COMPUT SCI] The ability of a device to handle data prepared or handled by other equipment, without alteration of the code or of the form of the data. { ə'kwip·mənt kəm¸pad·ə'bil·əd·ē }

equipment failure [COMPUT SCI] A fault in equipment that results in its improper behavior or prevents the execution of a job as scheduled. { ə'kwip·mənt ¸fāl·yər }

equipment misuse error [COMPUT SCI] An erroneous programming instruction, such as a read command addressed to a card punch. { ə¦kwip·mənt mis'yüs ¸er·ər }

equipment replacement study [IND ENG] A cost analysis based on estimates of operating costs over a stated time for the old facility compared with the new facility. { ə'kwip·mənt ri'plās·mənt ¸stəd·ē }

equipollent [MECH] Of two systems of forces, having the same vector sum and the same total torque about an arbitrary point. { ¦e·kwə¦päl·ənt }

equipotent [SCI TECH] Equal in capacity or effect. { ¦e·kwə¦pōt·ənt }

equipotent sets *See* equivalent sets. { ¸ek·wə¦pōt·ənt 'sets }

equipotential cathode *See* indirectly heated cathode. { ¦e·kwə·pə'ten·chəl 'kath¸ōd }

equipotential surface [ELEC] A surface on which the electric potential is the same at every point. [GEOPHYS] A surface characterized by the potential being constant everywhere on it for the attractive forces concerned. [MECH] A surface which is always normal to the lines of force of a field and on which the potential is everywhere the same. { ¦e·kwə·pə'ten·chəl 'sər·fəs }

equipressure contour [PETRO ENG] Within a reservoir, a plot or map of the equal isopressure flow network; used to locate sites for water-injection wells for flood coverage of an areal reservoir pattern. { ¦e·kwə¦presh·ər 'kän·túr }

Equisetales [BOT] The horsetails, a monogeneric order of the class Equisetopsida; the only living genus is *Equisetum*. { ¸e·kwə·sə'tā·lēz }

Equisetatae *See* Equisetopsida. { ¸e·kwə·sə'tā¸tē }

Equisetineae [BOT] The equivalent name for the Equisetophyta. { ¸e·kwə·sə'tin·ē¸ē }

Equisetophyta [BOT] A division of the subkingdom Embryobionta represented by a single living genus, *Equisetum*. { ¸e·kwə·sə'täf·ə·də }

Equisetopsida [BOT] A class of the division Equisetophyta whose members made up a major part of the flora, especially

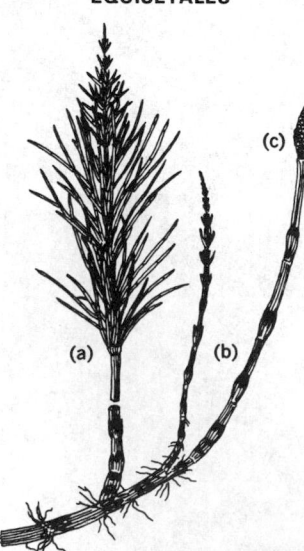

EQUISETALES

Equisetum arvense, (a) sterile shoot, (b) fertile shoot growing from underground rootstock, (c) cone. (From E. W. Sinnott and K. S. Wilson, Botany: Principles and Problems, 6th ed., McGraw-Hill, 1963)

in moist or swampy places, during the Carboniferous Period. { ¦e·kwə¦sə'täp·sə·də }

equisignal [COMMUN] **1.** Pertaining to two signals of equal intensity, used particularly with reference to the signals of a radio range station. **2.** Referring to a radio system in which two identifiable separate radio signals are received with the same intensity. { ¦e·kwə¦sig·nəl }

equisignal localizer [NAV] An aircraft guidance localizer in which the localizer on-course line is centered in a zone of equal amplitude of two transmitted signals; deviations from this zone are detectable as unbalance in the levels of the two signals. Also known as equisignal radio-range beacon; tone localizer. { ¦e·kwə¦sig·nəl 'lō·kə‚līz·ər }

equisignal radio-range beacon *See* equisignal localizer. { ¦e·kwə¦sig·nəl 'rād·ē·ō ‚rānj‚bē·kən }

equisignal surface [ELECTROMAG] Surface around an antenna formed by all points at which, for transmission, the field strength (usually measured in volts per meter) is constant. { ¦e·kwə¦sig·nəl ‚sər·fəs }

equisignal zone [NAV] That region in space where the difference in amplitude of two radio signals (usually emitted by a signal station) is indistinguishable. { ¦e·kwə¦sig·nəl ‚zōn }

equitangential curve *See* tractrix. { ‚ē·kwə·tan'jen·chəl 'kərv }

equitant [BOT] Of leaves, overlapping transversely at the base. { 'e·kwəd·ənt }

equitempered scale *See* equally tempered scale. { ¦e·kwə¦tem·pərd 'skāl }

equity crude [PETRO ENG] Crude produced which belongs to an oil company that owns a concession jointly with a host government. { 'e·kwəd·ē ‚krüd }

equivalence [MAP] In an equal-area map projection, the property of having the ratio between areas on the map the same as the ratio between corresponding areas on the earth's surface. [MATH] A logic operator having the property that if P, Q, R, etc., are statements, then the equivalence of P, Q, R, etc., is true if and only if all statements are true or all statements are false. { i'kwiv·ə·ləns }

equivalence classes [MATH] The collection of pairwise disjoint subsets determined by an equivalence relation on a set; two elements are in the same equivalence class if and only if they are equivalent under the given relation. { i'kwiv·ə·ləns ‚klas·əs }

equivalence element *See* equivalence gate. { i'kwiv·ə·ləns ‚el·ə·mənt }

equivalence gate [COMPUT SCI] A logic circuit that produces a binary output signal of 1 if its two binary input signals are the same, and an output signal of 0 if the input signals differ. Also known as biconditional gate; equality gate; equivalence element; exclusive-NOR gate; match gate. { i'kwiv·ə·ləns ‚gāt }

equivalence law of ordered sampling [STAT] If a random ordered sample of size s is drawn from a population of size N, then on any particular one of the s draws each of the N items has the same probability, 1/N, of appearing. { i'kwiv·ə·ləns ¦lō əv ¦òr·dərd 'sam·pliŋ }

equivalence point [CHEM] The point in a titration where the amounts of titrant and material being titrated are equivalent chemically. { i'kwiv·ə·ləns ‚pòint }

equivalence principle [RELAT] In general relativity, the principle that the observable local effects of a gravitational field are indistinguishable from those arising from acceleration of the frame of reference. Also known as Einstein's equivalency principle; principle of equivalence. { i'kwiv·ə·ləns ‚prin·sə·pəl }

equivalence relation [MATH] A relation which is reflexive, symmetric, and transitive. Also known as equals functions. { i'kwiv·ə·ləns ri'lā·shən }

equivalence transformation [MATH] A mapping which associates with each square matrix A the matrix B = SAT, where S and T are nonsingular matrices. Also known as equivalent transformation. { i'kwiv·ə·ləns ‚tranz·fər‚mā·shən }

equivalence zone *See* zone of optimal proportion. { i'kwiv·ə·ləns ‚zōn }

equivalent absorption area [ACOUS] Area of perfectly absorbing surface that will absorb sound energy at the same rate as the given object under the same conditions; the acoustic unit of equivalent absorption is the sabin. { i'kwiv·ə·lənt əb'sòrp·shən ‚er·ē·ə }

equivalent airspeed [AERO ENG] The product of the true airspeed and the square root of the density ratio; used in structural design work to designate various design conditions. { i'kwiv·ə·lənt 'er‚spēd }

equivalent angles [MATH] Two rotation angles that have the same measure. { i‚kwiv·ə·lənt 'aŋ·gəls }

equivalent annual rate [IND ENG] A measure used in setting up a monthly rate on a comparable basis for each of the months regardless of their variation in working days, or for making the rate comparable with an annual rate regardless of the variation in working days during each month. { i'kwiv·ə·lənt ¦an·yə·wəl 'rāt }

equivalent barotropic model [METEOROL] A model atmosphere characterized by frictionless and adiabatic flow and by hydrostatic quasigeostrophic equilibrium, and in which the vertical shear of the horizontal wind is assumed to be proportional to the horizontal wind itself. { i'kwiv·ə·lənt bar·ə'träp·ik 'mäd·əl }

equivalent bending moment [MECH] A bending moment which, acting alone, would produce in a circular shaft a normal stress of the same magnitude as the maximum normal stress produced by a given bending moment and a given twisting moment acting simultaneously. { i'kwiv·ə·lənt 'bend·iŋ ‚mō·mənt }

equivalent binary digits [COMPUT SCI] The number of binary positions required to enumerate the elements of a given set. { i'kwiv·ə·lənt 'bī‚ner·ē 'dij·əts }

equivalent blackbody temperature [THERMO] For a surface, the temperature of a blackbody which emits the same amount of radiation per unit area as does the surface. { i'kwiv·ə·lənt 'blak‚bäd·ē ‚tem·prə·chər }

equivalent circuit [ELEC] A circuit whose behavior is identical to that of a more complex circuit or device over a stated range of operating conditions. { i'kwiv·ə·lənt 'sər·kət }

equivalent conductance [PHYS CHEM] Property of an electrolyte, equal to the specific conductance divided by the number of gram equivalents of solute per cubic centimeter of solvent. { i'kwiv·ə·lənt kən'dək·təns }

equivalent continued fractions [MATH] Continued fractions whose values to n terms are the same for n = 1, 2, 3, ... { i¦kwiv·ə·lənt kən¦tin·yud 'frak·shənz }

equivalent diameter *See* nominal diameter. { i'kwiv·ə·lənt dī'am·əd·ər }

equivalent electrons [ATOM PHYS] Electrons in an atom which have the same principal and orbital quantum numbers, but not necessarily the same magnetic orbital and magnetic spin quantum numbers. { i'kwiv·ə·lənt i'lek‚tränz }

equivalent elements *See* associates. { ə¦kwiv·ə·lənt 'el·ə·məns }

equivalent equations [MATH] Equations that have the same set of solutions. { i¦kwiv·ə·lənt i'kwā·zhənz }

equivalent evaporation [FL MECH] The amount of water, usually in pounds per hour, evaporated from a temperature of 212°F (100°C) to saturated steam at the same temperature. { i'kwiv·ə·lənt i‚vap·ə'rā·shən }

equivalent focal length [OPTICS] The focal length of a thin lens which forms images that most nearly duplicate those of a given compound lens, thick lens, or system of lenses. { i'kwiv·ə·lənt 'fō·kəl ‚leŋkth }

equivalent footcandle *See* footlambert. { i'kwiv·ə·lənt 'fùt‚kan·dəl }

equivalent four-wire system [COMMUN] A transmission system in which multiplex techniques are used to carry on duplex operation over a single pair of wires. { i'kwiv·ə·lənt ¦fòr ¦wīr 'sis·təm }

equivalent head wind [NAV] A fictitious wind blowing along the track of an aircraft in the opposite direction to that of motion of the aircraft and of such speed that it would result in the same ground speed as that actually attained. { i'kwiv·ə·lənt 'hed ‚wind }

equivalent height *See* virtual height. { i'kwiv·ə·lənt 'hīt }

equivalent inequalities [MATH] Inequalities that have the same set of solutions. { i¦kwiv·ə·lənt ‚in·i'kwäl·əd·ēz }

equivalent magnetic length *See* effective magnetic length. { i¦kwiv·ə·lənt mag¦ned·ik 'leŋkth }

equivalent map projection *See* equal-area map projection. { i'kwiv·ə·lənt 'map prə‚jek·shən }

equivalent nitrogen pressure [MECH] The pressure that would be indicated by a device if the gas inside it were replaced by nitrogen of equivalent molecular density. { i'kwiv·ə·lənt 'nī·trə·jən ˌpresh·ər }

equivalent noise conductance [ELECTR] Spectral density of a noise current generator measured in conductance units at a specified frequency. { i'kwiv·ə·lənt 'nȯiz kənˌdək·təns }

equivalent noise pressure [ENG ACOUS] In an electroacoustic transducer or sound reception system, the root-mean-square sound pressure of a sinusoidal plane progressive wave, which when propagated parallel to the primary axis of the transducer, produces an open-circuit signal voltage equivalent to the root-mean-square of the inherent open-circuit noise voltage of the transducer in a transmission band with a bandwidth of 1 hertz and centered on the frequency of the plane sound wave. Also known as inherent noise pressure. { i'kwiv·ə·lənt 'nȯiz ˌpresh·ər }

equivalent noise resistance [ELECTR] Spectral density of a noise voltage generator measured in ohms at a specified frequency. { i'kwiv·ə·lənt 'nȯiz riˌzis·təns }

equivalent noise temperature [ELECTR] Absolute temperature at which a perfect resistor, of equal resistance to the component, would generate the same noise as does the component at room temperature. { i'kwiv·ə·lənt 'nȯiz ˌtem·prə·chər }

equivalent nuclei [PHYS CHEM] A set of nuclei in a molecule which are transformed into each other by rotations, reflections, or combinations of these operations, leaving the molecule invariant. { i'kwiv·ə·lənt 'nü·klē·ī }

equivalent orifice [MECH ENG] An expression of fan performance as the theoretical sharp-edge orifice area which would offer the same resistance to flow as the system resistance itself. { i'kwiv·ə·lənt 'ȯr·ə·fəs }

equivalent periodic line [ELEC] Of a uniform line, a periodic line having the same electrical behavior, at a given frequency, as the uniform line when measured at its terminals or at corresponding section junctions. { i'kwiv·ə·lənt pir·ē¦äd·ik 'līn }

equivalent potential temperature [METEOROL] The potential temperature corresponding to the adiabatic equivalent temperature. { i'kwiv·ə·lənt pə¦ten·chəl 'tem·prə·chər }

equivalent propositional functions [MATH] Propositional functions that have the same truth sets. { iˌkwiv·ə·lənt ˌpräp·ə¦zish·ən·əl 'fəŋk·shənz }

equivalent propositions [MATH] Two propositions, either of which is true if and only if the other is true. { iˌkwiv·ə·lənt ˌpräp·ə¦zish·ənz }

equivalent resistance [ELEC] Concentrated or lumped resistance that would cause the same power loss as the actual small resistance values distributed throughout a circuit. { i'kwiv·ə·lənt ri¦zis·təns }

equivalent round [ENG] The diameter of a circle whose circumference is equal to the circumference of a pipe whose cross section is not a perfect circle. { i¦kwiv·ə·lənt 'raund }

equivalent sets [MATH] Sets which have the same cardinal number; sets whose elements can be put into one-to-one correspondence with each other. Also known as equinumerable sets; equipotent sets. { i¦kwiv·ə·lənt 'sets }

equivalent sine wave [PHYS] A sine wave whose root-mean-square value and period are the same as that of a given periodic wave. { i¦kwiv·ə·lənt 'sīn ˌwāv }

equivalent tail wind [NAV] A fictitious wind blowing along the track of an aircraft in the same direction as that of motion of the aircraft and of such speed that it would result in the same ground speed as that actually attained. { i'kwiv·ə·lənt 'tāl ˌwind }

equivalent temperature [METEOROL] **1.** The temperature that an air parcel would have if all water vapor were condensed out at constant pressure, the latent heat released being used to heat the air. Also known as isobaric equivalent temperature. **2.** The temperature that an air parcel would have after undergoing the following theoretical process: dry-adiabatic expansion until saturated, pseudoadiabatic expansion until all moisture is precipitated out, and dry adiabatic compression to the initial pressure; this is the equivalent temperature as read from a thermodynamic chart and is always greater than the isobaric equivalent temperature. Also known as adiabatic equivalent temperature; pseudoequivalent temperature. [THERMO] A term used in British engineering for that temperature of a uniform enclosure in which, in still air, a sizable blackbody at 75°F (23.9°C) would lose heat at the same rate as in the environment. { i'kwiv·ə·lənt 'tem·prə·chər }

equivalent twisting moment [MECH] A twisting moment which, if acting alone, would produce in a circular shaft a shear stress of the same magnitude as the shear stress produced by a given twisting moment and a given bending moment acting simultaneously. { i'kwiv·ə·lənt 'twist·iŋ ˌmō·mənt }

equivalent vapor volume [PETRO ENG] The volume occupied by a barrel of oil if all the oil were to become a vapor; expressed as cubic foot per barrel at 60°F (15.6°C). { i'kwiv·ə·lənt 'vā·pər ˌväl·yəm }

equivalent viscous damping [MECH] An assumed value of viscous damping used in analyzing a vibratory motion, such that the dissipation of energy per cycle at resonance is the same for the assumed or the actual damping force. { i'kwiv·ə·lənt ¦vis·kəs 'damp·iŋ }

equivalent weight [CHEM] The number of parts by weight of an element or compound which will combine with or replace, directly or indirectly, 1.008 parts by weight of hydrogen, 8.00 parts of oxygen, or the equivalent weight of any other element or compound. { i'kwiv·ə·lənt 'wāt }

equivalent width [PHYS] A measure of the total absorption of radiant energy as indicated by an absorption line or absorption band. { i'kwiv·ə·lənt 'width }

equiviscous temperature [CHEM ENG] A measure of viscosity used in the tar industry, equal to the temperature in degrees Celsius at which the viscosity of tar is 50 seconds as measured in a standard tar efflux viscometer. Abbreviated EVT. { ¦e·kwə¦vis·kəs 'tem·prə·chər }

equivoluminal wave *See* S wave. { ¦e·kwə·və¦lüm·ə·nəl 'wāv }

Equl *See* Equuleus.

Equoidea [VERT ZOO] A superfamily of perissodactyl mammals in the suborder Hippomorpha comprising the living and extinct horses and their relatives. { ē'kwȯid·ē·ə }

Equuleus [ASTRON] A northern constellation near Aquarius, right ascension 21 hours, declination 10° north. Abbreviated Equl. Also known as Little Horse. { e'kwül·ē·əs }

Equus [VERT ZOO] The genus comprising the large, one-toed modern horses, including donkeys and zebras. { 'e·kwəs }

Er *See* erbium.

ER *See* electroreflectance; endoplasmic reticulum.

era [GEOL] A unit of geologic time constituting a subdivision of an eon and comprising one or more periods. { 'ir·ə }

ERA *See* electron ring accelerator.

eradiation *See* terrestrial radiation. { iˌrād·ē'ā·shən }

erasability of storage [COMPUT SCI] Ability to erase data that are recorded in a particular location and replace them with new data; storage media are said to be erasable (for example, magnetic tape) or nonerasable (for example, punched cards). { iˌrās·ə'bil·əd·ē əv 'stȯr·ij }

erasable programmable read-only memory [COMPUT SCI] A read-only memory in which stored data can be erased by ultraviolet light or other means and reprogrammed bit by bit with appropriate voltage pulses. Abbreviated EPROM. { i¦rās·ə·bəl prō¦gram·ə·bəl ¦rēd ˌōn·lē 'mem·rē }

erasable storage [COMPUT SCI] Any storage medium which permits new data to be written in place of the old, such as magnetic disk or tape, but not punched card or punched tape. { i¦rās·ə·bəl 'stȯr·ij }

erase [COMPUT SCI] To change all the binary digits in a digital computer storage device to binary zeros. [ELECTR] **1.** To remove recorded material from magnetic tape by passing the tape through a strong, constant magnetic field (dc erase) or through a high-frequency alternating magnetic field (ac erase). **2.** To eliminate previously stored information in a charge-storage tube by charging or discharging all storage elements. { i'rās }

erase character *See* ignore character. { i¦rās 'kar·ik·tər }

erase oscillator [ELECTR] The oscillator used in a magnetic recorder to provide the high-frequency signal needed to erase a recording on magnetic tape; the bias oscillator usually serves also as the erase oscillator. { i'rās ˌäs·ə¦lād·ər }

erasing head [ELECTR] A magnetic head used to obliterate material previously recorded on magnetic tape. { i'rās·iŋ ˌhed }

erasing speed [ELECTR] In charge-storage tubes, the rate of erasing successive storage elements. { i'rās·iŋ ˌspēd }

erathem [GEOL] A chronostratigraphic unit, below eonothem and above system, composed of rocks formed during an era of geologic time. { 'er·ə,them }

ERBE *See* Earth Radiation Budget Experiment. { 'ər,bē }

erbia *See* erbium oxide. { 'ər·bē·ə }

erbium [CHEM] A trivalent metallic rare-earth element, symbol Er, of the yttrium subgroup, found in euxenite, gadolinite, fergusonite, and xenotine; atomic number 68, atomic weight 167.26, specific gravity 9.051; insoluble in water, soluble in acids; melts at 1400–1500°C. { 'ər·bē·əm }

erbium-doped fiber amplifier [COMMUN] An optical-fiber amplifier whose fiber core is lightly doped with trivalent erbium ions which absorb light at pump wavelengths of 0.98 and 1.48 micrometers and emit it at a signal wavelength around 1.5 micrometers through stimulated emission. Abbreviated EDFA. { ¦ər·bē·əm ,dōpt ,fī·bər 'am·plə,fī·ər }

erbium halide [INV ZOO] A compound of erbium and one of the halide elements. { 'ər·bē·əm 'hal,īd }

erbium nitrate [INV ZOO] $Er(NO_3)_3 \cdot 5H_2O$ Pink crystals that are soluble in water, alcohol, and acetone; may explode if it is heated or shocked. { 'ər·bē·əm 'nī,trāt }

erbium oxalate [ORG CHEM] $Er_2(C_2O_4)_3 \cdot 10H_2O$ A red powder that decomposes at 575°C; used to separate erbium from common metals. { 'ər·bē·əm 'äk·sə,lāt }

erbium oxide [INV ZOO] Er_2O_3 Pink powder that is insoluble in water; used as an actuator for phosphors and in manufacture of glass that absorbs in the infrared. Also known as erbia. { 'ər·bē·əm 'äk,sīd }

erbium sulfate [INV ZOO] $Er_2(SO_4)_3 \cdot 8H_2O$ Red crystals that are soluble in water. { 'ər·bē·əm 'səl,fāt }

erbon [ORG CHEM] $C_{11}H_9Cl_5O_3$ A white solid with a melting point of 49–50°C; insoluble in water; used as a herbicide for perennial broadleaf weeds. { 'ər,bän }

erect image [OPTICS] An image in which directions are the same as those in the object, in contrast to an inverted image. { i'rekt 'im·ij }

erecting lens [OPTICS] An eyepiece sometimes used in Kepler telescopes that consists of four lenses and provides an erect image, which is more convenient for viewing terrestrial objects than the inverted image provided by simpler eyepieces. { i'rek·tiŋ ,lenz }

erecting prism [OPTICS] A system of prisms that converts the inverted image formed by most types of astronomical telescopes into an erect image. Also known as inverting prism. { i'rek·tiŋ ,priz·əm }

erection [CIV ENG] Positioning and fixing the frame of a structure. [PHYSIO] The enlarged state of erectile tissue when engorged with blood, as of the penis or clitoris. { i'rek·shən }

erection bolt [CIV ENG] A threaded rod with a head at one end, used to temporarily join parts of a structure during construction. { i'rek·shən ,bōlt }

erection stress [MECH] The internal forces exerted on a structural member during construction. { i'rek·shən ,stres }

erection tower [CIV ENG] A temporary framework built at a construction site for hoisting equipment. { i'rek·shən ,taů·ər }

erector [PHYSIO] Any muscle that produces erection of a part. { i'rek·tər }

erect stem [BOT] A stem that stands, having a vertical or upright habit. { i'rekt 'stem }

Eremascoideae [BOT] A monogeneric subfamily of ascosporogenous yeasts characterized by mostly septate mycelia, and spherical asci with eight oval to round ascospores. { er·ə·mə'skóid·ē,ē }

eremophobia [PSYCH] An abnormal fear of being lonely. { ə,rem·ə'fō·bē·ə }

Erethizontidae [VERT ZOO] The New World porcupines, a family of rodents characterized by sharply pointed, erectile hairs and four functional digits. { ,er·ə·thə'zänt·ə,dē }

ereuthophobia *See* erythrophobia. { ə,rüth·ə'fō·bē·ə }

erg [GEOGR] A large expanse of the earth's surface that is covered with sand, generally blown by wind into dune formations. [PHYS] A unit of energy or work in the centimeter-gram-second system of units, equal to the work done by a force of magnitude of 1 dyne when the point at which the force is applied is displaced 1 centimeter in the direction of the force. Also known as dyne centimeter (dyne-cm). { ərg }

ERG *See* electroretinogram.

Ergasilidae [INV ZOO] A family of copepod crustaceans in the suborder Cyclopoida in which the females are parasitic on aquatic animals, while the males are free-swimming. { ər·gə'sil·ə,dē }

ergastic [CYTOL] Pertaining to the nonliving components of protoplasm. { ər'gas·tik }

ergastoplasm [CYTOL] A cytoplasm component which shows an affinity for basic dyes; a form of the endoplasmic reticulum. { ,ər'gas·tə,plaz·əm }

ergodic [STAT] **1.** Property of a system or process in which averages computed from a data sample over time converge, in a probabilistic sense, to ensemble or special averages. **2.** Pertaining to such a system or process. { ər'gäd·ik }

ergodic theory [MATH] The study of measure-preserving transformations. [STAT MECH] Mathematical theory which attempts to show that the various possible microscopic states of a system are equally probable, and that the system is therefore ergodic. { ər'gäd·ik 'thē·ə·rē }

ergodic transformation [MATH] A measure-preserving transformation on X with the property that whenever X is written as a union of two disjoint invariant subsets, one of these must have measure zero. { ər'gäd·ik tranz·fər'mā·shən }

ergograph [ENG] An instrument with a recording device used to measure work capacity of muscles. { 'ər·gə,graf }

ergometer [ENG] An instrument with a recording device used to measure work performed by muscles under control conditions. { ər'gäm·əd·ər }

ergon [QUANT MECH] A quantum of energy; for any oscillator it is equal to the product of the oscillator's frequency and Planck's constant. { 'ər,gän }

ergonometrics [IND ENG] The application of various procedures for determining the time for an operator to perform a task satisfactorily, using the standard method in the usual environmental conditions, for example, time study or work sampling. Also known as work measurement. { ər,gän·ə'me,triks }

ergonomics [IND ENG] The study of human capability and psychology in relation to the working environment and the equipment operated by the worker. { ,ər·gə'näm·iks }

ergoregion *See* ergosphere. { 'ər·gō,rē·jən }

ergosphere [RELAT] The region outside the event horizon but inside the stationary limit of a Kerr black hole, within which no object can appear stationary to a distant observer. Also known as ergoregion. { 'ər·gə,sfir }

ergosterin *See* ergosterol. { ər'gäs·tə·rən }

ergosterol [BIOCHEM] $C_{28}H_{44}O$ A crystalline, water-insoluble, unsaturated sterol found in ergot, yeast, and other fungi, and which may be converted to vitamin D_2 on irradiation with ultraviolet light or activation with electrons. Also known as ergosterin. { ər'gäs·tə,ról }

ergot [MYCOL] The dark purple or black sclerotium of the fungus *Claviceps purpurea*. [ORG CHEM] Any of the five optically isomeric pairs of alkaloids obtained from this fungus; only the levorotatory isomers are physiologically active. { 'ər·gət }

ergotamine [ORG CHEM] $C_{33}H_{35}N_5O_5$ An alkaloid found in the fungal parasite ergot; causes smooth muscles in peripheral blood vessels to constrict, limiting blood flow; used to treat migraine headaches. { ər'gäd·ə,mēn }

ergotinine [ORG CHEM] An alkaloid and an isomer of ergotoxine that is a 1:1:1 mixture of ergocornine, ergocristine, and ergocryptine; crystallizes in long needles from acetone solutions, melting point 229°C, and soluble in chloroform, alcohol, and absolute ether. { ər'gät·ən,ēn }

ergotism [MED] Acute or chronic intoxication resulting from ingestion of grain infected with ergot fungus, or from chronic use of drugs containing ergot. { 'ər·gə,tiz·əm }

ergotoxine [ORG CHEM] An alkaloid and an isomer of ergotinine that is a 1:1:1 mixture of ergocornine, ergocristine, and ergocryptine; crystallizes in orthorhombic crystals, melts at 190°C, and is soluble in methyl alcohol, ethyl alcohol, acetone, and chloroform. { ,ər·gə'täk,sēn }

Eri *See* Eridanus.

Erian [GEOL] Middle Devonian geologic time; a North American provincial series. { 'i·rē·ən }

Erian orogeny [GEOL] One of the orogenies during Phanerozoic geologic time, at the end of the Silurian; the last part of the Caledonian orogenic era. Also known as Hibernian orogeny. { 'i·rē·ən ó'räj·ə·nē }

ERGOSTEROL

Structural diagram of ergosterol.

Ericaceae [BOT] A large family of dicotyledonous plants in the order Ericales distinguished by having twice as many stamens as corolla lobes. { ,er·ə'kās·ē,ē }

ericaceous mycorrhizal fungi [MYCOL] Mycorrhizal fungi that form symbiotic relationships with plants in the Ericaceae family; they are divided into three subgroups based on the presence or absence of a Hartig net, a fungal sheath, and fungal hyphae within the root cells. { er·ə¦kā·shəs ˌmī·kə¦rīz·əl 'fən·jī }

Ericales [BOT] An order of dicotyledonous plants in the subclass Dilleniidae; plants are generally sympetalous with unitegmic ovules and they have twice as many stamens as petals. { ,er·ə'kā·lēz }

Erichsen test [MET] A cupping test to measure the ductility of a piece of sheet metal and to determine its suitability for deep drawing. { 'er·ik·sən ,test }

Erichsen value [MET] The depth of impression in millimeters required to fracture a cupped sheet metal supported on a ring and deformed at the center by a spherically shaped tool. { 'er·ik·sən ,val·yü }

ericophyte [ECOL] A plant that grows on a heath or moor. { 'er·ək·ə,fīt }

Ericsson cycle [THERMO] An ideal thermodynamic cycle consisting of two isobaric processes interspersed with processes which are, in effect, isothermal, but each of which consists of an infinite number of alternating isentropic and isobaric processes. { 'er·ik·sən ,sī·kəl }

Erid See Eridanus.

Eridanus [ASTRON] A southern constellation made up of a long, crooked line of stars beginning near Rigel in the foot of Orion, and winding west and south to the first-magnitude star Achernar. Abbreviated Eri; Erid. Also known as River Po. { ,er·ə'dan·əs }

erigeron oil [MATER] A volatile oil, whose components are gallic and tannic acid, that is distilled from fleabane and horseweed. { ə'rij·ə·rən ,oil }

erikite [MINERAL] A brown mineral consisting of a silicate and phosphate of cerium metals; occurs in orthorhombic crystals. { 'er·ə,kīt }

Erinaceidae [VERT ZOO] The hedgehogs, a family of mammals in the order Insectivora characterized by dorsal and lateral body spines. { ,er·ə·nə'sē·ə,dē }

erineum [PL PATH] An abnormal growth of hairs induced on the epidermis of a leaf by certain mites. { ə'rin·ē·əm }

erinite [MINERAL] $Cu_5(OH)_4(AsO_4)_2$ Emerald-green mineral composed of basic copper arsenate. { 'er·ə,nīt }

Erinnidae [INV ZOO] A family of orthorrhaphous dipteran insects in the series Brachycera. { ə'rin·ə,dē }

Eriocaulaceae [BOT] The single family of the order Eriocaulales. { ,er·ē·ō,kȯ'lās·ē,ē }

Eriocaulales [BOT] An order of monocotyledonous plants in the subclass Commelinidae, having a perianth reduced or lacking and having unisexual flowers aggregated on a long peduncle. { ,er·ē·ō,kȯ'lā·lēz }

Eriococcinae [INV ZOO] A family of homopteran insects in the superfamily Coccoidea; adult females and late instar nymphs have an anal ring. { ,er·ē·ō'käk·sə,nē }

Eriocraniidae [INV ZOO] A small family of lepidopteran insects in the superfamily Eriocranioidea. { ,er·ē·ō·krə'nī·ə,dē }

Eriocranioidea [INV ZOO] A superfamily of lepidopteran insects in the suborder Homoneura comprising tiny moths with reduced, untoothed mandibles. { ,er·ē·ō,krə·nē'ȯid·ē·ə }

eriodictyol [ORG CHEM] $C_{15}H_{22}O_6$ A compound isolated from *Eriodictyon californicum* as needlelike crystals from a dilute alcohol solution, sparingly soluble in boiling water, hot alcohol, and glacial acetic acid; used in medicine as an expectorant. { ,er·ē·ō'dik·tē,ȯl }

eriometer [OPTICS] A device used to measure diameters of small particles or fibers by observing the diameter of the diffraction pattern produced by them in light coming from a small hole in a metal plate. { ,er·ē'äm·əd·ər }

erionite [MINERAL] A chabazite mineral of the zeolite family that contains calcium ions and crystallizes in the hexagonal system. { 'er·ē·ə,nīt }

Eriophyidae [INV ZOO] The bud mites or gall mites, a family of economically important plant-feeding mites in the suborder Trombidiformes. { ,er·ē·ō'fī·ə,dē }

eriophyllous [BOT] Having leaves covered by a cottony pubescence. { er·ē'äf·ə·ləs }

erlang [COMMUN] A unit of communication traffic load, equal to the traffic load whose calls, if placed end to end, will keep one path continuously occupied. { 'er,läŋ }

Erlang distribution See gamma distribution. { 'er,läŋ ,dis·trə,byü·shən }

Erlenmeyer flask [CHEM] A conical glass laboratory flask, with a broad bottom and a narrow neck. { 'ər·lən,mī·ər 'flask }

Erlenmeyer synthesis [ORG CHEM] Preparation of cyclic ethers by the condensation of an aldehyde with an α-acylamino acid in the presence of acetic anhydride and sodium acetate. { 'ər·lən,mī·ər 'sin·thə·səs }

eroding velocity [GEOL] The minimum average velocity required for eroding homogeneous material of a given particle size. { ə'rōd·iŋ və'läs·əd·ē }

Eros [ASTRON] The first asteroid to be orbited and landed on by a spacecraft, in 2000–2001; this elongated object's maximum diameter is 19.6 miles (31.6 kilometers); its closest approach to the earth is at about 14×10^6 miles (22.5×10^6 kilometers). { 'e,räs }

erose [BIOL] Having an irregular margin. { ē'rōs }

erosion [GEOL] **1.** The loosening and transportation of rock debris at the earth's surface. **2.** The wearing away of the land, chiefly by rain and running water. [MED] **1.** Surgical removal of tissues by scraping. **2.** Excision of a joint. { ə'rō·zhən }

erosional unconformity [GEOL] The surface that separates older, eroded rocks from younger, overlying sediments. { ə'rō·zhən·əl ,ən·kən'fȯr·məd·ē }

erosion-corrosion [MET] Attack on a metal surface resulting from the combined effects of erosion and corrosion. { ə¦rō·zhən kə¦rō·zhən }

erosion cycle [GEOL] A postulated sequence of conditions through which a new landmass proceeds as it wears down, classically the concept of youth, maturity, and old age, as stated by W.M. Davis; an original landmass is uplifted above base level, cut by canyons, gradually converted into steep hills and wide valleys, and is finally reduced to a flat lowland at or near base level. { ə'rō·zhən ,sī·kəl }

erosion pavement [GEOL] A layer of pebbles and small rocks that prevents the soil underneath from eroding. { ə'rō·zhən ,pāv·mənt }

erosion platform See wave-cut platform. { ə'rō·zhən ,plat,fȯrm }

erosion ridge [HYD] One of a group of ridges on the surface of snow; formed by the corrosive action of wind-blown snow. { ə'rō·zhən ,rij }

erosion scab [MET] A defect in the form of a solid mass of sand and metal that occurs when molten metal has been agitated or boiled or has partially eroded in the sand of a metal-casting mold. { ə'rō·zhən ,skab }

erosion surface [GEOL] A land surface shaped by agents of erosion. { ə'rō·zhən ,sər·fəs }

erotic [PSYCH] **1.** Pertaining to the libido or sexual passion. **2.** Moved by or arousing sexual desire. { ə'räd·ik }

erotophobia [PSYCH] An abnormal fear of love. { ə,räd·ə'fō·bē·ə }

Erotylidae [INV ZOO] The pleasing fungus beetles, a family of coleopteran insects in the superfamily Cucujoidea. { ,er·ə'tī·lə,dē }

ERP See effective radiated power. See emergency recorder plot.

Errantia [INV ZOO] A group of 34 families of polychaete annelids in which the anterior region is exposed and the linear body is often long and is dorsoventrally flattened. { ə'ran·chə }

erratic [GEOL] A rock fragment that has been transported a great distance, generally by glacier ice or floating ice, and differs from the bedrock on which it rests. { ə'rad·ik }

error [COMPUT SCI] An incorrect result arising from approximations used in numerical methods, rather than from a human mistake or computer malfunction. [SCI TECH] Any discrepancy between a computed, observed, or measured quantity and the true, specified, or theoretically correct value of that quantity. { 'er·ər }

error analysis [COMPUT SCI] In the solution of a problem on a digital computer, the estimation of the cumulative effect of rounding or truncation errors associated with basic arithmetic operations. { 'er·ər ə,nal·ə·səs }

error burst [COMPUT SCI] The condition when more than one bit is in error in a given number of bits. { 'er·ər ,bərst }

error character [COMPUT SCI] A character that indicates the existence of an error in the data being processed or transmitted, and usually specifies that a certain amount of preceding or following data is to be ignored. { 'er·ər ,kar·ik·tər }

error checking and recovery [COMPUT SCI] An automatic procedure which checks for parity and will proceed with the execution after error correction. { 'er·ər ,chek·iŋ ən ri'kəv·ə·rē }

error-checking code See self-checking code. { 'er·ər ,chek· iŋ ,kōd }

error code [COMPUT SCI] A specific character punched into a card or tape to indicate that a conscious error was made in the associated block of data; machines reading the error code may be programmed to throw out the entire block automatically. { 'er·ər ,kōd }

error coefficient [CONT SYS] The steady-state value of the output of a control system, or of some derivative of the output, divided by the steady-state actuating signal. Also known as error constant. { 'er·ər ,kō·i'fish·ənt }

error constant See error coefficient. { 'er·ər ,kän·stənt }

error-control procedures [COMMUN] Methods of detecting errors and correcting or recovering from those that occur in data transmission. { 'er·ər kən,trōl prə,sē·jərz }

error-correcting code [COMPUT SCI] Data representation that allows for error detection and error correction if the error is of a specific kind. { 'er·ər kə¦rek·tiŋ 'kōd }

error-correcting telegraph system [COMMUN] System employing an error-detecting code, and so conceived that any false signal initiates a repetition of the transmission of the character incorrectly received. { 'er·ər kə¦rek·tiŋ 'tel·ə,graf ,sis·təm }

error correction [COMMUN] Any system for reducing errors in an incoming message, such as sending redundant signals as a check. [COMPUT SCI] Computer device for automatically locating and correcting a machine error of dropping a bit or picking up an extraneous bit, without stopping the machine or having it go to a programmed recovery routine. [ELEC] Correction of time errors in interconnected alternating-current power systems resulting from deviations from normal frequency, in order to make all areas synchronous. { 'er·ər kə,rek·shən }

error correction routine [COMPUT SCI] A program which corrects specific error conditions in another program, routine, or subroutine. { 'er·ər kə,rek·shən rü,tēn }

error-detecting code See self-checking code. { 'er·ər di,tek· tiŋ ,kōd }

error-detecting system [COMPUT SCI] An automatic system which detects an error due to a lack of data, or erroneous data during transmission. { 'er·ər di,tek·tiŋ ,sis·təm }

error detection and feedback system [COMPUT SCI] An automatic system which retransmits a piece of data detected by the computer as being in error. { 'er·ər di,tek·shən ən 'fēd,bak ,sis·təm }

error detection routine See diagnostic routine. { 'er·ər di,tek· shən rü,tēn }

error diagnostic [COMPUT SCI] A computer printout of an instruction or data statement, pinpointing an error in the instruction or statement and spelling out the type of error involved. { 'er·ər ,dī·əg'näs·tik }

error equation [STAT] The equation of a normal distribution. { 'er·ər i,kwā·zhən }

error frequency limit [COMPUT SCI] The maximum number of single bit errors per unit of time that a computer will accept before a machine check interrupt is initiated. Abbreviated EFL. { 'er·ər ,frē·kwən·sē ,lim·ət }

error function [MATH] The real function defined as the integral from 0 to x of $e^{-t^2}\,dt$ or $e^{t^2}\,dt$, or the integral from x to ∞ of $e^{-t^2}\,dt$. { 'er·ər ,fəŋk·shən }

error handling [COMPUT SCI] The ability of a computer program to deal with errors automatically. { 'er·ər ,hand·liŋ }

error-indicating system [COMPUT SCI] Built-in circuits designed to indicate automatically that certain computational errors have occurred. { 'er·ər ,in·də,kād·iŋ ,sis·təm }

error interrupt [COMPUT SCI] The halt in execution of a program because of errors which the computer is not capable of correcting. { 'er·ər 'int·ə,rəpt }

error list [COMPUT SCI] A list generated by a compiler showing invalid or erroneous instructions in a source program. { 'er·ər ,list }

error log [COMPUT SCI] A file that is created during data processing to hold data known to contain errors, and that is usually printed after completion of processing so that the errors can be corrected. { 'er·ər ,läg }

error message [COMPUT SCI] A message indicating detection of an error. { 'er·ər ,mes·ij }

error of closure [ENG] Also known as angular error of closure. **1.** The amount by which the measurement of the azimuth of the first line of a traverse, made after completing the circuit, fails to equal the initial measurement. **2.** The amount by which the sum of the angles measured around the horizon differs from 360°. { 'er·ər əv 'klō·zhər }

error of perpendicularity [NAV] That error in the reading of a marine sextant due to nonperpendicularity of the index mirror to the frame. { 'er·ər əv ,pər·pən,dik·yə'lar·əd·ē }

error of the first kind See type I error. { ¦er·ər əv thə ¦fərst ,kīnd }

error of the second kind See type II error. { ¦er·ər əv thə ¦sekənd ,kīnd }

error range [COMPUT SCI] A range of values such that an error condition will result if a specified data item falls within it. [STAT] The difference between the highest and lowest error values; a measure of the uncertainty associated with a number. { 'er·ər ,rānj }

error rate [COMMUN] The number of erroneous bits or characters received for some fixed number of bits transmitted. { 'er·ər ,rāt }

error ratio [COMPUT SCI] The ratio of the number of erroneous items to the total number of bits or characters transmitted. { 'er·ər ,rā·shō }

error recovery routine [COMPUT SCI] A part of a computer program that attempts to handle errors without terminating the program. { 'er·ər ri¦kəv·ə·rē rü,tēn }

error report [COMPUT SCI] A list produced by a computer showing the error conditions, such as overflows and errors resulting from incorrect or unmatched data, that are generated during program execution. { 'er·ər ri,pórt }

error routine [COMPUT SCI] A routine which takes control of a program and initiates corrective actions when an error is detected. { 'er·ər rü,tēn }

error signal [CONT SYS] In an automatic control device, a signal whose magnitude and sign are used to correct the alignment between the controlling and the controlled elements. [ELEC] See error voltage. [ELECTR] A voltage that depends on the signal received from the target in a tracking system, having a polarity and magnitude dependent on the angle between the target and the center of the scanning beam. { 'er·ər ,sig·nəl }

error sum of squares [STAT] In analysis of variance, the sum of squares of the estimates of the contribution from the stochastic component. Also known as residual sum of squares. { ¦er·ər ¦səm əv ¦skwerz }

error tape [COMPUT SCI] The magnetic tape on which erroneous records are stored during processing. { 'er·ər ,tāp }

error voltage [ELEC] A voltage, usually obtained from a selsyn, that is proportional to the difference between the angular positions of the input and output shafts of a servosystem; this voltage acts on the system to produce a motion that tends to reduce the error in position. Also known as error signal. { 'er·ər ,vōl·tij }

ertor [METEOROL] The effective (radiational) temperature of the ozone layer (region). { 'ər,tór }

ERTS See earth resources technology satellite.

erucic acid [ORG CHEM] $C_{22}H_{42}O_2$ A monoethenoid acid that is the cis isomer of brassidic acid and makes up 40 to 50% of the total fatty acid in rapeseed, wallflower seed, and mustard seed; crystallizes as needles from alcohol solution, insoluble in water, soluble in ethanol and methanol. { ə'rüs·ik 'as·əd }

eruciform [INV ZOO] In certain insect larvae, having a soft cylindrical body with a well-defined head and usually short thoracic legs. { ə'rüs·ə,fórm }

eruption [GEOL] The ejection of solid, liquid, or gaseous material from a volcano. { i'rəp·shən }

eruptive prominence [ASTRON] A prominence on the sun

that is formed from active material above the chromosphere and reaches high altitudes on the sun at great speed. { i'rəp·tiv 'präm·ə·nəns }

eruptive rock [PETR] **1.** Rock formed from a volcanic eruption. **2.** Igneous rock that reaches the earth's surface in a molten condition. { ə'rəp·tiv 'räk }

eruptive star [ASTRON] A star that has a rapid change in its intensity because of the physical change it undergoes; examples are flare stars, recurrent novae, novae, supernovae, and nebular variables. { ə'rəp·tiv 'stär }

Erwinia [MICROBIO] A genus of motile, rod-shaped bacteria in the family Enterobacteriaceae; these organisms invade living plant tissues and cause dry necroses, galls, wilts, and soft rots. { ər'win·ē·ə }

Erwinieae [MICROBIO] Formerly a tribe of phytopathogenic bacteria in the family Enterobacteriaceae, including the single genus *Erwinia*. { ər'win·ē,ē }

erysipelas [MED] An acute, infectious bacterial disease caused by *Streptococcus pyogenes* and characterized by inflammation of the skin and subcutaneous tissues. { ,er·ə'sip·ə·ləs }

erysipeloid [MED] A bacterial infection caused by *Erysipelothrix rhuscopathiae* and occurring on the hands of people who handle infected meat or fish. { ,er·ə'sip·ə,lȯid }

Erysipelothrix [MICROBIO] A genus of gram-positive, rod-shaped bacteria of uncertain affiliation; cells have a tendency to form long filaments. { ,er·ə'sip·ə·lō,thriks }

Erysiphaceae [MYCOL] The powdery mildews, a family of ascomycetous fungi in the order Erysiphales with light-colored mycelia and conidia. { ,er·ə·sə'fās·ē,ē }

Erysiphales [MYCOL] An order of ascomycetous fungi which are obligate parasites of seed plants, causing powdery mildew and sooty mold. { ,er·ə·sə'fā·lēz }

erythema [MED] Localized redness of the skin in areas of variable size. { ,er·ə'thē·mə }

erythema migrans [MED] An expanding skin lesion characterized by a small red papule or macule that is a unique clinical marker for Lyme disease. { ,er·ə¦thē·mə 'mī,granz }

erythema multiforme [MED] An acute inflammatory skin disease characterized by red macules, papules, or tubercles on the extremities, neck, and face. { ,er·ə'thē·mə ,məl·tə'fȯr,mē }

erythema nodosum [MED] The occurrence of pink to blue, tender nodules on the anterior surfaces of the lower legs; more frequent in women than men. { ,er·ə'thē·mə nō'dō·səm }

erythorbate [FOOD ENG] A salt of erythorbic acid, an isomer of ascorbic acid; used in foods as an antioxidant. { ,er·ə'thȯr,bāt }

erythremia See erythrocytosis; polycythemia vera. { ,er·ə'thrē·mē·ə }

erythrine See erythrite. { 'er·ə,thrēn }

erythrite [MINERAL] Co₃(AsO₄)₂·8H₂O A crimson, peach, or pink-red secondary oxidized cobalt mineral that occurs in monoclinic crystals, in globular and reniform masses, or in earthy forms. Also known as cobalt bloom; cobalt ocher; erythrine; peachblossom ore; red cobalt. [ORG CHEM] See erythritol. { 'er·ə,thrīt }

erythritol [ORG CHEM] H(CHOH)₄H A tetrahydric alcohol; occurs as tetragonal prisms, melting at 121°C, soluble in water; used in medicine as a vasodilator. Also known as erythrite; erythrol. { ə'rith·rə,tȯl }

erythroblast [GISTOL] A nucleated cell occurring in bone marrow as the earliest recognizable cell of the erythrocytic series. { ə'rith·rə,blast }

erythroblastosis [MED] The abnormal presence of erythroblasts in the blood. [VET MED] A virus disease of birds; considered to be part of the avian leukosis complex in which there is an abnormal number of erythroblasts in the blood. { ə,rith·rə,bla'stō·səs }

erythroblastosis fetalis [MED] A form of hemolytic anemia affecting the fetus and newborn infant when a mother is Rh-negative and has developed antibodies against an Rh-positive fetus. Also known as hemolytic disease of newborn. { ə,rith·rə,bla'stō·səs fē'tal·əs }

erythrocruorin [BIOCHEM] Any of the iron-porphyrin protein respiratory pigments found in the blood and tissue fluids of certain invertebrates; corresponds to hemoglobin in vertebrates. { ə,rith·rə'krü·ə·rən }

erythrocyte [HISTOL] A type of blood cell that contains a

nucleus in all vertebrates but humans and that has hemoglobin in the cytoplasm. Also known as red blood cell. { ə'rith·rə,sīt }

erythrocytopoiesis See erythropoiesis. { ə,rith·rə,sīd·ə,pȯi'ē·səs }

erythrocytosis [MED] An increase in the number of circulating erythrocytes of more than two standard deviations above the mean normal, usually occurring secondary to hypoxia. Also known as erythremia; polycythemia. { ə,rith·rə,sī'tō·səs }

erythrodermia [MED] A skin condition in which the whole body surface is marked by an inflammatory blood vessel dilation. { ə,rith·rə'dər·mē·ə }

erythroidine [ORG CHEM] C₁₆H₁₉NO₃ An alkaloid existing in two forms: α-erythroidine and β-erythroidine, isolated from *Erythrina* species; β-erythroidine has an action similar to that of curare as a skeletal muscle relaxant. { ,er·ə'thrō·ə,dēn }

erythrol See erythritol. { 'er·ə,thrȯl }

erythromelalgia [MED] A cutaneous vasodilation of the feet or, more rarely, of the hands; characterized by redness, mottling, changes in skin temperature, and neuralgic pains. Also known as acromelalgia; Mitchell's disease. { ə,rith·rō·mə'lal·jē·ə }

erythromycin [MICROBIO] A crystalline antibiotic produced by *Streptomyces erythreus* and used in the treatment of gram-positive bacterial infections. { ə,rith·rə'mīs·ən }

D-erythropentose See ribulose. { ¦dē ə,rith·rə'pen,tōs }

erythrophilous [BIOL] Having an affinity for red dyes and other coloring matter. { ¦er·ə¦thräf·ə·ləs }

erythrophleine [ORG CHEM] C₂₄H₃₉NO₅ An alkaloid isolated from the bark of *Erythrophleum guineense;* used in medicine experimentally for its digitalislike action. { ə,rith·rə'flē·ən }

erythrophobia [PSYCH] **1.** An abnormal fear of red colors; may be associated with a fear of blood. **2.** Fear of blushing. Also known as ereuthophobia. { ə¦rith·rə'fō·bē·ə }

erythrophore [ZOO] A chromatophore containing a red pigment, especially a carotenoid. { ə'rith·rə,fȯr }

erythropia See erythropsia. { ,er·ə'thrō·pē·ə }

erythropoiesis [PHYSIO] The process by which erythrocytes are formed. Also known as erythrocytopoiesis. { ə,rith·rə,pȯi'ē·səs }

erythropoietin [BIOCHEM] A hormone, thought to be produced by the kidneys, that regulates erythropoiesis, at least in higher vertebrates. { ə,rith·rə'pȯi·ət·ən }

erythropsia [MED] An abnormality of vision in which all objects appear red; red vision. Also known as erythropia. { ,er·ə'thräp·sē·ə }

erythrose [ORG CHEM] HOCH₂(CHOH)₂CHO A tetrose sugar obtained from erythrol; a syrupy liquid at room temperature. { 'er·ə,thrōs }

erythrosiderite [MINERAL] K₂FeCl₅·H₂O Mineral composed of hydrous potassium iron chloride; occurs in lavas. { ə¦rith·rə'sid·ə,rīt }

erythrosin [ORG CHEM] C₁₃H₁₈O₆N₂ A red compound obtained by reacting tyrosine with nitric acid. { ə'rith·rə·sən }

erythrosis [MED] **1.** Overproliferation of erythropoietic tissue, as found in polycythemia. **2.** The unusual red skin color of individuals with polycythemia. { ə'rith·rə·səs }

Erythroxylaceae [BOT] A homogeneous family of dicotyledonous woody plants in the order Linales characterized by petals that are internally appendiculate, three carpels, and flowers without a disk. { ,er·ə,thräk·sə'lās·ē,ē }

erythrulose [BIOCHEM] C₄H₈O₄ A ketose sugar occurring as an oxidation product of erythritol due to the action of certain bacteria. { ə'rith·rə,lōs }

Erzgebirgian orogeny [GEOL] Diastrophism of the early Late Carboniferous. { 'erts·gə,bər·jən ȯ'räj·ə·nē }

Es See einsteinium.

Esaki tunnel diode See tunnel diode. { e'sä·kē ¦tən·əl 'dī,ōd }

Esbach's reagent [PATH] A solution of 1 gram of trinitrophenol and 2 grains of citric acid in 100 milliliters of water; used in determining albumin in urine. { 'es,bäks rē,ā·jənt }

ESCA See x-ray photoelectron spectroscopy.

escalation [IND ENG] Provision in actual or estimated costs for inflational increases in the costs of equipment, materials, labor, and so on, over those specified in an original contract. { ,es·kə'lā·shən }

escalator [MECH ENG] A continuously moving stairway and handrail. { 'es·kə,lād·ər }

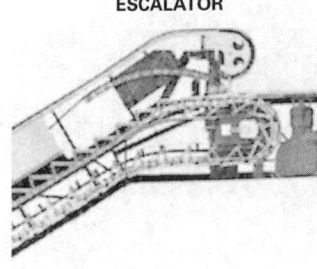

ESCALATOR

Cutaway view of escalator.
(Otis Elevator Co.)

E scan See E scope.   { 'ē ¦skan }

escape [COMPUT SCI] To exit from a program, routine, or mode.   { i'skāp }

escape character [COMPUT SCI] A character used to indicate that the succeeding character or characters are expressed in a code different from the code currently in use.   { ə'skāp ¦kar·ik·tər }

escape hatch [ENG] A hatch which permits persons to escape from a compartment, such as the interior of a submarine or aircraft, when normal means of exiting are blocked.   { ə'skāp ¦hach }

escapement [HOROL] A device in a timepiece consisting of a toothed wheel which engages alternate pallets attached to an oscillating member. [MECH ENG] A ratchet device that permits motion in one direction slowly.   { ə'skāp·mənt }

escape orbit [ASTRON] One of various paths that a body or particle escaping from a central force field must follow in order to escape.   { ə'skāp ¦ȯr·bət }

escape probability [NUCLEO] The proportion of neutrons produced in a reactor which eventually leaves the reactor without being absorbed.   { ə'skāp ¦präb·ə¦bil·əd·ē }

escape rocket [AERO ENG] A small rocket engine attached to the leading end of an escape tower, to provide additional thrust to the capsule in an emergency; it helps separate the capsule from the booster vehicle and carries it to an altitude where parachutes can be deployed.   { ə'skāp ¦räk·ət }

escape tower [AERO ENG] A trestle tower placed on top of a space capsule, connecting the capsule to the escape rocket on top of the tower; used for emergencies.   { ə'skāp ¦taú·ər }

escape trunk [NAV ARCH] An escape compartment in a submarine, designed to receive a rescue chamber or deep-submergence rescue vehicle.   { ə'skāp ¦trəŋk }

escape velocity [ASTRON] The minimum speed away from a parent body that a particle must acquire to escape permanently from the gravitational attraction of the parent.   { ə'skāp və¦läs·əd·ē }

escape wheel [HOROL] The final wheel in a timepiece train, designed to transmit impulses to a lever.   { ə'skāp ¦wēl }

escaping tendency [PHYS CHEM] The tendency of a solute species to escape from solution; related to the chemical potential of the solute.   { ə'skāp·iŋ ¦ten·dən·sē }

escar See esker.   { 'es·kər }

escarpment [GEOL] A cliff or steep slope of some extent, generally separating two level or gently sloping areas, and produced by erosion or by faulting. Also known as scarp. [ORD] The ground surrounding a fortified place which has been cut away nearly vertically to prevent an enemy's approach.   { ə'skärp·mənt }

eschar [GEOL] See esker. [MED] A dry crust or slough, especially one formed after a thermal or chemical burn.   { 'es·kər }

escharotic [MED] **1.** Caustic. **2.** Producing an eschar.   { ¦es·kə¦räd·ik }

Escherichia [MICROBIO] A genus of bacteria in the family Enterobacteriaceae; straight rods occurring singly or in pairs.   { ¦esh·ə¦rik·ē·ə }

Escherichia coli [MICROBIO] The type species of the genus, occurring as part of the normal intestinal flora in vertebrates. Also known as colon bacillus.   { ¦esh·ə¦rik·ē·ə 'kō¦lī }

Escherichia coli O157:H7 [MICROBIO] An unusually virulent food-borne pathogen that is found primarily in cattle and causes severe, sometimes life-threatening illness; symptoms include hemorrhagic colitis, hemolytic uremic syndrome, and thrombotic thrombocytopenic purpura.   { ¦es·kə¦rēk·ē·ə 'kō¦lī ¦ō¦wən¦fīv¦sev·ən 'āch¦sev·ən }

Escherichieae [MICROBIO] Formerly a tribe of bacteria in the family Enterobacteriaceae defined by the ability to ferment lactose, with the rapid production of acid and visible gas.   { ¦esh·ə·rə¦kī·ē¦ē }

Eschka mixture [ANALY CHEM] A mixture of two parts magnesium oxide and one part anhydrous sodium carbonate; used as a fusion mixture for determining sulfur in coal.   { 'esh·kə ¦miks·chər }

eschwegeite See tanteuxenite.   { ¦esh'vā·gē¦īt }

Eschweiler-Clarke modification [ORG CHEM] A modification of the Leuckart reaction, involving reductive alkylation of ammonia or amines (except tertiary amines) by formaldehyde and formic acid.   { ¦esh¦vīl·ər ¦klärk ¦mäd·ə·fə¦kā·shən }

eschynite [MINERAL] (Ce,Ca,Fe,Th)(Ti,Cb)$_2$O$_6$ A black

mineral, occurring in prismatic crystals; a rare oxide of cesium, titanium, and other metals, which is isomorphous with priorite.   { 'es·kə¦nīt }

Esclangon effect [OPTICS] Bending of a reflected light ray caused by movement of the mirror in a direction making an acute angle with its surface.   { e'skläŋ·gən i'fekt }

E scope [ELECTR] A cathode-ray scope on which signals appear as spots, with range as the horizontal coordinate and elevation angle or height as the vertical coordinate. Also known as E indicator; E scan.   { 'ē ¦skōp }

escort fighter [AERO ENG] A fighter designed or equipped for long-range missions, usually to accompany heavy bombers on raids.   { 'es¦kȯrt ¦fīd·ər }

escribed circle [MATH] For a triangle, a circle that lies outside of the triangle and is tangent to one side of the triangle and to the extensions of the other two sides. Also known as excircle.   { ə¦skrībd 'sər·kəl }

esculin [PHARM] C$_{15}$H$_{16}$O$_9$ A substance extracted from the leaves and bark of the horse chestnut tree; used as a skin protectant.   { 'es·kyə·lən }

esculoside See esculin.   { es'kyü·lə¦sīd }

escutcheon [DES ENG] An ornamental shield, flange, or border used around a dial, window, control knob, or other panel-mounted part. Also known as escutcheon plate.   { e'skəch·ən }

escutcheon plate See escutcheon.

ESD See external symbol dictionary.

ESDI See enhanced small device interface.   { 'ez¦dē }

ESEEM [SPECT] See electron spin echo envelope modulation.   { 'ē¦sēm or ¦ē¦es¦ē¦ē'em }

eserine See physostigmine.   { 'es·ə¦rēn }

E-set [ACOUS] The set of 10 English letters and digits that share the E sound and therefore tend to be more easily confused by speech recognition systems than other elements of an alpha-digit vocabulary: the letters B, C, D, E, G, P, T, V, and Z, and the digit 3.   { 'ē¦set }

Eshelby twist [SOLID STATE] A torsional deformation of a crystal whisker resulting from a screw dislocation along the whisker axis.   { 'esh·əl·bē 'twist }

eskar See esker.   { 'es·kər }

eskebornite [MINERAL] CuFeSe$_2$ The selenium analog of the mineral pyrrhotite (Fe$_{1-x}$S).   { ¦es·kə'bȯr¦nīt }

esker [GEOL] A sinuous ridge of constructional form, consisting of stratified accumulations, glacial sand, and gravel. Also known as asar; eschar; eskar; osar; serpent kame.   { 'es·kər }

Esocidae [VERT ZOO] The pikes, a family of fishes in the order Clupeiformes characterized by an elongated beaklike snout and sharp teeth.   { ə¦säs·ə¦dē }

Esocoidei [VERT ZOO] A small suborder of fresh-water fishes in the order Salmoniformes; includes the pikes, mudminnows, and pickerels.   { ¦es·ə¦kȯid·ē¦ī }

esophageal diverticulum [MED] An outpocketing of the wall of the esophagus, or of the pharynx just above the opening to the esophagus.   { ə¦säf·ə¦jē·əl ¦dī·vər¦tik·yə·ləm }

esophageal fistula [MED] Congenitally, an abnormal tube communicating between the esophagus and an internal organ, usually the trachea; an acquired esophageal fistula usually communicates between the esophagus and the skin through an external opening, or may communicate with internal organs.   { ə¦säf·ə¦jē·əl 'fis·chə·lə }

esophageal gland [ANAT] Any of the digestive glands within the submucosa of the esophagus; secretions are chiefly mucus and serve to lubricate the esophagus.   { ə¦säf·ə¦jē·əl 'gland }

esophageal hiatus [ANAT] The opening in the diaphragm for passage of the esophagus.   { ə¦säf·ə¦jē·əl hī'ād·əs }

esophageal teeth [VERT ZOO] The enamel-tipped hypapophyses of the posterior cervical vertebrae of certain snakes, which penetrate the esophagus and function to break eggshells.   { ə¦säf·ə¦jē·əl 'tēth }

esophagitis [MED] Inflammation of the esophagus.   { ə¦säf·ə'jīd·əs }

esophagogastrostomy [MED] Establishment, by surgery, of an anastomosis between the esophagus and the stomach; may be performed by the abdominal route or by transpleural operation.   { ə¦säf·ə·gō¦ga'sträs·tə·mē }

esophagoscopy [MED] Endoscopic examination of the interior of the esophagus.   { e¦säf·ə¦gäs·kə·pē }

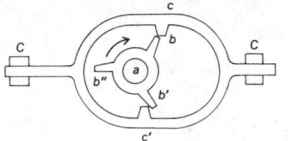

ESCAPEMENT

Simple escapement. Oscillating member *cc'* is an open bar arranged to slide longitudinally in bearings *CC*; wheel *a* turns continuously in direction of arrow; wheel has teeth *b*, *b'*, *b"*; tooth *b* is just ceasing to drive pallet *c* to the right.

ESKER

View along the crest of an esker in southeastern Wisconsin. *(W. C. Alden, USDG)*

esophagus [ANAT] The tubular portion of the alimentary canal interposed between the pharynx and the stomach. Also known as gullet. { ə'säf·ə·gəs }

esoteric name [COMPUT SCI] A symbolic name that is chosen in a computer program to designate a collection of devices. { es·ə'ter·ik 'nām }

esotropia [MED] Convergent strabismus, occurring when one eye fixes upon an object and the other deviates inward. Also known as cross-eye. { ˌes·ə'trō·pē·ə }

ESP See extrasensory perception.

espalier drainage See trellis drainage. { e'spal·yər ˌdrān·ij }

esparto wax [MATER] Vegetable wax extracted from esparto grass; it is hard, blends well, and is easy to emulsify; used to give smoothness to polishes. { e¦spär·dō ˌwaks }

esquisse [ARCH] A preliminary sketch or design drawing showing the basic features of a project. { e'skēs }

ESR See electron paramagnetic resonance.

Essen coefficient [ELEC] The torque exerted on the moving part of an electric rotating machine divided by the volume enclosed by the air gap. { 'es·ən ˌkō·i,fish·ənt }

essential amino acid [BIOCHEM] Any of eight of the 20 naturally occurring amino acids that are indispensable for optimum animal growth but cannot be formed in the body and must be supplied in the diet. { i'sen·chəl ə'mē·nō ˌas·əd }

essential bound [MATH] For a function f, a number A such that the set of points x for which the absolute value of f(x) is greater than A is of measure zero. { i¦sen·chəl 'baúnd }

essential constants [MATH] A set of constants in an equation that cannot be replaced by a smaller number of constants in another equation that has the same solutions. { i¦sen·chəl 'kän·stəns }

essential fatty acid [BIOCHEM] Any of the polyunsaturated fatty acids which are required in the diet of mammals; they are probably precursors of prostaglandins. { i'sen·chəl 'fad·ē ˌas·əd }

essential hypertension [MED] Elevation of the systemic blood pressure, of unknown origin. Also known as primary hypertension. { i'sen·chəl ˌhī·pər'ten·chən }

essentially bounded function [MATH] A function that has an essential bound. { i¦sen·chə·lē ¦baúnd·əd 'fəŋk·shən }

essential mapping [MATH] A mapping between topological spaces that is not homotopic to a mapping whose range is a single point. { i¦sen·chəl 'map·iŋ }

essential oil [MATER] Any of the odoriferous oily products of plant origin which are distillable; the principal constituents are terpenes, but benzenoid and aliphatic compounds may also be present. Also known as ethereal oil. { i'sen·chəl 'óil }

essential singularity [MATH] An isolated singularity of a complex function which is neither removable nor a pole. { i'sen·chəl sin·gyə'lar·əd·ē }

essential supremum [MATH] For an essentially bounded function, the greatest lower bound of the essential bounds. { i¦sen·chəl sə'prēm·əm }

essexite [PETR] A rock of igneous origin composed principally of plagioclase hornblende, biotite, and titanaugite. { 'e·sik,sīt }

established flow [FL MECH] The flow when the boundary layer of a fluid flowing in a duct completely fills the duct; that is, when the effect of the wall shearing stress extends completely across the duct. { i'stab·lisht 'flō }

establishment [OCEANOGR] The interval of time between the transit (upper or lower) of the moon and the next high water at a place. { i'stab·lish·mənt }

ester [ORG CHEM] The compound formed by the elimination of water and the bonding of an alcohol and an organic acid. { 'es·tər }

esterase [BIOCHEM] Any of a group of enzymes that catalyze the synthesis and hydrolysis of esters. { 'es·tə,rās }

ester gum [ORG CHEM] A compound obtained by forming an ester of a natural resin with a polyhydric alcohol; used in varnishes, paints, and cellulosic lacquers. Also known as rosin ester. { 'es·tər ˌgəm }

ester hydrolysis [ORG CHEM] A reaction in which an ester is converted into its alcohol and acid moieties. Also known as esterolysis. { 'e·stər hī'dräl·ə·səs }

esterification [ORG CHEM] A chemical reaction whereby esters are formed. { e,ster·ə·fə'kā·shən }

esterolysis See ester hydrolysis. { ˌe·stər'äl·ə·səs }

estersil [ORG CHEM] Hydrophobic silica powder, an ester of

—SiOH with a monohydric alcohol; used as a filler in silicone rubbers, plastics, and printing inks. { 'es·tər,sil }

esthacyte [INV ZOO] A simple sensory cell occurring in certain lower animals, such as sponges. Also spelled aesthacyte. { 'es·thə,sīt }

esthesia [PHYSIO] The capacity for sensation, perception, or feeling. Also spelled aesthesia. { es'thē·zhə }

esthesiometer [ENG] An instrument used to measure tactile sensibility by determining the distance by which two points pressed against the skin must be separated in order that they be felt as separate. Also spelled aesthesiometer. { es,thē·zē¦äm·əd·ər }

esthesioneuroblastoma See neuroepithelioma. { es¦thē·zē·ō¦nùr·ō,bla'stō·mə }

esthesioneuroepithelioma See neuroepithelioma. { es¦thē·zē·ō¦nùr·ō,ep·ə,thē·lē'ō·mə }

esthiomene [MED] The chronic ulcerative lesion of the vulva in lymphogranuloma venereum. { es'thī·ə,mēn }

estimate [SCI TECH] A statement of the value of a quantity or function based on incomplete data or evidence, or on a rough or approximate calculation. { 'es·tə,māt }

estimated exposure concentration [MED] Measured or calculated amount or mass concentration of a substance to which an organism is likely to be exposed. { ˌes·tə,mād·əd ik'spō·zhər ˌkəns·ən¦trā·shən }

estimated position [NAV] The most probable position of a craft determined from incomplete data or data of questionable accuracy. { 'es·tə,mād·əd pə'zish·ən }

estimated time [IND ENG] A predicted element or operation time. { 'es·tə,mād·əd 'tīm }

estimated time of arrival [NAV] The predicted time of reaching a destination or way point. Abbreviated ETA. { 'es·tə,mād·əd ¦tīm əv ə'rīv·əl }

estimated time of departure [NAV] The predicted time of leaving a place. Abbreviated ETD. { 'es·tə,mād·əd ¦tīm əv di'pär·chər }

estimated time of interception [NAV] The predicted time of intercepting a craft by another craft. { 'es·tə,mād·əd ¦tīm əv ˌin·tər'sep·shən }

estimation theory [STAT] A branch of probability and statistics concerned with deriving information about properties of random variables, stochastic processes, and systems based on observed samples. { ˌes·tə'mā·shən ˌthē·ə·rē }

estimator [STAT] A random variable or a function of it used to estimate population parameters. { 'es·tə,mād·ər }

estival [ASTRON] Of or pertaining to the summer. Also spelled aestival. { 'es·tə·vəl }

estivation [PHYSIO] **1.** The adaptation of certain animals to the conditions of summer, or the taking on of certain modifications, which enables them to survive a hot, dry summer. **2.** The dormant condition of an organism during the summer. { ¦es·tə'vā·shən }

estivoautumnal malaria See falciparum malaria. { ¦es·tə·vō,ó¦təm·nəl mə'ler·ē·ə }

estradiol [BIOCHEM] $C_{18}H_{24}O_2$ An estrogenic hormone produced by follicle cells of the vertebrate ovary; provokes estrus and proliferation of the human endometrium, and stimulates ICSH (interstitial-cell-stimulating hormone) secretion. { ˌes·trə'dī,ól }

estragole [ORG CHEM] $C_6H_4(C_3H_5)(OCH_3)$ A colorless liquid with the odor of anise, found in basil oil, estragon oil, and anise bark oil; used in perfumes and flavorings. { 'es·trə,gól }

estragon oil [MATER] Essential oil, colorless to yellowish green, with aniselike odor and aromatic flavor; distilled from flowering herb tarragon, *Artemisia dracunculus;* used chiefly as flavoring. Also known as tarragon oil. { 'es·trə,gän ˌóil }

estriol [BIOCHEM] $C_{18}H_{24}O_3$ A crystalline estrogenic hormone obtained from human pregnancy urine. { 'e,strī,ól }

estrogen [BIOCHEM] Any of various natural or synthetic substances possessing the biologic activity of estrus-producing hormones. { 'es·trə·jən }

estrogenic hormone [BIOCHEM] A hormone, found principally in ovaries and also in the placenta, which stimulates the accessory sex structures and the secondary sex characteristics in the female. { ¦es·trə¦jen·ik 'hór,mōn }

estrone [BIOCHEM] $C_{18}H_{22}O_2$ An estrogenic hormone produced by follicle cells of the vertebrate ovary; functions the same as estradiol. { 'e,strōn }

estrous cycle [PHYSIO] The physiological changes that take place between periods of estrus in the female mammal. { 'es·trəs ˌsī·kəl }

estrus [PHYSIO] The period in female mammals during which ovulation occurs and the animal is receptive to mating. { 'es·trəs }

estuarine circulation [OCEANOGR] In an estuary, the outflow (seaward) of low-salinity surface water over a deeper inflowing layer of dense, high-salinity water. { 'es·chə·wəˌrēn ˌsər·kyəˈlā·shən }

estuarine deposit [GEOL] A sediment deposited at the heads and floors of estuaries. { 'es·chə·wəˌrēn dəˈpäz·ət }

estuarine environment [OCEANOGR] The physical conditions and influences of an estuary. { 'es·chə·wəˌrēn enˈvī·rən·mənt }

estuarine oceanography [OCEANOGR] The study of the chemical, physical, biological, and geological properties of estuaries. { 'es·chə·wəˌrēn ˌō·shəˈnäg·rə·fē }

estuary [GEOGR] A semienclosed coastal body of water which has a free connection with the open sea and within which sea water is measurably diluted with fresh water. Also known as branching bay; drowned river mouth; firth. { 'es·chə ˌwer·ē }

esu See electrostatic units.

E.T. See ephemeris time.

ETA See estimated time of arrival.

etalon [OPTICS] **1.** Two adjustable parallel mirrors mounted so that either one may serve as one of the mirrors in a Michelson interferometer; used to measure distances in terms of wavelengths of spectral lines. **2.** An instrument similar to the Fabry-Pérot interferometer, except that the distance between the plates is fixed. Also known as Fabry-Pérot etalon. { 'ed·əlˌän }

eta meson [PART PHYS] Neutral pseudoscalar meson having zero isotopic spin and hypercharge, positive charge parity and G parity, and a mass of about 549 MeV; decays via electromagnetic interactions. { 'ād·ə 'māˌsän }

etamine [TEXT] Cotton material that is loosely woven and lightweight but has a coarse texture. { 'ad·əˌmēn }

eta-prime meson See chi meson. { ¦ad·ə ˌprīm ˌmāˌsän }

Etard reaction [ORG CHEM] Direct oxidation of an aromatic or heterocyclic bound methyl group to an aldehyde by utilizing chromyl chloride or certain metallic oxides. { āˈtär rēˌak·shən }

etch [GRAPHICS] To incise lines on a plate of metal, glass, or other material by covering it with an acid-resistant coating, scratching through the coating, and then permitting an acid bath to erode exposed parts of the plate. [MET] To corrode the surface of a metal in order to reveal its composition and structure. { ech }

etch cleaning [MET] Removing soil by electrolytic or chemical action; removes some of the surface metal along with the dirt. { 'ech ˌklēn·iŋ }

etch cracks [MET] Shallow cracks in the surface of hardened steel that result from reaction with an acid, causing hydrogen cracking. { 'ech ˌkraks }

etched circuit [ENG] A printed circuit formed by chemical or electrolytic removal of unwanted portions of a layer of conductive material bonded to an insulating base. { ¦echt 'sər·kət }

etch figures [MET] Minute, faceted pits or surfaces produced on a metal surface by chemical reaction with an etchant. [MINERAL] A minute pit produced by a solvent on the crystal face of a mineral which reveals its molecular structure. { 'ech ˌfig·yərz }

etch printing [GRAPHICS] Printing with a printing plate that has an image formed by chemical or electrolytic action. { 'ech ˌprint·iŋ }

ETD See estimated time of departure.

etesian climate See Mediterranean climate. { əˈtē·zhən 'klī·mət }

etesians [METEOROL] The prevailing northerly winds in summer in the eastern Mediterranean, and especially the Aegean Sea; basically similar to the monsoon and equivalent to the maestro of the Adriatic Sea. { əˈtē·zhənz }

ethamine See ethyl amine. { 'eth·əˌmēn }

ethane [ORG CHEM] CH_3CH_3 A colorless, odorless gas belonging to the alkane series of hydrocarbons, with freezing point of $-183.3°C$ and boiling point of $-88.6°C$; used as a fuel and refrigerant and for organic synthesis. { 'ethˌān }

1,2-ethanedithiol [ORG CHEM] $HSCH_2CH_2SH$ A liquid, freely soluble in alcohol and in alkalies; used as a metal complexing agent. { ¦wən ¦tü ¦eth·ānˈdī·əˌmēn }

ethanoic acid See acetic acid. { ¦eth·əˈnō·ikˌas·əd }

ethanol [ORG CHEM] C_2H_5OH A colorless liquid, miscible with water, boiling point 78.32°C; used as a reagent and solvent. Also known as ethyl alcohol; grain alcohol. { 'eth·əˌnȯl }

ethanolamine [ORG CHEM] $NH_2(CH_2)_2OH$ A colorless liquid, miscible in water; used in scrubbing hydrogen sulfide (H_2S) and carbon dioxide (CO_2) from petroleum gas streams, for dry cleaning, in paints, and in pharmaceuticals. { ˌethə'nälˌəˌmen }

ethanolurea [ORG CHEM] $NH_2CONHCH_2CH_2OH$ A white solid; its formaldehyde condensation products are thermoplastic and water-soluble. { ¦eth·əˌnȯlˈyu̇·rē·ə }

ethene See ethylene. { 'eˌthēn }

ethenol See vinyl alcohol. { 'eth·əˌnȯl }

ethephon [ORG CHEM] $C_2H_6ClO_3P$ A white solid with a melting point of 74.75°C; very soluble in water; used as a growth regulator for tomatoes, apples, cherries, and walnuts. Also known as CEPHA. { 'eth·əˌfän }

ether [ELECTROMAG] The medium postulated to carry electromagnetic waves, similar to the way a gas carries sound waves. [ORG CHEM] **1.** One of a class of organic compounds characterized by the structural feature of an oxygen linking two hydrocarbon groups (such as R-O-R). **2.** $(C_2H_5)_2O$ A colorless liquid, slightly soluble in water; used as a reagent, intermediate, anesthetic, and solvent. Also known as ethyl ether. { 'e·thər }

ether drag [ELECTROMAG] The hypothesis, advanced unsuccessfully to account for results of the Michelson-Morley experiment, that ether is dragged along with matter. { 'ē·thər ˌdrag }

ether drift [ELECTROMAG] Hypothetical motion of the ether relative to the earth. { 'ē·thər ˌdrift }

ethereal oil See essential oil. { ēˈthir·ē·əl ˌȯil }

etherification [ORG CHEM] The process of making an ether from an alcohol. { ēˌthir·ə·fəˈkā·shən }

etherize [MED] To produce anesthesia by administration of ether. { 'ē·thəˌrīz }

Ethernet [COMPUT SCI] A protocol for interconnecting computers and peripheral devices in a local area network. { 'ē·thərˌnet }

ether thermoscope [PHYS] A device for detecting radiant heat; consists of an evacuated U-shaped tube, with ether at the bottom of the tube and a bulb at each end, one bulb being blackened. { 'ē·thər 'thər·məˌskōp }

ethidine See ethylidene. { 'eth·əˌdēn }

ethidium bromide [ORG CHEM] $C_{21}H_{20}BrN_3$ Dark red crystals with a melting point of 238–240°C; used in treating trypanosomiasis in animals and as an inhibitor of deoxyribonucleic and ribonucleic acid synthesis. Also known as homidium bromide. { eˈthid·ē·əm 'brōˌmīd }

ethinyl [ORG CHEM] The $CH_3{:}C-$ radical from acetylene. Also known as acetenyl; acetylenyl; ethynyl. { eˈthīn·əl }

ethiolate [ORG CHEM] $C_7H_{15}ONS$ A yellow liquid with a boiling point of 206°C; used as a preemergence herbicide for corn. { əˈthī·əˌlāt }

ethionic acid [ORG CHEM] $HO·SO_2·CH_2·CH_2·SO_2OH$ An unstable diacid, known only in solution. Also known as ethylene sulfonic acid. { 'eth·ēˌän·ik 'as·əd }

ethionine [BIOCHEM] $C_5H_{13}O_2N$ An amino acid that is the ethyl analog of and the biological antagonist of methionine. { eˈthī·əˌnēn }

Ethiopian zoogeographic region [ECOL] A geographic unit of faunal homogeneity including all of Africa south of the Sahara. { ¦ē·thē'ō·pē·ən ¦zō·ō·jē·ə·ˈgraf·ik 'rē·jən }

ethmoid bone [ANAT] An irregularly shaped cartilage bone of the skull, forming the medial wall of each orbit and part of the roof and lateral walls of the nasal cavities. { 'ethˌmȯid ˌbōn }

ethmolith [GEOL] A downward tapering, funnel-shaped, discordant intrusion of igneous rocks. { 'eth·məˌlith }

ethmoturbinate [ANAT] Of or pertaining to the masses of ethmoid bone which form the lateral and superior portions of the turbinate bones in mammals. { ¦eth·mō'tər·bə·nāt }

ethnobotany [ARCH] The study of how cultures utilize plants and plant products. { ¦eth·nō'bät·ən·ē }

ethnogeography [ARCH] The scientific study of the geographic distribution of races, peoples, or cultural groups and

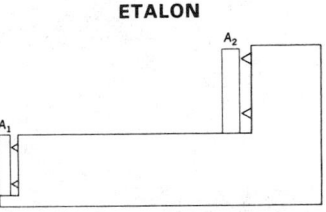

ETALON

An etalon with two adjustable mirrors A_1 and A_2 which are to be used with the Michelson interferometer in distance measurement.

their adaptation and relation to the environments in which they live. { ¦eth·nō·jē'äg·rə·fē }

ethnography [ARCH] The branch of ethnology that deals with the description of races or ethnic groups, without attempting to analyze or compare them. { eth'nä·grə·fē }

ethnology [ARCH] The science that deals with the study of the origin, distribution, and relations of races or ethnic groups of humankind. { eth'näl·ə·jē }

ethnopsychology [PSYCH] The study of the psychology of races or ethnic groups. { ¦eth·nō·sī'käl·ə·jē }

ethnozoology [ARCH] The study of how cultures use animals and animal by-products. { ¦eth·nō·zō'äl·ə·jē }

ethogram [ECOL] An extensive list, inventory, or description of the behavior of an organism. { 'ē·thə,gram }

ethohexadiol [ORG CHEM] $C_8H_{18}O_2$ A slightly oily liquid, used as an insect repellent. { ¦eth·ō,hek·sə'dī·ól }

ethological isolation See behavioral isolation. { ,ē·thə'läj·ə·kəl ī·sə'lā·shən }

ethology [VERT ZOO] The study of animal behavior in a natural context. { ē'thäl·ə·jē }

ethoprop [ORG CHEM] $C_8H_{19}O_2PS_2$ A pale yellow liquid compound, insoluble in water; used as an insecticide for soil insects and as a nematicide for plant parasitic nematodes. { 'ē·thō,präp }

ethoxide [ORG CHEM] A compound formed from ethanol by replacing the hydrogen of the hydroxy group by a monovalent metal. Also known as ethylate. { e'thäk,sīd }

ethoxy [ORG CHEM] The C_2H_5O- radical from ethyl alcohol. Also known as ethyoxyl. { e'thäk·sē }

2-ethoxyethanol See cellosolve. { ¦tü e¦thäk·sē'eth·ə,nól }

ethoxylation [CHEM ENG] A catalytic process which involves the direct addition of ethylene oxide to an alkyl phenol or to an aliphatic alcohol. { e,thäk·sə'lā·shən }

ethoxyquin [ORG CHEM] $C_{14}H_{19}NO$ A dark liquid, used as a growth regulator to protect apples and pears in storage.

ethyl [ORG CHEM] **1.** The hydrocarbon radical C_2H_5. **2.** Trade name for the tetraethyllead antiknock compound in gasoline. { 'eth·əl }

ethyl acetate [ORG CHEM] $CH_3COOC_2H_5$ A colorless liquid, slightly soluble in water; boils at 77°C; a medicine, reagent, and solvent. Also known as acetic ester; acetic ether; acetidin. { 'eth·əl 'as·ə,tāt }

ethyl acetoacetate [ORG CHEM] $CH_3COCH_2COOC_2H_5$ A colorless liquid, boiling at 181°C; used as a reagent, intermediate, and solvent. Also known as acetoacetic ester; diacetic ether. { 'eth·əl¦as·ə·to¦as·ə,tāt }

ethyl acetylene [ORG CHEM] Compound with boiling point 8.1°C; insoluble in water, soluble in alcohol; used in organic synthesis. { ə'sed·əl,ēn }

ethyl acrylate [ORG CHEM] $C_5H_8O_2$ A colorless liquid, boiling at 99°C; used to manufacture chemicals and resins. { 'eth·əl 'ak·rə,lāt }

ethyl alcohol See ethanol. { 'eth·əl 'al·kə,hól }

ethyl amine [ORG CHEM] A colorless liquid, boiling at 15°C, water-soluble; used as a solvent, as a dye intermediate, and in organic synthesis. Also known as aminoethane; ethamine. { 'eth·əl 'am,ēn }

ethyl-para-aminobenzoate [ORG CHEM] $C_6H_4NH_2CO_2-C_2H_5$ A white powder, melting point 88–92°C, slightly soluble in ethanol and ether, very slightly soluble in water; used as a local anesthetic. Also known as benzocaine. { 'eth·əl ¦par·ə ¦am·ə·nō'ben·zə,wāt }

ethyl amyl ketone [ORG CHEM] $C_8H_{16}O$ A colorless liquid, almost insoluble in water; used in perfumery. { 'eth·əl ¦am·əl 'ke,tōn }

ethylate See ethoxide. { 'eth·ə,lāt }

ethylation [ORG CHEM] Formation of a new compound by introducing the ethyl functional group (C_2H_5). { ,eth·ə'lā·shən }

ethyl benzene [ORG CHEM] $C_6H_5C_2H_5$ A colorless liquid that boils at 136°C, insoluble in water; used in organic synthesis, as a solvent, and in making styrene. { 'eth·əl 'ben,zēn }

ethyl benzoate [ORG CHEM] $C_6H_5COOCH_2CH_3$ Colorless, aromatic liquid, boiling at 213°C, insoluble in water; used as a solvent, in flavoring extracts, and in perfumery. { 'eth·əl 'ben·zə,wāt }

ethyl borate [ORG CHEM] $B(OC_2H_5)_3$ A salt of ethanol and boric acid; colorless, flammable liquid; used in antiseptics,

disinfectants, and fireproofing. Also known as boron triethoxide; triethylic borate. { 'eth·əl 'bór,āt }

ethyl bromide [ORG CHEM] C_2H_5Br A colorless liquid, boiling at 39°C; used as a refrigerant and in organic synthesis. { 'eth·əl 'brō,mīd }

2-ethylbutene [ORG CHEM] $CH_3CH_2(C_2H_5)CCH_2$ Colorless liquid, soluble in alcohol and organic solvents, insoluble in water; used in organic synthesis. { ¦tü ,eth·əl'byü,tān }

2-ethylbutyl acetate [ORG CHEM] $C_2H_5CH(C_2H_5)CH_2-O_2CCH_3$ Colorless liquid with mild odor; used as a solvent for resins, lacquers, and nitrocellulose. { ¦tü ¦eth·əl¦byüd·əl 'as·ə,tāt }

2-ethylbutyl alcohol [ORG CHEM] $(C_2H_5)_2CHCH_2OH$ A stable, colorless liquid, miscible in most organic solvents, slightly water-soluble; used as a solvent for resins, waxes, and dyes, and in the synthesis of perfumes, drugs, and flavorings. { ¦tü ¦eth·əl¦byüd·əl 'al·kə,hól }

ethyl butyl ketone [ORG CHEM] $C_2H_5COC_4H_9$ A colorless liquid, boiling at 147°C; used in solvent mixtures. Also known as 3-heptanone. { ¦eth·əl ¦byüd·əl 'kē,tōn }

ethyl butyrate [ORG CHEM] $C_3H_7COOC_2H_5$ A colorless liquid, boiling at 121°C; used in flavoring extracts and perfumery. { ¦eth·əl 'byüd·ə,rāt }

ethyl caprate [ORG CHEM] $CH_3(CH_2)_8COOC_2H_5$ A colorless liquid, used in the manufacture of wine bouquets and cognac essence. { ¦eth·əl 'ka,prāt }

ethyl caproate [ORG CHEM] $C_5H_{11}COOC_2H_5$ A colorless to yellow liquid, boiling at 167°C, soluble in ether and alcohol, and having a pleasant odor; used as a chemical intermediate and in the food industry as an artificial fruit essence. Also known as ethyl hexanoate; ethyl hexoate. { ¦eth·əl kə'prō,āt }

ethyl caprylate [ORG CHEM] $CH_3(CH_2)_6COOC_2H_5$ A clear, colorless liquid with a pineapple odor; used to make fruit ethers. Also known as ethyl octanoate. { ¦eth·əl 'kap·rə,lāt }

ethyl carbamate See urethane. { ¦eth·əl 'kär·bə,māt }

ethyl carbinol See propyl alcohol. { ¦eth·əl'kär·bə,nól }

ethyl carbonate See diethyl carbonate. { ¦eth·əl 'kär·bə,nāt }

ethyl cellulose [ORG CHEM] The ethyl ester of cellulose; it has film-forming properties and is inert to alkalies and dilute acids; used in adhesives, lacquers, and coatings. { ¦eth·əl'sel·yə,lōs }

ethyl chloride [ORG CHEM] C_2H_5Cl A colorless gas, liquefying at 12.2°C, slightly soluble in water; used as a solvent, in medicine, and as an intermediate. Also known as chloroethane. { ¦eth·əl 'klór,īd }

ethyl chloroacetate [ORG CHEM] $CH_2ClCOOC_2H_5$ A colorless liquid, boiling at 145°C; used as a poison gas, solvent, and chemical intermediate. { ¦eth·əl ,klór·ō'as·ə,tāt }

ethyl chlorophyllide [BIOCHEM] A compound formed by replacing the phytyl ($C_{20}H_{39}$) tail of the chlorophyll molecule with a short ethyl (C_2H_5) tail; crystallizes easily and has an absorption spectrum and electrochemical properties similar to those of chlorophyll. { ¦eth·əl ,klór·ō'fil,īd }

ethyl cinnamate [ORG CHEM] $C_6H_5CH=CHCOOC_2H_5$ An oily liquid with a faint cinnamon odor; used as a fixative for perfumes. Also known as ethyl phenylacrylate. { ¦eth·əl 'sin·ə,māt }

ethyl crotonate [ORG CHEM] $CH_3CHCHCO_2C_2H_5$ A compound with a pungent aroma; boiling point of 143–147°C, soluble in water, soluble in ether; one of two isomeric forms used as an organic intermediate, a solvent for cellulose esters, and as a plasticizer for acrylic resins. { ¦eth·əl 'krōt·ən,āt }

ethyl crotonic acid [ORG CHEM] $CH_3CHCC_2H_5COOH$ Colorless monoclinic crystals, subliming at 40°C; used as a peppermint flavoring. { ¦eth·əl krə'tän·ik 'as·əd }

ethyl cyanide [ORG CHEM] C_2H_5CN A colorless liquid that boils at 97.1°C; poisonous. { ¦eth·əl 'sī·ə,nīd }

S-ethyl-N,N-dipropylthiocarbamate [ORG CHEM] $C_9H_{19}-NOS$ An amber liquid soluble in water at 370 parts per million; used as a pre- and postemergence herbicide on vegetable crops. Abbreviated EDTC. { es ¦eth·əl ¦en ¦en dī¦prō·pəl,thī·ō'kär·bə,māt }

ethyl enanthate [ORG CHEM] $CH_3(CH_2)_5COOC_2H_5$ A clear, colorless oil with a boiling point of 187°C; soluble in alcohol, chloroform, and ether; taste and odor are fruity; used as a flavor for liqueurs and soft drinks. Also known as cognac oil; ethyl heptanoate; ethyl oenanthate. { ¦eth·əl ə'nan,thāt }

ethylene [ORG CHEM] C_2H_4 A colorless, flammable gas, boiling at $-102.7°C$; used as an agricultural chemical, in medicine, and for the manufacture of organic chemicals and polyethylene. Also known as ethene; olefiant gas. { 'eth·ə,lēn }

ethylene alkylation [CHEM ENG] A catalytic petroleum-refining process in which dry isobutane and ethylene react to form ethylene alkylate. { 'eth·ə,lēn ,al·kə'lā·shən }

ethylene bromide *See* ethylene dibromide. { 'eth·ə,lēn 'brō,mīd }

ethylene carbonate [ORG CHEM] $(CH_2O)_2CO$ Odorless, colorless solid with low melting point; soluble in water and organic solvents; used as a polymer and resin solvent, in solvent extraction, and in organic syntheses. { 'eth·ə,lēn 'kär·bə,nāt }

ethylene chloride [ORG CHEM] $ClCH_2CH_2Cl$ A colorless, oily liquid, boiling at 83.7°C; used as a solvent and fumigant, for organic synthesis, and for ore flotation. Also known as Dutch liquid; ethylene dichloride. { 'eth·ə,lēn 'klòr,īd }

ethylene chlorobromide [ORG CHEM] CH_2BrCH_2Cl Volatile, colorless liquid with chloroformlike odor; soluble in ether and alcohol but not in water; general-purpose solvent for cellulosics; used in organic synthesis. { 'eth·ə,lēn klòr·ə'brō,mīd }

ethylene chlorohydrin [ORG CHEM] $ClCH_2CH_2OH$ A colorless, poisonous liquid, boiling at 129°C; used as a solvent and in organic synthesis. Also known as chloroethyl alcohol. { 'eth·ə,lēn klòr·ə'hī·drən }

ethylene cyanide [ORG CHEM] $C_2H_4(CN)_2$ Colorless crystals, melting at 57°C; used in organic synthesis. Also known as succinonitrile. { 'eth·ə,lēn 'sī·ə,nīd }

ethylene cyanohydrin [ORG CHEM] C_3H_5ON A colorless liquid that is miscible with water and boils at 221°C. { 'eth·ə,lēn ,sī·ə·nō'hī·drən }

ethylene diacetate *See* ethylene glycol diacetate. { 'eth·ə,lēn dī'as·ə,tāt }

ethylenediamine [ORG CHEM] $NH_2CH_2CH_2NH_2$ Colorless liquid, melting at 8.5°C, soluble in water; used as a solvent, corrosion inhibitor, and resin and in adhesive manufacture. { 'eth·ə,lēn'dī·ə,mēn }

ethylenediaminetetraacetic acid [ORG CHEM] $(HOOC-CH_2)_2NCH_2CH_2N(CH_2COOH)_2$ White crystals, slightly soluble in water and decomposing above 160°C; the sodium salt is a strong chelating agent, reacting with many metallic ions to form soluble nonionic chelate. Abbreviated EDTA. { 'eth·ə·lēn'dī·ə,mēn,te·trə·ə'sēd·ik 'as·əd }

ethylene dibromide [ORG CHEM] $BrCH_2CH_2Br$ A colorless, poisonous liquid, boiling at 131°C; insoluble in water; used in medicine, as a solvent in organic synthesis, and in antiknock gasoline. Also known as ethylene bromide. { 'eth·ə,lēn dī'brō,mīd }

ethylene dichloride *See* ethylene chloride. { 'eth·ə,lēn dī'klòr,īd }

ethylene glycol *See* glycol. { 'eth·ə,lēn 'glī,kòl }

ethylene glycol bis(trichloroacetate) [ORG CHEM] $C_4H_4-Cl_6O_4$ A white solid with a melting point of 40.3°C; used as a herbicide for cotton and soybeans. Abbreviated EGT. { 'eth·ə,lēn 'glī,kòl ,bis·trī,klòr·ō'as·ə,tāt }

ethylene glycol diacetate [ORG CHEM] $CH_3COOCH_2-CH_2OOCCH_3$ A liquid used as a solvent for oils, cellulose esters, and explosives. Also known as ethylene diacetate; glycol diacetate. { 'eth·ə,lēn 'glī,kòl dī'as·ə,tāt }

ethyleneimine [ORG CHEM] C_2H_4NH Highly corrosive liquid, colorless and clear; miscible with organic solvents and water; used as an intermediate in fuel oil production, refining lubricants, textiles, and pharmaceuticals. Also known as aziridine. { ,eth·ə'lēn·ə,mīn }

ethylene nitrate [ORG CHEM] $(CH_2NO_3)_2$ An explosive yellow liquid, insoluble in water. Also known as glycol dinitrate. { 'eth·ə,lēn 'nī,trāt }

ethylene oxide [ORG CHEM] **1.** $(CH_2)_2O$ A colorless gas, soluble in organic solvents and miscible in water, boiling point 11°C; used in organic synthesis, for sterilizing, and for fumigating. Also known as 1,2-epoxyethane. *See* epoxide. *See* oxirane. { 'eth·ə,lēn 'äk,sīd }

ethylene-propylene terpolymer [MATER] An elastomer which is based on ethylene and propylene terpolymers with small amounts of a nonconjugated diene and which can be vulcanized; used for automotive parts, cable coating, hose, footwear, and other products. Abbreviated EPDM; EPT. { ,eth·ə,lēn ,prō·pə,lēn tər'pāl·ə·mər }

ethylene resin [ORG CHEM] A thermoplastic material composed of polymers of ethylene; the resin is synthesized by polymerization of ethylene at elevated temperatures and pressures in the presence of catalysts. Also known as polyethylene; polyethylene resin. { 'eth·ə,lēn 'rez·ən }

ethylene sulfonic acid *See* ethionic acid. { 'eth·ə,lēn səl'fän·ik 'as·əd }

ethylethanolamine [ORG CHEM] $C_2H_5NHCH_2CH_2OH$ Water-white liquid with amine odor; soluble in alcohol, ether, and water; used in dyes, insecticides, fungicides, and surface-active agents. { ¦eth·əl,eth·ə'näl·ə,mēn }

ethyl ether *See* ether. { ¦eth·əl 'ē·thər }

ethyl formate [ORG CHEM] $HCOOC_2H_5$ A colorless liquid, boiling at 54.4°C; used as a solvent, fumigant, and larvicide and in flavors, resins, and medicines. { ¦eth·əl 'fòr,māt }

ethyl hexanoate *See* ethyl caproate. { ¦eth·əl hek'san·ə,wāt }

ethyl hexoate *See* ethyl caproate. { ¦eth·əl 'hek·sə,wāt }

2-ethyl hexoic acid [ORG CHEM] $C_4H_9CH(C_2H_5)COOH$ A liquid that is slightly soluble in water, boils at 226.9°C, and has a mild odor; used as an intermediate to make metallic salts for paint and varnish driers, esters for plasticizers, and light metal salts for conversion of some oils to grease. { ¦tü ¦eth·əl hek'sō·ik 'as·əd }

2-ethylhexyl acetate [ORG CHEM] $CH_3COOCH_2CH-C_2H_5C_4H_9$ Water-white, stable liquid; used as a solvent for nitrocellulose, resins, and lacquers. Also known as octyl acetate. { ¦tü ¦eth·əl ¦hek·səl'as·ə,tāt }

2-ethylhexyl acrylate [ORG CHEM] $CH_2CHCOOCH_2CH-(C_2H_5)C_4H_9$ Pleasant-smelling liquid; used as monomer for plastics, protective coatings, and paper finishes. { ¦tü ¦eth·əl 'ak·rə,lāt }

2-ethylhexyl alcohol [ORG CHEM] $C_4H_9CH(C_2H_5)CH_2OH$ Colorless, slightly viscous liquid; used as a defoaming or wetting agent, as a solvent for protective coatings, waxes, and oils, and as a raw material for plasticizers. Also known as octyl alcohol. { ¦tü ¦eth·əl 'al·kə,hòl }

2-ethylhexylamine [ORG CHEM] $C_4H_9CH(C_2H_5)CH_2NH_2$ Water-white liquid with slight ammonia odor; slightly water-soluble; used to synthesize detergents, rubber chemicals, and oil additives. { ¦tü¦eth·əl·hek'sil·ə,mēn }

2-ethylhexyl bromide [ORG CHEM] $C_4H_9CH(C_2H_5)CH_2Br$ Water-white, water-insoluble liquid; used to prepare pharmaceuticals and disinfectants. { ¦tü ¦eth·əl¦hek·səl 'brō,mīd }

2-ethylhexyl chloride [ORG CHEM] $C_4H_9CH(C_2H_5)CH_2Cl$ Colorless liquid; used to synthesize cellulose derivatives, pharmaceuticals, resins, insecticides, and dyestuffs. { ¦tü ¦eth·əl¦hek·səl 'klòr,īd }

ethyl-*para*-hydroxybenzoate [ORG CHEM] $HOC_6H_4CO-OC_2H_5$ Crystals with a melting point of 116°C that are soluble in water, alcohol, and ether; used as a preservative for pharmaceuticals. Also known as ethylparaben. { ¦eth·əl ¦par·ə hī,dräk·sē'ben·zə,wāt }

1-ethyl-3-hydroxypiperidine [PHARM] $C_7H_{15}NO$ A liquid with a boiling point of 93–95°C; used as an antispasmodic drug. Also known as 1-ethyl-3-piperidinol. { ¦wan ¦eth·əl ¦thrē hī,dräk·sē·pə'per·ə,dēn }

ethyl-2-hydroxypropionate *See* ethyl lactate. { ¦eth·əl ¦tü hī,dräk·sē'prō·pē·ə,nāt }

ethylic compound [ORG CHEM] Generic term for ethyl compounds. { e'thil·ik 'käm,paúnd }

ethylic ether *See* diethyl ether. { e'thil·ik 'ē·thər }

ethylidine [ORG CHEM] The $CH_3·CH=$ radical from ethane, C_2H_5. Also known as ethidine. { e'thil·ə,dēn }

ethyl iodide [ORG CHEM] C_2H_5I A colorless liquid, boiling at 72.3°C; used in medicine and in organic synthesis. Also known as hydroiodic ether; iodoethane. { ¦eth·əl 'ī·ə,dīd }

ethyl isobutylmethane *See* 2-methylhexane. { ¦eth·əl ,ī·sō,byüd·əl'me,thān }

ethyl isovalerate [ORG CHEM] $(CH_3)_2CHCH_2COOC_2H_5$ A colorless, oily liquid with an apple odor, soluble in water and miscible with alcohol, benzene, and ether; used for flavoring beverages and confectioneries. { ¦eth·əl ,ī·sō'val·ə,rāt }

ethyl lactate [ORG CHEM] $CH_3CHOHCOOC_2H_5$ A colorless liquid that boils at 154°C, has a mild odor, and is miscible with water and organic solvents such as alcohols, ketones, esters, and hydrocarbons; used as a flavoring and as a solvent for cellulose compounds such as nitrocellulose, cellulose acetate, and cellulose ethers. Also known as ethyl-2-hydroxy-propionate. { ¦eth·əl 'lak,tāt }

ethyl malonate [ORG CHEM] $CH_2(COOC_2H_5)_2$ A colorless liquid, boiling at 198°C; used as an intermediate and a plasticizer. Also known as malonic ester. { ¦eth·əl 'mal·ə,nāt }

ethyl mercaptan [ORG CHEM] C_2H_5SH A colorless liquid, boiling at 36°C. Also known as ethyl sulfhydrate; thioethyl alcohol. { ¦eth·əl mər'kap·tan }

ethyl methacrylate [ORG CHEM] $CH_2CCH_3COOC_2H_5$ Colorless, easily polymerized liquid, water-insoluble; used to produce polymers and chemical intermediates. { ¦eth·əl me'thak·rə,lāt }

ethyl methyl ketone See methyl ethyl ketone. { ¦eth·əl ¦meth·əl 'kē,tōn }

ethyl nitrate [ORG CHEM] $C_2H_5NO_3$ A colorless, flammable liquid, boiling at 87.6°C; used in perfumes, drugs, and dyes and in organic synthesis. { ¦eth·əl 'nī,trāt }

ethyl nitrite [ORG CHEM] $C_2H_5NO_2$ A colorless liquid, boiling at 16.4°C; used in medicine and in organic synthesis. Also known as sweet spirits of niter. { ¦eth·əl 'nī,trīt }

ethyl octanoate See ethyl caprylate. { ¦eth·əl äk'tan·ə,wāt }

ethyl oenanthate See ethyl enanthate. { ¦eth·əl ē'nan,thāt }

ethyl oleate [ORG CHEM] $C_{20}H_{38}O_2$ A yellow oil, insoluble in water; used as a solvent, plasticizer, and lubricant. { ¦eth·əl 'ō·lē,āt }

ethyl orthosilicate See ethyl silicate. { ¦eth·əl ,òr·thō'sil·ə,kāt }

ethyl oxalate [ORG CHEM] $(COOC_2H_5)_2$ Oily, unstable, colorless liquid that is combustible; miscible with organic solvents, very slightly soluble in water; used as a solvent for cellulosics and resins, and as an intermediate for dyes and pharmaceuticals. { ¦eth·əl 'äk·sə,lāt }

ethyl oxide See diethyl ether. { ¦eth·əl 'äk,sīd }

ethylparaben See ethyl-*para*-hydroxybenzoate. { ¦eth·əl'par·ə·bən }

O-ethyl-O-para-nitrophenyl **phenylphosphonothioate** [ORG CHEM] $C_2H_5O_4NPS$ A yellow, crystalline compound with a melting point of 36°C; used as an insecticide and miticide on fruit crops. Abbreviated EPN. { ¦ō ¦eth·əl ¦ō ¦par·ə ,nī·trō'fen·əl,fen·əl·fäs¦fä·nō 'thī·ə,wāt }

ethyl phenylacrylate See ethyl cinnamate. { ¦eth·əl ,fen·əl'ak·rə,lāt }

N-ethyl-5-phenylisoxazolium-3′-sulfonate [ORG CHEM] $C_{11}H_{11}NO_4S$ Crystals that decompose at 207–208°C; used to form peptide bonds. Also known as Woodward's Reagent K. { ¦en ¦eth·əl ¦fīv ¦fen·əl,ī·säk·sə'zō·lē·əm ¦thrē,prīm 'səl·fə,nāt }

1-ethyl-3-piperidinol See 1-ethyl-3-hydroxypiperidine. { ¦wən ¦eth·əl ¦thrē ,pi'per·ə·də,nòl }

ethyl propionate [ORG CHEM] $C_2H_5COOC_2H_5$ A colorless liquid, slightly soluble in water, boiling at 99°C; used as solvent and pyroxylin cutting agent. Also known as propionic ether. { ¦eth·əl 'prō·pē·ə,nāt }

ethyl salicylate [ORG CHEM] $(HO)C_6H_4COOC_2H_5$ A clear liquid with a pleasant odor; used in commercial preparation of artificial perfumes. Also known as sal ethyl; salicylic acid ethyl ether; salicylic ether. { ¦eth·əl sə'lis·əl,āt }

ethyl silicate [ORG CHEM] $(C_2H_5)_4SiO_4$ A colorless, flammable liquid, hydrolyzed by water; used as a preservative for stone, brick, and masonry, in lacquers, and as a bonding agent. Also known as ethyl orthosilicate. { ¦eth·əl 'sil·ə,kāt }

ethyl sulfate See diethyl sulfate. { ¦eth·əl 'səl,fāt }

ethyl sulfhydrate See ethyl mercaptan. { ¦eth·əl ,səlf'hī,drāt }

ethyl sulfide [ORG CHEM] $(C_2H_5)_2S$ A colorless, oily liquid, boiling at 92°C; used as a solvent and in organic synthesis. Also known as diethyl sulfide; ethylthioethane. { ¦eth·əl 'səl,fīd }

ethylthioethane See ethyl sulfide. { ¦eth·əl,thī·ō'e,thān }

ortho-ethyl(O-2,4,5-trichlorophenyl)ethylphosphonothioate [ORG CHEM] $C_{10}H_{12}OPSCI$ An amber liquid with a boiling point of 108°C at 0.01 mmHg; solubility in water is 50 parts per million; used as an insecticide for vegetable crops and soil pests on meadows. Also known as trichloronate. { ¦òr·thō ¦eth·əl ¦ō ¦tü ¦fòr ¦fiv ¦trī,klòr·ō¦fen·əl¦eth·əl,fäs¦fan·ō'thī·ə,wāt }

ethyl urethane See urethane. { ¦eth·əl 'yùr·ə,thān }

ethyl vanillin [ORG CHEM] $C_2H_5O(OH)C_6H_3CHO$ A compound, crystallizing in fine white crystals that melt at 76.5°C, has a strong vanilla odor and four times the flavor of vanilla, soluble in organic solvents such as alcohol, chloroform, and ether; used in the food industry as a flavoring agent to replace or fortify vanilla. { ¦eth·əl və'nil·ən }

ethyne See acetylene. { 'e,thēn }

ethynyl See ethinyl. { 'eth·ə,nil }

ethynylation [ORG CHEM] Production of an acetylenic derivative by the condensation of acetylene with a compound such as an aldehyde; for example, production of butynediol from the union of formaldehyde with acetylene. { ,eth·ən·əl'ā·shən }

ethyoxyl See ethoxy. { ,eth·əl'äk·səl }

etioallocholane See androstane. { ¦ēd·ē·ō,al·ə'kō,lān }

etioblast [BOT] An immature chloroplast, containing prolamellar bodies. { 'ēd·ē·ō,blast }

etiolation [BOT] The yellowing or whitening of green plant parts grown in darkness. { ,ed·ē·ə'lā·shən }

etiology [MED] Any factors which cause disease. [SCI TECH] A branch of science dealing with the causes of phenomena. { ,ēd·ē'äl·ə·jē }

etiopathogenesis [MED] The cause and development of a disease or abnormal condition. { ,ē·tē·ō,path·ə'jen·ə·səs }

etioplast [BOT] The plastid of a dark-grown plant that contains crystalline prolamellar bodies. { 'ēd·ē·ō,plast }

etioporphyrin [ORG CHEM] $C_{31}H_{34}N_4$ A synthetic porphyrin that has four ethyl and four methyl groups in a red-pigmented compound whose crystals melt at 280°C. { ,ēd·ē·ō'pòr·fə·rən }

E transformer [ELECTROMAG] A transformer consisting of two coils wound around a laminated iron core in the shape of an E, with the primary and secondaries occupying the center and outside legs respectively. { 'ē tranz,fòr·mər }

E trisomy See trisomy 18 syndrome. { 'ē 'trī,sō·mē }

Ettingshausen coefficient [PHYS] A measure of the strength of the Ettingshausen effect, equal to the ratio of the temperature gradient to the product of the current density and magnetic field strength which produce this gradient. { 'ed·iŋz,haùz·ən ,kō·i,fish·ənt }

Ettingshausen effect [PHYS] The phenomenon that, when a metal strip is placed with its plane perpendicular to a magnetic field and an electric current is sent longitudinally through the strip, corresponding points on opposite edges of the strip have different temperatures. { 'ed·iŋz,haùz·ən i,fekt }

Ettingshausen-Nernst coefficient [PHYS] A measure of the strength of the Ettingshausen-Nernst effect, equal to the ratio of the electric field to the product of the temperature gradient and magnetic field strength which produce this field. { ¦ed·iŋz,haùz·ən ¦nərnst,kō·i,fish·ənt }

Ettingshausen-Nernst effect [PHYS] The phenomenon that, when a conductor or semiconductor is subjected to a temperature gradient and to a magnetic field perpendicular to the temperature gradient, an electric field arises perpendicular to both the temperature gradient and the magnetic field. { ¦ed·iŋz,haùz·ən ¦nərnst i,fekt }

ettringite [MINERAL] $Ca_6Al_2(SO_4)_3(OH)_{12}·26H_2O$ A mineral composed of hydrous basic calcium and aluminum sulfate. { 'e·triŋ,īt }

Eu See europium.

EU See expected value.

Eubacteriales [MICROBIO] Formerly an order of the class Schizomycetes; considered the true bacteria and characterized by simple, undifferentiated, rigid cells of either spherical or straight, rod-shaped form. { ,yü·bak,tir·ē'ā·lēz }

Eubacterium [MICROBIO] A genus of bacteria in the family Propionibacteriaceae; obligate anaerobes producing a mixture of organic acids (butyric, acetic, formic, and lactic) from carbohydrates and peptone. { ,yü·bak'tir·ē·əm }

Eubasidiomycetes [MYCOL] An equivalent name for Homobasidiomycetidae. { ,yü·bə,sid·ē·ō,mī'sēd·ēz }

Eubrya [BOT] A subclass of the mosses (Bryopsida); the leafy gametophytes arise from buds on the protonemata, which are nearly always filamentous or branched green threads attached to the substratum by rhizoids. { yü'brī·ə }

Eubryales [BOT] An order of mosses (Bryatae); plants have the sporophyte at the end of a stem, vary in size from small to robust, and generally grow in tufts. { ,yü,brī'ā·lēz }

eucairite [MINERAL] CuAgSe A white, native selenide that crystallizes in the isometric crystal system. { yü'kī,rīt }

eucalyptol [ORG CHEM] $C_{10}H_{18}O$ A colorless oil with a camphorlike odor; boiling point is 174–177°C; used in pharmaceuticals, perfumery, and flavoring. Also known as cajeputol; cineol. { ,yü·kə'lip,tòl }

Eucalyptus [BOT] A large genus of evergreen trees belonging to the myrtle family (Myrtaceae) and occurring in Australia and New Guinea. { ˌyü·kə'lip·təs }

eucalyptus gum [PHARM] The dried gummy exudate from *Eucalyptus longirostris* of Australia, composed of kinotannic acid, kino red, glucoside, catechol, and pyrocatechol; used in medicine as an astringent and antidiarrheal agent. Also known as eucalyptus kino; red gum. { ˌyü·kə'lip·təs ˌgəm }

eucalyptus kino *See* eucalyptus gum. { ˌyü·kə'lip·təs 'kē·nō }

eucalyptus oil [MATER] Any of various essential oils of various eucalyptus leaves, a yellow liquid miscible with alcohol; varies from peppermint to turpentine in odor; used in perfumery, medicine, and ore flotation. { ˌyü·kə'lip·təs ˌoil }

eucalyptus resin oil *See* eucalyptus tar. { ˌyü·kə'lip·təs 'rez·ən ˌoil }

eucalyptus tar [MATER] Residue from caustic treating of oil from distilled eucalyptus leaves; used to perfume soaps and as a disinfectant. Also known as eucalyptus resin oil. { ˌyü·kə'lip·təs ˌtär }

Eucarida [INV ZOO] A large superorder of the decapod crustaceans, subclass Malacostraca, including shrimps, lobsters, hermit crabs, and crabs; characterized by having the shell and thoracic segments fused dorsally and the eyes on movable stalks. { ˌyü'kar·ə·də }

Eucaryota [BIOL] Primitive, unicellular organisms having a well-defined nuclear membrane, chromosomes, and mitotic cell division. { yü·kar·ē'ōd·ə }

eucaryote *See* eukaryote. { yü'kar·ē·ˌōt }

eucatropine hydrochloride [PHARM] $C_{17}H_{25}O_3N \cdot HCl$ A white, odorless powder that melts at 183–186°C, is soluble in alcohol and chloroform, and insoluble in ether; used in medicine as a mydriatic. { yü'ka·trə·ˌpēn ˌhī·drə'klôr·ˌīd }

Eucestoda [INV ZOO] The true tapeworms, a subclass of the class Cestoda. { yü'ses·tə·də }

Eucharitidae [INV ZOO] A family of hymenopteran insects in the superfamily Chalcidoidea. { yü·kə'rid·ə·dē }

euchlorin [MINERAL] $(K,Na)_8Cu_9(SO_4)_{10}(OH)_6$ An emerald-green mineral consisting of a basic sulfate of potassium, sodium, and copper; found in lava at Vesuvius. { yü'klôr·ən }

euchroite [MINERAL] $Cu_2(AsO_4)(OH) \cdot 3H_2O$ An emerald green or leek green, orthorhombic mineral consisting of a hydrated basic copper arsenate. { 'yü·krō·ˌīt }

euchromatin [CYTOL] The portion of the chromosomes that stains with low intensity, uncoils during interphase, and condenses during cell division. { yü'krō·mə·tən }

Eucinetidae [INV ZOO] The plate thigh beetles, a family of coleopteran insects in the superfamily Dascilloidea. { yü·sə'ned·ə·dē }

euclase [MINERAL] $BeAlSiO_4(OH)$ A brittle, pale green, blue, yellow, or violet monoclinic mineral, occurring as prismatic crystals. { 'yü·klās }

Euclasterida [INV ZOO] An order of asteroid echinoderms in which the arms are sharply distinguished from a small, central disk-shaped body. { ˌyü·klə'ster·ə·də }

Eucleidae [INV ZOO] The slug moths, a family of lepidopteran insects in the suborder Heteroneura. { yü'klē·ə·dē }

euclidean algorithm [MATH] A method of finding the greatest common divisor of a pair of integers. { yü'klid·ē·ən 'al·gə·ˌrith·əm }

euclidean geometry [MATH] The study of the properties preserved by isometries of two- and three-dimensional euclidean space. { yü'klid·ē·ən jē'äm·ə·trē }

euclidean quantum field theory [QUANT MECH] A relativistic quantum field theory in which time is replaced by a purely formal imaginary time, resulting in replacement of Lorentz covariance by euclidean group covariance. { yü'klid·ē·ən ˌkwänt·əm 'fēld ˌthē·ə·rē }

Euclidean ring [MATH] A commutative ring, together with a function, f, from the nonzero elements of the ring to the nonnegative integers, such that (1) $f(xy) \geq f(x)$ if $xy \neq 0$, and (2) for any members of the ring, x and y, with $x \neq 0$, there are members q and r such that $y = qx + r$ and either $r = 0$ or $f(r) < f(x)$. { yü'klid·ē·ən 'riŋ }

euclidean space [MATH] A space consisting of all ordered sets $(x_1, \ldots, x_n)$ of n numbers with the distance between $(x_1, \ldots, x_n)$ and $(y_1, \ldots, y_n)$ being given by

$$\left[\sum_{i=1}^{n}(x_i - y_i)^2\right]^{1/2};$$

the number n is called the dimension of the space. { yü'klid·ē·ən 'spās }

Euclymeninae [INV ZOO] A subfamily of annelids in the family Maldonidae of the Sedentaria, having well-developed plaques and an anal pore within the plaque. { ˌyü·klə'men·ə·ˌnē }

Eucnemidae [INV ZOO] The false click beetles, a family of coleopteran insects in the superfamily Elateroidea. { yük'nem·ə·dē }

Eucoccida [INV ZOO] An order of parasitic protozoans in the subclass Coccidia characterized by alternating sexual and asexual phases; stages of the life cycle occur intracellularly in vertebrates and invertebrates. { yü'käk·sə·də }

Eucoelomata [ZOO] A large sector of the animal kingdom including all forms in which there is a true coelom or body cavity; includes all phyla above Aschelminthes. { yü·sē·lə'mäd·ə }

Eucommiales [BOT] A monotypic order of dicotyledonous plants in the subclass Hamamelidae; plants have two, unitegmic ovules and lack stipules. { yü·käm·ē'ā·lēz }

eucrite [MINERAL] An olivine-bearing gabbro containing unusually calcic plagiocase; a meteorite component. { 'yü·ˌkrīt }

eucryptite [MINERAL] $LiAlSiO_4$ A colorless or white lithium aluminum silicate mineral, crystallizing in the hexagonal system; specific gravity is 2.67. { yü'krip·ˌtīt }

Eudactylinidae [INV ZOO] A family of parasitic copepod crustaceans in the suborder Caligoida; found as ectoparasites on the gills of sharks. { ˌyü·dak·tə'lin·ə·dē }

eudialite [MINERAL] $(Na,Ca,Fe)_6ZrSi_6O_{18}(OH,Cl)$ Hexagonal-crystalline silicate chloride mineral; color is red to brown. { yü'dī·ə·ˌlīt }

eudidymite [MINERAL] $NaBeSi_3O_7(OH)$ A glassy white mineral composed of sodium beryllium silicate. { yü'did·ə·ˌmīt }

eudiometer [ENG] An instrument for measuring changes in volume during the combustion of gases, consisting of a graduated tube that is closed at one end and has two wires sealed into it, between which a spark may be passed. { ˌyü·dē'äm·əd·ər }

eudoxid *See* eudoxome. { yü'däk·ˌsīd }

eudoxome [INV ZOO] Cormidium of most calycophoran siphonophores which lead a free existence. Also known as eudoxid. { yü'däk·ˌsōm }

Euechinoidea [INV ZOO] A subclass of echinoderms in the class Echinoidea; distinguished by the relative stability of ambulacra and interambulacra. { yü·ek·ə'nóid·ē·ə }

Eugenia [ASTRON] An asteroid with a diameter of about 133 miles (215 kilometers), mean distance from the sun of 2.72 astronomical units, and F-type (C-like) surface composition; observed to have a small satellite orbiting around it. { yü'jēn·yə }

eugenics [GEN] The attempt to improve the phenotypes of future generations of the human population by fostering the reproduction of those with favorable phenotypes and genotypes and hampering or preventing breeding by those with "undesirable" phenotypes and genotypes. The concept is largely discredited. { yü'jen·iks }

eugenol [ORG CHEM] $CH_2CHCH_2C_6H_3(OCH_3)OH$ A colorless or yellowish aromatic liquid with spicy odor and taste, soluble in organic solvents, and extracted from clove oil; used in flavors, perfumes, medicines, and the manufacture of vanilla. { 'yü·jə·ˌnól }

eugeosyncline [GEOL] The internal volcanic belt of an orthogeosyncline. { ˌyü·jē·ō'sin·ˌklīn }

Euglena [BIOL] A genus of organisms with one or two flagella, chromatophores in most species, and a generally elongate, spindle-shaped body; classified as algae by botanists (Euglenophyta) and as protozoans by zoologists (Euglenida). { yü'glē·nə }

Euglenida [INV ZOO] An order of protozoans in the class Phytamastigophorea, including the largest green, noncolonial flagellates. { yü'glen·ə·də }

Euglenidae [INV ZOO] The antlike leaf beetles, a family of coleopteran insects in the superfamily Tenebrionoidea. { yü'glen·ə·dē }

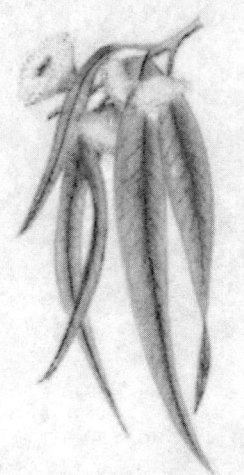

Eucalyptus globulus showing spray of mature foliage.

Euglenoidina [INV ZOO] The equivalent name for Euglenida. { ˌyü·glə·nȯi′dī·nə }

Euglenophyceae [BOT] The single class of the plant division Euglenophyta. { yü‚glē·nə′fīs·ē‚ē }

Euglenophyta [BOT] A division of the plant kingdom including one-celled, chiefly aquatic flagellate organisms having a spindle-shaped or flattened body, naked or with a pellicle. { ˌyü·glə′näf·əd·ə }

euglobulin [BIOCHEM] True globulin; a simple protein that is soluble in distilled water and dilute salt solutions. { yü′gläb·yə·lən }

Eugregarinida [INV ZOO] An order of protozoans in the subclass Gregarinia; parasites of certain invertebrates. { ˌyü·grə‚gar·ə′nīd·ə }

euhedral See automorphic. { yü′hē·drəl }

Eukaryotae [BIOL] A superkingdom that includes living and fossil organisms comprising all taxonomic groups above the primitive unicellular prokaryotic level. { yü‚kar·ē′ō‚tē }

eukaryote [BIOL] A cell with a definitive nucleus. Also spelled eucaryote. { yü′kar·ē‚ōt }

Eulamellibranchia [INV ZOO] The largest subclass of the molluscan class Bivalvia, having a heterodont shell hinge, leaf-like gills, and well-developed siphons. { ˌyü·lə‚mel·ə′braŋ‚kē·ə }

Eulenburg's disease See paramyotonia congenita. { ′ȯi·lən‚bərkz di‚zēz }

Euler angles [MECH] Three angular parameters that specify the orientation of a body with respect to reference axes. { ′ȯi·lər ‚aŋ·gəlz }

Euler characteristic of a topological space X [MATH] The number $\chi(X) = \Sigma(-1)^q \beta_q$, where β_q is the qth Betti number of X. { ′ȯi·lər ‚kar·ik·tə′ris·tik əv ə ‚täp·ə‚läj·i·kəl ′spās ′eks }

Euler diagram [MATH] A diagram consisting of closed curves, used to represent relations between logical propositions or sets; similar to a Venn diagram. { ′ȯi·lər ‚dī·ə‚gram }

Euler equation [MECH] Expression for the energy removed from a gas stream by a rotating blade system (as a gas turbine), independent of the blade system (as a radial- or axial-flow system). { ′ȯi·lər i‚kwā·zhən }

Euler equations of motion [MECH] A set of three differential equations expressing relations between the force moments, angular velocities, and angular accelerations of a rotating rigid body. { ′ȯi·lər i‚kwā·zhənz əv ′mō·shən }

Euler force [MECH] The greatest load that a long, slender column can carry without buckling, according to the Euler formula for long columns. { ′ȯi·lər ‚fȯrs }

Euler formula for long columns [MECH] A formula which gives the greatest axial load that a long, slender column can carry without buckling, in terms of its length, Young's modulus, and the moment of inertia about an axis along the center of the column. { ′ȯi·lər ‚fȯr·myə·lə fər ‚lȯŋ ′käl·əmz }

Eulerian coordinates [FL MECH] Any system of coordinates in which properties of a fluid are assigned to points in space at each given time, without attempt to identify individual fluid parcels from one time to the next; a sequence of synoptic charts is a Eulerian representation of the data. { ȯi′ler·ē·ən kō′ȯrd·ən‚ats }

Eulerian correlation [FL MECH] The correlation between the properties of a flow at various points in space at a single instant of time. Also known as synoptic correlation. { ȯi′ler·ē·ən ‚kä·rə′lā·shən }

Eulerian description See Euler method. { ȯi′ler·ē·ən di′skrip·shən }

Eulerian equation [FL MECH] A mathematical representation of the motions of a fluid in which the behavior and the properties of the fluid are described at fixed points in a coordinate system. { ȯi′ler·ē·ən i‚kwā·zhən }

Eulerian graph [MATH] A graph that has an Eulerian path. { ȯi′ler·ē·ən ′graf }

Eulerian nutation See Chandler wobble. { ȯi′ler·ē·ən nyü′tā·shən }

Eulerian path [MATH] A path that traverses each of the lines in a graph exactly once. { ȯi′ler·ē·ən ′path }

Eulerian wind [METEOROL] A wind motion only in response to the pressure force; the cyclostrophic wind is a special case of the Eulerian wind, which is limited in its meteorological applicability to those situations in which the Coriolis effect is negligible. { ȯi′ler·ē·ən ′wind }

Euler-Lagrange equation [MATH] A partial differential equation arising in the calculus of variations, which provides a necessary condition that $y(x)$ minimize the integral over some finite interval of $f(x,y,y')dx$, where $y' = dy/dx$; the equation is $(\delta f(x,y,y')/\delta y) - (d/dx)(\delta f(x,y,y')/\delta y') = 0$. Also known as Euler's equation. { ′ȯi·lər lə′gränj i‚kwā·zhən }

Euler-Maclaurin formula [MATH] A formula used in the numerical evaluation of integrals, which states that the value of an integral is equal to the sum of the value given by the trapezoidal rule and a series of terms involving the odd-numbered derivatives of the function at the end points of the interval over which the integral is evaluated. { ′ȯi·lər mə′klȯr·ən ‚fȯr·myə·lə }

Euler method [MATH] A method of obtaining an approximate solution of an ordinary differential equation of the form $dy/dx = f(x,y)$, where f is a specified function of x and y. Also known as Eulerian description. [MECH] A method of studying fluid motion and the mechanics of deformable bodies in which one considers volume elements at fixed locations in space, across which material flows; the Euler method is in contrast to the Lagrangian method. { ′ȯi·lər ‚meth·əd }

Euler number 1 [FL MECH] A dimensionless number used in the study of fluid friction in conduits, equal to the pressure drop due to friction divided by the product of the fluid density and the square of the fluid velocity. { ′ȯi·lər ‚nəm·bər ′wən }

Euler number 2 [FL MECH] A dimensionless number equal to two times the Fanning friction factor. { ′ȯi·lər ‚nəm·bər ′tü }

Euler-Rodrigues parameter [MECH] One of four numbers which may be used to specify the orientation of a rigid body; they are components of a quaternion. { ′ȯi·lər rə′drē·gəs pə‚ram·əd·ər }

Euler's constant [MATH] The limit as n approaches infinity, of $1 + 1/2 + 1/3 + \cdots + 1/n - \ln n$, equal to approximately 0.5772. Denoted γ. Also known as Mascheroni's constant. { ′ȯi·lərz ‚kän·stənt }

Euler's criterion [MATH] A criterion for the congruence $x^n \equiv a \pmod m$ to have a solution, namely that $a^{\phi/d} \equiv 1 \pmod m$, where $\phi = \phi(m)$ is Euler's phi function evaluated at m, and d is the greatest common divisor of ϕ and n. { ′ȯi·lərz krī′tir·ē·ən }

Euler's equation See Euler-Lagrange equation. { ′ȯi·lərz i‚kwā·zhən }

Euler's expansion [FL MECH] The transformation of a derivative (d/dt) describing the behavior of a moving particle with respect to time, into a local derivative $(\delta/\delta t)$ and three additional terms that describe the changing motion of a fluid as it passes around a fixed point. { ′ȯi·lərz ik′span·shən }

Euler's formula [MATH] The formula $e^{ix} = \cos x + i \sin x$, where $i = \sqrt{-1}$. { ′ȯi·lərz ‚fȯr·myə·lə }

Euler's numbers [MATH] The numbers E_{2n} defined by the equation

$$\frac{1}{\cos z} = \sum_{n=0}^{\infty} (-1)^n \frac{E_{2n}}{(2n)!} z^{2n}$$

{ ′ȯi·lərz ‚nəm·bərz }

Euler's phi function [MATH] A function ϕ, defined on the positive integers, whose value $\phi(n)$ is the number of integers equal to or less than n and relatively prime to n. Also known as indicator; phi function; totient. { ′ȯi·lərz ′fī ‚faŋk·shən }

Euler's spiral See Cornu's spiral. { ′ȯi·lərz ‚spī·rəl }

Euler's theorem [MATH] For any polyhedron, $V - E + F = 2$, where V, E, F represent the number of vertices, edges, and faces respectively. { ′ȯi·lərz ‚thir·əm }

Euler transformation [MATH] A method of obtaining from a given convergent series a new series which converges faster to the same limit, and for defining sums of certain divergent series; the transformation carries the series $a_0 - a_1 + a_2 - a_3 + \cdots$ into a series whose nth term is

$$\sum_{r=0}^{n-1} (-1)^r \binom{n-1}{r} a_r / 2^n$$

{ ′ȯi·lər ‚tranz·fər′mā·shən }

eulittoral [OCEANOGR] A subdivision of the benthic division of the littoral zone of the marine environment, extending from high-tide level to about 200 feet (60 meters), the lower limit for abundant growth of attached plants. { yü'lid·ə·rəl }

Eulophidae [INV ZOO] A family of hymenopteran insects in the superfamily Chalcidoidea including species that are parasitic on the larvae of other insects. { yü'läf·ə‚dē }

eulytine *See* eulytite. { 'yü·lə‚tēn }

eulytite [MINERAL] Bi₄Si₃O₁₂ A bismuth silicate mineral usually found as minute dark-brown or gray tetrahedral crystals; specific gravity is 6.11. Also known as agricolite; bismuth blende; eulytine. { 'yü·lə‚tīt }

Eumalacostraca [INV ZOO] A series of the class Crustacea comprising shrimplike crustaceans having eight thoracic segments, six abdominal segments, and a telson. { yü‚mal·ə'käs·trə·kə }

Eumetazoa [ZOO] A section of the animal kingdom that includes the phyla above the Porifera; contains those animals which have tissues or show some tissue formation and organ systems. { yü‚med·ə'zō·ə }

eumitosis [CYTOL] Typical mitosis. { ¦yü‚mī'tō·səs }

Eumycetes [MYCOL] The true fungi, a large group of microorganisms characterized by cell walls, lack of chlorophyll, and mycelia in most species; includes the unicellular yeasts. { ‚yü‚mī'sēd·ēz }

Eumycetozoida [INV ZOO] An order of protozoans in the subclass Mycetozoia; includes slime molds which form a plasmodium. { ¦yü‚mī‚sed·ə'zóid·ə }

Eumycophyta [MYCOL] An equivalent name for the Eumycetes. { ‚yü‚mī'käf·əd·ə }

Eumycota [MYCOL] An equivalent name for Eumycetes. { ‚yü‚mī'kōd·ə }

Eunicea [INV ZOO] A superfamily of polychaete annelids belonging to the Errantia. { yü'nis·ē·ə }

Eunicidae [INV ZOO] A family of polychaete annelids in the superfamily Eunicea having characteristic pharyngeal armature consisting of maxillae and mandibles. { yü'nis·əd·ē }

Eunomia [ASTRON] An asteroid with a diameter of about 161 miles (259 kilometers), mean distance from the sun of 2.64 astronomical units, and S-type surface composition. { yü'nō·mē·ə }

eunuch [MED] An individual who has undergone complete loss of testicular function. { 'yü·nik }

Euomphalacea [PALEON] A superfamily of extinct gastropod mollusks in the order Aspidobranchia characterized by shells with low spires, some approaching bivalve symmetry. { yü‚äm·fə'lās·ē·ə }

eupavarine [PHARM] C₂₀H₂₁NO₂ A crystalline alkaloid that melts at 214°C and is soluble in many organic solvents and hot water; used in medicine as an antispasmodic. { yü'pav·ə‚rēn }

eupelagic *See* pelagic. { yü·pə'laj·ik }

Eupelmidae [INV ZOO] A family of hymenopteran insects in the superfamily Chalcidoidea. { yü'pel·mə‚dē }

Euphausiacea [INV ZOO] An order of planktonic malacostracans in the class Crustacea possessing photophores which emit a brilliant blue-green light. { yü‚fó·zē'ās·ē·ə }

euphenics [GEN] The production of a satisfactory phenotype by means other than eugenics. { yü'fen·iks }

Eupheterochlorina [INV ZOO] A suborder of flagellate protozoans in the order Heterochlorida. { yü¦fed·ə·rō‚klə'rī·nə }

Euphorbiaceae [BOT] A family of dicotyledonous plants in the order Euphorbiales characterized by dehiscent fruit having more than one seed and by epitropous ovules. { yü‚fór·bē'ās·ē‚ē }

Euphorbiales [BOT] An order of dicotyledonous plants in the subclass Rosidae having simple leaves and unisexual flowers that are aggregated and reduced. { yü‚fór·bē'ā·lēz }

euphotic [OCEANOGR] Of or constituting the upper levels of the marine environment down to the limits of effective light penetration for photosynthesis. { yü'fäd·ik }

Euphrosinidae [INV ZOO] A family of amphinomorphan polychaete annelids with short, dorsolaterally flattened bodies. { yü‚frə'zin·ə‚dē }

Euphrosyne [ASTRON] An asteroid with a diameter of about 154 miles (248 kilometers), mean distance from the sun of 3.15 astronomical units, and B-type (C-like) surface composition. { yü'fräz·ən·ē }

Euplexoptera [INV ZOO] The equivalent name for Dermaptera. { yü‚plek'säp·tə·rə }

euploid [GEN] Having a chromosome complement that is an exact multiple of the haploid complement. { 'yü‚plóid }

eupnea [PHYSIO] Normal or easy respiration rhythm. { 'yüp·nē·ə }

Eupodidae [INV ZOO] A family of mites in the suborder Trombidiformes. { yü'päd·ə‚dē }

Euproopacea [PALEON] A group of Paleozoic horseshoe crabs belonging to the Limulida. { yü‚prō·ə'pās·ē·ə }

Euramerica [GEOL] The continent that was composed of Europe and North America during most of the Mesozoic Era. { ‚yür·ə'mer·ə·kə }

Eureka [ASTRON] An asteroid orbiting near the following Lagrangian point of the planet Mars, located on Mars's orbit, 60° behind the planet; the first Trojan asteroid of the sun-Mars system to be discovered, in 1990. { yə'rē·kə }

Europa [ASTRON] **1.** A satellite of Jupiter with a mean distance from Jupiter of 4.17 × 10⁵ miles (6.71 × 10⁵ kilometers), orbital period of 3.6 days, and diameter of about 1950 miles (3100 kilometers). Also known as Jupiter II. **2.** An asteroid with a diameter of about 183 miles (295 kilometers), mean distance from the sun of 3.09 astronomical units, and C-type surface composition. { yü'rō·pə }

Europe [GEOGR] A great western peninsula of the Eurasian landmass, usually called a continent; its eastern limits are arbitrary and are conventionally drawn along the water divide of the Ural Mountains, the Ural River, the Caspian Sea, and the Caucasus watershed to the Black Sea. { 'yùr·əp }

European boreal faunal region [ECOL] A zoogeographic region describing marine littoral faunal regions of the northern Atlantic Ocean between Greenland and the northwestern coast of Europe. { ¦yùr·ə¦pē·ən ¦bór·ē·əl 'fón·əl 'rē·jən }

European canker [PL PATH] **1.** A fungus disease of apple, pear, and other fruit and shade trees caused by *Nectria galligena* and characterized by cankers with concentric rings of callus on the trunk and branches. **2.** A fungus disease of poplars caused by *Dothichiza populea*. { ¦yùr·ə¦pē·ən 'kaŋ·kər }

European Geostationary Navigation Overlay System [NAV] A satellite-based augmentation system developed jointly by the European Union, European Space Agency, and EUCONTROL. Abbreviated EGNOS. { ‚yür·ə¦pē·ən jē·ō¦stā·shə·ner·ē ‚nav·ə¦gā·shən 'ō·vər‚lay ‚sis·təm }

European hornbeam [FOR] *Carpinus betulus.* A type of hornbeam native to Europe and Asia Minor that can be distinguished by its larger size, larger winter buds, and larger three-lobed, almost entire fruiting bracts, often cultivated in parks and estates. { ‚yür·ə¦pē·ən 'hórn‚bēm }

European porcelain *See* porcelain. { ¦yür·ə¦pē·ən 'pórs·lən }

europium [CHEM] A member of the rare-earth elements in the cerium subgroup, symbol Eu, atomic number 63, atomic weight 151.96, steel gray and malleable, melting at 1100–1200°C. { yu'rō·pe·əm }

europium halide [INV ZOO] Any of the compounds of the element europium and the halogen elements; for example, europium chloride, EuCl₃·xH₂O. { yu'rō·pē·əm 'ha‚līd }

europium oxide [INV ZOO] Eu₂O₃ A white powder, insoluble in water; used in red- and infrared-sensitive phosphors. { yu'rō·pē·əm 'äk‚sīd }

Eurotiaceae [MYCOL] A family of ascomycetous fungi of the order Eurotiales in which the asci are borne in cleistothecia or closed fruiting bodies. { yə‚rōd·ē'ās·ē‚ē }

Eurotiales [MYCOL] An order of fungi in the class Ascomycetes bearing ascospores in globose or broadly oval, delicate asci which lack a pore. { yə‚rōd·ē'ā·lēz }

Euryalae [INV ZOO] The basket fishes, a family of echinoderms in the subclass Ophiuroidea. { yə'rī·ə‚lē }

Euryalina [INV ZOO] A suborder of ophiuroid echinoderms in the order Phrynophiurida characterized by a leathery integument. { yùr·ē'a·lə·nə }

Euryapsida [PALEON] A subclass of fossil reptiles distinguished by an upper temporal opening on each side of the skull. { yùr·ē'ap·sə·də }

eurybathic [ECOL] Living at the bottom of a body of water. { ¦yùr·ə¦bath·ik }

eurycephalic [ARCH] Having a cephalic index of 80–84. { ¦yùr·ə·sə'fal·ik }

Eurychilinidae [PALEON] A family of extinct dimorphic ostracodes in the superfamily Hollinacea. { ‚yùr·ə·kə'lin·ə‚dē }

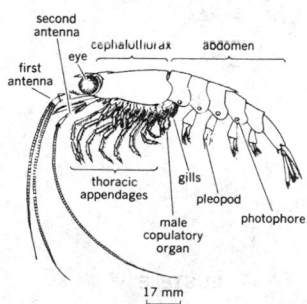

EUPHAUSIACEA

second antenna
first antenna
eye
cephalothorax
abdomen
thoracic appendages
gills
male copulatory organ
pleopod
photophore

17 mm

Diagram of a euphausiid crustacean.

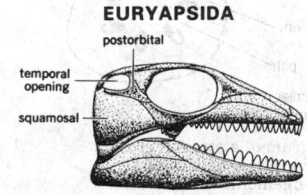

EURYAPSIDA

postorbital
temporal opening
squamosal

Lateral view of *Araeoscelis* skull. *(From A. S. Romer, Vertebrate Paleontology, 3d ed., University of Chicago Press, 1966)*

euryene [ARCH] Having a short or broad forehead or both with an upper facial index of 45–50. { ′yùr·ə‚ēn }

eurygamous [INV ZOO] Mating in flight, as in many insect species. { yù′rig·ə·məs }

euryhaline [ECOL] Pertaining to the ability of marine organisms to tolerate a wide range of saline conditions, and therefore a wide variation of osmotic pressure, in the environment. { ‚yùr·ə¦ha‚līn }

Eurylaimi [VERT ZOO] A monofamilial suborder of suboscine birds in the order Passeriformes. { ‚yùr·ə′lā‚mī }

Eurylaimidae [VERT ZOO] The broadbills, the single family of the avian suborder Eurylaimi. { ‚yùr·ə′lā·mə‚dē }

eurymeric [ARCH] Having a broad femur with a platymeric index of 85–100. { ¦yùr·ə¦mer·ik }

Eurymylidae [PALEON] A family of extinct mammals presumed to be the ancestral stock of the order Lagomorpha. { ‚yùr·ə′mil·ə‚dē }

euryon [ANAT] One of the two lateral points functioning as end points to measure the greatest transverse diameter of the skull. { ′yùr·ē‚än }

Euryphoridae [INV ZOO] A family of copepod crustaceans in the order Caligoida; members are fish ectoparasites. { ‚yùr·ə′fōr·ə‚dē }

euryplastic [BIOL] Referring to an organism with a marked ability to change and adapt to a wide spectrum of environmental conditions. { ¦yùr·ə¦plas·tik }

euryprosopic [ARCH] Having a short or broad face or both with a facial index of 80–85. { ¦yùr·ə·prə′säp·ik }

Eurypterida [PALEON] A group of extinct aquatic arthropods in the subphylum Chelicerata having elongate-lanceolate bodies encased in a chitinous exoskeleton. { ‚yùr·əp′ter·ə·də }

Eurypygidae [VERT ZOO] The sun bitterns, a family of tropical and subtropical New World birds belonging to the order Gruiformes. { ‚yùr·ə′pij·ə‚dē }

eurypylous [INV ZOO] Having a wide opening; applied to sponges with wide apopyles opening directly into excurrent canals, and wide prosopyles opening directly from incurrent canals. { ¦yùr·ə¦pī·ləs }

eurysome [ARCH] Having a broad, thickset body build. { ′yùr·ə‚sōm }

eurytherm [BIOL] An organism that is tolerant of a wide range of temperatures. { ′yùr·ə‚thərm }

Eurytomidae [INV ZOO] The seed and stem chalcids, a family of hymenopteran insects in the superfamily Chalcidoidea. { ‚yùr·ə′täm·ə‚dē }

eurytopic [ECOL] Referring to organisms which are widely distributed. { ‚yùr·ə′täp·ik }

Eusiridae [INV ZOO] A family of pelagic amphipod crustaceans in the suborder Gammaridea. { yü′sir·ə‚dē }

eusocial [ZOO] Pertaining to animal societies, such as those of certain insects, in which sterile individuals work on behalf of reproductive individuals. { ‚yü′sō·shəl }

eusporangiate [BOT] Having sporogenous tissue derived from a group of epidermal cells. { ¦yü·spə¦ran·jē‚āt }

eustachian tube [ANAT] A tube composed of bone and cartilage that connects the nasopharynx with the middle ear cavity. { yü′stā·shən ‚tüb }

eustachian valve See canal valve. { yü′stā·shən ‚valv }

eustacy [OCEANOGR] Worldwide fluctuations of sea level due to changing capacity of the ocean basins or the volume of ocean water. { ′yü′stə·sē }

eustele [BOT] A modified siphonostele containing collateral or bicollateral vascular bundles; found in most gymnosperm and angiosperm stems. { yü′stēl }

eusternum [INV ZOO] The anterior sternal plate in insects. { yü′stər·nəm }

Eustigmatophyceae [BOT] A small class of mostly nonmotile, photosynthetic, unicellular algae in the division Chromophycota, characterized by the unique organization of motile cells, photosynthetic pigments including chlorophyll *a*, betacarotene, and violaxanthin, and a single parietal yellow-green chloroplast; live chiefly in fresh water but also in marine and soil habitats. { ‚yüs·tig‚ma·də′fī·sē·ē }

Eusuchia [VERT ZOO] The modern crocodiles, a suborder of the order Crocodilia characterized by a fully developed secondary palate and procoelous vertebrae. { yü′sü·kē·ə }

Eusyllinae [INV ZOO] A subfamily of polychaete annelids in the family Syllidae having a thick body and unsegmented cirri. { yü′sil·ə‚nē }

Eutardigrada [INV ZOO] An order of tardigrades which lack both a sensory cephalic appendage and a club-shaped appendage. { yü‚tärd·ə′grād·ə }

eutaxite [PETR] A rock exhibiting eutaxitic structure. { yü′tak‚sīt }

eutaxitic [PETR] Referring to the banded appearance in certain extrusive rocks, resulting from the layering of different textures, materials, or colors. { ¦yü·tak′sid·ik }

eutectic [MET] The microstructure that results when a metal of eutectic composition solidifies. [PHYS CHEM] An alloy or solution that has the lowest possible constant melting point. { yü′tek·tik }

eutectic crystallization [MET] Simultaneous crystallization of the constituents of a eutectic alloy during cooling of the melt. { yü′tek·tik krist·əl·ə′zā·shən }

eutectic melting [MET] Melting of isolated, microscopic areas of an alloy that correspond to the location of the eutectic of the system. { yü′tek·tik ′melt·iŋ }

eutectic mixture See eutectic system. { yü¦tek·tik ′miks·chər }

eutectic point [PHYS CHEM] The point in the constitutional diagram indicating the composition and temperature of the lowest melting point of a eutectic. { yü′tek·tik ′pȯint }

eutectic system [PHYS CHEM] The particular composition and temperature of materials at the eutectic point. Also known as eutectic mixture. { yü′tek·tik ′sis·təm }

eutectic temperature [PHYS CHEM] The temperature at the lowest melting point of a eutectic. { yü′tek·tik ′tem·prə·chər }

eutectofelsite See eutectophyre. { yü¦tek·tō¦fel‚sīt }

eutectogenic system [PHYS CHEM] A multicomponent liquid-solid mixture in which pure solid phases of each component are in equilibrium with the remaining liquid mixture at a specific (usually minimum) temperature for a given composition, that is, the eutectic point. { yü¦tek·tə¦jen·ik ′sis·təm }

eutectoid [MET] A mixture of phases whose composition is determined by the eutectoid point in the solid region of an equilibrium diagram and whose constituents are formed by the eutectoid reaction. [PHYS CHEM] The point in an equilibrium diagram for a solid solution at which the solution on cooling is converted to a mixture of solids. { yü′tek‚tȯid }

eutectophyre [PETR] A light-colored tufflike igneous rock exhibiting a network of interlocking quartz and orthoclase crystals. Also known as eutectofelsite. { yü′tek·tə‚fīr }

eutely [BIOL] Having the body composed of a constant number of cells, as in certain rotifers. { ′yüd·əl·ē }

Euthacanthidae [PALEON] A family of extinct acanthodian fishes in the order Climatiiformes. { ‚yü·thə′kan·thə‚dē }

euthanasia [MED] The act or practice of putting to death or allowing the death, in a relatively painless way, of persons or animals with incurable or painful disease. { ‚yü·thə′nā·zhə }

euthenics [BIOL] The science that deals with the improvement of the future of humanity by changing the environment. { yü′then·iks }

Eutheria [VERT ZOO] An infraclass of therian mammals including all living forms except the monotremes and marsupials. { yü′thir·ē·ə }

Eutrichosomatidae [INV ZOO] Small family of hymenopteran insects in the superfamily Chalcidoidea. { yü¦trik·ə‚sō′mad·ə‚dē }

eutrophic [HYD] Pertaining to a lake containing a high concentration of dissolved nutrients; often shallow, with periods of oxygen deficiency. { yü′träf·ik }

eutrophication [ECOL] The process by which a body of water becomes, either by natural means or by pollution, excessively rich in dissolved nutrients, resulting in increased primary productivity that often leads to a seasonal deficiency in dissolved oxygen. { yü·trə·fə′kā·shən }

EUV radiation See vacuum ultraviolet radiation. { ¦ē¦yü¦vē ‚rād·ē′ā·shən }

euxenite [MINERAL] A brownish-black rare-earth mineral that crystallizes in the orthorhombic system, contains oxide of calcium, cerium, columbium, tantalum, titanium, and uranium, and has a metallic luster; hardness is 6.5 on Mohs scale, and specific gravity is 4.7–5.0. { ′yük·sə‚nīt }

euxinic [HYD] Of or pertaining to an environment of restricted circulation and stagnant or anaerobic conditions. { yük′sin·ik }

eV See electronvolt.

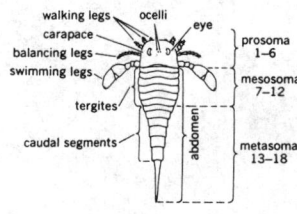

EURYPTERIDA

walking legs · ocelli · eye
carapace
balancing legs
swimming legs
tergites
caudal segments
prosoma 1–6
mesosoma 7–12
metasoma 13–18
abdomen

Dorsal view of features of a typical eurypterid.

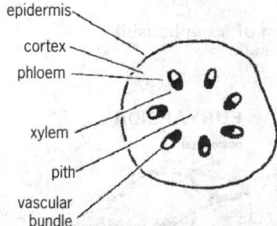

EUSTELE

epidermis
cortex
phloem
xylem
pith
vascular bundle

Diagram of a stem in cross section showing tissue arrangements of a eustele.

EV *See* expected value.

evacuate [SCI TECH] To remove something, especially gases and vapors, from an enclosure, such as from the envelope of an electron tube, or from a well. Also known as exhaust. { i'vak·yə,wāt }

Evaniidae [INV ZOO] The ensign flies, a family of hymenopteran insects in the superfamily Proctotrupoidea. { ,ev·ə'nī·ə,dē }

evansite [MINERAL] $Al_3(PO_4)(OH)_6·6H_2O$ A colorless to milky white mineral consisting of a hydrated basic aluminum phosphate; occurs in massive form and as stalactites. { 'ev·ən,zīt }

evaporable water [MATER] Water present in capillaries or held by surface forces in cement that has set; measured as the amount of water that can be removed by drying under specified conditions. { i'vap·rə·bəl 'wȯd·ər }

evaporated milk [FOOD ENG] A product made by removing enough moisture from whole milk that the final product contains no less than 7.9% fat or 25.9% total milk solids. { i'vap·ə,rād·əd 'milk }

evaporation [PHYS] Conversion of a liquid to the vapor state by the addition of latent heat. { i,vap·ə'rā·shən }

evaporation capacity *See* evaporative power. { i,vap·ə'rā·shən kə,pas·əd·ē }

evaporation current [OCEANOGR] An ocean current resulting from the accumulation of water through precipitation and river runoff at one point, and loss by evaporation at another point. { i,vap·ə'rā·shən ,kə·rənt }

evaporation gage *See* atmometer. { i,vap·ə'rā·shən ,gāj }

evaporation loss [CHEM ENG] The loss of a stored volatile liquid component or mixture by evaporation; controlled by temperature, pressure, and the presence or absence of vapor-recovery systems. { i,vap·ə'rā·shən ,lȯs }

evaporation pan [ENG] A type of atmometer consisting of a pan, used in the measurement of the evaporation of water into the atmosphere. { i,vap·ə'rā·shən ,pan }

evaporation power *See* evaporative power. { i,vap·ə'rā·shən ,paủ·ər }

evaporation tank [ENG] A tank used to measure the evaporation of water under controlled conditions. { i,vap·ə'rā·shən ,taŋk }

evaporative capacity *See* evaporative power. { i'vap·ə,rād·iv kə'pas·əd·ē }

evaporative condenser [MECH ENG] An apparatus in which vapor is condensed within tubes that are cooled by the evaporation of water flowing over the outside of the tubes. { i'vap·ə,rād·iv kən'den·sər }

evaporative control system [MECH ENG] A motor vehicle system that prevents escape of gasoline vapors from the fuel tank or carburetor to the atmosphere while the engine is not operating. { i'vap·ə,rād·iv kən'trōl ,sis·təm }

evaporative cooling [ENG] **1.** Lowering the temperature of a large mass of liquid by utilizing the latent heat of vaporization of a portion of the liquid. **2.** Cooling air by evaporating water into it. **3.** *See* vaporization cooling. { i'vap·ə,rād·iv 'kül·iŋ }

evaporative cooling tower *See* wet cooling tower. { i'vap·ə,rād·iv 'kül·iŋ ,taủ·ər }

evaporative heat regulation [PHYSIO] The composite process by which an animal body is cooled by evaporation of sensible perspiration; this avenue of heat loss serves as a physical means of regulating the body temperature. { i'vap·ə,rād·iv 'hēt ,reg·yə,lā·shən }

evaporative power [METEOROL] A measure of the degree to which the weather or climate of a region is favorable to the process of evaporation; it is usually considered to be the rate of evaporation, under existing atmospheric conditions, from a surface of water which is chemically pure and has the temperature of the lowest layer of the atmosphere. Also known as evaporation capacity; evaporation power; evaporative capacity; evaporativity; potential evaporation. { i'vap·ə,rād·iv 'paủ·ər }

evaporativity *See* evaporative power. { i,vap·ə·rə'tiv·əd·ē }

evaporator [CHEM ENG] A device used to vaporize part or all of the solvent from a solution; the valuable product is usually either a solid or concentrated solution of the solute. [MECH ENG] Any of many devices in which liquid is changed to the vapor state by the addition of heat, for example, distiller, still, dryer, water purifier, or refrigeration system element where evaporation proceeds at low pressure and consequent low temperature. { i'vap·ə,rād·ər }

evaporimeter *See* atmometer. { i,vap·ə'rim·əd·ər }

evaporite [GEOL] Deposits of mineral salts from sea water or salt lakes due to evaporation of the water. { i'vap·ə,rīt }

evaporite pond [IND ENG] Any containment area for brines or solution-mined effluents constructed to permit solar evaporation and harvesting of dewatered evaporite concentrates. { i'vap·ə,rīt ,pänd }

evapotranspiration [HYD] Discharge of water from the earth's surface to the atmosphere by evaporation from lakes, streams, and soil surfaces and by transpiration from plants. Also known as fly-off; total evaporation; water loss. { i,vap·ō,tranz·pə'rā·shən }

evapotranspirometer [ENG] An instrument which measures the rate of evapotranspiration; consists of a vegetation soil tank so designed that all water added to the tank and all water left after evapotranspiration can be measured. { i,vap·ō,tranz·pə'räm·əd·ər }

Evasé stack [CIV ENG] In tunnel engineering, an exhaust stack for air having a cross section that increases in the direction of airflow at a rate to regain pressure. { ā,vä,zā ,stak }

evection [ASTROPHYS] A perturbation of the moon in its orbit due to the attraction of the sun. { ē'vek·shən }

E vector [ELECTROMAG] Vector representing the electric field of an electromagnetic wave. { 'ē ,vek·tər }

even-even nucleus [NUC PHYS] A nucleus which has an even number of neutrons and an even number of protons. { 'ē·vən 'ē·vən ,nü·klē·əs }

even function [MATH] A function with the property that $f(x) = f(-x)$ for each number x. { 'ē·vən ,fəŋk·shən }

even harmonic [PHYS] A harmonic that is an even multiple of the fundamental frequency. { 'ē·vən här,män·ik }

evening gun [ORD] Firing of a gun as a signal for the lowering of the flag at retreat; the gun is fired after the sounding of the last note of the bugle call at retreat. Also known as retreat gun. { 'ev·niŋ 'gən }

evening star [ASTRON] A misnomer for a planet that can be seen without a telescope when it sets after the sun. { 'ev·niŋ 'stär }

evening twilight [ASTRON] The period of time between sunset and darkness. { 'ev·niŋ 'twī,līt }

even keel [NAV ARCH] Floating level, without inclination. { 'ē·vən 'kēl }

even number [MATH] An integer which is a multiple of 2. { 'ē·vən ,nəm·bər }

even-odd nucleus [NUC PHYS] A nucleus which has an even number of protons and an odd number of neutrons. { 'ē·vən 'äd 'nu·klē·əs }

even parity check [COMPUT SCI] A parity check in which the number of 0's or 1's in each word is expected to be even. { 'ē·vən 'par·əd·ē ,chek }

even permutation [MATH] A permutation which may be represented as a result of an even number of transpositions. { 'ē·vən pər·myə'tā·shən }

even pitch [DES ENG] The pitch of a screw in which the number of threads per inch is a multiple (or submultiple) of the threads per inch of the lead screw of the lathe on which the screw is cut. { 'ē·vən 'pich }

event [COMPUT SCI] The moment of time at which a specified change of state occurs; usually marks the completion of an asynchronous input/output operation. [GEOL] An incident of probable tectonic significance, but whose full implications are unknown. [IND ENG] A specified accomplishment in a program at a particular time; appears as a node in a graphic representation of an endeavor with a specific objective (project). [PHYS] A point in space-time. [STAT] A mathematical model of the result of a conceptual experiment; this model is a measurable subset of a probability space. { i'vent }

event-driven monitor [COMPUT SCI] A computer program that measures the performance of a computer system by counting the tasks performed by the system. { i'vent ,driv·ən 'män·əd·ər }

event horizon [RELAT] The boundary of a region of space-time from which it is not possible to escape to infinity. Symbolized $\mathscr{I}^+$. { i'vent hə,rīz·ən }

Eventognathi [VERT ZOO] The equivalent name for Cypriniformes. { ,e,ven'täg·nə,thī }

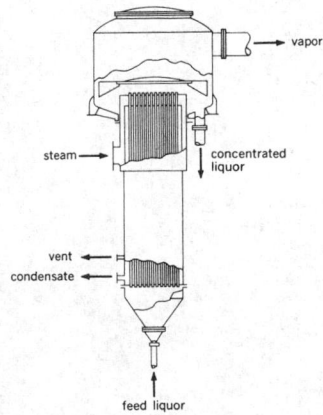

EVAPORATOR

vapor

steam

concentrated liquor

vent

condensate

feed liquor

Long-tube vertical evaporator.

EVOLUTIONARY PROGRAMMING

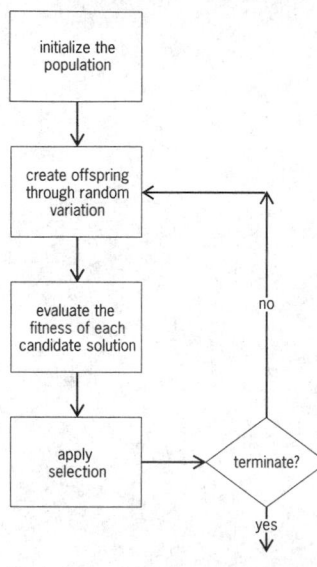

Typical flowchart for an evolutionary algorithm.

event recorder [ENG] A recorder that plots on-off information against time, to indicate when events start, how long they last, and how often they recur. { i'vent ri,kȯrd·ər }

event-related potential [NEUROSCI] Electrical activity produced by the brain in response to a sensory stimulus or associated with the execution of a motor, cognitive, or psychophysiologic task. { i¦vent ri,lād·əd pə'ten·chəl }

event tree [IND ENG] A graphical representation of the possible sequence of events that might occur following an event that initiates an accident. { i'vent ,trē }

eventually in [MATH] A net is eventually in a set if there is an element a of the directed system that indexes the net such that, if b is also an element of this directed system and $b \geq a$, then x_b (the element indexed by b) is in this set. { i'ven·chəl·ē ,in }

even vertex [MATH] A vertex whose degree is an even number. { ¦ēv·ən 'vər,teks }

even-word boundary [COMPUT SCI] A storage address that is an integral multiple of the computer's word length. { 'ēv·ən ¦wərd 'baún·drē }

Everett's interpolation formula [MATH] A formula for estimating the value of a function at an intermediate value of the independent variable, when its value is known at a series of equally spaced points (such as those that appear in a table), in terms of the central differences of the function of even order only and coefficients which are polynomial functions of the independent variable. { ¦ev·rəts ,in·tər·pə'lā·shən ,fȯr·myə·lə }

Everett-Wheeler interpretation [QUANT MECH] An interpretation of quantum mechanics which holds (at least according to some expositions) that the subjective impression of having observed one and only one outcome of a given experiment is an illusion and that there exist other parallel universes, said to be equally real, in which the unknown outcomes are realized. Also known as many-world interpretation, relative-state interpretation. { ¦ev·rit 'wēl·ər ,in·tər·prə,tā·shən }

everglade [ECOL] A type of wetland in southern Florida usually containing sedges and at least seasonally covered by slowly moving water. { 'ev·ər,glād }

evergreen [BOT] Pertaining to a perennially green plant. Also known as aiophyllous. { 'ev·ər,grēn }

Evershed effect [ASTRON] A displacement of spectral lines of sunspots near the sun's limb, caused by outward motion of gases from the center of the sunspot. { 'ev·ər,shed i,fekt }

Eve's constant [NUCLEO] A measure of a substance's intensity of radioactivity, equal to the number of ions produced per cubic centimeter per second in air by 1 gram of the substance at a distance of 1 centimeter. { 'ēvz ¦kän·stənt }

Evjen method [SOLID STATE] Method of calculating lattice sums in which groups of charges whose total charge is zero are taken together, so that the contribution of each group is small and the series rapidly converges. { 'ev·yən ,meth·əd }

evjite [PETR] A gabbro of hornblende in which the only light-colored mineral is labradorite or bytownite; hornblende must be primary, not uralitic. { 'ev,yīt }

evoked potential [NEUROSCI] Electrical response of any neuron to stimuli. { ē'vōkt pə'ten·chəl }

evolute [MATH] **1.** The locus of the centers of curvature of a curve. **2.** The two surfaces of center of a given surface. { 'ev·ə,lüt }

evolution [BIOL] The processes of biological and organic change in organisms by which descendants come to differ from their ancestors. { ,ev·ə'lü·shən }

evolutionarily significant unit [ECOL] A distinct local population within a species that has very different behavioral and phenological traits and thus harbors enough genetic uniqueness to warrant its own management and conservation agenda. Abbreviated ESU. { ,ev·ə,lü·shə¦ner·ə·lē sig¦nif·i·kənt 'yü·nət }

evolutionary computation See evolutionary programming. { ,ev·ə¦lü·shə,ner·ē ,kam·pyə'tā·shən }

evolutionary distance [EVOL] The number of base substitutions per homologous site that have occurred since the divergence of two deoxyribonucleic acid sequences. { ,ev·ə¦lü·sha,ner·ē 'dis·təns }

evolutionary divergence [EVOL] The degree of divergence, at the intra- and interspecific levels, of two or more populations, which presumably have evolved from a common ancestor. { ,ev·ə¦lü·shə,ner·ē də'vər·jəns }

evolutionary force [EVOL] Any factor that brings about changes in gene frequencies or chromosome frequencies in a population and is thus capable of causing evolutionary change. { ,ev·ə¦lü·shə,ner·ē 'fȯrs }

evolutionary operation [IND ENG] An iterative technique for optimizing a production process by systematically introducing small changes in the process and then observing and evaluating the results. { ¦ev·ə¦lü·shə,ner·ē ,äp·ə'rā·shən }

evolutionary plasticity [EVOL] The genetic adaptibility of populations or lines of descent. { ,ev·ə¦lü·shə,ner·ē plas'tis·əd·ē }

evolutionary programming [COMPUT SCI] Computer programming with genetic algorithms. Also known as evolutionary computation; genetic programming. { ,ev·ə¦lü·shə,ner·ē 'prō,gram·iŋ }

evolutionary progress [EVOL] The acquisition of new macromolecular and metabolic processes by which competitive superiority is achieved. { ,ev·ə¦lü·shə,ner·ē 'prä,gres }

evolutionary rate [EVOL] The amount of evolutionary change per unit of time. { ,ev·ə¦lü·shə,ner·ē 'rāt }

evolutionary strategy See genetic algorithm. { ,ev·ə¦lü·shə,ner·ē 'strad·ə·jē }

evolutionary tree [EVOL] **1.** A diagram that portrays the hypothesized genealogical ties and sequence of evolutionary relationships linking individual organisms, populations, or taxa. Also known as phylogenetic tree. **2.** A diagram that depicts the evolutionary relationship of protein or nucleic acid sequences. { ,ev·ə¦lü·shə,ner·ē 'trē }

evolutionary trend [EVOL] Any trend in the evolution of phyletic lines that is a consequence of genotypic cohesion. { ,ev·ə¦lü·shə,ner·ē 'trend }

evolution pressure [EVOL] The result of the combined action of mutation pressure, immigration and hybridization pressure, and selection pressure, giving rise to systematic changes in the gene frequency of a population. { ,ev·ə¦lü·shən ¦presh·ər }

evolved gas analysis [ANALY CHEM] An analytical technique in which the characteristics or the amount of volatile products released by a substance and its reaction products are determined as a function of temperature while the sample is subjected to a series of controlled temperature changes. Abbreviated EGA. { ē¦välvd 'gas ə,nal·ə·səs }

evolvon [EVOL] The operational unit in evolution, assumed to consist of a deoxyribonucleic acid master sequence with a series of redundant sequences that constitute a repository of genetic information. { 'ev·ə,län }

evorsion [GEOL] The process of pothole formation in riverbeds; plays an important role in denudation. { ē'vȯr·shən }

evorsion hollow See pothole. { ē'vȯr·shən ,häl·ō }

EVR See electronic video recording.

EVT See equiviscous temperature.

Ewald-Kornfeld method [SOLID STATE] An extension of the Ewald method to calculate Coulomb energies of dipole arrays. { ¦ē·valt ¦kȯrn,feld ,meth·əd }

Ewald method [SOLID STATE] Method of calculating lattice sums in which certain mathematical techniques are employed to make series converge rapidly. { 'ē·valt ,meth·əd }

Ewald sphere [SOLID STATE] A sphere superimposed on the reciprocal lattice of a crystal, used to determine the directions in which an x-ray or other beam will be reflected by a crystal lattice. { 'ē·valt ,sfir }

E wave See transverse magnetic wave. { 'ē ,wāv }

ewe [VERT ZOO] A mature female sheep, goat, or related animal, as the smaller antelopes. { yü }

Ewing's hysteresis tester [ENG] An instrument for determining the hysteresis loss of a specimen of magnetic material by measuring the deflection of a horseshoe magnet when the specimen is rapidly rotated between the poles of the magnet and the magnet is allowed to rotate about an axis that is aligned with the axis of rotation of the specimen. { ¦yü·iŋz ,his·tə'rē·səs ,tes·tər }

Ewing's sarcoma [MED] A primary malignant tumor of bone, usually arising as a central tumor in long bone. { 'yü·iŋz sär'kō·mə }

Ewing theory of ferromagnetism [SOLID STATE] Theory of ferromagnetic phenomena which assumes each atom is a permanent magnet which can turn freely about its center under the influence of applied fields and other magnets. { 'yü·iŋ ,thē·ə·rē əv ,fe·rō'mag·nə,tiz·əm }

exa- [SCI TECH] A prefix indicating 10^{18}. Abbreviated E. { 'ek·sə }

exact differential equation [MATH] A differential equation obtained by setting the total differential of some function equal to zero. { ig'zakt dif·ə'ren·chəl i‚kwā·zhən }

exact differential form [MATH] A differential form which is the differential of some other form. { ig'zakt dif·ə'ren·chəl ‚fórm }

exact division [MATH] Division wherein the remainder is zero. { ig'zakt di'vizh·ən }

exact divisor [MATH] A divisor that leaves a remainder of zero. { ig'zakt di'vī·zər }

exact sequence [MATH] A sequence of homomorphisms with the property that the kernel of each homomorphism is precisely the image of the previous homomorphism. { ig'zakt 'sē·kwəns }

EXAFS *See* extended x-ray absorption fine structure.

exalate [BOT] Being without winglike appendages. { 'ek·sə‚lāt }

exalbuminous *See* exendospermous. { ‚eks‚al'byü·mə·nəs }

exalted-carrier receiver [ELECTR] Receiver that counteracts selective fading by maintaining the carrier at a high level at all times; this minimizes the second harmonic distortion that would otherwise occur when the carrier drops out while leaving most of the sidebands at their normal amplitudes. { ig'zȯl·təd 'kar·ē·ər ri‚sēv·ər }

exanthema [MED] **1.** An eruption on the skin. **2.** Any disease or fever accompanied by a skin eruption. { ‚eg‚zan'thē·mə }

exanthem subitum [MED] A mild, sometimes epidemic viral disease of young children, with abrupt onset, high fever, and rash. Also known as roseola infantum. { eg'zan·thəm 'sü·həd·əm }

exarch [BOT] A vascular bundle in which the primary wood is centripetal. { 'ek‚särk }

exasperate [BIOL] Having a surface roughened by stiff elevations or bristles. { ig'zas·pə·rət }

exaspidean [VERT ZOO] Of the tarsal envelope of birds, being continuous around the outer edge of the tarsus. { ‚eg·zə'spid·ē·ən }

excavation [ARCHEO] Process of removing earth, stone, or other materials covering the remains of ancient civilizations. [CIV ENG] **1.** The process of digging a hollow in the earth. **2.** An uncovered cavity in the ground. { ‚ek·skə'vā·shən }

excavator [MECH ENG] A machine for digging and removing earth. { 'ek·skə‚vād·ər }

excelsior [MATER] Fine, curled wood shavings, used as packing material. { ek'sel·sē·ər }

excenter [MATH] The center of the escribed circle of a given triangle. { ‚ek'sen·tər }

except [MATH] A logical operator which has the property that if P and Q are two statements, then the statement "P except Q" is true only when P alone is true; it is false for the other three combinations (P false Q false, P false Q true, and P true Q true). { ek'sept }

except gate [ELECTR] A gate that produces an output pulse only for a pulse on one or more input lines and the absence of a pulse on one or more other lines. { ek'sept ‚gāt }

exceptional group [MATH] One of five Lie groups which leave invariant certain forms constructed out of the Cayley numbers; they are Lie groups with maximum symmetry in the sense that, compared with other simple groups with the same rank (number of independent invariant operators), they have maximum dimension (number of generators). { ek‚sep·shən·əl ‚grüp }

exceptional Jordan algebra [MATH] A Jordan algebra that cannot be written as a symmetrized product over a matrix algebra; used in formulating a generalization of quantum mechanics. { ek'sep·shən·əl ‚jȯrd·ən 'al·jə·brə }

exceptional space [QUANT MECH] A space used to describe a system with a finite number of degrees of freedom in a generalization of quantum mechanics; this generalization is achieved by reformulating quantum mechanics in terms of a Jordan algebra of observables and states, and then generalizing this to the exceptional Jordan algebra realized by the algebra of 3×3 Hermitian matrices over the Cayley numbers. { ek'sep·shən·əl 'spās }

exception handling [COMPUT SCI] Programming techniques for dealing with error conditions, generally without terminating execution of the program. [CONT SYS] The actions taken by a control system when unpredictable conditions or situations arise in which the controller must respond quickly. { ek'sep·shən ‚hand·liŋ }

exception-item encoding [COMPUT SCI] A technique which allows the uninterrupted flow of a process by the automatic shunting of erroneous records to an error tape for later corrections. { ek'sep·shən ‚īd·əm en'kōd·iŋ }

exception-principle system [COMPUT SCI] A technique which assumes no printouts except when an error is encountered. { ek'sep·shən ‚prin·sə·pəl ‚sis·təm }

exception reporting [COMPUT SCI] A form of programming in which only values that are outside predetermined limits, representing significant changes, are selected for printout at the output of a computer. { ek'sep·shən ri‚pȯrd·iŋ }

excess air [ENG] Amount of air in a combustion process greater than the amount theoretically required for complete oxidation. { ‚ek‚ses 'er }

excess coefficient [MECH ENG] The ratio $(A - R)/R$, where A is the amount of air admitted in the combustion of fuel and R is the amount required. { 'ek‚ses ‚kō·i‚fish·ənt }

excess conduction [SOLID STATE] Electrical conduction by excess electrons in a semiconductor. { 'ek‚ses kən'dək·shən }

excess electron [SOLID STATE] Electron introduced into a semiconductor by a donor impurity and available for conduction. { 'ek‚ses i'lek‚trän }

excess-fifty code [COMPUT SCI] A number code in which the number n is represented by the binary equivalent of $n + 50$. { 'ek‚ses 'fif·tē ‚kōd }

excessive precipitation [METEOROL] Precipitation (generally in the form of rain) of an unusually high rate of fall; although often used qualitatively, several meteorological services have adopted quantitative limits. { ek'ses·iv prə‚sip·ə'tā·shən }

excess of arc [NAV] The part of a sextant arc beinning at zero and extending in the direction opposite to that part usually considered positive. { 'ek‚ses əv 'ärk }

excess reactivity [NUCLEO] The amount of surplus reactivity over that needed to achieve criticality; it is built into a reactor (by using extra fuel) in order to compensate for fuel burnup and the accumulation of fission-product poisons during operation. { 'ek‚ses ‚rē·ak'tiv·əd·ē }

excess-three code [COMPUT SCI] A number code in which the decimal digit n is represented by the four-bit binary equivalent of $n + 3$. Also known as XS-3 code. { 'ek‚ses 'thrē ‚kōd }

exchange [COMMUN] **1.** A unit established by a telephone company for the administration of telephone service in a specified area, usually a town, a city, or a village and its environs, and consisting of one or more central offices together with the associated plant used in furnishing telephone service in that area. Also known as local exchange. **2.** Room or building equipped so telephone lines terminating there may be interconnected as required; equipment may include a switchboard or automatic switching apparatus. [COMPUT SCI] The interchange of contents between two locations. [QUANT MECH] **1.** Operation of exchanging the space and spin coordinates in a Schrödinger-Pauli wave function representing two identical particles; this operation must leave the wave function unchanged, except possibly for sign. **2.** Process of exchanging a real or virtual particle between two other particles. { iks‚chānj }

exchangeable disk storage [COMPUT SCI] A type of disk storage, used as a backing storage, in which the disks come in capsules, each containing several disks; the capsules can be replaced during operation of the computer and can be stored until needed. { iks‚chānj·ə·bəl 'disk‚stȯr·ij }

exchange adsorption [CHEM ENG] Ion exchange process in which the fluid phase contains (or consists of) two adsorbable components which together entirely saturate the surfaces of the adsorbent. { iks'chānj ad'sȯrp·shən }

exchange anisotropy [ELECTROMAG] Phenomenon observed in certain mixtures of magnetic materials under certain conditions, in which magnetization is favored in some direction (rather than merely along some axis); thought to be caused by exchange coupling across the interface between compounds when one is ferromagnetic and one is antiferromagnetic. { iks'chānj ¦an·ə'sä·trə·pē }

exchange broadening [SPECT] The broadening of a spectral line by some type of chemical or spin exchange process which limits the lifetime of the absorbing or emitting species and produces the broadening via the Heisenberg uncertainty principle. { iks'chānj 'bród·ən·iŋ }

exchange buffering [COMPUT SCI] An input/output buffering technique that avoids the internal moving of data. { iks'chānj ¦bəf·ə·riŋ }

exchange cable [ELEC] Lead covered, nonquadded, paper-insulated cable used within a given area to provide cable pairs between local subscribers and a central office. { iks'chānj ¦kā·bəl }

exchange capacity [GEOL] The ability of a soil material to participate in ion exchange as measured by the quantity of exchangeable ions in a given unit of the material. { iks'chānj kə,pas·əd·ē }

exchange coefficient [FL MECH] A coefficient of eddy flux in turbulent flow, defined in analogy to those coefficients of the kinetic theory of gases. Also known as austausch coefficient; eddy coefficient; interchange coefficient. { iks'chānj ¦kō·i,fish·ənt }

exchange current [ELEC] The magnitude of the current which flows through a galvanic cell when it is operating in a reversible manner. { iks'chānj ¦kə·rənt }

exchange degeneracy [PART PHYS] Coincidence of two Regge trajectories for particles having the same quantum numbers (except for parity, charge parity, and G parity) where one would have expected separate trajectories for alternate Regge recurrences. [QUANT MECH] An exchange process that leads back to the original configuration. Also known as exchange symmetry. { iks'chānj dē'jen·ə·rə·sē }

exchange force [QUANT MECH] The force arising in an exchange interaction. { iks'chānj ¦fórs }

exchange integral [QUANT MECH] Integral over the coordinates of two identical particles which can be thought of as the interaction between a given state and a second state in which the coordinates of the particles are exchanged. { iks'chānj 'in·tə·grəl }

exchange interaction [QUANT MECH] **1.** An interaction represented by a potential involving exchange of space or spin coordinates, or both, of the particles involved; can be visualized physically in terms of exchange of particles. **2.** Any interaction which can be looked upon as due to exchange of particles. { iks'chānj ¦int·ə'rak·shən }

exchange line [ELEC] Line joining a subscriber or switchboard to a commercial exchange. { iks'chānj ¦līn }

exchange message [COMPUT SCI] A device, placed between a communication line and a computer, in order to take care of certain communication functions and thereby free the computer for other work. { iks'chānj ¦mes·ij }

exchange narrowing [SPECT] The phenomenon in which, when a spectral line is split and thereby broadened by some variable perturbation, the broadening may be narrowed by a dynamic process that exchanges different values of the perturbation. { iks'chānj 'nar·ə·wiŋ }

exchange operator [QUANT MECH] An operator which exchanges the spatial coordinates of the particles in a wave function, or their spins, or both positions and spins. { iks'chānj ¦äp·ə,rād·ər }

exchange plant [COMMUN] Plant used to serve subscriber's local needs as distinguished from that used for long-distance communication. { iks'chānj ¦plant }

exchanger See heat exchanger. { iks'chānj·ər }

exchange reaction [CHEM] Reaction in which two atoms or ions exchange places either in two different molecules or in the same molecule. { iks'chānj rē,ak·shən }

exchange-repulsion [PHYS CHEM] A force that arises between neighboring molecules when they are close enough that their electron clouds overlap and, as a consequence of the Pauli exclusion principle, electrons are squeezed out from the region between the nuclei, which them repel each other. { iks'chānj ri,pəl·shən }

exchange sort [COMPUT SCI] A method of arranging records or other types of data into a specified order, in which adjacent pairs of records are exchanged until the correct order is achieved. { iks'chānj ¦sórt }

exchange symmetry See exchange degeneracy. { iks'chānj 'sim·ə·trē }

exchange transfusion [MED] The replacement of most or all of the recipient's blood in small amounts at a time by blood from a donor, a technique used particularly in cases of erythroblastosis fetalis, in certain types of poisoning such as salicylism, and occasionally in liver failure. Also known as replacement transfusion. { iks'chānj tranz,fyü·zhən }

exchange velocity [CHEM] In an ion-exchange process, the speed with which one ion is displaced from an exchanger in favor of another ion. { iks'chānj və'läs·əd·ē }

excimer [CHEM] An excited diatomic molecule where both atoms are of the same species and are dissociated in the ground state. { 'ek·sə·mər }

excimer laser [OPTICS] A laser containing a noble gas, such as helium or neon, which is based on a transition between an excited state in which a metastable bond exists between two gas atoms and a rapidly dissociating ground state. { 'ek·sə·mər ,lā·zər }

excipient [PHARM] Any inert substance combined with an active drug for preparing an agreeable or convenient dosage form. { ek'sip·ē·ənt }

exciplex [CHEM] An excited electron donor-acceptor complex which is dissociated in the ground state. { 'ek·sə,pleks }

Excipulaceae [MYCOL] The equivalent name for Discellaceae. { ,ek·sə·pə'lās·ē,ē }

excircle See escribed circle. { ¦ek'sər·kəl }

excision [GEN] Recombination involving removal of a genetic element. [MED] The cutting out of a part; removal of a foreign body or growth from a part, organ, or tissue. { ek'sizh·ən }

excision enzyme [BIOCHEM] A bacterial enzyme that removes damaged dimers from the deoxyribonucleic acid molecule of a bacterial cell following light or ultraviolet radiation or nitrogen mustard damage. { ek'sizh·ən 'en,zīm }

excitable [BIOL] Referring to a tissue or organism that exhibits irritability. { ek'sīd·ə·bəl }

excitation [ATOM PHYS] A process in which an atom or molecule gains energy from electromagnetic radiation or by collision, raising it to an excited state. [CONT SYS] The application of energy to one portion of a system or apparatus in a manner that enables another portion to carry out a specialized function; a generalization of the electricity and electronics definitions. [ELEC] The application of voltage to field coils to produce a magnetic field, as required for the operation of an excited-field loudspeaker or a generator. [ELECTR] **1.** The signal voltage that is applied to the control electrode of an electron tube. Also known as drive. **2.** Application of signal power to a transmitting antenna. [NEUROSCI] A change in the electrical state of a neuron leading to an action potential. [QUANT MECH] The addition of energy to a particle or system of particles at ground state to produce an excited state. { ,ek,sī'tā·shən }

excitation anode [ELECTR] An anode used to maintain a cathode spot on a pool cathode of a gas tube when output current is zero. { ,ek,sī'tā·shən ,an,ōd }

excitation curve [NUC PHYS] A curve showing the relative yield of a specified nuclear reaction as a function of the energy of the incident particles or photons. Also known as excitation function. { ,ek,sī'tā·shən ,kərv }

excitation energy [QUANT MECH] The minimum energy required to change a system from its ground state to a particular excited state. { ,ek,sī'tā·shən ,en·ər·jē }

excitation function [ATOM PHYS] The cross section for an incident electron to excite an atom to a particular excited state expressed as a function of the electron energy. See excitation curve. { ,ek,sī'tā·shən ,fəŋk·shən }

excitation index [SPECT] In emission spectroscopy, the ratio of intensities of a pair of extremely nonhomologous spectra lines; used to provide a sensitive indication of variation in excitation conditions. { ,ek,sī'tā·shən ,in,deks }

excitation loss See core loss. { ,ek,sī'tā·shən ,lós }

excitation potential [QUANT MECH] Electric potential which gives the excitation energy when multiplied by the magnitude of the electron charge. { ,ek,sī'tā·shən pə,ten·chəl }

excitation purity [ANALY CHEM] The ratio of the departure of the chromaticity of a specified color to that of the reference source, measured on a chromaticity diagram; used as a guide of the wavelength of spectrum color needed to be mixed with

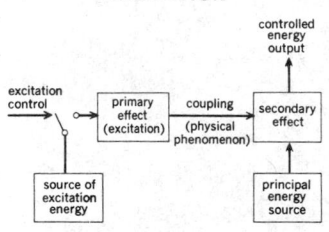

EXCITATION

Diagram showing function of excitation in a control system.

a reference color to give the specified color. { ‚ek‚sī'tā·shən ‚pyür·əd·ē }

excitation spectrum [SPECT] The graph of luminous efficiency per unit energy of the exciting light absorbed by a photoluminescent body versus the frequency of the exciting light. { ‚ek‚sī'tā·shən ‚spek·trəm }

excitation voltage [ELEC] Nominal voltage required for excitation of a circuit. { ‚ek‚sī'tā·shən ‚vōl·tij }

excitation volume [PHYS] In electron-probe microanalysis, the volume of the x-ray source used to penetrate and diffuse into the target sample. { ‚ek‚sī'tā·shən ‚väl·yəm }

excited state [QUANT MECH] A stationary state of higher energy than the lowest stationary state or ground state of a particle or system of particles. { ek'sīd·əd 'stāt }

excited-state effect [SOLID STATE] The motion of a crystal defect through a process in which the defect is first raised into an excited state and then decays, together with the surroundings, into a state in which motion of the defect readily occurs. { ek¦sīd·əd 'stāt i‚fekt }

excited-state maser [PHYS] A maser whose amplifying transition has a terminal level that is not appreciably populated at thermal equilibrium for the ambient temperature. { ck¦sīd·əd 'stāt 'mā·zər }

exciter [ELEC] 1. A small auxiliary generator that provides field current for an alternating-current generator. 2. See exciter lamp. [ELECTR] A crystal oscillator or self-excited oscillator used to generate the carrier frequency of a transmitter. [ELECTROMAG] 1. The portion of a directional transmitting antenna system that is directly connected to the transmitter. 2. A loop or probe extending into a resonant cavity or waveguide. { ek'sīd·ər }

exciter lamp [ELEC] A bright incandescent lamp having a concentrated filament, used to excite a phototube or photocell in sound movie and facsimile systems. Also known as exciter. { ek'sīd·ər ‚lamp }

exciter response [ELEC] In electrical rotating machinery, the rate of increase or decrease of the main exciter voltage when resistance is suddenly removed from or inserted in the main exciter field circuit. { ek'sīd·ər ri'späns }

exciting current See magnetizing current. { ek'sīd·iŋ ‚kə·rənt }

exciting line [SPECT] The frequency of electromagnetic radiation, that is, the spectral line from a noncontinuous source, which is absorbed by a system in connection with some particular process. { ek'sīd·iŋ ‚līn }

exciton [SOLID STATE] An excited state of an insulator or semiconductor which allows energy to be transported without transport of electric charge; may be thought of as an electron and a hole in a bound state. { 'ek·sə‚tän }

exciton-induced photoemission [ELECTR] A two-stage process that takes place in an ionic crystal in which color centers are present, in which photon absorption leads to the formation of an exciton, and the exciton then transfers enough energy to color centers to eject photoelectrons from the crystal. { ¦ek·si‚tän in‚düst 'fōd·ō·i‚mish·əm }

excitron [ELECTR] A single-anode mercury-pool tube provided with means for maintaining a continuous cathode spot. { 'ek·sə‚trän }

exclusion area [NUCLEO] The area around a nuclear operation (reactor, bomb test, and so on) where human habitation is restricted. { ik'sklü·zhən ‚er·ē·ə }

exclusion principle [ECOL] The principle according to which two species cannot coexist in the same locality if they have identical ecological requirements. [QUANT MECH] The principle that no two fermions of the same kind may simultaneously occupy the same quantum state. Also known as Pauli exclusion principle. { ik'sklü·zhən ‚prin·sə·pəl }

exclusive-NOR gate See equivalence gate. { ik¦sklü·siv 'nȯr ‚gāt }

exclusive or [COMPUT SCI] An instruction which performs the "exclusive or" operation on a bit-by-bit basis for its two operand words, usually storing the result in one of the operand locations. Abbreviated XOR. [MATH] A logic operator which has the property that if P is a statement and Q is a statement, then P exclusive or Q is true if either but not both statements are true, false if both are true or both are false. { ik¦sklü·siv 'ȯr }

exclusive segments [COMPUT SCI] Parts of an overlay program structure that cannot be resident in main memory simultaneously. { ik'sklü·siv 'seg·mənts }

exclusive species [ECOL] A species which is completely or nearly limited to one community. { ik'sklü·siv 'spē·shēz }

Excorallanidae [INV ZOO] A family of free-living and parasitic isopod crustaceans in the suborder Flabellifera which have mandibles and first maxillae modified as hooklike piercing organs. { ek‚skȯr·ə'lan·ə‚dē }

excoriation [MED] Abrasion of a portion of the skin. { ek‚skȯr·ē'ā·shən }

excrement [PHYSIO] An excreted substance; the feces. { 'ek·skrə·mənt }

excrescence [BIOL] 1. Abnormal or excessive increase in growth. 2. An abnormal outgrowth. { ek'skrē·səns }

excretion [PHYSIO] The removal of unusable or excess material from a cell or a living organism. { ek'skrē·shən }

excretory system [ANAT] Those organs concerned with solid, fluid, or gaseous excretion. { 'ek·skrə‚tȯr·ē ‚sis·təm }

excurrent [BIOL] Flowing out. [BOT] 1. Having an undivided main stem or trunk. 2. Having the midrib extending beyond the apex. { eks'kə·rənt }

excursion [NUCLEO] A sudden, very rapid rise in the power level of a nuclear reactor caused by supercriticality. { ik'skər·zhən }

excursion steamer [NAV ARCH] A steam-powered ship which carries passengers on recreational trips. { ik'skər·zhən ‚stē·mər }

executable module [COMPUT SCI] A file holding a computer program written in machine language so that it is ready to run. { ‚ek·sə'kyüd·ə·bəl 'māj·yül }

executable program [COMPUT SCI] A program that is ready to run on a computer. { ‚ek·sə¦kyüd·ə·bəl 'prō·grəm }

executable statement [COMPUT SCI] A program statement that causes the computer to carry out some operation, in contrast to a declarative statement. { ‚ek·sə'kyüd·ə·bəl 'stāt·mənt }

execute [COMPUT SCI] Usually, to run a compiled or assembled program on the computer; by extension, to compile or assemble and to run a source program. { 'ek·sə‚kyüt }

execute statement [COMPUT SCI] A program statement that indicates the beginning of a job statement in a job control language. { 'ek·sə‚kyüt ‚stāt·mənt }

execution control program [COMPUT SCI] The program delivered by the manufacturer which permits the computer to handle the programs fed to it. { ‚ek·sə¦kyü·shən kən'trōl ‚prō·grəm }

execution cycle [COMPUT SCI] The time during which an elementary operation takes place. { ‚ek·sə'kyü·shən ‚sī·kəl }

execution error detection [COMPUT SCI] The detection of errors which become apparent only during execution time. { ‚ek·sə¦kyü·shən 'er·ər di‚tek·shən }

execution time [COMPUT SCI] The time during which actual work, such as addition or multiplication, is carried out in the execution of a computer instruction. { ‚ek·sə'kyü·shən ‚tīm }

executive communications [COMPUT SCI] The routine information transmitted to the operator on the status of programs being executed and of the requirements made by these programs of the various components of the system. { ig‚zek·yəd·iv kə·‚myü·nə'kā·shənz }

executive control language [COMPUT SCI] The generic term for a finite set of instructions which enables the programmer to run a program more efficiently. { ig¦zek·yəd·iv kən'trōl ‚laŋ·gwij }

executive ego function [PSYCH] A psychoanalytic term for the ego's management of mental mechanisms in order to meet the needs of the individual. { ig¦zek·yəd·iv 'ē‚go ‚fəŋk·shən }

executive file-control system [COMPUT SCI] The assignment of intermediate storage devices performed by the computer, and over which the programmer has no control. { ig¦zek·yəd·iv 'fīl kən‚trōl ‚sis·təm }

executive guard mode [COMPUT SCI] A protective technique which prevents the programmer from accessing, or using, the executive instructions. { ig¦zek·yəd·iv 'gärd ‚mōd }

executive instruction [COMPUT SCI] Instruction to determine how a specially written computer program is to operate. { ig¦zek·yəd·iv in¦strək·shən }

executive logging [COMPUT SCI] The automatic bookkeeping of time utilization by programs of the various components of a computer system. { ig¦zek·yəd·iv 'läg·iŋ }

executive routine [COMPUT SCI] A digital computer routine designed to process and control other routines. Also known as master routine; monitor routine. { ig¦zek·yəd·iv rü¦tēn }

executive schedule maintenance [COMPUT SCI] The scheduling of jobs to be run according to priorities as established and maintained by the control executive or executive supervisor. { ig¦zek·yəd·iv 'sked·jəl ¦mān·tə·nəns }

executive supervisor [COMPUT SCI] The component of the computer system which controls the sequencing, setup, and execution of the jobs presented to it. { ig¦zek·yəd·iv 'sü·pər¸viz·ər }

executive system concurrency [COMPUT SCI] The capability of the executive of a computer system to handle more than one job at the same time if these jobs do not require the same components at the same time. { ig'zek·yəd·iv ¸sis·təm kən'kər·ən·sē }

executive system control [COMPUT SCI] The control exerted over the executive system by means of job control cards or commands issued at a terminal. { ig'zek·yəd·iv ¸sis·təm kən'trōl }

executive system utilities [COMPUT SCI] The set of programs, such as diagnostic programs or file utility programs, which enables the executive to handle the jobs efficiently and completely. { ig'zek·yəd·iv ¸sis·təm yü'til·əd·ēz }

exendospermous [BOT] Lacking endosperm. Also known as exalbuminous. { eks¦en·də¦spər·məs }

exercise physiology [MED] A field of sports medicine that involves the study of the body's response to physical stress; comprises the science of fitness, the preservation of fitness, and the role of fitness in the prevention and treatment of disease. { ¸ek·sər¸sīz ¸fiz·ē'äl·ə·jē }

exergonic [BIOCHEM] Of or pertaining to a biochemical reaction in which the end products possess less free energy than the starting materials; usually associated with catabolism. { ¸ek·sər'gän·ik }

exergy [THERMO] The portion of the total energy of a system that is available for conversion to useful work; in particular, the quantity of work that can be performed by a fluid relative to a reference condition, usually the surrounding ambient condition. { 'eks·ər·jē }

exfiltration [SCI TECH] A gradual escape of fluid, for example, through a membrane or a wall. { ¸eks¸fil'trā·shən }

exfoliation [GEOL] *See* sheeting. [MED] **1.** The separation of bone or other tissue in thin layers. **2.** A peeling and shedding of the horny layer of the skin. [MET] Peeling off or separation of metal at its surface in the form of thin, parallel scales or lamellae. [PETR] The breaking off of thin concentric shells, sheets, scales, plates, and so on, from a rock mass; measuring less than a centimeter to several meters in thickness, the loosened rock is spalled, peeled, or stripped. [SCI TECH] Flaking away or peeling off in scales. { eks¸fō·lē'ā·shən }

exfoliation corrosion [MET] A type of corrosion that progresses parallel to the metal surface in such a manner that underlying layers are gradually separated. { eks¸fō·lē'ā·shən kə¸rō·zhən }

exfoliation dome [GEOL] A large rounded dome-shaped structure produced in massive homogeneous coarse-grained rocks (usually igneous) by exfoliation. { eks¸fō·lē'ā·shən ¸dōm }

exfoliation joint *See* sheeting structure. { eks¸fō·lē'ā·shən ¸jȯint }

exfoliative cytology [PATH] The study of cells shed spontaneously from the body surfaces; used principally in the diagnosis of cancer. { eks'fō·lē¸ād·iv sī'tāl·ə·jē }

exfoliative erythroderma [MED] A type of dermatitis that is characterized by widespread warm redness and scaling; nail degeneration and loss, hair loss, fever, chills, and enlargement of the lymph nodes may also occur. { eks¸fōl·ē¸ād·iv ə¸rith·rə'dər·mə }

exhalation [GEOPHYS] The process by which radioactive gases escape from the surface layers of soil or loose rock, where they are formed by decay of radioactive salts. [PHYSIO] The giving off or sending forth in the form of vapor; expiration. { ¸eks·ə'lā·shən }

exhaust [MECH ENG] **1.** The working substance discharged from an engine cylinder or turbine after performing work on the moving parts of the machine. **2.** The phase of the engine cycle concerned with this discharge. **3.** A duct for the escape of gases, fumes, and odors from an enclosure, sometimes

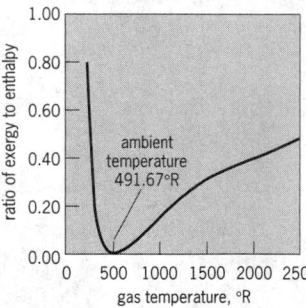

EXERGY

Ratio of exergy to enthalpy as a function of temperature. This ratio increases as the temperature difference between the working fluid and the ambient temperature increases (in both positive and negative directions).

equipped with an arrangement of fans. [SCI TECH] *See* evacuate. { ig'zȯst }

exhaust deflecting ring [MECH ENG] A type of jetavator consisting of a ring so mounted at the end of a nozzle as to permit it to be rotated into the exhaust stream. { ig'zȯst di¸flek·tiŋ ¸riŋ }

exhaust gas [MECH ENG] Spent gas leaving an internal combustion engine or gas turbine. { ig'zȯst ¸gas }

exhaust-gas analyzer [ENG] An instrument that analyzes the gaseous products to determine the effectiveness of the combustion process. { ig'zȯst ¸gas 'an·ə¸līz·ər }

exhaust head [ENG] A device placed on the end of an exhaust pipe to remove oil and water and to reduce noise. { ig'zȯst ¸hed }

exhaustion delirium [MED] Acute, confusional, delirious reactions brought about by extreme fatigue, long wasting illness, or prolonged insomnia. { ig'zȯs·chən də'lir·ē·əm }

exhaustion point [CHEM] In an ion-exchange process, the state of an adsorbent at which it no longer can produce a useful ion exchange. { ig'zȯs·chən ¸point }

exhaustion region [ELECTR] A layer in a semiconductor, adjacent to its contact with a metal, in which there is almost complete ionization of atoms in the lattice and few charge carriers, resulting in a space-charge density. { ig'zȯs·chən ¸rē·jən }

exhaust manifold [MECH ENG] A branched system of pipes to carry waste emissions away from the piston chambers of an internal combustion engine. { ig'zȯst ¸man·ə¸fōld }

exhaust nozzle [AERO ENG] The terminal portion of a jet engine tail pipe. { ig'zȯst ¸näz·əl }

exhaust pipe [MECH ENG] The duct through which engine exhaust is discharged. { ig'zȯst ¸pīp }

exhaust scrubber [ENG] A purifying device on internal combustion engines which removes noxious gases from engine exhaust. { ig'zȯst ¸skrəb·ər }

exhaust stream [AERO ENG] The stream of matter or radiation emitted from the nozzle of a rocket or other reaction engine. { ig'zȯst ¸strēm }

exhaust stroke [MECH ENG] The stroke of an engine, pump, or compressor that expels the fluid from the cylinder. { ig'zȯst ¸strōk }

exhaust suction stroke [MECH ENG] A stroke of an engine that simultaneously removes used fuel and introduces fresh fuel to the cylinder. { ig'zȯst ¸sək·shən ¸strōk }

exhaust trail [METEOROL] A visible condensation trail (contrail) that forms when the water vapor of an aircraft exhaust is mixed with and saturates (or slightly supersaturates) the air in the wake of the aircraft. { ig'zȯst ¸trāl }

exhaust valve [MECH ENG] The valve on a cylinder in an internal combustion engine which controls the discharge of spent gas. { ig'zȯst ¸valv }

exhaust velocity [FL MECH] The velocity of gaseous or other particles in the exhaust stream of the nozzle of a reaction engine, relative to the nozzle. { ig'zȯst və'läs·əd·ē }

exhibitionism [PSYCH] **1.** A sexual perversion in which pleasure is obtained by exposing the genitalia. **2.** In psychoanalysis, gratification of early sexual impulses in young children by physical activity, such as dancing. **3.** Any attracting of attention to oneself. { ¸ek·sə'bish·ə¸niz·əm }

exhumation [GEOL] The uncovering or exposure through erosion of a former surface, landscape, or feature that had been buried by subsequent deposition. { ¸eks·yü'mā·shən }

exhumed *See* resurrected. { ig'zyümd }

exine *See* exosporium. { 'ek¸sēn }

exinite [GEOL] A hydrogen-rich maceral group consisting of spore exines, cuticular matter, resins, and waxes; includes sporinite, cutinite, alginite, and resinite. Also known as liptinite. { 'ek·sə¸nīt }

existence doubtful [NAV] In marine operations, expression used principally on charts to indicate the possible existence of a rock, shoal, and so on, the actual existence of which has not been established. { ig'zis·təns 'daut·fəl }

existence proof [MATH] An argument that establishes the truth of an existence theorem. { ig'zis·təns ¸prüf }

existence theorem [MATH] The theorem that at least one object of a specified type exists. { ig'zis·təns ¸thir·əm }

existential psychiatry [PSYCH] A school of psychiatry that stresses the way in which a person experiences the phenomeno-logic world and takes responsibility for existence. { ‚eg‚zə¦sten·chəl sī'kī·ə·trē }

existential quantifier [MATH] A logical relation, often symbolized ∃, that may be expressed by the phrase "there is a" or "there exists"; if *P* is a predicate, the statement $(\exists x)P(x)$ is true if there exists at least one value of *x* in the domain of *P* for which *P*(*x*) is true, and is false otherwise. { ‚eg·zə¦sten·chəl 'kwän·tə‚fī·ər }

exit [COMPUT SCI] **1.** A way of terminating a repeated cycle of operations in a computer program. **2.** A place at which such a cycle can be stopped. [ENG] A door, passage, or place of egress. { 'eg·zət }

exite [INV ZOO] A movable appendage or lobe located on the external side of the limb of a generalized arthropod. { 'ek‚sīt }

exit pupil [OPTICS] The image of the aperture stop of an optical system formed in the image space by rays emanating from a point on the optical axis in the object space. { 'eg·zət ‚pyü·pəl }

exit region [METEOROL] The region of difluence at the downwind extremity of a jet stream. { 'eg·zət ‚rē·jən }

ex lighterage [IND ENG] Price quoted exclusive of lighterage fees. { ¦eks 'līd·ə·rij }

exline correction [FL MECH] Calculation of fluid-flow friction loss through annular sections with a correction for the flow eccentricity in the laminar-flow range. { 'eks‚līn kə'rek·shən }

exmeridian altitude [ASTRON] An altitude of a celestial body near the celestial meridian of the observer to which a correction is to be applied to determine the meridian altitude. Also known as circummeridian altitude. { eks·mə'rid·ē·ən 'al·tə‚tüd }

exmeridian observation [ASTRON] **1.** Measurement of the altitude of a celestial body near the celestial meridian of the observer, for conversion to a meridian altitude. **2.** The altitude so measured. { eks·mə'rid·ē·ən ‚äb·zər'vā·shən }

exo- [ORG CHEM] A conformation of carbon bonds in a six-membered ring such that the molecule is boat-shaped with one or more substituents directed outward from the ring. [SCI TECH] A prefix denoting outside or outer. { 'ek·sō }

exobiology [BIOL] The search for and study of extraterrestrial life. { ¦ek·sō·bī'äl·ə·jē }

exocarp *See* epicarp. { 'ek·sō‚kärp }

exocellular [CELL MOL] Referring to reactions or processes that are initiated inside a cell and take place outside it. { ‚ek·sə'sel·yə·lər }

exoccipital [ANAT] Lying to the side of the foramen magnum, as the exoccipital bone. { ¦eks·äk'sip·əd·əl }

exochorion [INV ZOO] The outer of two layers forming the covering of an insect egg. { ¦ek·sō'kór·ē‚än }

exocline [GEOL] An inverted anticline or syncline. { 'ek·sə‚klīn }

exocoel [INV ZOO] The space between pairs of adjacent mesenteries in anthozoan polyps. { 'ek·sə‚sēl }

Exocoetidae [VERT ZOO] The halfbeaks, a family of actinopterygian fishes in the order Atheriniformes. { ‚ek·sə'sēd·ə‚dē }

exocrine gland [PHYSIO] A structure whose secretion is passed directly or by ducts to its exterior surface, or to another surface which is continuous with the external surface of the gland. { 'ek·sə·krən ‚gland }

exocuticle [INV ZOO] The middle layer of the cuticle of insects. { ¦ek·sō'kyüd·ə·kəl }

exocyclic [CHEM] Pertaining to the outside of a ring structure. { ‚ek·sō'sī·klik }

exocyclic double bond [ORG CHEM] A double bond that is connected to and external to a ring structure. { ‚ek·sō¦sī·klik ¦dəb·əl 'bänd }

exocytosis [CYTOL] The extrusion of material from a cell. { ¦ek·sō·sī'tō·səs }

exodermis *See* hypodermis. { ‚ek·sō'dər·məs }

exoelectrons [PHYS] Electrons emitted from the surfaces of metals and certain ceramics after these surfaces have been freshly formed by a process such as abrasion or fracture; electrons obtain energy required for emission from processes such as establishment of surface films and rearrangement of disturbed atoms. { ¦ek·sō·i'lek‚tränz }

exoenzyme [BIOCHEM] An enzyme that functions outside the cell in which it was synthesized. { ¦ek·sō¦wen‚zīm }

exoergic *See* exothermic. { ¦ek·sō¦wər·jik }

exoergic collision *See* collision of the second kind. { ¦ek·sō¦wər·jik kə'lizh·ən }

exogamy [GEN] Union of gametes from organisms that are not closely related. Also known as outbreeding. { ek'säg·ə·mē }

exogastrula [EMBRYO] An abnormal gastrula that is unable to undergo invagination or further development because of a quantitative increase of presumptive endoderm. { ¦ek·sō'gas·trə·lə }

exogenote [GEN] The genetic fragment transferred from the donor to the recipient cell during the process of recombination in bacteria. { ‚ek'säj·ə‚nōt }

exogenous [BIOL] **1.** Due to an external cause; not arising within the organism. **2.** Growing by addition to the outer surfaces. [PHYSIO] Pertaining to those factors in the metabolism of nitrogenous substances obtained from food. { ‚ek'säj·ə·nəs }

exogenous electrification [ELEC] The separation of electric charge in a conductor placed in a preexisting electric field, especially applied to the charge separation observed on metal-covered aircraft, resulting from induction effects, and by itself does not create any net total charge on the conductor. { ‚ek'säj·ə·nəs i‚lek·trə·fə'kā·shən }

exogenous inclusion *See* xenolith. { ‚ek'säj·ə·nəs in'klü·zhən }

exogenous variables [MATH] In a mathematical model, the independent variables, which are predetermined and given outside the model. { ‚ek'säj·ə·nəs 'ver·ē·ə·bəlz }

exogeosyncline [GEOL] A parageosyncline that lies along the cratonal border and obtains its clastic sediments from erosion of the adjacent orthogeosynclinal belt outside the craton. Also known as delta geosyncline; foredeep; transverse basin. { ¦ek·sō‚jē·ō'sin‚klīn }

exognathite [INV ZOO] The external branch of an oral appendage of a crustacean. { ‚ek'säg·nə‚thīt }

Exogoninae [INV ZOO] A subfamily of polychaete annelids in the family Syllidae having a short, small body of few segments. { ¦ek·sō'gä·nə‚nē }

exogynous [BOT] Having the style longer than and exserted beyond the corolla. { ‚ek'säj·ə·nəs }

exomorphic zone *See* aureole. { ‚ek·sə'mór·fik ‚zōn }

exomorphism [PETR] A change in a rock mass caused by intrusion of external igneous material; in the usual sense, contact metamorphism. { ‚ek·sə'mór‚fiz·əm }

exon [GEN] The segment or segments of a gene which code for its final messenger ribonucleic acid. { 'ek‚sän }

exonephric [INV ZOO] Having the excretory organs discharge through the body wall. { ¦ek·sō'ne‚frik }

exon shuffling [GEN] In eukaryotic split genes, the creation of new genes by the addition or removal of exons through unequal crossing over within introns intervening between the exons of a split gene. { 'ek‚sän ‚shəf·liŋ }

exonuclease [BIOCHEM] Any of a group of enzymes which catalyze hydrolysis of single nucleotide residues from the end of a deoxyribonucleic acid chain. { ¦ek·sō'nü·klē‚ās }

exopathogen [PL PATH] An external, nonparasitic plant pathogen. { ¦ek·sō'path·ə·jən }

exopathogenesis [PL PATH] The external incitement of disease by a nonparasitic pathogen. { ¦ek·sō‚path·ə'jen·ə·səs }

exopeptidase [BIOCHEM] An enzyme that acts on the terminal peptide bonds of a protein chain. { ¦ek·sō'pep·tə‚dās }

exophoria [MED] A type of heterophoria in which the visual lines tend outward. { ‚ek·sə'fór·ē·ə }

exophthalmic goiter *See* hyperthyroidism. { ¦ek‚säf¦thal·mik 'góid·ər }

exophthalmos [MED] Abnormal protrusion of the eyeball from the orbit. { ‚ek‚säf'thal·məs }

exoplanet *See* extrasolar planet. { ‚ek·sō'plan·ət }

exopodite [INV ZOO] The outer branch of a biramous crustacean appendage. { ek'säp·ə‚dīt }

exoprosthesis [MED] An externally applied prosthesis. { ‚ek·sō·prəs'thē·səs }

Exopterygota [INV ZOO] A division of the insect subclass Pterygota including those insects which undergo a hemimetabolous metamorphosis. { ‚ek‚säp·ter·ə¦gōd·ə }

exorheic [GEOL] Referring to a basin or region characterized by external drainage. { ek·sə'rē·ik }

exoskeleton [INV ZOO] The external supportive covering of certain invertebrates, such as arthropods. [VERT ZOO] Bony or horny epidermal derivatives, such as nails, hoofs, and scales. { 'ek·sō'skel·ə·tən }

exosmosis [PHYSIO] Passage of a liquid outward through a cell membrane. { ¦ek·sō·äs'mō·səs }

exosphere [METEOROL] An outermost region of the atmosphere, estimated at 300–600 miles (500–1000 kilometers), where the density is so low that the mean free path of particles depends upon their direction with respect to the local vertical, being greatest for upward-traveling particles. Also known as region of escape. { 'ek·sō‚sfir }

exospore [MYCOL] An asexual spore formed by abstriction, as in certain Phycomycetes. { 'ek·sō‚spòr }

exosporium [BOT] The outer of two layers forming the wall of spores such as pollen and bacterial spores. Also known as exine. { ‚ek·sə'spòr·ē·əm }

exostome [BOT] The opening through the outer integument of a bitegmic ovule. { 'ek·sə‚stōm }

exostosis [MED] A benign cartilage-capped protuberance from the surface of long bones but also seen on flat bones, caused by chronic irritation as from infection, trauma, or osteoarthritis. { ‚ek·sə'tō·səs }

exotheca [INV ZOO] The tissue external to the theca of corals. { ‚ek·sə'thē·kə }

exotherm [CHEM ENG] The graphical plotting of heat rise and fall versus time for an exothermic reaction or process system. { 'ek·sə‚thərm }

exothermic [PHYS] Indicating liberation of heat. Also known as exoergic. { ¦ek·sō'thər·mik }

exotic [ECOL] Not endemic to an area. { ig'zäd·ik }

exotic atom [ATOM PHYS] A system in which either the proton that forms the nucleus of a hydrogen atom is replaced by another particle (such as a muon, to form muonium, or a positron, to form positronium), one electron in an ordinary atom is replaced by another particle (such as a muon, pion, or antiproton), or both substitutions are made (as in antihydrogen). { ik¦säd·ik 'ad·əm }

exotic four-space [MATH] A four-dimensional manifold that is homeomorphic, but not diffeomorphic, to four-dimensional Euclidean space. { ig¦zäd·ik 'fòr‚spās }

exotic fuels [MATER] The hydroborons which have higher calorific values than do the carbon-hydrogen fuels, once proposed as high-energy fuels for aircraft and missiles; include borane (BH₃), borobutane (B₄H₁₀), and borodecane (B₁₀H₁₄). { ig¦zäd·ik ¦fyülz }

exotic nucleus [NUC PHYS] An atomic nucleus in which the ratio of neutron number to proton number is much larger or much smaller than that of naturally occurring nuclei. { ig'zäd·ik 'nü·klē·əs }

exotic sphere [MATH] A smooth manifold that is homeomorphic, but not diffeomorphic, to a sphere. { ig¦zäd·ik 'sfir }

exotic stream [HYD] A stream that crosses a desert as it flows to the sea, or any stream which derives most of its water from the drainage system of another region. { ig'zäd·ik 'strēm }

exotic viral disease [MED] A viral disease that occurs only rarely in human populations of developed countries. { ig¦zäd·ik ¦vī·rəl diz‚ēz }

exotoxin [MICROBIO] A toxin that is excreted by a microorganism. { ¦ek·sə¦täk·sən }

exozodiacal dust [ASTRON] Dust that appears to be orbiting within a few astronomical units of a star and may be analogous to the zodiacal dust in the solar system. { ‚ek·sō·zō‚dī·ə·kəl 'dəst }

expandable space structure [AERO ENG] A structure which can be packaged in a small volume for launch and then erected to its full size and shape outside the earth's atmosphere. { ik¦span·də·bəl 'spās ‚strək·chər }

expanded batch [COMPUT SCI] A level of computer processing more complex than basic batch, in which computer programs perform complex computations and produce reports that analyze performance in addition to reporting it. { ik'spand·əd 'bach }

expanded clay [MATER] A material made from common brick clays by grinding, screening, and then feeding through a gas burner at about 2700°F (1482°C), thus changing the ferric oxide to ferrous oxide and causing the formation of bubbles. { ik'spand·əd 'klā }

expanded-flow bin [ENG] A bin formed by attaching

a mass-flow hopper to the bottom of a funnel-flow bin. { ik¦spand·əd 'flō ‚bin }

expanded foot [HYD] A broad, bulblike or fan-shaped ice mass formed where a valley glacier flows beyond its confining walls and extends onto an adjacent lowland at the bottom of a mountain slope. { ik'spand·əd 'fùt }

expanded metal [MET] An alloy which has expanded following cooling and solidification. { ik'spand·əd 'med·əl }

expanded notation [MATH] The representation of a number as the sum of a series of terms, each of which is written explicitly as the product of a digit and the base of the number system raised to some power. { ik'spand·əd nō'tā·shən }

expanded numeral [MATH] A number expressed in expanded notation. { ik¦spand·əd 'nüm·rəl }

expanded perlite [MATER] Perlite that has been finely ground and subjected to extreme heat, causing the particles to become considerably expanded and porous because of release of water. { ik'spand·əd 'pər‚līt }

expanded plastic [MATER] A light, spongy plastic made by introducing pockets of air or gas. Also known as foamed plastic; plastic foam. { ik'spand·əd 'plas·tik }

expanded position indicator display [ELECTR] Display of an expanded sector from a plan position indicator presentation. { ik'spand·əd pə'zish·ən ‚in·də‚kad·ər di‚splā }

expanded scope [ELECTR] Magnified portion of a given type of cathode-ray tube presentation. { ik'spand·əd 'skōp }

expanded slag [MATER] Slag formed by running slag from phosphate rock onto a forehearth at about 2000°F (1093°C) and then treating it with water, high-pressure steam, and air; used to make lightweight concrete blocks. { ik'spand·əd 'slag }

expanded sweep [ELECTR] A cathode-ray sweep in which the movement of the electron beam across the screen is speeded up during a selected portion of the sweep time. { ik'spand·əd 'swēp }

expander [ELECTR] A transducer that, for a given input amplitude range, produces a larger output range. { ik'spand·ər }

expander flange [ENG] A type of butt-welded flange designed with a tapered bore so that various pipe sizes can be matched. { ik'span·dər ‚flanj }

expanding [MET] A process used to increase the inside diameter of a hollow piece, such as a tube, cup, or shell. { ik'spand·iŋ }

expanding arm [ASTRON] A spiral arm of the Galaxy consisting of neutral hydrogen that lies between 2.5 and 4 kiloparsecs beyond the galactic center and is moving out from it at about 85 miles (135 kilometers) per second. { ik'spand·iŋ 'ärm }

expanding brake [MECH ENG] A brake that operates by moving outward against the inside rim of a drum or wheel. { ik'spand·iŋ 'brāk }

expanding population [ECOL] A population containing a large proportion of young individuals. { ik'spand·iŋ ‚päp·yə'lā·shən }

expanding square search See square search. { ik¦spand·iŋ ¦skwer ¦sərch }

expanding universe [ASTROPHYS] Explanation of the red shift observed in spectral lines from distant galaxies as due to a mutual recession of galaxies away from each other. [RELAT] A model of the universe describing the process defined in the astronomy definition, in which the universe is nonstatic, homogeneous, and isotropic; based on Einstein's field equations with a nonvanishing cosmical constant. { ik¦spand·iŋ 'yü·nə‚vərs }

expandor [ELECTR] The part of a compandor that is used at the receiving end of a circuit to return the compressed signal to its original form; attenuates weak signals and amplifies strong signals. { ik'spand·ər }

expansion [ELECTR] A process in which the effective gain of an amplifier is varied as a function of signal magnitude, the effective gain being greater for large signals than for small signals; the result is greater volume range in an audio amplifier and greater contrast range in facsimile. [MATH] The expression of a quantity as the sum of a finite or infinite series of terms, as a finite or infinite product of factors, or, in general, in any extended form. [MECH ENG] Increase in volume of working material with accompanying drop in pressure of a gaseous or vapor fluid, as in an internal combustion engine or

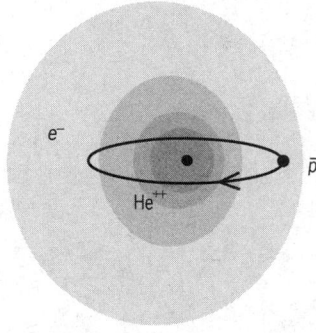

EXOTIC ATOM

Schematic diagram of an antiprotonic helium atom, with an electron (e^-) cloud around the helium nucleus (He^{++}), and an antiproton ($\bar{p}$) in orbit.

EXPANDER

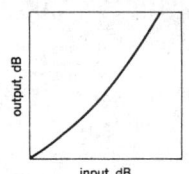

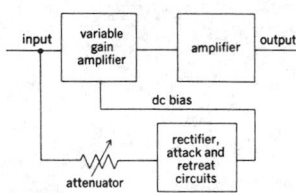

Input-output characteristics and schematic block diagram of an expander.

steam engine cylinder. [PHYS] Process in which the volume of a constant mass of a substance increases. { ik'span·shən }

expansion board [COMPUT SCI] A printed circuit board that can be plugged into a computer to provide it with additional peripherals or enhancements, such as increased memory or communications facilities. { ik'span·shən ,bȯrd }

expansion bolt [DES ENG] A bolt having an end which, when embedded into masonry or concrete, expands under a pull on the bolt, thereby providing anchorage. { ik'span·shən ,bōlt }

expansion bus [COMPUT SCI] The wiring and protocols that connect a computer's motherboard with the peripheral devices. { ik'span·shən ,bəs }

expansion chamber See cloud chamber. { ik'span·shən ,chām·bər }

expansion chucking reamer [DES ENG] A machine reamer with an expansion screw at the end which increases the diameter. { ik'span·shən 'chək·iŋ ,rē·mər }

expansion coefficient See coefficient of cubical expansion. { ik'span·shən kō·ə'fish·ənt }

expansion cooling [MECH ENG] Cooling of a substance by having it undergo adiabatic expansion. { ik'span·shən ,kül·iŋ }

expansion ellipsoid [SOLID STATE] An ellipsoid whose axes have lengths which are proportional to the coefficient of linear expansion in the corresponding direction in a crystal. { ik'span·shən ə'lip,sȯid }

expansion engine [MECH ENG] Piston-cylinder device that cools compressed air via sudden expansion; used in production of pure gaseous oxygen via the Claude cycle. { ik'span·shən ,en·jən }

expansion factor [NAV] In radio hyperbolic navigation, a factor representing the degree to which the hyperbolas of a given group of hyperbolic lines of position diverge as the distance from the base line increases. { ik'span·shən ,fak·tər }

expansion fissures [GEOL] A system of fissures which radiate randomly and pass through feldspars and other minerals adjacent to olivine crystals that have been replaced by serpentine. { ik'span·shən ,fish·ərz }

expansion fit [DES ENG] A condition of optimum clearance between certain mating parts in which the cold inner member is placed inside the warmer outer member and the temperature is allowed to equalize. { ik'span·shən ,fit }

expansion joint [CIV ENG] **1.** In masonry, a flexible bituminous fiber strip used to separate blocks or units of concrete to prevent cracking caused by thermally induced expansion and contraction. **2.** A union or gap between adjacent parts of a building, structure, or concrete work that permits the relative movement caused by temperature changes to occur without rupture or damage. [GEOL] See sheeting structure. [MECH ENG] **1.** A joint between parts of a structure or machine to avoid distortion when subjected to temperature change. **2.** A pipe coupling which, under temperature change, allows movement of a piping system without hazard to associated equipment. { ik'span·shən ,jȯint }

expansion loop [ENG] A complete loop installed in a pipeline to mitigate the effect of expansion or contraction of the line. { ik'span·shən 'lüp }

expansion opening [ENG] A chamber in line with a pipe or tunnel and of larger diameter than the conduit containing liquid or gas, to allow lowering of pressure within the conduit by expansion of the fluid. { ik'span·shən,ōp·ə·niŋ }

expansion ratio [FL MECH] For the calculation of the mass flow of a gas out of a nozzle or other expanding duct, the ratio of the nozzle exit section area to the nozzle throat area, or the ratio of final to initial volume. [MECH ENG] In a reciprocating piston engine, the ratio of cylinder volume with piston at bottom dead center to cylinder volume with piston at top dead center. { ik'span·shən ,rā·shō }

expansion reamer [ENG] A reamer whose diameter may be adjusted between limits by an expanding screw. { ik'span·shən ,rē·mər }

expansion rollers [CIV ENG] Rollers fitted to one support of a bridge or truss to allow for thermal expansion and contraction. { ik'span·shən ,rō·lərz }

expansion scab [MET] A defect in the form of a rough thin layer of metal that is partially separated from the body of a metal casting by a thin layer of sand and held in place by a thin vein of metal. { ik'span·chən ,skab }

expansion shield [DES ENG] An anchoring device that expands as it is driven into masonry or concrete, pressing against the sides of the hole. { ik'span·shən ,shēld }

expansion slot [COMPUT SCI] A location in a computer system where additional facilities, especially circuit boards, can be plugged in to extend the computer's capability. { ik'span·shən ,slät }

expansion system [PETRO ENG] Gas-liquid recovery system in which the refrigeration effect of rapidly depressurized well-stream effluent through a wellhead choke is used to obtain maximum removal of liquefiable hydrocarbons from the gas stream. { ik'span·shən ,sis·təm }

expansion valve [MECH ENG] A valve in which fluid flows under falling pressure and increasing volume. { ik'span·shən ,valv }

expansion wave [FL MECH] A pressure wave or shock wave that decreases the density of air as the air passes through it. { ik'span·shən ,wāv }

expansive bit [DES ENG] A bit in which the cutting blade can be set at various sizes. { ek'span·siv ,bit }

expansive cement [MATER] A type of hydraulic cement, usually of high sulfate and alumina content, that expands after hardening to compensate for drying shrinkage. { ek'span·siv si'ment }

expansivity See coefficient of cubical expansion. { ,ek,span'-siv·əd·ē }

expectation See expected value. { ,ek,spek'tā·shən }

expectation value [QUANT MECH] The average of the results of a large number of measurements of a quantity made on a system in a given state; in case the measurement disturbs the state, the state is reprepared before each measurement. { ,ek,spek'tā·shən ,val·yü }

expected approach clearance time [NAV] The anticipated time that an arriving aircraft will be cleared to commence approach for a landing. { ek'spek·təd ə'prōch ,klir·əns ,tīm }

expected utility See expected value. { ek'spek·təd yü'til·əd·ē }

expected value [MATH] **1.** For a random variable x with probability density function $f(x)$, this is the integral from $-\infty$ to ∞ of $xf(x)dx$. Also known as expectation. **2.** For a random variable x on a probability space (Ω, P), the integral of x with respect to the probability measure P. [SYS ENG] In decision theory, a measure of the value or utility expected to result from a given strategy, equal to the sum over states of nature of the product of the probability of the state times the consequence or outcome of the strategy in terms of some value or utility parameter. Abbreviated EV. Also known as expected utility (EU). { ek'spek·təd 'val·yü }

expectorant [PHARM] **1.** Tending to promote expectoration. **2.** An agent that promotes expectoration. { ik'spek·tə·rənt }

expectorate [PHYSIO] To eject phlegm or other material from the throat or lungs. { ik'spek·tə,rāt }

expendable pattern [MET] A pattern that is destroyed in the metal-casting process. { ik'spen·də·bəl 'pad·ərn }

expendable rocket See expendable vehicle. { ik'spen·də·bəl 'räk·ət }

expendable vehicle [AERO ENG] A rocket that is used only once to place a payload in orbit. Also known as expendable rocket. { ik'spen·də·bəl 've·ə·kəl }

experiment [SCI TECH] The test of a hypothesis under controlled conditions. { ik'sper·ə·mənt }

experimental breeder reactor [NUCLEO] A fast, heterogeneous nuclear reactor used for research and breeding; its core consists of enriched ^{235}U surrounded by a blanket of natural uranium. { ik,sper·ə'ment·əl 'brēd·ər rē,ak·tər }

experimental design [STAT] A pattern for setting up experiments and making observations about the relationship between several variables in which one attempts to obtain as much information as possible for a fixed expenditure level. { ik,sper·ə'ment·əl di'zīn }

experimental ecology [ECOL] The manipulation of organisms or their environments to discover the underlying mechanisms governing distribution and abundance. { ik,sper·ə,ment·əl ē'käl·ə·jē }

experimental petrology [PETR] A branch of petrology in which phenomena that occur during petrological processes are reproduced and studied in the laboratory. { ik,sper·ə'ment·əl pə'träl·ə·jē }

experimental psychology [PSYCH] The study of psychological phenomena by experimental methods. { ik,sper·ə'ment·əl sī'käl·ə·jē }

experimental reactor [NUCLEO] A reactor to test the design of a new reactor concept. { ik,sper·ə·'ment·əl rē'ak·tər }

expert control system [CONT SYS] A control system that uses expert systems to solve control problems. { 'ek,spərt kən'trōl ,sis·təm }

expert system [COMPUT SCI] A computer system composed of algorithms that perform a specialized, usually difficult professional task at the level of (or sometimes beyond the level of) a human expert. { 'ek,spərt ,sis·təm }

expiration date [MATER] The anticipated date when a material may go from usable to unusable. { ,ek·spə'rā·shən ,dāt }

expiratory reserve volume [PHYSIO] At the end of a normal expiration, the quantity of air that can be expelled by forcible expiration. { ek'spī·rə,tór·ē ri¦zərv ,väl·yəm }

expiratory standstill [PHYSIO] Suspension of action at the end of expiration. { ek'spī·rə,tór·ē 'stand,stil }

explant [CYTOL] An excised fragment of a tissue or an organ used to start a cell culture. { 'eks,plant }

explementary angles See conjugate angles. { ,ek·splə¦men·tə·rē 'aŋ·gəlz }

expletive [ENG] Any material used as fill, for example, a piece of masonry used to fill a cavity. { 'ek·spləd·iv }

explicit memory [PSYCH] A type of memory that has the self as the agent or experiencer of the event in question and requires conscious recall; for example, memory of facts, episodes, and images. Also known as declarative memory. { ik,splis·ət 'mem·rē }

explicit programming [CONT SYS] Robotic programming that employs detailed and exact descriptions of the tasks to be performed. { ik'splis·ət 'prō,gram·iŋ }

explicit symmetry breaking [PHYS] A phenomenon in which a system is not quite, but almost, the same for two configurations related by exact symmetry. { ik'splis·ət 'sim·ə·trē ,brāk·iŋ }

exploded file [COMPUT SCI] A file in which more data have been added to each record in order to adapt it to a new application. { ik'splōd·əd 'fīl }

exploded view [GRAPHICS] A drawing or picture of any article or piece of equipment in which the component parts are separated but so arranged to show their relationship to the whole. { ik'splōd·əd 'vyü }

exploding bridge wire [ENG] An initiator or system in which a very high energy electrical impulse is passed through a bridge wire, literally exploding the bridge wire and releasing thermal and shock energy capable of initiating a relatively insensitive explosive in contact with the bridge wire. { ik¦s·plōd·iŋ 'brij ,wīr }

exploitation [MIN ENG] The extraction from the earth and utilization of ore, gas, oil, and minerals found by exploration. { ,ek,splói'tā·shən }

exploration [MIN ENG] The search for economic deposits of minerals, ore, gas, oil, or coal by geological surveys, geophysical prospecting, boreholes and trial pits, or surface or underground headings, drifts, or tunnels. { ,ek·splə'rā·shən }

exploratory well [PETRO ENG] An oil well drilled for purposes of exploration for underlying petroleum. { ik'splór·ə,tór·ē 'wel }

exploring coil [ELECTROMAG] A small coil used to measure a magnetic field or to detect changes produced in a magnetic field by a hidden object; the coil is connected to an indicating instrument either directly or through an amplifier. Also known as magnetic test coil; search coil. { ik'splór·iŋ ,kóil }

explosimeter [PETRO ENG] An instrument for measuring the concentration of flammable vapors or gases in the air. Also known as gas sniffer. { ,ek·splō'sim·əd·ər }

explosion [CHEM] A chemical reaction or change of state which is effected in an exceedingly short space of time with the generation of a high temperature and generally a large quantity of gas. { ik'splō·zhən }

explosion breccia [PETR] Breccia resulting from volcanic eruption or a phreatic explosion. { ik'splō·zhən,brech·ə }

explosion crater [GEOL] A volcanic crater formed by explosion and commonly developed along rift zones on the flanks of large volcanoes. { ik'splō·zhən ,krād·ər }

explosion door [MECH ENG] A door in a furnace which is designed to open at a predetermined pressure. { ik'splō·zhən ,dór }

explosion method [THERMO] Method of measuring the specific heat of a gas at constant volume by enclosing the gas with an explosive mixture, whose heat of reaction is known, in a chamber closed with a corrugated steel membrane which acts as a manometer, and by deducing the maximum temperature reached on ignition of the mixture from the pressure change. { ik'splō·zhən ,meth·əd }

explosion rupture disk device [MECH ENG] A protective device used where the pressure rise in the vessel occurs at a rapid rate. { ik'splō·zhən 'rəp·chər ,disk di,vīs }

explosion tuff [GEOL] A tuff whose constituent ash particles are in the place they fell after being ejected from a volcanic vent. { ik'splō·zhən ,təf }

explosion welding [MET] A solid-state process wherein bonding is produced by a controlled detonation, resulting in rapid movement together of the members to be joined. Also known as explosive welding. { ik'splō·zhən ,weld·iŋ }

explosive [MATER] A substance, such as trinitrotoluene, or a mixture, such as gunpowder, that is characterized by chemical stability but may be made to undergo rapid chemical change without an outside source of oxygen, whereupon it produces a large quantity of energy generally accompanied by the evolution of hot gases. { ik'splō·siv }

explosive-actuated device [ENG] Any of various devices actuated by means of explosive; includes devices actuated either by high explosives or low explosives, whereas propellant-actuated devices include only the latter. { ik'splō·sive ,ak·chə,wād·əd di,vīs }

explosive bolt [AERO ENG] A bolt designed to contain a remote-initiated explosive charge which, upon detonation, will shear the bolt or cause it to fail otherwise; applicable to such uses as stage separation of rockets, jettison of expended fuel tanks, and ejection of parachutes. { ik'splō·siv 'bōlt }

explosive cladding [MET] Bonding of a metal coating or metal cladding to a base metal by using the force of an explosive charge. { ik'splō·siv 'klad·iŋ }

explosive decompression [AERO ENG] A sudden loss of pressure in a pressurized cabin, cockpit, or the like, so rapid as to be explosive, as when punctured by gunfire. { ik'splō·siv ,dē·kəm'presh·ən }

explosive disintegration [ENG] Explosive shattering when pressure is suddenly released on a pressured, permeable material (wood, mineral, and such) containing gas or liquid; the rupture of wood by this process is used to manufacture Masonite. { ik'splō·siv di,sin·tə'grā·shən }

explosive echo ranging [ENG] Sonar in which a charge is exploded underwater to produce a shock wave that serves the same purpose as an ultrasonic pulse; the elapsed time for return of the reflected wave gives target range. { ik'splō·siv 'ek·ō ,rānj·iŋ }

explosive evolution [EVOL] Rapid diversification of a group of fossil organisms in a short geological time. { ik'splō·siv ev·ə'lü·shən }

explosive filler [MATER] Main explosive charge contained in a projectile, missile, bomb, or the like. { ik'splō·siv 'fil·ər }

explosive fog signal [NAV] A fog signal produced by detonating an explosive charge. { ik'splō·siv 'fäg ,sig·nəl }

explosive forming [MET] Shaping metal parts in dies by using an explosive charge to generate forming pressure. { ik'splō·siv 'fór·miŋ }

explosive fuel [MATER] Any substance which combines with oxygen and other explosive ingredients to produce explosion energy, including aluminum, silicon, carbon, sulfur, glycerol, glucol, paraffin wax, diesel oil, and guar gum. { ik'splō·siv 'fyül }

explosive index [GEOL] The percentage of pyroclastics in the material from a volcanic eruption. { ik'splō·siv 'in,deks }

explosive limits [CHEM ENG] The upper and lower limits of percentage composition of a combustible gas mixed with other gases or air within which the mixture explodes when ignited. { ik'splō·siv 'lim·əts }

explosive nucleosynthesis [ASTRON] Nucleosynthetic processes that are believed to occur in novae and supernovae, and at the surfaces of neutron stars, such as the r-process and the rp-process. { ik'splō·siv ,nü·klē·ō'sin·thə·səs }

explosive ordnance [ORD] Ordnance materiel containing

or consisting of explosives, such as bombs, mines, torpedoes, missiles, or projectiles. { ik'splō·siv 'ȯrd·nəns }

explosive-ordnance disposal [ORD] The handling, disarming, or destroying of unexploded missiles and other explosive ordnance. Abbreviated EOD. { ik'splō·siv 'ȯrd·nəns di,spō·zəl }

explosive-ordnance disposal unit [ORD] Organization or personnel with special training or equipment who render safe explosive ordnance, make intelligence reports on such ordnance, and supervise the safe removal and disposal thereof. { ik'splō·siv 'ȯrd·nəns di,spō·zəl ,yü·nət }

explosive-ordnance reconnaissance [ORD] Act of reconnoitering to determine the presence of an unexploded missile, ascertaining its nature, applying all practicable protective measures for the protection of personnel, installations, and equipment, and, finally, reporting essential information to the authority directing explosive ordnance disposal operations. Abbreviated EOR. { ik'splō·siv 'ȯrd·nəns ri,kän·ə·səns }

explosive-ordnance reconnaissance agent [ORD] A person trained in explosive ordnance reconnaissance techniques who takes required actions and renders the needed reports so that the unexploded ordnance in question can be effectively neutralized by the military. Abbreviated EORA. { ik'splō·siv 'ȯrd·nəns ri,kän·ə·səns ,ā·jənt }

explosive oxidizer [MATER] Any substance which yields oxygen to combine with fuels or other explosive ingredients to produce explosive energy, such as nitrates, chlorates, and perchlorates. { ik'splō·siv 'äk·sə,dīz·ər }

explosive personality [PSYCH] A disorder of impulse control in which several episodes of serious outbursts of relatively unprovoked aggression lead to assault on others or the destruction of property. { ik'splō·siv ,pərs·ən'al·ə·dē }

explosive rivet [ENG] A rivet holding a charge of explosive material; when the charge is set off, the rivet expands to fit tightly in the hole. { ik'splō·siv 'riv·ət }

explosive train [ORD] A train of combustible and explosive elements arranged in order of decreasing sensitivity, inside a fuse, projectile, bomb, gun chamber, or the like; the function is to accomplish the controlled augmentation of a small impulse into one of suitable energy to cause the main charge of the munition to function. { ik'splō·siv 'trān }

explosive variable See cataclysmic variable. { ik'splō·siv 'ver·ē·ə·bəl }

explosive welding See explosion welding. { ik'splō·siv 'weld·iŋ }

exponent [MATH] A number or symbol placed to the right and above some given mathematical expression. { ik'spō·nənt }

exponential [MATH] For a bounded linear operator A on a Banach space, the sum of a series which is formally the exponential series in A. { ,ek·spə'nen·chəl }

exponential amplifier [ELECTR] An amplifier capable of supplying an output signal proportional to the exponential of the input signal. { ,ek·spə'nen·chəl 'am·plə,fī·ər }

exponential atmosphere See isothermal atmosphere. { ,ek·spə'nen·chəl 'at·mə,sfir }

exponential curve [MATH] A graph of the function $y = a^x$, where a is a positive constant. { ,ek·spə'nen·chəl 'kərv }

exponential decay [PHYS] The decrease of some physical quantity according to the exponential law $N(t) = N_0 e^{-t/\tau}$, where τ is a constant called the decay time. { ,ek·spə'n en·chəl di'kā }

exponential density function [MATH] A probability density function obtained by integrating a function of the form $\exp(-|x - m|/\sigma)$, where m is the mean and σ the standard deviation. { ,ek·spə'nen·chəl den·səd·ē ,fəŋk·shən }

exponential distribution [STAT] A continuous probability distribution whose density function is given by $f(x) = ae^{-ax}$, where $a > 0$ for $x > 0$, and $f(x) = 0$ for $x \le 0$; the mean and standard deviation are both $1/a$. { ,ek·spə'nen·chəl dis·trə'byü·shən }

exponential equation [MATH] An equation containing e^x (the Naperian base raised to a power) as a term. { ,ek·spə'nen·chəl i'kwā·zhən }

exponential experiment [NUCLEO] A nuclear experiment involving a subcritical assembly of fissionable and moderator material. { ,ek·spə'nen·chəl ik'sper·ə·mənt }

exponential function [MATH] The function $f(x) = e^x$, written $f(x) = \exp(x)$. { ,ek·spə'nen·chəl 'fəŋk·shən }

exponential generating function [MATH] A function, $G(x)$, corresponding to a sequence, $a_0, a_1, \ldots$, where $G(x) = a_0 + (a_1 x/1!) + (a_2 x^2/2!) + \cdots$. { ,eks·pə'nen·chəl ,jen·ə,rād·iŋ 'fəŋk·shən }

exponential growth [MICROBIO] The period of bacterial growth during which cells divide at a constant rate. Also known as logarithmic growth. [SCI TECH] The increase of a quantity x with time t according to the equation $x = Ka^t$, where K and a are constants, a is greater than 1, and K is greater than 0. { ,ek·spə'nen·chəl 'grōth }

exponential horn [ENG ACOUS] A horn whose cross-sectional area increases exponentially with axial distance. { ,ek·spə'nen·chəl 'hȯrn }

exponential integral [MATH] The function defined to be the integral from x to ∞ of $(e^{-t}/t) \, dt$ for x positive. { ,ek·spə'nen·chəl 'int·ə·grəl }

exponential law [MATH] See law of exponents. [PHYS] The principle that growth or decay of some physical quantity is at a rate such that its value at a certain time or place is the initial value times e raised to a power equal to a constant times some convenient coordinate, such as the elapsed time or the distance traveled by a wave; there is growth if the constant is positive, decay if it is negative. { ,ek·spə'nen·chəl 'lȯ }

exponential pile See exponential reactor. { ,ek·spə'nen·chəl 'pīl }

exponential pulse [PHYS] Variation of some quantity with time similar to the displacement of a critically damped harmonic oscillator which is initially given an impulse in its equilibrium position. { ,ek·spə'nen·chəl 'pəls }

exponential reactor [NUCLEO] A nuclear reactor which is designed specifically for exponential experiments, as well as for determining critical size. Also known as exponential pile. { ,ek·spə'nen·chəl rē'ak·tər }

exponential series [MATH] The Maclaurin series expansion of e^x, namely,

$$e^x = 1 + \sum_{n=1}^{\infty} \frac{x^n}{n!}$$

{ ,ek·spə'nen·chəl 'sir·ēz }

exponential smoothing [IND ENG] A mathematical-statistical method of forecasting used in industrial engineering which assumes that demand for the following period is some weighted average of the demands for the past periods. { ,ek·spə'nen·chəl 'smüth·iŋ }

exponential transmission line [ELEC] A two-conductor transmission line whose characteristic impedance varies exponentially with electrical length along the line. { ,ek·spə'nen·chəl tranz'mish·ən ,līn }

export kerosine [MATER] A grade of kerosine once used for export; has the darkest shade of the standard kerosine colors, namely standard white (also called export white). Also known as standard white kerosine. { 'ek,spȯrt 'ker·ə,sēn }

exposure [BUILD] The distance from the butt of one shingle to the butt of the shingle above it, or the amount of a shingle that is seen. [GRAPHICS] The act of permitting light to fall upon a photosensitive material. [MED] The state of being open to some action or influence that may affect detrimentally, as cold, disease, or wetness. [METEOROL] The general surroundings of a site, with special reference to its openness to winds and sunshine. [NUCLEO] **1.** The total quantity of radiation at a given point, measured in air. **2.** The cumulative amount of radiation exposure to which nuclear fuel has been subjected in a nuclear reactor; usually expressed in terms of the thermal energy produced by the reactor per ton of fuel initially present, as megawatt days per ton. [OPTICS] See light exposure. See radiant exposure. { ik'spō·zhər }

exposure dose [MED] A measure of the radiation in a certain place based upon its ability to produce ionization in air. { ik'spō·zhər ,dōs }

exposure factor [NUCLEO] A quantity f used to specify radiographic exposure equal to st/d^2, where s is the intensity of radioactive source, t is the time, and d is the source to film distance. { ik'spō·zhər ,fak·tər }

exposure index [GRAPHICS] A numerical speed rating of a photosensitive material. { ik'spō·zhər ,in,deks }

exposure latitude [GRAPHICS] The limits of film exposure within which there is no significant effect on the quality of the image. { ik'spō·zhər ‚lad·ə,tüd }

exposure limit [MED] The maximum radiation dose equivalent permitted under specified conditions. { ik'spō·zhər ‚lim·ət }

exposure meter [OPTICS] An instrument used to measure the intensity of light reflected from an object, for the purpose of determining proper camera exposure. { ik'spō·zhər ‚mēd·ər }

exposure pathway [MED] The mode of intake of a substance; for example, inhalation, ingestion, or absorption. { ik'spō·zhər ‚path·wā }

exposure rate [MED] Exposure dose per unit time. { ik'spō·zhər ‚rāt }

exposure time [CIV ENG] The time period of interest for seismic hazard calculations such as the design lifetime of a building or the time over which the numbers of casualties should be estimated. [PHYS] The amount of time a material is illuminated or irradiated. { ik'spō·zhər ‚tīm }

exposure voltage [ELEC] The voltage at which the document-illuminating lamps are operated during exposure. { ik'spō·zhər ‚vōl·tij }

expression [CHEM ENG] Separation of liquid from a two-phase solid-liquid system by compression under conditions that permit liquid to escape while the solid is retained between the compressing surfaces. Also known as mechanical expression. [COMPUT SCI] A mathematical or logical statement written in a source language, consisting of a collection of operands connected by operations in a logical manner. { ik'spresh·ən }

expression vector [MOL BIO] A cloning vector that promotes the expression of foreign gene inserts. { ik'spresh·ən ‚vek·tər }

expressway [CIV ENG] A limited-access, high-speed, divided highway having grade separations at points of intersection with other roads. Also known as limited-access highway. { ik'spres‚wā }

expulsion fuse See expulsion-fuse unit. { ik'spəl·shən ‚fyüz }

expulsion-fuse unit [ELEC] A vented fuse unit in which the arc is extinguished by the expulsion of gases generated by the arc and lining of the fuse holder, sometimes with the aid of a spring. Also known as expulsion fuse. { ik'spəl·shən ‚fyüz ‚yü·nət }

exradius [MATH] The radius of an escribed circle of a triangle. { ‚eks'rād·ē·əs }

exsecant [MATH] The trigonometric function defined by subtracting unity from the secant, that is exsec θ = sec θ − 1. { ‚ek'sē·kant }

exserted [BIOL] Protruding beyond the enclosing structure, such as stamens extending beyond the margin of the corolla. { ek'sərd·əd }

exsheath [INV ZOO] To escape from the residual membrane of a previous developmental stage, as pertaining to the larva of certain nematodes, microfilaria, and so on. { ek'shēth }

exsiccate [SCI TECH] To dry by driving off, or draining of, moisture. { 'ek·sə,kāt }

exsolution [GEOL] A phenomenon during which molten rock solutions separate when cooled. { ‚ek·sə'lü·shən }

exsolution lamellae [GEOL] Layers of sedimentary rock that solidify from solution by either precipitation or secretion. { ‚ek·sə'lü·shən lə'mel·ē }

exstipulate [BOT] Lacking stipules. { ek'stip·yə,lāt }

exstrophy [MED] Eversion; the turning inside out of a part. { 'ek·strə·fē }

exsurgence See resurgence. { ek'sər·jəns }

extended area [DES ENG] An engineering surface that has been extended areawise without increasing diameter, as by using pleats (as in filter cartridges) or fins (as in heat exchangers). { ik'stend·əd 'er·ē·ə }

extended-area service [COMMUN] Telephone exchange service, without toll charges, that extends over an area where there is a community of interest, often in return for a somewhat higher exchange service rate. { ik'stend·əd ‚er·ē·ə 'sər·vəs }

extended ASCII [COMMUN] An addition to the standard American Standard Code for Information Interchange, namely, characters 128 through 255; includes letters with diacritics, Greek letters, and special symbols. { ik'sten·dəd 'as,kē }

extended binary-coded decimal interchange code [COMPUT SCI] A computer code that uses eight binary positions to represent a single character, giving a possible maximum of 256

characters. Abbreviated EBCDIC. { ik'stend·əd 'bī,ner·ē ‚kōd·əd ‚des·məl 'int·ər,chānj ‚kōd }

extended channel status word [COMPUT SCI] Stored information which follows an input/output interrupt. Abbreviated ECSW. { ik'stend·əd 'chan·əl 'stad·əs ‚wərd }

extended close-coupling method [ATOM PHYS] A method of extending the close-coupling method to the case of ionizing collisions of an electron with a hydrogen atom by replacing the true continuum of ionized hydrogenic target states with a finite number of discrete, normalized, positive-energy pseudostates, while treating the incident electron with conventional, two-body scattering boundary conditions. { ik'stend·əd ‚klōs 'kəp·liŋ ‚meth·əd }

extended data out random-access memory [COMPUT SCI] A type of dynamic random-access memory that was optimized for the 66-megahertz bus but largely has been replaced by faster systems. Abbreviated EDO RAM. { ik'stend·əd ‚dad·ə ‚aut ‚ran·dəm 'ak,ses ‚mem·rē }

extended dislocation [CRYSTAL] A dislocation in a close-packed structure consisting of a strip of stacking fault edged by two lines across which slip through a fraction of a lattice constant, into one of the alternative stacking positions, has occurred. { ik'stend·əd ‚dis,lō'kā·shən }

extended-entry decision table [COMPUT SCI] A decision table in which the condition stub cites the identification of the condition but not the particular values, which are entered directly into the condition entries. { ik'stend·əd 'en·trē di'sizh·ən ‚tā·bəl }

extended forecast [METEOROL] In general, a forecast of weather conditions for a period extending beyond 2 days from the day of issue. Also known as long-range forecast. { ik'stend·əd 'fȯr,kast }

extended-interaction tube [ELECTR] Microwave tube in which a moving electron stream interacts with a traveling electric field in a long resonator; bandwidth is between that of klystrons and traveling-wave tubes. { ik'stend·əd int·ə'rak·shən ‚tüb }

extended mean-value theorem See second mean-value theorem. { ik'sten·dəd ‚mēn ‚val·yü 'thir·əm }

extended-precision word [COMPUT SCI] A piece of data of 16 bytes in floating-point arithmetic when additional precision is required. { ik'stend·əd prə'sizh·ən ‚wərd }

extended-range Dovap [NAV] A baseline extension of the Dovap system to provide a coherent reference to the ground transmitter and all Dovap receivers located beyond line-of-sight to the ground transmitter. Abbreviated Extradop. { ik'stend·əd ‚rānj 'dō,vap }

extended-range forecast See medium-range forecast. { ik'stend·əd ‚rānj 'fȯr,kast }

extended source [ASTRON] A radio source that has a large angular extent and is strongest at longer wavelengths, as distinguished from a compact source. [OPTICS] A source of radiation that can be resolved by the eye or a specified instrument into a geometrical image. { ik'stend·əd 'sȯrs }

extended state [QUANT MECH] A state of motion in which an electron may be found anywhere within a region of a material of linear extent equal to that of the material itself. { ik'stend·əd 'stāt }

extended stream [HYD] A stream lengthened by the extension of its downstream course; the course is through a newly emerged land such as a coastal plain. { ik'stend·əd 'strēm }

extended time scale See slow time scale. { ik'stend·əd 'tīm ‚skāl }

extended valley [GEOL] **1.** A valley that is lengthened downstream either by a regression of the sea or by uplift of the coastal region. **2.** A valley formed by or containing an extended stream. { ik'stend·əd 'val·ē }

extended x-ray absorption fine structure [PHYS] A variation in the x-ray absorption of a substance as a function of energy, at energies just above that required for photons to liberate core electrons into the continuum; it is due to interference between the outgoing photoelectron waves and electron waves backscattered from atoms adjacent to the absorbing atoms. Abbreviated EXAFS. { ik'stend·əd 'eks,rā əb'sȯrp·shən ‚fīn 'strək·chər }

extender [CHEM] A material used to dilute or extend or change the properties of resins, ceramics, paints, rubber, and so on. [ELEC] A male or female receptacle connected by a short cable to make a test point more conveniently accessible to a test probe. { ik'sten·dər }

extender plasticizer *See* secondary plasticizer. { ik'sten·dər 'plas·tə,sīz·ər }

extend flip-flop [COMPUT SCI] A special flag set when there is a carry-out of the most significant bit in the register after an addition or a subtraction. { ik'stend 'flip,fläp }

extending flow [HYD] A glacial flow pattern in which velocity increases as the distance downstream becomes greater. { ik'stend·iŋ ,flō }

extensibility [MATER] The extent to which a material can be stretched without causing it to tear or break. [MECH] The amount to which a material can be stretched or distorted without breaking. { ik,sten·sə'bil·əd·ē }

extensible language [COMPUT SCI] A programming language which can be modified by adding new features or changing existing ones. { ik'sten·sə·bəl 'laŋ·gwij }

Extensible Markup Language [COMMUN] A set of rules for writing markup languages which provides a robust, machine-readable information protocol that can handle complex objects. Abbreviated XML. { ik¦sten·sə·bəl 'märk,əp ,laŋ·gwij }

extensible system [COMPUT SCI] A computer system in which users may extend the basic system by implementing their own languages and subsystems and making them available for others to use. { ik'sten·sə·bəl 'sis·təm }

extension *See* extension fields. [PHYSIO] A movement which has the effect of straightening a limb. { ik'sten·chən }

extensional fault *See* tension fault. { ik'sten·chən·əl 'fólt }

extension bolt [DES ENG] A vertical bolt that can be slid into place by a long extension rod; used at the top of doors. { ik'sten·chən ,bōlt }

extension cord [ELEC] A line cord having a plug at one end and an outlet at the other end. { ik'sten·chən ,kórd }

extension field [MATH] An extension field of a given field *E* is a field *F* such that *E* is a subfield of *F*. Also known as extension. { ik'sten·chən ,fēld }

extension fracture [GEOL] A fracture that develops perpendicular to the direction of greatest stress and parallel to the direction of compression. { ik'sten·chən ,frak·chər }

extension jamb [BUILD] A jamb that extends past the head of a door or window. { ik'sten·chən ,jam }

extension joints [GEOL] Fractures that form parallel to a compressive force. { ik'sten·chən ,jóins }

extension ladder [DES ENG] A ladder of two or more nesting sections which can be extended to almost the combined length of the sections. { ik'sten·chən ,lad·ər }

extension map [MATH] An extension map of a map *f* from a set *A* to a set *L* is a map *g* from a set *B* to *L* such that *A* is a subset of *B* and the restriction of *g* to *A* equals *f*. { ik'sten·chən ,map }

extension mechanism [COMPUT SCI] One of the components of an extensible language which allows the definition of new language features in terms of the primitive facilities of the base language. { ik'sten·chən ,mek·ə,niz·əm }

extension ore *See* possible ore. { ik'sten·chən ,ór }

extension register [COMPUT SCI] A register that is combined with an accumulator register for calculations involving multiple precision arithmetic. { ik'sten·chən ,rej·ə·stər }

extension spring [DES ENG] A tightly coiled spring designed to resist a tensile force. { ik'sten·chən ,spriŋ }

extensive air shower *See* Auger shower. { ik¦sten·siv 'er ,shaü·ər }

extensive property [PHYS CHEM] A noninherent property of a system, such as volume or internal energy, that changes with the quantity of material in the system; the quantitative value equals the sum of the values of the property for the individual constituents. { ik'sten·siv 'präp·ərd·ē }

extensive shower *See* Auger shower. { ik'sten·siv 'shaür }

extensometer [ENG] **1.** A strainometer that measures the change in distance between two reference points separated 60–90 feet (20–30 meters) or more; used in studies of displacements due to seismic activities. **2.** An instrument designed to measure minute deformations of small objects subjected to stress. { ,ek,sten'säm·əd·ər }

extent [COMPUT SCI] The physical locations in a mass-storage device or volume allocated for use by a particular data set. { ik'stent }

exterior [MATH] **1.** For a set *A* in a topological space, the largest open set contained in the complement of *A*. **2.** For a

plane figure, the set of all points that are neither on the figure nor inside it. **3.** For an angle, the set of points that lie in the plane of the angle but not between the rays defining the angle. **4.** For a simple closed plane curve, one of the two regions into which the curve divides the plane according to the Jordan curve theorem, namely, the region that is not bounded. { ek'stir·ē·ər }

exterior algebra [MATH] An algebra whose structure is analogous to that of the collection of differential forms on a Riemannian manifold. Also known as Grassmann algebra. { ek'stir·ē·ər 'al·jə·brə }

exterior angle [MATH] **1.** An angle between one side of a polygon and the prolongation of an adjacent side. **2.** An angle made by a line (the transversal) that intersects two other lines, and either of the latter on the outside. { ek'stir·ē·ər 'aŋ·gəl }

exterior ballistics [MECH] The science concerned with behavior of a projectile after leaving the muzzle of the firing weapon. { ek'stir·ē·ər bə'lis·tiks }

exterior complex scaling [ATOM PHYS] A mathematical transformation, which has been used to simplify the boundary conditions on the wave functions in an electron-atom collision, in which the variable *r*, representing the distance of the electron from the nucleus, is replaced, at values of *r* greater than some constant *R*, by $R + (r - R)C$, where *C* is a complex number with unit modulus and positive imaginary part. { ik¦stir·ē·ər ¦käm,pleks ,skāl·iŋ }

exterior content *See* exterior Jordan content. { ek'stir·ē·ər 'kän,tent }

exterior Jordan content [MATH] Also known as exterior content. **1.** For a set of points on a line, the largest number *C* such that the sum of the lengths of a finite number of closed intervals that includes every point in the set is always equal to or greater than *C*. **2.** The exterior Jordan content of a set of points, *X*, in *n*-dimensional Euclidean space (where *n* is a positive integer) is the greatest lower bound on the hypervolume of the union of a finite set of hypercubes that contains *X*. { ek¦stir·ē·ər ¦jórd·ən 'kän,tent }

exterior measure *See* Lebesgue exterior measure. { ek¦stir·ē·ər 'mezh·ər }

extern [COMPUT SCI] A pseudoinstruction found in several assembly languages which explicitly tells an assembler that a symbol is external, that is, not defined in the program module. { ek'stərn }

external aileron [AERO ENG] An aileron offset from the wing; that is, not forming a part of the wing. { ek'stərn·əl 'ā·lə,rän }

external angle [MATH] The angle defined by an arc around the boundaries of an internal angle or included angle. { ek'stərn·əl 'aŋ·gəl }

external ankle height [ARCH] A measure of the vertical distance taken from the lower end of the fibula to the floor. { ek'stərn·əl 'aŋ·kəl ,hīt }

external armature [ELEC] Armature for a machine of special design in which the armature is a ring which rotates around the magnetic poles. { ek'stərn·əl 'är·mə·chər }

external auditory meatus [ANAT] The external passage of the ear, leading to the tympanic membrane in reptiles, birds, and mammals. { ek'stərn·əl 'ód·ə,tór·ē mē'ād·əs }

external beam [NUCLEO] A beam of particles which originate in a particle accelerator and are directed outside the accelerator so that they can be used for experiments with external apparatus. { ek'stərn·əl 'bēm }

external brake [MECH ENG] A brake that operates by contacting the outside of a brake drum. { ek'stərn·əl 'brāk }

external buffer [COMPUT SCI] A buffer storage located outside the computer's main storage, often within a control unit or other peripheral device. { ek'stərn·əl 'bəf·ər }

external burning [AGR] Combustion that can be established when a reactive fuel is injected onto the lateral surfaces of an airfoil or from the base of a projectile traveling at high speed, and, if properly controlled, can produce useful forces that can augment lift, provide attitude control, reduce or cancel drag, or produce thrust. { ik,stərn·əl 'bərn·iŋ }

external carotid artery [ANAT] An artery which originates at the common carotid and distributes blood to the anterior part of the neck, face, scalp, side of the head, ear, and dura mater. { ek'stərn·əl kə'räd·əd 'ärd·ə·rē }

external centerless grinding [MECH ENG] A process by which a metal workpiece is finished on its external surface by supporting the piece on a blade while it is advanced between a regulating wheel and grinding wheel. { ek¦stərn·əl 'sen·tər·ləs ,grīnd·iŋ }

external circuit [PHYS CHEM] All connecting wires, devices, and current sources which achieve desired conditions within an electrolytic cell. { ek¦stərn·əl 'sər·kət }

external combustion engine [MECH ENG] An engine in which the generation of heat is effected in a furnace or reactor outside the engine cylinder. { ek¦stərn·le kəm'bəs·chən ,en·jən }

external declaration [COMPUT SCI] A declarative statement in a computer program that specifies that a symbolic name used in the program is defined in another program. { ek¦stərn·əl ,dek·lə'rā·shən }

external delay [COMPUT SCI] Time during which a computer cannot be operated due to circumstances beyond the reasonable control of the operators and maintenance engineers, such as a failure of the public power supply. { ek¦stərn·əl di'lā }

external device [ENG] A piece of equipment that operates in conjunction with and under the control of a central system, such as a computer or control system, but is not part of the system itself. { ek¦stərn·əl di'vīs }

external-device address [COMPUT SCI] The address of a component such as a tape drive. { ek¦stərn·əl di¦vīs 'a,dres }

external-device control [COMPUT SCI] The capability of an external device to create an interrupt during the execution of a job. { ek¦stərn·əl di¦vīs kən,trōl }

external-device operands [COMPUT SCI] The part of an instruction referring to an external device such as a tape drive. { ek¦stərn·əl di¦vīs 'äp·ə,ranz }

external-device response [COMPUT SCI] The signal from an external device, such as a tape drive, that it is not busy. { ek¦stərn·əl di¦vīs ri,späns }

external dominating set See dominating vertex set. { ek¦stərn·əl ,däm·ə,nād·iŋ 'set }

external ear [ANAT] The portion of the ear that receives sound waves, including the pinna and external auditory meatus. { ek¦stərn·əl 'ēr }

external error [COMPUT SCI] An error sensed by the computer when this error occurs in a device such as a disk drive. { ek¦stərn·əl 'er·ər }

external fertilization [PHYSIO] Those processes involved in the union of male and female sex cells outside the body of the female. { ek¦stərn·əl ,fərd·əl·ə'zā·shən }

external force [MECH] A force exerted on a system or on some of its components by an agency outside the system. { ek¦stərn·əl 'fors }

external forcing [CLIMATOL] The influence on the earth system by solar radiation. { ik,stərn·əl 'fors·iŋ }

external galaxy [ASTRON] Any galaxy known to exist, besides the Milky Way. { ek¦stərn·əl 'gal·ik·sē }

external gas injection [PETRO ENG] Pressure-maintenance gas injection with wells located in the structurally higher positions of the reservoir, usually in the primary or secondary gas cap. Also known as crestal injection; gas-cap injection. { ek¦stərn·əl 'gas in,jek·shən }

external gill [ZOO] A gill that is external to the body wall, as in certain larval fishes and amphibians, and in many aquatic insects. { ek¦stərn·əl 'gil }

external grinding [MECH ENG] Grinding the outer surface of a rotating piece of work. { ek¦stərn·əl 'grīnd·iŋ }

external header [MECH ENG] Manifold connecting sections of a cast iron boiler. { ek¦stərn·əl 'hed·ər }

external interrupt [COMPUT SCI] Any interrupt caused by the operator or by some external device such as a tape drive. { ek¦stərn·əl 'int·ə,rəpt }

external-interrupt status word [COMPUT SCI] The content of a special register which indicates, among other things, the source of the interrupt. { ek¦stərn·əl 'int·ə,rəpt 'stad·əs ,wərd }

external label [COMPUT SCI] A reference to a variable not defined in a program segment. { ek¦stərn·əl 'lā·bəl }

external line [QUANT MECH] A component of a Feynman graph (in the diagrammatic presentation of perturbative quantum field theory) describing an incoming or outgoing particle in a scattering. { ek¦stərn·əl 'līn }

externally fired boiler [MECH ENG] A boiler that has refractory or cooling tubes surrounding its furnace. { ek¦stərn·əl·ē ¦fīrd 'bȯil·ər }

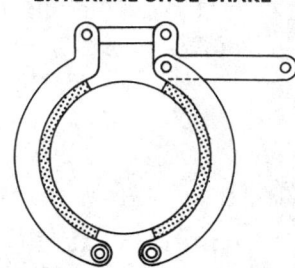

EXTERNAL SHOE BRAKE

External shoe brake; shoes are lined with frictional material (shaded area).

externally stored program [COMPUT SCI] A program achieved by wiring plugboards, as in some tabulating equipment. { ek¦stərn·əl·ē 'stȯrd 'prō·grəm }

externally tangent circles [MATH] Two circles, neither of which is inside the other, that have a single point in common. { ek¦stərn·əl·ē ¦tan·jənt 'sər·kəlz }

external memory [COMPUT SCI] Any storage device not an integral part of a computer system, such as a magnetic tape or a deck of cards. { ek¦stərn·əl 'mem·rē }

external-mix oil burner [ENG] A burner utilizing a jet stream of air to strike the liquid fuel after it has left the burner orifice. { ek¦stərn·əl ,miks 'ȯil,bərn·ər }

external operation [MATH] For a set *S*, a function of one or more independent variables such that at least one of the independent variables has values in *S* but either one or more of the independent variables or the dependent variable fails to have values in *S*. { ek¦stərn·əl ,äp·ə'rā·shən }

external phase See continuous phase. { ek¦stərn·əl 'fāz }

external photoelectric effect See photoemission. { ek¦stərn·əl ,fō·dō·i'lek·trik i,fekt }

external Q [ELECTR] The inverse of the difference between the loaded and unloaded Q values of a microwave tube. { ek¦stərn·əl 'kyü }

external reference [COMPUT SCI] In a computer program, a branch or call to a separate independent program or routine. { ek¦stərn·əl 'ref·rəns }

external respiration [PHYSIO] The processes by which oxygen is carried into living cells from the outside environment and by which carbon dioxide is carried in the reverse direction. { ek¦stərn·əl ,res·pə'rā·shən }

external sensor [CONT SYS] A device that senses information about the environment of a control system but is not part of the system itself. { ek¦stərn·əl 'sen·sər }

external shoe brake [MECH ENG] A friction brake operated by the application of externally contracting elements. { ek¦stərn·əl 'shü ,brāk }

external signal [COMPUT SCI] Any message to an operator for which no printout is required but which is self-explanatory, such as a light condition indicating whether the equipment is on or off. { ek¦stərn·əl 'sig·nəl }

external sorting [COMPUT SCI] The sorting of a list of items by a computer in which the list is too large to be brought into the memory at one time, and instead is brought into the memory a piece at a time so as to produce a collection of ordered sublists which are subsequently reordered by the computer to produce a single list. { ek¦stərn·əl 'sȯrd·iŋ }

external stability number See vertex domination number. { ek,stərn·əl stə'bil·əd·ē ,nəm·bər }

external storage [COMPUT SCI] Large-capacity, slow-access data storage attached to a digital computer and used to store information that exceeds the capacity of main storage. { ek¦stərn·əl 'stȯr·ij }

external symbol dictionary [COMPUT SCI] A list of external symbols and their relocatable addresses which allows the linkage editor to resolve interprogram references. Abbreviated ESD. { ek¦stərn·əl 'sim·bəl ,dik·shə,ner·ē }

external table [COMPUT SCI] A table whose data are located outside a computer program, usually in a separate file. { ek¦stərn·əl 'tā·bəl }

external tangent [MATH] For two circles, each exterior to the other, a line that is tangent to both circles such that both circles are on the same side of this line. { ek¦stərn·əl 'tan·jənt }

external thread [DES ENG] A screw thread cut on an outside surface. { ek¦stərn·əl 'thred }

external time [IND ENG] The time used to perform work by the operator outside the machine cycle, resulting in a loss of potential machine operating time. { ek¦stərn·əl 'tīm }

external upset [PETRO ENG] A thick wall on the exterior of the threaded end of a drill pipe or tubing. { ek¦stərn·əl 'əp,set }

external upset casing [PETRO ENG] Special oil- or gas-well casing designed for extreme conditions requiring greater than usual strength and leak resistance. Also known as extreme line casing. { ek¦stərn·əl 'əp,set,kās·iŋ }

external wave [FL MECH] **1.** A wave in fluid motion having its maximum amplitude at an external boundary such as a free surface. **2.** Any surface wave on the free surface of a homogeneous incompressible fluid is an external wave. { ek¦stərn·əl 'wāv }

external work [THERMO] The work done by a system in expanding against forces exerted from outside. { ek¦stərn·əl 'wərk }

external working environment [IND ENG] The workplace environment that is external to the human body; ranges from air quality to specific features such as clothing or tool handles. { ek¦stirn·əl ¦wərk·iŋ in'vī·rən·mənt }

exteroceptor [PHYSIO] Any sense receptor at the surface of the body that transmits information about the external environment. { ¦ek·stə·rō¦sep·tər }

extinction [ASTRON] The reduction in the apparent brightness of a celestial object due to absorption and scattering of its light by the atmosphere and by interstellar dust; it is greater at low altitudes. [EVOL] The worldwide death and disappearance of a specific organism or group of organisms. [HYD] The drying up of lake by either water loss or destruction of the lake basin. [OPTICS] Phenomenon in which plane polarized light is almost completely absorbed by a polarizer whose axis is perpendicular to the plane of polarization. [PHYS CHEM] See absorbance. [PSYCH] Decrease in frequency and elimination of a conditioned response if reinforcement of the response is withheld. { ek'stiŋk·shən }

extinction coefficient See absorptivity. { ek'stiŋk·shən ‚kō·i‚fish·ənt }

extinction meter [OPTICS] An exposure meter in which light intensity is measured by gradually attenuating the light by a known fraction until a selected design is just visible or disappears. { ek'stiŋk·shən ‚mēd·ər }

extinction voltage [ELECTR] The lowest anode voltage at which a discharge is sustained in a gas tube. { ek'stiŋk·shən ‚vōl·tij }

extirpate [BIOL] To uproot, destroy, make extinct, or exterminate. { 'ek·stər‚pāt }

extracellular [BIOL] Outside the cell. { ¦ek·strə'sel·yə·lər }

extracellular matrix [CELL MOL] A filamentous structure that is attached to the outer cell surface and provides anchorage, traction, and positional recognition to the cell. [HISTOL] A filamentous structure of glycoproteins and proteoglycans that is attached to the cell surface and provides cells with anchorage, traction for movement, and positional recognition. Abbreviated ECM. { ¦ek·strə'sel¦yə·lər 'mā·triks }

extrachromosomal inheritance See cytoplasmic inheritance. { ¦ek·strə‚krō·mə'sō·məl in'her·ət·əns }

extracorporeal shock-wave lithotripsy [MED] A treatment for renal calculi (kidney stones) in which powerful ultrasonic shock waves are focused on the stones, thereby breaking them into small fragments that can be excreted, thus avoiding surgery. Also known as lithotripsy. { ‚ek·strə·kór¦pòr·ē·əl ¦shäk‚wāv 'lith·ə‚trip·sē }

extract [CHEM] Material separated from liquid or solid mixture by a solvent. [COMPUT SCI] **1.** To form a new computer word by extracting and putting together selected segments of given words. **2.** To remove from a computer register or memory all items that meet a specified condition. [MET] To separate a metal or a mineral from an ore by various chemical or mechanical methods. [PHARM] **1.** A pharmaceutical preparation obtained by dissolving the active constituents of a drug with a suitable menstruum, evaporating the solvent, and adjusting to prescribed standards. **2.** A preparation, usually in a concentrated form, obtained by treating plant or animal tissue with a solvent to remove desired odiferous, flavorful, or nutritive components of the tissue. { 'ek‚strakt (noun) or ik'strakt (verb) }

extractant [CHEM] The liquid used to remove a solute from another liquid. { ik'strak·tənt }

extract a root [MATH] To determine a root of a given number, usually a positive real root, or a negative real odd root of a negative number. { ik'strakt ə 'rüt }

extracting agent [CHEM] In a liquid-liquid distribution, the reagent forming a complex or other adduct that has different solubilities in the two immiscible liquids of the extraction system. { ik'strak·tiŋ ‚ā·jənt }

extract instruction [COMPUT SCI] An instruction that requests the formation of a new expression from selected parts of given expressions. { ik'strakt in‚strək·shən }

extraction [CHEM] A method of separation in which a solid or solution is contacted with a liquid solvent (the two being essential mutually insoluble) to transfer one or more components into the solvent. [MED] The act or process of pulling out a tooth. { ik'strak·shən }

extraction column [CHEM ENG] Vertical-process vessel in which a desired product is separated from a liquid by counter-current contact with a solvent in which the desired product is preferentially soluble. { ik'strak·shən ‚käl·əm }

extraction parachute [AERO ENG] An auxiliary parachute designed to release and extract cargo from aircraft in flight and to deploy cargo parachutes. { ik'strak·shən 'par·ə‚shüt }

extraction plant [PETRO ENG] An installation for separating liquid components from casinghead gas or wet gas. { ik'strak·shən ‚plant }

extraction turbine [MECH ENG] A steam turbine equipped with openings through which partly expanded steam is bled at one or more stages. { ik'strak·shən 'tər‚bīn }

extractive distillation [CHEM ENG] A distillation process to separate components from eutectic mixtures; a solution of the mixture is cooled, causing one component to crystallize out and the other to remain in solution; used to separate *p*-xylene and *m*-xylene, using *n*-pentane as the solvent. { ik'strak·tiv ‚dis·tə'lā·shən }

extractive metallurgy [MET] Extraction of metals from ore by various chemical and mechanical methods. { ik'strak·tiv 'med·əl·ər·jē }

extract oil [MATER] The less desirable portion of the oil being solvent-refined; it is dissolved in and selectively removed by the solvent. { 'ek‚strakt ‚òil }

extractor [CHEM ENG] An apparatus for solvent-contact with liquids or solids for removal of specified components. [COMPUT SCI] See mask. [ENG] **1.** A machine for extracting a substance by a solvent or by centrifugal force, squeezing, or other action. **2.** An instrument for removing an object. [ORD] A device in the breech mechanism of a gun, rifle, or the like, for pulling an empty cartridge case or an unfired cartridge out of the chamber of a gun, rifle, or the like. { ik'strak·tər }

extractor groove [ORD] Groove machined in the base of a cartridge case, a short distance above the head, which receives the extractor of the breech mechanism and permits the case to be withdrawn by the extractor; used in automatic weapons, in preference to extractor rims (flanges) formed on the cartridge case base. { ik'strak·tər ‚grüv }

Extradop See extended-range Dovap. { 'ek·strə‚däp }

extrados [ARCH] The upper or outer curve of an arch. { 'ek·strə‚däs }

extraembryonic coelom [EMBRYO] The cavity in the extra-embryonic mesoderm; it is continuous with the embryonic coelom in the region of the umbilicus, and is obliterated by growth of the amnion. { ¦ek·strə‚em·brē'än·ik 'sē·ləm }

extraembryonic membrane See fetal membrane. { ¦ek·strə‚em·brē'än·ik 'mem‚brān }

extragalactic [ASTRON] Beyond the Milky Way. { ¦ek·strə·gə'lak·tik }

extragalactic background light [ASTRON] The contribution to the brightness of the sky from light coming from outside the Galaxy, chiefly distant galaxies, excluding resolved galaxies. { ¦ek·strə·gə¦lak·tik 'bak‚graund ‚līt }

extragalactic radio source [ASTROPHYS] A source of radio emission outside the Milky Way. { ¦ek·strə·gə'lak·tik 'rad·ē·ō ‚sòrs }

extra-high tension [ELECTR] British term for the high direct-current voltage applied to the second anode in a cathode-ray tube, ranging from about 4000 to 50,000 volts in various sizes of tubes. Abbreviated eht. { ¦ek·strə ¦hī 'ten·chən }

extra-high voltage [ELEC] A voltage above 345 kilovolts used for power transmission. Abbreviated ehv. { ¦ek·strə ¦hī 'vōl·tij }

extrajunctional receptor [PHYSIO] An acetylcholine receptor which occurs randomly over a muscle fiber surface outside the area of the neuromuscular junction. { ¦ek·strə¦jənk·shən·əl ri'sep·tər }

extraneous emission [ELECTR] Any emission of a transmitter or transponder, other than the output carrier fundamental, plus only those sidebands intentionally employed for the transmission of intelligence. { ik'strān·ē·əsə'mish·ən }

extraneous light [GRAPHICS] Undesired illumination striking photographic area or copy field. { ik'strān·ē·əs 'līt }

extraneous response [ELECTR] Any undersired response of a receiver, recorder, or other susceptible device, due to the desired signals, undersired signals, or any combination or interaction among them. { ik'strän·ē·əs ri'späns }

extraneous root [MATH] A root that is introduced into an equation in the process of solving another equation, but is not a solution of the equation to be solved. { ik¦strän·ē·əs 'rüt }

extranet [COMPUT SCI] A secure, Internet-based private network that allows organizations to share information with vendors, partners, customers, and so on; access requires either a password or digital encryption. { 'ek·strə,net }

extraordinary component *See* extraordinary wave. { ik'strȯr·dən,er·ē kəm'pō·nənt }

extraordinary index [OPTICS] The index of refraction of the extraordinary wave propagating in a direction perpendicular to the optical axis of a uniaxial crystal. { ik'strȯr·dən,er·ē 'in,deks }

extraordinary ray [OPTICS] One of two rays into which a ray incident on an anisotropic uniaxial crystal is split; its deviation at the crystal's surface depends on the orientation of the crystal, and it is deviated even in the case of normal incidence. { ik'strȯr·dən,er·ē 'rā }

extraordinary wave [GEOPHYS] Magneto-ionic wave component which, when viewed below the ionosphere in the direction of propagation, has clockwise or counterclockwise elliptical polarization respectively, accordingly as the earth's magnetic field has a positive or negative component in the same direction. Also known as X wave. [OPTICS] Component of electromagnetic radiation propagating in an anisotropic uniaxial crystal whose electric displacement vector lies in the plane containing the optical axis and the direction normal to the wavefront; it gives rise to the extraordinary ray. Also known as extraordinary compoment. { ik'strȯr·dən,er·ē 'wāv }

extrapolated boundary [NUCLEO] A hypothetical surface outside a medium sustaining a neutron chain reaction on which the neutron flux would be 0 if it could be extrapolated from the flux a few mean free paths inside the medium. { ik¦strap·ə,lād·əd 'baùn·drē }

extrapolation [MATH] Estimating a function at a point which is larger than (or smaller than) all the points at which the value of the function is known. { ik,strap·ə'lā·shən }

extrapolation distance [NUCLEO] The distance from the boundary of a medium sustaining a neutron chain reaction to the extrapolated boundary. { ik,strap·ə'lā·shən ,dis·təns }

extrapolation ionization chamber [NUCLEO] An ionization chamber so designed that volume, electrode separation, or some other factor can be varied in suitable steps for measurement purposes; the resulting measured values are plotted in appropriate form and the desired result is obtained by extrapolation of the curve. { ik,strap·ə'lā·shən ,ī·ə·nə'zā·shən ,chām·bər }

extrapyramidal system [NEUROSCI] Descending tracts of nerve fibers arising in the cortex and subcortical motor areas of the brain. { ¦ek·strə,pir·ə'mid·əl 'sis·təm }

extrasensory perception [PSYCH] The alleged phenomenon of perception or awareness of external events in the absence of any sensory stimulation arising from the events. Abbreviated ESP. { ¦ek·strə¦sen·sə·rē pər'sep·shən }

extrasolar planet [ASTRON] A planet in orbit about a star other than the sun. Also known as exoplanet. { ¦ek·strə¦sō·lər 'plan·ət }

extrasystole [MED] Premature beat of the heart. { ¦ek·strə'sis·tə·lē }

extraterrestrial intelligence [ASTRON] The potential existence beyond the earth of other advanced civilizations with a technology at least as developed as that on earth. { ¦ek·strə·tə'res·trē·əl in'tel·ə·jəns }

extraterrestrial noise [ELECTROMAG] Cosmic and solar noise; radio disturbances from sources other than those related to the earth. { ¦ek·strə·tə'res·trē·əl 'nȯiz }

extraterrestrial radiation [ASTROPHYS] Electromagnetic radiation which originates outside the earth or its atmosphere, as in the sun or stars. { ¦ek·strə·tə'res·trē·əl ,rād·ē·ā·shən }

extratropical cyclone [METEOROL] Any cyclone-scale storm that is not a tropical cyclone. Also known as extratropical low; extratropical storm. { ¦ek·strə¦träp·i·kəl 'sī,klōn }

extratropical low *See* extratropical cyclone. { ¦ek·strə¦träp·i·kəl 'lō }

extratropical storm *See* extratropical cyclone. { ¦ek·strə¦träp·i·kəl 'stȯrm }

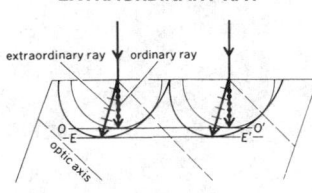

EXTRAORDINARY RAY

Huygens construction for a plane wave incident normally on transparent calcite showing the ordinary and extraordinary ray. *(From F. A. Jenkins and H. E. White, Fundamentals of Optics, 4th ed., McGraw-Hill, 1976)*

extrauterine pregnancy [MED] Gestation outside the uterus. { ¦ek·trə'yüd·ə,rēn 'preg·nən·sē }

extravasation [GEOL] The eruption of lava from a vent in the earth. [MED] The pouring out or eruption of a body fluid from its proper channel or vessel into the surrounding tissue. { ik,strav·ə'sā·shən }

extravehicular activity [AERO ENG] Activity conducted outside a spacecraft during space flight. { ¦ek·strə·və'hik·yə·lər ak'tiv·əd·ē }

extremals [MATH] For a variational problem in the calculus of variaitons entailing use of the Euler-Lagrange equation, the extremals are the solutions of this equation. { ek'strem·əlz }

extreme [CLIMATOL] The highest, and in some cases the lowest, value of a climatic element observed during a given period or during a given month or season of that period; if this is the whole period for which observations are available, it is the absolute extreme. [MATH] *See* extremum. { ek'strēm }

extreme and mean ratio *See* golden section. { ek'strēm ən 'mēn ,rā·shō }

extreme close-up [GRAPHICS] A picture taken at extremly close range; for example, an extreme close-up of a face might show only an eye. { ek¦strēm 'klō,səp }

extreme line casing *See* external upset casing. { ek¦strēm ¦līn 'kās·iŋ }

extreme line tubing [PETRO ENG] Special oil- or gas-well tubing designed for extreme conditions requiring greater than usual strength and leak resistance. { ek¦strēm ¦līn 'tüb·iŋ }

extremely high frequency [COMMUN] The frequency band from 30,000 to 300,000 megahertz in the radio spectrum. Abbreviated EHF. { ek¦strēm·lē 'hī 'frē·kwən·sē }

extremely low frequency [COMMUN] A frequency below 300 hertz in the radio spectrum. Appreviated ELF. { ek¦strēm·lē 'lō 'frē·kwən·sē }

extreme narrowing approximation [SPECT] A mathematical approximation in the theory of spectral-line shapes to the effect that the exchange narrowing of a perturbation is complete. { ek¦strēm 'nar·ə·wiŋ ə,präk·sə'mā·shən }

extreme point [MATH] **1.** A maximum or minimum value of a function. **2.** A point in a convex subset *K* of a vector space is called extreme if it does not lie on the interior of any line segment contained in *K*. { ek'strēm 'pȯint }

extreme-pressure lubricant *See* EP lubricant. { ek¦strēm ¦presh·ər 'lü·brə·kənt }

extreme range [ORD] Greatest range of a weapon; for example, the greatest distance a gun will shoot. { ek¦strēm 'rānj }

extreme relativistic limit [PHYS] Limit which a formula describing a particle's behavior approaches when the speed of the particle approaches the speed of light. { ek¦strēm ,rel·ə·tə'vis·tik 'lim·ət }

extreme spread [ORD] In a firing accuracy test, the distance between the two shots farthest from each other. { ek¦strēm 'spred }

extreme terms [MATH] The first and last terms in a proportion. { ek¦strēm 'tərmz }

extreme ultraviolet astronomy [ASTRON] Astronomical observations carried out in the region of the electromagnetic spectrum with wavelengths from approximately 10 nanometers to the ionization edge of hydrogen at 91.2 nanometers. { ik¦strēm ,əl·trə¦vī·lət ə'strän·ə·mē }

extreme ultraviolet radiation *See* vacuum ultraviolet radiation. { ek¦strēm ,əl·trə¦vī·lət ,rād·ē·ā·shən }

extreme value problem [MATH] A set of mathematical conditions which may be met by values that are less than or greater than an upper or a lower bound, that is, an extreme value. { ek¦strēm 'val·yü präb·ləm }

extremophiles [PHYSIO] Microorganisms belonging to the domains Bacteria and Archaea that can live and thrive in environments with extreme conditions such as high or low temperatures and pH levels, high salt concentrations, and high pressure. { ek'trem·ə,fīlz }

extremum [MATH] A maximum or minimum value of a function. Also known as extreme. { ek'strēm·əm }

extrinsic detector [ENG] A semiconductor detector of electromagnetic radiation that is doped with an electrical impurity and utilizes transitions of charge carriers from impurity states in the band gap to nearby energy bands. { ek¦strinz·ik di'tek·tər }

extrinsic factor *See* vitamin B₁₂. { ek¦strinz·ik ¦fak·tər }

extrinsic photoconductivity [ELECTR] Photoconductivity

that occurs for photon energies smaller than the band gap and corresponds to optical excitation from an occupied imperfection level to the conduction band, or to an unoccupied imperfection level from the valence band, of a material. { ek¦strinz·ik ¸fō·dō·kän·dək¦tiv·əd·ē }

extrinsic photoemission [ELECTR] Photoemission by an alkali halide crystal in which electrons are ejected directly from negative ion vacancies, forming color centers. Also known as direct ionization. { ek¦strin·sik ¸fōd·ō·i¦mish·ən }

extrinsic properties [ELECTR] The properties of a semiconductor as modified by impurities or imperfections within the crystal. { ek¦strinz·ik 'präp·ərd·ēz }

extrinsic protein See peripheral membrane protein. { ¦ek ¦strin·sik 'prō¸tēn }

extrinsic semiconductor [ELECTR] A semiconductor whose electrical properties are dependent on impurities added to the semiconductor crystal, in contrast to an intrinsic semiconductor, whose properties are characteristic of an ideal pure crystal. { ek¦strinz·ik 'sem·i·kən¸dək·tər }

extrinsic sol [PHYS CHEM] A colloid whose stability is attributed to electric charge on the surface of the colloidal particles. { ek¦strinz·ik 'säl }

extrinsic variable star [ASTRON] A variable star, such as an eclipsing variable, whose variation in apparent brightness is due to some external cause, rather than to actual variaiton in the amount of radiation emitted. { ek¦strinz·ik ¸ver·ē·ə·bəl 'stär }

extrophy [MED] Malformation of an organ. { 'ek·strə·fē }

extrorse [BIOL] Directed outward or away from the axis of growth. { ek'strórs }

extroversion [BIOL] A turning outward. [PSYCH] The turning to things and persons outside oneself rather than to one's own thoughts and feelings. { ¦ek·strə¦vər·zhən }

extrudate [ENG] Ductile metal, plastic, or other semisoft solid material that has been shaped into a continuous form (such as fiber, film, pipe, or wire coating) by forcing the semisolid material through a die opening of appropriate shape. { 'ek·strə¸dāt }

extruder [ENG] A device that forces ductile or semisoft solids through die openings of appropriate shape to produce a continuous film, strip, or tubing { ed'strüd·ər }

extrusion [ENG] A process in which a hot or cold semisoft solid material, such as metal or plastic, is forced through the orifice of a die to produce a continuously formed piece in the shape of the desired product. [GEOL] Emission of magma or magmatic materials at the surface of the earth. [TEXT] A process for making continuous-filament synthetic fibers by forcing a syruplike liquid through minute holes of a spinneret. { ek'strü·zhən }

extrusion billet [MET] A slug of heated metal that is forced through a die by a hydraulic ram in direct extrusion operations. { ek'strü·zhən ¸bil·ət }

extrusion coating [ENG] A process of placing resin on a substrate by extruding a thin film of molten resin and pressing it onto or into the substrates, or both, without the use of adhesives. { ek'strü·zhən ¸kōd·iŋ }

extrusion cooking [FOOD ENG] The process by which moistened, expansile materials are plasticized in a tube by combination of moisture, heat, pressure, and mechanical shear. { ek'strü·zhən ¸kúk·iŋ }

extrusion defect [MET] Impaired flow of an extrusion product due to surface oxidation of the ingot or billet. { ek'strü·zhən di¸fekt }

extrusion ingot [MET] A cylindrical casting used to form extruded products. { ek'strü·zhən ¸iŋ·gət }

extrusion metal [MET] Any of numerous nonferrous metals, alloys, and other materials used in extrusion operations. { ek'strü·zhən ¸med·əl }

extrusion pressing See cold extrusion. { ek'strü·zhən ¸pres·iŋ }

extrusive rock See volcanic rock. { ik'strü·siv 'räk }

exudate [MED] 1. A proteinaceous material that passes through blood vessel walls into the surrounding tissue in inflammation or a superficial lesion. 2. Any substance that is exuded. { 'ek·syü¸dāt }

exudation See sweating. { ¸ek·syə¦dā·shən }

exudation vein See segregated vein. { ¸ek·syə¦dā·shən¸vān }

exumbrella [INV ZOO] The outer, convex surface of the umbrella of jellyfishes. { ¸ek·səm'brel·ə }

eye [FOOD ENG] A hole formed in certain cheeses during ripening, such as in swiss cheese. [ZOO] A photoreceptive sense organ that is capable of forming an image in vertebrates and in some invertebrates such as the squids and crayfishes. { ī }

eye assay [MIN ENG] An estimate of the valuable mineral content of a core or ore sample as based on visual inspection. Also known as eyeball assay. { 'ī ¸as¸ā }

eyeball [ANAT] The globe of the eye. { 'ī¸ból }

eyeball assay See eye assay. { 'ī¸ból 'as¸ā }

eyeball potential [PHYSIO] Very small electrical potentials at the eyeball surface resulting from depolarization of muscles controlling eye position. { 'ī¸ból pə¸ten·chəl }

eyebar [DES ENG] A metal bar having a hole or eye through each enlarged end. { 'ī¸bär }

eyebolt [DES ENG] A bolt with a loop at one end. { 'ī¸bólt }

eye coal [GEOL] Coal characterized by small, circular or elliptic structural disks that reflect light and are arranged in parallel planes either in or normal to the bedding. Also known as augen kohle; circular coal. { 'ī ¸kōl }

eye-ear plane [ARCH] In craniometric study, a position for placing a human skull so that the lower margins of the orbits and the upper margin of the auditory meatus are on the same horizontal plane. Also known as Frankfurt horizontal. { 'ī ¦ēr ¸plān }

eyeglasses [OPTICS] Optical devices containing corrective lenses for defects of vision or for special purposes. { 'ī¸glas·əs }

eye lens [OPTICS] The lens in a two-lens eyepiece which is nearer to the eye. { 'ī ¸lenz }

eyelet [DES ENG] A small ring or barrel-shaped piece of metal inserted into a hole for reinforcement. { 'ī·lət }

eyeleting [ENG] Forming a lip around the rim of a hole. { 'ī·ləd·iŋ }

eyelid [ANAT] A movable, protective section of skin that covers and uncovers the eyeball of many terrestrial animals. { 'ī¸lid }

eyelights [GRAPHICS] Low-intensity light sources used to add sparkle to the eyes or teeth and reduce shadows on the face; usually placed at eye level. { 'ī¸līts }

eye of the storm [METEOROL] The center of a tropical cyclone, marked by relatively light winds, confused seas, rising temperature, lowered relative humidity, and often by clear skies. { 'ī əv thə 'stórm }

eye of the wind [METEOROL] The point or direction from which the wind is blowing. { 'ī əv thə 'wind }

eyepiece [OPTICS] A lens or optical system which offers to the eye the image originating from another system (the objective) at a suitable viewing distance. Also known as ocular. { 'ī¸pēs }

eyepoint [OPTICS] That point on the axis of a lens at which the brightest and sharpest visual image is obtained. { 'ī¸póint }

eye scanning [IND ENG] Scanning of the visual field by moving the eyeballs without rotation of the head. { 'ī ¸skan·iŋ }

eye screw [DES ENG] A screw with an open loop head. { 'ī ¸skrü }

eye socket See orbit. { 'ī ¸säk·ət }

eyespot [BOT] 1. A small photosensitive pigment body in certain unicellular algae. 2. A dark area around the hilum of certain seeds, as some beans. [INV ZOO] A simple organ of vision in many invertebrates consisting of pigmented cells overlying a sensory termination. [PL PATH] A fungus disease of sugarcane and certain other grasses which is caused by *Helminthosporium sacchari* and characterized by yellowish oval lesions on the stems and leaves. { 'ī¸spät }

eyestalk [INV ZOO] A movable peduncle bearing a terminal eye in decapod crustaceans. { 'ī¸stók }

eye wall [METEOROL] A zone at the periphery of the eye of the storm where winds reach their highest speed. { 'ī ¸wól }

Eykman formula [OPTICS] An empirical formula which relates the molal refraction of a liquid at a given optical frequency to its index of refraction, density, and molecular weight. { 'īk·mən ¸fór·myə·lə }

Eyring equation [PHYS CHEM] An equation, based on statistical mechanics, which gives the specific reaction rate for a chemical reaction in terms of the heat of activation, entropy

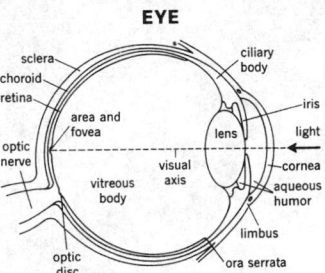

EYE

Cross section of human eye.

of activation, the temperature, and various constants. { 'ı̄·riŋ i,kwä·zhən }

Eyring formula [FL MECH] A formula, based on the Eyring theory of rate processes, which relates shear stress acting on a liquid and the resulting rate of shear. { 'ı̄·riŋ ,fȯr·myə·lə }

Eyring molecular system [FL MECH] Theory to account for liquid properties; assumes that each liquid molecule can move freely within a certain free volume. Also known as Eyring theory. { 'ı̄·riŋ mə'lek·yə·lər ,sis·təm }

Eyring theory See Eyring molecular system. { 'ı̄·riŋ ,thē·ə·rē }

e-zine [COMPUT SCI] A Web-published magazine. { 'ē,zēn }

E zone [COMMUN] One of the three zones into which the earth is divided to show the variations of the F_2 layer in respect to longitude when one is making frequency predictions; it roughly covers Asia, Australia, the Philippines, and Japan. { 'ē ,zōn }

F

f [PHYS] Notation representing the Coriolis parameter.

F *See* farad; fluorine.

F₁ *See* first filial generation.

F₂ [GEN] Notation for the progeny produced by intercrossing members of the first filial generation. Also known as second generation.

fA *See* femtoampere.

Fabales [BOT] An order of dicotyledonous plants whose members typically have stipulate, compound leaves, ten to many stamens which are often united by the filaments, and a single carpel which gives rise to a legume; many harbor symbiotic nitrogen-fixing bacteria in the roots. { fə'bā·lēz }

Faber flaw [SOLID STATE] A deformation in a superconducting material that acts as a nucleation center for the growth of a superconducting region. { 'fā·bər ‚flȯ }

Faber-Jackson relation [ASTRON] A relation between the spectral dispersion caused by the random motions of stars in an elliptical galaxy and the galaxy's intrinsic luminosity. { ¦fāb·ər 'jak·sən ri‚lā·shən }

Fabian system [MIN ENG] The free-fall drilling system from which all other free-fall systems have originated. { 'fā·bē·ən ‚sis·təm }

Fab region [IMMUNOL] Region of an antibody molecule that contains the antigen binding site; Fab is derived from the term antigen binding fragment. { 'fab ‚rē·jən }

fabric [ARCH] The framework of a building. [GEOL] The spatial orientation of the elements of a sedimentary rock. [PETR] The sum of all the structural and textural features of a rock. Also known as petrofabric; rock fabric; structural fabric. [SCI TECH] **1.** Arrangement or pattern of constituent parts. **2.** Materials used in fabrication. [TEXT] A thin, flexible material made of any combination of cloth, fiber, polymeric film, sheet, or foam. { 'fab·rik }

fabricable [MATER] Capable of being shaped, such as an alloy. { 'fab·rik·ə·bəl }

fabric analysis *See* structural petrology. { 'fab·rik ə‚nal·əs·əs }

fabricated food *See* food analog. { ¦fab·rə‚kād·əd 'füd }

fabrication [ENG] **1.** The manufacture of parts, usually structural or electromechanical parts. **2.** The assembly of parts into a structure. { ‚fab·ri'kā·shən }

fabric diagram [PETR] In structural petrology, a graphic representation of the data of fabric elements. Also known as petrofabric diagram. { 'fab·rik 'dī·ə‚gram }

fabric domain [PETR] A three-dimensional area or volume of uniform rock fabric delineated by boundaries such as structural or compositional discontinuities. { 'fab·rik də'mān }

fabric element [PETR] A surface or line of structural discontinuity in a rock fabric. { 'fab·rik 'el·ə·mənt }

Fabriciinae [INV ZOO] A subfamily of small to minute, colonial, sedentary polychaete annelids in the family Sabellidae. { ‚fa·brə'sī·ə‚nē }

fabric laminate [MATER] Layers of fabric alternating with plastic, used as insulation in electrical equipment. { 'fab·rik 'lam·ə‚nāt }

fabric-type dust collector [MIN ENG] A collector which removes dust particles from ore by means of a filter made of fabric. { 'fab·rik ¦tīp 'dəst kə‚lek·tər }

fabric weight [TEXT] The number of ounces per square yard of a fabric. { 'fab·rik ‚wāt }

Fabry-Barot method [OPTICS] Method of determining the index of refraction of a prism in which the prism is set up so that the incident beam is perpendicular to the emergent face, and the index of refraction is calculated from the angle of the prism and the angle of deviation. { fä'brē bə'rō ‚meth·əd }

Fabry-Pérot etalon *See* etalon. { fä'brē pə'rō 'ed·əl‚än }

Fabry-Pérot filter [OPTICS] An optical interference filter, similar to the Fabry-Pérot interferometer except that the space between the partially reflecting surfaces is only a few thousand angstroms. { fä'brē pə'rō 'fil·tər }

Fabry-Pérot fringes [OPTICS] Series of rings observed when a monochromatic light source is viewed through a Fabry-Pérot interferometer. { fä'brē pə'rō 'frin·jəz }

Fabry-Pérot interferometer [OPTICS] An interferometer having two parallel glass plates (whose separation of a few centimeters may be varied), silvered on their inner surfaces so that the incoming wave is multiply reflected between them and ultimately transmitted. { fä'brē pə'rō ‚int·ə·fə'räm·əd·ər }

facade [ARCH] The front of a building or a face of a building, given special architectural treatment. { fə'säd }

face [ANAT] The anterior portion of the head, including the forehead and jaws. [CIV ENG] **1.** The surface of the area that has been excavated in constructing a tunnel. **2.** In building construction, the exposed surface of a wall, masonry unit, or sheet of material. **3.** To install a surface layer of one material over another, such as laying brick on a wall built of concrete blocks. [CRYSTAL] *See* crystal face. [DES ENG] The surface of a flange on a pipe that is fitted against another flange. [ELECTR] *See* faceplate. [GEOL] **1.** The main surface of a landform. **2.** The original surface of a layer of rock. [GRAPHICS] **1.** A particular style or size of letter as distinguished from another style or size. Also known as typeface. **2.** The printing surface of a printing plate or the front surface of a piece of paper. [MATER] The veneer on the exposed surface of a sheet of plywood. [MATH] **1.** One of the plane polygons bounding a polyhedron. **2.** A face of a simplex is the subset obtained by setting one or more of the coordinates a_i, defining the simplex, equal to 0; for example, the faces of a triangle are its sides and vertices. **3.** The face of a half space is the plane that bounds it. **4.** One of the regions bounded by edges of a planar graph. [MIN ENG] A surface on which mining operations are being performed. Also known as breast. [TEXT] The side of a fabric which is more attractive than the other side because of features such as weave, luster, or finish. { fās }

face angle [MATH] An angle between two successive edges of a polyhedral angle. { 'fās ‚aŋ·gəl }

face area [MIN ENG] The working area toward the interior of the last open crosscut in an entry or room. { 'fās ‚er·ē·ə }

face belt conveyor [MIN ENG] A lightweight belt conveyor used at the working face in a mine. { ¦fās 'belt kən‚vā·ər }

face-bonding [ELECTR] Method of assembling hybrid microcircuits wherein semiconductor chips are provided with small mounting pads, turned facedown, and bonded directly to the ends of the thin-film conductors on the passive substrate. { 'fās ‚bänd·iŋ }

face boss [MIN ENG] A foreman in charge of operations at the working face in a bituminous coal mine. { 'fās ‚bȯs }

face brick [MATER] A brick of some esthetic quality to be used on the exposed surface of a building wall or other structure. { 'fās ‚brik }

face-bridging ligand [ORG CHEM] A ligand that forms a bridge over one triangular face of the polyhedron of a metal cluster structure. { 'fās ‚brij·iŋ 'lig·ənd }

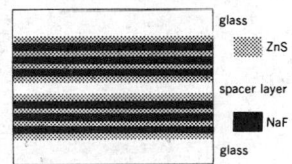

FABRY-PÉROT FILTER

glass
ZnS
spacer layer
NaF
glass

Schematic diagram of seven-layer solid Fabry-Pérot filter. *(From D. E. Gray, ed., American Institute of Physics Handbook, McGraw-Hill, 1957)*

face-centered cubic lattice [CRYSTAL] A lattice whose unit cells are cubes, with lattice points at the center of each face of the cube, as well as at the vertices. Abbreviated fcc lattice. { 'fās ˌsen·tərd ¦kyüb·ik 'lad·əs }

face-centered orthorhombic lattice [CRYSTAL] An orthorhombic lattice which has lattice points at the center of each face of a unit cell, as well as at the vertices. { 'fās ˌsen·tərd ¦ȯr·thō¦räm·bik 'lad·əs }

face conveyor [MIN ENG] Any type of mine conveyor used at and parallel to a working face. { 'fās kən¸vā·ər }

face-discharge bit [MECH ENG] A liquid-coolant bit designed for drilling in soft formations and for use on a double-tube core barrel, the inner tube of which fits snugly into a recess cut into the inside wall of the bit directly above the inside reaming stones; the coolant flows through the bit and is ejected at the cutting face. Also known as bottom-discharge bit; face-ejection bit. { ¦fās 'dis¸chärj ¸bit }

faced wall [BUILD] A wall whose masonry facing and backing are of different materials. { ¦fāst 'wȯl }

face-ejection bit *See* face-discharge bit. { ¦fās ē'jek·shən ¸bit }

face feed [MET] In brazing or soldering, the deposition of filler metal to the joint, usually by hand. { 'fās ¸fēd }

face gear [DES ENG] A gear having teeth cut on the face. { 'fās ¸gir }

face haulage *See* primary haulage. { 'fās ¸hȯl·ij }

face height [MIN ENG] The vertical distance between the top and toe of a quarry or opencast face. { 'fās ¸hīt }

facellite *See* kaliophilite. { fə'se¸līt }

faceman [MIN ENG] A coal miner who performs the duties involved in drilling underground openings into which explosives are charged and set off, to extract coal, slate, and rock. Also known as coal digger; coal getter. { 'fās·mən }

face mechanization [MIN ENG] The use of a cutter-loader on a longwall face. { ¦fās ¸mek·ə·nə'zā·shən }

face milling [MECH ENG] Milling flat surfaces perpendicular to the rotational axis of the cutting tool. { 'fās ¸mil·iŋ }

face mold [ENG] A pattern for cutting forms out of sheets of wood, metal, or other material. { 'fās ¸mōld }

face nailing [ENG] Nailing of facing wood to a base, leaving the nailheads exposed. { 'fās ¸nāl·iŋ }

faceplate [ELECTR] The transparent or semitransparent glass front of a cathode-ray tube, through which the image is viewed or projected; the inner surface of the face is coated with fluorescent chemicals that emit light when hit by an electron beam. Also known as face. [ENG] **1.** A disk fixed perpendicularly to the spindle of a lathe and used for attachment of the workpiece. **2.** A protective plate used to cover holes in machines or other devices. **3.** In scuba or skin diving, a glass or plastic window positioned over the face to provide an air space between the diver's eyes and the water. { 'fās¸plāt }

face sampling [MIN ENG] Taking random samples of ore and rock from exposed faces of ore and waste. { 'fās ¸sam·pliŋ }

face shield [ENG] A detachable wraparound guard fitted to a worker's helmet to protect the face from flying particles. { 'fās ¸shēld }

face signal [MIN ENG] A wire stretched along the face and connected to a panel near the main gate to control the running of a face conveyor. { 'fās ¸sig·nəl }

facet [ANAT] A small plane surface, especially on a bone or a hard body; may be produced by wear, as a worn spot on the surface of a tooth. [GEOGR] Any part of an intersecting surface that constitutes a unit of geographic study, for example, a flat or a slope. [INV ZOO] The surface of a simple eye in the compound eye of arthropods and certain other invertebrates. [MATER] The plane surface of a crystal, a cut precious stone, or other fractured surface. [MATH] A proper face of a convex polytope that is not contained in any larger face. { 'fas·ət }

faceted pebble [GEOL] A pebble with three or more faces naturally worn flat and meeting at sharp angles. { 'fas·əd·əd 'peb·əl }

faceted spur [GEOL] A spur or ridge with an inverted-V face resulting from faulting or from the trimming, beveling, or truncating motion of streams, waves, or glaciers. { 'fas·əd·əd 'spər }

face tile [MATER] Tile with one finished surface, intended for use on a face. { 'fās ¸tīl }

face timbering [MIN ENG] Positioning of safety posts at the

working portion of a coal face to support the roof of the mine. { 'fās 'tim·bə·riŋ }

face veneer [MATER] Wood veneer selected for its decorative qualities rather than its strength. { 'fās və¸nir }

facework [CIV ENG] Ornamental or otherwise special material on the front side or outside of a wall. { 'fās¸wərk }

face worker [MIN ENG] A miner who works regularly at the face. { 'fās ¸wərk·ər }

facial angle [ARCH] The angle formed by the union of a line connecting nasion and gnathion with the Frankfort horizontal plane of the head. { ¦fā·shəl 'aŋ·gəl }

facial artery [ANAT] The external branch of the external carotid artery. { ¦fā·shəl 'ärd·ə·rē }

facial bone [ANAT] The bone comprising the nose and jaws, formed by the maxilla, zygoma, nasal, lacrimal, palatine, inferior nasal concha, vomer, mandible, and parts of the ethmoid and sphenoid. { 'fā·shəl ¸bōn }

facial index [ARCH] The ratio of the breadth of the face to its length multiplied by 100. { 'fā·shəl 'in¸deks }

facial nerve [NEUROSCI] The seventh cranial nerve in vertebrates; a paired composite nerve, with motor elements supplying muscles of facial expression and with sensory fibers from the taste buds of the anterior two-thirds of the tongue and from other sensory endings in the anterior part of the throat. { 'fā·shəl ¸nərv }

facies [ANAT] Characteristic appearance of the face in association with a disease or abnormality. [ECOL] The makeup or appearance of a community or species population. [GEOL] Any observable attribute or attributes of a rock or stratigraphic unit, such as overall appearance or composition, of one part of the rock or unit as contrasted with other parts of the same rock or unit. { 'fā·shēz }

facies map [GEOL] A stratigraphic map indicating distribution of sedimentary facies within a specific geologic unit. { 'fā· shēz ¸map }

facilitated glucose transport [BIOCHEM] The movement of glucose across cell membranes that is driven by the glucose concentration gradient but assisted (facilitated) by carrier proteins. { fə'sil·ə¸tād·əd 'glü·kōs ¸tranz¸pȯrt }

facilitated transport [PHYSIO] The transport of certain materials across a cell membrane, down a concentration gradient, assisted by enzymelike carrier proteins embedded in the membrane and without the explicit provision of energy. { fə'sil·ə¸tād·əd 'trans¸pȯrt }

facility assignment [COMPUT SCI] The allocation of core memory and external devices by the executive as required by the program being executed. { fə'sil·əd·ē ə¸sīn·mənt }

facility dispersion [COMMUN] The distribution of circuits between two points over more than one physical or geographic route to reduce the likelihood of a trunk group being put completely out of service by facility damage or other circuit failure. { fə'sil·əd·ē di'spər·zhən }

facility security clearance [ORD] An administrative determination that, from a security viewpoint, a facility is eligible for access to classified information of a certain category (and all lower categories). { fə'sil·əd·ē sə'kyůr·əd·ē ¸klir·əns }

facing [CIV ENG] A covering or casting of some material applied to the outer face of embankments, buildings, and other structures. [MECH ENG] Machining the end of a flat rotating surface by applying a tool perpendicular to the axis of rotation in a spiral planar path. [MET] A fine molding sand applied to the face of a mold. { 'fās·iŋ }

facing-point lock [CIV ENG] A lock used on a railroad track, such as a switch track, which contains a plunger that engages a rod on the switch point to lock the device. { 'fās·iŋ ¸pȯint ¸läk }

facing wall [CIV ENG] Concrete lining against the earth face of an excavation; used instead of timber sheeting. { 'fās·iŋ ¸wȯl }

facsimile [COMMUN] **1.** A system of communication in which a transmitter scans a photograph, map, or other fixed graphic material and converts the information into signal waves for transmission by wire or radio to a facsimile receiver at a remote point. Also known as fax; phototelegraphy; radiophoto; telephoto; telephotography; wirephoto. **2.** A photograph transmitted by radio to a facsimile receiver. Also known as radiophoto. [GRAPHICS] An exact copy of a book, document, painting, or other material. { fak'sim·ə·lē }

facsimile chart [METEOROL] Any graphic form of weather

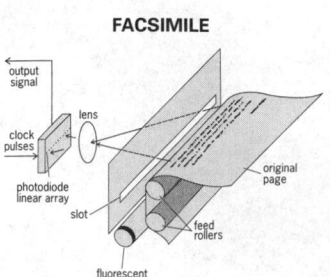

FACSIMILE

output signal
lens
clock pulses
photodiode linear array
slot
original page
feed rollers
fluorescent lamp

A photodiode facsimile scanner forms an image, similarly to a camera.

information, usually a type of synoptic chart, which has been reproduced by facsimile equipment.   Also known as fax chart; fax map.   { fak'sim·ə·lē ,chärt }

facsimile modulation [COMMUN] Process in which the amplitude, frequency, or phase of a transmitted wave is varied with time in accordance with a facsimile transmission signal.   { fak'sim·ə·lē ,mäj·ə·'lā·shən }

facsimile posting [COMPUT SCI] The process of transferring by a duplicating process a printed line of information from a report, such as a listing of transactions prepared on an accounting machine, to a ledger or other recorded sheet.   { fak'sim·ə·lē 'pōst·iŋ }

facsimile receiver [ELECTR] The receiver used to translate the facsimile signal from a wire or radio communication channel into a facsimile record of the subject copy.   { fak'sim·ə·lē ri'sē·vər }

facsimile recorder [ELECTR] The section of a facsimile receiver that performs the final conversion of electric signals to an image of the subject copy on the record medium.   { fak'-sim·ə·lē ri'kórd·ər }

facsimile signal [COMMUN] The picture signal produced by scanning the subject copy in a facsimile transmitter.   { fak'sim·ə·lē ,sig·nəl }

facsimile signal level [ELECTR] Maximum facsimile signal power or voltage (root mean square or direct current) measured at any point in a facsimile system.   { fak'sim·ə·lē 'sig·nəl ,lev·əl }

facsimile synchronizing [ELECTR] Maintenance of predetermined speed relations between the scanning spot and the recording spot within each scanning line.   { fak'sim·ə·lē 'siŋ·krə,niz·iŋ }

facsimile telegraph [COMMUN] A telegraph system designed to transmit pictures.   { fak'sim·ə·lē 'tel·ə,graf }

facsimile transmitter [ELECTR] The apparatus used to translate the subject copy into facsimile signals suitable for delivery over a communication system.   { fak'sim·ə·lē tranz'mid·ər }

factor [MATH] **1.** For an integer n, any integer which gives n when multiplied by another integer.   **2.** For a polynomial p, any polynomial which gives p when multiplied by another polynomial.   **3.** For a graph G, a spanning subgraph of G with at least one edge. [STAT] A quantity or a variable being studied in an experiment as a possible cause of variation.   { 'fak·tər }

factor I See fibrinogen.   { 'fak·tər 'wən }

factor II See prothrombin.   { 'fak·tər 'tü }

factor III See thromboplastin.   { 'fak·tər 'thrē }

factor IV [BIOCHEM] Calcium ions involved in the mechanism of blood coagulation.   { 'fak·tər 'fór }

factor V See proaccelerin.   { 'fak·tər 'fīv }

factor VII [BIOCHEM] A procoagulant, related to prothrombin, that is involved in the formation of a prothrombin-converting principle which transforms prothrombin to thrombin.   Also known as stable factor.   { 'fak·tər 'se·vən }

factor VIII See antihemophilic factor.   { 'fak·tər 'āt }

factor IX See Christmas factor.   { 'fak·tər 'nīn }

factor IX deficiency See Christmas disease.

factor X See Stuart factor.   { 'fak·tər 'ten }

factor XI [BIOCHEM] A procoagulant present in normal blood but deficient in hemophiliacs.   Also known as plasma thromboplastin antecedent (PTA).   { 'fak·tər ə'le·vən }

factor XII [BIOCHEM] A blood clotting factor effective experimentally only in vitro; deficient in hemophiliacs.   Also known as Hageman factor.   { 'fak·tər 'twelv }

factorable integer [MATH] An integer that has factors other than unity and itself.   { 'fak·trə·bəl 'int·ə·jər }

factorable polynomial [MATH] A polynomial which has polynomial factors other than itself.   { 'fak·tə·rə·bəl ,päl·ə'nō·mē· əl }

factor analysis [MATH] Given sets of variables which are related linearly, factor analysis studies techniques of approximating each set relative to the others; usually the variables denote numbers.   { 'fak·tər ə,nal·ə·səs }

factor comparison [IND ENG] A quantitative system of job evaluation in which jobs are given relative positions on a rating scale based on a comparison of factors composing the job with certain previously selected key jobs.   { 'fak·tər kəm,par·ə·sən }

factor group See quotient group.   { 'fak·tər ,grüp }

factorial [MATH] The product of all positive integers less than or equal to n; written $n!$; by convention $0! = 1$.   { fak'tór·ē·əl }

factorial design [STAT] A design for an experiment that allows the experimenter to find out the effect levels of each factor on levels of all the other factors.   { fak'tór·ē·əl di,zīn }

factorial moment [STAT] The nth factorial moment of a random variable X is the expected value of $X (X - 1) (X - 2) \cdots (X - n + 1)$.   { fak,tór·ē·əl 'mō·mənt }

factorial ring See unique-factorization domain.   { fak'tór·ē·əl ,riŋ }

factorial series [MATH] The series $1 + (1/1!) + (1/2!) + (1/3!) + \cdots$, whose $(n + 1)$st term is $1/n!$ for n = 1, 2, . . . ; its sum is the number e.   { fak,tór·ē·əl 'sir,ēz }

factoring [MATH] Finding the factors of an integer or polynomial.   { 'fak·tə·riŋ }

factoring of the secular equation [MATH] Factoring the polynomial that results from expanding the secular determinant of a matrix, in order to find the roots of this polynomial, which are the eigenvalues of the matrix.   { 'fak·tə·riŋ əv thə ,sek·yə· lər i'kwā·zhən }

factor model [STAT] Any one of the probability models which goes into the construction of a product model.   { 'fak·tər ,mäd·əl }

factor module [MATH] The factor module of a module M over a ring R by a submodule N is the quotient group M/N, where the product of a coset $x + N$ by an element a in R is defined to be the coset $ax + N$.   { 'fak·tər ,mä·jül }

factor of proportionality [MATH] Two quantities A and B are related by a factor of proportionality μ if either $A = \mu B$ or $B = \mu A$.   { 'fak·tər əv prə,pórsh·ən'al·əd·ē }

factor of safety [MECH] **1.** The ratio between the breaking load on a member, appliance, or hoisting rope and the safe permissible load on it.   Also known as safety factor.   **2.** See factor of stress intensity.   { 'fak·tər əv 'sáf·tē }

factor of stress concentration [MECH] Any irregularity producing localized stress in a structural member subject to load.   Also known as fatigue-strength reduction factor.   { 'fak·tər əv 'stres ,käns·ən,trā·shən }

factor of stress intensity [MECH] The ratio of the maximum stress to which a structural member can be subjected, to the maximum stress to which it is likely to be subjected.   Also known as factor of safety.   { 'fak·tər əv 'stres in,ten·səd·ē }

factor of subdivision [NAV ARCH] An arbitrary factor used in computing allowable floodable length of ships after damage, set up by regulations and international convention.   { 'fak·tər əv 'səb·də,vizh·ən }

factor-reversal test [STAT] A test for index numbers in which an index number of quantity, obtained if symbols for price and quantity are interchanged in an index number of price, is multiplied by the original price index to give an index of changes in total value.   { 'fak·tər ri'vər·səl ,test }

factor ring See quotient ring.   { 'fak·tər ,riŋ }

factor space See quotient space.   { 'fak·tər ,spās }

factor theorem of algebra [MATH] A polynomial $f(x)$ has $(x - a)$ as a factor if and only if $f(a) = 0$.   { 'fak·tər ,thir· əm əv 'al·jə·brə }

factory [IND ENG] A building or group of buildings where goods are manufactured.   { 'fak·trē }

factory-data collection [COMPUT SCI] The continuous input of data achieved in a working area by having the worker insert a precoded card into a device connected to a computer.   { 'fak·trē ,dad·ə kə,lek·shən }

factory farming [AGR] Raising livestock indoors under conditions of extremely restricted mobility.   { 'fak·trē ,fär·miŋ }

factory lumber [MATER] Softwood lumber graded and used in the factory for the manufacture of such items as doors, sashes, moldings, and so on.   { 'fak·trē ,ləm·bər }

factory ship [NAV ARCH] A ship equipped both to catch and to process fish into products such as frozen filet, frozen whole fish, and fish meal.   { 'fak·trē ,ship }

facula [ASTRON] Any of the large patches of bright material forming a veined network in the vicinity of sunspots; faculae appear to be more permanent than sunspots and are probably due to elevated clouds of luminous gas.   { 'fak·yə·lə }

facultative aerobe [MICROBIO] An anaerobic microorganism which can grow under aerobic conditions.   { 'fa·kəl,tād·iv 'er,ōb }

facultative anaerobe [MICROBIO] A microorganism that

grows equally well under aerobic and anaerobic conditions. { 'fak·əl,tād·iv 'an·ə,rōb }

facultative heterochromatin [GEN] Chromosomal material that may alternate in form and function between euchromatin and heterochromatin. { 'fak·əl,tād·iv ¦hed·ə,rō'krō·mə·tən }

facultative parasite [ECOL] An organism that can exist independently but may be parasitic on certain occasions, such as the flea. { 'fak·əl,tād·iv 'par·ə,sīt }

facultative photoheterotroph [MICROBIO] Any bacterium that usually grows anaerobically in light but can also grow aerobically in the dark. { 'fak·əl,tād·iv ¦fōd·ō¦hed·ə·rə,träf }

FAD See flavin adenine dinucleotide.

fade chart [ELECTROMAG] Graph on which the null areas of an air-search radar antenna are plotted as an aid to estimating target altitude. { 'fād ,chärt }

fade-in [COMMUN] A gradual increase in signal strength, as at the start of a radio or television program or when changing to a new scene, to make sound volume and picture brightness increase gradually. [GRAPHICS] In motion pictures, the gradual emergence of a screen image from black. { 'fād,in }

fade-out [COMMUN] A gradual and temporary loss of a received radio or television signal due to magnetic storms, atmospheric disturbances, or other conditions along the transmission path. { 'fād,aùt }

fader [ELECTR] A multiple-unit level control used for gradual changeover from one microphone, audio channel, or television camera to another. { 'fād·ər }

fading [COMMUN] Variations in the field strength of a radio signal that are caused by changes in the transmission medium. [GRAPHICS] The loss of density on a photographic image. [TEXT] A change in the color of a fabric produced by a natural or artificial agent, such as sunlight or an abrasive. { 'fād·iŋ }

fading margin [COMMUN] **1.** Number of decibels of attenuation which may be added to a specified radio-frequency propagation path before the signal-to-noise ratio of a specified channel falls below a specified minimum in order to avoid fading. **2.** Allowance made in radio system planning to accommodate estimated fading. { 'fād·iŋ ,mär·jən }

Fagaceae [BOT] A family of dicotyledonous plants in the order Fagales characterized by stipulate leaves, seeds without endosperm, female flowers generally not in catkins, and mostly three styles and locules. { fə'gās·ē,ē }

Fagales [BOT] An order of dicotyledonous woody plants in the subclass Hamamelidae having simple leaves and much reduced, mostly unisexual flowers. { fə'gā·lēz }

Fagergren cell [MIN ENG] A froth-flotation cell in which a squirrel-cage rotor is driven concentrically in a vertical stator, so that air is drawn down the rotor shaft and dispersed into the pulp. { 'fä·gər·grən ,sel }

Fagersta cut [MIN ENG] A cut drilled with handheld equipment in two steps, first as a pilot hole and then as an enlargement of this hole. { fə'gərs·tə ,kət }

fagopyrism [VET MED] Photosensitization of the skin and mucous membranes, accompanied by convulsions; produced especially in sheep and swine by feeding on the buckwheat plant, *Fagopyrum sagittatum*, or clovers and grasses containing flavin or carotene and xanthophyll. { ,fa·gō'pī,riz·əm }

fahlband [GEOL] A stratum containing metal sulfides; occurs in crystalline rock. { 'fäl,bänt }

fahlore See tetrahedrite. { 'fä,lór }

Fahnestock clip [ELEC] A spring-type terminal to which a temporary connection can readily be made. { 'fan,stäk ,klip }

Fahrenheit scale [THERMO] A temperature scale; the temperature in degrees Fahrenheit (°F) is the sum of 32 plus $^9/_5$ the temperature in degrees Celsius; water at 1 atmosphere (101,325 pascals) pressure freezes very near 32°F and boils very near 212°F. { 'far·ən,hīt ,skāl }

Fahrenheit's hydrometer [ENG] A type of hydrometer which carries a pan at its upper end in which weights are placed; the relative density of a liquid is measured by determining the weights necessary to sink the instrument to a fixed mark, first in water and then in the liquid being studied. { 'far·ən,hīts hī'dräm·əd·ər }

Fahrenholz's rule [ECOL] The rule that in groups of permanent parasites the classification of the parasites usually corresponds directly to the natural relationships of the hosts. { 'fär·ən,hōlt·səz ,rül }

failed hole [ENG] A drill hole loaded with dynamite which did not explode. Also known as missed hole. { 'fāld 'hōl }

failed star See brown dwarf. { 'fāld 'stär }

fail-safe system [ENG] A system designed so that failure of power, control circuits, structural members, or other components will not endanger people operating the system or other people in the vicinity. { 'fāl ¦sāf ,sis·təm }

failsafe tape See incremental dump tape. { 'fāl¦sāf ,tāp }

fail soft [ENG] A failure in the performance of a system component that neither results in immediate or major interruption of the system operation as a whole nor adversely affects the quality of its products. { 'fāl ¦sóft }

fail-soft system [COMPUT SCI] A computer system with automatic controls that allow function to continue after a malfunction and, if necessary, permit the shutdown of the system without loss of data. { 'fāl ¦sóft ,sis·təm }

failure [ENG] A permanent change in the volume of a powder or the stresses within it. [MECH] Condition caused by collapse, break, or bending, so that a structure or structural element can no longer fulfill its purpose. { 'fāl·yər }

failure logging [COMPUT SCI] The automatic recording of the state of various components of a computer system following detection of a machine fault; used to initiate corrective procedures, such as repeating attempts to read or write a magnetic tape, and to aid customer engineers in diagnosing errors. { 'fāl·yər ,läg·iŋ }

failure properties [ENG] The parameters that control the degree of the failure of a powder. { 'fāl·yər ,präp·ərd·ēz }

failure rate [ENG] The probability of failure per unit of time of items in operation; sometimes estimated as a ratio of the number of failures to the accumulated operating time for the items. { 'fāl·yər ,rāt }

fair [METEOROL] Generally descriptive of pleasant weather conditions, with regard for location and time of year; it is subject to popular misinterpretation, for it is a purely subjective description; when this term is used in forecasts of the U.S. Weather Bureau, it is meant to imply no precipitation, less than 0.4 sky cover of low clouds, and no other extreme conditions of cloudiness or windiness. { fer }

fairchildite [MINERAL] $K_2Ca(CO_3)_2$ A mineral composed of potassium calcium carbonate; occurs in partly burned trees. { 'fer,chīl,dīt }

fair curves [NAV ARCH] Curves which are smooth without sharp changes in direction over any portion of their length. { ¦fer 'kərvz }

faired cable [DES ENG] A trawling cable covered by streamlined surfaces to reduce hydrodynamic drag. { ¦ferd 'kā·bəl }

fairfieldite [MINERAL] $Ca_2Mn(PO_4)_2·2H_2O$ A white or pale-yellow mineral composed of hydrous calcium manganese phosphate and occurring in foliated or fibrous form. { 'fer,fēl,dīt }

fair game [MATH] A game in which all of the participants have equal expectation of gain. { ¦fer 'gām }

fairing [AERO ENG] A structure or surface on an aircraft or rocket that functions to reduce drag, such as the streamlined nose of a satellite-launching rocket. { 'fer·iŋ }

fairlead [AERO ENG] A guide through which an airplane antenna or control cable passes. [MECH ENG] A group of pulleys or rollers used in conjunction with a winch or similar apparatus to permit the cable to be reeled from any direction. [NAV ARCH] A block, ring, or other fitting through which passes a line or the running rigging on a ship to prevent chafing. { 'fer,lēd }

fair line [NAV ARCH] A line formed by the intersection of a plane with a ship's surface, which surface is smooth and is such as to minimize resistance to the ship's motion. { ¦fer 'līn }

fair tide [NAV] A tidal current setting in such a direction as to increase the speed of a vessel. { ¦fer 'tīd }

fairwater [NAV ARCH] A device for making fair the lines of an underwater fitting. { 'fer,wòd·ər }

fairway [NAV] Open water of sufficient depth for navigation; a marine thoroughfare. { 'fer,wā }

fairway buoy [NAV] A buoy indicating a fairway, having safe water on both sides. { 'fer,wā ,bói }

fair-weather cumulus See cumulus humilis cloud. { 'fer ,weth·ər 'kyü·myə·ləs }

fair wind See favorable wind. { ¦fer 'wind }

fairy ring spot [PL PATH] A fungus disease of carnations caused by *Heterosporium echinulatum*, producing bleached spots with concentric dark zones on the leaves. { 'fer·ē ,riŋ ,spät }

fairy stone *See* staurolite. { 'fer·ē ˌstōn }

faithful module [MATH] A module *M* over a commutative ring *R* such that if *a* is an element in *R* for which *am* = 0 for all *m* in *M*, then *a* = 0. { ¦fāth‚fül 'mä·jül }

faithful representation [MATH] A homomorphism *h* of a group onto some group of matrices or linear operators such that *h* is an injection. { ¦fāth‚fül ‚rep·rə·zen'tā·shən }

fake [NAV ARCH] To lay a rope or chain down in long bights side by side or in coils in regular order. { fāk }

fake set *See* false set. { ¦fāk ‚set }

falcate [ASTRON] Crescent-shaped; applied usually to the appearance of the moon, Venus, and Mercury during their crescent phases. [BIOL] Shaped like a sickle. { 'fal‚kat }

falciform [BIOL] Sickle-shaped. { 'fal·sə‚förm }

falciform ligament [ANAT] The ventral mesentery of the liver; its peripheral attachment extends from the diaphragm to the umbilicus and contains the round ligament of the liver. { 'fal·sə‚förm 'lig·ə·mənt }

falciger [INV ZOO] Seta with a distally blunt and curved tip. { 'fal·sə·gər }

falciparum malaria [MED] A severe form of malaria caused by *Plasmodium falciparum* and characterized by sudden attacks of chills, fever, and sweating at irregular intervals; the infecting organism usually localizes in a specific organ, causing capillary blockage. Also known as alged malaria; estivoautumnal malaria; malignant malaria; pernicious malaria. { fal'sip·ə·rəm mə'ler·ē·ə }

falcon [VERT ZOO] Any of the highly specialized diurnal birds of prey composing the family Falconidae; these birds have been captured and trained for hunting. { 'fal·kən }

Falcon [ORD] A U.S. Air Force air-to-air guided missile having either radar or infrared homing guidance, a speed of about Mach 2, and a range of about 5 miles (8 kilometers); can be carried in quantity by interceptor aircraft. { 'fal·kən }

Falconbridge process [MET] Recovery of nickel from a nickel-copper matte; the matte is first crushed and roasted to remove sulfur, and the copper is acid-leached, filtered off, and electrolyzed; the residual solids are melted, cast as anodes, and refined electrolytically to produce nickel. { 'fók·ən‚brij ‚präs· əs }

Falconidae [VERT ZOO] The falcons, a family consisting of long-winged predacious birds in the order Falconiformes. { fal'kän·ə‚dē }

Falconiformes [VERT ZOO] An order of birds containing the diurnal birds of prey, including falcons, hawks, vultures, and eagles. { fal‚kän·ə'för·mēz }

falculate [ZOO] Curved and with a sharp point. { 'fal·kyə‚lāt }

Fales-Stuart windmill [MECH ENG] A windmill developed for farm use from the two-blade airfoil propeller. Also known as Stuart windmill. { ¦fālz ¦stü·ərt 'wind‚mil }

Falk flexible coupling [MECH ENG] A spring coupling in which a continuous steel spring is threaded back and forth through axial slots in the periphery of two hubs on the shaft ends. { ¦fók ¦flek·sə·bəl 'kəp·liŋ }

Falkland Current [OCEANOGR] An ocean current flowing northward along the Argentine coast. { 'fók·lənd 'kə·rənt }

fall [ASTRON] **1.** Of a spacecraft or spatial body, to drop toward a spatial body under the influence of its gravity. **2.** *See* autumn. [ENG] The minimum slope that is required to facilitate proper drainage of liquid inside a pipe. [MECH ENG] The rope or chain of a hoisting tackle. [MIN ENG] A mass of rock, coal, or ore which has fallen from the roof or side in any subterranean working or gallery. { fól }

fallaway section [AERO ENG] A section of a rocket vehicle that is cast off from the vehicle during flight, especially such a section that falls back to the earth. { 'fól·ə‚wā ‚sek·shən }

fallback [COMPUT SCI] The system, electronic or manual, which is substituted for the computer system in case of breakdown. [GEOL] Fragmented ejecta from an impact or explosion crater during formation which partly refills the true crater almost immediately. [NUCLEO] That part of the material carried into the air by an atomic explosion which ultimately drops back to the earth or water at the site of the explosion. { 'fól‚bak }

fallback switch [COMMUN] A mechanical switch to transfer a communications path from a primary device to an identical standby device in the event of a primary device failure. { 'fól‚bak ‚swich }

fall block [MECH ENG] A pulley block that rises and falls with the load on a lifting tackle. { 'fól ‚bläk }

faller [MECH ENG] A machine part whose operation depends on a falling action. { 'fól·ər }

falling-ball viscometer *See* falling-sphere viscometer. { 'fól· iŋ ‚ból vi'skäm·əd·ər }

falling body [MECH] A body whose motion is accelerated toward the center of the earth by the force of gravity, other forces acting on it being negligible by comparison. { 'fól·iŋ 'bäd·ē }

falling disease [VET MED] A terminal manifestation of copper deficiency in which the animal collapses and dies because of heart failure. { 'fól·iŋ di‚zēz }

falling-drop method [PHYS] Technique for measurement of liquid densities in which the time of fall of a drop of the sample liquid through a reference liquid is measured. { 'fól·iŋ ‚dräp ‚meth·əd }

falling factorial polynomials [MATH] The polynomials $[x]_n = x(x-1)(x-2)\cdots(x-n+1)$. { ‚fól·iŋ fak‚tór· ē·əl ‚päl·ə'nō·mē·əlz }

falling film [FL MECH] A theoretical liquid film that moves downward in even flow on a vertical surface in laminar flow; the concept is used for heat-and mass-transfer calculations. { 'fól·iŋ ‚film }

falling-film cooler [ENG] Liquid cooling system in which the cooling liquid flows down vertical tube exterior surfaces in a thin film, and hot process fluid flows upward through the tubes. { 'fól·iŋ ‚film ‚kül·ər }

falling-film evaporator [ENG] Liquid evaporator system with heated vertical tubes; liquid to be evaporated flows down the inside tube surfaces as a film, evaporating as it flows. { 'fól·iŋ ‚film i'vap·ə‚rād·ər }

falling-film molecular still *See* falling-film still. { 'fól·iŋ ‚film mə¦lek·yə·lər 'stil }

falling-film still [CHEM ENG] Special molecular distillation apparatus designed for high evaporative and separation efficiency. Also known as falling-film molecular still. { 'fól·iŋ ‚film 'stil }

falling-sphere viscometer [ENG] A viscometer which measures the speed of a spherical body falling with constant velocity in the fluid whose viscosity is to be determined. Also known as falling-ball viscometer. { 'fól·iŋ ‚sfir vi'skäm· əd·ər }

falling tide *See* ebb tide. { 'fól·iŋ 'tīd }

fall line [GEOL] **1.** The zone or boundary between resistant rocks of older land and weaker strata of plains. **2.** The line indicated by the edge over which a waterway suddenly descends, as in waterfalls. { 'fól ‚līn }

falloff curve [PETRO ENG] Graphical representatation of bottom-hole pressure falloff for a shut-in well as the reservoir drainage area expands. { 'fól‚óf ‚kərv }

fall of ground [MIN ENG] The fall of rock from the roof into a mine opening. { 'fól əv 'graúnd }

Fallopian tube [ANAT] Either of the paired oviducts that extend from the ovary to the uterus for conduction of the ovum in mammals. { fə'lō·pē·ən 'tüb }

fallout [ELECTR] Failure of electronic components during burn-in. [NUCLEO] The material that descends to the earth or water well beyond the site of a surface or subsurface nuclear explosion. Also known as atomic fallout; radioactive fallout. { 'fól‚aút }

fallout area [NUCLEO] The area on which radioactive materials have settled out, or the area on which it is predicted from weather conditions that radioactive materials may settle out. { 'fól‚aút ‚er·ē·ə }

fallout shelter [CIV ENG] A structure that affords some protection against fallout radiation and other effects of nuclear explosion; maximum protection is in reinforced concrete shelters below the ground. Also known as radiation shelter. { 'fól‚aút ‚shel·tər }

fallout winds [METEOROL] Tropospheric winds that carry the radioactive fallout materials, observed by standard winds-aloft observation techniques. { 'fól‚aút ‚winz }

fallow [AGR] Pertaining to land normally used for crop production but left unsown for one or more growing seasons. { 'fal·ō }

fall-streak hole [METEOROL] A hole occurring in a cloud layer of supercooled water droplets; produced by the local

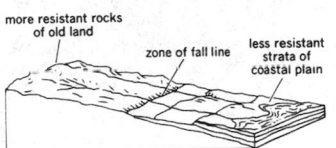

FALL LINE

more resistant rocks of old land

zone of fall line

less resistant strata of coastal plain

Diagram showing zone of a fall line.

freezing of some of the droplets and their coversion into fallout, frequently in a streak form. { 'fȯl ,strēk ,hōl }

fall streaks *See* virga. { 'fȯl ,strēks }

Fallstreifen *See* virga. { 'fäl,strīf·ən }

fall time [ELEC] Measure of time required for a circuit to change its output from a high level to a low level. { 'fȯl ,tīm }

fall velocity *See* settling velocity. { 'fȯl və,lä·səd·ē }

fall wind [METEOROL] A strong, cold, downslope wind, differing from a foehn in that the initially cold air remains relatively cold despite adiabatic warming upon descent, and from the gravity wind in that it is a larger-scale phenomenon prerequiring an accumulation of cold air at high elevations. { 'fȯl ,wind }

false acceptance [STAT] Accepting on the basis of a statistical test a hypothesis which is wrong. { ¦fȯls ak'sep·təns }

false alarm [ELECTR] In radar, an indication of a detected target even though one does not exist, due to noise or interference levels exceeding the set threshold of detection. { ¦fȯls ə'lärm }

false attic [BUILD] A section under a roof normally occupied by an attic, but which has no windows and does not enclose rooms. { ¦fȯls 'ad·ik }

false bedding [GEOL] An inclined bedding produced by currents. { ¦fȯls 'bed·iŋ }

false blossom [PL PATH] **1.** A fungus disease of the cranberry caused by *Exobasidium oxycocci;* erect flower buds are formed which produce malformed flowers that set no fruit. Also known as rosebloom. **2.** A similar virus disease of the cranberry transmitted by the leafhopper, *Scleroracus vaccinii.* Also known as Wisconsin false blossom. { ¦fȯls 'bläs·əm }

false body [PHYS CHEM] The property of certain colloidal substances, such as paints and printing inks, of solidifying when left standing. { ¦fȯls 'bäd·ē }

false bottom [CIV ENG] A temporary bottom installed in a caisson to add to its buoyancy. [MET] An insert put in either member of a die set to increase the strength and improve the life of the die. [MIN ENG] A flat, hexagonal or cylindrical iron die upon which ore is crushed in a stamp mill. { ¦fȯls 'bäd·əm }

false cirrus cloud [METEOROL] Cirrus composed of the debris of the upper frozen parts of a cumulonimbus cloud. { ¦fȯls 'sir·əs ,klaůd }

false cleavage [GEOL] **1.** A weak cleavage at an angle to the slaty cleavage. **2.** Spaced surfaces about a millimeter apart along which a rock splits. { ¦fȯls 'klēv·ij }

false color [OPTICS] Color assigned to frequency bands that are normally invisible to the human eye (such as infrared radiation) in an image in order to enhance contrasts or to display those colors. { ¦fȯls ¦kəl·ər }

false drop *See* false retrieval. { ¦fȯls 'dräp }

false drumlin *See* rock drumlin. { ¦fȯls 'drəm·lən }

false form *See* pseudomorph. { ¦fȯls 'fȯrm }

false galena *See* sphalerite. { ¦fȯls gə'lē·nə }

false header [CIV ENG] A half brick used to complete a visible bond; it is not a header. { ¦fȯls 'hed·ər }

false horizon [NAV] A line resembling the visible horizon but above or below it. { ¦fȯls hə'rīz·ən }

false ice foot [OCEANOGR] Ice that forms along a beach terrace and attaches to it just above the high-water mark; derived from water coming from melting snow above the terrace. { ¦fȯls 'īs ,fůt }

false lapis *See* lazulite. { ¦fȯls 'lap·əs }

false ligament [ANAT] Any peritoneal fold which is not a true supporting ligament. { ¦fȯls 'lig·ə·mənt }

false light [NAV] A light which is unavoidably exhibited by an aid to navigation and which is not intended to be a part of the proper characteristic of the light, such as reflections from storm panes. { ¦fȯls 'līt }

false ogive [ORD] A rounded or pointed hollow cup added to the nose of a projectile to improve streamlining. Also known as ballistic cap. { ¦fȯls 'ō,jīv }

false oolith *See* pseudo-oolith. { ¦fȯls 'ō,ō,līth }

false pyroelectricity *See* tertiary pyroelectricity. { ¦fȯls ¦pī·rō·i,lek'tri·səd·ē }

false rejection [STAT] Rejecting on the basis of a statistical test a hypothesis which is correct. { ¦fȯls ri'jek·shən }

false retrieval [COMPUT SCI] An item retrieved in an automatic library search which is unrelated or vaguely related to the subject of the search. Also known as false drop. { ¦fȯls ri'trē·vəl }

false rib [ANAT] A rib that is not attached to the sternum directly; any of the five lower ribs on each side in humans. { ¦fȯls 'rib }

false ring [BOT] A layer of wood that is less than a full season's growth and often does not form a complete ring. { ¦fȯls 'riŋ }

false set [MATER] Rapid hardening of freshly mixed cement paste, mortar, or concrete with minimum evolution of heat; plasticity can be restored by mixing without addition of water. [MIN ENG] A light, temporary lagging set of timber supporting the side and roof lagging until the drive is advanced sufficiently to allow the heavy permanent set to be put, at which time the false set is taken out and used again in advance of the next permanent set. Also known as fake set. { ¦fȯls 'set }

false smut [PL PATH] **1.** A fungus disease of palm caused by *Graphiola phoenicis* and characterized by small cylindrical protruding pustules, often surrounded by yellowish leaf tissue. **2.** *See* green smut. { ¦fȯls 'smət }

false sorts [COMPUT SCI] Entries irrelevant to the subject sought which are retrieved in a search. { ¦fȯls 'sȯrts }

false stull [MIN ENG] A stull so placed as to offer support or reinforcement for a stull, prop, or other timber. { ¦fȯls 'stəl }

false target [ELECTR] A nonexistent target which shows up on a radar scope as the result of time delay. { ¦fȯls 'tär·gət }

false-target generator [ELECTR] An electronic countermeasure device that generates a delayed return signal on an enemy radar frequency to give erroneous position information. { ¦fȯls ¦tär·gət 'jen·ə,rād·ər }

false topaz *See* citrine. { ¦fȯls 'tō,paz }

false twist [TEXT] A method by which certain synthetic yarns are given stretch characteristics; yarns are wound under heat to eliminate the twist, but because of the yarns' tendency to retain the twist, fabrics made from them have elasticity. Also known as memory twist. { ¦fȯls 'twist }

false warm sector [METEOROL] The sector, in a horizontal plane, between the occluded front and a secondary cold front of an occluded cyclone. { ¦fȯls 'wȯrm ,sek·tər }

false white rainbow *See* fogbow. { ¦fȯls ¦wīt 'rān,bō }

falsework [CIV ENG] A temporary support used until the main structure is strong enough to support itself. { 'fȯls,wərk }

faltung [MATH] A family of functions where the convolution of any two members of the family is also a member of the family. Also known as convolution family. { 'fäl,tůŋ }

falx [ANAT] A sickle-shaped structure. { falks }

famatinite [MINERAL] Cu_3SbS_4 A reddish-gray mineral composed of copper antimony sulfide. { ,fam·ə'tē,nīt }

familial [BIOL] Of, pertaining to, or occurring among the members of a family. { fə'mil·yəl }

familial aldosterone deficiency [MED] A hereditary metabolic disorder, probably due to a defect in the enzyme involved in dehydrogenation of 18-hydroxycorticosterone to aldosterone, characterized by growth retardation and hypoaldosteronism. { fə'mil·yəl al'däs·tə,rōn di,fish·ən·sē }

familial dysautonomia [MED] A hereditary disease transmitted as an autosomal recessive and characterized from infancy by evidence of autonomic nervous system dysfunction, including feeding difficulties, absence of overflow tears, indifference to pain, absent corneal reflexes and deep tendon reflexes, and absence of fungiform papillae on the tongue; most common in Jewish children. { fə'mil·yəl ,dis,ȯd·ə'nō·mē·ə }

familial Mediterranean fever [MED] A hereditary disease of unknown cause characterized by recurrent fever, abdominal and chest pain, arthralgia, and rash, sometimes terminating in renal failure. Abbreviated FMF. Also known as familial recurring polyserositis; periodic disease; periodic peritonitis. { fə'mil·yəl ¦med·ə·tə¦rā·nē·ən ¦fē·vər }

familial osteochondrodystrophy *See* Morquio's syndrome. { fə'mil·yəl ¦äs·tē·ō,kän·drə'dis·trə·fē }

familial polyposis [MED] A hereditary condition transmitted as an autosomal dominant and characterized by the appearance of polyps in the small intestine and colon; malignant degeneration is common. { fə'mil·yəl ,päl·ə'pō·səs }

familial recurring polyserositis *See* familial Mediterranean fever. { fə'mil·yəl ri'kər·iŋ ,päl·ē,ser·ə'sīd·əs }

familial splenic anemia *See* Gaucher's disease. { fə'mil·yəl 'splēn·ik ə'nē·myə }

family [CHEM] A group of elements whose chemical properties, such as valence, solubility of salts, and behavior toward reagents, are similar. [SYST] A taxonomic category based on the grouping of related genera. { 'fam·lē }

family mold [ENG] A multicavity injection mold where each cavity forms a component part of the finished product. { 'fam·lē ,mōld }

family of curves [MATH] A set of curves whose equations can be obtained by varying a finite number of parameters in a particular general equation. { ¦fam·lē əv 'kərvz }

family therapy [PSYCH] Treatment of more than one family member in the same therapeutic session. { ¦fam·i·lē 'ther·ə·pē }

FAMOS device See floating-gate avalanche-injection metal-oxide semiconductor device. { 'fā,mós di'vīs }

famphur [ORG CHEM] $C_{10}H_{16}NO_5PS_2$ A crystalline compound with a melting point of 55°C; slightly soluble in water; used as an insecticide for lice and grubs of reindeer and cattle. { 'fam·fər }

fan [AGR] A mechanical device used for winnowing grain. [BIOL] Any structure, such as a leaf or the tail of a bird, resembling an open fan. [ELECTROMAG] Volume of space periodically energized by a radar beam (or beams) repeatedly traversing an established pattern. [GEOL] A gently sloping, fan-shaped feature usually found near the lower termination of a canyon. [MECH ENG] **1.** A device, usually consisting of a rotating paddle wheel or an airscrew, with or without a casing, for producing currents in order to circulate, exhaust, or deliver large volumes of air or gas. **2.** A vane to keep the sails of a windmill facing the direction of the wind. { fan }

fan antenna [ELECTROMAG] An array of folded dipoles of different length forming a wide-band ultra-high-frequency or very-high-frequency antenna. { 'fan an,ten·ə }

fan beam [ELECTROMAG] **1.** A radio beam having an elliptically shaped cross section in which the ratio of the major to the minor axis usually exceeds 3 to 1; the beam is broad in the vertical plane and narrow in the horizontal plane. **2.** A radar beam having the shape of a fan. { 'fan ,bēm }

fan brake [MECH ENG] A fan used to provide a load for a driving mechanism. { 'fan ,brāk }

Fanconi's anemia [MED] An infantile anemia that resembles pernicious anemia; related to excessive chromosomal breakage and associated with the risk of developing leukemia. { fäŋ·kō·nēz ə'nē·myə }

Fanconi's syndrome See amino diabetes. { fäŋ·kō·nēz 'sin,drōm }

fan cut [ENG] A cut in which holes of equal or increasing length are drilled in a pattern on a horizontal plane or in a selected stratum to break out a considerable part of the plane or stratum before the rest of the round is fired. { 'fan ,kət }

fan drift [MIN ENG] The short tunnel connecting the upcast shaft with the exhaust fan. { 'fan ,drift }

fan-drift doors [MIN ENG] Isolation doors for each drift leading to each fan, when there are two fans at a mine. { 'fan ,drift ,dórz }

fan drilling [ENG] **1.** Drilling boreholes in different vertical and horizontal directions from a single-drill setup. **2.** A radial pattern of drill holes from a setup. { 'fan ,dril·iŋ }

fan efficiency [MECH ENG] The ratio obtained by dividing a fan's useful power output by the power input (the power supplied to the fan shaft); it is expressed as a percentage. { 'fan i,fish·ən·sē }

fan fold [GEOL] A fold of strata in which both limbs are overturned, forming a syncline or anticline. { 'fan ,fōld }

fanfold [COMPUT SCI] Continuous paper that is perforated at page boundaries and can be folded back and forth at the perforations to form a stack. { 'fan,fōld }

fang [ANAT] The root of a tooth. [VERT ZOO] A long, pointed tooth, especially one of a venomous serpent. { faŋ }

fang bolt [DES ENG] A bolt having a triangular nut with sharp projections at its corners; used to attach metal pieces to wood. { 'faŋ ,bōlt }

fanglomerate [GEOL] Coarse material in an alluvial fan, with the rock fragments being only slightly worn. { fan'gläm·ə·rət }

fan-in [ELECTR] The number of inputs that can be connected to a logic circuit. { 'fan,in }

fanjet [AERO ENG] A turbojet engine whose performance has been improved by the addition of a fan which operates in an annular duct surrounding the engine. { 'fan,jet }

fanlight [ARCH] A segmented semicircular window over a door or window. { 'fan·līt }

fan marker See fan-marker beacon. { 'fan ,märk·ər }

fan-marker beacon [NAV] A very-high frequency radio facility having a vertically directed fan beam interesecting a airway to provide a fix. Also known as fan marker; radio fan-marker beacon. { 'fan ,märk·ər ,bē·kən }

fanned-beam antenna [ELECTROMAG] Unidirectional antenna so designed that transverse cross sections of the major lobe are approximately elliptical. { ¦fand ¦bēm an,ten·ə }

fanning beam [ELECTROMAG] Narrow antenna beam which is repeatedly scanned over a limited arc. { 'fan·iŋ ,bēm }

Fanning friction factor [FL MECH] A dimensionless number used in studying fluid friction in pipes, equal to the pipe diameter times the drop in pressure in the fluid due to friction as it passes through the pipe, divided by the product of the pipe length and the kinetic energy of the fluid per unit volume. Symbolized f. { 'fan·iŋ 'frik·shən ,fak·tər }

fanning mill [FOOD ENG] A device consisting of two vibrating screens and utilizing an air blast to clean and separate grain. { 'fan·iŋ ,mil }

Fanning's equation [FL MECH] The equation expressing that frictional pressure drop of fluid flowing in a pipe is a function of the Reynolds number, rate of flow, acceleration due to gravity, and length and diameter of the pipe. { 'fan·iŋz i,kwā·zhən }

fanning strip [ELEC] Insulated board, often of wood, which serves to spread out the wires of a cable for distribution to a terminal board. { 'fan·iŋ ,strip }

Fanno flow [FL MECH] An ideal flow used to study the flow of fluids in long pipes; the flow obeys the same simplifying assumptions as Rayleigh flow except that the assumption there is no friction is replaced by the requirement the flow be adiabatic. { 'fan·ō ,flō }

Fano effect [ATOM PHYS] The spin polarization of photoelectrons from alkali atoms that is produced upon the atoms' absorption of circularly polarized light. { 'fan·ō i,fekt }

Fano plane [MATH] A projective plane in which the points of intersection of the three possible pairs of opposite sides of a quadrilateral are collinear. { 'fä·nō ,plān }

Fano's axiom [MATH] The postulate that the points of intersection of the three possible pairs of opposite sides of any quadrilateral in a given projective plane are noncollinear; thus a projective plane satisfying Fano's axiom is not a Fano plane, and a Fano plane does not satisfy Fano's axiom. { ¦fä·nōz 'ak·sē·əm }

fan-out [ELECTR] The number of parallel loads that can be driven from one output mode of a logic circuit. { 'fan,aút }

fan rating [MECH ENG] The head, quantity, power, and efficiency expected from a fan operating at peak efficiency. { 'fan ,rād·iŋ }

fan ring [DES ENG] Circular metallic collar encircling (but spaced away from) the tips of the fan blade in process equipment, such as air-cooled heat exchangers; ring design is critical to the efficiency of fan performance. { 'fan ,riŋ }

fan shaft [DES ENG] The spindle on which a fan impeller is mounted. [MIN ENG] The ventilating shaft to which a mine fan is connected. { 'fan,shaft }

fan-shaped delta See arcuate delta. { 'fan ,shāpt 'del·tə }

fan shooting [ENG] Seismic exploration in which seismometers are placed in a fan-shaped array to detect anomalies in refracted-wave arrival times indicative of circular rock structures such as salt domes. { 'fan ,shüd·iŋ }

fan static pressure [MECH ENG] The total pressure rise diminished by the velocity pressure in the fan outlet. { ¦fan ¦stad·ik ,presh·ər }

fantail [NAV ARCH] The area of the upper deck of a ship which is nearest the stern. { 'fan,tāl }

fantasy [PSYCH] An imagined image or series of images that serves to express unconscious conflicts, to gratify unconscious wishes, or to prepare for anticipated future events. { 'fan·tə·sē }

fan test [MECH ENG] Observations of the quantity, total pressure, and power of air circulated by a fan running at a known constant speed. { 'fan,test }

Fantl unit [BIOL] A unit for the standardization of thrombin. { 'fant·əl ,yü·nət }

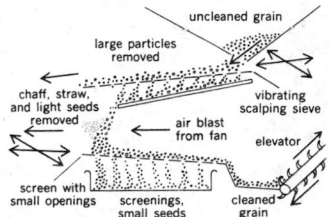

FANNING MILL

uncleaned grain
large particles removed
chaff, straw, and light seeds removed
vibrating scalping sieve
air blast from fan
elevator
screen with small openings
screenings, small seeds
cleaned grain

Fanning mill operation.

fan total head [MECH ENG] The sum of the fan static head and the velocity head at the fan discharge corresponding to a given quantity of airflow. { ¦fan ¦tōd·əl ¦hed }

fan total pressure [MECH ENG] The algebraic difference between the mean total pressure at the fan outlet and the mean total pressure at the fan inlet. { ¦fan ¦tōd·əl ¦presh·ər }

fan truss [CIV ENG] A truss with struts arranged as radiating lines. { 'fan ‚trəs }

fan vaulting [ARCH] Vaulting in which the ribs diverge like the rays of a fan. { 'fan ‚vȯlt·iŋ }

fan velocity pressure [MECH ENG] The velocity pressure corresponding to the average velocity at the fan outlet. { ¦fan və'läs·əd·ē ‚presh·ər }

FAQ See Frequently Asked Questions.

farad [ELEC] The unit of capacitance in the meter-kilogram-second system, equal to the capacitance of a capacitor which has a potential difference of 1 volt between its plates when the charge on one of its plates is 1 coulomb, there being an equal and opposite charge on the other plate. Symbolized F. { 'fa‚rad }

faradaic current See faradic current. { ‚far·ə¦dā·ik ¦kər·ənt }

faraday [PHYS] The electric charge required to liberate 1 gram-equivalent of a substance by electrolysis; experimentally equal to 96,485.3415 ± 0.0039 coulombs. Also known as Faraday constant. { 'far·ə‚dā }

Faraday birefringence [OPTICS] Difference in the indices of refraction of left and right circularly polarized light passing through matter parallel to an applied magnetic field; it is responsible for the Faraday effect. { 'far·ə‚dā ‚bī·ri'frin·jəns }

Faraday cage See Faraday shield. { 'far·ə‚dā ‚kāj }

Faraday constant See faraday. { 'far·ə‚dā ‚kän·stənt }

Faraday cylinder [ELEC] **1.** A closed, or nearly closed, hollow conductor, usually grounded, within which apparatus is placed to shield it from electrical fields. **2.** A nearly closed, insulated, hollow conductor, usually shielded by a second grounded cylinder, used to collect and detect a beam of charged particles. { 'far·ə‚dā ‚sil·ən·dər }

Faraday dark space [ELECTR] The relatively nonluminous region that separates the negative glow from the positive column in a cold-cathode glow-discharge tube. { 'far·ə‚dā 'därk ‚spās }

Faraday disk machine [ELECTROMAG] A device for demonstrating electromagnetic induction, consisting of a copper disk in which a radial electromotive force is induced when the disk is rotated between the poles of a magnet. Also known as Faraday generator. { 'far·ə‚dā 'disk mə‚shēn }

Faraday effect [OPTICS] Rotation of polarization of a beam of linearly polarized light when it passes through matter in the direction of an applied magnetic field; it is the result of Faraday birefringence. Also known as Faraday rotation; Kundt effect; magnetic rotation. { 'far·ə‚dā i'fekt }

Faraday generator See Faraday disk machine. { 'far·ə‚dā 'jen·ə‚rād·ər }

Faraday ice bucket experiment [ELEC] Experiment in which one lowers a charged metal body into a pail and observes the effect on an electroscope attached to the pail, with and without contact between body and pail; the experiment shows that charge resides on a conductor's outside surface. { 'far·ə‚dā 'īs ‚bək·ət ik‚sper·ə·mənt }

Faraday rotation See Faraday effect. { 'far·ə‚dā rō'tā·shən }

Faraday rotation experiment [ELECTROMAG] An experiment in which a wire dipping in a pool of mercury surrounding a magnet rotates around the magnet when a current passes through it, demonstrating the effect of a magnetic field on a current-carrying conductor. { ¦far·ə‚dā rō'tā·shən ik‚sper·ə·mənt }

Faraday rotation isolator See ferrite isolator. { 'far·ə‚dā rō'tā·shən 'īs·əl‚ād·ər }

Faraday screen See Faraday shield. { 'far·ə‚dā ‚skrēn }

Faraday shield [ELEC] Electrostatic shield composed of wire mesh or a series of parallel wires, usually connected at one end to another conductor which is grounded. Also known as Faraday cage; Faraday screen. { 'far·ə‚dā ‚shēld }

Faraday's law of electromagnetic induction [ELECTROMAG] The law that the electromotive force induced in a circuit by a changing magnetic field is equal to the negative of the rate of change of the magnetic flux linking the circuit. Also known as law of electromagnetic induction. { 'far·ə‚dāz 'lȯ əv i¦lek·trō‚mag¦ned·ik in'dək·shən }

Faraday's laws of electrolysis [PHYS CHEM] **1.** The amount of any substance dissolved or deposited in electrolysis is proportional to the total electric charge passed. **2.** The amounts of different substances dissolved or desposited by the passage of the same electric charge are proportional to their equivalent weights. { 'far·ə‚dāz ¦lȯz əv i‚lek'träl·ə·səs }

Faraday tube [ELEC] A tube of force for electric displacement which is of such size that the integral over any surface across the tube of the component of electric displacement perpendicular to that surface is unity. { 'far·ə‚dā ‚tüb }

faradic current Also spelled faradaic current. [CHEM] An electric current that corresponds to the reduction or oxidation of a chemical species. [ELEC] An intermittent and nonsymmetrical alternating current like that obtained from the secondary winding of an induction coil. { fə'rad·ik ‚kər·ənt }

faradization [BIOPHYS] Use of a faradic current to stimulate muscles and nerves. { ‚far·əd·ə'zā·shən }

farcy See glanders. { 'fär·sē }

far-end crosstalk [COMMUN] Crosstalk that travels along the disturbed circuit in the same direction as desired signals in that circuit. { ¦fär ¦end 'krȯs‚tȯk }

farewell buoy See sea buoy. { 'fer‚wel ‚bȯi }

Farey sequence [MATH] The Farey sequence of order n is the increasing sequence, from 0 to 1, of fractions whose denominator is equal to or less that n, with each fraction expressed in lowest terms. { 'far·ē ‚sē·kwəns }

far field See Fraunhofer region. { ¦fär ¦fēld }

farinaceous [BIOL] Having a mealy surface covering. [FOOD ENG] **1.** Containing starch or flour. **2.** Having the texture of meal. [GEOL] Of a rock or sediment, having a texture that is mealy, soft, and friable, for example, a limestone or a pelagic ooze. { ¦far·ə¦nā·shəs }

Farinales [BOT] An order that includes several groups regarded as orders of the Commelinidae in other systems of classification. { ‚far·ə'nā·lēz }

far-infrared maser [ENG] A gas maser that generates a beam having a wavelength well above 100 micrometers, and ranging up to the present lower wavelength limit of about 500 micrometers for microwave oscillators. { ¦fär in·frə'red 'mā·zər }

far-infrared radiation [ELECTROMAG] Infrared radiation the wavelengths of which are the longest of those in the infrared region, about 50–1000 micrometers; requires diffraction gratings for spectroscopic analysis. { ¦fär in·frə'red ‚rād·ē'ā·shən }

Farinosae [BOT] The equivalent name for Farinales. { ‚far·ə'nō·sē }

farinose [AGR] Yielding farina, a fine meal of vegetable matter. [BIOL] Covered with a white powdery substance. { 'far·ə‚nōs }

farm [AGR] A tract of land used for cultivating crops or raising animals. { färm }

Farmer dosimeter [NUCLEO] A small ionization chamber with an air wall, used for routine measurements of radiation. { ¦fär·mər dō'sim·əd·ər }

farmer's lung [MED] An acute pulmonary disorder caused by the inhalation of spores from moldy hay or straw. { 'fär·mərz ‚ləŋ }

farmer's year [CLIMATOL] In Great Britain, the 12-month period starting with the Sunday nearest March 1. { 'fär·mərz 'yir }

farming [AGR] The skills and practices of agriculture. { 'fär·miŋ }

farmstead [AGR] The whole area that constitutes a farm, including its land and buildings. { 'färm‚sted }

farnesol [BIOCHEM] $C_{15}H_{25}OH$ A colorless liquid extracted from oils of plants such as citronella, neroli, cyclamen, and tuberose; it has a delicate floral odor, and is an intermediate step in the biological synthesis of cholesterol from mevalonic acid in vertebrates; used in perfumery. { 'fär·nə‚sȯl }

Farnsworth image dissector tube See image dissector tube. { 'färnz‚wərth 'im·ij di‚sek·tər ‚tüb }

far point [OPTICS] The farthest point from an eye at which an object is distinctly seen; for a normal eye it is theoretically at infinity. Also known as punctum remotum. { ¦fär ¦pȯint }

far region See Fraunhofer region. { ¦fär ¦rē·jən }

farringtonite [MINERAL] $Mg_3(PO_4)_2$ A colorless, wax-white, or yellow phosphate mineral known only in meteorites. { 'far·iŋ·tə‚nīt }

farsightedness See hypermetropia. { 'fär¦sīd·əd·nəs }

far-ultraviolet radiation [ELECTROMAG] Ultraviolet radiation in the wavelength range of 200–300 nanometers; germicidal effects are greatest in this range. Abbreviated FUV radiation. { 'fär ˌəl·trə'vī·lət ˌrād·ē'ā·shən }

far vane [NAV] In marine operations, the instrument sighting vane on the opposite side of the instrument from the observer's eye. { ¦fär ¦vān }

far zone See Fraunhofer region. { ¦fär ¦zōn }

fascia [BUILD] A wide board fixed vertically on edge to the rafter ends or wall which carries the gutter around the eaves of a roof. [HISTOL] Layers of areolar connective tissue under the skin and between muscles, nerves, and blood vessels. { 'fā·shə }

fasciate [BOT] Having bands or stripes. { 'fa·shē‚āt }

fasciation [PL PATH] Malformation of plant parts resulting from disorganized tissue growth. { ‚fa·shē'ā·shən }

fascicle [BOT] A small bundle, as of fibers or leaves. { 'fas·i·kəl }

fasciculate [BOT] Arranged in tufts or fascicles. { fə'sik·yə·lət }

fasciculation potential [PHYSIO] An action potential which is quantitatively comparable to that of a motor unit and which represents spontaneous contraction of a bundle of muscle fibers. { fə‚sik·yə'lā·shən pə‚ten·chəl }

fasciculus [ANAT] A bundle or tract of nerve, muscle, or tendon fibers isolated by a sheath of connective tissues and having common origins, innervation, and functions. { fə'sik·yə·ləs }

fascine [CIV ENG] A cylindrical bundle of brushwood 1–3 feet (30–90 centimeters) in diameter and 10–20 feet (3–6 meters) long, used as a facing for seawalls on riverbanks, as a foundation mat, as a dam in an estuary, or to protect bridge, dike, and pier foundations from erosion. { fa'sēn }

Fasciola hepatica [INV ZOO] A digenetic trematode which parasitizes sheep, cattle, and occasionally humans. { fə·'sē·ə·lə he'pad·ə·kə }

fasciole [INV ZOO] A band of cilia on the test of certain sea urchins. { 'fas·ē‚ōl }

fascioliasis [MED] The infection of humans with Fasciola hepatica. { fə‚sē·ə'lī·ə·səs }

fasciolopsiasis [MED] The presence of the parasite Fasciolopsis buski in a person's small intestine. { fə‚sē·ə‚läp'sī·ə·səs }

Fasciolopsis buski [INV ZOO] A large, fleshy trematode, native to eastern Asia and the southwestern Pacific, which parasitizes humans. { fə‚sē·ə'läp·səs 'bəs·kē }

fascioscapulohumeral dystrophy [MED] A progressive hereditary form of muscular dystrophy involving atrophy of the muscles of the face, pectoral girdle, and upper arm. { ¦fa·sē·ō¦skap·yə·lō¦hyü·mə·rəl 'di·strə·fē }

Fas protein [CELL MOL] A cell-surface protein receptor expressed on essentially all cells of the body that when bound to its ligand (FasL) signals a caspase cascade, ultimately resulting in apoptosis (programmed cell death). { ¦ef¦ā·es ‚prō‚tēn }

fassaite [GEOCHEM] $Ca(Mg,Ti,Al)(Al,Si)_2O_6$ A mineral found in the millimeter-sized rocklets or refractory inclusions of carbonaceous chondrite meteorites. { 'fas·ə‚yīt }

fast [GRAPHICS] A relative term given to the speed of emulsion. { fast }

fast-access storage [COMPUT SCI] The section of a computer storage from which data can be obtained most rapidly. { ¦fast ¦ak·ses 'stōr·ij }

fast automatic gain control [ELECTR] Radar automatic gain control method characterized by a response time that is long with respect to a pulse width, and short with respect to the time on target. { ¦fast ‚ȯd·ə‚mad·ik 'gān kən‚trōl }

fast axis [OPTICS] The direction of the electrical displacement vector of light propagating in an anisotropic crystal with the greatest possible phase velocity corresponding to a specified direction of propagation. { ¦fast 'ak·səs }

fast break [MET] Interruption of the current in the magnetizing coil during nondestructive testing of magnetic particles; induces eddy currents and strong magnetization. { ¦fast 'brāk }

fast breeder reactor [NUCLEO] A type of fast reactor using highly enriched fuel in the core, fertile material in the blanket, and a liquid-metal coolant, such as sodium; high-speed neutrons fission the fuel in the compact core, and the excess neutrons convert fertile material to fissionable isotopes; the breeding ratio is 1.0 or larger. Abbreviated FBR. { ¦fast 'brēd·ər rē‚ak·tər }

fast-burst reactor [NUCLEO] A nuclear reactor that supplies microsecond pulses of fast neutrons for use in biomedical research. { ¦fast ¦bərst rē'ak·tər }

fast carbon-nitrogen-oxygen cycle See hot carbon-nitrogen-oxygen cycle. { ¦fast ¦kär·bən ¦nī·trə·jən ¦äks·ə·jən ‚sī·kəl }

fast chemical reaction [PHYS CHEM] A reaction with a half-life of milliseconds or less; such reactions occur so rapidly that special experimental techniques are required to observe their rate. { ¦fast ¦kem·ə·kəl rē'ak·shən }

fast coupling [MECH ENG] A flexible geared coupling that uses two interior hubs on the shafts with circumferential gear teeth surrounded by a casing having internal gear teeth to mesh and connect the two hubs. { ¦fast 'kəp·liŋ }

fast-delay detonation [ENG] The firing of blasts by means of a blasting timer or millisecond delay caps. { ¦fast di¦lā det·ən'ā·shən }

fast effect [NUCLEO] The reactivity change (increase in neutrons) due to fissions caused by fast neutrons in a thermal reactor. { ¦fast i'fekt }

fastener [DES ENG] **1.** A device for joining two separate parts of an article or structure. **2.** A device for holding closed a door, gate, or similar structure. { 'fas·nər }

fastening [DES ENG] A spike, bolt, nut, or other device to connect rails to ties. { 'fas·niŋ }

fastest mile [METEOROL] Over a specified period (usually the 24-hour observational day), the fastest speed, in miles per hour, of any mile of wind, with its accompanying direction. { ¦fas·təst 'mīl }

fast fission [NUC PHYS] Fission caused by fast neutrons. { ¦fast 'fish·ən }

fast-fission factor [NUCLEO] The ratio of the number of neutrons produced by nuclear fissions due to neutrons of all energies in an infinite medium, to the number of neutrons produced by nuclear fissions due to thermal neutrons only. { ¦fast 'fish·ən ‚fak·tər }

fast Fourier transform [MATH] A Fourier transform employing the Cooley-Tukey algorithm to reduce the number of operations. Abbreviated FFT. { ¦fast ‚fur·ē‚ā 'tranz‚fȯrm }

fast ice [HYD] Any type of sea, river, or lake ice attached to the shore (ice foot, ice shelf), beached (shore ice), stranded in shallow water, or frozen to the bottom of shallow waters (anchor ice). Also known as landfast ice. [OCEANOGR] Sea ice generally remaining in the position where originally formed and sometimes attaining a considerable thickness; it is attached to the shore or over shoals where it may be held in position by islands, grounded icebergs, or polar ice. Also known as coastal ice; coast ice. { ¦fast ¦īs }

fastigiate [BOT] **1.** Having erect branches that are close to the stem. **2.** Becoming narrower at the top. [ZOO] Arranged in a conical bundle. { fa'stij·ē·āt }

fast ion See small ion. { ¦fast 'ī‚än }

fast-joint [ENG] Pertaining to a joint with a permanently secured pin. { ¦fast 'jȯint }

fast line [PETRO ENG] In a drilling operation, the end of the drilling line that is attached to the drum or reel; it travels with greater velocity than any other part of the drilling line. { ¦fast ‚līn }

fast neutron [NUCLEO] A neutron having energy much greater than some arbitrary lower limit (that may be only a few thousand electronvolts). { ¦fast 'nü‚trän }

fast-neutron spectrometry [NUC PHYS] Neutron spectrometry in which nuclear reactions are produced by or yield fast neutrons; such reactions are more varied than in the slow-neutron case. { ¦fast 'nü‚trän spek'träm·ə·trē }

fast nova [ASTRON] A nova whose brightness rises quickly to a maximum, remains near maximum for a short time, and then decreases to the original value in a few years or less. { ¦fast 'nō·və }

fast pin [ENG] A pin that fastens immovably, particularly the pin in a fast joint. { ¦fast ¦pin }

fast powder [MATER] Any explosive having a high-speed detonation. { ¦fast 'pau̇d·ər }

fast pulsar [ASTROPHYS] A pulsar with a very short period, of the order of a millisecond. Also known as millisecond pulsar. { ¦fast 'pəl‚sär }

fast reactor [NUCLEO] A nuclear reactor in which most of the fissions are produced by fast neutrons, with little or no moderator to slow down the neutrons. { ¦fast rē'ak·tər }

fast sheave [PETRO ENG] In a drilling operation, the grooved pulley on the crown block sheave assembly over which the fast line is reeved. { 'fast 'shēv }

fast-spiral drill *See* high-helix drill. { ¦fast ¦spī·rəl 'dril }

fast time constant [ELEC] An electric circuit which combines resistance and capacitance to give a short time constant for capacitor discharge through the resistor. [ELECTR] Circuit with short time constant used to emphasize signals of short duration to produce discrimination against low-frequency components of clutter in radar. { 'fast 'tīm ,kän·stənt }

fast time scale [COMPUT SCI] In simulation by an analog computer, a scale in which the time duration of a simulated event is less than the actual time duration of the event in the physical system under study. { 'fast 'tīm ,skāl }

fast-vibration direction [OPTICS] The direction of the electric field vector of the ray of light that travels with the greatest velocity in an anisotropic crystal and therefore corresponds to the minimum refractive index. { ¦fast vī'brā·shən də,rek·shən }

fat [ANAT] Pertaining to an obese person. [BIOCHEM] Any of the glyceryl esters of fatty acids which form a class of neutral organic compounds. [PHYSIO] The chief component of fat cells and other animal and plant tissues. { fat }

FAT *See* file allocation table. { fat *or* ¦ef¦ā'tē }

fatal accident [MIN ENG] A coal mine accident in which less than five persons are killed and property damage is slight; excludes ignitions and mine fires. { ¦fād·əl 'ak·sə·dənt }

fatal error [COMPUT SCI] An error in a computer program which causes running of the program to be terminated. { ¦fād·əl 'er·ər }

Fata Morgana [OPTICS] A complex mirage characterized by multiple distortions of images, generally in the vertical, so that such objects as cliffs or cottages are distorted and magnified into fantastic castles. { ¦fäd·ə ,mȯr'gän·ə }

fat body [INV ZOO] A nutritional reservoir of fatty tissue surrounding the viscera or forming a layer beneath the integument in the immature larval stages of many insects. [VERT ZOO] A mass of adipose tissue attached to each genital gland in amphibians. { 'fat ,bäd·ē }

fat cell [GISTOL] The principal component of adipose connective tissue; two types are yellow fat cells and brown fat cells. { 'fat ,sel }

fat dye [MATER] A type of oil-soluble dye used in the coloring of candles and other wax products. { 'fat ,dī }

fate map [EMBRYO] A graphic scheme indicating the definite spatial arrangement of undifferentiated embryonic cells in accordance with their destination to become specific tissues. { 'fāt ,map }

fat embolus [MED] An embolus composed principally of fatty substances. { 'fat ,em·bə·ləs }

father file [COMPUT SCI] A copy of the master file from the cycle or generation that precedes the one being updated. { 'fäth·ər ,fīl }

fathom [OCEANOGR] The common unit of depth in the ocean, equal to 6 feet (1.8288 meters). { 'fath·əm }

fathom curve *See* isobath. { 'fath·əm ,kərv }

fatigue [ELECTR] The decrease of efficiency of a luminescent or light-sensitive material as a result of excitation. [MECH] Failure of a material by cracking resulting from repeated or cyclic stress. [PHYSIO] Exhaustion of strength or reduced capacity to respond to stimulation following a period of activity. { fə'tēg }

fatigue allowance [IND ENG] An adjustment to normal time to compensate for production time lost due to exhaustion of the worker. { fə'tēg ə,laü·əns }

fatigue factor [IND ENG] The element of physical and mental exhaustion in a time-motion study; the multiplier used to add the fatigue allowance to the normal time. { fə'tēg ,fak·tər }

fatigue life [MECH] The number of applied repeated stress cycles a material can endure before failure. { fə'tēg ,līf }

fatigue limit [MECH] The maximum stress that a material can endure for an infinite number of stress cycles without breaking. Also known as endurance limit. { fə'tēg ,lim·ət }

fatigue notch factor [MET] A notch, scratch, or other impairment on the surface of a metal resulting in premature failure of the metal. { fə'tēg 'näch ,fak·tər }

fatigue notch sensitivity [MET] A measure of the reduction

of fatigue strength of a metal resulting from a notch. { fə'tēg 'näch sen·sə,tiv·əd·ē }

fatigue ratio [MECH] The ratio of the fatigue limit or fatigue strength to the static tensile strength. Also known as endurance ratio. { fə'tēg ,rā·shō }

fatigue strength [MECH] The maximum stress a material can endure for a given number of stress cycles without breaking. Also known as endurance strength. { fə'tēg ,streŋkth }

fatigue-strength reduction factor *See* factor of stress concentration. { fə'tēg ,streŋkth ri'dək·shən ,fak·tər }

fatigue test [ENG] Test to determine the range of alternating stress which a material can withstand without breaking. { fə'tēg ,test }

fat liquoring agents [MATER] Oil-in-water emulsions used to replace oils in tanned (deoiled) leather hides. { 'fat ,lik·ə·riŋ ,ā·jəns }

fat-metabolizing hormone *See* ketogenic hormone. { 'fat mə¦tab·ə,līz·iŋ ,hȯr,mōn }

fat mortar [MATER] Mortar that adheres to the trowel. { ¦fat 'mȯr·dər }

fat necrosis [MED] Pathologic death of adipose tissue often accompanied by soap production from the hydrolyzed fat; associated with pancreatitis. { ¦fat nə'krō·səs }

fat oil [MATER] Enriched absorber oil that is drawn off from the absorber column after being saturated by hydrocarbon values stripped from a wet natural-gas stream. { 'fat ,ȯil }

Fatou-Lebesgue lemma [MATH] Given a sequence f_n of positive measurable functions on a measure space (X, μ), then

$$\int_X (\lim_{n \to \infty} \inf f_n) d\mu \leq \lim_{n \to \infty} \inf \int_X f_n d\mu$$

{ ,fä'tü lə'beg ¦lem·ə }

fatty acid [ORG CHEM] An organic monobasic acid of the general formula $C_nH_{2n+1}COOH$ derived from the saturated series of aliphatic hydrocarbons; examples are palmitic acid, stearic acid, and oleic acid; used as a lubricant in cosmetics and nutrition, and for soaps and detergents. { ¦fad·ē 'as·əd }

fatty acid peroxidase [BIOCHEM] An enzyme present in germinating plant seeds which catalyzes the oxidation of the carboxyl carbon of fatty acids to carbon dioxide. { ¦fad·ē 'as·əd pə'räk·sə,dās }

fatty-acid pitch *See* packing house pitch. { ¦fad·ē ,as·əd 'pich }

fatty acyl carnitine [BIOCHEM] Transport form of fatty acids which allows them to cross the mitochondrial membrane; formed by reaction of fatty acyl-coenzyme A with carnitine by employing the enzyme carnitine acyltransferase. Also known as acyl carnitine. { 'fad·ē 'as·əl 'kär·nə,tēn }

fatty acyl-coenzyme A [BIOCHEM] Activated form of fatty acids formed by the enzyme acyl-coenzyme A synthetase at the expense of adenosinetriphosphate. Also known as acyl-coenzyme A. { 'fad·ē 'as·əl kō¦en,zīm 'ā }

fatty alcohol [ORG CHEM] A high-molecular-weight, straight-chain primary alcohol derived from natural fats and oils; includes lauryl, stearyl, oleyl, and linoleyl alcohols; used in pharmaceuticals, cosmetics, detergents, plastics, and lube oils and in textile manufacture. { 'fad·ē 'al·kə,hȯl }

fatty amine [ORG CHEM] RCH_2NH_2 A normal aliphatic amine from oils and fats; used as a plasticizer, in medicine, as a chemical intermediate, and in rubber manufacture. { 'fad·ē 'am,ēn }

fatty ester [ORG CHEM] RCOOR′ A fatty acid in which the alkyl group (R′) of a monohydric alcohol replaces the active hydrogen; for example, $RCOOCH_3$ from reaction of RCOOH with methane. { 'fad·ē 'es·tər }

fatty infiltration [PHYSIO] Infiltration of an organ or tissue with excessive amounts of fats. { 'fad·ē ,in·fil'trā·shən }

fatty metamorphosis [MED] Fatty degeneration, fatty infiltration, or both. { 'fad·ē ,med·ə'mȯr·fə·səs }

fatty nitrile [ORG CHEM] RCN An ester of hydrogen cyanide derived from fatty acid; used in lube oil additives and plasticizers, and as a chemical intermediate. { 'fad·ē 'nī,trəl }

fatware [COMPUT SCI] Software that is overly laden with features or is inefficiently designed, so that it occupies inordinate space in disk storage and random-access memory, and requires an inappropriate share of microprocessor power. Also known as bloatware. { 'fat,wer }

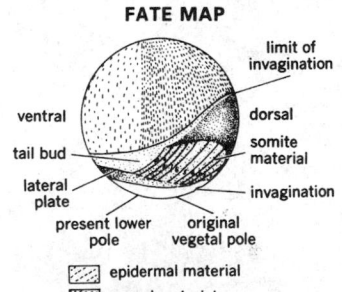

FATE MAP

limit of invagination
ventral
dorsal
tail bud
somite material
lateral plate
invagination
present lower pole
original vegetal pole

epidermal material
neural material
notochord

A fate map for the beginning gastrula of a urodele shown in lateral view.

faucal [BIOL] Of or pertaining to the fauces. [INV ZOO] The opening of a spiral shell. { 'fȯ·kəl }

fauces [ANAT] The passage in the throat between the soft palate and the base of the tongue. [BOT] The throat of a calyx, corolla, or similar part. { 'fȯ‚sēz }

faucet [ENG] A fixture through which water is drawn from a pipe or vessel. { 'fȯs·ət }

faucial tonsil *See* palatine tonsil. { 'fȯ·shəl 'tän·səl }

Faugeron kiln [ENG] A coal-fired tunnel kiln for firing feldspathic porcelain; the distinctive feature is the separation of the tunnel into a series of chambers by division walls on the cars and drop arches in the roof. { 'fō·zhə‚rän ‚kil }

faujasite [MINERAL] $(Na_2,Ca)Al_2Si_4O_{12}\cdot 6H_2O$ Zeolite mineral of the sodalite group, crystallizing in the cubic system. { 'fō·zhə‚sīt }

fault [ELEC] A defect, such as an open circuit, short circuit, or ground, in a circuit, component, or line. Also known as electrical fault; faulting. [ELECTR] Any physical condition that causes a component of a data-processing system to fail in performance. [GEOL] A fracture in rock along which the adjacent rock surfaces are differentially displaced. { fȯlt }

fault analysis [ENG] The detection and diagnosis of malfunctions in technical systems, in particular, by means of a scheme in which one or more computers monitor the technical equipment to signal any malfunction and designate the components responsible for it. { 'fȯlt ə‚nal·ə·səs }

fault basin [GEOL] A region depressed in relation to surrounding regions and separated from them by faults. { 'fȯlt ‚bās·ən }

fault block [GEOL] A rock mass that is bounded by faults; the faults may be elevated or depressed and not necessarily the same on all sides. { 'fȯlt ‚bläk }

fault-block mountain *See* block mountain. { 'fȯlt ‚bläk ‚maȯnt·ən }

fault breccia [GEOL] The assembly of angular fragments found frequently along faults. Also known as dislocation breccia. { 'fȯlt ‚brech·ə }

fault cliff *See* fault scarp. { 'fȯlt ‚klif }

fault current *See* fault electrode current. { 'fȯlt ‚kə·rənt }

fault detection and exclusion [NAV] The capability of a user of the Global Positioning System, once the presence of a fault in the system has been detected, to identify and exclude the malfunctioning satellite in order to continue navigating using the remaining satellites. Abbreviated FDE. { 'fȯlt di‚tek·shən and ik'sklü·zhən }

fault electrode current [ELEC] The current to an electrode under fault conditions, such as during arc-backs and load short circuits. Also known as fault current; surge electrode current. { 'fȯlt i'lek‚trōd ‚kə·rənt }

fault escarpment *See* fault scarp. { 'fȯlt e‚skärp·mənt }

fault finder [ENG] Test set for locating trouble conditions in communications circuits or systems. { 'fȯlt ‚fīnd·ər }

faulting [ELEC] *See* fault. [GEOL] The fracturing and displacement processes which produce a fault. { 'fȯl·tiŋ }

fault ledge *See* fault scarp. { 'fȯlt ‚lej }

fault line [GEOL] Intersection of the fault surface with the surface of the earth or any other horizontal surface of reference. Also known as fault trace. { 'fȯlt‚līn }

fault-line scarp [GEOL] A cliff produced when a soft rock erodes against hard rock at a fault. { 'fȯlt‚līn ‚skärp }

fault masking [COMPUT SCI] Any type of hardware redundancy in which faults are corrected immediately and the operations of fault detection, location, and correction are indistinguishable. { 'fȯlt ‚mask·iŋ }

fault monitoring [SYS ENG] A procedure for systematically checking for errors and malfunctions in the software and hardware of a computer or control system. { 'fȯlt ‚män·ə·triŋ }

fault plane [GEOL] A planar fault surface. { 'fȯlt ‚plān }

fault rock [GEOL] A rock often found along a fault plane and made up of fragments formed by the crushing and grinding which accompany a dislocation. { 'fȯlt ‚räk }

fault scarp [GEOL] A steep cliff formed by movement along one side of a fault. Also known as cliff of displacement; fault cliff; fault escarpment; fault ledge. { 'fȯlt ‚skärp }

fault separation [GEOL] Apparent displacement of a fault measured on the basis of disrupted linear features. { 'fȯlt ‚sep·ə‚rā·shən }

fault strike [GEOL] The angular direction, with respect to

north, of the intersection of the fault surface with a horizontal plane. { 'fȯlt ‚strīk }

fault system [GEOL] Two or more fault sets which interconnect. { 'fȯlt ‚sis·təm }

fault terrace [GEOL] A step on a slope, produced by displacement of two parallel faults. { 'fȯlt ‚ter·əs }

fault throw [GEOL] The amount of vertical displacement of rocks due to faulting. { 'fȯlt ‚thrō }

fault tolerance [SYS ENG] The capability of a system to perform in accordance with design specifications even when undesired changes in the internal structure or external environment occur. { 'fȯlt ‚täl·ə·rəns }

fault trace *See* fault line. { 'fȯlt ‚trās }

fault trap [GEOL] Oil or gas reservoir formed by a structural trap limited in one or more directions by subterranean geological faulting. { 'fȯlt ‚trap }

fault tree [IND ENG] A graphical representation of an undesired event caused by a combination of factors arising from equipment failure, human error, or environmental events. { 'fȯlt ‚trē }

fault-trough lake *See* sag pond. { 'fȯlt ‚trȯf ‚lāk }

fault vein [GEOL] A mineral vein deposited in a fault fissure. { 'fȯlt ‚vān }

fault wall [GEOL] The mass of rock on a particular side of a fault. { 'fȯlt ‚wȯl }

fault zone [GEOL] A fault expressed as an area of numerous small fractures. Also known as distributed fault. { 'fȯlt ‚zōn }

fauna [ZOO] **1.** Animals. **2.** The animal life characteristic of a particular region or environment. { 'fȯn·ə }

faunal extinction [EVOL] The worldwide death and disappearance of diverse animal groups under circumstances that suggest common and related causes. Also known as mass extinction. { 'fȯn·əl ik'stiŋk·shən }

faunal region [ECOL] A division of the zoosphere, defined by geographic and environmental barriers, to which certain animal communities are bound. { 'fȯn·əl ‚rē·jən }

faunizone [GEOL] A bed characterized by fossils of a particular assemblage of fauna. { 'fȯn·ə‚zōn }

faunule [PALEON] The localized stratigraphic and geographic distribution of a particular taxon. { 'fȯ‚nyül }

Faure storage battery [ELEC] A storage battery in which the plates consist of lead-antimony supporting grids covered with a lead oxide paste, immersed in weak sulfuric acid. Also known as pasted-plate storage battery. { 'fȯr 'stȯr·ij ‚bad·ə‚rē }

Faust jig [MIN ENG] A plunger-type jig, usually built with multiple compartments; distinguished by synchronized plungers on both sides of the screen plate, withdrawal of refuse through kettle valves in each compartment, and discharge of the hutch periodically by means of hand valves. { 'faȯst jig }

favism [MED] An acute hemolytic anemia, usually in persons of Mediterranean area descent, occurring when an individual with glucose-6-phosphate dehydrogenase deficiency of erythrocytes eats the beans or inhales the pollen of *Vicia faba*. { 'fä‚viz·əm }

favorable current [NAV] A current flowing in such a direction as to increase the speed of a vessel over the ground. { 'fāv·rə·bəl 'kə·rənt }

favorable wind [NAV] A wind which aids a craft in making progress in a desired direction; usually used in plural form and chiefly in connection with sailing vessels. Also known as fair wind. { 'fāv·rə·bəl 'wind }

Favorskii rearrangement [ORG CHEM] A reaction in which α-halogenated ketones undergo rearrangement in the presence of bases, with loss of the halogen and formation of carboxylic acids or their derivatives with the same number of carbon atoms. { fa'vȯr·skē ‚rē·ə'ränj·mənt }

Favositidae [PALEON] A family of extinct Paleozoic corals in the order Tabulata. { ‚fav·ə'sid·ə‚dē }

favus [MED] A fungal infection of the scalp, usually caused by *Trichophyton schoenleini*, characterized by round, yellow, cup-shaped crusts having a peculiar mousy odor. Also known as tinea favosa. { 'fā·vəs }

fax *See* facsimile. { faks }

fax chart *See* facsimile chart. { 'faks ‚chärt }

Faxen drag factor *See* drag factor. { 'fäk·sən 'drag ‚fak·tər }

fax map *See* facsimile chart. { 'faks ‚map }

fayalite [MINERAL] Fe_2SiO_4 A brown to black mineral of

F BAND

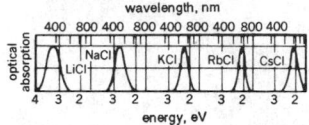

F bands in different alkali halide crystals.

FEATHER

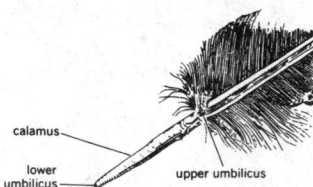

calamus

lower umbilicus   upper umbilicus

Detail of the base of a feather.

the olivine group, consisting of iron silicate and found either massive or in crystals; specific gravity is 4.1. { fə'yä,līt }

faying surface [ENG] The surfaces of materials in contact with each other and joined or about to be joined together. { 'fā·iŋ ,sər·fəs }

F band [SOLID STATE] The optical absorption band arising from *F* centers. { 'ef ,band }

FB data set [COMPUT SCI] A data set which has F-format logical records and whose physical records are all some multiple of the size of the logical record, except possibly for a few truncated blocks. Also known as blocked F-format data set. { ¦ef¦bē ,dad·ə ,set }

FBM data set [COMPUT SCI] An FB data set which has a machine-control (M) character in its first byte of information. { ¦ef¦bē¦em ,dad·ə ,set }

FBR *See* fast breeder reactor.

FBSA data set [COMPUT SCI] An FBS data set which has an ASCII (American Standard Code for Information Interchange) control (A) character in its first byte of information. { ¦ef¦bē¦es¦ā ,dad·ə ,set }

FBS data set [COMPUT SCI] An FB data set which has at most one truncated block, which must be the last one in the data set. Also known as standard blocked F-format data set. { ¦ef¦bē ¦es ,dad·ə ,set }

FCC *See* chlorofluorocarbon.

fcc lattice *See* face-centered cubic lattice. { ¦ef¦sē¦sē 'lad·əs }

F center [SOLID STATE] A color center consisting of an electron trapped by a negative ion vacancy in an ionic crystal, such as an alkali halide or an alkaline-earth fluoride or oxide. { ¦ef ¦sen·tər }

F' center [SOLID STATE] A color center that gives rise to a broad absorption band at longer wavelengths than the band of the F center; probably an F center that has trapped an additional electron. { ¦ef,prīm ,sen·tər }

F connector [ELECTR] A plug and socket for interconnecting coaxial cables; commonly used to interconnect television receivers, videocassette recorders, and cable or antenna sources. { 'ef kə,nek·tər }

F corona [ASTRON] The outer portion of the solar corona, consisting of sunlight that has been scattered from interplanetary dust between the sun and the earth. Also known as Fraunhofer corona. { 'ef kə,rō·nə }

Fc region [IMMUNOL] Region of an antibody molecule that binds to antibody receptors on the surface of cells such as macrophages and mast cells, and to complement protein; F_c is derived from the term crystallizable fragment. { ¦ef'sē ,rē jən }

FD&C color *See* food color. { ¦ef¦dē·ən'sē ,kəl·ər }

FDDI *See* fiber-optic data distribution interface.

FDE *See* fault detection and exclusion.

F display [ELECTR] A rectangular display in which a target appears as a centralized blip when the radar antenna is aimed at it; horizontal and vertical aiming errors are respectively indicated by the horizontal and vertical displacement of the blip. { 'ef di,splā }

F distribution [STAT] The ratio of two independent chi-square variables each divided by its degree of freedom; used to test hypotheses in the analysis of variance and hypotheses about whether or not two normal populations have the same variance. { 'ef ,dis·trə,byü·shən }

FDM *See* frequency-division multiplexing.

FDMA *See* frequency-division multiple access.

FDNR *See* frequency-dependent negative resistor.

Fe *See* iron.

fear [PSYCH] Emotional and physiologic response to recognized sources of danger. { fir }

feasibility study [SYS ENG] **1.** A study of applicability or desirability of any management or procedural system from the standpoint of advantages versus disadvantages in any given case. **2.** A study to determine the time at which it would be practicable or desirable to install such a system when determined to be advantageous. **3.** A study to determine whether a plan is capable of being accomplished successfully. { ,fēz·ə'bil·əd·ē ,stəd·ē }

feasibility test [SYS ENG] A test conducted to obtain data in support of a feasibility study or to demonstrate feasibility. { ,fēz·ə'bil·əd·ē ,test }

feasible flow [MATH] A flow on a directed network such

that the net flow at every intermediate vertex is zero. { ¦fē zə·bəl 'flō }

feasible ground [MIN ENG] Ground that is easy to work and yet will stand without the support of timber or boards. { 'fēz· ə·bəl 'graund }

feasible method *See* interaction prediction method. { 'fēz·ə bəl 'meth·əd }

feasible solution [COMPUT SCI] In linear programming, any set of values for the variables x_j, $j = 1, 2, \ldots, n$, that (1) satisfy the set of restrictions

$$\sum_{j=1}^{n} a_{ij}x_j \leq b_i, i = 1, 2, \ldots, m$$

$$\left(\text{alternatively,} \sum_{j=1}^{n} a_{ij}x_j \leq b_i, \text{ or } \sum_{j=1}^{n} a_{ij}x_j = b_i \right)$$

where the b_i are numerical constants known collectively as the right-hand side and the a_{ij} are coefficients of the variables x_j, and (2) satisfy the restrictions $x_j \geq 0$. { 'fēz·ə·bəl sə'lü·shən }

feather [MECH ENG] To change the pitch on a propeller in order to reduce drag and prevent windmilling in case of engine failure. *See* barb. [VERT ZOO] An ectodermal derivative which is a specialized keratinous outgrowth of the epidermis of birds; functions in flight and in providing insulation and protection. { 'feth·ər }

feather alum *See* alunogen; halotrichite. { 'feth·ər ,al·əm }

Feather analysis [NUCLEO] A technique for determining the range in aluminum of the beta rays of a species by comparing the absorption curve of that species with the absorption curve of a reference species. { 'feth·ər ə,nal·ə·səs }

feather cloth [TEXT] Fabric in which small feathers are added for softness and decoration. { 'feth·ər ,klóth }

featheredge [CIV ENG] The thin edge of a gravel-surfaced road. [DES ENG] A wood tool with a level edge used to straighten angles in the finish coat of plaster. { 'feth·ər,ej }

feathering [FOOD ENG] Flocculation of the cream (fat) in homogenized milk when added to hot coffee or tea due to a defect in the chemistry of the cream. [GRAPHICS] The diffusing or spreading of lines or photographic images. [MECH ENG] A pitch position in a controllable-pitch propeller; it is used in the event of engine failure to stop the windmilling action, and occurs when the blade angle is about 90° to the plane of rotation. Also known as full feathering. [VERT ZOO] Plumage. { 'feth·ə·riŋ }

feathering propeller [MECH ENG] A variable-pitch marine or airscrew propeller capable of increasing pitch beyond the normal high pitch value to the feathered position. { 'feth·ə· riŋ prə'pel·ər }

feather joint [ENG] A joint made by cutting a mating groove in each of the pieces to be joined and inserting a feather in the opening formed when the pieces are butted together. Also known as ploughed-and-tongued joint. [GEOL] One of a series of joints in a fault zone formed by shear and tension. Also known as pinnate joint. { 'feth·ər ,jòint }

feather ore *See* jamesonite. { 'feth·ər ,ór }

feather rot [PL PATH] A fungus rot of both dead and living tree trunks caused by *Poria subacida* and characterized by the white stringy or spongy nature of the rotted tissue. { 'feth· ər ,rät }

feature [COMPUT SCI] In automatic pattern recognition, a property of an image that is useful for its interpretation. { 'fē·chər }

feature extraction-classification model [COMPUT SCI] A method of automatic pattern recognition in which recognition is achieved by making measurements on the patterns to be recognized, and then deriving features from these measurements. { 'fē·chər ik¦strak·shən ,klas·ə·fə¦kā·shən ,mäd·əl }

febrile convulsion [MED] A type of convulsion that occurs in infants and young children in association with fever. { ¦fe,brīl kən'vəl·shən }

febrile disease [MED] Any disease associated with or characterized by fever. { 'feb·rəl di,zēz }

febriphobia [PSYCH] An abnormal fear of fever. { ,feb· rə'fō·bē·ə }

fecalith [MED] A hardened piece of fecal matter formed in the intestine or vermiform appendix. { 'fek·ə,lith }

fecal pellets [GEOL] Mainly the excreta of invertebrates occurring in marine deposits and as fossils in sedimentary rocks. Also known as castings. { 'fē·kəl 'pel·əts }

feces [PHYSIO] The waste material eliminated by the gastrointestinal tract. { 'fē·sēz }

Fechner color [OPTICS] A sensation of color caused by achromatic stimuli at intervals in time. { 'fek·nər ˌkäl·ər }

Fechner fraction [PHYSIO] The smallest difference in the brightness of two sources that can be detected by the human eye divided by the brightness of one of them. { 'fek·ner ˌfrak·shən }

Fechner law [PHYSIO] The intensity of a sensation produced by a stimulus varies directly as the logarithm of the numerical value of that stimulus. { 'fek·nər ˌlȯ }

fecundity [BIOL] The innate potential reproductive capacity of the individual organism, as denoted by its ability to form and separate from the body the mature germ cells. { fə'kən·dəd·ē }

Federal Telecommunications System [COMMUN] System of commercial telephone lines, leased by the government, for use between major government installations for official telecommunications. { 'fed·rəl ˌtel·ə·kə,myü·nə'kā·shənz ˌsis·təm }

Fedorov stage See universal stage. { fyȯ'dȯr·ȯf ˌstāj }

fedsim star [COMPUT SCI] The starlike shape that is characteristic of the Kiviat graph of a well-balanced computer system. { 'fed,sim ˌstär }

feed [AGR] Any crops or other food substances for livestock. [COMPUT SCI] **1.** To supply the material to be operated upon to a machine. **2.** A device capable of so feeding. [ELECTR] To supply a signal to the input of a circuit, transmission line, or antenna. [ELECTROMAG] The part of a radar antenna that is connected to or mounted on the end of the transmission line and serves to radiate radio-frequency electromagnetic energy to the reflector or receive energy therefrom. [ENG] **1.** Process or act of supplying material to a processing unit for treatment. **2.** The material supplied to a processing unit for treatment. **3.** A device that moves stock or workpieces to, in, or from a die. [FOOD ENG] The fermenting wort that is removed from the yeast troughs during brewing processes. [MECH ENG] Forward motion imparted to the cutters or drills of cutting or drilling machinery. { fēd }

feedback [CHEM] In a stepwise reaction, the formation of a substance in one step that affects the rate of a previous step. [ELECTR] The return of a portion of the output of a circuit or device to its input. [SCI TECH] The control of input as a function of output by returning a portion of the output to the input. { 'fēd,bak }

feedback admittance [ELECTR] Short-circuit transadmittance from the output electrode to the input electrode of an electron tube. { 'fēd,bak əd'mit·əns }

feedback amplifier [ELECTR] An amplifier in which a passive network is used to return a portion of the output signal to its input so as to change the performance characteristics of the amplifier. { 'fēd,bak 'am·plə,fī·ər }

feedback branch [CONT SYS] A branch in a signal-flow graph that belongs to a feedback loop. { 'fēd,bak ,branch }

feedback circuit [ELECTR] A circuit that returns a portion of the output signal of an electronic circuit or control system to the input of the circuit or system. { 'fēd,bak ,sər·kət }

feedback compensation [CONT SYS] Improvement of the response of a feedback control system by placing a compensator in the feedback path, in contrast to cascade compensation. Also known as parallel compensation. { 'fēd,bak ,käm·pən,sā·shən }

feedback control loop See feedback loop. { 'fēd,bak kən'trōl ˌlüp }

feedback control signal [CONT SYS] The portion of an output signal which is retransmitted as an input signal. { 'fēd,bak kən'trōl ,sig·nəl }

feedback control system [CONT SYS] A system in which the value of some output quantity is controlled by feeding back the value of the controlled quantity and using it to manipulate an input quantity so as to bring the value of the controlled quantity closer to a desired value. Also known as closed-loop control system. { 'fēd,bak kən'trōl ,sis·təm }

feedback factor [ELECTR] The fraction of the output voltage of an oscillator which is applied to the feedback network. { 'fēd,bak ,fak·tər }

feedback inhibition [BIOCHEM] A cellular control mechanism by which the end product of a series of metabolic reactions inhibits the activity of the first enzyme in the sequence. Also known as allosteric control. { 'fēd,bak ,in·ə,bish·ən }

feedback loop [CONT SYS] A closed transmission path or loop that includes an active transducer and consists of a forward path, a feedback path, and one or more mixing points arranged to maintain a prescribed relationship between the loop input signal and the loop output signal. Also known as feedback control loop. { 'fēd,bak ,lüp }

feedback oscillator [ELECTR] An oscillating circuit, including an amplifier, in which the output is fed back in phase with the input; oscillation is maintained at a frequency determined by the values of the components in the amplifier and the feedback circuits. { 'fēd,bak ,äs·ə,lād·ər }

feedback regulator [CONT SYS] A feedback control system that tends to maintain a prescribed relationship between certain system signals and other predetermined quantities. { 'fēd,bak ,reg·yə,lād·ər }

feedback transfer function [CONT SYS] In a feedback control loop, the transfer function of the feedback path. { 'fēd,bak 'tranz·fər ,faŋk·shən }

feedback winding [ELECTR] A winding to which feedback connections are made in a magnetic amplifier. { 'fēd,bak ,wīnd·iŋ }

feed chute [ORD] A chute or passage through which ammunition is guided into the breech mechanism of a machine gun. { 'fēd ,shüt }

feed-control valve [MECH ENG] A small valve, usually a needle valve, on the outlet of the hydraulic-feed cylinder on the swivel head of a diamond drill, used to control minutely the speed of the hydraulic piston travel and hence the rate at which the bit is made to penetrate the rock. { 'fēd kən,trōl ,valv }

feeder [ELEC] **1.** A transmission line used between a transmitter and an antenna. **2.** A conductor, or several conductors, connecting generating stations, substations, or feeding points in an electric power distribution system. **3.** A group of conductors in an interior wiring system which link a main distribution center with secondary or branch-circuit distribution centers. [GEOL] A small ore-bearing vein which merges with a larger one. [HYD] See tributary. [MECH ENG] **1.** A conveyor adapted to control the rate of delivery of bulk materials, packages, or objects, or a control device which separates or assembles objects. **2.** A device for delivering materials to a processing unit. [MET] A runner or riser so placed that it can feed molten metal to the contracting mass of the casting as it cools in its flask, therefore preventing formation of cavities or porous structure. [ORD] A device that supplies ammunition to a weapon, usually actuated by an automatic or semiautomatic mechanism. { 'fēd·ər }

feeder beach [GEOL] A beach that is artificially widened and nourishes downdrift beaches by natural littoral currents or forces. { 'fēd·ər ,bēch }

feeder-breaker [MECH ENG] A unit that breaks and feeds ore or crushed rock to a materials-handling system at a required rate. { 'fēd·ər 'brāk·ər }

feeder cable [COMMUN] In communications practice, a cable extending from the central office along a primary route (main feeder cable) or from a main feeder cable along a secondary route (branch feeder cable) and providing connections to one or more distribution cables. { 'fēd·ər ,kā·bəl }

feeder canal [CIV ENG] A canal serving to conduct water to a larger canal. { 'fēd·ər kə,nal }

feeder conveyor [MECH ENG] A short auxiliary conveyor designed to transport materials to another conveyor. Also known as stage loader. { 'fēd·ər kən,vā·ər }

feeder current [OCEANOGR] A current which flows parallel to the shore before converging with other such currents and forming the neck of a rip current. { 'fēd·ər ,kə·rənt }

feeder distribution center [COMMUN] Distribution center at which feeders or subfeeders are connected. { 'fēd·ər dis·trə'byü·shən ,sen·tər }

feeder panel [ELEC] The part of a switchboard in an electric power distribution system where feeder connections are made. { 'fēd·ər ,pan·əl }

feeder reactor [ELEC] A small inductor connected in series with a feeder in order to limit and localize the disturbances due to faults on the feeder. { 'fēd·ər rē,ak·tər }

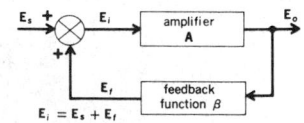

FEEDBACK CIRCUIT

$E_i = E_s + E_f$

Block diagram of a feedback circuit: E_s is sinusoidal input signal; E_i is actuating signal; E_o is output signal; E_f is feedback signal; A is amplifier gain; and β is feedback function.

feeder road [CIV ENG] A road that feeds traffic to a more important road. { 'fēd·ər ¦rōd }

feeder trough [MIN ENG] The trough connected to the conveyor pan line in a duckbill. { 'fēd·ər ¦tróf }

feeder yarn [TEXT] Yarn that is furnished to the throwing process. { 'fēd·ər ¦yärn }

feedforward control [CONT SYS] Process control in which changes are detected at the process input and an anticipating correction signal is applied before process output is affected. { 'fēd¦fór·wərd kən¦trōl }

feedhead [MET] A reservoir of molten metal that is left above a casting in order to supply additional metal as the casting solidifies and shrinks. Also known as riser; sinkhead. { 'fēd¦hed }

feed holes [COMPUT SCI] Holes along the edges of continuous-feed computer paper that are engaged by sprockets to move the paper and maintain alignment during printing. { 'fēd ¦hōlz }

feed horn [ELECTROMAG] A device located at the focus of a receiving paraboloidal antenna that acts as a receiver of radio waves which the antenna collects, focuses, and couples to transmission lines to the amplifier. { 'fēd ¦hórn }

feeding ground See drainage basin. { 'fēd·iŋ ¦graund }

feeding mechanism [ZOO] A mechanism by which an animal obtains and utilizes food materials. { 'fēd·iŋ ¦mek·ə¦niz·əm }

feeding rod [MET] A rod used by working up and down to keep the passage clear between riser and casting. { 'fēd·iŋ ¦räd }

feeding zone [CONT SYS] The area on the planar surface of a conveyor or pallet where the center of an object to be manipulated by a robotic system is placed. { 'fēd·iŋ ¦zōn }

feed lines [MET] The pattern produced on the surface of a piece of metal by machine grinding. { 'fēd ¦līnz }

feed materials [NUCLEO] Refined uranium or thorium metal or their pure compounds in a form suitable for use in nuclear reactor fuel elements or as feed for uranium-enrichment processes. { 'fēd mə¦tir·ē·əlz }

feed nut [MECH ENG] The threaded sleeve fitting around the feed screw on a gear-feed drill swivel head, which is rotated by means of paired gears driven from the spindle or feed shaft. { 'fēd ¦nət }

feed off [ENG] To lower the bit continuously or intermittently during a drilling operation by disengaging the drum brake. { 'fēd ¦óf }

feed pipe [MECH ENG] The pipe which conducts water to a boiler drum. { 'fēd ¦pīp }

feed pitch [DES ENG] The distance between the centers of adjacent feed holes in punched paper tape. { 'fēd ¦pich }

feed preparation unit [CHEM ENG] A processing unit (such as distillation or desulfurization units) providing feedstock for subsequent processing. { ¦fēd prep·ə¦rā·shən ¦yü·nət }

feed pressure [MECH ENG] Total weight or pressure, expressed in pounds or tons, applied to the drilling stem to make the drill bit cut and penetrate the geologic, rock, or ore formation. { 'fēd ¦presh·ər }

feed pump [MECH ENG] A pump used to supply water to a steam boiler. { 'fēd ¦pəmp }

feed rate See cutting speed. { 'fēd ¦rāt }

feed ratio [MECH ENG] The number of revolutions a drill stem and bit must turn to advance the drill bit 1 inch when the stem is attached to and rotated by a screw- or gear-feed type of drill swivel head with a particular pair of the set of gears engaged. Also known as feed speed. { 'fēd ¦rā·shō }

feed reel [ENG] The reel from which paper tape or magnetic tape is being fed. { 'fēd ¦rēl }

feed screw [MECH ENG] The externally threaded drill-rod drive rod in a screw- or gear-feed swivel head on a diamond drill; also used on percussion drills, lathes, and other machinery. { 'fēd ¦skrü }

feed shaft [MECH ENG] A short shaft or countershaft in a diamond-drill gear-feed swivel head which is rotated by the drill motor through gears or a fractional drive and by means of which the engaged pair of feed gears is driven. { 'fēd ¦shaft }

feed shelf [COMPUT SCI] **1.** A device for supporting documents for manual sensing. **2.** The first few feet of a tape reel, used to prime the tape drive. { 'fēd ¦shelf }

feed speed See feed ratio. { 'fēd ¦spēd }

feedstock [ENG] The raw material furnished to a machine or process. { 'fēd¦stäk }

feedstuff [AGR] Food, usually of lower quality, used for animals. { 'fēd¦stəf }

feed tank [ENG] A chamber that contains feedstock. { 'fēd ¦taŋk }

feed-tape [COMPUT SCI] A mechanism which will feed tape to be read or sensed. { 'fēd¦tāp }

feedthrough [ELEC] A conductor that connects patterns on opposite sides of a printed circuit board. Also known as interface connection. { 'fēd ¦thrü }

feedthrough capacitor [ELEC] A feedthrough terminal that provides a desired value of capacitance between the feedthrough conductor and the metal chassis or panel through which the conductor is passing; used chiefly for bypass purposes in ultra-high-frequency circuits. { 'fēd¦thrü kə¦pas·əd·ər }

feedthrough insulator See feedthrough terminal. { 'fēd¦thrü 'in·sə¦lād·ər }

feedthrough terminal [ELEC] An insulator designed for mounting in a hole in a panel, wall, or bulkhead, with a conductor in the center on the insulator to permit feeding electricity through the partition. Also known as feedthrough insulator. { 'fēd¦thrü 'tərm·ən·əl }

feed track [COMPUT SCI] The longitudinal channel on a paper tape that contains the feed holes. { 'fēd ¦trak }

feed travel [MECH ENG] The distance a drilling machine moves the steel shank in traveling from top to bottom of its feeding range. { 'fēd ¦trav·əl }

feed tray [CHEM ENG] For a tray-type distillation column, that tray on which fresh feedstock is introduced into the system. { 'fēd ¦trā }

feed trough [MECH ENG] A receptacle into which feedwater overflows from a boiler drum. { 'fēd ¦tróf }

feedwater [MECH ENG] The water supplied to a boiler or still. { 'fēd¦wód·ər }

feedwater heater [MECH ENG] An apparatus that utilizes steam extracted from an engine or turbine to heat boiler feedwater. { 'fēd¦wód·ər ¦hēd·ər }

feeler gage [MECH ENG] A tool with many blades of different thickness used to establish clearance between parts or for gapping spark plugs. { 'fēl·ər ¦gāj }

feeler pin [MECH ENG] A pin that allows a duplicating machine to operate only when there is a supply of paper. { 'fēl·ər ¦pin }

feel the bottom [NAV] The effect on a ship underway in shallow water which tends to reduce its speed, make it slow in answering the helm, and often make it sheer off course; the speed reduction is largely due to increased wave making resistance resulting from higher pressure differences due to restriction of flow around the hull; the increased velocity of the water flowing past the hull results in an increase in the squat. Also known as smell the bottom. { ¦fēl thə 'bäd·əm }

Fehling's reagent [ANALY CHEM] A solution of cupric sulfate, sodium potassium tartrate, and sodium hydroxide, used to test for the presence of reducing compounds such as sugars. { 'fāl·iŋz rē¦ā·jənt }

Feinc filter [MIN ENG] A vacuum-type drum filter in which a system of parallel strings is used to carry the filter cake away from the drum, instead of the usual filter cloth. { 'fīŋk ¦fil·tər }

Feit-Thompson theorem [MATH] The proposition that every group of odd order is solvable. { ¦fīt ¦täm·sən ¦thir·əm }

feldspar [MINERAL] A group of silicate minerals that make up about 60% of the outer 9 miles (15 kilometers) of the earth's crust; they are silicates of aluminum with the metals potassium, sodium, and calcium, and rarely, barium. { 'fel¦spär }

feldspathic graywacke [PETR] Sandstone containing less than 75% quartz and chert and 15–75% detrital clay matrix, and having feldspar grains in greater abundance than rock fragments. Also known as arkosic wacke; high-rank graywacke. { fel'spath·ik 'grā¦wak·ə }

feldspathic sandstone [PETR] Sandstone rich in feldspar; intermediate in composition between arkosic sandstone and quartz sandstone, made up of 10–25% feldspar and less than 20% matrix material. { fel'spath·ik 'san¦stōn }

feldspathic shale [PETR] A well-laminated shale with more than 10% feldspar in the silt size and with a finer matrix of kaolinitic clay minerals. { fel'spath·ik 'shāl }

feldspathization [GEOL] Formation of feldspar in a rock

usually as a result of metamorphism leading toward granitization. { ‚fel‚spa·thə'zā·shən }

feldspathoid [GEOL] Aluminosilicates of sodium, potassium, or calcium that are similar in composition to feldspars but contain less silica than the corresponding feldspar. { 'fel‚spa‚thȯid }

f electron [ATOM PHYS] An atomic electron that has an orbital angular momentum quantum number of 3 in the central field approximation. { 'ef i‚lek‚trän }

Felidae [VERT ZOO] The cats and saber-toothed cats, a family of mammals in the superfamily Feloidea. { 'fel·ə‚dē }

feline [VERT ZOO] **1.** Of or relating to the genus *Felis*. **2.** Catlike. { 'fē‚līn }

Felis [VERT ZOO] The type genus of the Felidae, comprising the true or typical cats, both wild and domestic. { 'fē·ləs }

fell [FOR] The timber cut in a given season. { fel }

fell-field [ECOL] A culture community of dwarfed, scattered plants or grasses above the timberline. { 'fel ‚fēld }

Fell system [CIV ENG] A method of traction intended for steep railroad slopes; a central rail is gripped between horizontal wheels on the locomotive. { 'fel ‚sis·təm }

Feloidea [VERT ZOO] A superfamily of catlike mammals in the order Carnivora. { fə'lȯid·ē·ə }

Felon's unit [BIOL] A unit for the standardization of antipneumococcic serum. { 'fel·ənz ‚yü·nət }

felsenmeer [GEOL] A flat or gently sloping veneer of angular rock fragments occurring on moderate mountain slopes above the timber line. { 'felz·ən‚mer }

felsic [MINERAL] A light-colored mineral. [PETR] Of an igneous rock, having a mode containing light-colored minerals. { 'fel·sik }

felsite [PETR] **1.** A light-colored, fine-grained igneous rock composed chiefly of quartz or feldspar. **2.** A rock characterized by felsitic texture. { 'fel‚sīt }

felsöbányaite [MINERAL] $Al_4(SO_4)(OH)_{10}·5H_2O$ A yellow to white, probably orthorhombic mineral consisting of a hydrated basic sulfate of aluminum; occurs as aggregates of lamellar crystals. { ‚fel·sȯ'ban·yə‚īt }

felsophyric *See* aphaniphyric. { ‚fel·sə'fir·ik }

felt [MATER] A fibrous, watertight heavy paper of organic or asbestos fibers impregnated with asphalt and used as an overlining or an underlining for roofs. Also known as felt paper. [TEXT] A compressed, densely matted unwoven fabric of wool, sometimes with rayon or hair. { felt }

felt side [MATER] The upper side of a sheet of paper which was not in contact with the wire in a papermaking machine. { 'felt ‚sīd }

felty [GEOL] Referring to a pilotaxitic texture in which the microlites are randomly oriented. { 'fel·tē }

Felty's syndrome [MED] A complex of symptoms involving rheumatoid arthritis, splenomegaly, lymphadenopathy, and anemia. { 'fel·tēz ‚sin‚drōm }

female [BOT] A flower lacking stamens. [ZOO] An individual that bears young or produces eggs. { 'fē‚māl }

female connector [ELEC] A connector having one or more contacts set into recessed openings; jacks, sockets, and wall outlets are examples. { 'fē‚māl kə'nek·tər }

female fitting [DES ENG] In a paired pipe or an electrical or mechanical connection, the portion (fitting) that receives, contrasted to the male portion (fitting) that inserts. { 'fē‚māl 'fid·iŋ }

female heterogamety [GEN] The production by females of two kinds of gametes differing in sex chromosome complement, as in birds and Lepidoptera. { 'fē‚māl ‚hed·ə·rō·gə'med·ē }

female homogamety [GEN] The production by females of a single type of gamete with respect to sex chromosome complement, as in mammals and *Drosophila*. { 'fē‚māl ‚hō·mō·gə'med·ē }

female pseudohermaphroditism *See* gynandry. { 'fē‚māl ‚sü·dō·hər‚maf·rə'dīd‚iz·əm }

feminizing syndrome [MED] Any of a number of symptom complexes in which males tend to take on feminine characteristics due to alterations of adrenocorticotropin output. { 'fem·ə‚niz·iŋ ‚sin‚drōm }

femitrons [ELECTR] Class of field-emission microwave devices. { 'fem·ə‚tränz }

femoral artery [ANAT] The principal artery of the thigh; originates as a continuation of the external iliac artery. { 'fem·ə·rəl 'ärd·ə·rē }

femoral hernia [MED] A hernia that occurs at the passage of the arteries and veins from the abdomen into the legs below the inguinal ligament. { 'fem·ə·rəl 'hər·nē·ə }

femoral nerve [NEUROSCI] A mixed nerve of the leg; the motor portion innervates muscles of the thigh, and the sensory portion innervates portions of the skin of the thigh, leg, hip, and knee. { 'fem·ə·rəl 'nərv }

femoral ring [ANAT] The abdominal opening of the femoral canal. { 'fem·ə·rəl 'riŋ }

femoral vein [ANAT] A vein accompanying the femoral artery. { 'fem·ə·rəl 'vān }

femorotibial index [ARCH] The ratio of the length of the femur to the length of the tibia multiplied by 100. { 'fem·ə·rō‚tib·ē·əl 'in‚deks }

femto- [SCI TECH] A prefix representing 10^{-15}, which is 0.000 000 000 000 001, or one-thousandth of a millionth of a millionth. { 'fem·tō }

femtoampere [ELEC] A unit of current equal to 10^{-15} ampere. Abbreviated fA. { 'fem·tō‚am·pir }

femtometer [MECH] A unit of length, equal to 10^{-15} meter; used particularly in measuring nuclear distances. Abbreviated fm. Also known as fermi. { 'fem·tō‚mēd·ər }

femtovolt [ELEC] A unit of voltage equal to 10^{-15} volt. Abbreviated fV. { 'fem·tō‚vōlt }

femur [ANAT] **1.** The proximal bone of the hind or lower limb in vertebrates. **2.** The thigh bone in humans, articulating with the acetabulum and tibia. { 'fē·mər }

fen [GEOGR] Peat land covered by water, especially in the upper regions of old estuaries and around lakes, that can be drained only artificially. { fen }

fenaminosulf [ORG CHEM] $C_8H_{10}N_3SO_3Na$ A yellow-brown powder, decomposing at 200°C; used as a fungicide for seeds and seedlings in crops. { ‚fen'am·ə·nō‚səlf }

fenazaflor [ORG CHEM] $C_{15}H_7Cl_2F_3N_2O_2$ A greenish-yellow, crystalline compound with a melting point of 103°C; used as an insecticide and miticide for spider mites and eggs. { fə'naz·ə‚flȯr }

fenbutatin oxide [ORG CHEM] $C_{60}H_{78}OSn_2$ A white, crystalline compound, insoluble in water; used to control mites in deciduous and citrus fruits. { fen'byüd·əd·ən 'äk‚sīd }

fence [AERO ENG] A stationary plate or vane projecting from the upper surface of an airfoil, substantially parallel to the airflow, used to prevent spanwise flow. [COMPUT SCI] *See* fence cell. [ENG] **1.** A line of data-acquisition or tracking stations used to monitor orbiting satellites. **2.** A line of radar or radio stations for detection of satellites or other objects in orbit. **3.** A line or network of early-warning radar stations. **4.** A concentric steel fence erected around a ground radar transmitting antenna to serve as an artificial horizon and suppress ground clutter that would otherwise drown out weak signals returning at a low angle from a target. **5.** An adjustable guide on a tool. { fens }

fence cell [COMPUT SCI] A criterion for dividing a list into two equal or nearly equal parts in the course of a binary search. Also known as fence. { 'fens ‚sel }

fenchol *See* fenchyl alcohol. { 'fen·chȯl }

fenchone [ORG CHEM] $C_{10}H_{16}O$ An isomer of camphor; a colorless oil that boils at 193°C and is soluble in ether; a constituent of fennel oil; used as a flavoring. { 'fen‚chōn }

fenchyl alcohol [ORG CHEM] $C_{10}H_{18}O$ A colorless solid or oily liquid, boiling at 198–204°C, isolated from pine oil and turpentine and also made synthetically; used as a solvent, an intermediate in organic synthesis, and as a flavoring. Also known as fenchol. { 'fen·chəl 'al·kə‚hȯl }

fender [CIV ENG] A timber, cluster of piles, or bag of rope placed along dock or bridge pier to prevent damage by docking ships or floating objects. [ENG] A cover over the upper part of a wheel of an automobile or other vehicle. [MIN ENG] A thin pillar of coal adjacent to the gob, left for protection while driving a lift through the mine pillar. [NAV ARCH] A padded device acting as a buffer to prevent damage between two ships or between a ship and dock. { 'fen·der }

Fenestellidae [PALEON] A family of extinct fenestrated, cryptostomatous bryozoans which abounded during the Silurian. { ‚fen·ə'stel·ə‚dē }

fenestra [ANAT] An opening in the medial wall of the middle ear. [MED] An opening in a bandage or plaster splint for examination or drainage. { fə'nes·trə }

fenestrated membrane [HISTOL] One of the layers of elastic tissue in the tunica media and tunica intima of large arteries. { 'fen·ə,sträd·əd 'mem,brān }

fenestration [ARCH] The arrangement of openings, especially windows, in the wall of a building. [BIOL] **1.** A transparent or windowlike break or opening in the surface. **2.** The presence of windowlike openings. { ,fen·ə'strā·shən }

fenitrothion [ORG CHEM] $C_9H_{12}NO_5PS$ A yellow-brown liquid, insoluble in water; used as a miticide and insecticide for rice, orchards, vegetables, cereals, and cotton, and for fly and mosquito control. { ,fen·ə·trō'thī,än }

fennel [BOT] *Foeniculum vulgare.* A tall perennial herb of the family Umbelliferae; a spice is derived from the fruit. { 'fen·əl }

fennel oil [MATER] The essential oil obtained from fennel; a colorless liquid with aromatic scent and bitter taste, insoluble in water and boiling at 160–220°C; used in medicine, perfumes, and liqueurs. Also known as oil of fennel. { 'fen·əl ,oil }

fen peat *See* low-moor peat. { 'fen ,pēt }

Fenske equation *See* Fenske-Underwood equation. { 'fenskē i,kwā·zhən }

Fenske-Underwood equation [CHEM ENG] Equation in plate-to-plate distillation-column calculations relating the number of theoretical plates needed at total reflux to overall relative volatility and the liquid-vapor composition ratios on upper and lower plates. Also known as Fenske equation. { ¦fen·skē 'ən·dər,wùd i,kwā·zhən }

fenster *See* window. { 'fen·stər }

fensulfothion [ORG CHEM] $C_{11}H_{17}S_2O_2P$ A brown liquid with a boiling point of 138–141°C; used as an insecticide and nematicide in soils. { ,fen,səl·fō'thī,än }

fentinacetate [ORG CHEM] $C_{20}H_{18}O_2Sn$ A yellow to brown, crystalline solid that melts at 124–125°C; used as a fungicide, molluscicide, and algicide for early and late blight on potatoes, sugarbeets, peanuts, and coffee. Also known as triphenyltinacetate. { ,fent·ən'as·ə,tāt }

fenuron [ORG CHEM] $C_9H_{12}N_2O$ A white, crystalline compound with a melting point of 133–134°C; soluble in water; used as a herbicide to kill weeds and bushes. { 'fen'yù,rän }

fenuron-TCA [ORG CHEM] $C_{11}H_{13}Cl_3N_2O_3$ A white, crystalline compound with a melting point of 65–68°C; moderately soluble in water; used as a herbicide for noncrop areas. { ,fen'yù,rän ¦tē¦sē¦ā }

FEP resin *See* fluorinated ethylene propylene resin. { ¦ef¦ē¦pē 'rez·ən }

ferbam [ORG CHEM] $C_9H_{18}FeN_3S_6$ [iron(III) dimethyldithiocarbamate] A fungicide for protecting fruits, vegetables, melons, and ornamental plants. { 'fər·bəm }

ferberite [MINERAL] $FeNO_4$ A black mineral of the wolframite solid-solution series occurring as monoclinic, prismatic crystals and having a submetallic luster; hardness is 4.5 on Mohs scale, and specific gravity is 7.5. { 'fər·bə,rīt }

ferghanite [MINERAL] $U_3(VO_4)_2·6H_2O$ Sulfur-yellow mineral composed of hydrated uranium vanadate, occurring in scales. { fər'gä,nīt }

fergusonite [MINERAL] $Y_2O_3·(Nb,Ta)_2O_5$ Brownish-black rare-earth mineral with a tetragonal crystal form; it is isomorphous with formanite. { 'fər·gə·sə,nīt }

Fermat numbers [MATH] The numbers of the form $F_n = (2^{(2^n)}) + 1$ for $n = 0, 1, 2, \ldots$ { 'fer·mä ,nəm·bərz }

Fermat's last theorem [MATH] The proposition, proven in 1995, that there are no positive integer solutions of the equation $x^n + y^n = z^n$ for $n \geq 3$. { fer'mäz ¦last 'thir·əm }

Fermat's principle [OPTICS] The principle that an electromagnetic wave will take a path that involves the least travel time when propagating between two points. Also known as least-time principle; stationary time principle. { fer'mäz 'prin·sə·pəl }

Fermat's spiral [MATH] A plane curve whose equation in polar coordinates (r,θ) is $r^2 = a^2\theta$, where a is a constant. { fer'mäz ,spī·rəl }

Fermat's theorem [MATH] The proposition that, if p is a prime number and a is a positive integer which is not divisible by p, then $a^{p-1} - 1$ is divisible by p. { 'fer,mäz ,thir·əm }

ferment [BIOCHEM] An agent that can initiate fermentation and other metabolic processes. { ¦fər¦ment }

fermentation [MICROBIO] An enzymatic transformation of organic substrates, especially carbohydrates, generally accompanied by the evolution of gas; a physiological counterpart of oxidation, permitting certain organisms to live and grow in the absence of air; used in various industrial processes for the manufacture of products such as alcohols, acids, and cheese by the action of yeasts, molds, and bacteria; alcoholic fermentation is the best-known example. Also known as zymosis. { ,fər·mən'tā·shən }

fermentation accelerator [MATER] Substance that speeds chemical fermentation (as for wines) without participating in the resulting chemical changes; can be an enzyme or other catalytic agent. { ,fər·mən'tā·shən ak'sel·ə,rād·ər }

fermentation tube [MICROBIO] A culture tube with a vertical closed arm to collect gas formed in a broth culture by microorganisms. { ,fər·mən'tā·shən ,tüb }

fermenter [FOOD ENG] A vessel used for fermenting, such as a vat for fermenting mash in brewing. { fər'ment·ər }

ferment oil [MATER] A volatile oil formed by the fermentation of plant material in which the oil was not present originally. { ¦fər¦ment ,oil }

fermi *See* femtometer. { 'fer·mē }

Fermi age [NUCLEO] The value calculated for the slowing-down area in the Fermi age model; it has the dimensions of area, not time. Also known as age; neutron age; symbolic age of neutrons. { 'fer·mē ,āj }

Fermi age equation [NUCLEO] An equation in the Fermi age model which states that the Laplacian of the slowing-down density equals the partial derivative of the slowing-down density with respect to the Fermi age. { 'fer·mē ¦āj i,kwā·zhən }

Fermi age model [NUCLEO] A model used in studying the slowing down of neutrons by elastic collisions; it is assumed that the slowing down takes place by a very large number of very small energy changes. { 'fer·mē ,āj ,mäd·əl }

Fermi beta-decay theory [NUC PHYS] Theory in which a nucleon source current interacts with an electron-neutrino field to produce beta decay, in a manner analogous to the interaction of an electric current with an electromagnetic field during the emission of a photon of electromagnetic radiation. { 'fer·mē ¦bād·ə di¦kā ,thē·ə·rē }

Fermi constant [NUC PHYS] A universal constant, introduced in beta-disintegration theory, that expresses the strength of the interaction between the transforming nucleon and the electron-neutrino field. { 'fer·mē ,kän·stənt }

Fermi derivative [RELAT] A generalization of covariant differentiation along a curve that reduces to covariant differentiation when the curve is geodesic; an orthonormal tetrad constructed at each point along a timelike curve such that the Fermi derivative of the tetrad along the curve is zero has (1) its timelike basis vector equal to the curve's unit tangent vector and (2) its spatial basis vectors nonrotating along the curve. { 'fer·mē də,riv·əd·iv }

Fermi-Dirac distribution function [STAT MECH] A function specifying the probability that a member of an assembly of independent fermions, such as electrons in a semiconductor or metal, will occupy a certain energy state when thermal equilibrium exists. { ¦fer·mē di¦rak ,dis·trə'byü·shən ,fəŋk·shən }

Fermi-Dirac gas *See* Fermi gas. { ¦fer·mē di¦rak ,gas }

Fermi-Dirac statistics [STAT MECH] The statistics of an assembly of identical half-integer spin particles; such particles have wave functions antisymmetrical with respect to particle interchange and satisfy the Pauli exclusion principle. { ¦fer·mē di¦rak stə'tis·tiks }

Fermi distribution [SOLID STATE] Distribution of energies of electrons in a semiconductor or metal as given by the Fermi-Dirac distribution function; nearly all energy levels below the Fermi level are filled, and nearly all above this level are empty. { 'fer·mē ,dis·trə,byü·shən }

Fermi energy [STAT MECH] **1.** The average energy of electrons in a metal, equal to three-fifths of the Fermi level. **2.** *See* Fermi level. { 'fer·mē ,en·ər·jē }

Fermi gas [STAT MECH] An assembly of independent particles that obey Fermi-Dirac statistics, and therefore obey the Pauli exclusion principle; this concept is used in the free-electron theory of metals and in one model of the behavior of the nucleons in a nucleus. Also known as Fermi-Dirac gas. { 'fer·mē ,gas }

Fermi hole [SOLID STATE] A region surrounding an electron in a solid in which the energy band theory predicts that the

FENNEL

Fennel (*Foeniculum vulgare*). (*USDA*)

probability of finding other electrons is less than the average over the volume of the solid. { 'fer·mē ˌhōl }

Fermi interaction [PART PHYS] The direct interaction between four Dirac fields at a single point in space-time, postulated in conventional theories of the weak interactions.

Fermi level [STAT MECH] The energy level at which the Fermi-Dirac distribution function of an assembly of fermions is equal to one-half. Also known as Fermi energy. { 'fer·mē ˌlev·əl }

Fermi liquid [CRYO] A liquid of particles which have Fermi-Dirac statistics; an example is the liquid phase of helium-3, in which the atoms belong to the isotope with mass number 3. { 'fer·mē ˌlik·wəd }

fermion [QUANT MECH] A particle, such as the electron, proton, or neutron, which obeys the rule that the wave function of several identical particles changes sign when the coordinates of any pair are interchanged; it therefore obeys the Pauli exclusion principle. { 'fer·mēˌän }

fermion field [QUANT MECH] An operator defined at each point in space-time that creates or annihilates a particular type of fermion and its antiparticle. { 'fer·mēˌän ˌfēld }

Fermi plot See Kurie plot. { 'fer·mē ˌplät }

Fermi-propagated [RELAT] A vector field is said to be Fermi-propagated along a curve γ when it is constructed so that its Fermi derivative along γ is 0. { 'fer·mē ˌpräp·əˌgād·əd }

Fermi resonance [PHYS CHEM] In a polyatomic molecule, the relationship of two vibrational levels that have in zero approximation nearly the same energy; they repel each other, and the eigenfunctions of the two states mix. { 'fer·mē ˌrez·ən·əns }

Fermi selection rules [NUC PHYS] Selection rules for beta decay in a Fermi transition; that is, there is no change in total angular momentum or parity of the nucleus in an allowed transition. { 'fer·mē si'lek·shən ˌrülz }

Fermi's golden rules [QUANT MECH] The equations giving the first-order (rule number 2) and second-order (rule number 1) contributions to the transition probability per unit time induced by a perturbation Hamiltonian, in terms of matrix elements of the perturbation Hamiltonian. { ˌfer·mēz ˌgōld·ən ˌrülz }

Fermi sphere [STAT MECH] The Fermi surface of an assembly of fermions in the approximation that the fermions are free particles. { 'fer·mē ˌsfir }

Fermi surface [SOLID STATE] A constant-energy surface in the space containing the wave vectors of states of members of an assembly of independent fermions, such as electrons in a semiconductor or metal, whose energy is that of the Fermi level. { 'fer·mē ˌsər·fəs }

Fermi temperature [STAT MECH] The energy of the Fermi level of an assembly of fermions divided by Boltzmann's constant, which appears as a parameter in the Fermi-Dirac distribution function. { 'fer·mē ˌtem·prə·chər }

Fermi transition [NUC PHYS] Beta decay subject to Fermi selection rules. { 'fer·mē tran'zish·ən }

fermium [CHEM] A synthetic radioactive element, symbol Fm, with atomic number 100; discovered in debris of the 1952 hydrogen bomb explosion, and now made in nuclear reactors. { 'fer·mē·əm }

fermorite [MINERAL] $(Ca,Sr)_5[(As,P)O_4]_3$ A white mineral composed of arsenate, phosphate, and fluoride of calcium and strontium, occurring in crystalline masses. { 'fər·məˌrīt }

fern [BOT] Any of a large number of vascular plants composing the division Polypodiophyta. { fərn }

fernandinite [MINERAL] A dull green mineral composed of hydrous calcium vanadyl vanadate. { ˌfər·nən'dēˌnīt }

fernico [MET] An iron-nickel-cobalt alloy used for metal-to-glass seals. { fər'nīˌkō }

Ferranti effect [ELEC] A rise in voltage occurring at the end of a long transmission line when its load is disconnected. { fə'ran·tē iˌfekt }

ferrate [INV ZOO] A multiple iron oxide with another oxide, for example, Na_2FeO_4. { 'feˌrāt }

ferredoxins [BIOCHEM] Iron-containing proteins that transfer electrons, usually at a low potential, to flavoproteins; the iron is not present as a heme. { ˌfer·ə'däkˌsənz }

ferreed switch [ELEC] A switch whose contacts are mounted on magnetic blades or reeds sealed into an evacuated tubular glass housing, the contacts being operated by external electromagnets or permanent magnets. { 'feˌrēd ˌswich }

Ferrers diagram [MATH] An array of dots associated with an integer partition $n = a_1 + \cdots + a_k$, whose ith row contains a_i dots. { 'fer·ərz ˌdī·əˌgram }

ferret [ORD] An aircraft, ship, or vehicle especially equipped for the detection, location, recording, and analyzing of electromagnetic radiation. [VERT ZOO] Mustela nigripes. The largest member of the weasel family, Mustelidae, and a relative of the European polecat; has yellowish fur with black feet, tail, and mask. { 'fer·ət }

ferriamphibole [MINERAL] The ferric ion equivalent of the amphibole group of minerals. { ˌfer·ē'am·fəˌbōl }

ferric [INV ZOO] The term for a compound of trivalent iron, for example, ferric bromide, $FeBr_3$. { 'fer·ik }

ferric acetate [ORG CHEM] $Fe_2(C_2H_3O_2)_3$ A brown compound, soluble in water; used as a tonic and dye mordant. { 'fer·ik 'as·əˌtāt }

ferric ammonium alum See ferric ammonium sulfate. { 'fer·ik ə'mōn·ē·əm 'al·əm }

ferric ammonium citrate [ORG CHEM] $Fe(NH_4)_3(C_6H_5O_7)_2$ Red, deliquescent scales or granules; odorless, water soluble, and affected by light; used in medicine and blueprint photography. { 'fer·ik ə'mōn·ē·əm 'sīˌtrāt }

ferric ammonium oxalate [ORG CHEM] $(NH_4)_3Fe(C_2O_4)_3 \cdot 3H_2O$ Green, crystalline material, soluble in water and alcohol, sensitive to light; used in blueprint photography. { 'fer·ik ə'mōn·ē·əm 'äk·səˌlāt }

ferric ammonium sulfate [INV ZOO] $FeNH_4(SO_4)_2 \cdot 12H_2O$ Efflorescent, water-soluble crystals; used in medicine, in analytical chemistry, and as a mordant in textile dyeing. Also known as ferric ammonium alum; iron ammonium sulfate. { 'fer·ik ə'mōn·ē·əm 'səlˌfāt }

ferric arsenate [INV ZOO] $FeAsO_4 \cdot 2H_2O$ A green or brown powder, insoluble in water, soluble in dilute mineral acids; used as an insecticide. { 'fer·ik 'ärs·ənˌāt }

ferric bromide [INV ZOO] $FeBr_3$ Red, deliquescent crystals that decompose upon heating; soluble in water, ether, and alcohol; used in medicine and analytical chemistry. Also known as ferric sesquibromide; ferric tribromide; iron bromide. { 'fer·ik 'brōˌmīd }

ferric chloride [INV ZOO] $FeCl_3$ Brown crystals, melting at 300°C, that are soluble in water, alcohol, and glycerol; used as a coagulant for sewage and industrial wastes, as an oxidizing and chlorinating agent, as a disinfectant, in copper etching, and as a mordant. Also known as anhydrous ferric chloride; ferric trichloride; flores martis; iron chloride. { 'fer·ik 'klȯrˌīd }

ferric citrate [ORG CHEM] $FeC_6H_5O_7 \cdot 3H_2O$ Red scales that react to light; soluble in water, insoluble in alcohol; used as a medicine for certain blood disorders, and for blueprint paper. Also known as iron citrate. { 'fer·ik 'sīˌtrāt }

ferric dichromate [INV ZOO] $Fe_2(CrO_4)_3$ A red-brown, granular powder, miscible in water; used as a mordant. { 'fer·ik dī'krōˌmāt }

ferric ferrocyanide [INV ZOO] $Fe_4[Fe(CN)_6]_3$ Dark-blue crystals, used as a pigment, and with oxalic acid in blue ink. Also known as iron ferrocyanide. { 'fer·ik ˌfer·ə'sī·əˌnīd }

ferric fluoride [INV ZOO] FeF_3 Green, rhombohedral crystals, soluble in water and acids; used in porcelain and pottery manufacture. Also known as iron fluoride. { 'fer·ik 'flu̇rˌīd }

ferrichrome [MICROBIO] A cyclic hexapeptide that is a microbial hydroxamic acid and is involved in iron transport and metabolism in microorganisms. { 'fer·əˌkrōm }

ferric hydrate See ferric hydroxide. { 'fer·ik 'hīˌdrāt }

ferric hydroxide [INV ZOO] $Fe(OH)_3$ A brown powder, insoluble in water; used as arsenic poisoning antidote, in pigments, and in pharmaceutical preparations. Also known as ferric hydrate; iron hydroxide. { 'fer·ik hī'dräkˌsīd }

ferric nitrate [INV ZOO] $Fe(NO_3)_3 \cdot 9H_2O$ Colorless crystals, soluble in water and decomposed by heat; used as a dyeing mordant, in tanning, and in analytical chemistry. Also known as iron nitrate. { 'fer·ik 'nīˌtrāt }

ferric oxalate [ORG CHEM] $Fe_2(COO)_3$ Yellow scales, soluble in water, decomposing when heated at about 100°C; used as a catalyst and in photographic printing papers. { 'fer·ik 'äk·səˌlāt }

ferric oxide [INV ZOO] Fe_2O_3 Red, hexagonal crystals or powder, insoluble in water and soluble in acids, melting at 1565°C; used as a catalyst and pigment for metal polishing, in

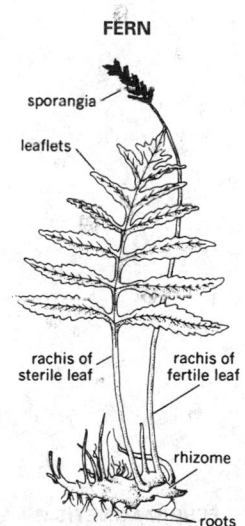

FERN

sporangia
leaflets
rachis of sterile leaf
rachis of fertile leaf
rhizome
roots

The sensitive fern (*Onoclea sensibilis*), a representative of the Polypodiospida.

FERRET

Black-footed ferret (*Mustela nigripes*).

metallurgy, and in medicine. Also known as ferric oxide red; jeweler's rouge; red ocher. { 'fer·ik 'äk,sīd }

ferric oxide red *See* ferric oxide. { 'fer·ik ,ak,sīd 'red }

ferric phosphate [INV ZOO] FePO$_4$·2H$_2$O Yellow, rhombohedral crystals, insoluble in water, soluble in acids; used in medicines and fertilizers. Also known as iron phosphate. { 'fer·ik 'fäs,fāt }

ferric resinate [ORG CHEM] Reddish-brown, water-insoluble powder; used as a drier for paints and varnishes. Also known as iron resinate. { 'fer·ik 'rez·ən,āt }

ferricrete [GEOL] A conglomerate of surficial sand and gravel held together by iron oxide resulting from percolating solutions of iron salts. { 'fer·ə,krēt }

ferric sesquibromide *See* ferric bromide. { 'fer·ik ,ses·kwə'brō,mīd }

ferric stearate [ORG CHEM] Fe(C$_{18}$H$_{35}$O$_2$)$_3$ A light-brown, water-insoluble powder; used as a varnish drier. Also known as iron stearate. { 'fer·ik 'stir,āt }

ferric sulfate [INV ZOO] Fe$_2$(SO$_4$)$_3$·9H$_2$O Yellow, water-soluble, rhombohedral crystals, decomposing when heated; used as a chemical intermediate, disinfectant, soil conditioner, pigment, and analytical reagent, and in medicine. Also known as iron sulfate. { 'fer·ik 'səl,fāt }

ferric tribromide *See* ferric bromide. { 'fer·ik trī'brō,mīd }

ferric trichloride *See* ferric chloride. { 'fer·ik trī'klór,īd }

ferric vanadate [INV ZOO] Fe(VO$_3$)$_3$ Grayish-brown powder, insoluble in water and alcohol; used in metallurgy. Also known as iron metavanadate. { 'fer·ik 'van·ə,dāt }

ferricyanic acid [INV ZOO] H$_3$Fe(CN)$_6$ A red-brown unstable solid. { ,fer·i·sī'an·ik 'as·əd }

ferricyanide [INV ZOO] A salt containing the radical Fe(CN)$_6^{3-}$. { fer·i'sī·ə,nīd }

ferrierite [MINERAL] (Na,K)$_2$MgAl$_3$Si$_{15}$O$_{36}$(OH)·9H$_2$O A zeolite mineral crystallizing in the orthorhombic system. { fə'rē·ə,rīt }

ferriferous [GEOL] Of a sedimentary rock, iron-rich. [MINERAL] Of a mineral, iron-bearing. { fə'rif·ə·rəs }

ferrihemoglobin [BIOCHEM] Hemoglobin in the oxidized state. Also known as methemoglobin. { ¦fe·ri¸hē·mə'glō·bən }

ferrimagnet *See* ferrimagnetic material. { 'fe·ri,mag·nət }

ferrimagnetic amplifier [ELECTR] A microwave amplifier using ferrites. { ,fe·ri·mag'ned·ik 'am·plə,fī·ər }

ferrimagnetic garnet [MATER] A synthetic ferrimagnetic oxide material that has the garnet structure and the chemical formula X$_3$Fe$_5$O$_{12}$, where the trivalent X ion is yttrium, or any of the rare-earth ions with an atomic number greater than 61. { ,fe·ri·mag'ned·ik 'gär·nət }

ferrimagnetic limiter [ELECTROMAG] Power limiter used in microwave systems to replace transmit-receive tubes; uses ferrimagnetic material (such as a piece of ferrite or garnet) that exhibits nonlinear properties. { ,fe·ri·mag'ned·ik 'lim·əd·ər }

ferrimagnetic material [SOLID STATE] A material displaying ferrimagnetism; the ferrites are the principal example. Also known as ferrimagnet. { ,fe·ri·mag'ned·ik mə'tir·ē·əl }

ferrimagnetic resonance [PHYS] Magnetic resonance of a ferrimagnetic material. { ,fe·ri·mag'ned·ik 'rez·ən·əns }

ferrimagnetism [SOLID STATE] A type of magnetism in which the magnetic moments of neighboring ions tend to align nonparallel, usually antiparallel, to each other, but the moments are of different magnitudes, so there is an appreciable resultant magnetization. Also known as Néel ferromagnetism. { ,fe·ri'mag·nə,tiz·əm }

ferrimolybdite [MINERAL] Fe$_2$(MoO$_4$)$_3$·8H$_2$O A colorless to canary yellow, probably orthorhombic mineral consisting of hydrated ferric molybdate; occurs in massive form, as crusts or aggregates. { fe·ri·mə'lib,dīt }

ferrimycin [MICROBIO] The representative antibiotic of the sideromycin group; a hydroxamic acid compound. { ,fe·ri'mīs·ən }

ferrinatrite [MINERAL] Na$_3$Fe(SO$_4$)$_3$·3H$_2$O A greenish or white mineral composed of sodium ferric iron double sulfate; usually occurs in spherical forms. { ,fe·ri'nā,trīt }

ferriporphyrin [BIOCHEM] A red-brown to black complex of iron and porphyrin in which the iron is in the 3+ oxidation state. { ¦fe·ri'pór·fə·rən }

ferrisicklerite [MINERAL] (Li,Fe,Mn)(PO$_4$) Mineral composed of phosphate of lithium, ferric iron, and manganese, more

iron being present than manganese; it is isomorphous with sicklerite. { ¦fe·ri'sik·lə,rīt }

ferristor [ELECTR] A miniature, two-winding, saturable reactor that operates at a high carrier frequency and may be connected as a coincidence gate, current discriminator, free-running multivibrator, oscillator, or ring counter. { fə'ris·tər }

ferrisulphas *See* ferrous sulfate. { ¦fe·ri'səl·fəs }

ferrite [INV ZOO] An unstable compound of a strong base and ferric oxide which exists in alkaline solution, such as NaFeO$_2$. [MET] Iron that has not combined with carbon in pig iron or steel. [PETR] Grains or scales of unidentifiable, generally transparent amorphous iron oxide in the matrix of a porphyritic rock. [SOLID STATE] Any ferrimagnetic material having high electrical resistivity which has a spinel crystal structure and the chemical formula XFe$_2$O$_4$, where X represents any divalent metal ion whose size is such that it will fit into the crystal structure. { 'fe,rīt }

ferrite attenuator *See* ferrite limiter. { 'fe,rīt ə'ten·yə,wād·ər }

ferrite banding [MET] The formation of faint bands (flow lines) of free ferrite in rolled steel, running in the direction of working. Also known as ferrite ghosts; ferrite streaks; ghost lines; ghost structure. { 'fe,rīt 'ban·diŋ }

ferrite bead [ELECTR] Magnetic information storage device consisting of ferrite powder mixtures in the form of a bead fired on the current-carrying wires of a memory matrix. { 'fe,rīt 'bēd }

ferrite circulator [ELECTROMAG] A combination of two dual-mode transducers and a 45° ferrite rotator, used with rectangular waveguides to control and switch microwave energy. Also known as ferrite phase-differential circulator. { 'fe,rīt 'sər·kyə,lād·ər }

ferrite core [ELECTR] A magnetic core made of ferrite material. Also known as dust core; powdered-iron core. { 'fe,rīt 'kór }

ferrite-core memory [ELECTR] A magnetic memory consisting of a matrix of tiny toroidal cores molded from a square-loop ferrite, through which are threaded the pulse-carrying wires and the sense wire. { 'fe,rīt ,kór 'mem·rē }

ferrite device [ELEC] An electrical device whose principle of operation is based upon the use of ferrites in powdered, compressed, sintered form, making use of their ferrimagnetism and their high electrical resistivity, which makes eddy-current losses extremely low at high frequencies. { 'fe,rīt di,vīs }

ferrite ghosts *See* ferrite banding. { 'fe,rīt 'gōsts }

ferrite isolator [ELECTROMAG] A device consisting of a ferrite rod, centered on the axis of a short length of circular waveguide, located between rectangular-waveguide sections displaced 45° with respect to each other, which passes energy traveling through the waveguide in one direction while absorbing energy from the opposite direction. Also known as Faraday rotation isolator. { 'fe,rīt 'ī·sə,lād·ər }

ferrite limiter [ELECTROMAG] A passive, low-power microwave limiter having an insertion loss of less than 1 decibel when operating in its linear range, with minimum phase distortion; the input signal is coupled to a single-crystal sample of either yttrium iron garnet or lithium ferrite, which is biased to resonance by a magnetic field. Also known as ferrite attenuator. { 'fe,rīt 'lim·əd·ər }

ferrite number [MET] The standard value assigned to austenitic stainless steel to denote a specific ferrite content; used instead of percent ferrite. { 'fe,rīt 'nəm·bər }

ferrite phase-differential circulator *See* ferrite circulator. { 'fe,rīt ¦fāz dif·ə¦ren·chəl 'sər·kyə,lād·ər }

ferrite-rod antenna [ELECTROMAG] An antenna consisting of a coil wound on a rod of ferrite; used in place of a loop antenna in radio receivers. Also known as ferrod; loopstick antenna. { 'fe,rīt ¦räd an'ten·ə }

ferrite rotator [ELECTROMAG] A gyrator consisting of a ferrite cylinder surrounded by a ring-type permanent magnet, inserted in a waveguide to rotate the plane of polarization of the electromagnetic wave passing through the waveguide. { 'fe,rīt 'rō,tād·ər }

ferrite streaks *See* ferrite banding. { 'fe,rīt 'strēks }

ferrite switch [ELECTROMAG] A ferrite device that blocks the flow of energy through a waveguide by rotating the electric field vector 90°; the switch is energized by sending direct

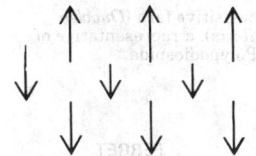

FERRIMAGNETISM

Schematic representation of arrangement of magnetic moments of neighboring ions in collinear ferrimagnetism.

current through its magnetizing coil; the rotated electromagnetic wave is then reflected from a reactive mismatch or absorbed in a resistive card. { 'fe,rīt 'swich }

ferrite-tuned oscillator [ELECTR] An oscillator in which the resonant characteristic of a ferrite-loaded cavity is changed by varying the ambient magnetic field, to give electronic tuning. { 'fe,rīt ¦tünd 'äs·ə,lād·ər }

ferritic stainless steel [MET] Any magnetic iron alloy containing more than 12% chromium having a body-centered cubic structure. Also known as stainless iron. { fe,rid·ik ¦stān·ləs 'stēl }

ferritin [BIOCHEM] An iron-protein complex occurring in tissues, probably as a storage form of iron. { 'fer·ət·ən }

ferritremolite [MINERAL] The ferric ion equivalent of the monoclinic amphibole, tremolite. { ¦fe·ri'trem·ə,līt }

ferritungstite [MINERAL] $Fe_2(WO_4)(OH)_4 \cdot 4H_2O$ A yellow ocher mineral composed of hydrous ferric tungstate, occurring as a powder. { ¦fer·ri'təŋ·,stīt }

ferroacoustic storage [ELECTR] A delay-line type of storage consisting of a thin tube of magnetostrictive material, a central conductor passing through the tube, and an ultrasonic driving transducer at one end of the tube. { ¦fe·rō·ə¦küs·tik 'stör·ij }

ferroalloy [MET] Any alloy containing iron, usually in major amount. Also known as ferrous alloy. { ¦fe·rō'al,öi }

ferroaluminum [MET] An alloy of iron and aluminum; added to molten steel as a deoxidizer or as an alloying component. { ¦fe·rō·ə'lü·mə·nəm }

ferroamphibole [MINERAL] The ferrous iron equivalent of the amphibole group of minerals. { ¦fe·rō'am·fə,bōl }

ferroan dolomite [MINERAL] A species of ankerite having less than 20% of the manganese positions occupied by iron. { 'fer·ə·wən 'dōl,mīt }

ferroaugite [MINERAL] A form of monoclinic pyroxene. { ¦fe·rō'ò,gīt }

ferroboron [MET] An alloy of iron and boron that is added to steel to form hardened special steels; two grades are used, 10% boron and 17% boron. { ¦fe·rō'bò,rän }

ferrocarbon titanium [MET] An alloy of iron with 15–20% titanium and 3–8% carbon; may be added to molten steel as a component of low-alloy steel. { ¦fe·rō'kär·bən ti'tān·ē·əm }

ferrocene [ORG CHEM] $(CH_2)_5Fe(CH_2)_5$ Orange crystals that are soluble in ether, melting point 174°C; used as a combustion control additive in fuels, and for heat stabilization in greases and plastics. { 'fer·ə,sēn }

ferrocerium [MET] An alloy of iron with a high percentage of cerium; used to make cigarette lighter flints. { ¦fe·rō'sir·ē·əm }

ferrochelatase [BIOCHEM] A mitochondrial enzyme which catalyzes the incorporation of iron into the protoporphyrin molecule. { ¦fe·rō'kel·ə,tās }

ferrochromium [MET] A crude ferroalloy containing chromium. { ¦fe·rō'krō·mē·əm }

ferrocolumbium [MET] An alloy of iron and columbium (niobium); used to add columbium to certain alloy steels. { ¦fe·rō·kə'ləm·bē·əm }

ferrocyanic acid [INV ZOO] $H_4Fe(CN)_6$ A white solid obtained by treating ferrocyanides with acid. { ¦fe·rō·sī'an·ik 'as·əd }

ferrocyanide [INV ZOO] A salt containing the radical $Fe(CN)_6^{4-}$. { ¦fe·rō'sī·ə,nīd }

ferrocyanide process [CHEM ENG] A regenerative chemical treatment for removal of mercaptans from petroleum fuels; uses caustic-sodium ferrocyanide reagent. { fe·rō'sī·ə,nīd ,präs·əs }

ferrod See ferrite-rod antenna. { 'fe,räd }

Ferrod [GEOL] A suborder of the soil order Spodosol that is well drained and contains an iron accumulation with little organic matter. { 'fe,räd }

ferrodolomite [MINERAL] $CaFe(CO_3)_2$ A mineral composed of calcium iron carbonate, isomorphous with dolomite, and occurring in ankerite. { ¦fe·rō'dō·lə,mīt }

ferroelectric [SOLID STATE] A crystalline substance displaying ferroelectricity, such as barium titanate, potassium dihydrogen phosphate, and Rochelle salt; used in ceramic capacitors, acoustic transducers, and dielectric amplifiers. Also known as Rochelle electric; Seignette electric. { ¦fe·rō·i'lek·trik }

ferroelectric Barkhausen effect [SOLID STATE] A series of

abrupt changes in the dielectric polarization of a ferroelectric material that occurs when the external electric field acting on the material is varied. { ¦fe·rō·i¦lek·trik 'bärk,haùz·ən i,fekt }

ferroelectric converter [ELEC] A converter that transforms thermal energy into electric energy by utilizing the change in the dielectric constant of a ferroelectric material when heated beyond its Curie temperature. { ¦fe·rō·i'lek·trik kən'vərd·ər }

ferroelectric crystal [SOLID STATE] A crystal of a ferroelectric material. { ¦fe·rō·i'lek·trik 'krist·əl }

ferroelectric domain [SOLID STATE] A region of a ferroelectric material within which the spontaneous polarization is constant. { ¦fe·rō·i'lek·trik də'mān }

ferroelectric hysteresis [ELEC] The dependence of the polarization of ferroelectric materials not only on the applied electric field but also on their previous history; analogous to magnetic hysteresis in ferromagnetic materials. Also known as dielectric hysteresis; electric hysteresis. { fe·rō·i'lek·trik ,his·tə'rē·səs }

ferroelectric hysteresis loop [ELEC] Graph of polarization or electric displacement versus applied electric field of a material displaying ferroelectric hysteresis. { ¦fe·rō·i'lek·trik ,his·tə'rē·səs ,lüp }

ferroelectricity [SOLID STATE] Spontaneous electric polarization in a crystal; analogous to ferromagnetism. { ¦fe·rō·i'lek·tris·əd·ē }

ferroelectric liquid-crystal display [ELECTR] An electronic display that employs a liquid crystal that is ferroelectric, such as smectic C*, which has two different stable molecular configurations; polarizers are positioned such that one state is optically transmissive while the other is dark. { ,fer·ō·i¦lek·trik ¦lik·wəd ¦kris·təl dis¦plā }

ferroelectric shutter [OPTICS] A shutter consisting of a slab of ferroelectric crystal located between polarizers whose planes are at right angles; opens to pass light when activated by a pulse of up to 100 volts. { ¦fe·rō·i'lek·trik 'shəd·ər }

ferrofluid [MATER] A colloidal suspension of ultramicroscopic magnetic particles in a carrier liquid, used as a lubricant or damping liquid. [PHYS CHEM] A colloidal suspension that becomes magnetized in a magnetic field because of a disperse phase consisting of ferromagnetic or ferrimagnetic particles. { 'fe·rō,flü·əd }

ferrogabbro [PETR] A gabbro rock in which the pyroxene and olivine constituents have an unusually high iron content. { ¦fe·rō'ga·brō }

ferrograph analyzer [ENG] An instrument used for ferrography; a pump delivers a small sample of the fluid to a microscope slide mounted above a magnet that generates a high-gradient magnetic field, causing particles to be deposited in a gradient of sizes along the slide. { 'fer·ə,graf 'an·ə,līz·ər }

ferrography [ENG] Wear analysis of machine bearing surfaces by collection of ferrous (or nonferrous) wear particles from lubricating oil in a ferrograph analyzer; the method can be applied to human joints by collecting fragments of cartilage, bone, or prosthetic materials from synovial fluid. { fe'räg·rə·fē }

ferrohydrodynamics [PHYS] The study of the motion of strongly magnetizable fluids subjected to magnetic fields. { ,fer·ō,hī·drə·dī'nam·iks }

ferromagnetic amplifier [ELECTR] A parametric amplifier based on the nonlinear behavior of ferromagnetic resonance at high radio-frequency power levels; incorrectly known as garnet maser. { ¦fe·rō·mag¦ned·ik 'am·plə,fī·ər }

ferromagnetic ceramic See ceramic magnet. { ¦fe·rō·mag¦ned·ik sə'ram·ik }

ferromagnetic crystal [SOLID STATE] A crystal of a ferromagnetic material. Also known as polar crystal. { ¦fe·rō·mag¦ned·ik 'krist·əl }

ferromagnetic domain [SOLID STATE] A region of a ferromagnetic material within which atomic or molecular magnetic moments are aligned parallel. Also known as magnetic domain. { ¦fe·rō·mag¦ned·ik də'mān }

ferromagnetic film See magnetic thin film. { ¦fe·rō·mag¦ned·ik 'film }

ferromagnetic material [SOLID STATE] A material displaying ferromagnetism, such as the various forms of iron, steel, cobalt, nickel, and their alloys. { ¦fe·rō·mag¦ned·ik mə'tir·ē·əl }

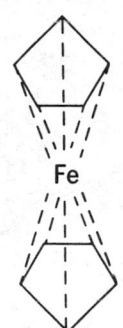

FERROCENE

Structural diagram of ferrocene.

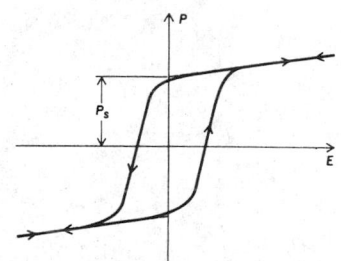

FERROELECTRIC HYSTERESIS LOOP

Hysteresis loop showing the relation between the resulting polarization P of the ferroelectric crystal and the externally applied electric field E; P_s is permanent or spontaneous magnetization.

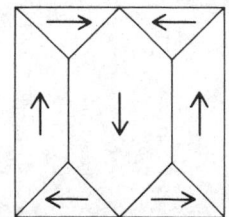

FERROMAGNETIC DOMAIN

Drawing of a uniaxial crystal showing the ferromagnetic domains. The arrows show the direction of the magnetic field in each domain.

ferromagnetic resonance [SOLID STATE] Magnetic resonance of a ferromagnetic material. { ¦fe·rō¦mag¦ned·ik 'rez·ən·əns }

ferromagnetics [ELECTR] The science that deals with the storage of binary information and the logical control of pulse sequences through the utilization of the magnetic polarization properties of materials. { ¦fe·rō¦mag¦ned·iks }

ferromagnetic tape [ELECTROMAG] A tape made of magnetic material for use in winding closed magnetic cores of toroids and transformers. { ¦fe·rō¦mag¦ned·ik 'tāp }

ferromagnetism [SOLID STATE] A property, exhibited by certain metals, alloys, and compounds of the transition (iron group) rare-earth and actinide elements, in which the internal magnetic moments spontaneously organize in a common direction; gives rise to a permeability considerably greater than that of vacuum, and to magnetic hysteresis. { ¦fe·rō'mag·nə¸tiz·əm }

ferromagnetography [GRAPHICS] A printing technique in which magnetic iron particles are applied to a latent image magnetized onto a metal sheet or drum, then transferred to ordinary paper and fixed in position by heat, pressure, or other means. { ¦fe·rō¸mag·nə'täg·rə·fē }

ferromanganese [MET] A ferroalloy containing about 80% manganese and used in steelmaking. { ¦fe·rō'maŋ·gə¸nēs }

ferrometer [ENG] An instrument used to make permeability and hysteresis tests of iron and steel. { fə'räm·əd·ər }

ferromolybdenum [MET] A molybdenum-iron alloy produced in the electric furnace or by a thermite process; used to introduce molybdenum into iron or steel alloys, and as a coating material on welding rods. { ¦fe·rō·mə'lib·də·nəm }

ferronickel [MET] A crude ferroalloy containing nickel. { ¦fe·rō¦nik·əl }

ferrophosphorus [MET] A by-product formed in the heating of iron, phosphate rock, silica, and coke; this alloy is used to increase fluidity in steel casting. { ¦fe·rō'fäs·fə·rəs }

ferroporphyrin [BIOCHEM] A red complex of porphyrin and iron in which the iron is in the 2+ oxidation state. { ¦fe·rō'pȯr·fə·rən }

ferroprussiate paper [MATER] A paper used in a blueprint process to reproduce plans and drawings. { ¦fe·rō'prəsh·ē¸āt ¸pā·pər }

ferroresonant circuit [ELECTR] A resonant circuit in which a saturable reactor provides nonlinear characteristics, with tuning being accomplished by varying circuit voltage or current. { ¦fe·rō'rez·ən·ənt 'sər·kət }

ferroresonant power supply [ELECTR] A transformer-based power supply, employed in high-current applications such as battery chargers, that uses nonlinear magnetic properties and a resonant circuit to regulate the output current. { ¸fe·rō¸rez·ən·ənt 'paú·ər sə¸plī }

ferroresonant static inverter [ELEC] A static inverter consisting of a simple square-wave inverter system and a tuned output transformer that performs filtering, voltage regulation, and current limiting. { ¦fe·rō¦rez·ən·ənt ¦stad·ik in'vərd·ər }

ferrosilicon [MET] A crude ferroalloy containing 15–95% silicon and used in steelmaking. { ¦fe·rō'sil·ə·kən }

ferrosilicon process See Pidgeon process. { ¦fe·rō'sil·ə·əkən ¸präs·əs }

ferrosilite [MINERAL] A mineral in the orthopyroxene group; the iron analog of enstatite; occurs in hypersthene, but is not found separately in nature. { 'fe·rō'si¸līt }

ferrospinel See hercynite. { ¦fe·rō·spə'nel }

ferrotitanium [MET] A ferroalloy containing 15–45% titanium and used in steelmaking. { ¦fe·rō·tī'tān·ē·əm }

ferrotremolite [MINERAL] The ferrous iron equivalent of the monoclinic amphibole, tremolite. { ¦fe·rō'tre·mə¸līt }

ferrotungsten [MET] A crude ferroalloy containing tungsten and used in steelmaking. { ¦fe·rō'təŋ·stən }

ferrotype [GRAPHICS] A photograph formed on a metal plate coated with collodion and sensitive salts. { 'fe·rə¸tīp }

ferrouranium [MET] An alloy of iron and uranium. { ¦fe·rō·yù'rän·ē·əm }

ferrous [CHEM] The term or prefix used to denote compounds of iron in which iron is in the divalent (2+) state. { 'fer·əs }

ferrous acetate [ORG CHEM] $Fe(CH_3COO)_2·4H_2O$ Soluble green crystals, soluble in water and alcohol, that are combustible and that oxidize to basic ferric acetate in air; used as textile dyeing mordant, as wood preservative, and in medicine. Also known as iron acetate. { 'fer·əs 'as·ə¸tāt }

ferrous alloy See ferroalloy. { 'fer·əs 'al¸ȯi }

ferrous ammonium sulfate [INV ZOO] $Fe(SO_4)·(NH_4)SO_4·6H_2O$ Light-green, water-soluble crystals; used in medicine, analytical chemistry, and metallurgy. Also known as iron ammonium sulfate; Mohr's salt. { 'fer·əs ə'mōn·ē·əm 'səl¸fāt }

ferrous arsenate [INV ZOO] $Fe_3(AsO_4)_2·6H_2O$ Water-insoluble, toxic green amorphous powder, soluble in acids; used in medicine and as an insecticide. Also known as iron arsenate. { 'fer·əs 'ärs·ən¸āt }

ferrous carbonate [INV ZOO] $FeCO_3$ Green rhombohedral crystals that are soluble in carbonated water and decompose when heated; used in medicine. { 'fer·əs 'kär·bə¸nāt }

ferrous chloride [INV ZOO] $FeCl_2·4H_2O$ Green, monoclinic crystals, soluble in water; used as a mordant in dyeing, for sewage treatment, in metallurgy, and in pharmaceutical preparations. Also known as iron chloride; iron dichloride. { 'fer·əs 'klȯr¸īd }

ferrous hydroxide [INV ZOO] $Fe(OH)_2$ A white, water-insoluble, gelatinous solid that turns reddish-brown as it oxidizes to ferric hydroxide. { 'fer·əs hī'dräk¸sīd }

ferrous oxalate [ORG CHEM] $Fe(COO)_2$ A water-soluble, yellow powder; used in photography and medicine. Also known as iron oxalate. { 'fer·əs 'äk·sə¸lāt }

ferrous oxide [INV ZOO] FeO A black powder, soluble in water, melting at 1419°C. Also known as black iron oxide; iron monoxide. { 'fer·əs 'äk¸sīd }

ferrous sulfate [INV ZOO] $FeSO_4·7H_2O$ Blue-green, water-soluble, monoclinic crystals; used as a mordant in dyeing wool, in the manufacture of ink, and as a disinfectant. Also known as copperas; ferrisulphas; green copperas; green vitriol; iron sulfate. { 'fer·əs 'səl¸fāt }

ferrous sulfide [INV ZOO] FeS Black crystals, insoluble in water, soluble in acids, melting point 1195°C; used to generate hydrogen sulfide in ceramics manufacture. Also known as iron sulfide. { 'fer·əs 'səl¸fīd }

ferrovanadium [MET] An iron alloy high in vanadium (35–55); used to add 0.1–2.5% vanadium during the manufacture of engineering steels and high-strength steels. { ¦fe·rō·və'nād·ē·əm }

ferruccite [MINERAL] $NaBF_4$ An orthorhombic boron mineral consisting of sodium fluoborate. { fə'rü¸chīt }

ferruginous [SCI TECH] **1.** Pertaining to or containing iron. **2.** Having the appearance or color of iron rust (ferric oxide). { fə'rü·jə·nəs }

ferrule [DES ENG] **1.** A metal ring or cap attached to the end of a tool handle, post, or other device to strengthen and protect it. **2.** A bushing inserted in the end of a boiler flue to spread and tighten it. [ENG] See stabilizer. { 'fer·əl }

ferrum [CHEM] Latin term for iron; derivation of the symbol Fe. { 'fer·əm }

ferry [NAV ARCH] A boat which carries people, automotive vehicles, or goods across a river or other body of water, usually traveling back and forth on a regular schedule. [ORD] **1.** To deliver aircraft or ships by operating them under their own power. **2.** To transport personnel and materiel by air. { 'fer·ē }

fersmanite [MINERAL] $(Na,Ca)_2(Ti,Cb)Si(O,F)_6$ A brown mineral composed of a silicate fluoride of sodium, calcium, titanium, and columbium. { 'fȯrz·mə¸nīt }

fersmite [MINERAL] $(Ca,Ce)(Cb,Ti)_2(O,F)_6$ A black mineral composed of an oxide and fluoride of calcium and columbium with cerium and titanium. { 'fȯrz¸mīt }

fertigation [AGR] The practice of fertilizing plants via a drip irrigation system. { ¸fər·tə'gā·shən }

fertile material [NUCLEO] A material, such as thorium-232 or uranium-238, which is capable of being transformed into a fissionable material by capture of a neutron. { 'fərd·əl mə'tir·ē·əl }

fertility [BIOL] The state of or capacity for abundant productivity. { fər'til·əd·ē }

fertility factor [GEN] An episomal bacterial sex factor which determines the role of a bacterium as either a male donor or as a female recipient of genetic material. Also known as F factor; sex factor. { fər'til·əd·ē ¸fak·tər }

fertilization [PHYSIO] The physicochemical processes

involved in the union of the male and female gametes to form the zygote. { 'fərd·əl·ə'zā·shən }

fertilization membrane [CYTOL] A membrane that separates from the surface of and surrounds many eggs following activation by the sperm; prevents multiple fertilization. { 'fərd·əl·ə'zā·shən ,mem,brān }

fertilizer [MATER] Material that is added to the soil to supply chemical elements needed for plant nutrition. { 'fərd·əl,īz·ər }

fertilizin [BIOCHEM] A mucopolysaccharide, derived from the jelly coat of an egg, that plays a role in sperm recognition and the stimulation of sperm motility and metabolic activity. { fər'til·ə·zən }

ferulic acid [ORG CHEM] $C_{10}H_{10}O_4$ A compound widely distributed in small amounts in plants, having two isomers: the cis form is a yellow oil, and the trans form is obtained from water solutions as orthorhombic crystals. { fə'rül·ik 'as·əd }

fervanite [MINERAL] $Fe_4V_4O_6\cdot5H_2O$ Golden-brown mineral composed of a hydrated iron vanadate; although itself not radioactive, it occurs with radioactive minerals. { 'fər·və,nīt }

fervenulin [MICROBIO] $C_7H_7N_5O_2$ An antibiotic from culture filtrates of *Streptomyces fervens;* yellow, orthorhombic crystals can be formed; melting point is 178–179°C. Also known as planomycin. { fər'ven·ə·lin }

Féry spectrograph [SPECT] A spectrograph whose only optical element consists of a back-reflecting prism with cylindrically curved faces. { ¦fãr·ē 'spek·trə,graf }

fescue [BOT] A group of grasses of the genus *Festuca*, used for both hay and pasture. { 'fes,kyü }

fescue foot [VET MED] A gangrenous condition of cattle feet caused by grazing tall fescue infected with the endophytic symbiotic fungus *Acremonium coenophialum.* { 'fes·kyü ,füt }

Feshbach resonance [QUANT MECH] A sharp resonance or peak which is seen when the cross section of an atomic or nuclear scattering process is plotted as a function of energy. It is associated with an energy threshold above which the scattering process can lead to a new result (such as excitation or ionization of one of the colliding objects), and it lies at an energy slightly below this threshold. { 'fesh,bäk ,rez·ən·əns }

Fessler compound [FOOD ENG] Flocculating-salt mixture used to remove copper and iron from wines. { 'fes·lər ,käm,paùnd }

Festuca arundinacea See tall fescue. { fe,stü·kə ə,rən·də'nãs·ē·ə }

FET See field-effect transistor.

fetal alcohol syndrome [MED] A spectrum of changes in the offspring of women who consume alcoholic beverages during pregnancy, ranging from mild mental changes to severe growth deficiency, mental retardation, and abnormal facial features. { ¦fēd·əl 'al·kə,hòl ,sin,drōm }

fetal asphyxia [MED] Deprivation of oxygen to the fetus due to interference with its blood supply. { 'fēd·əl əs'fik·sē·ə }

fetal fat-cell lipoma See liposarcoma. { 'fēd·əl ¦fat ¦sel li'pō·mə }

fetal hemoglobin [BIOCHEM] A normal embryonic hemoglobin having alpha chains identical to those of normal adult human hemoglobin, and gamma chains similar to adult beta chains. { 'fēd·əl 'hē·mə,glō·bən }

fetal membrane [EMBRYO] Any one of the membranous structures which surround the embryo during its development period. Also known as extraembryonic membrane. { 'fēd·əl 'mem,brān }

fetch [COMPUT SCI] To locate and load into main memory a requested load module, relocating it as necessary and leaving it in a ready-to-execute condition. [OCEANOGR] **1.** The distance traversed by waves without obstruction. **2.** An area of the sea surface over which seas are generated by a wind having a constant speed and direction. **3.** The length of the fetch area, measured in the direction of the wind in which the seas are generated. Also known as generating area. { fech }

fetch ahead See instruction lookahead. { ¦fech ə'hed }

fetch bit [COMPUT SCI] The fifth bit in a storage key; the value of the fetch bit can protect a stored block from destruction or from being accessed by unauthorized programs. { 'fech ,bit }

fetch cycle [COMPUT SCI] The period during which a

machine language instruction is read from memory into the control section of the central processing unit.

fetometamorphism [INV ZOO] A life cycle variation in the Cantharidae (Coleoptera); the larvae hatch prematurely as legless, immature prelarvae. { ¦fē·dō,med·ə'mór,fiz·əm }

fettling knife [GRAPHICS] A sharp instrument with a flexible blade tapering to a point; used in ceramics for carving of clay models, sgraffito, removing mold marks, and miscellaneous other purposes. { 'fet·liŋ ,nīf }

fetus [EMBRYO] **1.** The unborn offspring of viviparous mammals in the later stages of development. **2.** In human beings, the developing body in utero from the beginning of the ninth week after fertilization through the fortieth week of intrauterine gestation, or until birth. { 'fēd·əs }

Feulgen reaction [ANALY CHEM] An aldehyde specific reaction based on the formation of a purple-colored compound when aldehydes react with fuchsin-sulfuric acid; deoxyribonucleic acid gives this reaction after removal of its purine bases by acid hydrolysis; used as a nuclear stain. { 'fóil·gən rē,ak·shən }

fever [MED] An elevation in the central body temperature of warm-blooded animals caused by abnormal functioning of the thermoregulatory mechanisms. { 'fē·vər }

Feyliniidae [VERT ZOO] The limbless skinks, a family of reptiles in the suborder Sauria represented by four species in tropical Africa. { ,fā·lə'nī·ə,dē }

Feynman diagram [QUANT MECH] A diagram which gives an intuitive picture of a term in a perturbation expansion of a scattering matrix element or other physical quantity associated with interactions of particles; each line represents a particle, each vertex an interaction. { 'fīn·mən ,dī·ə,gram }

Feynman integral [QUANT MECH] A term in a perturbation expansion of a scattering matrix element; it is an integral over the Minkowski space of various particles (or over the corresponding momentum space) of the product of propagators of these particles and quantities representing interactions between the particles. { 'fīn·mən 'int·ə·grəl }

Feynman propagator [QUANT MECH] A factor $(\rho + m)/(\rho^2 - m^2 + i\epsilon)$ in a transition amplitude corresponding to a line that connects two vertices in a Feynman diagram, and that represents a virtual particle. { 'fīn·mən 'präp·ə,gād·ər }

Feynman's rules [QUANT MECH] Rules for carrying out perturbation expansions in quantum field theory codified by Feynman diagrams. { 'fīn·mənz ,rülz }

Feynman's superfluidity theory [CRYO] Microscopic theory of superfluid helium which accounts for the spectrum of elementary excitations assumed by Landau's superfluidity theory. { 'fīn·mənz ,sü·pər,flü'id·əd·ē ,thē·ə·rē }

F factor See fertility factor. { 'ef ,fak·tər }

F format [COMPUT SCI] **1.** In data management, a fixed-length logical record format. **2.** In FORTRAN, a real variable formatted as $F\mu.d$, where μ is the width of the field and d represents the number of digits to appear after the decimal point. { 'ef ,fòr·mat }

FFT See fast Fourier transform.

FG achromatism See actinic achromatism. { ¦ef¦jē ,ā'krō·mə,tiz·əm }

fiard See fjard. { fē'ärd }

fiber [BOT] **1.** An elongate, thick-walled, tapering plant cell that lacks protoplasm and has a small lumen. **2.** A very slender root. [MATH] The set of points in the total space of a bundle which are sent into the same element of the base of the bundle by the projection map. [MET] **1.** The characteristic of wrought metal that indicates directional properties as revealed by etching or by fracture appearance. **2.** The pattern of preferred orientation of metal crystals after a deformation process, usually wire-drawing. [OPTICS] A transparent threadlike object made of glass or clear plastic, used to conduct light along selected paths. [TEXT] An extremely long, pliable, cohesive natural or manufactured threadlike object from which yarns are spun to be woven into textiles. { 'fī·bər }

fiberboard [MATER] A hard isotropic board made by compressing wood chips or other vegetable fibers. { 'fī·bər ,bórd }

fiber bundle [MATH] A bundle whose total space is a *G*-space *X*, whose base is the homomorphic image of the orbit space of *X*, and whose fibers are isomorphic to the orbits of points in the base space under the action of *G*. [OPTICS] A flexible bundle of glass or other transparent fibers, parallel to

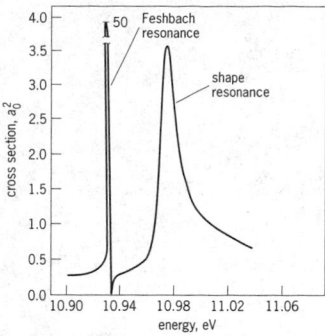
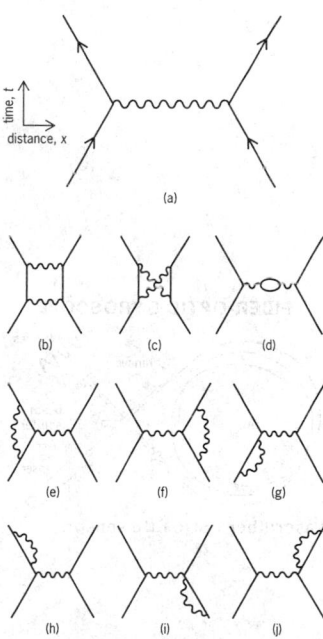

each other, used in fiber optics to transmit a complete image from one end of the bundle to the other. { 'fī·bər ˌbən·dəl }

fiber-composite material [MATER] A composite material in which a fibrous phase that retains its physical identity is dispersed in a continuous matrix phase. { 'fī·bər kəmˌpäz·ət məˌtir·ē·əl }

fiber crops [AGR] Plants, such as flax, hemp, jute, and sisal, cultivated for their content or yield of fibrous material. { 'fī·bər ˌkräps }

fiber diagram [SOLID STATE] The x-ray diffraction pattern of a collection of crystallites that have one crystallographic axis approximately parallel to a common direction but are otherwise randomly oriented. { 'fī·bər ˌdī·ə·gram }

fiber flax [BOT] The flax plant grown in fertile, well-drained, well-prepared soil and cool, humid climate; planted in the early spring and harvested when half the seed pods turn yellow; used in the manufacture of linen. { 'fī·bər ˌflaks }

fiberglass See glass fiber. { 'fī·bərˌglas }

fiber grease [MATER] Solid-base lubricating grease; contains soap fibers 1–1000 micrometers long and 0.1–1.0 micrometer wide, which stabilize lubricating action by immobilizing fluid lubricating components. { 'fī·bər ˌgrēs }

fiber gyro See fiber-optic gyroscope. { 'fī·bər 'jī·rō }

fiberizer [MIN ENG] A hammer mill which cracks open asbestos-bearing rock to yield a fibrous product. { 'fī·bəˌrīz·ər }

fiber metal [MET] Any material composed of metal fibers that are pressed or sintered together or infiltrated with resin or other material. { 'fī·bər ˌmed·əl }

fiber metallurgy [MET] A branch of metallurgy concerned with the study of metal fibers. { 'fī·bər 'med·əlˌər·jē }

fiber-optic circuit [COMMUN] A path for data transmission in which light acts as the information carrier and is transmitted by total internal reflection through a transparent optical waveguide. { 'fī·bər 'äp·tik 'sər·kət }

fiber-optic current sensor [ENG] An instrument for measuring currents on high-voltage lines, in which the magnetic field associated with the current changes the phase of light traveling through an optical fiber, and the phase change is measured in an interferometer. { 'fī·bər 'äp·tik 'kə·rənt ˌsen·sər }

fiber-optic data distribution interface [COMMUN] A set of standards for high-speed fiber-optic local-area networks. Abbreviated FDDI. { 'fī·bər 'äp·tik 'dad·ə ˌdis·trəˌbyü·shən 'in·tərˌfās }

fiber-optic gyroscope [ENG] An instrument for measuring rotation rate, in which light from a laser or light-emitting diode is split into two beams which travel in opposite directions around a coil of optical fiber and recombine to generate interference fringes whose shift is a measure of the rotation rate of the coil. Also known as fiber gyro; laser/fiber-optics gyroscope. { 'fī·bər 'äp·tik 'jī·rəˌskōp }

fiber-optic hydrophone See interferometric hydrophone. { 'fī·bər 'äp·tik 'hī·drəˌfōn }

fiber-optic imaging [OPTICS] The formation of optical images by transmission through precisely aligned bundles of optical fibers; each fiber transmits one element of the image. { 'fī·bər 'äp·tik 'im·əj·iŋ }

fiber-optic magnetometer [ENG] A magnetometer in which the deformation of a magnetostrictive body in the field causes phase changes in light traveling through an optical fiber wrapped around the body, and these phase changes are measured in an interferometer. { 'fī·bər 'äp·tik ˌmag·nəˈtäm·əd·ər }

fiber optics [OPTICS] The technique of transmitting light through long, thin, flexible fibers of glass, plastic, or other transparent materials; bundles of parallel fibers can be used to transmit complete images. { 'fī·bər ˌäp·tiks }

fiber-optic sensor See optical-fiber sensor. { 'fī·bər 'äp·tik 'sen·sər }

fiber-optic thermometer [ENG] A thermometer in which light from a mercury lamp is guided along an optical fiber to excite a tiny fluorescent crystal, whose light is in turn guided back along the fiber to an evaluation unit where the crystal temperature is determined from the ratios of the strengths of spectral lines in the fluorescent light or from the decay time of the fluorescence. { 'fī·bər 'äp·tik thərˈmäm·əd·ər }

fiber plaster [MATER] Gypsum plaster containing hair or wood fiber as a binder. { 'fī·bər ˌplas·tər }

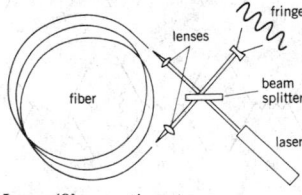

FIBER-OPTIC GYROSCOPE

fringes
lenses
fiber
beam splitter
laser

Laser/fiber-optic rate sensor.

fiber-reinforced concrete [MATER] A portland-cement concrete or mortar that is reinforced with dispersed, randomly oriented, unconnected fibers of metallic, mineral, or organic materials. { ˈfī·bər ˌrē·inˈfórst 'kanˌkrēt }

fiber-reinforced polymer [MATER] Lightweight, high-strength composite material composed of fibers (glass, carbon, or silicon carbide) embedded in a polymeric (epoxy, phenolic, or polyester) matrix. Abbreviated FRP. { 'fī·bər ˌrē·ənˈfórst 'päl·ə·mərz }

fiberscope [OPTICS] An arrangement of parallel glass fibers with an objective lens on one end and an eyepiece at the other; the assembly can be bent as required to view objects that are inaccessible for direct viewing. { 'fī·bərˌskōp }

fiber stress [MECH] **1.** The tensile or compressive stress on the fibers of a fiber metal or other fibrous material, especially when fiber orientation is parallel with the neutral axis. **2.** Local stress through a small area (a point or line) on a section where the stress is not uniform, as in a beam under bending load. { 'fī·bər ˌstres }

fiber waveguide See optical waveguide. { 'fī·bər 'wāvˌgīd }

Fibonacci number [MATH] A number in the Fibonacci sequence whose first two terms are $f_1 = f_2 = 1$. { ˌfib·əˈnä·chē 'nəm·bər }

Fibonacci sequence [MATH] The sequence 1,1,2,3,5, 8,13,21, ..., or any sequence where each entry is the sum of the two previous entries. { ˌfē·bəˈnäch·ē ˌsē·kwəns }

fibratus [METEOROL] A cloud species characterized by a fine hairlike or striated composition, the filaments of which are usually distinctly separated from each other; the extremities of these filaments are always thin and never terminated by tufts or hooks. Also known as filosus. { fiˈbräd·əs }

fibril [BIOL] A small thread or fiber, as a root hair or one of the structural units of a striated muscle. [MATER] One of the minute threadlike elements of a natural or synthetic fiber. { 'fī·brəl }

fibrillating [TEXT] A process for producing fibers by mechanically cracking sheets of polymer film that have been stretched and oriented along their direction of extrusion. { 'fib·rəˌlād·iŋ }

fibrillation [PHYSIO] An independent, spontaneous, local twitching of muscle fibers. [TEXT] A process in which yarn is extruded as a ribbon or a tape rather than as a fine filament. { ˌfib·rəˈlā·shən }

fibrillose [BIOL] Having fibrils. { 'fīb·rəˌlōs }

fibrin [BIOCHEM] The fibrous, insoluble protein that forms the structure of a blood clot; formed by the action of thrombin. { 'fī·brən }

fibrinase [BIOCHEM] An enzyme that catalyzes the formation of covalent bonds between fibrin molecules. Also known as fibrin-stabilizing factor. { 'fī·brəˌnās }

fibrinogen [BIOCHEM] A plasma protein synthesized by the parenchymal cells of the liver; the precursor of fibrin. Also known as factor I. { fiˈbrin·ə·jən }

fibrinogenopenia [MED] A congenital hemorrhagic diathesis in which there is a decrease in fibrinogen in the plasma. { fiˌbrin·ə·jēn·əˈpē·nē·ə }

fibrinoid [BIOCHEM] A homogeneous, refractile, oxyphilic substance occurring in degenerating connective tissue, as in term placentas, rheumatoid nodules, and Aschoff bodies, and in pulmonary alveoli in some prolonged pneumonitides. { 'fī·brəˌnóid }

fibrinolysin See plasmin. { ˌfī·brəˈnäl·ə·sən }

fibrinolysis [PHYSIO] Liquefaction of coagulated blood by the action of plasmin on fibrin. { ˌfī·brəˈnäl·ə·səs }

fibrinous pericarditis [MED] Inflammation of the pericardium involving the deposition of fibrin and leukocytes between the layers of the pericardium; seen in rheumatic carditis and acute infectious diseases. { 'fī·brə·nəs ˌper·ə·kärˈdīd·əs }

fibrin-stabilizing factor See fibrinase. { ˈfī·brən 'stā·bəˌlīz·iŋ ˌfak·tər }

Fibrist [GEOL] A suborder of the soil order Histosol, consisting mainly of recognizable plant residues or sphagnum moss and saturated with water most of the year. { 'fī·brəst }

fibroadenoma [MED] A benign tumor containing both fibrous and glandular elements. { ˈfī·brōˌad·ənˈō·mə }

fibroblast [HISTOL] A stellate connective tissue cell found in fibrous tissue. Also known as a fibrocyte. { 'fī·brəˌblast }

fibroblast growth factor [PHYSIO] A family of proteins

important in the development of the nervous and skeletal systems. { 'fī·brō‚blast 'grōth ‚fak·tər }

fibroblastic [PETR] Of a metamorphic rock, having a texture that is homeoblastic as a result of the development of minerals with a fibrous habit during recrystallization. { ‚fī·brə‚blas·tik }

fibrocartilage [HISTOL] A form of cartilage rich in dense, closely opposed bundles of collagen fibers; occurs in intervertebral disks, in the symphysis pubis, and in certain tendons. { ‚fī·brō'kärd·əl·ij }

fibrocyte See fibroblast. { 'fī·brə‚sīt }

fibroferrite [MINERAL] Fe(SO₄)(OH)·5H₂O A yellowish mineral composed of a hydrous basic ferric sulfate, occurring in fibrous form. { ‚fī·brō'fe‚rīt }

fibroid [HISTOL] Composed of fibrous tissue. { 'fī‚bróid }

fibroid deoxyribonucleic acid [MOL BIO] Sections of relatively uncoiled double-stranded deoxyribonucleic acid thought to be regions of specific base sequences for coding rather than gene control. { 'fī‚bróid dē¦äk·sē‚rī·bō·nü¦klē·ik 'as·əd }

fibroid tumor See fibroma. { 'fī‚bróid 'tü·mər }

fibroin [BIOCHEM] A protein secreted by spiders and silkworms which rapidly solidifies into strong, insoluble thread that is used to form webs or cocoons. { 'fī·brə·wən }

fibrolite See sillimanite. { 'fī·brə‚līt }

fibroma [MED] A benign tumor composed primarily of fibrous connective tissue. Also known as fibroid tumor. { fī'brō·mə }

fibroma molluscum See neurofibromatosis. { fī'brō·mə mə'ləs·kəm }

fibromatosis [MED] 1. The occurrence of multiple fibromas. 2. Localized proliferation of fibroblasts without apparent cause. { ‚fī·brō·mə'tō·səs }

fibromyoma [MED] A benign tumor, usually of smooth muscle, with a prominent fibrous stroma; commonly a uterine leiomyoma. { ‚fī·brō·mī'ō·mə }

fibromyosis See interstitial endometrosis. { ‚fī·bro·mī'ō·səs }

fibromyositis See myositis. { ‚fī·brō‚mī·ə'sīd·əs }

fibronectin [BIOCHEM] A type of large glycoprotein that is found on the surface of cells and mediates cellular adhesion, control of cell shape, and cell migration. { 'fī·brə'nek·tən }

fibroplasia [MED] The growth of fibrous tissue, as in the second phase of wound healing. { ‚fī·brə'plā·zhə }

fibrosarcoma [MED] A sarcoma composed of spindle cells that produce collagenous fibrils. { ‚fī·brō·sär'kō·mə }

fibrosing adenomatosis See sclerosing adenomatosis. { fī'brōs·iŋ ‚ad·ən‚ō·mə'tō·səs }

fibrosis [MED] Growth of fibrous connective tissue in an organ or part in excess of that naturally present. { fī'brō·səs }

fibrositis [MED] Inflammation of white fibrous connective tissue, usually in a joint region. { ‚fī brə'sīd·əs }

fibrous composite [MATER] A composite material consisting of fibers embedded in a matrix. { 'fī·brəs kəm'päz·ət }

fibrous dysplasia [MED] 1. Extensive formation of fibrous tissue and transformation of bony tissue in one or more bones. 2. Development of abnormal amounts of fibrous tissue in the mammary glands. { 'fī·brəs dis'plā·zhə }

fibrous fracture [MECH] Failure of a material resulting from a ductile crack; broken surfaces are dull and silky. Also known as ductile fracture. { 'fī·brəs 'frak·chər }

fibrous ice See acicular ice. { 'fī·brəs 'īs }

fibrous osteoma See ossifying fibroma. { 'fī·brəs ‚äs·tē'ō·mə }

fibrous plaster [MATER] Gypsum plaster reinforced or backed with sisal or canvas. { 'fī·brəs 'plas·tər }

fibrous protein [BIOCHEM] Any of a class of highly insoluble proteins representing the principal structural elements of many animal tissues. { 'fī·brəs 'prō‚tēn }

fibrous structure [MATER] A ropy surface on a fractured material. [MET] 1. A lamination on an etched section of a forging. 2. In wrought iron, a structure consisting of slag fibers embedded in ferrite. { 'fī·brəs 'strək·chər }

fibula [ANAT] The outer and usually slender bone of the hind or lower limb below the knee in vertebrates; it articulates with the tibia and astragalus in humans, and is ankylosed with the tibia in birds and some mammals. { 'fib·yə·lə }

ficin [ORG CHEM] A proteolytic enzyme obtained from fig latex or sap; hydrolyzes casein, meat, fibrin, and other proteinlike materials; used in the food industry and as a diagnostic aid in medicine. { 'fī·sən }

Fick's law [PHYS] The law that the rate of diffusion of matter across a plane is proportional to the negative of the rate of change of the concentration of the diffusing substance in the direction perpendicular to the plane. { 'fiks ‚lò }

fictitious [NAV] Pertaining to or measured from an arbitrary reference line as in fictitious equator, fictitious latitude, or fictitious longitude. { fik'tish·əs }

fictitious craft [NAV] In marine usage, an imaginary craft used in the solution of certain maneuvering problems, as when a ship to be intercepted is expected to change course or speed during the interception run. { fik'tish·əs 'kraft }

fictitious equator [NAV] A reference line serving as the origin for measurement of a fictitious latitude. { fik'tish·əs i'kwād·ər }

fictitious graticule [NAV] A graticule formed from a set of fictitious parallels and meridians to resemble the geographic graticule but offset from it; coordinates taken from the fictitious graticule are used with a transverse or oblique map projection or with a navigation grid. { fik'tish·əs 'grad·ə‚kyül }

fictitious latitude [NAV] Angular distance from the equator as shown on a fictitious graticule. { fik'tish·əs 'lad·ə‚tüd }

fictitious longitude [NAV] The arc of the equator between the prime meridian and any given meridian derived from the fictitious graticule. { fik'tish·əs 'län·jə‚tüd }

fictitious loxodrome See fictitious rhumb line. { fik'tish·əs 'läk·sə‚drōm }

fictitious loxodromic curve See fictitious rhumb line. { fik'tish·əs ‚läk·sə¦drōm·ik 'kərv }

fictitious meridian [NAV] One of the coordinates of a fictitious graticule. { fik'tish·əs mə'rid·ē·ən }

fictitious parallel [NAV] One of the coordinates derived from a fictitious graticule. { fik'tish·əs 'par·ə·lel }

fictitious pole [NAV] One of the two points 90° from the equator on a fictitious graticule; called transverse or oblique pole, depending upon the type of fictitious equator. { fik'tish·əs 'pōl }

fictitious rhumb line [NAV] A line on a fictitious graticule making the same oblique angle with all meridians. Also known as fictitious loxodrome; fictitious loxodromic curve. { fik'tish·əs 'rəm ‚lin }

fictitious ship [NAV] An imaginary ship serving as a fictitious craft. { fik'tish·əs 'ship }

fictitious vehicle [NAV] An imaginary vehicle serving as a fictitious craft. { fik'tish·əs 'vē·ə·kəl }

fictitious year [ASTRON] The period between successive returns of the fictitious mean sun to a sidereal hour angle of 80° (right ascension 18 hours 40 minutes; about January 1); the length of the fictitious year is the same as that of the tropical year, since both are based upon the position of the sun with respect to the vernal equinox. Also known as Besselian year. { fik'tish·əs 'yir }

Ficus [BOT] A genus of tropical trees in the family Moraceae including the rubber tree and the fig tree. [INV ZOO] A genus of gastropod mollusks having pear-shaped, spirally ribbed sculptured shells. { 'fī·kəs }

FID See free induction decay.

fiddley See fidley. { 'fid·lē }

fidelity [COMMUN] The degree to which a system accurately reproduces at its output the essential characteristics of the signal impressed on its input. { fə'del·əd·ē }

fidley [NAV ARCH] Also spelled fiddley. 1. A wide opening above a fire room through which pass ventilators. 2. A framework of iron about the ladder of a hatch in a ship's deck leading below the deck. { 'fid·lē }

FIDO [METEOROL] A system for artificially dissipating fog, in which gasoline or other fuel is burned at intervals along an airstrip to be cleared. Derived from fog investigation dispersal operations. { 'fī·dō }

fiducial inference [STAT] A type of inference whose purpose is to make probabilistic statements about values of unknown parameters; based on the distribution of population values about which the inference is to be made. { fə¦dü·shəl ‚in·tər'fir·əns }

fiducial limits [STAT] The boundaries within which a parameter is considered to be located; a concept in fiducial inference. { fə¦dü·shəl 'lim·əts }

fiducial point [OPTICS] A mark, or one of several marks, visible in the field of view of an optical instrument, used as a

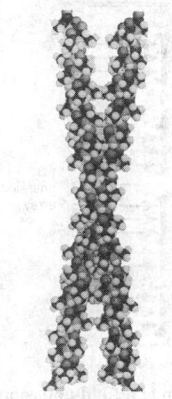

FIBROUS PROTEIN

α-Fibrous protein.

reference or for measurement. Also known as fiduciary point. { fə'dü·shəl ˌpȯint }

fiducial temperature [METEOROL] That temperature at which, in a specified latitude, the reading of a particular barometer does not require temperature or latitude correction. [THERMO] Any of the temperatures assigned to a number of reproducible equilibrium states on the International Practical Temperature Scale; standard instruments are calibrated at these temperatures. { fə'dü·shəl 'tem·prə·chər }

fiduciary point See fiducial point. { fə'dü·shē,er·ē ˌpȯint }

fiedlerite [MINERAL] $Pb_3(OH)_2Cl_4$ A colorless mineral composed of a hydroxychloride of lead, occurring as monoclinic crystals. { 'fēd·lə,rīt }

Fiedler's myocarditis See interstitial myocarditis. { 'fēd·lərz ,mī·ə,kär'dīd·əs }

field [COMPUT SCI] A specified area, such as a group of card columns or a set of bit locations in a computer word, used for a particular category of data. [ELEC] That part of an electric motor or generator which produces the magnetic flux which reacts with the armature, producing the desired machine action. [ELECTR] One of the equal parts into which a frame is divided in interlaced scanning for television; includes one complete scanning operation from top to bottom of the picture and back again. [GEOL] A region or area with a particular mineral resource, for example, a gold field. [GEOPHYS] That area or space in which a particular geophysical effect, such as gravity or magnetism, occurs and can be measured. [MATH] An algebraic system possessing two operations which have all the properties that addition and multiplication of real numbers have. [MED] The area in which surgery is taking place, bounded on all sides by sterilized tissue or drapes. Also known as sterile field. [OPTICS] See field of view. [PHYS] **1.** An entity which acts as an intermediary in interactions between particles, which is distributed over part or all of space, and whose properties are functions of space coordinates and, except for static fields, of time; examples include gravitational field, sound field, and the strain tensor of an elastic medium. **2.** The quantum-mechanical analog of this entity, in which the function of space and time is replaced by an operator at each point in space-time. { fēld }

field artillery [ORD] Artillery mounted on carriages, and mobile enough to accompany infantry or armored units in the field. { 'fēld är,til·ə·rē }

field artillery observer [ORD] A person who watches the effects of artillery fire, adjusts the center of impact of that fire onto a target, and reports results to the firing agency. { 'fēld är,til·ə·rē əb'zər·vər }

field artillery trainer [ORD] A small practice gun and carriage unit with a telescope and a mechanism for adjusting elevation and deflection; used in training field artillery personnel. { 'fēld är,til·ə·rē 'trān·ər }

fieldata code [COMMUN] A standardized military data transmission code, seven data bits plus one parity bit. { 'fēl,dad·ə ,kōd }

field brightness See adaptation luminance. { 'fēld ,brīt·nəs }

field capacity [HYD] The maximum amount of water that a soil can retain after gravitational water has drained away. { 'fēld kə,pas·əd·ē }

field changes [METEOROL] With regard to thunderstorm electricity, the rapid variations in the vertical component of the electric field strength at the earth's surface. { 'fēld ,chānj·əz }

field coil [ELECTROMAG] A coil used to produce a constant-strength magnetic field in an electric motor, generator, or excited-field loudspeaker; depending on the type of motor or generator, the field core may be on the stator or the rotor. Also known as field winding. { 'fēld ,kȯil }

field curvature See curvature of field. { 'fēld ,kər·və·chər }

field delimiter [COMPUT SCI] Any symbol, such as a slash, colon, tab, or space, which enables an assembler to recognize the end of a field. { 'fēld də,lim·əd·ər }

field designator [COMPUT SCI] A character generally placed at the beginning of a field to specify the nature of the data contained in it. { 'fēld ,dez·ig,nād·ər }

field desorption [SOLID STATE] A technique which tears atoms from a surface by an electric field applied at a sharp dip to produce very well-ordered, clean, plane surfaces of many crystallographic orientations. { 'fēld dē'sȯrp·shən }

field-desorption mass spectroscopy [SPECT] A technique for analysis of nonvolatile molecules in which a sample is deposited on a thin tungsten wire containing sharp microneedles of carbon on the surface; a voltage is applied to the wire, thus producing high electric-field gradients at the points of the needles, and moderate heating then causes desorption from the surface or molecular ions, which are focused into a mass spectrometer. { 'fēld dē'sȯrp·shən 'mas spek'trä·skə·pē }

field-desorption microscope [ELECTR] A type of field-ion microscope in which the tip specimen is imaged by ions that are field-desorbed or field-evaporated directly from the surface rather than by ions obtained from an externally supplied gas. { 'fēld dē,sȯrp·shən ,mī·krə,skōp }

field discharge [ELECTR] A spark discharge due to high potential across a gap. { 'fēld 'dis,chärj }

field-discharge switch [ELEC] A special type of switch that is connected in series with the field winding of an electrical machine, and that is operated to connect a resistor in parallel with the field winding before the main supply contacts are opened, in order to prevent the self-induced electromotive force in the field winding from reaching dangerous levels. { 'fēld 'dis,chärj ,swich }

field distortion [ELECTROMAG] Any alteration in the direction of an electric or magnetic field; in particular, distortion of the magnetic fields between the north and south poles of a generator due to the counter electromotive force in the armature winding. { 'fēld di,stȯr·shən }

field effect [ELECTR] The local change from the normal value that an electric field produces in the charge-carrier concentration of a semiconductor. { 'fēld i,fekt }

field-effect capacitor [ELECTR] A capacitor in which the effective dielectric is a region of semiconductor material that has been depleted or inverted by the field effect. { 'fēld i,fekt kə'pas·əd·ər }

field-effect device [ELECTR] A semiconductor device whose properties are determined largely by the effect of an electric field on a region within the semiconductor. { 'fēld i,fekt di,vīs }

field-effect diode [ELECTR] A semiconductor diode in which the charge carriers are of only one polarity. { 'fēld i,fekt 'dī,ōd }

field-effect display [OPTICS] A type of numerical display device in which a liquid-crystal cell is sandwiched between polarizers; the cell is treated so that it normally rotates light 90°, but ceases to rotate light when an electric field is applied to it, altering the transmission of the device. { 'fēld i,fekt di'splā }

field-effect phototransistor [ELECTR] A field-effect transistor that responds to modulated light as the input signal. { 'fēld i,fekt 'fōd·ō·tran'zis·tər }

field-effect tetrode [ELECTR] Four-terminal device consisting of two independently terminated semiconducting channels so displaced that the conductance of each is modulated along its length by the voltage conditions in the other. { 'fēld i,fekt 'te,trōd }

field-effect transistor [ELECTR] A transistor in which the resistance of the current path from source to drain is modulated by applying a transverse electric field between grid or gate electrodes; the electric field varies the thickness of the depletion layer between the gates, thereby reducing the conductance. Abbreviated FET. { 'fēld i,fekt tran'zis·tər }

field-effect-transistor resistor [ELECTR] A field-effect transistor in which the gate is generally tied to the drain; the resultant structure is used as a resistance load for another transistor. { 'fēld i,fekt tran'zis·tər ri'zis·tər }

field-effect varistor [ELECTR] A passive, two-terminal, nonlinear semiconductor device that maintains constant current over a wide voltage range. { 'fēld i,fekt və'ris·tər }

field emission [ELECTR] The emission of electrons from the surface of a metallic conductor into a vacuum (or into an insulator) under influence of a strong electric field; electrons penetrate through the surface potential barrier by virtue of the quantum-mechanical tunnel effect. Also known as cold emission. { 'fēld ə,mish·ən }

field-emission display [ELECTR] A flat-panel electronic display in which electrons are extracted from an array of cold-cathode emitters by applying a voltage between the cathode and a control electrode, and the electrons are then accelerated without deflection over a distance of less than 1 millimeter before colliding with a phosphor-coated flat faceplate. { 'fēld i,mish·ən di,splā }

FIELD COIL

A rotor of a synchronous motor showing the field coils. (*Allis-Chalmers*)

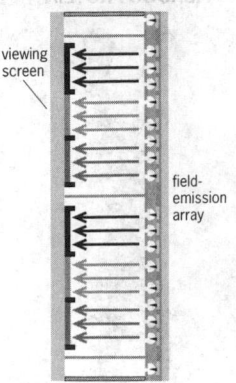

FIELD-EMISSION DISPLAY

viewing screen

field-emission array

Cross section of a field-emission display.

field-emission microscope [ELECTR] A device that uses field emission of electrons or of positive ions (field-ion microscope) to produce a magnified image of the emitter surface on a fluorescent screen. { 'fēld ə¦mish·ən 'mī·krə,skōp }

field-emission tube [ELECTR] A vacuum tube within which field emission is obtained from a sharp metal point; must be more highly evacuated than an ordinary vacuum tube to prevent contamination of the point. { 'fēld ə¦mish·ən ,tüb }

field-emitter array [ELECTR] An array of pyramidal silicon structures, with spacing on the order of 10 micrometers, designed for field emission of electrons into a vacuum. { 'fēld i¦mid·ər ə'rā }

field emplacement [ORD] Platform, support, or other position for artillery, machine guns, and so forth, in the field. { 'fēld em,plās·mənt }

field engineer [COMPUT SCI] A professional who installs computer hardware on customers' premises, performs routine preventive maintenance, and repairs equipment when it is out of order. Also known as field service representative. [ENG] **1.** An engineer who is in charge of directing civil, mechanical, and electrical engineering activities in the production and transmission of petroleum and natural gas. **2.** An engineer who operates at a construction site. { 'fēld en·jə,nir }

field-enhanced emission [ELECTR] An increase in electron emission resulting from an electric field near the surface of the emitter. { 'fēld in¦hanst i'mish·ən }

field excitation [MECH ENG] Control of the speed of a series motor in an electric or diesel-electric locomotive by changing the relation between the armature current and the field strength, either through a reduction in field current by shunting the field coils with resistance, or through the use of field taps. { 'fēld ,ek·sī'tā·shən }

field flattener [OPTICS] A thin planoconvex lens placed in front of the photographic plate in some telescopes that have a curved focal plane so as to focus light on the flat plate. { 'fēld ,flat·ən·ər }

field focus [GEOPHYS] The total area or volume occupied by an earthquake source. { 'fēld ,fō·kəs }

field fortification [ORD] Fortification constructed in the field to strengthen the natural defenses of the ground features, including foxholes, obstacles, trenches, gun emplacements, and so forth. { 'fēld fórd·ə·fə'kā·shən }

field-free emission current [ELECTR] Electron current emitted by a cathode when the electric field at the surface of the cathode is zero. Also known as zero-field emission. { 'fēld ,frē i'mish·ən ,kə·rənt }

field frequency [ELECTR] The number of fields transmitted per second in television; equal to the frame frequency multiplied by the number of fields that make up one frame. Also known as field repetition rate. { 'fēld ,frē·kwən·sē }

field galaxy [ASTRON] An isolated galaxy that does not belong to a cluster. { 'fēld ,gal·ik·sē }

field geology [GEOL] The study of rocks and rock materials in their environment and in their natural relations to one another. { 'fēld jē,äl·ə·jē }

field gradient [PHYS] **1.** A vector obtained by applying the del operator to a scalar field. **2.** A tensor obtained by dyadic multiplication of the del operator with a vector field. { 'fēld ,grād·ē·ənt }

field gun [ORD] Any artillery piece mounted on a carriage for use in the field, for example, a cannon. { 'fēld ,gən }

field index [NUCLEO] The constant n for a betatron in which the magnetic field strength at radius r is equal to $B_0 (r/R)^{-n}$, where R is the radius of equilibrium orbit of an electron, and B_0 is the corresponding magnetic field. Also known as n value. { 'fēld ,in,deks }

field intensity [COMMUN] In Federal Communications Commission regulations, the electric field intensity in the horizontal direction. [PHYS] *See* field strength. { 'fēld in,ten·səd·ē }

field investigation [SCI TECH] An investigation carried out in the field; usually applied to an investigation made by someone not domiciled at the site. { 'fēld ,in·ves·tə,gā·shən }

field ionization [ELECTR] The ionization of gaseous atoms and molecules by an intense electric field, often at the surface of a solid. { 'fēld ,ī·ən·ə'zā·shən }

field-ion microscope [ELECTR] A microscope in which atoms are ionized by an electric field near a sharp tip; the field then forces the ions to a fluorescent screen, which shows an enlarged image of the tip, and individual atoms are made visible; this is the most powerful microscope yet produced. Also known as ion microscope. { 'fēld ¦ī,än 'mī·krə,skōp }

field laboratory [SCI TECH] Usually a temporary or portable laboratory facility set up at the site of an operation to conduct chemical or physical evaluations. { 'fēld ,lab·rə,tór·ē }

field length [COMPUT SCI] The number of columns, characters, or bits in a specified field. { 'fēld ,leŋkth }

field lens [OPTICS] The lens in a two-lens eyepiece which is farther from the eye. { 'fēld ,lenz }

field-line annihilation *See* field-line reconnection. { 'fēld ,līn ə,nī·ə'lā·shən }

field-line reconnection [ASTRON] A topological rearrangement of the magnetic field lines surrounding an astronomical body, for example, the transfer of lines between open and closed configurations in the terrestrial magnetotail; a possible source of the energy released explosively in solar flares and magnetospheric substorms. Also known as field line annihilation; magnetic merging. { 'fēld ,līn ,rē·kə'nek·shən }

field luminance *See* adaptation luminance. { 'fēld ,lü·mə·nəns }

field magnet [ELECTROMAG] The magnet which creates a magnetic field in an electric machine or device. { 'fēld ,mag·nət }

field map [MAP] A map made in the field and bearing observations of various kinds upon which the final map is based. { 'fēld ,map }

field moisture [HYD] Water in the ground above the water table. { 'fēld ,mois·chər }

field of fire [ORD] The area which a weapon or group of weapons may cover effectively with fire from a given position. { 'fēld əv 'fīr }

field of planes on a manifold [MATH] A continuous assignment of a vector subspace of tangent vectors to each point in the manifold. Also known as plane field. { 'fēld əv 'plānz ón ə 'man·ə,fōld }

field of search [ELECTR] The space that a radar set or installation can cover effectively. { 'fēld əv 'sərch }

field of vectors on a manifold [MATH] A continuous assignment of a tangent vector to each point in the manifold. Also known as vector field. { 'fēld əv 'vek·tərz ón ə 'man·ə,fōld }

field of view [OPTICS] The area or solid angle which can be viewed through an optical instrument. Also known as field. { 'fēld əv 'vyü }

field operator [QUANT MECH] An operator function of space and time for the annihilation or creation of a particle. { 'fēld ,äp·ə,rād·ər }

field pattern *See* radiation pattern. { 'fēld ,pad·órn }

field piece [ORD] An artillery gun or howitzer mounted on a carriage for use in the field. { 'fēld ,pēs }

field pole [ELECTROMAG] A structure of magnetic material on which a field coil of a loudspeaker, motor, generator, or other electromagnetic device may be mounted. { 'fēld ,pōl }

field pressure [GEOL] The pressure of natural gas in the underground formations from which it is produced. { 'fēld ,presh·ər }

field processing [PETRO ENG] Treatment of oil and gas in the field prior to delivery to a refinery; includes separation into oil and gas fractions, separation of liquid hydrocarbons from the gas, and removal of water. { 'fēld ,präs,es·iŋ }

field-programmable gate array [ELECTR] A gate-array device that can be configured and reconfigured by the system manufacturer and sometimes by the end user of the system. { ¦fēld prō,gram·ə·bəl 'gāt ə,rā }

field-programmable logic array [ELECTR] A programmed logic array in which the internal connections of the logic gates can be programmed once in the field by passing high current through fusible links, by using avalanche-induced migration to short base-emitter junctions at desired interconnections, or by other means. Abbreviated FPLA. Also known as programmable logic array. { ¦fēld prō¦gram·ə·bəl 'läj·ik ə,rā }

field quenching [MET] The quench cooling and tempering of a heated metal object at the site of construction or operation by using portable equipment rather than fixed manufacturing facilities. [SOLID STATE] Decrease in the emission of light of a phosphor excited by ultraviolet radiation, x-rays, alpha particles, or cathode rays when an electric field is simultaneously applied. { 'fēld ,kwench·iŋ }

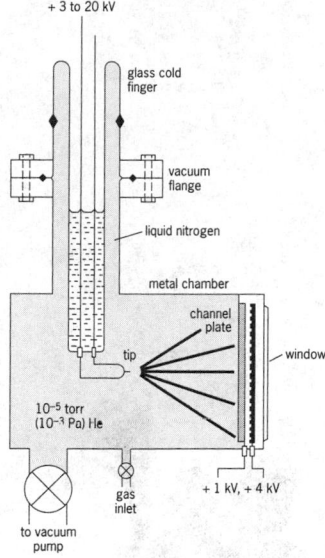

FIELD-EMISSION MICROSCOPE

Field ion microscope.

FIG

Fig (*Ficus carica*), details of branch and fruit.

field repetition rate See field frequency. { ¦fēld rep·ə'tish·ən ˌrāt }

field rheostat [ELEC] A rheostat used to adjust the current in the field winding of an electric machine. { 'fēld ¦rē·ə₁stat }

field scan [ELECTR] Television term denoting the vertical excursion of an electron beam downward across a cathode-ray tube face, the excursion being made in order to scan alternate lines. { 'fēld ₁skan }

field section [COMPUT SCI] A portion of a field, such as the section formed by the second and third character of a 10-character field. { 'fēld ₁sek·shən }

field separator [COMPUT SCI] A character that is used to mark the boundary between fields in a record. { 'fēld ₁sep·ə₁rād·ər }

field-sequential color television [COMMUN] A color television system in which the individual red, green, and blue primary colors are associated with successive fields. { ¦fēld sə¦kwen·chəl ¦kəl·ər 'tel·ə₁vizh·ən }

field service representative See field engineer. { 'fēld ₁sər·vəs ₁rep·rə₁zent·əd·iv }

field shift [NUC PHYS] The portion of the mass shift produced by the change in the size and shape of the nuclear charge distribution when neutrons are added to the nucleus. Also known as volume shift. { 'fēld ₁shift }

field squeeze [COMPUT SCI] In a mail merge operation, the elimination of extra blank spaces in a data field so that the data field is correctly printed within the text of the letter. { 'fēld ₁skwēz }

field stars [ASTRON] Background stars when a specific object is being observed. { 'fēld ₁stärz }

field stop [OPTICS] An opening, usually circular, in an opaque screen, whose edges determine the limits of the field of view of an optical instrument. { 'fēld ₁stäp }

field strength [PHYS] A vector characterizing a field. Also known as field intensity. { 'fēld ₁streŋkth }

field-strength meter [ENG] A calibrated radio receiver used to measure the field strength of radiated electromagnetic energy from a radio transmitter. { 'fēld ₁streŋkth ₁mēd·ər }

field-strip [ORD] To disassemble the major components of a machine gun, cannon, or other firearm for cleaning, inspection, or the like. { 'fēld ₁strip }

field telephone [COMMUN] A portable telephone designed for field or combat use. { 'fēld ₁tel·ə₁fōn }

field test [SCI TECH] A nonformal experiment, that is, one with fewer controls than a laboratory experiment, conducted under field conditions. { 'fēld ₁test }

field theory [MATH] The study of fields and their extensions. [PHYS] A theory in which the basic quantities are fields; classically the equations governing the fields may be given; in quantum field theory the commutation rules satisfied by the field operators also are specified. [PSYCH] A psychological theory that emphasizes the importance of interactions between events in an individual's environment. { 'fēld ₁thē·ə·rē }

field waveguide [ELECTROMAG] A single wire, threaded or coated with dielectric, which guides an electromagnetic field. Also known as G string. { 'fēld 'wāv₁gīd }

field weapons [ORD] Weapons designed or intended for actual use in the field. { 'fēld ₁wep·ənz }

field weld [MET] A weld made at the construction site. { 'fēld ₁weld }

field winding See field coil. { 'fēld ₁wīnd·iŋ }

field wire [ELEC] An insulated flexible wire or cable used in field telephone and telegraph systems. { 'fēld ₁wīr }

fieldwork [SCI TECH] Work done, such as surveying or making geological observations, in the field. { 'fēld₁wərk }

Fierz interference [NUC PHYS] Interference between the axial vector and tensor parts of the weak interaction of nucleon and lepton (electron-neutrino) fields in beta decay; measurements of the beta-particle energy spectrum indicate that it vanishes. { 'firts ₁int·ə'fir·əns }

fièvre boutonneuse [MED] A mild febrile rickettsial disease of humans caused by *Rickettsia conori;* characterized by a rash, tache noire (primary ulcer), and swollen lymph glands. Also known as boutonneuse fever; Marseilles fever. { 'fyev·rə bü·tə'nāz }

fife rail [NAV ARCH] A metal or wooden rail with holes for belaying pins. { 'fīf ₁rāl }

FIFO See first-in, first-out. { 'fī₁fō }

FILAMENT

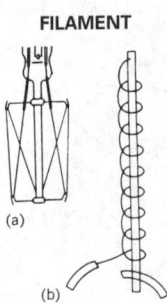

Two types of filaments: (*a*) in incandescent lamp; (*b*) in thermionic tube. (*General Electric Co.*)

fifteen-degrees calorie See calorie. { ¦fif·tēn di¦grēz ¦kal·ə·rē }

fifth-generation computer [COMPUT SCI] A computer that would use artificial intelligence techniques to learn, reason, and converse in natural languages resembling human languages. { 'fifth ₁jen·ə¦rā·shən kəm'pyüd·ər }

fifth sound [CRYO] A temperature oscillation which propagates in helium II contained in a porous material such as a tightly packed powder, where the normal component is immobilized by its viscosity. { ¦fifth ¦saúnd }

fifth wheel [MECH ENG] A coupling device in the form of two horizontal disks that rotate on each other positioned between a tractor and a semitrailer so that they can change direction independently. { ¦fifth ¦wēl }

fifty-percent zone [ORD] Area enclosing the center of dispersion or impact within which one half of all shots fired with the same setting will fall. { ¦fif·tē pər'sent ₁zōn }

fig [BOT] *Ficus carica.* A deciduous tree of the family Moraceae cultivated for its edible fruit, which is a syconium, consisting of a fleshy hollow receptacle lined with pistillate flowers. { fig }

fighter aircraft [AERO ENG] A military aircraft designed primarily to destroy other aircraft in the air; may also be used to bomb military targets; it is maneuverable and has a high rate of climb. { 'fīd·ər 'er₁kraft }

fighter bomber [AERO ENG] A fighter aircraft that is designed to have bombs, or rockets, added to it so that it may be used as a bomber. { 'fīd·ər ₁bäm·ər }

fighter interceptor [AERO ENG] A fighter aircraft designed to intercept and shoot down enemy aircraft. { 'fīd·ər ₁in·tər'sep·tər }

fighting compartment [ORD] A portion of a fighting vehicle in which the occupants service and fire the principal armament, occupying a part of the hull and all of the turret, if any. { 'fīd·iŋ kəm₁pärt·mənt }

Figitidae [INV ZOO] A family of hymenopteran insects in the superfamily Cynipoidea. { fə'jid·ə₁dē }

figurative constant [COMPUT SCI] A predefined constant in COBOL which does not require a description in data division, such as ZERO which stands for 0. { 'fig·yə₁rəd·iv 'kän₁stənt }

figure [MATER] The natural grain of wood, especially when it is cut as a veneer. { 'fig·yər }

figurehead [NAV ARCH] An ornament placed on the foremost edge of the stem just below the bowsprit. { 'fig·yər₁hed }

figure of merit [ANALY CHEM] A performance characteristic of an analytical chemical method that influences its choice for a specific type of determination, such as selectivity, sensitivity, detection limit, precision, and bias. [ELECTR] A performance rating that governs the choice of a device for a particular application; for example, the figure of merit of a magnetic amplifier is the ratio of usable power gain to the control time constant. { 'fig·yər əv 'mer·ət }

figure of the earth [GEOD] A precise geometric shape of the earth. { 'fig·yər əv thē 'ərth }

figures shift [COMMUN] A physical movement that permits a teletypewriter to print uppercase characters, numbers, symbols, and the like. { 'fig·yərz ₁shift }

figure stone See agalmatolite. { 'fig·yər ₁stōn }

figuring [OPTICS] Grinding or polishing of surfaces of optical components to remove aberrations. { 'fig·yər·iŋ }

Fiji disease [PL PATH] A virus disease of sugarcane; elongated swellings on the underside of leaves precede death of the plant. { 'fē₁jē di₁zēz }

Fijivirus [VIROL] A genus in the viral family Reoviridae that is the causative agent of Fiji disease in plants and insects. { 'fē·jē₁vī·rəs }

filament [ASTRON] A prominence, seen as a dark marking on the solar disk. [BOT] **1.** The stalk of a stamen which supports the anther. **2.** A chain of cells joined end to end, as in certain algae. [ELEC] Metallic wire or ribbon which is heated in an incandescent lamp to produce light, by passing an electric current through the filament. [ELECTR] A cathode made of resistance wire or ribbon, through which an electric current is sent to produce the high temperature required for emission of electrons in a thermionic tube. Also known as directly heated cathode; filamentary cathode; filament-type cathode. [INV ZOO] A single silk fiber in the cocoon of a silkworm. [MET] A long, flexible metal wire drawn very fine. [SCI TECH] A long, flexible object with a small cross section. [TEXT] A single continuous manufactured fiber which is extruded from a spinneret and joined with others to make a thread. { 'fil·ə·mənt }

filamentary cathode See filament. { ¦fil·ə'ment·ə·rē 'kath‚ōd }

filament current [ELECTR] The current supplied to the filament of an electron tube for heating purposes. { 'fil·ə·mənt ‚kə·rənt }

filament drawing [MET] Reducing the cross section of wire by pulling it through a die to form a filament. { 'fil·ə·mənt ‚dró·iŋ }

filament emission [ELECTR] Liberation of electrons from a heated filament wire in an electron tube. { 'fil·ə·mənt i'mish·ən }

filament lamp See incandescent lamp. { 'fil·ə·mənt ‚lamp }

filamentous bacteria [MICROBIO] Bacteria, especially in the order Actinomycetales, whose cells resemble filaments and are often branched. { ‚fil·ə'men·təs bak'tir·ē·ə }

filamentous bacteriophage [VIROL] Threadlike bacterial viruses that use pili to attach to the host. { ‚fil·ə¦men·təs bak'tir·ē·ə‚fāj }

filament saturation See temperature saturation. { 'fil·ə·mənt ‚sach·ə'rā·shən }

filament transformer [ELECTR] A small transformer used exclusively to supply filament or heater current for one or more electron tubes. { 'fil·ə·mənt tranz‚fór·mər }

filament-type cathode See filament. { 'fil·ə·mənt ‚tīp 'kath‚ōd }

filament winding [ELECTR] The secondary winding of a power transformer that furnishes alternating-current heater or filament voltage for one or more electron tubes. [ENG] A process for fabricating a composite structure in which continuous fiber reinforcement (glass, boron, silicon carbide), either previously impregnated with a matrix material or impregnated during winding, are wound under tension over a rotating core. { 'fil·ə·mənt ‚wīnd·iŋ }

filaria [INV ZOO] A parasitic filamentous nematode belonging to the order Filaroidea. { fə'lar·ē·ə }

filariasis [MED] A disease due to the presence of hairlike nematodes (filariae) in humans, including *Wuchereria bancrofti*, *W. pacifica*, and *Onchocerca volvulus*. { ‚fil·ə'rī·ə·səs }

Filarioidea [INV ZOO] An order of the class Nematoda comprising highly specialized parasites of humans and domestic animals. { fil·ə'rŏid·ē·ə }

filar micrometer [DES ENG] An instrument used to measure small distances in the field of an eyepiece by using two parallel wires, one of which is fixed while the other is moved at right angles to its length by means of an accurately cut screw. Also known as bifilar micrometer. { 'fī·lər mī'kräm·əd·ər }

filature [TEXT] A factory where silk is unwound from cocoons and the strands are collected into skeins. { 'fil·ə·chər }

filbert [BOT] Either of two European plants belonging to the genus *Corylus* and producing a thick-shelled, edible nut. Also known as hazelnut. { 'fil·bərt }

file [COMPUT SCI] A collection of related records treated as a unit. [DES ENG] A steel bar or rod with cutting teeth on its surface; used as a smoothing or forming tool. { fīl }

file allocation table [COMPUT SCI] A table stored on hard or removable disks used to locate files or sections of files if scattered about the disk. Abbreviated FAT. { ‚fīl ‚al·ə'kā·shən ‚tā·bəl }

file compression program See file compression utility. { 'fīl kəm‚presh·ən ‚prō·grəm }

file compression utility [COMPUT SCI] A utility program that encodes files so that they take up less space in storage. Also known as file compression program. { 'fīl kəm‚presh·ən yü‚til·əd·ē }

file control system [COMPUT SCI] Software package which handles the transfer of data from any device into any device. { 'fīl kən‚trōl ‚sis·təm }

file event [COMPUT SCI] A single access to any storage device for either input or output. { 'fīl i‚vent }

file gap [COMPUT SCI] An area in a data storage medium which is used mainly to indicate the end of a file and sometimes the beginning of another. { 'fīl ‚gap }

file-handling routine [COMPUT SCI] A part of a computer program that deals with reading and writing of data from and to a file. { 'fīl ‚hand·liŋ rü‚tēn }

file hardness [ENG] Hardness of a material as determined by testing with a file of standardized hardness; a material which cannot be cut with the file is considered as hard as or harder than the file. { 'fīl ‚härd·nəs }

file header [COMPUT SCI] A set of words comprising the file name and various characteristics of the file, found at the beginning of a file stored on magnetic tape or disk. { 'fīl ‚hed·ər }

file identification [COMPUT SCI] A device, such as a label or tag, used to identify, describe, or name a physical medium, such as a reel of digital magnetic tape or a box of punched cards, which contains data. { ¦fīl ī‚dent·ə·fə'kā·shən }

file layout [COMPUT SCI] A description of the arrangement of the data in a file. { 'fīl ‚lā‚aùt }

file locking [COMPUT SCI] A technique that prevents processing of a file by more than one program or user at a time, ensuring that a file in use by one user is made unavailable to others. { 'fīl ‚läk·iŋ }

file maintenance [COMPUT SCI] Data-processing operation in which a master file is updated on the basis of one or more transaction files. { 'fīl ‚mānt·ən·əns }

file management system [COMPUT SCI] Computer programs that control the space used for file storage and provide such services as input/output control and indexing. { 'fīl ¦man·ij·mənt ‚sis·təm }

file manager [COMPUT SCI] Software for managing data that works only with single files and lacks relational capability. { 'fīl ‚man·ə·jər }

file name [COMPUT SCI] The name given by the programmer to a specific set of data. { 'fīl ‚nām }

file opening [COMPUT SCI] The process, carried out by computer software, of identifying a file and comparing the file header with specifications in the program being run to ensure that the file corresponds. { 'fīl ‚ōp·ə·niŋ }

file organization [COMPUT SCI] The structure of a file meeting two requirements: to minimize the running time of the program, and to simplify the work involved in modifying the contents of the file. { ¦fīl ‚órg·ə·nə'zā·shən }

file organization routine [COMPUT SCI] A program which allocates data files into random-access storage devices. { 'fīl ‚órg·ə·nə'zā·shən rü‚tēn }

file-oriented system [COMPUT SCI] A computer configuration which considers a heavy, or exclusive, usage of data files. { 'fīl ‚ór·ē‚ent·əd ‚sis·təm }

file printout [COMPUT SCI] Output from a computer printer consisting of a copy of the contents of a file held in some storage device, usually to assist in debugging a program. { 'fīl ‚prin‚taùt }

file processing [COMPUT SCI] The job of updating, sorting, or validating a data file. { 'fīl ‚präs‚es·iŋ }

file protection [COMPUT SCI] A mechanical device or a computer command which prevents erasing of or writing upon a magnetic tape but allows a program to read the data from the tape. { 'fīl prə‚tek·shən }

file protection ring [COMPUT SCI] A ring that can be attached to, or detached from, the hub of a reel of magnetic tape, used to identify the reel's status and, in some computer systems, to prevent writing upon the tape when the ring is attached or detached. { 'fīl prə‚tek·shən ‚riŋ }

file reference [COMPUT SCI] An operation involving looking up and retrieving the information on file for a specified item or items. { 'fīl ‚ref·rəns }

file reorganization [COMPUT SCI] An activity performed periodically on files and data bases, involving such operations as deletion of unneeded records, in order to minimize space requirements of files and improve efficiency of processing. { 'fīl rē‚ór·gə·nə'zā·shən }

file search [COMPUT SCI] An operation involving looking through the file for information on all items falling in a specified category, extracting the information for any item where the information recorded meets certain criteria, and determining whether or not there exists a specified pattern of information anywhere in the file. { 'fīl ‚sərch }

file security See data security. { 'fīl sə‚kyür·əd·ē }

file server [COMPUT SCI] A mass storage device that holds programs and data that can be accessed and shared by the workstations connected to a local-area network. Also known as network server. { 'fīl ˌsər·vər }

file sharing [COMPUT SCI] The common use, by two or more users, of data and program files, usually located in a file server. { 'fīl ˌsher·iŋ }

FileSize metric [COMPUT SCI] A measure of computer program size, equal to the total number of characters in the source file of the program. { ¦fīl'sīz ¦me·trik }

file specification [COMPUT SCI] A designation that enables a file to be located on a disk and includes the disk drive, name of the directory/subdirectory, and name of file. { 'fīl ˌspes·ə·fəˌkā·shən }

file storage unit [COMPUT SCI] The component of a computer system that stores information required for reference. { 'fīl ˌstȯr·ij ˌyü·nət }

filet lace [TEXT] Lace with knotted square-mesh ground and square design in darned or woven effect. { fi'lā ˌlās }

file transfer [COMPUT SCI] The movement, under program control, of a file from one storage device to another. { 'fīl ˌtranz·fər }

file transfer access and management [COMPUT SCI] A standard communications protocol for transferring files between systems of different vendors. Abbreviated FTAM. { ¦fīl ¦tranz·fər ¦ak·ses ən 'man·ij·mənt }

file transfer protocol [COMPUT SCI] A set of standards that allows the user of any computer on the Internet to receive files from another computer, or to transmit files to another computer, after the user has specified a name and password for the other computer. Abbreviated FTP. { 'fīl ¦tranz·fər ˌprōd·ə·kȯl }

file transfer utility [COMPUT SCI] A computer program specifically designed to handle file transfers. { 'fīl ¦tranz·fər yü,til·əd·ē }

file virus [COMPUT SCI] A computer virus that infects application files such as spreadsheets, computer games, or accounting software. { 'fīl ˌvī·rəs }

filial generation [GEN] Any generation following the parental generation. { 'fil·ē·əl jen·əˌrā·shən }

Filicales [BOT] The equivalent name for Polypodiales. { ˌfil·ə'kā·lēz }

Filicineae [BOT] The equivalent name for Polypodiatae. { ˌfil·ə'sin·ē,ē }

Filicornia [INV ZOO] A group of hyperiid amphipod crustaceans in the suborder Genuina having the first antennae inserted anteriorly. { ˌfil·ə'kȯr·nē·ə }

filiform [BIOL] Threadlike or filamentous. { 'fil·əˌfȯrm }

filiform corrosion [MET] A random threadlike deterioration of a painted or lacquered metal caused by superficial corrosion of the base metal. { 'fil·əˌfȯrm kə'rō·zhən }

filiform lapilli See Pele's hair. { 'fil·əˌfȯrm lə'pil·ē }

filiform papilla [ANAT] Any one of the papillae occurring on the dorsum and margins of the oral part of the tongue, consisting of an elevation of connective tissue covered by a layer of epithelium. { 'fil·əˌfȯrm pə'pil·ə }

fill [CIV ENG] Earth used for embankments or as backfill. [MIN ENG] See pack. { fil }

fill characters [COMPUT SCI] Nondata characters or bits which are used to fill out a field on the left if data are right-justified or on the right if data are left-justified. { 'fil ˌkar·ik·tərz }

filled band [SOLID STATE] An energy band, each of whose energy levels is occupied by an electron. { ¦fild 'band }

filled composite [MATER] Mixture (composite) of thermoplastic or thermosetting resin and granular or short-strand fiber fill. { ¦fild kəm'päz·ət }

filled insulation [MATER] A loose insulating material that is poured or blown into walls. { ¦fild in·sə'lā·shən }

filled shell [PHYS] A set of energy levels in an atom or nucleus which have approximately the same energy and which are all occupied. { ˌfild 'shel }

filled stopes [MIN ENG] Stopes filled with barren stone, low-grade ore, sand, or tailings (mill waste) after the ore has been extracted. { ¦fild 'stōps }

filled-system thermometer [ENG] A thermometer which has a bourdon tube connected by a capillary tube to a hollow bulb; the deformation of the bourdon tube depends on the pressure of a gas (usually nitrogen or helium) or on the volume of a liquid filling the system. Also known as filled thermometer. { ¦fild ¦sis·təm thər'mäm·əd·ər }

filled thermometer See filled-system thermometer. { ¦fild thər'mäm·əd·ər }

filled thermoplastic [MATER] A thermoplastic resin material that has been extended (filled) with an inert filler powder or fibers before curing. { ¦fild ¦thər·mə¦plas·tik }

filled thermoset [MATER] Thermosetting resin material that has been extended (filled) with an inert filler powder or fibers before curing. { ¦fild 'thər·mə,set }

filler [COMPUT SCI] Storage space that does not contain significant data but is needed to comply with length requirements or is reserved to fulfill some future need. [MATER] **1.** An inert material added to paper, resin, bituminous material, and other substances to modify their properties and improve quality. **2.** A material used to fill holes in wood, plaster, or other surfaces before applying a coating such as paint or varnish. [MET] The rod used to deposit metal in a joint in brazing, soldering, or welding. Also known as filler metal. { 'fil·ər }

filler metal See filler. { 'fil·ər ˌmed·əl }

filler specks [MATER] In a cast plastic object, visible specks of a filler such as wood flour or asbestos that stand out in color contrast against the surface of the object. { 'fil·ər ˌspeks }

fillet [BUILD] A flat molding that separates rounded or angular moldings. [DES ENG] A concave transition surface between two otherwise intersecting surfaces. [ENG] **1.** Any narrow, flat metal or wood member. **2.** A corner piece at the juncture of perpendicular surfaces to lessen the danger of cracks, as in core boxes for castings. [FOOD ENG] A boneless slice of meat or fish. { 'fil·ət or, of food, fə'lā }

fillet gage [DES ENG] A gage for measuring convex or concave surfaces. { 'fil·ət ˌgāj }

fillet lightning See ribbon lightning. { 'fil·ət ˌlīt·niŋ }

fillet weld [MET] A weld joining two edges at right angles; cross-sectional configuration is approximately triangular. { 'fil·ət ˌweld }

fill factor [MECH ENG] The approximate load that the dipper of a shovel is carrying, expressed as a percentage of the rated capacity. { 'fil ˌfak·tər }

filling [ENG] The loading of trucks with any material. [METEOROL] An increase in the central pressure of a pressure system on a constant-height chart, or an analogous increase in height on a constant-pressure chart; the term is commonly applied to a low rather than to a high. [MIN ENG] Allowing a mine to fill with water. [TEXT] **1.** The yarn running perpendicular to the lengthwise, or warp, yarn in weaving. Also known as pick; weft; woof. **2.** In cloth finishing, a clay or starch used to add body and weight. { 'fil·iŋ }

filling factor [CRYO] The ratio of the electron density of a quantum Hall liquid to the density of magnetic flux quanta. { 'fil·iŋ ˌfak·tər }

filling in [GRAPHICS] A condition in letterpress or offset lithography in which the area between the halftone dots is filled with ink or ink is plugging up the type. { ¦fil·iŋ ¦in }

fill justification See justification. { ¦fil ˌjəs·tə·fə¦kā·shən }

fillowite [MINERAL] $H_2Na_6(Mn,Fe,Ca)_{14}(PO_4)_{12}·H_2O$ A brown, yellow, or colorless mineral composed of a hydrous phosphate of manganese, iron, sodium, and other metals. { 'fil·ə,wīt }

fill terrace See alluvial terrace. { 'fil ˌter·əs }

fill-up work See internal work. { 'fil,əp ˌwərk }

film [BIOL] A thin, membranous skin, such as a pellicle. [ELEC] The layer adjacent to the valve metal in an electrochemical valve, in which is located the high voltage drop when current flows in the direction of high impedance. [GRAPHICS] Plastic material, such as cellulose acetate or cellulose nitrate, coated with a light-sensitive emulsion, used to make negatives or transparencies in radiography or photography. [MATER] A flat section of material that is extremely thin in comparison to its other dimensions and has a nominal maximum thickness of 250 micrometers and a lower limit of thickness of about 25 micrometers. Also known as self-supported film. [MED] A pathological opacity, as of the cornea. [MET] Oxide coating on a metal. { film }

film analysis [IND ENG] A systematic detailed analysis of work from a motion picture film, usually derived from a memo-motion study. { 'film ə'nal·ə·səs }

film badge [NUCLEO] A device worn for the purpose of indicating the absorbed dose of radiation received by the wearer; usually made of metal, plastic, or paper and loaded with one or more pieces of x-ray film.　Also known as badge meter.　{ 'film ,baj }

film base [GRAPHICS] The celluloid component which supports the emulsion of photographic film.　{ 'film ,bās }

film boiling [PHYS CHEM] A stage in the boiling process in which the heater surface is totally covered by a film of vapor and the liquid does not contact the solid.　[THERMO] Boiling in which a continuous film of vapor forms at the hot surface of the container holding the boiling liquid, reducing heat transfer across the surface.　{ 'film ,bȯil·iŋ }

film coefficient [THERMO] For a fluid confined in a vessel, the rate of flow of heat out of the fluid, per unit area of vessel wall divided by the difference between the temperature in the interior of the fluid and the temperature at the surface of the wall.　Also known as convection coefficient.　{ 'film ,kō·i,fish·ənt }

film condensation [THERMO] The formation of a continuous film of liquid on a wall in contact with a vapor, when the wall is cooled below the local vapor saturation temperature and the liquid wets the cold surface.　{ 'film ,kän·dən,sā·shən }

film cooling [THERMO] The cooling of a body or surface, such as the inner surface of a rocket combustion chamber, by maintaining a thin fluid layer over the affected area.　{ 'film ,kül·iŋ }

film density [GRAPHICS] Degree of film image opacity.　{ 'film ,den·səd·ē }

film-development chromatography [ANALY CHEM] Liquid-analysis chromatographic technique in which the stationary phase (adsorbent) is a strip or layer, as in paper or thin-layer chromatography.　{ film di,vel·əp·mənt ,krō·mə'täg·rə·fē }

film dosimetry [NUCLEO] The determination of radiation dose by measurement of the darkening of a photographic film which is exposed to radiation and developed under controlled conditions.　{ ¦film dō'sim·ə·trē }

film integrated circuit [ELECTR] An integrated circuit whose elements are films formed in place on an insulating substrate.　{ ¦film int·ə¦grād·əd 'sər·kət }

film jacket [GRAPHICS] A transparent acetate single or multiple sleeve or pocket made to hold microfilm in flat strips.　{ 'film ,jak·ət }

film mottle [GRAPHICS] Cloudy or blotchy appearance of photographic film; the uneven density is generally caused by insufficient agitation.　{ 'film ,mäd·əl }

filmogen [MATER] The film-forming material or binder in paint which imparts continuity.　{ 'fil·mə,jen }

film opaque [GRAPHICS] Any of the liquids applied with small brushes to processed film to cover imperfections known as pinholes, very small transparent dots in an otherwise black emulsion area.　{ 'film ō,pāk }

film optical-sensing device [COMPUT SCI] A device capable of digitizing the information stored on a film.　{ 'film ¦äp·tə·kəl ¦sens·iŋ di,vīs }

film plane [GRAPHICS] The position in a camera occupied by the film or plate.　{ 'film ,plān }

film platen [ENG] A device which holds film in the focal plane during exposure.　{ 'film ,plat·ən }

film pressure [PHYS] The difference between the surface tension of a pure liquid and the surface tension of the liquid with a unimolecular layer of a given substance adsorbed on it.　Also known as surface pressure.　{ 'film ,presh·ər }

film reader [ELECTR] A device for converting a pattern of transparent or opaque spots on a photographic film into a series of electric pulses.　[OPTICS] A device for projecting or displaying microfilm so that an operator can read the data on the film; usually provided with equipment for moving or holding the film.　{ 'film ,rēd·ər }

film recorder [ELECTR] A device which places data, usually in the form of transparent and opaque spots or light and dark spots, on photographic film.　{ 'film ri,kȯrd·ər }

film resistor [ELEC] A fixed resistor in which the resistance element is a thin layer of conductive material on an insulated form; the conductive material does not contain binders or insulating material.　{ 'film ri,zis·tər }

film scanning [ELECTR] The process of converting motion picture film into corresponding electric signals that can be transmitted by a television system.　{ 'film ,skan·iŋ }

film size [GRAPHICS] Generally, an indication of film width, for example, 16 millimeter, or 5 inch, 8 inch, and so forth.　{ 'film ,sīz }

film sizing table [MIN ENG] A table used in ore dressing for sorting fine material by means of a film of flowing water.　{ 'film ,sīz·iŋ ,tā·bəl }

film strength [MATER] **1.** The measurement of a lubricant's ability to keep an unbroken film over surfaces.　**2.** The resistance to disruption by films of all types, such as plastic films or surface-coating films.　{ 'film ,streŋkth }

filmstrip [GRAPHICS] A continuous length of 35-millimeter film containing a number of still photographs, drawings, or charts, which are projected on a screen one at a time.　{ 'film,strip }

film tension [PHYS CHEM] The contractile force per unit length that is exerted by an equilibrium film in contact with a supporting substrate.　{ 'film ,ten·chən }

film theory [PHYS] A theory of the transfer of material or heat across a phase boundary, where one or both of the phases are flowing fluids, the main controlling factor being resistance to heat conduction or mass diffusion through a relatively stagnant film of the fluid next to the surface.　Also known as boundary-layer theory.　{ 'film ,thē·ə·rē }

film transport [MECH ENG] **1.** The mechanism for moving photographic film through the region where light strikes it in recording film tracks or sound tracks of motion pictures.　**2.** The mechanism which moves the film print past the area where light passes through it in reproduction of picture and sound.　{ 'film ,tranz,pȯrt }

film vault [ENG] A place for safekeeping of film.　{ 'film ,vȯlt }

film weld [GRAPHICS] A butt splice of photographic film made with a heat splicer.　{ 'film ,weld }

filoplume [VERT ZOO] A specialized feather that may be decorative, sensory, or both; it is always associated with papillae of contour feathers.　{ 'fil·ə,plüm }

filopodia [INV ZOO] Filamentous pseudopodia.　{ ,fil·ə'pōd·ē·ə }

filoreticulopodia [INV ZOO] Branched, filamentous pseudopodia.　{ ,fil·ə,red·ə,kyül·ə'pōd·ē·ə }

Filosia [INV ZOO] A subclass of the class Rhizopodea characterized by slender filopodia which rarely anastomose.　{ fī'lō·shə }

filosus See fibratus.　{ fī'lō·səs }

filter [COMPUT SCI] A device or program that separates data or signals in accordance with specified criteria.　[CONT SYS] See compensator.　[ELECTR] Any transmission network used in electrical systems for the selective enhancement of a given class of input signals.　Also known as electric filter; electric-wave filter.　[ENG] A porous article or material for separating suspended particulate matter from liquids by passing the liquid through the pores in the filter and sieving out the solids.　[ENG ACOUS] A device employed to reject sound in a particular range of frequencies while passing sound in another range of frequencies.　Also known as acoustic filter.　[MATH] A family of subsets of a set S: it does not include the empty set, the intersection of any two members of the family is also a member, and any subset of S containing a member is also a member.　[OPTICS] An optical element that partially absorbs incident electromagnetic radiation in the visible, ultraviolet, or infrared spectra, consisting of a pane of glass or other partially transparent material, or of films separated by narrow layers; the absorption may be either selective or nonselective with respect to wavelength.　Also known as optical filter.　[SCI TECH] In general, a selective device that transmits a desired range of matter or energy while substantially attenuating all other ranges.　{ 'fil·tər }

filterability [ENG] The adaptability of a liquid-solid system to filtration; system is not filterable if it is too viscous to be forced through a filter medium, or if the solids are too small to be stopped by the filter medium.　{ ,fil·trə'bil·əd·ē }

filterable virus [VIROL] Virus particles that remain in a fluid after passing through a diatomite or glazed porcelain filter with pores too minute to allow the passage of bacterial cells.　{ 'fil·trə·bəl 'vī·rəs }

filter aid [MATER] An inert powder or granules such as diatomaceous earth, fly ash, or sand added to a solution that is to be filtered in order to form a porous bed on the filter and increase the rate and improve the quality of filtration.　{ 'fil·tər ,ād }

filter base [MATH] A family of subsets of a given set with the property that it does not include the empty set, and the

FILOPLUME

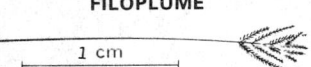

Drawing of a filoplume showing reduced hairlike vane.

FILTER

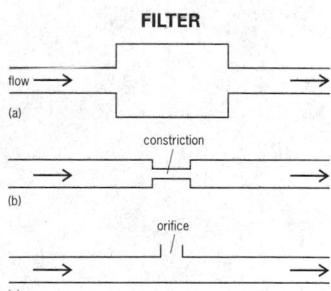

Acoustic filters. (*a*) Expansion-chamber filter. (*b*) Constriction filter. (*c*) High-pass filter.

intersection of any finite number of members of the family includes another member. { 'fil·tər ˌbās }

filter bauxite [MATER] Crushed, screened, and calcined bauxite, in particles that range from 20–60 and from 30–60 mesh grades; used in ore refineries for filtering. Also known as activated bauxite. { 'fil·tər 'bȯk,sīt }

filter bed [CIV ENG] A fill of pervious soil that provides a site for a septic field. [ENG] A contact bed used for filtering purposes. { 'fil·tər ˌbed }

filter cake [ENG] *See* mud cake. [MATER] A concentrated solid or semisolid material that is separated from a liquid and remains on the filter after pressure filtration. { 'fil·tər ˌkāk }

filter-cake washing [CHEM ENG] An operation performed at the end of a filtration, in which residual liquid impurities are washed out of the cake by the flow of another liquid through the cake. { 'fil·tər ˌkāk ˌwash·iŋ }

filter capacitor [ELEC] A capacitor used in a power-supply filter system to provide a low-reactance path for alternating currents and thereby suppress ripple currents, without affecting direct currents. { 'fil·tər kə,pas·əd·ər }

filter center [ORD] An information center at which all radar and other observed information concerning movements of friendly and enemy planes within a certain sector is screened and disseminated. { 'fil·tər ˌsen·tər }

filter choke [ELEC] An iron-core coil used in a power-supply filter system to pass direct current while offering high impedance to pulsating or alternating current. { 'fil·tər ˌchōk }

filter cloth [MATER] A fabric used as a medium for filtration. { 'fil·tər ˌklȯth }

filter crystal [ELECTR] Quartz crystal which is used in an electrical circuit designed to pass energy of certain frequencies. { 'fil·tər ˌkrist·əl }

filter design [ELECTR] The design of electrical networks in which the principle of electrical resonance is used to make the network accept wanted frequencies while rejecting unwanted ones. { 'fil·tər di,zīn }

filter discrimination [ELECTR] Difference between the minimum insertion loss at any frequency in a filter attenuation band and the maximum insertion loss at any frequency in the operating range of a filter transmission band. { 'fil·tər di,skrim·ə'nā·shən }

filtered-particle testing [ENG] A penetrant method of nondestructive testing by which cracks in porous objects (100 mesh or smaller) are indicated: a fluid containing suspended particles is sprayed on a test object; if a crack exists, particles are filtered out and concentrate at the surface as liquid flows into the crack. { ¦fil·tərd ¦pärd·ə·kəl ˌtest·iŋ }

filtered radar data [ELECTR] Radar data from which unwanted returns have been removed by mapping. { ¦fil·tərd 'rā,där ˌdad·ə }

filtered stock [MATER] Lubricating oil which has been filtered or refiltered to improve performance characteristics. { ¦fil·tərd 'stäk }

filter factor [OPTICS] The number of times the exposure must be increased when a filter is used on a camera, because the filter absorbs some of the light. { 'fil·tər ˌfak·tər }

filter feeder [INV ZOO] A microphagous organism that uses complex filtering mechanisms to trap particles suspended in water. { 'fil·tər ˌfēd·ər }

filter flask [CHEM] A flask with a side arm to which a vacuum can be applied; usually filter flasks have heavy side walls to withstand high vacuum. { 'fil·tər ˌflask }

filter impedance compensator [ELECTR] Impedance compensator which is connected across the common terminals of electric wave filters when the latter are used in parallel to compensate for the effects of the filters on each other. { ¦fil·tər im'ped·əns ˌkäm·pən'sād·ər }

filtering [ENG] The process of interpreting reported information on movements of aircraft, ships, and submarines in order to determine their probable true tracks and, where applicable, heights or depths. { 'fil·tə·riŋ }

filter leaf [CHEM ENG] The frame or structure in a filter press that holds the filter cloth or other filter medium; a number of leaves in series usually comprises a filter press. { 'fil·tər ˌlēf }

filter loss [SCI TECH] The amount of fluid passed through a permeable membrane in a given time. { 'fil·tər ˌlȯs }

filter medium [MATER] That portion of a filtration system that provides the liquid-solid separation, such as close-woven textiles or metal screens, papers, nonwoven fabrics, granular beds, or porous media. { 'fil·tər ˌmēd·ē·əm }

filter paper [MATER] Porous cellulose paper used for filtering, especially for quantitative purposes. { 'fil·tər ˌpā·pər }

filter pass band *See* filter transmission band. { ¦fil·tər 'pas ˌband }

filter photometer [ENG] A colorimeter in which the length of light is selected by the use of appropriate glass filters. { 'fil·tər fə'täm·əd·ər }

filter photometry [ANALY CHEM] **1.** Colorimetric analysis of solution colors with a filter applied to the eyepiece of a conventional colorimeter. **2.** Inspection of a pair of Nessler tubes through a filter. { 'fil·tər fə'täm·ə·trē }

filter press [ENG] A metal frame on which iron plates are suspended and pressed together by a screw device; liquid to be filtered is pumped into canvas bags between the plates, and the screw is tightened so that pressure is furnished for filtration. { 'fil·tər ˌpres }

filter-press cell [PHYS CHEM] An electrolytic cell consisting of several units in series, as in a filter press, in which each electrode, except the two end ones, acts as an anode on one side and a cathode on the other, and the space between electrodes is divided by porous asbestos diaphragms. { 'fil·tər ˌpres ˌsel }

filter pump [MECH ENG] An aspirator or vacuum pump which creates a negative pressure on the filtrate side of the filter to hasten the process of filtering. { 'fil·tər ˌpəmp }

filter reactor [ELEC] A reactor used for reducing the harmonic components of voltage in an alternating-current or direct-current circuit. { 'fil·tər rē,ak·tər }

filter sand [MATER] Graded sand used for filtering suspended matter from a flowing liquid stream. { 'fil·tər ˌsand }

filter screen [ENG] A fine-pored medium through which a liquid will pass and on which solids deposit; the medium may be a metal sieve screen or a woven fabric of metal or of natural or synthetic fibers. { 'fil·tər ˌskrēn }

filter section [ELEC] A simple *RC*, *RL*, or *LC* network used as a broad-band filter in a power supply, grid-bias feed, or similar device. { 'fil·tər ˌsek·shən }

filter slot [ELECTROMAG] Choke in the form of a slot designed to suppress unwanted modes in a waveguide. { 'fil·tər ˌslät }

filter spectrophotometer [SPECT] Spectrophotographic analyzer of spectral radiations in which a filter is used to isolate narrow portions of the spectrum. { 'fil·tər spek·trə·fə'täm·əd·ər }

filter thickener [ENG] Device that thickens a liquid-solid mixture by removing a portion of the liquid by filtration, rather than by settling. { ¦fil·tər 'thik·ə·nər }

filter transmission band [ELECTR] Frequency band of free transmission; that is, frequency band in which, if dissipation is neglected, the attenuation constant is zero. Also known as filter pass band. { ¦fil·tər tranz'mish·ən ˌband }

filter-type respirator [ENG] A protective device which removes dispersoids from the air by physically trapping the particles on the fibrous material of the filter. { 'fil·tər ˌtīp 'res·pə,rād·ər }

filtrate [SCI TECH] The discharge liquor in filtration. Also known as mother liquor; strong liquor. { 'fil,trāt }

filtration [SCI TECH] A process of separating particulate matter from a fluid, such as air or a liquid, by passing the fluid carrier through a medium that will not pass the particulates. { fil'trā·shən }

filtration sterilization [MICROBIO] The physical removal of microorganisms from liquid that may be destroyed by heat (such as blood serum, enzyme solutions, antibiotics, and some bacteriological media and medium constituents) by filtering through materials having relatively small pores. { fil'trā·shən ˌster·ə·lə,zā·shən }

fimbria *See* pilus. { 'fim·brē·ə }

fimbriate [BIOL] Having a fringe along the edge. { 'fim·brē,āt }

fin [AERO ENG] A fixed or adjustable vane or airfoil affixed longitudinally to an aerodynamically or ballistically designed body for stabilizing purposes. [DES ENG] A projecting flat plate or structure, as a cooling fin. [ENG] Material which remains in the holes of a molded part and which must be removed. [VERT ZOO] A paddle-shaped appendage on fish and other aquatic animals that is used for propulsion, balance, and guidance. { fin }

final amplifier [ELECTR] The transmitter stage that feeds the antenna. { ¦fīn·əl 'am·plə,fī·ər }

final boiling point *See* end point. { ¦fīn·əl 'bȯil·iŋ ,pȯint }

final common pathway *See* lower motor neuron. { ¦fīn·əl ¦käm·ən 'path,wā }

final cut [MIN ENG] In surface mining, the pit that remains after the final pass through the deposit. { 'fīn·əl 'kət }

final diameter [NAV] The diameter of the circle traversed by a vessel after turning through 360° and maintaining the same speed and rudder angle; it is always less than the tactical diameter, and is measured perpendicular to the original course and between the tangents at the points where 180° and 360° of the turn have been completed. { ¦fīn·əl dī'am·əd·ər }

final drawing [GRAPHICS] A completed mechanical drawing from which fabrication of an item can take place. { ¦fīn·əl 'dró·iŋ }

final filter *See* afterfilter. { ¦fīn·əl 'fil·tər }

final great-circle course [NAV] The direction, at the destination, of the great circle through that point and the point of departure, expressed as the angular distance from a reference direction, usually north, to that part of the great circle extending beyond the destination. { ¦fīn·əl ¦grāt 'sər·kəl ,kȯrs }

final heading [NAV] The aircraft heading at the end of a rating period while using gyro steering. { ¦fīn·əl 'hed·iŋ }

final lock mechanism [ORD] A device for locking the stroking member of a cartridge-actuated device in final position. { ¦fīn·əl 'läk ,mek·ə,niz·əm }

final mass [AERO ENG] The mass of a rocket after its propellants are consumed. { ¦fīn·əl 'mas }

final set [MATER] Hardening of a mixture of water and cement, concrete, or mortar to a greater degree than the hardening attained at the initial set; generally measured as the time required for the mixture to stiffen sufficiently to resist the penetration of a test needle. { ¦fīn·əl 'set }

final-value theorem [MATH] The theorem that if $f(t)$ is a function which has a Laplace transform $F(s)$, and if the derivative of $f(t)$ with respect to t is also Laplace transformable, and if the limit of $f(t)$ as t approaches infinity exists, then this limit is equal to the limit of $sF(s)$ as s approaches zero. { ¦fīn·əl ¦val·yü ¦thir·əm }

financial life *See* venture life. { fə'nan·chəl ¦līf }

financial planning system [COMPUT SCI] A decision-support system that allows the financial planner or manager to examine and evaluate many alternatives before making final decisions, and which employs the use of a model, usually a matrix of data elements which is constructed as a series of equations. { fī'nan·chəl 'plan·iŋ ,sis·təm }

fin assembly [ORD] An assembly of metal blades, usually mounted lengthwise on a sleeve, used on a missile, such as bomb or rifle grenade, to give directional stability. { 'fin ə,sem·blē }

finch [VERT ZOO] The common name for birds composing the family Fringillidae. { finch }

find [IND ENG] The therblig representing the mental reaction which occurs on recognizing an object at the end of the elemental motion search; now seldom used. { find }

finder [COMMUN] **1.** An optical or electronic device that shows the field of action covered by a television camera. **2.** Switch or relay group in telephone switching systems that selects the path which the call is to take through the system; operates under the instruction of the calling station's dial. [OPTICS] A small telescope having a wide-angle lens and low power, which is attached to a larger telescope and points in the same direction; used to locate objects that are to be viewed in the larger telescope. { 'fīnd·ər }

finder beam [COMPUT SCI] A beam of light projected by a light pen on the spot on the display screen where the light pen photodetector is focused, in order to aid the user in positioning the light pen. { 'fīnd·ər ,bēm }

finder light [GRAPHICS] Light beam projected to show outline of photographic field. { 'fīnd·ər ,līt }

F indicator *See* F scope. { 'ef ,in·də,kād·ər }

finding circuit *See* lockout circuit. { 'fīnd·iŋ ,sər·kət }

findspot [ARCHEO] The place where an archeological object has been found. { 'fīnd,spät }

fine admixture [GEOL] The smaller size grades of a sediment of mixed size grades. { ¦fīn 'ad,miks·chər }

fineblanking [ENG] A manufacturing process in which a part is fabricated to a shape very close to its final dimensions by use of high-precision tools that yield a final workpiece with smoothly sheared edges. { 'fīn,blaŋk·iŋ }

fine delay [NAV] A dial on a loran navigation system receiver, for controlling relatively small changes in the position of the B trace pedestal, and serving as a vernier for the coarse delay. { ¦fīn di'lā }

fine earth [GEOL] A soil which can be passed through a 2-millimeter sieve without grinding its primary particles. { ¦fīn 'ərth }

fin efficiency [ENG] In extended-surface heat-exchange equations, the ratio of the mean temperature difference from surface-to-fluid divided by the temperature difference from fin-to-fluid at the base or root of the fin. { 'fin ə,fish·ən·sē }

fine gold [MET] Almost pure gold; the value of bullion gold depends on its percentage of fineness. [MIN ENG] In placer mining, gold in exceedingly small particles. { ¦fīn 'gōld }

fine gravel [GEOL] Gravel consisting of particles with a diameter range of 1 to 2 millimeters. { ¦fīn 'grav·əl }

fine grinding [MECH ENG] Grinding performed in a mill rotating on a horizontal axis in which the material undergoes final size reduction, to −100 mesh. { ¦fīn 'grīnd·iŋ }

fine index [COMPUT SCI] The more specific of two indices consulted to gain access to a record. { ¦fīn 'in,deks }

fine-line printer [GRAPHICS] An accessory device mounted in a microfilmer to mark each filmed document. { 'fīn ,līn ,print·ər }

fineness [MATH] **1.** For a partition of a metric space, the least upper bound on distances between points in the same member of the partition. **2.** For a partition of an interval into subintervals, the length of the longest subinterval. Also known as mesh; norm. [MET] Degree of purity of gold or silver in parts per thousand. { 'fīn·nəs }

fineness modulus [ENG] A number denoting the fineness of a fine aggregate or other fine material such as sand or paint. { 'fīn·nəs 'mäj·ə·ləs }

fineness of grind [MATER] In inks and paint, the pigment particle size, size range, and population, which indirectly influence color strength, gloss, and rheology. { ¦fīn·nəs əv 'grīnd }

fineness ratio [AERO ENG] The ratio of the length of a streamlined body, as that of a fuselage or airship hull, to its maximum diameter. { 'fīn·nəs ,rā·shō }

finer [MATH] A partition P of a set is finer than another partition Q of the same set if each member of P is a subset of a member of Q. { 'fīn·ər }

fines [MATER] **1.** Particles smaller than average in a mixture of particles varying in size. **2.** Fine material which passes through a standard screen on which coarser fragments are retained. [MET] That portion of a metal powder consisting of particles smaller than a specified size. { fīnz }

fine sand [GEOL] Sand grains between 0.25 and 0.125 millimeter in diameter. { ¦fīn 'sand }

fine-screen halftone [GRAPHICS] An illustration (photograph or artwork) reproduced for printing in a continuous tone by photographing the illustration with a 120-line-to-the-inch screen between the illustration and the camera. { 'fīn ,skrēn 'haf,tōn }

fine silver [MET] Silver having a minimum fineness of 999; considered to be pure silver. { ¦fīn 'sil·vər }

fine structure [ATOM PHYS] The splitting of spectral lines in atomic and molecular spectra caused by the spin angular momentum of the electrons and the coupling of the spin to the orbital angular momentum. { ¦fīn 'strək·chər }

fine-structure constant [PHYS] A fundamental dimensionless constant, equal to $e^2/(4\pi\epsilon_0\hbar c)$ in International System (SI) units and to $e^2/(\hbar c)$ in centimeter-gram-second (cgs) electrostatic units, where e is the elementary charge, $\hbar$ is Planck's constant divided by 2π, c is the speed of light, and ϵ_0 is the electric constant; numerically, it is equal to 0.007 297 352 533 ± 0.000 000 000 027 or to 1/(137.035 999 76 ± 0.000 000 50); symbolized α. Also known as Sommerfeld fine-structure constant. { 'fīn ,strək·chər 'kän·stənt }

fin fold [EMBRYO] A median integumentary fold extending along the body of a fish embryo which gives rise to the dorsal, caudal, and anal fins. { 'fin ,fōld }

finger [ANAT] Any of the four digits on the hand other than the thumb. [GEOL] The tendency for gas which is displacing liquid hydrocarbons in a heterogeneous reservoir rock system to move forward irregularly (in fingers), rather than on a uniform front. [PETRO ENG] A pair or set of bracketlike projections placed at a strategic point in a drill tripod or derrick to keep a number of lengths of drill rods or casing in place when they are standing in the tripod or derrick. { 'fiŋ·gər }

finger bit [DES ENG] A steel rock-cutting bit having fingerlike, fixed or replaceable steel-cutting points. { 'fiŋ·gər ‚bit }

finger board [PETRO ENG] A board with projecting dowels or pipe fingers located in the upper part of the drill derrick or tripod to support stands of drill rod, drill pipe, or casing. { 'fiŋ·gər ‚bórd }

finger chute [MIN ENG] Steel rails hinged independently over an ore chute, to control rate of flow of rock. { 'fiŋ·gər ‚shüt }

finger coal *See* natural coke. { 'fiŋ·gər ‚kōl }

finger gripper [CONT SYS] A robot component that uses two or more joints for grasping objects. { 'fiŋ·gər ‚grip·ər }

finger lake [HYD] A long, comparatively narrow lake, generally glacial in origin; may occupy a rock basin in the floor of a glacial trough or be confined by a morainal dam across the lower end of the valley. { 'fiŋ·gər ‚lāk }

fingerprint [ANALY CHEM] Evidence for the presence or the identity of a substance that is obtained by techniques such as spectroscopy, chromatography, or electrophoresis. [FOREN SCI] **1.** A pattern of distinctive epidermal ridges on the bulbs of the inside of the end joints of fingers and thumbs. **2.** An impression of a human fingerprint. { 'fiŋ·gər‚print }

fingerprint pattern area [FOREN SCI] The part of a fingerprint that contains the cores, deltas, and ridges that are used for classification. { 'fiŋ·gər‚print 'pad·ərn ‚er·ē·ə }

finger raise [MIN ENG] Steeply sloping openings permitting caved ore to flow down raises through grizzlies to chutes on the haulage level. { 'fiŋ·gər ‚rāz }

finial [ARCH] An ornamental terminating or capping feature on an upper extremity, as over a door or on a pinnacle or gable. { 'fin·ē·əl }

fining [CHEM ENG] A process in which molten glass is cleared of bubbles, usually by the addition of chemical agents. [FOOD ENG] A process for clarifying a beverage, such as beer or wine, by causing the suspended matter to fall to the bottom. [MATER] A material such as gelatin, egg white, or bentonite that is used to clarify a liquid. Also known as clarifying agent. { 'fīn·iŋ }

finish [MATER] **1.** A chemical or other material applied to surfaces to protect them, to alter their appearances, or to modify their physical properties; finishes can be physically, chemically, or electrolytically applied and have value for fabrics and fibers, metals, paper products, plastics, woods, and so on. **2.** The ultimate quality, condition, or appearance of the surface of a material. { 'fin·ish }

finished goods [IND ENG] Manufactured products in inventory ready for packaging, shipment, or sale. { ¦fin·isht 'gùdz }

finished steel [MET] Steel that has undergone final processing and is ready for market. { ¦fin·isht 'stēl }

finisher [CIV ENG] A construction machine used to smooth the freshly placed surface of a roadway, or to prepare the foundation for a pavement. { 'fin·ish·ər }

finish grinding [MECH ENG] The last action of a grinding operation to achieve a good finish and accurate dimensions. { 'fin·ish ‚grīnd·iŋ }

finishing compound [MATER] A substance used to impart surface properties to textiles or leather, such as softness, flexibility, or fire resistance. { 'fin·ish·iŋ ‚käm‚paúnd }

finishing hardware [BUILD] Items, such as hinges, door pulls, and strike plates, made in attractive shapes and finishes, and usually visible on the completed structure. { 'fin·ish·iŋ ‚härd‚wer }

finishing hydrated lime [MATER] Any hydrated lime suitable for use in the finishing coat of plaster; characterized by a high degree of whiteness and plasticity. { 'fin·ish·iŋ 'hī‚drād·əd ‚līm }

finishing mill [MET] A rolling mill in which sheet, plate, and other mill products are subjected to final rolling operations. { 'fin·ish·iŋ ‚mil }

finishing nail [DES ENG] A wire nail with a small head that can easily be concealed. { 'fin·ish·iŋ ‚nāl }

finishing roll [MIN ENG] The last roll, or the one that does the finest crushing in ore dressing. { 'fin·ish·iŋ ‚rōl }

finishing temperature [MET] The temperature at which hot-working is completed. { 'fin·ish·iŋ ‚tem·prə·chər }

finish plate [DES ENG] A plate which covers and protects the cylinder setscrews; it is fastened to the underplate and forms part of the armored front for a mortise lock. { 'fin·ish ‚plāt }

finish turning [MECH ENG] The operation of machining a surface to accurate size and producing a smooth finish. { 'fin·ish ‚tərn·iŋ }

finite character [MATH] **1.** A property of a family C of sets such that any finite subset of a member of C belongs to C, and C includes any set all of whose finite subsets belong to C. **2.** A characteristic of a property of subsets of a set such that a subset S has the property if and only if all the nonempty finite subsets of S have the property. { 'fī‚nīt 'kar·ik·tər }

finite clipping [ELECTR] Clipping in which the threshold level is large but is below the peak input signal amplitude. { ¦fī‚nīt 'klip·iŋ }

finite closed aquifer [HYD] The part of a subterranean reservoir containing water (aquifer) in which the aquifer is limited (finite), with no water flow across the exterior reservoir boundary. { ¦fī‚nīt 'klōzd 'ak·wə·fər }

finite decimal *See* terminating decimal. { ¦fī‚nīt 'des·məl }

finite difference [MATH] The difference between the values of a function at two discrete points, used to approximate the derivative of the function. { ¦fī‚nīt 'dif·rəns }

finite-difference equations [MATH] Equations arising from differential equations by substituting difference quotients for derivatives, and then using these equations to approximate a solution. { ¦fī‚nīt 'dif·rəns i‚kwā·zhənz }

finite discontinuity [MATH] A discontinuity of a function that lies at the center of an interval on which the function is bounded. { 'fī‚nīt ‚dis·kän·tə'nü·əd·ē }

finite elasticity theory *See* finite strain theory. { ¦fī‚nīt i‚las'tis·əd·ē ‚thē·ə·rē }

finite element method [ENG] An approximation method for studying continuous physical systems, used in structural mechanics, electrical field theory, and fluid mechanics; the system is broken into discrete elements interconnected at discrete node points. { ¦fī‚nīt 'el·ə·mənt ‚meth·əd }

finite extension [MATH] An extension field F of a given field E such that F, viewed as a vector space over E, has finite dimension. { ¦fī‚nīt ik'sten·chən }

finite group [MATH] A group which contains a finite number of distinct elements. { ¦fī‚nīt 'grüp }

finite impulse response filter [ELECTR] An electric filter that will settle to a steady state within a finite amount of time after being exposed to a change in input. Abbreviated FIR filter. { ¦fī‚nīt ‚im‚pəls ri'späns ‚fil·tər }

finite intersection property of a family of sets [MATH] If the intersection of any finite number of them is nonempty, then the intersection of all the members of the family is nonempty. { ¦fī‚nīt ‚in·tər'sek·shən ‚präp·ərd·ē əv ə 'fam·lē əv 'sets }

finitely additive set function *See* additive set function. { ¦fī‚nīt·lē ¦ad·ə·div 'set ‚fəŋk·shən }

finitely generated extension [MATH] A finitely generated extension of a field k is the smallest field which contains k and some finite set of elements. { ¦fī‚nīt·lē 'gen·ə‚rād·əd ik'sten·chən }

finitely generated left module [MATH] A left module over a ring A that has a finite subset, $x_1, x_2, \ldots, x_n$, such that any member of the module has the form $a_1 x_1 + \cdots + a_n x_n$, where $a_1, \ldots, a_n$ are members of A. { ¦fī‚nīt·lē jen·ə‚rād·əd ¦left 'mäj·əl }

finitely representable [MATH] A Banach space A is said to be finitely representable in a Banach space B if every finite-dimensional subspace of A is nearly isometric to a subspace of B. { ¦fī‚nīt·lē ‚rep·rə'zen·tə·bəl }

finite mathematics [MATH] **1.** Those parts of mathematics which deal with finite sets. **2.** Those fields of mathematics which make no use of the concept of limit. Also known as discrete mathematics. { ¦fī‚nīt ‚math·ə'mad·iks }

finite matrix [MATH] A matrix with a finite number of rows and columns. { ¦fī‚nīt 'mā·triks }

finite measure space [MATH] A measure space in which the measure of the entire space is a finite number. { 'fī‚nīt ‚mezh·ər ‚spās }

finite moment theorem [MATH] The theorem that if $f(x)$ is a continuous function, and if the integral of $f(x) x^n$ over a finite interval is zero for all positive integers n, then $f(x)$ is identically zero in that interval. { ¦fī¸nīt 'mō·mənt ¸thir·əm }

finite plane [MATH] In projective geometry, a plane with a finite number of points and lines. { ¦fī¸nīt 'plān }

finite population [STAT] A population of finite individuals or elements. { ¦fī¸nīt ¸päp·yə'lā·shən }

finite precision number [COMPUT SCI] A number that can be represented by a finite set of symbols in a given numeration system. { ¦fī¸nīt prə¦sizh·ən 'nəm·bər }

finite quantity [MATH] Any bounded quantity. { ¦fī¸nīt 'kwän·əd·ē }

finite sequence [MATH] **1.** A listing of some finite number, n, of mathematical entities that is indexed by the first n positive integers, 1, 2, . . . , n. **2.** More precisely, a function whose domain is the first n positive integers. { ¦fī¸nīt 'sē·kwəns }

finite series [MATH] A series that has a limited number of terms. { 'fī¸nīt 'sir¸ēz }

finite set [MATH] A set whose elements can be indexed by integers 1,2,3, . . . , n inclusive. { ¦fī¸nīt 'set }

finite-state machine [COMPUT SCI] An automaton that has a finite number of distinguishable internal configurations. { 'fī¸nīt ¸stāt mə¸shēn }

finite strain theory [MECH] A theory of elasticity, appropriate for high compressions, in which it is not assumed that strains are infinitesimally small. Also known as finite elasticity theory. { 'fī¸nīt 'strān ¸thē·ə·rē }

fin keel [NAV ARCH] A metal plate or thin fairing attached to the keel of sailing craft to give resistance to lateral motion, and frequently having a cigar-shaped lead bulb at the bottom to give transverse stability. Also known as ballast fin. { 'fin ¸kēl }

Fink truss [CIV ENG] A symmetrical steel roof truss suitable for spans up to 50 feet (15 meters). { 'fiŋk ¸trəs }

finned surface [MECH ENG] A tubular heat-exchange surface with extended projections on one side. { ¦find 'sər·fəs }

finnemanite [MINERAL] $Pb_5Cl(AsO_3)_3$ A gray, olive-green, or black hexagonal mineral composed of arsenite and chloride of lead. { 'fin·ə·mə¸nīt }

fin-reinforcing assembly [ORD] An assemblage of components required to reinforce a bomb fin assembly. { 'fin rē·ən¸fórs·iŋ ə¸sem·blē }

fin rot [VET MED] A bacterial disease of hatchery fishes characterized by necrosis and erosion of the fin tissue. { 'fin ¸rät }

Finsen lamp [ELEC] A high-temperature carbon arc or mercury arc lamp that produces a mixture of blue, violet, and near-ultraviolet light; used to treat certain skin disorders and to test paints and other protective coatings. { 'fin·sən ¸lamp }

finsen unit [ELECTROMAG] A unit of intensity of ultraviolet radiation, equal to the intensity of ultraviolet radiation at a specified wavelength whose energy flux is 100,000 watts per square meter; the wavelength usually specified is 296.7 nanometers. Abbreviated FU. { 'fin·sən ¸yü·nət }

Finsler geometry [MATH] The study of the geometry of a manifold in terms of the various possible metrics on it by means of Finsler structures. { 'fin·slər jē¦äm·ə·trē }

Finsler structure on a manifold [MATH] A family of metrics varying continuously from point to point. { 'fin·slər ¸strək·chər ón ə 'man·ə¸fōld }

fin spine [VERT ZOO] A bony process that supports the fins of certain fishes. { 'fin ¸spīn }

fin stabilization [ORD] Method of stabilizing a projectile (as a rocket, bomb, or missile) during flight by the aerodynamic use of protruding fins. { 'fin ¸stā·bə·lə'zā·shən }

fin waveguide [ELECTROMAG] Waveguide containing a thin longitudinal metal fin that serves to increase the wavelength range over which the waveguide will transmit signals efficiently; usually used with circular waveguides. { ¦fin 'wāv¸gīd }

fiord See fjord. { fyórd }

fiorite See siliceous sinter. { fē'ór¸īt }

Fior process [MET] The prereduction of high-grade iron particles or concentrates in a hot gaseous reactor to produce low-oxygen fines for partially metallized briquettes suitable for electric-arc steelmaking furnaces. { 'fyór ¸präs·əs }

fir [BOT] The common name for any tree of the genus *Abies* in the pine family; needles are characteristically flat. { fər }

fire [CHEM] The manifestation of rapid combustion, or combination of materials with oxygen. [ENG] To blast with gunpowder or other explosives. [MIN ENG] A warning that a shot is being fired. [ORD] **1.** The discharge of a gun, launching of a missile, or the like. **2.** The projectiles or missiles fired. **3.** To discharge a weapon. { fīr }

fire adjustment [ORD] Correcting the elevation and direction of a weapon, or regulating the explosion time of its projectile, so that the projectile will strike or burst at the desired point; for automatic weapons, it is a continuous operation from the instant the first rounds reach the vicinity of the target until the command "cease firing" is given. { 'fīr ə¸jəs·mənt }

firearm [ORD] **1.** In a general sense, a gun. **2.** A small arm, as a pistol or rifle, designed to be carried and used by an individual. { 'fīr¸ärm }

fire assay [MATER] Analysis of a metal-bearing material, especially gold and silver, by heating a sample with a suitable flux and weghing the resulting metal beads. { ¦fīr 'as¸ā }

fireball [ASTRON] A bright meteor with luminosity equal to or exceeding that of the brightest planets. [NUCLEO] The luminous sphere of hot gases that forms a few millionths of a second after a nuclear explosion. { 'fīr¸ból }

fireball model [ASTROPHYS] A model of gamma-ray bursts according to which a black hole accelerates jets of gas or plasma to relativistic velocities, at about 0.9999 of the speed of light, and the huge amount of kinetic energy in these jets is dissipated in internal collisions within the jets, which produce the gamma-ray emission. { ¦fīr¸ból ¦mäd·əl }

fire blight [PL PATH] A bacterial disease of apple, pear, and related pomaceous fruit trees caused by *Erwinia amylovora*; leaves are blackened, cankers form on the trunk, and flowers and fruits become discolored. { 'fīr ¸blīt }

fireboat [NAV ARCH] A vessel similar to a tug but fitted with fire-fighting apparatus. { 'fīr¸bōt }

fire bomb [ORD] An item designed to be dropped from an aircraft to destroy or reduce the utility of a target by the effects of combustion. { 'fīr ¸bäm }

fire boss [MIN ENG] An individual who examines a mine for gas and other dangers. Also known as mine examiner. { 'fīr ¸bós }

firebox [MECH ENG] The furnace of a locomotive or similar type of fire-tube boiler. { 'fīr¸bäks }

firebreak [FOR] A cleared area of land intended to check the spread of forest or prairie fire. [MIN ENG] A strip across an area in which either no combustible material is employed or in which, if timber supports are used, sand is filled and packed tightly around them. { 'fīr¸brāk }

firebrick [MATER] A refractory brick, often made of fireclay, that is able to withstand high temperature (up to 1500–1600°C) without fusion; used to line furnaces, fireplaces, and chimneys. { 'fīr¸brik }

fire bridge [ENG] A low wall separating the hearth and the grate in a reverberatory furnace. { 'fīr ¸brij }

fireclay [GEOL] **1.** A clay that can resist high temperatures without becoming glassy. **2.** Soft, embedded, white or gray clay rich in hydrated aluminum silicates or silica and deficient in alkalies and iron. { 'fīr ¸klā }

fire control [ORD] Control over the direction, volume, and time of fire of guns or launchers by the use of certain electrical or electronic devices or aids, or by optical or mechanical systems. { 'fīr kən¸trōl }

fire-control circuit [ELECTR] An electric circuit in a fire-control system. { 'fīr kən¸trōl ¸sər·kət }

fire-control computer [ORD] A computer used to guide a fire-control system. { 'fīr kən¸trōl kəm¸pyüd·ər }

fire-control grid [ORD] A system of lines that divide a military map into squares, the distance between any two parallel lines representing 1000 yards or 1000 meters, depending on the type of map; maps using the fire-control grid are of sufficiently large scale to be useful in fire control. { 'fīr kən¸trōl ¸grid }

fire-control instrument [ORD] An aiming circle, range finder, compass, telescope, or other instrument used in fire control. { 'fīr kən¸trōl ¸in·strə·mənt }

fire-control quadrant [ORD] A chemical device with a scale to measure angles and a leveling adjustment, used to measure the elevation angles of a weapon for obtaining the horizontal

FIR

Cone and branches with flat needles of balsam fir (*Abies balsamea*).

range of a weapon; it is attached to the gun, gun mount, or gun carriage. { 'fīr kən,trōl ,kwäd·rənt }

fire-control radar [ORD] Radar equipment used in a fire-control system. { 'fīr kən,trōl ,rā,där }

fire-control sonar [ORD] Sonar used in determining the aiming of antisubmarine weapons. { 'fīr kən,trōl ,sō,när }

fire-control system [ORD] A setup for control of the aiming and firing of guns, rockets, or guided missiles. { 'fīr kən,trōl ,sis·təm }

fire coordination [ORD] The planning and execution of fire so that targets are adequately covered by a suitable weapon or groups of weapons. { 'fīr kō,órd·ən'ā·shən }

fire crack [ENG] A crack resulting from thermal stress which propagates on the heated side of a shell or header in a boiler or a heat transfer surface. { 'fīr ,krak }

firecracker [ENG] A cylindrically shaped item containing an explosive and a fuse; used to simulate the noise of an explosive charge. { 'fīr,krak·ər }

fire cut [BUILD] An angular cut made at the end of a joist which will rest on a brick wall. { 'fīr ,kət }

firedamp [MIN ENG] **1.** A gas formed in mines by decomposition of coal or other carbonaceous matter; consists chiefly of methane and is combustible. **2.** An airtight stopping to isolate an underground fire and to prevent the inflow of fresh air and the outflow of foul air. Also known as fire wall. { 'fīr,damp }

firedamp alarm [MIN ENG] An instrument which gives a warning signal when the methane content in the mine atmosphere exceeds a known value. { 'fīr,damp ə,lärm }

firedamp detector [MIN ENG] A portable device to detect the presence and determine the percentage of firedamp in mine air. { 'fīr,damp di,tek·tər }

firedamp drainage [MIN ENG] The collection of firedamp from coal strata, generally into pipes, with or without the use of suction. Also known as methane drainage. { 'fīr,damp ,drān·ij }

firedamp-drainage drill [MIN ENG] A heavy, compressed-air-operated, percussive, rotary or rotary-percussive drilling machine for putting up the boreholes in firedamp drainage. { 'fīr,damp ,drān·ij ,dril }

firedamp explosion [MIN ENG] An explosion of a mixture of firedamp and air. { 'fīr,damp ik'splō·zhən }

firedamp fringe [MIN ENG] The zone of contact between the coal gases and the ventilation air current at the face of the mine. { 'fīr,damp ,frinj }

firedamp layer [MIN ENG] An accumulation of firedamp under the roof of a mine roadway where the ventilation is insufficient to dilute and remove the gas. { 'fīr,damp ,lā·ər }

firedamp migration [MIN ENG] The movement of firedamp through the strata or coal of a mine. { 'fīr,damp mī'grā·shən }

firedamp pressure-chamber method [MIN ENG] A method of firedamp drainage in coal mines; pressure chambers built at the intake and the return of a worked-out area are used to trap firedamp, which is drawn off in pipes. { 'fīr,damp 'presh·ər ,chām·bər ,meth·əd }

firedamp probe [MIN ENG] A flexible rubber tube connected to a rod, which can be thrust into roof cavities and breaks so that a sample of the air may be transferred to a methanometer and its firedamp content determined. { 'fīr,damp ,prōb }

firedamp reforming process [CHEM ENG] A process in which methane (firedamp) is mixed with steam and passed over a nickel catalyst for conversion to a mixture of hydrogen and carbon monoxide; this mixture is blended with pure methane, and the result is a fuel of high calorific value. { 'fīr,damp ri'fór·miŋ ,präs·əs }

fire-danger meter [ENG] A graphical aid used in fire-weather forecasting to calculate the degree of forest-fire danger (or burning index): commonly in the form of a circular slide rule, it relates numerical indices of the seasonal stage of foliage, the cumulative effect of past precipitation or lack thereof (buildup index), the measured fuel moisture, and the speed of the wind in the woods; the fuel moisture is determined by weighing a special type of wooden stick that has been exposed in the woods, its weight being proportional to its contained water; the calculated burning index falls on a scale of 1 to 100: 1 to 11 is no fire danger; 12 to 35 medium danger; 40 to 100 high danger. { 'fīr ,dān·jər ,mēd·ər }

fire detector [ENG] A temperature-sensing device designed to sound an alarm, to turn on a sprinkler system, or to activate

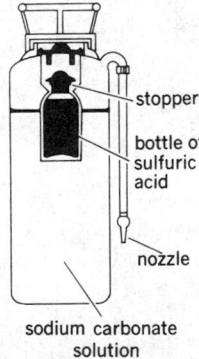

FIRE EXTINGUISHER

stopper

bottle of sulfuric acid

nozzle

sodium carbonate solution

Soda-acid fire extinguisher. When it is inverted, the stopper falls out of place allowing sulfuric acid to react with sodium carbonate to produce carbon dioxide, which propels the solution through the nozzle.

some other fire preventive measure at the first signs of fire. { 'fīr di,tek·tər }

fire direction [ORD] Tactical employment of fire power, exercise of tactical command of one or more units in the selection of targets, concentration or distribution of fire, and allocation of ammunition for each mission. { 'fīr də,rek·shən }

fire-direction net [COMMUN] A communication system linking observers, liaison officers, air observers, and firing batteries with the fire direction center for the purposes of fire control. { 'fīr də,rek·shən ,net }

fire disclimax [ECOL] A community that is perpetually maintained at an early stage of succession through recurrent destruction by fire followed by regeneration. { 'fīr dis'klī,maks }

fire distribution [ORD] **1.** The use of weapons of a unit to cover a target most effectively. **2.** The application of fire on targets, or on subdivisions of a target, in order of their importance. **3.** The systematic assignment of targets to the batteries of a number of vessels by message. { 'fīr dis·trə'byü·shən }

fire door [ENG] **1.** The door or opening through which fuel is supplied to a furnace or stove. **2.** A door that can be closed to prevent the spreading of fire, as through a building or mine. { 'fīr ,dór }

fired process equipment [ENG] Heaters, furnaces, reactors, incinerators, vaporizers, steam generators, boilers, and other process equipment for which the heat input is derived from fuel combustion (flames); can be direct-fired (flame in contact with the process stream) or indirect-fired (flame separated from the process fluid by a metallic wall). { ¦fīrd 'präs·əs i,kwip·mənt }

fired state [ELECTR] The "on" state of a silicon controlled rectifier or other semiconductor switching device, occurring when a suitable triggering pulse is applied to the gate. { ¦fīrd 'stāt }

fire effect [ORD] Result of firing on enemy personnel and materiel. { 'fīr i,fekt }

fire escape [BUILD] An outside stairway usually made of steel and used to escape from a building in case of fire. { 'fīr ə,skāp }

fire-exit bolt See panic exit device. { 'fīr ,eg·zət ,bōlt }

fire extinguisher [ENG] Any of various portable devices used to extinguish a fire by the ejection of a fire-inhibiting substance, such as water, carbon dioxide, gas, or chemical foam. { 'fīr ik,stiŋ·gwish·ər }

firefinder [ENG] An instrument consisting of a map and a sighting device; used in fire towers to locate forest fires. { 'fīr,fīn·dər }

fire flooding [PETRO ENG] A method to improve secondary recovery in an oil reservoir; a combustion process is started in the reservoir at an injection well by continued introduction of gas containing oxygen or other material to support combustion, and the combustion wave is driven through the reservoir toward the production well. { 'fīr ,fləd·iŋ }

firefly [INV ZOO] Any of various flying insects which produce light by bioluminescence. { 'fīr,flī }

fire foam [MATER] A colloidal solution of small gas bubbles produced by chemical reaction or mechanical agitation and used to extinguish hydrocarbon fire. { 'fīr ,fōm }

fire for adjustment [ORD] Fire delivered for the purpose of determining firing data that will place the center of impact or burst on the desired portion of the target. { 'fīr fər ə'jəs·mənt }

fire for effect [ORD] Fire which is delivered after the center of impact or burst is within the desired distance of the target or adjusting point. { 'fīr fər i'fekt }

fire-formed fragment See controlled fragment. { 'fīr ,fórmd 'frag·mənt }

fire fountain See lava fountain. { 'fīr ,faún·ən }

fire hook [ENG] **1.** A pole with a hooked metal head that is used in fire fighting to tear down walls or ceilings. Also known as pike pole. **2.** A hook used to rake a furnace fire. { 'fīr ,húk }

fire hose [ENG] A collapsible, flameproof hose that can be attached to a hydrant, standpipe, or similar outlet to supply water to extinguish a fire. { 'fīr ,hōz }

fire hydrant [CIV ENG] An outlet from a water main provided inside buildings or outdoors to which fire hoses can be connected. Also known as fire plug; hydrant. { 'fīr ,hī·drənt }

fire interrupter [ORD] For aircraft guns, a device (usually an electrical switch mechanically actuated) which interrupts firing of a weapon. { 'fīr ,int·ə,rəp·tər }

fire line [ENG] A pipework system dedicated to providing water for extinguishing fires. { 'fīr ,līn }

fire load [CIV ENG] The load of combustible material per square foot of floor space. { 'fīr ,lōd }

fire opal [MINERAL] A translucent or transparent, orangy-yellow, brownish-orange, or red variety of opal that gives out fiery reflections in bright light and that may have a play of colors. Also known as pyrophane; sun opal. { 'fīr ,ō·pəl }

fire partition [BUILD] A wall inside a building intended to retard fire. { 'fīr pər,tish·ən }

fire plug *See* fire hydrant. { 'fīr ,pləg }

fire point [CHEM] The lowest temperature at which a volatile combustible substance vaporizes rapidly enough to form above its surface an air-vapor mixture which burns continuously when ignited by a small flame. { 'fīr ,pȯint }

firepower [ORD] **1.** The capability to deliver fire. **2.** The fire itself, or the quantity or effectiveness of fire delivered. **3.** Any explosive or missile that wreaks damage upon the target against which it is directed; for example, guns or rockets, or aircraft units and so forth armed with guns or rockets. { 'fīr,pau̇·ər }

fireproof [BUILD] Having noncombustible walls, stairways, and stress-bearing members, and having all steel and iron structural members which could be damaged by heat protected by refractory materials. [MATER] The property of being relatively resistant to combustion. { 'fīr,prüf }

fireproofing compound *See* fire retardant. { 'fīr,prüf·iŋ ,käm,pau̇nd }

fire protection [CIV ENG] Measures for reducing injury and property loss by fire. { 'fīr prə,tek·shən }

fire pump [MECH ENG] A pump for fire protection purposes usually driven by an independent, reliable prime mover and approved by the National Board of Fire Underwriters. { 'fīr ,pəmp }

fire refining [MET] The refining of blister copper by treatment in a furnace under oxidizing conditions to remove the impurities and under reducing conditions to remove the excess oxygen. { 'fīr rī'fīn·iŋ }

fire-resistant [CIV ENG] Of a structural element, able to resist combustion for a specified time under conditions of standard heat intensity without burning or failing structurally. { 'fīr ri,zis·tənt }

fire retardant [MATER] A chemical used as a coating for or a component of a combustible material to reduce or eliminate a tendency to burn; used with textiles, plastics, rubbers, paints, and other materials. Also known as fireproofing compound. { 'fīr ri'tärd·ənt }

fire-retardant paint [MATER] A paint applied as a thin coating to reduce the rate of flame spread of a combustible material; based on silicone, casein, polyvinylchloride, or other substance. { 'fīr ri'tärd·ənt 'pānt }

fireroom [MECH ENG] That portion of a fossil fuel-burning plant which contains the furnace and associated equipment. { 'fīr,rüm }

fire scale [MET] Copper oxide remaining below the surface of silver-copper alloys after annealing and pickling. { 'fīr ,skāl }

fire sprinkling system *See* sprinkler system. { 'fīr 'spriŋk·liŋ ,sis·təm }

fire standpipe [CIV ENG] A high, vertical pipe or tank that holds water to assure a positive, relatively uniform pressure, particularly to provide fire protection to upper floors of tall buildings. { 'fīr 'stan,pīp }

firestone *See* flint. { 'fīr,stōn }

fire stop [BUILD] An incombustible, horizontal or vertical barrier, as of brick across a hollow wall or across an open room, to stop the spread of fire. Also known as draught stop. { 'fīr ,stäp }

fire storm [ORD] A fire engulfing a city, produced by an air raid with incendiary or fire bombs. { 'fīr ,stȯrm }

fire superiority [ORD] Fire with greater effect than that of the enemy because of its greater accuracy and volume, making possible advances against the enemy without heavy losses. { 'fīr sə,pir·ē'är·əd·ē }

fire support [ORD] The support or protection given forces in direct contact with the enemy by ground or naval guns or by aircraft engaging in close air support. { 'fīr sə,pȯrt }

fire tower [BUILD] A fireproof and smokeproof stairway compartment running the height of a building. [FOR] A tower used to watch for fires, especially forest fires. { 'fīr ,tau̇·ər }

fire-tube boiler [MECH ENG] A steam boiler in which hot gaseous products of combustion pass through tubes surrounded by boiler water. { 'fīr ,tüb ,bȯil·ər }

fire unit analyzer [ORD] An instrument for analyzing the effectiveness of an antiaircraft fire unit against hostile aircraft or missiles under the conditions stated on the face of the analyzer. { 'fīr ,yü·nət 'an·ə,līz·ər }

fire wall [CIV ENG] **1.** A fire-resisting wall separating two parts of a building from the lowest floor to several feet above the roof to prevent the spread of fire. **2.** A fire-resisting wall surrounding an oil storage tank to retain oil that may escape and to confine fire. [MIN ENG] *See* firedamp. { 'fīr ,wȯl }

firewall [COMPUT SCI] Hardware and software programs that protect the resources of a private network from users in other networks, controlling all traffic according to a predefined access policy. { 'fīr,wȯl }

fire weather [METEOROL] The state of the weather with respect to its effect upon the kindling and spreading of forest fires. { 'fīr ,weth·ər }

fire welding *See* forge welding. { 'fīr ,weld·iŋ }

firewire *See* IEEE 1394. { 'fīr,wīr }

FIR filter *See* finite impulse response filter. { 'fər ,fil·tər or ,ef'ī'är }

firing [ELECTR] **1.** The gas ionization that initiates current flow in a gas-discharge tube. **2.** Excitation of a magnetron or transmit-receive tube by a pulse. **3.** The transition from the unsaturated to the saturated state of a saturable reactor. [ENG] **1.** The act or process of adding fuel and air to a furnace. **2.** Igniting an explosive mixture. **3.** Treating a ceramic product with heat. { 'fīr,iŋ }

firing azimuth [ORD] Horizontal direction in which a gun or launcher is pointed for firing, expressed as an azimuth. { 'fīr,iŋ ,az·ə·məth }

firing base [ORD] Part of the mechanism in some cannon that supports the gun carriage when it is in position for firing. { 'fīr,iŋ ,bās }

firing box [ELEC] A boxlike item in which are mounted switches, cables, fuses, plugs, indicator lights, batteries, and the like, specifically designed for firing a rocket or guided missile from a remote position. { 'fīr,iŋ ,bäks }

firing button [ELEC] A button or switch for firing guns or rockets. { 'fīr·iŋ ,bət·ən }

firing cable *See* shot-firing cable. { 'fīr·iŋ ,kā·bəl }

firing chamber *See* combustion chamber. { 'fīr·iŋ ,chām·bər }

firing circuit [ELECTR] **1.** Circuit used with an ignitron to deliver a pulse of current of 5–50 amperes in the forward direction, from the igniter to the mercury, to start a cathode spot and to control the time of firing. **2.** By analogy, a similar control circuit of silicon-controlled rectifiers and like devices. { 'fīr·iŋ ,sər·kət }

firing data [ORD] All data necessary for firing a weapon at a given objective, which may be determined by computation and then transmitted as verbal commands, or may be applied electromechanically by one of the several types of directing devices. { 'fīr·iŋ ,dad·ə }

firing hammer [ORD] A metallic pivoted item, part of the firing mechanism of a firearm, designed to strike a firing pin or percussion cap and to fire a gun. { 'fīr·iŋ ,ham·ər }

firing interval [ORD] Period of time between firing one shot and the next. { 'fīr·iŋ ,int·ər·vəl }

firing jack [ORD] Adjustable device which stabilizes and levels certain mobile artillery weapons while the weapons are in firing position. { 'fīr·iŋ ,jak }

firing lanyard [ORD] A cord or cable of specific length, usually with a hook on one end and a handle on the opposite end, designed to be attached to a component of the firing mechanism of a gun, rocket launcher, or the like, and used to fire the weapon. { 'fīr·iŋ ,lan·yərd }

firing lock [ORD] A removable part of the firing mechanism in some weapons, incorporating the firing pin and the mechanism which drives it against the primer. { 'fīr·iŋ ,läk }

firing machine [ENG] An electric blasting machine. [MECH ENG] A mechanical stoker used to feed coal to a boiler furnace. { 'fīr·iŋ mə,shēn }

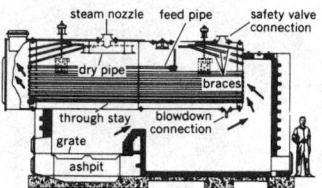

FIRE-TUBE BOILER

A horizontal-return-tube fire-tube boiler.

firing mechanism [ENG] A mechanism for firing a primer; the primer may be for initiating the propelling charge, in which case the firing mechanism forms a part of the weapon; if the primer is for the purpose of initiating detonation of the main charge, the firing mechanism is a part of the ammunition item and performs the function of a fuse. { 'fīr·iŋ ,mek·ə,niz·əm }

firing pin [ORD] A device used in the firing mechanism of a gun, mine, bomb, fuse, projectile, or the like, which strikes and detonates a sensitive explosive to initiate an explosive train or a propelling charge. { 'fīr·iŋ ,pin }

firing point [ELECTR] *See* critical grid voltage. [ORD] Location from which fire is delivered in target practice. { 'fīr·iŋ ,pȯint }

firing position [ORD] Position of a weapon ready for firing. { 'fīr·iŋ pə,zish·ən }

firing potential [ELECTR] Controlled potential at which conduction through a gas-filled tube begins. { 'fīr·iŋ pə,ten·chəl }

firing pressure [MECH ENG] The highest pressure in an engine cylinder during combustion. { 'fīr·iŋ ,presh·ər }

firing rate [MECH ENG] The rate at which fuel feed to a burner occurs, in terms of volume, heat units, or weight per unit time. { 'fīr·iŋ ,rāt }

firing table [ORD] A table or chart giving data needed for firing a gun accurately on a target under standard conditions, and also the corrections that must be made for special conditions such as winds or variations of temperature. { 'fīr·iŋ ,tā·bəl }

firing table elevation [ORD] The angle between the axis of the bore and the horizontal plane when the piece is laid to fire at a given range under conditions that are accepted as standard. { 'fīr·iŋ ,tā·bəl el·ə'vā·shən }

firing time [ORD] The period of time during which a weapon is fired. { 'fīr·iŋ ,tīm }

firmer chisel [DES ENG] A small hand chisel with a flat blade; used in woodworking. { 'fər·mər ,chiz·əl }

firming agent [FOOD ENG] A substance added prior to or during the processing of a foodstuff to protect and retain natural firmness. { 'firm·iŋ ,ā·jənt }

firm-joint caliper [DES ENG] An outside or inside caliper whose legs are jointed together at the top with a nut and which must be opened and closed by hand pressure. { 'fərm ,jȯint 'kal·ə·pər }

firmoviscosity [MECH] Property of a substance in which the stress is equal to the sum of a term proportional to the substance's deformation, and a term proportional to its rate of deformation. { ¦fər·mō·vis¦käs·əd·ē }

firmware [COMPUT SCI] A computer program or instruction, such as a microprogram, used so often that it is stored in a read-only memory instead of being included in software; often used in computers that monitor production processes. { 'fərm,wer }

firn [HYD] Material transitional between snow and glacier ice; it is formed from snow after existing through one summer melt season and becomes glacier ice when its permeability to liquid water drops to zero. Also known as firn snow. { fərn }

firn basin *See* firn field. { 'fərn ,bās·ən }

fir needle oil [MATER] An essential oil distilled from the needles and twigs of some trees of the genus *Abies;* used in perfumery, flavoring, and medicine. Also known as fir oil. { 'fər ,nēd·əl ,ȯil }

firn field [HYD] The accumulation area or upper region of a glacier where snow accumulates and firn is secreted. Also known as firn basin. { 'fərn ,fēld }

firn ice *See* iced firn. { 'fərn ,īs }

firnification [HYD] The process of firn formation from snow and of transformation of firn into glacier ice. { ,fər·nə·fə'kā·shən }

firn limit *See* firn line. { 'fərn ,lim·ət }

firn line [GEOL] **1.** The regional snow line on a glacier. **2.** The line that divides the ablation area of a glacier from the accumulation area. Also known as firn limit. { 'fərn ,līn }

firn snow *See* firn; old snow. { 'fərn ,snō }

fir oil *See* fir needle oil. { 'fər ,ȯil }

first answer print [GRAPHICS] The motion picture composite print, with the mixed sound track and the optical effects. { ¦fərst 'an·sər ,print }

first arrival [ENG] In exploration refraction seismology, the first seismic event recorded on a seismogram; it is noteworthy in that only first arrivals are considered in this usage. { ¦fərst ə'rī·vəl }

first bottom [GEOL] The floodplain of a river, below the first terrace. { ¦fərst 'bäd·əm }

first category [MATH] **1.** A set is of first category if it is a countable union of nowhere dense sets. **2.** A set *S* is of first category at a point *x* if there is a neighborhood of *x* whose intersection with *S* is of first category. { ¦fərst 'kad·ə,gȯr·ē }

first-class current [PART PHYS] A weak-interaction current whose charge symmetry (or G parity) properties are the same as those of currents which arise in the Fermi theory of beta decay. { ¦fərst ,klas 'kə·rənt }

first cost [IND ENG] The sum of the initial expenditures involved in capitalizing a property; includes items such as transportation, installation, preparation for service, as well as other related costs. { ¦fərst 'kȯst }

first countable topological space [MATH] A topological space in which every point has a countable number of open neighborhoods so that any neighborhood of this point contains one of these. { ¦fərst 'kaȯnt·ə·bəl ,täp·ə¦läj·ə·kəl 'spās }

first-degree burn [MED] A mild burn characterized by pain and reddening of the skin. { ¦fərst də¦grē 'bərn }

first derivative [MATH] The derivative of a function, considered as a function of the independent variable just as was the original function from which the derivative was taken. { ¦fərst də'riv·əd·iv }

first derived curve *See* derived curve. { ¦fərst də¦rīvd 'kərv }

first detector *See* mixer. { ¦fərst di'tek·tər }

first estimate-second estimate method [NAV] The method of determining time of meridian transit (especially local apparent noon) for a moving craft; the time of transit is computed for an estimated longitude of the craft, and the longitude estimate is then revised to agree with the time determined by the first estimate, and a second computation is made; the process is repeated as many times as necessary to obtain an answer of the desired precision. { ¦fərst ,es·tə·mət 'sek·ənd ,es·tə·mət ,meth·əd }

first filial generation [GEN] The first generation resulting from a cross between parents homozygous for different alleles at a locus; all members are heterozygous at the locus. Symbolized F_1. { ¦fərst ¦fil·ē·əl jen·ə'rā·shən }

first fire [ENG] The igniter used with pyrotechnic devices, consisting of first fire composition, loaded in direct contact with the main pyrotechnic charge; the ignition of the igniter or first fire is generally accomplished by fuse action. { ¦fərst 'fīr }

first fire composition [MATER] A pyrotechnic composition (readily ignitable and easily pressed into a strong, solid mass), compounded to produce a high temperature, preferably with creation of slag to give heat capacity. { ¦fərst ¦fīr ,käm·pə'zish·ən }

first Fresnel zone [ELECTROMAG] Circular portion of a wavefront transverse to the line between an emitter and a more distant point, where the resultant disturbance is being observed, whose center is the intersection of the front with the direct ray, and whose radius is such that the shortest path from the emitter through the periphery to the receiving point is one-half wavelength longer than the direct ray. { ¦fərst frə'nel ,zōn }

first-generation [COMPUT SCI] Denoting electronic hardware, logical organization, and software characteristic of a first-generation computer. { ¦fərst jen·ə'rā·shən }

first-generation computer [COMPUT SCI] A computer from the earliest stage of computer development, ending in the early 1960s, characterized by the use of vacuum tubes, the performance of one operation at a time in strictly sequential fashion, and elementary software, usually including a program loader, simple utility routines, and an assembler to assist in program writing. { ¦fərst jen·ə'rā·shən kəm'pyüd·ər }

first gust [METEOROL] The sharp increase in wind speed often associated with the early mature stage of a thunderstorm cell; it occurs with the passage of the discontinuity zone which is the boundary of the cold-air downdraft. { ¦fərst 'gəst }

first harmonic *See* fundamental. { ¦fərst här'män·ik }

first-in, first-out [IND ENG] An inventory cost evaluation method which transfers costs of material to the product in chronological order. Abbreviated FIFO. { ¦fərst 'in ¦fərst 'aȯt }

first-item list [COMPUT SCI] A series of records that is printed with descriptive information from only the first record of each group. { ¦fərst 'ī·dəm ˌlist }

first law of motion *See* Newton's first law. { ¦fərst ˌló əv 'mō·shən }

first law of the mean *See* mean value theorem. { ¦fərst ˌló əv thə 'mēn }

first law of the mean for integrals [MATH] The proposition that the definite integral of a continuous function over an interval equals the length of the interval multiplied by the value of the function at some point in the interval. { ¦fərst ¦ló əv thə ¦mēn fór 'int·ə·grəlz }

first law of thermodynamics [THERMO] The law that heat is a form of energy, and the total amount of energy of all kinds in an isolated system is constant; it is an application of the principle of conservation of energy. { ¦fərst ˌló əv ˌthər·mō· dī'nam·iks }

first-level address [COMPUT SCI] The location of a referenced operand. { ¦fərst ¦lev·əl ə'dres }

first-level controller [CONT SYS] A controller that is associated with one of the subsystems into which a large-scale control system is partitioned by plant decomposition, and acts to satisfy local objectives and constraints. Also known as local controller. { ¦fərst ¦lev·əl kən'trōl·ər }

first-level interrupt handler [COMPUT SCI] A software or hardware routine that is activated by interrupt signals sent by peripheral devices and decides, based on the relative importance of the interrupts, how they should be handled. Abbreviated FLIH. { ¦fərst ¦lev·əl 'int·ə,rəpt ,hand·lər }

first-level packaging [ELECTR] Electronic packaging which provides interconnection directly to the integrated circuit chip. { ˌfərst,lev·əl 'pak·ij·iŋ }

first light [NAV] The beginning of morning nautical twilight, that is, when the center of the morning sun is 12° below the horizon. { ¦fərst 'līt }

first lunar meridian [ASTRON] The great circle on the moon which passes through the poles and through the center of the side of the moon that faces the earth. { ¦fərst ¦lün·ər mə'rid· ē·ən }

first motion [AGR] The first indication of motion of a rocket, missile, or test vehicle from its launcher. { ¦fərst 'mō·shən }

first negative pedal *See* negative pedal. { ¦fərst 'neg·əd·iv 'ped·əl }

first-order climatological station [METEOROL] A meteorological station at which autographic records or hourly readings of atmospheric pressure, temperature, humidity, wind, sunshine, and precipitation are made, together with observations at fixed hours of the amount and form of clouds and notes on the weather. { ¦fərst ¦órd·ər klī·mə·tə¦läj·ə·kəl 'stā·shən }

first-order difference [MATH] A member of a sequence that is formed from a given sequence by subtracting each term of the original sequence from the next succeeding term. { ¦fərst ¦órd·ər 'dif·rəns }

first-order leveling [ENG] Spirit leveling of high precision and accuracy in which lines are run first forward to the objective point and then backward to the starting point. { ¦fərst ¦órd· ər 'lev·ə·liŋ }

first-order reaction [PHYS CHEM] A chemical reaction in which the rate of decrease of concentration of component A with time is proportional to the concentration of A. { ¦fərst ¦órd·ər rē'ak·shən }

first-order relief [GEOGR] Relief features on the largest scale, consisting of continental platforms and ocean basins. { ¦fərst ¦ór·dər 'ri·lēf }

first-order spectrum [SPECT] A spectrum, produced by a diffraction grating, in which the difference in path length of light from adjacent slits is one wavelength. { ¦fərst ¦órd·ər 'spek·trəm }

first-order station [METEOROL] After U.S. National Weather Service practice, any meteorological station that is staffed in whole or in part by National Weather Service (Civil Service) personnel, regardless of the type or extent of work required of that station. { ¦fərst ¦órd·ər 'stā·shən }

first-order subroutine [COMPUT SCI] A subroutine which is entered directly from a main routine or program and which leads back to that program. Also known as first-remove subroutine. { ¦fərst ¦órd·ər 'səb·rüˌtēn }

first-order theory [MATH] A logical theory in which predicates are not allowed to have other functions or predicates as arguments and in which predicate quantifiers and function

quantifiers are not permitted. [OPTICS] *See* Gaussian optics. [PHYS] A theory which takes into account only the most important terms, such as the term proportional to the independent variable in the series expansion of a function appearing in the theory. { ¦fərst ¦órd·ər 'the·ə·rē }

first-order transition [THERMO] A change in state of aggregation of a system accompanied by a discontinuous change in enthalpy, entropy, and volume at a single temperature and pressure. { ¦fərst ¦órd·ər trans'zish·ən }

first pedal curve *See* pedal curve. { ¦fərst ¦ped·əl ˌkərv }

first point of Aries *See* vernal equinox. { ¦fərst ¦póint əv 'er·ēz }

first point of Cancer *See* summer solstice. { ¦fərst ¦póint əv 'kan·sər }

first point of Capricorn *See* winter solstice. { ¦fərst ¦póint əv 'kap·rəˌkórn }

first point of Libra *See* autumnal equinox. { ¦fərst ¦póint əv 'lē·brə }

first positive pedal curve *See* pedal curve. { 'fərst ¦päz·əd· iv ¦ped·əl ˌkərv }

first quadrant [MATH] **1.** The range of angles from 0 to 90°. **2.** In a plane with a system of cartesian coordinates, the region in which the x and y coordinates are both positive. { ¦fərst 'kwäd·rənt }

first quarter [ASTRON] The phase of the moon when it is near east quadrature, when the western half of it is visible to an observer on the earth. { ¦fərst 'kwórd·ər }

first radiation constant [STAT MECH] A constant appearing in the Planck radiation formula; its value depends on the form of the formula used; in the formula for power emitted by a blackbody per unit area per unit wavelength interval, it is 2π times Planck's constant, times the square of the speed of light, or approximately 3.74177×10^{-16} watt (meter)^2. Symbolized c_1; C_1. { ¦fərst ¦rād·ē'ā·shən ˌkän·stənt }

first-remove subroutine *See* first-order subroutine. { 'fərst 'rə,müv 'səb·rüˌtēn }

first selector [ELECTR] Selector which immediately follows a line finder in a switch train and which responds to dial pulses of the first digit of the called telephone number. { 'fərst si'lek·tər }

first sound [CRYO] Ordinary sound in helium II, in which pressure and density variations are propagated; in contrast to second sound. { 'fərst ˌsaünd }

first species [MATH] The class of sets G_0 such that one of the sets G_n is the null set, where, in general, G_n is the derived set of G_{n-1}. { 'fərst 'spē,shēz }

first water [LAP] The highest quality or the purest luster of a gemstone, such as that of a diamond which is flawless, perfectly clear and transparent, and colorless or almost blue-white. { 'fərst 'wód·ər }

firth *See* estuary. { fərth }

fir wood oil *See* pine needle oil. { 'fər ˌwúd ˌóil }

Fischer ellipsoid of 1960 [GEOD] The reference ellipsoid of which the semimajor axis is 6,378,166.000 meters, the semiminor axis is 6,356,784.298 meters, and the flattening or ellipticity is 1/298.3. Also known as Fischer spheroid of 1960. { 'fish·ər ə'lip,sóid əv ¦nīn,tēn 'siks·tē }

Fischer-Hepp rearrangement [ORG CHEM] The rearrangement of a nitroso derivative of a secondary aromatic amine to a *p*-nitrosoarylamine; the reaction is brought about by an alcoholic solution of hydrogen chloride. { ¦fish·ər ¦hep rē· ə'rānj·mənt }

Fischer-Hinnen method [ELEC] Method of analysis of a complex waveform which has like loops above and below the time axis, in which the amplitude and phase of the *n*-th harmonic is determined from the ordinates of the resultant wave at a series of times which divide the half wave into 2*n* equal time intervals. { ¦fish·ər ¦hin·ən ,meth·əd }

Fischer indole synthesis [ORG CHEM] A reaction to form indole derivatives by means of a ring closure of aromatic hydrazones. { ˌfish·ər 'in,dōl ,sin·thə·səs }

fischerite [MINERAL] A green mineral composed of a basic aluminum phosphate; may be identical to wavellite. { 'fish· ə,rīt }

Fischer polypeptide synthesis [ORG CHEM] A synthesis of peptides in which α-amino acids or those peptides with a free amino group react with acid halides of α-haloacids, followed by amination with ammonia. { ¦fish·ər ¦päl·ē'pep,tīd ,sin· thə·səs }

FISHER

The fisher (*Martes pennanti*).

FISSIDENTALES

Fissidens adiantoides.

FISSION

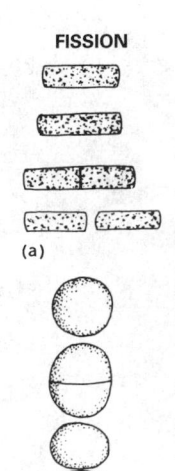

Fission in (*a*) a bacillus, (*b*) a coccus.

Fischer projection [ORG CHEM] A method for representing the spatial arrangement of groups around chiral carbon atoms; the four bonds to the chiral carbon are represented by a cross, with the assumption that the horizontal bonds project toward the viewer and the vertical bonds away from the viewer. { 'fish·ər prə‚jek·shən }

Fischer's distribution [STAT] Given data from a normal population with S_1^2 and S_2^2 two independent estimates of variance, the distribution $\frac{1}{2}$ log (S_1^2/S_2^2). { 'fish·ərz ‚dis·trə'byü·shən }

Fischer spheroid of 1960 See Fischer ellipsoid of 1960. { 'fish·ər 'sfir‚òid əv ‚nīn‚tēn 'siks·tē }

Fischer's salt See cobalt potassium nitrite. { ¦fish·ərz 'sòlt }

Fischer-Tropsch process [CHEM ENG] A catalytic process to synthesize hydrocarbons and their oxygen derivatives by the controlled reaction of hydrogen and carbon monoxide. { ¦fish·ər ¦tröpsh ‚präs·əs }

Fischer-Yates test [STAT] A test of independence of data arranged in a 2 × 2 contingency table. { ¦fish·ər ¦yāts ‚test }

fish [PETRO ENG] An object in the wellbore that must be removed before drilling or workover operations can be continued. [VERT ZOO] The common name for the cold-blooded aquatic vertebrates belonging to the groups Cyclostomata, Chondrichthyes, and Osteichthyes. { fish }

FISH [GEN] See fluorescent in situ hybridization. { fish or ¦ef‚ī‚es'äch }

fish-bone antenna [ELECTROMAG] 1. Antenna consisting of a series of coplanar elements arranged in collinear pairs, loosely coupled to a balanced transmission line. 2. Directional antenna in the form of a plane array of doublets arranged transversely along both sides of a transmission line. { 'fish ‚bōn an‚ten·ə }

fished joint [CIV ENG] A structural joint made with fish plates. { 'fisht ‚jòint }

fisher [VERT ZOO] *Martes pennanti*. An arboreal, carnivorous mammal of the family Mustelidae; a relatively large weasellike animal with dark fur, found in northern North America. { 'fish·ər }

fisheries conservation [ECOL] Those measures concerned with the protection and preservation of fish and other aquatic life, particularly in sea waters. { 'fish·ə·rēz ‚kän·sər'vā·shən }

Fisher-Irwin test [STAT] A method for testing the null hypothesis in an experiment with quantal response. { 'fish·ər ¦ər·wən ‚test }

Fisher's ideal index [STAT] The geometric mean of Laspeyres and Paasche index numbers. Also known as ideal index number. { ¦fish·ərz ¦ī‚dēl 'in‚deks }

Fisher's inequality [MATH] The inequality whereby the number *b* of blocks in a balanced incomplete block design is equal to or greater than the number *v* of elements arranged among the blocks. { ¦fish·ərz ‚in·i'kwäl·əd·ē }

fishery [ECOL] A place for harvesting fish or other aquatic life, particularly in sea waters. { 'fish·ə·rē }

Fishes See Pisces. { 'fish·əz }

fisheye [MATER] A small globular mass which has not blended completely into the surrounding material and is particularly evident in a transparent or translucent material, such as a plastic coating or surface coating. [MET] See flake. { 'fish‚ī }

fish-eye lens [OPTICS] A photographic lens that has a highly curved protruding front, enabling it to cover an angle of about 180°; provides a circular image with barrel distortion. { ¦fish ¦ī 'lenz }

fish-eye stone See apophyllite. { 'fish ‚ī ‚stōn }

fish gelatin See isinglass. { fish ‚jel·ət·ən }

fish glue [MATER] An adhesive obtained from the skin of certain fish, principally cod; used in gummed tape, letterpress printing plates, and blueprint paper. { fish ‚glü }

fishing [ENG] In drilling, the operation by which lost or damaged tools are secured and brought to the surface from the bottom of a well or drill hole. { 'fish·iŋ }

fishing boat [NAV ARCH] A ship having the necessary equipment to catch fish, such as a fathometer and fishing nets, but not the equipment to process fish. { 'fish·iŋ ‚bōt }

fishing grounds [NAV] Water areas in which fishing is frequently carried on. { 'fish·iŋ ‚graunz }

fishing space [CIV ENG] The space between base and head of a rail in which a joint bar is placed. { 'fish·iŋ ‚spās }

fishing tool [ENG] A device for retrieving objects from inaccessible locations. { 'fish·iŋ ‚tül }

fish ladder [CIV ENG] Contrivance that carries water around a dam through a series of stepped baffles or boxes and thus facilitates the migration of fish. Also known as fishway. { 'fish ‚lad·ər }

fish lead [ENG] A type of sounding lead used without removal from the water between soundings. { 'fish ‚led }

fish lice [INV ZOO] The common name for all members of the crustacean group Arguloida. { 'fish ‚līs }

fish liver oil [MATER] An oil extracted from certain fish livers and containing vitamin A; high-potency livers are obtained from cod, shark, and halibut; used in medicine and as a dietary supplement. { 'fish ‚liv·ər ‚òil }

fish meal [FOOD ENG] A protein-rich, dried food product produced from the inedible portions of fishes by dry or wet rendering. { 'fish ‚mēl }

fishmouthing See alligatoring. { 'fish‚mauth·iŋ }

fish net buoy [NAV] A buoy marking the limit of a fish net area. { 'fish ‚net ‚bòi }

fish oil [MATER] Oil obtained from fish such as menhaden, pilchard, herring, and sardine; used as a drying oil in paint and as a raw material for detergents, resins, margarine, and so on. { 'fish ‚òil }

fish paper [MATER] A type of fiber used in sheet form for insulating purposes where high mechanical strength is required, as in insulating transformer windings from the transformer core. { 'fish ‚pā·pər }

fish plate [CIV ENG] One of a pair of steel plates bolted to the sides of a rail or beam joint, to secure the joint. { 'fish ‚plāt }

fishpole antenna See whip antenna. { 'fish‚pōl an‚ten·ə }

fish protein concentrate [FOOD ENG] A dried protein-rich food product in the form of flour or paste prepared from whole fish; used as a dietary supplement. { 'fish ¦prō‚tēn 'käns·ən‚trāt }

fish screen [CIV ENG] 1. A screen set across a water intake canal or pipe to prevent fish from entering. 2. Any similar barrier to prevent fish from entering or leaving a pond. { 'fish ‚skrēn }

fish stakes [NAV] Poles or stakes placed in shallow water to outline fishing areas. { 'fish ‚stāks }

fish stock [ECOL] Any natural population of fish which is an isolated and self-perpetuating group of the same species. { 'fish ‚stäk }

fishtail [MET] Excess metal trailing on the end of a roll forging. { 'fish‚tāl }

fishtail bit [DES ENG] A drilling bit shaped like the tail of a fish. { 'fish‚tāl ‚bit }

fishtail burner [ENG] A burner in which two jets of gas impinge on each other to form a flame shaped like a fish's tail. { 'fish‚tāl ‚bərn·ər }

fishway See fish ladder. { 'fish‚wā }

Fissidentales [BOT] An order of the Bryopsida having erect to procumbent, simple or branching stems and two rows of leaves arranged in one plane. { ‚fis·ə‚den'tā·lēz }

fissile [GEOL] Capable of being split along the line of the grain or cleavage plane. [NUCLEO] See fissionable. { 'fis·əl }

fissiochemistry [CHEM] The process of producing chemical change by means of nuclear energy. { ¦fish·ō¦kem·ə·strē }

fission [BIOL] A method of asexual reproduction among bacteria, algae, and protozoans by which the organism splits into two or more parts, each part becoming a complete organism. [NUC PHYS] The division of an atomic nucleus into parts of comparable mass; usually restricted to heavier nuclei such as isotopes of uranium, plutonium, and thorium. Also known as atomic fission; nuclear fission. { 'fish·ən }

fissionable [NUCLEO] 1. A property of material whose nuclei are capable of undergoing fission. Also known as fissile. 2. A material capable of fission. { 'fish·nə·bəl }

fission barrier [NUC PHYS] One or more maxima in the plot of potential energy against nuclear deformation of a heavy nucleus, which inhibits spontaneous fission of the nucleus. { 'fish·ən ‚bar·ē·ər }

fission bomb See atomic bomb. { 'fish·ən ‚bäm }

fission chamber [NUCLEO] An ionization chamber used to detect slow neutrons; the inside wall has a thin coating of uranium, in which a slow neutron produces a fission; the resulting highly ionizing fission fragments produce a count in the chamber. Also known as fission counter. { 'fish·ən ,chām·bər }

fission counter *See* fission chamber. { 'fish·ən ,kau̇nt·ər }

fission cross section [NUC PHYS] The cross section for a bombarding neutron, gamma ray, or other particle to induce fission of a nucleus. { ¦fish·ən 'krȯ,sek·shən }

fission detector [NUCLEO] Device for detecting spontaneous fission, consisting of a mica or special glass which is placed near the sample and which is subsequently chemically etched, making fission tracks visible. { 'fish·ən di,tek·tər }

fission fraction [NUCLEO] The fraction of the total yield of a nuclear weapon that is due to fission; for thermonuclear weapons the average value is about 50. { 'fish·ən ,frak·shən }

fission fragments [NUCLEO] The nuclear species first produced when an atom such as uranium-238 or plutonium-239 undergoes fission. Also known as primary fission products. { 'fish·ən ,frag·məns }

fission fuel *See* nuclear fuel. { 'fish·ən ,fyül }

fission fungi [MICROBIO] A misnomer once used to describe the Schizomycetes. { ¦fish·ən 'fən,jī }

fission-fusion bomb [NUCLEO] An explosive device which derives its energy in comparable amounts from nuclear fission and nuclear fusion. { ¦fish·ən ¦fyü·zhən ,bäm }

fission isomer [NUC PHYS] A highly deformed nuclear state lying in the second well of a double-hump fission barrier. { ¦fish·ən 'ī·sə·mər }

fission neutron [NUC PHYS] A neutron emitted as a result of nuclear fission. { ¦fish·ən 'nü,trän }

fission product [NUC PHYS] Any radioactive or stable nuclide resulting from fission, including both primary fission fragments and their radioactive decay products. { 'fish·ən ,präd·əkt }

fission-product poisoning [NUCLEO] Inhibition of a nuclear chain reaction by fission products which have large cross sections for slow neutrons, and thus capture these neutrons before they can cause fission. { 'fish·ən ,präd·əkt ,pȯiz·niŋ }

fission reactor *See* nuclear reactor. { 'fish·ən rē,ak·tər }

fission spectrum [NUC PHYS] The energy distribution of neutrons arising from fission. { 'fish·ən ,spek·trəm }

fission spike [NUCLEO] A displacement spike produced by fission fragments. { 'fish·ən ,spīk }

fission threshold [NUC PHYS] The minimum kinetic energy of a bombarding neutron required to induce fission of a nucleus. { 'fish·ən ,thresh,hōld }

fission-track dating [GEOL] A method of dating geological specimens by counting the radiation-damage tracks produced by spontaneous fission of uranium impurities in minerals and glasses. { 'fish·ən ,trak ,dād·iŋ }

fission yield [NUCLEO] The amount of energy released by fission in a nuclear explosion, as distinct from that released by fusion. [NUC PHYS] The percent of fissions that gives a particular nuclide or group of isobars. { 'fish·ən ,yēld }

fissiped [VERT ZOO] **1.** Having the toes separated to the base. **2.** Of or relating to the Fissipeda. { 'fis·ə,ped }

Fissipeda [VERT ZOO] Former designation for a suborder of the Carnivora. { fə'sip·ə·də }

fissium [NUCLEO] An equilibrium mixture of fission products in reactor fuel that can improve the stability of uranium and uranium-plutonium fuel alloys under fast-neutron irradiation. { 'fish·yəm }

fissure [GEOL] **1.** A high, narrow cave passageway. **2.** An extensive crack in a rock. [MET] A small cracklike discontinuity with a slight opening or displacement of the fracture surfaces. { 'fish·ər }

Fissurellidae [INV ZOO] The keyhole limpets, a family of gastropod mollusks in the order Archeogastropoda. { ,fis·ə'rel·ə,dē }

fissure spring *See* artesian spring. { 'fish·ər ,spriŋ }

fissure system [GEOL] A group of fissures having the same age and generally parallel strike and dip. { 'fish·ər ,sis·təm }

fissure vein [GEOL] A mineral deposit in a cleft or crack in the rock material of the earth's crust. { 'fish·ər ,vān }

fistula [MED] An abnormal congenital or acquired communication between two surfaces such as a viscus or other hollow structure and the exterior. { 'fis·chə·lə }

Fistuliporidae [PALEON] A diverse family of extinct marine bryozoans in the order Cystoporata. { ,fis·chə·lə'pȯr·ə,dē }

fistulous withers [VET MED] A chronic inflammation of the withers of a horse accompanied by fluid discharge, which may be initiated by mechanical injury but depends on bacterial (*Brucella abortus*) infection for development. { 'fis·chə·ləs 'with·ərz }

fit [DES ENG] The dimensional relationship between mating parts, such as press, shrink, or sliding fit. { fit }

fitch [MATER] **1.** A thin sheet of wood, such as a veneer. **2.** A bundle of veneers arranged in the same order as they were cut from a log. { fich }

fitment [BUILD] A decorative or functional item or component in a room that is fixed in place but not actually built in. Also known as fitting. { 'fit·mənt }

fitness [GEN] A measure of reproductive success for a genotype, based on the average number of surviving progeny of this genotype as compared to the average number of other, competing genotypes. { 'fit·nəs }

fitness figure [METEOROL] In Great Britain, a measure of the "fitness" of the weather at an airport for the safe landing of aircraft; the figure F is computed on the basis of corrected values of visibility and cloud height; observed visibility is adjusted according to intensity of precipitation, and cloud height is corrected for height of nearby obstructions and cloud amount; further corrections are applied for the cross-runway component of the wind. Also known as fitness number. { 'fit·nəs ,fig·yər }

fitness number *See* fitness figure. { 'fit·nəs ,nəm·bər }

fitter [ENG] One who maintains, repairs, and assembles machines in an engineering shop. { 'fid·ər }

Fittig's synthesis [ORG CHEM] The synthesis of aromatic hydrocarbons by the condensation of aryl halides with alkyl halides, using sodium as a catalyst. { 'fid·iks ,sin·thə·səs }

fitting *See* fitment. [ENG] A small auxiliary part of standard dimensions used in the assembly of an engine, piping system, machine, or other apparatus. { 'fid·iŋ }

fitting out [NAV ARCH] The preparation of a ship for active service by placing on board equipment and consumables required by the allowance list or for operation. { ¦fid·iŋ 'au̇t }

FitzGerald-Lorentz contraction [RELAT] The contraction of a moving body in the direction of its motion when its speed is comparable to the speed of light. Also known as Lorentz contraction; Lorentz-FitzGerald contraction. { fits¦jer·əld lə'rens kən,trak·shən }

five-and-ten system [METEOROL] The most common system for representing wind speed, to the nearest 5 knots, in symbolic form on synoptic charts, consisting of drawing the appropriate number of half-barbs, barbs, and pennants from the end of the wind-direction shaft; in this system, a half-barb represents 5 knots, a barb 10 knots, a pennant 50 knots. { ¦fīv·ən 'ten ,sis·təm }

five-day forecast [METEOROL] A forecast of the average weather conditions and large-scale synoptic features in a 5-day period; a type of extended forecast. { 'fīv ,dā 'fȯr,kast }

five-dimensional space [MATH] A vector space whose basis has five vectors. { 'fīv də,men·chən·əl 'spās }

five-fourths power law [THERMO] The proposition that the rate of heat loss from a body by free convection is proportional to the five-fourths power of the difference between the temperature of the body and that of its surroundings. { ¦fīv ¦fȯrths 'pau̇·ər ,lȯ }

five-level code [COMPUT SCI] A code which uses five bits to specify each character. { 'fīv ,lev·əl 'kōd }

five-level start-stop operation [COMMUN] Simplex mode of operation used in teletypewriter circuits; each code character is divided into five electrical units; the machine distributor unit makes a positive start and stop for the transmission of each character. { 'fīv ,lev·əl 'stärt ¦stäp ,äp·ə,rā·shən }

five-limbed core [ELECTROMAG] The core of a core-form, three-limbed transformer to which additional core legs have been added at both ends of the core to carry the net flux resulting from unbalanced voltages, which could otherwise flow in the steel transformer tank and cause overheating. { ¦fīv ,limd 'kȯr }

five-spot well pattern [PETRO ENG] A symmetrical network pattern of five wells (one in center, four equally spaced in a square pattern) as used in water-injection pressure maintenance of reservoirs. { 'fīv ,spät 'wel ,pad·ərn }

five-wire line [ELEC] A transmission line which has four conductors, all in phase, at the corners of a square and a fifth conductor at the center of the square which is out of phase with the others. { ¦fīv ,wīr 'līn }

fix [BIOL] To kill, harden, or preserve a tissue, organ, or organism by immersion in dilute acids, alcohol, or solutions of coagulants. [COMPUT SCI] A piece of coding that is inserted in a computer program to correct an error. [NAV] A position of a vessel or craft determined by its master, pilot, or navigator through the use of some or all of the equipments and techniques available. { fiks }

fixation [PSYCH] A rigid habit developed as a consequence of repeated reinforcement, or of frustration. { fik'sā·shən }

fixative [MATER] **1.** A chemical or a mixture of chemicals used to treat biological specimens before preservation so as to retain a reasonable facsimile of their appearance when alive. **2.** A substance used to increase the durability of another substance; used to fix dye mordants, hold textile dyes and pigments, and slow the rate of perfume evaporation. Also known as fixing agent. { 'fik·səd·iv }

fixator [PHYSIO] A muscle whose action tends to hold a body part in a certain position or limit its movement. { 'fik,sād·ər }

fixed action pattern [PSYCH] An innate behavior that appears to be substantially complete the first time the organism encounters the relevant stimulus. { ¦fikst 'ak·shən ,pad·ərn }

fixed-active tooling [CONT SYS] Stationary equipment in a robotic system, such as numerical control equipment, sensors, cameras, conveying systems and parts feeders, that is activated and controlled by signals. { 'fikst ¦ak·tiv 'tül·iŋ }

fixed allele [GEN] An allele that is homozygous in all members of a population. { ¦fikst ə'lēl }

fixed ammunition [ORD] Ammunition with primer and propellant contained in a cartridge case, permanently crimped or attached to a projectile, and loaded into the weapon as a unit. { ¦fikst ,am·yə'nish·ən }

fixed and flashing light [NAV] A fixed light varied at regular intervals by one or more flashes of greater brilliance. { ¦fikst ən ¦flash·iŋ 'līt }

fixed and group flashing light [NAV] A fixed light varied at regular intervals by a group of two or more flashes of greater brilliance. { ¦fikst ən ¦grüp ¦flash·iŋ 'līt }

fixed arch [CIV ENG] A stiff arch having rotation prevented at its supports. { ¦fikst 'ärch }

fixed area [COMPUT SCI] That portion of the main storage occupied by the resident portion of the control program. { ¦fikst 'er·ē·ə }

fixed artillery [ORD] Artillery weapons permanently installed on land and sea frontiers for the protection of important areas. { ¦fikst är'til·ə·rē }

fixed attenuator See pad. { ¦fikst ə'ten·yə,wād·ər }

fixed-base index [STAT] In a time series, an index number whose base period for computing the index number is constant throughout the lifetime of the index. { ¦fikst ¦bās 'in,deks }

fixed-bed hydroforming [CHEM ENG] A cyclic petroleum process that utilizes a fixed bed of molybdenum oxide catalyst deposited on activated alumina. { ¦fikst ,bed 'hī·drə,fòr·miŋ }

fixed-bed operation [CHEM ENG] An operation in which the additive material (catalyst, absorbent, filter media, ion-exchange resin) remains stationary in the chemical reactor. { 'fikst ,bed ,äp·ə'rā·shən }

fixed bias [ELECTR] A constant value of bias voltage, independent of signal strength. { ¦fikst 'bī·əs }

fixed-bias transistor circuit [ELECTR] A transistor circuit in which a current flowing through a resistor is independent of the quiescent collector current. { ¦fikst ¦bī·əs tran'zis·tər ,sər·kət }

fixed-block [COMPUT SCI] Pertaining to an arrangement of data in which all the blocks of data have the same number of words or characters, as determined by either the hardware requirements of the computer or the programmer. { ¦fikst 'bläk }

fixed bridge [CIV ENG] A bridge having permanent horizontal or vertical alignment. { ¦fikst 'brij }

fixed capacitor [ELEC] A capacitor having a definite capacitance value that cannot be adjusted. { ¦fikst kə'pas·əd·ər }

fixed carbon [CHEM] Solid, combustible residue remaining after removal of moisture, ash, and volatile materials from coal, coke, and bituminous materials; expressed as a percentage. { ¦fikst 'kär·bən }

fixed-charge problem [IND ENG] A linear programming problem in which each variable has a fixed-charge coefficient in addition to the usual cost coefficient; the fixed charge (for example, a setup time charge) is a nonlinear function and is

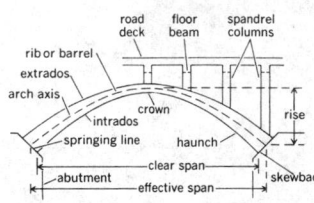

FIXED ARCH

An open-spandrel, concrete, fixed arch bridge.

incurred only when the variable appears in the solution with a positive level. { ¦fikst 'chärj ,präb·ləm }

fixed contact [ELEC] A relatively immovable contact that is engaged and disengaged by a moving contact to make and break a circuit, as in a switch or relay. { ¦fikst 'kän,takt }

fixed cost [IND ENG] A cost that remains unchanged during short-term changes in production level. Also known as overhead; overhead cost. { ¦fikst 'kóst }

fixed-cycle operation [COMPUT SCI] An operation completed in a specified number of regularly timed execution cycles. { ¦fikst 'sī·kəl ,äp·ə'rā·shən }

fixed disk [COMPUT SCI] A disk drive that permanently holds the disk platters. { ¦fikst 'disk }

fixed echo [ELECTR] An echo indication that remains stationary on a radar plan-position indicator display, indicating the presence of a fixed target. { ¦fikst 'ek·ō }

fixed-electrode method [ENG] A geophysical surveying method used in a self-potential system of prospecting in which one electrode remains stationary while the other is grounded at progressively greater distances from it. { ¦fikst i'lek,trōd ,meth·əd }

fixed emplacement [ORD] A fixed setting for a gun, usually made of reinforced concrete, with the base plate and base ring set in the concrete and bolted down. Also known as permanent emplacement. { ¦fikst em'plās·mənt }

fixed end [MECH] An end of a structure, such as a beam, that is clamped in place so that both its position and orientation are fixed. { 'fikst ,end }

fixed-end beam [CIV ENG] A beam that is supported at both free ends and is restrained against rotation and vertical movement. Also known as built-in beam; encastré beam. { 'fikst ,end 'bēm }

fixed-end column [CIV ENG] A column with the end fixed so that it cannot rotate. { 'fikst ,end 'käl·əm }

fixed end moment See fixing moment. { 'fikst ,end 'mō·mənt }

fixed-feed grinding [MECH ENG] Feeding processed material to a grinding wheel, or vice versa, in predetermined increments or at a given rate. { 'fikst ,fēd 'grind·iŋ }

fixed field [COMPUT SCI] A field in computers, film selection devices, or punched cards, or a given number of holes along the edge of a marginal punched card, set aside, or "fixed," for the recording of a given type of characteristic. { ¦fikst 'fēld }

fixed-field accelerator [NUCLEO] A circular particle accelerator whose magnetic fields do not vary with time, such as an ordinary cyclotron or a fixed-field, alternating-gradient synchrotron. { 'fikst ,fēld ak'sel·ə,rād·ər }

fixed-field method [COMPUT SCI] A method of data storage in which the same type of data is always placed in the same relative position. { 'fikst ,fēld 'meth·əd }

fixed fin [AGR] A nonadjustable vane or airfoil affixed longitudinally to an aerodynamically or ballistically designed body for stabilizing purposes. { ¦fikst 'fin }

fixed-focus lens [OPTICS] A lens whose focus is invariable, as on inexpensive cameras with no mechanism for adjusting focus but so designed that all objects from a few feet away to infinity are tolerably in focus. { 'fikst ,fō·kəs 'lenz }

fixed form coding [COMPUT SCI] Any method of coding a source language in which each part of the instruction appears in a fixed field. { 'fikst ,fòrm 'kōd·iŋ }

fixed gun [ORD] An aircraft machine gun mounted rigidly to the aircraft, and aimed by moving the aircraft. { ¦fikst 'gən }

fixed-head disk [COMPUT SCI] A disk storage device in which the read-write heads are fixed in position, one to a track, and the arms to which they are attached are immovable. { 'fikst ,hed 'disk }

fixed inductor [ELEC] An inductor whose coils are wound in such a manner that the turns remain fixed in position with respect to each other, and which either has no magnetic core or has a core whose air gap and position within the coil are fixed. { ¦fikst in'dək·tər }

fixed ion [ANALY CHEM] An ion in the lattice of a solid ion exchanger. [PHYS CHEM] One of a group of nonexchangeable ions in an ion exchanger that have a charge opposite to that of the counterions. { ¦fikst 'ī,än }

fixed-length field [COMPUT SCI] A field that always has the same number of characters, regardless of its content. { ¦fikst ,leŋkth 'fēld }

fixed-length operation [COMPUT SCI] A computer operation whose operands always have the same number of bits or characters. { ¦fikst ¦leŋkth ¦äp·ə'rā·shən }

fixed-length record [COMPUT SCI] One of a file of records, each of which must have the same specified number of data units, such as blocks, words, characters, or digits. { ¦fikst ¦leŋkth 'rek·ərd }

fixed-level chart *See* constant-height chart. { ¦fikst ¦lev·əl 'chärt }

fixed light [NAV] A light used in navigation having constant luminous intensity. [NAV ARCH] A thick, usually circular glass fitted in a frame and fixed in an opening in the side of a ship or deckhouse or in a bulkhead, to provide light. { ¦fikst 'līt }

fixed linkage system [IND ENG] Linkage formed between the skeletal elements of a human and a fixed machine in a human-machine system. { ¦fikst 'liŋk·ij ¸sis·təm }

fixed logic [COMPUT SCI] Circuit logic of computers or peripheral devices that cannot be changed by external controls; connections must be physically broken to arrange the logic. { ¦fikst 'läj·ik }

fixed medium [COMPUT SCI] A data storage device in which the reading and writing of data do not involve mechanical motion. { ¦fikst 'mē·dē·əm }

fixed memory [COMPUT SCI] Of a computer, a nondestructive readout memory that is only mechanically alterable. { ¦fikst 'mem·rē }

fixed mooring berth [CIV ENG] A marine structure consisting of dolphins for securing a ship and a platform to support cargo-handling equipment. { ¦fikst 'mu̇r·iŋ ¸bərth }

fixed-needle traverse [ENG] In surveying, a traverse with a compass fitted with a sight line which can be moved above a graduated horizontal circle, so that the azimuth angle can be read, as with a theodolite. { ¦fikst ¦nēd·əl trə'vərs }

fixed oil [MATER] A nonvolatile fatty oil of vegetable origin. { ¦fikst 'ȯil }

fixed-passive tooling [CONT SYS] Unpowered, accessory equipment in a robotic system, such as jigs, fixtures, and work-holding devices. { ¦fikst 'pas·iv 'tül·iŋ }

fixed point [ENG] A reproducible value, as for temperature, used to standardize measurements; derived from intrinsic properties of pure substances. [MATH] For a function *f* mapping a set *S* to itself, any element of *S* which *f* sends to itself. { ¦fikst 'pȯint }

fixed-point arithmetic [COMPUT SCI] **1.** A method of calculation in which the computer does not consider the location of the decimal or radix point because the point is given a fixed position. **2.** A type of arithmetic in which the operands and results of all arithmetic operations must be properly scaled so as to have a magnitude between certain fixed values. { ¦fikst ¸pȯint ə'rith·mə·tik }

fixed-point attractor [PHYS] An attractor that consists of a single point in phase space and describes a stationary state of a system. { ¦fikst ¦pȯint ə'trak·tər }

fixed-point calculation [COMPUT SCI] A calculation made with fixed-point arithmetic. { ¦fikst ¸pȯint ¸kal·kyə'lā·shən }

fixed-point computer [COMPUT SCI] A computer in which numbers in all registers and storage locations must have an arithmetic point that remains in the same fixed location. { ¦fikst ¸pȯint kəm'pyüd·ər }

fixed-point part *See* mantissa. { ¦fikst ¸pȯint 'pärt }

fixed-point representation [COMPUT SCI] Any method of representing a number in which a fixed-point convention is used. { ¦fikst ¸pȯint ¸rep·rə·zen'tā·shən }

fixed-point system [COMPUT SCI] A number system in which the location of the point is fixed with respect to one end of the numerals, according to some convention. { ¦fikst ¸pȯint 'sis·təm }

fixed-point theorem [MATH] Any theorem, such as the Brouwer theorem or Schauder's fixed-point theorem, which states that a certain type of mapping of a set into itself has at least one fixed point. { ¦fikst 'pȯint ¸thir·əm }

fixed-position addressing [COMPUT SCI] Direct access to an item in a data file on disk or drum, as opposed to a sequential search for this item starting with the first item in the file. { ¦fikst pə¸zish·ən ə'dres·iŋ }

fixed-position welding [MET] A welding operation in which the work is stationary. { ¦fikst ¸pȯint 'weld·iŋ }

fixed-product area [COMPUT SCI] The area in core memory

where multiplication takes place for certain types of computers. { ¦fikst ¸präd·əkt 'er·ē·ə }

fixed-program computer [COMPUT SCI] A special-purpose computer having a program permanently wired in. { ¦fikst ¸prō·grəm kəm'pyüd·ər }

fixed prosthodontics [MED] A subdivision of prosthodontics that focuses on the replacement of missing teeth by dental bridges. { ¸fikst ¸präs·thə'dän·tiks }

fixed radix notation [MATH] A form of positional notation in which successive digits are interpreted as coefficients of successive powers of an integer called the base or radix. { ¦fikst 'rā¸diks nō¸tā·shən }

fixed rent *See* minimum rent. { ¦fikst ¸'rent }

fixed resistor [ELEC] A resistor that has no provision for varying its resistance value. { ¦fikst ri'zis·tər }

fixed-satellite service [COMMUN] A radiocommunication service between earth stations at given positions that uses one or more satellites. Abbreviated FSS. { ¦fikst 'sad·əl¸īt ¸sər·vis }

fixed screen [MIN ENG] A stationary panel, commonly of wedge wire, used to remove a large proportion of water and fines from a suspension of coal in water. { ¦fikst ¸'skrēn }

fixed-sequence robot *See* fixed-stop robot. { ¦fikst ¦sē·kwəns 'rō¸bät }

fixed service [COMMUN] Service providing radio communications between fixed points. { ¦fikst 'sər·vəs }

fixed sonar [ENG] Sonar in which the receiving transducer is not constantly rotated, in contrast to scanning sonar. { ¦fikst 'sō¸när }

fixed square search *See* geographic square search. { ¦fikst ¸skwer 'sərch }

fixed star [ASTRON] A misnomer to indicate those stars which kept apparently the same position with respect to other stars, in contrast to the planets which were termed wandering stars. { ¦fikst 'stär }

fixed-stop robot [CONT SYS] A robot in which the motion along each axis has a fixed limit, but the motion between these limits is not controlled and the robot cannot stop except at these limits. Also known as fixed-sequence robot; limited-sequence robot; nonservo robot. { ¦fikst ¦stäp 'rō¸bät }

fixed storage [COMPUT SCI] A storage for data not alterable by computer instructions, such as magnetic-core storage with a lockout feature. { ¦fikst 'stȯr·ij }

fixed target [ORD] A nonmovable or immobilized target, such as a city, factory, or surrounded troops. { ¦fikst 'tär·gət }

fixed transmitter [ELECTR] Transmitter that is operated in a fixed or permanent location. { ¦fikst 'tranz¸mid·ər }

fixed word length [COMPUT SCI] The length of a computer machine word that always contains the same number of characters or digits. { ¦fikst 'wərd ¸leŋkth }

fixer network [NAV] A combination of radio or radar direction-finding installations which, operating in conjunction, are capable of plotting the position relative to the ground of an aircraft in flight. Also known as fixer system. { 'fik·sər ¸net¸wərk }

fixer system *See* fixer network. { 'fik·sər ¸sis·təm }

fixing agent *See* fixative. { 'fik·siŋ ¸ā·jənt }

fixing moment [MECH] The bending moment at the end support of a beam necessary to fix it and prevent rotation. Also known as fixed end moment. { 'fik·siŋ ¸mō·mənt }

fixity *See* continuity. { 'fik·səd·ē }

fixture [CIV ENG] An object permanently attached to a structure, such as a light or sink. [MECH ENG] A device used to hold and position a piece of work without guiding the cutting tool. { 'fiks·chər }

Fizeau effect [OPTICS] The change in the speed of light in a material medium that results from the motion of the medium relative to the source and to the observer. { 'fē'zō i¸fekt }

Fizeau fringes [OPTICS] **1.** Interference fringes of monochromatic light from interference in a geometrical situation other than plane parallel plates. Also known as fringes of equal thickness. **2.** Interference fringes in light from a Fizeau interferometer. { fē'zō ¸frin·jəz }

Fizeau interferometer [OPTICS] Interferometer in which light from a point source is collimated and multiply reflected between a plane mirror and the partially silvered inner surface of a parallel plane plate, and is viewed in reflection. { fē'zō ¸in·tər·fə'räm·əd·ər }

Fizeau toothed wheel [OPTICS] Rapidly rotating toothed

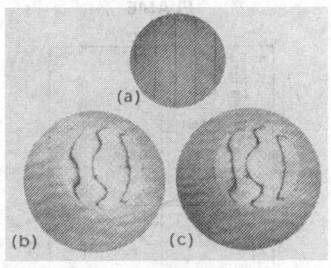

Fizeau multiple-beam fringe patterns (def. 2). (*a*) Narrow-gap plane mirrows. (*b*) Wide-gap test lens without end mirrors imaged on each other; (*c*) with mirrors imaged on each other.

wheel which was used to measure the speed of light by adjusting the rotation speed until light passing through one tooth opening and reflected from a distant mirror would pass through the next tooth opening on return. { fē′zō ¦tütht ′wēl }

fizelyite [MINERAL] A metallic, lead-gray mineral composed of a lead silver antimony sulfide, occurring as prisms. { fə′zä·lē͵īt }

fjard [GEOGR] A small, narrow, and irregular inlet of the sea with low banks on either side. Also spelled fiard. { fē′ärd }

fjord [GEOGR] A narrow, deep inlet of the sea between high cliffs or steep slopes. Also spelled fiord. { fyórd }

fjord valley [GEOGR] A deep, narrow channel occupied by the sea and extending inland about 50–100 miles (80–160 kilometers). { ′fyórd ͵val·ē }

flabellate [BIOL] Fan-shaped. { flə′bel·ət }

Flabellifera [INV ZOO] The largest and morphologically most generalized suborder of isopod crustaceans; the biramous uropods are attached to the sides of the abdomen and may form, with the last abdominal fragment, a caudal fan. { ͵flab·ə′lif·ə·rə }

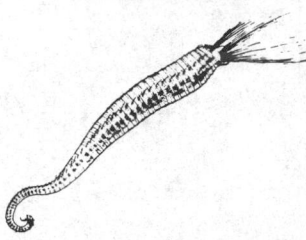

FLABELLIGERIDAE

Pheurusa, a cage worm, shown in right lateral view.

Flabelligeridae [INV ZOO] The cage worms, a family of spioniform worms belonging to the Sedentaria; the anterior part of the body is often concealed by a cage of setae arising from the first few segments. { flə͵bel·ə′jer·ə͵dē }

flabellum [INV ZOO] Any structure resembling a fan, as the epipodite of certain crustacean limbs. { flə′bel·əm }

flaccid [BOT] Deficient in turgor. [PHYSIO] Soft, flabby, or relaxed. { ′flas·əd *or* ′flak·səd }

flacherie [INV ZOO] A fatal bacterial disease of caterpillars, especially silkworms, marked by loss of appetite, dysentery, and flaccidity of the body; after death the body darkens and liquefies. { ͵flash·ə′rē }

Flacourtiaceae [BOT] A family of dicotyledonous plants in the order Violales having the characteristics of the more primitive members of the order. { flə͵kúrd·ē′ās·ē͵ē }

Flade potential [MET] The potential of a passive metal immediately preceding a final steep fall from the passive to the active region. { ′fläd pə′ten·chəl }

flag [COMPUT SCI] Any of various types of indicators used for identification, such as a work mark, or a character that signals the occurrence of some condition, such as the end of a word. [ELECTR] A small metal tab that holds the getter during assembly of an electron tube. [ENG] **1.** A piece of fabric used as a symbol or as a signaling or marking device. **2.** A large sheet of metal or fabric used to shield television camera lenses from light when not in use. { flag }

flag alarm [ENG] A semaphore-type flag in the indicator of an instrument to serve as a signal, usually to warn that the indications are unreliable. { ′flag ə͵lärm }

flag bridge [NAV ARCH] A separate structure built above the deck of a ship, designed for the use of a naval officer higher than the rank of captain; the visual control and communications center from which the officer exercises command of the tactical units and is separate from and at a different level from the navigation bridge. { ′flag ͵brij }

flagella [BIOL] Relatively long, whiplike, centriole-based locomotor organelles on some motile cells. { flə′jel·ə }

Flagellata [INV ZOO] The equivalent name for Mastigophora. { ͵flaj·ə′läd·ə }

flagellate [BIOL] **1.** Having flagella. **2.** An organism that propels itself by means of flagella. **3.** Resembling a flagellum. [INV ZOO] Any member of the protozoan superclass Mastigophora. { ′flaj·ə͵lāt }

flagellated chamber [INV ZOO] An outpouching of the wall of the central cavity in Porifera that is lined with choanocytes; connects with incurrent canals through prosophyles. { ¦flaj·ə͵lād·əd ′chäm·bər }

flagellation [BIOL] The arrangement of flagella on an organism. [PSYCH] Beating or whipping as a means of producing sexual gratification. { ͵flaj·ə′lā·shən }

flagellin [MICROBIO] The protein component of bacterial flagella. { flə′jel·ən }

flag flip-flop [COMPUT SCI] A one-bit register which indicates overflow, carry, or sign bit from past or current operations. { ′flag ′flip ͵fläp }

flag float [ENG] A pyrotechnic device that floats and burns upon the water, used for marking or signaling. { ′flag ͵flōt }

flaggy [GEOL] **1.** Of bedding, consisting of strata 4–40 inches (10–100 centimeters) in thickness. **2.** Of rock, tending

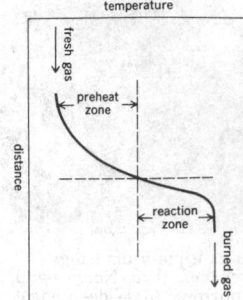

FLAME

Temperature distribution through a flame front.

to split into layers of suitable thickness (0.4–2 inches or 1–5 centimeters) for use as flagstones. { ′flag·ē }

flagilliflory [BOT] Of flowers, hanging down freely from ropelike twigs. { flə′jil·ə͵flór·ē }

flagman [CIV ENG] A range-pole carrier in a surveying party. { ′flag·mən }

flag operand [COMPUT SCI] A part of the instruction of some assembly languages denoting which elements of the object instruction will be flagged. { ′flag ′äp·ə͵rand }

flagpole [ENG] A single staff or pole rising from the ground and on which flags or other signals are displayed; on charts the term is used only when the pole is not attached to a building. { ′flag͵pōl }

flag smut [PL PATH] A smut affecting the leaves and stems of cereals and other grasses, characterized by formation of sori within the tissues, which rupture releasing black spore masses and causing fraying of the infected area. { ′flag ͵smət }

flagstaff [ENG] A pole or staff on which flags or other signals are displayed; on charts this term is used only when the pole is attached to a building. { ′flag͵staf }

flagstone [GEOL] **1.** A hard, thin-bedded sandstone, firm shale, or other rock that splits easily along bedding planes or joints into flat slabs. **2.** A piece of flagstone used for making pavement or covering the side of a house. { ′flag͵stōn }

flag tower [NAV] A scaffoldlike tower from which flags are displayed. { ′flag ͵taùr }

flail [AGR] A manual device for threshing consisting of a long wooden handle with a free-swinging stick attached to the end. { flāl }

flail tank [ORD] A specially constructed tank equipped with a flailing device, consisting of chain flails attached to a roller powered by the tank engine, employed to detonate antitank mines. { ′flāl ͵taŋk }

flair [CIV ENG] A gradual widening of the flangeway near the end of a guard line of a track or rail structure. { fler }

flajolotite [MINERAL] $4FeSbO_4 \cdot 3H_2O$ A claylike, lemon-yellow mineral composed of a hydrous iron antimonate, occurring in nodular masses. { ′flaj·ə′lō͵tīt }

flak [ORD] **1.** Explosive or exploding missile fired from anti-aircraft cannon. **2.** Antiaircraft cannon, as in flak battery or flak installation. { flak }

flak analysis [ORD] The examination and study of flak intelligence to determine the nature, effectiveness, or probable effectiveness of enemy antiaircraft defenses. { ′flak ə͵nal·ə·səs }

flake [MATER] **1.** Dry, unplasticized, cellulosic plastics base. **2.** Plastic chip used as feed in molding operations. **3.** A small, flat wood particle of predetermined dimensions and uniform thickness, with fiber direction essentially in the plane of the flake. [MET] **1.** Discontinuous, internal cracks formed in steel during cooling due usually to the release of hydrogen. Also known as fisheye; shattercrack; snowflake. **2.** Fish-scale, flat particles in powder metallurgy. Also known as flake powder. { flāk }

flakeboard [MATER] Particleboard composed of wood flakes. { ′flāk͵bórd }

flake powder *See* flake. { ′flak ͵paùd·ər }

flaking [CHEM ENG] Continuous process operation to remove heat from material in the liquid state to cause its solidification. [ENG] **1.** Reducing or separating into flakes. **2.** *See* frosting. [MIN ENG] Breaking small chips from the face of a refractory, particularly chrome ore containing refractories. { ′flāk·iŋ }

flaking mill [MECH ENG] A machine for converting material to flakes. { ′flāk·iŋ ͵mil }

flak jacket [ENG] A jacket or vest of heavy fabric containing metal, nylon, or ceramic plates, designed especially for protection against flak; usually covers the chest, abdomen, back, and genitals, leaving the arms and legs free. Also known as flak vest. { ′flak ͵jak·ət }

flak vest *See* flak jacket. { ′flak ͵vest }

flamboyant structure [GEOL] The optical continuity of crystals or grains as disturbed by a structure that is divergent. { flam′bói·ənt ′strək·chər }

flame [CHEM] A hot, luminous reaction front (or wave) in a gaseous medium into which the reactants flow and out of which the products flow. { flām }

flame annealing [MET] The careful heating of a metal part by flames, before or after working. { ′flām ə͵nēl·iŋ }

flame arc lamp [ELEC] An arc lamp in which carbon electrodes are impregnated with chemicals, such as calcium, barium, or titanium, which are more volatile than the carbon and radiate light when driven into the arc. { ¦flām ′ärk ‚lamp }

flame arrester [ENG] An assembly of screens, perforated plates, or metal-gauze packing attached to the breather vent on a flammable-product storage tank. { ′flām ə‚res·tər }

flame bucket [AERO ENG] A deep, cavelike construction built beneath a rocket launchpad, open at the top to receive the hot gases of the rocket, and open on one or three sides below, with a thick metal fourth side bent toward the open side or sides so as to deflect the exhaust gases. { ′flām ‚bək·ət }

flame bulb [INV ZOO] The enlarged terminal part of the flame cell of a protonephridium, consisting of a tuft of cilia. { ′flām ‚bəlb }

flame cell [INV ZOO] A hollow cell that contains the terminal branches of excretory vessels in certain flatworms and rotifers and some other invertebrates. { ′flām ‚sel }

flame cleaning [MET] Removing scale, rust, and dirt from metal surfaces by using a broad flame. { ′flām ‚klēn·iŋ }

flame coating See flame plating. { ′flām ‚kōd·iŋ }

flame collector [ENG] A device used in atmospheric electrical measurements for the removal of induction charge on apparatus; based upon the principle that products of combustion are ionized and will consequently conduct electricity from charged bodies. { ′flām kə‚lek·tər }

flame cultivator [AGR] A flamethrower for destroying weeds between rows of crops. { ′flām ‚kəl·tə‚vād·ər }

flame cutting [MET] Use of an oxyacetylene, oxyhydrogen, or oxycoal gas flame to cut thick metal sections. { ′flām ‚kəd·iŋ }

flame deflector [AERO ENG] **1.** In a vertical launch, any of variously designed obstructions that intercept the hot gases of the rocket engine so as to deflect them away from the ground or from a structure. **2.** In a captive test, an elbow in the exhaust conduit or flame bucket that deflects the flame into the open. { ′flām di‚flek·tər }

flame detector [MECH ENG] A sensing device which indicates whether or not a fuel is burning, or if ignition has been lost, by transmitting a signal to a control system. { ′flām di‚tek·tər }

flame emission spectroscopy [SPECT] A flame photometry technique in which the solution containing the sample to be analyzed is optically excited in an oxyhydrogen or oxyacetylene flame. { ′flām i‚mish·ən spek′träs·kə‚pē }

flame excitation [SPECT] Use of a high-temperature flame (such as oxyacetylene) to excite spectra emission lines from alkali and alkaline-earth elements and metals. { ¦flām ‚ek·sī′tā·shən }

flame gouging [MET] A form of oxygen cutting by means of a cutting torch with a slightly curved tip, enabling the flame to strike the metal surface at a low angle and making shallow cuts possible. Also known as oxygen gouging. { ′flām ‚gaủj·iŋ }

flame hardening [MET] A method for local surface hardening of steel by passing an oxyacetylene or similar flame over the work at a predetermined rate. { ′flām ‚härd·ən·iŋ }

flameholder [AERO ENG] A device that sustains combustion in a flowing mixture within the combustion chamber of some types of jet engines. { ′flām‚hōl·dər }

flame ionization detector [ANALY CHEM] A device in which the measured change in conductivity of a standard flame (usually hydrogen) due to the insertion of another gas or vapor is used to detect the gas or vapor. { ¦flām ‚ī·ə·nə′zā·shən di‚tek·tər }

flame laser [OPTICS] A molecular gas laser in which gases such as carbon disulfide and oxygen are mixed at low pressures and ignited; the flame is then self-sustaining and produces carbon monoxide laser emission. { ′flām ‚lā·zər }

flameout [AERO ENG] The extinguishing of the flame in a reaction engine, especially in a jet engine. { ′flām‚aủt }

flame photometer [SPECT] One of several types of instruments used in flame photometry, such as the emission flame photometer and the atomic absorption spectrophotometer, in each of which a solution of the chemical being analyzed is vaporized; the spectral lines resulting from the light source going through the vapors enters a monochromator that selects the band or bands of interest. { ′flām ‚fə′täm·əd·ər }

flame photometry [SPECT] A branch of spectrochemical analysis in which samples in solution are excited to produce line emission spectra by introduction into a flame. { ′flām fə′täm·ə·trē }

flame plate [ENG] One of the plates on a boiler firebox which are subjected to the maximum furnace temperature. { ′flām ‚plāt }

flame plating [MET] Coating a thin layer of refractory material on a surface by exploding a mixture of plating powder, oxygen, and acetylene. Also known as flame coating. { ′flām ‚plād·iŋ }

flameproofing [CHEM ENG] The process of treating materials chemically so that they will not support combustion. { ′flām‚prüf·iŋ }

flame propagation [CHEM] The spread of a flame in a combustible environment outward from the point at which the combustion started. { ′flām ‚präp·ə′gā·shən }

flame retardant [CHEM ENG] A substance that can suppress, reduce, or delay the propagation of a flame through a polymer material; may be inserted chemically into the polymer molecule or blended in after polymerization. { ′flām ri‚tärd·ənt }

flame-retarded resin [MATER] A resin which is compounded with certain chemicals to reduce or eliminate its tendency to burn. { ′flam ri‚tärd·əd ′rez·ən }

flame repellent [TEXT] Referring to a fabric that has been rendered resistant to flash or sustained burning by chemical treatment. { ′flām ri‚pel·ənt }

flame spectrometry [SPECT] A procedure used to measure the spectra or to determine wavelengths emitted by flame-excited substances. { ¦flām spek′träm·ə·trē }

flame spectrophotometry [SPECT] A method used to determine the intensity of radiations of various wavelengths in a spectrum emitted by a chemical inserted into a flame. { ¦flām ‚spek·trə·fə′täm·ə·trē }

flame spectrum [SPECT] An emission spectrum obtained by evaporating substances in a nonluminous flame. { ′flām ‚spek·trəm }

flame speed [CHEM] The rate at which combustion moves through an explosive mixture. { ′flām ‚spēd }

flame spraying [ENG] **1.** A method of applying a plastic coating onto a surface in which finely powdered fragments of the plastic, together with suitable fluxes, are projected through a cone of flame. **2.** Deposition of a conductor on a board in molten form, generally through a metal mask or stencil, by means of a spray gun that feeds wire into a gas flame and drives the molten particles against the work. { ′flām ‚sprā·iŋ }

flame straightening [MET] Correcting distorted structural metal to a straight form by local application of a gas-flame heat. { ′flām ‚strāt·ən·iŋ }

flamethrower [ENG] A device used to project ignited fuel from a nozzle so as to cause casualties to personnel or to destroy material such as weeds or insects. { ′flām‚thrō·ər }

flame trap [ENG] A device that prevents a gas flame from entering the supply pipe. { ′flām ‚trap }

flame treating [ENG] A method of rendering inert thermoplastic objects receptive to inks, lacquers, paints, or adhesives, in which the object is bathed in an open flame to promote oxidation of the surface. { ′flām ‚trēd·iŋ }

flamingo [VERT ZOO] Any of various long-legged and long-necked aquatic birds of the family Phoenicopteridae characterized by a broad bill resembling that of a duck but abruptly bent downward and rosy-white plumage with scarlet coverts. { flə′miŋ·gō }

flammability [CHEM] A measure of the extent to which a material will support combustion. Also known as inflammability. { ‚flam·ə′bil·əd·ē }

flammability limits [CHEM] The stoichiometric composition limits (maximum and minimum) of an ignited oxidizer-fuel mixture what will burn indefinitely at given conditions of temperature and pressure without further ignition. { ‚flam·ə′bil·əd·ē ‚lim·əts }

flammable [MATER] Of a material, capable of supporting combustion. { ′flam·ə·bəl }

flammable liquid [MATER] A liquid which gives off combustible vapors. { ¦flam·ə·bəl ′lik·wəd }

Flamsteed's number [ASTRON] A number sometimes used with the possessive form of the Latin name of the constellation to identify a star, for example, 72 Ophiuchi. { ′flam‚stēdz ‚nəm·bər }

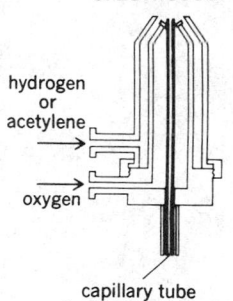

FLAME EMISSION SPECTROSCOPY

hydrogen or acetylene

oxygen

capillary tube for introduction of sample solution

Beckman aspirator burner used to aspirate the solution directly into oxyhydrogen or oxyacetylene flame.

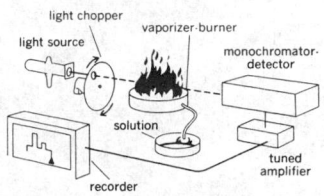

FLAME PHOTOMETER

light chopper
light source
vaporizer-burner
monochromator-detector
solution
recorder
tuned amplifier

Schematic diagram of atomic absorption spectrophotometer for determining metal concentrations.

flan [METEOROL] In Scotland, a sudden gust or squall of wind from land. { flan }

Flanders storm [METEOROL] In England, a heavy fall of snow coming with the south wind. { 'flan·dərz ˌstȯrm }

Flandrian transgression [OCEANOGR] The rapid rise of the North Sea between 8000 and 3000 B.C. from about 180 feet (55 meters) below to about 20 feet (6 meters) below its present level. { 'flan·drē·ən tranz'gresh·ən }

flange [SCI TECH] A projecting rim of an organism or mechanical part. { flanj }

flanged pipe [DES ENG] A pipe with flanges at the ends; can be bolted end to end to another pipe. { ¦flanjd 'pīp }

flange isolator See short waveguide isolator. { ¦flanj 'ī·sə¦lād·ər }

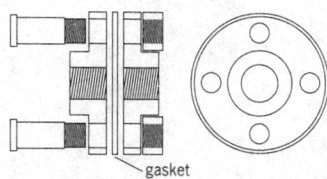

FLANGE UNION

A flange union in side view and cross section.

gasket

flange union [ENG] A pair of flanges that are screwed to the ends of pipes and then bolted or welded together to hold two pipes together. { 'flanj ˌyün·yən }

flangeway [CIV ENG] Open way through a rail or track structure that provides a passageway for the flange of a wheel. { 'flanj,wā }

flanging [ENG] A forming process in which the edge of a metal part is bent over to make a flange at a sharp angle to the body of the part. { 'flanj·iŋ }

flank [CIV ENG] The outer edge of a carriageway. [DES ENG] **1.** The end surface of a cutting tool, adjacent to the cutting edge. **2.** The side of a screw thread. [GEOL] See limb. [VERT ZOO] The part of a quadruped mammal between the ribs and the pelvic girdle. { flaŋk }

flank angle [DES ENG] The angle made by the flank of a screw thread with a line perpendicular to the axis of the screw. { 'flaŋk ˌaŋ·gəl }

flank hole [MIN ENG] **1.** A hole bored in advance of a working place when approaching old workings. **2.** A borehole driven from the side of an underground excavation, not parallel with the center line of the excavation, to detect water, gas, or other danger. { 'flaŋk ˌhōl }

flank observation [ORD] Observation of fire from a place on, or near, the flank of the target; the angle at the target between the gun and the observer is between 75 and 105°. { ¦flaŋk äb·zər'vā·shən }

flank wear [ENG] Loss of relief on the flank of a tool behind the cutting edge. { 'flaŋk ˌwer }

flannel [TEXT] A loosely woven, generally wool fabric with the weave concealed by a napped surface. { 'flan·əl }

flannelette [TEXT] Plain cotton weave finished with a nap on one side. Also known as kimono flannel. { ¦flan·əl¦et }

flap [AERO ENG] **1.** Any control surface, such as a speed brake, dive brake, or dive-recovery brake, used primarily to increase the lift or drag on an airplane, or to aid in recovery from a dive. **2.** Any rudder attached to a rocket and acting either in the air or within the jet stream. { flap }

flap attenuator [ELECTROMAG] A waveguide attenuator in which a contoured sheet of dissipative material is moved into the guide through a nonradiating slot to provide a desired amount of power absorption. Also known as vane attenuator. { ¦flap ə'ten·yə,wād·ər }

flaperon [AERO ENG] A control surface used both as a flap and as an aileron. { 'flap·ə,rän }

flap gate [CIV ENG] A gate that opens or closes by rotation around hinges at the top of the gate. Also known as pivot leaf gate. { 'flap ˌgāt }

flap hinge See backflap hinge. { 'flap ˌhiŋ }

flapping [MET] Striking through the surface of molten copper to hasten oxidation by increasing the exposure to air. { 'flap·iŋ }

flap trap [ENG] In plumbing, a trap fitted with a hinged flap that permits flow in one direction only, thus preventing backflow. { 'flap ˌtrap }

flap valve [MECH ENG] A valve fitted with a hinged flap or disk that swings in one direction only. { 'flap ˌvalv }

flare [AERO ENG] To descend in a smooth curve, making a transition from a relatively steep descent to a direction substantially parallel to the surface, when landing an aircraft. [ASTRON] A bright eruption from the sun's chromosphere; flares may appear within minutes and fade within an hour, cover a wide range of intensity and size, and tend to occur between sunspots or over their penumbrae. [CHEM ENG] A device for disposing of combustible gases from refining or chemical processes by burning in the open, in contrast to combustion in a furnace or closed vessel or chamber. [DES ENG] An expansion at the end of a cylindrical body, as at the base of a rocket. [ELECTR] A radar screen target indication having an enlarged and distorted shape due to excessive brightness. [ELECTROMAG] See horn antenna. [ENG] A pyrotechnic item designed to produce a single source of intense light for such purposes as target or airfield illumination. [NAV ARCH] A concave curve of a boat's or ship's sides away from the center line, above the waterline, normally at the bow. { fler }

flareback [ORD] A rearward escapement of flame or gas from a gun. { 'fler,bak }

flare chute [ENG] A flare attached to a parachute. { 'fler ,shüt }

flare factor [ENG ACOUS] Number expressing the degree of outward curvature of the horn of a loudspeaker. { 'fler ,fak·tər }

flare gas [CHEM ENG] Surplus gas that is disposed of by combustion in the open. { 'fler ,gas }

flareout [AERO ENG] That portion of the approach path of an aircraft in which the vertical component is modified to lessen the impact of landing. { 'fler,aùt }

flare spot [OPTICS] A small, diffuse, brightly illuminated region produced by multiple reflections of light from the various surfaces of an optical system. { 'fler ,spät }

flare stars See UV Ceti stars. { 'fler ,stärz }

flare-type bucket [MIN ENG] A dragline bucket that has flared sides to allow heaped loading. { 'fler ,tīp ,bək·ət }

flare-type burner [ENG] A circular burner which discharges flame in the form of a cone. { 'fler ,tīp ,bərn·ər }

flaring [MET] Increasing the diameter at the end of a pipe or tube. { 'fler·iŋ }

flaser [GEOL] Streaky layer of parallel, scaly aggregates that surrounds the lenticular bodies of granular material in flaser structure; caused by pressure and shearing during metamorphism. { 'flā·zər }

flaser gabbro [GEOL] A cataclastic gabbro that contains augen of feldspar or quartz surrounded by flakes of mica or chlorite. { 'flā·zər 'ga,brō }

flaser structure [GEOL] **1.** A metamorphic structure in which small lenses and layers of granular material are surrounded by a matrix of sheared, crushed material, resembling a crude flow structure. Also known as pachoidal structure. **2.** A primary sedimentary structure consisting of fine-sand or silt lenticles that are aligned and cross-bedded. { 'flā·zər ,strək·chər }

flash [ASTRON] A thermal instability that occurs in late stages of stellar evolution, according to numerical calculations. [ENG] In plastics or rubber molding or in metal casting, that portion of the charge which overflows from the mold cavity at the joint line. [MET] A fin of excess metal along the mold joint line of a casting, occurring between mating die faces of a forging or expelled from a joint in resistance welding. { flash }

flash arc [ELECTR] A sudden increase in the emission of large thermionic vacuum tubes, probably due to irregularities in the cathode surface. { 'flash ,ärk }

flashback See backfire. { 'flash,bak }

flashback arrester [ENG] A device which prevents a flashback from passing the point where the arrester is installed in a torch, thereby preventing damage. { 'flash,bak ə,res·tər }

flashback voltage [ELECTR] Inverse peak voltage at which ionization takes place in a gas tube. { 'flash,bak ,vōl·tij }

flash-bang [ORD] The time interval between visual observation of the flash of a weapon being fired and the auditory perception of the sound of the discharge. { ¦flash ¦baŋ }

flash barrier [ELEC] A fireproof structure between conductors of an electric machine, designed to minimize flashover or the damage caused by flashover. { 'flash ,bar·ē·ər }

flashboard [CIV ENG] A relatively low, temporary barrier constructed of a series of boards along the top of a dam spillway to increase storage capacity. { 'flash,bȯrd }

flash boiler [MECH ENG] A boiler with hot tubes of small capacity; designed to immediately convert small amounts of water to superheated steam. { 'flash ,bȯil·ər }

flash bomb [ENG] A bomb that illuminates the ground for night aerial photography. { 'flash ,bäm }

flash burn [MED] Tissue injury resulting from exposure to high-intensity radiant heat. { 'flash ,bərn }

flash butt welding [MET] Resistance welding to produce a butt joint by passing an electric current through two pieces of metal in light contact to create an arc which causes flashing and consequent heating; the weld is completed by applying pressure at the joint. { 'flash ,bət ,weld·iŋ }

flash carbonization [CHEM ENG] A carbonization process in which coal is subjected to a very brief residence time in the reactor in order to produce the largest possible yield of tar. { 'flash ,kär·bə·nə'zā·shən }

flash chamber [CHEM ENG] A conventional oil-and-gas separator operated at low pressure, with the liquid from a higher-pressure vessel being flashed into it. Also known as flash trap; flash vessel. { 'flash ,chām·bər }

flash coat [MET] A thin coating that is forced from a flash-welded joint after the abutting surfaces are forced together. { 'flash ,kōt }

flash coloration [ECOL] A type of protective coloration in which the prey is cryptic when at rest but reveals brilliantly colored parts while escaping. { ,flash kəl·ə'rā·shən }

flash depressor [MATER] A substance used to reduce the flash from a rocket motor. { 'flash di,pres·ər }

flash distillation See equilibrium flash vaporization. { 'flash ,dis·tə'lā·shən }

flash drum [CHEM ENG] A facility, such as a tower, which receives the products of a preheater or heat exchanger to release pressure; volatile components are vaporized and separated for further fractionation. { 'flash ,drəm }

flash dry [CHEM ENG] The rapid evaporation of moisture from a porous or granular solid by a sudden reduction in pressure or by placing the material in an updraft of warm air. { 'flash ,drī }

flasher [ELEC] A switch, generally either motor-driven or using a combination heater element and bimetallic strip, that turns lamps on and off rapidly. { 'flash·ər }

flash exposure [GRAPHICS] In halftone photography, additional exposure given to reinforce the dots in the shadow zones of negatives. { 'flash ik'spō·zhər }

flash factor [OPTICS] In photography using a photoflash lamp, a number dependent on the lamp and the film speed, equal to the product of the distance of the lamp from the subject and the correct f-number for that distance. { 'flash ,fak·tər }

flash flood [HYD] A sudden local flood of short duration and great volume; usually caused by heavy rainfall in the immediate vicinity. { 'flash ,fləd }

flash groove [ENG] **1.** A groove in a casting die so that excess material can escape during casting. **2.** See cutoff. { 'flash ,grüv }

flash hider [ORD] A metallic cone or flat disks attached to the muzzle of the gun to conceal the flash when the gun is fired and to prevent temporary blindness of the gun crew while firing. { 'flash ,hīd·ər }

flashing [BUILD] A strip of sheet metal placed at the junction of exterior building surfaces to render the joint watertight. [CHEM ENG] Vaporization of volatile liquids by either heat or vacuum. [ENG] Burning brick in an intermittent air supply in order to impart irregular color to the bricks. [MET] The violent expulsion of small metal particles due to arcing during flash butt welding. [OPTICS] The apparent filling of a curved mirror or lens with light when viewed from a distance, as a result of the production of a parallel beam by a light source at the focus. { 'flash·iŋ }

flashing block See raggle. { 'flash·iŋ ,bläk }

flashing flow [CHEM ENG] The condition when a liquid at its boiling point flows through a heated conduit and is further heated to cause partial vaporization (flashing), with a resultant two-phase (vapor-liquid) flow. { 'flash·iŋ ,flō }

flashing light [NAV] A light showing one or more flashes at regular intervals, the duration of the light period is less than that of the dark period; in particular, a light showing a single flash at regular intervals; distinctive names are used to indicate different combinations of flashes. { 'flash·iŋ ,līt }

flashing over [ELEC] Accidental formation of an arc over the surface of a rotating commutator from brush-to-brush; usually caused by faulty insulation between commutator segments. { 'flash·iŋ 'ō·vər }

flashing ring [ENG] A ring around a pipe that holds it in place as it passes through a partition such as a floor or wall. { 'flash·iŋ ,riŋ }

flash lamp [ELECTR] A gaseous-discharge lamp used in a photoflash unit to produce flashes of light of short duration and high intensity for stroboscopic photography. Also known as stroboscopic lamp. { 'flash ,lamp }

flashless [ORD] Of a propellant or a propelling charge that does not produce a muzzle flash in the weapon for which intended. { 'flash·ləs }

flash line [ENG] A raised line on the surface of a molding where the mold faces joined. { 'flash ,līn }

flash magnetization [ELECTROMAG] Magnetization of a ferromagnetic object by a current impulse of short duration. { 'flash ,mag·nə·tə'zā·shən }

flash memory [COMPUT SCI] A type of electrically erasable programmable read-only memory (EEPROM). While EPROM is reprogrammed bit-by-bit, flash memory is reprogrammed in blocks, making it faster. It is nonvolatile. { 'flash 'mem·rē }

flash message [COMMUN] A category of precedence reserved for initial enemy contact messages or operational combat messages of extreme urgency; brevity is mandatory. { 'flash 'mes·ij }

flash mold [ENG] A mold which permits excess material to escape during closing. { 'flash ,mōld }

flashover [ELEC] An electric discharge around or over the surface of an insulator. [ENG] A condition occurring during a fire in a building in which the surfaces of everything within a compartment or room seem to burst into flame simultaneously. { 'flash,ō·vər }

flashover voltage [ELECTR] The voltage at which an electric discharge occurs between two electrodes that are separated by an insulator; the value depends on whether the insulator surface is dry or wet. Also known as sparkover voltage. { 'flash,ō·vər ,vōl·tij }

flash pasteurization [MICROBIO] A pasteurization method in which a heat-labile liquid, such as milk, is briefly subjected to temperatures of 230°F (110°C). { 'flash pas·chə·rə'zā·shən }

flash photography See stroboscopic photography. { 'flash fə,täg·rə·fē }

flash photolysis [PHYS CHEM] A method of studying fast photochemical reactions in gas molecules; a powerful lamp is discharged in microsecond flashes near a reaction vessel holding the gas, and the products formed by the flash are observed spectroscopically. { 'flash fə,täl·ə·səs }

flash plating [MET] Electrodeposition of a thin film of metal. { 'flash ,plād·iŋ }

flash point [CHEM] The lowest temperature at which vapors from a volatile liquid will ignite momentarily upon the application of a small flame under specified conditions; test conditions can be either open- or closed-cup. { 'flash ,pȯint }

flash process [CHEM ENG] Liquid-vapor system in which the composition remains constant, but the proportion of gas and liquid phases changes as pressure or temperature change. { 'flash ,präs·əs }

flash radiography [GRAPHICS] A radiography technique employing very short exposure times, such as 1 microsecond, to give unblurred pictures of moving objects. { 'flash ,rād·ē'äg·rə·fē }

flash ranging [ORD] Finding the position of the burst of a projectile or of an enemy gun by observing its flash. { 'flash ,rān·jiŋ }

flash-ranging adjustment [ORD] Correcting friendly artillery fire on the basis of observation and location of the flash of shell bursts. { 'flash ,rān·jiŋ ə,jəs·mənt }

flash reconnaissance [ORD] Observation from ground posts or from aircraft to locate enemy gun positions by the flashes of enemy guns. { 'flash ri,kän·ə·səns }

flash ridge [ENG] The part of a flash mold along which the excess material escapes before the mold is closed. { 'flash ,rij }

flash roast [MIN ENG] Rapid removal of sulfur from ore by having finely divided sulfide mineral fall through a heated oxidizing atmosphere. Also known as suspension roast. { 'flash ,rōst }

flash separation [CHEM ENG] Process for separation of gas (vapor) from liquid components under reduced pressure; the liquid and gas remain in contact as the gas evolves from the liquid. { 'flash ,sep·ə'rā·shən }

flash set [MATER] Rapid hardening of freshly mixed cement paste, mortar, or concrete with considerable evolution of heat; plasticity cannot be restored. { 'flash 'set }

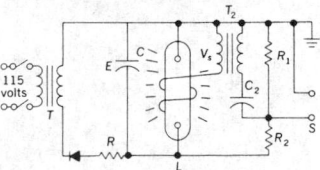

FLASH LAMP

Elementary circuit of electronic flash lamp L.

flash smelting [MET] Production of molten metal or matte in a vertical furnace in which concentrates are reacted with hot gases; the molten product is collected in a horizontal refractory-lined accumulator at the base of the furnace. { 'flash ,smel·tiŋ }

flash spectroscopy [SPECT] The study of the electronic states of molecules after they absorb energy from an intense, brief light flash. { ¦flash spek'träs·kə·pē }

flash spectrum [ASTRON] The emission spectrum of the sun's chromosphere, observed for a few seconds just before and just after a total solar eclipse. { 'flash ,spek·trəm }

flash steam [ENG] A mixture of steam and water that occurs when hot water under pressure moves to a region of lower pressure, such as in a flash boiler. { 'flash ,stēm }

flash tank [CHEM ENG] In a processing operation, a unit that is used to separate the liquid and gas phases. { 'flash ,taŋk }

flash test [ELEC] A method of testing insulation by applying momentarily a voltage much higher than the rated working voltage. { 'flash ,test }

flash trap See flash chamber. { 'flash ,trap }

flash vaporization [CHEM ENG] Rapid vaporization achieved by passing a volatile liquid through continuously heated coils. [ENG] A method used for withdrawing liquefied petroleum gas from storage in which liquid is first flashed into a vapor in an intermediate pressure system, and then a second stage regulator provides the low pressure required to use the gas in appliances. { ¦flash vā·pə·rə'zā·shən }

flash vessel See flash chamber. { 'flash ,ves·əl }

flash welding [MET] A form of resistance butt welding used to weld wide, thin members or members with irregular faces, and tubing to tubing. { 'flash ,weld·iŋ }

flask [CHEM] A long-necked vessel, frequently of glass, used for holding liquids. [MET] A frame used to hold molding sand in foundry work. { flask }

flat [ACOUS] A musical note that is a half step lower than a specified note. [ENG] A nonglossy painted surface. [GEOGR] A level tract of land. [GEOL] See mud flat. [GRAPHICS] **1.** The sheet of glass on which negative films are placed close together for printing on sensitized metal in the photoengraving process. **2.** An assemblage of negative or positive films used in preparing a photo-offset plate. [MINERAL] An inferior grade of rough diamonds. [NAV] **1.** A place covered with water too shallow for ordinary navigation. **2.** The area between high- and low-water marks along the edge of an arm of the sea, a bay, or tidal river; the term is usually used in the plural. [NAV ARCH] A partial deck below the main deck, constructed without any camber. [SCI TECH] **1.** A smooth, even surface. **2.** An object with a broad, shallow or thin form. { flat }

flat arch [ARCH] **1.** A straight horizontal arch consisting of mutually supportive wedge-shaped blocks. **2.** Any arch with a small rise-to-span ratio. { ¦flat 'ärch }

flat-back stope [MIN ENG] An overhand stoping method in which the ore is broken in slices parallel with the levels. { ¦flat ,bak 'stōp }

flatbed laser scanning [GRAPHICS] A method for the direct exposure of offset printing plates in which a laser beam is swept across a flat plate by an oscillation mirror or other scanner while being modulated by video signals from a similar reading system or by digital signals from a computer. { 'flat,bed 'lā·zər ,skan·iŋ }

flatbed plotter [ENG] A graphics output device that draws by moving a pen in both horizontal and vertical directions over a sheet of paper; the overall size of the drawing is limited by the height and width of this bed. { 'flat,bed 'pläd·ər }

flatbed press [GRAPHICS] A press whose printing plates for type and cuts are flat; it and the rotary press are the two major kinds of presses. { 'flat,bed 'pres }

flatbed truck [ENG] A truck whose body is in the form of a platform. { 'flat,bed 'trək }

flat belt [DES ENG] A power transmission belt, in the form of leather belting, used where high-speed motion rather than power is the main concern. { 'flat ,belt }

flat-belt conveyor [MECH ENG] A conveyor belt in which the carrying run is supported by flat-belt idlers or pulleys. { 'flat ,belt kən,vā·ər }

flat-belt pulley [DES ENG] A smooth, flat-faced pulley made of cast iron, fabricated steel, wood, and paper and used with a flat-belt drive. { 'flat ,belt ,pul·ē }

FLATFISH

A typical left-eye flounder, with both eyes on left side.

FLATHEAD RIVET

A flathead rivet, one of many standard types of rivet heads.

flat-blade turbine [MECH ENG] An impeller with flat blades attached to the margin. { 'flat ,blād 'tər,bīn }

flat-bottom crown See flat-face bit. { 'flat ,bäd·əm 'kraun }

flat cable [ELEC] A cable made of round or rectangular, parallel copper wires arranged in a plane and laminated or molded into a ribbon of flexible insulating plastic. { ¦flat 'kā·bəl }

flatcar [ENG] A railroad car without fixed walls or a cover. { 'flat,kär }

flat chisel [DES ENG] A steel chisel used to obtain a flat and finished surface. { 'flat 'chiz·əl }

flat-conductor cable [ELEC] A cable made of wide, flat conductors arranged side by side in a plane and protected by ribbons of insulating plastic. { 'flat kən,dək·tər 'kā·bəl }

flat crank [DES ENG] A crankshaft having one flat bearing journal. { 'flat ,kraŋk }

flat-crested weir [CIV ENG] A type of measuring weir whose crest is in the horizontal plane and whose length is great compared with the height of water passing over it. { 'flat ,krest·əd 'wer }

flat cut [MIN ENG] A manner of placing the boreholes, for the first shot in a tunnel, in which they are started about 2 or 3 feet (60 or 90 centimeters) above the floor and pointed downward so that the bottom of the hole will be about level with the floor. { 'flat ,kət }

flat die forging [MET] Die forging in which the metal is worked between simple contour dies. { 'flat 'dī ,fȯr·jiŋ }

flat drill [DES ENG] A type of rotary drill constructed from a flat piece of material. { 'flat ,dril }

flat edge trimmer [MECH ENG] A machine designed to trim the notched edges of metal shells. { 'flat ,ej 'trim·ər }

flat etching [GRAPHICS] The chemical reduction of the silver deposit on a continuous-tone or halftone plate by placing it in a tray which holds an etching solution. { ¦flat 'ech·iŋ }

flat-face bit [DES ENG] A diamond core bit whose face in cross section is square. Also known as flat-bottom crown; flat-nose bit; square-nose bit. { 'flat ,fās ,bit }

flat fading [COMMUN] Type of fading in which all components of the received radio signal fluctuate in the same proportion simultaneously. { ¦flat 'fād·iŋ }

flat file [COMPUT SCI] A two-dimensional array. { 'flat ,fīl }

flat fillet weld [MET] A fillet weld having a face that is relatively flat. { 'flat ,fil·ət 'weld }

flatfish [VERT ZOO] Any of a number of asymmetrical fishes which compose the order Pleuronectiformes; the body is laterally compressed, and both eyes are on the same side of the head. { 'flat,fish }

flat-flamed burner [ENG] A burner which emits a mixture of fuel and air in a flat stream through a rectangular nozzle. { 'flat ,flamd 'bərn·ər }

flat form tool [DES ENG] A tool having a square or rectangular cross section with the form along the end. { 'flat ,fȯrm ,tül }

flat grain [MATER] The grain pattern formed when soft wood is cut so that the tree's annular rings form an angle of 45° or less with the board's surface. { ¦flat 'grān }

flathead rivet [DES ENG] A small rivet with a flat manufactured head used for general-purpose riveting. { 'flat,hed 'riv·ət }

flat jack [CIV ENG] A hollow steel cushion which is made of two nearly flat disks welded around the edge and which can be inflated with oil or cement under controlled pressure; used at the arch abutments and crowns to relieve the load on the formwork at the moment of striking the formwork. { ¦flat 'jak }

flat line [ELECTROMAG] A radio-frequency transmission line, or part thereof, having essentially 1-to-1 standing wave ratio. { 'flat ,līn }

flat-lying [GEOL] Of mineral deposits and coal seams, having a relatively flat dip, up to 5°. { 'flat ,lī·iŋ }

flatness problem [ASTRON] The problem of explaining why, after 10^10 years of expansion, the density parameter is of the order of 1, whereas the standard big-bang theory suggests that once this parameter deviates even slightly from 1 it very quickly approaches an asymptotic value far away from 1 for open or closed universes. { 'flat·nəs ,präb·ləm }

flat nose [ORD] A missile used against submarines, designed to prevent ricocheting on water impact. { 'flat ,nōz }

flat-nose bit *See* flat-face bit.   { 'flat ‚nōz ‚bit }

flat of bottom [NAV ARCH] That portion of a ship's bottom without rise or having a rise with little or no curvature.   { 'flat əv 'bäd·əm }

flatpack [ELECTR] Semiconductor network encapsulated in a thin, rectangular package, with the necessary connecting leads projecting from the edges of the unit.   { 'flat‚pak }

flat-panel display *See* panel display.   { 'flat ‚pan·əl di'splā }

flat-plate collector [ENG] A solar collector consisting of a shallow metal box covered by a transparent lid.   { 'flat ‚plāt kə'lek·tər }

flat-position welding [MET] Welding above the joint with the face of the weld in the horizontal reference plane. Also known as downhand welding.   { 'flat pə‚zish·ən 'weld·iŋ }

flat roof [ARCH] A roof pitched only enough to allow water to drain off.   { ‚flat ‚rüf }

flat rope [DES ENG] A steel or fiber rope having a flat cross section and composed of a number of loosely twisted ropes placed side by side, the lay of the adjacent strands being in opposite directions to secure uniformity in wear and to prevent twisting during winding.   { ‚flat ‚rōp }

flats [GRAPHICS] Stage constructions, used in series, to produce a painted background, usually architectural details; they are canvas-covered frames and are used for all flat surfaces on the stage, such as room interiors or exterior walls of buildings.   { flats }

flat slab [CIV ENG] A flat plate of reinforced concrete designed to span in two directions.   { ‚flat ‚slab }

flat space [MATH] A Riemannian space for which a coordinate system exists such that the components of the metric tensor are constants throughout the space; equivalently, a space in which the Riemann-Christoffel tensor vanishes throughout the space.   { ‚flat ‚spās }

flat space-time [RELAT] Space-time in which the Riemann-Christoffel tensor vanishes; geometry is then equivalent to that of the Minkowski universe used in special relativity.   { ‚flat ‚spās ‚tīm }

flat spin [MECH] Motion of a projectile with a slow spin and a very large angle of yaw, happening most frequently in fin-stabilized projectiles with some spin-producing moment, when the period of revolution of the projectile coincides with the period of its oscillation; sometimes observed in bombs and in unstable spinning projectiles.   { ‚flat ‚spin }

flat spring *See* leaf spring.   { ‚flat ‚spriŋ }

flattening [GEOD] The ratio of the difference between the equatorial and polar radii of the earth; the flattening of the earth is the ellipticity of the spheroid; the magnitude of the flattening is sometimes expressed as the numerical value of the reciprocal of the flattening. Also known as compression. [MET] Straightening of metal sheet by passing it through special rollers which flatten it without changing its thickness. Also known as roll flattening.   { 'flat·ən·iŋ }

flattening test [MET] Quality test performed by flattening metal tubing between parallel plates that are a specified distance from each other.   { 'flat·ən·iŋ ‚test }

flatting agent [MATER] Additive substance for paints or varnishes to disperse incident light rays to give the dried surface a nonglossy matte finish.   { 'flat·ən·iŋ ‚ā·jənt }

flat-top antenna [ELECTROMAG] An antenna having two or more lengths of wire parallel to each other and in a plane parallel to the ground, each fed at or near its midpoint.   { 'flat ‚täp an‚ten·ə }

flat-top boom [ACOUS] A sonic boom whose pressure signature is shaped to reduce the perceived amplitude of the shocks by allowing plateaus following the bow shock and preceding the tail shock.   { 'flat ‚täp 'büm }

flat-top response *See* band-pass response.   { 'flat ‚täp ri'späns }

flat trajectory [MECH] A trajectory which is relatively flat, that is, described by a projectile of relatively high velocity.   { ‚flat trə'jek·trē }

flat tuning [ELECTR] Tuning of a radio receiver in which a change in frequency of the received waves produces only a small change in the current in the tuning apparatus.   { 'flat 'tün·iŋ }

flat-turret lathe [MECH ENG] A lathe with a low, flat turret on a power-fed cross-sliding headstock.   { 'flat ‚tə·rət 'lāth }

flatulence [MED] Excessive intestinal gas.   { 'flach·ə·ləns }

flatus [MED] Gas in the intestinal tract.   { 'flād·əs }

flatwood [ECOL] An almost-level zone containing mostly imperfectly drained, acid soils and vegetation consisting of wiregrass and saw palmetto at ground level, shrubs such as gallberry and waxmyrtle, and trees such as longleaf and slash pines.   { 'flat‚wud }

flatworm [INV ZOO] The common name for members of the phylum Platyhelminthes; individuals are dorsoventrally flattened.   { 'flat‚wərm }

flat yard [CIV ENG] A switchyard in which railroad cars are moved by locomotives, not by gravity.   { 'flat ‚yärd }

flavan [BIOCHEM] $C_{15}H_{14}O$ 2-Phenylbenzopyran, an aromatic heterocyclic compound from which all flavonoids are derived.   { 'fla·vən }

flavanol [BIOCHEM] Yellow needles with a melting point of 169°C, derived from flavanone; a flavanoid pigment used as a dye. Also known as 3-hydroxyflavone.   { 'fla·və‚nól }

flavanone [BIOCHEM] $C_{15}H_{12}O_2$ A colorless crystalline ketone, that often occurs in plants in the form of a glycoside.   { 'fla·və‚nōn }

flavescence [PL PATH] Yellowing or blanching of green plant parts due to diminution of chlorophyll accompanying certain virus disease.   { flə'ves·əns }

flavin [BIOCHEM] Any of several water-soluble yellow pigments occurring as coenzymes of flavoproteins.   { 'fla·vən }

flavin adenine dinucleotide [BIOCHEM] $C_{27}H_{33}N_9O_{15}P_2$ A coenzyme that functions as a hydrogen acceptor in aerobic dehydrogenases (flavoproteins). Abbreviated FAD.   { 'fla·vən 'ad·ən‚ēn dī'nü·klē·ə‚tīd }

flavin mononucleotide *See* riboflavin 5′-phosphate.   { 'fla·vən ‚mä·nō'nü·klē·ə‚tīd }

flavin phosphate *See* riboflavin 5′-phosphate.   { 'fla·vən 'fäs‚fāt }

Flavobacterium [MICROBIO] A genus of bacterium of uncertain affiliation; gram-negative coccobacilli or slender rods producing pigmented (yellow, red, orange, or brown) growth on solid media.   { ‚fla·vō·bak'tir·ē·əm }

flavone [BIOCHEM] **1.** Any of a number of ketones composing a class of flavonoid compounds. **2.** $C_{15}H_{10}O_2$ A colorless crystalline compound occurring as dust on the surface of many primrose plants.   { 'fla‚vōn }

flavonoid [BIOCHEM] Any of a series of widely distributed plant constituents related to the aromatic heterocyclic skeleton of flavan.   { 'fla·və‚nóid }

flavonol [BIOCHEM] **1.** Any of a class of flavonoid compounds that are hydroxy derivatives of flavone. **2.** $C_{16}H_{10}O_2$ A colorless, crystalline compound from which many yellow plant pigments are derived.   { 'fla·və‚nól }

flavoprotein [BIOCHEM] Any of a number of conjugated protein dehydrogenases containing flavin that play a role in biological oxidations in both plants and animals; a yellow enzyme.   { ‚fla·vō'prō‚tēn }

flavor [FOOD ENG] The set of characteristics of a food that causes a simultaneous reaction or sensation of taste on the tongue and odor in the olfactory center in the nose. [PART PHYS] A label used to distinguish different types of leptons (the electron, electron neutrino, muon, muon neutrino, and possibly others) and different color triplets of quarks (the up, down, strange, and charmed quarks, and possibly others).   { 'flā·vər }

flavor enhancer [FOOD ENG] A substance that accentuates the taste of a food to which it has been added without contributing a flavor of its own. Also known as potentiator.   { 'flā·vər in‚han·sər }

flaw [MATER] A discontinuity in a material beyond acceptable established limits. [METEOROL] An English nautical term for a sudden gust or squall of wind. [MINERAL] A faulty part of a gemstone, such as a crack, visible imperfect crystallization, or internal twinning or cleavage. [OCEANOGR] **1.** The seaward edge of fast ice. **2.** A shore lead just outside fast ice.   { fló }

flax [BOT] *Linum usitatissimum.* An erect annual plant with linear leaves and blue flowers; cultivated as a source of flaxseed and fiber.   { flaks }

flax rust [PL PATH] A disease of flax caused by the rust fungus *Melampsora lini.*   { 'flaks ‚rəst }

flaxseed [BOT] The seed obtained from the seed flax plant; a source of linseed oil. Also known as linseed.   { 'flak‚sēd }

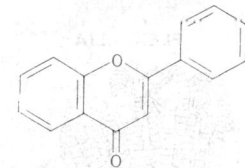

FLAVONE

Structural formula of flavone.

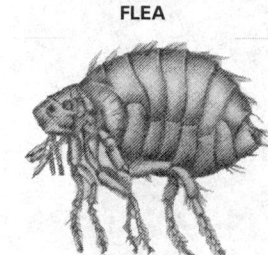

FLEA

The sticktight flea, a bird parasite.

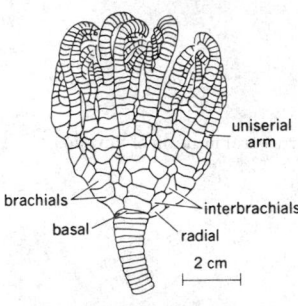

FLEXIBILIA

Talanterocrinus species.

flaxseed ore [GEOL] Iron ore composed of disk-shaped oauolites that have been partially flattened parallel to the bedding plane. { 'flak,sēd ,ȯr }

flax wilt [PL PATH] A fungus disease of flax caused by *Fusarium oxysporum lini;* diseased plants wilt, yellow, and die. { 'flaks ,wilt }

F layer [GEOPHYS] An ionized layer in the F region of the ionosphere which consists of the F_1 and F_2 layers in the day hemisphere, and the F_2 layer alone in the night hemisphere; it is capable of reflecting radio waves to earth at frequencies up to about 50 megahertz. { 'ef ,lā·ər }

F_1 layer [GEOPHYS] The ionosphere layer beneath the F_2 layer during the day, at a virtual height of 120–180 miles (200–300 kilometers), being closest to earth around noon; characterized by a distinct maximum of free-electron density, except at high latitudes during winter, when the layer is not detectable. { ¦ef 'wən ,lā·ər }

F_2 layer [GEOPHYS] The highest constantly observable ionosphere layer, characterized by a distinct maximum of free-electron density at a virtual height from about 135 miles (225 kilometers) in the polar winter to more than 240 miles (400 kilometers) in daytime near the magnetic equator. Also known as Appleton layer. { ¦ef 'tü ,lā·ər }

fl dr *See* fluid dram.

flea [INV ZOO] Any of the wingless insects composing the order Siphonaptera; most are ectoparasites of mammals and birds. { flē }

flea-allergy dermatitis [VET MED] Inflammation of the skin of small pets (dogs, cats, and ferrets) that results from an allergic reaction to protein substances deposited on or under the surface of the skin at the time of flea feeding. Also known as flea-bite dermatitis. { ¦flē,al·ər·jē ,dər·mə'tīd·əs }

flea-bite dermatitis *See* flea allergy dermatitis. { ¦flē,bīt ,dər·mə'tīd·əs }

flea-borne typhus *See* murine typhus. { 'flē ,bȯrn 'tī·fəs }

fleam [DES ENG] The angle of bevel of the edge of the teeth of a saw with respect to the plane of the blade. { flēm }

flechette [ORD] **1.** A small fin-stabilized missile, a large number of which can be loaded in an artillery canister or in a warhead. **2.** *See* aerial dart. { fle'shet }

flecnode [MATH] A node that is also a point of inflection of one of the two branches of the curve that cross at the node. { 'flek,nōd }

fleece [TEXT] A fabric with a deep, soft, napped surface. [VERT ZOO] Coat of wool shorn from sheep; usually taken off the animal in one piece. { flēs }

fleet [MECH ENG] Sidewise movement of a rope or cable when winding on a drum. [ORD] **1.** An organization of ships, aircraft, marine forces, and shore-based fleet activities, all under a commander who may exercise operational as well as administrative control. **2.** All naval operating forces. { flēt }

fleet angle [MECH ENG] In hoisting gear, the included angle between the rope, in its position of greatest travel across the drum, and a line drawn perpendicular to the drum shaft, passing through the center of the head sheave or lead sheave groove. { 'flēt ,aŋ·gəl }

fleet broadcast [COMMUN] The radio broadcast (in addition to the general broadcast) to all U.S. Navy ships and merchant ships in which storm warnings are given. { 'flēt ,brȯd,kast }

fleeting target [ORD] Moving or transient target that remains within observing or firing distance for such a short period that it affords little time for deliberate adjustment and fire against it; for example, aircraft, vehicles, marching troops, and so forth. { ¦flēd·iŋ 'tär·gət }

Flehmen response [ECOL] A courtship behavior displayed by the males of some mammalian species in which the upper lip is curled and the neck is extended, facilitating the reception of olfactory cues. { 'flā·mən ri,späns }

Fleming cracking process [CHEM ENG] An obsolete liquid-phase thermal cracking process for heavy petroleum fractions; the charge was heated under pressure in a vertical shell still. { 'flem·iŋ 'krak·iŋ ,präs·əs }

Fleming-Kennelly law [ELECTROMAG] The reluctivity of a ferromagnetic substance varies linearly with magnetic field strength at points near magnetic saturation. { ¦flem·iŋ 'ken·ə·lē ,lȯ }

Fleming's rule *See* left-hand rule; right-hand rule. { 'flem·iŋz ,rül }

Fleming's solution [MATER] A tissue fixative made up of a mixture of osmic, chromic and acetic acids. { 'flem·iŋ sə,lü·shən }

Fleming tube [ELECTR] The original diode, consisting of a heated filament and a cold metallic electrode in an evacuated glass envelope; negative current flows from the filament to the cold electrode, but not in the reverse direction. { 'flem·iŋ ,tüb }

Flemish bond [CIV ENG] A masonry bond consisting of alternating stretchers and headers in each course, laid with broken joints. { ¦flem·ish 'bänd }

Flemish garden wall bond [CIV ENG] A masonry bond consisting of headers and stretchers in the ratio of one to three or four in each course, with joints broken to give a variety of patterns. { ¦flem·ish 'gärd·ən ¦wȯl ,bänd }

flesh [ANAT] The soft parts of the body of a vertebrate, especially the skeletal muscle and associated connective tissue and fat. { flesh }

Flesh-Demag process [CHEM ENG] A gas-making process in which a cyclic water-gas apparatus is used for feeding and charring the coal charge and for gas generation, with periodic automatic removal of the resultant ash. { ¦flesh 'da·mäk ,präs·əs }

fleshing machine [ENG] A machine that removes flesh from hides in a tannery. { 'flesh·iŋ mə,shēn }

fleshy fruit [BOT] A fruit having a fleshy pericarp that is usually soft and juicy, but sometimes hard and tough. { ¦flesh·ē ¦früt }

Fletcher-Munson contour *See* equal loudness contour. { ¦flech·ər ¦mən·sən 'kän,tur }

Fletcher radial burner [ENG] A burner with gas jets arranged radially. { 'flech·ər ¦rād·ē·əl 'bərn·ər }

Flettner windmill [MECH ENG] An inefficient windmill with four arms, each consisting of a rotating cylinder actuated by a Savonius rotor. { 'flet·nər 'wind,mil }

fleuron [GRAPHICS] A typographical decorative symbol, generally in the shape of a flower or a leaf. { 'flu̇,rän }

flex [SCI TECH] To bend. { fleks }

Flexibacter [MICROBIO] A genus of bacteria in the family Cytophagaceae; cells are unsheathed rods or filaments and are motile; microcysts are not known. { ¦flek·sə¦bak·tər }

Flexibilia [PALEON] A subclass of extinct stalked or creeping Crinoidea; characteristics include a flexible tegmen with open ambulacral grooves, uniserial arms, a cylindrical stem, and five conspicuous basals and radials. { ,flek·sə'bil·ē·ə }

flexibility [MECH] The quality or state of being able to be flexed or bent repeatedly. { ,flek·sə'bil·əd·ē }

flexibilizer [MATER] An additive that gives an otherwise rigid plastic flexibility. Also known as plasticizer. { 'flek·sə·bə,līz·ər }

flexible circuit [ELECTR] A printed circuit made on a flexible plastic sheet that is usually die-cut to fit between large components. { ,flek·sə·bəl 'sər·kət }

flexible collodion [MATER] A collodion which has two additives (2% camphor and 3% castor oil) to make a pliable film. { ,flek·sə·bəl kə'lōd·ē·ən }

flexible coupling [ELECTROMAG] A coupling designed to allow a limited angular movement between the axes of two waveguides. [MECH ENG] A coupling used to connect two shafts and to accommodate their misalignment. { ,flek·sə·bəl 'kəp·liŋ }

flexible glue [MATER] A type of glue used for pliable molds and printers' rollers, for example, a mixture of glue, glycerol, and water. { ,flek·sə·bəl 'glü }

flexible gun [ORD] A gun, especially a machine gun, mounted in an aircraft turret or on a post, tripod, or other mount in such a manner that the gun may be swung in both a vertical and horizontal plane. { ,flek·sə·bəl 'gən }

flexible-joint pipe [ENG] Cast-iron pipe adapted to laying under water and capable of motion through several degrees without leakage. { ,flek·sə·bəl ¦jȯint 'pīp }

flexible manufacturing system [IND ENG] A form of computer-integrated manufacturing used to make small to moderate-sized batches of parts. { 'flek·sə·bəl ,man·yə·'fak·chə·riŋ ,sis·təm }

flexible mold [ENG] A coating mold made of flexible rubber or other elastomeric materials; used mainly for casting plastics. { ,flek·sə·bəl 'mōld }

flexible pavement [CIV ENG] A road or runway made of

bituminous material which has little tensile strength and is therefore flexible. { ‚flek·sə·bəl 'pav·mənt }

flexible resistor [ELEC] A wire-wound resistor having the appearance of a flexible lead; made by winding the Nichrome resistance wire around a length of asbestos or other heat-resistant cord, then covering the winding with asbestos and braided insulating covering. { ‚flek·sə·bəl ri'zis·tər }

flexible sandstone [GEOL] A variety of itacolumite that consists of fine grains and occurs in thin layers. { ‚flek·sə·bəl 'san‚stōn }

flexible shaft [MECH ENG] **1.** A shaft that transmits rotary motion at any angle up to about 90°. **2.** A shaft made of flexible material or of segments. **3.** A shaft whose bearings are designed to accommodate a small amount of misalignment. { ‚flek·sə·bəl 'shaft }

flexible ventilation ducting [MIN ENG] Flexible fabric tubes covered with rubber or polyvinyl chloride, used for auxiliary ventilation. { ‚flek·sə·bəl vent·əl'ā·shən ‚dək·tiŋ }

flexible waveguide [ELECTROMAG] A waveguide that can be bent or twisted without appreciably changing its electrical properties. { ‚flek·sə·bəl 'wāv‚gīd }

flexicoking [CHEM ENG] A continuous coke-making process that has a gasification section in which coke can be gasified to produce refinery fuel gas, allowing the production of both gas and coke in line with market requirements. { 'flek·sə‚kōk·iŋ }

flexion [BIOL] Act of bending, especially of a joint. { 'flek·shən }

flexional symbols [COMPUT SCI] Symbols in which the meaning of each component digit is dependent on those which precede it. { ¦flek·shən·əl ¦sim·bəlz }

flexion reflex [PHYSIO] An unconditioned, segmental reflex elicited by noxious stimulation and consisting of contraction of the flexor muscles of all joints on the same side. Also known as the nociceptive reflex. { 'flek·shən ‚rē‚fleks }

Flexithrix [MICROBIO] A genus of bacteria in the family Cytophagaceae; cells are usually sheathed filaments, and unsheathed cells are motile; microcysts are not known. { 'flek·sə‚thriks }

flexographic printing See flexography. { ‚flek·sə¦graf·ik 'print·iŋ }

flexography [GRAPHICS] Relief printing with plates fastened to a cylinder and with a single inking roller supplied with aniline ink from two rollers in the ink fountain. Also known as aniline printing; aniline process; flexographic printing. { flek'säg·rə·fē }

flexometer [ENG] An instrument for measuring the flexibility of materials. { flek'säm·əd·ər }

flexor [PHYSIO] A muscle that bends or flexes a limb or a part. { 'flek·sər }

flexowriter [COMPUT SCI] A typewriterlike device to read in manually or to read out information of a computer to which it is connected; it can also be used to punch paper tape. { 'flek·sə‚wrīd·ər }

flexuous [BIOL] **1.** Flexible. **2.** Bending in a zigzag manner. **3.** Wavy. { 'flek·shə·wəs }

flexural modulus [MECH] A measure of the resistance of a beam of specified material and cross section to bending, equal to the product of Young's modulus for the material and the square of the radius of gyration of the beam about its neutral axis. { 'flek·shə·rəl 'mäj·ə·ləs }

flexural rigidity [MECH] The ratio of the sideward force applied to one end of a beam to the resulting displacement of this end, when the other end is clamped. { 'flek·shə·rəl ri'jid·əd·ē }

flexural slip [GEOL] The slipping of sedimentary strata along bedding planes during folding, producing disharmonic folding and, when extreme, décollement. Also known as bedding-plane slip. { 'flek·shə·rəl 'slip }

flexural strength [MECH] Strength of a material in blending, that is, resistance to fracture. { 'flek·shə·rəl 'streŋkth }

flexure [EMBRYO] A sharp bend of the anterior part of the primary axis of the vertebrate embryo. [GEOL] **1.** A broad, domed structure. **2.** A fold. [MECH] **1.** The deformation of any beam subjected to a load. **2.** Any deformation of an elastic body in which the points originally lying on any straight line are displaced to form a plane curve. [VERT ZOO] The last joint of a bird's wing. { 'flek·shər }

flexure theory [MECH] Theory of the deformation of a prismatic beam having a length at least 10 times its depth and

consisting of a material obeying Hooke's law, in response to stresses within the elastic limit. { 'flek·shər ‚thē·ə·rē }

flicker [OPTICS] A visual sensation produced by periodic fluctuations in light at rates ranging from a few cycles per second to a few tens of cycles per second. { 'flik·ər }

flicker control [AERO ENG] Control of an aircraft, rocket, or such in which the control surfaces are deflected to their maximum degree with only a slight motion of the controller. { 'flik·ər kən‚trōl }

flicker effect [ELECTR] Random variations in the output current of an electron tube having an oxide-coated cathode, due to random changes in cathode emission. { 'flik·ər i‚fekt }

flicker fusion [PHYSIO] The tendency to perceive an oscillating or flickering sensory input signal as continuous when the frequency is above a specific threshold frequency. { 'flik·ər ‚fyü·zhən }

flicker photometer [OPTICS] A photometer in which a single field of view is alternately illuminated by the light sources to be compared, and the rate of alternation is such that color flicker is absent but brightness flicker is not; disappearance of flicker signifies equality of luminance. { 'flik·ər fə'täm·əd·ər }

flight [AERO ENG] The movement of an object through the atmosphere or through space, sustained by aerodynamic reaction or other forces. [CIV ENG] A series of stairs between landings or floors. [MECH ENG] Plain or shaped plates that are attached to the propelling mechanism of a flight conveyor. { flīt }

flight briefing See pilot briefing. { 'flīt ‚brēf·iŋ }

flight characteristic [AERO ENG] A characteristic exhibited by an aircraft, rocket, or the like in flight, such as a tendency to stall or to yaw, or an ability to remain stable at certain speeds. { 'flīt ‚kar·ik·tə‚ris·tik }

flight control system See vehicle control system. { 'flīt kən‚trōl ‚sis·təm }

flight conveyor [MECH ENG] A conveyor in which paddles, attached to single or double strands of chain, drag or push pulverized or granulated solid materials along a trough. Also known as drag conveyor. { 'flīt kən‚vā·ər }

flight deck [AERO ENG] In certain airplanes, an elevated compartment occupied by the crew for operating the airplane in flight. [NAV ARCH] The topmost complete deck of an aircraft carrier, used mainly for takeoff and landing of planes. { 'flīt ‚dek }

flight dynamics [AERO ENG] The study of the motion of an aircraft or missile; concerned with transient or short-term effects relating to the stability and control of the vehicle, rather than to calculating such performance as altitude or velocity. { 'flīt dī‚nam·iks }

flight envelope [AERO ENG] The boundary depicting, for a specific aircraft, the limits of speed, altitude, and acceleration which that aircraft cannot safely exceed. { 'flīt ‚en·və‚lōp }

flighter [FOOD ENG] A rotating vane used to cool brewers' wort. { 'flīd·ər }

flight feather [VERT ZOO] Any of the long contour feathers on the wing of a bird. Also known as remex. { 'flīt ‚feth·ər }

flight feeder [MECH ENG] Short-length flight conveyor used to feed solids materials to a process vessel or other receptacle at a preset rate. { 'flīt ‚fēd·ər }

flight forecast [METEOROL] An aviation weather forecast for a specific flight. { 'flīt ‚fȯr‚kast }

flight inspection [NAV] Flight investigation certification of certain operational performance characteristics of electronic and visual navigation facilities by an authorized inspector in conformance with the U.S. Standard Flight Inspection Manual. { 'flīt in‚spek·shən }

flight instrument [AERO ENG] An aircraft instrument used in the control of the direction of flight, attitude, altitude, or speed of an aircraft, for example, the artificial horizon, airspeed indicator, altimeter, compass, rate-of-climb indicator, accelerometer, turn-and-bank indicator, and so on. { 'flīt ‚in·strə·mənt }

flight level [AERO ENG] A surface of constant atmospheric pressure which is related to the standard pressure datum. { 'flīt ‚lev·əl }

flight log [NAV] **1.** A complete written record of a flight, normally showing flight planning information together with

FLIGHT FEATHER

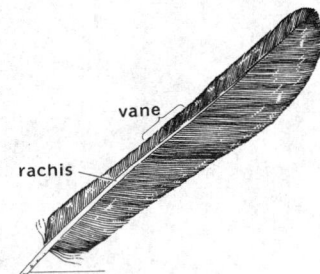

vane

rachis

Full view of a typical flight feather.

actual data recorded during the flight.   **2.** A device that automatically records on a screen or map the flight path flown by an aircraft.   { 'flīt ,läg }

flight-management computer   [NAV]   A computer carried on an aircraft to integrate the functions of navigation, guidance, and performance management.   { 'flīt ¦man·ij·mənt kəm,pyüd·ər }

flight of ideas   [PSYCH]   Verbal skipping from one idea to another in which the ideas bear only a superficial relation to one another and are fragmentary and often associated by chance.   { ¦flīt əv ī'dē·əz }

flight path   [AERO ENG]   The path made or followed in the air or in space by an aircraft, rocket, or such.   { 'flīt ,path }

flight-path angle   [AERO ENG]   The angle between the horizontal (or some other reference angle) and a tangent to the flight path at a point.   Also known as flight-path slope.   { 'flīt ,path ,aŋ·gəl }

flight-path computer   [COMPUT SCI]   A computer that includes all of the functions of a course-line computer and also provides means for controlling the altitude of an aircraft in accordance with a desired plan of flight.   { 'flīt ,path kəm'pyüd·ər }

flight-path deviation   [NAV]   The difference between the flight track of an aircraft and the flight path, expressed in terms of either angular or linear measurement.   { 'flīt ,path dē·vē'ā·shən }

flight-path deviation indicator   [NAV]   An instrument providing visual indication of deviation from a flight path.   { 'flīt ,path dē·vē'ā·shən ,in·də,kād·ər }

flight-path selector   [NAV]   An instrument used with a flight path computer to preset the values defining the flight path to a way point or terminal.   { 'flīt ,path si'lek·tər }

flight-path slope   *See* flight-path angle.   { 'flīt ,path ,slōp }

flight plan   [NAV]   Information provided to air-traffic service units, giving in detail the proposed plan of flight, including times, altitudes, way points, and so on, and submitted for approval.   { 'flīt ,plan }

flight profile   [AERO ENG]   A graphic portrayal or plot of the flight path of an aeronautical vehicle in the vertical plane.   { 'flīt ,prō,fīl }

flight recorder   [ENG]   Any instrument or device that records information about the performance of an aircraft in flight or about conditions encountered in flight, for future study and evaluation.   { 'flīt ri,kòrd·ər }

flight rules   [NAV]   Rules established by competent authority to govern flights; the type of flight involved determines whether instrument flight rules or visual flight rules apply.   { 'flīt ,rülz }

flight science   [AERO ENG]   The sum total of all knowledge that enables humans to accomplish flight; it is compounded of both science and engineering, and is concerned with airplanes, missiles, and crewed and crewless space vehicles.   { 'flīt ,sī·əns }

flight simulator   [AERO ENG]   A training device or apparatus that simulates certain conditions of actual flight or of flight operations.   { 'flīt ,sim·yə,lād·ər }

flight stability   [AERO ENG]   The property of an aircraft or missile to maintain its attitude and to resist displacement, and, if displaced, to tend to restore itself to the original attitude.   { 'flīt stə,bil·əd·ē }

flight technical error   [NAV]   The deviation of aircraft position, as reported by the navigation sensors, from the desired flight path.   { ¦flīt 'tek·nə·kəl ,er·ər }

flight test   [AERO ENG]   **1.** A test by means of actual or attempted flight to see how an aircraft, spacecraft, space-air vehicle, or missile flies.   **2.** A test of a component part of a flying vehicle, or of an object carried in such a vehicle, to determine its suitability or reliability in terms of its intended function by making it endure actual flight.   { 'flīt ,test }

flight time   [NAV]   The elapsed time from the moment an aircraft first moves under its own power for the purpose of taking off until the moment it comes to rest at the end of the flight.   { 'flīt ,tīm }

flight track   *See* track.   { 'flīt ,trak }

flight visibility   [METEOROL]   Average visibility in a forward direction from an aircraft in flight.   { 'flīt ,viz·ə,bil·əd·ē }

flight-weather briefing   *See* pilot briefing.   { 'flīt ,weth·ər ,brēf·iŋ }

FLIH   *See* first-level interrupt handler.

flinching   [IND ENG]   In inspection, failure to call a borderline defect a defect.   { 'flin·chiŋ }

Flinders bar   [NAV]   A bar of soft unmagnetized iron placed vertically near a magnetic compass to counteract deviation caused by magnetic induction in vertical soft iron of the craft.   { 'flin·dərz ,bär }

F line   [SPECT]   A green-blue line in the spectrum of hydrogen, at a wavelength of 486.133 nanometers.   { 'ef ,līn }

flinkite   [MINERAL] $Mn_3(AsO_4)(OH)_4$ Greenish-brown mineral composed of basic manganese arsenate, occurring in feathery forms.   { 'fliŋ,kīt }

flint   [MINERAL]   A black or gray, massive, hard, somewhat impure variety of chalcedony, breaking with a conchoidal fracture.   Also known as firestone.   { flint }

flint clay   [GEOL]   A hard, smooth, flintlike fireclay; when it is ground, it develops no plasticity, and it breaks with conchoidal fracture.   { 'flint ,klā }

flint glass   [MATER]   **1.** Heavy, colorless, brilliant glass that contains lead oxide.   **2.** Any high-quality glass.   { 'flint ,glas }

flint mill   [MECH ENG]   A mill employing pebbles to pulverize materials (for example, in cement manufacture).   { 'flint ,mil }

FLIP   *See* floating instrument platform.   { flip }

flip chip   [ELECTR]   A tiny semiconductor die having terminations all on one side in the form of solder pads or bump contacts; after the surface of the chip has been passivated or otherwise treated, it is flipped over for attaching to a matching substrate.   Also known as solder-ball flip chip.   { 'flip ,chip }

flip coil   [ELECTROMAG]   A small coil used to measure the strength of a magnetic field; it is placed in the field, connected to a ballistic galvanometer or other instrument, and suddenly flipped over 180°; alternatively, the coil may be held stationary and the magnetic field reversed.   { 'flip ,kòil }

flip-flop amplifier   *See* wall-attachment amplifier.   { 'flip,fläp ,am·plə,fī·ər }

flip-flop circuit   *See* bistable multivibrator.   { 'flip,fläp ,sər·kət }

flip-open cutout fuse   *See* dropout fuse.   { ¦flip ¦ō·pən 'kəd,aut ,fyüz }

flip-over process   *See* Umklapp process.   { 'flip,ō·vər ,präs·əs }

flipper   [VERT ZOO]   A broad, flat appendage used for locomotion by aquatic mammals and sea turtles.   { 'flip·ər }

FLIR imager   *See* forward-looking infrared imager.   { 'flir ,im·ij·ər }

flist   [METEOROL]   In Scotland, a keen blast or shower accompanied by a squall.   { flist }

flitch   [MATER]   **1.** A section that is cut from a log and subsequently manufactured into veneer or lumber.   **2.** Sheets of veneer that are stacked in sequence as they have been cut from a log.   **3.** A longitudinal section of a log, sometimes having bark on one or more edges.   { flich }

flitch beam   *See* flitch girder.   { 'flich ,bēm }

flitch girder   [BUILD]   A beam made of structural timbers bolted together with a steel plate between them.   Also known as flitch beam; sandwich beam.   { 'flich ,gərd·ər }

flitch plate   [CIV ENG]   The metal plate in a flitch beam or girder.   { 'flich ,plāt }

float   [AGR]   A device consisting of one or more blades used to level a seedbed.   [BIOL]   An air-filled sac in many pelagic flora and fauna that serves to buoy up the body of the organism.   [DES ENG]   A file which has a single set of parallel teeth.   [ENG]   **1.** A flat, rectangular piece of wood with a handle, used to apply and smooth coats of plaster.   **2.** A mechanical device to finish the surface of freshly placed concrete paving.   **3.** A marble-polishing block.   **4.** Any structure that provides positive buoyancy such as a hollow, watertight unit that floats or rests on the surface of a fluid.   **5.** *See* plummet.   [GEOL] An isolated, displaced rock or ore fragment.   [IND ENG]   *See* bank.   [TEXT]   **1.** A thread used to create patterns in fabric by passing over other threads.   **2.** A fabric defect caused by passing a thread over other threads where it should be interwoven.   { flōt }

floatability   [MIN ENG]   Response of a specific mineral to the flotation process.   { ,flōd·ə'bil·əd·ē }

float-and-sink analysis   [MIN ENG]   Use of a series of heavy liquids diminishing (or increasing) in density by accurately controlled stages in order to divide a sample of crushed coal

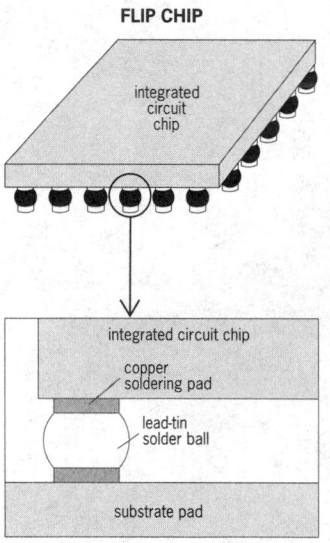

FLIP CHIP

integrated circuit chip

integrated circuit chip

copper soldering pad

lead-tin solder ball

substrate pad

Flip chip technique for connecting an integrated circuit chip to a substrate.

or other minerals or metals into fractions that are either equal-settling or equal-floating at each stage. { ¦flōt ən 'siŋk ə₁nal·ə·səs }

float barograph [ENG] A type of siphon barograph in which the mechanically magnified motion of a float resting on the lower mercury surface is used to record atmospheric pressure on a rotating drum. { 'flōt 'bar·ə₁graf }

float bowl [MECH ENG] A component of a carburetor that holds a small amount of liquid gasoline and serves as a constant-level reservoir of fuel that is metered into the passing flow of air. { 'flōt ₁bōl }

float chamber [ENG] A vessel in which a float regulates the level of a liquid. { 'flōt ₁chām·bər }

float coal [GEOL] Small, irregularly shaped, isolated deposits of coal embedded in sandstone or in siltstone. Also known as raft. { 'flōt ₁kōl }

float collar [PETRO ENG] A specialized coupling device inserted in a casing string just above the bottom that contains a check valve to permit downward passage of fluid but not upward passage through the casing. { 'flōt ₁käl·ər }

float control [ENG] Floating device used to transmit a liquid-level reading to a control apparatus, such as an on-off switch controlling liquid flow into and out of a storage tank. { 'flōt kən₁trōl }

float-cut file [DES ENG] A coarse file used on soft materials. { 'flōt ₁kət ₁fīl }

floater [OCEANOGR] See drift bottle. [PETRO ENG] A structure for offshore drilling that is secured to the sea floor only by anchors. { 'flōd·ər }

float finish [CIV ENG] A rough concrete finish, obtained by using a wooden float for finishing. { 'flōt ₁fin·ish }

float gage [ENG] Any one of several types of instruments in which the level of a liquid is determined from the height of a body floating on its surface, by using pulleys, levers, or other mechanical devices. { 'flōt ₁gāj }

float glass [MATER] Flat glass with a nearly true optical surface produced by floating a continuous sheet of molten glass on a bed of molten tin until the glass cools and hardens. { 'flōt ₁glas }

floating [ELECTR] The condition wherein a device or circuit is not grounded and not tied to an established voltage supply. { 'flōd·iŋ }

floating action [ENG] Controller action in which there is a predetermined relation between the deviation and the speed of a final control element; a neutral zone, in which no motion of the final control element occurs, is often used. { 'flōd·iŋ ₁ak·shən }

floating address [COMPUT SCI] The symbolic address used prior to its conversion to a machine address. { ¦flōd·iŋ ə'dres }

floating aid [NAV] A buoy serving as an aid to navigation, secured in its charted position by a mooring. { ¦flōd·iŋ 'ād }

floating arithmetic See floating-point arithmetic. { ¦flōd·iŋ ə'rith·mə·tik }

floating axle [MECH ENG] A live axle used to turn the wheels of an automotive vehicle; the weight of the vehicle is borne by housings at the ends of a fixed axle. { ¦flōd·iŋ 'ak·səl }

floating battery [ELEC] A storage battery connected permanently in parallel with another power source; the battery normally handles only small charging or discharging currents, but takes over the entire load upon failure of the main supply. { ¦flōd·iŋ 'bad·ə·rē }

floating block See traveling block. { ¦flōd·iŋ 'bläk }

floating carrier modulation See controlled carrier modulation. { ¦flōd·iŋ ₁kar·ē·ər ₁mäj·ə'lā·shən }

floating charge [ELEC] Application of a constant voltage to a storage battery, sufficient to maintain an approximately constant state of charge while the battery is idle or on light duty. { ¦flōd·iŋ 'chärj }

floating chase [ENG] A mold part that can move freely in a vertical plane, which fits over a lower member (such as a cavity or plug) and into which an upper plug can telescope. { ¦flōd·iŋ 'chās }

floating control [ENG] Control device in which the speed of correction of the control element (such as a piston in a hydraulic relay) is proportional to the error signal. Also known as proportional-speed control. { ¦flōd·iŋ kən'trōl }

floating crane [CIV ENG] A crane having a barge or scow for an undercarriage and moved by cables attached to anchors

set some distance off the corners of the barge; used for water work and for work on waterfronts. { ¦flōd·iŋ 'krän }

floating-decimal arithmetic See floating-point arithmetic. { ¦flōd·iŋ ₁des·məl ə'rith·mə·tik }

floating dock [CIV ENG] **1.** A form of dry dock for repairing ships; it can be partly submerged by controlled flooding to receive a vessel, then raised by pumping out the water so that the vessel's bottom can be exposed. Also known as floating dry dock. **2.** A barge or flatboat which is used as a wharf. { ¦flōd·iŋ 'däk }

floating dollar sign [COMPUT SCI] A dollar sign used with an edit mask, allowing the sign to be inserted before the nonzero leading digit of a dollar amount. { ¦flōd·iŋ 'däl·ər ₁sīn }

floating dry dock See floating dock. { ¦flōd·iŋ 'drī ₁däk }

floating floor [BUILD] A floor constructed so that the wearing surface is separated from the supporting structure by an insulating layer of mineral wool, resilient quilt, or other material to provide insulation against impact sound. { ¦flōd·iŋ 'flor }

floating foundation [CIV ENG] **1.** A reinforced concrete slab that distributes the concentrated load from columns; used on soft soil. **2.** A foundation mat several meters below the ground surface when it is combined with external walls. { ¦flōd·iŋ faùn'dā·shən }

floating-gate avalanche-injection metal-oxide semiconductor device [ELECTR] An erasable programmable read-only memory chip that holds its contents until they are erased by ultraviolet light. Abbreviated FAMOS device. { ¦flōd·iŋ ¦gāt ¦av·ə₁lanch in¦jek·shən ¦med·əl ¦äk₁sīd ₁sem·i·kən'dək·tər di₁vīs }

floating graphic [COMPUT SCI] A picture or graph that moves up or down on a page of a document as text is deleted or inserted above it. { ¦flōd·iŋ 'graf·ik }

floating grid [ELECTR] Vacuum-tube grid that is not connected to any circuit; it assumes a negative potential with respect to the cathode. Also known as free grid. { ¦flōd·iŋ 'grid }

floating ice [OCEANOGR] Any form of ice floating in water, including grounded ice and drifting land ice. { ¦flōd·iŋ 'īs }

floating input [ELEC] Isolated input circuit not connected to ground at any point. { ¦flōd·iŋ 'in₁pùt }

floating instrument platform [NAV ARCH] A crewed, spar-buoy-type oceanographic platform with marine instruments mounted on the submerged portions. Abbreviated FLIP. { ¦flōd·iŋ 'in·strə·mənt ₁plat₁form }

floating lever [MECH ENG] A horizontal brake lever with a movable fulcrum; used under railroad cars. { ¦flōd·iŋ 'lēv·ər }

floating neutral [ELEC] Neutral conductor whose voltage to ground is free to vary when circuit conditions change. { ¦flōd·iŋ 'nü·trəl }

floating pan [ENG] An evaporation pan in which the evaporation is measured from water in a pan floating in a larger body of water. { ¦flōd·iŋ 'pan }

floating platen [ENG] In a multidaylight press, a platen that is between the main head and the press table and can be moved independently of them. { ¦flōd·iŋ 'plat·ən }

floating plug [MET] A plug or mandrel attached to a rod and used in plug drawing. Also known as a plug die. { ¦flōd·iŋ 'pləg }

floating-point arithmetic [MATH] A method of performing arithmetical operations, used especially by automatic computers, in which numbers are expressed as integers multiplied by the radix raised to an integral power, as 87×10^{-4} instead of 0.0087. Also known as floating arithmetic; floating-decimal arithmetic. { ¦flōd·iŋ ¦point ə'rith·mə·tik }

floating-point calculation [COMPUT SCI] A calculation made with floating-point arithmetic. { ¦flōd·iŋ ¦point ₁kal·kyə'lā·shən }

floating-point coefficient See mantissa. { ¦flōd·iŋ ¦point ₁kō·i'fish·ənt }

floating-point package [COMPUT SCI] A program which enables a computer to perform arithmetic operations when such capabilities are not wired into the computer. Also known as floating-point routine. { ¦flōd·iŋ ¦point 'pak·ij }

floating-point processor [COMPUT SCI] A separate processor or a special section of a computer's main storage that is for the efficient handling of floating-point operations. { 'flōd·iŋ ¦point 'präs₁es·ər }

floating-point routine See floating-point package. { ¦flōd·iŋ ¦point rü'tēn }

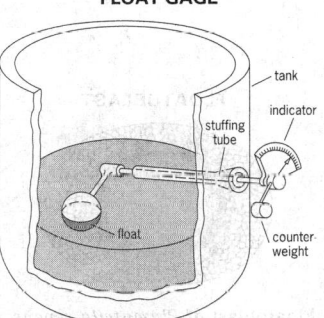

FLOAT GAGE

Internal float-and-lever mechanism.

floating-point system [COMPUT SCI] A number system in which the location of the point does not remain fixed with respect to one end of the numerals. { ¦flōd·iŋ ¦point 'sis·təm }

floating reticle [OPTICS] A reticle the image of which is movable within the field of view. { ¦flōd·iŋ 'red·ə·kəl }

floating rib [ANAT] One of the last two ribs in humans which have the anterior end free. { ¦flōd·iŋ 'rib }

floating roof [ENG] A type of tank roof (steel, plastic, sheet, or microballoons) which floats upon the surface of the stored liquid; used to decrease the vapor space and reduce the potential for evaporation. { ¦flōd·iŋ 'rüf }

floating sand [PETR] A single grain of quartz sand that does not appear to touch surrounding sand grains scattered throughout the finer-grained matrix of a sedimentary rock. { ¦flōd·iŋ 'sand }

floating scraper [MECH ENG] A balanced scraper blade that rests lightly on a drum filter; removes solids collected on the rotating drum surface by riding on the drum's surface contour. { ¦flōd·iŋ 'skrā·pər }

floating signal See differential signal. { ¦flōd·iŋ 'sig·nəl }

floating zone refining [MET] A variation of the zone-refining technique in which the molten zone is held in place by its own surface tension between two collinear rods; since no container is needed, contamination of the pure metal is avoided. { ¦flōd·iŋ 'zōn ri¸fīn·iŋ }

floatless level control [ENG] Any nonfloat device for measurement and control of liquid levels in storage tanks or process vessels; includes use of manometers, capacitances, electroprobes, nuclear radiation, and sonics. { ¦flōt·ləs 'lev·əl kən¸trōl }

float level [MECH ENG] The position of the float in a carburetor at which the needle valve closes the fuel inlet to prevent entry of additional fuel. { 'flōt ¸lev·əl }

float mineral [GEOL] Small ore fragments carried from the ore bed by the action of water or by gravity; a float mineral often leads to discovery of mines. { 'flōt ¸min·rəl }

floatoblast [INV ZOO] A free-floating statoblast having a float of air cells. { 'flōd·ə¸blast }

float switch [ENG] A switch actuated by a float at the surface of a liquid. { 'flōt ¸swich }

float-type rain gage [ENG] A class of rain gage in which the level of the collected rainwater is measured by the position of a float resting on the surface of the water; frequently used as a recording rain gage by connecting the float through a linkage to a pen which records on a clock-driven chart. { 'flōt ¸tīp 'rān ¸gāj }

float valve [ENG] A valve whose on-off action is controlled directly by the fall or rise of a float concurrent with the fall or rise of liquid level in a liquid-containing vessel. { 'flōt ¸valv }

floc [CHEM] Small masses formed in a fluid through coagulation, agglomeration, or biochemical reaction of fine suspended particles. { fläk }

floccose [BOT] Covered with tufts of woollike hairs. { 'flä¸kōs }

flocculant See flocculating agent. { 'fläk·yə·lənt }

flocculate [BIOL] Having small tufts of hairs. [CHEM] To cause to aggregate or coalesce into a flocculent mass. { 'fläk·yə¸lāt (adjective) or 'fläk·yə¸lāt (verb) }

flocculating agent [CHEM] A reagent added to a dispersion of solids in a liquid to bring together the fine particles to form flocs. Also known as flocculant. { 'fläk·yə¸lād·iŋ ¸ā·jənt }

flocculent [CHEM] Pertaining to a material that is cloudlike and noncrystalline. { 'fläk·yə·lənt }

flocculonodular lobes [ANAT] The pair of lateral cerebellar lobes in vertebrates which function to regulate vestibular reflexes underlying posture; referred to functionally as the vestibulocerebellum. { ¦fläk·yə·lō¦näj·ə·lər 'lōbz }

flocculus [ANAT] A prominent lobe of the cerebellum situated behind and below the middle cerebellar peduncle on each side of the median fissure. [ASTRON] A patch in the sun's surface seen in the light of calcium or hydrogen; the patch may be bright or dark and is usually in the vicinity of sunspots. { 'fläk·yə·ləs }

floccus [BOT] A tuft of woolly hairs. [METEOROL] A cloud species in which each element is a small tuft with a rounded top and a ragged bottom. { 'fläk·əs }

flock [TEXT] **1.** Pulverized wool, cotton, silk, or rayon fiber used to form velvety patterns on cloth. **2.** Woolen or cotton

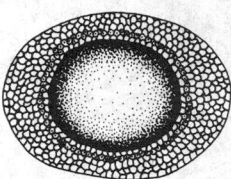

FLOATOBLAST

Floatoblast of *Plumatella repens*, in Phylactolaemata, a class of fresh-water bryozoans.

refuse reduced by machinery and used to stuff furniture. { fläk }

flocking [TEXT] A design or a surface finish made by spraying short fibers so that they adhere electrostatically to a material. { 'fläk·iŋ }

floc point [ANALY CHEM] The temperature at which wax or solids separate from kerosine and other illuminating oils as a definite floc. { 'fläk ¸point }

floc test [ANALY CHEM] A quantitative test applied to kerosine and other illuminating oils to detect substances rendered insoluble by heat. { 'fläk ¸test }

floe [OCEANOGR] A piece of floating sea ice other than fast ice or glacier ice; may consist of a single fragment or of many consolidated fragments, but is larger than an ice cake and smaller than an ice field. Also known as ice floe. { flō }

floeberg [OCEANOGR] A mass of hummocked ice formed by the piling up of many ice floes by lateral pressure; an extreme form of pressure ice; may be more than 50 feet (15 meters) high and resemble an iceberg. { 'flō¸bərg }

floe till [GEOL] **1.** A glacial till resulting from the intact deposition of a grounded iceberg in a lake bordering an ice sheet. **2.** A lacustrine clay with boulders, stones, and other glacial matter dropped into it by melting icebergs. Also known as berg till. { 'flō ¸til }

flokite See mordenite. { 'flō¸kīt }

flong [GRAPHICS] In stereotyping, a matrix, made from several paper sheets which have been moistened and pasted together, that receives the molten metal and becomes a duplicate of the type page after cooling. { fläŋ }

flood [ELECTR] To direct a large-area flow of electrons toward a storage assembly in a charge storage tube. [ENG] To cover or fill with fluid. [HYD] The condition that occurs when water overflows the natural or artificial confines of a stream or other body of water, or accumulates by drainage over low-lying areas. [MECH ENG] To supply an excess of fuel to a carburetor so that the level rises above the nozzle. [OCEANOGR] The highest point of a tide. { fləd }

floodable length [NAV ARCH] At any point in the length of a ship, the greatest part of the length centered at that point which can be flooded at the prescribed permeability without submerging the margin line. { ¦fləd·ə·bəl ¦leŋkth }

flood basalt See plateau basalt. { 'fləd bə¸sȯlt }

flood basin [GEOL] **1.** The tract of land actually submerged during the highest known flood in a specific region. **2.** The flat, wide area lying between a low, sloping plain and the natural levee of a river. { 'fləd ¸bās·ən }

flood control [CIV ENG] Use of levees, walls, reservoirs, floodways, and other means to protect land from water overflow. { 'fləd kən¸trōl }

flood coverage [PETRO ENG] The extent of subterranean coverage within an oil reservoir by the injection of pressure-maintenance (or water-drive) water. { 'fləd ¸kəv·rij }

flood current [OCEANOGR] The tidal current associated with the increase in the height of a tide. { 'fləd ¸kə·rənt }

flood dam [CIV ENG] A dam for storing floodwater, or for supplying a flood of water. { 'fləd ¸dam }

flooded stream See drowned stream. { ¦fləd·əd 'strēm }

flooded system [ENG] A system filled with so much tracer gas that probe testing for leaks suffers from a loss of sensitivity. { ¦fləd·əd 'sis·təm }

flood flow [HYD] Stream discharge during a flood. { 'fləd ¸flō }

flood fringe See pondage land. { 'fləd ¸frinj }

floodgate [CIV ENG] **1.** A gate used to restrain a flow or, when opened, to allow a flood flow to pass. **2.** The lower gate of a lock. { 'fləd¸gāt }

flood icing See icing. { 'fləd ¸īs·iŋ }

flooding [AGR] Filling of ditches or covering of land with water during the raising of crops; rice, for example, must have occasional flooding to grow properly. [CHEM ENG] Condition in a liquid-vapor counterflow device (such as a distillation column) in which the rate of vapor rise is such as to prevent liquid downflow, causing a buildup of the liquid (flooding) within the device. [PETRO ENG] Technique of increasing recovery of oil (secondary recovery) from a reservoir by injection of water into the formation to drive the oil toward producing wellholes. Also known as waterflooding. [PSYCH] A behavior therapy for phobias and other problems involving maladaptive anxiety, in which anxiety producers are presented in intense form (real or imagined) and continued until the stimuli no longer produce disabling anxiety. { 'fləd·iŋ }

flooding ice *See* icing. { 'fləd·iŋ ‚īs }

floodlight [ELEC] A light projector used for outdoor lighting of buildings, parking lots, sports fields, and the like, usually having a filament lamp or mercury-vapor lamp and a parabolic reflector. { 'fləd‚līt }

flood-out pattern [PETRO ENG] Pattern of subterranean water penetration and spread in an oil reservoir as a result of water injection. { 'fləd ‚aut ‚pad·ərn }

floodplain [GEOL] The relatively smooth valley floors adjacent to and formed by alluviating rivers which are subject to overflow. { 'fləd‚plān }

floodplain splay [GEOL] A small alluvial fan or other outspread deposit formed where an overloaded stream breaks through a levee (artificial or natural) and deposits its material (often coarse-grained) on the floodplain. Also known as channel splay. { 'fləd‚plān ‚splā }

flood plane [HYD] The position of a stream's water surface during a particular flood. { 'fləd ‚plān }

flood pot test [PETRO ENG] Laboratory simulation of an oil reservoir to appraise the residual reservoir saturation after waterflooding. { 'fləd ‚pät ‚test }

flood projection [COMMUN] In facsimile transmission, optical method in which the subject facsimile copy is illuminated and the scanning spot is delineated by an aperture between the subject copy and the light-sensitive device. { 'fləd prə‚jek·shən }

flood relief channel *See* bypass channel. { 'fləd ri‚lēf ‚chan·əl }

flood routing [HYD] The process of computing the progressive time and shape of a flood wave at successive points along a river. Also known as storage routing; streamflow routing. { 'fləd ‚rüd·iŋ }

Flood's equation [PHYS CHEM] A relation used to determine the liquidus temperature in a binary fused salt system. { 'flədz i‚kwā·zhən }

flood stage [HYD] The stage, on a fixed river gage, at which overflow of the natural banks of the stream begins to cause damage in any portion of the reach for which the gage is used as an index. { 'fləd ‚stāj }

flood tide [OCEANOGR] **1.** That period of tide between low water and the next high water. **2.** A tide at its highest point. { 'fləd ‚tīd }

flood tuff *See* ignimbrite. { 'fləd ‚təf }

flood wall [CIV ENG] A levee or similar wall for the purpose of protecting the land from inundation by flood waters. { 'fləd ‚wól }

floodway *See* bypass channel. { 'fləd‚wā }

floor [ENG] The bottom, horizontal surface of an enclosed space. [GEOL] **1.** The rock underlying a stratified or nearly horizontal deposit, corresponding to the footwall of more steeply dipping deposits. **2.** A horizontal, flat ore body. [MIN ENG] Boards laid at the heading to receive blasted rocks and to facilitate ore loading. [NAV ARCH] One of a series of vertical plates extending across the bottom of a ship at right angles to the center line and forming part of the bottom framing of the hull. { 'flór }

floor beam [BUILD] A beam used in the framing of floors in buildings. [CIV ENG] A large beam used in a bridge floor at right angles to the direction of the roadway, to transfer loads to bridge supports. { 'flór ‚bēm }

floorboard [MIN ENG] A thick wooden plank constituting part of a drill platform or other work platform. { 'flór‚bórd }

floor burst [MIN ENG] A type of outburst in longwall faces which is preceded by heavy weighting due to floor lift; gas evolved below the seam collects beneath an impervious layer of rock, and a gas blister forms beneath the face, giving the observed floor lift; later, the floor fractures and the firedamp escapes into the mine atmosphere. { 'flór ‚bərst }

floor collar [ENG] A relatively narrow upright structural part fitted around the periphery of a hole where a pipe passes through to prevent drainage water from entering the hole. { 'flór ‚käl·ər }

floor cut [MIN ENG] A machine-made cut in the floor dirt just below the coal seam. { 'flór ‚kət }

floor drain [CIV ENG] A pipe or channel to remove water from under a floor in contact with soil. { 'flór ‚drān }

floor framing [BUILD] Floor joists together with their strutting and supports. { 'flór ‚frām·iŋ }

flooring [MATER] Material suitable for use as a floor. { 'flór·iŋ }

flooring saw [DES ENG] A pointed saw with teeth on both edges; cuts its own entrance into a material. { 'flór·iŋ ‚só }

floor light [BUILD] A window set in a floor that is adapted for walking on and admitting light to areas below. { 'flór ‚līt }

floor outlet [ELEC] An electrical outlet whose face is level with or recessed into a floor. Also known as floor plug. { 'flór ‚aut·lət }

floor plan [ARCH] A diagram of a floor showing partitions, doors, windows, and other features. { 'flór ‚plan }

floor plate [BUILD] A flat board on a floor used to support wall studs. [ENG] A plate in a floor to which heavy work or machine tools can be bolted. { 'flór ‚plāt }

floor plug *See* floor outlet. { 'flór ‚pləg }

floor sill [MIN ENG] A large timber laid flat on the ground or in a level, shallow ditch, to which are fastened the drill-platform boards or planking, or which is used as the base for a full timber set. { 'flór ‚sil }

floor system [CIV ENG] The structural floor assembly between supporting beams or girders in buildings and bridges. { 'flór ‚sis·təm }

flop gate [MIN ENG] An automatic gate used in placer mining when there is a shortage of water; the gate closes a reservoir until it is filled with water, then automatically opens and allows the water to flow into the sluices; when the reservoir is empty, the gate closes, and the operation is repeated. { 'fläp ‚gāt }

flopover [ELECTR] A defect in television reception in which a series of frames move vertically up or down the screen, caused by lack of synchronization between the vertical and horizontal sweep frequencies. { 'fläp‚ō·vər }

floppy disk [COMPUT SCI] A flexible plastic disk coated with magnetic oxide and used for data entry to a computer; a slot in its protective envelope or housing, which remains stationary while the disk rotates, exposes the track positions for the magnetic read/write head of the drive unit. Also known as diskette. { 'fläp·ē 'disk }

flops [COMPUT SCI] A unit of computer speed, equal to one floating-point arithmetic operation per second. { fläps }

Floquet theorem [MATH] A second-order linear differential equation whose coefficients are periodic single-valued functions of an independent variable x has a solution of the form $e^{\mu x}P(x)$ where μ is a constant and $P(x)$ a periodic function. { flō'kā ‚thir·əm }

flora [BOT] **1.** Plants. **2.** The plant life characterizing a specific geographic region or environment. { 'flór·ə }

floral axis [BOT] A flower stalk. { 'flór·əl ‚ak·səs }

floral diagram [BOT] A diagram of a flower in cross section showing the number and arrangement of floral parts. { 'flór·əl 'dī·ə‚gram }

Florence oil *See* olive oil. { 'flär·əns ‚óil }

florencite [MINERAL] $CeAl_3(PO_4)_2(OH)_6$ Pale-yellow mineral composed of basic phosphate of cerium and aluminum. { 'flär·ən‚sīt }

florentium *See* promethium-147. { flō'ren·chəm }

flores [CHEM] A form of a chemical compound made by the process of sublimation. { 'flór·ēz }

flores martis *See* ferric chloride. { 'flór·ēz 'märd·əs }

floret [BOT] A small individual flower that is part of a compact group of flowers, such as the head of a composite plant or inflorescence. { 'flór·ət }

Florey unit [BIOL] A unit for the standardization of penicillin. { 'flór·ē ‚yü·nət }

floricome [INV ZOO] A type of branched hexaster spicule. { 'flór·ə‚kōm }

floriculture [AGR] A segment of horticulture concerned with commercial production, marketing, and retail sale of cut flowers and potted plants, as well as home gardening and flower arrangement. { 'flór·ə‚kəl·chər }

Florida Current [OCEANOGR] A fast current that sets through the Straits of Florida to a point north of Grand Bahama Island, where it joins the Antilles Current to form the Gulf Stream. { 'flär·ə·də ‚kə·rənt }

Florideophyceae [BOT] A class of red algae, division Rhodophyta, having prominent pit connections between cells. { flə‚rid·ē·ō'fīs·ē‚ē }

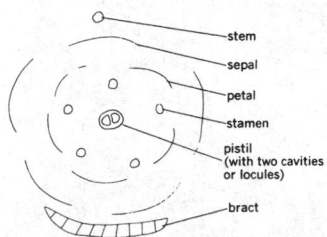

FLORAL DIAGRAM

stem
sepal
petal
stamen
pistil (with two cavities or locules)
bract

Graphic diagram of a cross section of a flower.

floriferous [BOT] Blooming freely, used principally of ornamental plants. { flȯ'rif·ə·rəs }

florigen [BIOCHEM] A plant hormone that stimulates buds to flower. { 'flȯr·ə·jen }

florivorous [ZOO] Feeding on flowers. { flȯ'riv·ə·rəs }

flor process [FOOD ENG] A technique used to make Spanish sherries involving the participation of a film-forming yeast growing on the surface of the wine in partially filled oak barrels; the yeast imparts the characteristic sherry flavor. { 'flȯr ‚präs·əs }

florula [ECOL] Plants which grow in a small, confined habitat, for example, a pond. { 'flȯr·yə·lə }

floscelle [INV ZOO] A flowerlike structure around the mouth of some echinoids. { flȯ'sel }

Flosculariacea [INV ZOO] A suborder of rotifers in the order Monogononta having a malleoramate mastax. { ‚fläs·kyə‚lar·ē'ās·ē·ə }

Flosculariidae [INV ZOO] A family of sessile rotifers in the suborder Flosculariacea. { ‚fläs·kyə·lə'rī·ə‚dē }

flosculous [BOT] **1.** Composed of florets. **2.** Of a floret, tubular in form. { 'fläs·kyə·ləs }

flosculus [BOT] A floret. { 'fläs·kyə·ləs }

flospinning [MET] Power-spinning or flowing metal over a rotating bar for shaping into cylindrical, conical, and curvilinear parts. { 'flō‚spin·iŋ }

floss [MET] Molten or solid slag floating on the surface of a metal melt. { fläs }

floss hole [MET] A small door or opening of the bottom of a smokestack or flue for removal of ash. { 'fläs ‚hōl }

flotation [ENG] A process used to separate particulate solids by causing one group of particles to float; utilizes differences in surface chemical properties of the particles, some of which are entirely wetted by water, others are not; the process is primarily applied to treatment of minerals but can be applied to chemical and biological materials; in mining engineering it is referred to as froth flotation. { flō'tā·shən }

flotation agent [CHEM] A chemical which alters the surface tension of water or which makes it froth easily. { flō'tā·shən ‚ā·jənt }

flotation analysis [PHYS] Technique to measure liquid density in which a float of known density is adjusted with weights to match that of the liquid. { flō'tā·shən ə‚nal·ə·səs }

flotation cell [MIN ENG] The device in which froth flotation of ores is performed. [PETRO ENG] A large tank for separating oil from contaminated water by rapidly bubbling gas through the water to scavenge the oil droplets. { flō'tā·shən ‚sel }

flotation collar [ENG] A buoyant bag carried by a spacecraft and designed so that it inflates and surrounds part of the outer surface if the spacecraft lands in the sea. { flō'tā·shən ‚käl·ər }

flotsam [ENG] Floating articles, particularly those that are thrown overboard to lighten a vessel in distress. { 'flät·səm }

flounder [VERT ZOO] Any of a number of flatfishes in the families Pleuronectidae and Bothidae of the order Pleuronectiformes. { 'flaün·dər }

flour [FOOD ENG] A powdery meal obtained by milling wheat and other cereal grains or dry food products such as potato or banana. { flaü·ər }

flour gold [MET] The finest-size gold dust, much of which will float on water. { 'flaü·ər ‚gōld }

flour mill [FOOD ENG] A machine or factory that processes cereal grains such as wheat and rye into flour. { 'flaü·ər ‚mil }

flow [COMPUT SCI] The sequence in which events take place or operations are carried out. [ENG] A forward movement in a continuous stream or sequence of fluids or discrete objects or materials, as in a continuous chemical process or solids-conveying or production-line operations. [FL MECH] The forward continuous movement of a fluid, such as gases, vapors, or liquids, through closed or open channels or conduits. [GEOL] Any rock deformation that is not instantly recoverable without permanent loss of cohesion. Also known as flowage; rock flowage. [MATH] A function from the set of arcs in an *s-t* network to the nonnegative integers whose value at each arc is equal to or less than the weight of the arc. [PHYS] The movement of electric charges, gases, liquids, or other materials or quantities. { flō }

flowability [FL MECH] Capability of a liquid or loose particulate solid to move by flow. { ‚flō·ə'bil·əd·ē }

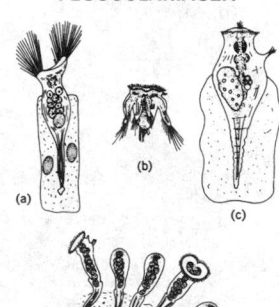

FLOSCULARIACEA

Sessile rotifers of the Flosculariacea. *(a) Floscularia mutabilis. (b) Pedalia mira. (c) Conchiloides species. (d) Lacinularia socialis.*

flowage [GEOL] *See* flow. [HYD] Flooding of water onto adjacent land. { 'flō·ij }

flowage line [GEOL] A contour line at the edge of a body of water, such as a reservoir, representing a given water level. { 'flō·ij ‚līn }

flow analysis [IND ENG] A detailed study of all aspects of the progressive travel by personnel or material from place to place during a particular operation or from one operation to another. { 'flō ə‚nal·ə·səs }

flow banding [GEOL] An igneous rock structure resulting from flowing of magmas or lavas and characterized by alternation of mineralogically unlike layers. { 'flō ‚band·iŋ }

flow bean [PETRO ENG] A plug containing a small hole placed in the flow line at the well head which serves to maintain oil flow at a proper rate. { 'flō ‚bēn }

flow birefringence [PHYS CHEM] Orientation of long, thin asymmetric molecules in the direction of flow of a solution forced to flow through a capillary tube. { 'flō ‚bī·rə'frin·jəns }

flow bog [ECOL] A peat bog with a surface level that fluctuates in accordance with rain and tides. { 'flō ‚bäg }

flow brazing [MET] A brazing process in which coalescence is produced by the heat of molten filler metal that is poured over a joint. { 'flō ‚brāz·iŋ }

flow breccia [GEOL] A breccia formed with the movement of lava flow while the flow is still in motion. { 'flō ‚brech·ə }

flow brightening [MET] In a soldering process, the melting of a chemical or mechanical metallic coating on the base metal to be soldered. { 'flō ‚brīt·ən·iŋ }

flow brush [ENG] A hollow tool for the continuous application of a broad coat of an adhesive. { 'flō ‚brəsh }

flow cast [PETR] One of a group of bedding plane structures formed in graywacke. { 'flō ‚kast }

flow chart [ENG] A graphical representation of the progress of a system for the definition, analysis, or solution of a data-processing or manufacturing problem in which symbols are used to represent operations, data or material flow, and equipment, and lines and arrows represent interrelationships among the components. Also known as control diagram; flow diagram; flow sheet. { 'flō ‚chärt }

flow-chart symbol [ENG] Any of the existing symbols normally used to represent operations, data or materials flow, or equipment in a data-processing problem or manufacturing-process description. { 'flō ‚chärt ‚sim·bəl }

flow cleavage [GEOL] Rock cleavage in which solid flow of rock accompanies recrystallization. Also known as slaty cleavage. { 'flō ‚klē·vij }

flow coat [ENG] A coating formed by pouring a liquid material over the object and allowing it to flow over the surface and drain off. { 'flō ‚kōt }

flow coefficient [FL MECH] An experimentally determined proportionality constant, relating the actual velocity of fluid flow in a pipe, duct, or open channel to the theoretical velocity expected under certain assumptions. [MECH ENG] A dimensionless number used in studying the power required by fans, equal to the volumetric flow rate through the fan divided by the product of the rate of rotation of the fan and the cube of the impeller diameter. { ‚flō ‚kō·i'fish·ənt }

flow cone [FL MECH] One of a collection of elements of conical shape that may be attached to an aerodynamic surface to visualize the fluid flow over the surface. { 'flō ‚kōn }

flow control [ENG] Any system used to control the flow of gases, vapors, liquids, slurries, pastes, or solid particles through or along conduits or channels. { 'flō kən‚trōl }

flow control valve [ENG] A valve whose flow opening is controlled by the rate of flow of the fluid through it; usually controlled by differential pressure across an orifice at the valve. Also known as rate-of-flow control valve. { 'flō kən‚trōl ‚valv }

flow counter *See* gas-flow counter tube. { 'flō ‚kaünt·ər }

flow curve [FL MECH] A graph of the total shear of a fluid as a function of time. [MECH] The stress-strain curve of a plastic material. { 'flō ‚kərv }

flow cytometry [CYTOL] A technique for optical analysis and separation of cells and metaphase chromosomes based on light scattering and fluorescence. { 'flō sī‚täm·ə·trē }

flow diagram *See* flow chart. { 'flō ‚dī·ə‚gram }

flow direction [ENG] The antecedent-to-successor relation, indicated by arrows or other conventions, between operations on a flow chart. { 'flō də‚rek·shən }

flow distribution *See* flow field. { 'flō ‚dis·trə‚byü·shən }

flow earth *See* solifluction mantle. { 'flō ‚ərth }

flow equation [FL MECH] Equation for the calculation of fluid (gas, vapor, liquid) flow through conduits or channels; consists of an interrelation of fluid properties (such as density or viscosity), environmental conditions (such as temperature or pressure), and conduit or channel geometry and conditions (such as diameter, cross-sectional shape, or surface roughness). { 'flō i‚kwā·zhən }

flower [BOT] The characteristic reproductive structure of a seed plant, particularly if some or all of the parts are brightly colored. { 'flaů·ər }

flowers of sulfur [PHARM] One of three forms of pharmaceutical sulfur, made by sublimation; the other two forms are precipitated sulfur and washed sulfur. Also known as sublimed sulfur. { 'flaů·ərz əv 'səl·fər }

flowers of tin *See* stannic oxide. { 'flaů·ərz əv 'tin }

flow field [FL MECH] The velocity and the density of a fluid as functions of position and time. Also known as flow distribution. { 'flō ‚fēld }

flow figure *See* strain figure. { 'flō ‚fig·yər }

flow fold [GEOL] Folding in beds, composed of relatively plastic rock, that assume any shape impressed upon them by the more rigid surrounding rocks or by the general stress pattern of the deformed zone; there are no apparent surfaces of slip. { 'flō ‚fōld }

flow graph [COMPUT SCI] A directed graph that represents a computer program, wherein a node in the graph corresponds to a block of sequential code and branches correspond to decisions taken in the program. [SYS ENG] *See* signal-low graph. { 'flō ‚graf }

flow-induced vibration [FL MECH] Structural and mechanical oscillations of structures immersed in or conveying fluid flow as a result of an interaction between the fluid-dynamic forces and the inertia, damping, and elastic forces in the structures. { ¦flō in‚düst vī'brā·shən }

flowing-film concentration [MIN ENG] A concentration based on the fact that liquid films in laminar flow possess a velocity which is not the same in all depths of the film; by this principle lighter particles of ore may be washed off while the heavier particles accumulate and are intermittently removed. { ¦flō·iŋ ¦film käns·ən'trā·shən }

flowing furnace [MET] A furnace from which molten metal can be tapped or drawn. { ¦flō·iŋ ¦fər·nəs }

flowing pressure [PETRO ENG] Pressure at the bottom of an oil-well bore (bottom-hole pressure) during normal oil production. { ¦flō·iŋ ‚presh·ər }

flowing-pressure gradient [PETRO ENG] The slope of decreasing pressure plotted against distance measured for upward liquid flow in a continuous-flow gas-lift oil well. { ¦flō·iŋ ‚presh·ər ‚grād·ē·ənt }

flowing-temperature factor [THERMO] Calculation correction factor for gases flowing at temperatures other than that for which a flow equation is valid, that is, other than 60°F (15.5°C). { ¦flō·iŋ ¦tem·prə·chər ‚fak·tər }

flowing well [PETRO ENG] Oil reservoir in which gas-drive pressure is sufficient to force oil flow up through and out of a wellhole. { ¦flō·iŋ ¦wel }

flow karyotype [CYTOL] A karyotype that is based on flow cytometry measurements. { ¦flō 'kar·ē·ə‚tīp }

flow layer [PETR] In an igneous rock, a layer which is different in composition or texture from adjacent layers. { 'flō ‚lā·ər }

flow line [ENG] **1.** The connecting line or arrow between symbols on a flow chart or block diagram. **2.** Mark on a molded plastic or metal article made by the meeting of two input-flow fronts during molding. Also known as weld line; weld mark. [HYD] A contour of the water level around a body of water. [PETR] In an igneous rock, any internal structure produced by parallel orientation of crystals, mineral streaks, or inclusions. [PETRO ENG] A pipeline that takes oil from a single well or a series of wells to a gathering center. { 'flō ‚līn }

flow marks [MATER] Wavy surface marks on a thermoplastic resin molding due to improper flow of material into the mold. { 'flō ‚märks }

flow measurement [ENG] The determination of the quantity of a fluid, either a liquid, a vapor, or a gas, that passes through a pipe, duct, or open channel. { 'flō ‚mezh·ər·mənt }

flowmeter [ENG] An instrument used to measure pressure, flow rate, and discharge rate of a liquid, vapor, or gas flowing in a pipe. Also known as fluid meter. { 'flō‚mēd·ər }

flow mixer [MECH ENG] Liquid-liquid mixing device in which the mixing action occurs as the liquids pass through it; includes jet nozzles and agitator vanes. Also known as line mixer. { 'flō ‚mik·sər }

flow net [FL MECH] A diagram used in studying the flow of a fluid through a permeable substance (such as water through a soil structure) having two nests of curves, one representing the flow lines, which follow the path of the fluid, and the other the equipotential lines, which connect points of equal head. { 'flō ‚net }

flow noise [ACOUS] **1.** Pressure variations associated with a turbulent flow field that do not propagate away from the turbulent source but are sensed as sound by a receiver in direct contact or close to the turbulent flow. Also known as near-field noise. **2.** More generally, any noise generated by turbulent fluid flow. { 'flō ‚nóiz }

flow nozzle [ENG] A flowmeter in a closed conduit, consisting of a short flared nozzle of reduced diameter inset into the inner diameter of a pipe; used to cause a temporary pressure drop in flowing fluid to determine flow rate via measurement of static pressures before and after the nozzle. { 'flō ‚näz·əl }

flow of variability [GEN] The movement of genetic variability within a population as a result of hybridization and segregation. { ¦flō əv ‚ver·ē·ə'bil·əd·ē }

flow pattern [FL MECH] Pattern of two-phase flow in a conduit or channel pipe, taking into consideration the ratio of gas to liquid and conditions of flow resistance and liquid holdup. { 'flō ‚pad·ərn }

flow process [ENG] System in which fluids or solids are handled in continuous movement during chemical or physical processing or manufacturing. { 'flō ‚präs·əs }

flow-programmed chromatography [ANALY CHEM] A chromatographic procedure in which the rate of flow of the mobile phase is periodically changed. { ¦flō ‚prō‚gramd ‚krō·mə'täg·rə·fē }

flow rate [FL MECH] Also known as rate of flow. **1.** Time required for a given quantity of flowable material to flow a measured distance. **2.** Weight or volume of flowable material flowing per unit time. { 'flō ‚rāt }

flow-rating pressure [MECH ENG] The value of inlet static pressure at which the relieving capacity of a pressure-relief device is established. { 'flō ‚rad·iŋ ‚presh·ər }

flow reactor [CHEM ENG] A dynamic reactor system in which reactants flow continuously into the vessel and products are continuously removed, in contrast to a batch reactor. { 'flō rē‚ak·tər }

flow regime [HYD] A range of streamflows having similar bed forms, flow resistance, and means of transporting sediment. { 'flō rə‚zhēm }

flow resistance [FL MECH] **1.** Any factor within a conduit or channel that impedes the flow of fluid, such as surface roughness or sudden bends, contractions, or expansions. **2.** *See* viscosity. { 'flō ri‚zis·təns }

flow rock [PETR] An igneous rock that had been liquid. { 'flō ‚räk }

flow separation *See* boundary-layer separation. { 'flō ‚sep·ə‚rā·shən }

flow sheet *See* flow chart. { 'flō ‚shēt }

flow shop [IND ENG] A manufacturing facility in which machine tools and robots are employed in the same manner on all jobs. { 'flō ‚shäp }

flow slide [GEOL] A slide of waterlogged material in which the slip surface is not well defined. { 'flō ‚slīd }

flow soldering [ENG] Soldering of printed circuit boards by moving them over a flowing wave of molten solder in a solder bath; the process permits precise control of the depth of immersion in the molten solder and minimizes heating of the board. Also known as wave soldering. { 'flō ‚säd·ə·riŋ }

flowstone [GEOL] Deposits of calcium carbonate that accumulated against the walls of a cave where water flowed on the rock. { 'flō‚stōn }

flow stress [MECH] The stress along one axis at a given value of strain that is required to produce plastic deformation. { 'flō ‚stres }

flow string [PETRO ENG] Total length of oil- or gas-well

Flutter of a polyethylene tube (garden hose) conveying fluid, an example of flow-induced vibration. The tube is illuminated by a stroboscope to show its motion.

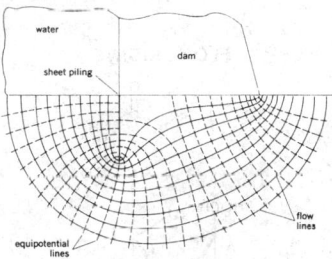

Flow net indicated under the cutoff wall of a dam. (*From D. P. Krynine, Soil Mechanics, 2d ed., McGraw-Hill, 1947*)

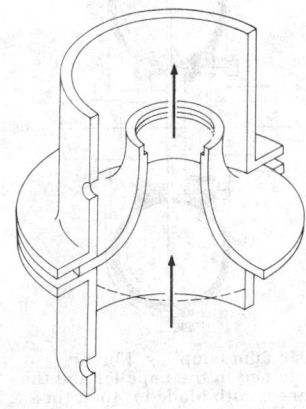

Diagram of a flow nozzle. Arrows indicate direction of fluid flow.

tubing made up of a string of interconnected tubing sections. { 'flō ˌstriŋ }

flow structure [GEOL] A primary sedimentary structure due to underwater slump or flow. { 'flō ˌstrək·chər }

flow tank [PETRO ENG] A tank which receives oil from the well and where gas and water may be separated before the oil passes into a stock tank. { 'flō ˌtaŋk }

flow texture [PETR] A pattern of an igneous rock that is formed when the stream or flow lines of a once-molten material have a subparallel arrangement of prismatic or tabular cyrstals or microlites. Also known as fluidal texture. { 'flō ˌteks·chər }

flow transmitter [ENG] A device used to measure the flow of liquids in pipelines and convert the results into proportional electric signals that can be transmitted to distant receivers or controllers. { 'flō tranzˌmid·ər }

flow value [MATH] For a feasible flow on an *s-t* network, the outflow from the source. { 'flō ˌval·yü }

flow valve [ENG] A valve that closes itself when the flow of a fluid exceeds a particular value. { 'flō ˌvalv }

flow velocity [GEOL] In soil, a vector point function used to indicate rate and direction of movement of water through soil per unit of time, perpendicular to the direction of flow. { 'flō və'läs·əd·ē }

flow visualization [ENG] Method of making visible the disturbances that occur in fluid flow, using the fact that light passing through a flow field of varying density exhibits refraction and a relative phase shift among different rays. { ¦flō vizh·ə·lə'zā·shən }

flow welding [MET] A welding process in which coalescence is produced by heating with molten filler metal, which is poured over the joint until the welding temperature is attained and the required amount of filler metal is added. { 'flō ˌweld·iŋ }

floxuridine [PHARM] $C_9H_{11}FN_2O_5$ Crystals that melt at 150–151°C; used as an antiviral drug and as an inhibitor of deoxyribonucleic acid synthesis. Abbreviated FUDR. { ˌfläks'yūr·əˌdēn }

fl oz *See* fluid ounce.

fluctuating current [ELEC] Direct current that changes in value but not at a steady rate. { ¦flək·chəˌwäd·iŋ 'kə·rənt }

fluctuation [OCEANOGR] **1.** Wavelike motion of water. **2.** The variations of water-level height from mean sea level that are not due to tide-producing forces. [SCI TECH] **1.** Variation, especially back and forth between successive values in a series of observations. **2.** Variation of data points about a smooth curve passing among them. { ˌflək·chə'wā·shən }

fluctuation noise *See* random noise. { ˌflək·chə'wā·shən ˌnȯiz }

fluctuation test [MICROBIO] A method of demonstrating that bacterial mutations preexist in a population before they are selected; a large parent population is divided into small parts which are grown independently and the number of mutants in each subculture determined; the number of mutants in the subculture will fluctuate because in some a mutant arises early (giving a large number of progeny), while in others the mutant arises late and gives few progeny. { ˌflək·chə'wā·shən ˌtest }

fluctuation theory [OPTICS] The theory proposed by M. von Smoluchowski and A. Einstein which states that the scattering of light occurs in pure water because random molecular motion causes density variations which effect changes in the refraction of light. { ˌflək·chə'wā·shən ˌthē·ə·rē }

fluctuation velocity *See* eddy velocity. { ˌflək·chə'wā·shən və'läs·əd·ē }

flue [ENG] A channel or passage for conveying combustion products from a furnace, boiler, or fireplace to or through a chimney. { flü }

flue dust [MET] Fine particles of metal or alloy emitted with the gases of a smelter or metallurgical furnace. { 'flü ˌdəst }

flue exhauster [ENG] A device installed as part of a vent in order to provide a positive induced draft. { 'flü igˌzȯs·tər }

flue gas [ENG] Gaseous combustion products from a furnace. { 'flü ˌgas }

flue gas analyzer [ENG] A device that monitors the composition of the flue gas of a boiler heating unit to determine if the mixture of air and fuel is at the proper ratio for maximum heat output. { 'flü ˌgas ˌan·ə,līz·ər }

flue gas expander [MECH ENG] In a petroleum processing system, a turbine for recovering energy at the point where combustion gases are discharged under pressure to the atmosphere; the reduction in pressure drives the turbine impeller. { 'flü ˌgas ik'spand·ər }

fluellite [MINERAL] $AlF_3 \cdot H_2O$ A colorless or white mineral composed of aluminum fluoride, occurring in crystals. { 'flü·əˌlīt }

fluence [ELECTROMAG] The total energy per unit area carried by a pulse of electromagnetic radiation. [PHYS] A measure of time-integrated particle flux, expressed in particles per square centimeter. { 'flü·əns }

fluent aphasia [PSYCH] Aphasia in which the facility of articulation, grammatical organization, and rate of speech are well preserved, while the comprehension of language and word choice are most affected; typically caused by lesions posterior to the rolandic fissure. { ¦flü·ənt ə'fā·zhə }

flufenamic acid [PHARM] $C_{14}H_{10}F_3NO_2$ Pale yellow needles with a melting point of 125°C; used as an anti-inflammatory drug or analgesic. { ¦flü·fə¦nam·ik 'as·əd }

fluid [PHYS] An aggregate of matter in which the molecules are able to flow past each other without limit and without fracture planes forming. { 'flü·əd }

fluidal texture *See* flow texture. { ¦flü·əd·əl 'teks·chər }

fluid amplifier [ENG] An amplifier in which all amplification is achieved by interaction between jets of fluid, with no electronic circuit and usually no moving parts. { ¦flü·əd 'am·plə,fī·ər }

fluid-bed process [CHEM ENG] A type of process based on the tendency of finely divided powders to behave in a fluidlike manner when supported and moved by a rising gas or vapor stream; used mainly for catalytic cracking of petroleum distillates. { ¦flü·əd ¦bed 'präs·əs }

fluid catalyst [CHEM ENG] Finely divided solid particles utilized as a catalyst in a fluid-bed process. { ¦flü·əd 'kad·əl,ist }

fluid catalytic cracking [CHEM ENG] An oil refining process in which the gas-oil is cracked by a catalyst bed fluidized by using oil vapors. { ¦flü·əd ¦kad·əl¦id·ik 'krak·iŋ }

fluid clutch *See* fluid drive. { ¦flü·əd ¦kləch }

fluid coal [MATER] Pulverized coal that, when mixed with air, can be transported through pipes. { ¦flü·əd ¦kōl }

fluid coating [FL MECH] The operation of depositing a liquid film on a solid, due to a relative motion between them. { 'flü·əd ˌkōd·iŋ }

fluid coefficient [PETRO ENG] A measure of the flow resistance to the leaking off of reservoir fracturing fluids into the formation during the fracturing operation. { ¦flü·əd ˌkōi'fish·ənt }

fluid coking [CHEM ENG] A thermal process utilizing the fluidized solids technique for continuous conversion of heavy, low-grade petroleum oils into petroleum coke and lighter hydrocarbon products. { ¦flü·əd ¦kōk·iŋ }

fluid-compressed [MET] Pertaining to steel that has been compressed while still fluid to remove gases and make the material more homogeneous. { 'flü·əd kəm,prest }

fluid computer [COMPUT SCI] A digital computer constructed entirely from air-powered fluid logic elements; it contains no moving parts and no electronic circuits; all logic functions are carried out by interaction between jets of air. { ¦flü·əd kəm¦pyüd·ər }

fluid contact [PETRO ENG] The site in a reservoir where there is gas-oil contact or oil-water contact. { ¦flü·əd 'kän,takt }

fluid-controlled valve [MECH ENG] A valve for which the valve operator is activated by a fluid energy, in contrast to electrical, pneumatic, or manual energy. { ¦flü·əd kən,trōld 'valv }

fluid coupling [MECH ENG] A device for transmitting rotation between shafts by means of the acceleration and deceleration of a fluid such as oil. Also known as hydraulic coupling. { ¦flü·əd ¦kəp·liŋ }

fluid density [FL MECH] The mass of a fluid per unit volume. { ¦flü·əd ¦den·səd·ē }

fluid die [MECH ENG] A die for shaping parts by liquid pressure; a plunger forces the liquid against the part to be shaped, making the part conform to the shape of a die. { ¦flü·əd ¦dī }

fluid distributor [ENG] Device for the controlled distribution of fluid feed to a process unit, such as a liquid-gas or liquid-solids contactor, reactor, mixer, burner, or heat exchanger; can be a simple perforated-pipe sparger, spray head, or such. { ¦flü·əd də'strib·yəd·ər }

fluid dram [MECH] Abbreviated fl dr. **1.** A unit of volume

FLOXURIDINE

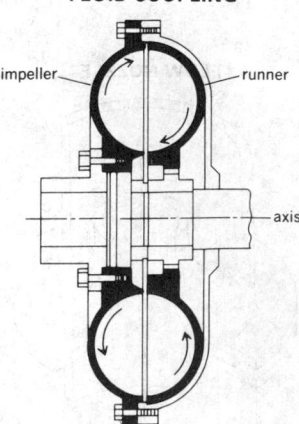

Structural formula of floxuridine.

FLUID COUPLING

impeller

runner

axis

Basic fluid coupling. Fluid is contained in the impeller and the runner, both bladed rotors; the impeller acts as a pump and the runner reacts as a turbine.

used in the United States for measurement of liquid substances, equal to $^1/_8$ fluid ounce, or $3.6966911953125 \times 10^{-6}$ cubic meter. **2.** A unit of volume used in the United Kingdom for measurement of liquid substances and occasionally of solid substances, equal to $^1/_8$ fluid ounce or $3.5516328125 \times 10^{-6}$ cubic meter. { ¦flü·əd 'dram }

fluid drive [MECH ENG] A power coupling operated on a hydraulic turbine principle in which the engine flywheel has a set of turbine blades which are connected directly to it and which are driven in oil, thereby turning another set of blades attached to the transmission gears of the automobile. Also known as fluid clutch; hydraulic clutch. { ¦flü·əd ¦drīv }

fluid duplicating [GRAPHICS] A system in which a master is produced by placing a special aniline-dye carbon paper under a sheet of coated paper; the carbon is placed face up, so that when the coated paper is typed or written on, a reversed copy is produced on the back of the sheet; a certain amount of dye from the carbon paper is transferred to the master, from which the copies are printed directly. { ¦flü·əd 'düp·lə‚kād·iŋ }

fluid dynamics [FL MECH] The science of fluids in motion. { ¦flü·əd dī'nam·iks }

fluid end [MECH ENG] In a fluid pump, the section that contains parts which are directly involved in moving the fluid. { 'flü·əd ‚end }

fluid-energy mill [ENG] A size-reduction unit in which grinding is achieved by collision between the particles being ground and the energy supplied by a compressed fluid entering the grinding chamber at high speed. Also known as jet mill. { ¦flü·əd 'en·ər·jē ‚mil }

fluid-film bearing [MECH ENG] An antifriction bearing in which rubbing surfaces are kept apart by a film of lubricant such as oil. { ¦flü·əd 'film ‚ber·iŋ }

fluid friction [FL MECH] Conversion of mechanical energy in fluid flow into heat energy. { ¦flü·əd 'frik·shən }

fluid fuel reactor [NUCLEO] A type of reactor (for example, a fused-salt reactor) whose fuel is in fluid form. { ¦flü·əd ‚fyül rē'ak·tər }

fluid geometry [GEOL] Fluid distribution in reservoir strata controlled by rock effective pore-size distribution, rock wettability characteristics in relation to the fluids present, method of producing saturation, and rock heterogeneity. { ¦flü·əd jē'äm·ə·trē }

fluid hydroforming [CHEM ENG] A type of fluid catalytic cracking process used by petroleum refineries to upgrade low-octane-number stocks. { ¦flü·əd 'hī·drə‚fòr·miŋ }

fluidic device [ENG] A device that operates by the interaction of streams of fluid. { flü¦id·ik di¦vīs }

fluidic flow sensor [ENG] A device for measuring the velocity of gas flows in which a jet of air or other selected gas is directed onto two adjacent small openings and is deflected by the flow of gas being measured so that the relative pressure on the two ports is a measure of gas velocity. Also known as deflected jet fluidic flowmeter. { flü¦id·ik 'flō ‚sen·sər }

fluidic oscillator meter [ENG] A flowmeter that measures the frequency with which a fluid entering the meter attaches to one of two opposite diverging side walls and then the other, because of the Coanda effect. { flü¦id·ik 'äs·ə‚lād·ər ‚mēd·ər }

fluidics [ENG] A control technology that employs fluid dynamic phenomena to perform sensing, control, information processing, and actuation functions without the use of moving mechanical parts. { flü'id·iks }

fluidic sensor [ENG] A proximity sensor that detects the presence of a nearby object from the back pressure created on an air jet when the object blocks the jet's exit area. { flü'id·ik 'sen·sər }

fluid inclusion [PETR] A tiny fluid-filled cavity in an igneous rock that forms by the entrapment of the liquid from which the rock crystallized. { ¦flü·əd in'klü·zhən }

fluid injection [PETRO ENG] The introduction of gases or liquids under pressure into a reservoir to force oil into producing wells. { ¦flü·əd in'jek·shən }

fluidity [FL MECH] The reciprocal of viscosity; expresses the ability of a substance to flow. { flü'id·ə·dē }

fluidization [CHEM ENG] A roasting process in which finely divided solids are suspended in a rising current of air (or other fluid), producing a fluidized bed; used in the calcination of various minerals, in Fischer-Tropsch synthesis, and in the coal industry. { ‚flü·ə·də'zā·shən }

fluidized adsorption [CHEM ENG] Method of vapor- or gas-fractionation (separation via adsorption-desorption cycles) in a fluidized bed of adsorbent material. { ¦flü·ə‚dīzd ad'sòrp·shən }

fluidized bed [ENG] A cushion of air or hot gas blown through the porous bottom slab of a container which can be used to float a powdered material as a means of drying, heating, quenching, or calcining the immersed components. { ¦flü·ə‚dīzd 'bed }

fluidized-bed coating [ENG] Method for plastic-coating of objects; the heated object is immersed into the fluidized bed of a thermoplastic resin that then fuses into a continuous uniform coating over the immersed object. { ¦flü·ə‚dīzd ¦bed 'kōd·iŋ }

fluidized-bed combustion [MECH ENG] A method of burning particulate fuel, such as coal, in which the amount of air required for combustion far exceeds that found in conventional burners; the fuel particles are continually fed into a bed of mineral ash in the proportions of 1 part fuel to 200 parts ash, while a flow of air passes up through the bed, causing it to act like a turbulent fluid. { ¦flü·ə‚dīzd ¦bed kəm'bəs·chən }

fluidized-bed reactor See fluidized reactor. { ¦flü·ə‚dīzd ¦bed rē'ak·tər }

fluidized reactor [NUCLEO] A nuclear reactor in which the fuel has been given the properties of a quasi-fluid, such as by suspension of fine fuel particles in a carrying gas or liquid. Also known as fluidized-bed reactor. { ¦flü·ə‚dīzd rē'ak·tər }

fluid level [PETRO ENG] In an oil well, the distance from the surface of the liquid in the tubing or the casing to ground level. { 'flü·əd ‚lev·əl }

fluid logic [ENG] The simulation of logical operations by means of devices that employ fluid dynamic phenomena to control the interactions between sets of gases or liquids. { ¦flü·əd ¦läj·ik }

fluid-loss agent [MATER] Material used to thicken or gel crude oil, light oil, and water or acid fracturing fluids to seal off pores and flow channels in the reservoir matrix. { ¦flü·əd ¦lòs ‚ā·jənt }

fluid-loss test [PETRO ENG] Measure of fracturing fluid loss versus time (spurt loss) before the fluid-loss agent forms a nonpermeable layer in the reservoir pore matrix. { ¦flü·əd ¦lòs ‚test }

fluid mechanics [MECH] The science concerned with fluids, either at rest or in motion, and dealing with pressures, velocities, and accelerations in the fluid, including fluid deformation and compression or expansion. { ¦flü·əd mə'kan·iks }

fluid meter See flowmeter. { 'flü·əd ‚mēd·ər }

fluid ounce [MECH] Abbreviated fl oz. **1.** A unit of volume that is used in the United States for measurement of liquid substances, equal to 1/16 liquid pint, or 231/128 cubic inches, or $2.95735295625 \times 10^{-5}$ cubic meter. **2.** A unit of volume used in the United Kingdom for measurement of liquid substances, and occasionally of solid substances, equal to 1/20 pint or $2.84130625 \times 10^{-5}$ cubic meter. { ¦flü·əd 'aùns }

fluid resistance [FL MECH] The force exerted by a gas or liquid opposing the motion of a body through it. Also known as resistance. { ¦flü·əd ri'zis·təns }

fluid saturation [GEOL] Measure of the gross void space in a reservoir rock that is occupied by a fluid. { ¦flü·əd ‚sach·ə'rā·shən }

fluid statics [FL MECH] The determination of pressure intensities and forces exerted by fluids at rest. { ¦flü·əd 'stad·iks }

fluid stress [MECH] Stress associated with plastic deformation in a solid material. { ¦flü·əd 'stres }

fluid ton [MECH] A unit of volume equal to 32 cubic feet or approximately 0.90614×10^{-2} cubic meter; used for many hydrometallurgical, hydraulic, and other industrial purposes. { ¦flü·əd 'tən }

fluid transmission [MECH ENG] Automotive transmission with fluid drive. { ¦flü·əd tranz'mish·ən }

fluid viscosity ratio [PETRO ENG] Ratio of viscosity of a displacing gas to that of oil in a gas-drive reservoir; used in unit displacement efficiency calculations. { ¦flü·əd vi'skäs·əd·ē ‚rā·shō }

fluing [ENG] A forming process in which a flange is formed around a hole in a sheet-metal part by pressing a cylindrical die through the hole. { 'flü·iŋ }

fluke [INV ZOO] The common name for more than 40,000 species of parasitic flatworms that form the class Trematoda.

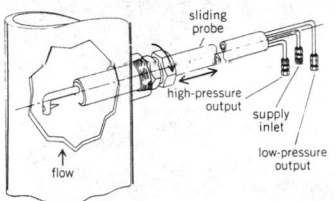

FLUIDIC FLOW SENSOR

Schematic drawing of a fluidic flow sensor.

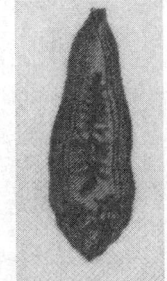

FLUKE

Liver fluke *Opisthorchis sinensis.*

[NAV ARCH] The broad end of each arm of an anchor. [VERT ZOO] A flatfish, especially summer flounder. { flük }

flume [ENG] **1.** An open channel constructed of steel, reinforced concrete, or wood and used to convey water to be utilized for power, to transport logs, and so on. **2.** To divert by a flume, as the waters of a stream, in order to lay bare the auriferous sand and gravel forming the bed. [GEOL] A ravine with a stream flowing through it. { flüm }

flumed [MIN ENG] In hydraulic mining, pertaining to the transportation of solids by suspension or flotation in flowing water. { flümd }

fluoborate See fluoroborate. { ‚flü·ə'bȯr‚āt }

fluoborite [MINERAL] $Mg_3(BO_3)(F,OH)_3$ A colorless mineral composed of magnesium fluoborate; occurs in hexagonal prisms. Also known as nocerite. { ‚flü·ə'bȯr‚īt }

fluocerite [MINERAL] $(Ce,La,Nd)F_3$ A reddish-yellow mineral composed of fluoride of cerium and related elements. { ¦flü·ə'se‚rīt }

fluolite See pitchstone. { 'flü·ə‚līt }

fluometuron [ORG CHEM] $C_{10}H_{11}F_3N_2O$ A white, crystalline solid with a melting point of 163–164.5°C; used as a herbicide for cotton and sugarcane. Also known as 1,1-dimethyl-3-(α,α,α-trifluoro-*meta*-tolyl)urea. { ¦flü·ō'me·chə ‚rän }

fluophor See luminophor. { 'flü·ə‚fȯr }

fluor See fluorite; luminophor. { 'flü‚ȯr }

fluoranthene [ORG CHEM] $C_{10}H_{10}$ A tetracyclic hydrocarbon found in coal tar fractions and petroleum, forming needlelike crystals, boiling point 250°C, and soluble in organic solvents such as ether and benzene. { flu'ran‚thēn }

fluorapatite [MINERAL] **1.** $Ca_5(PO_4)_3F$ A mineral of the solid-solution series of the apatite group; common accessory mineral in igneous rocks. **2.** An apatite mineral in which the fluoride member dominates. { flu·'rap·ə‚tīt }

fluorene [ORG CHEM] $C_{13}H_{10}$ A hydrocarbon chemical present in the middle oil fraction of coal tar; insoluble in water, soluble in ether and acetone, melting point 116–117°C; used as the basis for a group of dyes. Also known as 2,3-benzindene; diphenylenemethane. { 'flu‚rēn }

fluorescein [ORG CHEM] $C_{20}H_{12}O_5$ A yellowish to red powder, melts and decomposes at 290°C, insoluble in water, benzene, and chloroform, soluble in glacial acetic acid, boiling alcohol, ether, dilute acids, and dilute alkali; used in medicine, in oceanography as a marker in sea water, and in textiles to dye silk and wool. { ‚flu're·sē·ən }

fluorescence [ATOM PHYS] **1.** Emission of electromagnetic radiation that is caused by the flow of some form of energy into the emitting body and which ceases abruptly when the excitation ceases. **2.** Emission of electromagnetic radiation that is caused by the flow of some form of energy into the emitting body and whose decay, when the excitation ceases, is temperature-independent. [NUC PHYS] Gamma radiation scattered by nuclei which are excited to and radiate from an excited state. [OPTICS] See bloom. { flu'res·əns }

fluorescence analysis See fluorometric analysis. { flu'res·əns ə‚nal·ə·səs }

fluorescence digital imaging microscope [OPTICS] An instrument for observing images formed by emitted fluorescent light; consists of an electronic camera or a laser-photomultiplier scanning system, an analog-to-digital converter, and a computer that manipulates the digitized signal and converts it back to analog form for display. { flu'res·əns 'dij·əd·əl ¦im·ij·iŋ 'mī·krə‚skōp }

fluorescence method [GRAPHICS] A method of ultraviolet photography in which the subject is illuminated by ultraviolet light, and a filter is used on the camera to absorb the reflected ultraviolet light and permit only the visible fluorescence to reach the film. { flu'res·əns ‚meth·əd }

fluorescence microscope [OPTICS] A variation of the compound laboratory light microscope which is arranged to admit ultraviolet, violet, and sometimes blue radiations to a specimen, which then fluoresces. { flu'res·əns 'mī·krə‚skōp }

fluorescence spectra [SPECT] Emission spectra of fluorescence in which an atom or molecule is excited by absorbing light and then emits light of characteristic frequencies. { flu'res·əns ‚spek·trə }

fluorescence x-rays [ATOM PHYS] Characteristic x-rays emitted as the result of the absorption of x-rays of higher frequency. { flu'res·əns 'eks‚rāz }

fluorescence yield [ATOM PHYS] The probability that an atom in an excited state will emit an x-ray photon in its first transition rather than an Auger electron. { flu'res·əns ‚yēld }

fluorescent antibody [IMMUNOL] An antibody that is labeled by a fluorescent dye, such as fluorescein. { flu'res·ənt 'an·tē‚bäd·ē }

fluorescent antibody test [IMMUNOL] A clinical laboratory test based on the antigen used in the diagnosis of syphilis and lupus erythematosus and for identification of certain bacteria and fungi, including the tubercle bacillus. { flu'res·ənt 'an·tē‚bäd·ē ‚test }

fluorescent dye [CHEM] A highly reflective dye that serves to intensify color and add to the brilliance of a fabric. { flu¦res·ənt 'dī }

fluorescent in situ hybridization [GEN] A technique in which a deoxyribonucleic acid (DNA) probe is labeled with a fluorescent dye (that can be visualized under a fluorescent microscope) and then hybridized with target DNA, usually chromosome preparations on a microscopic slide. It is used to precisely map genes to a specific region of a chromosome in a prepared karyotype, or can enumerate chromosomes, or can detect chromosomal deletions, translocations, or gene amplifications in cancer cells. Abbreviated FISH. { flə¦res·ənt in¦sit‚chü ‚hī·brə·də'zā·shən }

fluorescent lamp [ELECTR] A tubular discharge lamp in which ionization of mercury vapor produces radiation that activates the fluorescent coating on the inner surface of the glass. { flu¦res·ənt 'lamp }

fluorescent minituft method [FL MECH] An adaptation of the tuft method in which the tuft material is treated with a fluorescent dye and the image is recorded with fluorescence photography, allowing reduction of the tuft size, so that there is less interference by the tufts with the fluid flow. { flə¦res·ənt 'min·ē‚təft ‚meth·əd }

fluorescent pigment [CHEM] A pigment capable of absorbing both visible and nonvisible electromagnetic radiations and releasing them quickly as energy of desired wavelength; examples are zinc sulfide or cadmium sulfide. { flu¦res·ənt 'pig·mənt }

fluorescent screen [ENG] A sheet of material coated with a fluorescent substance so as to emit visible light when struck by ionizing radiation such as x-rays or electron beams. { flu¦res·ənt 'skrēn }

fluorescent staining [CYTOL] The use of fluorescent dyes to mark specific cell structures, such as chromosomes. { flu¦res·ənt 'stān·iŋ }

fluoridation [ENG] The addition of the fluorine ion (F^-) to municipal water supplies in a final concentration of 0.8–1.6 parts per million to help prevent dental caries in children. [GEOCHEM] Formation in rocks of fluorine-containing minerals such as fluorite or topaz. { flur·ə'dā·shən }

fluoride [INV ZOO] A salt of hydrofluoric acid, HF, in which the fluorine atom is in the −1 oxidation state. { 'flur‚īd }

fluorimeter See fluorometer. { flu'rim·əd·ər }

fluorinated ethylene propylene resin [ORG CHEM] Copolymers of tetrafluoroethylene and hexafluoropropylene. Abbreviated FEP resin. { 'flur·ə‚nād·əd 'eth·ə‚lēn 'prō·pə‚lēn 'rez·ən }

fluorination [CHEM] A chemical reaction in which fluorine is introduced into a chemical compound. { ‚flur·ə'nā·shən }

fluorine [CHEM] A gaseous or liquid chemical element, symbol F, atomic number 9, atomic weight 18.998403; a member of the halide family, it is the most electronegative element and the most chemically energetic of the nonmetallic elements; highly toxic, corrosive, and flammable; used in rocket fuels and as a chemical intermediate. { 'flur‚ēn }

fluorite [MINERAL] CaF_2 A transparent to translucent, often blue or purple mineral, commonly found in crystalline cubes in veins and associated with lead, tin, and zinc ores; hardness is 4 on Mohs scale; the principal ore of fluorine. Also known as Derbyshire spar; fluor; fluorspar. { 'flur‚īt }

fluoroacetate [ORG CHEM] Acetate in which carbon-connected hydrogen atoms are replaced by fluorine atoms. { ¦flur·ō'as·ə‚tāt }

fluoroacetic acid [ORG CHEM] CH_2FCOOH A poisonous, crystalline compound obtained from plants, such as those of the Dichapetalaceae family, South Africa, soluble in water and alcohol, and burns with a green flame; the sodium salt is used as a water-soluble rodent poison. Also known as gifblaar poison. { ¦flur·ō·ə'sēd·ik 'as·əd }

fluoroalkane [ORG CHEM] Straight-chain, saturated hydrocarbon compound (or analog thereof) in which some of the hydrogen atoms are replaced by fluorine atoms. { ¦flür·ō'al‚kān }

para-fluoroaniline [ORG CHEM] $FC_6H_4NH_2$ A liquid that is an intermediate in the manufacture of herbicides and plant growth regulators. { ¦par·ə ‚flür·ō'an·ə‚lēn }

fluorobenzene [ORG CHEM] C_6H_5F A colorless liquid with a boiling point of 84.9°C; used as an insecticide intermediate. Also known as phenyl fluoride. { ¦flür·ō¦ben‚zēn }

fluoroborate [INV ZOO] **1.** Any of a group of compounds related to the borates in which one or more oxygens have been replaced by fluorine atoms. **2.** The BF_4^- ion, which is derived from fluoroboric acid, HBF_4. Also known as fluoborate. { ‚flür·ə'bȯr‚āt }

fluoroboric acid [INV ZOO] HBF_4 Colorless, clear, water-miscible acid; used for electrolytic brightening of aluminum and for forming stabilized diazo salts. { ¦flür·ə‚bȯr·ik 'as·əd }

fluorocarbon [ORG CHEM] A hydrocarbon in which part or all hydrogen atoms have been replaced by fluorine atoms, including chlorinated and brominated fluorocarbons. Also known as fluorohydrocarbon. { ¦flür·ō'kär·bən }

fluorocarbon-11 *See* trichlorofluoromethane. { ¦flür·ō'kär·bən ə'lev·ən }

fluorocarbon-21 *See* dichlorofluoromethane. { ¦flür·ō'kär·bən ‚twen·tē'wən }

fluorocarbon fiber [ORG CHEM] Fiber made from a fluorocarbon resin, such as polytetrafluoroethylene resin. { ¦flür·ō'kär·bən 'fī·bər }

fluorocarbon resin [ORG CHEM] Polymeric material made up of carbon and fluorine with or without other halogens (such as chlorine) or hydrogen; the resin is extremely inert and more dense than corresponding fluorocarbons such as Teflon. { ¦flür·ō'kär·bən 'rez·ən }

fluorochemical [CHEM] Any chemical compound containing fluorine; usually refers to the fluorocarbons. { ¦flür·ō'kem·ə·kəl }

fluorochlorocarbon *See* chlorofluorocarbon. { ¦flür·ō¦klȯr·ō'kär·bən }

fluorochromasia [CYTOL] The immediate appearance of fluorescence inside viable cells on exposure to a fluorogenic substrate. { ¦flür·ō·krə'mā·zhə }

fluorocummingtonite [MINERAL] Cummingtonite with a high content of fluorine. { ¦flür·ō'kəm·iŋ·tə‚nīt }

fluorod [NUCLEO] A rod made from silver-activated phosphate glass and used in solid-state dosimeters; under irradiation the rod absorbs ultraviolet light and emits orange fluorescent light; measurement of the intensity of the emitted light with a photomultiplier gives a measure of the absorbed dose of radiation. { 'flü‚räd }

fluorodichloromethane *See* dichlorofluoromethane. { ¦flür·ō·dī‚klȯr·ō'meth‚ān }

fluorodifen [ORG CHEM] $C_{13}H_7F_3N_2O_4$ A yellow, crystalline compound with a melting point of 93°C; used as a pre- and postemergence herbicide for food crops. { flü'räd·ə·fen }

1-fluoro-2,4-dinitrobenzene [ORG CHEM] $(NO_2)_2C_6H_3F$ Crystals that are soluble in benzene, propylene glycol, and ether; used as a reagent for labeling terminal amino acid groups and in the detection of phenols. Also known as Sanger's reagent. { ¦wən ¦flür·ō ¦tü ¦fȯr dī‚nī·trō'ben‚zēn }

fluoroelastomer [MATER] A partially fluorinated polymer or a copolymer; it is the most chemically resistant of the elastomers and has good mechanical properties at high and low temperatures. { ‚flür·ō·ī'las·tə·mər }

fluoroform [ORG CHEM] CHF_3 A colorless, nonflammable gas, boiling point 84°C at 1 atmosphere (101,325 pascals), freezing point 160°C at 1 atmosphere; used in refrigeration and as an intermediate in organic synthesis. Also known as propellant 23; refrigerant 23; trifluoromethane. { 'flür·ə‚fȯrm }

fluorogenic substrate [CHEM] A nonfluorescent material that is acted upon by an enzyme to produce a fluorescent compound. { 'flür·ə‚jen·ik 'səb‚strāt }

fluorograph [GRAPHICS] **1.** A photograph produced by the light of fluorescent materials. **2.** A photograph of an image produced on a fluorescent screen. { 'flür·ə‚graf }

fluorography [GRAPHICS] Photography of an image produced on a fluorescent screen. Also known as photofluorography. { flü'räg·rə·fē }

fluorohydrocarbon *See* fluorocarbon. { ¦flür·ō‚hī·drə'kär·bən }

fluorologging [ENG] A well-logging technique in which well cuttings are examined under ultraviolet light for fluorescence radiation related to trace occurrences of oil. { 'flür·ō‚läg·iŋ }

fluorometer [ENG] An instrument that measures the fluorescent radiation emitted by a sample which is exposed to monochromatic radiation, usually radiation from a mercury-arc lamp or a tungsten or molybdenum x-ray source that has passed through a filter; used in chemical analysis, or to determine the intensity of the radiation producing fluorescence. Also spelled fluorimeter. { flü'räm·əd·ər }

fluorometric analysis [ANALY CHEM] A method of chemical analysis in which a sample, exposed to radiation of one wavelength, absorbs this radiation and reemits radiation of the same or longer wavelength in about 10^{-9} second; the intensity of reemited radiation is almost directly proportional to the concentration of the fluorescing material. Also known as fluorescence analysis; fluorometry. { ¦flür·ə¦me·trik ə'nal·ə·səs }

fluorometry *See* fluorometric analysis. { flü'räm·ə·trē }

para-fluorophenylacetic acid [ORG CHEM] $FC_6H_4CH_2$-COOH Crystals with a melting point of 86°C; used as an intermediate in the manufacture of fluorinated anesthetics. { ¦par·ə ¦flü·rə‚fen·əl·ə'sēd·ik 'as·əd }

fluorophosphoric acid [INV ZOO] H_2PO_3F A colorless, viscous liquid that is miscible with water; used in metal cleaners and as a catalyst. { ¦flür·ō‚fäs'fȯr·ik 'as·əd }

fluoroplastics [MATER] A family of plastics based on fluorine replacement of hydrogen atoms in hydrocarbon molecules; includes polytetrafluoroethylene (PTFE), polychlorotrifluoroethylene (PCTFE), polyvinylidene fluoride, and fluorinated ethylene propylene (FEP). Also known as fluoropolymers. { ¦flür·ō¦plas·tiks }

fluoropolymers *See* fluoroplastics. { ¦flür·ō'päl·ə·mərz }

fluoroscope [ENG] A fluorescent screen designed for use with an x-ray tube to permit direct visual observation of x-ray shadow images of objects interposed between the x-ray tube and the screen. { 'flür·ə‚skōp }

fluoroscopic image intensifier [ELECTR] An electron-beam tube that converts a relatively feeble fluoroscopic image on the fluorescent input phosphor into a much brighter image on the output phosphor. { ‚flür·ə'skäp·ik 'im·ij in'ten·sə‚fī·ər }

fluoroscopy [ENG] Use of a fluoroscope for x-ray examination. { flü'räs·kə·pē }

fluorothene *See* chlorotrifluoroethylene polymer. { 'flür·ə‚thēn }

fluorotrichloromethane *See* trichlorofluoromethane. { ¦flür·ō·trī¦klȯr·ō'meth‚ān }

fluorouracil [PHARM] $C_4H_3FN_2O_2$ Crystals that decompose at 282–283°C; used as an antineoplastic drug and as an inhibitor of deoxyribonucleic acid synthesis. Also known as 2,4-dioxo-5-fluoropyrimidine; 5-FU. { ¦flür·ō'yùr·ə‚sil }

fluorspar *See* fluorite. { 'flür‚spär }

fluosilicate [INV ZOO] A salt derived from fluosilicic acid, H_2SiF_6, and containing the SiF_6^{-2} ion. { ¦flü·ə'sil·ə‚kāt }

fluosilicic acid [INV ZOO] H_2SiF_6 A colorless acid, soluble in water, which attacks glass and stoneware; highly corrosive and toxic; used in water fluoridation and electroplating. Also known as hydrofluorosilicic acid; hydrofluosilicic acid. { ¦flü·ə·sə'lis·ik 'as·əd }

fluosolids system [MET] In pyrometallurgy, a roasting method for finely divided solids, in which air under pressure is blown through a heated bed of mineral to keep it fluid. { ¦flü·ō¦säl·ədz 'sis·təm }

fluosulfonic acid [INV ZOO] HSO_3F Colorless, corrosive, fuming liquid; soluble in water with partial decomposition; used as organic synthesis catalyst and in electroplating. { ¦flü·ə·səl'fän·ik 'as·əd }

flurenol [ORG CHEM] $C_{18}H_{18}O_3$ A solid, crystalline compound with a melting point of 70–71°C; used as an herbicide for vegetables, cereals, and ornamental flowers. { 'flür·ə‚nȯl }

flurry [METEOROL] A brief shower of snow accompanied by a gust of wind, or a sudden, brief wind squall. { 'flər·ē }

flush [ECOL] An evergreen herbaceous or nonflowering vegetation growing in habitats where seepage water causes the surface to be constantly wet but rarely flooded. [ENG] Pertaining to separate surfaces that are on the same level.

FLUOROURACIL

Structural formula of fluorouracil.

[GRAPHICS] A printing term that means no indention; headings are often run flush left, that is, they align at the left margin; flush-right lines align at the right. { fləsh }

flush bead See quirk bead. { 'fləsh ‚bēd }

flush center See center-justify. { ¦fləsh 'sen·tər }

flush coat [CIV ENG] A coating of bituminous material, used to waterproof a surface. { 'fləsh ‚kōt }

flush cover [GRAPHICS] In bookbinding, a book cover that has been trimmed to the same size as the text pages inside. { ¦fləsh 'kəv·ər }

flushed-zone resistivity [PETRO ENG] Electrical resistivity of the reservoir area which surrounds a borehole to a distance of at least 3 inches (7.6 centimeters) and for which the original interstitial fluids have been flushed out by drilling-mud filtrate. { ¦fləsht ¦zōn rē‚zis'tiv·əd·ē }

flush gate [CIV ENG] A gate for flushing a channel that lies below the gate of a dam. { 'fləsh ‚gāt }

flushing [CIV ENG] The removal or reduction to a permissible level of dissolved or suspended contaminants in an estuary or harbor. [ENG] Removing lodged deposits of rock fragments and other debris by water flow at high velocity; used to clean water conduits and drilled boreholes. { 'fləsh·iŋ }

flushing oil [MATER] A solvent oil designed to remove used lubricating oil, decomposition products, and accumulated dirt from lubrication passages, crankcase surfaces, and lubricated moving parts of automotive engines. { 'fləsh·iŋ ‚oil }

flushing period [HYD] The interval of time required for a quantity of water equal to the volume of a lake to pass through the lake outlet; computed by dividing lake volume by mean flow rate of the outlet. { 'fləsh·iŋ ‚pir·ē·əd }

flush-joint casing [PETRO ENG] Lengths of casing that when connected end to end form a smooth joint flush with the outer diameter of the remainder of the section length. { ¦fləsh ¦joint 'kās·iŋ }

flush left See left-justify. { ¦fləsh 'left }

flushometer [ENG] A valve that discharges a fixed quantity of water when a handle is operated; used to flush toilets and urinals. { flə'shäm·əd·ər }

flush production [PETRO ENG] First yield from a flowing oil well during its most productive period. { 'fləsh prə‚dək·shən }

flush right See right-justify. { ¦fləsh 'rīt }

flush tank [CIV ENG] **1.** A tank in which water or sewage is retained for periodic release through a sewer. **2.** A small water-filled tank for flushing a water closet. { 'fləsh ‚taŋk }

flush valve [ENG] A valve used for flushing toilets. { 'fləsh ‚valv }

flute [DES ENG] A groove having a curved section, especially when parallel to the main axis, as on columns, drills, and other cylindrical or conical shaped pieces. [GEOL] **1.** A natural groove running vertically down the face of a rock. **2.** A groove in a sedimentary structure formed by the scouring action of a turbulent, sediment-laden water current, and having a steep upcurrent end. { flüt }

flute cast [GEOL] A raised, oblong, or subconical welt on the bottom surface of a siltstone or sandstone bed formed by the filling of a flute. { 'flüt ‚kast }

fluted chucking reamer [DES ENG] A machine reamer with a straight or tapered shank and with straight or spiral flutes; the ends of the teeth are ground on a slight chamfer for end cutting. { 'flüd·əd 'chək·iŋ ‚rēm·ər }

fluted coupling See stabilizer. { ¦flüd·əd¦kəp·liŋ }

flute length [DES ENG] On a twist drill, the length measured from the outside corners of the cutting lips to the farthest point at the back end of the flutes. { 'flüt ‚leŋkth }

flute storage [ELECTR] Ferrite storage consisting of a number of parallel lengths of fine prism-shaped tubing, each surrounding an insulated axial conductor that acts as a word line; the lengths of tubing are intersected at right angles by parallel sets of insulated wire bit lines that are displaced slightly from the word lines; each intersection stores one bit. { ¦flüt ¦stór·ij }

fluting [MECH ENG] A machining operation whereby flutes are formed parallel to the main axis of cylindrical or conical parts. { 'flüd·iŋ }

flutter [ACOUS] Distortion that occurs in sound reproduction as a result of undesired speed variations during the recording, duplicating, or reproducing process. [ELECTROMAG] A fast-changing variation in received signal strength, such as may be caused by antenna movements in a high wind or interaction

with a signal or another frequency. [ENG] The irregular alternating motion of the parts of a relief valve due to the application of pressure where no contact is made between the valve disk and the seat. [FL MECH] See aeronautical flutter. [MED] Rapid, regular contraction of the atrial muscle of the heart. { 'fləd·ər }

flutter echo [ACOUS] A multiple echo in which the reflections rapidly follow each other. [ELECTROMAG] A radar echo consisting of a rapid succession of reflected pulses resulting from a single transmitted pulse. { 'fləd·ər ‚ek·ō }

flutter valve [ENG] A valve that is operated by fluctuations in pressure of the material flowing over it; used in carburetors. { 'fləd·ər ‚valv }

fluvarium [ENG] A large aquarium in which the tanks contain flowing stream water maintained by gravity, not pumps. { flü'ver·ē·əm }

Fluvent [GEOL] A suborder of the soil order Entisol that is well-drained with visible marks of sedimentation and no identifiable horizons; occurs in recently deposited alluvium along streams or in fans. { 'flü·vənt }

fluvial [HYD] **1.** Pertaining to or produced by the action of a stream or river. **2.** Existing, growing, or living in or near a river or stream. { 'flü·vē·əl }

fluvial cycle of erosion See normal cycle. { 'flü·vē·əl 'sī·kəl əv ə'rō·zhən }

fluvial deposit [GEOL] A sedimentary deposit of material transported by or suspended in a river. { ¦flü·vē·əl di'päz·ət }

fluvial sand [GEOL] Sand laid down by a river or stream. { ¦flü·vē·əl 'sand }

fluvial soil [GEOL] Soil laid down by a river or stream. { ¦flü·vē·əl 'soil }

fluviatile [GEOL] Resulting from river action. { 'flü·vē·ə‚tīl }

fluviology [HYD] The science of rivers. { flü·vē'äl·ə·jē }

fluviomorphology See river morphology. { ¦flü·vē·ō·mór'fäl·ə·jē }

flux [ELECTROMAG] The electric or magnetic lines of force in a region. [MATER] **1.** In soldering, welding, and brazing, a material applied to the pieces to be united to reduce the melting point of solders and filler metals and to prevent the formation of oxides. **2.** A substance used to promote the fusing of minerals or metals. **3.** Additive for plastics composition to improve flow during physical processing. **4.** In enamel work, a substance composed of silicates and other materials that forms a colorless, transparent glass when fired. Also know as fondant. [NUCLEO] The product of the number of particles per unit volume and their average velocity; a special case of the physics definition. Also known as flux density. [PHYS] **1.** The integral over a given surface of the component of a vector field (for example, the magnetic flux density, electric displacement, or gravitational field) perpendicular to the surface; by definition, it is proportional to the number of lines of force crossing the surface. **2.** The amount of some quantity flowing across a given area (often a unit area perpendicular to the flow) per unit time; the quantity may be, for example, mass or volume of fluid, electromagnetic energy, or number of particles. { fləks }

fluxball [ELECTROMAG] A type of magnetic test coil in which the wire is wound into the form of a solid spherical winding by combining a series of coaxial cylindrical windings of different lengths; it gives accurate values of the magnetic flux density (or its variation) at its center, even in a nonuniform magnetic field. { 'fləks‚ból }

flux-closure domain See closure domain. { 'fləks ‚klō·zhər dō‚mān }

flux-compression generator [ELEC] A type of impulse generator in which megajoules of energy can be generated within microseconds by abruptly reducing the volume of a closed conducting cage that surrounds a region in which a magnetic field is established. Also known as magnetic cumulative generator. { ¦fləks kəm‚presh·ən 'jen·ə‚rād·ər }

flux-cored welding [MET] Welding with a metal electrode that has a flux core. { ¦fləks ¦kórd 'weld·iŋ }

flux density [NUCLEO] See flux. [PHYS] Any vector field whose flux is a significant physical quantity; examples are magnetic flux density, electric displacement, gravitational field, and the Poynting vector. { 'fləks ‚den·səd·ē }

flux density threshold See threshold illuminance. { 'fləks ‚den·səd·ē 'thresh‚hōld }

flux factor [MET] A factor for assessing the quality of steel-works-grade silica refractories. { 'fləks ,fak·tər }

flux gate [ENG] A detector that gives an electric signal whose magnitude and phase are proportional to the magnitude and direction of the external magnetic field acting along its axis; used to indicate the direction of the terrestrial magnetic field. { 'fləks ,gāt }

flux-gate magnetometer [ELECTROMAG] A magnetometer in which the degree of saturation of the core by an external magnetic field is used as a measure of the strength of the earth's magnetic field; the essential element is the flux gate. { 'fləks ,gāt ,mag·nə'täm·əd·ər }

flux guide [MET] A shaped piece of magnetic material used to guide magnetic flux in induction heating; may be used either to direct the flux to preferred locations or to prevent the flux from spreading beyond definite regions. { 'fləks ,gīd }

fluxing [MET] The development of the liquid phase in a ceramic body under heat treatment by the melting of low-fusion components; used in steel manufacture, metal smelting, and assaying. { 'fləks·iŋ }

fluxing ore [MET] An ore containing usually an appreciable amount of valuable metal, but smelted mainly because it contains fluxing agents which are required in the reduction of other ores. { 'fləks·iŋ ,ȯr }

fluxional compound [ORG CHEM] **1.** Any of a group of molecules which undergo rapid intramolecular rearrangements in which the component atoms are interchanged among equivalent structures. **2.** Molecules in which bonds are broken and reformed in the rearrangement process. { 'flək·shən·əl ,käm,paůnd }

flux jumping See Meissner effect. { 'fləks ,jəmp·iŋ }

flux lattice [SOLID STATE] The regular array of fluxoids in a type II superconductor in the mixed state, when the superconductor is sufficiently pure and free of defects. { 'fləks ,lad·əs }

flux-lattice melting [SOLID STATE] A phenomenon in high-temperature superconductors in the mixed state, in which the regular ordering of fluxoids breaks down above a certain temperature. { 'fləks,lad·əs ,melt·iŋ }

flux leakage [ELECTROMAG] Magnetic flux that does not pass through an air gap or other part of a magnetic circuit where it is required. { 'fləks ,lēk·ij }

flux line See fluxoid; line of force. { 'fləks ,līn }

flux-line pinning [SOLID STATE] The introduction, in a type II superconductor, of microscopic crystalline defects that suppress the motion of flux lines which would otherwise occur in the presence of a magnetic field. { 'fləks ,līn ,pin·iŋ }

flux linkage [ELECTROMAG] The product of the number of turns in a coil and the magnetic flux passing through the coil. Also known as linkage. { 'fləks ,liŋk·ij }

flux mapping [NUCLEO] The process of measuring the radiation flux at representative points within a nuclear reactor or around some other radiation source. { 'fləks ,map·iŋ }

fluxmeter [ENG] An instrument for measuring magnetic flux. { 'fləks,mēd·ər }

flux of energy [PHYS] The energy which passes through a surface per unit area per unit time. { 'fləks əv 'en·ər·jē }

fluxoid [SOLID STATE] One of the microscopic filaments of magnetic flux that penetrates a type II superconductor in the mixed state, consisting of a normal core in which the magnetic field is large, surrounded by a superconducting region in which flows a vortex of persistent supercurrent which maintains the field in the core. Also known as flux line; fluxon; vortex. { 'fluk,sȯid }

flux oil [MATER] An oil suitable for blending with bitumen or asphalt to form a product of greater fluidity or softer consistency. { 'fləks ,ȯil }

fluxon See fluxoid. { 'flək,sän }

flux oxygen cutting [MET] Oxygen cutting of metal with the aid of a flux to reduce the temperature. { 'fləks 'äk·sə·jən ,kəd·iŋ }

flux path [ELECTROMAG] A path which is followed by magnetic lines of force and in which the magnetic flux density is significant. { 'fləks ,path }

flux pump [CRYO] A cryogenic direct-current generator that converts a small alternating-current input to a large direct-current output when cooled to about 4 K; the output current builds up in a series of steps, much like the action of a pump. { 'fləks ,pəmp }

flux refraction [ELECTROMAG] The abrupt change in direction of magnetic flux lines at the boundary between two media having different permeabilities, or of the electric flux lines at the boundary between two media having different dielectric constants, when these lines are oblique to the boundary. { 'fləks ri,frak·shən }

flux unit [ASTROPHYS] A unit of energy flux density of radio-astronomical sources, equal to 10^{-26} watt per square meter per hertz. Abbreviated fu. { 'fləks ,yü·nət }

fly [INV ZOO] The common name for a number of species of the insect order Diptera characterized by a single pair of wings, antennae, compound eyes, and hindwings modified to form knoblike balancing organs, the halters. [MECH ENG] A fan with two or more blades used in timepieces or light machinery to govern speed by air resistance. { flī }

Fly See Musca. { flī }

fly ash [ENG] **1.** Fine particulate, essentially noncombustible refuse, carried in a gas stream from a furnace. **2.** Coal combustion residue. { 'flī ,ash }

flyback [ELECTR] The time interval in which the electron beam of a cathode-ray tube returns to its starting point after scanning one line or one field of a television picture or after completing one trace in an oscilloscope. Also known as retrace; return trace. [HOROL] The return to zero of the timing hand of a stopwatch or chronograph. { 'flī,bak }

flyback power supply [ELECTR] A high-voltage power supply used to produce the direct-current voltage of about 10,000–25,000 volts required for the second anode of a cathode-ray tube in a television receiver or oscilloscope. { 'flī,bak 'paůr sə,plī }

flyback transformer See horizontal output transformer. { 'flī,bak tranz,fȯr·mər }

flyby [AERO ENG] A close approach of a space vehicle to a target planet in which the vehicle does not impact the planet or go into orbit around it. Also known as swing-by. { 'flī,bī }

fly-by-wire system [AERO ENG] A flight control system that uses electric wiring instead of mechanical or hydraulic linkages to control the actuators for the ailerons, flaps, and other control surfaces of an aircraft. { 'flī bī 'wīr ,sis·təm }

fly cutter [MECH ENG] A cutting tool that revolves with the arbor of a lathe. { 'flī ,kəd·ər }

fly cutting [MECH ENG] Cutting with a milling cutter provided with only one tooth. { 'flī ,kəd·iŋ }

flying angle [AERO ENG] The acute angle between the longitudinal axis of an aircraft and the horizontal axis in normal level flight, or the angle of attack of a wing in normal level flight. { 'flī·iŋ ,aŋ·gəl }

flying-aperture scanner [ELECTR] An optical scanner, used in character recognition, in which a document is flooded with light, and light is collected sequentially spot by spot from the illuminated image. { 'flī·iŋ 'ap·ər·chər ,skan·ər }

flying boat [AERO ENG] A seaplane with a fuselage that acts as a hull and is the means of the plane's support on water. { 'flī·iŋ ,bōt }

flying bomb [ORD] Popularly, any explosive robot plane, guided missile, rocket bomb, or the like; specifically, the German V-1 explosive robot plane of World War II. { 'flī·iŋ 'bäm }

flying bridge [NAV ARCH] A narrow walkway or platform built at the level of the top of the pilothouse and containing a duplicate set of controls for steering gear and engine-room signals and for navigating instruments; the platform extends from one side of the ship to the other. Also known as navigating bridge. { 'flī·iŋ 'brij }

flying buttress [ARCH] A buttress connected by an arch to the building it supports. { 'flī·iŋ 'bə·trəs }

flying clock [HOROL] A cesium-controlled high-precision time standard that can be kept in continuous operation on commercial aircraft as well as in automobiles. { 'flī·iŋ 'kläk }

flying-crane helicopter [AERO ENG] A heavy-lift helicopter used in rapid loading and unloading of, for example, cargo ships. { 'flī·iŋ ,krān 'hel·ə,käp·tər }

flying fish [VERT ZOO] Any of about 65 species of marine fishes which form the family Exocoetidae in the order Atheriniformes; characteristic enlarged pectoral fins are used for gliding. { 'flī·iŋ 'fish }

Flying Fish See Volan. { 'flī·iŋ 'fish }

flying head [ELECTR] A read/write head used on magnetic disks and drums, so designed that it flies a microscopic distance

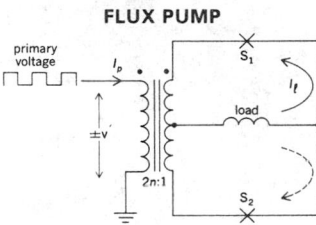

FLUX PUMP

A flux-pump circuit (simplified).

FLY

A common black fly (*Simulium*).

FLYING FISH

The four-winged flying fish (*Parexocoetus mesogaster*), with pectoral fins extended for gliding.

off the moving magnetic surface and is supported by a film of air. { ¦flī·iŋ 'hed }

flying hour rate *See* utilization rate. { ¦flī·iŋ ¦au̇r ¦rāt }

flying paster [GRAPHICS] In web printing, an automatic device that splices a new web of paper onto the roll that is being used up without interrupting the printing process. { ¦flī·iŋ 'pās·tər }

flying saucer *See* unidentified flying object. { ¦flī·iŋ 'sȯs·ər }

flying shear [MET] A machine which cuts lengths of rolled products and allows for continuous production by reciprocating with the product while cutting. { ¦flī·iŋ 'shir }

flying spot [ELECTR] A small point of light, controlled mechanically or electrically, which moves rapidly in a rectangular scanning pattern in a flying-spot scanner. { ¦flī·iŋ 'spät }

flying-spot microscope [OPTICS] A microscope in which a minute spot of light, produced in the lens system, passes through a specimen while sweeping over it systematically, and falls on a photocell; the image is produced on a cathode-ray tube that is scanned in synchronization with the spot. { ¦flī·iŋ ˌspät 'mī·krə,skōp }

flying-spot scanner [ELECTR] A scanner used for television film and slide transmission, electronic writing, and character recognition, in which a moving spot of light, controlled mechanically or electrically, scans the image field, and the light reflected from or transmitted by the image field is picked up by a phototube to generate electric signals. Also known as optical scanner. { ¦flī·iŋ ˌspät 'skan·ər }

flying switch [ENG] Disconnection of railroad cars from a locomotive while they are moving and switching them to another track under their own momentum. { ¦flī·iŋ 'swich }

flying veins [GEOL] A series of mineral-deposit veins which overlap or intersect in a branchlike pattern. { ¦flī·iŋ 'vānz }

fly-off *See* evapotranspiration. { 'flī,ȯf }

fly rock [ENG] The fragments of rock thrown and scattered during quarry or tunnel blasting. { 'flī ˌräk }

flysch [GEOL] Deposits of dark, fine-grained, thinly bedded sandstone shales and of clay, thought to be deposited by turbidity currents and originally defined as rock formations on the northern and southern borders of the Alps. { flīsh }

flyway [VERT ZOO] A geographic migration route for birds, including the breeding and wintering areas that it connects. { 'flī,wā }

flywheel [MECH ENG] A rotating element attached to the shaft of a machine for the maintenance of uniform angular velocity and revolutions per minute. Also known as balance wheel. { 'flī,wēl }

flywheel synchronization [ELECTR] Automatic frequency control of a scanning system by using the average timing of the incoming sync signals, rather than by making each pulse trigger the scanning circuit; used in high-sensitivity television receivers designed for fringe-area reception, when noise pulses might otherwise trigger the sweep circuit prematurely. { 'flī,wēl ˌsiŋ·krə·nə'zā·shən }

fm *See* femtometer.

Fm *See* fermium.

FM *See* frequency modulation.

FM/AM multiplier [ELECTR] Multiplier in which the frequency deviation from the central frequency of a carrier is proportional to one variable, and its amplitude is proportional to the other variable; the frequency-amplitude-modulated carrier is then consecutively demodulated for frequency modulation (FM) and for amplitude modulation (AM); the final output is proportional to the product of the two variables. { 'ef,em 'ā,em 'məl·tə,plī·ər }

F martingale [MATH] A stochastic process $\{X_t, t > 0\}$ such that the conditional expectation of X_t given F_s equals X_s whenever $s < t$, where $F = \{F_t, t \geq 0\}$ is an increasing family of sigma algebras that represents the amount of information increasing with time. { ¦ef 'mart·ən,gāl }

F meson [PARTIC PHYS] Former name of a charged meson that carries both strangeness and charm and has a mass of approximately 1.97 GeV, spin 0, and negative parity; now known as D_s meson. { 'ef ˌmā,sän }

FMF *See* familial Mediterranean fever.

FMN *See* riboflavin 5′-phosphate.

fnp *See* fusion point.

f number [OPTICS] A lens rating obtained by dividing the lens's focal length by its effective maximum diameter; the

larger the *f* number, the less exposure is given. Also known as focal ratio. Also known as stop number. { 'ef ,nəm·bər }

foal [VERT ZOO] A young horse, especially one under 1 year of age. { fōl }

foam [CHEM] An emulsionlike two-phase system where the dispersed phase is gas or air. [FL MECH] A collection of bubbles on the surface of a liquid, often stabilized by organic contaminants, as found at sea or along shore. Also known as froth. [GEOL] *See* pumice. { fōm }

foam-back fabric [TEXT] A fabric to the back of which is banded a thin layer of polyether or polyester foam. Also known as foam-laminated fabric. { ¦fōm ¦bak 'fab·rik }

foam blanketing [ENG] A technique for fighting fire within an oil tank or similar facility by generating foam that forms a coating inside the tank, thus depriving the fire of air. { ˌfōm 'blaŋ·kə·tiŋ }

foam booster [MATER] An additive used in detergents to increase suds production and stabilize lather. { 'fōm ,büs·tər }

foam crust [HYD] A snow surface feature that looks like small overlapping waves, like sea foam on a beach, occurring during the ablation of the snow surface and may further develop into a more pronounced wedge-shaped form, known as plowshares. { 'fōm ,krəst }

foam depresser [MATER] An additive used in detergents to suppress or inhibit the production of suds or to maintain suds at a low level. { 'fōm di,pres·ər }

foam drilling [MIN ENG] A method of dust suppression in which thick foam is forced through the drill by means of compressed air, and the foam-and-dust mixture emerges from the mouth of the hole in the form of a thick sludge. Also known as mist drilling. { 'fōm ,dril·iŋ }

foamed plastic *See* expanded plastic. { ¦fōmd 'plas·tik }

foam glass [MATER] A light, black, opaque, cellular glass made by adding powdered carbon to crushed glass and firing the mixture. { 'fōm ,glas }

foaminess [PHYS] The volume of foam produced in a liquid, in cubic centimeters, produced by passing air through it divided by the rate of flow of air, in cubic centimeters per second. { 'fō·mē·nəs }

foaming [ENG] Any of various processes by which air or gas is introduced into a liquid or solid to produce a foam material. [GRAPHICS] A froth in a photographic emulsion containing minute bubbles caused by excessive turbulence. { 'fōm·iŋ }

foaming agent *See* blowing agent. { 'fōm·iŋ ,ā·jənt }

foam-in-place [ENG] The deposition of reactive foam ingredients onto the surface to be covered, allowing the foaming reaction to take place upon that surface, as with polyurethane foam; used in applying thermal insulation for homes and industrial equipment. { ¦fōm in 'plās }

foam-laminated fabric *See* foam-back fabric. { ¦fōm ¦lam·ə,nād·əd 'fab·rik }

foam line [OCEANOGR] The front of a wave as it moves toward the shore, after the wave has broken. { 'fōm ,līn }

foam mark [GEOL] A surface sedimentary structure comprising a pattern of barely visible ridges and hollows formed where wind-driven sea foam passes over a surface of wet sand. { 'fōm ,märk }

foam metal [MET] Cast metal with finely divided gas bubbles evenly distributed throughout the body of the metal; an example is foam aluminum. { ¦fōm 'med·əl }

foam rubber *See* rubber sponge. { ¦fōm 'rəb·ər }

focal adhesion plaque [CYTOL] Points of attachment that form when cells attach to a substrate. { ¦fō·kəl əd'hē·zhən ,plak }

focal chord [MATH] For a conic, a chord that passes through a focus of the conic. { 'fō·kəl ¦kȯrd }

focal distance *See* focal length. { 'fō·kəl ,dis·təns }

focal infection [MED] Infection in a limited area, such as the tonsils, teeth, sinuses, or prostate. { ¦fō·kəl in¦fek·shən }

focal length [OPTICS] The distance from the focal point of a lens or curved mirror to the principal point; for a thin lens it is approximately the distance from the focal point to the lens. Also known as focal distance. { 'fō·kəl ,leŋkth }

focal lines *See* astigmatic foci. { 'fō·kəl ,līnz }

focal plane [OPTICS] A plane perpendicular to the axis of an optical system and passing through the focal point of the system. { 'fō·kəl ,plān }

focal-plane array [ELECTR] A photodetector that has up to a million photosensors on a single semiconductor silicon chip arranged in a rectangular grid matrix that is placed in the focal plane of an optical instrument. { ¦fō·kəl ¦plān ə'rä }

focal-plane shutter [OPTICS] A camera shutter consisting of a blind containing a slot; the blind is pulled rapidly across the film, exposing it through the slot. { 'fō·kəl ¦plān 'shəd·ər }

focal point [OPTICS] The point to which rays that are initially parallel to the axis of a lens, mirror, or other optical system are converged or from which they appear to diverge. Also known as principal focus. { 'fō·kəl ¦point }

focal power [OPTICS] A measure of the ability of a lens, mirror, prism, or optical system to converge a parallel beam of light; equals the reciprocal of the focal length. Also known as power. { 'fō·kəl ¦pau̇·ər }

focal property [MATH] **1.** The property of an ellipse or hyperbola whereby lines drawn from the foci to any point on the conic make equal angles with the tangent to the conic at that point. **2.** The property of a parabola whereby a line from the focus to any point on the parabola, and a line through this point parallel to the axis of the parabola, make equal angles with the tangent to the parabola at this point. { 'fō·kəl 'präp·ər·dē }

focal radius [MATH] For a conic, a line segment from a focus to any point on the conic. { 'fō·kəl 'rād·ē·əs }

focal ratio See f number. { ¦fō·kəl 'rā·shō }

focal seizure [MED] An epileptic manifestation of a restricted nature, usually without loss of consciousness, due to irritation of a localized area of the brain. { ¦fō·kəl 'sē·zhər }

focal spot [MET] In electron-beam or laser welding, the spot where the beam has the highest concentrated energy level. { 'fō·kəl ¦spät }

Foch space [QUANT MECH] An infinite-dimensional vector space in which the state of a quantum-mechanical system with a variable number of particles is represented by an infinite number of wave functions, each of which corresponds to a fixed number of particles. { 'fōsh ¦späs }

focometer [ENG] An instrument for measuring focal lengths of optical systems. { fō'käm·əd·ər }

focus [ELECTR] To control convergence or divergence of the electron paths within one or more beams, usually by adjusting a voltage or current in a circuit that controls the electric or magnetic fields through which the beams pass, in order to obtain a desired image or a desired current density within the beam. [GEOPHYS] The center of an earthquake and the origin of its elastic waves within the earth. [MATH] A point in the plane which together with a line (directrix) defines a conic section. [NUCLEO] To guide particles along a desired path in a particle accelerator by means of electric or magnetic fields. [OPTICS] **1.** The point or small region at which rays converge or from which they appear to diverge. **2.** To move an optical lens toward or away from a screen or film to obtain the sharpest possible image of a desired object. { 'fō·kəs }

focus control [ELECTR] A control that adjusts spot size at the screen of a cathode-ray tube to give the sharpest possible image; it may vary the current through a focusing coil or change the position of a permanent magnet. [OPTICS] A device to adjust a lens system to produce a sharp image. { 'fō·kəs kən,trōl }

focused collision sequence [PHYS] A cascade of interatomic collisions, initiated by the bombardment of a crystal with energetic particles, that propagates in a particular direction along a closely packed row of atoms in the crystal. { 'fō·kəst kə'lizh·ən ,sē·kwəns }

focused-current log [ENG] A resistivity log that is obtained by means of a multiple-electrode arrangement. { ¦fō·kəst ¦kə·rənt 'läg }

focusing anode [ELECTR] An anode used in a cathode-ray tube to change the size of the electron beam at the screen; varying the voltage on this anode alters the paths of electrons in the beam and thus changes the position at which they cross or focus. { 'fō·kəs·iŋ ¦an,ōd }

focusing coil [ELECTR] A coil that produces a magnetic field parallel to an electron beam for the purpose of focusing the beam. { 'fō·kəs·iŋ ¦kȯil }

focusing collector [ENG] A solar collector that uses semicircular aluminum reflectors to focus sunlight onto copper pipes containing circulating water. { 'fō·kəs·iŋ kə'lek·tər }

focusing electrode [ELECTR] An electrode to which a potential is applied to control the cross-sectional area of the electron beam in a cathode-ray tube. { 'fō·kəs·iŋ i,lek,trōd }

focusing glass [OPTICS] A magnifying glass designed to enlarge the image thrown on the ground glass of the viewfinder of a camera, to help achieve exact focusing. { 'fō·kəs·iŋ ,glas }

focusing magnet [ELECTR] A permanent magnet used to produce a magnetic field for focusing an electron beam. { 'fō·kəs·iŋ ,mag·nət }

focusing scale [OPTICS] A graduated scale to indicate appropriate lens-to-image plane positions for given lens-to-object plane distances. { 'fō·kəs·iŋ ,skāl }

focus lamp [ELEC] **1.** A lamp whose filament has a spiral or zigzag form in order to reduce its size, so that it can be brought into the focus of a lens or mirror. **2.** An arc lamp whose feeding mechanism is designed to hold the arc in a constant position with respect to an optical system that is used to focus its rays. { 'fō·kəs ,lamp }

focus projection and scanning [ELECTR] Method of magnetic focusing and electrostatic deflection of the electron beam of a hybrid vidicon; a transverse electrostatic field is used for beam deflection; this field is immersed with an axial magnetic field that focuses the electron beam. { ¦fō·kəs prə,jek·shən ən 'skan·iŋ }

focus wave mode [PHYS] A localized wave solution of the three-dimensional wave equation whose overall characteristics depend on a free parameter such that it resembles a transverse plane wave at one extreme and a narrow spatially transverse pulse at the other extreme. { ¦fō·kəs 'wāv ,mōd }

foehn [METEOROL] A warm, dry wind on the lee side of a mountain range, the warmth and dryness being due to adiabatic compression as the air descends the mountain slopes. Also spelled föhn. { fān }

foehn air [METEOROL] The warm, dry air associated with foehn winds. { 'fān ,er }

foehn cloud [METEOROL] Any cloud form associated with a foehn, but usually signifying only those clouds of the lenticularis species formed in the lee wave parallel to the mountain ridge. { 'fān ,klau̇d }

foehn cyclone [METEOROL] A cyclone formed (or at least enhanced) as a result of the foehn process on the lee side of a mountain range. { 'fān 'sī,klōn }

foehn island [METEOROL] An isolated area where the foehn has reached the ground, in contrast to the surrounding area where foehn air has not replaced colder surface air. { 'fān 'ī·lənd }

foehn nose [METEOROL] As seen on a synoptic surface chart, a typical deformation of the isobars in connection with a well-developed foehn situation; a ridge of high pressure is produced on the windward slopes of the mountain range, while a foehn trough forms on the lee side; the isobars "bulge" correspondingly, giving a noselike configuration. { 'fān ,nōz }

foehn pause [METEOROL] **1.** A temporary cessation of the foehn at the ground, due to the formation or intrusion of a cold air layer which lifts the foehn above the valley floor. **2.** The boundary between foehn air and its surroundings. { 'fān ,pȯz }

foehn period [METEOROL] The duration of continuous foehn conditions at a given location. { 'fān ,pir·ē·əd }

foehn phase [METEOROL] One of three stages to describe the development of the foehn in the Alps: the preliminary phase, when cold air at the surface is separated from warm dry air aloft by a subsidence inversion; the anticyclonic phase, when the warm air reaches a station as the result of the cold air flowing out from the plain; and the stationary phase or cyclonic phase, when the foehn wall forms and the downslope wind becomes appreciable. { 'fān ,fāz }

foehn sickness [MED] A phenomenon in humans in alpine regions, marked by adverse psychological and physiological effects during prolonged periods of foehn wind. { 'fān ,sik·nəs }

foehn storm [METEOROL] A type of destructive storm which frequently occurs in October in the Bavarian Alps. { 'fān ,stȯrm }

foehn trough [METEOROL] The dynamic trough formed in connection with the foehn. { 'fān ,trȯf }

foehn wall [METEOROL] The steep leeward boundary of flat, cumuliform clouds formed on the peaks and upper windward sides of mountains during foehn conditions. { 'fān ˌwol }

fog [GRAPHICS] A dark, hazy deposit or veil of uniform density over all or parts of a piece of film or paper; can be caused by light other than that forming the image, lens flare, aged materials, or chemical impurities. [METEOROL] Water droplets or, rarely, ice crystals suspended in the air in sufficient concentration to reduce visibility appreciably. { fäg }

fogbank [METEOROL] A fairly well-defined mass of fog observed in the distance, most commonly at sea. { 'fäg,baŋk }

fogbound [NAV] Surrounded by fog; the term refers to vessels which are unable to proceed because of the fog. { 'fäg,baund }

fogbow [OPTICS] A faintly colored circular arc similar to a rainbow but formed on fog layers containing drops whose diameters are of the order of 100 micrometers or less. Also known as false white rainbow; mistbow; white rainbow. { 'fäg,bō }

fog buoy See position buoy. { 'fäg ˌboi }

fog chamber See cloud chamber. { 'fäg ˌchām·bər }

fog climax [ECOL] A community that deviates from a climatic climax because of the persistent occurrence of a controlling fog blanket. { ¦fäg 'klī,maks }

fog deposit [HYD] The deposit of an ice coating on exposed surfaces by a freezing fog. { 'fäg dī,päz·ət }

fog detector light [NAV] A light fitted in certain lighthouses for the automatic detection of fog, in addition to the main light. { 'fäg di,tek·tər ˌlīt }

fog dispersal [METEOROL] Artificial dissipation of a fog by means such as seeding or heating. { 'fäg di,spərs·əl }

fog drip [HYD] Water dripping to the ground from trees or other objects which have collected the moisture from drifting fog; the dripping can be as heavy as light rain, as sometimes occurs among the redwood trees along the coast of northern California. { 'fäg ˌdrip }

fog drop [METEOROL] An elementary particle of fog, physically the same as a cloud drop. Also known as fog droplet. { 'fäg ˌdräp }

fog droplet See fog drop. { 'fäg ˌdräp·lət }

fog forest [ECOL] The dense, rich forest growth which is found at high or medium-high altitudes on tropical mountains; occurs when the tropical rain forest penetrates altitudes of cloud formation, and the climate is excessively moist and not too cold to prevent plant growth. { 'fäg ˌfär·əst }

fogged metal [MET] A metal whose luster has been highly reduced by corrosion products. { ¦fägd ¦med·əl }

fog gong [NAV] A gong used as a fog signal. { 'fäg ˌgäŋ }

fog gun [NAV] A gun used as a fog signal. { 'fäg ˌgən }

fog horizon [METEOROL] The top of a fog layer which is confined by a low-level temperature inversion so as to give the appearance of the horizon when viewed from above against the sky; the true horizon is usually obscured by the fog in such instances. { 'fäg hə,rīz·ən }

foghorn [NAV] A horn used as a fog signal. { 'fäg,horn }

fog quenching [MET] Rapid cooling of a metal piece in a fine vapor or mist. { 'fäg ,kwench·iŋ }

fog scale [METEOROL] A classification of fog intensity based on its effectiveness in decreasing horizontal visibility; such practice is not current in United States weather observing procedures. { 'fäg ,skāl }

fog signal [NAV] **1.** A warning signal transmitted by a vessel or aid to navigation during periods of low visibility. **2.** The device for producing such a signal. { 'fäg ,sig·nəl }

fog siren [NAV] A siren used as a fog signal. { 'fäg ,sī·rən }

fog track [NUCLEO] A line of condensation, produced in supersaturated water vapor by the passage of charged particles; used in studying the courses and collisions of particles in cloud chambers. { 'fäg ,trak }

fog trumpet [NAV] A trumpet used as a fog signal. { 'fäg ,trəm·pət }

fog whistle [NAV] A whistle used as a fog signal. { 'fäg ,wis·əl }

fog wind [METEOROL] Humid east wind which crosses the divide of the Andes east of Lake Titicaca and descends on the west in violent squalls; probably the same as puelche. { 'fäg ,wind }

föhn See foehn. { fān }

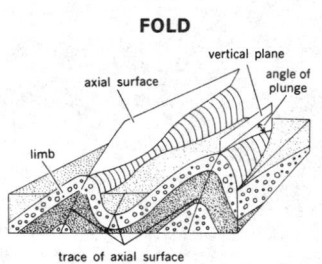

FOLD

Elements of folds.

foil [MET] A thin sheet of metal, usually less than 0.006 inch (0.15 millimeter) thick. { foil }

foil decorating [ENG] The molding of paper, textile, or plastic foil, printed with compatible inks, into a plastic part so that the foil is visible below the surface of the part as a decoration. { ¦foil 'dek·ə,rād·iŋ }

foil dosimeter [NUCLEO] A device for measuring the amount of radiation exposure by means of the degree of activation created in a metal foil inserted in the radiation field. { ¦foil dō'sim·əd·ər }

foil electret [ELEC] A thin film of strongly insulating material capable of trapping charge carriers, such as polyfluoroethylenepropylene, that is electrically charged to produce an external electric field; in the conventional design, charge carriers of one sign are injected into one surface, and a compensation charge of opposite sign forms on the opposite surface or an adjacent electrode. { ¦foil i'lek·trət }

Fokker-Planck equation [STAT MECH] An equation for the distribution function of a gas, analogous to the Boltzmann equation but applying where the forces are long-range and the collisions are not binary. { ¦fō·kər ¦pläŋk i'kwā·zhən }

fold [ANAT] A plication or doubling, as of various parts of the body such as membranes and other flat surfaces. [GEOL] A bend in rock strata or other planar structure, usually produced by deformation; folds are recognized where layered rocks have been distorted into wavelike form. [MET] See lap. { fōld }

fold belt See orogenic belt. { 'fōld ,belt }

folded cavity [ELECTR] Arrangement used in a klystron repeater to make the incoming wave act on the electron stream from the cathode at several places and produce a cumulative effect. { ¦fōld·əd 'kav·əd·ē }

folded dipole See folded-dipole antenna. { ¦fōld·əd 'dī,pōl }

folded-dipole antenna [ELECTROMAG] A dipole antenna whose outer ends are folded back and joined together at the center; the impedance is about 300 ohms, as compared to 70 ohms for a single-wire dipole; widely used with television and frequency-modulation receivers. Also known as folded dipole. { ¦fōld·əd 'dī,pōl an'ten·ə }

folded horn [ENG ACOUS] An acoustic horn in which the path from throat to mouth is folded or curled to give the longest possible path in a given volume. { ¦fōld·əd 'horn }

folded-plate roof [BUILD] A roof constructed of flat plates, usually of reinforced concrete, joined at various angles. { ¦fōld·əd ¦plāt 'rüf }

folding [COMPUT SCI] A method of hashing which consists of splitting the original key into two or more parts and then adding the parts together. [GEOL] Compression of planar structure in the formation of fold structures. { 'fōld·iŋ }

folding boards [MIN ENG] A shifting frame on which the cage rests in a mine. Also known as chairs; dogs; keeps; keps. { 'fōld·iŋ ,bordz }

folding door [ENG] A door in sections that can be folded back or can be moved apart by sliding. { 'fōld·iŋ ,dor }

folding fin [AERO ENG] A fin hinged at its base to lie flat, especially a fin on a rocket that lies flat until the rocket is in flight. { 'fōld·iŋ 'fin }

foldover [ELECTR] Picture distortion seen as a white line on the side, top, or bottom of a television picture; generally caused by nonlinear operation in either the horizontal or vertical deflection circuits of a receiver. { 'fōl,dō·vər }

fold system [GEOL] A group of folds with common trends and characteristics. { 'fōld ,sis·təm }

Foldy effect [ATOM PHYS] The interaction of the electric field that is produced by a neutron and results from the nonvanishing charge distribution in the neutron's interior with the electrons in an atom. { 'fōl·dē i,fekt }

Foley pits [ENG ACOUS] Open boxes that are used in ADR studios and contain various materials (such as water, sand, gravel, rice, and nails) for generating sound effects that could not be recorded well during filming or video recording. { 'fō·lē ,pits }

folia [PETR] Thin, leaflike layers that occur in gneissic or schistose rocks. { 'fō·lē·ə }

foliaceous [BOT] Consisting of or having the form or texture of a foliage leaf. [GEOL] Having a leaflike or platelike structure composed of thin layers of minerals. [ZOO] Resembling a leaf in growth form or mode. { ¦fō·lē'ā·shəs }

foliage [BOT] The leaves of a plant. { 'fō·lē·ij }

foliage leaf [BOT] The chief photosynthetic organ of most vascular plants. { 'fō·lē·ij ,lēf }

foliar [BOT] Of, pertaining to, or consisting of leaves. { 'fō·lē·ər }

foliated ice [HYD] Large masses of ice which grow in thermal contraction cracks in permafrost. Also known as ice wedge. { 'fō·lē,ād·əd 'īs }

foliate papilla [VERT ZOO] One of the papillae found on the posterolateral margin of the tongue of many mammals, but vestigial or absent in humans. { 'fō·lē·ət pə'pil·ə }

foliation [BOT] **1.** The process of developing into a leaf. **2.** The state of being in leaf. [GEOL] A laminated structure formed by segregation of different minerals into layers that are parallel to the schistosity. [MET] Beating metal into thin sheets. { ,fō·lē'ā·shən }

folic acid [BIOCHEM] $C_{19}H_{19}N_7O_6$ A yellow, crystalline vitamin of the B complex; it is slightly soluble in water, usually occurs in conjugates containing glutamic acid residues, and is found especially in plant leaves and vertebrate livers. Also known as pteroylglutamic acid (PGA). { 'fō·lik 'as·əd }

folic acid sodium salt See sodium folate. { ¦fō·lik ¦as·əd ¦sōd·ē·əm 'sōlt }

folie à deux [PSYCH] Identical or similar mental disorders affecting two individuals, usually members of the same family living together. Also known as shared paranoid delusion. { ¦fō·lē ä ¦dü }

foliferous [BOT] Producing leaves. { fə'lif·ə·rəs }

foliicolous [BIOL] Growing or parasitic upon leaves, as certain fungi. { fä·lē·ə'kə·ləs }

folimat [ORG CHEM] $C_5H_{12}NO_4PS$ An oily liquid that decomposes at 135°C; soluble in water; used as an insecticide and miticide on fruit and vegetable crops and on ornamental flowers. Also known as omethioate. { 'fä·lə,mat }

Folin solution [ANALY CHEM] An aqueous solution of 500 grams of ammonium sulfate, 5 grams of uranium acetate, and 6 grams of acetic acid in a volume of 1 liter; used to test for uric acid. { 'fō·lən sə,lü·shən }

folio [GRAPHICS] The page number as it will appear in printed text. { 'fō·lē,ō }

follobranchiate [VERT ZOO] Having leaflike gills. { ¦fō·lē·ō¦braŋ·kē,āt }

foliolate [BOT] Having leaflets. { 'fō·lē·ə,lāt }

Folist [GEOL] A suborder of the soil order Histosol, consisting of wet forest litter resting on rock or rubble. { 'fäl·əst }

folium [MATH] A plane curve that is a pedal curve (first positive pedal) of the deltoid. { 'fō·lē·əm }

folium of Descartes [MATH] A plane cubic curve whose equation in cartesian coordinates x and y is $x^3 + y^3 = 3axy$, where a is some constant. Also known as leaf of Descartes. { 'fō·lē·əm əv dā'kärt }

follicle [BIOL] A deep, narrow sheath or a small cavity. [BOT] A type of dehiscent fruit composed of one carpel opening along a single suture. { 'fäl·ə·kəl }

follicle cell [CELL MOL] A cell through which the developing ovum receives material for growth. { 'fäl·ə·kəl ,sel }

follicle-stimulating hormone [BIOCHEM] A protein hormone released by the anterior pituitary of vertebrates which stimulates growth and secretion of the Graafian follicle and also promotes spermatogenesis. Abbreviated FSH. { 'fäl·ə·kəl ¦stim·yə,lād·iŋ 'hȯr,mōn }

follicular lymphoma [MED] A premalignant lymphoma in which the lymph nodes show enlarged follicles composed predominantly of closely packed, large reticuloendothelial cells. { fə'lik·yə·lər lim'fō·mə }

folliculate [BIOL] Having or composed of follicles. { fə'lik·yə,lāt }

folliculitis [MED] Inflammation of a follicle or group of follicles. { fa,lik·yə'līd·əs }

folliculitis keloidalis See keloid acne. { fa,lik·yə'līd·əs ,kē·lȯi'dal·əs }

follow current [ELEC] The current at power frequency that passes through a surge diverter or other discharge path after a high-voltage surge has started the discharge. { 'fäl·ō ,kə·rənt }

follower [ENG] A drill used for making all but the first part of a hole, the first part being made with a drill of larger gage. { 'fäl·ə·wər }

follower rail [MIN ENG] The rail of a mine switch on the other side of the turnout corresponding to the lead rail. { 'fäl·ə·wər ,rāl }

following error [CONT SYS] The difference between commanded and actual positions in contouring control. { 'fäl·ə·wiŋ ,er·ər }

following limb [ASTRON] The half of the limb of a celestial body with an observable disk that appears to follow the body in its apparent motion across the field of view of a fixed telescope. { 'fäl·ə·wiŋ ,lim }

following sea [NAV] A sea in which the waves move in the general direction of the heading. { ¦fäl·ə·wiŋ 'sē }

following wind [METEOROL] **1.** A wind blowing in the direction of ocean-wave advance. **2.** See tailwind. [NAV] Wind blowing in the general direction of a vessel's course. { ¦fäl·ə·wiŋ 'wind }

follow spot [ELEC] A high-intensity spotlight used to follow action in arenas and stadiums and on large stages; it is equipped with adjustable iris and shutter controls, and its light source is either a carbon arc or an incandescent bulb. { 'fäl·ō ,spät }

follow-the-pointer indicator [ORD] Scale, on the mount of some types of artillery, that receives and registers firing data transmitted over a remote control system; the gun is kept properly aimed when its adjustment dials are matched with the readings on the indicator. { ¦fäl·ō thə ¦pȯint·ər 'in·də,kād·ər }

follow through [ORD] Material which follows the jet of a shaped charge through the hole formed in the target. { 'fäl·ō ,thrü }

folpet [ORG CHEM] $C_9H_4Cl_3NO_2S$ A buff or white, crystalline compound with a melting point of 177–178°C; insoluble in water; used as a fungicide on fruits, vegetables, and ornamental flowers. { 'fäl·pet }

Fomalhaut See α Piscis Australis. { 'fō·mə,lȯt }

fomite [MED] An inanimate object contaminated with an infectious organism (for example, a dish, clothing, towel, needle, or dust). { 'fō,mīt }

fondant See flux. { 'fän·dənt }

font [GRAPHICS] A particular typeface and size, including all the uppercase and lowercase letters, punctuation marks, numerals, and so forth. { fänt }

fontanelle [ANAT] A membrane-covered space between the bones of a fetal or young skull. [INV ZOO] A depression on the head of termites. { fänt·ən'el }

font cartridge [COMPUT SCI] A removable module that can be plugged into a slot in a printer and has one or more fonts stored in a read-only memory chip. { fänt ,kär·trij }

font compiler See font generator. { fänt kəm,pīl·ər }

Fontéchevade man [PALEON] A fossil man representing the third interglacial *Homo sapiens* and having browridges and a cranial vault similar to those of modern *Homo sapiens*. { fȯn·te·chə'väd ,man }

font generator [COMPUT SCI] A computer program that converts an outline font into the patterns of dots required for a particular size of font. Also known as font compiler. { fänt jen·ə,rād·ər }

food [BIOL] A material that can be ingested and utilized by the organism as a source of nutrition and energy. { füd }

food additive [FOOD ENG] A substance added to foods during processing to improve color, texture, flavor, or keeping qualities; examples are antioxidants, emulsifiers, thickeners, preservatives, and colorants. { 'füd ,ad·əd·iv }

food allergy [IMMUNOL] A hypersensitivity to certain foods. { 'füd ,al·ər·jē }

food analog [FOOD ENG] A manufactured food product designed to imitate a given food and frequently possessing characteristics equal or superior to that food. Also known as fabricated food; structured food; texturized food. { 'füd ,an·ə,läg }

food-borne disease [MED] Any disease transmitted by contaminated foods. { ¦füd ¦bȯrn di'zēz }

food chain [ECOL] The scheme of feeding relationships by trophic levels which unites the member species of a biological community. { 'füd ,chān }

food chemistry [FOOD ENG] A specialized phase of food technology concerned with an understanding of the fundamental changes of composition and the physical condition of foodstuffs which may occur during and subsequent to industrial processing. { 'füd ,kem·ə·strē }

food color [MATER] A colorant, either a dye (soluble) or a

FOLLICLE

Follicles of the milkweed plant.

lake (insoluble), permitted by the Food and Drug Administration for use in foods, drugs, and cosmetics. Also known as certified color; FD&C color. { 'füd ˌkəl·ər }

food engineering [ENG] The technical discipline involved in food manufacturing and processing. { 'füd ˌen·jə₁nir·iŋ }

food infection [MED] A type of bacterial food poisoning in which the host is infected by organisms carried by food. { 'füd in₁fek·shən }

food irradiation [FOOD ENG] The treatment of fresh or processed foods with ionizing radiation that inactivates biological contaminants (insects, molds, parasites, or bacteria), rendering foods safe to consume and extending their storage lifetime. { 'füd i₁rā·dē¦ā·shən }

food manufacturing [FOOD ENG] The commercial production and packaging of foods that are fabricated by processing, by combining various ingredients, or both. { 'füd ₁man·yə₁fak·chə·riŋ }

food microbiology [FOOD ENG] The science that deals with the microorganisms involved in the spoilage, contamination, and preservation of food. { 'füd ₁mī·krō·bī¦äl·ə·jē }

food poisoning [MED] Poisoning due to intake of food contaminated by bacteria or poisonous substances produced by bacteria. { 'füd ₁póiz·ən·iŋ }

food preservation [FOOD ENG] Processing designed to protect food from spoilage caused by microbes, enzymes, and autooxidation. { 'füd ₁prez·ər₁vā·shən }

food pyramid [ECOL] An ecological pyramid representing the food relationship among the animals in a community. { 'füd ₁pir·ə₁mid }

food science [FOOD ENG] The applied science which deals with the chemical, biochemical, physical, physiochemical, and biological properties of foods. { 'füd ₁sī·əns }

food spoilage [FOOD ENG] Deterioration in the color, flavor, odor, or consistency of a food product. { 'füd ₁spóil·ij }

foodstuff [AGR] Any substance that can be used to feed animals. { 'füd₁stəf }

food technology [FOOD ENG] The application of science and engineering to the refining, manufacturing, and handling of foods; many food technologists are food scientists rather than engineers. { 'füd tek₁näl·ə·jē }

food vacuole [CYTOL] A membrane-bound organelle in which digestion occurs in cells capable of phagocytosis. Also known as heterophagic vacuole; phagocytic vacuole. { 'füd ₁vak·yə₁wōl }

food web [ECOL] A modified food chain that expresses feeding relationships at various, changing trophic levels. { 'füd ₁web }

fool's gold See pyrite. { ¦fülz ¦gōld }

foot [ANAT] Terminal portion of a vertebrate leg. [BOT] In a fern, moss, or liverwort, the basal part of the young sporophyte that attaches it to the gametophyte. [INV ZOO] An organ for locomotion or attachment. [MECH] The unit of length in the British systems of units, equal to exactly 0.3048 meter. Abbreviated ft. { fút }

footage [ENG] The extent or length of a material expressed in feet. [MIN ENG] **1.** The number of feet of borehole drilled per unit of time, or that required to complete a specific project or contract. **2.** The payment of miners by the running foot of work. { 'fúd·ij }

foot-and-mouth disease [VET MED] A highly contagious virus disease of cattle, pigs, sheep, and goats that is transmissible to humans; characterized by fever, salivation, and formation of vesicles in the mouth and pharynx and on the feet. Also known as hoof-and-mouth disease. { ¦fút ən 'maúth di₁zēz }

foot block [ENG] Flat pieces of wood placed under props in tunneling to give a broad base and thus prevent the superincumbent weight from pressing the props down. { 'fút ₁bläk }

foot breadth [ARCH] The measure of the maximum distance across the left foot, when the subject stands with his or her weight evenly distributed on both feet. { 'fút ₁bredth }

foot bridge [CIV ENG] A bridge structure used only for pedestrian traffic. { 'fút ₁brij }

footcandle [OPTICS] A unit of illumination, equal to the illumination of a surface, 1 square foot in area, on which there is a luminous flux of 1 lumen uniformly distributed, or equal to the illumination of a surface all points of which are at a distance of 1 foot from a uniform point source of 1 candela; equal to approximately 10.7639 lux. Abbreviated ftc. { 'fút ₁kand·əl }

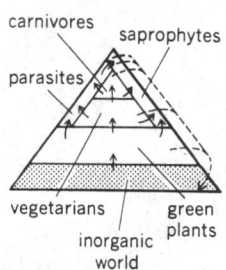

FOOD PYRAMID

Diagram of the food pyramid showing the general trends in food circulation.

footeite See connellite. { 'fút₁īt }

footer [GRAPHICS] Text that is printed at the bottom of every page of a document or report; may contain the page number. { 'fúd·ər }

foot gland [INV ZOO] A glandular structure which secretes an adhesive substance in many animals. Also known as pedal gland. { 'fút ₁gland }

foot guard [CIV ENG] A filler placed on the space between converging rails to prevent a foot from being wedged between the rails. { 'fút ₁gärd }

foothills [GEOGR] A region of relatively low, rounded hills at the base of, or on the periphery of, a mountain range. { 'fút₁hilz }

foot holes [MIN ENG] Holes cut in the sides of shafts or winzes to enable miners to climb up or down. { 'fút ₁hōlz }

footing [CIV ENG] The widened base or substructure forming the foundation for a wall or a column. { 'fúd·iŋ }

footlambert [OPTICS] A unit of luminance (photometric brightness), equal to $1/\pi$ candela per square foot, or to the uniform luminance of a perfectly diffusing surface emitting or reflecting light at the rate of 1 lumen per square foot; equal to approximately 3.42625 nit. Abbreviated ft-L. Also known as equivalent footcandle. { 'fút ¦lam·bərt }

foot length [ARCH] A measure of the distance from the heel to the longest toe of the left foot, when the subject stands with the weight evenly distributed on both feet. { 'fút ¦leŋkth }

footpad [AERO ENG] A somewhat flat base on the leg of a spacecraft to distribute weight and thereby minimize sinking into a surface. { 'fút₁pad }

footpoint [ASTRON] The intersection of tubes of magnetic field lines with the surface of the photosphere. { 'fút₁póint }

foot-pound [MECH] **1.** Unit of energy or work in the English gravitational system, equal to the work done by 1 pound of force when the point at which the force is applied is displaced 1 foot in the direction of the force; equal to approximately 1.355818 joule. Abbreviated ft-lb; ft-lbf. **2.** Unit of torque in the English gravitational system, equal to the torque produced by 1 pound of force acting at a perpendicular distance of 1 foot from an axis of rotation. Also known as pound-foot. Abbreviated lbf-ft. { 'fút ¦paúnd }

foot-poundal [MECH] **1.** A unit of energy or work in the English absolute system, equal to the work done by a force of magnitude 1 poundal when the point at which the force is applied is displaced 1 foot in the direction of the force; equal to approximately 0.04214011 joule. Abbreviated ft-pdl. **2.** A unit of torque in the English absolute system, equal to the torque produced by a force of magnitude 1 poundal acting at a perpendicular distance of 1 foot from the axis of rotation. Also known as poundal-foot. Abbreviated pdl-ft. { 'fút ¦paúnd·əl }

foot-pound-second system of units See British absolute system of units. { ¦fút ¦paúnd ¦sek·ənd ₁sis·təm əv 'yü·nəts }

footprint [BUILD] A description of the exact size, shape, and location of a building's foundation as the foundation has been installed on a specific site. Also known as building footprint. [COMMUN] The area of the earth's surface that can be covered by a communications satellite at any given time. [COMPUT SCI] The amount and shape of the area occupied by equipment, such as a terminal or microcomputer, on desktop, floor, or other surface area. { 'fút₁print }

foot rot [PL PATH] Any disease that involves rotting of the stem or trunk of a plant. [VET MED] See foul foot. { 'fút ₁rät }

foot screw [ENG] **1.** One of the three screws connecting the tribach of a theodolite or other level with the plate screwed to the tripod head. **2.** An adjusting screw that serves also as a foot. { 'fút ₁skrü }

foot section [MECH ENG] In both belt and chain conveyors that portion of the conveyor at the extreme opposite end from the delivery point. { 'fút ₁sek·shən }

foot's oil [CHEM] The oil sweated out of slack wax; it takes its name from the fact that it goes to the bottom, or foot, of the pan when sweated. { 'fúts ₁óil }

footstock [MECH ENG] A device containing a center which supports the workpiece on a milling machine; usually used in conjunction with a dividing head. { 'fút₁stäk }

foot valve [MECH ENG] A valve in the bottom of the suction pipe of a pump which prevents backward flow of water. { 'fút ₁valv }

footwall [GEOL] The mass of rock that lies beneath a fault, an ore body, or a mine working. Also known as heading side; heading wall; lower plate. { 'fut,wol }

footwall shaft See underlay shaft. { 'fut,wol ,shaft }

forage [AGR] A vegetable food for domestic animals. { 'fär·ij }

Foraky boring method [MIN ENG] A percussive boring system; a closed-in derrick contains the crown pulley, over which a steel rope with the boring tools is passed from a drum; the drum moves the tools, which are vibrated by a walking beam. { fə'räk·ē 'bȯr·iŋ ,meth·əd }

Foraky freezing process [MIN ENG] A method of shaft sinking through heavily watered sands by freezing the sands. { fə'räk·ē 'frēz·iŋ 'präs·əs }

foramen [BIOL] A small opening, orifice, pore, or perforation. { fə'rā·mən }

foramen magnum [ANAT] A large oval opening in the occipital bone at the base of the cranium that allows passage of the spinal cord, accessory nerves, and vertebral arteries. { fə'rā·mən 'mag·nəm }

foramen of Magendie [ANAT] The median aperture of the fourth ventricle of the brain. { fə'rā·mən əv mə,zhän'dē }

foramen of Monro See interventricular foramen. { fə'rā·mən əv mən'rō }

foramen of Winslow See epiploic foramen. { fə'rā·mən əv 'winz·lō }

foramen ovale [ANAT] An opening in the sphenoid for the passage of nerves and blood vessels. [EMBRYO] An opening in the fetal heart partition between the two atria. { fə'rā·mən ō'vä·lē }

foramen primum [EMBRYO] A temporary embryonic interatrial opening. { fə'rā·mən 'prī·məm }

Foraminiferida [INV ZOO] An order of dominantly marine protozoans in the subclass Granuloreticulosia having a secreted or agglutinated shell enclosing the ameboid body. { fə¦ram·ə·nə'fer·ə·də }

forb [BOT] A weed or broadleaf herb. { fȯrb }

Forbes bar [THERMO] A metal bar which has one end immersed in a crucible of molten metal and thermometers placed in holes at intervals along the bar; measurement of temperatures along the bar together with measurement of cooling of a short piece of the bar enables calculation of the thermal conductivity of the metal. { 'fȯrbz ,bär }

forbesite [MINERAL] H(Ni,Co)AsO$_4$·3^1/$_2$H$_2$O A grayish-white mineral composed of hydrous nickel cobalt arsenate; occurs in fibrocrystalline form. { 'fȯrb,zīt }

Forbes log [NAV] A log consisting essentially of a small rotator in a tube projecting below the bottom of a vessel, with suitable registering devices. { 'fȯrbz ,läg }

forbidden band [SOLID STATE] A range of unallowed energy levels for an electron in a solid. { fər¦bid·ən 'band }

forbidden-character code [COMPUT SCI] A bit code which exists only when an error occurs in the binary coding of characters. { fər¦bid·ən 'kar·ik·tər ,kōd }

forbidden-combination check [COMPUT SCI] A test for the occurrence of a nonpermissible code expression in a computer; used to detect computer errors. { fər¦bid·ən ,käm·bə'nā·shən ,chek }

forbidden line [ATOM PHYS] A spectral line associated with a transition forbidden by selection rules; optically this might be a magnetic dipole or electric quadrupole transition. { fər¦bid·ən 'līn }

forbidden transition [QUANT MECH] A transition between two states of a quantum-mechanical system which is considerably less probable than a competing allowed transition. { fər¦bid·ən tran'zish·ən }

Forbush decrease [ASTROPHYS] A sudden decrease in cosmic-ray intensity which occurs a day or two after a solar flare, and at the same time as the commencement of magnetic storms and auroral activity. { 'fȯr,bush di'krēs }

force [COMPUT SCI] To intervene manually in a computer routine and cause the computer to execute a jump instruction. [MECH] That influence on a body which causes it to accelerate; quantitatively it is a vector, equal to the body's time rate of change of momentum. { fȯrs }

force-balance meter [ENG] A flowmeter that measures a force, such as that associated with the air pressure in a small bellows, that is required to balance the net force created by the differential pressure, on opposite sides of a diaphragm or diaphragm capsule, generated by a differential-producing primary device. { 'fȯrs ,bal·əns ,mēd·ər }

force compensation [ENG] On an analytical balance, the weight force of a load that is held in equilibrium by a force of equal size which acts in the opposite direction. { 'fȯrs ,käm·pən,sā·shən }

force constant [MECH] The ratio of the force to the deformation of a system whose deformation is proportional to the applied force. [PHYS CHEM] An expression for the force acting to restrain the relative displacement of the nuclei in a molecule. { 'fȯrs ,kän·stənt }

force-controlled motion commands [CONT SYS] Robot control in which motion information is provided by computer software but sensing of forces or feedback is used by the robot to adapt this information to the environment. { 'fȯrs kən¦trōld 'mō·shən kə,manz }

forced-air heating [MECH ENG] A warm-air heating system in which positive air circulation is provided by means of a fan or a blower. { ¦fȯrst ,er 'hēd·iŋ }

forced auxiliary ventilation [MIN ENG] A system in which the duct delivers the intake air to the face. { ¦fȯrst ȯg¦zil·yə·rē ,vent·əl'ā·shən }

forced-caving system [MIN ENG] A stoping system in which the ore is broken down by large blasts into the stopes that are kept partly full of broken ore. { ¦fȯrst 'kāv·iŋ ,sis·təm }

forced circulation [MECH ENG] The use of a pump or other fluid-movement device in conjunction with liquid-processing equipment to move the liquid through pipes and process vessels; contrasted to gravity or thermal circulation. { ¦fȯrst ,sər·kyə'lā·shən }

forced-circulation boiler [MECH ENG] A once-through steam generator in which water is pumped through successive parts. { ¦fȯrst ,sər·kyə'lā·shən ,bȯil·ər }

forced convection [THERMO] Heat convection in which fluid motion is maintained by some external agency. { ¦fȯrst kən'vek·shən }

forced draft [MECH ENG] Air under positive pressure produced by fans at the point where air or gases enter a unit, such as a combustion furnace. { ¦fȯrst 'draft }

forced expiratory volume [MED] During the performance of a forced vital capacity measurement, the volume of exhaled gas over a specific time interval. { ¦fȯrst ik¦spī·rə,tȯr·ē 'väl·yəm }

forced-flow boiling [PHYS CHEM] Boiling of a liquid whose flow over a heater surface is imposed by external means. { ¦fȯrst ¦flō 'bȯil·iŋ }

forced oscillation [MECH] An oscillation produced in a simple oscillator or equivalent mechanical system by an external periodic driving force. Also known as forced vibration. { ¦fȯrst ,äs·ə'lā·shən }

forced programming See minimum-access programming. { ¦fȯrst 'prō,gram·iŋ }

forced ventilation [MECH ENG] A system of ventilation in which air is forced through ventilation ducts under pressure. { ¦fȯrst ,vent·əl'ā·shən }

forced vibration See forced oscillation. { ¦fȯrst vī'brā·shən }

forced vital capacity [MED] Maximum gas volume which can be expired, as quickly and forcibly as possible, after a maximum inspiration. { ¦fȯrst 'vīd·əl kə'pas·əd·ē }

forced wave [FL MECH] Any wave which is required to fit irregularities at the boundary of a system or satisfy some impressed force within the system; the forced wave will not in general be a characteristic mode of oscillation of the system. { ¦fȯrst 'wāv }

force feedback [CONT SYS] A method of error detection in which the force exerted on the effector is sensed and fed back to the control, usually by mechanical, hydraulic, or electric transducers. { 'fȯrs ¦fēd,bak }

force field method See molecular mechanics. { 'fȯrs ¦fēld ,meth·əd }

force fit See press fit. { 'fȯrs ,fit }

force gage [ENG] An instrument which measures the force exerted on an object. { 'fȯrs ,gāj }

force main [CIV ENG] The discharge pipeline of a pumping station. { 'fȯrs ,mān }

force piece See foreset. { 'fȯrs ,pēs }

force plate [ENG] A plate that carries the plunger or force plug of a mold and the guide pins on bushings. { 'fȯrs ,plāt }

force plug [ENG] A mold member that fits into the cavity block, exerting pressure on the molding compound. Also known as piston; plunger. { 'fȯrs ‚pləg }

force polygon [MECH] A closed polygon whose sides are vectors representing the forces acting on a body in equilibrium. { ¦fȯrs 'päl·ə‚gän }

forceps [DES ENG] A pincerlike instrument for grasping objects. [INV ZOO] A pair of curved, hard, movable appendages at the end of the abdomen of certain insects, for example, the earwig. [MED] A device with two blades or limbs opposite each other which is operated by handles or by direct force on the blades; used in surgery to grasp, compress, and hold tissue, a body part, or surgical substances. { 'fȯr·səps }

force pump [MECH ENG] A pump fitted with a solid plunger and a suction valve which draws and forces a liquid to a considerable height above the valve or puts the liquid under a considerable pressure. { 'fȯrs ‚pəmp }

force ratio See mechanical advantage. { 'fȯrs ‚rā·shō }

force-time [IND ENG] The product of an applied force and its time of application; used for quantitative determination of isometric work. { ¦fȯrs ¦tīm }

forcing [GRAPHICS] The attempt to bring out detail in an underexposed negative by extending the development time or by using an accelerator. { 'fȯrs·iŋ }

forcing cone [ORD] Tapered beginning of the lands at the origin of the rifling of a gun tube; the forcing cone allows the rotating band of the projectile to be gradually engaged by the rifling, thereby centering the projectile in the bore. { 'fȯrs·iŋ ‚kōn }

forcing fan [MIN ENG] A fan which forces the intake air into mine workings. Also known as blowing fan. { 'fȯrs·iŋ ‚fan }

forcipate [BIOL] Shaped like forceps; deeply forked. { 'fȯr·sə‚pāt }

forcipate trophus [INV ZOO] A type of masticatory apparatus in certain predatory rotifers which resembles forceps and is used for grasping. { 'fȯr·sə‚pāt 'trō·fəs }

Forcipulatida [INV ZOO] An order of echinoderms in the subclass Asteroidea characterized by crossed pedicellariae. { fȯr¦sip·ə'lad·ə·də }

ford [HYD] A shallow and usually narrow part of a stream, estuary, or other body of water that may be crossed; for example, by wading or by a wheeled land vehicle. { fȯrd }

Ford-Fulkerson theorem [MATH] The theorem that in any *s-t* network there exists a feasible flow and an *s-t* cut such that (1) the flow equals the weight of the cut, (2) on any arc belonging to the cut, this flow equals the weight of the arc, and (3) on any arc, that would belong to the cut if its orientation were reversed, the flow equals zero. Also known as max-flow min-cut theorem. { ¦fȯrd 'fül·kər·sən ‚thir·əm }

fording depth [ENG] Maximum depth at which a particular vehicle can operate in water. { 'fȯrd·iŋ ‚depth }

fore [NAV ARCH] **1.** The front part of a ship. **2.** In the direction of or toward the bow. { fȯr }

forearc [GEOL] The area between the trench and the volcanic arc of a subduction zone. { 'fȯr‚ärk }

forearm [ANAT] The part of the upper extremity between the wrist and the elbow. Also known as antebrachium. { 'fȯr‚ärm }

forearm circumference [ARCH] The measure of the circumference taken halfway between the elbow and the wrist. { ¦fȯr‚ärm sər'kəm·frəns }

forearm length [ARCH] A measure of the distance from the tip of the elbow to the tip of the middle finger, with the arm flexed at the elbow. { 'fȯr‚ärm ‚leŋkth }

forebay [CIV ENG] **1.** A small reservoir at the head of the pipeline that carries water to the consumer; it is the last free water surface of a distribution system. **2.** A reservoir feeding the penstocks of a hydro-power plant. { 'fȯr‚bā }

forebody [NAV ARCH] The part of a ship's hull in front of the amidships. { 'fȯr‚bäd·ē }

forebrain [EMBRYO] The most anterior expansion of the neural tube of a vertebrate embryo. [VERT ZOO] The part of the adult brain derived from the embryonic forebrain; includes the cerebrum, thalamus, and hypothalamus. { 'fȯr‚brān }

forebulge [GEOL] An uplift at the edge of a glacier caused by tilting of the lithosphere. { 'fȯr‚bəlj }

forecast [METEOROL] A statement of expected future meteorological occurrences. [STAT] To assess the magnitude that

a quantity will have at a specified time in the future. Also known as predict. { 'fȯr‚kast }

forecasting [COMMUN] The prediction of conditions of radio propagation for a period extending anywhere from a few hours to a few months. [METEOROL] Procedures for extrapolation of the future characteristics of weather on the basis of present and past conditions. { 'fȯr‚kast·iŋ }

forecastle deck [NAV ARCH] A deck extending from the stem aft over a forecastle erection. { 'fōk·səl ‚dek }

forecast period [METEOROL] The time interval for which a forecast is made. { 'fȯr‚kast ‚pir·ē·əd }

forecast-reversal test [METEOROL] A test used to evaluate the adequacy of a given method of forecast verification; the same verification method is applied, simultaneously, to a given forecast and to a fabricated forecast of opposite conditions; comparison of the verification scores gives an indication of the value of the verification system. { ¦fȯr‚kast ri'vər·səl ‚test }

forecast verification [METEOROL] Any process for determining the accuracy of a weather forecast by comparing the predicted weather with the observed weather of the forecast period; used to test forecasting skills and methods. { ¦fȯr‚kast ‚ver·ə·fə'kā·shən }

foredeep [GEOL] **1.** A long, narrow depression that borders an orogenic belt, such as an island arc, on the convex side. **2.** See exogeosyncline. { 'fȯr‚dēp }

fore drift [MIN ENG] That one of a pair of parallel headings which is kept a short distance in advance of the other. { 'fȯr ‚drift }

foredune [GEOL] A coastal dune or ridge that is parallel to the shoreline of a large lake or ocean and is stabilized by vegetation. { 'fȯr‚dün }

forefinger [ANAT] The index finger; the first finger next to the thumb. { 'fȯr‚fiŋ·gər }

forefoot [NAV ARCH] The extreme forward end of the bottom of a ship. [VERT ZOO] An anterior foot of a quadruped. { 'fȯr‚füt }

foreground [COMPUT SCI] A program or process of high priority that utilizes machine facilities as needed, with less critical, background work performed in otherwise unused time. { 'fȯr‚graùnd }

foregut [EMBRYO] The anterior alimentary canal in a vertebrate embryo, including those parts which will develop into the pharynx, esophagus, stomach, and anterior intestine. { 'fȯr ‚gət }

forehand welding [MET] Welding in which the flame is directed against the base metal ahead of the weld and is moved in the direction of welding. Also known as forward welding. { ¦fȯr‚hand 'weld·iŋ }

forehead [ANAT] The part of the face above the eyes. { 'fär·əd }

forehearth [MET] **1.** A bay in front of the hearth of a furnace. **2.** A receptacle in front of a hearth to receive the molten products. { 'fȯr‚härth }

foreign-body locator [ENG] A device for locating foreign metallic bodies in tissue by means of suitable probes that generate a magnetic field; the presence of a magnetic body within this field is indicated by a meter or a sound signal. { ¦fär·ən ¦bäd·ē 'lō‚kād·ər }

foreign element [IND ENG] A work element which is not a part of the normal work cycle, either because it is accidental or because it occurs only occasionally. { ¦fär·ən 'el·ə·mənt }

foreign inclusion [PETR] A fragmentary piece of country rock which is enclosed in an igneous intrusion. { ¦fär·ən in'klü·zhən }

foreign matter [SCI TECH] Any substance not belonging naturally in the place where found. { 'fär·ən ‚mad·ər }

foreland [GEOGR] An extensive area of land jutting out into the sea. [GEOL] **1.** A lowland area onto which piedmont glaciers have moved from adjacent mountains. **2.** A stable part of a continent bordering an orogenic or mobile belt. { 'fȯr·lənd }

foreland facies See shelf facies. { 'fȯr·lənd ‚fā·shēz }

forelimb [ANAT] An appendage (as a wing, fin, or arm) of a vertebrate that is, or is homologous to, the foreleg of a quadruped. { 'fȯr‚lim }

forellenstein See troctolite. { fə'rel·ən‚stīn }

Forel scale [OCEANOGR] A scale of yellows, greens, and blues for recording the color of sea water as seen against the white background of a Secchi disk. { fȯ'rel ‚skāl }

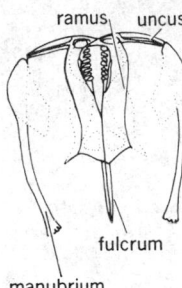

FORCIPATE TROPHUS

ramus   uncus

fulcrum

manubrium

Ventral view of *Dicranophorus* forcipate trophus.

forensic anthropology [FOREN SCI] The application of physical anthropology theory and techniques to answering legal questions involving human skeletal identification and analysis. { fə¦ren·sik ‚an·thrə'päl·ə·jē }

forensic biology [FOREN SCI] The analysis of the biological or genetic properties of evidence. { fə'ren·sik bī'äl·ə·jē }

forensic chemistry [FOREN SCI] The application of chemistry to the study of materials or problems in cases where the findings may be presented as technical evidence in a court of law. { fə'ren·sik 'kem·ə·strē }

forensic engineering [FOREN SCI] The application of accepted engineering practices and principles for discussion, debate, argumentative, or legal purposes. { fə'ren·sik ‚en·jə'nir·iŋ }

forensic entomology [FOREN SCI] The application of insect evidence to criminal investigations and civil cases. { fə¦ren·sik ‚en·tə'mä·lə·jē }

forensic medicine [FOREN SCI] Application of medical evidence or medical opinion for purposes of civil or criminal law. { fə'ren·sik 'med·ə·sən }

forensic odontology [FOREN SCI] A subspecialty of forensic medicine which focuses on the identification of deceased persons by dental examination, or of perpetrators by bite marks. { fə'ren·sik ‚ō·dän'täl·ə·jē }

forensic pathology [FOREN SCI] A subspecialty of forensic medicine which deals with the cause and manner of death. { fə'ren·sik pə'thäl·ə·jē }

forensic physics [FOREN SCI] The application of physics for discussion, debate, argumentative, or legal purposes. { fə'ren·sik 'fiz·iks }

forensic psychiatry [FOREN SCI] A branch of psychiatry dealing with legal issues related to mental disorders. { fə¦renz·ik sī'kī·ə·trē }

forensic science [FOREN SCI] The recognition, collection, identification, individualization, and interpretation of physical evidence, and the application of science and medicine for criminal and civil law, or regulatory purposes. Also known as criminalistics. { fə'ren·sik 'sī·əns }

forensic toxicology [FOREN SCI] An interdisciplinary field applying the methods of analytical chemistry, pharmacology, and toxicology to the analysis and interpretation of drugs and chemicals in biological samples for legal purposes. { fə'ren·sik ‚täk·sə'käl·ə·jē }

forepeak [NAV ARCH] A tank in the extreme forward end of a ship that is usually used for fresh water, but can carry ballast or bunkers. { 'for·pēk }

forepeak bulkhead [NAV ARCH] The partition wall nearest the stem; it forms the after-boundary of the forepeak tank. { 'for‚pēk 'bəlk‚hed }

forepoling [MIN ENG] A timbering method for a very weak roof in which a bench of timbers is set and boards or long wedges are placed above the header; as the next bench of timbers is placed at the inbye end of the wedges, other like wedges are driven in under the first wedges and over the second header. Also known as spiling. { 'for‚pōl·iŋ }

fore pump See backing pump. { 'for ‚pəmp }

forerunner [OCEANOGR] Low, long-period ocean swell which commonly precedes the main swell from a distant storm, especially a tropical cyclone. { 'for‚rən·ər }

foreset [MIN ENG] **1.** To place a prop under the coal-face end of a bar. **2.** Timber set used for roof support at the working face. Also known as force piece. { 'for‚set }

foreset bed [GEOL] One of a series of inclined symmetrically arranged layers of a cross-bedding unit formed by deposition of sediments that rolled down a steep frontal slope of a delta or dune. { 'for‚set ‚bed }

foreshaft sinking [MIN ENG] The first 150 feet (46 meters) of shaft sinking from the surface; the plant and services for the main shaft are installed during this step. { 'for‚shaft ‚siŋk·iŋ }

foreshock [GEOPHYS] A tremor which precedes a larger earthquake or main shock. { 'for‚shäk }

foreshore [GEOL] The zone that lies between the ordinary high- and low-watermarks and is daily traversed by the rise and fall of the tide. Also known as beach face. { 'for‚shor }

foresight [ENG] **1.** A sight or bearing on a new survey point, taken in a forward direction and made in order to determine its elevation. **2.** A sight on a previously established survey point, taken in order to close a circuit. **3.** A reading taken on a level rod to determine the elevation of the point on which the rod rests when read. Also known as minus sight. { 'for‚sīt }

forest [ECOL] An ecosystem consisting of plants and animals and their environment, with trees as the dominant form of vegetation. [MATH] See acyclic graph. { 'fär·əst }

forestaff See cross-staff. { 'for‚staf }

forestay [NAV ARCH] In sailing vessels, a stay extending from the foremast to the bow deck, bowsprit, or jibboom. { 'for‚stā }

forest climate See humid climate. { 'fär·əst ¦klī·mət }

forest conservation [ECOL] Those measures concerned with the protection and preservation of forest lands and resources. { 'fär·əst ‚kän·sər'vā·shən }

forest ecology [ECOL] The science that deals with the relationship of forest trees to their environment, to one another, and to other plants and to animals in the forest. { 'fär·əst i‚käl·ə·jē }

forest ecosystem [ECOL] The entire assemblage of forest organisms (trees, shrubs, herbs, bacteria, fungi, and animals, including people) together with their environmental substrate (the surrounding air, soil, water, organic debris, and rocks), interacting inside a defined boundary. { ¦fär·əst 'ek·ō‚sis·təm }

forest engineering [ENG] A branch of engineering concerned with the solution of forestry problems with regard to long-range environmental and economic effects. { 'fär·əst ‚en·jə‚nir·iŋ }

forest fire [FOR] Uncontrolled combustion of forest fuels. { 'fär·əst ‚fīr }

forest genetics [FOR] The study of variation and inheritance in forest trees; it provides the knowledge necessary to breed trees through traditional methods of selection and hybridization, and also through the newer biotechnologies. { ‚fär·əst jə'ned·iks }

forest management [FOR] Measures concerned with the effective organization of a forest to ensure continued production of its goods and services. { 'fär·əst ‚man·ij·mənt }

forest mapping [FOR] The branch of forestry dealing with the preparation of maps showing the distribution and conformation of individual forest stands. { 'fär·əst ‚map·iŋ }

forest measurement [FOR] The branch of forestry concerned with the measurement of standing trees, cut roundwood, and lumber products. { ¦fär·əst 'mezh·ər·mənt }

forest product [FOR] Any material afforded by a forest for commercial use, such as tree products and forage. { ¦fär·əst ¦präd·əkt }

forest resources [FOR] Forest land and the trees on it. { ¦fär·əst ri'sors·əs }

forestry [ECOL] The management of forest lands for wood, forages, water, wildlife, and recreation. { 'fär·ə·strē }

forest soil [FOR] The natural medium for growth of tree roots and associated forest vegetation. { 'fär·əst ‚soil }

forest stand [FOR] The basic unit of forest mapping; a group of trees that are more or less homogeneous with regard to species composition, density, size, and sometimes habitat. { 'fär·əst ‚stand }

forest-tundra [ECOL] A temperate and cold savanna which occurs at high altitudes and consists of scattered or clumped trees and a shrub layer of varying coverage. { ¦fär·əst 'tən·drə }

forest wind [METEOROL] A light breeze which blows from forests toward open country on calm clear nights. { ¦fär·əst 'wind }

forfeiture [MIN ENG] Loss of a mining claim by operation of the law, without regard to the intention of the locator, whenever he or she fails to preserve his or her right by complying with the conditions imposed by law. { 'for·fə·chər }

forge [MET] **1.** To form a metal, usually hot, into desirable shapes by employing compressive forces. **2.** A machine or place in which metal is formed hot, or where iron is produced from its ore. { forj }

forgeability [MET] Suitability of a material for forging. { 'for·jə'bil·əd·ē }

forge delay time [MET] The time between the start of weld time and the time when forging pressure is reached by the electrode force. { 'forj di'lā ‚tīm }

forge welding [MET] A group of welding processes in which the parts to be joined, usually iron, are heated to about 1000°C and then hammered or pressed together. Also known as fire welding. { 'fȯrj ,weld·iŋ }

forging [MET] **1.** Using compressive force to shape metal by plastic deformation; dies may be used. **2.** A piece of work made by forging. { 'fȯrj·iŋ }

forging brass [MET] Brass composed of 60% copper, 38% zinc, and 2% lead, used for hot forgings, hardware, and plumbing supplies; it is extremely plastic when hot, is corrosion-resistant, and has excellent mechanical properties. { 'fȯrj·iŋ ,bras }

forging hammer [MET] A hammer used to pound metal into forgings. { 'fȯrj·iŋ ,ham·ər }

forging plane [MET] The plane of the principal die face when oriented normal to the direction of ram travel. { 'fȯrj·iŋ ,plān }

forging press [MET] A press designed to operate dies in die forging. { 'fȯrj·iŋ ,pres }

forging range [MET] Optimum temperature range in which a metal can be forged. { 'fȯrj·iŋ ,rānj }

forging rolls [MET] A machine used in making forgings by rolling the metal. { 'fȯrj·iŋ ,rōlz }

forging stock [MET] A section or a piece of metal used to make a forging. { 'fȯrj·iŋ ,stäk }

forked lightning [GEOPHYS] A common form of lightning, in a cloud-to-ground discharge, which exhibits downward-directed branches from the main lightning channel. { ¦fȯrkt 'līt·niŋ }

fork-join model [COMPUT SCI] A method of programming on parallel machines in which one or more child processes branch out from the root task when it is time to do work in parallel, and end when the parallel work is done. { 'fȯrk ¦jȯin 'mäd·əl }

forklift [MECH ENG] A machine, usually powered by hydraulic means, consisting of two or more prongs which can be raised and lowered and are inserted under heavy materials or objects for hoisting and moving them. { 'fȯrk,lift }

forklift truck *See* fork truck. { 'fȯrk,lift ,trək }

fork oscillator [ELECTR] An oscillator that uses a tuning fork as the frequency-determining element. { ¦fȯrk ,äs·ə¦lād·ər }

fork pocket [MECH ENG] An opening in the base of a container or pallet for insertion of the prong of a forklift. { 'fȯrk ,pak·ət }

fork truck [MECH ENG] A vehicle equipped with a forklift. Also known as forklift truck. { 'fȯrk ,trək }

form [CIV ENG] Temporary boarding, sheeting, or pans of plywood, molded fiber glass, and so forth, used to give desired shape to poured concrete or the like. [GRAPHICS] Type and material that is secured in a chase and is ready for printing or for producing an electrotype plate. { fȯrm }

formability [MATER] Capability of a material to be shaped by plastic deformation. { ,fȯr·mə'bil·əd·ē }

formal charge [PHYS CHEM] The apparent charge of an element in a compound; for example, magnesium has a formal charge of +2 in MgO and oxygen has a charge of −2. { ¦fȯr·məl ¦chärj }

formaldehyde [ORG CHEM] HCHO The simplest aldehyde; a gas at room temperature, and a poisonous, clear, colorless liquid solution with pungent odor; used to make synthetic resins by reaction with phenols, urea, and melamine, as a chemical intermediate, as an embalming fluid, and as a disinfectant. Also known as formol; methanal; methylene oxide. { fȯr'mal·də,hīd }

formaldehyde sodium bisulfite [ORG CHEM] CH_3NaO_4S A compound used as a fixing agent for fibers containing keratin, in metallurgy for flotation of lead-zinc ores, and in photography. { fȯr'mal·də,hīd ¦sōd·ē·əm bī'səl,fīt }

formal derivative [MATH] For a polynomial, $a_n x^n + a_{n-1} x^{n-1} + \cdots + a_1 x + a_0$, where the coefficients $a_0, a_1, \ldots, a_n$ are elements of a ring, the formal derivative is the polynomial $n a_n x^{n-1} + (n-1) a_{n-1} x^{n-2} + \cdots + a_1$. { ,fȯrm·əl də'riv·əd·iv }

formalin [MATER] An aqueous solution of formaldehyde, usually 37% formaldehyde by weight. { 'fȯr·mə·lən }

formality [CHEM] A concentration scale that gives the number of formula weights of solute per liter of solution; designated by *F* preceded by a number to show solute concentration. { fȯr'mal·əd·ē }

formal language [COMPUT SCI] An abstract mathematical

object used to model the syntax of a programming or natural language. { ¦fȯr·məl 'laŋ·gwij }

formal logic [MATH] The study of the permissible relationships between propositions, a study that concerns the form rather than the content. { ¦fȯr·məl 'läj·ik }

formal power series [MATH] A power series whose convergence is disregarded, but which is subject to the operations of addition and multiplication with other such series. { ¦fȯr·məl 'paü·ər ,sir·ēz }

formamidase [BIOCHEM] An enzyme involved in tryptophane catabolism; catalyzes the conversion of *N*-formylkynurenine to kynurenine and formate. { fȯr'mam·ə,dās }

formamide [ORG CHEM] **1.** A compound containing the radical HCONH. **2.** $HCONH_2$ A clear, colorless hygroscopic liquid, boiling at 200–212°C; soluble in water and alcohol; used as a solvent, softener, and chemical intermediate. Also known as formylamine; methanamide. { 'fȯrm'am·əd }

formamidinesulfinic acid [ORG CHEM] $H_2NC(NH)SO_2H$ A reagent for the reduction of ketones to secondary alcohols. { fȯr¦mam·ə,dēn·səl'fin·ik 'as·əd }

formanite [MINERAL] A mineral composed of an oxide of uranium, zirconium, thorium, calcium, tantalum, and niobium with some rare-earth metals. { 'fȯr·mə,nīt }

formant [ACOUS] A set of resonances of a musical instrument or voice mechanism that form partials of sounds produced by the instrument, independent of the fundamental frequency, and give these sounds their quality. { 'fȯr·mənt }

format [COMPUT SCI] **1.** The specific arrangement of data on a printed page, display screen, or such, or in a record, data file, or storage device. **2.** To prepare a disk to store information by using a special program that divides the disk into storage units such as tracks and sectors. { 'fȯr,mat }

formate [ORG CHEM] A compound containing the HCOO− functional group. { 'fȯr,māt }

format effector *See* layout character. { 'fȯr,mat i,fek·tər }

formation [GEOL] Any assemblage of rocks which have some common character and are mappable as a unit. { fȯr'mā·shən }

formation bombing [ORD] Bombing by aircraft in formation. { fȯr'mā·shən ,bäm·iŋ }

formation damage [PETRO ENG] A reduction in the permeability of reservoir rock caused by penetration of drilling fluid and treating fluids into the section adjacent to the wellbore. { fȯr'mā·shən ,dam·ij }

formation factor [GEOCHEM] The ratio between the conductivity of an electrolyte and that of a rock saturated with the same electrolyte. Also known as resistivity factor. [GEOL] A function of the porosity and internal geometry of a reservoir rock system, expressed as $F = \phi^{-m}$, where ϕ is the fractional porosity of the rock, and m is the cementation factor (pore-opening reduction). { fȯr'mā·shən ,fak·tər }

formation fracturing [PETRO ENG] Method of applying hydraulic pressure to a reservoir formation to cause the rock to split open, that is, to fracture; used to increase oil production. { fȯr'mā·shən 'frak·chə·riŋ }

formation gas [PETRO ENG] The first gas produced from an underground reservoir. { fȯr'mā·shən ,gas }

formation microscanner [PETRO ENG] A downhole logging tool that provides a detailed image of the borehole wall on the basis of electrical contrasts. { fȯr¦mā·shə 'mīk·rō,skan·ər }

formation pressure *See* reservoir pressure. { fȯr'mā·shən ,presh·ər }

formation resistivity [GEOPHYS] Electrical resistivity of reservoir formations measured by electrical log sondes; used for clues to formation lithography and fluid content. { fȯr'mā·shən ri,zis'tiv·əd·ē }

formation solubility [PETRO ENG] Measure of formation rock solubility in oil-well acidizing solution (hydrochloric acid or hydrochloric-hydrofluoric acids). { fȯr'mā·shən ,säl·yə'bil·əd·ē }

formation tester [PETRO ENG] Device for retrieval of samples of fluid from an oil-reservoir formation. { fȯr'mā·shən ,test·ər }

formation water [HYD] Water present with petroleum or gas in reservoirs. Also known as oil-reservoir water. { fȯr'mā·shən ,wȯd·ər }

formatted tape [COMPUT SCI] A magnetic tape which employs a prerecorded timing track by means of which blocks

FORKLIFT

A high-lift forklift.

of data can be found after reference to a directory table. { 'fȯr¦mad·əd 'tāp }

formatting [COMPUT SCI] The preparation of a magnetic storage device to receive data structures; for example, the recording of track and sector information on a floppy disk. { 'fȯr,mad·iŋ }

form birefringence [OPTICS] Birefringence of a liquid caused by the orientation of rod-shaped particles in the liquid whose thickness and separation are much smaller than a wavelength of light. { ¦fȯrm ,bī·ri¦frin·jəns }

form clamp [CIV ENG] An adjustable metal clamp used to secure planks of wooden forms for concrete columns or beams. { 'fȯrm ,klamp }

form contour [MAP] A topographic contour determined either by stereoscopic examination of aerial photographs without the use of ground control or by some other method besides conventional surveying. { 'fȯrm ,kän,tür }

form cutter See formed cutter. { 'fȯrm ,kəd·ər }

form drag [FL MECH] **1.** The drag from all causes resulting from the particular shape of a body relative to its direction of motion, as of fuselage, wing, or nacelle. **2.** At supersonic speed, the drag caused by losses due to shock waves, exclusive of losses due to skin friction. { 'fȯrm ,drag }

formed cutter [MECH] A cutting tool shaped to make surfaces with irregular geometry. Also known as form cutter. { ¦fȯrmd 'kəd·ər }

form factor [ELEC] **1.** The ratio of the effective value of a periodic function, such as an alternating current, to its average absolute value. **2.** A factor that takes the shape of a coil into account when computing its inductance. Also known as shape factor. [MECH] The theoretical stress concentration factor for a given shape, for a perfectly elastic material. [PHYS] A function which describes the internal structure of a particle, allowing calculations to be made even though the structure is unknown. [QUANT MECH] An expression used in studying the scattering of electrons or radiation from atoms, nuclei, or elementary particles, which gives the deviation from point particle scattering due to the distribution of charge and current in the target. { 'fȯrm ,fak·tər }

form feed character [COMPUT SCI] A control character that determines when a printer or display device moves to the next page, form, or equivalent unit of data. { 'fȯrm ¦fēd ,kar·ik·tər }

form feeding [COMPUT SCI] The positioning of documents in order to move them past printing or sensing devices, either singly or in continuous rolls. { 'fȯrm ¦fēd·iŋ }

form feed printer [COMPUT SCI] A computer printer that accepts continuous forms or continuous sheets of paper. { 'fȯrm ¦fēd ,print·ər }

form function [ORD] The mathematical expression for the relationship between the fraction of the propellant burned and the distance that each burning surface has regressed. { 'fȯrm ,fəŋk·shən }

form grinding [MECH ENG] Grinding by use of a wheel whose cutting face is contoured to the reverse shape of the desired form. { 'fȯrm ,grīnd·iŋ }

formic acid [ORG CHEM] HCOOH A colorless, pungent, toxic, corrosive liquid melting at 8.4°C; soluble in water, ether, and alcohol; used as a chemical intermediate and solvent, in dyeing and electroplating processes, and in fumigants. Also known as methanoic acid. { ¦fȯr·mik 'as·əd }

Formicariidae [VERT ZOO] The antbirds, a family of suboscine birds in the order Passeriformes. { ,fȯr·mə·kə¦rī·ə,dē }

formication [PSYCH] An abnormal sensation as of insects crawling in or upon the skin; a common symptom in diseases of the spinal cord and the peripheral nerves; may be a hallucination. { ,fȯr·mə¦kā·shən }

formic ether See ethyl formate. { 'fȯr·mik 'ē·thər }

Formicidae [INV ZOO] The ants, social insects composing the single family of the hymenopteran superfamily Formicoidea. { fȯr'mis·ə,dē }

formicivorous [ZOO] Feeding on ants. { ,fȯr·mə'siv·ə·rəs }

Formicoidea [INV ZOO] A monofamilial superfamily of hymenopteran insects in the suborder Apocrita, containing the ants. { ,fȯr·mə'kȯid·ē·ə }

forming [ELEC] Application of voltage to an electrolytic capacitor, electrolytic rectifier, or semiconductor device to produce a desired permanent change in electrical characteristics as a part of the manufacturing process. [MECH ENG] A process for shaping or molding sheets, rods, or other pieces of hot glass, ceramic ware, plastic, or metal by the application of pressure. { 'fȯrm·iŋ }

forming die [ENG] A die like a drawing die, but without a blank holder. { 'fȯrm·iŋ ,dī }

forming press [MECH ENG] A punch press for forming metal parts. { 'fȯrm·iŋ ,pres }

forming rolls [MECH ENG] Rolls contoured to give a desired shape to parts passing through them. { 'fȯrm·iŋ ,rōlz }

forming tool [DES ENG] A nonrotating tool that produces its inverse form on the workpiece. { 'fȯrm·iŋ ,tül }

form line [MAP] An approximation of a contour line without a definite elevation value, as one derived by visual observation, sometimes supplemented by measured elevations but not in sufficient quantity to produce accurate results; used principally to indicate the appearance of terrain which has not been accurately surveyed. { 'fȯrm ,līn }

form of the physical store [COMPUT SCI] The code store considered as a physical structure, which can exhibit many different forms: discrete (cards), continuous (tapes), linear (tapes), cylindrical (drums), three-dimensional (array of cores), disks, strips, sheets, reels, and so on. { 'fȯrm əv thə ¦fiz·ə·kəl 'stȯr }

form oil [MATER] An oil utilized on the contact surface of wooden or metal concrete forms to prevent concrete from sticking. { 'fȯrm ,ȯil }

formol See formaldehyde. { 'fȯr,mȯl }

formonitrile See hydrocyanic acid. { ¦fȯr·mō¦nī·trəl }

form process chart [IND ENG] A graphic representation of the process flow of paperwork forms. Also known as forms analysis chart; functional forms analysis chart; information process analysis chart. { ¦fȯrm ¦präs·əs ,chärt }

form roller [GRAPHICS] An inking or dampening roller which is in direct contact with the plate on a printing press. { 'fȯrm ,rōl·ər }

forms [COMPUT SCI] Web pages that allow users to fill in and submit information, they are written in HTML and processed by CGI scripts. { fȯrmz }

forms analysis chart See form process chart. { ¦fȯrmz ə¦nal·ə·səs ,chärt }

form scabbing [CIV ENG] In placing of concrete using formwork, removal of the surface layer of concrete that adheres to the form when it is removed. { 'fȯrm ,skab·iŋ }

forms control buffer [COMPUT SCI] A reserved storage containing coordinates for a page position on the printer; earlier printers utilized a carriage control tape, allowing the page to be set at a specific position. { 'fȯrmz kən,trōl 'bəf·ər }

form stop [COMPUT SCI] A device which stops a machine when its supply of paper has run out. { 'fȯrm ,stäp }

formula [CHEM] **1.** A combination of chemical symbols that expresses a molecule's composition. **2.** A reaction formula showing the interrelationship between reactants and products. [MATH] An equation or rule relating mathematical objects or quantities. { 'fȯr·myə·lə }

formulation [CHEM] The particular mixture of base chemicals and additives required for a product. { ,fȯr·myə'lā·shən }

formula translation See FORTRAN. { ¦fȯr·myə·lə tranz¦lā·shən }

formula weight [CHEM] **1.** The gram-molecular weight of a substance. **2.** In the case of a substance of uncertain molecular weight such as certain proteins, the molecular weight calculated from the composition, assuming that the element present in the smallest proportion is represented by only one atom. { 'fȯr·myə·lə ,wāt }

formwork [CIV ENG] A temporary wooden casing used to contain concrete during its placing and hardening. Also known as shuttering. { 'fȯrm,wərk }

form-wound coil [ELEC] Armature coil that is formed or shaped over a fixture before being placed on the armature of a motor or generator. { 'fȯrm ¦waùnd ,kȯil }

formyl [ORG CHEM] The formic acid radical, HCO−; it is characteristic of aldehydes. { 'fȯr,mil }

formylamine See formamide. { ¦fȯr·məl'am,ēn }

formyl methionine [BIOCHEM] Formylated methionine; initiates peptide chain synthesis in bacteria. { 'fȯr,mil mə'thī·ə,nēn }

Fornax A [ASTRON] A peculiar giant elliptical galaxy on the periphery of the Fornax cluster which is a strong double radio source. { 'fȯr‚naks 'ā }

Fornax cluster [ASTRON] A cluster of galaxies with a few tens of members, about 30 megaparsecs (6×10^{19} miles or 9×10^{19} kilometers) distant. { 'fȯr‚naks ‚kləs·tər }

Fornax system [ASTRON] A dwarf elliptical galaxy in the Local Group, about 460,000 light-years (2.7×10^{18} miles or 4.4×10^{18} kilometers) distant, having a diameter of about 1600 light-years (9×10^{15} miles or 1.5×10^{16} kilometers) and a mass and luminosity about 7×10^6 that of the sun. { 'fȯr‚naks ‚sis·təm }

for-next loop [COMPUT SCI] In computer programming, a high-level logic statement which defines a part of a computer program that will be repeated a certain number of times. { ¦fȯr ¦nekst ‚lüp }

fornix [ANAT] A structure that is folded or arched. [BOT] A small scale, especially in the corolla tube of some plants. { 'fȯr‚niks }

Forrel cell [METEOROL] A type of atmospheric circulation in which air moves away from the thermal equator at low latitude levels and in the opposite direction in higher latitudes. { fə'rel ‚sel }

Forrester machine [MIN ENG] A pneumatic flotation cell in which pulp is aerated by low-pressure air, delivering a mineralized froth along the overflow, and tailings to the end weir. { 'fär·ə·stər mə‚shēn }

forril farina See rock milk. { 'fär·əl fə‚rēn·ē }

fors See G; gram-force. { fȯrs }

Forssman antibody [IMMUNOL] A heterophile antibody that reacts with Forssman antigen. { 'fȯrs·mən 'an·tə‚bäd·ē }

Forssman antigen [IMMUNOL] A heterophile antigen, occurring in a variety of unrelated animals, which elicits production of hemolysin (Forssman antibody) for sheep red blood cells. { 'fȯrs·mən 'an·tə‚jən }

FOR statement [COMPUT SCI] A statement in a computer program that is repeatedly executed a specified number of times, generally while a control variable takes on successive values over a specified range. { 'fȯr ‚stāt·mənt }

forsterite [MINERAL] Mg_2SiO_4 A whitish or yellowish, magnesium-rich variety of olivine. Also known as white olivine. { 'fȯr·stə‚rīt }

fort [ORD] **1.** Permanent post as opposed to a camp, which is a temporary installation. **2.** Land area within which harbor defense units are located. { fȯrt }

Forth [COMPUT SCI] A high-level programming language developed primarily for microcomputers and characterized by a number of features that make it highly adaptable and readily extensible, such as the ability to be used as an interpreter or an operating system. { fȯrth }

fortification [ORD] **1.** A structure or earthworks, usually heavily armed, constructed as a defense; a fortified place or position. **2.** The act or art of fortifying. { ‚fȯrd·ə·fə'kā·shən }

fortification agate See landscape agate. { ‚fȯrd·ə·fə'kā·shən 'ag·ət }

Fortin barometer [ENG] A type of cistern barometer; provision is made to increase or decrease the volume of the cistern so that when a pressure change occurs, the level of the cistern can be maintained at the zero of the barometer scale (the ivory point). { 'fȯrd·ən bə'räm·əd·ər }

fortnightly nutation [ASTRON] Nutation caused by the change in declination of the moon, having a displacement of up to 0.1 second of arc and a period of 15 days. { ‚fȯrt¦nīt·lē nü'tā·shən }

fortnightly tide [OCEANOGR] A tide occurring at intervals of one-half the period of oscillation of the moon, approximately 2 weeks. { ‚fȯrt¦nīt·lē 'tīd }

FORTRAN [COMPUT SCI] A family of procedure-oriented languages used mostly for scientific or algebraic applications; derived from formula translation. { 'fȯr‚tran }

Fortrat parabola [SPECT] Graph of wave numbers of lines in a molecular spectral band versus the serial number of the successive lines. { ‚fȯrträ pə'rab·ə·lə }

fortuitous distortion [COMMUN] Distortion in a telegraph system which includes effects that cannot be classified as bias or characteristic distortion; it is a departure (for one occurrence of a particular signal pulse) from the average combined effects of bias and characteristic distortion; the direct opposite of systematic distortion. { fȯr'tü·əd·əs di'stȯr·shən }

Fortuna [ASTRON] An asteroid with a diameter of about 130 miles (210 kilometers), mean distance from the sun of 2.44 astronomical units, and C-type surface composition. { fȯr'tü·nə }

forum See newsgroup. { 'fȯr·əm }

forty-four-type repeater [ELECTR] Type of telephone repeater employing two amplifiers and no hybrid arrangements; used in a four-wire system. { ¦fȯrd·ē‚fȯr ¦tīp ri'pēd·ər }

Forty Saints' storm [METEOROL] A southerly gale in Greece, occurring a little before the equinox in March. { ¦fȯrd·ē ¦sāns ‚stȯrm }

forward [NAV] In a direction nearer dead ahead than dead astern. { 'fȯr·wərd }

forward-acting regulator [ELECTR] Transmission regulator in which the adjustment made by the regulator does not affect the quantity which caused the adjustment. { 'fȯr·wərd ‚ak·tiŋ 'reg·yə‚lād·ər }

forward-backward counter [COMPUT SCI] A counter that has both an add and a subtract input so as to count in either an increasing or a decreasing direction. Also known as bidirectional counter. { ¦fȯr·wərd ¦bak·wərd 'kaunt·ər }

forward bias [ELECTR] A bias voltage that is applied to a pn-junction in the direction that causes a large current flow; used in some semiconductor diode circuits. { ¦fȯr·wərd 'bī·əs }

forward chaining [COMPUT SCI] In artificial intelligence, a method of reasoning which begins with a statement of all the relevant data and works toward the solution using the system's rules of inference. { ¦fȯr·wərd 'chān·iŋ }

forward compatibility See upward compatibility. { ¦fȯr·wərd kəm‚pad·ə'bil·əd·ē }

forward coupler [ELECTR] Directional coupler used to sample incident power. { ¦fȯr·wərd 'kəp·lər }

forward current [ELECTR] Current which flows upon application of forward voltage. { ¦fȯr·wərd 'kə·rənt }

forward difference [MATH] One of a series of quantities obtained from a function whose values are known at a series of equally spaced points by repeatedly applying the forward difference operator to these values; used in interpolation or numerical calculation and integration of functions. { ¦fȯr·wərd 'dif·rəns }

forward difference operator [MATH] A difference operator, denoted Δ, defined by the equation $\Delta f(x) = f(x + h) - f(x)$, where h is a constant indicating the difference between successive points of interpolation or calculation. { ¦fȯr·wərd ¦dif·rəns 'äp·ə‚rād·ər }

forward direction [ELECTR] Of a semiconductor diode, the direction of lower resistance to the flow of steady direct current. { ¦fȯr·wərd də'rek·shən }

forward drop [ELECTR] The voltage drop in the forward direction across a rectifier. { ¦fȯr·wərd 'dräp }

forward error analysis [COMPUT SCI] A method of error analysis based on the assumption that small changes in the input data lead to small changes in the results, so that bounds for the errors in the results caused by rounding or truncation errors in the input can be calculated. { ¦fȯr·wərd 'er·ər ə‚nal·ə·səs }

forward error correction [COMMUN] The location and correction of errors occurring in data communications by the receiver without retransmission of data. { ¦fȯr·wərd 'er·ər kə‚rek·shən }

forward extrusion [MET] A cold extrusion process in which a formed blank is placed in a die cavity and struck by a punch; the metal is extruded through an annular space between the die and the end of the punch, moving in the same direction as the punch. { ¦fȯr·wərd ik'strü·zhən }

forward-looking infrared imager [ENG] An infrared imaging device which employs an optomechanical system to make a two-dimensional scan, and produces a visible image corresponding to the spatial distribution of infrared radiation. Abbreviated FLIR imager. Also known as framing imager. { 'fȯr·wərd ¦luk·iŋ ‚in·frə‚red 'im·ij·ər }

forward of the beam [NAV] Any direction between broad on the beam and ahead. { 'fȯr·wərd əv thə 'bēm }

forward pass [ENG] In project management, scheduling from a known start date and calculating the finish date by proceeding from the first operation to the last. Also known as forward scheduling. { ¦fȯr·wərd 'pas }

forward path [CONT SYS] The transmission path from the

loop actuating signal to the loop output signal in a feedback control loop.   { 'fȯr·wərd ‚path }

forward perpendicular   [NAV ARCH]   A vertical line through the intersection of the design waterline and the forward end of a ship. Abbreviated FP.   { 'fȯr·wərd ‚pər·pən'dik·yə·lər }

forward propagation by ionospheric scatter   [COMMUN] Radio communications technique using the scattering phenomenon exhibited by electromagnetic waves in the 30–100-megahertz region when passing through the ionosphere at an elevation of about 50 miles (85 kilometers).   { ¦fȯr·wərd ‚präp·ə‚gā·shən bī ī'än·ə‚sfir·ik ‚skad·ər }

forward propagation by tropospheric scatter   [COMMUN] Radio communications technique using high transmitting power levels, large antenna arrays, and the scattering phenomenon of the troposphere to permit communications far beyond line-of-sight distances.   { ¦fȯr·wərd ‚präp·ə‚gā·shən bī 'träp·ə‚sfir·ik ‚skad·ər }

forward quarter   [NAV ARCH]   The portions of the sides of a ship immediately abaft the stem.   { 'fȯr·wərd 'kwärd·ər }

forward recovery time   [ELECTR]   Of a semiconductor diode, the time required for the forward current or voltage to reach a specified value after instantaneous application of a forward bias in a given circuit.   { 'fȯr·wərd ri'kəv·ə·re ‚tim }

forward reference   [COMPUT SCI]   Reference to a data element that has not yet been defined in the program being compiled.   { 'fȯr·wərd 'ref·rəns }

forward resistance   [ELECTR]   The resistance of a semiconductor diode to current flow in the forward direction.   { 'fȯr·wərd ri'zis·təns }

forward scatter   [COMMUN]   **1.** Propagation of electromagnetic waves at frequencies above the maximum usable high frequency through use of the scattering of a small portion of the transmitted energy when the signal passes from an unionized medium into a layer of the ionosphere.   **2.** Collectively, the very-high-frequency forward propagation by ionospheric scatter and ultra-high-frequency forward propagation by tropospheric scatter communications techniques.   [GEOPHYS]   The scattering of radiant energy into the hemisphere of space bounded by a plane normal to the direction of the incident radiation and lying on the side toward which the incident radiation was advancing.   { ¦fȯr·wərd 'skad·ər }

forward scattering   [PHYS]   **1.** Scattering in which there is no change in the direction of motion of the scattered particles. **2.** Scattering in which the angle between the initial and final directions of motion of the scattered particles is less than 90°.   { ¦fȯr·wərd 'skad·ə·riŋ }

forward-scatter propagation   See scatter propagation.   { 'fȯr·wərd ¦skad·ər präp·ə'gā·shən }

forward scheduling   See forward pass.   { 'fȯr·wərd 'skej·əl·iŋ }

forward shift operator   See displacement operator.   { 'fȯr·wərd ¦shift 'äp·ə‚rād·ər }

forward transfer function   [CONT SYS]   In a feedback control loop, the transfer function of the forward path.   { 'fȯr·wərd 'tranz·fər ‚fəŋk·shən }

forward voltage drop   See diode forward voltage.   { 'fȯr·wərd 'vōl·tij ‚dräp }

forward wave   [ELECTR]   Wave whose group velocity is the same direction as the electron stream motion.   { 'fȯr·wərd ¦wāv }

forward welding   See forehand welding.   { 'fȯr·wərd ¦weld·iŋ }

FOSDIC II   [COMPUT SCI]   An electronic scanner which reads filmed images of punched cards, searches for cards containing specified information, and copies the selected information onto new cards for computer input. Derived from film optical scanning device for input to computers.   { 'fäz·dik 'tü }

foshagite   [MINERAL] $Ca_5Si_3O_{10}(OH)_2 \cdot 2H_2O$ A white mineral composed of a basic hydrous calcium silicate.   { 'fō·shə‚gīt }

fossa   [ANAT]   A pit or depression.   [VERT ZOO]   *Cryptoprocta ferox.* A Madagascan carnivore related to the civets.   { 'fäs·ə }

fossil   [PALEON]   The organic remains, traces, or imprint of an organism preserved in the earth's crust since some time in the geologic past.   { 'fäs·əl }

fossil dune   [GEOL]   An ancient desert dune.   { 'fäs·əl 'dün }

fossil fuel   [GEOL]   Any hydrocarbon deposit that may be used for fuel; examples are petroleum, coal, and natural gas.   { ¦fäs·əl 'fyül }

fossil ice   [HYD]   **1.** Relatively old ground ice found in regions of permafrost.   **2.** Underground ice in regions where present-day temperatures are not low enough to have formed it.   { ¦fäs·əl 'īs }

fossil man   [PALEON]   Ancient human identified from prehistoric skeletal remains which are archeologically earlier than the Neolithic.   { ¦fäs·əl 'man }

fossil permafrost   See passive permafrost.   { ¦fäs·əl 'pər·mə‚frȯst }

fossil reef   [GEOL]   An ancient reef.   { ¦fäs·əl 'rēf }

fossil resin   [GEOL]   A natural resin in geologic deposits which is an exudate of long-buried plant life; for example, amber, retinite, and copal.   { ¦fäs·əl 'rez·ən }

fossil soil   See paleosol.   { ¦fäs·əl 'sȯil }

fossil turbulence   [METEOROL]   Inhomogeneities of temperature and humidity remaining in the air after the motion which produced them has subsided and the density has become uniform; causes scattering of radio waves, and lumpy clouds when air is rising.   { ¦fäs·əl 'tər·byə·ləns }

fossil wax   See ozocerite.   { ¦fäs·əl 'waks }

fossorial   [VERT ZOO]   Adapted for digging.   { fä'sȯr·ē·əl }

Foster-Seely discriminator   See phase-shift discriminator.   { 'fȯs·tər ¦sē·lē di'skrim·ə‚nād·ər }

Foster's formula   [MIN ENG]   The empirical formula $R = 3\sqrt{DT}$ for determining the radius R of a shaft pillar, where D = depth in feet and T = thickness of lode in feet.   { 'fȯs·tərz 'fȯr·myə·lə }

Foster's reactance theorem   [CONT SYS]   The theorem that the most general driving point impedance or admittance of a network, in which every mesh contains independent inductance and capacitance, is a meromorphic function whose poles and zeros are all simple and occur in conjugate pairs on the imaginary axis, and in which these poles and zeros alternate.   { 'fȯs·tərz rē'ak·təns ‚thir·əm }

Foucault current   See eddy current.   { fü'kō ¦kə·rənt }

Foucault knife-edge test   [OPTICS]   Test of a lens or a concave mirror in which a pinhole source is placed at twice the focal length behind the lens or at the mirror's center of curvature, the eye is placed at the image of the pinhole, and defects in the lens or mirror result in irregular darkening of the image when a knife edge is moved across the image immediately in front of the eye.   { fü'kō 'nīf ‚ej ‚test }

Foucault mirror   [OPTICS]   Experiment for measuring the speed of light in which light is reflected from a rapidly rotating mirror to a distant mirror and back, and the speed of light is deduced from the displacement of the beam after its second reflection from the rotating mirror, the angular speed of the rotating mirror, and the distance the light travels.   { fü'kō ¦mir·ər }

Foucault pendulum   [MECH]   A swinging weight supported by a long wire, so that the wire's upper support restrains the wire only in the vertical direction, and the weight is set swinging with no lateral or circular motion; the plane of the pendulum gradually changes, demonstrating the rotation of the earth on its axis.   { fü'kō 'pen·jə·ləm }

fougasse   [ORD]   A mine constructed so that upon explosion of the charge, pieces of metal, rock, gasoline, or other substances are blown in a predetermined direction.   { 'fü‚gas }

foulard   [TEXT]   Lightweight twill, often printed in a small pattern; originally silk, but frequently woven of rayon, wool, cotton, or nylon.   { fü'lärd }

foul berth   [NAV]   A berth in which a vessel at anchor is in danger of striking or fouling another vessel, the ground, or an obstruction.   { faul 'bərth }

foul bottom   [CIV ENG]   A hard, uneven, rocky or obstructed bottom having poor holding qualities for anchors, or one having rocks or wreckage that would endanger an anchored vessel.   [NAV ARCH]   Referring to the underwater portion of the hull when covered with foreign matter such as barnacles or grass.   { faul 'bäd·əm }

foulbrood   [INV ZOO]   The common name for three destructive bacterial diseases of honeybee larvae.   { faul‚brüd }

foul foot   [VET MED]   A feedlot disease of cattle and sheep marked by inflammation and ulceration of the feet; common in wet feedlots. Also called foot rot.   { faul 'füt }

Foulger's test   [ANALY CHEM]   A test for fructose in which urea, sulfuric acid, and stannous chloride are added to the solution to be tested, the solution is boiled, and in the presence of fructose a blue coloration forms. { 'fül‚jāz ‚test }

fouling [CHEM ENG] Deposition on the surface of a heat-transfer device of sediment in the form of scale derived from burned particles of the heated substance. [NAV ARCH] The adhesion of different marine organisms to the underwater parts of ships, causing the ships to lose speed. [ORD] The deposit that remains on the bore of a gun after firing. { 'faúl·iŋ }

fouling factor [CHEM ENG] In heat transfer, the lowering of clear-film transfer rates resulting from corrosion, dirt, or roughness of the surface of tube walls of heat exchangers. { 'faúl·iŋ ‚fak·tər }

fouling organism [ECOL] Any aquatic organism with a sessile adult stage that attaches to and fouls underwater structures of ships. { 'faúl·iŋ ‚ór·gə‚niz·əm }

fouling plates [ENG] Metal plates submerged in water to allow attachment of fouling organisms, which are then analyzed to determine species, growth rate, and growth pattern, as influenced by environmental conditions and time. { 'faúl·iŋ ‚plāts }

fouling point [CIV ENG] **1.** The point at a switch or turnout beyond which railroad cars must be placed so as not to interfere with cars on the main track. **2.** The location of insulated joints in a turnout on signaled tracks. { 'faúl·iŋ ‚póint }

foundation [CIV ENG] **1.** The ground that supports a building or other structure. **2.** The portion of a structure which transmits the building load to the ground. { faún'dā·shən }

foundation coefficient [GEOPHYS] A coefficient which expresses how much stronger the effect of an earthquake is on a given rock than it would be on an undisturbed crystalline rock under the same conditions. { faún'dā·shən ‚kō·i‚fish·ənt }

foundation engineering [CIV ENG] That branch of engineering concerned with evaluating the earth's ability to support a load and designing substructures to transmit the load of superstructures to the earth. { faún'dā·shən ‚en·jə‚nir·iŋ }

foundation mat See raft foundation. { faún'dā·shən ‚mat }

founder [GEOL] To sink under water either by depression of the land or by rise of sea level, especially in reference to large crustal masses, islands, or significant portions of continents. { 'faún·dər }

founder effect [GEN] The overrepresentation of a specific allele at one or more loci in a new population that arises from a small number of individuals whose small gene pool may be unrepresentative of the parental population initially or as a result of the ensuing genetic drift. { 'faún·dər i‚fekt }

founding [MET] The art and science of melting and casting metals. { 'faúnd·iŋ }

foundry [ENG] A building where metal or glass castings are produced. { 'faún·drē }

foundry alloy See master alloy. { 'faún·drē ‚al‚ói }

foundry core sand See core sand. { ¦faún·drē ¦kór ‚sand }

foundry engineering [ENG] The science and practice of melting and casting glass or metal. { 'faún·drē ‚en·jə‚nir·iŋ }

foundry facing [MET] A material applied to a sand mold to improve the surface quality of a casting. { 'faún·drē ‚fās·iŋ }

foundry proofs [GRAPHICS] Proofs of type pages that have been locked up within frames prior to casting; identified by black bands around each page of type, caused by bearers that lend support to the type page when stereotypes or electrotypes are made. { 'faún·drē ‚prüfs }

foundry return [MET] A scrapped casting that is returned to the furnace for remelting. { 'faún·drē ri‚tərn }

foundry sand [MET] Sand used in foundries to make molds for the casting of metal shapes. Also known as molding sand. { 'faún·drē ‚sand }

foundry type [GRAPHICS] Type cast as single characters. Also known as hand type. { 'faún·drē ‚tīp }

fountain [GRAPHICS] **1.** In printing, a container or reservoir on a press that contains an ink supply. **2.** In offset lithography, a fountain solution (usually a water-alcohol mixture) that wets the nonprinting areas of the plate. { 'faúnt·ən }

fountain effect [FL MECH] The effect occurring when two containers of superfluid helium are connected by a capillary tube and one of them is heated, so that helium flows through the tube in the direction of higher temperature. { 'faúnt·ən i‚fekt }

fourable [PETRO ENG] A section of drill pipe casing or tubing comprising four joints that are screwed together. { 'fór·ə·bəl }

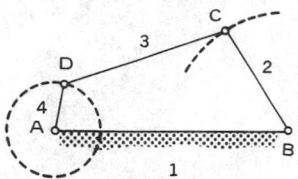

FOUR-BAR LINKAGE

Four-bar linkage, a common form of bar linkage. Bars join and pivot at points A, B, C, and D.

fourable board [PETRO ENG] A platform installed in an oil derrick at an elevation of 80–120 feet (24–37 meters) above the derrick floor to support the derrick operator while pipe is being raised or lowered. { 'fór·ə·bəl ‚bórd }

four-address [COMPUT SCI] Pertaining to an instruction address which contains four address parts. { 'fór ə‚dres }

four-ball tester [ENG] A machine designed to measure the efficiency of lubricants by driving one ball against three stationary balls clamped together in a cup filled with the lubricant; performance is evaluated by measuring wear-scar diameters on the stationary balls. { ¦fór ¦ból ‚tes·tər }

four-bar linkage [MECH ENG] A plane linkage consisting of four links pinned tail to head in a closed loop with lower, or closed, joints. { ¦fór ¦bär ‚liŋk·ij }

fourble [PETRO ENG] A section of drill pipe, casing, or tubing in which four joints are connected. { 'fór·bəl }

Fourcault process [ENG] A process for forming sheet glass in which the molten glass is drawn vertically upward. { für'kō ‚präs·əs }

four-channel sound system See quadraphonic sound system. { ¦fór ¦chan·əl 'saúnd ‚sis·təm }

fourchite [PETR] A monchiquite that lacks feldspar and olivine. { 'fúr‚shīt }

four-color printing [GRAPHICS] A method of reproducing full-color originals, such as paintings and color photographs, by overprinting a series of four plates in yellow, magenta, cyan, and black ink. { ¦fór ¦kəl·ər 'print·iŋ }

four-color problem [MATH] The problem of proving the statement that, given any map in the plane, it is possible to color the regions with four colors so that any two regions with a common boundary have different colors. { ¦fór 'kəl·ər ‚präb·ləm }

four-color separation process [GRAPHICS] Conversion of a color illustration into four negative films from which the four printing plates [yellow, magenta, cyan (blue), and black] that will be used in the printing process are made; the negative film for yellow is made by photographing the illustration through a blue filter, the magenta through a green filter, the cyan through a red filter, and the black through a yellow filter. { ¦fór ¦kəl·ər ‚sep·ə'rā·shən ‚präs·əs }

four-course radio range station [NAV] Radio navigation land station in the aeronautical radio navigation service providing radio equisignal zones. { ¦fór ¦kórs ¦rād·ē·ō 'ränj ‚stā·shən }

four-current density [RELAT] A four-vector whose three space components are those of the ordinary current density and whose time component is the charge density. { ¦fór ¦kə·rənt 'den·səd·ē }

4-D chart [METEOROL] A chart showing the field of D values (deviations of the actual altitudes along a constant-pressure surface from the standard atmosphere altitude of that surface) in terms of the three dimensions of space and one of time; it is a form of a four-dimensional display of pressure altitude; the space dimensions are represented by D-value contours, and the time dimension is provided by tau-value lines. { ¦fór ¦dē 'chärt }

four-degree calorie [CHEM] The heat needed to change the temperature of 1 gram of water from 3.5 to 4.5°C. { ¦fór di¦grē 'kal·ə·rē }

Fourdrinier machine [MECH ENG] A papermaking machine; a paper web is formed on an endless wire screen; the screen passes through presses and over dryers to the calenders and reels. { ‚for·drə'nir mə‚shēn }

four-factor formula [NUCLEO] The principle that the multiplication factor of a thermal reactor with no leakage is the product of the average number of fast neutrons emitted when a nucleus in the fuel material captures a thermal neutron, the fast fission factor, the fraction of neutrons which are not captured while being slowed down, and the number of thermal neutrons absorbed in the fuel divided by the total number of neutrons absorbed in the fuel and the moderator. { ¦fór ¦fak·tər 'fór·myə·lə }

four-force [RELAT] A four-vector equal to the product of the rest mass of a particle and the rate of change of its four-momentum with respect to its proper time. { ¦fór 'fórs }

four-frequency diplex telegraphy [COMMUN] Frequency-shift telegraphy in which each of the four possible signal combinations corresponding to two telegraph channels is represented by a separate frequency. { ¦fȯr ¦frē·kwən·sē ¦dī‚pleks tə'leg·rə·fē }

four-group [MATH] The only group of order 4 other than the cyclic group. { 'fȯr ‚grüp }

four-hundred-day clock *See* anniversary clock. { ¦fȯr ¦hən·drəd ¦dā 'kläk }

fourier *See* thermal ohm. { fùr·ē‚ā }

Fourier analysis [MATH] The study of convergence of Fourier series and when and how a function is approximated by its Fourier series or transform. { fùr·ē‚ā ə‚nal·ə·səs }

Fourier analyzer [ENG] A digital spectrum analyzer that provides push-button or other switch selection of averaging, coherence function, correlation, power spectrum, and other mathematical operations involved in calculating Fourier transforms of time-varying signal voltages for such applications as identification of underwater sounds, vibration analysis, oil prospecting, and brain-wave analysis. { fùr·ē‚ā 'an·ə‚līz·ər }

Fourier-Bessel integrals [MATH] Given a function $F(r,\theta)$ independent of θ where r,θ are the polar coordinates in the plane, these integrals have the form

$$\int_0^\infty u\,du \int_0^\infty F(r)J_m(ur)r\,dr$$

where J_m is a Bessel function order m. { fùr·ē‚ā ¦bes·əl 'int·ə·grəlz }

Fourier-Bessel series [MATH] For a function $f(x)$, the series whose mth term is $a_mJ_0(j_mx)$, where $j_1, j_2, \ldots$ are positive zeros of the Bessel function J_0 arranged in ascending order, and a_m is the product of $2/J_1{}^2(j_m)$ and the integral over t from 0 to 1 of $tf(t)J_0(j_mt)$; J_1 is a Bessel function. { fùr·ē‚ā ¦bes·əl ‚sir·ēz }

Fourier-Bessel transform *See* Hankel transform. { fùr·ē‚ā ¦bes·əl 'tranz‚fȯrm }

Fourier expansion *See* Fourier series. { fùr·ē‚ā ik'span·chən }

Fourier heat equation *See* Fourier law of heat conduction; heat equation. { fùr·ē‚ā 'hēt i‚kwā·zhən }

Fourier integrals [MATH] For a function $f(x)$ the Fourier integrals are

$$\frac{1}{\pi}\int_0^\infty du \int_{-\infty}^\infty f(t)\cos u(x-t)\,dt$$

$$\frac{1}{\pi}\int_0^\infty du \int_{-\infty}^\infty f(t)\sin u(x-t)\,dt$$

{ fùr·ē‚ā 'int·ə·grəlz }

Fourier kernel [MATH] Any kernel $K(x,y)$ of an integral transform which may be written in the form $K(x,y) = k(xy)$ and which is identical with the kernel of the inverse transform. { 'fȯr·ē‚ā ‚kər·nəl }

Fourier law of heat conduction [THERMO] The law that the rate of heat flow through a substance is proportional to the area normal to the direction of flow and to the negative of the rate of change of temperature with distance along the direction of flow. Also known as Fourier heat equation. { 'fùr·ē‚ā ‚lȯ əv 'hēt kən‚dək·shən }

Fourier-Legendre series [MATH] Given a function $f(x)$, the series from $n = 0$ to infinity of $a_nP_n(x)$, where $P_n(x)$, $n=0$, $1, 2, \ldots$, are the Legendre polynomials, and a_n is the product of $(2n + 1)/2$ and the integral over x from -1 to 1 of $f(x)P_n(x)$. { fùr·ē‚ā lə'zhän·drə ‚sir·ēz }

Fourier number [FL MECH] A dimensionless number used in unsteady-state flow problems, equal to the product of the dynamic viscosity and a characteristic time divided by the product of the fluid density and the square of a characteristic length. Symbolized Fo_f. [PHYS] A dimensionless number used in the study of unsteady-state mass transfer, equal to the product of the diffusion coefficient and a characteristic time divided by the square of a characteristic length. Symbolized N_{Fo_m}. [THERMO] A dimensionless number used in the study of unsteady-state heat transfer, equal to the product of the thermal conductivity and a characteristic time, divided by the product of the density, the specific heat at constant pressure, and the distance from the midpoint of the body through which heat is passing to the surface. Symbolized N_{Fo_h}. { fùr·ē‚ā ‚nəm·bər }

Fourier series [MATH] The Fourier series of a function $f(x)$ is

$$\tfrac{1}{2} a_0 + \sum_{n=1}^\infty (a_n \cos nx + b_n \sin nx)$$

with

$$a_n = \frac{1}{\pi}\int_{-\pi}^\pi f(x) \cos nx\,dx$$

$$b_n = \frac{1}{\Pi}\int_{-\infty}^\Pi f(x) \sin nx\,dx$$

Also known as Fourier expansion. { ‚fùr·ē‚ā ‚sir·ēz }

Fourier's half-range series [MATH] A Fourier series that either contains only terms that are even in the independent variable (the cosine series) or contains only terms that are odd (the sine series). { 'fȯr·ē‚āz ¦haf ¦rānj ‚sir‚ēz }

Fourier space [MATH] The space in which the Fourier transform of a function is defined. { ‚fùr·ē‚ā ‚spās }

Fourier spectrum [PHYS] A plot of the magnitude and phase of the Fourier transform of a function. { ‚fùr·ē‚ā ‚spek·trəm }

Fourier's theorem [MATH] If $f(x)$ satisfies the Dirichlet conditions on the interval $-\pi < x < \pi$, then its Fourier series converges to $f(x)$ for all values of x in this interval at which $f(x)$ is continuous, and approaches $1/2[f(x + 0) + f(x - 0)]$ at points at which $f(x)$ is discontinuous, where $f(x - 0)$ is the limit on the left of f at x and $f(x + 0)$ is the limit on the right of f at x. { ‚fùr·ē‚āz ‚thir·əm }

Fourier-Stieltjes series [MATH] For a function $f(x)$ of bounded variation on the interval $[0,2\pi]$, the series from $n = 0$ to infinity of $c_n \exp(inx)$, where c_n is $1/2\pi$ times the integral from $x = 0$ to $x = 2\pi$ of $\exp(-inx)df(x)$. { ‚fùr·ē‚ā 'stēl·yes ‚sir·ēz }

Fourier-Stieltjes transform [MATH] For a function $f(y)$ of bounded variation on the interval $(-\infty, \infty)$, the function $F(x)$ equal to $1/\sqrt{2\pi}$ times the integral from $y = -\infty$ to $y = \infty$ of $\exp(-ixy)df(y)$. { ‚fùr·ē‚ā 'stēl·yes ‚tranz‚fȯrm }

Fourier synthesis [MATH] The determination of a periodic function from its Fourier components. { ‚fùr·ē‚ā 'sin·thə·səs }

Fourier transform [MATH] For a function $f(t)$, the function $F(x)$ equal to $1/\sqrt{2\pi}$ times the integral over t from $-\infty$ to ∞ of $f(t) \exp(itx)$. { ‚fùr·ē‚ā 'tranz‚fȯrm }

Fourier transform spectroscopy [SPECT] A spectroscopic technique in which all pertinent wavelengths simultaneously irradiate the sample for a short period of time, and the absorption spectrum is found by mathematical manipulation of the Fourier transform so obtained. { ‚fùr·ē‚ā 'tranz‚fȯrm spek'träs·kə·pē }

four laws of black hole mechanics [RELAT] Four laws of general relativity theory describing black holes, which are closely analogous to the four laws of classical thermodynamics. { ¦fȯr ¦lȯz əv ¦blak ¦hōl mi'kan·iks }

four-layer device [ELECTR] A $pnpn$ semiconductor device, such as a silicon controlled rectifier, that has four layers of alternating p-and n-type material to give three pn junctions. { ¦fȯr ¦lā·ər di'vīs }

four-layer diode [ELECTR] A semiconductor diode having three junctions, terminal connections being made to the two outer layers that form the junctions; a Shockley diode is an example. { ¦fȯr ¦lā·ər 'dī‚ōd }

four-layer transistor [ELECTR] A junction transistor having four conductivity regions but only three terminals; a thyristor is an example. { ¦fȯr ¦lā·ər tran'zis·tər }

four-level laser [PHYS] A laser in which the lowest level for a laser transition is an excited state rather than the ground level. { ¦fȯr ¦lev·əl 'lā·zər }

four-limbed core [ELECTROMAG] The core of a core-form, three-limbed transformer to which an additional core leg has been added at one end of the core to carry the net flux resulting from unbalanced voltages, which could otherwise flow in the steel transformer tank and cause overheating. { ¦fȯr ‚limd 'kȯr }

fourmarierite [MINERAL] An orange-red to brown mineral composed of a hydrous oxide of lead and uranium. { fùr'mar·ē·ə‚rīt }

four-phase modulation [COMMUN] Modulation in which data are encoded on a carrier frequency as a succession of phase shifts that will be 45, 135, 225, or 315°; each phase shift contains 2 bits of information called dibits, as follows: 225° represents 00, 315° is 01, 45° is 11, and 135° is 10. { ¦fȯr ¦fāz ¦maj·ə¦lā·shən }

four-pi counter [ENG] An instrument which measures the radiation that a radioactive material emits in all directions. { ¦fȯr 'pī ¦kaün·tər }

four-piece set [MIN ENG] Squared timber frame used in underground driving to give all-around support to weak ground. { ¦fȯr ¦pēs 'set }

four-plus-one address [COMPUT SCI] An instruction that contains four operand addresses and a control address. { ¦fȯr ¦pləs ¦wən ə'dres }

four-point [MATH] A set of four points in a plane, no three of which are collinear. Also known as complete four-point. { 'fȯr ¦pȯint }

four-point bearing [NAV] A relative bearing of 045° or 315°. { 'fȯr ¦pȯint 'ber·iŋ }

four-pole double-throw [ELEC] A 12-terminal switch or relay contact arrangement that simultaneously connects two pairs of terminals to either of two other pairs of terminals. Abbreviated 4PDT. { ¦fȯr ¦pōl ¦dəb·əl 'thrō }

four-quadrant multiplier [COMPUT SCI] A multiplier in an analog computer in which both the reference signal and the number represented by the input may be bipolar, and the multiplication rules for algebraic sign are obeyed. Also known as quarter-square multiplier. { ¦fȯr ¦kwäd·rənt 'məl·tə¦plī·ər }

fourré See temperate and cold scrub; tropical scrub. { fu'rā }

four-stroke cycle [MECH ENG] An internal combustion engine cycle completed in four piston strokes; includes a suction stroke, compression stroke, expansion stroke, and exhaust stroke. { ¦fȯr ¦strōk 'sī·kəl }

four-tape [COMPUT SCI] To sort input data, supplied on two tapes, into incomplete sequences alternately on two output tapes; the output tapes are used for input on the succeeding pass, resulting in longer and longer sequences after each pass, until the data are all in one sequence on one output tape. { ¦fȯr ¦tāp }

fourth dimension [RELAT] Time in the theory of relativity, in which space and time are conceived as particular aspects of a four-dimensional world. { ¦fȯrth də'men·chən }

fourth-generation computer [COMPUT SCI] A type of general-purpose digital computer used in the 1970s and 1980s that is characterized by increasingly advanced very large-scale integrated circuits and increasing use of a hierarchy of memory devices. { ¦fȯrth ¦jen·ə¦rā·shən kəm'pyüd·ər }

fourth-generation language [COMPUT SCI] A higher-level programming language that automates many of the basic functions that must be spelled out in conventional languages, and can obtain results with an order-of-magnitude less coding because of its richer content of commands. { ¦fȯrth ¦jen·ə¦rā·shən 'laŋ·gwij }

fourth-power law See Stefan-Boltzmann law. { ¦fȯrth ¦paü·ər 'lȯ }

fourth proportional [MATH] For numbers a, b, and c, a number x such that $a/b = c/x$. { ¦fȯrth prə'pȯr·shən·əl }

fourth quadrant [MATH] **1.** The range of angles from 270 to 360°. **2.** In a plane with a system of cartesian coordinates, the region in which the x coordinate is positive and the y coordinate is negative. { ¦fȯrth 'kwäd·rənt }

fourth sound [CRYO] A pressure wave which propagates in helium II contained in a porous material such as a tightly packed powder, and which results entirely from motion of the superfluid component, the normal component being immobilized by its viscosity. { ¦fȯrth 'saünd }

four-track tape [ENG ACOUS] Magnetic tape on which two tracks are recorded for each direction of travel, to provide stereo sound reproduction or to double the amount of source material that can be recorded on a given length of ¹/₄-inch (0.635-centimeter) tape. { 'fȯr ¦trak 'tāp }

four-vector [RELAT] A set of four quantities which transform under a Lorentz transformation in the same way as the three space coordinates and the time coordinate of an event. Also known as Lorentz four-vector. { 'fȯr ¦vek·tər }

four-vector potential [ELECTROMAG] A four-vector whose space components are the magnetic vector potential and whose time component is the electric scalar potential. { 'fȯr ¦vek·tər pə'ten·chəl }

four-velocity [RELAT] A four-vector whose components are

the rates of change of the space and time coordinates of a particle with respect to the particle's proper time. { 'fȯr və¦läs·əd·ē }

four-way dip [GEOPHYS] In seismic prospecting, dip determined by an array of geophones which are set up at points in four directions from a shot point; three of the locations are essential and the fourth serves as a control point. { 'fȯr ¦wā 'dip }

four-way reinforcing [CIV ENG] A system of reinforcing rods in concrete slab construction in which the rods are placed parallel to two adjacent edges and to both diagonals of a rectangular slab. { 'fȯr ¦wā rē·ən'fȯrs·iŋ }

four-way switch [ELEC] An electric switch employed in house wiring, that makes it possible to turn a light on or off at three or more places. { 'fȯr ¦wā 'swich }

four-way valve [MECH ENG] A valve at the junction of four waterways which allows passage between any two adjacent waterways by means of a movable element operated by a quarter turn. { 'fȯr ¦wā 'valv }

four-wheel drive [MECH ENG] An arrangement in which the drive shaft acts on all four wheels of the automobile. { 'fȯr ¦wēl 'drīv }

four-wire circuit [COMMUN] A two-way circuit using two paths so arranged that communication currents are transmitted in one direction only on one path, and in the opposite direction on the other path; the transmission path may or may not employ four wires. { 'fȯr ¦wīr 'sər·kət }

four-wire line [ELECTROMAG] A transmission line in which four conductors lie at the corners of a rectangle, and each conductor is in phase with the conductor at the opposite corner and out of phase with the conductors at adjacent corners. { 'fȯr ¦wīr 'līn }

four-wire repeater [ELECTR] Telephone repeater for use in a four-wire circuit and in which there are two amplifiers, one serving to amplify the telephone currents in one side of the four-wire circuit, and the other serving to amplify the telephone currents in the other side of the four-wire circuit. { 'fȯr ¦wīr ri'pēd·ər }

four-wire subscriber line [COMMUN] Four-wire circuit connecting a subscriber directly to a switching center. { 'fȯr ¦wīr səb'skrīb·ər ¦līn }

four-wire terminating set [ELECTR] Hybrid arrangement by which four-wire circuits are terminated on a two-wire basis for interconnection with two-wire circuits. { 'fȯr ¦wīr 'ter·mə¦nād·iŋ ¦set }

fovea [BIOL] A small depression or pit. { 'fō·vē·ə }

fovea centralis [ANAT] A small, rodless depression of the retina in line with the visual axis, which affords acute vision. { 'fō·vē·ə sen'tral·əs }

foveal vision [PHYSIO] Vision achieved by looking directly at objects in the daylight so that the image falls on or near the fovea centralis. Also known as photopic vision. { ¦fō·vē·əl 'vizh·ən }

foveola [BIOL] A small pit, especially one in the embryonic gastric mucosa from which gastric glands develop. { fō'vē·ə·lə }

foveolate [BIOL] Having small depressions; pitted. { 'fō·vē·ə¦lāt }

fowan [METEOROL] A dry, scorching wind of Great Britain and the Isle of Man. { faü·ən }

fowl [AGR] A domestic cock or hen, especially an adult hen, such as among chickens or several other gallinaceous birds. { faül }

Fowler-DuBridge theory [SOLID STATE] Theory of photoelectric emission from a metal based on the Sommerfeld model, which takes into account the thermal agitation of electrons in the metal and predicts the photoelectric yield and the energy spectrum of photoelectrons as functions of temperature and the frequency of incident radiation. { ¦faül·ər dü'brij ¦thē·ə·rē }

Fowler function [SOLID STATE] A mathematical function used in the Fowler-DuBridge theory to calculate the photoelectric yield. { 'faülər ¦faŋk·shən }

fowlerite [MINERAL] A zinc-bearing variety of rhodonite. { 'faü·lə¦rīt }

fowl pox [VET MED] A disease of birds caused by a virus and characterized by wartlike nodules on the skin, particularly on the head. { 'faül ¦päks }

fox [COMPUT SCI] A name for the hexadecimal digit whose decimal equivalent is 15. [VERT ZOO] The common name

FOX

The gray fox (Urocynon cineroargenteus).

for certain members of the dog family (Canidae) having relatively short legs, long bodies, large erect ears, pointed snouts, and long bushy tails. { fäks }

Fox broadcast [COMMUN] Radio broadcast of messages for which receiving stations make no acknowledgment. { 'fäks ‚bród‚kast }

foxhole [ORD] A small pit used for cover, usually for one or two soldiers, and so constructed that an occupant can fire effectively from it. { 'fäks‚hōl }

fox lathe [MECH ENG] A lathe with chasing bar and leaders for cutting threads; used for turning brass. { 'fäks ‚lāth }

foyaite [PETR] A nepheline syenite composed chiefly of potassium feldspar. { 'fòi·yə‚īt }

fp See freezing point.

FP See forward perpendicular.

FPLA See field-programmable logic array.

F process [MATH] A stochastic process $\{X_t, t > 0\}$ whose value at time t is determined by the information up to time t; more precisely, the events $\{X_t \le a\}$ belong to F_t for every t and a, where $F = \{F_t, t \ge 0\}$ is an increasing family of sigma algebras that represents the amount of information increasing with time. { 'ef ‚präs·əs }

fps system of units See British absolute system of units. { ¦ef ¦pē¦es 'sis·təm əv 'yü·nəts }

Fr See francium. [ELEC] See statcoulomb.

fractal [MATH] A geometrical shape whose structure is such that magnification by a given factor reproduces the original object. { 'frakt·əl }

fractal dimensionality [MATH] A number D associated with a fractal which satisfies the equation $N = b^D$, where b is the factor by which the length scale changes under a magnification in each step of a recursive procedure defining the object, and N is the factor by which the number of basic units increases in each such step. Also known as Mandelbrot dimensionality. { 'frak·təl di‚men·shə'nal·əd·ē }

fractal persistence length [PHYS] A length L characterizing a solid that has a fractal material distribution at shorter length scales, such that the solid is homogeneous at length scales larger than L. { ¦frak·təl pər'sis·təns ‚leŋkth }

fraction [CHEM] One of the portions of a volatile liquid within certain boiling point ranges, such as petroleum naphtha fractions or gas-oil fractions. [MATH] An expression which is the product of a real number or complex number with the multiplicative inverse of a real or complex number. [MET] In powder metallurgy, that portion of sample that lies between two stated particle sizes. Also known as cut. [SCI TECH] A portion of a mixture which represents a discrete unit and can be isolated from the whole system. { 'frak·shən }

fractional condensation [CHEM] Separation of components of vaporized liquid mixtures by condensing the vapors in stages (partial condensation); highest-boiling-point components condense in the first condenser stage, allowing the remainder of the vapor to pass on to subsequent condenser stages. { ¦frak·shən·əl ‚kän·den'sā·shən }

fractional crystallization [PETR] Separation of a cooling magma into multiple minerals as the different minerals cool and congeal at progressively lower temperatures. Also known as crystallization differentiation; fractionation. { ¦frak·shən·əl ‚krist·əl·ə'zā·shən }

fractional distillation [CHEM] A method to separate a mixture of several volatile components of different boiling points; the mixture is distilled at the lowest boiling point, and the distillate is collected as one fraction until the temperature of the vapor rises, showing that the next higher boiling component of the mixture is beginning to distill; this component is then collected as a separate fraction. { ¦frak·shən·əl dis·tə'lā·shən }

fractional equation [MATH] 1. Any equation that contains fractions. 2. An equation in which the unknown variable appears in the denominator of one or more terms. { ¦frak·shən·əl i'kwā·zhən }

fractional factorial experiment [STAT] An experiment in which certain properly chosen levels of factors are left out. Also known as fractional replicate. { ¦frak·shən·əl fak¦tòr·ē·əl ik'sper·ə·mənt }

fractional gas-flow curve [PETRO ENG] Graph of the fraction of free injected gas flowing through a reservoir formation versus the liquid saturation of the gas for various parameter values of oil viscosity; used to calculate displacement efficiency during gas injection. { ¦frak·shən·əl 'gas ‚flō ‚kərv }

fractional horsepower motor [ELEC] Any motor built into a frame smaller than that for a motor having an open construction and a continuous rating of 1 horsepower (745.7 watts) at 1800 revolutions per minute. { ¦frak·shən·əl ¦hòrs‚paù·ər 'mōd·ər }

fractional ideal [MATH] A submodule of the quotient field of an integral domain. { ¦frak·shən·əl i'dēl }

fractional precipitation [ANALY CHEM] Method for separating elements or compounds with similar solubilities by a series of analytical precipitations, each one improving the purity of the desired element. { ¦frak·shən·əl prə‚sip·ə'tā·shən }

fractional quantum Hall effect [ELECTR] The version of the quantum Hall effect in which the Hall resistance becomes precisely equal to $h/(p/q)e^2$, where h is Planck's constant, e is the electronic charge, q is an odd integer, and p is an integer not divisible by q. { ¦frak·shən·əl ¦kwän·təm 'hòl i‚fekt }

fractional replicate See fractional factorial experiment. { ¦frak·shən·əl 'rep·lə·kət }

fractional sampling [MIN ENG] Mechanical selection of samples of uniformly graded material without segregation. { ¦frak·shən·əl 'sam‚pliŋ }

fractional sine wave [PHYS] A pulse train whose waveform is a truncated sine wave. { ¦frak·shən·əl 'sīn ‚wāv }

fractionating column [CHEM] An apparatus used widely for separation of fluid (gaseous or liquid) components by vapor-liquid fractionation or liquid-liquid extraction or liquid-solid adsorption. { 'frak·shə‚nād·iŋ ‚käl·əm }

fractionation [CHEM] Separation of a mixture in successive stages, each stage removing from the mixture some proportion of one of the substances, as by differential solubility in water-solvent mixtures. [NUCLEO] Alterations in the isotopic composition of substances found in nature or in radioactive weapon debris, which result from small differences in the physical and chemical properties of isotopes of an element. [PETR] See fractional crystallization. { ‚frak·shə'nā·shən }

fractionator [CHEM ENG] An apparatus used to separate a mixture by fractionation, especially by fractional distillation. { 'frak·shə‚nād·ər }

fraction defective [IND ENG] The number of units per 100 pieces which are defective in a lot; expressed as a decimal. { 'frak·shən di'fek·tiv }

fraction in lowest terms [MATH] A fraction from which all common factors have been divided out of the numerator and denominator. { 'frak·shən in ¦lō·əst 'tərmz }

fraction kill hypothesis [MED] A principle of chemotherapy that a uniform dose of a drug will destroy a constant fraction rather than a constant number of tumor cells regardless of the size of the tumor or the number of cells present. { 'frak·shən ‚kil hī‚päth·ə·səs }

fractoconformity [GEOL] The relation between conformable strata, where faulting of the older beds occurs at the same time as deposition of the newer beds. { ¦frak·tō·kən'fòr·məd·ē }

fractography [MET] The microscopic examination of fractured metal surfaces. { ‚frak'täg·rə·fē }

fracton [PHYS] 1. A quantum corresponding to a vibrational excitation of random fractal structure, which is localized in space. 2. A vibrational quantum of a fractal structure, either random or ordered, that may be localized or delocalized. 3. A quantum of a localized vibrational excitation of a disordered structure, whether or not it is fractal. { 'frak‚tän }

fracton dimension [PHYS] A parameter $\bar{d}$ associated with a fractal material distribution of fractal dimension D such that the average frequency ω of fractons of size l is given by a power law stating that $\omega^{\bar{d}}$ is proportional to l^{-D}, and the density of vibrational states is proportional to $\omega^{\bar{d}-1}$. Also known as spectral dimension. { 'frak·tən di‚men·shən }

fracture [GEOL] A crack, joint, or fault in a rock due to mechanical failure by stress. Also known as rupture. [MED] The breaking of bone, cartilage, or teeth. [MINERAL] A break in a mineral other than along a cleavage plane. [SCI TECH] 1. The act, process, or state of being broken. 2. The surface appearance of a freshly broken material. 3. The break produced by fracturing. { 'frak·shər }

fracture cleavage [GEOL] Cleavage that occurs in deformed but only slightly metamorphosed rocks along closely spaced, parallel joints and fractures. { 'frak·shər ‚klēv·ij }

fractured formation [PETRO ENG] Reservoir formation in which rock has been split by hydraulic pressure produced by injected fluids. { 'frak·shərd fôr'mā·shən }

fracture dome [MIN ENG] The zone of loose or semiloose rock which exists in the immediate hanging or footwall of a stope. { 'frak·shər ‚dōm }

fracture-plane inclination [GEOL] Gradient or inclination of the plane of fracture formed in a reservoir formation. { 'frak·shər ‚plān ‚in·klə'nā·shən }

fracture pressure [PETRO ENG] The pressure that must be exerted in a wellbore in order to crack a formation. { 'frak·shər ‚presh·ər }

fracture strength See fracture stress. { 'frak·shər ‚streŋkth }

fracture stress [MECH] The minimum tensile stress that will cause fracture. Also known as fracture strength. { 'frak·shər ‚stres }

fracture system [GEOL] A stress-related group of contemporaneous fractures. { 'frak·shər ‚sis·təm }

fracture test [ENG] **1.** Macro- or microscopic examination of a fractured surface to determine characteristics such as grain pattern, composition, or the presence of defects. **2.** A test designed to evaluate fracture stress. { 'frak·shər ‚test }

fracture wear [MECH] The wear on individual abrasive grains on the surface of a grinding wheel caused by fracture. { 'frak·shər ‚wer }

fracture zone [GEOL] An elongate zone on the deep-sea floor that is of irregular topography and often separates regions of different depths; frequently crosses and displaces the mid-oceanic ridge by faulting. { 'frak·shər ‚zōn }

fractus [METEOROL] A cloud species in which the cloud elements are irregular but generally small in size, and which presents a ragged, shredded appearance, as if torn; these characteristics change ceaselessly and often rapidly. { 'frak·təs }

fragile site [GEN] The chromosomal position of a deoxyribonucleic acid sequence predisposed to spontaneous or induced breakage; sometimes contains short repetitive sequences. { 'fraj·əl ‚sīt }

fragile X syndrome [MED] A hereditary condition resulting from a trinucleotide repeat at an inherited fragile site on the long arm of the X chromosome. Affected males usually have some characteristic facial features, enlarged testes, and mental retardation. Females with one fragile X chromosome and one normal X chromosome may have a lesser degree of mental retardation. { ‚fraj·əl 'eks ‚sin‚drōm }

fragility [SCI TECH] The state or quality of being fragile, that is, brittle or easily broken. { frə'jil·əd·ē }

fragility test [PATH] A measure of the resistance of red blood cells to osmotic hemolysis in hypotonic salt solutions of graded dilutions. { frə'jil·əd·ē ‚test }

fragipan [GEOL] A dense, natural subsurface layer of hard soil with relatively slow permeability to water, mostly because of its extreme density or compactness rather than its high clay content or cementation. { 'fraj·ə‚pan }

fragment [ORD] **1.** A piece of an exploding or exploded bomb, projectile, or the like. **2.** To break into fragments. { 'frag·mənt }

fragmental printing [GRAPHICS] A nonstandard typeface, used for printing large characters, in which the elements of a rectangular grid are either wholly filled, wholly empty, or half filled, with the portion on one side of a diagonal filled and the other half empty. { 'frag‚ment·əl 'print·iŋ }

fragmentation [COMPUT SCI] The tendency of files in disk storage to be divided up into many small areas scattered around the disk. [CELL MOL] Amitotic division; a type of asexual reproduction. [MIN ENG] The blasting of coal, ore, or rock into pieces small enough to load, handle, and transport without the need for hand-breaking or secondary blasting. [PSYCH] Disordered behavior and mental processes. { ‚frag·mən'tā·shən }

fragmentation ammunition [ORD] Ammunition that is primarily intended to produce a fragmentation effect. { ‚frag·mən'tā·shən am·yə‚nish·ən }

fragmentation bomb [ORD] An item designed to be dropped from aircraft to produce many small, high-velocity fragments when detonated. { ‚frag·mən'tā·shən ‚bäm }

fragmentation bomb cluster [ORD] Multiple fragmentation bombs suspended and dropped from a single station of a bomb rack on an airplane. { ‚frag·mən'tā·shən ¦bäm ‚kläs·tər }

fragmentation grenade [ORD] A hand grenade designed to fragment, an effective weapon against personnel; since the thrower needs protective cover, it is used primarily for defensive operations, and is often called a defensive grenade. { ‚frag·mən'tā·shən grə‚nād }

fragmentation nucleus [METEOROL] A tiny ice particle broken from a large ice crystal, serving as an ice nucleus; that is, a growth center for a new ice crystal. { ‚frag·mən'tā·shən ‚nü·klē·əs }

fragmentation protective body armor [ORD] Armor designed to provide fragmentation protection to vital areas of the body; usually provided in the form of garments which contain steel, nylon, or other resistant materials. { ‚frag·mən¦tā·shən prə¦tek·tiv ¦bäd·ē ‚är·mər }

fragmentation test [ORD] A test conducted to determine the number and weight distribution, and (where the method used permits) the velocity and spatial distribution of fragments produced by a projectile or other munition upon detonation. { ‚frag·mən'tā·shən ‚test }

fragment emission [ORD] In terminal ballistics, the pattern of the fragments upon leaving the exploded projectile or other munition, including the number of fragments and the direction, weight, and velocity of each fragment. { 'frag·mənt i‚mish·ən }

fragmenting [COMPUT SCI] The breaking up of a document into its various components. { 'frag‚ment·iŋ }

fragment simulator projectile [ORD] Projectile which simulates the action of a fragment; used in ballistic tests at the proving ground. { 'frag·mənt ‚sim·yə‚lād·ər prə‚jek·təl }

Frahm frequency meter See vibrating-reed frequency metery. { 'främ 'frē·kwən·sē ‚mēd·ər }

frambesia See yaws. { fram'bē·zhə }

framboid [GEOL] A microscopic aggregate of pyrite grains, often occurring in spheroidal clusters. { 'fram‚bȯid }

frame [BUILD] The skeleton structure of a building. Also known as framing. [COMMUN] **1.** One cycle of a regularly recurring series of pulses. **2.** An elementary block of data for transmission over a network or communications system. [COMPUT SCI] See main frame. [ELECTR] **1.** One complete coverage of a television picture. **2.** A rectangular area representing the size of copy handled by a facsimile system. [GRAPHICS] A single complete picture on motion picture film. { 'främ }

frame buffer [COMPUT SCI] A device that stores a television picture or frame for processing. { 'främ ‚bəf·ər }

frame dragging See dragging of inertial frames. { 'främ ‚drag·iŋ }

frame frequency [ELECTR] The number of times per second that the frame is completely scanned in television. Also known as picture frequency. { 'främ ‚frē·kwən·sē }

frame grabber [COMPUT SCI] An external device that digitizes standard television video images for storage or processing in a computer. { 'främ ‚grab·ər }

frame of reference [PHYS] A coordinate system for the purpose of assigning positions and times to events. Also known as reference frame. { ¦främ əv 'ref·rəns }

frame period [ELECTR] A time interval equal to the reciprocal of the frame frequency. { 'främ ‚pir·ē·əd }

framer [ELECTR] Device for adjusting facsimile equipment so the start and end of a recorded line are the same as on the corresponding line of the subject copy. { 'främ·ər }

frames [COMPUT SCI] Subdivisions of a browser window, with each section containing a separate Web page. { främz }

frame set [MIN ENG] The arrangement of the legs and cap or crossbar so as to provide support for the roof of an underground passage. Also known as framing; set. { 'främ ‚set }

frameshift mutation [GEN] The addition or deletion of nucleotides in numbers other than three, which shifts the translation reading frame so a new set of codons beyond the point of abnormality in the messenger ribonucleic acid is read. Also known as phase-shift mutation. { ¦främ‚shift myü'tā·shən }

frameshift suppression [GEN] Reversion of a frameshift mutation by a second frameshift mutation in the same gene. { 'främ‚shift sə‚presh·ən }

framework [ENG] The load-carrying frame of a structure; may be of timber, steel, or concrete. [GEOL] **1.** In a sediment or sedimentary rock, the rigid arrangement created by particles that support one another at contact points.   **2.** A fixed calcareous structure impervious to waves, built by sedentary organisms (for example, sponges, corals, and bryozoans) in a high-energy environment.   { 'frām,wərk }

framework silicate *See* tectosilicate.   { 'frām,wərk 'sil·ə·kət }

framework structure [SOLID STATE] A crystalline structure in which there are strong interatomic bonds which are not confined to a single plane, in contrast to a layer structure.   { 'frām,wərk ,strək·chər }

framing [BUILD] *See* frame. [ELECTR] **1.** Adjusting a television picture to a desired position on the screen of the picture tube.   **2.** Adjusting a facsimile picture to a desired position in the direction of line progression. Also known as phasing. [MIN ENG] *See* frame set.   { 'frām·iŋ }

framing anchor [BUILD] A metal device for joining elements such as studs, joists, and rafters in light wood-frame construction.   { 'frām·iŋ ,aŋk·ər }

framing camera [OPTICS] A motion picture camera that automatically controls the position of successive still photographs on the film so that, when the film is subsequently projected, the image will appear steady on the screen.   { 'frām·iŋ ,kam·rə }

framing control [ELECTR] **1.** A control that adjusts the centering, width, or height of the image on a television receiver screen.   **2.** A control that shifts a received facsimile picture horizontally.   { 'frām·iŋ kən,trō }

framing imager [ENG] *See* forward-looking infrared imageryy.   { 'frām·iŋ ,im·ij·ər }

framing plan [NAV ARCH] A diagram showing positions and type of construction of the framing members of a ship.   { 'frām·iŋ ,plan }

framing square [DES ENG] A graduated carpenter's square used for cutting off and making notches.   { 'frām·iŋ ,skwer }

framing table [MIN ENG] An inclined table on which ore slimes are separated by running water.   { 'frām·iŋ ,tā·bəl }

Francisella [MICROBIO] A genus of gram-negative, aerobic bacteria of uncertain affiliation; cells are small, coccoid to ellipsoidal, pleomorphic rods and can be parasitic on mammals, birds, and arthropods.   { ,fran·si'sel·ə }

Francis formula [FL MECH] An equation for the calculation of water flow rate over a rectangular weir in terms of length and head.   { 'fran·səs ,fòr·myə·lə }

Francis turbine [MECH ENG] A reaction hydraulic turbine of relatively medium speed with radial flow of water in the runner.   { 'fran·səs 'tər,bīn }

francium [CHEM] A radioactive alkali-metal element, symbol Fr, atomic number 87, atomic weight distinguished by nuclear instability; exists in short-lived radioactive forms, the chief isotope being francium-223.   { 'fran·sē·əm }

Franck-Condon principle [PHYS CHEM] The principle that in any molecular system the transition from one energy state to another is so rapid that the nuclei of the atoms involved can be considered to be stationary during the transition.   { 'fräŋk 'kän·dən ,prin·sə·pəl }

franckeite [MINERAL] A dark-gray or black massive mineral composed of lead antimony tin sulfide.   { 'fräŋ·kə,īt }

Franck-Hertz experiment [ELECTR] Experiment for measuring the kinetic energy lost by electrons in inelastic collisions with atoms; it established the existence of discrete energy levels in atoms, and can be used to determine excitation and ionization potentials.   { ¦fräŋk 'herts ik,sper·ə·mənt }

Franck-Rabinowitch hypothesis [PHYS CHEM] The hypothesis that the decreased quantum efficiencies of certain photochemical reactions observed in the dissolved or liquid state are due to the formation of a cage of solvent molecules around the molecule which has been excited by absorption of a photon.   { ¦fräŋk rə'bin·ə,wich hī,päth·ə·səs }

francolite [MINERAL] $Ca_5(PO_4,CO_3)_3(F,OH)$ Colorless fluoride-bearing carbonate-apatite.   { 'fraŋ·kə,līt }

Franconian [GEOL] A North American stage of geologic time; the middle Upper Cambrian.   { fraŋ'kō·nē·ən }

frangible [MECH] Breakable, fragile, or brittle.   { 'fran·jə·bəl }

frangible bullet [ORD] A brittle plastic or other nonmetallic bullet for firing practice which, upon striking a target, breaks into powder or small fragments without penetrating.   { ¦fran·jə·bəl 'bùl·ət }

frangible grenade [ORD] Improvised incendiary hand grenade consisting of a glass container filled with a flammable

liquid, with an igniter attached, and which breaks and ignites upon striking a resistant target, such as a tank.   { ¦fran·jə·bəl grə'nād }

frangula emodin *See* emodin.   { 'fraŋ·gyə·lə 'em·ə·dən }

frangulic acid *See* emodin.   { fraŋ'gyü·lik 'as·əd }

Frankfurt horizontal *See* eye-ear plane.   { 'fraŋk·fərt här·ə'zänt·əl }

Frankia [MICROBIO] The single genus of the family Frankiaceae.   { ,fraŋ'kē·ə }

Frankiaceae [MICROBIO] A family of bacteria in the order Actinomycetales; filamentous cells form true mycelia; they are symbiotic and found in active, nitrogen-fixing root nodules.   { ,fraŋ·kē'ās·ē,ē }

frankincense *See* olibanum.   { 'fraŋk·ən,sens }

frankincense oil *See* olibanum oil.   { 'fraŋk·ən,sens ,òil }

Frankland's method [ORG CHEM] Reaction of dialkyl zinc compounds with alkyl halides to form hydrocarbons; may be used to form paraffins containing a quaternary carbon atom.   { 'fraŋk·lənz ,meth·əd }

franklin *See* statcoulomb.   { 'fraŋk·lən }

franklin centimeter [ELEC] A unit of electric dipole moment, equal to the dipole moment of a charge distribution consisting of positive and negative charges of 1 statcoulomb separated by a distance of 1 centimeter.   { 'fraŋk·lən 'sent·ə,mēd·ər }

Franklin equation [ENG ACOUS] An equation for intensity of sound in a room as a function of time after shutting off the source, involving the volume and exposed surface area of the room, the speed of sound, and the mean sound-absorption coefficient.   { 'fraŋk·lən i,kwā·zhən }

franklinite [MINERAL] $ZnFe_2O_4$ Black, slightly magnetic mineral member of the spinel group; usually possesses extensive substitution of divalent manganese and iron for the divalent zinc, and limited trivalent manganese for the trivalent iron.   { 'fraŋ·klə,nīt }

Frank partial dislocation [CRYSTAL] A partial dislocation whose Burger's vector is not parallel to the fault plane, so that it can only diffuse and not glide, in contrast to a Schockley partial dislocation.   { 'fräŋk ¦pär·shəl ,dis·lō'kā·shən }

Frank-Read source [MET] **1.** The creation of dislocations by application of shear stress to an edge dislocation anchored terminally, causing formation of an unstable loop form followed by formation of a closed dislocation line and the establishment of the original condition.   **2.** One of the sources of dislocations in a plastically deforming metal.   { ¦fräŋk ¦rēd ,sors }

Franz-Keldysh effect [OPTICS] A shift to longer wavelength in the spectrum transmitted by a semiconductor when a strong electric field is applied.   { ¦fränts 'kel·dəsh i,fekt }

Frary metal [MET] Metal containing 97–98% lead alloyed with 1–2% barium and calcium; used for bearings.   { 'frer·ē ,med·əl }

Frasch process [MIN ENG] A process to remove sulfur from sulfur beds; superheated water is forced under pressure into the sulfur bed, and the molten sulfur is thus forced to the surface.   { 'fräsh ,präs·əs }

Fraser's air-sand process [MIN ENG] A process in which dry, specific-gravity separation of coal from refuse is achieved by utilizing a flowing dense medium intermediate in density between coal and refuse.   { 'frā·zərz ¦er 'sand ,präs·əs }

fraternal twins *See* dizygotic twins.   { frə¦tərn·əl 'twinz }

Fraude's reagent [INV ZOO] *See* perchloric acid.   { 'fròdz rē,ä·jənt }

fraunhofer [SPECT] A unit for measurement of the reduced width of a spectrum line such that a spectrum line's reduced width in fraunhofers equals 10^6 times its equivalent width divided by its wavelength.   { 'fraùn,hōf·ər }

Fraunhofer corona *See* F corona.   { 'fraùn,hōf·ər kə,rō·nə }

Fraunhofer diffraction [OPTICS] Diffraction of a beam of parallel light observed at an effectively infinite distance from the diffracting object, usually with the aid of lenses which collimate the light before diffraction and focus it at the point of observation.   { 'fraùn,hōf·ər di,frak·shən }

Fraunhofer lines [SPECT] The dark lines constituting the Fraunhofer spectrum.   { 'fraùn,hōf·ər ,līnz }

Fraunhofer region [ELECTROMAG] The region far from an antenna compared to the dimensions of the antenna and the wavelength of the radiation. Also known as far field; far region; far zone; radiation zone.   { 'fraùn,hōf·ər ,rē·jən }

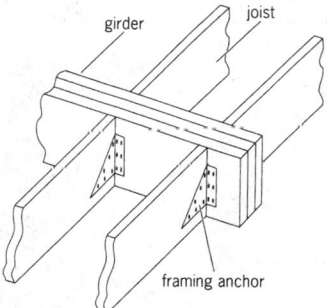

FRAMING ANCHOR

girder   joist

framing anchor

Diagram showing a framing anchor in a section of a wood frame.

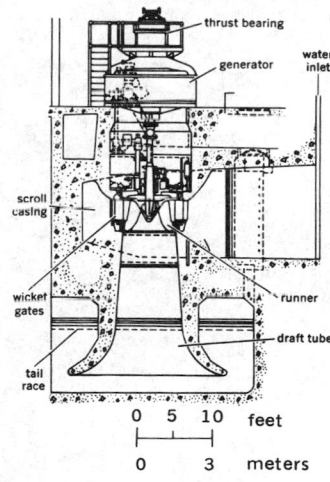

FRANCIS TURBINE

thrust bearing
generator
water inlet
scroll casing
wicket gates
runner
draft tube
tail race

0   5   10   feet
0   3   meters

Cross section of a Francis turbine installation.

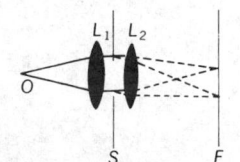

FRAUNHOFER DIFFRACTION

L_1 L_2
O
S   F

Diagram of Fraunhofer diffraction, with circular aperture. Light source O lies at the principal focus of lens L_1, which renders light parallel as it falls on aperture S. The second lens L_2 focuses parallel diffracted beams on observing screen F.

FRAUNHOFER SPECTRUM

Two sections of the Fraunhofer spectrum, showing bright continuum and dark absorption lines. The wavelength range covered by each strip is approximately 85 angstroms.

Fraunhofer spectrum [SPECT] The absorption lines in sunlight, due to the cooler outer layers of the sun's atmosphere. { 'fraún,hōf·ər ,spek·trəm }

Frazer-Brace extraction method [CHEM ENG] A method used to extract oil from citrus fruit; utilizes a machine which has abrasive carborundum rolls to rasp the peel from the fruit under a water spray; the water-and-peel mixture is screened and settled to allow oil separation. { 'frā·zər 'brās ik'strak·shən ,meth·əd }

frazil [HYD] Ice crystals which form in supercooled water that is too turbulent to permit coagulation of the crystals into sheet ice. { 'fra·zəl }

frazil ice [HYD] A spongy or slushy accumulation of frazil in a body of water. Also known as needle ice. { 'fra·zəl ,īs }

Fréchet space [MATH] **1.** A topological vector space that is locally convex, metrizable, and complete. **2.** A topological vector space that is metrizable and complete. **3.** See T1 space. { frā'shā ,spās }

freckle [MED] A pigmented macule resulting from focal increase in melanin, usually associated with exposure to sunlight, commonly on the face. { 'frek·əl }

Fredholm determinant [MATH] A power series obtained from the function $K(x,y)$ of the Fredholm equation which provides solutions to the equation under certain conditions. { 'fred,hōm di¦tər·mə·nənt }

Fredholm integral equations [MATH] Given functions $f(x)$ and $K(x,y)$, the Fredholm integral equations with unknown function y are

$$\text{type 1: } f(x) = \int_a^b K(x,t)y(t)dt$$

$$\text{type 2: } y(x) = f(x) + \lambda \int_a^b K(x,t)y(t)dt$$

{ 'fred,hōm ¦int·ə·grəl i'kwā·zhənz }

Fredholm operator [MATH] A linear operator between Banach spaces which has closed range, and both the Fredholm operator and its adjoint have finite dimensional null space. { 'fred,hōm ,äp·ə,rād·ər }

Fredholm theorem [MATH] A Fredholm equation of type 2 with continuous $f(x)$ has a unique continuous solution, or else the corresponding equation of type 1 has a positive number of linearly independent solutions. { 'fred,hōm ,thir·əm }

Fredholm theory [MATH] The study of the solutions of the Fredholm equations. { 'fred,hōm ,thē·ə·rē }

free admittance [ELEC] The reciprocal of the blocked impedance of a transducer. { ¦frē əd'mit·əns }

free air See free atmosphere. { ¦frē 'er }

free-air anomaly See free-air gravity anomaly. { 'frē ,er ə'näm·ə·lē }

free-air gravity anomaly [GEOPHYS] A measure of the mass excesses and deficiencies within the earth; calculated as the difference between the measured gravity and the theoretical gravity at sea level and a free-air coefficient determined by the elevation of the measuring station. Also known as free-air anomaly. { 'frē ,er 'grav·əd·ē ə,näm·ə·lē }

free-air ionization chamber [NUCLEO] An ionization chamber in which the ionizing radiation is limited by a diaphragm, so that ionization is detected in a region of accurately known volume, away from the electrodes and other internal parts of the equipment. { ¦frē ,er ,ī·ə·nə'zā·shən ,chām·bər }

free-air temperature [METEOROL] Temperature of the atmosphere, obtained by a thermometer located so as to avoid as completely as practicable the effects of extraneous heating. { 'frē ,er 'tem·prə·chər }

free ascent [ENG] Emergency ascent by a diver by floating to the surface through natural buoyancy or through assisted buoyancy with a life jacket. { ¦frē ə'sent }

free association [PSYCH] **1.** Spontaneous, consciously unrestricted association of ideas or mental images. **2.** A method used in psychoanalysis to gain an understanding of the organization of the content of the mind. { ¦frē ə,sō·sē'ā·shən }

free atmosphere [GEOPHYS] That portion of the earth's atmosphere, above the planetary boundary layer, in which the effect of the earth's surface friction on the air motion is negligible and in which the air is usually treated (dynamically) as an ideal fluid. Also known as free air. { ¦frē 'at·mə,sfir }

free atom [ATOM PHYS] An atom, as in a gas, whose properties, such as spectrum and magnetic moment, are not significantly affected by other atoms, ions, or molecules nearby. { ¦frē 'ad·əm }

free balloon [AERO ENG] A balloon that ascends without a tether, propulsion or guidance; it is made to descend by the release of gas. { ¦frē bə'lün }

freeboard [ANALY CHEM] The space provided above the resin bed in an ion-exchange column to allow for expansion of the bed during backwashing. [CHEM ENG] In a fluidized-bed reactor, the space between the top of the reaction bed and the top of the reactor. [CIV ENG] The height between normal water level and the crest of a dam or the top of a flume. [ENG] The vertical distance in a water tank between the maximum water level and the top of the tank. [NAV ARCH] The vertical distance from the intersection of the top of the freeboard deck amidships with the outer surface of the side plating to the upper edge of the summer load line, or in general terms, the distance from the waterline to the deck. { ¦frē,bord }

freeboard deck [NAV ARCH] The lowest exposed deck of a ship, below which all bulkheads must be made watertight. { ¦frē,bord ,dek }

free-burning coal See noncaking coal. { ¦frē ,bərn·iŋ 'kōl }

free carbon [MET] Elemental carbon present in a metal in an uncombined state. { ¦frē 'kär·bən }

free charge [ELEC] Electric charge which is not bound to a definite site in a solid, in contrast to the polarization charge. { ¦frē 'chärj }

free convection See natural convection. { ¦frē kən'vek·shən }

free convection number See Grashof number. { ¦frē kən'vek·shən ,nəm·bər }

free crushing [MIN ENG] Crushing under conditions of speed and feed so that there is ample room for the fine ore to fall away from the coarser material and thereby escape further crushing. { ¦frē 'krəsh·iŋ }

free-cutting steel [MET] Steel that contains a higher percentage of sulfur than carbon steel, making it very easy to machine. { ¦frē ,kəd·iŋ 'stēl }

free cyanide [CHEM] Cyanide not combined as part of an ionic complex. { ¦frē 'sī·ə,nīd }

free diving [ENG] Diving with the use of scuba equipment to allow freedom and maneuverability. { ¦frē 'dīv·iŋ }

freedom to mine [MIN ENG] The law by which anybody has the right to mine certain minerals when he or she has prospected for them and has filed a proper application for the right to mine them. { ¦frēd·əm tə 'mīn }

free-drop [ENG] To air-drop supplies or equipment without parachute. { 'frē ,dräp }

free electromagnetic field [ELECTROMAG] An electromagnetic field in empty space that does not interact with matter. { 'frē i,lek·trō·mag,ned·ik 'fēld }

free electron [PHYS] An electron that is not constrained to remain in a particular atom, and is therefore able to move in matter or in a vacuum when acted on by external electric or magnetic fields. { ¦frē i'lek,trän }

free-electron laser [OPTICS] A device in which a beam of relativistic electrons passes through a static periodic magnetic field to amplify a superimposed coherent optical wave and thereby produce a powerful beam of coherent light. { 'frē i¦lek,trän 'lā·zər }

free-electron paramagnetism [ELECTROMAG] Paramagnetism of certain metals that results from the magnetic moments of nearly free electrons in their conduction bands. Also known as Pauli paramagnetism. { ¦frē i¦lek,trän ,par·ə'mag·nə,tiz·əm }

free-electron theory of metals [SOLID STATE] A model of a metal in which the free electrons, that is, those giving rise to the conductivity, are regarded as moving in a potential (due to the metal ions in the lattice and to all the remaining free electrons) which is approximated as constant everywhere inside the metal. Also known as Sommerfeld model; Sommerfeld theory. { ¦frē i'lek,trän ,thē·ə·rē əv 'med·əlz }

free end See free face. { ¦frē 'end }

free energy [THERMO] **1.** The internal energy of a system minus the product of its temperature and its entropy. Also known as Helmholtz free energy; Helmholtz function; Helmholtz potential; thermodynamic potential at constant volume; work function. **2.** See Gibbs free energy. { ¦frē 'en·ər·jē }

free enthalpy See Gibbs free energy. { ¦frē 'en,thal·pē }

free face [GEOL] A vertical or steeply inclined layer of rock from which weathered material falls to form talus at its base. [MIN ENG] The exposed surface of a mass of rock or of coal. Also known as free end. { ¦frē 'fās }

free fall [MECH] The ideal falling motion of a body acted upon only by the pull of the earth's gravitational field. [PETRO ENG] In deep drilling, an arrangement by which the bit is permitted to fall freely to the bottom at each drop or down stroke. { 'frē ,fȯl }

free falling [MECH ENG] In ball milling, the peripheral speed at which part of the crop load breaks clear on the ascending side and falls clear to the toe of the charge. { 'frē ,fȯl·iŋ }

free-fed [MIN ENG] In comminution, pertaining to rolls fed only enough ore to maintain a ribbon of material between them. { 'frē ,fed }

free ferrite [MET] Relatively pure metallic iron phase present in steel or cast iron. { ¦frē 'fe,rīt }

free field [ACOUS] An isotropic, homogeneous sound field that is free from all bounding surfaces. [COMPUT SCI] A property of information retrieval devices which permits recording of information in the search medium without regard to preassigned fixed fields. [PHYS] A field in empty space not interacting with other fields or sources. { 'frē ,fēld }

free-field room See anechoic chamber. { 'frē ,fēld ,rüm }

free-field storage [COMPUT SCI] Data storage that allows recording of the data without regard for fixed or preassigned fields. { 'frē ,fēld ,stȯr·ij }

free fit [DES ENG] A fit between mating pieces where accuracy is not essential or where large variations in temperature may occur. { 'frē ,fit }

free flight [MECH] Unconstrained or unassisted flight. { 'frē ,flīt }

free-flight angle [MECH] The angle between the horizontal and a line in the direction of motion of a flying body, especially a rocket, at the beginning of free flight. { 'frē ,flīt ,aŋ·gəl }

free-flight melt spinning [MET] Rapid-quenching process in which the molten metal is forced through an orifice under pressure and the jet is solidified while in free flight; quench rates reach 1000 K per second. { 'frē ,flīt 'melt ,spin·iŋ }

free-flight trajectory [MECH] The path of a body in free fall. { 'frē ,flīt trə'jek·trē }

free float [IND ENG] The length of time, expressed as work units, that a specific activity may be delayed without delaying the start of another activity scheduled to follow immediately after. Also known as free slack. { ¦frē 'flōt }

free-floating anxiety [PSYCH] Severe, generalized, persistent anxiety not specifically ascribed to a particular object or event and often a precursor of panic. { ¦frē ¦flōd·iŋ aŋ'zī·ə·tē }

free foehn See high foehn. { ¦frē ¦fān }

freeform language [COMPUT SCI] A programming or command language that does not require rigid formatting. { ¦frē ,fȯrm 'laŋ·gwij }

freeform text [COMPUT SCI] A record, or a variable-length portion of a record, that stores plain, unformatted English. { 'frē,fȯrm 'tekst }

free-free absorption See inverse bremsstrahlung. { 'frē 'frē ab'sȯrp·shən }

free gas [PETRO ENG] A hydrocarbon that exists in the gaseous phase at reservoir pressure and temperature and remains a gas when produced under normal conditions. [PHYS] Any gas at any pressure not in solution, or mechanically held in the liquid hydrocarbon phase. { ¦frē ¦gas }

free-gas saturation [PETRO ENG] Proportion of oil-reservoir pore structure saturated by free (undissolved) gas. { ¦frē ¦gas ,sach·ə'rā·shən }

free gold [MET] Gold that is in the free state, that is, not combined with other substances. { ¦frē ¦gōld }

free grid See floating grid. { 'frē ,grid }

free group [MATH] A group whose generators satisfy the equation $x \cdot y = e$ (e is the identity element in the group) only when $x = y^{-1}$ or $y = x^{-1}$. { 'frē ,grüp }

free gyroscope [ENG] A gyroscope that uses the property of gyroscopic rigidity to sense changes in altitude of a machine, such as an airplane; the spinning wheel or rotor is isolated from the airplane by gimbals; when the plane changes from level flight, the gyro remains vertical and gives the pilot an artificial horizon reference. { ¦frē ¦jī·rə,skōp }

freehand grinding See offhand grinding. { ¦frē,hand 'grind·iŋ }

free hole [SOLID STATE] Any hole which is not bound to an impurity or to an exciton. { ¦frē ¦hōl }

free impedance [ELECTR] Impedance at the input of the transducer when the impedance of its load is made zero. Also known as normal impedance. { ¦frē im'pēd·əns }

free induction decay [SPECT] A type of electron paramagnetic resonance spectroscopy in which a material is exposed to a short high-power pulse (as short as 2 nanoseconds) of microwave radiation, and the response of the material is Fourier transformed into the normal spectrum. Abbreviated FID. { ¦frē in'dək·shən di,kā }

freeing port [NAV ARCH] An opening in a ship's side, at the level of the deck or in the bulwark, to allow water to escape. { 'frē·iŋ ,pȯrt }

free instruments [ENG] Instruments designed to initially sink to the ocean bottom, release their ballast, and then rise to the surface where they are retrieved with their acquired payload. { ¦frē 'in·strə·məns }

free ion [PHYS CHEM] An ion, such as found in an ionized gas, whose properties, such as spectrum and magnetic moment, are not significantly affected by other atoms, ions, or molecules nearby. { ¦frē 'ī,än }

free joint [MECH ENG] A robotic articulation that has six degrees of freedom. { ¦frē 'jȯint }

free-machining steel [MET] Steel to which impurities have been added to improve machinability. { 'frē mə¦shēn·iŋ 'stēl }

Freeman-Nichols roaster [MIN ENG] A unit in which pyrite flotation concentrates are flash-roasted. { ¦frē·mən ¦nik·əlz ,rōs·tər }

freemartin [VERT ZOO] An intersexual, usually sterile female calf twinborn with a male. { 'frē,märt·ən }

free-mass antenna [ENG] A detector of gravitational radiation that consists of suspended, almost inertial masses and a laser interferometer that detects their motions. { ,frē ,mas an'ten·ə }

free meander [HYD] A stream meander that displaces itself very easily by lateral corrasion. { ¦frē mē'an·dər }

free milling [MIN ENG] A process applied to ores which contain free gold or silver and can be reduced by crushing and amalgamation (by gravity or on blankets), without roasting or other chemical treatment. { ¦frē ,mil·iŋ }

free-milling gold [MET] Gold that has a clean surface so that it readily amalgamates with mercury (by gravity or on blankets) after liberation by comminution. { 'frē ,mil·iŋ 'gōld }

free-milling ore [MIN ENG] Ore containing gold which can be caught with mercury by a variety of gravity processes or on blankets. { 'frē ,mil·iŋ 'ȯr }

free module [MATH] A module which is a free group with respect to its additive group. { ¦frē ¦mäj·yül }

free moisture [MIN ENG] Water in a sample of coal that can be removed simply by drying in air. { 'frē 'mȯis·chər }

free molecule [PHYS CHEM] A molecule, as in a gas, whose properties, such as spectrum and magnetic moment, are not affected by other atoms, ions, and molecules nearby. { ¦frē 'mäl·ə,kyül }

free molecule flow [PHYS] Flow of a gas in which the mean free path of the molecules is long compared to a characteristic dimension of the flow field, such as the diameter of a tube through which gas is flowing. Also known as Knudsen flow. { ¦frē 'mäl·ə,kyül ,flō }

free motional impedance [ELECTR] Of a transducer, the complex remainder after the blocked impedance has been subtracted from the free impedance. { ¦frē ¦mō·shən·əl im'pēd·əns }

freenet [COMPUT SCI] A bulletin board system, based in a public library or other community or government organization, that provides access to useful resources. { 'frē,net }

free oscillation [PHYS] The oscillation of a physical system with no externally applied stimuli. Also known as free vibration. { ¦frē ,äs·ə'lā·shən }

free-piston engine [MECH ENG] A prime mover utilizing free-piston motion controlled by gas pressure in the cylinders. { ¦frē ,pis·tən 'en·jən }

free-piston gage [ENG] An instrument for measuring high fluid pressures in which the pressure is applied to the face of a small piston that can move in a cylinder and the force needed to keep the piston stationary is determined. Also known as piston gage. { ¦frē ¦pis·tən 'gāj }

free-point indicator [PETRO ENG] In a drilling operation in which a string of pipe has become stuck, a tool that measures the amount of stretch in the stuck string and indicates the deepest point at which the pipe is free. { 'frē ¦pȯint 'in·də̇,kād·ər }

free port [CIV ENG] An isolated, enclosed, and policed port in or adjacent to a port of entry, without a resident population. { 'frē ¦pȯrt }

free progressive wave [PHYS] A wave in a medium or in vacuum, free from boundary effects. Also known as free wave. { 'frē prə'gres·iv 'wāv }

free quark [PART PHYS] A hypothetical quark that is not bound together with other quarks within a hadron, and whose charge, mass, and other properties can therefore be measured individually. { 'frē 'kwärk }

free radical [CHEM] An atom or a diatomic or polyatomic molecule which possesses one unpaired electron. Also known as a radical. { ¦frē 'rad·ə·kəl }

free-radical reaction See homolytic cleavage. { ¦frē ¦rad·ə·kəl rē'ak·shən }

free recoil [ORD] The movement in recoil of the recoiling parts of a gun, if unimpeded by resistances such as springs or pneumatic pressure; it is a theoretical term used in recoil mechanism design. { ¦frē 'rē,kȯil }

free-recoil mount [ORD] A proving ground gun mount designed to closely approximate free-recoil conditions; used for determining design information. { ¦frē 'rē,kȯil ,maủnt }

free ribosome [CELL MOL] A ribosome located by itself or in a group (known as a polysome or polyribosome) in the cytosol, rather than bound to the endoplasmic reticulum; synthesizes soluble cytosolic proteins and most extrinsic membrane proteins. { ¦frē 'rī·bə,sōm }

free rocket [ORD] A rocket having fixed fins but no control surface, that is, no provision for guidance. { ¦frē 'räk·ət }

free run [ORD] As applied to guns, the travel of a projectile from its original position in the gun chamber until it engages with the rifling in the gun bore. { ¦frē 'rən }

free-running frequency [ELECTR] Frequency at which a normally driven oscillator operates in the absence of a driving signal. { ¦frē ,rən·iŋ ,frē·kwən·sē }

free-running multivibrator See astable multivibrator. { 'frē ,rən·iŋ məl·tə'vī,brād·ər }

free-running sweep [ELECTR] Sweep triggered continuously by an internal trigger generator. { 'frē ,rən·iŋ 'swēp }

free settling [MIN ENG] In classification, the free fall of particles through fluid media. { ¦frē 'sed·əl·iŋ }

free sheet [GRAPHICS] A type of paper free of mechanical pulp. { ¦frē 'shēt }

free slack See free float. { ¦frē 'slak }

free space [PHYS] A region of space in which there are no particles of matter and no electromagnetic or gravitational fields other than those whose behavior is under consideration. { ¦frē 'spās }

free-space field intensity [ELECTROMAG] Radio field intensity that would exist at a point in a uniform medium in the absence of waves reflected from the earth or other objects. { 'frē ,spās 'fēld in,ten·sə̇d·ē }

free-space loss [ELECTROMAG] The theoretical radiation loss, depending only on frequency and distance, that would occur if all variable factors were disregarded when transmitting energy between two antennas. { 'frē ,spās ,lȯs }

free-space propagation [ELECTROMAG] Propagation of electromagnetic radiation over a straight-line path in a vacuum or ideal atmosphere, sufficiently removed from all objects that affect the wave in any way. { 'frē ,spās ,präp·ə'gā·shən }

free-space radar equation [ELECTROMAG] Equation that governs a radar signal characteristic when it is propagated between a radar set and a reflecting object or target in otherwise empty space. { 'frē ,spās 'rā,där i,kwā·zhən }

free-space radiation pattern [ELECTROMAG] Radiation pattern that an antenna would have if it were in free space where there is nothing to reflect, refract, or absorb the radiated waves. { 'frē ,spās rād·ē'ā·shən ,pad·ərn }

free-space wave [ELECTROMAG] An electromagnetic wave propagating in a vacuum, free from boundary effects. { 'frē ,spās 'wāv }

freestone [BOT] A fruit stone to which the fruit does not cling, as in certain varieties of peach. [GEOL] Stone, particularly a thick-bedded, even-textured, fine-grained sandstone, that breaks freely and is able to be cut and dressed with equal facility in any direction without tending to split. { 'frē,stōn }

free streamline [FL MECH] A streamline separating fluid in motion from fluid at rest. { 'frē 'strēm,līn }

free-stream Mach number [AERO ENG] The Mach number of the total airframe (entire aircraft) as contrasted with the local Mach number of a section of the airframe. { 'frē ,strēm 'mäk ,nəm·bər }

free surface [FL MECH] A boundary between two homogeneous fluids. { ¦frē 'sər·fəs }

free-swelling index [ENG] A test for measuring the free-swelling properties of coal; consists of heating 1 gram of pulverized coal in a silica crucible over a gas flame under prescribed conditions to form a coke button, the size and shape of which are then compared with a series of standard profiles numbered 1 to 9 in increasing order of swelling. { 'frē ,swel·iŋ 'in,deks }

free symbol [COMPUT SCI] A contextual symbol preceded and followed by a space; it is always meaningful and always used to symbolize both grammatical and nongrammatical meaning; an example is the English "I." { ¦frē 'sim·bəl }

free symbol sequence [COMPUT SCI] A symbol sequence not preceded, or not followed, or neither preceded nor followed by space. { ¦frē 'sim·bəl 'sēk·wəns }

Free test [PATH] Demonstration of a delayed hypersensitivity reaction to Bedsonia antigen in the diagnosis of lymphogranuloma venereum. { 'frē ,test }

Freeth's nephroid [MATH] The strophoid of a circle with respect to a pole located at the center and a fixed point located on the circumference. Also known as nephroid of Freeth. { 'frāths 'nef,rȯid }

free-traveling wave See progressive wave. { 'frē ,trav·ə·liŋ 'wāv }

free tree [MATH] A tree graph in which there is no node which is distinguished as the root. { 'frē ,trē }

free turbine [MECH ENG] In a turbine engine, a turbine wheel that drives the output shaft and is not connected to the shaft driving the compressor. { ¦frē 'tər·bən }

free ultrafilter [MATH] An ultrafilter of a set, S, that contains any subset of S whose complement is finite. { ¦frē 'əl·trə,fil·tər }

free variable [MATH] In logic, a variable that has an occurrence which is not within the scope of a quantifier and thus can be replaced by a constant. { ¦frē 'ver·ē·ə·bəl }

free vector [MECH] A vector whose direction in space is prescribed but whose point of application and line of application are not prescribed. { ¦frē 'vek·tər }

free vibration See free oscillation. { ¦frē vī'brā·shən }

free volume [STAT MECH] In a lattice theory of a dense gas or liquid, the volume of the cage in which a given molecule is free to wander when its nearest neighbors are fixed at their lattice positions. { ¦frē 'väl·yəm }

free vortex [FL MECH] Two-dimensional fluid flow in which the fluid moves in concentric circles at speeds inversely proportional to the radii of the circles. { ¦frē 'vȯr,teks }

free wall [MIN ENG] The wall of an ore vein filling which scales off cleanly from the gouge. { ¦frē 'wȯl }

freeware [COMPUT SCI] Copyrighted software that is downloaded from the Internet for which there is no charge. { 'frē,wer }

free water [CHEM] The volume of water that is not contained in suspension in a vessel containing both water and a suspension of water and another liquid. [PETRO ENG] Water produced with oil from an oil well. { 'frē 'wȯd·ər }

free-water content See water content. { 'frē ,wȯd·ər ,kän·tent }

free-water elevation See water table. { 'frē ,wȯd·ər el·ə,vā·shən }

free-water knockout [PETRO ENG] A vessel into which oil or an oil-water emulsion is piped so that free water can settle out. { 'frē ,wȯd·ər 'näk,aủt }

free-water surface See water table. { 'frē ,wȯd·ər 'sər·fəs }

free wave See free progressive wave. { ¦frē 'wāv }

freeze [ENG] **1.** To permit drilling tools, casing, drivepipe, or drill rods to become lodged in a borehole by reason of caving walls or impaction of sand, mud, or drill cuttings, to the extent that they cannot be pulled out. Also known as bind-seize. **2.** To burn in a bit. Also known as burn-in. **3.** The premature setting of cement, especially when cement slurry hardens before it can be ejected fully from pumps or drill rods during a borehole cementation operation. **4.** The act or process of drilling a borehole by utilizing a drill fluid chilled to minus 30–40°F, (minus 34–40°C) as a means of consolidating, by freezing, the borehole wall materials or core as the drill penetrates a water-saturated formation, such as sand or gravel. [PHYS CHEM] To solidify a liquid by removal of heat. { frēz }

freeze drying [ENG] A method of drying materials, such as certain foods, that would be destroyed by the loss of volatile ingredients or by drying temperatures above the freezing point; the material is frozen under high vacuum so that ice or other frozen solvent will quickly sublime and a porous solid remain. { 'frēz ,drī·iŋ }

freeze etching [CRYO] A method using cryogenics to prepare specimens for study with a microscope. { 'frēz ,ech·iŋ }

freeze-fracture electron microscopy [CELL MOL] A technique used to visualize the inside of cellular membranes. Rapidly frozen cells are ruptured (fractured) so as to split open the membrane and expose the interior surfaces. A thin layer of carbon together with a metal (usually platinum) is then evaporated over the specimen to produce a replica of the surface, which is removed and examined in the electron microscope. { ¦frēz,frak·chər i,lek,trän mi'kräs·kə·pē }

freeze-out lake [HYD] A shallow lake which may be deeply frozen over for long periods of time. { 'frēz ,aút ,lāk }

freeze point [PETRO ENG] The depth in a drill hole at which the drill pipe, tubing, or casing has become stuck. { 'frēz ,póint }

freezer [MECH ENG] An insulated unit, compartment, or room in which perishable foods are quick-frozen and stored. { 'frēz·ər }

freeze sinking [MIN ENG] A method of shaft sinking in waterlogged strata by the use of cold brine circulating through a system of pipes until an ice wall is formed. Also known as freezing method. { 'frēz ,siŋk·iŋ }

freeze-up [HYD] The formation of a continuous ice cover on a body of water. [MECH ENG] Abnormal operation of a refrigerating unit because ice has formed at the expansion device. { 'frēz,əp }

freezing drizzle [METEOROL] Drizzle that falls in liquid form but freezes upon impact with the ground to form a coating of glaze. { 'frēz·iŋ ¦driz·əl }

freezing level [METEOROL] The lowest altitude in the atmosphere over a given location, at which the air temperature is 0°C; the height of the 0°C constant-temperature surface. { ¦frēz·iŋ ¦lev·əl }

freezing-level chart [METEOROL] A synoptic chart showing the height of the 0°C constant-temperature surface by means of contour lines. { 'frēz·iŋ ,lev·əl ,chärt }

freezing method See freeze sinking. { 'frēz·iŋ ,meth·əd }

freezing microtome [ENG] A microtome used to cut frozen tissue. { ¦frēz·iŋ 'mī·krə,tōm }

freezing mixture [PHYS CHEM] A mixture of substances whose freezing point is lower than that of its constituents. { 'frēz·iŋ ,miks·chər }

freezing nucleus [METEOROL] Any particle which, when present within a mass of supercooled water, will initiate growth of an ice crystal about itself. { 'frēz·iŋ ,nü·klē·əs }

freezing point [PHYS CHEM] The temperature at which a liquid and a solid may be in equilibrium. Abbreviated fp. { 'frēz·iŋ ,póint }

freezing-point depression [PHYS CHEM] The lowering of the freezing point of a solution compared to the pure solvent; the depression is proportional to the active mass of the solute in a given amount of solvent. { 'frēz·iŋ ,póint di,presh·ən }

freezing precipitation [METEOROL] Any form of liquid precipitation that freezes upon impact with the ground or exposed objects; that is, freezing rain or freezing drizzle. { 'frēz·iŋ prə,sip·ə'tā·shən }

freezing rain [METEOROL] Rain that falls in liquid form but freezes upon impact to form a coating of glaze upon the ground and on exposed objects. { ¦frēz·iŋ 'rān }

Fregatidae [VERT ZOO] Frigate birds or man-o'-war birds, a family of fish-eating birds in the order Pelecaniformes. { fre'gad·ə,dē }

F region [GEOPHYS] The general region of the ionosphere in which the F_1 and F_2 layers tend to form. { 'ef ,rē·jən }

F$_c$ region [IMMUNOL] Region of an antibody molecule that binds to antibody receptors on the surface of cells such as macrophages and mast cells, and to complement protein; F_c is

derived from the term crystallizable fragment. { ¦ef'sē ,rē·jən }

Frégoli syndrome [PSYCH] A misidentification syndrome characterized by the delusional confusion of an individual as a familiar person in disguise. { 'frā·gól·ē ,sin,drōm }

freibergite [MINERAL] A steel-gray, silver-bearing variety of tetrahedrite. { 'frī,bər,gīt }

freieslebenite [MINERAL] $Pb_3Ag_5Sb_5S_{12}$ A steel-gray to dark mineral composed of a sulfide of antimony, lead, and silver. { ¦frī·əs¦lā·bə,nīt }

freight car [ENG] A railroad car in or on which freight is transported. { 'frāt ,kär }

freighter [ENG] A ship or aircraft used mainly for carrying freight. { 'frād·ər }

freight ton See ton. { 'frāt ,tən }

freirinite [MINERAL] $Na_3Cu_3(AsO_4)_2(OH)_3 \cdot H_2O$ A lavender to turquoise-blue mineral composed of a basic hydrous arsenate of sodium and copper. { frā'rē,nīt }

fremontite See natromontebrasite. { 'frē·mən,tīt }

Frenatae [INV ZOO] The equivalent name for Heteroneura. { 'frē·nə,tē }

french [MECH] A unit of length used to measure small diameters, especially those of fiber optic bundles, equal to 1/3 millimeter. { french }

French chalk [MATER] Finely ground talc. { ¦french 'chók }

french coupling [DES ENG] A coupling having both right- and left-handed threads. { ¦french 'kəp·liŋ }

french curve [GRAPHICS] A guide, usually made of clear plastic, used for making regular, irregular, and reverse curves in mechanical drawings and illustrations. { ¦french 'kərv }

French division See Lyot division. { ¦french də'vizh·ən }

French drain [CIV ENG] An underground passage for water, consisting of loose stones covered with earth. { ¦french 'drān }

French measles See rubella. { ¦french 'mē·zəlz }

French polish [MATER] Shellac dissolved in methylated spirits. { ¦french 'päl·ish }

Frenet-Serret formulas [MATH] Formulas in the theory of space curves, which give the directional derivatives of the unit vectors along the tangent, principal normal and binormal of a space curve in the direction tangent to the curve. Also known as Serret-Frenet formulas. { fre'nā sə'rā ,fór·myə·ləz }

Frenkel defect [SOLID STATE] A crystal defect consisting of a vacancy and an interstitial which arise when an atom is plucked out of a normal lattice site and forced into an interstitial position. Also known as Frenkel pair. { 'freŋk·əl 'dē,fekt }

Frenkel exciton [SOLID STATE] A tightly bound exciton in which the electron and the hole are usually on the same atom, although the pair can travel anywhere in the crystal. { 'freŋk·əl 'ek·sə,tän }

Frenkel-Halsey-Hill isotherm equation [PHYS] An equation for the volume v of a gas adsorbed on a surface at a given temperature, $\ln (p/p_o) = k/v^s$, where p is the pressure of the gas, p_o is the vapor pressure, and k and s are constants. { 'freŋk·əl ¦hólz·ē ¦hil 'ī·sō,thərm i,kwā·zhən }

Frenkel pair See Frenkel defect. { 'freŋk·əl ,per }

frenulum [ANAT] **1.** A small fold of integument or mucous membrane. **2.** A small ridge on the upper part of the anterior medullary velum. [INV ZOO] A spine on most moths that projects from the hindwings and is held to the forewings by a clasp, thus coupling the wings together. { 'fren·yə·ləm }

frenum [ANAT] A fold of tissue that restricts the movements of an organ. { 'frē·nəm }

frequency [PHYS] The number of cycles completed by a periodic quantity in a unit time. [STAT] The number of times an event or item falls into or is expected to fall into a certain class or category. { 'frē·kwən·sē }

frequency agility [ORD] The ability to shift the frequency of a radar transmitter rapidly and continually in order to avoid jamming by the enemy, to reduce mutual interference with friendly sources, to enhance echoes from targets, or to provide the required patterns of electronic countermeasures or electronic counter-countermeasures radiation. { 'frē·kwən·sē ə,jil·əd·ē }

frequency allocation [COMMUN] Assignment of available frequencies in the radio spectrum to specific stations and for specific purposes, to give maximum utilization of frequencies with minimum interference between stations. { 'frē·kwən·sē ,al·ə'kā·shən }

FREQUENCY COUNTER

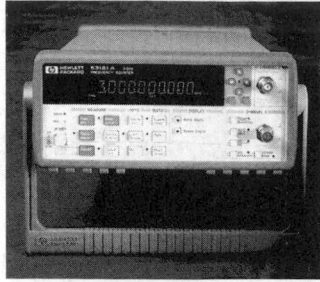

General-purpose frequency counter.
(Hewlett-Packard Co.)

FREQUENCY DIVIDER

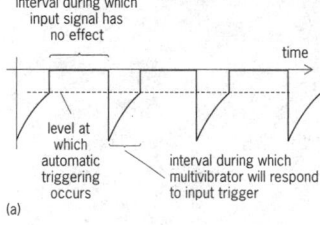

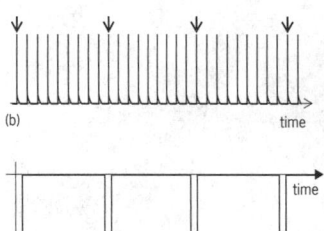

Waveform of frequency-dividing
astable multivibrator. (*a*) Waveform
at the base terminal of a free-running
astable multivibrator, showing how
the multivibrator would perform if
there were no input. (*b*) Input signals.
Arrows indicate triggering signals.
(*c*) Output signal, synchronized to
every *n*th input cycle.

frequency analysis [COMPUT SCI] A determination of the number of times certain parts of an algorithm are executed, indicating which parts of the algorithm consume large quantities of time and hence where efforts should be directed toward improving the algorithm. { ¦frē·kwən·sē ə¸nal·ə·səs }

frequency analyzer [ELECTR] A device which measures the intensity of many different frequency components in some oscillation, as in a radio band; used to identify transmitting sources. { 'frē·kwən·sē 'an·ə¸līz·ər }

frequency-azimuth intensity [ELECTR] Type of radar display in which frequency, azimuth, and strobe intensity are correlated. { ¦frē·kwən·sē ¦az·ə·məth in¸ten·səd·ē }

frequency band [PHYS] A continuous range of frequencies extending between two limiting frequencies. { 'frē·kwən·sē ¸band }

frequency bridge [ELECTR] A bridge in which the balance varies with frequency in a known manner, such as the Wien bridge; used to measure frequency. { 'frē·kwən·sē ¸brij }

frequency carrier system [COMMUN] A form of frequency division multiplex in which intelligence is carried on subcarriers. { 'frē·kwən·sē ¸kar·ē·ər ¸sis·təm }

frequency changer *See* frequency converter. { 'frē·kwən·sē ¸chānj·ər }

frequency-changer station [ELEC] An installation at which power is transmitted between two alternating-current electric power systems operating at different frequencies by a direct-current link. { 'frē·kwən·sē ¸chānj·ər ¸stā·shən }

frequency characteristic *See* frequency-response curve. { ¦frē·kwən·sē ¸kar·ik·tə'ris·tik }

frequency coding [NEUROSCI] A means by which the central nervous system analyzes the content of a receptor message; the frequency of discharged impulses is a function of the rate of rise of the generator current and indirectly of the stimulus strength; the greater the stimulus intensity, the higher the impulse frequency of the message. { 'frē·kwən·sē ¸kōd·iŋ }

frequency compensation *See* compensation. { ¦frē·kwən·sē ¸käm·pən'sā·shən }

frequency conversion [ELECTR] Converting the carrier frequency of a received signal from its original value to the intermediate frequency value in a superheterodyne receiver. { 'frē·kwən·sē kən¸vər·zhən }

frequency converter [ELEC] A circuit, device, or machine that changes an alternating current from one frequency to another, with or without a change in voltage or number of phases. Also known as frequency changer; frequency translator. { 'frē·kwən·sē kən¸vərd·ər }

frequency counter [ELECTR] An electronic counter used to measure frequency by counting the number of cycles in an electric signal during a preselected time interval. { 'frē·kwən·sē ¸kaůnt·ər }

frequency curve [STAT] A graphical representation of a continuous frequency distribution; the value of the variable is the abscissa and the frequency is the ordinate. { 'frē·kwən·sē ¸kərv }

frequency cutoff [ELECTR] The frequency at which the current gain of a transistor drops 3 decibels below the low-frequency gain value. { ¦frē·kwən·sē ¸kəd¸óf }

frequency-dependent selection [EVOL] A type of natural selection that decreases the frequency of more common phenotypes in a population and increases the frequency of less common phenotypes. { ¦frē·kwən·sē di¦pen·dənt si'lek·shən }

frequency deviation [COMMUN] The peak difference between the instantaneous frequency of a frequency-modulated wave and the carrier frequency. { ¦frē·kwən·sē ¸dē·vē'ā·shən }

frequency discriminator [ELECTR] A discriminator circuit that delivers an output voltage which is proportional to the deviations of a signal from a predetermined frequency value. { ¦frē·kwən·sē di'skrim·ə¸nād·ər }

frequency distortion [ELECTR] Distortion in which the relative magnitudes of the different frequency components of a wave are changed during transmission or amplification. Also known as amplitude distortion; amplitude-frequency distortion; waveform-amplitude distortion. { ¦frē·kwən·sē di'stór·shən }

frequency distribution [MATH] A function which measures the relative frequency or probability that a variable can take on a set of values. { ¦frē·kwən·sē ¸dis·trə'byü·shən }

frequency diversity [COMMUN] Diversity reception involving the use of carrier frequencies separated 500 hertz or more and having the same modulation, to take advantage of the fact that fading does not occur simultaneously on different frequencies. { ¦frē·kwən·sē də'vər·səd·ē }

frequency divider [ELECTR] A harmonic conversion transducer in which the frequency of the output signal is an integral submultiple of the input frequency. Also known as counting-down circuit. { 'frē·kwən·sē di¸vīd·ər }

frequency-division data link [COMMUN] Data link using frequency division techniques for channel spacing. { 'frē·kwən·sē di¸vizh·ən 'dad·ə ¸liŋk }

frequency-division multiple access [COMMUN] A technique by which multiple users who are geographically dispersed gain access to a communications channel to which they are assigned distinct and nonoverlapping sections of the electromagnetic spectrum. Abbreviated FDMA. { ¦frē·kwən·sē di¦vizh·ən ¸məl·tə·pəl 'ak¸ses }

frequency-division multiplexing [COMMUN] A multiplex system for transmitting two or more signals over a common path by using a different frequency band for each signal. Abbreviated fdm; FDM. Also known as frequency multiplexing. { 'frē·kwən·sē di¸vizh·ən 'məl·tə¸plek·siŋ }

frequency domain [COMMUN] A plane on which signal strength can be represented graphically as a function of frequency, instead of a function of time. [CONT SYS] Pertaining to a method of analysis, particularly useful for fixed linear systems in which one does not deal with functions of time explicitly, but with their Laplace or Fourier transforms, which are functions of frequency. { 'frē·kwən·sē də¸mān }

frequency-domain optical storage [COMPUT SCI] A technique whereby up to 1000 bits of information would be stored at each spatial location in an optical storage medium by using persistent spectral holeburning. { 'frē·kwən·sē də¸mān 'äp·tə·kəl 'stór·ij }

frequency-domain reflectometer [ELECTROMAG] A tuned reflectometer used for measuring reflection coefficients and impedance of waveguides over a wide frequency range, by sweeping a band of frequencies and analyzing the reflected returns. { 'frē·kwən·sē də¸mān ¸rē¸flek'täm·əd·ər }

frequency doubler [ELECTR] An amplifier stage whose resonant anode circuit is tuned to the second harmonic of the input frequency; the output frequency is then twice the input frequency. Also known as doubler. { 'frē·kwən·sē ¸dəb·lər }

frequency drift [ELECTR] A gradual change in the frequency of an oscillator or transmitter due to temperature or other changes in the circuit components that determine frequency. { 'frē·kwən·sē ¸drift }

frequency-exchange signaling [COMMUN] Signaling in which the change from one signaling condition to another is accompanied by decay in amplitude of one or more other frequencies. { 'frē·kwən·sē iks¸chānj ¸sig·nəl·iŋ }

frequency factor [PHYS CHEM] The constant A (or ν) in the Arrhenius equation, which is the relation between reaction rate and absolute temperature T; the equation is $k = Ae - (\Delta H_{act}/RT)$, where k is the specific rate constant, ΔH_{act} is the heat of activation, and R is the gas constant. { 'frē·kwən·sē ¸fak·tər }

frequency frogging [COMMUN] Interchanging of frequency allocations for carrier channels to prevent singing, reduce crosstalk, and reduce the need for equalization; modulators in each repeater translate a low-frequency group to a high-frequency group, and vice versa. { 'frē·kwən·sē ¸fräg·iŋ }

frequency function *See* probability density function. { 'frē·kwən·sē ¸fəŋk·shən }

frequency hopping [COMMUN] A spread-spectrum technique in which the frequency of the carrier changes pseudorandomly according to a pseudonoise code, with a consecutive group of code symbols defining a particular frequency. { 'frē·kwən·sē ¸häp·iŋ }

frequency interlace [COMMUN] Carrier chrominance signal frequency chosen so I and J sidebands are interwoven with luminance sidebands in the same bandwidth and in a manner that causes no mutual interference. { ¦frē·kwən·sē 'in·tər¸lās }

frequency locus [CONT SYS] The path followed by the frequency transfer function or its inverse, either in the complex plane or on a graph of amplitude against phase angle; used in determining zeros of the describing function.   { 'frē·kwən·sē ,lō·kəs }

frequency meter   [ENG]   **1.** An instrument for measuring the frequency of an alternating current; the scale is usually graduated in hertz, kilohertz, and megahertz.   **2.** A device cali-brated to indicate frequency of a radio wave.   { 'frē·kwən·sē ,mēd·ər }

frequency mixing   [OPTICS]   The combination of two or more electromagnetic waves in a nonlinear medium to form another wave whose frequency is a sum or difference of the frequencies of the incident waves.   { 'frē·kwən·sē 'mik·siŋ }

frequency-modulated carrier current telephony   [COMMUN]   Telephony involving the use of a frequency-modulated carrier signal transmitted over power-line wires or other wires.   { 'frē·kwən·sē ,mäj·ə,lād·əd 'kar·ē·ər 'kə·rənt tə'lef·ə·nē }

frequency-modulated cyclotron   See synchrocyclotron.   { 'frē·kwən·sē ,mäj·ə,lād·əd 'sī·klə,trän }

frequency-modulated jamming   [ELECTR]   Jamming technique consisting of a constant amplitude radio-frequency signal that is varied in frequency about a center frequency to produce a signal over a band of frequencies.   { 'frē·kwən·sē ,mäj·ə,lād·əd 'jam·iŋ }

frequency-modulated laser   [OPTICS]   A helium-neon or other laser in which an ultrasonic modulation cell is used to impress a frequency-modulated video signal on the output beam of the laser.   { 'frē·kwən·sē ,mäj·ə,lād·əd 'lā·zər }

frequency-modulated radar   [ENG]   Form of radar in which the radiated wave is frequency modulated, and the returning echo beats with the wave being radiated, thus enabling range to be measured.   { 'frē·kwən·sē ,mäj·ə,lād·əd 'rā,där }

frequency modulation   [COMMUN]   Modulation in which the instantaneous frequency of the modulated wave differs from the carrier frequency by an amount proportional to the instantaneous value of the modulating wave.   Abbreviated FM.   { 'frē·kwən·sē ,mäj·ə,lā·shən }

frequency-modulation broadcast band   [COMMUN]   The band of frequencies extending from 88 to 108 megahertz; used for frequency-modulation radio broadcasting in the United States.   { 'frē·kwən·sē ,mäj·ə,lā·shən 'brȯd,kast ,band }

frequency-modulation detector   [ELECTR]   A device, such as a Foster-Seely discriminator, for the detection or demodulation of a frequency-modulated wave.   { 'frē·kwən·sē ,mäj·ə,lā·shən di'tek·tər }

frequency-modulation Doppler   [ENG]   Type of radar involving frequency modulation of both carrier and modulation on radial sweep.   { 'frē·kwən·sē ,mäj·ə,lā·shən 'däp·lər }

frequency modulation-frequency modulation   [COMMUN]   System in which frequency-modulated subcarriers are used to frequency-modulate a second carrier.   { 'frē·kwən·sē ,mäj·ə,lā·shən 'frē·kwən·sē ,mäj·ə,lā·shən }

frequency-modulation laser   [OPTICS]   Conventional laser containing a phase modulator inside its Fabry-Pérot cavity; characterized by the lack of noise resulting from the random fluctuation in the phase in the various modes.   { 'frē·kwən·sē ,mäj·ə,lā·shən 'lā·zər }

frequency-modulation noise level on carrier   [COMMUN]   Residual frequency modulation resulting from disturbance produced in an aural transmitter operating within the band of 50 to 15,000 hertz.   { 'frē·kwən·sē ,mäj·ə,lā·shən 'nȯiz ,lev·əl ȯn 'kar·ē·ər }

frequency modulation-phase modulation   [COMMUN]   System in which the several frequency-modulated subcarriers are used to phase modulate a second carrier.   { 'frē·kwən·sē ,mäj·ə,lā·shən 'fāz ,mäj·ə,lā·shən }

frequency-modulation receiver   [ELECTR]   A radio receiver that receives frequency-modulated waves and delivers corresponding sound waves.   { 'frē·kwən·sē ,mäj·ə,lā·shən ri'sē·vər }

frequency-modulation receiver deviation sensitivity   [ELECTR]   Least frequency deviation that produces a specified output power.   { 'frē·kwən·sē ,mäj·ə,lā·shən ri'sē·vər dē·vē'ā·shən sen·sə'tiv·əd·ē }

frequency-modulation synthesis   [ENG ACOUS]   A method of synthesizing musical tones which, in its simplest form, is carried out using two digital oscillators, with the output of one adding to the frequency (or phase) control of the other.   { 'frē·kwən·sē ,mä·jə'lā·shən ,sin·thə·səs }

frequency-modulation transmitter   [ELECTR]   A radio transmitter that transmits a frequency-modulated wave.   { 'frē·kwən·sē ,mäj·ə,lā·shən tranz'mid·ər }

frequency-modulation tuner   [ELECTR]   A tuner containing a radio-frequency amplifier, converter, intermediate-frequency amplifier, and demodulator for frequency-modulated signals, used to feed a low-level audio-frequency signal to a separate audio-frequency amplifer and loudspeaker.   { 'frē·kwən·sē ,mäj·ə,lā·shən 'tün·ər }

frequency modulator   [ELECTR]   A circuit or device for producing frequency modulation.   { 'frē·kwən·sē 'mäj·ə,lād·ər }

frequency monitor   [ELECTR]   An instrument for indicating the amount of deviation of the carrier frequency of a transmitter from its assigned value.   { 'frē·kwən·sē ,män·əd·ər }

frequency multiplexing   See frequency-division multiplexing.   { 'frē·kwən·sē 'məl·tə,plek·siŋ }

frequency multiplier   [ELECTR]   A harmonic conversion transducer in which the frequency of the output signal is an exact integral multiple of the input frequency.   Also known as multiplier.   { 'frē·kwən·sē 'məl·tə,plī·ər }

frequency offset   [COMMUN]   **1.** Change in the frequency received on a transmission line from that which was transmitted, which can occur if suppressed-carrier multiplexing is used, and may result in errors.   **2.** A small difference in the carrier frequencies of television stations in adjacent cities operating on the same channel.   { 'frē·kwən·sē 'ȯf,set }

frequency-offset transponder   [ELECTR]   Transponder that changes the signal frequency by a fixed amount before retransmission.   { 'frē·kwən·sē 'ȯf·set tran'spän·dər }

frequency optimum traffic   See optimum working frequency.   { 'frē·kwən·sē ,äp·tə·məm 'traf·ik }

frequency polygon   [STAT]   A graph obtained from a frequency distribution by joining with straight lines points whose abscissae are the midpoints of successive class intervals and whose ordinates are the corresponding class frequencies.   { 'frē·kwən·sē 'päl·ə,gän }

frequency prediction chart   [COMMUN]   Graph showing curve for the maximum usable frequency, frequency optimum traffic, and lowest usable frequency between two specific points for various times throughout a 24-hour period.   { 'frē·kwən·se prə,dik·shən ,chärt }

frequency probabilities   See objective probabilities.   { 'frē·kwən·sē ,präb·ə,bil·əd·ēz }

frequency pulling   [ELECTR]   A change in the frequency of an oscillator due to a change in load impedance.   { 'frē·kwən·sē ,pul·iŋ }

frequency recorder   [ELEC]   An instrument which uses a frequency bridge to sense the frequency of an alternating current, and which makes a graphical record of this frequency as a function of time.   { 'frē·kwən·sē ri,kȯrd·ər }

frequency regulator   [ELEC]   A device that maintains the frequency of an alternating-current generator at a predetermined value.   { 'frē·kwən·sē ,reg·yə,lā·dər }

frequency relay   [ELECTR]   Relay which functions at a predetermined value of frequency; may be an over-frequency relay, an under-frequency relay, or a combination of both.   { 'frē·kwən·sē ,rē,lā }

frequency response   [ENG]   A measure of the effectiveness with which a circuit, device, or system transmits the different frequencies applied to it; it is a phasor whose magnitude is the ratio of the magnitude of the output signal to that of a sine-wave input, and whose phase is that of the output with respect to the input.   Also known as amplitude-frequency response; sine-wave response.   { 'frē·kwən·sē ri,späns }

frequency-response curve   [ENG]   A graph showing the magnitude or the phase of the freqency response of a device or system as a function of frequency.   Also known as frequency characteristic.   { 'frē·kwən·sē ri,späns ,kərv }

frequency-response equalization   See equalization.   { 'frē·kwən·sē ri,späns ,ē·kwə·lə'zā·shən }

frequency-response trajectory   [CONT SYS]   The path followed by the frequency-response phasor in the complex plane as the frequency is varied.   { 'frē·kwən·sē ri,späns trə'jek·trē }

frequency run   [ELECTR]   A series of tests made to determine the amplitude-frequency response characteristic of a transmission line, circuit, or device.   { 'frē·kwən·sē ,rən }

frequency scan antenna   [ELECTROMAG]   A radar antenna similar to a phased array antenna in which one dimensional scanning is accomplished through frequency variation.   { 'frē·kwən·sē ,skan an'ten·ə }

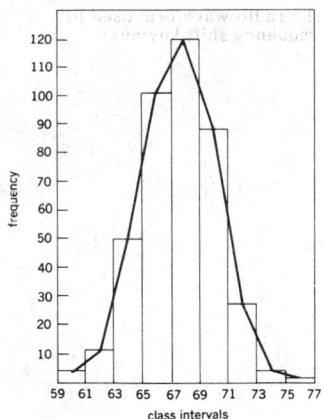

FREQUENCY POLYGON

The conversion of a histogram into a frequency polygon by connecting the midpoint value (at the top of each rectangle) with the adjacent midpoint value by straight lines.

frequency scanning [ELECTR] Type of system in which output frequency is made to vary at a mechanical rate over a desired frequency band. { 'frē·kwən·sē ,skan·iŋ }

frequency-selective device *See* electric filter. { 'frē·kwən·sē si,lek·tiv di,vīs }

frequency separation multiplier [ELECTR] Multiplier in which each of the variables is split into a low-frequency part and a high-frequency part that are multiplied separately, and the results added to give the required product; this system makes it possible to get high accuracy and broad bandwidth. { 'frē·kwən·sē ,sep·ə,rā·shən 'məl·tə,plī·ər }

frequency separator [ELECTR] The circuit that separates the horizontal and vertical synchronizing pulses in a monochrome or color television receiver. { 'frē·kwən·sē ,sep·ə,rād·ər }

frequency shift [ELECTR] A change in the frequency of a radio transmitter or oscillator. Also known as radio-frequency shift. { 'frē·kwən·sē ,shift }

frequency-shift converter [ELECTR] A device that converts a received frequency-shift signal to an amplitude-modulated signal or a direct-current signal. { 'frē·kwən·sē ,shift kən'vərd·ər }

frequency-shift keyer [ELECTR] A lever to effect a frequency shift, that is, a change in the frequency of a radio transmitter, oscillator, or receiver. { 'frē·kwən·sē ,shift 'kē·ər }

frequency-shift keying [COMMUN] A form of frequency modulation used especially in telegraph, data, and facsimile transmission, in which the modulating wave shifts the output frequency between predetermined values corresponding to the frequencies of correlated sources. Abbreviated FSK. Also known as frequency-shift modulation; frequency-shift transmission. { 'frē·kwən·sē ,shift 'kē·iŋ }

frequency-shift modulation *See* frequency-shift keying. { 'frē·kwən·sē ,shift ,mäj·ə'lā·shən }

frequency-shift transmission *See* frequency-shift keying. { 'frē·kwən·sē ,shift tranz'mish·ən }

frequency-slope modulation [COMMUN] Type of modulation in which the carrier signal is swept periodically over the entire width of the band, much as in chirp radar; modulation of the carrier with a voice or other communication signal changes the bandwidth of the system without affecting the uniform distribution of energy over the band. { 'frē·kwən·sē ,slōp ,mäj·ə'lā·shən }

frequency spectrum [PHYS] A plot of the distribution of the intensity of some type of electromagnetic or acoustic radiation as a function of frequency. [SYS ENG] In the analysis of a random function of time, such as the amplitude of noise in a system, the limit as T approaches infinity of $1/(2\pi T)$ times the ensemble average of the squared magnitude of the amplitude of the Fourier transform of the function from $-T$ to T. Also known as power-density spectrum; power spectrum; spectral density. { 'frē·kwən·sē ,spek·trəm }

frequency splitting [ELECTR] One condition of operation of a magnetron which causes rapid alternating from one mode of operation to another; this results in a similar rapid change in oscillatory frequency and consequent loss in power at the desired frequency. { 'frē·kwən·sē ,splid·iŋ }

frequency stability [ELECTR] The ability of an oscillator to maintain a desired frequency; usually expressed as percent deviation from the assigned frequency value. { 'frē·kwən·sē stə,bil·əd·ē }

frequency stabilization [COMMUN] Process of controlling the center or carrier frequency so that it differs from that of a reference source by not more than a prescribed amount. { 'frē·kwən·sē ,stā·bə·lə'zā·shən }

frequency standard [ELECTR] A stable oscillator, usually controlled by a crystal or tuning fork, that is used primarily for frequency calibration. { 'frē·kwən·sē ,stan·dərd }

frequency study *See* work sampling. { 'frē·kwən·sē ,stəd·ē }

frequency swing [COMMUN] **1.** Peak difference between the maximum and the minimum values of the instantaneous frequency. **2.** In frequency modulation, a term used to describe the change in frequency resulting from the modulation. { 'frē·kwən·sē ,swiŋ }

frequency synthesizer [ELECTR] A device that provides a choice of a large number of different frequencies by combining frequencies selected from groups of independent crystals, frequency dividers, and frequency multipliers. { 'frē·kwən·sē ¦sin·thə,sīz·ər }

frequency table [STAT] A tabular arrangement of the distribution of an event or item according to some specified category or class intervals. { 'frē·kwən·sē ,tā·bəl }

frequency telemetering [COMMUN] The transmittal of an alternating-current signal from a primary element by variations in the signal frequency, instead of amplitude. { ¦frē·kwən·sē ¦tel·ə¦mēd·ə·riŋ }

frequency theory [NEUROSCI] A theory of human hearing according to which every specific frequency of sound energy is represented by nerve impulses of the same frequency, and pitch differentiation and analysis are carried out by the brain centers. Also known as telephone theory. { 'frē·kwən·sē ,thē·ə·rē }

frequency-time-intensity [ELECTR] Type of radar display in which the frequency, time, and strobe intensity are correlated. { ¦frē·kwən·sē ¦tīm in'ten·səd·ē }

frequency tolerance [ELECTR] Of a radio transmitter, extent to which the carrier frequency of the transmitter may be permitted to depart from the frequency assigned. { 'frē·kwən·sē ,täl·ə·rəns }

frequency-to-voltage converter [ELECTR] A converter that provides an analog output voltage which is proportional to the frequency or repetition rate of the input signal derived from a flowmeter, tachometer, or other alternating-current generating device. Abbreviated F/V converter. { ¦frē·kwən·sē tə ¦vōl·tij kən'vərd·ər }

frequency transformation [CONT SYS] A transformation used in synthesizing a band-pass network from a low-pass prototype, in which the frequency variable of the transfer function is replaced by a function of the frequency. Also known as low-pass band-pass transformation. { ¦frē·kwən·sē ,tranz·fər'mā·shən }

frequency translation [COMMUN] Moving a modulated radio-frequency carrier signal to a new location in the frequency spectrum without disturbing the relationship of the carrier to its sidebands. { ¦frē·kwən·sē tranz'lā·shən }

frequency translator *See* frequency converter. { ¦frē·kwən·sē 'tranz,lād·ər }

frequency-type telemeter [ELECTR] Telemeter that employs frequency of an alternating current or voltage as the translating means. { 'frē·kwən·sē ,tīp 'tel·ə,mēd·ər }

frequency variation [ELECTR] The change over time of the deviation from assigned frequency of a radio-frequency carrier (or power supply system); usually tightly controlled because of national or industry standards. { ¦frē·kwən·sē ,ver·ē¦ā·shən }

Frequently Asked Questions [COMPUT SCI] Abbreviated FAQ **1.** A document containing answers to common questions about the subjects of other documents to which it is linked. **2.** In particular, a document associated with a Web site that contains answers to common questions about the site. { ,frē·kwənt·lē ,askt 'kwes·chənz }

frequently in [MATH] A net is frequently in a set if, for each element a of the directed system that indexes the set, there is an element b of the directed system such that $b \geq a$ and x_b (the element indexed by b) is in this set. { 'frē·kwənt·lē ,in }

fresco [GRAPHICS] Painting of two types, buon fresco and fresco secco; in buon fresco, dry pigments are ground with water and applied on a wet lime plaster wall, and as the plaster dries, the pigment is permanently bound to it; fresco secco utilizes pigments ground in glue, casein, or polymer emulsions and applied on a dry plaster wall that has been wetted with limewater. { 'fres·kō }

fresh [GEOL] Unweathered in reference to a rock or rock surface. [METEOROL] Pertaining to air which is stimulating and refreshing. { fresh }

fresh breeze [METEOROL] In the Beaufort wind scale, a wind whose speed is 17 to 21 knots (19 to 24 miles per hour, or 31 to 39 kilometers per hour). { ¦fresh 'brēz }

fresh-core technique [PETRO ENG] A method in which a core sample, fresh from the field, is subjected to waterflooding in the laboratory, and the resulting residual oil is determined; used to calculate total waterflood recovery of oil from a reservoir formation. { ¦fresh 'kòr tek,nēk }

freshet [HYD] **1.** The annual spring rise of streams in cold climates as a result of melting snow. **2.** A flood resulting from either rain or melting snow; usually applied only to small

FREQUENCY-SHIFT KEYING

The radio waveform used in frequency-shift keying.

streams and to floods of minor severity. **3.** A small freshwater stream. { 'fresh·ət }

fresh gale [METEOROL] In the Beaufort wind scale, a wind whose speed is from 34 to 40 knots (39 to 46 miles per hour, or 63 to 74 kilometers per hour). { ¦fresh 'gāl }

fresh ice See newly formed ice. { ¦fresh 'īs }

fresh water [HYD] Water containing no significant amounts of salts, such as in rivers and lakes. { ¦fresh 'wȯd·ər }

fresh-water ecosystem [ECOL] The living organisms and nonliving materials of an inland aquatic environment. { 'fresh ‚wȯd·ər 'ek·ō‚sis·təm }

fresnel [PHYS] A unit of frequency, equal to 10^{12} hertz. { frā'nel }

Fresnel-Arago laws [OPTICS] The three laws stating that two rays of polarized light interfere in the same way as ordinary light if they are polarized in the same plane, but do not interfere if they are polarized at right angles; two rays polarized from ordinary light at right angles do not interfere in the ordinary sense when they are brought into the same plane of polarization; and two rays polarized at right angles from plane polarized light, and then brought into the same polarization plane, interfere. { frā'nel ä·rä'gō ‚lóz }

Fresnel biprism [OPTICS] A very flat triangular prism which has two very acute angles and one very obtuse angle; used to observe the interference of light from a slit passing through the two halves of the prism. { frā'nel 'bī‚priz·əm }

Fresnel diffraction [OPTICS] Diffraction in which the source of light or the observing screen are at a finite distance from the aperture or obstacle. { frā'nel di'frak·shən }

Fresnel drag coefficient [OPTICS] The quantity $1 - (1/n^2)$, where n is the index of diffraction of a transparent medium, believed by Fresnel to be the ratio of the velocity with which ether was dragged along in the medium to the velocity of the medium itself. { frā'nel 'drag ‚kō·i‚fish·ənt }

Fresnel ellipsoid [OPTICS] An ellipsoid whose three perpendicular axes are proportional to the principal values of the wave velocity of light in an anisotropic medium. Also known as ray ellipsoid. { frā'nel i'lip‚sȯid }

Fresnel equations [OPTICS] Equations which give the intensity of each of the two polarization components of light which is reflected or transmitted at the boundary between two media with different indices of refraction. { frā'nel i‚kwā·zhənz }

Fresnel fringe [OPTICS] One of a series of light and dark bands that appear near the edge of a shadow in Fresnel diffraction. { frā'nel 'frinj }

Fresnel-Huygens principle See Huygens-Fresnel principle. { frā'nel 'hī·gənz ‚prin·sə·pəl }

Fresnel integrals [MATH] Given a parameter x, the integrals over t from 0 to x of $\sin t^2$ and of $\cos t^2$ or from x to ∞ of $(\cos t)/t^{1/2}$ and of $(\sin t)/t^{1/2}$. { frā'nel 'int·ə·grəlz }

Fresnel lens [OPTICS] A thin lens constructed with stepped setbacks so as to have the optical properties of a much thicker lens. { frā'nel 'lenz }

Fresnel mirrors [OPTICS] Two plane mirrors which are inclined to each other on the order of a degree and used to observe the interference of light which originates from a slit and is reflected from both mirrors. { frā'nel 'mir·ərz }

Fresnel ovaloid [OPTICS] For an anisotropic crystal, an ovaloid whose central section normal to the propagation direction of an electromagnetic wave gives the axes of polarization of the displacement vector and the associated wave velocities. { frā'nel 'ōv·ə‚lóid }

Fresnel reflection formula [OPTICS] The Fresnel equations for light reflected from a boundary. { frā'nel ri'flek·shən ‚fȯr·myə·lə }

Fresnel region [ELECTROMAG] The region between the near field of an antenna (close to the antenna compared to a wavelength) and the Fraunhofer region. { frā'nel ‚rē·jən }

Fresnel rhomb [OPTICS] A glass rhomb which has an acute angle of about 52°; light which is incident normal to the end of the rhomb undergoes two internal reflections, and if it is initially linearly polarized at an angle of 45° to the plane of incidence, it emerges circularly polarized. { frā'nel 'räm }

Fresnel spotlight [ELEC] A lighting instrument that is composed of a lamp and a Fresnel (stepped planoconvex) lens; the unit can be made with or without reflectors and has a system to adjust the spacing between the lamp and the lens so as to

control the light beam; models range from 100 to 5000 watts. { frā'nel 'spät‚līt }

Fresnel theory of double refraction [OPTICS] The theory which explains double refraction of a crystal in terms of nonspherical wave surfaces. { frā'nel ‚thē·ə·rē əv ‚dəb·əl ri'frak·shən }

Fresnel zones [ELECTROMAG] Circular portions of a wavefront transverse to a line between an emitter and a point where the disturbance is being observed; the nth zone includes all paths whose lengths are between $n-1$ and n half-wavelengths longer than the line-of-sight path. Also known as half-period zones. { frā'nel ‚zōnz }

Fresnian [GEOL] A North American stage of upper Eocene geologic time, above Narizian and below Refugian. { 'frez·nē·ən }

fretsaw [DES ENG] A narrow-bladed fine-toothed saw that is held under tension in a frame. { 'fret‚sȯ }

fretting corrosion [MET] Surface damage usually in an air environment between two surfaces, one or both of which are metals, in close contact under pressure and subject to a slight relative motion. Also known as chafing corrosion. { 'fred·iŋ kə'rō·zhən }

Freudianism [PSYCH] The psychoanalytic school of psychiatry founded by Sigmund Freud. { 'fród·ē·ən‚iz·əm }

Freudian slip [PSYCH] A verbal mistake that suggests some underlying motive, often sexual or aggressive in nature. { ¦fród·ē·ən 'slip }

Freundlich isotherm equation [PHYS] Equation which states that the volume of gas adsorbed on a surface at a given temperature is proportional to the pressure of the gas raised to a constant power. { 'fród·lik 'ī·sō‚thərm i‚kwā·zhən }

Freund method [ORG CHEM] A method for preparation of cycloparaffins in which dihalo derivatives of the paraffins are treated with zinc to produce the cycloparaffin. { 'fród ‚meth·əd }

Freund's adjuvant [IMMUNOL] A water-oil emulsion containing a killed microorganism (usually *Mycobacterium tuberculosis*) which enhances antigenicity. { 'fróinz 'a·jə·vənt }

friability [MATER] The ease with which a material is crumbled, pulverized, or reduced to powder. { ‚frī·ə'bil·əd·ē }

friable [MATER] Referring to the property of a substance capable of being easily rubbed, crumbled, or pulverized into powder. { 'frī·ə·bəl }

friagem [METEOROL] A period of cold weather in the middle and upper parts of the Amazon Valley and in eastern Bolivia, occurring during the dry season in the Southern Hemisphere winter. Also known as vriajem. { 'frī·ə‚jem }

friar's cloth See monk's cloth. { 'frī·ərz ‚klȯth }

friar's cowl See aconite. { 'frī·ərz ‚kaúl }

fricative [LING] A primary type of speech sound of the major languages that is produced by a partial constriction along the vocal tract which results in turbulence; for example, the fricatives in English may be illustrated by the initial and final consonants in the words vase, this, faith, hash. { 'frik·əd·iv }

Fricke dosimeter [NUCLEO] A radiation dosimeter in which the energy of ionizing radiation is determined from the amount of ferrous ions converted to ferric ions in an aerated acidic ferrous sulfate solution. { 'frik·ə dō'sim·əd·ər }

friction [MECH] A force which opposes the relative motion of two bodies whenever such motion exists or whenever there exist other forces which tend to produce such motion. { 'frik·shən }

frictional See cohesionless. { 'frik·shən·əl }

frictional electricity [ELEC] The electric charges produced on two different objects, such as silk and glass or catskin and ebonite, by rubbing them together. Also known as triboelectricity. { 'frik·shən·əl i‚lek'tri·səd·ē }

frictional grip [MECH] The adhesion between the wheels of a locomotive and the rails of the railroad track. { ¦frik·shən·əl 'grip }

frictional secondary flow See secondary flow. { ¦frik·shən·əl ¦sek·ən‚der·ē 'flō }

friction bearing [MECH ENG] A solid bearing that directly contacts and supports an axle end. { 'frik·shən ‚ber·iŋ }

friction blocks [PETRO ENG] Thin blocks with cylindrical surfaces that drag on the inside of the well casing to prevent rotation of the packer (seal between the outside of tubing and inside of casing). { 'frik·shən ‚bläks }

friction bonding [ENG] Soldering of a semiconductor chip

FRESNEL RHOMB

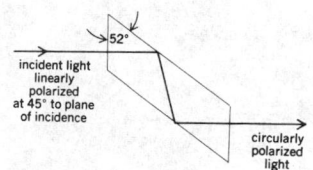

incident light
linearly
polarized
at 45° to plane
of incidence

52°

circularly
polarized
light

Graphic demonstration of the Fresnel rhomb.

to a substrate by vibrating the chip back and forth under pressure to create friction that breaks up oxide layers and helps alloy the mating terminals. { 'frik·shən ˌbänd·iŋ }

friction brake [MECH ENG] A brake in which the resistance is provided by friction. { 'frik·shən ˌbrāk }

friction calendering [ENG] Process wherein an elastomeric compound is forced into the interstices of woven or cord fabrics while passing between calender rolls. { 'frik·shən ˌkal·ən·driŋ }

friction catch [DES ENG] A catch consisting of a spring and plunger contained in a casing. { 'frik·shən ˌkach }

friction clutch [MECH ENG] A clutch in which torque is transmitted by pressure of the clutch faces on each other. { 'frik·shən ˌkləch }

friction coefficient See coefficient of friction. { 'frik·shən ˌkō·i'fish·ənt }

friction crack [GEOL] A short, crescent-shaped crack in glaciated rock produced by a localized increase in friction between rock and ice, oriented transverse to the direction of ice flow. { 'frik·shən ˌkrak }

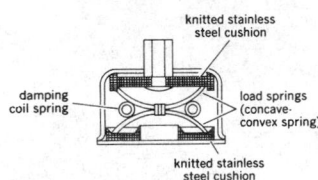

FRICTION DAMPING

knitted stainless steel cushion
damping coil spring
load springs (concave-convex spring)
knitted stainless steel cushion

A frictional damper, effective in vertical and horizontal directions.

friction damping [MECH] The conversion of the mechanical vibrational energy of solids into heat energy by causing one dry member to slide on another. { 'frik·shən ˌdamp·iŋ }

friction depth [OCEANOGR] The depth at which the velocity of wind-driven current becomes negligible compared to the surface velocity; sometimes referred to as the depth of the Ekman layer. { 'frik·shən ˌdepth }

friction drive [MECH ENG] A drive that operates by the friction forces set up when one rotating wheel is pressed against a second wheel. { 'frik·shən ˌdrīv }

friction factor [FL MECH] Any of several dimensionless numbers used in studying fluid friction in pipes, equal to the Fanning friction factor times some dimensionless constant. { 'frik·shən ˌfak·tər }

friction feed [MIN ENG] Longitudinal movement or advance of a drill stem and bit accomplished by friction devices in a diamond-drill swivel head, as opposed to a system consisting entirely of meshing gears. { 'frik·shən ˌfēd }

friction-feed printer [COMPUT SCI] A computer printer in which a roller is used to hold and advance the paper, much as in an ordinary typewriter. { 'frik·shən ˌfēd ˌprint·ər }

friction fit [DES ENG] A perfect fit between two parts. { 'frik·shən ˌfit }

friction flow [FL MECH] Fluid flow in which a significant amount of mechanical energy is dissipated into heat by action of viscosity. { 'frik·shən ˌflō }

friction force microscopy [ENG] The use of an atomic force microscope to measure the frictional forces on a surface. { ¦frik·shən ¦fors mī'krä·skə·pē }

friction gear [MECH ENG] Gearing in which motion is transmitted through friction between two surfaces in rolling contact. { 'frik·shən ˌgir }

friction head [FL MECH] The head lost by the flow in a stream or conduit due to frictional disturbances set up by the moving fluid and its containing conduit and by intermolecular friction. { 'frik·shən ˌhed }

friction horsepower [MECH ENG] Power dissipated in a machine through friction. { 'frik·shən ˌhors·pau·ər }

friction layer See surface boundary layer. { 'frik·shən ˌlā·ər }

frictionless flow See inviscid flow. { ¦frik·shən·ləs 'flō }

friction loss [MECH] Mechanical energy lost because of mechanical friction between moving parts of a machine. { 'frik·shən ˌlos }

friction pile [CIV ENG] A bearing pile surrounded by earth and supported entirely by friction; carries no load at its end. { 'frik·shən ˌpīl }

friction ridge [ANAT] One of the integumentary ridges on the plantar and palmar surfaces of primates. { 'frik·shən ˌrij }

friction saw [MECH ENG] A toothless circular saw used to cut materials by fusion due to frictional heat. { 'frik·shən ˌso }

friction sawing [MECH ENG] A burning process to cut stock to length by using a blade saw operating at high speed; used especially for the structural parts of mild steel and stainless steel. { 'frik·shən ˌso·iŋ }

friction shoe [ENG] An adjustable friction device that holds a window sash in any desired open position. { 'frik·shən ˌshü }

friction tape [MATER] Cotton tape impregnated with a sticky

moisture-repelling compound; used chiefly to hold rubber-tape insulation in position over a joint or splice. { 'frik·shən ˌtāp }

friction torque [MECH] The torque which is produced by frictional forces and opposes rotational motion, such as that associated with journal or sleeve bearings in machines. { 'frik·shən ˌtork }

friction-tube viscometer [ENG] Device to determine liquid viscosity by measurement of pressure drop through a friction tube with the liquid in viscous flow; gives direct solution to Poiseuille's equation. { 'frik·shən ˌtüb vi'skäm·əd·ər }

friction velocity [METEOROL] A reference wind velocity defined by the relation $u = \sqrt{|\tau/\rho|}$, where τ is the Reynolds stress, ρ the density, and u the friction velocity. { 'frik·shən və'läs·əd·ē }

friction welding [ENG] A welding process for metals and thermoplastic materials in which two members are joined by rubbing the mating faces together under high pressure. { 'frik·shən ˌweld·iŋ }

friction yielding prop See mechanical yielding prop. { 'frik·shən ¦yēld·iŋ ˌpräp }

Friedel-Crafts reaction [ORG CHEM] A substitution reaction, catalyzed by aluminum chloride in which an alkyl (R−) or an acyl (RCO−) group replaces a hydrogen atom of an aromatic nucleus to produce hydrocarbon or a ketone. { frē¦del 'krafs rē‚ak·shən }

friedelite [MINERAL] $Mn_8Si_6O_{18}(OH,Cl)_4·3H_2O$ A rose-red mineral composed of manganese silicate with chlorine. { 'frē·də‚līt }

Friedel's law [CRYSTAL] The law that x-ray or electron diffraction measurements cannot determine whether or not a crystal has a center of symmetry. { frē'delz ‚lo }

Friedlander's bacillus See Klebsiella pneumoniae. { 'frēd ‚lan·dərz bə‚sil·əs }

Friedlander's pneumonia [MED] Inflammation of the lungs caused by Klebsiella pneumoniae. { 'frēd‚lan·dərz nə'mō·nyə }

Friedlander synthesis [ORG CHEM] A synthesis of quinolines; the method is usually catalyzed by bases and consists of condensation of an aromatic o-amino-carbonyl derivative with a compound containing a methylene group in the alpha position to the carbonyl. { 'frēd‚lan·dər ‚sin·thə·səs }

Friedman solution [RELAT] A solution of Einstein's equations of general relativity with flat spatial sections describing a cosmological model. { 'frēd·mən sə‚lü·shən }

Friedman test [PATH] A pregnancy test in which a female rabbit is given an intravenous injection of urine from the patient; formation of corpora lutea in the ovaries indicates a positive test. { 'frēd·mən ‚test }

Friedmann universe [ASTRON] A nonstatic, homogeneous, isotropic model of the universe that displays expansion or contraction and has nonzero matter density. { 'frēd·mən 'yü·nə‚vərs }

Friedrich's ataxia [MED] A hereditary sclerosis of the spine with speech impairment, lateral curvature of the spine, and palsy of the lower limbs. Also known as hereditary spinal ataxia. { 'frēd·riks ā'tak·sē·ə }

friendship theorem [MATH] The proposition that, among a finite set of people, if every pair of people has exactly one common friend, then there is someone who knows everyone else. { 'fren‚ship ‚präb·ləm }

Fries rearrangement [ORG CHEM] The conversion of a phenolic ester into the corresponding o-and p-hydroxyketone by treatment with catalysts of the type of aluminum chloride. { 'frēz rē·ə'ränj·mənt }

Fries' rule [ORG CHEM] The rule that the most stable form of the bonds of a polynuclear compound is that arrangement which has the maximum number of rings in the benzenoid form, that is, three double bonds in each ring. { 'frēz ‚rül }

frieze [ARCH] A decorated band immediately below the cornice on an interior wall. [TEXT] Thick, heavyweight coating and upholstery fabric, with a rough, raised fibrous surface and a more or less hard feel. { frēz }

frigate [NAV ARCH] **1.** In the U.S. Navy, a ship larger than a destroyer and smaller than a cruiser, having a displacement of 4000–9000 tons, designed mainly as an escort ship for an attack aircraft carrier. **2.** In the British and Canadian navies, an escort ship larger than a corvette and smaller than a destroyer, having a displacement of 1200–2500 tons, corresponding in size to a destroyer escort in the U.S. Navy. { 'frig·ət }

frigid [PSYCH] Temperamentally, especially sexually, cold or unresponsive. { 'frij·id }

frigidoreceptor [PHYSIO] A cutaneous sense organ which is sensitive to cold. { ¦frij·ə·dō·ri¦sep·tər }

frigorie [THERMO] A unit of rate of extraction of heat used in refrigeration, equal to 1000 fifteen-degree calories per hour, or 1.16264 ± 0.00014 watts. { 'frig·ə·rē }

frigorimeter [ENG] A thermometer which measures low temperatures. { ¦frig·ə'rim·əd·ər }

frilling [GRAPHICS] Emulsion at the edges of a photographic film loosened from base of film. { 'fril·iŋ }

fringe [OPTICS] One of the light or dark bands produced by interference or diffraction of light. { frinj }

fringe area [COMMUN] An area just beyond the limits of the reliable service area of a television or radio transmitter, in which signals are weak and erratic. { 'frinj ,er·ē·ə }

fringe howl [ENG ACOUS] Squeal or howl heard when some circuit in a receiver is on the verge of oscillation. { 'frinj ,haül }

fringe joint [GEOL] A small-scale joint peripheral to, and usually at a 5–25° angle from the face of, the master joint. { 'frinj ,jóint }

fringe magnetic field [ELECTROMAG] The part of the magnetic field of a horseshoe magnet that extends outside the space between its poles. { 'frinj mag¦ned·ik 'fēld }

fringe ore [GEOL] Ore located on the outer boundary of a mineralization pattern or halo. Also known as halo ore. { 'frinj ,ór }

fringe region [METEOROL] The upper portion of the exosphere, where the conc of escape equals or exceeds 180°; in this region the individual atoms have so little chance of collision that they essentially travel in free orbits, subject to the earth's gravitation, at speeds imparted by the last collision. Also known as spray region. { 'frinj ,rē·jən }

fringes of equal thickness See Fizeau fringes. { 'frin·jəz əv ¦ē·kwəl 'thik·nəs }

fringe value [OPTICS] A quantity used in photoelastic work, equal to the stress which must be applied to a material, in pounds per square inch (1 pound per square inch equals approximately 6.89476 kilopascals), to produce a relative retardation of 1 wavelength between the components of a linearly polarized light beam when the light passes through a thickness of 1 inch (2.54 centimeters) in a direction perpendicular to the stress. { 'frinj ,val·yü }

Fringillidae [VERT ZOO] The finches, a family of oscine birds in the order Passeriformes. { frin'jil·ə,dē }

fringing fields [ELECTR] The electric fields produced by scattered electrons in an electron microscope. { 'frinj·iŋ ,fēlz }

fringing groove [ORD] A groove cut into a rotating band to collect metal from the band while it travels through the bore, to prevent its forming a fringe in rear of the rotating band; causes excess dispersion and short range. { 'frin·jiŋ ,grüv }

fringing reef [GEOL] A coral reef attached directly to or bordering the shore of an island or continental landmass. { ¦frin·jiŋ 'rēf }

Frise aileron [AERO ENG] A type of aileron having its leading edge projecting well ahead of the hinge axis. { 'frēz 'ā·lə,rän }

frisket [GRAPHICS] A mask used to protect the portions of photographs and artwork that are not be be airbrushed. { 'fris·kət }

frit [MATER] Fusible ceramic mixture used to make glazes and enamels for dinnerware and metallic surfaces, as on stoves and metal-base basins and tubs. { frit }

fritillary [BOT] The common name for plants of the genus Fritillaria. [INV ZOO] The common name for butterflies in several genera of the subfamily Nymphalinae. { frə'til·ə·rē }

frit seal [ENG] A seal made by fusing together metallic powders with a glass binder, for such applications as hermetically sealing ceramic packages for integrated circuits. { 'frit ,sēl }

fritting [ENG] Fusing materials for glass by application of heat. [MET] The pasty condition, usually occurring a little below the melting point, of the powdered ore, flux, and other reagents in fire assaying. { 'frid·iŋ }

Fritz process [FOOD ENG] Treatment of cream containing up to 40–50% fat by continuous feeding to a cooled horizontal cylindrical tank, where it is distributed in a film and beaten by blades at 3000 revolutions per minute; the formed mass falls into a trough, where buttermilk drains off, and butter is moved out for kneading and packing. { 'frits ,präs·əs }

Frobenius method [MATH] A method of finding a series solution near a point for a linear homogeneous ordinary differential equation. { frō'ben·yùs ,meth·əd }

Froehlich's syndrome See adiposogenital dystrophy. { 'frā liks ,sin,drōm }

frog [DES ENG] A hollow on one or both of the larger faces of a brick or block; reduces weight of the brick or block; may be filled with mortar. Also known as panel. [ENG] A device which permits the train or tram wheels on one rail of a track to cross the rail of an intersecting track. [VERT ZOO] The common name for a number of tailless amphibians in the order Anura; most have hindlegs adapted for jumping, scaleless skin, and large eyes. { fräg }

frogeye [PL PATH] Any of various leaf diseases characterized by formation of concentric rings around the diseased spots. { 'fräg,ī }

frogging repeater [ELECTR] Carrier repeater having provisions for frequency frogging to permit use of a single multipair voice cable without having excessive crosstalk. { ¦fräg·iŋ ri¦pēd·ər }

frog storm [METEOROL] The first bad weather in spring after a warm period. Also known as whippoorwill storm. { 'fräg ,stórm }

frohbergite [MINERAL] FeTe₂ A mineral composed of iron telluride; it is isomorphous with marcasite. { 'frō,bər,gīt }

Froissart bound [PART PHYS] A limit on the rate at which the cross section of a completely absorptive collision between hadrons can increase with energy, so that the interaction radius cannot increase more rapidly than the logarithm of the energy. { frwä'sär ,baünd }

from-to tester [ENG] Test equipment which checks continuity or impedance between points. { ¦främ,tü ,test·ər }

frond [BOT] **1.** The leaf of a palm or fern. **2.** A foliaceous thallus or thalloid shoot. { fränd }

frondelite [MINERAL] MnFe₄(PO₄)₅(OH)₅ A mineral composed of basic phosphate of manganese and iron; it is isomorphous with rockbridgeite. { 'frän·de,līt }

front [METEOROL] A sloping surface of discontinuity in the troposphere, separating air masses of different density or temperature. { frənt }

front abutment pressure [GEOPHYS] The release of energy in the superincumbent strata above the seam induced by the extraction of the seam. { 'frənt ə'bət·mənt ,presh·ər }

frontal-advance performance [PETRO ENG] The theory that, during the waterflood of a formation reservoir, displacement of oil causes a desaturation of the displaced fluids in accordance with relative-permeability relationships, and the displacement is linear. { ¦frənt·əl ad'vans pər,fòr·məns }

frontal angle [ARCH] The angle formed by the intersection of lines from the bregma and glabella to the auricular point. { ¦frənt·əl ¦aŋ·gəl }

frontal apron See outwash plain. { ¦frənt·əl 'ā·prən }

frontal bone [ANAT] Either of a pair of flat membrane bones in vertebrates, and a single bone in humans, forming the upper frontal portion of the cranium; the forehead bone. { 'frənt·əl ,bōn }

frontal contour [METEOROL] The line of intersection of a front (frontal surface) with a specified surface in the atmosphere, usually a constant-pressure surface; with respect to only one surface, this line is usually called the front. { ¦frənt·əl 'kän·túr }

frontal crest [ANAT] A median ridge on the internal surface of the frontal bone in humans. { 'frənt·əl ,krest }

frontal cyclone [METEOROL] Any cyclone associated with a front; often used synonymously with wave cyclone or with extratropical cyclone (as opposed to tropical cyclones, which are nonfrontal). { ¦frənt·əl 'sī,klōn }

frontal drive [PETRO ENG] In an oil reservoir with constant pressure (by gas injection or from a large gas-to-oil ratio), the driving of oil fluids into the wellbore by the free gas. { 'frənt·əl ,drīv }

frontal eminence [ANAT] The prominence of the frontal bone above each superciliary ridge in humans. { ¦frənt·əl 'em·ə·nəns }

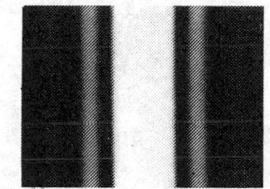

FRINGE

A Fraunhofer diffraction pattern, for a slit, or fringe, photographed with visible light. (*F. S. Harris*)

FROG

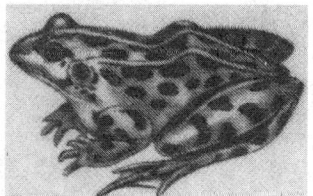

Common frog (*Ranan temporaria*).

frontal fire [ORD] Fire delivered at right angles to the front of the target. { ¦frənt·əl 'fīr }

frontal fog [METEOROL] Fog associated with frontal zones and frontal passages. { ¦frənt·əl 'fäg }

frontalia [INV ZOO] Paired sensory bristles on the frontal aspect of the head of gnathostomulids. { frən'tal·yə }

frontal index [ARCH] The ratio of the least to the greatest breadth of the forehead multiplied by 100. { ¦frənt·əl 'in,deks }

frontal inversion [METEOROL] A temperature inversion in the atmosphere, encountered upon vertical ascent through a sloping front (or frontal zone). { ¦frənt·əl in'vər·zhən }

frontal lifting [METEOROL] The forced ascent of the warmer, less-dense air at and near a front, occurring whenever the relative velocities of the two air masses are such that they converge at the front. { ¦frənt·əl 'lift·iŋ }

frontal lobe [NEUROSCI] The anterior portion of a cerebral hemisphere, bounded behind by the central sulcus and below by the lateral cerebral sulcus. { ¦frənt·əl ¦lōb }

frontal nerve [NEUROSCI] A somatic sensory nerve, attached to the ophthalmic nerve, which innervates the skin of the upper eyelid, the forehead, and the scalp. { ¦frənt·əl ¦nərv }

frontal occlusion See occluded front. { ¦frənt·əl ə'klü·zhən }

frontal passage [AERO ENG] The transit of an aircraft through a frontal zone. [METEOROL] The passage of a front over a point on the earth's surface. { ¦frənt·əl 'pas·ij }

frontal plain See outwash plain. { ¦frənt·əl 'plān }

frontal plane [ANAT] Any plane parallel with the long axis of the body and perpendicular to the sagittal plane. [MED] In electrocardiography and vectorcardiography, the projection of the vertical axis. { ¦frənt·əl ¦plān }

frontal precipitation [METEOROL] Any precipitation attributable to the action of a front; used mainly to distinguish this type from air-mass precipitation and orographic precipitation. { ¦frənt·əl prə,sip·ə'tā·shən }

frontal profile [METEOROL] The outline of a front as seen on a vertical cross section oriented normal to the frontal surface. { ¦frənt·əl 'prō,fīl }

frontal sinus [ANAT] Either of a pair of air spaces within the frontal bone above the nasal bridge. { ¦frənt·əl 'sī·nəs }

frontal strip [METEOROL] The presentation of a front, on a synoptic chart, as a frontal zone; that is, two lines, rather than a single line, are drawn to represent the boundaries of the zone; a rare usage. { 'frənt·əl ,strip }

frontal system [METEOROL] A system of fronts as they appear on a synoptic chart. { 'frənt·əl ,sis·təm }

frontal thunderstorm [METEOROL] A thunderstorm associated with a front; limited to thunderstorms resulting from the convection induced by frontal lifting. { 'frənt·əl 'thən·dər,stórm }

frontal wave [METEOROL] A horizontal, wavelike deformation of a front in the lower levels, commonly associated with a maximum of cyclonic circulations in the adjacent flow; it may develop into a wave cyclone. { 'frənt·əl ,wāv }

frontal zone [METEOROL] The three-dimensional zone or layer of large horizontal density gradient, bounded by frontal surfaces and surface front. { 'frənt·əl ,zōn }

front-end [COMPUT SCI] Of a minicomputer, under programmed instructions, performing data transfers and control operations to relieve a larger computer of these routines. { ¦frənt 'end }

front-end edit [COMPUT SCI] The process of checking and correcting data at the time it is entered into a computer system. { 'frənt ¦end 'ed·it }

front-end loader [MECH ENG] An excavator consisting of an articulated bucket mounted on a series of movable arms at the front of a crawler or rubber-tired tractor. { ¦frənt ¦end 'lōd·ər }

front-end processor [COMPUT SCI] A computer which connects to the main computer at one end and communications channels at the other, and which directs the transmitting and receiving of messages, detects and corrects transmission errors, assembles and disassembles messages, and performs other processing functions so that the main computer receives pure information. { ¦frənt ¦end ,präs,es·ər }

front-end volatility [CHEM ENG] The volatility of the lower-boiling fractions of gasoline, such as butanes. { ¦frənt ¦end väl·ə'til·əd·ē }

frontier [MATH] For a set in a topological space, all points in the closure of the set but not in its interior. Also known as boundary. { frən'tir əv ə 'set }

frontier orbitals [PHYS CHEM] Orbitals of two molecules that are spatially arranged so that a significant amount of overlap occurs between them. { frən¦tir 'òr·bə·təlz }

frontogenesis [METEOROL] **1.** The initial formation of a frontal zone or front. **2.** The increase in the horizontal gradient of an air mass property, mainly density, and the formation of the accompanying features of the wind field that typify a front. { ¦frən·tō¦jen·ə·səs }

frontogenetic function [METEOROL] A kinematic measure of the tendency of the flow in an air mass to increase the horizontal gradient of a conservative property. { ¦frən·tōjə¦ned·ik 'fəŋk·shən }

frontolysis [METEOROL] **1.** The dissipation of a front or frontal zone. **2.** In general, a decrease in the horizontal gradient of an air mass property, principally density, and the dissipation of the accompanying features of the wind field. { frən'täl·ə·səs }

front pinacoid [CRYSTAL] The {100} pinacoid in an orthorhombic, monoclinic, or triclinic crystal. Also known as macropinacoid; orthopinacoid. { ¦frənt 'pin·ə,kóid }

front porch [COMMUN] Portion of a composite picture signal which lies between the leading edge of the horizontal blanking pulse and the leading edge of the corresponding synchronizing pulse. { ¦frənt 'pórch }

front slagging [ENG] Skimming slag from the mixture of slag and molten metal as it flows through a taphole. { 'frənt ,slag·iŋ }

front slope See scarp slope. { 'frənt ,slōp }

front-to-back ratio [ELECTROMAG] Ratio of the effectiveness of a directional antenna, loudspeaker, or microphone toward the front and toward the rear. [SOLID STATE] Ratio of resistance of a crystal to current flowing in the normal direction to current flowing in the opposite direction. { ¦frənt tə ¦bak 'rā·shō }

frost [HYD] A covering of ice in one of its several forms, produced by the sublimation of water vapor on objects colder than 32°F (0°C). { fròst }

frost action [GEOL] **1.** The weathering process caused by cycles of freezing and thawing of water in surface pores, cracks, and other openings. **2.** Alternate or repeated cycles of freezing and thawing of water contained in materials; the term is especially applied to disruptive effects of this action. { 'fròst ,ak·shən }

frostbite [MED] Injury to skin and subcutaneous tissues, and in severe cases to deeper tissues also, from exposure to extreme cold. { 'fròst,bīt }

frost boil [GEOL] **1.** An accumulation of water and mud released from ground ice by accelerated spring thawing. **2.** A low mound formed by local differential frost heaving at a location most favorable for the formation of segregated ice and accompanied by the absence of an insulating cover of vegetation. { 'fròst ,bóil }

frost bursting See congelifraction. { 'fròst ,bərst·iŋ }

frost churning See congeliturbation. { 'fròst ,chərn·iŋ }

frost climate [CLIMATOL] The coldest temperature province in C. W. Thornthwaite's climatic classification: the climate of the ice cap regions of the earth, that is, those regions perennially covered with snow and ice. { 'fròst ,klī·mət }

frost cracks [BOT] Cracks in wood that have split outward from ray shakes. { 'fròst ,kraks }

frost day [METEOROL] An observational day on which frost occurs. { 'fròst ,dā }

frosted glass [MATER] Glass that has been etched with sand, or appears to have been so treated. { ¦fròs·təd 'glas }

frost feathers See ice feathers. { 'fròst ,feth·ərz }

frost flakes See ice fog. { 'fròst ,flāks }

frost flowers See ice flowers. { 'fròst ,flaù·ərz }

frost fog See ice fog. { 'fròst ,fäg }

frost hazard [METEOROL] The risk of damage by frost, expressed as the probability or frequency of killing frost on different dates during the growing season, or as the distribution of dates of the last killing frost of spring or the first of autumn. { 'fròst ,haz·ərd }

frost heaving [GEOL] The lifting and distortion of a surface due to internal action of frost resulting from subsurface ice formation; affects soil, rock, pavement, and other structures. { 'fròst ,hēv·iŋ }

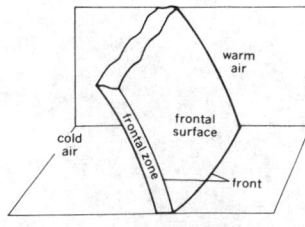

FRONTAL ZONE

Schematic diagram of the frontal zone (the angle with earth's surface is much exaggerated).

frost hollow [METEOROL] A small, low-lying zone which experiences frequent and severe frosts owing to the accumulation of cold night air; often severe where hills block the afternoon sunshine. { 'fròst ,häl·ō }

frosting [ENG] Decorating a scraped metal surface with a handscraper. Also known as flaking. { 'fròst·iŋ }

frostless zone [METEOROL] The warmest part of a slope above a valley floor, lying between the layer of cold air which forms over the valley floor on calm clear nights and the cold hill tops or plateaus; the air flowing down the slopes is warmed by mixing with the air above ground level, and to some extent also by adiabatic compression. Also known as green belt; verdant zone. { 'fròst·ləs ,zōn }

frost line [GEOL] **1.** The maximum depth of frozen ground during the winter. **2.** The lower limit of the permafrost. [MATER] In polyethylene film extrusion, a ring-shaped area with a frosty appearance at the point where the film reaches its final diameter. { 'fròst ,līn }

frost mound [GEOL] A hill and knoll associated with frozen ground in a permafrost region, containing a core of ice. Also known as soffosian knob; soil blister. { 'fròst ,maund }

frost pocket [METEOROL] A parcel of cold air in a hollow or at a valley floor, occurring when nighttime terrestrial radiation is greatest on valley slopes. { 'fròst ,päk·ət }

frost point [METEOROL] The temperature to which atmospheric moisture must be cooled to reach the point of saturation with respect to ice. { 'fròst ,pòint }

frost-point hygrometer [ENG] An instrument for measuring the frost point of the atmosphere; air under test is passed continuously across a polished surface whose temperature is adjusted so that a thin deposit of frost is formed which is in equilibrium with the air. { 'fròst ,pòint hī'gräm·əd·ər }

frost-point technique See 8D technique. { 'fròst ,pòint tek,nēk }

frost ring [BOT] A false annual growth ring in the trunk of a tree due to out-of-season defoliation by frost and subsequent regrowth of foliage. { 'fròst ,riŋ }

frost riving See congelifraction. { 'fròst ,rīv·iŋ }

frost shattering See congelifraction. { 'fròst ,shad·ə·riŋ }

frost smoke [METEOROL] **1.** A rare type of fog formed in the same manner as a steam fog, but at colder temperatures so that it is composed of ice particles instead of water droplets. **2.** See steam fog. { 'fròst ,smōk }

frost splitting See congelifraction. { 'fròst ,splid·iŋ }

frost stirring See congelifraction. { 'fròst ,stər·iŋ }

frost table [GEOL] An irregular surface in the ground which, at any given time, represents the penetration of thawing into seasonally frozen ground. { 'fròst ,tā·bəl }

frost thrusting [GEOL] Lateral dislocation of soil and rock materials by the action of freezing and resulting expansion of soil water. { 'fròst ,thrəst·iŋ }

frost weathering See congelifraction. { 'fròst ,weth·ə·riŋ }

frost wedging See congelifraction. { 'fròst ,wej·iŋ }

frosty mildew [PL PATH] A leaf spot caused by fungi of the genus *Cercosporella* and characterized by pale to white lesions. { 'fròs·tē 'mil,dü }

frost zone See seasonally frozen ground. { 'fròst ,zōn }

froth See foam. { fròth }

frother [CHEM] Substance used in flotation processes to make air bubbles sufficiently permanent, principally by reducing surface tension. { 'frò·thər }

froth flotation [ENG] A process for recovery of particles of ore or other material, in which the particles adhere to bubbles and can be removed as part of the froth. { ¦fròth flō'tā·shən }

frothing [ENG] The producing of relatively stable bubbles at an air-liquid interface as the result of agitation, aeration, ebulliation, or chemical reaction; it can be an undesired side effect, but in minerals beneficiation it is the basis of froth flotation. { 'frò·thiŋ }

frothing agent [MATER] A material that facilitates formation of bubbles in a froth flotation process. { 'fròth·iŋ ,ā·jənt }

frothing collector [MIN ENG] An ore collector which in addition produces a stable foam. { 'frò·thiŋ kə,lek·tər }

froth promoter [CHEM] A chemical compound used with a frothing agent. { 'fròth prə,mōd·ər }

frottage [GRAPHICS] A technique in which a material, such as paper, is placed on a rough or irregular surface and is rubbed with a pencil or paint; the approximate image of the peaks and valleys results; the method is used to copy bas-reliefs, tombstones, and bronzes. { frò'täzh }

Froude number 1 [FL MECH] A dimensionless number used in studying the motion of a body floating on a fluid with production of surface waves and eddies; equal to the ratio of the square of the relative speed to the product of the acceleration of gravity and a characteristic length of the body. Symbolized N_1. { ¦früd ,nəm·bər 'wən }

Froude number 2 [FL MECH] A dimensionless number, equal to the ratio of the speed of flow of a fluid in an open channel to the speed of very small gravity waves, the latter being equal to the square root of the product of the acceleration of gravity and a characteristic length. Symbolized N_2. { ¦früd ,nəm·bər 'tü }

frozen flux [PL PHYS] The lines of force of a frozen-in field. { ¦frōz·ən 'fləks }

frozen fog See ice fog. { ¦frōz·ən 'fäg }

frozen ground [GEOL] Soil having a temperature below freezing, generally containing water in the form of ice. Also known as gelisol; merzlota; taele; tjaele. { ¦frōz·ən 'graund }

frozen-in field [PL PHYS] A magnetic field in a plasma which has negligible electrical resistance; it can be shown that the lines of force of this field are constrained to move with the material. { ¦frōz·ən 'in ,fēld }

frozen pipe [PETRO ENG] A pipe that is immobilized in a borehole because caving has settled around the outside of the pipe. { ¦frōz·ən 'pīp }

frozen precipitation [METEOROL] Any form of precipitation that reaches the ground in frozen form; that is, snow, snow pellets, snow grains, ice crystals, ice pellets, and hail. { ¦frōz·ən prə,sip·ə'tā·shən }

frozen section [BIOL] A thin slice of material cut from a frozen sample of tissue or organ. { ¦frōz·ən 'sek·shən }

FRP See fiber-reinforced polymer.

fructescence [BOT] The period of fruit maturation. { frək'tes·əns }

fructification [BOT] **1.** The process of producing fruit. **2.** A fruit and its appendages. [MYCOL] A sporogenous structure. { 'frək·tə·fə'kā·shən }

fructivorous See frugivorous. { ¦frək'tiv·ə·rəs }

D-fructopyranose See fructose. { ¦dē ¦frək·tō'pī·rə,nōs }

fructose [BIOCHEM] $C_6H_{12}O_5$ The commonest of ketoses and the sweetest of sugars, found in the free state in fruit juices, honey, and nectar of plant glands. Also known as D-fructopyranose. { 'frük,tōs }

Frue vanner [MIN ENG] A side-shake type of ore-dressing apparatus consisting of an inclined rubber belt on which the material is washed by a constant flow of water. { 'frü ,van·ər }

frugivore [ZOO] A fruit-eating animal. { 'frü·ji,vòr }

frugivorous [ZOO] Fruit-eating. Also known as fructivorous. { frü'jiv·ə·rəs }

fruit [BOT] A fully matured plant ovary with or without other floral or shoot parts united with it at maturity. [NAV] Radar-beacon-system video display of a synchronous beacon return which results when several interrogator stations are located within the same general area; each interrogator receives its own interrogated reply as well as many synchronous replies resulting from interrogation of the airborne transponders by other ground stations. { früt }

fruit bud [BOT] A fertilized flower bud that matures into a fruit. { 'früt ,bəd }

fruit fly [INV ZOO] **1.** The common name for those acalypterate insects composing the family Tephritidae. **2.** Any insect whose larvae feed on fruit or decaying vegetable matter. { 'früt ,flī }

fruiting body [BOT] A specialized, spore-producing organ. { 'früd·iŋ ,bäd·ē }

fruiting myxobacteria See Myxobacterales. { 'früd·iŋ ,mik·sō·bak'tir·ē·ə }

frustrated internal reflectance See attenuated total reflectance. { 'frəs,trād·əd in,tərn·əl ri'flek·təns }

frustration [PSYCH] The experience of nonfulfillment of some wish or need. [SOLID STATE] In spin glasses, a phenomenon in which individual magnetic moments receive competing ordering instructions via different routes, because of the variation of the interaction between pairs of atomic moments with separation. { frəs'trā·shən }

FUCHSIN

Structural formula of fuchsin.

frustration threshold [PSYCH] The point at which an individual feels or shows frustration over inability to achieve an objective. { frəs'trā·shən ,thresh,hōld }

frustule [INV ZOO] **1.** The shell and protoplast of a diatom. **2.** A nonciliated planulalike bud in some hydrozoans. { 'frəs,chül }

frustum [MATH] The part of a solid between two cutting parallel planes. { 'frəs·təm }

frutescent [BIOL] *See* fruticose. [BOT] Shrublike in habit. { frü'tes·ənt }

fruticose [BIOL] Resembling a shrub; applied especially to lichens. Also known as frutescent. { 'früd·ə,kōs }

frying noise [ELEC] Noise in telephone transmission even when no conversation is taking place; caused by signal current flowing across a resistance element having multiple intermittent paths. Also known as transmitter noise. { 'frī·iŋ 'nóiz }

F scan *See* F scope. { 'ef ,skan }

F scope [ELECTR] A cathode-ray scope on which a single signal appears as a spot with bearing error as the horizontal coordinate and elevation angle error as the vertical coordinate, with cross hairs on the scope face to assist in bringing the system to bear on the target. Also known as F indicator; F scan. { 'ef ,skōp }

FSH *See* follicle-stimulating hormone.

FSK *See* frequency-shift keying.

f spot [ASTRON] One of a pair of sunspots that appears to follow the other across the face of the sun, or whose magnetic polarity is that normally found in such a sunspot during that sunspot cycle and in that hemisphere of the sun. { 'ef ,spät }

FSS *See* fixed-satellite service.

F star [ASTRON] A star whose spectral type is F; surface temperature is 7000 K, and color is yellowish. { 'ef ,stär }

f stop [OPTICS] An aperture setting for a camera lens; indicated by the f number. { 'ef ,stäp }

f-sum rule [ATOM PHYS] The rule that the sum of the *f* values (or oscillator strengths) of absorption transitions of an atom in a given state, minus the sum of the *f* values of the emission transitions in that state, equals the number of electrons which take part in these transitions. Also known as Thomas-Reiche-Kuhn sum rule. { 'ef ,səm ,rül }

ft *See* foot.

FTAM *See* file transfer access and management. { 'ef,tam }

ftc *See* footcandle.

F test *See* variance ratio test. { 'ef ,test }

ft-L *See* footlambert.

ft-lb *See* foot-pound.

ft-lbf *See* foot-pound.

FTP *See* file transfer protocol.

ft-pdl *See* foot-poundal.

fT value [NUC PHYS] The product of the half-period *T* of a nucleus that undergoes beta decay and an integral *f* that depends on the beta-decay energy and the type of transition; decays of a particular degree of forbiddenness have similar values of this product. { ¦ef¦tē ,val·yü }

fu *See* flux unit.

FU *See* finsen unit.

5-FU *See* fluorouracil.

Fubini's theorem [MATH] The theorem stating conditions under which

$$\iint f(u,v)dudv = \int du \int f(u,v)dv = \int dv \int f(u,v)du$$

{ fü'bē·nēz ,thir·əm }

Fucales [BOT] An order of brown algae composing the class Cyclosporeae. { fyü'kā·lēz }

Fuchsian differential equation [MATH] A homogeneous, linear differential equation whose coefficients are analytic functions whose only singularities, if any, are poles of order one. { ¦fyük·sē·ən ,dif·ə¦ren·chəl i'kwā·zhən }

Fuchsian group [MATH] A Kleinian group *G* for which there is a region *D* in the complex plane, consisting of either the interior of a circle or the portion of the plane on one side of a straight line, such that *D* is mapped onto itself by every element of *G*. { 'fyük·sē·ən ,grüp }

fuchsin [ORG CHEM] $C_{20}H_{19}N_3$ Brownish-red crystals, used as a dye or in the commercial preparation of other dyes, and as an antifungal drug. Also known as magenta; rosaniline. { 'fyük·sən }

fuchsinophile [BIOL] Having an affinity for the dye fuchsin. { 'fyük'sin·ə,fil }

fuchsite [MINERAL] A bright-green variety of muscovite rich in chromium. { 'fyük,sīt }

fucoid [GEOL] A tunnellike marking on a sedimentary structure identified as a trace fossil but not referred to a described genus. { 'fyü,kóid }

fucoidin [BIOCHEM] A gum composed of L-fucose and sulfate acid ester groups obtained from *Fucus* species and other brown algae. { fyü'kóid·ən }

Fucophyceae [BOT] A class of brown algae. { ,fyü·kə'fīs·ē,ē }

L-fucopyranose *See* L-fucose. { ¦el ,fyü·kō¦pī·rə,nōs }

L-fucose [BIOCHEM] $C_6H_{12}O_5$ A methyl pentose present in some algae and a number of gums and identified in the polysaccharides of blood groups and certain bacteria. Also known as 6-deoxy-L-galactose; L-fucopyranose; L-galactomethylose; L-rhodeose. { ¦el 'fyü,kōs }

fucoxanthin [BIOCHEM] $C_{40}H_{60}O_6$ A carotenoid pigment; a partial xanthophyll ester found in diatoms and brown algae. { ¦fyü·kō¦zan·thən }

Fucus [BOT] A genus of dichotomously branched brown algae; it is harvested in the kelp industry as a source of algin. { 'fyü·kəs }

FUDR *See* floxuridine.

fuel [MATER] A material that is burnt to release heat energy, for example, coal, oil, or uranium. { fyül }

fuel assembly [NUCLEO] A combination of fuel and structural materials, used in some nuclear reactors to facilitate assembly of the core. { 'fyül ə'sem·blē }

fuel bed [MECH ENG] A layer of burning fuel, as on a furnace grate or a cupola. { 'fyül ,bed }

fuel cell [PHYS CHEM] An electrochemical device in which the reaction between a fuel, such as hydrogen, and an oxidant, such as oxygen or air, converts the chemical energy of the fuel directly into electrical energy without combustion. { 'fyül ,sel }

fuel-cell catalyst [CHEM] A substance, such as platinum, silver, or nickel, from which the electrodes of a fuel cell are made, and which speeds the reaction of the cell; it is especially important in a fuel cell which does not operate at high temperatures. { 'fyül ,sel 'kad·ə,list }

fuel-cell electrolyte [CHEM] The substance which conducts electricity between the electrodes of a fuel cell. { 'fyül ,sel i'lek·trə,līt }

fuel-cell fuel [CHEM] A substance, such as hydrogen, carbon monoxide, sodium, alcohol, or a hydrocarbon, which reacts with oxygen to generate energy in a fuel cell. { 'fyül ,sel 'fyül }

fuel cycle *See* reactor fuel cycle. { 'fyül ,sī·kəl }

fuel decanner [NUCLEO] A machine used for removing the stainless steel or other metal cans that enclose the enriched uranium fuel rods of a nuclear reactor; the cans are removed by a chipless machining operation in which the tubing is sheared into a spiral strip. { 'fyül dē,kan·ər }

fuel element [NUCLEO] A rod, tube, plate, or other geometrical form into which nuclear fuel is fabricated for use in a reactor. { 'fyül ,el·ə·mənt }

fuel filter [ENG] A device, as in an internal combustion engine, that removes particles from the fuel. { 'fyül ,fil·tər }

fuel gas [MATER] A gaseous fuel used to provide heat energy when burned with oxygen. { 'fyül ,gas }

fuel injection [MECH ENG] The delivery of fuel to an internal combustion engine cylinder by pressure from a mechanical pump. { 'fyül in,jek·shən }

fuel injector [MECH ENG] A pump mechanism that sprays fuel into the cylinder of an internal combustion engine at the appropriate part of the cycle. { 'fyül in,jek·tər }

fuel oil [MATER] A liquid product burned to generate heat, exclusive of oils with a flash point below 100°F (38°C); includes heating oils, stove oils, furnace oils, bunker fuel oils. { 'fyül ,oil }

fuel pellet [NUCLEO] A small pellet of frozen deuterium and tritium that would be used as fuel in a laser-induced fusion power plant. { 'fyül ,pel·ət }

fuel plate [NUCLEO] A form of nuclear fuel element consisting of a flat or slightly curved sheet of fuel, which is usually a sandwich of uranium fuel protected by metallic cladding. { 'fyül ,plāt }

fuel pump [MECH ENG] A pump for drawing fuel from a storage tank and delivering it to an engine or furnace. { 'fyül ,pəmp }

fuel reprocessing [NUCLEO] The processing of nuclear reactor fuel to recover the unused fissionable material. { 'fyül rē'präs,es·iŋ }

fuel rod [NUCLEO] A long, rod-shaped fuel assembly. { 'fyül ,räd }

fuel seed *See* fuel spike. { 'fyül ,sēd }

fuel shutoff [AERO ENG] **1.** The action of shutting off the flow of liquid fuel into a combustion chamber or of stopping the combustion of a solid fuel. **2.** The event or time marking this action. { 'fyül ,shəd,óf }

fuel spike [NUCLEO] Nuclear fuel that is more highly enriched than the majority of the other fuel in a nuclear reactor. Also known as fuel seed. { 'fyül ,spīk }

fuel structure ratio *See* fuel-weight ratio. { 'fyül ,strək·chər ,rā·shō }

fuel system [MECH ENG] A system which stores fuel for present use and delivers it as needed. { 'fyül ,sis·təm }

fuel tank [MECH ENG] The operating, fuel-storage component of a fuel system. { 'fyül ,taŋk }

fuel-weight ratio [AERO ENG] The ratio of the weight of a rocket's fuel to the weight of the unfueled rocket. Also known as fuel structure ratio. { 'fyül 'wāt ,rā·shō }

fugacious [BOT] Lasting a short time; used principally to describe plant parts that fall soon after being formed. { fyü'gā·shəs }

fugacity [THERMO] A function used as an analog of the partial pressure in applying thermodynamics to real systems; at a constant temperature it is proportional to the exponential of the ratio of the chemical potential of a constituent of a system divided by the product of the gas constant and the temperature, and it approaches the partial pressure as the total pressure of the gas approaches zero. { fyü'gas·əd·ē }

fugacity coefficient [THERMO] The ratio of the fugacity of a gas to its pressure. { fyü'gas·əd·ē ,kō·ə,fish·ənt }

fugitive air [MIN ENG] Air which moves through the ventilation fan but never reaches the mine workings. { ¦fyü·jəd·iv 'er }

fugitive dye [CHEM] A dye that is unstable, that is, not fast; used in the textile processing for purposes of identity. { ¦fyü·jəd·iv 'dī }

fugue [PSYCH] A flight from reality, as in hysteria, during which an individual performs acts which later are not recollected. { fyüg }

Fulcher bands [SPECT] A group of bands in the spectrum of molecular hydrogen that are preferentially excited by a low-voltage discharge. { 'fəl·chər ,banz }

fulchronograph [ENG] An instrument for recording lightning strokes, consisting of a rotating aluminum disk with several hundred steel fins on its rim; the fins are magnetized if they pass between two coils when these are carrying the surge current of a lightning stroke. { 'fül'krän·ə,graf }

fulcrate [BIOL] Having a fulcrum. { 'fül,krāt }

fulcrate trophus [INV ZOO] A type of masticatory apparatus in certain rotifers characterized by an elongate fulcrum. { 'fül,-krāt 'trō·fəs }

fulcrum [MECH] The rigid point of support about which a lever pivots. { 'fül·krəm }

Fuld-Gross unit [BIOL] A unit for the standardization of trypsin. { ¦füld 'grōs ,yü·nət }

Fulgoroidea [INV ZOO] The lantern flies, a superfamily of homopteran insects in the series Auchenorrhyncha distinguished by the anterior and middle coxae being of equal length and joined to the body at some distance from the median line. { ,fül·gə'roid·ē·ə }

fulgurator [ENG] An atomizer used to spray salt solutions into a flame for analysis. { 'fül·gə,rād·ər }

fulgurite [GEOL] A glassy, rootlike tube formed when a lightning stroke terminates in dry sandy soil; the intense heating of the current passing down into the soil along an irregular path fuses the sand. { 'fül·gə,rīt }

full adder [ELECTR] A logic element which operates on two binary digits and a carry digit from a preceding stage, producing

as output a sum digit and a new carry digit. Also known as three-input adder. { ¦fül 'ad·ər }

full annealing [MET] Heating steel to a high temperature and then cooling to ambient or near-ambient temperatures. { ¦fül ə'nēl·iŋ }

full automatic [ORD] A weapon that provides continuous fire as long as the trigger is depressed; used to distinguish from semiautomatic. { ¦fül 'ȯd·ə'mad·ik }

full automatic plating [MET] Electroplating a piece of work that is carried through the full cycle automatically. { ¦fül ,ȯd·ə,mad·ik 'plād·iŋ }

full-cell process [ENG] A process of preservative treatment of wood that uses a pressure vessel and first draws a vacuum on the charge of wood and then introduces the preservative without breaking the vacuum. Also known as Bethell process. { ¦fül ¦sel 'präs·əs }

full duplex [COMMUN] Telegraph or other data channel able to operate in both directions simultaneously. [COMPUT SCI] The complete duplication of any data-processing facility. { ¦fül 'dü,pleks }

full-duplex operation [COMMUN] Simultaneous communications in both directions between two points. { ¦fül ¦dü,pleks ,äp·ə'rā·shən }

full-ended [NAV ARCH] The condition when the extremities of the waterlines in the vicinity of the load line are strongly convex to the surrounding water and the ends of the sectional area curve are full indicating that the displacement is carried well forward and aft toward the ends of the vessel. { ¦fül 'en·dəd }

fuller [MET] A die or portion of a die used in preliminary forging operations to reduce the cross section somewhere between the ends of a piece of stock. { 'fül·ər }

fullerene [CHEM] A large molecule composed entirely of carbon, with the chemical formula C_n, where n is any even number from 32 to over 100; believed to have the structure of a hollow spheroidal cage with a surface network of carbon atoms connected in hexagonal and pentagonal rings. { 'fül·ə,rēn }

fuller's earth [GEOL] A natural, fine-grained earthy material, such as a clay, with high adsorptive power; consists principally of hydrated aluminum silicates; used as an adsorbent in refining and decolorizing oils, as a catalyst, and as a bleaching agent. { ¦fül·ərz ¦ərth }

full-face firing [MIN ENG] Drilling of small-diameter holes from top to bottom of the face. { ¦fül ¦fās 'fīr·iŋ }

full-face tunneling [CIV ENG] A system of tunneling in which the tunnel opening is enlarged to desired diameter before extension of the tunnel face. { ¦fül ¦fās 'tən·əl·iŋ }

full feathering *See* feathering. { ¦fül 'feth·ə·riŋ }

full-featured software [COMPUT SCI] Software with the most advanced available functionality. { ¦fül ¦fē·chərd 'sȯft,wer }

full-gear [MECH ENG] The condition of a steam engine when the valve is operated to the maximum extent by the link motion. { ¦fül 'gir }

fulling [TEXT] A process in which a felt fabric is pounded with hammers in order to develop firmness. { 'fül·iŋ }

fulling agent [TEXT] A soap solution used during the fulling of wool. { 'fül·iŋ ,ā·jənt }

full justification *See* justification. { ¦fül ¦jəs·tə·fə'kā·shən }

full linear group [MATH] The group of all nonsingular linear transformations of a complex vector space whose group operation is composition. { ¦fül 'lin·ē·ər ,grüp }

full load [ELEC] The greatest load that a circuit or piece of equipment is designed to carry under specified conditions. { ¦fül 'lōd }

full-load current [ELEC] The greatest current that a circuit or piece of equipment is designed to carry under specified conditions. { ¦fül ,lōd 'kə·rənt }

full-mill [BUILD] A type of construction in which all vertical apertures open onto shafts of brick or other fireproof material; used for fire retardance. { ¦fül 'mil }

full moon [ASTRON] The moon at opposition, with a phase angle of 0°, when it appears as a round disk to an observer on the earth because the illuminated side is toward the observer. { ¦fül 'mün }

full-motion video adapter [COMPUT SCI] A video adapter

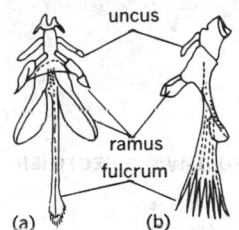

FULCRATE TROPHUS

Fulcrate trophus of *Seison*.
(*a*) Dorsal view. (*b*) Lateral view.

FULLERENE

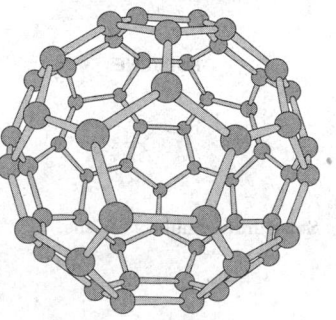

Model of the buckyball, C_{60}. In this highly symmetrical carbon molecule, all 12 pentagons are isolated from each other.

capable of displaying moving video images from a video cassette recorder, laser disk player, or camcorder on a computer screen. { ˌfu̇l ˈmō·shən ˈvid·ē·ō əˌdap·tər }

full-period allocated circuit [COMMUN] Communications link (allocated circuit) assigned for the exclusive use of previously defined users at two or more terminal points. { ˈfu̇l ˌpir·ē·əd ˌal·əˌkäd·əd ˈsər·kət }

full-pitch winding [ELEC] An armature winding in which the distance between two active conductors of a coil equals the pole pitch. { ˈfu̇l ˌpich ˌwīnd·iŋ }

full plate [HOROL] A watch with train wheels and escapement under a single plate; only the balance is exposed. { ˈfu̇l ˌplāt }

full pressure suit [AERO ENG] A pressure suit which completely encloses the body and in which a gas pressure sufficiently above ambient pressure for maintenance of function may be sustained. { ˈfu̇l ˈpresh·ər ˌsüt }

full-range fuse [ELEC] A high-voltage, current-limiting fuse that can safely interrupt any value of the fault current that causes the fuse elements (conductors) to melt. { ˈfu̇l ˌrānj ˈfyüz }

full-scale [ORD] Of an attack or other operation, all-out or maximum. { ˈfu̇l ˌskāl }

full-screen editor [COMPUT SCI] A computer program that allows the user to work with the computer in an interactive manner by using all or most of the area of a cathode-ray tube or similar electronic display. { ˈfu̇l ˌskrēn ˈed·əd·ər }

full-seam mining [MIN ENG] A mining system in which the entire section is dislodged together and the coal is separated from the rock outside the mine by the cleaning plant. { ˈfu̇l ˌsēm ˌmīn·iŋ }

full section filter [ELECTR] A filter network whose graphical representation has the shape of the Greek letter pi, connoting capacitance in the upright legs and inductance or reactance in the horizontal member. { ˈfu̇l ˌsek·shən ˈfil·tər }

full subsidence [MIN ENG] The greatest amount of subsidence occurring as a result of mine workings. { ˈfu̇l ˈsub·sə·dəns }

full subtracter [ELECTR] A logic element which operates on three binary input signals representing a minuend, subtrahend, and borrow digit, producing as output a different digit and a new borrow digit. Also known as three-input subtracter. { ˈfu̇l səbˈtrak·tər }

full-track vehicle [MECH ENG] A vehicle entirely supported, driven, and steered by an endless belt, or track, on each side; for example, a tank. { ˈfu̇l ˌtrak ˈvē·ə·kəl }

full trailer [MECH ENG] A towed vehicle whose weight rests completely on its own wheels. { ˈfu̇l ˈtrāl·ər }

full-wave amplifier [ELECTR] An amplifier without any clipping. { ˈfu̇l ˌwāv ˈam·pləˌfī·ər }

full-wave bridge [ELECTR] A circuit having a bridge with four diodes, which provides full-wave rectification and gives twice as much direct-current output voltage for a given alternating-current input voltage as a conventional full-wave rectifier. { ˈfu̇l ˌwāv ˈbrij }

full-wave control [ELECTR] Phase control that acts on both halves of each alternating-current cycle, for varying load power over the full range from 0 to the full-wave maximum value. { ˈfu̇l ˌwāv kənˈtrōl }

full-wave rectification [ELECTR] Rectification in which output current flows in the same direction during both half cycles of the alternating input voltage. { ˈfu̇l ˌwāv ˌrek·tə·fəˈkā·shən }

full-wave rectifier [ELECTR] A double-element rectifier that provides full-wave rectification; one element functions during positive half cycles and the other during negative half cycles. { ˈfu̇l ˌwāv ˈrek·təˌfī·ər }

full-wave vibrator [ELEC] A vibrator having an armature that moves back and forth between two fixed contacts so as to change the direction of direct-current flow through a transformer at regular intervals and thereby permit voltage stepup by the transformer; used in battery-operated power supplies for mobile and marine radio equipment. { ˈfu̇l ˌwāv ˈvīˌbrād·ər }

full width at half maximum [PHYS] The difference between the energies or frequencies on either side of a spectral line or resonance curve at which the line absorption or emission or the resonant quantity reaches half its maximum intensity. Abbreviated FWHM. { ˈfu̇l ˌwidth at ˈhaf ˈmak·sə·məm }

full wires [MATER] In wire-mesh cloth, wires running the short way of the cloth as woven. Also known as shute wires. { ˈfu̇l ˌwīrz }

full-word boundary [COMPUT SCI] In the IBM 360 system, any address which ends in 00, and is therefore a natural boundary for a four-byte machine word. { ˈfu̇l ˌwərd ˈbau̇n·drē }

fully arisen sea See fully developed sea. { ˈfu̇l·ē əˌriz·ən ˈsē }

fully developed mine [MIN ENG] In coal mining, a mine where all development work has reached the boundaries and further extraction will be done on the retreat. { ˈfu̇l·ē diˌvel·əpt ˈmīn }

fully developed nucleate boiling [PHYS CHEM] A stage in the boiling process in which vapor bubbles from neighboring sites on the heater surface merge, and the vapor appears to leave the heater in the form of jets, so that the vapor structures resemble mushrooms with multiple stems. { ˈfu̇l·ē diˈvel·əpt ˌnük·lē·ˌāt ˈbȯil·iŋ }

fully developed sea [OCEANOGR] The maximum ocean waves or sea state that can be produced by a given wind force blowing over sufficient fetch, regardless of duration. Also known as fully arisen sea. { ˈfu̇l·ē diˌvel·əpt ˈsē }

fully parenthesized notation [MATH] A method of writing arithmetic expressions in which parentheses are placed around each pair of operands and its associated operator. { ˈfu̇l·ē pəˌren·thəˌsīzd nōˈtā·shən }

fully populated board [COMPUT SCI] A printed circuit board on which no room remains to install additional chips or other electronic components that would provide additional capabilities. { ˈfu̇l·ē ˈpäp·yəˌläd·əd ˈbȯrd }

fulmar [VERT ZOO] Any of the oceanic birds composing the family Procellariidae; sometimes referred to as foul gulls because of the foul-smelling substance spat at intruders upon their nests. { ˈfu̇l·mər }

fulminate [MED] Of a disease, to come suddenly and follow a severe, intense, and rapid course. [ORG CHEM] **1.** A salt of fulminic acid. **2.** HgC₂N₂O₂ An explosive mercury compound derived from the fulminic acid; used for the caps or exploders by means of which charges of gunpowder, dynamite, and other explosives are fired. Also known as mercury fulminate. { ˈfu̇l·məˌnāt }

fulminic acid [ORG CHEM] CNOH An unstable isomer of cyanic acid, whose salts are known for their explosive characteristics. { fu̇lˈmin·ik ˈas·əd }

fulminuric acid [ORG CHEM] CN·CH(NO₂)·CONH₂ A trimer of cyanuric acid; a water-soluble compound, crystallizing in colorless needles, melting at 138°C, and exploding at 145°C. Also known as isocyanuric acid. { ˌfu̇l·məˌnu̇r·ik ˈas·əd }

fuloppite [MINERAL] Pb₃Sb₈S₁₅ A lead gray, monoclinic mineral consisting of lead antimony sulfide. { ˈfu̇l·əˌpīt }

fulvene [ORG CHEM] C₆H₆ A yellow oil, an isomer of benzene. { ˈfu̇l·vēn }

fumagillin [MICROBIO] C₂₆H₃₄O₇ An insoluble, crystalline antibiotic produced by a strain of the fungus *Aspergillus fumigatus.* { ˌfyü·məˈjil·ən }

fumarase [BIOCHEM] An enzyme that catalyzes the hydration of fumaric acid to malic acid, and the reverse dehydration. { ˈfyü·məˌrās }

Fumariaceae [BOT] A family of dicotyledonous plants in the order Papaverales having four or six stamens, irregular flowers, and no latex system. { fyüˌma·rēˈās·ēˌē }

fumaric acid [ORG CHEM] C₄H₄O₄ A dicarboxylic organic acid produced commercially by synthesis and fermentation; the trans isomer of maleic acid; colorless crystals, melting point 287°C; used to make resins, paints, varnishes, and inks, in foods, as a mordant, and as a chemical intermediate. Also known as boletic acid. { fyüˈmar·ik ˈas·əd }

fumarole [GEOL] A hole, usually found in volcanic areas, from which vapors or gases escape. { ˈfyü·məˌrōl }

fumble [IND ENG] An unintentional sensory-motor error that may be unavoidable. { ˈfəm·bəl }

fume hood [CHEM] A fume-collection device over an enclosed shelf or table, so that experiments involving poisonous or unpleasant fumes or gases may be conducted away from the experimental area. { ˈfyüm ˌhu̇d }

fumes [CHEM] Particulate matter consisting of the solid particles generated by condensation from the gaseous state, generally after volatilization from melted substances, and often accompanied by a chemical reaction, such as oxidation. { fyümz }

fumigant [CHEM] A chemical compound which acts in the gaseous state to destroy insects and their larvae and other

---

FULL-WAVE RECTIFIER

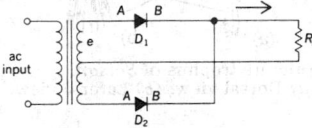

Circuit diagram of a full-wave diode rectifier.

FULVENE

Structural formula of fulvene.

pests; examples are dichlorethyl ether, *p*-dichlorobenzene, and ethylene oxide.   { 'fyü·mə,gənt }

fumigating [ENG] The use of a chemical compound in a gaseous state to kill insects, nematodes, arachnids, rodents, weeds, and fungi in confined or inaccessible locations; also used to control weeds, nematodes, and insects in the field.   { 'fyü·mə,gād·iŋ }

fuming nitric acid [INV ZOO] Concentrated nitric acid containing dissolved nitrogen dioxide; may be prepared by adding formaldehyde to concentrated nitric acid.   { ¦fyüm·iŋ 'nī,trik ,as·əd }

fuming sulfuric acid [INV ZOO] Concentrated sulfuric acid containing dissolved sulfur trioxide. Also known as oleum.   { ¦fyüm·iŋ səl'fyür·ik ,as·əd }

fumulus [METEOROL] A very thin cloud veil at any level, so delicate that it may be almost invisible.   { 'fyü·myə·ləs }

funal *See* sthène.   { 'fyün·əl }

Funariales [BOT] An order of mosses; plants are usually annual, are terrestrial, and have stems that are erect, short, simple, or sparingly branched.   { fyü,nar·ē'ā·lēz }

function [COMPUT SCI] In FORTRAN, a subroutine of a particular kind which returns a computational value whenever it is called. [MATH] A mathematical rule between two sets which assigns to each member of the first, exactly one member of the second.   { 'fəŋk·shən }

functional [COMPUT SCI] In a linear programming problem involving a set of variables x_n, $j = 1, 2, \ldots, n$, a function of the form $c_1 x_1 + c_2 x_2 + \cdots + c_n x_n$ (where the c_j are constants) which one wishes to optimize (maximize or minimize, depending on the problem) subject to a set of restrictions. [MATH] Any function from a vector space into its scalar field.   { 'fəŋk·shən·əl }

functional analysis [MATH] A branch of analysis which studies the properties of mappings of classes of functions from one topological vector space to another. [SYS ENG] A part of the design process that addresses the activities that a system, software, or organization must perform to achieve its desired outputs, that is, the transformations necessary to turn available inputs into the desired outputs.   { ¦'fəŋk·shən·əl ə'nal·ə·səs }

functional analysis diagram [SYS ENG] A representation of functional analysis and, in particular, the transformations necessary to turn available inputs into the desired outputs, the flow of data or items between functions, the processing instructions that are available to guide the transformation, and the control logic that dictates the activation and termination of functions.   { ¦fəŋk·shən·əl ə'nal·ə·səs ,dī·ə,gram }

functional anatomy [ANAT] The study of the human body and its parts with emphasis on those features that are directly involved in physiological function.   { 'fəŋk·shə·nəl ə'nad·ə·mē }

functional application [COMPUT SCI] A program or computer system, particularly a real-time system, that deals with the primary, ongoing operations of a business enterprise.   { 'fəŋk·shən·əl ,ap·lə'kā·shən }

functional bombing [ORD] The bombing of a specially selected class of key targets, such as dams or marshaling yards, that function within an industrial complex or transportation system, as opposed to the bombing of an entire system.   { 'fəŋk·shən·əl 'bäm·iŋ }

functional constraint [MATH] A mathematical equation which must be satisfied by the independent parameters in an optimization problem, representing some physical principle which governs the relationship among these parameters.   { 'fəŋk·shən·əl kən'stränt }

functional decomposition [CONT SYS] The partitioning of a large-scale control system into a nested set of generic control functions, namely the regulatory or direct control function, the optimizing control function, the adaptive control function, and the self-organizing function.   { 'fəŋk·shən·əl dē,käm·pə'zish·ən }

functional design [COMPUT SCI] A level of the design process in which subtasks are specified and the relationships among them defined, so that the total collection of subsystems performs the entire task of the system. [SYS ENG] The aspect of system design concerned with the system's objectives and functions, rather than its specific components.   { 'fəŋk·shən·əl di'zīn }

functional diagram [COMPUT SCI] A diagram that indicates the functions of the principal parts of a total system and also shows the important relationships and interactions among these parts.   { 'fəŋk·shən·əl 'dī·ə,gram }

functional disorder [MED] A disorder in which the performance of an organ or organ system is abnormal, but not as a result of known changes in structure.   { ¦fəŋk·shən·əl dis'ȯrd·ər }

functional electrical stimulation [MED] Therapeutic application of controlled amounts of electric current to muscles to make them contract in a nearly normal manner.   { 'fəŋk·shən·əl i¦lek·trə·kəl ,stim·yə'lā·shən }

functional error recovery [COMPUT SCI] A procedure whereby the operating system intervenes in certain common errors and attempts actions to allow execution of the computer program to continue.   { 'fəŋk·shən·əl 'er·ər ri,kəv·ə·rē }

functional failure [COMPUT SCI] Failure of a computer system to generate the correct results for a set of inputs.   { ,fəŋk·shən·əl 'fāl·yər }

functional forms analysis chart *See* form process chart.   { 'fəŋk·shən·əl 'fȯrmz ə,nal·ə·səs ,chärt }

functional generator *See* function generator.   { 'fəŋk·shən·əl 'jen·ə,rād·ər }

functional group [ORG CHEM] An atom or group of atoms, acting as a unit, that has replaced a hydrogen atom in a hydrocarbon molecule and whose presence imparts characteristic properties to this molecule; frequently represented as R—. Also known as functionality.   { ¦fəŋk·shən·əl 'grüp }

functional interleaving [COMPUT SCI] Alternating the parts of a number of sequences in a cyclic fashion, such as a number of accesses to memory followed by an access to a data channel.   { ¦fəŋk·shən·əl 'in·tər,lēv·iŋ }

functionalism [ARCH] The view that a social system is an expression of human biological and social needs.   { 'fəŋk·shə·nə,liz·əm }

functionality *See* functional group.   { ,fəŋk·shə'nal·əd·ē }

functional multiplier *See* function multiplier.   { ¦fəŋk·shən·əl ¦məl·tə,plī·ər }

functional paralysis *See* hysterical paralysis.   { ¦fəŋk·shən·əl pə'ral·ə·səs }

functional programming [COMPUT SCI] A type of computer programming in which functions are used to control the processing of logic.   { 'fəŋk·shən·əl 'prō,gram·iŋ }

functional requirement [COMPUT SCI] The documentation which accompanies a program and states in detail what is to be performed by the system.   { ¦fəŋk·shən·əl ri'kwīr·mənt }

functional residual capacity [PHYSIO] The volume of gas which remains within the lungs at expiratory standstill.   { ¦fəŋk·shən·əl ri¦zij·ə·wəl kə'pas·əd·ē }

functional specifications [COMPUT SCI] The documentation for the design of an information system, including the data base; the human and machine procedures; and the inputs, outputs, and processes for each data entry, query, update, and report program in the system.   { ¦fəŋk·shən·əl ,spes·ə·fə'kā·shənz }

functional switching circuit [ELECTR] One of a relatively small number of types of circuits which implements a Boolean function and constitutes a basic building block of a switching system; examples are the AND, OR, NOT, NAND, and NOR circuits.   { ¦fəŋk·shən·əl 'swich·iŋ ,sər·kət }

functional unit [COMPUT SCI] The part of the computer required to perform an elementary process such as an addition or a pulse generation.   { ¦fəŋk·shən·əl 'yü·nət }

function code [COMPUT SCI] Special code which appears on a medium such as a paper tape and which controls machine functions such as a carriage return.   { 'fəŋk·shən ,kōd }

function-evaluation routine [COMPUT SCI] A canned routine such as a log function or a sine function.   { ¦fəŋk·shən i,val·yə'wā·shən rü,tēn }

function failure safety [ENG] The capability of an electronic-mass measuring instrument to withhold the release of an incorrect measurement when there is a function failure.   { 'fəŋk·shən ¦fālyər ,sāf·tē }

function generator [ELECTR] **1.** Also known as functional generator. **2.** An analog computer device that indicates the value of a given function as the independent variable is increased. **3.** A signal generator that delivers a choice of a number of different waveforms, with provisions for varying the frequency over a wide range.   { 'fəŋk·shən 'jen·ə,rād·ər }

function hole *See* designation punch.   { 'fəŋk·shən ,hōl }

function key [COMPUT SCI] A special key on a keyboard

to control a mechanical function, initiate a specific computer operation, or transmit a signal that would otherwise require multiple key strokes. { 'fəŋk·shən ,kē }

function multiplier [ELECTR] An analog computer device that takes in the changing values of two functions and puts out the changing value of their product as the independent variable is changed. Also known as functional multiplier. { 'fəŋk·shən ¦məl·tə,plī·ər }

function space [MATH] A metric space whose elements are functions. { 'fəŋk·shən ,spās }

function switch [ELECTR] A network having a number of inputs and outputs so connected that input signals expressed in a certain code will produce output signals that are a function of the input information but in a different code. { 'fəŋk·shən ,swich }

function table [COMPUT SCI] **1.** Sets of computer information arranged so an entry in one set selects one or more entries in the other sets. **2.** A computer device that converts multiple inputs into a single output or encodes a single input into multiple outputs. [MATH] A table that lists the values of a function for various values of the variable. { 'fəŋk·shən ,tā·bəl }

function unit [COMPUT SCI] In computer systems, a device which can store a functional relationship and release it continuously or in increments. { 'fəŋk·shən ,yü·nət }

functor [COMPUT SCI] See logic element. [MATH] A function between categories which associates objects with objects and morphisms with morphisms. { 'fəŋk·tər }

fundamental [PHYS] The lowest frequency component of a complex wave. Also known as first harmonic; fundamental component. { ¦fən·də¦ment·əl }

fundamental affine connection [MATH] An affine connection whose coefficients arise from the covariant and contravariant metric tensors of a space. { ¦fən·də¦ment·əl ə'fīn kə'nek·shən }

fundamental circle See primary great circle. { ¦fən·də¦ment·əl 'sər·kəl }

fundamental complex [GEOL] An agglomeration of metamorphic rocks underlying sedimentary or unmetamorphosed rocks; specifically, an agglomeration of Archean rocks supporting a geological column. { ¦fən·də¦ment·əl 'käm,pleks }

fundamental component See fundamental. { ¦fən·də¦ment·əl kəm'pō·nənt }

fundamental constants [PHYS] The physical constants which play a fundamental role in the basic theories of physics, including the speed of light, electronic charge, electronic mass, Planck's constant, and the fine-structure constant. Also known as atomic constants; universal constants. { ¦fən·də¦ment·əl 'kän·stəns }

fundamental forms of a surface [MATH] Differential forms which express the area and curvature of the surface. { ¦fən·də¦ment·əl 'fȯrmz əv ə 'sər·fəs }

fundamental frequency [PHYS] **1.** The lowest frequency at which a system vibrates freely. **2.** The lowest frequency in a complex wave. { ¦fən·də¦ment·əl 'frē·kwən·sē }

fundamental frequency of the voice [PHYSIO] The rate of vibration of the vocal folds. Abbreviated F0. { ¦fən·də¦ment·əl ¦frē·kwən·sē əv thə 'vȯis }

fundamental group [COMMUN] In wire communications, a group of trunks that connect each local or trunk switching center to a trunk switching center of higher rank on which it homes; the term also applies to groups that interconnect zone centers. [MATH] For a topological space, the group of homotopy classes of all closed paths about a point in the space; this group yields information about the number and type of "holes" in a surface. { ¦fən·də¦ment·əl 'grüp }

fundamental interaction [PART PHYS] One of the fundamental forces that act between the elementary particles of matter. { ¦fən·də¦ment·əl ,in·tər'ak·shən }

fundamental interval [THERMO] **1.** The value arbitrarily assigned to the difference in temperature between two fixed points (such as the ice point and steam point) on a temperature scale, in order to define the scale. **2.** The difference between the values recorded by a thermometer at two fixed points; for example, the difference between the resistances recorded by a resistance thermometer at the ice point and steam point. { ¦fən·də¦men·təl 'int·ər·vəl }

fundamental jelly See ulmin. { ¦fən·də¦ment·əl 'jel·ē }

fundamental mode [ELECTROMAG] The waveguide mode having the lowest critical frequency. Also known as dominant mode; principal mode. [PHYS] The normal mode of vibration having the lowest frequency. { ¦fən·də¦ment·əl 'mōd }

fundamental motion See elemental motion. { ¦fən·də¦ment·əl 'mō·shən }

fundamental number [GEN] The number of chromosome arms of a karyotype. { ¦fən·də¦men·təl 'nəm·bər }

fundamental particle See elementary particle. { ¦fən·də¦ment·əl 'pärd·ə·kəl }

fundamental quantity See base quantity. { ¦fən·də¦ment·əl 'kwän·əd·ē }

fundamental region [MATH] Any region in the complex plane that can be mapped conformally onto all of the complex plane. { ¦fən·də¦ment·əl 'rē·jən }

fundamental sequence See Cauchy sequence. { ¦fən·də¦ment·əl 'sē·kwəns }

fundamental series [SPECT] A series occurring in the line spectra of many atoms and ions having one, two, or three electrons in the outer shell, in which the total orbital angular momentum quantum number changes from 3 to 2. { ¦fən·də¦ment·əl 'sir·ēz }

fundamental star places [ASTRON] The apparent right ascensions and declinations of 1535 standard comparison stars obtained by leading observatories and published annually under the sponsorship of the International Astronomical Union. { ¦fən·də¦ment·əl 'stär ,plās·əz }

fundamental strength [GEOPHYS] The maximum stress that a geological structure can withstand without creep under certain conditions but without reference to time. { ¦fən·də¦ment·əl 'streŋkth }

fundamental substance See ulmin. { ¦fən·də¦ment·əl 'səb·stəns }

fundamental tensor See metric tensor. { ¦fən·də¦ment·əl 'ten·sər }

fundamental theorem of algebra [MATH] Every polynomial of degree n with complex coefficients has exactly n roots counted according to multiplicity. { ¦fən·də¦ment·əl ¦thir·əm əv 'al·jə·brə }

fundamental theorem of arithmetic [MATH] Every positive integer greater than 1 can be factored uniquely into the form $P_1{}^{n_1} \ldots P_i{}^{n_i} \ldots P_k{}^{n_k}$, where the P_i are primes, the n_i positive integers. { ¦fən·də¦ment·əl ¦thir·əm əv ə'rith·mə·tik }

fundamental theorem of calculus [MATH] Given a continuous function $f(x)$ on the closed interval $[a,b]$ the functional

$$F(x) = \int_a^x f(t) \, dt$$

is differentiable on $[a,b]$ and $F'(x) = f(x)$ for every x in $[a,b]$, and if G is any function on $[a,b]$ such that $G'(x) = f(x)$ for all x in $[a,b]$, then

$$\int_a^b f(t) \, dt = G(b) - G(a)$$

{ ¦fən·də¦ment·əl ¦thir·əm əv 'kal·kyə·ləs }

fundamental tone [ACOUS] The component tone of lowest pitch in a complex tone. { ¦fən·də¦ment·əl 'tōn }

fundamental unit See base unit. { ¦fən·də¦ment·əl 'yü·nət }

fundamental wavelength [PHYS] Of an oscillatory device, that wavelength corresponding to its fundamental frequency. { ¦fən·də¦ment·əl 'wāv,leŋkth }

fundic gland [ANAT] Any of the glands of the corpus and fundus of the stomach. { 'fən·dik ,gland }

fundus [ANAT] The bottom of a hollow organ. { 'fən·dəs }

fungal ecology [ECOL] The subdiscipline in mycology and ecology that examines fungal community composition and structure; responses, activities, and interactions of single fungus species; and the functions of fungi in ecosystems. { ,fəŋ·gəl i'käl·ə·jē }

fungal sheath [MYCOL] A compact layer of fungal hyphae that surrounds the young root surface of the host plant and prevents direct contact between the root and the soil. { 'fəŋ·gəl ,shēth }

fungi [MYCOL] Nucleated, usually filamentous, sporebearing organisms devoid of chlorophyll. { 'fən,jī }

fungible [CHEM ENG] Pertaining to petroleum products whose characteristics are so similar they can be commingled. { 'fən·jə·bəl }

fungicide [MATER] An agent that kills or destroys fungi. { 'fən·jə,sīd }

fungiform [BIOL] Mushroom-shaped. { 'fən·jə,fòrm }

fungiform papilla [ANAT] One of the low, broad papillae scattered over the dorsum and margins of the tongue. { 'fən·jə,fòrm pə'pil·ə }

Fungi Imperfecti [MYCOL] A class of the subdivision Eumycetes; the name is derived from the lack of a sexual stage. { 'fən,jī ,im·pər'fek,tī }

fungi-proofing [ENG] Application of a protective chemical coating that inhibits growth of fungi. { 'fən,jī ,prúf·iŋ }

fungistat [MATER] A compound that inhibits or prevents growth of fungi. { 'fən·jə,stat }

Fungivoridae [INV ZOO] The fungus gnats, a family of orthorrhaphous dipteran insects in the series Nematocera; the larvae feed on fungi. { ,fən·jə'vòr·ə,dē }

fungivorous [ZOO] Feeding on or in fungi. { fən'jiv·ə·rəs }

fungus [MYCOL] Singular of fungi. { 'fəŋ·gəs }

fungus gall [PL PATH] A plant gall resulting from an attack of a parasitic fungus. { 'fəŋ·gəs ,gòl }

funicle See funiculus. { 'fyün·ə·kəl }

funicular See funicular railroad. { fə'nik·yə·lər }

funicular distribution [CHEM] The distribution of a two-phase, immiscible liquid mixture (such as oil and water, one a wetting phase, the other nonwetting) in a porous system when the wetting phase is continuous over the surface of the solids. { fə'nik·yə·lər dis·trə'byü·shən }

funicular polygon [MECH] **1.** The figure formed by a light string hung between two points from which weights are suspended at various points. **2.** A force diagram for such a string, in which the forces (weights and tensions) acting on points of the string from which weights are suspended are represented by a series of adjacent triangles. { fə'nik·yə·lər 'päl·ə,gän }

funicular railroad [ENG] A railroad system used primarily to ascend and descend mountains; the weight of the descending train helps to move the ascending train up the mountain. Also known as funicular. { fə'nik·yə·lər 'rāl,rōd }

funiculus [ANAT] **1.** Also known as funicle. **2.** Any structure in the form of a chord. **3.** A column of white matter in the spinal cord. [BOT] The stalk of an ovule. [INV ZOO] A band of tissue extending from the adoral end of the coelom to the adoral body wall in bryozoans. { fə'nik·yə·ləs }

Funkel effect [ELECTR] Fluctuations in the current from an oxide cathode, or any cathode that does not consist of pure metal, due to fluctuations in the work function resulting from changes with time in the cathode surface. { 'fəŋ·kəl i,fekt }

funnel [DES ENG] A tube with one conical end that sometimes holds a filter; the function is to direct flow of a liquid or, if a filter is present, to direct a flow that was filtered. [NAV ARCH] A smokestack of a ship. { 'fən·əl }

funnel chest [MED] A developmental deformity in which the sternum is depressed and the ribs and costal cartilages curve inward. { 'fən·əl ,chest }

funnel cloud [METEOROL] The popular term for the tornado cloud, often shaped like a funnel with the small end nearest the ground. { 'fən·əl ,klaùd }

funnel-flow bin [ENG] A bin in which solid flows toward the outlet in a channel that forms within stagnant material. { 'fən·əl ¦flō ,bin }

funneling [ASTRON] The convergence of the evolutionary paths of stars from different parts of the main sequence into the red giant region on a Hertzsprung-Russell diagram. { 'fən·əl·iŋ }

fur [MATER] The dressed pelt of a mammal. [VERT ZOO] The coat of a mammal. { fər }

2-furaldehyde See furfural. { ¦tü fə'ral·də,hīd }

furan [ORG CHEM] **1.** One of a group of organic heterocyclic compounds containing a diunsaturated ring of four carbon atoms and one oxygen atom. **2.** C_4H_4O The simplest furan type of molecule; a colorless, mildly toxic liquid, boiling at 32°C, insoluble in water, soluble in alcohol and ether; used as a chemical intermediate. Also known as furfuran; tetrol. { 'fyùr,an }

furancarboxylic acid See furoic acid. { ¦fyùr·ən,kar,bäk'sil·ik 'as·əd }

furan cement [MATER] A strong adhesive that is made from furfural-alcohol resins and is highly resistant to chemicals. { 'fyùr,an si,ment }

2,5-furandione See maleic anhydride. { ¦tü ¦fīv ¦fyùr·ən'dī,ōn }

furanones [BIOCHEM] Analogs of homoserine lactones that appear to interfere with the development of typical biofilm structure, leaving these organisms more susceptible to treatment with natural biocides. { 'fyür·ə,nōnz }

furanose [BIOCHEM] A sugar whose cyclic or ring structure resembles that of furan. { 'fyür·ə,nōs }

furanoside [ORG CHEM] A glycoside whose cyclic sugar component resembles that of furan. { fyə'ran·ə,sīd }

furan resin [ORG CHEM] A liquid, thermosetting resin in which the furan ring is an integral part of the polymer chain, made by the condensation of furfuryl alcohol; used as a cement and adhesive, casting resin, coating, and impregnant. { 'fyür,an ,rez·ən }

furca [INV ZOO] A forked process as the last abdominal segment of certain crustaceans, and as part of the spring in collembolans. { 'fər·kə }

furcate [BIOL] Forked. { 'fər,kāt }

furcocercous cercaria [INV ZOO] A free-swimming, digenetic trematode larva with a forked tail. { ¦fər·kō¦sər·kəs sər'kar·ē·ə }

furcula [ZOO] A forked structure, especially the wishbone of fowl. { 'fər·kyə·lə }

furfural [ORG CHEM] C_4H_3OCHO When pure, a colorless liquid, soluble in organic solvents, slightly soluble in water; used as a lube oil-refining solvent, in cellulosic formulations, in making resins, as a weed killer, as a fungicide, and as a chemical intermediate. Also known as 2-furaldehyde; furfuraldehyde; furfurol; furol. { 'fər·fə,ral }

furfuraldehyde See furfural. { 'fər·fə'ral·də,hīd }

furfural extraction [CHEM ENG] Process for the refining of lubricating oils and other organic materials by contact with furfural. { 'fər·fə,ral ik'strak·shən }

furfuran See furan. { 'fər·fə,ran }

furfurol See furfural. { 'fər·fə,rōl }

furfuryl [ORG CHEM] The functional group C_5H_6O- from furfural. { 'fər·fə,ril }

furfuryl alcohol [ORG CHEM] $C_5H_6O_2$ A liquid with a faint burning odor and bitter taste, soluble in alcohol and ether, usually prepared from furfural; used as a solvent in the manufacturing of wetting agents and resins. { 'fər·fə,ril 'al·kə,hòl }

furiani [METEOROL] A southwest wind that blows in the vicinity of the Po River, Italy, and is vehement and short-lived, followed by a gale from the south or southeast. { fù·rē'ä·nē }

Furipteridae [VERT ZOO] The smoky bats, a family of mammals in the order Chiroptera having a vestigial thumb and small ears. { ,fù·rip'ter·ə,dē }

furlong [MECH] A unit of length, equal to 1/8 mile, 660 feet, or 201.168 meters. { 'fər,lòŋ }

furnace [ENG] An apparatus in which heat is liberated and transferred directly or indirectly to a solid or fluid mass for the purpose of effecting a physical or chemical change. { 'fər·nəs }

furnace black [CHEM] A carbon black formed by partial combustion of liquid and gaseous hydrocarbons in a closed furnace with a deficiency of oxygen; used as a reinforcing filler for synthetic rubber. { 'fər·nəs ,blak }

furnace brazing [MET] Joining two metals by mechanical union of the filler metal and joint, then heating the composite in a furnace. { 'fər·nəs ,brāz·iŋ }

furnace cupola See cupola. { 'fər·nəs ,kyü·pə·lə }

furnace lining [ENG] The interior part of a furnace in contact with a molten charge and hot gases; constructed of heat-resistant material. { 'fər·nəs ,līn·iŋ }

furnace oil [MATER] Distillate fuel oil intended primarily for domestic central heating systems; usually No. 1 fuel oil. { 'fər·nəs ,òil }

furnace refining [MET] Purification of molten metal by treatment in a reverberatory furnace. { ¦fər·nəs ri'fīn·iŋ }

furnace soldering [MET] Soldering by heating clamped members to the appropriate temperature in a furnace. { ¦fər·nəs 'säd·ə·riŋ }

Furnariidae [VERT ZOO] The oven birds, a family of perching birds in the superfamily Furnarioidea. { ,fər·nə'rī·ə,dē }

Furnarioidea [VERT ZOO] A superfamily of birds in the order Passeriformes characterized by a predominance of gray, brown, and black plumage. { fər,nar·ē'óid·ē·ə }

FURFURAL

Structural formula of furfural.

furnish [CHEM ENG] In papermaking, the raw materials placed in a beater for producing paper pulp. { 'fər·nish }

furniture [GRAPHICS] Wood or metal blocks which are used to fill the blank spaces in a form in the lockup process of letterpress printing. { 'fər·nə·chər }

furoic acid [ORG CHEM] $C_5H_4O_3$ Long monoclinic prisms crystallized from the water solution, soluble in ether and alcohol; used as a preservative and bactericide. Also known as furancarboxylic acid; pyromucic acid. { fyü'rō·ik 'as·əd }

furol See furfural. { 'fyü,rōl }

furred ceiling [BUILD] A ceiling in which the furring units are attached directly to the structural units of the building. { ¦fərd 'sē·liŋ }

furring [BUILD] Thin strips of wood or metal fastened to joists, studs, ceilings, or inner walls of a building to provide a level surface or air space over which the finished surface can be applied. Also known as batten; furring strip. { 'fər·iŋ }

furring brick [MATER] Hollow brick grooved for plastering. { 'fər·iŋ ,brik }

furring strip See furring. { 'fər·iŋ ,strip }

furring tile [MATER] Non-load-bearing clay tile used for lining interior walls. { 'fər·iŋ ,tīl }

furrow [ENG] A trench plowed in the ground. { 'fər·ō }

furrow irrigation [AGR] Irrigation via furrows between rows of crops. { 'fər·ō ,ir·ə'gā·shən }

furrow press [AGR] A device that firms the earth in a furrow after plowing. { 'fər·ō ,pres }

Furry theorem [QUANT MECH] In quantum electrodynamics, the theorem that the contribution of a Feynman diagram, consisting of a closed polygon of fermion lines connected to an odd number of photon lines, vanishes. { 'fər·ē ,thir·əm }

furuncle [MED] A small cutaneous abscess, usually resulting from infection of a hair follicle by *Staphylococcus aureus*. Also known as boil. { 'fyür,əŋ·kəl }

furunculosis [MED] A condition marked by numerous furuncles, or the recurrence of furuncles following healing of a preceding crop. { fyü,rəŋ·kyə'lō·səs }

fusain [GEOL] The local lithotype strands or patches, characterized by silky luster, fibrous structure, friability, and black color. Also known as mineral charcoal; mother-of-coal. { 'fyü,zān }

Fusarium [MYCOL] A genus of fungi in the family Tuberculariaceae having sickle-shaped, multicelled conidia; includes many important plant pathogens. { fyü'za·rē·əm }

Fusarium oxysporum [MYCOL] A pathogenic fungus causing a variety of plant diseases, including cabbage yellows and wilt of tomato, flax, cotton, peas, and muskmelon. { fyü'za·rē·əm ,äk·sə'spȯr·əm }

Fusarium solani [MYCOL] A pathogenic fungus implicated in root rot and wilt diseases of several plants, including sisal and squash. { fyü'za·rē·əm sō'lan·ē }

fuse [ELEC] An expendable device for opening an electric circuit when the current therein becomes excessive, containing a section of conductor which melts when the current through it exceeds a rated value for a definite period of time. Also known as electric fuse. [ENG] Also spelled fuze. **1.** A device with explosive components designed to initiate a train of fire or detonation in an item of ammunition by an action such as hydrostatic pressure, electrical energy, chemical energy, impact, or a combination of these. **2.** A nonexplosive device designed to initiate an explosion in an item of ammunition by an action such as continuous or pulsating electromagnetic waves or acceleration. { fyüz }

fuse alarm [ELEC] Circuit that produces a visual or audible signal to indicate a blown fuse. { 'fyüz ə,lärm }

fuse blasting cap [ENG] A small copper cylinder closed at one end and charged with a fulminate. { 'fyüz 'blast·iŋ ,kap }

fuse block [ELEC] An insulating base on which are mounted fuse clips or other contacts for fuses. Also known as fuseboard. { 'fyüz ,bläk }

fuseboard See fuse block. { 'fyüz,bȯrd }

fuse body [ENG] The part of a fuse contributing the major portion of the total weight, and which houses the majority of the functioning parts, and to which smaller parts are attached. { 'fyüz ,bäd·ē }

fuse box See cutout box. { 'fyüz ,bäks }

fuse clip [ELEC] A spring contact used to hold and make connection to a cartridge-type fuse. { 'fyüz ,klip }

fuse cutout [ELEC] Assembly of a fuse support and a fuse holder which may or may not include the fuse link. { 'fyüz 'kə,daut }

fused aromatic ring [ORG CHEM] A molecular structure in which two or more aromatic rings have two carbon atoms in common. { ¦fyüzd ar·ə¦mad·ik 'riŋ }

fused-electrolyte battery See thermal battery. { ¦fyüzd i'lek·trə,līt ,bad·ə·rē }

fuse diode [ELECTR] A diode that opens under specified current surge conditions. { 'fyüz ,dī,ōd }

fuse disconnecting switch [ELEC] Disconnecting switch in which a fuse unit forms a part of the blade. { 'fyüz dis·kə'nek·tiŋ ,swich }

fused junction See alloy junction. { ¦füzd 'jəŋk·shən }

fused-junction diode See alloy-junction diode. { ¦fyüzd ¦jəŋk·shən 'dī,ōd }

fused-junction transistor See alloy-junction transistor. { ¦fyüzd ¦jəŋk·shən tran'zis·tər }

fused potassium sulfide See potassium sulfide. { ¦fyüzd pə'tas·ē·əm 'səl,fīd }

fused quartz [MATER] A glasslike insulating material made by melting crushed crystals of natural quartz or a certain type of quartz sand. { ¦fyüzd 'kwȯrts }

fused-salt electrolysis [PHYS CHEM] Electrolysis with use of purified fused salts as raw material and as an electrolyte. { ¦fyüzd ¦sȯlt i,lek'trä·lə·səs }

fused-salt reactor See molten-salt reactor. { ¦fyüzd ¦sȯlt rē'ak·tər }

fused semiconductor [ELECTR] Junction formed by recrystallization on a base crystal from a liquid phase of one or more components and the semiconductor. { ¦fyüzd 'sem·i·kən,dək·tər }

fused silica See silica glass. { ¦fyüzd 'sil·ə·kə }

fused silver nitrate See lunar caustic. { ¦fyüzd 'sil·vər 'nī,trāt }

fused spray deposit [MET] In thermal spraying, deposit which is sprayed on a preheated substrate and has the capability to coalesce within itself as well as to the substrate. { ¦fyüzd ¦sprā di'päz·ət }

fusee [HOROL] In a timepiece, a conical pulley with grooves in a spiral configuration from which a cord or chain unwinds onto a barrel containing the spring; the increasing diameter of the pulley compensates for the lessening power of the spring. [VET MED] A bony growth occurring on a horse's leg. Also spelled fuzee. { fyü'zē }

fuse gage [ENG] An instrument for slicing time fuses to length. { 'fyüz ,gāj }

fusehead [ENG] That part of an electric detonator consisting of twin metal conductors, bridged by fine resistance wire, and surrounded by a bead of igniting compound which burns when the firing current is passed through the bridge wire. { 'fyüz,hed }

fuselage [AERO ENG] In an airplane, the central structure to which wings and tail are attached; it accommodates flight crew, passengers, and cargo. { 'fyü·sə,läzh }

fuse lighter [ENG] A device for facilitating the ignition of the powder core of a fuse. { 'fyüz ,līd·ər }

fuse link [ELEC] Part of a fuse that carries the current of the circuit and all or part of which melts when the current exceeds a predetermined value. { 'fyüz ,liŋk }

fusel oil [MATER] A volatile, poisonous mixture of isoamyl, butyl, propyl, and heptyl alcohols produced as by-products in alcoholic fermentation of starches, grains, or fruits to produce ethyl alcohol. { 'fyü·zəl ,ȯil }

fuse PROM [COMPUT SCI] A programmable read-only memory in which the programming is carried out either by blowing open microscopic fuse links to define a logic one or zero for each cell in the memory array, or by causing metal to short out base-emitter transistor junctions to program the ones or zeros into the memory. { 'fyüz ,präm }

fuse wire [ELEC] Wire made from an alloy that melts at a relatively low temperature and overheats to this temperature when carrying a particular value of overload current. { 'fyüz ,wīr }

fusibility [THERMO] The quality or degree of being capable of being liquefied by heat. { ,fyü·zə'bil·əd·ē }

fusible alloy [MET] A low melting alloy, usually of bismuth, tin, cadmium, and lead, which melts at temperatures as low as 70°C (160°F). { ¦fyü·zə·bəl 'al,ȯi }

fusible plug See safety plug. { ¦fyü·zə·bəl 'pləg }

fusible resistor [ELEC] A resistor designed to protect a circuit against overload; its resistance limits current flow and thereby protects against surges when power is first applied to a circuit; its fuse characteristic opens the circuit when current drain exceeds design limits. { ¦fyü·zə·bəl ri'zis·tər }

fusicoccin [PL PATH] A nonselective pathotoxin with growth-regulator properties that is produced by *Fusicoccum amygdale* and causes a wilt disease of peach and almond trees. { ¸fyüz·i'käk·sən }

Fusicoccum amygdali [MYCOL] A fungal pathogen that produces fusicoccin, the cause of wilt disease in peach and almond trees. { ¸fyüz·i¸käk·əm ə'mig·də¸lē }

fusiform [BIOL] Spindle-shaped; tapering toward the ends. { 'fyü·zə¸form }

fusiform bacillus [MICROBIO] A bacillus having one blunt and one pointed end, as *Fusobacterium fusiforme*. { 'fyü·zə¸form bə'sil·əs }

fusiform initial cell [BOT] A cell type of the vascular cambium that gives rise to all cells in the vertical system of secondary xylem and phloem. { 'fyü·zə¸form ə¦nish·əl 'sel }

fusimotoneuron [NEUROSCI] One of the small motor fibers, composing about 30% of the fibers in the ventral root of the spinal cord, which innervate intrafusal fibers. { ¦fyü·zē¸mō·dō'nü¸rän }

fusing disk [MECH ENG] A rapidly spinning disk that cuts metal by melting it. { 'fyüz·iŋ ¸disk }

fusinite [GEOL] The micropetrological constituent of fusain which consists of carbonized woody tissue. { 'fyüz·ən¸īt }

fusinization [GEOL] The process of formation of fusain in coal. { ¸fyüz·ən·ə'zā·shən }

fusion [NUC PHYS] Combination of two light nuclei to form a heavier nucleus (and perhaps other reaction products) with release of some binding energy. Also known as atomic fusion; nuclear fusion. [PHYS CHEM] A change of the state of a substance from the solid phase to the liquid phase. Also known as melting. { 'fyü·zhən }

fusion bomb [ORD] A bomb that depends upon nuclear fusion for release of energy. Also known as fusion weapon. { 'fyü·zhən ¸büm }

fusion crust [GEOL] A thin, glassy coating, usually black and rarely more than 1 millimeter thick, which is formed by ablation on the surface of a meteorite. { 'fyü·zhən ¸krəst }

fusion face [MET] That portion of a metal surface that will be fused in a welding operation. { 'fyü·zhən ¸fās }

fusion frequency [PHYSIO] The frequency of a series of retinal images above which their differences in luminosity or color (that is, flicker) can no longer be perceived. { 'fyü·zhən ¸frē·kwən·sē }

fusion fuel [NUCLEO] A substance which may generate energy in a fusion reaction, such as deuterium, deuterium and tritium, or deuterium and helium-3. { 'fyü·zhən ¸fyül }

fusion nucleus [BOT] The triploid, or 3*n*, nucleus which results from double fertilization and which produces the endosperm in some seed plants. { ¦fyü·zhən ¦nü·klē·əs }

fusion piercing [ENG] A method of producing vertical blastholes by virtually burning holes in rock. Also known as piercing. { 'fyü·zhən ¸pir·siŋ }

fusion-piercing drill [ENG] A machine designed to use the fusion-piercing mode of producing holes in rock. Also known as det drill; jet-piercing drill; Linde drill. { 'fyü·zhən ¸pirs·iŋ ¸dril }

fusion point [NUCLEO] The temperature of a plasma above which the rate of energy generation by nuclear fusion reactions exceeds the rate of energy loss from the plasma, so that the fusion reaction can be self-sustaining. Abbreviated fnp. { 'fyü·zhən ¸point }

fusion reactor [NUCLEO] A proposed device in which controlled, self-sustaining nuclear fusion reactions would be carried out in order to produce useful power. { 'fyü·zhən rē ¸ak·tər }

fusion tube [ANALY CHEM] Device used for the analysis of the elements in a compound by fusing them with another compound; for example, analysis of nitrogen in organic compounds by fusing the compound with sodium and analyzing for sodium cyanide. { 'fyü·zhən ¸tüb }

fusion weapon See fusion bomb. { 'fyü·zhən ¸wep·ən }

fusion welding [MET] Any welding operation involving melting of the base or parent metal. { 'fyü·zhən ¸weld·iŋ }

fusion zone [MET] The volume of base or parent metal melted during a welding operation. { 'fyü·zhən ¸zōn }

fusospirochetosis See Vincent's infection. { ¦fyü·zō¸spī·rə·kē'tō·səs }

fusula [INV ZOO] A spindle-shaped, terminal projection of the spinneret of a spider. { 'fyü·zə·lə }

Fusulinacea [PALEON] A superfamily of large, marine extinct protozoans in the order Foraminiferida characterized by a chambered calcareous shell. { ¸fyü·zə·lə'nās·ē·ə }

Fusulinidae [PALEON] A family of extinct protozoans in the superfamily Fusulinacea. { ¸fyü·zə'lin·ə¸dē }

Fusulinina [PALEON] A suborder of extinct rhizopod protozoans in the order Foraminiferida having a monolamellar, microgranular calcite wall. { ¸fyü·zə·lə'nī·nə }

future [RELAT] For an event in space-time, those events that can be reached by a signal that is emitted at the original event and moves at a speed less than or equal to the speed of light in a vacuum. { 'fyü·chər }

future address patch [COMPUT SCI] A computer output containing the address of a symbol and the address of the last reference to that symbol. { 'fyü·chər ə'dres ¸pach }

future asymptotically predictable [RELAT] A mathematical restriction on the global nature of an asymptotically flat space-time such that Cauchy data set on a spacelike surface (partial Cauchy surface) *S* will determine the evolution of the space-time to the future of *S*; naked singularities to the future of *S* are thereby ruled out. { ¦fyü·chər ¸ā¸sim¦täd·ə·klē prə'dik·tə·bəl }

future Cauchy development [RELAT] The set of points *p* relative to a surface *S* in a space-time such that every past-directed inextendible timelike or null curve through *p* intersects *S*. Symbolized $D^+(s)$. Also known as domain of dependence. { ¦fyü·chər 'kō·shē di¸vel·əp·mənt }

future horismos [RELAT] The set of points *p* relative to a surface *S* that can be causally affected by events in *S*; that is, the set of points in the future of *S* which can be reached from *S* by future-directed timelike curves. { ¦fyü·chər hə'riz·mōs }

future label [COMPUT SCI] An address referenced in the operand field of an instruction, but which has not been previously defined. { 'fyü·chər ¸lā·bəl }

future light cone [RELAT] The set of all points in space time that are reached by signals traveling at the speed of light from a specified point. { 'fyü·chər 'līt ¸kōn }

future trapped set [RELAT] A set of points in a space-time such that no two points of the set have timelike separation and the future horismos is compact. { ¦fyü·chər ¦trapt 'set }

FUV radiation See far-ultraviolet radiation. { ¦ef¦yü¦vē ¸rād·ē'ā·shən }

fuze See fuse. { fyüz }

fuzee See fusee. { fyü'zē }

fuzz [MATER] Fibers which protrude from the surface of a sheet of paper. { fəz }

fuzzy algorithm [COMPUT SCI] An ordered set of instructions, comprising fuzzy assignment statements, fuzzy conditional statements, and fuzzy unconditional action statements, that, upon execution, yield an approximate solution to a specified problem. { ¦fəz·ē 'al·gə¸rith·əm }

fuzzy assignment statement [COMPUT SCI] An instruction in a fuzzy algorithm that assigns a possibly fuzzy value to a variable. { ¦fəz·ē ə'sīn·mənt ¸stāt·mənt }

fuzzy conditional statement [COMPUT SCI] An instruction in a fuzzy algorithm that assigns a possibly fuzzy value to a variable or causes an action to be executed, provided that a fuzzy condition holds. { ¦fəz·ē kən'dish·ən·əl ¸stāt·mənt }

fuzzy controller [CONT SYS] An automatic controller in which the relation between the state variables of the process under control and the action variables, whose values are computed from observations of the state variables, is given as a set of fuzzy implications or as a fuzzy relation. { ¦fəz·ē kən'trōl·ər }

fuzzy logic [MATH] The logic of approximate reasoning, bearing the same relation to approximate reasoning that two-valued logic does to precise reasoning. { ¦fəz·ē 'läj·ik }

fuzzy mathematics [MATH] A methodology for systematically handling concepts that embody imprecision and vagueness. { ¦fəz·ē ¸math·ə'mad·iks }

fuzzy model [MATH] A finite set of fuzzy relations that form an algorithm for determining the outputs of a process from some finite number of past inputs and outputs. { ¦fəz·ē ¸mäd·əl }

fuzzy relation [MATH] A fuzzy subset of the cartesian product $X \times Y$, denoted as a relation from a set *X* to a set *Y*. { ¦fəz·ē ri'lā·shən }

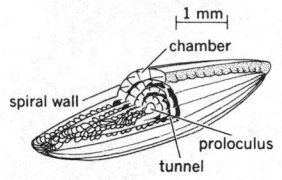

FUSULINIDAE

1 mm

chamber

spiral wall

proloculus

tunnel

A representative of the Fusulinidae shown in a cutaway diagram.

fuzzy relational equation [MATH] An equation of the form $A \circ R = B$, where A and B are fuzzy sets, R is a fuzzy relation, and $A \circ R$ stands for the composition of A with R. { ¦fəz·ē ri¦lā·shən·əl i'kwā·zhən }

fuzzy set [MATH] An extension of the concept of a set, in which the characteristic function which determines membership of an object in the set is not limited to the two values 1 (for membership in the set) and 0 (for nonmembership), but can take on any value between 0 and 1 as well. { 'fəz·ē 'set }

fuzzy-structure acoustics [ACOUS] A class of conceptual viewpoints in the study of vibrations of large structure in which precise, computationally intensive models of the overall structure are replaced by nonprecise analytical models. { ¦fəz·ē ¦strək·chər ə'küs·tiks }

fuzzy system [SYS ENG] A process that is too complex to be modeled by using conventional mathematical methods, and that gives rise to data that are, in general, soft, with no precise boundaries; examples are large-scale engineering complex systems, social systems, economic systems, management systems, medical diagnostic processes, and human perception. { ¦fəz·ē 'sis·təm }

fuzzy unconditional action statement [COMPUT SCI] An instruction in a fuzzy algorithm that specifies a possibly fuzzy mathematical operation or an action to be executed. { ¦fəz·ē ən·kən¦dish·ən·əl 'ak·shən ‚stāt·mənt }

fuzzy value [MATH] A membership function of a fuzzy set that serves as the value assigned to a variable. { ¦fəz·ē 'val‚yü }

fV See femtovolt.

f value See oscillator strength. { 'ef ‚väl·yü }

F/V converter See frequency-to-voltage converter. { ¦ef¦vē kən'vərd·ər }

FWHM See full width at half maximum.

g *See* gram.

G [ELEC] *See* conductance. [MECH] A unit of acceleration equal to the standard acceleration of gravity, 9.80665 meters per second per second, or approximately 32.1740 feet per second per second. Also known as fors; grav. [SCI TECH] *See* giga-.

Ga *See* gallium.

GaAs FET *See* gallium arsenide field-effect transistor. { 'gas,fet }

GABA *See* γ-aminobutyric acid. { 'ga·bə *or* ,jē,ā,bē'ā }

gabardine [TEXT] Steep twill fabric made of closely woven single or two-ply fibers of wool, rayon, polyester, or other yarn. { 'gab·ər,dēn }

gabbro [PETR] A group of dark-colored, intrusive igneous rocks with granular texture, composed largely of basic plagioclase and clinopyroxene. { 'gab·rō }

gabion [ENG] A bottomless basket of wickerwork or metal iron filled with earth or stones; used in building fieldworks or as revetments in mining. Also known as pannier. { 'gā·bē·ən }

gable [ARCH] The upper, triangular portion of the terminal wall of a building under the ridge of a sloped roof. { 'gā·bəl }

gableboard *See* vergeboard. { 'gā·bəl,bórd }

gable roof [ARCH] A sloping peaked roof that forms a gable at each end. { 'gā·bəl ,rüf }

GABOB *See* 4-amino-3-hydroxybutyric acid. { 'ga,bäb *or* ,jē,ā,bē,ō'bē }

Gabor trolley [ENG] A small three-wheel trolley with knife-edge wheels, used in constructing trajectories of charged particles in an electric field. { 'gä,bór ,trä·lē }

Gabriel's synthesis [ORG CHEM] A synthesis of primary amines by the hydrolysis of N-alkylphthalimides; the latter are obtained from potassium phthalimide and alkyl halides. { 'gā·brē·əlz ,sin·thə·səs }

gad [MIN ENG] **1.** A heavy steel wedge, 6 or 8 inches (15 or 20 centimeters) long, with a narrow chisel point used in mining to cut samples, break out pieces of loose rock, and so on. **2.** A small iron punch with a wooden handle used to break up ore. { gad }

gadder [MIN ENG] A small car or platform with a drilling machine attached, to make a straight line of holes along its course in getting out dimension stone. Also known as gadding car; gadding machine. { 'gad·ər }

gadding car *See* gadder. { 'gad·iŋ ,kär }

gadding machine *See* gadder. { 'gad·iŋ mə,shēn }

Gadidae [VERT ZOO] A family of fishes in the order Gadiformes, including cod, haddock, pollock, and hake. { 'ga·də,dē }

Gadiformes [VERT ZOO] An order of actinopterygian fishes that lack fin spines and a swim bladder duct and have cycloid scales and many-rayed pelvic fins. { ,gad·ə'fór,mēz }

gadoleic acid [ORG CHEM] $C_{20}H_{38}O_2$ A fatty acid derived from cod liver oil, and melting at 20°C. { ¦gad·ə¦lē·ik 'as·əd }

gadolinite [MINERAL] $Be_2FeY_2Si_2O_{10}$ A black, greenish-black, or brown rare-earth mineral; hardness is 6.5–7 on Mohs scale, and specific gravity is 4–4.5. { 'gad·əl·ə,nīt }

gadolinium [CHEM] A rare-earth element, symbol Gd, atomic number 64, atomic weight 157.25; highly magnetic, especially at low temperatures. { ,gad·əl'in·ē·əm }

gage Also spelled gauge. [CIV ENG] The distance between the inner faces of the rails of railway track; standard gage in the United States is 4 feet 8½ inches (1.44 meters). [DES ENG] **1.** A device for determining the relative shape or size of an object. **2.** The thickness of a metal sheet, a rod, or a wire. [ENG] The minimum sieve size through which most (95% or more) of an aggregate will pass. [ORD] The interior diameter of the barrel of a shotgun expressed by the number of spherical lead bullets fitting it that are required to make a pound. [TEXT] A measure of the density of knit cloth, given in the number of stitches in 1.5 inches (3.8 centimeters). { gāj }

gage block [DES ENG] A chrome steel block having two flat, parallel surfaces with the parallel distance between them being the size marked on the block to a guaranteed accuracy of a few millionths of an inch; used as the standard of precise lineal measurement for most manufacturing processes. Also known as precision block; size block. { 'gāj ,bläk }

gage cock [ENG] A valve located on a water column of a boiler drum. { 'gāj ,käk }

gage complete penetration [ORD] Penetration in which a hole of sufficient size is made through the plate so as to fully admit a plug gage of a designated percentage of diameter; if the projectile remains in the plate and prevents the complete insertion of the plug gage through the hole, the round is disregarded. { ¦gāj kəm¦plēt pen·ə'trā·shən }

gaged brick [MATER] Brick which has been ground or otherwise produced to accurate dimensions. { ¦gājd ¦brik }

gage glass [ENG] A glass, plastic, or metal tube, usually equipped with shutoff valves, that is connected by a suitable fitting to a tank or vessel, for the measurement of liquid level. { 'gāj ,glas }

gageite [MINERAL] $(Mn,Mg,Zn)_8Si_3O_{14}\cdot 2H_2O$ (or $3H_2O$) A mineral composed of a hydrous silicate of manganese, magnesium, and zinc. { 'gā,jīt }

gage length [ENG] Original length of the portion of a specimen measured for strain, length changes, and other characteristics. { 'gāj ,leŋkth }

gage loss [MIN ENG] The diametrical reduction in the size of a bit or reaming shell caused by wear through use. { 'gāj ,lós }

gage plate [CIV ENG] A plate inserted between the parallel rails of a railroad track to maintain the gage. { 'gāj ,plāt }

gage point [DES ENG] A point used to position a part in a jig, fixture, or qualifying gage. { 'gāj ,póint }

gage pressure [MECH ENG] The amount by which the total absolute pressure exceeds the ambient atmospheric pressure. { 'gāj ,presh·ər }

gager [PETRO ENG] An oil-field worker who gathers oil samples, tests them to determine their gravity and freedom from water, and measures the quantity of oil that is run from the producer's tank to the pipeline. { 'gā·jər }

gagger [MET] An irregular-shaped piece of metal used in a sand mold to reinforce and support a metal casting. { 'gag·ər }

gaging [NUCLEO] The measurement of the thickness, density, or quantity of material by the amount of radiation it absorbs; this is the most common use of radioactive isotopes in industry. Also spelled gauging. { 'gā·jiŋ }

gaging hatch [ENG] An opening in a tank or other vessel through which measuring and sampling can be performed. { 'gāj·iŋ ,hach }

gaging table *See* strapping table. { 'gāj·iŋ ,tā·bəl }

gaging tape [ENG] A metal measuring tape used to determine the depth of liquid in a tank. { 'gāj·iŋ ,tāp }

gahnite [MINERAL] $ZnAl_2O_4$ A usually dark-green, but

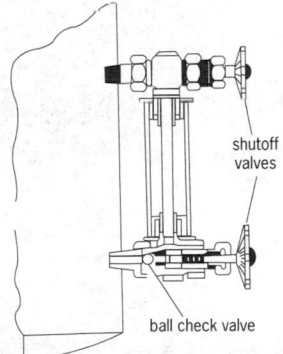

GAGE GLASS

Low-pressure type of gage glass. (*From D. M. Considine, ed., Process Instruments and Controls Handbook, McGraw-Hill, 1985*)

sometimes yellow, gray, or black spinel mineral consisting of an oxide of zinc and aluminum. Also known as zinc spinel. { 'gä,nīt }

gaign [METEOROL] A cross-mountain wind that causes clouds to form on the crests of mountains in Italy. { gān }

gain [ELECTR] **1.** The increase in signal power that is produced by an amplifier; usually given as the ratio of output to input voltage, current, or power, expressed in decibels. Also known as transmission gain. **2.** *See* antenna gain. [ENG] A cavity in a piece of wood prepared by notching or mortising so that a hinge or other hardware or another piece of wood can be placed on the cavity. { gān }

gain asymptotes [CONT SYS] Asymptotes to a logarithmic graph of gain as a function of frequency. { 'gān 'as·əm,tōts }

gain-bandwidth product [ELECTR] The midband gain of an amplifier stage multiplied by the bandwidth in megacycles. { ¦gān ¦band,width ,präd·əkt }

gain control [ELECTR] A device for adjusting the gain of a system or component. { 'gān kən,trōl }

gain-crossover frequency [CONT SYS] The frequency at which the magnitude of the loop ratio is unity. { ¦gān ¦krós,ō·vər ,frē·kwən·sē }

gain margin [CONT SYS] The reciprocal of the magnitude of the loop ratio at the phase crossover frequency, frequently expressed in decibels. { 'gān ,mär·jən }

gain reduction [ELECTR] Diminution of the output of an amplifier, usually achieved by reducing the drive from feed lines by use of equalizer pads or reducing amplification by a volume control. { 'gān ri,dək·shən }

gain scheduling [CONT SYS] A method of eliminating influences of variations in the process dynamics of a control system by changing the parameters of the regulator as functions of auxiliary variables which correlate well with those dynamics. { 'gān ,skej·ə·liŋ }

gain sensitivity control *See* differential gain control. { ¦gān ,sen·sə'tiv·əd·ē kən,trōl }

gain turndown [ELEC] A receiver gain control incorporated in a transponder to protect the transmitter from overload. { ¦gān 'tərn,daùn }

gain twist [ORD] A type of rifling in which the twist is more rapid at the muzzle than at the breech, thus gradually increasing the rotation of the projectile. { 'gān ,twist }

gait analysis [PHYSIO] An aspect of kinesiology that involves the study of walking or other types of ambulation. { 'gāt ə,nal·ə·səs }

gal [MECH] **1.** The unit of acceleration in the centimeter-gram-second system, equal to 1 centimeter per second squared; commonly used in geodetic measurement. Formerly known as galileo. Symbolized Gal. **2.** *See* gallon. { gal }

Gal *See* gal. { gal }

galactan [BIOCHEM] Any of a number of polysaccharides composed of galactose units. Also known as galactosan. { gə'lak·tən }

galactaric acid *See* mucic acid. { ¦gal·ək¦tar·ik 'as·əd }

galactic bulge [ASTRON] A spheroidal distribution of stars that is centered on the nucleus of the Milky Way Galaxy and extends to a distance of about 3 kiloparsecs from the center. { gə¦lak·tik 'bəlj }

galactic center [ASTRON] The gravitational center of the Milky Way Galaxy; the sun and other stars of the Galaxy revolve about this center. { gə'lak·tik 'sen·tər }

galactic circle *See* galactic equator. { gə'lak·tik 'sər·kəl }

galactic cluster *See* open cluster. { gə'lak·tik 'kləs·tər }

galactic concentration [ASTRON] A measure of the increasing density of stars toward the galactic plane, equal to the ratio of the density of stars of a given magnitude at the galactic plane to that at the galactic poles. { gə'lak·tik ,käns·ən'trā·shən }

galactic coordinates *See* galactic system. { gə'lak·tik kō'órd·ən·əts }

galactic corona [ASTRON] A low-density gaseous region extending away from the dense gas of the disk of the Milky Way Galaxy into the halo for distances estimated to be at least 3000 parsecs. { gə¦lak·tik kə'rōn·ə }

galactic disk [ASTRON] The flat distribution of stars and interstellar matter in the spiral arms and plane of the Milky Way Galaxy. { gə'lak·tik 'disk }

galactic equator [ASTRON] A great circle of the celestial sphere, inclined 62° to the celestial equator, coinciding approximately with the center line of the Milky Way, and constituting the primary great circle for the galactic system of coordinates; it is everywhere 90° from the galactic poles. Also known as galactic circle. { gə'lak·tik i'kwäd·ər }

galactic halo [ASTRON] The spherical distribution of oldest stars that are centered about the galactic center of the Milky Way Galaxy. { gə'lak·tik 'hā·lō }

galactic latitude [ASTRON] Angular distance north or south of the galactic equator; the arc of a great circle through the galactic poles, between the galactic equator and a point on the celestial sphere, measured northward or southward from the galactic equator through 90° and labeled N or S to indicate the direction of measurement. { gə'lak·tik 'lad·ə,tüd }

galactic light [ASTRON] The part of the illumination of the night sky that is due to light emitted from stars but diffused through interstellar space. { gə'lak·tik 'līt }

galactic longitude [ASTRON] Angular distance east of sidereal hour angle 94.4° along the galactic equator; the arc of the galactic equator or the angle at the galactic pole between the great circle through the intersection of the galactic equator and the celestial equator in Sagittarius (SHA 94.4°) and a great circle through the galactic poles, measured eastward from the great circle through SHA 94.4° through 360°. { gə'lak·tik 'län·jə,tüd }

galactic nebula [ASTRON] A nebula that is in or near the galactic system known as the Milky Way. { gə'lak·tik 'neb·yə·lə }

galactic noise [ASTRON] Radio-frequency noise that originates outside the solar system; it is similar to thermal noise and is strongest in the direction of the Milky Way. { gə'lak·tik 'nóiz }

galactic nova [ASTRON] One of the novae that are concentrated largely in a band 10° on each side of the plane of the galaxy and are most frequent toward the center of the galaxy. { gə'lak·tik 'nō·və }

galactic nucleus [ASTRON] The center area in the galaxy about which there is a large spherical distribution of stars and from which the spiral arms emanate. { gə'lak·tik 'nü·klē·əs }

galactic plane [ASTRON] The plane that may be drawn through the galactic equator; the plane of the Milky Way Galaxy. { gə'lak·tik 'plān }

galactic pole [ASTRON] On the celestial sphere, either of the two points 90° from the galactic equator. { gə'lak·tik 'pōl }

galactic radiation [ASTROPHYS] Radiation emanating from the Milky Way Galaxy. { gə'lak·tik rād·ē'ā·shən }

galactic radio waves [ELECTROMAG] Radio waves emanating from the Milky Way Galaxy. { gə'lak·tik 'rād·ē·ō ,wāvz }

galactic rotation [ASTRON] The rotation of the Milky Way about an axis through the center and perpendicular to the plane of the Galaxy; the rotation is apparent from the highly flattened shape and from relative stellar motion. { gə'lak·tik rō'tā·shən }

galactic system [ASTRON] An astronomical coordinate system using latitude measured north and south from the galactic equator, and longitude measured in the sense of increasing right ascension from 0 to 360°. Also known as galactic coordinates. { gə'lak·tik 'sis·təm }

galactic windows [ASTROPHYS] The regions near the equator of the Milky Way where there is low absorption of light by interstellar clouds so that some distant external galaxies may be seen through them. { gə'lak·tik 'win,dōz }

galactocele [MED] **1.** A retention cyst caused by obstruction of one or more of the mammary ducts. **2.** A hydrocele with milky contents. { gə'lak·tə,sēl }

galactogen [BIOCHEM] A polysaccharide, in snails, that yields galactose on hydrolysis. { gə'lak·tə·jən }

galactoglucomannan [BIOCHEM] Any of a group of polysaccharides which are prominent components of coniferous woods; they are soluble in alkali and consist of D-glucopyranose and D-mannopyranose units. { gə¦lak·tō,glü·kə'man·ən }

galactokinase [BIOCHEM] An enzyme which reacts D-galactose with adenosine triphosphate to give D-galactose-1-phosphate and adenosine diphosphate. { gə¦lak·tə'kī,nās }

galactolipid *See* cerebroside. { gə¦lak·tə'lip·id }

galactomannan [BIOCHEM] Any of a group of polysaccharides which are composed of D-galactose and D-mannose units, are soluble in water, and form highly viscous solutions; they are plant mucilages existing as reserve carbohydrates in the endosperm of leguminous seeds. { gə¦lak·tō'man·ən }

L-galactomethylose See L-fucose. { ¦el gə¦lak·tə'meth·ə,lōs }

galactonic acid [BIOCHEM] $C_6H_{12}O_7$ A monobasic acid derived from galactose, occurring in three optically different forms, and melting at 97°C. Also known as pentahydroxyhexoic acid. { ga,lak¦tän·ik 'as·əd }

galactophore [ANAT] A duct that carries milk. { gə'lak·tə,fòr }

galactopoiesis [PHYSIO] Formation of the components of milk by the cells composing the lobuloalveolar glandular structure. { gə,lak·tə,pòi'ē·səs }

galactorrhea [MED] Excessive flow of milk. { gə,lak·tə'rē·ə }

galactosamine [BIOCHEM] $C_6H_{14}O_5N$ A crystalline amino acid derivative of galactose; found in bacterial cell walls. { gə,lak'tō·sə,mēn }

galactosan See galactan. { gə'lak·tə,san }

galactose [BIOCHEM] $C_6H_{12}O_6$ A monosaccharide occurring in both levo and dextro forms as a constituent of plant and animal oligosaccharides (lactose and raffinose) and polysaccharides (agar and pectin). Also known as cerebrose. { gə'lak,tōs }

galactosemia [MED] A congenital metabolic disorder caused by an enzyme deficiency and marked by high blood levels of galactose. { gə,lak·tō'sē·mē·ə }

galactosidase [BIOCHEM] An enzyme that hydrolyzes galactosides. { gə,lak·tə'sī,dās }

galactoside [BIOCHEM] A glycoside formed by the reaction of galactose with an alcohol; yields galactose on hydrolysis. { gə'lak·tə,sīd }

galactosuria [MED] Passage of urine containing galactose. { gə,lak·tə'sur·ē·ə }

galactosyl ceramide [BIOCHEM] A type of glycolipid that enriches brain tissue and is a major component of the myelin sheaths around nerves. { gə,lak·tə,sil 'ser·ə,mīd }

galacturonic acid [BIOCHEM] The monobasic acid resulting from oxidation of the primary alcohol group of D-galactose to carboxyl; it is widely distributed as a constituent of pectins and many plant gums and mucilages. { gə¦lakt·yə¦rän·ik 'as·əd }

galago See bushbaby. { gə'lä·gō }

galatea [TEXT] Strong warp-effect twill cotton fabric. { ,gal·ə'tē·ə }

Galatea [ASTRON] A satellite of Neptune orbiting at a mean distance of 38,500 miles (62,000 kilometers) with a period of 10.3 hours, and a diameter of about 90 miles (150 kilometers). { ,gal·ə'tē·ə }

Galatheidea [INV ZOO] A group of decapod crustaceans belonging to the Anomura and having a symmetrical abdomen bent upon itself and a well developed tail fan. { ,gal·ə·thē'ī·dē·ə }

Galaxioidei [VERT ZOO] A suborder of mostly small, freshwater fishes in the order Salmoniformes. { gə,lak·sē'òid·ē,ī }

galaxite [MINERAL] $MnAl_2O_4$ A black mineral of the spinel series composed of an oxide of manganese and aluminum. { 'gä·lak,sīt }

galaxy [ASTRON] A large-scale aggregate of stars, gas, and dust; the aggregate is a separate system of stars covering a mass range from 10^7 to 10^{12} solar masses and ranging in diameter from 1500 to 300,000 light-years. { 'gal·ik·sē }

Galaxy See Milky Way Galaxy. { 'gal·ik·sē }

galaxy cluster [ASTRON] A collection of from two to several hundred galaxies which are much more densely distributed than the average density of galaxies in space. { 'gal·ik·sē ,kləs·tər }

galbanum [MATER] A yellowish to brownish gum resin derived from *Ferula galbaniflua*, a perennial herb of western Asia; used in medicine. { 'gal·bə·nəm }

Galbulidae [VERT ZOO] The jacamars, a family of highly iridescent birds of the order Piciformes that resemble giant hummingbirds. { ,gal'bul·ə,dē }

gale [METEOROL] **1.** An unusually strong wind. **2.** In storm-warning terminology, a wind of 28–47 knots (52–87 kilometers per hour). **3.** In the Beaufort wind scale, a wind whose speed is 28–55 knots (52–102 kilometers per hour). { gāl }

galea [ANAT] The epicranial aponeurosis linking the occipital and frontal muscles. [BIOL] A helmet-shaped structure. [BOT] A helmet-shaped petal near the axis. [INV ZOO] **1.** The endopodite of the maxilla of certain insects. **2.** A spinning organ on the movable digit of chelicerae of pseudoscorpions. { 'gā·lē·ə }

galeate [BIOL] **1.** Shaped like a helmet. **2.** Having a galea. { 'ga·lē,āt }

galena [MINERAL] PbS A bluish-gray to lead-gray mineral with brilliant metallic luster, specific gravity 7.5, and hardness 2.5 on Mohs scale; occurs in cubic or octahedral crystals, in masses, or in grains. Also known as blue lead; lead glance. { gə'lē·nə }

galenic [MINERAL] Containing galena. Also known as galenical. { gə'len·ik }

galenical [MINERAL] See galenic. [PHARM] A medicinal preparation containing one or several active plant ingredients and produced so that inert constituents and other undesirable contents of the plant remain undissolved. { gə'len·i·kəl }

galenobismutite [MINERAL] $PbBi_2S_4$ A lead-gray or tin-white mineral consisting of bismuth sulfide; specific gravity is 6.9. { gə¦lē·nō'biz·mə,tīt }

Galen's vein [ANAT] One of the two veins running along the roof the third ventricle that drain the interior of the brain. { 'gā·lənz 'vān }

Galeritidae [PALEON] A family of extinct exocyclic Euechinoidea in the order Holectypoida, characterized by large ambulacral plates with small, widely separated pore pairs. { ,ga·lə'rid·ə,dē }

galerna See galerne. { gə'lər·nə }

galerne [METEOROL] A squally northwesterly wind that is cold, humid, and showery, occurring in the rear of a low-pressure area over the English Channel and off the Atlantic coast of France and northern Spain. Also known as galerna; galerno; giboulee. { gə'lər·nə }

galerno See galerne. { gə'lər·nō }

gale warning [METEOROL] A storm warning for marine interests of impending winds from 28 to 47 knots (52–87 kilometers per hour), signaled by two triangular red pennants by day, and a white lantern over a red lantern by night. { gāl ,wòrn·iŋ }

Galilean glass See Galilean telescope. { ,gal·ə¦lē·ən 'glas }

Galilean satellites [ASTRON] The four largest and brightest satellites of Jupiter (Io, Europa, Ganymede, and Callisto). { ,gal·ə¦lē·ən 'sad·əl,īts }

Galilean telescope [OPTICS] A refracting telescope whose objective is a converging (convex) lens and whose eyepiece is a diverging (concave) lens; it forms erect images. Also known as Galilean glass. { ,gal·ə¦lē·ən 'tel·ə,skōp }

Galilean transformation [MECH] A mathematical transformation used to relate the space and time variables of two uniformly moving (inertial) reference systems in nonrelativistic kinematics. { ,gal·ə¦lē·ən tranz·fər'mā·shən }

galileo See gal. { ,gal·ə'lē·ō }

Galileo number [FL MECH] A dimensionless number used in studying the circulation of viscous liquids, equal to the cube of a characteristic dimension, times the acceleration of gravity, times the square of the liquid's density, divided by the square of its viscosity. Symbol N_{Gal}. { ,gal·ə'lē·ō ,nəm·bər }

Galileo's law of inertia See Newton's first law. { ,gal·ə'lē·ōz 'lò əv i'nər·shə }

galipol [ORG CHEM] $C_{15}H_{26}O$ A terpene alcohol derived from the oil of the angostura bark; colorless crystals that melt at 89°C. { 'gal·ə,pòl }

Galitzin pendulum [MECH] A massive horizontal pendulum that is used to measure variations in the direction of the force of gravity with time, and thus serves as the basis of a seismograph. { gä¦lit·sən 'pen·jə·ləm }

gall [MED] A sore on the skin that is caused by chafing. [MET] Damage to metal surfaces resulting from friction and improper lubrication. [PHYSIO] See bile. [PL PATH] A large swelling on plant tissues caused by the invasion of parasites, such as fungi or bacteria, following puncture by an insect; insect oviposit and larvae of insects are found in galls. { gòl }

gallacetophenone [ORG CHEM] $C_8H_8O_4$ A white to brownish-gray, crystalline powder, melting at 173°C, soluble in water, alcohol, and ether; used as an antiseptic. { ¦gòl¦as·ə·tä·fə'nōn }

gallbladder [ANAT] A hollow, muscular organ in humans and most vertebrates which receives dilute bile from the liver, concentrates it, and discharges it into the duodenum. { 'gòl,blad·ər }

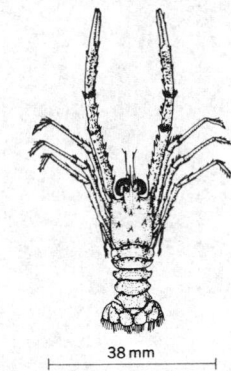

GALATHEIDEA

Munida evermanni.

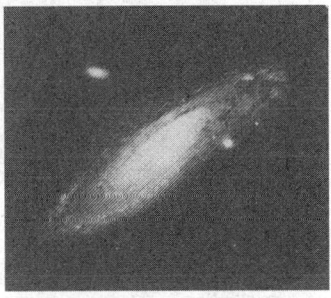

GALAXY

Great Spiral NGC 224 in Andromeda, which resembles the Milky Way Galaxy. Above it is elliptical galaxy NGC 205. (*Hale Observatories*)

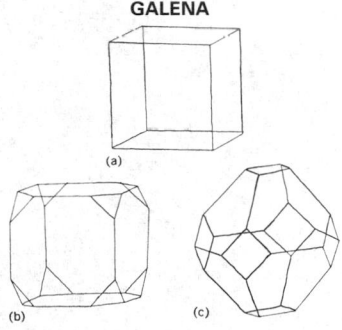

GALENA

Forms of galena crystals; usually (*a*) cubic but sometimes modified to (*b*, *c*) octahedral truncations. (*From C. S. Hurlbut, Jr., Dana's Manual of Mineralogy, 17th ed., Wiley, 1959*)

gallego [METEOROL] A cold, piercing, northerly wind in Spain and Portugal. { gə'yä·gō }

gallein [ORG CHEM] $C_{20}H_{10}O_7$ A brown powder or green scales, broken down by heat; used as a pH indicator in the analysis of phosphates in urine and as an intermediate in the manufacture of dyes. Also known as anthracene violet; gallin; pyrogallolphthalein. { 'gal·ē·ən }

galleria forest [ECOL] A modified tropical deciduous forest occurring along stream banks. { ˌgal·ə'rē·ə ˌfär·əst }

Galleriinae [INV ZOO] A monotypic subfamily of lepidopteran insects in the family Pyralididae; contains the bee moth or wax worm (*Galleria mellonella*), which lives in beehives and whose larvae feed on beeswax. { ˌgal·ə'rī·ə,nē }

gallery [GEOL] **1.** A horizontal, or nearly horizontal, underground passage. **2.** A subsidiary passage in a cave at a higher level than the main passage. [MIN ENG] *See* drift; level. { 'gal·rē }

gallery deck [NAV ARCH] A partial deck below a flight deck on an aircraft carrier. { 'gal·rē ˌdek }

gallery forest [FOR] A forest occurring on both banks of a river in a region that is otherwise treeless. { 'gal·rē ˌfär·əst }

gallery practice ammunition [ORD] Small arms ammunition with a reduced charge, used for practice shooting in a gallery and also for guard purposes. { 'gal·rē ˌprak·təs ˌam·yə'nish·ən }

gallery testing [MIN ENG] A method of testing explosives; a test condition is achieved by firing light charges without any stemming, and heavier charges with only 1 inch (2.5 centimeters) of stemming. { 'gal·rē ˌtest·iŋ }

galley [ENG] The kitchen of a ship, airplane, or trailer. [GRAPHICS] A flat, oblong, open-ended tray into which the letters assembled by hand in a composing stick are transferred after the composing stick is full. { 'gal·ē }

galley proof [GRAPHICS] A reproduction taken from type while the type is still in the galley; the reproduction is reviewed for errors. { 'gal·ē ˌprüf }

gallic acid [ORG CHEM] $C_7H_6O_5$ A crystalline compound that forms needles from solutions of absolute methanol or chloroform, dissolves in water, alcohol, ether, and glycerol; obtained from nutgall tannins or from *Penicillium notatum* fermentation; used to make antioxidants and ink dyes and in photography. { 'gal·ik 'as·əd }

gallicolous [BIOL] Producing or inhabiting galls. { gə'lik·ə·ləs }

Galliformes [VERT ZOO] An order of birds that includes important domestic and game birds, such as turkeys, pheasants, and quails. { ˌgal·ə'fȯr,mēz }

gallin *See* gallein. { 'gal·ən }

gallinaceous [VERT ZOO] Of, pertaining to, or resembling birds of the order Galliformes. { ˌgal·ə'nā·shəs }

galling [MET] Surface damage on mating, moving metal parts due to friction caused by local welding of high spots. { 'gȯl·iŋ }

Gallionella [MICROBIO] A genus of appendaged bacteria; cells are kidney-shaped or rounded and occur on stalks; reproduce by binary fission. { ˌgal·yə'nel·ə }

gallium [CHEM] A chemical element, symbol Ga, atomic number 31, atomic weight 69.72. [MET] A silvery-white metal, melting at 29.7°C, boiling at 1983°C. { 'gal·ē·əm }

gallium arsenide [INV ZOO] GaAs A crystalline material, melting point 1238°C; frequently alloys of this material are formed with gallium phosphide or indium arsenide. { 'gal·ē·əm 'ärs·ən,īd }

gallium arsenide field-effect transistor [ELECTR] A field-effect transistor in which current between the ohmic source and drain contacts is carried by free electrons in a channel consisting of *n*-type gallium arsenide, and this current is modulated by a Schottky-barrier rectifying contact called the gate that varies the cross-sectional area of the channel. Abbreviated GaAs FET. { 'gal·ē·əm 'ärs·ən,īd 'fēld i,fekt tran'zis·tər }

gallium arsenide laser [OPTICS] A laser that emits light at right angles to a junction region in gallium arsenide, at a wavelength of 9000 angstroms (900 nanometers); can be modulated directly at microwave frequencies; cryogenic cooling is required. { 'gal·ē·əm 'ärs·ən,īd 'lā·zər }

gallium arsenide semiconductor [SOLID STATE] A semiconductor having a forbidden-band gap of 1.4 electronvolts

and a maximum operating temperature of 400°C when used in a transistor. { 'gal·ē·əm 'ärs·ən,īd 'sem·i·kən,dək·tər }

gallium halide [INV ZOO] A compound formed by bonding of gallium to either chlorine, bromine, iodine, fluorine, or astatine. { 'gal·ē·əm 'ha,līd }

gallium phosphide [INV ZOO] GaP Transparent crystals made by reacting phosphorus and gallium suboxide at low temperature. { 'gal·ē·əm 'fäs,fīd }

gallium phosphide semiconductor [SOLID STATE] A semiconductor having a forbidden-band gap of 2.4 electronvolts and a maximum operating temperature of 870°C when used in a transistor. { 'gal·ē·əm 'fäs,fīd 'sem·i·kən,dək·tər }

gallivorous [ZOO] Feeding on the tissues of galls, especially certain insect larvae. { gȯ'liv·ə·rəs }

gallnut [PL PATH] A gall resembling a nut. { 'gȯl,nət }

gallocyanine [ORG CHEM] $C_{15}H_{13}ClN_2O_5$ Green crystals soluble in alcohol, glacial acetic acid, alkali carbonates, and concentrated hydrochloric acid; used as a dye and as a reagent for the determination of lead. { ˌga·lō'sī·ə,nēn }

gallodesoxycholic acid *See* chenodeoxycholic acid. { ¦ga·lō·de,zäk·sē'käl·ik 'as·əd }

gallogen *See* ellagic acid. { 'gal·ə·jən }

gallon [MECH] Abbreviated gal. **1.** A unit of volume used in the United States for measurement of liquid substances, equal to 231 cubic inches, or to 3.785 411 784 × 10⁻³ cubic meter, or to 3.785 411 784 liters; equal to 128 fluid ounces. **2.** A unit of volume used in the United Kingdom for measurement of liquid and solid substances, usually the former; equal to 4.54609 × 10⁻³ cubic meter, or to 4.54609 liters; equal to 160 fluid ounces. Also known as imperial gallon. { 'gal·ən }

galloping [FL MECH] Large-amplitude oscillations of a wire or cable in a strong wind, which may become destructive. { 'gal·əp·iŋ }

gallop rhythm [MED] A three-sound sequence resulting from the intensification of the normal third or fourth heart sounds, occurring usually with a rapid ventricular rate. { 'gal·əp ,rith·əm }

gallotannic acid *See* tannic acid. { ¦ga·lō¦tan·ik 'as·əd }

gallotannin *See* tannic acid. { ¦ga·lō¦tan·ən }

Galloway sinking and walling stage *See* sinking-and-walling scaffold. { 'gal·ə,wā ¦siŋk·iŋ ən 'wȯl·iŋ ,stāj }

Galloway stage [MIN ENG] A platform of several decks suspended near the shaft during the sinking operation. { 'gal·ə,wā ,stāj }

gallows frame [MIN ENG] *See* headframe. [NAV ARCH] A deck-mounted frame consisting of one U-shaped structural member or two legs and a header to provide large spacing at the head of the frame; accommodates securing points for a number of head blocks that can swing transversely without interference from each other. { 'gal·ōz ,frām }

gallstone [PATH] A nodule formed in the gallbladder or biliary tubes and composed of calcium, cholesterol, or bilirubin, or a combination of these. { 'gȯl,stōn }

galmei *See* hemimorphite. { 'gäl'mī }

Galofaro [OCEANOGR] A whirlpool in the Strait of Messina, between Sicily and Italy; formerly called Charybdis. { ˌgäl·ə'fä·rō }

Galois field [MATH] A type of field extension obtained from considering the coefficients and roots of a given polynomial. Also known as root field; splitting field. { 'gal,wä ,fēld }

Galois group [MATH] A group of isomorphisms of a particular field extension associated with a polynomial's roots. { 'gal,wä ,grüp }

Galois theory [MATH] The study of the Galois field and Galois group corresponding to a polynomial. { 'gal,wä ,thē·ə·rē }

Galtonian curve [STAT] A graph showing the variation of any quantity from its normal value. { gȯl'tō·nē·ən 'kərv }

Galton whistle [ENG ACOUS] A short cylindrical pipe with an annular nozzle, which is set into resonant vibration in order to generate ultrasonic sound waves. { 'gȯl·tən ,wis·əl }

Galumnidae [INV ZOO] A family of oribatid mites in the suborder Sarcoptiformes. { gə'ləm·nə,dē }

galvanic [ELEC] Pertaining to electricity flowing as a result of chemical action. { gal'van·ik }

galvanic battery [ELEC] A galvanic cell, or two or more such cells electrically connected to produce energy. { gal'van·ik 'bad·ə·rē }

galvanic cell [ELEC] An electrolytic cell that is capable of producing electric energy by electrochemical action. { gal'-van·ik 'sel }

galvanic corrosion [MET] Electrochemical corrosion associated with the current in a galvanic cell, caused by dissimilar metals in an electrolyte because of the difference in potential (emf) of the two metals. Also known as contact corrosion. { gal'van·ik kə'rō·zhən }

galvanic couple [ELEC] A pair of unlike substances, such as metals, which generate a voltage when brought in contact with an electrolyte. { gal'van·ik 'kəp·əl }

galvanic current [ELEC] A steady direct current. { gal'-van·ik 'kə·rənt }

galvanic series [CHEM] The relative hierarchy of metals arranged in order from magnesium (least noble) at the anodic, corroded end through platinum (most noble) at the cathodic, protected end. { gal'van·ik 'sir·ēz }

galvanic skin response [PHYSIO] The electrical reactions of the skin to any stimulus as detected by a sensitive galvanometer; most often used experimentally to measure the resistance of the skin to the passage of a weak electric current. Also known as electrodermal response. { gal'van·ik 'skin ri¦späns }

galvanism [BIOL] The use of a galvanic current for medical or biological purposes. { 'gal·və₁niz·əm }

galvanize [MET] To deposit zinc on the surface of iron or steel by the processes of hot dipping, sherardizing, or sometimes electroplating. { 'gal·və₁nīz }

galvanoluminescence [PHYS] Light emission which may occur when electrodes of certain metals, such as aluminum or tantalum, are immersed in suitable electrolytes and current is passed between them. { gal·və·nō₁lü·mə'nes·əns }

galvanomagnetic effect [ELECTROMAG] One of the electrical or thermal phenomena occurring when a current-carrying conductor or semiconductor is placed in a magnetic field; examples are the Hall effect, Ettingshausen effect, transverse magnetoresistance, and Nernst effect. Also known as magnetogalvanic effect. { ¦gal·və·nō₁mag¦ned·ik i'fekt }

galvanometer [ENG] An instrument for indicating or measuring a small electric current by means of a mechanical motion derived from electromagnetic or electrodynamic forces produced by the current. { ₁gal·və'näm·əd·ər }

galvanometer constant [ELEC] Number by which a certain function of the reading of a galvanometer must be multiplied to obtain the current value in ordinary units. { ₁gal·və'näm·əd·ər 'kän·stənt }

galvanometer recorder [ENG ACOUS] A sound recorder in which the audio signal voltage is applied to a coil suspended in a magnetic field; the resulting movements of the coil cause a tiny attached mirror to move a reflected light beam back and forth across a slit in front of a moving photographic film. { ₁gal·və'näm·əd·ər ri'kórd·ər }

galvanometer shunt [ELEC] Resistor connected in parallel with a galvanometer to increase its range under certain conditions; it allows only a known fraction of the current to pass through the galvanometer. { ₁gal·və'näm·əd·ər ₁shənt }

galvanostat [ELEC] A device to deliver constant current from a high-voltage battery. { gal'van·ə₁stat }

galvanotaxis [BIOL] Movement of a free-living organism in response to an electrical stimulus. { ¦gal·və·nō¦tak·səs }

galvanotropism [BIOL] Response of an organism to electrical stimulation. { ₁gal·və'nä·trə₁piz·əm }

galvo *See* metal fume fever. { 'gal₁vō }

gambir [MATER] The yellowish extract from the twigs and leaves of the Malayan wood vine *Uncaria gambir* (Rubiaceae); used for tanning and dyeing and as an astringent. Also known as pale catechu; terra japonica; white cutch. { 'gam₁bir }

gambler's ruin [STAT] A game of chance which can be considered to be a series of Bernoulli trials at which each player wins a specified sum of money for every success and loses another sum for every failure; play goes on until the initial capital is lost and the player is ruined. { ¦gam·blərz 'rü·ən }

gambrel roof [BUILD] A roof with two sloping sides stepped at different angles on each side of the center ridge; the lower slope is steeper than the upper slope. { 'gam·brəl 'rüf }

game [MATH] A mathematical model expressing a contest between two or more players under specified rules. { gām }

game bird [BIOL] A bird that is legal quarry for hunters. { 'gām ₁bərd }

Gamella [MICROBIO] A genus of bacteria in the family Streptococcaceae; spherical cells occurring singly or in pairs with flattened adjacent sides; ferment glucose. { gə'mel·ə }

gamene *See* madder. { 'ga₁mēn }

gametangial copulation [MYCOL] Direct fusion of certain fungal gametangia without differentiation of the gametes. { ₁ga·mə'tan·jē·əl ₁käp·yə'la·shən }

gametangium [BIOL] A cell or organ that produces sex cells; occurs in algae, fungi, and plants. { ₁gam·ə'tan·jē·əm }

gamete [BIOL] A cell which participates in fertilization and development of a new organism. Also known as germ cell; sex cell. { 'ga₁mēt }

gamete intrafallopian transfer [MED] A variation of in vitro fertilization in which the spermatozoa and oocytes are placed directly into the fimbriated end of the Fallopian tube during the laparoscopy. Abbreviated GIFT. { ¦ga₁mēt ₁in·trə·fə¦lō·pē·ən 'tranz·fər }

game theory [MATH] The mathematical study of games or abstract models of conflict situations from the viewpoint of determining an optimal policy or strategy. Also known as theory of games. { 'gām ₁thē·ə·rē }

gametic copulation [MYCOL] The fusion of pairs of differentiated, uninucleate sexual cells or gametes formed in specialized gametangia. { gə'med·ik ₁käp·yə'la·shən }

gametoblast [BOT] An archespore that has not yet undergone differentiation. { gə'mēd·ə₁blast }

gametocyte [HISTOL] An undifferentiated cell from which gametes are produced. { gə'mēd·ə₁sīt }

gametogenesis [BIOL] The formation of gametes, or reproductive cells such as ova or sperm. { gə₁mēd·ə'jen·ə·səs }

gametophore [BOT] A branch that bears gametangia. { gə'mēd·ə₁fòr }

gametophyte [BOT] **1.** The haploid generation producing gametes in plants exhibiting metagenesis. **2.** An individual plant of this generation. { gə'mēd·ə₁fīt }

game tree [MATH] A tree graph used in the analysis of strategies for a game, in which the vertices of the graph represent positions in the game, and a given vertex has as its successors all vertices that can be reached in one move from the given position. Also known as lookahead tree. { 'gām ₁trē }

gamma [CHEM] The gamma position (the third carbon atom in an aliphatic carbon chain) on a chemical compound. [ELECTROMAG] A unit of magnetic field strength, equal to 10 microoersteds, or 0.00001 oersted. [GRAPHICS] A numerical indication of the degree of contrast in a television or photographic image; equal to the slope of the straight-line portion of the H and D curve for the emulsion or screen. [MECH] A unit of mass equal to 10^{-6} gram or 10^{-9} kilogram. { 'gam·ə }

gamma-absorption gage *See* gamma gage. { ¦gam·ə əb'sòrp·shən ₁gāj }

gamma acid [ORG CHEM] $C_{10}H_5NH_2OHSO_3H$ White crystals, slightly soluble in water; an intermediate in dyestuff manufacture. Also known as 2-amino-8-naphthol-6-sulfonic acid; 7-amino-1-naphthol-3-sulfonic acid; 2,5-naphthylamine sulfonic acid; 3-sulfonic acid; 6-sulfonic acid. { 'gam·ə 'as·əd }

gamma camera [ENG] An instrument consisting of a large, thin scintillation crystal or array of photomultiplier tubes, a multichannel collimator, and circuitry to analyze the pulses produced by the photomultipliers; used to visualize the distribution of radioactive compounds in the human body. { 'gam·ə ₁kam·rə }

gamma counter [ENG] A device for detecting gamma radiation, primarily through the detection of fast electrons produced by the gamma rays; it either yields information about integrated intensity within a time interval or detects each photon separately. { 'gam·ə ₁kaủnt·ər }

gamma cross section [NUC PHYS] The cross section for absorption or scattering of gamma rays by a nucleus or atom. { ¦gam·ə 'krò ₁sek·shən }

gamma decay *See* gamma emission. { 'gam·ə di₁kā }

gamma distribution [STAT] A normal distribution whose frequency function involves a gamma function. Also known as Erlang distribution. { ¦gam·ə ₁dis·trə'byü·shən }

gamma emission [NUC PHYS] A quantum transition between two energy levels of a nucleus in which a gamma ray is emitted. Also known as gamma decay. { 'gam·ə i'mish·ən }

gamma flux density [NUC PHYS] The number of gamma rays passing through a unit area in a unit time. { ¦gam·ə 'flòks ₁den·səd·ē }

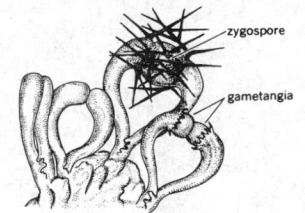

GAMETANGIAL COPULATION

Gametangial copulation, a sexual mechanism in fungi.

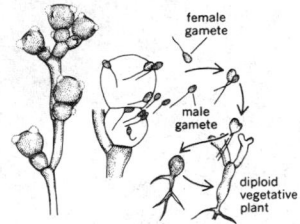

GAMETIC COPULATION

Gametic copulation, a sexual mechanism in fungi.

gamma function [MATH] The complex function $\Gamma(z)$ given by the integral with respect to t from 0 to ∞ of $e^{-t}t^{z-1}$; this function helps determine the general solution of Gauss' hypergeometric equation. { 'gam·ə ‚fəŋk·shən }

gamma gage [NUCLEO] A penetration-type thickness gage that measures the thickness or density of a sample by measuring its absorption of gamma rays. Also known as gamma-absorption gage. { 'gam·ə ‚gāj }

gamma globulin [IMMUNOL] Any of the serum proteins with antibody activity. { 'gam·ə 'gläb·yə·lən }

gamma heating [NUCLEO] Heating resulting from absorption of gamma-ray energy by a material. { 'gam·ə ‚hēd·iŋ }

gamma hemolysis [MICROBIO] The absence of activity in the area surrounding a bacterial colony growing on blood agar. Also known as nonhemolysis. { ‚gam·ə hi'mäl·ə·səs }

Gammaherpesvirinae [VIROL] A subfamily of animal, double-stranded linear DNA viruses of the family Herpesviridae, which are enveloped by a lipid bilayer and several glycoproteins. Also known as lymphoproliferative virus group. { ‚gam·ə‚hər‚pēz'vī·rə‚nē }

gamma iron [MET] Iron having a face-centered cubic lattice structure, stable between 910 and 1400°C. { 'gam·ə ‚ī·ərn }

gamma irradiation [NUCLEO] Exposure of a material to gamma rays. { 'gam·ə i‚rād·ē'ā·shən }

gamma logging [ENG] Obtaining, by means of a gamma-ray probe, a record of the intensities of gamma rays emitted by the rock strata penetrated by a borehole. { 'gam·ə ‚läg·iŋ }

gamma matrix See Dirac matrix. { 'gam·ə ‚mā·triks }

gamma radiation [NUCLEO] Radiation of gamma rays. { 'gam·ə ‚rād·ē'ā·shən }

gamma radiography [NUCLEO] Radiography by means of gamma rays. { 'gam·ə ‚rād·ē·ā‚gra·fē }

gamma random variable [MATH] A random variable that has a gamma distribution. { ¦gam·ə ¦ran·dəm 'ver·ē·ə·bəl }

gamma ray [NUC PHYS] A high-energy photon, especially as emitted by a nucleus in a transition between two energy levels. { 'gam·ə ‚rā }

gamma-ray altimeter [ENG] An altimeter, used at altitudes under several hundred feet, that measures the photon backscatter from the earth resulting from the transmission of photons to earth from a cobalt-60 gamma source in the plane. { 'gam·ə ‚rā al'tim·əd·ər }

gamma-ray astronomy [ASTRON] The study of gamma rays from extraterrestrial sources, especially gamma-ray bursts. { 'gam·ə ‚rā ə'strän·ə·mē }

gamma-ray bursts [ASTRON] Intense blasts of soft gamma rays of unknown origin, which range in duration from a tenth of a second to tens of seconds and occur several times a year from sources widely distributed over the sky. { 'gam·ə ‚rā ‚bərsts }

gamma-ray detector [ENG] An instrument that registers the presence of gamma rays. { 'gam·ə ‚rā di'tek·tər }

gamma-ray laser [PHYS] A hypothetical device which would generate coherent radiation in the range 0.005–0.5 nanometer by inducing isomeric radiative transitions between isomeric nuclear states. Also known as graser. { 'gam·ə ‚rā 'lā·zər }

gamma-ray level indicator [ENG] A level indicator in which the rising level of the liquid or other material reduces the amount of radiation passing from a gamma-ray source through the container to a Geiger counter or other radiation detector. { 'gam·ə ‚rā ‚lev·əl 'in·də‚kād·ər }

gamma-ray probe [ENG] A gamma-ray counter built into a watertight case small enough to be lowered into a borehole. { 'gam·ə ‚rā ‚prōb }

gamma-ray scattering See Compton scattering. { 'gam·ə ‚rā 'skad·ə·riŋ }

gamma-ray source [NUCLEO] A quantity of radioactive material that emits gamma radiation and is in a form convenient for radiology. { 'gam·ə ‚rā ‚sòrs }

gamma-ray spectrometry [NUCLEO] **1.** Determination of the energy distribution of gamma rays emitted by nuclei. Also known as gamma-ray spectroscopy. **2.** In particular, a variation of neutron activation analysis in which the induced radiation from the sample is gamma rays instead of neutrons. { 'gam·ə ‚rā spek'träm·ə·trē }

gamma-ray spectroscopy See gamma-ray spectrometry. { 'gam·ə ‚rā spek'träs·kə·pē }

gamma-ray spectrum [SPECT] The set of wavelengths or energies of gamma rays emitted by a given source. { 'gam·ə ‚rā ‚spek·trəm }

gamma-ray telescope [ASTRON] Any device for detecting and determining the directions of extraterrestrial gamma rays, using coincidence or anticoincidence circuits with scintillation or semiconductor detectors to obtain directional discrimination. { 'gam·ə ‚rā 'tel·ə‚skōp }

gamma-ray tracking [ENG] Use of three tracking stations, located at the three corners of a triangle centered on a missile about to be launched, to obtain accurate azimuthal tracking of a cobalt-60 gamma source in the tail. { 'gram·ə ‚rā 'trak·iŋ }

gamma-ray transformation [NUC PHYS] A radioactive decay in which gamma rays are emitted. { 'gam·ə ‚rā ‚tranz·fər‚mā·shən }

gamma-ray well logging [ENG] Measurement of gamma-ray intensity versus depth down the wellbore; used to identify rock strata, their position, and their thicknesses. { 'gam·ə ‚rā 'wel ‚läg·iŋ }

Gammaridea [INV ZOO] The scuds or sand hoppers, a suborder of amphipod crustaceans; individuals are usually compressed laterally, are poor walkers, and lack a carapace. { ‚gam·ə'rid·ē·ə }

gamma scanning [NUCLEO] The scanning of a fuel rod in a nuclear reactor for gamma activity by moving the rod past a slit in a lead block; photons emerging from the slit are detected by a scintillation spectrometer and recorded as a function of rod position. { 'gam·ə ‚skan·iŋ }

gamma structure [SOLID STATE] A Hume-Rothery designation for structurally analogous phases or intermetallic phases having 21 valence electrons to 13 atoms, analogous to the γ-brass structure. { 'gam·ə ‚strək·chər }

gamma taxonomy [SYST] The level of taxonomic study concerned with biological aspects of taxa, including intraspecific populations, speciation, and evolutionary rates and trends. { 'gam·ə tak'sän·ə·mē }

gamma transition See glass transition. { 'gam·ə tran'zish·ən }

gammeter [ENG] A template fashioned of transparent material and marked with a calibrated scale; when positioned on a sensitometric curve it is used to determine the slope of the straight-line portion. { 'ga‚mēd·ər }

gammil [CHEM] A unit of concentration, equal to a concentration of 1 milligram of solute in 1 liter of solvent. Also known as micril; microgammil. { 'gam·əl }

gamodeme [ECOL] An isolated breeding community. { 'gam·ə‚dēm }

gamogony [INV ZOO] Spore formation by multiple fission in sporozoans. [ZOO] Sexual reproduction. { gə'mäg·ə·nē }

gamone [PHYSIO] Any substance released by a gamete that facilitates fertilization processes. { ga'mōn }

gamont [INV ZOO] The gametocyte of sporozoans. { 'ga‚mänt }

gamopetalous [BOT] Having petals united at their edges. Also known as sympetalous. { ¦gam·ə¦ped·əl·əs }

gamophobia [PSYCH] An abnormal fear of marriage. { ‚gam·ə'fō·bē·ə }

gamophyllous [BOT] Having the leaves of the perianth united. { ¦gam·ə¦fil·əs }

gamosepalous [BOT] Having sepals united at their edges. Also known as synsepalous. { ‚ga·mō¦sep·ə·ləs }

Gamow barrier [NUC PHYS] The potential barrier which retards the escape of alpha particles from the nucleus according to the Gamow-Condon-Gurney theory. { 'ga‚mòf ‚bar·ē·ər }

Gamow-Condon-Gurney theory [NUC PHYS] An early quantum-mechanical theory of alpha-particle decay according to which the alpha particle penetrates a potential barrier near the surface of the nucleus by a tunneling process. { ¦ga‚mòf ¦känd·ən 'gər‚nē ‚thē·ə·rē }

Gamow-Teller interaction [NUC PHYS] Interaction between a nucleon source current and a lepton field which has an axial vector or tensor form. { ¦ga‚mòf 'tel·ər ‚in‚tər'ak·shən }

Gamow-Teller selection rules [NUC PHYS] Selection rules for beta decay caused by the Gamow-Teller interaction; that is, in an allowed transition there is no parity change of the nuclear state, and the spin of the nucleus can either remain unchanged or change by ±1; transitions from spin 0 to spin 0 are excluded, however. { ¦ga‚mòf 'tel·ər si'lek·shən ‚rülz }

Gampsonychidae [PALEON] A family of extinct crustaceans in the order Palaeocaridacea. { ‚gam·sə'nī·kə‚dē }

gang [ELEC] A mechanical connection of two or more circuit devices so that they can be varied at the same time. { gaŋ }

gang capacitor [ELEC] A combination of two or more variable capacitors mounted on a common shaft to permit adjustment by a single control. { 'gaŋ kə'pas·əd·ər }

gang chart [IND ENG] A multiple-activity process chart used for groups of men on materials-handling operations. { 'gaŋ ‚chärt }

gang drill [MECH ENG] A set of drills operated together in the same machine; used in rock drilling. { 'gaŋ ‚dril }

ganged control [ELECTR] Controls of two or more circuits mounted on a common shaft to permit simultaneous control of the circuits by turning a single knob. { 'gaŋd kən'trōl }

gangliated cord [NEUROSCI] One of the two main trunks of the sympathetic nervous system, one trunk running along each side of the spinal column. { 'gaŋ·glē‚ād·əd 'kórd }

ganglioma [MED] A form of ganglioneuroma in which neuronal and glial elements appear in about equal proportions. { ‚gaŋ·glē'ō·mə }

ganglion [NEUROSCI] A group of nerve cell bodies, usually located outside the brain and spinal cord. { 'gaŋ·glē·ən }

ganglioneuroma [MED] A tumor composed of sympathetic ganglion cells and sheathed nerve fibers. { ‚gaŋ·glē·ō·nü'rō·mə }

ganglionitis [MED] Inflammation of a ganglion. { gaŋ·glē·ə'nīd·əs }

ganglioside [BIOCHEM] One of a group of glycosphingolipids found in neuronal surface membranes and spleen; they contain an *N*-acyl fatty acid derivative of sphingosine linked to a carbohydrate (glucose or galactose); they also contain *N*-acetylglucosamine or *N*-acetylgalactosamine, and *N*-acetylneuraminic acid. { 'gaŋ·glē·ō‚sīd }

gang milling [ENG] Rolling of material by means of a composite machine with numerous cutting blades. { 'gaŋ ‚mil·iŋ }

gangosa [MED] Destructive lesions of the nose and hard palate, sometimes more extensive, considered to be the tertiary stage of yaws. { gaŋ'gō·sə }

gangplank [NAV ARCH] A long, narrow, movable bridge or plank from a ship to a pier or to another ship alongside. { 'gaŋ‚plaŋk }

gangplow [AGR] A plow with two or more cutters that turn parallel furrows. { 'gaŋ‚plaù }

gangrene [MED] A form of tissue death usually occurring in an extremity due to insufficient blood supply. { gaŋ'grēn }

gangrenous stomatitis *See* noma. { 'gaŋ·grə·nəs ‚stō·mə'tīd·əs }

gang saw [MECH ENG] A steel frame in which thin, parallel saws are arranged to operate simultaneously in cutting logs. { 'gaŋ ‚só }

gang switch [ELEC] A combination of two or more switches mounted on a common shaft to permit operation by a single control. Also known as deck switch. { 'gaŋ ‚swich }

gangue [GEOL] The valueless rock or aggregates of minerals in an ore. { gaŋ }

gangway [MIN ENG] **1.** A principal underground haulage road. **2.** A passageway into or out of an underground mine. [NAV ARCH] An opening in the rail or bulwarks of a ship through which one can enter or leave it. { 'gaŋ‚wā }

ganister [MATER] A fine mixture of quartz and fireclay which is used to line certain furnaces for metallurgical processes. [PETR] A fine, hard quartzose sandstone; used to make refractory silica brick to line furnace reactors. { 'gan·ə·stər }

Ganoderma lucidum [MYCOL] A mushroom found throughout the United States, Europe, South America, and Asia that appears to have antiallergic, anti-inflammatory, antibacterial, antioxidant, antitumor, and immunostimulating activity. Also known as ling-zhi; reishi mushroom. { gen·ə‚dər·mə 'lüs·ə·dəm }

ganoid scale [VERT ZOO] A structure having several layers of enamellike material (ganoin) on the upper surface and laminated bone below. { ‚ga‚nóid 'skāl }

ganoin [VERT ZOO] The enamellike covering of a ganoid scale. { 'gan·ə·wən }

ganomalite [MINERAL] $(Ca_2)Pb_3Si_3O_{11}$ A colorless to gray silicate of lead with calcium crystallizing in the tetragonal system. { gə'näm·ə‚līt }

ganophyllite [MINERAL] $(Na,K)(Mn,Fe,Al)_5(Si,Al)_6O_{15}$-$(OH)_5 \cdot 2H_2O$ A brown, prismatic crystalline or foliated mineral composed of a hydrous silicate of manganese and aluminum. { ‚gan·ə'fi‚līt }

gantlet [CIV ENG] A stretch of overlapping railroad track, with one rail of one track being between the two rails of another track; used over narrow bridges and passes. { 'gónt·lət }

gantry [ENG] A frame erected on side supports so as to span an area and support and hoist machinery and heavy materials. { 'gan·trē }

gantry crane [MECH ENG] A bridgelike hoisting machine having fixed supports or arranged for running along tracks on ground level. { 'gan·trē ‚krān }

gantry-type robot [CONT SYS] A continuous-path, cartesian-coordinate robot constructed in a bridge shape that uses rails to move along a single horizontal axis or along either of two perpendicular horizontal axes. { 'gan·trē ‚tīp 'rō‚bät }

Gantt chart [IND ENG] In production planning and control, a type of bar chart depicting the work planned and done in relation to time; each division of space represents both a time interval and the amount of work to be done during that interval. { 'gant ‚chärt }

Gantt task and bonus plan [IND ENG] A wage incentive plan in which high task efficiency is maintained by providing a percentage bonus as a reward for production in excess of standard. { 'gant ‚task ən 'bō·nəs ‚plan }

Ganymede [ASTRON] A satellite of Jupiter orbiting at a mean distance of 664,000 miles (1,071,000 kilometers). Also known as Jupiter III. { 'gan·ə‚mēd }

gap [COMMUN] A region not adequately covered by the main lobes of a radar antenna. [COMPUT SCI] A uniformly magnetized area in a magnetic storage device (tape, disk), used to indicate the end of an area containing information. [ELEC] The spacing between two electric contacts. [ELECTROMAG] A break in a closed magnetic circuit, containing only air or filled with a nonmagnetic material. [GEN] A short region that is missing in one strand of a double-stranded deoxyribonucleic acid. [GEOGR] Any sharp, deep notch in a mountain ridge or between hills. [MET] An opening at the point of closest approach between faces of members in a weld joint. { gap }

gap coding [COMMUN] A process for conveying information by inserting gaps or periods of nontransmission in a system that normally transmits continuously. { 'gap ‚kōd·iŋ }

gap digit [COMPUT SCI] A digit in a machine word that does not represent data or instructions, such as a parity bit or a digit included for engineering purposes. { 'gap ‚dij·it }

gape [ANAT] The margin to margin distance between open jaws. [INV ZOO] The space between the margins of a closed mollusk valve. { gāp }

gap factor [ELECTR] Ratio of the maximum energy gained in volts to the maximum gap voltage in a tube employing electron accelerating gaps, that is, a traveling-wave tube. { 'gap ‚fak·tər }

gap-filler radar [ENG] Radar used to fill gaps in radar coverage of other radar. { 'gap ‚fil·ər 'rā‚där }

gap filling [ELECTROMAG] Electrical or mechanical rearrangement of an antenna array, or the use of a supplementary array, to produce lobes where gaps previously occurred. { 'gap ‚fil·iŋ }

gap-framepress [MECH ENG] A punch press whose frame is open at bed level so that wide work or strip work can be inserted. { 'gāp 'frām‚pres }

gap-graded aggregate [MATER] Aggregate in which certain size particles are entirely or substantially absent. { 'gap ‚grād·əd 'ag·rə‚gət }

gap junction [NEUROSCI] An intercellular junction composed of cylindrical channels connecting adjacent cells; considered to be a low-resistance pathway for intercellular communication. Also known as communicating junction; nexus. { 'gap ‚jəŋk·shən }

gap lathe [MECH ENG] An engine lathe with a sliding bed providing enough space for turning large-diameter work. { 'gap ‚lāth }

gapless tape [COMPUT SCI] A magnetic tape upon which raw data is recorded in a continuous manner; the data are streamed onto the tape without the word gaps; the data still may contain signs and end-of-record marks in the gapless form. { 'gap·ləs 'tāp }

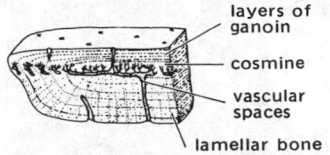

layers of ganoin

cosmine

vascular spaces

lamellar bone

Ganoid scale, a type of primitive vertebrate scale. (*From A. S. Romer, The Vertebrate Body, Saunders, 1962*)

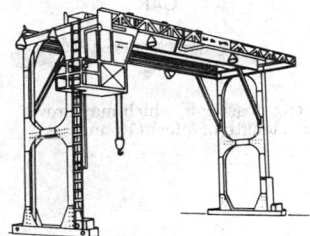

An overhead-type traveling gantry crane.

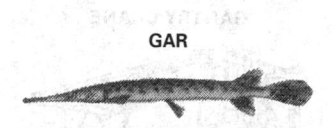

GAR

Long-nosed gar, which may grow to a length of 5 feet (1.5 meters).

gap markers [ORD] In land mine warfare, markers used to identify a minefield gap; gap markers at the entrance to and exit from the gap are referenced to a landmark or intermediate marker. { 'gap ¦märk·ərz }

gapped tape [COMPUT SCI] A magnetic tape upon which blocked data has been recorded; it contains all of the flag bits and format to be read directly into a computer for immediate use. { ¦gapt 'tāp }

gap scanning [ENG] In ultrasonic testing, a coupling technique in which a sound beam is projected through a short fluid column that flows through a nozzle on an ultrasonic search unit. { 'gap ¦skan·iŋ }

gap scatter [COMPUT SCI] The deviation from the exact distance required between read/write heads and the magnetized surface. { 'gap ¦skad·ər }

gar [VERT ZOO] The common name for about seven species of bony fishes in the order Semionotiformes having a slim form, an elongate snout, and close-set ganoid scales. { gar }

garbage See hash. { 'gär·bij }

garbage collection [COMPUT SCI] In a computer program with dynamic storage allocation, the automatic process of identifying those memory cells whose contents are no longer useful for the computation in progress and then making them available for some other use. { 'gär·bij kə¦lek·shən }

garbage in, garbage out [COMPUT SCI] A phrase often stressed during introductory courses in computer utilization as a reminder that, regardless of the correctness of the logic built into the program, no answer can be valid if the input is erroneous. Abbreviated GIGO. { ¦gär·bij 'in ¦gär·bij 'aút }

garbage pitch [MATER] Dark-brown to black pitch material obtained as a by-product residue from the burning of garbage; properties are analogous to complex hydrocarbons; used to make paints, varnishes, tarred paper, and waterproofing compound. { 'gär·bij ¦pich }

garbet rod See garbutt rod. { 'gär·bət ¦räd }

garbin [METEOROL] A sea breeze; in southwest France it refers to a southwesterly sea breeze which sets in about 9 a.m., reaches its maximum toward 2 p.m., and ceases about 5 p.m. { gär'ba }

garble [COMMUN] To alter a message intentionally or unintentionally so that it is difficult to understand. { 'gär·bəl }

garboard strake [NAV ARCH] The strake of shell plating adjacent to the keel; this row of plates acts in conjunction with the keel, and the plates are made heavier than the other bottom plates. { ¦gär·bərd 'strāk }

garbot rod See garbutt rod. { 'gär·bət ¦räd }

garbutt rod [PETRO ENG] A device used to pull the standing valve out of a tubing-type oil-well sucker-rod pump. Also known as garbet rod; garbot rod. { 'gär·bət ¦räd }

gardening [ASTRON] A phenomenon in which the lunar regolith is constantly churning at a very slow rate because of successive impacts; the result is that bottom material is brought up to the top and surface material is buried. { 'gärd·ən·iŋ }

Garden path problem [PSYCH] A problem that induces performance that has the false appearance of success. { 'gärd·ən 'path ¦präb·ləm }

Gardner crusher [MIN ENG] A swing-and-hammer crusher; the U-shaped hammers are thrown by a revolving shaft against the feed and a heavy anvil inside the housing. { 'gärd·nər 'krəsh·ər }

Gardner's syndrome [MED] A hereditary disorder transmitted as an autosomal dominant; manifested in childhood by multiple neoplasms, including bony and mesenteric tumors, fatty and fibrous skin, and intestinal polyps. { 'gärd·nərz ¦sin¦drōm }

gargoylism See Hurler's syndrome. { 'gär·gói¦liz·əm }

garigue [ECOL] A low, open scrubland restricted to limestone sites in the Mediterranean area; characterized by small evergreen shrubs and low trees. { gə'rēg }

garland [MIN ENG] A channel fixed around a shaft in order to catch the water draining down the walls and conduct it to a lower level. Also known as water curb; water garland; water ring. { 'gär·lənd }

garlic [BOT] Allium sativum. A perennial plant of the order Liliales grown for its pungent, edible bulbs. { 'gär·lik }

garlic oil [MATER] An essential oil obtained from steam distillation of garlic; contains chiefly a mixture of terpenes, with organic sulfides also present. { 'gär·lik ¦oil }

garner [AGR] **1.** A building or section of a building in which grain is stored. **2.** A bin for weighing grain. { 'gär·nər }

garnet [MINERAL] A generic name for a group of mineral silicates that are isometric in crystallization and have the general chemical formula $A_3B_2(SiO_4)_3$, where A is Fe^{2+}, Mn^{2+}, Mg, or Ca, and B is Al, Fe^{3+}, Cr^{3+}, or Ti^{3+}; used as a gemstone and as an abrasive. { 'gär·nət }

garnet hinge [DES ENG] A hinge with a vertical bar and horizontal strap. { 'gär·nət ¦hinj }

garnet maser [ELECTR] A name incorrectly applied to a ferromagnetic amplifier. { 'gär·nət 'mā·zər }

garnet paper [MATER] Paper with a layer of crushed garnet on one side; used as an abrasive or polisher. { 'gär·nət ¦pā·pər }

garnett [TEXT] **1.** A machine for removing foreign materials from fiber before it is carded. **2.** A machine for converting textile waste to fiber. **3.** The waste products of a garnett machine. { 'gär·net }

garnetting [TEXT] The reduction of textile waste materials to fiber. { 'gär·ned·iŋ }

garnierite [MINERAL] $(Ni,Mg)_3Si_2O_5(OH)_4$ An apple-green or pale-green, monoclinic serpentine; a gemstone and an ore of nickel. Also known as nepuite; noumeite. { 'gär·nē·ə¦rīt }

garret [BUILD] The part of a house just under the roof. { 'gar·ət }

garronite [MINERAL] $Na_2Ca_5Al_{12}Si_{20}O_{64}\cdot27H_2O$ A zeolite mineral belonging to the phillipsite group; crystallizes in the tetragonal system. { 'ga·rə¦nīt }

garter spring [DES ENG] A closed ring formed of helically wound wire. { 'gärd·ər ¦spriŋ }

Gartner's duct [ANAT] The remnant of the embryonic Wolffian duct in the adult female mammal. { 'gärt·nərz ¦dəkt }

garúa [METEOROL] A dense fog or drizzle from low stratus clouds on the west coast of South America, creating a raw, cold atmosphere that may last for weeks in winter, and supplying a limited amount of moisture to the area. Also known as camanchaca. { gä'rü·ə }

Garvey-Kelson mass relations [NUC PHYS] A set of equations relating the masses of atomic nuclei with slightly different numbers of neutrons and protons. { 'gär·vē 'kel·sən 'mas ri'lā·shənz }

gas [MATER] See gasoline. [ORD] To expose to a war gas. [PHYS] A phase of matter in which the substance expands readily to fill any containing vessel; characterized by relatively low density. { gas }

gas absorption operation [CHEM ENG] The recovery of solute gases present in gaseous mixtures of noncondensables; this recovery is generally achieved by contacting the gas stream with a liquid that offers specific or selective solubility for the solute gas to be recovered, or with an adsorbent (for example, synthetic or natural zeolite) that accepts only specific molecule sizes or shapes. { gas əb¦sòrp·shən ¦äp·ə¦rā·shən }

gas-activated battery [ELEC] A reserve battery which is activated by introducing a gas which reacts with a material between the electrodes of the battery to form an electrolyte. { ¦gas ak·tə¦vād·əd 'bad·ə·rē }

gas adsorption [PHYS CHEM] The concentration of a gas upon the surface of a solid substance by attractive forces between the surface and the gas molecules. { ¦gas ad'sorp·shən }

gas alarm [MIN ENG] A signal system which warns mine workers of dangerous concentration of firedamp. { 'gas ə¦lärm }

gas amplification [NUCLEO] The ratio of the charge collected to the charge liberated by the initial ionizing event in a radiation-counter tube. { ¦gas ¦am·plə·fə'kā·shən }

gas analysis [ANALY CHEM] Analysis of the constituents or properties of a gas (either pure or mixed); composition can be measured by chemical adsorption, combustion, electrochemical cells, indicator papers, chromatography, mass spectroscopy, and so on; properties analyzed for include heating value, molecular weight, density, and viscosity. { 'gas ə¦nal·ə·səs }

gas anchor [PETRO ENG] A downhole gas separator used to reduce gas-in-oil froth before the pump to increase pump efficiency. { 'gas ¦aŋ·kər }

gas and mist sampler [MIN ENG] An instrument for automatic collection of one sample per hour of airborne contaminants such as sulfur dioxide or ammonia. { ¦gas ən 'mist ¦sam·plər }

gas bag [ENG] A bag made of gas-impermeable material and designed for insertion into a pipeline followed by inflation to halt the flow of gas. { 'gas ¦bag }

gas bearing [MECH ENG] A journal or thrust bearing lubricated with gas. Also known as gas-lubricated bearing. { 'gas ¦ber·iŋ }

gas bell [GRAPHICS] A bubble resulting from a chemical reaction in the vicinity of the emulsion which causes separation of the emulsion from the support and the formation of minute holes in the negative. { 'gas ¦bel }

gas black [CHEM] Fine particles of carbon formed by partial combustion or thermal decomposition of natural gas; used to reinforce rubber products such as tires. Also known as carbon black; channel black. { 'gas ¦blak }

gas bomb [ORD] A bomb designed to produce casualties among personnel and to contaminate an area; a burster charge splits the bomb case and disperses the gas filler over the area. { 'gas ¦bäm }

gas-bounded nebula [ASTRON] An emission nebula whose central star is hot enough, or in which the density of the cloud is small enough, to ionize the entire cloud. { 'gas ¦baùnd·əd 'neb·yə·lə }

gas brazing See gas-flame brazing. { 'gas ¦brāz·iŋ }

gas-bubble protective device See Buchholz protective device. { 'gas ¦bəb·əl prə'tek·tiv di¸vīs }

gas buoy [NAV] A metal buoy having a gas light. { 'gas ¦bȯi }

gas burner [ENG] A hole or a group of holes through which a combustible gas or gas-air mixture flows and burns. { 'gas ¦bər·nər }

gas cap [GEOPHYS] The gas immediately in front of a meteoroid as it travels through the atmosphere. [PETRO ENG] Gas occurring above liquid hydrocarbons in a reservoir under such trap conditions as the presence of water which prevents downward migration or the abutment of an impermeable formation against the reservoir. { 'gas ¦kap }

gas capacitor [ELEC] A capacitor consisting of two or more electrodes separated by a gas, other than air, that serves as a dielectric. { 'gas kə'pas·əd·ər }

gas-cap drive [PETRO ENG] Driving liquid hydrocarbons through a porous reservoir and toward well holes by utilizing the pressure of gas overlying the liquid pool. { 'gas ¦kap ¸drīv }

gas-cap expansion [PETRO ENG] Process of reservoir-liquids displacement by the natural expansion of the reservoir gas cap to fill the voids vacated by recovered liquids. { 'gas ¦kap ik'span·shən }

gas-cap gas See associated gas. { 'gas ¸kap 'gas }

gas-cap injection See external gas injection. { 'gas ¸kap in¸jek·shən }

gas-cap reservoir [PETRO ENG] Two-phase reservoir in which a free area of gas (a gas cap) is underlain by an oil or liquid phase. { 'gas ¦kap 'rez·əv¸wär }

gas carburizing [MET] Surface hardening by heating a metal in gas of high carbon content in order to introduce carbon into the surface layers. { 'gas 'kär·bə¸rīz·iŋ }

gas cell [ELEC] Cell in which the action depends on the absorption of gases by the electrodes. { 'gas ¸sel }

gas-cell frequency standard [ATOM PHYS] An atomic frequency standard in which the frequency-determining element is a gas cell containing rubidium, cesium, or sodium vapor. { 'gas ¸sel 'frē·kwən·sē ¸stan·dərd }

gas centrifuge process [NUCLEO] A method of isotope separation in which a mixture of isotopes in the gaseous state is spun at high speeds in a centrifuge, and centrifugal forces cause a concentration of heavy isotopes near the walls and light isotopes near the center. { 'gas 'sen·trə¸fyüj ¸präs·əs }

gas check [ORD] Device in a gun that prevents escape of gas through the breech. { 'gas ¸chek }

gas chromatograph [ANALY CHEM] The instrument used in gas chromatography to detect volatile compounds present; also used to determine certain physical properties such as distribution or partition coefficients and adsorption isotherms, and as a preparative technique for isolating pure components or certain fractions from complex mixtures. { 'gas krō'mad·ə¸graf }

gas chromatography [ANALY CHEM] A separation technique involving passage of a gaseous moving phase through a column containing a fixed adsorbent phase; it is used principally as a quantitative analytical technique for volatile compounds. { 'gas ¸krō·mə'täg·rə·fē }

gas clathrate See gas hydrate. { 'gas ¸'klath¸rāt }

gas cleaning [ENG] Removing ingredients, pollutants, or contaminants from domestic and industrial gases. { 'gas ¸klēn·iŋ }

gas column [GEOL] The difference in elevation between the highest and lowest parts of the various producing zones of a gas-producing formation. { 'gas ¸ka·ləm }

gas-compression cycle [MECH ENG] A refrigeration cycle in which hot, compressed gas is cooled in a heat exchanger, then passes into a gas expander which provides an exhaust stream of cold gas to another heat exchanger that handles the sensible-heat refrigeration effect and exhausts the gas to the compressor. { 'gas kəm¦presh·ən ¸sī·kəl }

gas compressor [MECH ENG] A machine that increases the pressure of a gas or vapor by increasing the gas density and delivering the fluid against the connected system resistance. { 'gas kəm¸pres·ər }

gas-condensate liquid [ORG CHEM] A hydrocarbon, such as propane, butane, and pentane, obtained as condensate when wet natural gas is compressed or refrigerated. { 'gas 'känd·ən¸sät ¸lik·wəd }

gas-condensate reservoir [GEOL] Hydrocarbon reservoir in which conditions of temperature and pressure have resulted in the condensation of the heavier hydrocarbon constituents from the reservoir gas. { 'gas 'känd·ən¸sät ¸rez·əv¸wär }

gas-condensate well [PETRO ENG] A well producing hydrocarbons from a gas-condensate reservoir. { 'gas 'känd·ən¸sät ¸wel }

gas coning [PETRO ENG] The tendency of gas in a gas-drive reservoir to push oil downward in an inverse cone contour toward the casing perforations; at the extreme of coning, gas, not oil, will be produced from the well. { 'gas 'kōn·iŋ }

gas constant [THERMO] The constant of proportionality appearing in the equation of state of an ideal gas, equal to the pressure of the gas times its molar volume divided by its temperature. Also known as gas-law constant; universal gas constant. { 'gas 'kän·stənt }

gas-cooled reactor [NUCLEO] A nuclear reactor in which a gas, such as air, carbon dioxide, or helium, is used as a coolant. { 'gas ¸küld rē'ak·tər }

gas counter [NUCLEO] A counter in which the radioactive sample is prepared in the form of a gas and introduced into the counter tube. { 'gas ¸kaùnt·ər }

gas current [ELECTR] A positive-ion current produced by collisions between electrons and residual gas molecules in an electron tube. Also known as ionization current. { 'gas ¸kə·rənt }

gas cut [PETRO ENG] A foamy mixture of gas and drilling mud recovered in testing. { 'gas ¸kət }

gas cutting [MET] Cutting metal with the heat of an oxyacetylene flame. { 'gas ¸kəd·iŋ }

gas cyaniding See carbonitriding. { 'gas 'sī·ə¸nīd·iŋ }

gas cycle [THERMO] A sequence in which a gaseous fluid undergoes a series of thermodynamic phases, ultimately returning to its original state. { 'gas ¸sī·kəl }

gas cycling [PETRO ENG] A petroleum enhanced-recovery process which injects the gas produced with oil back into the oil sand to help produce more oil. { 'gas ¸sī·kliŋ }

gas cylinder [MECH ENG] The chamber in which a piston moves in a positive displacement engine or compressor. { 'gas ¸sil·ən·dər }

gas dehydrator [CHEM ENG] A device or system to remove moisture vapor from a gas stream, usually incorporates desiccant-type packed towers. { 'gas dē'hī¸drād·ər }

gas depletion drive See internal gas drive. { 'gas di'plē·shən ¸drīv }

gas detector [MIN ENG] A device which indicates the existence of firedamp or other combustible or noxious gas in a mine. { 'gas di¸tek·tər }

gas-deviation factor See compressibility factor. { 'gas ¸dē·vē'ā·shən ¸fak·tər }

gas discharge [ELECTR] Conduction of electricity in a gas, due to movements of ions produced by collisions between electrons and gas molecules. { 'gas 'dis¸chärj }

GAS CHROMATOGRAPH

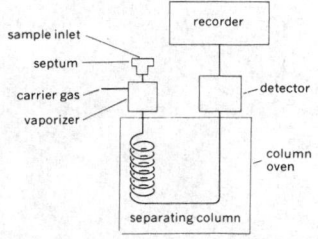

Diagram of a gas chromatograph.

gas-discharge centrifuge [NUCLEO] A type of plasma centrifuge in which gas between two coaxial electrodes is partially ionized by a radial current between the electrodes and is caused to rotate by the Lorentz force on the radial current. { 'gas ¦dis,chärj 'sen·trə,fyüj }

gas-discharge display [ELECTR] A display in which seven or more cathode elements form the segments of numerical or alphameric characters when energized by about 160 volts direct current; the segments are vacuum-sealed in a neon-mercury gas mixture. { 'gas 'dis,chärj di,splā }

gas-discharge lamp See discharge lamp. { ¦gas 'dis,chärj ,lamp }

gas-discharge laser [OPTICS] A gas laser in which optical pumping is caused by nonequilibrium processes in a gas discharge. { ¦gas 'dis,chärj ,lā·zər }

gas doping [ELECTR] The introduction of impurity atoms into a semiconductor material by epitaxial growth, by using streams of gas that are mixed before being fed into the reactor vessel. { 'gas ,dōp·iŋ }

gas-drive reservoir [PETRO ENG] An oil reservoir in which gas (either natural or reinjected) provides the driving force to sweep liquids through the formation and into the wellbore. { 'gas ,drīv 'rēz·əv,wär }

gas dynamic laser [OPTICS] A gas laser that converts thermal energy directly into coherent radiation at an efficiency high enough to offer promise of wireless power transmission. { ¦gas dī¦nam·ik 'lā·zər }

gas dynamics [PHYS] The study of the motion of gases, and of its causes, which takes into account thermal effects generated by the motion. { ¦gas dī'nam·iks }

gas embolus [MED] An embolus composed of a gas resulting from trauma or other causes. { 'gas ,em·bə·ləs }

gas emission [MIN ENG] The release of gas from the strata into the mine workings. { 'gas i,mish·ən }

gas engine [MECH ENG] An internal combustion engine that uses gaseous fuel. { 'gas ,en·jən }

gaseous conduction analyzer [ENG] A device to detect organic vapors in air by measuring the change in current that flows between a heated platinum anode and a concentric platinum cathode. { ¦gash·əs kən¦dək·shən 'an·ə,līz·ər }

gaseous diffusion [CHEM ENG] **1.** Pressure-induced free-molecular transfer of gas through microporous barriers as in the process of making fissionable fuel. **2.** Selective solubility diffusion of gas through nonporous polymers by absorption and solution of the gas in the polymer matrix. { ¦gash·əs di'fyü·zhən }

gaseous diffusion plant [NUCLEO] A facility where the gaseous diffusion process is used in the separation of uranium isotopes in order to provide fissionable fuel for nuclear power plants and weapons. { ¦gash·əs di'fyü·zhən ,plant }

gaseous diffusion process [NUCLEO] A method of separating isotopes in which an isotopic mixture of gases is allowed to diffuse through a porous wall; the lighter molecules pass through the porous wall more readily than the heavier molecules. { ¦gash·əs di'fyü·zhən ,präs·əs }

gaseous fuel [MATER] A combustible gas that can be burned in a furnace or an engine. { ¦gash·əs 'fyül }

gaseous nebulae [ASTRON] Clouds of gas, such as the Network Nebula in Cygnus, that are members of the Milky Way galactic system and are small compared with its overall dimensions. { ¦gash·əs 'neb·yə·lē }

gas etching [ENG] The removal of material from a semiconductor circuit by reaction with a gas that forms a volatile compound. { 'gas ,ech·iŋ }

gas explosion [MIN ENG] An explosion of firedamp in a coal mine; coal dust apparently does not play a significant part. { 'gas ik,splō·zhən }

GasFET [ENG] A gas sensor based on changes, upon exposure to hydrogen, in the surface part of the work function of a palladium component that serves as the gate contact of a metal oxide semiconductor field-effect transistor (MOSFET). { 'gas,fet }

gas field [PETRO ENG] An area underlain with little or no interruption by one or more reservoirs of commercially valuable gas. { 'gas ,fēld }

gas-filled cable [ELEC] A coaxial or other cable containing gas under pressure to serve as insulation and keep out moisture. { 'gas ,fild 'kā·bəl }

gas-filled diode [ELECTR] A gas tube which is a diode, such as a cold-cathode rectifier or phanotron. { 'gas ,fild 'dī,ōd }

gas-filled porosity [GEOL] A reservoir formation in which the pore space is filled by gas instead of liquid hydrocarbons. { 'gas ,fild pə'räs·əd·ē }

gas-filled radiation counter [NUCLEO] A gas tube used to detect radiation by means of gas ionization. { 'gas ,fild ,rād·ē'ā·shən ,kaunt·ər }

gas-filled rectifier See cold-cathode rectifier. { 'gas ,fild 'rek·tə,fī·ər }

gas-filled thermometer [ENG] A thermometer which uses a gas (usually nitrogen or hydrogen), that approximately follows the ideal gas law. { 'gas ,fild thər'mäm·əd·ər }

gas-filled triode [ELECTR] A gas tube which has a grid or other control element, such as a thyratron or ignitron. { 'gas ,fild 'trī,ōd }

gas filter [CHEM ENG] A device used to remove liquid or solid particles from a flowing gas stream. { 'gas fil·tər }

gas-flame brazing [MET] A brazing process for which the heat is supplied by a gas flame. Also known as gas brazing. { 'gas ,flām 'brāz·iŋ }

gas floor [GEOL] In a sedimentary basin, the depth below which there is no economic accumulation of gaseous hydrocarbons. { 'gas ,flór }

gas-flow counter tube [NUCLEO] A radiation-counter tube in which an appropriate atmosphere is maintained by a flow of gas through the tube. Also known as flow counter; gas-flow radiation counter. { 'gas ,flō 'kaunt·ər ,tüb }

gas-flow radiation counter See gas-flow counter tube. { 'gas ,flō ,rād·ē'ā·shən ,kaunt·ər }

gas focusing [ELECTR] A method of concentrating an electron beam by utilizing the residual gas in a tube; beam electrons ionize the gas molecules, forming a core of positive ions along the path of the beam which attracts beam electrons and thereby makes the beam more compact. Also known as ionic focusing. { 'gas ,fō·kəs·iŋ }

gas furnace [ENG] An enclosure in which a gaseous fuel is burned. { 'gas ,fər·nəs }

gas gangrene [MED] A localized, but rapidly spreading, necrotizing bacterial wound infection characterized by edema, gas production, and discoloration; caused by several species of Clostridium. { 'gas 'gaŋ,grēn }

gas generator [CHEM] A device used to generate gases in the laboratory. [CHEM ENG] A chemical plant for producing gas from coal, for example, water gas. [MECH ENG] An apparatus that supplies a high-pressure gas flow to drive compressors, airscrews, and other machines. { 'gas 'jen·ə,rād·ər }

gas gland [VERT ZOO] A structure inside the swim bladder of many teleosts which secretes gas into the bladder. { 'gas ,gland }

gas grenade [ORD] Popular name for a chemical grenade designed to release a war gas, usually tear gas or other irritants. { 'gas grə,nād }

gas gun [ORD] Automatic gun operated solely by gas, utilizing a portion of the gas pressure from firing a bullet to act on some form of piston-and-cylinder arrangement. { 'gas ,gən }

gas heater [MECH ENG] A unit heater designed to supply heat by forced convection, using gas as a heat source. { 'gas ,hēd·ər }

gash fracture [GEOL] Open gashes that are formed diagonally to a fault or fault zone. { 'gash ,frak·chər }

gas holder [ENG] Gas storage container with vertically free top section that moves up or down to adjust to the volume of gas held. { 'gas ,hōl·dər }

gas hole [ENG] A cavity formed in a casting as a result of cavitation. { 'gas ,hōl }

gas hydrate [GEOCHEM] A naturally occurring solid composed of crystallized water (ice) molecules, forming a rigid lattice of cages (a clathrate) with most of the cages containing a molecule of natural gas, mainly methane. Also known as clathrate hydrate, gas clathrate. { ¦gas 'hī,drāt }

gash vein [GEOL] A mineralized fissure that extends a short distance vertically. { 'gash ,vān }

gasification [CHEM ENG] Any chemical or heat process used to convert a substance to a gas; coal is converted by the Hygas process to a gaseous fuel. [MIN ENG] A method for exploiting poor-quality coal and thin coal seams by burning the coal in place to produce combustible gas which can be burned to generate power or processed into chemicals and fuels. Also known as underground gasification. { 'gas·ə·fə'kā·shən }

gasifier [CHEM ENG] A unit for producing gas, particularly synthesis gas from coal. { 'gas·ə,fī·ər }

gas ignition [MIN ENG] The setting on fire of an accumulation of firedamp in a coal mine. { 'gas ig'nish·ən }

gas injection [MECH ENG] Injection of gaseous fuel into the cylinder of an internal combustion engine at the appropriate part of the cycle. [PETRO ENG] The injection of gas into a reservoir to maintain formation pressure and to drive liquid hydrocarbons toward the wellbores. { 'gas in,jek·shən }

gas-injection well [PETRO ENG] A well hole in a reservoir into which pressurized gas is injected to maintain formation pressure or to drive liquid hydrocarbons toward other well holes. Also known as gas-input well. { 'gas in,jek·shən ,wel }

gas-input well See gas-injection well. { 'gas ,in,pút ,wel }

gas-insulated substation [ELEC] An electric power substation in which all live equipment and busbars are housed in grounded metal enclosures sealed and filled with sulfur hexafluoride gas. { 'gas ,in·sə,lād·əd 'səb,stā·shən }

gas ionization [ELECTR] Removal of the planetary electrons from the atoms of gas filling an electron tube, so that the resulting ions participate in current flow through the tube. { 'gas ,ī·ə·nə'zā·shən }

gasket [ENG] A packing made of deformable material, usually in the form of a sheet or ring, used to make a pressure-tight joint between stationary parts. Also known as static seal. { 'gas·kət }

gas kinematics [FL MECH] The motion of a gas considered by itself, without regard for the causes of motion. { 'gas ,kin·ə'mad·iks }

Gaskin's theorem [MATH] A theorem in projective geometry which states that if a circle circumscribes a triangle which is identical with its conjugate triangle with respect to a given conic, then the tangent to the circle at either of its intersections with the director circle of the conic is perpendicular to the tangent to the director circle at the same intersection. { 'gas·kinz ,thir·əm }

gas laser [OPTICS] A laser in which the active medium is a discharge in a gas contained in a glass or quartz tube with a Brewster-angle window at each end; the gas can be excited by a high-frequency oscillator or direct-current flow between electrodes inside the tube; the function of the discharge is to pump the medium, to obtain population inversion. { 'gas ,lā·zər }

gas law [THERMO] Any law relating the pressure, volume, and temperature of a gas. { 'gas ,lò }

gas-law constant See gas constant. { 'gas ,lò ,kän·stənt }

gas lens [OPTICS] An optical lens formed by a flow of gas which gives rise to gradients of refractive index that bring about the focusing of light. { 'gas ,lenz }

gas lift [CHEM ENG] Solids movement operation in which an upward-flowing gas stream in a closed conduit or vessel is used to lift and move powdered or granular solid material. [PETRO ENG] The injection of gas near the bottom of an oil well to aerate and lighten the column of oil to increase oil production from the well. { 'gas ,lift }

gas-lift mandrel [PETRO ENG] In a gas-lift well, a component installed in the tubing string to support the gas-lift valve. { 'gas ,lift 'man·drəl }

gas-lift valve [PETRO ENG] In a gas-lift well, a device on the tubing string that allows gas to be injected into the fluid, causing it to rise to the surface. { 'gas ,lift 'valv }

gas-liquid chromatography [ANALY CHEM] A form of gas chromatography in which the fixed phase (column packing) is a liquid solvent distributed on an inert solid support. Abbreviated GLC. Also known as gas-liquid partition chromatography. { 'gas ,lik·wəd ,krō·mə'täg·rə·fē }

gas-liquid partition chromatography See gas-liquid chromatography. { 'gas ,lik·wəd pär'tish·ən ,krō·mə'täg·rə·fē }

gas logging [PETRO ENG] Hot-wire-detector or gas-chromatographic analysis and record of gas contained in the mud stream and cuttings for a well being drilled; a common way of detecting subsurface oil and gas shows. { 'gas ,läg·iŋ }

gas-lubricated bearing See gas bearing. { 'gas ,lü·brə,kād·əd 'ber·iŋ }

gas magnification [ELECTR] Increase in current through a

phototube due to ionization of the gas in the tube. { 'gas ,mag·nə·fə'kā·shən }

gas making [CHEM ENG] Making water gas or air gas by the action of steam and air upon hot coke. { 'gas ,māk·iŋ }

gasman [MIN ENG] An underground official who examines the mine for firedamp and has charge of its removal. { 'gas,mən }

gas manometer [ENG] A gage for determining the difference in pressure of two gases, usually by measuring the difference in height of liquid columns in the two sides of a U-tube. { 'gas mə'näm·əd·ər }

gas maser [PHYS] A maser in which the microwave electromagnetic radiation interacts with the molecules of a gas such as ammonia; used chiefly in highly stable oscillator applications, as in atomic clocks. { 'gas ,mā·zər }

gas mask [ENG] A device to protect the eyes and respiratory tract from noxious gases, vapors, and aerosols, by removing contamination with a filter and a bed of adsorbent material. { 'gas ,mask }

gas mechanics [FL MECH] The action of forces on gases. { 'gas mə,kan·iks }

gas-metal arc welding [MET] A welding procedure that employs an electric arc to heat the joint to fusion, and an inert gas to prevent oxidation of the weld. { 'gas 'med·əl 'ärk ,weld·iŋ }

gas meter [ENG] An instrument for measuring and recording the amount of gas flow through a pipe. { 'gas ,mēd·ər }

gas munition [ORD] A munition such as a bomb, projectile, pot, candle, or spray tank containing a war gas and a means of release. { 'gas myü'nish·ən }

gasogene [MATER] A fuel gas formed by incomplete combustion of charcoal; a European development as a substitute for gasoline. Also spelled gazogene. { 'gas·ə,jēn }

gasohol [MATER] A liquid fuel made from a blend of gasoline and ethanol. { 'gas·ə,hòl }

gas oil [MATER] A petroleum distillate boiling within the general range 450–800°F (232–426°C); usually includes kerosine, diesel fuel, heating oils, and light fuel oils. { 'gas ,òil }

gas-oil contact [PETRO ENG] In a petroleum reservoir, that point or plane where the bottom of a gas sand is in contact with the top of an oil sand. { 'gas 'òil 'kän,takt }

gas-oil ratio [PETRO ENG] Approximation of oil-reservoir composition, expressed in cubic feet of gas per barrel of liquid at 14.7 psia and 60°F (15.6°C). { 'gas 'òil ,rā·shō }

gas-oil separator [PETRO ENG] An oil-field stock tank or series of tanks in which wellhead pressure is reduced so that the dissolved gas associated with reservoir oil is flashed off or separated as a separate phase. Also known as gas separator; oil-field separator; oil-gas separator; oil separator; separator. { 'gas 'òil 'sep·ə,rād·ər }

gasoline [MATER] A fuel for internal combustion engines consisting essentially of volatile flammable liquid hydrocarbons; derived from crude petroleum by processes such as distillation reforming, polymerization, catalytic cracking, and alkylation; the common name is gas. Also known as petrol. { 'gas·ə,lēn }

gasoline alkylate [MATER] A gasoline component usually formed by union of an olefin with isobutane by a refinery process known as alkylation; the product has high octane value and is blended with motor and aviation gasoline to improve antiknock value. { 'gas·ə,lēn 'al·kə,lāt }

gasoline engine [MECH ENG] An internal combustion engine that uses a mixture of air and gasoline vapor as a fuel. { 'gas·ə,lēn 'en·jən }

gasoline gel See gelatinized gasoline. { 'gas·ə,lēn 'jel }

gasoline pool [MATER] A concept that considers gasolines of different qualities as a single group for the purpose of blending to meet final product specifications. { 'gas·ə,lēn ,pül }

gasoline pump [MECH ENG] A device that pumps and measures the gasoline supplied to a motor vehicle, as at a filling station. { 'gas·ə,lēn ,pəmp }

gasometer [ENG] A piece of equipment that holds and measures gas; may be used in analytical chemistry to measure the quantity of gas evolved in a reaction. { ga'säm·əd·ər }

gasometric method [ANALY CHEM] An analytical technique for gases; the gas may be measured by instrumental methods or through chemical reactions with specific reagents. { ,gas·ə'me·trik 'meth·əd }

gas packing [IND ENG] Packing a material such as food

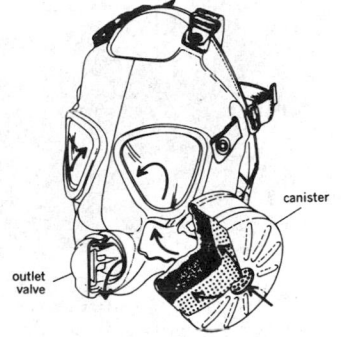

GAS MASK

A gas mask with arrows showing airflow direction.

canister

outlet valve

in an atmosphere consisting of an oxygen-free gas. { 'gas ‚pak·iŋ }

gaspar [MATER] A mixture of finely ground glass and quartz; a feldspar substitute in some applications and a hard-rubber filler. { 'ga‚spär }

gaspeite [MINERAL] $NaCO_3$ An anhydrous normal carbonate mineral with calcite structure. { ga'spē‚īt }

gas phototube [ELECTR] A phototube into which a quantity of gas has been introduced after evacuation, usually to increase its sensitivity. { 'gas 'fōd·ō‚tüb }

gas pliers [DES ENG] Pliers for gripping round objects such as pipes, tubes, and circular rods. { 'gas ‚plī·ərz }

gas pocket [GEOL] A gas-filled cavity in rocks, especially above an oil pocket. [MET] A cavity in a metal which contains trapped gases. { 'gas ‚päk·ət }

gas port [ORD] An opening for the passage of gas in the cylinder of a gas-operated automatic weapon to admit some of the propellant gases from the gun barrel. { 'gas ‚pȯrt }

gas-pressure maintenance [PETRO ENG] The maintenance of oil-reservoir gas pressure, usually by gas injection, to increase hydrocarbon recovery and to improve reservoir production characteristics. { 'gas ‚presh·ər ‚mānt·ən·əns }

gas producer [CHEM ENG] A device for complete gasification of coal by utilizing simultaneously the air and water-gas reactions. { 'gas prə‚düs·ər }

gas reservoir [GEOL] An accumulation of natural gas found with or near accumulations of crude oil in the earth's crust. { ¦gas ¦rez·əv‚wär }

gas reversion [CHEM ENG] A process which combines thermal cracking or reforming of naphtha with thermal polymerization or alkylation of hydrocarbon gases carried out in the same reaction zone. { 'gas ri'vər‚zhən }

gas rig [MIN ENG] A borehole drill, either rotary or churn type, driven by a combustion-type engine energized by a combustible liquid, such as gasoline, or a combustible gas, such as bottle gas. { 'gas ‚rig }

gas sand [GEOL] A stratum of sand or porous sandstone from which natural gas may be extracted. { 'gas ‚sand }

gas scattering [ELECTR] The scattering of electrons or other particles in a beam by residual gas in the vacuum system. { 'gas ‚skad·ə·riŋ }

gas scrubbing [CHEM ENG] Removal of gaseous or liquid impurities from a gas by the action of a liquid; the gas is contacted with the liquid which removes the impurities by dissolving or by chemical combination. { 'gas ‚skrəb·iŋ }

gas seal [ENG] A seal which prevents gas from leaking to or from a machine along a shaft. { 'gas ‚sēl }

gas sendout [PETRO ENG] The total gas that is produced, purchased, or withdrawn from underground storage in a certain interval of time. { ¦gas 'sen‚daut }

gas-sensitive field-effect transistor [ELECTR] A field-effect transistor whose gate electrode is composed of a material, such as palladium, that is sensitive to a particular gas, such as hydrogen, so that the gain of the transistor depends on the concentration of this gas. { 'gas ¦sen·səd·iv 'fēld i‚fekt tran‚zis·tər }

gas separator *See* gas-oil separator. { ¦gas 'sep·ə‚rād·ər }

Gasserian ganglion [NEUROSCI] A group of nerve cells of the sensory root of the trigeminal nerve. Also known as semilunar ganglion. { ga'ser·ē·ən 'gaŋ·glē·ən }

gas-shielded arc welding [MET] Use of a gas atmosphere to shield the molten metal from air in arc welding. { 'gas ‚shēl·dəd 'ärk ‚weld·iŋ }

gassiness [ELECTR] Presence of unwanted gas in a vacuum tube, usually in relatively small amounts, caused by the leakage from outside or evolution from the inside walls or elements of the tube. { 'gas·ē·nəs }

gassing [ELEC] The evolution of gas in the form of small bubbles in a storage battery when charging continues after the battery has been completely charged. [ENG] **1.** Absorption of gas by a material. **2.** Formation of gas pockets in a material. **3.** Evolution of gas from a material during a process or procedure. [TEXT] *See* singeing. { 'gas·iŋ }

gas slippage [FL MECH] Phenomenon of gas bypassing liquids that occurs when the diameter of capillary openings approaches the mean free path of the gas; occurs not only in capillary tubing, but in porous oil-reservoir formations. { ¦gas ¦slip·ij }

gas sniffer *See* explosimeter. { 'gas ‚snif·ər }

gas-solid chromatography [ANALY CHEM] A form of gas chromatography in which the moving phase is a gas and the stationary phase is a surface-active sorbent (charcoal, silica gel, or activated alumina). Abbreviated GSC. { ¦gas ¦säl·əd ‚krō·mə'täg·rə·fē }

gas solubility [PHYS CHEM] The extent that a gas dissolves in a liquid to produce a homogeneous system. { ¦gas ‚säl·yə'bil·əd·ē }

gas-solubility factor [PETRO ENG] The number of standard cubic feet of gas liberated under specified gas-oil separator conditions that are in solution in one barrel of stock-tank oil at reservoir temperature and pressure. { ¦gas ‚säl·yə'bil·əd·ē ‚fak·tər }

gas spurt [GEOL] An accumulation of organic matter on certain strata caused by escaping gas. { 'gas ‚spȯrt }

gas sterilization [MICROBIO] Sterilization of heat and liquid-labile materials by means of gaseous agents, such as formaldehyde, ethylene oxide, and β-propiolactone. { ¦gas ‚ster·ə·lə'zā·shən }

gas stimulation [PETRO ENG] The detonation of a nuclear explosive in the strata of a natural-gas field to make the gas flow more freely. { ¦gas ‚stim·yə'lā·shən }

gas storage [FOOD ENG] Storage in a low-oxygen atmosphere to delay ripening. { ¦gas ¦stȯr·ij }

gassy [MIN ENG] A coal mine rating by the U.S. Bureau of Mines, applicable when an ignition occurs or if a methane content exceeding 0.25% can be detected; work must be halted if the methane exceeds 1.5% in a return airway. { 'gas·ē }

gassy tube [ELECTR] A vacuum tube that has not been fully evacuated or has lost part of its vacuum due to release of gas by the electrode structure during use, so that enough gas is present to impair operating characteristics appreciably. Also known as soft tube. { 'gas·ē 'tüb }

gas tank [ENG] A tank for storing gas or gasoline. { 'gas ‚taŋk }

Gasteromycetes [MYCOL] A group of basidiomycetous fungi in the subclass Homobasidiomycetidae with enclosed basidia and with basidiospores borne symmetrically on long sterigmata and not forcibly discharged. { ¦gas·tə·rō‚mī'sēd·ēz }

Gasterophilidae [INV ZOO] The horse bots, a family of myodarian cyclorrhaphous dipteran insects in the subsection Calypteratae, including individuals that cause myiasis in horses. { ¦gas·tə·rō'fil·ə‚dē }

Gasterophilus [INV ZOO] A large genus of botflies in the family Gasterophilidae. { ‚gas·tə'rä·fə·ləs }

Gasterosteidae [VERT ZOO] The sticklebacks, a family of actinopterygian fishes in the order Gasterosteiformes. { ‚gas·tə·rō'stē·ə‚dē }

Gasterosteiformes [VERT ZOO] An order of actinopterygian fishes characterized by a ductless swim bladder, a pelvic fin that is abdominal to subthoracic in position, and an elongate snout. { ‚gas·tə‚rä·stē·ə'fȯr‚mēz }

Gasteruptiidae [INV ZOO] A family of hymenopteran insects in the superfamily Proctotrupoidea. { ‚gas·tə‚rəp'tī·ə‚dē }

gas tetrode *See* tetrode thyratron. { ¦gas 'te‚trōd }

gas thermometer [ENG] A device to measure temperature by measuring the pressure exerted by a definite amount of gas enclosed in a constant volume; the gas (preferably hydrogen or helium) is enclosed in a glass or fused-quartz bulb connected to a mercury manometer. Also known as constant-volume gas thermometer. { ¦gas thər'mäm·əd·ər }

gas thermometry [ENG] Measurement of temperatures with a gas thermometer; used with helium down to about 1 K. { ¦gas thər'mäm·ə·trē }

gas thermostatic switch [ELEC] A thermostatic switch in which heat causes the pressure of gas in a sealed metal bellows to increase, thereby moving the bellows and closing the contacts of a switch. { ¦gas 'thər·mə‚stad·ik 'swich }

gas tracer [MIN ENG] Dust clouds, chemical smoke, or gaseous or radioactive tracers used to detect slowly moving air currents in a mine. { ¦gas 'trā·sər }

gastralium [INV ZOO] A microsclere located just beneath the inner cell layer of hexactinellid sponges. [VERT ZOO] One of the riblike structures in the abdomen of certain reptiles. { ga'strā·lē·əm }

gas trap [CIV ENG] A bend or chamber in a drain or sewer pipe that prevents sewer gas from escaping. { 'gas ‚trap }

gas-treating system [CHEM ENG] A process system to remove nonhydrocarbon impurities (such as water vapor, hydrogen sulfide, or carbon dioxide) from wellhead gas. { 'gas ,trēd·iŋ ,sis·təm }

gastrectomy [MED] Surgical removal of all or part of the stomach. { ga'strek·tə·mē }

gastric acid [BIOCHEM] Hydrochloric acid secreted by parietal cells in the fundus of the stomach. { 'gas·trik 'as·əd }

gastric cecum [INV ZOO] One of the elongated pouchlike projections of the upper end of the stomach in insects. { 'gas·trik 'sē·kəm }

gastric enzyme [BIOCHEM] Any digestive enzyme secreted by cells lining the stomach. { 'gas·trik 'en,zīm }

gastric filament [INV ZOO] In scyphozoans, a row of filaments on the surface of the gastric cavity which function to kill or paralyze live prey taken into the stomach. Also known as phacella. { 'gas·trik 'fil·ə·mənt }

gastric gland [ANAT] Any of the glands in the wall of the stomach that secrete components of the gastric juice. { 'gas·trik ,gland }

gastric hypothermia [MED] Cooling of the upper digestive tract; useful in the management of bleeding disorders. { 'gas·trik ,hī·pō'thər·mē·ə }

gastric juice [PHYSIO] The digestive fluid secreted by gastric glands; contains gastric acid and enzymes. { 'gas·trik ,jüs }

gastric mill [INV ZOO] A grinding apparatus consisting of calcareous or chitinous plates in the pharynx or stomach of certain invertebrates. { 'gas·trik ,mil }

gastric ostium [INV ZOO] The opening into the gastric pouch in scyphozoans. { 'gas·trik 'äs·tē·əm }

gastric pouch [INV ZOO] One of the pouchlike diversions of a scyphozoan stomach. { 'gas·trik 'paůch }

gastric shield [INV ZOO] A thickening of the stomach wall in some mollusks for mixing the contents. { 'gas·trik 'shēld }

gastric ulcer [MED] An ulcer of the mucous membrane of the stomach. { 'gas·trik 'əl·sər }

gastrin [BIOCHEM] A polypeptide hormone secreted by the pyloric mucosa which stimulates the pancreas to release pancreatic fluid and the stomach to release gastric acid. { 'gas·trən }

gastritis [MED] Inflammation of the stomach. { ga'strīd·əs }

gastroanastomosis [MED] The surgical formation of a communication between the two pouches of the stomach. Also known as gastrogastrostomy. { ,ga·strō·ə,nas·tə'mō·səs }

gastroblast [INV ZOO] A feeding zooid of a tunicate colony. { 'ga·strə,blast }

gastrocnemius [ANAT] A large muscle of the posterior aspect of the leg, arising by two heads from the posterior surfaces of the lateral and medial condyles of the femur, and inserted with the soleus muscle into the calcaneal tendon, and through this into the back of the calcaneus. { ga,sträk'nē·mē·əs }

gastrocoele See archenteron. { 'ga·strə,sēl }

gastrodermis [INV ZOO] The cellular lining of the digestive cavity of certain invertebrates. { ,ga·strō'dər·məs }

gastroduodenitis [MED] Inflammation of the stomach and duodenum. { ,ga·strō·dü,äd·ən'īd·əs }

gastroenteritis [MED] Inflammation of the mucosa of the stomach and intestine. { ,ga·strō,ent·ə'rīd·əs }

gastroenterology [MED] The branch of medicine concerned with study of the stomach and intestine. { ,ga·strō,ent·ə'räl·ə·jē }

gastroenterostomy [MED] Surgical formation of a connection between the stomach and small intestine. { ,ga·strō,ent·ə'räs·tə·mē }

gastroepiploic artery [ANAT] Either of two arteries arising from the gastroduodenal and splenic arteries respectively and forming an anastomosis along the greater curvature of the stomach. { ,ga·strō,ep·ə,plóik 'ärd·ə·rē }

gastrogastrostomy See gastroanastomosis. { ,ga·strō ,ga'sträs·tə·mē }

gastrointestinal hormone [BIOCHEM] Any hormone secreted by the gastrointestinal system. { ,ga·strō,in'tes·tən·əl 'hór,mōn }

gastrointestinal system [ANAT] The portion of the digestive system including the stomach, intestine, and all accessory organs. { ,ga·strō,in'tes·tən·əl 'sis·təm }

gastrointestinal tract [ANAT] The stomach and intestine. { ,ga·strō,in'tes·tən·əl 'trakt }

gastrojejunostomy [MED] Surgical establishment of an anastomosis between the jejunum and the anterior or posterior wall of the stomach. { ,ga·strō,jē·jə'nä·stə·mē }

gastrolith [VERT ZOO] A pebble swallowed by certain animals and retained in the gizzard or stomach, where it serves to grind food. { 'ga·strə,lith }

gastrolysis [MED] The breaking up of adhesions between the stomach and adjacent organs. { ga'sträl·ə·səs }

Gastromyzontidae [VERT ZOO] A small family of actinopterygian fishes of the suborder Cyprinoidei found in southeastern Asia. { ,ga·strō,mī'zän·tə,dē }

gastropexy [MED] The fixation of a prolapsed stomach in its normal position by suturing it to the abdominal wall or other structure. { 'ga·strə,pek·sē }

gastroplication [MED] An operation for relief of chronic dilation of the stomach by suturing a large horizontal fold in the stomach wall. { ,ga·strō·plə'kā·shən }

Gastropoda [INV ZOO] A large, morphologically diverse class of the phylum Mollusca, containing the snails, slugs, limpets, and conchs. { ga'sträp·ə·də }

gastropore [INV ZOO] A pore containing a gastrozooid in hydrozoan corals. { 'ga·strə,pór }

gastroptosis [MED] Prolapse or downward displacement of the stomach. { ,ga,sträp'tō·səs }

gastroscope [MED] A hollow, tubular instrument used to examine the inside of the stomach by passage through the mouth and esophagus. { 'ga·strə,skōp }

gastrosplenic ligament [ANAT] The fold of peritoneum passing from the stomach to the spleen. Also known as gastrosplenic omentum. { ,ga·strō'splen·ik 'lig·ə·mənt }

gastrosplenic omentum See gastrosplenic ligament. { ,ga·strō'splen·ik ō'men·təm }

gastrostome [INV ZOO] The opening of a gastropore. { 'ga·strə,stōm }

gastrostomy [MED] The establishment of a fistulous opening into the stomach, with an external opening in the skin; used for artificial feeding. { ga'sträs·tə·mē }

gastrostyle [INV ZOO] A spiculated projection that extends into the gastrozooid from the base of the gastropore. { 'ga·strə,stīl }

Gastrotricha [INV ZOO] A group of microscopic, pseudocoelomate animals considered either to be a class of the Aschelminthes or to constitute a separate phylum. { ,ga'strä·trə·kə }

gastrozooid [INV ZOO] A nutritive polyp of colonial coelenterates, characterized by having tentacles and a mouth. { ,ga·strə'zō,óid }

gastrula [EMBRYO] The stage of development in animals in which the endoderm is formed and invagination of the blastula has occurred. { 'ga·strə·lə }

gastrulation [EMBRYO] The process by which the endoderm is formed during development. { ,ga·strə'lā·shən }

gas tube [ELECTR] An electron tube into which a small amount of gas or vapor is admitted after the tube has been evacuated; ionization of gas molecules during operation greatly increases current flow. { 'gas ,tüb }

gas-tube boiler See waste-heat boiler. { 'gas ,tüb 'bóil·ər }

gas-tungsten arc welding [MET] Gas-metal arc welding in which tungsten is the agent of fusion. { 'gas 'təŋ·stən 'ärk ,weld·iŋ }

gas turbine [MECH ENG] A heat engine that converts the energy of fuel into work by using compressed, hot gas as the working medium and that usually delivers its mechanical output power either as torque through a rotating shaft (industrial gas turbines) or as jet power in the form of velocity through an exhaust nozzle (aircraft gas engines). Also known as combustion turbine. { 'gas ,tər·bən }

gas-turbine nozzle [MECH ENG] The component of a gas turbine in which the hot, high-pressure gas expands and accelerates to high velocity. { 'gas ,tər·bən ,näz·əl }

gas vacuole [BIOL] A membrane-bound, gas-filled cavity in some algae and protozoans; thought to control buoyancy. { 'gas 'vak·yə,wól }

gas vacuum breakdown [ELECTR] Ionization of residual gas in a vacuum, causing reverse conduction in an electron tube. { 'gas ,vak·yəm 'brāk,daůn }

gas valve [ENG] An exhaust valve, held shut by rubber

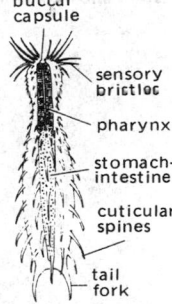

GASTROTRICHA

buccal capsule

sensory bristles

pharynx

stomach-intestine

cuticular spines

tail fork

Chaetonotus, a freshwater gastrotrich.

springs, used to discharge gas from the extreme top of a balloon. { 'gas ‚valv }

gas vent [ENG] A pipe or hole that allows gas to pass off. { 'gas ‚vent }

gas viscosity [FL MECH] The internal fluid function of a gas. { 'gas vi‚skäs·əd·ē }

gas welding [MET] A welding process in which metals are joined by the heat of an oxyacetylene flame. { 'gas ‚weld·iŋ }

gas well [PETRO ENG] A well drilled for extraction of natural gas from a gas reservoir. { 'gas ‚wel }

gas zone [GEOL] A rock formation containing gas under a pressure large enough to force the gas out if tapped from the surface. { 'gas ‚zōn }

gat [NAV] A natural or artificial passage or channel extending inland through shoals or steep banks. { gat }

GAT See Greenwich apparent time.

gate [CIV ENG] A movable barrier across an opening in a large barrier, a fence, or a wall. [ELECTR] **1.** A circuit having an output and a multiplicity of inputs and so designed that the output is energized only when a certain combination of pulses is present at the inputs. **2.** A circuit in which one signal, generally a square signal, serves to switch another signal on and off. **3.** One of the electrodes in a field-effect transistor. **4.** An output element of a cryotron. **5.** To control the passage of a pulse or signal. **6.** In radar, an electric waveform which is applied to the control point of a circuit to alter the mode of operation of the circuit at the time when the waveform is applied. Also known as gating waveform. [ENG] **1.** A device, such as a valve or door, for controlling the passage of materials through a pipe, channel, or other passageway. **2.** A device for positioning the film in a camera, printer, or projector. [GRAPHICS] The area or component in which the film is held at a fixed relationship to a lens. [MET] The opening in a casting mold through which molten metal enters the cavity. Also known as in-gate. [NAV] The position on the extension of the axis of a runway in use above which an aircraft heading toward that runway is required to pass at a time assigned by proper control authority. [NUCLEO] A movable barrier of shielding material used for closing a hole in a nuclear reactor. [ORD] A metal part in the rear of the cylinder of old-pattern revolvers that was turned out to expose the cylinder for loading. { gāt }

gate-array device [ELECTR] An integrated logic circuit that is manufactured by first fabricating a two-dimensional array of logic cells, each of which is equivalent to one or a few logic gates, and then adding final layers of metallization that determine the exact function of each cell and interconnect the cells to form a specific network when the customer orders the device. { gāt ə‚rā di‚vīs }

gate-controlled rectifier [ELECTR] A three-terminal semiconductor device, such as a silicon controlled rectifier, in which the unidirectional current flow between the rectifier terminals is controlled by a signal applied to a third terminal called the gate. { gāt kən‚trōld 'rek·tə‚fī·ər }

gate-controlled switch [ELECTR] A semiconductor device that can be switched from its nonconducting or "off" state to its conducting or "on" state by applying a negative pulse to its gate terminal and that can be turned off at any time by applying reverse drive to the gate. Abbreviated GCS. { gāt kən‚trōld 'swich }

gate conveyor [MIN ENG] A conveyor that carries coal from one source or face only, that is, from a single-unit or double-unit face. { 'gāt kən‚vā·ər }

gated-beam tube [ELECTR] A pentode electron tube having special electrodes that form a sheet-shaped beam of electrons; this beam may be deflected away from the anode by a relatively small voltage applied to a control electrode, thus giving extremely sharp cutoff of anode current. { ‚gād·əd ‚bēm ‚tüb }

gated sweep [ELECTR] Sweep in which the duration as well as the starting time is controlled to exclude undesired echoes from the indicator screen. { ‚gād·əd 'swēp }

gate equivalent circuit [ELECTR] A unit of measure for specifying relative complexity of digital circuits, equal to the number of individual logic gates that would have to be interconnected to perform the same function as the digital circuit under evaluation. { gāt i‚kwiv·ə·lənt ‚sər·kət }

gate generator [ELECTR] A circuit used to generate gate pulses; in one form it consists of a multivibrator having one stable and one unstable position. { gāt jen·ə·rād·ər }

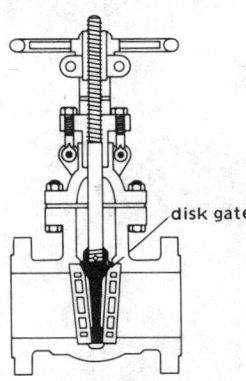

GATE VALVE

disk gate

Gate valve with disk gate; rising threaded stem shows when valve is open.

gate interlock [MIN ENG] A system that prevents movement of shaft conveyances or transmission of action signals until all shaft gates are closed. { gāt 'in·tər‚läk }

gate multivibrator [ELECTR] Rectangular-wave generator designed to produce a single positive or negative gate voltage when triggered and then to become inactive until the next trigger pulse. { gāt ‚məl·ti'vī‚brād·ər }

gate pulse [ELECTR] A pulse that triggers a gate circuit so it will pass a signal. { gāt ‚pəls }

gate road bunker [MIN ENG] An appliance for coal storage from the face conveyors during peaks of production or during a stoppage of the outby transport. { gāt 'rōd ‚bəŋ·kər }

Gates crusher [MECH ENG] A gyratory crusher which has a cone or mantle that is moved eccentrically by the lower bearing sleeve. { 'gāts 'krəsh·ər }

gate turnoff [ELECTR] A pnpn switching device comparable to a silicon-controlled rectifier, but having a more complex gate structure that permits easy and fast turnoff as well as turn-on from its gate input terminal, at frequencies up to 100 kilohertz. { ‚gāt ‚tər‚nóf }

gate-turnoff silicon-controlled rectifier [ELECTR] A silicon-controlled rectifier that can be turned off by applying a current to its gate; used largely for direct-current switching, because turnoff can be achieved in a fraction of a microsecond. { ‚gāt ‚tər‚nóf 'sil·ə·kən kən‚trōld 'rek·tə‚fī·ər }

gate valve [MECH ENG] A valve with a disk-shaped closing element that fits tightly over an opening through which water passes. { 'gāt ‚valv }

gateway [COMMUN] A point of entry and exit to another system, such as the connection point between a local-area network and an external-communications network. { 'gāt‚wā }

gate winding [ELECTR] A winding used in a magnetic amplifier to produce on-off action of load current. { 'gāt ‚wīnd·iŋ }

gather [GRAPHICS] In bookbinding, to arrange in sequence the folded sheets or signatures of a book. [MIN ENG] **1.** To assemble loaded cars from several production points and deliver them to main haulage for transport to the surface or pit bottom. **2.** To drive a heading through disturbed or faulty ground so as to meet the seam of coal at a convenient level or point on the opposite side. { 'gath·ər }

gathering area [PETRO ENG] The area, usually down the regional dip from a hydrocarbon trap, from which the oil or gas may have migrated updip into the trap. { 'gath·ə·riŋ ‚er·ē·ə }

gathering arm loader [MIN ENG] A machine for loading loose rock or coal; has a tractor-mounted chassis and carries a chain conveyor whose front end is built into a wedge-shaped blade. { 'gath·ə·riŋ ‚ärm ‚lōd·ər }

gathering conveyor [MIN ENG] Any conveyor which gathers coal from other conveyors and delivers it either into mine cars or onto another conveyor. { 'gath·ə·riŋ kən‚vā·ər }

gathering ground See drainage basin. { 'gath·ə·riŋ ‚graúnd }

gathering iron [ENG] A rod used to collect molten glass for glassblowing. { 'gath·ə·riŋ ‚ī·ərn }

gathering line [PETRO ENG] A special pipeline, frequently small in diameter, used to transport crude oil or gas from the field to the main pipeline. { 'gath·ə·riŋ ‚līn }

gathering locomotive See gathering motor. { 'gath·ə·riŋ ‚lō·kə‚mōd·iv }

gathering mine locomotive See gathering motor. { 'gath·ə·riŋ 'mīn ‚lō·kə‚mōd·iv }

gathering motor [MIN ENG] A lightweight type of electric locomotive used to haul loaded cars from the working places to the main haulage road and to replace them with empties. Also known as electric gathering locomotive; gathering locomotive; gathering mine locomotive. { 'gath·ə·riŋ ‚mōd·ər }

gathering motorman [MIN ENG] In bituminous coal mining, one who operates a mine locomotive to haul loaded mine cars from working places to sidings. { 'gath·ə·riŋ ‚mōd·ər·mən }

gathering mule [MIN ENG] The mule used to collect the loaded cars from the separate working places and to return empties. { 'gath·ə·riŋ ‚myül }

gathering pallet [HOROL] A finger that moves the rack in a striking timepiece for each blow struck. { 'gath·ə·riŋ ‚pal·ət }

gathering pump [MIN ENG] A portable or semiportable pump that is required for removing water encountered while opening a new mine, for extending headings or entries in an operating mine, for pump rooms or rib sections lying in the dip, for collecting water from local pools, or for sinking a shaft. { 'gath·ə·riŋ ‚pəmp }

gathering ring [ENG] A clay ring placed on molten glass to collect impurities and thus permit high-quality glass to be taken from the center. { 'gath·ə·riŋ ‚riŋ }

gathering system [PETRO ENG] The array of gathering lines and supporting equipment needed to transport gas and oil from the fields to the main pipeline. { 'gath·ə·riŋ ‚sis·təm }

gather way [NAV] To attain headway. { 'gath·ər 'wā }

gather write [COMPUT SCI] An operation that creates a single output record from data items gathered from nonconsecutive locations in main memory. { ‚gath·ər 'wrīt }

gating [ELECTR] The process of selecting those portions of a wave that exist during one or more selected time intervals or that have magnitudes between selected limits. [ENG] A network of connecting channels, including sprues, runners, gates, and cavities, which conduct molten metal to the mold. { 'gād·iŋ }

gating waveform See gate. { 'gād·iŋ 'wāv‚fȯrm }

Gatling gun [ORD] Type of machine gun using multibarrels which fire in rotation; it permits very rapid fire because it reduces the problem of overheating in a single barrel. { 'gat·liŋ ‚gən }

Gatterman-Koch synthesis [ORG CHEM] A synthesis of aldehydes; aldehydes form when an aromatic hydrocarbon is heated in the presence of hydrogen chloride, certain metallic chloride catalysts, and either carbon monoxide or hydrogen cyanide. { 'gäd·ər‚män 'kōk ‚sin·thə·səs }

Gatterman reaction [ORG CHEM] **1.** Reaction of a phenol or phenol ester, and hydrogen chloride or hydrogen cyanide, in the presence of a metallic chloride such as aluminum chloride to form, after hydrolysis, an aldehyde. **2.** Reaction of an aqueous ethanolic solution of diazonium salts with precipitated copper powder or other reducing agent to form diaryl compounds. { 'gäd·ər‚män rē‚ak·shən }

Gaucher's cells [PATH] Abnormal macrophages associated with Gaucher's disease. { gō'shāz ‚selz }

Gaucher's disease [MED] A rare chronic, probably hereditary disease in which cells loaded with cerebrosides become localized in reticuloendothelial tissue and eventually cause tissue destruction; manifestations include enlargement of the spleen, bronzing of the skin, and anemia. Also known as cerebroside lipoidosis; familial splenic anemia. { gō'shāz di‚zēz }

gaufrage See plaiting. { gō'fräzh }

gauge [ELECTROMAG] One of the family of possible choices for the electric scalar potential and magnetic vector potential, given the electric and magnetic fields. { gāj }

gauge boson [PHYS] A massless spin-1 particle, such as the photon and gluons, whose existence is required by gauge invariance in a gauge theory; such particles can acquire mass through spontaneous symmetry breaking, as in the case of intermediate vector bosons. Also known as gauge particle. { 'gāj ‚bō‚sän }

gauge-fixing term [QUANT MECH] A term added to a Lagrangian in quantum field theory that breaks gauge invariance. { 'gāj ‚fiks·iŋ ‚tərm }

gauge group [PHYS] The group of gauge transformations in a gauge theory. { 'gāj ‚grüp }

gauge invariance [ELECTROMAG] The invariance of electric and magnetic fields and electrodynamic interactions under gauge transformations. [PHYS] The invariance of any field theory under gauge transformations. [QUANT MECH] An invariance of a Lagrangian based on an internal gauge group, such as U(1) for electromagnetism or U(1) × SU(2) for the Weinberg-Salam unified model of weak and electromagnetic interactions. { 'gāj in'ver·ē·əns }

gauge particle See gauge boson. { 'gāj ‚pärd·ə·kəl }

gauge theory [PHYS] Any field theory in which, as the result of the conservation of some quantity, it is possible to perform a transformation in which the phase of the fields is altered by a function of space and time without altering any measurable physical quantity, so that the fields obtained by any such transformation give a valid description of a given physical situation. { 'gāj ‚thē·ə·rē }

gauge transformation [ELECTROMAG] The addition of the gradient of some function of space and time to the magnetic vector potential, and the addition of the negative of the partial derivative of the same function with respect to time, divided by the speed of light, to the electric scalar potential; this procedure gives different potentials but leaves the electric and magnetic fields unchanged. [PHYS] An alteration of the phase of the fields of a gauge theory as a function of space and time which does not alter the value of any measurable physical quantity. { 'gaj tranz·fər'mā·shən }

gauging See gaging. { 'gāj·iŋ }

Gaultheria [BOT] A genus of upright or creeping evergreen shrubs (Ericaceae). { gȯl'thir·ē·ə }

gaultheria oil See methyl salicylate. { gȯl'thir·ē·ə ‚ȯil }

Gause's principle [ECOL] A statement that two species cannot occupy the same niche simultaneously. Also known as competitive-exclusion principle. { 'gau̇z·əz ‚prin·sə·pəl }

gauss [ELECTROMAG] Unit of magnetic induction in the electromagnetic and Gaussian systems of units, equal to 1 maxwell per square centimeter, or 10^{-4} weber per square meter. Also known as abtesla (abt). { gau̇s }

Gauss A position See end-on position. { 'gau̇s ‚ā pə‚zish·ən }

Gauss-Bonnet theorem [MATH] The theorem that the Euler characteristic of a compact Riemannian surface is $1/(2\pi)$ times the integral over the surface of the Gaussian curvature. { ‚gau̇s bə'nā ‚thir·əm }

Gauss B position See broadside-on position. { ‚gau̇s ‚bē pə‚zish·ən }

Gauss-Codazzi equations [MATH] Equations dealing with the components of the fundamental tensor and Riemann-Christoffel tensor of a surface. { ‚gau̇s kō'dat·sē i‚kwā·zhənz }

Gauss' error curve See normal distribution. { 'gau̇s 'er·ər ‚kərv }

Gauss eyepiece [OPTICS] A Ramsden eyepiece which has a thin glass plate between the two lenses, making an angle of 45° with the optical axis; used to set a telescope perpendicular to a plane reflecting surface. { 'gau̇s 'ī‚pēs }

Gauss formulas [MATH] Formulas dealing with the sine and cosine of angles in a spherical triangle. Also known as Delambre analogies. { 'gau̇s ‚fȯr·myə·ləz }

Gauss' hypergeometric equation [MATH] The differential equation, arising in many physical contexts, $x(1 - x)y'' + [c - (a + b + 1)x]y' - aby = 0$. { 'gau̇s ‚hī·pər‚jē·ə'me·trik i'kwā·zhən }

Gaussian beam [ELECTROMAG] A beam of electromagnetic radiation whose wave front is approximately spherical at any point along the beam and whose transverse field intensity over any wave front is a Gaussian function of the distance from the axis of the beam. { 'gau̇s·ē·ən 'bēm }

Gaussian complex integers [MATH] Complex numbers whose real and imaginary parts are both integers. { ‚gau̇·sē·ən ‚käm‚pleks 'int·ə·jərz }

Gaussian constant [ASTRON] The acceleration caused by the attraction of the sun at the mean distance of the earth from the sun. { ‚gau̇·sē·ən 'kän·stənt }

Gaussian curvature [MATH] The invariant of a surface specified by Gauss' theorem. Also known as total curvature. { ‚gau̇·sē·ən 'kər·və·chər }

Gaussian curve [STAT] The bell-shaped curve corresponding to a population which has a normal distribution. Also known as normal curve. { ‚gau̇·sē·ən 'kərv }

Gaussian distribution See normal distribution. { ‚gau̇·sē·ən ‚dis·trə'byü·shən }

Gaussian elimination [MATH] A method of solving a system of n linear equations in n unknowns, in which there are first $n - 1$ steps, the mth step of which consists of subtracting a multiple of the mth equation from each of the following ones so as to eliminate one variable, resulting in a triangular set of equations which can be solved by back substitution, computing the nth variable from the nth equation, the $(n - 1)$st variable from the $(n - 1)$st equation, and so forth. { ‚gau̇·sē·ən ə‚lim·ə'nā·shən }

Gaussian error [NAV] Deviation of a magnetic compass due to transient magnetism which remains in a vessel's structure for short periods after the inducing force has been removed, usually appearing after a vessel has been on the same heading for a considerable time. { ‚gau̇·sē·ən 'er·ər }

Gaussian integer [MATH] A complex number whose real and imaginary parts are both ordinary (real) integers. Also known as complex integer. { ‚gau̇s·ē·ən 'int·ə·jər }

Gaussian noise [COMMUN] Noise that has a frequency distribution which follows the Gaussian curve. [MATH] *See* Wiener process. { ¦gaů·sē·ən 'nȯiz }

Gaussian noise generator [ELECTR] A signal generator that produces a random noise signal whose frequency components have a Gaussian distribution centered on a predetermined frequency value. { ¦gaů·sē·ən 'nȯiz ‚jen·ə‚rād·ər }

Gaussian optics [OPTICS] An approximation which describes rays which are very close to the axis of an optical system and are nearly parallel to this axis, so that only the linear terms of Taylor series for the distance of a point from the axis or the angle which a ray makes with the axis need be considered. Also known as first-order theory. { ¦gaů·sē·ən 'äp·tiks }

Gaussian pulse [PHYS] A pulse for which the graph of intensity as a function of time is a Gaussian curve. { ¦gaů·sē·ən 'pəls }

Gaussian reduction [MATH] A procedure of simplification of the rows of a matrix which is based upon the notion of solving a system of simultaneous equations. Also known as Gauss-Jordan elimination. { ¦gaů·sē·ən ri'dək·shən }

Gaussian representation *See* spherical image. { ¦gaůs·ē·ən ‚rep·rə·zen'tā·shən }

Gaussian system [ELECTROMAG] A combination of the electrostatic and electromagnetic systems of units (esu and emu), in which electrostatic quantities are expressed in esu and magnetic and electromagnetic quantities in emu, with appropriate use of the conversion constant c (the speed of light) between the two systems. Also known as Gaussian units. { ¦gaů·sē·ən ‚sis·təm }

Gaussian units *See* Gaussian system. { ¦gaů·sē·ən 'yü·nəts }

Gaussian weighing method [ENG] A method used to determine the accuracy of equal-arm balances and to test standard weights in which the sample is placed on one pan and the comparative weights on the other, and then the weights are interchanged in a second weighing. { 'gaůs·ē·ən 'wā·iŋ ‚meth·əd }

Gaussian year [ASTRON] The period, according to Kepler's laws, of a body of negligible mass traveling in an orbit about the sun whose semimajor axis is 1 astronomical unit; equal to about 365.2569 days. { 'gaůs·ē·ən 'yir }

Gauss image point [OPTICS] A point through which pass all paraxial rays from a specified point source in an optical system. { ¦gaůs 'im·ij ‚pȯint }

Gauss-Jordan elimination *See* Gaussian reduction. { ¦gaůs ¦jȯrd·ən ə‚lim·ə'nā·shən }

Gauss' law of flux [ELEC] The law that the total electric flux which passes out from a closed surface equals (in rationalized units) the total charge within the surface. { ¦gaůs ‚lȯ əv 'fləks }

Gauss' law of the arithmetic mean [MATH] The law that a harmonic function can attain its maximum value only on the boundary of its domain of definition, unless it is a constant. { 'gaůs ‚lȯ əv thə ‚a·rith¦med·ik 'mēn }

Gauss-Legendre rule [MATH] An approximation technique of definite integrals by a finite series which uses the zeros and derivatives of the Legendre polynomials. { ¦gaůs lə'zhän·drə ‚rül }

Gauss lens system *See* Celor lens system. { ¦gaůs 'lenz ‚sis·təm }

Gauss' mean value theorem [MATH] The value of a harmonic function at a point in a planar region is equal to its integral about a circle centered at the point. { ¦gaůs 'mēn ‚val·yü ‚thir·əm }

gaussmeter [ENG] A magnetometer whose scale is graduated in gauss or kilogauss, and usually measures only the intensity, and not the direction, of the magnetic field. { 'gaůs‚mēd·ər }

Gauss method of weighing *See* double weighing. { ¦gaůs ¦meth·əd əv 'wā·iŋ }

Gauss objective lens *See* Celor lens system. { ¦gaůs əb¦jek·tiv 'lenz }

Gauss point *See* cardinal point. { ¦gaůs ‚pȯint }

Gauss positions [ELECTROMAG] The Gauss A and B positions; that is, a point on the axis of a bar magnet is in Gauss A position, and a point on the magnetic equator of the magnet is in Gauss B position, with respect to the magnet. { 'gaůs pə‚zish·ənz }

Gauss' principle of least constraint [MECH] The principle that the motion of a system of interconnected material points subjected to any influence is such as to minimize the constraint on the system; here the constraint, during an infinitesimal period of time, is the sum over the points of the product of the mass of the point times the square of its deviation from the position it would have occupied at the end of the time period if it had not been connected to other points. { 'gaůs 'prin·sə·pəl əv ¦lēst kən'strānt }

Gauss-Seidel method *See* Seidel method. { ¦gaůs 'zīd·əl ‚meth·əd }

Gauss test [MATH] In an infinite series with general term a_n, if $a_{n+1}/a_n = 1 - (x/n) - [f(n)/n^\lambda]$ where x and λ are greater than 1, and $f(n)$ is a bounded integer function, then the series converges. { 'gaůs ‚test }

Gauss' theorem [MATH] **1.** The assertion, under certain light restrictions, that the volume integral through a volume V of the divergence of a vector function is equal to the surface integral of the exterior normal component of the vector function over the boundary surface of V. Also known as divergence theorem; Green's theorem in space; Ostrogradski's theorem. **2.** At a point on a surface the product of the principal curvatures is an invariant of the surface, called the Gaussian curvature. { 'gaůs ‚thir·əm }

gauze [MATER] **1.** A sheer, loosely woven textile fabric similar to cheesecloth; used for surgical dressings. **2.** A plastic or wire mesh. { gȯz }

gauze tones *See* howling tones. { 'gȯz ‚tōnz }

gavage [MED] The administration of nourishment through a stomach tube. { gə'väzh }

gavial [VERT ZOO] The name for two species of reptiles composing the family Gavialidae. { 'gā·vē·əl }

Gavialidae [VERT ZOO] The gavials, a family of reptiles in the order Crocodilia distinguished by an extremely long, slender snout with an enlarged tip. { ‚gā·vē'al·ə‚dē }

Gaviidae [VERT ZOO] The single, monogeneric family of the order Gaviiformes. { gə'vī·ə‚dē }

Gaviiformes [VERT ZOO] The loons, a monofamilial order of diving birds characterized by webbed feet, compressed, bladelike tarsi, and a heavy, pointed bill. { ‚gā·vē·ə'fȯr‚mēz }

Gay-Lussac acid [MATER] Product of a Gay-Lussac tower in the chamber process for sulfuric acid manufacture; a mixture of sulfuric acid and nitrogen oxides. { ‚gā·lů‚säk 'as·əd }

Gay-Lussac's first law *See* Charles' law. { ‚gā·lů‚säks 'fərst ‚lȯ }

Gay-Lussac's law of volumes *See* combining-volumes principle. { ‚gā·lů‚säks ‚lȯ əv 'väl·yəmz }

Gay-Lussac's second law [THERMO] The law that the internal energy of an ideal gas is independent of its volume. { ‚gā·lů‚säks 'sek·ənd ‚lȯ }

Gay-Lussac tower [CHEM ENG] A component part in the chamber process for sulfuric acid production that absorbs nitrogen oxides to form nitrous vitriol. { ‚ga·lů‚säk 'taů·ər }

gaylussite [MINERAL] $Na_2Ca(CO_3)_2 \cdot 5H_2O$ A translucent, yellowish-white hydrous carbonate mineral, with a vitreous luster, crystallizing in the monoclinic system; found in dry lakes. { 'gā·lə‚sīt }

gazogene *See* gasogene. { 'gaz·ə‚jēn }

gc *See* gigahertz.

g-cal *See* calorie. { 'jē‚kal }

GCA radar *See* ground-controlled approach radar. { ¦jē¦sē¦ā 'rā‚där }

gcd *See* greatest common divisor.

G center *See* N center. { 'jē ‚sen·tər }

GCI radar *See* ground-controlled intercept radar. { ¦jē¦sē¦ī 'rā‚där }

g-cm *See* gram-centimeter.

g-completeness *See* geodesic completeness. { ¦jē kəm'plēt·nəs }

g constant [SOLID STATE] The ratio of the induced electric field in a piezoelectric material to the applied force that produces this field. { 'jē ‚kän·stənt }

GCS *See* gate-controlled switch.

Gd *See* gadolinium.

GDCH *See* α-dichlorohydrin.

GDC survey [MIN ENG] A special density log that includes simultaneous gamma-ray, density, and caliper logs. { ¦jē‚d·ē¦sē 'sər‚vā }

GDG *See* generation data group.

GAVIAL

Indian gavial.

G display [ELECTR] A rectangular radar display in which horizontal and vertical aiming errors are indicated by horizontal and vertical displacement, respectively, and range is indicated by the length of wings appearing on the blip, with length increasing as range decreases. { 'jē di,splā }

gDNA *See* genomic deoxyribonucleic acid.

G drift [NAV] A drift component in gyros and accelerometers which is proportional to the nongravitational acceleration and which is caused by torques generated by mass unbalance. { 'jē ,drift }

G² drift [NAV] A drift component in gyros and accelerometers which is proportional to the square of the nongravitational acceleration and which is generated by the anisoelasticity of the rotor support. { 'jē,skwerd ,drift }

Ge *See* germanium.

geanticline [GEOL] A broad land uplift; refers to the land mass from which sediments in a geosyncline are derived. { jē'ant·i,klīn }

gear [DES ENG] A toothed machine element used to transmit motion between rotating shafts when the center distance of the shafts is not too large. [MECH ENG] **1.** A mechanism performing a specific function in a machine. **2.** An adjustment device of the transmission in a motor vehicle which determines mechanical advantage, relative speed, and direction of travel. { gir }

gearbox *See* transmission. { 'gir,bäks }

gear case [MECH ENG] An enclosure, usually filled with lubricating fluid, in which gears operate. { 'gir ,kās }

gear cutter [MECH ENG] A machine or tool for cutting teeth in a gear. { 'gir ,kəd·ər }

gear cutting [MECH ENG] The cutting or forming of a uniform series of toothlike projections on the surface of a workpiece. { 'gir ,kəd·iŋ }

gear down [MECH ENG] To arrange gears so the driven part rotates at a slower speed than the driving part. { ¦gir ¦daủn }

gear drive [MECH ENG] Transmission of motion or torque from one shaft to another by means of direct contact between toothed wheels. { 'gir ,drīv }

geared turbine [MECH ENG] A turbine connected to a set of reduction gears. { ¦gird 'tər·bən }

gear forming [MECH ENG] A method of gear cutting in which the desired tooth shape is produced by a tool whose cutting profile matches the tooth form. { 'gir ,for·miŋ }

gear generating [MECH ENG] A method of gear cutting in which the tooth is produced by the conjugate or total cutting action of the tool plus the rotation of the workpiece. { 'gir ,jen·ə,rād·iŋ }

gear grinding [MECH ENG] A gear-cutting method in which gears are shaped by formed grinding wheels and by generation; primarily a finishing operation. { 'gir ,grīnd·iŋ }

gear hobber [MECH ENG] A machine that mills gear teeth; the rotational speed of the hob has a precise relationship to that of the work. { 'gir ,häb·ər }

gearing [MECH ENG] A set of gear wheels. { 'gir·iŋ }

gearing chain [MECH ENG] A continuous chain used to transmit motion from one toothed wheel, or sprocket, to another. { 'gir·iŋ ,chān }

gearksutite [MINERAL] CaAl(OH)F₄·H₂O A clayey mineral composed of hydrous calcium aluminum fluoride, occurring with cryolite. { jē'ärk·sə,tīt }

gearless traction [MECH ENG] Direct drive, without reduction gears. { ¦gir·ləs 'trak·shən }

gear level [MECH ENG] To arrange gears so that the driven part and driving part turn at the same speed. { 'gir ,lev·əl }

gear loading [MECH ENG] The power transmitted or the contact force per unit length of a gear. { 'gir ,lōd·iŋ }

gear meter [ENG] A type of positive-displacement fluid quantity meter in which the rotating elements are two meshing gear wheels. { 'gir ,mēd·ər }

gearmotor [MECH ENG] A motor combined with a set of speed-reducing gears. { 'gir,mōd·ər }

gear oil [MATER] A lubricating oil for use in transmissions, most types of differential gears, and gears in gear boxes. { 'gir ,oil }

gear pump [MECH ENG] A rotary pump in which two meshing gear wheels contrarotate so that the fluid is entrained on one side and discharged on the other. { 'gir ,pəmp }

gear ratio [MECH ENG] The ratio of the angular speed of the driving member of a gear train or similar mechanism to that

of the driven member; specifically, the number of revolutions made by the engine per revolution of the rear wheels of an automobile. { 'gir ,rā·shō }

gear shaper [MECH ENG] A machine that makes gear teeth by means of a reciprocating cutter that rotates slowly with the work. { 'gir ,shāp·ər }

gear-shaving machine [MECH ENG] A finishing machine that removes excess metal from machined gears by the axial sliding motion of a straight-rack cutter or a circular gear cutter. { 'gir ,shāv·iŋ mə,shēn }

gearshift [MECH ENG] A device for engaging and disengaging gears. { 'gir,shift }

gear streaks [GRAPHICS] A series of parallel streaks on a printed sheet which appear at intervals equal to those on the gear teeth of the printing cylinder. { 'gir ,strēks }

gear teeth [DES ENG] Projections on the circumference or face of a wheel which engage with complementary projections on another wheel to transmit force and motion. { 'gir ,tēth }

gear train [MECH ENG] A combination of two or more gears used to transmit motion between two rotating shafts or between a shaft and a slide. { 'gir ,trān }

gear up [MECH ENG] To arrange gears so that the driven part rotates faster than the driving part. { ¦gir ¦əp }

gear wheel [MECH ENG] A wheel that meshes gear teeth with another part. { 'gir ,wēl }

gebli *See* ghibli. { 'geb·lē }

Gecarcinidae [INV ZOO] The true land crabs, a family of decapod crustaceans belonging to the Brachygnatha. { jē·kär'sin·ə,dē }

gecko [VERT ZOO] The common name for more than 300 species of arboreal and nocturnal reptiles composing the family Gekkonidae. { 'gek·ō }

Geco sampler [MIN ENG] Straight-line cutter designed to traverse a falling stream of ore or pulp at regular intervals, so as to divert a representative sample to a holding vessel. { ¦gek·ō 'sam·plər }

gedanite [MINERAL] A brittle, wine-yellow variety of amber containing little succinic acid; found on the shore of the Baltic Sea. { 'ged·ən,īt }

Gedanken experiment [PHYS] A hypothetical ("thought") experiment which is possible in principle and is analyzed (but not performed) to test some hypothesis. Also known as thought experiment. { ge'däŋk·ən ik,sper·ə·mənt }

gedrite [MINERAL] An aluminous variety of the mineral anthophyllite. { 'je,drīt }

Gee [NAV] An electronic navigational system that establishes hyperbolic lines of position similar to those produced by loran, but employing a radio frequency of 22–30 and 40–85 megahertz, whereas standard loran operates in the 1700–2000-megahertz band. { jē }

Gee chart [NAV] A chart showing the hyperbolic lines of position which are produced by the Gee navigation system. { 'jē ,chärt }

geepound *See* slug. { 'jē,paủnd }

geg [METEOROL] A desert dust whirl of China and Tibet. { geg }

Gegenbauer polynomials [MATH] A family of polynomials solving a special case of the Gauss hypergeometric equation. Also known as ultraspherical polynomials. { 'gāg·ən,baủr ,päl·i'nō·mē·əlz }

gegenschein [ASTRON] A round or elongated, faint, ill-defined spot of light in the sky at a point 180° from the sun. Also known as counterglow; zodiacal counterglow. { 'gāg·ən,shīn }

gehlenite [MINERAL] Ca₂Al₂SiO₇ A mineral of the melilite group that crystallizes in the tetragonal crystal system and is isomorphous with akermanite; a green, resinous material found with spinel. { 'gā·lə,nīt }

Geiger-Briggs rule [NUCLEO] The rule that the range of an alpha ray in dry air, at 15°C and 1 atmosphere, above 5 centimeters, is proportional to its initial velocity raised to the 3.26 power; an improvement on the Geiger formula. { ¦gī·gər 'brigz ,rül }

Geiger counter *See* Geiger-Müller counter. { 'gī·gər ,kaủnt·ər }

Geiger counter tube *See* Geiger-Müller tube. { 'gī·gər ,kaủnt·ər ,tüb }

Geiger formula [NUCLEO] A formula which states that the range of an alpha particle in dry air, at 15°C and 1 atmosphere,

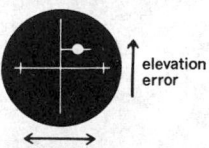

G DISPLAY

Type-G radar display, a three-dimensional display for use in an airplane cockpit.

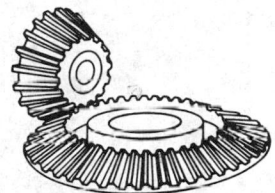

GEAR DRIVE

Bevel gears, a type of gear drive.

GECKO

Banded gecko, which is brown with broad yellow bands.

is proportional to the cube of its initial velocity. { 'gī·gər ,fȯr·myə·lə }

Geiger-Müller counter [NUCLEO] **1.** A radiation counter that uses a Geiger-Müller tube in appropriate circuits to detect and count ionizing particles; each particle crossing the tube produces ionization of gas in the tube which is roughly independent of the particle's nature and energy, resulting in a uniform discharge across the tube. Abbreviated GM counter. Also known as Geiger counter. **2.** See Geiger-Müller tube. { 'gī·gər 'myül·ər ,kau̇nt·ər }

Geiger-Müller counter tube See Geiger-Müller tube. { 'gī·gər 'myül·ər ,kau̇nt·ər ,tüb }

Geiger-Müller probe [ENG] A Geiger-Müller counter in a watertight container, lowered into a borehole to log the intensity of the gamma rays emitted by radioactive substances in traversed rock. Also known as electronic logger; Geiger probe. { ¦gī·gər 'myül·ər ,prōb }

Geiger-Müller region See Geiger region. { ¦gī·gər 'myül·ər ,rē·jən }

Geiger-Müller tube [NUCLEO] A radiation-counter tube operated in the Geiger region; it usually consists of a gas-filled cylindrical metal chamber containing a fine-wire anode at its axis. Also known as Geiger counter tube; Geiger-Müller counter; Geiger-Müller counter tube. { 'gī·gər 'myül·ər ,tüb }

Geiger-Nutall rule [NUC PHYS] The rule that the logarithm of the decay constant of an alpha emitter is linearly related to the logarithm of the range of the alpha particles emitted by it. { 'gī·gər 'nəd,ȯl ,rül }

Geiger probe See Geiger-Müller probe. { 'gī·gər ,prōb }

Geiger region [NUCLEO] The range of operating voltages of a radiation counter tube within which the output charge per count does not depend on the nature of the initial ionizing event. Also known as Geiger-Müller region. { 'gī·gər ,rē·jən }

Geiger threshold [NUCLEO] The lower limit of the Geiger region. { 'gī·gər ,thresh,hōld }

geikielite [MINERAL] MgTiO₃ A bluish-black or brownish-black mineral that crystallizes in the rhombohedral system and occurs in the form of rolled pebbles; it is isomorphous with ilmenite. { 'gē·kē,līt }

Geissler pump [ENG] A type of air pump that uses the principle of the Torricellian vacuum, and in which the vacuum is produced by the flow of mercury back and forth between a vertically adjustable and a fixed reservoir. { 'gīs·lər ,pəmp }

Geissler tube [ELECTR] An experimental discharge tube with two electrodes at opposite ends, used to demonstrate and study the luminous effects of electric discharges through various gases at low pressures. { 'gīs·lər ,tüb }

geitonogamy [BOT] Pollination and fertilization of one flower by another on the same plant. { ,gīt·ən'äg·ə·mē }

geking [OCEANOGR] Obtaining measurements of ocean movements with a geomagnetic electrokinetograph (GEK). { 'jē·kiŋ }

Gekkonidae [VERT ZOO] The geckos, a family of small lizards in the order Squamata distinguished by a flattened body, a long sensitive tongue, and adhesive pads on the toes of many species. { ge'kän·ə,dē }

gel [CHEM] A two-phase colloidal system consisting of a solid and a liquid in more solid form than a sol. { jel }

Gelastocoridae [INV ZOO] The toad bugs, a family of tropical and subtropical hemipteran insects in the subdivision Hydrocorisae. { je,la·stō'kȯr·ə,dē }

gelatin [MATER] See gelatin dynamite. [ORG CHEM] A protein derived from the skin, white connective tissue, and bones of animals; used as a food and in photography, the plastics industry, metallurgy, and pharmaceuticals. { 'jel·ət·ən }

gelatinase [BIOCHEM] An enzyme, found in some yeasts and molds, that liquefies gelatin. { 'jel·ə·tə,nās }

gelatin duplicating [GRAPHICS] A gelatin process in which the item to be reproduced is typed or drawn on a hard-surfaced master with inks containing aniline dyes; the master is then placed face down on a gelatin surface, and the dye is transferred to the gelatin within seconds; when blank paper is pressed against the gelatin, a small amount of the dye is released and a copy made. { 'jel·ət·ən 'dü·plə,kād·iŋ }

gelatin dynamite [MATER] A high explosive consisting mainly of a jellylike mass of nitroglycerin, with sodium nitrate, meal, collodion cotton, and sodium carbonate; commonly used by drillers to shatter boulders encountered in driving pipe

through overburden, especially in water-filled or saturated ground. Also known as gelatin; gelignite; nitrogelatin. { 'jel·ət·ən 'dī·nə,mīt }

gelatin effect See Ross effect. { 'jel·ət·ən i,fekt }

gelatin extra [MATER] An explosive in which some of the nitroglycerin is replaced with ammonium nitrate. { 'jel·ət·ən 'ek·strə }

gelatinize [ENG] To coat or treat with a solution of gelatin. [MATER] To convert into a gelatinous form or into a gel. { jə'lat·ən,īz }

gelatinized gasoline [MATER] Gasoline treated with a thickening agent; used in napalm bombs and flamethrowers. Also known as gasoline gel; jellied gasoline; thickened fuel. { jə'lat·ən,īzd 'gas·ə,lēn }

gelatinizing agent [MATER] In manufacture of propellants, a material which softens the nitrocellulose, permitting the mixture to be processed and formed. Also known as gelling agent. { jə'lat·ən,iz·iŋ ,ā·jənt }

gelatin liquefaction [MICROBIO] Reduction of a gelatin culture medium to the liquid state by enzymes produced by bacteria in a stab culture; used in identifying bacteria. { 'jel·ət·ən ,lik·wə'fak·shən }

gelatin model See electrolytic model. { 'jel·ət·ən ,mäd·əl }

gelatinobromide [MATER] A preparation of gelatin silver bromide that is light-sensitive and used in photography. { jə'lat·ən·ō¦brō,mīd }

gelatin process [GRAPHICS] A process for reproducing a direct image by putting the image (writing, typed matter, or drawn images) on a master paper using a special process, pressing it into a gelatin pad, then transferring it from the gelatin to duplicator paper. { 'jel·ət·ən ,präs·əs }

gelation [CHEM] **1.** The act or process of freezing. **2.** Formation of a gel from a sol. { jə'lā·shən }

gelation model [PETRO ENG] Electrolytic analog of a reservoir; used to investigate the areal sweep movement of water from multiple injection wells; operates by movement of copper ammonium and zinc ammonium ions through the gelatin in a flat tray that simulates the reservoir. { jə'lā·shən ,mäd·əl }

gelation time [CHEM ENG] In the manufacture of a thermosetting resin, the time interval between the addition of the catalyst into a liquid adhesive system and the formation of a gel. { jə'lā·shən ,tīm }

gel cement [MATER] Cement containing a small percentage of bentonite, which makes the mixture more homogeneous, increases the water-cement ratio, and reduces loss of water to the formation. { 'jel si'ment }

gel coat [MATER] A resin applied to the surface of a mold and gelled prior to placing plastic material in the mold in position for production; the gel coat becomes an integral part of the finished laminate and improves surface appearance. { 'jel ,kōt }

Geld-Benussi phenomenon [PSYCH] An effect whereby an observer who is presented consecutively with three spatially equidistant light sources lying on a line perceives additional light stimuli at other locations and times. Also known as tau phenomenon. { 'gelt bə'nüs·ē fə,näm·ə·nən }

Gelechiidae [INV ZOO] A large family of minute to small moths in the lepidopteran superfamily Tineoidea, generally having forewings and trapezoidal hindwings. { jel·ə'kī·ə,dē }

gel electrophoresis [CHEM] Electrophoresis performed in silica gel, which is a porous, inert medium. { 'jel i,lek·trō·fə'rē·səs }

gel filtration [ANALY CHEM] A type of column chromatography which separates molecules on the basis of size; higher-molecular-weight substances pass through the column first. Also known as molecular exclusion chromatography; molecular sieve chromatography. { 'jel fil'trā·shən }

Gelfond-Schneider theorem [MATH] The theorem that if a and b are algebraic numbers, where a is not equal to 0 or 1, and b is not a rational number, then a^b is a transcendental number. { 'gel,fänd 'shnīd·ər ,thir·əm }

gelfoam [BIOCHEM] Absorbable gelatin sponge partially insolubilized by crosslinking, used for arresting hemorrhage during surgery. { 'jel ,fōm }

gelifluction [GEOL] The slow, continuous downslope movement of rock debris and water-saturated soil that occurs above frozen ground, as in most polar regions and in many high

GEIGER-MÜLLER TUBE

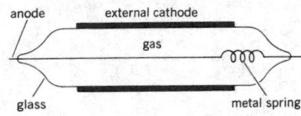

Cylindrical external-cathode Geiger-Müller tube. Metal spring keeps central wire taut.

mountain ranges. Also known as congelifluction; gelisolifluction. { ¦jel·ə¦flək·shən }

gelifraction *See* congeliturbation. { ¦jel·ə¦frak·shən }

gelignite *See* gelatin dynamite. { 'jel·əg‚nīt }

gelisol *See* frozen ground. { 'jel·ə‚sȯl }

gelisolifluction *See* gelifluction. { jə‚las·ə'fiək·shən }

geliturbation *See* congeliturbation. { ¦jel·ə‚ter'bāsh·ən }

gelivation *See* congelifraction. { ¦jel·ə¦vā·shən }

gelled cell [ELEC] A lead-acid cell with a nonspillable gelled electrolyte for portable use. { ¦jeld ¦sel }

gelling agent *See* gelatinizing agent. { 'jel·iŋ ‚ā·jənt }

Gell-Mann-Nishijima scheme [PART PHYS] A classification of elementary particles according to hypercharge, total isotopic spin, and its third component (which distinguishes between members of an isospin multiplet). { 'gel¦män ‚nē·shē·jē·mə ‚skēm }

Gell-Mann-Okubo mass formula [PART PHYS] A formula, based on SU(3) symmetry, giving the masses of the members of a unitary multiplet of mesons or baryons in terms of their total isotopic spin, hypercharge, and coefficients characteristic of the multiplet considered. { 'gel¦män ‚ō·ku̇·bō ¦mas ¦fȯr·myə·lə }

Gell-Mann relation [PART PHYS] A relation, derived from the Gell-Mann-Okubo mass formula, between the masses of the pi, eta, and K mesons of a meson octet. { 'gel¦män ri‚lā·shən }

gel mineral *See* mineraloid. { 'jel ‚min·rəl }

Gelocidae [PALEON] A family of extinct pecoran ruminants in the superfamily Traguloidea. { jə'läs·ə‚dē }

gelose *See* ulmin. { 'je‚lōs }

gel paint [MATER] Paint formulation made thixotropic by the reaction of a small amount of polyamide resin with an alkyd resin vehicle. { 'jel ‚pānt }

gel permeation chromatography [ANALY CHEM] Analysis by chromatography in which the stationary phase consists of beads of porous polymeric material such as a cross-linked dextran carbohydrate derivative sold under the trade name Sephadex; the moving phase is a liquid. { 'jel ‚pər·mē'ā·shən ‚krō·mə'täg·rə·fē }

gel point [PHYS CHEM] Stage at which a liquid begins to exhibit elastic properties and increased viscosity. { 'jel ‚pȯint }

gelsemine [PHARM] $C_{20}H_{22}O_2N_2$ A white, crystalline alkaloid with a melting point of 178°C; derived from *Gelsemium sempervirens*; soluble in alcohol, ether, and dilute acids; used in medicine as a central nervous system stimulant. { 'jel·sə‚mēn }

gel strength [FL MECH] Of a colloid, the ability or a measure of its ability to form gels. { 'jel ‚streŋkth }

gem [MINERAL] A natural or artificially produced mineral or other material that has sufficient beauty and durability for use as a personal adornment. { jem }

GEM *See* air-cushion vehicle.

gem cutting [LAP] The polishing of rough gem materials into faceted or rounded forms for use in jewelry. { 'jem ‚kəd·iŋ }

geminal [ORG CHEM] Referring to like atoms or groups attached to the same atom in a molecule. { 'jem·ə·nəl }

geminate [BIOL] Growing in pairs or couples. { 'jem·ə·nət }

Geminga [ASTRON] A relatively nearby neutron star, about 150 parsecs (450 light-years) distant, that emits pulsed x-rays and gamma rays (making it an x-ray and gamma-ray pulsar), steady optical radiation, and possible unconfirmed radio and optical pulsations. The term is derived from Gemini gamma-ray source. { 'jem·iŋ·gə }

Gemini [ASTRON] A northern constellation; right ascension 7 hours, declination 20°N. Also known as Twins. { 'jem·ə‚nē }

Geminids [ASTRON] A meteor shower that reaches maximum about December 13. { 'jem·ə·nidz }

geminiflorous [BOT] Having flowers in pairs. { ‚jem·ə'nif·lə·rəs }

gemma [BOT] A small, multicellular, asexual reproductive body of some liverworts and mosses. { 'jem·ə }

gemmation *See* budding. { je'mā·shən }

gemmho [ELEC] A unit of conductance, equal to 10^{-6} mho, being the conductance of a substance which has a resistance of 10^6 ohms. { 'je‚mō }

gemmiform [BOT] Resembling a gemma or bud. { 'jem·ə‚fȯrm }

gemmiparous [BIOL] Producing a bud or reproducing by a bud. { je'mip·ə·rəs }

gem mounting [LAP] Setting gems and gemstones in rings or other jewelry pieces made of precious metals. { 'jem ‚mau̇nt·iŋ }

gemmule [BIOL] Any bud formed by gemmation. [INV ZOO] A cystlike, asexual reproductive structure of many Porifera that germinates when proper environmental conditions exist; it is a protective, overwintering structure which germinates the following spring. [NEUROSCI] A minute dendritic process functioning as a synaptic contact point. { 'je‚myül }

Gemolite [OPTICS] A binocular magnifier with dark-field illumination, used to distinguish natural from synthetic gem materials. { 'jem·ə‚līt }

gemology [MINERAL] The science concerned with the identification, grading, evaluation, fashioning, and other aspects of gemstones. { je'mäl·ə·jē }

Gempylidae [VERT ZOO] The snake mackerels, a family of the suborder Scombroidei comprising compressed, elongate, or eel-shaped spiny-rayed fishes with caniniform teeth. { jem'pil·ə‚dē }

gem stick [LAP] A stick on which a gem is mounted before it is cut. { 'jem ‚stik }

gemstone [GEOL] A mineral or petrified organic matter suitable for use in jewelry. { 'jem‚stōn }

Gemuendinoidei [PALEON] A suborder of extinct raylike placoderm fishes in the order Rhenanida. { je¦myü·ən·də¦nȯid·ē‚ī }

gen [COMPUT SCI] To install an operating system or a systems software package for a particular configuration of computer equipment. Abbreviation for generate. { jen }

gena [ANAT] Cheek, or side of the head. { 'jē·nə }

gender [ELEC] The classification of a connector as female or male. { 'jen·dər }

gender changer [ELEC] A small passive device that is placed between two connectors of the same gender to enable them to be joined. Also known as cable matcher. { 'jen·dər ‚chān·jər }

gender identity [PSYCH] The sum of those aspects of personal appearance and behavior culturally attributed to masculinity or femininity. { 'jen·dər ī¦dent·əd·ē }

gending [METEOROL] A local dry wind in the northern plains of Java that resembles the foehn, caused by a wind crossing the mountains near the south coast and pushing between the volcanoes. { 'jen·diŋ }

gene [GEN] The basic unit of inheritance; composed of a deoxyribonucleic acid (DNA) sequence that contains the elements required for transcription of a complementary ribonucleic acid (RNA) which is sometimes the functional gene product but more often is converted into messenger RNAs that specify the amino acid sequence of a protein product. { jēn }

gene action [GEN] The functioning of a gene in determining the phenotype of an individual. { jēn ‚ak·shən }

genealogy [GEN] A record of the descent of a family, group, or person from an ancestor or ancestors. { jē·nē'äl·ə·jē }

gene amplification [CELL MOL] Any process by which a deoxyribonucleic acid sequence is disproportionately duplicated in comparison with the parent genome. [GEN] Repeated replication in a single cell cycle of the deoxyribonucleic acid (DNA) in a limited portion of the genome, resulting in an increase in the number of copies of a particular gene or DNA segment. { jēn ‚am·plə·fə‚kā·shən }

gene assignment [CELL MOL] The physical or functional localization of specific genes to individual chromosomes. { 'jēn ə‚sīn·mənt }

gene bank *See* gene library. { 'jēn ‚baŋk }

gene cluster [GEN] Any group of two or more closely linked genes that encode for the same or similar products. { 'jēn ‚kləs·tər }

genecology [BIOL] The study of species and their genetic subdivisions, their place in nature, and the genetic and ecological factors controlling speciation. { ‚jēn·ə'käl·ə·jē }

gene control region [GEN] The portion of a eukaryotic gene containing promoter and regulatory sequences of deoxyribonucleic acid that control transcription. { 'jēn kən‚trōl ‚rē·jən }

gene conversion [GEN] A situation in which gametocytes of an individual that is heterozygous for a pair of alleles undergo

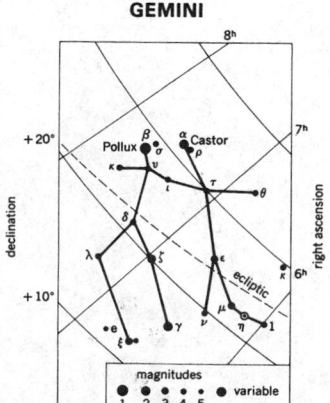

GEMINI

Line pattern of the constellation Gemini. Grid lines represent coordinates of the sky. Apparent brightness, or magnitude, of stars is shown by sizes of dots graded by appropriate numbers as indicated.

meiosis, and the gametes produced are in a 3:1 ratio rather than the expected 2:2 ratio, implying that one allele was converted to the other. { 'jēn kən‚vər·zhən }

gene duplication [GEN] The reduction in fitness of a diploid population due to new mutant genes and those already in the gene pool. { 'jēn düp·lə'kā·shən }

gene escape [MOL BIO] The movement of genetic material from a genetically engineered organism to another population or another species. { 'jēn ə‚skāp }

gene expression [GEN] The transcription, translation, and phenotypic manifestation of a gene. { 'jēn ik‚spresh·ən }

gene family [GEN] A group of related genes of similar sequence and function resulting from multiple duplications and subsequent mutational variation among the different copies. { 'jēn ‚fam·lē }

gene flow [GEN] The passage and establishment of alleles characteristic of one breeding population into the gene pool of another population through hybridization and backcrossing. { 'jēn ‚flō }

gene frequency See allele frequency. { 'jēn ‚frē·kwən·sē }

gene library [GEN] A random collection of cloned deoxyribonucleic acid fragments in a vector; includes all the genetic information of the species. Also known as gene bank. { 'jēn ‚lī·brēr·ē }

gene loss [GEN] Gene elimination from differentiating cells in some protozoans, insects, and crustaceans. { 'jēn ‚lòs }

gene overlap [GEN] The ability of a sequence of deoxyribonucleic acid to code for more than one protein by use of different reading frames. { 'jēn 'ō·vər‚lap }

gene penetrance See penetrance. { 'jēn ‚pen·ə·trəns }

gene pool [GEN] The totality of the genes of a specific population at a given time. { 'jēn ‚pül }

general address [COMMUN] Group of characters included in the heading of a message that causes the message to be routed to all addresses included in the general address category. { 'jen·rəl ə'dres }

general anesthesia [MED] Loss of sensation with loss of consciousness, produced by administration of anesthetic drugs. { 'jen·rəl ‚an·əs'thē·zhə }

general anesthetic [PHARM] An agent that produces general anesthesia. { 'jen·rəl ‚an·əs'thed·ik }

general arrangement plans [NAV ARCH] Plans showing the quarters, spaces, and compartments of a ship. { 'jen·rəl ə'rānj·mənt ‚planz }

general aviation [AERO ENG] All aviation activity not associated with either certificated air carriers or the military, including business uses, commuter airlines, air taxi operators, various commercial applications, and personal flying. { 'jen·rəl ‚ā·vē'ā·shən }

general broadcast [COMMUN] The radio broadcast, to all U.S. Navy ships and merchant ships, in which storm warnings, area forecasts, map analyses, surface weather reports, and upper air, upper wind, and aircraft reports are given. { 'jen·rəl 'bród‚kast }

general chart [NAV] A nautical chart intended for offshore coastwise navigation, using scales ranging from about 1:100,000 to 1:600,000, which are smaller than those of a coast chart, but larger than those of a sailing chart. { 'jen·rəl 'chärt }

general circulation [METEOROL] The complete statistical description of atmospheric motions over the earth. Also known as planetary circulation. { 'jen·rəl ‚sər·kyə'lā·shən }

general continuum hypothesis [MATH] A generalization of the continuum hypothesis which asserts that the smallest cardinal number greater than the cardinal number of an infinite set, S, is the cardinal number of the set of subsets of S. { 'jen·rəl kən'tin·yə·wəm hī‚päth·ə·səs }

general depot [ORD] Large supply establishment for receiving, storing, and issuing supplies for more than one technical service. { 'jen·rəl 'dep·ō }

general formula [CHEM] A formula that can apply not only to one specific compound but to a series of related compounds; for example, the general formula for an aldehyde RCHO, where R is hydrogen in formaldehyde (the simplest aldehyde) and is a hydrocarbon radical for other aldehydes in the series such as CH_3 for acetaldehyde and C_2H_5 for proprionaldehyde. { 'jen·rəl 'fòr·myə·lə }

general integral See general solution. { 'gen·rəl 'int·ə·grəl }

generalized anxiety disorder [PSYCH] A disorder characterized by excessive worry and apprehension concerning health,

daily problems, or other anxiety-provoking situations. { ‚jen·rə‚līzd ang'zī·əd·ē dis‚ór·der }

generalized binomial trials model [STAT] A product model in which the nth factor model has two simple events with probabilities p_n and $q_n = 1 - p_n$. Also known as Poisson binomial trials model. { 'jen·rə‚līzd bī'nō·mē·əl 'trīlz ‚mäd·əl }

generalized coordinates [MECH] A set of variables used to specify the position and orientation of a system, in principle defined in terms of cartesian coordinates of the system's particles and of the time in some convenient manner; the number of such coordinates equals the number of degrees of freedom of the system Also known as Lagrangian coordinates. { 'jen·rə‚līzd kō'órd·ən·əts }

generalized Euclidean space See inner-product space. { ‚jen·rə‚līzd yü‚klid·ē·ən 'spās }

generalized feasible flow [MATH] A feasible flow in a generalized s-t network such that the outflow at any intermediate vertex does not exceed the weight of that vertex. { ‚jen·rə‚līzd ‚fēz·ə·bəl 'flō }

generalized force [MECH] The generalized force corresponding to a generalized coordinate is the ratio of the virtual work done in an infinitesimal virtual displacement, which alters that coordinate and no other, to the change in the coordinate. { 'jen·rə‚līzd 'fòrs }

generalized function See distribution. { 'jen·rə‚līzd 'fəŋk·shən }

generalized hydrostatic equation [GEOPHYS] The vertical component of the vector equation of motion in natural coordinates when the acceleration of gravity is replaced by the virtual gravity; for most purposes it is identical to the hydrostatic equation. { 'jen·rə‚līzd ‚hī·drə‚stad·ik i'kwā·zhən }

generalized max-flow min-cut theorem [MATH] The theorem that in a generalized s-t network the maximum possible flow value of a generalized feasible flow equals the minimum possible weight of a generalized s-t cut. { ‚jen·rə‚līzd ‚maks‚flō ‚min'kət ‚thir·əm }

generalized mean-value theorem See second mean-value theorem. { ‚jen·rə‚līzd 'mēn ‚val·yü ‚thir·əm }

generalized momentum See conjugate momentum. { 'jen·rə‚līzd mə'ment·əm }

generalized permutation [MATH] Any ordering of a finite set of elements that are not necessarily distinct. { ‚jen·rə‚līzd ‚pər·myə'tā·shən }

generalized Poincaré conjecture [MATH] The question as to whether every closed n-manifold which has the homotopy type of the n-sphere is homeomorphic to the n-sphere. { 'jen·rə‚līzd 'pwän·ka‚rā kən'jek·chər }

generalized power [MATH] For a positive number a and an irrational number x, the number a^x defined by the equation $a^x = e^{x \log a}$, where e is the base of the natural logarithms and $\log a$ is taken to that base. { 'jen·rə‚līzd 'pau·ər }

generalized ratio test See d'Alembert's test for convergence. { ‚jen·rə‚līzd 'rā·shō ‚test }

generalized routine [COMPUT SCI] A routine which can process a wide variety of jobs; for example, a generalized sort routine which will sort in ascending or descending order on any number of fields whether alphabetic or numeric, or both, and whether binary coded decimals or pure binaries. { 'jen·rə‚līzd rü'tēn }

generalized s-t cut [MATH] A set of arcs and vertices in a generalized s-t network such that any directed path from the source to the terminal includes at least one element of this set. { ‚jen·rə‚līzd 'es‚tē 'kət }

generalized s-t network [MATH] An s-t network on which is defined a weight function from the vertices of the network to the nonnegative integers. { ‚jen·rə‚līzd 'es‚tē 'net‚wərk }

generalized system [COMPUT SCI] A computer system developed for a broad range of users. { 'jen·rə‚līzd 'sis·təm }

generalized transmission function [GEOPHYS] In atmospheric-radiation theory, a set of values, variable with wavelength, each one of which represents an average transmission coefficient for a small wavelength interval and for a specified optical path through the absorbing gas in question. { 'jen·rə‚līzd tranz'mish·ən ‚fəŋk·shən }

generalized velocity [MECH] The derivative with respect to time of one of the generalized coordinates of a particle. Also known as Lagrangian generalized velocity. { 'jen·rə‚līzd və'läs·əd·ē }

general manager [IND ENG] The person of general authority who performs all reasonable tasks in conducting the usual and customary business of the principal head or owner. { ¦jen·rəl 'man·ə·jər }

general paralysis [MED] A chronic, progressive form of syphilis caused by inflammatory and degenerative changes in the brain and meninges and characterized by physical, emotional, and intellectual deterioration; occurs at least 10 to 15 years after initial infection. { ¦jen·rəl pə'ral·ə·səs }

general paresis [MED] An inflammatory and degenerative disease of the brain caused by infection with *Treponema pallidum*. Also known as syphilitic meningoencephalitis. { ¦jen·rəl pə'rē·səs }

general precession [ASTRON] The resultant motion of the components causing precession of the equinoxes westward along the ecliptic at the rate of about 50.3″ per year. { ¦jen·rəl prē'sesh·ən }

general program [COMPUT SCI] A computer program designed to solve a specific type of problem when values of appropriate parameters are supplied. { ¦jen·rəl 'prō·grəm }

general-purpose automatic test system [ELECTR] Modular, computer-type, automatic electronic checkout system capable of finding faults in electronic equipment at the system, subsystem, line replaceable unit, module, and piece part levels. { ¦jen·rəl ¦pər·pəs ¦od·ə¸mad·ik 'test ¸sis·təm }

general-purpose bomb [ORD] An item designed to be dropped from an aircraft to destroy or reduce the utility of a target by explosive effect. { ¦jen·rəl ¦pər·pəs 'bäm }

general-purpose computer [COMPUT SCI] A device that manipulates data without detailed, step-by step control by human hand and is designed to be used for many different types of problems. { ¦jen·rəl ¦pər·pəs kəm'pyüd·ər }

general-purpose function generator [COMPUT SCI] A function generator which can be adjusted to generate many different functions, rather than being designed for a particular function. Also known as arbitrary function generator. { ¦jen·rəl ¦pər·pəs 'fəŋk·shən jen·ə¸rād·ər }

general-purpose language [COMPUT SCI] A computer programming language whose use is not restricted to a particular type of computer or a specialized application. { 'jen·rəl ¦pər·pəs 'laŋ·gwij }

general-purpose systems simulation See GPSS. { ¦jen·rəl ¦pər·pəs 'sis·təmz ¸sim·yə¸lā·shən }

general-purpose vehicle [ORD] Motor vehicle designated to be used interchangeably for movement of personnel, supplies, ammunition, or equipment and for towing artillery carriages, trailers, or semitrailers; used without modification to body or chassis to satisfy general automotive transport needs. { ¦jen·rəl ¦pər·pəs 've·ə·kəl }

general register See local register. { ¦jen·rəl 'rej·ə·stər }

general relativistic collapse [RELAT] Process in which a star undergoing gravitational collapse cannot release its kinetic energy to the outside universe, but continues to collapse into a general relativistic singularity of infinite density. { ¦jen·rəl ¸rel·ə·tə¦vis·tik kə'laps }

general relativity [RELAT] The theory of Einstein which generalizes special relativity to noninertial frames of reference and incorporates gravitation, and in which events take place in a curved space. { ¦jen·rəl ¸rel·ə'tiv·əd·ē }

general research [SCI TECH] Pure or applied research which is not directly concerned with an applied research program. { ¦jen·rəl 'rē·sərch }

general reserve artillery [ORD] All artillery units not integral to divisions and corps, comprising a pool of reserve artillery available to the field forces commander for assignment or attachment to subordinate commands as required by combat conditions. { ¦jen·rəl ri¦zərv är'til·ə·rē }

general routine [COMPUT SCI] In computers, a routine, or program, applicable to a class of problems; it provides instructions for solving a specific problem when appropriate parameters are supplied. { ¦jen·rəl rü'tēn }

general solution [MATH] For an nth-order differential equation, a function of the independent variables of the equation and of n parameters such that assignment of any numerical values to the parameters yields a solution to the equation. Also known as general integral. { ¦jen·rəl sə'lü·shən }

general supplies [ORD] Intraservice classification applied to ordnance, quartermaster, and transportation supplies; ordnance general supplies include all ordnance supplies, with the exception of ammunition, required for the maintenance of an organization. { ¦jen·rəl sə'plīz }

general-support artillery [ORD] Artillery which executes the fire directed by the commandeer of the unit to which it organically belongs or is attached; it fires in support of a specific subordinate unit. { ¦jen·rəl sə¦pòrt är'til·ə·rē }

general term [MATH] The general term of a sequence or series is an expression subscripted by an integer which determines any desired entry. { ¦jen·rəl 'tərm }

general topology [MATH] The branch of topology that studies the relationships between the basic topological properties that spaces may possess. Also known as point-set topology. { ¦jen·rəl tə'päl·ə·jē }

general transcription factor [CELL MOL] In eukaryotes, a protein that binds to ribonucleic acid polymerase to form a preinitiation complex that is necessary to begin transcription. { jen·rəl tranz'kript·shən ¸fak·tər }

generate [COMPUT SCI] **1.** To create a particular program by selecting parts of a general-program skeleton (or outline) and specializing these parts into a cohesive entity. **2.** See gen. { 'jen·ə¸rāt }

generate and test [COMPUT SCI] A computer problem-solving method in which a sequence of candidate solutions is generated, and each is tested to determine if it is an appropriate solution. { 'jen·ə¸rāt ən 'test }

generated address [COMPUT SCI] An address calculated or determined by instructions contained in a computer program for subsequent use by that program. Also known as calculated address; synthetic address. { jen·ə¸rād·əd ə'dres }

generating area See fetch. { 'jen·ə¸rād·iŋ ¸er·ē·ə }

generating flow [FL MECH] For a liquid allowed to flow smoothly into a duct, the flow while the boundary layer, which starts at the entrance and grows until it fills the duct, is growing. { 'jen·ə¸rād·iŋ ¸flō }

generating function [MATH] **1.** A function $g(x,y)$ corresponding to a family of orthogonal polynomials $f_0(x)$, $f_1(x)$, . . . , where a Taylor series expansion of $g(x,y)$ in powers of y will have the polynomial $f_n(x)$ as the coefficient for the term y^n. **2.** A function, $g(y)$, corresponding to a sequence a_0, a_1, . . . , where $g(y) = a_0 + a_1 y + a_2 y^2 + \cdots$. Also known as ordinary generating function. { 'jen·ə¸rād·iŋ ¸fəŋk·shən }

generating magnetometer [ENG] A magnetometer in which a coil is rotated in the magnetic field to be measured with the resulting generated voltage being proportional to the strength of the magnetic field. { 'jen·ə¸rād·iŋ mag·nə'täm·əd·ər }

generating plant See generating station. { 'jen·ə¸rād·iŋ ¸plant }

generating routine See generator. { 'jen·ə¸rād·iŋ rü¸tēn }

generating station [MECH ENG] A stationary plant containing apparatus for large-scale conversion of some form of energy (such as hydraulic, steam, chemical, or nuclear energy) into electrical energy. Also known as generating plant; power station. { 'jen·ə¸rād·iŋ ¸stā·shən }

generation [BIOL] A group of organisms having a common parent or parents and comprising a single level in line of descent. [COMPUT SCI] **1.** Any one of three groups used to historically classify computers according to their electronic hardware components, logical organization and software, or programming techniques; computers are thus known as first-, second-, or third-generation; a particular computer may possess characteristics of all generations simultaneously. **2.** One of a family of data sets, related to one another in that each is a modification of the next most recent data set. { jen·ə'rā·shən }

generation data group [COMPUT SCI] A collection of files, each a modification of the previous one, with the newest numbered 0, the next −1, and so forth, and organized so that each time a new file is added the oldest is deleted. Abbreviated GDG. { ¸jen·ə'rā·shən 'dad·ə ¸grüp }

generation number [COMPUT SCI] A number contained in the file label of a reel of magnetic tape that indicates the generation of the data set of the tape. { jen·ə'rā·shən ¸nəm·bər }

generation rate [ELECTR] In a semiconductor, the time rate of creation of electron-hole pairs. { jen·ə'rā·shən ¸rāt }

generation time [MICROBIO] The time interval required for a bacterial cell to divide or for the population to double.

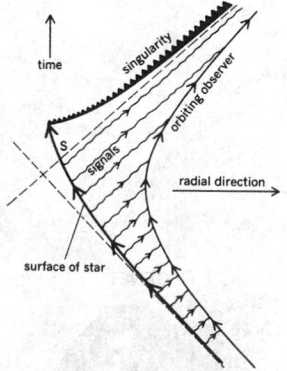

GENERAL RELATIVISTIC COLLAPSE

Space-time diagram for the collapse of a star past its Schwarzschild radium *(S)* and into a general relativistic singularity of infinite density.

[NUCLEO] The mean time required for a neutron arising from a fission to produce a new fission. { jen·ə'rā·shən ˌtīm }

generative grammar [COMPUT SCI] A set of rules that describes the valid expressions in a formal language on the basis of a set of the parts of speech (formally called the set of metavariables or phrase names) and the alphabet or character set of the language. { 'jen·rəd·iv 'gram·ər }

generative nucleus [BOT] A haploid nucleus in a pollen grain that produces two sperm nuclei by mitosis. { 'jen·rəd·iv 'nü·klē·əs }

generator [COMPUT SCI] A program that produces specific programs as directed by input parameters. Also known as generating routine. [ELEC] A machine that converts mechanical energy into electrical energy; in its commonest form, a large number of conductors are mounted on an armature that is rotated in a magnetic field produced by field coils. Also known as dynamo; electric generator. [ELECTR] **1.** A vacuum-tube oscillator or any other nonrotating device that generates an alternating voltage at a desired frequency when energized with direct-current power or low-frequency alternating-current power. **2.** A circuit that generates a desired repetitive or nonrepetitive waveform, such as a pulse generator. [MATH] **1.** One of the set of elements of an algebraic system such as a group, ring, or module which determine all other elements when all admissible operations are performed upon them. **2.** *See* generatrix. { 'jen·ə,rād·ər }

generator field control [ELEC] Method of regulating the output voltage of a generator by controlling the voltage which excites the field of the generator. { 'jen·ə,rād·ər 'fēld kən,trōl }

generator lock [ELECTR] Circuitry that synchronizes two video signals so that they can be mixed. Abbreviated genlock. { 'jen·ə,rād·ər ,läk }

generator reactor [ELEC] A small inductor connected between power-plant generators and the rest of an electric power system in order to limit and localize the effects of voltage transients. { 'jen·ə,rād·ər rē,ak·tər }

generator resistance [ELEC] The resistance of the current source in a network; usually much smaller than the load but taken into account in some network calculations. { 'jen·ə,rād·ər ri,zis·təns }

generator set [ENG] The aggregate of one or more generators together with the equipment and plant for producing the energy that drives them. { 'jen·ə,rād·ər ,set }

generatrix [MATH] The straight line generating a ruled surface. Also known as generator. { 'jen·ə,rā·triks }

gene redundancy [GEN] The presence of additional copies of a gene within a cell. { 'jēn ri'dən·dən·sē }

gene regulatory protein [CELL MOL] The general name for a protein that controls gene expression via its ability to bind to specific deoxyribonucleic acid sequences during transcription. { 'jēn 'reg·yə·lə,tór·ē 'prō,tēn }

generic [BIOL] Pertaining to or having the rank of a biological genus. [SCI TECH] **1.** A term applied to or descriptive of an entire group or a class. **2.** In general use, that which is nonproprietary. { jə'ner·ik }

gene scanning [CELL MOL] A method by which mutations are inserted at specific sites on a deoxyribonucleic acid (DNA) segment to determine those DNA sequences needed for gene activity. { 'jēn ,skan·iŋ }

gene sharing [GEN] The acquisition and maintenance of a second function for a gene without duplication and without loss of primary function. { 'jēn ,sher·iŋ }

genesis rocks [GEOL] Rocks that have retained their character from nearly 4.6×10^9 years ago, when planets were still occulting out of the cloud of dust and gas referred to as the solar nebula; examples are meteorites and asteroids. { 'jen·ə·səs ,räks }

gene splicing *See* recombinant technology. { 'jēn ,splīs·iŋ }

gene substitution [GEN] The replacement of an allele with a mutant allele. { 'jēn ,səb·stə,tü·shən }

gene suppression [GEN] The development of a normal phenotype in a mutant individual or cell due to a second mutation either in the same gene or in a different gene. { 'jēn sə,presh·ən }

genet [VERT ZOO] The common name for nine species of small, arboreal African carnivores in the family Viverridae. { 'jen·ət }

gene targeting [GEN] Replacement, by genetic engineering

GENET

Spotted genet (*Genetta genetta*).

and homologous recombination, of a mutant (or wild-type) gene by a wild-type (or mutant) copy that may also contain a reporter gene. { 'jen ,tär·gəd·iŋ }

gene therapy [GEN] An experimental technique in which a normal gene is inserted into an organism to correct a genetic defect. { 'jēn ,ther·ə·pē }

genetic algorithm [COMPUT SCI] A search procedure based on the mechanics of natural selection and genetics. Also known as evolutionary strategy. { jə,ned·ik 'al·gə,rith·əm }

genetic block [GEN] The reduction in enzyme activity due to a gene mutation. { jə,ned·ik 'bläk }

genetic carrier [GEN] An individual who is heterozygous for a recessive gene that predisposes for a hereditary disease. { jə,ned·ik kar·ē·ər }

genetic code [CELL MOL] The genetic information in the nucleotide sequences in deoxyribonucleic acid represented by a four-letter alphabet that makes up a vocabulary of 64 three-nucleotide sequences, or codons; a sequence of such codons (averaging about 100 codons) constructs a message for a polypeptide chain. { jə'ned·ik 'kōd }

genetic colonization [GEN] Natural introduction of genetic material into the deoxyribonucleic acid of a host cell; for example, transmission of a tumor-inducing plasmid into a plant cell by the bacterium *Agrobacterium tumefaciens*. { jə,ned·ik ,käl·ə·nə'zā·shən }

genetic death [GEN] **1.** Preferential elimination of genotypes that are carriers of alleles that reduce the adaptive value or fitness of those genotypes. **2.** The death of an individual without reproducing. { jə,ned·ik 'deth }

genetic differentiation [GEN] The accumulation of differences in allelic frequencies between completely or partially isolated populations due to evolutionary forces such as selection or genetic drift. { jə,ned·ik ,dif·ə,ren·chē'ā·shən }

genetic distance [GEN] **1.** A measure of the allelic substitutions per locus that have occurred during the separate evolution of two populations or species. **2.** The distance between linked genes in terms of recombination frequency or map units. { jə,ned·ik 'dis·təns }

genetic drift [GEN] The random fluctuation of gene frequencies from generation to generation that occurs in small populations. { jə,ned·ik 'drift }

genetic engineering [GEN] The intentional production of new genes and alteration of genomes by the substitution or addition of new genetic material. Also known as biogenetics. { jə,ned·ik en·jə'nir·iŋ }

genetic equilibrium [GEN] In a population, the condition in which the frequencies of allelic genes are maintained at the same values from generation to generation. { jə,ned·ik ,ē·kwə'lib·rē·əm }

genetic facies [GEOL] An ancient deposit of rocks which have been formed by similar sedimentary processes. { jə,ned·ik 'fā·shēz }

genetic fingerprinting [FOREN SCI] A forensic identification technique that enables virtually 100% discrimination between individuals from small samples of blood or semen, using probes for hypervariable minisatellite deoxyribonucleic acid. Also known as DNA fingerprinting. [CELL MOL] Identification of chemical entities in animal tissues as indicative of the presence of specific genes. { jə,ned·ik 'fiŋ·gər,print·iŋ }

genetic homeostasis [GEN] The tendency of Mendelian populations to maintain a constant genetic composition. { jə,ned·ik ,hō·mē·ō'stā·səs }

genetic identity [GEN] A measure of the proportion of genes that are identical in two populations. { jə,ned·ik ī'den·əd·ē }

genetic induction [GEN] Gene activation by a molecule that inactivates a repressor protein and thereby activates transcription of one or more structural genes. [MOL BIO] Gene activation by a chemical inducer; results in transcription of structural genes. { jə,ned·ik in'dək·shən }

genetic isolation [GEN] The absence of genetic exchange between populations or species as a result of geographic separation or of mechanisms that prevent reproduction. { jə,ned·ik īs·əl'ā·shən }

genetic load [GEN] The reduction in fitness of a diploid population due to new mutant genes and those already in the gene pool. { jə,ned·ik 'lōd }

genetic map [GEN] A graphic presentation of the linear arrangement of genes on a chromosome; gene positions are determined by percentages of recombination in linkage experiments. Also known as chromosome map. { jə¦ned·ik 'map }

genetic marker [GEN] A gene whose phenotypic expression is easily discerned and thereby can be used to identify an individual or a cell that carries it, or as a probe to mark a nucleus, chromosome, or locus. { jə¦ned·ik 'märk·ər }

genetic material [GEN] The nuclear (chromosomal) and cytoplasmic (mitochondrial and chloroplast) material that plays a fundamental role in determining the nature of all cell substances, cell structures, and cell effects; the genes have properties of self-propagation and variation. { jə¦ned·ik mə'tir·ē·əl }

genetic programming See evolutionary programming. { jə¦ned·ik 'prō¸gram·iŋ }

genetics [BIOL] The science that is concerned with the study of biological inheritance. { jə'ned·iks }

genetic system [GEN] For a given species, the organization of genetic material and the ways in which the genetic material is transmitted. { jə¦ned·ik 'sis·təm }

genetic transformation See transformation. { jə¦ned·ik ¸tranz·fər'mā·shən }

genetic variance [GEN] The phenotypic variance in a population that is due to genetic heterogeneity. { jə¦ned·ik ¸ver·ē·əns }

Geneva stop [HOROL] A device that transmits power only from the middle portion of a watch mainspring to provide an even force. { jə¦nē·və 'stäp }

Geneva system [ORG CHEM] An international system of nomenclature for organic compounds based on hydrocarbon derivatives; names correspond to the longest straight carbon chain in the molecule. { jə¦nē·və 'sis·təm }

genial See mental. { 'jēn·yəl }

genic hybrid sterility [GEN] Sterility resulting from the interaction of genes in a hybrid to cause disturbances of sex-cell formation or meiosis. { 'jen·ik 'hī·brəd stə'ril·əd·ē }

genicide [ORG CHEM] $C_{13}H_8O_2$ A compound with needle-like crystals and a melting point of 174°C; insoluble in water; used as an insecticide, miticide, and ovicide. Also known as oxoxanthone; 9-xanthenone; xanthone. { 'jen·ə¸sīd }

geniculate [SCI TECH] Bent abruptly at an angle, as a bent knee. { jə'nik·yə·lət }

geniculate body [NEUROSCI] Any of the four oval, flattened prominences on the posterior inferior aspect of the thalamus; functions as the synaptic center for fibers leading to the cerebral cortex. { jə'nik·yə·lət ¸bäd·ē }

geniculate ganglion [NEUROSCI] A mass of sensory and sympathetic nerve cells located along the facial nerve. { jə'nik·yə·lət 'gaŋ·glē·ən }

geniculum [ANAT] **1.** A small, kneelike, anatomical structure. **2.** A sharp bend in any small organ. { je'nik·yə·ləm }

genioglossus [ANAT] An extrinsic muscle of the tongue, arising from the superior mental spine of the mandible. { ¦jē·nē·ō¦glä·səs }

Geniohyidae [PALEON] A family of extinct ungulate mammals in the order Hyracoidea; all members were medium to large-sized animals with long snouts. { ¦jē·nē·ō'hī·ə¸dē }

genistin [ORG CHEM] $C_{21}H_{20}O_{10}$ A pale-yellow glucoside derived from soybean meal, crystallizes from 80% methanol solution, melting point 256°C, soluble in hot 80% ethanol, hot 80% methanol, and hot acetone. Also known as 7-D-glucoside. { jə'nis·tən }

genital atrium [ZOO] A common chamber receiving openings of male, female, and accessory organs. { 'jen·ət·əl 'ā·trē·əm }

genital coelom [INV ZOO] In mollusks, the lumina of the gonads. { 'jen·ət·əl 'sē·ləm }

genital cord [EMBRYO] A mesenchymal shelf bridging the coeloms in mammalian embryos, produced by fusion of the caudal part of the urogenital folds; fuses with the urinary bladder in the male, and is the primordium of the broad ligament and the uterine walls in the female. [INV ZOO] Strands of cells located in the genital canal which are primordial sex cells in crinoids. Also known as genital rachis. { 'jen·ət·əl 'kȯrd }

genitalia [ANAT] The organs of reproduction, especially those which are external. { ¸jen·ə'tāl·yə }

genital orifice See genital pore. { 'jen·ət·əl 'ȯr·ə·fəs }

genital pore [INV ZOO] A small opening on the side of the head in some gastropods through which the penis is protruded. Also known as genital orifice. { 'jen·ət·əl 'pȯr }

genital rachis See genital cord. { 'jen·ət·əl 'rā·kəs }

genital recess [INV ZOO] A depression between the calyx surface and anal cone in entoprocts which serves as a brood chamber. { 'jen·ət·əl 'rē¸ses }

genital ridge [EMBRYO] A medial ridge or fold on the ventromedial surface of the mesonephros in the embryo, produced by growth of the peritoneum; the primordium of the gonads and their ligaments. { 'jen·ət·əl 'rij }

genital scale [INV ZOO] Any of the small calcareous plates in ophiuroids associated with the buccal shields. { 'jen·ət·əl 'skāl }

genital segment See gonosomite. { 'jen·ət·əl 'seg·mənt }

genital shield [INV ZOO] In ophiuroids, a support of a bursal slit in the arms located near the arm base. { 'jen·ət·əl 'shēld }

genital stolon [INV ZOO] Part of the axial complex in ophiuroids. { 'jen·ət·əl 'stō·lən }

genital sucker [INV ZOO] In some trematodes, a suckerlike structure surrounding the gonopore. { 'jen·ət·əl 'sək·ər }

genital tract [ANAT] The ducts of the reproductive system. { 'jen·ət·əl ¸trakt }

genital tube [INV ZOO] A blood lacuna in crinoids, connected with the subtegminal plexus and suspended in the genital canal. { 'jen·ət·əl 'tüb }

genitourinary system See urogenital system. { ¦jen·ə·tō'yür·ə¸ner·ē ¸sis·təm }

genlock See generator lock. { 'gen¸läk }

Genoa cyclone [METEOROL] A cyclone, or low, which appears to have formed or developed in the vicinity of the Gulf of Genoa. Also known as Genoa low. { 'jen·ə·wə 'sī¸klōn }

Genoa low See Genoa cyclone. { 'jen·ə·wə 'lō }

Genocchi number [MATH] An integer of the form $G_n = 2(2^{2n} - 1)B_n$, where B_n is the nth Bernoulli number. { gə'näk·ē¸näm·bər }

genome [GEN] **1.** The genetic endowment of a species. **2.** The haploid set of chromosomes. { 'jē¸nōm }

genomic deoxyribonucleic acid [CELL MOL] Fragments of deoxyribonucleic acid (DNA) that are produced by the action of restriction enzymes on the DNA of a cell or organism. Abbreviated gDNA. { jə¦nōm·ik dē¸äk·sē¸rī·bō·nü¦klē·ik 'as·əd }

genomic imprinting See parental imprinting. { jə¦nō·mik im¸print·iŋ }

genomics [GEN] The study of the entire deoxyribonucleic sequence of an organism. { jə'nō·miks }

genomic stress [GEN] Any influence that may disrupt the stability of the genome by fostering chromosome damage or mutation, such as environmental factors or altered genetic background. { jə¦nōm·ik 'stres }

genophobia [PSYCH] An abnormal fear of sex. { ¦jē·nə'fō·hē·ə }

genotoxant [BIOCHEM] An agent that induces toxic, lethal, or heritable effects to nuclear and extranuclear genetic material in cells. { ¸jēn·ə'täk·sənt }

genotype [GEN] The genetic constitution of an organism, usually in respect to one gene or a few genes relevant in a particular context. [SYST] The type species of a genus. { 'jē·nə¸tīp }

genotype frequency [GEN] The proportion or frequency of any particular genotype among the individuals of a population. { 'jēn·ə¸tīp ¸frē·kwən·sē }

genotypic cohesion [EVOL] The process whereby balanced and superior gene combinations are held together under the force of genetic recombination, thus reducing the frequency of deleterious recombinants, and with it the genetic load. { ¦jēn·ə¸tip·ik kō'hē·zhən }

genotypic distance [EVOL] For two individuals A and B, the probability that the genotype of A is not the same as that of B for a given locus; the distance is zero when the genotype is the same at the particular locus. { ¦jēn·ə¸tip·ik 'dis·təns }

genotypic structure [GEN] The set of the genotype frequencies of a population. { ¦jēn·ə¸tip·ik 'strək·chər }

gentamicin [MICROBIO] A broad-spectrum antibiotic produced by a species of Micromonospora. { ¦jent·ə¦mīs·ən }

Gentianaceae [BOT] A family of dicotyledonous herbaceous plants in the order Gentianales distinguished by lacking stipules and having parietal placentation. { ¸jen·chə'nās·ē¸ē }

Gentianales [BOT] A family of dicotyledonous plants in the subclass Asteridae having well-developed internal phloem and opposite, simple, mostly entire leaves. { ¸jen·chə'nā·lēz }

gentianic acid *See* gentisic acid. { ¦jen·chē¦an·ik 'as·əd }

gentian violet *See* methyl violet. { 'jen·chən 'vī·lət }

gentisic acid [ORG CHEM] $C_7H_6O_4$ A crystalline compound that forms monoclinic prisms from a water solution, sublimes at 200°C, melts at 250°C, and is soluble in water, alcohol, ether, sodium, and salt; used in medicine. Also known as gentianic acid. { jen'tis·ik 'as·əd }

gentle breeze [METEOROL] In the Beaufort wind scale, a wind whose speed is from 7 to 10 knots (13–19 kilometers per hour). { ¦jent·əl 'brēz }

gentnerite [MINERAL] $Cu_8Fe_3Cr_{11}S_{18}$ A sulfide mineral known only in meteorites. { 'jent·nə‚rīt }

genu *See* knee. { 'ge·nü }

genus [MATH] An integer associated to a surface which measures the number of holes in the surface. [SYST] A taxonomic category that includes groups of closely related species; the principal subdivision of a family. { 'jē·nəs }

genu valgum [MED] Inward or medial curving of the knee; knock-knee. { 'ge·nü 'val·gəm }

geo [GEOGR] A narrow coastal inlet bordered by steep cliffs. Also spelled gio. { 'gyō }

GEO *See* geosynchronous orbit. { ¦jē¦ē¦ō *or* 'jē·ō }

geoacoustics [ACOUS] Study of the acoustic properties of rock, mainly to study possible use of the rock system as a carrier of seismic signals in a communications system. { ¦jē·ō·ə'küs·tiks }

geobotanical prospecting [GEOL] The use of the distribution, appearance, and growth anomalies of plants in locating ore deposits. { ¦jē·ō·bə¦tan·ə·kəl 'präs·pek·tiŋ }

geobotany [BOT] The study of plants as related to their geologic environment. { ¦jē·ō'bät·ən·ē }

geocentric [ASTRON] Relative to the earth as a center; that is measured from the center of the earth. { jē·ō¦sen·trik }

geocentric coordinates [ASTRON] Coordinates that define the position of a point with respect to the center of the earth; can be either cartesian (*x*, *y*, and *z*) or spherical (latitude, longitude, and radial distance). Also known as geocentric coordinate system; geocentric position. { jē·ō¦sen·trik kō'órd·ən·əts }

geocentric coordinate system *See* geocentric coordinates. { ¦jē·ō¦sen·trik kō'órd·ən·ət ‚sis·təm }

geocentric latitude [ASTRON] The latitude of a celestial body from the center of the earth. [GEOD] Of a position on the earth's surface, the angle between a line to the center of the earth and the plane of the equator. { ¦jē·ō¦sen·trik 'lad·ə‚tüd }

geocentric longitude [ASTRON] The celestial longitude of the position of a body projected on the celestial sphere when the body is viewed from the center of the earth. [GEOD] At a position on the earth's surface, the angle between the plane of the reference meridian and a plane through the polar axis and a line from the position in question to the center of mass of the earth. { jē·ō¦sen·trik 'län·jə‚tüd }

geocentric parallax [ASTRON] The difference in the apparent direction or position of a celestial body, measured in seconds of arc, as determined from the center of the earth and from a point on its surface; this varies with the body's altitude and distance from the earth. Also known as diurnal parallax. { ¦jē·ō¦sen·trik 'par·ə‚laks }

geocentric position *See* geocentric coordinates. { ¦jē·ō¦sen·trik pə'zish·ən }

geocentric vertical [GEOD] The direction of the radius vector drawn from the center of the earth through the location of the observer. Also known as geometric vertical. { jē·ō¦sen·trik 'verd·ə·kəl }

geocentric zenith [ASTRON] The point where a line from the center of the earth through a point on its surface meets the celestial sphere. { ¦jē·ō¦sen·trik 'zē·nith }

geocerite [MINERAL] A white, waxy mineral composed of carbon, oxygen, and hydrogen, occurring in brown coal. { jē·ō'si‚rīt }

geochemical anomaly [GEOCHEM] Above-average concentration of a chemical element in a sample of rock, soil, vegetation, stream, or sediment; indicative of nearby mineral deposit. { ¦jē·ō¦kem·ə·kəl ə'näm·ə·lē }

geochemical balance [GEOCHEM] The proportional distribution, and the migration rate, in the global fractionation of elements, minerals, or compounds; for example, the distribution of quartz in igneous rocks, its liberation by weathering, and its

redistribution into sediments and, in solution, into lakes, rivers, and oceans. { ¦jē·ō¦kem·ə·kəl 'bal·əns }

geochemical cycle [GEOCHEM] During geologic changes, the sequence of stages in the migration of elements between the lithosphere, hydrosphere, and atmosphere. { ¦jē·ō¦kem·ə·kəl 'sī·kəl }

geochemical evolution [GEOCHEM] **1.** A change in any constituent of a rock beyond that amount present in the parent rock. **2.** A change in chemical composition of a major segment of the earth during geologic time, as the oceans. { ¦jē·ō¦kem·ə·kəl ‚ev·ə'lü·shən }

geochemical prospecting [ENG] The use of geochemical and biogeochemical principles and data in the search for economic deposits of minerals, petroleum, and natural gases. { ¦jē·ō¦kem·ə·kəl 'prä‚spek·tiŋ }

geochemical well logging [ENG] Well logging dependent on geochemical analysis of the data. { ¦jē·ō¦kem·ə·kəl 'wel ‚läg·iŋ }

geochemistry [GEOL] The study of the chemical composition of the various phases of the earth and the physical and chemical processes which have produced the observed distribution of the elements and nuclides in these phases. { ¦jē·ō¦kem·ə·strē }

geochron *See* isochron. { 'jē·ə‚krän }

geochronology [GEOL] **1.** The dating of the events in the earth's history. **2.** A system of dating developed for the purposes of study of the earth's history. { ¦jē·ō·krə'näl·ə·jē }

geochronometry [GEOL] The study of the absolute age of the rocks of the earth based on the radioactive decay of isotopes, such as ^{238}U, ^{235}U, ^{232}Th, ^{87}Rb, ^{40}K, and ^{14}C, present in minerals and rocks. { ¦jē·ō·krə'näm·ə·trē }

Geocorisae [INV ZOO] A subdivision of hemipteran insects containing those land bugs with conspicuous antennae and an ejaculatory bulb in the male. { jē·ə'kór·ə‚sē }

geocorona [GEOPHYS] The outermost part of the earth's atmosphere, consisting of extremely attenuated hydrogen extending to perhaps 15 earth radii, that emits Lyman-alpha radiation under the action of sunlight. { jē·ō·kə'rō·nə }

geocosmogony [GEOL] The study of the origin of the earth. { ¦jē·ō‚käz'mäj·ə·nē }

geocronite [MINERAL] $Pb_5(Sb,As)_2S_3$ A mineral composed of lead-gray lead antimony arsenic sulfide. { jē'äk·rə‚nīt }

geode [GEOL] A roughly spheroidal, hollow body lined inside with inward-projecting, small crystals; found frequently in limestone beds but may occur in shale. { 'jē‚ōd }

Geodermatophilus [MICROBIO] A genus of bacteria in the family Dermatophilaceae; the mycelium is rudimentary and a muriform thallus is produced; motile spores are elliptical to lanceolate. { ‚jē·ó‚dər·mə'täf·ə·ləs }

geodesic [MATH] A curve joining two points in a Riemannian manifold which has minimum length. { ¦jē·ə¦des·ik }

geodesic circle [MATH] The locus of all points on a given surface whose geodesic distance from a given point on the surface (called the center of the circle) is a given constant. { ¦jē·ə¦des·ik 'sər·kəl }

geodesic completeness [RELAT] Property of a space-time wherein all timelike and null geodesics can be extended to arbitrary values of their affine parameter. Also known as g-completeness. { ¦jē·ə¦des·ik kəm'plēt·nəs }

geodesic coordinates [RELAT] Coordinates in the neighborhood of a point *P* such that the gradient of the metric tensor is zero at *P*. { ¦jē·ə¦des·ik kō'órd·ən·əts }

geodesic curvature [MATH] For a point on a curve lying on a surface, the curvature of the orthogonal projection of the curve onto the tangent plane to the surface at the point; it measures the departure of the curve from a geodesic. Also known as tangential curvature. { ¦jē·ə¦des·ik 'kərv·ə·chər }

geodesic distance [MATH] For two points in a Riemannian manifold, the length of a geodesic connecting them. { ¦jē·ə¦des·ik 'di·stəns }

geodesic dome [ARCH] A dome constructed of many light, straight structural elements in tension, arranged in a framework of triangles to reduce stress and weight. { ¦jē·ə¦des·ik 'dōm }

geodesic ellipse [MATH] The locus of all points on a given surface at which the sum of geodesic distances from a fixed pair of points is a constant. { ¦jē·ə¦des·ik i'lips }

geodesic hyperbola [MATH] The locus of all points on a given surface at which the difference between the geodesic

GEODE

Geode, lined with quartz crystals, Keokuk, Iowa (*Brooks Museum University of Virginia*).

distances to two fixed points is a constant. { ¦jē·ə¦des·ik hī'pər·bə·lə }

geodesic incompleteness [RELAT] Property of a space-time wherein there exists at least one timelike or null geodesic that cannot be extended to arbitrarily large values of its affine parameter; such a space-time contains a singularity. Also known as g-incompleteness. { ¦jē·ə¦des·ik ¸iŋ·kəm'plēt·nəs }

geodesic line [MATH] The shortest line between two points on a mathematically derived surface. { ¦jē·ə¦des·ik 'līn }

geodesic motion [RELAT] Motion of a particle along a geodesic path in the four dimensional space-time continuum; according to general relativity, this is the motion which occurs in the absence of nongravitational forces. { ¦jē·ə¦des·ik 'mō·shən }

geodesic parallels [MATH] Two curves on a given surface such that the lengths of geodesics between the curves that intersect both curves orthogonally is a constant. { ¦jē·ə¦des·ik 'par·ə¸lelz }

geodesic parameters [MATH] Coordinates u and v of a surface such that the curves obtained by setting u equal to various constants form a family of geodesic parallels, while the curves obtained by setting v equal to various constants form the corresponding orthogonal family, of length $u_2 - u_1$ between the points (u_1, v) and (u_2, v). { ¦jē·ə¦des·ik pə'ram·əd·ərz }

geodesic polar coordinates [MATH] Coordinates u and v of a surface such that the curves obtained by setting u equal to various constants are geodesic circles with a common center P and geodesic radius u, and the curves obtained by setting v equal to various constants are geodesics passing through P such that v_0 is the angle between the tangents at P to the lines v = 0 and $v = v_0$. { ¦jē·ə¦des·ik ¦pōl·ər kō'órd·ə·nəts }

geodesic radius [MATH] For a geodesic circle on a surface, the geodesic distance from the center of a circle to the points on the circle. { ¦jē·ə¦des·ik 'rād·ē·əs }

geodesic torsion [MATH] **1.** For a given point on a surface and a given direction, the torsion of the geodesic on the surface through the point and in the given direction. **2.** For a given curve on a surface at a given point, the torsion of the geodesic through the point in the same direction as the given curve. { ¦jē·ə¦des·ik 'tór·shən }

geodesic triangle [MATH] The figure formed by three geodesics joining three points on a given surface. { ¦jē·ə¦des·ik 'trī¸aŋ·gəl }

geodesy [GEOPHYS] A subdivision of geophysics which includes determination of the size and shape of the earth, the earth's gravitational field, and the location of points fixed to the earth's crust in an earth-referred coordinate system. { jē'äd·ə·sē }

geodetic astronomy [GEOD] The branch of geodesy which utilizes astronomic observations to extract geodetic information. { ¦jē·ə¦ded·ik ə'strän·ə·mē }

geodetic coordinates [GEOD] The quantities latitude, longitude, and elevation which define the position of a point on the surface of the earth with respect to the reference spheroid. { ¦jē·ə¦ded·ik kō'órd·ən¸əts }

geodetic datum [GEOD] A datum consisting of five quantities: the latitude and longitude of an initial point, the azimuth of a line from this point, and two constants necessary to define the terrestrial spheroid. { ¦jē·ə¦ded·ik 'dad·əm }

geodetic equator [GEOD] The great circle midway between the poles of revolution of the earth, connecting points of 0° geodetic latitude. { ¦jē·ə¦ded·ik i'kwād·ər }

geodetic gravimetry [GEOD] Worldwide relative measurements of gravitational acceleration used in geodetic studies of the earth. { ¦jē·ə¦ded·ik grə'vim·ə·trē }

geodetic latitude [GEOD] Angular distance between the plane of the equator and a normal to the spheroid; a geodetic latitude differs from the corresponding astronomical latitude by the amount of the meridional component of station error. Also known as geographic latitude; topographical latitude. { ¦jē·ə¦ded·ik 'lad·ə¸tüd }

geodetic line [GEOD] The shortest line between any two points on the surface of the spheroid. { ¦jē·ə¦ded·ik 'līn }

geodetic longitude [GEOD] The angle between the plane of the reference meridian and the plane through the polar axis and the normal to the spheroid; a geodetic longitude differs from the corresponding astronomical longitude by the amount of the prime-vertical component of station error divided by the cosine of the latitude. Also known as geographic longitude. { ¦jē·ə¦ded·ik 'län·jə·tüd }

geodetic meridian [GEOD] A line on a spheroid connecting points of equal geodetic longitude. Also known as geographic meridian. { ¦jē·ə¦ded·ik mə'rid·ē·ən }

geodetic parallel [GEOD] A line connecting points of equal geodetic latitude. Also known as geographic parallel. { ¦jē·ə¦ded·ik 'par·ə¸lel }

geodetic position [GEOD] **1.** A point on the earth, the coordinates of which have been determined by triangulation from an initial station, whose location has been established as a result of astronomical observations, the coordinates depending upon the reference spheroid used. **2.** A point on the earth, defined in terms of geodetic latitude and longitude. { ¦jē·ə¦ded·ik pə'zish·ən }

geodetic precession [RELAT] The precession of a gyroscope in orbit about the earth due to the curvature of space of the Schwarzschild metric. { ¦jē·ə¦ded·ik prē'sesh·ən }

geodetic satellite [AERO ENG] An artificial earth satellite used to obtain data for geodetic triangulation calculations. { ¦jē·ə¦ded·ik 'sad·əl¸īt }

geodetic survey [ENG] A survey in which the figure and size of the earth are considered; it is applicable for large areas and long lines and is used for the precise location of basic points suitable for controlling other surveys. { ¦jē·ə¦ded·ik 'sər¸vā }

geodetic triangle See spheroidal triangle. { ¦jē·ə¦ded·ik 'trī¸aŋ·gəl }

geodynamic height See dynamic height. { ¦jē·ō·dī¦nam·ik 'hīt }

geodynamics [GEOPHYS] The branch of geophysics concerned with measuring, modeling, and interpreting the configuration and motion of the crust, mantle, and core of the earth. { ¦jē·ō·dī¦nam·iks }

geodynamo [GEOPHYS] The self-sustaining process responsible for maintaining the earth's magnetic field in which the kinetic energy of convective motion of the earth's liquid core is converted into magnetic energy. { jē·ō'dī·nə·mō }

geoeconomy [GEOGR] The study of economic conditions that are influenced by geographic factors. { ¦jē·ō·i¦kän·ə·mē }

geoelectricity See terrestrial electricity. { ¦jē·ō·i¸lek'tris·əd·ē }

geoengineering [SCI TECH] Artificial modification of earth systems to counteract anthropogenic effects, such as increasing carbon dioxide uptake by fertilizing ocean surface waters or screening out sunlight with orbiting mirrors. { jē·ō¸en·jə'nir·iŋ }

geoflex See orocline. { 'jē·ə¸fleks }

geognosy [GEOL] The science dealing with the solid body of the earth as a whole, occurrences of minerals and rocks, and the origin of these and their relations. { jē'äg·nə·sē }

geographical botany See plant geography. { ¦jē·ə¦graf·ə·kəl 'bät·ən·ē }

geographical coordinates [GEOGR] Spherical coordinates, designating both astronomical and geodetic coordinates, defining a point on the surface of the earth, usually latitude and longitude. Also known as terrestrial coordinates. { ¦jē·ə¦graf·ə·kəl kō'órd·ən·əts }

geographical cycle See geomorphic cycle. { ¦jē·ə¦graf·ə·kəl 'sī·kəl }

geographical mile [MECH] The length of 1 minute of arc of the Equator, or 6087.08 feet (1855.34 meters), which approximates the length of the nautical mile. { ¦jē·ə¦graf·ə·kəl 'mīl }

geographical plot [NAV] A plot of the movements of one or more craft relative to the surface of the earth. { ¦jē·ə¦graf·ə·kəl 'plät }

geographical position [ASTRON] That point on the earth at which a given celestial body is in the zenith at a specified time. [GEOGR] Any position on the earth defined by means of its geographical coordinates, either astronomical or geodetic. { ¦jē·ə¦graf·ə·kəl pə'zish·ən }

Geographic Information Systems [GEOGR] Computer-based technologies for the storage, manipulation, and analysis of geographically referenced information. { jē·ō¦graf·ik ¸in·fər'mā·shən ¸sis·təmz }

geographic latitude See geodetic latitude. { ¦jē·ə¦graf·ik 'lad·ə¸tüd }

geographic longitude See geodetic longitude. { ¦jē·ə¦graf·ik 'län·jə¸tüd }

geographic meridian *See* geodetic meridian. { ¦jē·ə¦graf·ik mə'rid·ē·ən }

geographic number [NAV] The number assigned to an aid to navigation for identification purposes in accordance with the lateral system of numbering. { ¦jē·ə¦graf·ik 'nəm·bər }

geographic parallel *See* geodetic parallel. { ¦jē·ə¦graf·ik 'par·ə‚lel }

geographic position [GEOGR] The position of a point on the surface of the earth expressed in terms of geographical coordinates either geodetic or astronomical. { ¦jē·ə¦graf·ik pə'zish·ən }

geographic range [GEOD] The extreme distance at which an object or light can be seen when limited only by the curvature of the earth and the heights of the object and the observer. { ¦jē·ə¦graf·ik 'rānj }

geographic search [NAV] An orderly arrangement of course lines in an area in which the area is defined in relation to one or more geographical points on the earth. { ¦jē·ə¦graf·ik 'sərch }

geographic sector search [NAV] An orderly arrangement of course lines consisting of three legs, the turning points being at equal distances along radial lines from a fixed point, the first leg being along the first radial, the second leg along the straight line connecting the equidistant points on the two radials, and the third leg along the course line intersecting a fixed or moving base at the time of return. { ¦jē·ə¦graf·ik 'sek·tər ‚sərch }

geographic speciation [EVOL] Evolution of two or more species from a single species following geographic isolation. { ¦jē·ə¦graf·ik ‚spē·shē'ā·shən }

geographic square search [NAV] An orderly arrangement of course lines (a search) consisting of a series of lines of increasing length, each change of course being 90° in the same direction (right or left), so that the pattern of the search is an expanding square relative to a geographic point. Also known as fixed square search. { ¦jē·ə¦graf·ik 'skwer ‚sərch }

geographic vertical [GEOD] A line perpendicular to the surface of the geoid; it is the direction in which the force of gravity acts. [MAP] The direction of a line normal to the surface of the geoid. Also known as map vertical. { ¦jē·ə¦graf·ik 'vərd·ə·kəl }

Geographos [ASTRON] An asteroid whose orbit has a semimajor axis of 1.25 astronomical units and an eccentricity of 0.335, giving it a perihelion inside the earth's orbit. { jē'äg·rə‚fōs }

geography [SCI TECH] The study of all aspects of the earth's surface, comprising its natural and political divisions, the differentiation of areas, and, sometimes people in relationship to the environment. { jē'äg·rə·fē }

geohydrology [HYD] The science dealing with underground water, often referred to as hydrogeology. { ¦jē·ō‚hī'dräl·ə·jē }

geoid [GEOD] The figure of the earth considered as a sea-level surface extended continuously over the entire earth's surface. { 'jē‚ȯid }

geoidal horizon [ASTRON] That circle of the celestial sphere formed by the intersection of the celestial sphere and a plane tangent to the sea-level surface of the earth at the zenith-nadir line. { jē'ȯid·əl hə'rīz·ən }

geoisotherm [GEOPHYS] The locus of points of equal temperature in the interior of the earth; a line in two dimensions or a surface in three dimensions. Also known as geotherm; isogeotherm. { ¦jē·ō'ī·sə‚thərm }

geolith *See* rock-stratigraphic unit. { 'jē·ə‚lith }

geologic age [GEOL] **1.** Any great time period in the earth's history marked by special phases of physical conditions or organic development. **2.** A formal geologic unit of time that corresponds to a stage. **3.** An informal geologic time unit that corresponds to any stratigraphic unit. { ¦jē·ə¦läj·ik 'āj }

geological oceanography [GEOL] The study of the floors and margins of the oceans, including descriptions of topography, composition of bottom materials, interaction of sediments and rocks with air and sea water, the effects of movements in the mantle on the sea floor, and action of wave energy in the submarine crust of the earth. Also known as marine geology; submarine geology. { ¦jē·ə¦läj·ə·kəl ‚ō·shə'näg·rə·fē }

geological steering [PETRO ENG] In oil mining, the use of geological measurements to place the wellbore accurately in the reservoir relative to a fluid contact, a geological marker, or a reservoir structural feature. { ¦jē·ə¦läj·ə·kəl 'stir·iŋ }

geological survey [GEOL] **1.** An organization making geological surveys and studies. **2.** A systematic geologic mapping of a terrain. { ¦jē·ə¦läj·ə·kəl 'sər‚vā }

geological transportation [GEOL] Shifting of material by the action of moving water, ice, or air. { ¦jē·ə¦läj·ə·kəl ‚tranz·pər'tā·shən }

geologic climate *See* paleoclimate. { ¦jē·ə¦läj·ik 'klī·mət }

geologic column [GEOL] **1.** The vertical sequence of strata of various ages found in an area or region. Also known as column. **2.** The geologic time scale as represented by rocks. { ¦jē·ə¦läj·ik 'käl·əm }

geologic erosion *See* normal erosion. { ¦jē·ə¦läj·ik ə'rō·zhən }

geologic log [GEOL] A graphic presentation of the lithologic or stratigraphic units or both traversed by a borehole; used in petroleum and mining engineering as well as geological surveys. { ¦jē·ə¦läj·ik 'läg }

geologic map [GEOL] A representation of the geologic surface or subsurface features by means of signs and symbols and with an indicated means of orientation; includes nature and distribution of rock units, and the occurrence of structural features, mineral deposits, and fossil localities. { ¦jē·ə¦läj·ik 'map }

geologic noise [GEOPHYS] Disturbances in observed data caused by random inhomogeneities in surface and near-surface material. { ¦jē·ə¦läj·ik'nȯiz }

geologic province [GEOL] An area in which geologic history has been the same. { ¦jē·ə¦läj·ik 'präv·əns }

geologic section [GEOL] Any succession of rock units found at the surface or below ground in an area. Also known as section. { ¦jē·ə¦läj·ik 'sek·shən }

geologic structure [GEOL] The total structural features in an area. { ¦jē·ə¦läj·ik ‚strək·chər }

geologic thermometer *See* geothermometer. { ¦jē·ə¦läj·ik thər'mäm·əd·ər }

geologic thermometry *See* geothermometry. { ¦jē·ə¦läj·ik thər'mäm·ə·trē }

geologic time [GEOL] The period of time covered by historical geology, from the end of the formation of the earth as a separate planet to the beginning of written history. { ¦jē·ə¦läj·ik 'tīm }

geologic time scale [GEOL] The relative age of various geologic periods and the absolute time intervals. { ¦jē·ə¦läj·ik 'tīm ‚skāl }

geologist [GEOL] An individual who specializes in the geological sciences. { jē'äl·ə·jəst }

geolograph [ENG] A device that records the penetration rate of a bit during the drilling of a well. { jē'äl·ə‚graf }

geology [SCI TECH] The study or science of the earth, its history, and its life as recorded in the rocks; includes the study of geologic features of an area, such as the geometry of rock formations, weathering and erosion, and sedimentation. { jē'äl·ə·jē }

geomagnetic coordinates [GEOPHYS] A system of spherical coordinates based on the best fit of a centered dipole to the actual magnetic field of the earth. { ¦jē·ō·mag¦ned·ik kō'ȯrd·ən·əts }

geomagnetic cutoff [GEOPHYS] The minimum energy of a cosmic-ray particle able to reach the top of the atmosphere at a particular geomagnetic latitude. { ¦jē·ō·mag¦ned·ik 'kə‚dȯf }

geomagnetic dipole [GEOPHYS] The magnetic dipole caused by the earth's magnetic field. { ¦jē·ō·mag¦ned·ik 'dī‚pōl }

geomagnetic electrokinetograph [ENG] An instrument that can be suspended from the side of a ship to measure the direction and speed of ocean currents while the ship is under way by measuring the voltage induced in the moving conductive seawater by the magnetic field of the earth. { ¦jē·ō·mag¦ned·ik i¦lek·trə·kə'ned·ə‚graf }

geomagnetic equator [GEOPHYS] That terrestrial great circle which is 90° from the geomagnetic poles. { ¦jē·ō·mag¦ned·ik i'kwād·ər }

geomagnetic field [GEOPHYS] The earth's magnetic field. { ¦jē·ō·mag¦ned·ik 'fēld }

geomagnetic field reversal [GEOPHYS] Reversed magnetization in sedimentary and igneous rock, that is, polarized opposite to the mean geomagnetic field. { ¦jē·ō·mag¦ned·ik 'fēld ‚ri‚vər·səl }

geomagnetic latitude [GEOPHYS] The magnetic latitude

that a location would have if the field of the earth were to be replaced by a dipole field closely approximating it. { ¦jē·ō·mag¦ned·ik 'lad·ə,tüd }

geomagnetic longitude [GEOPHYS] Longitude that is determined around the geomagnetic axis instead of around the rotation axis of the earth. { ¦jē·ō·mag¦ned·ik 'län·jə,tüd }

geomagnetic meridian [GEOPHYS] A semicircle connecting the geomagnetic poles. { ¦jē·ō·mag¦ned·ik mə'rid·ē·ən }

geomagnetic noise [COMMUN] Interference in radio communications arising from terrestrial magnetism. [GEOPHYS] Unwanted frequencies caused by fluctuations in the geomagnetic field of the earth. { ¦jē·ō·mag¦ned·ik 'nȯiz }

geomagnetic pole [GEOPHYS] Either of two antipodal points marking the intersection of the earth's surface with the extended axis of a powerful bar magnet assumed to be located at the center of the earth and having a field approximating the actual magnetic field of the earth. { ¦jē·ō·mag¦ned·ik 'pōl }

geomagnetic reversal [GEOPHYS] Reversed magnetization of the earth's magnetic dipole. { ¦jē·ō·mag¦ned·ik ri'vər·səl }

geomagnetic secular variation See secular variation. { ¦jē·ō·mag¦ned·ik ¦sek·yə·lər ver·ē'a·shən }

geomagnetic storm See magnetic storm. { ¦jē·ō·mag¦ned·ik 'stȯrm }

geomagnetic variation [GEOPHYS] Temporal changes in the geomagnetic field, both long-term (secular) and short-term (transient). { ¦jē·ō·mag¦ned·ik ver·ē'ā·shən }

geomagnetism [GEOPHYS] 1. The magnetism of the earth. Also known as terrestrial magnetism. 2. The branch of science that deals with the earth's magnetism. { ¦jē·ō'mag·nə,tiz·əm }

geomembrane [CIV ENG] Any impermeable membrane (usually made of synthetic polymers in sheets) used with soils, rock, earth, or other geotechnical material in order to block the migration of fluids. { jē·ō'mem,brān }

geometrical acoustics See ray acoustics. { ¦jē·ə¦me·trə·kəl ə'kü·stiks }

geometric albedo [OPTICS] The ratio of the light flux received from an object to that which would be received from a perfectly reflecting, perfectly diffusing disk of the same size at the same distance at zero phase angle. { ¦jē·ə¦me·trik al'bē·dō }

geometrical dip [GEOD] The vertical angle, at the eye of an observer, between the horizontal and a straight line tangent to the surface of the earth. { ¦jē·ə¦me·trə·kəl 'dip }

geometrical distortion [COMPUT SCI] A discrepancy between the horizontal and vertical dimensions of the picture elements on an electronic display, causing, for example, circles to appear as ovals unless corrected for in software. { ¦jē·ə¦me·trə·kəl di'stȯr·shən }

geometrical factor [NAV] In loran and similar navigation systems, the ratio of the linear distance between a point on a loran line of position (hyperbola) and the nearest point on an adjacent line of the same family of lines to the interval between the corresponding time difference. { ¦jē·ə¦me·trə·kəl 'fak·tər }

geometrical horizon [GEOD] The intersection of the celestial sphere and an infinite number of straight lines radiating from the eye of the observer and tangent to the earth's surface. { ¦jē·ə¦me·trə·kəl hə'rīz·ən }

geometrical isomerism [PHYS CHEM] The phenomenon in which isomers contain atoms attached to each other in the same order and with the same bonds but with different spatial, or geometrical, relationships; the explicit geometry imposed upon a molecule by, say, a double bond between carbon atoms makes possible the existence of these isomers. { ¦jē·ə¦me·trə·kəl ī'sä·mə,riz·əm }

geometrical optics [OPTICS] The geometry of paths of light rays and their imagery through optical systems. { ¦jē·ə¦me·trə·kəl 'äp·tiks }

geometrical pitch [AERO ENG] The distance a component of an airplane propeller would move forward in one complete turn of the propeller if the path it was moving along was a helix that had an angle equal to an angle between a plane perpendicular to the axis of the propeller and the chord of the component. { ¦jē·ə¦me·trə·kəl 'pich }

geometrical similarity [FL MECH] Property of two fluid flows for which a simple alteration of scales of length and velocity transforms one into the other. { ¦jē·ə¦me·trə·kəl ¦sim·ə'lar·əd·ē }

geometric attenuation [NUCLEO] That part of the reduction in the intensity of ionizing radiation with distance from a source which is associated with spreading out of the radiation and is independent of the interaction of the radiation with matter. { ¦jē·ə¦me·trik ə,ten·yə'wā·shən }

geometric average See geometric mean. { ¦jē·ə¦me·trik 'av·rij }

geometric complex See simplicial complex. { ¦jē·ə¦me·trik 'käm,pleks }

geometric construction [ENG] Construction that employs only straightedge and compasses or is carried out by drawing only straight lines and circles. { ¦jē·ə¦me·trik kən'strək·shən }

geometric distribution [STAT] A discrete probability distribution whose probability function is given by the equation $P(x) = p(1 - p)^{x-1}$ for x any positive integer, $p(x) = 0$ otherwise, when $0 \leq p \leq 1$; the mean is $1/p$. { ¦jē·ə¦me·trik ,dis·trə'byü·shən }

geometric duals [MATH] Two polyhedra such that the vertices of one are in unique correspondence with the faces of the other. { ¦jē·ə¦me·trik 'dülz }

geometric latitude See reduced latitude. { ¦jē·ə¦me·trik 'lad·ə,tud }

geometric mean [MATH] The geometric mean of n given quantities is the nth root of their product. Also known as geometric average. { ¦jē·ə¦me·trik 'mēn }

geometric moment of inertia [MATH] The geometric moment of inertia of a plane figure about an axis in or perpendicular to the plane is the integral over the area of the figure of the square of the distance from the axis. Also known as second moment of area. { ¦jē·ə¦me·trik ¦mō·mənt əv i'nər·shə }

geometric number theory [MATH] The branch of number theory studying relationships among numbers by examining the geometric properties of ordered pair sets of such numbers. { ¦jē·ə¦me·trik 'nəm·bər ,thē·ə·rē }

geometric phase [PHYS] A unifying mathematical concept that describes the relation between the history of internal states of a system and the system's resulting orientation in space. { ¦jē·ə¦me·trik 'fāz }

geometric programming [SYS ENG] A nonlinear programming technique in which the relative contribution of each of the component costs is first determined; only then are the variables in the component costs determined. { ¦jē·ə¦me·trik 'prō ,gram·iŋ }

geometric progression [MATH] A sequence which has the form $a, ar, ar^2, ar^3, \ldots$ { ¦jē·ə¦me·trik prə'gresh·ən }

geometric sequence [MATH] A sequence in which the ratio of a term to its predecessor is the same for one term as for any other. { ¦jē·ə,me·trik 'sē·kwəns }

geometric series [MATH] An infinite series of the form $a + ar + ar^2 + ar^3 + \cdots$ { ¦jē·ə¦me·trik 'sir·ēz }

geometric shadow [PHYS] That region which a given type of radiation would not reach, because of the presence of an object, if the effects of diffraction and interference could be neglected. { ¦jē·ə¦me·trik 'shad·ō }

geometric vertical See geocentric vertical. { ¦jē·ə¦me·trik 'vərd·ə·kəl }

Geometridae [INV ZOO] A large family of lepidopteran insects in the superfamily Geometroidea that have slender bodies and relatively broad wings; includes measuring worms, loopers, and cankerworms. { jē·ə'me·trə,dē }

geometrodynamics [RELAT] A theory involving only geometry which attempts to combine gravitational and electromagnetic theory; characterized by a multiply connected space-time manifold containing structures, descriptively called wormholes, associated with electric charge. { ¦jē·ō,me·trə·dī'nam·iks }

Geometroidea [INV ZOO] A superfamily of lepidopteran insects in the suborder Heteroneura comprising small to large moths with reduced maxillary palpi and tympanal organs at the base of the abdomen. { jē·ə·mə'trȯid·ē·ə }

geometry [MATH] The qualitative study of shape and size. { jē'äm·ə·trē }

geomorphic cycle [GEOL] The cycle of change in the surface configuration of the earth. Also known as cycle of erosion; geographical cycle. { ¦jē·ō'mȯr·fik 'sī·kəl }

geomorphology [GEOL] The study of the origin of secondary topographic features which are carved by erosion in the primary elements and built up of the erosional debris. { ¦jē·ō·mȯr'fäl·ə·jē }

Geomyidae [VERT ZOO] The pocket gophers, a family of rodents characterized by fur-lined cheek pouches which open outward, a stout body with short legs, and a broad, blunt head. { jē·ō'mī·ə‚dē }

geon [PHYS] A hypothetical electromagnetic field that is held together by its own gravitational attraction. { 'jē‚än }

geonavigation [NAV] Plotting a course by computation based on information obtained by reference to other places on earth. { jē·ō‚nav·ə'gā·shən }

geonium [ATOM PHYS] A microscopic system consisting of a single electron in a Penning trap, forming a kind of synthetic atom. Also known as monoelectron oscillator. { jē'ō·nē·əm }

geopetal [PETR] Pertaining to the top-to-bottom relations in rocks at the time of formation. { jē·ə¦ped·əl }

geopetal fabric [PETR] The internal structure of a rock indicating the original orientation of the top-to-bottom strata. { ¦jē·ə¦ped·əl 'fab·rik }

geophagia [ZOO] Soil ingestion by animals. { jē·ə'fā·jə }

geophagous [ZOO] Feeding on soil, as certain worms. { jē'äf·ə·gəs }

Geophilomorpha [INV ZOO] An order of centipedes in the class Chilopoda including specialized forms that are blind, epimorphic, and dorsoventrally flattened. { jē·ō‚fil·ə'mór·fə }

geophilous [ECOL] Living or growing in or on the ground. { jē'äf·ə·ləs }

geophone [ELECTR] A transducer, used in seismic work, that responds to motion of the ground at a location on or below the surface of the earth. { 'jē·ə‚fōn }

geophysical engineering [ENG] A branch of engineering that applies scientific methods for locating mineral deposits. { ¦jē·ə¦fiz·ə·kəl ‚en·jə'nir·iŋ }

geophysical fluid dynamics [GEOPHYS] The study of the naturally occurring, large-scale flows in the atmosphere and oceans, such as in weather patterns, atmospheric fronts, ocean currents, coastal upwelling, and the El Niño phenomenon. { jē·ō¦fiz·ə·kəl ¦flü·əd dī'nam·iks }

geophysical prospecting [ENG] Application of quantitative concepts and principles of physics and mathematics in geologic explorations to discover the character of and mineral resources in underground rocks in the upper portions of the earth's crust. { ¦jē·ə¦fiz·ə·kəl 'prä‚spek·tiŋ }

geophysicist [GEOPHYS] An individual who specializes in geophysics. { jē'fiz·ə·sist }

geophysics [GEOL] The physics of the earth and its environment, that is, earth, air, and (by extension) space. { ¦jē·ə'fiz·iks }

geophyte [ECOL] A perennial plant that is deeply embedded in the soil substrata. { 'jē·ə‚fīt }

geopotential [PHYS] The potential energy of a unit mass relative to sea level, numerically equal to the work that would be done in lifting the unit mass from sea level to the height at which the mass is located, against the force of gravity. { ¦jē·ō·pə'ten·chəl }

geopotential height [GEOPHYS] The height of a given point in the atmosphere in units proportional to the potential energy of unit mass (geopotential) at this height, relative to sea level. { ¦jē·ō·pə'ten·chəl 'hīt }

geopotential number [GEOPHYS] The numerical value C that is assigned to a given geopotential surface when expressed in geopotential units (1 gpu = 1 meter × 1 kilogal). { ¦jē·ō·pə'ten·chəl 'nəm·bər }

geopotential surface [GEOPHYS] A surface of constant geopotential, that is, a surface along which a parcel of air could move without undergoing any changes in its potential energy. Also known as equigeopotential surface; level surface. { ¦jē·ō·pə'ten·chəl 'sər·fəs }

geopotential thickness [GEOPHYS] The difference in the geopotential height of two constant-pressure surfaces in the atmosphere, proportional to the appropriately defined mean air temperature between the two surfaces. { ¦jē·ō·pə'ten·chəl 'thik·nəs }

geopotential topography [GEOPHYS] The topography of any surface as represented by lines of equal geopotential; these lines are the contours of intersection between the actual surface and the level surfaces (which everywhere are normal to the direction of the force of gravity), and are spaced at equal

intervals of dynamic height. Also known as absolute geopotential topography. { ¦jē·ō·pə'ten·chəl tə'päg·rə·fē }

geopotential unit [GEOPHYS] A unit of gravitational potential used in describing the earth's gravitational field; it is equal to the difference in gravitational potential of two points separated by a distance of 1 meter when the gravitational field has a strength of 10 meters per second squared and is directed along the line joining the points. Abbreviated gpu. { ¦jē·ō·pə'ten·chəl 'yü·nət }

geopressure [GEOPHYS] An unusually high pressure exerted by a subsurface formation. { 'jē·ō‚presh·ər }

geopressurized geothermal system [GEOL] A geothermal system dominated by the presence of hot fluids under high pressure (brine plus methane) and having higher-than-normal temperatures because of their low thermal conductivity, the presence of interbedded shale layers, or the existence of local, exothermic chemical reactions. { ¦jē·ō'presh·ə‚rīzd ¦jē·ō¦thər·məl 'sis·təm }

georef See world geographic reference system. { 'jē·ō‚ref }

georef grid [MAP] The grid system used on U.S. Air Force aeronautical charts for identifying the location of any point or area in the world. { 'jē·ō‚ref ‚grid }

Georges Banks [GEOL] An elevation beneath the sea east of Cape Cod, Massachusetts. { ¦jór·jəz 'baŋks }

georgette [TEXT] Heavy sheer crepe made of yarn twisted both ways in the weave; the same yarn is usually used in warp and filling. { jór'jet }

georgiadesite [MINERAL] $Pb_3(AsO_4)Cl_3$ A white or brownish-yellow mineral composed of lead chloroarsenate, occurring in orthorhombic crystals. { jór'jäd·ə‚sīt }

georgiaite [GEOL] Any of a group of North American tektites, 134 million years of age, found in Georgia. { 'jór·jə‚īt }

Georyssidae [INV ZOO] The minute mud-loving beetles, a family of coleopteran insects belonging to the Polyphaga. { jē·ə'ris·ə‚dē }

geoscience See earth science. { 'jē·ō‚sī·əns }

geosensing [BOT] The sensing or detecting of gravity by a plant relative to its longitudinal axis. { 'jē·ō‚sens·iŋ }

geosere [GEOL] A series of ecological climax communities following each other in geologic time and changing in response to changing climate and physical conditions. { 'jē·ō‚sir }

Geosiridaceae [BOT] A monotypic family of monocotyledonous plants in the order Orchidales characterized by regular flowers with three stamens that are opposite the sepals. { jē·ə‚sir·ə'dās·ē‚ē }

geosol [GEOL] A body of sediment or rock composed of one or more soil horizons. { 'jē·ə‚sōl }

geosophy [GEOGR] The study of the nature and expression of geographical knowledge, both past and present. { je'äs·ə·fē }

geosphere [GEOL] **1.** The solid mass of earth, as distinct from the atmosphere and hydrosphere. **2.** The lithosphere, hydrosphere, and atmosphere combined. { 'jē·ō‚sfir }

Geospizinae [VERT ZOO] Darwin finches, a subfamily of perching birds in the family Fringillidae. { jē·ō'spiz·ə‚nē }

geostatic pressure See ground pressure. { ¦jē·ō¦stad·ik 'presh·ər }

geostationary satellite [AERO ENG] A satellite that follows a circular orbit in the plane of the earth's equator from west to east at such a speed as to remain fixed over a given place on the equator at an altitude of 22,280 miles (35,860 kilometers). { ¦jē·ō¦stā·shə‚ner·ē 'sad·əl‚īt }

geostatistics [GEOL] A branch of applied statistics that focuses on mathematical description and analysis of geological observations. { jē·ō·stə'tis·tiks }

geostrophic [GEOPHYS] Pertaining to deflecting force resulting from the earth's rotation. { ¦jē·ō¦sträf·ik }

geostrophic approximation [GEOPHYS] The assumption that the geostrophic current can represent the actual horizontal current. Also known as geostrophic assumption. { ¦jē·ō¦sträf·ik ə¦präk·sə'mā·shən }

geostrophic assumption See geostrophic approximation. { ¦jē·ō¦sträf·ik ə'səm·shən }

geostrophic current [GEOPHYS] A current defined by assuming the existence of an exact balance between the horizontal pressure gradient force and the Coriolis force. { ¦jē·ō¦sträf·ik 'kə·rənt }

geostrophic departure [METEOROL] A vector representing the difference between the real wind and the geostrophic wind.

GEOPHILOMORPHA

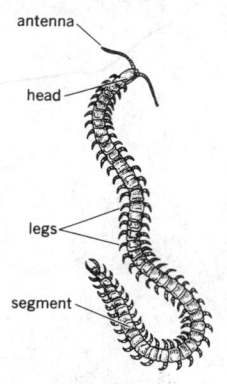

antenna

head

legs

segment

An epimorphic chilopod, an unidentified species 40 millimeters long. *(From R. E. Snodgrass, A Textbook of Arthropod Anatomy, copyright 1952 by Cornell University Press; used by permission)*

Also known as ageostrophic wind; geostrophic deviation. { ¦jē·ō¦sträf·ik di'pär·chər }

geostrophic deviation *See* geostrophic departure. { ¦jē·ō¦sträf·ik ‚dē·vē'ā·shən }

geostrophic distance [METEOROL] The distance (in degrees latitude) along a constant-pressure surface over which the change in height (in feet) is equal to the geostrophic wind speed (in knots). { ¦jē·ō¦sträf·ik 'dis·təns }

geostrophic equation [GEOPHYS] An equation, used to compute geostrophic current speed, which represents a balance between the horizontal pressure gradient force and the Coriolis force. { ¦jē·ō¦sträf·ik i'kwā·shən }

geostrophic equilibrium [GEOPHYS] A state of motion of a nonviscous fluid in which the horizontal Coriolis force exactly balances the horizontal pressure force at all points of the field so described. { ¦jē·ō¦sträf·ik ‚ē·kwə'lib·rē·əm }

geostrophic flow [GEOPHYS] A form of gradient flow where the Coriolis force exactly balances the horizontal pressure force. { ¦jē·ō¦sträf·ik 'flō }

geostrophic flux [METEOROL] The transport of an atmospheric property by means of the geostrophic wind. { ¦jē·ō¦sträf·ik 'fləks }

geostrophic vorticity [METEOROL] The vorticity of the geostrophic wind. { ¦jē·ō¦sträf·ik vór'tis·əd·ē }

geostrophic wind [METEOROL] That horizontal wind velocity for which the Coriolis acceleration exactly balances the horizontal pressure force. { ¦jē·ō¦sträf·ik 'wind }

geostrophic-wind level [METEOROL] The lowest level at which the wind becomes geostrophic in the theory of the Ekman spiral. Also known as gradient-wind level. { ¦jē·ō¦sträf·ik 'wind ‚lev·əl }

geostrophic-wind scale [METEOROL] A graphical device used for the determination of the speed of the geostrophic wind from the isobar or contour-line spacing on a synoptic chart; it is a nomogram representing solutions of the geostrophic-wind equation. { ¦jē·ō¦sträf·ik 'wind ‚skāl }

geosynchronous orbit [AERO ENG] A satellite orbit that has a period of one sidereal day (23 hours, 56 minutes, 4 seconds). Abbreviated GEO. { ¦jē·ō¦siŋ·krə·nəs 'ór·bət }

geosynchronous satellite [AERO ENG] An earth satellite that makes one revolution in one sidereal day (23 hours, 56 minutes, 4 seconds), synchronous with the earth's rotation; the orbit can have arbitrary eccentricity and arbitrary inclination to the earth's equator. Also known as synchronous satellite. { ¦jē·ō¦siŋ·krə·nəs 'sad·əl‚īt }

geosynclinal couple *See* orthogeosyncline. { ¦jē·ō‚sin'klīn·əl 'kəp·əl }

geosynclinal cycle *See* tectonic cycle. { ¦jē·ō‚sin'klīn·əl 'sī·kəl }

geosynclinal facies [GEOL] A sedimentary facies marked by great thickness, a generally argillaceous character, and few carbonate rocks. { ¦jē·ō‚sin'klīn·əl 'fā·shēz }

geosyncline [GEOL] A linear part of the earth's crust, hundreds of kilometers long and tens of kilometers wide, that subsided during millions of years as it received thousands of meters of sedimentary and volcanic accumulations. { ¦jē·ō'sin‚klīn }

geosynthetic [CIV ENG] Any synthetic material used in geotechnical engineering, such as geotextiles and geomembranes. { ¦jē·ō·sin'thed·ik }

geotaxis [PHYSIO] Movement of a free-living organism in response to the stimulus of gravity. { ¦jē·ō¦tak·səs }

geotechnics [CIV ENG] The application of scientific methods and engineering principles to civil engineering problems through acquiring, interpreting, and using knowledge of materials of the crust of the earth. { ¦jē·ō¦tek·niks }

geotechnology [ENG] Application of the methods of engineering and science to exploitation of natural resources. { ¦jē·ō·tek'näl·ə·jē }

geotectogene *See* tectogene. { ¦jē·ō¦tek·tə‚jēn }

geotectonic cycle *See* orogenic cycle. { ¦jē·ō·tek'tän·ik 'sī·kəl }

geotectonics *See* tectonics. { ¦jē·ō·tek'tän·iks }

geotextiles [CIV ENG] Woven or nonwoven fabrics used with foundations, soils, rock, earth, or other geotechnical material as an integral part of a manufactured project, structure, or system. Also known as civil engineering fabrics; erosion control cloth; filter fabrics; support membranes. { ¦jē·ō¦tek‚stīlz }

geotherm *See* geoisotherm. { ¦jē·ō‚thərm }

geothermal [GEOPHYS] Pertaining to heat within the earth. { ¦jē·ō¦thər·məl }

geothermal energy [GEOPHYS] Thermal energy contained in the earth; can be used directly to supply heat or can be converted to mechanical or electrical energy. { ¦jē·ō‚thərm·əl 'en·ər·jē }

geothermal gradient [GEOPHYS] The change in temperature with depth of the earth. { ¦jē·ō¦thər·məl 'grād·ē·ənt }

geothermal prospecting [ENG] Exploration for sources of geothermal energy. { ¦jē·ō¦thər·məl 'prä‚spek·tiŋ }

geothermal system [GEOL] Any regionally localized geological setting where naturally occurring portions of the earth's internal heat flow are transported close enough to the earth's surface by circulating steam or hot water to be readily harnessed for use; examples are the Geysers Region of northern California and the hot brine fields in the Imperial Valley of southern California. { ¦jē·ō¦thər·məl 'sis·təm }

geothermal well logging [ENG] Measurement of the change in temperature of the earth by means of well logging. { ¦jē·ō¦thər·məl 'wel ‚läg·iŋ }

geothermometer [ENG] A thermometer constructed to measure temperatures in boreholes or deep-sea deposits. [GEOL] A mineral that yields information about the temperature range within which it was formed. Also known as geologic thermometer. { ¦jē·ō·thər'mäm·əd·ər }

geothermometry [GEOL] Measurement of the temperatures at which geologic processes occur or occurred. Also known as geologic thermometry. { ¦jē·ō·thər'mäm·ə·trē }

geotropism [BOT] Response of a plant to the force of gravity. Also known as gravitropism. { jē'ä·trə‚piz·əm }

Gephyrea [INV ZOO] A class of burrowing worms in the phylum Annelida. { jə'fir·ē·ə }

gephyrocercal [VERT ZOO] Having the dorsal and anal fins coming together smoothly at the aborted end of the vertebral column of a fish's tail. { ¦jef·ə·rō¦sər·kəl }

Geraniaceae [BOT] A family of dicotyledonous plants in the order Geraniales in which the fruit is beaked, styles are usually united, and the leaves have stipules. { jə‚rā·nē'ās·ē‚ē }

geranial *See* citral. { jə'rā·nē·əl }

geranialdehyde *See* citral. { jə‚rā·nē'al·də‚hīd }

Geraniales [BOT] An order of dicotyledonous plants in the subclass Rosidae comprising herbs or soft shrubs with a superior ovary and with compound or deeply cleft leaves. { jə‚rā·nē'ā·lēz }

geraniol [ORG CHEM] $(CH_3)_2CCH(CH_2)_2C(CH_3)CHCH_2OH$ A colorless to pale-yellow liquid, an alcohol and a terpene, boiling point 230°C; soluble in alcohol and ether, insoluble in water; used in perfumery and flavoring. { jə'rā·nē‚ól }

Geranium [BOT] A genus of plants in the family Geraniaceae characterized by regular flowers, and glands alternating with the petals. { jə'rā·nē·əm }

geranium oil [MATER] A pale-yellow or green liquid distilled from the herb of several Pelargonium species; chief known constituents are citronellol and geraniol, both used in perfumery and as flavoring agents. { jə'rā·nē·əm ‚óil }

geranyl [ORG CHEM] $C_{10}H_{17}$ The functional group from geraniol, $(CH_3)_2$:$CHCH_2CH_2$·$CHCH_3$:CH·CH_2OH. { jə'ran·əl }

Gerardiidae [INV ZOO] A family of anthozoans in the order Zoanthidea. { ‚jər·är'dī·ə‚dē }

Gerard reagent [CHEM] The quaternary ammonium compounds, acethydrazide-pyridinium chloride and trimethylacethydrazide ammonium chloride; used to separate aldehydes and ketones from oily or fatty natural materials and to extract sex hormones from urine. { jə'rärd rē‚ā·jənt }

gerber beam [CIV ENG] A long, straight beam that functions essentially as a cantilevered beam by the insertion of two hinges in alternate spans. { 'gər·bər ‚bēm }

Gerber test [FOOD ENG] A volumetric measurement of the quantity of fat in a sample of milk; the European counterpart of the Babcock test. { 'gər·bər ‚test }

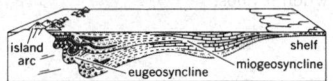

GEOSYNCLINE

Diagrammatic section of Cordilleran geosyncline in southeastern Alaska and British Columbia at the close of the Permian. Volcanic deposits are indicated in black. (*After A. J. Eardley, J. Geol., 55:319–342, 1947*)

GERANIALES

A common eastern United States species of geranium (*Geranium maculatum*), Which is characteristic of the order Geraniales. (*Courtesy of A. W. Ambler, form the National Audubon Society*)

GERBIL

A gerbil, with long tail for balance when it hops.

gerbil [VERT ZOO] The common name for about 100 species of African and Asian rodents composing the subfamily Gerbillinae. { 'jər·bəl }

Gerbillinae [VERT ZOO] The gerbils, a subfamily of rodents in the family Muridae characterized by hindlegs that are longer than the front ones, and a long, slightly haired, usually tufted tail. { jər'bil·ə‚nē }

gerhardtite [MINERAL] Cu₂(NO₃)(OH)₃ An emerald-green mineral composed of basic copper nitrate. { 'ger‚härd‚īt }

Gerhardt's test [PATH] A test for acetoacetic acid in urine. { 'ger‚härts ‚test }

geriatrics [MED] The study of the biological and physical changes and the diseases of old age. { ‚jer·ē'a·triks }

germ [BIOL] A primary source, especially one from which growth and development are expected. [MICROBIO] General designation for a microorganism. { jərm }

German cupellation [MET] A refining method using a large reverberatory furnace with a fixed bed and a movable roof; the bullion to be cupelled is all charged at once, and the silver is not refined in the same furnace where the cupellation is carried on. { ¦jər·mən ‚kyü·pə'lā·shən }

germane [INV ZOO] **1.** A hydride of germanium whose general formula is GeₙH_{2n+2}. **2.** The compound GeH₄, a hydride of germanium, a colorless gas that is combustible in air and burns with a blue flame. { ¦jər¦mān }

germanide [INV ZOO] A compound of an alkaline earth or alkali metal with germanium; an example is magnesium germanide, Mg₂Ge; the germanides are reactive with water. { 'jər·mə‚nīd }

germanite [MINERAL] Cu₃(Ge,Ga,Fe)(S,As)₄ Reddish-gray mineral occurring in massive form; an important source of germanium. { 'jər·mə‚nīt }

germanium [CHEM] A brittle, water-insoluble, silvery-gray metallic element in the carbon family, symbol Ge, atomic number 32, atomic weight 72.59, melting at 959°C. [MET] A rare metal used in semiconductors, alloys, and glass. { jər'mān·ē·əm }

germanium diode [ELECTR] A semiconductor diode that uses a germanium crystal pellet as the rectifying element. Also known as germanium rectifier. { jər'mān·ē·əm 'dī‚ōd }

germanium halide [INV ZOO] A dihalide or tetrahalide of fluorine, chlorine, bromine, or iodine with germanium. { jər'mān·ē·əm 'ha‚līd }

germanium oxide [INV ZOO] The monoxide GeO or dioxide GeO₂; a study of GeO indicates it exists in polymeric form; GeO₂ is a white powder, soluble in alkalies; used in special glass and in medicine. { jər'mān·ē·əm 'äk‚sīd }

germanium rectifier See germanium diode. { jər'mān·ē·əm 'rek·tə‚fī·ər }

germanium transistor [ELECTR] A transistor in which the semiconductor material is germanium, to which electric contacts are made. { jər'mān·ē·əm tran'zis·tər }

German measles See rubella. { ¦jər·mən 'mē·zəlz }

German R unit [NUCLEO] A unit of radiation dose rate due to x-rays, equal to approximately 2.5 Solomon R units, or approximately 1.5 roentgens per second. Also known as R unit. { ¦jər·mən 'är ‚yü·nət }

German silver See nickel silver. { ¦jər·mən 'sil·vər }

German tubbing [MIN ENG] A form of tubbing, with internal flanges and bolts, for lining circular shafts sunk through heavily watered strata. { ¦jər·mən 'təb·iŋ }

germarium [INV ZOO] The egg-producing portion of an ovary and the sperm-producing portion of a testis in Platyhelminthes and Rotifera. { jər'mar·ē·əm }

germ ball [INV ZOO] A group of cells in digenetic trematode miracidial larvae which are embryos. { 'jərm ‚ból }

germ cell See gamete. { 'jərm ‚sel }

germfree animal [MICROBIO] An animal having no demonstrable, viable microorganisms living in intimate association with it. { 'jərm‚frē 'an·ə·məl }

germfree isolator [MICROBIO] An apparatus that provides a mechanical barrier surrounding the area in which germfree vertebrates and accessory equipment are housed. { 'jərm‚frē 'ī·sə‚lād·ər }

germicide [MATER] An agent that destroys germs. { 'jər·mə‚sīd }

germinal epithelium [EMBRYO] The region of the dorsal coelomic epithelium lying between the dorsal mesentery and the mesonephros. { 'jər·mən·əl ‚ep·ə'thē·lē·əm }

germinal spot [CELL MOL] The nucleolus of an egg cell. { 'jər·mən·əl ‚spät }

germinal vesicle [CELL MOL] The enlarged nucleus of the primary oocyte before reduction divisions are complete. { 'jər·mən·əl 'ves·ə·kəl }

germination [BOT] The beginning or the process of development of a spore or seed. [PETR] See grain growth. { ‚jər·mə'nā·shən }

germ layer [EMBRYO] One of the primitive cell layers which appear in the early animal embryo and from which the embryo body is constructed. { 'jərm ‚lā·ər }

germ-layer theory [EMBRYO] The theory that three primary germ layers, ectoderm, mesoderm, and endoderm, are established in the early embryo and all organs and structures are derived from a specific germ layer. { 'jərm ‚lā·ər ‚thē·ə·rē }

germline [BIOL] A lineage of cells from which gametes are derived. Also known as germ track. { 'jərm ‚līn }

germline mutation [GEN] A mutation within a lineage of cells that form gametes. { ¦jərm‚līn myü'tā·shən }

germovitellarium [INV ZOO] A sex gland which differentiates into a yolk-producing or egg-producing region. { ¦jər‚mō‚vīd·əl'a·rē·əm }

germ plasm [BIOL] The genetic material contained within a germ cell. { 'jərm ‚plaz·əm }

germ theory [MED] The theory that contagious and infectious diseases are caused by microorganisms. { 'jərm ‚thē·ə·rē }

germ track See germ line. { 'jərm ‚trak }

geroderma [MED] The skin of old age, showing atrophy, loss of fat, and loss of elasticity. { ‚jer·ō'dər·mə }

gerontology [PHYSIO] The scientific study of aging processes in biological systems, particularly in humans. { ‚jer·ən'täl·ə·jē }

Gerrhosauridae [VERT ZOO] A small family of lizards in the suborder Sauria confined to Africa and Madagascar. { ‚jer·ō'sór·ə‚dē }

Gerridae [INV ZOO] The water striders, a family of hemipteran insects in the subdivision Amphibicorisae having long middle and hind legs and a median scent gland opening on the metasternum. { 'jer·ə‚dē }

Gerroidea [INV ZOO] The single superfamily of the hemipteran subdivision Amphibicorisae; all members have conspicuous antennae and hydrofuge hairs covering the body. { jə'róid·ē·ə }

gersdorffite [MINERAL] NiAsS A silver-white to steel-gray mineral, crystallizing in the isometric system; resembles cobaltite and may contain some iron and cobalt. Also known as nickel glance. { 'gerz‚dór‚fīt }

Gershgorin's method [MATH] A method of obtaining bounds on the eigenvalue of a matrix, based on the fact that the absolute value of any eigenvalue is equal to or less than the maximum over the rows of the matrix of the sum of the absolute values of the entries in a row, and is also equal to or less than the maximum over the columns of the matrix of the sum of the absolute values of the entries in a column. { gərsh'gór·ənz ‚meth·əd }

Gerstner wave [FL MECH] A rotational gravity wave of finite amplitude. { 'gerst·nər ‚wāv }

Gesneriaceae [BOT] A family of dicotyledonous plants in the order Scrophulariales characterized by parietal placentation, mostly opposite or whorled leaves, and a well-developed embryo. { ge‚snir·ē'ās·ē‚ē }

gesso [MATER] A material made from chalk and gelatin or casein glue; painted on panels to furnish a surface for tempera work or for polymer-based paints. { 'je·sō }

gestalt [METEOROL] A complex of weather elements occurring in a familiar form, and though not necessarily referring to basic hydrodynamical or thermodynamical quantities, may persist for an appreciable length of time and is often considered to be an entity in itself. { ge'shtält }

gestalt psychology [PSYCH] A school of psychology that views and examines the person as a whole. { ge'shtält sī'käl·ə·jē }

gestalt vision [PHYSIO] The visual perception and retention in the memory of an object in terms of its geometric shape. { gə¦shtält ‚vizh·ən }

gestate [EMBRYO] To carry the young in the uterus from conception to delivery. { 'je,stāt }

gestation period [EMBRYO] The period in mammals from fertilization to birth. { jə'stā·shən ,pir·ē·əd }

get [COMPUT SCI] An instruction in a computer program to read data from a file. [IND ENG] A combination of two or more of the elemental motions of search, select, grasp, transport empty, and transport loaded; applied to time-motion studies. { get }

getmain [COMPUT SCI] An instruction used in some programming languages to request dynamic allocation of additional storage space to the program. { 'get,mān }

getter [CHEM] See scavenger. [PHYS CHEM] **1.** A substance, such as thallium, that binds gases on its surface and is used to maintain a high vacuum in a vacuum tube. **2.** A special metal alloy that is placed in a vacuum tube during manufacture and vaporized after the tube has been evacuated; when the vaporized metal condenses, it absorbs residual gases. Also known as degasser. { 'ged·ər }

getter-ion pump [ENG] A high-vacuum pump that employs chemically active metal layers which are continuously or intermittently deposited on the wall of the pump, and which chemisorb active gases while inert gases are "cleaned up" by ionizing them in an electric discharge and drawing the positive ions to the wall, where the neutralized ions are buried by fresh deposits of metal. Also known as sputter-ion pump. { 'ged·ər 'ī,än ,pəmp }

getter sputtering [ELECTR] The deposition of high-purity thin films at ordinary vacuum levels by using a getter to remove contaminants remaining in the vacuum. { 'gəd·ər ,spəd·ə·riŋ }

GeV See gigaelectronvolt.

gewel hinge [DES ENG] A hinge consisting of a hook inserted in a loop. { 'jü·əl ,hinj }

geyser [HYD] A natural spring or fountain which discharges a column of water or steam into the air at more or less regular intervals. { 'gī·zər }

geyserite See siliceous sinter. { 'gī·zə,rīt }

gf See gram-force.

g factor See Landé g factor.

G factor See G value. { 'jē 'fak·tər }

g force [PHYS] A force such that a body subjected to it would have the acceleration of gravity at sea level; used as a unit of measurement for bodies undergoing the stress of acceleration. { 'jē ,fors }

GFP See green fluorescent protein.

gf-value [ATOM PHYS] The product of the oscillator strength *f* of an atomic transition and the statistical weight *g* of the lower level. Also known as weighted oscillator strength. { 'jē'ef ,val·yü }

GH See growth hormone.

GHA See Greenwich hour angle.

gharbi [METEOROL] A fresh westerly wind of oceanic origin in Morocco. { 'gär·bē }

gharra [METEOROL] Hard squalls from the northeast in Libya and Africa that are sudden and frequent, and are accompanied by heavy rain and thunder. { 'gä·rä }

ghatti gum [MATER] A water-soluble gum from the tree *Anogeissus latifolia*, forming a viscous glue in water; used as an emulsifier. { 'gäd·ē ,gəm }

ghedda wax [MATER] A beeswax that is obtained from African and Indian bees. { 'ged·ə ,waks }

ghee [FOOD ENG] In India, a common food fat produced from boiled buffalo milk; it can be kept for years, without refrigeration, and has a more intense flavor than butter or butter oil. { gē }

ghibli [METEOROL] A hot, dust-bearing, desert wind in North Africa, similar to the foehn. Also known as chibli; gebli; gibleh; gibli; kibli. { 'gib·lē }

Ghirardi-Rimini-Weber-Pearle theory See GRWP theory. { ,gi,rär·dē ,rim·i·nē ,vā·bər ,pərl ,thē·ə·rē }

Ghon complex [MED] The combination of a focus of subpleural tuberculosis with associated hilar and mediastinal lymph node tuberculosis. { 'gōn ,käm,pleks }

ghost [COMPUT SCI] To display a menu option in a dimmed, fuzzy typeface to indicate that this option is no longer available. [ORD] In passive detection, one of the intersection points of lines of position which do not represent actual targets but are only crossover points of multiple plotted lines of position from two or more detection stations. [PETR] The discernible outline of the shape of a former crystal or of another rock structure that has been partly obliterated and has as its boundaries inclusions, bubbles, or other foreign matter. Also known as phantom. { gōst }

ghost algebraic manipulation language [COMPUT SCI] An algebraic manipulation language which externally gives the appearance of manipulating quite general mathematical expressions, although internally it is functioning with canonically represented data, much like the simpler seminumerical languages. { gōst al·jə'brā·ik mə,nip·yə'lā·shən ,laŋ·gwij }

ghost crystal See phantom crystal. { gōst ,krist·əl }

ghost image [ELECTR] An undesired duplicate image near the desired image on a television receiver or computer display screen. **2.** See ghost pulse. [OPTICS] An undesired image appearing at the image plane of an optical system; it may be a false image of the object or an out-of-focus image of a bright source of light in the field of the optical system. [SPECT] A false image of a spectral line produced by irregularities in the ruling of a diffraction grating. { gōst ,im·ij }

ghost layer [CYTOL] A single layer of cultivated animal cells that has been treated with a nonionic detergent in order to disrupt the membranes. { gōst ,lā·ər }

ghost lines See ferrite banding. { gōst ,līnz }

ghost mode [ELECTROMAG] Waveguide mode having a trapped field associated with an imperfection in the wall of the waveguide; a ghost mode can cause trouble in a waveguide operating close to the cutoff frequency of a propagation mode. { gōst ,mōd }

ghost pulse [ELECTR] An unwanted signal appearing on the screen of a radar indicator and caused by echoes which have a basic repetition frequency differing from that of the desired signals. Also known as ghost image; ghost signal. { gōst ,pəls }

ghost signal [ELECTR] **1.** The reflection-path signal that produces a ghost image on a television receiver. Also known as echo. **2.** See ghost pulse. { gōst ,sig·nəl }

ghost spot [PL PATH] A disease of tomato characterized by small white rings on the fruit. { gōst ,spät }

ghost structure See ferrite banding. { gōst ,strək·chər }

GHz See gigahertz.

Giacobinids [ASTRON] A meteor shower that reaches maximum about October 10, associated with Comet P/Giacobini-Zinner. { jə'kä·bə,nidz }

giant See hydraulic monitor. { 'jī·ənt }

giant branch [ASTRON] A grouping of stars on the Hertzsprung-Russell diagram that extends upwards and to the right of the main sequence; it represents the first stage of giant-star evolution in which hydrogen fuses to helium in a shell surrounding the core where hydrogen fusion has been exhausted. { 'jī·ənt 'branch }

giant-cell arteritis [MED] Inflammation of the arteries, particularly the carotid branches, characterized by the appearance of multinucleate giant cells in the exudate. Also known as temporal arteritis. { 'jī·ənt ,sel ,ärd·ə'rīd·əs }

giant granite See pegmatite. { 'jī·ənt 'gran·ət }

giant magnetoresistance [SOLID STATE] A very large decrease in electrical resistance upon application of a magnetic field in certain structures composed of alternating layers of magnetic and nonmagnetic metals. { 'jī·ənt ,mag·nēd·ō·ri'zis·təns }

giant nuclear resonance [NUC PHYS] A systematic excitation of the atomic nucleus which occurs with great strength in a concentrated energy region. { 'jī·ənt 'nü·klē·ər 'rez·ən·əns }

giant planets [ASTRON] The planets Jupiter, Saturn, Uranus, and Neptune. { 'jī·ənt 'plan·əts }

giant powder [MATER] A blasting powder made of nitroglycerin, sodium nitrate, sulfur, and rosin, sometimes with kieselguhr. { 'jī·ənt 'paud·ər }

giant pulse laser See Q-switched laser. { 'jī·ənt 'pəls ,lā·zər }

giant's cauldron See giant's kettle. { 'jī·əns 'kol·drən }

giant's kettle [GEOL] A cylindrical hole bored in bedrock beneath a glacier by water falling through a deep moulin or by boulders rotating in the bed of a meltwater stream. Also known as giant's cauldron; moulin pothole; potash kettle. { 'jī·əns 'ked·əl }

giant star [ASTRON] One of a class of stars that is 20 or 30 or more times larger than the sun and over 100 times more luminous. { 'jī·ənt 'stär }

GEYSER

Old Faithful, Yellowstone Park, Wyoming. (*National Park Service, U. S. Department of the Interior*)

Giaque-Debye method See adiabatic demagnetization. { ¦zhyäk də'bī ¸meth·əd }

Giaque's temperature scale [THERMO] The internationally accepted scale of absolute temperature, in which the triple point of water is defined to have a temperature of 273.16 K. { ¦zhyäks 'tem·prə·chər ¸skāl }

Giardia [INV ZOO] A genus of zooflagellates that inhabit the intestine of numerous vertebrates, and may cause diarrhea in humans. { je'ärd·ē·ə }

giardiasis [MED] Presence of the protozoon *Giardia lamblia* in the human small intestine. { jē¸är'dī·ə·səs }

gib [ENG] A removable plate designed to hold other parts in place or act as a bearing or wear surface. [MIN ENG] **1.** A temporary support at the face to prevent coal from falling before the cut is complete, either by hand or by machine. **2.** A prop put in the holing of a seam while being undercut. **3.** A piece of metal often used in the same hole with a wedge-shaped key for holding pieces together. { gib }

Gibberella fujikuroi [MYCOL] A fungal pathogen that causes bakanae disease, a seed-borne disease of rice that is characterized by the growth of excessively long internodes, through its production of plant growth hormones called gibberellins. { ¸jib·ə¸rel·ə ¸fü·jē'kü¸roi }

gibberellic acid [BIOCHEM] $C_{18}H_{22}O_6$ A crystalline acid occurring in plants that is similar to the gibberellins in its growth-promoting effects. { ¦jib·ə¦rel·ik 'as·əd }

gibberellin [BIOCHEM] Any member of a family of naturally derived compounds which have a gibbane skeleton and a broad spectrum of biological activity but are noted as plant growth regulators. { jib·ə'rel·ən }

gibberish See hash. { 'jib·rish }

gibbon [VERT ZOO] The common name for seven species of large, tailless primates belonging to the genus *Hylobates*; the face and ears are hairless, and the arms are longer than the legs. { 'gib·ən }

gibbous [MATH] Bounded by convex curves. { 'jib·əs }

gibbous moon [ASTRON] The shape of the moon's visible surface when the sun is illuminating more than half of the side facing the earth. { 'jib·əs 'mün }

gibbs [PHYS] A unit of amount of adsorption, equal to a surface concentration of 10^{-6} mole per square meter. { gibz }

Gibbs adsorption equation [PHYS CHEM] A formula for a system involving a solvent and a solute, according to which there is an excess surface concentration of solute if the solute decreases the surface tension, and a deficient surface concentration of solute if the solute increases the surface tension. { 'gibz ad'sorp·shən i¸kwā·zhən }

Gibbs adsorption isotherm [PHYS CHEM] An equation for the surface pressure of surface monolayers,

$$\phi = RT \int_0^p \Gamma d(\ln p)$$

where ϕ is surface pressure, T is absolute temperature, R is the gas constant, Γ is the number of molecules adsorbed per gram per unit surface area, and p is the pressure of the gas. { 'gibz ad'sorp·shən 'ī·sō¸thərm }

Gibbs apparatus [ENG] A compressed-oxygen breathing apparatus used by miners in the United States. { 'gibz ¸ap·ə'rad·əs }

Gibbs diaphragm cell [CHEM ENG] A type of electrolytic diaphragm cell for chlorine production, with graphite electrodes and a cylindrical shape. { 'gibz 'dī·ə¸fram ¸sel }

Gibbs-Donnan equilibrium See Donnan equilibrium. { ¦gibz 'dän·ən ē·kwə'lib·rē·əm }

Gibbs-Duhem equation [PHYS CHEM] A relation that imposes a condition on the composition variation of the set of chemical potentials of a system of two or more components,

$$SdT - VdP + \sum_{i=1}^r n_i d\mu_i = 0$$

where S is entropy, T absolute temperature, P pressure, n_i the number of moles of the ith component, and μ_i is the chemical potential of the ith component. Also known as Duhem's equation. { ¦gibz 'dü·əm i¸kwä·zhən }

Gibbs elasticity [PHYS] The elasticity of a film of liquid, equal to twice the product of the surface area and the derivative of the surface tension with respect to surface area. { 'gibz i¸las'tis·əd·ē }

Gibbs free energy [THERMO] The thermodynamic function $G = H - TS$, where H is enthalpy, T absolute temperature, and S entropy. Also known as free energy; free enthalpy; Gibbs function. { 'gibz ¦frē 'en·ər·jē }

Gibbs function See Gibbs free energy. { 'gibz ¸fəŋk·shən }

Gibbs-Helmholtz equation [PHYS CHEM] An expression for the influence of temperature upon the equilibrium constant of a chemical reaction, $(d \ln K^0/dT)_P = \Delta H^0/RT^2$, where K^0 is the equilibrium constant, ΔH^0 the standard heat of the reaction at the absolute temperature T, and R the gas constant. [THERMO] **1.** Either of two thermodynamic relations that are useful in calculating the internal energy U or enthalpy H of a system; they may be written $U = F - T(\partial F/\partial T)_V$ and $H = G - T(\partial G/\partial T)_P$, where F is the free energy, G is the Gibbs free energy, T is the absolute temperature, V is the volume, and P is the pressure. **2.** Any of the similar equations for changes in thermodynamic potentials during an isothermal process. { 'gibz 'helm¸hōlts i¸kwā·zhən }

gibbsite [MINERAL] $Al(OH)_3$ A white or tinted mineral, crystallizing in the monoclinic system; a principal constituent of bauxite. Also known as hydrargillite. { 'gib¸zit }

Gibbs paradox [STAT MECH] The paradox in which there is an increase in entropy when two separate volumes of gases of the same kind, at the same temperature and pressure, are mixed. { 'gibz 'par·ə¸däks }

Gibbs phase rule [PHYS CHEM] A relationship used to determine the number of state variables F, usually chosen from among temperature, pressure, and species composition in each phase, which must be specified to fix the thermodynamic state of a system in equilibrium: $F = C - P - M + 2$, where C is the number of chemical species presented at equilibrium, P is the number of phases, and M is the number of independent chemical reactions. Also known as Gibbs rule; phase rule. { 'gibz 'fāz ¸rül }

Gibbs' phenomenon [MATH] A convergence phenomenon occurring when a function with a discontinuity is approximated by a finite number of terms from a Fourier series. { 'gibz fə¸näm·ə¸nän }

Gibbs-Poynting equation [PHYS CHEM] An expression relating the effect of the total applied pressure P upon the vapor pressure p of a liquid, (dp/dP):$yT = V_l/V_g$, where V_l and V_g are molar volumes of the liquid and vapor. { ¦gibz 'póint·iŋ i¸kwā·zhən }

Gibbs rule See Gibbs phase rule. { 'gibz ¸rül }

Gibbs system [STAT MECH] **1.** A hypothetical replica of a physical system. **2.** A set of such replicas forming an ensemble. { 'gibz ¸sis·təm }

gibleh See ghibli. { 'gib·lə }

gibli See ghibli. { 'gib·lē }

giboulee See galerne. { jə'bü·lē }

Gibraltar stone See onyx marble. { jə'bròld·ər ¸stōn }

Gibrat's distribution [MATH] The distribution of a variable whose logarithm has a normal distribution. { zhē'bräz di·strə'byü·shən }

gid [VET MED] A chronic brain disease of sheep, less frequently of cattle, characterized by forced movements of circling or rolling, caused by the larval form of the tapeworm *Multiceps multiceps*. { gid }

Giegy-Hardisty process [CHEM ENG] The production of sebacic acid from castor oil or its acids by reaction of the acid at a high temperature with caustic alkali. { 'gē·gē 'här·də·stē ¸präs·əs }

Giemsa stain [CHEM] A stain for hemopoietic tissue and hemoprotozoa consisting of a stock glycerol methanol solution of eosinates of Azure B and methylene blue with some excess of the basic dyes. { 'gēm·sə ¸stān }

Giesler coal test [ENG] A plastometric method for estimating the coking properties of coals. { 'gēs·lər 'kōl ¸test }

GIF See graphics interchange format. { gif }

gifblaar poison See fluoroacetic acid. { 'gif¸blär ¸pói·zən }

GIFT See gamete intrafallopian transfer. { gift or ¦jē¦ī¦ef'tē }

giga- [SCI TECH] A prefix representing 10^9, which is 1,000,000,000, or a billion. Abbreviated G. Also known as kilomega- (deprecated usage). { 'gig·ə }

gigabit [COMMUN] One billion bits, or 1,000,000,000 bits. { 'gig·ə¸bit }

gigacycle See gigahertz. { 'gig·ə¸sī·kəl }

GIBBON

Hylobates lar, which is a typical gibbon.

gigaelectronvolt [PHYS] A unit of energy, used primarily in high-energy physics, equal to 10^9 electronvolts or approximately 1.602×10^{-10} joule. Abbreviated GeV. { ˈgig·ə·iˈlek,trän,vōlt }

gigaflops [COMPUT SCI] A unit of computer speed, equal to 10^9 flops. { ˈgig·ə,fläps }

gigahertz [COMMUN] Unit of frequency equal to 10^9 hertz. Abbreviated GHz. Also known as gigacycle (gc); kilomegacycle; kilomegahertz. { ˈgig·ə,hərts }

gigantism [MED] Abnormal largeness of the body due to hypersecretion of growth hormone. { jīˈgan,tiz·əm }

Giganturoidei [VERT ZOO] A suborder of small, mesopelagic actinopterygian fishes in the order Cetomimiformes having large mouths and strong teeth. { jī,gan·təˈroid·ē,ī }

gigawatt [ELEC] One billion watts, or 10^9 watts. Abbreviated GW. { ˈgig·ə,wät }

gigging [TEXT] Passing a fabric across rollers equipped with teasels to produce a nap on the surface. { ˈgig·iŋ }

gig mill [TEXT] A textile mill employing rotary wire cylinders for napping. { ˈgig ,mil }

GIGO See garbage in, garbage out. { ˈgī,gō }

gigohm [ELEC] One thousand megohms, or 10^9 ohms. { ˈgig,ōm }

Gila monster [VERT ZOO] The common name for two species of reptiles in the genus *Heloderma* (Helodermatidae) distinguished by a rounded body that is covered with multicolored beaded tubercles, and a bifid protrusible tongue. { ˈhē·lə ,män·stər }

gilbert [ELECTROMAG] The unit of magnetomotive force in the electromagnetic system, equal to the magnetomotive force of a closed loop of one turn in which there is a current of $1/(4\pi)$ abamp. { ˈgil·bərt }

Gilbert circuit [ELECTR] A circuit that compensates for nonlinearities and instabilities in a monolithic variable-transconductance circuit by using the logarithmic properties of diodes and transistors. { ˈgil·bərt ,sər·kət }

Gilbert-Varshamov bound [COMPUT SCI] In the theory of quantum computation, a sufficient condition for an algorithm that encodes N logical qubits into N' carrier qubits (with N' larger than N) to correct any error on any M carrier qubits; namely, that N/N' be smaller than

$$1 - 2[-x \log_2 x - (1 - x) \log_2 (1 - x)], \text{ where } x = 2M/N'$$

{ ˈgil·bərt ,värˈsha·məv ˈbaund }

Gilbrethian variables [IND ENG] A system of three sets of variables that are considered to be intrinsic to every task: variables involving the response of the worker to anatomic and psychological factors, environmental variables, and variables of motion; used in analyzing and designing work systems. { gilˈbreth·ē·ən ˈver,ē·ə·həlz }

Gilbreth's micromotion study [IND ENG] A time and motion study based on the concept that all work is performed by using a relatively few basic operations in varying combinations and sequence; basic elements (therbligs) include grasp, search, move, reach, and hold. { ˈgil·brəths ˈmī·krōˈmō·shən ,stəd·ē }

gilding [GRAPHICS] Overlaying material with a thin layer of gold. { ˈgild·iŋ }

gilding metal [MET] A copper alloy (about 90% copper, 10% zinc) used to jacket small-arms bullets, to form detonator or primer cups, and to form rotating bands for artillery projectiles; it can be readily engraved by the lands as the projectile moves down the bore. { ˈgild·iŋ ,med·əl }

gill [MECH] **1.** A unit of volume used in the United States for the measurement of liquid substances, equal to $1/4$ U.S. liquid pint, or to $1.1829411825 \times 10^{-4}$ cubic meter. **2.** A unit of volume used in the United Kingdom for the measurement of liquid substances, and occasionally of solid substances, equal to $1/4$ U.K. pint, or to $1.420653125 \times 10^{-4}$ cubic meter. [MYCOL] A structure consisting of radially arranged rows of tissue that hang from the underside of the mushroom cap of certain basidiomycetes. [VERT ZOO] The respiratory organ of water-breathing animals. Also known as branchia. { gil }

gill cover [VERT ZOO] The fold of skin providing external protection for the gill apparatus of most fishes; it may be stiffened by bony plates and covered with scales. { ˈgil kəv·ər }

Gilles de la Tourette syndrome See Tourette's syndrome. { ,zhēl də lä tüˈret ,sinˌdrōm }

Gillespie equilibrium still [ANALY CHEM] A recirculating equilibrium distillation apparatus used to establish azeotropic properties of liquid mixtures. { gəˈles·pē ,ē·kwəˈlibrē·əm ,stil }

gillespite [MINERAL] $BaFeSi_4O_{10}$ A micalike mineral composed of barium and iron silicate. { gəˈle,spīt }

Gilliland correlation [CHEM ENG] Approximation method for distillation-column calculations; correlates reflux ratio and number of plates for the column as functions of minimum reflux and minimum plates. { gəˈlil·ənd ,kä·rə,lā·shən }

gill net [ENG] A net that entangles the gill covers of fish. { ˈgil ,net }

gill raker [VERT ZOO] One of the bony processes on the inside of the branchial arches of fishes which prevents the passage of solid substances through the branchial clefts. { ˈgil ,rāk·ər }

Gilmour heat-exchange method [ENG] Thermal design method for heat exchangers by solution of five unique equations containing a minimum number of variables and involving tubeside, shell side, tube-wall, and dirt resistance. { ˈgil·mòr ˈhēt iks,chānj ,meth·əd }

gilsonite [MINERAL] A variety of asphalt; it has black color, brilliant luster, brown streaks, and conchoidal fracture. { ˈgil·sə,nīt }

gimbal [ENG] **1.** A device with two mutually perpendicular and intersecting axes of rotation, thus giving free angular movement in two directions, on which an engine or other object may be mounted. **2.** In a gyro, a support which provides the spin axis with a degree of freedom. **3.** To move a reaction engine about on a gimbal so as to obtain pitching and yawing correction moments. **4.** To mount something on a gimbal. { ˈgim·bəl }

gimbaled inertial system [NAV] An inertial guidance system that makes use of a three-gimbal mounting whose inner gimbal is a stable platform on which three gyroscopes and accelerometers are mounted; the gyroscopes sense any rotation of the vehicle and drive the gimbals in the opposite direction, so that the inner platform remains fixed in inertial space. { ˈgim·bəld iˈnər·shəl ,sis·təm }

gimbaled motor [AERO ENG] A rocket engine mounted on a gimbal. { ˈgim·bəld ˈmōd·ər }

gimbaled nozzle [MECH ENG] A nozzle supported on a gimbal. { ˈgim·bəld ˈnäz·əl }

gimbal freedom [ENG] Of a gyro, the maximum angular displacement about the output axis of a gimbal. { ˈgim·bəl ,frē·dəm }

gimbaling error [NAV] That error introduced in a gyro compass by the tilting of its gimbal mounting system due to horizontal acceleration caused by motion of the vessel, such as rolling. { ˈgim·bə·liŋ ,er·ər }

gimballess inertial navigation equipment See strapped-down inertial navigation equipment. { ˈgim·bə·ləs iˈnər·shəl ,nav·əˈgā·shən i,kwip·mənt }

gimbal lock [ENG] A condition of a two-degree-of-freedom gyro wherein the alignment of the spin axis with an axis of freedom deprives the gyro of a degree-of-freedom and therefore its useful properties. { ˈgim·bəl ,läk }

gimlet [DES ENG] A small tool consisting of a threaded tip, grooved shank, and a cross handle; used for boring holes in wood. { ˈgim·lət }

gimlet bit [DES ENG] A bit with a threaded point and spiral flute; used for drilling small holes in wood. { ˈgim·lət ,bit }

gimmick [ELEC] Length of twisted two-conductor cable, used as a variable capacitive load, in which the capacitance is varied by untwisting and separating the individual conductors. { ˈgim·ik }

gimp [TEXT] **1.** Ornamental cord made of various materials, often with a wire, and sometimes with a coarse cord, running through it. **2.** Cord used to outline the design in lace. **3.** A braid or tape used to hide upholstery tacks in places where the covering fabric is attached to exposed woodwork. { gimp }

gin [AGR] **1.** A machine used to separate cotton fiber from the seed and waste. **2.** To thus separate cotton fiber. [FOOD ENG] An alcoholic beverage made from distilled spirits flavored with an extract of the juniperberry or other flavoring botanicals. [MECH ENG] A hoisting machine in the form of a tripod with a windlass, pulleys, and ropes. { jin }

GILA MONSTER

Gila monster (*Heloderma suspectum*), about 20 inches (50 cenitmeters) long.

GINSENG

Panax quinquefolius, a ginseng, showing shoot and root,

GIRAFFE

The giraffe of Africia.

gin block [NAV ARCH] A block with a metal frame and a single large-diameter sheave, used to facilitate overhauling, and to hoist numerous packages, such as in cargo hoisting. { 'jin ˌbläk }

g-incompleteness *See* geodesic incompleteness. { 'jē ˌiŋ·kəm'plēt·nəs }

G indicator *See* G scope. { jē ˌin·də,kād·ər }

gingelly oil *See* sesame oil. { 'jin·jə·lē ˌoil }

ginger [BOT] *Zingiber officinale*. An erect perennial herb of the family Zingiberaceae having thick, scaly branched rhizomes; a spice oleoresin is made by an organic solvent extraction of the ground dried rhizome. { 'jin·jər }

ginger-grass oil [MATER] A type of citronella oil derived from sofia grass that contains about 50% geraniol; used in perfumery. { 'jin·jər,gras ˌoil }

ginger oil [MATER] A thick, yellowish essential oil, soluble in most organic solvents and insoluble in water; distilled from dried ginger; main components are citral, borneol, and phellandrene; used as flavoring in liqueurs and soft drinks. { 'jin·jər ˌoil }

gingham [TEXT] Plain-weave cotton, or cotton-polyester blends, in checked or striped patterns or plain colors. { 'giŋ·əm }

ginging [MIN ENG] **1.** Lining a shaft with masonry or brick. **2.** The brick or masonry of a shaft lining. { 'gin·jiŋ }

gingiva [ANAT] The mucous membrane surrounding the teeth sockets. { 'jin·jə·və }

gingival crevice [ANAT] The space between the free margin of the gingiva and the surface of a tooth. Also known as gingival sulcus. { 'jin·jə·vəl 'krev·əs }

gingival sulcus *See* gingival crevice. { 'jin·jə·vəl 'səl·kəs }

gingivectomy [MED] Excision of a portion of the gingiva. { ˌjin·jə'vek·tə·mē }

gingivitis [MED] Inflammation of the gingiva. { ˌjin·jə'vīd·əs }

gingivostomatitis [MED] An inflammation of the gingiva and oral mucosa. { ˌjin·jə·vō,stō·mə'tīd·əs }

gingko tree *See* ginkgo. { 'giŋ·kō }

ginglymoarthrodia [ANAT] A composite joint consisting of one hinged and one gliding element. { ˌjiŋ·glə·mō,är'thrō·dē·ə }

Ginglymodi [VERT ZOO] An equivalent name for Semionotiformes. { ˌjiŋ·glə'mō,dī }

ginglymus [ANAT] A type of diarthrosis permitting motion only in one plane. Also known as hinge joint. { 'jiŋ·glə·məs }

ginkgo [BOT] A dioecious tree, commonly known as the maidenhair tree (*Ginkgo biloba*), that is native to China and is cultivated as a shade tree, it is the only surviving species of the class Ginkgoatae and is considered a living fossil. Also known as gingko tree. { 'giŋ·kō }

Ginkgoales [BOT] An order of gymnosperms composing the class Ginkgoopsida with one living species, the dioecious maidenhair tree (*Ginkgo biloba*). { ˌgiŋ·kō'ā·lēz }

Ginkgoatae *See* Ginkgoopsida. { ˌgiŋ'kō·ə,tē }

Ginkgoopsida [BOT] A class of the subdivision Pinicae containing the single, monotypic order Ginkgoales. { ˌgiŋ·kō'äp·sə·də }

Ginkgophyta [BOT] The equivalent name for Ginkgoopsida. { ˌgiŋ'käf·əd·ə }

ginko tree *See* ginkgo. { 'giŋ·kō ˌtrē }

ginning [AGR] The separation of lint from cottonseed. { 'jin·iŋ }

ginorite [MINERAL] $Ca_2B_{14}O_{23} \cdot 8H_2O$ A white monoclinic mineral composed of hydrous borate of calcium. { 'jin·ə,rīt }

gin pit [MIN ENG] A shallow mine, the hoisting from which is done by a gin. { 'jin ˌpit }

gin pole [MECH ENG] A hand-operated derrick which has a nearly vertical pole supported by guy ropes; the load is raised on a rope that passes through a pulley at the top and over a winch at the foot. Also known as guyed-mast derrick; pole derrick; standing derrick. { 'jin ˌpōl }

ginseng [BOT] The common name for plants of the genus *Panax*, a group of perennial herbs in the family Araliaceae; the aromatic root of the plant has been used medicinally in China. { 'jin,seŋ }

gin tackle [MECH ENG] A tackle made for use with a gin. { 'jin ˌtak·əl }

Ginzburg-Landau theory [CRYO] A phenomenological theory of superconductivity which accounts for the coherence length; the ordered state of a superconductor is described by a complex order parameter which is similar to a Schrödinger wave function, but describes all the condensed superelectrons, rather than a single charged particle. Also known as Landau-Ginzburg theory. { 'ginz·bərg 'lan·daù ,thē·ə·rē }

Ginzburg-London superconductivity theory [SOLID STATE] A modification of the London superconductivity theory to take into account the boundary energy. { 'ginz·bərg 'lən·dən 'sü·pər,kän,dək'tiv·əd·ē ,thē·ə·rē }

gio *See* geo. { gyō }

giobertite *See* magnesite. { 'jō·bər,tīt }

Giorgi system *See* meter-kilogram-second-ampere system. { ˌgyór·gē ,sis·təm }

gipsy head *See* gypsy head. { 'jip·sē ,hed }

gipsy winch [MIN ENG] A small winch that may be attached to a post and operated by a rotary motion or the reciprocating action of a handle having a pair of pawls and a ratchet. { 'jip·sē ,winch }

giraffe [VERT ZOO] *Giraffa camelopardalis*. An artiodactyl mammal in the family Giraffidae characterized by extreme elongation of the neck vertebrae, and two prominent horns on the head. { jə'raf }

Giraffe *See* Camelopardalis. { jə'raf }

Giraffidae [VERT ZOO] A family of pecoran ruminants in the superfamily Bovoidea including giraffe, okapi, and relatives. { jə'raf·ə,dē }

Girbotal process [CHEM ENG] A regenerative absorption process to remove carbon dioxide, hydrogen sulfide, and other acid impurities from natural gas, using mono-, di-, or triethanolamine as the reagent. { 'gər·bə,tól ,präs·əs }

girder [CIV ENG] A large beam made of metal or concrete, and sometimes of wood. { 'gər·dər }

girder clamp *See* beam clip. { 'gərd·ər ,klamp }

girder clip *See* beam clip. { 'gərd·ər ,klip }

girdle [ANAT] Either of the ringlike groups of bones supporting the forelimbs (arms) and hindlimbs (legs) in vertebrates. [INV ZOO] **1.** Either of the hooplike bands constituting the sides of the two valves of a diatom. **2.** The peripheral portion of the mantle in chitons. [LAP] The periphery of a cut gemstone that is usually grasped by the setting or mounting. [PETR] With reference to a fabric diagram or equal-area projection net, a belt showing concentration of points which is approximately coincident with a great circle of the net and which represents orientation of the fabric elements. { 'gərd·əl }

girt [CIV ENG] **1.** A timber in the second-floor corner posts of a house to serve as a footing for roof rafters. **2.** A horizontal member to stiffen the framework of a building frame or trestle. [ENG] A brace member running horizontally between the legs of a drill tripod or derrick. [MIN ENG] In square-set timbering, a horizontal brace running parallel to the drift. { gərt }

girth [NAV ARCH] The distance measured along a frame line from a waterline on one side around the hull to the corresponding point on the opposite side. { gərth }

gismondite [MINERAL] $CaAl_2Si_2O_8 \cdot 4H_2O$ A light-colored mineral composed of hydrous calcium aluminum silicate, occurring in pyramidal crystals. { jiz'män,dīt }

gitogenin [PHARM] $C_{27}H_{44}O_4$ A crystalline compound obtained from gitonin by heating with dilute hydrochloric acid, forms leaflets from benzene solutions, soluble in organic solvents such as chloroform, hot alcohol, and ether; used in medicine for treating coronary disease. { jə'täj·ə·nən }

gitonin [ORG CHEM] The gitogenin tetraglycoside in *Digitalis purpurea* seed; resembles digitonin. { jə'tōn·ən }

gitoxigenin [PHARM] $C_{23}H_{34}O_5$ Platelike crystals, used as a cardiotonic drug. { jə,täk·sə'jen·ən }

gitoxin [PHARM] $C_{41}H_{64}O_{14}$ A secondary glycoside derived from *Digitalis purpurea*, crystallizes in stout prisms from chloroform methanol solution, soluble in a mixture of chloroform and alcohol; used in medicine for coronary disease. { jə'täk·sən }

gitter cell [PATH] A compound granule cell that is characteristic of certain brain lesions. { 'gid·ər ,sel }

give-and-take lines [MATH] Straight lines which are used to approximate the boundary of an irregular, curvilinear figure for the purpose of approximating its area; they are placed so that small portions excluded from the area under consideration are balanced by other small portions outside the boundary. { 'giv ən 'tāk ,līnz }

Givens's method [MATH] A transformation method for finding the eigenvalues of a matrix, in which each of the orthogonal transformations that reduce the original matrix to a triple-diagonal matrix makes one pair of elements, a_{ij} and a_{ji}, lying off the principal diagonal and the diagonals immediately above and below it, equal to zero, without affecting zeros obtained earlier. { 'giv·ənz·əz ,meth·əd }

given-year method See Paasche's index. { ¦giv·ən 'yir ,meth·əd }

gizzard [VERT ZOO] The muscular portion of the stomach of most birds where food is ground with the aid of ingested pebbles. { 'giz·ərd }

GKS See graphical kernel system.

glabella [ANAT] The bony prominence on the frontal bone joining the supraorbital ridges. { glə'bel·ə }

glabello-occipital length [ARCH] The distance between the glabella and the opisthocranion. { glə¦bel·ō·äk¦sip·əd·əl 'leŋkth }

glabrous [BIOL] Having a smooth surface; specifically, having the epidermis devoid of hair or down. { 'glab·rəs }

glacial [GEOL] Pertaining to an interval of geologic time which was marked by an equatorward advance of ice during an ice age; the opposite of interglacial; these intervals are variously called glacial periods, glacial epochs, glacial stages, and so on. [HYD] Pertaining to ice, especially in great masses such as sheets of land ice or glaciers. { 'glā·shəl }

glacial abrasion [GEOL] Alteration of portions of the earth's surface as a result of glacial flow. { ¦glā·shəl ə'brā·zhən }

glacial accretion [GEOL] Deposition of material as a result of glacial flow. { ¦glā·shəl ə'krē·shən }

glacial acetic acid [ORG CHEM] CH_3COOH Pure acetic acid (containing less than 1% water); a clear, colorless, caustic hygroscopic liquid, boiling at 118°C, soluble in water, alcohol, and ether, and crystallizing readily; used as a solvent for oils and resins. { ¦glā·shəl ə¦sēd·ik 'as·əd }

glacial advance [GEOL] **1.** Increase in the thickness and area of a glacier. **2.** A time period equal to that increase. { ¦glā·shəl əd'vans }

glacial anticyclone [METEOROL] A type of semipermanent anticyclone which overlies the ice caps of Greenland and Antarctica. Also known as glacial high. { ¦glā·shəl ,an·ti'sī,klōn }

glacial boulder [GEOL] A boulder moved to a point distant from its original site by a glacier. { ¦glā·shəl 'bōl·dər }

glacial deposit [GEOL] Material carried to a point beyond its original location by a glacier. { ¦glā·shəl di'päz·ət }

glacial drift [GEOL] All rock material in transport by glacial ice, and all deposits predominantly of glacial origin made in the sea or in bodies of glacial meltwater, including rocks rafted by icebergs. { ¦glā·shəl 'drift }

glacial epoch [GEOL] **1.** Any of the geologic epochs characterized by an ice age; thus, the Pleistocene epoch may be termed a glacial epoch. **2.** Generally, an interval of geologic time which was marked by a major equatorward advance of ice; the term has been applied to an entire ice age or (rarely) to the individual glacial stages which make up an ice age. { ¦glā·shəl 'ep·ək }

glacial erosion [GEOL] Movement of soil or rock from one point to another by the action of the moving ice of a glacier. Also known as ice erosion. { ¦glā·shəl ə'rō·zhən }

glacial flour See rock flour. { ¦glā·shəl 'flaů·ər }

glacial flow See glacier flow. { ¦glā·shəl 'flō }

glacial geology [GEOL] The study of land features resulting from glaciation. { ¦glā·shəl jē'äl·ə·jē }

glacial high See glacial anticyclone. { ¦glā·shəl 'hī }

glacial ice [HYD] Ice that is flowing or that exhibits evidence of having flowed. { ¦glā·shəl 'īs }

glacial lake [GEOL] A lake that exists because of the effects of the glacial period. { ¦glā·shəl 'lāk }

glacial lobe [HYD] A tonguelike projection from a continental glacier's main mass. { ¦glā·shəl 'lōb }

glacial maximum [GEOL] The time or position of the greatest extent of any glaciation; most frequently applied to the greatest equatorward advance of Pleistocene glaciation. { ¦glā·shəl 'mak·sə·məm }

glacial mill See moulin. { ¦glā·shəl 'mil }

glacial outwash See outwash. { ¦glā·shəl 'aůt,wäsh }

glacial period [GEOL] **1.** Any of the geologic periods which embraced an ice age; for example, the Quaternary period may

be called a glacial period. **2.** Generally, an interval of geologic time which was marked by a major equatorward advance of ice. { ¦glā·shəl 'pir·ē·əd }

glacial plucking See plucking. { ¦glā·shəl 'plək·iŋ }

glacial retreat [GEOL] A condition occurring when backward melting at the front of a glacier takes place at a rate exceeding forward motion. { ¦glā·shəl ri'trēt }

glacial scour [GEOL] Erosion resulting from glacial action, whereby the surface material is removed and the rock fragments carried by the glacier abrade, scratch, and polish the bedrock. Also known as scouring. { ¦glā·shəl 'skaůr }

glacial striae [GEOL] Scratches, commonly parallel, on smooth rock surfaces due to glacial abrasion. { ¦glā·shəl 'strī,ī }

glacial till See till. { ¦glā·shəl 'til }

glacial trough [GEOL] A deep U-shaped valley with steep sides that leads down from a cirque and was excavated by a glacier. { ¦glā·shəl 'trôf }

glacial varve See varve. { ¦glā·shəl 'värv }

glaciated terrain [GEOL] A region that once bore great masses of glacial ice; a distinguishing feature is marks of glaciation. { 'glā·shē,ād·əd tə'rān }

glaciation [GEOL] Alteration of any part of the earth's surface by passage of a glacier, chiefly by glacial erosion or deposition. [METEOROL] The transformation of cloud particles from waterdrops to ice crystals, as in the upper portion of a cumulonimbus cloud. { ,glā·shē'ā·shən }

glaciation limit [GEOPHYS] For a given locality, the lowest altitude at which glaciers can develop. { ,glā·shē'ā·shən ,lim·ət }

glacier [HYD] A mass of land ice, formed by the further recrystallization of firn, flowing slowly (at present or in the past) from an accumulation area to an area of ablation. { 'glā·shər }

glacieret See snowdrift ice. { ¦glā·shə¦ret }

glacier flow [HYD] The motion that exists within a glacier's body. Also known as glacial flow. { 'glā·shər ,flō }

glacier front [HYD] The leading edge of a glacier. { 'glā·shər ,frənt }

glacier ice [HYD] Any ice that is or was once a part of a glacier, consolidated from firn by further melting and refreezing and by static pressure; for example, an iceberg. { 'glā·shər ,īs }

glacier mill See moulin. { 'glā·shər ,mil }

glacier pothole See moulin. { 'glā·shər 'pät,hōl }

glacier table [GEOL] A stone block supported by an ice pedestal above the surface of a glacier. { 'glā·shər ,tā·bəl }

glacier well See moulin. { 'glā·shər ,wel }

glacier wind [METEOROL] A shallow gravity wind along the icy surface of a glacier, caused by the temperature difference between the air in contact with the glacier and free air at the same altitude. { 'glā·shər ,wind }

glacioeustasy [GEOL] Changes in sea level due to storage or release of water from glacier ice. { ¦glās·ē·ō'yü·stə·sē }

glaciofluvial [GEOL] Pertaining to streams fed by melting glaciers, or to the deposits and landforms produced by such streams. { ¦glā·shē·ō¦flü·vē·əl }

glacioisostasy [GEOL] Lithospheric depression or rebound due to the weight or melting of glacier ice. { ,glā·sē·ō·ī'säs·tə·sē }

glaciolacustrine [GEOL] Pertaining to lakes fed by melting glaciers, or to the deposits forming therein. { ¦glā·shē·ō·lə'kəs·trən }

glaciology [GEOL] A broad field encompassing all aspects of the study of ice: glaciers, the largest ice masses on earth; ice that forms on rivers, lakes, and the sea; ice in the ground, including both permafrost and seasonal ice such as that which disrupts roads; ice that crystallizes directly from the air on structures such as airplanes and antennas; and all forms of snow research, including hydrological and avalanche forecasting. { ,glā·shē'äl·ə·jē }

glaçon [OCEANOGR] A piece of sea ice which is smaller than a medium-sized floe. { gla'sōn }

gladiate [BOT] Sword-shaped. { 'glad·ē,āt }

gladiolus See mesosternum. { ,glad·ē'ō·ləs }

Gladiolus [BOT] A genus of chiefly African plants in the family Iridaceae having erect, sword-shaped leaves and spikes of brightly colored irregular flowers. { ,glad·ē'ō·ləs }

GLADIOLUS

A specimen of the genus *Gladious* showing the erect sword-shaped leaves and spikes of flowers.

gladite [MINERAL] $PbCuBi_5S_9$ A lead gray mineral consisting of lead and copper bismuth sulfide; occurs as prismatic crystals. { 'gla,dīt }

gladius *See* pen. { 'glād·ē·əs }

Gladstone-Dale constant [OPTICS] The ratio $(n-1)/\rho$, where n is the index of refraction of a gas and ρ is its density. { 'glad·stŏn 'dāl ,kän·stənt }

Gladstone-Dale law [OPTICS] A law for the variation of the index of refraction n of a substance, according to which $n-1$ is proportional to its density. { 'glad·stŏn 'dāl ,lò }

glair [MATER] A sizing liquid made of egg white beaten with vinegar; used to prepare a surface of a book binding for gilding. { gler }

glance pitch [GEOL] A variety of asphaltite having brilliant conchoidal fracture, and resembling gilsonite but having higher specific gravity and percentage of fixed carbon. { 'glans ,pich }

glancing angle [PHYS] The angle between a surface and a beam of particles or radiation incident upon it; it is the complement of the angle of incidence. { 'glans·iŋ ,aŋ·gəl }

gland [ANAT] A structure which produces a substance essential and vital to the existence of the organism. [ENG] **1.** A device for preventing leakage at a machine joint, as where a shaft emerges from a vessel containing a pressurized fluid. **2.** A movable part used in a stuffing box to compress the packing. { gland }

glanders [VET MED] A bacterial disease of equines caused by *Actinobacillus mallei*; involves the respiratory system, skin, and lymphatics. Also known as farcy. { 'glan·dərz }

glands of Brunner *See* Brunner's glands. { ¦glanz əv 'brün·ər }

glands of Leydig [VERT ZOO] Unicellular, epidermal structures of urodele larvae and the adult *Necturus* that secrete a substance which digests the egg capsule and permits hatching. { ¦glanz əv 'lī·dig }

glandular fever *See* infectious mononucleosis. { ¦glan·jə·lər 'fē·vər }

glandulomuscular [ANAT] Of or pertaining to glands and muscles. { ¦glan·jə·lə'məs·kyə·lər }

glans [ANAT] The conical body forming the distal end of the clitoris or penis. { glanz }

glare [COMMUN] The interference that arises when an attempt is made to place a telephone call just as an incoming call is arriving; in the case of data transmission under the control of a computer, this can render the line or even the computer temporarily inoperative. [OPTICS] **1.** Discomfort produced in an observer by one or more visible sources of light. Also known as discomfort glare. **2.** Visual disability caused by visible sources or areas of luminance which are in an observer's field of view but do not assist in viewing. Also known as disability glare. **3.** Dazzling brightness of the atmosphere, caused by excessive reflection and scattering of light by particles in the line of sight. { gler }

glare filter [ENG] A screen that is placed over the face of a cathode-ray tube to reduce glare from ambient and overhead light. { 'gler ,fil·tər }

glare ice [HYD] Ice with a smooth, shiny surface. { 'gler ,īs }

Glareolidae [VERT ZOO] A family of birds in the order Charadriiformes including the ploverlike coursers and the swallowlike pratincoles. { ,gla·rē'ä·lə,dē }

glareous [ECOL] Growing in gravelly soil; refers specifically to plants. { 'gla·rē·əs }

glaserite *See* arcanite. { 'gla·zə,rīt }

glass [MATER] A hard, amorphous, inorganic, usually transparent, brittle substance made by fusing silicates, sometimes borates and phosphates, with certain basic oxides and then rapidly cooling to prevent crystallization. [METEOROL] In nautical terminology, a contraction for "weather glass" (a mercury barometer). { glas }

glass armor [ORD] Any of several special-purpose protective barrier materials composed of glass or containing glass. { ¦glas ¦är·mər }

glassblowing [ENG] Shaping a mass of viscid glass by inflating it with air introduced through a tube. { 'glas,blō·iŋ }

glass-bonded mica [MATER] An insulating material made by compressing a mixture of powdered glass and powdered natural or synthetic mica at high temperatures. { 'glas ,bän·dəd 'mī·kə }

glass brick [MATER] A hollow block of translucent glass with patterns molded on the faces; used in partitions. { ¦glas 'brik }

glass capacitor [ELEC] A capacitor whose dielectric material is glass. { ¦glas kə'pas·əd·ər }

glass-ceramic [MATER] Hard, strong, nucleated glass with a nonporous crystalline structure; has high flexural strength and shock resistance; used for coatings, molded mechanical and electrical parts, heat-exchanger tubes, missile cones, and cookware and dinnerware. { ¦glas sə'ram·ik }

glass cockpit *See* electronic flight instrument system. { ¦glas 'käk,pit }

glass cutter [ENG] A tool equipped with a steel wheel or a diamond point used to cut glass. { 'glas,kəd·ər }

glass dosimeter [NUCLEO] A dosimeter using as its radiation-sensing element a fluorod of special glass that fluoresces under ultraviolet light following gamma irradiation. { 'glas dō'sim·əd·ər }

glassed steel [CHEM ENG] Process piping or vessels lined with glass; a glass-steel composite has structural strength of steel and corrosion resistance of glass. { ¦glast ¦stēl }

glass electrode [PHYS CHEM] An electrode or half cell in which potential measurements are made through a glass membrane, which acts as a cation-exchange membrane; thus, the potential arises from phase-boundary and diffusion potentials which, depending on the composition of the glass, are logarithmic functions of the activity of the cations such as H^+, Na^+, or K^+ of the solutions in which the electrode is immersed. { ¦glas i'lek,trŏd }

Glasser's disease [VET MED] A generalized bacterial infection of swine caused by *Mycoplasma hyorhinis*. { 'glas·ərz di,zēz }

glass fiber [MATER] A glass thread less than a thousandth of an inch (25 micrometers) thick, used loosely or in woven form as an acoustic, electrical, or thermal insulating material and as a reinforcing material in laminated plastics. Also known as fiberglass. { 'glas ¦fī·bər }

glass film plates [GRAPHICS] Film plates made of glass sheets coated with a sensitized emulsion; they have been largely replaced by sheet film, but are still used in some critical work because of their dimensional stability. { 'glas 'film ,plāts }

glass fission detector [NUCLEO] A piece of glass in which fission fragments, flying apart with high energy, can create narrow but continuous, submicroscopic trails of altered material, which can be seen in an ordinary microscope after the altered material has been dissolved by a chemical reagent. { ¦glas 'fish·ən di,tek·tər }

glass former [MATER] **1.** An oxide that can readily form a glass. **2.** An oxide that can contribute to the network of a silica glass. { 'glas ,fòrm·ər }

glass furnace [ENG] A large, covered furnace or tank for melting large batches of glass, in which heat is supplied by a flame playing over the glass surface, and regenerative heating of combustion air and gas is usually employed. Also known as glass tank. { 'glas ,fər·nəs }

glass guide [GRAPHICS] In microfilm photography, a transparent bar of optical glass employed to guide documents through the photographic field. { 'glas 'gīd }

glass heat exchanger [ENG] Any heat exchanger in which glass replaces metal, such as shell-and-tube, cascade, double-pipe, bayonet, and coil exchangers. { ¦glas 'hēt iks,chān·jər }

glassine [MATER] A thin, dense, transparent, supercalendered paper from highly refined sulfite pulp, used for envelope windows, for sanitary wrapping, and as an insulating paper between layers of iron-core transformer windings. { ¦gla¦sēn }

glass insulator [MATER] An insulator for a power transmission line made of annealed or toughened (tempered) glass. { ¦glas 'in·sə,lād·ər }

glass ionomer [MED] The only dental restorative material that forms a durable chemical bond to dentin; it is formed by the reaction of aluminosilicate glass with polyacrylic acid. { ¦glas ī'än·ə·mər }

glassivation [ELECTR] Method of transistor passivation by a pyrolytic glass-deposition technique, whereby silicon semiconductor devices, complete with metal contact systems, are fully encapsulated in glass. { ,glas·ə'vā·shən }

glass laser [OPTICS] A solid laser in which glass serves as the host for laser ions of such materials as erbium, holmium,

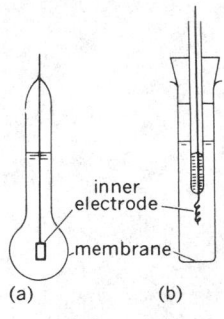

GLASS ELECTRODE

inner electrode

membrane

(a)   (b)

Two types of glass electrode: *(a)* bulb type; *(b)* flat-membrane type.

neodymium, and ytterbium. Also known as amorphous laser. { ¦glas 'lā·zər }

glassmakers' soap [MATER] A substance such as manganese dioxide added to glass to remove the green color created by iron salts. { 'glas₁māk·ərz ₁sōp }

glass paper [MATER] **1.** Paper with a layer of pulverized glass; used as an abrasive. **2.** Paper made of glass fibers. { 'glas ₁pā·pər }

glass-plate capacitor [ELEC] High-voltage capacitor in which the metal plates are separated by sheets of glass serving as the dielectric, with the complete assembly generally immersed in oil. { 'glas ₁plāt kə'pas·əd·ər }

glass porphyry See vitrophyre. { 'glas 'pȯr·fə·rē }

glass pot [ENG] A crucible used for making small amounts of glass. { 'glas ₁pät }

glass resistor [ELEC] A glass tube with a helical carbon resistance element painted on it. { 'glas ri'zis·tər }

glass sand [MATER] High-quartz sand used in glassmaking; contains small amounts of aluminum oxide, iron oxide, calcium oxide, and magnesium oxide. { 'glas ₁sand }

glass schorl See axinite. { 'glas ₁shȯrl }

glass screen See crossline screen. { 'glas ₁skrēn }

glass seal [ENG] An airtight seal made by molten glass. { 'glas ₁sēl }

glass sponge [INV ZOO] A siliceous sponge belonging to the class Hyalospongiae. { 'glas ₁spənj }

glass switch [ELECTR] An amorphous solid-state device used to control the flow of electric current. Also known as ovonic device. { ¦glas 'swich }

glass tank See glass furnace. { 'glas ₁taŋk }

glass textile [TEXT] Fabric woven from glass fibers; used for electrical insulation, as filter cloth, and in plastic laminates. { 'glas 'tek·stīl }

glass-to-metal seal [ELECTR] An airtight seal between glass and metal parts of an electron tube, made by fusing together a special glass and special metal alloy having nearly the same temperature coefficients of expansion. { ¦glas tə ¦med·əl 'sēl }

glass transition [PHYS] The transition that occurs when a liquid is cooled to an amorphous or glassy solid. [PHYS CHEM] The change in an amorphous region of a partially crystalline polymer from a viscous or rubbery condition to a hard and relatively brittle one; usually brought about by changing the temperature. Also known as gamma transition; glassy transition. { 'glas tran₁zish·ən }

glass transition temperature [PHYS CHEM] The temperature at which a liquid changes to an amorphous or glassy solid. { 'glas ₁tran'zish·ən ₁tem·prə·chər }

glass-tube manometer [ENG] A manometer for simple indication of difference of pressure, in contrast to the metallic-housed mercury manometer, used to record or control difference of pressure or fluid flow. { 'glas ₁tüb mə'näm·əd·ər }

glassware [MATER] Articles, especially tableware, made of glass. { 'glas₁wer }

glass wool [MATER] A mass of glass fibers resembling wool and used as insulation, packing, and air filters. { 'glas 'wu̇l }

glassy alloy [MET] An alloy having an amorphous or glassy structure. Also known as metallic glass. { 'glas·ē 'al₁ȯi }

glassy feldspar See sanidine. { ¦glas·ē 'fel₁spär }

glassy state See vitreous state. { ¦glas·ē 'stāt }

glassy transition See glass transition. { ¦glas·ē tran'zish·ən }

glauberite [MINERAL] $Na_2Ca(SO_4)_2$ A brittle, gray-yellow monoclinic mineral having vitreous luster and saline taste. { 'glau̇·bə₁rīt }

Glauber's salt [INV ZOO] $Na_2SO_4·10H_2O$ Crystalline hydrated sodium sulfate; loses water when exposed to air; water soluble, alcohol insoluble; used in textile dyeing and medicine. { 'glau̇·bərz ₁sȯlt }

glaucocerinite [MINERAL] A mineral composed of a hydrous basic sulfate of copper, zinc, and aluminum. { ₁glȯ·kō'se·rə₁nīt }

glaucochroite [MINERAL] $CaMnSiO_4$ A bluish-green mineral that is related to monticellite, is composed of calcium manganese silicate, and occurs in prismatic crystals. { ₁glȯ·kə'krō₁īt }

glaucodot [MINERAL] $(Co,Fe)AsS$ A grayish-white, metallic-looking mineral composed of cobalt iron sulfarsenide, occurring in orthorhombic crystals. { 'glȯ·kə₁dät }

glaucoma [MED] A disease of the eye characterized by increased fluid pressure within the eyeball. { glau̇'kō·mə }

glauconite [MINERAL] $K_{15}(Fe,Mg,Al)_{4-6}(Si,Al)_8O_{20}(OH)_4$ A type of clay mineral; it is dioctohedral and occurs in flakes and as pigmentary material. { 'glȯ·kə₁nīt }

glauconitic sandstone [PETR] A quartz sandstone or an arkosic sandstone that has many glauconite grains. { ¦glȯ·kə¦nid·ik 'san₁stōn }

glaucophane [MINERAL] $Na_2Mg_3Al_2Si_8$ A blue to black monoclinic sodium amphibole; blue to black coloration with marked pleochroism. { 'glȯ·kə₁fān }

glaucophane schist [PETR] Metamorphic schist that contains glaucophane. { 'glȯ·kə₁fān ₁shist }

glaucous [BOT] Having a white or grayish powdery coating that gives a frosty appearance and rubs off easily. { 'glȯ·kəs }

glave See glaves. { 'glä·və }

glaves [METEOROL] A foehnlike wind of the Faroe Islands. Also known as glave; glavis. { 'glä·vəs }

glavis See glaves. { 'glä·vəs }

glaze [ENG] A glossy coating. Also known as enamel. [HYD] A coating of ice, generally clear and smooth but usually containing some air pockets, formed on exposed objects by the freezing of a film of supercooled water deposited by rain, drizzle, or fog, or possibly condensed from supercooled water vapor. Also known as glaze ice; glazed frost; verglas. { glāz }

glazed [MECH ENG] Pertaining to an abrasive surface that has become smooth and cannot abrade efficiently. [TEXT] Referring to a polished surface finish given to cotton fabric by treating it with starch, glue, paraffin, or shellac and then running it throught a hot friction roller. { glāzd }

glazed frost See glaze. { ¦glāzd 'frȯst }

glaze ice See glaze. { 'glāz ₁īs }

glaze stain [INV ZOO] Colorant for ceramic glazes; made of a finely ground calcined oxide, such as of cobalt, copper, manganese, or iron. { 'glāz ₁stān }

glazier's point [ENG] A small piece of sheet metal, usually shaped like a triangle, used to hold a pane of glass in place. Also known as sprig. { 'glā·zərz ₁pȯint }

glazing [ENG] **1.** Cutting and fitting panes of glass into frames. **2.** Smoothing the lead of a wiped pipe joint by passing a hot iron over it. { 'glāz·iŋ }

glazing bar See sash bar. { 'glāz·iŋ ₁bär }

glazing compound [MATER] A caulking compound used to seal and support a glass pane in place. { 'glāz·iŋ ₁käm₁pau̇nd }

glb See greatest lower bound.

GLC See gas-liquid chromatography.

Gleason bevel gear system [DES ENG] The standard for bevel gear designs in the United States; employs a basic pressure angle of 20° with long and short addenda for ratios other than 1:1 to avoid undercut pinions and to increase strength. { 'glēs·ən ¦bev·əl ¦gir ₁sis·təm }

gleba [MYCOL] The central, sporogenous tissue of the sporophore in certain basidiomycetous fungi. { 'glē·bə }

Gledhill disk [ASTRON] The outer magnetosphere of Jupiter, which forms a disk-shaped region of hot plasma near the plane of Jupiter's magnetic equator. { 'gled₁hil ₁disk }

gleet [MED] The chronic stage of gonorrheal urethritis, characterized by a slight mucopurulent discharge. { glēt }

Gleissberg cycle [ASTRON] An 80-year cycle in the amplitude of the 11-year sunspot cycle. { 'glīs₁bərg ₁sī·kəl }

glenoid [ANAT] A smooth, shallow, socketlike depression, particularly of the skeleton. { 'gle₁nȯid }

glenoid cavity [ANAT] The articular surface on the scapula for articulation with the head of the humerus. { ¦gle₁nȯid 'kav·əd·ē }

glessite [GEOL] Fossil resin similar to amber. { 'gle₁sīt }

gley [GEOL] A sticky subsurface layer of clay in some water-logged soils. { glā }

glial cell [NEUROSCI] See neuroglia. { 'glē·əl ₁sel }

glide [AERO ENG] Descent of an aircraft at a normal angle of attack, with little or no thrust. [CRYSTAL] See slip. { glīd }

glide angle See gliding angle. { 'glīd ₁aŋ·gəl }

glide bomb [ORD] A bomb fitted with airfoils to provide lift, released in the direction of a target by an airplane. { 'glīd ₁bäm }

glide fold See shear fold. { 'glīd ₁fōld }

glide path [AERO ENG] **1.** The flight path of an aeronautical vehicle in a glide, seen from the side. Also known as glide trajectory. **2.** The path used by an aircraft or spacecraft in a landing approach procedure. { 'glīd ,path }

glide path indicator [NAV] A system which provides signals for indicating vertical guidance of an aircraft along an inclined surface. { 'glīd ,path ¦in·də,kād·ər }

glide plane [CRYSTAL] A lattice plane in a crystal on which translation or twin gliding occurs. Also known as slip plane. [NAV] *See* glide slope. { 'glīd ,plān }

glider [AERO ENG] A fixed-wing aircraft designed to glide, and sometimes to soar; usually does not have a power plant. { 'glīd·ər }

glide rocket [AERO ENG] A rocket that is kept within or near the sensible atmosphere so that it assumes a flat, gliding attitude when power is shut off. { 'glīd ,räk·ət }

glide slope [AERO ENG] *See* gliding angle. [NAV] An inclined electromagnetic surface which is generated by instrument-landing approach facilities and which includes a glide path supplying guidance in the vertical plane. Also known as glide plane. { 'glīd ,slōp }

glide slope facility [NAV] An instrument approach landing facility furnishing vertical guidance information to aircraft from its altitude down to the runway. { 'glīd ¦slōp fə,sil·əd·ē }

glide slope sector [NAV] A vertical sector containing that portion of the glide slope within which the cooperating aircraft equipment provides a quantitative measurement of deviations above or below the glide slope. { 'glīd ,slōp ,sek·tər }

glide trajectory *See* glide path. { 'glīd trə,jek·trē }

gliding angle [AERO ENG] The angle between the horizontal and the glide path of an aircraft. Also known as glide angle; glide slope. { 'glīd·iŋ ,aŋ·gəl }

gliding bacteria [MICROBIO] The descriptive term for members of the orders Beggiatoales and Myxobacterales; they are motile by means of creeping movements. { ¦glīd·iŋ bak'tir·ē·ə }

gliding joint *See* arthrodia. { ¦glīd·iŋ 'jóint }

gliding motility [MICROBIO] A means of bacterial self-propulsion by slow gliding or creeping movements on the surface of a substrate. { ¦glīd·iŋ mō'til·əd·ē }

glime [HYD] An ice coating with a consistency intermediate between glaze and rime. { glīm }

glimmer ice [HYD] Ice newly formed within the cracks or holes of old ice, or on the puddles on old ice. { 'glim·ər ,īs }

G line [ELECTROMAG] A single dielectric-coated, round wire used for transmitting microwave energy. { 'jē,līn }

glint [ELECTR] **1.** Pulse-to-pulse variation in amplitude of a reflected radar signal, owing to the reflection of the radar from a body that is rapidly changing its reflecting surface, for example, a spinning airplane propeller. **2.** The use of this effect to degrade tracking or seeking functions of an enemy weapons system. [OPTICS] A small region designed to strongly reflect light from a target. { glint }

glioma [MED] A malignant tumor derived from the supporting tissue of the central nervous system. { glī'ō·mə }

gliosis [MED] Proliferation of neuroglia in the brain or spinal cord, either as a replacement process or in response to a low-grade inflammation. { glī'ō·səs }

gliotoxin [MICROBIO] $C_{13}H_{14}O_4N_2S_2$ A heat-labile, bacteriostatic antibiotic produced by species of *Trichoderma* and *Cliocladium* and by *Aspergillus fumigatus*. { ¦glī·ō,täk·sən }

Gliridae [VERT ZOO] The dormice, a family of mammals in the order Rodentia. { 'glir·ə,dē }

glisette [MATH] A curve, such as Watt's curve, traced out by a point attached to a curve which moves so that it always touches two fixed curves, or the envelope of any line or curve attached to the moving curve. { gli'set }

glissile dislocation *See* Shockley partial dislocation. { 'glis·əl ,dis·lō'kā·shən }

Glisson's capsule [ANAT] The membranous sheet of collagenous and elastic fibers covering the liver. { 'glis·ənz ,kap·səl }

glitch [ASTRON] A sudden change in the period of a pulsar, believed to result from a phenomenon analogous to an earthquake that changes the pulsar's moment of inertia. [ELECTR] **1.** An undesired transient voltage spike occurring on a signal being processed. **2.** A minor technical problem arising in electronic equipment. { glich }

glitter [MATER] A group of decorative materials consisting of flakes large enough so that each flake produces a plainly visible sparkle or reflection; incorporated into plastic during

compounding. [OPTICS] The spots of light reflected from a point source by the surface of the sea or wave facets, that is, specular reflection. { 'glid·ər }

Glivenko-Cantelli lemma [MATH] The empirical distribution functions of a random variable converge uniformly in probability to the distribution function of the random variable. { gli'veŋ·kō kan'tel·ē 'lem·ə }

g load [PHYS] The numerical ratio of any applied force to the gravitational force at the earth's surface. { 'jē ,lōd }

global climate change [CLIMATOL] The periodic fluctuations in global temperatures and precipitation, such as the glacial (cold) and interglacial (warm) cycles. { ¦glō·bəl 'klī·mət ,chānj }

global format [COMPUT SCI] A choice of label alignment or numeric format in a spreadsheet program that applies to all the cells of the spreadsheet. { ¦glō·bəl 'fòr,mat }

globally hyperbolic [RELAT] Property of a space-time M that satisfies certain causality conditions ensuring that the solution to the wave equation for a delta function source at a point p in M is unique and vanishes outside the causal future of p. { ¦glō·bə·lē ,hī·pər'bäl·ik }

global memory [COMPUT SCI] Computer storage that can be used by a number of processors connected together in a multiprocessor system. { 'glō·bəl 'mem·rē }

global orbiting navigation satellite system *See* GLONASS. { ¦glō·bəl ¦órb·əd·iŋ ,nav·ə¦gā·shən 'sad·ə,līt ,sis·təm }

Global Positioning System [NAV] A positioning or navigation system designed to use 24 satellites, each carrying atomic clocks, to provide a receiver anywhere on earth with extremely accurate measurements of its three-dimensional position, velocity, and time. Abbreviated GPS. { 'glō·bəl pə'zish·niŋ ,sis·təm }

global property [MATH] A property of an object (such as a space, function, curve, or surface) whose specification requires consideration of the entire object, rather than merely the neighborhoods of certain points. { ¦glō·bəl 'präp·ərd·ē }

global radiation [GEOPHYS] The total of direct solar radiation and diffuse sky radiation received by a horizontal surface of unit area. { 'glō·bəl ,rād·ē'ā·shən }

global resource sharing [COMPUT SCI] The ability of all of the users of a local-area network to share any of the resources (storage devices, input/output devices, and so forth) connected to the network. { 'glō·bəl ri'sórs ,sher·iŋ }

global sea [OCEANOGR] All the seawater of the earth considered as a single ocean constantly intermixing. { 'glō·bəl 'sē }

global search and replace [COMPUT SCI] A text-editing function of a word-processing system in which text is scanned for a given combination of characters, and each such combination is replaced by another set of characters. { ¦glō·bəl ¦sərch ən ri'plās }

global symmetry [PHYS] A type of symmetry of the Hamiltonian of a physical system, specifying that the system must obey some general conservation law. { ,glō·bəl 'sim·ə·trē }

global system for mobile communications *See* GSM. { ¦glō·bəl 'sis·təm fər ,mō·bəl kə,myü·nə'kā·shənz }

global variable [COMPUT SCI] A variable which can be accessed (used or changed) throughout a computer program and is not confined to a single block. { ¦glō·bəl 'ver·ē·ə·bəl }

global warming potential [METEOROL] The ratio of global warming or radiative forcing from 1 kilogram of a greenhouse gas to 1 kilogram of carbon dioxide over 100 years, expressed per mole or per kilogram; it provides a way to calculate the contribution of each greenhouse gas to the annual increase in radiative forcing. { ¦glō·bəl 'wórm·iŋ pə,ten·chəl }

Globar lamp [ELEC] A lamp whose illuminating element is a silicon carbide rod which gives off blackbody radiation when heated. { 'glō,bär ,lamp }

globe [MAP] A sphere on the surface of which is a map of the world. { glōb }

globe lightning *See* ball lightning. { 'glōb ,līt·niŋ }

globe valve [MECH ENG] A device for regulating flow in a pipeline, consisting of a movable disk-type element and a stationary ring seat in a generally spherical body. { 'glōb ,valv }

GLOBAL POSITIONING SYSTEM

Constellation of operational GPS spacecraft.

GLOBE VALVE

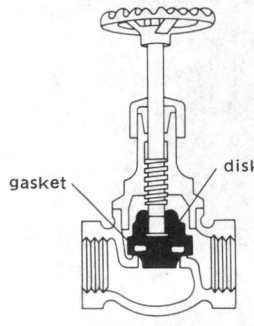

A globe valve with the gasket in the disk.

Globigerinacea [INV ZOO] A superfamily of foraminiferan protozoans in the suborder Rotaliina characterized by a radial calcite test with bilamellar septa and a large aperture. { glō,bij·ə·rə'nās·ē·ə }

globigerina ooze [GEOL] A pelagic sediment consisting of than 30% calcium carbonate in the form of foraminiferal tests of which *Globigerina* is the dominant genus. { glō,bij·ə'rī·nə ,üz }

globin [BIOCHEM] Any of a class of soluble histone proteins obtained from animal hemoglobins. { 'glō·bən }

globin zinc insulin [PHARM] A preparation of insulin modified by the addition of globin (derived from the hemoglobin of beef blood) and zinc chloride; it has intermediate duration of action. { ¦glō·bən ¦ziŋk 'in·sə·lən }

globoside [BIOCHEM] A glycoside of ceramide containing several sugar residues, but not neuraminic acid; obtained from human, sheep, and hog erythrocytes. { 'glō·bə,sīd }

globular *See* spherulitic. { 'gläb·yə·lər }

globular deoxyribonucleic acid [CELL MOL] A compact arrangement of double-stranded deoxyribonucleic acid formed by additional twisting of the fiber into a helical double helix. { 'gläb·yə·lər dē¦äk·sē,rī·bō,nü¦klē·ik 'as·əd }

globular projection [MAP] A projection, in perspective, of a hemisphere upon a plane parallel to the base of the hemisphere. { 'gläb·yə·lər prə'jek·shən }

globular protein [BIOCHEM] Any protein that is readily soluble in aqueous solvents. { 'gläb·yə·lər 'prō,tēn }

globular star cluster [ASTRON] A group of many thousands of stars that are much closer to each other than the stars around the group and that are traveling through space together; a globular cluster has a slightly flattened spheroidal shape. { 'gläb·yə·lər 'stär ,kləs·tər }

globular transfer [MET] In electric arc welding, the transfer of weld metal across the arc in large drops. { 'gläb·yə·lər 'tranz·fər }

globule [ASTRON] A black volume of cosmic dust viewed against the brighter background of bright nebulae. { 'gläb·yəl }

globulin [BIOCHEM] A heat-labile serum protein precipitated by 50% saturated ammonium sulfate and soluble in dilute salt solutions. { 'gläb·yə·lən }

globulite [GEOL] A small, isotropic, globular of spherulelike crystallite; usually dark in color and found in glassy extrusive rocks. { 'gläb·yə,līt }

globulomaxillary cyst [MED] A cyst in the alveolar process between the upper lateral incisor and canine teeth. { ¦gläb·yə·lō¦mak·sə,ler·ē 'sist }

glochid *See* glochidium. { 'glō,kòid }

glochidium [BOT] A barbed hair. Also known as glochid. [INV ZOO] The larva of fresh-water mussels in the family Unionidae. { glō'kid·ē·əm }

glockerite [MINERAL] A brown, ocher yellow, black, or dull green mineral consisting of a hydrated basic sulfate of ferric iron; occurs in stalactitic, encrusting, or earthy forms. { 'glä·kə,rīt }

gloea [INV ZOO] An adhesive mucoid substance secreted by certain protozoans and other lower organisms. { 'glē·ə }

glomerulonephritis [MED] Inflammation of the kidney, primarily involving the glomeruli. { glə¦mər·yə·lō·nə¦frīd·əs }

glomerulosclerosis [MED] Fibrosis of the renal glomeruli. { glə¦mər·yə·lō·sklə¦rō·səs }

glomerulus [ANAT] A tuft of capillary loops projecting into the lumen of a renal corpuscle. { glə'mər·yə·ləs }

glomus [ANAT] **1.** A fold of the mesothelium arising near the base of the mesentery in the pronephros and containing a ball of blood vessels. **2.** A prominent portion of the choroid plexus of the lateral ventricle of the brain. { 'glō·məs }

glomus aorticum *See* paraaortic body. { 'glō·məs ā'órd·ə·kəm }

glomus caroticum *See* carotid body. { 'glō·məs kə'räd·ə·kəm }

GLONASS [NAV] A worldwide Russian navigation system designed to use 24 satellites in three uniformly spaced orbital planes to provide three-dimensional position and velocity data to equipped users on or above the earth's surface. Acronym for global orbiting navigation satellite system. { 'glō,nas }

gloom [METEOROL] The condition existing when daylight is very much reduced by dense cloud or smoke accumulation above the surface, the surface visibility not being materially reduced. { glüm }

glory [OPTICS] A set of concentric, colored rings of light around the shadow cast by an observer or his head onto a cloud or fog bank. { 'glō·rē }

glory hole [CIV ENG] A funnel-shaped, fixed-crest spillway. [ENG] A furnace for resoftening or fire polishing glass during working, or an entrance in such a furnace. [MIN ENG] An opening formed by the removal of soft or broken ore through an underground passage. [NUCLEO] *See* beam hole. { 'glō·rē ,hōl }

glory hole system *See* chute system. { 'glō·rē ,hōl ,sis·təm }

gloss [OPTICS] The ratio of the light specularly reflected from a surface to the total light reflected. { gläs }

glossa [INV ZOO] A tongue or tonguelike structure in insects, especially the median projection of the labium. { 'gläs·ə }

glossalgia [MED] Pain in the tongue. { glä'sal·jə }

glossary [COMPUT SCI] A file of commonly used phrases that can be retrieved in a word-processing program, usually through use of a command and a keyword. { 'gläs·ə·rē }

glossate [INV ZOO] Having a glossa or tonguelike structure. { 'glä,sāt }

glossimeter [ENG] An instrument, often photoelectric, for measuring the ratio of the light reflected from a surface in a definite direction to the total light reflected in all directions. Also known as glossmeter. { glä'sim·əd·ər }

Glossinidae [INV ZOO] The tsetse flies, a family of cyclorrhaphous dipteran insects in the section Pupipara. { glä'sin·ə,dē }

Glossiphoniidae [INV ZOO] A family of small leeches with flattened bodies in the order Rhynchobdellae. { ¦glä·sə·fə'nī·ə,dē }

glossitis [MED] Inflammation of the tongue. { glä'sīd·əs }

glossmeter *See* glossimeter. { 'gläs,mēd·ər }

gloss oil [MATER] Low-grade varnish composed of rosin dissolved in solvent naphtha. { 'gläs ,òil }

glossolalia [PSYCH] Gibberishlike speech; unintelligible jargon. { ,gläs·ō'lä·lē·ə }

glossopalatine nerve [NEUROSCI] The intermediate branch of the facial nerve. { 'gläs·ō'pal·ə,tēn ,nərv }

glossopharyngeal nerve [NEUROSCI] The ninth cranial nerve in vertebrates; a paired mixed nerve that supplies autonomic innervation to the parotid gland and contains sensory fibers from the posterior one-third of the tongue and the anterior pharynx. { ¦gläs·ō·fə¦rin·jē·əl ,nərv }

glossopterid flora [PALEOBOT] Permian and Triassic fossil ferns of the genus *Glossopteris*. { glä'säp·tə·rəd 'flór·ə }

glossopyrosis [MED] Burning sensation of the tongue. { ,gläs·ō,pī'rō·səs }

glossy [OPTICS] Pertaining to a surface from which much more light is specularly reflected than is diffusely reflected. { 'gläs·ē }

glossy print [GRAPHICS] A photograph dried on a ferrotype plate or drum; the surface has a glazed appearance and is not easily scratched or soiled in handling. { 'gläs·ē ,print }

glost firing [CHEM ENG] The process of glazing and firing ceramic ware which has previously been fired at a higher temperature. { 'glóst ,fīr·iŋ }

glottis [ANAT] The opening between the margins of the vocal folds. { 'gläd·əs }

gloup [GEOL] An opening in the roof of a sea cave. { glüp }

glove-and-stocking anesthesia [MED] Loss or diminution of sensation in the hands and feet, corresponding to the areas covered by gloves and stockings. { ¦gləv ən 'stäk·iŋ ,an·əs,thē·zhə }

glove anesthesia [MED] Loss or diminution of sensation in the hands, corresponding to the area covered by gloves. { 'gləv ,an·əs,thē·zhə }

glove box [ENG] A sealed box with gloves attached and passing through openings into the box, so that workers can handle materials in the box; used to handle certain radioactive and biologically dangerous materials and to prevent contamination of materials and objects such as germfree rats or lunar rocks. { 'gləv ,bäks }

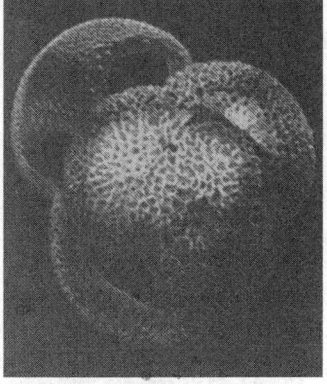
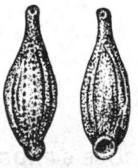

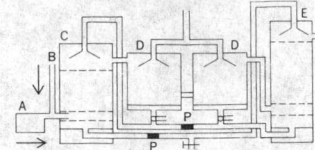

GLOVER TOWER

The lead chamber process. A, sulfur or pyrite burners; B, inlet for nitrogen oxides; C, Glover tower; D, lead chambers; E, Gay-Lussac tower; P, pumps.

GLUCOSE-6-PHOSPHATE

Structural formula for glucose-6-phosphate.

Glover tower [CHEM ENG] A tower in the lead chamber process for manufacturing sulfuric acid; in this tower the nitrogen oxide, sulfur dioxide, and air mixture is passed upward and sprayed with a sulfuric acid-nitrosyl sulfuric acid mixture. { 'gləv·ər ˌtaů·ər }

glow discharge [ELECTR] A discharge of electricity through gas at relatively low pressure in an electron tube, characterized by several regions of diffuse, luminous glow and a voltage drop in the vicinity of the cathode that is much higher than the ionization voltage of the gas. Also known as cold-cathode discharge. { 'glō ˌdis·chärj }

glow-discharge cold-cathode tube See glow-discharge tube. { 'glō ˌdis·chärj ˌkōld 'kath·ōd ˌtüb }

glow-discharge microphone [ENG ACOUS] Microphone in which the action of sound waves on the current forming a glow discharge between two electrodes causes corresponding variations in the current. { 'glō ˌdis·chärj 'mī·krəˌfōn }

glow-discharge tube [ELECTR] A gas tube that depends for its operation on the properties of a glow discharge. Also known as glow-discharge cold-cathode tube; glow tube. { 'glō ˌdis·chärj ˌtüb }

glow-discharge voltage regulator [ELECTR] Gas tube that varies in resistance, depending on the value of the applied voltage; used for voltage regulation. { 'glō ˌdis·chärj 'vōl·tij ˌreg·yəˌlād·ər }

glowing avalanche See ash flow. { 'glō·iŋ 'av·əˌlanch }

glowing cloud See nuée ardente. { 'glō·iŋ 'klaůd }

glowing combustion [CHEM ENG] A reaction between oxygen or an oxidizer and the surface of a solid fuel so that there is emission of heat and light without a flame. Also known as surface burning. { 'glō·iŋ kəm'bəs·chən }

glow lamp [ELECTR] A two-electrode electron tube containing a small quantity of an inert gas, in which light is produced by a negative glow close to the negative electrode when voltage is applied between the electrodes. { 'glō ˌlamp }

glow plug [MECH ENG] A small electric heater, located inside a cylinder of a diesel engine, that preheats the air and aids the engine in starting. { 'glō ˌpləg }

glow potential [ELECTR] The potential across a glow discharge, which is greater than the ionization potential and less than the sparking potential, and is relatively constant as the current is varied across an appreciable range. { 'glō pəˌten·chəl }

glow tube See glow-discharge tube. { 'glō ˌtüb }

glow-tube oscillator [ELECTR] A circuit using a glow-discharge tube which functions as a simple relaxation oscillator, generating a fixed-amplitude periodic sawtooth waveform. { 'glō ˌtüb 'äs·əˌlād·ər }

glucagon [BIOCHEM] The protein hormone secreted by α-cells of the pancreas which plays a role in carbohydrate metabolism. Also known as hyperglycemic factor; hyperglycemic glycogenolytic factor. { 'glü·kəˌgän }

glucamine [BIOCHEM] $C_6H_{15}O_4N$ An amine formed by reduction of glucosylamine or of glucose oxime. { 'glü·kəˌmēn }

glucan [BIOCHEM] A polysaccharide composed of the hexose sugar D-glucose. { 'glüˌkan }

glucinium [CHEM] The former name for the element beryllium, coined because the salts of beryllium are sweet-tasting. { glü'sin·ē·əm }

glucoamylase See amyloglucosidase. { ˌglü·kō'am·əˌlās }

glucocerebrosidase [BIOCHEM] An enzyme that removes the glucose from glycosyl ceramide and is defective or missing in Gaucher's disease. { ˌglü·kōˌser·ə'bräs·iˌdās }

glucocerebroside [BIOCHEM] A glycoside of ceramide that contains glucose. { ˌglü·kōˌsə'rēb·rəˌsīd }

glucochloral See chloralose. { ˌglü·kō'klȯr·əl }

glucochloralose See chloralose. { ˌglü·kō'klȯr·əˌlōs }

α-D-glucochloralose See chloralose. { ˌal·fə ˌdē ˌglü·kō'klȯr·əˌlōs }

glucocorticoid [BIOCHEM] A corticoid that affects glucose metabolism; secreted principally by the adrenal cortex. { ˌglü·kō'kȯrd·əˌkȯid }

glucogenesis [BIOCHEM] Formation of glucose within the animal body from products of glycolysis. { ˌglü·kō'jen·əˌsəs }

glucokinase [BIOCHEM] An enzyme that catalyzes the phosphorylation of D-glucose to glucose-6-phosphate. { ˌglü·kō'kīˌnās }

glucolipid [BIOCHEM] A glycolipid that yields glucose on hydrolysis. { ˌglü·kō'lip·əd }

glucomannan [BIOCHEM] A polysaccharide composed of D-glucose and D-mannose; a prominent component of coniferous trees. { ˌglü·kō'man·ən }

gluconate [ORG CHEM] A salt of gluconic acid. { 'glü·kəˌnāt }

gluconeogenesis [BIOCHEM] Formation of glucose within the animal body from substances other than carbohydrates, particularly proteins and fats. { ˌglü·kōˌnē·ō'jen·ə·səs }

gluconic acid [ORG CHEM] $C_6H_{12}O_7$ A crystalline acid obtained from glucose by oxidation; used in cleaning metals. { glü'kän·ik 'as·əd }

gluconic acid sodium salt See sodium gluconate. { glü'kän·ik 'as·əd 'sōd·ē·əm 'sȯlt }

Gluconobacter [MICROBIO] A genus of bacteria in the family Pseudomonadaceae; ellipsoidal to rod-shaped cells having three to eight polar flagella and occurring singly, in pairs, or in chains. { glü'kän·əˌbak·tər }

D-glucopyranose See glucose. { ˌdē ˌglü·kə'pir·əˌnōs }

glucopyranoside [BIOCHEM] Any glucoside that contains a six-membered ring. { ˌglü·kō·pir'an·əˌsīd }

glucosamine [BIOCHEM] $C_6H_{13}O_5$ An amino sugar; the most abundant in nature, occurring in glycoproteins and chitin. { 'glü·kōs·əˌmēn }

glucose [BIOCHEM] $C_6H_{12}O_6$ A monosaccharide; occurs free or combined and is the most common sugar. Also known as cerelose; D-glucopyranose. { 'glüˌkōs }

glucose-6-phosphatase [BIOCHEM] An enzyme found in liver which catalyzes the hydrolysis of glucose-6-phosphate to free glucose and inorganic phosphate. { 'glüˌkōs ˌsiks 'fäsˌfās }

glucose phosphate [BIOCHEM] A phosphoric derivative of glucose, as glucose-1-phosphate. { 'glüˌkōs 'fäsˌfāt }

glucose-1-phosphate [BIOCHEM] $C_6H_{12}O_8P$ An ester of glucopyranose in which a phosphate group is attached to carbon atom 1; there are two types: α-D- and β-D-glucose-1-phosphates. Also known as Cori ester. { 'glüˌkōs ˌwən 'fäsˌfāt }

glucose-6-phosphate [BIOCHEM] $C_6H_{13}O_9P$ An ester of glucose with phosphate attached to carbon atom 6. Also known as Robisonester. { 'glüˌkōs ˌsiks 'fäsˌfāt }

glucose-6-phosphate dehydrogenase [BIOCHEM] The mammalian enzyme that catalyzes the oxidation of glucose-6-phosphate by TPN+ (triphosphopyridine nucleotide). { 'glüˌkōs ˌsiks 'fäsˌfāt dēˌhī'drä·jəˌnās }

glucose tolerance test [PATH] A test to measure the ability of the liver to convert glucose to glycogen. { 'glüˌkōs ˌtäl·ə·rəns ˌtest }

glucosidase [BIOCHEM] An enzyme that hydrolyzes glucosides. { glü'kō·səˌdās }

glucoside [BIOCHEM] One of a group of compounds containing the cyclic forms of glucose, in which the hydrogen of the hemiacetal hydroxyl has been replaced by an alkyl or aryl group. { 'glü·kəˌsīd }

7-D-glucoside See genistin. { ˌsev·ən ˌdē 'glü·kəˌsīd }

glucosulfone sodium See sodium glucosulfone. { 'glüˌkō ˌsəlˌfōn 'sōd·ē·əm }

glucosyl ceramide [BIOCHEM] A type of glycolipid that is present in the cell membranes of many cell types and is abundant in serum. { ˌglü·kəˌsil 'ser·əˌmīd }

glucosyltransferase [BIOCHEM] An enzyme that catalyzes the glucosylation of hydroxymethyl cytosine; a constituent of bacteriophage deoxyribonucleic acid. { ˌglü·kəˌsil'tranz·fəˌräs }

glucuronic acid [BIOCHEM] $C_6H_{10}O_7$ An acid resulting from oxidation of the CH_2OH radical of D-glucose to COOH; a component of many polysaccharides and certain vegetable gums. Also known as glycuronic acid. { ˌglü·kyə'rän·ik 'as·əd }

glucuronidase [BIOCHEM] An enzyme that catalyzes hydrolysis of glucuronides. Also known as glycuronidase. { ˌglü·kyə'rän·əˌdās }

glucuronide [BIOCHEM] A compound resulting from the interaction of glucuronic acid with a phenol, an alcohol, or an acid containing a carboxyl group. Also known as glycuronide. { 'glü·kyùr·əˌnīd }

D-glucuronolactone [BIOCHEM] $C_6H_8O_6$ A water-soluble crystalline compound found in plant gums in polymers with other carbohydrates, and an important structural component of almost all fibrous and connective tissues in animals; used in medicine as an antiarthritic. { ¦de glü¦kyùr·ə·nō'lak‚tōn }

glue [MATER] A crude, impure, amber-colored form of commercial gelatin of unknown detailed composition produced by the hydrolysis of animal collagen; gelatinizes in aqueous solutions and dries to form a strong, adhesive layer. { glü }

glueball [PART PHYS] A hadron consisting entirely of gluons, without any quarks. Also known as bound glue state; gluonia. { 'glü‚bȯl }

glue block See angle block. { 'glü ‚bläk }

glue cell See adhesive cell. { 'glü ‚sel }

glue-joint ripsaw [MECH ENG] A heavy-gage ripsaw used on straight-line or self-feed rip machines; the cut is smooth enough to permit gluing of joints from the saw. { 'glü ‚jȯint 'rip‚sȯ }

glue-line heating [ENG] Dielectric heating in which the electrodes are designed to give preferential heating to a thin film of glue or other relatively high-loss material located between layers of relatively low-loss material such as wood. { 'glü ‚līn 'hēd·iŋ }

glug [MECH] A unit of mass, equal to the mass which is accelerated by 1 centimeter per second per second by a force of 1 gram-force, or to 980.665 grams. { gləg }

glulam [MATER] A material fabricated by joining two or more layers of wood with an adhesive so that the grains of all the layers are approximately parallel. { 'glü‚lam }

glume [BOT] One of two bracts at the base of a spikelet of grass. { glüm }

glumiferous [BOT] Bearing glumes. { glü'mif·ə·rəs }

Glumiflorae [BOT] An equivalent name for Cyperales. { ‚glü·mə'flȯr‚ē }

gluon [PART PHYS] One of eight hypothetical massless particles with spin quantum number and negative parity that mediate strong interactions between quarks. { 'glü‚än }

gluonia See glueball. { glü'ō·nyə }

gluside See saccharin. { 'glü‚sīd }

glutamate [BIOCHEM] A salt or ester of glutamic acid. { 'glüd·ə‚māt }

glutamic acid [BIOCHEM] $C_5H_9O_4N$ A dicarboxylic amino acid of the α-ketoglutaric acid family occurring widely in proteins. { glü'tam·ik 'as·əd }

glutaminase [BIOCHEM] The enzyme which catalyzes the conversion of glutamine to glutamic acid and ammonia. { glü'tam·ə‚nās }

glutamine [BIOCHEM] $C_5H_{10}O_3N_2$ An amino acid; the monamide of glutamic acid; found in the juice of many plants and essential to the development of certain bacteria. { 'glüd·ə‚mēn }

glutamine synthetase [BIOCHEM] An enzyme which catalyzes the formation of glutamine from glutamic acid and ammonia, using adenosine triphosphate as a source of energy. { 'glüd·ə‚mēn 'sin·thə‚tās }

glutaraldehyde [ORG CHEM] $OHC(CH_2)_3CHO$ A liquid with a boiling point of 188°C; soluble in water and alcohol; used as a biological solution (50) and for leather tanning. { ¦glüd·ə'ral·də‚hīd }

glutarate [BIOCHEM] The salt or ester of glutaric acid. { 'glüd·ə‚rāt }

glutaric acid [BIOCHEM] $C_5H_5O_4$ A water-soluble, crystalline acid that occurs in green sugarbeets and in water extracts of crude wool. { glü'tar·ik 'as·əd }

glutathione [BIOCHEM] $C_{10}H_{17}O_6N_3S$ A widely distributed tripeptide that is important in plant and animal tissue oxidation reactions. { ¦glüd·ə'thī‚ōn }

glutelin [BIOCHEM] A class of simple, heat-labile proteins occurring in seeds of cereals; soluble in dilute acids and alkalies. { 'glüd·əl·ən }

gluten [BIOCHEM] **1.** A mixture of proteins found in the seeds of cereals; gives dough elasticity and cohesiveness. **2.** An albuminous element of animal tissue. { 'glüt·ən }

gluten enteropathy [MED] A malabsorption disease characterized by inflammation and loss of the normal architecture of the small intestine following ingestion of some proteins. Also known as celiac sprue, nontropical sprue. { ‚glüt·ən ‚ent·ə'räp·ə·thē }

glutenin [BIOCHEM] A glutelin of wheat. { 'glüt·ən·ən }

glutethimide [PHARM] $C_{13}H_{15}NO_2$ A minor or sedative antianxiety tranquilizer that acts as a central nervous system depressant. { glü'teth·ə·məd }

gluteus maximus [ANAT] The largest and most superficial muscle of the buttocks. { 'glüd·ē·əs 'mak·sə·məs }

gluteus medius [ANAT] The muscle of the buttocks lying between the gluteus maximus and gluteus minimus. { 'glüd·ē·əs 'mēd·ē·əs }

gluteus minimus [ANAT] The smallest and deepest muscle of the buttocks. { 'glüd·ē·əs 'min·ə·məs }

glutinant nematocyst [INV ZOO] A nematocyst characterized by an open, sticky tube used for anchoring the cnidarian when walking on its tentacles. { 'glüt·ən·ənt nə'mad·ə‚sist }

glutinous [BOT] Having a sticky surface. { 'glüt·ən·əs }

glycal [BIOCHEM] A monosaccharide having a double bond between carbons 1 and 2, and no hydroxyl group on either of those carbons. { 'glī‚kal }

glycan See polysaccharide. { 'glī‚kan }

glycemia [PHYSIO] The presence of glucose in the blood. { glī'sē·mē·ə }

glycemic index [MED] A ranking of foods based on how they affect blood glucose (sugar) levels in the 2–3 hours after eating; foods with carbohydrates that break down quickly during digestion have the highest glycemic indexes. { glī'sēm·ik 'in‚deks }

glyceraldehyde [BIOCHEM] $CH_2OHCHOHCHO$ A colorless solid, isomeric with dehydroxyacetone; soluble in water and insoluble in organic solvents; an important intermediate in carbohydrate metabolism; used as a chemical intermediate in biochemical research and nutrition. { ‚glis·ə'ral·də‚hīd }

glycerate [BIOCHEM] A salt or ester of glyceric acid. { 'glis·ə‚rāt }

glyceric acid [BIOCHEM] $C_3H_6O_4$ A hydroxy acid obtained by oxidation of glycerin. { glə'ser·ik 'as·əd }

Glyceridae [INV ZOO] A family of polychaete annelids belonging to the Errantia and characterized by an enormous eversible proboscis. { glə'ser·ə‚dē }

glyceride [BIOCHEM] An ester of glycerin and an organic acid radical; fats are glycerides of certain long-chain fatty acids. { 'glis·ə‚rīd }

glycerin See glycerol. { 'glis·ə·rən }

glycerine See glycerol. { 'glis·ə·rən }

glycerinated vaccine virus See smallpox vaccine. { 'glis·ə·rə‚nād·əd 'vak‚sēn ‚vī·rəs }

glycerokinase [BIOCHEM] An enzyme that catalyzes the phosphorylation of glycerol to glycerophosphate during microbial fermentation of propionic acid. { ¦glis·ə·rō'kī‚nās }

glycerol [ORG CHEM] $CH_2OHCHOHCH_2OH$ The simplest trihedric alcohol; when pure, it is a colorless, odorless, viscous liquid with a sweet taste; it is completely soluble in water and alcohol but only partially soluble in common solvents such as ether and ethyl acetate; used in manufacture of alkyd resins, explosives, antifreezes, medicines, inks, perfumes, cosmetics, soaps, and finishes. Also known as glycerin; glycerine; glycyl alcohol. { 'glis·ə‚rȯl }

glycerophosphate [BIOCHEM] Any salt of glycerophosphoric acid. { ‚glis·ə·rō'fäs‚fāt }

glycerophosphoric acid [BIOCHEM] $C_3H_5(OH)_2OPO_3H_2$ Either of two pale-yellow, water-soluble, isomeric dibasic acids occurring in nature in combined form as cephalin and lecithin. { ¦glis·ə·rō‚fäs'fȯr·ik 'as·əd }

glyceryl [ORG CHEM] $OCH_2OCHOCH_2\equiv$ The functional group from glycerol, $(CH_2OH)_2CHOH$. { 'glis·ə·rəl }

glyceryl diacetate See diacetin. { 'glis·ə·rəl dī'as·ə‚tāt }

glyceryl tristearate See stearin. { 'glis·ə·rəl trī'stir‚āt }

glycidic acid [ORG CHEM] $C_2H_3O·CO_2H$ A volatile liquid. Also known as epoxy-propionic acid. { glə'sid·ik 'as·əd }

glycidol [ORG CHEM] $C_3H_6O_2$ A colorless, liquid epoxide that boils at 162°C and is miscible with water; used in organic synthesis. Also known as epihydrin alcohol. { 'glis·ə‚dȯl }

glycin [ORG CHEM] $C_8H_9NO_3$ A crystalline compound that forms shiny leaflets from water solution, melts at 245–247°C, and is soluble in alkalies and mineral acids; used as a photographic developer and in the analytical determination of iron, phosphorus, and silicon. Also known as photoglycine. { 'glī·sən }

glycine [BIOCHEM] $C_2H_5O_2N$ A white, crystalline amino acid found as a constituent of many proteins. Also known as aminoacetic acid. { 'glī‚sēn }

glyco- [ORG CHEM] Chemical prefix indicating sweetness, or relating to sugar or glycine. { 'glī·kō }

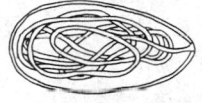

GLUTAMIC ACID

$$COOH$$
$$|$$
$$CH_2$$
$$|$$
$$CH_2$$
$$|$$
$$C$$
$$H_2N \quad H \quad COOH$$

Structural formula of glutamic acid.

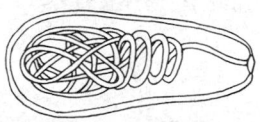

GLUTINANT NEMATOCYST

The glutinant nematocyst of *Hydra*. (*After Schulze, from T. I. Storer and R. L. Usinger, General Zoology, 3d ed., McGraw-Hill, 1957*)

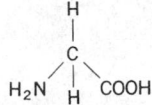

GLYCINE

$$H$$
$$|$$
$$C$$
$$H_2N \quad H \quad COOH$$

Structural formula of glycine.

glycocalyx [CELL MOL] The outer component of a cell surface, outside the plasmalemma; usually contains strongly acidic sugars, hence it carries a negative electric charge. { 'glī·kō,kā,liks }

glycocholic acid [BIOCHEM] $C_{26}H_{43}NO_6$ A bile obtained by the conjugation of cholic acid with glycine. { ¦glī·kō¦käl·ik 'as·əd }

glycocyamine [BIOCHEM] $C_3H_7N_3O_2$ A product of interaction of aminocetic acid and arginine, which on transmethylation with methionine is converted to creatine. Also known as guanidine-acetic acid. { ¦glī·kō'sī·ə,mēn }

glycogen [BIOCHEM] A nonreducing, white, amorphous polysaccharide found as a reserve carbohydrate stored in muscle and liver cells of all higher animals, as well as in cells of lower animals. { 'glī·kə·jən }

glycogenesis [BIOCHEM] The metabolic formation of glycogen from glucose. { 'glī·kə¦jen·ə·səs }

glycogenolysis [BIOCHEM] The metabolic breakdown of glycogen. { ¦glī·kə·jə¦näl·ə·səs }

glycogenosis [MED] One of several inborn errors in the metabolism of glycogen, classified on the basis of the enzyme deficiency and clinical findings as von Gierke's disease, Pompe's disease, limit dextrinosis, amylopectinosis, McArdle's disease, or Hers' disease. { ¦glī·kō·jə'nō·səs }

glycogen storage disease *See* von Gierke's disease. { 'glī·kə·jən ,stȯr·ij di,zēz }

glycogen synthetase [BIOCHEM] An enzyme that catalyzes the synthesis of the amylose chain of glycogen. { 'glī·kə·jən 'sin·thə,tās }

glycol [ORG CHEM] **1.** $C_nH_2(OH)_2$ An organic chemical with two hydroxyl groups on an essentially aliphatic carbon chain. Also known as dihydroxy alcohol. **2.** $HOCH_2CH_2OH$ A colorless dihydroxy alcohol used as an antifreeze, in hydraulic fluids, and in the manufacture of dynamites and resins. Also known as ethlene glycol. { 'glī,kȯl }

glycol dehydrator [CHEM ENG] Processing equipment for removing all or most of the water from a wet gas by contacting with glycol. { 'glī,kȯl dē'hī,drād·ər }

glycol diacetate *See* ethylene glycol diacetate. { 'glī,kȯl dī'as·ə,tāt }

glycoldinitrate *See* ethylene nitrate. { 'glī,kȯl dī'nī,trāt }

glycol ester [ORG CHEM] Chemical compound composed of the reaction products of a glycol, $C_nH_{2n}(OH)_2$, and an organic acid; an example is ethylene glycol diacetate, the product of ethylene glycol and acetic acid. { 'glī,kȯl 'es·tər }

glycol ether [ORG CHEM] A colorless liquid used as a solvent, in detergents, and as a diluent; a typical example is ethylene glycol diethyl ether, $C_2H_5OCH_2CH_2OC_2H_5$. { 'glī,kȯl 'ē·thər }

glycolic acid [ORG CHEM] $CH_2OHCOOH$ Colorless, deliquescent leaflets, decomposing about 78°C; soluble in water, alcohol, and ether; used as a chemical intermediate in fabric dyeing. Also known as hydroxyacetic acid. { glī'käl·ik 'as·əd }

glycolipid [BIOCHEM] Any of a class of complex lipids which contain carbohydrate residues. { ¦glī·kō¦lip·əd }

glycolysis [BIOCHEM] The enzymatic breakdown of glucose or other carbohydrate, with the formation of lactic acid or pyruvic acid and the release of energy in the form of adenosinetriphosphate. { ¦glī'käl·ə·səs }

glycolythiourea *See* 2-thiohydantoin. { ¦gli·kȯl¦thī·ō·yu'rē·ə }

glycolytic pathway [BIOCHEM] The principal series of phosphorylative reactions involved in pyruvic acid production in phosphorylative fermentations. Also known as Embden-Meyerhof pathway; hexose diphosphate pathway. { ¦glī·kə¦lid·ik 'path,wā }

glycolyurea *See* hydantoin. { ¦glī,kȯl·yu'rē·ə }

glyconeogenesis [BIOCHEM] The metabolic process of glycogen formation from noncarbohydrate precursors. { ¦glī·kō¦nē·ō'jen·ə·səs }

glycopeptide *See* glycoprotein. { ¦glī·kō'pep,tīd }

glycophyte [BOT] A plant requiring more than 0.5% sodium chloride solution in the substratum. { 'gli·kə,fīt }

glycoprotein [BIOCHEM] Any of a class of conjugated proteins containing both carbohydrate and protein units. Also known as glycopeptide. { ¦glī·kō'prō,tēn }

glycose [BIOCHEM] A simple sugar whose structure is in the form either of an open-chain aldehyde or ketone or of a cyclic hemiacetal. { 'glī,kōs }

glycosidase [BIOCHEM] An enzyme that hydrolyzes a glycoside. { glī'kō·sə,dās }

glycoside [BIOCHEM] A compound that yields on hydrolysis a sugar (glucose, galactose) and an aglycon; many of the glycosides are therapeutically valuable. { 'glī·kə,sīd }

glycosphingolipid [BIOCHEM] A glycoside of ceramide that is the most abundant and structurally diverse type of glycolipid in animals. { ¦glī·kō,sfiŋ·gō'lip·id }

glycosuria [MED] Presence of sugar in the urine. { ¦glī·kō'shur·ē·ə }

glycosyl [BIOCHEM] A univalent functional group derived from the cyclic form of glycose by removal of the hemiacetal hydroxyl group. { 'glī·kə,sil }

glycosylation [BIOCHEM] A chemical reaction in which glycosyl groups are added to a protein to produce a glycoprotein. { glī,käs·ə'lā·shən }

glycosyl glyceride [BIOCHEM] A class of glycolipids structurally analogous to phospholipids; they are the major glycolipids of plants and microorganisms but are rare in animals. { glī,kōs·əl 'glis·ə,rīd }

glycosyl phosphatidylinositol [BIOCHEM] Any of a class of glycolipids that serve as membrane anchors for a multitude of proteins in organisms ranging from yeast to protozoa to humans. { glī¦kōs·əl fäs,fäd·ə,dil·i'nos·ə,tól }

glycosyltransferase [BIOCHEM] Any of a group of enzymes that participate in the biosynthesis of glycoproteins by transferring one sugar at a time from suitable donors to particular acceptors. { ¦glī·kō·sil'tranz·fə,rās }

glycotropic [BIOCHEM] Acting to antagonize the action of insulin. { ¦glī·kō'träp·ik }

glycuresis [PHYSIO] Excretion of sugar seen normally in urine. { ,glik·yə'rē·səs }

glycuronic acid *See* glucuronic acid. { ¦glik·yə¦rän·ik 'as·əd }

glycuronidase *See* glucuronidase. { ,glik·yə'rän·ə,dās }

glycuronide *See* glucuronide. { gli'kyúr·ə,nīd }

glycyl [ORG CHEM] NH_2CH_2COO- or $NHCH_2COO=$ The radical from glycine, NH_2CH_2COOH; found in peptides. { 'glī·səl }

glycyl alcohol *See* glycerol. { 'glī·səl 'al·kə,hól }

glyoxal [ORG CHEM] $(CHO)_2$ Colorless, deliquescent powder or liquid with mild odor, melting point 15°C, boiling point 51°C; used to insolubilize starches, cellulosic materials, and proteins, in embalming fluids, for leather tanning, and for rayon shrinkproofing. { glī'äk,sal }

glyoxalase [BIOCHEM] An enzyme present in various body tissues which catalyzes the conversion of methylglyoxal into lactic acid. { glī'äk·sə,lās }

glyoxalic acid [ORG CHEM] $CHOCOOH$ Colorless crystals that are soluble in water, forming glyoxylic acid. { ¦glī,äk¦sal·ik 'as·əd }

glyoxylate cycle [BIOCHEM] A sequence of biochemical reactions related to respiration in germinating fatty seeds by which acetyl coenzyme A is converted to succinic acid and then to hexose. { glī'äk·sə,lāt ,sī·kəl }

glyoxylic acid [BIOCHEM] $CH(OH)_2COOH$ An aldehyde acid found in many plant and animal tissues, especially unripe fruit. { ¦glī,äk¦sil·ik 'as·əd }

glyoxysome [BOT] A specialized type of peroxisome found in plant tissues that is bounded by a single membrane and contains a broad spectrum of enzymes, including those of the glyoxylate cycle and the β-oxidation cycle in addition to catalase and oxidase. { glī'äk·sə,sōm }

glyphs [GRAPHICS] All typographic images, from letters and symbols to Chinese and Japanese characters. { glifz }

Glyphocyphidae [PALEON] A family of extinct echinoderms in the order Temnopleuroida comprising small forms with a sculptured test, perforate crenulate tubercles, and diademoid ambulacral plates. { ,glif·ō'sīf·ə,dē }

glyphosate [ORG CHEM] $C_3H_8NO_5P$ A white solid with a melting point of 200°C; slight solubility in water; used as a herbicide in postharvest treatment of crops. { 'glif·ə,sāt }

glyphosine [ORG CHEM] $C_4H_{11}NO_8P_2$ A white solid with a melting point of 203°C; quite soluble in water; used as a growth regulator in sugarcane. { 'glif·ə,sēn }

glyptal resin [ORG CHEM] A phthalic anhytxride glycerol made from an emulsion of an alkyd resin; used in lacquers and insulation. { 'glipt·əl 'rez·ən }

Glyptocrinina [PALEON] A suborder of extinct crinoids in the order Monobathrida. { ¦glip·tō·krə'nī·nə }

glyptolith See ventifact. { 'glip·tə,lith }

gm See gram.

GM counter See Geiger-Müller counter. { ¦jē'em ¸kaȯnt·ər }

gmelinite [MINERAL] (Na₂Ca)Al₂Si₄O₁₂·6H₂O Zeolite mineral that is colorless or lightly colored and crystallizes in the hexagonal system. { gə'mel·ə¸nīt }

Gmelin's test [PATH] A qualitative test for the pigments in bile; test solution is mixed with nitric acid containing nitrous acid; reaction is positive if color appears at the acid-solution junction. { gə'mäl·ənz ¸test }

GMT See Greenwich mean time.

gnab gib See big crunch. { gə'näb ¸gib }

gnat [INV ZOO] The common name for a large variety of biting insects in the order Diptera. { nat }

gnathic index [ARCH] The ratio of the distance from the nasion to the basion to that from the basion to the alveolar point multiplied by 100. { 'nath·ik 'in¸deks }

Gnathiidea [INV ZOO] A suborder of isopod crustaceans characterized by a much reduced second thoracomere, short antennules and antennae, and a straight pleon. { nā'thī·ə·də }

gnathion [ARCH] The midpoint of the lower margin of the mandible in humans. [VERT ZOO] The most anterior point of the premaxillae on or near the middle line in certain lower mammals. { 'nā·thē¸än }

gnathite [INV ZOO] A mouth appendage in arthropods. { 'nā¸thīt }

Gnathobdellae [INV ZOO] A suborder of leeches in the order Arhynchobdellae having jaws and a conspicuous posterior sucker; it contains most of the important blood-sucking leeches of humans and other warm-blooded animals. { ¦nā¸thäb 'del·ē }

Gnathobelodontinae [PALEON] A subfamily of extinct elephantoid proboscideans containing the shovel-jawed forms of the family Gomphotheriidae. { nā¦thäb·ə·lō'dän·tə¸nē }

gnathocephalon [INV ZOO] The part of the insect head lying behind the protocephalon; bears the maxillae and mandibles. { ¸nā·thō'sef·ə¸län }

gnathochilarium [INV ZOO] The lower lip of certain arthropods; thought to be fused maxillae. { ¸nā·thō¸kī'lar·ē·əm }

Gnathodontidae [PALEON] A family of extinct conodonts having platforms with large, cup-shaped attachment scars. { ¸nā·thō'dän·tə¸dē }

gnathopod [INV ZOO] Any of the crustacean paired thoracic appendages modified for manipulation of food but sometimes functioning in copulatory amplexion. { 'nā·thə¸päd }

gnathopodite [INV ZOO] A segmental, modified appendage which serves as a jaw in arthropods. { nā'thä·pə¸dīt }

gnathos [INV ZOO] A mid-ventral plate on the ninth tergum in lepidopterans. { 'nā¸thōs }

gnathostegite [INV ZOO] One of a pair of broad plates formed from the outer maxillipeds of some crustaceans, which function to cover other mouthparts. { nə'thäs·ə¸jīt }

Gnathostomata [INV ZOO] A suborder of echinoderms in the order Echinoidea characterized by a rigid, exocyclic test and a lantern or jaw apparatus. [VERT ZOO] A group of the subphylum Vertebrata which possess jaws and usually have paired appendages. { ¸nā·thə'stō·məd·ə }

Gnathostomidae [INV ZOO] A family of parasitic nematodes in the superfamily Spiruroidea; sometimes placed in the superfamily Physalopteroidea. { ¸nā·thə'stō·məd·ē }

Gnathostomulida [INV ZOO] Microscopic marine worms of uncertain systematic relationship, mainly characterized by cuticular structures in the pharynx and a monociliated skin epithelium. { nə¦thäs·tə'myül·ə·də }

gnathothorax [INV ZOO] The thorax and part of the head bearing feeding organs in arthropods, regarded as a primary region of the body. { ¸nā·thō'thȯr¸aks }

gnd See ground.

gneiss [PETR] A variety of rocks with a banded or coarsely foliated structure formed by regional metamorphism. { nīs }

gneissic granodiorites [PETR] Granodiorite rocks with gneissic characteristics. { 'nīs·ik ¦gra·nō'dī·ə¸rīts }

Gnetales [BOT] A monogeneric order of the subdivision Gneticae; most species are lianas with opposite, oval, entire-margined leaves. { nə'tā·lēz }

Gnetatae See Gnetopsida. { nə'täd¸ē }

Gneticae [BOT] A subdivision of the division Pinophyta

characterized by vessels in the secondary wood, ovules with two integuments, opposite leaves, and an embryo with two cotyledons. { 'ned·ə¸sē }

Gnetophyta [BOT] The equivalent name for Gnetopsida. { nə'täf·əd·ə }

Gnetopsida [BOT] A class of gymnosperms comprising the subdivision Gneticae. { nə'täp·səd·ə }

gnomic [ASTRON] Pertaining to the gnomon of a sundial. { 'nō·mik }

gnomon [ENG] On a sundial, the inclined plate or pin that casts a shadow. Also known as style. [MATH] A geometric figure formed by removing from a parallelogram a similar parallelogram that contains one of its corners. { 'nō·mən }

gnomonic chart [MAP] A chart on the gnomonic projection where great circles project as straight lines. Also known as great-circle chart. { nō'män·ik 'chärt }

gnomonic projection [CRYSTAL] A projection for displaying the poles of a crystal in which the poles are projected radially from the center of a reference sphere onto a plane tangent to the sphere. [MAP] A projection on a plane tangent to the surface of a sphere having the point of projection at the center of the sphere. { nō'män·ik prə'jek·shən }

Gnostidae [INV ZOO] An equivalent name for Ptinidae. { 'näs·tə¸dē }

gnotobiology [BIOL] That branch of biology dealing with known living forms; the study of higher organisms in the absence of all demonstrable, viable microorganisms except those known to be present. { ¸nō·dō·bī'äl·ə·jē }

gnotobiote [MICROBIO] An individual (host) living in intimate association with another known species (microorganism). **2.** The known microorganism living on a host. { ¸nō·dō'bī¸ōt }

gnotobiotics [MICROBIO] The science involved with maintaining a microbiologically controlled environment, and with the knowledge necessary to obtain and use biological specimens in this environment. { ¸nōd·ə·bī'äd·iks }

gnu [VERT ZOO] Any of several large African antelopes of the genera *Connochaetes* and *Gorgon* having a large oxlike head with horns that characteristically curve downward and outward and then up, with the bases forming a frontal shield in older individuals. { nü }

GNU [COMPUT SCI] Freely distributed software for producing and distributing nonproprietary software that is compatible with Unix, but is not Unix. { gə'nü }

goaf [MIN ENG] **1.** That part of a mine from which the coal has been worked away and the space more or less filled up. **2.** The refuse or waste left in the mine. Also known as gob. { gōf }

goal coordination method [CONT SYS] A method for coordinating the subproblem solutions in plant decomposition, in which Lagrange multipliers enter into the subsystem cost functions as shadow prices, and these are adjusted by the second-level controller in an iterative procedure which culminates (if the method is applicable) in the satisfaction of the subsystem coupling relationships. Also known as interaction balance method; nonfeasible method. { 'gōl kȯ¸ȯrd·ə·nā·shən ¸meth·əd }

goat [VERT ZOO] The common name for a number of artiodactyl mammals in the genus *Capra*; closely related to sheep but differing in having a lighter build and hollow, swept-back, sometimes spiral or twisted horns. { gōt }

gob See goaf. { gäb }

gobi [GEOL] Sedimentary deposits in a synclinal basin. { 'gō·bē }

Gobiatheriinae [PALEON] A subfamily of extinct herbivorous mammals in the family Uintatheriidae known from one late Eocene genus; characterized by extreme reduction of anterior dentition and by lack of horns. { gō¦bī·ə·thə'rī·ə¸nē }

Gobiesocidae [VERT ZOO] The single family of the order Gobiesociformes. { gō¦bī·ə¦säs·ə¸dē }

Gobiesociformes [VERT ZOO] The clingfishes, a monofamilial order of scaleless bony fishes equipped with a thoracic sucking disk which serves for attachment. { gō¦bī·ə¸säs·ə'fȯr¸mēz }

Gobiidae [VERT ZOO] A family of perciform fishes in the suborder Gobioidei characterized by pelvic fins united to form a sucking disk on the breast. { gō'bī·ə¸dē }

Gobioidei [VERT ZOO] The gobies, a suborder of morphologically diverse actinopterygian fishes in the order Perciformes; all lack a lateral line. { ¸gō·bē'ȯid·ē¸ī }

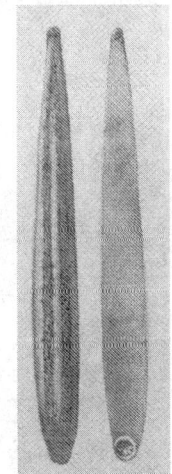

Haemopis grandis, dorsal and ventral view showing the conspicuous posterior sucker.

GNOMONIC PROJECTION

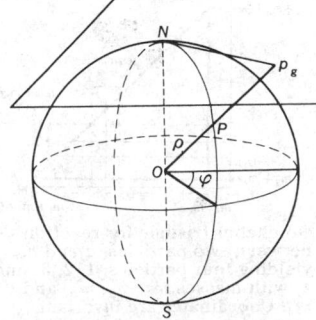

Gnomonic projection of *P.* Line *OP* cuts planes tangent to the sphere at *N* at point *p*_g. Coordinates φ and ρ define pole *P.*

goblet cell [HISTOL] A unicellular, mucus-secreting intraepithelial gland that is distended on the free surface. Also known as chalice cell. [INV ZOO] Any of the unicellular choanocytes of the genus *Monosiga*. { 'gäb·lət ,sel }

gobo [ENG] A panel used to shield a television camera lens from direct light. [ENG ACOUS] A sound-absorbing shield used with a microphone to block unwanted sounds. { 'gō,bō }

gob stink [MIN ENG] **1.** The odor from the burning coal given off by an underground fire. **2.** The odor given off by the spontaneous heating of coal, not necessarily in the gob. Also known as stink. { 'gäb ,stiŋk }

Gödel numbering See arithmetization. { 'gərd·əl ,nəm·bə·riŋ }

Gödel's proof [MATH] Any formal arithmetical system is incomplete in the sense that, given any consistent set of arithmetical axioms, there are true statements in the resulting arithmetical system that cannot be derived from these axioms. { 'gərd·əlz 'prüf }

Gödel's second theorem [MATH] The theorem that any formal arithmetical system is incomplete in the sense that, if it is consistent, it cannot prove its own consistency. { 'gərd·əlz ,sek·ənd 'thir·əm }

Gödel's universe [RELAT] An exact solution of the nonvacuum equations of general relativity with matter in the form of dust; there are closed timelike lines in this solution. { 'gərd·əlz 'yü·nə,vərs }

godet [TEXT] A roller used in stretching synthetic fiber filaments. { gō'det }

go-devil [ENG] **1.** A device inserted in a pipe or hole for purposes such as cleaning or for detonating an explosive. **2.** A sled for moving logs or cultivating. **3.** A large rake for gathering hay. **4.** A small railroad car used for transporting workers and materials. { 'gō ,dev·əl }

Goertler parameter [FL MECH] A dimensionless number used in studying boundary-layer flow on curved surfaces, equal to the Reynolds number, where the characteristic length is the boundary-layer momentum thickness, times the square root of this thickness, divided by the square root of the surface's radius of curvature. { 'gərt·lər pə'ram·əd·ər }

goethite [MINERAL] FeO(OH) A yellow, red, or dark-brown mineral crystallizing in the orthorhombic system, although it is usually found in radiating fibrous aggregates; a common constituent of natural rust or limonite. Also known as xanthosiderite. { 'gə,tīt }

go gage [DES ENG] A test device that just fits a part if it has the proper dimensions (often used in pairs with a "no go" gage to establish maximum and minimum dimensions). { 'gō ,gāj }

goggles [ENG] Spectacle-like eye protectors having shields at the sides and short, projecting eye tubes. { 'gäg·əlz }

going [CIV ENG] On a staircase, the distance between the faces of two successive risers. { 'gō·iŋ }

going barrel [HOROL] A mainspring barrel with a toothed edge for driving the train; it is mounted on an arbor that is stationary except while the timepiece is being wound. { 'gō·iŋ ,bar·əl }

going train [HOROL] Gears that drive the hands in a striking timepiece. { 'gō·iŋ ,trān }

goiter [MED] An enlargement of all or part of the thyroid gland; may be accompanied by a hormonal dysfunction. { 'gȯid·ər }

Golay cell [ENG] A radiometer in which radiation absorbed in a gas chamber heats the gas, causing it to expand and deflect a diaphragm in accordance with the amount of radiation. { gə'lā ,sel }

Golay code [COMMUN] A linear, block-based error-correcting code that is particularly suited to applications where short code word length and low latency are important. { gə'lā ,kōd }

gold [CHEM] A chemical element, symbol Au, atomic number 79, atomic weight 196.9665; soluble in aqua regia; melts at 1065°C. [MET] The native metallic element; a deep-yellow, very dense, soft, isometric metal, usually found alloyed with silver or copper; used in jewelry, dentistry, gilding, anodes, alloys, and solders. { gōld }

gold-198 [NUC PHYS] The radioisotope of gold, atomic mass-number 198 and half-life 2.7 days; used in medical treatment of tumors by injecting it in colloidal form directly into tumor tissue. { |gōld |wən|nīn·tē'āt }

gold alloy [MET] Any alloy containing gold. { |gōld 'al,ȯi }

Goldbach conjecture [MATH] The unestablished conjecture that every even number except the number 2 is the sum of two primes. { 'gōl,bäk kən,jek·chər }

goldbeater's skin [MATER] The treated outside membrane of the large intestine of cattle; used between leaves of metal in goldbeating, and sometimes in hygrometers. { 'gōl,bēd·ərz ,skin }

goldbeater's-skin hygrometer [ENG] A hygrometer using goldbeater's skin as the sensitive element; variations in the physical dimensions of the skin caused by its hygroscopic character indicate relative atmospheric humidity. { 'gōl,bēd·ərz ,skin hī'gräm·əd·ər }

goldbeating [MET] The process of producing gold leaf. { 'gōl,bēd·iŋ }

Goldberg-Mohn friction [FL MECH] A force proportional to the velocity of a current and the density of the medium; used as a first approximation in estimating frictional effects in the atmosphere and the ocean. { 'gōlt·berk 'mōn 'frik·shən }

gold beryl See chrysoberyl. { 'gōld ,ber·əl }

gold bronze [MET] A powdered alloy of copper used to simulate gold, or an alloy of copper containing 3–5% aluminum. { |gōld 'bränz }

gold chloride [INV ZOO] AuCl₃ A red, soluble compound made by reaction of gold and chlorine or by reaction of HAuCl₄ with chlorine; decomposes by heat; soluble in water, alcohol, and ether; used in photography, plating, inks, medicine, and ceramics. { |gōld 'klȯr,īd }

gold doping [ELECTR] A technique for controlling the lifetime of minority carriers in a transistor; gold is diffused into the base and collector regions to reduce storage time in transistor circuits. { 'gōl ,dōp·iŋ }

golden algae [BOT] The common name for members of the class Chrysophyceae. { 'gol·dən 'al·jē }

golden antimony sulfide See antimony pentasulfide. { 'gōl·dən 'ant·ə,mō·nē 'səl,fīd }

golden-brown algae [BOT] The common name for members of the division Chrysophyta. { ,gōl·dən 'braún 'al·jē }

golden mean See golden section. { ,gōld·ən 'mēn }

golden ratio See golden section. { ,gōld·ən 'rā·shō }

goldenrod paper [GRAPHICS] A specially coated masking paper used in lithography for assembling and positioning negatives for exposure onto plates. { 'gōl·dən,räd 'pā·pər }

golden rectangle [MATH] A rectangle that can be divided into a square and another rectangle similar to itself; its sides have the ratio $(1 + \sqrt{5})/2$. { 'gōl·dən 'rek,taŋ·gəl }

golden section [MATH] The division of a line so that the ratio of the whole line to the larger interval equals the ratio of the larger interval to the smaller. Also known as divine proportion; extreme and mean ratio; golden mean; golden ratio. { 'gōl·dən 'sek·shən }

goldenseal See Hydrastis canadensis. { 'gōl·dən ,sēl }

golden-section search [COMPUT SCI] A dichotomizing search in which, in each step, the remaining items are divided as closely as possible according to the golden section. { 'gōl·dən 'sek·shən ,sərch }

gold-filled [MET] Covered by a layer of gold alloy. { |gōld 'fild }

goldfish [VERT ZOO] *Crassius auratus*. An orange cyprini-form fish of the family Cyprinidae that can grow to over 18 inches (46 centimeters); closely related to the carps. { 'gōl,fish }

gold foil [MET] A thin sheet of gold, thicker than gold leaf, formed by rolling or hammering. { |gōld 'fȯil }

Goldhaber triangle [PART PHYS] A plot describing a high-energy reaction leading to four or more particles; its coordinates are the invariant masses of two intermediate-state quasi-particle composites, and its kinematical limits form a right-angled isosceles triangle; resonances in the quasi-particle composites appear as horizontal and vertical bands. { 'gōlt,hä·bər ,trī,aŋ·gəl }

gold hydroxide [INV ZOO] Au(OH)₃ A yellow-brown, light-sensitive, water-insoluble powder; dissolves in most acids; easily reduced to metallic gold; used in medicine, porcelain, gold plating, and daguerreotypes. { 'gōld hī'dräk,sīd }

gold leaf [MET] Gold beaten or rolled into extremely thin sheets or leaves (10^{-6} inch or 25 nanometers thick); leaves are stored in books (a book consists of 25 leaves), the paper of

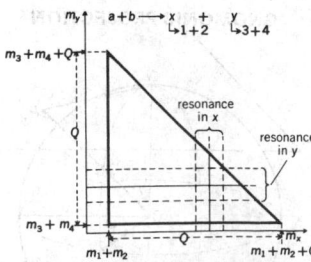

GOLDHABER TRIANGLE

Goldhaber triangle for reaction between two particles, a and b, yielding four particles, 1, 2, 3, and 4, with masses m_1, m_2, m_3, and m_4. Coordinates are invariant masses m_x and m_y of intermediate-state quasi-particle composites x and y which decay into two particles each as indicated at top of figure. Q is total center of mass energy of a and b minus sum of m_1, m_2, m_3, and m_4.

which is rubbed with chalk to keep the leaves from sticking. { ¦gōld 'lēf }

gold-leaf electroscope [ELEC] An electroscope in which two narrow strips of gold foil or leaf suspended in a glass jar spread apart when charged; the angle between the strips is related to the charge. { 'gōld ¦lēf i'lek·trə‚skōp }

gold metallurgy [MET] The science and technology of gold recovery and refining. { ¦gōld 'med·əl‚ər·jē }

gold number [ANALY CHEM] A measure of the amount of protective colloid which must be added to a standard red gold sol mixed with sodium chloride solution to prevent the solution from causing the sol to coagulate, as manifested by a change in color from red to blue. { 'gōld ‚nəm·bər }

gold oxide [INV ZOO] Au₂O₃ Water-insoluble, heat-decomposable, brownish-black powder; soluble in hydrochloric acid; used to gild, in medicine and porcelain, and for daguerreotypes. Also known as auric oxide; gold trioxide. { 'gōld 'äk‚sīd }

gold plate [MET] Gold which has been electroplated on a material in a thin layer of controlled thickness; used on electric contacts for corrosion resistance and solderability and on jewelry and ornaments. { ¦gōld 'plāt }

gold point [THERMO] The temperature of the freezing point of gold at a pressure of 1 standard atmosphere (101,325 pascals); used to define the International Temperature Scale of 1990, on which it is assigned a value of 1337.33 K or 1064.18°C. { 'gōld ‚póint }

gold potassium chloride See potassium gold chloride. { ¦gōld pə'tas·ē·əm 'klōr‚īd }

gold potassium cyanide See potassium gold cyanide. { ¦gōld pə'tas·ē·əm 'sī·ə‚nīd }

gold salt See sodium gold chloride. { 'gōld ‚sólt }

goldschmidtine See stephanite. { 'gōl‚shmid‚ēn }

goldschmidtite See sylvanite. { 'gōl‚shmid‚īt }

Goldschmidt process [MET] **1.** The thermite process of welding. **2.** A process by which dry chlorine is employed to remove tin from scrap tinplate. { 'gōl‚shmit ‚präs·əs }

Goldschmidt's law [SOLID STATE] The law that crystal structure is determined by the ratios of the numbers of the constituents, the ratios of their sizes, and their polarization properties. { 'gōl‚shmits ‚lò }

Goldschmidt's mineralogical phase rule [GEOL] The rule that the probability of finding a system with degrees of freedom less than two is small under natural rock-forming conditions. { 'gōl‚shmits ‚min·ə·rə'läj·ə·kəl ¦fāz ‚rül }

gold size [CHEM] A solution of white and red lead and yellow ocher in linseed oil; used to seal permanently microscopical preparations. [MATER] An adhesive used to fix gold leaf to a surface. { ¦gōld ¦sīz }

Gold slide [ENG] A slide rule used on British ships to compute barometric corrections and reduction of pressure to sea level; it includes the effects of temperature, latitude, index correction, and barometric height above sea level. { 'gōld ‚slīd }

gold sodium chloride See sodium gold chloride. { ¦gōld 'sōd·ē·əm 'klōr‚īd }

gold sodium cyanide See sodium gold cyanide. { ¦gōld 'sōd·ē·əm 'sī·ə‚nīd }

gold sodium thiomalate [PHARM] C₄H₃AuNa₂O₄S A crystalline compound, soluble in water; used as an antirheumatic in medicine. Also known as sodium aurothiomalate. { ¦gōld 'sōd·ē·əm ‚thī·ō'ma‚lāt }

gold sodium thiosulfate [PHARM] Na₃Au(S₂O₃)₂·2H₂O A white crystalline compound that is freely soluble in water; used for treatment of rheumatoid arthritis and lupus erythematosus. { ¦gōld 'sōd·ē·əm ‚thī·ō'səl‚fāt }

gold solder [MET] Solder composed of 60% gold, 20% silver, and 20% copper. { ¦gōld 'säd·ər }

Goldstone bosons [PHYS] Particles with zero mass and zero spin which accompany spontaneous breaking of exact fundamental symmetries. { 'gōl‚stōn 'bō‚sänz }

gold tin precipitate See gold tin purple. { 'gōld ¦tin prə'sip·ə‚tāt }

gold tin purple [ORG CHEM] A brown powder which is a mixture of gold chloride and brown tin oxide, soluble in ammonia; used in coloring enamels, manufacturing ruby glass, and painting porcelain. Also known as gold tin precipitate; purple of Cassius. { 'gōld ¦tin 'pər·pəl }

gold trioxide See gold oxide. { ¦gōld trī'äk‚sīd }

golfada [METEOROL] A heavy gale of the Mediterranean. { gól'fäd·ə }

golf ball [ENG] A printing element used on some typewriters and serial printers, consisting of a rotating, spherically shape, removable typehead that skims across the printed line while the typewriter or printer carriage does not move. { 'gälf ‚ból }

Golgi apparatus [CELL MOL] A cellular organelle that is part of the cytoplasmic membrane system; it is composed of regions of stacked cisternae and it functions in secretory processes. { 'gól·jē ‚ap·ə‚rad·əs }

Golgi cell [NEUROSCI] **1.** A nerve cell with long axons. **2.** A nerve cell with short axons that branch repeatedly and terminate near the cell body. { 'gól·jē ‚sel }

Golgi-Mazzoni's corpuscle [ANAT] A small sensory lamellar corpuscle located in the parietal pleura. { 'gól·jē mät'sō‚nēz ‚kòr·pə·səl }

Golgi stack [CELL MOL] The central structure of the Golgi apparatus consisting of flattened membrane-bounded cisternae. Formerly known as dictyosome. { 'gōl·jē ‚stak }

Golgi tendon organ [PHYSIO] Any of the kinesthetic receptors situated near the junction of muscle fibers and a tendon which act as muscle-tension recorders. { 'gól·jē 'ten·dən ‚ór·gən }

Gomberg-Bachmann-Hey reaction [ORG CHEM] Production of diaryl compounds by adding alkali to a mixture of a diazonium salt and a liquid aromatic hydrocarbon or a derivative. { 'góm‚berk 'bäk‚män 'hī rē‚ak·shən }

Gomberg reaction [ORG CHEM] The production of free radicals by reaction of metals with triarylmethyl halides. { 'góm‚berk rē‚ak·shən }

Gompertz curve [STAT] A curve similar to the exponential curve except that the constant a is raised to the b^x power instead of the x power; used in fitting a trend line to a nonlinear time series. { 'gäm‚pərts ‚kərv }

Gomphidae [INV ZOO] A family of dragonflies belonging to the Anisoptera. { 'gäm·fə‚dē }

gomphosis [ANAT] An immovable articulation, as that formed by the insertion of teeth into the bony sockets. { gäm'fō·səs }

Gomphotheriidae [PALEON] A family of extinct proboscidean mammals in the suborder Elephantoidea consisting of species with shoveling or digging specializations of the lower tusks. { ‚gam·fō·thə'rī·ə‚dē }

Gomphotheriinae [PALEON] A subfamily of extinct elephantoid proboscideans in the family Gomphotheriidae containing species with long jaws and bunomastodont teeth. { ‚gäm·fō·thə'rī·ə‚nē }

gon See grade. { gän }

gonad [ANAT] A primary sex gland; an ovary or a testis. { 'gō‚nad }

gonadal agenesis [MED] Failure of the gonad to develop, or retrogression of the gonad at very early stages. Also known as gonadal dysgenesis. { gō'nad·əl ā'jēn·ə·səs }

gonadal dysgenesis See gonadal agenesis. { gō'nad·əl dis'·jen·ə·səs }

gonadectomy [MED] Surgical removal of a gonad. { ‚gō·na'dek·tə·mē }

gonadotropic hormone [BIOCHEM] Either of two adenohypophyseal hormones, FSH (follicle-stimulating hormone) or ICSH (interstitial-cell-stimulating hormone), that act to stimulate the gonads. { gō¦nad·ə¦träp·ik 'hòr‚mōn }

gonadotropin [BIOCHEM] A substance that acts to stimulate the gonads. { gō‚nad·ə'trō·pən }

gonapophysis [INV ZOO] A paired, modified appendage of the anal region in insects that functions in copulation, oviposition, or stinging. { gän·ə'päf·ə·səs }

gondola car [ENG] A flat-bottomed railroad car which has no top, fixed sides, and often removable ends, in which steel, rock, or heavy bulk commodities are transported. { 'gän·də·lə ‚kär }

Gondwana [GEOL] The ancient continent that is supposed to have fragmented and drifted apart during the Triassic to form eventually the present continents. Also known as Gondwanaland. { gänd'wä·nə }

Gondwanaland See Gondwana. { gän'dwän·ə‚land }

gong buoy [NAV] A buoy similar in construction to a bell buoy, but sounding a distinctive note because of the use of sets of gongs, each gong having a different tone. { 'gäŋ ‚bói }

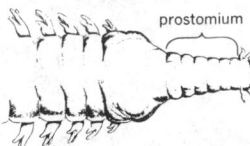

GONIADIDAE

prostomium

Multiannulate prostomium of *Goniada* in dorsal view.

GOOSE

Canada goose (*Branta canadensis*).

GOOSEBERRY

Gooseberry branch bearing leaves and fruits. (*USDA*)

gonia [HISTOL] Primordial sex cells, such as oogonia and spermatogonia. { 'gō·nē·ə }

Goniadidae [INV ZOO] A family of marine polychaete annelids belonging to the Errantia. { ,gō·nē'ad·ə,dē }

gonidium [BIOL] An asexual reproductive cell or group of cells arising in a special organ on or in a gametophyte. { gō'nid·ē·əm }

gonimoblast [BOT] A filament arising from the fertilized carpogonium of most red algae. { 'gän·ə·mō,blast }

goniometer [ELECTROMAG] An instrument for determining the direction of maximum response to a received radio signal, or selecting the direction of maximum radiation of a transmitted radio signal; consists of two fixed perpendicular coils, each attached to one of a pair of loop antennas which are also perpendicular, and a rotatable coil which bears the same space relationship to the coils as the direction of the signal to the antennas. [ENG] **1.** An instrument used to measure the angles between crystal faces. **2.** An instrument which uses x-ray diffraction to measure the angular positions of the axes of a crystal. **3.** Any instrument for measuring angles. { ,gō·nē'äm·əd·ər }

goniometric locator [ELECTROMAG] In radio direction finding, a rotating device that samples signals from orthogonal fixed antennas. { ¦gō·nē·ō¦me·trik 'lō,kād·ər }

goniophotometer [OPTICS] A photometer designed to measure the intensity of light reflected from a surface at various angles. { ¦gō·nē·ō·fə'täm·əd·ər }

gonioscope [MED] A special optical instrument for studying in detail the angle of the anterior chamber of the eye. { 'gō·nē·ə,skōp }

gonitis [MED] Inflammation of the knee joint. { gō'nīd·əs }

gonnardite [MINERAL] $Na_2CaAl_5Si_6O_{20}\cdot7H_2O$ Zeolite mineral occurring in fibrous, radiating spherules; specific gravity is 2.3. { 'gän·ər,dīt }

gonococcal arthritis [MED] A blood-borne joint infection by *Neisseria gonorrhoeae* occurring as a manifestation of gonorrhea. { ¦gän·ə¦käk·əl är'thrīd·əs }

gonococcal epididymitis [MED] Inflammation of the epididymitis due to infection by *Neisseria gonorrhoeae*; a secondary manifestation of gonorrhea. { ¦gän·ə¦käk·əl ,ep·ə,did·ə'mīd·əs }

gonococcus *See* Neisseria gonorrhoeae. { ,gän·ō'käk·əs }

Gonodactylidae [INV ZOO] A family of mantis shrimp in the order Stomatopoda. { ,gän·ō·dak'til·ə,dē }

gonodendrum [INV ZOO] A branched structure which bears clusters of gonophores. { ,gän·ə'den·drəm }

go/no-go detector [ENG] An instrument having only two operating states, such as a common fuse which is either intact or melted. { 'gō 'nō ,gō di,tek·tər }

go/no-go test [ENG] A test based on the measurement of one or more parameters but which can have only one of two possible results, to pass or reject the device under test. { 'gō 'nō ,gō ,test }

gonomery [EMBRYO] In some insect embryos, grouping of maternal and paternal chromosomes separately during the first couple of mitotic divisions after fertilization. { gō'näm·ə·rē }

gonopalpon [INV ZOO] Tentacle-like, sensitive structures associated with cnidarian gonophores. { ,gän·ə'pal·pən }

gonophore [BOT] An elongation of the receptacle extending between the stamens and corolla. [INV ZOO] Reproductive zooid of a hydroid colony. { 'gän·ə,fōr }

gonopodium [VERT ZOO] Anal fin modified as a copulatory organ in certain fishes. { ,gän·ə'pōd·ē·əm }

Gonorhynchiformes [VERT ZOO] A small order of soft-rayed teleost fishes having fusiform or moderately compressed bodies, single short dorsal and anal fins, a forked caudal fin, and weak toothless jaws. { gän·ə,riŋ·kə'fȯr,mēz }

gonorrhea [MED] A bacterial infection of humans caused by the gonococcus (*Neisseria gonorrhoeae*) which invades the mucous membrane of the urogenital tract. { ,gän·ə'rē·ə }

gonorrheal urethritis [MED] Inflammation of the urethra, particularly in males, as the result of gonorrhea. { ,gän·ə'rē·əl ,yur·ē'thrīd·əs }

gonorrheal vulvovaginitis [MED] Inflammation of the vulva and vagina as the result of gonorrhea. { ,gän·ə'rē·əl ¦vəl·vō,va·jə'nīd·əs }

gonosome [INV ZOO] Aggregate of gonophores in a hydroid colony. { 'gän·ə,sōm }

gonosomite [INV ZOO] The ninth segment of the abdomen of the male insect. Also known as genital segment. { ,gän·ə'sō,mīt }

gonostome [INV ZOO] The part of the genital duct of a coelomate invertebrate known as the coelomic funnel. Also known as coelomostome. { 'gän·ə,stōm }

gonostyle [INV ZOO] Gonapophysis of dipteran insects. { 'gän·ə,stīl }

gonotocont *See* auxocyte. { gə'näd·ə,känt }

gonotome [EMBRYO] The part of an embryonic somite involved in gonad formation. { 'gän·ə,tōm }

gonozooid [INV ZOO] A zooid of bryozoans and tunicates which produces gametes. { ¦gän·ə'zō,ȯid }

gonyautoxin [BIOCHEM] One of a group of saxitoxin-related compounds that are produced by the dinoflagellates *Gonyaulax catenella* and *G. tamarensis*. { ¦gō·nē·ō¦täk·sən }

gonys [VERT ZOO] A ridge along the mid-ventral line of the lower mandible of certain birds. { 'gō·nəs }

Gooch crucible [ANALY CHEM] A ceramic crucible with a perforated base; in analysis it is used for filtration through asbestos or glass. { 'güch ,krü·sə·bəl }

good geometry [NUC PHYS] An arrangement of source and detecting equipment such that little error is produced by the finite sizes of the source and the detector aperture. { ¦gud jē'äm·ə·trē }

Goodman duckbill loader [MIN ENG] A gathering and loading assembly for coal, consisting of a shovel trough with a shovel head fitting inside the feeder trough, an operating carrier which controls the connection between these troughs, a sliding shoe which moves to and fro on the floor of the seam, a swivel trough, and a pendulum jack; it loads the coal into the shaker conveyor pan column. { 'gud·mən ¦dək·bil 'lōd·ər }

Goodman loader [MIN ENG] **1.** An electrohydraulic power shovel designed for loading coal where the seams are 6 feet (1.8 meters) or more in thickness. **2.** A loader designed for loading coal from thin seams; has a telescoping fan-shaped apron that extends from the entry of the room to the working face. { 'gud·mən ,lōd·ər }

goodness of fit [STAT] The degree to which the observed frequencies of occurrence of events in an experiment correspond to the probabilities in a model of the experiment. Also known as best fit. { ¦gud·nəs əv 'fit }

good oil *See* raffinate. { 'gud ,ȯil }

Goodpasture's syndrome [MED] A complex of symptoms associated with acute glomerulonephritis and pulmonary hemorrhage. { 'gud,pas·chərz ,sin,drōm }

good seamanship [NAV] Any precaution which may be required by the ordinary practice of seamen. { ¦gud 'sē·mən·ship }

googol [MATH] A name for 10 to the power 100. { 'gü,gȯl }

googolplex [MATH] A name for 10 to the power googol. { 'gü,gȯl,pleks }

goongarrite [MINERAL] $Pb_4Bi_2S_7$ A mineral composed of a sulfide of lead and bismuth. { gün'ga,rīt }

goop [MATER] A compound in paste form containing finely divided magnesium, used as a constituent of certain incendiary bomb fillings. { güp }

goose [VERT ZOO] The common name for a number of waterfowl in the subfamily Anatinae; they are intermediate in size and features between ducks and swans. { güs }

gooseberry [BOT] The common name for about six species of thorny, spreading bushes of the genus *Ribes* in the order Rosales, producing small, acidic, edible fruit. { 'güs,ber·ē }

gooseberry stone *See* grossularite. { 'güs,ber·ē ,stōn }

goosefoot oil *See* chenopodium oil. { 'güs,fut ,ȯil }

gooseneck [DES ENG] **1.** A pipe, bar, or other device having a curved or bent shape resembling that of the neck of a goose. **2.** *See* water swivel. [NAV ARCH] An iron hook which joins a spar to a mast. { 'güs,nek }

gooseneck barnacle [INV ZOO] Any stalked barnacle, especially of the genus *Lepas*. { 'güs,nek 'bär·ni·kəl }

Goos-Hähnchen effect [OPTICS] A shift of a few wavelengths in the position of a light beam that undergoes total internal reflection from a surface. { 'gos 'hench·ən i,fekt }

goovoo [COMPUT SCI] A file within a generation data group, so called because of the notation used in some systems in which, for example, G003 V001 is volume 1, generation −3 of a generation data group. { 'gü,vü }

gopher [VERT ZOO] The common name for North American rodents composing the family Geomyidae. Also known as pocket gopher. { 'gō·fər }

Gopher [COMPUT SCI] A menu-based program for browsing the Internet and finding and gaining access to files, programs, definitions, and other Internet resources. { 'gō·fər }

gopher hole [ENG] Horizontal T-shaped opening made in rock in preparation for blasting. Also known as coyote hole. [MIN ENG] An irregular pitting hole made during prospecting. { 'gō·fər ˌhōl }

gopher-hole blasting See coyote blasting. { 'gō·fər ˌhōl ˈblast·iŋ }

gophering [MIN ENG] A method of breaking up a sandy medium-hard overburden where usual blastholes tend to cave in, by firing an explosive charge in each of a series of shallow holes; debris is cleared, and holes are made deeper for further charges, until they are deep enough to take enough explosives to break up the deposit. { 'gō·fə·riŋ }

gorceixite [MINERAL] $BaAl_3(PO_4)_2(OH)_5 \cdot H_2O$ A brown mineral composed of a hydrous basic phosphate of barium and aluminum. { 'gȯr·sək,sīt }

Gordiidae [INV ZOO] A monogeneric family of worms in the order Gordioidea distinguished by a smooth cuticle. { gȯr'dī·ə,dē }

Gordioidea [INV ZOO] An order of worms belonging to the Nematomorpha in which there is one ventral epidermal cord, a body cavity filled with mesenchymal tissue, and paired gonads. { ˌgȯr·dē'ȯid·ē·ə }

gordioid larva [INV ZOO] The developmental stage of nematomorphs, free-living for a short time. { 'gȯr·dē,ȯid 'lär·və }

gordonite [MINERAL] $MgAl_2(PO_4)_2(OH)_2 \cdot 8H_2O$ A colorless mineral composed of a hydrous basic phosphate of magnesium and aluminum. { 'gȯrd·ən,īt }

Gordon's formula [CIV ENG] An empirical formula which gives the collapsing load of a column in terms of its cross-sectional area, length, and least diameter. { 'gȯrd·ənz ˌfȯr·myə·lə }

gore [CIV ENG] A small triangular parcel of land. { gȯr }

gorge [ARCH] The entrance to a bastion. [GEOGR] A narrow passage between mountains or the walls of a canyon, especially one with steep, rocky walls. [OCEANOGR] A collection of solid matter obstructing a channel or a river, as an ice gorge. { gȯrj }

gorge wind See canyon wind. { 'gȯrj ˌwind }

Gorgonacea [INV ZOO] The horny corals, an order of the cnidarian subclass Alcyonaria; colonies are fanlike or featherlike with branches spread radially or oppositely in one plane. { ˌgȯr·gə'nās·ē·ə }

gorgonin [BIOCHEM] The protein, frequently containing iodine and bromine, composing the horny skeleton of members of the Gorgonacea; contains iodine and bromine. { 'gȯr·gə·nən }

Gorgonocephalidae [INV ZOO] A family of ophiuroid echinoderms in the order Phrynophiurida in which the individuals often have branched arms. { ˌgȯr·gə,nō·sə'fal·ə,dē }

gorilla [VERT ZOO] *Gorilla gorilla.* An anthropoid ape, the largest living primate; the two African subspecies are the lowland gorilla and the mountain gorilla. { gə'ril·ə }

gorlic acid [ORG CHEM] $C_5H_7(C_{12}H_{22})COOH$ An unsaturated acid derived from sapucainha oil, obtained from the seeds of a tree in the Amazon Valley. { 'gȯr·lik 'as·əd }

Gorsky effect [MET] For hydrogen dissolved in a metal, its migration from the compressed side to the stretched side when the metal is bent. { 'gȯr·skē i,fekt }

goslarite [MINERAL] $ZnSO_4 \cdot 7H_2O$ A white mineral composed of hydrous zinc sulfate. { 'gäs·lə,rīt }

gosling blast [METEOROL] A sudden squall of rain or sleet in England. Also known as gosling storm. { 'gäz·liŋ ,blast }

gosling storm See gosling blast. { 'gäz·liŋ ,stȯrm }

gossamer [TEXT] A filmy gauze fabric. { 'gäs·ə·mər }

gossan [GEOL] A rusty, ferruginous deposit filling the upper regions of mineral veins and overlying a sulfide deposit; formed by oxidation of pyrites. Also known as capping; gozzan; iron hat. { 'gas·ən }

gossypose See raffinose. { 'gäs·ə,pōs }

Gotlandian [GEOL] A geologic time period recognized in Europe to include the Ordovician; it appears before the Devonian. { gät'lan·dē·ən }

GOTO-less programming [COMPUT SCI] The writing of computer programs without the use of GOTO statements. { ˌgō·tü,les 'prō,gram·iŋ }

Goto pair [ELECTR] Two tunnel diodes connected in series in such a way that when one is in the forward conduction region, the other is in the reverse tunneling region; used in high-speed gate circuits. { 'gō·dō ,per }

GOTO statement [COMPUT SCI] A statement in a computer program that provides for the direct transfer of control to another statement with the identifier that is the argument of the GOTO statement. { 'gō,tü ,stāt·mənt }

gouge [DES ENG] A curved chisel for wood, bone, stone, and so on. [GEOL] Soft, pulverized mixture of rock and mineral material found along shear (fault) zones and produced by the differential movement across the plane of slippage. [MIN ENG] A layer of soft material along the wall of a vein which favors miners by enabling them, after gouging it out with a pick, to attack the solid vein from the side. { gau̇j }

gouging [ENG] The removal of material by electrical, mechanical, or manual means for the formation of a groove. { 'gau̇j·iŋ }

Gould's belt [ASTRON] A belt of bright stars inclined about 20° to the Milky Way and including most of the bright stars in Orion, Scorpio, Carina, and Centaurus, apparently resulting from a slight tilt of the spiral arm of the Milky Way Galaxy containing the sun, with respect to the galactic plane. { 'gülz ,belt }

gout [MED] A condition of purine metabolism resulting in increased blood levels of uric acid with ultimate deposition as urates in soft tissues around joints. { gau̇t }

gouy [PHYS CHEM] An electrokinetic unit equal to the product of the electrokinetic potential and the electric displacement divided by 4π times the polarization of the electrolyte. { 'gü·ē }

Gouy balance [ANALY CHEM] Device for measurement of diamagnetic and paramagnetic susceptibilities of samples (solid, liquid, solution). { 'gō·ē ,bal·əns }

government frequency bands [COMMUN] Radio-frequency bands which are allotted to various departments and services of the federal government. { 'gəv·ər·mənt ,frekwən·sē ,banz }

governor [MECH ENG] A device, especially one actuated by the centrifugal force of whirling weights opposed by gravity or by springs, used to provide automatic control of speed or power of a prime mover. { 'gəv·ə·nər }

gowk storm [METEOROL] In England, a storm or gale occurring at about the end of April or the beginning of May. { 'gau̇k ,stȯrm }

goyazite [MINERAL] $SrAl_3(PO_4)_2(OH)_5 \cdot H_2O$ A granular, yellowish-white mineral composed of a hydrous strontium aluminum phosphate. { 'gȯi·ə,zīt }

gozzan See gossan. { 'gäz·ən }

g parameter [ELECTR] One of a set of four transistor equivalent-circuit parameters; they are the inverse of the h parameters. { 'jē pə,ram·əd·ər }

G parity [PART PHYS] The eigenvalue of a system under the operation of inversion in isotopic spin space; it is conserved by the strong interactions. Also known as isotopic parity. { 'jē ,par·əd·ē }

GPI See ground point of intercept.

G protein [CELL MOL] See GTP-binding protein. { 'jē ,prō,tēn }

G-protein-linked receptor [CELL MOL] A cell surface receptor that consists of a polypeptide chain threaded across the membrane seven times and that, when activated by the binding of a ligand, in turn activates a cytosolic G-protein molecule, which then initiates a cascade of reactions effecting the intracellular response to the extracellular signal (the ligand). { 'jē ,prō,tēn ,liŋkt ri'sep·tər }

GPS See Global Positioning System.

GPSS [COMPUT SCI] A problem-oriented programming language designed to assist the user in developing models. Acronym for general-purpose systems simulation.

gpu See geopotential unit.

gr See grain.

Graafian follicle [HISTOL] The mature mammalian ovum with its surrounding epithelial cells. { 'gräf·ē·ən 'fäl·ə·kəl }

grab [ENG] An instrument for extricating broken boring tools from a borehole. { grab }

grabbing crane [MECH ENG] An excavator made up of a

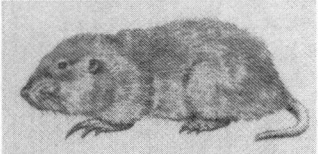

GOPHER

Pocket gopher of North America.

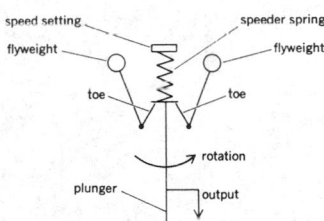

GOVERNOR

speed setting　　　　　speeder spring
flyweight　　　　　　　　flyweight
toe　　　　　　　　　　　toe
rotation
plunger
output

Ballhead of governor containing whirling flyweights driven at a speed proportional to the speed of the prime mover.

GRABEN

Diagram of simple graben. *(From A. K. Lobeck, Geomorphology, McGraw-Hill, 1939)*

crane carrying a large grab or bucket in the form of a pair of half scoops, hinged to dig into the earth as they are lifted. { 'grab·iŋ ,krän }

grab bucket [MECH ENG] A bucket with hinged jaws or teeth that is hung from cables on a crane or excavator and is used to dig and pick up materials. { 'grab ,bək·ət }

grab dredger [MECH ENG] Dredging equipment comprising a grab or grab bucket that is suspended from the jib head of a crane. Also known as grapple dredger. { 'grab ,drej·ər }

graben [GEOL] A block of the earth's crust, generally with a length much greater than its width, that has dropped relative to the blocks on either side. { 'grä·bən }

grabhook [DES ENG] A hook used for grabbing, as in lifting blocks of stone, in which case the hooks are used in pairs connected with a chain, and are so constructed that the tension of the chain causes them to adhere firmly to the rock. { 'grab,hùk }

grab sample [MIN ENG] A random mode of sampling; the samples may be taken from the pile broken in the process of mining, or from a truck or car of ore or coal. { 'grab ,sam·pəl }

graceful degradation [COMPUT SCI] A programming technique to prevent catastrophic system failure by allowing the machine to operate, though in a degraded mode, despite failure or malfunction of several integral units or subsystems. { 'grās·fùl ,deg·rə'dā·shən }

graceful exit [COMPUT SCI] The ability to escape from a problem situation in a computer program without having to reboot the computer. { 'grās·fəl 'eg·zət }

Gracilariidae [INV ZOO] A family of small moths in the superfamily Tineoidea; both pairs of wings are lanceolate and widely fringed. { ,gra·sil·ə'rī·ə,dē }

gracilis [ANAT] A long slender muscle on the medial aspect of the thigh. { 'gras·ə·ləs }

gradation [GEOL] **1.** The leveling of the land, or the bringing of a land surface or area to a uniform or nearly uniform grade or slope through erosion, transportation, and deposition. **2.** Specifically, the bringing of a stream bed to a slope at which the water is just able to transport the material delivered to it. { grā'dā·shən }

gradation period [GEOL] The time during which the base level of the sea remains in one position. Also known as base-leveling epoch. { grā'dā·shən ,pir·ē·əd }

grade [AGR] An individual having one parent (usually the sire) that is a purebred, and one parent a grade or scrub. [CIV ENG] **1.** To prepare a roadway or other land surface of uniform slope. **2.** A surface prepared for the support of rails, a road, or a conduit. **3.** The elevation of the finished surface of an engineering project. [COMMUN] One of two types of television service, designated grade A and grade B, each having a specified signal strength, that of grade A being several times larger than B. [ENG] The degree of strength of a high explosive. [EVOL] A stage of evolution in which a similar level of organization is reached by one or more species in the development of a structure, physiological process, or behavioral character. [GEOL] The slope of the bed of a stream, or of a surface over which water flows, upon which the current can just transport its load without either eroding or depositing. [MATER] Any of the various purity standards for chemicals and chemical products that have been established for specific applications. [MATH] A unit of plane angle, equal to 0.01 right angle, or $\pi/200$ radians, or 0.9°. Also known as gon. [MIN ENG] **1.** A classification of ore according to recoverable amount of a valuable metal. **2.** To sort and classify diamonds. { grād }

gradeability [MECH ENG] The performance of earthmovers on various inclines, measured in percent grade. { ,grād·ə'bil· əd·ē }

grade beam [CIV ENG] A reinforced concrete beam placed directly on the ground to provide the foundation for the superstructure. { 'grād ,bēm }

grade correction See slope correction. { 'grād kə,rek·shən }

grade crossing [CIV ENG] The intersection of roadways, railways, pedestrian walks, or combinations of these at grade. { 'grād ,krȯs·iŋ }

graded [GEOL] Brought to or established at grade. { 'grād·əd }

graded bedding [GEOL] A stratification in which each stratum displays a gradation in the size of grains from coarse below to fine above. { 'grād·əd 'bed·iŋ }

graded index lens See gradient index lens. { 'grād·əd 'in,deks ,lenz }

graded-junction transistor See rate-grown transistor. { 'grād·əd 'jəŋk·shən tran'zis·tər }

graded Lie algebra [MATH] A generalization of a Lie algebra in which both commutators and anticommutators occur. { 'grād·əd 'lē 'al·jə·brə }

graded periodicity technique [ELECTR] A technique for modifying the response of a surface acoustic wave filter by varying the spacing between successive electrodes of the interdigital transducer. { 'grād·əd ,pir·ē·ə'dis·əd·ē tek,nēk }

graded profile See profile of equilibrium. { 'grād·əd 'prō,fīl }

graded refractive index rod lens [OPTICS] A type of gradient index lens used in optical fiber components, consisting of a rod with a refractive index that has its maximum value on the axis, and decreases approximately as the square of the distance. Abbreviated GRIN-rod lens. { 'grād·əd ri'frak·tiv 'in,deks ,räd ,lenz }

graded stream [HYD] A stream in which, over a period of years, slope is adjusted to yield the velocity required for transportation of the load supplied from the drainage basin. { 'grād·əd 'strēm }

graded topocline [ECOL] A topocline having a wide range, or ranging into different kinds of environment, thus subjecting its members to differential selection so that divergence between local races may become sufficient to warrant creation of varietal, or even specific, names. { 'grād·əd 'täp·ə,klīn }

grade line [CIV ENG] A line or slope used as a longitudinal reference for a railroad or highway. { 'grād 'līn }

grade of coal [MIN ENG] A classification of coal based on the amount and nature of the ash and sulfur content. { 'grād əv 'kōl }

grader [MECH ENG] A high-bodied, wheeled vehicle with a leveling blade mounted between the front and rear wheels; used for fine-grading relatively loose and level earth. { 'grād·ər }

grade scale [GEOL] A continuous scale of particle sizes divided into a series of size classes. { 'grād ,skāl }

grade separation [CIV ENG] A grade crossing employing an underpass and overpass. { 'grād sep·ə,rā·shən }

grade slab [CIV ENG] A reinforced concrete slab placed directly on the ground to provide the foundation for the superstructure. { 'grād ,slab }

grade stake [CIV ENG] A stake used as an elevation reference. { 'grād ,stāk }

gradient [GEOL] The rate of descent or ascent (steepness of slope) of any topographic feature, such as streams or hillsides. [MATH] A vector obtained from a real function $f(x_1, x_2, \ldots, x_n)$ whose components are the partial derivatives of f; this measures the maximum rate of change of f in a given direction. [NAV] A slope expressed in feet per mile as a ratio of the horizontal to the vertical distance; for example, 40:1 means 40 feet horizontally to 1 foot vertically. { 'grād·ē·ənt }

gradient coupling [PART PHYS] A hypothetical interaction of particles in which the interaction Hamilton depends explicitly on first derivatives of wave functions associated with the particles with respect to position and time. { 'grād·ē·ənt ,kəp·liŋ }

gradient current [OCEANOGR] A current defined by assuming that the horizontal pressure gradient in the sea is balanced by the sum of the Coriolis and bottom frictional forces; at some distance from the bottom the effect of friction becomes negligible, and above this the gradient and geostrophic currents are equivalent. Also known as slope current. { 'grād·ē·ənt ,kə·rənt }

gradient elution analysis [ANALY CHEM] A form of gas-liquid chromatography in which the eluting solvent is changed with time, either by gradually mixing a second solvent of greater eluting power with the first, less powerful solvent, or by a gradual change in pH or other property. { 'grād·ē·ənt i'lü·shən ə,nal·ə·səs }

gradienter [ENG] An attachment placed on a surveyor's transit to measure angle of inclination in terms of the tangent of the angle. { 'grād·ē,en·tər }

gradient flow [METEOROL] Horizontal frictionless flow in which isobars and streamlines coincide, or equivalently, in which the tangential acceleration is everywhere zero; the balance of normal forces (pressure force, Coriolis force, centrifugal force) is then given by the gradient wind equation. { 'grād·ē·ənt ,flō }

gradient index lens [OPTICS] An optical element within

which the refractive index is a smooth, but not constant, function of position and, as a result, the ray paths are curved. Also known as graded index lens. { 'grād·ē·ənt ¦in‚deks ‚lenz }

gradient method [MATH] A finite iterative procedure for solving a system of *n* equations in *n* unknowns. { 'grād·ē·ənt ‚meth·əd }

gradient microphone [ENG ACOUS] A microphone whose electrical response corresponds to some function of the difference in pressure between two points in space. { 'grād·ē·ənt 'mī·krə‚fōn }

gradient of equal traction [MIN ENG] The gradient at which the tractive force necessary to pull an empty tram inby (slightly uphill) is equal to that force required to pull a loaded tram outby. { 'grād·ē·ənt əv ¦ē·kwəl 'trak·shən }

gradient projection method [MATH] Computational method used in nonlinear programming when constraint functions are linear. { 'grād·ē·ənt prə'jek·shən ‚meth·əd }

gradient tints [MAP] A series of color tints used on maps or charts to indicate relative heights or depths. Also known as elevation tints; hypsometric tints. { 'grād·ē·ənt ‚tins }

gradient wind [METEOROL] A wind for which Coriolis acceleration and the centripetal acceleration exactly balance the horizontal pressure force. { 'grād·ē·ənt ‚wind }

gradient-wind level See geostrophic-wind level. { 'grād·ē·ənt ‚wind ‚lev·əl }

grading [GEOL] The gradual reduction of the land to a level surface; for example, erosion of land to base level by streams. [IND ENG] Segregating a product into a number of adjoining categories which often form a spectrum of quality. Also known as classification. { 'grād·iŋ }

grading-up [AGR] The process of breeding purebred sires to grade females and their female offspring for generation after generation. { 'grād·iŋ 'əp }

gradiometer [ENG] Any instrument that measures the gradient of some physical quantity, such as certain types of magnetometers which are designed to measure the gradient of magnetic field, or the Eötvös torsion balance and related instruments which measure the gradient of gravitational field. { ‚grād·ē'äm·əd·ər }

gradualism [EVOL] A model of evolution in which change is slow, steady, and on the whole ameliorative, resulting in a gradual and continuous increase in biological diversity. Also known as phyletic gradualism. { 'graj·ə‚wə‚liz·əm }

gradual phase [ASTROPHYS] The second phase of a solar flare, characterized by emission of relatively low-energy (soft) x-radiation which appears soon after the beginning of the impulsive phase, grows in intensity as the impulsive bursts wane, and lasts up to several hours. Also known as thermal phase. { 'graj·ə·wəl ‚fāz }

graduate [CHEM] A cylindrical vessel that is calibrated in fluid ounces or milliliters or both; used to measure the volume of liquids. { 'graj·ə·wət }

graduated coating [MET] In thermal spraying, a deposit consisting of a number of layers which vary in material composition. { ¦graj·ə‚wād·əd 'kōd·iŋ }

graduator [ENG] An evaporation unit in which liquid is forced to flow over large surfaces which are subjected to air currents. { 'graj·ə‚wād·ər }

Graebe-Ullman reaction [ORG CHEM] 1. Production of fluorenone by boiling 2-benzoylbenzenediazonium salts in dilute acid solution. 2. Reaction of 2-aminodiphenylamines with nitrous acid to form a benzotriazole which on heating loses nitrogen to form a carbazole. { ¦gre·bə 'ùl·mən rē‚ak·shən }

Graeffe's method [MATH] A method of solving algebraic equations by means of squaring the exponents and making appropriate substitutions. { 'gref·əz ‚meth·əd }

Graetz number [THERMO] A dimensionless number used in the study of streamline flow, equal to the mass flow rate of a fluid times its specific heat at constant pressure divided by the product of its thermal conductivity and a characteristic length. Also spelled Grätz number. Symbolized N_{Gz}. { 'grets ‚nəm·bər }

Graetz problem [FL MECH] The problem of determining the steady-state temperature field in a fluid flowing in a circular tube when the wall of the tube is held at a uniform temperature and the fluid enters the tube at a different uniform temperature. { 'grets ‚präb·ləm }

graft [BIOL] 1. To unite to form a graft. 2. A piece of tissue transplanted from one individual to another or to a different

place on the same individual. 3. An individual resulting from the grafting of parts. [BOT] To unite a scion to an understock in such manner that the two grow together and continue development as a single plant without change in scion or stock. { graft }

graft copolymer [ORG CHEM] Any high polymer composed of two or more different polymeric entities chemically united. { ¦graft kō'päl·ə·mər }

graftonite [MINERAL] $(Fe,Mn,Ca)_3(PO_4)_2$ A salmon-pink mineral, crystallizing in the monoclinic system, and found as laminated intergrowths of triphylite; hardness is 5 on Mohs scale, and specific gravity is 3.7. { 'graf‚tə‚nīt }

Graham-Cole test See cholecystography. { ¦grā·əm 'kōl ‚test }

Grahamella [MICROBIO] A genus of the family Bartonellaceae; intracellular parasites in red blood cells of rodents and other mammals. { ‚grā·ə'mel·ə }

grahamite [GEOL] See mesosiderite. [MINERAL] A solid, jet-black hydrocarbon that occurs in veinlike masses; soluble in carbon disulfide and chloroform. { 'grā·ə‚mīt }

Graham's law of diffusion [FL MECH] The law that the rate of diffusion of a gas is inversely proportional to the square root of its density. { 'grā·əmz ¦lò əv di'fyü·zhən }

Graham's pendulum [DES ENG] A type of compensated pendulum having a hollow bob containing mercury whose thermal expansion balances the thermal expansion of the pendulum rod. { 'gramz 'pen·jə·ləm }

grain [BOT] 1. A rounded, granular prominence on the back of a sepal. 2. See cereal. 3. See drupelet. [GEOL] The particles or discrete crystals that make up a sediment or rock. [GRAPHICS] A small particle of metallic silver remaining in a photographic emulsion after developing and fixing; these grains together form the dark areas of a photographic image. [HYD] The particles which make up settled snow, firn, and glacier ice. [MATER] 1. The appearance and texture of wood due to the arrangement of constituent fibers. 2. The woodlike appearance or texture of a rock, metal, or other material. 3. The direction in which most fibers lie in a sample of paper, which corresponds with the way the paper was made on the manufacturing machine. [MECH] A unit of mass in the United States and United Kingdom, common to the avoirdupois, apothecaries', and troy systems, equal to 1/7000 of a pound, or to 6.479891×10^{-5} kilogram. Abbreviated gr. [ORD] A single piece of solid propellant, regardless of size or shape, used in a gun or rocket; a rocket grain is often very large and shaped to fit its requirements. [TEXT] The direction in a piece of fabric which is parallel with the selvage. { grān }

grain alcohol See alcohol. { 'grān 'al·kə‚hól }

grain boundary [SOLID STATE] An interface between individual crystals in a polycrystalline solid. { 'grān ‚baùn·drē }

grain diminution See degradation recrystallization. { 'grān dim·yə'nish·ən }

grain direction [COMPUT SCI] In character recognition, the arrangement of paper fibers in relation to a document's travel through a character reader. { 'grān də‚rek·shən }

grain drill [AGR] A drill used to sow small grains or seeds. { 'grān ‚dril }

grainer process [CHEM ENG] A salt production method in which salt is produced by surface evaporation of brine in open-air flat pans. { 'grān·ər ‚präs·əs }

grainer salt [MATER] Sodium chloride produced by the grainer process. { 'grān·ər ‚sólt }

grain fineness number [MATER] Average grain size of a granular material. { ¦grān 'fīn·nəs ‚nəm·bər }

grain flow [MET] The fiber patterns appearing on the surface of a forging and resulting from the alignment of the crystalline structure of the base metal in the direction of working. { 'grān ‚flō }

grain growth [MET] Enlargement of grains in a metal, usually through heat treatment. [PETR] Enlargement of some individual crystals in a monomineralic rock, producing a coarser texture. Also known as germination. { 'grān ‚grōth }

graininess [GRAPHICS] A mottled effect in film caused by clumping of the silver particles. { 'grān·ē·nəs }

graining [ENG] Simulating a grain such as wood or marble on a painted surface by applying a translucent stain, then working it into suitable patterns with tools such as special combs, brushes, and rags. [GRAPHICS] A technique in which lithographic metal plates are abraded for greater water retention and coating adhesion. { 'grān·iŋ }

grain leather [MATER] Leather made from the grain side of a skin. { 'grān ‚leth·ər }

grain neutral spirits *See* neutral spirits. { ¦grān ¦nü·trəl 'spir·əts }

grain size [GEOL] Average size of mineral particles composing a rock or sediment. [GRAPHICS] Average size of silver halide grains in a photosensitive material. [MET] Average size of grains in a metal expressed as average diameter, or grains per unit area or volume. { 'grān ‚sīz }

grain sorghum [BOT] *Sorghum bicolor.* A grass plant cultivated for its grain and to a lesser extent for forage. Also known as nonsaccharine sorghum. { ¦grān 'sȯr·gəm }

grain spacing [DES ENG] Relative location of abrasive grains on the surface of a grinding wheel. { 'grān ‚spās·iŋ }

grainstone [PETR] A mud-free (micrite-free) limestone. { 'grān‚stōn }

gram [MECH] The unit of mass in the centimeter-gram-second system of units, equal to 0.001 kilogram. Abbreviated g; gm. { gram }

gramagrass [BOT] Any grass of the genus *Bouteloua*; pasture grass. { 'gram·ə‚gras }

gram-atomic weight [CHEM] The atomic weight of an element expressed in grams, that is, the atomic weight on a scale on which the atomic weight of carbon-12 isotope is taken as 12 exactly. { ¦gram ə¦tam·ik 'wāt }

gram-calorie *See* calorie. { ¦gram ¦kal·ə·rē }

gram-centimeter [MECH] A unit of energy in the centimeter-gram-second gravitational system, equal to the work done by a force of magnitude 1 gram force when the point at which the force is applied is displaced 1 centimeter in the direction of the force. Abbreviated g-cm. { ¦gram 'sent·ə‚mēd·ər }

Gram determinant [MATH] The Gram determinant of vectors $v_1, \ldots, v_n$ from an inner product space is the determinant of the $n \times n$ matrix with the inner product of v_i and v_j as entry in the *i*th column and *j*th row; its vanishing is a necessary and sufficient condition for linear dependence. { 'gram di'tərm·ə·nənt }

gramenite *See* nontronite. { 'gra·mə‚nīt }

gram-equivalent weight [CHEM] The equivalent weight of an element or compound expressed in grams on a scale in which carbon-12 has an equivalent weight of 3 grams in those compounds in which its formal valence is 4. { ¦gram i¦kwiv·ə·lənt 'wāt }

gram-force [MECH] A unit of force in the centimeter-gram-second gravitational system, equal to the gravitational force on a 1-gram mass at a specified location. Abbreviated gf. Also known as fors; gram-weight; pond. { 'gram ‚fȯrs }

gramicidin [MICROBIO] A polypeptide antibacterial antibiotic produced by *Bacillus brevis*; active locally against gram-positive bacteria. { ‚gram·ə'sīd·ən }

Graminales [BOT] The equivalent name for Cyperales. { ‚gram·ə'nā·lēz }

Gramineae [BOT] The grasses, a family of monocotyledonous plants in the order Cyperales characterized by distichously arranged flowers on the axis of the spikelet. { grə'min·ē‚ē }

graminicolous [ECOL] Living upon grass. { ¦gram·ə¦nik·ə·ləs }

graminivorous [ZOO] Feeding on grasses. { ¦gram·ə¦niv·ə·rəs }

graminoid [BOT] Of or resembling the grasses. { 'gram·ə‚nȯid }

gram-molecular volume [CHEM] The volume occupied by a gram-molecular weight of a chemical in the gaseous state at 0°C and 760 millimeters of pressure (101,325 pascals). { ¦gram mə¦lek·yə·lər 'väl·yəm }

gram-molecular weight [CHEM] The molecular weight of compound expressed in grams, that is, the molecular weight on a scale on which the atomic weight of carbon-12 isotope is taken as 12 exactly. { ¦gram mə¦lek·yə·lər 'wāt }

gram-negative [MICROBIO] Of bacteria, decolorizing and staining with the counterstain when treated with Gram's stain. { 'gram¦neg·əd·iv }

gram-negative diplococci [MICROBIO] Three bacteriologic genera composing the family Neisseriaceae: *Gemella*, *Veillonella*, and *Neisseria*. { 'gram¦neg·əd·iv ‚dip·lə'käk·sē }

gram-positive [MICROBIO] Of bacteria, holding the color of the primary stain when treated with Gram's stain. { 'gram ¦päs·əd·iv }

gram-rad [NUCLEO] A unit of integral absorbed dose of radiation, equal to 100 ergs. { ¦gram ¦rad }

gram-roentgen [NUCLEO] A unit of energy conversion, equal to a dose of 1 roentgen delivered to 1 gram of air. { 'gram ¦rent·gən }

Gram-Schmidt orthogonalization process [MATH] A process by which an orthogonal set of vectors is obtained from a linearly independent set of vectors in an inner product space. { ¦gram 'shmit ‚ȯr¦thäg·ən·əl·ə'zā·shən ‚präs·əs }

Gram's stain [MICROBIO] A differential bacteriological stain; a fixed smear is stained with a slightly alkaline solution of basic dye, treated with a solution of iodine in potassium iodide, and then with a neutral decolorizing agent, and usually counterstained; bacteria stain either blue (gram-positive) or red (gram-negative). { 'gramz ‚stān }

Gram's theorem [MATH] A set of vectors are linearly dependent if and only if their Gram determinant vanishes. { 'gramz ‚thir·əm }

gram-variable [MICROBIO] Pertaining to staining inconsistently with Gram's stain. { 'gram ¦ver·ē·ə·bəl }

gram-weight *See* gram-force. { 'gram¦wāt }

grana [CELL MOL] A multilayered membrane unit formed by stacks of the lobes or branches of a chloroplast thylakoid. { 'grän·ə }

Granby car [MIN ENG] An automatically dumped car for hand loading or power-shovel loading; a wheel attached to its side engages an inclined track at the dumping point. { 'gran·bē ‚kär }

Grand Banks [GEOGR] Banks off southeastern Newfoundland, important for cod fishing. { ¦grand ¦baŋks }

grand canonical ensemble [STAT MECH] A collection of systems of particles used to describe an individual system which is allowed to exchange both energy and particles with its environment. { 'grand kə¦nän·ə·kəl än'säm·bəl }

grandfather [COMPUT SCI] A data set that is two generations earlier than the data set under consideration. { 'gran‚fath·ər }

grandfather cycle [COMPUT SCI] The period during which records are kept but not used except to reconstruct other records which are accidentally lost. { 'gran‚fath·ər ‚sī·kəl }

grandite [MINERAL] A garnet that is intermediate in chemical composition between grossular and androdite. { 'gran‚dīt }

grand mal [MED] A complete epileptic seizure involving sudden loss of consciousness and tonic convulsion of the skeletal musculature followed by clonic muscular spasms. { 'gran ¦mäl }

grandrelle yarn [TEXT] Yarn of blended or mixed colors made by twisting together two contrasting single yarns. { gran'drel ‚yärn }

grand unified field theory [PART PHYS] A theory in which the strong, electromagnetic, and weak interactions become aspects of one interaction. { ¦grand ¦yü·nə‚fīd ¦fēld ‚thē·ə·rē }

granellare [INV ZOO] In xenophyophores, that portion of the body consisting of the multinucleate plasmodium and its enclosing, branching organic tube. { 'gran·əl‚är }

granin [BIOCHEM] Any of a group of proteins that localize to secretory vesicles, are secreted by a regulated pathway, are posttranslationally glycosylated, and are typically responsive to hormones. { 'gran·ən }

granite [PETR] A visibly crystalline plutonic rock with granular texture; composed of quartz and alkali feldspar with subordinate plagioclase and biotite and hornblende. { 'gran·ət }

granite cloth [TEXT] A fabric having a hard finish with an irregular pebbled surface made by weaving tightly twisted wool or blended yarns. { 'gran·ət ‚klȯth }

granite-gneiss [PETR] A banded metamorphic rock derived from igneous or sedimentary rocks mineralogically equivalent to granite. { 'gran·ət ¦nīs }

granite moss [BOT] The common name for a group of the class Bryatae represented by two Arctic genera and distinguished by longitudinal splitting of the mature capsule into four valves. { 'gran·ət ‚mós }

granite pegmatite *See* pegmatite. { 'gran·ət 'peg·mə‚tīt }

granite porphyry *See* quartz porphyry. { 'gran·ət 'pȯr·fə·rē }

granite series [GEOL] A sequence of products that evolve continuously during crustal fusion; earlier products tend to be deep-seated, syntectonic, and granodioritic, and later products

GRANA

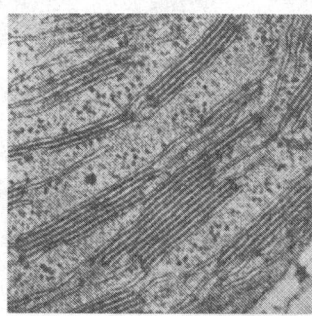

Part of oat leaf chloroplast at high magnification. Grana are seen in side view, revealing profiles of membranes. Thylakoids that interconnect grana are shown, with components of stroma, or ground substance, in which thylakoids lie. The dense granules are choroplast ribosomes.

GRANITE MOSS

sporophytes

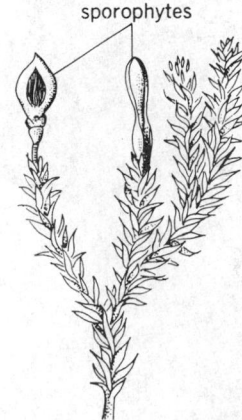

Granite moss (*Andreaea rupestris*), showing gametophyte with sporophytes. (*From G. M. Smith, Crytogamic Botany, vol. 2, 2d ed., McGraw-Hill, 1955*)

tend to be shallower, late syntectonic, or postsyntectonic, and more potassic. { 'gran·ət ,sir·ēz }

granite wash [GEOL] Material eroded from granites and redeposited, forming a rock with the same major mineral constituents as the original rock. { 'gran·ət ,wäsh }

granitic batholith [GEOL] A granitic shield mass intruded as the fusion of older formations. { grə'nid·ik 'bath·ə,lith }

granitic layer See sial. { grə'nid·ik 'lā·ər }

granitic magma [PETR] A coarse-grained igneous rock. { grə'nid·ik 'mag·mə }

granitization [PETR] A process whereby various types of rock may be converted to granite or closely related material. { ,gran·əd·ə'zā·shən }

granoblastic fabric [PETR] The texture of metamorphic rocks composed of equidimensional elements formed during recrystallization. { gra·nō¦blas·tik 'fab·rik }

granodiorite [PETR] A visibly crystalline plutonic rock composed chiefly of sodic plagioclase, alkali feldspar, quartz, and subordinate dark-colored minerals. { gra·nō'dī·ə,rīt }

granofels [PETR] A medium-to coarse-grained metamorphic rock possessing a granoblastic fabric and either lacking foliation or lineation entirely or exhibiting such characteristics only indistinctly. { 'gran·ə,felz }

granogabbro [PETR] Plutonic rock composed of quartz, basic plagioclase, potash-feldspar, and at least one ferromagnesian mineral; intermediate between a granite and a gabbro, and in a strict sense, a granodiorite with more than 50% boric plagioclase. { gra·nō'ga·brō }

granophyre [PETR] A quartz porphyry or fine grained porphyritic granite. { 'gran·ə,fīr }

Grantiidae [INV ZOO] A family of calcareous sponges in the order Sycettida. { gran'tī·ə,dē }

granular [SCI TECH] Having a grainy texture. { 'gran·yə·lər }

granular-bed separator [ENG] Vessel or chamber in which a bed of granular material is used to remove dust from a dust-laden gas as it passes through the bed. { 'gran·yə·lər ,bed 'sep·ə,rād·ər }

granular component [CELL MOL] The component of the nucleolus that contains the cleaved preribosomal particles. { ¦gran·yə·lər kəm¦pō·nənt }

granular fracture [MET] Grain-like or crystalline surface appearance of a broken metal. { 'gran·yə·lər 'frak·chər }

granular gland [PHYSIO] A gland that produces and secretes a granular material. { 'gran·yə·lər 'gland }

granular ice [HYD] Ice composed of many tiny, opaque, white or milky pellets or grains frozen together and presenting a rough surface; this is the type of ice deposited as rime and compacted as névé. { 'gran·yə·lər 'īs }

granularity [GRAPHICS] The distribution of grains in a portion of photographic material that has been uniformly exposed and processed. [PETR] The feature of rock texture relating to the size of the constituent grains or crystals. [SYS ENG] The degree to which a system can be broken down into separate components, making it customizable and flexible. { ,gran·yə'lar·əd·ē }

granular leukocyte See granulocyte. { 'gran·yə·lər 'lü·kə,sīt }

granular powder [MET] Equidimensional metal particles that are not spherical. { 'gran·yə·lər 'paùd·ər }

granular snow See snow grains. { 'gran·yə·lər 'snō }

granular structure [MATER] Nonuniform appearance of molded or compressed material due to presence of particles of composition, either within the material or on the surface. { 'gran·yə·lər 'strək·chər }

granulate [CHEM] To form or crystallize into grains, granules, or small masses. { 'gran·yə,lāt }

granulated metal [MET] Small pellets produced by pouring molten metal through a screen or similar device and chilling the droppings in water. { 'gran·yə,lād·əd 'med·əl }

granulation [ASTRON] The small "rice grain" markings on the sun's photosphere. Also known as photospheric granulation. [MED] **1.** Tiny red granules made of capillary loops in the base of an ulcer. **2.** Process of granular tissue formation around a focus of inflammation. [PL PATH] Dry, tasteless condition of citrus fruit due to hardening of the juice sacs when fruit is left on trees too late in the season. [SCI TECH] The state or process of reducing a material to grains or small particles. { ,gran·yə'lā·shən }

granulator [FOOD ENG] A revolving cylinder in which sugar is dried and granulated. { 'gran·yə,lād·ər }

granule [ASTRON] A convective cell in the solar photosphere, about 600 miles (1000 kilometers) in diameter. [GEOL] A somewhat rounded rock fragment ranging in diameter from 2 to 4 millimeters; larger than a coarse sand grain and smaller than a pebble. { 'gran·yül }

granulite [PETR] **1.** Granite that contains muscovite. **2.** A relatively coarse, granuloblastic rock formed at the high temperatures and pressures of the granulite facies. { 'gran·yə,līt }

granulite facies [PETR] A group of gneissic rocks characterized by a granoblastic fabric and formed by regional dynamothermal metamorphism at temperatures above 650°C and pressures of 3000–12,000 bars. { 'gran·yə,līt 'fā·shēz }

granuloblastosis [VET MED] An avian leukosis characterized by the presence of excessive numbers of immature granulocytes in the blood of affected birds. { ¦gran·yə·lō¦bla'stō·səs }

granulocyte [HISTOL] A leukocyte containing granules in the cytoplasm. Also known as granular leukocyte; polymorph; polymorphonuclear leukocyte. { 'gran·yə·lō,sīt }

granulocytic leukemia [MED] A blood disease involving neoplastic transformation of granulocytes, principally the neutrophilic series. Also known as myelogenous leukemia; myeloid leukemia. { ¦gran·yə·lō¦sid·ik lü'kē·mē·ə }

granulocytopenia [MED] A deficiency of granulocytes in circulating blood. Also known as granulopenia. { ¦gran·yə·lō,sīd·ə'pēn·yə }

granulocytosis [MED] An increase in the number of granulocytes in the circulation. { ,gran·yə·lō,sī'tō·səs }

granuloma [MED] A discrete nodular lesion of inflammatory tissue in which granulation is significant. { ,gran·yə'lō·mə }

granuloma inguinale [MED] An infectious, chronic, destructive granulomatous lesion of humans most frequently localized in the genital and inguinal regions; caused by Donovan bodies (*Donovania granulomatis*). { ,gran·yə'lō·mə ,iŋ·gwə'nä·lē }

granuloma pyogenicum [MED] A hemangioma with superimposed inflammation on the skin or other epithelial surfaces. { ,gran·yə'lō·mə ,pī·ō'jen·ə·kəm }

granulomatosis [MED] Any disease characterized by multiple granulomas. { ,gran·yə,lō·mə'tō·səs }

granulometry [PETR] Measurement of grain sizes of sedimentary rock. { ,gran·yə'läm·ə·trē }

granulopenia See granulocytopenia. { ,gran·yə·lō'pēn·yə }

Granuloreticulosia [INV ZOO] A subclass of the protozoan class Rhizopodea characterized by reticulopodia which often fuse into networks. { ¦gran·yə·lō·re,tik·yə'lō·shə }

granulosis [INV ZOO] A virus disease of lepidopteran larvae characterized by the accumulation of small granular inclusion bodies (capsules) in the infected cells. { ,gran·yə'lō·səs }

Granville wilt [PL PATH] A bacterial wilt of tobacco caused by *Pseudomonas solanacearum*. { 'gran·vəl 'wilt }

grape [BOT] The common name for plants of the genus *Vitis* characterized by climbing stems with cylindrical-tapering tendrils and polygamodioecious flowers; grown for the edible, pulpy berries. { grāp }

grapefruit [BOT] *Citrus paradisi*. An evergreen tree with a well-rounded top cultivated for its edible fruit, a large, globose citrus fruit characterized by a yellow rind and white, pink, or red pulp. { 'grāp,früt }

grapefruit oil [MATER] Pale-yellow, volatile liquid with grapefruit aroma; soluble in oils, insoluble in glycerin; derived from fresh rind of *Citrus paradisi*; used in flavors and toiletries. Also known as oil of grapefruit; oil of shaddock. { 'grāp,früt ,óil }

grapefruit seed oil [MATER] Reddish-brown oil expressed from grapefruit seeds; has nutlike aroma and bitter taste; solidifies at −10°C; used as lubricant for textile fibers and leather. { 'grāp,früt ,sēd ,óil }

grapestone [GEOL] A cluster of sand-size grains, such as calcareous pellets, held together by incipient cementation shortly after deposition; the outer surface is lumpy, resembling a bunch of grapes. { 'grāp,stōn }

grape sugar See dextrose. { 'grāp ,shùg·ər }

grapevine drainage See trellis drainage. { 'grāp,vīn ,drān·ij }

grapevine stopper [NAV ARCH] A device that allows temporary grip on a large-diameter cable at high tension; consists of a webbed sleeve that contracts under tension. { 'grāp,vīn ,stäp·ər }

GRAPE

Foliage, tendrils, fruit, and stem characteristics of the Concord grape (*Vitis labruscana*).

graph [MATH] **1.** The planar object, formed from points and line segments between them, used in the study of circuits and networks. **2.** The graph of a function f is the set of all ordered pairs $[x, f(x)]$, where x is in the domain of f. **3.** *See* graphical representation. **4.** The set of all points that satisfy a particular equation, inequality, or system of equations or inequalities. { graf }

graph component [MATH] A particular type of maximal connected subgraph of a graph. { 'graf kəm'pō·nənt }

graphechon [ELECTR] A storage tube having two electron guns, one for writing and the other for reading and simultaneous erasing, on opposite sides of the storage medium, which consists of an insulator or semiconductor deposited on a thin substratum of metal supported by a fine mesh. { 'graf·ə‚kän }

grapheme [COMMUN] A pictorial representation of a semanteme, such as X-reference for cross-reference. { 'gra‚fēm }

graph follower *See* curve follower. { 'graf ‚fäl·ə·wər }

graphical analysis [MATH] The study of interdependent phenomena by analyzing graphical representations. { ¦graf·ə·kəl ə'nal·ə·səs }

graphical design [ELECTR] Methods of obtaining operating data for an electron tube or semiconductor circuit by using graphs which plot the relationship between two variables, such as plate voltage and grid voltage, while another variable, such as plate current, is held constant. { ¦graf·ə·kəl di'zīn }

graphical formula [CHEM] A chemical formula that suggests a three-dimensional representation of the structure of a molecule by rendering chemical bonds within the plane of the paper as straight lines, those above the plane of the paper as wedge-shaped bonds, and those below the plane of the paper either as broken lines or broken-line wedges. { ¦graf·ə·kəl 'fȯr·myə·lə }

graphical kernel system [COMPUT SCI] A standard system and language for creating two- and three-dimensional master graphics images on many types of display devices. Abbreviated GKS. { ¦graf·i·kəl 'kər·nəl ‚sis·təm }

graphical representation [MATH] The plot of the points in the plane which constitute the graph of a given real function or a pictorial diagram depicting interdependence of variables. Also known as graph. { ¦graf·ə·kəl ‚rep·rə·zen'tā·shən }

graphical statics [MECH] A method of determining forces acting on a rigid body in equilibrium, in which forces are represented on a diagram by straight lines whose lengths are proportional to the magnitudes of the forces. { ¦graf·ə·kəl 'stad·iks }

graphical symbol [ELEC] A true symbol, rather than a coarse picture, representing an element in an electrical diagram. { ¦graf·ə·kəl 'sim·bəl }

graphical user interface [COMPUT SCI] A user interface in which program features are represented by icons that the user can access and manipulate with a pointing device. Abbreviated GUI. { ¦graf·ə·kəl ‚yü·zər 'in·tər‚fās }

graphical vector [MATH] A finite, nonincreasing sequence of nonnegative integers that is the degree vector of some simple graph. { ¦graf·ə·kəl 'vek·tər }

graphical visual display device [COMPUT SCI] A computer input-output device which enables the user to manipulate graphic material in a visible two-way, real-time communication with the computer, and which consists of a light pen, keyboard, or other data entry devices, and a visual display unit monitored by a controller. Also known as graphoscope. { ¦graf·ə·kəl ¦vizh·ə·wəl di'spla di‚vīs }

graphic arts [GRAPHICS] Those methods of applied arts used to form a visual end product that conveys information or is a decoration; methods include drawing, painting, photography, all types of printing, and bookmaking. { ¦graf·ik 'ärts }

graphic display [ELECTR] The display of data in graphical form on the screen of a cathode-ray tube. { 'graf·ik di'splā }

graphic equalizer [ENG ACOUS] A device that allows the response of audio equipment to be modified independently in several frequency bands through the use of a bank of slide controls whose positions form a graph of the frequency response. { ¦graf·ik 'ē·kwə‚lī·zər }

graphic granite [PETR] A distinct type of pegmatite in which quartz and orthoclase crystals grew together along a parallel axis. Also known as Hebraic granite; runite. { ¦graf·ik'gran·ət }

graphic intergrowth [PETR] An intergrowth of crystals, commonly feldspar and quartz, that produces a type of poikilitic texture in which the larger crystals have a fairly regular geometric outline and orientation, and resemble cuneiform writing. { ¦graf·ik 'in·tər‚grōth }

graphic methods [GRAPHICS] Pencil-and-paper methods that employ the geometry of a plane to express mathematical relationships and to carry out mathematical operations in analog form. { ¦graf·ik ¦meth·ədz }

graphic panel [CONT SYS] A master control panel which indicates the status of equipment and operations in a system, and their relationships. { ¦graf·ik 'pan·əl }

graphic recording instrument [ENG] An instrument that makes a graphic record of one or more quantities as a function of another variable, usually time. { 'graf·ik ri‚kȯrd·iŋ ‚in·strə·mənt }

graphics [COMMUN] **1.** In communications systems, an information mode in which a graphic system is used to reproduce intelligence; a variation of facsimile. **2.** Nonvoice analog information devices and modes such as facsimile, photographics, and television. [SCI TECH] **1.** The graphic media. **2.** The art of drawing a three-dimensional object on a two-dimensional surface according to mathematical rules of projection. { 'graf·iks }

graphics-based molecular modeling *See* molecular graphics. { 'graf·iks ¦bāst mə'lek·yə·lər 'mäd·əl·iŋ }

graphic scale [MAP] A graduated line that indicates the length of miles or kilometers as they appear on a map; the line has the advantage of remaining true after the map is enlarged or reduced in reproduction. Also known as bar scale. { ¦graf·ik ¦skāl }

graphics driver [COMPUT SCI] A series of instructions that activates a graphics device, such as a display screen or plotter. { 'graf·iks ‚drīv·ər }

graphics engine [COMPUT SCI] A specialized processor that carries out graphics processing independently of the main central processing unit. Also known as graphics processor. { 'graf·iks ¦en·jən }

graphics interface [COMPUT SCI] A user interface that displays icons to represent objects. { 'graf·iks ¦in·tər‚fās }

graphics interchange format [COMPUT SCI] Common file format for compressed graphic images on the World Wide Web that is limited to 256 colors. Abbreviated GIF. { ¦graf·iks 'in·tər‚chānj ‚fȯr‚mat }

graphics primitive [COMPUT SCI] A basic building block for graphic images, such as a dot, line, or curve. { 'graf·iks ¦prim·əd·iv }

graphics processor *See* graphics engine. { 'graf·iks ¦prä·ses·ər }

graphics program [COMPUT SCI] A program for the generation of images, ranging in complexity from simple line drawings to realistically shaded pictures that resemble photographs. { 'graf·iks ‚prō·grəm }

graphics tablet [COMPUT SCI] A padlike peripheral device which is designed so that shapes appear on the monitor's screen when the tablet is drawn upon with a pointed device. { 'graf·iks ‚tab·lət }

graphics terminal [COMPUT SCI] An input/output device that can accept and display picture images. { 'graf·iks ¦tər·mən·əl }

graphic tellurium *See* sylvanite. { ¦graf·ik te'lür·ē·əm }

graphic texture [GEOL] A pattern of rocks that is similar to cuneiform characters. { 'graf·ik ‚teks·chər }

Graphidaceae [BOT] A family of mosses formerly grouped with lichenized Hysteriales but now included in the order Lecanorales; individuals have true paraphyses. { ‚graf·ə'dās·ē‚ē }

Graphiolaceae [MYCOL] A family of parasitic fungi in the order Ustilaginales in which teleutospores are produced in a cuplike fruiting body. { ‚graf·ē·ō'lās·ē‚ē }

graphite [MINERAL] A mineral consisting of a low-pressure allotropic form of carbon; it is soft, black, and lustrous and has a greasy feeling; it occurs naturally in hexagonal crystals or massive or can be synthesized from petroleum coke; hardness is 1–2 on Mohs scale, and specific gravity is 2.09–2.23; used in pencils, crucibles, lubricants, paints, and polishes. Also known as black lead; plumbago. { 'gra‚fīt }

graphite anode [CHEM ENG] One of the electrodes of graphite used in a mercury cell to produce chlorine by electrolysis. [ELECTR] **1.** The rod of graphite which is inserted into the mercury-pool cathode of an ignitron to start current flow.

2. The collector of electrons in a beam power tube or other high-current tube. { 'gra,fīt 'an,ōd }

graphite flake [MET] A curved graphite particle in gray cast iron. { 'gra,fīt ¦flāk }

graphite grease [MATER] A lubricating grease that contains 2–10% amorphous graphite; used for bearings, especially in damp places. { 'gra,fīt ¦grēs }

graphite-moderated reactor [NUCLEO] A nuclear reactor in which graphite is the principal moderating material. { 'gra,fīt ¦mäd·ə,rād·əd rē'ak·tər }

graphite oil [MATER] A deflocculated suspension of graphite in oil; used as a lubricant. { 'gra,fīt ¦óil }

graphite resistor [ELEC] A resistor made of carbon for resistance heating. { 'gra,fīt ri'zis·tər }

graphite rosette [MET] Graphite flakes which extend radially outward from the centers of crystallization. { 'gra,fīt rō'zet }

graphitic carbon [MET] Carbon in iron or steel present in the form of graphite. { grə'fid·ik 'kär·bən }

graphitic corrosion [MET] Corrosion of gray cast iron in which the metallic iron constituent is converted into corrosion products which cement together the residual graphite. { grə'fid·ik kə'rō·zhən }

graphitic steel [MET] Alloy steel containing graphitic carbon. { grə'fid·ik 'stēl }

graphitization [ORG CHEM] The formation of graphitelike material from organic compounds. { ,graf·əd·ə'zā·shən }

graphitizing [MET] Annealing cast iron to convert all or some of the combined carbon to graphitic carbon. { 'graf·ə,tiz·iŋ }

graphoscope See graphical visual display device. { 'graf·ə,skōp }

graph theory [MATH] **1.** The mathematical study of the structure of graphs and networks. **2.** The body of techniques used in graphing functions in the plane. { 'graf ,thē·ə·rē }

grapnel [DES ENG] An implement with claws used to recover a lost core, drill fittings, and junk from a borehole or for other grappling operations. Also known as grapple. [NAV ARCH] An anchor with four or five hooks used for dragging the bottom or for anchorage. { 'grap·nəl }

grappier cement [MATER] A cement composed of finely ground lumps of leftover underburned and overburned slaked lime. { 'grap·ē,ā si,ment }

grapple See grapnel. { 'grap·əl }

grapple dredger See grab dredger. { 'grap·əl ,drej·ər }

grapple hook [DES ENG] An iron hook used on the end of a rope to snag lines, to hold one ship alongside another, or as a fishing tool. Also known as grappling iron. { 'grap·əl ,húk }

grappling iron See grapple hook. { 'grap·liŋ ,ī·ərn }

Grapsidae [INV ZOO] The square-backed crabs, a family of decapod crustaceans in the section Brachyura. { 'grap·sə,dē }

graptolites See graptolithina. { 'grap·tə,līts }

graptolite shale [GEOL] Shale containing an abundance of extinct colonial marine organisms known as graptolites. { 'grap·tə,līt 'shāl }

Graptolithina [PALEON] A class of extinct colonial animals believed to be related to the class Pterobranchia of the Hemichordata. Also known as graptolites. { ,grap·tə·lə'thīn·ə }

Graptoloidea [PALEON] An order of extinct animals in the class Graptolithina including branched, planktonic forms described from black shales. { ,grap·tə'lóid·ē·ə }

Graptozoa [PALEON] The equivalent name for Graptolithina. { ,grap·tə'zō·ə }

GRAS [FOOD ENG] Referring to food ingredients Generally Regarded As Safe in the United States by virtue of a history of safe use or scientific procedures. { gras or 'gē¦är¦ā'es }

graser See gamma-ray laser. { 'grā·zər }

Grashof formula [FL MECH] A formula, $m = 0.0165A_2p_1^{0.97}$, used to express the discharge m of saturated steam in pounds per second, where A_2 is the area of the orifice in square feet, and p_1 is reservoir pressure in pounds per square foot. { 'gräs ,hóf ,fór·myə·lə }

Grashof number [FL MECH] A dimensionless number used in the study of free convection of a fluid caused by a hot body, equal to the product of the fluid's coefficient of thermal expansion, the temperature difference between the hot body and the fluid, the cube of a typical dimension of the body and the square of the fluid's density, divided by the square of

the fluid's dynamic viscosity. Also known as free convection number. { 'gräs,hóf ,nəm·bər }

grasp [IND ENG] A basic element (therblig) in time-motion study; a useful element that accomplishes work. { grasp }

grass [BOT] The common name for all members of the family Gramineae; moncotyledonous plants having leaves that consist of a sheath which fits around the stem like a split tube, and a long, narrow blade. [ELECTR] Clutter due to circuit noise in a radar receiver, seen on an A scope as a pattern resembling a cross section of turf. Also known as hash. { gras }

grass crops [AGR] Plants cultivated as forage and grain to be consumed by domestic livestock. { 'gras ,kräps }

grasserie [INV ZOO] A polyhedrosis disease of silkworms characterized by spotty yellowing of the skin and internal liquefaction. Also known as jaundice. { 'gras·ə·rē }

grasshopper [INV ZOO] The common name for a number of plant-eating orthopteran insects composing the subfamily Saltatoria; individuals have hindlegs adapted for jumping, and mouthparts adapted for biting and chewing. { 'gras,häp·ər }

grasshopper fuse [ELEC] Small fuse incorporating a spring which, upon release by the fusing wire, connects an auxiliary circuit to operate an alarm. { 'gras,häp·ər ,fyüz }

grasshopper linkage [MECH ENG] A straight-line mechanism used in some early steam engines. { 'gras,häp·ər ,liŋ·kij }

grassland [ECOL] Any area of herbaceous terrestrial vegetation dominated by grasses and graminoid species. { 'gras,land }

grassland climate See subhumid climate. { 'gras,land ,klī·mət }

Grassmann algebra See exterior algebra. { 'gräs·mən ,al·jə·brə }

Grassmannian See Grassmann manifold. { 'gräs¦man·ē·ən }

Grassmann manifold [MATH] The differentiable manifold whose points are all k-dimensional planes passing through the origin in n-dimensional euclidean space. Also known as Grassmannian. { 'gräs·mən 'man·ə,fōld }

Grassmann's laws [ANALY CHEM] Seven laws of color identification and mixing that form the basis of modern analytical colorimetry. { 'gräs·mənz ,lóz }

grass minimum [METEOROL] The minimum temperature shown by a minimum thermometer exposed in an open situation with its bulb on the level of the tops of the grass blades of short turf. { 'gras 'min·ə·məm }

Grassot fluxmeter [ENG] A type of fluxmeter in which a light coil of wire is suspended in a magnetic field in such a way that it can rotate; the ends of the suspended coil are connected to a search coil of known area penetrated by the magnetic flux to be measured; the flux is determined from the rotation of the suspended coil when the search coil is moved. { ,grä,sō 'fləks,mēd·ər }

grass-roots deposit [MIN ENG] A deposit that is discovered in surface croppings, is easily exploited, and can pay for its own development while in progress. { 'gras ,rüts di'päz·ət }

grass-roots mining [MIN ENG] Also known as mining on a shoestring. **1.** Inadequately financed mining operation, with catch-as-catch-can practices. **2.** Mining from surface down to bedrock. { 'gras ,rüts 'mīn·iŋ }

grass-roots plant [CHEM ENG] A complete plant erected on a virgin site. { 'gras ,rüts 'plant }

grass sickness [VET MED] A disease of horses occurring mainly in Scotland; thought to be caused by a virus similar to the one that causes poliomyelitis in humans. { 'gras ,sik·nəs }

grass temperature [METEOROL] The temperature registered by a thermometer with its bulb at the level of the tops of the blades of grass in short turf. { 'gras ,tem·prə·chər }

grass tetany [VET MED] A magnesium-deficiency disease of cows. Also known as hypomagnesemia. { 'gras ,tet·ən·ē }

grate [ENG] A support for burning solid fuels; usually made of closely spaced bars to hold the burning fuel, while allowing combustion air to rise up to the fuel from beneath, and ashes to fall away from the burning fuel. { grāt }

graticule [MAP] A network of lines representing the earth's parallels of latitude and meridians of longitude on a map, chart, or plotting sheet. [OPTICS] A scale at the focal plane of an optical instrument to aid in the measurement of objects. { 'grad·ə,kyül }

grating [ELECTROMAG] **1.** An arrangement of fine, parallel

GRASSHOPPER

Melanoplus mexicanus, a grasshopper.

GRASSHOPPER LINKAGE

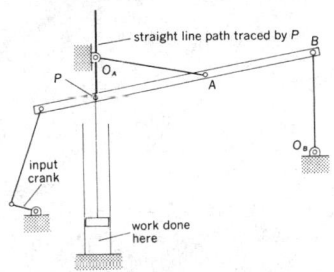

Grasshopper linkage used in the early steam engine. As input crank rotates, the cranks AO_A and BO_B pivot about O_A and O_B respectively, and point P moves approximately in a straight line.

wires used in waveguides to pass only a certain type of wave. **2.** An arrangement of crossed metal ribs or wires that acts as a reflector for a microwave antenna and offers minimum wind resistance. [SPECT] *See* diffraction grating. { 'grād·iŋ }

grating constant [OPTICS] The distance between consecutive diffraction centers of a ultrasonic wave which is producing a light diffraction spectrum. [SPECT] The distance between consecutive grooves of a diffraction grating. { 'grād·iŋ ,kän·stənt }

grating spectrograph [SPECT] A grating spectroscope provided with a photographic camera or other device for recording the spectrum. { 'grād·iŋ 'spek·trə,graf }

grating spectroscope [SPECT] A spectroscope which employs a transmission or reflection grating to disperse light, and usually also has a slit, a mirror or lenses to collimate the light sent through the slit and to focus the light dispersed by the grating into spectrum lines, and an eyepiece for viewing the spectrum. { 'grād·iŋ 'spek·trə,skōp }

gratonite [MINERAL] $Pb_9As_4S_{15}$ A mineral composed of lead arsenic sulfide, occurring in rhombohedral crystals. { 'grat·ən,īt }

Grätz number *See* Graetz number. { 'grets ,nəm·bər }

Gratz rectifier [ELECTR] Three-phase, full-wave rectifying circuit using six rectifiers connected in a bridge circuit. { 'grats ,rek·tə,fī·ər }

graupel *See* snow pellets. { 'graù·pəl }

grav *See* G. { grav }

gravel [GEOL] A loose or unconsolidated deposit of rounded pebbles, cobbles, or boulders. { 'grav·əl }

gravel bank [GEOL] A natural mound or exposed face of gravel, particularly such a place from which gravel is dug. { 'grav·əl ,baŋk }

gravel desert *See* reg. { 'grav·əl ¦dez·ərt }

gravel mine [MIN ENG] A mine extracting gold from sand or gravel. Also known as placer mine. { 'grav·əl ,mīn }

gravel pack [PETRO ENG] A packing of very fine gravel surrounding a slotted or perforated liner in well. { 'grav·əl ,pak }

gravel pump [MECH ENG] A centrifugal pump with renewable impellers and lining, used to pump a mixture of gravel and water. { 'grav·əl ,pəmp }

gravel stop [BUILD] Metal flashing placed at the edge of a roof to prevent gravel from falling off. { 'grav·əl ,stäp }

Grave's disease *See* hyperthyroidism. { 'grāvz di,zēz }

graveyard *See* burial ground. { 'grāv,yärd }

graveyard shift [IND ENG] The shift of workers that begins at or around midnight; the last shift of the day. { 'grāv,yärd ,shift }

graviceptor *See* gravireceptor. { 'grav·i,sep·tər }

gravid [ZOO] **1.** Of the uterus, containing a fetus. **2.** Pertaining to female animals when carrying young or eggs. { 'grav·əd }

Gravigrada [VERT ZOO] The sloths, a group of herbivorous xenarthran mammals in the order Edentata; members are completely hairy and have five upper and four lower prismatic teeth without enamel. { grə'vig·rə·də }

gravimeter [ENG] A highly sensitive weighing device used for relative measurement of the force of gravity by detecting small weight differences of a constant mass at different points on the earth. Also known as gravity meter. { grə'vim·əd·ər }

gravimetric absorption method [ANALY CHEM] A method of measuring the moisture content of a gas in which a known volume of gas is passed through a suitable desiccant, such as phosphorus pentoxide or silica gel, and the change in weight of the desiccant is observed. { grav·ə'me·trik əb'sórp·shən ,meth·əd }

gravimetric analysis [ANALY CHEM] That branch of quantitative analytical chemistry in which a desired constituent is converted, usually by precipitation or combustion, to a pure compound or element, of definite known composition, and is weighed; in a few cases a compound or element is formed which does not contain the constituent but bears a definite mathematical relationship to it. { grav·ə'me·trik ə'nal·ə·səs }

gravimetric geodesy [GEOD] The science that utilizes measurements and characteristics of the earth's gravity field, as well as theories regarding this field, to deduce the shape of the earth and, in combination with arc measurements, the earth's size. { grav·ə'me·trik jē'äd·ə·sē }

gravimetry [ENG] Measurement of gravitational force. { grə'vim·ə·trē }

graving dock [CIV ENG] A form of dry dock consisting of an artificial basin fitted with a gate or caisson, into which a vessel can be floated and the water pumped out to expose the vessel's bottom. { 'grāv·iŋ ,däk }

graviperception [BOT] The magnitude of acceleration required to induce a directed growth response in plants carried aboard spacecraft. { 'grav·ə·pər,sep·shən }

graviportal [BIOL] Weight-bearing. { 'grav·ə¦pórd·əl }

gravireceptor [NEUROSCI] One of the highly specialized nerve endings and receptor organs located in the skeletal muscles, tendons, joints, and inner ear that furnishes information to the brain with respect to body position, equilibrium, and the direction of gravitational forces. Also known as graviceptor. { 'grav·ə·ri,sep·tər }

gravitation [PHYS] The mutual attraction between all masses in the universe. Also known as gravitational attraction. { grav·ə'tā·shən }

gravitational acceleration [PHYS] The acceleration imparted to a body by the attraction of the earth; approximately equal to 980.7 cm/s², or 32.2 ft/s². { grav·ə'tā·shən·əl ak,sel·ə'rā·shən }

gravitational astronomy *See* celestial mechanics. { grav·ə'tā·shən·əl ə'strän·ə·mē }

gravitational attraction *See* gravitation. { grav·ə'tā·shən·əl ə'trak·shən }

gravitational bremsstrahlung [RELAT] The emission of gravitational radiation by two massive objects that pass each other at a high relative velocity and deflect each other slightly. Also known as relativistic bremsstrahlung. { grav·ə'tā·shən·əl 'brem,shträ·lən }

gravitational clustering [ASTRON] A theory that attributes the hierarchy structure of the universe to growth of density fluctuations in a statistically uniform and isotropic universe. { grav·ə'tā·shən·əl 'kləs·tə,riŋ }

gravitational collapse [ASTRON] The implosion of a star or other astronomical body from an initial size to a size hundreds or thousands of times smaller. { grav·ə'tā·shən·əl kə'laps }

gravitational constant [MECH] The constant of proportionality in Newton's law of gravitation, equal to the gravitational force between any two particles times the square of the distance between them, divided by the product of their masses. Also known as constant of gravitation. { grav·ə'tā·shən·əl 'kän·stənt }

gravitational convection *See* thermal convection. { grav·ə'tā·shən·əl kən'vek·shən }

gravitational displacement [MECH] The gravitational field strength times the gravitational constant. Also known as gravitational flux density. { grav·ə'tā·shən·əl dis'plās·mənt }

gravitational encounter [ASTRON] An approach of two massive bodies in which the directions of motion of both bodies are altered by their mutual gravitational attraction. { grav·ə'tā·shən·əl in'kaùnt·ər }

gravitational energy *See* gravitational potential energy. { grav·ə'tā·shən·əl 'en·ər·jē }

gravitational equilibrium [ASTROPHYS] The condition of a star in which the weight of overlying layers at each point is balanced by the total pressure at that point. { grav·ə'tā·shən·əl ,ē·kwə'lib·rē·əm }

gravitational field [MECH] The field in a region in space in which a test particle would experience a gravitational force; quantitatively, the gravitational force per unit mass on the particle at a particular point. { grav·ə'tā·shən·əl 'fēld }

gravitational-field theory [RELAT] A theory in which gravity is treated as a field, as opposed to a theory in which the force acts instantaneously at a distance. { grav·ə'tā·shən·əl 'fēld ,thē·ə·rē }

gravitational flux density *See* gravitational displacement. { grav·ə'tā·shən·əl 'fləks ,den·səd·ē }

gravitational force [MECH] The force on a particle due to its gravitational attraction to other particles. { grav·ə'tā·shən·əl 'fórs }

gravitational geon [PHYS] A hypothetical gravitational field that is held together by its own gravitational attraction. { grav·ə¦tā·shən·əl 'jē,än }

gravitational gradient guidance [NAV] A system for maintaining the desired orientation of a space vehicle in the vicinity of the earth (or any large celestial body) which utilizes the

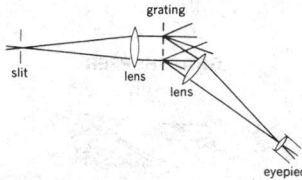
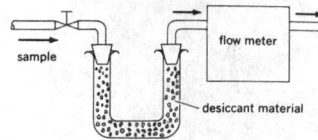

differences in the pull of gravity between points on the craft having different distances from the earth. { ¦grav·ə'tā·shən·əl 'grād·ē·ənt ¦gīd·əns }

gravitational instability [MECH] Instability of a dynamic system in which gravity is the restoring force. { ¦grav·ə'tā·shən·əl ¦in·stə'bil·əd·ē }

gravitational lens [ASTRON] A massive galaxy or other massive object whose gravitational field focuses light from a distant quasar near or along its line of sight, giving a double or multiple image of the quasar. { ¦grav·ə'tā·shən·əl 'lenz }

gravitational mass [PHYS] The mass of a particle as it determines the force it experiences in a gravitational field; equal to inertial mass according to the equivalence principle. { ¦grav·ə'tā·shən·əl 'mas }

gravitational potential [MECH] The amount of work which must be done against gravitational forces to move a particle of unit mass to a specified position from a reference position, usually a point at infinity. { ¦grav·ə'tā·shən·əl pə'ten·chəl }

gravitational potential energy [MECH] The energy that a system of particles has by virtue of their positions, equal to the work that must be done against gravitational forces to assemble the particles from some reference configuration, such as mutually infinite separation. Also known as gravitational energy. { ¦grav·ə'tā·shən·əl pə'ten·chəl 'en·ər·jē }

gravitational pressure *See* hydrostatic pressure. { ¦grav·ə'tā·shən·əl 'presh·ər }

gravitational radiation *See* gravitational wave. { ¦grav·ə'tā·shən·əl ¦rād·ē'ā·shən }

gravitational radius *See* Schwarzschild radius. { ¦grav·ə'tā·shən·əl 'rād·ē·əs }

gravitational redshift [RELAT] A displacement of spectral lines toward the red when the gravitational potential at the observer of the light is greater than at its source. { ¦grav·ə'tā·shən·əl 'red ¦shift }

gravitational repulsion [PHYS] Hypothetical repulsion of matter and antimatter; however, experimental results indicate that matter and antimatter attract according to the same laws as matter and matter. { ¦grav·ə'tā·shən·əl ri'pəl·shən }

gravitational settling [GEOL] A movement of sediment resulting from gravitational forces. { ¦grav·ə'tā·shən·əl 'set·liŋ }

gravitational sliding [GEOL] Extensive sliding of strata down a slope of an uplifted area. Also known as sliding. { ¦grav·ə'tā·shən·əl 'slīd·iŋ }

gravitational systems of units [MECH] Systems in which length, force, and time are regarded as fundamental, and the unit of force is the gravitational force on a standard body at a specified location on the earth's surface. { ¦grav·ə'tā·shən·əl ¦sis·təmz əv 'yü·nəts }

gravitational tide [OCEANOGR] An atmospheric tide due to gravitational attraction of the sun and moon. { ¦grav·ə'tā·shən·əl 'tīd }

gravitational water [HYD] Soil water of a temporary character that results from prolonged infiltration from above and which moves downward to the groundwater zone in response to gravity. { ¦grav·ə'tā·shən·əl 'wód·ər }

gravitational wave [RELAT] A propagating gravitational field predicted by general relativity, which is produced by some change in the distribution of matter; it travels at the speed of light, exerting forces on masses in its path. Also known as gravitational radiation. { ¦grav·ə'tā·shən·əl 'wāv }

gravitino [PART PHYS] A hypothetical counterpart of the graviton; postulated to exist in supersymmetry theories. { ¦grav·ə'tēn·ō }

gravitoelectric field [RELAT] In general relativity, those components of the gravitational field that are analogous to the electric-field components of the electromagnetic field and normally provide the dominant contribution. { ¦grav·ə·tō·i¦lek·trik 'fēld }

gravitomagnetic field [RELAT] In general relativity, those components of the gravitational field that are analogous to the magnetic-field components of the electromagnetic field. { ¦grav·ə·tō·mag¦ned·ik 'fēld }

gravitometer *See* densimeter. { ¦grav·ə'täm·əd·ər }

graviton [PHYS] A theoretically deduced particle postulated as the quantum of the gravitational field, having a rest mass and charge of zero and a spin of 2. { 'grav·ə¦tän }

gravitropism *See* geotropism. { grə'vi·trə¦piz·əm }

gravity [MECH] The gravitational attraction at the surface of a planet or other celestial body. { 'grav·əd·ē }

gravity anomaly [GEOPHYS] The difference between the observed gravity and the theoretical or predicted gravity. { 'grav·əd·ē ə¦näm·ə·lē }

gravity bar [MIN ENG] A 5-foot (1.5-meter) length of heavy half-round rod forming the link between the wedge-oriented coupling and the drill-rod swivel coupling on an assembled Thompson retrievable borehole-deflecting wedge. { 'grav·əd·ē ¦bär }

gravity bed [ENG] A moving body of solids in which particles (granules, pellets, beads, or briquets) flow downward by gravity through a vessel, while process fluid flows upward; the moving-bed technique is used in blast and shaft furnaces, petroleum catalytic cracking, pellet dryers, and coolers. { 'grav·əd·ē ¦bed }

gravity cell [PHYS CHEM] An electrolytic cell in which two ionic solutions are separated by means of gravity. { 'grav·əd·ē ¦sel }

gravity chute [ENG] A gravity conveyor in the form of an inclined plane, trough, or framework that depends on sliding friction to control the rate of descent. { 'grav·əd·ē ¦shüt }

gravity classification [MIN ENG] The grading of ores, or the separation of waste from coal, by the differences in specific gravities of the substances. { 'grav·əd·ē ¦klas·ə·fə¦kā·shən }

gravity-collapse structure *See* collapse structure. { 'grav·əd·ē kə¦laps ¦strək·chər }

gravity concentration [ENG] **1.** Any of various methods for separating a mixture of particles, such as minerals, based on the differences in density of the various species and on the resistance to relative motion exerted upon the particles by the fluid or semifluid medium in which separation takes place. **2.** The separation of liquid-liquid dispersions based on settling out of the dense phase by gravity. { 'grav·əd·ē ¦käns·ən'trā·shən }

gravity conveyor [ENG] Any unpowered conveyor such as a gravity chute or a roller conveyor, which uses the force of gravity to move materials over a downward path. { 'grav·əd·ē kən'vā·ər }

gravity corer [ENG] Any type of corer that achieves bottom penetration solely as a result of gravitational force acting upon its mass. { 'grav·əd·ē ¦kór·ər }

gravity dam [CIV ENG] A dam which depends on its weight for stability. { 'grav·əd·ē ¦dam }

gravity drainage [HYD] Withdrawal of water from strata as a result of gravitational forces. { 'grav·əd·ē 'drān·ij }

gravity drainage reservoir [GEOL] A reservoir in which production is significantly affected by gas, oil, and water separating under the influence of gravity while production takes place. { 'grav·əd·ē 'drān·ij ¦rez·əv¦wär }

gravity erosion *See* mass erosion. { 'grav·əd·ē i¦rō·zhən }

gravity fault *See* normal fault. { 'grav·əd·ē ¦fólt }

gravity feed [ENG] Movement of materials from one location to another using the force of gravity. { 'grav·əd·ē ¦fēd }

gravity flow [HYD] A form of glacier movement in which the flow of the ice results from the downslope gravitational component in an ice mass resting on a sloping floor. { 'grav·əd·ē ¦flō }

gravity-flow gathering system [PETRO ENG] The use of gravity (downhill flow) through pipelines to transport and collect liquid at a central location; used for gathering of waste water from waterflooding operations for treatment prior to reuse or disposal. { 'grav·əd·ē ¦flō 'gath·ə·riŋ ¦sis·təm }

gravity-gradient attitude control [AERO ENG] A device that regulates automatically attitude or orientation of an aircraft or spacecraft by responding to changes in gravity acting on the craft. { 'grav·əd·ē ¦grād·ē·ənt 'ad·ə¦tüd kən¦trōl }

gravity gradiometry [PHYS] The study and measurement of variations in the acceleration due to gravity. { 'grav·əd·ē ¦grād·ē'äm·ə·trē }

gravity haulage [MIN ENG] A type of haulage system in which the set of full cars is lowered at the end of a rope, and gravity force pulls up the empty cars, the rope being passed around a sheave at the top of the incline. Also known as self-acting incline. { 'grav·əd·ē ¦hól·ij }

gravity incline [MIN ENG] An opening made in the direction of, and along the same gradient as, the dip of the deposit. { 'grav·əd·ē 'in¦klīn }

gravity map [GEOPHYS] A map of gravitational variations

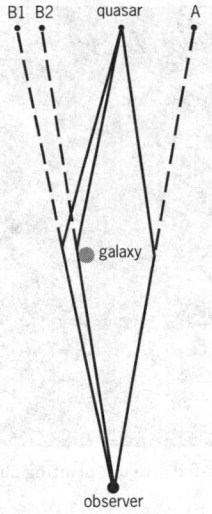

GRAVITATIONAL LENS

The angles are exaggerated for clarity. Here, the lens action produces three images of a quasar (A, B1, and B2), since light from the image can travel along three different curved paths and still reach the observer.

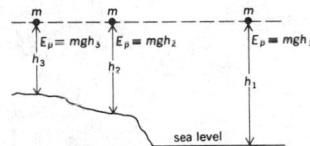

GRAVITATIONAL POTENTIAL ENERGY

The gravitational potential energy E_p of a body of weight mg (m = mass, g = force of gravity) for different reference levels (h_1, h_2, h_3).

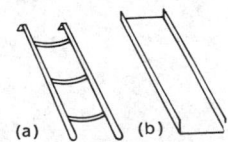

GRAVITY CHUTE

Two types of gravity chutes. (*a*) Barrel skid. (*b*) Plain chute.

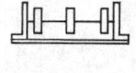
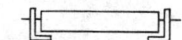

GRAVITY CONVEYOR

Wheels and roller used in some types of gravity conveyors.

GRAVURE PRINTING

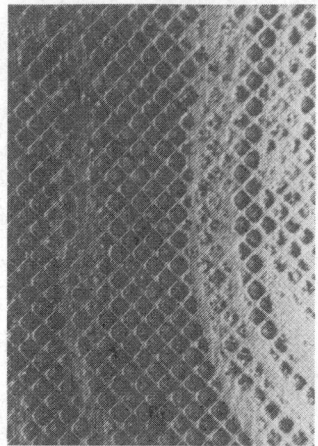

Sheet-fed gravure printing surface.

in an area displaying gravitational highs and lows. { 'grav·əd·ē ,map }

gravity meter [ENG] **1.** U-tube-manometer type of device for direct reading of solution specific gravities in semimicro quantities. **2.** An electrical device for measuring variations in gravitation through different geologic formations; used in mineral exploration. **3.** *See* gravimeter. { 'grav·əd·ē ,mēd·ər }

gravity pendulum *See* pendulum. { 'grav·əd·ē 'pen·jə·ləm }

gravity prospecting [ENG] Identifying and mapping the distribution of rock masses of different specific gravity by means of a gravity meter. { 'grav·əd·ē 'präs,pek·tiŋ }

gravity railroad [ENG] A cable railroad in which cars descend a slope by gravity and are hauled back up the slope by a stationary engine, or there may be two tracks with cars so connected that cars going down may help to raise the cars going up and thus conserve energy. { 'grav·əd·ē 'rāl,rōd }

gravity segregation [ENG] Tendency of immiscible liquids or multicomponent granular mixtures to separate into distinct layers in accordance with their respective densities. { 'grav·əd·ē ,seg·rə'gā·shən }

gravity separation [ENG] Separation of immiscible phases (gas-solid, liquid-solid, liquid-liquid, solid-solid) by allowing the denser phase to settle out under the influence of gravity; used in ore dressing and various industrial chemical processes. { 'grav·əd·ē ,sep·ə'rā·shən }

gravity settling chamber [ENG] Chamber or vessel in which the velocity of heavy particles (solids or liquids) in a fluid stream is reduced to allow them to settle downward by gravity, as in the case of a dust-laden gas stream. { 'grav·əd·ē 'set·liŋ ,chām·bər }

gravity simulation [AERO ENG] The spinning of part or all of a space vehicle so that the centripetal force on bodies within the vehicle near the outer periphery mimics the force of gravity on objects at the earth's surface. { 'grav·əd·ē ,sim·yə,lā·shən }

gravity slope [GEOL] The relatively steep slope on a hillside above the wash slope; usually situated at the angle of repose of the material eroded from it. { 'grav·əd·ē ,slōp }

gravity spring [HYD] A spring that issues under the influence of gravity, not internal pressure. { 'grav·əd·ē ,spriŋ }

gravity stamp [MIN ENG] Unit in a stamp battery which directs a heavy falling weight onto a die on which rock is crushed. { 'grav·əd·ē ,stamp }

gravity station [ENG] The site of installation of gravimeters. { 'grav·əd·ē ,stā·shən }

gravity survey [ENG] The measurement of the differences in gravity force at two or more points. { 'grav·əd·ē 'sər,vā }

gravity tide [GEOPHYS] Cyclic motion of the earth's surface caused by interaction of gravitational forces of the moon, sun, and earth. { 'grav·əd·ē ,tīd }

gravity vector [MECH] The force of gravity per unit mass at a given point. Symbolized **g**. { 'grav·əd·ē ,vek·tər }

gravity wall [CIV ENG] A retaining wall which is kept upright by the force of its own weight. { 'grav·əd·ē ,wöl }

gravity wave [FL MECH] **1.** A wave at a gas-liquid interface which depends primarily upon gravitational forces, surface tension and viscosity being of secondary importance. **2.** A wave in a fluid medium in which restoring forces are provided primarily by buoyancy (that is, gravity) rather than by compression. { 'grav·əd·ē ,wāv }

gravity wheel conveyor [MECH ENG] A downward-sloping conveyor trough with closely spaced axle-mounted wheel units on which flat-bottomed containers or objects are conveyed from point to point by gravity pull. { 'grav·əd·ē ,wel kən'vā·ər }

gravity wind [METEOROL] A wind (or component thereof) directed down the slope of an incline and caused by greater air density near the slope than at the same levels some distance horizontally from the slope. Also known as drainage wind; katabatic wind. { 'grav·əd·ē ,wind }

gravity yard *See* hump yard. { 'grav·əd·ē ,yärd }

gravure coating *See* engraved-roll coating. { grə'vyür ,kōd·iŋ }

gravure printing [GRAPHICS] A process that prints from sunken or depressed surfaces or cups on a plate, in contrast to the raised printing surfaces of letterpress; the depth (and the area) of the depressed areas varies, thus yielding more or less ink on the paper. { grə'vyür ,print·iŋ }

Grawitz's tumor *See* renal-cell carcinoma. { 'grä·vits·əz ,tü·mər }

gray [NUCLEO] The International System unit of absorbed dose, equal to the energy imparted by ionizing radiation to a mass of matter corresponding to 1 joule per kilogram. Symbolized Gy. { grā }

gray antimony *See* antimonite; jamesonite. { ¦grā 'ant·ə,mō·nē }

gray blight [PL PATH] A fungus disease of tea caused by *Pestalotia* (*Pestalozzia*) *theae*, which invades the tissues and causes the formation of black dots on the leaves. { ¦grā 'blīt }

graybody [THERMO] An energy radiator which has a blackbody energy distribution, reduced by a constant factor, throughout the radiation spectrum or within a certain wavelength interval. Also known as nonselective radiator. { 'grā ,bäd·ē }

gray casting [MET] A casting of gray iron. { 'grā ,kast·iŋ }

Gray clay treating [CHEM ENG] A fixed-bed, vapor-phase treating process used to polymerize selectively unsaturated gum-forming constituents (diolefins); a fixed bed is used of 30- to 60-mesh fuller's earth. { 'grā 'klā ,trēd·iŋ }

gray cobalt *See* cobaltite. { ¦grā 'kō,bölt }

Gray code [COMMUN] A modified binary code in which sequential numbers are represented by expressions that differ only in one bit, to minimize errors. Also known as reflective binary code. { 'grā ,kōd }

gray copper ore *See* tetrahedrite. { grā 'käp·ər ,ör }

gray filter *See* neutral-density filter. { 'grā ,fil·tər }

gray goods [TEXT] Cloths of any color that have been woven in a loom but have not been otherwise treated in dry- or wet-finishing operations. Also known as greige goods. { 'grā ,gùdz }

gray hematite *See* specularite. { ¦grā 'hē·mə,tīt }

gray iron [MET] Pig or cast iron in which the carbon not contained in pearlite is present in the form of graphitic carbon. { 'grā ,ī·ərn }

gray leaf spot [PL PATH] A fungus disease of tomatoes caused by *Stemphylium solani* and characterized by water-soaked brown spots on the leaves that become gray with age. { 'grā 'lēf 'spät }

Grayloc tubing joint [PETRO ENG] Special wellbore-tubing joint that has greater leak resistance and strength than standard API tubing joints. { 'grā·läk 'tüb·iŋ ,jóint }

gray manganese ore *See* manganite. { ¦grā 'maŋ·gə,nēs ,ör }

gray matter [NEUROSCI] The part of the central nervous system composed of nerve cell bodies, their dendrites, and the proximal and terminal unmyelinated portions of axons. { 'grā ,mad·ər }

gray mold [PL PATH] Any fungus disease characterized by a gray surface appearance of the affected part. { 'grā ,mōld }

gray scab [PL PATH] A fungus disease of willow caused by *Sphaceloma murrayae* and characterized by irregular raised leaf spots having grayish-white centers and dark-brown margins. { 'grā ,skab }

gray scale [OPTICS] A series of achromatic tones having varying proportions of white and black, to give a full range of grays between white and black; a gray scale is usually divided into 10 steps; however, electronic scanners can typically differentiate 16 to 256 levels. { 'grā ,skāl }

gray sour *See* lime sour. { 'grā ,saùr }

gray speck [PL PATH] A manganese-deficiency disease of oats characterized by light-green to grayish leaf spots, and later by the buff or light-brown discoloration of the blades. { 'grā ,spek }

graywacke [PETR] An argillaceous sandstone characterized by an abundance of unstable mineral and rock fragments and a fine-grained clay matrix binding the larger, sand-size detrital fragments. { 'grā,wak·ə }

graywall [PL PATH] A disease of tomatoes thought to be caused by excess sunlight and characterized by translucent grayish-brown streaks or blotches on the fruit and browning of the vascular strands. { 'grā,wöl }

graze [ORD] **1.** To pass close to the surface, as a shot that follows a path nearly parallel to the ground and low enough to strike a standing person. **2.** Burst of a projectile at the instant of impact with the ground. Also known as graze burst. In time fire, a burst on impact with the ground or other material object on a level with or below the target. [VERT ZOO] To feed by browsing on, cropping, and eating grass. { grāz }

graze burst *See* graze. { 'grāz ‚bərst }

grazing angle [PHYS] A very small glancing angle. { 'grāz·iŋ ‚aŋ·gəl }

grazing fire [ORD] Small-arms fire which is approximately parallel to the ground and does not rise above the height of a standing person. { 'grāz·iŋ ‚fīr }

grazing food web [ECOL] A trophic web that is based on the consumption of the tissues of living organisms. { 'grāz·iŋ ‚füd ‚web }

grazing incidence [PHYS] Incidence at a small glancing angle. { 'grāz·iŋ ‚in·sə·dəns }

grazing-incidence telescope [ASTRON] An instrument for forming images of celestial x-ray or gamma-ray sources in which the total external reflection of the x-rays or gamma rays from a surface at sufficiently shallow angles of incidence is used to focus them. { 'grāz·iŋ ‚in·səd·əns 'tel·ə‚skōp }

grease [MATER] **1.** Rendered, inedible animal fat that is soft at room temperature and is obtained from lard, tallow, bone, raw animal fat, and other waste products. **2.** A lubricant in the form of a solid to semisolid dispersion of a thickening agent in a fluid lubricant, such as petroleum oil thickened with metallic soap. { grēs }

grease cup [ENG] A receptacle used to apply a solid or semifluid lubricant to a bearing; the receptacle is packed with grease and the cap forces the grease to the bearing. { 'grēs ‚kəp }

greased-deck concentration [MIN ENG] A separation process based on selective adhesion of certain grains (diamonds) to quasi-solid grease. { 'grēst ‚dek käns·ən‚trā·shən }

grease gun [ENG] A small hand-operated device that pumps grease under pressure into bearings. { 'grēs ‚gən }

grease ice [HYD] A kind of slush with a greasy appearance, formed from the congelation of ice crystals in the early stages of freezing. Also known as ice fat; lard ice. { 'grēs ‚īs }

greasepaint [MATER] Makeup made of melted tallow or grease used by theatrical performers. { 'grēs‚pānt }

grease seal [ENG] **1.** Type of seal used on floating pistons of some hydropneumatic recoil systems to prevent leakage past the piston of gas or oil; also used in cylinders of some hydropneumatic equilibrators. **2.** Seal used to retain grease in a case or housing, as on an axle shaft. { 'grēs‚sēl }

grease spot [PL PATH] A fungus disease of turf grasses caused by *Pythium aphanidermatum* and characterized by spots that have a greasy border of blackened leaves and intermingled cottony mycelia. Also known as spot blight. { 'grēs‚spät }

grease spot photometer [OPTICS] A photometer in which the light sources whose intensities are to be compared illuminate a thin sheet of opaque paper with a translucent spot at the center. { 'grēs‚spät fə'täm·əd·ər }

grease table [MIN ENG] An apparatus for concentrating minerals, such as diamonds, which adhere to grease; usually a shaking table coated with grease or wax over which an aqueous pulp is flowed. { 'grēs ‚tā·bəl }

grease trap [CIV ENG] A trap in a drain or waste pipe to stop grease from entering a sewer system. { 'grēs ‚trap }

greasewood [BOT] Any plant of the genus *Sarcobatus*, especially *S. vermiculatus*, which is a low shrub that grows in alkali soils of the western United States. { 'grēs‚wůd }

grease wool [TEXT] Raw wool as shorn form the sheep, containing original natural oils. { 'grēs‚wůl }

greasy quartz *See* milky quartz. { 'grē·sē 'kwôrts }

Great Attractor [ASTRON] A great supercluster of galaxies and dark matter, approximately 150×10^6 light-years distant, whose existence has been hypothesized to account for the peculiar motions of galaxies, including the Milky Way Galaxy. { 'grāt ə'trak·tər }

Great Basin high [METEOROL] A high-pressure system centered over the Great Basin of the western United States; it is a frequent feature of the surface chart in the winter season. { 'grāt ‚bas·ən 'hī }

great circle [GEOD] A circle, or near circle, described on the earth's surface by a plane passing through the center of the earth. [MATH] The circle on the two-sphere produced by a plane passing through the center of the sphere. { 'grāt ‚sər·kəl }

great-circle bearing [NAV] The initial direction along a great circle through two terrestrial points, expressed as angular distance from a reference direction. { 'grāt ‚sər·kəl 'ber·iŋ }

great-circle chart *See* gnomonic chart. { 'grāt ‚sər·kəl 'chärt }

great-circle course [NAV] Course along the direction of the great circle through the departure point and the destination; expressed as the angular distance from a reference direction, usually north to the direction of the great circle; the angle varies from point to point along the great circle. { 'grāt ‚sər·kəl 'kórs }

great-circle direction [GEOD] Horizontal direction of a great circle, expressed as angular distance from a reference direction. { 'grāt ‚sər·kəl di'rek·shən }

great-circle distance [GEOD] The length of the shorter arc of the great circle joining two points. { 'grāt ‚sər·kəl 'dis·təns }

great-circle sailing [NAV] Any method of solving the various problems involving courses, distance, and so on, as they are relating to a great-circle track. { 'grāt ‚sər·kəl 'sāl·iŋ }

great-circle track [NAV] The track of a craft following a great circle, or the approximate great circle which a craft intends to follow. { 'grāt ‚sər·kəl 'trak }

great cluster [ASTRON] A galaxy cluster containing thousands of member galaxies and having a radius of 5×10^6 to 20×10^6 light-years. { 'grāt ‚kləs·tər }

great diurnal range [OCEANOGR] The difference in height between mean higher high water and mean lower low water. Also known as diurnal range. { 'grāt dī‚ərn·əl 'rānj }

Greater Dog *See* Canis Major. { 'grād·ər ‚dòg }

greater ebb [OCEANOGR] The stronger of two ebb currents occurring during a tidal day. { 'grād·ər ‚eb }

greater flood [OCEANOGR] The stronger of two flood currents occurring during a tidal day. { 'grād·ər ‚flod }

greater omentum [ANAT] A fold of peritoneum that is attached to the greater curvature of the stomach and hangs down over the intestine and fuses with the mesocolon. { 'grād·ər ō'men·təm }

greatest common divisor [MATH] The greatest common divisor of integers $n_1, n_2, \ldots, n_k$ is the largest of all integers that divide each n_i. Abbreviated gcd. Also known as highest common factor (hcf). { 'grād·əst ‚käm·ən di'vīz·ər }

greatest elongation [ASTRON] The maximum angular distance of a body of the solar system from the sun, as observed from the earth. { 'grād·əst ‚ē‚lòŋ'gā·shən }

greatest lower bound [MATH] The greatest lower bound of a set of numbers S is the largest number among the lower bounds of S. Abbreviated glb. Also known as infimum (inf). { 'grād·əst ‚lō·ər 'baund }

greatest-lower-bound axiom [MATH] The statement that any set of real numbers that has a lower bound also has a greatest lower bound. { 'grād·əst ‚lō·ər‚baund 'ak·sē·əm }

great galago *See* thick-tailed bushbaby. { 'grāt gə'lä·gō }

Great Ice Age [GEOL] The Pleistocene epoch. { 'grāt 'īs ‚āj }

Great Nebula of Orion *See* Orion Nebula. { 'grāt 'neb·yə·lə əv ə'rī·ən }

Great Rift [ASTRON] An apparent break in the Milky Way between Cygnus and Sagittarius caused by a series of large, dark overlapping clouds about 100 parsecs (2×10^{15} miles or 3×10^{15} kilometers) distant in the equatorial plane of the Galaxy. { 'grāt 'rift }

great soil group [GEOL] A group of soils having common internal soil characteristics; a subdivision of a soil order. { 'grāt 'sóil ‚grüp }

great tropic range [OCEANOGR] The difference in height between tropic higher high water and tropic lower low water. { 'grāt 'träp·ik ‚rānj }

Great Wall [ASTRON] A layer of several thousand galaxies, estimated to extend for about 500×10^6 by 200×10^6 light-years but to be less than 15×10^6 light-years thick, constituting the largest known structure in the universe. { 'grāt 'wòl }

great year [ASTRON] The period of one complete cycle of the equinoxes around the ecliptic, about 25,800 years. Also known as platonic year. { 'grāt 'yir }

GREAT CIRCLE

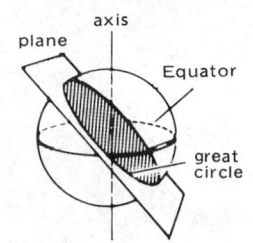

Diagram of a great circle described by a plane through the center of the earth.

grebe [VERT ZOO] The common name for members of the family Podicipedidae; these birds have legs set far posteriorly, compressed bladelike tarsi, individually broadened and lobed toes, and a rudimentary tail. { grēb }

greco [METEOROL] An Italian name for the northeast wind. { 'grek·ō }

Greco-Latin square [STAT] An arrangement of combinations of two sets of letters (one set Greek, the other Roman) in a square array, in such a way that no letter occurs more than once in the array. Also known as orthogonal Latin square. { ¦grek·ō¦lat·ən 'skwer }

Greeffiellidae [INV ZOO] A family of free-living nematodes in the superfamily Desmoscolecoidea. { grē·fē'el·ə,dē }

Greeffielloidea [INV ZOO] A superfamily of primarily marine, free-living nematodes in the order Desmoscolecida, distinguished by a prominent nonencrusted annulation that bears a ring of elongate spines or short scales, and large subdorsal and subventral tubular setae along the body. { grē·fə'lóid·ē·ə }

Greek group [ASTRON] The group of Trojan planets which lies near the Lagrangian point 60° ahead of Jupiter. Also known as Achilles group. { 'grēk ¦grüp }

greeking [COMPUT SCI] The display of the format of a document without displaying the characters. { 'grēk·iŋ }

green [MET] Pertaining to an unsintered powder. [OPTICS] The hue evoked in an average observer by monochromatic radiation having a wavelength in the approximate range from 492 to 577 nanometers; however, the same sensation can be produced in a variety of other ways. { grēn }

green algae [BOT] The common name for members of the plant division Chlorophyta. { 'grēn ¦al·jē }

green belt See frostless zone. { 'grēn ¦belt }

Greenburg-Smith impinger [MIN ENG] A dust-sampling apparatus based on the principle of impingement of the dust-carrying air at high velocity upon a wetted glass surface; also involves bubbling the air through a liquid medium. { 'grēn·bərg 'smith im'pin·jər }

green chalcedony See chrysoprase. { ¦grēn kal'sed·ən·ē }

green chemistry [CHEM] The use of chemical products and processes that reduce or eliminate substances hazardous to human health or the environment. { ¦grēn 'kem·ə·strē }

green concrete [MATER] Concrete that has set but not hardened. { ¦grēn 'kaŋ,krēt }

green copperas See ferrous sulfate. { ¦grēn 'käp·ə·rəs }

green design See industrial ecology. { ¦grēn di'zīn }

green flash [ASTRON] A brilliant green coloration of the upper limb of the sun occasionally observed just as the sun's apparent disk is about to sink below a distant clear horizon. Also known as blue flash; blue-green flame; green segment; green sun. { 'grēn ¦flash }

green fluorescent protein [CELL MOL] A protein that is produced by the bioluminescent jellyfish *Aequorea victoria*; used to trace the synthesis, location, and movement of proteins of interest in cell biology research. Abbreviated GFP. { ¦grēn flə¦res·ənt 'prō,tēn }

green forming [MATER] The ceramic fabrication step in which powders are formed into useful shapes; used in casting, extrusion, die pressing, tape casting, and injection molding. { 'grēn fórm·iŋ }

green gland See antennal gland. { 'grēn ¦gland }

green glass [MATER] Glass given a blue-green hue by substituting cupric oxide for the chromium compound used in ordinary glass. { 'grēn ¦glas }

green gold [MET] A greenish alloy of gold obtained by using silver, silver and cadmium, or silver and copper as the alloying metal. { 'grēn ¦gōld }

greenheart [MATER] Wood from *Octolea rodioei*; resistant to fungi and termites and used for shipbuilding, docks, and marine planking. { 'grēn,härt }

greenhouse [BOT] Glass-enclosed, climate-controlled structure in which young or out-of-season plants are cultivated and protected. { 'grēn,hàus }

greenhouse effect [METEOROL] The effect created by the earth's atmosphere in trapping heat from the sun; the atmosphere acts like a greenhouse. { 'grēn,hàus i,fekt }

greenhouse gases [METEOROL] Gases whose concentration is small and varies, mostly due to anthropogenic factors; they absorb heat from incoming solar radiation but do not allow long-wave radiation to reflect back into space. They include carbon dioxide, methane, and nitrous oxide, as well as water vapor, carbon monoxide, nitrogen oxide, nitrogen dioxide, and ozone. { 'grēn,hàus ,gas·əz }

Greenland anticyclone [METEOROL] The glacial anticyclone which is supposed to overlie Greenland; analogous to the Antarctic anticyclone. { 'grēn·lənd ,ant·i'sī,klōn }

greenlandite See columbite. { 'grēn·lən,dīt }

Greenland spar See cryolite. { 'grēn·lənd 'spär }

green laser [OPTICS] A gas laser using mercury and argon to generate a green line at 5225 angstroms, corresponding to the wavelength that is most readily transmitted through seawater. { ¦grēn ¦lā·zər }

green lead ore See pyromorphite. { ¦grēn 'led ,ór }

green lumber [MATER] Freshly sawed lumber, before drying. { 'grēn ¦ləm·bər }

green manure [AGR] Herbaceous plant material plowed into the soil while still green. { ¦grēn mə'nü·ər }

green mold [MYCOL] Any fungus, especially *Penicillium* and *Aspergillus* species, that is green or has green spores. { ¦grēn ¦mōld }

green mortar [MATER] Mortar that has set but not dried. { ¦grēn ¦mórd·ər }

green mud [GEOL] **1.** A fine-grained, greenish terrigenous mud or oceanic ooze found near the edge of a continental shelf at depths of 300–7500 feet (90–2300 meters). **2.** A deep-sea terrigenous deposit characterized by the presence of a considerable proportion of glauconite and calcium carbonate. { ¦grēn 'məd }

green muscardine [INV ZOO] A disease of the European corn borer, the wheat cockchafer, and other insects caused by the fungus *Metarhizium anisopliae*. { ,grēn 'məs·kər,dēn }

green nickel oxide See nickel oxide. { 'grēn ¦nik·əl 'äk,sīd }

green oats [AGR] Oat crops before separating and grading. { 'grēn ¦ōts }

greenockite [MINERAL] CdS A green or orange mineral that crystallizes in the hexagonal system; occurs as an earthy encrustation and is dimorphous with hawleyite. Also known as cadmium blende; cadmium ocher; xanthochroite. { 'grē·nə,kīt }

green oil [MATER] In the Scottish shale-oil industry, once-run crude shale oil. { 'grēn ¦óil }

green roof [MIN ENG] A mine roof which has not broken down or which shows no sign of taking weight. { 'grēn ¦rüf }

green rosette [PL PATH] A virus disease of the peanut characterized by bunching and yellowing of the leaves with severe stunting of the plant. { ¦grēn rō'zet }

green rot [PL PATH] A decay of fallen deciduous trees in which the wood is colored a malachite green by the fungus *Peziza aeruginosa*. { 'grēn ,rät }

green salt See uranium tetrafluoride. { 'grēn ¦sólt }

greensand [GEOL] A greenish sand consisting principally of grains of glauconite and found between the low-water mark and the inner mud line. [MATER] A highly siliceous sand that contains small amounts of magnesia and alumina, with about 8% of its bulk in powdered coal or charcoal; dampened with water to make foundry molds. [PETR] Sandstone composed of greensand with little or no cement. { 'grēn,sand }

greenschist [PETR] A schistose metamorphic rock with abundant chlorite, epidote, or actinolite present, giving it a green color. { 'grēn,shist }

greenschist facies [PETR] Any schistose rock containing an abundance of green minerals and produced under conditions of low to intermediate temperatures (300–500°C) and low to moderate hydrostatic pressures (3000–8000 bars). { 'grēn,shist 'fā·shēz }

Green's dyadic [MATH] A vector operator which plays a role analogous to a Green's function in a partial differential equation expressed in terms of vectors. { 'grēnz dī'ad·ik }

green segment See green flash. { 'grēn ¦seg·mənt }

Green's function [MATH] A function, associated with a given boundary value problem, which appears as an integrand for an integral representation of the solution to the problem. { 'grēnz ,fəŋk·shən }

Green's identities [MATH] Formulas, obtained from Green's theorem, which relate the volume integral of a function and its gradient to a surface integral of the function and its partial derivatives. { 'grēnz i'den·ə,dēz }

green sky [METEOROL] A greenish tinge to part of the sky, supposed by seamen to herald wind or rain, or in some cases, a tropical cyclone. { 'grēn ¦skī }

green smut [PL PATH] A fungus disease of rice characterized by enlarged grains covered with a green powder consisting of conidia, and caused by *Ustilaginoidea virens*. Also known as false smut. { 'grēn ¦smət }

green snow [HYD] A snow surface that has attained a greenish tint as a result of the growth within it of certain microscopic algae. { 'grēn ¦snō }

Green's theorem [MATH] Under certain general conditions, an integral along a closed curve C involving the sum of functions $P(x,y)$ and $Q(x,y)$ is equal to a surface integral, over the region D enclosed by C, of the partial derivatives of P and Q; namely,

$$\int_C P \, dx + Q \, dy = \iint_D \left(\frac{\partial Q}{\partial x} - \frac{\partial P}{\partial y} \right) dx \, dy$$

{ 'grēnz ¦thir·əm }

Green's theorem in space *See* Gauss' theorem. { ¦grēnz ¦thir·əm in 'spās }

greenstick fracture [MED] An incomplete fracture of a long bone, seen in children; the bone is bent but splintered only on the convex side. { 'grēn¦stik ¦frak·chər }

greenstone [MINERAL] *See* nephrite. [PETR] Any altered basic igneous rock which is green due to the presence of chlorite, hornblende, or epidote. { 'grēn¦stōn }

greenstone belts [GEOL] Oceanic and island arclike sequences that are similar to, and run to the south and north of, the Swaziland System. { 'grēn¦stōn ¦belts }

greenstone schist [PETR] Greenstone with a foliated structure. { 'grēn¦stōn ¦shist }

green strength [MET] The mechanical strength which a compacted powder must have in order to withstand mechanical operations to which it is subjected after pressing and before sintering, without damaging its fine details and sharp edges. { 'grēn ¦streŋkth }

green sulfur bacteria [MICROBIO] A physiologic group of green photosynthetic bacteria of the Chloraceae that are capable of using hydrogen sulfide (H_2S) and other inorganic electron donors. { 'grēn ¦səl·fər bak'tir·ē·ə }

green sun *See* green flash. { 'grēn ¦sən }

green vitriol *See* ferrous sulfate. { 'grēn ¦vi·trē,òl }

greenware [MATER] Ceramic ware that has not yet been fired. { 'grēn,wer }

Greenwell formula [MIN ENG] A formula used for calculating the thickness of tubing and involving the required thickness of tubing in feet, the vertical depth in feet, the diameter of the shaft in feet, and an allowance for possible flaws or corrosion. { 'grēn,wel ¦fòr·myə·lə }

Greenwich apparent noon [ASTRON] Local apparent noon at the Greenwich meridian; twelve o'clock Greenwich apparent time, or the instant the apparent sun is over the upper branch of the Greenwich meridian. { 'gren·ich ə'par·ənt ¦nün }

Greenwich apparent time [ASTRON] Local apparent time at the Greenwich meridian. Abbreviated GAT. { 'gren·ich ə'par·ənt ¦tīm }

Greenwich civil time *See* Greenwich mean time. { 'gren·ich 'siv·əl ¦tīm }

Greenwich hour angle [ASTRON] Angular distance west of the Greenwich celestial meridian; the arc of the celestial equator, or the angle at the celestial pole, between the upper branch of the Greenwich celestial meridian and the hour circle of a point on the celestial sphere, measured westward from the Greenwich celestial meridian through 360°. Abbreviated GHA. { 'gren·ich ¦aur ¦aŋ·gəl }

Greenwich interval [ASTRON] An interval based on the moon's transit of the Greenwich celestial meridian, as distinguished from a local interval based on the moon's transit of the local celestial meridian. { 'gren·ich 'in·tər·vəl }

Greenwich lunar time [ASTRON] Local lunar time at the Greenwich meridian; the arc of the celestial equator, or the angle at the celestial pole, between the lower branch of the Greenwich celestial meridian and the hour circle of the moon, measured westward from the lower branch of the Greenwich celestial meridian through 24 hours. { 'gren·ich 'lün·ər ¦tīm }

Greenwich mean noon [ASTRON] Local mean noon at the Greenwich meridian, twelve o'clock Greenwich mean time, or the instant the mean sun is over the upper branch of the Greenwich meridian. { 'gren·ich 'mēn ¦nün }

Greenwich mean time [ASTRON] Mean solar time at the meridian of Greenwich. Abbreviated GMT. Also known as Greenwich civil time; universal time; Z time; zulu time. { 'gren·ich 'mēn ¦tīm }

Greenwich meridian [GEOD] The meridian passing through Greenwich, England, and serving as the reference for Greenwich time; it also serves as the origin of measurement of longitude. { 'gren·ich mə'rid·ē·ən }

Greenwich sidereal time [ASTRON] Local sidereal time at the Greenwich meridian. Abbreviated GST. { 'gren·ich sī'dir·ē·əl ¦tīm }

gregale [METEOROL] The Maltese and best-known variant of a term for a strong northeast wind in the central and western Mediterranean and adjacent European land areas; it occurs either with high pressure over central Europe or the Balkans and low pressure over Libya, when it may continue for up to 5 days, or with the passage of a depression to the south or southeast, when it lasts only 1 or 2 days; it is most frequent in winter. { grā'gä·lā }

Gregarinia [INV ZOO] A subclass of the protozoan class Telosporea occurring principally as extracellular parasites of invertebrates. { greg·ə'rin·ē·ə }

Gregorian calendar [ASTRON] The calendar used for civil purposes throughout the world, replacing the Julian calendar and closely adjusted to the tropical year. { grə'gór·ē·ən 'kal·ən·dər }

Gregorian telescope [OPTICS] A reflecting telescope having a paraboloidal mirror with a hole in the center and a small secondary (concave ellipsoidal) mirror placed beyond the focus of the primary mirror; light is reflected to the secondary mirror and back to an eyepiece at the hole; the telescope produces an erect image but a small field of view. { grə'gór·ē·ən 'tel·ə,skōp }

Gregory formula [MATH] A formula used in the numerical evaluation of integrals derived from the Newton formula. { 'greg·ə·rē ¦fòr·myə·lə }

greige goods *See* gray goods. { 'grāzh ¦gùdz }

greisen [PETR] A pneumatolytically altered granite consisting of mainly quartz and a light-green mica. { 'grīz·ən }

grenade [ORD] A small explosive or chemical missile, originally designed to be thrown by hand, but now also designed to be projected from special grenade launchers, usually fitted to rifles or carbines. { grə'nād }

grenade cartridge extractor [ORD] An extractor designed for pulling an empty cartridge case or unfired cartridge out of the chamber of a grenade launcher. { grə'nād ¦kär·trij ik,s-trak·tər }

grenade launcher [ORD] A device designed to hold a grenade in position for firing but does not itself propel the missile. { grə'nād ¦lòn·chər }

grenade net [ORD] A net of chicken wire or some similar material placed over a trench, for example, as a protection against grenades. { grə'nād ¦net }

grenade pit [ORD] A pit, usually at the bottom of an incline, to catch hand grenades and limit the effects of their explosion. { grə'nād ¦pit }

grenadine [TEXT] A gauzelike fabric made on a jacquard loom. { 'gren·ə,dēn }

grenatite *See* leucite; staurolite. { 'gren·ə,tīt }

Greninger chart [CRYSTAL] A chart that enables angular relations between planes and zones in a crystal to be read directly from an x-ray diffraction photograph. { 'gren·iŋ·ər ,chärt }

Grenville orogeny [GEOL] A Precambrian mountain-forming epoch. { 'gren·vəl ò'räj·ə·nē }

grenz ray [NUCLEO] An x-ray produced at the long-wavelength end of the x-ray spectrum, involving wavelengths of the order of 1 to 10 angstroms (0.1 to 1 nanometer), by using special x-ray tubes that operate at voltages from only 5000 to 15,000 volts. { 'grents ,rā }

gressorial [VERT ZOO] Adapted for walking, as certain birds' feet. { gre'sór·ē·əl }

greyhound [PETRO ENG] A stand of drill pipe comprising one or more lengths with a tool joint at each end; used to fashion lengths of less than one regular stand during drilling. { 'grā,haùnd }

grid [COMPUT SCI] In optical character recognition, a system of two groups of parallel lines, perpendicular to each other, used to measure or specify character images. [DES ENG] A network of equally spaced lines forming squares, used for determining permissible locations of holes on a printed circuit board or a chassis. [ELEC] **1.** A metal plate with holes or ridges, used in a storage cell or battery as a conductor and a support for the active material. **2.** Any systematic network, such as of telephone lines or power lines. [ELECTR] An electrode located between the cathode and anode of an electron tube, which has one or more openings through which electrons or ions can pass, and serves to control the flow of electrons from cathode to anode. [MAP] A system of uniformly spaced perpendicular lines and horizontal lines running north and south, and east and west on a map, chart, or aerial photograph; used in locating points. *See* Potter-Bucky grid. [MIN ENG] Imaginary line used to divide the surface of an area when following a checkerboard placement of boreholes. { grid }

grid amplitude [NAV] Amplitude relative to grid east or west. { 'grid ˌam·plə₁tüd }

grid-anode transconductance *See* transconductance. { 'grid ¦an₁ōd ˌtranz·kən¦dək·təns }

grid azimuth [NAV] In grid navigation, the angle in the plane of projection measured clockwise between a straight line and the central meridian of a plane-rectangular coordinate system. { 'grid ¦az·ə·məth }

grid battery *See* C battery. { 'grid ₁bad·ə·rē }

grid bearing [NAV] In grid navigation, the angle in the plane of the projection between a line and a north-south grid line. { 'grid ¦ber·iŋ }

grid bias [ELECTR] The direct-current voltage applied between the control grid and cathode of an electron tube to establish the desired operating point. Also known as bias; C bias; direct grid bias. { 'grid ₁bī·əs }

grid-bias cell *See* bias cell. { 'grid ₁bī·əs ₁sel }

grid blocking [ELECTR] **1.** Method of keying a circuit by applying negative grid bias several times cutoff value to the grid of a tube during key-up conditions; when the key is down, the blocking bias is removed and normal current flows through the keyed circuit. **2.** Blocking of capacitance-coupled stages in an amplifier caused by the accumulation of charge on the coupling capacitors due to grid current passed during the reception of excessive signals. { 'grid ₁bläk·iŋ }

grid blocking capacitor *See* grid capacitor. { 'grid ¦bläk·iŋ kə₁pas·əd·ər }

grid cap [ELECTR] A top-cap terminal for the control grid of an electron tube. { 'grid ₁kap }

grid capacitor [ELECTR] A small capacitor used in the grid circuit of an electron tube to pass signal current while blocking the direct-current anode voltage of the preceding stage. Also known as grid blocking capacitor; grid condenser. { 'grid kə₁pas·əd·ər }

grid-cathode capacitance [ELECTR] Capacitance between the grid and the cathode in a vacuum tube. { 'grid ¦kath₁ōd kə₁pas·əd·əns }

grid characteristic [ELECTR] Relationship of grid current to grid voltage of a vacuum tube. { 'grid ¦kar·ik·tə₁ris·tik }

grid circuit [ELECTR] The circuit connected between the grid and cathode of an electron tube. { 'grid ₁sər·kət }

grid condenser *See* grid capacitor. { 'grid kən₁den·sər }

grid conductance *See* electrode conductance. { 'grid kən'dək·təns }

grid control [ELECTR] Control of anode current of an electron tube by variation (control) of the control grid potential with respect to the cathode of the tube. { 'grid kən₁trōl }

grid-controlled mercury-arc rectifier [ELECTR] A mercury-arc rectifier in which one or more electrodes are employed exclusively to control the starting of the discharge. Also known as grid-controlled rectifier. { 'grid kən₁trōld 'mər·kyə·re ¦ärk 'rek·tə₁fī·ər }

grid-controlled rectifier *See* grid-controlled mercury-arc rectifier. { 'grid kən₁trōld 'rek·tə₁fī·ər }

grid control tube [ELECTR] Mercury-vapor-filled thermionic vacuum tube with an external grid control. { 'grid kən₁trōl ₁tüb }

grid convergence *See* grid declination. { 'grid kən₁vər·jəns }

grid course [NAV] In grid navigation, the horizontal direction expressed as the angular distance from grid north. { 'grid ¦kȯrs }

grid current [ELECTR] Electron flow to a positive grid in an electron tube. { 'grid ₁kə·rənt }

grid declination [NAV] In grid navigation, the angular difference in direction between grid north and true north; it is measured east or west from true north. Also known as declination of grid north; grid convergence. { 'grid ₁dek·lə¦nā·shən }

grid-dip meter [ELECTR] A multiple-range electron-tube oscillator incorporating a meter in the grid circuit to indicate grid current; the meter reading dips (reads lower grid current) when an external resonant circuit is tuned to the oscillator frequency. Also known as grid-dip oscillator. { 'grid ₁dip ₁mēd·ər }

grid-dip oscillator *See* grid-dip meter. { 'grid ₁dip ¦äs·ə₁lād·ər }

grid direction [NAV] Horizontal direction expressed as angular distance from grid north, measured from grid north, clockwise through 360°. { 'grid di¦rek·shən }

grid drive [ELECTR] A signal applied to the grid of a transmitting tube. { 'grid ₁drīv }

grid driving power [ELECTR] Average product of the instantaneous value of the grid current and of the alternating component of the grid voltage over a complete cycle; this comprises the power supplied to the biasing device and to the grid. { 'grid ¦drīv·iŋ ₁paủ·ər }

grid element [ELEC] A sinuous resistor used to heat a furnace, made of heavy wire, strap, or casting and suspended from refractory or stainless supports built into the furnace walls, floor, and roof. { 'grid ₁el·ə·mənt }

grid equator [NAV] A line perpendicular to a prime meridian, at the origin; for the usual orientation in polar regions, the grid equator is the 90°W–90°E meridian forming the basic grid parallel, from which grid latitude is measured. { 'grid i¦kwād·ər }

grid-glow tube [ELECTR] A glow-discharge tube in which one or more control electrodes initiate but do not limit the anode current except under certain operating conditions. { 'grid ¦glō ₁tüb }

grid heading [NAV] In grid navigation, the heading relative to grid north. { 'grid ¦hed·iŋ }

gridiron pendulum [HOROL] A compensation pendulum in which the unequal thermal expansion of two metals is utilized to keep the length constant. { 'grid₁ī·ərn 'pen·jə·ləm }

gridistor [ELECTR] Field-effect transistor which uses the principle of centripetal striction and has a multichannel structure, combining advantages of both field effect transistors and minority carrier injection transistors. { gri'dis·tər }

grid latitude [NAV] In grid navigation, the angular distance from a grid equator. { 'grid ¦lad·ə₁tüd }

grid leak [ELECTR] A resistor used in the grid circuit of an electron tube to provide a discharge path for the grid capacitor and for charges built up on the control grid. { 'grid ₁lēk }

grid-leak detector [ELECTR] A detector in which the desired audio-frequency voltage is developed across a grid leak and grid capacitor by the flow of modulated radio-frequency current; the circuit provides square-law detection on weak signals and linear detection on strong signals, along with amplification of the audio-frequency signal. { 'grid ₁lēk di₁tek·tər }

grid limiter [ELECTR] Limiter circuit which operates by limiting positive grid voltages by means of a large ohmic value resistor; as the exciting signal moves in a positive direction with respect to the cathode, current through the resistor causes an *IR* drop which holds the grid voltage essentially at cathode potential; during negative excursions no current flows in the grid circuit, so no voltage drop occurs across the resistor. { 'grid ₁lim·əd·ər }

grid locking [ELECTR] Defect of tube operation in which the grid potential becomes continuously positive due to excessive grid emission. { 'grid ₁läk·iŋ }

grid longitude [NAV] In grid navigation, the angular distance between a prime grid and any designated grid meridian. { 'grid ¦län·jə₁tüd }

grid magnetic angle [NAV] In grid navigation, the angular difference in direction between grid north and magnetic north; it is measured east or west from grid north. { 'grid mag¦ned·ik 'aŋ·gəl }

grid meridian [NAV] One of the grid lines extending in a grid north-south direction. { 'grid mə'rid·ē·ən }

grid metal [MET] An alloy of lead with 5–12% antimony

and sometimes with a smaller amount of tin; used for grids in storage batteries. { 'grid ‚med·əl }

grid modulation [ELECTR] Modulation produced by feeding the modulating signal to the control-grid circuit of any electron tube in which the carrier is present. { 'grid ‚mäj·ə'lā·shən }

grid navigation [NAV] A navigation system related to a reference grid instead of the true north for the measurement of angles. { 'grid ‚nav·ə'gā·shən }

grid nephosope [ENG] A nephoscope constructed of a grid work of bars mounted horizontally on the end of a vertical column and rotating freely about the vertical axis; the observer rotates the grid and adjusts the position until some feature of the cloud appears to move along the major axis of the grid; the azimuth angle at which the grid is set is taken as the direction of the cloud motion. { 'grid 'nef·ə‚skōp }

grid neutralization [ELECTR] Method of amplifier neutralization in which a portion of the grid-cathode alternating-current voltage is shifted 180° and applied to the plate-cathode circuit through a neutralizing capacitor. { 'grid ‚nü·trə·lə'zā·shən }

grid north [NAV] In grid navigation, the northerly or zero direction indicated by the grid datum of directional reference. { 'grid ‚nòrth }

grid parallel [NAV] A line parallel to a grid equator, connecting all points of equal grid latitude. { 'grid 'par·ə·lel }

grid-plate capacitance [ELECTR] Direct capacitance between the grid and the plate in a vacuum tube. { 'grid ‖plāt kə'pas·əd·əns }

grid-plate transconductance See transconductance. { 'grid ‖plāt tranz·kən'dək·təns }

grid-pool tube [ELECTR] An electron tube having a mercury-pool cathode, one or more anodes, and a control electrode or grid that controls the start of current flow in each cycle; the excitron and ignitron are examples. { 'grid 'pül ‚tüb }

grid prime vertical [NAV] The vertical circle through the grid east and west points of the horizon. { 'grid 'prīm 'vərd·ə·kəl }

grid pulse modulation [ELECTR] Modulation produced in an amplifier or oscillator by applying one or more pulses to a grid circuit. { 'grid 'pəls ‚mäj·ə'lā shən }

grid pulsing [ELECTR] Circuit arrangement of a radio-frequency oscillator in which the grid of the oscillator is biased so negatively that no oscillation takes place even when full plate voltage is applied; pulsing is accomplished by removing this negative bias through the application of a positive pulse on the grid. { 'grid ‚pəls·iŋ }

grid-rectification meter [ENG] A type of vacuum-tube voltmeter in which the grid and cathode of a tube act as a diode rectifier, and the rectified grid voltage, amplified by the tube, operates a meter in the plate circuit. { 'grid ‚rek·tə·fə‖kā·shən ‚mēd·ər }

grid resistor [ELECTR] A general term used to denote any resistor in the grid circuit. { 'grid ri‚zis·tər }

grid return [ELECTR] External conducting path for the return grid current to the cathode. { 'grid ri‚tərn }

grid rhumb line [NAV] A line making the same oblique angle with all grid meridians; grid parallels and meridians may be considered special cases of the grid rhumb line. { 'grid ‖rəm ‚līn }

grid spectrometer [SPECT] A grating spectrometer in which a large increase in light flux without loss of resolution is achieved by replacing entrance and exit slits with grids consisting of opaque and transparent areas, patterned to have large transmittance only when the entrance grid image coincides with that of the exit grid. { 'grid spek'träm·əd·ər }

grid suppressor [ELECTR] Resistor of low ohmic value inserted in the grid circuit of a radio-frequency amplifier to prevent low-frequency parasitic oscillations. { 'grid sə‚pres·ər }

grid swing [ELECTR] Total variation in grid-cathode voltage from the positive peak to the negative peak of the applied signal voltage. { 'grid ‚swiŋ }

grid track [NAV] The direction of the track relative to grid north. { 'grid ‚trak }

grid transformer [ELECTR] Transformer to supply an alternating voltage to a grid circuit or circuits. { 'grid tranz‚fōr·mər }

grid-type level detector [ELECTR] A detector using a vacuum tube with input applied to a grid. { 'grid ‚tip 'lev·əl di‚tek·tər }

grid variation [NAV] In grid navigation, the angular difference between grid north and magnetic north. { 'grid ‚ver·ē‖ā·shən }

grid voltage [ELECTR] The voltage between a grid and the cathode of an electron tube. { 'grid ‚vōl·tij }

Griebe-Schiebe method [SOLID STATE] A method of observing the piezoelectric behavior of small crystals, in which the crystals are placed between two electrodes connected to the resonant circuit of an oscillator, and tuning of the resonant circuit results in jumps in the oscillator frequency which produce clicks in headphones or a loudspeaker attached to the plate circuit of the oscillator. { 'grē·bə 'shē·bə ‚meth·əd }

Griebhard's rings [ELEC] A method of producing lines of constant color on a copper plate, coinciding with the equipotential lines of an electric field. { 'grēb·härts ‚riŋz }

Griess reagent [ANALY CHEM] A reagent used to test for nitrous acid; it is a solution of sulfanilic acid, α-naphthylamine and acetic acid in water. { 'grēs rē‚ā·jənt }

Griffith crack [MET] Any small flaw in a metal theorized as creating a low order of fracture strength. { 'grif·əth ‚krak }

griffithite [MINERAL] A micalike mineral containing magnesium, iron, calcium, and aluminosilicate. { 'grif·ə‚thīt }

Griffith's criterion [MECH] A criterion for the fracture of a brittle material under biaxial stress, based on the theory that the strength of such a material is limited by small cracks. { 'grif·əths krī‚tir·ē·ən }

Griffiths' method [THERMO] A method of measuring the mechanical equivalent of heat in which the temperature rise of a known mass of water is compared with the electrical energy needed to produce this rise. { 'grif·əths ‚meth·əd }

Griffith's white See lithopone. { 'grif·əths ‖wīt }

Grifola frondosa [MYCOL] A type of mushroom found in parts of the eastern United States, Europe, and Asia, growing in masses at the base of stumps and on roots; has an anticancer effect in patients with lung and stomach cancers or leukemia. Also known as maitake mushroom. { grə‚fō·lə frän'dōs·ə }

Grignard reaction [ORG CHEM] A reaction between an alkyl or aryl halide and magnesium metal in a suitable solvent, usually absolute ether, to form an organometallic halide. { grin'yär rē‚ak·shən }

Grignard reagent [ORG CHEM] RMgX The organometallic halide formed in the Grignard reaction; an example is C_2H_5MgCl; it is useful in organic synthesis. { grin'yär rē‚ā·jənt }

Grignard synthesis [ORG CHEM] Use of the Grignard reagent in any one of a vast number of reactions, usually condensations; typical syntheses involve formation of a hydrocarbon, acid, ketone, or secondary or tertiary alcohol. { grin'yär ‚sin·thə·səs }

grike [GEOL] A vertical fissure developed along a joint in limestone by dissolution of some of the rock. Also spelled gryke. { grīk }

grillage [CIV ENG] A footing that consists of two or more tiers of closely spaced structural steel beams resting on a concrete block, each tier being at right angles to the one below. { grē'yazh }

grille [ENG] A grating or openwork barrier that is used to conceal or protect an opening in a floor, wall, or pavement. [ENG ACOUS] An arrangement of wood, metal, or plastic bars placed across the front of a loudspeaker in a cabinet for decorative and protective purposes. { gril }

grille cloth [ENG ACOUS] A loosely woven cloth stretched across the front of a loudspeaker to keep out dust and provide protection without appreciably impeding sound waves. { 'gril ‚klòth }

Grimmiales [BOT] An order of mosses commonly growing upon rock in dense tufts or cushions and having hygroscopic, costate, usually lanceolate leaves arranged in many rows on the stem. { ‚grim·ē‚ā·lēz }

grindability [MATER] Relative ease with which a material can be ground. { ‚grīn·də'bil·əd·ē }

grindability index [MATER] A numerical indication of the capacity of a material to be ground. { ‚grīn·də'bil·əd·ē ‚in‚deks }

grinder [MECH ENG] Any device or machine that grinds, such as a pulverizer or a grinding wheel. { 'grīn·dər }

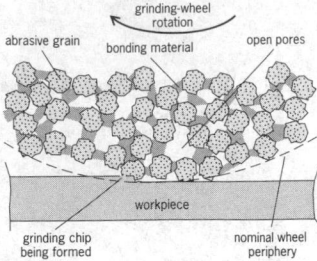

GRINDING WHEEL

grinding-wheel rotation

abrasive grain

bonding material

open pores

workpiece

grinding chip being formed

nominal wheel periphery

Action of a grinding wheel, showing how individual grains of the wheel produce chips from a workpiece.

GRISEOFULVIN

Structural formula for griseofulvin.

GROMIIDA

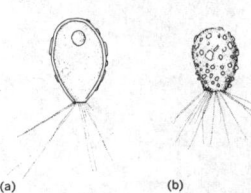

(a)   (b)

Representative Gromiida.
(a) Plagiophrys parvipunctata.
(b) Pseudodifflugia fluva (after Penard). (From R. P. Hall, Protozoology, Prentice-Hall, 1953)

grind gage *See* grindometer. { 'grīnd ¦gāj }

grinding [ELECTR] **1.** A mechanical operation performed on silicon substrates of semiconductors to provide a smooth surface for epitaxial deposition or diffusion of impurities. **2.** A mechanical operation performed on quartz crystals to alter their physical size and hence their resonant frequencies. [MECH ENG] **1.** Reducing a material to relatively small particles. **2.** Removing material from a workpiece with a grinding wheel. [MIN ENG] The act or process of continuing to drill after the bit or core barrel is blocked, thereby crushing and destroying any core that might have been produced. { 'grīn·diŋ }

grinding aid [ENG] An additive to the charge in a ball mill or rod mill to accelerate the grinding process. { 'grīn·diŋ ¦ād }

grinding burn [MECH ENG] Overheating a localized area of the work in grinding operations. { 'grīn·diŋ ¦bərn }

grinding cracks [MATER] Cracks in a workpiece resulting from grinding. { 'grīn·diŋ ¦kraks }

grinding fluid [MATER] Cutting fluid used in metal-grinding operations. { 'grīn·diŋ ¦flü·əd }

grinding medium [ENG] Any material including balls and rods, used in a grinding mill. { 'grīn·diŋ ¦mēd·ē·əm }

grinding mill [LAP] A lathe designed for lapidary work. [MECH ENG] A machine consisting of a rotating cylindrical drum, that reduces the size of particles of ore or other materials fed into it; three main types are ball, rod, and tube mills. { 'grīn·diŋ ¦mil }

grinding pebbles [ENG] Pebbles, of chert or quartz, used for grinding in mills, where contamination with iron has to be avoided. { 'grīn·diŋ ¦peb·əlz }

grinding ratio [MECH ENG] Ratio of the volume of ground material removed from the workpiece to the volume removed from the grinding wheel. { 'grīn·diŋ ¦rā·shō }

grinding relief [MET] A groove at the edge of a metal surface which permits overhang of the corner of the grinding wheel. { 'grīn·diŋ ri¦lēf }

grinding sensitivity [MATER] Susceptibility of a material to the formation of grinding cracks. { 'grīn·diŋ ¦sen·sə·tiv·əd·ē }

grinding stress [MECH] Residual tensile or compressive stress, or a combination of both, on the surface of a material due to grinding. { 'grīn·diŋ ¦stres }

grinding-type resin [ORG CHEM] Vinyl or other resin that requires grinding before dispersal into plastisols or organosols. { 'grīn·diŋ ¦tīp ¦rez·ən }

grinding wheel [DES ENG] A wheel or disk having an abrasive material such as alumina or silicon carbide bonded to the surface. { 'grīn·diŋ ¦wēl }

grindle *See* bowfin. { 'grind·əl }

grindometer [MATER] A device used to measure the size of pigment particles in the inks and paint. Also know as grind gage. { grīn'däm·əd·ər }

grindstone [ENG] A stone disk on a revolving axle, used for grinding, smoothing, and shaping. { 'grīnd¦stōn }

GRIN-rod lens *See* graded refractive index rod lens. { 'grin ¦räd ¦lenz }

grip [ORD] One of a pair of wooden or plastic items designed to be attached by threaded fasteners to the two sides of the frame of a weapon, such as a revolver or bayonet; it is shaped to fit the hand and to provide a formed surface to hold the weapon. { grip }

griphite [MINERAL] $(Na,Al,Ca,Fe)_6Mn_4(PO_4)_5(OH)_4$ Mineral composed of a basic phosphate of sodium, calcium, iron, aluminum, and manganese. { 'gri¦fīt }

gripper [CONT SYS] A component of a robot that grasps an object, generally through the use of suction cups, magnets, or articulated mechanisms. [GRAPHICS] One of the metal fingers on a printing press that clamps the paper into position and controls its flow as it passes through. { 'grip·ər }

gripper edge [GRAPHICS] **1.** The leading edge of the paper as it passes through a printing press. **2.** The front edge of a lithographic or wraparound plate that is secured to the front clamp of the plate cylinder. { 'grip·ər ¦ej }

gripper margin [GRAPHICS] The blank edge of paper which is unprinted because it is held by the grippers. { 'grip·ər ¦mär·jən }

gripping zone [CONT SYS] The area in which the center of an object must be located in order for the object to be properly handled by the gripper of a robot. { 'grip·iŋ ¦zōn }

grip safety [ORD] Safety mechanism that prevents a gun from being fired unless the stock is firmly grasped while the trigger is pulled; used mainly on automatic pistols. { 'grip ¦sāf·tē }

grip vector [CONT SYS] A vector from a point on the wrist socket of a robot to the point where the end effector grasps an object; describes the orientation of the object in space. { 'grip ¦vek·tər }

griquaite [PETR] A hypabyssal rock that contains garnet and diopside, and sometimes olivine or phlogopite, and is found in kimberlite pipes and dikes. { 'grē·kwə¦īt }

grisaille [GRAPHICS] **1.** A technique of painting to imitate a bas-relief and done in shades of gray. **2.** All methods of painting in which full modeling is done in black and white or other contrasting tones, and then finished by the application of transparent glazes. [TEXT] A poplin-type fabric in salt-and-pepper gray with printed warp and coarse filling that imparts a texture. { grə'zī }

grisein [MICROBIO] $C_{40}H_{61}O_{20}N_{10}SFe$ A red, crystalline, water-soluble antibiotic produced by strains of *Streptomyces griseus*. { grə'zēn }

griseofulvin [MICROBIO] $C_{17}H_{17}O_6Cl$ A colorless, crystalline antifungal antibiotic produced by several species of *Penicillium*. Also known as curling factor. { ¦griz·ē·ō'ful·vən }

griseolutein [MICROBIO] Either of two fractions, A or B, of broad-spectrum antibiotics produced by *Streptomyces griseoluteus*; more active against gram-positive than gram-negative microorganisms. { ¦griz·ē·ō'lüd·ē·ən }

griseomycin [MICROBIO] A white, crystalline antibiotic produced by an actinomycete resembling *Streptomyces griseolus*. { ¦griz·ē·ō'mīs·ən }

grism [SPECT] A combination of a diffraction grating and a prism, wherein the grating spreads light into colors and the prism moves the spectrum's position to the point in an image where the observed object appears. { 'griz·əm }

grist [AGR] Grain for grinding or the products and by-products of grinding. [FOOD ENG] Ground malt for brewing. { grist }

gristmill [FOOD ENG] A mill that grinds grains. { 'grist¦mil }

grit [GEOL] **1.** A hard, sharp granule, as of sand. **2.** A coarse sand. [MATER] An abrasive material composed of angular grains. [PETR] A sandstone composed of angular grains of different sizes. { grit }

grit chamber [CIV ENG] A chamber designed to remove sand, gravel, or other heavy solids that have subsiding velocities or specific gravities substantially greater than those of the organic solids in waste water. { 'grit ¦chām·bər }

grits [FOOD ENG] Coarsely ground hominy (corn kernel endosperm) that is boiled and served as a breakfast cereal. { grits }

grit size [DES ENG] Size of the abrasive particles on a grinding wheel. { 'grit ¦sīz }

grizzly [ENG] **1.** A coarse screen used for rough sizing and separation of ore, gravel, or soil. **2.** A grating to protect chutes, manways, and winzes, in mines, or to prevent debris from entering a water inlet. { 'griz·lē }

grizzly bear [VERT ZOO] The common name for a number of species of large carnivorous mammals in the genus *Ursus*, family Ursidae. { 'griz·lē ¦ber }

grizzly chute [MIN ENG] A chute equipped with grizzlies which separate fine from coarse material as it passes through the chute. { 'griz·lē ¦shüt }

grizzly crusher [MECH ENG] A machine with a series of parallel rods or bars for crushing rock and sorting particles by size. { 'griz·lē ¦krəsh·ər }

grizzly worker [MIN ENG] In metal mining, a laborer who works underground at a grizzly. { 'griz·lē ¦wərk·ər }

Groeberiidae [PALEON] A family of extinct rodentlike marsupials. { ¦grə·bə'rī·ə¦dē }

grog [FOOD ENG] Liquor diluted with water and served hot. [MATER] Fired refractory material that is used in the manufacture of products which must withstand extreme heat. { gräg }

groin [ANAT] Depression between the abdomen and the thigh. [ARCH] The projecting edge at the intersection of two vaults. [CIV ENG] A barrier built out from a seashore or riverbank to protect the land from erosion and sand movements, among other functions. Also known as groyne; jetty; spur dike; wing dam. { gröin }

Gromiida [INV ZOO] An order of protozoans in the subclass

Filosia; the test, which is chitinous in some species and thin and somewhat flexible in others, is reinforced with sand grains or siliceous particles. { grə'mī·ə·də }

grommet [ENG] **1.** A metal washer or eyelet. **2.** A piece of fiber soaked in a packing material and used under bolt and nut heads to preserve tightness. [ORD] Device made of rope, plastic, rubber, or metal to protect the rotating band of projectiles. { 'gräm·ət }

grommet nut [DES ENG] A blind nut with a round head; used with a screw to attach a hinge to a door. { 'gräm·ət ,nət }

groove [BIOCHEM] Any of a group of depressions in the double helix of deoxyribonucleic acid that are believed to be sites occupied by nuclear proteins. [DES ENG] A long, narrow channel in a surface. [GEOL] Glaciated marks of large size on rock. [ORD] One of the spiral depressions in the rifling of a gun to impart a spinning motion to a projectile which stabilizes it in flight. { grüv }

groove casts [GEOL] Rounded or sharp, crested, rectilinear ridges that are a few millimeters high and a few centimeters long; found on the undersurfaces of sandstone layers lying on mudstone. { 'grüv ,kasts }

grooved drum [DES ENG] Drum with a grooved surface to support and guide a rope. { 'grüvd ¦drəm }

groove diameter [ORD] One of the two diameters of a rifled barrel; it is measured between two grooves that are diametrically opposed. { 'grüv dī,am·əd·ər }

groove face [MIN ENG] The portion of a surface of a member that is included in a groove. { 'grüv ,fās }

groover [ENG] A tool for forming grooves in a slab of concrete not yet hardened. { 'grüv·ər }

groove sample [MIN ENG] A sample of coal or ore obtained by cutting appropriate grooves along or across the road exposures. Also known as channel sample. { 'grüv ,sam·pəl }

groove weld [MET] A weld in the groove between a pair of members. { 'grüv ,weld }

grooving saw [MECH ENG] A circular saw for cutting grooves. { 'grüv·iŋ ,sȯ }

grosgrain [TEXT] A ribbed material, usually with a dull finish, made by using cotton in the filling. { 'grō,grān }

Grosch's law [COMPUT SCI] The law that the processing power of a computer is proportional to the square of its cost. { 'grȯsh·əz ,lȯ }

gros point See rose point lace. { 'grō ,pȯint }

gross anatomy [ANAT] Anatomy that deals with the naked-eye appearance of tissues. { 'grȯs ə'nad·ə·mē }

gross area [BUILD] Sum of the areas of all stories included within the outside face of the exterior walls of a building. { 'grōs ¦er·ē·ə }

gross-austausch [METEOROL] The exchange of air mass properties and the associated momentum and energy transports produced on a worldwide scale by the migratory large-scale disturbances of middle latitudes. { 'grōs 'aȯs,taȯsh }

gross errors [STAT] Errors that occur when a measurement process is subject occasionally to large inaccuracies. { 'grōs 'er·ərz }

gross index [COMPUT SCI] The first of two indexes consulted to gain access to a record. { 'grōs 'in,deks }

gross information content [COMMUN] Measure of the total information, redundant or otherwise, contained in a message; expressed as the number of bits, nits, or Hartleys required to transmit the message with specified accuracy over a noiseless medium without coding. { 'grōs ,in·fər'mā·shən ,kän·tent }

gross porosity [MET] Gas pockets or pores of undesirable size and quantity in a casting or a weld metal. { 'grōs pə'räs·əd·ē }

gross primary production [ECOL] The incorporation of organic matter or biocontent by a grassland community over a given period of time. { 'grōs 'prī,mer·ē prə'dək·shən }

gross production rate [ECOL] The speed of assimilation of organisms belonging to a specific trophic level. { 'grōs prə'dək·shən ,rāt }

gross recoverable value [MIN ENG] The part of the total metal recovered from an ore multiplied by the price. { 'grōs ri¦kəv·rə·bəl 'val·yü }

gross rubber [CHEM ENG] In rubber manufacturing, the total weight of salable product, including elastomer, carbon black, extender oils, and other materials used in compounding the rubber. { 'grōs 'rəb·ər }

gross sample [ANALY CHEM] One or more increments taken from a larger quantity of a material that is to be analyzed. Also known as bulk sample; lot sample. { ¦grōs 'sam·pəl }

gross stress reaction [PSYCH] A transient personality disorder in which, under conditions of great or unusual stress, a normal person utilizes neurotic mechanisms to deal with danger. { ¦grōs 'stres rē,ak·shən }

gross ton See ton. { ¦grōs ¦tən }

gross tonnage [NAV ARCH] The total volume of the interior of a ship, measured in tons (units of 100 cubic feet). { ¦grōs 'tən·ij }

grossular See grossularite. { 'gräs·yə·lər }

grossularite [MINERAL] $Ca_3Al_2(SiO_4)_3$ The colorless or green, yellow, brown, or red end member of the garnet group, often occurring in contact-metamorphosis impure limestones. Also known as gooseberry stone; grossular. { 'gräs·yə·lə,rīt }

gross unit value [MIN ENG] The weight of metal per long or short ton as determined by assay or analysis, multiplied by the market price of the metal. { ¦grōs 'yü·nət ,val·yü }

gross vehicle weight [IND ENG] A truck rating based on the combined weight of the vehicle and its load. Abbreviated gvw. { ¦grōs 'vē·ə·kəl ,wāt }

gross weight [IND ENG] The weight of a vehicle or container when it is loaded with goods. Abbreviated gr wt. { ¦grōs 'wāt }

grothite See sphene. { 'grō,thīt }

Grotthus' chain theory [PHYS CHEM] An early theory used to explain the conductivity of an electrolyte, in which it was assumed that the cathode and anode attract hydrogen and oxygen respectively, and the molecules of the electrolyte are stretched out in chains between the electrodes, with decomposition occurring in molecules closest to the electrodes. { 'grōt·hus 'chān ,thē·ə·rē }

ground [AERO ENG] To forbid (an aircraft or individual) to fly, usually for a relatively short time. [ELEC] **1.** A conducting path, intentional or accidental, between an electric circuit or equipment and the earth, or some conducting body serving in place of the earth. Abbreviated gnd. Also known as earth (British usage); earth connection. **2.** To connect electrical equipment to the earth or to some conducting body which serves in place of the earth. [GEOL] **1.** Any rock or rock material. **2.** A mineralized deposit. **3.** Rock in which a mineral deposit occurs. [NAV] To touch bottom or run aground; in a serious grounding, the vessel is said to strand. { graȯnd }

ground absorption [ELECTROMAG] Loss of energy in transmission of radio waves, due to dissipation in the ground. { 'graȯnd əb,sȯrp·shən }

ground-air [ORD] **1.** Of missiles, signals, and so on, from the ground to the air. **2.** Of a military force, made up of army or marine and air force units. **3.** Of actions, conducted by both ground and air units. { 'graȯnd ¦er }

ground anchor See anchor log. { 'graȯnd ,aŋ·kər }

ground area [BUILD] The area of a building at ground level. { 'graȯnd ,er·ē·ə }

ground-based navigation aid [NAV] That portion of a navigation system which is located on the ground and emits signals or receives them from crafts or vehicles; these signals provide the navigation information. { 'graȯnd ,bāst ,nav·ə'gā·shən ,ād }

ground bed [MET] In cathodic protection, a device that consists of an interconnected group of impressed-current anodes which absorbs the damage caused by generated electric current. { 'graȯnd ,bed }

ground block [CIV ENG] A pulley fastened to the anchor log which changes a horizontal pull to a vertical pull on a wire line. { 'graȯnd ,bläk }

ground cable [ELEC] A heavy cable connected to earth for the purpose of grounding electric equipment. { 'graȯnd ,kā·bəl }

ground chain [NAV ARCH] A chain that keeps the anchor, when it is being pulled up from the ocean mooring, away from the side of the ship; it is attached along the first length of the anchor chain. { 'graȯnd ,chān }

ground check [ENG] **1.** A procedure followed prior to the release of a radiosonde in order to obtain the temperature and humidity corrections for the radiosonde system. **2.** Any instrumental check prior to the ground launch of an airborne experiment. Also known as base-line check. { 'graȯnd ,chek }

ground-check chamber [ENG] A chamber that is used to

check the sensing elements of radiosonde equipment and that houses sources of heat and water vapor plus instruments for measuring temperature, humidity, and pressure, and in which air circulation is maintained by a motor-driven fan. { 'graund ‚chek ‚chām·bər }

ground circuit [ELEC] A telephone or telegraph circuit part of which passes through the ground. { 'graund ‚sər·kət }

ground clutter [ELECTROMAG] Clutter on a ground or airborne radar due to reflection of signals from the ground or objects on the ground. Also known as ground flutter; ground return; land return; terrain echoes. { 'graund ‚kləd·ər }

ground coal [MIN ENG] The bottom of a coal seam. { 'graund ‚kōl }

ground conductivity [ELEC] The effective conductivity of the ground, used in calculating the attenuation of radio waves. { 'graund ‚kän·dək¦tiv·əd·ē }

ground control [CIV ENG] Supervision or direction of all airport surface traffic, except an aircraft landing or taking off. [ENG] The marking of survey, triangulation, or other key points or system of points on the earth's surface so that they may be recognized in aerial photographs. { 'graund kən‚trōl }

ground-controlled approach [NAV] Technique or procedures by which a ground controller directs an aircraft to reach a point from which a landing may be made; the controller utilizes information from various sensors, including radar. { 'graund kən‚trōld ə‚prōch }

ground-controlled approach minimums [NAV] The minimum ceiling or visibility under which an aircraft may be landed with the use of ground-controlled approach (GCA); the pilot must be qualified to use GCA (and the aircraft must be equipped with the necessary radio equipment) before GCA minimums are applicable to the particular flight; at most airports GCA minimums are less than other minimums because of the double check on the "glide path" position of the aircraft. { 'graund kən‚trōld ə‚prōch 'min·ə‚məmz }

ground-controlled approach radar [ENG] A ground radar system providing information by which aircraft approaches may be directed by radio communications. Abbreviated GCA radar. { 'graund kən‚trōld ə‚prōch 'rā‚där }

ground-controlled interception [ORD] In air defense, an interception in which the interceptor weapon is vectored to the target by instructions transmitted from the ground. { 'graund kən‚trōld ‚in·tər'sep·shən }

ground-controlled intercept radar [ENG] A radar system by means of which a controller may direct an aircraft to make an interception of another aircraft. Abbreviated GCI radar. { 'graund kən‚trōld 'in·tər‚sept ‚rā‚där }

ground controller [ENG] Aircraft controller stationed on the ground; a generic term, applied to the controller in ground-controlled approach, ground-controlled interception, and so on. { 'graund kən‚trōl·ər }

ground cover [BOT] Prostrate or low plants that cover the ground instead of grass. [FOR] All forest plants except trees. { 'graund ‚kəv·ər }

ground current *See* earth current. { 'graund ‚kə·rənt }

ground data equipment [ENG] Any device located on the ground that aids in obtaining space-position or tracking data (including computation function); reads out data telemetry, video, and so on, from payload instrumentation, or is capable of transmitting command and control signals to a satellite or space vehicle. { 'graund ‚dad·ə i‚kwip·mənt }

ground-derived navigation data [NAV] That data used in navigation and obtained by systems or parts of systems that are located on the ground or sea. { 'graund də‚rīvd ‚nav·ə'gā‚shən ‚dad·ə }

ground detector [ELEC] An instrument or equipment used for indicating the presence of a ground on an ungrounded system. Also known as ground indicator. { 'graund di‚tek·tər }

ground dielectric constant [ELEC] Dielectric constant of the earth at a given location. { 'graund di·ə¦lek·trik 'kän·stənt }

ground discharge *See* cloud-to-ground discharge. { 'graund ‚dis‚chärj }

ground distance [NAV] The great-circle distance between two ground positions, as contrasted with slant distance or slant range, the straight-line distance between two points. Also known as ground range. { 'graund ‚dis·təns }

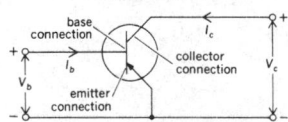

GROUNDED-EMITTER CONNECTION

Schematic diagram of grounded-emitter connection of an alloy junction transistor.

grounded-anode amplifier *See* cathode follower. { ¦graund·əd 'an‚ōd ‚am·plə‚fī·ər }

grounded-base amplifier [ELECTR] An amplifier that uses a transistor in a grounded-base connection. { ¦graund·əd 'bās ‚am·plə‚fī·ər }

grounded-base connection [ELECTR] A transistor circuit in which the base electrode is common to both the input and output circuits; the base need not be directly connected to circuit ground. Also known as common-base connection. { ¦graund·əd 'bās kə‚nek·shən }

grounded-cathode amplifier [ELECTR] Electron-tube amplifier with a cathode at ground potential at the operating frequency, with input applied between control grid and ground, and with the output load connected between plate and ground. { ¦graund·əd 'kath‚ōd ‚am·plə‚fī·ər }

grounded-collector connection [ELECTR] A transistor circuit in which the collector electrode is common to both the input and output circuits; the collector need not be directly connected to circuit ground. Also known as common-collector connection. { ¦graund·əd kə'lek·tər kə‚nek·shən }

grounded-emitter amplifier [ELECTR] An amplifier that uses a transistor in a grounded-emitter connection. { ¦graund·əd i'mid·ər ‚am·plə‚fī·ər }

grounded-emitter connection [ELECTR] A transistor circuit in which the emitter electrode is common to both the input and output circuits; the emitter need not be directly connected to circuit ground. Also known as common-emitter connection. { ¦graund·əd i'mid·ər kə‚nek·shən }

grounded-gate amplifier [ELECTR] Amplifier that uses thin-film transistors in which the gate electrode is connected to ground; the input signal is fed to the source electrode and the output is obtained from the drain electrode. { ¦graund·əd 'gāt ‚am·plə‚fī·ər }

grounded-grid amplifier [ELECTR] An electron-tube amplifier circuit in which the control grid is at ground potential at the operating frequency; the input signal is applied between cathode and ground, and the output load is connected between anode and ground. { ¦graund·əd 'grid ‚am·plə‚fī·ər }

grounded-grid-triode circuit [ELECTR] Circuit in which the input signal is applied to the cathode and the output is taken from the plate; the grid is at radio-frequency ground and serves as a screen between the input and output circuits. { ¦graund·əd ¦grid ‚trī‚ōd 'sər·kət }

grounded-grid-triode mixer [ELECTR] Triode in which the grid forms part of a grounded electrostatic screen between the anode and cathode, and is used as a mixer for centimeter wavelengths. { ¦graund·əd ¦grid ‚trī‚ōd 'mik·sər }

grounded ice *See* stranded ice. { ¦graund·əd ¦īs }

grounded-plate amplifier *See* cathode follower. { ¦graund·əd 'plāt ‚am·plə‚fī·ər }

grounded system [ELEC] Any conducting apparatus connected to ground. Also known as earthed system. { ¦graund·əd 'sis·təm }

ground effect [AERO ENG] Increase in the lift of an aircraft operating close to the ground caused by reaction between high-velocity downwash from its wing or rotor and the ground. [COMMUN] The effect of ground conditions on radio communications. { 'graund i‚fekt }

ground-effect machine *See* air-cushion vehicle. { 'graund i‚fekt mə‚shēn }

ground electrode [ELEC] A conductor buried in the ground, used to maintain conductors connected to it at ground potential and dissipate current conducted to it into the earth, or to provide a return path for electric current in a direct-current power transmission system. Also known as earth electrode; grounding electrode. { 'graund i¦lek‚trōd }

ground environment [ENG] **1.** Environment that surrounds and affects a system or piece of equipment that operates on the ground. **2.** System or part of a system, as of a guidance system, that functions on the ground; the aggregate of equipment, conditions, facilities, and personnel that go to make up a system, or part of a system, functioning on the ground. [ORD] In air defense, the portion of the air defense system that provides for the detection, surveillance, and control of airborne objects; normally these are ground-based facilities, but also include overwater facilities, such as picket vessels, and airborne early warning and control aircraft. { 'graund in¦vī·ərn·mənt }

ground equalizer inductors [ELECTROMAG] Coils, having relatively low inductance, inserted in the circuit to one or more of the grounding points of an antenna to distribute the current to the various points in any desired manner. { 'graund 'ē·kwə,līz·ər in,dok·tərz }

ground fallout plot [NUCLEO] Time lines plotted within a radioactive fallout plot that approximate the distances to which local fallout will have spread at the end of succeeding hours; these lines are determined by using the rate of fall of the radioactive material and the mean wind vector (fallout wind) in the layer of air through which the material is falling. { 'graund 'fȯl,aut ,plät }

ground fault [ELEC] Accidental grounding of a conductor. { 'graund ,fȯlt }

ground fault interrupter [ELEC] A fast-acting circuit breaker that also senses very small ground fault currents such as might flow through the body of a person standing on damp ground while touching a hot alternating-current line wire. { 'graund ,fȯlt ,int·ə,rəp·tər }

ground fire [ORD] Fire, such as antiaircraft fire, that originates from the ground. { 'graund ,fīr }

ground flutter *See* ground clutter. { 'graund ,fləd·ər }

ground fog [METEOROL] A fog that hides less than 0.6 of the sky and does not extend to the base of any clouds that may lie above it. { 'graund ,fäg }

ground frost [METEOROL] In British usage, a freezing condition injurious to vegetation, which is considered to have occurred when a minimum thermometer exposed to the sky at a point just above a grass surface records a temperature (grass temperature) of 30.4°F (−0.9°C) or below. { 'graund ,frȯst }

ground glass [OPTICS] A sheet of matte-surfaced glass on the back of a view camera or process camera so that the image of the subject can be focused on it; it is exactly in the film plane. { 'graund 'glas }

ground handling equipment *See* ground support equipment. { 'graund ,hand·liŋ i,kwip·mənt }

groundhog *See* barney. { 'graund,häg }

ground ice [HYD] **1.** A body of clear ice in frozen ground, most commonly found in more or less permanently frozen ground (permafrost), and may be of sufficient age to be termed fossil ice. Also known as stone ice; subsoil ice; subterranean ice; underground ice. **2.** *See* anchor ice. { 'graund ,īs }

ground ice mound [GEOL] A frost mound containing bodies of ice. Also known as ice mound. { 'graund ,īs ,maund }

ground indicator *See* ground detector. { 'graund ,in·də,kād·ər }

grounding [ELEC] Intentional electrical connection to a reference conducting plane, which may be earth, but which more generally consists of a specific array of interconnected electrical conductors referred to as the grounding conductor. { 'graund·iŋ }

grounding conductor [ELEC] An array of interconnected electric conductors at a uniform potential, to which electrical connections are made for the purpose of grounding. { 'graund·iŋ kən,dək·tər }

grounding electrode *See* ground electrode. { 'graund·iŋ i,lek,trōd }

grounding plate [ELEC] An electrically grounded metal plate on which a person stands to discharge static electricity picked up by his body, or a similar plate buried in the ground to act as a ground rod. { 'graund·iŋ ,plāt }

grounding reactor [ELEC] A reactor sometimes used in a grounded alternating-current system which joins a conductor or neutral point to ground and serves to limit ground current in case of a fault. Also known as earthing reactor (British usage). { 'graund·iŋ rē,ak·tər }

grounding receptacle [ELEC] A receptacle which has an extra contact that accepts the third round or U-shaped prong of a grounding attachment plug and is connected internally to a supporting strap, providing a ground both through the outlet box and the grounding conductor, armor, or raceway of the wiring system. { 'graund·iŋ ri,sep·tə·kəl }

grounding transformer [ELEC] Transformer intended primarily for the purpose of providing a neutral point for grounding purposes. { 'graund·iŋ tranz,fȯr·mər }

ground instrumentation *See* spacecraft ground instrumentation. { 'graund ,in·strə·mən'tā·shən }

ground inversion *See* surface inversion. { 'graund in,vər·zhən }

ground joint [CIV ENG] A closely fitted masonry joint, usually set without mortar. [MECH ENG] A machined metal joint

that makes a tight fit without packing or a gasket. { 'graund ,jȯint }

ground junction *See* grown junction. { 'graund ,jəŋk·shən }

ground layer *See* surface boundary layer. { 'graund ,lā·ər }

ground lead *See* work lead. { 'graund ,lēd }

ground log [NAV] A device for determining the course and speed made good over the ground in shallow water, consisting of a lead or weight attached to a line and thrown overboard to the bottom; the course being made good is indicated by the direction the line tends, and the speed by the amount of line paid out in unit time. { 'graund läg }

ground loop [AERO ENG] A sharp, uncontrollable turn made by an aircraft on the ground during landing, taking off, or taxiing. [COMMUN] Return currents or magnetic fields from relatively high-powered circuits or components which generate unwanted noisy signals in the common return of relatively low-level signal circuits. [ELEC] Potentially detrimental loop formed when two or more points in an electric system that are nominally at ground potential are connected by a conducting path. { 'graund ,lüp }

ground lug [ELEC] A lug that connects a grounding conductor to a grounding electrode. { 'graund ,ləg }

ground magnetic survey [ENG] A determination of the magnetic field at the surface of the earth by means of ground-based instruments. { 'graund mag,ned·ik 'sər,vā }

groundman [ENG] **1.** A person employed in digging or excavating. **2.** *See* mucker. { 'graund,mən }

groundmass *See* matrix. { 'graund,mas }

ground meristem [BOT] Partially differentiated meristematic tissue derived from the apical meristem that gives rise to ground tissue. { 'graund 'mer·ə,stem }

ground mine [ORD] An underwater mine with negative buoyancy, intended to rest on the bottom of relatively shallow water. { 'graund ,mīn }

ground moraine [GEOL] Rock material carried and deposited in the base of a glacier. Also known as bottom moraine; subglacial moraine. { 'graund mə,rān }

ground noise [ENG ACOUS] The residual system noise in the absence of the signal in recording and reproducing; usually caused by inhomogeneity in the recording and reproducing media, but may also include tube noise and noise generated in resistive elements in the amplifier system. [GEOPHYS] In seismic exploration, disturbance of the ground due to some cause other than the shot. { 'graund ,nȯiz }

ground outlet [ELEC] Outlet equipped with a receptacle of the polarity type having, in addition to the current-carrying contacts, one grounded contact which can be used for the connection of an equipment-grounding conductor. { 'graund ,aut·let }

ground-penetrating radar [ENG] *See* ground-probing radar. { 'graund ,pen·ə,trad·iŋ 'rā,där }

ground plane [ELEC] A grounding plate, aboveground counterpoise, or arrangement of buried radial wires required with a ground-mounted antenna that depends on the earth as the return path for radiated radio-frequency energy. { 'graund ,plān }

ground-plane antenna [ELECTROMAG] Vertical antenna combined with a grounded horizontal disk, turnstile element, or similar ground-plane simulation; such antennas may be mounted several wavelengths above the ground, and provide a low radiation angle. { 'graund ,plān an'ten·ə }

groundplasm [CELL MOL] A polyphasic system in which the resolvable elements of the cytoplasm are suspended, including the larger organelles, enzymes of intermediate cell metabolism, contractile protein molecules, and the main cellular pool of soluble precursors. { 'graund,plaz·əm }

ground plate [ELEC] A plate of conducting material embedded in the ground to act as a ground electrode. { 'graund ,plāt }

ground plot *See* dead-reckoning plot. { 'graund ,plät }

ground point of intercept [NAV] In terminal approach procedures, a point in the vertical plane on the runway center line by which it is assumed that the straight-line extension of the slide slope intercepts the runway approach surface base line. Abbreviated as GPI. { 'graund ,point əv 'in·tər,sept }

ground position [NAV] A position on the spherical or spheroidal earth, particularly such a position vertically below an aircraft. { 'graund pə,zish·ən }

ground-position indicator [NAV] An airborne computing

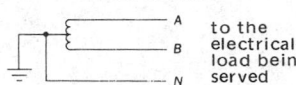
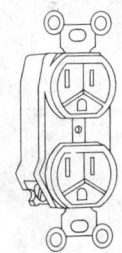

mechanism that provides a continuous indication of position through data derived from instruments which give heading, airspeed, elapsed time, and drift. { 'graůnd pə,zish·ən 'in·də,kād·ər }

ground potential [ELEC] Zero potential with respect to the ground or earth. { 'graůnd pə,ten·chəl }

ground pressure [GEOPHYS] The pressure to which a rock formation is subjected by the weight of the superimposed rock and rock material or by diastrophic forces created by movements in the rocks forming the earth's crust. Also known as geostatic pressure; lithostatic pressure; rock pressure. { 'graůnd ,presh·ər }

GROUND-PROBING RADAR

Portable ground-probing radar equipment with a 500-megahertz antenna and a compact industrial personal computer that displays scrolling vertical slices of the soil in real time. *(ERA Technology UK)*

ground-probing radar [ENG] A nondestructive technique using electromagnetic waves to locate objects or interfaces buried beneath the earth's surface or located within a visually opaque structure. Also known as ground-penetrating radar; subsurface radar; surface-penetrating radar. { 'graůnd ,prōb·iŋ 'rā,där }

ground protection [ELEC] Protection provided a circuit by a device which opens the circuit when a fault to ground occurs. { 'graůnd prə,tek·shən }

ground proximity warning system [NAV] An airborne computer that provides a pilot with visual and audible warnings when it senses the aircraft may inadvertently make contact with the ground. { 'graůnd präk¦sim·əd·ē 'wórn·iŋ ,sis·təm }

ground range *See* ground distance. { 'graůnd ,rānj }

ground recharge [ELEC] The flow of electrons from the ground, in reference to lightning effects. { 'graůnd 'rē,chärj }

ground-referenced navigation data [NAV] Navigation data which are referenced to points on the earth's surface. { 'graůnd 'ref·rənst ,nav·ə'gā·shən ,dad·ə }

ground-reflected wave [ELECTROMAG] Component of the ground wave that is reflected from the ground. { 'graůnd ri¦flek·təd 'wāv }

ground resistance [ELEC] Opposition of the earth to the flow of current through it; its value depends on the nature and moisture content of the soil, on the material, composition, and nature of connections to the earth, and on the electrolytic action present. { 'graůnd ri,zis·təns }

ground return [ELEC] Use of the earth as the return path for a transmission line. [ELECTROMAG] **1.** An echo received from the ground by an airborne radar set. **2.** *See* ground clutter. { 'graůnd ri,tərn }

ground-return circuit [ELEC] Circuit which has a conductor (or two or more in parallel) between two points and which is completed through the ground or earth. { 'graůnd ri,tərn ,sər·kət }

ground rod [ELEC] A rod that is driven into the earth to provide good grounding. { 'graůnd ,räd }

groundscatter propagation [COMMUN] Multihop ionospheric radio propagation along other than the great-circle path between transmitting and receiving stations; radiation from the transmitter is first reflected back to earth from the ionosphere, then scattered in many directions from the earth's surface. { 'graůnd,skad·ər ,präp·ə'gā·shən }

ground sluice [MIN ENG] A channel through which goldbearing earth is passed in placer mining. { 'graůnd ,slüs }

ground speed [NAV] Speed of an aircraft relative to the surface of the earth. { 'graůnd ,spēd }

ground start [AERO ENG] A propulsion starting sequence of a rocket or missile that is initiated and carried through to ignition of the main-stage engines on the ground. { 'graůnd ,stärt }

ground state [QUANT MECH] The stationary state of lowest energy of a particle or a system of particles. { 'graůnd ,stāt }

ground-state maser [PHYS] A maser whose amplifying transition has a terminal level that is appreciably populated at thermal equilibrium for the ambient temperature. { 'graůnd ,stāt 'mā·zər }

ground strafing [ORD] Attack upon ground troops by lowflying aircraft using bombs, machine gun, and cannon. { 'graůnd ,sträf·iŋ }

ground streamer [METEOROL] An upward advancing column of high-ion density which rises from a point on the surface of the earth toward which a stepped leader descends at the start of a lightning discharge. { 'graůnd ,strē·mər }

ground substance *See* matrix. { 'graůnd ,səb·stəns }

ground support equipment [AERO ENG] That equipment on the ground, including all implements, tools, and devices

(mobile or fixed), required to inspect, test, adjust, calibrate, appraise, gage, measure, repair, overhaul, assemble, disassemble, transport, safeguard, record, store, or otherwise function in support of a rocket, space vehicle, or the like, either in the research and development phase or in an operational phase, or in support of the guidance system used with the missile, vehicle, or the like. Abbreviated GSE. Also known as ground handling equipment. { 'graůnd sə,pórt i,kwip·mənt }

ground surveillance radar [ENG] **1.** A surveillance radar operated at a fixed point on the earth's surface for observation and control of the position of aircraft or other vehicles in the vicinity. **2.** A radar system capable of detecting objects on the ground from points on the ground. { 'graůnd sər,vā·ləns ,rā,där }

ground swell [OCEANOGR] A swell passing through shallow water, characterized by a marked increase in height in water shallower than one-tenth wavelength. { 'graůnd ,swel }

ground system [ELECTROMAG] The portion of an antenna that is closely associated with an extensive conducting surface, which may be the earth itself. { 'graůnd ,sis·təm }

ground tackle [NAV ARCH] The anchoring equipment on a ship. { 'graůnd ,tak·əl }

ground tissue [BOT] In leaves and young roots and stems, any tissue other than the epidermis and vascular tissues. { 'graůnd ,tish·ü }

ground-to-cloud discharge [GEOPHYS] A lightning discharge in which the original streamer processes start upward from an object located on the ground. { 'graůnd tə ¦klaůd 'dis,chärj }

ground trace [ENG] The theoretical mark traced upon the surface of the earth by a flying object, missile, or satellite as it passes over the surface, the mark being made vertically from the object making the trace. { 'graůnd ,trās }

ground truth measurements [GEOPHYS] Measurements of various properties, such as temperature and land utilization, which are conducted on the ground to calibrate observations made from satellites or aircraft. { 'graůnd ,trüth ,mezh·ər·məns }

ground-up read-only memory [COMPUT SCI] A read-only memory which is designed from the bottom up, and for which all fabrication masks used in the multiple mask process are custom-generated. { 'graůnd ¦əp ¦rēd ¦ōn·lē 'mem·rē }

ground visibility [METEOROL] In aviation terminology, the horizontal visibility observed at the ground, that is, surface visibility or control-tower visibility. { 'graůnd ,viz·ə'bil·əd·ē }

groundwater [HYD] All subsurface water, especially that part that is in the zone of saturation. { 'graůnd,wȯd·ər }

groundwater decrement *See* groundwater discharge. { 'graůnd,wȯd·ər ¦dek·rə·mənt }

groundwater depletion curve [HYD] A recession curve of streamflow, so adjusted that the slope of the curve represents the runoff (depletion rate) of the groundwater; it is formed by the observed hydrograph during prolonged periods of no precipitation. Also known as groundwater recession. { 'graůnd,wȯd·ər di'plē·shən ,kərv }

groundwater discharge [HYD] **1.** Water released from the zone of saturation. **2.** Release of such water. Also known as groundwater decrement; phreatic-water discharge. { 'graůnd,wȯd·ər 'dis,chärj }

groundwater flow [HYD] That portion of the precipitation that has been absorbed by the ground and has become part of the groundwater. { 'graůnd,wȯd·ər ,flō }

groundwater hydrology [HYD] The study of the occurrence, circulation, distribution, and properties of any liquid water residing beneath the surface of the earth. { ,graůnd ,wȯd·ər hī'dräl·ə·jē }

groundwater increment *See* recharge. { 'graůnd,wȯd·ər ¦iŋ·krə·mənt }

groundwater level [HYD] **1.** The level below which the rocks and subsoil are full of water. **2.** *See* water table. { 'graůnd,wȯd·ər ,lev·əl }

groundwater recession *See* groundwater depletion curve. { 'graůnd,wȯd·ər ri'sesh·ən }

groundwater recharge *See* recharge. { 'graůnd,wȯd·ər 'rē,chärj }

groundwater replenishment *See* recharge. { 'graůnd,wȯdər ri'plen·ish·mənt }

groundwater surface *See* water table. { 'graünd,wȯd·ər ¦sər·fəs }

groundwater table *See* water table. { 'graünd,wȯd·ər ¦tā·bəl }

ground wave [COMMUN] A radio wave that is propagated along the earth and is ordinarily affected by the presence of the ground and the troposphere; includes all components of a radio wave over the earth except ionospheric and tropospheric waves. Also known as surface wave. [ORD] One of the waves formed in the ground by an explosion. { 'graünd ,wāv }

ground ways [CIV ENG] Supports, usually made of heavy timbers, which are placed on the ground on either side of the keel of a ship under construction, providing a track for launching, and supporting the sliding ways. Also known as standing ways. { 'graünd ,wāz }

ground wire [CIV ENG] A small-gage, high-strength steel wire used to establish line and grade for air-blown mortar or concrete. Also known as alignment wire; screed wire. [ELEC] A conductor used to connect electric equipment to a ground rod or other grounded object. { 'graünd ,wīr }

groundwood pulp *See* mechanical pulp. { 'graünd,wu̇d ¦pəlp }

ground zero [ORD] The point on the surface of the earth (including water) at which, above which, or below which an atomic detonation has actually occurred. Also known as surface zero. { 'graünd ¦zir·ō }

group [ASTRON] A number of stars moving in the same direction with the same speed. [CHEM] **1.** A family of elements with similar chemical properties. **2.** A combination of bonded atoms that behave as a unit under certain conditions, for example, the sulfate group, $SO_4{}^{2-}$. [COMMUN] A communications transmission subdivision containing a number of voice channels, either within a supergroup or separately, normally comprised of up to 12 voice channels occupying the frequency band 60–108 kilohertz; each voice channel may be multiplexed for teletypewriter operation, if required; the number of voice channels which may be simultaneously multiplexed for teletypewriter operation will vary according to equipment design. [GEOL] A lithostratigraphic material unit comprising several formations. [MATH] A set G with an associative binary operation where $g_1 \cdot g_2$ always exists and is an element of G, each g has an inverse element g^{-1}, and G contains an identity element. { grüp }

group A kits [ELECTR] Normally those items of electronic equipment which may be permanently or semipermanently installed in an aircraft for supporting, securing, or interconnecting the components and controls of the equipment, and which will not in any manner compromise the security classification of the equipment. { ¦grüp 'ā ,kits }

group B kits [ELECTR] Normally, the operating or operable component of the electronic equipment in an aircraft which, when installed on or in connection with group A parts, constitute the complete operable equipment. { 'grüp 'bē ,kits }

group bus [ELEC] A scheme of electrical connections for a generating station in which more than two feeder lines are supplied by two bus-selector circuit breakers which lead to a main bus and an auxiliary bus. { 'grüp ¦bəs }

group busy tone [COMMUN] High tone connected to the jack sleeves of an outgoing trunk group as an indication that all trunks in the group are busy. { 'grüp ¦biz·ē ¦tōn }

group code *See* systematic error-checking code. { 'grüp ,kōd }

group-coded record [COMPUT SCI] A method of recording data on magnetic tape with eight tracks of data and one parity track, in which every eighth byte in conjunction with the parity track is used for detection and correction of all single-bit errors. { 'grüp ¦kōd·əd 'rek·ərd }

group communications software *See* groupware. { ,grüp kə,myü·nə,kā·shənz 'sȯf,wer }

group decision support system [COMPUT SCI] A computer-based system that enables members of a work group to meet face to face in a conference room or via a computer network and to analyze problems, discuss issues, develop plans, and evaluate alternative sources of action via specialized software that is simultaneously available to everyone in the group. { ¦grüp di,sizh·ən sə¦pȯrt ,sis·təm }

group diffusion method [NUCLEO] An approximation used in studying the diffusion of neutrons in a nuclear reactor, in which the range of neutron energies, from source energy to thermal energy, is divided into a finite number of intervals, or groups, and the neutrons in each group are assumed to diffuse with no loss of energy, until they have undergone the average number of collisions needed to reduce their energy to that of the next lower group. { 'grüp di¦fyü·zhən ,meth·əd }

group dynamics [PSYCH] A branch of social psychology which studies problems involving the structure of a group. { 'grüp dī'nam·iks }

grouped-frequency operation [COMMUN] Use of different frequency bands for channels in opposite directions in a two-wire carrier system. { 'grüpt ¦frē·kwən·sē ,äp·ə,rā·shən }

grouped records [COMPUT SCI] Two or more records placed together and identified by a single key, to save storage space or reduce access time. { 'grüpt 'rek·ərdz }

group flashing light [NAV] A light showing groups of flashes at regular intervals, the duration of light being less than that of darkness. { 'grüp 'flash·iŋ ,līt }

group frequency [ELECTROMAG] Frequency corresponding to group velocity of propagated waves in a transmission line or waveguide. { 'grüp 'frē·kwən·sē }

group incentive [IND ENG] Any wage incentive applied to more than one employee who is engaged in group work characterized by interdependent relationship between operations with consequent physical proximity and unification of interest. { 'grüp in'sen·tiv }

group-indicate [COMPUT SCI] To print indicative information from only the first record of a group. { 'grüp ¦in·də,kāt }

grouping [COMMUN] Periodic error in the spacing of recorded lines in a facsimile system. { 'grüp·iŋ }

grouping circuits [COMMUN] Circuits used to interconnect two or more switchboard positions together, so that one operator may handle the several switchboard positions from one operator's set. { 'grüp·iŋ ,sər·kəts }

grouping of records [COMPUT SCI] Placing records together in a group to either conserve storage space or reduce access time. { 'grüp·iŋ əv 'rek·ərdz }

group mark [COMPUT SCI] A character signaling the beginning or end of a group of data. { 'grüp ¦märk }

group modulation [COMMUN] Process by which a number of channels, already separately modulated to a specific frequency range, are again modulated to shift the group to another range. { 'grüp ,mäj·ə'lā·shən }

group number [OCEANOGR] The first two numbers in the argument number in A. T. Doodson's scheme for predicting tides. { 'grüp ,nəm·bər }

group occulting light [NAV] A light having groups of eclipses at regular intervals; the duration of the light period is equal to or greater than that of the dark period. { 'grüp ə'kəl·tiŋ ,līt }

groupoid [MATH] A set having a binary relation everywhere defined. { 'grü,pȯid }

group printing [COMPUT SCI] The printing of information summarizing the data on a group of cards or other records when a key change occurs. { 'grüp ¦print·iŋ }

group psychotherapy [PSYCH] Therapy given to a group of people by a therapist relying on the group effect on the individual and the person's interactions with the group. { 'grüp ,sī·kō'ther·ə·pē }

group selection [EVOL] Selection in which changes in gene frequency are brought about by the differential extinction and proliferation of the local population. { 'grüp si¦lek·shən }

group technology [IND ENG] A manufacturing system that uses a classification and coding scheme to group parts into families based on similar manufacturing requirements, and specifies parts characteristics, process plans, setups, and manufacturing sequences. { 'grüp tek'näl·ə·jē }

group theory [MATH] The study of the structure of groups which especially deals with the classification of finite groups. { 'grüp ,thē·ə·rē }

group therapy [PSYCH] Application of psychotherapeutic techniques to a group, including utilization of interactions of members of the group. { 'grüp 'ther·ə·pē }

group velocity [PHYS] The velocity of the envelope of a group of interfering waves having slightly different frequencies and phase velocities. { ¦grüp və'läs·əd·ē }

groupware [COMPUT SCI] Multiuser software that supports information sharing through digital media, such as electronic mail and messaging, electronic meeting systems and audio conferencing, group calendaring and scheduling, workflow

GROUSE

The ruffed grouse (*Bonasa umbellus*).

GROWTH CURVE

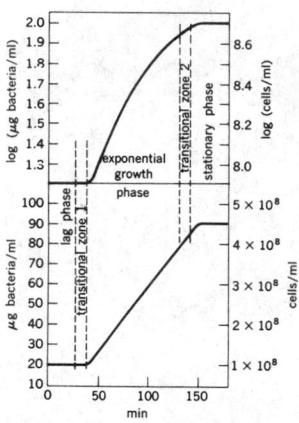

Growth curve of a typical bacterium. Lag phase = 25 minutes, doubling time = 45 minutes, and the specific growth rate = 0.154 min^{-1}.

process diagramming and analysis tools, and group document handling including group editing. { 'grüp,wer }

group without small subgroups [MATH] A topological group in which there is a neighborhood of the identity element that contains no subgroup other than the subgroup consisting of the identity element alone. { ¦grüp with¦aut ¦smól 'səb,grüps }

grouse [VERT ZOO] Any of a number of game birds in the family Tetraonidae having a plump body and strong, feathered legs. { graús }

grouser [ENG] A temporary pile or a heavy, iron-shod pole driven into the bottom of a stream to hold a drilling or dredging boat or other floating object in position. Also known as spud. { 'graús·ər }

grout [MATER] **1.** A fluid mixture of cement and water, or a mixture of cement, sand, and water. **2.** Waste material of all sizes obtained in quarrying stone. { graút }

grout curtain [ENG] A row of vertically drilled holes filled with grout under pressure to form the cutoff wall under a dam, or to form a barrier around an excavation through which water cannot seep or flow. { 'graút ,kərt·ən }

grout hole [ENG] **1.** One of the holes in a grout curtain. **2.** Any hole into which grout is forced under pressure to consolidate the surrounding earth or rock. { 'graút ,hōl }

grouting [ENG] The act or process of applying grout or of injecting grout into grout holes or crevices of a rock. { 'graúd·iŋ }

grout injector [ENG] A machine that mixes the dry ingredients for a grout with water and injects it, under pressure, into a grout hole. { 'graút in,jek·tər }

groutite [MINERAL] HMnO₂ A mineral of the diaspore group, composed of manganese, hydrogen, and oxygen; it is polymorphous with manganite. { 'graú,tīt }

grout pipe [ENG] A pipe that transports grout under pressure for injection into a grout hole or a rock formation. { 'graút ,pīp }

grove cell [ELEC] Primary cell, having a platinum electrode in an electrolyte of nitric acid within a porous cup, outside of which is a zinc electrode in an electrolyte of sulfuric acid; it normally operates on a closed circuit. { 'grōv ,sel }

Grover's algorithm [COMPUT SCI] An algorithm for finding an item in a database of 2N items, using a quantum computer, in a time of order $2^{N/2}$ steps instead of order 2^N steps. { ¦grō·vərz 'al·gə,rith·əm }

Grove's synthesis [ORG CHEM] Production of alkyl chlorides by passing hydrochloric acid into an alcohol in the presence of anhydrous zinc chloride. { 'grōvz 'sin·thə·səs }

grower's year [CLIMATOL] In Great Britain, the 12-month period starting November 6, and referring to the cycle of seasonal change of weather. { 'grō·ərz ,yir }

growing season [AGR] The period of the year when climatic conditions are favorable for plant growth, common to a place or an area. { 'grō·iŋ ,sēz·ən }

growl [GEOPHYS] Noise heard when strata are subjected to great pressure. { graúl }

growler [ELEC] An electromagnetic device consisting essentially of two field poles arranged as in a motor, used for locating short-circuited coils in the armature of a generator or motor and for magnetizing or demagnetizing objects; a growling noise indicates a short-circuited coil. [OCEANOGR] A small piece of floating sea ice, usually a fragment of an iceberg or floeberg; it floats low in the water, and its surface often is heavily pitted; it often appears greenish in color. Also known as bergy-bit. { 'graúl·ər }

grown-diffused transistor [ELECTR] A junction transistor in which the final junctions are formed by diffusion of impurities near a grown junction. { ¦grōn di¦fyüzd tran'zis·tər }

grown junction [ELECTR] A junction produced by changing the types and amounts of donor and acceptor impurities that are added during the growth of a semiconductor crystal from a melt. Also known as ground junction. { ¦grōn ¦jəŋk·shən }

grown-junction photocell [ELECTR] A photodiode consisting of a bar of semiconductor material having a *pn* junction at right angles to its length and an ohmic contact at each end of the bar. { ¦grōn ¦jəŋk·shən 'fōd·ō,sel }

grown-junction transistor [ELECTR] A junction transistor in which different impurities are placed in the melt in sequence as the silicon or germanium seed crystal is slowly withdrawn, to produce the alternate *pn* and *np* junctions. { ¦grōn ¦jəŋk·shən tran'zis·tər }

growth [MED] Any abnormal, localized increase in cells, such as a tumor. [PHYSIO] Increase in the quantity of metabolically active protoplasm, accompanied by an increase in cell number or cell size, or both. { grōth }

growth cone [NEUROSCI] A specialized structure at the end of a growing nerve fiber that guides the fiber to its destination during the development of the nervous system by means of interaction with signaling molecules in its surroundings and its own motile mechanism. { 'grōth ,kōn }

growth curve [MICROBIO] A graphic representation of the growth of a bacterial population in which the log of the number of bacteria or the actual number of bacteria is plotted against time. [NUCLEO] A curve showing how some quantity associated with a radioactive transformation or induced nuclear reaction increases with time. { 'grōth ,kərv }

growth fabric [PETR] Orientation of fabric elements independent of the influences of stress and resultant movement. { 'grōth ,fab·rik }

growth factor [AERO ENG] The additional weight of fuel and structural material required by the addition of 1 pound (0.45 kilogram) of payload to the original payload. [PHYSIO] Any factor, genetic or extrinsic, which affects growth. { 'grōth ,fak·tər }

growth form [ECOL] The habit of a plant determined by its appearance of branching and periodicity. { 'grōth ,fórm }

growth hormone [BIOCHEM] **1.** A polypeptide hormone secreted by the anterior pituitary which promotes an increase in body size. Abbreviated GH. **2.** Any hormone that regulates growth in plants and animals. { 'grōth ¦hór,mōn }

growth index [MATH] For a function of bounded growth f, the smallest real number a such that for some positive real constant M the quantity Me^{ax} is greater than the absolute value of $f(x)$ for all positive x; for a function that is not of bounded growth, the quantity $+\infty$. { 'grōth ,in,deks }

growth lattice [GEOL] The rigid, reef-building, inplace framework of an organic reef, consisting of skeletons of sessile organisms and excluding reef-flank and other associated fragmental deposits. Also known as organic lattice. { 'grōth ,lad·əs }

growth rate [MICROBIO] Increase in the number of bacteria in a population per unit time. { 'grōth ,rāt }

growth regulator [BIOCHEM] A synthetic substance that produces the effect of a naturally occurring hormone in stimulating plant growth; an example is dichlorophenoxyacetic acid. { 'grōth ,reg·yə,lād·ər }

growth spiral [CRYSTAL] A structure on a crystal surface, observed after growth, consisting of a growth step winding downward and outward in an Archimedean spiral which may be distorted by the crystal structure. { 'grōth ,spī·rəl }

growth step [CRYSTAL] A ledge on a crystal surface, one or more lattice spacings high, where crystal growth can take place. { 'grōth ,step }

groyne *See* groin. { gróin }

GR-S rubber [ORG CHEM] Former designation for general-purpose synthetic rubbers formed by copolymerization of emulsions of styrene and butadiene; used in tires and other rubber products; previously also known as Buna-S, currently known as SBR (styrene-butadiene rubber). { ¦jē¦är'es ,rəb·ər }

grub [INV ZOO] The larva of certain insects; commonly used in reference to beetles. { grəb }

grubbing [CIV ENG] Clearing stumps and roots. { 'grəb·iŋ }

grub screw [DES ENG] A headless screw with a slot at one end to receive a screwdriver. { 'grəb ,skrü }

grubstake [MIN ENG] In the United States, supplies or money furnished to a mining prospector for a share in his discoveries. { 'grəb,stāk }

grubstake contract [MIN ENG] An agreement between two or more persons to locate mines upon the public domain by their joint aid, effort, labor, or expense, with each to acquire by virtue of the act of location such an interest in the mine as agreed upon in the contract. { 'grəb,stāk ,kän,trakt }

gruenlingite [MINERAL] Bi₄TeS₃ A mineral composed of sulfide and telluride of bismuth. { 'grün·liŋ,īt }

Gruidae [VERT ZOO] The cranes, a family of large, tall, cosmopolitan wading birds in the order Gruiformes. { 'grü·ə,dē }

Gruiformes [VERT ZOO] A heterogeneous order of generally cosmopolitan birds including the rails, coots, limpkins, button quails, sun grebes, and cranes. { 'grü·ə'fór,mēz }

Grüneisen constant [SOLID STATE] Three times the bulk modulus of a solid times its linear expansion coefficient, divided by its specific heat per unit volume; it is reasonably constant for most cubic crystals. Also known as Grüneisen gamma. { 'grü·nīz·ən ˌkän·stənt }

Grüneisen gamma See Grüneisen constant. { 'grü·nīz·ən ˌgam·ə }

Grüneisen relation [SOLID STATE] The relation stating that the electrical resistivity of a very pure metal is proportional to a mathematical function which depends on the ratio of the temperature to a characteristic temperature. { 'grü·nīz·ən ri̇ˌlā·shən }

grunerite [MINERAL] $(Mg,Fe)_7Si_8O_{22}(OH)_2$ Variety of amphibole; forms monoclinic crystals. { 'grün·əˌrīt }

grus See gruss. { grüs }

Grus [ASTRON] A constellation, right ascension 22 hours, declination 45°S. Also known as Crane. { grüs }

gruss [GEOL] A loose accumulation of fragmental products formed from the weathering of granite. Also spelled grus. { grüs }

GRWP theory [QUANT MECH] A theory that attempts to resolve the quantum measurement paradox by postulating the existence of new laws whose corrections to quantum mechanics become significant over time periods of t_0/N, where t_0 is a characteristic time of the order of the age of the universe and N is the number of particles in the system in question. Derived from Ghirardi-Rimini-Weber-Pearle theory. { 'jē¦är¦dəb·əlˌyü'pē ˌthē·ə·rē }

gr wt See gross weight.

gryke See grike. { grīk }

Gryllidae [INV ZOO] The true crickets, a family of orthopteran insects in which individuals are dark-colored and chunky with long antennae and long, cylindrical ovipositors. { 'gril·əˌdē }

Grylloblattidae [INV ZOO] A monogeneric family of crickets in the order Orthoptera; members are small, slender, wingless insects with hindlegs not adapted for jumping. { ˌgril·ō'blad·əˌdē }

Gryllotalpidae [INV ZOO] A family of North American insects in the order Orthoptera which live in sand or mud; they eat the roots of seedlings growing in moist, light soils. { gril·ō'tal·pəˌdē }

Gs See stimulatory G protein.

GSC See gas-solid chromatography.

G scan See G scope. { 'jē ˌskan }

G scope [ELECTR] A cathode-ray scope on which a single signal appears as a spot on which wings grow as the distance to the target is decreased, with bearing error as the horizontal coordinate and elevation angle error as the vertical coordinate. Also known as G indicator; G scan. { 'jē ˌskōp }

GSE See ground support equipment.

g service [COMMUN] A Federal Aviation Administration service pertaining to aural and visual monitoring of radio aids to air navigation and of the landlines and radio communications systems, to detect faulty operation. { 'jē ˌsər·vəs }

GSM [COMMUN] A digital cellular telephone technology that is based on time-division multiple access; it operates on the 900-megahertz and 1.8-gigahertz bands in Europe, where it is the predominant cellular system, and on the 1.9-gigahertz band in the United States. Derived from global system for mobile communications.

G space [MATH] A topological space X together with a topological group G and a continuous function on the cartesian product of X and G to X such that if the values of this function at (x,g) are denoted by xg, then $x(g_1g_2) = (xg_1)g_2$ and $xe = x$ where e is the identity in G and g_1,g_2 are elements in G. { 'jē ˌspās }

GST See Greenwich sidereal time.

G star [ASTRON] A star of spectral type G; many metallic lines are seen in the spectra, with hydrogen and potassium being strong; G stars are yellow stars, with surface temperatures of 4200–5500 K for giants, 5000–6000 K for dwarfs. { 'jē ˌstär }

G string See field waveguide. { 'jē ˌstriŋ }

g suit [ENG] A suit that exerts pressure on the abdomen and lower parts of the body to prevent or retard the collection of blood below the chest under positive acceleration. Also known as anti-g suit. { 'jē ˌsüt }

GTP [CELL MOL] See guanosine 5'-triphosphate.

GTPase [CELL MOL] One of a family of monomeric GTP-binding proteins. { 'jē·tēˌpās }

GTP-binding protein [CELL MOL] One of a large family of heterotrimeric or monomeric proteins that bind GTP (guanosine 5'-triphosphate) as intermediaries in intracellular signaling pathways. Also known as G protein. { 'jē·tēˌpē 'bīn·diŋ ˌprō·tēn }

Guadalupian [GEOL] A North American provincial series in the Lower and Upper Permian, above the Leonardian and below the Ochoan. { ˌgwäd·əlˌü·pē·ən }

guaiac [MATER] A resin obtained from the trees *Guaiacum santum* and *G. officinale;* soluble in alcohol, ether, and chloroform; used in medicine and varnish. { 'gwīˌak }

guaiacol [ORG CHEM] $C_6H_4(OH)OCH_3$ A colorless, crystalline compound, soluble in water; used as a reagent to determine the presence of such substances as lignin, narceine, and nitrous acid. { 'gwī·əˌkȯl }

guaiazulene [PHARM] $C_{15}H_{18}$ A blue oil with a boiling point of 165–170°C; used as an anti-inflammatory drug. { ˌgwī'az·əˌlēn }

guanajuatite [MINERAL] Bi_2Se_3 Bluish-gray mineral composed of bismuth selenide, occurring in crystals or masses. { ˌgwän·ə'hwäˌtīt }

guanidine [BIOCHEM] CH_5N_2 Aminomethanamidine, a product of protein metabolism found in urine. { 'gwän·əˌdēn }

guanidine-acetic acid See glycocyamine. { 'gwän·əˌdēn əˌsēd·ik 'as·əd }

guanine [BIOCHEM] $C_5H_5ON_5$ A purine base; occurs naturally as a fundamental component of nucleic acids. { 'gwäˌēn }

guano [MATER] Phosphate- and nitrogen-rich, partially decomposed excrement of seabirds; used as a fertilizer. { 'gwän·ō }

guanophore See iridocyte. { 'gwän·əˌfȯr }

guanosine [BIOCHEM] $C_{10}H_{13}O_5N_5$ Guanine riboside, a nucleoside composed of guanine and ribose. Also known as vernine. { 'gwän·əˌsēn }

guanosine 5'-triphosphate [CELL MOL] A nucleoside triphosphate that is instrumental in many cellular processes, including microtubule assembly, protein synthesis, and cell signaling, due to the energy it releases upon removal of its terminal phosphate group (producing guanosine 5'-diphosphate). Abbreviated GTP. { 'gwän·əˌsēn ¦fīvˌprīm trī'fäsˌfat }

guanosine tetraphosphate [BIOCHEM] A nucleotide which participates in the regulation of gene transcription in bacteria by turning off the synthesis of ribosomal ribonucleic acid. { 'gwän·əˌsēn te·trə'fäsˌfat }

guanylic acid [BIOCHEM] A nucleotide composed of guanine, a pentose sugar, and phosphoric acid and formed during the hydrolysis of nucleic acid. Abbreviated GMP. Also known as guanosine monophosphate; guanosine phosphoric acid. { gwə'nil·ik 'as·əd }

guar [AGR] *Cyanopsis tetragonaloba.* A leguminous crop adapted to semiarid regions of the southwestern United States and Mexico. Also known as cluster bean. { gwär }

guard [ENG] A shield or other fixture designed to protect against injury. [MIN ENG] A support in front of a roll train to guide the bar into the groove. { gärd }

guard ammunition [ORD] Ammunition specifically designed for use by guards, usually containing a reduced propelling charge. { 'gärd ˌam·yə¦nish·ən }

guard arm [ELEC] **1.** Crossarm placed across and in line with a cable to prevent damage to the cable. **2.** Crossarm located over wires to prevent foreign wires from falling into them. { 'gärd ˌärm }

guard band [ELECTR] A narrow frequency band provided between adjacent channels in certain portions of the radio spectrum to prevent interference between stations. { 'gärd ˌband }

guard cell [BOT] Either of two specialized cells surrounding each stoma in the epidermis of plants; functions in regulating stoma size. { 'gärd ˌsel }

guard circle [DES ENG] The closed loop at the end of a grooved record. { 'gärd ˌsər·kəl }

guarded command [COMPUT SCI] A program statement within a group of such statements that determines whether the other statements will be executed by the computer. { 'gärd·əd kə'mand }

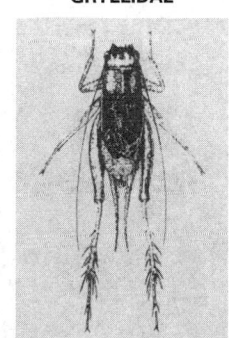

GRYLLIDAE

Grass cricket (*Nemobius fasciatus*).

GUANINE

Structural formula of guanine.

guard-electrode system [PETRO ENG] System of extra electrodes used during electrical logging of reservoir formations to confine the surveying current from the measuring electrode to a generally horizontal path. { 'gärd i‚lek‚trōd ‚sis·təm }

guarding [ELEC] A method of eliminating surface-leakage effects from measurements of electrical resistance which employs a low-resistance conductor in the vicinity of one of the terminals or a portion of the measuring circuit. { 'gärd·iŋ }

guard lock [CIV ENG] *See* entrance lock. [ENG] An auxiliary lock that must be opened before the key can be turned in a main lock. { 'gärd ‚läk }

guard magnet [MIN ENG] A magnet employed in a crushing system to remove or arrest tramp iron ahead of the machinery. { 'gärd ‚mag·nət }

guardrail [CIV ENG] **1.** A handrail. **2.** A rail made of posts and a metal strip used on a road as a divider between lines of traffic in opposite directions or used as a safety barrier on curves. **3.** A rail fixed close to the outside of the inner rail on railway curves to hold the inner wheels of a railway car on the rail. Also known as check rail; safety rail; slide rail. { 'gärd ‚rāl }

guard relay [ELEC] Used in the linefinder circuit to make sure that only one linefinder can be connected to any line circuit when two or more line relays are operated simultaneously. { 'gärd ‚rē‚lā }

guard ring [ELEC] A ring-shaped auxiliary electrode surrounding one of the plates of a parallel-plate capacitor to reduce edge effects. [ELECTR] A ring-shaped auxiliary electrode used in an electron tube or other device to modify the electric field or reduce insulator leakage; in a counter tube or ionization chamber a guard ring may also serve to define the sensitive volume. [THERMO] A device used in heat flow experiments to ensure an even distribution of heat, consisting of a ring that surrounds the specimen and is made of a similar material. { 'gärd ‚riŋ }

guard screen *See* oversize control screen. { 'gärd ‚skrēn }

guard shield [ELECTR] Internal floating shield that surrounds the entire input section of an amplifier; effective shielding is achieved only when the absolute potential of the guard is stabilized with respect to the incoming signal. { 'gärd ‚shēld }

guard signal [COMPUT SCI] A signal used in digital-to-analog converters, analog-to-digital converters, or other converters which permits values to be read or converted only when the values are not changing, usually to avoid ambiguity error. { 'gärd ‚sig·nəl }

guard wire [ELEC] A grounded conductor placed beneath an overhead transmission line in order to ground the line, in case it breaks, before reaching the ground. { 'gärd ‚wīr }

guar gum [MATER] A mucilage formed from seeds of the guar plant; light-gray powder dispersible in water; used as a thickening agent in paper, foods, pharmaceuticals, and cosmetics. { 'gwär ‚gəm }

Guarnieri body [PATH] Eosinophilic cytoplasmic inclusion bodies found in the epidermal cells of patients with smallpox or chickenpox. { gwär'nyer·ē ‚bäd·ē }

GUAVA

Psidium guajava, showing a branch with leaves and two berries and a berry cut in half.

guava [BOT] *Psidium guajava.* A shrub or low tree of tropical America belonging to the family Myrtaceae; produces an edible, aromatic, sweet, juicy berry. { 'gwäv·ə }

guayule [BOT] *Parthenium argentatum.* A subshrub of the family Compositae that is native to Mexico and the southwestern United States; it has been cultivated as a source of rubber. { wī'yü·lē }

guba [METEOROL] In New Guinea, a rain squall on the sea. { 'gü·bä }

gubernaculum [ANAT] A guiding structure, as the fibrous cord extending from the fetal testes to the scrotal swellings. [INV ZOO] **1.** A posterior flagellum of certain protozoans. **2.** A sclerotized structure associated with the copulatory spicules of certain nematodes. { ‚gü·bər'nak·yə·ləm }

Gudden-Pohl effect [ELECTR] The momentary illumination produced when an electric field is applied to a phosphor previously excited by ultraviolet radiation. { 'gùd·ən 'pōl i‚fekt }

Gudermannian [MATH] The function *y* of the variable *x* satisfying tan *y* = sinh *x* or sin *y* = tanh *x*; written gd*x*. { 'gùd·ər‚män·ē·ən }

gudgeon [ENG] **1.** A pivot. **2.** A pin for fastening stone blocks. [NAV ARCH] Metal fittings on the sternpost of a boat

or on the rudderpost of a ship on which the rudder is hung; the gudgeon forms the pivot point. { 'gəj·ən }

gudmundite [MINERAL] FeSbS A silver-white to steel-gray orthorhombic mineral composed of a sulfide and antimonide of iron. { 'gùd·mən‚dīt }

Guerbet reaction [ORG CHEM] A condensation of alcohols at high temperatures through the action of sodium alkoxides. { ‚ger'bā rē‚ak·shən }

guerrilla warfare [ORD] Operations carried on by independent or semi-independent forces in the rear of the enemy; these operations usually are conducted by irregular forces acting either separately from, or in conjunction with, regular forces but may at times be conducted entirely with regular troops. { gə'ril·ə 'wär‚fer }

guess-warp [NAV ARCH] A line used in conjunction with a Jacob's ladder on the thimble of which boat crews reeve their bowlines. { 'ges‚wärp }

guest [CHEM] Cationic, anionic, or neutral organic, inorganic, or biological substance, bound by means of various interactions (electrostatic, hydrogen bonding, van der Waals, donor-acceptor) within a crystalline or molecular structure. Also known as guest molecule; guest substance. { gest }

guest computer [COMPUT SCI] A computer that operates under the control of another computer (the host). { 'gest kəm‚pyüd·ər }

guest element *See* trace element. { 'gest ‚el·ə·mənt }

guest molecule *See* guest. { 'gest ‚mäl·ə‚kyül }

guest substance *See* guest. { 'gest ‚səb·stəns }

Guest unit [BIOL] A unit for the standardization of plasmin. { 'gest ‚yü·nət }

Guggenheim process [CIV ENG] A method of chemical precipitation which employs ferric chloride and aeration to prepare sludge for filtration. { 'gùg·ən·hīm ‚präs·əs }

GUI *See* graphical user interface. { 'gü‚ē *or* ‚jē‚yü'ī }

Guiana Current [OCEANOGR] A current flowing northwestward along the northeastern coast of South America. { gī'an·ə ‚kə·rənt }

guidance [NAV] The process of directing the movements of any vehicle, especially an aeronautical vehicle or space vehicle, with particular reference to the selection of a course or flight path. { 'gīd·əns }

guidance site [ENG] Specific location of high-order geodetic accuracy containing equipment and structures necessary to provide guidance services or a given launch rate; it may be an integrated part of a launch site, or it may be a remote facility. { 'gīd·əns ‚sīt }

guidance station equipment [ENG] The ground-based portion of the missile guidance system necessary to provide guidance during missile flight; it specifically includes the tracking radar, the rate measuring equipment, the data link equipment, and the computer, test, and maintenance equipment integral to these items. { 'gīd·əns ‚stā·shən i‚kwip·mənt }

guidance system [AERO ENG] The control devices used in guidance of an aircraft or spacecraft. [NAV] Apparatus for generating and detecting the path along which a vehicle or craft is guided, often remotely and automatically. { 'gīd·əns ‚sis·təm }

guide bearing [MECH ENG] A plain bearing used to guide a machine element in its lengthwise motion, usually without rotation of the element. { 'gīd ‚ber·iŋ }

guide bracket [MIN ENG] A steel bracket fixed to a bunton to secure rigid guides in a shaft. { 'gīd ‚brak·ət }

guide coat [MATER] A thin coat of paint applied to a surface over a sealer or filler to indicate the locations of bumps or imperfections and thereby to serve as a guide for removing them. { 'gīd ‚kōt }

guide coupling [MIN ENG] A short coupling with a projecting reamer guide or pup to which is attached a reaming bit, which it couples to a reaming barrel. { 'gīd ‚kəp·liŋ }

guided bend test [MET] A bend test in which the specimen is bent to a predetermined shape. { 'gīd·əd 'bend ‚test }

guided bomb [ORD] An aerial bomb that is guided in range or azimuth, or both, during its drop. { 'gīd·əd 'bäm }

guided missile [ORD] An uncrewed self-propelled vehicle, with or without a warhead, which is designed to move in a trajectory or flight path all or partially above the earth's surface and whose trajectory or course, while in flight, is capable of being directed by remote control, by homing systems, or by inertial or programmed guidance from within; excludes drones, torpedoes, and rockets and other vehicles whose trajectory or course cannot be controlled while in flight. { 'gīd·əd ¦mis·əl }

guided-missile control [ORD] The guidance or direction exercised over a guided missile during its flight to a target. { 'gīd·əd ¦mis·əl kən,trōl }

guided-missile exercise head [ORD] An item designed to simulate a guided missile warhead, with or without telemetering devices. { 'gīd·əd ¦mis·əl ¦ek·sər,sīz ,hed }

guided-missile ship [NAV ARCH] A warship equipped with guided-missile launchers, and sometimes with a gun battery, long-range sonar, and antisubmarine warfare weapons. Also known as missile ship. { 'gīd·əd ¦mis·əl 'ship }

guided-missile submarine [NAV ARCH] A submarine designed to have an additional capability to launch guided-missile attacks from surface condition. { 'gīd·əd ¦mis·əl ¦səb·mə'rēn }

guided propagation [COMMUN] Type of radio-wave propagation in which radiated rays are bent excessively by refraction in the lower layers of the atmosphere; this bending creates an effect much as if a duct or waveguide has been formed in the atmosphere to guide part of the radiated energy over distances far beyond the normal range. Also known as trapping. { 'gīd·əd präp·ə'gā·shən }

guided rocket [ORD] A guided missile having rocket propulsion. { 'gīd·əd 'räk·ət }

guided wave [ELECTROMAG] A wave whose energy is concentrated near a boundary or between substantially parallel boundaries separating materials of different properties and whose direction of propagation is effectively parallel to these boundaries; waveguides transmit guided waves. { 'gīd·əd 'wāv }

guide fossil [PETR] A fossil used for rock correlation and age determination. { 'gīd ,fäs·əl }

guide frame [MIN ENG] A frame held rigidly in place by roof jacks or timbers, with provisions for attaching a shaker conveyor pan line to the movable portion of the frame; prevents jumping or side movement of the pan line. { 'gīd ,frām }

guide idler [MECH ENG] An idler roll with its supporting structure mounted on a conveyor frame to guide the belt in a defined horizontal path, usually by contact with the edge of the belt. { 'gīd ,īd·lər }

guide key *See* home key. { 'gīd ,kē }

guideline [IND ENG] A document containing recommendations for methods that should be used to achieve a desired goal. { 'gīd,līn }

guide mill [MET] A hand rolling mill with a series of stands and guides at the entrance to the rolls. { 'gīd ,mil }

guide number [GRAPHICS] A number that relates the output of a light source (such as a flashbulb) to the sensitivity of a particular film; when the number is divided by the distance in feet to the subject, it gives the T stop at which the lens should be set. { 'gīd ,nəm·bər }

guidepath [ENG] The path over which an automated guided vehicle travels; often contains some means of communication with the guidance system, such as a guidewire. { 'gīd,path }

guide pin [ENG] A pin used to line up a tool or die with the work. { 'gīd ,pin }

guide post [CIV ENG] A post along a road that bears direction signs or guide boards. { 'gīd ,pōst }

guide rail [CIV ENG] A track or rail that serves to guide movement, as of a sliding door, window, or similar element. { 'gīd ,rāl }

guide ribonucleic acid [CELL MOL] A ribonucleic acid sequence that provides a template for the alignment of splice junctions. { 'gīd ,rī·bō·nü¦klē·ik 'as·əd }

guide ring [MIN ENG] A longitudinally grooved, annular ring made almost full borehole size, which is fitted to an extension coupling between the core barrel and the first drill rod. { 'gīd ,riŋ }

guide rod [MIN ENG] A heavy drill rod coupled to and having the same diameter as a core barrel on which it is used; gives additional rigidity to the core barrel and helps to prevent deflection of the borehole. Also known as core barrel rod; oversize rod. { 'gīd ,räd }

guides [MECH ENG] **1.** Pulleys to lead a driving belt or rope in a new direction or to keep it from leaving its desired direction. **2.** Tracks that support and determine the path of a skip bucket and skip bucket bail. **3.** Tracks guiding the chain or buckets of a bucket elevator. **4.** The runway paralleling the path of

the conveyor which limits the conveyor or parts of a conveyor to movement in a defined path. [MIN ENG] **1.** Steel, wood, or steel-wire rope conductors in a mine shaft to guide the movement of the cages. **2.** Timber, rope, or metal tracks in a hoisting shaft, which are engaged by shoes on the cage or skip so as to steady it in transit. **3.** The holes in a crossbeam through which the stems of the stamps in a stamp mill rise and fall. { gīdz }

guide wavelength [ELECTROMAG] Wavelength of electromagnetic energy conducted in a waveguide; guide wavelength for all air-filled guides is always longer than the corresponding free-space wavelength. { 'gīd ¦wāv,leŋkth }

guidewire [ENG] A wire embedded in the surface of the path traveled by an electromagnetically guided automated guided vehicle. { 'gīd,wīr }

guiding center [ELECTROMAG] A slowly moving point about which a charged particle rapidly revolves; this is used in an approximation for the motion of a charged particle in slowly varying electric and magnetic fields. { 'gīd·iŋ ¦sent·ər }

guiding telescope [OPTICS] A telescope that is mounted so that it remains parallel to a photographic telescope and is used by a person observing through it to supplement the clock motion in keeping the image of a celestial body motionless on a photographic plate. { 'gīd·iŋ ¦tel·ə,skōp }

guild [ECOL] A group of species that utilize the same kinds of resources, such as food, nesting sites, or places to live, in a similar manner. { gild }

guildite [MINERAL] $(Cu,Fe)_3(Fe,Al)_4(SO_4)_7(OH)_4 \cdot 15H_2O$ A dark-brown mineral composed of a basic hydrated sulfate of copper, iron, and aluminum. { 'gil,dīt }

Guillain-Barré syndrome *See* Landry Guillain-Barré syndrome. { ¦gē·yan bə'rā ,sin,drōm }

Guillemin effect [ELECTROMAG] The tendency of a bent magnetostrictive rod to straighten in a magnetic field parallel to its length. { gē·yə'ma i,fekt }

Guillemin line [ELECTR] A network or artificial line used in high-level pulse modulation to generate a nearly square pulse, with steep rise and fall; used in radar sets to control pulse width. { gē·yə'ma ,līn }

guillotine [GRAPHICS] A heavy steel knife used to cut printing plates made of copper, zinc, or magnesium and to trim excess metal from electrotype or stereotype casts. { ,gē·yə'tēn }

guillotine factor [ASTROPHYS] A quantity that expresses the sharp reduction in the opacity of a gas which occurs when its temperature becomes sufficiently high to ionize the atoms down to their K shells. { 'gē·ə,tēn ,fak·tər }

guillotine shears [ENG] A cutting tool fitted with vertically mounted blades, the bottom blade being fixed in position and the top blade mounted on a movable ram. { 'gē·ə,tēn ,shirz }

Guinea Current [OCEANOGR] A current flowing eastward along the southern coast of northwestern Africa into the Gulf of Guinea. { 'gin·ē ¦kə·rənt }

guinea fowl [VERT ZOO] The common name for plump African game birds composing the family Numididae; individuals have few feathers on the head and neck, but may have a crest of feathers and various fleshy appendages. { 'gin·ē ,faül }

guinea pig [VERT ZOO] The common name for several species of wild and domestic hystricomorph rodents in the genus *Cavia*, family Caviidae; individuals are stocky, short-eared, short-legged, and nearly tailless. { 'gin·ē ,pig }

guinea worm [INV ZOO] *Dracunculus medinensis.* A parasitic nematode that infects the subcutaneous tissues of humans and other mammals. { 'gin·ē ,wərm }

Guinier-Preston zones [MET] The initially formed zones of a precipitate as it comes out of solid solution. { gēn'yä 'pres·tən ,zōnz }

guipure [TEXT] A machine-made lace with a heavy gimp motif connected by bars and without mesh grounds. { gē'pyür }

guitarfish [VERT ZOO] The common name for fishes composing the family Rhinobatidae. { gə'tär,fish }

guitermanite [MINERAL] $Pb_{10}Ar_6S_{19}$ A bluish-gray mineral composed of lead, arsenic, and sulfur, occurring in compact masses. { 'gid·ər·mə,nīt }

Gukhman number [THERMO] A dimensionless number used in studying convective heat transfer in evaporation, equal to $(t_0 - t_m)/T_0$, where t_0 is the temperature of a hot gas stream,

GUINEA FOWL

The common guinea fowl (*Numida meleagris*), with red wattles and pearl-green plumage.

GUINEA PIG

The guinea pig, with short legs and large angular claws.

t_m is the temperature of a moist surface over which it is flowing, and T_0 is the absolute temperature of the gas stream. Symbolized Gu; N_{Gu}. { 'gük·mən ‚nəm·bər }

gular [ANAT] Of, pertaining to, or situated in the gula or upper throat. [VERT ZOO] A horny shield on the plastron of turtles. { 'gyü·lər }

gulch [GEOGR] A gulley, sometimes occupied by a torrential stream. { gəlch }

gulch claim [MIN ENG] A claim laid upon and along the bed of an unnavigable stream winding through a canyon, with precipitous, nonmineral, and uncultivable bands wherein have accumulated placer deposits. { 'gəlch ‚klām }

Guldberg and Waage law See mass action law. { 'gült·berk and 'väg·ə ‚lo }

Guldberg-Waage group [CHEM ENG] A dimensionless number used in studying chemical reactions in blast furnaces; it is given by an equation relating volumes of reacting gases and reacting products. Symbolized N_{GW}. { 'gült·berk 'väg·ə ‚grüp }

gulf [GEOGR] **1.** An abyss or chasm. **2.** A large extension of the sea partially enclosed by land. { gəlf }

Gulfian [GEOL] A North American provincial series in Upper Cretaceous geologic time, above the Comanchean and below the Paleocene of the Tertiary. { 'gəlf·ē·ən }

Gulf Stream [OCEANOGR] A relatively warm, well-defined, swift, relatively narrow, northward-flowing ocean current which originates north of Grand Bahama Island where the Florida Current and the Antilles Current meet, and which eventually becomes the eastward-flowing North Atlantic Current. { 'gəlf ‚strēm }

Gulf Stream Countercurrent [OCEANOGR] **1.** A surface current opposite to the Gulf Stream, one current component on the Sargasso Sea side and the other component much weaker, on the inshore side. **2.** A predicted, but as yet unobserved, large current deep under the Gulf Stream but opposite to it. { 'gəlf ‚strēm 'kaunt·ər‚kə·rənt }

Gulf Stream eddy [OCEANOGR] A cutoff meander of the Gulf Stream. { 'gəlf ‚strēm 'ed·ē }

Gulf Stream front [OCEANOGR] The pronounced horizontal temperature gradient that defines a cross section of the Gulf Stream. { 'gəlf ‚strēm 'frənt }

Gulf Stream meander [OCEANOGR] One of the changeable, winding bends in the Gulf Stream; such bends intensify as the Gulf Stream merges into North Atlantic Drift and break up into detached eddies at times, at about 40°N. { 'gəlf ‚strēm mē'and·ər }

Gulf Stream system [OCEANOGR] The Florida Current, Gulf Stream, and North Atlantic Current, collectively. { 'gəlf ‚strēm 'sis·təm }

gulfweed [BOT] Brown algae of the genus *Sargassum*. { 'gəlf‚wēd }

gull [VERT ZOO] The common name for a number of long-winged swimming birds in the family Laridae having a stout build, a thick, somewhat hooked bill, a short tail, and webbed feet. { gəl }

gullet [ANAT] See esophagus. [INV ZOO] A canal between the cytostome and reservoir that functions in food intake in ciliates. { 'gəl·ət }

gull wing [AERO ENG] An airplane wing that slants upward from the fuselage for a short distance and then levels out. { 'gəl ‚wiŋ }

gull-wing door [DES ENG] A door on an automotive vehicle that is hinged at the top, opens upward, and, in the open position, resembles an airplane gull wing. { 'gəl ‚wiŋ 'dor }

gully [GEOGR] A narrow ravine. { 'gəl·ē }

gully erosion [GEOL] Erosion of soil by running water. { 'gəl·ē i‚rō·zhən }

gully-squall [METEOROL] A nautical term for a violent squall of wind from mountain ravines on the Pacific side of Central America. { 'gəl·ē ‚skwol }

gulp [COMPUT SCI] A series of bytes considered as a unit. { gəlp }

gum [MATER] A hydrophilic plant polysaccharide or derivative that swells to produce a viscous dispersion or solution when added to water. Also known as hydrocolloid. [PETRO ENG] Any one of the partially oxidized high-molecular-weight hydrocarbons that can form in gasoline stored without the addition of an oxidation inhibitor. { gəm }

gum accroides See acaroid resin. { 'gəm ə'kroi·dēz }

gum arabic [MATER] A water-soluble gum obtained from acacia trees in Africa and Australia; produced commercially as a white powder; used in the manufacture of inks and adhesives, in textile finishing, and as the principal binder in water-color and gouache. Also known as acacia gum; gum Kordofan; gum Senegal. { 'gəm 'ar·ə·bik }

gum benzoin See benzoin. { 'gəm 'ben·zə·wən }

gumbo [BOT] See okra. [GEOL] A soil that forms a sticky mud when wet. { 'gəm·bō }

gumbotil [GEOL] Deoxidized, leached clay that contains siliceous stones. { 'gəm·bō‚til }

gum dammar See dammar. { 'gəm 'da‚mär }

gum Kordofan See gum arabic. { 'gəm ‚kor·də'fan }

gumme [PATH] A mass of rubberlike necrotic tissue found in any of various organs and tissues in tertiary syphilis. { 'gəm·ə }

gummed paper [MATER] **1.** A variety of colored, patterned papers with adhesive on one side for easy application. **2.** Coated or uncoated book paper, gummed on one side, used for stickers, labels, stamps, seals, and tapes. { 'gəmd ‚pā·pər }

gummite [MINERAL] Any of various yellow, orange, red, or brown secondary minerals containing hydrous oxides of uranium, thorium, and lead. Also known as uranium ocher. { 'gə‚mīt }

gummosis [PL PATH] Production of gummy exudates in diseased plants as a result of cell degeneration. { ‚gə'mō·səs }

Gum Nebula [ASTRON] A giant nebula about 250 parsecs (5×10^{15} miles or 8×10^{15} kilometers) in diameter, with its near edge about 300 parsecs (6×10^{15} miles or 9×10^{15} kilometers) distant, which is both an old supernova remnant and an H II region. { 'gəm 'neb·yə·lə }

gum resin [MATER] A group of oleoresinous substances obtained from plants; mixtures of true gums and resins less soluble in alcohol than natural resins; examples are rubber, gutta-percha, gamboge, myrrh, and olibanum; used to make certain pharmaceuticals. { 'gəm 'rez·ən }

gum Senegal See gum arabic. { 'gəm ‚sen·ə·gəl }

gum test [CHEM ENG] A standard American Society for Testing and Materials test to determine the amount of gums in gasolines. { 'gəm ‚test }

gum thus See olibanum. { 'gəm ‚thəs }

gum vein [BOT] Local accumulation of resin occurring as a wide streak in certain hardwoods. { 'gəm ‚vān }

gun [ORD] A piece of ordnance, consisting essentially of a tube or barrel, for throwing projectiles by force, usually the force of an explosive but sometimes that of compressed gas, a spring, or so on. { gən }

gunbarrel [CHEM ENG] An atmospheric vessel used for treatment of waterflood waste water. { 'gən‚bar·əl }

gunboat [NAV ARCH] A small, moderate-speed, heavily armed vessel for general patrol and escort duty; usually unarmored and of less than 2000 tons. { 'gən‚bōt }

gun-bomb-rocket sight [ORD] A computing sight used in a fighter aircraft in which different data may be set for use in aiming gunfire, bombs, or rockets. { 'gən ‚bäm 'räk·ət ‚sīt }

gun book [ORD] Log that records the history of the operations and inspections of a particular gun. { 'gən ‚buk }

gun bore line [ORD] The extended bore axis of a gun. { 'gən ‚bor ‚līn }

gun breech [ORD] The mass of metal at the rear end of a cannon, from the front slope to the rear face, exclusive of the breech mechanism; in this metal are formed the seat for the breech mechanism, the powder chamber, and the slopes connecting the latter with the rifled portion of the bore. { 'gən ‚brēch }

gun burner [ENG] A burner which sprays liquid fuel into a furnace for combustion. { 'gən ‚bər·nər }

gun camera [OPTICS] **1.** A camera used in gunnery training that records the image of the target at which it is aimed on a strip of motion-picture film. **2.** A camera synchronized to a gun to record the results of firing, with the film revealing if the camera was correctly sighted on the target. { 'gən ‚kam·rə }

gun carriage [ORD] Mobile or fixed support for a gun; sometimes includes the elevating and traversing mechanisms. { 'gən ‚kar·ij }

gun charger [ORD] A mechanism on a gun that operates to retract the breech mechanism or bolt to the rear and to insert a charge into the chamber. { 'gən ‚chär·jər }

guncotton [MATER] Any of various nitrocellulose explosives of high nitration (13.35–13.4% nitrogen) made by treating cotton with nitric and sulfuric acids; used principally in the manufacture of single-base and double-base propellants. { 'gən¦kät·ən }

gundeck [NAV ARCH] A deck on old-time warships on which the ship's guns were carried. { 'gən¦dek }

gun deflection board See deflection board. { 'gən di'flek·shən ¦bȯrd }

gun-directing radar [ORD] Radar used to direct antiaircraft artillery or similar fire. { 'gən di¦rek·tiŋ ¦rā·där }

gun displacement [ORD] **1.** Distance from a gun to the directing point or the base piece of a battery. **2.** Movement of a gun to a new firing position. { 'gən di¦splās·mənt }

gun emplacement [ORD] Firing location of a gun together with necessary installations, such as camouflage, and ammunition supply. { 'gən em¦plās·mənt }

gunfire [ORD] Use of artillery, rifles, and small arms as distinguished from the use of bayonets, swords, torpedoes, and bombs. { 'gən ¦fīr }

gun fore-end [ORD] A wood or plastic piece that is usually designed with a semicircular groove to fit under the barrel of a gun and is attached by metal fastening devices; it is shaped to fit the hand and is used to steady the weapon during firing. { 'gən ¦fȯr‚end }

gun group [ORD] Major parts of a gun, considered as a unit distinct from its mount. { 'gən ¦grüp }

gun handguard [ORD] A wood or plastic piece that is usually designed with a semicircular groove to fit over the top of the barrel of a carbine or rifle and is attached by metal fastening devices; it is shaped to fit the hand and to protect it during firing; it excludes the gun fore-end. { ¦gən 'han‚gärd }

gun hoist [ORD] A device placed near the breech of a gun for lifting propellant and projectiles. { 'gən ‚hȯist }

gunite [CIV ENG] A mixture of cement, sand, and water that is sprayed on a surface for repairing portions of existing structures, lining reservoirs, and encasing steel for fireproofing. { 'gə‚nīt }

gun jack [ORD] A jack for forcing a tube of a gun out of battery. { 'gən jak }

gunk See rod dope. { gəŋk }

gun launcher [ORD] A gun adapted to launching guided missiles or rockets. { 'gən ¦lȯnch·ər }

gun-laying radar [ENG] Radar equipment specifically designed to determine range, azimuth, and elevation of a target and sometimes also to automatically aim and fire antiaircraft artillery or other guns. { 'gən ‚lā·iŋ ¦rā·där }

gun line [ORD] For a single aircraft gun, the axis of the bore extended; for multiple guns, it is the mean gun line. { 'gən ‚līn }

gunlock [ORD] Fastening device used to prevent a weapon from being fired. { 'gən‚läk }

gunmetal [MET] **1.** Bronze composed of copper and tin in proportions of 9:1, formerly used to make cannons. **2.** Any metal or alloy from which guns are made. **3.** Any metal or alloy treated to give the appearance of black, tarnished copper-alloy gunmetal. { 'gən‚med·əl }

gun mount [ORD] An item designed to support a gun. { 'gən ‚mau̇nt }

Gunn amplifier [ELECTR] A microwave amplifier in which a Gunn oscillator functions as a negative-resistance amplifier when placed across the terminals of a microwave source. { 'gən ¦am·plə‚fī·ər }

Gunn diode See Gunn oscillator. { 'gən ¦dī‚ōd }

Gunn effect [ELECTR] Development of a rapidly fluctuating current in a small block of a semiconductor (perhaps *n*-type gallium arsenide) when a constant voltage above a critical value is applied to contacts on opposite faces. { 'gən i‚fekt }

gunnel See gunwale. { 'gən·əl }

Gunneraceae [BOT] A family of dicotyledonous terrestrial herbs in the order Haloragales, distinguished by two to four styles, a unilocular bitegmic ovule, large inflorescences with no petals, and drupaceous fruit. { ‚gən·ə'rās·ē‚ē }

gunner's quadrant [ENG] Mechanical device having scales graduated in mils, with fine micrometer adjustments and leveling or cross-leveling vials; it is a separate, unattached instrument for hand placement on a reference surface. { ¦gən·ərz 'kwäd·rənt }

gunnery [ORD] The art or practice of using machine guns or cannon either on the ground or in the air. { 'gən·ə·rē }

Gunn oscillator [ELECTR] A microwave oscillator utilizing the Gunn effect. Also known as Gunn diode. { 'gən ¦äs·ə‚läd·ər }

gunny See burlap. { 'gən·ē }

gun parallax [ORD] The difference in azimuth between the line from the directing point to the target and the line from the gun to the target. { 'gən ¦par·ə‚laks }

gun pendulum [ENG] A device used to determine the initial velocity of a projectile fired from a gun in which the gun is mounted as a pendulum and its excursion upon firing is measured. { 'gən ¦pen·jə·ləm }

gunpowder [MATER] A black or brown explosive mixture of potassium nitrate, charcoal, and sulfur; originally, it was made in powder form, now generally in grains of various sizes. { 'gən‚pau̇d·ər }

gunpowder paper [MATER] Paper with an explosive on it that is rolled up for use in loading. { 'gən‚pau̇d·ər ‚pā·pər }

gun pressure [ORD] Pressure within a gun tube or barrel, as used in design practices. { 'gən ‚presh·ər }

gun reaction [MECH] The force exerted on the gun mount by the rearward movement of the gun resulting from the forward motion of the projectile and hot gases. Also known as recoil. { 'gən rē‚ak·shən }

gunsight [ORD] A mechanical or optical device for aiming a firearm or for placing a gun or rocket launcher in position; based on the principle that two points (the observer's eye and a suitable mark in the instrument) in fixed relation to each other may be brought in line with a third (the target). { 'gən‚sīt }

gunsight computer [ORD] The component of a computing gunsight that calculates for variables in gunnery and computes a prediction angle. { 'gən‚sīt kəm‚pyüd·ər }

gun silencer [ORD] An item specifically designed to silence the explosive report caused by the discharge of cartridges by small arms weapon; it incorporates integral chambers or baffles which allow the gases to expand gradually. Also known as silencer. { 'gən ‚sī·lən·sər }

gun slide [ORD] Portion of a gun which rests on the cradle guides. { 'gən ‚slīd }

gunstock [ORD] The wooden stock in which the barrel and mechanism of a gun are fixed. { 'gən‚stäk }

gun stoppage [ORD] A condition which prevents a gun from being fired, usually in automatic weapons as the result of a cartridge not feeding properly from the magazine to the breech of the gun. { 'gən ¦stäp·ij }

Gunter's chain [ENG] A chain 66 feet (20.1168 meters) long, consisting of 100 steel links, each 7.92 inches (20.1168 centimeters) long, joined by rings, which is used as the unit of length for surveying public lands in the United States. Also known as chain. { 'gən·tərz ¦chān }

gun turret [ORD] Dome-shaped or cylindrical armored structure containing one or more guns, located on forts, warships, airplanes, or tanks. { 'gən ‚tər·ət }

gun-type burner [ENG] An oil burner that uses a nozzle to atomize the fuel. { 'gən‚tīp ¦bər·nər }

gunwale [NAV ARCH] The upper edge of the side of a boat. Also spelled gunnel. { 'gən·əl }

Günz [GEOL] A European stage of geologic time, in the Pleistocene (above Astian of Pliocene, below Mindel); it is the first stage of glaciation of the Pleistocene in the Alps. { gints }

Günzberg reagent [ANALY CHEM] A solution of 2 grams of vanillin and 4 grams of phloroglucinol in 80 milliliters of 95% alcohol; used as a test reagent for determining free hydrochloric acid in gastric juice. { 'gints‚berk rē‚ā·jənt }

Günz-Mindel [GEOL] The first interglacial stage of the Pleistocene in the Alps, between Günz and Mindel glacial stages. { 'gints 'mind·əl }

Gurevich effect [SOLID STATE] An effect observed in electric conductors in which phonon-electron collisions are important, in the presence of a temperature gradient, in which phonons carrying a thermal current tend to drag the electrons with them from hot to cold. Also known as phonon-drag effect. { 'gür·ə·vich i‚fekt }

Gurney formulas [ORD] A series of formulas proposed by R.W. Gurney for determining initial fragment velocity as a function of the type of explosive and the ratio of explosive charge to metal weight; each formula corresponds to a particular shape of container. { 'gər·nē ‚fȯr·myə·ləz }

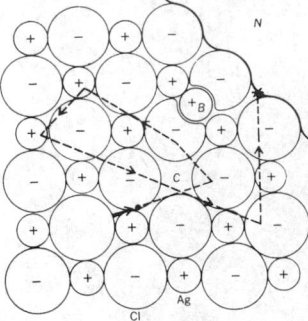

Schematic representation of Gurney-Mott theory. Broken line shows path of photoelectron which is eventually trapped in vicinity of a silver speck *N*. The interstitial silver ion *B* has come up to neutralize the charge of the electron and thus to add to the silver speck.

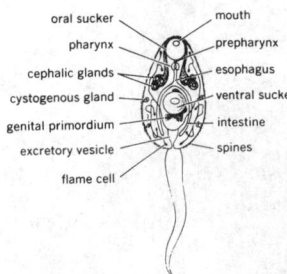

Anatomy of gymnocephalous cercaria. (*From R. M. Cable, An Illustrated Laboratory Manual of Parasitology, Burgess, 1958*)

Gurney-Mott theory [CHEM] A theory of the photographic process that proposes a two-stage mechanism; in the first stage, a light quantum is absorbed at a point within the silver halide gelatin, releasing a mobile electron and a positive hole; these mobile defects diffuse to trapping sites (sensitivity centers) within the volume or on the surface of the grain; in the second stage, trapped (negatively charged) electron is neutralized by an interstitial (positively charged) silver ion, which combines with the electron to form a silver atom; the silver atom is capable of trapping a second electron, after which the process repeats itself, causing the silver speck to grow. { 'gər·nē 'mät ,thē·ə·rē }

gusset [CIV ENG] A plate that is used to strengthen truss joints. [MIN ENG] A V-shaped cut in the face of a heading. { 'gəs·ət }

gusset plate [CIV ENG] A rectangular or triangular steel plate that connects members of a truss. { 'gəs·ət ,plāt }

gust [METEOROL] A sudden, brief increase in the speed of the wind; it is of a more transient character than a squall and is followed by a lull or slackening in the wind speed. { gəst }

gustation [PHYSIO] The act or the sensation of tasting. { gə'stā·shən }

gustatoreceptor [ANAT] A taste bud. [PHYSIO] Any sense organ that functions as a receptor for the sense of taste. { ¦gəs·tə·tō ri'sep·tər }

gust-gradient distance [AERO ENG] The horizontal distance along an aircraft flight path from the "edge" of a gust to the point at which the gust reaches its maximum speed. { 'gəst¦grād·ē·ənt ,dis·təns }

gustiness [METEOROL] A quality of airflow characterized by gusts. { 'gəs·tē·nəs }

gustiness components [METEOROL] **1.** The ratios, to the mean wind speed, of the average magnitudes of the component fluctuations of the wind along three mutually perpendicular axes. **2.** The ratios of the root-mean-squares of the eddy velocities to the mean wind speed. Also known as intensity of turbulence. { 'gəs·tē·nəs kəm,pō·nəns }

gustiness factor [METEOROL] A measure of the intensity of wind gusts; it is the ratio of the total range of wind speeds between gusts and the intermediate periods of lighter wind to the mean wind speed, averaged over both gusts and lulls. { 'gəs·tē·nəs ,fak·tər }

gust load [MECH] The wind load on an antenna due to gusts. { 'gəst ,lōd }

gustsonde [ENG] An instrument dropped from high altitude by a stable parachute, to measure the vertical component of turbulence aloft; consists of an accelerometer and radio telemetering equipment. { 'gəst,sänd }

gust tunnel [AERO ENG] A type of wind tunnel that has an enclosed space and is used to test the effect of gusts on an airplane model in free flight to determine how atmospheric gusts affect the flight of an airplane. { 'gəst ,tən·əl }

gut [ANAT] The intestine. [EMBRYO] The embryonic digestive tube. [GEOL] **1.** A narrow water passage such as a strait. **2.** A channel deeper than the surrounding water; generally formed by water in motion. { gət }

Guthrie test [PATH] A screening test for the detection of phenylketonuria in which the inhibition of growth of a strain of *Bacillus subtilis* by a phenylalanine analog is reversed by L-phenylalanine, as found in elevated concentration in the plasma of patients with phenylketonuria. { 'gəth·rē ,test }

guti weather [METEOROL] In Rhodesia, a dense stratocumulus overcast, frequently with drizzle, occurring mainly in early summer, and associated with easterly winds that invade the interior, bringing in cool and stable maritime air when an anticyclone moves eastward south of Africa. { 'güd·ē ,weth·ər }

Gutschoven's curve See kappa curve. { 'güt,shō·fənz ,kərv }

gutta-balata See balata. { ¦gəd·ə·bə'läd·ə }

gutta-percha [MATER] A leathery, thermoplastic substance consisting of gutta hydrocarbon with some resin obtained from the latex of certain Malaysian sapotaceous trees; used as insulation for submarine cables, and in golf balls and other products. { ¦gəd·ə'pər·chə }

guttation [BOT] The discharge of water from a plant surface, especially from a hydathode. { gə'tā·shən }

gutter [BUILD] A trough along the edge of the eaves of a building to carry off rainwater. [CIV ENG] A shallow trench provided beside a canal, bordering a highway, or elsewhere, for surface drainage. [GRAPHICS] In the pages of a book,

the unprinted space or inner margin between the printed area and the binding. [MET] A groove along the periphery of a die impression to allow for excess flash during forging. [MIN ENG] A drainage trench cut along the side of a mine shaft to conduct the water back into a lodge or sump. { 'gəd·ər }

guttering [ENG] A process of quarrying stone in which channels, several inches wide, are cut by hand tools, and the stone block is detached from the bed by pinch bars. [MIN ENG] The process of cutting gutters in a mine shaft. { 'gəd·ə·riŋ }

Guttiferae [BOT] A family of dicotyledonous plants in the order Theales characterized by extipulate leaves and conspicuous secretory canals or cavities in all organs. { gə'tif·ə,rē }

guttra [METEOROL] In Iran, sudden squalls in May. { 'gü·trə }

Guttulinaceae [MICROBIO] A family of microorganisms in the Acrasiales characterized by simple fruiting structures with only slightly differentiated component cells containing little or no cellulose. { ,gùd·əl·ə'nās·ē,ē }

Gutzeit test [ANALY CHEM] A test for arsenic; zinc and dilute sulfuric acid are added to the substance, which is then covered with a filter paper moistened with mercuric chloride solution; a yellow spot forms on the paper if arsenic is in the sample. { 'güt,sīt ,test }

Gutzkow's process [MET] A modification of the sulfuric acid parting process for bullion containing large amounts of copper; a large excess of acid is used, and the silver sulfate is then reduced with charcoal, or, in the original process, ferrous sulfate. { 'güts,kōz ,präs·əs }

guxen [METEOROL] A cold wind of the Alps in Switzerland. { 'gúk·sən }

guy [ENG] A rope or wire securing a pole, derrick, or similar temporary structure in a vertical position. { gī }

guy derrick [MECH ENG] A derrick having a vertical pole supported by guy ropes to which a boom is attached by rope or cable suspension at the top and by a pivot at the foot. { 'gī ,der·ik }

guyot [GEOL] A seamount, usually deeper than 100 fathoms (180 meters), having a smooth platform top. Also known as tablemount. { gē'ō }

guzzle [METEOROL] In the Shetland Islands, an angry blast of wind, dry and parching. { 'gəz·əl }

G value [NUCLEO] The number of molecules produced or destroyed for each 100 electronvolts absorbed by a substance from ionizing radiation. Also known as G factor. { 'jē ,val·yü }

gvw See gross vehicle weight.

GW See gigawatt.

Gy See gray.

Gymnarchidae [VERT ZOO] A monotypic family of electrogenic fishes in the order Osteoglossiformes in which individuals lack pelvic, anal, and caudal fins. { jim'närk·ə,dē }

Gymnarthridae [PALEON] A family of extinct lepospondylous amphibians that have a skull with only a single bone representing the tabular and temporal elements of the primitive skull roof. { jim'närth·rə,dē }

gymnite See deweylite. { 'jim,nīt }

Gymnoascaceae [MYCOL] A family of ascomycetous fungi in the order Eurotiales including dermatophytes and forms that grow on dung, soil, and feathers. { ¦jim·nō·ə'skās·ē,ē }

Gymnoblastea [INV ZOO] A suborder of cnidarians in the order Hydroida comprising hydroids without protective cups around the hydranths and gonozooids. { ¦jim·nə'blas·tē·ə }

gymnoblastic [INV ZOO] Having naked medusa buds, referring to anthomedusan hydroids. { ¦jim·nə¦bla·stik }

gymnocarpous [BOT] Having the hymenium uncovered on the surface of the thallus or fruiting body of lichens or fungi. { ¦jim·nə¦kär·pəs }

gymnocephalous cercaria [INV ZOO] A type of digenetic trematode larva. { ¦jim·nə¦sef·ə·ləs sər'kar·ē·ə }

Gymnocerata [INV ZOO] An equivalent name for Hydrocorisae. { ¦jim·nō'ser·əd·ə }

Gymnocodiaceae [PALEOBOT] A family of fossil red algae. { ¦jim·nō,kō·dē'as·ē,ē }

Gymnodinia [INV ZOO] A suborder of flagellate protozoans in the order Dinoflagellida that are naked or have thin pellicles. { ¦jim·nə¦din·ē·ə }

gymnogynous [BOT] Having a naked ovary. { jim'näj·ə·nəs }

Gymnolaemata [INV ZOO] A class of ectoproct bryozoans

possessing lophophores which are circular in basal outline and zooecia which are short, wide, and vaselike or boxlike. { ¦jim·nə'lē·məd·ə }

Gymnonoti [VERT ZOO] An equivalent name for Cypriniformes. { ¦jim·nə'nōd·ī }

Gymnophiona [VERT ZOO] An equivalent name for Apoda. { ¦jim·nə'fī·ə·nə }

gymnoplast [BOT] In angiosperms, a cell without a cell wall. { 'jim·nə‚plast }

Gymnopleura [INV ZOO] A subsection of brachyuran decapod crustaceans including the primitive burrowing crabs with trapezoidal or elongate carapaces, the first pereiopods subchelate, and some or all of the remaining pereiopods flattened and expanded. { ¦jim·nə'plůr·ə }

gymnosperm [BOT] The common name for members of the division Pinophyta; seed plants having naked ovules at the time of pollination. { 'jim·nə‚spərm }

Gymnospermae [BOT] The equivalent name for Pinophyta. { ¦jim·nə'spər·mē }

Gymnosporangium juniperi-virginianae [MYCOL] A heteroecious fungal pathogen that is the cause of apple-cedar rust. { ¦jim·nō·spə¦ran·jē·əm ¦jü·ni‚per·ē ‚vər‚jin·ē'a‚nī }

Gymnostomatida [INV ZOO] An order of the protozoan subclass Holotrichia containing the most primitive ciliates, distinguished by the lack of ciliature in the oral area. { ¦jim·nə'stō·məd·ə }

Gymnotidae [VERT ZOO] The single family of the suborder Gymnotoidei; eel-shaped fishes having numerous vertebrae, and anus located far forward, and lacking pelvic and developed dorsal fins. { jim'näd·ə‚dē }

Gymnotoidei [VERT ZOO] A monofamilial suborder of actinopterygian fishes in the order Cypriniformes. { ¦jim·nə'tóid‚ē‚ī }

gynaecandrous [BOT] Having staminate and pistillate flowers on the same spike. { ¦gī·nə¦kan·drəs }

gynander [BIOL] A mosaic individual composed of diploid female portions derived from both parents and haploid male portions derived from an extra egg or sperm nucleus. { gīn'an·dər }

gynandromorph [BIOL] An individual of a dioecious species made up of a mosaic of tissues of male and female genotypes. { gī'nan·drə‚mòrf }

gynandry [PHYSIO] A form of pseudohermaphroditism in which the external sexual characteristics are partly or wholly of the male aspect, but internal female genitalia are present. Also known as female pseudohermaphroditism; virilism. { gīn'an·drē }

gynecology [MED] The branch of medicine dealing with diseases of women, particularly those affecting the sex organs. { ‚gīn·ə'käl·ə·jē }

gynecomastia [MED] Abnormal enlargement of the mammary glands in the male. { ‚gīn·ə·kō'mas·tē·ə }

gynephobia [PSYCH] An abnormal fear of women. { ‚gīn·ə'fō·bē·ə }

gynobase [BOT] A gynoecium-bearing elongation of the receptacle in certain plants. { 'gīn·ō‚bās }

gynodioecious [BOT] Dioecious but with some perfect flowers on a plant bearing pistillate flowers. { ¦gīn·ō·dī'ē·shəs }

gynoecious [BOT] Pertaining to plants that have only female flowers. { gī'nē·shəs }

gynoecium [BOT] The aggregate of carpels in a flower. { gī'nē·sē·əm }

gynogenesis [EMBRYO] Development of a fertilized egg through the action of the egg nucleus, without participation of the sperm nucleus. { ‚gīn·ō'jen·ə·səs }

gynogenetic diploid [GEN] A diploid organism having two maternal haploid chromosome sets; lethal in mammals because of imprinting. { gī·nō·jə¦ned·ik 'dip‚lóid }

gynomerogony [EMBRYO] Development of a fragment of a fertilized egg containing the haploid egg nucleus. { ¦gīn·ō·mə'rag·ə·nē }

gynomonoecious [BOT] Having complete and pistillate flowers on the same plant. { ¦gīn·ō·mä'nē·shəs }

gynophore [BOT] **1.** A stalk that bears the gynoecium. **2.** An elongation of the receptacle between pistil and stamens. { 'gīn·ə‚fòr }

gynostemium [BOT] The column composed of the united gynoecia and androecium. { ¦gīn·ō'stē·mē·əm }

gyotaku [GRAPHICS] The art of Japanese fish printing, a modern application of stone-rubbing technique. { 'gyō·tä‚kù }

gypcrete [GEOL] A type of duricrust composed of hydrous calcium sulfate. { 'jip‚krēt }

gypsite [GEOL] A variety of gypsum consisting of dirt and sand; found as an efflorescent deposit in arid regions, overlying gypsum. Also known as gypsum earth. { 'jip‚sīt }

gypsophilous [ECOL] Flourishing on a gypsum-rich substratum. { jip'säf·ə·ləs }

gypsum [MINERAL] $CaSO_4 \cdot 2H_2O$ A mineral, the commonest sulfate mineral; crystals are monoclinic, clear, white to gray, yellowish, or brownish in color, with well-developed cleavages; luster is subvitreous to pearly, hardness is 2 on Mohs scale, and specific gravity is 2.3; it is calcined at 190–200°C to produce plaster of paris. { 'jip·səm }

gypsum board [MATER] A plaster board covered with paper. { 'jip·səm ‚bórd }

gypsum cement See gypsum plaster. { 'jip·səm si¦ment }

gypsum earth See gypsite. { 'jip·səm ¦ərth }

gypsum lath [MATER] Lath consisting of a core of set gypsum surfaced with paper that is treated to receive plaster. { 'jip·səm ¦lath }

gypsum plank [MATER] A structural precast unit consisting of gypsum core reinforced with welded galvanized steel mesh and bounded on all four edges with a tongue-and-groove steel form; used as the roof deck of steel-frame buildings, and sometimes for the floor system. { 'jip·səm ¦plaŋk }

gypsum plaster [MATER] Plaster made principally from gypsum. Also known as gypsum cement. { 'jip·səm ¦plas·tər }

gypsum wallboard [MATER] Wallboard consisting of a core of set gypsum surfaced with paper or other fibrous material suitable to receive paint or paper. { 'jip·səm ¦wól‚bòrd }

gypsy [NAV ARCH] A small auxiliary drum fitted to one or both ends of a winch or windlass. { 'jip·sē }

gypsy head [NAV ARCH] A small auxiliary drum at the end of a windlass or capstan, used to handle lines. Also spelled gipsy head. { 'jip·sē ‚hed }

gypsy moth [INV ZOO] *Porthetria dispar.* A large lepidopteran insect of the family Lymantriidae that was accidentally imported into New England from Europe in the late 19th century; larvae are economically important as pests of deciduous trees. { 'jip·sē ¦móth }

Gyracanthididae [PALEON] A family of extinct acanthodian fishes in the suborder Diplacanthoidei. { ¦jī·rə‚kan'thid·ə‚dē }

gyration tensor [SOLID STATE] A tensor characteristic of an optically active crystal, whose product with a unit vector in the direction of propagation of a light ray gives the gyration vector. { ji'rā·shən ¦ten·sər }

gyration vector [OPTICS] For light propagating in an optically active medium, a vector whose cross product with the time derivative of the electric displacement vector gives a negative contribution to the electric field. { ji'rā·shən ¦vek·tər }

gyrator [ELECTROMAG] A waveguide component that uses a ferrite section to give zero phase shift for one direction of propagation and 180° phase shift for the other direction; in other words, it causes a reversal of signal polarity for one direction of propagation but not for the other direction. Also known as microwave gyrator. { 'jī‚rād·ər }

gyrator filter [ELECTR] A highly selective active filter that uses a gyrator which is terminated in a capacitor so as to have an inductive input impedance. { 'jī‚rād·ər ‚fil·tər }

gyratory breaker See gyratory crusher. { 'jī·rə‚tòr·ē 'brāk·ər }

gyratory crusher [MECH ENG] A primary breaking machine in the form of two cones, an outer fixed cone and a solid inner erect cone mounted on an eccentric bearing. Also known as gyratory breaker. { 'jī·rə‚tòr·ē 'krəsh·ər }

gyratory screen [MECH ENG] Boxlike machine with a series of horizontal screens nested in a vertical stack with downward-decreasing mesh-opening sizes; near-circular motion causes undersized material to sift down through each screen in succession. { 'jī·rə‚tòr·ē 'skrēn }

gyre [OCEANOGR] A closed circulatory system that is larger than a whirlpool or eddy. { jīr }

Gyrinidae [INV ZOO] The whirligig beetles, a family of large coleopteran insects in the suborder Adephaga. { jə'rin·ə‚dē }

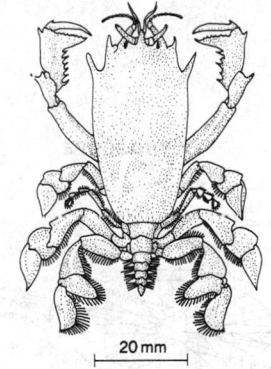

GYMNOPLEURA

20 mm

Gymnopleuran crab, *Raninoides louisianensis.*

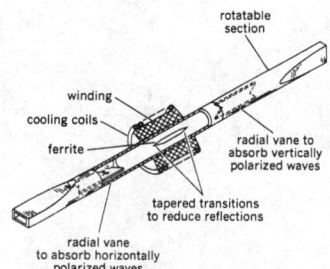

GYRATOR

rotatable section

winding

cooling coils

ferrite

radial vane to absorb vertically polarized waves

tapered transitions to reduce reflections

radial vane to absorb horizontally polarized waves

A practical nonreciprocal gyrator.

Gyrinocheilidae [VERT ZOO] A monogeneric family of cypriniform fishes in the suborder Cyprinoidei. { ¦jir·ə·nō·kī'lī·ə‚dē }

gyro *See* gyrocompass; gyroscope. { 'jī·rō }

gyrocompass [NAV] A north-seeking form of gyroscope used as a vehicle's or craft's directional reference. Also known as gyro; gyroscopic compass. { 'jī·rō‚käm·pəs }

gyrocompass alignment [NAV] Automatic alignment of a gyrocompass with the celestial meridian. { 'jī·rō‚käm·pəs ə'līn·mənt }

gyrocompassing [NAV] In inertial navigation equipment, a process of self-alignment in azimuth accomplished without any external equipment, consisting basically of using the inertial platform's accelerometers to slave it to the local gravity vector, and employing one of the gyros to seek true north by sensing the rotation of the earth in the manner of a gyro compass. { 'jī·rō‚käm·pəs·iŋ }

Gyrocotylidea [INV ZOO] An order of tapeworms of the subclass Cestodaria; species are intestinal parasites of chimaeroid fishes and are characterized by an anterior eversible proboscis and a posterior ruffled adhesive organ. { ¦jī·rō·kä'til·əd·ē }

Gyrocotyloidea [INV ZOO] A class of trematode worms according to some systems of classification. { ¦jī·rō‚käd·əl'óid·ē·ə }

Gyrodactyloidea [INV ZOO] A superfamily of ectoparasitic trematodes in the subclass Monogenea; the posterior holdfast is solid and is armed with central anchors and marginal hooks. { ¦jī·rō‚dak·tə'lóid·ē·ə }

gyrodynamics [MECH] The study of rotating bodies, especially those subject to precession. { ¦jī·rō·dī'nam·iks }

gyro error [NAV] The error in the reading of the gyro compass, expressed in degrees east or west to indicate the direction in which the axis of the compass is offset from the north. { 'jī·rō ‚er·ər }

gyro flux-gate compass [NAV] A compass in which a flux gate, horizontally stabilized by a gyroscope, senses the horizontal component of the earth's magnetic field; being fixed with respect to the aircraft, the compass reacts to each change in heading with a change in current. { 'jī·rō 'fləks ‚gāt ‚käm·pəs }

gyrofrequency *See* cyclotron frequency. { 'jī·rō‚frē·kwən‚sē }

gyrogonite [PALEON] A minute, ovoid body that is the residue of the calcareous encrustation about the female sex organs of a fossil stonewort. { 'jī·räg·ə‚nīt }

gyro graph [NAV] A graph for recording and indicating drift of a directional gyro; usually a graph of drift from the desired heading versus time. { 'jī·rō ‚graf }

gyroklystron [ELECTR] A microwave tube which, like the gyrotron, is based on cyclotron resonance coupling between microwave fields and an electron beam in vacuum, but which employs two or more cavities, and in which electrons give up their energy to an alternating electric field in a circuit separate from the one that supports the field that bunches the electrons. { jī·rə'klī‚strän }

gyro log [NAV] **1.** A written record of directional gyro drift. **2.** A written record of the performance of a gyro compass. { 'jī·rō ‚läg }

gyromagnetic compass [NAV] A magnetic compass in which gyroscopic stabilization is used to indicate direction. { ¦jī·rō·mag'ned·ik ‚käm·pəs }

gyromagnetic coupler [ELECTR] A coupler in which a single-crystal yig (yttrium iron garnet) resonator provides coupling at the required low signal levels between two crossed stripline resonant circuits. { ¦jī·rō·mag'ned·ik 'kəp·lər }

gyromagnetic effect [ELECTROMAG] The rotation induced in a body by a change in its magnetization, or the magnetization resulting from a rotation. { ¦jī·rō·mag'ned·ik i'fekt }

gyromagnetic radius *See* Larmor radius. { ¦jī·rō·mag'ned·ik 'rād·ē·əs }

gyromagnetic ratio [PHYS] **1.** The ratio of the magnetic dipole moment to the angular momentum for a classical, atomic,

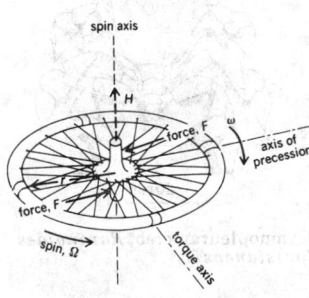

GYROSCOPE

Illustration of gyroscope principle. Bicycle wheel with high spin velocity Ω has angular momentum $H = mr^2 \Omega$, where m and r are mass and radius of wheel. Torque T resulting from force F produces precession with small angular velocity $\omega = T/H$ about axis perpendicular to both spin axis and torque axis.

or nuclear system. **2.** Occasionally, the reciprocal of the quantity in the first definition. { ¦jī·rō·mag'ned·ik 'rā·shō }

gyromagnetics [ELECTROMAG] The study of the relation between the angular momentum and the magnetization of a substance as exhibited in the gyromagnetic effect. { ¦jī·rō·mag'ned·iks }

gyropendulum [MECH ENG] A gravity pendulum attached to a rapidly spinning gyro wheel. { ¦jī·rō¦pen·jə·ləm }

Gyropidae [INV ZOO] A family of biting lice in the order Mallophaga; members are ectoparasites of South American rodents. { jə'räp·ə‚dē }

gyroplane [AERO ENG] A rotorcraft whose rotors are not power-driven. { 'jī·rō‚plān }

gyrorepeater [ENG] That part of a remote indicating gyro compass system which repeats at a distance the indications of the master gyro compass system. { 'jī·rō·ri'pēd·ər }

gyroscope [ENG] An instrument that maintains an angular reference direction by virtue of a rapidly spinning, heavy mass; all applications of the gyroscope depend on a special form of Newton's second law, which states that a massive, rapidly spinning body rigidly resists being disturbed and tends to react to a disturbing torque by precessing (rotating slowly) in a direction at right angles to the direction of torque. Also known as gyro. { 'jī·rə‚skōp }

gyroscopic-clinograph method [ENG] A method used in borehole surveying which measures time, temperature, and temperature on 16-millimeter film while a gyroscope maintains the casing on a fixed bearing. { jī·rə'skäp·ik 'klīn·ə‚graf ‚meth·əd }

gyroscopic compass *See* gyrocompass. { jī·rə'skäp·ik 'käm·pəs }

gyroscopic/Coriolis-type mass flowmeter [ENG] An instrument consisting of a C-shaped pipe and a T-shaped leaf-spring tuning fork which is excited by an electromagnetic forcer, resulting in an angular deflection of the pipe which is directly proportional to the mass-flow rate within the pipe. { jī·rə'skäp·ik ‚kór·ē'ō·ləs ¦tīp ¦mas 'flō‚mēd·ər }

gyroscopic couple [MECH ENG] The turning moment which opposes any change of the inclination of the axis of rotation of a gyroscope. { jī·rə'skäp·ik 'kəp·əl }

gyroscopic drift [NAV] The horizontal rotation of the spin axis of a gyroscope about the drift axis. { jī·rə'skäp·ik 'drift }

gyroscopic horizon [NAV] A gyroscopic instrument that indicates the lateral and longitudinal attitude of an aircraft by simulating the natural horizon. { jī·rə'skäp·ik hə'rīz·ən }

gyroscopic mass flowmeter [ENG] An instrument in which the torque on a rotating pipe of suitable shape, through which a fluid is made to flow, is measured to determine the mass flow through the pipe. { jī·rə'skäp·ik ¦mas 'flō‚mēd·ər }

gyroscopic precession [MECH] The turning of the axis of spin of a gyroscope as a result of an external torque acting on the gyroscope; the axis always turns toward the direction of the torque. { jī·rə'skäp·ik prē'sesh·ən }

gyroscopics [MECH] The branch of mechanics concerned with gyroscopes and their use in stabilization and control of ships, aircraft, projectiles, and other objects. { jī·rə'skäp·iks }

gyrosextant [NAV] A sextant provided with a gyroscope for the purpose of determining the horizontal. { ¦jī·rō'sek·stənt }

gyrostabilizer [ENG] A gyroscope used to stabilize ships and airplanes. { ¦jī·rō'stā·bə‚līz·ər }

gyrotron [ELECTR] **1.** A device that detects motion of a system by measuring the phase distortion that occurs when a vibrating tuning fork is moved. **2.** A type of microwave tube in which microwave amplification or generation results from cyclotron-resonance coupling between microwave fields and an electron beam in vacuum. Also known as cyclotron-resonance maser. { 'jī·rə‚trän }

gyro wheel [MECH ENG] The rapidly spinning wheel in a gyroscope, which resists being disturbed. { 'jī·rō ‚wēl }

gyrus [ANAT] One of the convolutions (ridges) on the surface of the cerebrum. { 'jī·rəs }

gyttja [GEOL] A fresh-water anaerobic mud containing an abundance of organic matter; capable of supporting aerobic life. { 'yi‚chä }

H

h *See* hour; Planck's constant.

H *See* henry; hydrogen.

Ĥ [MED] A symbol used in electrocardiography for the longitudinal axis of the heart as projected on the frontal plane.

ha *See* hectare.

Ha *See* hahnium.

HAA *See* height above airport.

Haanel depth rule [GEOPHYS] A rule for estimating the depth of a magnetic body, provided the body may be considered magnetically equivalent to a single pole; the depth of the pole is then equal to the horizontal distance from the point of maximum vertical magnetic intensity to the points where the intensity is one-third of the maximum value. { 'hän·əl 'depth ,rül }

haar [METEOROL] A wet sea fog or very fine drizzle which drifts in from the sea in coastal districts of eastern Scotland and northeastern England; it occurs most frequently in summer. { här }

Haar measure [MATH] A measure on the Borel subsets of a locally compact topological group whose value on a Borel subset *U* is unchanged if every member of *U* is multiplied by a fixed element of the group. { 'här ,mezh·ər }

Haas effect [ACOUS] A phenomenon whereby sound produced by the second of two loudspeakers cannot be detected if it is delayed relative to sound from the other loudspeaker by a time interval between 1 and 30 milliseconds and has an intensity less that 10 decibels above that of the primary sound. { 'häs ,fekt }

Haase system [MIN ENG] Shaft sinking in loose ground or quicksand by piles in the form of iron tubes connected by webs; downward movement is facilitated by water forced down the tubes to wash away loose material beneath their points. { 'hä·sə ,sis·təm }

habenula [ANAT] **1.** Stalk of the pineal body. **2.** A ribbonlike structure. { hə'ben·yə·lə }

habenular commissure [ANAT] The commissure connecting the habenular ganglia in the roof of the diencephalon. { hə'ben·yə·lər 'käm·ə,shur }

habenular ganglia [NEUROSCI] Olfactory centers anterior to the pineal body. { hə'ben·yə·lər 'gaŋ·glē·ə }

habenular nucleus [NEUROSCI] Either of a pair of nerve centers that are located at the base of the pineal body on either side and serve as an olfactory correlation center. { hə'ben·yə·lər 'nü·klē·əs }

Haber-Bosch process [CHEM ENG] Early nitrogen-fixation process for production of ammonia from hydrogen and nitrogen, catalyzed by iron; now replaced by more efficient ammonia synthesis processes. Also known as Haber process. { 'hä·bər ,bȯsh ,prä·səs }

Haber process *See* Haber-Bosch process. { 'hä·bər ,prä·səs }

habit [CRYSTAL] *See* crystal habit. [PSYCH] A repetitive behavior pattern. { 'hab·ət }

habitat [ECOL] The part of the physical environment in which a plant or animal lives. { 'hab·ə,tat }

habit plane [CRYSTAL] The crystallographic plane or system of planes along which certain phenomena such as twinning occur. { 'hab·ət ,plān }

habitual abortion [MED] Recurring, successive spontaneous abortion. { hə'bich·ə·wəl ə,bȯr·shən }

habituation [MED] **1.** A condition of tolerance to the effects of a drug or a poison, acquired by its continued use; marked by a psychic craving for it when the drug is withdrawn.

2. Mild drug addiction in which withdrawal symptoms are not severe. { hə,bich·ə'wā·shən }

habitus [BIOL] General appearance or constitution of an organism. { 'hab·ə·təs }

haboob [METEOROL] A strong wind and sandstorm or duststorm in the northern and central Sudan, especially around Khartum, where the average number is about 24 haboobs a year. { hə'büb }

habutai [TEXT] A smooth, lightweight silk fabric, originally hand-woven in Japan. { 'ha·bə·tī }

HACCP *See* hazard analysis critical control points.

hachure [MAP] A short line used to denote slopes of the ground, as on maps. { ha'shur }

H acid [ORG CHEM] $H_2NC_{10}H_4(OH)(SO_3H)_2$ A gray powder or crystalline substance that is soluble in water, ether, and alcohol; used as a dye intermediate. { 'ach 'as·əd }

hackberry [BOT] **1.** *Celtis occidentalis.* A tree of the eastern United States characterized by corky or warty bark, and by alternate, long-pointed serrate leaves unequal at the base; produces small, sweet, edible drupaceous fruit. **2.** Any of several other trees of the genus *Celtis.* { 'hak,ber·ē }

hacker [COMPUT SCI] A person who uses a computer system without a specific, constructive purpose or without proper authorization. { 'hak·ər }

hacking [COMPUT SCI] Use of a computer system without a specific, constructive purpose, or without proper authorization. [ENG] The technique of roughening a surface by striking it with a tool. [LAP] A system of grooves in a lap that hold diamond powder for cutting and polishing gems. { 'hak·iŋ }

hacking knife [ENG] A tool for removing old putty from a window frame prior to reglazing. Also known as hacking-out tool. { 'hak·iŋ ,nīf }

hacking-out tool *See* hacking knife. { 'hak·iŋ ,aut ,tül }

hackle [TEXT] A board studded with long, thin wire brushes and used for hackling. { 'hak·əl }

hackling [TEXT] Passing a comb through flax to clean and straighten the fibers. { 'hak·liŋ }

hackly fracture [MINERAL] A break in a mineral characterized by jagged irregular surfaces with sharp edges. { 'hak·lē 'frak·chər }

hackmanite [MINERAL] A mineral of the sodalite family containing a small amount of sulfur; fluoresces orange or red in ultraviolet light. { 'hak·mə,nīt }

hackmarack *See* tamarack. { 'hak·mə,rak }

hacksaw [ENG] A hand or power tool consisting of a fine-toothed blade held in tension in a bow-shaped frame; used for cutting metal, wood, and other hard materials. { 'hak,sȯ }

hack watch [HOROL] A watch having a device for stopping the balance so that the hour, minute, and second hands may be reset precisely; used for timing observations of celestial bodies and regulating ship clocks. { 'hak ,wäch }

hadal [OCEANOGR] Pertaining to the environment of the ocean trenches, over 4 miles (6.5 kilometers) in depth. { 'häd·əl }

Hadamard's conjecture [MATH] The conjecture that any partial differential equation that is essentially different from the wave equation fails to satisfy Huygens' principle. { 'had·ə,märd kən'jek·chər }

Hadamard's inequality [MATH] An inequality that gives an upper bound for the square of the absolute value of the determinant of a matrix in terms of the squares of the matrix entries; the upper bound is the product, over the rows of the matrix,

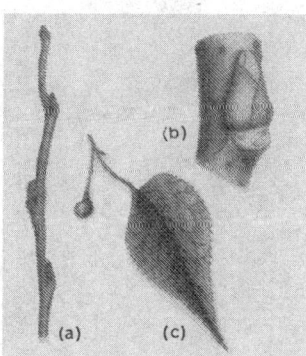

HACKBERRY

Common hackberry (*Celtis occidentalis*). (*a*) Twig. (*b*) Lateral bud. (*c*) Leaf.

of the sum of the squares of the absolute values of the entries in a row. { 'had·ə,märdz ,in·ə¦kwäl·əd·ē }

Hadamard's three-circle theorem [MATH] The theorem that if the complex function $f(z)$ is analytic in the ring $a < |z| < b$, and if $m(r)$ denotes the maximum value of $|f(z)|$ on the circle $|z| = r$ with $a < r < b$, then log $m(r)$ is a convex function of log r. { 'had·ə,märdz ¦thrē ¦sər·kəl 'thir·əm }

Hadar See β Centauri. { ha'där }

haddock [VERT ZOO] *Melanogrammus aeglefinus.* A fish of the family Gadidae characterized by a black lateral line and a dark spot behind the gills. { 'had·ək }

hade [GEOL] **1.** The angle of inclination of a fault as measured from the vertical. **2.** The inclination angle of a vein or lode. { hād }

Hadean [GEOL] The period (more than 3800 million years ago) extending for several hundred millions of years from the end of the accretion of the earth to the formation of the oldest recognized rocks. { 'hā·dē·ən }

Hadfield manganese steel [MET] Austenitic steel (face-centered cubic structure) containing 11–14% manganese; resistant to shock and wear. Also known as austenitic manganese steel; manganese steel. { 'had,fēld ¦maŋ·gə,nēs 'stēl }

Hadley cell [METEOROL] A direct, thermally driven, and zonally symmetric circulation first proposed by George Hadley as an explanation for the trade winds; it consists of the equatorward movement of the trade winds between about latitude 30° and the equator in each hemisphere, with rising wind components near the equator, poleward flow aloft, and finally descending components at about latitude 30° again. { 'had·lē ,sel }

Hadromerina [INV ZOO] A suborder of sponges in the class Clavaxinellida having monactinal megascleres, usually with a terminal knob at one end. { ,had·rō·mə'rī·nə }

hadromycosis [PL PATH] Any plant disease resulting from infestation of the xylem by a fungus. { ¦had·rō,mī'kō·səs }

hadron [PART PHYS] An elementary particle which has strong interactions. { 'had,rän }

hadron era [ASTRON] The period in the early universe when the physical forces (gravity, weak, strong, and electromagnetic) diverged from a condition of rough equivalence to increasing disparity, roughly the time between about 10^{-43} and 10^{-4} second after the big bang. { 'had,rän ,ir·ə }

hadronic atom [ATOM PHYS] An atom consisting of a negatively charged, strongly interacting particle orbiting around an ordinary nucleus. { ha'drän·ik 'ad·əm }

hadrosaur [PALEON] A duck-billed dinosaur. { 'had·rə,sór }

Hadsel mill [MIN ENG] An early autogenous grinding mill in which comminution was caused by the fall of ore on ore that was rotating in a large-diameter horizontal cylinder. { 'hads·əl ,mil }

Haeckel's law See recapitulation theory. { 'hek·əlz ,lo }

haematodocha [INV ZOO] A sac in the palpus of male spiders that fills with hemolymph and becomes distended during pairing. { ,hē·məd·ō'dō·kə }

Haemobartonella [MICROBIO] A genus of bacteria in the family Anaplasmataceae; parasites in or on red blood cells of many vertebrates. { ¦hē·mō,bart·ən'el·ə }

Haemophilus [MICROBIO] A genus of gram-negative coccobacilli or rod-shaped bacteria of uncertain affiliation; cells may form threads and filaments and are aerobic or facultatively anaerobic; strictly blood parasites. { hē'mä·fə·ləs }

Haemophilus aegyptius [MICROBIO] A pathogenic bacterium associated with acute contagious forms of conjunctivitis and Brazilian purpuric fever. { hē,mäf·ə·ləs ə'jip·tē·əs }

Haemophilus ducreyi [MICROBIO] A bacterial pathogen that causes the sexually transmitted disease soft chancre, or chancroid. { hē¦mäf·ə·ləs dü'krā,ī }

Haemophilus (para) gallinarum [MICROBIO] A bacterial pathogen that causes infectious coryza in chickens and some birds. { hē,mäf·ə·ləs ,par·ə ,gal·ə'när·əm }

Haemophilus parasuis [MICROBIO] A bacterial pathogen that frequently inhabits the normal upper respiratory tract, can cause secondary pneumonias and, in young or otherwise susceptible animals, generalized illness with arthritis, meningitis, pleuritis, and peritonitis. { hē,mäf·ə·ləs ,par·ə'sü·is }

Haemosporina [INV ZOO] A suborder of sporozoan protozoans in the subclass Coccidia; all are parasites of vertebrates,

and human malarial parasites are included. { ,hē·mō'spór·ə·nə }

haerangium [MYCOL] The fruiting body of *Fugascus* and *Ceratostomella.* { hē'ran·je·əm }

HAF black See high-abrasion furnace black. { ¦āch¦ā¦ef 'blak }

haff [GEOGR] A freshwater lagoon separated from the sea by a sandbar. { haf }

Hafnia [MICROBIO] A genus of bacteria in the family Enterobacteriaceae; motile rods that can utilize citrate as the only source of carbon. { 'haf·ne·ə }

hafnium [CHEM] A metallic element, symbol Hf, atomic number 72, atomic weight 178.49; melting point 2000°C, boiling point above 5400°C. { 'haf·nē·əm }

hafnium carbide [INV ZOO] HfC Gray powder, melting at 3887°C; used in the control rods of nuclear reactors. { 'haf·nē·əm 'kär,bīd }

Haftplatte [INV ZOO] An adhesive plate or disk in some turbellarians; it is a glanduloepidermal organ. { 'häft,pläd·ə }

Hageman factor See factor XII. { 'häg·ə·män ,fak·tər }

Hagen-Poiseuille law [FL MECH] In the case of laminar flow of fluid through a circular pipe, the loss of head due to fluid friction is 32 times the product of the fluid's viscosity, the pipe length, and the fluid velocity, divided by the product of the acceleration of gravity, the fluid density, and the square of the pipe diameter. { 'häg·ən pwä·zói ,lo }

Hagen-Rubens relation [OPTICS] An equation for the reflectivity of a solid surface in terms of the frequency of radiation of the conductivity of the solid; it applies at wavelengths long enough that the product of the frequency and the relaxation time is much less than unity. { 'häg·ən ¦rü·bənz ri'lā·shən }

hagfish [VERT ZOO] The common name for the jawless fishes composing the order Myxinoidea. { 'hag,fish }

Haggenmacher equation [CHEM] Equation to calculate latent heats of vaporizations of pure compounds by using critical conditions with Antoine constants. { ¦häg·ən¦mäk·ər i,kwā·zhən }

H agglutinin [IMMUNOL] An antibody that is type-specific for the flagella of cells or microorganisms. { ¦āch ə'glüt·ən·ən }

Hahn-Banach extension theorem [MATH] The theorem that every continuous linear functional defined on a subspace or linear manifold in a normed linear space X may be extended to a continuous linear functional defined on all of X. { ¦hän ¦bän·äk̲ ek'sten·chən ,thir·əm }

Hahn decomposition [MATH] The Hahn decomposition of a measurable space X with signed measure m consists of two disjoint subsets A and B of X such that the union of A and B equals X, A is positive with respect to m, and B is negative with respect to m. { ¦hän dē,käm·pə'zish·ən }

Hahnemannism See homeopathy. { 'hän·ə·mä,niz·əm }

hahnium [CHEM] The name suggested by workers in the United States for element 105. Symbolized Ha. { 'hän·ē·əm }

Hahn technique [SOLID STATE] A method of studying changes in solids under various treatments that involves incorporating small amounts of radium into the solid and measuring the emanating power. { 'hän tek,nēk }

Haidinger brushes [OPTICS] Faint yellow, brushlike patterns that are observed when a bright surface is viewed through a polarizer such as a rotating Nicol prism or sheet of Polaroid film; believed to be caused by birefringence of fibers at the fovea of the eye. { 'hī·diŋ·ər ,brəsh·əz }

Haidinger fringes [OPTICS] Interference fringes produced by nearly normal incidence of light on thick, flat plates. Also known as constant-angle fringes; constant-deviation fringes. { 'hī·diŋ·ər ,friŋ·jəz }

haidingerite [MINERAL] $HCaAsO_4 \cdot H_2O$ A white mineral composed of hydrous calcium arsenate. { 'hī·diŋ·ə,rīt }

hail [METEOROL] Precipitation composed of lumps of ice formed in strong updrafts in cumulonimbus clouds, having a diameter of at least 0.2 inch (5 millimeters); most hailstones are spherical or oblong, some are conical, and some are bumpy and irregular. { hāl }

hail stage [METEOROL] Thermodynamic process of freezing of suspended water drops in adiabatically rising air with temperature below the freezing point, under the assumption that release of latent heat of fusion maintains constant temperature until all water is frozen. { 'hāl ,stāj }

hailstone [METEOROL] A single unit of hail, ranging in size from that of a pea to that of a grapefruit, or from less than $^1/_4$ inch (6 millimeters) to more than 5 inches (13 centimeters) diameter; may be spheroidal, conical, or generally irregular in shape. { 'hāl₁stōn }

hair [ZOO] **1.** A threadlike outgrowth of the epidermis of animals, especially a keratinized structure in mammalian skin. **2.** The hairy coat of a mammal, or of a part of the animal. { her }

hair ball *See* lake ball. { 'her ₁bȯl }

hair cell [HISTOL] The basic sensory unit of the inner ear of vertebrates; a columnar, polarized structure with specialized cilia on the free surface. { 'her ₁sel }

haircloth [TEXT] A stiff, wiry fabric made with a weft of hair, especially of horsehair or camel's hair, and a warp of cotton, linen, or wool. { 'her₁klȯth }

hair copper *See* chalcotrichite. { 'her ₁käp·ər }

haircord [TEXT] **1.** An English dress muslin made with thick warp cords. **2.** An English bleached cotton fabric with colored warp cords; similar to dimity but heavier. { 'her₁kȯrd }

hair cracks [MATER] Fine, random cracks in the surface of the top coat of paint or other coating material. { 'her ₁kraks }

hair cycle [PHYSIO] The formation and growth of a new hair, followed by a resting stage, and ending with growth of another new hair from the same follicle. { 'her ₁sī·kəl }

hair felt [MATER] Felt made of cattle hair; used as insulation in buildings. { 'her ₁felt }

hair follicle [ANAT] An epithelial ingrowth of the dermis that surrounds a hair. { 'her ₁fäl·ə·kəl }

hair gland [ANAT] Sebaceous gland associated with hair follicles. { 'her ₁gland }

hair hygrometer [ENG] A hygrometer in which the sensing element is a bundle of human hair, which is held under slight tension by a spring and which expands and contracts with changes in the moisture of the surrounding air or gas. { 'her hī'gräm·əd·ər }

hairline [ENG] *See* air line. [GRAPHICS] **1.** A reference line that has very little apparent width. **2.** A fine, undesirable line on a negative, for example, a scratch. { 'her₁līn }

hairpin loop [MOL BIO] Any double-stranded region of single-stranded deoxyribonucleic acid or ribonucleic acid formed by base-pairing between complementary base sequences on the same strand. { 'her₁pin ₁lüp }

hairpin tube [DES ENG] A boiler tube bent into a hairpin, or U, shape. { 'her₁pin ₁tüb }

hair pyrites *See* millerite. { 'her ₁pī₁rits }

hair salt *See* alunogen. { 'her ₁sȯlt }

hairspring *See* balance spring. { 'her₁spriŋ }

hairstone [GEOL] Quartz embedded with hairlike crystals of rutile, actinolite, or other mineral. { 'her₁stōn }

hair trigger [ORD] The trigger of a firearm when it is delicately balanced so as to fire easily. { 'her ₁trig·ər }

hairworm [INV ZOO] The common name for about 80 species of worms composing the class Nematomorpha. { 'her₁wərm }

hairy cell leukemia *See* leukemic reticuloendotheliosis. { ₁her·ē ₁sel lü'kē·mē·ə }

hairy tongue [MED] Hyperplasia of the papillae forming hairlike projections on the tongue. { ₁her·ē 'təŋ }

Halacaridae [INV ZOO] A family of marine arachnids in the order Acarina. { ₁hal·ə'kar·ə₁dē }

halation [ELECTR] An area of glow surrounding a bright spot on a fluorescent screen, due to scattering by the phosphor or to multiple reflections at front and back surfaces of the glass faceplate. [OPTICS] A halo on a photographic image of a bright object caused by light reflected from the back of the film or plate. { hā'lā·shən }

halazone [ORG CHEM] $COOHC_6H_4SO_2NCl_2$ White crystals, with strong chlorine aroma; slightly soluble in water and chloroform; used as water disinfectant. { 'hal·ə₁zōn }

Halbach array [ELECTROMAG] An array of permanent magnets that produces a strong, concentrated, spatially periodic magnetic field; used on the moving object in the Inductrack magnetic levitation system. { 'häl₁bäk ə₁rā }

halcyon days [METEOROL] A period of fine weather. { 'hal·sē·ən ₁dāz }

Haldane's rule [GEN] The rule that if one sex in a first generation of hybrids between species is rare, absent, or sterile, then it is the heterogametic sex. { 'hȯl₁dānz ₁rül }

haldenhang *See* wash slope. { 'hal·dən₁haŋ }

Halecomorphi [VERT ZOO] The equivalent name for Amiiformes. { ₁hal·ə·kō'mȯr·fī }

Halecostomi [VERT ZOO] The equivalent name for Pholidophoriformes. { ₁hal·ə'käst·ə·mē }

Hale cycle [ASTRON] The variation of the sun's magnetic field over a period of approximately 22 years, during which the field reverses and is restored to its original polarity; one such cycle comprises two successive sunspot cycles. { 'hāl ₁sī·kəl }

half-adder [ELECTR] A logic element which operates on two binary digits (but no carry digits) from a preceding stage, producing as output a sum digit and a carry digit. { ₁haf ₁ad·ər }

half-adjust [COMPUT SCI] A rounding process in which the least significant digit is dropped and, if the least significant digit is one-half or more of the number base, one is added to the next more significant digit and all carries are propagated. { ₁haf ə₁jəst }

half-and-half solder [MET] Solder composed of tin and lead in equal parts. { ₁haf ən ₁haf 'säd·ər }

half-angle formulas [MATH] In trigonometry, formulas that express the trigonometric functions of half an angle in terms of trigonometric functions of the angle. { ₁haf ₁aŋ·gəl ₁fȯr·myə·ləz }

half-arc angle [METEOROL] The elevation angle of that point which a given observer regards as the bisector of the arc from his zenith to his horizon; a measure of the apparent degree of flattening of the dome of the sky. { ₁haf ₁ärk 'aŋ·gəl }

half bat [MATER] One half of a brick, cut across the length. { ₁haf ₁bat }

half block [COMPUT SCI] The unit of transfer between main storage and the buffer control unit; it consists of a column of 128 elements, each element 16 bytes long. { ₁haf ₁bläk }

half-breadth plan [NAV ARCH] A plan of a ship showing outlines of horizontal sections or waterlines at various levels, from the main deck to the keel; only the left or right half is shown. { 'haf ₁bredth ₁plan }

half-bridge [ELEC] A bridge having two power supplies, located in two of the bridge arms, to replace the single power supply of a conventional bridge. { ₁haf ₁brij }

half carry [COMPUT SCI] A flag used in the central processing unit of some computers to indicate that a carry has occurred from the low-order N bits of a $2N$-bit number to the high-order N bits. { ₁haf ₁kar·ē }

half-cell [PHYS CHEM] A single electrode immersed in an electrolyte. { ₁haf ₁sel }

half-cell potential [PHYS CHEM] In electrochemical cells, the electrical potential developed by the overall cell reaction; can be considered, for calculation purposes, as the sum of the potential developed at the anode and the potential developed at the cathode, each being a half-cell. { ₁haf ₁sel pə₁ten·chəl }

half chronometer [HOROL] **1.** A watch in which the lever and chronometer escapements are combined. **2.** A watch adjusted for temperature. { 'haf krə'näm·əd·ər }

half cock [ORD] Position of the hammer of a small arm weapon when it is held by the first cocking notch, with the trigger locked and the weapon relatively safe. { 'haf ₁käk }

half column [ARCH] A column projecting from a wall by about half its diameter. { 'haf ₁käl·əm }

half-convergency *See* conversion angle. { 'haf kən'vər·jən·sē }

half-course [MIN ENG] The drift or opening driven at an angle of about 45° to the strike and in the plane of the seam. { 'haf ₁kȯrs }

half cycle [ENG] The time interval corresponding to half a cycle, or 180°, at the operating frequency of a circuit or device. { 'haf ₁sī·kəl }

half-cycle transmission [COMMUN] Data transmission and control system that uses synchronized sources of 60-hertz power at the transmitting and receiving ends; either of two receiver relays can be actuated by choosing the appropriate half-cycle polarity of the 60-hertz transmitter power supply. { 'haf ₁sī·kəl tranz'mish·ən }

half-dog setscrew [DES ENG] A setscrew with a short, blunt point. { 'haf ₁dȯg 'set₁skrü }

half-duplex circuit [COMMUN] A circuit designed for half-duplex operation. Abbreviated HDX. { 'haf ₁dü₁pleks ₁sər·kət }

half-duplex operation [COMMUN] Operation of a telegraph

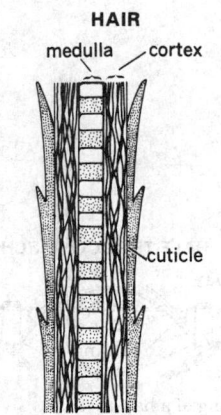

HAIR

Longitudinal section of a hair shaft.

HALF-THROUGH ARCH

roadway
hangers
column
pier hinges

Drawing of a half-through, two-hinged, crescent-rib arch. *(From G. A. Hool and W. S. Kinne, Movable and Long-span Steel Bridges, 2d ed., McGraw-Hill 1943)*

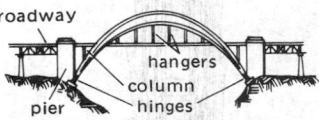

HALFTONE

Greatly enlarged detail of photoengraved halftone plate, showing formation of dots of different sizes. *(American Museum of Photography)*

system in either direction over a single channel, but not in both directions simultaneously. { 'haf ¦dü,pleks äp·ə'rā·shən }

half-duplex repeater [ELECTR] Duplex telegraph repeater provided with interlocking arrangements which restrict the transmission of signals to one direction at a time. { 'haf ¦dü,pleks ri'pēd·ər }

half-hard [MET] A rolled-metal product of intermediate hardness or temper. { 'haf ,härd }

half-hardy plant [BOT] A plant that can withstand relatively low temperatures but cannot survive severe freezing in cold climates unless carefully protected. { 'haf ¦här·dē 'plant }

half-header [MIN ENG] A large cap piece; used by sawing a header in two and placing, generally, two timbers under the half header on one side of the haulage, with the end extending over the haulage. { 'haf ¦hed·ər }

half-height drive [COMPUT SCI] A personal-computer disk drive whose height is half that of earlier disk drives. { ¦haf ¦hīt 'drīv }

half-life [CHEM] The time required for one-half of a given material to undergo chemical reactions. [NUCLEO] The average time interval required for one-half of any quantity of identical radioactive atoms to undergo radioactive decay. Also known as half-value period; radioactive half-life. { 'haf ,līf }

half line *See* ray. { 'haf ,līn }

half-loaded [ORD] In automatic arms, the condition wherein the belt or magazine is inserted and the receiver is charged, but without the first cartridge actually in the chamber. { 'haf 'lōd·əd }

half-moon [ASTRON] The moon as seen in the first quarter and the last quarter. { 'haf ,mün }

half nut [DES ENG] A nut split lengthwise so that it can be clamped around a screw. { 'haf ,nət }

half-period zones *See* Fresnel zones. { 'haf ,pir·ē·əd ,zōnz }

half plane [MATH] The portion of a plane lying on one side of some line in the plane; in particular, all points of the complex plane either above or below the real axis. { 'haf ¦plān }

half-power beamwidth [ELECTROMAG] The angle across the main lobe of an antenna pattern between the two directions at which the antenna's sensitivity is half its maximum value at the center of the lobe. Abbreviated HPBW. { 'haf ¦pau̇·ər 'bēm,width }

half-power frequency [ELECTR] One of the two values of frequency, on the sides of an amplifier response curve, at which the voltage is $1/\sqrt{2}$ (70.7%) of a midband or other reference value. Also known as half-power point. { 'haf ¦pau̇·ər 'frē·kwən·sē }

half-power point [ELECTR] **1.** A point on the graph of some quantity in an antenna, network, or control system, versus frequency, distance, or some other variable at which the power is half that of a nearby point at which power is a maximum. **2.** *See* half-power frequency. { 'haf ¦pau̇·ər ,pȯint }

half-pulse-repetition-rate delay [ELECTR] In the loran navigation system, an interval of time equal to half the pulse repetition rate of a pair of loran transmitting stations, introduced as a delay between transmission of the master and slave signals, to place the slave station signal on the B trace when the master station signal is mounted on the A trace pedestal. { 'haf ,pəls ,rep·ə'tish·ən ,rāt di,lā }

half-round file [DES ENG] A file that is flat on one side and convex on the other. { 'haf ¦rau̇nd ,fīl }

half section [GRAPHICS] View of a section of a symmetrical workpiece or assembly on an engineering drawing, terminating at an axis of symmetry. { 'haf ¦sek·shən }

half set [MIN ENG] In mine timbering, one leg piece and a collar. { 'haf ,set }

half-shade plate [OPTICS] A half-wave plate that is placed near the polarizer of a polariscope, between it and the analyzer. { 'haf ¦shād ,plāt }

half-shift register [ELECTR] Logic circuit consisting of a gated input storage element, with or without an inverter. { 'haf ,shift ,rej·ə·stər }

half-side formulas [MATH] In trigonometry, formulas that express the tangents of one-half of each of the sides of a spherical triangle in terms of its angles. { 'haf ¦sīd ,fȯr·myə·ləz }

half-silvered surface [OPTICS] A surface covered with metallic film of a thickness such that approximately half the light falling on it at normal incidence is reflected and half is transmitted. { 'haf ¦sil·vərd ,sər·fəs }

half space [BUILD] A broad step between two half flights of a stair. [MATH] A space bounded only by an infinite plane. { 'haf ,spās }

half step *See* semitone. { 'haf ,step }

half-subtracter [ELECTR] A logic element which operates on two digits from a preceding stage, producing as output a difference digit and a borrow digit. Also known as one-digit subtracter; two-input subtracter. { 'haf səb'trak·tər }

half tap [ELEC] Bridge placed across conductors without disturbing their continuity. { 'haf ,tap }

half thickness [PHYS] The thickness of a sheet of material which reduces the intensity of a beam of radiation passing through it to one-half its initial value. Also known as half-value layer; half-value thickness. { 'haf ¦thik·nəs }

half-through arch [CIV ENG] A bridge arch having the roadway running through it at an elevation midway between the base and the crown. { 'haf ,thrü 'ärch }

half tide [OCEANOGR] The condition when the tide is at the level between any given high tide and the following or preceding low tide. Also known as mean tide. { 'haf ,tīd }

half-tide basin [CIV ENG] A lock of very large size and usually of irregular shape, the gates of which are kept open for several hours after high tide so that vessels may enter as long as there is sufficient depth over the sill; vessels remain in the half-tide basin until the ensuing flood tide, when they may pass through the gate to the inner harbor; if entry to the inner harbor is required before this time, water must be admitted to the half-tide basin from some external source. { 'haf ,tīd ,bās·ən }

half-tide level [OCEANOGR] The level midway between mean high water and mean low water. { 'haf ,tīd ,lev·əl }

half-timbered [BUILD] Pertaining to a timber frame building with brickwork, plaster, or wattle and daub filling the spaces between the timbers. { 'haf ¦tim·bərd }

half time [NUCLEO] The time during which half the radioactive material resulting from a nuclear explosion remains in the atmosphere. { 'haf ,tīm }

halftone [GRAPHICS] An engraving used in printing to reproduce photographs and drawings that contain continuous tones, that is, grays (middle tones or halftones) in addition to black and white; preparation involves photographing the artwork through a screen. { 'haf,tōn }

halftone characteristic [COMMUN] Fidelity of the recorded density shadings compared with the subject copy transmitted; the term may also be used to express the relationship between the facsimile signal and the subject copy or recorded copy. { 'haf,tōn ,kar·ik·tə'ris·tik }

halftone contact screen [GRAPHICS] A screen consisting of a regular pattern of dots that is used for making halftone negatives in a camera. { 'haf,tōn 'kän,tak ,skrēn }

half-track [MECH ENG] **1.** A chain-track drive system for a vehicle; consists of an endless metal belt on each side of the vehicle driven by one of two inside sprockets and running on bogie wheels; the revolving belt lays down on the ground a flexible track of cleated steel or hard-rubber plates; the front end of the vehicle is supported by a pair of wheels. **2.** A motor vehicle equipped with half-tracks. { 'haf ,trak }

half-track tape recorder *See* double-track tape recorder. { 'haf ,trak 'tāp ri,kȯrd·ər }

half-value layer *See* half thickness. { 'haf ¦val·yü ,lā·ər }

half-value period *See* half-life. { 'haf ¦val·yü ,pir·ē·əd }

half-value thickness *See* half thickness. { 'haf ¦val·yü ,thik·nəs }

half-wave [ELEC] Pertaining to half of one cycle of a wave. [ELECTROMAG] Having an electrical length of a half wavelength. { 'haf ¦wāv }

half-wave amplifier [ELECTR] A magnetic amplifier whose total induced voltage has a frequency equal to the power supply frequency. { 'haf ¦wāv 'am·plə,fī·ər }

half-wave antenna [ELECTROMAG] An antenna whose electrical length is half the wavelength being transmitted or received. { 'haf ¦wāv an'ten·ə }

half-wave dipole *See* dipole antenna. { 'haf ¦wāv 'dī,pōl }

half-wavelength [ELECTROMAG] The distance corresponding to an electrical length of half a wavelength at the operating frequency of a transmission line, antenna element, or other device. { 'haf ¦wāv,leŋkth }

half-wave plate [OPTICS] A thin section of a doubly refracting crystal, of a thickness such that the ordinary and extraordinary components of normally incident light emerge from it with a phase difference corresponding to an odd number of half wavelengths. { 'haf ¦wāv 'plāt }

half-wave rectification [ELECTR] Rectification in which current flows only during alternate half cycles. { 'haf ¦wāv ‚rek·tə·fə'kā·shən }

half-wave rectifier [ELECTR] A rectifier that provides half-wave rectification. { 'haf ¦wāv 'rek·tə‚fī·ər }

half-wave transmission line [ELECTROMAG] Transmission line which has an electrical length equal to one-half the wavelength of the signal being transmitted or received. { 'haf ¦wāv tranz'mish·ən ‚līn }

half-wave vibrator [ELEC] A vibrator having only one pair of contacts; interrupts the flow of direct current through the primary of a power transformer, but does not reverse the current. { 'haf ¦wāv 'vī‚brād·ər }

half-width [MATH] For a function which has a maximum and falls off rapidly on either side of the maximum, the difference between the two values of the independent variable for which the dependent variable has one-half its maximum value. { 'haf ¦width }

half-word I/O buffer [COMPUT SCI] A buffer, the upper half being used to store the upper half of a word for both input and output characters, the lower half of the buffer being used for purposes such as the storage of constants. { 'haf ‚wərd ¦ī ¦ō ‚bəf·ər }

halibut [VERT ZOO] Either of two large species of flatfishes in the genus *Hippoglossus;* commonly known as a right-eye flounder. { 'hal·ə·bət }

halibut liver oil [MATER] Fishy-tasting oil, pale yellow to dark red, with slightly fishy smell; soluble in alcohol, ether, chloroform, and carbon disulfide, insoluble in water; derived from halibut livers; used as medicine, as vitamin A and D source, and to dress leather. { 'hal·ə·bət 'liv·ər ‚oil }

Halichondrida [INV ZOO] A small order of sponges in the class Demospongiae with a skeleton of diactinal or monactinal, siliceous megascleres (or both), a skinlike dermis, and small amounts of spongin. { ‚hal·ə'kän‚drē·də }

Halictidae [INV ZOO] The halictid and sweat bees, a family of hymenopteran insects in the superfamily Apoidea. { hə'lik·tə‚dē }

halide [CHEM] A compound of the type MX, where X is fluorine, chlorine, iodine, bromine, or astatine, and M is another element or organic radical. { 'ha‚līd }

Halimeda [BOT] A genus of small, bushy green algae in the family Codiaceae composed of thick, leaflike segments; important as a fossil and as a limestone builder. { ‚hal·ə'mē·də }

Halimond tube [MIN ENG] A miniature pneumatic flotation cell used for examination of small ore samples under closely controllable conditions. { 'hal·ə·mənd ‚tüb }

Haliotis [INV ZOO] A genus of gastropod mollusks commonly known as the abalones. { ‚hal·ē'ōd·əs }

Haliplidae [INV ZOO] The crawling water beetles, a family of coleopteran insects in the suborder Adephaga. { hə'lip·lə‚dē }

halite [MINERAL] NaCl Native salt; an evaporite mineral occurring as isometric crystals or in massive, granular, or compact form. Also known as common salt; rock salt. { 'ha‚līt }

Halitheriinae [PALEON] A subfamily of extinct sirenian mammals in the family Dugongidae. { hə‚lith·ə'rī·ə‚nē }

Hall accelerator [PL PHYS] A plasma accelerator based on the Hall effect. { 'hol ak'sel·ə‚rād·ər }

Hall angle [ELECTROMAG] The electric field, resulting from the Hall effect, perpendicular to a current, divided by the electric field generating the current. { 'hol ‚aŋ·gəl }

Hall coefficient [ELECTROMAG] A measure of the Hall effect, equal to the transverse electric field (Hall field) divided by the product of the current density and the magnetic induction. Also known as Hall constant. { 'hol ‚kō·i'fish·ənt }

Hall constant *See* Hall coefficient. { 'hol ‚kän·stənt }

Hall cyclic thermal reforming [CHEM ENG] A gas-making process that uses component parts of carbureted-water gas apparatus to generate high-Btu gas from feedstocks ranging from naphtha to Bunker C. { 'hol 'sī·klik ‚thər·məl rē'for·miŋ }

Hall effect [ELECTROMAG] The development of a transverse electric field in a current-carrying conductor placed in a magnetic field; ordinarily the conductor is positioned so that the magnetic field is perpendicular to the direction of current flow and the electric field is perpendicular to both. { 'hol i‚fekt }

Hall-effect gaussmeter [ENG] A gaussmeter that consists of a thin piece of silicon or other semiconductor material which is inserted between the poles of a magnet to measure the magnetic field strength by means of the Hall effect. { 'hol i‚fekt 'gaus‚mēd·ər }

Hall-effect isolator [ELECTROMAG] An isolator that makes use of the Hall effect in a semiconductor plate mounted in a magnetic field, to provide greater loss in one direction of signal travel through a waveguide than in the other direction. { 'hol i‚fekt 'ī·sə‚lād·ər }

Hall-effect modulator [ELECTR] A Hall-effect multiplier used as a modulator to give an output voltage that is proportional to the product of two input voltages or currents. { 'hol i‚fekt 'mäj·ə‚lād·ər }

Hall-effect multiplier [ELECTR] A multiplier based on the Hall effect, used in analog computers to solve such problems as finding the square root of the sum of the squares of three independent variables. { 'hol i‚fekt 'məl·tə‚plī·ər }

Hall-effect switch [ELECTR] A magnetically activated switch that uses a Hall generator, trigger circuit, and transistor amplifier on a silicon chip. { 'hol i‚fekt ‚swich }

Haller's organ [INV ZOO] A chemoreceptor on the tarsus of certain ticks. { 'hal·ərz ‚or·gən }

Hallett table [MIN ENG] A Wilfley-type concentrating table having the tops of the riffles in the same plane as the cleaning planes, and riffles inclined toward the waste water side. { 'hal·ət ‚tā·bəl }

Halley's Comet [ASTRON] A member of the solar system, with an orbit and a period of about 76 years; its nucleus is about 9 miles (15 kilometers) in diameter; next due to appear in 2061. Also known as Comet Halley. { 'hal·ēz ¦käm·ət }

Hall generator [ELECTROMAG] A generator using the Hall effect to give an output voltage proportional to magnetic field strength. { 'hol 'jen·ə‚rād·ər }

Hallian [GEOL] A North American stage of Pleistocene geologic time, above the Wheelerian and below the Recent. { 'hol·ē·ən }

halliard *See* halyard. { 'hal·yərd }

Hallinger shield [MIN ENG] A tunneling shield valuable for working in very soft ground; incorporates a mechanical excavator and does not entail the use of timbering to protect the miners. { 'hal·in·jər ‚shēld }

Hall mobility [SOLID STATE] The product of conductivity and the Hall constant for a conductor or semiconductor; a measure of the mobility of the electrons or holes in a semiconductor. { 'hol mō'bil·əd·ē }

halloysite [MINERAL] $Al_2Si_2O_5(OH)_4 \cdot 2H_2O$ Porcelainlike clay mineral whose composition is like that of kaolinite but contains more water and is structurally distinct; varieties are known as metahalloysites. { hə'loi‚sīt }

Hall plasma thruster [AERO ENG] A type of electromagnetic propulsion system in which electrons drawn through an annular discharge chamber feel the effect of a radial magnetic field between the central and outside pole pieces bounding the discharge region, and thereby follow azimuthal cyclotron (Hall-effect) trajectories that increase the probability that they will have ionizing collisions with propellant atoms; the positive ions produced in these collisions are accelerated downstream, thereby producing thrust. { 'hol 'plaz·mə‚ thrəst·ər }

Hall-plate device [ENG] A sensor that uses the Hall effect to measure magnetic field strength. { 'hol 'plāt di‚vīs }

Hall process [MET] An electrolytic recovery process for aluminum employing a fused-bauxite (aluminum oxide), cryolite electrolyte. { 'hol ‚prä·səs }

Hall resistance [ELECTR] The ratio of the transverse voltage developed across a current-carrying conductor, due to the Hall effect, to the current itself. { 'hol ri‚zis·təns }

Hall's theorem *See* marriage theorem. { 'holz ‚thir·əm }

hallucination [PSYCH] A perception without an appropriate stimulus. { hə‚lüs·ən'ā·shən }

hallucinogen [PHARM] A substance, such as LSD, that induces hallucinations. { hə'lüs·ən·ə‚jən }

hallucinogenic [PHARM] Of or pertaining to a hallucinogen. [PSYCH] Referring to any stimulus that creates the impression of experiencing a hallucination. { hə‚lüs·ən·ə'jen·ik }

hallucinosis [PSYCH] The condition of being possessed by more or less persistent hallucinations, especially while fully conscious. { hə‚lüs·ən'ō·səs }

hallux [ANAT] The first digit of the hindlimb; the big toe of a human. { 'hal·əks }

hallux valgus [MED] A deformity of the great toe, in which the head of the first metatarsal deviates away from the second metatarsal, that is, toward the outside of the foot, and the phalanges are deviated toward the second toe, causing prominence of the metatarsophalangeal joint. { 'hal·əks 'val·gəs }

Hall voltage [ELECTR] The no-load voltage developed across a semiconductor plate due to the Hall effect, when a specified value of control current flows in the presence of a specified magnetic field. { 'hȯl ˌvōl·tij }

Hallwachs effect [ELECTR] The discharge of a negatively charged metal plate caused by photoemission when the plate is exposed to ultraviolet light. [PHYS] The ability of ultraviolet radiation to discharge a negatively charged body in a vacuum. { 'häl‚väks iˌfekt }

halmeic [GEOL] Referring to minerals or sediments derived directly from sea water. Also known as halmyrogenic; halogenic. { hal'mē·ik }

halmophagous [ZOO] Pertaining to organisms which infest and eat stalks or culms of plants. { hal'mäf·ə·gəs }

halmyrogenic *See* halmeic. { hal‚mī·rə'jen·ik }

halmyrolysis [GEOCHEM] Postdepositional chemical changes that occur while sediment is on the sea floor. { ‚hal·mə'räl·ə·səs }

halo [ASTRON] A type of ray system in which many short, filamentary streaks form a complex network of bright matter surrounding the lunar crater. Also known as nimbus. [ELECTR] An undesirable bright or dark ring surrounding an image on the fluorescent screen of a television cathode-ray tube; generally due to overloading or maladjustment of the camera tube. [GEOL] A ring or crescent surrounding an area of opposite sign; it is a diffusion of a high concentration of the sought mineral into surrounding ground or rock; it is encountered in mineral prospecting and in magnetic and geochemical surveys. [METEOROL] Any one of a large class of atmospheric optical phenomena which appear as colored or whitish rings and arcs about the sun or moon when seen through an ice crystal cloud or in a sky filled with falling ice crystals. [OPTICS] A ring around the photographic image of a bright source caused by light scattering in any one of a number of possible ways. { 'hā·lō }

haloalkane [ORG CHEM] Halogenated aliphatic hydrocarbon. { ‚ha·lō'al‚kān }

Halobacteriaceae [MICROBIO] A family of gram-negative, aerobic rods and cocci which require high salt (sodium chloride) concentrations for maintenance and growth. { ‚hal·ə‚bak·tir·ē'ās·ē‚ē }

Halobacterium [MICROBIO] A genus of bacteria in the family Halobacteriaceae; single, rod-shaped cells which may be pleomorphic when media are deficient. { ‚hal·ə·bak'tir·ē·əm }

halo blight [PL PATH] A bacterial blight of beans and sometimes other legumes caused by *Pseudomonas phaseolicola* and characterized by water-soaked lesions surrounded by a yellow ring on the leaves, stems, and pods. { 'hā·lō ‚blīt }

halocarbon [ORG CHEM] A compound of carbon and a halogen, sometimes with hydrogen. { ‚ha·lō'kär·bən }

halocarbon plastic [ORG CHEM] Plastic made from halocarbon resins. { ‚ha·lō'kär·bən 'plas·tik }

halocarbon resin [ORG CHEM] Resin produced by the polymerization of monomers made of halogenated hydrocarbons, such as tetrafluoroethylene, C_2F_4, and trifluorochloroethylene, C_2F_3Cl. { ‚ha·lō'kär·bən 'rez·ən }

halocline [OCEANOGR] A well-defined vertical gradient of salinity in the oceans and seas. { 'hal·ə‚klīn }

Halococcus [MICROBIO] A genus of bacteria in the family Halobacteriaceae; nonmotile cocci which occur in pairs, tetrads, or clusters of tetrads. { ‚hal·ə'käk·əs }

Halocypridacea [INV ZOO] A superfamily of ostracods in the suborder Myodocopa; individuals are straight-backed with a very thin, usually calcified carapace. { ‚hal·ə‚sip·rə'dās·ē·ə }

halo effect [IND ENG] A tendency when rating a person in regard to a specific trait to be influenced by a general impression or by another trait of the person. { 'hā·lō iˌfekt }

haloform [ORG CHEM] CHX_3 A compound made by reaction of acetaldehyde or methyl ketones with NaOX, where X

is a halogen; an example is iodoform, HCI_3, or bromoform, $HCBr_3$ or chloroform, $HCCl_3$. { 'hal·ə‚fȯrm }

haloform reaction [ORG CHEM] Halogenation of acetaldehyde or a methyl ketone in aqueous basic solution; the reaction is characteristic of compounds containing a CH_3CO group linked to a hydrogen or to another carbon. { 'hal·ə‚fȯrm rē‚ak·shən }

halogen [CHEM] Any of the elements of the halogen family, consisting of fluorine, chlorine, bromine, iodine, and astatine. { 'hal·ə·jən }

halogen acid [INV ZOO] A compound composed of hydrogen bonded to a halogen element, for example, hydrochloric acid. { 'hal·ə·jən ‚as·əd }

halogenated hydrocarbon [ORG CHEM] One of a group of halogen derivatives of organic hydrogen- and carbon-containing compounds; the group includes monohalogen compounds (alkyl or aryl halides) and polyhalogen compounds that contain the same or different halogen atoms. { 'hal·ə·jə‚nād·əd ‚hī·drə'kär·bən }

halogenation [ORG CHEM] A chemical process or reaction in which a halogen element is introduced into a substance, generally by the use of the element itself. { ‚hal·ə·jə'nā·shən }

halogen counter [NUCLEO] A Geiger counter in which the self-quenching action is provided by a halogen gas, such as chlorine or bromine. { 'hal·ə·jən ‚kaunt·ər }

halogenic *See* halmeic. { ‚hal·ə'jen·ik }

halogen mineral [MINERAL] Any of the naturally occurring compounds containing a halogen as the sole or principal anionic constituent. { 'hal·ə·jən ‚min·rəl }

halohydrin [ORG CHEM] A compound with the general formula $X-R-OH$ where X is a halide such as Cl^-; an example is chlorohydrin. { ‚hal·ə'hī·drən }

halokinesis *See* salt tectonics. { ‚hal·ə·kə'nē·səs }

halomorphic [GEOCHEM] Referring to an intrazonal soil whose features have been strongly affected by either neutral or alkali salts, or both. { ‚hal·ə'mȯr·fik }

halon [ORG CHEM] A fluorocarbon that has one or more bromine atoms in its molecule. { 'ha‚län }

halonate [MYCOL] Pertaining to a spore surrounded by a colored circle. [PL PATH] A leaf spot surrounded by a halo. { 'hal·ə‚nāt }

halo of 22° [OPTICS] A halo phenomenon in the form of a prismatically colored circle of 22° angular radius around the sun or moon, exhibiting coloration from red on the inside to blue on the outside. { 'hā·lō əv ‚twen·tē'tü di'grēz }

halo of 46° [OPTICS] A halo phenomenon in the form of a prismatically colored circle or incomplete arc thereof, centered on the sun or moon and having an angular radius of about 46°; the coloration is red on the inner edge to blue on the outer edge. { 'hā·lō əv ‚fȯrd·ē'siks di'grēz }

halo of 90° *See* Hevelian halo. { 'hā·lō əv 'nīn·tē di'grēz }

halo of Hevelius *See* Hevelian halo. { 'hā·lō əv hə'vel·yəs }

halo ore *See* fringe ore. { 'hā·lō ‚ȯr }

halophile [BIOL] An organism that requires high salt concentrations for growth and maintenance. { 'hal·ə‚fīl }

halophilism [BIOL] The phenomenon of demand for high salt concentrations for growth and maintenance. { ‚hal·ə'fil‚iz·əm }

halophone [ENG] A device that records patterns in time in a manner analogous to the way that optical holograms record space. { 'hal·ə‚fōn }

halophyte [ECOL] A plant or microorganism that grows well in soils having a high salt content. { 'hal·ə‚fīt }

halo population *See* population II. { 'hā·lō ‚päp·yə‚lā·shən }

Haloragaceae [BOT] A family of dicotyledonous plants in the order Haloragales distinguished by an apical ovary of 2–4 loculi, small inflorescences, and small, alternate or opposite or whorled, exstipulate leaves. { ‚hal·ə·rə'gās·ē‚ē }

Haloragales [BOT] An order of dicotyledonous plants in the subclass Rosidae containing herbs with perfect or often unisexual, more or less reduced flowers, and a minute or vestigial perianth. { ‚hal·ə·rə'gā·lēz }

Halosauridae [VERT ZOO] A family of mostly extinct deep-sea teleost fishes in the order Notacanthiformes. { ‚hal·ə'sȯr·ə‚dē }

halosere [ECOL] The series of communities succeeding one another, from the pioneer stage to the climax, and commencing in salt water or on saline soil. { 'hal·ə‚sir }

halothane [PHARM] $C_2HBrClF_3$ A colorless, nonflammable liquid used as a general anesthetic, by inhalation. { 'hal·ə,thān }

halotrichite [MINERAL] **1.** $FeAl_2(SO_4)_4 \cdot 22H_2O$ A mineral composed of hydrous sulfate of iron and aluminum. Also known as butter rock; feather alum; iron alum; mountain butter. **2.** Any sulfate mineral resembling halotrichite in structure and habit. { ha'lä·trə,kīt }

Halsey premium plan [IND ENG] A wage-incentive plan which sets a guaranteed daily rate to an employee and provides for predetermined compensation for superior performance. { 'hȯl·zē 'prē·mē·əm ,plan }

halt [COMPUT SCI] The cessation of the execution of the sequence of operations in a computer program resulting from a halt instruction, hang-up, or interrupt. { hȯlt }

haltere [INV ZOO] Either of a pair of capitate filaments representing rudimentary hindwings in Diptera. Also known as balancer. { 'hȯl,tir }

halving [NAV] The process of adjusting magnetic compass correctors so as to remove half of the deviation on the opposite cardinal or adjacent intercardinal headings to those on which adjustment was originally made when all deviation was removed; this action is taken to equalize the error on opposite headings. { 'hav·iŋ }

halyard [NAV ARCH] A rope or tackle used to raise or lower a flag, sail, or spar. Also spelled halliard. { 'hal·yərd }

Halysitidae [PALEON] A family of extinct Paleozoic corals of the order Tabulata. { ,hal·ə'sid·ə,dē }

hamada [GEOL] A barren desert surface composed of consolidated material usually consisting of exposed bedrock, but sometimes of consolidated sedimentary material. { hə'mä·də }

Hamal [ASTRON] A second-magnitude star in the constellation Aries; the star α Ari. { hə'mäl }

Hamamelidaceae [BOT] A family of dicotyledonous trees or shrubs in the order Hamamelidales characterized by united carpels, alternate leaves, perfect or unisexual flowers, and free filaments. { ,ha·mə,mel·ə'dās·ē,ē }

Hamamelidae [BOT] A small subclass of plants in the class Magnoliopsida having strongly reduced, often unisexual flowers with poorly developed or no perianth. { ,ha·mə'mel·ə,dē }

Hamamelidales [BOT] A small order of dicotyledonous plants in the subclass Hamamelidae characterized by vessels in the wood and a gynoecium consisting either of separate carpels or of united carpels that open at maturity. { ,ha·mə,mel·ə'dā·lēz }

hamartoma [MED] An abnormal condition resulting in the formation of a mass of tissue of disproportionate size and distribution but composed of the normal tissue of the region. { ,ha·mər'tō·mə }

hamartophobia [PSYCH] An abnormal fear of error or sin. { ha,mard·ə'fō·bē·ə }

hamate [BIOL] Hook-shaped or hooked. { 'hā,māt }

hambergite [MINERAL] Be_2BO_3OH A grayish-white or colorless mineral composed of beryllium borate and occurring as prismatic crystals; hardness is 7.5 on Mohs scale, and specific gravity is 2.35. { 'ham·bər,gīt }

Hamel basis [MATH] For a normed space, a collection of vectors with every finite subset linearly independent, while any vector of the space is a linear combination of at most countably many vectors from this subset. { 'ham·əl ,bā·səs }

Hamilton-Cayley theorem See Cayley-Hamilton theorem. { 'ham·əl·tən 'kā·lē ,thir·əm }

Hamiltonian circuit See Hamiltonian path. { ,ham·əl,tō·nē·ən 'sər·kət }

Hamiltonian cycle See Hamiltonian path. { ,ham·əl'tō·nē·ən ,sī·kəl }

Hamiltonian function [MECH] A function of the generalized coordinates and momenta of a system, equal in value to the sum over the coordinates of the product of the generalized momentum corresponding to the coordinate, and the coordinate's time derivative, minus the Lagrangian of the system; it is numerically equal to the total energy if the Lagrangian does not depend on time explicitly; the equations of motion of the system are determined by the functional dependence of the Hamiltonian on the generalized coordinates and momenta. { ,ham·əl'tō·nē·ən 'fəŋk·shən }

Hamiltonian graph [MATH] A graph which has a Hamiltonian path. { ,ham·əl'tō·nē·ən ,graf }

Hamiltonian operator See energy operator. { ,ham·əl'tō·nē·ən ¦äp·ə,rād·ər }

Hamiltonian path [MATH] A path along the edges of a graph that traverses every vertex exactly once and terminates at its starting point. Also known as Hamiltonian circuit; Hamiltonian cycle. { ,ham·əl'tō·nē·ən ,path }

Hamilton-Jacobi equation [MATH] A particular partial differential equation useful in studying certain systems of ordinary equations arising in the calculus of variations, dynamics, and optics: $H(q_1, \ldots, q_n, \partial\phi/\partial q_1, \ldots, \partial\phi/\partial q_n, t) + \partial\phi/\partial t = 0$, where $q_1, \ldots, q_n$ are generalized coordinates, t is the time coordinate, H is the Hamiltonian function, and ϕ is a function that generates a transformation by means of which the generalized coordinates and momenta may be expressed in terms of new generalized coordinates and momenta which are constants of motion. { 'ham·əl·tən jə'kō·bē i,kwā·zhən }

Hamilton-Jacobi theory [MATH] The study of the solutions of the Hamilton-Jacobi equation and the information they provide concerning solutions of the related systems of ordinary differential equations. [MECH] A theory that provides a means for discussing the motion of a dynamic system in terms of a single partial differential equation of the first order, the Hamilton-Jacobi equation. { 'ham·əl·tən jə'kō·bē ,thē·ə·rē }

Hamilton's equations of motion [MECH] A set of first-order, highly symmetrical equations describing the motion of a classical dynamical system, namely $\dot{q}_j = \partial H/\partial p_j$, $p_j = -\partial H/\partial q_j$; here q_j ($j = 1, 2, \ldots$) are generalized coordinates of the system, p_j is the momentum conjugate to q_j, and H is the Hamiltonian. Also known as canonical equations of motion. { 'ham·əl·tənz i¦kwā·zhənz əv 'mō·shən }

Hamilton's principle [MECH] A variational principle which states that the path of a conservative system in configuration space between two configurations is such that the integral of the Lagrangian function over time is a minimum or maximum relative to nearby paths between the same end points and taking the same time. { 'ham·əl·tənz ¦prin·sə·pəl }

hammarite [MINERAL] $Pb_2Cu_2Bi_4S_9$ A monoclinic mineral whose color is a steel gray with red tone; consists of lead and copper bismuth sulfide. { 'ham·ə,rīt }

hammer [DES ENG] **1.** A hand tool used for pounding and consisting of a solid metal head set crosswise on the end of a handle. **2.** An arm with a striking head for sounding a bell or gong. [GRAPHICS] The mechanism of an impact printer that presses the typeface against the ribbon and paper, or presses the paper against the ribbon and typeface. [MECH ENG] A power tool with a metal block or a drill for the head. [ORD] A metallic pivoted item in a firing mechanism designed to strike a firing pin or percussion cap and thus fire a gun. { 'ham·ər }

hammer drill [MECH ENG] Any of three types of fast-cutting, compressed-air rock drills (drifter, sinker, and stoper) in which a hammer strikes rapid blows on a loosely held piston, and the bit remains against the rock in the bottom of the hole, rebounding slightly at each blow, but does not reciprocate. { 'ham·ər ,dril }

hammer forging [MET] Forging by means of repeated blows of a hammer. { 'ham·ər ¦fȯrj·iŋ }

hammer gun [ORD] Gun with an outside, visible hammer. { 'ham·ər ,gən }

hammerhead [DES ENG] The striking part of a hammer. { 'ham·ər,hed }

hammerhead crane [MECH ENG] A crane with a horizontal jib that is counterbalanced. { 'ham·ər,hed ,krān }

hammerless gun [ORD] Gun with a wholly enclosed hammer and firing mechanism. { 'ham·ər,ləs ,gən }

hammer mill [MECH ENG] **1.** A type of impact mill or crusher in which materials are reduced in size by hammers revolving rapidly in a vertical plane within a steel casing. Also known as beater mill. **2.** A grinding machine which pulverizes feed and other products by several rows of thin hammers revolving at high speed. { 'ham·ər ,mil }

hammer milling [MECH ENG] Crushing or fracturing materials in a hammer mill. { 'ham·ər ,mil·iŋ }

hammer pick [MIN ENG] A pneumatic hand-held machine used to break up the harder rocks in a mine; consists of a pick which is driven by a hammer set in a cylinder which receives compressed air. { 'ham·ər ,pik }

hammer test [MET] An impact test conducted by dropping weights from increasing heights until a specified deflection of the weight is produced. { 'ham·ər ,test }

hammertoe [MED] A condition of the toe, usually the second, in which the proximal phalanx is extremely extended while the two distal phalanges are flexed. { 'ham·ər,tō }

hammer track [NUCLEO] A hammer or T-shaped track in a nuclear emulsion that is formed by a nucleus that comes to rest in the emulsion and decays into two fragments that travel in opposite directions. { 'ham·ər ,trak }

hammer welding [MET] Forge welding by means of a hammer. { 'ham·ər ,weld·iŋ }

Hammett acidity function [CHEM] An expression for the acidity of a medium, defined as $h_0 = K_{BH^+}[BH^+]/[B]$, where K_{BH^+} is the dissociation constant of the acid form of the indicator, and $[BH^+]$ and $[B]$ are the concentrations of the protonated base and the unprotonated base respectively. { 'ham·ət ə'sid·əd·ē ,fəŋk·shən }

Hamming code [COMMUN] An error-correcting code used in data transmission. { 'ham·iŋ ,kōd }

hamming distance See signal distance. { 'ham·iŋ ,dis·təns }

hammock See hummock. { 'ham·ək }

Hamoproteidae [INV ZOO] A family of parasitic protozoans in the suborder Haemosporina; only the gametocytes occur in blood cells. { ,ham·ō·prə'tē·ə,dē }

hamper [NAV ARCH] Articles of outfit, especially spars, rigging, and so on, above the deck. { 'ham·pər }

Hampson process [CRYO] A process for liquefying gases which resembles the Linde process except that the Joule-Thomson expansion reduces the gas pressure to approximately atmospheric pressure. { 'ham·sən ,prä·səs }

ham radio See amateur radio. { ¦ham 'rād·ē·ō }

ham sandwich [MATH] **1.** The theorem that if the functions f and h have the same limit L, and if the value of a third function g is greater to or equal than that of f and less than or equal to than that of h for all values of the independent variable, then g also has the limit L. **2.** The theorem that there is a plane that cuts each of three bounded, connected, open sets in space into two sets of equal volume. { ¦ham 'san,wich }

hamster [VERT ZOO] The common name for any of 14 species of rodents in the family Cricetidae characterized by scent glands in the flanks, large cheek pouches, and a short tail. { 'ham·stər }

hamstring muscles [ANAT] The biceps femoris, semitendinosus, and semimembranosus collectively. { 'ham,striŋ ,məs·əlz }

hamulus [VERT ZOO] A hooklike process, especially a small terminal hook on the barbicel of a feather. { 'ham·yə·ləs }

hamus [BIOL] A hook or a curved process. { 'hä·məs }

hancockite [MINERAL] A complex silicate mineral containing lead, calcium, strontium, and other minerals; it is isomorphous with epidote. { 'han,kä,kīt }

Hancock jig [MIN ENG] A moving-screen jig used to treat lead-zinc ores; the material is jigged in a tank of water, and the heavy layer settles through slots. { 'han,käk ,jig }

hand [ANAT] The terminal part of the upper extremity modified for grasping. [CONT SYS] See end effector. [TEXT] The quality or feel of a fabric. { hand }

hand-and-foot counter See hand-and-foot monitor. { ¦hand ən 'fût ,kaůnt·ər }

hand-and-foot monitor [NUCLEO] An instrument routinely used to monitor the hands and feet of atomic energy workers as they leave locations in which radioactive materials are handled. Also known as hand-and-foot counter. { ¦hand ən 'fût ,män·əd·ər }

hand arm See hand weapon. { 'hand ,ärm }

hand auger [DES ENG] A hand tool resembling a large carpenters' bit or comprising a short cylindrical container with cutting lips attached to a rod; used to bore shallow holes in the soil to obtain samples of it and other relatively unconsolidated near-surface materials. { 'hand ¦óg·ər }

handbarrow [ENG] A flat, rectangular frame with handles at both ends, carried by two persons to transport objects. Also known as barrow. { 'hand,bar·ō }

hand brake [MECH ENG] A manually operated brake. { 'hand ,brāk }

hand breadth [ANTHRO] The distance between the outside projections of the distal ends of the second and fifth metacarpals of the right hand, with the fingers extended and together. { 'hand ,bredth }

hand cable [MIN ENG] A flexible cable for electrical connection between a mining machine and a truck carrying a reel of

HAMSTER

Hamster, *Cricetus cricetus*, with a body 8–12 inches (20–30 centimeters) long.

portable cable. Also known as butt cable; head cable. { 'hand ,kā·bəl }

handcar [MECH ENG] A small, four-wheeled, hand-pumped car used on railroad tracks to transport workers and equipment for construction or repair work; other cars for the same purpose are motor-operated. { 'hand,kär }

hand drill [DES ENG] A small, portable drilling machine which is operated by hand. { 'hand ,dril }

handedness [PHYS] A division of objects, such as coordinate systems, screws, and circularly polarized light beams, into two classes (right and left), which distinguishes an object from a mirror image but not from a rotated object. { 'han·dəd·nəs }

hand electric lamp [MIN ENG] In mining, a portable, battery operated hand lamp with a tungsten-filament light source that forms a self-contained unit. { 'hand i¦lek·trik 'lamp }

hand feed [ENG] A drill machine in which the rate at which the bit is made to penetrate the rock is controlled by a hand-operated ratchet and lever or a hand-turned wheel meshing with a screw mechanism. { 'hand ,fēd }

hand float [ENG] A wooden tool used to fill in and smooth a plaster surface in order to produce a level base coat or a textured finish coat. { 'hand ,flōt }

hand-foot-and-mouth disease [MED] An infectious disease of humans caused by a coxsackie virus and characterized by maculopapular and vesicular eruptions in the mouth and on the hands and feet. { ¦hand ¦fût ən 'maůth di,zēz }

hand forging [MET] Plastic deformation of a metal by manual force. { 'hand ¦fórj·iŋ }

hand generator [ELEC] A manually cranked dynamo or alternator, usually used as the prime mover for emergency radio transmitters. { 'hand ¦jen·ə,rād·ər }

hand grenade [ORD] A grenade designed to be thrown by hand. { 'hand grə,nād }

handguard [ORD] **1.** A wooden part on a rifle to protect the shooter's hand from the hot barrel. **2.** The part of a sword or dagger designed to protect the hand. { 'hand,gärd }

hand hammer drill [ENG] A hand-held rock drill. { 'hand 'ham·ər ,dril }

hand-held computer [COMPUT SCI] A small, battery-powered mobile computer for personal or business use. Also known as palmtop; personal digital assistant (PDA). { ¦hand ,held kəm'pyüd·ər }

hand-held scanner [ENG] An image-reading device that is held and operated by a person. { ¦hand ,held 'skan·ər }

handhole [ENG] A shallow access hole large enough for a hand to be inserted for maintenance and repair of machinery or equipment. { 'hand,hōl }

handie-talkie [ELECTR] Two-way radio communications unit small enough to be carried in the hand. { 'hand·dē 'tók·ē }

hand jig [MIN ENG] A moving-screen jig operated by hand and used to treat small batches of ore; the jig box is attached to a rocking beam and moved in a tank of water. { 'hand ,jig }

hand lance [ENG] A hand-held pipe with a nozzle through which steam or air is discharged; used to remove soot deposits from the external surfaces of boiler tubes. { 'hand ,lans }

handle [COMPUT SCI] **1.** One of several small squares that appear around a selected object in an object-oriented computer-graphics program, and can be dragged with a mouse to move, enlarge, reduce, or change the shape of the object. **2.** In particular, one of the two interior points on a Bézier curve that can be dragged to alter its shape. Also known as control handle. [MECH ENG] The arm connecting the bucket with the boom in a dipper shovel or hoe. { 'han·dəl }

hand lead [ENG] A light sounding lead (7–14 pounds or 3–6 kilograms) usually having a line not more than 25 fathoms (46 meters) in length. { 'hand ,led }

hand length [ANTHRO] The distance measured from the end of the small wrist bone at the base of the thumb to the tip of the middle finger of the right hand, palm turned up, with the fingers extended and together. { 'hand ,leŋkth }

hand lens See simple microscope. { 'hand ,lenz }

handler [COMPUT SCI] A computer program developed to perform one particular function, such as control of input from, and output to, a specific peripheral device. { ,hand·lər }

hand level [ENG] A hand-held surveyor's level, basically a telescope with a bubble tube attached so that the position of the bubble can be seen when looking through the telescope. { 'hand ,lev·əl }

handling time [IND ENG] The time needed to transport parts or materials to or from a work area. { 'hand·liŋ ,tīm }

hand loader [MIN ENG] A miner who uses a shovel, rather than a machine to load coal. { 'hand ,lōd·ər }

hand-operating device [ORD] Mechanism on certain automatic firing weapons that permits the piece to be prepared by hand for firing. { 'hand ,äp·ə,rād·iŋ di,vīs }

handpicking [MIN ENG] Manual removal of a selected fraction of coarse run-of-mine ore, after washing and screening away waste. { 'hand¦pik·iŋ }

hand punch [DES ENG] A hand-held device for punching holes in paper or cards. { 'hand ,pənch }

handrail [ENG] A narrow rail to be grasped by a person for support. { 'hand,rāl }

hand-rammed [ORD] Indicating that the cartridge is intended to be rammed into the gun by hand rather than by power. { 'hand,ramd }

hand-reset [ELEC] Pertaining to a relay in which the contacts must be reset manually to their original positions when normal conditions are resumed. { 'hand 'rē,set }

hand rule See right-hand rule. { 'hand ,rül }

hand sampling [MIN ENG] Using manual methods to detach and reduce in size representative samples of ore; one of the major methods in sampling small batches of ore, others being grab sampling, trench or channel sampling, fractional selection, coning and quartering, and pipe sampling. { 'hand ,sam·pliŋ }

handsaw [DES ENG] A saw operated by hand, with a backward and forward arm movement. { 'hand,sȯ }

Hand-Schüller-Christian disease [MED] A childhood syndrome characterized by exopthalmos, diabetes insipidus, and softened or punched-out areas in the bones. { ¦hänt ¦shil·ər 'kris·chən di,zēz }

handset [DES ENG] A combination of a telephone-type receiver and transmitter, designed for holding in one hand. { 'hand,set }

handset bit [DES ENG] A bit in which the diamonds are manually set into holes that are drilled into a malleable-steel bit blank and shaped to fit the diamonds. { 'hand,set ,bit }

handshaking [COMMUN] The establishment of synchronization between sending and receiving equipment by means of exchange of specific character configurations. { 'hand,shak·iŋ }

handshaking lemma [MATH] The result that the sum of the degrees of a graph is twice the number of its edges. { 'han ,shāk·iŋ ,lem·ə }

hand-shoulder syndrome See shoulder-hand syndrome. { ¦hand ¦shōl·dər 'sin,drōm }

handspike [ORD] Handle attached to the trails of certain firearms for ease in handling. { 'hand,spīk }

hand sugar refractometer [ANALY CHEM] Portable device to read refractive indices of sugar solutions. Also known as proteinometer. { 'hand 'shu̇g·ər ,rē,frak'täm·əd·ər }

hand-tight [ENG] The extent of tightening of screwed fittings that can be accomplished without mechanical assistance. { ¦hand ¦tīt }

hand time [IND ENG] The time necessary to complete a manual element. Also known as manual time. { 'hand ,tīm }

hand tool [ENG] Any implement used by hand. { 'hand ,tül }

hand tramming [MIN ENG] Manual pushing of mine cars; limited to small mines with low output. { 'hand ¦tram·iŋ }

hand truck [ENG] **1.** A manually operated, two-wheeled truck consisting of a rectangular frame with handles at the top and a plate at the bottom to slide under the load. **2.** Any of various small, manually operated, multiwheeled platform trucks for transporting materials. { 'hand ,trək }

hand type See foundry type. { 'hand ,tīp }

hand viewer [OPTICS] An optical magnifying device small enough for hand use. { 'hand ,vyü·ər }

hand weapon [ORD] A weapon, such as a pistol, knife, or sword, used by one hand. Also known as hand arm. { 'hand ,wep·ən }

hand winch [MECH ENG] A winch that is operated by hand. { 'hand ,winch }

hang [ORD] To lock the receiver or bolt of a gun in an open position. { 'haŋ }

hangar [CIV ENG] A building at an airport specially designed in height and width to enable aircraft to be stored or maintained in it. { 'haŋ·ər }

hangar deck [NAV ARCH] A deck, below the flight deck of a carrier, where aircraft are parked and serviced. { 'haŋ·ər ,dek }

hanger [CIV ENG] An iron strap which lends support to a joist beam or pipe. [GEOL] See hanging wall. [PETRO ENG] **1.** A device to seat in the bowl of a lowermost casing head to suspend the next-smaller casing string and form a seal between the two. Also known as casing hanger. **2.** A device to provide a seal between the tubing and the tubing head. Also known as tubing hanger. { 'haŋ·ər }

hanger bolt [DES ENG] A bolt with a machine-screw thread on one end and a lag-screw thread on the other. { 'haŋ·ər ,bōlt }

hanger plug [PETRO ENG] A pressure-tight sealing device placed in the casing below the blowout preventer. { 'haŋ·ər ,pləg }

hangfire [ENG] Delay in the explosion of a charge. { 'haŋ,fīr }

hang glider [AERO ENG] A flexible, deployable, steerable, kitelike glider from which a harnessed rider hangs during flight. { 'haŋ ,glīd·ər }

hanging [GEOL] See hanging wall. [MET] Sticking or wedging of part of the charge in a blast furnace. { 'haŋ·iŋ }

hanging bolt [MIN ENG] A bolt used to suspend wall plates in shaft construction. { 'haŋ·iŋ ,bōlt }

hanging coal [MIN ENG] A portion of the coal seam which, by undercutting, has had its natural support removed. { 'haŋ·iŋ ¦kōl }

hanging compass See inverted compass. { 'haŋ·iŋ ¦käm·pəs }

hanging-drop atomizer [MECH ENG] An atomizing device used in gravitational atomization; functions by quasi-static emission of a drop from a wetted surface. Also known as pendant atomizer. { 'haŋ·iŋ ,dräp 'ad·ə,mīz·ər }

hanging-drop preparation [MICROBIO] A technique used in microscopy in which a specimen is placed in a drop of a suitable fluid on a cover slip and the cover slip is inverted over a concavity on a slide. { 'haŋ·iŋ ,dräp ,prep·ə'rā·shən }

hanging glacier [HYD] A glacier lying above a cliff or steep mountainside; as the glacier advances, calving can cause ice avalanches. { 'haŋ·iŋ ¦glā·shər }

hanging load [MECH ENG] **1.** The weight that can be suspended on a hoist line or hook device in a drill tripod or derrick without causing the members of the derrick or tripod to buckle. **2.** The weight suspended or supported by a bearing. { 'haŋ·iŋ ¦lōd }

hanging scaffold [CIV ENG] A movable platform suspended by ropes and pulleys; used by workers for above-ground building construction and maintenance. { 'haŋ·iŋ ¦ska,fȯld }

hanging sets [MIN ENG] Timbers from which cribs are suspended in working through soft strata. { 'haŋ·iŋ ¦sets }

hanging sheave [MIN ENG] The grooved wheel or pulley which is suspended from the drill tripod clevis or from the roof or side of a haulage road, and over which the hoist line runs to minimize friction. { 'haŋ·iŋ ¦shēv }

hanging side See hanging wall. { 'haŋ·iŋ ¦sīd }

hanging valley [GEOL] A valley whose floor is higher than the level of the shore or other valley to which it leads. { 'haŋ·iŋ ¦val·ē }

hanging wall [GEOL] The rock mass above a fault plane, vein, lode, ore body, or other structure. Also known as hanger; hanging; hanging side. { 'haŋ·iŋ ¦wȯl }

hanging-wall drift [MIN ENG] A horizontal gallery that is driven in the hanging wall of a vein. { 'haŋ·iŋ ,wȯl ,drift }

hangline [PETRO ENG] A single length of wire rope attached to the crown block at the top of a derrick for suspending the traveling block when it is not being used. Also known as hang-off line. { 'haŋ,līn }

hang-off line See hangline. { 'haŋ ,ȯf ,līn }

hangover [COMMUN] **1.** In television, overlapping and blurring of successive frames opposite to direction of subject motion, due to improper adjustment of transient response. **2.** In facsimile, distortion produced when the signal changes from maximum to minimum conditions at a slower rate than required, resulting in tailing on the lines in the recorded copy. [MED] After effect of excessive intake of alcohol or certain drugs, such as barbiturates. { 'haŋ,ō·vər }

HAND TRUCK

curved handles

legs

nose

straight frame

straight crossmember

An Eastern type of hand truck.

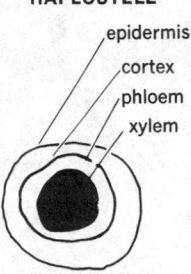

HAPLOSTELE

epidermis
cortex
phloem
xylem

Cross section of a haplostele.

hang-up [COMPUT SCI] A nonprogrammed stop in a computer routine caused by a human mistake or a computer malfunction. [ENG] A virtual leak resulting from the release of entrapped tracer gas from a leak detector vacuum system. [MIN ENG] Blockage of the movement of ore by rock in an underground chute. { 'haŋ,əp }

hangwire [ORD] A length of wire connecting the fuse assembly of an aerial flare or bomb to the structure of an aircraft; it removes the safety and arms the fuse at the beginning of the drop; it may also open a parachute or stabilizing sleeve. { 'haŋ,wīr }

hank [TEXT] A unit of measure of yarn or thread equal to 840 yards (768.096 meters) for cotton, and 560 yards (512.064 meters) for first-grade wool. { haŋk }

Hankel functions [MATH] The Bessel functions of the third kind, occurring frequently in physical studies. { 'häŋk·əl ,faŋk·shənz }

Hankel transform [MATH] The Hankel transform of order m of a real function $f(t)$ is the function $F(s)$ given by the integral from 0 to ∞ of $f(t)tJ^m(st)dt$, where J^m denotes the mth-order Bessel function. Also known as Bessel transform; Fourier-Bessel transform. { 'häŋk·əl ,tranz,form }

hanksite [MINERAL] $Na_{22}K(SO_4)_9(CO_3)_2Cl$ A white or yellow mineral crystallizing in the hexagonal system; found in California. { 'haŋk,sīt }

Hanle effect [OPTICS] A reduction in the polarization of light that is emitted from atoms excited by linearly polarized light when a magnetic field is applied in the direction of observation. { 'hän·lē i,fekt }

hannayite [MINERAL] $Mg_3(NH_4)_2H_2(PO_4)_4\cdot8H_2O$ Mineral composed of hydrous acid ammonium magnesium phosphate; occurs as yellow crystals in guano. { 'ha·nē,īt }

Hansa yellow [ORG CHEM] Group of organic azo pigments with strong tinting power, but poor opacity in paints; used where nontoxicity is important. { 'hän·sə 'yel·ō }

Hansen's disease [MED] An infectious disease of humans thought to be caused by *Mycobacterium leprae*; common manifestations are cutaneous and neural lesions. Also known as leprosy. { 'han·sənz di,zēz }

H antigen [MICROBIO] A general term for microbial flagellar antigens; former designation for species-specific flagellar antigens of *Salmonella*. { 'āch ,ant·i·jən }

Hantzsch synthesis [ORG CHEM] The reaction whereby a pyrrole compound is formed when a β-ketoester, chloroacetone, and a primary amine condense. { 'hänsh ,sin·thə·səs }

Hanus solution [ANALY CHEM] Iodine monobromide in glacial acetic acid; used to determine iodine values in oils containing unsaturated organic compounds. { 'han·əs sə,lü·shən }

HAP See hazardous air pollutants. { hap *or* 'āch'ā'pē }

haplobiont [BOT] A plant that produces only sexual haploid individuals. { 'ha·plō'bī,änt }

haplocaulescent [BOT] Having a simple axis with the reproductive organs on the principal axis. { 'ha·plō·kȯ'les·ənt }

Haplodoci [VERT ZOO] The equivalent name for Batrachoidiformes. { hə'pläd·ə,sī }

haploid [GEN] Having half of the diploid complement of chromosomes, that is, one complete set, as in mature gametes. { 'ha,plȯid }

haploidization [MYCOL] In certain fungi, the transformation of a diploid into a haploid cell by progressive loss of chromosomes due to nondisjunction. { ,hap,lȯid·i'zā·shən }

haplo-insufficiency [GEN] In a diploid organism, the presence of an abnormal phenotype in which only one of the two copies of a particular gene yields its normal product, and a reduced amount rather than a qualitative change in the gene product leads to the phenotype. It is responsible for some, but not all, autosomal dominant disorders. { ,hap,lō 'in·sə,fish·ən·sē }

Haplolepidae [PALEON] A family of Carboniferous chondrostean fishes in the suborder Palaeoniscoidei having a reduced number of fin rays and a vertical jaw suspension. { ,ha·plō'lep·ə,dē }

Haplomi [VERT ZOO] An equivalent name for Salmoniformes. { ha'plō,mī }

haplomitosis [CYTOL] Type of primitive mitosis in which the nuclear granules form into threadlike masses rather than clearly differentiated chromosomes. { ,ha·plō,mī'tō·səs }

haplont [BOT] A plant with only haploid somatic cells; the zygote is diploid. { 'ha,plänt }

haplophase [BIOL] Haploid stage in the life cycle of an organism. { 'ha·plō,fāz }

haplopore [PALEON] Any randomly distributed pore on the surface of fossil cystoid echinoderms. { 'ha·plō,pȯr }

Haplosclerida [INV ZOO] An order of sponges in the class Demospongiae including species with a skeleton made up of siliceous megascleres embedded in spongin fibers or spongin cement. { ,ha·plō'skler·ə·də }

haplosis [CELL MOL] Reduction of the chromosome number to half during meiosis. { ha'plō·səs }

Haplosporea [INV ZOO] A class of Protozoa in the subphylum Sporozoa distinguished by the production of spores lacking polar filaments. { ,ha·plō'spȯr·ē·ə }

Haplosporida [INV ZOO] An order of Protozoa in the class Haplosporea distinguished by the production of uninucleate spores that lack both polar capsules and filaments. { ,ha·plō'spȯr·ə·də }

haplostele [BOT] A type of protostele with the core of xylem characterized by a smooth outline. { 'ha·plō,stēl }

haplo-sufficient gene [GEN] Any gene that allows the production of viable adults even when one copy of the gene in diploids is mutant or deleted from one of the homologous chromosomes. { ,hap,lō sə,fish·ənt 'jēn }

haplotype [GEN] A set of alleles of closely linked loci on a chromosome that tend to be inherited together; commonly used in reference to the linked genes of the major histocompatibility complex. { 'hap·lə,tīp }

hapten [IMMUNOL] An incomplete antigen that cannot induce antibody formation by itself but can do so by coupling with a larger carrier molecule. { 'hap·tən }

haptephobia [PSYCH] An abnormal fear of being touched. { ,hap·tə'fō·bē·ə }

hapteron [BOT] A disklike holdfast on the stem of certain algae. { 'hap·tə,rän }

haptic interface [COMPUT SCI] A device that allows a user to interact with a computer by receiving tactile feedback; for example, glove or pen devices that allow users to touch and manipulate three-dimensional virtual objects. { ,hap·tik 'in·tər,fās }

haptics [COMPUT SCI] The study of the use of touch in order to produce computer interfaces that will allow users to interact with digital objects by means of force feedback and tactile feedback. { 'hap·tiks }

haptochlamydeous [BOT] Having the sporophylls protected by rudimentary perianth leaves. { ,hap·tō·klə'mid·ē·əs }

haptoglobin [BIOCHEM] An alpha globulin that constitutes 1–2% of normal blood serum; contains about 5% carbohydrate. { 'hap·tə,glō·bən }

Haptophyceae [BOT] A class of the phylum Chrysophyta that contains the Coccolithophorida. { ,hap·tə'fīs·ē,ē }

haptor [INV ZOO] The posterior organ of attachment in certain monogenetic trematodes characterized by multiple suckers and the presence of hooks. { 'hap·tər }

haptotropism [BIOL] Movement of sessile organisms in response to contact, especially in plants. { hap'tä·trə,piz·əm }

harassing agent [ORD] A chemical agent that forces troops to wear gas masks and to cut down their efficiency; examples are an irritating gas or smoke. { 'har·ə·siŋ ,ā·jənt }

harassing fire [ORD] Rifle or artillery fire designed to inflict losses, or to disturb the enemy troops by the threat of losses; the fire restricts movement and lowers troop morale. { 'har·ə·siŋ ,fīr }

harbor [GEOGR] Any body of water of sufficient depth for ships to enter and find shelter from storms or other natural phenomena. Also known as port. { 'här·bər }

harbor chart [NAV] A nautical chart for navigation and anchorage in small waterways and harbors. { 'här·bər ,chärt }

harbor engineering [CIV ENG] Planning and design of facilities for ships to discharge or receive cargo and passengers. { 'här·bər ,en·jə'nir·iŋ }

harbor line [CIV ENG] The line beyond which wharves and other structures cannot be extended. { 'här·bər ,līn }

harbor reach [GEOGR] The stretch of a river or estuary which leads directly to the harbor. { 'här·bər ,rēch }

hard [MATER] Quality of a material that is compact, solid, and difficult to deform. { härd }

hard acid [CHEM] A Lewis acid of low polarizability, small size, and high positive oxidation state; it does not have easily excitable outer electrons; some examples are H^+, Li^+, and Al^+. { 'härd 'as·əd }

hard automation [IND ENG] Automation that makes use of specially designed equipment for production. { 'härd ˌöd·ə'mā·shən }

hard base [CHEM] A Lewis base (electron donor) that has high polarizability and low electronegativity, is easily oxidized, or possesses lowlying empty orbitals; some examples are H_2O, HO^-, OCH_3^-, and F^-. [ORD] A launching base that is protected against a nuclear explosion. { 'härd ˌbās }

hard beach [CIV ENG] A portion of a beach especially prepared with a hard surface extending into the water, employed for the purpose of loading or unloading directly into or from landing ships or landing craft. { 'härd ˌbēch }

hardboard [MATER] A fiberboard formed to a density of 30–50 pounds per cubic foot (480–800 kilograms per cubic meter) and having one textured and one smooth face. { 'härd,bȯrd }

hard bottom [MIN ENG] A condition encountered in some opencut mines wherein the rock occasionally does not break down to grade because of an extra-hard streak of ground or because insufficient powder is used. { 'härd ˌbäd·əm }

hard bronze [MET] A high-tensile-strength alloy containing 88% copper, 7% tin, 3% zinc, and 2% lead. { 'härd ˌbränz }

hard-burned brick [MATER] A brick that has been fired and sintered at high temperature. { 'härd ˌbȯrnd 'brik }

hardcap [MIN ENG] In bauxite mining, the uppermost foot or two (0.3–0.6 meter) deposit of bauxite; usually serves as a roof during mining. { 'härd,kap }

hard chromium [MET] A thick coating of electrodeposited chromium on a base metal; increases wear resistance. { 'härd ˌkrō·mē·əm }

hard cider [FOOD ENG] Fermented apple cider that contains less than 10% alcohol. { 'härd ˌsīd·ər }

hard coal *See* anthracite. { 'härd ˌkōl }

hard-coal plough [MIN ENG] A plough-type cutter-loader consisting of rigid or swiveling kerfing bits which precut the coal in hard-coal seams. { 'härd,kōl ˌplau }

hard code [COMPUT SCI] Program statements that are written into the computer program itself, in contrast to external tables and files to hold values and parameters used by the program. { 'härd 'kōd }

hard-coded program [COMPUT SCI] A software program or program subroutine that is designed to perform a specific task and is not easily modified. { 'härd ˌkōd·əd 'prō·grəm }

hard copy [COMPUT SCI] Human-readable typewritten or printed characters produced on paper at the same time that information is being keyboarded in a coded machine language, as when punching cards or paper tape. [GRAPHICS] The typewritten or paper copy of material keyboarded into a computer, word processor, or typesetter. { 'härd ˌkäp·ē }

hard cosmic ray [NUCLEO] A cosmic-radiation component that penetrates a moderate thickness of an absorber, such as 4 inches (10 centimeters) of lead. { 'härd ˌkäz·mik 'rā }

hard crash [COMPUT SCI] An abrupt halting of operations by a computer due to a malfunction, allowing the users or operators of the computer little or no time to minimize its effects. { 'härd 'krash }

hard data [SCI TECH] Data in the form of numbers or graphs, as opposed to qualitative information. { 'härd ˌdad·ə }

hard detergent [CHEM] A nonbiodegradable detergent. { 'härd di'tər·jənt }

hard disk [COMPUT SCI] A magnetic disk made of rigid material, providing high-capacity random-access storage. { 'härd ˌdisk }

hard-drawn wire [MET] Cold-drawn metal wire, usually of high tensile strength. { 'härd ˌdrȯn 'wīr }

hard edit [COMPUT SCI] The process of checking and correction that causes data containing errors to be rejected by a computer system. { 'härd 'ed·it }

hardenability [MET] In a ferrous alloy, the property that determines the depth and distribution of hardness induced by quenching from elevated temperatures. { ˌhärd·ən·ə'bil·əd·ē }

hardened circuit [ELECTR] A circuit that uses components whose tolerance to radiation released by a nuclear explosion has been increased by various radiation-hardening procedures. { 'härd·ənd ˌsər·kət }

hardened links [COMMUN] Transmission links that require special construction or installation to assure a high probability of survival under nuclear attack. { 'härd·ənd ˌliŋks }

hardened site [ORD] An underground missile-launching site, control center, or other facility that can withstand in varying degrees the effects of an enemy nuclear blast. { 'härd·ənd ˌsīt }

hardened steel [MET] Steel hardened by quenching from high temperatures. { 'härd·ənd ˌstēl }

hardener [MET] A master alloy added to a melt to control hardness. [ORG CHEM] Compound reacted with a resin polymer to harden it, such as the amines or anhydrides that react with epoxides to cure or harden them into plastic materials. Also known as curing agent. { 'härd·ən·ər }

hardener bath [GRAPHICS] A fixing bath containing chemicals such as formalin and chrome alum to toughen the emulsion of a film. { 'härd·ən·ər ˌbath }

hardening [MET] **1.** Imparting hardness to carbon steel by abrupt cooling (quenching) through a critical temperature range. **2.** Heat-treating an age-hardening or precipitation-hardening alloy at intermediate temperatures. [BOT] Treatment of plants designed to increase their resistance to extremes in temperature or drought. { 'härd·ən·iŋ }

Harderian gland [VERT ZOO] An accessory lacrimal gland associated with lower eyelid structures in all vertebrates except land mammals. { här'dir·ē·ən ˌgland }

hard error [COMPUT SCI] Any error that results from malfunctioning of hardware, including storage devices and data transmission equipment. { 'härd 'er·ər }

hard-face [MET] To apply a layer of hard, abrasion-resistant metal to a less resistant metal part by plating, welding, spraying, or other techniques. Also known as hard-surface. { 'härd ˌfās }

hard failure [COMPUT SCI] Equipment failure that requires repair by a person with specialized knowledge before the equipment can be put back into operation. { 'härd 'fāl·yər }

hard fiber [BOT] A heavily lignified leaf fiber used in making cordage, twine, and textiles. [MATER] Indicating vulcanization with zinc chloride; used of paper or boards. { 'härd ˌfī·bər }

hard freeze [HYD] A freeze in which seasonal vegetation is destroyed, the ground surface is frozen solid underfoot, and heavy ice is formed on small water surfaces such as puddles and water containers. { 'härd ˌfrēz }

hard frost *See* black frost. { 'härd ˌfrȯst }

hard glass [MATER] A potash-lime glass with a high silica content, used for making brilliant glassware. Also known as Bohemian glass. { 'härd ˌglas }

hard goods *See* durable goods. { 'härd ˌgudz }

hard grease [MATER] A lubricating grease that flows at a temperature of about 90°C. { 'härd ˌgrēs }

hard ground [MIN ENG] Ground that is difficult to work. { 'härd ˌgraund }

Hardgrove grindability index [ENG] The relative grindability of ores and minerals in comparison with standard coal, chosen as 100 grindability, as determined by a miniature ball-ring pulverizer. Also known as Hardgrove number. { 'här ˌgrōv ˌgrīn·də'bil·əd·ē ˌin,deks }

Hardgrove number *See* Hardgrove grindability index. { 'här ˌgrōv ˌnəm·bər }

hard hat [ENG] A safety hat usually having a metal crown; used by construction workers and miners. { 'härd ˌhat }

hardhead [MET] A hard white deposit formed during tin refining by liquation. { 'härd,hed }

Hardinge feeder-weigher [MECH ENG] A pivoted, short belt conveyor which controls the rate of material flow from a hopper by weight per cubic foot. { 'här·diŋ ˌfēd·ər ˌwā·ər }

Hardinge mill [MECH ENG] A tricone type of ball mill; the cones become steeper from the feed end toward the discharge end. { 'här·diŋ ˌmil }

Hardinge thickener [ENG] A machine for removing the maximum amount of liquid from a mixture of liquid and finally divided solids by allowing the solids to settle out on the bottom as sludge while the liquid overflows at the top. { 'här·diŋ 'thik·ən·ər }

hard iron [MET] Iron or steel which is not readily magnetized by induction, but which retains a high percentage of the magnetism acquired. { 'härd ¦ī·ərn }

hard lac [MATER] Solvent-extracted shellac. Also known as hard-lac resin. { 'härd ¦lak }

hard-lac resin *See* hard lac. { 'härd ¦lak ‚rez·ən }

hard-laid [DES ENG] Pertaining to rope with strands twisted at a 45° angle. { 'härd ¦lād }

hard landing [AERO ENG] A landing made without deceleration, as by impact on the moon. { 'härd ¦land·iŋ }

hard lead [MET] Lead alloy with reduced malleability due to the presence of impurities, usually antimony. { 'härd ¦led }

hard-limiting [COMMUN] Limiting condition for which there is little variation in the output signal over the input signal range where the input is subject to limiting. { 'härd ‚lim·əd·iŋ }

hard magnetic material [MET] A metal having a high coercive force which gives a large magnetic hysteresis. { 'härd mag¦ned·ik mə'tir·ē·əl }

hardness [CHEM] The amount of calcium carbonate dissolved in water, usually expressed as parts of calcium carbonate per million parts of water. [ELECTROMAG] That quality which determines the penetrating ability of x-rays; the shorter the wavelength, the harder and more penetrating the rays. [ENG] Property of an installation, facility, transmission link, or equipment that will prevent an unacceptable level of damage. [MATER] Resistance of a metal or other material to indentation, scratching, abrasion, or cutting. { 'härd·nəs }

hardness number [ENG] A number representing the relative hardness of a mineral, metal, or other material as determined by any of more than 30 different hardness tests. { 'härd·nəs ‚nəm·bər }

hardness test [ANALY CHEM] A test to determine the calcium and magnesium content of water. [ENG] A test to determine the relative hardness of a metal, mineral, or other material according to one of several scales, such as Brinell, Mohs, or Shore. { 'härd·nəs ‚test }

hard page [COMPUT SCI] A page break that is inserted in a document by the user, and whose location is not changed by the addition, deletion, or reformatting of text. { 'härd ‚pāj }

hard pad [VET MED] A disease of dogs, probably associated with the canine distemper virus; often characterized by encephalitis and hardening of the foot pads. { 'härd ‚pad }

hard palate [ANAT] The anterior portion of the roof of the mouth formed by paired palatine processes of the maxillary bones and by the horizontal part of each palate bone. { 'härd ¦pal·ət }

hardpan *See* caliche. { 'härd‚pan }

hard paste porcelain *See* porcelain. { 'härd ‚pāst 'pórs·lən }

hard patch [COMPUT SCI] A modification of a computer program, generally to repair a software error, which is applied to a stored copy of the program in machine language, so that recompilation of the source program is unnecessary and the change is permanent. { 'härd ‚pach }

hard porcelain [MATER] A ceramic material having good resistance to thermal shock. { 'härd ¦pórs·lən }

hard radiation [PHYS] Radiation whose particles or photons have a high energy and, as a result, readily penetrate all kinds of materials, including metals. { 'härd ‚rād·ē'ā·shən }

hard return [COMPUT SCI] A control code that is entered into a document by pressing the enter key. { 'härd ri‚tərn }

hard rime [HYD] Opaque, granular masses of rime deposited chiefly on vertical surfaces by a dense super-cooled fog; it is more compact and amorphous than soft rime, and may build out into the wind as glazed cones or feathers. { 'härd ¦rīm }

hard rock [GEOL] Rock which needs drilling and blasting for removal. { 'härd ¦räk }

hard-rock driller [MIN ENG] A worker who operates a drill in a mine where the rocks are generally igneous or metamorphosed and considered hard, such as rocks in which coal and salt are generally found. { 'härd ¦räk 'dril·ər }

hard-rock mine [MIN ENG] A mine located in hard rock, especially a mine difficult to drill, blast, and square up. { 'härd ¦räk 'mīn }

hard-rock miner [MIN ENG] A worker competent to mine in hard rock, usually an expert miner. { 'härd ¦räk 'mīn·ər }

hard-rock tunnel boring [MIN ENG] A tunneling method utilizing a mole to cut out 7-foot-diameter (2.1-meter) drifts in hard rock at an average rate of 5 feet (1.5 meters) per hour. { 'härd ¦räk 'tən·əl ‚bór·iŋ }

hard rot [PL PATH] **1.** Any plant disease characterized by

lesions with hard surfaces. **2.** A fungus disease of gladiolus caused by *Septoria gladioli* which produces hard-surfaced lesions on the leaves and corms. { 'härd ‚rät }

hard rubber [MATER] Rubber that has been vulcanized at high temperatures and pressures to give hardness; used as an electrical insulating material and in tool handles. Also known as ebonite. { 'härd ‚rəb·ər }

hard-sectored disk [COMPUT SCI] A disk whose sectors are set up during manufacture. { 'härd ¦sek·tərd 'disk }

hard silk [TEXT] Silk that has not been degummed, so that the finished fabric contains sericin, or silk gum. { 'härd ¦silk }

hard site [ORD] Underground missile-launching site, control center, or other facility that can withstand in varying degrees the effects of an enemy nuclear blast. { 'härd ¦sīt }

hard solder *See* brazing alloy. { 'härd ¦säd·ər }

hard-solder *See* braze. { 'härd ‚säd·ər }

hard-sphere collision theory [PHYS CHEM] A theory for calculating reaction rate constants for biomolecular gas-phase reactions in which the molecules are considered to be colliding, hard spheres. { 'härd ‚sfir kə'lizh·ən ‚thē·ə·rē }

hardstand [CIV ENG] **1.** A paved or stabilized area where vehicles or aircraft are parked. **2.** Open ground area having a prepared surface and used for storage of material. { 'härd‚stand }

hard superconductor [CRYO] A superconductor that requires a strong magnetic field, over 1000 oersteds (79,577 amperes per meter), to destroy superconductivity; niobium and vanadium are examples. { 'härd 'sü·pər·kən‚däk·tər }

hard-surface [CIV ENG] To treat a ground surface in order to prevent muddiness. [MET] *See* hard-face. { 'härd ¦sər·fəs }

hard tube *See* high-vacuum tube. { 'härd ‚tüb }

hardware [COMPUT SCI] The physical, tangible, and permanent components of a computer or a data-processing system. [ENG] Items made of metal, such as tools, fittings, fasteners, and appliances. [ORD] Metal military items for use in combat. { 'härd‚wer }

hardware check *See* machine check. { 'härd‚wer ‚chek }

hardware compatibility [COMPUT SCI] Property of two computers such that the object code from one machine can be loaded and executed on the other to produce exactly the same results. { 'härd‚wer kəm‚pad·ə'bil·əd·ē }

hardware control [COMPUT SCI] The control of, and communications between, the various parts of a computer system. { 'härd‚wer kən‚trōl }

hardware description language [COMPUT SCI] A computer language that facilitates the documentation, design, and manufacturing of digital systems, particularly very large-scale integrated circuits, and combines program verification techniques with expert system design methodologies. { 'här‚dwer di 'skrip·shən ‚laŋ·gwij }

hardware diagnostic [COMPUT SCI] A computer program designed to determine whether the components of a computer are operating properly. { 'härd‚wer dī·əg'näs·tik }

hardware division [COMPUT SCI] Mathematical division performed by electronic circuitry on a large computer as a result of a single machine instruction. { 'här‚dwer di‚vizh·ən }

hardware floating point [COMPUT SCI] Complex circuitry within a central processing unit that carries out floating-point arithmetic. { 'här‚dwer 'flōd·iŋ 'póint }

hardware key *See* dongle. { 'här‚dwer ‚kē }

hardware monitor [COMPUT SCI] A system used to evaluate the performance of computer hardware; it collects information such as central processing unit usage from voltage level sensors that are attached to the circuitry and measure the length of time or the number of times various signals occur, and displays this information or stores it on a medium that is then fed into a special data-reduction program. { 'härd‚wer ‚män·əd·ər }

hardware multiplexing [COMPUT SCI] A procedure in which a servicing unit interleaves its attention among a family of serviced units in such a way that the serviced units appear to be receiving constant attention. { 'härd‚wer 'məl·tə‚plek·siŋ }

hardware multiplication [COMPUT SCI] Multiplication performed by electronic circuitry on a large computer as a result of a single machine instruction. { 'här‚dwer ‚məl·tə·plə‚kā·shən }

hard water [CHEM] Water that contains certain salts, such as those of calcium or magnesium, which form insoluble deposits in boilers and form precipitates with soap. { 'härd ¦wȯd·ər }

Hardwick conveyor loader head [MIN ENG] A dust collector for belt conveyors used at the loading station; a scraper chain runs at the bottom of a coal hopper and collects underbelt fines. { 'härd,wik kən¦vā·ər 'lōd·ər ¦hed }

hard-wire [ELEC] To connect electric components with solid, metallic wires as opposed to radio links and the like. { 'härd ¦wīr }

hard-wired [COMPUT SCI] Having a fixed wired program or control system built in by the manufacturer and not subject to change by programming. { 'härd ¦wīrd }

hard-wire telemetry See wire-link telemetry. { 'härd ,wīr tə'lem·ə·trē }

hardwood [MATER] Dense, close-grained wood of an angiospermous tree, such as oak, walnut, cherry, and maple. { 'härd,wud }

hardwood bearing [MECH ENG] A fluid-film bearing made of lignum vitae which has a natural gum, or of hard maple which is impregnated with oil, grease, or wax. { 'härd,wud ¦ber·iŋ }

hardwood forest [ECOL] **1.** An ecosystem having deciduous trees as the dominant form of vegetation. **2.** An ecosystem consisting principally of trees that yield hardwood. { 'härd ,wud ¦fär·əst }

hard x-ray [ELECTR] An x-ray having high penetrating power. { 'härd ¦eks,rā }

Hardy plankton indicator [ENG] Metal-shrouded net sampler designed to collect specimens of plankton during normal passage of a ship. { 'härd·ē 'plaŋk·tən ,in·də,kād·ər }

hardy plant [BOT] A plant able to withstand low temperatures without artificial protection. { ¦här·dē ¦plant }

Hardy-Schulz rule [PHYS CHEM] An increase in the charge of ions results in a large increase in their flocculating power. { 'härd·ē 'shults ,rül }

hardystonite [MINERAL] Ca₂ZnSi₂O₇ A white mineral composed of zinc calcium silicate. { 'här·dē·stə,nīt }

Hardy-Weinberg law [GEN] The concept that frequencies of both genes and genotypes will remain constant from generation to generation in an idealized population where mating is random and evolutionary forces (such as mutation, migration, selection, or genetic drift) are absent. { 'här·dē 'wīn,bərg ,lȯ }

hare [VERT ZOO] The common name for a number of lagomorphs in the family Leporidae; they differ from rabbits in being larger with longer ears, legs, and tails. { her }

Hare See Lepus. { her }

harelip [MED] A congenital defect, sometimes hereditary, marked by an abnormal cleft between the upper lip and the base of the nose. Also known as cleft lip. { 'her,lip }

Hare's hygrometer [ENG] A type of hydrometer in which the ratio of the densities of two liquids is determined by measuring the heights to which they rise in two vertical glass tubes, connected at their upper ends, when suction is applied. { 'herz hī'gräm·əd·ər }

Hargreaves process [CHEM ENG] A process for the manufacture of salt cake (sodium sulfate) by passing a mixture of sulfur dioxide and air through sodium chloride brine in a countercurrent manner. { 'här·grēvz ,prä·səs }

Haring cell [PHYS CHEM] An electrolytic cell with four electrodes used to measure electrolyte resistance and polarization of electrodes. { 'her·iŋ ,sel }

Harker diagram See variation diagram. { 'härk·ər ,dī·ə,gram }

Harker-Kasper inequalities [SOLID STATE] Inequalities used in the analysis of crystal structure by x-ray diffraction which relate the structure factors and help to determine their phase factors. { 'härk·ər 'kas·pər ,in·i·'kwäl·əd·ēz }

Harkin's rule [PHYS] An empirical rule for the calculation of the nuclear abundances of an element's isotopes stating that isotopes with an odd mass number are less abundant than their even-mass-number neighbors. { 'här·kənz ,rül }

Harlechian [GEOL] A European stage of geologic time: Lower Cambrian. { här'lek·ē·ən }

HARM See high-aspect-ratio micromachining.. { ¦āch¦ā¦är'em or härm }

harman [ORG CHEM] C₁₂H₁₀N₂ Crystals that melt at 237–238°C; inhibits growth of molds and certain bacteria. Also known as arabine; loturine; passiflorin. { 'här·mən }

harmatan See harmattan. { ,här·mə'tan }

harmattan [METEOROL] A dry, dust-bearing wind from the northeast or east which blows in West Africa especially from late November until mid-March; it originates in the Sahara as a desert wind and extends southward to about 5°N in January and 18°N in July. Also spelled harmatan; harmetan; hermitan. { ,här·mə'tan }

harmetan See harmattan. { ,här·mə'tan }

harmful interference [COMMUN] Radiation, emission, or induction which endangers the functioning of a radionavigation broadcasting service or of a safety broadcasting service, or obstructs or repeatedly interrupts a radio service operating in accordance with the appropriate regulations. { 'härm·ful ,int·ə'fir·əns }

harmless-depth theory [MIN ENG] Formerly, the hypothesis that there was a certain depth below which mining could be carried on without risk of damage to the surface. { 'härm·ləs ¦depth ,thē·ə·rē }

harmonic [ACOUS] One of a series of sounds, each of which has a frequency which is an integral multiple of some fundamental frequency. [MATH] A solution of Laplace's equation which is separable in a specified coordinate system. [PHYS] A sinusoidal component of a periodic wave, having a frequency that is an integral multiple of the fundamental frequency. Also known as harmonic component. { här'män·ik }

harmonica bug [ELECTR] A surreptitious interception technique applied to telephone lines; the target instrument is modified so that a tuned relay bypasses the switch hook and ringing circuit when a 500-hertz tone is received; this tone was originally generated by use of a harmonica. { här'män·ə·kə ,bəg }

harmonic analysis [MATH] A study of functions by attempting to represent them as infinite series or integrals which involve functions from some particular well-understood family; it subsumes studying a function via its Fourier series. [PHYS] Any method of identifying and evaluating the harmonics that make up a complex waveform of sound pressure, voltage, current, or some other varying quantity. { här'man·ik ə'nal·ə·səs }

harmonic analyzer [ELECTR] An instrument that measures the strength of each harmonic in a complex wave. Also known as harmonic wave analyzer. { här'man·ik 'an·ə,līz·ər }

harmonic antenna [ELECTROMAG] An antenna whose electrical length is an integral multiple of a half-wavelength at the operating frequency of the transmitter or receiver. { här'män·ik an'ten·ə }

harmonic attenuation [ELECTR] Attenuation of an undesired harmonic component in the output of a transmitter. { här'män·ik ə,ten·yə'wā·shən }

harmonic average See harmonic mean. { här,män·ik 'av·rij }

harmonic component See harmonic. { här'män·ik kəm'pō·nənt }

harmonic conjugates [MATH] **1.** Two points, P_3 and P_4, that are collinear with two given points, P_1 and P_2, such that P_3 lies in the line segment P_1P_2 while P_4 lies outside it; and, if x_1, x_2, x_3, and x_4 are the abscissas of the points, $(x_3 - x_1)/(x_3 - x_2) = -(x_4 - x_1)/(x_4 - x_2)$. **2.** A pair of harmonic functions, u and v, such that $u + iv$ is an analytic function, or, equivalently, u and v satisfy the Cauchy-Riemann equations. { här'män·ik 'kän·jə·gəts }

harmonic content [PHYS] The components remaining after the fundamental frequency has been removed from a complex wave. { här'män·ik 'kän·tent }

harmonic conversion transducer [ELECTR] A conversion transducer of which the useful output frequency is a multiple or a submultiple of the input frequency. { här'män·ik kən¦vər·zhən tranz,dü·sər }

harmonic decline [PETRO ENG] One of three types of decline in oil or gas production rate (the others are constant-percentage and hyperbolic), in which the nominal decline in production rate per unit of time expressed as a fraction of the production rate is proportional to the production rate itself. { här'män·ik di'klīn }

harmonic detector [ELECTR] Voltmeter circuit so arranged as to measure only a particular harmonic of the fundamental frequency. { här'män·ik di'tek·tər }

harmonic distortion [ELECTR] Nonlinear distortion in which undesired harmonics of a sinusoidal input signal are generated because of circuit nonlinearity. { här'män·ik di'stȯr·shən }

harmonic division [MATH] The division of a line segment externally and internally in the same ratio; that is, the division of a line segment by the harmonic conjugates of its end points. { här¦män·ik di'vizh·ən }

harmonic drive [MECH ENG] A drive system that uses inner and outer gear bands to provide smooth motion. { här'män·ik 'drīv }

harmonic echo [ACOUS] An echo that appears to be higher in pitch than the original sound, due to enhancement of harmonics in the original complex tone. { här¦män·ik 'ek·ō }

harmonic fields [ELECTROMAG] The sinusoidal Fourier components of a magnetic or other field confined to a finite region of space; their half-wavelengths are integral divisors of the length of the space in which the field is confined. { här¦män·ik 'fēlz }

harmonic filter [ELECTR] A filter that is tuned to suppress an undesired harmonic in a circuit. { här'män·ik 'fil·tər }

harmonic folding [GEOL] Folding in the earth's surface, with no sharp changes with depth in the form of the folds. { här¦män·ik 'fōld·iŋ }

harmonic frequency [PHYS] An integral multiple of the fundamental frequency of a periodic wave. { här'män·ik 'frē·kwən·sē }

harmonic function [MATH] **1.** A function of two real variables which is a solution of Laplace's equation in two variables. **2.** A function of three real variables which is a solution of Laplace's equation in three variables. { här'män·ik 'fəŋk·shən }

harmonic generator [ELECTR] A generator operated under conditions such that it generates strong harmonics along with the fundamental frequency. { här'män·ik 'jen·ə,rād·ər }

harmonic interference [COMMUN] Interference due to the presence of harmonics in the output of a radio transmission. { här'män·ik ,in·tər'fir·əns }

harmonic law [ASTRON] The third of Kepler's laws, which states that the squares of the periods of revolution of any two planets are proportional to the cubes of their mean distances from the sun. { här'män·ik 'lȯ }

harmonic loss [ELECTROMAG] Energy loss in a generator due to space harmonics of the magnetomotive force produced by armature current, especially losses resulting from the fifth and seventh harmonics. { här'män·ik 'lȯs }

harmonic mean [MATH] For n positive numbers $x_1, x_2, \ldots, x_n$ their harmonic mean is the number $n/(1/x_1 + 1/x_2 + \cdots + 1/x_n)$. Also known as harmonic average. { här'män·ik 'mēn }

harmonic measure [MATH] Let D be a domain in the complex plane bounded by a finite number of Jordan curves Γ, and let Γ be the disjoint union of α and β, where α and β are Jordan arcs; the harmonic measure of α with respect to D is the harmonic function on D which assumes the value 1 on α and the value 0 on β. { här¦män·ik 'mezh·ər }

harmonic motion [MECH] A periodic motion that is a sinusoidal function of time, that is, motion along a line given by the equation $x = a \cos(kt + \theta)$, where t is the time parameter, and a, k, and θ are constants. Also known as harmonic vibration; simple harmonic motion (SHM). { här'män·ik 'mō·shən }

harmonic oscillator [ELECTR] *See* sinusoidal oscillator. [MECH] Any physical system that is bound to a position of stable equilibrium by a restoring force or torque proportional to the linear or angular displacement from this position. [PHYS] Anything which has equations of motion that are the same as the system in the mechanics definition. Also known as linear oscillator; simple oscillator. { här'män·ik 'äs·ə,lād·ər }

harmonic pencil [MATH] The configuration of four lines, passing through a single point, such that any line that is not parallel to one of the four cuts the four lines at points which are harmonic conjugates. { här¦män·ik 'pen·səl }

harmonic prediction [OCEANOGR] A method used in predicting the tides and tidal currents by combining the harmonic constituents into a single tide curve. { här'män·ik prə'dik·shən }

harmonic producer [ELECTR] Tuning-fork controlled oscillator device capable of producing odd and even harmonics of the fundamental tuning-fork frequency; used to provide carrier frequencies for broad-band carrier systems. { här'män·ik prə,dü·sər }

harmonic progression [MATH] A sequence of numbers whose reciprocals form an arithmetic progression. Also known as harmonic sequence. { här¦män·ik prə'gresh·ən }

harmonic range [MATH] The configuration of four collinear points which are harmonic conjugates. { här¦män·ik 'rānj }

harmonic ratio [MATH] A cross ratio that is equal to −1. { här¦män·ik 'rā·sho }

harmonic selective ringing [COMMUN] Selective ringing which employs currents of several frequencies and ringers, each tuned mechanically or electrically to the frequency of one of the ringing currents, so that only the desired ringer responds. { här'män·ik si¦lek·tiv 'riŋ·iŋ }

harmonic sequence *See* harmonic progression. { här,män·ik 'sē·kwəns }

harmonic series [MATH] A series whose terms form a harmonic progression. { här¦män·ik 'sir,ēz }

harmonic speed changer [MECH ENG] A mechanical-drive system used to transmit rotary, linear, or angular motion at high ratios and with positive motion. { här'män·ik 'spēd ,chän·jər }

harmonic synthesizer [MECH] A machine which combines elementary harmonic constituents into a single periodic function; a tide-predicting machine is an example. { här'män·ik 'sin·thə,sīz·ər }

harmonic telephone ringer [ELECTR] Telephone ringer which responds only to alternating current within a very narrow frequency band. { här'män·ik 'tel·ə,fōn ,riŋ·ər }

harmonic tide plane *See* Indian spring low water. { här'män·ik 'tīd ,plān }

harmonic vibration *See* harmonic motion. { här'män·ik vī'brā·shən }

harmonic vibration-rotation band [SPECT] A vibration-rotation band of a molecule in which the harmonic oscillator approximation holds for the vibrational levels, so that the vibrational levels are equally spaced. { här'män·ik vī'brā·shən rō'tā·shən ,band }

harmonic wave [PHYS] A transverse waveform obtained by mapping onto a time base the periodic up and down excursions of simple harmonic motion. { här'män·ik 'wāv }

harmonic wave analyzer *See* harmonic analyzer. { här'män·ik 'wāv ,an·ə,līz·ər }

harmonize [ORD] **1.** To align the sights of a gun so that the curving path of the projectile will meet the straight line of sight at the target. **2.** To adjust or align the gunsight, guns, rocket launchers, or gun camera of a fighter aircraft so that an accurate aim is obtained at a given range, or so that the guns will produce a desired pattern of fire. { 'här·mə,nīz }

harmotome [MINERAL] $(K,Ba)(Al,Si)_2(Si_6O_{16})\cdot 6H_2O$ A zeolite mineral with ion-exchange properties that forms cruciform twin crystals. Also known as cross-stone. { 'här·mə,tōm }

Harnack's first convergence theorem [MATH] The theorem that if a sequence of functions harmonic in a common domain of three-dimensional space and continuous on the boundary of the domain converges uniformly on the boundary, then it converges uniformly in the domain to a function which is itself harmonic; the sequence of any partial derivative of the functions in the original sequence converges uniformly to the corresponding partial derivative of the limit function in every closed subregion of the domain. { 'här·naks ¦fərst kən'vər·jəns ,thir·əm }

Harnack's second convergence theorem [MATH] The theorem that if a sequence of functions is harmonic in a common domain of three-dimensional space and their values are monotonically decreasing at any point in the domain, then convergence of the sequence at any point in the domain implies uniform convergence of the sequence in every closed subregion of the domain to a function which is itself harmonic. { 'här·naks ¦sek·ənd kən'vər·jəns ,thir·əm }

harness [AERO ENG] **1.** Straps arranged to hold an occupant of a spacecraft or aircraft in the seat. **2.** Straps worn by a parachutist or used to suspend a load from a parachute. [ELEC] Wire and cables so arranged and tied together that they may be inserted and connected, or may be removed after disconnection, as a unit. [TEXT] One of two or more frames on a loom which are raised to separate the warp from the filler yarns to allow the shuttle to pass between them. { 'här·nəs }

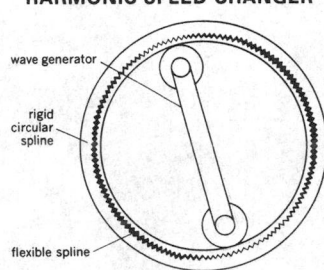

HARMONIC SPEED CHANGER

wave generator
rigid circular spline
flexible spline

Rotary-to-rotary harmonic speed changer that effects a speed reduction of 1:66.

Harpacticoida [INV ZOO] An order of minute copepod crustaceans of variable form, but generally being linear and more or less cylindrical. { här‚pak·tə'kȯid·ə }

harpago [INV ZOO] Part of the clasper on the copulatory organ of certain male insects. { 'här·pə‚gō }

Harpidae [INV ZOO] A family of marine gastropod mollusks in the order Neogastropoda. { 'här·pə‚dē }

harpoon [DES ENG] A barbed spear used to catch whales. { här'pün }

harpoon log [ENG] A log which consists essentially of a rotator and distance registering device combined in a single unit, and towed through the water; it has been largely replaced by the taffrail log; the two types of logs are similar except that the registering device of the taffrail log is located at the taffrail and only the rotator is in the water. { här'pün ‚läg }

Harris flow [ELECTR] Electron flow in a cylindrical beam in which a radial electric field is used to overcome space charge divergence. { 'har·əs ‚flō }

Harrison's gridiron pendulum [DES ENG] A type of compensated pendulum that has five iron rods and four brass rods arranged so that the effects of their thermal expansion cancel. { ‚har·ə·sənz 'grid‚ī·ərn ‚pen·jə·ləm }

Harris process [MET] A method for refining lead in which the liquid bullion is sprayed through molten caustic soda and molten sodium nitrate; arsenic, antimony, and tin are oxidized, converted into sodium salts, and skimmed from the bath. { 'har·əs ‚prä·səs }

harrow [AGR] An implement that is pulled over plowed soil to break clods, level the surface, and destroy weeds. { 'här·ō }

harrowing [AGR] Cultivation of the soil with a harrow. { 'här·ə·wiŋ }

harstigite [MINERAL] $Be_2Ca_3Si_3O_{11}$ A mineral composed of silicate of beryllium and calcium. { 'härs·tə‚git }

Hartford loop [MECH ENG] A condensate return arrangement for low-pressure, steam-heating systems featuring a steady water line in the boiler. { 'härt·fərd ‚lüp }

Hartig net [MYCOL] A complex network of fungal hyphae that is the site of nutrient exchange between the fungus and the host plant. { 'här·tig ‚net }

hartite [GEOL] A white, crystalline, fossil resin that is found in lignites. Also known as bombiccite; branchite; hofmannite; josen. { 'här‚tīt }

hartley [COMMUN] A unit of information content, equal to the designation of 1 of 10 possible and equally likely values or states of anything used to store or convey information. { 'härt·lē }

Hartley formula [COMMUN] The relationship expressing that as the time function is made narrower, the frequency spectrum must become broader. { 'härt·lē ‚fȯr·myə·lə }

Hartley oscillator [ELECTR] A vacuum-tube oscillator in which the parallel-tuned tank circuit is connected between grid and anode; the tank coil has an intermediate tap at cathode potential, so the grid-cathode portion of the coil provides the necessary feedback voltage. { 'härt·lē ‚äs·ə‚lād·ər }

Hartley principle [COMMUN] The principle that the total number of bits of information that can be transmitted over a channel in a given time is proportional to the product of channel bandwidth and transmission time. { 'härt·lē ‚prin·sə·pəl }

Hartley transform [MATH] An analog of the Fourier transform for finite, real-valued data sets; for a function f defined at N data values, 0, 1, 2, ..., $N - 1$, the Hartley transform is a function, F, also defined on the set (0, 1, 2, ..., $N - 1$), whose value at n is the sum over the variable r, from 0 through $N - 1$, of the quantity $N^{-1}f(r)$ cas $(2\pi nr/N)$, where cas $\theta =$ cos $\theta +$ sin θ. { 'härt·lē ‚tranz‚fȯrm }

Hartmann diaphragm [ANALY CHEM] Comparison device for positive-element-identification readings from emission spectra. { 'härt·män ‚dī·ə‚fram }

Hartmann dispersion formula [OPTICS] A semiempirical formula relating the index of refraction n and wavelengths λ; $n = n_0 + a/(\lambda - \lambda_0)$, where n_0, a, and λ_0 are empirical constants. Also known as Cornu-Hartmann formula. { 'härt·män di'spər·zhən ‚fȯr·myə·lə }

Hartmann flow [PL PHYS] The steady flow of an electrically conducting fluid between two parallel plates when there is a uniform applied magnetic field normal to the plates. { 'härt·män ‚flō }

Hartmann generator [ENG ACOUS] A device in which shock waves generated at the edges of a nozzle by a supersonic gas jet resonate with the opening of a small cylindrical pipe, placed opposite the nozzle, to produce powerful ultrasonic sound waves. { 'härt·mən ‚jen·ə‚rād·ər }

Hartmann lines *See* Lüders' lines. { 'härt·män ‚līnz }

Hartmann number [PL PHYS] A dimensionless number which gives a measure of the relative importance of drag forces resulting from magnetic induction and viscous forces in Hartmann flow, and determines the velocity profile for such flow. { 'härt·män ‚näm·bər }

Hartmann test [OPTICS] A test for telescope mirrors in which the mirror is covered with a screen with regularly spaced holes, and a photographic plate is placed near the focus; for a perfect mirror, this results in regularly spaced dots on the plate. [SPECT] A test for spectrometers in which light is passed through different parts of the entrance slit; any resulting changes of the spectrum indicate a fault in the instrument. { 'härt·män ‚test }

Hartman's solution [ANALY CHEM] Solution of thymol, ethyl alcohol, and sulfuric ether; used for selective dentin analysis. { 'härt·mənz sə‚lü·shən }

Hart-Park virus [VIROL] A ribonucleic acid-containing animal virus of the rhabdovirus group. { 'härt ‚pärk ‚vī·rəs }

hartree [ATOM PHYS] A unit of energy used in studies of atomic spectra and structure, equal to $2R_\infty hc$ or $\alpha^2 mc^2$, where R_∞ is the Rydberg constant, h is Planck's constant, c is the speed of light, α is the fine-structure constant, and m is the mass of the electron; numerically, it is approximately 27.21 electronvolts or 4.360×10^{-18} joule. Also known as Hartree energy. { 'här‚trē }

Hartree energy *See* hartree. { 'här‚trē ‚en·ər·jē }

Hartree equation [ELECTR] An equation which gives the lowest anode voltage at which it is theoretically possible to maintain oscillation in the different modes of a magnetron. { 'här‚trē i‚kwa·zhən }

Hartree-Fock approximation [QUANT MECH] A refinement of the Hartree method in which one uses determinants of single-particle wave functions rather than products, thereby introducing exchange terms into the Hamiltonian. { 'här‚trē ‚fäk ə‚präk·sə‚mā·shən }

Hartree method [QUANT MECH] An iterative variational method of finding an approximate wave function for a system of many electrons, in which one attempts to find a product of single-particle wave functions, each one of which is a solution of the Schrödinger equation with the field deduced from the charge density distribution due to all the other electrons. Also known as self-consistent field method. { 'här‚trē ‚meth·əd }

Hartree units [ATOM PHYS] A system of units in which the unit of angular momentum is Planck's constant divided by 2π, the unit of mass is the mass of the electron, and the unit of charge is the charge of the electron. Also known as atomic units. { 'här‚trē ‚yü·nəts }

hartshorn oil *See* bone oil. { 'härts‚hȯrn ‚ȯil }

hartshorn salts *See* aromatic spirits of ammonia. { 'härts‚hȯrn ‚sȯls }

Harvard-Draper sequence [ASTRON] A system of classification of stellar spectra based on features that are found to vary in a smooth way from one star to another, and on the star's color. Also known as Harvard sequence { 'här·vərd 'drā·pər ‚sē·kwəns }

Harvard sequence *See* Harvard-Draper sequence. { 'här·vərd ‚sē·kwəns }

harvester [AGR] A machine used to reap field crops. { 'här·və·stər }

harvester-thresher [AGR] A machine that combines the harvesting and threshing of grain crops. { 'här·və·stər 'thresh·ər }

harvesting [AGR] The gathering of mature field crops. { 'här·və·stiŋ }

harvest moon [ASTRON] A full moon that is seen nearest the autumnal equinox. { 'här·vəst ‚mün }

harzburgite [PETR] A peridotite consisting principally of olivine and orthopyroxene. { 'härts·bər‚gīt }

Harz jig [MIN ENG] A device used to separate coal and foreign matter which gives pulsion intermittently with suction. { 'härts ‚jig }

Hasche process [CHEM ENG] A thermal reforming process for hydrocarbon fuels; it is a noncatalytic regenerative method in which a mixture of hydrocarbon gas or vapor and air is passed through a regenerative mass that is progressively hotter

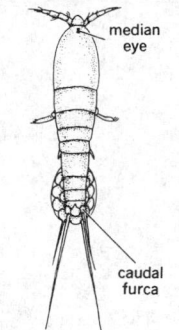

HARPACTICOIDA

Harpacticus chelifer, a typical harpacticoid.

HARVESTER

Self-propelled combine. (*Sherry New Holland, Division of Sperry Rand Corp.*)

in the direction of the gas flow; partial combustion occurs, liberating heat to crack the remaining hydrocarbons in a combustion zone. { 'häsh·ə ˌprä·səs }

Hasenclever turntable [MIN ENG] A turntable that is made to rotate by the friction between the positively driven pulley, the car, and the table; used as an alternative to the shunt-back or the traverser for changing the direction of mine cars or tubs, either on the surface or underground. { 'häz·ənˌklev·ər ˌtərnˌtā·bəl }

hash [COMPUT SCI] Data which are obviously meaningless, caused by human mistakes or computer malfunction. Also known as garbage; gibberish. [ELEC] Electric noise produced by the contacts of a vibrator or by the brushes of a generator or motor. [ELECTR] *See* grass. { hash }

hash coding *See* hashing. { 'hash ˌkōd·iŋ }

Hashimoto's disease *See* struma lymphomatosa. { ˌhä·shi'mō·dōz diˌzēz }

Hashimoto's struma *See* struma lymphomatosa. { ˌhä·shi'mō·dōz 'strü·mə }

hashing [COMPUT SCI] **1.** A method for converting representations of values within fields, usually keys, to a more compact form. **2.** An addressing technique that uses keys to store and retrieve data in a file. { 'hash·iŋ }

hashish [PHARM] A narcotic drug derived from the plant *Cannabis sativa;* can be smoked, chewed, or drunk. { 'hashˌēsh }

hash total [COMPUT SCI] A sum obtained by adding together numbers having different meanings; the sole purpose is to ensure that the correct number of data have been read by the computer. { 'hash ˌtōd·əl }

hasp [DES ENG] A two-piece fastening device having a loop on one piece and a hinged plate that fits over the loop on the other. { hasp }

HASP [COMPUT SCI] A technique used on some types of larger computers to control input and output between a computer and its peripheral devices by utilizing mass-storage devices to temporarily store data. Acronym for Houston Automatic Spooling Processor. { hasp }

Hassal's body *See* thymic corpuscle. { 'has·əlz ˌbäd·ē }

Hasse diagram [MATH] A representation of a partially ordered set as a directed graph, in which elements of the set are represented by vertices of the graph, and there is a directed arc from *x* to *y* if and only if *y* covers *x*. { 'häs·ə ˌdī·əˌgram }

hassium [CHEM] A chemical element, symbolized Hs, atomic number 108, a synthetic element; the sixteenth transuranium element. { 'hä·sē·əm }

hastate [BIOL] Shaped like an arrowhead with divergent barbs. { 'haˌstāt }

haster [METEOROL] In England, a violent rain storm. { 'has·tər }

hastingsite [MINERAL] $NaCa_2(Fe,Mg)_5Al_2Si_6O_{22}(OH)_2$ A mineral of the amphibole group crystallizing in the monoclinic system and composed chiefly of sodium, calcium, and iron, but usually with some potassium and magnesium. { 'häs·tiŋˌzīt }

hasty mine field [ORD] Field of mines quickly laid as a protection against an enemy attack; when practicable, it is laid in a definite pattern, as is a deliberate field, but measurements are approximate rather than exact. { 'häs·tē 'mīn ˌfēld }

HAT *See* height above touch-down.

hatch [ENG] A door or opening, especially on an airplane, spacecraft, or ship. { hach }

hatch battens [NAV ARCH] Flat bars used to fasten and make tight the edges of a tarpaulin covering a hatch. { 'hach ˌbat·ənz }

hatch beam [ENG] A heavy, portable beam which supports a hatch cover. { 'hach ˌbēm }

hatch carlings [NAV ARCH] Fore and aft girders running under the coamings of hatches, to which the partial or half deck beams are attached. { 'hach ˌkär·liŋz }

hatch coaming [NAV ARCH] A raised frame around a hatch; it forms a support for the hatch cover and strengthens the edges of the opening. { 'hach ˌkōm·iŋ }

hatch cover [ENG] A steel or wooden cover for a hatch. { 'hach ˌkəv·ər }

hatch end beam [NAV ARCH] The deck beam at the fore and aft end of a hatch. { 'hach ˌend ˌbēm }

hatchet [DES ENG] A small ax with a short handle and a hammerhead in addition to the cutting edge. { 'hach·ət }

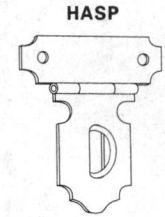

HASP

An example of a hasp.

hatchettine *See* hatchettite. { 'ha·chədˌēn }

hatchettite [MINERAL] $C_{38}H_{78}$ A yellow-white mineral paraffin wax, melting at 55–65°C in the natural state and 79°C in the pure state; occurs in masses in ironstone nodules or in cavities in limestone. Also known as adipocerite; adipocire; hatchettine; mineral tallow; mountain tallow; naphthine. { 'ha·chədˌīt }

hatchettolite *See* ellsworthite. { 'ha·ched·ōˌlīt }

hatching [GRAPHICS] Parallel lines drawn on sections of plans for buildings or machines to distinguish between different materials. { 'hach·iŋ }

hatchite [MINERAL] A lead-gray mineral composed of sulfide of lead and arsenic; occurs in triclinic crystals. { 'haˌchīt }

Hatch-Slack pathway [BIOCHEM] A metabolic cycle involved in the non-light-requiring phase of photosynthesis in certain plants having specific metabolic and anatomical modifications in their mesophyll and bundle sheath cells which facilitate the temporary fixation of carbon dioxide (CO_2) into four-carbon organic acid; these acids are next broken down to three-carbon organic acids plus CO_2 in bundle sheath cells, where this freed CO_2 is then fixed into carbohydrates in a normal Calvin cycle pathway. { 'hach 'slak 'pathˌwā }

hatted code [COMMUN] Randomized code consisting of an encoding section; the plain text groups are arranged in alphabetical or other significant order, accompanied by their code groups arranged in a nonalphabetical or random order. { 'had·əd ˌkōd }

H attenuator *See* H network. { 'āch ə'ten·yəˌwād·ər }

haud [METEOROL] In Scotland, a squall. { hȯd }

hauerite [MINERAL] MnS_2 A reddish-brown or brownish-black mineral composed of native manganese sulfide; occurs massive or in octahedral or pyritohedral crystals. { 'haȯ·əˌrīt }

haughtonite [PETR] A black variety of biotite that is rich in iron. { 'hȯt·ənˌīt }

haul [ENG] A single tow of a net or dredge. [NAV] To change the course of a vessel so as to bring the wind farther forward. { hȯl }

haulage [MIN ENG] The movement, in cars or otherwise, of men, supplies, ore, and waste, underground and on the surface. { 'hȯl·ij }

haulage conveyor [MIN ENG] A conveyor used to transport material between the gathering conveyor and the outside. { 'hȯl·ij kənˌvā·ər }

haulage curve [MIN ENG] A bend in a haulage road in any direction. { 'hȯl·ij ˌkərv }

haulage drum [MIN ENG] A cylinder on which steel haulage rope is coiled. { 'hȯl·ij ˌdrəm }

haulage level [MIN ENG] An underground level, either along and inside the ore body or closely parallel to it, and usually in the footwall, in which mineral is loaded into trams and moved out to the hoisting shaft. { 'hȯl·ij ˌlev·əl }

haulage stage [MIN ENG] A mine roadway along which a load is moved by one form of haulage without coupling or uncoupling of cars. { 'hȯl·ij ˌstāj }

haulageway [MIN ENG] The gangway, entry, or tunnel through which loaded or empty mine cars are hauled by animal or mechanical power. { 'hȯl·ijˌwā }

haul-cycle time [MIN ENG] The time required for the scraper to haul a load to the dumping area and to return to its position in the loading area. { 'hȯl ˌsī·kəl ˌtīm }

haul road [MIN ENG] A road built to carry heavily loaded trucks at a good speed; the grade is limited and usually kept to less than 17% of climb. { 'hȯl ˌrōd }

Hausdorff maximal principle [MATH] The principle that every partially ordered set has a linearly ordered subset S which is maximal in the sense that S is not a proper subset of another linearly ordered subset. { 'haȯsˌdȯrf 'mak·sə·məl ˌprin·sə·pəl }

Hausdorff paradox [MATH] The theorem that a sphere can be represented as the union of four disjoint sets, A, B, C, and D, where D is a countable set, and A is congruent to each of the three sets B, C, and the union of B and C. { 'haȯsˌdȯrf ˌpar·əˌdäks }

Hausdorff space [MATH] A topological space where each pair of distinct points can be enclosed in disjoint open neighborhoods. Also known as T_2 space. { 'haȯsˌdȯrf ˌspās }

hausmannite [MINERAL] Mn_3O_4 Brownish-black, opaque mineral composed of manganese tetroxide. { 'haus·mä,nīt }

haustellum [INV ZOO] A proboscis modified for sucking. { hȯ'stel·əm }

haustoria [MYCOL] Specialized branches of hyphae that penetrate host cells and absorb nutrients from them. { hau'stȯr·ē·ə }

haustorial [MYCOL] Pertaining to fungi that have food-absorbing cells in the host. { hau'stȯr·ē·əl }

haustorium [BOT] **1.** An outgrowth of certain parasitic plants which serves to absorb food from the host. **2.** Food-absorbing cell of the embryo sac in nonparasitic plants. { hȯ'stȯr·ē·əm }

haustrum [ANAT] An outpocketing or pouch of the colon. { 'hȯ·strəm }

Hauterivian [GEOL] A European stage of geologic time, in the Lower Cretaceous, above Valanginian and below Barremian. { ō·trə'vē·ən }

Haüy law [CRYSTAL] The law that for a given crystal there is a set of ratios such that the ratios of the intercepts of any crystal plane on the crystal axes are rational fractions of these ratios. { ä'wē ,lȯ }

haüyne [MINERAL] $(Na,Ca)_{4-8}(Al_6Si_6O_{24})(SO_4S)_{1-2}$ An isometric silicate mineral of the sodalite group occurring as grains embedded in various igneous rocks; hardness is 5.5–6 on Mohs scale, and specific gravity is 2.4–2.5. Also known as haüynite. { ä'wēn }

haüynite See haüyne. { ä'wē,nīt }

Hauzeur furnace [MET] A double furnace for the distillation of zinc wherein waste heat from one set of retorts is utilized for heating the second set. { ō'zúr ,fər·nəs }

hav See haversine.

Havelock's law [OPTICS] The law that in a substance displaying the Kerr effect, $n_p - n = 2(n_s - n)$, where n is the index of refraction in the absence of an electric field, and n_p and n_s are the indices of refraction of light whose magnetic vector is parallel and perpendicular to the applied electric field. { 'hav,läks ,lȯ }

haven [NAV] A place of safety for vessels; it is accessible in all states of weather and tides. { 'hā·vən }

Haverhill fever [MED] An acute bacterial infection caused by *Streptobacillus moniliformis*, usually acquired by rat bite, and characterized by acute onset, intermittent fever, erythematous rash, and polyarthritis. Also known as streptobacillary fever. { 'hāv·ə·rəl ,fē·vər }

Haversian canal [HISTOL] The central, longitudinal channel of an osteon containing blood vessels and connective tissue. { hə'vər·zhən kə,nal }

Haversian lamella [HISTOL] One of the concentric layers of bone composing a Haversian system. { hə'vər·zhən lə 'mel·ə }

Haversian system See osteon. { hə'vər·zhən ,sis·təm }

haversine [MATH] The haversine of an angle A is half of the versine of A, or is $\frac{1}{2}(1 - \cos A)$. Abbreviated hav. { 'ha·vər,sīn }

havgul See havgull. { 'hav·gəl }

havgula See havgull. { 'hav·gəl·ə }

havgull [METEOROL] Cold, damp wind blowing from the sea during summer in Scotland and Norway. Also known as havgul; havgula. { 'hav·gəl }

HA virus See hemadsorption virus. { 'āch¦ā ,vī·rəs }

hawk [ENG] A board with a handle underneath used by a workman to hold mortar. [VERT ZOO] Any of the various smaller diurnal birds of prey in the family Accipitridae; some species are used for hunting hare and partridge in India and other parts of Asia. { hȯk }

Hawk [ORD] A U.S. Army surface-to-air guided missile that has a range of about 25 miles (40 kilometers), a maximum speed of about Mach 3, and a ceiling of about 45,000 feet (14,800 meters); originally guided by radio for attacking low-flying enemy aircraft, but newer models are radar-guided. { hȯk }

hawse [NAV ARCH] **1.** The area in the bow of the ship where the hawsepipes are located. **2.** The distance between the bow of a ship and the anchors. { hȯz }

hawse bolster [NAV ARCH] A steel ring or fairing at the end of or around a hawsepipe to aid the motion of a cable through the pipe, prevent its chafing, and guide the anchor into stowed position. Also known as anchor bolster. { 'hȯz ¦bōl·stər }

hawsepipe [NAV ARCH] A pipe, made of heavy cast iron or steel, through which the anchor chain runs; placed in the ship's bow on each side of the stem, or in some cases also at the stern when a stern anchor is used. { 'hȯz,pīp }

hawser [NAV ARCH] A large rope or cable, usually over 5 inches (13 centimeters) in diameter, generally used to tow or moor a ship or secure it at a dock. { 'hȯz·ər }

hay [AGR] Forage plants cut and dried for animal feed. { hā }

Hayashi track [ASTRON] A vertical track on the Hertz-sprung-Russell diagram along which a star of small mass descends during its early stages of formation, when convective heat transport prevails over most of the star. { ha'ya·shē ,trak }

Hay bridge [ELEC] A four-arm alternating-current bridge used to measure inductance in terms of capacitance, resistance, and frequency; bridge balance depends on frequency. { 'hā ,brij }

haycock [HYD] An isolated ice cone created on land ice or shelf ice because of pressure or ice movement. { 'hā,käk }

Hayden process [MET] A method of electrolytic copper refining; anodes of unrefined copper are suspended in an acid electrolyte, and one side of each then acts as an anode and the other as a cathode. { 'hād·ən ,prä·səs }

haydite [MATER] Expanded shale, slate, or clay characterized by low unit weight and satisfactory structural properties; used as an aggregate to produce lightweight structural concrete. { 'hā,dīt }

hay fever [MED] An allergic disorder of the nasal membranes and related structures due to sensitization by certain plant pollens. Also known as allergic rhinitis; pollinosis. { 'hā ,fē·vər }

Hayflick limit [PHYSIO] The finite replicative capacity of normal somatic cells. { 'hā,flik ,lim·ət }

haymaker [AGR] A machine for curing hay. { 'hā,māk·ər }

hay section [PETRO ENG] A compartment in a heater or a heater-treater which is filled with fibrous material for filtering oil-and-water emulsions. { 'hā ,sek·shən }

Hay's test [PATH] A test for bile salts; the salts lower the surface tension of water, and therefore a light powder such as flowers of sulfur will not float in a solution containing a high concentration of the salts. { 'hāz ,test }

Hayward grab bucket [MECH ENG] A clamshell type of grab bucket used for handling coal, sand, gravel, and other flowable materials. { 'hā·wərd 'grab ,bək·ət }

Hayward orange peel [MECH ENG] A grab bucket that operates like the clamshell type but has four blades pivoted to close. { 'hā·wərd 'ä·rənj ,pēl }

hazard [IND ENG] Any risk to which a worker is subject as a direct result (in whole or in part) of his being employed. { 'haz·ərd }

hazard analysis critical control points [FOOD ENG] A food safety technique that examines points in food production (ingredients, processing, and so on) that are most likely to cause food-borne illness; these points are then heavily monitored during production. Abbreviated HACCP. { ¦haz·ərd ə,nal·ə·səs ,krid·i·kəl kən'trōl ,poins }

hazard beacon See obstruction beacon. { 'haz·ərd ,bē·kən }

hazardous air pollutants [ENG] Chemicals that are known or suspected to cause cancer or other serious health effects, such as reproductive effects or birth defects, or adverse environmental effects. Listed hazardous air pollutants include benzene, found in gasoline; perchlorethylene, emitted from some dry cleaning facilities; and methylene chloride, used as a solvent and paint stripper in industry; as well as dioxin, asbestos, toluene, and metals such as cadmium, mercury, chromium, and lead compounds. Also known as air toxics. Abbreviated HAP. { ,haz·ər·dəs 'er pə,lüt·əns }

hazardous material [MATER] A poison, corrosive agent, flammable substance, explosive, radioactive chemical, or any other material which can endanger human health or well-being if handled improperly. { 'haz·ərd·əs mə'tir·ē·əl }

hazardous waste [CHEM ENG] Any solid, liquid, or gaseous waste material that, if improperly managed or disposed, may pose substantial hazards to human health and the environment; such waste exhibits ignitability, corrosivity, reactivity, or toxicity. { ,haz·ər·dəs 'wāst }

haze [METEOROL] Fine dust or salt particles dispersed

HAWK

Rough-legged hawk (*Buteo lagopus*).

through a portion of the atmosphere; the particles are so small that they cannot be felt, or individually seen with the naked eye, but they diminish horizontal visibility and give the atmosphere a characteristic opalescent appearance that subdues all colors. [OPTICS] The degree of cloudiness in a solution, cured plastic material, or coating material. { hāz }

haze droplet [METEOROL] Any small liquid droplet contributing to an atmospheric haze condition. { 'hāz ,dräp·lət }

haze factor [METEOROL] The ratio of the luminance of a mist or fog through which an object is viewed to the luminance of the object. { 'hāz ,fak·tər }

haze horizon [METEOROL] The top of a haze layer which is confined by a low-level temperature inversion and has the appearance of the horizon when viewed from above against the sky. { 'hāz hə,rīz·ən }

haze layer [METEOROL] A layer of haze in the atmosphere, usually bounded at the top by a temperature inversion and frequently extending downward to the ground. { 'hāz lā·ər }

haze level See haze line. { 'hāz ,lev·əl }

haze line [METEOROL] The boundary surface in the atmosphere between a haze layer and the relatively clean, transparent air above the top of a haze layer. Also known as haze level. { 'hāz ,līn }

hazelnut See filbert. { 'ha·zəl,nət }

hazel sandstone [GEOL] An arkosic, iron-bearing redbed sandstone from the Precambrian found in western Texas. { 'ha·zəl 'san,stōn }

hazemeter See transmissometer. { 'hāz,mēd·ər }

h-bar [QUANT MECH] A fundamental constant equal to $h/2\pi$, where h is Planck's constant. Symbolized $\hbar$. Also known as Dirac h; h-line. { 'āch ,bär }

H beacon [NAV] Nondirectional homing beacon which has a power output of 50 to 2000 watts. { 'āch ,bē·kən }

H beam [CIV ENG] A beam similar to the I beam but with longer flanges. Also known as wide-flange beam. { 'āch ,bēm }

H bend See H-plane bend. { 'āch ,bend }

HBFC See hydrobromofluorocarbon.

H bit [DES ENG] A core bit manufactured and used in Canada having inside and outside diameters of 2.875 and 3.875 inches (73.025 and 98.425 millimeters), respectively; the matching reaming shell has an outside diameter of 3.906 inches (99.2124 millimeters). { 'āch ,bit }

H bomb See hydrogen bomb. { 'āch ,bäm }

HBT See heterojunction bipolar transistor.

H-B virus [VIROL] A subgroup-A picornavirus associated with diseases in rodents. { 'āch 'bē ,vī·rəs }

H carrier system [COMMUN] Low-frequency carrier system which provides one carrier channel, using frequencies up to about 10 kilohertz, by means of effective four-wire transmission on a single open-wire pair. { 'āch 'kar·ē·ər ,sis·təm }

HCB See hexachlorobenzene.

hcf See greatest common divisor.

HCFC See hydrochlorofluorocarbon.

HCG See human chorionic gonadotropin.

H chain See heavy chain. { 'āch ,chān }

hcp structure See hexagonal close-packed structure. { 'āch 'sē'pē 'strək·chər }

HDA See head/disk assembly.

HdC star [ASTRON] A type of hydrogen-deficient supergiant carbon star that resembles the R Coronae Borealis stars but does not display significant variability. { ,āch,dē'sē ,stär }

H-D curve See Hurter and Driffield characteristic curve. { 'āch'dē 'kərv }

H display [NAV] In radar, a B display modified to include indication of angle of elevation. { 'āch di,splā }

HDL See high-density lipoprotein.

HDLC See high-level data-link control..

HDPE See high-density polyethylene. { 'āch'dē'pē'dē }

HDTV See high-definition television.

HDX See half-duplex circuit.

He See helium.

head [ANAT] **1.** The region of the body consisting of the skull, its contents, and related structures. **2.** Proximal end of a long bone. [ASTRON] See coma. [BOT] A dense cluster of nearly sessile flowers on a very short stem. [BUILD] The upper part of the frame on a door or window. [COMPUT SCI] A device that reads, records, or erases data on a storage medium such as a drum or tape; examples are a small electromagnet or

a sensing or punching device. [ELECTR] The photoelectric unit that converts the sound track on motion picture film into corresponding audio signals in a motion picture projector. [ENG] **1.** The end section of a plastics blow-molding machine in which a hollow parison is formed from the melt. **2.** The section of a shell-and-tube heat exchanger from which fluid from the tube bundle is discharged. [ENG ACOUS] See cutter. [FL MECH] See pressure head. [GEOGR] See headland. { hed }

headache [MED] A deep form of pain, with a characteristic aching quality, localized in the head. { 'hed,āk }

headache post [MECH ENG] A post installed on a cable-tool rig for supporting the end of the walking beam when the rig is not operating. { 'hed,āk ,pōst }

headblock [MIN ENG] **1.** A stop at the head of a slope or shaft to keep cars from going down the shaft or slope. **2.** A cap piece. { 'hed,bläk }

headboard [MIN ENG] **1.** A wooden wedge placed against the hanging wall, and against which one end of the stull is jammed. **2.** A board in the roof of a heading, contacting the earth above and supported by a headtree on each side. { 'hed,bȯrd }

headbox [ENG] A device for controlling the flow of a suspension of solids into a machine. { 'hed,bäks }

head breadth [ANTHRO] Greatest horizontal breadth of the head above the ear openings, at whichever level is found, with moderate pressure applied on the caliper points. { 'hed ,bredth }

head bulb [INV ZOO] A structure armed with spines behind the lips of spiruroid nematodes. { 'hed ,bəlb }

head cable See hand cable. { 'hed ,kā·bəl }

head circumference [ANTHRO] Maximum of three measurements taken above the eyebrows. { 'hed sər'kəm·frəns }

head crash [COMPUT SCI] The collision of the read-write head and the magnetic recording surface of a hard disk. Also known as disk crash. { 'hed ,krash }

head/disk assembly [COMPUT SCI] An airtight assembly including a disk pack and read/write heads. Abbreviated HDA. { 'hed 'disk ə'sem·blē }

header [BUILD] A framing beam positioned between trimmers and supported at each end by a tail beam. [CIV ENG] Brick or stone laid in a wall with its narrow end facing the wall. [COMMUN] The first section of a message, which contains information such as the addressee, routing, data, and origination time. [COMPUT SCI] See header label. [ELEC] A mounting plate through which the insulated terminals or leads are brought out from a hermetically sealed relay, transformer, transistor, tube, or other device. [ENG] A pipe, conduit, or chamber which distributes fluid from a series of smaller pipes or conduits; an example is a manifold. [GRAPHICS] Text that is printed at the top of every page of a document or report; may contain the page number. [MECH ENG] A machine used for gathering or upsetting materials; used for screw, rivet, and bolt heads. [MIN ENG] **1.** An entry-boring machine that bores the entire section of the entry in one operation. **2.** A rock that heads off or delays progress. **3.** A blasthole at or above the head. { 'hed·ər }

header bond [CIV ENG] A masonry bond consisting of header courses exclusively. { 'hed·ər ,bänd }

header course [CIV ENG] A masonry course of bricks laid as headers. { 'hed·ər ,kȯrs }

header label [COMPUT SCI] A block of data at the beginning of a magnetic tape file containing descriptive information to identify the file. Also known as header. { 'hed·ər ,lā·bəl }

head erosion See headward erosion. { 'hed i,rō·zhən }

header record [COMPUT SCI] Computer input record containing common, constant, or identifying information for records that follow. { 'hed·ər ,rek·ərd }

header-type boiler See straight-tube boiler. { 'hed·ər ,tīp ,bȯil·ər }

head fold [EMBRYO] A ventral fold formed by rapid growth of the head of the embryo over the embryonic disk, resulting in the formation of the foregut accompanied by anteroposterior reversal of the anterior part of the embryonic disk. { 'hed ,fōld }

headframe [MIN ENG] **1.** The frame at the top of a shaft, on which is mounted the hoisting pulley. Also known as gallows; gallows frame; headstock; hoist frame. **2.** The shaft frame, sheaves, hoisting arrangements, dumping gear, and connected works at the top of a shaft. Also known as headgear. { 'hed,frām }

head gap [COMPUT SCI] The space between the read/write head and the recording medium, such as a disk in a computer. { 'hed ,gap }

head gate [CIV ENG] **1.** A gate on the upstream side of a lock or conduit. **2.** A gate at the starting point of an irrigation ditch. [PETRO ENG] The gate valve that is installed nearest the pump or compressor on oil or gas lines. { 'hed ,gāt }

headgear *See* headframe. { 'hed,gir }

head height [ANTHRO] Average perpendicular distance between the tragion point and the mid-longitudinal line on top of the head, measured on both sides, when the head is positioned so that the line between the tragion point and the bottom of the bony orbit is horizontal. { 'hed ,hīt }

headhouse [MIN ENG] **1.** A timber framing located at the top of a shaft and receiving the shaft guides that carry the cage or elevator. **2.** A structure that houses the headframe. { 'hed ,haús }

heading [CIV ENG] In tunnel construction, one or more small tunnels excavated within a large tunnel cross section that will later be enlarged to full section. [NAV] **1.** The horizontal direction in which a ship actually points or heads at any instant, expressed in angular units from a reference direction, usually from 0° at the reference direction clockwise through 360°. **2.** In air navigation, the horizontal direction in which an aircraft points or heads, that is the direction of the longitudinal axis, measured as in the first definition. [PETRO ENG] An intermittent flow from an oil well. { 'hed·iŋ }

heading-and-bench mining [MIN ENG] A stoping method used in thicker ore where it is customary first to take out a slice or heading 7 or 8 feet (2.1 or 2.4 meters) high directly under the top of the ore, and then to bench or stope down the ore between the bottom of the heading and the bottom of the ore or floor of the level. { 'hed·iŋ ən 'bench ,mīn·iŋ }

heading angle [NAV] Heading measured from 0° at the reference direction clockwise or counterclockwise through 90 or 180° it is labeled with the reference direction as a prefix and the direction of measurement from the reference as a suffix; for example, heading angle S 107° E is 107° east of south, or heading 073°. { 'hed·iŋ ,aŋ·gəl }

heading blasting *See* coyote blasting. { 'hed·iŋ ,blast·iŋ }

heading joint [BUILD] **1.** A joint between two pieces of timber which are joined in a straight line, end to end. **2.** A masonry joint formed between two stones in the same course. { 'hed·iŋ ,jóint }

heading line [NAV] The line extending in the direction of a heading. { 'hed·iŋ ,līn }

heading-overhand bench method [MIN ENG] A tunneling method in which the heading is the lower part of the section and is driven at least a round or two in advance of the upper part (bench), which is taken out by overhand excavating. Also known as inverted heading and bench method. { 'hed·iŋ 'ō·vər,hand 'bench ,meth·əd }

heading side *See* footwall. { 'hed·iŋ ,sīd }

heading-upward plan position indicator [ELECTR] A plan position indicator in which the heading of the craft appears at the top of the indicator at all times. { 'hed·iŋ 'əp·wərd 'plan pə,zish·ən 'ind·ə,kād·ər }

heading wall *See* footwall. { 'hed·iŋ ,wól }

headland [GEOGR] **1.** A high, steep-faced promontory extending into the sea. Also known as head; mull. **2.** High ground surrounding a body of water. { 'hed·lənd }

head length [ANTHRO] The distance measured between glabella and opisthocranion, with application of moderate pressure. { 'hed ,leŋkth }

headlight [ELEC] A lamp, usually fitted with a reflector and a special lens, that is mounted on the front of a locomotive or automotive vehicle to illuminate the road ahead. { 'hed,līt }

headline [MIN ENG] In dredging, the line which is anchored ahead of the dredge pond and holds the dredge up to its digging front. { 'hed,līn }

head loss [FL MECH] The drop in the sum of pressure head, velocity head, and potential head between two points along the path of a flowing fluid, due to causes such as fluid friction. { 'hed ,lós }

headman [MIN ENG] **1.** A person who brings coal to the

tramway from the workings. **2.** One who engages or disengages grips on mine cars at the top of a haulage slope. { 'hed·mən }

head margin [GRAPHICS] The unprinted space left above the first line of printing on a page. { 'hed ,mär·jən }

head mast [MIN ENG] The tower carrying the working lines of a cable excavator. { 'hed ,mast }

head meter [ENG] A flowmeter that is dependent upon change of pressure head to operate. { 'hed ,mēd·ər }

head motion [MECH ENG] The vibrator on a reciprocating table concentrator which imparts motion to the deck. { 'hed ,mō·shən }

head-mounted display [COMPUT SCI] A tracking device incorporating liquid-crystal displays or miniature cathode-ray tubes worn on a user's head to simulate a virtual environment (a three-dimensional sensation of depth) and to provide information on head movements for updating visual images. { 'hed ,maúnt·əd di'splā }

head organ [INV ZOO] One of the bulbous structures in the prohaptor of monogenetic trematodes which are openings for adhesive glands. { 'hed ,ór·gən }

head parking [COMPUT SCI] The positioning of the read/write head of a hard disk over the landing zone to ensure against head crashes. { 'hed ,pärk·iŋ }

head-per-track [COMPUT SCI] An arrangement having one read/write head for each magnetized track on a disk or drum to eliminate the need to move a single head from track to track. { 'hed pər 'trak }

headphone [ENG ACOUS] An electroacoustic transducer designed to be held against an ear by a clamp passing over the head, for private listening to the audio output of a communications, radio, or television receiver or other source of audio-frequency signals. Also known as phone. { 'hed,fōn }

head process [EMBRYO] The notochord or notochordal plate formed as an axial outgrowth of the primitive node. { 'hed ,prä·səs }

head pulley [MECH ENG] The pulley at the discharge end of a conveyor belt; may be either an idler or a drive pulley. { 'hed ,púl·ē }

head-pulley-drive conveyor [MECH ENG] A conveyor having the belt driven by the head pulley without a snub pulley. { 'hed ,púl·ē ,drīv kən'vā·ər }

headroom [MIN ENG] **1.** Distance between the drill platform and the bottom of the sheave wheel. **2.** Height between the floor and the roof in a mine opening. { 'hed,rüm }

headrope [MIN ENG] In rope haulage, that rope used to pull the loaded transportation device toward the discharge point. { 'hed ,rōp }

heads [MIN ENG] **1.** Material removed from the ore in the treatment plant and containing the valuable metallic constituents. **2.** The feed to a concentrating system in ore dressing. { hedz }

head scanning [IND ENG] Scanning of the visual field by using movement of both the head and the eyeballs. { 'hed ,skan·iŋ }

head section [COMMUN] *See* end section. [ENG] That part of belt conveyor which consists of a drive pulley, a head pulley which may or may not be a drive pulley, belt idlers if included, and the necessary framing. { 'hed ,sek·shən }

headset [ENG ACOUS] A single headphone or a pair of headphones, with a clamping strap or wires holding them in position. { 'hed,set }

head shaft [MECH ENG] The shaft driven by a chain and mounted at the delivery end of a chain conveyor; it serves as the mount for a sprocket which drives the drag chain. { 'hed ,shaft }

head sheave [MIN ENG] Pulley in the headgear of a winding shaft over which the hoisting rope runs. { 'hed ,shēv }

head shield [INV ZOO] A conspicuous structure arching over the lips of certain nematodes. { 'hed ,shēld }

headsill [BUILD] A horizontal beam at the top of the frame of a door or window. { 'hed,sil }

head smut [PL PATH] A fungus disease of corn and sorghum caused by *Sphacelotheca reiliana* which destroys the head of the plant. { 'hed ,smət }

headspace [ORD] Distance between the face of the locked bolt or breechblock of a gun and some specified point in the chamber; in guns using rimless bottlenecked cartridges, the space between the boltface and a specified point on the shoulder of the cartridge. { 'hed,spās }

head stepping rate [COMPUT SCI] The rate at which the read/write head of a disk drive moves from one track to another on the disk surface. { 'hed ¦step·iŋ ,rāt }

headstock [MECH ENG] **1.** The device on a lathe for carrying the revolving spindle. **2.** The movable head of certain measuring machines. **3.** The device on a cylindrical grinding machine for rotating the work. **4.** Also known as workhead. { 'hed ,stäk }

head tide [NAV] A tidal current setting in a direction approximately opposite to the heading of a vessel. { 'hed ,tīd }

headtree [MIN ENG] The horizontal timber placed at each side of a rectangular heading to support the headboard. { 'hed,trē }

head up [ENG] To tighten bolts on a hatch cover or access hole plate to prevent leakage from or into an operating vessel. { 'hed ,əp }

head-up display [OPTICS] A device that enables an aircraft pilot to view the instrument panel while looking out the cockpit window, by projecting an image of the panel in the direction of the window and forming the image at infinity. { 'hed ,əp di'splā }

head value [MIN ENG] Assay value of the feed to a concentrating system. { 'hed ,val·yü }

headwall [CIV ENG] A retaining wall at the outlet of a drain or culvert. [GEOL] The steep cliff at the back of a cirque. { 'hed,wól }

headward erosion [GEOL] Erosion caused by water flowing at the head of a valley. Also known as head erosion; headwater erosion. { 'hed·wərd i'rō·zhən }

headwater erosion See headward erosion. { 'hed,wòd·ər i'rō·zhən }

headwaters [HYD] The source and upstream waters of a stream. { 'hed,wòd·ərz }

headway [MIN ENG] See cross heading. [NAV] Motion in a forward direction. { 'hed,wā }

headworks [CIV ENG] Any device or structure at the head or diversion point of a waterway. { 'hed,wərks }

health [MED] A state of dynamic equilibrium between an organism and its environment in which all functions of mind and body are normal. { helth }

health physics [NUCLEO] The study of the protection of personnel from harmful effects of ionizing radiation by such means as routine radiation surveys, area and personnel monitoring, and protective equipment and procedures. { 'helth ,fiz·iks }

heaped capacity [MIN ENG] In scraper loading, the volume of heaped material that a scraper will hold. { ¦hēpt kə'pas·əd·ē }

heap leaching [MET] A process used for the recovery of copper from weathered ore and material from mine dumps; material is laid to a thickness of 20 feet (6 meters) in alternately fine and coarse beds and treated with water at intervals during which oxidation occurs; liquor that runs off is treated with scrap iron to precipitate copper. { 'hēp ,lēch·iŋ }

heap roasting [MIN ENG] A process in which ore with a high sulfur content is roasted by the combustion of the sulfur. { 'hēp ,rōst·iŋ }

heap sampling [MIN ENG] Method of ore sampling in which the material is shoveled into a conical heap which is then flattened with a spade and shoveled into four equal heaps, two of which are retained, crushed, mixed, and formed into another, smaller cone; the process is repeated until the required small sample is produced. { 'hēp ,sam·pliŋ }

hearing [PHYSIO] The general perceptual behavior and the specific responses made in relation to sound stimuli. { 'hir·iŋ }

hearing aid [ENG ACOUS] A miniature, portable sound amplifier for persons with impaired hearing, consisting of a microphone, audio amplifier, earphone, and battery. { 'hir·iŋ ,ād }

heart [ANAT] The hollow muscular pumping organ of the cardiovascular system in vertebrates. { härt }

heartbeat [PHYSIO] Pulsation of the heart coincident with ventricular systole. { 'härt,bēt }

heart block [MED] The cardiac condition resulting from defective transmission of impulses from atrium to ventricle. { 'härt ,bläk }

heart bond [CIV ENG] A masonry bond in which two header stones meet in the middle of the wall, their joint being covered by another stone; no headers stretch across the wall. { 'härt ,bänd }

heartburn [MED] A burning sensation emanating from the esophagus below the sternum. { 'härt,bərn }

heart cam [HOROL] A heart-shaped cam used in stopwatches and chronographs to return the recording hand to zero. { 'härt ,kam }

heart failure See cardiac failure. { 'härt ,fāl·yər }

heart-failure cells [PATH] Macrophages containing hemosiderin granules found in the lung in certain heart disorders. { 'härt ,fāl·yər ,selz }

hearth [BUILD] **1.** The floor of a fireplace or brick oven. **2.** The projection in front of a fireplace, made of brick, stone, or cement. [MET] The floor of a reverberatory, open-hearth, cupola, or blast furnace; it is made of refractory material able to support the charge and to collect the molten products. { härth }

hearth furnace [MET] A furnace designed to heat the charge, resting on the hearth, by passing hot gases over it. { 'härth ,fər·nəs }

hearth roasting [MIN ENG] A process in which ore or concentrate enters at the top of a multiple hearth roaster and drops from hearth until it is discharged at the bottom. { 'härth ,rōst·iŋ }

heart-lung machine [MED] A machine through which blood is shunted to maintain circulation during heart surgery. { 'härt 'ləŋ mə,shēn }

heart murmur See cardiac murmur. { 'härt 'mər·mər }

heart pacemaker See pacemaker. { 'härt 'pās,māk·ər }

heart rate [PHYSIO] The number of heartbeats per minute. { 'härt ,rāt }

heartrot [PL PATH] **1.** A rot involving disintegration of the heartwood of a tree. **2.** A fungus disease of beets and rutabagas caused by *Mycosphaerella tabifica* which results in decay of the central tissues of the plant. **3.** A boron-deficiency disease of sugarbeets that causes rot. **4.** A fatal disease of palms associated with a trypanosomatid flagellate *Phytomonas*. Also known as cedros wilt; marchitez sorpresiva. { 'härt,rät }

heart valve [ANAT] Flaps of tissue that prevent reflux of blood from the ventricles to the atria or from the pulmonary arteries or aorta to the atria. { 'härt ,valv }

heartwater disease [VET MED] A septicemic infectious disease of cattle, sheep, and goats in Africa caused by the rickettsial microorganism *Cowdria ruminantium*. { 'härt,wòd·ər di ,zēz }

heartwood [BOT] Xylem of an angiosperm. { 'härt,wùd }

heart worm [INV ZOO] *Dirofilaria immitis*. A filarial nematode parasitic on dogs and other carnivores. { 'härt ,wərm }

heat [THERMO] Energy in transit due to a temperature difference between the source from which the energy is coming and a sink toward which the energy is going; other types of energy in transit are called work. { hēt }

heat-activated battery See thermal battery. { 'hēt ¦ak·tə,vād·əd 'bad·ə·rē }

heat-affected zone [MET] The zone within a base metal that undergoes structural changes but does not melt during welding, cutting, or brazing. { 'hēt ə¦fek·təd 'zōn }

heat balance [GEOPHYS] The equilibrium which exists on the average between the radiation received by the earth and atmosphere from the sun and that emitted by the earth and atmosphere. [MET] The calculation used in fluidization roasting so that the addition or removal of heat can be controlled to maintain the optimum temperature in the reacting vessel. [THERMO] The equilibrium which is known to exist when all sources of heat gain and loss for a given region or body are accounted for. { 'hēt ,bal·əns }

heat barrier See thermal barrier. { 'hēt ,bar·ē·ər }

heat budget [GEOPHYS] Amount of heat needed to raise a lake's water from the winter temperature to the maximum summer temperature. [THERMO] The statement of the total inflow and outflow of heat for a planet, spacecraft, biological organism, or other entity. { 'hēt ,bəj·ət }

heat capacity [THERMO] The quantity of heat required to raise a system one degree in temperature in a specified way, usually at constant pressure or constant volume. Also known as thermal capacity. { 'hēt kə,pas·əd·ē }

heat check [MET] Parallel surface cracks forming a pattern on the surface of a metal as a result of thermal fatigue. { 'hēt ‚chek }

heat coil [ELEC] Protective device which uses a mechanical element which is allowed to move when the fusible substance that holds it in place is heated above a predetermined temperature by current in the circuit. { 'hēt ‚kóil }

heat conduction [THERMO] The flow of thermal energy through a substance from a higher- to a lower-temperature region. { 'hēt kən‚dək·shən }

heat conductivity See thermal conductivity. { 'hēt ‚kän·dək'tiv·əd·ē }

heat content See enthalpy. { 'hēt ¦kän·tent }

heat convection [THERMO] The transfer of thermal energy by actual physical movement from one location to another of a substance in which thermal energy is stored. Also known as thermal convection. { 'hēt kən¦vek·shən }

heat cramps [MED] Painful voluntary-muscle spasm and cramps following strenuous exercise, usually in persons in good physical condition, due to loss of sodium chloride and water from excessive sweating. { 'hēt ‚kramps }

heat cycle See thermodynamic cycle. { 'hēt ‚sī·kəl }

heat death [THERMO] The condition of any isolated system when its entropy reaches a maximum, in which matter is totally disordered and at a uniform temperature, and no energy is available for doing work. { 'hēt ‚deth }

heat development [GRAPHICS] A method that employs heat absorption to form the image. { 'hēt di‚vel·əp·mənt }

heat dissipation See heat loss. { 'hēt ‚dis·ə¦pā·shən }

heat distortion point [ENG] The temperature at which a standard test bar (American Society for Testing and Materials test) deflects 0.010 inch (0.254 millimeter) under a load of either 66 or 264 pounds per square inch (4.55×10^5 or 18.20×10^5 pascals), as specified. { 'hēt di‚stór·shən ‚póint }

heat dump See heatsink. { 'hēt ‚dəmp }

heat energy See internal energy. { 'hēt ‚en·ər·jē }

heat engine [MECH ENG] A machine that converts heat into work (mechanical energy). [THERMO] A thermodynamic system which undergoes a cyclic process during which a positive amount of work is done by the system; some heat flows into the system and a smaller amount flows out in each cycle. { 'hēt ‚en·jən }

heat equation [THERMO] A parabolic second-order differential equation for the temperature of a substance in a region where no heat source exists: $\partial t/\partial \tau = (k/\rho c)(\partial^2 t/\partial x^2 + \partial^2 t/\partial y^2 + \partial^2 t/\partial z^2)$, where x, y, and z are space coordinates, τ is the time, $t(x,y,z,\tau)$ is the temperature, k is the thermal conductivity of the body, ρ is its density, and c is its specific heat; this equation is fundamental to the study of heat flow in bodies. Also known as Fourier heat equation; heat flow equation. { 'hēt i‚kwā·zhən }

heat equator [METEOROL] **1.** The line which circumscribes the earth and connects all points of highest mean annual temperature for their longitudes. **2.** The parallel of latitude of 10°N, which has the highest mean temperature of any latitude. Also known as thermal equator. { 'hēt i¦kwäd·ər }

heater [ELECTR] An electric heating element for supplying heat to an indirectly heated cathode in an electron tube. Also known as electron-tube heater. [ENG] A contrivance designed to give off heat. { 'hēd·ər }

heater oil See heating oil. { 'hēd·ər ‚óil }

heater-treater [PETRO ENG] A unit for heating an oil-and-water emulsion and then removing the water and gas. { 'hēd·ər ‚trēd·ər }

heater-type cathode See indirectly heated cathode. { 'hēd·ər ‚tīp ‚kath‚ōd }

heat exchange [CHEM ENG] A unit operation based on heat transfer which functions in the heating and cooling of fluids with or without phase change. { 'hēt iks‚chānj }

heat exchanger [ENG] Any device, such as an automobile radiator, that transfers heat from one fluid to another or to the environment. Also known as exchanger. { 'hēt iks‚chānj·ər }

heat exhaustion [MED] A heat-exposure syndrome characterized by weakness, vertigo, headache, nausea, and peripheral vascular collapse, usually precipitated by physical exertion in a hot environment. { 'hēt ig‚zós·chən }

heat filter [OPTICS] Special glass in condenser lens systems to keep heat from film. { 'hēt ‚fil·tər }

heat flow [THERMO] Heat thought of as energy flowing from

one substance to another; quantitatively, the amount of heat transferred in a unit time. Also known as heat transmission. { 'hēt ‚flō }

heat flow equation See heat equation. { 'hēt ¦flō i‚kwā·zhən }

heat flow province [GEOPHYS] A geographic area in which the heat flow and heat production are linearly related. { 'hēt ¦flō ‚präv·əns }

heat flux [THERMO] The amount of heat transferred across a surface of unit area in a unit time. Also known as thermal flux. { 'hēt ‚fləks }

heat gain [ENG] The increase of heat within a given space as a result of direct heating by solar radiation and of heat radiated by other sources such as lights, equipment, or people. { 'hēt ‚gān }

heath See temperate and cold scrub. { hēth }

heather [BOT] *Calluna vulgaris.* An evergreen heath of northern and alpine regions distinguished by racemes of small purple-pink flowers. { 'heth·ər }

heating chamber [ENG] The part of an injection mold in which cold plastic feed is changed into a hot melt. { 'hēd·iŋ ‚chām·bər }

heating coils [NAV ARCH] A system of piping in the bottom of an oil tanker which carries steam to heat high-pour-point liquid cargoes to a viscosity suitable for pumping. { 'hēd·iŋ ‚kóilz }

heating degree-day [METEOROL] A form of degree-day used as an indication of fuel consumption; in United States usage, one heating degree-day is given for each degree that the daily mean temperature departs below the base of 65°F (where the Celsius scale is used, the base is usually 19°C). { 'hēd·iŋ di'grē ‚dā }

heating element [ELEC] The part of a heating appliance in which electrical energy is transformed into heat. { 'hēd·iŋ ‚el·ə·mənt }

heating fuel See heating oil. { 'hēd·iŋ ‚fyül }

heating load [CIV ENG] The quantity of heat per unit time that must be provided to maintain the temperature in a building at a given level. { 'hēd·iŋ ‚lōd }

heating oil [MATER] No. 2 fuel oil; used in domestic heating units. Also known as heater oil; heating fuel. { 'hēd·iŋ ‚óil }

heating plant [CIV ENG] The whole system for heating an enclosed space. Also known as heating system. { 'hēd·iŋ ‚plant }

heating surface [ENG] The surface for the absorption and transfer of heat from one medium to another. { 'hēd·iŋ ‚sər·fəs }

heating system See heating plant. { 'hēd·iŋ ‚sis·təm }

heating value See heat of combustion. { 'hēd·iŋ ‚val·yü }

heat insulator [MATER] A substance having relatively low heat conductivity. { 'hēt ‚ins·ə‚lād·ər }

heat island effect [METEOROL] In urban areas with tall buildings, an atmospheric condition in which heat and pollutants create a haze dome that prevents warm air from rising and being cooled at a normal rate, especially in the absence of strong winds. { 'hēt ‚ī·lənd i‚fekt }

heat lamp [ELEC] An infrared lamp used for brooders in farming, for drying paint or ink, for keeping food warm, and for therapeutic and other applications requiring heat with or without some visible light. { 'hēt ‚lamp }

heat lightning [GEOPHYS] Nontechnically, the luminosity observed from ordinary lightning too far away for its thunder to be heard. { 'hēt ‚līt·niŋ }

heat loss [PHYS] Energy or power transmitted out of a system in the form of heat. Also known as heat dissipation. { 'hēt ‚lós }

heat-loss flowmeter [ENG] Any of various instruments that determine gas velocities or mass flows from the cooling effect of the flow on an electrical sensor such as a thermistor or resistor; a second sensor is used to compensate for the temperature of the fluid. Also known as thermal-loss meter. { 'hēt ‚lós 'flō‚mēd·ər }

heat low See thermal low. { 'hēt ¦lō }

heat of ablation [THERMO] A measure of the effective heat capacity of an ablating material, numerically the heating rate input divided by the mass loss rate which results from ablation. { 'hēt əv ə'blā·shən }

heat of activation [PHYS CHEM] The increase in enthalpy when a substance is transformed from a less active to a more reactive form at constant pressure. { 'hēt əv ‚ak·tə'vā·shən }

heat of adsorption [THERMO] The increase in enthalpy when 1 mole of a substance is adsorbed upon another at constant pressure. { 'hēt əv ad'sórp·shən }

heat of aggregation [THERMO] The increase in enthalpy when an aggregate of matter, such as a crystal, is formed at constant pressure. { 'hēt əv ˌag·rə'gā·shən }

heat of association [PHYS CHEM] Increase in enthalpy accompanying the formation of 1 mole of a coordination compound from its constituent molecules or other particles at constant pressure. { 'hēt əv əˌsō·sē'ā·shən }

heat of atomization [PHYS CHEM] The change in enthalpy accompanying the conversion of 1 mole of an element or a compound at 298 K (77°F) and 1 atmosphere (10^5 pascals) into free atoms. { 'hēt əv ˌad·ə·mə'zā·shən }

heat of combustion [PHYS CHEM] The amount of heat released in the oxidation of 1 mole of a substance at constant pressure, or constant volume. Also known as heat value; heating value. { 'hēt əv kəm'bəs·chən }

heat of compression [THERMO] Heat generated when air is compressed. { 'hēt əv kəm'presh·ən }

heat of condensation [THERMO] The increase in enthalpy accompanying the conversion of 1 mole of vapor into liquid at constant pressure and temperature. { 'hēt əv ˌkänd·ən'sā·shən }

heat of cooling [THERMO] Increase in enthalpy during cooling of a system at constant pressure, resulting from an internal change such as an allotropic transformation. { 'hēt əv 'kül·iŋ }

heat of crystallization [THERMO] The increase in enthalpy when 1 mole of a substance is transformed into its crystalline state at constant pressure. { 'hēt əv ˌkrist·əl·ə'zā·shən }

heat of decomposition [PHYS CHEM] The change in enthalpy accompanying the decomposition of 1 mole of a compound into its elements at constant pressure. { 'hēt əv dē,käm·pə'zish·ən }

heat of dilution [PHYS CHEM] **1.** The increase in enthalpy accompanying the addition of a specified amount of solvent to a solution of constant pressure. Also known as integral heat of dilution; total heat of dilution. **2.** The increase in enthalpy when an infinitesimal amount of solvent is added to a solution at constant pressure. Also known as differential heat of dilution. { 'hēt əv də'lü·shən }

heat of dissociation [PHYS CHEM] The increase in enthalpy at constant pressure, when molecules break apart or valence linkages rupture. { 'hēt əv di,sō·sē'ā·shən }

heat of emission [ELECTR] Additional heat energy that must be supplied to an electron-emitting surface to maintain it at a constant temperature. { 'hēt əv i'mish·ən }

heat of evaporation See heat of vaporization. { 'hēt əv iˌvap·ə'rā·shən }

heat of formation [PHYS CHEM] The increase in enthalpy resulting from the formation of 1 mole of a substance from its elements at constant pressure. { 'hēt əv fòr'mā·shən }

heat of fusion [THERMO] The increase in enthalpy accompanying the conversion of 1 mole, or a unit mass, of a solid to a liquid at its melting point at constant pressure and temperature. Also known as latent heat of fusion. { 'hēt əv 'fyü·zhən }

heat of hydration [PHYS CHEM] The increase in enthalpy accompanying the formation of 1 mole of a hydrate from the anhydrous form of the compound and from water at constant pressure. { 'hēt əv hī'drā·shən }

heat of ionization [PHYS CHEM] The increase in enthalpy when 1 mole of a substance is completely ionized at constant pressure. { 'hēt əv ˌī·ən·ə'zā·shən }

heat of linkage [PHYS CHEM] The bond energy of a particular type of valence linkage between atoms in a molecule, as determined by the energy required to dissociate all bonds of the type in 1 mole of the compound divided by the number of such bonds in a compound. { 'hēt əv 'liŋk·ij }

heat of mixing [THERMO] The difference between the enthalpy of a mixture and the sum of the enthalpies of its components at the same pressure and temperature. { 'hēt əv 'mik·siŋ }

heat of reaction [PHYS CHEM] **1.** The negative of the change in enthalpy accompanying a chemical reaction at constant pressure. **2.** The negative of the change in internal energy accompanying a chemical reaction at constant volume. { 'hēt əv rē'ak·shən }

heat of solidification [THERMO] The increase in enthalpy when 1 mole of a solid is formed from a liquid or, less commonly, a gas at constant pressure and temperature. { 'hēt əv sə,lid·ə·fə'kā·shən }

heat of solution [PHYS CHEM] The enthalpy of a solution minus the sum of the enthalpies of its components. Also known as integral heat of solution; total heat of solution. { 'hēt əv sə'lü·shən }

heat of sublimation [THERMO] The increase in enthalpy accompanying the conversion of 1 mole, or unit mass, of a solid to a vapor at constant pressure and temperature. Also known as latent heat of sublimation. { 'hēt əv ˌsəb·lə'mā·shən }

heat of transformation [THERMO] The increase in enthalpy of a substance when it undergoes some phase change at constant pressure and temperature. { 'hēt əv ˌtranz·fər'mā·shən }

heat of vaporization [THERMO] The quantity of energy required to evaporate 1 mole, or a unit mass, of a liquid, at constant pressure and temperature. Also known as enthalpy of vaporization; heat of evaporation; latent heat of vaporization. { 'hēt əv ˌvā·pə·rə'zā·shən }

heat of wetting [THERMO] **1.** The heat of adsorption of water on a substance. **2.** The additional heat required, above the heat of vaporization of free water, to evaporate water from a substance in which it has been absorbed. { 'hēt əv 'wed·iŋ }

heat pipe [ENG] A heat-transfer device consisting of a sealed metal tube with an inner lining of wicklike capillary material and a small amount of fluid in a partial vacuum; heat is absorbed at one end by vaporization of the fluid and is released at the other end by condensation of the vapor. { 'hēt ˌpīp }

heat pump [MECH ENG] A device which transfers heat from a cooler reservoir to a hotter one, expending mechanical energy in the process, especially when the main purpose is to heat the hot reservoir rather than refrigerate the cold one. { 'hēt ˌpəmp }

heat quantity [THERMO] A measured amount of heat; units are the small calorie, normal calorie, mean calorie, and large calorie. { 'hēt ˌkwän·əd·ē }

heat radiation [THERMO] The energy radiated by solids, liquids, and gases in the form of electromagnetic waves as a result of their temperature. Also known as thermal radiation. { 'hēt ˌrād·ē'ā·shən }

heat rash See miliaria. { 'hēt ˌrash }

heat rate [MECH ENG] An expression of the conversion efficiency of a thermal power plant or engine, as heat input per unit of work output; for example, Btu/kWh. { 'hēt ˌrāt }

heat reactor [NUCLEO] A nuclear reactor designed primarily to supply heat for industrial purposes. { 'hēt rē'ak·tər }

heat release [THERMO] The quantity of heat released by a furnace or other heating mechanism per second, divided by its volume. { 'hēt ri,lēs }

heat resistance See thermal resistance. { 'hēt ri,zis·təns }

heat-resistant alloy [MET] An oxidation-resistant alloy. { 'hēt ri,zis·tənt 'al,ói }

heat-resistant glass [MATER] Glass, such as borosilicate glass, that is heat-treated or leached to remove alkali so that it withstands high heat and sudden cooling without shattering. { 'hēt ri,zis·tənt 'glas }

heat run [ELEC] A series of temperature measurements made on an electric device during operating tests under various conditions. { 'hēt ˌrən }

heat seal [ENG] A union between two thermoplastic surfaces by application of heat and pressure to the joint. { 'hēt ˌsēl }

heatseeker [ORD] A guided missile incorporating an infrared device for homing on heat-radiating machines or installations, such as an aircraft engine or a blast furnace. { 'hēt ˌsēk·ər }

heat set [TEXT] A process to fix or set a crimp or texture in yarn by use of heat. { 'hēt ˌset }

heat shield [MATER] Any protective layer that gives protection from heat; used on the front of a reentry capsule. { 'hēt ˌshēld }

heat shock protein [CELL MOL] Any of a group of proteins that are synthesized in the cytoplasm of cells as part of the heat shock response and act to protect the chromosomes from damage. { 'hēt ˌshäk 'prō,tēn }

heat shock response [CELL MOL] A cellular reaction to a stimulus such as elevated temperatures or abrupt environmental changes, in which there is cessation or slowdown of normal protein synthesis and activation of previously inactive genes, resulting in the production of heat shock proteins. { 'hēt ,shäk ri,späns }

heat-shrinkable tubing [MATER] A type of plastic tubing that can be heated and shrink-fitted over terminals and other objects of varying sizes and shapes, for insulating and other purposes. { 'hēt 'shriŋk·ə·bəl 'tüb·iŋ }

heat shunt [MET] A heatsink placed in contact with the lead of a delicate component to prevent overheating during soldering. { 'hēt ,shənt }

heatsink [AERO ENG] 1. A type of protective device capable of absorbing heat and used as a heat shield. 2. In nuclear propulsion, any thermodynamic device, such as a radiator or condenser, that is designed to absorb the excess heat energy of the working fluid. Also known as heat dump. [ELEC] A mass of metal that is added to a device for the purpose of absorbing and dissipating heat; used with power transistors and many types of metallic rectifiers. Also known as dissipator. [THERMO] Any (gas, solid, or liquid) region where heat is absorbed. { 'hēt,siŋk }

heatsink cooling [ENG] Cooling a body or system by allowing heat to be absorbed from it by another body. { 'hēt ,siŋk ,kül·iŋ }

heat source [THERMO] Any device or natural body that supplies heat. { 'hēt ,sòrs }

heat sterilization [ENG] An act of destroying all forms of life on and in bacteriological media, foods, hospital supplies, and other materials by means of moist or dry heat. { 'hēt ,ster·ə·lə'zā·shən }

heat storage [OCEANOGR] The tendency of the ocean to act as a heat reservoir; results in smaller daily and annual variations in temperature over the sea. { 'hēt ,stòr·ij }

heat stress index [PHYSIO] Relation of the amount of evaporation or perspiration required for particular job conditions as related to the maximum evaporative capacity of an average person. Abbreviated HSI. { 'hēt ,stres 'in,deks }

heatstroke [MED] A heat-exposure syndrome characterized by hyperpyrexia and prostration due to diminution or cessation of sweating, occurring most commonly in persons with underlying disease. { 'hēt,strōk }

heat thunderstorm [METEOROL] In popular terminology, a thunderstorm of the air mass type which develops near the end of a hot, humid summer day. { 'hēt ,thən·dər,stòrm }

heat time [MET] Duration of a single current impulse in pulsation welding. { 'hēt ,tīm }

heat tinting [MET] Oxidation of a polished metal surface by heating to reveal the microstructure. { 'hēt ,tint·iŋ }

heat transfer [THERMO] The movement of heat from one body to another (gas, liquid, solid, or combinations thereof) by means of radiation, convection, or conduction. { 'hēt ,tranz·fər }

heat-transfer coefficient [THERMO] The amount of heat which passes through a unit area of a medium or system in a unit time when the temperature difference between the boundaries of the system is 1 degree. { 'hēt ,tranz·fər ,kō·i'fish·ənt }

heat-transfer oil [MATER] An oil used to transport heat or cold between two areas of process-equipment surface, and especially compounded to avoid heat degradation in the temperature range of application. { 'hēt ,tranz·fər ,òil }

heat transmission See heat flow. { 'hēt tranz,mish·ən }

heat transport [THERMO] Process by which heat is carried past a fixed point or across a fixed plane, as in a warm current. { 'hēt ,tranz,pórt }

heat-treatable alloy [MET] An alloy that can be hardened by thermal treatment. { 'hēt ,trēd·ə·bəl 'al,òi }

heat-treating film [MET] An oxide coating formed on a metal surface by heat treating. { 'hēt ,trēd·iŋ ,film }

heat treatment [MET] Heating and cooling a metal or alloy to obtain desired properties or conditions. { 'hēt ,trēt·mənt }

heat value See heat of combustion. { 'hēt ,val·yü }

heat wave [ELECTROMAG] Infrared radiation, much higher in frequency than radio waves. [METEOROL] A period of abnormally and uncomfortably hot and usually humid weather; the condition must prevail at least 1 day to be a heat wave, but conventionally the term is reserved for periods of several days to several weeks. Also known as hot wave; warm wave. { 'hēt ,wāv }

heat wheel [MECH ENG] In a ventilating system, a device to condition incoming air by causing it to approach thermal equilibrium with the exiting air; hot incoming air is cooled, and cold incoming air is warmed. { 'hēt ,wēl }

heave [GEOL] The horizontal component of the slip, measured at right angles to the strike of the fault. [MIN ENG] 1. A rising of the floor of a mine caused by its being too soft to resist the weight on the pillars. 2. A predominantly upward movement of the surface of the soil due to expansion or displacement. [OCEANOGR] The motion imparted to a floating body by wave action. { hēv }

heave compensator [PETRO ENG] A motion compensator on a floating offshore drilling rig that moves with vertical motion to maintain a constant pressure on the drilling bit. { 'hēv käm·pən,sād·ər }

heavenly body See celestial body. { 'hev·ən·lē 'bäd·ē }

heaves [VET MED] Chronic emphysema in horses marked by labored breathing due to overdistension of the alveoli. Also known as broken wind. { hēvz }

heave to [NAV] To bring a ship into such a position that there is no headway. { ¦hēv 'tü }

heavier-than-air craft [AERO ENG] Any aircraft weighing more than the air it displaces. { 'hev·ē·ər thən 'er ¦kraft }

heaving [NAV ARCH] Vertical motion of a ship, as distinguished from pitching. [PETRO ENG] Partial or total collapse of drill hole walls resulting from internal pressures mainly due to swelling from hydration or formation gas pressures. { 'hēv·iŋ }

heaving plug [PETRO ENG] A plug at the bottom of an oil well which stops unconsolidated sand from mixing with the oil. { 'hēv·iŋ ,pləg }

Heaviside calculus [MATH] A type of operational calculus that is used to completely analyze a linear dynamical system which represents some vibrating physical system. { 'hev·ē,sīd ,kal·kyə·ləs }

Heaviside layer See F layer. { 'hev·ē,sīd ,lā·ər }

Heaviside-Lorentz system [ELECTROMAG] A system of electrical units which is the same as the Gaussian system except that the units of charge and current are smaller by a factor of $1/\sqrt{4\pi}$, and those of electric and magnetic field are larger by a factor by $\sqrt{4\pi}$. Also known as Lorentz-Heaviside system. { 'hev·ē,sīd lò'rents ,sis·təm }

Heaviside's expansion theorem [MATH] A theorem providing an infinite series representation for the inverse Laplace transforms of functions of a particular type. { 'hev·ē,sīdz ik 'span·chən ,thir·əm }

Heaviside unit function [MATH] The real function $f(x)$ whose value is 0 if x is negative and whose value is 1 otherwise. { 'hev·ē,sīd 'yü·nət ¦fəŋk·shən }

heavy acid See phosphotungstic acid. { ¦hev·ē 'as·əd }

heavy alloy [MET] A tungsten-nickel alloy produced by pressing and sintering the metallic powders; used for screens for x-ray tubes and radioactivity units and for contact surfaces of circuit breakers. { 'hev·ē 'al,òi }

heavy antiaircraft artillery [ORD] Conventional antiaircraft artillery pieces larger than 90-millimeter, the weight of which in a trailed mount is greater than 40,000 pounds (18,000 kilograms). { 'hev·ē ,an·tē'er,kraft är'til·ə·rē }

heavy artillery [ORD] Artillery other than antiaircraft artillery; consists of howitzers and longer-barreled cannon not classified as medium artillery. { 'hev·ē är'til·ə·rē }

heavy bombardment [ORD] A bombardment of great intensity, especially one with large aerial bombs or other missiles. { 'hev·ē bäm'bärd·mənt }

heavy bomber [AERO ENG] Any large bomber considered to be relatively heavy, such as a bomber having a gross weight, including bomb load, of 250,000 pounds (113,000 kilograms) or more, as the B-36 and the B-52. { 'hev·ē 'bäm·ər }

heavy chain [IMMUNOL] The heavier of the two types of polypeptide chains occurring in immunoglobulin molecules, its molecular weight range being 50,000–70,000. Also known as A chain; H chain. { 'hev·ē 'chān }

heavy concrete [MATER] Concrete in which some or all rock aggregate is replaced by metal aggregate. { 'hev·ē kän'krēt }

heavy crude [PETRO ENG] Crude oil having a high proportion of viscous, high-molecular-weight hydrocarbons, and often having a high sulfur content. { 'hev·ē 'krüd }

heavy cruiser [NAV ARCH] A warship designed to operate with strike, antisubmarine-warfare, or amphibious forces against air and surface threats. { 'hev·ē 'krüz·ər }

heavy drop [ORD] An airdrop in which heavy articles, such as trucks or artillery pieces, are dropped by parachute. { 'hev·ē ¦dräp }

heavy-duty [ENG] Designed to withstand excessive strain. { ¦hev·ē ¦düd·ē }

heavy-duty car [MECH ENG] A railway motorcar weighing more than 1400 pounds (635 kilograms), propelled by an engine of 12–30 horsepower (8900–22,400 watts), and designed for hauling heavy equipment and for hump-yard service. { ¦hev·ē ¦düd·ē 'kär }

heavy-duty oil [MATER] Lubricating oil with good oxidation stability and corrosion-preventive and detergent-dispersant characterisitics; used in high-speed diesel and gasoline engines under heavy-duty service conditions. { ¦hev·ē ¦düd·ē 'ȯil }

heavy-duty tool block See open-side tool block. { 'hev·ē ¦düd·ē 'tül ¦bläk }

heavy ends [MATER] The highest boiling portion of a petroleum fraction. { 'hev·ē 'enz }

heavy-fermion superconductor [SOLID STATE] A superconductor in which the superconducting electrons have unusually large effective masses, more than 100 times the mass of a free electron. { ¦hev·ē 'fər¸mē¸än 'sü·pər·kən¸dək·tər }

heavy-fermion system [SOLID STATE] A lanthanide-based or actinide-based intermetallic compound in which the low-energy excitations (quasiparticles) of the conduction electron system have effective masses at low temperatures that are several hundred times the free-electron mass. { ¸hev·ē 'fər¸mē¸än ¸sis·təm }

heavy field artillery [ORD] Field artillery of the largest calibers, such as the 155-millimeter gun, the 6-inch (152.4-millimeter) gun, the 8-inch (203.2-millimeter) howitzer, and the 240-millimeter howitzer. { 'hev·ē 'fēld är¸til·ə·rē }

heavy floe [OCEANOGR] A mass of sea ice that is more than 10 feet (3 meters) thick. Also known as heavy ice. { 'hev·ē 'flō }

heavy force fit [DES ENG] A fit for heavy steel parts or shrink fits in medium sections. { 'hev·ē 'fȯrs ¸fit }

heavy fraction [PETRO ENG] The final products retrieved from crude oil during the process of distillation. Also known as end cut. { 'hev·ē 'frak·shən }

heavy ground [MIN ENG] Dangerous hanging wall requiring vigilance against possible rock fall. { 'hev·ē 'graůnd }

heavy howitzer [ORD] A complete projectile-firing weapon, with a medium muzzle velocity and a curved trajectory; the bore diameter is larger than 200 millimeters. { 'hev·ē 'haů·it·sər }

heavy hydrogen [NUC PHYS] Hydrogen consisting of isotopes whose mass number is greater than one, namely deuterium or tritium. { 'hev·ē 'hī·drə·jən }

heavy ice See heavy floe. { 'hev·ē 'īs }

heavy ion See large ion. { 'hev·ē 'ī¸än }

heavy-ion linac See heavy-ion linear accelerator. { 'hev·ē 'ī¸än 'lin¸ak }

heavy-ion linear accelerator [NUCLEO] A linear accelerator which produces a beam of heavy particles of high intensity and sharp energy; used to produce transuranic elements and short-lived isotopes, and to study nuclear reactions, nuclear spectroscopy, and the absorption of heavy ions in matter. Also known as heavy-ion linac; hilac. { 'hev·ē ¦ī¸än ¦lin·ē·ər ak'sel·ə¸rād·ər }

heavy-ion source [ELECTR] Any source of ionized molecules or atoms of elements heavier than helium. { 'hev·ē ¦ī¸än 'sȯrs }

heavy-lift ship [NAV ARCH] A ship capable of loading and discharging large individual cargo units of up to 1000 metric tons. { 'hev·ē 'lift ¸ship }

heavy liquid [MATER] Any of a group of heavy organic liquids, inorganic solutions, and fused salts used for determination of specific gravity of mineral particles or for separation of minerals having lower and higher specific gravities than the liquids; examples are methylene iodide and bromoform. { 'hev·ē 'lik·wəd }

heavy-liquid bubble chamber [NUCLEO] A bubble chamber which contains deuterium or an organic liquid such as propane or Freon. { 'hev·ē ¦lik·wəd 'bəb·əl ¸chām·bər }

heavy-liquid separation [MIN ENG] A laboratory technique for separating ore particles by allowing them to settle through, or float above, a fluid of intermediate density. { 'hev·ē ¦lik·wəd ¸sep·ə'rā·shən }

heavy machine gun [ORD] **1.** Any machine gun of relatively heavy weight, including caliber-.30 water-cooled machine guns and caliber-.50 machine guns. **2.** Any aircraft machine gun above caliber .30. { 'hev·ē mə'shēn ¸gən }

heavy-media separation [MIN ENG] A series of processes for the concentration of ore developed at one time, but now used in coal cleaning; uses suspensions of magnetic materials such as magnetite. { 'hev·ē ¸mēd·ē·ə ¸sep·ə'rā·shən }

heavy meromyosin [BIOCHEM] The larger of two fragments obtained from the muscle protein myosin following limited proteolysis by trypsin or chymotrypsin. { 'hev·ē ¸mer·ə¦mī·ə¸sin }

heavy metal [MET] A metal whose specific gravity is approximately 5.0 or higher. { 'hev·ē 'med·əl }

heavy-metal star [ASTRON] A member of a class of peculiar giants that includes the barium stars and S stars, characterized by unusually strong lines of heavy metals, including barium and zirconium. { 'hev·ē 'med·əl 'stär }

heavy mineral [MINERAL] A mineral with a density above 2.9, which is the density of bromoform, the liquid used to separate the heavy from the light minerals. { 'hev·ē 'min·rəl }

heavy-mineral prospecting [MIN ENG] Locating the source of an economic mineral by determining the relative amounts of the mineral in stream sediments and tracing the drainage upstream. { 'hev·ē ¦min·rəl 'präs¸pek·tiŋ }

heavy naphtha [MATER] A dark amber to red liquid that is a mixture of xylene and higher homologs; it is flammable; used as a solvent for asphalt and in production of coumarone resins. Also known as high-flash naphtha. { 'hev·ē 'naf·thə }

heavy oil [MATER] The high-boiling, relatively viscous fractions of petroleum or coal tar oils. { 'hev·ē 'ȯil }

heavy oxygen See oxygen-18. { 'hev·ē 'äk·sə·jən }

heavy particle See baryon. { 'hev·ē 'pärd·ə·kəl }

heavy resin oil [MATER] Reddish-brown, high-boiling, heavy coal tar oils. { 'hev·ē ¦rez·ən 'ȯil }

heavy section car [MECH ENG] A railway motorcar weighing 1200–1400 pounds (544–635 kilograms) and propelled by an 8–12 horsepower (6000–8900 watts) engine. { 'hev·ē 'sek·shən ¸kär }

heavy tank [ORD] A full-track combat tank with a weight of 56–85 tons. { 'hev·ē 'taŋk }

heavy water [INV ZOO] A compound of hydrogen and oxygen containing a higher proportion of the hydrogen isotope deuterium than does naturally occurring water. Also known as deuterium oxide. { 'hev·ē 'wȯd·ər }

heavy-water reactor [NUCLEO] A nuclear reactor in which heavy water serves as moderator and sometimes also as coolant. { 'hev·ē ¦wȯd·ər rē'ak·tər }

heavy weapon [ORD] Any weapon such as a howitzer, mortar, heavy machine gun, and recoilless rifle that is usually part of infantry equipment. { 'hev·ē 'wep·ən }

heazelwoodite [MINERAL] Ni_3S_2 A meteorite mineral consisting of a sulfide of nickel. { 'hē¸zəl¸wů¸dīt }

Hebe [ASTRON] An asteroid with a diameter of about 126 miles (202 kilometers), mean distance from the sun of 2.42 astronomical units, and S-type surface composition. { 'hē¸bē }

hebephrenia [PSYCH] A type of schizophrenia marked by disorganized thinking, mannerisms, and regressive caricaturing that is seen in some adolescents, such as silliness, unpredictable giggling, and posturing. { ¸hē·bə'frē·nē·ə }

Heberden-Rosenbach node See Heberden's node. { 'heb·ərd·ən 'rōz·ən¸bäk ¸nōd }

Heberden's arthritis (Obsolete) [MED] Degenerative joint disease of the terminal joints of the fingers, producing enlargement (Heberden's nodes) and flexion deformities. { 'heb·ərd·ənz är'thrīd·əs }

Heberden's node [MED] Nodose deformity of the fingers in degenerative joint disease. Also known as Heberden-Rosenbach node. { 'heb·ərd·ənz ¸nōd }

Hebraic granite See graphic granite. { hē'brā·ik ¦gran·ət }

Hebrovellidae [INV ZOO] A family of hemipteran insects in the subdivision Amphibicorisae. { ¸heb·rō'vel·ə¸dē }

hecatolite See moonstone. { hə'kat·əl¸īt }

hectare [MECH] A unit of area in the metric system equal to 100 ares or 10,000 square meters. Abbreviated ha. { 'hek¸tar }

hecto- [SCI TECH] A prefix representing 10^2 or 100. { 'hek·tō }

hectocotylus [INV ZOO] A specialized appendage of male cephalopods adapted for the transference of sperm. { ˌhek·tōˈkäd·əl·əs }

hectogram [MECH] A unit of mass equal to 100 grams. Abbreviated hg. { ˈhek·təˌgram }

hectoliter [MECH] A metric unit of volume equal to 100 liters or to 0.1 cubic meter. Abbreviated hl. { ˈhek·təˌlēd·ər }

hectometer [MECH] A unit of length equal to 100 meters. Abbreviated hm. { ˈhek·təˌmēd·ər }

hectometric wave [COMMUN] A radio wave between the wavelength limits of 100 and 1000 meters, corresponding to the frequency range of 3000 to 300 kilohertz. { ˌhek·təˈmetrik ˈwāv }

hectorite [MINERAL] $(Mg,Li)_3Si_4O_{10}(OH)_2$ A trioctohedral clay mineral of the montmorillonite group composed of a hydrous silicate of magnesium and lithium. { ˈhek·təˌrīt }

heddle [TEXT] **1.** A twisted wire with an eye, attached to the harness to guide the warp threads. **2.** One of the sets of parallel cords or wires composing the harness of a loom. { ˈhed·əl }

hedenbergite [MINERAL] $CaFeSi_2O_6$ A black mineral consisting of calcium-iron pyroxene and occurring at the contacts of limestone with granitic masses. { ˈhed·ənˌbərˌgīt }

hedeoma oil [MATER] A yellowish essential oil distilled from the leaves of *Hedeoma pulegioides;* soluble in two or more parts of 70% alcohol, ether, and chloroform; used in medicine and perfumery. Also known as American pennyroyal oil; pulegium oil. { ˌhēˈdēˈō·məˌoil }

hedgehog [ORD] **1.** A portable obstacle, made of crossed poles laced with barbed wire, in the general shape of an hourglass. **2.** A beach obstacle, usually made of steel bars or channel iron, imbedded in concrete and used to interfere with beach landings. **3.** A concentration of troops securely entrenched or fortified, with arms and defenses facing all directions. [VERT ZOO] The common name for members of the insectivorous family Erinaceidae characterized by spines on their back and sides. { ˈhejˌhäg }

hedgehog round [ORD] A small, mortarlike, antisubmarine projectile. { ˈhejˌhägˌraủnd }

hedleyite [MINERAL] A mineral composed of an alloy of bismuth and tellurium. { ˈhedˈlēˌīt }

hedonic gland [VERT ZOO] One of the mucus-secreting scent glands in many urodeles; functions in courtship. { hēˈdän·ik ˌgland }

hedonism [PSYCH] The doctrine that every act is motivated by the desire for pleasure or the aversion from pain and unpleasantness. { ˈhēd·ənˌiz·əm }

hedonophobia [PSYCH] An abnormal fear of pleasure. { ˌhēd·ən·ōˈfō·bē·ə }

hedreocraton [GEOL] A craton that influenced later continental development. { ˌhedˌrē·ōˈkrāˌtän }

Hedvall effect I [SOLID STATE] A discontinuous change in the temperature dependence of the chemical reaction rate of certain substances at the Curie temperture. { ˈhedˌvȯl iˌfekt ˈwən }

Hedvall effect II [SOLID STATE] A discontinuous change in the activation energy of certain substances at the Curie temperature. { ˈhedˌvȯl iˌfekt ˈtü }

Hedwigiaceae [BOT] A family of mosses in the order Isobryales. { ˌhedˌvig·ēˈāsˌēˌē }

hedyphane [MINERAL] $(Ca,Pb)_5Cl(AsO_4)_3$ Yellowish-white mineral composed of lead and calcium arsenate and chloride; occurs in monoclinic crystals. { ˈhed·əˌfān }

HEED See high-energy electron diffraction.

heel [MECH ENG] See heel block. [MET] **1.** A quantity of molten metal remaining in the ladle after pouring a metal casting. **2.** A quantity of metal retained in an induction furnace during a stand-by period. [NAV] Of a ship, to incline or to be inclined to one side. [ORD] Upper corner of the butt of a rifle stock held in firing position. { hēl }

heel block [MECH ENG] A block or plate that is usually fixed on the die shoe to minimize deflection of a punch or cam. Also known as heel. { ˈhēl ˌbläk }

heeling adjuster [ENG] A dip needle with a sliding weight that can be moved along one of its arms to balance the magnetic force; used to determine the correct position of a heeling magnet. Also known as heeling error instrument; vertical force instrument. { ˈhēl·iŋ əˌjəs·tər }

heeling error [NAV] The change in the indication of a magnetic compass when a craft heels, due to the change in the position of the magnetic influences on the craft relative to the earth's magnetic field and to the compass. { ˈhēl·iŋ ˌer·ər }

heeling error instrument See heeling adjuster. { ˈhēl·iŋ ˌer·ər ˌinˈstrə·mənt }

heeling magnet [ENG] A permanent magnet placed vertically in a tube under the center of a marine magnetic compass, to correct for heeling error. { ˈhēl·iŋ ˌmag·nət }

heel of a shot [ENG] **1.** In blasting, the front or face of a shot farthest from the charge. **2.** The distance between the mouth of the drill hole and the corner of the nearest free face. **3.** That portion of a drill hole which is filled with the tamping. [MIN ENG] That portion of the coal to be fractured which is outside the powder. { ˈhēl əv ə ˈshät }

heel plate [CIV ENG] A plate at the end of a truss. { ˈhēl ˌplāt }

heel post [CIV ENG] A post to which are secured the hinges of a gate or door.

heel spur [MED] A bony growth produced by excessive musculoskeletal tension at the heel. Also known as calcaneal exostosis. { ˈhēl ˌspər }

Hefner candle [OPTICS] A luminous intensity standard, formerly used in Germany, equal to 0.9 international candle; produced by a Hefner lamp burning under standard conditions. Abbreviated HK. Also known as Hefnerkerze. { ˈhef·nər ˌkand·əl }

Hefnerkerze See Hefner candle. { ˈhef·nərˌkert·sə }

Hefner lamp [CHEM] A flame lamp that burns amyl acetate. { ˈhef·nər ˌlamp }

Hegeler furnace [MET] A muffle furnace having seven tiers of hearths; lower hearths are heated by gas burned in flues beneath them. { ˈheg·lər ˌfər·nəs }

Hehner number [ANALY CHEM] Weight percent of water-insoluble fatty acids in fats and oils. { ˈhān·ər ˌnəm·bər }

Heidelberg capsule [ELECTR] A radio pill for telemetering pH values of gastric acidity. { ˈhīd·əl·bərg ˌkap·səl }

Heidelberg man [PALEON] An early type of European fossil man known from an isolated lower jaw; considered a variant of *Homo erectus* or an early stock of Neanderthal man. { ˈhīd·əl·bərg ˌman }

heifer [VERT ZOO] A female cow less than 3 years of age that has not borne a calf. { ˈhef·ər }

hei function [MATH] A function that is expressed in terms of Hankel functions in a manner similar to that in which the bei function is expressed in terms of Bessel functions. { ˈhī ˌfȯŋkˈshən }

height [MATH] **1.** The perpendicular distance between horizontal lines or planes passing through the top and bottom of an object. **2.** The height of a rational number q is the maximum of $|m|$ and $|n|$, where m and n are relatively prime integers such that $q = m/n$. { hīt }

height above airport [NAV] In air operations, the height of the minimum descent altitude above the published airport elevation. Abbreviated HAA. { ˈhīt əˌbav ˈerˌpȯrt }

height above touch-down [NAV] In air operations, the height of the minimum descent altitude above the highest elevation in the touch-down zone. Abbreviated HAT. { ˈhīt əˌbav ˈtəchˌdaủn }

height-change chart [METEOROL] A chart indicating the change in height of a constant-pressure surface over a specified previous time interval; comparable to a pressure-change chart. { ˈhīt ˌchānj ˌchärt }

height-change line [METEOROL] A line of equal change in height of a constant-pressure surface over a specified previous interval of time; the lines drawn on a height-change chart. Also known as contour-change line; isallohypse. { ˈhīt ˌchānj ˌlīn }

height control [ELECTR] The television receiver control that adjusts picture height. { ˈhīt kənˌtrōl }

height equivalent of theoretical plate [CHEM ENG] In a packed fractionating column, a height of packing that makes a separation equivalent to that of a theoretical plate; used in sorption and distillation calculations. Abbreviated HETP. { ˈhīt iˌkwiv·ə·lənt əv ˌthē·əˈred·ə·kəl ˈplāt }

height finder [ENG] A radar equipment, used to determine height of aerial targets. { ˈhīt ˌfīn·dər }

height finding [ENG] Determination of the height of an airborne object. { ˈhīt ˌfīnd·iŋ }

HEDGEHOG

Hedgehog (*Erinaceus europaeus*).

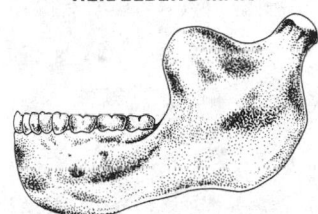

HEIDELBERG MAN

Reconstruction of the Heidelberg jaw. *(From M. F. Ashley Montagu, An Introduction to Physical Anthropology, 2d ed., Charles C. Thomas, 1951)*

height-finding radar [ENG] A radar set that measures and determines the height of an airborne object. { 'hīt ¦fīnd·iŋ 'rā,där }

height gage [ENG] A gage used to measure heights by either a micrometer or a vernier scale. [GRAPHICS] A C-shaped device for measuring foot-to-face height of printing type or mounted printing plates. { 'hīt ,gāj }

height gain [ELECTR] A radio-wave interference phenomenon which results in a more or less periodic signal strength variation with height; this specifically refers to interference between direct and surface-reflected waves; maxima or minima in these height-gain curves occur at those elevations at which the direct and reflected waves are exactly in phase or out of phase respectively. { 'hīt ,gān }

height input [ELECTR] Radar height information on target received by a computer from height finders and relayed via ground-to-ground data link or telephone. { 'hīt 'in,pu̇t }

height of burst [ORD] Vertical distance from the ground, or target, to the point of burst. { 'hīt əv 'bərst }

height-of-eye correction [NAV] A correction applied to sextant altitude to compensate for the dip of the horizon. Also known as dip correction. { 'hīt əv 'ī kə¦rek·shən }

height of instrument [ENG] **1.** In survey leveling, the vertical height of the line of collimation of the instrument over the station above which it is centered, or above a specified datum level. **2.** In spirit leveling, the vertical distance from datum to line of sight of the instrument. **3.** In stadia leveling the height of center of transit above the station stake. **4.** In differential leveling, the elevation of the line of sight of the telescope when the instrument is leveled. { 'hīt əv 'in·strə·mənt }

height of tide [OCEANOGR] Vertical distance from the chart datum to the level of the water at any time; it is positive if the water level is higher than the chart datum. { 'hīt əv 'tīd }

height of transfer unit [CHEM ENG] A dimensionless parameter used to calculate countercurrent sorption tower operations; it is proportional to the apparent resident time of the fluid. Abbreviated HTU. { 'hīt əv 'tranz·fər ,yü·nət }

height overlap coverage [ELECTR] Height-finder coverage within which there is an area of overlapping coverage from adjacent height finders or other radar stations. { 'hīt 'ō·vər,lap ,kəv·rij }

height pattern [METEOROL] The general geometric characteristics of the distribution of height of a constant-pressure surface as shown by contour lines on a constant-pressure chart. Also known as baric topography; isobaric topography; pressure topography. { 'hīt ,pad·ərn }

height-position indicator [ELECTR] Radar display which shows simultaneously angular elevation, slant range, and height of objects detected in the vertical sight plane. { 'hīt pə¦zish·ən 'in·də,kād·ər }

height-range indicator [ELECTR] **1.** Radar display which shows an echo as a bright spot on a rectangular field, slant range being indicated along the X axis, height above the horizontal plane being indicated (on a magnified scale) along the Y axis, and height above the earth being shown by a cursor. **2.** Cathode-ray tube from which altitude and range measurements of flightborne objects may be viewed. { 'hīt ¦rānj 'in·də,kād·ər }

heiligenschein [OPTICS] A diffuse white ring surrounding the shadow cast by the observer's head upon a dew-covered lawn when the solar elevation is low and, therefore, the distance from observer to shadow is great. { 'hī·lə·gən,shīn }

Heine-Borel theorem [MATH] The theorem that the only compact subsets of the real line are those which are closed and bounded. { 'hī·nə bȯ'rel ,thir·əm }

Heine-Medin disease See poliomyelitis. { 'hī·nə 'med·ən di,zēz }

Heinz bodies [PATH] Refractile spots seen in erythrocytes in hemolytic anemia that may represent denatured globulin. { 'hīnts ,bäd·ēz }

Heisenberg algebra [QUANT MECH] The Lie algebra formed by the operators of position and momentum. { 'hīz·ən·bərg ,al·jə·brə }

Heisenberg equation of motion [QUANT MECH] An equation which gives the rate of change of an operator corresponding to a physical quantity in the Heisenberg picture. { 'hīz·ən·bərg i¦kwā·zhən əv 'mō·shən }

Heisenberg exchange coupling [SOLID STATE] The exchange forces between electrons in neighboring atoms which give rise to ferromagnetism in the Heisenberg theory. { 'hīz·ən·bərg iks'chānj ,kəp·liŋ }

Heisenberg force [NUC PHYS] A force between two nucleons derivable from a potential with an operator which exchanges both the positions and the spins of the particles. { 'hīz·ən·bərg ,fȯrs }

Heisenberg picture [QUANT MECH] A mode of description of a system in which dynamic states are represented by stationary vectors and physical quantities are represented by operators which evolve in the course of time. Also known as Heisenberg representation. { 'hīz·ən·bərg ,pik·chər }

Heisenberg representation See Heisenberg picture. { 'hīz·ən·bərg ,re·prə,zen'tā·shən }

Heisenberg theory of ferromagnetism [SOLID STATE] A theory in which exchange forces between electrons in neighboring atoms are shown to depend on relative orientations of electron spins, and ferromagnetism is explained by the assumption that parallel spins are favored so that all the spins in a lattice have a tendency to point in the same direction. { 'hīz·ən·bərg 'thē·ə·rē əv ,fer·ō'mag·nə,tiz·əm }

Heisenberg uncertainty principle See uncertainty principle. { 'hīz·ən·bərg ən'sərt·ən·tē ,prin·sə·pəl }

Heisenberg uncertainty relation See uncertainty relation. { 'hīz·ən·bərg ən'sərt·ən·tē ri,lā·shən }

Heising modulation See constant-current modulation. { 'hī·ziŋ ,mäj·ə,lā·shən }

Heitler-London covalence theory [PHYS CHEM] A calculation of the binding energy and the distance between the atoms of a diatomic hydrogen molecule, which assumes that the two electrons are in atomic orbitals about each of the nuclei, and then combines these orbitals into a symmetric or antisymmetric function. { 'hīt·lər 'lən·dən kō'vā·ləns ,thē·ə·rē }

hekistotherm [ECOL] Plant adapted for conditions of minimal heat; can withstand long dark periods. { he'kis·tō,thərm }

Hektor [ASTRON] An asteroid, believed to be the largest of the Trojan planets, which circles the sun in the orbit of and approximately 60° ahead of Jupiter; it has an elongated shape, about 186 × 93 miles (300 × 150 kilometers) and D-type surface composition. { 'hek·tər }

HeLa cells [PATH] Human cancer cells maintained in tissue culture since 1953, originally excised from the cervical carcinoma of a patient named Helen Lane. { 'hel·ə ,selz }

Helaletidae [PALEON] A family of extinct perissodactyl mammals in the superfamily Tapiroidea. { ,hel·ə'led·ə,dē }

Helcionellacea [PALEON] A superfamily of extinct gastropod mollusks in the order Aspidobranchia. { ¦hel·sē·ō·nə'las·ē·ə }

Helderbergian [GEOL] A North American stage of geologic time, in the lower Lower Devonian. { ,hel·dər'bərg·ē·ən }

held in common [MIN ENG] Pertaining to a claim whereof there is more than one owner. { ¦held in 'kam·ən }

Heleidae [INV ZOO] The biting midges, a family of orthorrhaphous dipteran insects in the series Nematocera. { hə'lē·ə,dē }

heliacal rising [ASTRON] The rising of a celestial body at the same time or just before that of the sun. { hi'lī·ə·kəl 'rīz·iŋ }

heliacal setting [ASTRON] The setting of a celestial body at the same time or just after that of the sun. { hi'lī·ə·kəl 'sed·iŋ }

Heliarc welding See inert gas-shielded arc welding. { 'hēl·ē,ärk ,weld·iŋ }

Heliasteridae [INV ZOO] A family of echinoderms in the subclass Asteroidea lacking pentameral symmetry but structurally resembling common asteroids. { ,hēl·ē·ə'ster·ə,dē }

helical [MATH] Pertaining to a cylindrical spiral, for example, a screw thread. { 'hel·ə·kəl }

helical angle [MECH] In the study of torsion, the angular displacement of a longitudinal element, originally straight on the surface of an untwisted bar, which becomes helical after twisting. { 'hel·ə·kəl 'aŋ·gəl }

helical antenna [ELECTROMAG] An antenna having the form of a helix. Also known as helix antenna. { 'hel·ə·kəl an'ten·ə }

helical conveyor [MECH ENG] A conveyor for the transport of bulk materials which consists of a horizontal shaft with helical paddles or ribbons rotating inside a stationary tube. { 'hel·ə·kəl kən'vā·ər }

helical-fin section [CHEM ENG] Helical-shaped, extended-surface addition for the external surfaces of process-fluid tubes to increase heat-exchange efficiency; used for gas heating and cooling and in fuel oil residuum exchangers. { 'hel·ə·kəl 'fin ,sek·shən }

helical-flow turbine [MECH ENG] A steam turbine in which the steam is directed tangentially and radially inward by nozzles against buckets milled in the wheel rim; the steam flows in a helical path, reentering the buckets one or more times. Also known as tangential helical-flow turbine. { 'hel·ə·kəl ¦flō 'tər·bən }

helical gear [MECH ENG] Gear wheels running on parallel axes, with teeth twisted oblique to the gear axis. { 'hel·ə·kəl 'gir }

helical line [ELECTROMAG] A transmission line with a helical inner conductor. { 'hel·ə·kəl 'līn }

helical milling [MECH ENG] Milling in which the work is simultaneously rotated and translated. { 'hel·ə·kəl 'mil·iŋ }

helical potentiometer [ELEC] A multiturn precision potentiometer in which a number of complete turns of the control knob are required to move the contact arm from one end of the helically wound resistance element to the other end. { 'hel·ə·kəl pə,ten·chē'äm·əd·ər }

helical rake angle [DES ENG] The angle between the axis of a reamer and a plane tangent to its helical cutting edge; also applied to milling cutters. { 'hel·ə·kəl 'rāk ,aŋ·gəl }

helical repeat [MOL BIO] The number of base pairs in one turn of a deoxyribonucleic acid helix. { 'hel·ə·kəl ri'pēt }

helical resonator [ELECTROMAG] A cavity resonator with a helical inner conductor. { 'hel·ə·kəl 'rez·ən,ād·ər }

helical scanning [COMMUN] A method of facsimile scanning in which a single-turn helix rotates against a stationary bar to give horizontal movement of an elemental area. [ELECTR] A method of recording on videotape and digital audio tape in which the tracks are recorded diagonally from top to bottom by wrapping the tape around the rotating-head drum in a helical path. [ENG] A method of radar scanning in which the antenna beam rotates continuously about the vertical axis while the elevation angle changes slowly from horizontal to vertical, so that a point on the radar beam describes a distorted helix. { 'hel·ə·kəl 'skan·iŋ }

helical-spline broach [MECH ENG] A broach used to produce internal helical splines having a straight-sided or involute form. { 'hel·ə·kəl 'splīn ,brōch }

helical spring [DES ENG] A bar or wire of uniform cross section wound into a helix. { 'hel·ə·kəl 'spriŋ }

helical steel support [MIN ENG] A continuous, screw-shaped steel joist lining used for staple shafts. { 'hel·ə·kəl 'stēl sə,pórt }

helical traveling-wave tube See helix tube. { 'hel·ə·kəl ¦trav·ə·liŋ 'wāv ,tüb }

helicase [BIOCHEM] An enzyme that is capable of unwinding the deoxyribonucleic acid double helix at a replication fork. { 'hel·ə,kās }

helicate [ORG CHEM] Any member of a group of synthetic, helical arrays of molecules formed by the chemical recognition and organization of metals and organic bases. { 'hel·i,kāt }

helicin See salicylaldehyde. { 'hel·ə·sən }

Helicinidae [INV ZOO] A family of gastropod mollusks in the order Archeogastropoda containing tropical terrestrial snails. { ,hel·ə'sin·ə,dē }

helicity [QUANT MECH] The component of the spin of a particle along its momentum. { he'lis·əd·ē }

helicoid [INV ZOO] Of a gastropod shell, shaped like a flat coil or flattened spiral. [MATH] A surface generated by a curve which is rotated about a straight line and also is translated in the direction of the line at a rate that is a constant multiple of its rate of rotation. { 'hel·ə,kóid }

helicoid cyme [BOT] A type of determinate inflorescence having a coiled cluster, with flowers on only one side of the axis. { 'hel·ə,kóid 'sīm }

helicon [ELECTROMAG] A low-frequency, circularly polarized electromagnetic wave that is propagated in a metal in the presence of an external magnetic field. { 'hel·ə,kän }

Heliconiaceae [BOT] A family of monocotyledonous plants in the order Zingiberales characterized by perfect flowers with a solitary ovule in each locule, schizocarpic fruit, and capitate stigma. { ,hel·ə,kän·ē'ās·ē,ē }

Helicoplacoidea [PALEON] A class of free-living, spindle- or pear-shaped, plated echinozoans known only from the Lower Cambrian of California. { ,hel·ə·kō·plə'kóid·ē·ə }

helicopter [AERO ENG] An aircraft fitted to sustain itself by motor-driven horizontal rotating blades (rotors) that accelerate the air downward, providing a reactive lift force, or accelerate the air at an angle to the vertical, providing lift and thrust. { 'hel·ə,käp·tər }

helicopter yarding [FOR] A technique that uses a helicopter to bring logs from a forest to a wide level spot along a road where the logs can be loaded onto trucks for hauling to a sawmill; used where steep terrain and unstable ground makes it impossible to build roads without damaging watersheds. { 'hel·ə,käp·tər ¦yard·iŋ }

Helicosporae [MYCOL] A spore group of the Fungi Imperfecti characterized by spirally coiled, septate spores. { ,hel·ə'kä·spə,rē }

helicospore [INV ZOO] Mature spore of the Helicosporida characterized by a peripheral spiral filament. { 'hel·ə·kə,spór }

Helicosporida [INV ZOO] An order of protozoans in the class Myxosporidea characterized by production of spores with a relatively thick, single intrasporal filament and three uninucleate sporoplasms. { ,hel·ə·kō¦spór·ə·də }

helicotrema [ANAT] The opening at the apex of the cochlea through which the scala tympani and the scala vestibuli communicate with each other. { ,hel·ə·kō¦trē·mə }

helictite [GEOL] A speleothem whose origin is similar to that of a stalactite or stalagmite but that angles or twists in an irregular fashion. { 'hē·lik,tīt }

Heligmosomidae [INV ZOO] A family of parasitic roundworms belonging to the Strongyloidea. { hə,lig·mō'sō·mə,dē }

helimagnet [SOLID STATE] A metal, alloy, or salt that possesses helimagnetism. { 'hel·ə,mag·nət }

helimagnetism [SOLID STATE] A property possessed by some metals, alloys, and salts of transition elements or rare earths, in which the atomic magnetic moments, at sufficiently low temperatures, are arranged in ferromagnetic planes, the direction of the magnetism varying in a uniform way from plane to plane. { ,hel·ə'mag·nə,tiz·əm }

heliocentric [ASTRON] Relative to the sun as a center. { ¦hē·lē·ō¦sen·trik }

heliocentric coordinates [ASTRON] A coordinate system relative to the sun as a center. { ¦hē·lē·ō¦sen·trik kō'órd·ən·əts }

heliocentric Julian date [ASTRON] The Julian date corrected to the time at which light from the celestial object in question reaches the sun (rather than the earth). Abbreviated HJD. { ¦hē·lē·ō¦sen·trik 'jül·yən 'dāt }

heliocentric latitude [ASTRON] Sun-centered coordinate of angular distance perpendicular to the ecliptic plane. { ¦hē·lē·ō¦sen·trik 'lad·ə,tüd }

heliocentric longitude [ASTRON] The angular distance east or west from a given point on the sun's equator. { ¦hē·lē·ō¦sen·trik 'län·jə,tüd }

heliocentric orbit [ASTRON] An orbit relative to the sun as a center. { ¦hē·lē·ō¦sen·trik 'ór·bət }

heliocentric parallax See annual parallax. { ¦hē·lē·ō¦sen·trik 'par·ə,laks }

Heliodinidae [INV ZOO] A family of lepidopteran insects in the suborder Heteroneura. { ,hē·lē·ə'din·ə,dē }

heliogram [COMMUN] A message transmitted on a heliograph. { 'hē·lē·ə,gram }

heliograph [COMMUN] An instrument for sending telegraphic messages by reflecting the sun's rays from a mirror. [ENG] An instrument that records the duration of sunshine and gives a qualitative measure of its amount by action of sun's rays on blueprint paper. { 'hē·lē·ə,graf }

heliographic latitude [ASTRON] On the sun, angular distance north or south of its equator. { ,hē·lē·ə'graf·ik 'lad·ə,tüd }

heliographic longitude [ASTRON] On the sun, angular distance east or west from given point on the equator of the sun. { ,hē·lē·ə'graf·ik 'lan·jə,tüd }

heliolite See sunstone. { 'hē·lē·ə,līt }

Heliolitidae [PALEON] A family of extinct corals in the order Tabulata. { ,hē·lē·ō'lid·ə,dē }

heliometer [OPTICS] A split-lens telescope used to measure

HELICAL GEAR

A helical gear. *(Fellows Corp.)*

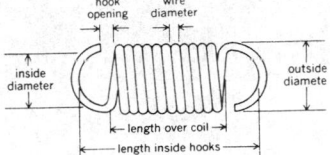

HELICAL SPRING

A helical spring wound tight to extend under axial tension.

HELICOID CYME

Drawing of a helicoid cyme.

the sun's diameter as well as small distances between stars or other celestial bodies. { ‚hē·lē·'äm·əd·ər }

helion [NUC PHYS] The nucleus of a helium-3 atom, consisting of two protons and one neutron. { 'hē·lē‚än }

heliopause [ASTRON] A shock front about 50 astronomical units from the sun, where the thermal pressure of the interstellar gas which surrounds the solar system overcomes the pressure of the solar wind, which is mostly ram pressure due to its high-velocity flow. { 'hē·lē·ə‚póz }

heliophilous [ECOL] Attracted by and adapted for a high intensity of sunlight. { ‚hē·lē‚äf·ə·ləs }

heliophobe [MED] An individual who is extremely sensitive to the sun's rays. [PSYCH] One who has an abnormal fear of the sun's rays. { 'hē·lē·ə‚fōb }

heliophobia [PSYCH] An abnormal fear of exposure to the sun's rays. { ‚hē·lē·ə'fō·bē·ə }

heliophyllite [MINERAL] $Pb_6As_2O_7Cl_4$ A yellow to greenish-yellow, orthorhombic mineral consisting of an oxychloride of lead and arsenic; occurs in massive and tabular form and as crystals. { ‚hē·lē·ō'fi‚līt }

heliophyte [ECOL] A plant that thrives in full sunlight. { 'hē·lē·ə‚fīt }

Heliornithidae [VERT ZOO] The lobed-toed sun grebes, a family of pantropical birds in the order Gruiformes. { ‚hē·lē‚ór'nith·ə‚dē }

helioscope [OPTICS] A telescope for observing the sun that protects the observer's eyes from the sun's glare. { 'hē·lē·ə‚skōp }

helioseismology [ASTRON] The analysis of wave motions of the solar surface to determine the structure of the sun's interior. { ‚hē·lē·ō‚sīz'mäl·ə·jē }

heliosphere [ASTRON] The region surrounding the sun where the solar wind dominates the interstellar medium. Also known as solar cavity. [GEOPHYS] The region in the ionosphere where helium ions are predominant (sometimes there may be no region in which helium ions dominate). { 'hē·lē·ə‚sfir }

heliostat [ENG] A clock-driven instrument mounting which automatically and continuously points in the direction of the sun; it is used with a pyrheliometer when continuous direct solar radiation measurements are required. { 'hē·lē·ə‚stat }

heliotaxis [BIOL] Orientation movement of an organism in response to the stimulus of sunlight. { ‚hē·lē·ō'tak·səs }

heliotrope [BOT] A plant whose flower or stem turns toward the sun. [ENG] An instrument that reflects the sun's rays over long distances; used in geodetic surveys. [MINERAL] *See* bloodstone. { 'hē·lē·ə‚trōp }

heliotropic wind [METEOROL] A subtle, diurnal component of the wind velocity leading to a diurnal shift of the wind or turning of the wind with the sun, produced by the east-to-west progression of daytime surface heating. { ‚hē·lē·ə'träp·ik 'wind }

heliotropin *See* piperonal. { ‚hē·lē·ə'trō·pən }

heliotropism [BIOL] Growth or orientation movement of a sessile organism or part, such as a plant, in response to the stimulus of sunlight. { ‚hē·lē'ä·trə‚piz·əm }

heliotype *See* photogelatin printing plate. { 'hē·lē·ə‚tīp }

heliox [MATER] A mixture of helium and a few percent of oxygen used for breathing during deep dives. { 'hēlē‚äks }

Heliozoia [INV ZOO] A subclass of the protozoan class Actinopodea; individuals lack a central capsule and have either axopodia or filopodia. { ‚hē·lē·ə'zói·ə }

heliozooid [BIOL] Ameboid, but with distinct filamentous pseudopodia. { ‚hē·lē·ə'zō‚óid }

helipad [CIV ENG] The launch and landing area of a heliport. Also known as pad. { 'hel·ə‚pad }

heliport [CIV ENG] A place built for helicopter takeoffs and landings. { 'hel·ə‚pórt }

helitron [ELECTR] An electrostatically focused, low-noise backward-wave oscillator; the microwave output signal frequency can be swept rapidly over a wide range by varying the voltage applied between the cathode and the associated radio-frequency circuit. { 'hel·ə‚trän }

helium [CHEM] A gaseous chemical element, symbol He, atomic number 2, and atomic weight 4.0026; one of the noble gases in group 0 of the periodic table. { 'hē·lē·əm }

helium I [CRYO] The phase of liquid helium-4 which is stable at temperatures above the lambda point (about 2.2 K) and has

the properties of a normal liquid, except low density. { 'hē·lē·əm 'wən }

helium II [CRYO] The phase of liquid helium-4 which is stable at temperatures between absolute zero and the lambda point (about 2.2 K), and has many remarkable properties such as vanishing viscosity, extremely high heat conductivity, and the fountain effect. { 'hē·lē·əm 'tü }

helium-3 [NUC PHYS] The isotope of helium with mass number 3, constituting approximately 1.3 parts per million of naturally occurring helium. { 'hē·lē·əm 'thrē }

helium-4 [NUC PHYS] The isotope of helium with mass number 4, constituting nearly all naturally occurring helium. { 'hē·lē·əm 'fór }

helium burning [NUC PHYS] The synthesis of elements in stars through the fusion of three alpha particles to form a carbon-12 nucleus, followed by further captures of alpha particles. { 'hē·lē·əm ‚bərn·iŋ }

helium-cadmium laser [OPTICS] A metal-vapor ion laser in which cadmium vapor, produced by heat or other means, migrates through a high-voltage glow discharge in helium, generating a continuous laser beam at wavelengths in the ultraviolet and blue parts of the spectrum (about 0.3 to 0.5 micrometer). { 'hē·lē·əm 'kad·mē·əm 'lā·zər }

helium film [CRYO] A superfluid film that covers any surface in contact with helium II. Also known as Rollin film. { 'hē·lē·əm ‚film }

helium flash [ASTRON] The onset of runaway helium burning in the degenerate core of a red giant star and the resulting expansion of the core. { 'hē·lē·əm ‚flash }

helium-like ion [ATOM PHYS] An atom from which all the electrons except two have been removed. { 'hē·lē·əm‚līk ‚i·ən }

helium liquefier [CRYO] Any one of several machines which liquefy helium by causing it to undergo adiabatic expansion and to do external work. { 'hē·lē·əm 'lik·wə‚fī·ər }

helium magnetometer [PHYS] A device for measuring magnetic fields by observing the Zeeman effect in the lowest triplet level of helium atoms subjected to the field. { 'hē·lē·əm mag·nə'täm·əd·ər }

helium-3 maser [PHYS] A gas maser in which the gas used is helium-3. { 'hē·lē·əm ‚thrē 'mā·zər }

helium-neon laser [OPTICS] An atomic gas laser in which a combination of helium and neon gases is used. { 'hē·lē·əm ‚nē‚än 'lā·zər }

helium-oxygen diving [ENG] Diving operations employing a breathing mixture of helium and oxygen. { 'hē·lē·əm ‚äk·sə·jən 'dīv·iŋ }

helium refrigerator [MECH ENG] A refrigerator which uses liquid helium to cool substances to temperatures of 4 K or less. { 'hē·lē·əm ri'frij·ə‚rād·ər }

helium spectrometer [SPECT] A small mass spectrometer used to detect the presence of helium in a vacuum system; for leak detection, a jet of helium is applied to suspected leaks in the outer surface of the system. { 'hē·lē·əm spek'träm·əd·ər }

helium stars [ASTRON] The class B stars. { 'hē·lē·əm ‚stärz }

helix [CELL MOL] A spiral structure with a repeating pattern that characterizes many biological polymers, for example, double-stranded nucleic acids and proteins. [ELEC] A spread-out, single-layer coil of wire, either wound around a supporting cylinder or made of stiff enough wire to be self-supporting. [MATH] A curve traced on a cylindrical or conical surface where all points of the surface are cut at the same angle. { 'hē‚liks }

Helix [INV ZOO] A genus of pulmonate land mollusks including many of the edible snails; individuals have a coiled shell with a low conical spire. { 'hē‚liks }

helix angle [DES ENG] That angle formed by the helix of the thread at the pitch-diameter line and a line at right angles to the axis. [MATH] The constant angle between the tangent to a helix and a generator of the cylinder upon which the helix lies. { 'hē‚liks ‚aŋ·gəl }

helix antenna *See* helical antenna. { 'hē‚liks an'ten·ə }

helix-destabilizing protein [BIOCHEM] Any of a group of proteins that bind to single-stranded regions of duplex deoxyribonucleic acid and cause unwinding of the helix. { 'hē‚liks di'stā·bə‚līz·iŋ ‚prō·tēn }

Helix Nebula [ASTRON] A planetary nebula in Aquarius about 140 parsecs (2.7×10^{15} miles or 4.3×10^{15} kilometers) distant that has a high helium abundance and the largest known diameter of any planetary nebula. { 'hē,liks 'neb·yə·lə }

helix tube [ELECTR] A traveling-wave tube in which the electromagnetic wave travels along a wire wound in a spiral about the path of the beam, so that the wave travels along the tube at a velocity approximately equal to the beam velocity. Also known as helical traveling-wave tube. { 'hē·liks ,tüb }

hellandite [MINERAL] Mineral composed of silicate of metals in the cerium group with aluminum, iron, manganese, and calcium. { 'hel·ən,dīt }

Hellas [ASTRON] The largest impact basin on Mars, approximately 1240 miles (2000 kilometers) across and 2.5 miles (4 kilometers) deep, appearing as a bright circular region in earth-based telescopes. { 'hel·əs }

hellbender [VERT ZOO] *Cryptobranchus alleganiensis.* A large amphibian of the order Urodela which is the most primitive of the living salamanders, retaining some larval characteristics. { 'hel¦ben·der }

Heller's syndrome *See* childhood disintegrative disorder. { 'hel·ərz ,sin,drōm }

Heller's test [PATH] A test for albumin in urine; presence of albumin is indicated by formation of a white ring at the junction of the solution and a concentrate solution of nitric acid. { 'hel·ərz ,test }

Hellespontus [ASTRON] A surface region of the planet Mars between the regions Hellas and Noachis. { hel·əs¦pän·təs }

Hellman-Feynman theorem [QUANT MECH] A theorem which states that in the Born-Oppenheimer approximation the forces on nuclei in molecules or solids are those which would arise electrostatically if the electron probability density were treated as a static distribution of negative electric charge. { ¦hel·mən ¦fīn·mən ,thir·əm }

Hell-Volhard-Zelinsky reaction [ORG CHEM] Preparation of an ester or α-halo substituted acid (chloro or bromo) by reacting the halogen on the acid in the presence of phosphorus or phosphorus halide, and then followed by hydrolysis or alcoholysis of the haloacyl halide resulting. { ¦hel ¦fōl,härt zə'lins·kē rē,ak·shən }

Helly's theorem [MATH] The theorem that there is a point that belongs to each member of a collection of bounded closed convex sets in an *n*-dimensional Euclidean space if the collection has at least *n* + 1 members and any *n* + 1 members of the collection has a common point. { 'hel·ēz ,thir·əm }

helm [NAV ARCH] **1.** The tiller or wheel controlling a ship's rudder. **2.** The entire apparatus for steering a ship. { helm }

Helmert's formula [GEOPHYS] A formula for the acceleration due to gravity in terms of the latitude and the altitude above sea level. { 'hel·mərts ,fȯr·myə·lə }

helmet [ENG] A globe-shaped head covering made of copper and supplied with air pumped through a hose; attached to the breastplate of a diving suit for deep-sea diving. { 'hel·mət }

helmet-mounted display [ELECTR] An electronic display that presents, on a combining glass within the visor of the helmet of a helicopter gunner, primary information for directing firepower; the angular direction of the helmet is sensed and used to control weapons to point in the same direction as the gunner is looking. Also known as visually coupled display. { 'hel·mət ¦maȯnt·əd di'splā }

helmholtz [ELEC] A unit of dipole moment per unit area, equal to 1 Debye unit per square angstrom, or approximately 3.335 × 10⁻¹⁰ coulomb per meter. { 'helm,hōlts }

Helmholtz coils [ELECTROMAG] A pair of flat, circular coils having equal numbers of turns and equal diameters, arranged with a common axis, and connected in series; used to obtain a magnetic field more nearly uniform than that of a single coil. { 'helm,hōlts ,kȯilz }

Helmholtz double layer [PHYS] An electrical double layer of positive and negative charges one molecule thick which occurs at a surface where two bodies of different materials are in contact, or at the surface of a metal or other substance capable of existing in solution as ions and immersed in a dissociating solvent. { 'helm,hōlts ¦dəb·əl 'lā·ər }

Helmholtz equation [MATH] A partial differential equation obtained by setting the Laplacian of a function equal to the function multiplied by a negative constant. [OPTICS] An equation which relates the linear and angular magnifications of a spherical refracting interface. Also known as Lagrange-Helmholtz equation. [PHYS CHEM] The relationship stating that the emf (electromotive force) of a reversible electrolytic cell equals the work equivalent of the chemical reaction when charge passes through the cell plus the product of the temperature and the derivative of the emf with respect to temperature. { 'helm,hōlts i,kwā·zhən }

Helmholtz flow [FL MECH] Flow with free streamlines or vortex sheets. { 'helm,hōlts ,flō }

Helmholtz free energy *See* free energy. { 'helm,hōlts ¦frē 'en·ər·jē }

Helmholtz function *See* free energy. { 'helm,hōlts ,fəŋk·shən }

Helmholtz instability [FL MECH] The hydrodynamic instability arising from a shear, or discontinuity, in current speed at the interface between two fluids in two-dimensional motion; the perturbation gains kinetic energy at the expense of that of the basic currents. Also known as shearing instability. { 'helm,hōlts ,in·stə'bil·əd·ē }

Helmholtz-Keteller formula [OPTICS] A dispersion formula in which the difference between the square of the index of refraction and unity is set equal to a sum of terms each of which is associated with a resonant wavelength of the medium. { 'helm,hōlts 'kcd·əl·ər ,fȯr·myə·lə }

Helmholtz potential *See* free energy. { 'helm,hōlts pə,ten·chəl }

Helmholtz resonator [ENG ACOUS] An enclosure having a small opening consisting of a straight tube of such dimensions that the enclosure resonates at a single frequency determined by the geometry of the resonator. { 'helm,hōlts ,rez·ən,ād·ər }

Helmholtz's theorem [ELEC] *See* Thévenin's theorem. [FL MECH] The theorem that in the isentropic flow of a nonviscous fluid which is not subject to body forces, individual vortices always consist of the same fluid particles. [MATH] The theorem determining a general class of vector fields as being everywhere expressible as the sum of an irrotational vector with a divergence-free vector. { 'helm,hōlt·səz ,thir·əm }

Helmholtz theory *See* Young-Helmholtz theory. { 'helm,hōlts ,thē·ə·rē }

Helmholtz wave [FL MECH] An unstable wave in a system of two homogeneous fluids with a velocity discontinuity at the interface. { 'helm,hōlts ,wāv }

helminth [INV ZOO] Any parasitic worm. { 'hel,minth }

helminthemesis [MED] Vomiting of worms. { ¦hel·mən'them·ə·səs }

helminthiasis [MED] Any disease caused by the presence of parasitic worms in the body. { ,hel·mən'thī·ə·səs }

helminthic abscess [MED] An abscess caused by worms. { hel'min·thik 'ab,ses }

helminthogogue [PHARM] An anthelminthic. { hel'min·thə,gäg }

helminthoid [BIOL] Resembling a helminth. { hel'min·thȯid }

helminthologist [BIOL] An individual who studies helminths. { ,hel,mən'thäl·ə·jəst }

helminthophobia [PSYCH] An abnormal fear of worms or becoming infested with worms. { hel,min·thə'fō·bē·ə }

helminthosporin [BIOCHEM] $C_{15}H_{10}O_5$ A maroon, crystalline pigment formed by certain fungi growing on a sugar substrate. { hel,min·thə'spȯr·ən }

Helminthosporium [MYCOL] A genus of parasitic fungi of the family Dematiaceae having conidiophores which are more or less irregular or bent and bear conidia successively on new growing tips. { hel,min·thə'spȯr·ē·əm }

Helminthosporium leaf spot [PL PATH] Any leaf spot, commonly of grasses, caused by *Helminthosporium* species. { hel,min·thə'spȯr·ē·əm 'lēf ,spät }

Helminthosporium victoriae [MYCOL] A fungal pathogen that produces victorin, the cause of Victoria blight of oats. { hel,min·thə,spȯr·ē·əm vik'tȯr·ē,ī }

helm roof [ARCH] A steeply pitched roof with four faces rising from gables to a point. { 'helm ¦rüf }

helm wind [METEOROL] A strong, cold northeasterly wind blowing down into the Eden valley from the western slope of the Crossfell Range in northern England. { 'helm ¦wind }

Helobiae [BOT] The equivalent name for Helobiales. { he'lō·bē,ē }

Helobiales [BOT] An order embracing most of the Alismatidae in certain systems of classification. { he,lō·bē'ā·lēz }

Heloderma [VERT ZOO] The single genus in the reptilian family Helodermatidae; contains the only known poisonous

HELLBENDER

The hellbender, with a maximum length of 18 inches (46 centimeters).

HELMHOLTZ RESONATOR

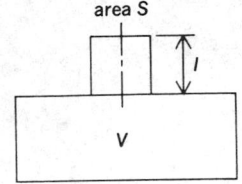

Schematic drawing of a Helmholtz resonator. Here l = length of straight tube, S = cross-sectional area of tube, V = closed volume. Resonant frequency is $f_0 = 1/2\pi \sqrt{c^2 S/l_e V}$ where c = speed of sound and l_e = effective length of tube, approximately $l + 0.8 \sqrt{S}$.

lizards, the Gila monster (*H. suspectum*) and the beaded lizard (*H. horridum*). { ˌhē·lōˈdər·mə }

Helodermatidae [VERT ZOO] A family of lizards in the suborder Sauria. { ˌhē·lō·dərˈmadˌə‚dē }

Helodidae [INV ZOO] The marsh beetles, a family of coleopteran insects in the superfamily Dascilloidea. { həˈlōd·ə‚dē }

Helodontidae [PALEON] A family of extinct ratfishes conditionally placed in the order Bradyodonti. { ‚he·lōˈdäntˌə‚dē }

heloma *See* corn. { hēˈlōm·ə }

Helomyzidae [INV ZOO] The sun flies, a family of myodarian cyclorrhaphous dipteran insects in the subsection Acalypteratae. { ‚he·lōˈmīzˌə‚dē }

helophyte [ECOL] A marsh plant; buds overwinter underwater. { ˈhe·lə‚fīt }

helophytia [ECOL] Differences in ecological control by fluctuations in water level such as in marshes. { ˌhe·lōˈfī·shə }

Heloridae [INV ZOO] A family of hymenopteran insects in the superfamily Proctotrupoidea. { həˈlòr·ə‚dē }

Helotiales [MYCOL] An order of fungi in the class Ascomycetes. { hə‚lō·shēˈā·lēz }

Helotidae [INV ZOO] The metallic sap beetles, a family of coleopteran insects in the superfamily Cucujoidea. { həˈlädˌə‚dē }

helotism [ECOL] Symbiosis in which one organism is a slave to the other, as between certain species of ants. { ˈhel·ə‚tiz·əm }

Helotrephidae [INV ZOO] A family of true aquatic, tropical hemipteran insects in the subdivision Hydrocorisae. { ‚he·ləˈtref·ə‚dē }

helper grade [MIN ENG] A grade on which helper engines are required to assist road locomotives. Also known as pusher grade. { ˈhel·pər ‚grād }

helper set [MIN ENG] A set of timbers to reinforce the normal set of timbers in a mine. { ˈhel·pər ‚set }

helper virus [VIROL] A virus that, by its infection of a cell, enables a defective virus to multiply by supplying one or more functions that the defective virus lacks. { ˈhel·pər ‚vī·rəs }

help screen [COMPUT SCI] Instructions that explain how to use the software of a computer system and that can be presented on the screen of a video display terminal at any time. { ˈhelp ‚skrēn }

help-yourself system [IND ENG] A tool-crib system for temporary issue of tools employed in small shops; employees have access to tools in the crib and help themselves. { ˈhelp yūrˈself ‚sis·təm }

helve hammer [MET] A belt-driven trip hammer with the hammer face or swage carried on the end of a beam; used for welding, forging, plating, drawing, and other metal-working operations. { ˈhelv ‚ham·ər }

helvine *See* helvite. { ˈhel‚vēn }

helvite [MINERAL] $(Mn,Fe,Zn)_4Be_3(SiO_4)_3S$ A silicate mineral isomorphous with danalite and genthelvite. Also known as helvine. { ˈhel‚vīt }

hem-, hema-, hemo-, haem- [HISTOL] Combining form for blood. { hēm, ˈhēm·ə, ˈhēm·o, hēm }

hemacytometer *See* hemocytometer. { ‚hē·mə·sīˈtäm·əd·ər }

hemadsorption virus [VIROL] A descriptive term for myxoviruses that agglutinate red blood cells and cause the cells to adsorb to each other. Abbreviated HA virus. { ‚hēmˈadˈsòrp·shən ‚vī·rəs }

hemafibrite [MINERAL] $Mn_3(AsO_4)(OH)_3·H_2O$ A brownish to garnet-red mineral composed of basic manganese arsenate. { ˌhē·məˈfī‚brīt }

hemagglutination [IMMUNOL] Agglutination of red blood cells. { ‚hē·mə‚glüd·ənˈā·shən }

hemagglutination-inhibition test [IMMUNOL] A test to identify a virus antigen or to quantitate an antibody by adding virus-specific antibody to a mixture of agglutinating virus and red blood cells. { ‚hē·mə‚glüd·ənˈā·shən ‚in·əˈbish·ən ‚test }

hemagglutinin [IMMUNOL] An erythrocyte-agglutinating antibody. { ‚hē·məˈglüd·ən·ən }

hemal arch [ANAT] **1.** A ventral loop on the body of vertebrate caudal vertebrae surrounding the blood vessels. **2.** In humans, the ventral vertebral process formed by the centrum together with the ribs. { ˈhē·məl ˈärch }

hemal ring [INV ZOO] A vessel in certain echinoderms, variously located, associated with the coelom and axial gland. { ˈhē·məl ˈriŋ }

hemal sinus [INV ZOO] The two principal lacunae along the digestive tube in certain echinoderms. { ˈhē·məl ˈsī·nəs }

hemal tuft [INV ZOO] Series of fine vessels in echioderms arising from the axial gland. { ˈhē·məl ˈtəft }

hemangioendothelioma [MED] A malignant tumor composed of anoplastic endothelial cells. Also known as hemangiosarcoma. { hēˈmän·jē·ō‚en·dō‚thē·lē ˈō·mə }

hemangioma [MED] A tumor composed of blood vessels. Also known as capillary angioma. { hēˈmän·jēˈō·mə }

hemangiopericytoma [MED] A tumor composed of endothelium-lined tubes or cords of cells surrounded by spherical cells with supporting reticulin network. { hēˈmän·jē·ō‚per·ə·sīˈtō·mə }

hemangiosarcoma *See* hemangioendothelioma. { hēˈmän·jē·ō·särˈkō·mə }

hemapodium [INV ZOO] The dorsal lobe of a parapodium. { ˌhē·məˈpō·dē·əm }

hemarthrosis [MED] Passage of blood into a joint. { ˌhē·märˈthrō·səs }

hematein [BIOCHEM] $C_{16}H_{12}O_6$ A brownish stain and chemical indicator obtained by oxidation of hematoxylin. { ‚hē·məˈtē·ən }

hematemesis [MED] Vomiting of blood. { ‚hē·məˈtem·ə·səs }

hematidrosis [MED] The appearance of blood or blood products in sweat gland secretions. { hē‚mad·əˈdrō·səs }

hematin [ORG CHEM] $C_{34}H_{33}O_5N_4Fe$ The hydroxide of ferriheme derived from oxidized heme. { ˈhē·məd·ən }

hematite [MINERAL] Fe_2O_3 An iron mineral crystallizing in the rhombohedral system; the most important ore of iron, it is dimorphous with maghemite, occurs in black metallic-looking crystals, in reniform masses or fibrous aggregates, or in reddish earthy forms. Also known as bloodstone; red hematite; red iron ore; red ocher; rhombohedral iron ore. { ˈhē·mə‚tīt }

hematoblast [HISTOL] An immature erythrocyte. { ˈhe·məd·ō‚blast }

hematocele [MED] Collection of blood in a body part. { ˈhe·məd·ō‚sēl }

hematochrome [BIOCHEM] A red pigment occurring in green algae, especially when plants are exposed to intense light on subaerial habitats. { hiˈmad·ə‚krōm }

hematocrit [PATH] The volume, after centrifugation, occupied by the cellular elements of blood, in relation to the total volume. { hiˈmad·ə‚krit }

hematodocha [INV ZOO] In some spiders, a thin sac on the male that is distended during copulation. { hi‚mad·əˈdō·kə }

hematogenous [PHYSIO] **1.** Pertaining to the production of blood or of its fractions. **2.** Carried by way of the bloodstream. **3.** Originating in blood. { ˈhēm·əˈtäj·ə·nəs }

hematoidin crystals [PATH] Yellow to brown crystals in the feces following gastrointestinal hemorrhage. { ‚hēm·əˈtòid·ən ‚krist·əlz }

hematolite [MINERAL] $(Mn,Mg)_4Al(AsO_4)(OH)_8$ A brownish-red mineral composed of aluminum manganese arsenate; occurs in rhombohedral crystals. { ˈhe·məd·ō‚līt }

hematologic disorder [MED] A disorder marked by aberrations in structure or function of the blood cells or the blood-clotting mechanism. { ‚hē·mə·tə‚läj·ik disˈòrd·ər }

hematologist [MED] A specialist in the study of blood. { ‚hē·məˈtäl·ə·jəst }

hematology [MED] The science of the blood, its nature, functions, and diseases. { ‚hē·məˈtäl·ə·jē }

hematoma [MED] A localized mass of blood in tissue; usually it clots and becomes encapsulated by connective tissue. { ‚hē·məˈtō·mə }

hematometra [MED] Accumulation of blood in the uterus. { ‚he·məd·ōˈme·trə }

hematomyelia [MED] Hemorrhage into the spinal cord. { ‚he·məd·ō‚mīˈe·lē·ə }

hematopathology *See* hemopathology. { ‚he·məd·ō·pəˈthäl·ə·jē }

hematophagous [ZOO] Feeding on blood. { ˈhē·məˈtäf·ə·gəs }

hematophanite [MINERAL] $Pb_5Fe_4O_{10}(Cl,OH)_2$ A mineral composed of oxychloride lead and iron. { ‚hē·məˈtäf·ə‚nīt }

hematopoiesis [PHYSIO] The process by which the cellular elements of the blood are formed. Also known as hemopoiesis. { ‚he·məd·ō·pói'ē·səs }

hematopoietic system *See* reticuloendothelial system. { ¦he·məd·ō·pȯi¦ed·ik 'sis·təm }

hematopoietic tissue [HISTOL] Blood-forming tissue, consisting of reticular fibers and cells. Also known as hemopoietic tissue. { ¦he·məd·ō·pȯi¦ed·ik 'tish·ü }

hematopoietin [BIOCHEM] A substance which is produced by the juxtaglomerular apparatus in the kidney and controls the rate of red cell production. Also known as hemopoietin. { ¦he·məd·ō'pȯi·ət·ən }

hematoporphyrin [BIOCHEM] $C_{34}H_{38}O_6N$ Iron-free heme, a porphyrin obtained by treating hemoglobin with sulfuric acid in vitro. Also known as hemoporphyrin. { ¦he·məd·ō'pȯr·fə·rən }

hematorrhachis [MED] Hemorrhage into the spinal meninges, producing irritative phenomena. { ¦he·mə'tȯr·ə·kəs }

hematosalpinx [MED] Accumulation of blood in a Fallopian tube. Also known as hemosalpinx. { ¦he·məd·ō'sal¸piŋks }

hematoxylin [ORG CHEM] $C_{16}H_{14}O_6$ A colorless, crystalline compound occurring in hematoxylon; upon oxidation, it is converted to hematein which forms deeply colored lakes with various metals; used as a stain in microscopy. { ¦hē·mə'täk·sə·lən }

hematoxylon [BOT] The heartwood of *Haematoxylon campechianum*. Also known as logwood. { ¦hē·mə'täk·sə¸län }

hematuria [MED] A pathological condition in which the urine contains blood. { ¦he·mə'tùr·ē·ə }

heme [BIOCHEM] $C_{34}H_{32}O_4N_4Fe$ An iron-protoporphyrin complex associated with each polypeptide unit of hemoglobin. { hēm }

hemeralopia [MED] Day blindness. { ¸hem·ə·rə'lō·pē·ə }

hemerythrin *See* hemoerythrin. { ¸hem·ə'rith·rən }

heme synthetase [BIOCHEM] An enzyme which combines protoporphyrin IX, ferrous iron, and globin to form the intact hemoglobin molecule. { ¸hēm 'sin·thə¸tās }

hemi- [BIOL] **1.** Prefix for half. **2.** Prefix denoting one side of the body. { 'he·mē }

hemiacetal [ORG CHEM] A class of compounds that have the grouping $\geq$C(OH)−(OR) and that result from the reaction of an aldehyde and alcohol. { ¦he·mē'as·ə¸tal }

hemianesthesia [MED] Loss of sensation on one side of the body. { ¦he·mē¸an·əs'thē·zha }

hemianopsia [MED] Bilateral or unilateral blindness in one-half of the field of vision. { ¸he·mē·ə'näp·sē·ə }

Hemiascomycetes [MYCOL] The equivalent name for Hemiascomycetidae. { ¦he·mē¸as·kō¸mī'sēd·ēz }

Hemiascomycetidae [MYCOL] A subclass of fungi in the class Ascomycetes. { ¦he·mē¸as·kō¸mī'sed·ə¸dē }

hemiazygous vein [ANAT] A vein on the left side of the vertebral column which drains blood from the left ascending lumbar vein to the azygos vein. { ¦he·mē'az·ə·gəs 'vān }

hemiballismus [MED] Sudden, violent, spasmodic movements involving particularly the proximal portions of the extremities of one side of the body; caused by a destructive lesion of the contralateral subthalamic nucleus or its neighboring structures or pathways. { ¦he·mē·bə'liz·məs }

Hemibasidiomycetes [MYCOL] The equivalent name for Heterobasidiomycetidae. { ¦he·mē·bə¸sid·ē·ō¸mī's ēd·ēz }

hemicellulase [FOOD ENG] An enzyme that hydrolyzes pentosan (a type of hemicellulose); mainly used in the baking industry to improve the quality of dough, also used in food, beverage, animal feed, and pharmaceutical industries. { ¸hem·i'sel·yə¸lās }

hemicellulose [BIOCHEM] $(C_6H_{10}O_5)_n$ A type of polysaccharide found in plant cell walls in association with cellulose and lignin; it is soluble in and extractable by dilute alkaline solutions. Also known as hexosan. { ¦he·mē'sel·yə¸lōs }

hemicephaly [MED] Congenital absence of the cerebrum. { ¦he·mē'sef·ə·lē }

Hemichordata [SYST] A group of marine animals categorized as either a phylum of deuterostomes or a subphylum of chordates; includes the Enteropneusta, Pterobranchia, and Graptolithina. { ¦he·mē·kȯr'däd·ə }

Hemicidaridae [PALEON] A family of extinct Echinacea in the order Hemicidaroida distinguished by a stirodont lantern, and ambulacra abruptly widened at the ambitus. { ¦he·mē·si'där·ə¸dē }

Hemicidaroida [PALEON] An order of extinct echinoderms in the superorder Echinacea characterized by one very large

tubercle on each interambulacral plate. { ¦he·mē¸sid·ə'rȯid·ə }

hemic murmur [MED] Blowing or rasping sound heard in the heart or vessels, usually in association with systole, in abnormal conditions of increased velocity of blood flow. { 'hē·mik 'mər·mər }

hemicolectomy [MED] Surgical removal of a portion of the colon. { ¦he·mē·kə'lek·tə·mē }

hemicone *See* alluvial cone. { 'he·mē¸kōn }

hemicryptophyte [ECOL] A plant having buds at the soil surface and protected by scales, snow, or litter. { ¦he·mē'k-rip·tə¸fīt }

hemicrystalline *See* hypocrystalline. { ¦he·mē'krist·əl·ən }

hemicycle [MATH] A curve in the form of a semicircle. { 'he·mē¸sī·kəl }

hemicyclic [BOT] Of flowers, having the floral leaves arranged partly in whorls and partly in spirals. { 'he·mē'sī·dik }

hemidesmosome [HISTOL] A structure similar to a desmosome that joins a cell to its basilar membrane rather than to another cell. { ¸hem·ē'dez·mə¸sōm }

hemidiaphragm [ANAT] A lateral half of a diaphragm. [MED] Diaphragm with normal muscle development only on one side. { ¦he·mē'dī·ə¸fram }

Hemidiscosa [INV ZOO] An order of sponges in the subclass Amphidiscophora distinguished by birotulates that are hemidiscs with asymmetrical ends. { ¦he·mē·dis'kō·sə }

hemihedral symmetry [CRYSTAL] The possession by a crystal of only half of the elements of symmetry which are possible in the crystal system to which it belongs. { ¦he·mē¦hē·drəl 'sim·ə·trē }

hemiholohedral [CRYSTAL] Of hemihedral form but with half of the octants having the full number of planes. { ¦he·mē¸hō·lə'hē·drəl }

hemihydrate [ORG CHEM] A hydrate with a 2:1 molecular ratio of anhydrous compound to water; plaster of paris is the hemihydrate of calcium sulfate, composition $CaSO_4 \cdot 1/2 H_2O$. { ¸hem·ē'hī¸drate }

hemiketal [ORG CHEM] A carbonyl compound that results from the addition of an alcohol to the carbonyl group of a ketone, with the general formula (R)(R')C(OH)(OR). { ¦he·mē'ked·əl }

Hemileia vastatrix [MYCOL] A fungus of the order Uredinales which is the causative agent of orange coffee rust. { ¸hem·ə¸lē·yə 'vas·tə¸triks }

Hemileucinae [INV ZOO] A subfamily of lepidopteran insects in the family Saturnidae consisting of the buck moths and relatives. { ¦he·mē'lüs·ən¸ē }

hemimellitic acid [ORG CHEM] $C_6H_3(COOH)_3$ A compound crystallizing in colorless needles; melting point 196°C; slightly soluble in water. { ¦he·mē·mə¦lid·ik 'as·əd }

Hemimetabola [INV ZOO] A division of the insect subclass Pterygota; members are characterized by hemimetabolous metamorphosis. { ¦he·mē·me'tab·ə·lə }

hemimetabolous metamorphosis [INV ZOO] An incomplete metamorphosis; gills are present in aquatic larvae, or naiads. { ¦he·mē·me'tab·ə·ləs ¸med·ə'mȯr·fə·səs }

hemimorphic crystal [CRYSTAL] A crystal with no transverse plane of symmetry and no center of symmetry; composed of forms belonging to only one end of the axis of symmetry. { ¦he·mē¦mȯr·fik 'krist·əl }

hemimorphite [MINERAL] $Zn_4Si_2O_7(OH)_2 \cdot H_2O$ A white, colorless, pale-green, blue, or yellow mineral having an orthorhombic crystal structure; an ore of zinc. Also known as calamine; electric calamine; galmei. { ¸he·mē'mȯr¸fīt }

hemin [BIOCHEM] $C_{34}H_{32}O_4N_4FeCl$ The crystalline salt of ferriheme, containing iron in the ferric state. { 'hē·mən }

hemiparasite [ECOL] A parasite capable of a saprophytic existence, especially certain parasitic plants containing some chlorophyll. Also known as semiparasite. { ¦he·mē'par·ə¸sīt }

hemiparesis [MED] Muscle weakness on one side of the body. { ¦he·mē·pə'rē·səs }

hemipelagic [ECOL] Of the biogeographic environment of the hemipelagic region with both neritic and pelagic qualities. { ¦he·mē·pə'laj·ik }

hemipelagic region [OCEANOGR] The region of the ocean extending from the edge of a shelf to the pelagic environment; roughly corresponds to the bathyal zone, in which the bottom is 660 to 3300 feet (200 to 1000 meters) below the surface. { ¦he·mē·pə'laj·ik 'rē·jən }

hemipelagic sediment [GEOL] Deposits containing terrestrial material and the remains of pelagic organisms, found in the ocean depths. { ¦he·mē·pə'laj·ik 'sed·ə·mənt }

hemipelagite [OCEANOGR] Deep-sea mud deposits in which more than 25% of the fraction of particles coarser than 5 micrometers is of terrigenous, volcanogenic, or neritic origin. { ‚hem·ē'pel·ə‚jīt }

hemipenis [VERT ZOO] Either of a pair of nonerectile, evertible sacs that lie on the floor of the cloaca in snakes and lizards; used as intromittent organs. { ¦he·mē'pē·nəs }

Hemipeplidae [INV ZOO] An equivalent name for Cucujidae. { ¦he·mē'pep·lə‚dē }

hemiplegia [MED] Unilateral paralysis of the body. { ‚hē·mə'plē·jə }

hemiprism [CRYSTAL] A pinacoid that cuts two crystallographic axes. { 'he·mē‚priz·əm }

Hemiprocnidae [VERT ZOO] The crested swifts, a family comprising three species of perching birds found only in southeastern Asia. { ‚he·mē'präk·nə‚dē }

Hemiptera [INV ZOO] The true bugs, an order of the class Insecta characterized by forewings differentiated into a basal area and a membranous apical region. { he'mip·tə·rə }

Hemisphaeriales [MYCOL] A group of ascomycetous fungi characterized by the wall of the fruit body being a stroma. { ¦he·mē‚sfir·ē'ā·lēz }

hemisphere [GEOGR] A half of the earth divided into north and south sections by the equator, or into an east section containing Europe, Asia, and Africa, and a west section containing the Americas. [MATH] One of the two pieces of a sphere divided by a great circle. { 'he·mē‚sfir }

hemispherical candlepower [OPTICS] Luminous intensity of a hemispherical light source. { ‚he·mē'sfir·ə·kəl 'kand·əl‚pau̇·ər }

hemispherical pyrheliometer [ENG] An instrument for measuring the total solar energy from the sun and sky striking a horizontal surface, in which a thermopile measures the temperature difference between white and black portions of a thermally insulated horizontal target within a partially evacuated transparent sphere or hemisphere. { ‚he·mē'sfir·ə·kəl ‚pīr‚hē·lē'äm·əd·ər }

hemispheric wave number See angular wave number. { ‚he·mē'sfir·ik ¦wāv ‚nəm·bər }

hemispheroid [MATH] One of the halves into which a spheroid is divided by a plane of symmetry. { ‚he·mē'sfir‚ȯid }

Hemist [GEOL] A suborder of the soil order Histosol, consisting of partially decayed plant residues and saturated with water most of the time. { 'he·mist }

hemithorax [ANAT] One side of the chest. { ¦he·mē'thȯr‚aks }

hemitropic [CRYSTAL] Pertaining to a twinned structure in which, if one part were rotated 180°, the two parts would be parallel. { ¦he·mē¦träp·ik }

Hemizonida [PALEON] A Paleozoic order of echinoderms of the subclass Asteroidea having an ambulacral groove that is well defined by adambulacral ossicles, but with restricted or undeveloped marginal plates. { ¦he·mē'zän·ə·də }

hemizygous [GEN] In diploid organisms, the presence of a single copy of a gene; it may be a result of deletion or chromosome loss, or simply may reflect the presence of a single copy of a sex chromosome, such as the X in male mammals. { ¦he·mē¦zī·gəs }

hemlock [BOT] The common name for members of the genus Tsuga in the pine family characterized by two white lines beneath the flattened, needlelike leaves. { 'hem‚läk }

hemlock oil See spruce oil. { 'hem‚läk ‚ȯil }

hemming [MECH ENG] Forming of an edge by bending the metal back on itself. { 'hem·iŋ }

hemobilirubin [BIOCHEM] Bilirubin in normal blood serum before passage through the liver. { ¦hē·mō‚bil·i'rü·bən }

hemoblast See hemocytoblast. { 'hē·mə‚blast }

hemochorial placenta [EMBRYO] A type of placenta having the maternal blood in direct contact with the chorionic trophoblast. Also known as labyrinthine placenta. { ¦hē·mō'kȯr·ē·əl plə'sen·tə }

hemochromatosis [MED] A disorder of iron metabolism characterized by excessive accumulation of iron in the liver and

other tissues and by development of severe cirrhosis. { ¦hē·mō‚krō'mə'tō·səs }

hemocoel [INV ZOO] An expanded portion of the blood system in arthropods that replaces a portion of the coelom. { 'hē·mə‚sēl }

hemoconcentration [MED] An increase in the concentration of blood cells resulting from the loss of plasma or water from the bloodstream. { ¦hē·mō‚käns·ən'trā·shən }

hemoconia [BIOCHEM] Round or dumbbell-shaped, refractile, colorless particles found in blood plasma. { ‚hē·mə'kō·nē·ə }

hemoconiosis [MED] Condition of having an abnormal amount of hemoconia in the blood. { ‚hē·mō‚kō·nē'ō·səs }

hemocyanin [BIOCHEM] A blue respiratory pigment found only in mollusks and in arthropods other than insects. { ‚hē·mō'sī·ə·nən }

hemocyte [INV ZOO] A cellular element of blood, especially in invertebrates. { 'hē·mə‚sīt }

hemocytoblast [HISTOL] A pluripotential blast cell thought to be capable of giving rise to all other blood cells. Also known as hemoblast; stem cell. { ‚hē·mə'sīd·ə‚blast }

hemocytolysis [PHYSIO] The dissolution of blood cells. { ‚hē·mə‚sī'täl·ə·səs }

hemocytometer [PATH] A specifically designed, ruled and calibrated glass slide used with a microscope to count red and white blood cells. Also spelled hemacytometer. { ‚hē·mə‚sī'täm·əd·ər }

hemodialysis [MED] The filtering of toxic solutes and excess fluid from the blood via an external membranous coil placed between the blood and a rinsing solution. { ‚hē·mō·dī'al·ə·səs }

hemodichorial placenta [EMBRYO] A placenta with a double trophoblastic layer. { ¦hē·mə·də'kȯr·ē·əl plə'sen·tə }

hemodynamics [PHYSIO] A branch of physiology concerned with circulatory movements of the blood and the forces involved in circulation. { ¦hē·mō·dī'nam·iks }

hemoendothelial placenta [EMBRYO] A placenta having the endothelium of vessels of chorionic villi in direct contact with the maternal blood. { ¦hē·mō‚en·də'thē·lē·əl plə'sen·tə }

hemoerythrin [BIOCHEM] A red respiratory pigment found in a few annelid and sipunculid worms and in the brachiopod Lingula. Also known as hemerythrin. { ‚hē·mō·ə'rith·rən }

hemoflagellate [INV ZOO] A parasitic, flagellate protozoan that lives in the blood of the host. { ‚hē·mə'flaj·ə·lət }

hemoglobin [BIOCHEM] The iron-containing, oxygen-carrying molecule of the red blood cells of vertebrates comprising four polypeptide subunits in a heme group. { 'hē·mə‚glō·bən }

hemoglobin A [BIOCHEM] The type of hemoglobin found in normal adults, which moves as a single component in an electrophoretic field, is rapidly denatured by highly alkaline solutions, and contains two titratable sulfhydryl groups per molecule. { 'hē·mə‚glō·bən 'ā }

hemoglobin C [PATH] A slow-moving abnormal hemoglobin associated with intraerythrocytic crystal formation, target cells, and chronic hemolytic anemia. { ‚hē·mə‚glō·bən 'sē }

hemoglobin E [PATH] An abnormal hemoglobin found in people of Southeast Asia, migrating slightly faster than hemoglobin C; in the homozygous form it causes a mild hemolytic anemia with normochromic target cells. { 'hē·mə‚glō·bən 'ē }

hemoglobinemia [MED] The presence of hemoglobin in the blood plasma. { ‚hē·mə‚glō·bə'nē·mē·ə }

hemoglobin H [PATH] An abnormal hemoglobin migrating more rapidly than normal hemoglobin on electrophoresis, and usually associated with thalassemia. { 'hē·mə‚glō·bən 'āch }

hemoglobin M [PATH] An abnormal hemoglobin associated with hereditary methemoglobinemia, differing from normal hemoglobin in its electrophoretic mobility by the starch-block method. { 'hē·mə‚glō·bən 'em }

hemoglobinopathy [MED] Any blood dyscrasia resulting from the genetically determined alteration of the chemical nature of hemoglobin. { ‚hē·mə‚glō·bə'näp·ə·thē }

hemoglobin S See sickle-cell hemoglobin. { 'hē·mə‚glō·bən 'es }

hemoglobinuria [MED] A pathological condition in which the urine contains hemoglobin. { ‚hē·mə‚glō·bə'nu̇r·ē·ə }

hemoglobinuric nephrosis See lower nephron nephrosis. { ‚he·mə‚glō·bə'nu̇r·ik nə'frō·səs }

hemogram [PATH] 1. Erythrocyte and leukocyte count per

HEMLOCK

Branch and cone of Eastern hemlock (Tsugo canadensis).

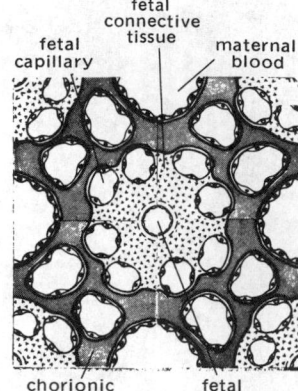

HEMOCHORIAL PLACENTA

fetal connective tissue
fetal capillary
maternal blood
chorionic trophoblast
fetal artery

Diagram of a hemochorial placenta found in rodents, many insectivores, and tarsiers.

cubic millimeter of blood plus the differential leukocyte count and hemoglobin level in grams per 100 milliliters of blood. **2.** The differential leukocyte count. { 'hē·mə‚gram }

hemohistioblast [HISTOL] The hypothetical reticuloendothelial cell from which all the cells of the blood are eventually differentiated. { ‚hē·mō'his·tē·ō‚blast }

hemolymph [INV ZOO] The circulating fluid of the open circulatory systems of many invertebrates. { 'hē·mə‚limf }

hemolysin [IMMUNOL] A substance that lyses erythrocytes. { ‚hē·mə'līs·ən }

hemolysis [PHYSIO] The lysis, or destruction, of erythrocytes with the release of hemoglobin. { hē'mäl·ə·səs }

hemolytic anemia [MED] A decrease in the blood concentration of hemoglobin and the number of erythrocytes, due to the inability of the mature erythrocytes to survive in the circulating blood. { ¦hē·mə¦lid·ik ə'nē·mē·ə }

hemolytic disease of newborn *See* erythroblastosis fetalis. { ¦hē·mə¦lid·ik di¦zēz əv 'nü‚bórn }

hemolytic jaundice [MED] Accumulation of bile pigments in the plasma as a result of excessive hemolysis. { ¦hē·mə¦lid·ik 'jón·dəs }

hemolytic uremic syndrome [MED] An illness characterized by the abrupt onset of decreased urine production, loss of kidney function, and anemia. It may be accompanied by edema, hypertension, blood-clotting disorders, and seizures. It is often caused by infection with *Escherichia coli* O157:H7 but has also been associated with *Salmonella* and *Shigella*. { hē·mə¦lid·ik yə¦rēm·ik 'sin‚drōm }

hemomonochorial placenta [EMBRYO] A placenta with a single trophoblastic layer. { ¦hē·mō‚män·ə'kór·ē·əl plə'sen·tə }

hemoparasite [INV ZOO] A parasitic animal that lives in the blood of a vertebrate. { ¦hē·mō'par·ə‚sīt }

hemopathology [MED] A branch of medicine dealing with blood diseases. Also known as hematopathology. { ¦hē·mō· pə'thäl·ə·jē }

hemopathy [MED] Any disease of the blood. { hē'mäp· ə·thē }

hemopericardium [MED] The presence of blood or bloody effusion in the pericardial sac. { ¦hē·mō‚per·ə'kärd·ē·əm }

hemoperitoneum [MED] An effusion of blood in the peritoneal cavity. { ¦hē·mō‚per·ə·tə'nē·əm }

hemopexin [BIOCHEM] A heme-binding protein in human plasma that may be a regulator of heme and drug metabolism, and a distributor of heme. { ¦hē·mə'pek·sən }

hemophilia [MED] A rare, hereditary blood disorder marked by a tendency toward bleeding and hemorrhages due to a deficiency of factor VIII. { ¦hē·mə¦fil·ē·ə }

hemophilic bacteria [MICROBIO] Bacteria of the genera *Hemophilus, Bordetella,* and *Moraxella;* all are small, gramnegative, nonmotile, parasitic rods, dependent upon blood factors for growth. { ¦hē·mə¦fil·ik bak'tir·ē·ə }

hemophilioid disease [MED] Any hemophilialike disease; it is the same as hemophilia clinically, but caused by a deficiency of factors IX, X, and XII. { ¦hē·mə¦fil·ē‚óid di‚zēz }

Hemophilus [MICROBIO] A genus of hemophilic bacteria in the family Brucellaceae requiring hemin and nicotinamide nucleoside for growth. { hə'mäf·ə·ləs }

hemophobia [PSYCH] An abnormal fear of the sight of blood. { ‚hē·mə'fō·bē·ə }

hemopoiesis *See* hematopoiesis. { ‚hē·mō‚pói'ē·səs }

hemopoietic tissue *See* hematopoietic tissue. { ‚hē·mō ‚pói'ed·ik 'tish·ü }

hemopoietin *See* hematopoietin. { ‚hē·mō‚pói'ēt·ən }

hemoporphyrin *See* hematoporphyrin. { ‚hē·mō'pór·fə·rən }

hemoptysis [MED] Discharge of blood from the larynx, trachea, bronchi, or lungs. { hē'mäp·tə·səs }

hemorrhage [MED] The escape of blood from the vascular system. { 'hem·rij }

hemorrhagic colitis [MED] An acute disease characterized by overtly bloody diarrhea that is caused by infection with the enterohemorrhagic strain of *Escherichia coli* (EC O157:H7). { ‚hem·ə¦raj·ik kə'līd·əs }

hemorrhagic diathesis [MED] Any condition marked by abnormal bleeding tendency. { ‚hem·ə¦raj·ik dī'ath·ə·səs }

hemorrhagic fever virus [VIROL] Any of several arboviruses causing acute infectious human diseases characterized by fever, prostration, vomiting, and hemorrhage. { ‚hem·ə¦raj·ik ¦fē·vər ‚vī·rəs }

hemorrhagic measles [MED] A grave variety of measles with a hemorrhagic eruption and severe constitutional symptoms. Also known as black measles. { ‚hem·ə¦raj·ik 'mēz· əlz }

hemorrhagic pericarditis [MED] Inflammation of the pericardium accompanied by the hemorrhagic appearance of the exudate imparted by the presence of red cells. { ‚hem·ə¦raj· ik ‚per·ə‚kär'dīd·əs }

hemorrhagic pleuritis [MED] Inflammation of the pleura characterized by the presence of bloody fluid in the pleural cavity. { ‚hem·ə¦raj·ik plə'rīd·əs }

hemorrhagic septicemia [VET MED] An infectious bacterial disease of fowl, rabbit, buffalo, and other animals caused by *Pasteurella mulfocida.* Also known as pasteurellosis. { ‚hem·ə¦raj·ik ‚sep·tə'sē·mē·ə }

hemorrhagic unit [BIOL] A unit for the standardization of snake venom. { ‚hem·ə¦raj·ik 'yü·nət }

hemorrhoid [MED] A varicosity of the external hemorrhoidal veins, causing painful swelling in the anal region. { 'hem‚róid }

hemorrhoidectomy [MED] Surgical removal of hemorrhoids. { ‚hem‚rói'dek·tə·mē }

hemosalpinx *See* hematosalpinx. { ‚hē·mō'sal‚piŋks }

hemosiderin [BIOCHEM] An iron-containing glycoprotein found in most tissues and especially in liver. { ‚hē·mō'sid· ə·rən }

hemosiderosis [PHYSIO] Deposition of hemosiderin in body tissues without tissue damage, reflecting an increase in body iron stores. { ‚hē·mō‚sid·ə'rō·səs }

hemostasis [MED] **1.** The arrest of a flow of blood or hemorrhage. **2.** The stopping or slowing of circulation. { ‚hē· mə'stā·səs }

hemostat [MED] An instrument to compress a bleeding vessel. { 'hē·mə‚stat }

hemostatic [MED] An agent that arrests or checks bleeding, especially by shortening clotting time. { ¦hē·mə¦stad·ik }

hemothorax [MED] Accumulation of blood in the pleural cavity. { ¦hē·mō'thór‚aks }

hemotrichorial placenta [EMBRYO] A placenta with a triple trophoblastic layer. { ‚hē·mō·trə'kór·ē·əl plə'sen·tə }

hemotrophe [BIOCHEM] The nutritive substance supplied via the placenta to embryos of viviparous animals. { hē'mä· trə·fē }

hemp-core cable *See* standard wire rope. { 'hemp ‚kór ‚kā· bəl }

hempseed oil [MATER] A fatty oil, light green or brownish yellow when dry, obtained from hempseed; used in soft soap, paints, and varnishes, and in Asia in foods. { 'hemp‚sēd óil }

HEMT *See* high-electron-mobility transistor.

hen [VERT ZOO] The female of several bird species, especially gallinaceous species. { hen }

henbane [BOT] *Hyoscyamus niger.* A poisonous herb containing the toxic alkaloids hyoscyamine and hyoscine; extracts have properties similar to belladonna. { 'hen‚bān }

hendecanal *See* undecanal. { hen'dek·ə·nəl }

hendecane *See* undecane. { 'hen·də‚kān *or* hen'de‚kān }

hendecyl *See* undecyl. { hen'des·əl }

Henderson equation for pH [PHYS CHEM] An equation for the pH of an acid during its neutralization: $pH = pK_a + \log [salt]/[acid]$, where pK_a is the logarithm to base 10 of the reciprocal of the dissociation constant of the acid; the equation is found to be useful for the pH range 4–10, providing the solutions are not too dilute. { 'hen·dər·sən i¦kwā·zhən fər ¦pē'āch }

Henderson process [MET] The treatment of copper sulfide ores by roasting with salt to form chlorides, which are then leached out and precipitated. { 'hen·dər·sən ‚prä·səs }

heneicosane [ORG CHEM] $C_{21}H_{44}$ Saturated hydrocarbon of the methane series; the crystals melt at 40°C and boil at 215°C (at 15 mm Hg). { hen'ī·kə‚sān }

henequen [MATER] A hard plant fiber, obtained from the leaves of the American agave (*Agave fourcroydes*) and other agave species; used to make rope, twine, and cord. { 'hen· ə·kən }

Hengstebeck approximation [CHEM ENG] A method of calculation to estimate the distribution of non-key components in distillation column products. { 'heŋg·stə·bek ə‚präk· sə‚mā·shən }

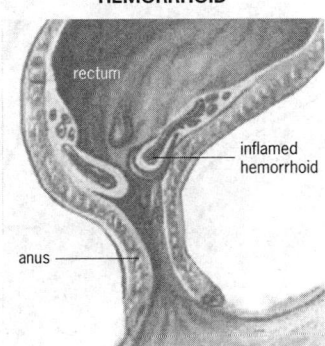

HEMORRHOID

Cross section of the human lower colon with a hemorrhoidal condition.

HENBANE

Drawing of henbane (*Hyoscyamus niger*).

Henicocephalidae [INV ZOO] A family of hemopteran insects of uncertain affinities. { ‚hen·ə·kō·sə'fal·ə‚dē }

Henle's loop *See* loop of Henle. { 'hen·lēz ‚lüp }

henna [BOT] *Lawsonia inermis.* An Old World plant having small opposite leaves and axillary panicles of white flowers; a reddish-brown dye extracted from the leaves is used in hair dyes. Also known as Egyptian henna. { 'hen·ə }

henry [ELECTROMAG] The mks unit of self and mutual inductance, equal to the self-inductance of a circuit or the mutual inductance between two circuits if there is an induced electromotive force of 1 volt when the current is changing at the rate of 1 ampere per second. Symbolized H. { 'hen·rē }

Henry's law [PHYS CHEM] The law that at sufficiently high dilution in a liquid solution, the fugacity of a nondissociating solute becomes proportional to its concentration. { 'hen·rēz ‚lò }

Hensen's node [EMBRYO] Thickening formed by a group of cells at the anterior end of the primitive streak in vertebrate gastrulas. { 'hen·səns ‚nōd }

hentriacontane [ORG CHEM] $C_{31}H_{64}$ A hydrocarbon; a crystalline material melting at 68°C and boiling at 302°C (at 15 mmHg); derived from roots of *Oenanthe crocata* and found in beeswax. { ‚hen‚trī·ə'kän‚tān }

Henyey track [ASTRON] An almost horizontal track on the Hertzsprung-Russell diagram that a star of small mass follows in an early stage of evolution after leaving the Hayashi track and before reaching the main sequence, during which the star is almost wholly in radiative equilibrium. { 'hen·yē ‚trak }

HEPA filter *See* high-efficiency particulate air filter. { 'hep·ə ‚fil·tər }

hepar calcies *See* calcium sulfide. { 'hē‚pär 'kal‚sēz }

heparin [BIOCHEM] An acid mucopolysaccharide acting as an antithrombin, antithromboplastin, and antiplatelet factor to prolong the clotting time of whole blood; occurs in a variety of tissues, most abundantly in liver. { 'hep·ə·rən }

hepar lobatum [MED] The liver in syphilitic cirrhosis, having a nodular lobulated appearance. { 'hē‚pär lō'bäd·əm }

hepar sulfuris *See* potassium sulfide. { 'hē‚pär səl'fyùr·əs }

HEPAT charge *See* high-explosive plastic antitank charge. { 'hē‚pat ‚chärj }

hepatectomy [MED] Surgical removal of the liver or a part of it. { ‚hep·ə'tek·tə·mē }

Hepaticae [BOT] The equivalent name for Marchantiatae. { he'pad·ə‚sē }

hepatic artery [ANAT] A branch of the celiac artery that carries blood to the stomach, pancreas, great omentum, liver, and gallbladder. { he'pad·ik 'ärd·ə·rē }

hepatic cecum [INV ZOO] A hollow outpocketing of the foregut of *Branchiostoma*; receives veins from the intestine. { he'pad·ik 'sē·kəm }

hepatic coma [MED] Unconscious state associated with advanced liver disease. { he'pad·ik 'kō·mə }

hepatic duct [ANAT] The common duct draining the liver. Also known as common hepatic duct. { he'pad·ik 'dəkt }

hepatic duct system [ANAT] The biliary tract including the hepatic ducts, gallbladder, cystic duct, and common bile duct. { he'pad·ik 'dəkt ‚sis·təm }

hepatic encephalopathy [MED] Behavioral, psychological, and neurological changes associated with advanced liver disease. { he'pad·ik en‚sef·ə'läp·ə·thē }

hepatic glycogenosis *See* von Gierke's disease. { he'pad·ik ‚glī·kə·jə'nō·səs }

hepatic plexus [NEUROSCI] Nerve network accompanying the hepatic artery to the liver. { he'pad·ik 'plek·səs }

hepatic portal system [ANAT] A system of veins in vertebrates which collect blood from the digestive tract and spleen and pass it through capillaries in the liver. { he'pad·ik 'pòrd·əl ‚sis·təm }

hepatic vein [ANAT] A blood vessel that drains blood from the liver into the inferior vena cava. { he'pad·ik 'vān }

hepatitis [MED] Inflammation of the liver; commonly of viral origin but also occurring in association with syphilis, typhoid fever, malaria, toxemias, and parasitic infestations. { ‚hep·ə'tīd·əs }

hepatitis virus [VIROL] Any of several viruses causing hepatitis in humans and lower mammals. { 'hep·ə'tīd·əs 'vī·rəs }

hepatization [PATH] The conversion of tissue into a liverlike substance, as of the lungs during the exudative stage of pneumonia. { ‚hep·əd·ə'zā·shən }

hepatocyte [HISTOL] An epithelial cell constituting the major cell type in the liver. { hə'pad·ə‚sīt }

hepatolenticular degeneration *See* Wilson's disease. { he¦pad·ō‚len¦tik·yə·lər di‚jen·ə'rā·shən }

hepatoma [MED] A usually malignant neoplasm arising from parenchymal cells of the liver. { ‚hep·ə'tō·mə }

hepatomegaly [MED] Enlargement of the liver. { ‚hep·əd·ō'meg·ə·lē }

hepatopancreas [INV ZOO] A gland in crustaceans and certain other invertebrates that combines the digestive functions of the liver and pancreas of vertebrates. { ¦hep·əd·ō'paŋ‚krē·əs }

hepatorenal syndrome [MED] A complex of syndromes due to hepatic and renal failure, including hyperpyrexia, oliguria, and coma. Also known as Heyd's syndrome. { ¦hep·əd·ō'rēn·əl 'sin‚drōm }

hepatoscopy [MED] Inspection of the liver, as by laparotomy or peritoneoscopy. { ‚hep·ə'täs·kə·pē }

hepatotoxin [PHARM] An agent capable of damaging the liver. { ¦hep·əd·ō'täk·sən }

Hepialidae [INV ZOO] A family of lepidopteran insects in the superfamily Hepialoidea. { ‚hep·ē'al·ə‚dē }

Hepialoidea [INV ZOO] A superfamily of lepidopteran insects in the suborder Homoneura including medium- to large-sized moths which possess rudimentary mouthparts. { ‚hep·ē·ə'lòid·ē·ə }

Hepplewhite-Gray lamp [MIN ENG] A lamp which drew its air from the top down through four tubular pillars into the base, where the air fed the flame through a gauze ring; the outlet was through a metal chimney closed by a gauze disk. { 'hep·əl‚wīt 'grā ‚lamp }

HEP projectile *See* high-explosive plastic projectile. { 'hep prə‚jek·təl }

Hepsogastridae [INV ZOO] A family of parasitic insects in the order Mallophaga. { ‚hep·sə'gas·trə‚dē }

heptachlor [ORG CHEM] $C_{10}H_7Cl_7$ An insecticide; a white to tan, waxy solid; insoluble in water, soluble in alcohol and xylene; melts at 95–96°C. { 'hep·tə‚klòr }

heptacosane [ORG CHEM] $C_{27}H_{56}$ A hydrocarbon; water-insoluble crystals melting at 60°C and boiling at 270°C (at 15 mmHg); soluble in alcohol; found in beeswax. { hep'tāk·ə‚sān *or* ‚hep·tə'kō‚sān }

heptad [SCI TECH] A group of seven. { 'hep‚tad }

heptadecane [ORG CHEM] $C_{17}H_{36}$ A hydrocarbon; water-insoluble, alcohol-soluble solid melting at 23°C and boiling at 303°C; used as a chemical intermediate. { ‚hep·tə'de‚kān }

***n*-heptadecanoic acid** [ORG CHEM] $CH_3(CH_2)_{15}COOH$ A fatty acid that is saturated; soluble in ether and alcohol, insoluble in water; colorless crystals melt at 61°C. Also known as margaric acid. { ¦en ¦hep·tə¦dek·ə¦nō·ik 'as·əd }

heptadecanol [ORG CHEM] $C_{17}H_{35}OH$ An alcohol; colorless liquid boiling at 309°C; slightly soluble in water; used as a chemical intermediate, as a perfume fixative, in cosmetics and soaps, and to manufacture surfactants. { ¦hep·tə¦dek·ə‚nól }

heptadione-2,3 *See* acetyl valeryl. { ‚hep·tə'dī‚ōn ¦tü ¦thrē }

heptagon [MATH] A seven-sided polygon. { 'hep·tə‚gän }

heptahedron [MATH] A polyhedron with seven faces. { ‚hep·tə'hē·drən }

heptakaidecagon [MATH] A polygon with 17 sides. { ‚hep·tə‚kī'dek·ə‚gän }

heptaldehyde [ORG CHEM] $C_6H_{13}CHO$ An aldehyde; ether-soluble, colorless oil with fruity aroma; slightly soluble in water; boils at 153°C; used as a chemical intermediate and for perfumes and pharmaceuticals. Also known as heptanal. { ‚hep·'tal·də‚hīd }

heptanal *See* heptaldehyde. { 'hep·tə‚nal }

heptane [ORG CHEM] $CH_3(CH_2)_5CH_3$ A hydrocarbon; water-insoluble, flammable, colorless liquid boiling at 98°C; soluble in alcohol, chloroform, and ether; used as an anesthetic, solvent, and chemical intermediate, and in standard octane-rating tests. { 'hep‚tān }

heptanoic acid [ORG CHEM] $CH_3(CH_2)_5COOH$ Clear oil boiling at 223°C; soluble in alcohol and ether, insoluble in water; used as a chemical intermediate. { ¦hep·tə¦nō·ik 'as·əd }

1-heptanol [ORG CHEM] $C_7H_{15}OH$ An alcohol; a fragrant, colorless liquid boiling at 174°C; soluble in water, ether, or alcohol; used as a chemical intermediate, as a solvent, and in cosmetics. Also known as heptyl alcohol. { ¦wən 'hep·tə‚nȯl }

3-heptanol [ORG CHEM] $CH_3CH_2CH(OH)C_4H_9$ An alcohol; a liquid boiling at 156°C; used as a coating, solvent, and diluent, as a chemical intermediate, and as a flotation frother. { ¦thrē 'hep·tə‚nȯl }

2-heptanone See methyl-*n*-amyl ketone. { ¦tü 'hep·tə‚nōn }

3-heptanone See ethyl butyl ketone. { ¦thrē 'hep·tə‚nōn }

4-heptanone [ORG CHEM] $(CH_3CH_2CH_2)_2CO$ A colorless liquid that is stable and has a pleasant odor; boils at approximately 98°C; used to put nitrocellulose and raw and blown oils into solution, and used in lacquers and as a flavoring in foods. { ¦fȯr 'hep·tə‚nōn }

heptene [ORG CHEM] $C_{17}H_{14}$ A liquid that is a mixture of isomers; boils at 189.5°C; used as an additive in lubricants, as a catalyst, and as a surface active agent. Also known as heptylene. { 'hep‚tēn }

heptode [ELECTR] A seven-electrode electron tube containing an anode, a cathode, a control electrode, and four additional electrodes that are ordinarily grids. Also known as pentagrid. { 'hep‚tōd }

heptomino [MATH] One of the 108 plane figures that can be formed by joining seven unit squares along their sides. { hep'täm·ə‚nō }

heptose [BIOCHEM] Any member of the group of monosaccharides containing seven carbon atoms. { 'hep‚tōs }

heptoxide [CHEM] An oxide whose molecule contains seven atoms of oxygen. { hep'täk‚sīd }

heptulose [BIOCHEM] The generic term for a ketose formed from a seven-carbon monosaccharide. { 'hep·tə‚lōs }

heptyl [ORG CHEM] $CH_3(CH_2)_6$— The functional group from heptane, $CH_3(CH_2)_5CH_3$. { 'hep·təl }

heptyl alcohol See 1-heptanol. { 'hep·təl 'al·kə‚hȯl }

heptylene See heptene. { 'hep·tə‚lēn }

herb [BOT] **1.** A seed plant that lacks a persistent, woody stem aboveground and dies at the end of the season. **2.** An aromatic plant or plant part used medicinally or for food flavoring. { hərb }

herbaceous [BOT] **1.** Resembling or pertaining to a herb. **2.** Pertaining to a stem with little or no woody tissue. { hər'bā·shəs }

herbaceous dicotyledon [BOT] A type of dicotyledon in which the primary vascular cylinder forms an ectophloic siphonostele with widely separated vascular strands. { hər'bā·shəs ‚dī‚käd·əl'ēd·ən }

herbaceous monocotyledon [BOT] A type of monocotyledon with a vascular system composed of widely spaced strands arranged in one of four ways. { hər'bā·shəs ‚män·ə‚käd·əl'ēd·ən }

herbarium [BOT] **1.** A collection of plant specimens, pressed and mounted on paper or placed in liquid preservatives, and systematically arranged. **2.** A building where a herbarium is housed. { hər'ber·ē·əm }

Herbert cloudburst test [MET] A hardness test in which a shower of steel balls, dropped from a predetermined height, dulls the surface of a hardened part in proportion to its softness and thus reveals defective areas. { 'hər·bərt 'klaud‚bərst ‚test }

Herbert pendulum method [MET] Hardness testing in which a 1-millimeter steel or jewel ball resting on the surface to be tested acts as the fulcrum for a 4-kilogram compound pendulum of 10-second period; the swinging of the pendulum causes a rolling indentation in the material, and several hardness factors, such as work hardenability, are determined. { 'hər·bərt 'pen·jə·ləm ‚meth·əd }

herbicide [MATER] A chemical agent that destroys or inhibits plant growth. { 'her·bə‚sīd }

herbicolous [ECOL] Living on herbs. { hər'bik·əl·əs }

Herbig emission star [ASTRON] A relatively massive star in early stages of formation, still surrounded by a nebula which makes it variable in luminosity and renders its spectrum very peculiar. { 'hər·bik i¦mish·ən ‚stär }

Herbig-Haro object [ASTRON] A bright patch on the surface of a dark cloud of gas and dust, consisting of light that has been scattered and reflected from a newborn star embedded in the cloud. { ¦hər·big 'hä·rō ‚äb·jəkt }

herbivore [VERT ZOO] An animal that eats only vegetation. { 'hər·bə‚vȯr }

herbivory [ECOL] The consumption of plants without killing them. { hər'biv·ə·rē }

Herbst corpuscle [VERT ZOO] A cutaneous sense organ found in the mucous membrane of the tongue of the duck. { 'hərbst ‚kȯr·pə·səl }

hercularc lining [MIN ENG] A German method of lining mine roadways subjected to heavy pressures; a closed circular arch of wedge-shaped precast concrete blocks made in two sizes are erected so that alternate blocks offer their wedge action in opposite direction, the larger blocks toward the center of the roadway and the smaller outward. { 'hər·kyə‚lärk ‚līn·iŋ }

Hercules [ASTRON] A constellation with no stars brighter than third magnitude; right ascension 17 hours, declination 30° north. { 'hər·kyə‚lēz }

Hercules cluster [ASTRON] A cluster of about 75 bright galaxies with a recession velocity of 6200 miles (10,000 kilometers) per second. { 'hər·kyə‚lēz ‚kləs·tər }

Hercules stone See lodestone. { 'hər·kyə‚lēz ‚stōn }

Hercules superclusters [ASTRON] A pair of superclusters with recession velocities around 10,000 kilometers (6200 miles) per second, one of which contains the Hercules cluster. { ¦hərk·yə‚lēz 'sü·pər‚kləs·tərz }

Hercules trap [ANALY CHEM] Water-measuring liquid trap used in aquametry when the material collected is heavier than water. { 'hər·kyə‚lēz ‚trap }

Hercules X-1 [ASTROPHYS] A source of x-rays that pulses with a period of 1.237 seconds, and is eclipsed for 6 of every 42 hours, associated with a variable star, designated HZ Herculis, that also has a period of 42 hours and faint 1.237-second pulsations; believed to be a binary star whose invisible member is a rotating neutron star. Abbreviated Her X-1. { 'hər·kyə‚lēz ¦eks 'wən }

Hercynian geosyncline [GEOL] A principal area of geosynclinal sediment accumulation in Devonian time; found in south-central and southern Europe and northern Africa. { hər'sin·ē·ən ‚jē·ō'sin‚klīn }

Hercynian orogeny See Variscan orogeny. { hər'sin·ē·ən ō'räj·ə·nē }

hercynite [MINERAL] $(Fe,Mg)Al_2O_4$ A black mineral of the spinel group; crystallizes in the isometric system. Also known as ferrospinel; iron spinel. { 'hərs·ən‚īt }

herd [VERT ZOO] A number of one kind of wild, semidomesticated, or domesticated animals grouped or kept together under human control. { hərd }

herderite [MINERAL] $CaBe(PO_4)(F,OH)$ A colorless to pale-yellow or greenish-white mineral consisting of phosphate and fluoride of calcium and beryllium; hardness is 7.5–8 on Mohs scale, and specific gravity is 3.92. { 'hər·də‚rīt }

herd immunity [IMMUNOL] Immunity of a sufficient number of individuals in a population such that infection of one individual will not result in an epidemic. { 'hərd i‚myü·nəd·ē }

herd instinct [PSYCH] Psychic need for identification with a group. { 'hərd ‚in‚stiŋkt }

hereditary deforming chondrodysplasia See multiple hereditary exostoses. { hə'red·ə‚ter·ē di'fȯr·miŋ ¦kän·drō‚dis'plā·zhə }

hereditary determinant [MOL BIO] A nuclear or extranuclear genetically functional unit that is replicated with conservation of specificity. { hə¦red·ə‚ter·ē di'tər·mə·nənt }

hereditary disease [MED] A genetically determined illness transmitted from parent to child. { hə'red·ə‚ter·ē di‚zēz }

hereditary hemorrhagic telangiectasia [MED] An inherited disease characterized by dilatation of groups of capillaries and a tendency to hemorrhage. Also known as Osler-Rendu-Weber disease. { hə'red·ə‚ter·ē ‚hem·ə'raj·ik tə¦lan·jē·ek'tā·zhə }

hereditary hypophosphatemic rickets [MED] Sex-linked syndrome involving defective bone growth and decreased concentrations of phosphates in the serum. { hə'red·ə‚ter·ē ¦hī·pō‚fäs·fə¦tē·mik 'rik·əts }

hereditary mechanics [MECH] A field of mechanics in which quantities, such as stress, depend not only on other quantities, such as strain, at the same instant but also on integrals involving the values of such quantities at previous times. { hə'red·ə‚ter·ē mi'kan·iks }

hereditary nephritis [MED] A familial disease characterized by recurrent attacks of interstitial inflammation of the kidneys

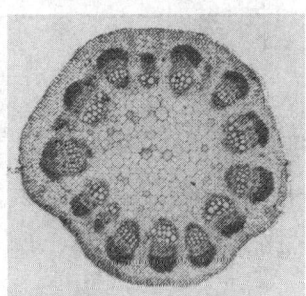

HERBACEOUS DICOTYLEDON

Cross section through the stem of the sunflower. (*From J. B. Hill, L. O. Overholts, and H. W. Popp, Botany: A Textbook for Colleges, 2d ed., McGraw-Hill, 1950*)

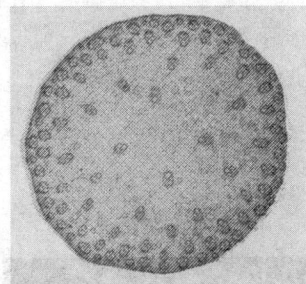

HERBACEOUS MONOCOTYLEDON

Photomicrograph of transverse section through corn stem. (*From J. B. Hill, L. O. Overholts, and H. W. Popp, Botany: A Textbook for Colleges, 2d ed., McGraw-Hill, 1950*)

and discharge of blood in the urine. { hə'red·ə,ter·ē ne 'frīd·əs }

hereditary spherocytosis [MED] A chronic congenital disorder of the erythrocytopoietic system characterized by a preponderance of spherical erythrocytes, increased osmotic fragility, hemolytic anemia, and splenomegaly. { hə'red·ə,ter·ē ,sfer·ō,sī'tō·səs }

hereditary spinal ataxia See Friedrich's ataxia. { hə'red·ə,ter·ē ¦spīn·əl ə'tak·sē·ə }

heredity [GEN] The transmission of phenotypes and alleles from one generation to the next. The sum of genetic endowment obtained from the parents. { hə'red·əd·ē }

heredofamilial [MED] Referring to a disease or disorder having a familial pattern of occurrence and thought to be hereditary. { hə¦red·ō·fə¦mil·yəl }

her function [MATH] A function that is expressed in terms of Hankel functions in a manner similar to that in which the ber function is expressed in terms of Bessel functions. { 'her ,fəŋk·shən }

Hering theory [PHYSIO] A theory of color vision which assumes that three qualitatively different processes are present in the visual system, and that each of the three is capable of responding in two opposite ways. { 'her·iŋ ,thē·ə·rē }

heritability [GEN] In a population, the ratio of the total genetic variance to the total phenotypic variance. { ,her·əd·ə'bil·əd·ē }

heritable change [GEN] A nonlethal genetic change that is passed on to the descendants. { ¦her·əd·ə·bəl 'chānj }

Hermann-Mauguin symbols [CRYSTAL] Symbols representing the 32 symmetry classes, consisting of series of numbers giving the multiplicity of symmetry axes in descending order, with other symbols indicating inversion axes and mirror planes. { 'her¦män 'mō,gan ,sim·bəlz }

hermaphrodite [BIOL] An individual animal or plant exhibiting hermaphroditism. { hər'maf·rə,dīt }

hermaphrodite caliper [DES ENG] A layout tool having one leg pointed and the other like that of an inside caliper; used to locate the center of irregularly shaped stock or to lay out a line parallel to an edge. { hər'maf·rə,dīt ¦kal·ə·pər }

hermaphroditic See monoecious. { hər,ma·frə'did·ik }

hermaphroditic connector [ELEC] A connector in which both mating parts are exactly alike at their mating surfaces. Also known as sexless connector. { hər,ma·frə'did·ik kə'nek·tər }

hermaphroditism [PHYSIO] An abnormal condition, especially in humans and other higher vertebrates, in which both male and female reproductive organs are present in the individual. { hər'ma·frə·dīd,iz·əm }

hermatype See hermatypic coral. { 'hər·mə,tīp }

hermatypic coral [INV ZOO] Reef-building coral characterized by the presence of symbiotic algae within their endodermal tissue. Also known as hermatype. { ,hər·mə'tip·ik 'kär·əl }

Hermes [ASTRON] A very small asteroid which passed within 485,000 miles (780,000 kilometers) of the earth in 1937, the closest known approach of a celestial body other than the moon. { 'hər,mēz }

hermetic seal [ENG] An airtight seal. { hər'med·ik 'sēl }

hermit crab [INV ZOO] The common name for a number of marine decapod crustaceans of the families Paguridae and Parapaguridae; all lack right-sided appendages and have a large, soft, coiled abdomen. { 'hər·mət ,krab }

Hermite polynomials [MATH] A family of orthogonal polynomials which arise as solutions to Hermite's differential equation, a particular case of the hypergeometric differential equation. { er'mēt ,päl·ə'nō·mē·əlz }

Hermite's differential equation [MATH] A particular case of the hypergeometric equation; it has the form $w'' - 2zw' + 2nw = 0$, where n is an integer. { er'mēts dif·ə¦ren·chəl i'kwā·zhən }

Hermitian conjugate [MATH] For a matrix A, the transpose of the complex conjugate of A. Also known as adjoint; associate matrix. { er'mish·ən 'kän·jə·gət }

Hermitian conjugate operator See adjoint operator. { er'mish·ən 'kän·jə·gət 'äp·ə,rād·ər }

Hermitian form [MATH] **1.** A polynomial in n real or complex variables where the matrix constructed from its coefficients is Hermitian. **2.** More generally, a sesquilinear form g such that $g(x, y) = \overline{g(y, x)}$ for all values of the independent variables

x and y, where $\overline{g(x, y)}$ is the image of $g(x, y)$ under the automorphism of the underlying ring. { er'mish·ən 'fȯrm }

Hermitian inner product See inner product. { er'mish·ən 'in·ər ¦präd·əkt }

Hermitian kernel [MATH] A kernel $K(x, t)$ of an integral transformation or integral equation is Hermitian if $K(x, t)$ equals its adjoint kernel, $K^*(t, x)$. { er'mish·ən 'kər·nəl }

Hermitian matrix [MATH] A matrix which equals its conjugate transpose matrix, that is, is self-adjoint. { er'mish·ən 'mā·triks }

Hermitian operator [MATH] A linear operator A on vectors in a Hilbert space, such that if x and y are in the range of A then the inner products (Ax, y) and (x, Ay) are equal. { er'mish·ən 'äp·ə,rād·ər }

Hermitian scalar product See inner product. { er'mish·ən 'skāl·ər ¦präd·əkt }

Hermitian space See inner product space. { er'mish·ən 'spās }

hermit point See isolated point. { 'hər·mit ,pȯint }

hernia [MED] Abnormal protrusion of an organ or other body part through its containing wall. Also called rupture. { 'hər·nē·ə }

hernial sac [MED] A pouch of peritoneum containing a herniated organ or other body part. { 'hər·nē·əl 'sak }

herniated disk [MED] An intervertebral disk in which the pulpy center has pushed through the fibrocartilage. Also known as slipped disk. { 'hər·nē,ad·əd 'disk }

herniation of the nucleus pulposus [MED] Occurrence of a herniated disk. { ,hər·nē'ā·shən əv th̲ə 'nü·klē·əs pəl'pō·səs }

herniorrhaphy [MED] Operation for repair and suturing of a hernia. { ,hər·nē'ȯr·ə·fē }

herniotomy [MED] An operation for the relief of irreducible hernia, by cutting through the neck of the sac. { ,hər·nē'äd·ə·mē }

heroin [PHARM] $C_{21}H_{23}O_5N$ A white, crystalline powder made from morphine; the hydrochloride compound is used as a sedative and narcotic. { 'her·ə·wən }

heron [VERT ZOO] The common name for wading birds composing the family Ardeidae characterized by long legs and neck, a long tapered bill, large wings, and soft plumage. { 'her·ən }

Heron's formula See Hero's formula. { 'her·ənz ,fȯr·myə·lə }

Hero's formula [MATH] A formula expressing the area of a triangle in terms of the sides a, b, and c as

$$\sqrt{s(s - a)(s - b)(s - c)}$$

where $s = \sqrt{(1/2)(a + b + c)}$

Also known as Heron's formula. { 'hir·ōz ,fȯr·myə·lə }

herpangia [MED] A mild viral disease of humans caused by a coxsackie virus and characterized by fever, anorexia, and grayish papules surrounded by a red areola in the mouth. { hər'pan·jē·ə }

herpes [MED] An acute inflammation of the skin or mucous membranes, characterized by the development of groups of vesicles on an inflammatory base. { 'hər,pēz }

herpes simplex [MED] An acute vesicular eruption of the skin or mucous membranes caused by a virus, commonly seen as cold sores or fever blisters. { 'hər,pēz 'sim,pleks }

herpes simplex encephalitis See herpetic encephalitis. { ¦hər,pēz 'sim,pleks in,sef·ə'līd·əs }

herpes simplex virus [VIROL] Either of two types of subgroup A herpesviruses that are specific for humans; given the binomial designation *Herpesvirus hominis*. { 'hər,pēz 'sim ,pleks 'vī·rəs }

Herpesviridae [VIROL] A family of deoxyribonucleic acid (DNA)-containing viruses characterized by enveloped virions containing one molecule of double-stranded linear DNA wrapped around an associated spool-shaped protein inside an icosahedron. It includes subfamilies Alphaherpesvirinae (herpes simplex virus group), Betaherpesvirinae (cytomegalovirus group), and Gammaherpesvirinae (lymphoproliferative virus group). { ,hər·pēz'vir·ə,dī }

herpesvirus [VIROL] A major group of deoxyribonucleic acid-containing animal viruses, distinguished by a cubic capsid, enveloped virion, and affinity for the host nucleus as a site of maturation. { 'hər,pēz,vī·rəs }

HERMAPHRODITE CALIPER

Drawing of a hermaphrodite caliper. *(From R. J. Sweeney, Measurement Techniques in Mechanical Engineering, Wiley, 1953)*

HERMIT CRAB

Representative species of an anomuran, the hermit crab.

herpes zoster [MED] A systemic virus infection affecting spinal nerve roots, characterized by vesicular eruptions distributed along the course of a cutaneous nerve. Also known as shingles; zoster. { ¦hər‚pēz 'zäs·tər }

herpetic encephalitis [MED] A type of meningoencephalitis characterized by large intranuclear inclusion bodies in the brain. Also known as herpes simplex encephalitis. { hər'ped·ik in‚sef·ə'līd·əs }

herpetic stomatitis [MED] Inflammation of the soft tissues of the mouth characterized by fever blisters. { hər'ped·ik ‚stō·mə'tīd·əs }

herpetic tonsillitis [MED] Acute inflammation of the tonsils characterized by fever and vesicles caused by a herpesvirus. { hər'ped·ik tän·sə'līd·əs }

Herpetosiphon [MICROBIO] A genus of bacteria in the family Cytophagaceae; cells are unbranched, sheathed rods or filaments; unsheathed segments are motile; microcysts are not known. { ¦hər·pəd·ō'sī·fən }

herpolhode [MECH] The curve traced out on the invariable plane by the point of contact between the plane and the inertia ellipsoid of a rotating rigid body not subject to external torque. { ¦hər·pəl'hōd }

herpolhode cone See space cone. { ¦hər·pəl'hōd ‚kōn }

Herreshoff furnace [MET] **1.** A rectangular-shaft blast furnace for smelting copper ore. **2.** A mechanical, cylindrical, multiple-deck muffle furnace of the McDougall type. { 'her·əs‚hóf ‚fər·nəs }

herring [VERT ZOO] The common name for fishes composing the family Clupeidae; fins are soft-rayed and have no supporting spines, there are usually four gill clefts, and scales are on the body but absent on the head. { 'her·iŋ }

Herring body [HISTOL] Any of the distinct colloid masses in the vertebrate pituitary gland, possibly representing greatly dilated endings of nerve fibers. { 'her·iŋ ‚bäd·ē }

herringbone gear [MECH ENG] The equivalent of two helical gears of opposite hand placed side by side. { 'her·iŋ‚bōn ‚gir }

herringbone pattern [ELECTR] An interference pattern sometimes seen on television receiver screens, consisting of a horizontal band of closely spaced V- or S-shaped lines. { 'her·iŋ‚bōn ‚pad·ərn }

herringbone stoping [MIN ENG] Method used in flattish Rand stope panels 500–1000 feet (150–300 meters) long for breaking and moving the ore; the stope is divided into 20-foot (6-meter) panels, and a different gang works each panel; a tramming system delivers the cut rock to a central scraper system. { 'her·iŋ‚bōn ¦stōp·iŋ }

herringbone timbering [MIN ENG] A method of timber support in a roadway with a weak roof and strong sides, using neither arms nor side uprights; the crossbar is notched into the sides and supported at its center by a bar under it and parallel with the roadway; the bar is supported by struts notched into the sides at about half height. { 'her·iŋ‚bōn ‚tim·bə·riŋ }

herringbone weave [TEXT] Broken-twill weave giving a zigzag effect produced by alternating the direction of the twill, like the skeleton of a herring. { 'her·iŋ‚bōn ‚wēv }

herring oil [MATER] Pale-yellow to dark-red liquid obtained from herring; soluble in carbon disulfide, chloroform, and ether, insoluble in water; used in soaps, in leather dressing, and as machinery lubricant. { 'her·iŋ ‚oil }

Herschel-Cassegrain telescope [OPTICS] A modification of a Cassegrain telescope in which the primary paraboloidal mirror is slightly inclined to the optical axis, and both the secondary hyperboloidal mirror and the eyepiece are located off the axis, so that it is not necessary to pierce the primary. { 'hər·shəl 'kas·gran ‚tel·ə‚skōp }

Herschel effect [GRAPHICS] **1.** The reaction whereby a photographic image on silver chloride is destroyed by exposure to red light. Also known as visual Herschel effect. **2.** The reaction whereby a latent photographic image in gelatin emulsion which is not dye-sensitized is destroyed by exposure to red light. Also known as latent Herschel effect. { 'hər·shəl i‚fekt }

Herschel-Quincke tube [ACOUS] A device for demonstrating the interference of sound in which sound waves from a common source travel through two tubes of different lengths and recombine, producing reinforcement or cancellation of sound depending on the difference in path length; used to demonstrate the interference of sound and as a wave filter. Also known as Quincke tube. { ¦hər·shəl 'kwiŋ·kə ‚tüb }

Herschel-type venturi tube [ENG] A type of venturi tube in which the converging and diverging sections are cones, the throat section is relatively short, the diverging cone is long, and the pressures preceding the inlet cone and in the throat are transferred through multiple openings into annular openings, called piezometer rings. { 'hər·shəl ‚tīp ven'tùr·ē ‚tüb }

hertz [PHYS] Unit of frequency; a periodic oscillation has a frequency of n hertz if in 1 second it goes through n cycles. Also known as cycle per second (cps). Symbolized Hz. { hərts }

Hertz antenna [ELECTROMAG] An ungrounded half-wave antenna. { 'hərts an‚ten·ə }

Hertz effect [ELECTR] Increase in the length of a spark induced across a spark gap when the gap is irradiated with ultraviolet light. [ELECTROMAG] A dependence of the attenuation of a linearly polarized electromagnetic wave passing through a grating of metal rods on the angle between the electric vector and the rod direction, with the attenuation being a minimum when the two are perpendicular. { 'hərts i‚fekt }

Hertzian oscillator [ELECTROMAG] **1.** A generator of electric dipole radiation; consists of two capacitors joined by a conducting rod having a small spark gap; an oscillatory discharge occurs when the two halves of the oscillator are raised to a sufficiently high potential difference. **2.** A dumbbell-shaped conductor in which electrons oscillate from one end to the other, producing electric dipole radiation. { 'hərt·sē·ən 'äs·ə‚lād·ər }

Hertzian wave See electric wave. { 'hərt·sē·ən 'wāv }

Hertz's law [MECH] A law which gives the radius of contact between a sphere of elastic material and a surface in terms of the sphere's radius, the normal force exerted on the sphere, and Young's modulus for the material of the sphere. { 'hərt·səs ‚ló }

Hertzsprung gap [ASTRON] A gap on the Hertzsprung-Russell diagram between giant stars of spectral types A0 and G0, caused by the fact that the movement of stars across this region occupies a relatively brief time. { 'hert·sprùŋ ‚gap }

Hertzsprung-Russell diagram [ASTRON] A plot showing the relation between the luminosity and surface temperature of stars; other related quantities frequently used in plotting this diagram are the absolute magnitude for luminosity, and spectral type or color index for the surface temperatures. Abbreviated H-R diagram. Also known as Russell diagram. { 'hert·sprùŋ 'rəs·əl 'dī·ə‚gram }

Hertz vector See polarization potential. { 'hərts ‚vek·tər }

hervidero See mud volcano. { ¦hər·və‚der·ō }

Her X-1 See Hercules X-1.

Herzau-Ogle phenomenon [PSYCH] An effect whereby identical geometrical figures seen with stereoscopic vision are perceived as double when there is unequal enlargement in the two eyes, or when corresponding retinal positions do not correspond to the same geometrical points. { 'härt‚saù 'ō·gəl fə'näm·ə‚nän }

Hesionidae [INV ZOO] A family of small polychaete worms belonging to the Errantia. { ‚hes·ē'än·ə‚dē }

hesitation [COMPUT SCI] A brief automatic suspension of the operations of a main program in order to perform all or part of another operation, such as rapid transmission of data to or from a peripheral unit. { ‚hez·ə'tā·shən }

hesperidium [BOT] A modified berry, with few seeds, a leathery rind, and membranous extensions of the endocarp dividing the pulp into chambers; an example is the orange. { ‚hes·pə'rid·ē·əm }

Hesperiidae [INV ZOO] The single family of the superfamily Hesperioidea comprising butterflies known as skippers because of their rapid, erratic flight. { hes·pə'rī·ə‚dē }

Hesperioidea [INV ZOO] A monofamilial superfamily of lepidopteran insects in the suborder Heteroneura including heavy-bodied, mostly diurnal insects with clubbed antennae that are bent, curved, or reflexed at the tip. { he‚spir·ē'óid·ē·ə }

Hesperornithidae [PALEON] A family of extinct North American birds in the order Hesperornithiformes. { ‚hes·pər‚ór'nith·ə‚dē }

Hesperornithiformes [PALEON] An order of ancient extinct birds; individuals were large, flightless, aquatic diving birds with the shoulder girdle and wings much reduced and the legs

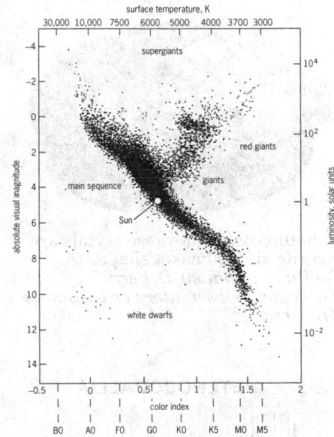

HERTZSPRUNG-RUSSELL DIAGRAM

Diagram for about 15,000 stars within a sphere of radius 100 parsecs, taken from the *Hipparcos Catalogue*. The color index and absolute visual magnitude scales are directly measured. The spectral class, surface temperature, and luminosity (in terms of solar luminosity) are approximate relationships appropriate for the main sequence.

specialized for strong swimming. { ,hes·pə,ròr,nith·ə'fòr,mēz }

Hesperus [ASTRON] Greek name for the planet Venus as an evening star. { 'hes·prəs }

Hess diagram [ASTRON] A diagram showing the frequencies of occurrence of stars at various positions on the Hertzsprung-Russell diagram. { 'hes ,dī·ə,gram }

Hesser's variation [COMPUT SCI] A variation of a Kiviat graph in which all variables are arranged so that their plots approach the circumference of the graph as the system being evaluated approaches saturation, and the scales on the various axes may not cover the full 0–100% range, or may be in units other than percent. { 'hes·ərz ,ver·ē,ā·shən }

Hesse's theorem [MATH] A theorem in projective geometry which states that, from the three pairs of lines containing the two pairs of opposite sides and the diagonals of a quadrilateral, if any two pairs are conjugate lines with respect to a given conic, then so is the third. { 'hes·əz ,thir·əm }

hessian *See* burlap. { 'hesh·ən }

Hessian [MATH] For a function $f(x_1, \ldots, x_n)$ of n real variables, the real-valued function of $(x_1, \ldots, x_n)$ given by the determinant of the matrix with entry $\partial^2 f/\partial x_i \partial x_j$ in the ith row and jth column; used for analyzing critical points. { 'hesh·ən }

hessite [MINERAL] Ag_2Te A lead-gray sectile mineral crystallizing in the isometric system; usually massive and often auriferous. { 'he,sīt }

Hess's law [PHYS CHEM] The law that the evolved or absorbed heat in a chemical reaction is the same whether the reaction takes one step or several steps. Also known as the law of constant heat summation. { 'hes·əz ,lò }

hetaerolite [MINERAL] $ZnMn_2O_4$ A black mineral consisting of zinc-manganese oxide found with chalcophanite. { hə'tir·ə,līt }

Heteractinida [PALEON] A group of Paleozoic sponges with calcareous spicules; probably related to the Calcarea. { ,hed·ə·rak'tin·əd·ə }

Heterakidae [INV ZOO] A group of nematodes assigned either to the suborder Oxyurina or the suborder Ascaridina. { ,hed·ə'rak·ə,dē }

Heterakoidea [INV ZOO] A superfamily of parasitic nematodes of the order Ascaridida, characterized by small, well-developed lips with paired sensilla in the labial region, an infundibular stoma, and a rarely cylindrical esophagus divided into three parts. { ,hed·ə·rə'kóid·ē·ə }

heterandrous [BOT] Having stamens differing from each other in length or form. { ¦hed·ə¦ran·drəs }

heterauxesis *See* allometry. { ¦hed·ər,òg'zē·səs }

hetero- [CHEM] Prefix meaning different; for example, a heterocyclic compound is one in which the ring is made of more than one kind of atom. { 'hed·ə·rō }

heteroagglutinin [IMMUNOL] An antibody in normal blood serum capable of agglutinating foreign particles and erythrocytes of other species. { ¦hed·ə·rō·ə'glüd·ən·ən }

heteroallele [GEN] One of two or more alternative forms of a gene that differ at nonidentical mutation sites. { ,hed·ə·rō·ə'lēl }

heteroatom [ORG CHEM] In an organic compound, any atom other than carbon or hydrogen. { 'hed·ə·rō,ad·əm }

heteroauxin [BIOCHEM] $C_{10}H_9O_2N$ A plant growth hormone with an indole skeleton. { ¦hed·ə·rō'òk·sən }

heteroazeotrope [CHEM] Liquid mixture that is not completely miscible in all proportions in the liquid phase, yet does not form an azeotrope. Also known as heterogeneous zeotrope. { ¦hed·ə·rō·ā'zē·ə,tròp }

Heterobasidiomycetidae [MYCOL] A class of fungi in which the basidium either is branched or is divided into cells by cross walls. { ¦hed·ə·rō·bə,sid·ē·ō,mī¦sed·ə,dē }

heteroblastic [EMBRYO] Arising from different tissues or germ layers, in referring to similar organs in different species. [PETR] Pertaining to rocks in which the essential constituents are of two distinct orders of magnitude of size. { ¦hed·ə·rō¦blas·tik }

Heterocapsina [BOT] An order of green algae in the class Xanthophyceae. [INV ZOO] A suborder of yellow-green to green flagellate protozoans in the order Heterochlorida. { ¦hed·ə·rō'kap·sə·nə }

heterocarpous [BOT] Producing two distinct types of fruit. { ¦hed·ə·rō'kär·pəs }

heterocentric chromosome [GEN] A dicentric chromosome whose centromeres are of unequal strength. { ¦hed·ə·rə,sen·trik 'krō·mə,sōm }

Heterocera [INV ZOO] A formerly recognized suborder of Lepidoptera including all forms without clubbed antennae. { ,hed·ə'räs·ə·rə }

heterocercal [VERT ZOO] Pertaining to the caudal fin of certain fishes and indicating that the upper lobe is larger, with the vertebral column terminating in this lobe. { ¦hed·ə·rō¦sər·kəl }

Heteroceridae [INV ZOO] The variegated mud-loving beetles, a family of coleopteran insects in the superfamily Dryopoidea. { ¦hed·ə·rō¦ser·ə,dē }

Heterocheilidae [INV ZOO] A family of parasitic roundworms in the superfamily Ascaridoidea. { ¦hed·ə·rō¦kī·lə,dē }

heterochlamydeous [BOT] Having the perianth differentiated into a distinct calyx and a corolla. { ¦hed·ə·rō·klə'mid·ē·əs }

Heterochlorida [INV ZOO] An order of yellow-green to green flagellate oraganisms of the class Phytamastigophorea. { ¦hed·ə·rō'klòr·ə·də }

heterochromatic photometry [OPTICS] The branch of photometry concerned with comparing the illuminating powers of light sources with different colors. { ,hed·ə·rō·krə¦mad·ik fō'täm·ə·trē }

heterochromatin [CELL MOL] Specialized chromosome material which remains tightly coiled even in the nondividing nucleus and stains darkly in interphase. { ¦hed·ə·rō'krō·məd·ən }

heterochromia [PHYSIO] A condition in which the two irises of an individual have different colors, or in which one iris has two colors. { ,hed·ə·rō'krō·mē·ə }

heterochronic mutation [GEN] A mutation that perturbs the relative timing of events during postembryonic development. { ¦hed·ə·rə,krä·nik myü'tā·shən }

heterochronism [EMBRYO] Deviation from the normal sequence of organ formation; a factor in evolution. [GEOL] A phenomenon in which two similar geologic deposits may not be of the same age even though they underwent like processes of formation. { ,hed·ə'räk·rə,niz·əm }

heterochrony [EVOL] An evolutionary phenomenon that involves changes in the rate and timing of species development. { ,hed·ə'rä·krə·nē }

heterochthonous [SCI TECH] Not indigenous to the area of present occurrence. { ,hed·ə'räk·thə·nəs }

heterococcolith [BIOL] A coccolith with crystals arranged into boat, trumpet, or basket shapes. { ,hed·ə·rō'käk·ə,lith }

heterocoelous [ANAT] Pertaining to vertebrae with centra having saddle-shaped articulations. { ¦hed·ə·rō'sē·ləs }

Heterocorallia [PALEON] An extinct small, monofamilial order of fossil corals with elongate skeletons; found in calcareous shales and in limestones. { ,hed·ə·rō·kə'ral·ē·ə }

Heterocotylea [INV ZOO] The equivalent name for Monogenea. { ,hed·ə·rō·kə'til·ē·ə }

heterocyclic compound [ORG CHEM] Compound in which the ring structure is a combination of more than one kind of atom; for example, pyridine, C_5H_5N. { ,hed·ə·rō'sī·klik ¦käm,pȧund }

heterocyst [BOT] Clear, thick-walled cell occurring at intervals along the filament of certain blue-green algae. { 'hed·ə·rə,sist }

heterodactylous [VERT ZOO] Having the first two toes turned backward. { ¦hed·ə·rō¦dakt·əl·əs }

Heterodera [INV ZOO] The cyst nematodes, a genus of phytoparasitic worms that live in the internal root systems of many plants. { ,hed·ə'räd·ə·rə }

heterodesmic [CRYSTAL] Pertaining to those atoms bonded in more than one way in crystals. { ,hed·ə·rō'dez·mik }

heterodimer [BIOCHEM] A protein made of paired polypeptides that differ in their amino acid sequences. { ,hed·ə·rō'dī·mər }

heterodisomy [GEN] A type of uniparental disomy in which two different homologous chromosomes and their frequently different alleles are inherited from the same parent. { ,hed·ə,rō·dī'sō·mē }

heterodont [ANAT] Having teeth that are variable in shape and differentiated into incisors, canines, and molars. [INV ZOO] In bivalves, having two types of teeth on one valve

---

HETEROAUXIN

$$CH_2COOH$$

Structural formula of heteroauxin.

HETEROCOCCOLITH

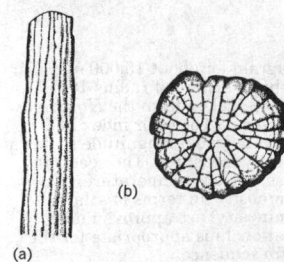

A heterococcolith whose crytals are arranged in a basket shape. (*A. McIntyre, Lamont-Doherty Geological Observatory or Columbia Unversity*)

HETEROCORALLIA

(a)

(b)

Heterocorallian skeleton. (*a*) External view. (*b*) Transverse thin section. (*From R. C. Moore, ed.,* Treatise on Invertebrate Paleontology, *pt. F, Geological Society of America, University of Kansas Press, 1956*)

which fit into depressions on the other valve. { 'hed·ə·rə‚dänt }

Heterodonta [INV ZOO] An order of bivalve mollusks in some systems of classification; hinge teeth are few in number and variable in form. { ‚hed·ə·rə'dän·tə }

Heterodontoidea [VERT ZOO] A suborder of sharks in the order Selachii which is represented by the single living genus *Heterodontus*. { ‚hed·ə·rō‚dän'tóid·ē·ə }

heteroduplex [GEN] A double-stranded deoxyribonucleic molecule comprising strands from different individuals. [CELL MOL] A double-stranded molecule of deoxyribonucleic acid in which the two strands show noncomplementary sections. { ‚hed·ə·rō'dü‚pleks }

heterodyne [ELECTR] To mix two alternating-current signals of different frequencies in a nonlinear device for the purpose of producing two new frequencies, the sum of and difference between the two original frequencies. { 'hed·ə·rə‚dīn }

heterodyne analyzer [ENG ACOUS] A type of constant-bandwidth analyzer in which the electric signal from a microphone beats with the signal from an oscillator, and one of the side bands produced by this modulation is then passed through a fixed filter and detected. { 'hed·ə·rə‚dīn 'an·ə‚līz·ər }

heterodyne conversion transducer See converter. { 'hed·ə·rə‚dīn kən‚vər·zhən tranz‚dü·sər }

heterodyne detector [ELECTR] A detector in which an unmodulated carrier frequency is combined with the signal of a local oscillator having a slightly different frequency, to provide an audio-frequency beat signal that can be heard with a loudspeaker or headphones; used chiefly for code reception. { 'hed·ə·rə‚dīn di'tek·tər }

heterodyne frequency [COMMUN] Either of the two new frequencies resulting from heterodyne action between the two input frequencies of a heterodyne detector. { 'hed·ə·rə‚dīn 'frē·kwən·sē }

heterodyne frequency meter [ELECTR] A frequency meter in which a known frequency, which may be adjustable or fixed, is heterodyned with an unknown frequency to produce a zero beat or an audio-frequency signal whose value is measured by other means. Also known as heterodyne wavemeter. { 'hed·ə·rə‚dīn 'frē·kwən·sē ‚mēd·ər }

heterodyne interference See heterodyne whistle. { 'hed·ə·rə‚dīn ‚in·tər'fir·əns }

heterodyne measurement [ELECTR] A measurement carried out by a type of harmonic analyzer which employs a highly selective filter, at a frequency well above the highest frequency to be measured, and a heterodyning oscillator. { 'hed·ə·rə‚dīn 'mezh·ər·mənt }

heterodyne modulator See mixer. { 'hed·ə·rə‚dīn 'mäj·ə‚lād·ər }

heterodyne oscillator [ELECTR] **1.** A separate variable-frequency oscillator used to produce the second frequency required in a heterodyne detector for code reception. **2.** See beat-frequency oscillator. { 'hed·ə·rə‚dīn 'äs·ə‚lād·ər }

heterodyne reception [ELECTR] Radio reception in which the incoming radio-frequency signal is combined with a locally generated rf signal of different frequency, followed by detection. Also known as beat reception. { 'hed·ə·rə‚dīn ri'sep·shən }

heterodyne repeater [ELECTR] A radio repeater in which the received radio signals are converted to an intermediate frequency, amplified, and reconverted to a new frequency band for transmission over the next repeater section. { 'hed·ə·rə‚dīn ri'pēd·ər }

heterodyne wavemeter See heterodyne frequency meter. { 'hed·ə·rə‚dīn 'wāv‚mēd·ər }

heterodyne whistle [COMMUN] A steady, high-pitched audio tone heard in an ordinary amplitude-modulation radio receiver under certain conditions when two signals that differ slightly in carrier frequency enter the receiver and heterodyne to produce an audio beat. Also known as heterodyne interference. { 'hed·ə·rə‚dīn 'wis·əl }

heteroecious [BIOL] Pertaining to forms that pass through different stages of a life cycle in different hosts. { ‚hed·ə'rē·shəs }

heteroerotism [PSYCH] Sexual desire directed away from one's self or sex. { ‚hed·ə·rō·ə'räd·ə‚siz·əm }

heterogamete [BIOL] A gamete that differs in size, appearance, structure, or sex chromosome content from the gamete of the opposite sex. Also known as anisogamete. { ‚hed·ə·rō'ga‚mēt }

heterogametic sex [GEN] That sex of some species which produces two or more different kinds of gametes that differ in their sex chromosome content. { ‚hed·ə·rō·gə'med·ik 'seks }

heterogamety [GEN] The production of different kinds of gametes by one sex of a species. { ‚hed·ə·rō'gam·əd·ē }

heterogamous [BIOL] Of or pertaining to heterogamy. { ‚hed·ə'räg·ə·məs }

heterogamy [BIOL] **1.** Alternation of a true sexual generation with a parthenogenetic generation. **2.** Sexual reproduction by fusion of unlike gametes. Also known as anisogamy. [BOT] Condition of producing two kinds of flowers. { ‚hed·ə'räg·ə·mē }

heterogeneity [BIOL] The condition or state of being different in kind or nature. [SCI TECH] The condition of a sample of matter that is composed of particles or aggregates of different substances of dissimilar composition. { ‚hed·ə·rə·jə'nē·əd·ē }

heterogeneous [CHEM] Pertaining to a mixture of phases such as liquid-vapor, or liquid-vapor-solid. [MATH] Pertaining to quantities having different degrees or dimensions. [SCI TECH] Composed of dissimilar or nonuniform constituents. { ‚hed·ə'räj·ə·nəs }

heterogeneous catalysis [CHEM] A chemical process in which the catalyst is in a separate phase; usually the reactants and products are in gaseous or liquid phases and the catalyst is a solid, and the catalytic reaction occurs on the surface of the solid. { ‚hed·ə·rə'jē·nē·əs kə'tal·ə·səs }

heterogeneous chemical reaction [CHEM] Chemical reaction system in which the reactants are of different phases; for example, gas with liquid, liquid with solid, or a solid catalyst with liquid or gaseous reactants. { ‚hed·ə·rə'jē·nē·əs 'kem·ə·kəl rē'ak·shən }

heterogeneous fluid [FL MECH] A fluid within which the density varies from point to point; for most purposes the atmosphere must be treated as heterogeneous, particularly with regard to the decrease of density with height. { ‚hed·ə·rə'jē·nē·əs 'flü·əd }

heterogeneous nucleation [PHYS CHEM] The formation of vapor bubbles on cavities or scratches of a surface bounding a superheated liquid. { ‚hed·ə·rə‚jē·nē·əs ‚nü·klē'ā·shən }

heterogeneous radiation [PHYS] Radiation having a number of different frequencies, different particles, or different particle energies. { ‚hed·ə·rə'jē·nē·əs ‚rād·ē'ā·shən }

heterogeneous reactor [NUCLEO] A nuclear reactor in which fissionable material and moderator are arranged in a regular pattern of discrete bodies with dimensions such that a nonhomogeneous medium is presented to neutrons. { ‚hed·ə·rə'jē·nē·əs rē'ak·tər }

heterogeneous reservoir [GEOL] Formation with two or more noncommunicating sand members, each possibly with different specific- and relative-permeability characteristics. { ‚hed·ə·rə'jē·nē·əs 'rez·əv‚wär }

heterogeneous ribonucleic acid [MOL BIO] A large molecule of ribonucleic acid that is believed to be the precursor of messenger ribonucleic acid. Abbreviated H-RNA. { ‚hed·ə·rə'jē·nē·əs ‚rī·bō·nü‚klē·ik 'as·əd }

heterogeneous strain [MECH] A strain in which the components of the displacement of a point in the body cannot be expressed as linear functions of the original coordinates. { ‚hed·ə·rə'jē·nē·əs 'strān }

Heterogeneratae [BOT] A class of brown algae distinguished by a heteromorphic alteration of generations. { ‚hed·ə·rō‚ji'ner·əd·ē }

heterogenesis [BIOL] Alternation of generations in a complete life cycle, especially the alternation of a dioecious generation with one or more parthenogenetic generations. { ‚hed·ə·rō'jen·ə·səs }

heterogenetic antigen See heterophile antigen. { ‚hed·ə·rō·jə‚ned·ik 'ant·i·jən }

heterogenite [MINERAL] CoO(OH) A black cobalt mineral, sometimes with some copper and iron, found in mammillary masses. Also known as stainierite. { ‚hed·ə·rə'räj·ə‚nīt }

heterogenous [BIOL] Not originating within the body of the organism. { ‚hed·ə·rə'räj·ə·nəs }

heterogeneous nuclear ribonuclear protein [CELL MOL] Any of a large family of proteins involved in the processing of pre-messenger ribonucleic acid (mRNA). { ¦hed·ə·rə¸jēn·ē·əs ¦nü·klē·ər ¦rī·bō¸nü·klē·ər 'prō¸tēn }

heterogenous vaccine [IMMUNOL] A vaccine derived from a source other than the patient. { ¦hed·ə'räj·ə·nəs vak'sēn }

Heterognathi [VERT ZOO] An equivalent name for Cypriniformes. { ¸hed·ə'räg·nə¸thī }

heterogony [BIOL] **1.** Alteration of generations in a complete life cycle, especially of a dioecious and hermaphroditic generation. **2.** *See* allometry. [BOT] Having heteromorphic perfect flowers with respect to lengths of the stamens or styles. { ¸hed·ə'räg·ə·nē }

heterograft [IMMUNOL] A tissue or organ obtained from an animal of one species and transplanted to the body of an animal of another species. Also known as heterologous graft. { 'hed·ə·rō¸graft }

heterohemolysin [IMMUNOL] Hemolytic amboceptor against the erythrocytes of a species different from that used to obtain the amboceptor. { ¦hed·ə·rō·hə'mäl·ə·sən }

heterojunction [ELECTR] The boundary between two different semiconductor materials, usually with a negligible discontinuity in the crystal structure. { ¦hed·ə·rō'jəŋk·shən }

heterojunction bipolar transistor [ELECTR] A bipolar transistor that has two or more materials making up the emitter, base, and collector regions, giving it a much higher maximum frequency than a silicon bipolar transistor. Abbreviated HBT. { ¦hed·ə·rə¸jəŋk·shən 'bī¸pōl·ər tran¸zis·tər }

heterojunction field-effect transistor *See* high-electron-mobility transistor. { ¦hed·ə·rə¸jəŋk·shən 'fēld i¸fekt tran¸zis·tər }

heterokaryon [GEN] Cell with two or more nuclei originating from different cell types or species. [MYCOL] A bi- or multinucleate cell having genetically different kinds of nuclei. { ¦hed·ə·rō'kar·ē¸än }

heterokaryosis [MYCOL] The condition of a bi- or multinucleate cell having nuclei of genetically different kinds. { ¸hed·ə·rō¸kar·ē'ō·səs }

heterokaryotype [GEN] A karyotype that is heterozygous for a chromosome mutation. { ¸hed·ə·rə'kar·ē·ə¸tīp }

heterokont [BIOL] An individual, especially among certain algae, having unequal flagella. { 'hed·ə·rō¸känt }

heterolactic fermentation [MICROBIO] A type of lactic acid fermentation by which small yields of lactic acid are produced and much of the sugar is converted to carbon dioxide and other products. { ¦hed·ə·rō'lak·tik ¸fər·mən'tā·shən }

heterolalia [PSYCH] Saying one thing and meaning another as by an unconscious mechanism or in motor aphasia. { ¦hed·ə·rō'lal·ē·ə }

heterolateral [ANAT] Of, pertaining to, or located on the opposite side. { ¸hed·ə·rō'lad·ə·rəl }

heterolecithal [CELL MOL] Of an egg, having the yolk distributed unevenly throughout the cytoplasm. { ¸hed·ə·rō'les·ə·thəl }

heterologous graft *See* heterograft. { ¸hed·ə'räl·ə·gəs 'graft }

heterologous stimulus [PHYSIO] A form of energy capable of exciting any sensory receptor or form of nervous tissue. { ¸hed·ə'räl·ə·gəs 'stim·yə·ləs }

heterologous tumor [MED] A neoplasm composed of tissues that differ from those of the organ at the site of the tumor. { ¸hed·ə'räl·ə·gəs 'tü·mər }

heterolysis *See* heterolytic cleavage. { ¸hed·ə'räl·ə·səs }

heterolytic bond dissociation energy [PHYS CHEM] The change in enthalpy that occurs when a chemical bond undergoes heterolytic cleavage. { ¸hed·ə·rō¦lid·ik ¸bänd ¸dis·ə¸sō·sē'ā·shən ¸en·ər·jē }

heterolytic cleavage [ORG CHEM] The breaking of a single (two-electron) chemical bond in which both electrons remain on one of the atoms. Also known as heterolysis. { ¸hed·ə·rō¦lid·ik 'klēv·ij }

heteromedusoid [INV ZOO] A styloid type of sessile gonophore. { ¦hed·ə·rō'med·yə¸sóid }

Heteromera [INV ZOO] The equivalent name for Tenebrionoidea. { ¸hed·ə¦räm·ə·rə }

heteromerous [BOT] Of a flower, having one or more whorls made up of a different number of members than the remaining whorls. { ¦hed·ə¦räm·ə·rəs }

heterometaplasia [MED] Change in the character of an autograft. { ¦hed·ə·rō¦med·ə'plā·zhə }

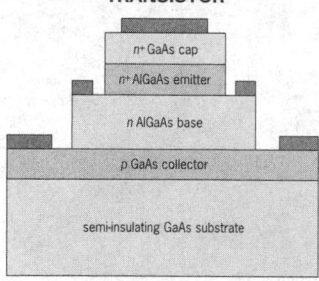

HETEROJUNCTION BIPOLAR TRANSISTOR

n+ GaAs cap
n+ AlGaAs emitter
n AlGaAs base
p GaAs collector
semi-insulating GaAs substrate

Materials composition for a heterojunction bipolar transistor.

Heteromi [VERT ZOO] An equivalent name for Notacanthiformes. { ¸hed·ə'rō¸mī }

heteromixis [MYCOL] In Fungi, sexual reproduction which involves the fusion of genetically different nuclei, each from a different thallus. { ¸hed·ə·rō'mik·səs }

heteromorphic [CELL MOL] Having synoptic or sex chromosomes that differ in size or form. [MED] Differing from the normal in size or morphology. [ZOO] Having a different form at each stage of the life history. { ¸hed·ə·rō'mór·fik }

heteromorphic transformation [THERMO] A change in the values of the thermodynamic variables of a system in which one or more of the component substances also undergo a change of state. { ¸hed·ə·rə¦mór·fik ¸tranz·fər'mā·shən }

heteromorphite [MINERAL] $Pb_7Sb_8S_{19}$ An iron black, monoclinic mineral consisting of lead antimony sulfide. { ¦hed·ə·rō'mór¸fīt }

heteromorphosis [BIOL] Regeneration of an organ or part that differs from the original structure at the site. [EMBRYO] Formation of an organ at an abnormal site. Also known as homoeosis. { ¸hed·ə·rō¦mór·fə·səs }

Heteromyidae [VERT ZOO] A family of the mammalian order Rodentia containing the North American kangaroo mice and the pocket mice. { ¸hed·ə'räm·ə¸dē }

Heteromyinae [VERT ZOO] The spiny pocket mice, a subfamily of the rodent family Heteromyidae. { ¸hed·ə·rō'mī·ə¸nē }

Heteromyota [INV ZOO] A monospecific order of wormlike animals in the phylum Echiurida. { ¸hed·ə·rō'mī·ə·tə }

Heterondontidae [VERT ZOO] The Port Jackson sharks, a family of aberrant modern elasmobranchs in the suborder Heterodontoidea. { ¸hed·ə·rō¦dänt·ə¸dē }

Heteronemertini [INV ZOO] An order of the class Anopla; individuals have a middorsal blood vessel and a body wall composed of three muscular layers. { ¦hed·ə·rō¸nem·ər'tī·nī }

Heteroneura [INV ZOO] A suborder of Lepidoptera; individuals are characterized by fore- and hindwings that differ in shape and venation and by sucking mouthparts. { ¸hed·ə·rō'nür·ə }

heteronuclear culture [CELL MOL] A cell culture showing a marked variation in chromosome complement among the cells. { ¦hed·ə·rə¸nü·klē·ər 'kəl·chər }

heteronuclear molecule [CHEM] A diatomic molecule having atoms of different elements. { ¦hed·ə·rə¸nü·klē·ər 'mäl·ə¸kyül }

heteropelmous [VERT ZOO] Having bifid flexor tendons of the toes. { ¦hed·ə·rō¦pel·məs }

heterophagic vacuole *See* food vacuole. { ¸hed·ə·rō'faj·ik 'vak·yə·wōl }

heterophile agglutination test [PATH] A test for the presence of heterophile antibodies in the serum produced in infectious mononucleosis; agglutination of sheep red cells is a positive test. Also known as heterophile antibody test; Paul-Bunnell test. { 'hed·ə·rə¸fīl ə¸glüd·ən'ā·shən ¸test }

heterophile antibody [IMMUNOL] Substance that will react with heterophile antigen; found in the serum of patients with infectious mononucleosis. { 'hed·ə·rə¸fīl 'an·ti·bäd·ē }

heterophile antibody test *See* heterophile agglutination test. { 'hed·ə·rə¸fīl 'an·ti¸bäd·ē ¸test }

heterophile antigen [IMMUNOL] A substance that occurs in unrelated species of animals but has similar serologic properties among them. Also known as heterogenetic antigen. { 'hed·ə·rə¸fīl 'ant·i·jən }

heterophile leukocyte [HISTOL] A neutrophil of vertebrates other than humans. { 'hed·ə·rə¸fīl 'lü·kə¸sīt }

heterophyiasis [MED] Presence of the minute intestinal fluke *Heterophyes heterophyes* in the small intestine of humans. { ¦hed·ə·rō'fī·ə·səs }

Heterophyllidae [PALEON] The single family of the extinct cnidarian order Heterocorallia. { ¦hed·ə·rō'fil·ə¸dē }

heterophyllous [BOT] Having more than one form of foliage leaves on the same plant or stem. { ¸hed·ə'räf·ə·ləs }

heterophyte [BOT] A plant that depends upon living or dead plants or their products for food materials. { 'hed·ə·rə¸fīt }

Heteropiidae [INV ZOO] A family of calcareous sponges in the order Sycettida. { ¦hed·ə·rō'pī·ə¸dē }

heteroplasia [MED] **1.** The presence of a tissue in an abnormal location. **2.** A process whereby tissues are displaced to or developed in locations foreign to their normal habitats. { ¸hed·ə·rō'plā·zhə }

heteroplastic [BIOL] Pertaining to transplantation between individuals of different species within the same genus. { ‚hed·ə·rō‚plas·tik }

heteroplastidy [BOT] The condition of having two kinds of plastids, chloroplasts and leukoplasts. { ‚hed·ə·rō‚plas·təd·ē }

heteroploidy [GEN] A chromosome complement in which one or more chromosomes, or parts of chromosomes, are present in number different from the numbers of the rest. { ‚hed·ə·rō‚plȯid·ē }

heteropolar bond [PHYS CHEM] A covalent bond whose total dipole moment is not 0. { ‚hed·ə·rə‚pō·lər ‚bänd }

heteropolar generator [ELECTROMAG] A generator whose active conductors successively pass through magnetic fields of opposite direction. { ‚hed·ə·rō‚pō·lər ‚jen·ə·rā‚d·ər }

heteropoly acid [INV ZOO] Complex acids of metals, whose specific gravity is greater than 4, with phosphoric acid; an example is phosphomolybdic acid. { ‚hed·ə‚räp·ə·lē ‚as·əd }

heteropoly compound [INV ZOO] Polymeric compounds of molybdates with anhydrides of other elements such as phosphorus; the yellow precipitate $(NH_4)_3P(Mo_3O_{10})_4$ is such a compound. { ‚hed·ə‚räp·ə·lē ‚käm‚paůnd }

heteropolymer [CHEM] A compound comprising two or more molecules that are different from one another. { ‚hed·ə·rə‚päl·ə·mər }

heteropolysaccharide [BIOCHEM] A polysaccharide which is a polymer consisting of two or more different monosaccharides. { ‚hed·ə·rō‚päl·ē‚sak·ə‚rīd }

Heteroporidae [INV ZOO] A family of trepostomatous bryozoans in the order Cyclostomata. { ‚hed·ə·rō‚pór·ə‚dē }

Heteroptera [INV ZOO] The equivalent name for Hemiptera. { ‚hed·ə‚räp·tə·rə }

heteropycnosis [CYTOL] Differential condensation of certain chromosomes, such as sex chromosomes, or chromosome parts. { ‚hed·ə·rō‚pik'nō·səs }

heterosexuality [PSYCH] Having sexual feeling toward members of the opposite sex. { ‚hed·ə·rō‚sek·shə'wal·əd·ē }

heterosis [GEN] The increase in size, yield, and performance found in some hybrid plants and animals, especially if the parents are from inbred stocks. Also known as hybrid vigor. { ‚hed·ə'rō·səs }

heterosite [MINERAL] A mineral composed of phosphate of iron and manganese; it is isomorphous with purpurite. { 'hed·ə·rə‚sīt }

Heterosomata [VERT ZOO] The equivalent name for Pleuronectiformes. { ‚hed·ə·rō'sō·məd·ə }

Heterosoricinae [PALEON] A subfamily of extinct insectivores in the family Soricidae distinguished by a short jaw and hedgehoglike teeth. { ‚hed·ə·rō·sə'ris·ə‚nē }

heterosphere [METEOROL] The upper portion of a two-part division of the atmosphere (the lower portion is the homosphere) according to the general homogeneity of atmospheric composition; characterized by variation in composition, and in mean molecular weight of constituent gases; starts at 50–62 miles (80–100 kilometers) above the earth and therefore closely coincides with the ionosphere and the thermosphere. { 'hed·ə·rə‚sfir }

Heterospionidae [INV ZOO] A monogeneric family of spioniform worms found in shallow and abyssal depths of the Atlantic and Pacific oceans. { ‚hed·ə·rō‚spī'än·ə‚dē }

heterospory [BOT] Development of more than one type of spores, especially relating to the microspores and megaspores in ferns and seed plants. { ‚hed·ə'räs·pə·rē }

heterostatic [ELEC] Pertaining to the measurement of one electrostatic potential by means of a different potential. { ‚hed·ə·rō'stad·ik }

heterostatic connection [ELEC] An arrangement of a quadrant electrometer in which the vane is maintained at a high potential with respect to one of the quadrant pairs and the deflection of the vane is linearly proportional to the unknown voltage applied across the quadrant pairs. { ‚hed·ə·rə‚stad·ik kə'nek·shən }

heterostemony [BOT] Presence of two or more different types of stamens in the same flower. { ‚hed·ə·rō'stem·ə·nē }

Heterostraci [PALEON] An extinct group of ostracoderms, or armored, jawless vertebrates; armor consisted of bone lacking cavities for bone cells. { ‚hed·ə'räs·trə‚sī }

heterostyly [BOT] Condition or state of flowers having unequal styles. { ‚hed·ə'räst·əl·ē }

heterosuggestion [PSYCH] A suggestion from a source outside an individual's mind. { ‚hed·ə·rō·səg'jes·chən }

Heterotardigrada [INV ZOO] An order of the tardigrades exhibiting wide morphologic variations. { ‚hed·ə·rō‚tär'dig·rə·də }

heterotaxia [ANAT] The reversed polarity of one or more individual organs with respect to the left-right axis. { ‚hed·ə·rə'tak·sē·ə }

heterothallic [BOT] Pertaining to a mycelium with genetically incompatible hyphae, therefore requiring different hyphae to form a zygospore; refers to fungi and some algae. { ‚hed·ə·rō'thal·ik }

heterotherm [ECOL] An animal that is endothermic part of the time but can reduce metabolic heat production and lower body temperature when conservation of food energy supplies is necessary. { 'hed·ə·rə‚thərm }

heterotic superstring theory [PART PHYS] The most promising version of superstring theory, with only closed strings, 10 dimensions, and 496 massless gauge bosons. { ‚hed·ə‚räd·ik 'sü·pər‚striŋ ‚thē·ə·rē }

heterotopia [ECOL] An abnormal habitat. [MED] Displacement of an organ or other body part from its natural position. { ‚hed·ə·rō'tō·pē·ə }

heterotopic [BIOL] Pertaining to transplantation of tissue from one site to another on the same organism. [MED] Occurring in an abnormal anatomic location. { ‚hed·ə·rō‚täp·ik }

heterotopic epithelium [MED] Intestinal epithelium and goblet cells occurring in the stomach. { ‚hed·ə·rō‚täp·ik ‚ep·ə'thē·lē·əm }

heterotopic faces [ORG CHEM] On molecules, faces of double bonds where addition gives rise to isomeric structures. { ‚hed·ə·rō‚täp·ik 'fās·əz }

heterotopic ligands [CHEM] Constitutionally identical ligands whose separate replacement by a different ligand gives rise to isomeric structures. { ‚hed·ə·rō‚täp·ik 'līg·ənz }

heterotopic pregnancy [MED] Double pregnancy with one fetus within and the other outside the uterus. { ‚hed·ə·rō‚täp·ik 'preg·nən·sē }

heterotopic transplantation [BIOL] A graft transplanted to an abnormal anatomical location on the host. { ‚hed·ə·rō‚täp·ik ‚tranz·plən'tā·shən }

Heterotrichida [INV ZOO] A large order of large ciliates in the protozoan subclass Spirotrichia; buccal ciliature is well developed and some species are pigmented. { ‚hed·ə·rō'trik·ə·də }

Heterotrichina [INV ZOO] A suborder of the protozoan order Heterotrichida. { ‚hed·ə·rō'trik·ə·nə }

heterotrichous [BOT] In certain algae, a body that is divided into both prostrate and erect parts. { ‚hed·ə‚rä·trə·kəs }

heterotroph [BIOL] An organism that obtains nourishment from the ingestion and breakdown of organic matter. { 'hed·ə·rō‚träf }

heterotrophic ecosystem [ECOL] An ecosystem that depends upon preformed organic matter that is imported from autotrophic ecosystems elsewhere. { ‚hed·ə·rə‚träf·ik 'ek·ō‚sis·təm }

heterotrophic effect [BIOCHEM] The interaction between different ligands, such as the effect of an inhibitor or activator on the binding of a substrate by an enzyme. { ‚hed·ə·rō‚träf·ik i'fekt }

heterotrophic succession [ECOL] A type of ecological succession that involves decomposer organisms. { ‚hed·ə·rə‚träf·ik sək'sesh·ən }

heterotropia See strabismus. { ‚hed·ə·rō'trō·pē·ə }

heterotropic enzyme [BIOCHEM] A type of allosteric enzyme in which a small molecule other than the substrate serves as the allosteric reflector. { ‚hed·ə·rō‚träp·ik 'en‚zīm }

heteroxenous [BIOL] Requiring more than one host to complete a life cycle. { ‚hed·ə‚räk·sə·nəs }

heterozooid [INV ZOO] Any of the specialized, nonfeeding zooids in a bryozoan colony. { ‚hed·ə·rō'zō‚ȯid }

heterozygote [GEN] An individual that has different alleles at one or more loci and therefore produces gametes of two or more different kinds with respect to their loci. { ‚hed·ə·rō'zī‚gōt }

heterozygote advantage See heterozygote superiority. { ‚hed·ə·rō‚zī‚gōt ad'van·tij }

heterozygote superiority [GEN] The greater fitness of an

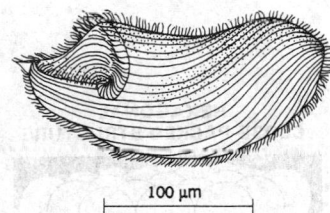

HETEROTRICHIDA

100 μm

Climacostomum, an example of a heterotrich.

organism that is heterozygous at a given genetic locus as compared with either homozygote. Also known as heterozygote advantage. { ‚hed·ə·rō‚zī‚gōt sü‚pir·ē'ôr·əd·ē }

heterozygous [GEN] Of or pertaining to a heterozygote. { ‚hed·ə·rō‚zī'gəs }

HETP *See* height equivalent of theoretical plate.

Hettangian [GEOL] A stage of Lower Jurassic geologic time. { he'tan·jē·ən }

heulandite [MINERAL] $CaAl_2Si_6O_{16}·5H_2O$ A zeolite mineral that crystallizes in the monoclinic system; often occurs as foliated masses or in crystal form in cavities of decomposed basic igneous rocks. { 'hyü·lən‚dīt }

Heulinger equations [PHYS] Equations which relate the values of various quantities, such as the Hall coefficient, thermoelectric power, electrical resistivity, and thermal conductivity, in isothermal and adiabatic thermoelectric and thermomagnetic effects. { 'hói·liŋ·ər i‚kwā·zhənz }

heuristic algorithm *See* dynamic algorithm. { hyü'ris·tik 'al·gə‚rith·əm }

heuristic method [MATH] A method of solving a problem in which one tries each of several approaches or methods and evaluates progress toward a solution after each attempt. { hyü'ris·tik 'meth·əd }

heuristic program [COMPUT SCI] A program in which a computer tries each of several methods of solving a problem and judges whether the program is closer to solution after each attempt. Also known as heuristic routine. { hyü'ris·tik 'prō·grəm }

heuristic routine *See* heuristic program. { hyü'ris·tik rü'tēn }

heuristics [PSYCH] The study of the mental processes involved in problem solving. { hyü'ris·tiks }

Heusler alloy [MET] Any of a group of ferromagnetic nonferrous alloys typically composed of 18–25% manganese, 10–25% aluminum, and the balance copper. { 'hóis·lər ‚al‚ói }

Hevea [BOT] The rubber tree genus of the order Euphoriales from which the largest volumes of latex are harvested for use in the manufacture of natural rubber. { 'hē·vē·ə }

hevea rubber [MATER] Rubber made from latex obtained from the rubber tree (*Hevea brasiliensis*); used for electrical insulation. { 'hē·vē·ə ‚rəb·ər }

Hevelian halo [OPTICS] A faint, white halo with an angular radius of 90°, centered on the sun or moon, and only occasionally seen; it is a member of the class of halos reported but not yet fully explained. Also known as halo of Hevelius; halo of 90°. { he'vāl·yən 'hā·lō }

Hevelius's parhelia [OPTICS] Bright spots infrequently observed on the parhelic circle halfway around from the sun to the anthelion; these two brighter areas on the parhelic circle are probably a result of superposition of luminosity of the parhelic circle and the Hevelian halo. { he'vel·yəs·əz pär 'hēl·yə }

hewettite [MINERAL] $CaV_6O_{16}·9H_2O$ A deep-red mineral composed of hydrated calcium vanadate; found in silky orthorhombic crystal aggregates in Colorado, Utah, and Peru. { 'hyü·ə‚tīt }

hexacanth [INV ZOO] Having six hooks; refers specifically to the embryo of certain tapeworms. { 'hek·sə‚kanth }

hexachlorobenzene [ORG CHEM] C_6Cl_6 Colorless, needlelike crystals with a melting point of 231°C; used in organic synthesis and as a fungicide. Abbreviated HCB. { ‚hek·sə‚klór·ō'ben‚zēn }

hexachlorobutadiene [ORG CHEM] $Cl_2C:CClCCl:CCl_2$ A colorless liquid with mild aroma, boiling at 210–220°C; soluble in alcohol and ether, insoluble in water; used as solvent, heat-transfer liquid, and hydraulic fluid. { ‚hek·sə‚klór·ō‚byüd·ə'dī‚ēn }

1,2,3,4,5,6-hexachlorocyclohexane [ORG CHEM] $C_6H_6Cl_6$ A white or yellow powder or flakes with a musty odor; a systemic insecticide toxic to flies, cockroaches, aphids, and boll weevils. Abbreviated TBH. { ‚wən‚tü‚thrē‚fór‚fīv‚siks ‚hek·sə‚klór·ō‚sī·klō'hek‚sān }

hexachloroethane [ORG CHEM] Cl_3CCCl_3 Colorless crystals with a camphorlike odor, melting point 185°C, toxic; used in organic synthesis, as a retarding agent in fermentation, and as a rubber accelerator. { ‚hek·sə‚klór·ō'e‚thān }

hexachlorophene [ORG CHEM] $(C_6HCl_3OH)_2CH_2$ A white powder melting at 161°C; soluble in alcohol, ether, acetone, and chloroform, insoluble in water; bacteriostat used in

antiseptic soaps, cosmetics, and dermatologicals. { ‚hek·sə'klór·ə‚fen }

hexachloropropylene [ORG CHEM] $CCl_3CCl:CCl_2$ A water-white liquid boiling at 210°C, soluble in alcohol, ether, and chlorinated solvents, insoluble in water; used as a solvent, plasticizer, and hydraulic fluid. { ‚hek·sə‚klór·ō'prō·pə‚lēn }

hexacontane [ORG CHEM] $C_{60}H_{122}$ Solid, saturated hydrocarbon of the methane series; melts at 101°C. { ‚hek·sə'kän‚tān }

Hexacorallia [INV ZOO] The equivalent name for Zoantharia. { ‚hek·sə·kə'ral·ē·ə }

hexacosane [ORG CHEM] $C_{26}H_{54}$ Saturated hydrocarbon of the methane series; colorless crystals melting at 57°C. { ‚hek·sə'kō‚sān }

hexacosanoic acid *See* cerotic acid. { ‚hek·sə‚kō·sə'nō·ik 'as·əd }

hexactin [INV ZOO] A spicule, especially in Porifera, having six equal rays at right angles to each other. { hek'sak‚tən }

Hexactinellida [INV ZOO] A class of the phylum Porifera which includes sponges with a skeleton made up basically of hexactinal siliceous spicules. { ‚hek‚sak·tə'nel·ə·də }

Hexactinosa [INV ZOO] An order of sponges in the subclass Hexasterophora; parenchymal megascleres form a rigid framework and consist of simple hexactins. { hek‚sak·tə'nō·sə }

hexad axis [CRYSTAL] A rotation axis whose multiplicity is equal to 6. { 'hek‚sad ‚ak·səs }

***n*-hexadecane** [ORG CHEM] $C_{16}H_{34}$ A colorless, solid hydrocarbon, melting point 20°C; a standard reference fuel in determining the ignition quality (cetane number) of diesel fuels. Also known as cetane. { ‚en‚hek·sə·de‚kān }

1-hexadecene [ORG CHEM] $CH_3(CH_2)_{13}CH:CH_2$ A colorless liquid made by treating cetyl alcohol with phosphorus pentoxide; boils at 274°C; soluble in organic solvents such as alcohol, ether, and petroleum; used as an intermediate in organic synthesis. { ‚wən ‚hek·sə·de‚sēn }

hexadecimal [MATH] Pertaining to a number system using the base 16. Also known as sexadecimal. { ‚hek·sə'des·məl }

hexadecimal notation [COMPUT SCI] A notation in the scale of 16, using decimal digits 0 to 9 and six more digits that are sometimes represented by *A, B, C, D, E,* and *F.* { ‚hek·sə'des·məl nō'tā·shən }

hexadecimal number system [MATH] A digital system based on powers of 16, as compared with the use of powers of 10 in the decimal number system. Also known as sexadecimal number system. { ‚hek·sə'des·məl 'nəm·bər ‚sis·təm }

hexadentate ligand [INV ZOO] A chelating agent having six groups capable of attachment to a metal ion. Also known as sexadentate ligand. { ‚hek·sə'den‚tāt 'līg·ənd }

hexadiene [ORG CHEM] C_6H_{10} A group of unsaturated hydrocarbons with two double bonds; some members of the group are 1,4-hexadiene, 1,5-hexadiene, and 2,4-hexadiene. { ‚hek·sə'dī‚ēn }

hexafoil [MATH] A multifoil consisting of six congruent arcs of a circle arranged around a regular hexagon. { 'hek·sə‚fóil }

hexagon [MATH] A six-sided polygon. { 'hek·sə‚gän }

hexagonal boron nitride [MATER] A synthetic material composed of boron and nitrogen with the atoms combined in a hexagonal lattice such that it has a structure similar to graphite. Also known as white graphite. { hek‚sag·ən·əl ‚bór·än 'nī‚trīd }

hexagonal close-packed structure [CRYSTAL] Close-packed crystal structure characterized by the regular alternation of two layers; the atoms in each layer lie at the vertices of a series of equilateral triangles, and the atoms in one layer lie directly above the centers of the triangles in neighboring layers. Abbreviated hcp structure. { hek'sag·ə·nəl ‚klōs ‚pakt 'strək·chər }

hexagonal column [METEOROL] One of the many forms in which ice crystals are found in the atmosphere; this crystal habit is characterized by hexagonal cross-section in a plane perpendicular to the long direction (principal axis, optic axis, or *c* axis) of the columns; it differs from that found in hexagonal platelets only in that environmental conditions have favored growth along the principal axis rather than perpendicular to that axis. { hek'sag·ə·nəl 'käl·əm }

hexagonal-head bolt [DES ENG] A standard wrench head bolt with a hexagonal head. { hek'sag·ə·nəl ‚hed ‚bōlt }

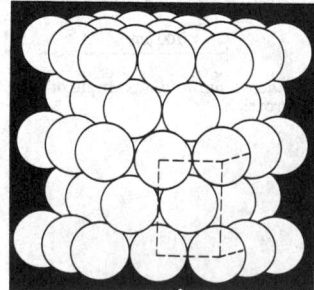

HEXAGONAL CLOSE-PACKED STRUCTURE

Hexagonal close packing of spheres that simulates the arrangement of the atoms in a crystal that is a hexagonal close-packed structure.

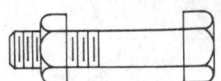

HEXAGONAL-HEAD BOLT

A drawing of a heavy-weight standard bolt that has a hexagonal head.

hexagonal lattice [CRYSTAL] A Bravais lattice whose unit cells are right prisms with hexagonal bases and whose lattice points are located at the vertices of the unit cell and at the centers of the bases. { hek′sag·ə·nəl ′lad·əs }

hexagonal nipple [DES ENG] A nipple for joining pipe with a hexagonal configuration around the center of the exterior surface to permit tightening with a spanner. { hek¦sag·ən·əl ′nip·əl }

hexagonal nut [DES ENG] A plain nut in hexagon form. { hek′sag·ə·nəl ′nət }

hexagonal platelet [METEOROL] A small ice crystal of the hexagonal tabular form; the distance across the crystal from one side of the hexagon to the opposite side may be as large as about 1 millimeter, and the thickness perpendicular to this dimension is of the order of one-tenth as great; this crystal form is usually formed at temperatures of −10 to −20°C by sublimation; at higher temperatures the apices of the hexagon grow out and develop dendritic forms. { hek′sag·ə·nəl ′plāt·lət }

hexagonal system [CRYSTAL] A crystal system that has three equal axes intersecting at 120° and lying in one plane; a fourth, unequal axis is perpendicular to the other three. { hek′sag·ə·nəl ′sis·təm }

hexahedrite [GEOL] An iron meteorite composed of single crystals or aggregates of kamacite, usually containing 4–6% nickel in the metal phase. { ‚hek·sə′he‚drīt }

hexahedron [MATH] A polyhedron with six faces. { ‚hek·sə′hē·drən }

hexahydric alcohol [ORG CHEM] A member of the mannitol-sorbitol-dulcitol sugar group; isomer of $C_6H_8(OH)_6$. { ¦hek·sə¦hī·drik ′al·kə‚hól }

hexahydrite [MINERAL] $MgSO_4·6H_2O$ A white or greenish-white monoclinic mineral composed of hydrous magnesium sulfate. { ‚hek·sə′hī‚drīt }

hexahydrotoluene See methyl cyclohexane. { ¦hek·sə¦hī·drō′täl·yə‚wēn }

n-hexaldehyde [ORG CHEM] $CH_3(CH_2)_4CHO$ Colorless liquid with sharp aroma, boiling at 128.6°C; used as an intermediate for plasticizers, dyes, insecticides, resins, and rubber chemicals. { ¦en ‚heks′al·də‚hīd }

hexametapol [ORG CHEM] $C_6H_{18}N_3OP$ A liquid used as a solvent in organic synthesis, as a deicing additive for jet engine fuel, and as an insect pest chemosterilant and chemical mutagen. { ‚hek·sə′med·ə‚pól }

hexamethonium [PHARM] One of a homologous series of polymethylene bis(trimethylammonium) ions, of the general formula $[(CH_3)_3N(CH_2)_6N(CH_3)_3]^{2+}$, in which n is 6; possesses potent ganglion-blocking action, effecting reduction in blood pressure; used clinically as a salt, commonly bromide or iodide. { ‚hek·sə·mə′thō·nē·əm ′klór‚īd }

hexamethylene See cyclohexane. { ‚hek·sə′meth·ə‚lēn }

hexamethylenediamine [ORG CHEM] $H_2N(CH_2)_6NH_2$ Colorless solid boiling at 205°C; slightly soluble in water, alcohol, and ether; used to make nylon and other high polymers. { ‚hek·sə′meth·ə‚lēn′dī·ə‚mēn }

hexamethylene tetramine See cystamine. { ‚hek·sə′meth·ə‚lēn ′te·trə‚mēn }

hexamethylphosphoric triamide See bempa. { ‚hek·sə‚meth·əl′fás′fór·ik trī′am·əd }

Hexanchidae [VERT ZOO] The six- and seven-gill sharks, a group of aberrant modern elasmobranchs in the suborder Notidanoidea. { ‚hek¦saŋ·kə‚dē }

hexane [ORG CHEM] C_6H_{14} Water-insoluble, toxic, flammable, colorless liquid with faint aroma; forms include: n-hexane, a straight-chain compound boiling at 68.7°C and used as a solvent, paint diluent, alcohol denaturant, and polymerization-reaction medium; isohexane, a mixture of hexane isomers boiling at 54–61°C and used as a solvent and freezing-point depressant; and neohexane. { ′hek‚sān }

1,6-hexanediol [ORG CHEM] $HO(CH_2)_6OH$ A crystalline substance, soluble in water and alcohol; used in gasoline refining, as an intermediate in nylon manufacturing, and in making polyesters and polyurethanes. { ¦wən ‚siks ‚hek‚sān′dī‚ól }

hexanitrodiphenyl amine [ORG CHEM] $(NO_2)_3C_6H_2NHC_6H_2(NO_2)_3$ Explosive, yellow solid melting at 238–244°C; insoluble in water, ether, alcohol, or benzene; soluble in alkalies and acetic and nitric acids; used as an explosive and in potassium analysis. { ‚hek·sə‚nī·trō·dī′fen·əl ′am‚ēn }

hexapetalous [BOT] Having or being a perianth comprising six petaloid divisions. { ¦hek·sə′ped·əl·əs }

hexaphenylethane [ORG CHEM] $(C_6H_5)_3CC(C_6H_5)_3$ The dimer of triphenylmethyl radical. { ¦hek·sə¦fen·əl′eth‚ān }

hexapod [CONT SYS] A robot that uses six leglike appendages to stride over a surface. { ′hek·sə‚päd }

Hexapoda [INV ZOO] An equivalent name for Insecta. { hek′säp·əd·ə }

hexaprismane [CHEM] $C_{12}H_{12}$ A highly strained saturated hydrocarbon cage structure which consists of two flat cyclohexanes fused by six cyclobutanes. { ‚hek·sə′priz‚mān }

hexaster [INV ZOO] A type of hexactin with branching rays that form star-shaped figures. { ′hek‚sas·tər }

Hexasterophora [INV ZOO] A subclass of sponges of the class Hexactinellida in which parenchymal microscleres are typically hexasters. { ‚hek‚sas·tə′räf·ə·rə }

hexatic phase [PHYS] A phase of matter that is intermediate between the normal solid and the isotropic liquid phases, and corresponds to a two-dimensional fluid with sixfold orientational order but no positional order. { hek¦sad·ik ′fāz }

hexenbesen See witches′-broom disease. { ′heks·ən‚baz·ən }

1-hexene [ORG CHEM] $CH_3(CH_2)_3HC:CH_2$ Colorless, olefinic hydrocarbon boiling at 64°C; soluble in alcohol, acetone, ether, and hydrocarbons, insoluble in water; used as a chemical intermediate and for resins, drugs, and insecticides. Also known as hexylene. { ¦wən ′hek‚sēn }

hex nut [DES ENG] A nut in the shape of a hexagon. { ′heks ‚nət }

hexobarbital [PHARM] $C_{12}H_{16}N_2O_3$ A sedative and hypnotic of short duration of action; used also, as the sodium derivative, to induce surgical anesthesia. { ¦hek·sə′bär·bə‚tól }

hexoctahedron [CRYSTAL] A cubic crystal form that has 48 equal triangular faces, each of which cuts the three crystallographic axes at different distances. { ‚hek′säk·tə‚hē·drən }

hexode [ELECTR] A six-electrode electron tube containing an anode, a cathode, a control electrode, and three additional electrodes that are ordinarily grids. { ′hek‚sōd }

hexoglycan [BIOCHEM] A polysaccharide that yields hexose monosaccharides on hydrolysis. { ‚hek·sə′glī‚kan }

hexokinase [BIOCHEM] Any enzyme that catalyzes the phosphorylation of hexoses. { ¦hek·sō′kī‚nās }

hexomino [MATH] One of the 35 plane figures that can be formed by joining six unit squares along their sides. { hek′säm·ə‚nō }

hexone See methyl isobutyl ketone. { ′hek‚sōn }

hexosamine [BIOCHEM] A primary amine derived from a hexose by replacing the hydroxyl with an amine group. { ‚hek′säs·ə‚mēn }

hexosaminidase A [BIOCHEM] An enzyme which catalyzes the hydrolysis of the N-acetylgalactosamine residue from certain gangliosides. { hek‚säs·ə′min·ə‚dās ′ā }

hexosan See hemicellulose. { ′hek·sō‚san }

hexose [BIOCHEM] Any monosaccharide that contains six carbon atoms in the molecule. { ′hek‚sōs }

hexose diphosphate pathway See glycolytic pathway. { ′hek‚sōs dī′fäs‚fāt ′path‚wā }

hexose monophosphate cycle [BIOCHEM] A pathway for carbohydrate metabolism in which one molecule of hexose monophosphate is completely oxidized. { ′hek‚sōs ¦män·ō¦fäs‚fāt ′sī·kəl }

hexose phosphate [BIOCHEM] Any one of the phosphoric acid esters of a hexose, notably glucose, formed during the metabolism of carbohydrates by living organisms. { ′hek‚sōs ′fäs‚fāt }

hextetrahedron [CRYSTAL] A 24-faced form of crystal in the tetrahedral group of the isometric system. { ‚heks‚te·trə′hē·drən }

hexulose [BIOCHEM] A ketose made from a six-carbon-chain monosaccharide. { ′heks·yə‚lōs }

n-hexyl acetate [ORG CHEM] $CH_3COOC_6H_{13}$ Colorless liquid boiling at 169°C; soluble in alcohol and ether, insoluble in water; used as a solvent for resins and cellulosic esters. { ¦en ′hek·səl ′as·ə‚tāt }

hexyl alcohol [ORG CHEM] $CH_3(CH_2)_4CH_2OH$ Colorless liquid boiling at 156°C; soluble in alcohol and ether, slightly

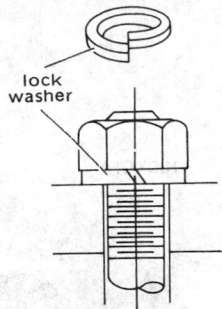

HEXAGONAL NUT

A hexagonal nut used with a lock washer.

HEXASTEROPHORA

A representative hexasterophoran, *Polylophus*.

soluble in water; used as a chemical intermediate for pharmaceuticals, perfume esters, and antiseptics. { 'hek·səl 'al·kə‚hȯl }

hexylamine [ORG CHEM] $CH_3(CH_2)_5NH_2$ Poisonous, water-white liquid with amine aroma; boils at 129°C; a ptomaine base from the autolysis of protoplasm. { hek'sil·ə‚mēn }

hexylene *See* 1-hexene. { 'hek·sə‚lēn }

hexylene glycol [ORG CHEM] $C_6H_{14}O_2$ Water-miscible, colorless liquid boiling at 198°C; used in hydraulic brake fluids, in printing inks, and in textile processing. { 'hek·sə‚lēn 'glī‚kȯl }

n-hexyl ether [ORG CHEM] $C_6H_{13}OC_6H_{13}$ Faintly colored liquid with a characteristic odor, only slightly water-soluble; used in solvent extraction and in the manufacture of collodion and various cellulosic products. { ‚en 'hek·səl 'ē·thər }

hexylresorcinol [ORG CHEM] $C_6H_{13}C_6H_3(OH)_2$ Sharp-tasting, white to yellowish crystals melting at 64°C; slightly soluble in water, soluble in glycerin, vegetable oils, and organic solvents; used in medicine. { ‚hek·səl·ri'sȯrs·ən‚ȯl }

1-hexyne [ORG CHEM] C_4H_9CCH A colorless, water-white liquid, either *n*-butylacetylene, boiling at 71.5°C, or methylpropylacetylene, boiling at 84°C. { ‚wən 'hek‚sēn }

Heyd's syndrome *See* hepatorenal syndrome. { 'hīdz ‚sin‚drōm }

Heyn's reagent [MET] Double chloride of copper and ammonia; used in microanalysis of carbon steels. { 'hīnz rē‚ā·jənt }

Hf *See* hafnium.

HF *See* high frequency.

HF akylation [CHEM ENG] Petroleum refinery alkylation process in which olefins (C_3, C_4, C_5) are reacted with isobutane in the presence of hydrofluoric acid catalyst. { ‚ach‚ef ‚al·kə'lā·shən }

HFC *See* hydrofluorocarbon.

HFET *See* high-electron-mobility transistor. { 'ach‚fet }

HFIR *See* high-flux isotope reactor.

Hfr *See* high-frequency recombination.

hfs *See* hyperfine structure.

HFS *See* type II superconductor.

hg *See* hectogram.

Hg *See* mercury.

HG [MED] A symbol used in electrocardiography for the longitudinal axis of the heart as projected on the frontal plane.

HH beacon [NAV] Nondirectional radio homing beacon which has a power output of 2000 watts or greater. { ‚ach'ach ‚bē·kən }

H hour [ORD] The hour at which some specified operation will begin, as an attack, amphibious assault, or movement. { 'ach ‚au̇r }

hiascent [BIOL] Gaping. { hī'ā·shənt }

hiatus [ANAT] A space or a passage through an organ. [GEOL] A gap in a rock sequence due to a lack of deposition of a bed or to erosion of beds. { hī'ād·əs }

hiatus hernia [MED] Hernia through the esophageal hiatus, usually of a portion of the stomach. { hī'ād·əs 'hər·nē·ə }

hibernaculum [BIOL] A winter shelter for plants or dormant animals. [BOT] A winter bud or other winter plant part. [INV ZOO] A winter resting bud produced by a few freshwater bryozoans which grows into a new colony in the spring. { ‚hī·bər'nak·yə·ləm }

hibernal [METEOROL] Of or pertaining to winter. { hī'bərn·əl }

hibernation [PHYSIO] **1.** Condition of dormancy and torpor found in cold-blooded vertebrates and invertebrates. **2.** *See* deep hibernation. { ‚hī·bər'nā·shən }

Hibernian orogeny *See* Erian orogeny. { hī'bər·nē·ən ȯ'räj·ə·nē }

Hibiscus cannabinus *See* kenaf. { hī‚bis·kəs kə'nab·ə·nəs }

hibonite [MINERAL] $CaAl_{12}O_{19}$ Common mineral found in carbonaceous chondrite meteorites; occurs only rarely on earth. { 'hib·ə‚nīt }

hiccup *See* singultus. { 'hik·əp }

hickey [ELEC] A threaded coupling for attaching an electrical fixture to an outlet box, used when wires from the fixture come out of the end of a stem on the fixture, rather than through an opening in the side of the stem. [GRAPHICS] A spot or imperfection in a printed item caused by dirt, dried ink membrane, or such. { 'hik·ē }

hickory [BOT] The common name for species of the genus *Carya* in the order Fagales; tall deciduous tree with pinnately compound leaves, solid pith, and terminal, scaly winter buds. { 'hik·ə·rē }

Hidalgo [ASTRON] The asteroid with the second largest known mean distance from the sun, about 5.8 astronomical units. { hi'däl·gō }

hidden file [COMPUT SCI] A disk file that does not appear in a directory listing and cannot be displayed, changed, or deleted. { ‚hid·ən 'fīl }

hiddenite [MINERAL] A transparent green or yellowish-green spodumene mineral containing chromium and valued as a gem. { 'hid·ən‚īt }

hidden line [GRAPHICS] For a three-dimensional object, a line that cannot be seen because view of it is obstructed by part of the object or of another object. { 'hid·ən 'līn }

hidden Markov model [MATH] A finite-state machine that is also a doubly stochastic process involving at least two levels of uncertainty: a random process associated with each state, and a Markov chain, which characterizes the probabilistic relationship among the states in terms of how likely one state is to follow another. { ‚hid·ən 'mär·kəf ‚mäd·əl }

hidden variables [QUANT MECH] Hypothetical additional variables or parameters which would supplement quantum mechanics, making it possible to unambiguously predict the result of a single measurement on a single microscopic system. { 'hid·ən 'ver·ē·ə·bəlz }

hidden-variable theory of the first kind [QUANT MECH] A theory postulating the existence of hidden variables and constructed so as to be self-consistent and to reproduce all the statistical predictions of quantum mechanics when the hidden variables are in an equilibrium distribution. { 'hid·ən ‚ver·ē·ə·bəl ‚thē·ə·rē əv thə 'fərst ‚kīnd }

hidden-variable theory of the second kind [QUANT MECH] A theory, postulating the existence of hidden variables, that predicts deviations from the statistical predictions of quantum mechanics, even for the equilibrium situations; such theories are required to satisfy a locality condition. Also known as local hidden-variable theory. { 'hid·ən ‚ver·ē·ə·bəl ‚thē·ə·rē əv thə 'sek·ənd 'kīnd }

hide [MATER] A raw or dressed pelt, especially from a large, adult animal such as a cow. [VERT ZOO] Outer covering of an animal. { hīd }

hide glue [MATER] A strong, light-brown glue made from animal hides. { 'hīd ‚glü }

hidradenitis [MED] Inflammation of sweat glands. { ‚hī‚drad·ən'īd·əs }

hidradenoma papilliferum [MED] Benign tumor of sweat glands, usually on the vulva or perineum. { ‚hī‚drad·ən'ō·mə ‚pap·ə'lif·ə·rəm }

hidrosis [MED] Abnormally profuse sweat. [PHYSIO] The formation and excretion of sweat. { hī'drō·səs }

hiemal climate [CLIMATOL] Climate pertaining to winter. { 'hī·ə·məl ‚klī·mət }

hierarchical control [CONT SYS] The organization of controllers in a large-scale system into two or more levels so that controllers in each level send control signals to controllers in the level below and feedback or sensing signals to controllers in the level above. Also known as control hierarchy. { ‚hī·ər‚är·kə·kəl kən'trōl }

hierarchical distributed processing system [COMPUT SCI] A type of distributed processing system in which processing functions are distributed outward from a central computer to intelligent terminal controllers or satellite information processors. Also known as host-centered system; host/satellite system. { ‚hī·ər‚är·kə·kəl di‚strib·yəd·əd 'prä‚ses·iŋ ‚sis·təm }

hierarchical file [COMPUT SCI] A file with a grandfather-father-son structure. { ‚hī·ər‚är·kə·kəl 'fīl }

hierarchical level I [ELEC] The level of reliability evaluation of an electric power system that is concerned only with the generation facilities. { ‚hī·ər‚ärk·ə·kəl ‚lev·əl 'wən }

hierarchical level II [ELEC] The level of reliability evaluation of an electric power system that is concerned only with the generation and transmission facilities. { ‚hī·ər'ark·ə·kəl ‚lev·əl 'tü }

hierarchical level III [ELEC] The level of reliability evaluation of an electric power system that is concerned with all three functional zones of the system, that is, generation, transmission, and distribution facilities. { ‚hī·ər'ärk·ə·kəl ‚lev·əl 'thrē }

HICKORY

Twig, bud, and leaf of the shagbark hickory, *Carya ovata*.

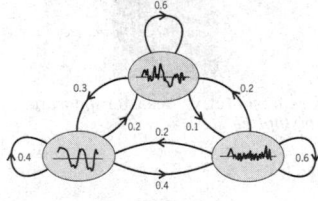

HIDDEN MARKOV MODEL

Three-state hidden Markov model used in implementing speech-recognition systems.

hierarchical storage management [COMPUT SCI] A method of managing large amounts of data in which files are assigned to various storage media based on how soon or how frequently they will be needed. { ¦hī·ər¦är·kə·kəl 'stȯr·ij ˌman·ij·mənt }

hieratite [MINERAL] K_2SiF_6 A grayish mineral composed of potassium fluosilicate; occurs as deposits in volcanic holes. { 'hī·ər·əˌtīt }

hieroglyph [GEOL] Any sort of sedimentary mark or structure occurring on a bedding plane. { 'hī·rəˌglif }

hierophobia [PSYCH] An abnormal fear of sacred objects. { ˌhī·rə'fō·bē·ə }

hi-fi See high fidelity. { 'hī'fī }

Hi-Fix [NAV] A Decca radio navigation system utilizing lightweight, portable stations, and used in either of two configurations; in the hyperbolic configuration, the relative phases of continuous-wave signals from the master and slave stations ashore, as measured at the receiver at the mobile station, provide hyperbolic lines of position; in the range configuration, the master station is located at the mobile receiver; phase comparison of continuous-wave signals from one shore station and the master provides range to the shore station. { 'hī'fiks }

Higbie model [CHEM ENG] Mass-transfer theory for packed absorption towers, stating that liquid flows across each packing piece in laminar flow and is mixed with other liquids meeting it at the points of discontinuity between packing elements. { 'hig·bē ˌmäd·əl }

Higgs bosons [PART PHYS] Massive scalar mesons whose existence is predicted by certain unified gage theories of the weak and electromagnetic interactions; they are not eliminated by the Higgs mechanism. { 'higz ¦bō¦sänz }

Higgs mechanism [PART PHYS] The feature of the spontaneously broken gage symmetries that the Goldstone bosons do not appear as physical particles, but instead constitute the zero helicity states of gage vector bosons of nonzero mass (such as the intermediate vector boson). [QUANT MECH] A mathematical procedure in which particles in a field theory gain or lose mass due to spontaneous breakdown of symmetry. { 'higz ˌmek·ə·niz·əm }

high [METEOROL] An area of high pressure, referring to a maximum of atmospheric pressure in two dimensions (closed isobars) in the synoptic surface chart, or a maximum of height (closed contours) in the constant-pressure chart; since a high is, on the synoptic chart, always associated with anticyclonic circulation, the term is used interchangeably with anticyclone. { hī }

high-abrasion furnace black [MATER] A variety of carbon black made by burning oil in a deficiency of air and then quenching to stop the reaction short of equilibrium; particles are 26–30 nanometers in diameter; used in tires and mechanical rubber goods. Abbreviated HAF black. { 'hī ə,brā·zhən 'fər·nəs ˌblak }

high-alloy steel [MET] Steel containing large percentages of elements other than carbon. { ¦hī 'al,ȯi ¦stēl }

high aloft See upper-level anticyclone. { ¦hī ə'lȯft }

high-altitude bombing [ORD] Bombing, especially horizontal bombing, from a high altitude, such as over 15,000 feet (4500 meters). { 'hī ¦al·tə,tüd 'bäm·iŋ }

high-altitude disease See mountain sickness. { 'hī ¦al·tə,tüd di,zēz }

high-altitude erythremia See mountain sickness. { 'hī ¦al·tə,tüd ,er·ə'thrē·mē·ə }

high-altitude method [NAV] The establishing of a circular line of position from the observation of the altitude of a celestial body; the geographical position and zenith distance of the body are used in constructing this line of position. { 'hī ¦al·tə,tüd 'meth·əd }

high-altitude radio altimeter See radar altimeter. { 'hī ¦al·tə,tüd 'rād·ē·ō al'tim·əd·ər }

high-altitude station [METEOROL] A weather observing station at a sufficiently high elevation to be nonrepresentative of conditions near sea level; 6500 feet (about 2000 meters) has been given as a reasonable lower limit. { 'hī ¦al·tə,tüd 'stā·shən }

high-alumina brick [MATER] Refractory brick made from raw materials rich in alumina, such as diaspore and bauxite; when well fired, they contain a large amount of mullite; used for unusually severe temperature or load conditions. { ¦hī ə'lüm·ə·nə ¦brik }

high-alumina cement See aluminate cement. { ¦hī ə'lüm·ə·nə si'ment }

high-angle fault [GEOL] A fault with a dip greater than 45°. { ¦hī ,aŋ·gəl 'fȯlt }

high-angle fire [ORD] **1.** Fire delivered at elevations greater than the elevation of maximum range. **2.** Fire whose range decreases as the angle of elevation is increased; typical of mortar and howitzer fire. { ¦hī ,aŋ·gəl 'fīr }

high-angle gun [ORD] A cannon, such as an antiaircraft cannon, capable of firing at a high angle of elevation. { ¦hī ,aŋ·gəl 'gən }

high-angle strafing [ORD] Strafing from an airplane at a comparatively high dive angle; for example, approximately 45°. { ¦hī ,aŋ·gəl 'sträf·iŋ }

high-aspect-ratio micromachining [ENG] Microfabrication processes that produce tall microstructures with vertical sidewalls. Abbreviated HARM. { ¦hī ¦as,pekt ,rā·shō ,mī·krō·mə'shēn·iŋ }

high boiler [MATER] A solvent added to lacquer thinner to slow the rate of evaporation; boiling point 150–200°C. { 'hī ,bȯil·ər }

high boost See high-frequency compensation. { 'hī ,büst }

high brass [MET] The most common commercial wrought brass containing about 35% zinc and 65% copper. { 'hī ,bras }

high-burst ranging [ORD] Adjustment of gunfire by observation of airbursts. { 'hī ,bərst 'rānj·iŋ }

high-capacity projectile [ORD] A projectile with thin walls and high explosive loading, for use where no special penetrative qualities are required. { 'hī kə,pas·əd·ē prə'jek·təl }

high-carbon chromium [MET] Chromium containing at least 86% chromium, 8–11% carbon, and a maximum of 0.5% each of iron and silicon. { 'hī ,kär·bən 'krō·mē·əm }

high-carbon steel [MET] A cast or forged steel containing more than 0.5% carbon. { 'hī ,kär·bən 'stēl }

high clouds [METEOROL] Types of clouds whose mean lower level is above 20,000 feet (6100 meters); principal clouds in this group are cirrus, cirrocumulus, and cirrostratus. { 'hī ¦klaudz }

high contrast [GRAPHICS] That area where the degree of difference between black and white approaches the maximum. { 'hī ¦kän,trast }

high core [COMPUT SCI] The locations with higher addresses in a computer's main storage, usually occupied by the operating system. { 'hī 'kȯr }

high-current rectifier [ELECTR] A solid-state device, gas tube, or vacuum tube used to convert alternating to direct current for powering low-impedance loads. { 'hī ,kə·rənt 'rek·tə,fī·ər }

high-current switch [ELEC] A switch used to redirect heavy current flow; usually has a make-before-break feature to prevent excessive arcing. { 'hī ,kə·rənt 'swich }

high definition [COMMUN] Television or facsimile equivalent of high fidelity, in which the reproduced image contains such a large number of accurately reproduced elements that picture details approximate those of the original scene. { 'hī ,def·ə'nish·ən }

high-definition television [COMMUN] A television system with a resolution of more than 1000 scan lines, as compared to 525–625 scan lines in conventional systems. Abbreviated HDTV. { ¦hī def·ə¦nish·ən 'tel·ə,vizh·ən }

high-density disk [COMPUT SCI] A diskette that holds two or more times as much data per unit area as a double-density disk of the same size. { ¦hī ¦den·səd·ē 'disk }

high-density drive [COMPUT SCI] A disk drive that accepts both high-density and double-density disks. { ¦hī ¦den·səd·ē 'drīv }

high-density lipoprotein [BIOCHEM] A lipoprotein containing more proteins than lipids that transports excess cholesterol from tissues to the liver for excretion. Abbreviated HDL. { ¦hī ,den·səd·ē ,lī·pō'prō,tēn }

high-density polyethylene [ORG CHEM] A thermoplastic polyolefin with a density of 0.941–0.960 gram per cubic centimeter (0.543–0.555 ounce per cubic inch). Abbreviated HDPE. { ,hī ¦den·səd·ē ,päl·ē'eth·ə,lēn }

high-efficiency particulate air filter [MECH ENG] An air filter capable of reducing the concentration of solid particles (0.3 millimeter in diameter or larger) in the airstream by 99.97%. Also known as HEPA filter. { ,hī i¦fish·ən·sē pər,tik·yə·lət 'er ,fil·tər }

high-electron-mobility transistor [ELECTR] A type of field-effect transistor consisting of gallium arsenide and gallium aluminum arsenide, with a Schottky metal contact on the gallium aluminum arsenide layer and two ohmic contacts penetrating into the gallium arsenide layer, serving as the gate, source, and drain respectively. Abbreviated HEMT. Also known as heterojunction field-effect transistor (HFET); modulation-doped field-effect transistor (MODFET); selectively doped heterojunction transistor (SDHT); two-dimensional electron gas field-effect transistor (TEGFET). { 'hī i'lek‚trän mōˌbil·əd·ē tran‚zis·tər }

high enema [MED] An enema injected into the colon. { 'hī 'en·ə·mə }

high-energy astrophysics [ASTROPHYS] A science concerned with studies of acceleration of charged particles to high energies in space, cosmic rays, radio galaxies, pulsars, and quasi-stellar sources. { 'hī ‚en·ər·jē as·trəˈfiz·iks }

high-energy bond [PHYS CHEM] Any chemical bond yielding a decrease in free energy of at least 5 kilocalories per mole. { 'hī ‚en·ər·jē 'bänd }

high-energy electron diffraction [PHYS] The diffraction of electrons with high energies, usually in the range of 30,000–70,000 electronvolts, mainly to study the structure of atoms and molecules in gases and liquids. Abbreviated HEED. { 'hī ‚en·ər·jē i'lek‚trän di‚frak·shən }

high-energy environment [GEOL] An aqueous sedimentary environment which features a high energy level and turbulent motion, created by waves, currents, or surf, which prevents the settling and piling up of fine-grained sediment. { 'hī ‚en·ər·jē in'vī·ərn·mənt }

high-energy fuel [MATER] Fuel that upon combustion provides greater energy than that from conventional carbonaceous fuels; specifically, a hydroboron. { 'hī ‚en·ər·jē 'fyül }

high-energy neutron-proton scattering [NUC PHYS] A collision of a neutron having an energy greater than 440 megaelectronvolts with a proton; the collision is inelastic because of pion production. { 'hī ‚en·ər·jē ¦nü‚trän 'prō‚tän ‚skad·ər·iŋ }

high-energy particle [PART PHYS] An elementary particle having an energy of hundreds of megaelectronvolts or more. { 'hī ‚en·ər·jē 'pärd·ə·kəl }

high-energy physics See particle physics. { 'hī ‚en·ər·jē 'fiz·iks }

high-energy proton-proton scattering [NUC PHYS] A collision of a proton having an energy greater than 440 megaelectronvolts with another proton; the collision is inelastic because of pion production. { 'hī ‚en·ər·jē ¦prō‚tän 'prō‚tän ‚skad·ər·iŋ }

high-energy-rate forging [MET] The production of a forging by the use of a machine which utilizes a sudden surge of kinetic energy from compressed gas against a piston, causing a high ram velocity against the work. { 'hī ‚en·ər·jē ‚rāt 'forj·iŋ }

high-energy scattering [PART PHYS] Collisions of particles having energies of hundreds of megaelectronvolts or more, sufficient to produce new particles. { 'hī ‚en·ər·jē 'skad·ə·riŋ }

high-epithermal neutron range [NUCLEO] The neutron energy range of 1000 to 100,000 electronvolts. { 'hī ‚ep·ə‚thər·məl 'nü‚trän ‚rānj }

higher harmonic control [AERO ENG] An active control concept for reducing helicopter vibration levels, in which the control system senses vibrations originating with aerodynamic forces acting on the rotor blades, and applies high-frequency pitch motion to the rotor blades at very small angles to suppress this aerodynamic excitation. { ¦hī·ər här'män·ik kən‚trōl }

higher high water [OCEANOGR] The higher of two high tides occurring during a tidal day. { 'hī·ər ¦hī ‚wòd·ər }

higher-level language See high-level language. { 'hī·ər ‚lev·əl ‚laŋ·gwij }

higher low water [OCEANOGR] The higher of two low tides occurring during a tidal day. { 'hī·ər ¦lō ‚wòd·ər }

higher mode [ELECTROMAG] A waveguide mode whose frequency is higher than the lowest one. { 'hī·ər ‚mōd }

higher-order language See high-level language. { 'hī·ər ‚or·dər ‚laŋ·gwij }

higher-order software [COMPUT SCI] Software for designing and documenting an information system by decomposing the system into elementary components that are mathematically correct and error-free. Abbreviated HOS. { ¦hī·ər ‚or·dər 'sòft‚wer }

higher pair [MECH ENG] A link in a mechanism in which the mating parts have surface (instead of line or point) contact. { 'hī·ər 'per }

higher plane curve [MATH] Any algebraic curve whose degree exceeds 2. { 'hī·ər ‚plän ¦kərv }

higher than high-level language [COMPUT SCI] A programming language, such as an application development language, report program, or financial planning language, that is oriented toward a particular application and is much easier to use for that application than a conventional programming language. { 'hī·ər ‚than 'hī ‚lev·əl ‚laŋ·gwij }

highest common factor See greatest common divisor. { 'hī·əst 'käm·ən 'fak·tər }

highest occupied molecular orbital [PHYS CHEM] The highest-energy molecular orbital that is occupied by electrons. Abbreviated HOMO. { ‚hī·əst ¦äk·yə·pīd məˌlek·yə·lər 'òr·bə·təl }

high etch See dry-relief offset. { 'hī 'ech }

high-expansion alloy [MET] An alloy possessing a high coefficient of expansion. { 'hī ik'span·chən 'al‚òi }

high-expansion foam [MATER] Noncombustible foam made from ammonium lauryl sulfate; used in underground mine fire fighting. { 'hī ik'span·chən 'fōm }

high explosive [MATER] An explosive with a nitroglycerin base requiring a detonator; the explosion is violent and practically instantaneous. { 'hī ik'splō·siv }

high-explosive bomb [ORD] **1.** Any aerial bomb charged with a high explosive. **2.** Any bomb chiefly dependent upon its explosion or blast effect to create damage. { 'hī ik‚splō·siv 'bämb }

high-explosive plastic [MATER] High-explosive substance or mixture which, within normal ranges of atmospheric temperature, is capable of being molded into desired shapes. Also known as plastic explosive. { 'hī ik‚splō·siv 'plas·tik }

high-explosive plastic antitank charge [ORD] A shaped charge coupled with a high-explosive plastic charge, intended to produce jet penetration followed by a detonation of the plastic charge. Abbreviated HEPAT charge. { 'hī ik‚splō·siv 'plas·tik an·tē'taŋk ‚chärj }

high-explosive plastic projectile [ORD] A thin-walled projectile, filled with plastic explosive; designed to squash against an armored target before detonation, and to defeat the armor by producing spalls which are detached with considerable velocity from the back of the target plate. Abbreviated HEP projectile. { 'hī ik‚splō·siv 'plas·tik prə'jek·təl }

high-explosive projectile [ORD] Projectile with a bursting charge of high explosive. { 'hī ik'splō·siv prə'jek·təl }

high fidelity [ENG ACOUS] Audio reproduction that closely approximates the sound of the original performance. Also known as hi-fi. { ¦hī fi'del·əd·ē }

high-field superconductor See type II superconductor. { 'hī ‚fēld 'sü·pər·kən‚dək·tər }

high-flash naphtha See heavy naphtha. { 'hī ‚flash 'naf·thə }

high-flux isotope reactor [NUCLEO] A thermal research reactor at the Oak Ridge National Laboratory, used mainly in the production of elements with atomic numbers greater than that of plutonium. Abbreviated HFIR. { 'hī ‚fləks ¦ī·sə‚tōp rē‚ak·tər }

high foehn [METEOROL] The occurrence of warm, dry air above the level of the general surface, accompanied by clear skies, resembling foehn conditions; it is due to subsiding air in an anticyclone, above a cold surface layer; in such circumstances the mountain peaks may be warmer than the lowlands. Also known as free foehn. { 'hī 'fān }

high fog [METEOROL] The frequent fog on the slopes of the coastal mountains of California, especially applied when the fog overtops the range and extends as stratus over the leeward valleys. { ¦hī 'fäg }

high frequency [COMMUN] Federal Communications Commission designation for the band from 3 to 30 megahertz in the radio spectrum. Abbreviated HF. { 'hī ¦frē·kwən·sē }

high-frequency carrier telegraphy [COMMUN] Form of carrier telegraphy in which the carrier currents have their frequencies above the range transmitted over a voice-frequency telephone channel. { 'hī ¦frē·kwən·sē 'kar·ē·ər tə'leg·rə·fē }

high-frequency compensation [ELECTR] Increasing the amplification at high frequencies with respect to that at low

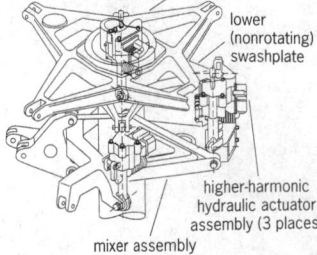

and middle frequencies in a given band, such as in a video band or an audio band. Also known as high boost. { 'hī ¦frē‧kwən‧sē ‚käm‧pən'sā‧shən }

high-frequency furnace [ENG] An induction furnace in which the heat is generated within the charge, within the walls of the containing crucible, or within both, by currents induced by high-frequency magnetic flux produced by a surrounding coil. Also known as coreless-type induction furnace; high-frequency heater. { 'hī ¦frē‧kwən‧sē 'fər‧nəs }

high-frequency heater *See* high-frequency furnace. { 'hī ¦frē‧kwən‧sē 'hēd‧ər }

high-frequency heating *See* electronic heating. { 'hī ¦frē‧kwən‧sē 'hēd‧iŋ }

high-frequency propagation [COMMUN] Propagation of radio waves in the high-frequency band, which depends entirely on reflection from the ionosphere. { 'hī ¦frē‧kwən‧sē ‚präp‧ə'gā‧shən }

high-frequency recombination [MICROBIO] A bacterial cell type, especially *Escherichia coli,* having an integrated F factor and characterized by a high frequency of recombination. Abbreviated Hfr. { 'hī ¦frē‧kwən‧sē ‚rē‚käm‧bə'nā‧shən }

high-frequency resistance [ELEC] The total resistance offered by a device in an alternating-current circuit, including the direct-current resistance and the resistance due to eddy current, hysteresis, dielectric, and corona losses. Also known as alternating-current resistance; effective resistance; radio-frequency resistance. { 'hī ¦frē‧kwən‧sē ri'zis‧təns }

high-frequency titration [ANALY CHEM] A conductimetric titration in which two electrodes are mounted on the outside of the beaker or vessel containing the solution to be analyzed and an alternating current source in the megahertz range is used to measure the course of a titration. { 'hī ¦frē‧kwən‧sē tī'trā‧shən }

high-frequency transformer [ELECTR] A transformer which matches impedances and transmits a frequency band in the carrier (or higher) frequency ranges. { 'hī ¦frē‧kwən‧sē tranz'fór‧mər }

high-frequency triode [ELECTR] A triode designed for operation at high frequency, having small spacings between the grid and the cathode and anode, large emission and power densities, and low active and inactive capacitances. { 'hī ¦frē‧kwən‧sē 'trī‧ōd }

high-frequency voltmeter [ELECTR] A voltmeter designed to measure currents alternating at high frequencies. { 'hī ¦frē‧kwən‧sē 'vōlt‚mēd‧ər }

high-frequency welding [MET] Resistance welding in which the heat is produced by the current flow induced by a high-frequency electromagnetic field. Also known as radio-frequency welding. { 'hī ¦frē‧kwən‧sē 'weld‧iŋ }

high-front shovel [MECH ENG] A power shovel with a dipper stick mounted high on the boom for stripping and overburden removal. { 'hī ‚frənt 'shəv‧əl }

high-grade [MIN ENG] To steal or pilfer ore or gold from a mine or miner. { 'hī ‚grād }

high-grade dynamite [MATER] Dynamite of 40% strength or over. { 'hī ‚grād 'dī‧nə‚mīt }

high-gradient magnetic separation [ENG] A magnetic separation technique applicable to weakly paramagnetic compounds and to particle sizes down to the colloidal domain. { 'hī ‚grād‧ē‧ənt mag'ned‧ik ‚sep‧ə'rā‧shən }

high hat [ENG] A very low tripod head resembling a formal top hat in shape. { 'hī ‚hat }

high heat [THERMO] Heat absorbed by the cooling medium in a calorimeter when products of combustion are cooled to the initial atmospheric (ambient) temperature. { 'hī ‚hēt }

high-heat cement [MATER] A type of cement which releases a large amount of heat during curing. { 'hī ‚hēt si'ment }

high-helix drill [DES ENG] A two-flute twist drill with a helix angle of 35–40°; used for drilling deep holes in metals, such as aluminum, copper, hard brass, and soft steel. Also known as fast-spiral drill. { 'hī ‚hē‧liks ‚dril }

high-impedance voltmeter [ELEC] A voltage-measuring device with a high-impedance input to reduce load on the unit under test; a vacuum-tube voltmeter is one type. { 'hī im‚pēd‧əns 'vōlt‚mēd‧ər }

high index [METEOROL] A relatively high value of the zonal index which, in middle latitudes, indicates a relatively strong westerly component of wind flow and the characteristic weather features attending such motion; a synoptic circulation pattern

of this type is commonly called a high-index situation. { 'hī ‚in‚deks }

high-information-content display [ELECTR] An electronic display that has a sufficient number of pixels (75,000 to 2,000,000) to show standard or high-definition television images or comparable computer images. { ¦hī ‚in‧fər'mā‧shən ‚kän‚tent di‚splā }

high-intensity atomizer [MECH ENG] A type of atomizer used in electrostatic atomization, based on stress sufficient to overcome tensile strength of the liquid. { 'hī in‚ten‧səd‧ē 'ad‧ə‚miz‧ər }

high-K capacitor [ELEC] A capacitor whose dielectric material is a ferroelectric having a high dielectric constant, up to about 6000. { 'hī ‚kā kə'pas‧əd‧ər }

high key [GRAPHICS] A negative or print that has an overall light effect, with very few dark tones or shadows. { 'hī ‚kē }

high-key lighting [GRAPHICS] In photography, lighting that produces tones that fall mostly between white and gray, with very few dark-gray or black tones. { 'hī ‚kē 'līd‧iŋ }

highland [GEOGR] **1.** Any relatively large area of elevated or mountainous land standing prominently above adjacent low areas. **2.** The higher land of a region. [GEOL] **1.** A lofty headland, cliff, or other high platform. **2.** A dissected mountain region composed of old folded rocks. { 'hī‧lənd }

highland climate *See* mountain climate. { 'hīlənd ‚klī‧mət }

highland glacier [HYD] A semicontinuous ice cap or glacier that covers the highest or central portion of a mountainous area and partly reflects irregularities of the land surface lying beneath it. Also known as highland ice. { 'hī‧lənd ‚glā‧shər }

highland halftone [GRAPHICS] A halftone with elimination of the residual dot formation in the pure highlights, done either by spotting the highlight areas on the negative with opaque or by etching the printing plate. { 'hī‧lənd ‚haf‚tōn }

highland ice *See* highland glacier. { 'hī‧lənd ‚īs }

high-lead bronze [MET] Bronze containing high percentages of lead to give a soft matrix alloy. { 'hī ‚led 'bränz }

high level [ELECTR] The more positive of the two logic levels or states in a binary digital logic system. { 'hī ‚lev‧əl }

high-level anticyclone *See* upper-level anticyclone. { 'hī ‚lev‧əl an‚tē'sī‚klōn }

high-level cyclone *See* upper-level cyclone. { 'hī ‚lev‧əl 'sī‚klōn }

high-level data-link control Abbreviated HDLC. [COMMUN] A bit-oriented protocol for managing information flow in a data communications channel that supports both full-duplex and half-duplex transmission, and both point-to-point and multipoint communications using synchronous data transmission. [COMPUT SCI] A communications protocol that allows devices from different manufacturers to interface with each other and standardizes the transmission of packets of information between them. { 'hī ‚lev‧əl 'dad‧ə ‚liŋk kən‚trōl }

high-level index [COMPUT SCI] The first part of a file name, which frequently specifies the category of data to which it belongs. { 'hī ‚lev‧əl 'in‚deks }

high-level language [COMPUT SCI] A computer language whose instructions or statements each correspond to several machine language instructions, designed to make coding easier. Also known as higher-level language; higher-order language. { 'hī ‚lev‧əl 'laŋ‧gwij }

high-level modulation [COMMUN] Modulation produced at a point in a system where the power level approximates that at the output of the system. { 'hī ‚lev‧əl ‚mäj‧ə'lā‧shən }

high-level ridge *See* upper-level ridge. { 'hī ‚lev‧əl 'rij }

high-level thunderstorm [METEOROL] Generally, a thunderstorm based at a comparatively high altitude in the atmosphere, roughly 8000 feet (2400 meters) or higher. { 'hī ‚lev‧əl 'thən‧dər‚stórm }

high-level trough *See* upper-level trough. { 'hī ‚lev‧əl 'tróf }

high-lift truck [MECH ENG] A forklift truck with a fixed or telescoping mast to permit high elevation of a load. { 'hī ‚lift 'trək }

highlights [ELECTR] Bright areas occurring in a television image. [GRAPHICS] The bright parts of a subject that show up as the densest part of the negative and the lightest or whitest part of a print. { 'hī‚līts }

high-low bias test [ELECTR] A routine maintenance procedure that tests equipment over and under normal operating conditions in order to detect defective units. { ¦hī ¦lō 'bī·əs ˌtest }

highly skewed propeller [NAV ARCH] A marine propeller whose blades are in the form of scimitars, typically with the tip of one blade aligning radially with the root of the following blade. { 'hī·lē ˌskyüd prə'pel·ər }

high-mass x-ray binary [ASTRON] A binary system consisting of a massive (greater than 5 solar masses), early-type star and a neutron star or black hole that accretes material through a stellar wind or Roche-lobe overflow, resulting in the emission of hard x-rays. Abbreviated HMXRB. { 'hī¦mas ¦eks¦rā'bī¦ner·ē }

high-modulus furnace black [MATER] A variety of carbon black made by burning oil in a deficiency of air; particle diameter is 49–60 nanometers; production is now negligible, but it was used in tire carcasses. Abbreviated HMF black. { 'hī ˌmäj·ə·ləs 'fər·nəs ˌblak }

highmoor bog [ECOL] A bog whose surface is covered by sphagnum mosses and is not dependent upon the water table. { 'hīˌmür 'bäg }

high-mu tube [ELECTR] A tube having a very high amplification factor. { 'hī ˌmyü 'tüb }

high-octane [MATER] Having an octane number in the middle or high 90s and therefore having good antiknock properties. { 'hī ¦äk¦tān }

high-order [COMPUT SCI] Pertaining to a digit location in a numeral, the leftmost digit being the highest-order digit. { 'hī ¦ȯr·dər }

high-pass filter [ELECTR] A filter that transmits all frequencies above a given cutoff frequency and substantially attenuates all others. { 'hī ˌpas 'fil·tər }

high-performance liquid chromatography [ANALY CHEM] A type of column chromatography in which the solvent is conveyed through the column under pressure. Abbreviated HPLC. { ¦hī pər'fȯrm·əns ˌlik·wəd ˌkrō·mə'täg·rə·fē }

high plain [GEOGR] A large area of level land situated above sea level. { 'hī ¦plān }

high polymer [ORG CHEM] A large molecule (of molecular weight greater than 10,000) usually composed of repeat units of low-molecular-weight species; for example, ethylene or propylene. { 'hī ¦päl·ə·mər }

high-positive indicator [COMPUT SCI] A component in some computers whose status is "on" if the number tested is positive and nonzero. { 'hī ¦päz·əd·iv 'in·də¦kād·ər }

high-potting [ELEC] Testing with a high voltage, generally on a production line. { 'hī ¦päd·iŋ }

high-precision shoran See hiran. { 'hī prə¦sizh·ən 'shȯˌran }

high-pressure area See anticyclone. { 'hī ¦presh·ər 'er·ē·ə }

high-pressure chemistry [PHYS CHEM] The study of chemical reactions and phenomena that occur at pressures exceeding 10,000 bars (a bar is nearly equivalent to a kilogram per square centimeter), mainly concerned with the properties of the solid state. { 'hī ¦presh·ər 'kem·ə·strē }

high-pressure cloud chamber [NUCLEO] A cloud chamber in which the gas is maintained at high pressure to reduce the range of high-energy particles and thereby increase the probability of observing events. { 'hī ¦presh·ər 'klaud ˌchäm·bər }

high-pressure gage glass [ENG] A gage glass consisting of a metal tube with thick glass windows. { 'hī ¦presh·ər 'gāj ˌglas }

high-pressure gas injection [PETRO ENG] Oil reservoir pressure maintenance by injection of gas at pressures higher than those used in conventional equilibrium gas drives. { 'hī ¦presh·ər 'gas inˌjek·shən }

high-pressure laminate [MATER] A plastic-substrate laminate molded and cured at pressures of, commonly, 1200–2000 pounds per square inch ($8-14 \times 10^6$ pascals). { 'hī ¦presh·ər 'lam·əˌnāt }

high-pressure mercury-vapor lamp [ELECTR] A discharge tube containing an inert gas and a small quantity of liquid mercury; the initial glow discharge through the gas heats and vaporizes the mercury, after which the discharge through mercury vapor produces an intensely brilliant light. { 'hī ¦presh·ər 'mər·kyə·rē ¦vā·pər 'lamp }

high-pressure phenomena [PHYS] Natural conditions and processes occurring at high pressures, and their duplication in the laboratory. { 'hī ¦presh·ər fə'näm·ə·nä }

high-pressure physics [PHYS] The study of the effects of high pressure on the properties of matter. { 'hī ¦presh·ər 'fiz·iks }

high-pressure process [CHEM ENG] A chemical process operating at elevated pressure; for example, phenol manufacture at 330 atmospheres (1 atmosphere = 101,325 pascals), ethylene polymerization at 2000 atm, ammonia synthesis at 100–1000 atm, and synthetic-diamond manufacture up to 100,000 atm. { 'hī ¦presh·ər 'präˌsəs }

high-pressure separator [PETRO ENG] A horizontal vessel through which a low-temperature, high-pressure gas stream is fed, and in which free liquids separate out from the gas. { 'hī ¦presh·ər 'sep·əˌrād·ər }

high-pressure torch [ENG] A type of torch in which both acetylene and oxygen are delivered to the mixing chamber under pressure. { 'hī ¦presh·ər 'tȯrch }

high-pressure well [PETRO ENG] A well with a shut-in wellhead pressure of more than 2000 pounds per square inch absolute (1.4×10^7 pascals absolute). { 'hī ¦presh·ər 'wel }

high Q [ELECTR] A characteristic wherein a component has a high ratio of reactance to effective resistance, so that its Q factor is high. { ¦hī 'kyü }

high-Q cavity [ELECTROMAG] A cavity resonator which has a large Q factor, and thus has a small energy loss. Also known as high-Q resonator. { 'hī ˌkyü 'kav·əd·ē }

high-Q resonator See high-Q cavity. { 'hī ˌkyü 'rez·ənˌād·ər }

high quartz [MINERAL] Quartz that was formed at high temperatures. { 'hī ¦kwȯrts }

high-rank coal [GEOL] Coal consisting of less than 4% moisture when air-dried, or more than 84% carbon. { 'hī ˌraŋk 'kōl }

high-rank graywacke See feldspathic graywacke. { 'hī ˌraŋk 'grāˌwak·ə }

high relief [GRAPHICS] Sculpture projecting from the background block to the extent of at least half of the natural depth of the modeled forms. { 'hī ri¦lēf }

high-residual-phosphorus copper [MET] Deoxidized copper having reduced conductivity due to the presence of residual phosphorus, usually less than 0.1. { ¦hī ri¦zij·yə·wəl 'fäs·fə·rəs 'käp·ər }

high-resistance voltmeter [ELEC] A voltmeter having a resistance considerably higher than 1000 ohms per volt, so that it draws little current from the circuit in which a measurement is made. { 'hī ri¦zis·təns 'vōltˌmēd·ər }

high-resolution electron energy loss spectroscopy [SPECT] A type of electron energy loss spectroscopy in which electron scattering is performed by using a monoenergetic beam and electron energy analyzers to achieve a resolution of 5 to 10 millielectronvolts. Abbreviated HREELS. { 'hī ˌrez·ə'lü·shən i¦lek¦trän 'en·ər·jē ˌlȯs spek'träs·kə·pē }

high-resolution electron microscope [ELECTR] An electron microscope in which lens aberrations are minimized and lens currents and the accelerating voltage are maintained with a high degree of stability, in order to achieve extremely high resolution. { 'hī ˌrez·ə'lü·shən i¦lek¦trän 'mī·krəˌskōp }

high-resolution radar [ENG] A radar system which can discriminate between two close targets. { 'hī ˌrez·ə'lü·shən 'rāˌdär }

high-rise building See tall building. { ¦hī ¦rīz 'bild·iŋ }

high side [COMPUT SCI] The part of a remote device that communicates with a computer. { 'hī ˌsīd }

high-side capacitance coupling [ELECTR] Taking the output of an oscillator or amplifier from a point of high potential, using a capacitor to block direct current flow. { 'hī ˌsīd kə'pas·əd·əns ˌkəp·liŋ }

high-solvency naphtha [MATER] Any of the petroleum-based solvents that boil in the naphtha range (95–650°F; 35–343°C) and have a high aromatic-chemical content to give high-solvency power for nitrocelluloses, dry paints, and certain resins. { 'hī ˌsäl·vən·sē 'naf·thə }

high-speed carry [COMPUT SCI] A technique in parallel addition to speed up the propagation of carries. { 'hī ˌspēd'karˌē }

high-speed cement See aluminate cement. { 'hī ˌspēd si'ment }

high-speed data acquisition system [COMPUT SCI] A system which collects and transmits data rapidly to a monitoring and controlling center. { 'hī ˌspēd 'dad·ə ¦ak·wəˌzish·ən ˌsis·təm }

high-speed excitation system [ELEC] Excitation system capable of changing its voltage rapidly in response to a change in the excited generator field circuit. { 'hī 'spēd ˌek·sə'tā·shən ˌsis·təm }

high-speed machine [MECH ENG] A diamond drill capable of rotating a drill string at a minimum of 2500 revolutions per minute, as contrasted with the normal maximum speed of 1600–1800 revolutions per minute attained by the average diamond drill. { 'hī ˌspēd mə'shēn }

high-speed oscilloscope [ELECTR] An oscilloscope with a very fast sweep, capable of observing signals with rise times or periods on the order of nanoseconds. { 'hī ˌspēd ä'sil·ə·ˌskōp }

high-speed photography [GRAPHICS] Photography to record movement or events that occur too quickly to be observed by usual visual or photographic means; motion pictures may be shot at high speeds (50–500 frames per second) and projected at normal rates, so that the action of the subject is slowed to a point where it can be observed; or a series of individual still photos may be produced. { 'hī ˌspēd fə'täg·rə·fē }

high-speed printer [COMPUT SCI] A printer which can function at a high rate, relative to the state of the art; 600 lines per minute is considered high speed. Abbreviated HSP. { 'hī ˌspēd 'print·ər }

high-speed reader [COMPUT SCI] The fastest input device existing at a particular time in the state of the technology. { 'hī ˌspēd 'rēd·ər }

high-speed relay [ELECTR] A relay specifically designed for short operate time, short release time, or both. { 'hī ˌspēd 'rē·lā }

high-speed steel [MET] An alloy steel that remains hard and tough at red heat. { 'hī ˌspēd 'stēl }

high-speed storage See rapid storage. { 'hī ˌspēd 'stòr·ij }

high-strength alloy See high-tensile alloy. { 'hī ˌstreŋkth 'al·ói }

high-strength low-alloy steel [MET] Steel containing small amounts of niobium or vanadium and having higher strength, better low-temperature impact toughness, and, in some grades, better atmospheric corrosion resistance than carbon steel. Abbreviated HSLA steel. { 'hī ˌstreŋkth ˌlō ˌal·ói 'stēl }

high-strength steel See high-tensile steel. { 'hī ˌstreŋkth 'stēl }

high-technology robot [CONT SYS] A robot equipped with feedback, vision, real-time data acquisition, and powerful controllers. { 'hī tek'näl·ə·jē 'rō·ˌbät }

high-temperature alloy [MET] An alloy suitable for use at temperatures of 500°C and above. { 'hī ˌtem·prə·chər 'al·ói }

high-temperature cement [MATER] A cement that resists fusing, softening, or spalling at elevated temperatures; used to bond refractory materials. { 'hī ˌtem·prə·chər si'ment }

high-temperature chemistry [PHYS CHEM] The study of chemical phenomena occurring above about 500 K. { 'hī ˌtem·prə·chər 'kem·ə·strē }

high-temperature coke [MATER] Coke produced at temperatures of 900–1150°C; used mainly for metallurgical purposes. { 'hī ˌtem·prə·chər 'kōk }

high-temperature fuel cell [ELEC] A fuel cell which operates at temperatures above about 550°C, can use inexpensive hydrocarbon fuels, and usually uses a molten salt as an electrolyte. { 'hī ˌtem·prə·chər 'fyül ˌsel }

high-temperature gas-cooled reactor [NUCLEO] A prototype gas-cooled nuclear reactor in which the coolant is pressurized helium gas with an inlet temperature of about 325°C and an outlet temperature of about 750°C, and the fuel consists of fully enriched uranium and thorium. Abbreviated HTGR. { 'hī ˌtem·prə·chər ˌgas ˌküld rē'ak·tər }

high-temperature material [MATER] A material with high-temperature capability, including the superalloys, refractory alloys, and ceramics; used in structures such as spacecraft subjected to extreme thermal environments. { 'hī ˌtem·prə·chər mə'tir·ē·əl }

high-temperature phenomena [PHYS] Phenomena occurring at temperatures above about 500 K. { 'hī ˌtem·prə·chər fə'näm·ə·nə }

high-temperature reactor [NUCLEO] A nuclear power reactor in which the temperature is high enough for efficient generation of mechanical power. { 'hī ˌtem·prə·chər rē'ak·tər }

high-temperature superconductor [SOLID STATE] A ceramic material, consisting of an oxide of a rare-earth element, barium, and copper, which displays superconductivity at temperatures of 90 K (−298°F) or more. { 'hī 'tem·prə·chər 'sü·pər·kən·ˌdək·tər }

high-temperature water boiler [MECH ENG] A boiler which provides hot water, under pressure, for space heating of large areas. { 'hī ˌtem·prə·chər 'wòd·ər ˌbòil·ər }

high-tensile alloy [MET] An alloy having a high tensile strength. Also known as high-strength alloy. { 'hī ˌten·səl 'al·ói }

high-tensile bolt [ENG] A bolt that is adjusted to a carefully controlled tension by means of a calibrated torsion wrench; used in place of a rivet. Also known as high-tension bolt. { 'hī ˌten·səl 'bōlt }

high-tensile steel [MET] Low-alloy steel having a yield strength range of 50,000 to 100,000 pounds per square inch (3.4–6.9 × 10⁸ pascals). Also known as high-strength steel. { 'hī ˌten·səl 'stēl }

high tension See high voltage. { 'hī ˌten·chən }

high-tension bolt See high-tensile bolt. { ˌhī ˌten·chən 'bōlt }

high-tension detonator [ENG] A detonator requiring an electric potential of about 50 volts for firing. { 'hī ˌten·chən 'det·ən·ˌād·ər }

high-tension separation See electrostatic separation. { 'hī ˌten·chən ˌsep·ə'rā·shən }

high-test chain [ENG] Chain made from heat-treatable plain-carbon steel, usually with a carbon content of 0.15–0.20; used for load binding, tie-downs, and other applications where failure would be costly. { 'hī ˌtest 'chān }

high tide [OCEANOGR] The maximum height reached by a rising tide. Also known as high water. { 'hī ˌtīd }

high-tier system [COMMUN] A wireless telephone system that supports base stations with large coverage areas and low traffic densities, but provides low-quality voice service and has limited data-service capabilities with high delays. { ˌhī 'tir ˌsis·təm }

high vacuum [PHYS] A vacuum with a pressure between 1 × 10⁻³ and 1 × 10⁻⁶ mmHg (0.1333224 and 0.0001333 pascal). { 'hī ˌvak·yüm }

high-vacuum insulation [CHEM ENG] High vacuum between the walls of double-wall vessels to serve as thermal insulation at ultralow (cryogenic) temperatures, such as in Dewar vessels. { 'hī ˌvak·yüm ˌin·sə'lā·shən }

high-vacuum rectifier [ELECTR] Vacuum-tube rectifier in which conduction is entirely by electrons emitted from the cathode. { 'hī ˌvak·yüm 'rek·tə·ˌfī·ər }

high-vacuum switching tube [ELECTR] A microwave transmit-receive (TR) tube of the high-vacuum variety, as contrasted with gas-tube or semiconductor devices. { 'hī ˌvak·yüm 'swich·iŋ ˌtüb }

high-vacuum tube [ELECTR] Electron tube evacuated to such a degree that its electrical characteristics are essentially unaffected by gaseous ionization. Also known as hard tube. { 'hī ˌvak·yüm ˌtüb }

high-velocity cloud [ASTRON] A rapidly moving interstellar cloud with a radial velocity greater than about 12 miles (20 kilometers) per second, consisting primarily of neutral atomic hydrogen, observed in the ultraviolet. { 'hī və'läs·əd·ē 'klaud }

high-velocity star [ASTRON] A star that moves across the galactic track along which the majority of the stars execute their galactic rotation, thus exhibiting high velocity with respect to the sun and low velocity with respect to the galactic center. { 'hī və'läs·əd·ē 'stär }

high-volatile bituminous coal [GEOL] A bituminous coal composed of more than 31% volatile matter. { 'hī ˌväl·əd·əl bə'tü·mə·nəs 'kōl }

high voltage [ELEC] A voltage on the order of thousands of volts. Also known as high tension. { 'hī ˌvōl·tij }

high-voltage direct current [ELEC] A long-distance direct-current power transmission system that uses direct-current voltages up to about 1 megavolt to keep transmission losses down. Abbreviated HVDC. { 'hī ˌvōl·tij di·ˌrekt 'kə·rənt }

high-voltage electron microscope [ELECTR] An electron microscope whose accelerating voltage is on the order of 10⁶ volts, as compared with 40–100 kilovolts for an ordinary electron microscope; it has the advantages of increased specimen penetration, reduced specimen damage, better theoretical resolution, and more efficient dark-field operation. { 'hī ˌvōl·tij i·ˌlek·ˌträn 'mī·krə·ˌskōp }

high-voltage insulation [ELEC] Electrical insulation designed to prevent breakdown in a circuit operating at high voltages. { 'hī ¦vōl·tij ‚in·sə'lā·shən }

highwall [MIN ENG] The unexcavated face of exposed overburden and coal or ore in an opencast mine or the face or bank of the uphill side of a contour strip-mine excavation. { 'hī‚wȯl }

high water *See* high tide. { 'hī ¦wȯd·ər }

high-water full and change [GEOPHYS] The average interval of time between the transit (upper or lower) of the full or new moon and the next high water at a place. Also known as common establishment; vulgar establishment. { 'hī ¦wȯd·ər 'fùl ·ən 'chānj }

high-water inequality [OCEANOGR] The difference between the heights of the two high tides during a tidal day. { 'hī ¦wȯd·ər ‚in·ə'kwäl·əd·ē }

high-water line [OCEANOGR] The intersection of the plane of mean high water with the shore. { 'hī ¦wȯd·ər ‚līn }

high-water lunitidal interval [GEOPHYS] The interval of time between the transit (upper or lower) of the moon and the next high water at a place. { 'hī ¦wȯd·ər ¦lün·ə¦tīd·əl 'in·tər·vəl }

high-water mark [COMPUT SCI] The maximum number of jobs that are in a queue awaiting execution by a large computer system during a specified period of observation. { 'hī ¦wȯd·ər ‚märk }

high-water platform *See* wave-cut bench. { 'hī ¦wȯd·ər 'plat‚fȯrm }

high-water quadrature [OCEANOGR] The average high-water interval when the moon is at quadrature. { 'hī ¦wȯd·ər'kwäd·rə·chər }

high-water springs *See* mean high-water springs. { 'hī ¦wȯd·ər 'spriŋz }

high-water stand [OCEANOGR] The condition at high tide when there is no change in the height of the water. { 'hī ¦wȯd·ər 'stand }

highway [CIV ENG] A public road where traffic has the right to pass and to which owners of adjacent property have access. { 'hī‚wā }

highway engineering [CIV ENG] A branch of civil engineering dealing with highway planning, location, design, and maintenance. { 'hī‚wā ‚en·jə'nir·iŋ }

Hikojima serotype [IMMUNOL] An immunologically distinct group of *Vibrio* somatic O antigens. { ‚hē·kō'jē·mə 'ser·ə‚tīp }

hilac *See* heavy-ion linear accelerator. { 'hī‚lak }

Hilbert cube [MATH] The topological space which is the cartesian product of a countable number of copies of I, the unit interval. { 'hil·bərt ‚kyüb }

Hilbert parallelotope [MATH] **1.** A subset of an infinite-dimensional Hilbert space with coordinates $x_1, x_2, \ldots,$ for which the absolute value of x_n is equal to or less that $(1/2)^n$ for each n. **2.** The subset of this space for which the absolute value of x_n is equal to or less that $1/n$ for each n. { 'hil·bərt ‚par·ə'lel·ə‚tōp }

Hilbert-Schmidt theory [MATH] A body of theorems which investigates the kernel of an integral equation via its eigenfunctions, and then applies these functions to help determine solutions of the equation. { ¦hil·bərt 'shmit ‚thē·ə·rē }

Hilbert space [MATH] A Banach space which also is an inner-product space with the inner product of a vector with itself being the same as the square of the norm of the vector. { 'hil·bərt ¦spās }

Hilbert's theorem [MATH] The proposition that the ring of polynomials with coefficients in a commutative Noetherian ring is itself a Noetherian ring. { 'hil‚bərts ‚thir·əm }

Hilbert transform [MATH] The transform $F(y)$ of a function $f(x)$ realized by taking the Cauchy principal value of the integral over the real numbers of $(1/\pi) f(x)[1/(x-y)] \, dx$. { 'hil·bərt ¦tranz‚fȯrm }

Hilbert transformer [ELECTR] An electric filter whose gain is $-j$ for positive frequencies and j for negative frequencies, where j is the square root of -1. { 'hil·bərt tranz‚fȯr·mər }

Hilda group [ASTRON] A group of asteroids whose periods of revolution about the sun are approximately two-thirds that of Jupiter, and whose motions are in resonance with Jupiter. { 'hil·də ‚grüp }

Hildebrand function [THERMO] The heat of vaporization of a compound as a function of the molal concentration of the

vapor; it is nearly the same for many compounds. { 'hil·də‚brand ‚faŋk·shən }

hilgardite [MINERAL] $Ca_8(B_6O_{11})_3Cl_4 \cdot H_2O$ Colorless mineral composed of hydrous borate and chloride of calcium; occurs as monoclinic domatic crystals. { 'hil‚gär‚dīt }

hill [GEOGR] A land surface feature characterized by strong relief; it is a prominence smaller than a mountain. { hil }

hill-and-dale recording *See* vertical recording. { ¦hil ən ¦dāl ri'kȯrd·iŋ }

hill bandwidth [ELECTR] The difference between the upper and lower frequencies at which the gain of an amplifier is 3 decibels less than its maximum value. { ¦hil 'band‚width }

hill-climbing [MATH] Any numerical procedure for finding the maximum or maxima of a function. [MECH ENG] Adjustment, either continuous or periodic, of a self-regulating system to achieve optimum performance. { 'hil ‚klim·iŋ }

hill creep [GEOL] Slow gravity movement of rock and soil waste down a steep hillside. Also known as hillside creep. { 'hil ‚krēp }

hillebrandite [MINERAL] $Ca_2SiO_3(OH)_2$ A white mineral composed of hydrous calcium silicate; occurs in masses. { 'hil·ə‚bran‚dīt }

hillock [GEOL] A small, low hill. { 'hil·ək }

Hill plot [BIOCHEM] A graphic representation of the Hill reaction. { 'hil ‚plät }

Hill reaction [BIOCHEM] The release of molecular oxygen by isolated chloroplasts in the presence of a suitable electron receptor, such as ferricyanide. [ORG CHEM] Production of substituted phenylacetic acids by the oxidation of the corresponding alkylbenzene by potassium permanganate in the presence of acetic acid. { 'hil rē‚ak·shən }

hill shading [MAP] Also known as hillwork; plastic shading; relief shading; shading. **1.** A method of showing relief on a map by simulating shadows by assuming oblique sunlight from the northwest so that slopes facing south and east are shaded. **2.** The pictorial effect created by this method. { 'hil ‚shād·iŋ }

hillside creep *See* hill creep. { 'hil‚sīd ‚krēp }

hillside quarry [MIN ENG] A quarry cut along a hillside. { 'hil‚sīd ¦kwä·rē }

hillwork *See* hill shading. { 'hil‚wərk }

Hiltner-Hall effect [ASTRON] The polarization of the light received from distant stars; this effect is thought to take place in interstellar space. { ¦hilt·nər 'hȯl i‚fekt }

Hilt's law [GEOL] The law that in a small area the deeper coals are of higher rank than those above them. { 'hilts ‚lȯ }

hilum [ANAT] *See* hilus. [BOT] Scar on a seed marking the point of detachment from the funiculus. { 'hī·ləm }

hilus [ANAT] An opening or recess in an organ, usually for passage of a vessel or duct. Also known as hilum. { 'hī·ləs }

Himalia [ASTRON] A small satellite of Jupiter with a diameter of about 35 miles (56 kilometers), orbiting at a mean distance of 7.12×10^6 miles (11.46×10^6 kilometers). Also known as Jupiter VI. { hi'mäl·ē·ə }

Himantandraceae [BOT] A family of dicotyledonous plants in the order Magnoliales characterized by several, uniovulate carpels and laminar stamens. { hə¦mant·ən'drās·ē‚ē }

himantioid [MYCOL] Pertaining to a mycelium arranged in spreading fanlike cords. { hə'man·tē‚ȯid }

Himantopterinae [INV ZOO] A subfamily of lepidopteran insects in the family Zygaenidae including small, brightly colored moths with narrow hindwings, ribbonlike tails, and long hairs covering the body and wings. { hə¦man·tō'ter·ə‚nē }

hindbrain *See* rhombencephalon. { 'hīn‚brān }

hindered contraction [MET] Thermal contraction of a casting that is hindered locally due to the particular geometry. { 'hin·dərd kən'trak·shən }

hindered settling [MIN ENG] Settling of particles in a thick suspension in water through which their fall is hindered by rising water. { 'hin·dərd 'set·liŋ }

hindered-settling ratio [MIN ENG] The ratio of the specific gravity of a mineral to that of the suspension of ore raised to a power between one-half and unity. { 'hin·dərd 'set·liŋ 'rä·shō }

hindgut [EMBRYO] The caudal portion of the embryonic digestive tube in vertebrates. { 'hīn‚gət }

H indicator *See* H scope. { 'āch ¦in·də‚kād·ər }

Hindley screw [DES ENG] An endless screw or worm of hourglass shape that fits a part of the circumference of a worm

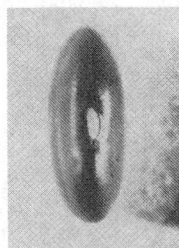

HILUM

A mature kidney bean showing the hilum.

wheel so as to increase the bearing area and thus diminish wear. Also known as hourglass screw; hourglass worm. { 'hind·lē ‚skrü }

hindrance factor *See* drag factor. { 'hin·drəns ‚fak·tər }

Hind's Nebula [ASTRON] A reflection nebula illuminated by the star T Tauri that undergoes marked changes in brightness. { 'hīnz 'neb·yə·lə }

Hindu-Arabic numerals *See* arabic numerals. { ‚hin·dü ‚ar·ə·bik 'nüm·rəlz }

hinge [DES ENG] A pair of metal leaves forming a jointed device on which a swinging part turns. { hinj }

hinged arch [CIV ENG] A structure that can rotate at its supports or in the center or at both places. { 'hinjd 'ärch }

hinged bar [MIN ENG] A steel extension bar placed in contact with the mine roof perpendicular to the longwall face and supported by yielding steel props. Also known as link bar. { 'hinjd ‚bär }

hinge fault [GEOL] A fault whose movement is an angular or rotational one on a side of an axis that is normal to the fault plane. { 'hinj ‚fȯlt }

hinge joint *See* ginglymus. { 'hinj ‚jȯint }

hinge line [GEOL] **1.** The line separating the region in which a beach has been thrust upward from that in which it is horizontal. **2.** A line in the plane of a hinge fault separating the part of a fault along which thrust or reverse movement occurred from that having normal movement. { 'hinj ‚līn }

hinge moment [AERO ENG] The tendency of an aerodynamic force to produce motion about the hinge line of a control surface. { 'hinjd ‚mō·mənt }

hinge plate [INV ZOO] **1.** In bivalve mollusks, the portion of a valve that supports the hinge teeth. **2.** The socket-bearing part of the dorsal valve in brachiopods. { 'hinj ‚plāt }

hinge tooth [INV ZOO] A projection on a valve of a bivalve mollusk near the hinge line. { 'hinj ‚tüth }

Hinsberg test [ANALY CHEM] A test to distinguish between primary and secondary amines; it involves reaction of an amine with benzene disulforyl chloride in alkaline solution; primary amines give sulfonamides that are soluble in basic solution; secondary amines give insoluble derivatives; tertiary amines do not react with the reagent. { 'hinz·bərg ‚test }

hinsdalite [MINERAL] $(Pb,Sr)Al_3(PO_4)(SO_4)(OH)_6$ Dark-gray or greenish rhombohedral mineral composed of basic lead and strontium aluminum sulfate and phosphate; occurs in coarse crystals and masses. { 'hinz‚dā‚līt }

hinterland [GEOL] **1.** The region behind the coastal district. **2.** The terrain on the back of a folded mountain chain. **3.** The moving block which forces geosynclinal sediments toward the foreland. { 'hin·tər‚land }

Hiodontidae [VERT ZOO] A family of tropical, fresh-water actinopterygian fishes in the order Osteoglossiformes containing the mooneyes of North America. { ‚hī·ə'dänt·ə‚dē }

hip [ANAT] **1.** The region of the junction of thigh and trunk. **2.** The hip joint, formed by articulation of the femur and hip-bone. [BUILD] **1.** The external angle formed by the junction of two sloping roofs or the sides of a roof. **2.** A rafter that is positioned at the junction of two sloping roofs or the sides of a roof. [CIV ENG] *See* hip joint. { hip }

HIP *See* hot isostatic pressing. { hip *or* ‚āch‚ī'pē }

hipbone [ANAT] A large broad bone consisting of three parts, the ilium, ischium, and pubis; makes up a lateral half of the pelvis in mammals. Also known as innominate. { 'hip‚bōn }

hip joint [CIV ENG] The junction of an inclined head post and the top chord of a truss. Also known as hip. { 'hip ‚jȯint }

hi pot [ELEC] High potential voltage applied across a conductor to test the insulation or applied to an etched circuit to burn out tenuous conducting paths that might later fail in service. { 'hī ‚pät }

Hippidea [INV ZOO] A group of decapod crustaceans belonging to the Anomura and including cylindrical or squarish burrowing crustaceans in which the abdomen is symmetrical and bent under the thorax. { hi'pid·ē·ə }

Hippoboscidae [INV ZOO] The louse flies, a family of cyclorrhaphous dipteran insects in the section Pupipara. { ‚hip·ə'bäs·kə‚dē }

hippocampal sulcus [ANAT] A fissure on the brain situated between the para hippocampal gyrus and the fimbria hippocampi. Also known as dentate fissure. { ‚hip·ə‚kam·pəl 'səl·kəs }

hippocampus [ANAT] A ridge that extends over the floor of the descending horn of each lateral ventricle of the brain. { ‚hip·ə'kam·pəs }

hippocampus [VERT ZOO] A genus of marine fishes in the order Gasterosteiformes which contains the sea horses. { ‚hip·ə'kam·pəs }

Hippocrateaceae [BOT] A family of dicotyledonous plants in the order Celastrales distinguished by an extrastaminal disk, mostly opposite leaves, seeds without endosperm, and a well-developed latex system. { ‚hip·ə‚krād·ē'ās·ē‚ē }

hippocrepiform [BIOL] Horseshoe-shaped. { ‚hip·ə'krep·ə‚fȯrm }

Hippoglossidae [VERT ZOO] A family of actinopterygian fishes in the order Pleuronectiformes composed of the flounders and plaice. { ‚hip·ə'gläs·ə‚dē }

Hippomorpha [VERT ZOO] A suborder of the mammalian order Perissodactyla containing horses, zebras, and related forms. { ‚hip·ə'mȯr·fə }

hippopede [MATH] A plane curve whose equation in polar coordinates r and θ is $r^2 = 4b(a - b\sin^2\theta)$, where a and b are positive constants. Also known as horse fetter. { 'hip·ə‚pēd }

Hippopotamidae [VERT ZOO] The hippopotamuses, a family of palaeodont mammals in the superfamily Anthracotherioidea. { ‚hip·ə·pəd'am·ə‚dē }

hippopotamus [VERT ZOO] The common name for two species of artiodactyl ungulates composing the family Hippopotamidae. { ‚hip·ə'päd·ə·məs }

Hipposideridae [VERT ZOO] The Old World leaf-nosed bats, a family of mammals in the order Chiroptera. { ‚hi‚pō·sə'der·ə‚dē }

hippuric acid [ORG CHEM] $C_6H_5CONHCH_2{\cdot}COOH$ Colorless crystals melting at 188°C; soluble in hot water, alcohol, and ether; used in medicine and as a chemical intermediate. { hi'pyu̇r·ik 'as·əd }

hip rafter [BUILD] A diagonal rafter extending from the plate to the ridge of a roof. { 'hip ‚raf·tər }

hip roof [ARCH] A roof having four slopes; the two shorter ends are triangular. { 'hip ‚rüf }

hiran [NAV] A modification of the shoran system; special operating techniques permit distance measurements of an accuracy comparable to first-order triangulation. Derived from high-precision shoran. { 'hī‚ran }

Hirayama family [ASTRON] A clustering of asteroids whose orbits have similar values of semimajor axis, eccentricity, and inclination; over 100 such families have been tabulated. { ‚hi·rä'yä·mä ‚fam·lē }

Hirschback method [MIN ENG] A method for draining firedamp from coal seams by means of superjacent entries located 80–138 feet (24–42 meters) above the seams and supplemented by boreholes drilled at right angles to the entry walls. Also known as superjacent roadway system. { 'hərsh‚bak ‚meth·əd }

Hirschsprung's disease [MED] A disease caused by absence of the myenteric ganglion cells in a segment of rectum or distal colon, resulting in spasm of the affected part and dilation of the bowel proximal to the defect. { 'hərsh‚pru̇nz di‚zēz }

hirsute [BIOL] Shaggy; hairy. { 'hər‚süt }

hirsutism [MED] An abnormal condition characterized by growth of hair in unusual places and in unusual amounts. { 'hər·sə‚tiz·əm }

Hirudinea [INV ZOO] A class of parasitic or predatory annelid worms commonly known as leeches; all have 34 body segments and terminal suckers for attachment and locomotion. { ‚hi·rə'din·ē·ə }

Hirudinidae [VERT ZOO] The swallows, a family of passeriform birds in the suborder Oscines. { ‚hi·rə'din·ə‚dē }

hisingerite [MINERAL] $Fe_2^{3-}Si_2O_5(OH)_4{\cdot}2H_2O$ A black, amorphous mineral composed of hydrous ferric silicate; an iron ore. { 'hi·siŋ·ə‚rīt }

***his* operon** [GEN] A sequence of nine contiguous genes in the bacterial chromosome in various species; these code for all the enzymes of histidine biosynthesis. { 'his 'äp·ə‚rän }

hispid [BIOL] Having a covering of bristles or minute spines. { 'his·pəd }

hispidulous [BIOL] Hispid to a minute degree. { his'pij·ə·ləs }

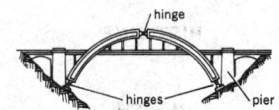

HINGED ARCH

A three-hinged bridge arch, which has the advantage that it is statically determinate, and no stresses result from temperature, shrinkage, or rib-shortening effects.

HIPPOPOTAMUS

The great African hippopotamus.

HIRUDINEA

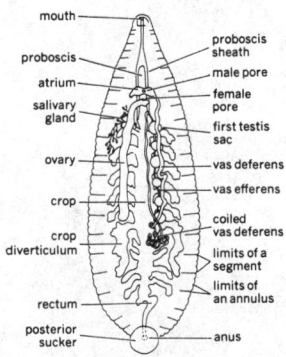

General structure of a leech. Male reproductive system is shown on the right, the female on the left.

HISTIDINE

Structural formula of histidine.

hiss [COMMUN] Random noise in the audio-frequency range, similar to prolonged sibilant sounds. { his }

histamine [BIOCHEM] $C_5H_9N_3$ An amine derivative of histadine which is widely distributed in human tissues. { 'his·tə,mēn }

Histeridae [INV ZOO] The clown beetles, a large family of coleopteran insects in the superfamily Histeroidea. { hi'ster·ə,dē }

Histeroidea [INV ZOO] A superfamily of coleopteran insects in the suborder Polyphaga. { ,his·tə'rȯid·ē·ə }

histidase [BIOCHEM] An enzyme found in the liver of higher vertebrates that catalyzes the deamination of histidine to urocanic acid. { 'his·tə,dās }

histidine [BIOCHEM] $C_6H_9O_2N_3$ A crystalline basic amino acid present in large amounts in hemoglobin and resulting from the hydrolysis of most proteins. { 'his·tə,dēn }

histidine-kinase-associated receptor [CELL MOL] A class of enzyme-linked cell-surface receptors found in bacteria, yeast, and plant cells that transduce intracellular signals via a two-component pathway that results in the phosphorylation of an intracellular messenger protein. { ¦his·tə,dēn ¦kī,nās ə,sō·sē,ād·əd ri'sep·tər }

histidinemia [MED] An asymptomatic, hereditary metabolic disorder involving a deficiency of histidase with high blood and urine levels of histidine, urocanic acid, and sometimes alanine. { ¦his·tə·də'nē·ē·ə }

histiocyte See macrophage. { 'his·tē·ə,sīt }

histiocytoma [MED] 1. Benign tumor composed of histiocytes. 2. Dermatofibroma. { ,his·tē·ō,sī'tō·mə }

histiocytosis [MED] Abnormal proliferation of histiocytes, especially in hematopoietic tissues. { ,his·tē·ō,sī'tō·səs }

Histioteuthidae [INV ZOO] A family of cephalopod mollusks containing several species of squids. { ,his·tē·ō'tü·thə,dē }

histochemistry [BIOCHEM] A science that deals with the distribution and activities of chemical components in tissues. { ¦hi·stō'kem·ə·strē }

histocompatibility [IMMUNOL] The capacity to accept or reject a tissue graft. { ¦hi·stō·kəm¦pad·ə'bil·əd·ē }

histocompatibility gene [GEN] In mammals, any of the genes, especially those of the major histocompatibility complex, that influence immunological properties of cellular antigens. { ,his·tə·kəm¦pad·ə,bil·əd·ē ¦jēn }

histodifferentiation [EMBRYO] Differentiation of cell groups into tissues. { ¦hi·stō·dif·ə,ren·chē'ā·shən }

histogen [BOT] A clearly delimited region or primary tissue from which the specific parts of a plant organ are thought to be produced. { 'his·tə·jən }

histogenesis [EMBRYO] The developmental process by which the definite cells and tissues which make up the body of an organism arise from embryonic cells. { ,his·tə'jen·ə·səs }

histogram [STAT] A graphical representation of a distribution function by means of rectangles whose widths represent intervals into which the range of observed values is divided and whose heights represent the number of observations occurring in each interval. { 'his·tə,gram }

histoincompatibility [IMMUNOL] The condition in which a recipient rejects a tissue graft. { ¦hi·stō,in·kəm,pad·ə'bil·əd·ē }

histologist [ANAT] An individual who specializes in histology. { hi'stäl·ə·jəst }

histology [ANAT] The study of the structure and chemical composition of animal and plant tissues as related to their function. { hi'stäl·ə·jē }

histolysis [PATH] Disintegration of organic tissue. { hi'stäl·ə·səs }

histomycosis [MED] Infection of deep tissues by a fungus. { ,his·tə,mī'kō·səs }

histone [BIOCHEM] Any of the strong, soluble basic proteins of cell nuclei that are precipitated by ammonium hydroxide. { 'hi,stōn }

histopathology [PATH] A branch of pathology that deals with tissue changes associated with disease. { ,hi·stō·pə'thäl·ə·jē }

histophysiology [PHYSIO] The science of tissue functions. { ,hi·stō,fiz·ē'äl·ə·jē }

Histoplasma [MYCOL] A genus of parasitic fungi. { ,his·tə'plaz·mə }

Histoplasma capsulatum [MYCOL] The parasitic fungus that causes histoplasmosis in humans. { ,his·tə'plaz·mə ,kap·sə'läd·əm }

histoplasmin [PHARM] A standardized liquid concentrate of soluble growth factors developed by the fungus *Histoplasma capsulatum*. { ,his·tə'plaz·mən }

histoplasmin test [IMMUNOL] Skin test for hypersensitivity reaction to *Histoplasma capsulatum* products in the diagnosis of histoplasmosis. { ,his·tə'plaz·mən ,test }

histoplasmoma [MED] A tumorlike swelling caused by an inflammatory reaction to *Histoplasma capsulatum*. { ,his·tə,plaz'mō·mə }

histoplasmosis [MED] An infectious fungus disease of the lungs of humans caused by *Histoplasma capsulatum*. { ,his·tə,plaz'mō·səs }

historadiography [BIOPHYS] A technique for taking x-ray pictures of cells, tissues, or sometimes whole small organisms. { ,his·tə,rād·ē'äg·rə·fē }

historical biogeography [ECOL] The study of how species' distributions have changed over time in relationship to the history of landforms, ocean basins, and climate, as well as how those changes have contributed to the evolution of biotas. { his,tär·i·kəl ,bī·ō·jē'ag·rə·fē }

historical climate [CLIMATOL] A climate of the historical period (the past 7000 years). { hi'stär·ə·kəl 'klī·mət }

historical data [COMPUT SCI] Any data that is not actively maintained by a computer system and cannot be readily revised or updated. { hi'stär·ə·kəl 'dad·ə }

historical geology [GEOL] A branch of geology concerned with the systematic study of bedded rocks and their relations in time and the study of fossils and their locations in a sequence of bedded rocks. { hi'stär·ə·kəl jē'äl·ə·jē }

historical zoogeography [ECOL] The study of animal distributions in terms of evolutionary history. { his,tär·i·kəl ,zō·ō·jē'äg·rə·fē }

Histosol [GEOL] An order of wet soils consisting mostly of organic matter, popularly called peats and mucks. { 'his·tə,sȯl }

histotome [BIOL] A microtome used to cut tissue sections for microscopic examination. { 'his·tə,tōm }

Histriobdellidae [INV ZOO] A small family of errantian polychaete worms that live as ectoparasites on crayfishes. { ,his·trē·əb'del·ə,dē }

histrionism [PSYCH] Dramatic affect, associated with some psychoneuroses and psychoses. { 'his·trē·ə,niz·əm }

hit [COMPUT SCI] The obtaining of a correct answer in a mechanical information-retrieval system. [ELEC] A momentary electrical disturbance on a transmission line. [ORD] 1. A blow or impact on a target by a bullet, bomb, or other projectile. 2. An instance of striking something with a bomb or the like. { hit }

hitch [GEOL] 1. A fault of strata common in coal measures, accompanied by displacement. 2. A minor dislocation of a vein or stratum not exceeding in extent the thickness of the vein or stratum. [MIN ENG] 1. A step cut in the rock face to hold timber support in an underground working. 2. A hole cut in side rock solid enough to hold the cap of a set of timbers, permitting the leg to be dispensed with. { hich }

Hitchcock transportation problem [MATH] The problem in linear programming of minimizing the cost of moving ships between two configurations in both of which there is a specified number of ships in each of a finite number of ports, when the costs of moving one ship from each of the ports in the first configuration to each of the ports in the second are specified. { ¦hich,käk ,tranz·pər'tā·shən ,präb·ləm }

hitcher [MIN ENG] 1. The worker who runs trams into or out of the cages, gives the signals, and attends at the shaft when workers are riding in the cage. 2. A worker at the bottom of a haulage slope or plane who engages the clips or grips by means of which mine cars are attached to a hoisting cable or chain. Also known as hitcher-on. { 'hich·ər }

hitcher-on See hitcher. { 'hich·ər¦ȯn }

hitchhiking effect [GEN] The increase in frequency of a neutral allele at a locus closely linked to a selectively favored allele at a different locus. { 'hich,hīk·iŋ i,fekt }

hitch timbering [MIN ENG] Installing timbers in hitches either cut or drilled in the rock. { 'hich ,tim·bə·riŋ }

hit-on-the-fly system [COMPUT SCI] A printer in a computer system where either the print roller or the paper is in continuous motion. { ¦hid ȯn t̲h̲ə 'flī ,sis·təm }

hit probability [ORD] The probability of hits being made on a target out of a given number of projectiles directed at the target. { 'hit ˌpräb·əˈbil·əd·ē }

hit rate [COMPUT SCI] The ratio of the number of records found and processed during a particular processing run, to the total number of records available. { 'hit ˌrāt }

Hittorf dark space *See* cathode dark space. { 'hi·dȯrf 'därk ˌspās }

Hittorf method [PHYS CHEM] A procedure for determining transference numbers in which one measures changes in the composition of the solution near the cathode and near the anode of an electrolytic cell, due to passage of a known amount of electricity. { 'hi·dȯrf ˌmeth·əd }

Hittorf principle [ELECTR] The principle that a discharge between electrodes in a gas at a given pressure does not necessarily occur between the closest points of the electrodes if the distance between these points lies to the left of the minimum on a graph of spark potential versus distance. Also known as short-path principle. { 'hi·dȯrf ˌprin·sə·pəl }

HIV *See* human immunodeficiency virus.

hive *See* beehive. { hīv }

hives *See* urticaria. { hīvz }

HJD *See* heliocentric Julian date.

hjelmite [MINERAL] A black mineral containing yttrium, iron, manganese, uranium, calcium, columbium, tantalum, tin, and tungsten oxide; often occurs with crystal structure disrupted by radiation. { 'yel,mīt }

Hjelmslev plane *See* affine Hjelmslev plane. { 'hyelm,slev ˌplän }

HK *See* Hefner candle.

hl *See* hectoliter.

HLA *See* human leukocyte antigen.

HLA complex [IMMUNOL] The major histocompatibility complex of humans. { ˈāchˈelˈā ˌkäm,pleks }

h-line *See* h-bar. { 'āch ˌlīn }

hm *See* hectometer.

H mode *See* transverse electric mode. { 'āch,mōd }

H Monel [MET] A Monel containing 3.2% silicon available in cast form; it is harder and stronger than Monel. { 'āch mō,nel }

HMPA *See* hexametapol.

HMXRB *See* high-mass x-ray binary.

H network [ELECTR] An attenuation network composed of five branches and having the form of the letter H. Also known as H attenuator; H pad. { 'āch ˌnet,wərk }

Ho *See* holmium.

hoar crystal [HYD] An individual ice crystal in a deposit of hoarfrost; always grows by sublimation. { 'hȯr ˌkrist·əl }

hoarding behavior [VERT ZOO] The carrying of food to the home nest for storage, in quantities exceeding daily need. { 'hȯrd·iŋ bi,hāv·yər }

hoarfrost [HYD] A deposit of interlocking ice crystals formed by direct sublimation on objects. Also known as white frost. { 'hȯr,frȯst }

hoarhound *See* marrubiumum. { 'hȯr,haůnd }

hoarse [MED] Having a harsh, discordant voice, caused by an abnormal condition of the larynx or throat. { hȯrs }

hoary [BOT] Having grayish or whitish color, referring to leaves. { 'hȯr·ē }

hob [DES ENG] A master model made from hardened steel which is used to press the shape of a plastics mold into a block of soft steel. [MECH ENG] A rotary cutting tool with its teeth arranged along a helical thread; used for generating gear teeth. { häb }

hobber *See* hobbing machine. { 'häb·ər }

hobbing [DES ENG] In plastics manufacturing, the act of creating multiple mold cavities by pressing a hob into soft metal cavity blanks. [MECH ENG] Cutting evenly spaced forms, such as gear teeth, on the periphery of cylindrical workpieces. { 'häb·iŋ }

hobbing machine [MECH ENG] A machine for cutting gear teeth in gear blanks or for cutting worm, spur, or helical gears. Also known as hobber. { 'häb·iŋ mə,shēn }

hobbing steel [MET] A high-speed steel used to make gear teeth cutters. { 'häb·iŋ ˌstēl }

hobnail [DES ENG] A short, large-headed, sharp-pointed nail; used to attach soles to heavy shoes. { 'häb,nāl }

hobo connection [ENG] A parallel electrical connection used in blasting. { 'hō·bō kə,nek·shən }

hod [CIV ENG] A tray fitted with a handle by which it can be carried on the shoulder for transporting bricks or mortar. { häd }

Hodge conjecture [MATH] The $2p$-dimensional rational cohomology classes in an n-dimensional algebraic manifold M which are carried by algebraic cycles are those with dual cohomology classes representable by differential forms of bidegree $(n\text{-}p, n\text{-}p)$ on M. { 'häj kən,jek·chər }

Hodgkin's disease [MED] A disease characterized by a neoplastic proliferation of atypical histiocytes in one or several lymph nodes. Also known as lymphogranulomatosis. { 'häj,känz di,zēz }

hodgkinsonite [MINERAL] $MnZnSiO_5 \cdot H_2O$ A pink to reddish-brown mineral composed of hydrous zinc manganese silicate; occurs as crystals. { 'häj·kən·sə,nīt }

Hodgson number [CHEM ENG] Method of predicting the metering error during pulsating gas flow when a surge tank is located between the pulsation source (pump or compressor) and the meter (orifice, nozzle, or venturi). { 'häj·sən ˌnəm·bər }

hodograph [PHYS] **1.** The curve traced out in the course of time by the tip of a vector representing some physical quantity. **2.** In particular, the path traced out by the velocity vector of a given particle. { 'häd·ə,graf }

hodograph method [FL MECH] A method for studying two-dimensional steady fluid flow in which the independent variables are taken as the components of the velocity with respect to cartesian or polar coordinates, rather than the coordinates themselves. { 'häd·ə,graf ,meth·əd }

hodoscope [NUCLEO] An array of small Geiger counters, scintillation counters, or other radiation counters used in tracing paths of high-energy particles in experiments with particle accelerators or in cosmic rays. [OPTICS] *See* conoscope. { 'häd·ə,skōp }

Hodotermitidae [INV ZOO] A family of lower (primitive) termites in the order Isoptera. { ˌhäd·ō·tərˈmid·ə,dē }

hoe [DES ENG] An implement consisting of a long handle with a thin, flat, straight-edged blade attached transversely to the end; used for cultivating and weeding. { hō }

hoegbomite [MINERAL] $Mg(Al,Fe,Ti)_4O_7$ A black mineral composed of an oxide of magnesium, aluminum, iron, and titanium. Also spelled högbomite. { 'hāg·bə,mīt }

Hoepfner process *See* Hopfner process. { 'hepf·nər ,präs·əs }

hoernesite [MINERAL] $Mg_3As_2O_8 \cdot H_2O$ A white, monoclinic mineral composed of hydrous magnesium arsenate; occurs as gypsumlike crystals. { 'hər·nə,sīt }

hoe scraper [MIN ENG] A scraper-loader consisting of a box-sided hoe pulled by cables and used in mining to gather and transport severed rock. { 'hō ˌskrāp·ər }

hoe shovel [MECH ENG] A revolving shovel with a pull-type bucket rigidly attached to a stick hinged on the end of a live boom. { 'hō ˌshəv·əl }

Hofbauer cell [HISTOL] A large, possibly phagocytic cell found in chorionic villi. { 'hȯf·baůr ,sel }

Hofmann amine separation [ORG CHEM] A technique to separate a mixture of primary, secondary, and tertiary amines; they are heated with ethyl oxalate; there is no reaction with tertiary amines, primary amines form a diamide, and the secondary amines form a monoamide; when the reaction mixture is distilled, the mixture is separated into components. { 'häf·mən 'am,ēn ,sep·ə,rā·shən }

Hofmann degradation [ORG CHEM] The action of bromine and an alkali on an amide so that it is converted into a primary amine with one less carbon atom. { 'häf·mən deg·rə'dā·shən }

Hoffmann electrometer [ENG] A variant of the quadrant electrometer that has two sections instead of four. { ˈhäf·mən i,lek'träm·əd·ər }

Hofmann exhaustive methylation reaction [ORG CHEM] The thermal decomposition of quaternary ammonium hydroxide compounds to yield an olefin and water; an exception is tetramethylammonium hydroxide, which decomposes to give an alcohol. { 'häf·mən ig'zȯs·tiv ,meth·ə'lā·shən rē,ak·shən }

hofmannite *See* hartite. { 'häf·mə,nīt }

Hofmann mustard-oil reaction [ORG CHEM] Preparation of alkylisothiocyanates by heating together a primary amine, mercuric chloride, and carbon disulfide. { 'häf·mən 'məs·tərd ,oil rē,ak·shən }

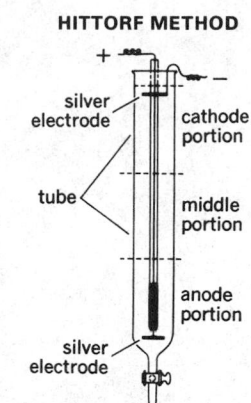

HITTORF METHOD

silver electrode — cathode portion

tube — middle portion

— anode portion

silver electrode

The cell used in the Hittorf method.

Hofmann reaction [ORG CHEM] A reaction in which amides are degraded by treatment with bromine and alkali (caustic soda) to amines containing one less carbon; used commercially in the production of nylon. { 'häf·mən rē,ak·shən }

Hofmann rearrangement [ORG CHEM] A chemical rearrangement of the hydrohalides of N-alkylanilines upon heating to give aminoalkyl benzenes. { 'häf·mən ,rē·ə'rānj·mənt }

Hofmeister series [CHEM] An arrangement of anions or cations in order of decreasing ability to produce coagulation when their salts are added to lyophilic sols. Also known as lyotopic series. { 'hōf,mīs·tər ,sir·ēz }

hog [AGR] A domestic swine. [COMPUT SCI] A computer program that uses excessive computer resources, such as memory or processing power, or requires excessive time to execute. { häg }

hogback [GEOL] Alternate ridges and ravines in certain areas of mountains, caused by erosive action of mountain torrents. { 'häg,bak }

högbomite See hoegbomite. { 'häg·bə,mīt }

hog cholera [VET MED] A fatal infectious virus disease of swine characterized by fever, diarrhea, and inflammation and ulceration of the intestine; secondary infection by *Salmonella cholerae suis* is common. Also known as African swine fever. { 'häg ,käl·ə·rə }

hog frame [NAV ARCH] A rigid framework of beams, usually extending the length of a ship and above the deck, used to prevent hogging and increase longitudinal stiffness. { 'häg ,frām }

hogged fuel [MATER] Sawmill refuse that has been fed through a disintegrator, or hog, by which the various sizes and forms are reduced to a practically uniform size of chips or shreds. { 'hägd ¦fyül }

hogging [ENG] Mechanical chipping of wood waste for fuel. [NAV ARCH] Sagging of the bow and stern of a ship with respect to the amidships section. { 'häg·iŋ }

hog gum [MATER] A sticky gum, an exudate from various species of *Sterculia* trees, whose chief constituent is galactan; the gum is dried and marketed either as a powder or as white flakes; used in the food, cosmetic, and textile industry. { 'häg ,gəm }

hoghorn antenna See horn antenna. { 'häg,hörn an'ten·ə }

hohlraum See blackbody. { 'hōl,raüm }

hohmannite [MINERAL] $Fe_2(SO_4)_2(OH)_2·7H_2O$ A chestnut brown to burnt orange and amaranth red, triclinic mineral consisting of a hydrated basic sulfate of iron. { 'hō·mə,nīt }

Hohmann orbit [AERO ENG] A minimum-energy-transfer orbit. { 'hō·mən ,ör·bət }

Hohmann trajectory [AERO ENG] The minimum-energy trajectory between two planetary orbits, utilizing only two propulsive impulses. { 'hō·mən trə,jek·tə·rē }

hoist [MECH ENG] **1.** To move or lift something by a rope-and-pulley device. **2.** A power unit for a hoisting machine, designed to lift from a position directly above the load and therefore mounted to facilitate mobile service. Also known as winding engine. { höist }

hoist back-out switch [MECH ENG] A protective switch that permits hoist operation only in the reverse direction in case of overwind. { 'höist ¦bak,aüt ,swich }

hoist cable [MECH ENG] A fiber rope, wire rope, or chain by means of which force is exerted on the sheaves and pulleys of a hoisting machine. { 'höist ,kā·bəl }

hoist frame See headframe. { 'höist ,frām }

hoist hook [DES ENG] A swivel hook attached to the end of a hoist cable for securing a load. { 'höist ,hùk }

hoisting [MECH ENG] **1.** Raising a load, especially by means of tackle. **2.** Either of two power-shovel operations: the raising or lowering of the boom, or the lifting or dropping of the dipper stick in relation to the boom. { 'höist·iŋ }

hoisting compartment [MIN ENG] The section of a shaft used for hoisting the mined mineral to the surface. { 'höist·iŋ kəm,pärt·mənt }

hoisting cycle [MIN ENG] The periods of acceleration, uniform speed, retardation, and rest; the deeper the shaft, the greater is the ratio of the time of full-speed hoisting to the whole cycle. { 'höist·iŋ ,sī·kəl }

hoisting drum See drum. { 'höist·iŋ ,drəm }

hoisting machine [MECH ENG] A mechanism for raising and

lowering material with intermittent motion while holding the material freely suspended. { 'höist·iŋ mə,shen }

hoisting power [MECH ENG] The capacity of the hoisting mechanism on a hoisting machine. { 'höist·iŋ ,paù·ər }

hoistman [ENG] One who operates steam or electric hoisting machinery to lower and raise cages, skips, or instruments into a mine or an oil or gas well. Also known as hoist operator; winch operator. { 'höist·mən }

hoist operator See hoistman. { 'höist ,äp·ə,rād·ər }

hoist overspeed device [MECH ENG] A device used to prevent a hoist from operating at speeds greater than predetermined values by activating an emergency brake when the predetermined speed is exceeded. { 'höist ¦ō·vər,spēd di,vīs }

hoist overwind device [MECH ENG] A device which can activate an emergency brake when a hoisted load travels beyond a predetermined point into a danger zone. { 'höist ¦ō·vər,wīnd di,vīs }

hoist slack-brake switch [MECH ENG] A device that automatically cuts off power to the hoist motor and sets the brake if the links in the brake rigging require tightening or if the brakes require relining. { 'höist ¦slak ,brāk ,swich }

hoist tower [CIV ENG] A temporary shaft of scaffolding used to hoist materials for building construction. { 'höist ,taù·ər }

hoistway [MECH ENG] A shaft for one or more elevators, lifts, or dumbwaiters. { 'höist,wā }

hoja blanca [PL PATH] A major virus disease of rice in Cuba and Venezuela. { 'ō·hä 'bläŋ·kä }

holandric trait [GEN] Any trait appearing only in males. { hə¦lan,drik 'trāt }

holarctic zoogeographic region [ECOL] A major unit of the earth's surface extending from the North Pole to 30–45°N latitude and characterized by faunal homogeneity. { hō'lärd·ik ,zō·ō,jē·ə¦graf·ik 'rē·jən }

Holasteridae [INV ZOO] A family of exocyclic Euechinoidea in the order Holasteroida; individuals are oval or heart-shaped, with fully developed pore pairs. { ,häl·ə'ster·ə,dē }

Holasteroida [INV ZOO] An order of exocyclic Euechinoidea in which the apical system is elongated along the antero-posterior axis and teeth occur only in juvenile stages. { ,häl·ə·stə'röid·ə }

holcodont [VERT ZOO] Having the teeth fixed in a long, continuous groove. { 'häl·kə,dänt }

hold [AERO ENG] A scheduled or unscheduled pause in a testing or launching sequence or countdown of a missile or space vehicle. [COMPUT SCI] To retain information in a computer storage device for further use after it has been initially utilized. [ELECTR] To maintain storage elements at equilibrium voltages in a charge storage tube by electron bombardment. [ENG] The interior of a ship or plane, especially the cargo department. [IND ENG] A therblig, or basic operation, in time-and-motion study in which the hand or other body member maintains an object in a fixed position and location. [MECH ENG] A machine motion that is halted by an operator or interlock until it is restarted. { hōld }

hold back [GRAPHICS] In printing of photos, to shade certain portions of the image while printing. { 'hōl 'bak }

holdback [MECH ENG] A brake on an inclined-belt conveyor system which is automatically activated in the event of power failure, thus preventing the loaded belt from running downward. { 'hōl,bak }

holdbeam [NAV ARCH] Any one of a tier of beams which go from side to side, spanning the hold from frame to frame, and upon which a deck is not attached. { 'hōl,bēm }

hold circuit [ELECTR] A circuit in a sampled-data control system that converts the series of impulses, generated by the sampler, into a rectangular function, in order to smooth the signal to the motor or plant. { 'hōld ,sər·kət }

hold control [ELECTR] A manual control that changes the frequency of the horizontal or vertical sweep oscillator in a television receiver, so that the frequency more nearly corresponds to that of the incoming synchronizing pulses. { 'hōld kən,trōl }

holddown [MET] A device that holds the outer portion of a metal sheet in place during deep-drawing operations in order to keep it from becoming wrinkled. [PETRO ENG] A device to anchor an oil well rod pump in its position. { 'hōl,daùn }

holddown groove [ENG] A groove in the side wall of the molding surface which assists in holding the molded plastic article in place when the mold opens. { 'hōl,daùn ,grüv }

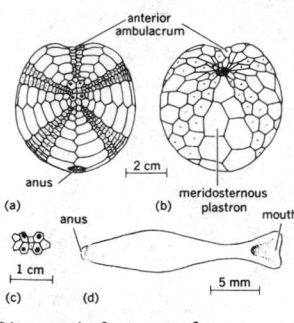

HOLASTEROIDA

Diagnostic features of holasteroids. *(a)* Aboral aspect. *(b)* Adoral aspect. *(c)* Apical system. *(d) Echinosigra paradoxa*, a deep-sea species of Pourtalesiidae, aboral aspect.

holdenite [MINERAL] A red, orthorhombic mineral composed of basic manganese zinc arsenate with a small amount of calcium, magnesium, and iron. { 'hōl·də,nīt }

holder [ELEC] A device that mechanically and electrically accommodates one or more crystals, fuses, or other components in such a way that they can readily be inserted or removed. { 'hōl·dər }

Hölder condition [MATH] **1.** A function $f(x)$ satisfies the Hölder condition in a neighborhood of a point x_0 if

$$|f(x) - f(x_0)| \le c|(x - x_0)|^n$$

where c and n are constants. **2.** A function $f(x)$ satisfies a Hölder condition in an interval or in a region of the plane if

$$|f(x) - f(y)| \le c|x - y|^n$$

for all x and y in the interval or region, where c and n are constants. { 'hel·dər kən,dish·ən }

Hölder's inequality [MATH] Generalization of the Schwarz inequality: for real functions

$$\left| \int f(x)g(x)dx \right| \le \left(\int |f(x)|^p dx \right)^{1/p} \cdot \left(\int |g(x)|^q dx \right)^{1/q}$$

where $1/p + 1/q = 1$. { 'hel·dərz ,in·i'kwäl·əd·ē }

Hölder summation [MATH] A method of attributing a sum to certain divergent series in which a new series is formed, each of whose partial sums is the average of the first n partial sums of the original series, and this process is repeated until a stage is reached where the limit of this average exists. { 'hel·dər sə,mā·shən }

hold facility [COMPUT SCI] The ability of a computer to operate in a hold mode. { 'hōld fə,sil·əd·ē }

holdfast [BOT] **1.** A suckerlike base which attaches the thallus of certain algae to the support. **2.** A disklike terminal structure on the tendrils of various plants used for attachment to a flat surface. [INV ZOO] An organ by which parasites such as tapeworms attach themselves to the host. { 'hōl,fast }

holding anode [ELECTR] A small auxiliary anode used in a mercury-pool rectifier to keep a cathode spot energized during the intervals when the main-anode current is zero. { 'hōl·diŋ ,an,ōd }

holding beam [ELECTR] A diffused beam of electrons used to regenerate the charges stored on the dielectric surface of a cathode-ray storage tube. { 'hōl·diŋ ,bēm }

holding coil [ELECTR] A separate relay coil that is energized by contacts which close when a relay pulls in, to hold the relay in its energized position after the orginal operating circuit is opened. { 'hōl·diŋ ,kȯil }

holding current [ELECTR] The minimum current required to maintain a switching device in a closed or conducting state after it is energized or triggered. { 'hōl·diŋ ,kə·rənt }

holding furnace [MET] A heated reservoir to hold molten metal preparatory to casting. { 'hōl·diŋ ,fər·nəs }

holding ground [NAV] The bottom ground of an anchorage; the term is usually preceded by an adjective to indicate the quality of the holding power of the material constituting the bottom. { 'hōl·diŋ ,graund }

holding magnet See lifting magnet. { 'hōl·diŋ ,mag·nət }

holding pattern [NAV] A course over which an aircraft is instructed to fly repeatedly while waiting for clearance to land. { 'hōl·diŋ ,pad·ərn }

holding point [NAV] An identifiable point designated by an air-traffic control center as the position in the vicinity of which an aircraft is instructed to remain. { 'hōl·diŋ ,pȯint }

holding time [COMMUN] Period of time a trunk or circuit is in use on a call, including operator's time in connecting and subscriber's or user's conversation time. { 'hōl·diŋ ,tīm }

hold lamp [ELEC] Indicating lamp which remains lighted while a telephone connection is being held. { 'hōld ,lamp }

hold mode [COMPUT SCI] The state of an analog computer in which its operation is interrupted without altering the values of the variables it is handling, so that computation can continue when the interruption is over. Also known as interrupt mode. { 'hōld ,mōd }

hold-over command [COMPUT SCI] A command punched at the end of each card to cause machines to treat the several cards as if they were one continuous record. { 'hōl,dō·vər kə,mand }

hold queue [COMPUT SCI] A queue consisting of jobs that have been submitted for execution by a large computer system and are waiting to be run. { 'hōld ,kyü }

hold time [MET] In resistance welding, the time that pressure is applied to the electrodes after the welding current is cut off. { 'hōl ,tīm }

holdup [CHEM ENG] **1.** Volume of material held or contained in a process vessel or line. **2.** Liquid held up (suspended) in a vertical process vessel or line by rising gas or vapor streams. { 'hōl,dəp }

hole [SOLID STATE] A vacant electron energy state near the top of an energy band in a solid; behaves as though it were a positively charged particle. Also known as electron hole. { hōl }

hole burning [PHYS] Saturation of attenuation or gain that is confined to a narrow range of frequencies (hole) within an inhomogeneously broadened transition when the saturating radiation is confined to frequencies within this range. { 'hōl ,bərn·iŋ }

hole-burning spectroscopy [SPECT] A method of observing extremely narrow line widths in certain ions and molecules embedded in crystalline solids, in which broadening produced by crystal-site-dependent statistical field variations is overcome by having a monochromatic laser temporarily remove ions or molecules at selected crystal sites from their absorption levels, and observing the resulting dip in the absorption profile with a second laser beam. { 'hōl ,bərn·iŋ spek'träs·kə·pē }

hole conduction [ELECTR] Conduction occurring in a semiconductor when electrons move into holes under the influence of an applied voltage and thereby create new holes. { 'hōl kən¦dək·shən }

Holectypidae [PALEON] A family of extinct exocyclic Euechinoidea in the order Holectypoida; individuals are hemispherical. { hō,lek'tip·ə,dē }

Holectypoida [INV ZOO] An order of exocyclic Euechinoidea with keeled, flanged teeth, with distinct genital plates, and with the ambulacra narrower than the interambulacra on the adoral side. { hō,lek·tə'pȯid·ə }

hole deviation [ENG] The change in the course or direction that a borehole follows. { 'hōl ,dō·vē,ā·shən }

hole director [MIN ENG] A steel framework used in underground tunneling to set the angle at which holes for a blasting round are to be drilled. { 'hōl di,rek·tər }

hole injection [ELECTR] The production of holes in an n-type semiconductor when voltage is applied to a sharp metal point in contact with the surface of the material. { 'hōl in,jek·shən }

hole layout [MIN ENG] In quarrying, an arrangement of vertical and horizontal holes. { 'hōl 'lā,aut }

hole mobility [ELECTR] A measure of the ability of a hole to travel readily through a semiconductor, equal to the average drift velocity of holes divided by the electric field. { 'hōl mō,bil·əd·ē }

hole saw See crown saw. { 'hōl ,sȯ }

hole site [COMPUT SCI] The area on a punch card where a hole may be punched. { 'hōl ,sīt }

hole theory [QUANT MECH] A theory about the significance of negative energy states in the Dirac theory which leads to the prediction of the existence of the positron and, by extension, to that of other antiparticles. { 'hōl ,thē·ə·rē }

hole-through [MIN ENG] The meeting of two approaching tunnel heads. { 'hōl ,thrü }

hole trap [ELECTR] A semiconductor impurity capable of releasing electrons to the conduction or valence bands, equivalent to trapping a hole. { 'hōl ,trap }

holiday [ENG] An undesirable discontinuity or break in the anticorrosion protection on pipe or tubing. { 'häl·ə,dā }

holiday detector [ENG] An electrical device used to determine the location of a gap or void in the anticorrosion coating of a metal surface. { 'häl·ə,dā di,tek·tər }

holing [MIN ENG] **1.** The working of a lower part of a bed of coal to bring down the upper mass. **2.** The final act of connecting two workings underground. **3.** The meeting of two mine roadways driven to intersect. Also known as thirling. { 'hōl·iŋ }

holism [BIOL] The view that the whole of a complex system, such as a cell or organism, is functionally greater than the sum of its parts. Also known as organicism. { 'hō,liz·əm }

holistic masks [COMPUT SCI] In character recognition, that set of characters which resides within a character reader and theoretically represents the exact replicas of all possible input characters. { hō'lis·tik 'masks }

hollander [MECH ENG] An elongate tube with a central midfeather and a cylindrical beater roll; formerly used for stock preparation in paper manufacture. { 'häl·ən·dər }

Holland formula [ENG] A formula used to calculate the height of a plume formed by pollutants emitted from a stack in terms of the diameter of the stack exit, the exit velocity and heat emission rate of the stack, and the mean wind speed. { 'häl·ənd ,fȯr·myə·lə }

hollandite [MINERAL] $Ba(Mn^{2+},Mn^{4+})_8O_{16}$ A silvery-gray to black mineral composed of manganate of barium and manganese; occurs as crystals. { 'hä·lən,dīt }

Hollerith code [COMPUT SCI] A code used to represent letters, numbers, or special symbols to be punched in standard 80-column punch cards. { 'häl·ə·rəth ,kōd }

Hollerith string [COMPUT SCI] A sequence of characters preceded by an H and a character count in FORTRAN, as 4HSTOP. { 'häl·ə·rəth ,striŋ }

Hollinacea [PALEON] A dimorphic superfamily of extinct ostracods in the suborder Beyrichicopina including forms with sulci, lobation, and some form of velar structure. { ,häl·ə'nās·ē·ə }

Hollinidae [PALEON] An extinct family of ostracodes in the superfamily Hollinacea distinguished by having a bulbous third lobe on the valve. { hə'lin·ə,dē }

hollow [SCI TECH] **1.** Having a concave surface. **2.** Having an interior cavity. { 'häl·ō }

hollow atom [ATOM PHYS] An atom whose electrons are in highly excited states, leaving the states of lower energy (in which the electron is, on average, closer to the nucleus) vacant. { 'häl·ō 'ad·əm }

hollow block See hollow tile. { 'häl·ō 'bläk }

hollow cathode [ELECTR] A cathode which is hollow and closed at one end in a discharge tube filled with inert gas, designed so that radiation is emitted from the cathode glow inside the cathode. { 'häl·ō 'kath,ōd }

hollow-core construction [BUILD] Panel construction with wood faces bonded to a framed-core assembly of elements which support the facing at spaced intervals. { 'häl·ō ¦kȯr kən'strək·shən }

hollow drill [DES ENG] A drill rod or stem having an axial hole for the passage of water or compressed air to remove cuttings from a drill hole. Also known as hollow rod; hollow stem. { 'häl·ō 'dril }

hollow ended [NAV ARCH] Pertaining to when the extremities of the waterlines in the neighborhood of the designed load line are concave to the surrounding water, and when the sectional area curve at the ends is fine, indicating relatively small displacement in these locations. { 'häl·ō ,end·əd }

hollow gravity dam [CIV ENG] A fixed gravity dam, usually of reinforced concrete, constructed of inclined slabs or arched sections supported by transverse buttresses. { 'häl·ō 'grav·əd·ē ,dam }

hollow mill [MECH ENG] A milling cutter with three or more cutting edges that revolve around the cylindrical workpiece. { 'häl·ō ,mil }

hollow-pipe waveguide [ELECTROMAG] A waveguide consisting of a hollow metal pipe; electromagnetic waves are transmitted through the interior and electric currents flow on the inner surfaces. { 'häl·ō ,pīp 'wāv,gīd }

hollow-plunger pump [MIN ENG] A pump for mining and quarrying in gritty and muddy water. { 'häl·ō ,plən·jər 'pəmp }

hollow reamer [ENG] A tool or bit used to correct the curvature in a crooked borehole. { 'häl·ō 'rēm·ər }

hollow rod See hollow drill. { 'häl·ō 'räd }

hollow-rod churn drill [MECH ENG] A churn drill with hollow rods instead of steel wire rope. { 'häl·ō ,räd 'chərn ,dril }

hollow-rod drilling [ENG] A modification of wash boring in which a check valve is introduced at the bit so that the churning action may be also used to pump the cuttings up the drill rods. { 'häl·ō ,räd 'dril·iŋ }

hollow shafting [MECH ENG] Shafting made from hollowed-out rods or hollow tubing to minimize weight, allow internal support, or permit other shafting to operate through the interior. { 'häl·ō 'shaft·iŋ }

hollow stalk [PL PATH] Any plant disease characterized by deterioration of the pith in the stalk. { 'häl·ō 'stȯk }

hollow stem See hollow drill. { 'häl·ō 'stem }

hollow tile [MATER] A hollow building block of concrete or burnt clay used for making partitions, exterior walls, or suspended floors or roofs. Also known as hollow block. { 'häl·ō 'tīl }

hollow wall [BUILD] A masonry wall provided with an air space between the inner and outer wythes. { 'häl·ō 'wȯl }

holly [BOT] The common name for the trees and shrubs composing the genus *Ilex*; distinguished by spiny leaves and small berries. { 'häl·ē }

hollyhock rust [PL PATH] A fungal disease of the hollyhock caused by the rust fungus *Puccinia malvacearum;* characterized by the formation of orange-brown pustules that bear rust-colored teliospores on the undersides of the leaves. { 'häl·ē,äk ,rəst }

hollywood lignumvitae See lignumvitae. { 'häl·ē,wu̇d ,lig·nəm'vī·dē }

Holman counterbalanced drill rig [MIN ENG] A drill rig consisting of a rail-track carriage with a counterbalanced boom 10 feet (3 meters) long. { 'hōl·mən ¦kau̇nt·ər,bal·ənst 'dril ,rig }

Holman dust extractor [MIN ENG] A dust-trapping system in which the dust and chippings from percussive drilling operations are drawn back through the hollow drill rod and along a hose to a metal container with filter elements. { 'hōl·mən 'dəst ik,strak·tər }

Holman stamp [MIN ENG] A crushing stamp raised by a crank and accelerated in its fall by compressed air. { 'hōl·mən ,stamp }

Holmberg radius [ASTRON] The radius of an external galaxy at which the surface brightness is such that the light emitted from one square arc-second equals that from a star of magnitude 26.6. { 'hōm,bərg ,rād·ē·əs }

Holme mud sampler [ENG] A scooplike device which can be lowered by cable to the ocean floor to collect sediment samples. { 'hōm 'məd ,sam·plər }

holmium [CHEM] A rare-earth element belonging to the yttrium subgroup, symbol Ho, atomic number 67, atomic weight 164.9304, melting point 1400–1525°C. { 'hōl·mē·əm }

holmquistite [MINERAL] $(Na,K,Ca)Li(Mg,Fe)_3Al_2Si_8O_{22}$-$(OH)_2$ A bluish-black, orthorhombic mineral composed of alkali and silicate of iron, magnesium, lithium, and aluminum. { 'hōm,kwi,stīt }

holoaxial [CRYSTAL] Having all possible axes of symmetry. { ¦häl·ō'ak·sē·əl }

Holobasidiomycetes [MYCOL] An equivalent name for Homobasidiomycetidae. { ¦häl·ō·bə,sid·ē·ō·mī' sēd·ēz }

holoblastic [EMBRYO] Pertaining to eggs that undergo total cleavage due to the absence of a mass of yolk. { ¦häl·ə¦blas·tik }

holobranch [VERT ZOO] A gill with a row of filaments on each side of the branchial arch. { 'häl·ə,braŋk }

holocarpic [BOT] **1.** Having the entire thallus developed into a fruiting body or sporangium. **2.** Lacking rhizoids and haustoria. { ¦häl·ō¦kär·pik }

holocellulose [BIOCHEM] The total polysaccharide fraction of wood, straw, and so on, that is composed of cellulose and all of the hemicelluloses and that is obtained when the extractives and the lignin are removed from the natural material. { ¦häl·ō'sel·yə,lōs }

Holocene [GEOL] An epoch of the Quaternary Period from the end of the Pleistocene, around 10,000 years ago, to the present. Also known as Postglacial; Recent. { 'hō·lə,sēn }

holocentric chromosome [GEN] A chromosome with centromeric activity spread along its entire length. { ,hō·lə,sen·trik 'krō·mə,sōm }

Holocentridae [VERT ZOO] A family of nocturnal beryciform fishes found in shallow tropical and subtropical reefs; contains the squirrelfishes and soldierfishes. { ,häl·ə'sen·trə,dē }

Holocephali [VERT ZOO] The chimaeras, a subclass of the Chondrichthyes, distinguished by four pairs of gills and gill arches, an erectile dorsal fin and spine, and naked skin. { ¦häl·ō¦sef·ə,lī }

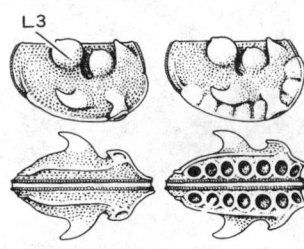

HOLLINIDAE

Abditoloculina pulchra, male (left) and female (right) carapaces showing bulbous L3 (lobe 3) typical of the Hollinidae.

holochrome [BIOCHEM] A colored chromophore bound to an apoprotein. { 'hōl·ə,krōm }

holoclastic [PETR] Being or belonging to ordinary (sedimentary) clastic rock. { ¦häl·ō¦klas·tik }

holococcolith [BIOL] A coccolith with simple rhombic or hexagonal crystals arranged like a mosaic. { ¦häl·ō¦käk·ə,lith }

holocoenosis [ECOL] The nature of the action of the environment on living organisms. { ¦häl·ō·sə¦nō·səs }

holocrine gland [PHYSIO] A structure whose cells undergo dissolution and are entirely extruded, together with the secretory product. { 'häl·ə·krən ,gland }

holocrystalline [PETR] Pertaining to igneous rocks that are entirely crystallized minerals, without glass. { ¦häl·ō'krist·əl·ən }

holoechinate [INV ZOO] Having the whole body covered with spines. { ¦hō·lō·ē'kī,nāt }

holoenzyme [BIOCHEM] A complex, fully active enzyme, containing an apoenzyme and a coenzyme. { ¦häl·ō'en,zīm }

hologamy [BIOL] Condition of having gametes similar in size and form to somatic cells. [BOT] Condition of having the whole thallus develop into a gametangium. { lɪə'läg·ə·mē }

hologony [INV ZOO] Condition of having the germinal area extend the full length of a gonad; refers specifically to certain nematodes. { hə'läg·ə·nē }

hologram [OPTICS] The special photographic plate used in holography; when this negative is developed and illuminated from behind by a coherent gas-laser beam, it produces a three-dimensional image in space. Also known as hologram interferometer. { 'häl·ə,gram }

hologram interferometer See hologram. { 'häl·ə,gram ,in·tər·fə'räm·əd·ər }

holographic interferometry [OPTICS] The study of the formation and interpretation of the fringe pattern which appears when a wave, generated at some earlier time and stored in a hologram, is later reconstructed and caused to interfere with a comparison wave. { ¦häl·ə'graf·ik ,in·tər·fə'räm·ə·trē }

holographic memory [COMPUT SCI] A memory in which information is stored in the form of holographic images on thermoplastic or other recording films. { ¦häl·ə'graf·ik 'mem·rē }

holographic multiplexing See quasi-acoustical holography. { ¦häl·ə'graf·ik 'məl·tə,pleks·iŋ }

holographic optical element [OPTICS] A hologram that is used to control transmitted light beams, rather than to display images, based on the principles of diffraction. { ¦häl·ə'graf·ik 'äp·tə·kəl 'el·ə·mənt }

holographic storage [COMMUN] A form of data storage in which bits of information are distributed throughout the storage volume and recorded interferometrically, rather than being stored at discrete locations in the medium. { ¦häl·ə,graf·ik 'stór·ij }

holography [PHYS] A technique for recording, and later reconstructing, the amplitude and phase distributions of a wave disturbance; widely used as a method of three-dimensional optical image formation, and also with acoustical and radio waves; in optical image formation, the technique is accomplished by recording on a photographic plate the pattern of interference between coherent light reflected from the object of interest, and light that comes directly from the same source or is reflected from a mirror. { hə'läg·rə·fē }

hologynic trait [GEN] Any trait appearing only in females. { ¦hō·lə,gī'nik 'trāt }

holohedral [CRYSTAL] Pertaining to a crystal structure having the highest symmetry in each crystal class. Also known as holosymmetric; holosystemic. { ¦häl·ō¦hē·drəl }

holohedron [CRYSTAL] A crystal form of the holohedral class, having all the faces needed for complete symmetry. { ¦häl·ō¦hē·drən }

holohyaline [PETR] Pertaining to an entirely glassy rock. { ¦häl·ō¦hī·ə·lən }

Holometabola [INV ZOO] A division of the insect subclass Pterygota whose members undergo holometabolous metamorphosis during development. { ¦häl·ō·mə'tab·ə·lə }

holometabolous metamorphosis [INV ZOO] Complete metamorphosis, during which there are four stages; the egg, larva, pupa, and imago or adult. { ¦häl·ō·mə'tab·ə·ləs ,med·ə'mór·fə·səs }

holomicrography [OPTICS] The use of holography to produce three-dimensional images with various types of microscopes. { ¦häl·ō·mī'kräg·rə·fē }

holomictic lake [HYD] A lake whose water circulates completely from top to bottom. { ¦häl·ō¦mik·tik 'lāk }

holomorphic function See analytic function. { ¦häl·ō¦mór·fik 'fəŋk·shən }

holomorphosis [BIOL] Complete regeneration of a lost body structure. { ¦häl·ō'mór·fə·səs }

holomyarian [INV ZOO] Having zones of muscle layers but no muscle cells; refers specifically to certain nematodes. { ¦häl·ō·mī¦a·rē·ən }

holonephros [VERT ZOO] Type of kidney having one nephron beside each somite along the entire length of the coelom; seen in larvae of myxinoid cyclostomes. { ¦häl·ō'ne,frōs }

holonomic constraints [MECH] An integrable set of differential equations which describe the restrictions on the motion of a system; a function relating several variables, in the form $f(x_1, \ldots, x_n) = 0$, in optimization or physical problems. { ¦häl·ə¦näm·ik kən'strans }

holonomic system [MECH] A system in which the constraints are such that the original coordinates can be expressed in terms of independent coordinates and possibly also the time. { ¦häl·ə¦näm·ik 'sis·təm }

holophotal [OPTICS] 1. Pertaining to a holophote. 2. Reflecting all the light from a source in one direction. { ¦häl·ə¦fōd·əl }

holophote [OPTICS] An optical system consisting of lenses or reflectors that collect a large amount of the light from a source (such as the lamp of a lighthouse) and send it in a desired direction. { 'häl·ə,fōt }

holophyte [BIOL] An organism that obtains food in the manner of a green plant, that is, by synthesis of organic substances from inorganic substances using the energy of light. { 'häl·ə,fīt }

holoplankton [ZOO] Organisms that live their complete life cycle in the floating state. { ,häl·ō'plaŋk·tən }

Holoptychidae [PALEON] A family of extinct lobefin fishes in the order Osteolepiformes. { ,häl·əp·tə'kī·ə,dē }

holopulping process [CHEM ENG] A process for making paper pulp by alkaline oxidation of extremely thin wood chips at low temperature and pressure and then solubilization of the lignin fraction. { ¦häl·ō¦pəl·piŋ ,präs·əs }

holorhinal [VERT ZOO] Among birds, having a rounded anterior margin on the nasal bones. { ¦häl·ō'rīn·əl }

holospondyly [VERT ZOO] The condition in which the vertebral centra and spines are single-pieced or fused. { ,häl·ō'span·də·lē }

Holostei [VERT ZOO] An infraclass of fishes in the subclass Actinopterygii descended from the Chondrostei and ancestral to the Teleostei. { hə'läs·tē,ī }

holostome [INV ZOO] A type of adult digenetic trematode having a portion of the ventral surface modified as a complex adhesive organ. { 'häl·ə,stōm }

holostratotype [GEOL] The originally defined stratotype. { ¦häl·ō'strad·ə,tīp }

holosymmetric See holohedral. { ¦häl·ō·si'me·trik }

holosystemic See holohedral. { ¦häl·ō·si'stem·ik }

Holothuriidae [VERT ZOO] A family of aspidochirotacean echinoderms in the order Aspidochirotida possessing tentacular ampullae and only the left gonad. { ,häl·ō·thə'rī·ə,dē }

Holothuroidea [INV ZOO] The sea cucumbers, a class of the subphylum Echinozoa characterized by a cylindrical body and smooth, leathery skin. { ,häl·ō,thù·rē'ȯid·ē·ə }

Holothyridae [INV ZOO] The single family of the acarine suborder Holothyrina. { ,häl·ō'thī·rə,dē }

Holothyrina [INV ZOO] A suborder of mites in the order Acarina which are large and hemispherical with a deep-brown, smooth, heavily sclerotized cuticle. { ,häl·ō'thī·rə·nə }

Holotrichia [INV ZOO] A major subclass of the protozoan class Ciliatea; body ciliation is uniform with cilia arranged in longitudinal rows. { ,häl·ō'trik·ē·ə }

holotype [SYST] A nomenclatural type for the single specimen designated as "the type" by the original author at the time of publication of the original description. { 'häl·ə,tīp }

holozoic [ZOO] Obtaining food in the manner of most animals, by ingesting complex organic matter. { ¦häl·ə¦zō·ik }

holster [ORD] A pocket-type device with a single compartment designed to be worn on a belt or shoulder harness, which

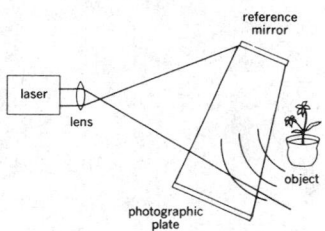

HOLOGRAPHY

Interference pattern between light from laser source reflected from object of interest and light from same source reflected from mirror is recorded on photographic plate which, when developed, forms a hologram.

HOLOSTOME

Diagram of ventral surface of adult holostome showing adhesive organs. (*From R. M. Cable, An Illustrated Laboratory Manual of Parasitology, Burgess, 1940*)

HOLOTHUROIDEA

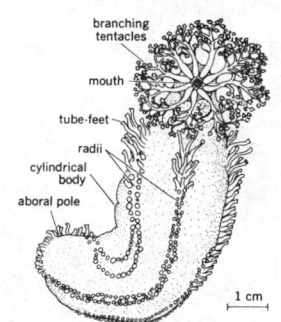

branching tentacles
mouth
tube feet
radii
cylindrical body
aboral pole
1 cm

Cucumaria, a representative holothurian.

may be furnished with it, and used to carry a pistol, revolver, or the like. { 'hōl·stər }

Holtz machine *See* Toepler-Holtz machine. { 'hōlts mə,shēn }

Holuridae [PALEON] A group of extinct chondrostean fishes in the suborder Palaeoniscoidei distinguished in having lepidotrichia of all fins articulated but not bifurcated, fins without fulcra, and the tail not cleft. { hə'lùr·ə,dē }

holystone [MATER] A soft sandstone that is used in scrubbing a ship's deck. { 'hō·lē,stōn }

Holzer's method [MECH] A method of determining the shapes and frequencies of the torsional modes of vibration of a system, in which one imagines the system to consist of a number of flywheels on a massless flexible shaft and, starting with a trial frequency and motion for one flywheel, determines the torques and motions of successive flywheels. { 'hōt·sərz ,meth·əd }

Homacodontidae [PALEON] A family of extinct palaeodont mammals in the superfamily Dichobunoidea. { ,häm·ə·kō'dänt·ə,dē }

Homalopteridae [VERT ZOO] A small family of cypriniform fishes in the suborder Cyprinoidei. { ¦häm·ə·läp'ter·ə,dē }

Homalorhagae [INV ZOO] A suborder of the class Kinorhyncha having a single dorsal plate covering the neck and three ventral plates on the third zonite. { ,häm·ə'lȯr·ə,gē }

Homalozoa [INV ZOO] A subphylum of echinoderms characterized by the complete absence of radial symmetry. { hə¦mal·ə¦zō·ə }

Homan's sign [MED] Pain in the calf and popliteal area on passive dorsiflexion of the foot, indicating deep venous thrombosis of the calf. Also known as dorsiflexion sign. { 'hō·mənz ,sīn }

Homaridae [INV ZOO] A family of marine decapod crustaceans containing the lobsters. { hō'mar·ə,dē }

Homarus [INV ZOO] A genus of the family Homaridae comprising most species of lobsters. { 'häm·ə·rəs }

homatropine [ORG CHEM] $C_{16}H_{21}O_3N$ An alkaloid that causes pupil dilation and paralysis of accommodation. { hōm'a·trə,pēn }

homaxial [BIOL] Having all axes equal. { hōm'ak·sē·əl }

home [COMPUT SCI] The location at the upper left-hand corner of an electronic display. [ELEC] To return to the starting position, as in a stepping relay or turning motor. [NAV] To navigate toward a point by maintaining constant some navigational parameter other than altitude. [ORD] To travel to a target by guidance of heat radiation, radar echoes, radio waves, sound waves, laser beams, or other phenomena reflected from or originating in the target, as in homing missiles and homing torpedoes. { hōm }

home address [COMPUT SCI] A technique used to identify each disk track uniquely by means of a 9-byte record immediately following the index marker; the record contains a flag (good or defective track), cylinder number, head number, cyclic check, and bit count appendage. { ¦hōm 'ad,res }

home key [ENG] One of the eight keys on a keyboard on which the typist's fingers normally rest in the starting position for touch typing. Also known as guide key. { 'hōm ,kē }

homenergic flow [THERMO] Fluid flow in which the sum of kinetic energy, potential energy, and enthalpy per unit mass is the same at all locations in the fluid and at all times. { 'häm·ə,nər·jik 'flō }

homentropic flow [FL MECH] Fluid flow in which the entropy per unit mass is the same at all locations in the fluid and at all times. { 'häm·ən,träp·ik 'flō }

homeoblastic [PETR] Of a metamorphic crystalloblastic texture, having constituent minerals of approximately the same size. { ¦hō·mē·ō¦blas·tik }

homeobox [CELL MOL] A highly conserved sequence of deoxyribonucleic acid (DNA) that occurs in the coding region of development-controlling regulatory genes and codes for a protein domain that is similar in structure to certain DNA-binding proteins and is thought to be involved in the control of gene expression during morphogenesis and development. { 'hō·mē·ə,bäks }

homeohydric [PHYSIO] Pertaining to the ability to restrict cellular water loss regardless of environmental conditions. { ¦hō·mē·o 'hī,drik }

homeomorph [BIOL] An organism which exhibits a superficial resemblance to another organism even though they have different ancestors. [CRYSTAL] A crystal that displays a form similar to that of a crystal with a different chemical composition. { 'hō·mē·ə,mȯrf }

homeomorphic spaces [MATH] Two topological spaces with a homeomorphism existing between them; intuitively one can be obtained from the other by stretching, twisting, or shrinking. { ¦hō·mē·ə¦mȯr·fik 'spās·əz }

homeomorphism [MATH] A continuous map between topological spaces which is one-to-one, onto, and its inverse function is continuous. Also known as bicontinuous function; topological mapping. { ¦hō·mē·ə¦mȯr,fiz·əm }

home-on-jam [ELECTR] A feature that permits radar to track a jamming source in angle. { 'hōm ,ȯn ,jam }

homeopathy [MED] A system of medicine expounded by Samuel Hahnemann that treats disease by administering to the patient small doses of drugs which produce the signs and symptoms of the disease in a healthy person. Also known as Hahnemannism. { ,hō·mē'äp·ə·thē }

homeosis *See* heteromorphosis. { ,hō·mē'ō·səs }

homeostasis [BIOL] In higher animals, the maintenance of an internal constancy and an independence of the environment. { ,hō·mē·ō'stā·səs }

homeotherm [PHYSIO] An endotherm that maintains a constant body temperature, as do most mammals and birds. { 'hō·mē·ə,thərm }

homeothermia [PHYSIO] The condition of being warm-blooded. { ¦hō·mē·ə¦thər·mē·ə }

homeotic mutation [GEN] A gene whose mutation in *Drosophila* can lead to the substitution of one body part for another, such as a leg for an antenna. { 'hō·mē,äd·ik myü'tā·shən }

home page [COMPUT SCI] A document in a hypertext system that serves as the point of entry to a web of related documents, and generally contains introductory information and hyperlinks to other documents in the web. Also known as welcome page. { ,hōm 'pāj }

homer [NAV] A ground-based direction-finding station that utilizes radio transmissions from aircraft to determine their bearing, then guides the aircraft toward the station by voice communications. [ORD] *See* target seeker. { 'hō·mər }

home range [ECOL] The physical area of an organism's normal activity. { 'hōm ¦rānj }

home record [COMPUT SCI] The first record in the chaining method of file organization. { 'hōm¦rek·ərd }

home row [ENG] The row on a keyboard that contains the home keys. { 'hōm ,rō }

home signal [CIV ENG] A signal at the beginning of a block of railroad track that indicates whether the block is clear. { 'hōm ¦sig·nəl }

homespun [TEXT] **1.** A general term for cloth handwoven at home instead of in a mill such as the linsey-woolsey, butternut, and coarse flannels made on handlooms by early American settlers. **2.** Coarse fabrics woven from linen or cotton, for example, drapery fabrics. { 'hōm,spən }

hometaxial-base transistor [ELECTR] Transistor manufactured by a single-diffusion process to form both emitter and collector junctions in a uniformly doped silicon slice; the resulting homogeneously doped base region is free from accelerating fields in the axial (collector-to-emitter) direction, which could cause undesirable high current flow and destroy the transistor. { 'häm·ə,tak·sē·əl ,bās tran'zis·tər }

homeward bound [NAV] Of a vessel or person, headed for the home port or country. { 'hōm·wərd 'baùnd }

homichlophobia [PSYCH] An abnormal fear of fog. { ,häm·ə·klə'fō·bē·ə }

homidium bromide *See* ethidium bromide. { hə'mid·ē·əm 'brō,mīd }

homilite [MINERAL] $Ca_2(Fe,Mg)B_2Si_2O_{10}$ A black or blackish brown mineral composed of iron calcium borosilicate. { 'hä·mə,līt }

homing [NAV] A process in navigation by which the destination is approached by keeping some parameter constant. { 'hōm·iŋ }

homing adapter [NAV] A device which, when used with an aircraft radio receiver, produces aural or visual signals which indicate the direction of a transmitting radio station with respect to the heading of the aircraft. { 'hōm·iŋ ə,dap·tər }

homing antenna [ELECTROMAG] A directional antenna array used in flying directly to a target that is emitting or reflecting radio or radar waves. { 'hōm·iŋ an,ten·ə }

homing beacon [NAV] A radio beacon, either airborne or

on the ground, toward which an aircraft can fly if equipped with a radio compass or homing adapter. Also known as radio homing beacon. { 'hōm·iŋ ˌbē·kən }

homing device [ELECTR] A control device that automatically starts in the correct direction of motion or rotation to achieve a desired change, as in a remote-control tuning motor for a television receiver. [ENG] A device incorporated in a guided missile or the like to home it on a target. [NAV] A transmitter, receiver, or adapter used for homing aircraft or used by aircraft for homing purposes. { 'hōm·iŋ di‚vīs }

homing guidance [ENG] A guidance system in which a missile directs itself to a target by means of a self-contained mechanism that reacts to a particular characteristic of the target. { 'hōm·iŋ ‚gīd·əns }

homing range [NAV] The maximum distance from a target or homing beacon at which a homing device is effective. { 'hōm·iŋ ‚rānj }

homing relay [ELEC] A stepping relay that returns to a specified starting position before each operating cycle. { 'hōm·iŋ ‚rē‚lā }

homing station [NAV] A station at which a beacon emits signals that may be used for homing. { 'hōm·iŋ ‚stā·shən }

homing system [NAV] The two sets of equipment, one on a vehicle and the other on the ground or shore, that act cooperatively to keep some navigational parameter constant in the homing process. { 'hōm·iŋ ‚sis·təm }

homing torpedo [ORD] A torpedo having homing guidance, designed for homing on a surface vessel or a submerged submarine. { 'hōm·iŋ tȯr'pē·dō }

homing transponder [NAV] A small acoustic transponder used in navigation of a submersible vehicle, which can be carried by the vehicle and quickly dropped when an area of interest is reached; a single transponder and a dead reckoning system are used. { 'hōm·iŋ tranz'pän·dər }

hominid [ANTHRO] Any of the bipedal primates of the family Hominidae (modern or extinct); contains the genera *Ardipithecus*, *Australopithecus*, and *Homo*. { 'häm·ə·nid }

Hominidae [VERT ZOO] A family of primates in the superfamily Hominoidea containing one living species, *Homo sapiens*. { hä'min·ə‚dē }

hominoid [ANTHRO] A member of the biological superfamily Hominoidea, including humans, the apes (great and lesser apes), and a number of their extinct ancestors and relatives. { 'häm·ə‚nȯid }

Hominoidea [VERT ZOO] A superfamily of the order Primates comprising apes and humans. { hä·mə'nȯid·ē·ə }

hominy [FOOD ENG] Whole corn kernels that do not contain the bran and germ (which are removed by bleaching or crushing and sifting), that is, corn kernel endosperm. { 'häm·ə·nē }

homo- [ORG CHEM] **1.** Indicating the homolog of a compound differing in formula from the latter by an increase of one CH_2 group. **2.** Indicating a homopolymer made up of a single type of monomer, such as polyethylene from ethylene. **3.** Indicating that a skeletal atom has been added to a well-known structure. [SCI TECH] Indicating the same or similar. { 'hō·mō }

Homo [VERT ZOO] The genus of human beings, including modern humans and many extinct species. { 'hō·mō }

HOMO *See* Highest Occupied Molecular Orbital. { ¦āch¦ō¦em¦ō or 'hō‚mō }

homoallele [GEN] One of two or more alternative forms of a gene that differ at identical mutation sites. { ‚hō·mō·ə'lēl }

Homobasidiomycetidae [MYCOL] A subclass of basidiomycetous fungi in which the basidium is not divided by cross walls. { ¦hä·mō·bə‚sid·ē·ō‚mī'sed·ə‚dē }

homobront *See* isobront. { 'häm·ə¦bränt }

homocentric [OPTICS] Pertaining to rays which have the same focal point, or which are parallel. Also known as stigmatic. { ‚häm·ə'sen·trik }

homocercal [VERT ZOO] Pertaining to the caudal fin of certain fishes which has almost equal lobes, with the vertebral column terminating near the middle of the base. { ‚häm·ə'sər·kəl }

homochiral *See* enantiomerically pure. { ‚hō·mə'kī·rəl }

homochlamydeous [BIOL] Having all members of the perianth similar or not differentiated into calyx or corolla. { ¦hä·mō·klə'mid·ē·əs }

homochromy [ZOO] A form of protective coloration

whereby the individual blends into the background. { hə'mäk·rə·mē }

homocline [GEOL] Any rock unit in which the strata exhibit the same dip. { 'hä·mə‚klīn }

homocyclic compound [ORG CHEM] A ring compound that has one type of atom in its structure; an example is benzene. { ¦hä·mə'sī·klik 'käm‚paund }

homocysteine [BIOCHEM] $C_4H_9O_2NS$ An amino acid formed in animals by demethylation of methionine. { ¦hä·mə'sis·tēn }

homocystinuria [MED] A hereditary disease characterized by a deficiency of the enzyme serine dehydratase causing incompletely dislocated lenses after the age of 10, thromboembolisms, and usually mental retardation. { ‚hō·mō‚sis·tə'nur·ē·ə }

homodesmic [CRYSTAL] Of a crystal, having atoms bonded in a single way. { ‚hä·mə'dez·mik }

homodimer [BIOCHEM] A protein made of paired identical polypeptides. { ‚hō·mō'dī·mər }

homodont [VERT ZOO] Having all teeth similar in form; characteristic of nonmammalian vertebrates. { 'hä·mə‚dänt }

homoduplex [CELL MOL] A deoxyribonucleic acid duplex in which the nitrogenous bases of the two strands are precisely complementary. { ‚hō·mō'dü‚pleks }

homodynamic [INV ZOO] Developing through continuous successive generations without a diapause; applied to insects. { ‚hä·mə‚dī'näm·ik }

homodyne reception [ELECTR] A system of radio reception for suppressed-carrier systems of radiotelephony, in which the receiver generates a voltage having the original carrier frequency and combines it with the incoming signal. Also known as zero-beat reception. { 'hä·mə‚dīn ri'sep·shən }

homoecious [BIOL] Having one host for all stages of the life cycle. { hō'mē·shəs }

homoeomerous [BOT] Having algae distributed uniformly throughout the thallus of a lichen. { ‚hō·mē¦äm·ə·rəs }

Homo erectus [PALEON] A type of fossil human from the Pleistocene of Java and China representing a specialized side branch in human evolution. { 'hō·mō ə'rek·təs }

homocrotism [PSYCH] Sexual desire directed toward a member of the same sex; usually sublimated and not expressed. { ‚hō·mō'er·ə‚tiz·əm }

homofermentative lactobacilli [MICROBIO] Bacteria that produce a single end product, lactic acid, from fermentation of carbohydrates. { ¦hō·mō·fər'men·tə·tiv ‚lak·tō bə‚sil·ē }

homogametic sex [GEN] The sex that produces one type of gamete with respect to sex chromosome content, such as female mammals and male birds. { ¦hä·mō·gə¦med·ik 'seks }

homogamous [BIOL] Of or pertaining to homogamy. { hə'mäg·ə·məs }

homogamy [BIOL] Inbreeding due to isolation. [BOT] Condition of having all flowers alike. { hə'mäg·ə·mē }

homogenate [BIOL] A tissue that has been finely divided and mixed. { hə'mäj·ə·nət }

homogeneity [PHYS] Quality of a substance whose properties are independent of position. [STAT] Equality of the distribution functions of several populations. { ‚hō·mə·jə'nē·əd·ē }

homogeneous [CHEM] Pertaining to a substance having uniform composition or structure. [MATH] Pertaining to a group of mathematical symbols of uniform dimensions or degree. [SCI TECH] Uniform in structure or composition. { ¦hä·mə'jē·nē·əs }

homogeneous atmosphere [METEOROL] A hypothetical atmosphere in which the density is constant with height. { ‚hä·mə'jē·nē·əs 'at·mə‚sfir }

homogeneous catalysis [CHEM] Catalysis occurring within a single phase, usually a gas or liquid. { ‚hä·mə'jē·nē·əs kə'tal·ə·səs }

homogeneous chemical reaction [CHEM] Chemical reaction system in which all constitutents (reactants and catalyst) are of the same phase. { ‚hä·mə'jē·nē·əs ¦kem·i·kəl rē'ak·shən }

homogeneous coordinates [MATH] To a point in the plane with cartesian coordinates (x,y) there corresponds the homogeneous coordinates (x_1, x_2, x_3), where $x_1/x_3 = x$, $x_2/x_3 = y$; any polynomial equation in cartesian coordinates becomes homogeneous if a change into these coordinates is made. { ‚hä·mə'jē·nē·əs kō'ȯrd·ən·əts }

homogeneous differential equation [MATH] A differential

equation where every scalar multiple of a solution is also a solution. { ¦hä·mə¦jē·nē·əs ¦dif·ə¦ren·chəl i¦kwā·zhən }

homogeneous equation [MATH] An equation that can be rewritten into the form having zero on one side of the equal sign and a homogeneous function of all the variables on the other side. { ¦hä·mə¦jē·nē·əs i'kwā·zhən }

homogeneous function [MATH] A real function $f(x_1, x_2, \ldots, x_n)$ is homogeneous of degree r if $f(ax_1, ax_2, \ldots, ax_n) = a^r f(x_1, x_2, \ldots, x_n)$ for every real number a. { ¦hä·mə¦jē·nē·əs 'fəŋk·shən }

homogeneous integral equation [MATH] An integral equation where every scalar multiple of a solution is also a solution. { ¦hä·mə¦jē·nē·əs ¦int·ə·grəl i¦kwā·zhən }

homogeneous line-broadening [OPTICS] An increase beyond the natural linewidth of an absorption or emission line which results from a disturbance (such as collisions or lattice vibrations) that is the same for all the source emitters. { ¦hō·mə¦jē·nē·əs ¦līn ¦bröd·ən·iŋ }

homogeneously staining region [CYTOL] In human chromosomes, an extended chromosomal segment that has a banding pattern and represents a site of gene amplification; found mostly in cancer cells. [GEN] An unusual chromosomal region of variable length that fails to show chromosome banding, the result of amplification of a complex short deoxyribonucleic acid sequence. { ¦hō·mə¦jēn·ē·əs·lē ¦stān·iŋ ¦rē·jən }

homogeneous network [COMPUT SCI] A computer network consisting of fairly similar computers from a single manufacturer. { ¦hō·mə¦jē·nē·əs 'net¦wərk }

homogeneous nucleation [PHYS CHEM] The process of creation of vapor bubble nuclei in a superheated liquid away from bounding walls and in the absence of any foreign material. { ¦hō·mə¦jē·nē·əs ¦nü·klē'a·shən }

homogeneous polynomial [MATH] A polynomial all of whose terms have the same total degree; equivalently it is a homogeneous function of the variables involved. { ¦hä·mə¦jē·nē·əs ¦päl·ə'nō·mē·əl }

homogeneous radiation [PHYS] Radiation having an extremely narrow band of frequencies, or a beam of monoenergetic particles of a single type, so that all components of the radiation are alike. { ¦hä·mə¦jē·nē·əs ¦rād·ē'ā·shən }

homogeneous reactor [NUCLEO] A nuclear reactor in which fissionable material and moderator (if used) are intimately mixed to form an effectively homogeneous medium for neutrons. { ¦hä·mə¦jē·nē·əs rē'ak·tər }

homogeneous space [MATH] A topological space having a group of transformations acting upon it, that is, a transformation group, where for any two points x and y some transformation from the group will send x to y. { ¦hä·mə¦jē·nē·əs 'spās }

homogeneous strain [MECH] A strain in which the components of the displacement of any point in the body are linear functions of the original coordinates. { ¦hō·mə¦jē·nē·əs 'strān }

homogeneous transformation See linear transformation. { ¦hä·mə¦jē·nē·əs ¦tranz·fər'mā·shən }

homogeneously staining region [GEN] An unusual chromosomal region of variable length that fails to show chromosome banding, the result of amplification of a complex short deoxyribonucleic acid sequence. { ¦hä·mə¦jē·nē·əs·lē ¦stān·iŋ ¦rē·jən }

homogenize [MET] To hold metal at a high temperature long enough to eliminate by diffusion any chemical segregation of the components. { hə'mäj·ə¦nīz }

homogenizer [MECH ENG] A machine that blends or emulsifies a substance by forcing it through fine openings against a hard surface. { hə'mäj·ə¦nīz·ər }

homogentisase [BIOCHEM] The enzyme that catalyzes the conversion of homogentisic acid to fumaryl acetoacetic acid. { ¦hä·mə¦jen·tə¦sās }

homogentisic acid [BIOCHEM] $C_8H_8O_4$ An intermediate product in the metabolism of phenylalanine and tyrosine; found in excess in persons with phenylketonuria and alkaptonuria. { ¦hä·mə¦jen¦tiz·ik 'as·əd }

homogony [BOT] Condition of having one type of flower, with stamens and pistil of uniform length. { hə'mäg·ə·nē }

homograft rejection [IMMUNOL] An immunologic process by which an individual destroys and casts off a tissue transplanted from a donor of the same species. { 'hä·mə¦graft ri'jek·shən }

homographic transformations See Möbius transformations. { ¦hä·mə¦graf·ik ¦tranz·fər'mā·shənz }

homoiochlamydeous [BOT] Having perianth leaves alike, not differentiated into sepals and petals. { hō¦mòi·ō·klə'mid·ē·əs }

homoiogenetic [EMBRYO] Of a determined part of an embryo, capable of inducing formation of a similar part when grafted into an undetermined field. { hō¦mòi·ō·jə'ned·ik }

homoiothermal [PHYSIO] Referring to an organism which maintains a constant internal temperature which is often higher than that of the environment; common among birds and mammals. Also known as warm-blooded. { hō¦mòi·ō¦thər·məl }

Homoistela [PALEON] A class of extinct echinoderms in the subphylum Homalozoa. { hō¦mòi·stə·lə }

homojunction bipolar transistor [ELECTR] Any bipolar transistor that is composed entirely of one type of semiconductor. { ¦hō·mō¦jəŋk·shən bī¦pō·lər tran'zis·tər }

homokaryon [MYCOL] A bi- or multinucleate cell having nuclei all of the same kind. { ¦hä·mə'kar·ē¦än }

homokaryosis [MYCOL] The condition of a bi- or multinucleate cell having nuclei all of the same kind. { ¦hä·mə¦kar·ē'ō·səs }

homokaryotype [GEN] A karyotype that is homozygous for a chromosome mutation. { ¦hō·mə'kar·ē·ə¦tīp }

homolateral [MED] Situated on the same side. Also known as ipsilateral. { ¦hä·mə'lad·ə·rəl }

homolecithal [CELL MOL] Referring to eggs having small amounts of evenly distributed yolk. Also known as isolecithal. { ¦hä·mə'les·ə·thəl }

homolog [GEN] One of a pair of homologous chromosomes. Also spelled homologue. { 'häm·ə¦läg }

homologation [ORG CHEM] A type of hydroformylation in which carbon monoxide reacts with certain saturated alcohols to yield either aldehydes or alcohols (or a mixture of both) containing one more carbon atom than the parent. { hə¦mal·ə'gā·shən }

homological algebra [MATH] The study of the structure of modules, particularly by means of exact sequences; it has application to the study of a topological space via its homology groups. { ¦hä·mə¦läj·ə·kəl 'al·jə·brə }

homologous [BIOL] Pertaining to a structural relation between parts of different organisms due to evolutionary development from the same or a corresponding part, such as the wing of a bird and the pectoral fin of a fish. [GEOL] **1.** Referring to strata, in separated areas, that are correlatable (contemporaneous) and are of the same general character or facies, or occupy analogous structural positions along the strike. **2.** Pertaining to faults, in separated areas, that have the same relative position or structure. { hə'mäl·ə·gəs }

homologous chromosomes [GEN] A pair of chromosomes, one inherited from each parent, that have corresponding gene sequences and that pair during meiosis. { hə¦mäl·ə·gəs 'krō·mə¦sōmz }

homologous motion [IND ENG] A motion produced by one set of muscles that can be substituted for an essentially similar motion performed by another set of muscles; the substitution is usually made in order to reduce the stress needed to perform a work task. { hə¦mäl·ə·gəs 'mō·shən }

homologous recombination [GEN] Deoxyribonucleic acid exchange between identical chromosome regions on homologous chromosomes that occurs naturally during meiosis. { hə¦mäl·ə·gəs ¦rē·käm·bə'nā·shən }

homologous stimulus [PHYSIO] A form of energy to which a specific sensory receptor is most sensitive. { hə¦mäl·ə·gəs 'stim·yə·ləs }

homologous transformation [ASTRON] A mathematical transformation in the study of stellar models. { hə¦mäl·ə·gəs ¦tranz·fər'mā·shən }

homologous tumor [MED] A neoplasm composed of tissue identical with those of the organ at the site of the tumor. { hə¦mäl·ə·gəs 'tü·mər }

homology [BIOL] A fundamental similarity between structures or processes in different organisms that usually results from their having descended from a common ancestor. [CHEM] The relation among elements of the same group, or family, in the periodic table. [ORG CHEM] That state, in a series of organic compounds that differ from each other by a CH_2 such as the methane series C_nH_{2n+2}, in which there is a similarity between the compounds in the series and a graded change of their properties. { hə'mäl·ə·jē }

homology group [MATH] Associated to a topological space

X, one of a sequence of Abelian groups $H_n(X)$ that reflect how n-dimensional simplicial complexes can be used to fill up X and also help determine the presence of n-dimensional holes appearing in X. Also known as Betti group. { hə'mäl·ə·jē ,grüp }

homology theory [MATH] Theory attempting to compare topological spaces and investigate their structures by determining the algebraic nature and interrelationships appearing in the various homology groups. { hə'mäl·ə·jē ,thē·ə·rē }

homolysis *See* homolytic cleavage. { hə'mäl·ə·sis }

homolytic cleavage [ORG CHEM] The breaking of a single (two-electron) bond in which one electron remains on each of the atoms. Also known as free-radical reaction; homolysis. { ¦häm·ə¸lid·ik 'klēv·ij }

homometric pair [CRYSTAL] A pair of crystal structures whose x-ray diffraction patterns are identical. { ¦hä·mə¦me·trik 'per }

homomorphism [BOT] Having perfect flowers consisting of only one type. [MATH] A function between two algebraic systems of the same type which preserves the algebraic operations. { ¦hä·mə'mȯr¸fiz·əm }

homomorphous transformation [THERMO] A change in the values of the thermodynamic variables of a system in which none of the component substances undergoes a change of state. { ¸hō·mə¦mȯr·fəs ¸tranz·fər'mā·shən }

homomorphs [CHEM] Chemical molecules that are similar in size and shape, but not necessarily having any other characteristics in common. { 'hä·mə¸mȯrfs }

Homoneura [INV ZOO] A suborder of the Lepidoptera with mandibulate mouthparts, and fore- and hindwings that are similar in shape and venation. { ¦hä·mə'nu̇r·ə }

homonomous hemianopsia [MED] Partial blindness affecting the inner half of one field of vision or the outer half of the other; caused by optic nerve lesions posterior to the chiasma. { hə'män·ə·məs ¦hē·mē·ə'näp·sē·ə }

homonuclear molecule [CHEM] A diatomic molecule, both of whose atoms are of the same element. { ¦hō·mō¦nü·klē·ər 'mäl·ə¸kyül }

homopause [GEOPHYS] The level of transition between the homosphere and the heterosphere; it lies about 50 to 56 miles (80 to 90 kilometers) above the earth. Also known as turbopause. { 'hä·mə¸pȯz }

homopetalous [BOT] Having all petals identical. { ¦hä·mə¦ped·əl·əs }

homoplastic [BIOL] Pertaining to transplantation between individuals of the same species. { ¸hō·mō'plas·tik }

homoplastidy [BOT] The condition of having one kind of plastid. { ¸hō·mō'plas·təd·ē }

homoplasy [BIOL] Correspondence between organs or structures in different organisms acquired as a result of evolutionary convergence or of parallel evolution. { 'hä·mə¸plā·sē }

homopolar [ELEC] **1.** Electrically symmetrical. **2.** Having equal distribution of charge. { ¦hä·mə'pō·lər }

homopolar bond [PHYS CHEM] A covalent bond whose total dipole moment is zero. { ¦hä·mə'pō·lər 'bänd }

homopolar crystal [SOLID STATE] A crystal in which the bonds are all covalent. { ¦hä·mə'pō·lər 'krist·əl }

homopolar generator [ELECTR] A direct-current generator in which the poles presented to the armature are all of the same polarity, so that the voltage generated in active conductors has the same polarity at all times; a pure direct current is thus produced, without commutation. Also known as acyclic machine; homopolar machine; unipolar machine. { ¦hä·mə'pō·lər 'jen·ə¸rād·ər }

homopolar machine *See* homopolar generator. { ¦hä·mə'pō·lər mə'shēn }

homopolymer [ORG CHEM] A polymer formed from a single monomer; an example is polyethylene, formed by polymerization of ethylene. { ¸hä·mō'päl·ə·mər }

homopolymer tail [MOL BIO] A segment that contains only one sort of nucleotide at the 3′ end of a deoxyribonucleic acid or ribonucleic acid molecule. { ¸hō·mə¦päl·i·mər 'tāl }

homopolysaccharide [BIOCHEM] A polysaccharide which is a polymer of one kind of monosaccharide. { ¦hä·mō¸päl·ē'sak·ə¸rīd }

Homoptera [INV ZOO] An order of the class Insecta including a large number of sucking insects of diverse forms. { hō'mäp·tə·rə }

Homo sapiens [VERT ZOO] Modern human species; a large, erect, omnivorous terrestrial biped of the primate family Hominidae. { ¦hō·mō 'sap·ē·ənz }

homoscedastic [STAT] **1.** Pertaining to two or more distributions whose variances are equal. **2.** Pertaining to a variate in a bivariate distribution whose variance is the same for all values of the other variate. { ¸hä·mō·skə¦das·tik }

Homosclerophorida [INV ZOO] An order of primitive sponges of the class Demospongiae with a skeleton consisting of equirayed, tetraxonid, siliceous spicules. { ¸hä·mō¦skler·ə'fȯr·ə·də }

homoserine [BIOCHEM] $C_4H_9O_3N$ An amino acid formed as an intermediate product in animals in the metabolic breakdown of cystathionine to cysteine. { hə'mäs·ə¸rēn }

homosexual [BIOL] Of, pertaining to, or being the same sex. [PSYCH] **1.** Of, pertaining to, or exhibiting homosexuality. **2.** One who practices homosexuality. { ¦hō·mə'sek·shə·wəl }

homosexuality [PSYCH] **1.** State of being sexually attracted to members of the same sex. **2.** A form of homoerotism involving sexual interest without genital expression. { ¦hō·mə¸sek·shə'wal·əd·ē }

homosexual panic [PSYCH] An acute syndrome that comes as a climax of prolonged tension from unconscious homosexual conflicts or sometimes bisexual tendencies. { ¦hō·mə'sek·shə·wəl 'pan·ik }

homosphere [METEOROL] The lower portion of a two-part division of the atmosphere (the upper portion is the heterosphere) according to the general homogeneity of atmospheric composition; the region in which there is no gross change in atmospheric composition, that is, all of the atmosphere from the earth's surface to about 50 to 62 miles (80–100 kilometers). { 'hä·mə¸sfir }

homospory [BOT] Production of only one kind of asexual spore. { hə'mäs·pə·rē }

homothallic [MYCOL] Having genetically compatible hyphae and therefore forming zygospores from two branches of the same mycelium. { ¸hä·mə'thal·ik }

homothetic center [MATH] The fixed point through which pass lines joining corresponding points of homothetic figures. Also known as center of similitude; ray center. { ¸häm·ə¸thed·ik 'sen·tər }

homothetic curves [MATH] For a given point, a set of curves such that any straight line through the point intersects all the curves in the set at the same angle. { ¦häm·ə¸thed·ik 'kərvz }

homothetic figures [MATH] Similar figures which are placed so that lines joining corresponding points pass through a common point and are divided in a constant ratio by this point. Also known as radially related figures. { ¦häm·ə¸thed·ik 'fig·yərz }

homothetic ratio *See* ratio of similitude. { ¦hō·mə¦thed·ik 'rā·shō }

homothetic transformation [MATH] A transformation that leaves the origin of coordinates fixed and multiplies the distance between any two points by the same fixed constant. Also known as transformation of similitude. { ¦häm·ə¸thed·ik ¸tranz·fər'mā·shən }

homotopy [MATH] Between two mappings of the same topological spaces, a continuous function representing how, in a step-by-step fashion, the image of one mapping can be continuously deformed onto the image of the other. { hō'mäd·ə·pē }

homotopy groups [MATH] Associated to a topological space X, the groups appearing for each positive integer n, which reflect the number of different ways (up to homotopy) than an n-dimensional sphere may be mapped to X. { hō'mäd·ə·pē ¸grüps }

homotopy theory [MATH] The study of the topological structure of a space by examining the algebraic properties of its various homotopy groups. { hō'mäd·ə·pē ¸thē·ə·rē }

homotropic enzyme [BIOCHEM] A type of allosteric enzyme in which the substrate serves as the allosteric reflector. { ¦hä·mə¦träp·ik 'en¸zīm }

homotropous [BOT] Having the radicle directed toward the hilum. { hō'mä·trə·pəs }

homotype [SYST] A taxonomic type for a specimen which has been compared with the holotype by another than the author of the species and determined by him to be conspecific with it. { 'hä·mə,tīp }

homotypy [BIOL] Protective device based on resemblance of shape to the background. { ¦hä·mə¦tī·pē }

homovanillic acid [BIOCHEM] $HOC_6H_3(OCH_3)CH_2CO_2H$ A major metabolite of 3-*O*-methyl dopa; used in enzyme determination. Abbreviated HVA. { ¦hä·mə·və¦nil·ik 'as·əd }

homozeotrope [CHEM] Mixture in which the liquid components are miscible in all proportions in the liquid phase, and may be separated by ordinary distillation. { ¦hä·mə'zē·ə,trōp }

homozygote [GEN] An individual who has identical alleles at one or more loci and therefore produces identical gametes with respect to these loci. { ¦hō·mə'zī,gōt }

homozygous [GEN] Of or pertaining to a homozygote. { ,hō·mə'zī·gəs }

hone [MATER] A fine-grit stone that is used for sharpening a cutting tool. [MECH ENG] A machine for honing that consists of a holding device containing several oblong stones arranged in a circular pattern. { hōn }

honed-bore tube [DES ENG] Tubing manufactured to very close tolerances and having a very smooth surface in the bore. { ¦hōnd ¦bȯr 'tüb }

honey [FOOD ENG] The sweet, viscous secretion composed principally of levulose and dextrose that is deposited in the honeycomb by the honeybee. { 'hən·ē }

honeybee [INV ZOO] *Apis mellifera*. The bee kept for the commercial production of honey; a member of the dipterous family Apidae. { 'hən·ē,bē }

honeycomb [INV ZOO] A mass of wax cells in the form of hexagonal prisms constructed by honeybees for their brood and honey. { 'hən·ē,kōm }

honeycomb coil [ELECTROMAG] A coil wound in a crisscross manner to reduce distributed capacitance. Also known as duolateral coil; lattice-wound coil. { 'hən·ē,kōm ,kȯil }

honeycomb coral [PALEON] The common name for members of the extinct order Tabulata; has prismatic sections arranged like the cells of a honeycomb. { 'hən·ē,kōm ¦kär·əl }

honeycomb formation [GEOL] A rock stratum containing large cavities or caverns. { 'hən·ē,kōm fȯr,mā·shən }

honeycombing [MATER] **1.** Internal fiber separation in drying timber. **2.** Local roughness and weakening on the face of a concrete wall due to segregation of the concrete, with the result that there is little sand to fill in between the stone aggregate. { 'hən·ē,kōm·iŋ }

honeycomb lung [MED] **1.** Condition of the lung in emphysema. **2.** A lung containing small pus-filled cavities. { 'hən·ē,kōm ,ləŋ }

honeycomb radiator [MECH ENG] A heat-exchange device utilizing many small cells, shaped like a bees' comb, for cooling circulating water in an automobile. { 'hən·ē,kōm 'rād·ē,ād·ər }

honeycomb wall [BUILD] A brick wall having openings created either by allowing gaps between stretchers or by omitting bricks and used to support floor joists and provide ventilation under floors. { 'hən·ē,kōm ,wȯl }

honeycomb weave See waffle weave. { 'hən·ē,kōm ,wēv }

honeydew [INV ZOO] The viscous secretion deposited on leaves by many aphids and scale insects; an attractant for ants. { 'hən·ē,dü }

Honey Dew melon [BOT] A variety of muskmelon (*Cucumis melo*) belonging to the Violales; fruit is large, oval, smooth, and creamy yellow to ivory, without surface markings. { 'hən·ē ,dü ,mel·ən }

honey tube [INV ZOO] Either of a pair of cornicles on the dorsal aspect of one abdominal segment in certain aphids. { 'hən·ē ,tüb }

Honigmann process [MIN ENG] A method of shaft sinking through water-bearing sand; the shaft is bored in stages, increasing in size from the pilot hole, about 4 feet (1.2 meters) in diameter to the final size. { 'hän·ik·mən ,präs·əs }

honing [MECH ENG] The process of removing a relatively small amount of material from a cylindrical surface by means of abrasive stones to obtain a desired finish or extremely close dimensional tolerance. { 'hōn·iŋ }

honing gage [ENG] A device for keeping a chisel steady at the proper angle while it is sharpened on a flat stone. { 'hōn·iŋ ,gāj }

hood [DES ENG] An opaque shield placed above or around

the screen of a cathode-ray tube to eliminate extraneous light. [ENG] **1.** Close-fitting, rubber head covering that leaves the face exposed; used in scuba diving. **2.** A protective covering, usually providing special ventilation to carry away objectionable fumes, dusts, and gases, in which dangerous chemical, biological, or radioactive materials can be safely handled. { hu̇d }

hoodmold [ARCH] The molding that is projected above the arch over a door or window. Also known as drip; label. { 'hu̇d,mōld }

hood test [ENG] A leak detection method in which the vessel under test is enclosed by a metallic casing so that a dynamic leak test may be carried out on a large portion of the external surface. { 'hu̇d ,test }

hoof [VERT ZOO] **1.** Horny covering for terminal portions of the digits of ungulate mammals. **2.** A hoofed foot, as of a horse. { hu̇f }

hoof-and-mouth disease See foot-and-mouth disease. { ¦hu̇f ən 'mȧu̇th di,zēz }

hoof oil See neatsfoot oil. { 'hu̇f ,ȯil }

hook [COMPUT SCI] A modification of a computer program to add instructions to an existing part of the program. [DES ENG] A piece of hard material, especially metal, formed into a curve for catching, holding, or pulling something. [ELECTR] A circuit phenomenon occurring in four-zone transistors, wherein hole or electron conduction can occur in opposite directions to produce voltage drops that encourage other types of conduction. [GEOGR] The end of a spit of land that is turned toward shore. Also known as hooked spit; recurved spit. { hu̇k }

hookah [ENG] An air supply device used in free diving, comprising a demand regulator worn by the diver and a hose extending to a compressed air supply at the surface. { 'hü·kə }

hook-and-eye hinge [DES ENG] A hinge consisting of a hook (usually attached to a gate post) over which an eye (usually attached to the gate) is placed. { ¦hu̇k ən 'ī ,hinj }

hook bolt [DES ENG] A bolt with a hook or L band at one end and threads at the other to fit a nut. { 'hu̇k ,bōlt }

hook collector transistor [ELECTR] A transistor in which there are four layers of alternating *n*- and *p*-type semiconductor material and the two interior layers are thin compared to the diffusion length. Also known as hook transistor; *pn* hook transistor. { ¦hu̇k kə'lek·tər tran,zis·tər }

Hookean deformation [MECH] Deformation of a substance which is proportional to the force applied to it. { 'hu̇k·ē·ən ,def·ər'mā·shən }

Hookean solid [MECH] An ideal solid which obeys Hooke's law exactly for all values of stress, however large. { 'hu̇k·ē·ən 'säl·əd }

hooked spit See hook. { ¦hu̇kt 'spit }

Hooke number See Cauchy number. { 'hu̇k ,nəm·bər }

hooker [MIN ENG] A worker who detaches empty downcoming buckets and hook-loads buckets or cans onto the hoisting rope. { 'hu̇k·ər }

Hooker diaphragm cell [CHEM ENG] A device used in industry for the electrolysis of brine (sodium chloride) to make chlorine and caustic soda (sodium hydroxide) or caustic potash (potassium hydroxide); saturated purified brine fed around the anode passes through the diaphragm to the cathode; chlorine is formed at the anode and hydrogen released at the cathode, leaving sodium hydroxide and residual sodium chloride in the cell liquor; the diaphragm prevents the products from mixing. { 'hu̇k·ər 'dī·ə,fram ,sel }

Hookeriales [BOT] An order of the mosses with irregularly branched stems and leaves that appear to be in one plane. { hu̇,kir·ē'ā·lēz }

Hooker process [MET] A forming process in which pierced slugs or cups are punched through a die to produce tubing and other shapes. { 'hu̇k·ər ,präs·əs }

Hooke's joint [MECH ENG] A simple universal joint; consists of two yokes attached to their respective shafts and connected by means of a spider. Also known as Cardan joint. { 'hu̇ks ,jȯint }

Hooke's law [MECH] The law that the stress of a solid is directly proportional to the strain applied to it. { 'hu̇ks ,lȯ }

hook gage [ENG] An instrument used to measure changes in the level of the water in an evaporation pan; it consists of a pointed metal hook, mounted in the vertical, whose position with respect to its supporting member may be adjusted by

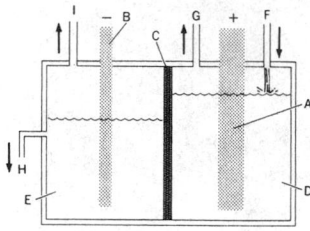

HOOKER DIAPHRAGM CELL

Diagram of diaphragm cell for making chlorine and caustic soda. A = graphic anode, B = iron screen cathode, C = asbestos diaphragm, D = anode compartment for brine and chlorine, E = cathode compartment for NaOHNaCl cell liquor, F = brine inlet, G = chlorine outlet, H = cell liquor (caustic soda) outlet, I = hydrogen outlet.

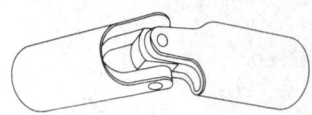

HOOKE'S JOINT

A drawing of a Hooke's joint.

means of a micrometer arrangement; the gage is placed on the still well, and a measurement is taken when the point of the hook just breaks above the surface of the water. { 'hȯk ˌgāj }

hook tender [MIN ENG] In bituminous coal mining, a worker who attaches the hook of a hoisting cable to the link of a trip of cars to be hauled up or lowered down an incline. Also known as rope cutter. { 'hȯk ˌten·dər }

hook transistor See hook collector transistor. { 'hȯk tranˌzis·tər }

hookup [ELEC] An arrangement of circuits and apparatus for a particular purpose. { 'hȯkˌəp }

hookup wire [ELEC] Tinned and insulated solid or stranded soft-drawn copper wire used in making circuit connections. { 'hȯkˌəp ˌwīr }

hook-wall packer [PETRO ENG] Fluid-proof seal between the outside of oil well tubing and the inside of the casing; hooks hold it in place. { 'hȯk ˌwȯl ˌpak·ər }

hookworm [INV ZOO] The common name for parasitic roundworms composing the family Ancylostomidae. { 'hȯkˌwərm }

hookworm disease [MED] Microcytic hypochromic anemia in humans produced by the nematodes *Necator americanus* or *Ancylostoma duodenale* in the the intestine. { 'hȯkˌwərm diˌzēz }

hook wrench [DES ENG] A wrench with a hook for turning a nut or bolt. { 'hȯk ˌrench }

hoop [CIV ENG] A ring-shaped binder placed around the main reinforcement in a reinforced concrete column. { hüp }

hooped column [CIV ENG] A column of reinforced concrete with hoops around the main reinforcements. { 'hüpt ˌkäl·əm }

Hooper jig [MIN ENG] A pneumatic jig, used when water is scarce or the ore must be kept dry, to concentrate values from sands. { 'hüp·ər ˌjig }

hoot stop [COMPUT SCI] A closed loop that generates an audible signal; usually employed to signal an error or for operating convenience. { 'hüt ˌstäp }

hop [BOT] *Humulus lupulus.* A dioecious liana of the order Urticales distinguished by herbaceous vines produced from a perennial crown; the inflorescence, a catkin, of the female plant is used commercially for beer production. [COMMUN] A single reflection of a radio wave from the ionosphere back to the earth in traveling from one point to another. { häp }

hopeite [MINERAL] $Zn_3(PO_4)_2·4H_2O$ A gray, orthorhombic mineral composed of hydrous phosphate of zinc; specific gravity is 2.76–2.85; dimorphous with parahopeite. { 'hōˌpīt }

Hope's apparatus [THERMO] An apparatus consisting of a vessel containing water, a freezing mixture in a tray surrounding the vessel, and thermometers inserted in the water at points above and below the freezing mixture; used to show that the maximum density of water lies at about 4°C. { 'hōps ˌap·əˌrad·əs }

Hopfner process [MET] A process for the recovery of copper from its sulfide ores by leaching with a solution of cuprous chloride in sodium or calcium chloride and electrolyzing the resulting solution in tanks that are protected by diaphragms. Also spelled Hoepfner process. { 'häpf·nər ˌprä·səs }

hophornbeam [BOT] Any tree of the genus *Ostrya* in the birch family recognized by its very scaly bark and the fruit which closely resembles that of the hopvine. { 'häp 'hȯrnˌbēm }

Hopkins-Cole reaction [ANALY CHEM] The appearance of a violet ring when concentrated sulfuric acid is added to a mixture that includes a protein and glyoxylic acid; however, gelatin and zein do not show the reaction. { 'häpˌkənz'kōl rēˌak·shən }

Hopkinson effect [ELECTROMAG] A phenomenon in which the permeability of a ferromagnetic material at low field strengths, measured as a function of temperature, reaches a maximum at a temperature a little below the Curie temperature. { 'häp·kən·sən iˌfekt }

Hopkinson's coefficient [ELECTROMAG] The average magnetic flux per turn of an induction coil divided by the average flux per turn of another coil linked with it. { 'häp·kən·sənz ˌkō·i'fish·ənt }

Hoplocarida [INV ZOO] A superorder of the class Crustacea with the single order Stomatopoda. { ˌhäp·lō'kar·ə·də }

Hoplonemertini [INV ZOO] An order of unsegmented, ribbonlike worms in the class Enopla; all species have an armed proboscis. { ˌhäp·lōˌne·mər'tīˌnī }

Hoplophoridae [INV ZOO] A family of prawns containing numerous bathypelagic representatives. { ˌhäp·lə'fȯr·əˌdē }

hopper [ENG] A funnel-shaped receptacle with an opening at the top for loading and a discharge opening at the bottom for bulk-delivering material such as grain or coal. { 'häp·ər }

hopperburn [PL PATH] A disease of potato and peanut plants caused by a leafhopper which secretes a toxic substance on the leaves, causing browning and shriveling. { 'häp·ər ˌbərn }

hopper car [ENG] A freight car with a permanent roof and a hinged floor sloping to one or more hoppers for discharging contents by gravity. { 'häp·ər ˌkär }

hopper dryer [ENG] In extrusion and injection molding of plastics, a combined feeding and drying device in which hot air flows through the hopper. { 'häp·ər ˌdrī·ər }

hoppit [MIN ENG] A large bucket, usually up to about 80 cubic feet (2.3 cubic meters), used in shaft sinking for hoisting men, rock, materials, and tools. Also known as bowk; kibble; sinking bucket. { 'häp·ət }

hopsacking [TEXT] A fabric with an open, plied-yarn, coarse basket weave, similar to sacking used to gather hops. { 'häpˌsak·iŋ }

hops oil [MATER] Greenish-yellow essential oil with strong aroma; soluble in alcohol, ether, and chloroform, insoluble in water; distilled from strobiles of the hop (*Humulus lupulus*); main components are humulene, geraniol, and terpenes; used to aromatize beer and tobacco. { 'häps ˌȯil }

horadiam drilling [MIN ENG] The drilling of a number of horizontal boreholes radiating outward from a common center. Also known as horizontal-ring drilling. { ˌhȯr·ə'dī·am 'dril·iŋ }

hordeolum [MED] A furuncular inflammation of the connective tissue of the eyelids near a hair follicle. Also known as sty. { hȯr'dē·ə·ləm }

Hordeum [BOT] A genus of the order Cyperales containing all species of barley. { 'hȯr·dē·əm }

horehound See marrubium. { 'hȯrˌhau̇nd }

horizon [ASTRON] **1.** The apparent boundary line between the sky and the earth or sea. Also known as apparent horizon. **2.** The distance a light ray could have traveled since the big-bang explosion at any given epoch in the evolution of the universe. [GEOL] **1.** The surface separating two beds. **2.** One of the layers, each of which is a few inches to a foot thick, that make up a soil. { hə'rīz·ən }

horizon bar [NAV] The gyro-stabilized line or bar in an artificial-horizon flight instrument. { hə'rīz·ən ˌbär }

horizon distance [GEOD] The distance, at any given azimuth, to the point on the earth's surface constituting the horizon for some specified observer. { hə'rīz·ən ˌdis·təns }

horizon glass [NAV] That glass of a marine sextant, through which the horizon is observed. { hə'rīz·ən ˌglas }

horizon lights [NAV] Lights arranged to provide a ground reference for pilots taking off. { hə'rīz·ən ˌlīts }

horizon mining [MIN ENG] A system of mining suitable for inclined, and perhaps faulted, coal seams; main stone headings are driven, at predetermined levels, from the winding shaft to intersect the seams to be developed. { hə'rīz·ən ˌmīn·iŋ }

horizon mirror [NAV] On a marine sextant, the mirror part of the horizon glass; the term is sometimes used loosely to refer to the horizon glass. { hə'rīz·ən ˌmir·ər }

horizon prism [NAV] A prism which can be inserted in the optical path of an instrument, such as a bubble sextant, to permit observation of the visible horizon. { hə'rīz·ən ˌpriz·əm }

horizon problem [ASTRON] The problem of explaining the observed uniformity of the universe, and in particular of the cosmic background radiation, when, according to the standard big-bang theory, sources of radiation coming from opposite directions in the sky were separated by manyfold the horizon distance at the time of emission, and thus could not possibly have been in physical contact. { hə'rīz·ən ˌpräb·ləm }

horizon scanner See horizon tracker. { hə'rīz·ən ˌskan·ər }

horizon sensor [ENG] A passive infrared device that detects the thermal discontinuity between the earth and space; used in establishing a stable vertical reference for control of the attitude or orientation of a missile or satellite in space. { hə'rīz·ən ˌsen·sər }

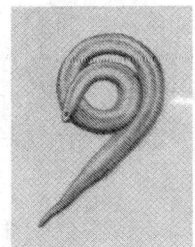

HOOKWORM

A typical hookworm.

HOP

Hop, female inflorescences. *(USDA)*

horizon system of coordinates [ASTRON] A set of celestial coordinates based on the celestial horizon as the primary great circle. { hə'rīz·ən ¦sis·təm əv kō'örd·ən·əts }

horizontal [SCI TECH] Being in a plane perpendicular to the gravitational field, that is, perpendicular to a plumb line, at a given point on the earth's surface. { ¸här·ə'zänt·əl }

horizontal auger [MECH ENG] A rotary drill, usually powered by a gasoline engine, for making horizontal blasting holes in quarries and opencast pits. { ¸här·ə'zänt·əl 'óg·ər }

horizontal base-line method [ORD] Method of locating targets or other points by intersection from two observation stations located at opposite ends of a base line. { ¸här·ə'zänt·əl 'bās ¸līn ¸meth·əd }

horizontal blanking [ELECTR] Blanking of a television picture tube during the horizontal retrace. { ¸här·ə'zänt·əl 'blaŋk·iŋ }

horizontal blanking pulse [ELECTR] The rectangular pulse that forms the pedestal of the composite television signal between active horizontal lines and causes the beam current of the picture tube to be cut off during retrace. Also known as line-frequency blanking pulse. { ¸här·ə'zänt·əl 'blaŋk·iŋ ¸pəls }

horizontal boiler [MECH ENG] A water-tube boiler having a main bank of straight tubes inclined toward the rear at an angle of 5 to 15° from the horizontal. { ¸här·ə'zänt·əl 'bóil·ər }

horizontal bombing [ORD] Bombing from an airplane in horizontal flight, as distinguished from dive bombing, glide bombing, toss bombing, and the like. Also known as level bombing. { ¸här·ə'zänt·əl 'bäm·iŋ }

horizontal borer [MIN ENG] A machine that makes holes 2–6 inches (5–15 centimeters) in diameter, used for drilling at opencut coal mines. { ¸här·ə'zänt·əl 'bór·ər }

horizontal boring machine [MECH ENG] A boring machine adapted for work not conveniently revolved, for milling, slotting, drilling, tapping, boring, and reaming long holes and for making interchangeable parts that must be produced without jigs and fixtures. { ¸här·ə'zänt·əl 'bór·iŋ mə,shēn }

horizontal branch [ASTRON] A region in the Hertzsprung-Russell diagram of a typical globular cluster that extends in the blue direction from the giant branch at an absolute bolometric magnitude of 0.3 and consists of stars that are burning helium in their cores and hydrogen in their surrounding envelopes. { ¸här·ə'zänt·əl 'branch }

horizontal broaching machine [MECH ENG] A pull-type broaching machine having the broach mounted on the horizontal plane. { ¸här·ə'zänt·əl 'brōch·iŋ mə,shēn }

horizontal cells [NEUROSCI] Interneurons located in the outer plexiform layer of the vertebrate retina that influence retinal signal processing in response to visual stimuli at the level of contact between the photoreceptor cells and the bipolar cells. { ¸här·ə,zänt·əl 'sels }

horizontal centering control [ELECTR] The centering control provided in a television receiver or cathode-ray oscilloscope to shift the position of the entire image horizontally in either direction on the screen. { ¸här·ə'zänt·əl 'sen·tə·riŋ kən,trōl }

horizontal chromatography [ANALY CHEM] Paper chromatography in which the chromatogram is horizontal instead of vertical. { ¸här·ə'zänt·əl ¸krō·mə'täg·rə·fē }

horizontal circle [ENG] A graduated disk affixed to the base of a transit or theodolite which is used to measure horizontal angles. { ¸här·ə'zänt·əl 'sər·kəl }

horizontal control [MAP] A system of points whose horizontal positions and interrelationships have been accurately determined for use as fixed references in positioning and correlating map features. { ¸här·ə'zänt·əl kən'trōl }

horizontal convergence control [ELECTR] The control that adjusts the amplitude of the horizontal dynamic convergence voltage in a color television receiver. { ¸här·ə'zänt·əl kən'vər·jəns kən,trōl }

horizontal crosscut See horizontal drive. { ¸här·ə'zänt·əl 'krós,kət }

horizontal crusher [MECH ENG] Rotary size reducer in which the crushing cone is supported on a horizontal shaft; needs less headroom than vertical models. { ¸här·ə'zänt·əl 'krəsh·ər }

horizontal danger angle [NAV] The maximum or minimum angle between two points on a chart, as observed from a craft, indicating the limit of safe approach to an off-lying danger. Also known as danger angle. { ¸här·ə'zänt·əl 'dān·jər ¸aŋ·gəl }

horizontal definition See horizontal resolution. { ¸här·ə'zänt·əl ¸def·ə'nish·ən }

horizontal deflection electrode [ELECTR] One of a pair of electrodes that move the electron beam horizontally from side to side on the fluorescent screen of a cathode-ray tube employing electrostatic deflection. { ¸här·ə'zänt·əl di'flek·shən i'lek ,trōd }

horizontal deflection oscillator [ELECTR] The oscillator that produces, under control of the horizontal synchronizing signals, the sawtooth voltage waveform that is amplified to feed the horizontal deflection coils on the picture tube of a television receiver. Also known as horizontal oscillator. { ¸här·ə'zänt·əl di'flek·shən 'äs·ə,lād·ər }

horizontal displacement See strike slip. { ¸här·ə'zänt·əl dis'plās·mənt }

horizontal distributed processing system [COMPUT SCI] A type of distributed system in which two or more computers which are logically equivalent are connected together, with no hierarchy or master/slave relationship. { ¸här·ə'zänt·əl di ¦strib·yəd·əd 'prä,ses·iŋ ¸sis·təm }

horizontal drilling machine [MECH ENG] A drilling machine in which the drill bits extend in a horizontal direction. { ¸här·ə'zänt·əl 'dril·iŋ mə,shēn }

horizontal drive [MIN ENG] An opening with a small inclination directed toward the shaft to drain the water and facilitate hauling of full cars to the shaft. Also known as horizontal crosscut. { ¸här·ə'zänt·əl 'drīv }

horizontal drive control [ELECTR] The control in a television receiver, usually at the rear, that adjusts the output of the horizontal oscillator. Also known as drive control. { ¸här·ə'zänt·əl 'drīv kən,trōl }

horizontal earth rate [NAV] The rate at which the spin axis of a gyroscope must be tilted about the horizontal axis to remain parallel to the earth's surface to compensate for the effect of earth rate; horizontal earth rate is maximum at the equator, zero at the poles, and varies as the cosine of the latitude. { ¸här·ə'zänt·əl 'ərth ¸rāt }

horizontal engine [MECH ENG] An engine with horizontal stroke. { ¸här·ə'zänt·əl 'en·jən }

horizontal error [ORD] The error in range, deflection, or radius which a weapon may be expected to exceed as often as not. { ¸här·ə'zänt·əl 'er·ər }

horizontal evolution See coincidental evolution. { ¸hä·ri¦zänt·əl ¸ev·ə'lü·shən }

horizontal field balance [ENG] An instrument that measures the horizontal component of the magnetic field by means of the torque that the field component exerts on a vertical permanent magnet. { ¸här·ə'zänt·əl 'fēld ¸bal·əns }

horizontal field-strength diagram [ELECTROMAG] Representation of the field strength at a constant distance from an antenna and in a horizontal plane; unless otherwise specified, this plane is that passing through the antenna. { ¸här·ə'zänt·əl 'fēld ¸streŋkth ¸dī·ə,gram }

horizontal firing [MECH ENG] The firing of fuel in a boiler furnace in which the burners discharge fuel and air into the furnace horizontally. { ¸här·ə'zänt·əl 'fīr·iŋ }

horizontal flow chart [COMPUT SCI] A graphical representation of the movement of forms, punch cards, and other recording media through an organization, showing the movement of each medium from the time it is first used to the time it is destroyed. { ¸här·ə'zänt·əl 'flō ¸chärt }

horizontal flyback [ELECTR] Flyback in which the electron beam of a television picture tube returns from the end of one scanning line to the beginning of the next line. Also known as horizontal retrace. { ¸här·ə'zänt·əl 'flī,bak }

horizontal fold See nonplunging fold. { ¸här·ə'zänt·əl 'fōld }

horizontal force instrument [ENG] An instrument used to make a comparison between the intensity of the horizontal component of the earth's magnetic field and the magnetic field at the compass location on board a craft; basically, it consists of a magnetized needle pivoted in a horizontal plane, as a dry-card compass; it settles in some position which indicates the direction of the resultant magnetic field; if the needle is started swinging, it damps down with a certain period of oscillation dependent upon the strength of the magnetic field. Also known as horizontal vibrating needle. { ¸här·ə'zänt·əl ¦förs 'in·strə·mənt }

horizontal frequency *See* line frequency. { ˌhär·ə'zänt·əl 'frē·kwən·sē }

horizontal hold control [ELECTR] The hold control that changes the free-running period of the horizontal deflection oscillator in a television receiver, so that the picture remains steady in the horizontal direction. { ˌhär·ə'zänt·əl 'hōld kən,trōl }

horizontal instruction [COMPUT SCI] An instruction in machine language to carry out independent operations on various operands in parallel or in a well-defined time sequence. { ˌhär·ə'zänt·əl in'strək·shən }

horizontal intensity [GEOPHYS] The strength of the horizontal component of the earth's magnetic field. { ˌhär·ə'zänt·əl in'ten·səd·ē }

horizontal intensity variometer [ENG] Essentially a declination variometer with a larger, stiffer fiber than in the standard model; there is enough torsion in the fiber to cause the magnet to turn 90° out of the magnetic meridian; the magnet is aligned with the magnetic prime vertical to within 0.5° so it does not respond appreciably to changes in declination. Also known as H variometer. { ˌhär·ə'zänt·əl in'ten·səd·ē ‚ver·ē'äm·əd·ər }

horizontal lathe [MECH ENG] A horizontally mounted lathe with which longitudinal and radial movements are applied to a workpiece that rotates. { ˌhär·ə'zänt·əl 'lāth }

horizontal linearity control [ELECTR] A linearity control that permits narrowing or expanding of the width of the left half of a television receiver image, to give linearity in the horizontal direction so that circular objects appear as true circles. { ˌhär·ə'zänt·əl ‚lin·ē'ar·əd·ē kən,trōl }

horizontal line frequency *See* line frequency. { ˌhär·ə'zänt·əl 'līn ‚frē·kwən·sē }

horizontal magnetometer [ENG] A measuring instrument for ascertaining changes in the horizontal component of the magnetic field intensity. { ˌhär·ə'zänt·əl ‚mag·nə'täm·əd·ər }

horizontal milling machine [MECH ENG] A knee-type milling machine with a horizontal spindle and a swiveling table for cutting helices. { ˌhär·ə'zänt·əl 'mil·iŋ mə,shēn }

horizontal obstacle sonar [NAV] A constant-transmission, frequency-modulated sonar used in underwater navigation to detect objects ahead of a submersible vehicle and to determine the bearing and range of specially designed transponders. Abbreviated HOS. { ˌhär·ə'zänt·əl ‚äb·stə·kəl 'sō,när }

horizontal oscillator *See* horizontal deflection oscillator. { ˌhär·ə'zänt·əl 'äs·ə,lād·ər }

horizontal output stage [ELECTR] The television receiver stage that feeds the horizontal deflection coils of the picture tube through the horizontal output transformer; may also include a part of the second-anode power supply for the picture tube. { ˌhär·ə'zänt·əl 'aut,put ‚stāj }

horizontal output transformer [ELECTR] A transformer used in a television receiver to provide the horizontal deflection voltage, the high voltage for the second-anode power supply of the picture tube, and the filament voltage for the high-voltage rectifier tube. Also known as flyback transformer; horizontal sweep transformer. { ˌhär·ə'zänt·əl 'aut,put tranz,fòr·mər }

horizontal parallax [ASTRON] The geocentric parallax of a celestial object when it is rising or setting. [GRAPHICS] *See* absolute stereoscopic parallax. { ˌhär·ə'zänt·əl 'par·ə,laks }

horizontal parity check *See* longitudinal parity check. { ˌhär·ə'zänt·əl 'par·əd·ē ‚chek }

horizontal pendulum [MECH] A pendulum that moves in a horizontal plane, such as a compass needle turning on its pivot. { ˌhär·ə'zänt·əl 'pen·jə·ləm }

horizontal plane [ANAT] A transverse plane at right angles to the longitudinal axis of the body. { ˌhär·ə'zänt·əl 'plān }

horizontal polarization [COMMUN] Transmission of linear polarized radio waves whose electric field vector is parallel to the earth's surface. { ˌhär·ə'zänt·əl ‚pō·lə·rə'zā·shən }

horizontal-position welding [MET] **1.** Making a fillet weld on the upper side of the intersection of a vertical surface and a horizontal surface. **2.** Making a horizontal groove weld on a vertical surface. { ˌhär·ə'zänt·əl pə‚zish·ən ‚weld·iŋ }

horizontal pressure force [GEOPHYS] The horizontal pressure gradient per unit mass, $-\alpha\nabla_H p$, where α is the specific volume, p the pressure, and ∇_H the horizontal component of the del operator; this force acts normal to the horizontal isobars toward lower pressure; it is one of the three important forces appearing in the horizontal equations of motion, the others being the Coriolis force and friction. { ˌhär·ə'zänt·əl 'presh·ər ‚fòrs }

horizontal projection [GRAPHICS] A drawing of a structure as it would appear if projected on a horizontal plane. { ˌhär·ə'zänt·əl prə'jek·shən }

horizontal range [ORD] The distance measured horizontally between a gun and its target; specifically in antiaircraft gunnery, the distance between a gun and a spot on the ground directly beneath the target. { ˌhär·ə'zänt·əl 'rānj }

horizontal resolution [ELECTR] The number of individual picture elements or dots that can be distinguished in a horizontal scanning line of a television or facsimile image. Also known as horizontal definition. { ˌhär·ə'zänt·əl ‚rez·ə'lü·shən }

horizontal retort [MET] An intermittent unit made from a siliceous fireclay and formerly used for zinc smelting. { ˌhär·ə'zänt·əl 'rē,tòrt }

horizontal retort process [MET] A zinc-smelting process that employs vast, honeycomblike batteries of fireclay or silicon-carbide retorts set horizontally in a gas- or coal-fired furnace. Also known as Belgian retort process. { ˌhär·ə'zänt·əl 'rē,tòrt ‚prä·səs }

horizontal retrace *See* horizontal flyback. { ˌhär·ə'zänt·əl 'rē,trās }

horizontal return tubular boiler [MECH ENG] A fire-tube boiler having tubes within a cylindrical shell that are attached to the end closures; products of combustion are transported under the lower half of the shell and back through the tubes. { ˌhär·ə'zänt·əl ri'tərn ‚tü·byə·lər 'bòil·ər }

horizontal-ring drilling *See* horadiam drilling. { ˌhär·ə'zänt·əl ‚riŋ 'dril·iŋ }

horizontal-rolled-position welding [MET] Top-side welding of a butt joint connecting two horizontal pieces of rotating pipe. { ˌhär·ə'zänt·əl ‚rōld·pə‚zish·ən 'weld·iŋ }

horizontal scanning [ENG] In radar scanning, rotating the antenna in azimuth around the horizon or in a sector. Also known as searching lighting. { ˌhär·ə'zänt·əl 'skan·iŋ }

horizontal scanning frequency [ELECTR] The number of horizontal lines scanned by the electron beam in a television receiver in 1 second. { ˌhär·ə'zänt·əl 'skan·iŋ ‚frē·kwən·sē }

horizontal screen [MECH ENG] Shaking screen with horizontal plates. { ˌhär·ə'zänt·əl 'skrēn }

horizontal separation *See* strike slip. { ˌhär·ə'zänt·əl ‚sep·ə'rā·shən }

horizontal separator [PETRO ENG] Horizontal tank used to separate free oil well gas from liquid hydrocarbons. { ˌhär·ə'zänt·əl 'sep·ə,rād·ər }

horizontal silo [AGR] A trench or bunker for storing silage. { ˌhär·ə'zänt·əl 'sī·lō }

horizontal sweep [ELECTR] The sweep of the electron beam from left to right across the screen of a cathode-ray tube. { ˌhär·ə'zänt·əl 'swēp }

horizontal sweep transformer *See* horizontal output transformer. { ˌhär·ə'zänt·əl 'swēp tranz,fòr·mər }

horizontal synchronizing pulse [ELECTR] The rectangular pulse transmitted at the end of each line in a television system, to keep the receiver in line-by-line synchronism with the transmitter. Also known as line synchronizing pulse. { ˌhär·ə'zänt·əl 'siŋ·krə,niz·iŋ ‚pəls }

horizontal system [COMPUT SCI] A programming system in which instructions are written horizontally, that is, across the page. { ˌhär·ə'zänt·əl 'sis·təm }

horizontal transmission [GEN] Passage of genetic information by invasive processes between individual organisms or cells of different species. { ˌhär·ə‚zänt·əl tranz'mish·ən }

horizontal-tube evaporator [MECH ENG] A horizontally mounted tube-and-shell type of liquid evaporator, used most often for preparation of boiler feedwater. { ˌhär·ə'zänt·əl ‚tüb i'vap·ə,rād·ər }

horizontal vee [ELECTROMAG] An antenna consisting of two linear radiators in the form of the letter V, lying in a horizontal plane. { ˌhär·ə'zänt·əl 'vē }

horizontal vibrating needle *See* horizontal force instrument. { ˌhär·ə'zänt·əl ‚vī,brād·iŋ 'nēd·əl }

horizon tracker [NAV] A device for establishing a vertical reference in a navigation system by precisely tracking the visible horizon. Also known as horizon scanner. { ˌhär·ə'zänt·əl 'trak·ər }

hormesis [BIOL] Providing stimulus by nontoxic amounts of a toxic agent. { 'hòr·mə·səs }

HORN

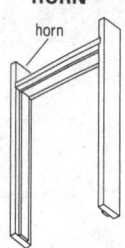

A horn on a right-angle wood framing joint.

HORN ANTENNA

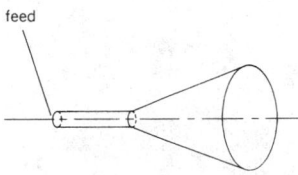

Horn antenna, a type of direct-aperture antenna.

HORNED TOAD

Horned toad (*Phrynosoma*).

HORN-LOADED SPEAKER

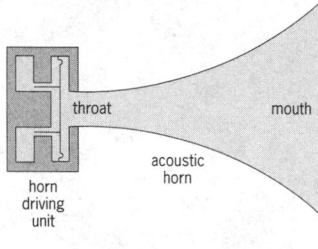

Speaker with the driver connected to an acoustic horn.

HORSEHEAD NEBULA

Horsehead Nebula in Orion. (*Hole Observatories*)

hormogonium [BOT] Portion of a filament between heterocysts in certain algae; detaches as a reproductive body. { ˌhȯr·mə'gō·nē·əm }

hormone [BIOCHEM] **1.** A chemical messenger produced by endocrine glands and secreted directly into the bloodstream to exert a specific effect on a distant part of the body. **2.** An organic compound that is synthesized in minute quantities in one part of a plant and translocated to another part, where it influences a specific physiological process. { 'hȯr·mōn }

hormone-responsive element [CELL MOL] Specific regulatory nucleotide sequences in most hormonally regulated genes that are located near the promoter and mediate the action of steroid hormones. { 'hȯrˌmōn ri·spän·siv ˌel·ə·mənt }

horn [BUILD] A section projecting from the end of one of the members of a right-angle wood framing joint. [ELECTROMAG] *See* horn antenna. [ENG ACOUS] A tube whose cross-sectional area increases from one end to the other, used to radiate or receive sound waves and to intensify and direct them. Also known as acoustic horn. [GEOL] A topographically high, sharp, pyramid-shaped mountain peak produced by the headward erosion of mountain glaciers; the Matterhorn is the classic example. [MET] Holding arm for the electrode of a resistance spot-welding machine. { hȯrn }

horn angle [MATH] A geometric figure formed by two tangent plane curves that lie on the same side of their mutual tangent line in the neighborhood of the point of tangency. { 'hȯrn ˌaŋ·gəl }

horn antenna [ELECTROMAG] A microwave antenna produced by flaring out the end of a circular or rectangular waveguide into the shape of a horn, for radiating radio waves directly into space. Also known as electromagnetic horn; flare (British usage); hoghorn antenna (British usage); horn; horn radiator. { 'hȯrn an'ten·ə }

horn arrester [ELEC] A lightning arrester in which the spark gap has thick wire horns that spread outward and upward; the arc forms at the narrowest bottom part of the gap, travels upward, and extinguishes itself when it reaches the widest part of the gap. Also known as horn lightning arrester. { 'hȯrn ə'res·tər }

hornbeam [BOT] Any tree of the genus *Carpinus* in the birch family distinguished by doubly serrate leaves and by small, pointed, angular winter buds with scales in four rows. { 'hȯrnˌbēm }

hornblende [MINERAL] A general name given to the monoclinic calcium amphiboles that form an extensive solid-solution series between the various metals in the generalized formula $(Ca,Na)_2(Mg,Fe,Al)_5(Al,Si)_8O_{22}(OH,F)_2$. { 'hȯrnˌblend }

hornblendite [PETR] A plutonic rock consisting mainly of hornblende. { 'hȯrn·blenˌdīt }

horn buoy [NAV] A buoy provided with a horn. { 'hȯrn ˌbȯi }

horned dinosaur [PALEON] Common name for extinct reptiles of the suborder Ceratopsia. { ˌhȯrnd 'dīn·ə·sȯr }

horned liverwort [BOT] The common name for bryophytes of the class Anthocerotae. Also known as hornwort. { ˌhȯrnd 'liv·ərˌwȯrt }

horned toad [VERT ZOO] The common name for any of the lizards of the genus *Phrynosoma;* a reptile that resembles a toad but is less bulky. { ˌhȯrnd 'tōd }

horned-toad dinosaur [PALEON] The common name for extinct reptiles composing the suborder Ankylosauria. { ˌhȯrndˌtōd 'dīn·ə·sȯr }

horn equation [ACOUS] A second-order partial differential equation for the velocity potential as a function of time and of distance along an acoustic horn. { 'hȯrn iˌkwā·zhən }

Horner's method [MATH] A technique for approximating the real roots of an algebraic equation; a root is located between consecutive integers, then a successive search is performed. { 'hȯrn·ərz ˌmeth·əd }

Horner's syndrome [MED] A complex of symptoms due to unilateral destruction of the cervical sympathetics. { 'hȯrn·ərz ˌsin·drōm }

hornet [INV ZOO] The common name for a number of large wasps in the hymenopteran family Vespidae. { 'hȯr·nət }

hornfels [PETR] A common name for a class of metamorphic rocks produced by contact metamorphism and characterized by equidimensional grains without preferred orientation. { 'hȯrˌfelz }

hornfels facies [PETR] Rock formed at depths in the earth's crust not exceeding 6.2 miles (10 kilometers) at temperatures of 250–800°C; includes albite-epidote hornfels facies, pyroxene-hornfels facies, and hornblende-hornfels facies. { 'hȯrˌfelz ˌfā·shēz }

horn gap [ELEC] Type of spark gap which is provided with divergent electrodes. { 'hȯrn ˌgap }

horn-gap switch [ELEC] Form of air switch provided with arcing horns. { 'hȯrn ˌgap ˌswich }

horn lead *See* phosgenite. { 'hȯrn ˌled }

horn lightning arrester *See* horn arrester. { 'hȯrn 'līt·niŋ əˌres·tər }

horn-loaded speaker [ENG ACOUS] A loudspeaker that has an acoustic horn between the diaphragm and the air load. { ˌhȯrn ˌlōd·əd 'spēk·ər }

horn loudspeaker [ENG ACOUS] A loudspeaker in which the radiating element is coupled to the air or another medium by means of a horn. { 'hȯrn ˌlaudˌspēk·ər }

horn quicksilver *See* calomel. { 'hȯrn 'kwikˌsil·vər }

horn radiator *See* horn antenna. { 'hȯrn 'rād·ēˌād·ər }

horn silver *See* cerargyrite. { 'hȯrn ˌsil·vər }

horn socket [DES ENG] A cone-shaped fishing tool especially designed to recover lost collared drill rods, drill pipe, or tools in bored wells. { 'hȯrn ˌsäk·ət }

horn spacing [MET] Unobstructed work space in a resistance-welding machine between horns at right angles to the throat depth. { 'hȯrn ˌspās·iŋ }

horn spoon [MIN ENG] A troughlike section cut from a cow horn and scraped thin; used for washing auriferous gravel and pulp when exacting tests are to be performed. { 'hȯrn ˌspün }

hornstone *See* chert. { 'hȯrnˌstōn }

hornwort *See* horned liverwort. { 'hȯrnˌwȯrt }

horny coral [INV ZOO] The common name for cnidarian members of the order Gorgonacea. { 'hȯr·nē 'kär·əl }

horography [HOROL] The making of watches and clocks. { hə'räg·rə·fē }

Horologium [ASTRON] A constellation with right ascension 3 hours, declination 60° south. Also known as Clock. { ˌhȯr·ə'lō·jē·əm }

Horologium superclusters [ASTRON] Two superclusters in approximately the same direction, with recession velocities around 12,000 kilometers (7500 miles) and 18,000 kilometers (11,200 miles) per second. { ˌhȯr·ə·lō·gē·əm 'sü·pərˌkləs·tərz }

horology [SCI TECH] The science of measuring time and the technology of constructing instruments for this measurement. { hə'räl·ə·jē }

horse [GEOL] A large rock caught along a fault. [MIN ENG] *See* horseback. [VERT ZOO] *Equus caballus.* A herbivorous mammal in the family Equidae; the feet are characterized by a single functional digit. { hȯrs }

horseback [GEOL] A low and sharp ridge of sand, gravel, or rock. [MIN ENG] **1.** Shale or sandstone occurring in a channel that was cut by flowing water in a coal seam. Also known as cutout; horse; roll; swell; symon fault; washout. **2.** To move or raise a heavy piece of machinery or timber by using a pinch bar as a lever. Also known as pinch. { 'hȯrsˌbak }

horse chestnut [BOT] *Aesculus hippocastanum.* An ornamental buckeye tree in the order Sapindales, usually with seven leaflets per leaf and resinous buds. { 'hȯrs ˌches·nət }

horse fetter *See* hippopede. { 'hȯrs ˌfed·ər }

horsehair blight [PL PATH] A fungus disease of tea and certain other tropical plants caused by *Marasmius equicrinis* and characterized by black festoons of mycelia hanging from the branches. { 'hȯrsˌher ˌblīt }

horsehead [MIN ENG] Timbers or steel joists used to support planks in tunneling through loose ground. [PETRO ENG] A curved section of the walking beam of a beam pumping unit installed in an oil well. { 'hȯrsˌhed }

Horsehead Nebula [ASTRON] A cloud of obscuring particles between the earth and a gaseous emission nebula in the constellation Orion. { 'hȯrsˌhed 'neb·yə·lə }

horse latitudes [METEOROL] The belt of latitudes over the oceans at approximately 30–35°N and S where winds are predominantly calm or very light and weather is hot and dry. { 'hȯrs ˌlad·əˌtüdz }

horsepower [MECH] The unit of power in the British engineering system, equal to 550 foot-pounds per second, approximately 745.7 watts. Abbreviated hp. { 'hȯrsˌpau̇·ər }

horsepox [VET MED] A disease of horses such as pseudotuberculosis, contagious pustular stomatitis, or a vesicular exanthema. { 'hȯrs,päks }

horseradish [BOT] *Armoracia rusticana.* A perennial crucifer belonging to the order Capparales and grown for its pungent roots, used as a condiment. { 'hȯrs'rad·ish }

horse serum [IMMUNOL] Immune serum obtained from the blood of horses. { 'hȯrs ,sir·əm }

horseshoe bend *See* oxbow. { 'hȯr,shü ,bend }

horseshoe crab [INV ZOO] The common name for arthropods composing the subclass Xiphosurida, especially the subgroup Limulida. { 'hȯr,shü ,krab }

horseshoe kidney [MED] Congenital fusion of two kidneys at one pole. { 'hȯr,shü ,kid·nē }

horseshoe lake *See* oxbow lake. { 'hȯr,shü ,lāk }

horseshoe magnet [ELECTROMAG] A permanent magnet or electromagnet in which the core is horseshoe-shaped or U-shaped, to bring the two poles near each other. { 'hȯr,shü ,mag·nət }

horsetail [BOT] The common name for plants of the genus *Equisetum* composing the order Equisetales. Also known as scouring rush. { 'hȯrs,tāl }

horsetail ore [GEOL] An ore occurring in fractures which diverge from a larger fracture. { 'hȯrs,tāl ,ȯr }

horsfordite [MINERAL] Cu₅Sb A silver-white mineral composed of copper-antimony alloy. { 'hȯrs·fər,dīt }

horst [GEOL] **1.** A block of the earth's crust uplifted along faults relative to the rocks on either side. **2.** A mass of the earth's crust limited by faults and standing in relief. **3.** One of the older mountain masses limiting the Alps on the west and north. **4.** A knobby ledge of limestone beneath a thin soil mantle. { hȯrst }

horticultural crop [AGR] Any food-producing plant. { ¦hȯrd·ə¦kəlch·ə·rəl 'kräp }

horticulture [BOT] The art and science of growing plants. { 'hȯrd·ə,kəl·chər }

Horton number [HYD] A dimensionless number that is formed by the product of runoff intensity and erosion proportionality factor; expresses the relative intensity of erosion on the slopes of a drainage basin. { 'hȯrt·ən ,nəm·bər }

hortonolite [MINERAL] (Fe,Mg,Mn)₂SiO₄ A dark mineral composed of silicate of iron, magnesium, and manganese; a member of the olivine series. { hȯr'tän·əl,īt }

Hortvet sublimator [ANALY CHEM] Device for the determination of the condensation temperature (sublimation point) of sublimed solids. { 'hȯrt¦vet 'səb·lə,mād·ər }

HOS *See* higher-order software; horizontal obstacle sonar.

hose [DES ENG] Flexible tube used for conveying fluids. { hōz }

hose clamp [DES ENG] Band or brace to attach the raw end of a hose to a water outlet. { 'hōz ,klamp }

hose coupling [DES ENG] Device to interconnect two or more pieces of hose. { 'hōz ,kəp·liŋ }

hose fitting [DES ENG] Any attachment or accessory item for a hose. { 'hōz ,fid·iŋ }

Hoskold formula [MIN ENG] A two-rate valuation formula formerly used to determine present value of mining properties or shares, with redemption of invested capital. { 'häs,kōld ,fȯr·myə·lə }

hospital information system [COMPUT SCI] The collection, evaluation or verification, storage, and retrieval of information about a patient. { 'häs,pid·əl ,in·fər'mā·shən ,sis·təm }

hospital ship [NAV ARCH] An unarmed ship equipped as a hospital, in particular, one assigned only to assist sick and wounded persons in time of war and protected from attack by international law. { 'häs,pid·əl ,ship }

host [BIOL] **1.** An organism on or in which a parasite lives. **2.** The dominant partner of a symbiotic or commensal pair. [CHEM] A crystalline lattice or receptor molecule for the strong and selective binding of a cationic, anionic, or neutral organic, inorganic, or biological substance (guest) by means of electrostatic, hydrogen-bonding, van der Waals, or donor-acceptor interactions. Examples include clathrates, crown ethers, cryptands, cyclodextrins, calixarenes, cavitands, cyclophanes, and cryptophanes. Also known as host structure; host substance. { hōst }

host-based system [COMMUN] A communications system that is controlled by a central computer system. { 'hōst ,bāst ,sis·təm }

host-centered system *See* hierarchical distributed processing system. { 'hōst ,sen·tərd ,sis·təm }

host computer [COMPUT SCI] **1.** The central or controlling computer in a time-sharing or distributed-processing system. **2.** The computer upon which depends a specialized computer handling the input/output functions in a real-time system. **3.** A computer that can function as the source or recipient of data transfers on a network. { 'hōst kəm¦pyüd·ər }

host-guest complexation chemistry [ORG CHEM] The design, synthesis, and study of highly structured organic molecular complexes that mimic biological complexes. { 'hōst 'gest ,käm·plek'sā·shən ,kem·ə·strē }

hostile-environment machine [MECH ENG] A robot capable of operating in extreme conditions of temperature, vibration, moisture, pollution, or electromagnetic or nuclear radiation. { 'häs·təl in'vī·rən·mənt mə,shēn }

hostile identification [PSYCH] The assumption by an individual, particularly a child, of socially undesirable characteristics of an older person important to the child so as to gain some special recognition from that person. { 'häst·əl ī,dent·ə·fə'kā·shən }

hostility [PSYCH] The feeling or display of anger, antagonism, or resistance toward an individual or group. { häs'til·əd·ē }

host language data-base management system [COMPUT SCI] A data-base management system that, from a programmer's point of view, represents an extension of an existing programming language. { 'hōst ,laŋ·gwij 'dad·ə ,bās 'man·ij·mənt ,sis·təm }

host processor [COMPUT SCI] The central computer in a hierarchical distributed processing system, which is typically located at some central site where it serves as a focal point for the collection of data, and often for the provision of services which cannot economically be distributed. { 'hōst 'prä,ses·ər }

host rock [GEOL] Rock which serves as a host for other rocks or for mineral deposits. { 'hōst ,räk }

host/satellite system *See* hierarchical distributed processing system. { 'hōst 'sad·əl,īt ,sis·təm }

host structure *See* host. { ¦hōst 'strək·chər }

host substance *See* host. { ¦hōst 'səb·stəns }

hot [ELEC] *See* energized. [PHYS] Having or charged with high energy, such as high thermal energy or a high level of radioactivity. { hät }

hot-air engine [MECH ENG] A heat engine in which air or other gases, such as hydrogen, helium, or nitrogen, are used as the working fluid, operating on cycles such as the Stirling or Ericsson. { 'hä¦der 'en·jən }

hot-air furnace [MECH ENG] An encased heating unit providing warm air to ducts for circulation by gravity convection or by fans. { 'hä¦der 'fər·nəs }

hot-air soldering [MET] Soldering with a narrow blast of air whose temperature is closely controlled at the value required for soldering individual joints on printed circuit boards. { 'hä¦der 'säd·ə·riŋ }

hot-air sterilization [ENG] A method of sterilization using dry heat for glassware and other heat-resistant materials which need to be dry after treatment; temperatures of 160–165°C are generated for at least 2 hours. { 'hä¦der ,ster·ə·lə'zā·shən }

hot atom [NUCLEO] An atom that has high internal or kinetic energy as a result of a nuclear process such as beta decay or neutron capture. { 'häd ,ad·əm }

hotbed [AGR] A bed of soil enclosed by a low frame with glass panels and heated by fermented manure or electric cables; used for forcing tender plants to grow out of season or to protect tender exotic plants. [MET] An area where hot-rolled metal is placed to cool. Also known as cooling table. { 'hät,bed }

hot belt [CLIMATOL] The belt around the earth within which the annual mean temperature exceeds 20°C. { 'hät ,belt }

hot-blast stove [MET] A retracting device for preheating the incoming air in an iron blast furnace by using heat from the burning gases of the furnace. { 'hät ¦blast ,stōv }

hot-bulb [MECH ENG] Pertaining to an ignition method used in semidiesel engines in which the fuel mixture is ignited in a separate chamber kept above the ignition temperature by the heat of compression. { 'hät ,bəlb }

hot carbon-nitrogen-oxygen cycle [NUC PHYS] A modification of the carbon-nitrogen cycle that occurs at the high

HORST

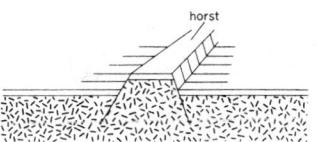

A simple horst with associated faults. (*From A. K. Lobeck, Geomorphology, McGraw-Hill, 1939*)

temperatures and high densities encountered in stellar explosions such as novae and supernovae, in which a nitrogen-13 nucleus captures a proton to form oxygen-14 before it can undergo beta decay to carbon-13. Also known as fast carbon-nitrogen-oxygen cycle. { 'hät ¦kär·bən ¦nī·trə·jən ¦äk·sə·jən ¦sī·kəl }

hot carrier [ELECTR] A carrier, which may be either an electron or a hole, that has relatively high energy with respect to the carriers normally found in majority-carrier devices such as thin-film transistors. { 'hät ¦kar·ē·ər }

hot-carrier diode *See* Schottky barrier diode. { 'hät ¦kar·ē·ər 'dī‚ōd }

hot cathode [ELECTR] A cathode in which electron or ion emission is produced by heat. Also known as thermionic cathode. { 'hät 'kath‚ōd }

hot-cathode gas-filled tube *See* thyratron. { 'hät ¦kath‚ōd ¦gas ¦fild 'tüb }

hot-cathode tube *See* thermionic tube. { 'hät ¦kath‚ōd ‚tüb }

hot cell *See* cave. { 'hät ‚sel }

hot-chamber die casting [ENG] A die-casting process in which a piston is driven through a reservoir of molten metal and thereby delivers a quantity of molten metal to the die cavity. { 'hät ¦chām·bər 'dī ‚kast·iŋ }

Hotchkiss drive [MECH ENG] An automobile rear suspension designed to take torque reactions through longitudinal leaf springs. { 'häch‚kis ‚drīv }

Hotchkiss superdip [ENG] A sensitive dip needle consisting of a freely rotating magnetic needle about a horizontal axis and a nonmagnetic bar with a counterweight at the end which is attached to the pivot point of the needle. { 'häch‚kis 'sü·pər‚dip }

hot dark matter [ASTRON] A hypothetical type of dark matter consisting of entities that were not in thermal equilibrium in the early universe; possibilities include massive neutrinos and cosmic strings. { ¦hät ¦därk 'mad·ər }

hot-die steel [MET] A high-temperature, shock-resistant alloy steel used in forging machines. { 'hät ¦dī 'stēl }

hot dipping [MET] Coating metal components by immersion in a molten metal bath, such as tin or zinc. { 'hät 'dip·iŋ }

hot-draw [ENG] To draw a material while it is hot. { 'hät ¦dró }

hot editing [CONT SYS] A method for detecting errors in the programming of a robot in which as many errors as possible are identified and resolved during testing, without setting the robotic program to its starting condition. { 'hät 'ed·əd·iŋ }

hot electron [ELECTR] An electron that is in excess of the thermal equilibrium number and, for metals, has an energy greater than the Fermi level; for semiconductors, the energy must be a definite amount above that of the edge of the conduction band. { 'hät i'lek‚trän }

hot-electron transistor [ELECTR] A transistor in which electrons tunnel through a thin emitter-base barrier ballistically (that is, without scattering), traverse a very narrow base region, and cross a barrier at the base-collector interface whose height, controlled by the collector voltage, determines the fraction of electrons coming to the collector. { ¦hät i'lek‚trän ‚tran'zis·tər }

hot-electron triode [ELECTR] Solid-state, evaporated thin-film structure directly equivalent to a vacuum triode. { 'hät i‚lek‚trän 'trī‚ōd }

hot extrusion [MET] The process of extruding metal at very high temperatures. { 'hät ik'strü·zhən }

hot-filament ionization gage [ELECTR] An ionization gage in which electrons emitted by an incandescent filament, and attracted toward a positively charged grid electrode, collide with gas molecules to produce ions which are then attracted to a negatively charged electrode; the ion current is a measure of the number of gas molecules. { 'hät ¦fil·ə·mənt ‚ī·ə·nə'zā·shən ‚gāj }

hot flash [PHYSIO] A sudden transitory sensation of heat, often involving the whole body, due to cessation of ovarian function; a symptom of the climacteric. { 'hät ‚flash }

hot forming [MET] Shaping operations performed at temperatures above the recrystallization temperature of the metal. { 'hät ¦fór·miŋ }

hot-gas welding [ENG] Joining of thermoplastic materials by softening first with a jet of hot air, then joining at the softened points. { 'hät ‚gas 'weld·iŋ }

hot hole [ELECTR] A hole that can move at much greater velocity than normal holes in a semiconductor. { 'hät ‚hōl }

hothouse [ENG] A greenhouse heated to grow plants out of season. { 'hät‚haùs }

hot isostatic pressing [ENG] A process in which a ceramic or metal powder is consolidated by heating and compressing the powder equally from all directions inside a sealed flexible mold. Abbreviated HIP. { ¦hät ‚ī·sō¦stad·ik 'pres·iŋ }

hot junction [ELECTR] The heated junction of a thermocouple. { ¦hät ¦jəŋk·shən }

hot key [COMPUT SCI] A computer key or key combination that causes a specified action to occur, regardless of what else the computer is currently doing. { 'hät ‚kē }

hot laboratory [NUCLEO] A laboratory designed for research with radioactive materials that have such high strengths that special handling precautions are required. { 'hät 'lab·rə ‚tór·ē }

hot line [COMMUN] Direct circuit between two points, available for immediate use without patching or switching. { 'hät 'līn }

hot link [COMPUT SCI] A linking of information in two documents so that modification of the information in the source document results in the same change in the destination document. { 'hät 'liŋk }

hot nucleus [NUC PHYS] An excited nucleus whose energy is shared among its many degrees of freedom. { ¦hät 'nü·klē·əs }

hot patching [ENG] Repair of a hot refractory lining in a furnace, usually by spraying with a refractory slurry. { 'hät 'pach·iŋ }

hot plate [MET] A heated surface on which joints are brought to soldering temperature. { 'hät ‚plāt }

hot press forge [MET] A press in which metal parts are formed by forcing hot metal into dies under high pressure. { 'hät ‚pres 'fórj }

hot pressing [ENG] **1.** Forming a metal-powder compact or a ceramic shape by applying pressure and heat simultaneously at temperatures high enough for sintering to occur. **2.** Fabrication of a composite material through joining the reinforcement and the matrix by means of heat and pressure, usually in a hydraulically actuated press. { 'hät 'pres·iŋ }

hot-press printing *See* hot stamping. { 'hät ‚pres 'print·iŋ }

hot-pressure welding [MET] A pressure welding process in which macrodeformation of the base material to produce coalescence results from the application of heat and pressure. { 'hät ‚presh·ər 'weld·iŋ }

hot-quenching [MET] Quenching from high temperatures into a medium of lower but still high temperature. { 'hät ¦kwench·iŋ }

hot rolling [MET] Rolling of metal bars or sheets when hot. { 'hät 'rōl·iŋ }

hot-runner mold [ENG] A plastics mold in which the runners are kept hot by insulation from the chilled cavities. { 'hät ‚rən·ər 'mōld }

hot saw [MECH ENG] A power saw used to cut hot metal. { 'hät ‚só }

hot shortness [MET] Brittleness, usually of steel or wrought iron, when the metal is hot, due to a high sulfur content. { 'hät 'shórt·nəs }

hotshot wind tunnel [AERO ENG] A wind tunnel in which electrical energy is discharged into a pressurized arc chamber, increasing the temperature and pressure in the arc chamber so that a diaphragm separating the arc chamber from an evacuated chamber is ruptured, and the heated gas from the arc chamber is then accelerated in a conical nozzle to provide flows with Mach numbers of 10 to 27 for durations of 10 to 100 milliseconds. { 'hät¦shät 'win ‚tən·əl }

hot-solder coating [ENG] The application of a protective finish to a printed circuit board by dip soldering in a solder bath. { 'hät ¦säd·ər 'kōd·iŋ }

hot spot [CHEM ENG] An area or point within a reaction system at which the temperature is appreciably higher than in the bulk of the reactor; usually locates the reaction front. [COMPUT SCI] A word in a multiprocessor memory that several processors attempt to access simultaneously, creating a conflict or bottleneck. [ENG] An area in a pipeline that is subject to excessive corrosion. [FOR] A forest region where fires occur at frequent intervals. [GRAPHICS] A region of excessive illumination on a photo. [MOL BIO] A site in a gene at which

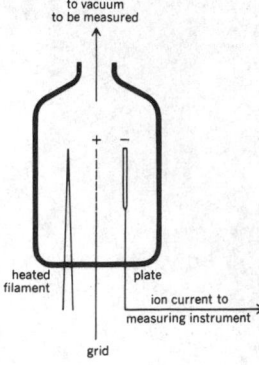

**HOT-FILAMENT
IONIZATION GAGE**

to vacuum
to be measured

heated
filament

plate

ion current to
measuring instrument

grid

Diagram of hot-filament ionization gage.

there is an unusually high frequency of mutation. [NUCLEO] **1.** A surface area of higher than average radioactivity. **2.** A part of a reactor fuel surface element that has become overheated. [PHYS] A localized region with temperature higher than the surroundings. { 'hät ,spät }

hot spraying [ENG] A paint-spraying technique in which paint viscosity is reduced by heat rather than a solvent. { 'hät ,sprā·iŋ }

hot spring [HYD] A thermal spring whose water temperature is above 98°F (37°C). { 'hät ,spriŋ }

hot stamp [ENG] An impression on a forging made in a heated condition. { 'hät ,stamp }

hot stamping [GRAPHICS] A method of printing in which heated type or stamping dies are pressed against a thin leaf of gold or other metal, or against pigment upon a surface such as paper, board, plastic, or leather; the thin leaf contains a sizing agent that is set by heat; this system is sometimes used for making movie titles and artwork for slides. Also known as hot-press printing. { 'hät ,stamp·iŋ }

hot strength See tensile strength. { 'hät ,streŋkth }

hot-swage [MET] To reduce the cross section of a hot metal tube or rod. { 'hät ,swāj }

hot tear [MET] A separation either internally or externally in a casting due to loadings or internal stresses or both; it results from improper solidification, and shrinkage near the temperature at which the casting is completely solid. { 'hät ,ter }

hot trim [MET] Removal of flash in a heated forging. { 'hät ,trim }

hot-water heating [MECH ENG] A heating system for a building in which the heat-conveying medium is hot water and the heat-emitting means are radiators, convectors, or panel coils. Also known as hydronic heating. { 'hät ,wòd·ər ,hēd·iŋ }

hot wave See heat wave. { 'hät ,wāv }

hot well [MECH ENG] A chamber for collecting condensate, as in a steam condenser serving an engine or turbine. { 'hät ,wel }

hot wind [METEOROL] General term for winds characterized by intense heat and low relative humidity, such as summertime desert winds or an extreme foehn. { 'hät ,wind }

hot wire [ELEC] **1.** A resistive wire in an electric relay that expands when heated and contracts when cooled. **2.** An electrical lead that has an electric potential with respect to the ground. { 'hät 'wīr }

hot-wire ammeter [ENG] An ammeter which measures alternating or direct current by sending it through a fine wire, causing the wire to heat and to expand or sag, deflecting a pointer. Also known as thermal ammeter. { 'hät ,wīr 'a,med·ər }

hot-wire anemometer [ENG] An anemometer used in research on air turbulence and boundary layers; the resistance of an electrically heated fine wire placed in a gas stream is altered by cooling by an amount which depends on the fluid velocity. { 'hät ,wīr ,an·ə'mäm·əd·ər }

hot-wire instrument [ENG] An instrument that depends for its operation on the expansion by heat of a wire carrying a current. { 'hät ,wīr 'in·strə·mənt }

hot-wire microphone [ENG ACOUS] A velocity microphone that depends for its operation on the change in resistance of a hot wire as the wire is cooled by varying particle velocities in a sound wave. { 'hät ,wīr 'mī·krə,fōn }

hot work [IND ENG] A task that requires working on, or in proximity to, exposed energized electrical equipment or wiring. { 'hät ,wərk }

hot working [MET] Plastic deformation of a metal at a rate and temperature such that strain hardening cannot occur. { 'hät 'wərk·iŋ }

Houben-Hoesch synthesis [ORG CHEM] Condensation of cyanides with polyhydric phenols in the presence of hydrogen chloride and zinc chloride to yield phenolic ketones. { 'hü·bən 'hərsh 'sin·thə·səs }

Houdry butane dehydrogenation [CHEM ENG] A catalytic process for dehydrogenating light hydrocarbons from crude oil to their corresponding mono- or diolefins; chromia-alumina catalysts with inert material are used in pellet form. { 'hü·drē byü,tān dē,hī·drə·jə'nā·shən }

Houdry fixed-bed catalytic cracking [CHEM ENG] A cyclic, regenerable process for cracking of petroleum distillates to produce high-octane gasoline from higher-boiling petroleum

fractions; synthetic or natural bead catalysts of activated hydrosilicate of alumina may be used. Also known as Houdry process. { 'hü·drē ¦fixt ¦bed ,kad·əl¦id·ik 'krak·iŋ }

Houdry hydrocracking [CHEM ENG] A catalytic process combining cracking and desulfurization of crude petroleum oil in the presence of hydrogen; catalysts may be nickel oxide or nickel sulfide on silica alumina, and cobalt molybdate on alumina. { 'hü·drē ¦hī·drō¦krak·iŋ }

Houdry process See Houdry fixed-bed catalytic cracking. { 'hü·drē ,prä·səs }

hour [MECH] A unit of time equal to 3600 seconds. Abbreviated h; hr. { aùr }

hour angle [ASTRON] Angular distance west of a celestial meridian or hour circle; the arc of the celestial equator, or the angle at the celestial pole, between the upper branch of a celestial meridian or hour circle and the hour circle of a celestial body or the vernal equinox, measured westward through 360°. { 'aùr ,aŋ·gəl }

hour-angle difference See meridian angle difference. { 'aùr ,aŋ·gəl 'dif·rəns }

hour circle [ASTRON] An imaginary great circle passing through the celestial poles on the celestial sphere above which declination is measured. Also known as circle of declination; circle of right ascension. { 'aùr ,sər·kəl }

hourglass [HOROL] An instrument for measuring time, consisting of two somewhat funnel-shaped vessels joined at their narrow points by a thin, hollow neck; one hour is required for a given quantity of sand to fall from the upper to the lower vessel. { 'aùr,glas }

hourglass screw See Hindley screw. { 'aùr,glas ,skrü }

hourglass stomach [MED] A stomach with an equatorial constriction usually caused by formation of scar tissue around an ulcer. { 'aùr,glas 'stəm·ək }

hourglass worm See Hindley screw. { 'aùr,glas ,wərm }

hour hand [HOROL] The pointer that indicates the hour on a timepiece. { 'aùr ,hand }

hourly observation See record observation. { 'aù·ər·lē ,äb·sər'vā·shən }

hour-out line See time front. { 'aùr ¦aùt ,līn }

hour wheel [HOROL] The wheel carrying the hour hand of a timepiece. { 'aùr ,wel }

houseboat [NAV ARCH] A barge-type craft with cabins on its deck that is used as a dwelling and for leisurely cruising. { 'haùs,bōt }

housed joint See dado joint. { 'haùzd ,jóint }

house drain [CIV ENG] Horizontal drain in a basement receiving waste from stacks. { 'haùs ,drān }

housefly [INV ZOO] Musca domestica. A dipteran insect with lapping mouthparts commonly found near human habitations; a vector in the transmission of many disease pathogens. { 'haùs,flī }

Householder's method [MATH] A transformation method for finding the eigenvalues of a symmetric matrix, in which each of the orthogonal transformations that reduce the original matrix to a triple-diagonal matrix reduces one complete row to the required form. { 'haùs,hōl·dərz ,meth·əd }

housekeeping [COMPUT SCI] Those operations or routines which do not contribute directly to the solution of a computer program, but rather to the organization of the program. { 'haùs,kēp·iŋ }

housekeeping gene See constitutive gene. { 'haùs,kēp·iŋ ¦jēn }

housekeeping run [COMPUT SCI] The performance of a program or routine to maintain the structure of files, such as sorting, merging, addition of new records, or deletion or modification of existing records. { 'haùs,kēp·iŋ ,rən }

house physician [MED] A physician employed by a hospital. { 'haùs fə'zish·ən }

house sewer [CIV ENG] Connection between house drain and public sewer. { 'haùs ,sü·ər }

housing [ENG] A case or enclosure to cover and protect a structure or a mechanical device. { 'haù·ziŋ }

Houskeeper seal [ENG] A vacuum-tight seal made between copper and glass by bringing the copper to a flexible feather edge before fusing it to the glass; the copper then flexes as the glass shrinks during cooling. { 'haùs,kēp·ər ,sēl }

Houston Automatic Spooling Processor See HASP. { hyüs·tən ód·ə¦mad·ik 'spül·iŋ ,prä,ses·ər }

hovercraft See air-cushion vehicle. { 'həv·ər,kraft }

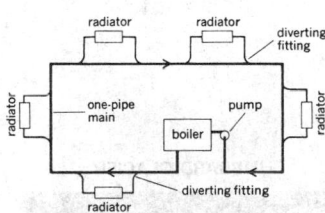

HOT-WATER HEATING

One pipe hot-water heating system.

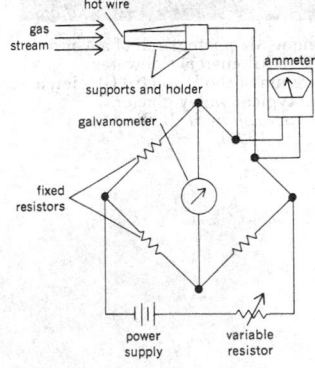

HOT-WIRE ANEMOMETER

Constant-resistance hot-wire anemometer.

howardite [GEOL] An achondritic stony meteorite composed chiefly of calcic plagioclase and orthopyroxene. { 'haů·ər,dīt }

Howell-Bunger valve *See* cone valve. { 'haů·əl 'bəŋ·gər ,valv }

Howell-Jolly bodies [PATH] Small, round basophilic inclusions of nuclear material in erythrocytes of splenectomized persons. { 'haů·əl 'jäl·ē ,bäd·ēz }

Howe truss [CIV ENG] A truss for spans up to 80 feet (24 meters) having both vertical and diagonal members; made of steel or timber or both. { 'haů ,trəs }

howitzer [ORD] A complete projectile-firing weapon with a bore diameter greater than 30 millimeters, and a length shorter than a gun of the same caliber; maximum angle of velocity is about 65°; used to deliver curved fire with projectiles of lower muzzle velocities than those from the gun. { 'haů·ət·sər }

howl [ENG ACOUS] Undesirable prolonged sound produced by a radio receiver or audio-frequency amplifier system because of either electric or acoustic feedback. { haůl }

howler [COMMUN] In telephone practice, an associated unit by which the test desk operator may connect a high tone of varying loudness to a subscriber's line to call the subscriber's attention to the fact that the phone receiver is off the hook. [ELECTR] An audio device used to warn a radar operator that signals are appearing on a radar screen. { 'haůl·ər }

howling tones [ACOUS] The sounds produced by a howling tube. Also known as gauze tones. { 'haůl·iŋ ,tōnz }

howling tube [ACOUS] A vertical open tube with a piece of gauze in the lower half which is placed over a flame to make the tube produce powerful sound waves with many overtones. { 'haůl·iŋ ,tüb }

howlite [MINERAL] $Ca_2Bi_5SiO_9(OH)_5$ A white mineral occurring in nodular or earthy form. { 'haů,līt }

howl repeater [COMMUN] Condition in telephone repeater operation where more energy is returned than sent, resulting in an oscillation being set up on the circuit. { 'haůl ri'pēd·ər }

Howship's lacuna [HISTOL] Minute depressions in the surface of a bone undergoing resorption. { 'haů·shəps lə'kü·nə }

Hoyer method of prestressing *See* pretensioning. { 'hȯi·yər ,meth·əd əv prē'stres·iŋ }

hp *See* horsepower.

H pad *See* H network. { 'āch ,pad }

h parameter [ELECTR] One of a set of four transistor equivalent-circuit parameters that conveniently specify transistor performance for small voltages and currents in a particular circuit. Also known as hybrid parameter. { 'āch pə,ram·əd·ər }

HPBW *See* half-power beamwidth.

H pile [CIV ENG] A steel pile that is H-shaped in section. { 'āch ,pīl }

H plane [ELECTROMAG] The plane of an antenna in which lies the magnetic field vector of linearly polarized radiation. { 'āch ,plān }

H-plane bend [ELECTROMAG] A rectangular waveguide bend in which the longitudinal axis of the waveguide remains in a plane parallel to the plane of the magnetic field vector throughout the bend. Also known as H bend. { 'āch ,plān ,bend }

H-plane T junction [ELECTROMAG] Waveguide T junction in which the change in structure occurs in the plane of the magnetic field. Also known as shunt T junction. { 'āch ,plān 'tē ,jəŋk·shən }

HPLC *See* high-performance liquid chromatography.

HPV *See* human papillomavirus.

hr *See* hour.

H-R diagram *See* Hertzsprung-Russell diagram. { 'āch 'är ,dī·ə,gram }

HREELS *See* high-resolution electron energy loss spectroscopy. { 'āch ,rēlz }

H I region [ASTRON] A region of interstellar space where neutral hydrogen is present. { 'āch 'wən ,rē·jən }

H II region [ASTRON] A region of interstellar space occupied by gas that is largely atomic hydrogen and mostly ionized. { 'āch 'tü ,rē·jən }

H-RNA *See* heterogeneous ribonucleic acid.

H rod [DES ENG] A drill rod having an outside diameter of 3-1/2 inches (8.89 centimeters). { 'āch ,räd }

H scan *See* H scope. { 'āch ,skan }

H scope [NAV] A modified form of B scope on which the signal appears as two dots, the slope of the line joining them

indicating elevation angle; the horizontal coordinate indicates bearing angle and the vertical coordinate range, with respect to the left-hand dot. Also known as H indicator; H scan. { 'āch ,skōp }

HSI *See* heat stress index.

HSLA steel *See* high-strength low-alloy steel. { 'āch,es,el,ā 'stēl }

HSP *See* high-speed printer.

H substance [BIOCHEM] An agent similar to histamine and believed to play a role in local blood vessel response in tissue damage. { 'āch ,səb·stəns }

H system [NAV] A radar navigation system that uses two ground radar beacons, in conjunction with airborne equipment which gives the direction and distance to each beacon. { 'āch ,sis·təm }

HTGR *See* high-temperature gas-cooled reactor.

H theorem of Boltzmann *See* Boltzmann H theorem. { 'āch ,thir·əm əv 'bōlts,män }

HTLV *See* human T-lymphotropic virus.

HTML *See* Hypertext Markup Language.

HTTP *See* Hypertext Transfer Protocol.

HTU *See* height of transfer unit.

huangho deposit [GEOL] A coastal-plain deposit comprising alluvium spread over a level surface (such as a floodplain) but extending into marine beds of equivalent age. { 'hwäŋ·hō di,päz·ət }

huarizo [VERT ZOO] A hybrid offspring of a male llama and a female alpaca, bred for its fine fleece. { wä'rē·zō }

hub [BUILD] The core section of a building from which corridors extend. [COMPUT SCI] An electric socket in a plugboard into which one may insert or connect leads or may plug wires. [DES ENG] **1.** The cylindrical central part of a wheel, propeller, or fan. **2.** A piece in a lock that is turned by the knob spindle, causing the bolt to move. **3.** A short coupling that joins plumbing pipes. [ENG] In surveying, a stake that marks the position of a theodolite. [MET] A steel punch used in making a working die for a coin or medal. { həb }

Hubbard Glacier [GEOGR] A valley glacier which reaches tidewater from a source area of Mount St. Elias of Alaska and the Yukon. { 'həb·ərd 'glā·shər }

Hubbard tank [MED] A large, specially designed tank in which a patient may be immersed for various therapeutic underwater exercises. { 'həb·ərd ,taŋk }

hubbing [MET] Forcing a male die into a blank to form a female die. { 'həb·iŋ }

hubble [ASTRON] A unit of astronomical distance equal to 10^9 light-years or 9.4605×10^{24} meters. { 'həb·əl }

Hubble constant [ASTROPHYS] The rate at which the velocity of recession of the galaxies increases with distance; the value is about 70 kilometers per second per megaparsec (or 2.3×10^{-18} s^{-1}) with a relative uncertainty of about ± 10%. { 'həb·əl ,kän·stənt }

Hubble effect *See* redshift. { 'həb·əl i,fekt }

Hubble flow [ASTRON] The mutual recession of celestial objects from each other by virtue of the cosmological expansion of the universe. { 'həb·əl ,flō }

Hubble law [ASTRON] The principle that the distance of external galaxies from the earth is proportional to their redshift. { 'həb·əl ,lȯ }

Hubble Space Telescope [OPTICS] An astronomical reflecting telescope with a mirror 94.5 inches (2.4 meters) in diameter; placed in orbit above the earth's atmosphere in April 1990. { 'həb·əl 'spās ,tel·ə,skōp }

Hubble's Variable Nebula [ASTRON] A variable-brightness nebula associated with variable stars and fan-shaped in appearance. { 'həb·əlz ,ver·ē·ə·bəl 'neb·yə·lə }

Hubble time [ASTRON] The reciprocal of the Hubble constant. { 'həb·əl ,tīm }

hub cannon [ORD] A cannon mounted through a propeller hub of an aircraft. { 'həb ,kan·ən }

hubcap [DES ENG] A metal cap fastened or clamped to the end of an axle, as on motor vehicles. { 'həb,kap }

Huber's reagent [ANALY CHEM] Aqueous solution of ammonium molybdate and potassium ferrocyanide used as a reagent to detect free mineral acid. { 'hyü·bərz rē,ā·jənt }

Hubl's reagent [ANALY CHEM] Solution of iodine and mercuric chloride in alcohol; used to determine the iodine content of oils and fats. { 'həb·əlz rē,ā·jənt }

Hübner rhomb [OPTICS] A glass rhombohedron used in

photometry to compare two illuminated surfaces. { 'hyib·nər ,räm }

hub ring [COMPUT SCI] A thin plastic ring placed around the center hole of a floppy disk to prevent the disk from warping and damaging its contents if it is improperly inserted in a disk drive. { 'həb ,riŋ }

Hückel's 4n + 2 rule [ORG CHEM] Aromatic (ring) compounds must have $4n + 2$ pi-bonding electrons, where n is a whole number and generally limited to $n = 0$ to 5. When $n = 1$, for example, there are six pi-electrons, as for benzene. Also known as Hückel's rule. { 'hük·əlz ¦fȯr ¸en pləs 'tü ¸rül }

Hückel's rule See Hückel's $4n + 2$ rule. { 'hük·əlz ¸rül }

huckleberry [BOT] The common name for shrubs of the genus *Gaylussacia* in the family Ericaceae distinguished by an ovary with 10 locules and 10 ovules; the dark-blue berries are edible. { 'hək·əl¸ber·ē }

Hudsonian life zone [ECOL] A zone comprising the climate and biotic communities of the northern portions of North American coniferous forests and the peaks of high mountains. { ¦həd¦sō·nē·ən 'līf ¸zōn }

hudsonite See cortlandite. { 'həd·sə¸nīt }

hue [OPTICS] The name of a color, such as red, yellow, green, blue, or purple, as perceived subjectively. { hyü }

huebnerite [MINERAL] $MnWO_4$ A brownish-red to black manganese member of the wolframite series, occurring in short, monoclinic, prismatic crystals; isomorphous with ferberite. { 'hēb·nə¸rīt }

hue control [ELECTR] A control that varies the phase of the chrominance signals with respect to that of the burst signal in a color television receiver, in order to change the hues in the image. Also known as phase control. { 'hyü kən¸trōl }

Huffman method [COMPUT SCI] A data compression technique in which a bit representation for each character is determined that is as close as possible to the character's predicted information content, based on its frequency of occurrence. { 'həf·mən ¸meth·əd }

Huggenberger tensometer [ENG] A type of extensometer having a short gage length (10 to 20 millimeters) and employing a compound lever system that gives a magnification of about 1200. { 'həg·ən¸bərg·ər ten'säm·əd·ər }

Hughes effect [ELECTROMAG] An asymmetry in the hysteresis curves of laminated cores made of certain magnetic materials such as permalloy or mu metal in alternating magnetic fields. { 'yüz i¸fekt }

Hughes plane [MATH] A finite projective plane with nine points on each line that can be represented by a nonlinear ternary ring generated by a four-point in the plane. { 'hyüz ¸plān }

Hugoniot function [PHYS] A function specifying the locus of states which are possible immediately after the passage of a shock front; gives the state's pressure as a function of its specific volume. { yü'gōn·yō ¸fəŋk·shən }

hühnerkobelite [MINERAL] $(Na,Ca)(Fe,Mn)_2(PO_4)_2$ A mineral composed of phosphate of sodium, calcium, iron, and manganese; it is isomorphous with varulite. { ¦hyü·nər¦kō·bə¸līt }

Huhner's test [PATH] An examination of seminal fluid obtained from the vaginal fornix and cervical canal after a specified interval following coitus; used in fertility studies to evaluate spermatozoal survival and activity in the female lower genital tract. { 'hyü·nərz ¸test }

hull [BOT] The outer, usually hard, covering of a fruit or seed. [FOOD ENG] **1.** To remove husks from fruits and seeds, as from ears of corn, nuts, or peas. **2.** To remove the shell of a crustacean or mollusk, as an oyster. [MATH] See span. [NAV ARCH] The body or shell of a ship. [ORD] **1.** The outer casing of a rocket, guided missile, or the like. **2.** Massive armored body of a tank, exclusive of tracks, motor, turret, and armament. { həl }

Hull cell [PHYS CHEM] An electrodeposition cell that operates within a simultaneous range of known current densities. { 'həl ¸sel }

hulless buckwheat See tartary buckwheat. { 'həl·əs 'bək¸wēt }

hulsite [MINERAL] $(Fe^{2+},Mg)_2(Fe^{3+},Sn)(BO_3)O_2$ A black mineral composed of iron calcium magnesium tin borate. { 'həl¸sīt }

hum [ELEC] A sound produced by an iron core of a transformer due to loose laminations or magnetostrictive effects; the frequency of the sound is twice the power line frequency.

[ELECTR] An electrical disturbance occurring at the power supply frequency or its harmonics, usually 60 or 120 hertz in the United States. { həm }

human biogeography [ECOL] The science concerned with the distribution of human populations on the earth. { 'hyü·mən ¦bī·ō·jē¦äg·rə·fē }

human chorionic gonadotropin [BIOCHEM] A gonadotropic and luteotropic hormone secreted by the chorionic vesicle. Abbreviated HCG. Also known as chorionic gonadotropin. { 'hyü·mən kȯr·ē¦än·ik gō¸nad·ə'trō·pən }

human community [ECOL] That portion of a human ecosystem composed of human beings and associated plant and animal species. { 'hyü·mən kə'myün·əd·ē }

human-computer interaction [COMPUT SCI] The processes through which human users work with interactive computer systems. { ¦yü·mən kəm¦pyüd·ər ¸in·tər'ak·shən }

human ecology [ECOL] The branch of ecology that considers the relations of individual persons and of human communities with their particular environment. { 'hyü·mən ē'käl·ə·jē }

human engineering See human-factors engineering. { 'hyü·mən ¸en·jə'nir·iŋ }

human-factors engineering [ENG] The area of knowledge dealing with the capabilities and limitations of human performance in relation to design of machines, jobs, and other modifications of the human's physical environment. Also known as human engineering. { 'hyü·mən ¦fak·tərz ¸en·jə'nir·iŋ }

human geography [GEOGR] The study of the characteristics and phenomena of the earth's surface that relate directly to or are due to human activities. Also known as anthropogeography. { ¦yü·mən jē¦äg·rə·fē }

human immunodeficiency virus [VIROL] The retrovirus that causes acquired immune deficiency syndrome. Abbreviated HIV. { 'hyü·mən ¦im·yə·nō·di'fish·ən·sē ¸vī·rəs }

human leukocyte antigen [IMMUNOL] Any of a group of antigens present on the surface of nucleated body cells that are coded for by the major histocompatibility complex of humans and thus allow the immune system to distinguish self and nonself. Abbreviated HLA. { ¦yü·mən ¦lü·kə¸sīt 'ant·i·jən }

human-machine chart [IND ENG] A two-column, multiple-activity process chart listing the steps performed by an operator and the operations performed by a machine and showing the corresponding idle times for each. Also known as man-machine chart. { ¦yü·mən mə¦shēn 'chärt }

human-machine system [ENG] A system in which the functions of the worker and the machine are interrelated and necessary for the operation of the system. Also known as man-machine system. { ¦yü·mən mə¦shēn 'sis·təm }

human measles immune serum [IMMUNOL] Serum from the blood of a person who has recovered from measles. { 'hyü·mən ¦mē·zəlz I'myun ¸sir·əm }

human myeloma protein [IMMUNOL] Any of a group of the first structurally discrete immunoglobulins known, resulting from the malignant expansion of a normal clone of antibody-producing cells. { ¸hyü·mən ¸mī·ə¸lō·mə 'prō¸tēn }

human papillomavirus [MED] One of a family of more than 100 different viruses, most commonly spread via sexual contact, that cause warts on the hands and feet and in the genital area; several types are associated with premalignant and malignant changes in the cervix. Abbreviated HPV. { ¦yü·mən ¸pap·ə'lō·mə ¸vī·rəs }

human threadworm See pinworm. { 'hyü·mən 'thred¸wərm }

human T-lymphotropic virus type 1 [VIROL] A retrovirus associated with a rare form of T-cell leukemia occurring primarily in the Caribbean area, Africa, and Japan. Also known as human T-cell lymphotropic virus type 1. { 'hyü·mən 'tē ¦lim·fə¦träp·ik ¸vī·rəs ¦tīp 'wən }

human T-lymphotropic virus type 2 [VIROL] A retrovirus associated with at least one form of leukemia. Also known as human T-cell lymphotropic virus type 2. { 'hyü·mən 'tē ¦lim·fə¦träp·ik ¸vī·rəs ¦tīp 'tü }

human T-lymphotropic virus type 3 [VIROL] A former designation for the human immunodeficiency virus. Also known as human T-cell lymphotropic virus type 3. { 'hyü·mən 'tē ¦lim·fə¦träp·ik ¸vī·rəs ¦tīp 'thrē }

hum bar [ELECTR] A dark horizontal band extending across a television picture due to excessive hum in the video signal applied to the input of the picture tube. { 'həm ¸bär }

Humble gage [PETRO ENG] Device to measure oil well bottom-hole pressure; a piston acts through a stuffing box against a helical spring in tension. { 'həm·bəl ˌgāj }

Humble relation [PETRO ENG] Equation used by oil companies to estimate porosity of the oil-bearing formation from measurements made with a contact resistivity device, such as a microlog. { 'həm·bəl riˌlā·shən }

Humboldt Current *See* Peru Current. { 'həm,bōlt 'kə·rənt }

Humboldt Glacier [HYD] The largest Arctic iceberg, at latitude 79°, with a seaward front extending 65 miles (105 kilometers). { 'həm,bōlt 'glā·shər }

humboldtine [MINERAL] $FeC_2O_4 \cdot 2H_2O$ A mineral composed of hydrous ferrous oxalate. Also known as humboldtite; oxalite. { 'həm,bōl,tēn }

humboldtite *See* humboldtine. { 'həm,bōl,tīt }

Humboldt jig [MIN ENG] An ore jig with a movable screen. { 'həm,bōlt ˌjig }

hum-bucking coil [ENG ACOUS] A coil wound on the field coil of an excited-field loudspeaker and connected in series opposition with the voice coil, so that hum voltage induced in the voice coil is canceled by that induced in the hum-bucking coil. { 'həm ˌbək·iŋ ˌkȯil }

humectant [CHEM] A substance which absorbs or retains moisture; examples are glycerol, propylene glycol, and sorbitol; used in preparing confectioneries and dried fruit. { hyü 'mek·tənt }

humeral [ANAT] Of or pertaining to the humerus or the shoulder region. { 'hyüm·ə·rəl }

humeroglandular effector system [PHYSIO] Glandular effector system in which the activating agent is a blood-borne chemical. { ¦hyüm·ə·rō¦glan·jə·lər i'fek·tər ˌsis·təm }

humeromuscular effector system [PHYSIO] Muscular effector system in which the activating agent is a blood-borne chemical. { ¦hyüm·ə·rō¦məs·kyə·lər i'fek·tər ˌsis·təm }

Hume-Rothery compound *See* electron compound. { 'hyüm 'rȯth·ə·rē ˌkäm,pau̇nd }

Hume-Rothery rule [SOLID STATE] The rule that the ratio of the number of valence electrons to the number of atoms in a given phase of an electron compound depends only on the phase, and not on the elements making up the compounds. { 'hyüm 'rȯth·ə·rē ˌrül }

humerus [ANAT] The proximal bone of the forelimb in vertebrates; the bone of the upper arm in humans, articulating with the glenoid fossa and the radius and ulna. { 'hyüm·ə·rəs }

humic [GEOL] Pertaining to or derived from humus. { 'hyü·mik }

humic acid [ORG CHEM] Any of various complex organic acids obtained from humus; insoluble in acids and organic solvents. { 'hyü·mik 'as·əd }

humic-cannel coal *See* pseudocannel coal. { 'hyü·mik ¦kan·əl ˌkōl }

humic coal [GEOL] A coal whose attritus is composed mainly of transparent humic degradation material. { 'hyü·mik ˌkōl }

humicolous [ECOL] Of or pertaining to plant species inhabiting medium-dry ground. { 'hyü·mik·ə·ləs }

humid climate [CLIMATOL] A climate whose typical vegetation is forest. Also known as forest climate. { 'hyü·məd ¦klī·mət }

humidification [ENG] The process of increasing the water vapor content of a gas. { yü,mid·i·fə'kā·shən }

humidifier [MECH ENG] An apparatus for supplying moisture to the air and for maintaining desired humidity conditions. { yü'mid·ə,fī·ər }

humidistat [ENG] An instrument that measures and controls relative humidity. Also known as hydrostat. { yü'mid·ə,stat }

humidity [METEOROL] Atmospheric water vapor content, expressed in any of several measures, such as relative humidity. { hyü'mid·əd·ē }

humidity capacitor [ELECTR] A device for measuring ambient relative humidity by sensing a change in capacitance. { hyü'mid·əd·ē kə'pas·əd·ər }

humidity coefficient [METEOROL] A measure of the precipitation effectiveness of a region; it recognizes the exponential relationship of temperature versus plant growth and is expressed as humidity coefficient = $P/(1.07)^t$, where P is the precipitation in centimeters, and t is the mean temperature in degrees Celsius

for the period in question; the denominator approximately doubles with each 10°C rise in temperature. { hyü'mid·əd·ē ˌkō·i'fish·ənt }

humidity element [ENG] The transducer of any hygrometer, that is, that part of a hygrometer that quantitatively senses atmospheric water vapor. { hyü'mid·əd·ē ˌel·ə·mənt }

humidity index [CLIMATOL] An index of the degree of water surplus over water need at any given station; it is calculated as humidity index = $100s/n$, where s (the water surplus) is the sum of the monthly differences between precipitation and potential evapotranspiration for those months when the normal precipitation exceeds the latter, and where n (the water need) is the sum of monthly potential evapotranspiration for those months of surplus. { hyü'mid·əd·ē ˌin,deks }

humidity indicator [INORG CHEM] Cobalt salt (for example, cobaltous chloride) that changes color as the surrounding humidity changes; changes from pink when hydrated, to greenish-blue when anhydrous. { hyü'mid·əd·ē ˌin·də,kād·ər }

humidity mixing ratio [METEOROL] The amount of water vapor mixed with one unit mass of dry air, usually expressed as grams of water vapor per kilogram of air. { hyü'mid·əd·ē 'mik·siŋ ˌrā·shō }

humidity province [CLIMATOL] A region in which the precipitation effectiveness of its climate produces a definite type of biological consequence, in particular the climatic climax formations of vegetation (rain forest, tundra, and the like). { hyü'mid·əd·ē ˌpräv·əns }

humidity strip [ENG] The humidity transducing element in a Diamond-Hinman radiosonde; it consists of a flat plastic strip bounded by electrodes on two sides and coated with a hygroscopic chemical compound such as lithium chloride; the electrical resistance of this coating is a function of the amount of moisture absorbed from the atmosphere and the temperature of the strip. Also known as electrolytic strip. { hyü'mid·əd·ē ˌstrip }

humidity test [MET] Corrosion test in which a specimen is exposed to an environment of controlled humidity and temperature. { hyü'mid·əd·ē ˌtest }

humid transition life zone [ECOL] A zone comprising the climate and biotic communities of the northwest moist coniferous forest of the north-central United States. { 'hyü·məd tran¦zish·ən 'līf ˌzōn }

humification [GEOL] Formation of humus. { ˌhyü·mə·fə'kā·shən }

humifuse [BIOL] Spread over the ground surface. { 'hyü·mə,fyüs }

humin [GEOL] *See* ulmin. [ORG CHEM] An insoluble pigment formed in the acid hydrolysis of a protein that contains tryptophan. { 'hyü·mən }

Humiriaceae [BOT] A family of dicotyledonous plants in the order Linales characterized by exappendiculate petals, usually five petals, flowers with an intrastaminal disk, and leaves lacking stipules. { hyü,mir·ē'ās·ē,ē }

humite [MINERAL] **1.** A humic coal mineral. **2.** A series of magnesium neosilicate minerals closely related in crystal structure and chemical composition. { 'hyü,mīt }

humivore [ECOL] An organism that feeds on humus. { 'hyü·mə,vȯr }

hummer screen [MIN ENG] An electrically vibrated ore screen for sizing moderately small material. { 'həm·ər ˌskrēn }

hummingbird [VERT ZOO] The common name for members of the family Trochilidae; fast-flying, short-legged, weak-footed insectivorous birds with a tubular, pointed bill and a fringed tongue. { 'həm·iŋ,bərd }

hummock [ECOL] A rounded or conical knoll frequently formed of earth and covered with vegetation. [GEOL] A rounded or conical knoll, mound, hillock, or other small elevation, generally of equal dimensions and not ridgelike. Also known as hammock. [HYD] A mound, hillock, or pile of broken floating ice, either fresh or weathered, that has been forced upward by pressure, as in an ice field or ice floe. { 'həm·ək }

hummocked ice [OCEANOGR] Pressure ice, characterized by haphazardly arranged mounds or hillocks; it has less definite form, and show the effects of greater pressure, than either rafted ice or tented ice, but in fact may develop from either of those. { 'həm·əkt 'īs }

HUMIDISTAT

plastic coating

to relay amplifier

conductors

Electronic humidistat. *(Honeywell Inc.)*

HUMMINGBIRD

Hummingbird, capable of hovering in flight.

hummocky [GEOL] Any topographic surface characterized by rounded or conical mounds. { 'həm·ə·kē }

hum modulation [ELECTR] Modulation of a radio-frequency signal or detected audio-frequency signal by hum; heard in a radio receiver only when a station is tuned in. { 'həm ,mäj·ə,lā·shən }

Humod [GEOL] A suborder of the soil order Spodosol having an accumulation of humus, and of aluminum but not iron. { 'hyü,mäd }

humodurite *See* translucent attritus. { ¦hyü·mə¦dù,rīt }

humogelite *See* ulmin. { hyü'mäj·əl,īt }

humor [PHYSIO] A fluid or semifluid part of the body. { 'hyü·mər }

humoral immunity [IMMUNOL] Immunity in which immune responses are mediated by immunoglobulins. Also known as antibody-mediated immunity; immunoglobulin-mediated immunity. { ¦hyüm·ə·rəl i'myü·nəd·ē }

Humox [GEOL] A suborder of the soil order Oxisol that is high in organic matter, well drained but moist all or nearly all year, and restricted to relatively cool climates and high altitudes for Oxisols. { 'hyü,mäks }

humpback *See* kyphosis. { 'həmp,bak }

Humphrey gas pump [MECH ENG] A combined internal combustion engine and pump in which the metal piston has been replaced by a column of water. { 'həm·frē 'gas ,pəmp }

Humphreys series [SPECT] A series of lines in the infrared spectrum of atomic hydrogen whose wave numbers are given by $R_H[(1/36) - (1/n^2)]$, where R_H is the Rydberg constant for hydrogen, and n is any number greater than 6. { 'həm·frēz ,sir·ēz }

Humphrey's spiral [MIN ENG] An ore concentrator consisting of a stationary spiral trough through which ore pulp gravitates; heavy particles stay on the inside and lighter ones climb to the outside. { 'həm·frēz ,spī·rəl }

Humphries equation [THERMO] An equation which gives the ratio of specific heats at constant pressure and constant volume in moist air as a function of water vapor pressure. { 'həm·frēz i,kwā·zhən }

hump yard [CIV ENG] A switch yard in a railway system that has a hump or steep incline down which freight cars can coast to prescheduled locations. Also known as gravity yard. { 'həmp ,yärd }

Humult [GEOL] A suborder of the soil order Ultisol, well drained with a moderately thick surface horizon; formed under conditions of high rainfall distributed evenly over the year; common in southeastern Brazil. { 'hyü·məlt }

humus [GEOL] The amorphous, ordinarily dark-colored, colloidal matter in soil; a complex of the fractions of organic matter of plant, animal, and microbial origin that are most resistant to decomposition. { 'hyü·məs }

Hund coupling cases [ATOM PHYS] Five ways of combining the electron-spin angular momentum, electron-orbital angular momentum, and nuclear-rotation angular momentum to form the total angular momentum of a molecule. { 'hənd ¦kəp·liŋ ,kā·səz }

hundred-percent rectangle [ORD] A rectangle whose length is eight probable errors in range, and whose breadth is eight probable errors in direction; its center is the center of dispersion; it is the area in which all shots (except wild shots) will fall when fired with the same data under identical conditions. Also known as rectangle of dispersion. { ¦hən·drəd pər'sent 'rek,taŋ·gəl }

hundredweight [SCI TECH] Abbreviated cwt. **1.** A unit of weight, in common use in the United States, equal to 100 pounds or the weight of 45.359237 kilograms. Also known as cental; centner; kintal; quintal; short hundredweight. **2.** A unit of weight in common use in the United Kingdom equal to 112 pounds or to the weight of 50.80234544 kilograms. Also known as long hundredweight. **3.** A unit of weight in troy measure equal to 100 troy pounds or the weight of 37.32417216 kilograms. { 'hən·drəd,dwāt }

Hund rules [ATOM PHYS] Two rules giving the order in energy of atomic states formed by equivalent electrons: of the terms given by equivalent electrons, the ones with greatest multiplicity have the least energy, and of these the one with greatest orbital angular momentum is lowest; the state of a multiplet with lowest energy is that in which the total angular

momentum is the least possible, if the shell is less than half-filled, and the greatest possible, if more than half filled. { 'hənd ,rülz }

Hundsdieke reaction [ORG CHEM] Production of an alkyl halide by boiling a silver carboxylate with an equivalent weight of bromine in carbon tetrachloride. { 'hənz,dēk·ə rē,ak·shən }

hunger [PSYCH] The need for food and the physiological and psychological mechanisms regulating food intake. { 'həŋ·gər }

hungry joint *See* starved joint. { 'həŋ·grē 'jòint }

hung shot [ENG] A shot whose explosion is delayed after detonation or ignition. { 'həŋ ,shät }

Hunner's ulcer [MED] A chronic ulcer of the urinary bladder, frequently in association with interstitial cystitis. Also known as elusive ulcer. { 'hən·ərz 'əl·sər }

hunt [AERO ENG] **1.** Of an aircraft or rocket, to weave about its flight path, as if seeking a new direction or another angle of attack; specifically, to yaw back and forth. **2.** Of a control surface, to rotate up and down or back and forth without being detected by the pilot. { hənt }

Hunt and Douglas process [MET] Smelting process involving the roasting of matte carrying copper, lead, gold, and silver to form copper sulfate and oxide (but not silver sulfate); this product is leached with sulfuric acid for copper; the resulting solution is treated with calcium chloride by passing sulfur dioxide through it; the cuprous chloride is then reduced to cuprous oxide by milk of lime, (regenerating calcium chloride), and the cuprous oxide is smelted. { ¦hənt ən 'dəg·ləs ,prä·səs }

Hunt continuous filter [MIN ENG] A continuous-vacuum filter consisting of a horizontally revolving, annular filter bed on which pulp is washed and then vacuum-dried. { 'hənt kən¦tin·yə·wəs 'fil·tər }

hunter's moon [ASTRON] The full moon next following the harvest moon. { 'hən·tərz ,mün }

Hunter syndrome [MED] An X-linked recessive disease in which a deficiency of the enzyme iduronate sulfatase leads to the accumulation of mucopolysaccharides in various body tissues, resulting in developmental abnormalities, skeletal deformations, mental retardation, and, in severe cases, early death. Also known as mucopolysaccharidosis II. { 'hənt·ər ,sin,drōm }

hunting [CONT SYS] Undesirable oscillation of an automatic control system, wherein the controlled variable swings on both sides of the desired value. [ELECTR] Operation of a selector in moving from terminal to terminal until one is found which is idle. [MECH ENG] Irregular engine speed resulting from instability of the governing device. { 'hənt·iŋ }

hunting circuit *See* lockout circuit. { 'hənt·iŋ ,sər·kət }

Hunting Dogs *See* Canes Venatici. { 'hənt·iŋ ,dògz }

Huntington-Heberlein process [MIN ENG] A sink-float process employing a galena medium, which is recovered by froth flotation. { 'hənt·iŋ·tən 'hā·bər,līn ,prä·səs }

Huntington's chorea [MED] A rare hereditary disease of the basal ganglia and cerebral cortex resulting in choreiform (dancelike) movements, intellectual deterioration, and psychosis. { 'hənt·iŋ·tənz kə'rē·ə }

hunting tooth [DES ENG] An extra tooth on the larger of two gear wheels so that the total number of teeth will not be an integral multiple of the number on the smaller wheel. { 'hənt·iŋ ,tüth }

huntite [MINERAL] $CaMg_3(CO_3)_4$ A white mineral consisting of calcium magnesium carbonate. { 'hən,tīt }

hurdle sheet [MIN ENG] A brattice-cloth screen across a roadway below a roof cavity or at the ripping lip to divert air current upward, thus diluting and removing firedamp. { 'hərd·əl ,shēt }

hureaulite [MINERAL] $Mn_5H_2(PO_4)_4 \cdot 4H_2O$ A monoclinic mineral of varying colors consisting of a hydrated acid phosphate of manganese. { 'hyù·rō,līt }

Hurler's syndrome [MED] Mucopolysaccharoidosis I, a hereditary condition transmitted as an autosomal recessive in which there is excessive chondroitin sulfate B and heparin sulfate in the urine and tissues, and which is marked clinically by a complex of symptoms including grotesque skeletal and facial deformities, skin and cardiac changes, clouding of the cornea, and mental deficiency. Also known as gargoylism; lipochondrodystrophy. { 'hər·lərz ,sin,drōm }

HURRICANE

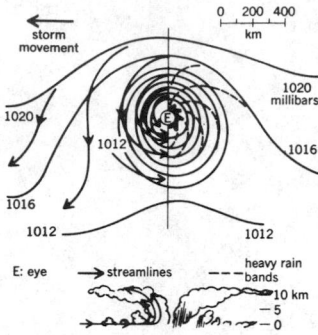

Simplified model diagram of circulation of a Northern Hemisphere hurricane. Isobars (in millibars) have been omitted near the cyclone center. Below is a vertical section through the center, showing clouds and vertical circulation.

HUYGENS EYEPIECE

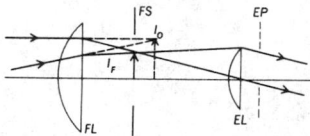

Arrangement of lenses in Huygens eyepiece. *FL* = field lens; *EL* = eye lens; *FS* = field stop; *EP* = exit pupil or eye point; I_O = image formed by preceding system; I_F = image formed by preceding system and the field lens.

HUYGENS' PRINCIPLE

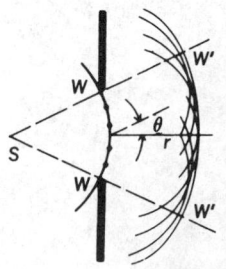

The construction for a spherical wave. *WW* = wave originating at *S*; *W'W'* = envelope of secondary waves; θ = angle between the normal to the original wavefront and any point on the secondary wave; *r* = radius representing the distance wave would travel in time *t*.

Huronian [GEOL] The lower system of the restricted Proterozoic. { hyü'rō·nē·ən }

hurricane [METEOROL] A tropical cyclone of great intensity; any wind reaching a speed of more than 73 miles per hour (117 kilometers per hour) is said to have hurricane force. { 'hər·ə‚kān }

hurricane air stemmer [MIN ENG] A mechanical device for rapidly tamping shotholes; consists of a funnel connected by a T piece to the charge tube, with a connection to a compressed-air column; sand is put in the funnel and injected into the shothole as the charge tube is withdrawn. { 'hər·ə‚kān 'er ‚stem·ər }

hurricane band See spiral band. { 'hər·ə‚kān ‚band }

hurricane beacon [ENG] An air-launched balloon designed to be released in the eye of a tropical cyclone, to float within the eye at predetermined levels, and to transmit radio signals. { 'hər·ə‚kān ‚bē·kən }

hurricane deck See awning deck; promenade deck. { 'hər·ə‚kān ‚dek }

hurricane-force wind [METEOROL] In the Beaufort wind scale, a wind whose speed is 64 knots (117 kilometers per hour) or higher. { 'hər·ə‚kān ‚fȯrs ‚wind }

hurricane lamp [ENG] An oil lamp with a glass chimney and perforated lid to protect the flame, or a candle with a glass chimney. { 'hər·ə‚kān ‚lamp }

hurricane monitoring buoy [METEOROL] A free-floating automatic weather station designed as an expendable instrument in connection with hurricane and typhoon monitoring and forecasting services. { 'hər·ə‚kān 'män·ə·triŋ ‚bȯi }

hurricane radar band See spiral band. { 'hər·ə‚kān 'rā‚där ‚band }

hurricane surge See hurricane wave. { 'hər·ə‚kān ‚sərj }

hurricane tide See hurricane wave. { 'hər·ə‚kān ‚tīd }

hurricane tracking [ENG] Recording of the movement of individual hurricanes by means of airplane sightings and satellite photography. { 'hər·ə‚kān ‚trak·iŋ }

hurricane warning [METEOROL] A warning of impending winds of hurricane force; for maritime interests, the storm warning signals for this condition are two square red flags with black centers by day, and a white lantern between two red lanterns by night. { 'hər·ə‚kān ‚wȯrn·iŋ }

hurricane watch [METEOROL] An announcement for a specific area that hurricane conditions pose a threat; residents are cautioned to take stock of their preparedness needs but, otherwise, are advised to continue normal activities. { 'hər·ə‚kān ‚wäch }

hurricane wave [OCEANOGR] As experienced on islands and along a shore, a sudden rise in the level of the sea associated with a hurricane. Also known as hurricane surge; hurricane tide. { 'hər·ə‚kān ‚wāv }

hurricane wind [METEOROL] In general, the severe wind of an intense tropical cyclone (hurricane or typhoon); the term has no further technical connotation, but is easily confused with the strictly defined hurricane-force wind. Also known as typhoon wind. { 'hər·ə‚kān ‚wind }

Hurst formula [PETRO ENG] Relationship used in reservoir material-balance analysis; interrelates field pressure and production data at a number of different times. { 'hərst ‚fȯr·myə·lə }

Hurst method [PETRO ENG] A calculation method for the bottom-hole static pressure of a well; uses graphical extrapolation of pressure buildup over a short period of time. { 'hərst ‚meth·əd }

Hurter and Driffield characteristic curve [GRAPHICS] A graph showing the dependence of the optical density of a photographic emulsion on the logarithm of its exposure to light. Abbreviated H-D curve. { 'hərd·ər ən 'dri‚fēld ‚kar·ik·tə‚ris·tik 'kərv }

Hürthle cells [PATH] Enlarged epithelial cells of the thyroid follicles containing acidophilic cytoplasm, seen most frequently in adenomas. { 'hirt·lē ‚selz }

Hurtley's test [PATH] A test for the presence of acetoacetic acid in urine: take 10 cubic centimeters of a urine specimen and add 2.5 cm³ of concentrated hydrochloric acid and 1 cm³ of a freshly prepared 1% solution of sodium nitrite, then shake and let stand for 2 minutes; add 15 cm³ of 0.880 ammonia and 5 cm³ of 10% ferrous sulfate, then shake and pour into a large boiling tube, allowing it to stand; a violet or purple color confirms the presence of acetoacetic acid. { 'hərt·lēz ‚test }

Hurwitz polynomial [MATH] A polynomial whose zeros all have negative real parts. { 'hər·vits ‚päl·ə'nō·mē·əl }

Hurwitz's criterion [MATH] A criterion that determines whether a polynomial is a Hurwitz polynomial, based on the signs of a set of determinants formed from the polynomial's coefficients. { 'hər‚wit·səz krī‚tir·ē·ən }

husk [BOT] The outer coat of certain seeds, particularly if it is a dry, membranous structure. { həsk }

Hutchinson-Gilford syndrome See progeria. { 'həch·ən·sən 'gil·fərd 'sin‚drōm }

hutchinsonite [MINERAL] $(Pb,Tl)_2(Cu,Ag)As_5S_{10}$ Red mineral composed of sulfide of lead, copper, and arsenic, with varying amounts of thallium and silver, occurring in small orthorhombic crystals. { 'həch·ən·sə‚nīt }

Hutchinson's freckle See melanotic freckle. { 'həch·ən·sənz ‚frek·əl }

Hutchinson's teeth [MED] Deformity of permanent incisor teeth associated with congenital syphilis; the crown of the incisor is wider in the cervical portion than at the incisal edge, and the incisal edge has a characteristic crescent-shaped notch. { 'həch·ən·sənz ‚tēth }

hutch product [MIN ENG] In ore dressing, the fine, heavy materials that pass through the screen of a jig and collect in the bottom compartment called the hutch. { 'həch ‚präd·əkt }

Huttig equation [THERMO] An equation which states that the ratio of the volume of gas adsorbed on the surface of a nonporous solid at a given pressure and temperature to the volume of gas required to cover the surface completely with a unimolecular layer equals $(1 + r) c^r/(1 + c^r)$, where r is the ratio of the equilibrium gas pressure to the saturated vapor pressure of the adsorbate at the temperature of adsorption, and c is the product of a constant and the exponential of $(q - q_l)/RT$, where q is the heat of adsorption into a first layer molecule, q_l is the heat of liquefaction of the adsorbate, T is the temperature, and R is the gas constant. { 'həd·ik i‚kwā·zhən }

huttonite [MINERAL] $ThSiO_4$ A colorless to pale-green monoclinic mineral composed of silicate of thorium; it is dimorphous with thorite. { 'hət·ən‚īt }

Huwood loader [MIN ENG] A machine consisting of a number of horizontal rotating flight bars working near the floor of the seam which push prepared coal up a ramp onto a low, bottom-loaded coveyor belt. { 'hü‚wu̇d ‚lōd·ər }

Huxley's anastomosis [HISTOL] Polyhedral cells forming the middle layer of a hair root sheath. { 'həks·lēz ə‚nas·tə'mō·səs }

Huygens' approximation [MATH] The length of a small circular arc is approximately $\frac{1}{3}(8c' - c)$, where c is the chord of the arc and c' is the chord of half the arc. { 'hī·gənz ə‚prak·sə‚mā·shən }

Huygens eyepiece [OPTICS] An eyepiece in which there are two plano-convex lenses, and the plane sides of both lenses face the eye. { 'hī·gənz 'ī‚pēs }

Huygens-Fresnel principle [OPTICS] A modification of Huygens' principle according to which the amplitude of secondary waves falls off in proportion to the cosine of the angle between the normals to the original and secondary waves, and the secondary waves interfere with each other according to the principle of superposition. Also known as Fresnel-Huygens principle. { 'hī·gənz frā'nel ‚prin·sə·pəl }

Huygens' principle [OPTICS] The principle that each point on a light wavefront may be regarded as a source of secondary waves, the envelope of these secondary waves determining the position of the wavefront at a later time. { 'hī·gənz ‚prin·sə·pəl }

Huygens wavelet [OPTICS] A secondary wave as used in Huygens' principle. { 'hī·gənz ‚wāv·lət }

HVA See homovanillic acid.

HVAC [CIV ENG] The abbreviation for heating, ventilation, and air conditioning systems, used in building design and construction. { ‚āch‚vē‚ā'sē or 'āch‚vak }

HVAP See hypervelocity armor-piercing.

H variometer See horizontal intensity variometer. { 'āch ‚ver·ē'äm·əd·ər }

HVDC See high-voltage direct current.

H vector [ELECTROMAG] A vector that is the magnetic field. For a plane wave in free space, it is perpendicular to the E vector and to the direction of propagation. { 'āch ‚vek·tər }

H wave See transverse electric wave. { 'āch ‚wāv }

hyacinth See zircon. { 'hī·ə‚sinth }

Hyades [ASTRON] A V-shaped open star cluster about 150 light-years from the sun, which appears in the constellation Taurus near the star Aldebaran. { 'hī·ə,dēz }

Hyaenidae [VERT ZOO] A family of catlike carnivores in the superfamily Feloidea including the hyenas and aardwolf. { hī'e·nə,dē }

Hyaenodontidae [PALEON] A family of extinct carnivorous mammals in the order Deltatheridia. { hī,ē·nə'dänt·ə,dē }

Hyalellidae [INV ZOO] A family of amphipod crustaceans in the suborder Gammaridea. { ,hī·ə'lel·ə,dē }

hyaline [BIOCHEM] A clear, homogeneous, structureless material found in the matrix of cartilage, vitreous body, mucin, and glycogen. [GEOL] Transparent and resembling glass. { 'hī·ə·lən }

hyaline cartilage [HISTOL] A translucent connective tissue comprising about two-thirds clear, homogeneous matrix with few or no collagen fibrils. { 'hī·ə·lən 'kärt·lij }

hyaline cast [PATH] A clear, structureless mass of proteinaceous material found in the urine in association with certain kidney diseases. { 'hī·ə·lən ,kast }

hyaline degeneration [PATH] Degenerative change involving tissues and cells so that they become clear, structureless, and homogeneous. { 'hī·ə·lən dē,jen·ə'rā·shən }

hyaline membrane [HISTOL] **1.** A basement membrane. **2.** A membrane of a hair follicle between the inner fibrous layer and the outer root sheath. { 'hī·ə·lən 'mem,brān }

hyaline membrane disease [MED] A disease occurring during the first few days of neonatal life, characterized by respiratory distress due to formation of a hyaline-like membrane within the alveoli. { 'hī·ə·lən 'mem,brān di,zēz }

hyaline test [INV ZOO] A translucent wall or shell of certain foraminiferans composed of layers of calcite interspersed with separating membranes. { 'hī·ə·lən ,test }

hyalinization [PATH] Replacement of the normal components of cells or tissues by a hyaline material. { ,hī·ə·lə·nə'zā·shən }

hyalinocrystalline [PETR] Of porphyritic rock texture, having the phenocrysts lying in a glassy ground mass. { hī,al·ə·nō'krist·əl·ən }

hyalite [MINERAL] A colorless, clear or translucent variety of opal occurring as globular concretions or botryoidal crusts in cavities or cracks of rocks. Also known as Müller's glass; water opal. { 'hī·ə,līt }

hyalobasalt See tachylite. { hī·ə·lō·bə'sȯlt }

hyaloclastite [GEOL] A tufflike deposit formed by the flowing of basalt under water and ice and its consequent fragmentation. Also known as aquagene tuff. { hī·ə·lō'kla,stīt }

Hyalodictyae [MYCOL] A subdivision of the spore group Dictyosporae characterized by hyaline spores. { ,hi·ə·lō'dik·tē,ē }

Hyalodidymae [MYCOL] A subdivision of the spore group Didymosporae characterized by hyaline spores. { ,hi·ə·lō'did·ə,mē }

hyalography [GRAPHICS] The art of engraving on glass, with diamond or emery cutting tools or with hydrofluoric acid or other etching solutions. { ,hī·ə'läg·rə·fē }

Hyalohelicosporae [MYCOL] A subdivision of the spore group Helicosporae characterized by hyaline spores. { hī·ə·lō,hel·ə'käs·pə,rē }

hyaloid membrane [ANAT] The limiting membrane surrounding the vitreous body of the eyeball, and forming the suspensory ligament. { 'hī·ə,lȯid'mem,brān }

hyaloophitic [PETR] Of the texture of igneous rocks, being composed principally of a glassy ground mass with little interstitial texture. { hī·ə·lō,ō'fid·ik }

hyalophane [MINERAL] $BaAl_2Si_2O_8$ A colorless feldspar mineral crystallizing in the monoclinic system; isomorphous with adularia. Also known as baryta feldspar. { hī'al·ə,fān }

hyalophobia [PSYCH] An abnormal fear of glass. { ,hī·ə·lō'fō·bē·ə }

Hyalophragmiae [MYCOL] A subdivision of the spore group Phragmosporae characterized by hyaline spores. { ,hī·ə·lō'frag·mē,ē }

hyaloplasm [CYTOL] The optically clear, viscous to gelatinous ground substance of cytoplasm in which formed bodies are suspended. { hī'al·ə,plaz·əm }

hyalopsite See obsidian. { ,hī·ə'läp,sīt }

Hyaloscolecosporae [MYCOL] A subdivision of the spore

group Scalecosporae characterized by hyaline spores. { ,hī·ə·lō,skäl·ə'käs·pə,rē }

Hyalospongia [PALEON] A class of extinct glass sponges, equivalent to the living Hexactinellida, having siliceous spicules made of opaline silica. { ,hi·ə·lō'spən·jē·ə }

Hyalosporae [MYCOL] A subdivision of the spore group Amerosporae characterized by hyaline spores. { hī·ə'läs·pə·rē }

Hyalostaurosporae [MYCOL] A subdivision of the spore group Staurosporae characterized by hyaline spores. { 'hī·ə·lō·stȯ'räs·pə,rē }

hyalotekite [MINERAL] $(Pb,Ca,Ba)_4BSi_6O_{17}(OH,F)$ A white gray mineral composed of borosilicate and fluoride of lead, barium, and calcium, occurring in crystalline masses. { hī·ə·lō'tek,tīt }

hyaluronate [BIOCHEM] A salt or ester of hyaluronic acid. { ,hī·ə'lùr·ə,nāt }

hyaluronate lyase See hyaluronidase. { ,hī·ə'lùr·ə,nāt 'lī,ās }

hyaluronic acid [BIOCHEM] A polysaccharide found as an integral part of the gellike substance of animal connective tissue. { hī·ə·lù'rän·ik 'as·əd }

hyaluronidase [BIOCHEM] Any one of a family of enzymes which catalyze the breakdown of hyaluronic acid. Also known as hyaluronate lyase; spreading factor. { ,hī·ə·lù'rän·ə,dās }

Hybinette process [MET] A process used for refining of crude nickel anodes; anodes are placed in asphalt-lined, reinforced concrete tanks and dissolved electrochemically so that impurities such as copper and iron pass into solution while pure nickel electrolyte is continuously added. { 'hī·bə,net ,prä·səs }

Hybodontoidea [PALEON] An ancient suborder of extinct fossil sharks in the order Selachii. { ,hī·bə,dän'tȯid·ē·ə }

hybrid [GEN] The offspring of parents of different species or varieties. [PETR] Pertaining to a rock formed by the assimilation of two magmas. [SCI TECH] Having two or more different characteristics or types of structure. { 'hī·brəd }

hybrid algebraic manipulation language [COMPUT SCI] The most ambitious type of algebraic manipulation language, which accepts the broadest spectrum of mathematical expressions but possesses, in addition, special representations and special algorithms for particular special classes of expressions. { 'hī·brəd ,al·jə'brā·ik mə,nip·yə'lā·shən ,laŋ·gwij }

hybrid-arrested translation [MOL BIO] A method for identifying the proteins coded for by a cloned deoxyribonucleic acid (DNA) sequence by depending on the ability of the cloned DNA to form a base pair with its messenger ribonucleic acid and thereby inhibit its translation. { 'hī·brəd ə,res·təd tranz'lā·shən }

hybrid balance [ELEC] Loss between two conjugate sides of a hybrid set less the same loss when one of the other sides is open or shorted. { 'hī·brəd 'bal·əns }

hybrid beam [ENG] A metal beam with flanges fabricated from a material that differs from that of the web plate and has a different minimum yield strength. { 'hī·brəd 'bēm }

hybrid circuit [ELEC] A circuit in which two or more basically different types of components, such as tubes and transistors, performing similar functions are used together. { 'hī·brəd 'sər·kət }

hybrid coil See hybrid transformer. { 'hī·brəd 'kȯil }

hybrid composite [MATER] A composite material in which two or more high-performance reinforcements are combined. { 'hī·brəd kəm'päz·ət }

hybrid computer [COMPUT SCI] A computer designed to handle both analog and digital data. Also known as analog-digital computer; hybrid system. { 'hī·brəd kəm'pyüd·ər }

hybrid distributed processing system [COMPUT SCI] A distributed processing system that includes both horizontal and hierarchical distribution. { 'hī·brəd di,strib·yəd·əd 'prä,ses·iŋ ,sis·təm }

hybrid dysgenesis [GEN] A syndrome of abnormal traits that appears in the hybrids between certain strains of the fruit fly *Drosophila melanogaster*, and includes such traits as partial sterility and greatly elevated genetic mutations and chromosome rearrangements. Its cause is mobilization of a transposable element. { 'hī,brid dis'jen·ə·səs }

hybrid electromagnetic wave [ELECTROMAG] A wave

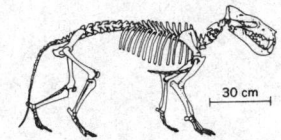

HYAENODONTIDAE

The deltatheridian *Hyaenodon*, original about 4 feet (1.2 meters) long. (*From A. S. Romer, Vertebrate Paleontology, 3d ed., University of Chicago Press, 1966*)

HYALURONIC ACID

Repeating unit in hyaluronic acid molecule.

which has both transverse and longitudinal components of displacement.   { 'hī·brəd iˌlek·trō·magˌned·ik 'wāv }

hybrid enzyme   [BIOCHEM]   A form of polymeric enzyme occurring in heterozygous individuals that shows a hybrid molecular form made up of subunits differing in one or more amino acids.   { 'hī·brəd 'en,zīm }

hybrid gene   [CELL MOL]   Any gene constructed by recombinant deoxyribonucleic acid technology that contains segments derived from different parents.   { 'hī·brəd jēn }

hybrid hardware control   [COMPUT SCI]   The control of and communication between the various parts of a hybrid computer.   { 'hī·brəd 'härd,wer kən,trōl }

hybrid inflation   [ASTRON]   A version of the inflationary universe cosmology that relies on the evolution of several interacting fields.   { 'hī·brid in,flā·shən }

hybrid inlet noise reduction   [ENG ACOUS]   A method of reducing the noise from the inlet of a jet engine, which involves the use of both high-Mach-number flows to retard or block the passage of sound waves and acoustic treatment of the walls of the inlet.   { 'hī·brəd ˌin·lət 'nóiz ri,dək·shən }

hybrid input/output   [COMPUT SCI]   The routines required to handle inputs and outputs from a computer system comprising digital and analog computers.   { 'hī·brəd ˌin,pút ˌaút,pút }

hybrid integrated circuit   [ELECTR]   A circuit in which one or more discrete components are used in combination with integrated-circuit construction.   { 'hī·brəd ˌint·ə,grād·əd 'sər·kət }

hybrid interface   [COMPUT SCI]   A device that joins a digital to an analog computer, converting digital signals transmitted serially by the digital computer into analog signals that are transmitted simultaneously to the various units of the analog computer, and vice versa.   { 'hī·brəd 'in·tər,fās }

hybridization   [CYTOL]   The production of viable hybrid somatic cells following experimentally induced cell fusion. [GEN]   **1.** Production of a hybrid by pairing complementary ribonucleic acid and deoxyribonucleic acid (DNA) strands.   **2.** Production of a hybrid by pairing complementary DNA single strands. [PHYS CHEM]   The mixing together on the same atom of two or more orbitals that have similar energies, forming a hybrid orbital.   { ˌhī·brəd·ə'zā·shən }

hybridization probe   [CELL MOL]   A small molecule of deoxyribonucleic acid or ribonucleic acid that is radioactively labeled and used to identify complementary nucleic acid sequences by hybridization.   { ˌhī·brəd·ə'zā·shən 'prōb }

hybridized orbital   [PHYS CHEM]   A molecular orbital which is a linear combination of two or more orbitals of comparable energy (such as $2s$ and $2p$ orbitals), is concentrated along a certain direction in space, and participates in formation of a directed valence bond.   { 'hī·brəd,īzd 'ór·bəd·əl }

hybrid junction   [ELECTR]   A transformer, resistor, or waveguide circuit or device that has four pairs of terminals so arranged that a signal entering at one terminal pair divides and emerges from the two adjacent terminal pairs, but is unable to reach the opposite terminal pair.   Also known as bridge hybrid.   { 'hī·brəd 'jəŋk·shən }

hybrid magnet   [CRYO]   A type of superconducting magnet consisting of a large-bore NbTi (niobium-titanium) external coil, which provides an external field of about 5 teslas, and an inner Nb_3Sn (niobium-tin) coil which provides additional field strength. [ELECTROMAG]   An air-cooled magnet consisting of a large-volume superconducting magnet surrounding a water-cooled normal-conductor magnet that operates at the highest field.   { 'hī·brəd 'mag·nət }

hybrid merogony   [EMBRYO]   The fertilization of cytoplasmic fragments of the egg of one species by the sperm of a related species.   { 'hī·brəd mə'räj·ə·nē }

hybrid microcircuit   [ELECTR]   Microcircuit in which thin-film, thick-film, or diffusion techniques are combined with separately attached semiconductor chips to form the circuit.   { 'hī·brəd 'mī·krō,sər·kət }

hybrid molecule   [BIOCHEM]   A single molecule, usually protein, peculiar to heterozygotes and containing two structurally different polypeptide chains determined by two different alleles.   { 'hī·brəd 'mäl·ə,kyül }

hybrid network   [COMMUN]   Nonhomogeneous communications network required to operate with signals of dissimilar characteristics (such as analog and digital modes).   { 'hī·brəd 'net,wərk }

hybridoma   [IMMUNOL]   A hybrid myeloma formed by fusing myeloma cells with lymphocytes that produce a specific antibody; the individual cells can be cloned, and each clone produces large amounts of identical (monoclonal) antibody.   { ˌhī·brə'dō·mə }

hybrid orbital   [PHYS CHEM]   An orbital formed by the combination of two or more atomic orbitals on a single atom.   { ˌhī·brəd 'ór·bəd·əl }

hybrid parameter   See h parameter.   { 'hī·brəd pə'ram·əd·ər }

hybrid plasmid   [MOL BIO]   In recombinant deoxyribonucleic acid (DNA) technology, any plasmid chimera containing inserted DNA sequences.   { 'plaz·mid }

hybrid problem analysis   [COMPUT SCI]   The determination of the parts of a problem best suited for the digital computer.   { 'hī·brəd ˌpräb·ləm ə'nal·ə·səs }

hybrid programming   [COMPUT SCI]   Hybrid system routines that handle timing, function generation, and simulation.   { 'hī·brəd 'prō,gram·iŋ }

hybrid propellant   [MATER]   A propellant using a combination of liquid and solid materials to provide propulsion energy and working fluid.   { 'hī·brəd prə'pel·ənt }

hybrid propulsion   [AERO ENG]   Propulsion utilizing energy released by a liquid propellant with a solid propellant in the same rocket engine.   { 'hī·brəd prə'pəl·shən }

hybrid redundancy   [COMPUT SCI]   A synthesis of triple modular redundancy and standby replacement redundancy, consisting of a triple modular redundancy system (or, in general, an N-modular redundancy system) with a bank of spares so that when one of the units in the triple modular redundancy system fails it is replaced by a spare unit.   { 'hī·brəd ri'dən·dən·sē }

hybrid relay   [ELEC]   A relay in which solid-state elements are combined with moving contacts.   { 'hī·brəd 'rē,lā }

hybrid repeater   See hybrid transformer.   { 'hī·brəd ri'pēd·ər }

hybrid rocket   [AERO ENG]   A rocket with an engine utilizing a liquid propellant with a solid propellant in the same rocket engine.   { 'hī·brəd 'räk·ət }

hybrid set   [ELEC]   Two or more transformers interconnected to form a hybrid junction.   Also known as transformer hybrid.   { 'hī·brəd 'set }

hybrid simulation   [COMPUT SCI]   The use of a hybrid computer for purposes of simulation.   { 'hī·brəd ˌsim·yə'lā·shən }

hybrid sterility   [GEN]   Inability to form functional gametes in a hybrid due to disturbances in sex-cell development or in meiosis, caused by incompatible genetic constitution.   { 'hī·brəd stə'ril·əd·ē }

hybrid swarm   [GEN]   A collection of hybrids produced when there is a breakdown of isolating barriers between two species whose areas of distribution overlap.   { 'hī·brəd 'swórm }

hybrid system   [COMPUT SCI]   **1.** A computer system that performs two or more functions, such as data processing and word processing.   **2.** See hybrid computer.   { 'hī,brid ˌsis·təm }

hybrid system checkout   [COMPUT SCI]   The static check of a hybrid system and of the digital program and analog wiring required to solve a problem.   { 'hī·brəd ˌsis·təm 'chek,aút }

hybrid tee   [ELECTROMAG]   A microwave hybrid junction composed of an E-H tee with internal matching elements; it is reflectionless for a wave propagating into the junction from any arm when the other three arms are match-terminated.   Also known as magic tee.   { 'hī·brəd 'tē }

hybrid thin-film circuit   [ELECTR]   Microcircuit formed by attaching discrete components and semiconductor devices to networks of passive components and conductors that have been vacuum-deposited on glazed ceramic, sapphire, or glass substrates.   { 'hī·brəd 'thin ˌfilm 'sər·kət }

hybrid transformer   [ELEC]   A single transformer that performs the essential functions of a hybrid set.   Also known as bridge transformer; hybrid coil; hybrid repeater.   { 'hī·brəd tranz'fór·mər }

hybrid vigor   See heterosis.   { 'hī·brəd 'vig·ər }

hybrid wave function   [QUANT MECH]   A linear combination of wave functions of one problem used as an approximation to the wave function in another problem; for example, a linear combination of atomic orbitals used to represent a molecular bond.   { 'hī·brəd 'wāv ,fəŋk·shən }

hybrid zone   [ECOL]   A geographic zone in which two populations hybridize after the breakdown of the geographic barrier that separated them.   { 'hī·brəd ˌzōn }

HYCATS *See* Hydrofoil Collision Avoidance and Tracking System. { 'hī₁kats }

hydantoin [ORG CHEM] $C_3N_2O_2H$ A white, crystalline compound, melting point 220°C; used as an intermediate in certain pharmaceutical manufacturing and as a textile softener and lubricant. Also known as glycolyurea. { hī'dant·ə·wən }

Hydatellales [BOT] An order of monocotyledonous flowering plants, division Magnoliophyta, of the class Liliopsida, characterized by small, submersed or partly submersed aquatic annuals with greatly simplified internal anatomy; consists of a single family with five species native to Australia, New Zealand, and Tasmania. { ₁hī·dad·ə'lā·lēz }

hydathode [BOT] An opening of the epidermis of higher plants specialized for exudation of water. { 'hīd·ə₁thōd }

hydatid [MED] **1.** A cyst formed in tissues due to growth of the larval stage of *Echinococcus granulosus.* **2.** A cystic remnant of an embryonal structure. { 'hīd·ə·dəd }

hydatid disease *See* echinococcosis. { 'hīd·ə·dəd di₁zēz }

hydatid of Morgagni *See* appendix testis. { 'hīd·ə·dəd əv mòr'gan·yē }

hydatidosis *See* echinococcosis. { ₁hīd·ə₁ti'dō·səs }

hydatiform mole [MED] A benign placental tumor formed as a cystic growth of the chorionic villi. Also known as hydatiform tumor. { hī'dad·ə₁fòrm 'mōl }

hydatiform tumor *See* hydatiform mole. { hī'dad·ə₁fòrm 'tü·mər }

hydatogenesis [GEOL] Crystallization and deposition of minerals from aqueous solutions. { ₁hīd·ə·tō'jen·ə·səs }

hydatosis [MED] Multiple hydatid cysts. { ₁hīd·ə'tō·səs }

hydnocarpic acid [ORG CHEM] $C_{16}H_{28}O_2$ A nonedible fat and oil isolated from chaulmoogra oil, forming white crystals that melt at 60°C; used to treat Hansen's disease. { ₁hīd·nə₁kär·pik 'as·əd }

hydra [INV ZOO] Any species of *Hydra* or related genera, consisting of a simple, tubular body with a mouth at one end surrounded by tentacles, and a foot at the other end for attachment. { 'lī·drə }

Hydra [ASTRON] A large constellation of the Southern Hemisphere, right ascension 10 hours, declination 20° south. Also known as Water Monster. [INV ZOO] A common genus of cnidarians in the suborder Anthomedusae. { 'hī·drə }

Hydra-Centaurus-Pavo supercluster [ASTRON] The nearest supercluster outside the local supercluster; includes the Centaurus cluster, the Hydra I cluster, and a number of smaller clusters in the constellation Pavo. { ₁hī·drə sen₁tòr·əs ₁pä·vō 'sü·pər₁kləs·tər }

Hydrachnellae [INV ZOO] A family of generally fresh-water predacious mites in the suborder Trombidiformes, including some parasitic forms. { ₁hī₁drak'ne·lē }

Hydra I cluster [ASTRON] A large cluster of galaxies with recession velocities around 3500 kilometers (2200 miles) per second, part of the Hydra-Centaurus-Pavo supercluster. { ₁hī·drə ₁wən 'kləs·tər }

hydracrylic acid [ORG CHEM] $CH_2OH·CH_2COOH$ An oily liquid that is an isomer of lactic acid and that breaks down on heating to acrylic acid. { ₁hī·drə₁kril·ik 'as·əd }

Hydraenidae [INV ZOO] The equivalent name for Limnebiidae. { hī'drē·nə₁dē }

hydragogue [MED] Causing the discharge of watery fluid, especially from the bowel. { 'hī·drə₁gäg }

hydralazine [PHARM] $C_8H_8N_4$ An antihypertensive drug; used as the hydrochloride salt. { hī'dral·ə₁zēn }

hydramnios *See* polyhydramnios. { hī'dram·nē₁äs }

hydranencephaly [MED] A congenital anomaly in which there are vestiges of a cerebellum, occipital lobes, and basal nuclei; frontal and parietal lobes are replaced by a cyst; and the neurocranium is undeveloped. { ₁hī·drə₁en'sef·ə·lē }

hydrant *See* fire hydrant. { 'hī·drənt }

hydranth [INV ZOO] Nutritive individual in a hydroid colony. { 'hī₁dranth }

hydrargillite *See* gibbsite. { hī'drär·jə₁līt }

hydrargyrism *See* mercurialism. { hī'drär·jə₁riz·əm }

hydrargyrum *See* mercury. { hī'drär·jə·rəm }

hydrarthrosis [MED] An accumulation of fluid in a joint. { ₁hī'drär'thrō·səs }

hydrase [BIOCHEM] An enzyme that catalyzes removal or addition of water to a substrate without hydrolyzing it. { 'hī₁drās }

hydrastine [ORG CHEM] $C_{21}H_{21}NO_6$ An alkaloid isolated from species of the family Ranunculaceae and from *Hydrastis canadensis;* orthorhombic prisms crystallize from alcohol solution, melting point 132°C; highly soluble in acetone and benzene, soluble in chloroform, less soluble in ether and alcohol. { 'hī₁dra₁stēn }

hydrastinine [ORG CHEM] $C_{11}H_{13}O_3N$ A compound formed by the decomposition of hydrastine; crystallizes as needles from petroleum-ether solution, soluble in organic solvents such as alcohol, chloroform, and ether; used in medicine as a stimulant in coronary disease and as a hemostatic in uterine hemorrhage. { hī'dras·tə₁nēn }

Hydrastis canadensis [BOT] A perennial medicinal herb in the family Ranunculaceae from which the alkaloid hydrastine is derived. Also known as goldenseal. { hī₁dras·tis ₁kan·ə'den·sis }

hydrate [CHEM] **1.** A form of a solid compound which has water in the form of H_2O molecules associated with it; for example, anhydrous copper sulfate is a white solid with the formula $CuSO_4$, but when crystallized from water a blue crystalline solid with formula $CuSO_4·5H_2O$ results, and the water molecules are an integral part of the crystal. **2.** A crystalline compound resulting from the combination of water and a gas; frequently a constituent of natural gas that is under pressure. { 'hī₁drāt }

hydrate aluminum oxide *See* alumina trihydrate. { 'hī₁drāt ə'lü·mə·nəm 'äk₁sīd }

hydrated alumina *See* alumina trihydrate. { 'hī₁drād·əd ə'lü·mə·nə }

hydrated cellulose *See* hydrocellulose. { 'hī₁drād·əd 'sel·yə₁lōs }

hydrated chloral *See* chloral hydrate. { 'hī₁drād·əd 'klòr·əl }

hydrated electron [PHYS CHEM] An electron released during ionization of a water molecule by water and surrounded by water molecules oriented so that the electron cannot escape. Also known as aqueous electron. { 'hī₁drād·əd i'lek₁trän }

hydrated grease [MATER] Grease made with a soap containing a hydrated alkali. { 'hī₁drād·əd 'grēs }

hydrated halloysite *See* endellite. { 'hī₁drād·əd hə'lòi₁sīt }

hydrated lime *See* calcium hydroxide. { 'hī₁drād·əd 'līm }

hydrated manganic hydroxide *See* manganic hydroxide. { 'hī₁drād·əd maŋ'gan·ik hī'drāk₁sīd }

hydrated mercurous nitrate [INORG CHEM] $Hg_2(NO_3)_2·2H_2O$ Poisonous, light-sensitive crystals, soluble in warm water, decomposes at 70°C; used as an analytical reagent and in cosmetics and medicine. { 'hī₁drād·əd mər'kyùr·əs 'nī₁trāt }

hydrated silica *See* silicic acid. { 'hī₁drād·əd 'sil·ə·kə }

hydrate inhibitor [CHEM] A material (such as alcohol or glycol) added to a gas stream to prevent the formation and freezing of gas hydrates in low-temperature systems. { 'hī₁drāt in₁hib·əd·ər }

hydration [CHEM] The incorporation of molecular water into a complex molecule with the molecules or units of another species; the complex may be held together by relatively weak forces or may exist as a definite compound. { hī'drā·shən }

hydraucone [DES ENG] A conical, spreading type of draft tube used on hydraulic turbine installations. { 'hī₁drò₁kōn }

hydraulic [ENG] Operated or effected by the action of water or other fluid of low viscosity. { hī'drò·lik }

hydraulic accumulator [MECH ENG] A hydraulic flywheel that stores potential energy by accumulating a quantity of pressurized hydraulic fluid in a suitable enclosed vessel. { hī'drò·lik ə'kyü·myə₁lād·ər }

hydraulic actuator [MECH ENG] A cylinder or fluid motor that converts hydraulic power into useful mechanical work; mechanical motion produced may be linear, rotary, or oscillatory. { hī'drò·lik 'ak·chə₁wād·ər }

hydraulic air compressor [MECH ENG] A device in which water falling down a pipe entrains air which is released at the bottom under compression to do useful work. { hī'drò·lik 'er kəm₁pres·ər }

hydraulic amplifier [CONT SYS] A device which increases the power of a signal in a hydraulic servomechanism or other system through the use of fixed and variable orifices. Also known as hydraulic intensifier. { hī'drò·lik 'am·plə₁fī·ər }

hydraulic analog table [FL MECH] An experimental facility based on the hydraulic analogy; the water flows over a smooth

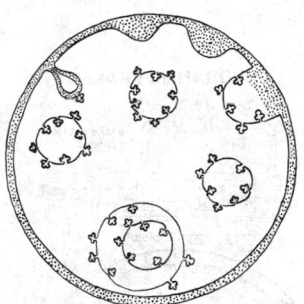

HYDATID

A drawing of the hydatid cyst showing large numbers of attached and free-floating brood capsules, each with one or more scoleces.

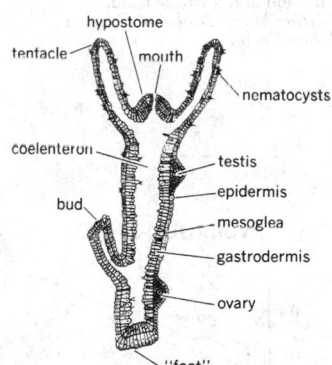

HYDRA

Longitudinal section of *Hydra.* (From T. I. Storer and R. L. Usinger, General Zoology, 4th ed., McGraw-Hill, 1965)

horizontal surface and is bounded by vertical walls geometrically similar to the boundaries of the corresponding compressible gas flow; flow patterns are easily observed, and boundary changes may be made rapidly and inexpensively during exploratory studies. { hī'drȯ·lik 'an·ə,läg ,tä·bəl }

hydraulic analogy [FL MECH] The analogy between the flow of a shallow liquid and the flow of a compressible gas; various phenomena such as shock waves occur in both systems; the analogy requires neglect of vertical accelerations in the liquid, and restrictions on the ratio of specific heats for the gas. { hī'drȯ·lik ə'nal·ə·jē }

hydraulic backhoe [MECH ENG] A backhoe operated by a hydraulic mechanism. { hī'drȯ·lik 'bak,hō }

hydraulic blasting [MIN ENG] Fracturing coal by means of a hydraulic cartridge. { hī'drȯ·lik 'blast·iŋ }

hydraulic bottom-hole pump [PETRO ENG] Liquid power-operated oil production pump; the liquid is oil, piped under pressure to the bottom of the well to operate the engine that drives the pump. { hī'drȯ·lik 'bäd·əm ,hōl ,pəmp }

hydraulic brake [MECH ENG] A brake in which the retarding force is applied through the action of a hydraulic press. { hī'drȯ·lik 'brāk }

hydraulic cartridge [MIN ENG] A device used in mining to split coal or rock and having 8–12 small hydraulic rams in the sides of a steel cylinder. { hī'drȯ·lik 'kär·trij }

hydraulic cement [MATER] Cement that hardens in the presence of water. { hī'drȯ·lik si'ment }

hydraulic chock [MIN ENG] A steel face-support structure consisting of one to four hydraulic legs mounted in a steel frame with a large head and base plate. { hī'drȯ·lik 'chäk }

hydraulic circuit [MECH ENG] A circuit whose operation is analogous to that of an electric circuit except that electric currents are replaced by currents of water or other fluids, as in a hydraulic control. { hī'drȯ·lik 'sər·kət }

hydraulic circulating system [MIN ENG] A method used to drill a borehole wherein water or a mud-laden liquid is circulated through the drill string. { hī'drȯ·lik 'sər·kyə,lād·iŋ ,sis·təm }

hydraulic classification [ENG] Classification of particles in a tank by specific gravity, utilizing the action of rising water currents. { hī'drȯ·lik ,klas·ə·fə'kā·shən }

hydraulic classifier [MECH ENG] A classifier in which particles are sorted by specific gravity in a stream of hydraulic water that rises at a controlled rate; heavier particles gravitate down and are discharged at the bottom, while lighter ones are carried up and out. Also known as hydrosizer. { hī'drȯ·lik 'klas·ə,fī·ər }

hydraulic clutch See fluid drive. { hī'drȯ·lik 'kləch }

hydraulic computer [COMPUT SCI] A computer in which electric current and gates are replaced by fluid and valves. { hī'drȯ·lik kəm'pyüd·ər }

hydraulic conductivity See permeability coefficient. { hī'drȯ·lik ,kän,dək'tiv·əd·ē }

hydraulic conveyor [MECH ENG] A system for handling material, such as ash from a coal-fired furnace; refuse is flushed from a hopper or slag tank to a grinder which discharges to a pump for conveying to a disposal area or a dewatering bin. { hī'drȯ·lik kən'vā·ər }

hydraulic coupling See fluid coupling. { hī'drȯ·lik 'kəp·liŋ }

hydraulic current [OCEANOGR] A current in a channel, due to a difference in the water level at the two ends. { hī'drȯ·lik 'kə·rənt }

hydraulic cylinder [MECH ENG] The cylindrical chamber of a positive displacement pump. { hī'drȯ·lik 'sil·ən·dər }

hydraulic discharge [HYD] The direct discharge of groundwater from the zone of saturation upon the land or into a body of surface water. { hī'drȯ·lik 'dis,chärj }

hydraulic dredge [MECH ENG] A dredge consisting of a large suction pipe which is mounted on a hull and supported and moved about by a boom, a mechanical agitator or cutter head which churns up earth in front of the pipe, and centrifugal pumps mounted on a dredge which suck up water and loose solids. { hī'drȯ·lik 'drej }

hydraulic drill [MECH ENG] A rotary drill powered by hydrodynamic means and used to make shot-firing holes in coal or rock, or to make a well hole. { hī'drȯ·lik 'dril }

hydraulic drive [MECH ENG] A mechanism transmitting motion from one shaft to another, the velocity ratio of the

shafts being controlled by hydrostatic or hydrodynamic means. { hī'drȯ·lik 'drīv }

hydraulic ejector [ENG] A pipe for removing excavated material from a pneumatic caisson. { hī'drȯ·lik i'jek·tər }

hydraulic elevator [MECH ENG] An elevator operated by water pressure. Also known as hydraulic lift. { hī'drȯ·lik 'el·ə,vād·ər }

hydraulic engineering [CIV ENG] A branch of civil engineering concerned with the design, erection, and construction of sewage disposal plants, waterworks, dams, water-operated power plants, and such. { hī'drȯ·lik ,en·jə'nir·iŋ }

hydraulic excavation See hydraulicking. { hī'drȯ·lik ,eks·kə'vā·shən }

hydraulic excavator digger [MECH ENG] An excavation machine which employs hydraulic pistons to actuate mechanical digging elements. { hī'drȯ·lik ,eks·kə'vād·ər 'dig·ər }

hydraulic extraction See hydraulicking. { hī'drȯ·lik ik'strak·shən }

hydraulic filling [MIN ENG] The use of water to wash waste material into stopes in order to prevent failure of rock walls and subsidence. { hī'drȯ·lik 'fil·iŋ }

hydraulic fluid [MATER] A low-viscosity fluid used in operating a hydraulic mechanism. { hī'drȯ·lik 'flü·əd }

hydraulic flume transport [MIN ENG] The transport of coal, pulp, or minerals in water flowing in semicircular or rectangular channels. { hī'drȯ·lik 'flüm ,tranz,pȯrt }

hydraulic fracturing [PETRO ENG] A method in which sand-water mixtures are forced into underground wells under pressure; the pressure splits the petroleum-bearing sandstone, thereby allowing the oil to move toward the wells more freely. { hī'drȯ·lik 'frak·chə·riŋ }

hydraulic friction [FL MECH] Resistance to flow which is exerted on the surface of contact between a stream and its conduit and which induces a loss of energy. { hī'drȯ·lik 'frik·shən }

hydraulic giant See hydraulic monitor. { hī'drȯ·lik 'jī·ənt }

hydraulic grade line [FL MECH] **1.** In a closed channel, a line joining the elevations that water would reach under atmospheric pressure. **2.** The free water surface in an open channel. { hī'drȯ·lik 'grād ,līn }

hydraulic gradient [FL MECH] With regard to an aquifer, the rate of change of pressure head per unit of distance of flow at a given point and in a given direction. [HYD] The slope of the hydraulic grade line of a stream. { hī'drȯ·lik 'grād·ē·ənt }

hydraulic intensifier See hydraulic amplifier. { hī'drȯ·lik in'ten·sə,fī·ər }

hydraulic jack [MECH ENG] A jack in which force is applied through the mechanism of a hydraulic press. { hī'drȯ·lik 'jak }

hydraulic jetting [ENG] Use of high-pressure water forced through nozzles to clean tube interiors and exteriors in heat exchangers and boilers. { hī'drȯ·lik 'jed·iŋ }

hydraulic jump [FL MECH] A steady-state, finite-amplitude disturbance in a channel, in which water passes turbulently from a region of (uniform) low depth and high velocity to a region of (uniform) high depth and low velocity; when applied to hydraulic jumps, the usual hydraulic formulas governing the relations of velocity and depth do not conserve energy. { hī'drȯ·lik 'jəmp }

hydraulicking [MIN ENG] Excavating alluvial or other mineral deposits by means of high-pressure water jets. Also known as hydraulic excavation; hydraulic extraction; hydroextraction. { hī'drȯ·lə·kiŋ }

hydraulic lift See hydraulic elevator. { hī'drȯ·lik 'lift }

hydraulic lime [MATER] A type of limestone which has been heated and pulverized, and absorbs water without swelling or heating, yielding a cement that hardens under water. { hī'drȯ·lik 'līm }

hydraulic limestone [MATER] Limestone, containing silica and alumina, which produces lime that hardens in water. { hī'drȯ·lik 'līm,stōn }

hydraulic loading [MIN ENG] The flushing of coal or other material broken down by jets of water along the mine floor and into flumes. { hī'drȯ·lik 'lōd·iŋ }

hydraulic locomotive [MIN ENG] A diesel locomotive in which traction wheels are driven by hydraulic motors powered by a hydraulic system on the unit; used in mine haulage. { hī'drȯ·lik ,lok·ə'mōd·iv }

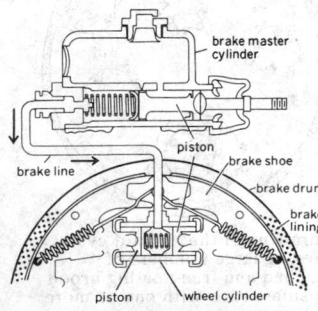

HYDRAULIC BRAKE

Schematic view of hydraulic braking system for one wheel on an automobile. Brakes are shown applied. Arrows indicate direction of motion of hydraulic fluid. (*Pontiac Motor Division, General Motors Corp.*)

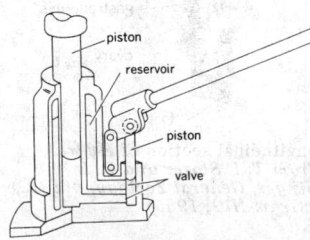

HYDRAULIC JACK

Hydraulic jack showing working parts.

hydraulic loss [FL MECH] The loss in fluid power due to flow friction within the system. { hī'drȯ·lik 'lȯs }

hydraulic machine [MECH ENG] A machine powered by a motor activated by the confined flow of a stream of liquid, such as oil or water under pressure. { hī'drȯ·lik mə'shēn }

hydraulic mine [MIN ENG] A placer mine worked by means of a water stream directed against a bank. { hī'drȯ·lik 'mīn }

hydraulic monitor [MIN ENG] A device for directing a high-pressure jet of water in hydraulicking; essentially, a swivel-mounted, counterweighted nozzle attached to a tripod or other type of stand and so designed that one worker can easily control and direct the vertical and lateral movements of the nozzle. Also known as giant; hydraulic giant; monitor. { hī'drȯ·lik 'män·əd·ər }

hydraulic motor [MECH ENG] A motor activated by water or other liquid under pressure. { hī'drȯ·lik 'mōd·ər }

hydraulic nozzle [MECH ENG] An atomizing device in which fluid pressure is converted into fluid velocity. { hī'drȯ·lik 'näz·əl }

hydraulic packer holddown [PETRO ENG] A pressure-actuated anchor located below a production packer (the seal between tubing and casing) to prevent well pressure from forcing the packer upward in the casing. { hī'drȯ·lik 'pak·ər 'hōl,daủn }

hydraulic packing [ENG] Packing material that resists the effects of water even under high pressure. { hī'drȯ·lik 'pak·iŋ }

hydraulic power oil [MATER] Well production oil from which corrosive and abrasive impurities have been removed; used to power downhole hydraulic pumps. { hī'drȯ·lik 'paủ·ər 'ȯil }

hydraulic power system [MECH ENG] A power transmission system comprising machinery and auxiliary components which function to generate, transmit, control, and utilize hydraulic energy. { hī'drȯ·lik 'paủ·ər ,sis·təm }

hydraulic press [MECH ENG] A combination of a large and a small cylinder connected by a pipe and filled with a fluid so that the fluid pressure created by a small force acting on the small-cylinder piston will result in a large force on the large piston. Also known as hydrostatic press. { hi'drȯ·lik 'pres }

hydraulic profile [HYD] A vertical section of an aquifer's potentiometric surface. { hi'drȯl·lik 'prō,fīl }

hydraulic prop [MIN ENG] A supporting device consisting of two telescoping steel cylinders extended by hydraulic pressure provided by a built-in hand pump. { hi'drȯ·lik 'präp }

hydraulic pump See hydraulic ram. { hi'drȯ·lik 'pəmp }

hydraulic radius [FL MECH] The ratio of the cross-sectional area of a conduit in which a fluid is flowing to the inner perimeter of the conduit. { hi'drȯ·lik 'rād·ē·əs }

hydraulic ram [MECH ENG] A device for forcing running water to a higher level by using the kinetic energy of flow; the flow of water in the supply pipeline is periodically stopped so that a small portion of water is lifted by the velocity head of a larger portion. Also known as hydraulic pump. { hi'drȯ·lik 'ram }

hydraulic ratio [GEOL] The weight of a heavy mineral multiplied by 100 and divided by the weight of a hydraulically equivalent light mineral. { hi'drȯ·lik 'rā·shō }

hydraulic robot [CONT SYS] A robot that is powered by hydraulic actuators, usually controlled by servovalves and analog resolvers. { hī'drȯl·ik 'rō,bät }

hydraulic scale [MECH ENG] An industrial scale in which the load applied to the load-cell piston is converted to hydraulic pressure. { hī'drȯ·lik 'skāl }

hydraulic rope-geared elevator [MECH ENG] An elevator hoisted by a system of ropes and sheaves attached to a piston in a hydraulic cylinder. { hi'drȯ·lik 'rōp ,gird 'el·ə,vād·ər }

hydraulics [FL MECH] The branch of engineering that focuses on the practical problems of collecting, storing, measuring, transporting, controlling, and using water and other liquids. { hī'drȯ·liks }

hydraulic separation [MECH ENG] Mechanical classification using a hydraulic classifier. { hī'drȯ·lik ,sep·ə'rā·shən }

hydraulic shovel [MECH ENG] A revolving shovel in which hydraulic rams or motors are substituted for drums and cables. { hī'drȯ·lik 'shəv·əl }

hydraulic sprayer [MECH ENG] A machine that sprays large quantities of insecticide or fungicide on crops. { hī'drȯ·lik 'sprā·ər }

hydraulic spraying See airless spraying. { hi'drȯ·lik 'sprā·iŋ }

hydraulic stacker [MECH ENG] A tiering machine whose carriage is raised or lowered by a hydraulic cylinder. { hi'drȯ·lik 'stak·ər }

hydraulic swivel head [MECH ENG] In a drill machine, a swivel head equipped with hydraulically actuated cylinders and pistons to exert pressure on and move the drill rod string longitudinally. { hi'drȯ·lik 'swiv·əl ,hed }

hydraulic telemetry [COMMUN] A form of mechanical telemetry in which signals are transmitted in the form of sound or other waves through water or some other liquid. { hi'drȯ·lik tə'lem·ə·trē }

hydraulic transport [ENG] Movement of material by water. { hī'drȯ·lik 'tranz,pȯrt }

hydraulic turbine [MECH ENG] A machine which converts the energy of an elevated water supply into mechanical energy of a rotating shaft. { hī'drȯ·lik 'tər·bən }

hydrazide [INORG CHEM] An acyl hydrazine; a compound of the formula

$$R-\overset{\overset{O}{\|}}{C}-NH=NH_2$$

where R may be an alkyl group. { 'hī·drə,zīd }

hydrazine [INORG CHEM] H_2NNH_2 A colorless, hygroscopic liquid, boiling point 114°C, with an ammonialike odor; it is reducing, decomposable, basic, and bifunctional; used as a rocket fuel, in corrosion inhibition in boilers, and in the synthesis of biologically active materials, explosives, antioxidants, and photographic chemicals. { 'hī·drə,zēn }

hydrazine hydrate [ORG CHEM] $H_2NNH_2OH_2O$ A colorless, fuming liquid that boils at 119.4°C; used as a component in jet fuels and as an intermediate in organic synthesis. { 'hī·drə,zēn 'hī,drāt }

hydrazinobenzene See phenylhydrazine. { hi'draz·ə·nō 'ben,zēn }

2-hydrazinoethanol See 2-hydroxyethylhydrazine. { 'tü hī'draz·ə·nō'e·thə,nȯl }

hydrazobenzene [ORG CHEM] $C_{12}H_{12}N_2$ A colorless, crystalline compound, melts at 132°C, slightly soluble in water, soluble in alcohol; used as an intermediate in the synthesis of benzidine. { 'hī·drə·zō'ben,zēn }

hydrazoic acid [INORG CHEM] NHN:N Explosive liquid, a strong protoplasmic poison boiling at 37°C. { 'hī·drə'zō·ik 'as·əd }

hydrazone [ORG CHEM] A compound containing the grouping −NH·N:C−, and obtained from a condensation reaction involving hydrazines with aldehydes or ketones; has been used as an exotic fuel. { 'hī·drə,zōn }

hydremia [MED] An excessive amount of water in the blood; disproportionate increase in plasma volume as compared with red blood cell volume. { hī'drē·mē·ə }

hydric [ECOL] Characterized by or thriving in abundance of moisture. { 'hī·drik }

hydride [INV ZOO] A compound containing hydrogen and another element; examples are H_2S, which is a hydride although it may be properly called hydrogen sulfide, and lithium hydride, LiH. { 'hī,drīd }

hydride descaling [MET] Removing surface deposits of oxides from a metal by immersion in molten alkali that contains hydrides. { 'hī,drīd dē'skāl·iŋ }

hydrindantin [ORG CHEM] $C_{18}H_{10}O_6$ A compound used as a reagent for the photometric determination of amino acids. { ,hī·drən'dant·ən }

hydriodic acid [INORG CHEM] A yellow liquid that is a water solution of the gas hydrogen iodide; a solution of 59% hydrogen iodide produces a liquid that is constant-boiling; it is a strong acid used in organic synthesis and as a reagent in analytical chemistry. { 'hī·drē,äd·ik 'as·əd }

hydriodic acid gas See hydrogen iodide. { 'hī·drē,äd·ik 'as·əd 'gas }

hydroacoustics See underwater acoustics. { ,hī·drō·ə'kü·stiks }

hydrobasaluminite [MINERAL] $Al_4(SO_4)(OH)_{10}\cdot36H_2O$ Mineral composed of a hydrous sulfate and hydroxide of aluminum. { 'hī·drō'bas·ə'lüm·ə,nīt }

Hydrobatidae [VERT ZOO] The storm petrels, a family of

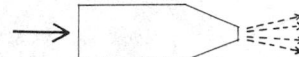

HYDRAULIC NOZZLE

A simple jet, a type of hydraulic nozzle.

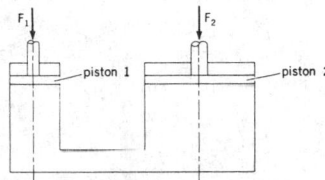

HYDRAULIC PRESS

Principle of hydraulic press. F_1 and F_2 are forces applied to pistons 1 and 2.

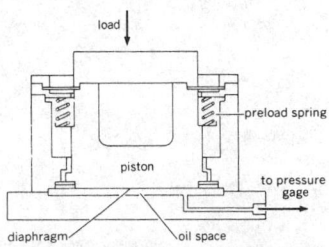

HYDRAULIC SCALE

Hydraulic scale, in which load cell is used. (From D. M. Considine, ed., Process Instruments and Controls Handbook, McGraw-Hill, 1957)

oceanic birds in the order Procellariiformes. { 'hī·drō'bad·ə,dē }

hydrobenzoin [ORG CHEM] $C_{14}H_{14}O_2$ A colorless, crystalline compound formed by action of sodium amalgam on benzaldehyde, melts at 136°C, and is slightly soluble in water. { ¦hī·drō'ben·zə·wən }

hydrobiotite [MINERAL] A light-green, trioctahedral clay mineral of mixed layers of biotite and vermiculite. { ¦hī·drō'bī·ə,tīt }

hydroboracite [MINERAL] $CaMgB_6O_{11}·6H_2O$ A white mineral composed of hydrous calcium magnesium borate, occurring in fibrous and foliated masses. { ¦hī·drō'bȯr·ə,sīt }

hydroboration [ORG CHEM] The process of producing organoboranes by the addition of a compound with a B-H bond to an unsaturated hydrocarbon; for example, the reaction of diborane ion with a carbonyl compound. Also known as boration. { ¦hī·drō·bə'rā·shən }

hydrobromic acid [INORG CHEM] HBr A solution of hydrogen bromide in water, usually 40%; a clear, colorless liquid; used in medicine, analytical chemistry, and synthesis of organic compounds. { ¦hī·drə'brō·mik 'as·əd }

hydrobromofluorocarbon [ORG CHEM] A compound consisting of hydrogen, bromine, fluorine, and carbon. Abbreviated HBFC. { ¦hī·drə,brō·mō,flùr·ō'kär·bən }

hydrocalumite [MINERAL] $Ca_2Al(OH)_7·3H_2O$ A colorless to light-green mineral composed of a hydrous hydroxide of calcium and aluminum. { ¦hī·drō'kal·yə,mīt }

hydrocarbon [ORG CHEM] One of a very large group of chemical compounds composed only of carbon and hydrogen; the largest source of hydrocarbons is from petroleum crude oil. { ¦hī·drə'kär·bən }

hydrocarbon blending value [ENG] Octane number rating for a 20% blend of a hydrocarbon with a 60:40 mixture of isooctane:*n*-heptane, which has been recalculated for a hypothetical 100% concentration of the tested hydrocarbon. { ¦hī·drə'kär·bən 'blend·iŋ ,val·yü }

hydrocarbon mud log [PETRO ENG] Record of oil, gas, or cuttings released into mud during rock drilling; used to detect the presence of hydrocarbon-bearing strata. { ¦hī·drə'kär·bən ¦məd ,läg }

hydrocarbon pore volume [PETRO ENG] The pore volume in a reservoir formation available to hydrocarbon intrusion. { ¦hī·drə'kär·bən ¦pȯr ,väl·yəm }

hydrocarbon resins [ORG CHEM] Brittle or gummy materials prepared by the polymerization of several unsaturated constituents of coal tar, rosin, or petroleum; they are inexpensive and find uses in rubber and asphalt formulations and in coating and caulking compositions. { ¦hī·drə'kär·bən 'rez·ənz }

hydrocarbon stabilization [PETRO ENG] The stepwise pressure reduction of a well stream to allow the release of dissolved gases until the liquid is stable at storage-tank conditions. { ¦hī·drə'kär·bən ,stā·bə·lə'zā·shən }

hydrocast [OCEANOGR] A series of water samplers on a single hydrographic wire which obtain samples simultaneously. { 'hī·drə,kast }

hydrocaulus [INV ZOO] Branched, upright stem of a hydroid colony. { 'hī·drō'kȯl·əs }

hydrocele [MED] Accumulation of fluid in the membranes surrounding the testis. { 'hī·drə,sēl }

hydrocellulose [MATER] A gelatinous mass formed from the reaction of cellulose with water either by grinding cellulose and mixing with water, or by using strong salt solutions, acids, or alkalies; used in the manufacture of artifical fibers such as rayon, mercerized cotton, paper, and vulcanized fiber. Also known as hydrated cellulose. { ¦hī·drō'sel·yə,lōs }

hydrocephaly [MED] Increased volume of cerebrospinal fluid in the skull. { ¦hī·drə'sef·ə·lē }

hydrocerussite [MINERAL] $Pb_3(OH)_2(CO_3)_2$ A colorless mineral composed of basic lead carbonate, occurring as crystals in thin hexagonal plates. { ¦hī·drō·sə'rə,sīt }

Hydrocharitaceae [BOT] The single family of the order Hydrocharitales, characterized by an inferior, compound ovary with laminar placentation. { ¦hī·drō,kar·ə'tās·ē,ē }

Hydrocharitales [BOT] A monofamilial order of aquatic monocotyledonous plants in the subclass Alismatidae. { ¦hī·drō,kar·ə'tā·lēz }

hydrochinone See hydroquinone. { ¦hī·drə·kə'nōn }

hydrochloric acid [INORG CHEM] HCl A solution of hydrogen chloride gas in water; a poisonous, pungent liquid forming a constant-boiling mixture at 20% concentration in water; widely used as a reagent, in organic synthesis, in acidizing oil wells, ore reduction, food processing, and metal cleaning and pickling. Also known as muriatic acid. { ¦hī·drə'klȯr·ik 'as·əd }

hydrochlorofluorocarbon [ORG CHEM] A compound composed of hydrogen, chlorine, fluorine, and carbon atoms. Also known as HCFC. { ¦hī·drə¦klȯr·ə'flùr·ə,kär·bən }

hydrochlorthiazide [PHARM] $C_7H_8ClN_3O_4S_2$ An orally effective diuretic and antihypertensive drug. { ¦hī·drō,klȯr'thī·ə,zīd }

hydrocholeresis [MED] Choleresis characterized by an increase of water output, or of a bile relatively low in specific gravity, viscosity, and content of total solids. { ¦hī·drō,kō·lə'rē·səs }

hydrochory [BIOL] Dispersal of disseminules by water. { 'hī·drə,kȯr·ē }

hydrocinnamic acid [ORG CHEM] $C_6H_5CH_2CH_2COOH$ A compound whose crystals have a floral odor (hyacinth-rose) and melt at 46°C; used in perfumes and flavoring. { ¦hī·drō·si'nam·ik 'as·əd }

hydrocinnamic alcohol See phenylpropyl alcohol. { ¦hī·drō·si'nam·ik 'al·kə,hȯl }

hydrocinnamic aldehyde See phenylpropyl aldehyde. { ¦hī·drō·si'nam·ik 'al·də,hīd }

hydrocladium [INV ZOO] Branchlet of a hydrocaulus. { ¦hī·drə'klād·ē·əm }

hydroclone [CHEM ENG] A device for separating a solid-liquid mixture during an industrial process by using a conical vortex and centrifugal force. { 'hī·drə,klōn }

hydrocoele [INV ZOO] 1. Water vascular system in Echinodermata. 2. Embryonic precursor of the system. { 'hī·drə,sel }

hydrocolloid See gum. { ,hī·drə'käl,ȯid }

hydrocooling [FOOD ENG] Removing heat from freshly harvested fruits and vegetables by immersion in ice water. { ¦hī·drō'kül·iŋ }

Hydrocorallina [INV ZOO] An order in some systems of classification set up to include the cnidarian groups Milleporina and Stylasterina. { ,hī·drō,kȯr·ə'lī·nə }

Hydrocorisae [INV ZOO] A subdivision of the Hemiptera containing water bugs with concealed antennae and without a bulbus ejaculatorius in the male. { ,hī·drō'kȯr·ə,sē }

hydrocortisone [BIOCHEM] $C_{21}H_{30}O_5$ The generic name for 17-hydroxycorticosterone; an adrenocortical steroid occurring naturally and prepared synthetically; its effects are similar to cortisone, but it is more active. Also known as cortisol. { ¦hī·drə'kȯrd·ə,zōn }

hydrocrackate [ORG CHEM] The product of a hydrocracker. { ¦hī·drō'kra,kāt }

hydrocracker [CHEM ENG] A high-pressure processing unit that cracks long hydrocarbon molecules under a high-hydrogen-content atmosphere. { 'hī·drō,krak·ər }

hydrocracking [CHEM ENG] A catalytic, high-pressure petroleum refinery process that is flexible enough to produce either high-octane gasoline or aviation jet fuel; the two main reactions are the adding of hydrogen to petroleum-derived molecules too massive and complex for gasoline and then the cracking of them to the required fuels; the catalyst is an acidic solid and a hydrogenating metal component. { 'hī·drō,krak·iŋ }

hydrocyanic acid [INV ZOO] HCN A highly toxic liquid that has the odor of bitter almonds and boils at 25.6°C; used to manufacture cyanide salts, acrylonitrile, and dyes, and as a fumigant in agriculture. Also known as formonitrile; hydrogen cyanide; prussic acid. { ¦hī·drō·sī'an·ik 'as·əd }

hydrocyanite See chalcocyanite. { ,hī·drə'sī·ə,nīt }

hydrocyclone [MECH ENG] A cyclone separator in which granular solids are removed from a stream of water and classified by centrifugal force. { 'hī·drō,sī,klōn }

hydrocystoma [MED] A group of clear vesicles, usually located around the eyes, composed of cystic sweat glands. { ,hī·drə·si'stō·mə }

Hydrodamalinae [VERT ZOO] A monogeneric subfamily of sirenian mammals in the family Dugongidae. { ,hī·drō·də'mal·ə,nē }

hydrodealkylation [CHEM ENG] A petroleum refining operation in which heat and pressure are used to remove methyl groups or larger alkyl groups from hydrocarbons, or to change positions of these groups on the molecule; used to upgrade low-value products. { ˌhī·drō·dē͟‚al·kə'lā·shən }

hydrodesulfurization [CHEM ENG] A catalytic process in which the petroleum feedstock is reacted with hydrogen to reduce the sulfur content in the oil. { ˌhī·drō·dē͟‚səl·fə·rə'zā·shən }

hydrodist [NAV] A navigation positioning system in which two separate tellurometers are mounted side by side on board a vessel and aimed at their respective remote units. { 'hī·drə‚dist }

hydrodynamic equations [FL MECH] Three equations which express the net acceleration of a unit water particle as the sum of the partial accelerations due to pressure gradient force, frictional force, earth's deflecting force, gravitational force, and other factors. { ˌhī·drō·dī'nam·ik i'kwā·zhənz }

hydrodynamic oscillator [ENG ACOUS] A transducer for generating sound waves in fluids, in which a continuous flow through an orifice is modulated by a reciprocating valve system controlled by acoustic feedback. { 'hī·drō·dī'nam·ik 'äs·ə‚lād·ər }

hydrodynamic pressure [FL MECH] The difference between the pressure of a fluid and the hydrostatic pressure; this concept is useful chiefly in problems of the steady flow of an incompressible fluid in which the hydrostatic pressure is constant for a given elevation (as when the fluid is bounded above by a rigid plate), so that the external force field (gravity) may be eliminated from the problem. { ˌhī·drō·dī'nam·ik 'presh·ər }

hydrodynamics [FL MECH] The study of the motion of a fluid and of the interactions of the fluid with its boundaries, especially in the incompressible inviscid case. { ˌhī·drō·dī'nam·iks }

hydroecium [INV ZOO] The closed, funnel-shaped tube at the upper end of cnidarians belonging to the Siphonophora. { hi'drē·shəm }

hydroelasticity [FL MECH] **1.** Theory of elasticity of a fluid. **2.** The interaction between the flow of water or other liquid and the elastic behavior of a body immersed in it. { ˌhī·drō‚ē·las'tis·əd·ē }

hydroelectric generator [MECH ENG] An electric rotating machine that transforms mechanical power from a hydraulic turbine or water wheel into electric power. { 'hī·drō·i'lek·trik 'jen·ə‚rād·ər }

hydroelectricity [ELEC] Electric power produced by hydroelectric generators. Also known as hydropower. { 'hī·drō‚i‚lek'tris·əd·ē }

hydroelectric plant [MECH ENG] A facility at which electric energy is produced by hydroelectric generators. Also known as hydroelectric power station. { 'hī·drō·i'lek·trik 'plant }

hydroelectric power station See hydroelectric plant. { 'hī·drō·i'lek·trik 'pau·ər ‚stā·shən }

hydroexcavation [MIN ENG] The use of high-pressure waterjets for cutting and removing material. { ˌhī·drō‚eks·kə'vā·shən }

hydroextraction See hydraulicking. { ˌhī·drō·ik'strak·shən }

hydrofining [CHEM ENG] A fixed-bed catalytic process to desulfurize and hydrogenate a wide range of charge stocks, from gases through waxes; the catalyst comprises cobalt oxide and molybdenum oxide on an extruded alumina support and may be regenerated in place by air and steam or flue gas. { 'hī·drə‚fīn·iŋ }

hydrofluoric acid [INORG CHEM] An aqueous solution of hydrogen fluoride, HF; colorless, fuming, poisonous liquid; extremely corrosive, it is a weak acid as compared to hydrochloric acid, but will attack glass and other silica materials; used to polish, frost, and etch glass, to pickle copper, brass, and alloy steels, to clean stone and brick, to acidize oil wells, and to dissolve ores. { 'hī·drə‚flùr·ik 'as·əd }

hydrofluorocarbon [ORG CHEM] A compound consisting of hydrogen, fluorine, and carbon. Abbreviated HFC. { 'hī·drə‚flùr·ə'kär·bən }

hydrofluorosilicic acid See fluosilicic acid. { 'hī·drō‚flùr·ō·sə'lis·ik 'as·əd }

hydrofoil [NAV ARCH] **1.** A flat or airfoil-shaped plate attached to a ship to stabilize it in roll. **2.** Foils attached by struts to the bottom of a hydrofoil boat to lift the hull out of the water as its speed is increased. { 'hī·drə‚fòil }

hydrofoil boat [NAV ARCH] A marine craft of generally conventional hull form which is supported above the water, when the craft is traveling at high speed, by dynamic lift produced by winglike surfaces (hydrofoils) below the hull. { 'hī·drə‚fòil ‚bōt }

Hydrofoil Collision Avoidance and Tracking System [NAV] A computer-based system designed to automate target tracking and navigation functions in order to increase the safety of high-speed ships; it makes use of the superposition of chart data over radar data, allowing precise positioning to be accomplished. Abbreviated HYCATS. { 'hī·drə‚fòil kə'lish·ən ə‚vòid·əns ən 'trak·iŋ ‚sis·təm }

hydroforming [CHEM ENG] A petroleum-refinery process in which naphthas are passed over a catalyst at elevated temperatures and moderate pressures in the presence of added hydrogen or hydrogen-containing gases, to form high-octane BTX aromatics for motor fuels or chemical manufacture. { 'hī·drə‚fòr·miŋ }

hydroformylation [CHEM ENG] The reaction of adding hydrogen and the —CHO group to the carbon atoms across a double bond to yield oxygenated derivatives; an example is in the oxo process where the term hydroformylation applies to those reactions brought about by treating olefins with a mixture of hydrogen and carbon monoxide in the presence of a cobalt catalyst. { 'hī·drə‚fòr·mə'lā·shən }

hydrofuge [ZOO] Of a structure, shedding water, as the hair on certain animals. { 'hī·drə‚fyüj }

hydrogarnet [MINERAL] One of a group of minerals having the general formula $A_3B_2(SiO_4)_{3-x}(OH)_{4x}$; isomorphous with certain garnets. { 'hī·drō'gär·nət }

hydrogasification [CHEM ENG] A technique to manufacture synthetic pipeline gas from coal; pulverized coal is reacted with hot, raw, hydrogen-rich gas containing a substantial amount of steam at 1000 pounds per square inch gage (6.9×10^6 pascals, gage) to form methane. { ˌhī·drə‚gas·ə·fə'kā·shən }

hydrogel [CHEM] The formation of a colloid in which the disperse phase (colloid) has combined with the continuous phase (water) to produce a viscous jellylike product; for example, coagulated silicic acid. { 'hī·drə‚jel }

hydrogen [CHEM] The first chemical element, symbol H, in the periodic table, atomic number 1, atomic weight 1.00797; under ordinary conditions it is a colorless, odorless, tasteless gas composed of diatomic molecules, H_2; used in manufacture of ammonia and methanol, for hydrofining, for desulfurization of petroleum products, and to reduce metallic oxide ores. { 'hī·drə·jən }

hydrogenase [BIOCHEM] Enzyme that catalyzes the oxidation of hydrogen. { hī'draj·ə‚nās }

hydrogenated oil [ORG CHEM] Unsaturated liquid vegetable oil that has had hydrogen catalytically added so as to convert the oil to a hydrogen-saturated solid. { 'hī·drə·jə‚nād·əd 'òil }

hydrogenation [CHEM ENG] Saturation of diolefin impurities in gasolines to form a stable product. [ORG CHEM] Catalytic reaction of hydrogen with other compounds, usually unsaturated; for example, unsaturated cottonseed oil is hydrogenated to form solid fats. { hī‚draj·ə'nā·shən }

hydrogen bacteria [MICROBIO] Bacteria capable of obtaining energy from the oxidation of molecular hydrogen. { 'hī·drə·jən bak'tir·ē·ə }

hydrogen blistering [MET] Cracks or blisters caused when atomic hydrogen penetrates steel via submicroscopic discontinuities or voids and becomes molecular hydrogen and develops internal pressures. { 'hī·drə·jən 'blis·tə·riŋ }

hydrogen bomb [ORD] A device in which heavy hydrogen nuclei, under intense heat and pressure, undergo an uncontrolled, self-sustaining fusion reaction to produce an explosion. Also known as H bomb. { 'hī·drə·jən 'bäm }

hydrogen bond [PHYS CHEM] A type of bond formed when a hydrogen atom bonded to atom A in one molecule makes an additional bond to atom B either in the same or another molecule; the strongest hydrogen bonds are formed when A and B are highly electronegative atoms, such as fluorine, oxygen, or nitrogen. { 'hī·drə·jən 'bänd }

hydrogen brazing [MET] Brazing in an atmosphere rich in hydrogen. { 'hi·drə·jən 'brāz·iŋ }

hydrogen bromide [INV ZOO] HBr A hazardous, toxic gas used as a chemical intermediate and as an alkylation catalyst; forms hydrobromic acid in aqueous solution. { 'hī·drə·jən 'brō‚mīd }

hydrogen-bubble method [FL MECH] A method of flow

HYDROFOIL BOAT

Jetfoil *Kamehameha*, a ferry with fully submerged foils. (*Boeing Airplane Co.*)

visualization in which hydrogen bubbles are generated by electrolysis of water using wire electrodes, and are made to form either time lines (by applying a pulsating voltage to a bare cathode) or streamlines (by applying a continuous voltage to a kink in an insulated cathode). { 'hī·drə·jən 'bəb·əl ,meth·əd }

hydrogen burning [ASTROPHYS] Thermonuclear reactions occurring in the cores of main-sequence stars, in which nuclei of hydrogen fuse to form helium nuclei. { 'hī·drə·jən ,bərn·iŋ }

hydrogen chloride [INV ZOO] HCl A fuming, highly toxic, colorless gas soluble in water, alcohol, and ether; used in the production of vinyl chloride and alkyl chloride, and in polymerization, isomerization, and other reactions. { 'hī·drə·jən 'klȯr,īd }

hydrogen cyanide See hydrocyanic acid. { 'hī·drə·jən 'sī·ə,nīd }

hydrogen cyanide laser [OPTICS] A gas laser using hydrogen cyanide, which emits infrared radiation at wavelengths of 311 and 337 micrometers. { 'hī·drə·jən ¦sī·ə,nīd 'lā·zər }

hydrogen cycle [CHEM] The complete process of a cation-exchange operation in which the adsorbent is used in the hydrogen or free acid form. { 'hī·drə·jən ,sī·kəl }

hydrogen damage [MET] Corrosion, common in boilers, caused by diffusion of hydrogen through steel reacting with carbon to form methane, which builds up local stresses at the interfaces between grains, forming voids that ultimately produce failure. { 'hī·drə·jən ,dam·ij }

hydrogen discharge lamp [ELECTR] A discharge lamp containing hydrogen and used as a source of ultraviolet radiation. { 'hī·drə·jən 'dis,chärj ,lamp }

hydrogen disulfide See hydrogen sulfide. { 'hī·drə·jən dī'səl,fīd }

hydrogen electrode [PHYS CHEM] A noble metal (such as platinum) of large surface area covered with hydrogen gas in a solution of hydrogen ion saturated with hydrogen gas; metal is used in a foil form and is welded to a wire sealed in the bottom of a hollow glass tube, which is partially filled with mercury; used as a standard electrode with a potential of zero to measure hydrogen ion activity. { 'hī·drə·jən i'lek,trōd }

hydrogen embrittlement See acid brittleness. { 'hī·drə·jən em'brid·əl·mənt }

hydrogen equivalent [CHEM] The number of replaceable hydrogen atoms or hydroxyl groups in a molecule of an acid or a base. { 'hī·drə·jən i'kwiv·ə·lənt }

hydrogen fluoride [INV ZOO] HF The hydride of fluoride; anhydrous HF is a mobile, colorless, liquid that fumes in air, melts at −83°C, boils at 19.8°C; used to make fluorine-containing refrigerants (such as Freon) and organic fluorocarbon compounds, as a catalyst in alkylate gasoline manufacture, as a fluorinating agent, and in preparation of hydrofluoric acid. { 'hī·drə·jən 'flȯr,īd }

hydrogenic ion [ATOM PHYS] An atom from which all but one of the electrons have been removed. { ¦hī·drə¦jen·ik 'ī,än }

hydrogenic rock See aqueous rock. { 'hī·drə,jen·ik 'räk }

hydrogen iodide [INV ZOO] HI A water-soluble, colorless gas that may be used in organic synthesis and as a reagent. Also known as hydriodic acid gas. { 'hī·drə·jən 'ī·ə,dīd }

hydrogen ion See hydronium ion. { 'hī·drə·jən 'ī,än }

hydrogen ion concentration [CHEM] The normality of a solution with respect to hydrogen ions, H⁺; it is related to acidity measurements in most cases by pH = log $\frac{1}{2}$ [1/(H⁺)], where (H⁺) is the hydrogen ion concentration in gram equivalents per liter of solution. { 'hī·drə·jən 'ī,än ,käns·ən,trā·shən }

hydrogen ion exponent [CHEM] A way of expressing pH; namely, pH = −log c_H, where c_H = hydrogen ion concentration. { 'hī·drə·jən 'ī,än ik'spō·nənt }

hydrogen laser [OPTICS] A molecular gas laser in which hydrogen is used to generate coherent wavelengths near 0.6 micrometer in the vacuum ultraviolet region. { 'hī·drə·jən 'lā·zər }

hydrogen-like ion [ATOM PHYS] An atom from which all the electrons except one have been removed. { 'hī·drə·jən ,līk ,ad·əm }

hydrogen line [SPECT] A spectral line emitted by neutral hydrogen having a frequency of 1420 megahertz and a wavelength of 21 centimeters; radiation from this line is used in radio astronomy to study the amount and velocity of hydrogen in the Galaxy. { 'hī·drə·jən ,līn }

hydrogen loss [MET] Loss of weight by a compact or a metal powder when heated in a hydrogen atmosphere; used as a measure of oxygen content of the sample. { 'hī·drə·jən ,lȯs }

hydrogen maser [PHYS] A maser in which hydrogen gas is the basis for providing an output signal with a high degree of stability and spectral purity. { 'hī·drə·jən 'mā·zər }

hydrogenolysis [CHEM] A reaction in which hydrogen gas causes a chemical change that is similar to the role of water in hydrolysis. { hī·drə·jə'näl·ə·səs }

hydrogenosome [MICROBIO] A membrane-bound organelle found in some anaerobic fungi and protozoa that is involved in the production of hydrogen and carbon dioxide. { ,hī·drə'jen·ə,sōm }

hydrogenous [CHEM] Of, pertaining to, or containing hydrogen. { hī'dräj·ə·nəs }

hydrogen overvoltage [MET] An overvoltage occurring at an electrode as a result of the liberation of hydrogen gas. { 'hī·drə·jən ¦ō·vər¦vōl·tij }

hydrogen oxide See water. { 'hī·drə·jən 'äk,sīd }

hydrogen peroxide [INORG CHEM] H_2O_2 Unstable, colorless, heavy liquid boiling at 158°C; soluble in water and alcohol; used as a bleach, chemical intermediate, rocket fuel, and antiseptic. Also known as peroxide. { 'hī·drə·jən pə'räk,sīd }

hydrogen phosphide See phosphine. { 'hī·drə·jən 'fäs,fīd }

hydrogen-reduced powder [MET] Metal powder produced by hydrogen-reduction of a metal, metallic compound, or surface-contaminated metal particles. { 'hī·drə·jən ri¦düst 'paúd·ər }

hydrogen selenide [INORG CHEM] H_2Se A toxic, colorless gas, soluble in water, carbon disulfide, and phosgene; used to make metallic selenides and organoselenium compounds and in the preparation of semiconductor materials. { 'hī·drə·jən 'sel·ə,nīd }

hydrogen star [ASTROPHYS] A star of spectral class A, a white star with a surface temperature of 8000 to 11,000 K. { 'hī·drə·jən ,stär }

hydrogen sulfide [INV ZOO] H_2S Flammable, toxic, colorless gas with offensive odor, boiling at −60°C; soluble in water and alcohol; used as an analytical reagent, as a sulfur source, and for purification of hydrochloric and sulfuric acids. Also known as hydrogen disulfide. { 'hī·drə·jən 'səl,fīd }

hydrogen tellurate See telluric acid. { 'hī·drə·jən 'tel·yə,rāt }

hydrogen thyratron [ELECTR] A thyratron containing hydrogen instead of mercury vapor to give freedom from effects of changes in ambient temperature; used in radar pulse circuits and stroboscopic photography. { 'hī·drə·jən 'thī·rə,trän }

hydrogeochemistry [GEOCHEM] The study of the chemical characteristics of ground and surface waters as related to areal and regional geology. { ¦hī·drō,jē·ō'kem·ə·strē }

hydrogeology [HYD] The science dealing with the occurrence of surface and ground water, its utilization, and its functions in modifying the earth, primarily by erosion and deposition. { ¦hī·drō·jē'äl·ə·jē }

hydrograph [HYD] A graphical representation of stage, flow, velocity, or other characteristics of water at a given point as a function of time. { 'hī·drə,graf }

hydrographic basin See drainage basin. { 'hī·drə'graf·ik 'bās·ən }

hydrographic chart [MAP] A map designed from data obtained by hydrographic surveys for purposes of navigation. { 'hī·drə'graf·ik 'chärt }

hydrographic cruise [OCEANOGR] Exploration of a body of water for hydrographic surveys. { 'hī·drə'graf·ik 'krüz }

hydrographic sextant [ENG] A surveying sextant similar to those used for celestial navigation but smaller and lighter, constructed so that the maximum angle that can be read is slightly greater than that on the navigating sextant; usually the angles can be read only to the nearest minute by means of a vernier; it is fitted with a telescope with a large object glass and field of view. Also known as sounding sextant; surveying sextant. { 'hī·drə'graf·ik 'seks·tənt }

hydrographic sonar [ENG] An echo sounder used in mapping ocean bottoms. { 'hī·drə'graf·ik 'sō,när }

hydrographic survey [OCEANOGR] Survey of a water area with particular reference to tidal currents, submarine relief, and any adjacent land. { 'hī·drə'graf·ik 'sər,vā }

hydrographic table [OCEANOGR] Tabular arrangement of

data relating sea-water density to salinity, temperature, and pressure. { 'hī·drə'graf·ik 'tā·bəl }

hydrography [GEOGR] Science which deals with the measurement and description of the physical features of the oceans, lakes, rivers, and their adjoining coastal areas, with particular reference to their control and utilization. [NAV] Measurement of the tides and currents as an aid to navigation. { hī'dräg·rə·fē }

hydrohalite [MINERAL] $Na_2Cl \cdot 2H_2O$ A mineral composed of hydrated sodium chloride, formed only from salty water cooled below 0°C. { ¦hī·drə¦ha,līt }

hydrohalloysite *See* endellite. { ¦hī·drō·hə'lòi,zīt }

hydrohetaerolite [MINERAL] $Zn_2Mn_4O_8 \cdot H_2O$ A dark brown to brownish-black mineral consisting of a hydrated oxide of zinc and manganese; occurs in massive form. { ¦hī·drō·hə'tir·ə,līt }

hydroid [INV ZOO] **1.** The polyp form of a hydrozoan cnidarian. Also known as hydroid polyp; hydropolyp. **2.** Any member of the Hydroida. { 'hī,dròid }

Hydroida [INV ZOO] An order of cnidarians in the class Hydrozoa including usually colonial forms with a well-developed polyp stage. { hī'dròi·də }

hydroid polyp *See* hydroid. { 'hī,dròid 'päl·əp }

hydroiodic ether *See* ethyl iodide. { ¦hī·drói¦äd·ik 'ē·thər }

hydrokaolin *See* endellite. { ¦hī·drə'kā·ə·lən }

hydrokinematics [FL MECH] The study of the motion of a liquid apart from the cause of motion. { 'hī·drə,kin·ə'mad·iks }

hydrokinetics [FL MECH] The study of the forces produced by a liquid as a consequence of its motion. { 'hī·drə·kə'ned·iks }

hydrolaccolith [GEOL] A frost mound, 0.3–20 feet (0.1–6 meters) in height, having a core of ice and resembling a laccolith in section. Also known as cryolaccolith. { ¦hī·drə'lak·ə,lith }

HYDROLANT [NAV] An urgent notice of dangers to navigation in the Atlantic Ocean, originated by the U.S. Naval Oceanographic Office and disseminated for the immediate safeguarding of shipping. { 'hī·drə,lant }

hydrolase [BIOCHEM] Any of a class of enzymes which catalyze the hydrolysis of proteins, nucleic acids, starch, fats, phosphate esters, and other macromolecular substances. { 'hī·drə,lās }

hydrolith [PETR] **1.** A chemically precipitated aqueous rock, such as rock salt. **2.** A rock that is free of organic material. { 'hī·drə,lith }

hydrologic accounting [HYD] A systematic summary of the terms (inflow, outflow, and storage) of the storage equation as applied to the computation of soil-moisture changes, groundwater changes, and so forth; an evaluation of the hydrologic balance of an area. Also known as basin accounting; water budget. { ¦hī·drə¦läj·ik ə'kaùnt·iŋ }

hydrological civilization [ANTHRO] A civilization that depends on sophisticated water management to sustain its agriculture. { ¦hī·drə¦läj·ə·kəl ,siv·ə·lə'zā·shən }

hydrologic cycle [HYD] The complete cycle through which water passes, from the oceans, through the atmosphere, to the land, and back to the ocean. Also known as water cycle. { ¦hī·drə¦läj·ik 'sī·kəl }

hydrologic sequence [GEOL] A series of soil sections from differentiated parent material that shows increasing lack of drainage downslope. { ¦hī·drə¦läj·ik 'sē·kwəns }

hydrologist [HYD] An individual who specializes in hydrology. { hī'dräl·ə·jəst }

hydrology [GEOPHYS] The science that treats the occurrence, circulation, distribution, and properties of the waters of the earth, and their reaction with the environment. { hī'dräl·ə·jē }

hydrolysis [CHEM] **1.** Decomposition or alteration of a chemical substance by water. **2.** In aqueous solutions of electrolytes, the reactions of cations with water to produce a weak base or of anions to produce a weak acid. { hī'dräl·ə·səs }

hydrolytic enzyme [BIOCHEM] A catalyst that acts like a hydrolase. { ¦hī·drə¦lid·ik 'en,zīm }

hydrolytic process [CHEM] A reaction of both organic and inorganic chemistry wherein water effects a double decomposition with another compound, hydrogen going to one compound and hydroxyl to another. { ¦hī·drə¦lid·ik 'prä·səs }

hydrolyzate [GEOL] A sediment characterized by elements such as aluminum, potassium, or sodium which are readily hydrolyzed. { hī'dräl·ə,zāt }

hydromagnesite [MINERAL] $Mg_4(OH)_2(CO_3)_3 \cdot 3H_2O$ A white, earthy mineral crystallizing in the monoclinic system and found in small crystals, amorphous masses, or chalky crusts. { ¦hī·drō'mag·nə,zīt }

hydromagnetic instability *See* magnetohydrodynamic instability. { ¦hī·drō·mag¦ned·ik ,in·stə'bil·əd·ē }

hydromagnetics *See* magnetohydrodynamics. { ¦hī·drō·mag¦ned·iks }

hydromagnetic stability *See* magnetohydrodynamic stability. { ¦hī·drō·mag¦ned·ik stə'bil·əd·ē }

hydromagnetic wave *See* magnetohydrodynamic wave. { ¦hī·drō·mag¦ned·ik 'wāv }

hydromechanics [FL MECH] The study of liquids, traditionally water, as a medium for the transmission of forces. { ¦hī·drō·mi'kan·iks }

hydrometallurgy [MET] Treatment of metals and metal-containing materials by wet processes. { ¦hī·drō¦med·əl,ər·jē }

hydrometamorphism [GEOL] Alteration of rocks by material carried in solution by water without the influence of high temperature or pressure. { ¦hī·drə,med·ə'mòr·fiz·əm }

hydrometeor [HYD] **1.** Any product of condensation or sublimation of atmospheric water vapor, whether formed in the free atmosphere or at the earth's surface. **2.** Any water particles blown by the wind from the earth's surface. { ¦hī·drō 'mēd·ē·ər }

hydrometeorology [METEOROL] That part of meteorology of direct concern to hydrologic problems, particularly to flood control, hydroelectric power, irrigation, and similar fields of engineering and water resources. { ¦hī·drō,mēd·ē·ə'räl·ə·jē }

hydrometer [ENG] A direct-reading instrument for indicating the density, specific gravity, or some similar characteristic of liquids. { hī'dräm·əd·ər }

Hydrometridae [INV ZOO] The marsh treaders, a family of hemipteran insects in the subdivision Amphibicorisae. { ,hī·drə'me·trə,dē }

hydrometrograph [ENG] An instrument that measures and records the rate of water discharge from a pipe or an orifice. { ¦hī·drə¦me·trə,graf }

hydrometry [FL MECH] The science and technology of measuring specific gravities, particularly of liquids. { hī'dräm·ə·trē }

hydromica [GEOL] Any of several varieties of muscovite, especially illite, which are less elastic than mica, have a pearly luster, and sometimes contain less potash and more water than muscovite. Also known as hydrous mica. { ¦hī·drō'mī·kə }

hydromorphic [GEOL] Referring to an intrazonal soil with characteristics that were developed in the presence of excess water all or part of the time. { ¦hī·drə'mòr·fik }

hydronephrosis [MED] Accumulation of urine in and distension of the renal pelvis and calyces due to obstructed outflow. { ,hī·drə·nə'frō·səs }

hydronic heating *See* hot-water heating. { hī'drän·ik 'hēd·iŋ }

hydronic radiation [COMMUN] A form of electromagnetic radiation used for communication underwater. { hī'drän·ik ,rād·ē'ā·shən }

hydronium ion [INV ZOO] H_3O^+ An oxonium ion consisting of a proton combined with a molecule of water; found in pure water and in all aqueous solutions. Also known as hydrogen ion. { hī'drō·nē·əm ,ī,än }

HYDROPAC [NAV] An urgent notice of dangers to navigation in the Pacific Ocean, originated by the U.S. Naval Oceanographic Office and disseminated for the immediate safeguarding of shipping and life at sea. { 'hī·drə,pak }

hydropathy [MED] The system of internal and external use of water in attempting to cure disease. { hī'drä·pə·thē }

hydropericardium [MED] Accumulation of serous fluid in the pericardial sac. { ¦hī·drō,per·ə'kärd·ē·əm }

Hydrophiidae [VERT ZOO] A family of proglyphodont snakes in the suborder Serpentes found in Indian-Pacific oceans. { ¦hī·drō'fī·ə,dē }

hydrophile-lipophile balance [ORG CHEM] The relative simultaneous attraction of an emulsifier for two phases of an emulsion system; for example, water and oil. { 'hī·drə,fīl 'lip·ə,fīl ,bal·əns }

hydrophilic [CHEM] Having an affinity for, attracting, adsorbing, or absorbing water. { ,hī·drə'fil·ik }

HYDROMETER

A plain hydrometer. *(Taylor Instrument Co.)*

HYDROPHILIDAE

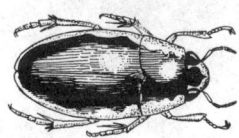

A drawing of a water scavenger beetle. *(From T. I. Storer and R. L. Usinger, General Zoology, 3d ed., McGraw-Hill, 1957)*

Hydrophilidae [INV ZOO] The water scavenger beetles, a large family of coleopteran insects in the superfamily Hydrophiloidea. { ˌhī·drə'fil·ə‚dē }

hydrophilite *See* chlorocalcite. { hī'draf·ə‚līt }

Hydrophiloidea [INV ZOO] A superfamily of coleopteran insects in the suborder Polyphaga. { ˌhī·drə·fə'lȯid·ē·ə }

hydrophilous [ECOL] Inhabiting moist places. { hī'draf·ə·ləs }

hydrophobia [MED] *See* rabies. [PSYCH] An abnormal fear of water. { ‚hī·drə'fō·bē·ə }

hydrophobic [CHEM] Lacking an affinity for, repelling, or failing to adsorb or absorb water. [MED] Of, pertaining to, or suffering from hydrophobia. { hī·drə'fō·bik }

hydrophobophobia [PSYCH] An abnormal fear of hydrophobia. { ˌhī·drə‚fōb·ə'fō·bē·ə }

hydrophone [ENG ACOUS] A device which receives underwater sound waves and converts them to electric waves. { 'hī·drə‚fōn }

hydrophone array [COMMUN] A group of two or more hydrophones which feed into a common receiver. { 'hī·drə‚fōn ə‚rā }

hydrophone noise [ELEC] Any unwanted disturbance in the electric waves delivered by a hydrophone. { 'hī·drə‚fōn ‚nȯiz }

hydrophone response [ELEC] The electric waves delivered by a hydrophone in response to waterborne sound waves. { 'hī·drə‚fōn ri'späns }

hydrophotometer [OPTICS] An instrument for measuring the attenuation coefficient of collimated light in sea water, in which light from a collimated source is directed through a column of sea water and is measured by a photocell or other electronic device at the other end of the column. { ‚hī·drə·fə'täm·əd·ər }

Hydrophyllaceae [BOT] A family of dicotyledonous plants in the order Polemoniales distinguished by two carpels, parietal placentation, and generally imbricate corolla lobes in the bud. { ˌhī·drō·fə'lās·ē‚ē }

hydrophyllium [INV ZOO] A transparent body partly covering the spore sacs of siphonophoran cnidarians. { ‚hī·drə'fil·ē·əm }

hydrophyte [BOT] **1.** A plant that grows in a moist habitat. **2.** A plant requiring large amounts of water for growth. Also known as hygrophyte. { 'hī·drə‚fīt }

hydroplane [NAV ARCH] **1.** A boat which when operated at high speed planes on the surface of the water; the bottom of such a craft is normally a prismatic surface. Also known as planing boat. **2.** *See* diving rudder. { 'hī·drə‚plān }

hydroplanula [INV ZOO] A cnidarian larval stage between the planula and actinula stages. { ‚hī·drə'plan·yə·lə }

hydropneumatic [ENG] Operated by both water and air power. { ‚hī·drō·nü'mad·ik }

hydropneumatic recoil system [MECH ENG] A recoil mechanism that absorbs the energy of recoil by the forcing of oil through orifices and returns the gun to battery by compressed gas. { ‚hī·drō·nü'mad·ik 'rē‚kȯil ‚sis·təm }

hydropolyp *See* hydroid. { ‚hī·drō'päl·əp }

hydroponics [BOT] Growing of plants in a nutrient solution with the mechanical support of an inert medium such as sand. { ˌhī·drə'pän·iks }

hydropore [INV ZOO] In certain asteroids and echinoids, an opening on the aboral surface of a canal which extends from the ring canal in one of the interradii. { 'hī·drə‚pȯr }

hydropower *See* hydroelectricity. { 'hī·drə‚pau̇·ər }

hydroquinol *See* hydroquinone. { ‚hī·drō'kwi‚nȯl }

hydroquinone [ORG CHEM] $C_6H_4(OH)_2$ White crystals melting at 170°C and boiling at 285°C; soluble in alcohol, ether, and water; used in photographic dye chemicals, in medicine, as an antioxidant and inhibitor, and in paints, varnishes, and motor fuels and oils. Also known as hydrochinone; hydroquinol; quinol. { ‚hī·drə·kwə'nōn }

hydroquinone dimethyl ether [ORG CHEM] $C_6H_4(OCH_3)_2$ White flakes with a melting point of 56°C; used as a weathering agent in paint, as a flavoring, and in dyes and cosmetics. { ‚hī·drə·kwə'nōn dī‚meth·əl 'ē·thər }

hydroquinone monomethyl ether [ORG CHEM] $CH_3O\text{-}C_6H_4OH$ A white, waxy solid with a melting point of 52.5°C; soluble in benzene, acetone, and alcohol; used for antioxidants, pharmaceuticals, and dyestuffs. { ‚hī·drə·kwə'nōn ‚män·ō‚meth·əl 'ē·thər }

HYDROQUINONE

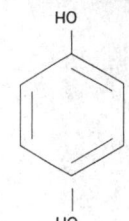

Structural formula of hydroquinone.

hydrorhiza [INV ZOO] Rootlike structure of a hydroid colony. { ‚hī·drō'rī·zə }

hydrosalpinx [MED] A distension of a Fallopian tube with fluid. { ‚hī·drō'sal‚piŋks }

Hydroscaphidae [INV ZOO] The skiff beetles, a small family of coleopteran insects in the suborder Myxophaga. { ‚hī·drə'skaf·ə‚dē }

hydroscope [OPTICS] An instrument designed to observe objects an appreciable distance below the surface of water, consisting of a series of mirrors enclosed in a steel tube. { 'hī·drə‚skōp }

hydroseparator [MECH ENG] A separator in which solids in suspension are agitated by hydraulic pressure or stirring devices. { ‚hī·drō'sep·ə‚rād·ər }

hydrosere [ECOL] Community in which the pioneer plants invade open water, eventually forming some kind of soil such as peat or muck. { 'hī·drə‚sir }

hydrosilylation [ORG CHEM] The addition of a Si-H bond to a C-C double bond of an olefin. { ‚hī·drō‚sil·ē'ā·shən }

hydrosizer *See* hydraulic classifier. { 'hī·drə‚sīz·ər }

hydroskeleton [INV ZOO] Water contained within the coelenteron and serving a skeletal function in most cnidarian polyps. { ‚hī·drō'skel·ət·ən }

hydrosol [CHEM] A colloidal system in which the dispersion medium is water, and the dispersed phase may be a solid, a gas, or another liquid. Also known as aquasol. { 'hī·drə‚sȯl }

hydrospace detection [ORD] Detection of underwater targets, as with sonar. { 'hī·drə‚spās di'tek·shən }

hydrosphere [HYD] The water portion of the earth as distinguished from the solid part (lithosphere) and from the gaseous outer envelope (atmosphere). { 'hī·drə‚sfir }

hydrospire [INV ZOO] Either of a pair of flattened tubes composing part of the respiratory system in blastoids. { 'hī·drə‚spīr }

hydrospore [INV ZOO] Opening into the hydrocoele on the right side in echinoderm larvae. { 'hī·drə‚spȯr }

hydrostat *See* humidistat. { 'hī·drə‚stat }

hydrostatic analogy [PHYS] An analogy between the relations among current, potential difference, and resistance in an electric circuit and the relations among corresponding quantities describing water flowing under a hydrostatic head. { ‚hī·drə'stad·ik ə'nal·ə·jē }

hydrostatic approximation [METEOROL] The assumption that the atmosphere is in hydrostatic equilibrium. { ‚hī·drə'stad·ik ə‚präk·sə'mā·shən }

hydrostatic assumption [GEOPHYS] **1.** The assumption that the pressure of seawater increases by 1 atmosphere (101,325 pascals) over approximately 33 feet (10 meters) of depth, the exact value depending on the water density and the local acceleration of gravity. **2.** Specifically, the assumption that fluid is not undergoing vertical accelerations, hence the vertical component of the passive gradient force per unit mass is equal to g, the local acceleration due to gravity. { ‚hī·drə'stad·ik ə'səm·shən }

hydrostatic balance [MECH] An equal-arm balance in which an object is weighed first in air and then in a beaker of water to determine its specific gravity. { ‚hī·drə'stad·ik i'kwä·zhən }

hydrostatic bearing [MECH ENG] A sleeve bearing in which high-pressure oil is pumped into the area between the shaft and the bearing so that the shaft is raised and supported by an oil film. { ‚hī·drə'stad·ik 'ber·iŋ }

hydrostatic equation [PHYS] The form assumed by the vertical component of the vector equation of motion when all Coriolis force, earth curvature, frictional, and vertical acceleration terms are considered negligible compared with those involving the vertical pressure force and the force of gravity. { ‚hī·drə'stad·ik i'kwä·zhən }

hydrostatic equilibrium [PHYS] The state of a fluid whose surfaces of constant mass (or density) coincide and are horizontal throughout; complete balance exists between the force of gravity and the pressure force; the relation between the pressure and the geometric height is given by the hydrostatic equation. { ‚hī·drə'stad·ik ‚ē·kwə'lib·rē·əm }

hydrostatic forging [MET] Forging a metal part by using pressure supplied by a liquid. { ‚hī·drə'stad·ik 'fȯrj·iŋ }

hydrostatic fuse [ORD] A fuse used with a depth bomb to cause an underwater explosion at a predetermined depth;

initiation is caused by the pressure of the water as the depth bomb sinks.　{ ‚hī·drə'stad·ik 'fyüz }

hydrostatic modulus　See bulk modulus of elasticity.　{ ‚hī·drə'stad·ik 'mäj·ə·ləs }

hydrostatic press　See hydraulic press.　{ ‚hī·drə'stad·ik 'pres }

hydrostatic pressing　[ENG]　Compacting ceramic or metal powders by packing them in a rubber bag which is subjected to pressure from a hydraulic press.　{ ‚hī·drə'stad·ik 'pres·iŋ }

hydrostatic pressure　[FL MECH]　**1.** The pressure at a point in a fluid at rest due to the weight of the fluid above it.　Also known as gravitational pressure.　**2.** The negative of the stress normal to a surface in a fluid.　{ ‚hī·drə'stad·ik 'presh·ər }

hydrostatic roller conveyor　[MECH ENG]　A portion of a roller conveyor that has rolls weighted with liquid to control the speed of the moving objects.　{ ‚hī·drə'stad·ik rō·lər kən‚vā·ər }

hydrostatics　[FL MECH]　The study of liquids at rest and the forces exerted on them or by them.　{ ‚hī·drə'stad·iks }

hydrostatic stability　See static stability.　{ ‚hī·drə'stad·ik stə'bil·əd·ē }

hydrostatic strength　[MECH]　The ability of a body to withstand hydrostatic stress.　{ ‚hī·drə'stad·ik 'streŋkth }

hydrostatic stress　[MECH]　The condition in which there are equal compressive stresses or equal tensile stresses in all directions, and no shear stresses on any plane.　{ ‚hī·drə'stad·ik 'stres }

hydrostatic test　[ENG]　Test of strength and leak-resistance of a vessel, pipe, or other hollow equipment by internal pressurization with a test liquid.　{ ‚hī·drə'stad·ik 'test }

hydrostatic weighing　[FL MECH]　A method of determining the density of a sample in which the sample is weighed in air, and then weighed in a liquid of known density; the volume of the sample is equal to the loss of weight in the liquid divided by the density of the liquid.　{ ‚hī·drə'stad·ik 'wā·iŋ }

hydrosulfide　[CHEM]　A compound that has the SH− radical; for example, sulfhydrates, sulfhydryls, thioalcohols, thiols, sulfur alcohols, and mercaptans.　{ ‚hī·drə'səl‚fīd }

hydrotalcite　[MINERAL]　$Mg_6Al_2(OH)_{16}(CO_3)·4H_2O$　Pearly-white mineral composed of hydrous aluminum and magnesium hydroxide and carbonate.　{ ‚hī·drə'tal‚sīt }

hydrotheca　[INV ZOO]　Cup-shaped portion of the perisarc in some cnidarians that serves to hold and protect a withdrawn hydranth.　{ ‚hī·drō'thē·kə }

hydrotherapy　[MED]　Treatment of disease by external application of water.　{ ‚hī·drə'ther·ə·pē }

hydrothermal　[GEOL]　Of or pertaining to heated water, to its action, or to the products of such action.　{ ‚hī·drə'thər·məl }

hydrothermal alteration　[GEOL]　Rock or mineral phase changes that are caused by the interaction of hydrothermal liquids and wall rock.　{ ‚hī·drə'thər·məl ‚ól·tə'rā·shən }

hydrothermal crystal growth　[CHEM ENG]　Formation of simple crystals of quartz at elevated temperatures and pressures in an autoclave with an alkaline solution.　{ ‚hī·drə'thər·məl 'krist·əl ‚gróth }

hydrothermal deposit　[GEOL]　A mineral deposit precipitated from a hot, aqueous solution.　{ ‚hī·drə'thər·məl di'päz·ət }

hydrothermal solution　[GEOL]　Hot, residual watery fluids derived from magmas during the later stages of their crystallization and commonly containing large amounts of dissolved metals which are deposited as ore veins in fissures along which the solutions often move.　{ ‚hī·drə'thər·məl sə'lü·shən }

hydrothermal synthesis　[GEOL]　Mineral synthesis in the presence of heated water.　{ ‚hī·drə'thər·məl 'sin·thə·səs }

hydrothermal vent　[OCEANOGR]　A hot spring on the ocean floor, found mostly along mid-oceanic ridges, where heated fluids exit from cracks in the earth's crust. Iron, sulfur, and other materials precipitate from these waters to form dark clouds.　Also known as black smoker.　{ ‚hī·drə'thərm·əl 'vent }

hydrothorax　[MED]　Collection of serous fluid in the pleural spaces.　{ ‚hī·drə'thór‚aks }

hydrotreating　[CHEM ENG]　Oil refinery catalytic process in which hydrogen is contacted with petroleum intermediate or product streams to remove impurities, such as oxygen, sulfur, nitrogen, or unsaturated hydrocarbons.　{ 'hī·drō‚trēd·iŋ }

hydrotroilite　[MINERAL]　$FeS·nH_2O$　A black, finely divided colloidal material reported in many muds and clays; thought to be formed by bacteria on bottoms of marine basins.　{ ‚hī·drō'trói‚līt }

hydrotrope　[CHEM]　Compound with the ability to increase the solubilities of certain slightly soluble organic compounds.　{ 'hī·drə‚trōp }

hydrotropism　[BIOL]　Orientation involving growth or movement of a sessile organism or part, especially plant roots, in response to the presence of water.　{ hī'drä·trə‚piz·əm }

hydrotungstite　[MINERAL]　$H_2WO_4·H_2O$　A mineral composed of hydrous tungstic acid.　{ ‚hī·drō'təŋz‚tīt }

hydroureter　[MED]　Accumulation of urine in and distention of the ureter due to obstructed outflow.　{ ‚hī·drō'yùr·əd·ər }

hydrous　[CHEM]　Indicating the presence of an indefinite amount of water.　[MINERAL]　Indicating a definite proportion of combined water.　{ 'hī·drəs }

hydrous mica　See hydromica.　{ 'hī·drəs 'mī·kə }

hydrowire　[ENG]　A wire to which equipment is clamped so that it can be lowered over the side of the ship into the water.　{ 'hī·drō‚wīr }

hydroxamic acid　[ORG CHEM]　An organic compound that contains the group −C(=O)NHOH.　{ ‚hī‚drak‚sam·ik 'as·əd }

hydroxide　[CHEM]　Compound containing the OH^- group; the hydroxides of metals are usually bases and those of nonmetals are usually acids; a hydroxide can be organic or inorganic.　{ hī'drak‚sīd }

hydroximino　See nitroso.　{ ‚hī‚drak'sim·ə·nō }

hydroxisoxazole　[ORG CHEM]　$C_4H_5NO_2$　A colorless, crystalline compound with a melting point of 86–87°C; used as a fungicide in soil and as a growth regulator for seeds.　Also known as 3-hydroxy-5-methylisoxazole; hymexazol.　{ hī‚drak·sə‚säk·sə‚zōl }

hydroxy-　[ORG CHEM]　Chemical prefix indicating the OH^- group in an organic compound, such as hydroxybenzene for phenol, C_6H_5OH; the use of just oxy- for the prefix is incorrect.　Also spelled hydroxyl-.　{ hī'drak·sē }

hydroxyacetic acid　See glycolic acid.　{ hī‚drak·sə·ə'sēd·ik 'as·əd }

hydroxy acid　[ORG CHEM]　Any organic acid, with an OH^- group, such as hydroxyacetic acid.　{ hī'drak·sē 'as·əd }

hydroxyamphetamine　[PHARM]　C_9H_3ON　A sympathetic amine used as the hydrobromide salt orally as a drug and locally as a mydriatic and nasal decongestant.　{ hī‚drak·sə·am‚fed·ə‚mēn }

hydroxybenzoic acid　[ORG CHEM]　$C_7H_6O_3$　Any one of three crystalline derivatives of benzoic acid: ortho, meta, and para forms; the ester of the para compound is used as a bacteriostatic agent.　{ hī‚drak·sē·ben'zō·ik 'as·əd }

para-hydroxybenzoic acid　[ORG CHEM]　$C_6H_4(OH)$-$COOH·2H_2O$　Colorless crystals melting at 210°C; soluble in alcohol, water, and ether; used as a chemical intermediate and for synthetic drugs.　{ ‚par·ə hī‚drak·sē·ben'zō·ik 'as·əd }

2-hydroxybiphenyl　See phenylphenol.　{ ‚tü hī‚drak·sē·bī'fen·əl }

β-hydroxybutyric dehydrogenase　[BIOCHEM]　The enzyme that catalyzes the conversion of L-β-hydroxybutyric acid to acetoacetic acid by dehydrogenation.　{ ‚bād·ə hī‚drak·sē·byü'tir·ik dē·hī'draj·ə‚nās }

hydroxycarbamide　See hydroxyurea.　{ hī‚drak·sē'kär·bə‚mīd }

hydroxycarbonyl compound　[ORG CHEM]　A compound possessing one or more hydroxy (−OH) groups and one or more carbonyl (=C=O) groups.　{ hī‚drak·sē'kär·bə·nəl 'käm‚paùnd }

hydroxychloroquine　[PHARM]　$C_{18}H_{26}ClN_3O$　A drug used as the sulfate salt for the treatment of malaria, lupus erythematosus, and rheumatoid arthritis.　{ hī‚drak·sē'klór·ə‚kwīn }

hydroxycholine　See muscarine.　{ hī‚drak·sē'kō‚lēn }

hydroxycinchonine　See cupreine.　{ hī‚drak·sē'siŋ·kə‚nēn }

hydroxycitronellal　[ORG CHEM]　$C_{10}H_{20}O_2$　A colorless or light yellow, viscous liquid with a boiling range of 94–96°C; soluble in 50% alcohol and fixed oils; used in perfumery and flavoring.　Also known as citronellal hydrate.　{ hī‚drak·sē‚sī·trə'nel·əl }

2-(hydroxydiphenyl)methane　[ORG CHEM]　$C_6H_5CH_2$-C_6H_4OH　A crystalline substance with a melting point of 20.2–20.9°C, or a liquid; used as a germicide, preservative, and antiseptic.　{ ‚tü hī‚drak·sē·dī'fen·əl 'meth‚ān }

hydroxyethylcellulose　[MATER]　A white powder made

from cellulose, used for textile finishes and as a thickener for water-base paints. { ¦hī¦drak·sē¦eth·əl'sel·yə,lōs }

2-hydroxyethylhydrazine [ORG CHEM] $HOCH_2CH_2-NHNH_2$ A colorless, slightly viscous liquid with a melting point of $-70°C$; soluble in lower alcohols; used as an abscission agent in fruit. Also known as 2-hydrazinoethanol. { ¦tü hī¦drak·sē¦eth·əl'hī·drə,zēn }

3-hydroxyflavone See flavanol. { ¦thrē hī¦drak·sē'fla,vōn }

hydroxyine [PHARM] $C_{21}H_{27}ClN_2O_2$ A tranquilizer, also possessing antiemetic and antihistaminic effects; used as the hydrochloride salt. { hī'drak·sə,lēn }

hydroxyl- See hydroxy-. { hī'drak·səl }

hydroxylamine [INORG CHEM] NH_2OH A colorless, crystalline compound produced commercially by acid hydrolysis of nitroparaffins, decomposes on heating, melts at 33°C; used in organic synthesis and as a reducing agent. { ¦hī,drak'sil·ə,mēn }

hydroxylamine hydrochloride [ORG CHEM] $(NH_2OH)Cl$ A crystalline substance with a melting point of 151°C; soluble in glycerol and propylene glycol; used as a reducing agent in photography and in synthetic and analytic chemistry, as an antioxidant in fatty acids and soaps, and as a reagent for enzyme reactivation. { ¦hī,drak'sil·ə,mēn ¦hī·drə'klór,īd }

ortho-**hydroxylaniline** [ORG CHEM] $C_6H_4NH_2OH$ White crystals that turn brownish upon standing for some time; melts at 172–173°C, and will sublime upon more heating; soluble in cold water and benzene; used as a dye for hair and furs, and as a dye intermediate. Also known as *ortho*-aminophenol; oxammonium. { ¦ór,thō ¦hī¦drak·səl'an·əl·ən }

hydroxylapatite [MINERAL] $Ca_5(PO_4)_3OH$ A rare form of the apatite group that crystallizes in the hexagonal system. { hī¦drak·səl'ap·ə,tīt }

hydroxylase [BIOCHEM] Any of several enzymes that catalyze certain hydroxylation reactions involving atomic oxygen. { hī'drak·sə,lās }

hydroxylation reaction [ORG CHEM] One of several types of reactions used to introduce one or more hydroxyl groups into organic compounds; an oxidation reaction as opposed to hydrolysis. { hī,drak·sə'lā·shən rē,ak·shən }

hydroxylherderite [MINERAL] $CaBe(PO_4)(OH)$ A monoclinic mineral composed of a phosphate and hydroxide of calcium and beryllium; isomorphous with herderite. { hī¦drak·səl'hər·də,rīt }

β-hydroxynaphthoic acid [ORG CHEM] $C_{10}H_6OHCOOH$ A yellow solid that is soluble in ether and alcohol and melts at about 218°C; used as a dye and a pigment. { ¦bād·ə hī¦drak·sē·naf'thō·ik 'as·əd }

4-hydroxy-3-nitrobenzenearsonic acid [ORG CHEM] $HOC_6H_3(NO_2)AsO(OH)_2$ Crystals used as a reagent for zirconium; also used to control enteric infections and to improve growth and feed efficiency in animals. Also known as roxarsone. { ¦fór hī¦drak·sē ¦thrē ,nī·trō¦ben,zēn·är'sän·ik 'as·əd }

hydroxyproline [BIOCHEM] $C_5H_9O_3N$ An amino acid that is essentially limited to structural proteins of the collagen type. { hī¦drak·sə'prō,lēn }

para-**hydroxypropiophenone** [PHARM] $HOC_6H_4COC_2H_5$ A crystalline substance with a melting point of 149°C; soluble in alcohol and ether; used as an inhibitor of pituitary gonadotropic hormone. { ¦par·ə hī¦drak·sē,prō·pē'ä·fə,nōn }

8-hydroxyquinoline [ORG CHEM] C_9H_6NOH White crystals or powder that darken on exposure to light, slightly soluble in water, soluble in benzene, melting at 73–75°C; used in preparing fungicides and in the separation of metals by acting as a precipitating agent. Also known as oxine; oxyquinoline; 8-quinolinol. { ¦āt hī¦drak·sē'kwin·ə·lən }

8-hydroxyquinoline sulfate [PHARM] $C_{18}H_{16}N_2O_6S$ A pale yellow, crystalline powder with a melting point of 175–178°C; soluble in water; used as an antiseptic, deodorant, and antiperspirant. { ¦āt hī¦drak·sē'kwin·ə·lən 'səl,fāt }

5-hydroxytryptamine See serotonin. { ¦fīv hī¦drak·sē'trip·tə,mēn }

5-hydroxytryptophan [BIOCHEM] $C_{11}H_{12}N_2O_3$ Minute rods or needlelike crystals; the biological precursor of serotonin. { ¦fīv hī¦drak·sē'trip·tə,fan }

3-hydroxytyramine hydrobromide [ORG CHEM] $(HO)_2-C_6H_3CH_2CH_2NH_2·HBr$ A source of dopamine for the synthesis of catecholamine analogs. { ¦thrē hī¦drak·sē'tī·rə,mēn ,hī·drə'brō,mīd }

hydroxyurea [PHARM] $HONHCONH_2$ Needlelike crystals with a melting point of 133–136°C; used as an antineoplastic agent. Also known as hydroxycarbamide. { hī¦drak·sē·yü'rē·ə }

hydrozincite [MINERAL] $Zn_5(OH)_5(CO_3)_2$ A white, grayish, or yellowish mineral composed of basic zinc carbonate, occurring as masses or crusts. { ,hī·drō'zin,kīt }

Hydrozoa [INV ZOO] A class of the phylum Cnidaria which includes the fresh-water hydras, the marine hydroids, many small jellyfish, a few corals, and the Portuguese man-of-war. { ,hī·drə'zō·ə }

Hydrus [ASTRON] A southern constellation, right ascension 2 hours, declination 75°S. Also known as Water Snake. { 'hī·drəs }

hyena [VERT ZOO] An African carnivore represented by three species of the family Hyaenidae that resemble dogs but are more closely related to cats. { hī'ē·nə }

Hyeniales [PALEOBOT] An order of Devonian plants characterized by small, dichotomously forked leaves borne in whorls. { ,hī·ə'nā·lēz }

Hyeniatae See Hyeniopsida. { ,hī·ə'nī·ə,tē }

Hyeniopsida [PALEOBOT] An extinct class of the division Equisetophyta. { ¦hī·ə·nē'äp·sə·də }

hyetal coefficient See pluviometric coefficient. { 'hī·əd·əl ,kō·i'fish·ənt }

hyetal equator [CLIMATOL] A line (or transition zone) which encircles the earth (north of the geographical equator) and lies between two belts that typify the annual time distribution of rainfall in the lower latitudes of each hemisphere; a form of meteorological equator. { 'hī·əd·əl i'kwād·ər }

hyetal region [CLIMATOL] A region in which the amount and seasonal variation of rainfall are of a given type. { 'hī·əd·əl ,rē·jən }

hyetograph [CLIMATOL] A map or chart displaying temporal or areal distribution of precipitation. { hī'ed·ə,graf }

hyetography [CLIMATOL] The study of the annual variation and geographic distribution of precipitation. { ,hī·ə'täg·rə·fē }

hyetology [METEOROL] The science which treats of the origin, structure, and various other features of all the forms of precipitation. { ,hī·ə'täl·ə·jē }

Hygiea [ASTRON] The fourth largest asteroid, with a diameter of about 260 miles (419 kilometers), mean distance from the sun of 3.14 astronomical units, and C-type surface composition. { hī'jē·ə }

hygiene [MED] The science that deals with the principles and practices of good health. { 'hī,jēn }

hygristor [ELECTR] A resistor whose resistance varies with humidity; used in some types of recording hygrometers. { hī'gris·tər }

Hygrobiidae [INV ZOO] The squeaker beetles, a small family of coleopteran insects in the suborder Adephaga. { hī·grə'bī·ə,dē }

hygrodeik [ENG] A form of psychrometer with wet-bulb and dry-bulb thermometers mounted on opposite edges of a specially designed graph of the psychrometric tables, arranged so that the intersections of two curves determined by the wet-bulb and dry-bulb readings yield the relative humidity, dew-point, and absolute humidity. { 'hī·grə,dīk }

hygrogram [ENG] The record made by a hygrograph. { 'hī·grə,gram }

hygrograph [ENG] A recording hygrometer. { 'hī·grə,graf }

hygrokinematics [METEOROL] The descriptive study of the motion of water substances in the atmosphere. { ¦hī·grə,kin·ə'mad·iks }

hygrology [METEOROL] The study which deals with the water vapor content (humidity) of the atmosphere. { hī'gräl·ə·jē }

hygroma [MED] A congenital disorder in which a lymph-filled cystic cavity is formed from distended lymphatics. { hī'grō·mə }

hygrometer [ENG] An instrument for giving a direct indication of the amount of moisture in the air or other gas, the indication usually being in terms of relative humidity as a percentage which the moisture present bears to the maximum amount of moisture that could be present at the location temperature without condensation taking place. { hī'gräm·əd·ər }

hygrometry [ENG] The study which treats of the measurement of the humidity of the atmosphere and other gases. { hī'gräm·ə·trē }

hygromycin [MICROBIO] $C_{25}H_{33}O_{12}N$ A weakly acidic, soluble antibiotic with a fairly broad spectrum, produced by a strain of *Streptomyces hygroscopicus*. { ‚hī·grə'mīs·ən }

hygrophobia [PSYCH] An abnormal fear of liquids or moisture. { ‚hī·grə'fō·bē·ə }

hygrophyte *See* hydrophyte. { 'hī·grə‚fīt }

hygroscopic [BOT] Being sensitive to moisture, such as certain tissues. [CHEM] **1.** Possessing a marked ability to accelerate the condensation of water vapor; applied to condensation nuclei composed of salts which yield aqueous solutions of a very low equilibrium vapor pressure compared with that of pure water at the same temperature. **2.** Pertaining to a substance whose physical characteristics are appreciably altered by effects of water vapor. **3.** Pertaining to water absorbed by dry soil minerals from the atmosphere; the amounts depend on the physicochemical character of the surfaces, and increase with rising relative humidity. { ‚hī·grə‚skäp·ik }

hygroscopic coefficient [HYD] The percentage of water that a soil will absorb and hold in equilibrium in a saturated atmosphere. { ‚hī·grə‚skäp·ik ‚kō·i'fish·ənt }

hygroscopic depression [CHEM] The measure of a desiccant's capacity to take on water. { ‚hī·grə‚skäp·ik di'presh·ən }

hygroscopic water [HYD] The component of soil water that is held adsorbed on the surface of soil particles and is not available to vegetation. { ‚hī·grə‚skäp·ik 'wòd·ər }

hygrothermograph [ENG] An instrument for recording temperature and humidity on a single chart. { ‚hī·grə'thər·mə‚graf }

hyl *See* metric-technical unit of mass.

hylaea *See* tropical rainforest. { hī·lē·ə }

Hylidae [VERT ZOO] The tree frogs, a large amphibian family in the suborder Procoela; many are adapted to arboreal life, having expanded digital disks. { 'hī·lə‚dē }

Hylleraas coordinates [ATOM PHYS] Coordinates for two particles used in studying the helium atom; they comprise the distance between the two particles, the sum of the distances of the particles from the origin, and the difference of the distances of the particles from the origin. { 'hil·ə‚räs kō‚órd·ən·əts }

Hylobatidae [VERT ZOO] A family of anthropoid primates in the superfamily Hominoidea including the gibbon and the siamang of southeastern Asia. { ‚hī·lō'bad·ə‚dē }

hylophagous [ZOO] Feeding on wood, as termites. { hī‚läf·ə·gəs }

hylophobia [PSYCH] An abnormal fear of forests. { ‚hī·lə'fō·bē·ə }

hylotomous [ZOO] Cutting wood, as wood-boring insects. { hī‚läd·ə·məs }

hymecromone [ORG CHEM] $C_{10}H_8O_3$ A crystalline substance with a melting point of 194–195°C; soluble in methanol and glacial acetic acid; used as choleretic and antispasmodic drugs and as a standard for the fluorometric determination of enzyme activity. { hī'mek·rə‚mōn }

hymen [ANAT] A mucous membrane partly closing off the vaginal orifice. Also known as maidenhead. { 'hī·mən }

hymenium [MYCOL] The outer, sporebearing layer of certain fungi or their fruiting bodies. { hī'mē·nē·əm }

hymenolepiasis [MED] Intestinal infection by tapeworms of the genus *Hymenolepis*. { ‚hī·mə·nō·lə'pī·ə·səs }

Hymenolepis [INV ZOO] A genus of tapeworms parasitic in humans, birds, and mammals. { ‚hī·mə'näl·ə·pəs }

Hymenomycetes [MYCOL] A group of the Homobasidiomycetidae including forms such as mushrooms and pore fungi in which basidia are formed in an exposed layer (hymenium) and basidiospores are borne asymmetrically on slender stalks. { ‚hī·mə·nō‚mī'sēd·ēz }

hymenophore [MYCOL] Portion of a sporophore that bears the hymenium. { hī'men·ə‚fòr }

hymenopodium [MYCOL] **1.** Tissue beneath the hymenium in certain fungi. **2.** A genus of the Moniliales. { ‚hī·mə·nō'pōd·ē·əm }

Hymenoptera [INV ZOO] A large order of insects including ants, wasps, bees, sawflies, and related forms; head, thorax and abdomen are clearly differentiated; wings, when present, and legs are attached to the thorax. { ‚hī·mə'näp·trə }

Hymenostomatida [INV ZOO] An order of ciliated protozoans in the subclass Holotrichia having fairly uniform ciliation and a definite buccal ciliature. { ‚hī·mə·nō·stō'mad·ə·də }

hymenotomy [MED] Surgical incision of the hymen. { ‚hī·mə'näd·ə·mē }

hymexazol *See* hydroxisoxazole. { hī'mek·sə‚zól }

Hynobiidae [VERT ZOO] A family of salamanders in the suborder Cryptobranchoidea. { ‚hī·nō'bī·ə‚dē }

hyobranchium [VERT ZOO] A Y-shaped bone supporting the tongue and tongue muscles in a snake. { ‚hī·ō'braŋ·kē·əm }

Hyocephalidae [INV ZOO] A monospecific family of hemipteran insects in the superfamily Pentatomorpha. { ‚hī·ō·sə'fal·ə‚dē }

hyoglossus [ANAT] An extrinsic muscle of the tongue arising from the hyoid bone. { ‚hī·ō'glä·səs }

hyoid [ANAT] **1.** A bone or complex of bones at the base of the tongue supporting the tongue and its muscles. **2.** Of or pertaining to structures derived from the hyoid arch. { 'hī‚óid }

hyoid arch [EMBRYO] Either of the second pair of pharyngeal segments or gill arches in vertebrate embryos. { 'hī‚óid ‚ärch }

hyoid tooth [VERT ZOO] One of a number of teeth on the tongue of fishes. { 'hī‚óid ‚tüth }

hyomandibular [VERT ZOO] The upper portion of the hyoid arch in fishes. { ‚hī·ō‚man'dib·yə·lər }

hyomandibular cleft [EMBRYO] The space between the hyoid arch and the mandibular arch in the vertebrate embryo. { ‚hī·ō‚man'dib·yə·lər 'kleft }

hyomandibular pouch [EMBRYO] A portion of the endodermal lining of the pharyngeal cavity which separates the paired hyoid and mandibular arches in vertebrate embryos. { ‚hī·ō‚man'dib·yə·lər 'paùch }

Hyopssodontidae [PALEON] A family of extinct mammalian herbivores in the order Condylarthra. { ‚hī·äp·sə'dänt·ə‚dē }

hyoscyamine [ORG CHEM] $C_{17}H_{23}O_3N$ A white, crystalline alkaloid isolated from henbane, belladonna, and other plants of the family Solanaceae, which is freely soluble in alcohol and dilute acids; used in medicine as an anticholinergic. { ‚hī·ə'sī·ə‚mēn }

hyostylic [VERT ZOO] Having the jaws and cranium connected by the hyomandibular, as certain fishes. { ‚hī·ō'stī·lik }

hypabyssal rock [PETR] Those igneous rocks that rose from great depths as magmas but solidified as minor intrusions before reaching the surface. { ‚hip·ə'bis·əl 'räk }

hypacusia [MED] Impairment of hearing. { ‚hip·ə'kyü·zhə }

hypalgesia [PHYSIO] Diminished sensitivity to pain. { ‚hip·al'jē·zhə }

hypandrium [INV ZOO] A plate covering the genitalia on the ninth abdominal segment of certain male insects. { hi'pan·drē·əm }

hypanthium [BOT] Expanded receptacle margin to which the sepals, petals, and stamens are attached in some flowers. { hi'pan·thē·əm }

hypantrum [VERT ZOO] In reptiles, a notch on the anterior portion of the neural arch that articulates with the hyposphene. { hi'pan·trəm }

hypautomorphic *See* hypidiomorphic. { hi‚pód·ə'mòr·fik }

hypaxial musculature [ANAT] The ventral portion of the axial musculature of vertebrates including subvertebral flank and ventral abdominal muscle groups. { hi'pak·sē·əl 'məs·kyə·lə‚chər }

hypengyophobia [PSYCH] An abnormal fear of responsibility. { hī'pen·jē·ə'fō·bē·ə }

hyperacid [PHYSIO] Containing more than the normal concentration of acid in the gastric juice. { ‚hī·pər'as·əd }

hyperacoustic zone [GEOPHYS] The region in the upper atmosphere, between 62 and 100 miles (100 and 160 kilometers), where the distance between the rarefied air molecules roughly equals the wavelength of sound, so that sound is transmitted with less volume than at lower levels. { ‚hī·pər·ə'kü·stik ‚zōn }

hyperactivity [PHYSIO] Excessive or pathologic activity. { ‚hī·pər·ak'tiv·əd·ē }

hyperacusia [MED] A hearing impairment characterized by an acute sense of hearing. { ‚hī·pər·ə'kyü·zhə }

HYLIDAE

Green tree frog, *Hyla cinera* (left) and gray tree frog, *H. versicolor* (right). (*American Museum of Natural History photograph*)

HYMENOSTOMATIDA

(a)

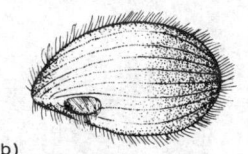

(b)

Examples of Hymenostomatida. (*a*) *Paramecium*. (*b*) *Tetrahymena*.

hyperadrenalism [MED] Hypersecretion of adrenal hormones marked by increased basal metabolism, decreased sugar tolerance, and glycosuria. Also known as hypercorticism. { ¦hī·pər·ə'dren·əl͵iz·əm }

hyperadrenocorticism [MED] Hypersecretion of adrenocortical hormones resulting in Cushing's syndrome, or virilism. { 'hī·pər·ə͵drē·nō'kórd·ə͵siz·əm }

hyperaldosteronism [MED] Hypersecretion of aldosterone by the adrenal cortex. { ¦hī·pər͵al·dō'ster·ə͵niz·əm }

hyperalgesia [PHYSIO] Increased or heightened sensitivity to pain stimulation. { ¦hī·pər·al'jē·zhə }

hyperbaric [MED] Pertaining to an anesthetic solution with a specific gravity greater than that of the cerebrospinal fluid. { ¦hī·pər¦bar·ik }

hyperbaric chamber [ENG] A specially equipped pressure vessel used in medicine and physiological research to administer oxygen at elevated pressures. { ¦hī·pər¦bar·ik 'chām·bər }

hyperbaric medicine [PHARM] Any agent having a specific gravity greater than that of spinal fluid; used for spinal anesthesia. { ¦hī·pər¦bar·ik 'med·ə·sən }

hyperbaric oxygenation [MED] The administration of oxygen, under greater than atmospheric pressure, by placing the patient in a room or chamber especially designed for the purpose. { ¦hī·pər¦bar·ik ͵äk·sə·jə'nā·shən }

hyperbilirubinemia [MED] **1.** Excessive amounts of bilirubin in the blood. **2.** A severe, prolonged physiologic jaundice. { ¦hī·pər͵bil·ə͵rü·bə'nē·mē·ə }

hyperbola [MATH] The plane curve obtained by intersecting a circular cone of two nappes with a plane parallel to the axis of the cone. { hī·pər·bə·lə }

hyperbolic amplitude [COMMUN] Excursion of a signal measured along hyperbolic rather than Cartesian coordinates. { ¦hī·pər¦bäl·ik 'am·plə͵tüd }

hyperbolic antenna [ELECTROMAG] A radiator whose reflector in cross section describes a half hyperbola. { ¦hī·pər¦bäl·ik an'ten·ə }

hyperbolic cosecant [MATH] A function whose value is equal to the reciprocal of the value of the hyperbolic sine. Abbreviated csch. { ¦hī·pər¦bäl·ik kō'sē͵kant }

hyperbolic cosine [MATH] A function whose value for the complex number z is one-half the sum of the exponential of z and the exponential of −z. Abbreviated cosh. { ¦hī·pər¦bäl·ik 'kō͵sīn }

hyperbolic cotangent [MATH] A function whose value is equal to the value of the hyperbolic cosine divided by the value of the hyperbolic sine. Abbreviated coth. { ¦hī·pər¦bäl·ik kō'tan·jənt }

hyperbolic cylinder [MATH] A cylinder whose directrix is a hyperbola. { ¦hī·pər¦bäl·ik 'sil·ən·dər }

hyperbolic decline [PETRO ENG] One of three types of decline in oil or gas production rate (the others are constant-percentage and harmonic decline). { ¦hī·pər¦bäl·ik di'klīn }

hyperbolic differential equation [MATH] A general type of second-order partial differential equation which includes the wave equation and has the form

$$\sum_{i,j=1}^{n} A_{ij}(\partial^2 u/\partial x_i \partial x_j) + \sum_{i=1}^{n} B_i(\partial u/\partial x_i) + Cu + F = 0$$

where the A_{ij}, B_i, C, and F are suitably differentiable real functions of $x_1, x_2, \ldots, x_n$, and there exists at each point $(x_1, x_2, \ldots, x_n)$ a real linear transformation on the x_i which reduces the quadratic form

$$\sum_{i,j=1}^{n} A_{ij} x_i x_j$$

to a sum of n squares not all of the same sign. { ¦hī·pər¦bäl·ik ͵dif·ə'ren·chəl i'kwā·zhən }

hyperbolic distance [ELECTROMAG] A function of pairs of points within a unit circle, where the interior of this circle is a conformal or projective representation of a hyperbolic space used in transmission line theory and waveguide analysis. { ¦hī·pər¦bäl·ik 'dis·təns }

hyperbolic Dovap [NAV] System using four or more Dovap stations with a common reference signal that is not coherent with the interrogation signal. { ¦hī·pər¦bäl·ik 'dō͵vap }

hyperbolic fix [NAV] A fix established by means of hyperbolic lines of position. { ¦hī·pər¦bäl·ik 'fiks }

hyperbolic flareout [AERO ENG] A flareout obtained by changing the glide slope from a straight line to a hyperbolic curve at an appropriate distance from touchdown at an airport. { ¦hī·pər¦bäl·ik 'fler͵aút }

hyperbolic form [MATH] A nondegenerate, symmetric or alternating form on a vector space E such that E is a hyperbolic space under this form. { ¦hī·pər¦bäl·ik 'fórm }

hyperbolic functions [MATH] The real or complex functions sinh (x), cosh (x), tanh (x), coth (x), sech (x), csch (x); they are related to the hyperbola in somewhat the same fashion as the trigonometric functions are related to the circle, and have properties analogous to those of the trigonometric functions. { ¦hī·pər¦bäl·ik 'fəŋk·shənz }

hyperbolic geometry See Lobachevski geometry. { ¦hī·pər¦bäl·ik jē'äm·ə·trē }

hyperbolic guidance [ORD] Missile guidance in which the difference in the arrival times of radio signals transmitted simultaneously from two ground stations is used to control the position of the missile. { ¦hī·pər¦bäl·ik 'gīd·əns }

hyperbolic horn [ENG] Horn whose equivalent cross-sectional radius increases according to a hyperbolic law. { ¦hī·pər¦bäl·ik 'hórn }

hyperbolic line of position [NAV] A line of position in the shape of a hyperbola, determined by measuring the difference in the phase or time of transit of radiations from fixed points; measurement may be made through the use of radio, sound, or light. { ¦hī·pər¦bäl·ik 'līn əv pə'zish·ən }

hyperbolic logarithm See logarithm. { ¦hī·pər¦bäl·ik 'läg·ə͵rith·əm }

hyperbolic navigation [NAV] Navigation by maintaining constant the indication of two parameters; the parameters can have any reasonable ratio to each other. { ¦hī·pər¦bäl·ik ͵nav·ə'gā·shən }

hyperbolic orbit [ASTRON] The path of a body moving along a hyperbola, such as a body that is subject to the gravitational attraction of another body from which it has sufficient energy to escape, and that is otherwise undisturbed. { ¦hī·pər͵bäl·ik 'ór·bət }

hyperbolic paraboloid [MATH] A surface which can be so situated that sections parallel to one coordinate plane are parabolas while those parallel to the other plane are hyperbolas. { ¦hī·pər¦bäl·ik pə'rab·ə͵lóid }

hyperbolic plane [MATH] A two-dimensional vector space E on which there is a nondegenerate, symmetric or alternating form $f(x,y)$ such that there exists a nonzero element w in E for which $f(w,w) = 0$. { ¦hī·pər¦bäl·ik 'plān }

hyperbolic point [FL MECH] A singular point in a streamline field which constitutes the intersection of a convergence line and a divergence line; it is analogous to a col in the field of a single-valued scalar quantity. Also known as neutral point. [MATH] A point on a surface where the Gaussian curvature is strictly negative. { ¦hī·pər¦bäl·ik 'póint }

hyperbolic Riemann surface See hyperbolic type. { ¦hī·pər¦bäl·ik 'rē͵män ͵sər·fəs }

hyperbolic secant [MATH] A function whose value is equal to the reciprocal of the value of the hyperbolic cosine. Abbreviated sech. { ¦hī·pər¦bäl·ik 'sē͵kant }

hyperbolic sine [MATH] A function whose value for the complex number z is one-half the difference between the exponential of z and the exponential of −z. Abbreviated sinh. { ¦hī·pər¦bäl·ik 'sīn }

hyperbolic space [MATH] A space described by hyperbolic rather than cartesian coordinates. { ¦hī·pər¦bäl·ik 'spās }

hyperbolic spiral [MATH] A plane curve for which the radius vector is inversely proportional to the polar angle. Also known as reciprocal spiral. { ¦hī·pər¦bäl·ik 'spi·rəl }

hyperbolic sweep generator [ELECTR] A sweep generator that generates a waveform resembling a hyperbola. { ¦hī·pər¦bäl·ik 'swēp ͵jen·ə͵rād·ər }

hyperbolic tangent [MATH] A function whose value is equal to the value of the hyperbolic sine divided by the value of the hyperbolic cosine. Abbreviated tanh. { ¦hī·pər¦bäl·ik 'tan·jənt }

hyperbolic trajectory [AERO ENG] A trajectory entered by a spacecraft when its velocity exceeds the escape velocity of a planet, satellite, or star. { ¦hī·pər¦bäl·ik trə'jek·trē }

hyperbolic type [MATH] A type of simply connected Riemann surface that can be mapped conformally on the interior of the unit circle. Also known as hyperbolic Riemann surface. { ¦hī·pər¦bäl·ik 'tīp }

hyperbolic waveform [ELECTR] A waveform which is an approximate hyperbola. { ¦hī·pər¦bäl·ik 'wāv‚fórm }

hyperboloid [MATH] A quadric surface given by an equation of the form $(x^2/a^2) \pm (y^2/b^2) - (z^2/c^2) = 1$; in certain cases it is a hyperboloid of revolution, which can be realized by rotating the pieces of a hyperbola about an appropriate axis. { hī'pər·bə‚lóid }

hyperboloid of one sheet [MATH] A surface whose equation in standard form is $(x^2/a^2) + (y^2/b^2) - (z^2/c^2) = 1$, so that it is in one piece, and cuts planes perpendicular to the x or y axes in hyperbolas and planes perpendicular to the z axis in ellipses. { hī'pər·bə‚lóid əv 'wən ‚shēt }

hyperboloid of revolution [MATH] A surface generated by rotating a hyperbola about one of its axes. { hī'pər·bə‚lóid əv ‚rev·ə'lü·shən }

hyperboloid of two sheets [MATH] A surface whose equation in standard form is $(x^2/a^2) - (y^2/b^2) - (z^2/c^2) = 1$, so that it is in two pieces, and cuts planes perpendicular to the y and z axis in hyperbolas and planes perpendicular to the x axis in ellipses, except for the interval $-a < x < a$, where there is no intersection. { hī'pər·bə‚lóid əv 'tü ‚shēts }

hyperbrachycephalic [ANTHRO] Having a round or broad head with a cephalic index of more than 85. { ¦hī·pər‚brak·ə·sə'fal·ik }

hyperbrachycranial [ANTHRO] Having a round or broad head with a cranial index of 85 to 90. { hī'pər‚brak·ə'krā·nē·əl }

hyperbrachyskelic [ANTHRO] Having the legs 75% of the length of the trunk with a skelic index below 75. { hī·pər‚brak·ə'skel·ik }

hypercalcemia [MED] Excessive amounts of calcium in the blood. Also known as calcemia. { ‚hī·pər‚kal'sē·mē·ə }

hypercapnia [MED] Excessive amount of carbon dioxide in the blood. { ‚hī·pər'kap·nē·ə }

hyperchamaerrhine [ANTHRO] Having a short broad nose with a nasal index of 58 or over. { ¦hī·pər'kam·ə‚rīn }

hypercharge [PART PHYS] A quantum number conserved by strong interactions, equal to twice the average of the charges of the members of an isospin multiplet. { 'hī·pər‚chärj }

hyperchlorhydria [MED] Excessive secretion of hydrochloric acid in the stomach. { ¦hī·pər‚klór'hī·drē·ə }

hypercholesteremia [MED] Elevated cholesterol levels in the blood. { ¦hī·pər‚kə‚les·tə'rē·mē·ə }

hyperchromatic [BIOL] Staining more intensely than normal. { ¦hī·pər·krō'mad·ik }

hyperchromatism [PATH] **1.** Excessive pigment formation in the skin. **2.** A condition in which cells or parts of cells stain more intensely than is normal. { ‚hī·pər'krō·mə‚tiz·əm }

hyperchromic [PATH] Pertaining to increased hemoglobin content in erythrocytes due to increased cell thickness, not increased hemoglobin concentration. { ¦hī·pər'krō·mik }

hyperchromic anemia [MED] Any of several blood disorders in which erythrocytes show an increase in hemoglobin and a reduction in number. { ¦hī·pər¦krō·mik ə'nē·mē·ə }

hyperchromicity [PHYS CHEM] An increase in the absorption of ultraviolet light by polynucleotide solutions due to a loss of the ordered secondary structure. { ‚hī·pər·krō'mis·əd·ē }

hypercircle method [MATH] A geometric method of obtaining approximate solutions of linear boundary value problems of mathematical physics that cannot be solved exactly, in which a correspondence is made between physical variables and vectors in a function space. { ¦hī·pər¦sər·kəl ‚meth·əd }

hypercoagulability [MED] Coagulation of blood more rapidly than normal. { ‚hī·pər·kō‚ag·yə·lə'bil·əd·ē }

hypercomplex number [MATH] **1.** An element of a division algebra. **2.** See quaternion. { ¦hī·pər¦käm‚pleks 'nəm·bər }

hypercomplex system See algebra. { ¦hī·pər¦käm‚pleks 'sis·təm }

hyperconjugation [PHYS CHEM] An arrangement of bonds in a molecule that is similar to conjugation in its formulation and manifestations, but the effects are weaker; it occurs when a CH_2 or CH_3 group (or in general, an AR_2 or AR_3 group where A may be any polyvalent atom and R any atom or radical) is adjacent to a multiple bond or to a group containing an atom with a lone π-electron, π-electron pair or quartet, or π-electron

vacancy; it can be sacrificial (relatively weak) or isovalent (stronger). { ‚hī·pər‚kän·jə'gā·shən }

hypercoracoid [VERT ZOO] The upper of two bones at the base of the pectoral fin in teleosts. { ¦hī·pər'kór·ə‚kóid }

hypercorticism See hyperadrenalism. { ‚hī·pər'kórd·ə‚siz·əm }

hypercube [COMPUT SCI] A configuration of parallel processors in which the locations of the processors correspond to the vertices of a mathematical hypercube and the links between them correspond to its edges. [MATH] The analog of a cube in n dimensions ($n = 2, 3, \ldots.$), with 2^n vertices, $n2^{n-1}$ edges, and $2n$ cells; for an object with edges of length $2a$, the coordinates of the vertices are $(\pm a, \pm a, \ldots, \pm a)$. { 'hī·pər ‚kyüb }

hyperdisk [COMPUT SCI] A mass-storage technique which uses a large-capacity storage and a disk for overflow. { 'hī·pər‚disk }

hyperdolichocephalic [ANTHRO] Having a long narrow head with a cephalic index of less than 70. { ¦hī·pər‚däl·ə·kō·sə'fal·ik }

hyperdolichocranial [ANTHRO] Having a long narrow skull with a cranial index of 65 to 70. { ¦hī·pər‚däl·ə·kō'krā·nē·əl }

hyperdynamic ileus See spastic ileus. { ¦hī·pər·dī'nam·ik 'il·ē·əs }

hyperemesis [MED] Excessive vomiting. { hī·pər'em·ə·səs }

hyperemesis gravedorium [MED] Pernicious vomiting in pregnancy. { hī·pər'em·ə·səs ‚grav·ə'dór·ē·əm }

hyperemia [MED] An excess of blood within an organ or tissue caused by blood vessel dilation or impaired drainage, especially of the skin. { ‚hī·pə're·mē·ə }

hyperergia [IMMUNOL] An altered state of reactivity to antigenic materials, in which the response is more marked than usual; one form of allery or pathergy. { ‚hī·pər'ər·jē·ə }

hyperesthesia [PHYSIO] Increased sensitivity or sensation. { ‚hī·pər·əs'thē·zhə }

hypereuryene [ANTHRO] Having a high wide forehead with an upper facial index less than 45. { ¦hī·pər'yùr·ə‚ēn }

hypereutectic alloy [MET] Any binary alloy whose composition lies to the right of the eutectic on an equilibrium diagram and which contains some eutectic structure. { ¦hī·pər·yü'tek·tik 'al‚ói }

hypereutectoid steel [MET] Steel containing more than 0.8% carbon. { ¦hī·pər·yü'tek‚tóid 'stēl }

hyperfine enhanced nuclear cooling [CRYO] A version of adiabatic demagnetization in which a sample containing magnetic ions, embedded in a suitable crystal which quenches the hyperfine fields, is cooled in a moderate external field which reinduces the hyperfine fields, and is then thermally isolated and removed from the external field. { 'hī·pər‚fīn in'hanst ¦nü·klē·ər 'kül·iŋ }

hyperfine structure [SPECT] A splitting of spectral lines due to the spin of the atomic nucleus or to the occurrence of a mixture of isotopes in the element. Abbreviated hfs. { 'hī·pər‚fīn 'strək·chər }

hyperfocal distance [OPTICS] The distance from the camera lens to the nearest object in acceptable focus when the lens is focused on infinity. { ¦hī·pər¦fō·kəl 'dis·təns }

hyperforming [CHEM ENG] A catalytic, petroleum-refinery hydrogenation process to improve naphtha octane number by removal of sulfur and nitrogen compounds; the catalyst is cobalt molybdate on a silica-alumina base. { 'hī·pər‚fór·miŋ }

hyperfrequency waves [ELECTROMAG] Microwaves having wavelengths in the range from 1 centimeter to 1 meter. { ‚hī·pər‚frē·kwən·sē 'wāvz }

hypergammaglobulinemia [MED] Increased blood levels of gamma globulin, usually associated with hepatic disease. { ¦hī·pər‚gam·ə¦gläb·yə·lə'nē·mē·ə }

hypergene See supergene. { 'hī·pər‚jēn }

hypergeometric differential equation See Gauss' hypergeometric equation. { ‚hī·pər‚jē·ə'me·trik ‚dif·ə‚ren·chəl i'kwā·zhən }

hypergeometric distribution [STAT] The distribution of the number D of special items in a random sample of size s drawn from a population of size N that contains r of the special items:

$$P(D = d) = \binom{r}{d}\binom{N-r}{s-d} \Big/ \binom{N}{s}$$

{ ‚hī·pər‚jē·ə'me·trik ‚dis·trə'byü·shən }

hypergeometric function [MATH] A function which is a solution to the hypergeometric equation and obtained as an infinite series expansion. { ˌhī·pərˌjē·ə'me·trik 'fəŋk·shən }

hypergeometric series [MATH] A particular infinite series which in certain cases is a solution to the hypergeometric equation, and having the form:

$$1 + \frac{ab}{c}z + \frac{1}{2!}\frac{a(a+1)b(b+1)}{c(c+1)}z^2 + \cdots$$

{ ˌhī·pərˌjē·ə'me·trik 'sir·ēz }

hypergiant star [ASTRON] A member of the brightest known class of stars, with absolute visual magnitude around −10, about 10^6 times as bright as the sun. { 'hī·pərˌjī·ənt 'stär }

hyperglobulinemia [MED] Increased blood levels of globulin. { ˌhī·pərˌgläb·yə·lə'nē·mē·ə }

hyperglycemia [MED] Excessive amounts of sugar in the blood. { ˌhī·pərˌglī'sē·mē·ə }

hyperglycemic factor See glucagon. { ˌhī·pərˌglī'sē·mik 'fak·tər }

hyperglycemic glycogenolytic factor See glucagon. { ˌhī·pərˌglī'sē·mik ˌglī·kō·jen·ə'lid·ik 'fak·tər }

hyperglycinenemia [MED] A hereditary metabolic disorder of males in which blood levels of glycine are excessive, resulting in vomiting, dehydration, osteoporosis, and mental retardation. { ˌhī·pərˌglīs·ən'ē·mē·ə }

hypergolic [CHEM] Capable of igniting spontaneously upon contact. { ˌhī·pər'gäl·ik }

hypergolic fuel [MATER] A combination of fuel and oxidizer that ignite spontaneously on contact, such as methanol and hydrogen peroxide; used as rocket propellant. { ˌhī·pərˌgäl·ik 'fyül }

hyperhidrosis [MED] Excessive sweating, which may be localized or generalized, chronic or acute, and often accumulating in visible drops on the skin. Also known as ephidrosis; polyhidrosis; sudatoria. { ˌhī·pərˌhī'drō·səs }

Hyperiidea [INV ZOO] A suborder of amphipod crustaceans distinguished by large eyes which cover nearly the entire head. { ˌhī·pə·rī'id·ē·ə }

hyperimmune antibody [IMMUNOL] An antibody having the characteristics of a blocking antibody. { ˌhī·pər·ə'myün 'antˌiˌbäd·ē }

hyperimmune serum [IMMUNOL] An antiserum that provides a very high degree of immunity due to a high antibody titer. { ˌhī·pər·ə'myün 'sir·əm }

hyperinsulinism [MED] Condition caused by abnormally high levels of insulin in the blood. { ˌhī·pər'in·sə·lə,niz·əm }

Hyperion [ASTRON] A satellite of Saturn approximately 300 miles (480 kilometers) in diameter. { hī'pir·ē·ən }

hyperkalemia See hyperpotassemia. { ˌhī·pər·kə'lē·mē·ə }

hyperkeratosis [MED] **1.** Hypertrophy of the cornea. **2.** Hypertrophy of the horny layer of the skin. { ˌhī·pər·ker·ə'tō·səs }

hyperkinesia [MED] Excessive and usually uncontrollable muscle movement. { ˌhī·pər·kə'nē·zhə }

hyperleptene [ANTHRO] Having a high narrow forehead with an upper facial index of 60 or more. { ˌhī·pər'lep,tēn }

hyperleptoprosopic [ANTHRO] Having a long narrow face with a facial index of 93 or more on the living and of 95 or more on the skull. { ˌhī·pərˌlep·tō·prə'säp·ik }

hyperleptinemia [MED] Increased serum leptin level. { ˌhī·pərˌlep·tə'nē·mē·ə }

hyperleptorrhine [ANTHRO] Having a long narrow nose with a nasal index of 40 to 55. { ˌhī·pər'lep·tə,rīn }

hyperlink [COMPUT SCI] A highlighted word, phrase, or image in the display of a computer document which, when chosen, connects the user to another part of the same document or to a different document (text, image, audio, video, or animation). In electronic documents, these cross references can be followed by a mouse click, and the target of the hyperlink may be on a physically distant computer connected by a network or the Internet. { 'hī·pərˌliŋk }

hyperlipemia [MED] Excessive amounts of fat in the blood. { ˌhī·pər·lə'pē·mē·ə }

hypermakroskelic [ANTHRO] Having long legs in proportion to the length of the trunk with a skelic index of 100 or more. { ˌhī·pərˌmak·rō'skel·ik }

Hypermastigida [INV ZOO] An order of the multiflagellate protozoans in the class Zoomastigophorea; all inhabit the alimentary canal of termites, cockroaches, and woodroaches. { ˌhī·pərˌma'stij·ə·də }

hypermedia [COMPUT SCI] Hypertext-based systems that combine data, text, graphics, video, and sound. { 'hī·pərˌmē·dē·ə }

hypermenorrhea See menorrhagia. { ˌhī·pərˌmen·ə'rē·ə }

hypermetabolism [MED] Any state in which there is an increase in basal metabolic rate. { ˌhī·pər·mə'tab·ə,liz·əm }

hypermetamorphism [INV ZOO] Type of embryological development in certain insects in which one or more stages have been interpolated between the full-grown larva and the adult. { ˌhī·pərˌmed·ə'mór,fiz·əm }

hypermetropia [MED] A defect of vision resulting from too short an eyeball so that unaccommodated rays focus behind the retina. Also known as farsightedness; hyperopia. { ˌhī·pər·mə'trō·pē·ə }

hypermotility [MED] Increased motility, as of the stomach or intestine. { ˌhī·pərˌmō'til·əd·ē }

hypernatremia [MED] Excessive amounts of sodium in the blood. { ˌhī·pər·nə'trē·mē·ə }

hypernephroma See renal-cell carcinoma. { ˌhī·pər·nə'frō·mə }

hypernucleus [NUC PHYS] A nucleus that consists of protons, neutrons, and one or more strange particles, such as lambda particles. { ˌhī·pər'nü·klē·əs }

Hyperoartii [VERT ZOO] A superorder in the subclass Monorhina distinguished by the single median dorsal nasal opening leading into a blind hypophyseal sac. { ˌhī·pə·rō'är·shē,ī }

hyperoid axle [MECH ENG] A type of rear-axle drive gear set which generally carries the pinion 1.5–2 inches (38–51 millimeters) or more below the centerline of the gear. { 'hī·pəˌroid 'ak·səl }

hyperon [PARTIC PHYS] **1.** An elementary particle which has baryon number $B = +1$, that is, which can be transformed into a nucleon and some number of mesons or lighter particles, and which has nonzero strangeness number. **2.** A hyperon (as in the first definition) which is semistable (the lifetime is much longer than 10^{-22} second). { 'hī·pəˌrän }

hyperopia See hypermetropia. { ˌhī·pə'rō·pē·ə }

hyperorthognathous [ANTHRO] Having a face that is flat in profile with a facial angle of 93° or more. { ˌhī·pər·ór'thäg·nə·thəs }

hyperosmia [MED] An abnormally acute sense of smell. { ˌhī·pə'räz·mē·ə }

hyperostosis [MED] Hypertrophy of bony tissue. { ˌhī·pə·rä'stō·səs }

Hyperotreti [VERT ZOO] A suborder in the subclass Monorhina distinguished by the nasal opening which is located at the tip of the snout and communicates with the pharynx by a long duct. { ˌhī·pə·rō'trēd,ī }

hyperoxemia [MED] Extreme acidity of the blood. { ˌhī·pər,äk'sē·mē·ə }

hyperparasite [ECOL] An organism that is parasitic on other parasites. { ˌhī·pər'par·ə,sīt }

hyperparathyroidism [MED] Condition caused by increased functioning of the parathyroid glands. { ˌhī·pərˌpar·ə'thī,róid,iz·əm }

hyperpathia [MED] An exaggerated or excessive perception of or response to any stimulus as being disagreeable or painful. { ˌhī·pər'path·ē·ə }

hyperperistalsis [MED] An increase in the rate and depth of the peristaltic waves. { ˌhī·pərˌper·ə'stäl·səs }

hyperphagia See bulimia. { ˌhī·pər'fā·jə }

hyperphosphaturia [MED] An excess of phosphates in the urine. Also known as phosphaturia. { ˌhī·pərˌfäs·fə'túr·ē·ə }

hyperpigmentation [MED] Increased pigmentation. { ˌhī·pərˌpig·mən'tā·shən }

hyperpituitarism [MED] Any abnormal condition resulting from overactivity of the anterior pituitary. { ˌhī·pər·pi'tü·ə·tə,riz·əm }

hyperplane [MATH] A hyperplane is an $(n − 1)$-dimensional subspace of an n-dimensional vector space. { ˌhī·pərˌplān }

hyperplane of support [MATH] Relative to a convex body in a normed vector space, a hyperplane whose distance from the body is zero, and which separates the normed vector space into two halves, one of which contains no points of the convex body. { 'hī·pər‚plān əv sə'pòrt }

hyperplasia [MED] Increase in cell number causing an increase in the size of a tissue or organ. { ‚hī·pər'plā·zhə }

hyperplatycnemic [ANTHRO] Having a shinbone that is laterally flattened with a platycnemic index below 55. { ‚hī·pər‚plad·ik‚nē·mik }

hyperploid [GEN] Having one or more chromosomes or parts of chromosomes in excess of the haploid number, or of a whole multiple of the haploid number. { 'hī·pər‚plȯid }

hyperpnea [MED] Increase in depth and rate of respiration. { ‚hī·pər'nē·ə }

hyperpotassemia [MED] Excessive amounts of potassium in the blood. Also known as hyperkalemia. { ‚hī·pə‚päd·ə'sē·mē·ə }

hyperprognathous [ANTHRO] Having prominent jaws with a facial profile angle less than 70°. { ‚hī·pər‚präg·nə·thəs }

hyperproteinemia [MED] Excessive protein levels in the blood. { ‚hī·pər‚prōd·ə'nē·mē·ə }

hyperpure germanium detector [ELECTR] A variant of the lithium-drifted germanium crystal which uses high-purity germanium, making it possible to store the detector at room temperature rather than liquid nitrogen temperature. { ‚hī·pər‚pyùr jər‚mā·nē·əm di'tek·tər }

hyperpycnal inflow [HYD] A denser inflow that occurs when a sediment-laden fluid flows down the side of a basin and along the bottom as a turbidity current. { ‚hī·pər‚pik·nəl 'in‚flō }

hyperpyrexia [MED] Extremely high fever. { ‚hī·pər‚pī'rek·sē·ə }

hyper-Raman effect [OPTICS] The phenomenon whereby, when light is scattered from an intense laser beam with frequency ν_0, the scattered light has components not only with frequency $2\nu_0$ but also with frequencies $2\nu_0 \pm \nu_m$, where ν_m is the frequency of a transition in the scattering molecules. { ‚hī·pər‚rä‚män i‚fekt }

hyperreal numbers *See* nonstandard numbers. { ‚hī·pər‚rēl 'nəm·bərz }

hyperreflexia [MED] A condition of abnormally increased reflex action. { ‚hī·pər·ri'flek·sē·ə }

hyperresonance [MED] Exaggeration of normal resonance on percussion of the chest; heard chiefly in pulmonary emphysema and pneumothorax. { ‚hī·pə'rez·ən·əns }

hypersaline [GEOL] Geologic material with high salinity. { ‚hī·pər'sā‚lēn }

hypersensitivity [IMMUNOL] The state of being abnormally sensitive, especially to allergens; responsible for allergic reactions. { ‚hī·pər‚sen·sə'tiv·əd·ē }

hypersensitization [GRAPHICS] Any of various techniques for increasing the sensitivity of a photographic plate; includes chemical treatments, removal of oxygen and water, baking, soaking in hydrogen gas, preflashing, and push development. [IMMUNOL] The process of producing hypersensitivity. { ‚hī·pər‚sen·sə·tə'zā·shən }

hypersensor [ELECTR] Single-component, resettable circuit breaker which operates as a majority-carrier tunneling device, and is used for overcurrent or overvoltage protection of integrated circuits. { ‚hī·pər‚sen·sər }

hypersomnia [MED] Excessive sleepiness. { ‚hī·pər'säm·nē·ə }

hypersonic [ACOUS] Pertaining to frequencies above 500 megahertz. [FL MECH] Pertaining to hypersonic speeds, or air currents moving at hypersonic speeds. { ‚hī·pər'sän·ik }

hypersonic flight [AERO ENG] Flight at speeds well above the local velocity of sound; by convention, hypersonic regime starts at about five times the speed of sound and extends upward indefinitely. { ‚hī·pər'sän·ik 'flīt }

hypersonic flow [FL MECH] Flow of a fluid over a body at hypersonic speeds, and in which shock waves start at a finite distance from the surface of the body. { ‚hī·pər'sän·ik 'flō }

hypersonic glider [AERO ENG] An unpowered vehicle, specifically a reentry vehicle, designed to fly at hypersonic speeds. { ‚hī·pər'sän·ik 'glīd·ər }

hypersonic inlet [FL MECH] An entrance or orifice for admission of fluids at hypersonic speeds. { ‚hī·pər'sän·ik 'in‚let }

hypersonic nozzle [FL MECH] A supersonic nozzle designed to accelerate a fluid to hypersonic speeds. { ‚hī·pər'sän·ik 'näz·əl }

hypersonics [ACOUS] Production and utilization of sound waves of frequencies above 500 megahertz. { ‚hī·pər'sän·iks }

hypersonic speed [FL MECH] A speed of an object greater than about five times the speed of sound in the fluid through which the object is moving. { ‚hī·pər'sän·ik 'spēd }

hypersonic wind tunnel [ENG] A wind tunnel in which air flows at speeds roughly in the range from 5 to 15 times the speed of sound. { ‚hī·pər'sän·ik 'win ‚tən·əl }

hypersorption [CHEM ENG] Process with recirculating bed of activated-carbon adsorbent for continuous recovery of ethylene from methane and other low-molecular-weight gases. { ‚hī·pər‚sȯrp·shən }

hyperspectral imaging system [ENG] An infrared imaging system that has more than 30 spectral channels with relatively fine spectral resolution, allowing imaging spectroscopy to be carried out. { ‚hī·pər‚spek·trəl 'im·ij·iŋ ‚sis·təm }

hypersplenism [MED] Condition caused by abnormal spleen activity. { ‚hī·pər‚sple‚niz·əm }

hyperstereoscopy [MAP] Stereoscopic viewing of a map in which the scale (usually vertical) along the line of sight is exaggerated in comparison with the scale perpendicular to the line of sight. Also known as appearance ratio. { ‚hī·pər‚ste·rē'äs·kə·pē }

hypersthene [MINERAL] $(Mg,Fe)SiO_3$ A grayish, greenish, black, or dark-brown rock-forming mineral of the orthopyroxene group, with bronzelike luster on the cleavage surface. { 'hī·pər‚sthēn }

hypersthenfels *See* norite. { ‚hī·pər'sthēn‚felz }

hypersurface [MATH] The analog of a surface in n-dimensional Euclidean space, where n is a positive integer; the set of points, $(x_1, x_2, \ldots, x_n)$, satisfying an equation of the form $f(x_1, \ldots, x_n) = 0$. { 'hī·pər‚sər·fəs }

hypertape control unit *See* tape control unit. { 'hī·pər‚tāp kən'trōl ‚yü·nət }

hypertape drive *See* cartridge tape drive. { 'hī·pər‚tāp ‚drīv }

hypertelorism [ANAT] An unusually large distance between paired body parts or organs. { ‚hī·pər'tel·ə‚riz·əm }

hypertely [EVOL] An extreme overdevelopment of an organ or body part during evolution that is disadvantageous to the organism. [ZOO] An extreme degree of imitative coloration, beyond the aspect of utility. { hī'pərd·əl·ē }

hypertensin *See* angiotensin. { ‚hī·pər'ten·sən }

hypertension [MED] Abnormal elevation of blood pressure, generally regarded to be levels of 165 systolic and 95 diastolic. { ‚hī·pər'ten·chən }

hypertext [COMPUT SCI] A data structure in which there are links between words, phrases, graphics, or other elements and associated information so that selection of a key object can activate a linkage and reveal the information. { 'hī·pər‚tekst }

Hypertext Markup Language [COMPUT SCI] The language used to specifically encode the content and format of a document and to link documents on the World Wide Web. Abbreviated HTML. { ‚hī·pər‚tekst 'märk‚əp ‚laŋ·gwij }

Hypertext Transfer Protocol [COMPUT SCI] The communication protocol for transmitting linked documents between computers; it is the basis for the World Wide Web and follows the TCP/IP protocol for the client-server model of computing. Abbreviated HTTP. { 'hī·pər‚tekst 'tranz·fər ‚prōd·ə‚kȯl }

hyperthecosis [PATH] Abnormal thickening of the inner layer of the Graafian follicle with increased leutein formation. { ‚hī·pər·thə'kō·səs }

hyperthermia [PHYSIO] A condition of elevated body temperature. { ‚hī·pər'thər·mē·ə }

hyperthermophile [MICROBIC] An extremophile that thrives in high-temperature (above 60°C or 140°F) environments. { ‚hī·pər'thər·mə‚fīl }

hyperthyroidism [MED] The constellation of signs and symptoms caused by excessive thyroid hormone in the blood, either from exaggerated functional activity of the thyroid gland or from excessive administration of thyroid hormone, and manifested by thyroid enlargement, emaciation, sweating, tachycardia, exophthalmos, and tremor. Also known as exophthalmic goiter; Grave's disease; thyrotoxicosis; toxic goiter. { ‚hī·pər'thī‚ròid‚iz·əm }

hyperthyrotropinism [MED] Excessive thyrotropic hormone secretion by the adenohypophysis. { ‚hī·pər‚thī·rō'träp·ə‚niz·əm }

hypertonia [MED] Abnormal increase in muscle tonicity. { ‚hī·pər'tō·nē·ə }

hypertonic [PHYSIO] **1.** Excessive or above normal in tone or tension, as a muscle. **2.** Having an osmotic pressure greater than that of physiologic salt solution or of any other solution taken as a standard. { ‚hī·pər'tän·ik }

hypertonic bladder [MED] Hypertonia of the urinary bladder. { ‚hī·pər'tän·ik 'blad·ər }

hypertonic contracture [MED] Prolonged muscular spasms in spastic paralysis. { ‚hī·pər'tän·ik kən'trak·chər }

Hypertragulidae [PALEON] A family of extinct chevrotainlike pecoran ruminants in the superfamily Traguloidea. { ‚hī·pər‚trə'gyül·ə‚dē }

hypertrophic arthritis See degenerative joint disease. { ‚hī·pər‚trä·fik är'thrīd·əs }

hypertrophic gastritis [MED] Chronic inflammation of the stomach with hypertrophy of the mucosa and rugae. { ‚hī·pər‚trä·fik ga'strīd·əs }

hypertrophy [PATH] Increase in cell size causing an increase in the size of an organ or tissue. { hī'pər·trə·fē }

hyperuricemia [MED] Abnormally high level of uric acid in the blood. Also known as lithemia. { ‚hī·pər‚yür·ə'sē·mē·ə }

hypervalent atom [CHEM] A central atom in a single-bonded structure that imparts more than eight valence electrons in forming covalent bonds. { ‚hī·pər'vā·lənt 'ad·əm }

hypervalent compounds [CHEM] Stable compounds of the main group elements in the third row of the periodic table, such as silicon, phosphorus, and sulfur, that have more than eight valence shell electrons; for example, PF_5 and SF_4. { ‚hī·pər‚vā·lənt 'käm‚paúnz }

hypervariable minisatellite [GEN] One of a number of tandem repetitive regions of deoxyribonucleic acid dispersed throughout the human and other genomes that display polymorphism associated with the allelic variation in the number of repetitive copies of each minisatellite. { ‚hī·pər'ver·ē·ə·bəl ‚min·ē'sad·əl‚īt }

hypervelocity [MECH] **1.** Muzzle velocity of an artillery projectile of 3500 feet per second (1067 meters per second) or more. **2.** Muzzle velocity of a small-arms projectile of 5000 feet per second (1524 meters per second) or more. **3.** Muzzle velocity of a tank-cannon projectile in excess of 3350 feet per second (1021 meters per second). { ‚hī·pər·və'läs·əd·ē }

hypervelocity armor-piercing [ORD] Designating a type of artillery projectile consisting of a core of extremely hard, high-density material, such as tungsten carbide, contained within a lightweight carrier called a sabot. Abbreviated HVAP. { ‚hī·pər·və'läs·əd·ē 'är·mər ‚pirs·iŋ }

hypervelocity wind tunnel [ENG] A wind tunnel in which higher airspeeds and temperatures can be attained than in a hypersonic wind tunnel. { ‚hī·pər·və'läs·əd·ē 'win ‚tən·əl }

hyperventilation [MED] Increase in air intake or of the rate or depth of respiration. { ‚hī·pər‚vent·əl'ā·shən }

hypervisor [COMPUT SCI] A control program enabling two operating systems to share a common computing system. { 'hī·pər‚vīz·ər }

hypervitaminosis [MED] Condition caused by intake of toxic amounts of a vitamin. { ‚hī·pər‚vīd·ə·mə'nō·səs }

hypervolume [MATH] **1.** The hypervolume of the direct product of open or closed intervals in each of the coordinates of a Euclidean n-space (where n is a positive integer) is the product of the lengths of the intervals. **2.** The Jordan content of any set in Euclidean n-space whose exterior Jordan content equals its interior Jordan content. { 'hī·pər‚väl·yəm }

hypesthesia [MED] Reduced or subnormal tactile sensibility. { ‚hī·pə'sthē·zhə }

hypha [MYCOL] One of the filaments composing the mycelium of a fungus. { 'hī·fə }

hyphenation zone [COMPUT SCI] In word processing, the area adjacent to the right margin consisting of those positions at which words may be hyphenated. { 'hī·fə‚nā·shən ‚zōn }

hyphidium [MYCOL] A sterile hymenial structure of hyphal origin. { hī'fid·ē·əm }

Hyphochytriales [MYCOL] An order of aquatic fungi in the class Phycomycetes having a saclike to limited hyphal thallus and zoospores with two flagella. { ‚hī·fō·ki‚trī'ā·lēz }

Hyphochytridiomycetes [MYCOL] A class of the true fungi; usually grouped with other classes under the general term Phycomycetes. { ‚hī·fō·ki‚trid·ē·ō‚mī'sēd‚ēz }

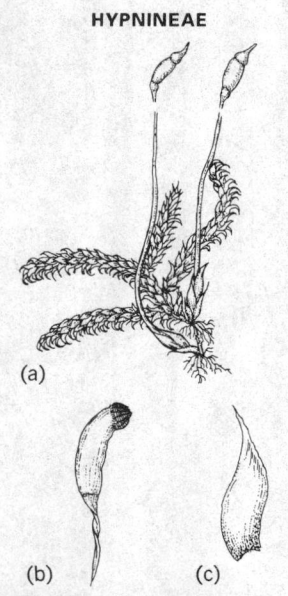

HYPNINEAE

Hypnum reptile. (*a*) Portion of plant, with stems shortened. (*b*) Urn and peristome. (*c*) Faintly bicostate leaf. (*From W. H. Welch, Mosses of Indiana, Indiana Department of Conservation, 1957*)

hyphoid [MYCOL] Hypha-like. { 'hī‚fóid }

Hyphomicrobiaceae [MICROBIO] Formerly a family of bacteria in the order Hyphomicrobiales; cells occurring in free-floating groups with individual cells attached to each other by a slender filament. { ‚hī·fō‚mī‚krō·bē'ās·ē‚ē }

Hyphomicrobiales [MICROBIO] Formerly an order of bacteria in the class Schizomycetes containing forms that multiply by budding. { ‚hī·fō‚mī‚krō·bē'ā·lēz }

Hyphomicrobium [MICROBIO] A genus of prosthecate bacteria; they reproduce by budding at hyphal tips; cells are rod-shaped with pointed ends, ovoid, or bean-shaped; hyphae are not septate. { ‚hī·fō‚mī'krō·bē·əm }

Hyphomonas [MICROBIO] A genus of prosthecate bacteria; cells are oval or pear-shaped, and reproduction is by budding of the hyphae or by direct budding of cells. { hi·fō'mō·nəs }

hyphopodium [MYCOL] Hypha with a haustorium in certain ectoparasitic fungi. { ‚hī·fō'pō·dē·əm }

hypidiomorphic [PETR] Of the texture of igneous rocks, having the crystals bounded partly by the crystal faces characteristic of the mineral species. Also known as hypautomorphic; subidiomorphic. { hī‚pid·ē·ō'mór·fik }

hypnagogic hallucination [PSYCH] Mental images occurring normally while falling asleep. { ‚hip·nə‚gäj·ik hə‚lüs·ən'ā·shən }

hypnic myoclonia [PSYCH] Startle-like muscle jerks often preceded by the sensation of having to fall. { ‚hip·nik ‚mī·ə'klō·nē·ə }

Hypnineae [BOT] A suborder of the Hypnobryales characterized by complanate, glossy plants with ecostate or costate leaves and paraphyllia rarely present. { hip'nī·nē‚ē }

hypnoanalysis [PSYCH] Technique used in psychotherapy combining hypnosis with psychoanalysis. { ‚hip·nō·ə'nal·ə·səs }

Hypnobryales [BOT] An order of mosses composed of procumbent and pleurocumbent plants with usually symmetrical leaves arranged in more than two rows. { ‚hip·nō‚brī'ā·lēz }

hypnophobia [PSYCH] An abnormal fear of sleeping. { ‚hīp·nə'fō·bē·ə }

hypnosis [PSYCH] An altered state of consciousness in which the individual is more susceptible to suggestion and in which regressive behavior may spontaneously occur. { hip'nō·səs }

hypnotherapy [MED] Treatment of disease by means of hypnotism. { ‚hip·nō'ther·ə·pē }

hypnotic [PHARM] A drug which induces sleep. Also known as somnificant; soporific. { hip'näd·ik }

hypnotism [PSYCH] The practice or study of inducing hypnosis. { 'hip·nə‚tiz·əm }

hypnotize [PSYCH] To induce a state of hypnosis. { 'hip·nə‚tīz }

hypnotoxin [BIOCHEM] A supposed hormone produced by brain tissue and inducing sleep. { ‚hip·nə'täk·sən }

hypo [GRAPHICS] In photography, the common fixing agent sodium thiosulfate, which formerly was incorrectly called hyposulfate of soda. [INV ZOO] See sodium thiosulfate. [PSYCH] Informal term for a hypochondriac or hypochondria. { 'hī·pō }

hypoadrenalism See hypoadrenia. { ‚hī·pō·ə'dren·əl‚iz·əm }

hypoadrenia [MED] Reduced functioning of the adrenal glands. Also known as hypoadrenalism. { ‚hī·pō·ə'drē·nē·ə }

hypoadrenocorticism [MED] Lowered or subnormal adrenal cortex activity. { ‚hī·pō·ə‚drē·nō'kórd·ə‚siz·əm }

hypoalbuminemia [MED] Abnormally low levels of albumin in the blood. { ‚hī·pō·al‚byü·mə'nē·mē·ə }

hypoallergenic [PHARM] Having a low tendency to induce allergic reactions; used particularly for formulated dermatologic preparations. { ‚hī·pō‚al·ər'jen·ik }

hypobaric [MED] Pertaining to an anesthetic solution of specific gravity lower than the cerebrospinal fluid. [PHYS] Having less weight or pressure. { ‚hī·pō'bar·ik }

hypobasal [BOT] Located posterior to the basal wall. { ‚hī·pō‚bā·səl }

hypoblast See endoderm. { 'hī·pō‚blast }

hypobranchial musculature [ANAT] The ventral musculature in vertebrates extending from the pectoral girdle forward to the hyoid arch, chin, and tongue. { ‚hī·pō'braŋ‚kē·əl 'məs·kyə·lə‚chər }

hypocalcemia [MED] Condition in which there are reduced levels of calcium in the blood. { ‚hī·pō‚kal'sē·mē·ə }

hypocalcification [MED] Reduction of normal amounts of mineral salts in calcified tissue. { ‚hī·pō‚kal·sə·fə'kā·shən }

hypocalciuria [MED] Decreased excretion of calcium in the urine. { ‚hī·pō‚kal·sē'yùr·ē·ə }

hypocapnia [MED] Reduced or subnormal blood levels of carbon dioxide. { ‚hī·pō'kap·nē·ə }

hypocenter [GEOPHYS] The point along a fault where an earthquake is initiated. { 'hī·pə‚sent·ər }

hypochil [BOT] Lower portion of the lip in orchids. Also known as hypochillium. { 'hī·pə‚kil }

Hypochilidae [INV ZOO] A family of true spiders in the order Araneida. { ‚hī·pə'kil·ə‚dē }

hypochillium See hypochil. { ‚hī·pə'kil·ē·əm }

Hypochilomorphae [INV ZOO] A monofamilial suborder of spiders in the order Araneida. { ‚hī·pə‚kil·ə'mòr·fē }

hypochloremia [MED] Reduction in the amount of blood chlorides. { ‚hī·pō·klō'rē·mē·ə }

hypochlorhydria [MED] Reduction in the hydrochloric acid content of gastric juice. { ‚hī·pō‚klòr'hī·drē·ə }

hypochlorite [INV ZOO] ClO⁻ A negative ion derived from hypochlorous acid, HClO; the ion is an oxidizing agent and a constituent of bleaching agents. { ‚hī·pə'klòr‚īt }

hypochlorite sweetening [CHEM ENG] A petroleum refinery process to oxidize gasoline mercaptans by agitation with an aqueous, alkaline hypochlorite solution. { ‚hī·pə'klòr‚īt 'swet·ən·iŋ }

hypochlorization [MED] Reduction of the dietary intake of sodium chloride. { ‚hī·pə‚klòr·ə'zā·shən }

hypochlorous acid [INV ZOO] HOCl Weak, unstable acid existing in solution only; its salts (such as calcium hypochlorite) are used as bleaching agents. { ‚hī·pə'klòr·əs 'as·əd }

hypochnoid [MYCOL] Having generally compacted hyphae. { hī'päk‚nòid }

hypocholesterolemia [MED] Subnormal levels of serum cholesterol. { ‚hī·pō·kə‚les·tə·rə'lē·mē·ə }

hypochondria See hypochondriasis. { ‚hī·pə'kän·drē·ə }

hypochondriac [PSYCH] A person affected with hypochondriasis. { ‚hī·pə‚kän·drē‚ak }

hypochondriac region [ANAT] The upper, lateral abdominal region just below the ribs on each side of the body. { ‚hī·pə‚kän·drē‚ak ‚rē·jən }

hypochondriasis [PSYCH] A chronic condition in which the patient is morbidly concerned with his own health and believes himself suffering from grave bodily diseases. Also known as hypochondria. { ‚hī·pə·kən'drī·ə·səs }

hypochromia [PATH] Lack of complete saturation of the erythrocyte stroma with hemoglobin, as judged by pallor of the unstained or stained erythrocytes when examined microscopically. { ‚hī·pə'krō·mē·ə }

hypochromicity [PHYS CHEM] A decrease in the absorption of ultraviolet light by polynucleotide solutions due to the formation of an ordered secondary structure. { ‚hī·pə·krə'mis·əd·ē }

hypochromic microcytic anemia [MED] An anemia associated with erythrocytes of reduced size and hemoglobin content. { ‚hī·pə'krō·mik ‚mī·krə‚sid·ik ə‚nē·mē·ə }

hypocleidium [ANAT] The median, ventral bone between clavicles. [VERT ZOO] Median process on the wishbone of birds. { ‚hī·pə'klīd·ē·əm }

hypocone [ANAT] The posterior inner cusp of an upper molar. [INV ZOO] Region of a dinoflagellate posterior to the girdle. { 'hī·pə‚kōn }

Hypocopridae [INV ZOO] A small family of coleopteran insects in the superfamily Cucujoidea. { ‚hī·pə'käp·rə‚dē }

hypocoracoid [VERT ZOO] The lower of two bones at the base of the pectoral fin in teleosts. { ‚hī·pō'kòr·ə‚kòid }

hypocotyl [BOT] The portion of the embryonic plant axis below the cotyledon. { 'hī·pə‚käd·əl }

hypocrateriform [BIOL] Saucer-shaped. { ‚hī·pə·krə'ter·ə‚fòrm }

Hypocreales [MYCOL] An order of fungi belonging to the Ascomycetes and including several entomophilic fungi. { ‚hī·pō‚krē'ā·lēz }

hypocrystalline [PETR] Pertaining to the texture of igneous rock characterized by crystalline components in an amorphous groundmass. Also known as hemicrystalline; hypohyaline;

merocrystalline; miocrystalline; semicrystalline. { ‚hī·pō'krist·əl·ən }

hypocycloid [MATH] The curve which is traced in the plane as a given point fixed on a circle moves while this circle rolls along the inside of another circle. { ‚hī·pō'sī‚klòid }

Hypodermatidae [INV ZOO] The warble flies, a family of myodarian cyclorrhaphous dipteran insects in the subsection Calypteratae. { ‚hī·pō·dər'mad·ə‚dē }

hypodermic needle [MED] A hollow needle, with a slanted open point, used for subcutaneous and intramuscular injections of fluid. { ‚hī·pə'dər·mik 'nēd·əl }

hypodermic syringe See syringe. { ‚hī·pə'dər·mik sə'rinj }

hypodermis [BOT] The outermost cell layer of the cortex of plants. Also known as exodermis. [INV ZOO] The layer of cells that underlies and secretes the cuticle in arthropods and some other invertebrates. { ‚hī·pə'dər·mis }

hypodermoclysis [MED] Subcutaneous injections of large quantities of fluid for therapeutic purposes. { ‚hī·pə·dər'mäk·lə·səs }

hypodontia [MED] The congenital absence of teeth. Also known as anodontia; oligodontia. { ‚hī·pə'dän·chə }

hypo eliminator [GRAPHICS] A solution used to facilitate the removal of fixer solutions during the wash bath. { 'hī·pō ə‚lim·ə‚nād·ər }

hypoergia [IMMUNOL] A state of less than normal reactivity to antigenic materials, in which the response is less marked than usual; one form of allergy or pathergy. { ‚hī·pō‚er·jē·ə }

hypoeutectic alloy [MET] Any binary alloy whose composition lies to the left of the eutectic. { ‚hī·pō·yü'tek·tik 'al‚ói }

hypoeutectoid steel [MET] Steel containing less than 0.8% carbon. { ‚hī·pō·yü'tek‚tòid 'stēl }

hypofibrinogenemia [MED] A decrease in plasma fibrogen level. { ‚hī·pō·fī‚brin·ə·jə'nē·mē·ə }

hypogammaglobulinemia [MED] Reduced blood levels of gamma globulin. { ‚hī·pō‚gam·ə‚gläb·yə·lə'nē·mē·ə }

hypogeal See hypogeous. { ‚hī·pə'jē·əl }

hypogene [GEOL] **1.** Of minerals or ores, formed by ascending waters. **2.** Of geologic processes, originating within or below the crust of the earth. { 'hī·pə‚jen }

hypogeous [BIOL] Living or maturing below the surface of the ground. Also known as hypogeal. { ‚hī·pə'jē·əs }

hypoglossal nerve [NEUROSCI] The twelfth cranial nerve; a paired motor nerve in tetrapod vertebrates innervating tongue muscles; corresponds to the hypobranchial nerve in fishes. { ‚hī·pə'gläs·əl 'nərv }

hypoglossal nucleus [NEUROSCI] A long nerve nucleus throughout most of the length of the medulla oblongata; cells give rise to the hypoglossal nerve fibers. { ‚hī·pə'gläs·əl 'nü·klē·əs }

hypoglycemia [MED] Condition caused by low levels of sugar in the blood. { ‚hī·pō·glī'sē·mē·ə }

hypogonadism [MED] Reduced hormonal secretion by the testes or ovaries. { ‚hī·pō‚gō·na‚diz·əm }

hypogynium [BOT] Structure that supports the ovary in plants such as sedges. { ‚hī·pə‚jin·ē·əm }

hypogynous [BOT] Having all flower parts attached to the receptacle below the pistil and free from it. { ‚hī'päj·ə·nəs }

hypohidrosis [MED] Deficient perspiration. { ‚hī·pō·hī'drō·səs }

hypohyaline See hypocrystalline. { ‚hī·pō'hī·ə·lən }

hypohalous acid [INORG CHEM] An oxyacid of a halogen (fluorine, chlorine, bromine, iodine, or astatine) possessing the general chemical formula HOX, where X is the halogen atom. { hī'päf·ə·ləs 'as·əd }

hypoid gear [MECH ENG] Gear wheels connecting nonparallel, nonintersecting shafts, usually at right angles. { 'hī‚pòid 'gir }

hypoid generator [MECH ENG] A gear-cutting machine for making hypoid gears. { 'hī‚pòic 'jen·ə‚rād·ər }

hypoiodous acid [INORG CHEM] HIO A very weak unstable acid that occurs as the result of the weak hydrolysis of iodine in water. { ‚hī·pō‚ī'ōd·əs 'as·əd }

hypokalemia [PHYSIO] A reduction in the normal amount of potassium in the blood. { ‚hī·pō·kə'lē·mē·ə }

hypokinesia [MED] Subnormal muscular movements. { ‚hī·pō·ki'nē·zhə }

hypokinetic syndrome [MED] General decrease in motor functions due to a form of minimal brain dysfunction. { ‚hī·pō·ki'ned·ik 'sin‚drōm }

HYPOGYNOUS

pistil — stamen
petal — sepal

receptacle

Flower arrangement and parts on receptacle in hypogynous flowers.

HYPOID GEAR

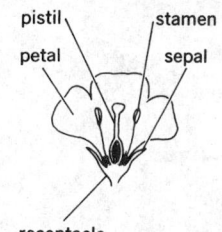

Picture of hypoid gear. (*Gleason Works*)

hypolimnion [HYD] The lower level of water in a stratified lake, characterized by a uniform temperature that is generally cooler than that of other strata in the lake. { ¦hī·pō'lim·nē,än }

hypomagma [GEOL] Relatively immobile, viscous lava that forms at depth beneath a shield volcano, is undersaturated with gases, and initiates volcanic activity. { ¦hī·pō¦mag·mə }

hypomagnesemia See grass tetany. { ,hī·pō,mag·nə'sē·mē·yə }

hypomenorrhea [MED] A deficient amount of menstrual flow at the regular period. { ¦hī·pō,men·ə'rē·ə }

hypomere [EMBRYO] The lateral or lower mesodermal plate zone in vertebrate embryos. [INV ZOO] The basal portion of certain sponges that contain no flagellated chambers. { 'hī·pə,mir }

hypometabolism [MED] Metabolism below the normal rate. { ¦hī·pō·mə'tab·ə,liz·əm }

hypomorphic allele [GEN] An allele that has reduced levels of gene activity. { ¦hī·pə¦mȯr·fik ə'lēl }

hypomotility [MED] Decreased motility, especially of the gastrointestinal tract. { ¦hī·pō·mō'til·əd·ē }

hyponasty [BOT] A nastic movement involving inward and upward bending of a plant part. { 'hī·pə,nas·tē }

hyponatremia [MED] Subnormal or reduced blood sodium levels. { ,hī·pə·nə'trē·mē·ə }

hyponeural sinus [INV ZOO] Tubular portion of the coelom containing hemal vessels and motor nerves in certain echinoderms. { ¦hī·pə'nu̇r·əl 'sī·nəs }

hyponychium [HISTOL] The thickened stratum corneum of the epidermis, which lies under the free edge of the nail. { hī'päŋ·kē·əm }

hypoovarianism [MED] Decrease in ovarian endocrine activity. { ¦hī·pō,ō'ver·ē·ə,niz·əm }

hypoparathyroidism [MED] Condition caused by insufficient functioning of the parathyroid gland. { ,hī·pō,par·ə'thī,rȯi,diz·əm }

hypophagia [MED] Undereating. { ,hī·pə'fā·jē·ə }

hypophalangism [MED] Congenital absence of one or more phalanges in a finger or toe. { ,hī·pō·fə'lan,jiz·əm }

hypopharynx [ANAT] See laryngopharynx. [INV ZOO] A sensory, tonguelike structure on the floor of the mouth of many insects; sometimes modified for piercing. { ¦hī·pō'far·iŋks }

hypophosphatasia [MED] **1.** Alkaline phosphatase deficiency. **2.** A hereditary metabolic disorder characterized by subnormal amounts of alkaline phosphatase in the tissues. { ,hī·pō,fas·fə'tā·zhə }

hypophrenia [PSYCH] Mental retardation. { ,hī·pə'frē·nē·ə }

hypophyseal cachexia See Simmonds' disease. { hī¦päf·ə¦sē·əl kə'kek·sē·ə }

hypophyseal duct tumor [MED] Any tumor derived from epithelial remnants of Rathke's pouch. { hī¦päf·ə¦sē·əl ¦dəkt 'tü·mər }

hypophysectomy [MED] Surgical removal of the pituitary gland. { hī¦päf·ə¦sek·tə,mē }

hypophysis [ANAT] A small rounded endocrine gland which lies in the sella turcica of the sphenoid bone and is attached to the floor of the third ventricle of the brain in all craniate vertebrates. Also known as pituitary gland. { hī'päf·ə·səs }

hypopituitarism [MED] Condition caused by insufficient secretion of pituitary hormones, especially of the adenohypophysis. Also known as panhypopituitarism. { ,hī·pō·pə'tü·ə·tə,riz·əm }

hypopituitary cachexia See Simmonds' disease. { hī·pō·pə'tü·ə,ter·ē kə'kek·sē·ə }

hypoplankton [BIOL] Forms of marine life whose swimming ability lies somewhere between that of the plankton and the nekton; includes some mysids, amphipods, and cumacids. { ¦hī·pō¦plaŋk·tən }

hypoplasia [MED] Failure of a tissue or organ to achieve complete development. { ,hī·pō'plā·zhə }

hypoplastic dwarf [MED] A normally proportioned individual of subnormal size. { ¦hī·pō¦plas·tik 'dwȯrf }

hypoplastron [VERT ZOO] Either of the third pair of lateral bony plates in the plastron of most turtles. { ¦hī·pō¦plas·trən }

hypopleura [INV ZOO] Sclerite above and in front of the hind coxa in Diptera. { ¦hī·pō¦plu̇r·ə }

hypoploid [GEN] Having a deficit of one or more chromosomes, or parts of chromosomes, from a whole multiple of the haploid number. { 'hī·pō,plȯid }

hypoploidy [GEN] The condition or state of being hypoploid. { 'hī·pə,plȯid·ē }

hypoproliferative anemia [MED] Decreased concentration of hemoglobin and number of red blood cells due to subnormal numbers of erythrocyte primordial cells in relation to the stimulus of anemia. { ¦hī·pō·prə'lif·rəd·iv ə'nē·mē·ə }

hypoproteinemia [MED] Abnormally low levels of protein in the blood. { ¦hī·pō,prō·də'nē·mē·ə }

hypoprothrombinemia [MED] Deficiency of prothrombin in the blood. { ¦hī·pō,prō¦thräm·bə'nē·mē·ə }

hypopus [INV ZOO] The resting larval stage of certain mites. { 'hī·pə·pəs }

hypopycnal inflow [HYD] Flowing water of lower density than the body of water into which it flows. { ,hī·pō'pik·nəl 'in,flō }

hypopygium [INV ZOO] A modified ninth abdominal segment together with the copulatory apparatus in Diptera. { ,hī·pə'pij·ē·əm }

hypopyon [MED] A collection of pus in the anterior chamber of the eye. { hə'pō·pē,än }

hyporeactive [MED] Characterized by decreased responsiveness to stimuli. { ¦hī·pō·rē'ak·tiv }

hyporeflexia [MED] A condition in which reflexes are below normal, due to a variety of causes. { ¦hī·pō·ri'flek·sē·ə }

hyporheic zone [ECOL] The saturated sediment environment below a stream that exchanges water, nutrients, and fauna with surface flowing waters. { ,hī·pə'rē·ik ,zōn }

hyposensitivity [MED] Condition marked by diminished sensitivity to stimuli. { ,hī·pō,sen·sə'tiv·əd·ē }

hyposensitization See desensitization. { ¦hī·pō,sen·səd·ə'zā·shən }

hyposmia [MED] Decreased olfactory sensitivity. { hī'päz·mē·ə }

hypospadias [MED] **1.** Congenital anomaly in which the urethra opens on the ventral surface of the penis or in the perineum. **2.** Congenital anomaly in which the urethra opens into the vagina. { ,hī·pō'spād·ē·əs }

hypospermatogenesis [MED] Decreased sperm production. { ¦hī·pō·spər,mad·ə'jen·ə·səs }

hypostasis [MED] A condition involving settling of blood in dependent parts of an organ. { hī'päs·tə·səs }

hypostatic [GEN] Subject to being suppressed, as a gene that can be suppressed by a nonallelic gene. { ¦hī·pō'stad·ik }

hyposthenia [MED] Weakness; subnormal strength. { ¦hī·pə'sthē·nē·ə }

hyposthenuria [MED] The secretion of urine of low specific gravity. { hī,päs·thə'nu̇r·ē·ə }

hypostome [INV ZOO] **1.** Projection surrounding the oral aperture in many cnidarian polyps. **2.** Anteroventral part of the head in Diptera. **3.** Median ventral mouthpart in ticks. **4.** Raised area on the posterior oral margin in crustaceans. { 'hī·pə,stōm }

hypostracum [INV ZOO] The innermost layer of the cuticle of ticks lying above the hypodermis. { ¦hī·pō¦strak·əm }

hyposulculus [INV ZOO] A groove of the siphonoglyph below the pharynx in anthozoans. { ,hī·pō¦səl·kyə·ləs }

hyposynergia [MED] Defective coordination. { ¦hī·pō·sə'nər·jē·ə }

hypotarsus [VERT ZOO] A process on the tarsometatarsal bone in birds. { ¦hī·pō¦tär·səs }

hypotelorism [MED] Decrease in distance between two organs or body parts. { ,hī·pō'tel·ə,riz·əm }

hypotension [MED] Abnormally low blood pressure, commonly considered to be levels below 100 diastolic and 40 systolic. { ¦hī·pō'ten·chən }

hypotenuse [MATH] On a right triangle, the side opposite the right angle. { hī'pät·ən,üs }

hypo test [GRAPHICS] A method of checking the washing efficiency of processed film or paper. { 'hī·pō ,test }

hypothalamic center [NEUROSCI] Any of the neural centers which regulate autonomic functions. { ¦hī·pō·thə'lam·ik ¦sen·tər }

hypothalamic releasing factor [NEUROSCI] Any of the hormones secreted by the hypothalamus which travel by way of nerve fibers to the anterior pituitary, where they cause selective release of specific pituitary hormones. { ¦hī·pō·thə'lam·ik ri'lēs·iŋ ,fak·tər }

hypothalamoneurohypophyseal system [PHYSIO] The

hormones and neurosecretory structures involved in the endocrine activity of the adenohypophysis, neurohypophysis, and hypothalamus. { ¦hī·pō·thə¦lam·ə¦nur·ō,hī¦päf·ə'sē·əl 'sis·təm }

hypothalamoneurohypophyseal tract [NEUROSCI] A bundle of nerve fibers connecting the supraoptic and paraventricular neurons of the hypothalamus with the infundibular stem and neurohypophysis. { ¦hī·pō·thə¦lam·ə¦nur·ō,hī¦päf·ə'sē·əl 'trakt }

hypothalamus [NEUROSCI] The floor of the third brain ventricle; site of production of several substances that act on the adenohypophysis. { ¦hī·pō'thal·ə·məs }

hypotheca [INV ZOO] **1.** The lower valve of a diatom frustule. **2.** Covering on the hypocone in dinoflagellates. { ¦hī·pō'thē·kə }

hypothenar [ANAT] Of or pertaining to the prominent portion of the palm above the base of the little finger. { hī'päth·ə,när }

hypothermal [GEOL] Referring to the high-temperature (300–500°C) environment of hypothermal deposits. { ¦hī·pō'thər·məl }

hypothermal deposit [MINERAL] Mineral deposit formed at great depths and high (300–500°C) temperatures. { ¦hī·pō'thər·məl di'päz·ət }

hypothermia [PHYSIO] Condition of reduced body temperature in homeotherms. { ,hī·pō'thər·mē·ə }

hypothesis [SCI TECH] **1.** A proposition which is assumed to be true in proving another proposition. **2.** A proposition which is thought to be true because its consequences are found to be true. [STAT] A statement which specifies a population or distribution, and whose truth can be tested by sample evidence. { hī'päth·ə·səs }

hypothesis testing [STAT] The branch of statistics which considers the problem of choosing between two actions on the basis of the observed value of a random variable whose distribution depends on a parameter, the value of which would indicate the correct action. { hī'päth·ə·səs ,test·iŋ }

hypothetical parallax See dynamic parallax. { ¦hī·pə¦thed·ə·kəl 'par·ə,laks }

hypothyroidism [MED] Condition caused by deficient secretion of the thyroid hormone. { ¦hī·pō'thī,roi,diz·əm }

hypotonia [MED] Decrease of normal tonicity or tension, especially diminution of intraocular pressure or of muscle tone. { ¦hī·pə'tō·nē·ə }

hypotonic [PHYSIO] **1.** Pertaining to subnormal muscle strength or tension. **2.** Referring to a solution with a lower osmotic pressure than physiological saline. { ¦hī·pə'tän·ik }

Hypotrichida [INV ZOO] An order of highly specialized protozoans in the subclass Spirotrichia characterized by cirri on the ventral surface and a lack of ciliature on the dorsal surface. { ,hī·pə'trik·ə·də }

hypotrochoid [MATH] A curve traced by a point rigidly attached to a circle at a point other than the center when the circle rolls without slipping on the inside of a fixed circle. { ¦hī·pō'trō,kóid }

hypotype [SYST] A specimen of a species, which, though not a member of the original type series, is known from a published description or listing. { 'hī·pə,tīp }

hypovitaminosis [MED] Condition due to deficiency of an essential vitamin. { ,hī·pə,vīd·ə·mə'nō·səs }

hypovolemia [MED] Low blood volume. { ,hī·pō,vä'lē·mē·ə }

hypovolemic shock [MED] Shock caused by reduced blood volume which may be due to loss of blood or plasma as in burns, the crush syndrome, perforating gastrointestinal wounds, or other trauma. Also known as wound shock. { ¦hī·pō,vä¦lē·mik 'shäk }

hypoxanthine [BIOCHEM] $C_5H_4ON_4$ An intermediate product derived from adenine in the hydrolysis of nucleic acid. { ¦hī·pō'zan,thēn }

hypoxemia See hypoxia. { ,hī,päk'sē·mē·ə }

hypoxia [ECOL] A condition characterized by a low level of dissolved oxygen in an aquatic environment. [MED] Oxygen deficiency; any state wherein a physiologically inadequate amount of oxygen is available to or is utilized by tissue, without respect to cause or degree. Also known as hypoxemia. { hī'päk·sē·ə }

hypoxic encephalopathy [MED] Brain damage syndrome caused by hypoxia. { hī'päk·sik en,sef·ə'läp·ə,thē }

hypozygal [INV ZOO] In comatulids, the proximal member of adjacent brachials in an articulation. { ,hī·pō'zīg·əl }

hypsicephalic [ANTHRO] Having a high forehead with a length-height index of 62.6 or more. { ¦hip·sə·sə¦fal·ik }

hypsiconch [ANTHRO] Having high orbits with an orbital index of 89 or more. { 'hip·sə,kaŋk }

hypsicranial [ANTHRO] Having a high skull with a length-height index of 75 or more { ¦hip·sə¦krā·nē·əl }

hypsidolichocephalic [ANTHRO] Having a head that is high and narrow, high and long, or high, long, and narrow. { ¦hip·sə däl·ə·kō·sə¦fal·ik }

hypsistenocephalic [ANTHRO] Having a very high and narrow head. { ¦hip·sə,sten·ō·sə¦fal·ik }

Hypsithermal See Altithermal. { ¦hip·sə'thərm·əl }

hypsodont [VERT ZOO] Of teeth, having crowns that are high or deep and roots that are short. { 'hip·sə,dänt }

hypsographic map [MAP] A chart showing topographic relief in reference to a given datum, usually sea level. { ¦hip·sə¦graf·ik 'map }

hypsography [GEOGR] The science of measuring or describing elevations of the earth's surface with reference to a given datum, usually sea level. { hip'säg·rə·fē }

hypsometer [ENG] **1.** An instrument for measuring atmospheric pressure to ascertain elevations by determining the boiling point of liquids. **2.** Any of several instruments for determining tree heights by triangulation. { hip'säm·əd·ər }

hypsometric [ENG] Pertaining to hypsometry. { ,hip·sə'me·trik }

hypsometric formula [GEOPHYS] A formula, based on the hydrostatic equation, for either determining the geopotential difference or thickness between any two pressure levels, or for reducing the pressure observed at a given level to that at some other level. { ,hip·sə'me·trik 'fόr·myə·lə }

hypsometric map [MAP] In topographic surveying, a map giving elevations by contours, or sometimes by means of shading, tinting, or batching. { ,hip·sə'me·trik 'map }

hypsometric tinting [MAP] A technique of showing relief on maps and charts by coloring in different shades, those parts which lie between different levels. Also known as altitude tints. { ,hip·sə'me·trik 'tint·iŋ }

hypsometric tints See gradient tints. { ,hip·sə'me·trik 'tins }

hypsometry [ENG] The measuring of elevation with reference to sea level. { hip'säm·ə trē }

hypsophobia [PSYCH] An abnormal fear of being at a great height. { ,hip·sə'fō·bē·ə }

hypural [VERT ZOO] Of or pertaining to the bony structure formed by fusion of the hemal spines of the last few vertebrae in most teleost fishes. { hī'pyůr·əl }

Hyracodontidae [PALEON] The running rhinoceroses, an extinct family of perissodactyl mammals in the superfamily Rhinoceratoidea. { ,hī·rə·kō'dänt·ə,dē }

Hyracoidea [VERT ZOO] An order of ungulate mammals represented only by the conies of Africa, Arabia, and Syria. { ,hī·rə'kóid·ē·ə }

hyster- [MED] A combining form that denotes a relation to or a connection with the uterus. [PSYCH] A combining form indicating a relation to hysteria. { 'his·tər }

hysterectomy [MED] Surgical removal of all or part of the uterus. { ,his·tə'rek·tə·mē }

hysteresimeter [ENG] A device for measuring hysteresis. { his,ter·ə'sim·əd·ər }

hysteresis [ELECTR] An oscillator effect wherein a given value of an operating parameter may result in multiple values of output power or frequency. [ELECTROMAG] See magnetic hysteresis. [NUCLEO] A temporary change in the counting-rate-voltage characteristic of a radiation counter tube, caused by its previous operation. [PHYS] The dependence of the state of a system on its previous history, generally in the form of a lagging of a physical effect behind its cause. { ,his·tə'rē·səs }

hysteresis clutch [MECH ENG] A clutch in which torque is produced by attraction between induced poles in a magnetized iron ring and the control field. { ,his·tə'rē·səs ,kləch }

hysteresis coefficient [PHYS] A constant, characteristic of a particular material, in a formula for hysteresis loss. { ,his·tə rē·səs ,kō·i'fish·ənt }

hysteresis damping [MECH] Damping of a vibration due to energy lost through mechanical hysteresis. { ,his·tə'rē·səs 'dam·piŋ }

HYPOTRICHIDA

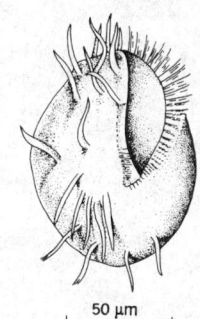

Euplotes, an example of Hypotrichida.

HYPOXANTHINE

Structural formula of hypoxanthine.

HYSTERESIS LOOP

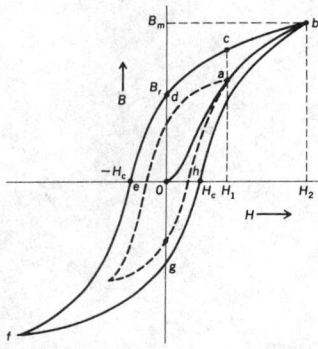

Hysteresis loop. H = magnetizing force; B = flux density; Oab = normal magnetization curve; H_c = coercive force; B_r = retentivity; B_m = maximum value of flux density; $bdefghb$ = the hysteresis loop.

HYSTRICHOSPHERIDA

75 μm

A drawing of a hystrichospherid.

hysteresis error [PHYS] The maximum separation due to hysteresis between upscale-going and downscale-going indications of a measured variable. { ¦his·tə¦rē·səs ¦er·ər }

hysteresis heating [PHYS] **1.** Supply of heat to a material through hysteresis loss. **2.** In particular, supply of a controlled amount of heat to a thermally isolated paramagnetic sample at temperatures below 1 kelvin by taking it through a magnetic hysteresis loop. { ¦his·tə¦rē·səs ¦hēd·iŋ }

hysteresis loop [PHYS] The closed curve followed by a material displaying hysteresis (such as a ferromagnet or ferroelectric) on a graph of a driven variable (such as magnetic flux density or electric polarization) versus the driving variable (such as magnetic field or electric field). { ¦his·tə'rē·səs ¦lüp }

hysteresis loss [PHYS] The energy converted to heat in a material because of magnetic or other hysteresis, accompanying cyclic variation of the magnetic field or other driving variable. { ¦his·tə¦rē·səs ¦lós }

hysteresis motor [ELEC] A synchronous motor without salient poles and without direct-current excitation which utilizes the hysteresis and eddy-current losses induced in its hardened-steel rotor to produce rotor torque. { ¦his·tə'rē·səs ¦mōd·ər }

hysteretic damping [MECH] Damping of a vibrating system in which the retarding force is proportional to the velocity and inversely proportional to the frequency of the vibration. { ¦his·tə'red·ik ¦damp·iŋ }

hysteria [PSYCH] A type of neurosis characterized by extreme emotionalism involving disorders of somatic and psychological functions; the conversion type is associated with neuromuscular and sensory symptoms such as paralysis, tremors, seizures, or blindness, whereas the dissociative displays disorders of consciousness such as amnesia, somnolence, and multiple personalities. { hi'ster·ē·ə }

hysteriaceous [MYCOL] Of, belonging to, or resembling the Hysteriales. { hi¸ster·ē'ā·shəs }

Hysteriales [BOT] An order of lichens in the class Ascolichenes including those species with an ascolocular structure. { hi¸ster·ē'ā·lēz }

hysterical anesthesia [MED] A loss of cutaneous pain sensation accompanying hysteria. { hi'ster·ə·kəl ¸an·əs'thē·zhə }

hysterical paralysis [MED] Muscle weakness or paralysis without loss of reflex activity, in which no organic nerve lesion can be demonstrated, but which is due to psychogenic factors. Also known as functional paralysis. { hi'ster·ə·kəl pə'ral·ə·səs }

hysterics [PSYCH] **1.** Attack of hysteria. **2.** Extreme display of emotions. { hi'ster·iks }

hysterochroic [MYCOL] Having fruiting bodies which discolor progressively from base to apex with age. { ¦his·tə·rō¦krō·ik }

hysterogram [MED] A roentgenogram with opacification of the cavity of the uterus by the injection of contrast medium. { 'his·tə·rō¸gram }

hysterography [MED] Roentgenologic examination of the uterus after the introduction of a contrast medium. { ¸his·tə'räg·rə·fē }

hysteromania *See* nymphomania. { ¸his·tə·rə'mā·nē·ə }

hystero-oophorectomy [MED] Surgical removal of the uterus and ovaries. { ¦his·tə·rō¸ō¸ō·fə'rek·tə·mē }

hysteropexy [MED] Fixation of the uterus by a surgical operation to correct displacement. { his¸te·rō'pek·sē }

hysterorrhaphy [MED] The closure of a uterine incision by suture. { ¸his·ter'ór·ə·fē }

hysterorrhexis [MED] Rupture of the uterus. { ¸his·tə·rō'rek·səs }

hysterosalpingectomy [MED] Surgical removal of the uterus and oviducts. { ¸his·tə·rō¸sal·pin'jek·tə·mē }

hysterosalpingography [MED] Roentgenographic examination of the uterus and oviducts after injection of a radiopaque substance. { ¸his·tə·rō¸sal·piŋ'gäg·rə·fē }

hysterosalpingo-oophorectomy [MED] The excision of the uterus, oviducts, and ovaries. { ¦his·tə·rō·sal¦piŋ·gō¸ō¸ō·fə'rek·tə·mē }

hysteroscope [MED] A uterine speculum with a reflector. { 'his·tə·rə¸skōp }

hysterosoma [INV ZOO] A body division of an acarid mite composed of the metapodosoma and opisthosoma. { ¦his·tə·rə'sō·mə }

hysterotomy [MED] **1.** Incision into the uterine wall. **2.** Cesarean section. { ¸his·tə'räd·ə·mē }

Hystrichospherida [PALEON] A group of protistan microfossils. { ¸his·trə·kō'sfer·ə·də }

Hystricidae [VERT ZOO] The Old World porcupines, a family of Rodentia ranging from southern Europe to Africa and eastern Asia and into the Philippines. { hi'stris·ə¸dē }

Hystricomorpha [VERT ZOO] A superorder of the class Rodentia. { ¸his·trə·kō'mór·fə }

Hz *See* hertz.

HZE particles [NUC PHYS] A component of cosmic radiation consisting of energetic heavy nuclei (atomic number 3 or greater); so named for their high atomic number (Z) and energy (E). { ¦āch¦zē'ē ¸pärd·i·kəlz }

H zone [HISTOL] The central portion of an A band in a sarcomere; characterized by the presence of myosin filaments. { 'āch ¸zōn }

I *See* iodine.

IA *See* international angstrom.

IAA *See* indoleacetic acid.

IAC *See* international analysis code.

ianthinite [MINERAL] $2UO_2 \cdot 7H_2O$ A violet mineral composed of hydrous uranium dioxide, occurring as orthorhombic crystals. { ē′an·thə‚nīt }

Iapetus [ASTRON] A satellite of Saturn that orbits at a mean distance of 2.207×10^6 miles (3.560×10^6 kilometers) and has a diameter of about 900 miles (1500 kilometers). { ‚yap· əd·əs }

IAS *See* indicated airspeed.

IAT *See* international atomic time.

iatrogenic [MED] Pertaining to an abnormal state or condition produced by a physician in a patient by inadvertent or incorrect treatment. { ī‚a·trə′jen·ik }

IBA *See* indolebutyric acid.

IBAC *See* in-band/adjacent-channel. { ′ī‚bē‚ā′sē *or* ′ī‚bak }

Iballidae [INV ZOO] A small family of hymenopteran insects in the superfamily Cynipoidea. { ī′bal·ə‚dē }

I band [HISTOL] The band on either side of a Z line; encompasses portions of two adjacent sarcomeres and is characterized by the presence of actin filaments. { ′ī ‚band }

IBD *See* inflammatory bowel disease.

I beam [CIV ENG] A rolled iron or steel joist having an I section, with short flanges. { ′ī ‚bēm }

Ibe wind [METEOROL] A local strong wind which blows through the Dzungarian Gate (western China), a gap in the mountain ridge separating the depression of Lakes Balkash and Ala Kul from that of Lake Ebi Nor; the wind resembles the foehn and brings a sudden rise of temperature, in winter from about −15 to about 30°F (−20 to −1°C). { ‚ē·be ‚wind }

IBIB *See* isobutyl isobutyrate.

ibis [VERT ZOO] The common name for wading birds making up the family Threskiornithidae and distinguished by a long, slender, downward-curving bill. { ′ī·bəs }

I blood group [IMMUNOL] The erythrocyte antigens defined by reactions with anti-I and anti-i antibodies, which occur both in acquired hemolytic anemia and naturally in normal persons of the rare phenotype i. { ′ī ′bləd ‚grüp }

IBOC *See* in-band/on-channel. { ′ī‚bē′ō′sē *or* ′ī‚bäk }

ibogaine [ORG CHEM] $C_{26}H_{32}O_2N_2$ An alkaloid isolated from the stems and leaves of the shrub *Tabernanthe iboga*, crystallizing from absolute ethanol as prismatic needles, melting at 152–153°C, soluble in ethanol, ether, and chloroform; used in medicine. { ə′bō·gə‚ēn }

IC *See* integrated circuit.

IC₅₀ *See* incapacitating concentration 50. { ′ī‚sē ′fif′tē }

Icacinaceae [BOT] A family of dicotyledonous plants in the order Celastrales characterized by haplostemonous flowers, pendulous ovules, stipules wanting or vestigial, a polypetalous corolla, valvate petals, and usually one (sometimes three) locules. { i‚ka·sə‚nās·ē‚ē }

Icarus [ASTRON] An asteroid with a highly eccentric orbit (eccentricity of 0.827) that crosses the earth's orbit and takes the asteroid to only 0.187 astronomical units from the sun, closer than Mercury. { ik·ə·rəs }

ICBM *See* intercontinental ballistic missile.

ice [PHYS CHEM] **1.** The dense substance formed by the freezing of water to the solid state; has a melting point of 32°F (0°C) and commonly occurs in the form of hexagonal crystals. **2.** A layer or mass of frozen water. { īs }

ice accretion [HYD] The process by which a layer of ice builds up on solid objects which are exposed to freezing precipitation or to supercooled fog or cloud droplets. { ′īs ə‚krē·shən }

ice-accretion indicator [ENG] An instrument used to detect the occurrence of freezing precipitation, usually consisting of a strip of sheet aluminum about $1\,^1/_2$ inches (4 centimeters) wide, and is exposed horizontally, face up, in the free air a few meters above the ground. { ′īs ə‚krē·shən ‚ind·ə‚kād·ər }

ice age [GEOL] A major interval of geologic time during which extensive ice sheets (continental glaciers) formed over many parts of the world. { ′īs ‚āj }

Ice Age *See* Pleistocene. { ′īs ‚āj }

ice anchor [NAV ARCH] An anchor, usually with only one fluke, used for securing a vessel to ice. { ′īs ‚aŋ·kər }

ice apron [CIV ENG] A wedge-shaped structure which protects a bridge pier from floating ice. [HYD] **1.** The snow and ice attached to the walls of a cirque. **2.** The ice that is flowing from an ice sheet over the edge of a plateau. **3.** A piedmont glacier's lobe. **4.** Ice that adheres to a wall of a valley below a hanging glacier. { ′īs ‚ā·prən }

ice atlas [NAV] A publication containing a series of ice charts showing geographic distribution of ice, usually by seasons or months; used in navigation. { ′īs ‚at·ləs }

ice band [HYD] A layer of ice in firn or snow. { ′īs ‚band }

ice barrier [HYD] The periphery of the Antarctic ice sheet; or used generally for any ice dam. { ′īs ‚bar·ē·ər }

ice bay [OCEANOGR] A bay-like recess in the edge of a large ice floe or ice shelf. Also known as ice bight. { ′īs ‚bā }

ice belt [OCEANOGR] A band of fragments of sea ice in otherwise open water. Also known as ice strip. { ′īs ‚belt }

iceberg [OCEANOGR] A large mass of glacial ice broken off and drifted from parent glaciers or ice shelves along polar seas; it is distinguished from polar pack ice, which is sea ice, and from frozen seawater, whose rafted or hummocked fragments may resemble small icebergs. { ′īs‚bərg }

ice bight *See* ice bay. { ′īs ‚bīt }

ice blink [METEOROL] A relatively bright, usually yellowish-white glare on the underside of a cloud layer, produced by light reflected from an ice-covered surface such as pack ice; used in polar regions with reference to the sky map; ice blink is not as bright as snow blink, but much brighter than water sky or land sky. { ′īs‚bliŋk }

icebound [NAV] Surrounded so closely by ice as to be incapable of proceeding. { ′īs‚baúnd }

ice boundary [HYD] At any given time, the boundary between fast ice and pack ice or between areas of different concentrations of pack ice. { ′īs ‚baún·drē }

icebreaker [NAV ARCH] A vessel designed for operating in and breaking heavy ice, having a special-shaped, reinforced bow, protected propellers, and powerful engines. { ′īs‚brāk·ər }

ice bridge [OCEANOGR] Surface river ice of sufficient thickness to impede or prevent navigation. { ′īs ‚brij }

ice buoy [ENG] A sturdy buoy, usually a metal spar, used to replace a more easily damaged buoy during a period when heavy ice is anticipated. { ′īs ‚bȯi }

ice cake [HYD] A single, usually relatively flat piece of ice of any size in a body of water. { ′īs ‚kāk }

ice calorimeter *See* Bunsen ice calorimeter. { ′īs ‚kal·ə′rim· əd·ər }

ice calving *See* calving. { ′īs ‚kav·iŋ }

Arctic iceberg, eroded to form a valley or drydock type.

ice canopy See pack ice. { 'īs ‚kam·ə·pē }

ice cap [HYD] **1.** A perennial cover of ice and snow in the shape of a dome or plate on the summit area of a mountain through which the mountain peaks emerge. **2.** A perennial cover of ice and snow on a flat land mass such as an Arctic island. { 'īs ‚kap }

ice-cap climate See perpetual frost climate. { 'īs ‚kap ‚klīm·ət }

ice cascade See icefall. { 'īs ka'skād }

ice cave [GEOL] A cave that is cool enough to hold ice through all or most of the warm season. [HYD] A cave in ice such as a glacier formed by a stream of melted water. { 'īs ‚kāv }

ice chart [NAV] A chart showing prevalence of ice, usually with reference to navigable waters. { 'īs ‚chärt }

ice clearing See polyn'ya. { 'īs ‚klir·iŋ }

ice color See azoic dye. { 'īs ‚kəl·ər }

ice concentration [OCEANOGR] In sea ice reporting, the ratio of the areal extent of ice present and the total areal extent of ice and water. Also known as ice cover. { 'īs ‚käns·ən'trā·shən }

ice-contact delta [GEOL] A delta formed by a stream flowing between a valley slope and the margin of glacial ice. Also known as delta moraine; morainal delta. { 'īs ‚kän‚tak ‚del·tə }

ice cover See ice concentration. { 'īs ‚kəv·ər }

ice crust [HYD] A type of snow crust; a layer of ice, thicker than a film crust, upon a snow surface, formed by the freezing of meltwater or rainwater which has flowed onto it. { 'īs ‚krəst }

ice crystal [PHYS CHEM] Any one of a number of macroscopic crystalline forms in which ice appears, including hexagonal columns, hexagonal platelets, dendritic crystals, ice needles, and combinations of these forms; although the crystal lattice of ice is hexagonal in its symmetry, varying conditions of temperature and vapor pressure can lead to growth of crystalline forms in which the simple hexagonal pattern is almost undiscernible. { 'īs ‚krist·əl }

ice-crystal cloud [METEOROL] A cloud consisting entirely of ice crystals, such as cirrus (in this sense distinguished from water clouds and mixed clouds), and having a diffuse and fibrous appearance quite different from that typical of water droplet clouds. { 'īs ‚krist·əl ‚klaůd }

ice-crystal fog See ice fog. { 'īs ‚krist·əl ‚fäg }

ice-crystal haze [METEOROL] A type of very light ice fog composed only of ice crystals and at times observable to altitudes as great as 20,000 feet (6100 meters), and usually associated with precipitation of ice crystals. { 'īs ‚krist·əl ‚hāz }

ice-crystal theory See Bergeron-Findeisen theory. { 'īs ‚krist·əl ‚thē·ə·rē }

ice day [CLIMATOL] A day on which the maximum air temperature in a thermometer shelter does not rise above 32°F (0°C), and ice on the surface of water does not thaw. { 'īs ‚dā }

ice desert [CLIMATOL] Any polar area permanently covered by ice and snow, with no vegetation other than occasional red snow or green snow. { 'īs ‚dez·ərt }

iced firn [HYD] A mixture of glacier ice and firn; firn permeated with meltwater and then refrozen. Also known as firn ice. { 'īst ‚fərn }

ice erosion [GEOL] **1.** Erosion due to freezing of water in rock fractures. **2.** See glacial erosion. { 'īs i'rō·zhən }

icefall [HYD] That portion of a glacier where a sudden steepening of descent causes a chaotic breaking up of the ice. Also known as ice cascade. { 'īs ‚fól }

ice fat See grease ice. { 'īs ‚fat }

ice feathers [HYD] A type of hoarfrost formed on the windward side of terrestrial objects and on aircraft flying from cold to warm air layers. Also known as frost feathers. { 'īs ‚feth·ərz }

ice field [HYD] A mass of land ice resting on a mountain region and covering all but the highest peaks. [OCEANOGR] A flat sheet of sea ice that is more than 5 miles (8 kilometers) across. { 'īs ‚fēld }

ice floe See floe. { 'īs ‚flō }

ice flowers [HYD] **1.** Formations of ice crystals on the surface of a quiet, slowly freezing body of water. **2.** Delicate tufts of hoarfrost that occasionally form in great abundance on an ice or snow surface. Also known as frost flowers. **3.** Frost crystals resembling a flower, formed on salt nuclei on the surface of sea ice as a result of rapid freezing of sea water. Also known as salt flowers. { 'īs ‚flaů·ərz }

ice fog [METEOROL] A type of fog composed of suspended particles of ice, partly ice crystals 20–100 micrometers in diameter but chiefly, especially when dense, droxtals 12–20 micrometers in diameter; occurs at very low temperatures and usually in clear, calm weather in high latitudes. Also known as frost flakes; frost fog; frozen fog; ice-crystal fog; pogonip; rime fog. { 'īs ‚fäg }

ice foot [OCEANOGR] Sea ice firmly frozen to a polar coast at the high-tide line and unaffected by tide; this fast ice is formed by the freezing of seawater during ebb tide, and of spray, and it is separated from the floating sea ice by a tide crack. { 'īs ‚fůt }

ice-free [HYD] **1.** Referring to a harbor, river, estuary, and so on, when there is not sufficient ice present to interfere with navigation. **2.** Descriptive of a water surface completely free of ice. { 'īs ‚frē }

ice fringe [HYD] An ice deposit on plant surfaces, not of hoarfrost from atmospheric water vapor, but of moisture exuded from the stems of plants and appearing as frosted fringes or ribbons. Also known as ice ribbon. [OCEANOGR] A belt of sea ice extending a short distance from the shore. { 'īs ‚frinj }

ice front [HYD] The floating vertical cliff forming the seaward face or edge of an ice shelf or other glacier that enters water. { 'īs ‚frənt }

ice gland [HYD] A column of ice in the granular snow at the top of a glacier. { 'īs ‚gland }

ice gruel [HYD] A type of slush formed by the irregular freezing together of ice crystals. { 'īs ‚grül }

ice island [OCEANOGR] A large tabular fragment of shelf ice found in the Arctic Ocean and having an irregular surface, thickness of 15–50 meters (50–165 feet), and an area between a few thousand square meters and 500 square kilometers (200 square miles) or more. { 'īs ‚ī·lənd }

ice-island iceberg [OCEANOGR] An iceberg having a conical or dome-shaped summit, often mistaken by mariners for ice-covered islands. { 'īs ‚ī·lənd 'īs‚bərg }

ice jam [HYD] **1.** An accumulation of broken river ice caught in a narrow channel, frequently producing local floods during a spring breakup. **2.** Fields of lake or sea ice thawed loose from the shores in early spring, and blown against the shore, sometimes exerting great pressure. { 'īs ‚jam }

ice-laid drift See till. { 'īs ‚lād ‚drift }

Iceland agate See obsidian. { 'īs·lənd 'ag·ət }

Iceland crystal See Iceland spar. { 'īs·lənd 'krist·əl }

Iceland disease See epidemic neuromyasthenia. { 'īs·lənd di‚zēz }

Icelandic low [METEOROL] **1.** The low-pressure center located near Iceland (mainly between Iceland and southern Greenland) on mean charts of sea-level pressure. **2.** On a synoptic chart, any low centered near Iceland. { 'īs'land·ik 'lō }

Iceland spar [MINERAL] A pure, transparent form of calcite found particularly in Iceland; easily cleaved to form rhombohedral crystals that are doubly refracting. Also known as Iceland crystal. { 'īs·lənd 'spär }

ice layer [HYD] An ice crust covered with new snow; when exposed at a glacier front or in crevasses, the ice layers viewed in cross section are termed ice bands. { 'īs ‚lā·ər }

ice line [THERMO] A graph of the freezing point of water as a function of pressure. { 'īs ‚līn }

I-cell disease See inclusion-cell disease. { 'ī‚sel diz‚ēz }

ice load [ENG] The weight of glaze deposited on an overhead wire in a power supply system; standard safety codes require allowance for ½-inch (12.7-millimeter) radial thickness in heavy loading districts and ¼-inch (6.35-millimeter) in medium. { 'īs ‚lōd }

ice mantle See ice sheet. { 'īs ‚mant·əl }

ice mine [ORD] A waterproof mine placed in or under the ice, detonated by a pressure device on the surface to break up river or lake ice. { 'īs ‚mīn }

ice mound See ground ice mound. { 'īs ‚maůnd }

ice needle [PHYS CHEM] A long, thin ice crystal whose cross section perpendicular to its long dimension is typically hexagonal. Also called ice spicule. { 'īs ‚nēd·əl }

ice nucleus [METEOROL] Any particle which may act as a nucleus in formation of ice crystals in the atmosphere. { 'īs ‚nü·klē·əs }

ice pack See pack ice. { 'īs ‚pak }

ice pellets [METEOROL] A type of precipitation consisting of transparent or translucent pellets of ice 0.2 inch (5 millimeters) or less in diameter; may be spherical, irregular, or (rarely) conical in shape. { 'īs ,pel·əts }

ice period [CLIMATOL] The interval between the first appearance and the final dissipation of ice during any year in a given locale. { 'īs ,pir·ē·əd }

ice pick [DES ENG] A hand tool for chipping ice. { 'īs ,pik }

ice pillar [HYD] A column of glacial ice covered with stones or debris which tend to protect the ice from melting. { 'īs ,pil·ər }

ice point [PHYS CHEM] The true freezing point of water; the temperature at which a mixture of air-saturated pure water and pure ice may exist in equilibrium at a pressure of 1 standard atmosphere (101,325 pascals). { 'īs ,póint }

ice pole [GEOGR] The approximate center of the most consolidated portion of the arctic pack ice, near 83 or 84°N and 160°W. Also known as pole of inaccessibility. { 'īs ,pōl }

ice push [GEOL] Lateral pressure that is caused by expansion of shoreward-moving ice on a lake or a bay of the sea and that follows a rise in temperature. Also known as ice shove; ice thrust. { 'īs ,pùsh }

icequake [HYD] The crash or concussion that accompanies the breakup of ice masses, frequently owing to contraction from the extreme cold. { 'īs,kwāk }

ice-rafting [GEOL] The transporting of rock and other minerals, of all sizes, on or within icebergs, ice floes, river drift, or other forms of floating ice. { 'īs ,raf·tiŋ }

ice ribbon See ice fringe. { 'īs ,rib·ən }

ice rind [HYD] A thin but hard layer of sea ice, river ice, or lake ice, which is either a new encrustation upon old ice or a single layer of ice usually found in bays and fiords, where fresh water freezes on top of slightly colder sea water. { 'īs ,rīnd }

ice run [HYD] The initial stage in the spring or summer breakup of river ice, being an exceedingly rapid process, seldom taking more than 1 day. { 'īs ,rən }

ice sheet [HYD] A thick glacier, more than 19,300 square miles (50,000 square kilometers) in area, forming a cover of ice and snow that is continuous over a land surface and moving outward in all directions. Also known as ice mantle. { 'īs ,shēt }

ice shelf [OCEANOGR] A thick sheet of ice with a fairly level or undulating surface, formed along a polar coast and in shallow bays and inlets, fastened to the shore along one side but mostly afloat and nourished by annual accumulation of snow and by the seaward extension of land glaciers. { 'īs ,shelf }

ice shove See ice push. { 'īs ,shəv }

ice spar See sanidine. { 'īs ,spär }

ice spicule See ice needle. { 'īs ,spik·yəl }

ice splinters [PHYS CHEM] Minute, electrically charged fragments of ice which have been observed under laboratory conditions to be torn away from dendritic crystals or spatial aggregates exposed to moving air. { 'īs ,splin·tərz }

ice stone See cryolite. { 'īs ,stōn }

ice storm [METEOROL] A storm characterized by a fall of freezing precipitation, forming a glaze on terrestrial objects that creates many hazards. Also known as silver storm. { 'īs ,stórm }

ice stream [HYD] A current of ice flowing in an ice sheet or ice cap; usually moves toward an ocean or to an ice shelf. { 'īs ,strēm }

ice strip See ice belt. { 'īs ,strip }

ice thrust See ice push. { 'īs ,thrəst }

ice tongs [DES ENG] Tongs for handling cubes or blocks of ice. { 'īs ,täŋz }

ice tongue [HYD] Any narrow extension of a glacier or ice shelf, such as a projection floating in the sea or an outlet glacier of an ice cap. { 'īs ,təŋ }

ice wall [HYD] A cliff of ice forming the seaward margin of a glacier that is not afloat. { 'īs ,wól }

ice wedge See foliated ice. { 'īs ,wej }

ich [VET MED] A dermatitis of fresh-water fishes caused by the invasion of the skin by the ciliated protozoan *Ichthyophthirius multifiliis*. { ik }

ichneumon [INV ZOO] The common name for members of the family Ichneumonidae. { ik nü·mən }

Ichneumonidae [INV ZOO] The ichneumon flies, a large family of cosmopolitan, parasitic wasps in the superfamily Ichneumonoidea. { ik·n i'män·ə,dē }

Ichneumonoidea [INV ZOO] A superfamily of hymenopteran insects; members are parasites of other insects. { ik,nü·mə'nóid·ē·ə }

ichnite [PALEON] An ichnofossil of the footprint or track of an organism. Also known as ichnolite. { 'ik,nīt }

ichnofacies [GEOL] A recurrent assemblage of ichnofossils that represent certain environmental conditions. { 'ik·rō̇,fā,shēz }

ichnofossil See trace fossil. { 'ik·nə,fäs·əl }

ichnography [GRAPHICS] **1.** The art of making decorative drawings by use of compass and rule, in a sense revived in Op art (optical art) techniques. **2.** The art of tracing plans and illustrations. **3.** A horizontal section, for example, of a structure, that shows the true dimensions according to a geometric scale. { ik'näg·rə·fē }

ichnolite [PALEON] **1.** A rock containing a fossilized track or footprint. **2.** See ichnite. { 'ik·nə,līt }

ichnology [PALEON] The study of ichnofossils, especially fossil footprints. { ik'näl·ə·jē }

ichor [GEOL] A fluid rich in mineralizers. [MED] (Obsolete) An acrid, watery or blood-tinged discharge from an ulcer or wound. { 'ī,kór }

ichthammol [PHARM] A viscid fluid obtained by the destructive distillation of certain bituminous schists, followed by sulfonation of the distillate and neutralization with ammonia; used as a weak antiseptic and stimulant in skin diseases, and occasionally as an expectorant. { 'ik·thə·mól }

ichthyism [MED] Food poisoning caused by eating spoiled fish. { 'ik·thē,iz·əm }

Ichthyobdellidae [INV ZOO] A family of leeches in the order Rhynchobdellae distinguished by cylindrical bodies with conspicuous, powerful suckers. { ,ik·thē·äb'del·ə,dē }

ichthyocolla See isinglass. { ,ik·thē·ə'käl·ə }

Ichthyodectidae [PALEON] A family of Cretaceous marine osteoglossiform fishes. { ,ik·thē·ə'dek·tə,dē }

ichthyolith [PALEON] Fossil fish remains. { 'ik·thē·ə,lith }

ichthyology [VERT ZOO] A branch of vertebrate zoology that deals with the study of fishes. { ,ik·thē'äl·ə·jē }

ichthyophagous [ZOO] Subsisting on a diet of fish. { ,ik·thē'äf·ə·gəs }

ichthyophobia [PSYCH] An abnormal fear of fish. { ,ik·thē·ə'fo·bē·ə }

Ichthyopterygia [PALEON] A subclass of extinct Mesozoic reptiles composed of predatory fish-finned and sea-swimming forms with short necks and a porpoiselike body. { ,ik·thē,äp·tə'rij·ē·ə }

Ichthyornis [PALEON] The type genus of Ichthyornithidae. { ik·thē'ór·nəs }

Ichthyornithes [PALEON] A superorder of fossil birds of the order Ichthyornithiformes according to some systems of classification. { ,ik·thē'ór·nə,thēz }

Ichthyornithidae [PALEON] A family of extinct birds in the order Ichthyornithiformes. { ,ik·thē,ór'nith·ə,dē }

Ichthyornithiformes [PALEON] An order of ancient fossil birds including strong flying species from the Upper Cretaceous that possessed all skeletal characteristics of modern birds. { ,ik·thē,ór·nə·thə'fór,mēz }

ichthyosarcotoxism [MED] Poisoning caused by eating the flesh of fish containing toxic substances. { ,ik·thē·ō,sär·kə'täk,siz·əm }

Ichthyosauria [PALEON] The only order of the reptilian subclass Ichthyopterygia, comprising the extinct predacious fish-lizards; all were adapted to a sea life in having tail flukes, paddles, and dorsal fins. { ,ik·thē·ə'sór·ē·ə }

ichthyosis [MED] A congenital skin disease characterized by dryness and scales, especially on the extensor surfaces of the extremities. { ,ik·thē'ō·səs }

ichthyosis congenita [MED] A severe form of ichthyosis characterized by cracked, thickened skin and mucous membranes. { ,ik·thē'ō·səs kən'jen·əd·ə }

ichthyosis simplex [MED] A common, childhood form of ichthyosis characterized by large, papery scales on the skin. { ,ik·thē'ō·səs 'sim,pleks }

Ichthyostega [PALEON] Four-legged vertebrates that

ICHNEUMON

Long-tailed ichneumon (*Megarhyssa lunator*), about 6 inches (15 centimeters) long.

ICHTHYOSAURIA

A Jurassic ichthyosaur.

evolved from their lobe-finned fish ancestors during the later Devonian Period (400–350 million years ago). { ˌik·thē·ə'steg·ə }

Ichthyostegalia [PALEON] An extinct Devonian order of labyrinthodont amphibians, the oldest known representatives of the class. { ˌik·thē·ō·stə'gal·ē·ə }

Ichthyotomidae [INV ZOO] A monotypic family of errantian annelids in the superfamily Eunicea. { ˌik·thē·ō'täm·ə‚dē }

icicle [HYD] Ice shaped like a narrow cone, hanging point downward from a roof, fence, or other sheltered or heated source from which water flows and freezes in below-freezing air. { ˈī‚sik·əl }

icing [HYD] **1.** Any deposit or coating of ice on an object, caused by the impingement and freezing of liquid (usually supercooled) hydrometeors. **2.** A mass or sheet of ice formed on the ground surface during the winter by successive freezing of sheets of water that may seep from the ground, from a river, or from a spring. Also known as flood icing; flooding ice. { ˈī‚siŋ }

icing level [METEOROL] The lowest level in the atmosphere at which an aircraft in flight does, or could, encounter aircraft icing conditions over a given locality. { ˈī‚siŋ ‚lev·əl }

icing-rate meter [ENG] An instrument for the measurement of the rate of ice accretion on an unheated body. { ˈī‚siŋ ‚rāt ‚mēd·ər }

ICL See lifting condensation level.

ICNI See integrated communications-navigation-identification.

icon [COMPUT SCI] A symbolic representation of a computer function that appears on an electronic display and makes it possible to command this function by selecting the symbol. { ˈī‚kän }

iconic imagery [PSYCH] Fleeting images due perhaps to continuing activity of sensory organs following stimulation. { i'kän·ik 'im·ij·rē }

iconocenter [ELECTROMAG] The image of the reflection coefficient of a matched load as plotted on an Argand diagram. { ī'kän·ə‚sen·tər }

iconometer [OPTICS] **1.** An instrument used to find the size of an object of known distance or the distance of an object of known size by measurement of the image of it produced by a lens whose focal length is known. **2.** A direct viewfinder with a metal frame. { ‚ī·kə'näm·əd·ər }

iconoscope [ELECTR] A television camera tube in which a beam of high-velocity electrons scans a photoemissive mosaic that is capable of storing an electric charge pattern corresponding to an optical image focused on the mosaic. Also known as storage camera; storage-type camera tube. { ī'kän·ə‚skōp }

icosahedral group [MATH] The group of motions of three-dimensional space that transform a regular icosahedron into itself. { ī‚käs·ə¦hē·drəl 'grüp }

icosahedral virus [VIROL] A virion in the form of an icosahedron. { ī¦kä·sə¦hē·drəl 'vī·rəs }

icosahedron [MATH] A 20-sided polyhedron. { ī¦kä·sə¦hē·drən }

icositetrahedron See trapezohedron. { ī¦kä·sə‚te·trə'hē·drən }

Icosteidae [VERT ZOO] The ragfishes, a family of perciform fishes in the suborder Stromateoidei found in high seas. { ‚ī‚kä'stē·ə‚dē }

icotype [SYST] A typical, accurately identified specimen of a species, but not the basis for a published description. { ˈī‚kə‚tīp }

ICP-AES See inductively coupled plasma-atomic emission spectroscopy.

ICSH See luteinizing hormone.

ICS system See intercarrier sound system. { ī¦sē'es ‚sis·təm }

ICT See International Critical Tables.

icteric [MED] Pertaining to or characterized by jaundice. { ik'ter·ik }

Icteridae [VERT ZOO] The troupials, a family of New World perching birds in the suborder Oscines. { ik'ter·ə‚dē }

icteroanemia [VET MED] A disease of swine characterized by jaundice, anemia, and erythrocytolysis. { ¦ik·tə·rō·ə'nē·mē·ə }

icterogenic [MED] Causing icterus, or jaundice. { ¦ik·tə·rō¦jen·ik }

icterohematuria [VET MED] A disease of sheep caused by the protozoan *Babesia ovis* and characterized by hemolysis of

erythrocytes accompanied by jaundice. { ¦ik·tə·rō‚hē·mə'tùr·ē·ə }

icterohemorrhagic fever See Weil's disease. { ¦ik·tə·rō‚hēm·ə'raj·ik 'fē·vər }

icterus See jaundice. { 'ik·tə·rəs }

icterus gravis [MED] Acute yellow atrophy of the liver marked by jaundice and nervous system dysfunctions. { 'ik·tə·rəs 'grav·əs }

icterus gravis neonatorum [MED] Icterus gravis in the newborn caused by physiologic jaundice, erythroblastosis fetalis, and severe jaundice. { 'ik·tə·rəs 'grav·əs nē‚än·ə'tòr·əm }

icterus index [PATH] Measure of serum bilirubin levels by comparing the yellow blood serum from a jaundiced patient with the colors of standard potassium dichromate solutions. { 'ik·tə·rəs ‚in‚deks }

Ictidosauria [PALEON] An extinct order of mammallike reptiles in the subclass Synapsida including small carnivorous and herbivorous terrestrial forms. { ik'tid·ə'sòr·ē·ə }

ICW See interrupted continuous wave.

id [PSYCH] The primitive, psychic energy source of the unconscious. { id }

ID See inside diameter.

ID₅₀ See infective dose 50.

IDA See iminodiacetic acid.

iddingsite [MINERAL] A reddish-brown mixture of silicates, forming patches in basic igneous rocks. { 'id·iŋ‚zīt }

idea [PSYCH] **1.** A mental impression or thought. **2.** An experience or thought not directly due to an external sensory stimulation. { ī'dē·ə }

ideal [MATH] A subset I of a ring R where $x - y$ is in I for every x,y in I and either rx is in I for every r in R and x in I or xr is in I for every r in R and x in I; in the first case I is called a left ideal, and in the second a right ideal; an ideal is two-sided if it is both a left and a right ideal. { ī'dēl }

ideal aerodynamics [FL MECH] A branch of aerodynamics that deals with simplifying assumptions that help explain some airflow problems and provide approximate answers. Also known as ideal fluid dynamics. { ī'dēl ‚er·ō·dī'nam·iks }

ideal bunching [ELECTR] Theoretical condition in which the bunching of electrons in a velocity-modulated tube would give a single infinitely large current peak during each cycle. { ī'dēl 'bən·chiŋ }

ideal crystal See perfect crystal. { ¦ī‚dēl 'krist·əl }

ideal dielectric [ELEC] Dielectric in which all the energy required to establish an electric field in the dielectric is returned to the source when the field is removed. Also known as perfect dielectric. { ī'dēl ‚dī·i'lek·trik }

ideal exhaust velocity [FL MECH] The theoretical maximum velocity, relative to the nozzle, of the gas flow as it passes from a given nozzle inlet temperature and pressure to a given ambient pressure, when the combustion gas has a given mean molecular weight. { ī'dēl ig'zóst və‚läs·əd·ē }

ideal flow [FL MECH] **1.** Fluid flow which is incompressible, two-dimensional, irrotational, steady, and nonviscous. **2.** See inviscid flow. { ī'dēl 'flō }

ideal fluid [FL MECH] **1.** A fluid which has ideal flow. **2.** See inviscid fluid. { ī'dēl 'flü·əd }

ideal fluid dynamics See ideal aerodynamics. { ī'dēl ¦flü·əd dī'nam·iks }

ideal gas [THERMO] Also known as perfect gas. **1.** A gas whose molecules are infinitely small and exert no force on each other. **2.** A gas that obeys Boyle's law (the product of the pressure and volume is constant at constant temperature) and Joule's law (the internal energy is a function of the temperature alone). { ī'dēl 'gas }

ideal gas law [THERMO] The equation of state of an ideal gas which is a good approximation to real gases at sufficiently high temperatures and low pressures; that is, $PV = RT$, where P is the pressure, V is the volume per mole of gas, T is the temperature, and R is the gas constant. { ī'dēl 'gas ‚ló }

ideal grain [ORD] Propellant granulation which produces the maximum obtainable velocity of a projectile without exceeding the permissible pressure at any point along the bore. { ī'dēl 'grān }

ideal index number See Fisher's ideal index. { ī'dēl 'in‚deks ‚nəm·bər }

idealization [PSYCH] A conscious or unconscious defense

mechanism in which a person overestimates an admired aspect or attribute of another person. { ī‚dēl·ə'zā·shən }

ideal line [MATH] The collection of all ideal points, each corresponding to a given family of parallel lines. Also known as line at infinity. { ī'dēl 'līn }

ideal network [ELECTR] An interconnection of lumped, constant electrical quantities analyzed without consideration of noise and distributed parameters that would exist in actual settings. { ī'dēl 'net‚wərk }

ideal point [MATH] In projective geometry, all lines parallel to a given line are hypothesized to meet at a point at infinity, called an ideal point. Also known as point at infinity. { ī'dēl 'pȯint }

ideal productivity index [PETRO ENG] Theoretical straight-line relationship between oil production from a reservoir and the resultant pressure drop within that reservoir. { ī'dēl ‚präd·ək'tiv·əd·ē ‚in‚deks }

ideal propeller [FL MECH] A propeller which is considered as acting alone on an inviscid, incompressible fluid stream. { ī'dēl prə'pel·ər }

ideal radiator See blackbody. { ī'dēl 'rād·ē‚ād·ər }

ideal rocket [AERO ENG] A rocket motor or rocket engine that would have a velocity equal to the velocity of its jet gases. { ī'dēl 'räk·ət }

ideal solution [CHEM] A solution that conforms to Raoult's law over all ranges of temperature and concentration and shows no internal energy change on mixing and no attractive force between components. { ī'dēl sə'lü·shən }

ideal theory [MATH] The branch of algebra studying the properties of ideals. { ī'dēl 'thē·ə·rē }

ideal transducer [ELEC] Hypothetical passive transducer which transfers the maximum possible power from the source to the load. { ī'dēl tranz'dü·sər }

ideal transformer [ELEC] A hypothetical transformer that neither stores nor dissipates energy, has unity coefficient of coupling, and has pure inductances of infinitely great value. { ī'dēl tranz'fȯr·mər }

ideas of influence [PSYCH] A clinical manifestation of certain psychotic disorders in which the patients may believe that their thoughts are read, that their limbs move without their consent, or that they are under the control of someone else or some external force or influence. { ī'dē·əz əv 'in·flü·əns }

ideas of reference [PSYCH] A symptom complex in which, through the mechanism of projection, an individual incorrectly believes himself or herself to be the direct object of casual remarks or incidents or of external events. { ī'dē·əz əv 'ref·rəns }

ideation [PSYCH] Conceptualization of an idea. { ‚id·ē·ā'shən }

ideational apraxia [MED] Inability to perform meaningful motor functions or to use objects properly due to mental confusion caused by diffuse brain disease. { ‚id·ē·ā'shən·əl a'prak·sē·ə }

idem factor [MATH] The dyadic $I = ii + jj + kk$ such that scalar multiplication of I by any vector yields that vector. { 'ī‚dem ‚fak·tər }

I demodulator [ELECTR] Stage of a color television receiver which combines the chrominance signal with the color oscillator output to restore the I signal. { 'ī dē'mäj·ə‚lād·ər }

idempotent [MATH] **1.** An element x of an algebraic system satisfying the equation $x^2 = x$. **2.** An algebraic system in which every element x satisfies $x^2 = x$. { ‚ī‚dem'pōt·ənt }

idempotent law [MATH] A law which states that an element x of an algebraic system satisfies $x^2 = x$. { ‚ī‚dem'pōt·ənt 'lȯ }

idempotent matrix [MATH] A matrix E satisfying the equation $E^2 = E$. { ‚ī‚dem'pōt·ənt 'mā·triks }

identical twins See monozygotic twins. { ī'dent·ə·kəl 'twinz }

identifiable flying object [SCI TECH] A reported event or sighting identifiable as such; a thing as a balloon, meteor, planet, aircraft, or mirage reflection. Abbreviated IFO. { ī'dent·ə‚fī·ə·bəl 'flī·iŋ 'äb·jəkt }

identification [CONT SYS] The procedures for deducing a system's transfer function from its response to a step-function input or to an impulse. [PSYCH] **1.** The tendency of children to model their behavior after that of one or more selected adults. **2.** A defense mechanism in which a person likens himself or herself to someone else. { ī‚dent·ə·fə'kā·shən }

identification and authentication [COMPUT SCI] The process of determining with high assurance the identity of a person who is seeking access to a computing system. { ī‚den·tə·fə'kā·shən and ə‚then·tə'kā·shən }

identification division [COMPUT SCI] The section of a program, written in the COBOL language, which contains the name of the program and the name of the programmer. { ī‚dent·ə·fə'kā·shən di‚vizh·ən }

identification, friend or foe [ENG] A system using pulsed radio transmissions to which equipment carried by friendly forces automatically responds, by emitting a pulse code, thereby identifying themselves from enemy forces; a method of determining the friendly or unfriendly character of aircraft, ships, and army units by other aircraft, ships, or ground force units. Abbreviated IFF. { ī‚dent·ə·fə'kā·shən 'frend ər 'fō }

identifier [COMPUT SCI] A symbol whose purpose is to specify a body of data. { ī'dent·ə‚fī·ər }

identifier word [COMPUT SCI] A full-length computer word associated with a search function. { ī'dent·ə‚fī·ər ‚wərd }

identity [MATH] **1.** An equation satisfied for all possible choices of values for the variables involved. **2.** See identity element. { ī'den·ə‚dē }

identity crisis [PSYCH] The critical period in emotional maturation and personality development, occurring usually during adolescence, which involves the reworking and abandonment of childhood identifications and the integration of new personal and social identifications. { ī'den·ə‚dē ‚krī·səs }

identity element [MATH] The unique element e of a group where $g·e = e·g = g$ for every element g of the group. Also known as identity. { ī'den·ə‚dē ‚el·ə·mənt }

identity function [MATH] The function of a set to itself which assigns to each element the same element. Also known as identity operator. { ī'den·ə‚dē ‚fəŋk·shən }

identity gate See identity unit. { ī'den·ə‚dē ‚gāt }

identity matrix [MATH] The square matrix all of whose entries are zero except along the principal diagonal where they all are 1. { ī'den·ə‚dē ‚mā·triks }

identity operator See identity function. { ī'den·ə‚dē ‚äp·ə‚rād·ər }

identity unit [COMPUT SCI] A logic element with several binary input signals and a single binary output signal whose value is 1 if all the input signals have the same value and 0 if they do not. Also known as identity gate. { ī'den·ə‚dē ‚yü·nət }

ideogenous See syngenetic. { id·ē'äj·ə·nəs }

ideokinetic apraxia See ideomotor apraxia. { ‚id·ē·ō·kə'ned·ik a'prak·sē·ə }

ideomotor [PHYSIO] **1.** Pertaining to involuntary movement resulting from or accompanying some mental activity, as moving the lips while reading. **2.** Pertaining to both ideation and motor activity. { 'id·ē·ə‚mōd·ər }

ideomotor apraxia [MED] A nervous system disorder caused by lesions of the cerebral cortex in which simple single acts can be performed but not a sequence of associated acts. Also known as ideokinetic apraxia. { 'id·ē·ə‚mōd·ər a'pak·sē·ə }

ideotype [SYST] A specimen identified as belonging to a specific taxon but collected from other than the type locality. { 'id·ē·ō‚tīp }

ID grinding [MET] The grinding of the inner diameter of a piece of tubing or piping. { 'ī‚ce ‚grīn·diŋ }

idioblast [BOT] A plant cell that differs markedly in shape or function from neighboring cells within the same tissue. [GEOL] A mineral constituent of a metamorphic rock formed by recrystallization which is bounded by its own crystal faces. { 'īd·ē·ō‚blast }

idiochromatic [MINERAL] Having characteristic color, usually applied to minerals. { ‚id·ē·ō·krō'mad·ik }

idiochromatin [CYTOL] The portion of the nuclear chromatin thought to function as the physical carrier of genes. { ‚id·ē·ō·krō'mə·tən }

idiocy [PSYCH] The lowest grade of mental deficiency; the individual's mental age is less than 3 years. { 'id·ē·ə·sē }

idioglossia [PSYCH] Speech or other vocalizations unique to an individual and generally incomprehensible to others; may be normal, as during childhood development, or representative of a pathological process. { ‚id·ē·ə'gläs·ē·ə }

idiogram [GEN] A diagram of chromosome morphology,

especially the banding pattern, that is used to compare karyotypes of different cells, individuals, or species. { 'id·ē·ə,gram }

idiomorphic *See* automorphic. { ¦id·ē·ō¦mȯr·fik }

idiomuscular [PHYSIO] Pertaining to any phenomenon occurring in a muscle which is independent of outside stimuli. { ¦id·ē·ō¦məs·kyə·lər }

idiopathic arthritis *See* Reiter's syndrome. { ¦id·ē·ə¦path·ik är'thrīd·əs }

idiopathic colitis [MED] Any form of colitis for which the causative agent is not identified. { ¦id·ē·ə¦path·ik kə'līd·əs }

idiopathic eunuchoidism [MED] A primary type of eunuchoidism in which there is no mammary gland enlargement and testes remain at a prepubertal stage of development. { ¦id·ē·ə¦path·ik ,yü·nə'kȯi,diz·əm }

idiopathic familial jaundice [MED] A familial form of obstructive jaundice of unknown cause in which there is decreased ability to excrete conjugated bilirubin into the bile ducts. { ¦id·ē·ə¦path·ik fə'mil·yəl 'jȯn·dəs }

idiopathic hypercholesterolemia [MED] A genetic derangement of fat metabolism characterized by high blood, cell, and plasma levels of cholesterol. { ¦id·ē·ə¦path·ik ¦hī·pər·kə,les·tə·rə'lē·mē·ə }

idiopathic megacolon [MED] Hypertrophy and dilation of the colon. { ¦id·ē·ə¦path·ik ¦meg·ə¦kō·lən }

idiopathic pulmonary hemisiderosis [MED] A disease of unknown etiology characterized by recurrent hemorrhaging from pulmonary capillaries. { ¦id·ē·ə¦path·ik 'pu̇l·mə,ner·ē ¦hem·ə,sid·ə'rō·səs }

idiopathic steatorrhea *See* celiac syndrome. { ¦id·ē·ə¦path·ik ,stē·əd·ə'rē·ə }

idiopathic thrombocytopenic purpura [MED] Thrombocytopenic purpura of unknown causes. { ¦id·ē·ə¦path·ik ¦thräm·bō,sīd·ə'pē·nik 'pər·pyə·rə }

idiopathic ulcerative colitis [MED] A form of ulcerative colitis of unknown cause. { ¦id·ē·ə¦path·ik ,əl·sə'rād·iv kə'līd·əs }

idiopathy [MED] **1.** A primary disease; one not a result of any other disease, but of spontaneous origin. **2.** Disease for which no cause is known. { ,id·ē'äp·ə·thē }

idiophase [MICROBIO] A period in culture growth during which secondary metabolites are produced. { 'id·ē·ə,fāz }

idiosome [CYTOL] **1.** A hypothetical unit of a cell, such as the region of modified cytoplasm surrounding the centriole or centrosome. **2.** A sex chromosome. { 'id·ē·ə,sōm }

idiostatic connection [ELEC] An arrangement of a quadrant electrometer in which the vane is electrically connected to one of the quadrant pairs and the deflection of the vane is proportional to the square of the unknown voltage applied across the quadrant pairs. { ,id·ē·ə,stad·ik kə'nek·shən }

Idiostolidae [INV ZOO] A small family of hemipteran insects in the superfamily Pentatomorpha. { ,id·ē·ə'stäl·ə,dē }

idiosyncrasy [MED] A peculiarity of constitution that makes an individual react differently from most persons to drugs, diet, treatment, or other situations. [PSYCH] Any special or peculiar characteristic or temperament by which a person differs from other persons. { ,id·ē·ə'siŋ·krə·sē }

idiot [PSYCH] A person afflicted with idiocy and requiring custodial or protective care. { 'id·ē·ət }

idiot tape [GRAPHICS] Paper or magnetic tape used in computerized phototypesetting that has not been hyphenated or justified. { 'id·ē·ət ,tāp }

idiotype [IMMUNOL] The unique amino acid sequence and corresponding three-dimensional structure of the variable region of an immunoglobulin molecule that determines its antigenic specificity. { 'id·ē·ə,tīp }

I display [ELECTR] A radarscope display in which a target appears as a complete circle when the radar antenna is correctly pointed at it, the radius of the circle being proportional to target distance; when the antenna is not aimed at the target, the circle reduces to a circle segment. { 'ī di,splā }

idle [MECH ENG] To run without a load. { 'īd·əl }

idle component *See* reactive component. { 'īd·əl kəm'pō·nənt }

idle current *See* reactive current. { 'īd·əl ,kə·rənt }

idler arm [MECH ENG] In an automotive steering system, a link that supports the tie rod and transmits steering motion to both wheels through the ends of the tie rod. { 'īd·lər ,ärm }

idler frequency [ELECTR] Of a parametric device, a sum or difference frequency generated within the parametric device other than the input, output, or pump frequencies which require specific circuit consideration to achieve the desired device performance; it is called an idler frequency since, in conventional parametric amplifiers, it is more or less a useless by-product of the parametric process. { 'īd·lər ,frē·kwən·sē }

idler gear [MECH ENG] A gear situated between a driving gear and a driven gear to transfer motion, without any change of direction or of gear ratio. { 'īd·lər ,gir }

idler pulley [MECH ENG] A pulley used to guide and tighten the belt or chain of a conveyor system. { 'īd·lər ,pu̇l·ē }

idler wheel [MECH ENG] **1.** A wheel used to transmit motion or to guide and support something. **2.** A roller with a rubber surface used to transfer power by frictional means in a sound-recording or sound-reproducing system. { 'īd·lər ,wēl }

idle-stop solenoid [MECH ENG] An electrically operated plunger in a carburetor that provides a predetermined throttle setting at idle and closes the throttle completely when the ignition switch is turned off. Also known as antidieseling solenoid. { 'īd·əl ,stäp 'sō·lə,nȯid }

idle time [COMPUT SCI] The time during which a piece of hardware in good operating condition is unused. [IND ENG] A period of time during a regular work cycle when a worker is not active because of waiting for materials or instruction. Also known as waiting time. { 'īd·əl ,tīm }

idle trunk lamp [ELEC] Signal lamp associated with an outgoing trunk to indicate that the trunk is not busy. { 'īd·əl 'trəŋk ,lamp }

idling jet [MECH ENG] A carburetor part that introduces gasoline during minimum load or speed of the engine. { 'īd·liŋ ,jet }

idling system [MECH ENG] A system to obtain adequate metering forces at low airspeeds and small throttle openings in an automobile carburetor in the idling position. { 'īd·liŋ ,sis·təm }

idocrase *See* vesuvianite. { 'ī·dō,krās }

Idoteidae [INV ZOO] A family of isopod crustaceans in the suborder Valvifera having a flattened body and seven pairs of similar walking legs. { ,ī·dō'tē·ə,dē }

IDP *See* integrated data processing.

id reaction [MED] Papular and vesicular eruptions on the skin, occurring suddenly following exacerbation of foci of some cutaneous fungus infections. { 'id rē'ak·shən }

idrialite [MINERAL] A mineral composed of crystalline hydrocarbon, $C_{22}H_{14}$. { 'id·rē·ə,līt }

IDU *See* 5-iodo-2′-deoxyuridine.

IDUR *See* 5-iodo-2′-deoxyuridine.

IEEE 1394 [COMPUT SCI] The standard for connecting storage, digital audio and video, and other peripheral devices to personal computers at data transfer rates up to 400 million bits per second. Also known as firewire.

i-f *See* intermediate frequency.

i-f amplifier *See* intermediate-frequency amplifier. { 'ī'ef 'am·plə,fī·ər }

if and only if operation *See* biconditional operation. { ¦if ən 'ōn·lē ¦if ,äp·ə,rā·shən }

IF canceler [ELECTR] In radar, a moving-target indicator canceler that operates at intermediate frequencies. { 'ī'ef 'kans·lər }

IFF *See* identification, friend or foe.

IFO *See* identifiable flying object.

IFR *See* instrument flight rules.

IFR terminal minimums [METEOROL] The operational weather limits concerned with minimum conditions of ceiling and visibility at an airport under which aircraft may legally approach and land under instrument flight rules; these minimums frequently are in the form of a sliding scale, and also vary with aircraft type, pilot experience, and from airport to airport. { 'ī'ef'är ¦tər·mən·əl 'min·ə·məmz }

IFR weather *See* instrument weather. { 'ī'ef'är 'weth·ər }

IF statement *See* conditional jump. { 'if ,stāt·mənt }

if then else [COMPUT SCI] A logic statement in a high-level programming language that defines the data to be compared and the actions to be taken as the result of a comparison. { ¦if ¦then 'els }

if-then operation *See* implication.

i-f transformer *See* intermediate-frequency transformer. { 'ī'ef tranz'fȯr·mər }

Ig *See* immunoglobulin entries.

IDOTEIDAE

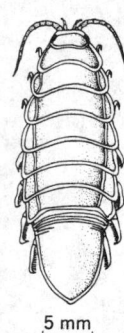

5 mm

Idotea neglecta, showing flattened body and seven pairs of legs. *(From G. O. Sars, An Account of the Crustacea of Norway, vol. 2, 1899)*

IGBT *See* insulated-gate bipolar transistor.

IGES *See* initial graphics exchange specification.

Igewsky's solution [MET] Etchant used to prepare carbon steels for microscopic analysis; consists of 5% picric acid in absolute alcohol. { ē'gev·skēz sə,lü·shən }

IGFET *See* metal oxide semiconductor field-effect transistor. { 'ig,fet }

igneous [PETR] Pertaining to rocks which have congealed from a molten mass. { 'ig·nē·əs }

igneous complex [PETR] An assemblage of igneous rocks that are intimately associated and roughly contemporaneous. { 'ig·nē·əs ¦käm,pleks }

igneous facies [PETR] A part of an igneous rock differing in structure, texture, or composition from the main mass. { 'ig·nē·əs 'fā·shēz }

igneous meteor [GEOPHYS] A visible electric discharge in the atmosphere; lightning is the most common and important type, but types of corona discharge are also included. { 'ig·nē·əs ,mēd·ē·ər }

igneous mineral [MINERAL] Mineral material forming igneous rock. { 'ig·nē·əs ¦min·rəl }

igneous petrology [PETR] The study of igneous rocks, their occurrence, composition, and origin. { 'ig·nē·əs pi'träl·ə·jē }

igneous province *See* petrographic province. { 'ig·nē·əs präv·əns }

igneous theory *See* volcanic theory. { 'ig·nē·əs ,thē·ə·rē }

ignimbrite [PETR] A rock deposit (welded or not) resulting from one or more ground-hugging flows of hot volcanic fragments and particles commonly produced during explosive eruptions (pyroclastic flows and tephra fall). Most ignimbrites have a sheet-like shape, cover many thousands of square kilometers, and have chemical compositions that span the range commonly exhibited by igneous rocks (basaltic to rhyolitic). Also known as ash-flow tuff; pyroclastic-flow deposit; welded tuff. { 'ig·nəm,brīt }

ignite [CHEM] To start a fuel burning. { ig'nīt }

igniter [ENG] **1.** A device for igniting a fuel mixture. **2.** A charge, as of black powder, to facilitate ignition of a propelling or bursting charge. { ig'nīd·ər }

igniter cord [ENG] A cord which passes an intense flame along its length at a uniform rate to light safety fuses in succession. { ig'nīd·ər ,kōrd }

igniter pad [ORD] A black powder charge in the form of a thin pad of cartridge cloth attached to separate-loading propelling charge to facilitate complete and uniform ignition. { ig'nīd·ər ,pad }

igniter train [ORD] Step-by-step arrangement of charges in pyrotechnic munitions by which the initial fire from the primer is transmitted and intensified until it reaches and sets off the main charge. Also known as burning train. { ig'nīd·ər ,trān }

igniting fuse [ORD] A fuse designed to initiate its main munition by an igniting action, as compared to the detonating action of a detonating fuse, and suitable only for munitions using a main charge of low explosive or other readily ignitable material. { ig'nīd·iŋ ,fyüz }

igniting primer [ORD] An auxiliary primer that carries the fire from a primer to the propelling charge in certain subcaliber tubes. { ig'nīd·iŋ ,prī·mər }

ignition [CHEM] The process of starting a fuel mixture burning, or the means for such a process. { ig'nish·ən }

ignition charge [MATER] A small quantity of explosive, usually composed of black powder, that facilitates the firing of the main charge. { ig'nish·ən ,chärj }

ignition coil [ELECTROMAG] A coil in an ignition system which stores energy in a magnetic field relatively slowly and releases it suddenly to ignite a fuel mixture. { ig'nish·ən ,kȯil }

ignition delay *See* ignition lag. { ig'nish·ən di,lā }

ignition interference [COMMUN] Radio interference due to the spark discharges in an automotive or other ignition system. { ig'nish·ən ,int·ə'fir·əns }

ignition lag [MECH ENG] In the internal combustion engine, the time interval between the passage of the spark and the inflammation of the air-fuel mixture. Also known as ignition delay. { ig'nish·ən ,lag }

ignition point *See* ignition temperature. { ig'nish·ən ,pȯint }

ignition quality [CHEM ENG] The property of a fuel that ignites when injected into the compressed-air charge in a diesel engine cylinder; measurement is given in terms of cetane number. { ig'nish·ən ,kwäl·əd·ē }

ignition reserve [ELEC] In an ignition system for an internal combustion engine, the difference between the minimum voltage available and the maximum voltage required by the system. { ig'nish·ən ri,zərv }

ignition system [MECH ENG] The system in an internal combustion engine that initiates the chemical reaction between fuel and air in the cylinder charge by producing a spark. { ig'nish·ən ,sis·təm }

ignition temperature [CHEM] The lowest temperature at which combustion begins and continues in a substance when it is heated in air. Also known as autogenous ignition temperature; ignition point. [PL PHYS] The lowest temperature at which the fusion energy generated in a plasma exceeds the energy lost through bremsstrahlung radiation. { ig'nish·ən ,tem·prə·chər }

ignition timing [ELEC] The time of delivery of the spark from the coil to the spark plug in relation to the time the piston reaches the correct position for the power stroke in an internal combustion engine. { ig'nish·ən ,tīm·iŋ }

ignitor [ELECTR] **1.** An electrode used to initiate and sustain the discharge in a switching tube. Also known as keep-alive electrode (deprecated). **2.** A pencil-shaped electrode, made of carborundum or some other conducting material that is not wetted by mercury, partly immersed in the mercury-pool cathode of an ignitron and used to initiate conduction at the desired point in each alternating-current cycle. { ig'nīd·ər }

ignitron [ELECTR] A single-anode pool tube in which an ignitor electrode is employed to initiate the cathode spot on the surface of the mercury pool before each conducting period. { 'ig·nə,trän }

ignitron contactor [ELECTR] A circuit containing an ignitron and control contacts that serves as a heavy-duty switch in the primary of a resistance-welding transformer. { 'ig·nə,trän 'kän,tak·tər }

ignorable coordinate *See* cyclic coordinate. { ig'nȯr·ə·bəl kō'ȯrd·ən·ət }

ignore character [COMPUT SCI] Also known as erase character. **1.** A character indicating that no action whatever is to be taken, that is, a character to be ignored; often used to obliterate an erroneous character. **2.** A character indicating that the preceding or following character is to be ignored, as specified. **3.** A character indicating that some specified action is not to be taken. { ig'nȯr ,kar·ik·tər }

iguana [VERT ZOO] The common name for a number of species of herbivorous, arboreal reptiles in the family Iguanidae characterized by a dorsal crest of soft spines and a dewlap; there are only two species of true iguanas. { i'gwän·ə }

iguanid [VERT ZOO] The common name for members of the reptilian family Iguanidae. { i'gwän·əd }

Iguanidae [VERT ZOO] A family of reptiles in the order Squamata having teeth fixed to the inner edge of the jaws, a nonretractile tongue, a compressed body, five clawed toes, and a long but rarely prehensile tail. { i'gwän·ə,dē }

Iguanodon [PALEON] A herbivorous ornithopod dinosaur, 30 feet (9 meters) long and weighing 5 tons, that appeared during the Early Cretaceous Period. { i'gwän·ə,dän }

IGY *See* International Geophysical Year.

I-head cylinder [MECH ENG] The internal combustion engine construction having both inlet and exhaust valves located in the cylinder head. { 'Ī,hed ,sil·ən·dər }

ihleite *See* copiapite. { 'ē·lə,īt }

ihp *See* indicated horsepower.

IIR filter *See* infinite impulse response filter. { ¦Ī¦Ī'är ,fil·tər }

I indicator *See* I scope. { 'Ī ,in·də,kād·ər }

ijolite [PETR] A plutonic rock of nepheline and 30–60% mafic materials, generally sodic pyroxene, with accessory apatite, sphene, calcite, and titaniferous garnet. { 'ē·ə,līt }

IL *See* interleukin entries.

I²L *See* integrated injection logic.

ilang-ilang oil [MATER] An oil made synthetically or derived from the flowers of the tree *Canangium odorata*, containing pinene, geraniol, and linalol. Also known as cananga oil; ylang-ylang oil. { 'ē,läŋ 'ē,läŋ ,ȯil }

Ilarvirus [VIROL] A genus of the family Bromoviridae that is characterized by three types of particles, which are quasi-isometric, different in diameter, and contain four types of ribonucleic acid (1.1, 0.9, 0.7, 0.3 × 10⁶) separately encapsidated,

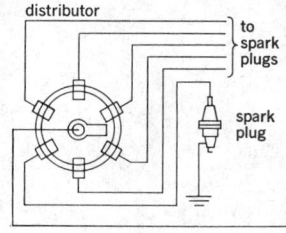

IGNITION SYSTEM

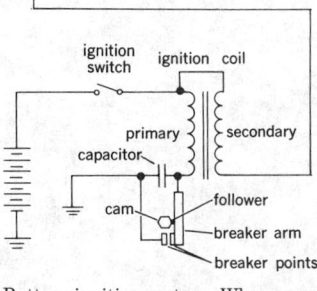

Battery ignition system. When switch and breaker parts are closed, current flows through primary, inducing magnetic field in core of ignition coil. When breaker points open, as engine driven cam rotates, primary current falls rapidly and magnetic field collapses, inducing very high voltage in secondary, which is led to spark plugs, in proper sequence, by distributor. *(Modified from A. R. Rogowski, Elements of Internal-combustion Engines, McGraw-Hill, 1953)*

except for the 3×10^5 species which may be found in combination with others; the type species is tobacco streak virus.   Also known as tobacco streak virus group.   { 'ī·lə̩vī·rəs }

ileitis   [MED]   Inflammation of the ileum.   { ‚il·ē'īd·əs }

ileocecal valve   [ANAT]   A muscular structure at the junction of the ileum and cecum which prevents reflex of the cecal contents.   { ‚il·ē·ō̩sē·kəl 'valv }

ileocolic artery   [ANAT]   A branch of the superior mesenteric artery that supplies blood to the terminal part of the ileum and the beginning of the colon.   { ‚il·ē·ō̩käl·ik 'ärd·ə·rē }

ileocolic intussusception   [MED]   Slipping of the ileum through the ileocecal valve into the colon.   { ‚il·ē·ō̩käl·ik ‚in·tə·sə'sep·shən }

ileocolitis   [MED]   Inflammation of both the ileum and the colon.   { ‚il·ē·ō·kə'līd·əs }

ileocolostomy   [MED]   Surgical formation of a bypass channel between the ileum and the colon.   { ‚il·ē·ō·kə'lä·stə·mē }

ileocutaneous   [MED]   Pertaining to a joining of the ileum and the skin resulting from a fistulous connection.   { ‚il·ē·ō̩kyü'tā·nē·əs }

ileostomy   [MED]   Surgical formation of an artificial anus through the abdominal wall into the ileum.   { ‚il·ē'ä·stə·mē }

ilesite   [MINERAL] $(Mn,Zn,Fe)SO_4·4H_2O$ A green mineral composed of hydrous manganese zinc iron sulfate.   { 'īl̩zīt }

ileum   [ANAT]   The last portion of the small intestine, extending from the jejunum to the large intestine.   { 'il·ē·əm }

ileus   [MED]   Acute intestinal obstruction of neurogenic origin.   { 'il·ē·əs }

Ilex opaca   See American holly.   { ‚ī‚leks ō'päk·ə }

ILF   See infralow frequency.

iliac artery   [ANAT]   Either of the two large arteries arising by bifurcation of the abdominal aorta and supplying blood to the lower trunk and legs (or hind limbs in quadrupeds).   Also known as common iliac artery.   { 'il·ē̩ak 'ärd·ə·rē }

iliac crest height   [ANTHRO]   A measure of the vertical distance from the top of the iliac crest to the floor while the subject stands.   { 'il·ē̩ak ̩krest ̩hīt }

iliac fascia   [ANAT]   The fascia covering the pelvic surface of the iliacus muscle.   { 'il·ē̩ak 'fā·shə }

iliac index   [ANTHRO]   The ratio multiplied by 100 of the distance between the iliac spines and the distance between the lower margin of the acetabulum and the topmost crest of the ilium.   { 'il·ē̩ak 'in‚deks }

iliac region   See inguinal region.   { 'il·ē̩ak ̩rē·jən }

iliacus   [ANAT]   The portion of the iliopsoas muscle arising from the iliac fossa and sacrum.   { i'lī·ə·kəs }

iliac vein   [ANAT]   Any of the three veins on each side of the body which correspond to and accompany the iliac artery.   { 'il·ē̩ak ̩vān }

iliocostalis   [ANAT]   The lateral portion of the erector spinal muscle that extends the vertebral column and assists in lateral movements of the trunk.   { ‚il·ē·ō̩kä'sta·ləs }

iliofemoral ligament   [ANAT]   A strong band of dense fibrous tissue extending from the anterior inferior iliac spine to the lesser trochanter and the intertrochanteric line.   Also known as Y ligament.   { ‚il·ē·ō̩fem·ə·rəl ̩lig·ə·mənt }

iliolumbar ligament   [ANAT]   A fibrous band that radiates laterally from the transverse processes of the fourth and fifth lumbar vertebrae and attaches to the pelvis by two main bands.   { ‚il·ē·ō̩ləm̩bär ̩lig·ə·mənt }

iliopsoas   [ANAT]   The combined iliacus and psoas muscles.   { ‚il·ē·ō̩sō·əs }

iliotibial tract   [ANAT]   A thickened portion of the fascia lata extending from the lateral condyle of the tibia to the iliac crest.   { ‚il·ē·ō̩tib·ē·əl ̩trakt }

ilium   [ANAT]   Either of a pair of bones forming the superior portion of the pelvis bone in vertebrates.   { 'il·ē·əm }

Ilkovič equation   [ANALY CHEM]   Mathematical relationship between diffusion current, diffusion coefficient, and active-substance concentration; used for polarographic analysis calculations.   { 'il·kə̩vich i̩kwä·zhən }

ill-conditioned problem   [COMPUT SCI]   A problem in which a small error in the data or in subsequent calculation results in much larger errors in the answers.   { ‚il kən̩dish·ənd 'präb·ləm }

illegal character   [COMPUT SCI]   A character or combination of bits that is not accepted as a valid representation by a computer or by a specific routine; commonly detected and used as an indication of a machine malfunction.   { i'lē·gəl 'kar·ik·tər }

illegal operation   [COMPUT SCI]   An operation specified by a program instruction that cannot be carried out by the computer.   { i'lē·gəl äp·ə'rā·shən }

illepé fat   See mowrah fat.   { i·lə·pē ‚fat }

Illiciales   [BOT]   An order of dicotyledonous flowering plants, division Magnoliophyta, of the class Magnoliopsida, characterized by having woody plants with scattered spherical cells containing volatile oils.   { i‚lis·ē'ā·lēz }

illinium   See promethium-147.   { ə'lin·ē·əm }

Illinoian   [GEOL]   The third glaciation of the Pleistocene in North America, between the Yarmouth and Sangamon interglacial stages.   { ‚il·ə̩nói·ən }

illite   [MINERAL]   A group of gray, green, or yellowish-brown micalike clay minerals found in argillaceous sediments; intermediate in composition and structure between montmorillonite and muscovite.   { 'i‚līt }

illness   [MED]   **1.** The state of being sick.   **2.** A sickness, disease, or disorder.   { 'il·nəs }

ill-posed problem   [MATH]   A problem which may have more than one solution, or in which the solutions depend discontinuously upon the initial data.   Also known as improperly posed problem.   { 'il ̩pōzd 'präb·ləm }

illuminance   [OPTICS]   The density of the luminous flux on a surface.   Also known as illumination; luminous flux density.   { ə'lü·mə·nəns }

illuminating   [GRAPHICS]   The hand decoration of books, as done in medieval times, with drawings and miniature paintings, or with embellishments, usually in red, blue, and gold, added to initial letters and borders.   { ə'lü·mə̩nād·iŋ }

illuminating gas   [MATER]   Flammable mixture of gases suitable for illuminating purposes; contains hydrogen, methane, ethane, carbon monoxide, and some nitrogen and oxygen.   { ə'lü·mə̩nād·iŋ ̩gas }

illuminating grenade   [ORD]   Hand grenade or rifle grenade designed to provide illumination by a burning action; used also as a trip flare or as an incendiary device.   { ə'lü·mə̩nād·iŋ grə̩nād }

illuminating oil   [MATER]   An oil such as kerosine suitable for burning to provide illumination.   { ə'lü·mə̩nād·iŋ ̩ói l }

illuminating projectile   [ORD]   Projectile, with a time fuse, that releases a parachute flare at any desired height; used for lighting up an area.   Also known as star shell.   { ə'lü·mə̩nād·iŋ prə̩jek·təl }

illumination   [ELECTROMAG]   **1.** The geometric distribution of power reaching various parts of a dish reflector in an antenna system.   **2.** The power distribution to elements of an antenna array.   [OPTICS]   **1.** The science of the application of lighting.   **2.** See illuminance.   { ə̩lü·mə̩nā·shən }

illumination climate   [METEOROL]   Also known as light climate.   **1.** The worldwide distribution of natural light from the sun and sky (direct solar radiation plus diffuse sky radiation) as received on a horizontal surface.   **2.** The character of total illumination at any given place.   { ə̩lü·mə̩nā·shən ̩klī·mət }

illumination control   [ELECTR]   A photoelectric control that turns on lights when outdoor illumination decreases below a predetermined level.   { ə̩lü·mə̩nā·shən kən̩trōl }

illumination design   [ENG]   Design of sources of lighting and of systems which distribute light in order to effect a comfortable and satisfactory environment for seeing.   { ə̩lü·mə̩nā·shən di̩zīn }

illumination distribution   [OPTICS]   The manner in which light is dispersed on a surface.   { ə̩lü·mə̩nā·shən ̩di·strə'byü·shən }

illuminometer   [OPTICS]   A portable photometer which is used in the field or outside the laboratory and yields results of lower accuracy than a laboratory photometer.   { ə̩lü·mə'näm·əd·ər }

illusion   [PSYCH]   A false interpretation of a real sensation; a perception that misinterprets the object perceived.   { ə'lü·zhən }

illusory correlation   See nonsense correlation.   { i̩lü·zə·rē ̩kä·rə'lā·shən }

illustration board   [GRAPHICS]   A board made by mounting good drawing paper on a stiff backing, usually a filled pulpboard; surfaces vary from smooth to rough.   { il·ə'strā·shən ̩bórd }

illustration program   See drawing program.   { ‚il·ə'strā·shən ̩prō·grəm }

illuvial [GEOL] Pertaining to a region or material characterized by the accumulation of soil by the illuviation of another zone or material. { i'lü·vē·əl }

illuvial horizon *See* B horizon. { i'lü·vē·əl hə'rīz·ən }

illuviation [GEOL] The deposition of colloids, soluble salts, and small mineral particles in an underlying layer of soil. { i,lü·vē'ā·shən }

illuvium [GEOL] Material leached by chemical or other processes from one soil horizon and deposited in another. { i'lü·vē·əm }

ilmenite [MINERAL] FeTiO₃ An iron-black, opaque, rhombohedral mineral that is the principal ore of titanium. Also known as mohsite; titanic iron ore. { il·mə,nīt }

ILS *See* instrument landing system.

ilsemannite [MINERAL] A black, blue-black, or blue mineral composed of hydrous molybdenum oxide or perhaps sulfate, occurring in earthy massive form. { 'il·sə·mə,nīt }

ILS reference point [NAV] A point that is on the centerline of a runway served by an ILS and that is designated as the optimum point of contact for the landing aircraft. Derived from instrument landing system reference point. { ī'el¦es 'ref·rəns ,pȯint }

image [ACOUS] *See* acoustic image. [COMMUN] **1.** One of two groups of side bands generated in the process of modulation; the unused group is referred to as the unwanted image. **2.** The scene reproduced by a television or facsimile receiver. [COMPUT SCI] A copy of the information contained in one medium recorded on a different data medium. [ELEC] *See* electric image. [ELECTROMAG] The input reflection coefficient corresponding to the reflection coefficient of a specified load when the load is placed on one side of a waveguide junction and a slotted line is placed on the other. [MATH] **1.** For a point x in the domain of a function f, the point $f(x)$. **2.** For a subset A of the domain of a function f, the set of all points that are equal to $f(x)$ for some point x in A. [OPTICS] An optical counterpart of a self-luminous or illuminated object formed by the light rays that traverse an optical system; each point of the object has a corresponding point in the image from which rays diverge or appear to diverge. [PHYS] Any reproduction of an object produced by means of focusing light, sound, electron radiation, or other emanations coming from the object or reflected by the object. [PSYCH] A representation of a sensory experience, occurring in the brain. { 'im·ij }

image admittance [ELECTR] The reciprocal of image impedance. { 'im·ij ad,mit·əns }

image antenna [ELECTROMAG] A fictitious electrical counterpart of an actual antenna, acting mathematically as if it existed in the ground directly under the real antenna and served as the direct source of the wave that is reflected from the ground by the actual antenna. { 'im·ij an,ten·ə }

image attenuation constant [ELECTR] The real part of the image transfer constant. { 'im·ij ə,ten·yə'wā·shən ,kän·stənt }

image converter [ELECTR] *See* image tube. [OPTICS] A converter that uses a fiber optic bundle to change the form of an image, for more convenient recording and display or for the coding of secret messages. { 'im·ij kən,vərd·ər }

image converter camera [ELECTR] A camera consisting of an image tube and an optical system which focuses the image produced on the phosphorescent screen of the tube onto photographic film. { 'im·ij ,kən,vərd·ər ,kam·rə }

image dissection photography [ELECTR] A method of high-speed photography in which an image is split in any one of various ways into interlaced space and time elements which can be unscrambled or played back through the system either to be viewed or to give a master negative. { 'im·ij di,sek·shən fə,täg·rə·fē }

image dissector [COMPUT SCI] In optical character recognition, a device that optically examines an input character for the purpose of breaking it down into its prescribed elements. { 'im·ij di,sek·tər }

image dissector tube [ELECTR] A television camera tube in which an electron image produced by a photoemitting surface is focused in the plane of the defining aperture and is scanned past that aperture. Also known as Farnsworth image dissector tube. { 'im·ij di,sek·tər ,tüb }

image effect [ELECTROMAG] Effect produced on the field of an antenna due to the presence of the earth; electromagnetic

waves are reflected from the earth's surface, and these reflections often are accounted for by an image antenna at an equal distance below the earth's surface. { 'im·ij i,fekt }

image enhancement [COMPUT SCI] Improvement of the quality of a picture, with the aid of a computer, by giving it higher contrast or making it less blurred or less noisy. { 'im·ij in'hans·mənt }

image force [ELEC] The electrostatic force on a charge in the neighborhood of a conductor, which may be thought of as the attraction to the charge's electric image. { 'im·ij ,fȯrs }

image frequency [ELECTR] An undesired carrier frequency that differs from the frequency to which a superheterodyne receiver is tuned by twice the intermediate frequency. { 'im·ij ,frē·kwən·sē }

image iconoscope [ELECTR] A camera tube in which an optical image is projected on a semitransparent photocathode, and the resulting electron image emitted from the other side of the photocathode is focused on a separate storage target; the target is scanned on the same side by a high-velocity electron beam, neutralizing the elemental charges in sequence to produce the camera output signal at the target. Also known as superemitron camera (British usage). { 'im·ij ī'kän·ə,skōp }

image impedance [ELECTR] One of the impedances that, when connected to the input and output of a transducer, will make the impedances in both directions equal at the input terminals and at the output terminals. { 'im·ij im,pēd·əns }

image intensifier *See* light amplifier. { 'im·ij in'ten·sə,fī·ər }

image interference [COMMUN] Interference occurring in a superheterodyne receiver when a station broadcasting on the image frequency is received along with the desired station. { 'im·ij ,in·tər'fir·əns }

image isocon [ELECTR] A television camera tube which is similar to the image orthicon but whose return beam consists of scanning beam electrons that are scattered by positive stored charges on the target. { 'im·ij 'ī·sə,kän }

image load [ELECTR] Load parameters reflected back to the source by line discontinuities. { 'im·ij ,lōd }

image orthicon [ELECTR] A television camera tube in which an electron image is produced by a photoemitting surface and focused on one side of a separate storage tube that is scanned on its opposite side by a beam of low-velocity electrons; electrons that are reflected from the storage tube, after positive stored charges are neutralized by the scanning beam, form a return beam which is amplified by an electron multiplier. { 'im·ij 'ȯr·thə,kän }

image parameter design [ELECTR] A method of filter design using image impedance and image transfer functions as the fundamental network functions. { 'im·ij pə¦ram·əd·ər di,zīn }

image parameter filter [ELECTR] A filter constructed by image parameter design. { 'im·ij pə¦ram·əd·ər ,fil·tər }

image phase constant [ELECTR] The imaginary part of the image transfer constant. { 'im·ij 'fāz ,kän·stənt }

image plane [OPTICS] The plane in which an image produced by an optical system is formed; if the object plane is perpendicular to the optical axis, the image plane will ordinarily also be perpendicular to the axis. { 'im·ij ,plān }

image potential [ELEC] The potential set up by an electric image. { 'im·ij pə,ten·chəl }

image processing [COMPUT SCI] A technique in which the data from an image are digitized and various mathematical operations are applied to the data, generally with a digital computer, in order to create an enhanced image that is more useful or pleasing to a human observer, or to perform some of the interpretation and recognition tasks usually performed by humans. Also known as picture processing. { 'im·ij ,präses·iŋ }

image ratio [ELECTR] In a heterodyne receiver, the ratio of the image frequency signal input at the antenna to the desired signal input for identical outputs. { 'im·ij ,rā·shō }

image reject mixer [ELECTR] Combination of two balanced mixers and associated hybrid circuits designed to separate the image channel from the signal channels normally present in a conventional mixer; the arrangement gives image rejection up to 30 decibels without the use of filters. { 'im·ij 'rē,jekt ,mik·sər }

image response [ELECTR] The response of a superheterodyne receiver to an undesired signal at its image frequency. { 'im·ij ri,späns }

image restoration [COMPUT SCI] Operation on a picture with a digital computer to make it more closely resemble the original object. { 'im·ij ˌres·tə'rā·shən }

imagery [PSYCH] Mental images that are collectively recalled. { 'im·ij·rē }

image space [OPTICS] The region of space where real or virtual images are formed by an optical system. { 'im·ij ˌspās }

image spacing [GRAPHICS] In microfilm technology, the area between the trailing edge of one image and the leading edge of the next. { 'im·ij ˌspās·iŋ }

image-storage array [ELECTR] A solid-state panel or chip in which the image-sensing elements may be a metal oxide semiconductor or a charge-coupled or other light-sensitive device that can be manufactured in a high-density configuration. { 'im·ij ˌstȯr·ij əˌrā }

image surface [OPTICS] A surface on which lie images of points on a given plane perpendicular to the axis of an optical system. { 'im·ij ˌsər·fəs }

image table [CONT SYS] A data table that contains the status of all inputs, registers, and coils in a programmable controller. { 'im·ij ˌtā·bəl }

image transfer constant [ELECTR] One-half the natural logarithm of the complex ratio of the steady-state apparent power entering and leaving a network terminated in its image impedance. { 'im·ij 'tranz·fər ˌkän·stənt }

image tube [ELECTR] An electron tube that reproduces on its fluorescent screen an image of the optical image or other irradiation pattern incident on its photosensitive surface. Also known as electron image tube; image converter. { 'im·ij ˌtüb }

imaginal disk [INV ZOO] Any of the thickened areas within the sac of the body wall in holometabolous insects which give rise to specific organs in the adult. { ə'maj·ən·əl ˌdisk }

imaginary axis [MATH] All complex numbers $x + iy$ where $x = 0$; the vertical coordinate axis for the complex plane. { ə'maj·əˌner·ē 'ak·səs }

imaginary circle [MATH] The set of points in the x-y plane that satisfy the equation $x^2 + y^2 = -r^2$, or $(x - h)^2 + (y - k)^2 = -r^2$, where r is greater than zero, and $x, y, h,$ and k are allowed to be complex numbers. { i'maj·əˌner·ē 'sər·kəl }

imaginary number [MATH] A complex number of the form $a + bi$, with b not equal to zero, where a and b are real numbers, and $i = \sqrt{-1}$; some mathematicians require also that $a = 0$. Also known as imaginary quantity. { ə'maj·əˌner·ē 'nəm·bər }

imaginary part [MATH] For a complex number $x + iy$ the imaginary part is the real number y. { ə'maj·əˌner·ē 'pärt }

imaginary quantity See imaginary number. { ə'maj·əˌner·ē 'kwän·əd·ē }

imaging [PHYS] The formation of images of objects. { 'im·i·jiŋ }

imaging radar [ENG] Radar carried on aircraft which forms images of the terrain. { 'im·i·jiŋ 'rāˌdär }

imago [INV ZOO] The sexually mature, usually winged stage of insect development. [PSYCH] An unconscious mental picture, usually idealized, of a parent or loved person important in the early development of an individual and carried into adulthood. { ə'mäˌgō }

IMAP See Internet Mail Access Protocol. { 'īˌmap }

imbecile [PSYCH] A person of middle-grade mental deficiency; the individual's mental age is between 3 and 7 years. { 'im·bə·səl }

imbed See embed. { im'bed }

imbedding [MATH] A homeomorphism of one topological space to a subspace of another topological space. { im'bed·iŋ }

imbibition [PHYS CHEM] Absorption of liquid by a solid or a semisolid material. { ˌim·bə'bish·ən }

imbricate [BIOL] Having overlapping edges, such as scales, or the petals of a flower. { 'im·brə·kət }

imbricate structure [GEOL] **1.** A sedimentary structure characterized by shingling of pebbles all inclined in the same direction with the upper edge of each leaning downstream or toward the sea. Also known as shingle structure. **2.** Tabular masses that overlap one another and are inclined in the same direction. Also known as schuppen structure; shingle-block structure. { 'im·brə·kət ˌstrək·chər }

imbrication [GEOL] Formation of an imbricate structure. Also known as shingling. { ˌim·brə'kā·shən }

imerinite [MINERAL] $Na_2(Mg,Fe)_6Si_8O_{22}(O,OH)_2$ A colorless to blue mineral composed of a basic silicate of sodium, iron, and magnesium, occurring as acicular crystals. { ˌim·ə'rēˌnīt }

Imhoff cone [CIV ENG] A graduated glass vessel for measuring settled solids in testing the composition of sewage. { 'imˌhȯf ˌkōn }

Imhoff tank [CIV ENG] A sewage treatment tank in which digestion and settlement take place in separate compartments, one below the other. { 'imˌhȯf ˌtaŋk }

imidazole [ORG CHEM] $C_3H_4N_2$ One of a group of organic heterocyclic compounds containing a five-membered diunsaturated ring with two nonadjacent nitrogen atoms as part of the ring; the particular compound imidazole is a member of the group. { ˌim·ə'daˌzōl }

imidazolyl [ORG CHEM] $C_3H_3N_2$· A free radical derived from imidazole. { ˌim·ə'daˌzəˌlil }

imide [ORG CHEM] **1.** A compound derived from acid anhydrides by replacing the oxygen (O) with the =NH group. **2.** A compound that has either the =NH group or a secondary amine in which R is an acyl functional group, as R_2NH. { 'iˌmīd }

imine [ORG CHEM] A class of compounds that are the product of condensation reactions of aldehydes or ketones with ammonia or amines; they have the NH radical attached to the carbon with the double bond, as R−HC=NH; an example is benzaldimine. { 'iˌmēn }

imino acid [ORG CHEM] Organic acid in which the =NH group is attached to one or two carbons; for example, acetic acid, $NH(CH_2COOH)_2$. { 'im·əˌnō 'as·əd }

imino compound [ORG CHEM] A compound that has the =NH radical attached to one or two carbon atoms. { 'im·əˌnō ˌkämˌpau̇nd }

iminodiacetic acid [ORG CHEM] $C_4H_7NO_4$ A crystalline substance used as an intermediate in the manufacture of chelating agents, surface-active agents, and complex salts. Abbreviated IDA. Also known as diglycine; iminodiethanoic acid. { ˌim·əˌnō·dī·ə'sēd·ik 'as·əd }

imino nitrogen [ORG CHEM] Nitrogen combined with hydrogen in the imino group. { 'im·əˌnō 'nī·trə·jən }

imitative deception [ELECTR] Introduction of electromagnetic radiations into enemy channels which imitate their own emissions, in order to mislead them. { 'im·əˌtād·iv di'sep·shən }

immarginate [BIOL] Lacking a clearly defined margin. { i'mär·jəˌnāt }

immature soil See azonal soil. { ˌim·ə'chu̇r 'sȯil }

immediate-access [COMPUT SCI] **1.** Pertaining to an access time which is relatively brief, or to a relatively fast transfer of information. **2.** Pertaining to a device which is directly connected with another device. { i'mē·dē·ət 'ak·ses }

immediate address [COMPUT SCI] The value of an operand contained in the address part of an instruction and used as data by this instruction. { i'mē·dē·ət 'aˌdres }

immediate data [COMPUT SCI] Data that appears in an instruction exactly as it is to be processed. { i'mēd·ē·ət 'dad·ə }

immediate hypersensitivity [IMMUNOL] A type of hypersensitivity in which the response rapidly occurs following exposure of a sensitized individual to the antigen. { i'mē·dē·ət ˌhī·pər·sen·sə'tiv·əd·ē }

immediate instruction [COMPUT SCI] A computer program instruction, part of which contains the actual data to be operated upon, rather than the address of that data. { i'mēd·ē·ət in'strək·shən }

immediate operand [COMPUT SCI] An operand contained in the instruction which specifies the operation. { i'mē·dē·ət 'äp·əˌrand }

immediate processing See demand processing. { i'mē·dē·ət 'präs·es·iŋ }

immersion [ASTRON] The disappearance of a celestial body either by passing behind another or passing into another's shadow. [MATH] A mapping f of a topological space X into a topological space Y such that for every $x \in X$ there exists a neighborhood N of x, such that f is a homeomorphism of N onto $f(N)$. [SCI TECH] Placement into or within a fluid, usually water. { ə'mər·zhən }

IMIDAZOLE

Structural formula of imidazole.

immersion cleaning [MET] Removing surface dirt from metal by dipping into a cleaning liquid. { ə'mər·zhən ‖klēn·iŋ }

immersion coating [ENG] Applying material to the surface of a metal or ceramic by dipping into a liquid. { ə'mər·zhən ‖kōd·iŋ }

immersion electron lens [ELECTR] An electron lens in which the object, usually the cathode, lies deep within the electric field so that the index of refraction varies rapidly in its vicinity. { i‖mər·zhən i'lk‚trän ‚lenz }

immersion electron microscope [ELECTR] An emission electron microscope in which the specimen is a flat conducting surface which may be heated, illuminated, or bombarded by high-velocity electrons or ions so as to emit low-velocity thermionic, photo-, or secondary electrons; these are accelerated to a high velocity in an immersion objective or cathode lens and imaged as in a transmission electron microscope. { ə'mər·zhən i‖lek‚trän 'mī·krə‚skōp }

immersion electrostatic lens See bipotential electrostatic lens. { ə'mər·zhən i‖lek·trə‚stad·ik 'lenz }

immersion foot [MED] A serious and disabling condition of the feet due to prolonged immersion in seawater at 60°F (15.6°C) or lower, but not at freezing temperature. { ə'mər·zhən ‚fút }

immersion heater [ELEC] An electric device for heating a liquid by direct immersion in the liquid. { ə'mər·zhən ‚hēd·ər }

immersion lens See immersion objective. { ə'mər·zhən ‚lenz }

immersion objective [OPTICS] A high-power microscope objective designed to work with the space between the objective and the cover glass over the object filled with an oil whose index of refraction is nearly the same as that of the objective and the cover glass, in order to reduce reflection losses and increase the index of refraction of the object space. Also known as immersion lens. { ə'mər·zhən əb‚jek·tiv }

immersion plating [MET] Applying an adherent layer of more-noble metal to the surface of a metal object by dipping in a solution of more-noble metal ions; a replacement reaction. Also known as dip plating; metal replacement. { ə'mər·zhən ‚plād·iŋ }

immersion proof [ORD] An item of equipment when ready for field transport can be submerged for 2 hours in salt or fresh water to a covering depth of 3 feet (90 centimeters), and be capable of operating at normal effectiveness immediately after being removed from the water. { ə'mər·zhən ‚prüf }

immersion refractometer [OPTICS] Device to measure refractive indices by immersing the prism portion in the sample being checked. Also known as dipping refractometer. { ə'mər·zhən ‚rē‚frak'täm·əd·ər }

immersion sampling [ANALY CHEM] Collection of a liquid sample for laboratory or other analysis by immersing a container in the liquid and filling it. { ə'mər·zhən ‚sam·pliŋ }

immersion scanning [ENG] Ultrasonic scanning in which both the ultrasonic transducer and the object being scanned are both immersed in water or some other liquid that provides good coupling while the transducer is being moved around the object. { ə'mər·zhən ‚skan·iŋ }

immersive simulation See virtual reality. { i‖mər·siv sim·yə'lā·shən }

immigrant [ECOL] An organism that settles in a zone where it was previously unknown. { 'im·ə·grənt }

immigration [ECOL] The one-way inward movement of individuals or their disseminules into a population or population area. [GEN] Gene flow from one population into another by interbreeding between members of the populations. { ‚im·ə'grā·shən }

immiscible [CHEM] Pertaining to liquids that will not mix with each other. { i'mis·ə·bəl }

immittance [ELEC] A term used to denote both impedance and admittance, as commonly applied to transmission lines, networks, and certain types of measuring instruments. { i'mit·əns }

immittance bridge [ELECTROMAG] A modification of an admittance bridge which compares the output current of a four-terminal device with admittance standards in a T configuration in order to measure transfer admittance by a null method. { i'mit·əns ‚brij }

immobilize [MED] To render motionless or to fix in place, as by splints or surgery. { i'mō·bə‚līz }

immobilized catalyst [CHEM] A molecular catalyst that is bound without substantial change in its structure to an insoluble solid to prevent solution of the catalyst in the contacting liquid. Also known as anchored catalyst. { i‚mō·bə‚līzd 'kad·ə‚list }

immortalization [CELL MOL] The process whereby a cell line gains the ability to undergo continuous cell division. { i‚mòrd·əl·ə'zā·shən }

immune [IMMUNOL] 1. Safe from attack; protected against a disease by an innate or an acquired immunity. 2. Pertaining to or conferring immunity. { i'myün }

immune body See antibody. { i'myün ‚bäd·ē }

immune complex disease [MED] A disease that results from deposition of antigen-antibody complexes in tissues. { i‖myün ‚käm‚pleks di‚zēz }

immune hemolysin [IMMUNOL] A substance formed in blood in response to an injection of erythrocytes from another species. { i'myün hē·mə'līs·ən }

immune horse serum [IMMUNOL] Serum obtained from the blood of an immunized horse. { i'myün 'hòrs ‚sir·əm }

immune lysin [IMMUNOL] An antibody that will disrupt a particular type of cell in the presence of complement and cofactors, such as magnesium or calcium ions. { i'myün 'līs·ən }

immune opsonin [IMMUNOL] A substance produced in blood in response to an infection or to inoculation with dead cells of the infecting species of bacteria. { i'myün 'äp·sə·nən }

immune precipitation [IMMUNOL] A method of isolating a protein from mixtures by using a specific antibody as the precipitating agent. { i'myün prə‚sip·ə'tā·shən }

immune protein [IMMUNOL] Any antibody. { i'myün 'prō‚tēn }

immune response [IMMUNOL] The physiological responses stemming from activation of the immune system by antigens, consisting of a primary response in which the antigen is recognized as foreign and eliminated, and a secondary response to subsequent contact with the same antigen. { i'myün ri‚späns }

immune response gene [IMMUNOL] Any of a group of genes in the major histocompatibility complex that determines the degree of immune response. Abbreviated IR gene. { i‖myün ri'späns ‚jēn }

immune serum [IMMUNOL] Blood serum obtained from an immunized individual and carrying antibodies. { i'myün ‚sir·əm }

immunity [IMMUNOL] The condition of a living organism whereby it resists and overcomes an infection or a disease. [MET] The ability of metal to resist corrosion as a result of thermodynamic stability. { i'myü·nəd·ē }

immunization [IMMUNOL] Rendering an organism immune to a specific communicable disease. { ‚im·yə·nə'zā·shən }

immunization therapy [MED] The use of vaccines or antiserums to produce immunity against a specific disease. { ‚im·yə·nə'zā·shən ‚ther·ə‚pē }

immunoassay [IMMUNOL] A laboratory detection method that uses antibodies to react with specific substances. { ‚im·yə·nō'a‚sāy }

immunochemistry [IMMUNOL] A branch of science dealing with the chemical changes associated with immunity factors. { ‖im·yə·nō'kem·ə·strē }

immunocompromised [IMMUNOL] Having an impaired or weakened immune system (usually due to drugs or illness). { ‖im·yə·nō'käm·prə‚mīzd }

immunodeficiency [IMMUNOL] Any defect of antibody function or cell-mediated immunity. { ‖im·yə‚nō·də'fish·ən·sē }

immunodiffusion [IMMUNOL] A serological procedure in which antigen and antibody solutions are permitted to diffuse toward each other through a gel matrix; interaction is manifested by a precipitin line for each system. { ‖im·yə·nō·də'fyü·zhən }

immunoelectrophoresis [IMMUNOL] A serological procedure in which the components of an antigen are separated by electrophoretic migration and then made visible by immunodiffusion of specific antibodies. { ‖im·yə·nō·i‚lek·trə·fə'rē·səs }

immunofluorescence [IMMUNOL] Fluorescence as the result of, or identifying, an immune response; a specifically stained antigen fluoresces in ultraviolet light and can thus be easily identified with a homologous antigen. { ¦im·yə·nō·flə'res·əns }

immunogen [IMMUNOL] A substance which stimulates production of specific antibody or of cellular immunity, and which can react with these products. { ə'myü·nə·jən }

immuno-gene therapy [MED] A type of gene therapy used in cancer treatment that aims to enhance the immune response against tumor cells. { 'im·yə·nō‚jēn ¦ther·ə·pē }

immunogenetics [MED] A branch of immunology dealing with the relationships between immunity and genetic factors in disease. { ¦im·yə·nō·jə'ned·iks }

immunogenic [IMMUNOL] Producing immunity. { ¦im·yə·nō¦jen·ik }

immunogenic peptides [IMMUNOL] Peptides that are recognized by T cells primed by immunization or transplantation. { ‚im·yə·nō‚jen·ik 'pep‚tīdz }

immunoglobulin [IMMUNOL] Any of a set of serum glycoproteins which have the ability to bind other molecules with a high degree of specificity. Abbreviated Ig. { ¦im·yə·nō'glä·byə·lən }

immunoglobulin A [IMMUNOL] A class of immunoglobulins that inhibits the binding of microorganisms to mucosal surfaces; large amounts are found in breast milk, saliva, and gastrointestinal secretions. Abbreviated IgA. { ‚im·yə·nō‚gläb·yə·lən 'ā }

immunoglobulin A deficiency [IMMUNOL] An immune system disorder in which lower than normal amounts of immunoglobulin A are produced, resulting in increased susceptibility to infections such as chronic sinusitis, chronic pulmonary infections, and digestive problems. { ‚im·yə·nō‚gläb·yə·lən ¦ā di'fish·ən‚sē }

immunoglobulin D [IMMUNOL] A class of immunoglobulins that are found on the surface of B cells and in minute amounts in normal human serum. { ‚im·yə·nō‚gläb·yə·lən 'dē }

immunoglobulin domain [IMMUNOL] The basic structural unit of immunoglobulins. { ‚im·yə·nō‚gläb·yə·lən də'mān }

immunoglobulin E [IMMUNOL] A class of immunoglobulins present in minute amounts in normal human serum that is active against parasites and acts as a mediator of immediate hypersensitivity. Abbreviated IgE. Also known as reagin. { ‚im·yə·nō‚gläb·yə·lən 'ē }

immunoglobulin G [IMMUNOL] The most abundant immunoglobulin class in human serum; it is associated with complement fixation, opsonization, fixation to macrophages, and membrane transport. Abbreviated IgG. { ‚im·yə·nō‚gläb·yə·lən 'jē }

immunoglobulin M [IMMUNOL] The first immunoglobulin class to appear during the primary immune response; it is mainly contained in the bloodstream, where it can readily neutralize agents attempting to gain entrance through the blood. Abbreviated IgM. { ‚im·yə·nō‚gläb·yə·lən 'em }

immunoglobulin-mediated immunity See humoral immunity. { ‚im·yə·nō¦gläb·yə·lən ¦mē·dē‚ād·əd i'myü·nəd·ē }

immunogold electron microscopy [CELL MOL] A technique in which cellular components are visualized with an electron microscope by using gold particles as antibody/protein labels. { ¦im·yə·nə‚gold i‚lek‚trän mī'kräs·kə·pē }

immunogranulomatous disease [MED] A condition in which a deviation from the standard immune mechanisms is considered to be associated with widespread granulomatosis. { ¦im·yə·nō‚gran·yə'läm·əd·əs di‚zēz }

immunological deficiency [IMMUNOL] A state wherein the immune mechanisms are inadequate in their ability to perform their normal function, that is, the elimination of foreign materials (usually infectious agents such as bacteria, viruses, and fungi). { ‚im·yə·nə‚läj·ə·kəl di'fish·ən‚sē }

immunological memory [IMMUNOL] The capacity of the immune system to respond more rapidly and vigorously to the second contact with a specific antigen than to the primary contact. { ‚im·yə·nə¦läj·ə·kəl 'mem‚rē }

immunological ontogeny [IMMUNOL] The origin and development of the lymphocyte system, from its earliest stages to the two major populations of mature lymphocytes: the thymus-dependent or T lymphocytes, and the thymus-independent or B lymphocytes. { ‚im·yə·nə‚läj·ə·kəl än'täj·ə·nē }

immunological paralysis See acquired immunological tolerance. { ¦im·yə·nə¦läj·ə·kəl pə'ral·ə·səs }

immunological phylogeny [IMMUNOL] The study of immunology and the immune system in evolution. { ¦im·yə·nə‚läj·i·kəl fī'läj·ə·nē }

immunologic cytotoxicity [IMMUNOL] The mechanism by which the immune system destroys or damages foreign or abnormal cells. { ¦im·yə·nə¦läj·ik ‚sīd·ō·täk'sis·əd·ē }

immunologic suppression [IMMUNOL] The use of x-irradiation, chemicals, corticosteroid hormones, or antilymphocyte antisera to suppress antibody production, particularly in graft transplants. Also known as immunosuppression. { ¦im·yə·nə¦läj·ik sə'presh·ən }

immunologic tolerance [IMMUNOL] **1.** A condition in which an animal will accept a homograft without rejection. **2.** A state of specific unresponsiveness to an antigen or antigens in adult life as a consequence of exposure to the antigen in utero or in the neonatal period. { ¦im·yə·nə¦läj·ik 'täl·ə·rəns }

immunologist [IMMUNOL] A person who specializes in immunology. { ‚im·yə'näl·ə·jəst }

immunology [BIOL] A branch of biological science concerned with the native or acquired resistance of higher animal forms and humans to infection with microorganisms. { ‚im·yə'näl·ə·jē }

immunonephelometry [IMMUNOL] The application of nephelometry to the quantification of antigen or antibody. { ¦im·yə·nō‚nef·ə'läm·ə·trē }

immunopathology [MED] The study of various human and animal diseases in which humoral and cellular immune factors seem important in causing pathological damage to cells, tissues, and the host. { ¦im·yə·nō·pə'thäl·ə·jē }

immunopotentiation [IMMUNOL] Enhancement of an immune response by a variety of adjuvants. { ‚im·yə·nō·pə‚ten·chē'ā·shən }

immunoprecipitation [IMMUNOL] A protein purification method which involves the formation of an antibody-protein complex to separate out the protein of interest. { ‚im·yə·nō·prə‚sip·ə'tā·shən }

immunoselective adsorption [IMMUNOL] A process that exploits biospecific interactions between antibodies and their corresponding antigens in order to separate pure high-value bioactive products from biological sources (for example, blood, food materials, and products of genetically engineered organisms). { ‚im·yə·nō·si‚lek·tiv ad'sȯrp·shən }

immunosuppression See immunologic suppression. { ¦im·yə·nō·sə'presh·ən }

immunosuppressive [PHARM] Any drug or agent used to suppress antibody production. { ¦im·yə·nō·sə'pres·iv }

immunotherapy [MED] **1.** Therapy utilizing immunosuppressives. **2.** See serotherapy. { ¦im·yə·nō'ther·ə·pē }

immunotoxicity [IMMUNOL] Adverse effects on the normal functioning of the immune system, caused by exposure to a toxic chemical. The result can be higher rates of infectious diseases or cancer, more severe cases of such autoimmune disease, or allergic reactions. { ‚im·yə·nō·täk'sis·əd·ē }

immunotoxin [IMMUNOL] Conjugate of antibody and toxic protein such that the specificity of the antibody molecule is combined with the cytotoxic property of the toxin. { ‚im·yə·nō'täk·sən }

impact [MECH] A forceful collision between two bodies which is sufficient to cause an appreciable change in the momentum of the system on which it acts. Also known as impulsive force. { 'im‚pakt }

impact area [ENG] An area with designated boundaries within which all objects that travel over a range are to make contact with the ground. { 'im‚pakt ‚er·ē·ə }

impact avalanche and transit time diode See IMPATT diode. { 'im‚pakt ‚av·ə‚lanch ən 'tran·zit ‚tīm 'dī‚ōd }

impact bar [ENG] Specimen used to test the relative susceptibility of a plastic material to fracture by shock. { 'im‚pakt ‚bär }

impact breaker [MECH ENG] A device that utilizes the energy from falling stones in addition to power from massive impellers for complete breaking up of stone. Also known as double impeller breaker. { 'im‚pakt ‚brāk·ər }

impact cast See prod cast. { 'im‚pakt ‚kast }

impact crater [GEOL] A crater formed on a planetary surface by the impact of a projectile. { 'im‚pakt ‚krād·ər }

impact crusher [MECH ENG] A machine for crushing large chunks of solid materials by sharp blows imposed by rotating hammers, or steel plates or bars; some crushers accept lumps as large as 28 inches (about 70 centimeters) in diameter, reducing them to $1/4$ inch (6 millimeters) and smaller. { 'im,pakt ,krəsh·ər }

impact energy [MECH] The energy necessary to fracture a material. Also known as impact strength. { 'im,pakt ,en·ər·jē }

impact excitation [ELEC] Starting of damped oscillations in a radio circuit by a sudden surge, such as that produced by a spark discharge. { 'im,pakt ,ek·sə'tā·shən }

impact extrusion [MET] A cold extrusion process for producing tubular components by striking a slug of the metal, which has been placed in the cavity of the die, with a punch moving at high velocity. { 'im,pakt ik,strü'zhən }

impact force See set forward force. { 'im,pakt ,fórs }

impact forging [MET] Plastic deformation of a metal using an impactive force. { 'im,pakt ,fórj·iŋ }

impact fuse [ORD] A fuse that detonates upon striking an object. Also known as contact fuse. { 'im,pakt ,fyüz }

impact grinding [MECH ENG] A technique used to break up particles by direct fall of crushing bodies on them. { 'im,pakt ,grīn·diŋ }

impaction [MED] **1.** The state of being lodged and retained in a body part. **2.** Confinement of a tooth in the jaw so that its eruption is prevented. **3.** A condition in which one fragment of bone is firmly driven into another fragment so that neither can move against the other. { im'pak·shən }

impact ionization [ELECTR] Ionization produced by the impact of a high-energy charge carrier on an atom of semiconductor material; the effect is an increase in the number of charge carriers. { 'im,pakt ,ī·ə·nə'zā·shən }

impactite [GEOL] Glassy fused rock or meteor fragments resulting from heat of impact of a meteor on the earth. { 'im,pak,tīt }

impact law [PHYS] The relationship of fluid density, particle density, and fluid viscosity in the settling velocity of large particles in a given liquid: settling velocity is directly proportional to the square root of the particle diameter. { 'im,pakt ,ló }

impact load [ENG] A force delivered by a blow, as opposed to a force applied gradually and maintained over a long period. { 'im,pakt ,lōd }

impact loss [FL MECH] Loss of head in a flowing stream due to the impact of water particles upon themselves or some bounding surface. { 'im,pakt ,lós }

impact mark See prod mark. { 'im,pakt ,märk }

impact microphone [ENG ACOUS] An instrument that picks up the vibration of an object impinging upon another, used especially on space probes to record the impact of small meteoroids. { 'im,pakt 'mī·krə,fōn }

impact mill [MECH ENG] A unit that reduces the size of rocks and minerals by the action of rotating blades projecting the material against steel plates. { 'im,pakt ,mil }

impact modifier [MATER] A material added to a substance during manufacture to improve resistance to deformation or breaking. { 'im,pakt ,mäd·ə,fī·ər }

impact-noise analyzer [ENG] An analyzer used with a sound-level meter to evaluate the characteristics of impact-type sounds and electric noise impulses that cannot be measured accurately with a noise meter alone. { 'im,pakt ,nóiz 'an·ə,līz·ər }

impactometer See impactor. { ,im,pak'täm·əd·ər }

impactor [ENG] A general term for instruments which sample atmospheric suspensoids by impaction; such instruments consist of a housing which constrains the air flow past a sensitized sampling plate. Also known as impactometer. [MECH ENG] A machine or part whose operating principle is striking blows. [MIN ENG] A rotary hammermill which crushes ore by impacting it against crushing plates or elements. { im'pak·tər }

impact parameter [NUC PHYS] In a nuclear collision, the perpendicular distance from the target nucleus to the initial line of motion of the incident particle. { 'im,pakt pə,ram·əd·ər }

impact predictor [AERO ENG] A device which takes information from a trajectory measuring system and continuously computes the point (in real time) at which the rocket will strike the earth. { 'im,pakt prə,dik·tər }

impact pressure See dynamic pressure. { 'im,pakt ,presh·ər }

impact printer [GRAPHICS] A line printer that has one or more character fonts, a ribbon or other inking device, a paper transport, and some means of impacting desired characters or character elements on the paper. { 'im,pakt ,print·ər }

impact roll [MECH ENG] An idler roll protected by a covering of a resilient material from the shock of the loading of material onto a conveyor belt, so as to reduce the damage to the belt. { 'im,pakt ,rōl }

impact screen [MECH ENG] A screen designed to swing or rock forward when loaded and to stop abruptly by coming in contact with a stop. { 'im,pakt ,skrēn }

impact strength [MECH] **1.** Ability of a material to resist shock loading. **2.** See impact energy. { 'im,pakt ,streŋkth }

impact stress [MECH] Force per unit area imposed on a material by a suddenly applied force. { 'im,pakt ,stres }

impact test [ENG] Determination of the degree of resistance of a material to breaking by impact, under bending, tension, and torsion loads; the energy absorbed is measured in breaking the material by a single blow. { 'im,pakt ,test }

impact theory [ASTRON] A theory which holds that most features of the moon's surface were formed by the impact of meteorites. Also known as meteoric theory; meteoritic theory. { 'im,pakt ,thē·ə·rē }

impact tube See pitot tube. { 'im,pakt ,tüb }

impact velocity [MECH] The velocity of a projectile or missile at the instant of impact. Also known as striking velocity. { 'im,pakt və'läs·əd·ē }

impact wrench [MECH ENG] A compressed-air or electrically operated wrench that gives a rapid succession of sudden torques. { 'im,pakt ,rench }

imparipinnate See odd-pinnate. { ,im·par·ə'pi,nāt }

impasto [GRAPHICS] **1.** The thick, heavy application of oil paint to a canvas, often with a palette knife; impasto sections stand out in considerable relief. **2.** The thick application of polymer or other paint to any surface. { im'pä·stō }

IMPATT amplifier [ELECTR] A diode amplifier that uses an IMPATT diode; operating frequency range is from about 5 to 100 gigahertz, primarily in the C and X bands, with power output up to about 20 watts continuous-wave or 100 watts pulsed. { 'im,pat ,am·plə,fī·ər }

IMPATT diode [ELECTR] A pn junction diode that has a depletion region adjacent to the junction, through which electrons and holes can drift, and is biased beyond the avalanche breakdown voltage. Derived from impact avalanche and transit time diode. { 'im,pat ,dī,ōd }

impedance [ELEC] See electrical impedance. [PHYS] **1.** The ratio of a sinusoidally varying quantity to a second quantity which measures the response of a physical system to the first, both being considered in complex notation; examples are electrical impedance, acoustic impedance, and mechanical impedance. Also known as complex impedance. **2.** The ratio of the greatest magnitude of a sinusoidally varying quantity to the greatest magnitude of a second quantity which measures the response of a physical system to the first; equal to the magnitude of the quantity in the first definition. { im'pēd·əns }

impedance-admittance matrix [ELECTR] A four-element matrix used to describe analytically a transistor in terms of impedances or admittances. { im'pēd·əns ad'mit·əns 'mā·triks }

impedance bridge [ELEC] A device similar to a Wheatstone bridge, used to compare impedances which may contain inductance, capacitance, and resistance. { im'pēd·əns ,brij }

impedance coil [ELEC] A coil of wire designed to provide impedance in an electric circuit. { im'pēd·əns ,kóil }

impedance compensator [ELEC] Electric network designed to be associated with another network or a line with the purpose of giving the impedance of the combination a desired characteristic with frequency over a desired frequency range. { im'pēd·əns 'käm·pən,sād·ər }

impedance component [ELEC] **1.** Resistance or reactance. **2.** A device such as a resistor, inductor, or capacitor designed to provide impedance in an electric circuit. { im'pēd·əns kəm,pō·nənt }

impedance coupling [ELEC] Coupling of two signal circuits with an impedance. { im'pēd·əns ,kəp·liŋ }

impedance drop [ELEC] The total voltage drop across a component or conductor of an alternating-current circuit, equal to the phasor sum of the resistance drop and the reactance drop. { im'pēd·əns ,dräp }

impedance irregularity [ELEC] A discontinuity or abrupt

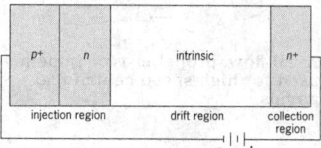

IMPATT DIODE

IMPATT p^+nin^+ diode structure and DC bias (Read design).

change which results from a junction between unlike sections of a transmission line or an irregularity on a line. { im'pēd·əns i‚reg·yə'lar·əd·ē }

impedance magnetometer [ENG] An instrument for determining local variations in magnetic field by measuring the change in impedance of a high-permeability nickel-iron wire. { im'ped·əns ‚mag·nə'täm·əd·ər }

impedance match [ELEC] The condition in which the external impedance of a connected load is equal to the internal impedance of the source or to the surge impedance of a transmission line, thereby giving maximum transfer of energy from source to load, minimum reflection, and minimum distortion. { im'pēd·əns ‚mach }

impedance-matching network [ELEC] A network of two or more resistors, coils, or capacitors used to couple two circuits in such a manner that the impedance of each circuit will be equal to the impedance into which it locks. Also known as line-building-out network. { im'pēd·əns ¦mach·iŋ ‚net‚wərk }

impedance matrix [ELEC] A matrix Z whose elements are the mutual impedances between the various meshes of an electrical network; satisfies the matrix equation $V = ZI$, where V and I are column vectors whose elements are the voltages and currents in the meshes. { im'pēd·əns ‚mā·triks }

impedance meter See electrical impedance meter. { im'pēd·əns ‚mēd·ər }

impedance plethysmography [MED] A technique by which changes in the volume of certain segments of the body can be determined by impedance measurements. These changes are related to factors involving the mechanical activity of the heart, to conditions of the circulatory system, or to respiratory flow and other physiological functions. { im‚pēd·əns ‚pleth·ə'smäg·rə·fē }

impedometer [ELECTROMAG] An instrument used to measure impedances in waveguides. { ‚im·pə'däm·əd·ər }

impeller [MECH ENG] The rotating member of a turbine, blower, fan, axial or centrifugal pump, or mixing apparatus. Also known as rotor. { im'pel·ər }

impeller pump [MECH ENG] Any pump using a mechanical agency to provide continuous power to move liquids. { im'pel·ər ‚pəmp }

Impennes [VERT ZOO] A superorder of birds for the order Sphenisciformes in some systems of classification. { im'pen·ēz }

imperative language [COMPUT SCI] A programming language in which programs largely consist of a series of commands to assign values to objects. { im'per·əd·iv ‚laŋ·gwij }

imperative statement [COMPUT SCI] A statement in a symbolic program which is translated into actual machine-language instructions by the assembly routine. { im'per·əd·iv ‚stāt·mənt }

imperfect crystal [CRYSTAL] A crystal in which the regular, periodic structure is interrupted by various defects. { im'pər‚fekt 'krist·əl }

imperfect flower [BOT] A flower lacking either stamens or carpels. { im'pər·fikt 'flau·ər }

imperfect gas See real gas. { im'pər·fikt 'gas }

imperforate [BIOL] Lacking a normal opening. [GRAPHICS] In printing, no perforations between repeated designs, for example, stamps or labels, which are to be used separately. { im'pər·fə·rət }

imperforate anus [MED] A congenital malformation in which the large intestine ends blindly. { im'pər·fə·rət 'ā·nəs }

imperial gallon See gallon. { im'pir·ē·əl 'gal·ən }

imperial pint See pint. { im'pir·ē·əl 'pīnt }

imperial red [INORG CHEM] Any of the red varieties of ferric oxide used as pigment. { im'pir·ē·əl 'red }

imperial smelting process [MET] A pyrometallurgical process which treats a complex concentrated feed to a single furnace to recover zinc, copper, lead, cadmium, silver, gold, and other metals in one pass. Abbreviated ISP. { im'pir·ē·əl 'smel·tiŋ ‚präs·əs }

impermeable [SCI TECH] Not permitting water or other fluid to pass through. Also known as impervious. { im'pər·mē·ə·bəl }

impermeable junction See tight junction. { im'pər‚mē·ə·bəl 'jəŋk·shən }

impersonal micrometer [ENG] An instrument consisting of a vertical wire that is mounted in the focal plane of a transit circle and can be moved across the field of view to follow a

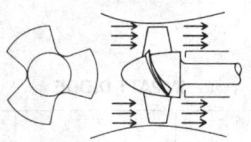

IMPELLER

Axial-flow, propeller-type impeller used for high-speed centrifugal pumps.

---

star, and instrumentation to record the position of the wire as a function of time; used to reduce systematic observational errors. { im'pərs·ən·əl mī'kräm·əd·ər }

impervious See impermeable. { im'pər·vē·əs }

impervious carbon [MATER] Carbon compressed with bituminous binder, then carbonized by sintering to produce a dense, impervious material; used as brick to line chemical process and storage vessels. { im'pər·vē·əs 'kär·bən }

impetigo [MED] An acute, contagious, inflammatory skin disease caused by streptococcal or staphylococcal infections and characterized by vesicular or pustular lesions. { ‚im·pə'tī‚gō }

impingement [ENG] Removal of liquid droplets from a flowing gas or vapor stream by causing it to collide with a baffle plate at high velocity, so that the droplets fall away from the stream. Also known as liquid knockout. { im'pinj·mənt }

impingement attack [MET] Accelerated corrosive attack on a metal by moving liquids, resulting usually from erosion of a protective surface layer. { im'pinj·mənt ə¦tak }

impinger [ENG] A device used to sample dust in the air that draws in a measured volume of dusty air and directs it through a jet to impact on a wetted glass plate; the dust particles adhering to the plate are counted. { im'pin·jər }

implant [MED] **1.** A quantity of radioactive material in a suitable container, intended to be embedded in a tissue or tumor for therapeutic purposes. **2.** A tissue graft placed in depth in the body. { 'im‚plant }

implantation [MED] **1.** Placement of a tissue transplant in depth in the body. **2.** Placement in the body of a device for mechanical repair, such as for a ventral hernia or a fracture. **3.** Embedding of an embryo into the endometrium. { ‚im‚plan'tā·shən }

implanted atom [ELECTR] An atom introduced into semiconductor material by ion implantation. { im'plant·əd 'ad·əm }

implanted device [ELECTR] A resistor or other device that is fabricated within a silicon or other semiconducting substrate by ion implantation. [MED] A heart pacemaker or other medical electronic device that is surgically placed in the body. { im'plant·əd di'vīs }

implementation [COMPUT SCI] **1.** The installation of a computer system or an information system. **2.** The use of software on a particular computer system. { ‚im‚plə‚men'tā·shən }

implexed [INV ZOO] In insects, having the integument infolded for muscle attachment. { 'im‚plekst }

implication [MATH] **1.** The logical relation between two statements p and q, usually expressed as "if p then q." **2.** A logic operator having the characteristic that if p and q are statements, the implication of p and q is false if p is true and q is false, and is true otherwise. Also known as conditional implication; if-then operation; material implication. { ‚im‚plə'kā·shən }

implicit differentiation [MATH] The process of finding the derivative of one of two variables with respect to the other by differentiating all the terms of a given equation in the two variables and solving the resulting equation for this derivative. { im'plis·it ‚dif·ə‚ren·chē'ā·shən }

implicit enumeration [MATH] A method of solving integer programming problems, in which tests that follow conceptually from using implied upper and lower bounds on variables are used to eliminate all but a tiny fraction of the possible values, with implicit treatment of all other possibilities. { im'plis·ət i‚nü·mə'rā·shən }

implicit function [MATH] A function defined by an equation $f(x,y) = 0$, when x is considered as an independent variable and y, called an implicit function of x, as a dependent variable. { im'plis·ət 'fəŋk·shən }

implicit function theorem [MATH] A theorem that gives conditions under which an equation in variables x and y may be solved so as to express y directly as a function of x; it states that if $F(x,y)$ and $\partial F(x,y)/\partial y$ are continuous in a neighborhood of the point (x_0,y_0) and if $F(x,y) = 0$ and $\partial F(x,y)/\partial y \neq 0$, then there is a number $\epsilon > 0$ such that there is one and only one function $f(x)$ that is continuous and satisfies $F[x, f(x)] = 0$ for $|x - x_0| < \epsilon$, and satisfies $f(x_0) = y_0$. { im'plis·ət 'fəŋk·shən ‚thir·əm }

implicit memory [PSYCH] A type of memory that is expressed through performance, rather than conscious recall, such as information acquired during skill learning, habit formation, classical conditioning, emotional learning, and priming. Also known as nondeclarative memory. { im‚plis·ət 'mem·rē }

implicit programming [CONT SYS] Robotic programming that uses descriptions of the tasks at hand which are less exact than in explicit programming. { im'plis·ət 'prō‚gram·iŋ }

implosion [CHEM] The sudden reduction of pressure by chemical reaction or change of state which causes an inrushing of the surrounding medium. [PHYS] A bursting inward, as in the inward collapse of an evacuated container (such as the glass envelope of a cathode-ray tube) or the compression of fissionable material by ordinary explosives in a nuclear weapon. { im'plō·zhən }

implosion weapon [ORD] A nuclear weapon in which a quantity of fissionable material, less than a critical mass in its untriggered configuration, has its volume suddenly reduced by compression with ordinary explosives, so that it becomes supercritical and a nuclear explosion can occur. { im'plō·zhən ‚wep·ən }

imposed date [IND ENG] An assignment of a date to an activity that represents either the earliest or the latest date at which the activity can be either started or finished. { im'pōzd 'dāt }

imposed load [CIV ENG] Any load which a structure must sustain, other than the weight of the structure itself. { im'pōzd 'lōd }

imposition [GRAPHICS] The pattern of arranging pages for a signature of a book so that the pages will be in sequence when folding occurs. { ‚im·pə'zish·ən }

impost [ARCH] The highest part of a column, pillar, pier, or wall upon which the end of an arch rests. { 'im‚pōst ‚ärch }

impotence [MED] **1.** Inability in the male to perform the sexual act. **2.** Lack of sexual vigor. { 'im·pəd·əns }

impound [CIV ENG] To collect water for irrigation, flood control, or similar purpose. { im'paùnd }

impounding reservoir [CIV ENG] A reservoir with outlets controlled by gates that release stored surface water as needed in a dry season; may also store water for domestic or industrial use or for flood control. Also known as storage reservoir. { im'paùnd·iŋ ‚rez·əv‚wär }

impregnate [ENG] To force a liquid substance into the spaces of a porous solid in order to change its properties, as the impregnation of turquoise gems with plastic to improve color and durability, the impregnation of porous tungsten with a molten barium compound to manufacture a dispenser cathode, or the impregnation of wood with creosote to preserve its integrity against water damage. [MED] To fertilize or cause to become pregnant. { im'preg‚nāt }

impregnated bit [DES ENG] A sintered, powder-metal matrix bit with fragmented bort or whole diamonds of selected screen sizes uniformly distributed throughout the entire crown section. { im'preg‚nād·əd 'bit }

impregnated timber [MATER] Timber which has been made flame-resistant, fungi-resistant, or insect-proof by forcing into it under vacuum or pressure a flame retardant or a fungal or insect poison. { im'preg‚nād·əd 'tim·bər }

impressed current [ELEC] Direct current supplied by an external power source in a cathodic protection installation. { im'prest 'kər·ənt }

impressed voltage [ELEC] Voltage applied to a circuit or device. { im'prest 'vōl·tij }

impression [GEOL] A form left on a soft soil surface by plant parts; the soil hardens and usually the imprint is a concave feature. [GRAPHICS] **1.** A print made from an engraved plate. **2.** A press run or printing of a book. [MET] A machined cavity in a forging die for production of a specific geometric shape in the workpiece. { im'presh·ən }

impression block [PETRO ENG] A block with wax or lead on the bottom run into a well and allowed to rest on a lost tool or other object so that an examination of the resultant impression is revelatory concerning the size, shape, or position of the object. { im'presh·ən ‚blak }

impression cylinder [GRAPHICS] A cylinder onto which an inked image is pressed, so that this image can be transferred to paper in the offset duplicating process. { im'presh·ən ‚sil·ən·dər }

imprint See overprint. { 'im‚print }

imprinter [GRAPHICS] Any device for entering markings

onto a form, including, but not limited to, printing presses, typewriters, pressure imprinting devices such as those used with credit cards and address plates, pencils, pens, cash registers, adding machines, and bookkeeping machines. { im'print·ər }

imprinting [PSYCH] The very rapid development of a response or learning pattern to a stimulus at an early and usually critical period of development; particularly characteristic of some species of birds. { im'print·iŋ }

imprisoned incompleteness [RELAT] The property of incomplete geodesics in a space-time being confined to a compact neighborhood. { im'priz·ənd ‚in·kəm'plēt·nəs }

improper divisor [MATH] An improper divisor of an element x in a commutative ring with identity is any unit of the ring or any associate of x. { im'präp·ər di'vī·zər }

improper face [MATH] For a convex polytope, either the empty set or the polytope itself. { ‚im'präp·ər 'fās }

improper fraction [MATH] **1.** In arithmetic, the quotient of two integers in which the numerator is greater than or equal to the denominator. **2.** In algebra, the quotient of two polynomials in which the degree of the numerator is greater than or equal to that of the denominator. { im'präp·ər 'frak·shən }

improper integral [MATH] Any integral in which either the integrand becomes unbounded on the domain of integration, or the domain of integration is itself unbounded. { im'präp·ər int·ə·grəl }

improperly posed problem See ill-posed problem. { im'präp·ər·lē ‚pōzd 'präb·ləm }

improper orthogonal transformation [MATH] An orthogonal transformation such that the determinant of its matrix is −1. { im'präp·ər ór'thäg·ə·nəl ‚tranz·fər'mā·shən }

improvement factor See noise improvement factor. { im'prüv·mənt ‚fak·tər }

improvement threshold [COMMUN] The condition of unity for the ratio of peak carrier voltage to peak noise voltage after selection and before any nonlinear process such as amplitude limiting. { im'prüv·mənt ‚thresh‚hōld }

improvised grenade [ORD] Any nonstandard type of grenade which is prepared by the user, for example, frangible grenades and fragmentation grenades composed of nails, cartridge cases, or other fragments taped to the sides of a trinitrotoluene block, with suitable detonating device. { 'im·prə‚vīzd grə'nād }

improvised mine [ORD] A mine manufactured of available materials because standard mines are either unavailable or are incapable of producing the desired result. { 'im·prə‚vīzd 'mīn }

impsonite [GEOL] A black, asphaltic pyrobitumen with a high fixed-carbon content derived from the metamorphosis of petroleum. { 'im·sə‚nīt }

impulse [MECH] The integral of a force over an interval of time. [MET] A single pulse or several pulses in welding current used in resistance welding. [PHYS] A pulse which lasts for so short a time that its duration can be thought of as infinitesimal. [PSYCH] A sudden psychogenic urge to act. { 'im‚pəls }

impulse approximation [PHYS] An approximation for studying the collision of an incident particle with a bound target particle, in which the binding forces on the target particle during the collision are ignored. { ‚im‚pəls ə'präk·sə‚mā·shən }

impulse excitation See shock excitation. { 'im‚pəls ‚ek·sə'tā·shən }

impulse face [HOROL] Lifting surface of a club tooth on an escape wheel, or the surface of a pallet stone engaged by such a wheel. { 'im‚pəls ‚fās }

impulse function [MATH] An idealized or generalized function defined not by its values but by its behavior under integration, such as the (Dirac) delta function. { 'im‚pəls ‚fəŋk·shən }

impulse generator [ELEC] An apparatus which produces very short surges of high-voltage or high-current power by discharging capacitors in parallel or in series. Also known as pulse generator. { 'im‚pəls ‚jen·ə‚rād·ər }

impulse modulation [CONT SYS] Modulation of a signal in which it is replaced by a series of impulses, equally spaced in time, whose strengths (integrals over time) are proportional to the amplitude of the signal at the time of the impulse. { 'im‚pəls ‚mäj·ə‚lā·shən }

impulse movement [HOROL] Clock movement in which the

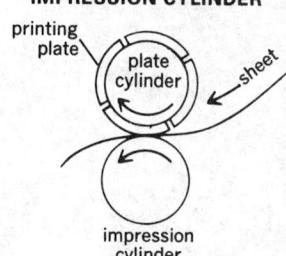

IMPRESSION CYLINDER

The plate and impression cylinders in a rotary press.

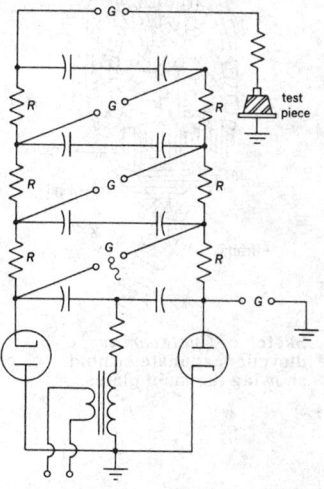

IMPULSE GENERATOR

Typical four-stage Marx impulse generator circuit. R = resistor, G = spark gap.

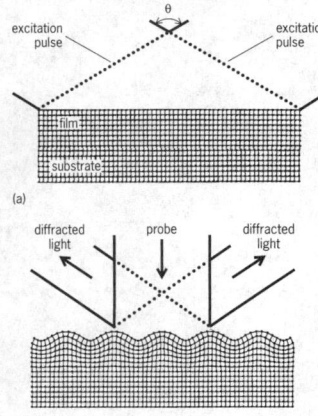

(a)

(b)

In this experiment, *(a)* crossed laser pulses induce coherent, monochromatic acoustic and thermal motions in a thin film. *(b)* Continuous-wave probing laser diffracts from the ripple on the sample surface. Measuring the diffracted light intensity with a fast detector and transient recorder reveals the time dependence of the motions.

INADUNATA

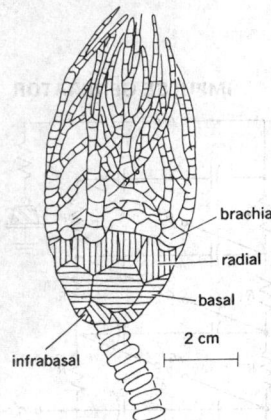

brachial

radial

basal

2 cm

infrabasal

Sketch of *Cyathocrinus*, a dicyclic inadunate crinoid, showing the main plates.

hands are driven by electromagnet impulses from the pendulum. { 'im‚pəls ‚müv·mənt }

impulse noise [ELEC] Noise characterized by transient short-duration disturbances distributed essentially uniformly over the useful passband of a transmission system. { 'im‚pəls ‚nóiz }

impulse pallet [HOROL] Pallet stone in a chronometer balance, driven by impulses from the escape wheel. { 'im‚pəls ‚pal·ət }

impulse period See pulse period. { 'im‚pəls ‚pir·ē·əd }

impulse relay [ELEC] A relay that stores the energy of a short pulse, to operate the relay after the pulse ends. { 'im‚pəls ‚rē‚lā }

impulse response [CONT SYS] The response of a system to an impulse which differs from zero for an infinitesimal time, but whose integral over time is unity; this impulse may be represented mathematically by a Dirac delta function. { 'im‚pəls ri‚späns }

impulse sealing [ENG] Heat-sealing of plastic materials by applying a pulse of intense thermal energy to the sealing area for a very short time, followed immediately by cooling. { 'im‚pəls ‚sēl·iŋ }

impulse separator [ELECTR] In a television receiver, the circuit that separates the horizontal synchronizing impulses in the received signal from the vertical synchronizing impulses. { 'im‚pəls ‚sep·ə‚rād·ər }

impulse signaling [COMMUN] Conveying information by means of on-off conditions transmitted down a line or over free space. { 'im‚pəls ‚sig·nə·liŋ }

impulse solenoid [ELECTROMAG] A solenoid that operates on pulse power, at speeds up to several hundred strokes per second. { 'im‚pəls ‚sō·lə‚nóid }

impulse strength [ELEC] Voltage breakdown of insulation under voltage surges on the order of microseconds in duration. { 'im‚pəls ‚streŋkth }

impulse tachometer [ENG] A tachometer in which each rotation of a shaft generates an electric pulse and the time rate of pulses is then measured; classified as capacitory-current, inductory, or interrupted direct-current tachometer. { 'im‚pəls tə'käm·əd·ər }

impulse train [CONT SYS] An input consisting of an infinite series of unit impulses, equally separated in time. { 'im‚pəls ‚trān }

impulse transmission [COMMUN] Form of signaling which employs impulses of either or both polarities for transmission to indicate the occurrence of transitions in the signals; used principally to reduce the effects of low-frequency interference; the impulses are generally formed by suppressing the low-frequency components, including direct current, of the signals. { 'im‚pəls tranz'mish·ən }

impulse turbine [MECH ENG] A prime mover in which fluid under pressure enters a stationary nozzle where its pressure (potential) energy is converted to velocity (kinetic) energy and absorbed by the rotor. { 'im‚pəls ‚tər‚bən }

impulse-type telemeter [COMMUN] A telemeter that employs electric impulses as the translating means. { 'im‚pəls ‚tīp tə'lem·əd·ər }

impulse voltage [ELEC] A unidirectional voltage that rapidly rises to a peak value and then drops to zero more or less rapidly. Also known as pulse voltage. { 'im‚pəls ‚vōl·tij }

impulse welding [ENG] A welding process in which two layers of thermoplastic film are heated and fused to form a welded seam by clamping them together in close contact with a shielded electric heating element. { 'im‚pəls 'weld·iŋ }

impulsive force See impact. { im'pəl·siv 'fórs }

impulsive phase [ASTROPHYS] The first phase of a solar flare, in which x-radiation rises to a maximum in a few seconds or minutes, and can then vary rapidly for several minutes in bursts of decreasing amplitude with rise times as short as 10 milliseconds. { im'pəl·siv 'fāz }

impulsive sound [ACOUS] A sound that lasts for a short period of time and includes frequencies over a large portion of the acoustic spectrum, such as a hammer blow or hand clap. { im'pəl·siv ‚saund }

impulsive sound equation [ACOUS] An equation which states that the total sound energy produced by a short burst of sound in a room is an exponentially decreasing function of time, whose decay constant depends on the speed of sound,

the sound absorption coefficient, and the volume and surface area of the room. { im'pəl·siv 'saund i‚kwā·zhən }

impulsive stimulated thermal scattering [ENG] An optical, noncontacting method for characterizing the high-frequency acoustic behavior of surfaces, thin membrane, coatings, and multilayer assemblies, in which picosecond pulses of light from an excitation laser stimulate motions which are then detected with a continuous-wave probing laser. Abbreviated ISTS. Also known as transient grating photoacoustics. { im'pəl·siv ‚stim·yə‚lād·əd ‚thərm·əl 'skad·ər·iŋ }

impunctate [BIOL] Lacking pores. { im'pəŋk‚tāt }

impurity [SCI TECH] An undesirable foreign material in a pure substance. [SOLID STATE] A substance that, when diffused into semiconductor metal in small amounts, either provides free electrons to the metal or accepts electrons from it. { im'pyür·əd·ē }

impurity band [SOLID STATE] The impurity levels in a semiconductor, occupying a certain range of energies. { im'pyür·əd·ē ‚band }

impurity level [SOLID STATE] An energy level in the band gap of a semiconductor that results from the presence of an impurity atom. { im'pyür·əd·ē ‚lev·əl }

impurity scattering [SOLID STATE] Scattering of electrons by holes or phonons in the crystal. { im'pyür·əd·ē ‚skad·ə‚riŋ }

impurity semiconductor [SOLID STATE] A semiconductor whose properties are due to impurity levels produced by foreign atoms. { im'pyür·əd·ē ‚sem·i·kən‚dək·tər }

IMVIC test [MICROBIO] A group of four cultural tests used to differentiate genera of bacteria in the family Enterobacteriaceae and to distinguish them from other bacteria; tests are indole, methyl red, Voges-Proskauer, and citrate. { 'im‚vik ‚test }

In See indium.

in. See inch.

inactivated vaccine See killed vaccine. { in‚ak·tə‚vād·əd vak‚sēn }

inactive front [METEOROL] A front, or portion thereof, that produces very little cloudiness and no precipitation, as opposed to an active front. Also known as passive front. { in'ak·tiv 'frənt }

inactive tartaric acid See racemic acid. { in'ak·tiv tär'tär·ik 'as·əd }

inadequate personality [PSYCH] An individual showing no obvious mental or physical defect, but characterized by inappropriate or inadequate response to intellectual, emotional, and physical demand, and whose behavior pattern shows inadaptability, ineptitude, poor judgment, lack of stamina, and social incompatibility. { in'ad·ə·kwət pərs·ən'al·əd·ē }

Inadunata [PALEON] An extinct subclass of stalked Paleozoic Crinozoa characterized by branched or simple arms that were free and in no way incorporated into the calyx. { i‚nä·jə‚näd·ə }

inadunate [BIOL] Not united. [INV ZOO] In crinoids, having the arms free from the calyx. { i'näj·ə·nət }

in-and-out bond [CIV ENG] Masonry bond composed of vertically alternating stretchers and headers. { ‚in ən ‚aut 'bänd }

inanimate [MED] Lacking consciousness or life. [SCI TECH] A lifeless object. { in'an·ə·mət }

inanition [MED] The exhausted, pathologic condition resulting from starvation. { ‚in·ə'nish·ən }

inaperturate [BIOL] Lacking apertures. { in'ap·ə‚chür·ət }

inarching [BOT] A kind of repair grafting in which two plants growing on their own roots are grafted together and one plant is severed from its roots after the graft union is established. { 'in'ärch·iŋ }

Inarticulata [INV ZOO] A class of the phylum Brachiopoda; valves are typically not articulated and are held together only by soft tissue of the living animal. { ‚in·är‚tik·yə'läd·ə }

in-band/adjacent-channel [COMMUN] The characteristic of radio or television broadcasting with a new type of signal (such as in digital audio broadcasting) using the same frequency channel adjacent to the one which that station uses to broadcast signals with conventional analog modulation. Abbreviated IBAC. { ‚in band ə‚jās·ənt 'chan·əl }

in-band/on-channel [COMMUN] The characteristic of radio or television broadcasting with a new type of signal (such as in digital audio broadcasting) using the same frequency channel as that station uses to broadcast signals with conventional analog modulation. Abbreviated IBOC. { ¦in ¦band ¦ōn 'chan·əl }

inboard [ENG] Toward or close to the longitudinal axis of a ship or aircraft. { 'in,bȯrd }

inboard profile [NAV ARCH] A plan having a longitudinal section through the center of a ship, showing all dimensions, positions of bulkheads, location of spaces, machinery, and fittings. { 'in,bȯrd 'prō,fīl }

inbond [CIV ENG] Pertaining to bricks or stones laid as headers across a wall. { 'in,bänd }

inborn [BIOL] Of or pertaining to a congenital or hereditary characteristic. { 'in,bȯrn }

inbred strain [GEN] Animal strain that results when individuals that are more closely related to each other than randomly chosen individuals mate together for many generations. { 'in,bred ,strān }

inbreeding [GEN] Reproduction behavior between closely related individuals; self-fertilization, as in some plants, is the most extreme form. { 'in,brēd·iŋ }

inbreeding coefficient [GEN] A measure of the rate of inbreeding or the degree to which an individual is inbred. Also known as Wright's inbreeding coefficient. { 'in,brēd·iŋ ,kō·ə,fish·ənt }

inbreeding depression [GEN] A decrease in fitness and vigor as a result of inbreeding. { 'in,brēd·iŋ di,presh·ən }

inby [MIN ENG] Away from the shaft or mine entrance and therefore toward the working face. { 'in¦bī }

incandescence [OPTICS] The emission of visible radiation by a hot body. { ,in·kən'des·əns }

incandescent lamp [ELEC] An electric lamp that produces light when a metallic filament is heated white-hot in a vacuum by passing an electric current through the filament. Also known as filament lamp; light bulb. { ,in·kən'des·ənt 'lamp }

incandescent readout [ELECTR] A readout in which each character is formed by energizing an appropriate combination of seven bar-shaped incandescent lamps. { ,in·kən'des·ənt 'red,aȯt }

incandescent tuff flow See ash flow. { ,in·kən'des·ənt 'təf ,flō }

incapacitating concentration 50 [PHYSIO] The concentration of gas or smoke that will incapacitate 50% of the test animals within a given time of exposure. Abbreviated IC$_{50}$. { ,in·kə¦pas·ə,tād·iŋ ,kans·ən¦trā·shən 'fif·tē }

incapsidation [VIROL] The construction of a capsid around the genetic material of a virus. { in,kap·sə'dā·shən }

incarbonization See coalification. { in,kär·bə·nə'zā·shən }

incarcerated hernia [MED] A hernia in which the intestinal loop is permanently trapped in the hernia sac. { in'kär,sə,rād·əd 'hər·nē·ə }

incendiary [ORD] Ammunition equipped so that an incendiary effect at the target occurs. { in'sen·dē,er·ē }

incendiary bomb [ORD] A bomb designed to be dropped from an aircraft to destroy or reduce the utility of a target by the effects of combustion. { in'sen·dē,er·ē 'bäm }

incendiary file destroyer [ORD] An incendiary device designed for use in destroying combustible file material. { in'sen·dē,er·ē 'fīl di,stroi·ər }

incendiary grenade [ORD] Hand grenade designed to be filled with incendiary materials, or used primarily for incendiary purposes. { in'sen·dē,er·ē grə'nād }

incendiary rocket [ORD] A rocket with a warhead designed to produce an incendiary effect at the target. { in'sen·dē,er·ē 'räk·ət }

incenter [MATH] The center of the inscribed circle of a given triangle. { ¦in¦sen·tər }

incentive operator [IND ENG] An employee whose wage is based on the quantity or quality of output. { in'sen·tiv ,äp·ə,rād·ər }

incentive wage system See wage incentive plan. { in'sen·tiv 'wāj ,sis·təm }

Inceptisol [GEOL] A soil order characterized by soils that are usually moist, with pedogenic horizons of alteration of parent materials but not of illuviation. { in'sep·tə,sȯl }

incertae sedis [SYST] Placed in an uncertain taxonomic position. { in¦kər,tī 'sā·dəs }

inch [MECH] A unit of length in common use in the United States and Great Britain, equal to $1/12$ foot or 2.54 centimeters. Abbreviated in. { inch }

inching See jogging. { 'inch·iŋ }

inch of mercury [MECH] The pressure exerted by a 1-inch-high (2.54-centimeter) column of mercury that has a density of 13.5951 grams per cubic centimeter when the acceleration of gravity has the standard value of 9.80665 m/s^2 or approximately 32.17398 ft/s^2; equal to 3386.388640341 pascals; used as a unit in the measurement of atmospheric pressure. { 'inch əv 'mər·kyə·rē }

incidence angle See angle of incidence. { 'in·səd·əns ,aŋ·gəl }

incidence function [MATH] The function that assigns a pair of vertices to each edge of a graph. { 'in·səd·əns ,fəŋk·shən }

incidence matrix [MATH] In a graph, the $p \times q$ matrix (b_{ij}) for which $b_{ij} = 1$ if the ith vertex is an end point of the jth edge, and $b_{ij} = 0$ otherwise. { 'in·səd·əns ,mā·triks }

incidence plane See plane of incidence. { 'in·səd·əns ,plān }

incidental element See irregular element. { ¦in·sə¦dent·əl 'el·ə·mənt }

incident field intensity [ELECTROMAG] Field strength of a sky wave without including the effects of earth reflections at the receiving location. { 'ir·sə·dənt 'fēld in,ten·səd·ē }

incident light [OPTICS] The direct light that falls on a surface. { 'in·sə·dənt 'līt }

incident power [ELEC] Product of the outgoing current and voltage, from a transmitter, traveling down a transmission line to the antenna. { 'in·sə·dənt ¦paȯ·ər }

incident wave [ELECTR] A current or voltage wave that is traveling through a transmission line in the direction from source to load. [PHYS] A wave that impinges on a discontinuity, particle, or body, or on a medium having different propagation characteristics. { 'in·sə·dənt ¦wāv }

incineration [CHEM] The process of burning a material so that only ashes remain. { in,sin·ə'rā·shən }

incinerator [ENG] A furnace or other container in which materials are burned. { in'sin·ə,rād·ər }

incipient species [EVOL] Populations that are in the process of diverging to the point of speciation but still have the potential to interbreed. { in'sip·ē·ənt ¦spē·shēz }

incircle See inscribed circle. { ¦in¦sər·kəl }

Incirrata [INV ZOO] A suborder of cephalopod mollusks in the order Octopoda. { ¦in·sə¦räd·ə }

incised [BIOL] Having a deeply and irregularly notched margin. [MED] Made by cutting, as a wound. { in'sīzd }

incised meander [GEOL] A deep, tortuous valley cut by a meandering stream that was rejuvenated. { in'sīzd mē'an·dər }

incision [MED] A cut or wound of the body tissue, as an abdominal incision or a vertical or oblique incision. { in'sizh·ən }

incisional hernia [MED] Abnormal protrusion of an organ through an operative or accidental incision. Also known as postoperative hernia; posttraumatic hernia. { in'sizh·ən·əl 'hər·nē·ə }

incisive canal [ANAT] The bifurcated bony passage from the floor of the nasal cavity to the incisive fossa. { in'sī·siv kə¦nal }

incisive foramen [ANAT] One of the two to four openings of the incisive canal on the floor of the incisive fossa. { in'sī·siv fə'rā·mən }

incisive fossa [ANAT] **1.** A bony pit behind the upper incisors into which the incisive canals open. **2.** A depression on the maxilla at the origin of the depressor muscle of the nose. **3.** A depression of the mandible at the origin of the mentalis muscle. { in'sī·siv 'fäs·ə }

incisor [ANAT] A tooth specialized for cutting, especially those in front of the canines on the upper jaw of mammals. { in'sīz·ər }

inclination [GEOL] The angle at which a geological body or surface deviates from the horizontal or vertical; often used synonymously with dip. [GEOPHYS] In magnetic inclination, the dip angle of the earth's magnetic field. Also known as magnetic dip. [MATH] **1.** The inclination of a line in a plane is the angle made with the positive x axis. **2.** The inclination of a line in space with respect to a plane is the smaller angle the line makes with its orthogonal projection in the plane. **3.** The inclination of a plane with respect to a given plane is the smaller of the dihedral angles which it makes with the given plane. [SCI TECH] **1.** Angular deviation of a direction or surface from the true vertical or horizontal.

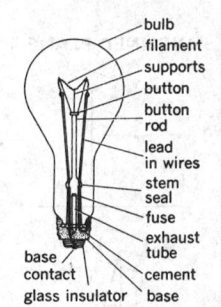

INCANDESCENT LAMP

bulb
filament
supports
button
button rod
lead in wires
stem
seal
fuse
exhaust tube
base contact
glass insulator
cement
base

Parts of an incandescent lamp. *(General Electric Co.)*

2. The angle which a direction or surface makes with the vertical or horizontal. **3.** A surface which deviates from the vertical or horizontal. { ‚iŋ·klə'nā·shən }

inclination of axis [ASTRON] The angle between a planet's axis of rotation and the perpendicular to the plane of its orbit. { ‚iŋ·klə'nā·shən əv 'ak·səs }

inclination of planetary orbits [ASTRON] The angle between the plane of the orbit and the plane of the ecliptic, which is the plane of the earth's orbit. { ‚iŋ·klə'nā·shən əv ‚plan·ə‚ter·ē 'òr·bəts }

inclination of the wind [METEOROL] The angle between the direction of the wind and the isobars. { ‚iŋ·klə'nā·shən əv the 'wind }

incline [SCI TECH] An upward-or downward-sloping surface. { 'in‚klīn }

inclined bedding [GEOL] A type of bedding in which the strata dip in the direction of current flow. { in'klīnd 'bed·iŋ }

inclined cableway [MECH ENG] A monocable arrangement in which the track cable has a slope sufficiently steep to allow the carrier to run down under its own weight. { in'klīnd 'kā·bəl‚wā }

inclined contact [GEOL] A contact plane of gas or oil with water underlying, in which the plane slopes or is inclined. { in'klīnd 'kän‚takt }

inclined drilling [ENG] The drilling of blastholes at an angle with the vertical. { in'klīnd 'dril·iŋ }

inclined extinction [OPTICS] Extinction in which the vibration directions are inclined to a crystal axis or direction of cleavage. Also known as oblique extinction. { in'klīnd ik'stiŋk·shən }

inclined orbit [AERO ENG] A satellite orbit which is inclined with respect to the earth's equator. { in'klīnd 'òr·bət }

inclined plane [MECH] A plane surface at an angle to some force or reference line. { in‚klīnd 'plān }

inclined skip hoist [MIN ENG] A skip hoist that operates on steeply inclined rails placed on a mine pit slope or wall. { in'klīnd 'skip ‚hòist }

inclined-tube manometer [ENG] A glass-tube manometer with the leg inclined from the vertical to extend the scale for more minute readings. { in'klīnd ‚tüb mə'näm·əd·ər }

incline shaft [MIN ENG] A shaft which has been dug at an angle to the vertical to follow the depth of the lode. { 'in‚klīn ‚shaft }

inclining experiment [NAV ARCH] An experimental determination of the weight of a ship and of the position of its center of gravity, in which a known weight, already aboard, is moved a measured distance perpendicular to the ship's centerline plane, and the resulting angle of list is measured. { in'klīn·iŋ ik‚sper·ə·mənt }

inclinometer [ENG] **1.** An instrument that measures the attitude of an aircraft with respect to the horizontal. **2.** An instrument for measuring the angle between the earth's magnetic field vector and the horizontal plane. Also known as dip circle. **3.** An apparatus used to ascertain the direction of the magnetic field of the earth with reference to the plane of the horizon. { ‚in·klə'näm·əd·ər }

inclusion [CELL MOL] A visible product of cellular metabolism within the protoplasm. [CRYSTAL] **1.** A crystal or fragment of a crystal found in another crystal. **2.** A small cavity filled with gas or liquid in a crystal. [MET] An impure particle, such as sand, trapped in molten metal during solidification. [PETR] A fragment of older rock enclosed in an igneous rock. { in'klü·zhən }

inclusion blennorrhea See inclusion conjunctivitis. { in'klü·zhən ‚blen·ə'rē·ə }

inclusion body [VIROL] Any of the abnormal structures appearing within the cell nucleus or the cytoplasm during the course of virus multiplication. { in'klü·zhən ‚bäd·ē }

inclusion-cell disease [MED] A rare genetic disorder in which lysosomal hydrolases are transported out of the cell into the blood, rather than into the lysosome, resulting in the accumulation of undigested macromolecules within the lysosome. Abbreviated I-cell disease. { iŋ'klü·zhən ‚sel ‚diz‚ēz }

inclusion complex [CHEM] An unbonded association in which the molecules of one component are contained wholly or partially within the crystal lattice of the other component. { in'klü·zhən 'käm‚pleks }

inclusion compound See clathrate. { in'klü·zhən 'käm‚paund }

inclusion conjunctivitis [MED] An acute inflammation of the conjunctiva with pus formation caused by a virus and identified from epithelial-cell inclusion bodies in conjunctival scrapings. Also known as inclusion blennorrhea; paratrachoma; swimmer's conjunctivitis; swimming-pool conjunctivitis. { in'klü·zhən kən‚jəŋk·tə'vīd·əs }

inclusion cyst [MED] A cyst formed by the implantation of epithelial tissue into another structure. { in'klü·zhən ‚sist }

inclusion encephalitis [MED] A chronic inflammation of the brain in which large inclusion bodies occur in the nuclei of oligodendria and sometimes in nerve cells. { in'klü·zhən en‚sef·ə'līd·əs }

inclusion-exclusion principle [MATH] The principle that, if A and B are finite sets, the number of elements in the union of A and B can be obtained by adding the number of elements in A to the number of elements in B, and then subtracting from this sum the number of elements in the intersection of A and B. { ‚in‚klü·zhən 'eks‚klü·zhən ‚prin·sə·pəl }

inclusion relation [MATH] **1.** A set theoretic relation, usually denoted by the symbol ⊂, such that, if A and B are two sets, A ⊂ B if and only if every element of A is an element of B. **2.** Any relation on a Boolean algebra which is reflexive, antisymmetric, and transitive. { iŋ'klü·zhən ri‚lā·shən }

inclusive or See or. { in'klü·siv 'òr }

incoalation See coalification. { ‚in·kō'lā·shən }

incoherence [MED] Lack of coherence, relevance, or continuity of ideas or language. { ‚in·kō'hir·əns }

incoherent [GEOL] Pertaining to a rock or deposit that is loose or unconsolidated, or that is unable to hold together firmly or solidly. { ‚in·kō'hir·ənt }

incoherent light [OPTICS] Electromagnetic radiant energy not all of the same phase, and possibly also consisting of various wavelengths. { ‚in·kō'hir·ənt 'līt }

incoherent scattering [PHYS] Scattering of particles or photons in which the scattering elements act independently of one another, so that there are no definite phase relationships among the different parts of the scattered beam. { ‚in·kō'hir·ənt 'skad·ə·riŋ }

incoherent waves [PHYS] Waves having no fixed phase relationship. { ‚in·kō'hir·ənt 'wāvz }

incoming first selector [ELEC] Connects incoming calls from outlying dial offices to local second selectors. { 'in‚kəm·iŋ ‚fərst si'lek·tər }

incoming selector [ELEC] Selector associated with trunk circuits from another central office. { 'in‚kəm·iŋ si'lek·tər }

incommensurable line segments [MATH] Two line segments the ratio of whose lengths is irrational. { ‚in·kə‚mens·ə·rə·bəl 'līn ‚seg·məns }

incommensurable numbers [MATH] Two numbers whose ratio is irrational. { ‚in·kə'mens·ə·rə·bəl ‚nəm·bərz }

incompatibility [IMMUNOL] Genetic or antigenic differences between donor and recipient tissues that result in a rejection response. { ‚in·kəm‚pad·ə'bil·əd·ē }

incompatible equations [MATH] Two or more equations that are not satisfied by any set of values for the variables appearing. Also known as inconsistent equations. { ‚in·kəm'pad·ə·bəl i‚kwā·zhənz }

incompatible inequalities [MATH] Two or more inequalities that are not satisfied by any set of values of the variables involved. Also known as inconsistent inequalities. { ‚in·kəm'pad·ə·bəl ‚in·ə'kwäl·əd·ēz }

incompetence [FOREN SCI] Inability to function within the law, as the incompetence of an individual to drive when under the influence of alcohol. [MED] Insufficiency or inadequacy in performing natural functions. { in'käm·pəd·əns }

incompetent bed [GEOL] A bed not combining sufficient firmness and flexibility to transmit a thrust and to lift a load by bending. { in'käm·pəd·ənt 'bed }

incompetent rock [ENG] Soft or fragmented rock in which an opening, such as a borehole or an underground working place, cannot be maintained unless artificially supported by casing, cementing, or timbering. { in'käm·pəd·ənt 'räk }

incomplete abortion [MED] Expulsion of only part of the product of conception, with some of the membranes or placenta remaining in the uterus. { ‚in·kəm'plēt ə'bòr·shən }

incomplete antibody [IMMUNOL] An antibody that cannot directly agglutinate saline-suspended red blood cells; it needs

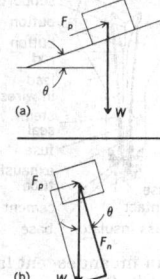

INCLINED PLANE

Weight resting on an inclined plane (a) with principal forces applied, and (b) with their resolution into normal force. θ is angle of inclination of plane, W is weight of body, F_p is force parallel to the surface, F_n is force normal to the surface.

INCLINED-TUBE MANOMETER

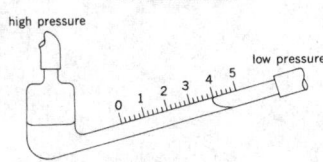

high pressure   low pressure

Drawing of inclined-tube manometer.

an additive to complete the agglutination. { ˌin·kəmˌplēt 'ant·iˌbäd·ē }

incomplete beta function [MATH] The function $\beta_x(p,q)$ defined by

$$\beta_x(p,q) = \int_0^x t^{p-1}(1-t)^{q-1}\,dt$$

where $0 \le x \le 1$, $p > 0$, and $q > 0$. { ˌin·kəmˌplēt 'bād·ə ˌfəŋk·shən }

incomplete combustion [CHEM] Combustion in which oxidation of the fuel is incomplete. { ˌin·kəm'plēt kəm'bəs·shən }

incomplete dominance [GEN] Expression of alleles such that the phenotype of the heterozygote is intermediate between that of the two homozygotes. { ˌin·kəmˌplēt 'däm·ən·əns }

incomplete flower [BOT] A flower lacking one or more modified leaves, such as petals, sepals, pistils, or stamens. { ˌin·kəm'plēt 'flaů·ər }

incomplete fusion See deep inelastic collision. { ˌin·kəm'plēt 'fyü·zhən }

incomplete gamma function [MATH] Either of the functions $\gamma(a,x)$ and $\Gamma(a,x)$ defined by

$$\gamma(a,x) = \int_0^x t^{a-1}e^{-t}\,dt$$

$$\Gamma(a,x) = \int_x^\infty t^{a-1}e^{-t}\,dt$$

where $0 \le x \le \infty$ and $a > 0$. { ˌin·kəm'plēt 'gam·ə ˌfəŋk·shən }

incomplete Latin square See Yonden square. { ˌin·kəm'plēt ˌlat·ən 'skwer }

incomplete lubrication [MECH ENG] Lubrication that takes place when the load on the rubbing surfaces is carried partly by a fluid viscous film and partly by areas of boundary lubrication; friction is intermediate between that of fluid and boundary lubrication. { ˌin·kəm'plēt ˌlü·brə'kā·shən }

incompressibility [MECH] Quality of a substance which maintains its original volume under increased pressure. { ˌin·kəmˌpres·ə'bil·əd·ē }

incompressibility condition [FL MECH] The condition prevailing when the time rate of change of the density of a fluid is zero; this is a valid assumption for most problems in dynamic oceanography. { ˌin·kəmˌpres·ə'bil·əd·ē kən·dish·ən }

incompressible flow [FL MECH] Fluid motion without any change in density. { ˌin·kəm'pres·ə·bal 'flō }

incompressible fluid [FL MECH] A fluid which is not reduced in volume by an increase in pressure. { ˌin·kəm'pres·ə·bal 'flü·əd }

incongruous [GEOL] Of a drag fold, having an axis and axial surface not parallel to the axis and axial surface of the main fold to which it is related. { in'käŋ·grü·əs }

inconsistent axioms [MATH] A set of axioms from which both a proposition and its negation can be deduced. { ˌin·kənˌsis·tənt 'ak·sē·əmz }

inconsistent equations See incompatible equations. { ˌin·kənˌsis·tənt i'kwā·zhənz }

inconsistent inequalities See incompatible inequalities. { ˌin·kənˌsis·tənt ˌin·ə'kwäl·əd·ēz }

incontinence [MED] Inability to control the natural evacuations, as the feces or the urine; specifically, involuntary evacuation from organic causes. { in'känt·ən·əns }

incorporate [COMPUT SCI] To place in storage. { in'kȯr·pəˌrāt }

incrassate [BIOL] State of being swollen or thickened. { in'kraˌsāt }

increaser [ENG] An adapter for connecting a small-diameter pipe to a larger-diameter pipe. { in'krēs·ər }

increasing function [MATH] A function, f, of a real variable, x, whose value gets larger as x gets larger; that is, if $x < y$, then $f(x) < f(y)$. Also known as strictly increasing function. { in'krēs·iŋ ˌfəŋk·shən }

increasing sequence [MATH] A sequence of real numbers in which each term is greater than the preceding term. { in'krēs·iŋ 'sē·kwəns }

increment [ANALY CHEM] An individual portion of material of a group of samples collected by a single operation of a sampling device from parts of a lot that are separated in time or space. [HYD] See recharge. [MATH] A change in the argument or values of a function, usually restricted to being a small positive or negative quantity. [SCI TECH] A small change in the value of a variable. { 'iŋ·krə·mənt }

incremental compiler [COMPUT SCI] A compiler that generates code for a statement, or group of statements, which is independent of the code generated for other statements. { ˌiŋ·krə'ment·əl kəm'pīl·ər }

incremental computer [COMPUT SCI] A special-purpose computer designed to process changes in variables as well as absolute values; for instance, a digital differential analyzer. { ˌiŋ·krə'ment·əl kəm'pyüd·ər }

incremental cost [IND ENG] **1.** The difference between the costs and the revenues between two alternative procedures. **2.** The cost of the last unit produced at a given level of production. { ˌiŋ·krəˌment·əl 'kȯst }

incremental digital recorder [COMPUT SCI] Magnetic tape recorder in which the tape advances across the recording head step by step, as in a punched-paper-tape recorder; used for recording an irregular flow of data economically and reliably. { ˌiŋ·krə'ment·əl ˌdij·əd·ə ri'kȯrd·ər }

incremental dump tape [COMPUT SCI] A safety technique used in time-sharing which consists in copying all files (created or modified by a user during a day) on a magnetic tape; in case of system failure, the file storage can then be reconstructed. Also known as failsafe tape. { ˌiŋ·krə'ment·əl 'dəmp ˌtāp }

incremental frequency shift [COMMUN] Method of superimposing incremental intelligence on another intelligence by shifting the center frequency of an oscillator a predetermined amount. { ˌiŋ·krə'ment·əl 'frē·kwən·sē ˌshift }

incremental hysteresis loss [ELECTROMAG] Hysteresis loss when a magnetic material is subjected to a pulsating magnetizing force. { ˌiŋ·krə'ment·əl ˌhis·tə'rē·səs ˌlȯs }

incremental induction [ELECTROMAG] The quantity lying between the highest and lowest value of a magnetic induction at a point in a polarized material, when subjected to a small cycle of magnetization. { ˌiŋ·krə'ment·əl in'dək·shən }

incremental mode [COMPUT SCI] The plotting of a curve on a cathode-ray tube by illuminating a fixed number of points at a time. { ˌiŋ·krə'ment·əl ˌmōd }

incremental permeability [ELECTROMAG] The ratio of a small cyclic change in magnetic induction to the corresponding cyclic change in magnetizing force when the average magnetic induction is greater than zero. { ˌiŋ·krə'ment·əl ˌpər·mē·ə'bil·əd·ē }

incremental printer [GRAPHICS] A printer, such as a computer-controlled electric typewriter, that prints sequentially, character by character, on each line. { ˌiŋ·krə'ment·əl ˌprint·ər }

incremental representation [COMPUT SCI] A way of representing variables used in incremental computers, in which changes in the variables are represented instead of the values of the variables themselves. { ˌiŋ·krə'ment·əl ˌrep·rə·sən'tā·shən }

increment borer [FOR] An augerlike instrument with a hollow bit, used to extract thin radial cylinders of wood from trees to determine age and growth rate. { 'iŋ·krə·mənt ˌbȯr·ər }

incretion [PHYSIO] An internal secretion. { in'krē·shən }

incross [GEN] Mating between individuals from the same inbred line. { 'inˌkrȯs }

incubation [CHEM] Maintenance of chemical mixtures at specified temperatures for varying time periods to study chemical reactions, such as enzyme activity. [MED] The phase of an infectious disease process between infection by the pathogen and appearance of symptoms. [VERT ZOO] The act or process of brooding. { ˌiŋ·kyə'bē·shən }

incubation period [MED] The period of time required for the development of symptoms of a disease after infection, or of altered reactivity after exposure to an allergen. [VERT ZOO] The brooding period required to bring an egg to hatching. { ˌiŋ·kyə'bē·shən pir·ē·əd }

incubator [AGR] A device for the artificial hatching of eggs. [MED] A small chamber with controlled oxygen, temperature, and humidity for newborn infants requiring special care. [MICROBIO] A laboratory cabinet with controlled temperature for the cultivation of bacteria, or for facilitating biologic tests. { 'iŋ·kyəˌbād·ər }

incubator oil [MATER] Special grade of long-burning petroleum heating oil used to heat farm incubators. { 'iŋ·kyə‚bäd·ər ‚öil }

incubatory carrier [MED] A person infected with a certain microorganism but in such an early stage of disease that clinical manifestations are not apparent. { 'iŋ·kyə·bə‚tòr·ē ‚kar·ē·ər }

incubous [BOT] The juxtaposition of leaves such that the anterior margins of older leaves overlap the posterior margins of younger leaves. { 'iŋ·kyə·bəs }

incudate [BIOL] Of, pertaining to, or having an incus. { 'iŋ·kyə‚dāt }

incumbent [BIOL] Lying on or down. [GEOL] Lying above, said of a stratum that is superimposed or overlies another stratum. [ECOL] Referring to the occupation and utilization of resources to the exclusion of other species. { in'kəm·bənt }

incunabula printing *See* cradle printing. { ‚in·kyə'nab·yə·lə ¦print·iŋ }

incurrent canal [INV ZOO] A canal through which water enters a sponge. { in'kər·ənt kə'nal }

incurrent siphon *See* inhalant siphon. { in'kə·rənt 'sī·fən }

Incurvariidae [INV ZOO] A family of lepidopteran insects in the superfamily Incurvarioidea; includes yucca moths and relatives. { ‚in‚kər·və'rī·ə‚dē }

Incurvarioidea [INV ZOO] A monofamilial superfamily of lepidopteran insects in the suborder Heteroneura having wings covered with microscopic spines, a single genital opening in the female, and venation that is almost complete. { ‚in‚kər‚var·ē'òīd·ē·ə }

incus [ANAT] The middle one of three ossicles in the middle ear. Also known as anvil. [METEOROL] A supplementary cloud feature peculiar to cumulonimbus capillatus; the spreading of the upper portion of cumulonimbus when this part takes the form of an anvil with a fibrous or smooth aspect. Also known as anvil; thunderhead. { 'iŋ·kəs }

indamine [ORG CHEM] HN:C₆H₄:N·C₆H₄NH₂ An unstable dye obtained by the reaction of *para*-phenylenediamine and aniline. Also known as phenylene blue. { 'in·də‚mēn }

indan [ORG CHEM] C₆H₄(CH₂)₃ Colorless liquid boiling at 177°C; soluble in alcohol and ether, insoluble in water; derived from coal tar. { 'in‚dan }

indanthrone [ORG CHEM] C₂₈H₁₄N₂O₄ A blue pigment or vat dye soluble in dilute base solutions; used in cotton dyeing and as a pigment in paints and enamels. { in'dan‚thrōn }

indeciduate placenta [EMBRYO] A placenta having the maternal and fetal elements associated but not fused. { ¦in·də¦sij·ə·wət plə'sent·ə }

indefinite ceiling [METEOROL] After United States weather observing practice, the ceiling classification applied when the reported ceiling value represents the vertical visibility upward into surface-based, atmospheric phenomena (except precipitation), such as fog, blowing snow, and all of the lithometeors. Formerly known as ragged ceiling. { in'def·ə·nət 'sēl·iŋ }

indefinite integral [MATH] An indefinite integral of a function $f(x)$ is a function $F(x)$ whose derivative equals $f(x)$. Also known as antiderivative; integral. { in'def·ə·nət 'int·ə·grəl }

indegree [MATH] For a vertex, v, in a directed graph, the number of arcs directed from other vertices to v. { 'in·di‚grē }

indehiscent [BOT] **1.** Remaining closed at maturity, as certain fruits. **2.** Not splitting along regular lines. { ¦in·də'his·ənt }

indelible ink [MATER] An ink that cannot be removed, for example, India ink. { in'del·ə·bəl 'iŋk }

indene [ORG CHEM] C₉H₈ A colorless, liquid, polynuclear hydrocarbon; boils at 181°C and freezes at −2°C; derived from coal tar distillates; copolymers with benzofuran have been manufactured on a small scale for use in coatings and floor coverings. { 'in‚dēn }

indent [SCI TECH] To form a depression by forcing inward. { in'dent }

indentation hardness [MET] The resistance of a metal surface to indention when subjected to pressure by a hard pointed or rounded tool. Also known as penetration hardness. { ‚in‚den'tā·shən ¦hard·nəs }

indented bolt [DES ENG] A type of anchor bolt that has indentations to hold better in cemented grout. { in'den·təd 'bòlt }

independence number [MATH] For a graph, the largest possible number of vertices in an independent set. Also

INDENE

Structural formula of indene.

known as internal stability number. { ‚in·di'pen·dəns ‚nəm·bər }

independent assortment [GEN] In meiosis, the random assortment of the alleles at two or more loci because they are on different chromosome pairs or far apart on the same chromosome pair. { ‚in·də'pen·dənt ə'sòrt·mənt }

independent axiom [MATH] A member of a set of axioms that cannot be deduced as a consequence of the other axioms in the set. { ‚in·di‚pen·dənt 'ak·sē·əm }

independent chuck [DES ENG] A chuck for holding work by means of four jaws, each of which is moved independently of the others. { ‚in·də'pen·dənt 'chək }

independent contractor [ENG] One who exercises independent control over the mode and method of operations to produce the results demanded by the contract. { ‚in·də'pen·dənt 'kän‚trak·tər }

independent edge set *See* matching. { ‚in·di‚pen·dənt 'ej ‚set }

independent equations [MATH] A system of equations such that no one of them is necessarily satisfied by a solution to the rest. { ‚in·də'pen·dənt i'kwā·zhənz }

independent events [STAT] Two events in probability such that the occurrence of one of them does not affect the probability of the occurrence of the other. { ‚in·də'pen·dənt i'vens }

independent footing [CIV ENG] A footing that supports a concentrated load, such as a single column. { ‚in·də'pen·dənt 'fùd·iŋ }

independent functions [MATH] A set of functions such that knowledge of the values obtained by all but one of them at a point is insufficient to determine the value of the remaining function. { ‚in·də'pen·dənt 'fəŋk·shənz }

independent line of sighting [ORD] A system for laying a gun, whereby the angle of site and the angle of elevation (range) mechanisms work independently of each other. { ‚in·də'pen·dənt ¦līn əv 'sīd·iŋ }

independent migration law [ANALY CHEM] The law that each ion in a conductiometric titration contributes a definite amount to the total conductance, irrespective of the nature of the other ions in the electrolyte. { ‚in·də'pen·dənt mī'grā·shən ‚lò }

independent-particle model [ATOM PHYS] A model of an atomic system in which the electrons are assumed to move independently of each other in the average field generated by the nucleus and the other electrons. { ‚in·də¦pen·dənt 'pard·i·kəl ‚mäd·əl }

independent random variables [STAT] The discrete random variables $X_1, X_2, \ldots, X_n$ are independent if for arbitrary values $x_1, x_2, \ldots, x_n$ of the variables the probability that $X_1 = x_1$ and $X_2 = x_2$, etc., is equal to the product of the probabilities that $X_i = x_i$ for $i = 1, 2, \ldots, n$; random variables which are unrelated. { ‚in·də'pen·dənt ¦ran·dəm ‚ver·ē·ə·bəls }

independent recoil system [ORD] A recoil mechanism for artillery that has an independent recuperator, that is, the recuperator is entirely independent of the recoil brake in the recoil mechanism. { ‚in·də'pen·dənt 'rē‚kòil ‚sis·təm }

independent sector [COMPUT SCI] A device on some punched-card tabulators that allows only the first of a series of similar data items to be printed and prevents printing of the rest. { ‚in·də'pen·dənt 'sek·tər }

independent set [MATH] A set of vertices in a simple graph such that no two vertices in this set are adjacent. Also known as internally stable set. { ‚in·di‚pen·dənt 'set }

independent-sideband modulation [COMMUN] Modulation in which the radio-frequency carrier is reduced or eliminated and two channels of information are transmitted, one on an upper and one on a lower sideband. Abbreviated ISB modulation. { ‚in·də'pen·dənt ¦sīd‚band ‚maj·ə'lā·shən }

independent-sideband receiver [ELECTR] A radio receiver designed for the reception of independent-sideband modulation, having provisions for restoring the carrier. { ‚in·də'pen·dənt ¦sīd‚band ri'sē·vər }

independent-sideband transmitter [ELECTR] A transmitter which produces independent-sideband modulated signals. { ‚in·də'pen·dənt ¦sīd‚band tranz'mid·ər }

independent suspension [MECH ENG] In automobiles, a system of springs and guide links by which wheels are mounted independently on the chassis. { ‚in·də'pen·dənt sə'spen·chən }

independent variable [MATH] In an equation $y = f(x)$, the input variable *x*. Also known as argument. { ‚in·də'pen·dənt 'ver·ē·ə·bəl }

independent wire-rope core [DES ENG] A core of steel in a wire rope made in accordance with the best practice and design, either bright (uncoated) galvanized or drawn galvanized wire. { ‚in·də'pen·dənt 'wīr ‚rōp ‚kȯr }

inderborite [MINERAL] $CaMgB_6O_{11} \cdot 11H_2O$ A monoclinic mineral composed of hydrous calcium and magnesium borate. { ‚in·dər'bȯ‚rīt }

inderite [MINERAL] $Mg_2B_6O_{11} \cdot 15H_2O$ A hydrated borate mineral. { 'in·də‚rīt }

indeterminacy principle *See* uncertainty principle. { ‚in·də'tərm·ə·nə·sē ‚prin·sə·pəl }

indeterminate cleavage [EMBRYO] Cleavage in which all the early cells have the same potencies with respect to development of the entire zygote. { ‚in·də'tərm·ə·nət 'klē·vij }

indeterminate equations [MATH] A set of equations possessing an infinite number of solutions. { ‚in·də'tərm·ə·nət i'kwā·zhənz }

indeterminate forms [MATH] Products, quotients, differences, or powers of functions which are undefined when the argument of the function has a certain value, because one or both of the functions are zero or infinite; however, the limit of the product, quotient, and so on as the argument approaches this value is well defined. { ‚in·də'tərm·ə·nət 'fȯrmz }

indeterminate growth [BOT] Growth of a plant in which the axis is not limited by development of a reproductive structure, and therefore growth continues indefinitely. { ‚in·də'tərm·ə·nət 'grōth }

indeterminate truss [CIV ENG] A truss having redundant bars. { ‚in·də'tərm·ə·nət 'trəs }

index [COMPUT SCI] **1.** A list of record surrogates arranged in order of some attribute expressible in machine-orderable form. **2.** To produce a machine-orderable set of record surrogates, as in indexing a book. **3.** To compute a machine location by indirection, as is done by index registers. **4.** The portion of a computer instruction which indicates what index register (if any) is to be used to modify the address of an instruction. [MATH] **1.** Unity of a logarithmic scale, as the C scale of a slide rule. **2.** A subscript or superscript used to indicate a specific element of a set or sequence. **3.** The number above and to the left of a radical sign, indicating the root to be extracted. **4.** For a subgroup of a finite group, the order of the group divided by the order of the subgroup. **5.** For a continuous complex-valued function defined on a closed plane curve, the change in the amplitude of the function when traversing the curve in a counterclockwise direction, divided by 2π. **6.** For a quadratic or Hermitian form, the number of terms with positive coefficients when the form is reduced by a linear transformation to a sum of squares or a sum of squares of absolute values. **7.** For a symmetric or Hermitian matrix, the number of positive entries when the matrix is transformed to diagonal form. **8.** *See* winding number. [PHYS] A numerical quantity, usually dimensionless, denoting the magnitude of some physical effect, such as the refractive index. { 'in‚deks }

index arithmetic unit [COMPUT SCI] A section of some computers that performs addition or subtraction operations on address parts of instructions for the purpose of indexing, boundary tests for memory protection, and so forth. { 'in‚deks ə'rith·mə‚tik ‚yü·nət }

index arm [NAV] On a marine sextant, a slender bar carrying the index; the bar pivots at the center of curvature of the arc of the sextant and carries the index and the vernier or micrometer. { 'in‚deks ‚ärm }

index bed *See* key bed. { 'in‚deks ‚bed }

index catalog [ASTRON] A supplement to the New General Catalog of nebulae. { 'in‚deks ‚kad·əl‚äg }

index center [MECH ENG] One of two machine-tool centers used to hold work and to rotate it by a fixed amount. { 'in‚deks ‚sen·tər }

index chart [MECH ENG] **1.** A chart used in conjunction with an indexing or dividing head, which correlates the index plate, hole circle, and index crank motion with the desired angular subdivisions. **2.** A chart indicating the arrangement of levers in a machine to obtain desired output speed or fuel rate. [NAV] In marine operations, an outline chart showing the limits and identifying designations of navigational charts, volumes of sailing directions, and so on. { 'in‚deks ‚chärt }

index counter [ENG] A counter indicating revolutions of the

tape supply reel, making it possible to index selections within a reel of tape. { 'in‚deks ‚kaún·tər }

index crank [MECH ENG] The crank handle of an index head used to turn the spindle. { 'in‚deks ‚krank }

index cycle [METEOROL] A roughly cyclic variation in the zonal index. { 'in‚deks ‚sī·kəl }

indexed address [COMPUT SCI] An address which is modified, generally by means of index registers, before or during execution of a computer instruction. { 'in‚dekst ə'dres }

indexed array [COMPUT SCI] An array of data items in which the individual items can be accessed by specifying their position through use of a subscript. { 'in‚dekst ə'rā }

indexed sequential data set [COMPUT SCI] A collection of related data items that are stored sequentially on a key, but are also accessible through index tables maintained by the system. { 'in‚dekst si‚kwen·chəl 'dad·ə ‚set }

indexed sequential organization [COMPUT SCI] A sequence of records arranged in collating sequence used with direct-access devices. { 'in‚dekst si‚kwen·chəl ‚ȯr·gə·nə'zā·shən }

index ellipsoid [OPTICS] An ellipsoid whose three perpendicular axes are proportional in length to the principal values of the index of refraction of light in an anisotropic medium and point in the direction of the corresponding electric vector. Also known as ellipsoid of wave normals; indicatrix; optical indicatrix; polarizability ellipsoid; reciprocal ellipsoid. { 'in‚deks ə'lip‚sȯid }

index error [ENG] An error caused by the misalignment of the vernier and the graduated circle (arc) of an instrument. { 'in‚deks ‚er·ər }

index forest [FOR] A forest reaching the highest average in a given locality for density, volume, and increment. { 'in‚deks ‚fär·əst }

index fossil [PALEON] The ancient remains and traces of an organism that lived during a particular geologic time period and that geologically date the containing rocks. { 'in‚deks ‚fäs·əl }

index glass *See* index mirror. { 'in‚deks ‚glas }

index head [MECH ENG] A headstock that can be affixed to the table of a milling machine, planer, or shaper; work may be mounted on it by a chuck or centers, for indexing. { 'in‚deks ‚hed }

indexing [MECH ENG] The process of providing discrete spaces, parts, or angles in a workpiece by using an index head. { 'in‚dek·siŋ }

indexing fixture [MECH ENG] A fixture that changes position with regular steplike movements. { 'in‚dek·siŋ ‚fiks·chər }

index line *See* isopleth. { 'in‚deks ‚līn }

index liquid [OPTICS] A liquid whose index of refraction is known, used to find the index of refraction of powdered substances with a microscope. { 'in‚deks ‚lik·wəd }

index marker [COMPUT SCI] The beginning (and end) of each track in a disk, which is recognized by a special sensing device within the disk mechanism. { 'in‚deks ‚märk·ər }

index mineral [PETR] A mineral whose first appearance in passing from low to higher grades of metamorphism indicates the outer limit of a zone. { 'in‚deks ‚min·rəl }

index mirror [NAV] The mirror attached to the index arm of a marine sextant. Also known as index glass. { 'in‚deks ‚mir·ər }

index number [STAT] A number indicating change in magnitude, as of cost or of volume of production, as compared with the magnitude at a specified time, usually taken as 100; for example, if production volume in 1970 was two times as much as the volume in 1950 (taken as 100), its index number is 200. { 'in‚deks ‚nəm·bər }

index of absorption *See* absorption index. { 'in‚deks əv əb'sȯrp·shən }

index of aridity [CLIMATOL] A measure of the precipitation effectiveness or aridity of a region, given by the following relationship: index of aridity = $P/(T + 10)$, where P is the annual precipitation in centimeters, and T the annual mean temperature in degrees Celsius. { 'in‚deks əv ə'rid·əd·ē }

index of cooperation [COMMUN] In rectilinear scanning or recording, the product of the total length of a scanning or recording line by the number of scanning or recording lines per unit length divided by pi. { 'in‚deks əv kō‚äp·ə'rā·shən }

index of modulation *See* modulation factor. { 'in‚deks əv ‚mäj·ə'lā·shən }

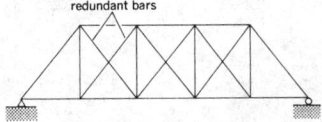

INDETERMINATE TRUSS

Redundant bars in an indeterminate truss.

index of precision [STAT] The constant h in the normal curve $y = K \exp[-h^2(x - u)^2]$; a large value of h indicates a high precision, or small standard deviation. { 'in,deks əv prə'sizh·ən }

index of refraction [OPTICS] The ratio of the phase velocity of light in a vacuum to that in a specified medium. Also known as absolute index of refraction; absolute refractive constant; refractive constant; refractive index. { 'in,deks əv ri'frak·shən }

index of unsaturation [ORG CHEM] A numerical value that represents the number of rings or double bonds in a molecule; a triple bond is considered to have the numerical value of 2. { 'in,deks əv ¦ən,sach·ə'ra·shən }

index of work tolerance [IND ENG] A measure of the period of time during which an individual can perform a given task with the required efficiency while maintaining appropriate levels of physiological and emotional well-being. { ¦in,deks əv 'wərk ,täl·ə·rəns }

index plane [GEOL] A surface used as a reference point in determining geological structure. { 'in,deks ,plān }

index plate [DES ENG] A plate with circular graduations or holes arranged in circles, each circle with different spacing; used for indexing on machines. { 'in,deks ,plāt }

index point [COMPUT SCI] A hardware reference mark on a disk or drum for use in timing. { 'in,deks ,pȯint }

index prism [NAV] A sextant prism which can be rotated to any angle corresponding to altitudes between established limits; the bubble or pendulum sextant counterpart of the index mirror of a marine sextant. { 'in,deks ,priz·əm }

index ratio [ELECTROMAG] The ratio of the radius of a conductor used in induction heating to its skin depth at the frequency used. { 'in,deks ,rā·shō }

index register [COMPUT SCI] A hardware element which holds a number that can be added to (or, in some cases, subtracted from) the address portion of a computer instruction to form an effective address. Also known as base register; B box; B line; B register; B store; modifier register. { 'in,deks ,rej·ə·stər }

index stock [GRAPHICS] A stiff paper receptive to writing ink and to printing; available in smooth and antique finishes and used for index records and business cards. { 'in,deks ,stäk }

index thermometer [ENG] A thermometer in which steel index particles are carried by mercury in the capillary and adhere to the capillary wall in the high and low positions, thus indicating minimum and maximum inertial scales. { 'in,deks thər'mäm·əd·ər }

index word See modifier. { 'in,deks ,wərd }

india ink [MATER] A permanent black ink made of lampblack and blue binder; some varieties are waterproof. Also known as Chinese ink; sumi ink. { 'in·dē·ə 'iŋk }

indialite [MINERAL] $Mg_2Al_4Si_5O_{18}$ A hexagonal cordierite mineral; it is isotypic with beryl. { 'in·dē·ə,līt }

Indian See Indus. { 'in·dē·ən }

indianaite [MINERAL] A white porcelainlike clay mineral; a variety of halloysite found in Indiana. { ,in·dē'a·nə,īt }

Indiana limestone See spergenite. { ,in·dē'a·nə 'līm,stōn }

Indian balsam See Peru balsam. { 'in·dē·ən 'bȯl·səm }

Indian grass oil See palmarosa oil. { 'in·dē·ən 'gras ,ȯil }

Indian gum [MATER] Any of the gums, such as ghatti gum and sterculia gum, with mucilage consistency from trees in the forests in India and Ceylon. { 'in·dē·ən 'gəm }

Indian Ocean [GEOGR] The smallest and geologically the most youthful of the three oceans, whose surface area is 29,300,000 square miles (75,900,000 square kilometers); it is bounded on the north by India, Pakistan, and Iran; on the east by the Malay Peninsula; on the south by Antarctica; and on the west by the Arabian peninsula and Africa. { 'in·dē·ən 'ō·shən }

Indian red [MATER] Iron-oxide-base, maroon pigment; used to polish gold, silver, and other metals. Also known as iron saffron. { 'in·dē·ən 'red }

Indian spring low water [OCEANOGR] An arbitrary tidal datum approximating the level of the mean of the lower low waters at spring time, first used in waters surrounding India. Also known as harmonic tide plane; Indian tide plane. { 'in·dē·ən ¦spriŋ ¦lō 'wȯd·ər }

Indian summer [CLIMATOL] A period, in mid-or late autumn, of abnormally warm weather, generally clear skies, sunny but hazy days, and cool nights; in New England, at least one killing frost and preferably a substantial period of normally cool weather must precede this warm spell in order for it to be considered a true Indian summer; it does not occur every year, and in some years there may be two or three Indian summers; the term is most often heard in the northeastern United States, but its usage extends throughout English-speaking countries. { 'in·dē·ən 'səm·ər }

Indian tide plane See Indian spring low water. { 'in·dē·ən 'tīd ,plān }

Indian tragacanth See karaya gum. { 'in·dē·ən 'trag·ə,kanth }

Indian yellow [MATER] A yellow pigment which may be aureolin, made of cobalt and potassium nitrates; or puree, the impure basic magnesium salt of euxanthic acid; or the synthetic dye primuline. { 'in·dē·ən 'yel·ō }

indican [BIOCHEM] C_8H_6NOSOK The potassium salt of indoxylsulfate found in urine as a result of bacterial action on tryptophan in the bowel. [ORG CHEM] $C_{14}H_{17}O_6N$ A glucoside of indoxyl occurring in the indigo plant; on hydrolysis indican gives rise to indoxyl, which is oxidized to indigo by air. { 'in·də,kan }

indicated airspeed [AERO ENG] The airspeed as shown by a differential-pressure airspeed indicator, uncorrected for instrument and installation errors; a simple computation for altitude and temperature converts indicated airspeed to true airspeed. Abbreviated IAS. { 'in·də,kād·əd 'er,spēd }

indicated air temperature [METEOROL] The uncorrected reading from the free air temperature gage. Also known as outside air temperature. { 'in·də,kād·əd 'er ,tem·prə·chər }

indicated altitude [AERO ENG] The uncorrected reading of a barometric altimeter. { 'in·də,kād·əd 'al·tə,tüd }

indicated horsepower [MECH ENG] The horsepower delivered by an engine as calculated from the average pressure of the working fluid in the cylinders and the displacement. Abbreviated ihp. { 'in·də,kād·əd 'hȯrs,paů·ər }

indicated ore [MIN ENG] A known mineral deposit for which quantitative estimates are made partly from inference and partly from specific sampling. Also known as probable ore. { 'in·də,kād·əd 'ȯr }

indicating gage [ENG] A gage consisting essentially of a case and mounting, a spindle carrying the contact point, an amplifying mechanism, a pointer, and a graduated dial; used to amplify and measure the displacement of a movable contact point. { 'in·də,kād·iŋ ,gāj }

indicating instrument [ENG] An instrument in which the present value of the quantity being measured is visually indicated. { 'in·də,kād·iŋ ,in·strə·mənt }

indication [ENG] In ultrasonic testing, determination of the presence of a flaw by detection of a reflected ultrasonic beam. { ,in·də'kā·shən }

indicative data [COMPUT SCI] Data which describe a specific item. { in'dik·əd·iv 'dad·ə }

indicator [CHEM] See chemical indicator. [COMPUT SCI] A device announcing an error or failure. [ELECTR] A cathode-ray tube or other device that presents information transmitted or relayed from some other source, as from a radar receiver. [ENG] An instrument for obtaining a diagram of the pressure-volume changes in a running positive-displacement engine, compressor, or pump cylinder during the working cycle. [MATH] See Euler's phi function. { 'in·də,kād·ər }

indicator card [ENG] A chart on which an indicator diagram is produced by an instrument called an engine indicator which traces the real-performance cycle diagram as the machine is running. { 'in·də,kād·ər ,kärd }

indicator diagram [ENG] A pressure-volume diagram representing and measuring the work done by or on a fluid while performing the work cycle in a reciprocating engine, pump, or compressor cylinder. { 'in·də,kād·ər 'dī·ə,gram }

indicator element [ELECTR] A component whose variability under conditions of manufacture or use is likely to cause the greatest variation in some measurable parameter. { 'in·də,kād·ər 'el·ə·mənt }

indicator function See characteristic function. { 'in·də,kād·ər ,fəŋk·shən }

indicator gate [ELECTR] Rectangular voltage waveform which is applied to the grid or cathode circuit of an indicator cathode-ray tube to sensitize it or desensitize it during a desired portion of the operating cycle. { 'in·də,kād·ər ,gāt }

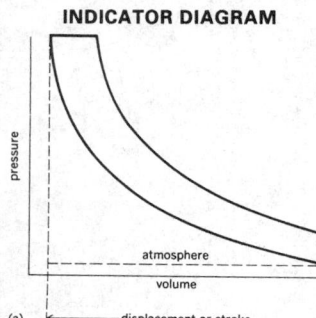

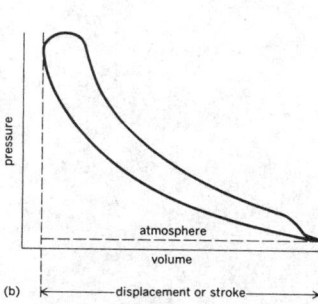

INDICATOR DIAGRAM

Pressure-volume diagrams for a diesel engine. *(a)* Theoretical ideal conditions; *(b)* actual two-cycle engine card. The areas outlined by the curve represent the effective work of the cycle.

indicator lamp [ELEC] A neon lamp whose on-off condition is used to convey information. { 'in·də,kad·ər ,lamp }

indicator medium [MICROBIO] A usually solid culture medium containing substances capable of undergoing a color change in the vicinity of a colony which has effected a particular chemical change, such as fermenting a certain sugar. { 'in·də,kad·ər ,mē·dē·əm }

indicator plant [BOT] A plant used in geobotanical prospecting as an indicator of a certain geological phenomenon. { 'in·də,kad·ər ,plant }

indicator species [ECOL] A species whose presence is directly related to a particular quality in its environment at a given location. { 'in·də,kad·ər ,spē·shēz }

indicator tube [ELECTR] An electron-beam tube in which useful information is conveyed by the variation in cross section of the beam at a luminescent target. { 'in·də,kad·ər ,tüb }

indicator unit [ENG] An instrument which detects the presence of an electrical quantity without necessarily measuring it. { 'in·də,kad·ər ,yü·nət }

indicatrix See index ellipsoid. { in'dik·ə,triks }

indicolite [MINERAL] An indigo-blue variety of tourmaline that is used as a gemstone. Also known as indigolite. { in'dik·ə,līt }

indifferent equilibrium See neutral equilibrium. { in'dif·ərnt ,ē·kwə'lib·rē·əm }

indifferent stability See neutral stability. { in'dif·ərnt stə'bil·əd·ē }

indigenous [SCI TECH] Existing and having originated naturally in a particular region or environment. { in'dij·ə·nəs }

indigenous coal See autochthonous coal. { in'dij·ə·nəs kōl }

indigenous limonite [MINERAL] Sulfide-derived limonite that remains fixed at the site of the parent sulfide. { in'dij·ə·nəs 'lī·mə,nīt }

indigo [ORG CHEM] **1.** A blue dye extracted from species of the *Indigofera* bush. **2.** See indigo blue. { 'in·də·gō }

indigo blue [ORG CHEM] $C_{16}H_{10}O_2N_2$ A component of the dye indigo, crystallizing as dark-blue rhomboids that break down at 30°C, that are soluble in hot aniline and hot chloroform, and that are also made synthetically; used as a reagent and a dye. Also known as indigo. { 'in·də·gō 'blü }

indigo carmine [ORG CHEM] $C_{16}H_8N_2Na_2O_8S_2$ A dark blue powder with coppery luster; used as a dye in testing kidney function and as a reagent in detecting chlorate and nitrate. Also known as soluble indigo blue. { 'in·də·gō 'kär·mən }

indigo copper See covellite. { 'in·də·gō 'kap·ər }

indigoid dye [ORG CHEM] Any of the vat dyes with $C_{16}H_{10}O_2N_2$ (indigo) or $C_{16}H_8S_2O_2$ (thioindigo) groupings; used to dye cotton and rayon, sometimes silk. { 'in·də,gòid ,dī }

indigolite See indicolite. { 'in·də,gō,līt }

indigo red [ORG CHEM] $C_{16}H_{10}O_2N_2$ A red isomer of indigo obtained in the manufacture of indigo. Also known as indirubin. { 'in·də·gō 'red }

indirect address [COMPUT SCI] An address in a computer instruction that indicates a location where the address of the referenced operand is to be found. Also known as multilevel address. { ,in·də'rekt ə'dres }

indirect addressing [COMPUT SCI] A programming device whereby the address part of an instruction is not the address of the operand but rather the location in storage where the address of the operand may be found. { ,in·də'rekt ə'dres·iŋ }

indirect-arc furnace [ENG] A refractory-lined furnace in which the burden is heated indirectly by the radiant heat from an electric arc. { ,in·də'rekt 'ärk 'fər·nəs }

indirect-band-gap semiconductor [SOLID STATE] A semiconductor material in which the state of minimun energy in the conduction band and the state of maximum energy in the valence band have different momenta, and consequently optical transitions between free electrons and holes are forbidden. { ,in·də'rekt 'band ,gap 'sem·i·kən,dək·tər }

indirect cell [METEOROL] A closed circulation in a vertical plane in which the rising motion occurs at lower potential temperature than the descending motion, thus forming an energy sink. { ,in·də'rekt 'sel }

indirect control [COMPUT SCI] The control of one peripheral unit by another through some sequence of events that involves human intervention. { ,in·də'rekt kən'trōl }

indirect Coombs test See Rh blocking test. { ,in·də'rekt 'kümz ,test }

indirect cost [IND ENG] A cost that is not readily indentifiable with or chargeable to a specific product or service. { ,in·də'rekt kóst }

indirect cycle [NUCLEO] A nuclear reactor cycle in which a heat exchanger transfers heat from the reactor coolant to a second fluid, which then drives a prime mover. { ,in·də'rekt 'sī·kəl }

indirect developing test See Rh blocking test. { ,in·də'rekt di'vel·ə·piŋ ,test }

indirect-drive approach [NUCLEO] An approach to inertial-confinement fusion in which the laser or particle beam energy is first converted to heat radiation of very high temperature, in the x-ray range, in materials of high atomic number, and this radiation then compresses the fuel pellets. { ,in·də'rekt 'drīv ə,prōch }

indirect effect [PHYS CHEM] A chemical effect of ionizing radiation on a dilute solution caused by the interaction of solute molecules with highly reactive transient molecules or ions formed by reaction of the radiation with the solvent. { ,in·də'rekt i'fekt }

indirect extrusion [MET] An extrusion process in which the billet remains stationary while a hollow die stand forces the die back into the cylinder. { ,in·də'rekt ik'strü·zhən }

indirect fire [ORD] Gunfire delivered at a target that cannot be seen from the gun position or firing ship. { ,in·də'rekt 'fīr }

indirect heater [ENG] A vessel containing equipment in which heat generated by a primary source is transferred to a fluid or solid which then serves as the heating medium. { ,in·də'rekt 'hēd·ər }

indirect hernia [MED] A form of inguinal hernia that passes out of the abdomen through the inguinal canal. { ,in·də'rekt 'hər·nē·ə }

indirect immunofluorescence [IMMUNOL] The use of a labeled indicator antibody which reacts with an unlabeled detector antibody that has previously reacted with an antigen. { ,in·də'rekt ,im·yə·nō·flü'res·əns }

indirect labor [IND ENG] Labor not directly engaged in the actual production of the product or performance of a service. { ,in·də'rekt 'lā·bər }

indirect laying [ORD] Aiming a gun either by sighting at a fixed object called the aiming point instead of at the target or by using a means of pointing other than a sight, such as a gun director, when the target cannot be seen from the gun position. { ,in·də'rekt 'lā·iŋ }

indirect lighting [ENG] A system of lighting in which more than 90% of the light from luminaires is distributed upward toward the ceiling, from which it is diffusely reflected. { ,in·də'rekt 'līd·iŋ }

indirectly heated cathode [ELECTR] A cathode to which heat is supplied by an independent heater element in a thermionic tube; this cathode has the same potential on its entire surface, whereas the potential along a directly heated filament varies from one end to the other. Also known as equipotential cathode; heater-type cathode; unipotential cathode. { ,in·də'rek·lē ,hēd·əd 'kath,ōd }

indirect material [IND ENG] Any material used in the manufacture of a product which does not itself become a part of the product and whose cost is indirect. { ,in·də'rekt mə'tir·ē·əl }

indirect proof [MATH] **1.** A proof of a proposition in which another theorem is first proven from which the given theorem follows. **2.** See reductio ad absurdum. { ,in·də,rekt 'prüf }

indirect stratification See secondary stratification. { ,in·də'rekt ,strad·ə·fə'kā·shən }

indirect stroke [ELEC] A lightning stroke that induces a voltage in a power or communications system without actually striking it. { ,in·də'rekt 'strōk }

indirect vision See peripheral vision. { ,in·də'rekt 'vizh·ən }

indirect wave [PHYS] Any radio wave which arrives by an indirect path, having undergone an abrupt change of direction by refraction or reflection. { ,in·də'rekt 'wāv }

indirubin See indigo red. { in·də'rü·bər }

indiscreet topology See trivial topology. { ,in·də,skrēt tə'päl·ə·jē }

indium [CHEM] A metallic element, symbol In, atomic number 49, atomic weight 114.82; soluble in acids; melts at 156°C, boils at 1450°C. [MET] A ductile, silver-white, shiny metal that resists tarnishing and is used in precious-metal alloys for jewelry and dentistry, in glass-sealing alloys, lubricants, and bearing metals, and as an atomic-pile neutron indicator. { 'in·dē·əm }

indium antimonide [INORG CHEM] InSb Crystals that melt at 535°C; an intermetallic compound having semiconductor properties and the highest room-temperature electron mobility of any known material; used in Hall-effect and magnetoresistive devices and as an infrared detector. { 'in·dē·əm ,an'tim·ə,nīd }

indium arsenide [INORG CHEM] InAs Metallic crystals that melt at 943°C; an intermetallic compound having semiconductor properties; used in Hall-effect devices. { 'in·dē·əm 'ärs·ən,īd }

indium chloride [INORG CHEM] InCl₃ Hygroscopic white powder, soluble in water and alcohol. { 'in·dē·əm 'klȯr,īd }

indium phosphide [INORG CHEM] InP A metallic mass that is brittle and melts at 1070°C; an intermetallic compound having semiconductor properties. { 'in·dē·əm 'fäs,fīd }

indium sulfate [INORG CHEM] In₂(SO₄)₃ Deliquescent, water-soluble, grayish powder; decomposes when heated. { 'in·dē·əm 'səl,fāt }

individual distributed numerical control [CONT SYS] A form of distributed numerical control involving only a few machines, each of which operates independently of the others and is unaffected by their failures. { ,in·də'vij·ə·wəl di'strib·yəd·əd nü'mer·ə·kəl kən'trōl }

individual line [COMMUN] Subscriber line arranged to serve only one main station, although additional stations may be connected to the line as extensions; an individual line is not arranged for discriminatory ringing with respect to the stations on that line. { ¦in·də¦vij·ə·wəl 'līn }

individual psychology [PSYCH] A system of psychology in which traits of an individual are compared in terms of striving for superiority and then restated in the form of a composite of this single tendency. { ¦in·də¦vij·ə·wəl sī'käl·ə·jē }

individuation [EMBRYO] The process whereby, through induction, a spatially organized tissue, organ, or embryo develops. { ,in·di,vij·ə'wā·shən }

indogen [ORG CHEM] The functional group C₆H₄(NH)COC=; it occurs, for example, in the molecule indigo. { 'in·də·jən }

indogenide [ORG CHEM] A compound containing the function group C₆H₄(NH)·CO·C= from indogen. { 'in·də·jə,nīd }

indole Also known as 2,3-benzopyrrole. [BIOCHEM] C₆H₄(CHNH)CH A decomposition product of tryptophan formed in the intestine during putrefaction and by certain cultures of bacteria. [ORG CHEM] Carcinogenic, white to yellowish scales with unpleasant aroma; soluble in alcohol, ether, hot water, and fixed oils; melt at 52°C; used as a chemical reagent and in perfumery and medicine. { 'in,dōl }

indoleacetic acid [BIOCHEM] C₁₀H₉O₂N A decomposition product of tryptophan produced by bacteria and occurring in urine and feces; used as a hormone to promote plant growth. Abbreviated IAA. { ¦in,dōl·ə¦sēd·ik 'as·əd }

indolebutyric acid [ORG CHEM] C₁₂H₁₃O₂N A crystalline acid similar to indoleacetic acid in auxin activity. Abbreviated IBA. { ¦in,dōl·byü'tir·ik 'as·əd }

indolent [MED] **1.** Of or relating to a slow-growing, non-painful neoplasm. **2.** Slowness in the process of healing. { 'in·də·lənt }

indole test [MICROBIO] A test for the production of indole from tryptophan by microorganisms; a solution of para-dimethylaminobenzaldehyde, amyl alcohol, and hydrochloric acid added to the incubated culture of bacteria shows a red color in the alcoholic layer if indole is present. { 'in,dōl ,test }

Indo-Pacific faunal region [ECOL] A marine littoral faunal region extending eastward from the east coast of Africa, passing north of Australia and south of Japan, and ending in the east Pacific south of Alaska. { ¦in·dō·pə'sif·ik 'fȯn·əl ,rē·jən }

indoxyl [ORG CHEM] (C₈H₆N)OH A yellow crystalline glycoside, used as an intermediate in the manufacture of indigo. { in'däk·səl }

Indriidae [VERT ZOO] A family of Madagascan prosimians containing wholly arboreal vertical clingers and leapers. { in'drī·ə,dē }

induced angle of attack [AERO ENG] The downward vertical angle between the horizontal and the velocity (relative to the wing of an aircraft) of the airstream passing over the wing. { in'düst ¦aŋ·gəl əv ə'tak }

induced anisotropy [SOLID STATE] A type of uniaxial anisotropy in a magnetic material produced by annealing the magnetic material in a magnetic field. { in'düst ,an·ə'sä·trə·pē }

induced capacity See absolute permeability. { in'düst kə'pas·əd·ē }

induced current [ELECTROMAG] A current produced in a conductor by a time-varying magnetic field, as in induction heating. { in'düst 'kə·rənt }

induced dipole [ELEC] An electric dipole produced by application of an electric field. { in'düst 'dī,pōl }

induced angle of attack [AERO ENG] The downward vertical angle between the horizontal and the velocity (relative to the wing of an aircraft) of the airstream passing over the wing. { in'düst ¦aŋ·gəl əv ə'tak }

induced draft [MECH ENG] A mechanical draft produced by suction stream jets or fans at the point where air or gases leave a unit. { in'düst 'draft }

induced-draft cooling tower [MECH ENG] A structure for cooling water by circulating air where the load is on the suction side of the fan. { in'düst ¦draft 'kül·iŋ ,taů·ər }

induced drag [FL MECH] That part of the drag caused by the downflow or downwash of the airstream passing over the wing of an aircraft, equal to the lift times the tangent of the induced angle of attack. { in'düst 'drag }

induced electromotive force [ELECTROMAG] An electromotive force resulting from the motion of a conductor through a magnetic field, or from a change in the magnetic flux that threads a conductor. { in'düst i,lek·trə¦mōd·iv 'fȯrs }

induced emission See stimulated emission. { in'düst i'mish·ən }

induced fission [NUC PHYS] Fission which takes place only when a nucleus is bombarded with neutrons, gamma rays, or other carriers of energy. { in'düst 'fish·ən }

induced magnetism [ELECTROMAG] The magnetism acquired by magnetic material while it is in a magnetic field. { in'düst 'mag·nə,tiz·əm }

induced magnetization [GEOPHYS] That component of a rock's magnetization which is proportional to, and has the same direction as, the ambient magnetic field. { in'düst ,mag·nə·tə'zā·shən }

induced moment [ELEC] The average electric dipole moment per molecule which is produced by the action of an electric field on a dielectric substance. { in'düst 'mō·mənt }

induced movement [PSYCH] The perceived movement of a stationary object when its frame of reference moves. { in,düst 'müv·mənt }

induced orientation [MATH] An orientation of a face of a simplex S opposite a vertex p_i obtained by deleting p_i from the ordering defining the orientation of S. { in¦düsd ,ȯr·ē·ən'tā·shən }

induced potential See induced voltage. { in'düst pə'ten·chəl }

induced precession See real precession. { in'düst pri'sesh·ən }

induced radioactivity [NUCLEO] Radioactivity created by bombarding a substance with radiation. Also known as artificial radioactivity. { in'düst ,rād·ē·ō·ak'tiv·əd·ē }

induced subgraph See vertex-induced subgraph. { in,düst 'səb,graf }

induced voltage [ELECTROMAG] A voltage produced by electromagnetic or electrostatic induction. Also known as induced potential. { in'düst 'vōl·tij }

inducer [EMBRYO] The cell group that functions as the acting system in embryonic induction by controlling the mode of development of the reacting system. Also known as inductor. { in'dü·sər }

inducible enzyme [BIOCHEM] An enzyme which is present in trace quantities within a cell but whose concentration increases dramatically in the presence of substrate molecules. { in'dü·sə·bəl 'en,zīm }

inductance [ELECTROMAG] **1.** That property of an electric circuit or of two neighboring circuits whereby an electromotive force is generated (by the process of electromagnetic induction) in one circuit by a change of current in itself or in the other. **2.** Quantitatively, the ratio of the emf (electromotive force) to the rate of change of the current. **3.** See coil. { in'dək·təns }

inductance bridge [ELECTROMAG] **1.** A device, similar to a Wheatstone bridge, for comparing inductances. **2.** A four-coil alternating-current bridge circuit used for transmitting a mechanical movement to a remote location over a three-wire circuit; half of the bridge is at each location. { in'dək·təns ,brij }

inductance coil *See* coil. { in'dək·təns ,kȯil }

inductance measurement [ELECTROMAG] The determination of the self-inductance of a circuit or the mutual inductance of two circuits. { in'dək·təns ,mezh·ər·mənt }

inductance meter [ELECTROMAG] A device which measures the self-inductance of a circuit or the mutual inductance of two circuits. { in'dək·təns ,mēd·ər }

inductance standards [ELECTROMAG] Two equal, multilayer coils, wound on toroidal cores of nonmagnetic materials, connected in series and located so that their interactions with external fields tend to cancel one another. { in'dək·təns ,stan·dərdz }

induction [ELEC] *See* electrostatic induction. [ELECTROMAG] *See* electromagnetic induction. [EMBRYO] *See* embryonic induction. [MED] The period from administration of the anesthetic to loss of consciousness by the patient. [SCI TECH] The act or process of causing. { in'dək·shən }

induction accelerator [NUCLEO] A cyclic or linear device for accelerating electrons to high energies by using the inductive electric field produced by a time-varying magnetic field. { in'dək·shən ak'sel·ə,rād·ər }

induction brazing [MET] A brazing process in which coalescence is produced by heat generated within the work by an induced electric current. { in'dək·shən ,brāz·iŋ }

induction burner [ENG] Fuel-air burner into which the fuel is fed under pressure to entrain needed air into the combustion nozzle area. { in'dək·shən ,bər·nər }

induction charging [ELEC] Production of electric charge on a body by means of electrostatic induction. { in'dək·shən ,chär·jiŋ }

induction coil [ELECTROMAG] A device for producing high-voltage alternating current or high-voltage pulses from low-voltage direct current, in which interruption of direct current in a primary coil, containing relatively few turns of wire, induces a high voltage in a secondary coil, containing many turns of wire wound over the primary. { in'dək·shən ,kȯil }

induction disk relay [ELECTROMAG] A unit widely used in regulating and protective relays, in which alternating current applied to a coil produces torque to rotate a disk. { in'dək·shən ¦disk 'rē¦lā }

induction-electrical survey [ENG] Study of subterranean formations by combined induction and electrical logging. { in'dək·shən i¦lek·trə·kəl 'sər,vā }

induction field [ELECTROMAG] A component of an electromagnetic field associated with an alternating current in a loop, coil, or antenna which carries energy alternately away from and back into the source, with no net loss, and which is responsible for self-inductance in a coil or mutual inductance with neighboring coils. { in'dək·shən ,fēld }

induction flowmeter [ENG] An instrument for measuring the flow of a conducting liquid passing through a tube, in which the tube is placed in a transverse magnetic field and the induced electromotive force between electrodes at opposite ends of a diameter of the tube perpendicular to the field is measured. { in'dək·shən ,flō,mēd·ər }

induction force [PHYS CHEM] A type of van der Waals force resulting from the interaction of the dipole moment of a polar molecule and the induced dipole moment of a nonpolar molecule. Also known as Debye force. { in'dək·shən ,fȯrs }

induction frequency converter [ELEC] Slip-ring induction machine which is driven by an external source of mechanical power and whose primary circuits are connected to a source of electric energy having a fixed frequency; the secondary circuits deliver energy at a frequency proportional to the relative speed of the primary magnetic field and the secondary member. { in'dək·shən 'frē·kwən·sē kən,vərd·ər }

induction furnace [ENG] An electric furnace in which heat is produced in a metal charge by electromagnetic induction. { in'dək·shən ,fər·nəs }

induction generator [ELEC] A nonsynchronous alternating-current generator whose construction is identical to that of an ac motor, and which is driven above synchronous speed by external sources of mechanical power. { in'dək·shən ¦jen·ə·rād·ər }

induction hardening [MET] A quench-hardening technique in which the required elevated temperature is obtained by electromagnetic induction. { in'dək·shən ,härd·ən·iŋ }

induction heating [ENG] Increasing the temperature in a material by induced electric current. Also known as eddy-current heating. { in'dək·shən ¦hēd·iŋ }

induction inclinometer *See* earth inductor. { in'dək·shən ,in·klə'näm·əd·ər }

induction inoculation [PL PATH] Repeated inoculation of plants to induce a maximum level of systemic resistance to disease. { in'dək·shən in,äk·yə,lā·shən }

induction instrument [ENG] Meter that depends for its operation on the reaction between magnetic flux set up by current in fixed windings, and other currents set up by electromagnetic induction in conducting parts of the moving system. { in'dək·shən ,in·strə·mənt }

induction log [ENG] An electric log of the conductivity of rock with depth obtained by lowering into an uncased borehole a generating coil that induces eddy currents on the rocks and these are detected by a receiver coil. { in'dək·shən ,läg }

induction loudspeaker [ENG ACOUS] Loudspeaker in which the current which reacts with the steady magnetic field is induced in the moving member. { in'dək·shən ¦laȯd,spēk·ər }

induction machine [ELEC] An asynchronous alternating-current machine, such as an induction motor or induction generator, in which the windings of two electric circuits rotate with respect to each other and power is transferred from one circuit to the other by electromagnetic induction. { in'dək·shən mə,shēn }

induction melting [MET] Converting a solid metal to the molten state in an induction furnace. { in'dək·shən ¦melt·iŋ }

induction method [GEOPHYS] In studies of the radioactivity of the atmosphere, a technique for estimating the concentration of the radioactive gases by exposing a negatively charged wire to the air and then using an ionization chamber to count the activity of the radioactive deposit formed on the wire. { in'dək·shən ,meth·əd }

induction motor [ELEC] An alternating-current motor in which a primary winding on one member (usually the stator) is connected to the power source, and a secondary winding on the other member (usually the rotor) carries only current induced by the magnetic field of the primary. { in'dək·shən ,mōd·ər }

induction noise [ACOUS] The noise caused by the periodic inrush of intake air into the cylinder of an automobile engine or an air compressor as the intake valve opens and the piston moves on the intake stroke. { in'dək·shən ,nȯiz }

induction period [PHYS CHEM] A time of acceleration of a chemical reaction from zero to a maximum rate. { in'dək·shən ,pir·ē·əd }

induction problem [ELECTROMAG] An effect of potentials and currents induced in conductors of a telephone system by paralleling power facilities or power lines. { in'dək·shən ,präb·ləm }

induction pump [MECH ENG] Any pump operated by electromagnetic induction. { in'dək·shən ,pəmp }

induction regulator [ELEC] A transformer in which the voltage produced in a secondary winding is varied by changing the position of the primary winding. { in'dək·shən 'reg·yə,lād·ər }

induction salinometer [ENG] A device for measuring salinity by taking voltage readings of the current in seawater. { in'dək·shən ,sal·ə'näm·əd·ər }

induction silencer [ENG] A device for reducing engine induction noise, which consists essentially of a low-pass acoustic filter with the inertance of the air-entrance tube and the acoustic compliance of the annular and central volumes providing acoustic filtering elements. { in'dək·shən ¦sī·lən·sər }

induction soldering [MET] A soldering process in which the metals are heated by an induced electric current. { in'dək·shən ¦säd·ə·riŋ }

induction valve *See* inlet valve. { in'dək·shən ,valv }

induction voltage regulator [ELECTROMAG] A type of transformer having a primary winding connected in parallel with a circuit and a secondary winding in series with the circuit; the relative positions of the primary and secondary windings

---

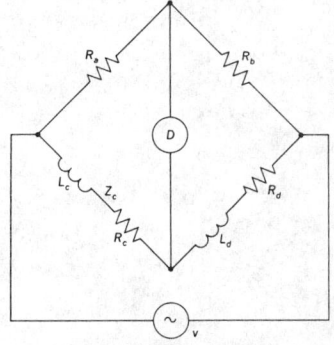

INDUCTANCE BRIDGE

Circuit diagram for a general inductance bridge (def. 1); R_{a-d} are resistors; L_{c-d} are inductances; Z_c is impedance; V is source of ac voltage; and D is a null detector.

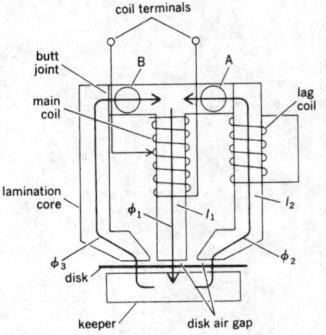

INDUCTION DISK RELAY

Schematic of the induction disk type of relay. I_1 = current in main coil, I_2 = current in lag coil, ϕ_1 = flux in main coil, ϕ_2 = flux in lag coil, ϕ_3 = differential induction flux, A = flux field about ϕ_3, B = flux field about ϕ_2.

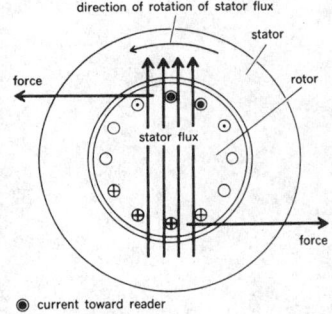

INDUCTION MOTOR

◉ current toward reader
⊕ current away from reader
weight of ● or + indicates magnitude of current

Forces on the rotor winding in an induction motor.

are changed to vary the voltage or phase relations in the circuit. { in'dək·shən 'vōl·tij ˌreg·yə¦lād·ər }

induction watthour meter [ELEC] A watthour meter used with alternating current; the energy taken by a circuit over a period of time is proportional to the rotation in that period of a light aluminum disk, in which a driving torque is developed by the joint action of the alternating magnetic flux produced by the potential circuit and by the load current. { in'dək·shən 'wät,aủr ˌmēd·ər }

induction welding [MET] A process of welding by means of heat generated within the work by induced electric currents. { in'dək·shən ¦weld·iŋ }

inductive capacities [ELECTROMAG] The permeability and permittivity of a substance. { in¦dək·tiv kə'pas·əd·ēz }

inductive charge [ELEC] The charge that exists on an object as a result of its being near another charged object. { in'dək·tiv 'chärj }

inductive circuit [ELEC] A circuit containing a higher value of inductive reactance than capacitive reactance. { in'dək·tiv 'sər·kət }

inductive coordination [ELECTROMAG] Measures to reduce induction problems. { in'dək·tiv kō,ȯrd·ən'ā·shən }

inductive coupler [ELEC] A mutual inductance that provides electrical coupling between two circuits; used in radio equipment. { in'dək·tiv 'kəp·lər }

inductive coupling [ELEC] Coupling of two circuits by means of the mutual inductance provided by a transformer. Also known as transformer coupling. { in'dək·tiv 'kəp·liŋ }

inductive divider [ELECTROMAG] A device for incorporating a desired fraction of an inductance into a circuit, usually consisting of an autotransformer with an intermediate tap. { in'dək·tiv di'vīd·ər }

inductive effect [PHYS CHEM] In a molecule, a shift of electron density due to the polarization of a bond by a nearby electronegative or electropositive atom. { in'dək·tiv ə'fekt }

inductive fault analysis [ELECTR] A method of analyzing the effects of defects on an integrated circuit, in which a computer simulates an electron that scatters at random faults in the form of additional or missing areas of material on the set of drawings of the masks from which the circuits are fabricated. { in,dək·tiv 'fȯlt ə,nal·ə·səs }

inductive feedback [ELECTR] **1.** Transfer of energy from the plate circuit to the grid circuit of a vacuum tube by means of induction. **2.** Transfer of energy from the output circuit to the input circuit of an amplifying device through an inductor, or by means of inductive coupling. { in'dək·tiv 'fēd·bak }

inductive filter [ELECTR] A low-pass filter used for smoothing the direct-current output voltage of a rectifier; consists of one or more sections in series, each section consisting of an inductor on one of the pair of conductors in series with a capacitor between the conductors. Also known as LC filter. { in'dək·tiv 'fil·tər }

inductive grounding [ELEC] Use of grounding connections containing an inductance in order to reduce the magnitude of short-circuit currents created by line-to-ground faults. { in'dək·tiv 'graủnd·iŋ }

inductive interference [COMMUN] Effect arising from the characteristics and inductive relations of electric supply and communications systems of such character and magnitude as would prevent the communications circuits from rendering service satisfactorily and economically if methods of inductive coordination were not applied. { in'dək·tiv ,in·tər'fir·əns }

inductive line pair [COMMUN] A telephone line displaying induction whose effects are of consequence, as in crosstalk; opposed to twisted pair. { in'dək·tiv 'līn ,per }

inductive load [ELEC] A load that is predominantly inductive, so that the alternating load current lags behind the alternating voltage of the load. Also known as lagging load. { in'dək·tiv 'lōd }

inductively coupled plasma-atomic emission spectroscopy [SPECT] A type of atomic spectroscopy in which the light emitted by atoms and ions in an inductively coupled plasma is observed. Abbreviated ICP-AES. { in'dək·tiv·lē ¦kəp·əld ¦plaz·mə ə'täm·ik i¦mish·ən spek'träs·kə·pē }

inductively coupled plasma discharge [PL PHYS] A high-temperature (8000–10,000 K) discharge generated by inducing a magnetic field in a flowing conducting gas, usually argon or argon and nitrogen, by means of a water-cooled copper coil which surrounds tubes through which the gas flows. { in'dək·tiv·lē ¦kəp·əld 'plaz·mə 'dis,chärj }

inductive neutralization [ELECTR] Neutralizing an amplifier whereby the feedback susceptance due to an interelement capacitance is canceled by the equal and opposite susceptance of an inductor. Also known as coil neutralization; shunt neutralization. { in'dək·tiv ,nü·trə·lə'zā·shən }

inductive-output tube [ELECTR] A tube in which output energy is obtained from the electron stream by electric induction between a cylindrical output electrode and the electron stream that flows through but does not touch the electrode. { in¦dək·tiv 'aủt,pủt ,tüb }

inductive post [ELECTROMAG] Metal post or screw extending across a waveguide parallel to the E field, to add inductive susceptance in parallel with the waveguide for tuning or matching purposes. { in'dək·tiv 'pōst }

inductive pressure transducer [ELECTROMAG] A type of pressure transducer in which changes in pressure cause a bourdon tube or other pressure-sensing element to move a magnetic core, and this results in a change in inductance of one or more windings of a coil surrounding the core. { in'dək·tiv 'presh·ər tranz,dü·sər }

inductive reactance [ELEC] Reactance due to the inductance of a coil or circuit. { in'dək·tiv rē'ak·təns }

inductive relay [ELECTROMAG] A relay displaying inductance, as opposed to one wound to be noninductive. { in'dək·tiv 'rē,lā }

inductive spacing [ELECTROMAG] Spacing of parallel transmission lines so that there is transfer of energy by mutual inductance. { in'dək·tiv 'spās·iŋ }

inductive superconducting fault-current limiter See shielded-core superconducting fault-current limiter. { in¦dək·tiv ,sü·pər·kən¦dək·tiŋ 'fȯlt ,cər·ənt ,lim·əd·ər }

inductive surge [ELECTROMAG] A surge in voltage caused by sudden interruption of current in an inductive circuit. { in'dək·tiv 'sərj }

inductive susceptance [ELEC] In a circuit containing almost no resistance, the part of the susceptance due to inductance. { in'dək·tiv sə'sep·təns }

inductive tuning [ELECTR] Tuning involving the use of a variable inductance. { in'dək·tiv 'tün·iŋ }

inductive voltage divider [ELEC] An autotransformer that has its winding subdivided into 10 equal turn sections so that when an alternating voltage V is applied to the whole winding the voltage across each section is nominally V/10; used as a ratio standard for electrical measurements. { in'dək·tiv 'vōl·tij di,vīd·ər }

inductive waveform [ELEC] A graph or trace of the effect of current buildup across an inductive network; proportional to the exponential of the product of a negative constant and the time. { in'dək·tiv 'wāv,fȯrm }

inductive window [ELECTROMAG] Conducting diaphragm extending into a waveguide from one or both sidewalls of the waveguide, to give the effect of an inductive susceptance in parallel with the waveguide. { in'dək·tiv 'win,dō }

inductometer [ELECTROMAG] A coil of wire of known inductance; the inductance may be fixed as in the case of primary standards, adjustable by means of switches, or continuously variable by means of a movable-coil construction. { ,in,dək'täm·əd·ər }

inductor See coil; inducer. { in'dək·tər }

inductor alternator [ELEC] A synchronous generator in which the field winding is fixed in magnetic position relative to the armature conductors. { in'dək·tər 'ȯl·tər,nād·ər }

inductor generator [ELEC] An alternating-current generator in which all the windings are fixed, and the flux linkages are varied by rotating an appropriately toothed ferromagnetic rotor; sometimes used for generating high power at frequencies up to several thousand hertz for induction heating. { in'dək·tər 'jen·ə,rād·ər }

inductor microphone [ENG ACOUS] Moving-conductor microphone in which the moving element is in the form of a straight-line conductor. { in'dək·tər 'mī·krə,fōn }

inductor tachometer [ENG] A type of impulse tachometer in which the rotating member, consisting of a magnetic material, causes the magnetic flux threading a circuit containing a magnet and a pickup coil to rise and fall, producing pulses in the circuit which are rectified for a permanent-magnet, movable-coil instrument. { in'dək·tər tə'käm·əd·ər }

inductosyn [CONT SYS] A resolver whose output phase is proportional to the shaft angle. { in'dək·tə,sin }

Inductrack [ENG] A magnetic levitation concept for trains and other moving objects that uses special arrays of permanent magnets to achieve levitation forces, and is inherently stable. { in'dək,trak }

inductura [INV ZOO] A layer of lamellar shell material along the inner lip of the aperture in gastropods. { in'dək·chə·rə }

indumentum [BOT] A covering, such as one that is woolly. [MYCOL] A covering of hairs. [VERT ZOO] The plumage covering a bird. { ,in·də'men·təm }

induplicate [BOT] Having the edges turned or rolled inward without twisting or overlapping; applied to the leaves of a bud. { in'dü·plə·kət }

induration [BIOL] The process of hardening, especially by increasing the fibrous elements. [GEOL] **1.** The hardening of a rock material by the application of heat or pressure or by the introduction of a cementing material. **2.** A hardened mass formed by such processes. **3.** The hardening of a soil horizon by chemical action to form a hardpan. [MED] Hardening of a tissue or organ due to an accumulation of blood, inflammation, or neoplastic growth. { ,in·də'rā·shən }

Indus [ASTRON] A constellation, right ascension 21 hours, declination 55° south. Also known as Indian. { 'in·dəs }

indusium [ANAT] A covering membrane such as the amnion. [BOT] An epidermal outgrowth covering the sori in many ferns. [MYCOL] The annulus of certain fungi. { in'dü·zē·əm }

industrial alcohol [ORG CHEM] Ethyl alcohol that has been denatured by acetates, ketones, gasoline, or other additives to make it unfit for beverage purposes. { in'dəs·trē·əl 'al·kə,hȯl }

industrial anthropometry [IND ENG] Application of the knowledge of physical anthropology to the design and construction of equipment for human use, such as automobiles. { in'dəs·trē·əl ,an·thrə'päm·ə·trē }

industrial car [IND ENG] Any of various narrow-gage railcars used for indoor or outdoor handling of bulk and package materials. { in'dəs·trē·əl 'kär }

industrial climatology [CLIMATOL] A type of applied climatology which studies the effect of climate and weather on industry's operations; the goal is to provide industry with a sound statistical basis for all administrative and operational decisions which involve a weather factor. { in'dəs·trē·əl ,klī·mə'täl·ə·jē }

industrial cost control [IND ENG] A specific system or procedure used to keep manufacturing costs in line. Also known as cost control. { in'dəs·trē·əl 'kȯst kən,trōl }

industrial crop [AGR] Any crop that provides materials for industrial processes and products such as soybeans, cotton (lint and seed), flax, and tobacco. { in'dəs·trē·əl ‚kräp }

industrial diamond [MINERAL] Diamond that is too hard or too radial-grained to be used for jewel cutting. { in'dəs·trē·əl 'dī·mənd }

industrial ecology [IND ENG] The development and use of industrial processes that result in products based on simultaneous consideration of product functionality and competitiveness, natural-resource conservation, and environmental preservation. Also known as design for environment; green design. [SYS ENG] The multidisciplinary study of industrial and economic systems and their linkages with fundamental natural systems. { in'dəs·trē·əl ē'käl·ə·jē }

industrial engineering [ENG] A branch of engineering concerned with the design, improvement, and installation of integrated systems of people, materials, and equipment. Also known as management engineering. { in'dəs·trē·əl ,en·jə'nir·iŋ }

industrial frequency bands [COMMUN] The radio-frequency bands allocated in the United States for land mobile communications of private industries other than transportation. { in'dəs·trē·əl 'frē·kwən·sē ,banz }

industrial geography [GEOGR] A branch of geography that deals with location, raw materials, products, and distribution, as influenced by geography. { in'dəs·trē·əl jē'äg·rə·fē }

industrial glass [MATER] Any glass molded into shapes for product parts, for example lime glass and lead glass. { in'dəs·trē·əl ‚glas }

industrial hygiene [MED] The science that deals with the anticipation and control of unhealthy conditions in workplaces in order to prevent illness among employees. { in'dəs·trē·əl 'hī,jēn }

industrial jewel [MINERAL] A hard stone, such as ruby or sapphire, used for bearings and impulse pins in instruments and for recording needles. { in'dəs·trē·əl 'jül }

industrial meteorology [METEOROL] The application of meteorological information and techniques to industrial problems. { in'dəs·trē·əl ,mē·dē·ə'räl·ə·jē }

industrial microbiology [MICROEIO] The study, utilization, and manipulation of those microorganisms capable of economically producing desirable substances or changes in substances, and the control of undesirable microorganisms. { in'dəs·trē·əl ,mī·krō·bī'äl·ə·jē }

industrial microorganism [MICROBIO] Any microorganism utilized for industrial microbiology. { in'dəs·trē·əl ,mī·krō'ȯr·gə,niz·əm }

industrial mobilization [IND ENG] Transformation of industry and other productive facilities and contributory services from their peacetime activities to the fulfillment of the munitions program necessary to support a military effort. { in'dəs·trē·əl ,mō·bə·lə'zā·shən }

industrial psychology [PSYCH] Psychology applied to problems in industry, dealing chiefly with the selection, efficiency, and mental health of personnel. { in'dəs·trē·əl sī'käl·ə·jē }

industrial railway [IND ENG] **1.** A usually short feeder line that is either owned or controlled and wholly operated by an industrial firm. **2.** Narrow-gage rail lines used on construction jobs or around industrial plants. { in'dəs·trē·əl 'rāl,wā }

industrial revolution [IND ENG] A widespread change in industrial or production methods, toward production by machine and away from manual labor. { in'dəs·trē·əl ,rev·ə'lü·shən }

industrial security [IND ENG] The portion of internal security which refers to the protection of industrial installations, resources, utilities, materials, and classified information essential to protection from loss or damage. { in'dəs·trē·əl si'kyür·əd·ē }

industrial television [COMMUN] Closed-circuit television used for remote viewing of industrial processes and operations. Abbreviated ITV. { in'dəs·trē·əl 'te·ə,vizh·ən }

industrial truck [ENG] A manually propelled or powered wheeled vehicle for transporting materials over level or slightly inclined running surfaces in a manufacturing or warehousing facility. { in'dəs·trē·əl 'trək }

industrial waste [ENG] Worthless materials remaining from industrial operations. { in'dəs·trē·əl 'wāst }

industrial yeast [MICROBIO] Any yeast used for the production of fermented foods and beverages, for baking, or for the production of vitamins, proteins, alcohol, glycerol, and enzymes. { in'dəs·trē·əl ‚yēst }

indwelling catheter [MED] A thin tube communicating to the body surface, inserted into the vascular system and positioned to permit pressure measurements or blood sampling over a long period of time. { 'in,dwel·iŋ kath·əd·ər }

ineffective time [COMPUT SCI] Time during which a computer can operate normally but which is not used effectively because of mistakes or inefficiency in operating the installation or for other reasons. { ,in·i'fek·tiv 'tīm }

inelastic [MECH] Not capable of sustaining a deformation without permanent change in size or shape { ,in·ə'las·tik }

inelastic buckling [MECH] Sudden increase of deflection or twist in a column when compressive stress reaches the elastic limit but before elastic buckling develops. { ,in·ə'las·tik 'bək·liŋ }

inelastic collision [MECH] A collision in which the total kinetic energy of the colliding particles is not the same after the collision as before it. { ,in·ə'las·tik kə'lizh·ən }

inelastic cross section [PHYS] The cross section for an inelastic collision. { ,in·ə'las·tik 'krȯs ,sek·shən }

inelastic neutron scattering *See* slow-neutron spectroscopy. { ,in·ə'las·tik 'nü,trän ,skad·ə·r·iŋ }

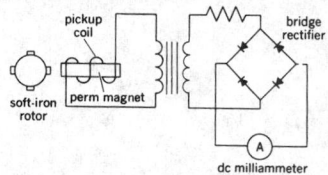

Circuit for an inductor form of tachometer.

INDUCTRACK

Elements of the system. An end view of the Halbach array on a moving car is shown above the upper conductors of the shorted levitation circuits in the track. Straight arrows show magnetization of permanent magnets in the array. Curved arrows show the periodic magnetic field below the array.

inelastic scattering [PHYS] Scattering that results from inelastic collisions. { ‚in·ə'las·tik 'skad·ə·riŋ }

inelastic stress [MECH] A force acting on a solid which produces a deformation such that the original shape and size of the solid are not restored after removal of the force. { ‚in·ə'las·tik 'stres }

inequality [MATH] A statement that one quantity is less than, less than or equal to, greater than, or greater than or equal to another quantity. { ‚in·i'kwäl·əd·ē }

inequality of Clausius See Clausius inequality. { ‚in·i'kwäl·əd·ē əv 'klau·zē·əs }

inequilateral [BIOL] Having the two sides or ends unequal, as the ends of a bivalve mollusk on either side of a line from umbo to gape. { in¦ē·kwə'lad·ə·rəl }

inermous [BIOL] Lacking mechanisms for defense or offense, especially spines. { i'nər·məs }

inert [SCI TECH] Lacking an activity, reactivity, or effect. { i'nərt }

inert atmosphere [CHEM ENG] A nonreactive gas atmosphere, such as nitrogen, carbon dioxide, or helium; used to blanket reactive liquids in storage, to purge process lines and vessels of reactive gases and liquids, and to cover a reaction mix in a partially filled vessel. { i'nərt 'at·mə‚sfir }

inert gas See noble gas. { i'nərt 'gas }

inert-gas blanketing [ENG] Purging the air from a unit of a heat exchanger by using an inert gas as the unit is being shut down. { i¦nərt ‚gas 'blaŋ·kəd·iŋ }

inert-gas cutting [MET] Cutting of metal while inert gas flows around the cutting area to prevent oxidation. { i'nərt ¦gas 'kəd·iŋ }

inert gas-shielded arc welding [MET] An arc-welding process in which the weld area is shielded by an inert gas to prevent oxidation. Also known as Heliarc welding. { i'nərt ¦gas ¦shēld·əd 'ärk ‚weld·iŋ }

inertia [MECH] That property of matter which manifests itself as a resistance to any change in the momentum of a body. [MED] Sluggishness, especially of muscular activity. { i'nər·shə }

inertia currents [OCEANOGR] Currents resulting after the cessation of wind in a generating area or after the water movement has left the generating area; circular currents with a period of one-half pendulum day. { i'nər·shə ‚kə·rəns }

inertia ellipsoid [MECH] An ellipsoid used in describing the motion of a rigid body; it is fixed in the body, and the distance from its center to its surface in any direction is inversely proportional to the square root of the moment of inertia about the corresponding axis. Also known as Poinsot ellipsoid. { i'nər·shə i'lip‚sȯid }

inertia governor [MECH ENG] A speed-control device utilizing suspended masses that respond to speed changes by reason of their inertia. { i'nər·shə ‚gəv·ə·nər }

inertial circle [METEOROL] A loop in the path of an air parcel in inertial flow, which is approximately circular if the latitudinal displacement is small. Also known as circle of inertia. [OCEANOGR] The circle described by inertial motion in a body of ocean water and having a radius $R = C/f$, where C is the particle velocity in a given direction and f is the Coriolis parameter. { i'nər·shəl ‚sər·kəl }

inertial confinement [NUCLEO] The rapid implosion of a high-density pellet, under bombardment by laser or charged-particle beams, to produce a core that is heated to extremely high temperatures before it can fly apart; proposed as a method for generating power from controlled thermonuclear reactions. { in'ər·shəl kən'fīn·mənt }

inertial-confinement fusion See pellet fusion. { i¦nər·shəl kən'fīn·mənt ‚fyü·zhən }

inertial coordinate system See inertial reference frame. { i'nər·shəl kō'ȯrd·ən‚at ‚sis·təm }

inertial flow [FL MECH] Flow in which no external forces are exerted on a fluid. [GEOPHYS] Frictionless flow in a geopotential surface in which there is no pressure gradient; the centrifugal and Coriolis accelerations must therefore be equal and opposite, and the constant inertial wind speed V_i is given by $V_i = fR$, where f is the Coriolis parameter and R the radius of curvature of the path. { i'nər·shəl 'flō }

inertial force [MECH] The fictitious force acting on a body as a result of using a noninertial frame of reference; examples are the centrifugal and Coriolis forces that appear in rotating

coordinate systems. Also known as effective force. { i'nər·shəl 'fȯrs }

inertial guidance [NAV] **1.** Guidance by means of accelerations measured and integrated within the craft. **2.** Guidance by the use of an inertial navigation system. { i'nər·shəl 'gīd·əns }

inertial instability [FL MECH] **1.** Generally, instability in which the only form of energy transferred between the steady state and the disturbance in the fluid is kinetic energy. **2.** The hydrodynamic instability arising in a rotating fluid mass when the velocity distribution is such that the kinetic energy of a disturbance grows at the expense of kinetic energy of the rotation. Also known as dynamic instability. { i'nər·shəl ‚in·stə'bil·əd·ē }

inertial mass [MECH] The mass of an object as determined by Newton's second law, in contrast to the mass as determined by the proportionality to the gravitational force. { i'nər·shəl 'mas }

inertial navigation system [NAV] A self-contained system that can automatically determine the position, velocity, and attitude of a moving vehicle by means of the double integration of the outputs of accelerometers that are either strapped to the vehicle or stabilized with respect to inertial space. Also known as inertial navigator. { in'ər·shəl ‚nav·ə'gā·shən ‚sis·təm }

inertial navigator See inertial navigation system. { i'nər·shəl 'nav·ə‚gād·ər }

inertial orbit [ASTRON] The path described by an object that is subject only to gravitational forces, such as a celestial body or a spacecraft that is not under any type of propulsive power. { in'ər·shəl 'ȯr·bət }

inertial platform [NAV] In an inertial navigator a platform that maintains sensing instruments in a precise known angular orientation in space. { i'nər·shəl 'plat‚fȯrm }

inertial reference frame [MECH] A coordinate system in which a body moves with constant velocity as long as no force is acting on it. Also known as inertial coordinate system. { i'nər·shəl 'ref·rəns ‚frām }

inertial size See aerodynamic size. { i'nər·shəl 'sīz }

inertial space [NAV] A coordinate system or frame of reference defined with respect to the stars whose apparent positions relative to surrounding stars appear to be fixed or unvarying for long periods of time. { i'nər·shəl 'spās }

inertial theory [OCEANOGR] The theory associated with the motion of an ocean current under the influences of inertia and the Coriolis force, which cause it to take a circular path. { i'nər·shəl 'thē·ə·rē }

inertia matrix [MECH] A matrix **M** used to express the kinetic energy T of a mechanical system during small displacements from an equilibrium position, by means of the equation $T = \frac{1}{2}\dot{\mathbf{q}}^T\mathbf{M}\dot{\mathbf{q}}$, where $\dot{\mathbf{q}}$ is the vector whose components are the derivatives of the generalized coordinates of the system with respect to time, and $\dot{\mathbf{q}}^T$ is the transpose of $\dot{\mathbf{q}}$. { i'nər·shə ‚mā·triks }

inertia of energy [RELAT] The principle that the inertial properties of matter both determine and are determined by its total energy content. { i'nər·shə əv 'en·ər·jē }

inertia period [OCEANOGR] The time required for a given particle to complete an inertia circle. { i'nər·shə ‚pir·ē·əd }

inertia starter [MECH ENG] A device utilizing inertial principles to start the rotator of an internal combustion engine. { i'nər·shə ¦stärd·ər }

inertia switch [ELEC] A switch that is actuated by an abrupt change in the velocity of the item on which it is mounted. { i'nər·shə ‚swich }

inertia tensor [MECH] A tensor associated with a rigid body whose product with the body's rotation vector yields the body's angular momentum. { i'nər·shə ‚ten·sər }

inertia wave [FL MECH] **1.** Any wave motion in which no form of energy other than kinetic energy is present; in this general sense, Helmholtz waves, barotropic disturbances, Rossby waves, and so forth, are inertia waves. **2.** More restrictedly, a wave motion in which the source of kinetic energy of the disturbance is the rotation of the fluid about some given axis; in the atmosphere a westerly wind system is such a source, the inertia waves here being, in general, stable. { i'nər·shə ‚wāv }

inertia welding [MET] A form of friction welding which utilizes kinetic energy stored in a flywheel system to supply the power required for all of the heating and much of the forging. { i'nər·shə 'weld·iŋ }

inertinite [GEOL] A carbon-rich maceral group, which includes micrinite, sclerotinite, fusinite, and semifusinite. { i'nərt·ən‚īt }

inert primer [ENG] A cylinder which enshrouds a detonator but does not interfere with the detonation of the explosive charge. { i'nərt 'prī·mər }

inert retarder [CIV ENG] A braking device built into a railroad track and operating without an external source of power that reduces car speed by means of brake shoes applied to the lower sides of the wheels. { i‚nərt ri'tär·dər }

inesite [MINERAL] Ca₂Mn₇Si₁₀O₂₈(OH)₂·5H₂O A pale-red mineral composed of hydrous manganese calcium silicate, occurring in small prismatic crystals or massive. { 'in·ə‚sīt }

inessential mapping [MATH] A mapping between topological spaces that is homotopic to a mapping whose range is a single point. { ‚in·ə‚sen·chəl 'map·iŋ }

inevitable abortion [MED] An abortion that has progressed to a stage where termination of the pregnancy cannot be avoided. { i'nev·əd·ə·bəl ə'bör·shən }

inextensional deformation [MECH] A bending of a surface that leaves unchanged the length of any line drawn on the surface and the curvature of the surface at each point. { in‚ek 'sten·chən·əl ‚def·ər'mā·shən }

in extremis [MED] At the point of death. { ‚in ik'strā·məs }

inf See greatest lower bound.

inface See scarp slope. { 'in‚fās }

infall process [ASTROPHYS] A process in which gas falls upon a very compact object such as a neutron star or black hole, reaching a high velocity and forming a hot plasma; postulated as a model for x-ray sources such as Centaurus X-1 and Hercules X-1. { 'in‚föl ‚prä·səs }

infall zone [ASTRON] The region that forms between the tidal radius of a planet in formation and the actual surface of the planet when the planet contracts from the tidal radius, so that any matter that enters this region falls to the planet's surface. { 'in‚föl ‚zōn }

infancy [GEOL] The initial (youthful) or very early stage of the cycle of erosion characterized by smooth, nearly level erosional surfaces dissected by narrow stream gorges, numerous depressions filled with marshy lakes and ponds, and shallow streams. Also known as topographic infancy. { 'in·fən·sē }

infant [ANTHRO] **1.** A baby; child under 2 years of age. **2.** An individual under legal age. { 'in·fənt }

infant botulism [MED] Botulism that involves ingestion of *Clostridium botulinum* spores with subsequent germination and toxin production in the gastrointestinal tract, found mostly in children aged 6 months or younger. { ‚in·fənt 'bäch·ə‚liz·əm }

infantile amaurotic familial idiocy See Tay-Sachs disease. { 'in·fən‚tīl ‚a‚mö¦räd·ik fə¦mil·yəl 'id·ē·ə·sē }

infantile autism [PSYCH] A disorder of children characterized by extreme withdrawal and introspection without regard for reality. { 'in·fən‚tīl 'ö‚tiz·əm }

infantile celiac disease [MED] Celiac syndrome of infants and young children. { 'in·fən‚tīl 'sē·lē‚ak di‚zēz }

infantile cortical hyperostosis [MED] A condition occurring during the first 3 months of life in which there is fever and painful swelling of the soft tissue of the lower jaw, characterized by periosteal proliferation of the mandible. { 'in·fən‚tīl 'körd·ə·kəl ‚hī·pə‚rä'stō·səs }

infantile diarrhea [MED] An acute gastrointestinal disease in infants resulting from damage of the intestinal mucosa by an infectious organism. { 'in·fən‚tīl ‚dī·ə'rē·ə }

infantile eczema [MED] An allergic inflammation of the skin in young children, usually due to common antigens such as food or inhalants. { 'in·fən‚tīl 'ek·sə·mə }

infantile genitalia [ANAT] The genital organs of an infant. [MED] Underdeveloped genitals in an adult. { 'in·fən‚tīl jen·ə'tāl·yə }

infantile neuroaxonal dystrophy [MED] A familial disease of the central nervous system occurring early in life and characterized by axonal swellings, arrested development, atrophy of the optic nerves, and eventual blindness. { 'in·fən‚tīl ¦nù·rō¦ak·sən·əl 'dis·trə·fē }

infantile paralysis See poliomyelitis. { 'in·fən‚tīl pə'ral·ə·səs }

infantile scurvy [MED] Acute scurvy of infants and young

children characterized by periosteal hemorrhage and swelling, especially of long bones. Also known as Cheadle's disease; Moeller-Barlow disease. { 'in·fən‚tīl 'skər·vē }

infantile sexuality [PSYCH] An infant's or child's capacity for and enjoyment of activities and experiences that are essentially sexual. { 'in·fər‚tīl ‚sek·shə'wal·əd·ē }

infantile spasm [MED] A type of seizure seen in infants and young children, characterized by a sudden, brief, massive myoclonic jerk. { 'in·fən‚tīl 'spaz·əm }

infantilism [MED] Persistence of physical, behavioral, or mental infantile characteristics into childhood, adolescence, or adult life. { 'in·fən·tə‚iz·əm }

infant respiratory distress syndrome [MED] A disorder usually affecting prematurely born infants and characterized by a rapid breathing rate, respiratory muscle retraction during expiration, and blood gas values reflecting oxygen deficiency, excessive carbon dioxide and acidosis. { 'in·fənt 'res·prə‚tör·ē di'stres ‚sin‚drōm }

infarct [MED] Localized death of tissue that is caused by obstructed inflow of arterial blood. Also known as infarction. { 'in‚färkt }

infarction [MED] **1.** Condition or process leading to the development of an infarct. **2.** See infarct. { in'färk·shən }

infauna [ZOO] Aquatic animals which live in the bottom sediment of a body of water. { in'fön·ə }

infect [MED] To cause an infection, as by contamination with or invasion by a pathogen. [MICROBIO] To cause a phage infection of bacteria. { in'fekt }

infection [MED] **1.** Invasion of the body by a pathogenic organism, with or without disease manifestation. **2.** Pathologic condition resulting from invasion of a pathogen. { in'fek·shən }

infectious [MED] Caused by infection. { in'fek·shəs }

infectious abortion See contagious abortion. { in'fek·shəs ə'bör·shən }

infectious anemia [VET MED] A virus disease of horses and mules characterized by intermittent fever, weakness, jaundice, and hemorrhages of mucous membranes; it is often fatal. Also known as swamp fever. { in'fek·shəs ə'nē·mē·ə }

infectious arthritis [MED] An inflammatory joint disease caused by microbial invasion of the articular tissue. { in'fek·shəs är'thrīd·əs }

infectious bronchitis [VET MED] A highly contagious respiratory viral disease of chickens. { in'fek·shəs brän'kīd·əs }

infectious canine hepatitis [VET MED] An acute inflammatory liver disease of dogs caused by a virus. { in'fek·shəs 'kā‚nīn ‚hep·ə'tīd·əs }

infectious chlorosis [PL PATH] A virus disease of plants characterized by yellowing of the green parts. { in'fek·shəs klə'rō·səs }

infectious conjunctivitis [MED] Conjunctivitis due to invasion by a microorganism. { in'fek·shəs kən‚jəŋk·tə'vīd·əs }

infectious disease [MED] Any disease caused by invasion by a pathogen which subsequently grows and multiplies in the body. { in'fek·shəs di'zēz }

infectious drug resistance [MICROBIO] A type of drug resistance that is transmissible from one bacterium to another by infectivelike agents referred to as resistance factors. { in'fek·shəs 'drəg ri‚zis·təns }

infectious endocarditis [MED] Inflammation of the endocardium due to an infectious microorganism. { in'fek·shəs ‚en·dō‚kär'dīd·əs }

infectious hepatitis [MED] Type A viral hepatitis, an acute infectious virus disease of the liver associated with hepatic inflammation and characterized by fever, liver enlargement, and jaundice. Also known as catarrhal jaundice; epidemic hepatitis; epidemic jaundice; virus hepatitis. { in'fek·shəs ‚hep·ə'tīd·əs }

infectious laryngotracheitis [VET MED] A highly contagious respiratory disease of viral etiology affecting chickens. { in'fek·shəs lə‚rin·jō‚trā·kē'īd·əs }

infectious mononucleosis [MED] A disorder of unknown etiology characterized by irregular fever, pathology of lymph nodes, lymphocytosis, and high serum levels of heterophil antibodies against sheep erythrocytes. Also known as acute benign lymphoblastosis; glandular fever; kissing disease; lymphocytic angina; monocytic angina; Pfeiffer's disease. { in'fek·shəs ‚män·ō‚nü·klē'ō·səs }

infectious myocarditis [MED] Inflammation of the myocardium due to an infectious microorganism. { in'fek·shəs ˌmī·ō‚kär'dīd·əs }

infectious myxomatosis [VET MED] An infectious virus disease of rabbits characterized by myxomatous lesions. { in'fek·shəs ˌmik·sō·mə'tō·səs }

infectious nucleic acid [VIROL] Purified viral nucleic acid capable of infecting a host cell and causing the production of viral progeny. { in¦fek·shəs nü¦klē·ik 'as·əd }

infectious papillomatosis [VET MED] A virus disease of cattle characterized by the appearance of warts on the body. { in'fek·shəs ˌpap·ə‚lō·mə'tō·səs }

infectious rhinitis [MED] Inflammation of the nasal mucous membrane due to an infectious microorganism. { in'fek·shəs rī'nīd·əs }

infectious transfer [GEN] The rapid spread of extrachromosomal episomes from donor to recipient cells in a bacterial population. { in¦fek·shən 'tranz‚fər }

infectious unit [VIROL] The smallest number of virus particles that will cause a lytic infection in a susceptible cell. { in'fek·shəs ˌyü·nət }

infectious uroarthritis See Reiter's syndrome. { in'fek·shəs ¦yùr·ō·är'thrīd·əs }

infective dose 50 [MICROBIO] The dose of microorganisms required to cause infection in 50% of the experimental animals; a special case of the median effective dose. Abbreviated ID₅₀. Also known as median infective dose. { in'fek·tiv ¦dōs 'fif·tē }

in-feed centerless grinding [MECH ENG] A metal-cutting process by which a cylindrical workpiece is ground to a prescribed surface smoothness and diameter by the insertion of the workpiece between a grinding wheel and a canted regulating wheel; the rotation of the regulating wheel controls the rotation and feed rate of the workpiece. { 'in‚fēd ¦sen·tər‚les 'grīnd·iŋ }

inference control [COMPUT SCI] A method of preventing data about specific individuals from being inferred from statistical information in a data base about groups of people. { 'in·frəns kən‚trōl }

inference program [COMPUT SCI] A computer program that uses certain facts provided as input to reach conclusions. { 'in·frəns ‚prō·grəm }

inferential flow meter [ENG] A flow meter in which the flow is determined by measurement of a phenomenon associated with the flow, such as a drop in static pressure at a restriction in a pipe, or the rotation of an impeller or rotor, rather than measurement of the actual mass flow. { ¦in·fə¦ren·chəl 'flō ‚mēd·ər }

inferential liquid-level meter [ENG] A liquid-level meter in which the level of a liquid is determined by measurement of some phenomenon associated with this level, such as the buoyancy of a solid partly immersed in the liquid, the pressure at a certain level, the conductance of the liquid, or its absorption of gamma radiation, rather than by direct measurement. { ¦in·fə¦ren·chəl ¦lik·wəd 'lev·əl ‚mēd·ər }

inferior [BIOL] The lower of two structures. { in'fir·ē·ər }

inferior alveolar artery [ANAT] A branch of the internal maxillary artery supplying the mucous membrane of the mouth and teeth of the lower jaw. { in'fir·ē·ər al¦vē·ə·lər 'ärd·ə·rē }

inferior alveolar nerve [NEUROSCI] A branch of the mandibular nerve that innervates the teeth of the lower jaw. { in'fir·ē·ər al¦vē·ə·lər 'nərv }

inferior cerebellar peduncle [NEUROSCI] A large bundle of nerve fibers running from the medulla oblongata to the cerebellum. Also known as restiform body. { in'fir·ē·ər ¦ser·ə¦bel·ər 'pē‚dəŋ·kəl }

inferior colliculus [NEUROSCI] One of the posterior pair of rounded eminences arising from the dorsal portion of the mesencephalon. { in'fir·ē·ər kə'lik·yə·ləs }

inferior conjunction [ASTRON] A type of configuration in which two celestial bodies have their least apparent separation; the smaller body is nearer the observer than the larger body, about which it orbits; for example, Venus is closest to the earth at its inferior conjunction. { in'fir·ē·ər kən'jəŋk·shən }

inferior fruit See accessory fruit. { in‚fir·ē·ər 'früt }

inferior ganglion [NEUROSCI] **1.** The lower sensory ganglion in the glossopharyngeal nerve. **2.** The lower sensory ganglion on the vagus. { in'fir·ē·ər 'gaŋ·glē·ən }

inferior hypogastric plexus [NEUROSCI] A network of nerves in the pelvic fascia containing autonomic nerve elements. { in'fir·ē·ər ˌhī·pə'gas·trik 'plek·səs }

inferiority [MED] An organic or psychic state or condition of being inferior or less adequate. { in‚fir·ē'är·əd·ē }

inferiority complex [PSYCH] Repressed unconsious fears and feelings of physical or social inadequacy or both, which may result in excessive anxiety, inability to function, or actual failure. { in‚fir·ē'är·əd·ē 'käm‚pleks }

inferior mesenteric ganglion [NEUROSCI] A sympathetic ganglion within the inferior mesenteric plexus at the origin of the inferior mesenteric artery. { in'fir·ē·ər ˌmez·ən'ter·ik 'gaŋ·glē·ən }

inferior mirage [OPTICS] A spurious image of an object formed below the true position of that object by abnormal refraction conditions along the line of sight; one of the most common types of mirage, and the opposite of a superior mirage. { in'fir·ē·ər mə'räzh }

inferior planet [ASTRON] A planet that circles the sun in an orbit that is smaller than the earth's. { in'fir·ē·ər 'plan·ət }

inferior temporal gyrus [NEUROSCI] A convolution on the temporal lobe of the cerebral hemispheres lying below the middle temporal sulcus and extending to the inferior sulcus. { in'fir·ē·ər ¦tem·pə·rəl 'jī·rəs }

inferior vena cava [ANAT] A large vein which drains blood from the iliac veins, lower extremities, and abdomen into the right atrium. { in'fir·ē·ər ‚vē·nə 'kā·və }

inferior vena cava syndrome [MED] Edema and venous distention of the abdomen and legs due to obstruction of the inferior vena cava. { in'fir·ē·ər ¦vē·nə ¦kā·və ‚sin‚drōm }

inferior vermis [NEUROSCI] The inferior portion of the vermis of the brain. { in'fir·ē·ər 'vər·məs }

inferior vestibular nucleus [NEUROSCI] The terminal nucleus for the spinal vestibular nerve tract. { in'fir·ē·ər və'stib·yə·lər 'nü·klē·əs }

infernal machine [ORD] Disguised or cleverly concealed explosive device, usually intended for sabotage. { in'fərn·əl mə‚shēn }

inferred ore [MIN ENG] An ore whose estimate of tonnage and grade is based largely on knowledge of the deposit's geological character and to a lesser degree on samples and other data. { in'fərd 'ȯr }

infertility [MED] Involuntary reduction in reproductive ability. { ˌin·fər'til·əd·ē }

infest [MED] To live on or within the host's body. { in'fest }

infestation [MED] The state or condition of having animal parasites on or in the body. { ‚in·fe'stā·shən }

infilling well [PETRO ENG] A well drilled between producing wells for the purpose of more efficient recovery of petroleum from the reservoir. { 'in‚fil·iŋ ‚wel }

infiltrating lipoma See liposarcoma. { in'fil‚trād·iŋ li'pō·mə }

infiltration [ENG] Leakage of outdoor air into a building by natural forces, for example, by seepage through cracks or other openings. [GEOL] Deposition of mineral matter among the pores or grains of a rock by permeation of water carrying the matter in solution. [HYD] Movement of water through the soil surface into the ground. [MET] **1.** Filling the pores of a metal powder compact with metal having a lower melting point. **2.** Movement of molten metal into the pores of a fiber or foam metal. { ‚in·fil'trā·shən }

infiltration capacity [HYD] The maximum rate at which water enters the soil or other porous material in a given condition. { ‚in·fil'trā·shən kə'pas·əd·ē }

infiltration gallery [CIV ENG] A large, horizontal underground conduit of perforated or porous material with openings on the sides for collecting percolating water by infiltration. { ‚in·fil'trā·shən ‚gal·rē }

infiltration vein [GEOL] Vein deposited in rock by percolating water. { ‚in·fil'trā·shən vān }

infimum See greatest lower bound. { 'in·fə·məm }

infinite [MATH] Larger than any fixed number. { 'in·fə·nət }

infinite aquifer [HYD] The portion of a formation that contains water, and for which the exterior boundary is at an effectively infinite distance from the oil reservoir. { 'in·fə·nət 'ak·wə·fər }

infinite baffle [ENG ACOUS] A loudspeaker baffle which prevents interaction between the front and back radiation of the loudspeaker. { 'in·fə·nət 'baf·əl }

infinite-capacity loading [CONT SYS] The deliberate overloading of a robotic work center with excessive force or weight in order to determine the overload protection necessary to maintain proper load conditions. { 'in·fə·nət kə'pas·əd·ē ,lōd·iŋ }

infinite discontinuity [MATH] A discontinuity of a function for which the absolute value of the function can have arbitrarily large values arbitrarily close to the discontinuity. { 'in·fə·nit ,dis,känt·ən'ü·əd·ē }

infinite extension [MATH] An extension field F of a given field E such that F, viewed as a vector space over E, has infinite dimension. { 'in·fi·nit ik'sten·chən }

infinite group [MATH] A group that contains an infinite number of distinct elements. { 'in·fə·nit 'grüp }

infinite impulse response filter [ELECTR] An electronic filter that will continue oscillating in a decaying manner forever after being exposed to a change in input. Abbreviated IIR filter. { ,in·fə·nət ¦im,pəls ri¦späns ,fil·tər }

infinite integral [MATH] An integral at least one of whose limits of integration is infinite. { 'in·fə·nət 'int·ə·grəl }

infinite multiplication factor [NUCLEO] The multiplication factor of a theoretical system from which there is no leakage of neutrons, that is, a reactor of infinite size. { 'in·fə·nət ,məl·tə·plə'kā·shən ,fak·tər }

infinite population [STAT] A universe which contains an infinite number of elements; it can be continuous or discrete. { 'in·fi·nit ,päp·yə'lā·shən }

infinite reservoir [PETRO ENG] In reservoir unsteady-state liquid-diffusion calculations, a reservoir in which the outer boundary is considered to be effectively at an infinite distance from the inner boundary at the well of an aquifer. { 'in·fə·nət 'rez·əv,wär }

infinite root [MATH] An equation $f(x) = 0$ is said to have an infinite root if the equation $f(1/y) = 0$ has a root at $y = 0$. { ,in·fə·nət 'rüt }

infinite sequence See sequence. { 'in·fə·nət 'sē·kwəns }

infinite series [MATH] An indicated sum of an infinite sequence of quantities, written $a_1 + a_2 + a_3 + \cdots$, or

$$\sum_{k=1}^{\infty} a_k$$

{ 'in·fə·nət 'sir·ēz }

infinite set [MATH] A set with more elements than any fixed integer; such a set can be put into a one to one correspondence with a proper subset of itself. { 'in·fə·nət 'set }

infinitesimal [MATH] A function whose value approaches 0 as its argument approaches some specified limit.

infinitesimal generator [MATH] A closed linear operator defined relative to some semigroup of operators and which uniquely determines that semigroup. { ¦in,fin·ə¦tes·ə·məl 'jen·ə,rād·ər }

infinity [COMPUT SCI] Any number larger than the maximum number that a computer is able to store in any register. [MATH] The concept of a value larger than any finite value. { in'fin·əd·ē }

infinity method [OPTICS] Method of adjusting two lines of sight to make them parallel; lines are adjusted on an object at great distance, for example, a star. { in'fin·əd·ē ,meth·əd }

infinity transmitter [ELECTR] A device used to tap a telephone; the telephone instrument is so modified that an interception device can be actuated from a distant source without the caller's becoming aware. { in'fin·əd·ē tranz'mid·ər }

infix notation [MATH] A method of forming mathematical or logical expressions in which operators are written between the operands on which they act. { 'in,fiks nō,tā·shən }

infix operation [COMPUT SCI] An operation carried out within an operation, as the addition of a and b prior to the multiplication by c or division by d in the operation $(a+b)c/d$. { 'in,fiks ,äp·ə,rā·shən }

inflammability See flammability. { in,flam·ə'bil·əd·ē }

inflammation [MED] Local tissue response to injury characterized by redness, swelling, pain, and heat. { ,in·flə'mā·shən }

inflammatory arthritis [MED] A type of arthritis that is characterized by inflammation of tissues associated with joints; rheumatoid arthritis is the most common variety. { in¦flam·ə,tȯr·ē ärth'rīd·əs }

inflammatory bowel disease [MED] A general term for two closely related conditions, ulcerative colitis and regional enteritis or Crohn's disease. { in,flam·ə,tȯr·ē 'baul diz,ēz }

inflammatory carcinoma [MED] A carcinoma, usually of the breast, associated with inflammation and characterized by rapid metastasis. { in'f am·ə,tȯr·ē ,kärs·ən'ō·mə }

inflammatory response [IMMUNOL] A nonspecific defensive reaction of the body to invasion by a foreign substance or organism that involves phagocytosis by white blood cells and is often accompanied by accumulation of pus and an increase in the local temperature. { in¦flam·ə,tȯre ri'späns }

inflammatory tissue [MED] Tissue characterized by exudation or cell proliferation caused by trauma. { in'flam·ə,tȯr·ē 'tish·ü }

inflatable gasket [DES ENG] A gasket whose seal is activated by inflation with compressed air. { in¦flād·ə·bəl 'gas·kət }

inflatable packer [PETRO ENG] A packer (downhole pressure seal between tubing and casing) set and held in place by an element that is inflated with hydraulic pressure. { in'flād·ə·bəl 'pak·ər }

inflated [BIOL] **1.** Distended, applied to a hollow structure. **2.** Open and enlarged. [ENG] Filled or distended with air or gas. { in'flād·əd }

inflationary universe cosmology [ASTRON] A theory of the evolution of the early universe which asserts that at some early time the observable universe underwent a period of exponential expansion, during which the scale of the universe increased by at least 28 orders of magnitude. { in'flā·shə,ner·ē 'yü·nə,vərs käz'mäl·ə·jē }

inflected [BOT] Curved or bent sharply inward, downward, or toward the axis. Also known as inflexed. { in'flek·təd }

inflected arch See inverted arch. { in'flek·təd 'ärch }

inflectional tangent [MATH] A tangent to a curve at a point of inflection. { in¦flek·shə·nəl 'tan·jənt }

inflection point See point of inflection. { in'flek·shən ,point }

inflexed See inflected. { in'flekst }

inflight start [AERO ENG] An engine ignition sequence that takes place after takeoff and during flight. { 'in¦flīt 'stärt }

inflorescence [BOT] A flower cluster segregated from any other flowers on the same plant, together with the stems and bracts (reduced leaves) associated with it. { ,in·flə'res·əns }

inflorescence structure [BOT] The way that the flowers are clustered or arranged on a flowering branch. { ,in·flə¦res·əns 'strək·chər }

inflow [MATH] The inflow to a vertex in an s-t network is the sum of the flows of all the arcs that terminate at that vertex. { 'in,flō }

influence diagram [SYS ENG] A graph-theoretic representation of a decision, which may include four types of nodes (decision, chance, value, and deterministic), directed arcs between the nodes (which identify dependencies between them), a marginal or conditional probability distribution defined at each chance node, and a mathematical function associated with each of the other types of node. { 'in,flü·əns ,dī·ə,gram }

influence factor See telephone influence factor. { 'in,flü·əns ,fak·tər }

influence function [PETRO ENG] Mathematical statement of the influence on pressure and production of each oil reservoir pool in a multipool aquifer { 'in,flü·əns ,fəŋk·shən }

influence fuse See proximity fuse. { 'in,flü·əns ,fyüz }

influence line [MECH] A graph of the shear, stress, bending moment, or other effect of a movable load on a structural member versus the position of the load. { 'in,flü·əns ,līn }

influence mine [ORD] A mine that is detonated by methods that are different than target contact. { 'in,flü·əns ,mīn }

influent [ECOL] An organism that disturbs the ecological balance of a community. [SCI TECH] An input stream of a fluid, as water into a reservoir, or liquid into a process vessel. { 'in,flü·ənt }

influent stream [HYD] A stream that contributes water to the zone of saturation of groundwater and develops bank storage. Also known as losing stream. { 'in,flü·ənt 'strēm }

influenza [MED] An acute virus disease of the respiratory system characterized by headache, muscle pain, fever, and prostration. { ,in·flü'en·zə }

influenzal meningitis [MED] Inflammation of the meninges caused by Hemophilus influenzae. { ,in·flü'enz·əl ,men·ən'jīd·əs }

influenzal pneumonia [MED] Pneumonia resulting from

infection by *Hemophilus influenzae*. { ‚in·flü'enz·əl nə 'mōn·yə }

influenza vaccine [IMMUNOL] A vaccine prepared from formaldehyde-attenuated mixtures of strains of influenza virus. { ‚in·flü'en·zə vak'sēn }

influenza virus [VIROL] Any of three immunological types, designated A, B, and C, belonging to the myxovirus group which cause influenza. { ‚in·flü'en·zə ‚vī·rəs }

influx *See* mouth. { 'in‚fləks }

information [COMMUN] Data which has been recorded, classified, organized, related, or interpreted within a framework so that meaning emerges. { ‚in·fər'mā·shən }

information architecture [COMPUT SCI] The organization of large bodies of content, as well as the organization and labeling (tagging) of content at the document level to make information easy to search, navigate, and manage. { ‚in·fər‚mā·shən 'är·kə‚tek·chər }

information bit [COMMUN] Bit that is generated by the data source but is not used by the data-transmission system. { ‚in·fər'mā·shən ‚bit }

information center [COMMUN] Center designed specifically for storing, processing, and retrieving information for dissemination at regular intervals, on demand or selectively, according to express needs of users. { ‚in·fər'mā·shən ‚sen·tər }

information channel [COMMUN] A facility used to transmit information between data-processing terminals separated by large distances. { ‚in·fər'mā·shən ‚chan·əl }

information content [COMMUN] A numerical measure of the information generated in selecting a specific symbol (or message), equal to the negative of the logarithm of the probability of the symbol (or message) selected. Also known as negentropy. { ‚in·fər'mā·shən ‚kän‚tent }

information engineering [COMPUT SCI] The process of networking, collecting, analyzing, and reporting information, as well as controlling business, manufacturing, or service operations. { ‚in·fər'mā·shən ‚en·jə‚nir·iŋ }

information feedback system [COMMUN] An information transmission system in which a return transmission is used to verify the accuracy of the sent transmission. { ‚in·fər'mā·shən 'fēd‚bak ‚sis·təm }

information float [COMPUT SCI] Information that is not located in a file or data base but is traveling between systems or is not assigned to a particular computer system. { ‚in·fər'mā·shən ‚flōt }

information flow [COMPUT SCI] The graphic representation of data collection, data processing, and report distribution throughout an organization. { ‚in·fər'mā·shən ‚flō }

information flow control [COMPUT SCI] A restriction on the use of information generated by a computer system that is consistent with the access controls on the resources of the system itself. { ‚in·fər'mā·shən 'flō kən‚trōl }

information function of a partition [MATH] If ξ is a finite partition of a probability space, the information function of ξ is a step function whose sets of constancy are the elements of ξ and whose value on an element of ξ is the negative of the logarithm of the probability of this element. { ‚in·fər'mā·shən ‚fəŋk·shən əv ə pär'tish·ən }

information interchange [COMMUN] The exchange of information between machines. { ‚in·fər'mā·shən 'in·tər‚chānj }

information link *See* data link. { ‚in·fər'mā·shən ‚liŋk }

information management [COMMUN] The science that deals with definitions, uses, value and distribution of information that is processed by an organization, whether or not it is handled by a computer. { ‚in·fər'mā·shən 'man·ij·mənt }

information network [COMPUT SCI] A service that provides a variety of information services to subscribers on a dial-up basis. Also known as subscription data base. { ‚in·fər'mā·shən 'net‚wərk }

information precedence relation [COMPUT SCI] A statement that some specified piece of data is required for the production of another piece of data. { ‚in·fər'mā·shən 'pres·ə·dəns ri‚lā·shən }

information process analysis chart *See* form process chart. { ‚in·fər'mā·shən ‚prä·ses ə¦nal·ə·səs ‚chärt }

information processing [COMPUT SCI] **1.** The manipulation of data so that new data (implicit in the original) appear in a useful form. **2.** *See* data processing. [PSYCH] The coding,

retrieval, and combination of information in perceptual recognition, learning, remembering, thinking, problem solving, and performance of sensory-motor acts. { ‚in·fər'mā·shən 'prä·ses·iŋ }

Information Processing Language *See* IPL. { ‚in·fər'mā·shən 'prä·ses·iŋ 'laŋ·gwij }

information rate [COMMUN] The information content generated per symbol or per second by an information source. { ‚in·fər'mā·shən ‚rāt }

information redundancy [COMPUT SCI] The use of more information than is absolutely necessary, such as the application of error-detection and error-correction codes, in order to increase the reliability of a computer system. { ‚in·fər'mā·shən rə'dən·dən·sē }

information requirements [COMPUT SCI] Actual or anticipated questions which may be posed to an information retrieval system. { ‚in·fər'mā·shən rə'kwīr·məns }

information resources management [COMPUT SCI] A concept for processing information that focuses on the information and places data-processing technology (software and hardware) in a secondary role. { ‚in·fər'mā·shən ri'sór·səz ‚man·ij·mənt }

information retrieval [COMPUT SCI] The technique and process of searching, recovering, and interpreting information from large amounts of stored data. { ‚in·fər'mā·shən ri‚trē·vəl }

information science [SCI TECH] The theory and techniques relating to the collection, classification, storage, retrieval, and dissemination of recorded knowledge. { ‚in·fər¦mā·shən ¦sī·əns }

information selection systems [COMPUT SCI] A class of information processing systems which carry out a sequence of operations necessary to locate in storage one or more items assumed to have certain specified characteristics and to retrieve such items directly or indirectly, in whole or in part. { ‚in·fər'mā·shən si'lek·shən ‚sis·təmz }

information separator [COMPUT SCI] A character that separates items or fields of information in a record, especially a variable-length record. { ‚in·fər'mā·shən 'sep·ə‚rād·ər }

information source [COMMUN] A system which produces messages by making successive selections from a group of symbols. { ‚in·fər'mā·shən ‚sórs }

information system [COMMUN] Any means for communicating knowledge from one person to another, such as by simple verbal communication, punched-card systems, optical coincidence systems based on coordinate indexing, and completely computerized methods of storing, searching, and retrieving of information. { ‚in·fər'mā·shən ‚sis·təm }

information system architecture [COMPUT SCI] The study of the structure of both computer systems and the organizations that use them, in order to develop computer systems that support the objectives of the organizations more effectively. { ‚in·fər¦mā·shən ¦sis·təm 'är·kə‚tek·chər }

information systems engineering [ENG] The discipline concerned with the design, development, testing, and maintenance of information systems. { ‚in·fər¦mā·shən ¦sis·təmz ‚en·jə'nir·iŋ }

information technology [COMPUT SCI] The collection of technologies that deal specifically with processing, storing, and communicating information, including all types of computer and communications systems as well as reprographics methodologies. { ‚in·fər'mā·shən tek¦näl·ə·jē }

information theory [COMMUN] A branch of theory which is devoted to problems in communications, and which provides criteria for comparing different communications systems on the basis of signaling rate, using a numerical measure of the amount of information gained when the content of a message is learned. [MATH] The branch of probability theory concerned with the likelihood of the transmission of messages, accurate to within specified limits, when the bits of information composing the message are subject to possible distortion. { ‚in·fər'mā·shən ‚thē·ə·rē }

information unit [COMMUN] A unit of information content, equal to a bit, nit, or hartley, according to whether logarithms are taken to base 2, *e*, or 10. { ‚in·fər'mā·shən ‚yü·nət }

information utility [COMPUT SCI] An information network that specializes in supplying information to businesses and other organizations. { ‚in·fər'mā·shən yü‚til·əd·ē }

information word *See* data word. { ‚in·fər'mā·shən ‚wərd }

informosome [CELL MOL] A type of cellular particle that is thought to be a complex of messenger ribonucleic acid with ribonucleoprotein. { in'fŏr·mə,sōm }

infrabasal [BIOL] Inferior to a basal structure. { ¦in·frə'bā·səl }

infrabranchial [VERT ZOO] Situated below the gills. { ¦in·frə'braŋ·kē·əl }

Infracambrian See Eocambrian. { ¦in·frə'kam·brē·ən }

infracentral [ANAT] Located below the centrum. { ¦in·frə'sen·trəl }

infracerebral gland [INV ZOO] A structure lying ventral to the brain in annelids which is thought to produce a hormone that inhibits maturation of the gametes. { ¦in·frə·sə'rē·brəl ¦gland }

infraciliature [INV ZOO] The neuromotor apparatus, silverline system, or neuroneme system of ciliates. { ¸in·frə'sil·yə·chər }

infraclass [SYST] A subdivision of a subclass; equivalent to a superorder. { 'in·frə,klas }

infraclavicle [VERT ZOO] A bony element of the shoulder girdle located below the cleithrum in some ganoid and crossopterygian fishes. { in·frə'klav·ə·kəl }

infradyne receiver [ELECTR] A superheterodyne receiver in which the intermediate frequency is higher than the signal frequency, so as to obtain high selectivity. { 'in·frə,dīn ri'sē·vər }

infrafoliar [BOT] Located below the leaves. { ¸in·frə'fō·lē·ər }

infraglenoid [ANAT] Below the glenoid cavity of the scapula. { ¦in·frə'gle,nóid }

infraglenoid tubercle [ANAT] A rough impression below the glenoid cavity, from which the long head of the triceps muscle arises. { ¦in·frə'gle,nóid 'tü·bər·kəl }

infragravity wave [FL MECH] A gravity wave whose period ranges from 30 seconds to 5 minutes. { ¸in·frə'grav·əd·ē ,wāv }

infralateral tangent arcs [METEOROL] Two oblique, colored arcs, convex toward the sun and tangent to the halo of 46° at points below the altitude of the sun, produced by refraction (90° effective prism angle) in hexagonal columnar ice crystals whose principal axes are horizontal but randomly directed in azimuth; if the sun's elevation exceeds about 68°, the arcs cannot appear. { ¦in·frə¦lad·ə·rəl 'tan·jənt ,ärks }

infralow frequency [COMMUN] A designation for the band from 0.3 to 3 kilohertz in the radio spectrum. Abbreviated ILF. { 'in·frə,lō 'frē·kwən·sē }

infraorbital [ANAT] Located beneath the orbit. { ¦in·frə'ór·bəd·əl }

infrared [ELECTROMAG] Pertaining to infrared radiation. { ¦in·frə¦red }

infrared absorption [ELECTROMAG] The taking up of energy from infrared radiation by a medium through which the radiation is passing. { ¦in·frə¦red əb'sórp·shən }

infrared array [ENG] A collection of several thousand infrared detector elements arranged in a grid pattern and connected to readout electronics to display infrared images focused on the array by an astronomical telescope. { ¦in·frə¦red ə'rā }

infrared astronomy [ASTROPHYS] The study of electromagnetic radiation in the spectrum between 0.75 and 1000 micrometers emanating from astronomical sources. { ¦in·frə¦red ə'strän·ə·mē }

infrared beacon [NAV] A source of infrared radiation used to establish a geographical reference point, the bearings of which may be determined. { ¦in·frə¦red 'bē·kən }

infrared binoculars [OPTICS] An instrument for viewing an enlarged infrared image with both eyes; it has two infrared telescopes whose lens systems are similar to those of ordinary binoculars. { ¦in·frə¦red bə'näk·yə·lərz }

infrared bolometer [ELECTR] A bolometer adapted to detecting infrared radiation, as opposed to microwave radiation. { ¦in·frə¦red bə'läm·əd·ər }

infrared brazing [MET] A brazing process in which coalescence is produced by heat generated by infrared radiation. { ¦in·frə¦red 'brāz·iŋ }

infrared catastrophe [QUANT MECH] The logarithmic divergence in the cross section (which one would expect to be finite) for the emission of low-energy photons in bremsstrahlung and in the double Compton effect, according to quantum electrodynamics; the difficulty is resolved by taking radiative

corrections to elastic scattering into account. Also known as infrared problem. { ¦ir·¦frə¦red kə'tás·trə·fē }

infrared communications set [ELECTR] Components required to operate a two-way electronic system using infrared radiation to carry intelligence. { ¦in·frə¦red kə,myü·nə'kā·shənz ,set }

infrared detector [ELECTR] A device responding to infrared radiation, used in detecting fires, or overheating in machinery, planes, vehicles, and people, and in controlling temperature-sensitive industrial processes. { ¦in·frə¦red di'tek·tər }

infrared dome See irdome. { ¦in·frə¦red 'dōm }

infrared dwarf See brown dwarf. { ¦in·frə,red 'dwórf }

infrared emission [PHYS] The act of emitting infrared waves. { ¦in·frə¦red i'mish·ən }

infrared-emitting diode [ELECTR] A light-emitting diode that has maximum emission in the near-infrared region, typically at 0.9 micrometer for *pn* gallium arsenide. { ¦in·frə¦red ¦mid·iŋ 'dī,ōd }

infrared film [GRAPHICS] Film that is sensitive to wavelengths in the near-infrared region. { ¦in·frə¦red 'film }

infrared filter [OPTICS] A substance or device which is highly transparent to infrared radiation at certain wavelengths while absorbing other types of electromagnetic radiation. { ¦in·frə¦red 'fil·tər }

infrared galaxy [ASTRON] A galaxy or quasar whose nucleus emits enormous amounts of infrared radiation, in some cases more than 1000 times the output of the entire Milky Way Galaxy at all wavelengths. { ¦in·frə¦red 'gal·ik·sē }

infrared heating [ENG] Heating by means of infrared radiation. { ¦in·frə¦red 'hēd·iŋ }

infrared heterodyne detector [ELECTR] A heterodyne detector in which both the incoming signal and the local oscillator signal frequencies are in the infrared range and are combined in a photodetector to give an intermediate frequency in the kilohertz or megahertz range for conventional amplification. { ¦in·frə¦red 'hed·ə·rə,dīn di'tek·tər }

infrared homing [ENG] Homing in which the target is tracked by means of its emitted infrared radiation. { ¦in·frə¦red 'hōm·iŋ }

infrared image converter [ELECTR] A device for converting an invisible infrared image into a visible image, consisting of an infrared-sensitive, semitransparent photocathode on one end of an evacuated envelope and a phosphor screen on the other, with an electrostatic lens system between the two. Also known as infrared image tube. { ¦in·frə¦red 'im·ij kən,vərd·ər }

infrared image tube See infrared image converter. { ¦in·frə¦red 'im·ij ,tüb }

infrared imaging device [ENG] Any device which converts an invisible infrared image into a visible image. { ¦in·frə¦red 'im·ə·jiŋ di,vīs }

infrared jamming [ELECTR] An attempt to confuse heat-seeking missiles by emissions which overload their inputs or misdirect them. { ¦in·frə¦red 'jam·iŋ }

infrared lamp [ELEC] An incandescent lamp which operates at reduced voltage with a filament temperature of 4000°F (2200°C) so that it radiates electromagnetic energy primarily in the infrared region. { ¦in·frə¦red 'lamp }

infrared laser [PHYS] A laser which emits infrared radiation, especially in the near- and intermediate-infrared regions. { ¦in·frə¦red 'lā·zər }

infrared mapping [MAP] Mapping in which a sensitive infrared detector is mounted on an infrared scanner, and the resulting thermal image is translated into varying shades of gray on photographic film, to give a line-pattern image much like that seen on a television screen. { ¦in·frə¦red 'map·iŋ }

infrared maser [PHYS] A laser which emits infrared radiation, especially in the far-infrared region, or which is pumped with radiation at infrared frequencies and emits radiation at millimeter wavelengths. { ¦in·frə¦red 'mā·zər }

infrared microscope [OPTICS] A type of reflecting microscope which uses radiation of wavelengths greater than 700 nanometers and is used to reveal detail in materials that are opaque to light, such as molybdenum, wood, corals, and many red-dyed materials. { ¦in·frə¦red 'mī·krə,skōp }

infrared optical material [ELECTROMAG] A material which is transparent to infrared radiation. { 'in·frə¦red 'äp·tə·kəl mə,tir·ē·əl }

infrared phosphor [SOLID STATE] A phosphor which, when

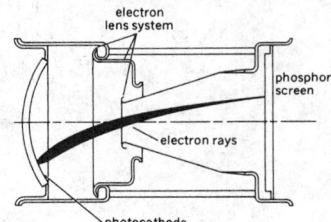

INFRARED IMAGE CONVERTER

electron lens system / phosphor screen / electron rays / photocathode

Components of the infrared image converter.

exposed to infrared radiation during or even after decay of luminescence resulting from its usual or dominant activator, emits light having the same spectrum as that of the dominant activator; sulfide and selenide phosphors are the most important examples. { ¦in·frə¦red 'fäs·fər }

infrared photoconductor [ELECTR] A conductor whose conductivity increases when it is exposed to infrared radiation. { ¦in·frə¦red ¦fōd·ō·kən¦dək·tər }

infrared photography [GRAPHICS] Photography in which an infrared optical system projects an image directly on infrared film, to provide a record of point-to-point variations in temperature of a scene. { ¦in·frə¦red fə'täg·rə·fē }

infrared problem *See* infrared catastrophe. { ¦in·frə¦red 'präb·ləm }

infrared radiation [ELECTROMAG] Electromagnetic radiation whose wavelengths lie in the range from 0.75 or 0.8 micrometer (the long-wavelength limit of visible red light) to 1000 micrometers (the shortest microwaves). { ¦in·frə¦red ¦räd·ē·ā·shən }

infrared reflectography [ANALY CHEM] In art conservation, a nondestructive digital imaging technique used to investigate underdrawings (below the painted surface) of paintings. { ¦in·frə¦red ¦rē¦flek'täg·rə·fē }

infrared receiver [ELECTR] A device that intercepts or demodulates infrared radiation that may carry intelligence. Also known as nancy receiver. { ¦in·frə¦red ri'sē·vər }

infrared scanner [ELECTR] An infrared detector mounted on a motor-driven platform which causes it to scan a field of view line by line, much as in television. { ¦in·frə¦red 'skan·ər }

infrared searchlight [OPTICS] A device for illuminating a scene with infrared radiation so that it may be viewed through an infrared image converter tube, although it is invisible to the unaided eye. { ¦in·frə¦red 'sərch,līt }

infrared soldering [MET] Soldering in which infrared radiation furnishes the required heat. { ¦in·frə¦red 'säd·ə·riŋ }

infrared spectrometer [SPECT] An instrument used to identify and measure the concentration of chemical compounds (gases, nonaqueous liquids, and solids) with electromagnetic radiation from 800 nanometers to 1 millimeter. { ¦in·frə¦red spek'träm·əd·ər }

infrared spectrophotometry [SPECT] Spectrophotometry in the infrared region, usually for the purpose of chemical analysis through measurement of absorption spectra associated with rotational and vibrational energy levels of molecules. { ¦in·frə¦red ¦spek·trə·fə'täm·ə·trē }

infrared spectroscopy [SPECT] The study of the properties of material systems by means of their interaction with infrared radiation; ordinarily the radiation is dispersed into a spectrum after passing through the material. { ¦in·frə¦red spek'träs·kə·pē }

infrared spectrum [ELECTROMAG] **1.** The range of wavelengths of infrared radiation. **2.** A display or graph of the intensity of infrared radiation emitted or absorbed by a material as a function of wavelength or some related parameter. { ¦in·frə¦red 'spek·trəm }

infrared star [ASTROPHYS] A star that emits a large amount of radiant energy in the infrared portion of the electromagnetic spectrum. { ¦in·frə¦red 'stär }

infrared telescope [OPTICS] An instrument that converts an invisible infrared image into a visible image and enlarges this image, consisting of an infrared image converter tube, an objective lens for imaging the scene to be viewed onto the photocathode of the tube, and an ocular for viewing the phosphor screen of the tube. { ¦in·frə¦red 'tel·ə,skōp }

infrared thermistor [ELECTR] A thermistor used to measure the power of infrared radiation. { ¦in·frə¦red thər'mis·tər }

infrared thermography [ENG] A method of measuring surface temperatures by observing the infrared emission from the surface. { ¦in·frə¦red thər'mäg·rə·fē }

infrared thermometer [ENG] An instrument that focuses and detects the infrared radiation emitted by an object in order to determine its temperature. { ¦in·frə·'red thər'mäm·əd·ər }

infrared transmitter [ELECTR] A transmitter that emits energy in the infrared spectrum; may be modulated with intelligence signals. { ¦in·frə¦red tranz'mid·ər }

infrared-transparent material [MATER] An optical material

that transmits infrared radiation; examples include sodium chloride (0.25 to 16 micrometers), cesium iodide (1 to 50 micrometers), and high-density polyethylene (16 to 300 micrometers). { ¦in·frə¦red tranz¦par·ənt mə,tir·ē·əl }

infrared vidicon [ELECTR] A vidicon whose photoconductor surface is sensitive to infrared radiation. { ¦in·frə¦red 'vid·ə,kän }

infrared window [GEOPHYS] A frequency region in the infrared where there is good transmission of electromagnetic radiation through the atmosphere. { ¦in·frə¦red 'win·dō }

infrasonic [ACOUS] Pertaining to signals, equipment, or phenomena involving frequencies below the range of human hearing, hence below about 15 hertz. Also known as subsonic (deprecated usage). { ¦in·frə¦sän·ik }

infrasound [ACOUS] Vibrations of the air at frequencies too low to be perceived as sound by the human ear, below about 15 hertz. { 'in·frə,saund }

infraspinous [ANAT] Below the spine of the scapula. { ¦in·frə'spī·nəs }

infraspinous fossa [ANAT] The recess on the posterior surface of the scapula occupied by the infraspinatus muscle. { ¦in·frə'spī·nəs 'fäs·ə }

infratemporal [ANAT] Situated below the temporal fossa. { ¦in·frə'tem·prəl }

infratemporal fossa [ANAT] An irregular space situated below and medial to the zygomatic arch, behind the maxilla and medial to the upper part of the ramus of the mandible. { ¦in·frə'tem·prəl 'fäs·ə }

infructescence [BOT] An inflorescence's fruiting stage. { ,in,frək'tes·əns }

infundibular canal [INV ZOO] A pathway from the mantle cavity through the funnel for water in cephalopods. { ,in·fən'dib·yə·lər kə'nal }

infundibular ganglion [INV ZOO] A branch of pedal ganglion which supplies the funnel in cephalopods. { ,in·fən'dib·yə·lər 'gaŋ·glē·ən }

infundibular process [ANAT] The distal portion of the neural lobe of the pituitary. { ,in·fən'dib·yə·lər 'prä·səs }

infundibulum [ANAT] **1.** A funnel-shaped passage or part. **2.** The stalk of the neurohypophysis. { ,in·fən'dib·yə·ləm }

infusion [CHEM] The aqueous solution of a soluble constituent of a substance as the result of the substance's steeping in the solvent for a period of time. [MED] The slow injection of a solution into a vein or into subcutaneous or other tissue of the body. { in'fyü·zhən }

infusorial earth [GEOL] Formerly, and incorrectly, a soft rock or an earthy substance composed of siliceous remains of diatoms. { ,in·fyə'sór·ē·əl 'ərth }

infusoriform larva [INV ZOO] The final larval stage, arising from germ cells within the infusorigen, in the life cycle of dicyemid mesozoans. { ¦in·fyə¦zór·ə,fórm 'lär·və }

infusorigen [INV ZOO] An individual that gives rise to the infusoriform larva in dicyemid mesozoans. { ,in·fyə'zór·ə·jən }

in-gate *See* gate. { 'in,gāt }

Ingen-Hausz apparatus [THERMO] An apparatus for comparing the thermal conductivities of different conductors; specimens consisting of long wax-coated rods of equal length are placed with one end in a tank of boiling water covered with a radiation shield, and the lengths along the rods from which the wax melts are compared. { 'iŋ·gən 'haus ,ap·ə,rad·əs }

ingesta [BIOL] Food and other substances taken into an animal body. { in'jes·tə }

ingestion [BIOL] The act or process of taking food and other substances into the animal body. { in'jes·chən }

Ingolfiellidea [INV ZOO] A suborder of amphipod crustaceans in which both abdomen and maxilliped are well developed and the head often bears a separate ocular lobe lacking eyes. { in¦gäl·fē·ə¦lid·ē·ə }

ingot [MET] **1.** A solid metal casting suitable for remelting or working. **2.** A bar of gold or silver. { 'iŋ·gət }

ingot iron [MET] Relatively pure iron produced in an open-hearth furnace. { 'iŋ·gət ,ī·ərn }

ingrain color *See* azoic dye. { 'in,grān ,kəl·ər }

ingress [ASTRON] The entrance of the moon into the shadow of the earth in an eclipse, of a planet into the disk of the sun, or of a satellite (or its shadow) onto the disk of the parent planet. [SCI TECH] The act of entering, as of air into the lungs or a liquid into an orifice. { 'in,gres }

ingrown [MED] Of a hair or nail, grown inward so that the normally free end is embedded in or under the skin. { 'in,grōn }

ingrown meander [GEOL] A meander of a stream with an undercut bank on one side and a gentle slope on the other. { 'in,grōn mē'an·dər }

inguinal canal [ANAT] A short, narrow passage between the abdominal ring and the inguinal ring in which lies the spermatic cord in males and the round ligament in females. { 'iŋ·gwən·əl kə'nal }

inguinal fold [EMBRYO] A fold of embryonic tissue on the urogenital ridge in which the gubernaculum testis develops. { 'iŋ·gwən·əl 'fōld }

inguinal gland [ANAT] Any of the superficial lymphatic glands in the groin. { 'iŋ·gwən·əl ‚gland }

inguinal hernia [MED] Protrusion of the abdominal viscera through the inguinal canal. { 'iŋ·gwən·əl 'her·nē·ə }

inguinal ligament [ANAT] The thickened lower portion of the aponeurosis of the external oblique muscle extending from the anterior superior spine of the ileum to the tubercle of the pubis and the pectineal line. Also known as Poupart's ligament. { 'iŋ·gwən·əl 'lig·ə·mənt }

inguinal lymphadenopathy [MED] Tenderness and swelling of the lymph nodes of the groin. { ‚iŋ·gwən·əl lim‚fad·ən'ä·pə·thē }

inguinal region [ANAT] The abdominal region occurring on each side of the body as a depression between the abdomen and the thigh. Also known as iliac region. { 'iŋ·gwən·əl ‚rē·jən }

inhabited building distance [ENG] The minimum distance permitted between an ammunition or explosive location and any building used for habitation or where people are accustomed to assemble, except operating buildings or magazines. { in'hab·əd·əd ‚bil·diŋ ‚dis·təns }

inhalant canal [INV ZOO] The incurrent canal in sponges and mollusks. { in'hā·lənt kə,nal }

inhalant siphon [INV ZOO] A channel for water intake in the mantle of bivalve mollusks. Also known as incurrent siphon. { in'hā·lənt 'sī·fən }

inhalation [PHYSIO] The process of breathing in. { ‚in·ə'lā·shən }

inhalator [MED] A device for facilitating the inhalation of a gas or spray, as for providing oxygen or oxygen-carbon dioxide mixtures for respiration in resuscitation. { 'in·ə,lād·ər }

inhaler [MED] **1.** A device containing a solid medication through which air is drawn into the air passages. **2.** An atomizer containing a liquid medication. [MIN ENG] An apparatus, of different forms, for permitting the supply of fresh air to a miner. { in'hā·lər }

inhaul cable [MECH ENG] In a cable excavator, the line that pulls the bucket to dig and bring in soil. Also known as digging line. { 'in,hól ‚kā·bəl }

inherent bursts [MIN ENG] Rock bursts that occur in development. { in'hir·ənt 'bərsts }

inherent damping [MECH ENG] A method of vibration damping which makes use of the mechanical hysteresis of such materials as rubber, felt, and cork. { in'hir·ənt 'dam·piŋ }

inherent noise pressure See equivalent noise pressure. { in'hir·ənt 'nóiz ‚presh·ər }

inherent storage [COMPUT SCI] Any type of storage in which the storage medium is part of the hardware of the computer medium. { in'hir·ənt 'stór·ij }

inheritance [COMPUT SCI] A feature of object-oriented programming that allows a new class to be defined simply by stating how it differs from an existing class. [GEN] **1.** The acquisition of characteristics by transmission of particular alleles from ancestor to descendant. **2.** The sum total of characteristics dependent upon the constitution of the sperm-fertilized ovum. { in'her·əd·əns }

inherited error [COMPUT SCI] The error existing in the data supplied at the beginning of a step in a step-by-step calculation as executed by a program. { in'her·əd·əd 'er·ər }

inhibited acid [PETRO ENG] Acid used in an acid fracturing operation that has been treated chemically to diminish its corrosive effect on the piping while retaining its effectiveness in the fracturing process. { in'hib·əd·əd 'as·əd }

inhibited mud [PETRO ENG] A chemically treated drilling fluid that prevents swelling of clay particles in the formation, thus protecting the permeability of a productive zone. { in'hib·əd·əd 'məd }

inhibit-gate [ELECTR] Gate circuit whose output is energized only when certain signals are present and other signals are not present at the inputs. { in'hib·ət‚gāt }

inhibiting antibody [IMMUNOL] A substance sometimes produced in the blood of immunized persons which is thought to prevent the expected antigen-reagin reaction. { in'hib·əd·iŋ 'ant·i‚bäd·ē }

inhibiting input [ELECTR] A gate input which, if in its prescribed state, prevents any output which might otherwise occur. { in'hib·əd·iŋ 'in,put }

inhibiting pigment [MATER] A paint additive that inhibits or prevents rust and corrosion of metals or the formation of mildew, for example, lead chromate. { in'hib·əd·iŋ 'pig‚mənt }

inhibiting signal [ELECTR] A signal, which when entered into a specific circuit will prevent the circuit from exercising its normal function; for example, an inhibit signal fed into an AND gate will prevent the gate from yielding an output when all normal input signals are present. { in'hib·əd·iŋ ‚sig·nəl }

inhibition [PSYCH] An unconscious mechanism for restraining an impulse by means of an opposing impulse. [SCI TECH] The act of repressing or restraining a physical or chemical action. { ‚in·ə'bish·ən }

inhibition-action balance [PSYCH] Relative balance maintained in every individual between his experiencing of emotional feelings and his outward behavior in response to them. { ‚in·ə'bish·ən 'ak·shən ‚bal·əns }

inhibition immunonephelometry [IMMUNOL] A technique that uses a constant amount of antibody and a predetermined quantity of the same reactant as that assayed, chemically coupled to a macromolecular carrier; used for the assay of substances of low molecular weight. { ‚in·ə‚bish·ən ‚im·yə·nō‚ne·fə'läm·ə·trē }

inhibition index [BIOCHEM] The amount of antimetabolite that will overcome the biological effect of a unit weight of metabolite. { ‚in·ə'bish·ən ‚in‚deks }

inhibitor [AERO ENG] A substance bonded, taped, or dipdried onto a solid propellant to restrict the burning surface and to give direction to the burning process. [CHEM] A substance which is capable of stopping or retarding a chemical reaction; to be technically useful, it must be effective in low concentration. { in'hib·əd·ər }

inhibitor sweetening [CHEM ENG] Petroleum-refinery treating process to sweeten gasoline (convert mercaptans to disulfides) of low mercaptan content; uses a phenylenediamine inhibitor, air, and caustic. { in'hib·əd·ər ‚swēt·ən·iŋ }

inhibitory G protein [CELL MOL] A guanine nucleotide-binding protein that lowers cellular levels of the second messenger cyclic adenosine monophosphate by inhibiting adenyl cyclase, the enzyme that catalyzes its conversion from adenosine triphosphate. { in‚hib·ə‚tór·ē 'jē ‚prō‚tēn }

inhibitory postsynaptic potential [NEUROSCI] A transient, graded hyperpolarization of the postsynaptic membrane, mediated by a chemical neurotransmitter, in response to action potentials arriving at the endings of the presynaptic neurons. { in'hib·ə·tór·ē pōst·sə'nap·tik pə'ten·chəl }

inhibit pulse [ELECTR] A drive pulse that tends to prevent flux reversal of a magnetic cell by certain specified drive pulses. { in'hib·ət ‚pəls }

inhomogeneous line-broadening [OPTICS] An increase beyond the natural linewidth in the width of an absorption or emission line that results from a disturbance (such as strains or imperfections) that can differ from one source emitter to another. { ‚in,hä·mə'jē·nē·əs lin ‚bród·ən·iŋ }

inhour [NUCLEO] A unit of reactivity of a reactor; 1 inhour is the reactivity that will give the reactor a period of 1 hour. Derived from inverse hour. { ‚in,aúr }

inhour equation [NUCLEO] An equation relating the reactivity of a nuclear reactor to the parameters of the delayed-neutron emitters and the neutron lifetime of the reactor. { ‚in,aúr i'kwā·zhən }

in-house [IND ENG] Pertaining to an operation produced or carried on within a plant or organization, rather than done elsewhere under contract. { ‚in,haús }

in-house scrap [MET] Metal that has been shaved, cropped, or slit off in various stages of casting and rolling of ingots into sheets. Also known as runaround scrap. { ‚in,haús 'skrap }

Iniomi [VERT ZOO] An equivalent name for Salmoniformes. { ¦in·ē'ō¦mī }

inion [ANTHRO] The external occipital protuberance of the skull. { 'in·ē¸än }

initial aiming point [ORD] Point on which a gun is sighted to establish a reference line from which direction angles for targets are measured; from this reference line, other aiming points that give the direction of the targets are measured off. { i'nish·əl 'ām·iŋ ¸póint }

initial boiling point [CHEM ENG] According to American Society for Testing and Materials petroleum-analysis distillation procedures, the recorded temperature when the first drop of distilled vapor is liquefied and falls from the end of the condenser. { i'nish·əl 'bóil·iŋ ¸póint }

initial condition [COMPUT SCI] See entry condition. [METEOROL] A prescription of the state of a dynamical system at some specified time; for all subsequent times the equations of motion and boundary conditions determine the state of the system; the appropriate synoptic weather charts, for example, constitute a (discrete) set of initial conditions for a forecast; in many contexts, initial conditions are considered as boundary conditions in the dimension of time. { i'nish·əl kən'dish·ən }

initial condition mode See reset mode. { i'nish·əl kən'dish·ən ¸mōd }

initial detention See surface storage. { i'nish·əl di'ten·chən }

initial dip See primary dip. { i'nish·əl 'dip }

initial free space [MECH] In interior ballistics, the portion of the effective chamber capacity not displaced by propellant. { i'nish·əl ¦frē 'spās }

initial graphics exchange specification [COMPUT SCI] A standard graphics file format for three-dimensional wire-frame models. Abbreviated IGES. { i¦nish·əl ¦graf·iks iks¸chānj ¸spes·ə·fə'kā·shən }

initial great-circle course [NAV] The direction, at the point of departure, of the great circle through that point and the destination, expressed as the angular distance from a reference direction, usually north, to that part of the great circle extending toward the destination. Also known as initial great-circle direction. { i'nish·əl 'grāt ¸sər·kəl ¸kórs }

initial great-circle direction See initial great-circle course. { i'nish·əl 'grāt ¸sər·kəl di¸rek·shən }

initial heading [NAV] The aircraft heading at the beginning of a rating period while using gyro steering. { i'nish·əl 'hed·iŋ }

initial instructions [COMPUT SCI] A routine stored in a computer to aid in placing a program in memory. Also known as initial orders. { i'nish·əl in'strək·shənz }

initial inverse voltage [ELECTR] Of a rectifier tube, the peak inverse anode voltage immediately following the conducting period. { i'nish·əl ¦in¸vərs 'vōl·tij }

initialize [COMPUT SCI] **1.** To set counters, switches, and addresses to zero or other starting values at the beginning of, or at prescribed points in, a computer routine. **2.** To begin an operation, and more specifically, to adjust the environment to the required starting configuration. { i'nish·ə¸līz }

initial landform [GEOL] A landform that is produced directly by epeirogenic, orogenic, or volcanic activity, and whose original features are only slightly modified by erosion. { i'nish·əl 'land¸fórm }

initial lead [ORD] The amount a gun is pointed in front of, above, or below a moving target when opening fire; this amount allows for the distance the target will travel while the projectile is in flight. { i'nish·əl 'lēd }

initial line [MATH] One of the two rays that form an angle and that may be regarded as remaining stationary while the other ray (the terminal line) is rotated about a fixed point on it to form the angle. { i¦nish·əl 'līn }

initial lock mechanism [ORD] Device for preventing inadvertent motion of stroking member in a cartridge-actuated device prior to firing. { i'nish·əl 'läk ¸mek·ə¸niz·əm }

initial mass [AERO ENG] The mass of a rocket missile at the beginning of its flight. { i'nish·əl 'mas }

initial mass function [ASTRON] The distribution of the masses of stars at the time of their formation. { i'nish·əl 'mas ¸fəŋk·shən }

initial nuclear radiation [NUCLEO] Radiation emitted from the fireball of a nuclear explosive during the first minute (an arbitrary time interval) after detonation. { i'nish·əl ¦nü·klē·ər ¸rād·ē'ā·shən }

initial orders See initial instructions. { i'nish·əl 'ór·dərz }

initial permeability [ELECTROMAG] The limit of the normal permeability as the magnetic induction and magnetic field strength approach 0. { i¦nish·əl ¸pər·mē·ə'bil·əd·ē }

initial potential [PETRO ENG] The early production of an oil well as recorded following testing operations and recovery of load oil; indicates the production ability of the well. { i'nish·əl pə'ten·chəl }

initial program load [COMPUT SCI] A routine, used in starting up a computer, that loads the operating system from a direct-access storage device, usually a disk or diskette, into the computer's main storage. Abbreviated IPL. { i'nish·əl 'prō·grəm ¸lōd }

initial program load button See bootstrap button. { i'nish·əl 'prō·grəm ¸lōd ¦bət·ən }

initial saturation [PETRO ENG] A reservoir's initial relative content (saturation) of water, oil, and gas. { i'nish·əl ¸sach·ə'rā·shən }

initial set [MATER] The onset of hardening after water has been added to concrete, cement, or plaster. { i'nish·əl 'set }

initial shot start pressure [MECH] In interior ballistics, the pressure required to start the motion of the projectile from its initial loaded position; in fixed ammunition, it includes pressure required to separate projectile and cartridge case and to start engraving the rotating band. { i'nish·əl 'shät ¦stärt ¸presh·ər }

initial surge voltage [ELEC] A spike of voltage experienced when a noncompensated load is first connected to a generator. { i'nish·əl 'sərj ¸vōl·tij }

initial-value problem [FL MECH] A dynamical problem whose solution determines the state of a system at all times subsequent to a given time at which the state of the system is specified by given initial conditions; the initial-value problem is contrasted with the steady-state problem, in which the state of the system remains unchanged in time. Also known as transient problem. [MATH] An nth-order ordinary or partial differential equation in which the solution and its first $(n − 1)$ derivatives are required to take on specified values at a particular value of a given independent variable. { i'nish·əl ¦val·yü ¸präb·ləm }

initial-value theorem [MATH] The theorem that, if a function $f(t)$ and its first derivative have Laplace transforms, and if $g(s)$ is the Laplace transform of $f(t)$, and if the limit of $sg(s)$ as s approaches infinity exists, then this limit equals the limit of $f(t)$ as t approaches zero. { i'nish·əl ¦val·yü ¸thir·əm }

initial velocity [PHYS] The velocity of anything at the beginning of a specific phase of its motion. { i'nish·əl və'läs·əd·ē }

initial yaw [MECH] The yaw of a projectile the instant it leaves the muzzle of a gun. { i'nish·əl 'yó }

initiate See trigger. { i'nish·ē¸āt }

initiating agent [MATER] An explosive material which has the necessary sensitivity to heat, friction, or percussion to make it suitable for use as the initial element in an explosive train. { i'nish·ē¸ād·iŋ ¸ā·jənt }

initiation [ORD] **1.** As applied to an explosive item, the beginning of the deflagration or detonation of the explosive. **2.** The first action in a fuse which occurs as a direct result of the action of the functioning medium. **3.** In a time fuse, the starting of the action which is terminated in the functioning of the fused munition. { i¸nish·ē'ā·shən }

initiation codon [GEN] A codon that signals the first amino acid in a protein sequence; usually *AUG*, but sometimes *GUG*. Also known as start codon. { i¸nish·ē'ā·shən 'kō¸dän }

initiation complex [CELL MOL] An intermediate of protein synthesis consisting of messenger ribonucleic acid, initiator codons, initiation factors, and initiator transfer ribonucleic acid. { i¸nish·ē'ā·shən ¸käm¸pleks }

initiation factor [CELL MOL] Any protein required for the initiation of protein synthesis. { i¸nish·ē'ā·shən ¸fak·tər }

initiation step [CHEM] The reaction that causes a chain reaction to begin but is not itself the principal source of products. { i¸nish·ē'ā·shən ¸step }

initiator [CHEM] The substance or molecule (other than reactant) that initiates a chain reaction, as in polymerization; an example is acetyl peroxide. [COMPUT SCI] A part of an operating system of a large computer that runs several jobs at the same time, setting up the job, monitoring its progress, and performing any necessary cleanup after the job's completion. [ORD] A device used as the first element of an explosive train, such as a detonator or squib, which upon receipt of the proper mechanical or electrical impulse produces a burning or detonating action; it generally contains a small quantity of a sensitive explosive. { i′nish·ē‚ād·ər }

initiator codon [CELL MOL] A codon that acts as a start signal for the synthesis of a polypeptide. { i′nish·ē‚ād·ər ′kō‚dän }

initiator ribonucleic acid [CELL MOL] An oligoribonucleotide that primes the initiation of Okazaki fragments during deoxyribonucleic acid synthesis. { i‚nish·ē‚ād·ər ‚rī·bō·nü‚klē·ik ′as·əd }

initiator tRNA [CELL MOL] A special type of transfer ribonucleic acid (RNA) that initiates protein synthesis by binding to the amino acid methionine and delivering it to the small ribosomal subunit. { i‚nish·ē‚ād·ər ‚tē‚ār‚en′ā }

injected [PETR] Pertaining to intrusive igneous rock or other mobile rock that has erupted through rock walls to neighboring older rocks. { in′jek·təd }

injected gas [PETRO ENG] Gas that has been pumped into an oil-producing reservoir to provide a gas-drive for increased oil production. { in′jek·təd ′gas }

injected hole [MIN ENG] A borehole into which a cement slurry or grout has been forced by high-pressure pumps and allowed to harden. { in′jek·təd ′hōl }

injection [AERO ENG] The process of placing a spacecraft into a specific trajectory, such as an earth orbit or an encounter trajectory to Mars. Also known as insertion. [ELECTR] **1.** The method of applying a signal to an electronic circuit or device. **2.** The process of introducing electrons or holes into a semiconductor so that their total number exceeds the number present at thermal equilibrium. [GEOL] Also known as intrusion; sedimentary injection. **1.** A process by which sedimentary material is forced under abnormal pressure into a preexisting rock or deposit. **2.** A structure formed by an injection process. [MATH] A mapping f from a set A into a set B which has the property that for any element b of B there is at most one element a of A for which $f(a) = b$. Also known as injective mapping; one-to-one mapping; univalent function. [MECH ENG] The introduction of fuel, fuel and air, fuel and oxidizer, water, or other substance into an engine induction system or combustion chamber. [MED] **1.** Introduction of a fluid into the skin, vessels, muscle, subcutaneous tissue, or any cavity of the body. **2.** The substance injected. [MIN ENG] The introduction under pressure of a liquid or plastic material into cracks, cavities, or pores in a rock formation. { in′jek·shən }

injection blow molding [ENG] Plastics molding process in which a hollow-plastic tube is formed by injection molding. { in′jek·shən ′blō ‚mōl·diŋ }

injection carburetor [MECH ENG] A carburetor in which fuel is delivered under pressure into a heated part of the engine intake system. Also known as pressure carburetor. { in′jek·shən ′kär·bə‚rād·ər }

injection chimera [BIOL] A chimera produced experimentally by inserting embryonic cells of different genetic makeup into the preimplantation blastocyst. { in′jek·shən kī‚mir·ə }

injection efficiency [ELECTR] A measure of the efficiency of a semiconductor junction when a forward bias is applied, equal to the current of injected minority carriers divided by the total current across the junction. { in′jek·shən ə‚fish·ən·sē }

injection electroluminescence [ELECTR] Radiation resulting from recombination of minority charge carriers injected in a *pn* or *pin* junction that is biased in the forward direction. Also known as Lossev effect; recombination electroluminescence. { in′jek·shən i‚lek·trō‚lü·mə′nes·əns }

injection fluid [PETRO ENG] Gas or water, depending on the nature of the reservoir and its fluid content, for injection into the formation to increase hydrocarbon production. { in′jek·shən ‚flü·əd }

injection-fluid front [PETRO ENG] The moving interfacial contact between an injected fluid (gas or water) and the natural fluid content of the reservoir formation. { in′jek·shən ‚flü·əd ′frənt }

injection gas-fluid ratio [PETRO ENG] The ratio of gas injected into a reservoir formation to the fluid hydrocarbons produced by the resultant gas lift. { in′jek·shən ′gas ‚flü·əd ′rā·shō }

injection gneiss [PETR] A composite rock with banding entirely or partly caused by layer-by-layer injection of granitic magma into rock layers. { in′jek·shən ′nīs }

injection grid [ELECTR] Grid introduced into a vacuum tube in such a way that it exercises control over the electron stream without causing interaction between the screen grid and control grid. { in′jek·shən ‚grid }

injection laser [OPTICS] A laser in which a forward-biased gallium arsenide diode converts direct-current input power directly into coherent light, without optical pumping. { in′jek·shən ‚lā·zər }

injection locking [ELECTR] The capture or synchronization of a free-running oscillator by a weak injected signal at a frequency close to the natural oscillator frequency or to one of its subharmonics; used for frequency stabilization in IMPATT or magnetron microwave oscillators, gas-laser oscillators, and many other types of oscillators. { in′jek·shən ‚läk·iŋ }

injection luminescent diode [ELECTR] Gallium arsenide diode, operating in either the laser or the noncoherent mode, that can be used as a visible or near-infrared light source for triggering such devices as light-activated switches. { in′jek·shən ‚lü·mə‚nes·ənt ′dī‚ōd }

injection mold [ENG] A plastics mold into which the material to be formed is introduced from an exterior heating cylinder. { in′jek·shən ‚mōld }

injection molding [ENG] Molding metal, plastic, or nonplastic ceramic shapes by injecting a measured quantity of the molten material into dies. { in′jek·shən ′mōl·diŋ }

injection pressure [PETRO ENG] Pressure of fluid injected into oil formations for waterflood (water) or pressure maintenance (gas). { in′jek·shən ′presh·ər }

injection pump [MECH ENG] A pump that forces a measured amount of fuel through a fuel line and atomizing nozzle in the combustion chamber of an internal combustion engine. { in′jek·shən ′pəmp }

injection ram [ENG] In injection molding, the ram that applies pressure to the feed plunger in the process of either injection or transfer molding. { in′jek·shən ‚ram }

injection signal [ENG ACOUS] The sawtooth frequency-modulated signal which is added to the first detector circuit for mixing with the incoming target signal. { in′jek·shən ‚sig·nəl }

injection temperature [NAV ARCH] The temperature of the sea water as measured at the sea-water intakes in the engine room of a vessel. { in′jek·shən ‚tem·prə·chər }

injection well [HYD] *See* recharge well. [PETRO ENG] In secondary recovery of petroleum, a well in which a fluid such as gas or water is injected to provide supplemental energy to drive the oil remaining in the reservoir to the vicinity of production wells. { in′jek·shən ‚wel }

injection-well plugging [PETRO ENG] Plugging of the sand face of an injection well because of lubricant or corrosion-product carryover from surface lines or well equipment. { in′jek·shən ‚wel ‚pləg·iŋ }

injective mapping *See* injection. { in′jek·tiv ′map·iŋ }

injectivity index [PETRO ENG] The number of barrels per day of gross liquid pumped into an injection well per psi (pound per square inch) pressure differential between the mean injection pressure and the mean formation pressure. { in‚jek′tiv·əd·ē ‚in‚deks }

injectivity test [PETRO ENG] A test series of reservoir water injection rates at different pressures to predict the performance of an injection well. { in‚jek′tiv·əd·ē ‚test }

injector [ELECTR] An electrode through which charge carriers (holes or electrons) are forced to enter the high-field region in a spacistor. [MECH ENG] **1.** An apparatus containing a nozzle in an actuating fluid which is accelerated and thus entrains a second fluid, so delivering the mixture against a pressure in excess of the actuating fluid. **2.** A plug with a valved nozzle through which fuel is metered to the combustion chambers in diesel- or full-injection engines. **3.** A jet through which feedwater is injected into a boiler, or fuel is injected into a combustion chamber. { in′jek·tər }

injector torch *See* low-pressure torch. { in′jek·tər ‚tȯrch }

injury [MED] **1.** A structural or functional stress or trauma that induces a pathologic process. **2.** Damage resulting from the stress. { ′in·jə‚rē }

injury current *See* injury potential. { ′in·jə‚rē ‚kə·rənt }

injury potential [PHYSIO] The potential difference observed between the injured and the noninjured regions of an injured tissue or cell. Also known as demarcation potential; injury current. { ′in·jə‚rē pə‚ten·chəl }

ink [MATER] A dispersion of a pigment or a solution of a dye in a carrier vehicle, yielding a fluid, paste, or powder to be applied to and dried on a substrate; writing, marking, drawing, and printing inks are applied by several methods to paper, metal, plastic, wood, glass, fabric, or other substrate. { 'iŋk }

ink bleed [COMPUT SCI] In character recognition, the capillary extension of ink beyond the original edges of a printed or handwritten character. { 'iŋk ‚blēd }

ink disease [PL PATH] A fungus disease of the chestnut in Europe caused by *Phytophthora cambivora* which produces black cankers and a black exudate in the trunk. Also known as black canker. { 'iŋk di‚zēz }

ink fountain [GRAPHICS] A device on a printing press for storing ink and supplying it to the rollers. { 'iŋk ‚faůnt·ən }

ink-jet printer [GRAPHICS] A nonimpact printer that uses electrostatic acceleration and deflection of ink particles emerging from nozzles to form characters on plain paper in a dot-matrix format. { 'iŋk ‚jet ‚print·ər }

ink knife [GRAPHICS] An instrument resembling a large spatula, with a handle and a blade with square or rounded end; used for handling paste inks. Also known as ink slice. { 'iŋk ‚nīf }

ink-mist recording *See* ink-vapor recording. { 'iŋk ‚mist ri'kórd·iŋ }

inkometer [ENG] An instrument for measuring adhesion of liquids by rotating drums in contact with the liquid. { iŋ'käm·əd·ər }

ink recorder [GRAPHICS] A recorder that employs an ink-filled pen or capillary tube to produce the graphic record. { 'iŋk ri‚kórd·ər }

ink sac [INV ZOO] An organ attached to the rectum in many cephalopods which secretes and ejects an inky fluid. { 'iŋk ‚sak }

ink slice *See* ink knife. { 'iŋk ‚slīs }

ink smudge [COMPUT SCI] In character recognition, the overflow of ink beyond the original edges of a printed or handwritten character. { 'iŋk ‚sməj }

ink squeezeout [COMPUT SCI] In character recognition, the overflow of ink from the stroke centerline to the edges of a printed or handwritten character. { 'iŋk ‚skwē‚zaůt }

ink-vapor recording [GRAPHICS] A type of recording in which vaporized ink particles are directly deposited on the record sheet. Also known as ink-mist recording. { 'iŋk ‚vā·pər ri'kórd·iŋ }

inland [GEOGR] Interior land, not bordered by the sea. { 'in·lənd }

inland ice [HYD] Ice composing the inner portion of a continental glacier or large ice sheet; applied particularly to Greenland ice. { 'in·lənd 'īs }

inland rules of the road [NAV] Rules to be followed by all vessels while navigating upon certain inland waters of the United States. { 'in·lənd ‚rülz əv thə 'rōd }

inland sea *See* epicontinental sea. { 'in·lənd 'sē }

inland water [GEOGR] **1.** A lake, river, or other body of water wholly within the boundaries of a state. **2.** An interior body of water not bordered by the sea. { 'in·lənd 'wód·ər }

inlay [GRAPHICS] A picture or ornament made by inserting a material such as metal into a space in metal, stone, or wood; the material (such as wire) may be burnished, heated, or fused. { 'in‚lā }

inlay cladding [MET] A mechanical process in which a groove, $^7/_{100}$–$^1/_8$ inch (1.778–3.175 millimeters) wide, is cut into a base metal and filled with cladding metal; mechanical bonding of the metals is accomplished by passing them through the pressure rolls of a bonding mill. { 'in‚lā 'klad·iŋ }

inlet [ENG] An entrance or orifice for the admission of fluid. [GEOGR] **1.** A short, narrow waterway connecting a bay or lagoon with the sea. **2.** A recess or bay in the shore of a body of water. **3.** A waterway flowing into a larger body of water. { 'in‚let }

inlet box [MECH ENG] A closure at the fan inlet or inlets in a boiler for attachment of the fan to the duct system. { 'in‚let ‚bäks }

inlet of the pelvis [ANAT] The space within the brim of the pelvis. { 'in‚let əv thə 'pel·vəs }

inlet valve [MECH ENG] The valve through which a fluid is drawn into the cylinder of a positive-displacement engine, pump, or compressor. Also known as induction valve. { 'in‚let ‚valv }

INKOMETER

The inkometer, used to measure ink tack.

IN-LINE LINKAGE

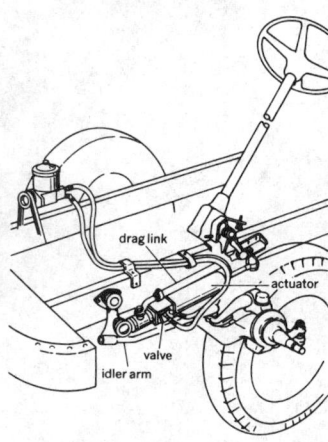

drag link

actuator

valve

idler arm

In-line linkage power steering.

inlier [GEOL] A circular or elliptical area of older rocks surrounded by strata that are younger. { 'in‚lī·ər }

in line [ENG] **1.** Over the center of a borehole and parallel with its long axis. **2.** Of a drill motor, mounted so that its drive shaft and the drive rod in the drill swivel head are parallel, or mounted so that the shaft driving the drill-swivel-head bevel gear and the drill-motor drive shaft are centered in a direct line and parallel with each other. **3.** Having similar units mounted together in a line. { 'in ‚līn }

in-line assembly machine [IND ENG] An assembly machine that inserts components into a wiring board one at a time as the board is moved from station to station by a conveyor or other transport mechanism. { 'in ‚līn ə‚sem·blē mə‚shēn }

in-line coding [COMPUT SCI] Any group of instructions within the main body of a program. { 'in ‚līn 'kōd·iŋ }

in-line engine [MECH ENG] A multiple-cylinder engine with cylinders aligned in a row. { 'in ‚līn 'en·jən }

in-line equipment [ENG] **1.** A sequence of equipment or processing items mounted along the same vertical or horizontal plane. **2.** Equipment mounted within a process line, such as an in-line pump, pressure-drop flowmeter, or nozzle mixer. { 'in ‚līn i'kwip·mənt }

in-line guns [ELECTR] An arrangement of three electron guns in a horizontal line; used in color picture tubes that have a slot mask in front of vertical color phosphor stripes. { 'in ‚līn 'gənz }

in-line linkage [MECH ENG] A power-steering linkage which has the control valve and actuator combined in a single assembly. { 'in ‚līn 'liŋ·kij }

in-line procedure [COMPUT SCI] A short body of coding or instruction which accomplishes some purpose. { 'in ‚līn prə'sē·jər }

in-line processing [COMPUT SCI] The processing of data in random order, not subject to preliminary editing or sorting. { 'in ‚līn 'prä·ses·iŋ }

in-line subroutine [COMPUT SCI] A subroutine which is an integral part of a program. { 'in ‚līn 'səb·rü‚tēn }

in-line tuning [ELECTR] Method of tuning the intermediate-frequency strip of a superheterodyne receiver in which all the intermediate-frequency amplifier stages are made resonant to the same frequency. { 'in ‚līn 'tün·iŋ }

innage [ENG] The volume or the measured height of liquid introduced into a tank or container. { 'in·ij }

innate [BIOL] Pertaining to a natural or inborn character dependent on genetic constitution. [BOT] Positioned at the apex of a supporting structure. [MYCOL] Embedded in, especially of an organ such as the fruiting body embedded in the thallus of some fungi. { i'nāt }

inner automorphism [MATH] An automorphism h of a group where $h(g) = g_0^{-1} \cdot g \cdot g_0$, for every g in the group with g_0 some fixed group element. { ‚in·ər ‚ód·ō'mór‚fiz·əm }

inner barrel *See* inner tube. { ‚in·ər 'bar·əl }

inner bottom [NAV ARCH] A watertight plating laid over the frames and longitudinals, forming the inner layer of the double bottom of the hull. { ‚in·ər 'bäd·əm }

inner bremsstrahlung [NUC PHYS] The emission of a photon during beta decay or electron capture by a nucleus. Also known as internal bremsstrahlung. { ‚in·ər 'brem‚shträ·ləŋ }

inner cell mass [EMBRYO] The cells at the animal pole of a blastocyst which give rise to the embryo and certain extraembryonic membranes. { ‚in·ər 'sel ‚mas }

inner core [GEOL] The central part of the earth's core, extending from a depth of 3160 miles (5100 kilometers) to the center of the earth. Also known as siderosphere. { ‚in·ər 'kór }

inner ear [ANAT] The part of the vertebrate ear concerned with labyrinthine sense and sound reception; consists generally of a bony and a membranous labyrinth, made up of the vestibular apparatus, three semicircular canals, and the cochlea. Also known as internal ear. { ‚in·ər 'ir }

inner function [MATH] A continuous open mapping of a topological space X into a topological space Y where the inverse image of each point in Y is zero dimensional. { ‚in·ər 'fəŋk·shən }

inner harbor [GEOGR] The part of a harbor more remote from the sea, as contrasted with the outer harbor; this expression is normally used only in a harbor that is clearly divided into parts, by a narrow passageway or artificial structure; the inner harbor generally has additional protection and is often the principal berthing area. { ¦in·ər ¦här·bər }

inner hearth See back hearth. { ¦in·ər ¦härth }

inner keel [NAV ARCH] The inner plate of a double, flat plate keel. { ¦in·ər ¦kēl }

inner Lagrangian point [ASTRON] A Lagrangian point that lies between two primary bodies on the line passing through their centers of mass, and through which mass transfer may occur between them. Also known as conical point. { ¦in·ər lə'gran·jē·ən ˌpóint }

inner mantle See lower mantle. { ¦in·ər ¦mant·əl }

inner marker [NAV] A 75-megahertz marker beacon normally used with the instrument landing system (ILS) to indicate that the aircraft is over the boundary of the airport. { ¦in·ər 'mär·kər }

inner measure See Lebesgue interior measure. { ˌin·ər 'mezh·ər }

inner planet [ASTRON] Any of the four planets (Mercury, Venus, Earth, and Mars) in the solar system whose orbits are closest to the sun. { ¦in·ər 'plan·ət }

inner potential [SOLID STATE] The average value of the electrostatic potential, taken over the volume of a crystal. { ¦in·ər pə'ten·chəl }

inner product [MATH] **1.** A scalar valued function of pairs of vectors from a vector space, denoted by (x,y) where x and y are vectors, and with the properties that (x,x) is always positive and is zero only if $x = 0$, that $(ax + by,z) = a(x,z) + b(y,z)$ for any scalars a and b, and that $(x,y) = (y,x)$ if the scalars are real numbers, $(x,y) = \overline{(y,x)}$ if the scalars are complex numbers. Also known as Hermitian inner product; Hermitian scalar product. **2.** The inner product of vectors $(x_1, \ldots, x_n)$ and $(y_1, \ldots, y_n)$ from n-dimensional euclidean space is the sum of $x_i y_i$ as i ranges from 1 to n. Also known as dot product; scalar product. **3.** The inner product of two functions f and g of a real or complex variable is the integral of $f(x)\overline{g(x)}dx$, where $\overline{g(x)}$ denotes the conjugate of $g(x)$. **4.** The inner product of two tensors is the contracted tensor obtained from their product by means of pairing contravariant indices of one with covariant indices of the other. { ¦in·ər 'präd·əkt }

inner product space [MATH] A vector space that has an inner product defined on it. Also known as generalized Euclidean space; Hermitian space; pre-Hilbert space. { ¦in·ər 'präd·əkt ˌspās }

inner quantum number [ATOM PHYS] A quantum number J which gives an atom's total angular momentum, excluding the nuclear spin. { ¦in·ər 'kwänt·əm ˌnəm·bər }

inner strake [NAV ARCH] The inner part of an in and out system of shell plating; the strakes adjacent to the molded frame line. { ¦in·ər 'strāk }

inner tube [ENG] A rubber tube used inside a pneumatic tire casing to hold air under pressure. Also known as tube. [MIN ENG] The inside tube which acts as the core container of a double-tube core barrel; used to obtain core samples for analysis of an ore formation. Also known as inner barrel. { 'in·ər ˌtüb }

inner-tube extension See lifter case. { 'in·ər ˌtüb ik'sten·shən }

innervation [ANAT] The distribution of nerves to a part. [PHYSIO] The amount of nerve stimulation received by a part. { ˌin·ər'vā·shən }

innominate See hipbone. { i¦näm·ə·nət }

innominate artery [ANAT] The first artery branching from the aortic arch; distributes blood to the head, neck, shoulder, and arm on the right side of the body. { i¦näm·ə·nət 'ärd·ə·rē }

inoculant [MET] A substance which augments a melt, usually in the latter part of the melting operation, thus altering the solidification structure of the cast metal, as in grain refinement of aluminum alloys. { i'näk·yə·lənt }

inoculation [BIOL] Introduction of a disease agent into an animal or plant to produce a mild form of disease and render the individual immune. [MET] Treating a molten material with another material before casting in order to nucleate crystals. [MICROBIO] Introduction of microorganisms onto or into a culture medium. { i¦näk·yə'lā·shən }

inoculum [MICROBIO] A small amount of substance containing bacteria from a pure culture which is used to start a new culture or to infect an experimental animal. { i'näk·yə·ləm }

inoperculate [BIOL] Lacking an operculum. { ¦in·ä'pər·kyə·lət }

inorganic [INORG CHEM] Pertaining to or composed of chemical compounds that do not contain carbon as the principal element (excepting carbonates, cyanides, and cyanates), that is, matter other than plant or animal. { ¦in·ór¦gan·ik }

inorganic acid [INORG CHEM] A compound composed of hydrogen and a nonmetal element or radical; examples are hydrochloric acid, HCl, sulfuric acid, H_2SO_4, and carbonic acid, H_2CO_3. { ¦in·ór¦gan·ik 'as·əd }

inorganic biochemistry See bioinorganic chemistry. { ˌin·ór¦gan·ik ¦bī·ō'kem·ə·strē }

inorganic chemistry [CHEM] The study of chemical reactions and properties of all the elements and their compounds, with the exception of hydrocarbons, and usually including carbides, oxides of carbon, metallic carbonates, carbon-sulfur compounds, and carbon-nitrogen compounds. { ¦in·ór¦gan·ik 'kem·ə·strē }

inorganic chert [PETR] Chert derived from siliceous colloids precipitated from silica-saturated waters. { ¦in·ór¦gan·ik 'chərt }

inorganic liquid laser [OPTICS] A liquid laser in which an inorganic liquid such as neodymium-selenium oxychloride or neodymium-doped phosphorus chloride is used as the active material. Also known as neodymium liquid laser. { ¦in·ór¦gan·ik ¦lik·wəd 'lā·zər }

inorganic peroxide [INORG CHEM] An inorganic compound containing an element at its highest state of oxidation (such as perchloric acid, $HClO_4$), or having the peroxy group, $-O-O-$ (such as perchromic acid, $H_3CrO_8 \cdot 2H_2O$). { ¦in·ór¦gan·ik pə'räk·ˌsīd }

inorganic pigment [INORG CHEM] A natural or synthetic metal oxide, sulfide, or other salt used as a coloring agent for paints, plastics, and inks. { ¦in·ór¦gan·ik 'pig·mənt }

inorganic polymer [INORG CHEM] Large molecules, usually linear or branched chains with atoms other than carbon in their backbone; an example is glass, an inorganic polymer made up of rings and chains of repeating silicate units. { ¦in·ór¦gan·ik 'päl·ə·mər }

inosculation See anastomosis. { in¦äs·kyə'lā·shən }

inosilicate [GEOL] A class or structural type of silicate in which the SiO_4 tetrahedrons are linked together by the sharing of oxygens to form linear chains of indefinite length. { ¦in·ō'sil·ə·ˌkāt }

inosine [BIOCHEM] $C_{10}H_{12}N_4C_5$ A compound occurring in muscle; a hydrolysis product of inosinic acid. { 'in·ə·ˌsēn }

inosinic acid [BIOCHEM] $C_{10}H_{13}N_4O_8P$ A nucleotide constituent of muscle, formed by deamination of adenylic acid; on hydrolysis it yields hypoxanthine and D-ribose-5-phosphoric acid. { 'in·ə¦sin·ik 'as·əd }

inositol [ORG CHEM] $C_6H_6(OH)_6 \cdot 2H_2O$ A water-soluble alcohol often grouped with the vitamins; there are nine stereoisomers of hexahydroxycyclohexane, and the only one of biological importance is optically inactive *meso*-inositol, comprising white crystals, widely distributed in animals and plants; it serves as a growth factor for animals and microorganisms. { i¦näs·ə·ˌtól }

Inoviridae [VIROL] A family of nontailed bacterial viruses (bacteriophages) characterized by a nonenveloped rod-shaped virion containing a single-stranded circular deoxyribonucleic acid genome. { ē·nō'vir·ə·dē }

Inovirus [VIROL] A genus of bacterial viruses of the family Inoviridae that are characterized by semiflexible filamentous virions with helical symmetry. { 'ī·nə·ˌvī·rəs }

in phase [PHYS] Having waveforms that are of the same frequency and that pass through corresponding values at the same instant. { 'in ˌfāz }

in-phase component [ELEC] The component of the phasor representing an alternating current which is parallel to the phasor representing voltage. { 'in ˌfāz kəm'pō·nənt }

in-phase rejection See common-mode rejection. { 'in ˌfāz ri'jek·shən }

in-phase signal See common-mode signal. { 'in ˌfāz 'sig·nəl }

in-pile [NUCLEO] Referring to experiments or equipment inside a reactor. { 'in ˌpīl }

in-pile loop [NUCLEO] An experiment inserted directly in a nuclear reactor (pile) incorporating a closed circuit (loop) of fluid usually for cooling purposes. { 'in ˌpīl ˌlüp }

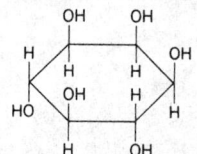

INOSITOL

Structural formula for inositol.

in-place stress field *See* ambient stress field. { ˌin ˈplās ˈstres ˌfēld }

in-place value [IND ENG] The site value of property, that is, the market value of equipment plus costs of transportation to the site and subsequent installation. { ˈinˌplās ˈval·yü }

input [COMPUT SCI] The information that is delivered to a data-processing device from the external world, the process of delivering this data, or the equipment that performs this process. [ELECTR] **1.** The power or signal fed into an electrical or electronic device. **2.** The terminals to which the power or signal is applied. [SCI TECH] Those resources and other environmental factors converted by a system. { ˈinˌpu̇t }

input admittance [ELEC] The admittance measured across the input terminals of a four-terminal network with the output terminals short-circuited. { ˈinˌpu̇t ədˌmit·əns }

input area [COMPUT SCI] A section of internal storage reserved for storage of data or instructions received from an input unit such as cards or tape. Also known as input block; input storage. { ˈinˌpu̇t ˌer·ē·ə }

input block [COMPUT SCI] **1.** A block of data read or transferred into a computer. **2.** *See* input area. { ˈinˌpu̇t ˌbläk }

input capacitance [ELECTR] The short-circuited transfer capacitance that exists between the input terminals and all other terminals of an electron tube (except the output terminal) connected together. { ˈinˌpu̇t kəˈpas·əd·əns }

input data [COMPUT SCI] Data employed as input. { ˈinˌpu̇t ˌdad·ə }

input equipment [COMPUT SCI] **1.** The equipment used for transferring data and instructions into an automatic data-processing system. **2.** The equipment by which an operator transcribes original data and instructions to a medium that may be used in an automatic data-processing system. { ˈinˌpu̇t iˌkwip·mənt }

input gap [ELECTR] An interaction gap used to initiate a variation in an electron stream; in a velocity-modulated tube it is in the buncher resonator. { ˈinˌpu̇t ˌgap }

input impedance [ELEC] The impedance across the input terminals of a four-terminal network when the output terminals are short-circuited. { ˈinˌpu̇t imˌpēd·əns }

input-limited [COMPUT SCI] Pertaining to a system or operation whose speed or efficiency depends mainly on the speed of input into the machine rather than the speed of the machine itself. { ˈinˌpu̇t ˌlim·əd·əd }

input magazine [COMPUT SCI] A part of a card-handling device which supplies the cards to the processing portion of the machine. Also known as magazine. { ˈinˌpu̇t ˌmag·əˌzēn }

input/output [COMPUT SCI] Pertaining to all equipment and activity that transfers information into or out of a computer. Abbreviated I/O. { ˈinˌpu̇t ˈau̇tˌpu̇t }

input/output adapter [COMPUT SCI] A circuitry which allows input/output devices to be attached directly to the central processing unit. { ˈinˌpu̇t ˈau̇tˌpu̇t əˌdap·tər }

input/output area [COMPUT SCI] A portion of computer memory that is reserved for accepting data from input devices and holding data for transfer to output devices. { ˈinˌpu̇t ˈau̇tˌpu̇t ˌer·ē·ə }

input/output bound [COMPUT SCI] Pertaining to a system or condition in which the time for input and output operation exceeds other operations. Also known as input/output limited. { ˈinˌpu̇t ˈau̇tˌpu̇t ˌbau̇nd }

input/output buffer [COMPUT SCI] An area of a computer memory used to temporarily store data and instructions transferred into and out of a computer, permitting several such transfers to take place simultaneously with processing of data. { ˈinˌpu̇t ˈau̇tˌpu̇t ˌbəf·ər }

input/output channel [COMPUT SCI] The physical link connecting the computer to an input device or to an output device. { ˈinˌpu̇t ˈau̇tˌpu̇t ˌchan·əl }

input/output controller [COMPUT SCI] An independent processor which provides the data paths between input and output devices and main memory. { ˈinˌpu̇t ˈau̇tˌpu̇t kənˌtrōl·ər }

input/output control system [COMPUT SCI] A set of flexible routines that supervise the input and output operations of a computer at the detailed machine-language level. Abbreviated IOCS. { ˈinˌpu̇t ˈau̇tˌpu̇t kənˌtrōl ˌsis·təm }

input/output control unit [COMPUT SCI] The piece of hardware which controls the operation of one or more of a type of devices such as tape drives or disk drives; this unit is frequently

an integral part of the input/output device itself. { ˈinˌpu̇t ˈau̇tˌpu̇t kənˈtrōl ˌyü·nət }

input/output device *See* peripheral device. { ˈinˌpu̇t ˈau̇t ˌpu̇t diˌvīs }

input/output generation [COMPUT SCI] A procedure involved in installing an operating system on a large computer, in which addresses and attributes of peripheral equipment under the computer's control are described in a language that can be read by the operating system. Abbreviated IOGEN. { ˈinˌpu̇t ˈau̇tˌpu̇t ˌjen·əˌrā·shən }

input/output instruction [COMPUT SCI] An instruction in a computer program that causes transfer of data between peripheral devices and main memory, and enables the central processing unit to control the peripheral devices connected to it. { ˈinˌpu̇t ˈau̇tˌpu̇t inˌstrək·shən }

input/output interrupt [COMPUT SCI] A technique by which the central processor needs only initiate an input/output operation and then handle other matters, while other units within the system carry out the rest of the operation. { ˈinˌpu̇t ˈau̇tˌpu̇t ˈint·əˌrəpt }

input/output interrupt identification [COMPUT SCI] The ascertainment of the device and channel taking part in the transfer of information into or out of a computer that causes a particular input/output interrupt, and of the status of the device and channel. { ˈinˌpu̇t ˈau̇tˌpu̇t ˈint·əˌrəpt īˌdent·ə·fəˌkā·shən }

input/output interrupt indicator [COMPUT SCI] A device which registers an input/output interrupt associated with a particular input/output channel; it can be used in input/output interrupt identification. { ˈinˌpu̇t ˈau̇tˌpu̇t ˈint·əˌrəpt ˌin·dəˌkād·ər }

input/output library [COMPUT SCI] A set of programs which take over the job from the programmer of creating the required instructions to access the various peripheral devices. Also known as input/output routines. { ˈinˌpu̇t ˈau̇tˌpu̇t ˌlīˌbrer·ē }

input/output limited *See* input/output bound. { ˈinˌpu̇t ˈau̇t ˌpu̇t ˌlim·əd·əd }

input/output order [COMPUT SCI] A procedure of transferring data between main memory and peripheral devices which is assigned to and performed by an input/output controller. { ˈinˌpu̇t ˈau̇tˌpu̇t ˌȯr·dər }

input/output processor [COMPUT SCI] A hardware device or software processor whose sole function is to handle input and output operations. { ˈinˌpu̇t ˈau̇tˌpu̇t ˌpräˌses·ər }

input/output referencing [COMPUT SCI] The use of symbolic names in a computer program to indicate data on input/output devices, the actual devices allocated to the program being determined when the program is executed. { ˈinˌpu̇t ˈau̇tˌpu̇t ˌref·rənˌsiŋ }

input/output register [COMPUT SCI] Computer register that provides the transfer of information from inputs to the central computer, or from it to output equipment. { ˈinˌpu̇t ˈau̇tˌpu̇t ˌrej·əˌstər }

input/output relation [SYS ENG] The relation between two vectors whose components are the inputs (excitations, stimuli) of a system and the outputs (responses) respectively. { ˈinˌpu̇t ˈau̇tˌpu̇t riˌlā·shən }

input/output routines *See* input/output library. { ˈinˌpu̇t ˈau̇t ˌpu̇t rüˌtēnz }

input/output statement [COMPUT SCI] A statement in a computer language that summons data or stores data in a peripheral device. { ˈinˌpu̇t ˈau̇tˌpu̇t ˌstāt·mənt }

input/output switching [COMPUT SCI] A technique in which a number of channels can connect input and output devices to a central processing unit; each device may be assigned to any available channel, so that several different channels may service a particular device during the execution of a program. { ˈinˌpu̇t ˈau̇tˌpu̇t ˌswich·iŋ }

input/output traffic control [COMPUT SCI] The coordination, by both hardware and software facilities, of the actions of a central processing unit and the input, output, and storage devices under its control, in order to permit several input/output devices to operate simultaneously while the central processing unit is processing data. { ˈinˌpu̇t ˈau̇tˌpu̇t ˌtraf·ik kənˌtrōl }

input/output wedge [COMPUT SCI] The characteristic shape of a Kiviat graph of a system which is approaching complete input/output boundedness. { ˈinˌpu̇t ˈau̇tˌpu̇t ˌwej }

input program *See* data entry program. { ˈinˌpu̇t ˌprōˌgrəm }

input record [COMPUT SCI] **1.** A record that is read from an input device into a computer memory during the performance of a program or routine. **2.** A record that has been stored in an input area and is ready to be processed. { 'in,pùt ,rek·ərd }

input register [COMPUT SCI] A register that accepts input information from a computer at one speed and supplies the information to the central processing unit at another speed, usually much greater. { 'in,pùt ,rej·ə·stər }

input resistance *See* transistor input resistance. { 'in,pùt ri,zis·təns }

input resonator *See* buncher resonator. { 'in,pùt 'rēz·ən,ād·ər }

input routine [COMPUT SCI] A routine which controls the loading and reading of programs, data, and other routines into a computer for storage or immediate use. Also known as loading routine. { 'in,pùt rü,tēn }

input section [COMPUT SCI] **1.** The part of a program which controls the reading of data into a computer memory from external devices. **2.** *See* input area. { 'in,pùt ,sek·shən }

input station [COMPUT SCI] A terminal in an in-plant communications system at which data can be entered into the system directly as events take place, enabling files to be immediately updated. { 'in,pùt ,stā·shən }

input storage *See* input area. { 'in,pùt ,stòr·ij }

inquartation [MET] A step in bullion assay that uses nitric acid to dissolve silver from associated gold. Also known as quartation. { ,in,kwòr'tā·shən }

inquiline [ZOO] An animal that inhabits the nest of another species. { 'in·kwə,līn }

inquiry [COMPUT SCI] A request for the retrieval of a particular item or set of items from storage. { in'kwī·ə·rē }

inquiry and communications system [COMPUT SCI] A computer system in which centralized records are maintained with data transmitted to and from terminals at remote locations or in an in-plant system, and which immediately responds to inquiries from remote terminals. { in'kwī·ə·rē ən kə,myü·nə'kā·shənz ,sis·təm }

inquiry and subscriber display [COMPUT SCI] An inquiry display unit that is distant from its computer and communicates with it over wire lines. { in'kwī·ə·rē ən səb'skrīb·ər di,splā }

inquiry display terminal [COMPUT SCI] A cathode-ray-tube terminal which allows the user to query the computer through a keyboard, the answer appearing on the screen. { in'kwī·ə·rē di,splā ,tər·mən·əl }

inquiry station [COMPUT SCI] A remote terminal from which an inquiry may be sent to a computer over wire lines. { in'kwī·ə·rē ,stā·shən }

inquiry unit [COMPUT SCI] Any terminal which enables a user to query a computer and get a hard-copy answer. { in'kwī·ə·rē ,yü·nət }

inradius [MATH] The radius of the inscribed circle of a triangle. { 'in,rād·ē·əs }

inrolling [BOT] Inward rolling of the corolla of a flower, a physical process associated with senescence. { 'in,rōl·iŋ }

insanity [FOREN SCI] In forensic psychiatry, a mental disorder which prevents one from managing one's affairs, impairs one's ability to distinguish right from wrong, or renders one harmful to oneself or others. [PSYCH] Term previously used to indicate mental disorder; no longer used in medical contexts. { in'san·əd·ē }

inscribe [COMPUT SCI] To rewrite data on a document in a form which can be read by an optical or magnetic ink character recognition machine. { in'skrīb }

inscribed circle [MATH] A circle that lies within a given triangle and is tangent to each of its sides. Also known as incircle. { in'skrībd 'sər·kəl }

inscribed polygon [MATH] A polygon that lies within a given circle or curve and whose vertices all lie on the circle or curve. { in'skrībd 'päl·ə,gän }

insect [INV ZOO] **1.** A member of the Insecta. **2.** An invertebrate that resembles an insect, such as a spider, mite, or centipede. { 'in,sekt }

Insecta [INV ZOO] A class of the Arthropoda typically having a segmented body with an external, chitinous covering, a pair of compound eyes, a pair of antennae, three pairs of mouthparts, and two pairs of wings. { in'sek·tə }

insect attractant [MATER] A chemical agent, usually associated with an insect's sexual drive, which may be used to attract pests to poisoned bait or for insect surveys. { 'in,sekt ə,trak·tənt }

insect control [ECOL] Regulation of insect populations by biological or chemical means. { 'in,sekt kən,trōl }

insecticide [MATER] A chemical agent that destroys insects. { in'sek·tə,sīd }

insectistasis [ECOL] The use of pheromones to trap, confuse, or inhibit insects in order to hold populations below a level where they can cause significant economic damage. { in'sek·tə,stā·səs }

Insectivora [VERT ZOO] An order of mammals including hedgehogs, shrews, moles, and other forms, most of which have spines. { in,sek'tiv·ə·rə }

insectivorous [BIOL] Feeding on a diet of insects. { in,sek'tiv·ə·rəs }

insectivorous plant [BOT] A plant that captures and digests insects as a source of nutrients by using specialized leaves. Also known as carnivorous plant. { in,sek'tiv·ə·rəs 'plant }

insect pathology [INV ZOO] A biological discipline embracing the general principles of pathology as applied to insects. { 'in,sekt pə'thäl·ə·jē }

insect physiology [INV ZOO] The study of the functional properties of insect tissues and organs. { 'in,sekt ,fiz·ē'äl·ə·je }

inselberg [GEOL] A large, steep-sided residual hill, knob, or mountain, generally rocky and bare, rising abruptly from an extensive, nearly level lowland erosion surface in arid or semiarid regions. Also known as island mountain. { 'in·səl,bərg }

insemination [BIOL] The planting of seed. [PHYSIO] **1.** The introduction of sperm into the vagina. **2.** Impregnation. { in,sem·ə'nā·shən }

insensitive time *See* dead time. { in'sen·sə·tiv ,tīm }

inseparable degree [MATH] Let E be a finite extension of a field F; the inseparable degree of E over F is the dimension of E viewed as a vector space over F divided by the separable degree of E over F. { in'sep·rə·bəl di'grē }

insequent stream [HYD] A stream that has developed on the present surface, but not consequent upon it, and seemingly not controlled or adjusted by the rock structure and surface features. { in'sē·kwənt 'strēm }

insert [MET] **1.** The part of a die or mold that can be removed. **2.** A part, usually metal, which is placed in a mold and appears as an integral part of the final casting. Also known as bowl. { 'in,sərt }

insert bit [DES ENG] A bit into which inset cutting points of various preshaped pieces of hard metal (usually a sintered tungsten carbide-cobalt powder alloy) are brazed or hand-peened into slots or holes cut or drilled into a blank bit. Also known as slug bit. { 'in,sərt ,bit }

inserted [BIOL] United or attached to the supporting structure by natural growth. { ir'sərd·əd }

inserted-tooth cutter [DES ENG] A milling cutter in which the teeth can be replaced. { in'sərd·əd ,tüth 'kəd·ər }

insert film [GRAPHICS] A microfilm strip cut in lengths to fit a jacket channel length. { 'in,sərt ,film }

insertion [AERO ENG] *See* injection. [ANAT] The point at which a muscle is attached to a bone that moves when the muscle contracts; it is the distal end of the muscle. [CELL MOL] The addition of an extranumerary base pair to double-stranded deoxyribonucleic acid; causes errors in transcription. { in'sər·shən }

insertion gain [ELECTR] The ratio of the power delivered to a part of the system following insertion of an amplifier, to the power delivered to that same part before insertion of the amplifier; usually expressed in decibels. { in'sər·shən ,gān }

insertion loss [ELECTR] The loss in load power due to the insertion of a component or device at some point in a transmission system; generally expressed as the ratio in decibels of the power received at the load before insertion of the apparatus, to the power received at the load after insertion. { in'sər·shən ,lòs }

insertion meter [ENG] A type of flowmeter which measures the rotation rate of a small propeller or turbine rotor mounted at right angles to the end of a support rod and inserted into the flowing stream or closed pipe. { in'sər·shən ,mēd·ər }

insertion mutagenesis [CELL MOL] Gene alteration due to insertion of unusual nucleotide sequences from sources such as transposons, viruses, or synthetic deoxyribonucleic acid. { in'sər·shən myüd·ə,jen·ə·səs }

insertion site [CELL MOL] **1.** In a cloning vector molecule of deoxyribonucleic acid (DNA), a restriction site into which foreign DNA can be inserted. **2.** The position at which a transposable genetic element is integrated. { in'sər·shən ˌsīt }

insertion switch [COMPUT SCI] Process by which information is inserted into the computer by an operator who manually operates switches. { in'sər·shən ˌswich }

insert pump *See* rod pump. { 'in,sərt ˌpəmp }

inshore [GEOGR] **1.** Located near the shore. **2.** Indicating a shoreward position. { 'in'shór }

inshore current [OCEANOGR] The horizontal movement of water inside the surf zone, including longshore and rip currents. { 'in,shór 'kə·rənt }

inshore zone [GEOL] The zone of variable width extending from the shoreline at low tide through the breaker zone. { 'in,shór 'zōn }

inside caliper [DES ENG] A caliper that has two legs with feet that turn outward; used to measure inside dimensions, as the diameter of a hole. { 'in,sīd 'kal·ə·pər }

inside diameter [DES ENG] The length of a line which passes through the center of a hollow cylindrical or spherical object, and whose end points lie on the inner surface of the object. Abbreviated ID. { 'in,sīd dī'am·əd·ər }

inside face [DES ENG] That part of the bit crown nearest to or parallel with the inside wall of an annular or coring bit. { 'in,sīd ˌfās }

inside gage [DES ENG] The inside diameter of a bit as measured between the cutting points, such as between inset diamonds on the inside-wall surface of a core bit. { 'in,sīd ˌgāj }

inside micrometer [DES ENG] A micrometer caliper with the points turned outward for measuring the internal dimensions of an object. { 'in,sīd mī'kräm·əd·ər }

inside work *See* internal work. { 'in,sīd ˌwərk }

insight therapy [PSYCH] Treatment of a personality disorder by attempting to uncover the deep causes of the individual's problem and to help eliminate defense mechanisms. { 'in,sīt ˌther·ə·pē }

in situ [SCI TECH] In the original location. { in 'si·chü }

in situ combustion [PETRO ENG] A method of driving high-viscosity, low-gravity ore otherwise unrecoverable from a formation by setting fire to the oil sand and thereby heating the oil in the horizon to increase its mobility by reducing its viscosity. { in 'si·chü kəm'bəs·chən }

in situ foaming [ENG] Depositing of the ingredients of a foamable plastic onto the location where foaming is to take place; for example, in situ foam insulation on equipment or walls. { in 'si·chü 'fōm·iŋ }

in situ hybridization [CELL MOL] A technique permitting identification of particular deoxyribonucleic acid or ribonucleic acid sequences while these sequences remain in their original location in the cell. A cell or tissue is treated with a fixative and then exposed to a labeled (by radioactivity or fluorescence) single-stranded nucleic acid probe that hybridizes with the targeted nucleic acid sequence, revealing its location on a chromosome band when the hybridized sequence is analyzed microscopically. { ,in ,si·chü ,hī·brəd·ə·zā·shən }

insol *See* insoluble. { 'in,säl }

insolation [ASTRON] **1.** Exposure of an object to the sun. **2.** Solar energy received, often expressed as a rate of energy per unit horizontal surface. { ,in·sō'lā·shən }

insoluble [CHEM] Incapable of being dissolved in another material; usually refers to solid-liquid or liquid-liquid systems. Abbreviated insol. { in'säl·yə·bəl }

insoluble anode [CHEM] An anode that resists dissolution during electrolysis. { in'säl·yə·bəl 'an,ōd }

insoluble residue [GEOL] Material remaining after a geological specimen is dissolved in hydrochloric or acetic acid. { in'säl·yə·bəl 'rez·ə,dü }

insomnia [MED] Sleeplessness; disturbed sleep; prolonged inability to sleep. { in'säm·nē·ə }

insomniac [MED] A person who is susceptible to insomnia. { in'säm·nē,ak }

insonify *See* ensonify. { in'sän·ə·fī }

inspect [IND ENG] To examine an object to determine whether it conforms to standards; may employ sight, hearing, touch, odor, or taste. { in'spekt }

inspection [IND ENG] The critical examination of a product to determine its conformance to applicable quality standards or specifications. { in'spek·shən }

inspection by variables [IND ENG] A quality-control

inspection method in which the sampled articles are evaluated on the basis of quantitative criteria. { in,spek·shən bī 'ver·ē·ə·bəlz }

inspector [MIN ENG] One employed to make examinations of and to report upon mines and surface plants relative to compliance with mining laws, rules and regulations, and safety methods. { in'spek·tər }

inspiration [PHYSIO] The drawing in of the breath. { ,in·spə'rā·shən }

inspiratory capacity [PHYSIO] Commencing from expiratory standstill, the maximum volume of gas which can be drawn into the lungs. { in'spī·rə,tór·ē kə'pas·əd·ē }

inspiratory reserve volume [PHYSIO] The amount of air that can be inhaled by forcible inspiration after completion of a normal inspiration. { in'spī·rə,tór·ē ri,zərv ,väl·yəm }

inspirometer [MED] An instrument for measuring the amount of air inspired. { ,in·spə'räm·əd·ər }

inspissation [CHEM] The process of thickening a liquid by evaporation. [GEOCHEM] Thickening of an oil deposit by evaporation or oxidation, resulting, for example, after long exposure in pitch or gum formation. { ,in·spi'sā·shən }

instability [CONT SYS] A condition of a control system in which excessive positive feedback causes persistent, unwanted oscillations in the output of the system. [PHYS] A property of the steady state of a system such that certain disturbances or perturbations introduced into the steady state will increase in magnitude, the maximum perturbation amplitude always remaining larger than the initial amplitude. { ,in·stə'bil·əd·ē }

instability line [METEOROL] Any nonfrontal line or band of convective activity in the atmosphere; this is the general term and includes the developing, mature, and dissipating stages; however, when the mature stage consists of a line of active thunderstorms, it is properly termed a squall line; therefore, in practice, instability line often refers only to the less active phases. { ,in·stə'bil·əd·ē ,līn }

instability strip [ASTRON] A portion of the Hertzsprung-Russell diagram occupied by pulsating stars; stars traverse this region at least once after they leave the main sequence. { ,in·stə'bil·əd·ē ,strip }

installation [ENG] Procedures for setting up equipment for use or service. { ,in·stə'lā·shən }

installation kit [ORD] A collection of items, which are of a supplementary nature to a major component or equipment; the items within the collection are used to establish and install an accessory or equipment to an operational condition or to the component or equipment. { ,in·stə'lā·shən ,kit }

installation processing control [COMPUT SCI] A system that automatically schedules the processing of jobs by a computer installation, in order to minimize waiting time and time taken to prepare equipment for operation. { ,in·stə'lā·shən 'präs,es·iŋ kən,trōl }

installation specification [COMPUT SCI] The criteria defined by a computer manufacturer for specifying correct physical installation. { ,in·stə'lā·shən ,spes·ə·fə,kā·shən }

installation tape number [COMPUT SCI] A number that is permanently assigned to a reel of magnetic tape to identify it. { ,in·stə'lā·shən ,tāp ,nəm·bər }

installed capacity [ELEC] The maximum runoff of a hydroelectric facility that can be constantly maintained and utilized by equipment. { in'stóld kə'pas·əd·ē }

install program [COMPUT SCI] A computer program that adapts a software package for use on a particular computer system. { in'stól ,prō·grəm }

instance variable [COMPUT SCI] The data in an object of an object-oriented program. { in'stəns ,ver·ē·ə·bəl }

instantaneous automatic gain control [ELECTR] Portion of a radar system that automatically adjusts the gain of an amplifier for each pulse to obtain a substantially constant output-pulse peak amplitude with different input-pulse peak amplitudes; the circuit is fast enough to act during the time a pulse is passing through the amplifier. { ¦in·stən¦tā·nē·əs ,ód·ə¦mad·ik 'gān kən,trōl }

instantaneous axis [MECH] The axis about which a rigid body is carrying out a pure rotation at a given instant in time. { ¦in·stən¦tā·nē·əs 'ak·səs }

instantaneous carrying current [ELEC] The maximum value of current which a switch, circuit breaker, or similar

apparatus can carry instantaneously. 　{ ¦in·stən¦tā·nē·əs 'kar·ē·iŋ ˌkə·rənt }

instantaneous center [MECH] A point about which a rigid body is rotating at a given instant in time. Also known as instant center. 　{ ¦in·stən¦tā·nē·əs 'sen·tər }

instantaneous companding [ELECTR] Companding in which the effective gain variations are made in response to instantaneous values of the signal wave. 　{ ¦in·stən¦tā·nē·əs kəm'pan·diŋ }

instantaneous condition [PHYS] The condition of a system at a particular instant in time. 　{ ¦in·stən¦tā·nē·əs kən'dish·ən }

instantaneous cut [ENG] A cut that is set off by instantaneous detonators to be certain that all charges in the cut go off at the same time; the drilling and ignition are carried out so that all the holes break smaller top angles. 　{ ¦in·stən¦tā·nē·əs 'kət }

instantaneous description [COMPUT SCI] For a Turing machine, the set of machine conditions at a given point in the computation, including the contents of the tape, the position of the read-write head on the tape, and the internal state of the machine. 　{ ¦in·stən¦tā·nē·əs di'skrip·shən }

instantaneous detonator [ENG] A type of detonator that does not have a delay period between the passage of the electric current through the detonator and its explosion. 　{ ¦in·stən¦tā·nē·əs 'det·ən,ād·ər }

instantaneous effects [COMMUN] Impairment of telephone or telegraph transmission caused by instantaneous changes in phase or amplitude of the wave in a transmission line. 　{ ¦in·stən¦tā·nē·əs i'feks }

instantaneous field of view [OPTICS] The solid angle within which radiation is detected by an imaging system employing some form of scanning mechanism, at a given instant of time. 　{ ¦in·stən¦tā·nē·əs ¦fēld əv 'vyü }

instantaneous frequency [COMMUN] The time rate of change of the angle of an angle-modulated wave. 　{ ¦in·stən¦tā·nē·əs 'frē·kwən·sē }

instantaneous frequency-indicating receiver [ELECTR] A radio receiver with a digital, cathode-ray, or other display that shows the frequency of a signal at the instant it is picked up anywhere in the band covered by the receiver. 　{ ¦in·stən¦tā·nē·əs ¦frē·kwən·sē ˌin·də,kād·iŋ ri'sē·vər }

instantaneous fuse [ENG] A fuse with an ignition rate of several thousand feet per minute; an example is PETN. 　{ ¦in·stən¦tā·nē·əs 'fyüz }

instantaneous power [ELEC] The product of the instantaneous voltage and the instantaneous current for a circuit or component. 　{ ¦in·stən¦tā·nē·əs 'paů·ər }

instantaneous readout [COMMUN] Readout by a radio transmitter instantaneous with the computation of data to be transmitted. 　{ ¦in·stən¦tā·nē·əs 'rēd,aůt }

instantaneous recording [ENG ACOUS] A recording intended for direct reproduction without further processing. 　{ ¦in·stən¦tā·nē·əs ri'kórd·iŋ }

instantaneous recovery [MECH] The immediate reduction in the strain of a solid when a stress is removed or reduced, in contrast to creep recovery. 　{ ¦in·stən¦tā·nē·əs ri'kəv·ə·rē }

instantaneous sample [COMMUN] One of a sequence of instantaneous values of a wave taken at regular intervals. 　{ ¦in·stən¦tā·nē·əs 'sam·pəl }

instantaneous strain [MECH] The immediate deformation of a solid upon initial application of a stress, in contrast to creep strain. 　{ ¦in·stən¦tā·nē·əs 'strān }

instantaneous value [PHYS] The value of a sinusoidal or otherwise varying quantity at a particular instant. 　{ ¦in·stən¦tā·nē·əs 'val·yü }

instant center See instantaneous center. 　{ 'in·stənt 'sen·tər }

instantiation [COMPUT SCI] **1.** An external declaration or a reference to another program or subprogram in the Ada programming language. **2.** The deduction of omitted values in a set of data from the known values. **3.** The creation of an object of a specific class in an object-oriented program. 　{ in,stan·chē'ā·shən }

instantizing [FOOD ENG] Redrying a wet agglomerate of nonfat dry milk powder to render the product more easily reconstituted. 　{ 'in·stən,tīz·iŋ }

instanton [PARTIC PHYS] A hypothetical pseudoparticle which provides solutions to equations describing the gauge fields of quantum chromodynamics, and represents large vacuum fluctuations in these fields that would exert forces on quarks. 　{ 'in·stən,tän }

instant-on switch [ELECTR] A switch that applies a reduced filament voltage to all tubes in a television receiver continuously, so the picture appears almost instantaneously after the set is turned on. 　{ 'in·stənt 'ón ,swich }

instant replay See video replay. 　{ 'in·stənt 'rē,plā }

instar [INV ZOO] A stage between molts in the life of arthropods, especially insects. 　{ 'in,stär }

instep [ANAT] The arch on the medial side of the foot. 　{ 'in,step }

instinct [PSYCH] A primary tendency or inborn drive, as toward life, sexual reproduction, and death. [ZOO] A precise form of behavior in which there is an invariable association of a particular series of responses with specific stimuli; an unconditioned compound reflex. 　{ 'in,stiŋkt }

instinctive behavior [ZOO] Any species-typical pattern of responses not clearly acquired through training. 　{ in'stiŋk·tiv bi'hā·vyər }

instinctual [PSYCH] Pertaining to an emotional, impulsive, and generally unreasoned behavior or mental process which is a function of the id. [ZOO] Of or pertaining to instincts. 　{ in'stiŋk·chə·wəl }

instruction [COMPUT SCI] A pattern of digits which signifies to a computer that a particular operation is to be performed and which may also indicate the operands (or the locations of operands) to be operated on. 　{ in'strək·shən }

instruction address [COMPUT SCI] The address of the storage location in which a given instruction is stored. 　{ in'strək·shən ə'dres }

instruction address register [COMPUT SCI] A special storage location, forming part of the program controller, in which addresses of instructions are stored in order to control their sequential retrieval from memory during the execution of a program. 　{ in'strək·shən ə·dres 'rej·ə·stər }

instruction area [COMPUT SCI] A section of storage used for storing program instructions. 　{ in'strək·shən ,er·ē·ə }

instruction card [IND ENG] A written description of the standard method used by a worker, to guide his activities. 　{ in'strək·shən ,kärd }

instruction code [COMPUT SCI] That part of an instruction which distinguishes it from all other instructions and specifies the action to be performed. 　{ in'strək·shən ,kōd }

instruction constant [COMPUT SCI] A dummy instruction of the type K = 1, where K is irrelevant to the program. 　{ in'strək·shən ,kän·stənt }

instruction counter [COMPUT SCI] A counter that indicates the location of the next computer instruction to be interpreted. Also known as location counter; program counter; sequence counter. 　{ in'strək·shən ,kaůnt·ər }

instruction cycle [COMPUT SCI] The steps involved in carrying out an instruction. 　{ in'strək·shən ,sī·kəl }

instruction format [COMPUT SCI] Any rule which assigns various functions to the various digits of an instruction. 　{ in'strək·shən ,fór,mat }

instruction length [COMPUT SCI] The number of bits or bytes (eight bits per byte) which defines an instruction. 　{ in'strək·shən ,leŋkth }

instruction lookahead [COMPUT SCI] A technique for speeding up the process of fetching and decoding instructions in a computer program, and of computing addresses of required operands and fetching them, in which the control unit fetches any unexecuted instructions on hand, to the extent this is feasible. Also known as fetch ahead. 　{ in'strək·shən 'lùk·ə,hed }

instruction mix [COMPUT SCI] The proportion of various types of instructions that appear in a particular computer program, or in a benchmark representing a class of programs. 　{ in'strək·shən ,miks }

instruction modification [COMPUT SCI] A change, carried out by the program, in an instruction so that, upon being repeated, this instruction will perform a different operation. 　{ in'strək·shən ,mäd·ə·fə'kā·shən }

instruction pointer [COMPUT SCI] **1.** A component of a task descriptor that designates the next instruction to be executed by the task. **2.** An element of the control component of the stack model of block structure execution, which points to the current instruction. 　{ in'strək·shən ,póint·ər }

instruction register [COMPUT SCI] A hardware element that receives and holds an instruction as it is extracted from memory; the register either contains or is connected to circuits that interpret the instruction (or discover its meaning). Also known as current-instruction register. { in'strək·shən ,rej·ə·stər }

instruction repertory *See* instruction set. { in'strək·shən ,rep·ə,tór·ē }

instruction set [COMPUT SCI] Also known as instruction repertory. **1.** The set of instructions which a computing or data-processing system is capable of performing. **2.** The set of instructions which an automatic coding system assembles. { in'strək·shən ,set }

instruction time [COMPUT SCI] The time required to carry out an instruction having a specified number of addresses in a particular computer. { in'strək·shən ,tīm }

instruction transfer [COMPUT SCI] An instruction which transfers control to one or another subprogram, depending upon the value of some operation. { in'strək·shən ,tranz·fər }

instruction word [COMPUT SCI] A computer word containing an instruction rather than data. Also known as coding line. { in'strək·shən ,wərd }

instrument [ENG] A device for measuring and sometimes also recording and controlling the value of a quantity under observation. { 'in·strə·mənt }

instrumental analysis [ENG] The use of an instrument to measure a component, to detect the completion of a quantitative reaction, or to detect a change in the properties of a system. { ,in·strə'ment·əl ə'nal·ə·səs }

instrumental conditioning *See* operant conditioning. { ,in·strə'ment·əl kən'dish·ə·niŋ }

instrument approach chart [NAV] An aeronautical chart designed for use under instrument flight conditions, for making instrument approach and letdown to contact flight conditions in the vicinity of an aerodrome. { 'in·strə·mənt ə¦prōch ,chärt }

instrument approach procedure [NAV] A series of predetermined maneuvers for the orderly transfer of an aircraft under instrument flight conditions from the initial approach to a landing, or to a point from which a landing can be made visually. { 'in·strə·mət ə¦prōch prə,sē·jər }

instrument approach system [NAV] An aircraft navigation system that furnishes guidance in the vertical and horizontal planes to aircraft during descent from an initial-approach altitude to a point near the landing area; completion of a landing requires guidance to touchdown by visual or other means. { 'in·strə·mənt ə¦prōch ,sis·təm }

instrumentation [ENG] Designing, manufacturing, and utilizing physical instruments or instrument systems for detection, observation, measurement, automatic control, automatic computation, communication, or data processing. { ,in·strə·men'tā·shən }

instrumentation amplifier [ELECTR] An amplifier that accepts a voltage signal as an input and produces a linearly scaled version of this signal at the output; it is a closed-loop fixed-gain amplifier, usually differential, and has high input impedance, low drift, and high common-mode rejection over a wide range of frequencies. { ,in·strə·men'tā·shən 'am·plə,fī·ər }

instrument correction [ENG] A correction of measurements made on a unit under test for either inaccuracy of the instrument or eroding effect of the instrument. { 'in·strə·mənt kə,rek·shən }

instrumented buoy [OCEANOGR] An uncrewed floating structure for the mounting, operation, data collection, and transmission of meteorological and oceanographic parameter-measuring systems. { ¦in·strə¦men·təd 'bói }

instrument flight [NAV] A flight in which the navigation of the aircraft is controlled solely by reference to instruments. { 'in·strə·mənt ,flīt }

instrument flight rules [NAV] Regulations governing flying when weather conditions are below the minimum for visual flight rules. Abbreviated IFR. { 'in·strə·mənt ¦flīt ,rülz }

instrument housing [ENG] A case or enclosure to cover and protect an instrument. { 'in·strə·mənt ,haú·ziŋ }

instrument landing [NAV] A landing made through the use of a system of electronic beacons and radar. { 'in·strə·mənt ,lan·diŋ }

instrument landing system [NAV] A system of radio navigation which provides lateral and vertical guidance, as well as other navigational parameters required by a pilot in a low approach or a landing. Abbreviated ILS. { 'in·strə·mənt ¦lan·diŋ ,sis·təm }

instrument landing system localizer [NAV] System of horizontal guidance embodied in the instrument landing system which indicates the horizontal deviation of the aircraft from its optimum path of descent along the axis of the runway. { 'in·strə·mənt ¦lan·diŋ ,sis·təm ,lō·kə,liz·ər }

instrument landing system reference point *See* ILS reference point. { 'in·strə·mənt ¦lan·diŋ ,sis·təm 'ref·rəns ,póint }

instrument multiplier [ELEC] A highly accurate resistor used in series with a voltmeter to extend its voltage range. Also known as voltage multiplier; voltage-range multiplier. { 'in·strə·mənt ¦məl·tə,plī·ər }

instrument oil [MATER] Special grade of lubricating oil that has been refined to have oxidation resistance and gum resistance, that has compatibility with electrical insulation, and that prevents tarnish or oxidation of contacted metal surfaces; used to lubricate instruments and other intricate mechanisms. { 'in·strə·mənt ,óil }

instrument panel [ENG] A panel or board containing indicating meters. { 'in·strə·mənt ,pan·əl }

instrument reading time [ENG] The time, after a change in a measured quantity, which it takes for the indication of an instrument to come and remain within a specified percentage of its final value. { 'in·strə·mənt 'rēd·iŋ ,tīm }

instrument resistor [ELEC] A high-accuracy, four-terminal resistor used to bypass the major portion of currents around the low-current elements of an instrument, such as a direct-current ammeter. { 'in·strə·mənt ri,zis·tər }

instrument science [ENG] The systematically organized body of general concepts and principles underlying the design, analysis, and application of instruments and instrument systems. { 'in·strə·mənt ,sī·əns }

instrument shelter [ENG] A boxlike structure designed to protect certain meteorological instruments from exposure to direct sunshine, precipitation, and condensation, while providing adequate ventilation. Also known as thermometer screen; thermometer shelter; thermoscreen. { 'in·strə·mənt ,shel·tər }

instrument shunt [ELEC] A resistor designed to be connected in parallel with an ammeter to extend its current range. { 'in·strə·mənt ,shənt }

instrument system [ENG] A system which integrates one or more instruments with auxiliary or associated devices for detection, observation, measurement, automatic control, automatic computation, communication, or data processing. { 'in·strə·mənt ,sis·təm }

instrument transformer [ELEC] A transformer that transfers primary current, voltage, or phase values to the secondary circuit with sufficient accuracy to permit connecting an instrument to the secondary rather than the primary; used so only low currents or low voltages are brought to the instrument. { 'in·strə·mənt tranz,fór·mər }

instrument-type relay [ELEC] A relay constructed like a meter, with one adjustable contact mounted on the scale and the other contact mounted on the pointer. Also known as contact-making meter. { 'in·strə·mənt ,tīp 'rē,lā }

instrument weather [METEOROL] Route or terminal weather conditions of sufficiently low visibility to require the operation of aircraft under instrument flight rules (IFR). Also known as IFR weather. { 'in·strə·mənt ,weth·ər }

insulated [ELEC] Separated from other conducting surfaces by a nonconducting material. { 'in·sə,lād·əd }

insulated conductor [ELEC] A conductor surrounded by insulation to prevent current leakage or short circuits. Also known as insulated wire. { 'in·sə,lād·əd kən'dək·tər }

insulated-gate bipolar transistor [ELECTR] A power semiconductor device that combines low forward voltage drop, gate-controlled turnoff, and high switching speed. It structurally resembles a vertically diffused MOSFET, featuring a double diffusion of a *p*-type region and an *n*-type region, but differs from the MOSFET in the use of a *p*+ substrate layer (in the case of an *n*-channel device) for the drain. The effect is to change the transistor into a bipolar device, as this *p*-type region injects holes into the *n*-type drift region. Abbreviated IGBT. { ¦in·sə,lād·əd,gāt bī,pō·lər tran'zis·tər }

insulated-gate field-effect transistor *See* metal oxide semiconductor field-effect transistor. { 'in·sə,lād·əd ¦gāt ¦fēld i,fekt tran'zis·tər }

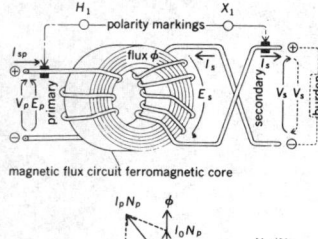

INSTRUMENT TRANSFORMER

A simple instrument transformer.
(General Electric Co.)

insulated-return power system [ELEC] A system for distributing electric power to trains or other vehicles, in which both the outgoing and return conductors are insulated, in contrast to a track-return system. { 'in·sə‚lād·əd ri¦tərn 'paů·ər ‚sis·təm }

insulated-substrate monolithic circuit [ELECTR] Integrated circuit which may be either an all-diffused device or a compatible structure so constructed that the components within the silicon substrate are insulated from one another by a layer of silicon dioxide, instead of reverse-biased *pn* junctions used for isolation in other techniques. { 'in·sə‚lād·əd ¦səb‚strāt ¦män·ə‚lith·ik 'sər·kət }

insulated wire *See* insulated conductor. { 'in·sə‚lād·əd 'wīr }

insulating board [MATER] Any board used in a wall or ceiling to provide insulation. { 'in·sə‚lād·iŋ ‚bórd }

insulating compound [MATER] A liquid, at low temperatures, which is poured into joint boxes and allowed to solidify; as a poor conductor of heat and electricity, it provides good insulation. { 'in·sə‚lād·iŋ ‚käm‚paůnd }

insulating concrete [MATER] Concrete with insulating properties, often made with asbestos fibers and in the form of blocks, corrugated slabs, or sheathing. { 'in·sə‚lād·iŋ ¦käŋ‚krēt }

insulating oil [MATER] A chlorinated hydrocarbon, such as trichlorobenzene, mixed with fluorinated hydrocarbons, whose high dielectric strength and high flash point allow it to be used in switches, circuit breakers, and transformers as an insulator and cooling medium. Also known as electrical oil. { 'in·sə‚lād·iŋ ‚óil }

insulating paper [MATER] A standard material for insulating electrical equipment, usually consisting of bond or kraft paper coated with black or yellow insulating varnish on both sides. Also known as electrical insulating paper; varnish paper. { 'in·sə‚lād·iŋ ‚pā·pər }

insulating strength [ELEC] Measure of the ability of an insulating material to withstand electric stress without breakdown; it is defined as the voltage per unit thickness necessary to initiate a disruptive discharge; usually measured in volts per centimeter. { 'in·sə‚lād·iŋ ‚streŋkth }

insulating tape [MATER] Tape impregnated with insulating material, and usually adhesive; used to cover joints in insulated wires or cables. Also known as electrical tape. { 'in·sə‚lād·iŋ ‚tāp }

insulation [BUILD] Material used in walls, ceilings, and floors to retard the passage of heat and sound. [ELEC] A material having high electrical resistivity and therefore suitable for separating adjacent conductors in an electric circuit or preventing possible future contact between conductors. Also known as electrical insulation. { ‚in·sə'lā·shən }

insulation coordination [ELEC] Steps taken to ensure that electric equipment is not damaged by overvoltages and that flashovers are localized in regions where no damage results from them. { ‚in·sə'lā·shən kō'órd·ən‚ā·shən }

insulation porcelain [MATER] Any of the various insulating materials consisting of molded silica, molded steatite, or specially compounded ceramics, often containing zirconia or beryllia. Also known as electrical porcelain. { ‚in·sə'lā·shən ¦pór·slən }

insulation protection [ELEC] Use of devices to protect insulators of power transmission lines from damage by heavy arcs. { ‚in·sə'lā·shən prə¦tek·shən }

insulation resistance [ELEC] The electrical resistance between two conductors separated by an insulating material. { ‚in·sə'lā·shən ri¦zis·təns }

insulation sampler [ENG] A device for collecting deep water which prevents any significant conduction of heat from the water sample so that it maintains its original temperature as it is hauled to the surface. { ‚in·sə'lā·shən ‚sam·plər }

insulation testing set [ENG] An instrument for measuring insulation resistance, consisting of a high-range ohmmeter having a hand-driven direct-current generator as its voltage source. { ‚in·sə'lā·shən 'test·iŋ ‚set }

insulator [ELEC] A device having high electrical resistance and used for supporting or separating conductors to prevent undesired flow of current from them to other objects. Also known as electrical insulator. [MATER] A material that is a poor conductor of heat, sound, or electricity. [SOLID STATE] A substance in which the normal energy band is full and is separated from the first excitation band by a forbidden band that can be penetrated only by an electron having an energy

of several electronvolts, sufficient to disrupt the substance. { 'in·sə‚lād·ər }

insulator arc-over [ELEC] Discharge of power current in the form of an arc, following a surface discharge over an insulator. { 'in·sə‚lād·ər 'ärk‚ō·vər }

insulator arrangement [ELECTROMAG] The placement of insulators on a transmission mast. { 'in·sə‚lād·ər ə‚rānj·mənt }

insulin [BIOCHEM] A protein hormone produced by the beta cells of the islets of Langerhans which participates in carbohydrate and fat metabolism. { 'in·sə·lən }

insulinase [BIOCHEM] An enzyme produced by the liver which is able to inactivate insulin. { 'in·sə·lə‚nās }

insulin-like growth factor *See* somatomedin. { 'in·sə·lən ‚līk 'grōth ‚fak·tər }

insulinoma *See* islet-cell tumor. { ‚in·sə·lə'nō·mə }

insulin shock [MED] Clinical manifestation of hypoglycemia due to excess amounts of insulin in the blood. { 'in·sə·lən ‚shäk }

insulin shock therapy [MED] Administration of large doses of insulin to induce hypoglycemic comas, followed by administration of glucose, in the treatment of certain psychotic disorders. { 'in·sə·lən ‚shäk 'ther·ə·pē }

insuloma *See* islet-cell tumor. { ‚in·sə'lō·mə }

intaglio [LAP] A type of carved gemstone in which the figure is engraved on the surface of the stone rather than left in relief by cutting away the background, as in a cameo. { in'tal·yō }

intaglio plate [GRAPHICS] A metal surface into which the printing elements are formed in intaglio printing. { in'tal·yō ‚plāt }

intaglio printing [GRAPHICS] A printing method in which the printing elements are all below the plate surface, having been cut, scratched, engraved, or etched into the metal to form ink-retaining grooves or cups; surplus ink on the surface must be wiped or scraped off after each inking and before each printing impression. { in'tal·yō 'prin·iŋ }

intake [ENG] **1.** An entrance for air, water, fuel, or other fluid, or the amount of such fluid taken in. **2.** A main passage for air in a mine. [HYD] *See* recharge. { 'in‚tāk }

intake area *See* recharge area. { 'in‚tāk ‚er·ē·ə }

intake chamber [CIV ENG] A large chamber that gradually narrows to an intake tunnel; designed to avoid undesirable water currents. { 'in‚tāk ‚chām·bər }

intake gate [CIV ENG] A movable partition for opening or closing a water intake opening. { 'in‚tāk ‚gāt }

intake manifold [MECH ENG] A system of pipes which feeds fuel to the various cylinders of a multicylinder internal combustion engine. { 'in‚tāk ‚man·ə‚fōld }

intake stroke [MECH ENG] The fluid admission phase or travel of a reciprocating piston and cylinder mechanism as, for example, in an engine, pump, or compressor. { 'in‚tāk ‚strōk }

intake valve [MECH ENG] The valve which opens to allow air or an air-fuel mixture to enter an engine cylinder. { 'in‚tāk ‚valv }

intarsia [GRAPHICS] Decorative designs of inlaid wood in a background of wood; often used in furniture making. Also known as tarsia. [TEXT] A pattern in several colors, usually geometrical, in a knitted fabric in which both sides of the fabric are alike. { in'tär·sē·ə }

integer [MATH] Any positive or negative counting number or zero. { 'int·ə·jər }

integer constant [COMPUT SCI] A constant that uses the values 0, 1, . . . , 9 with no decimal point in FORTRAN. { 'int·ə·jər ‚kän·stənt }

integer data type [COMPUT SCI] A scalar date type which is used to represent whole numbers, that is, values without fractional parts. { 'int·ə·jər 'dad·ə ‚tāp }

integer partition [MATH] For a positive integer n, a nonincreasing sequence of positive integers whose sum equals n. { 'int·ə·jər pär'tish·ən }

integer programming [SYS ENG] A series of procedures used in operations research to find maxima or minima of a function subject to one or more constraints, including one which requires that the values of some or all of the variables be whole numbers. { 'int·ə·jər 'prō‚gram·iŋ }

integer spin [QUANT MECH] Property of a particle whose spin angular momentum is a whole number times Planck's constant divided by 2π; bosons have this property; in contrast, fermions have half-integer spin. { 'int·ə·jər ‚spin }

integer variable [COMPUT SCI] A variable in FORTRAN whose first character is normally I, J, K, L, M, or N. { 'int·ə·jər 'ver·ē·ə·bəl }

integrable differential equation [MATH] A differential equation that either is exact or can be transformed into an exact differential equation by multiplying each equation term by a common factor. { ¦int·i·grə·bəl ¦dif·ə¸ren·chəl i'kwā·zhən }

integrable function [MATH] A function whose integral, defined in a specific manner, exists and is finite. { ¦int·i·grə·bəl 'fəŋk·shən }

integrable system [MECH] A dynamical system whose motion is governed by an integrable differential equation. { ¦int·i·grə·bəl ¦sis·təm }

integral [MATH] **1.** A solution of a differential equation is sometimes called an integral of the equation. **2.** An element *a* of a ring *B* is said to be integral over a ring *A* contained in *B* if it is the root of a polynomial with coefficients in *A* and with leading coefficient 1. **3.** *See* definite Riemann integral; 'int·ə·grəl }

integral absorbed dose *See* integral dose. { 'int·ə·grəl əb¦sȯrbd 'dōs }

integral action [CONT SYS] A control action in which the rate of change of the correcting force is proportional to the deviation. { 'int·ə·grəl ¸ak·shən }

integral calculus [MATH] The study of integration and its applications to finding areas, volumes, or solutions of differential equations. { 'int·ə·grəl 'kal·kyəl·ləs }

integral closure [MATH] The integral closure of a subring *A* of a ring *B* is the set of all elements in *B* that are integral over *A*. { 'int·ə·grəl 'klō·zhər }

integral compensation [CONT SYS] Use of a compensator whose output changes at a rate proportional to its input. { 'int·ə·grəl ¸käm·pən'sā·shən }

integral control [CONT SYS] Use of a control system in which the control signal changes at a rate proportional to the error signal. { 'int·ə·grəl kən¸trōl }

integral curvature [MATH] For a given region on a surface, the integral of the Gaussian curvature over the region. { 'int·ə·grəl 'kərv·ə·chər }

integral curves [MATH] A family of curves that satisfy a particular differential equation. { 'int·ə·grəl 'kərvz }

integral discriminator [ELECTR] A circuit which accepts only pulses greater than a certain minimum height. { 'int·ə·grəl di'skrim·ə¸nād·ər }

integral domain [MATH] A commutative ring with identity where the product of nonzero elements is never zero. Also known as entire ring. { 'int·ə·grəl dō'mān }

integral dose [NUCLEO] The total energy imparted to an irradiated body by an ionizing radiation; usually expressed in gram-rads or gram-roentgens. Also known as integral absorbed dose; volume dose. { 'int·ə·grəl 'dōs }

integral equation [MATH] An equation where the unknown function occurs under an integral sign. { 'int·ə·grəl i'kwā·zhən }

integral extension [MATH] An integral extension of a commutative ring *A* is a commutative ring *B* containing *A* such that every element of *B* is integral over *A*. { 'int·ə·grəl ik'sten·chən }

integral function [MATH] **1.** A function taking on integer values. **2.** *See* entire function. { 'int·ə·grəl ¸fəŋk·shən }

integral-furnace boiler [MECH ENG] A type of steam boiler which incorporates furnace water-cooling in the circulatory system. { 'int·ə·grəl ¦fər·nəs ¸bȯil·ər }

integral heat of dilution *See* heat of dilution. { 'int·ə·grəl ¦hēt əv də'lü·shən }

integral heat of solution *See* heat of solution. { 'int·ə·grəl ¦hed əv sə'lü·shən }

integral hologram [OPTICS] A type of hologram that is automatically synthesized from a large collection of photographs, each taken from a slightly different position. { 'int·ə·grəl 'häl·ə¸gram }

integral-joint casing [PETRO ENG] Oil well casing lengths on whose ends the connection joints are integrally formed. { 'int·ə·grəl ¦jȯint ¸kās·iŋ }

integrally closed ring [MATH] An integral domain which is equal to its integral closure in its quotient field. { in¦teg·rə·lē ¦klōzd 'riŋ }

integral map [MATH] A homomorphism from a commutative ring *A* into a commutative ring *B* such that *B* is an integral extension of *f*(*A*). { 'int·ə·grəl ¸map }

integral membrane protein [CELL MOL] A protein that is firmly anchored in the plasma membrane via interactions between its hydrophobic domains and the membrane phospholipids. Also known as intrinsic protein. { ¦int·i·grəl 'mem¸brān ¸prō¸tēn }

integral-mode controller [CONT SYS] A controller which produces a control signal proportional to the integral of the error signal. { 'int·ə·grəl ¦mōd kən¸trōl·ər }

integral modem [COMMUN] A modem built directly into a machine to enable it to communicate over a telephone line. { 'int·ə·grəl 'mō¸dem }

integral network [CONT SYS] A compensating network which produces high gain at low input frequencies and low gain at high frequencies, and is therefore useful in achieving low steady-state errors. Also known as lagging network; lag network. { 'int·ə·grəl 'net¸wərk }

integral operator [MATH] A rule for transforming one function into another function by means of an integral; this often is in context a linear transformation on some vector space of functions. { 'int·ə·grəl 'äp·ə¸rād·ər }

integral photography [OPTICS] A type of three-dimensional photography in which the photographic medium is placed at the focal plane of a microlens array, and the developed image is viewed through the same lens array, allowing the object to be reconstructed in full parallax. { ¸int·i·grəl fə'täg·rə·fē }

integral procedure decomposition temperature [PHYS CHEM] Decomposition temperatures derived from graphical integration of the thermogravimetric analysis of a polymer. { 'int·ə·grəl prə¦sē·jər dē¸käm·pə'zish·ən ¸tem·prə·chər }

integral quantum Hall effect [ELECTR] The version of the quantum Hall effect in which the Hall resistance becomes precisely equal to (h/e^2)/*n*, where *h* is Planck's constant, *e* is the electronic charge, and *n* is an integer. { 'int·ə·grəl ¦kwän·təm 'hȯl i¸fekt }

integral square error [CONT SYS] A measure of system performance formed by integrating the square of the system error over a fixed interval of time; this performance measure and its generalizations are frequently used in linear optimal control and estimation theory. { 'int·ə·grəl ¦skwer ¸er·ər }

integral test [MATH] If *f*(*x*) is a function that is positive and decreasing for positive *x*, then the infinite series with *n*th term *f*(*n*) and the integral of *f*(*x*) from 1 to ∞ are either both convergent (finite) or both infinite. { 'int·ə·grəl ¸test }

integral transform *See* integral transformation. { 'int·ə·grəl 'tranz¸fȯrm }

integral transformation [MATH] A transform of a function *F*(*x*) given by the function

$$f(y) = \int_a^b K(x,y)F(x)\, dx$$

where *K*(*x*,*y*) is some function. Also known as integral transform. { 'int·ə·grəl ¸tranz·fər'mā·shən }

integral-type flange [DES ENG] A flange which is forged or cast with, or butt-welded to, a nozzle neck, pressure vessel, or piping wall. { 'int·ə·grəl ¦tīp 'flanj }

integral waterproofing [ENG] Waterproofing concrete by adding the waterproofing material to the cement or to the mixing water. { 'int·ə·grəl 'wȯd·ər¸prüf·iŋ }

integrand [MATH] The function which is being integrated in a given integral. { 'int·ə¸grand }

integraph [ENG] A device used for completing a mathematical integration by graphical methods. { 'int·ə¸graf }

integrase [BIOCHEM] An enzyme that facilitates prophage integration into or excision from a bacterial chromosome. { 'int·ə¸grās }

integrated circuit [ELECTR] An interconnected array of active and passive elements integrated with a single semiconductor substrate or deposited on the substrate by a continuous series of compatible processes, and capable of performing at least one complete electronic circuit function. Abbreviated IC. Also known as integrated semiconductor. { 'int·ə¸grād·əd 'sər·kət }

integrated-circuit capacitor [ELECTR] A capacitor that can

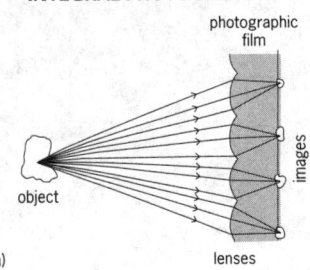

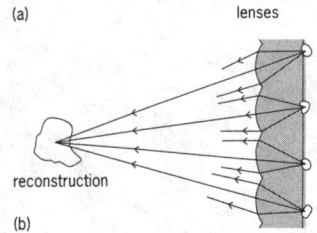

INTEGRAL PHOTOGRAPHY

photographic film

object

(a)

lenses

images

reconstruction

(b)

Diagrams of (*a*) recording of an object by a series of microcameras with a range of viewpoints, and (*b*) reconstruction, when the recording is illuminated, of an object that has full parallax.

be produced in a silicon substrate by conventional semiconductor production processes. { 'int·ə,grād·əd ¦sərkət kə'pas·əd·ər }

integrated-circuit filter [ELECTR] An electronic filter implemented as an integrated circuit, rather than by interconnecting discrete electrical components. { ¦int·i,grād·əd 'sər·kət ,fil·tər }

integrated-circuit memory See semiconductor memory. { 'int·ə,grād·əd ¦sərkət 'mem·rē }

integrated-circuit resistor [ELECTR] A resistor that can be produced in or on an integrated-circuit substrate as part of the manufacturing process. { 'int·ə,grād·əd ¦sərkət ri'zis·tər }

integrated communications - navigation - identification [NAV] The concept of coordinating the electronic units of a system to improve the efficiency of providing communications, navigation, and identification for civilian and military aircraft. Abbreviated ICNI. { 'int·ə,grād·əd kə,myü·nə'kā·shənz ,nav·ə'gā·shən ī,dent·ə·fə'kā·shən }

integrated communications system [COMMUN] Communications system on either a unilateral or joint basis in which a message can be filed at any communications center in that system and be delivered to the addressee by any other appropriate communications center in that system without reprocessing enroute. { 'int·ə,grād·əd kə,myü·nə'kā·shənz ,sis·təm }

integrated console [COMMUN] Computer control console that is capable of controlling the operation of the switching center equipment of an integrated communications system. { 'int·ə,grād·əd 'kän,sōl }

integrated data dictionary [COMPUT SCI] An index or catalog of information about a data base that is physically and logically integrated into the data base. { 'in·tə,grād·əd 'dad·ə 'dik·shə,ner·ē }

integrated data processing [COMPUT SCI] Data processing that has been organized and carried out as a whole, so that intermediate outputs may serve as inputs for subsequent processing with no human copying required. Abbreviated IDP. { 'in·tə,grād·əd 'dad·ə 'prä·ses·iŋ }

integrated data retrieval system [COMPUT SCI] A section of a data-processing system that provides facilities for simultaneous operation of several video-data interrogations in a single line and performs required communications with the rest of the system; it provides storage and retrieval of both data subsystems and files and standard formats for data representation. { 'in·tə,grād·əd 'dad·ə ri'trē·vəl ,sis·təm }

integrated drainage [HYD] Drainage resulting after folding and faulting of a surface under arid conditions; the streams by working headward have joined basins across intervening mountains or ridges. { 'in·tə,grād·əd 'drān·ij }

integrated ducted propeller [NAV ARCH] A coaxial duct fitted in two halves ahead of a previously existing propeller and tapered from top to bottom to increase efficiency. { 'int·ə,grād·əd 'dək·təd prə'pel·ər }

integrated electric propulsion [NAV ARCH] The use of a single electric system to provide both ship service and propulsion power. { ¦in·tə,grād·əd i¦lek·trik prə'pəl·shən }

integrated electronics [ELECTR] A generic term for that portion of electronic art and technology in which the interdependence of material, device, circuit, and system-design consideration is especially significant; more specifically, that portion of the art dealing with integrated circuits. { 'in·tə,grād·əd i,lek'trän·iks }

integrated fire control system [ORD] A system which combines target acquisition and tracking data computation, and weapon laying and firing, primarily using electronic means assisted by electromechanical devices. { 'in·tə,grād·əd 'fīr ,kən,trōl ,sis·təm }

integrated inertial navigation system [NAV] An air navigation system in which an inertial navigation or inertial reference subsystem is interconnected with all the other subsystems associated with avionics, so that data from all subsystems are automatically blended and interpreted to provide the control data needed to operate aircraft efficiently and safely. { 'in·tə,grād·əd in'ər·shəl ,nav·ə'gā·shən ,sis·təm }

integrated information processing [COMPUT SCI] System of computers and peripheral systems arranged and coordinated to work concurrently or independently on different problems at the same time. { 'in·tə,grād·əd ,in·fər'mā·shən ,präs·es·iŋ }

integrated information system [COMMUN] An expansion of a basic information system achieved through system design

of an improved or broader capability by functionally or technically relating two or more information systems, or by incorporating a portion of the functional or technical elements of one information system into another. { 'in·tə,grād·əd ,in·fər'mā·shən ,sis·təm }

integrated injection logic [ELECTR] Integrated-circuit logic that uses a simple and compact bipolar transistor gate structure which makes possible large-scale integration on silicon for logic arrays, memories, watch circuits, and various other analog and digital applications. Abbreviated I^2L. Also known as merged-transistor logic. { 'in·tə,grād·əd in'jek·shən 'läj·ik }

integrated neutron flux [NUCLEO] A measure of radiation exposure, equal to the product of the number of free neutrons per unit volume, the average speed of neutrons, and the exposure time. { 'in·tə,grād·əd 'nü,trän ,fləks }

integrated optics [OPTICS] A thin-film device containing tiny lenses, prisms, and switches to transmit very thin laser beams, and serving the same purposes as the manipulation of electrons in thin-film devices of integrated electronics. { 'in·tə,grād·əd 'äp·tiks }

integrated profile See mean profile. { 'in·tə,grād·əd 'prō,fīl }

integrated radiation [OPTICS] The integral of the radiance over the duration of exposure. { 'in·tə,grād·əd ,rād·ē'ā·shən }

integrated reflection [PHYS] The intensity of a beam of x-rays or neutrons reflected from a given atomic plane of a crystal, integrated over a small range of angles about the general direction of the beam. { 'in·tə,grād·əd ri'flek·shən }

integrated semiconductor See integrated circuit. { 'in·tə,grād·əd ,sem·i·kən'dək·tər }

integrated sensor [ENG] A very small device in which the sensing of some physical quantity is integrated with the functions of signal processing and information processing. { ¦in·tə,grād·əd 'sen·sər }

integrated services digital network [COMMUN] A public end-to-end digital communications network which has capabilities of signaling, switching, and transport over facilities such as wire pairs, coaxial cables, optical fibers, microwave radio, and satellites, and which supports a wide range of services, such as voice, data, video, facsimile, and music, over standard interfaces. Abbreviated ISDN. { 'in·tə,grād·əd ¦sər·vəs·əz 'dij·əd·əl 'net,wərk }

integrated software [COMPUT SCI] **1.** A collection of computer programs designed to work together to handle an application, either by passing data from one to another or as components of a single system. **2.** A collection of computer programs that work as a unit with a unified command structure to handle several applications, such as word processing, spreadsheets, data-base management, graphics, and data communications. { 'in·tə,grād·əd 'sóft,wər }

integrated thermionic circuit [ELECTR] A circuit fabricated from subminiature thin-film metal patterns on two planar substrates separated by an evacuated space about 1 millimeter in thickness to form miniature planar, thermionic, vacuum-tube devices, with densities approaching those of conventional integrated circuits. { 'in·tə,grād·əd ¦thər·mē¦än·ik 'sər·kət }

integrated train [MIN ENG] A long string of cars, permanently coupled together, that shuttles endlessly between one mine and one generating plant, not even stopping to load and unload, since rotary couplers permit each car to be flipped over and dumped as the train moves slowly across a trestle. { 'in·tə,grād·əd 'trān }

integrating accelerometer [ENG] A device whose output signals are proportional to the velocity of the vehicle or to the distance traveled (depending on the number of integrations) instead of acceleration. { 'in·tə,grād·əd ak,sel·ə'räm·əd·ər }

integrating amplifier [ELECTR] An operational amplifier with a shunt capacitor such that mathematically the waveform at the output is the integral (usually over time) of the input. { 'int·ə,grād·iŋ 'am·plə,fī·ər }

integrating detector [ELECTR] A frequency-modulation detector in which a frequency-modulated wave is converted to an intermediate-frequency pulse-rate modulated wave, from which the original modulating signal can be recovered by use of an integrator. { 'int·ə,grād·iŋ di'tek·tər }

integrating factor [MATH] A factor which when multiplied into a differential equation makes the portion involving derivatives an exact differential. { int·ə,grād·iŋ 'fak·tər }

integrating filter [ELECTR] A filter in which successive

INTEGUMENT

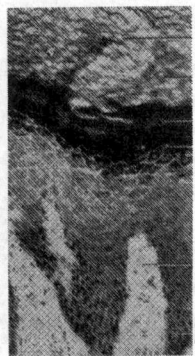

duct of
sweat gland
stratum
corneum
stratum
lucidum
stratum
granulosum

papilla
of corium

stratum
germinativum

The characteristic strata of thick
skin of the human finger as seen in
cross section at high magnification.
*(From J. F. Nonidez and W. F. Windle,
Textbook of Histology, McGraw-Hill,
1949)*

pulses of applied voltage cause cumulative buildup of charge
and voltage on an output capacitor. { 'int·ə‚grād·iŋ 'fil·tər }

integrating frequency meter [ENG] An instrument that
measures the total number of cycles through which the alternating voltage of an electric power system has passed in a given
period of time, enabling this total to be compared with the
number of cycles that would have elapsed if the prescribed
frequency had been maintained. Also known as master frequency meter. { 'int·ə‚grād·iŋ 'frē·kwən·sē ‚mēd·ər }

integrating galvanometer [ENG] A modification of the
d'Arsonval galvanometer which measures the integral of current over time; it is designed to be able to measure changes of
flux in an exploring coil which last over periods of several
minutes. { 'int·ə‚grād·iŋ ‚gal·və'näm·əd·ər }

integrating gyroscope [ENG] A gyroscope that senses the
rate of angular displacement and measures and transmits the
time integral of this rate. { 'int·ə‚grād·iŋ 'jī·rə‚skōp }

integrating ionization chamber [NUCLEO] An ionization
chamber in which the collected charge is stored on a capacitor
for subsequent measurement. { 'int·ə‚grād·iŋ ‚ī·ə·nə'zā-
shən ‚chām·bər }

integrating meter [ENG] An instrument that totalizes electric energy or some other quantity consumed over a period of
time. { 'int·ə‚grād·iŋ 'mēd·ər }

integrating network [ELECTR] A circuit or network whose
output waveform is the time integral of its input waveform.
Also known as integrator. { 'int·ə‚grād·iŋ 'net‚wərk }

integrating-sphere photometer [OPTICS] An instrument
for measuring the total luminous flux of a lamp or luminaire;
the source is placed inside a sphere whose inside surface has
a diffusely reflecting white finish, and the light reflected from
this surface onto a window is measured by an ordinary photometer. Also known as sphere photometer. { 'int·ə‚grād·iŋ ‚sfir
fə'täm·əd·ər }

integrating water sampler [ENG] A water sampling device
comprising a cylinder with a free piston whose movement is
regulated by the evacuation of a charge of fresh water. { 'int·
ə‚grād·iŋ 'wȯd·ər ‚sam·plər }

integration [GEN] Recombination involving insertion of a
genetic element. [MATH] The act of taking a definite or
indefinite integral. [SYS ENG] The arrangement of components in a system so that they function together in an efficient
and logical way. { ‚int·ə'grā·shən }

integration by parts [MATH] A technique used to find the
integral of the product of two functions by means of an identity
involving another simpler integral; for functions of one variable
the identity is

$$\int_a^b fg'\, dx + \int_a^b gf'\, dx = f(b)g(b) - f(a)g(a);$$

for functions of several variables the technique is tantamount
to using Stokes' theorem or the divergence theorem. { ‚int·
ə'grā·shən bī 'parts }

integration constant *See* constant of integration. { ‚int·ə'grā·
shən ‚kän·stənt }

integration efficiency [MOL BIO] The frequency with which
a segment of foreign deoxyribonucleic acid is incorporated into
a host bacterial genome. { ‚int·ə'grā·shən i‚fish·ən·sē }

integration test [COMPUT SCI] A stage in testing a computer
system in which a collection of modules in the system is tested
as a group. { ‚int·ə'grā·shən ‚test }

integrator [ELECTR] **1.** A computer device that approximates the mathematical process of integration. **2.** *See* integrating network. { 'int·ə‚grād·ər }

integrin [CELL MOL] A heterodimeric transmembrane receptor protein of animal cells that binds to components of the
extracellular matrix on the outside of a cell and to the cytoskeleton on the inside of the cell, functionally connecting the cell
interior to its exterior; in blood cells, integrins are also involved
in cell-cell adhesion. { in'teg·rən }

integrity [COMPUT SCI] Property of data which can be recovered in the event of its destruction through failure of the
recording medium, user carelessness, program malfunction, or
other mishap. [NAV] The ability of a navigation system to
inform the user in a timely manner of a latent failure that may
cause a hazardous condition. { in'teg·rəd·ē }

integrodifferential equation [MATH] An equation relating

a function, its derivatives, and its integrals. { in‚teg·rō‚dif·
ə‚ren·chəl i'kwä·zhən }

integument [ANAT] An outer covering, especially the skin,
together with its various derivatives. { in'teg·yə·mənt }

integumentary musculature [VERT ZOO] Superficial skeletal muscles which are spread out beneath the skin and are
inserted into it in some terrestrial vertebrates. { in‚teg·yə‚men·
trē 'məs·kyə·lə·chər }

integumentary pattern [ANAT] Any of the features of the
skin and its derivatives that are arranged in designs, such as
scales, epidermal ridges, feathers coloration, or hair. { in‚teg·
yə‚men·trē 'pad·ərn }

integumentary system [ANAT] A system encompassing the
integument and its derivatives. { in‚teg·yə‚men·trē 'sis·təm }

intelligence [COMMUN] Data, information, or messages that
are to be transmitted. [PSYCH] **1.** The intellect or astuteness
of the mind. **2.** Ability to recognize and understand qualities
and attributes of the physical world and of humankind. **3.**
Ability to solve problems and engage in abstract thought processes. { in'tel·ə·jəns }

intelligence quotient [PSYCH] The numerical designation
for intelligence expressed as a ratio of an individual's performance on a standardized test to the average performance
according to age. Abbreviated IQ. { in'tel·ə·jəns ‚kwōsh·
ənt }

intelligence test [PSYCH] A series of standardized tasks or
problems presented to an individual to measure his innate
capacity to think, conceive, or reason; examples are the Stanford-Binet test and the Wechsler-Bellevue intelligence test.
{ in'tel·ə·jəns ‚test }

intelligent agent [IND ENG] A computing hardware- or soft-
ware-based system that operates without the direct intervention
of humans or other agents; examples include robots, smart
sensors, and Web-search software agents. { in‚tel·ə·jənt 'ā·
jənt }

intelligent cable [COMMUN] A multiline communications
cable that is equipped with a microprocessor to analyze or
convert signals. { in'tel·ə·jənt ‚kā·bəl }

intelligent controller [COMPUT SCI] A peripheral control
unit whose operation is controlled by a built-in microprocessor.
{ in'tel·ə·jənt kən'trō·lər }

intelligent database [COMPUT SCI] **1.** A database that can
respond to queries in a high-level, interactive language. **2.** A
database that can store validation criteria with each item of
data, so that all programs entering or updating the data must
conform to these criteria. { in'tel·ə·jənt 'dad·ə‚bās }

intelligent machine [COMPUT SCI] A machine that uses sensors to monitor the environment and thereby adjust its actions
to accomplish specific tasks in the face of uncertainty and
variability. [ENG] Any machine that can accomplish its specific task in the presence of uncertainty and variability in its
environment. { in'tel·ə·jənt mə'shēn }

intelligent manufacturing [IND ENG] **1.** The use of production process technology that can automatically adapt to changing environments and varying process requirements, with the
capability of manufacturing various products with minimal
supervision and assistance from operators. **2.** The development and implementation of artificial intelligence in manufacturing. { in'tel·ə·jənt ‚man·ə‚fak·chər·iŋ }

intelligent robot [CONT SYS] A robot that functions as an
intelligent machine, that is, it can be programmed to take actions
or make choices based on input from sensors. { in'tel·ə·
jənt 'rō‚bät }

intelligent sensor *See* smart sensor. { in‚tel·ə·jənt 'sen·sər }

intelligent terminal [COMPUT SCI] A computer input/output
device with its own memory and logic circuits which can perform certain operations normally carried out by the computer.
Also known as smart terminal. { in'tel·ə·jənt 'ter·mən·əl }

intelligent transportation systems [CIV ENG] The application of advanced technologies to surface transportation problems, including traffic and transportation management, travel
demand management, advanced public transportation management, electronic payment, commercial vehicle operations,
emergency services management, and advanced vehicle control
and safety systems. Previously known as intelligent vehicle
highway systems. { in‚tel·ə·jənt ‚tranz·pər'tā·shən ‚sis·
təmz }

intelligent vehicle highway systems *See* intelligent transportation systems. { in‚tel·ə·jənt ‚vē·ə·kəl 'hī‚wā ‚sis·təmz }

intelligent work station [COMPUT SCI] A work station that has an intelligent terminal to carry out a variety of functions independently. { in'tel·ə·jənt 'wərk ˌstā·shən }

intelligibility [COMMUN] The percentage of speech units understood correctly by a listener in a communications system; customarily used for regular messages where the context aids the listener, in distinction to articulation. Also known as speech intelligibility. { in‚tel·ə·jə'bil·əd·ē }

intelligible crosstalk [COMMUN] Crosstalk which is sufficiently understandable under pertinent circuit and room noise conditions that meaningful information can be obtained by more sensitive listeners. { in'tel·ə·jə·bəl 'kròs‚tòk }

Intelsat [COMMUN] A satellite network, formerly under international control, used for global communication by more than 100 countries; the system uses geostationary satellites over the Atlantic, Pacific, and Indian oceans and directional antennas at earth stations. Derived from International Telecommunications Satellite. { in'tel‚sat }

intensification [MATH] An operation that increases the value of the membership function of a fuzzy set if the value is equal to or greater than 0.5, and decreases it if it is less than 0.5. { in‚tens·ə·fə'kā·shən }

intensifier [GRAPHICS] In photography, a means used to strengthen the image of either a negative or a positive; usually, metal is added to the silver image, and the increase in density is proportional to the existing density; thus when a negative is intensified, the highlights are always intensified more than the shadows. [PETRO ENG] Hydrofluoric acid added to hydrochloric acid for oil well acidizing; the fluoride destroys silica films that are insoluble by hydrochloric acid. { in'ten·sə ‚fī·ər }

intensifier electrode [ELECTR] An electrode used to increase the velocity of electrons in a beam near the end of their trajectory, after deflection of the beam. Also known as post-accelerating electrode; post-deflection accelerating electrode. { in'ten·sə‚fī·ər i‚lek‚tröd }

intensifier image orthicon [ELECTR] An image orthicon combined with an image intensifier that amplifies the electron stream originating at the photocathode before it strikes the target. { in'ten·sə‚fī·ər ‚im·ij 'òr·thə‚kän }

intensifying screen [GRAPHICS] A layer of material, such as a salt screen or a metal screen, that is placed next to an x-ray film to increase the effect of x-rays on the film. { in'ten·sə‚fī·iŋ ‚skrēn }

intensity [PHYS] **1.** The strength or amount of a quantity, as of electric field, current, magnetization, radiation, or radioactivity. **2.** The power transmitted by a light or sound wave across a unit area perpendicular to the wave. { in'ten·səd·ē }

intensity control See brightness control. { in'ten·səd·ē kən‚tröl }

intensity level [PHYS] The logarithm of the ratio of two intensities, powers or energies, usually expressed in decibels. { in'ten·səd·ē ‚lev·əl }

intensity modulation [ELECTR] Modulation of electron beam intensity in a cathode-ray tube in accordance with the magnitude of the received signal. { in'ten·səd·ē ‚mäj·ə'lā·shən }

intensity of magnetization See intrinsic induction. { in'ten·səd·ē əv ‚mag·nəd·ə'zā·shən }

intensity of turbulence See gustiness components. { in'ten·səd·ē əv 'tər·byə·ləns }

intensive properties [CHEM] Properties independent of the quantity or shape of the substance under consideration; for example, temperature, pressure, or composition. { in'ten·siv 'präp·ərd·ēz }

intention tremor [MED] A clinical manifestation of certain diseases of the nervous system characterized by involuntary trembling of the limbs brought on by voluntary movements, and ceasing on rest. { in'ten·chən ‚trem·ər }

interacting boson model [NUC PHYS] A model of nuclear structure which assumes that the particles making up a complex nucleus are bosons, each of which corresponds to a pair of correlated nucleons. { ‚in·tər'ak·tiŋ 'bō‚sän ‚mäd·əl }

interaction [FL MECH] With respect to wave components, the nonlinear action by which properties of fluid flow (such as momentum, energy, vorticity), are transferred from one portion of the wave spectrum to another, or viewed in another manner, between eddies of different size-scales. [PHYS] A process in which two or more bodies exert mutual forces on each other. [STAT] The phenomenon which causes the response to applying two treatments not to be the simple sum of the responses to each treatment. { ‚in·tə‚rak·shən }

interaction balance method See goal coordination method. { ‚in·tə‚rak·shən 'bal·əns ‚meth·əd }

interaction picture [QUANT MECH] A mode of description of a system in which the time dependence is carried partly by the operators and partly by the state vectors, the time dependence of the state vectors being due entirely to that part of the Hamiltonian arising from interactions between particles. Also known as interaction representation. { ‚in·tə‚rak·shən ‚pik·chər }

interaction prediction method [CONT SYS] A method for coordinating the subproblem solutions in plant decomposition, in which the interaction variables are specified by the second-level controller according to overall optimality conditions, and the subproblems are solved to satisfy local optimality conditions constrained by the specified values of the interaction variables. Also known as feasible method. { ‚in·tə‚rak·shən prə'dik·shən ‚meth·əd }

interaction representation See interaction picture. { ‚in·tə‚rak·shən ‚rep·ri·zen'tā·shən }

interaction space [ELECTR] A region of an electron tube in which electrons interact with an alternating electromagnetic field. { ‚in·tə‚rak·shən ‚spās }

interactive graphical input [COMPUT SCI] Information which is delivered to a computer by using hand-held devices, such as writing styli used with electronic tablets and lightpens used with cathode-ray tube displays, to sketch a problem description in an on-line interactive mode in which the computer acts as a drafting assistant with unusual powers, such as converting rough freehand motions of a pen or stylus to accurate picture elements. { ‚in·tə‚rak·tiv 'graf·ə·kəl 'in‚pút }

interactive information system [COMPUT SCI] An information system in which the user communicates with the computing facility through a terminal and receives rapid responses which can be used to prepare the next input. { ‚in·tə‚rak·tiv ‚in·fər'mā·shən ‚sis·təm }

interactive language [COMPUT SCI] A programming language designed to operate in an environment in which the user and computer communicate as transactions are being processed. { ‚in·tə‚rak·tiv 'laŋ·gwij }

interactive processing [COMPUT SCI] Computer processing in which the user can modify the operation appropriately while observing results at critical steps. { ‚in·tə‚rak·tiv 'prä·ses·iŋ }

interactive television [COMMUN] A form of television in which the signal is personalized and the viewer can control its various parameters. { ‚in·tə‚rak·tiv 'tel·ə‚vizh·ən }

interactive terminal [COMPUT SCI] A computer terminal designed for two-way communication between operator and computer. { ‚in·tə‚rak·tiv 'tər·mən·əl }

interambulacrum [INV ZOO] In echinoderms, an area between two ambulacra. { ‚in·ər‚am·byə'la·krəm }

Interamnia [ASTRON] An asteroid with a diameter of about 203 miles (327 kilometers), mean distance from the sun of 3.06 astronomical units, and F-type (C-like) surface composition. { ‚in·tər'am·nē·ə }

interarticular [ANAT] Situated between articulating surfaces. { ‚in·tər‚är'tik·yə·lər }

interatrial [ANAT] Located between the atria of the heart. { ‚in·tər'ā·trē·əl }

interatrial septal defect [MED] A congenital malformation of the septum between the atria of the heart. { ‚in·tər'ā·trē·əl ‚sept·əl 'dē‚fekt }

interatrial septum See atrial septum. { ‚in·tər'ā·trē·əl 'sep·təm }

interaxillary [BOT] Located within or between the axils of leaves. { ‚in·tər'ak·sə‚ler·ē }

interbase current [ELECTR] The current that flows from one base connection of a junction tetrode transistor to the other, through the base region. { 'in·tər‚bās 'kə·rənt }

interbedded [GEOL] Having beds lying between other beds with different characteristics. { ‚in·tər'bed·əd }

interblock [COMPUT SCI] A device or system that prevents one part of a computing system from interfering with another. { 'in·tər‚bläk }

interblock gap [COMPUT SCI] A space separating two blocks of data on a magnetic tape. { 'in·tər‚bläk ‚gap }

intercalary [BOT] Referring to growth occurring between

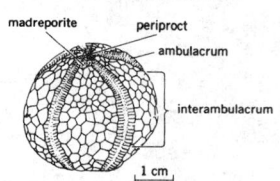

INTERAMBULACRUM

madreporite · periproct · ambulacrum · interambulacrum

1 cm

Morphological features of *Aulechinus grayae*, an echinocystitoid example from the Upper Ordovician of Scotland.

the apex and leaf. [SCI TECH] Inserted between two original components. { in'tər·kə,ler·ē }

intercalary day [ASTRON] A day inserted or introduced among others in a calendar, as February 29 during leap years. { in'tər·kə,ler·ē 'dā }

intercalary meristem [BOT] A meristem that is forming between regions of permanent or mature meristem. { in'tər·kə,ler·ē 'mer·ə,stem }

intercalated disc [HISTOL] A dense region at the junction of cellular units in cardiac muscle. { in'tər·kə,lād·əd 'disk }

intercalated graphite [MATER] An electrically conductive material made by impregnating graphite fiber or powder with metal-rich compounds that lodge between the stacked layers of the graphite. { ,in·tər'ka,lād·əd 'gra,fīt }

intercalated nucleus [ANAT] A nucleus of the medulla oblongata in the central gray matter of the ventricular floor located between the hypoglossal nucleus and the dorsal motor nucleus of the vagus. { in'tər·kə,lād·əd 'nü·klē·əs }

intercalating agent [CELL MOL] A chemical substance that can insert itself between base pairs in a deoxyribonucleic acid molecule. { ,in·tər'ka,lād·iŋ ,ā·jənt }

intercalation [GEOL] A layer located between layers of different character. { in,tər·kə'lā·shən }

intercalibration [ANALY CHEM] A state achieved by a group of laboratories engaged in a monitoring program in which they produce and maintain compatible data outputs. { ,in·tər,kal·ə'brā·shən }

intercapillary [ANAT] Located between capillaries. { ,in·tər'kap·ə,ler·ē }

intercapillary glomerulosclerosis [PATH] Nodular eosinophilic hyalin deposits on the periphery of glomeruli in individuals with diabetes. Also known as diabetic glomerulosclerosis; Kimmelstiel-Wilson disease. { ,in·tər'kap·ə,ler·ē glə,mer·yə·lō·sklə'rō·səs }

intercardinal heading [NAV] A heading in the direction of any of the intercardinal points. { ,in·tər'kärd·ən·əl 'hed·iŋ }

intercardinal point [GEOD] Any of the four directions midway between the cardinal points, that is, northeast, southeast, southwest, and northwest. Also known as quadrantal point. { ,in·tər'kärd·ən·əl ,point }

intercardinal rolling error [NAV] Quadrantal error of a gyro compass. { ,in·tər'kärd·ən·əl ,rōl·iŋ ,er·ər }

intercarpal [ANAT] Located between the carpal bones. { ,in·tər'kärp·əl }

intercarrier channel [COMMUN] A carrier telegraph channel in the available frequency spectrum between carrier telephone channels. { ,in·tər'kar·ē·ər ,chan·əl }

intercarrier noise suppression [ELECTR] Means of suppressing the noise resulting from increased gain when a high-gain receiver with automatic volume control is tuned between stations; the suppression circuit automatically blocks the audio-frequency input of the receiver when no signal exists at the second detector. Also known as interstation noise suppression. { ,in·tər'kar·ē·ər ,noiz sə,presh·ən }

intercarrier sound system [ELECTR] A television receiver arrangement in which the television picture carrier and the associated sound carrier are amplified together by the video intermediate-frequency amplifier and passed through the second detector, to give the conventional video signal plus a frequency-modulated sound signal whose center frequency is the 4.5 megahertz difference between the two carrier frequencies. Abbreviated ICS system. { ,in·tər'kar·ē·ər 'saund ,sis·təm }

intercavernous sinuses [ANAT] Venous sinuses located on the median line of the dura mater, connecting the cavernous sinuses of each side. { ,in·tər'kav·ər·nəs 'sī·nə·səz }

intercellular [HISTOL] Of or pertaining to the region between cells. { ,in·tər'sel·yə·lər }

intercellular cement [HISTOL] A substance bonding epithelial cells together. { ,in·tər'sel·yə·lər si'ment }

intercellular junction [CELL MOL] Any specialized region of contact between the membranes of adjacent cells. { ,in·tər'səl·yə·lər 'jəŋk·shən }

intercellular plexus [NEUROSCI] A network of neuronal processes surrounding the cells in a sympathetic ganglion. { ,in·tər'sel·yə·lər 'plek·səs }

intercellular space [HISTOL] A space between adjacent cells. { ,in·tər'sel·yə·lər ,spās }

intercellular substance [HISTOL] Tissue component that lies between cells. { ,in·tər'sel·yə·lər ,səb·stəns }

intercentrum [VERT ZOO] A type of crescentic intervertebral structure between successive centra in certain reptilian and mammalian tails. { ,in·tər'sen·trəm }

intercept [CRYSTAL] One of the distances cut off a crystal's reference axis by planes. [MAP] See altitude difference. [MATH] The point where a straight line crosses one of the axes of a cartesian coordinate system. { ,in·tər'sept }

intercept call [COMMUN] In telephone practice, routing of a call placed to a disconnected or nonexisting telephone number, to an operator, to a machine answering device, or to a tone. { ,in·tər·sept ,kol }

intercepting sewer [CIV ENG] A sewer that receives flow from transverse sewers and conducts the water to a treatment plant or disposal point. { ,in·tər'sep·tiŋ 'sü·ər }

interception [COMMUN] Tapping or tuning in to a telephone or radio message not intended for the listener. [HYD] **1.** The process by which precipitation is caught and retained on vegetation or structures and subsequently evaporated without reaching the ground. **2.** That part of the precipitation intercepted by vegetation. [METEOROL] **1.** The loss of sunshine, a part of which may be intercepted by hills, trees, or tall buildings. **2.** The depletion of part of the solar spectrum by atmospheric gases and suspensoids; this commonly refers to the absorption of ultraviolet radiation by ozone and dust. [ORD] Meeting or interrupting the course of a moving vessel, aircraft, or missile. { ,in·tər'sep·shən }

intercept method [MET] A method for determining grain size or the quantity of a phase in a microstructure by measuring the number of grains or phase particles per unit length intersected by straight lines. [NAV] See Saint Hilaire method. { 'in·tər,sept ,meth·əd }

interceptometer [ENG] A rain gage which is placed under trees or in foliage to determine the rainfall in that location; by comparing this catch with that from a rain gage set in the open, the amount of rainfall which has been intercepted by foliage is found. { ,in·tər,sep'täm·əd·ər }

interceptor [AERO ENG] A manned aircraft utilized for the identification or engagement of airborne objects. { ,in·tər·sep·tər }

intercept station [COMMUN] Provides service for subscribers whereby calls to disconnected stations or dead lines are either routed to an intercept operator for explanation, or the calling party receives a distinctive tone that informs the party that such a call has been made. { 'in·tər,sept ,stā·shən }

intercept tape [COMMUN] A tape used for temporary storage of messages for trunk channels and tributary stations that are having equipment or circuit trouble. { 'in·tər,sept ,tāp }

intercept trunk [COMMUN] Trunk to which a call for a vacant number, a changed number, or a line out of order is connected for action by an operator. { 'in·tər,sept ,trəŋk }

interchange [CIV ENG] A junction of two or more highways at a number of separate levels so that traffic can pass from one highway to another without the crossing at grade of traffic streams. [ELEC] The current flowing into or out of a power system which is interconnected with one or more other power systems. { 'in·tər,chānj }

interchangeability [ENG] The ability to replace the components, parts, or equipment of one manufacturer with those of another, without losing function or suitability. { ,in·tər,chānj·ə'bil·əd·ē }

interchangeable lens [OPTICS] A lens which can be used in place of another, generally of different magnification. { ,in·tər,chānj·ə·bəl 'lenz }

interchange coefficient See exchange coefficient. { 'in·tər,chānj ,kō·i'fish·ənt }

interchannel crosstalk [COMMUN] Crosstalk between channels in a multiplex system. { ,in·tər,chan·əl 'krós,tók }

interclavicle [VERT ZOO] A membrane bone in front of the sternum and between the clavicles in monotremes and most reptiles. { ,in·tər,klav·ə·kəl }

intercloud discharge See cloud-to-cloud discharge. { 'in·tər,klaud 'dis,chärj }

intercluster medium [ASTRON] A hot x-ray-emitting gas that pervades the space between the members of a galaxy cluster. { ,in·tər,kləs·tər 'mēd·ē·əm }

intercolumnation [ARCH] Distance between columns, measured between the bottoms of shafts, just above the apophyge, and expressed in terms of the lower diameter of the column. { ,in·tər,kál·əm'nā·shən }

intercom *See* intercommunicating system. { 'in·tər,käm }

intercombination line [ASTRON] A spectral line emitted in a transition between energy levels that have different multiplicities, that is, different values of the total spin quantum number. { ¦in·tər,käm·bə'nā·shən ,līn }

intercommunicating porosity [MET] The type of porosity in a sintered metal powder compact that allows fluid to pass from pore to pore. { ¦in·tər·kə'myü·nə,kād·iŋ pə'räs·əd·ēe }

intercommunicating system Also known as intercom. [COMMUN] **1.** A telephone system providing direct communication between telephones on the same premises. **2.** A two-way communication system having a microphone and loudspeaker at each station and providing communication within a limited area. { ¦in·tər·kə'myü·nə,kād·iŋ ,sis·təm }

intercommunication [PETRO ENG] Flow interconnection between the reservoir areas being drained by adjacent wells. { ¦in·tər·kə,myü·nə'kā·shən }

intercondenser [MECH ENG] A condenser between stages of a multistage steam jet pump. { ¦in·tər·kən'den·sər }

interconnected multiple processor [COMPUT SCI] A collection of computers that are physically separated but linked by communication channels to handle distributed data processing. { ¦in·tər·kə·nek·təd 'məl·tə·pəl 'präs,es·ər }

interconnection [ELEC] A link between power systems enabling them to draw on one another's reserves in time of need and to take advantage of energy cost differentials resulting from such factors as load diversity, seasonal conditions, time-zone differences, and shared investment in larger generating units. { ¦in·tər·kə'nek·shən }

intercontinental ballistic missile [ORD] A missile flying a ballistic trajectory after guided powered flight, usually over ranges in excess of 4000 miles (6500 kilometers). Abbreviated ICBM. { ¦in·tər,kant·ən'ent·əl bə¦lis·tik 'mis·əl }

intercontinental sea [GEOGR] A large body of salt water extending between two continents. { ¦in·tər,kant·ən'ent·əl 'sē }

interconversion [COMMUN] Changing the representation of information from one code to another, as from six-bit to ASCII. { ¦in·tər·kən'vər·zhən }

intercooler [MECH ENG] A heat exchanger for cooling fluid between stages of a multistage compressor with consequent saving in power. { ¦in·tər¦kül·ər }

intercostal [ANAT] Situated or occurring between adjacent ribs. [NAV ARCH] Situated or fitted between adjacent members of a ship's frame. { ¦in·tər¦käst·əl }

intercostal floor [NAV ARCH] A ship's floor constructed of a range of plates fitted between and clipped to longitudinal or side keelsons. { ¦in·tər¦käst·əl ¦flōr }

intercostal muscles [ANAT] Voluntary muscles between adjacent ribs. { ¦in·tər¦käst·əl ¦məs·əlz }

intercostal nerve [NEUROSCI] Any of the branches of the thoracic nerves in the intercostal spaces. { ¦in·tər¦käst·əl ¦nerv }

intercourse *See* coitus. { 'in·tər,kórs }

intercrescence [BIOL] A growing together of tissues. { ¦in·tər'krēs·əns }

intercropping [AGR] A form of multiple cropping in which two or more crops simultaneously occupy the same field. { ¦in·tər'kräp·iŋ }

intercrystalline corrosion [MET] Localized attack occurring along the crystal boundaries of a metal or alloy. Also known as intergranular corrosion. { ¦in·tər'krist·əl·ən kə'rō·zhən }

intercycle [COMPUT SCI] A cycle of operation of a punched-card tabulator or other punched-card machine during which card feeding is stopped to permit calculation and printing of control totals or to effect a change in control. { 'in·tər,sī·kəl }

interdendritic attack *See* interdendritic corrosion. { ¦in·tər,den'drid·ik ə'tak }

interdendritic corrosion [MET] Preferential corrosion of the metal immediately surrounding dendrites in unworked or slightly worked alloys caused by composition gradients. Also known as interdendritic attack. { ¦in·tər,den'drid·ik kə'rō·zhən }

interdiction [ORD] The prevention or destruction of, or interference with, enemy movements, communications, and lines of communication, as by gunfire, shelling, or bombing. { ,in·tər'dik·shən }

interdiction fire [ORD] Gunfire delivered in the process of interdiction; for example, firing on a specified place, as a rail yard, assembly area, crossroad, or the like, to prevent effective use of that area. { ,in·tər'dik·shən ¦fīr }

interdiffusion [PHYS CHEM] The self-mixing of two fluids, initially separated by a diaphragm. { ¦in·tər·də'fyü·zhən }

interdigital magnetron [ELECTR] Magnetron having axial anode segments around the cathode, alternate segments being connected together at one end, remaining segments connected together at the opposite end. { ,in·tər'dij·əd·əl 'mag·nə,trän }

interdigital structure [ELECTR] A structure in which the length of the region between two electrodes is increased by an interlocking-finger design for metallization of the electrodes. Also known as interdigitated structure. { ,in·tər'dij·əd·əl ,strək·chər }

interdigital transducer [ELECTR] Two interlocking comb-shaped metallic patterns applied to a piezoelectric substrate such as quartz or lithium niobate, used for converting microwave voltages to surface acoustic waves, or vice versa. { ,in·tər'dij·əd·əl tranz'dü·sər }

interdigitated structure *See* interdigital structure. { ,in·tər'dij·ə,tād·əd ,strək·chər }

interelectrode capacitance [ELECTR] The capacitance between one electrode of an electron tube and the next electrode on the anode side. Also known as direct interelectrode capacitance. { ,in·tər·i'lek,trōd kə'pas·əd·əns }

interelectrode transit time [ELECTR] Time required for an electron to traverse the distance between the two electrodes. { ,in·tər·i'lek,trōd 'tran·zət ,tīm }

interface [COMPUT SCI] **1.** Some form of electronic device that enables one piece of gear to communicate with or control another. **2.** A device linking two otherwise incompatible devices, such as an editing terminal of one manufacturer to typesetter of another. [GEOPHYS] *See* seismic discontinuity. [PHYS CHEM] The boundary between any two phases; among the three phases (gas, liquid, and solid), there are five types of interfaces: gas-liquid, gas-solid, liquid-liquid, liquid-solid, and solid-solid. [SCI TECH] A shared boundary; it may be a piece of hardware used between two pieces of equipment, a portion of computer storage accessed by two or more programs, or a surface that forms the boundary between two types of materials. { 'in·tər,fās }

interface adapter [COMMUN] A device that connects a terminal or computer to a network. { 'in·tər,fās ə,dap·tər }

interface card [COMPUT SCI] A card containing circuits that allow a device to interface with other devices. { 'in·tər,fās ,kärd }

interface connection *See* feedthrough. { 'in·tər,fās kə,nek·shən }

interface control module [COMPUT SCI] Relocatable modularized compiler allowing for efficient operation and easy maintenance. { 'in·tər,fās kən'trōl ,mäj·ül }

interface mixing [PHYS CHEM] The mixing of two immiscible or partially miscible liquids at the plane of contact (interface). { 'in·tər,fās ,mik·siŋ }

interface resistance [THERMO] **1.** Impairment of heat flow caused by the imperfect contact between two materials at an interface. **2.** Quantitatively, the temperature difference across the interface divided by the heat flux through it. { 'in·tər,fās ri'zis·təns }

interfacial angle [CRYSTAL] The angle between two crystal faces. { 'in·tər,fā·shəl ,aŋ·gəl }

interfacial energy [PHYS] The free energy of the surfaces at an interface, resulting from differences in the tendencies of each phase to attract its own molecules; equal to the surface tension. Also known as surface energy. { 'in·tər,fā·shəl ,en·ər·jē }

interfacial force *See* surface tension. { 'in·tər,fā·shəl ,fórs }

interfacial layer [PHYS CHEM] A one- or two-molecules-thick boundary between any two bulk phases (gas, liquid, or solid) in contact where the properties differ from the properties of the bulk phases. { ,in·tər,fā·shəl 'lā·ər }

interfacial polarization [ELEC] *See* space-charge polarization. [OPTICS] Polarization of light by reflection from the surface of a dielectric at Brewster's angle. { 'in·tər,fā·shəl ,pō·lə·rə'zā·shən }

interfacial tension *See* surface tension. { 'in·tər,fā·shəl 'ten·chən }

INTERFACE RESISTANCE

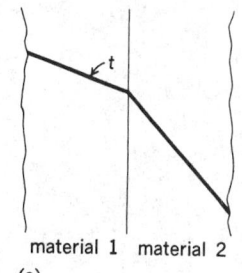

material 1   material 2
(a)

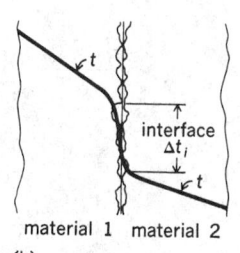

material 1   material 2
(b)

Distribution of temperature t through composite wall. (*a*) With perfect interface contact. (*b*) For typical actual surfaces. Δt_i is temperature difference across interface. Interface resistance equals Δt_i divided by heat flux.

interfacility transfer trunk [COMMUN] Trunk interconnecting switching centers of two different facilities. { ˌin·tər·fəˈsil·əd·ē ˈtranz·fər ˌtrəŋk }

interfacing [TEXT] A woven or nonwoven fabric used between the outer fabric and lining to reinforce or stiffen a feature or detail on a garment. { ˈin·tərˌfās·iŋ }

interfascicular cambium [BOT] The vascular cambium that develops between vascular bundles. { ¦in·tər·faˈsik·yə·lər ˈkam·bē·əm }

interference [ANALY CHEM] A systematic error in measurement that occurs when concomitants are present in the sample being analyzed. [COMMUN] Any undesired energy that tends to interfere with the reception of desired signals. Also known as electrical interference; radio interference. [PHYS] The variation with distance or time of the amplitude of a wave which results from the superposition (algebraic or vector addition) of two or more waves having the same, or nearly the same, frequency. Also known as wave interference. { ˌin·tərˈfir·əns }

interference analyzer [ELECTR] An instrument that discloses the frequency and amplitude of unwanted input. { ˌin·tərˈfir·əns ¦an·əˌlīz·ər }

interference blanker [ELECTR] Device that permits simultaneous operation of two or more pieces of radio or radar equipment without confusion of intelligence, or that suppresses undesired signals when used with a single receiver. { ˌin·tərˈfir·əns ¦blaŋ·kər }

interference colors [OPTICS] Colors formed by interference of a beam of light passed through a thin section of a mineral placed in a polarizing microscope. { ˌin·tərˈfir·əns ˌkəl·ərz }

interference control [ORD] The monitoring of radio frequencies assigned to a missile range to detect interfering radiations that could cause missile-borne equipment to malfunction, and the concerted effort to locate and terminate the source of such radiations. { ˌin·tərˈfir·əns kənˌtrōl }

interference fading [COMMUN] Fading of the signal produced by different wave components traveling slightly different paths in arriving at the receiver (often termed multipath). { ˌin·tərˈfir·əns ˌfād·iŋ }

interference figure [OPTICS] A pattern of light and dark areas observed with a conoscope when a birefringent crystal is placed in a convergent beam of linearly polarized light. { ˌin·tərˈfir·əns ˌfig·yər }

interference filter [ELECTR] **1.** A filter used to attenuate artificial interference signals entering a receiver through its power line. **2.** A filter used to attenuate unwanted carrier-frequency signals in the tuned circuits of a receiver. [OPTICS] An optical filter in which the wavelengths that are not transmitted are removed by interference phenomena rather then by absorbtion or scattering. { ˌin·tərˈfir·əns ¦fil·tər }

interference fit [DES ENG] A fit wherein one of the mating parts of an assembly is forced into a space provided by the other part in such a way that the condition of maximum metal overlap is achieved. { ˌin·tərˈfir·əns ˌfit }

interference fringes [OPTICS] A series of light and dark bands produced by interference of light waves. { ˌin·tərˈfir·əns ¦frin·jəz }

interference microscope [OPTICS] A microscope used for visualizing and measuring differences in phase or optical paths in transparent or reflecting specimens; it differs from the phase contrast microscope in that the incident and diffracted waves are not separated, but interference is produced between the transmitted wave and another wave which originates from the same source. { ˌin·tərˈfir·əns ¦mī·krəˌskōp }

interference pattern [ELECTR] Pattern produced on a radarscope by interference signals. [PHYS] Resulting space distribution of pressure, particle density, particle velocity, energy density, or energy flux when progressive waves of the same frequency and kind are superimposed. { ˌin·tərˈfir·əns ¦pad·ərn }

interference phenomenon [VIROL] Inhibition by a virus of the simultaneous infection of host cells by some other virus. { ˌin·tərˈfir·əns fəˈnäm·əˌnän }

interference prediction [ELECTR] Process of estimating the interference level of a particular equipment as a function of its future electromagnetic environment. { ˌin·tərˈfir·əns prəˈdik·shən }

interference range [GEN] The smallest genetic distance that

is large enough for two crossing-over events not to interfere with each other. { ˌin·tərˈfir·əns ˌranj }

interference reduction [ELECTR] Reduction of interference from such causes as power lines and equipment, radio transmitters, and lightning, usually through the use of electric filters. Also known as interference suppression. { ˌin·tərˈfir·əns riˈdək·shən }

interference region [COMMUN] That region in space in which interference between wave trains occurs; in microwave propagation, it refers to the region bounded by the ray path and the surface of the earth which is above the radio horizon. { ˌin·tərˈfir·əns ˌrē·jən }

interference rejection [ELECTR] Use of a filter to reject (to bypass to ground) unwanted input. { ˌin·tərˈfir·əns riˈjek·shən }

interference ripple mark [GEOL] A pattern resulting from two sets of symmetrical ripples formed by waves crossing at right angles. { ˌin·tərˈfir·əns ˈrip·əl ˌmärk }

interference source suppression [ELECTR] Techniques applied at or near the source to reduce its emission of undesired signals. { ˌin·tərˈfir·əns ˈsōrs səˌpresh·ən }

interference spectrum [ELECTR] Frequency distribution of the jamming interference in the propagation medium external to the receiver. [SPECT] A spectrum that results from interference of light, as in a very thin film. { ˌin·tərˈfir·əns ¦spek·trəm }

interference suppression *See* interference reduction. { ˌin·tərˈfir·əns səˈpresh·ən }

interference test [PETRO ENG] Test of pressure interrelationships (interference) between wells serving the same formation. { ˌin·tərˈfir·əns ˌtest }

interference theory [PSYCH] The concept that forgetting occurs when one memory replaces or becomes confused with another memory. { ˌin·tərˈfir·əns ˌthē·ə·rē }

interference time [IND ENG] Idle machine time occurring when a machine operator, assigned to two or more semiautomatic machines, is unable to service a machine requiring attention. { ˌin·tərˈfir·əns ˌtīm }

interference wave [COMMUN] A radio wave reflected by the lower atmosphere which produces an interference pattern when combined with the direct wave. { ˌin·tərˈfir·əns ˌwāv }

interference wiggler *See* undulator. { ˌin·tərˈfir·əns ¦wig·lər }

interferogram [GRAPHICS] A photograph of a shock wave or other fluid flow, produced with the aid of an interferometer, in which the displacement of interference fringes is proportional to the density change across the fluid. [SPECT] A graph of the variation of the output signal from an interferometer as the condition for interference within the interferometer is varied. { ˌin·təˈfir·əˌgram }

interferometer [OPTICS] An instrument in which light from a source is split into two or more beams, which are subsequently reunited after traveling over different paths and display interference. { ˌin·təˈfəˈräm·əd·ər }

interferometer systems [ELECTR] Method of determining the position of a target in azimuth by using an interferometer to compare the phases of signals at the output terminals of a pair of antennas receiving a common signal from a distant source. { ˌin·təˈfəˈräm·əd·ər ˌsis·təmz }

interferometric hydrophone [ENG] A hydrophone in which pressure changes act directly or indirectly to deform an optical fiber and thus produce a phase change in light from a laser or light-emitting diode; the phase change is detected in an interferometer. Also known as fiber-optic hydrophone. { ˌin·tərˈfir·əˌme·trik ˌhī·drəˌfōn }

interferometry [OPTICS] The design and use of optical inferometers; uses include precise measurement of wavelength, measurement of very small distances and thicknesses, study of hyperfine structure of spectral lines, precise measurement of indices of refraction, and determination of separations of binary stars and diameters of very large stars. { ˌin·təˈfəˈräm·ə·trē }

interferon [BIOCHEM] A protein produced by intact animal cells when infected with viruses; acts to inhibit viral reproduction and to induce resistance in host cells. { ˌin·tərˈfirˌän }

interferon-alpha [IMMUNOL] A low-molecular-weight protein produced by leukocytes in response to viral infection. { ˌin·tərˌfirˌän ˈal·fə }

interferon-beta [IMMUNOL] An interferon produced by

fibroblasts in response to viruses or foreign nucleic acids. { ˌin·tərˈfir‚än ˈbäd·ə }

interferon-gamma [IMMUNOL] An interferon produced by T lymphocytes and large granular lymphocytes in response to foreign macromolecules. { ˌin·tərˈfir‚än ˈgam·ə }

interferonogen [VIROL] A preparation made of inactivated virus particles used as an inoculant to stimulate formation of interferon. { ¦in·tə·fəˈrän·ə·jən }

interfit [ENG] The distance extended by the ends of one bit cone into the grooves of an adjacent one in a roller cone bit. Also known as intermesh. { ˈin·tərˌfit }

interfix [COMPUT SCI] A technique for describing relationships of key words in an item or document in a way which prevents crosstalk from causing false retrievals when very specific entries are made. { ˈin·tərˌfiks }

interflow [HYD] The water, derived from precipitation, that infiltrates the soil surface and then moves laterally through the upper layers of soil above the water table until it reaches a stream channel or returns to the surface at some point downslope from its point of infiltration. { ˈin·tərˌflō }

interfluve [GEOL] The area of land between two rivers, usually an upland or ridge between two adjacent valleys that contain streams flowing in approximately the same direction. { ˈin·tərˌflüv }

interfoliaceous [BOT] Between a pair of leaves, such as between those which are opposite or verticillate. { ¦in·tərˌfōˈlēˈā·shəs }

intergalactic [ASTRON] Pertaining to the space between the galaxies. { ¦in·tər·gəˈlak·tik }

intergalactic matter [ASTRON] The material between the galaxies. { ˌin·tər·gəˌlak·tik ˈmad·ər }

intergelisol See pereletok. { ¦in·tərˈjel·əˌsȯl }

intergenic crossing-over [CELL MOL] Recombination between distinct genes or cistrons. { ¦in·tərˈjen·ik ¦krȯs·iŋˈō·vər }

intergenic suppression [GEN] The restoration of a suppressed function or character by a second mutation that is located in a different gene than the original or first mutation. { ¦in·tərˈjen·ik səˈpresh·ən }

intergenote [CELL MOL] In hybrid bacteria, a chromosome with integrated deoxyribonucleic acid of foreign origin. { ˌin·tərˈjēˌnōt }

interglacial [GEOL] Pertaining to or formed during a period of geologic time between two successive glacial epochs or between two glacial stages. { ¦in·tərˈglā·shəl }

intergranular [SCI TECH] Occurring between grains. { ¦in·tərˈgran·yə·lər }

intergranular corrosion See intercrystalline corrosion. { ¦in·tərˈgran·yə·lər kəˈrō·zhən }

intergranular fracture [MET] Propagation of a crack along the grain boundaries of a metal or alloy. { ¦in·tərˈgran·yə·lər ˈfrak·chər }

intergranular pressure See effective stress. { ¦in·tərˈgran·yə·lər ˈpresh·ər }

intergrowth [MINERAL] A state of interlocking of different mineral crystals because of simultaneous crystallization. { ˈin·tərˌgrōth }

interhalogen [INORG CHEM] Any of the compounds formed from the elements of the halogen family that react with each other to form a series of binary compounds; for example, iodine monofluoride. { ¦in·tərˈhal·ə·jən }

interhemispheric integration [PSYCH] The process by which information is exchanged between the cerebral hemispheres. { ¦in·tərˌhem·əˈsfir·ik ˌint·əˈgrā·shən }

interionic attraction [PHYS CHEM] The coulomb attraction between ions of opposite sign in a solution. { ˌin·tir·ēˈän·ik əˈtrak·shən }

interior [MATH] **1.** For a set A in a topological space, the set of all interior points of A. **2.** For a plane figure, the set of all points inside the figure. **3.** For an angle, the set of points that lie in the plane of the angle and between the rays defining the angle. **4.** For a simple closed plane curve, one of the two regions into which the curve divides the plane according to the Jordan curve theorem, namely, the region that is bounded. { inˈtir·ē·ər }

interior angle [MATH] **1.** An angle between two adjacent sides of a polygon that lies within the polygon. **2.** For a line (called the transversal) that intersects two other lines, an angle between the transversal and one of the two lines that lies within the space between the two lines. { inˈtir·ē·ər ˈaŋ·gəl }

interior ballistics [MECH] The science concerned with the combustion of powder, development of pressure, and movement of a projectile in the bore of a gun. { inˈtir·ē·ər bəˈlis·tiks }

interior content See interior Jordan content. { inˈtir·ē·ər ˈkänˌtent }

interior distribution [ELEC] Distribution of electric power within a building or plant. { inˈtir·ē·ər ˌdi·strəˈbyü·shən }

interior Jordan content [MATH] Also known as interior content. **1.** For a set a points on a line, the smallest number C such that the sum of the lengths of a finite number of open, nonoverlapping intervals that are completely contained in the set is always equal to or less than C. **2.** The interior Jordan content of a set of points, X, in n-dimensional Euclidean space (where n is a positive integer) is the least upper bound on the hypervolume of the union of a finite set of hypercubes that is contained in X. { inˌtir·ē ər ˈjȯrd·ən ¦känˌtent }

interior label [COMPUT SCI] A label attached to the data that it identifies. { inˈtir·ē·ər ˈlā·bəl }

interior measure See Lebesgue interior measure. { inˈtir·ē·ər ˈmezh·ər }

interior point [MATH] A point p in a topological space is an interior point of a set S if there is some open neighborhood of p which is contained in S. { inˈtir·ē·ər ˈpȯint }

interkinesis See interphase. { ¦in·tər·kəˈnē·səs }

interlabium [INV ZOO] A small lobe situated between the lips in certain nematodes. { ¦in·tərˈlā·bē·əm }

interlace [COMPUT SCI] To assign successive memory location numbers to physically separated locations on a storage tape or magnetic drum of a computer, usually to reduce access time. { ¦in·tərˈlās }

interlaced scanning [ELECTR] A scanning process in which the distance from center to center of successively scanned lines is two or more times the nominal line width, so that adjacent lines belong to different fields. Also known as line interlace. { ¦in·tərˈlāst ˈskan·iŋ }

interlace operation [COMPUT SCI] System of computer operation where data can be read out or copied into memory without interfering with the other activities of the computer. { ¦in·tərˈlās ˌäp·əˈrā·shən }

interlacing arches [ARCH] Arches with intersecting curved moldings that appear to be interlaced. { ¦in·tərˈlās·iŋ ˈär·chəz }

interlaminated [SCI TECH] **1.** Having insertions between laminae. **2.** Arranged in alternate laminae. { ¦in·tərˈlam·əˌnād·əd }

interleave [COMPUT SCI] **1.** To alternate parts of one sequence with parts of one or more other sequences in a cyclic fashion such that each sequence retains its identity. **2.** To arrange the members of a sequence of memory addresses in different memory modules of a computer system, in order to reduce the time taken to access the sequence. { ˌin·tərˈlēv }

interleaved windings [ELEC] An arrangement of winding coils around a transformer core in which the coils are wound in the form of a disk, with a group of disks for the low-voltage windings stacked alternately with a group of disks for the high-voltage windings. { ¦in·tərˈlēvd ˈwīn·diŋz }

interleukin [IMMUNOL] Any of a class of proteins that are secreted mostly by macrophages and T lymphocytes and induce growth and differentiation of lymphocytes and hematopoietic stem cells. { ˌin·tərˈlü·kən }

interleukin-1 [IMMUNOL] A cytokine produced by macrophages, endothelial cells, lymphocytes, and epidermal cells that plays roles in the inflammatory process and in the immune response. Abbreviated IL-1. { ˌin·tərˈlük·ən ˈwən }

interleukin-2 [IMMUNOL] A lymphokine secreted mostly by helper T lymphocytes that promotes the growth of T lymphocytes. Abbreviated IL-2. { ˌin·tərˈlük·ən ˈtü }

interleukin-3 [IMMUNOL] A cytokine produced by a subset of helper T cells as well as by nonlymphoid cells; it is a growth factor for multiple lineages of hematopoietic cells and can act as an immunoregulatory factor. Abbreviated IL-3. { ˌin·tərˈlük·ən ˈthrē }

interleukin-4 [IMMUNOL] A cytokine that is capable of a variety of activities, such as induction of proliferation by T cells, mast cells, megakarocytes, and erythroid precursors; induction of antibody secretion by B cells; and potentiation

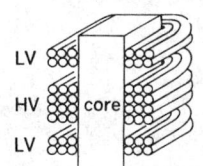

INTERLEAVED WINDINGS

Interleaved windings arrangement. LV = low voltage, HV = high voltage.

of the proliferation of mast cells. Abbreviated IL-4. { ‚in·tər¦lük·ən ¦fŏr }

interleukin-5 [IMMUNOL] A cytokine produced by helper T cells that promotes the development of B cells and stimulates them to produce IgA antibodies, induces proliferation and differentiation of eosinophils, and acts with IL-4 to enhance the production of IgE antibodies. Abbreviated IL-5. { ‚in·tər¦lük·ən ¦fīv }

interleukin-6 [IMMUNOL] A cytokine derived from activated T lymphocytes that has many functions, including induction of B-cell growth; induction of B-cell differentiation and antibody production; induction of differentiation and proliferation of T cells; synergistic induction with IL-3 of hematopoietic cell growth; and induction of hepatocyte secretion of acute-phase inflammatory proteins. Abbreviated IL-6. { ‚in·tər¦lük·ən ¦siks }

interleukin-7 [IMMUNOL] A cytokine that acts as a growth factor for precursors of B cells and T cells and also enhances the generation of interleukin-2–activated nonspecific killer cells and interleukin-2–activated antigen-specific cytotoxic T lymphocytes. Abbreviated IL-7. { ‚in·tər¦lük·ən ¦sev·ən }

interleukin-8 [IMMUNOL] A group of peptides produced by a variety of cell types, which activate and recruit polymorphonuclear leukocytes in the inflammatory process, and are probably involved in initiation of labor and delivery in pregnant women. Abbreviated IL-8. { ‚in·tər¦lük·ən ¦āt }

interleukin-9 [IMMUNOL] A cytokine produced by a type of activated helper T cells; it stimulates hemopoietic cells that develop into red blood cells. Abbreviated IL-9. { ‚in·tər¦lük·ən ¦nīn }

interleukin-10 [IMMUNOL] An immunoregulatory cytokine produced by a subset of helper T cells as well as by B lymphocytes and some cells of the uterus during pregnancy; it inhibits secretion (and function) of cytokine by macrophages and the second population of helper T cells called Th1. Abbreviated IL-10. { ‚in·tər¦lük·ən ¦ten }

interleukin-11 [IMMUNOL] A pleiotropic cytokine produced by bone marrow-derived fibroblasts which supports the growth of certain cell types, such as B cells, neutrophils, and platelet-producing megakaryocytes. Abbreviated IL-11. { ‚in·tər¦lük·ən i'lev·ən }

interleukin-12 [IMMUNOL] A heterodimeric cytokine that stimulates nonspecific cytotoxic natural killer–type cells to produce gamma interferon. Abbreviated IL-12. { ‚in·tər¦lük·ən ¦twelv }

interleukin-13 [IMMUNOL] A cytokine produced by activated T lymphocytes that inhibits inflammatory cytokine production induced by bacterial endotoxin and stimulates gamma-interferon production by natural killer cells, enhancing the effect of interleukin-2. Abbreviated IL-13. { ‚in·tər¦lük·ən ¦thər‚tēn }

interlobate moraine *See* intermediate moraine. { ¦in·tər'lō‚bāt mə'rān }

interlock [COMPUT SCI] **1.** A mechanism, implemented in hardware or software, to coordinate the activity of two or more processes within a computing system, and to ensure that one process has reached a suitable state such that the other may proceed. **2.** *See* deadlock. [ENG] A switch or other device that prevents activation of a piece of equipment when a protective door is open or some other hazard exists. { 'in·tər‚läk }

interlocking cutter [DES ENG] A milling cutter assembly consisting of two mating sections with uniform or alternate overlapping teeth. { ¦in·tər¦läk·iŋ ¦kəd·ər }

interlocking ring structures [ASTRON] Two lunar craters that overlap, but both have their walls intact, indicating that they were formed at the same time. { ¦in·tər¦läk·iŋ ¦riŋ ‚strək·chərz }

interlock relay [ELEC] A relay composed of two or more coils, each with its own armature and associated contacts, so arranged that movement of one armature or the energizing of its coil is dependent on the position of the other armature. { 'in·tər‚läk ‚rē‚lā }

interlock switch [ELEC] A switch designed for mounting on a door, drawer, or cover so that it opens automatically when the door or other part is opened. { 'in·tər‚läk ‚swich }

interlude [COMPUT SCI] A small routine or program which is designed to carry out minor preliminary calculations or housekeeping operations before the main routine begins to

operate, and which can usually be overwritten after it has performed its function. { 'in·tər‚lüd }

intermediary metabolism [BIOCHEM] Intermediate steps in the chemical synthesis and breakdown of foodstuffs within body cells. { ¦in·tər¦mēd·ē‚er·ē me'tab·ə‚liz·əm }

intermediate [CHEM] A precursor to a desired product; ethylene is an intermediate for polyethylene, and ethane is an intermediate for ethylene. [GRAPHICS] That print which is used as a master for further reproduction. { ‚in·tər'mēd·ē·ət }

intermediate annealing [MET] Softening of a metal by heat treatment at one or more stages during cold working and before final treatment. { ‚in·tər'mēd·ē·ət ə'nēl·iŋ }

intermediate control change [COMPUT SCI] A change of function that is of average or relatively moderate magnitude of importance such as the printing of intermediate totals by a punched-card tabulator, resulting from a change in intermediate control data between one card and the next. { ‚in·tər'mēd·ē·ət kən'trōl ‚chānj }

intermediate control data [COMPUT SCI] Control data at a level which is neither the most nor the least significant, or which is used to sort records into groups that are neither the largest nor the smallest used; for example, if control data are used to specify state, town, and street, then the data specifying town would be intermediate control data. { ‚in·tər'mēd·ē·ət kən'trōl ‚dad·ə }

intermediate distributing frame [ELEC] Frame in a local telephone central office, the primary purpose of which is to cross-connect the subscriber line multiple to the subscriber line circuit; in a private exchange, the intermediate distributing frame is for similar purposes. { ‚in·tər'mēd·ē·ət di'strib·yəd·iŋ ‚frām }

intermediate-energy neutron-proton scattering [NUC PHYS] An elastic collision of a neutron having an energy from 10 to 440 megaelectronvolts with a proton (usually the nucleus of a hydrogen atom). { ‚in·tər¦mē·dē·ət 'en·ər·jē ‚nü‚trän ¦prō‚tän ‚skad·ər·iŋ }

intermediate-energy proton-proton scattering [NUC PHYS] An elastic collision of a proton having an energy from 10 to 440 megaelectronvolts with another proton (usually the nucleus of a hydrogen atom). { ‚in·tər¦mē·dē·ət 'en·ər·jē ¦prō‚tän ¦prō‚tän ‚skad·ər·iŋ }

intermediate filament [CELL MOLL] Any of several classes of cell-specific cytoplasmic filaments of 8–12 nanometers in diameter; protein composition varies from one cell type to another. { ‚int·ər'mēd·ē·ət ¦fil·ə·mənt }

intermediate flux [MET] A flux consisting of organic halide compounds whose residues are decomposed by the heat of soldering; fluxing action approaches that of corrosive flux. { ‚in·tər'mēd·ē·ət ¦fləks }

intermediate frequency [ELECTR] The frequency produced by combining the received signal with that of the local oscillator in a superheterodyne receiver. Abbreviated i-f. { ‚in·tər'mēd·ē·ət ¦frē·kwən·sē }

intermediate-frequency amplifier [ELECTR] The section of a superheterodyne receiver that amplifies signals after they have been converted to the fixed intermediate-frequency value by the frequency converter. Abbreviated i-f amplifier. { ‚in·tər'mēd·ē·ət ¦frē·kwən·sē 'am·plə‚fī·ər }

intermediate-frequency jamming [ELECTR] Form of continuous wave jamming that is accomplished by transmitting two continuous wave signals separated by a frequency equal to the center frequency of the radar receiver intermediate-frequency amplifier. { ‚in·tər'mēd·ē·ət ¦frē·kwən·sē 'jam·iŋ }

intermediate-frequency response ratio [ELECTR] In a superheterodyne receiver, the ratio of the intermediate-frequency signal input at the antenna to the desired signal input for identical outputs. Also known as intermediate-interference ratio. { ‚in·tər'mēd·ē·ət ¦frē·kwən·sē ri'späns ‚rā·shō }

intermediate-frequency signal [ELECTR] A modulated or continuous-wave signal whose frequency is the intermediate-frequency value of a superheterodyne receiver and is produced by frequency conversion before demodulation. { ‚in·tər'mēd·ē·ət ¦frē·kwən·sē ‚sig·nəl }

intermediate-frequency stage [ELECTR] One of the stages in the intermediate-frequency amplifier of a superheterodyne receiver. { ‚in·tər'mēd·ē·ət ¦frē·kwən·sē ‚stāj }

intermediate-frequency strip [ELECTR] A receiver subassembly consisting of the intermediate-frequency amplifier stages, installed or replaced as a unit. { ‚in·tər'mēd·ē·ət ¦frē·kwən·sē ‚strip }

intermediate-frequency transformer [ELECTR] The transformer used at the input and output of each intermediate-frequency amplifier stage in a superheterodyne receiver for coupling purposes and to provide selectivity. Abbreviated i-f transformer. { ‚in·tər'mēd·ē·ət ¦frē·kwən·sē tranz'fōr·mər }

intermediate ganglion [NEUROSCI] Any of certain small groups of nerve cells found along communicating branches of spinal nerves. { ‚in·tər'mēd·ē·ət 'gaŋ·glē·ən }

intermediate gear [MECH ENG] An idler gear interposed between a driver and driven gear. { ‚in·tər'mēd·ē·ət ¦gir }

intermediate haulage See relay haulage. { ‚in·tər'mēd·ē·ət ¦hȯl·ij }

intermediate haulage conveyor [MIN ENG] A type of conveyor, usually 500 to 3000 feet (150 to 900 meters) in length, that transports material between the gathering conveyor and the main haulage conveyor. { ‚in·tər'mēd·ē·ət ¦hȯl·ij kən‚vā·ər }

intermediate horizon [ELECTROMAG] Screening object (hill, mountain, ridge, building, and so on) similar to the radar horizon, but not the most distant. { ‚in·tər'mēd·ē·ət hə'rīz·ən }

intermediate host [BIOL] The host in which a parasite multiplies asexually. { ‚in·tər'mēd·ē·ət 'hōst }

intermediate-infrared radiation [ELECTROMAG] Infrared radiation having a wavelength between about 2.5 micrometers and about 50 micrometers; this range includes most molecular vibrations. Also known as mid-infrared radiation. { ‚in·tər'mēd·ē·ət ¦in·frə¦red ‚rād·ē'ā·shən }

intermediate-interference ratio See intermediate-frequency response ratio. { ‚in·tər'mēd·ē·ət ‚in·tər'fir·əns ‚rā·shō }

intermediate ion [METEOROL] An atmospheric ion of size and mobility intermediate between the small ion and the large ion. { ‚in·tər'mēd·ē·ət 'ī‚än }

intermediate language level [COMPUT SCI] A computer program that has been converted by a compiler into a form that does not resemble the original program but that still requires further processing by an interpreter at run time before it can be executed. { ‚in·tər'mēd·ē·ət 'laŋ·gwij ‚lev·əl }

intermediate layer See sima. { ‚in·tər'mēd·ē·ət ¦lā·ər }

intermediate lobe [ANAT] The intermediate portion of the adenohypophysis. { ‚in·tər'mēd·ē·ət ¦lōb }

intermediate material [IND ENG] A manufactured product that requires additional processing before it becomes finished goods. { ‚in·tər'mēd·ē·ət mə'tir·ē·əl }

intermediate memory storage [COMPUT SCI] An electronic device for holding working figures temporarily until needed and for releasing final figures to the output. { ‚in·tər'mēd·ē·ət 'mem‚rē ‚stȯr·ij }

intermediate moraine [GEOL] A type of lateral moraine formed at the junction of two adjacent glacial lobes. Also known as interlobate moraine. { ‚in·tər'mēd·ē·ət mə'rān }

intermediate neutron [NUCLEO] A neutron having energy in a range from about 100 to 100,000 electronvolts. { ‚in·tər'mēd·ē·ət 'nü‚trän }

intermediate phase [MET] In an alloy system, a distinct phase whose composition ranges do not extend to any of the pure constituents of the system. { ‚in·tər'mēd·ē·ət ¦fāz }

intermediate polar [ASTRON] A member of a class of cataclysmic variable stars whose x-ray and optical light curves display large pulses on time scales of minutes. Also known as DQ Herculis star. { ‚in·tər'mēd·ē·ət ¦pōl·ər }

intermediate-range ballistic missile [ORD] A missile flying a ballistic trajectory after guided powered flight and having a range of about 200 to 1500 miles (300 to 2500 kilometers). Abbreviated IRBM. { ‚in·tər'mēd·ē·ət ¦ranj bə'lis·tik ‚mis·əl }

intermediate reactor [NUCLEO] A reactor in which the chain reaction is sustained mainly by intermediate neutrons. { ‚in·tər'mēd·ē·ət rē'ak·tər }

intermediate repeater [ELECTR] Repeater for use in a trunk or line at a point other than an end. { ‚in·tər'mēd·ē·ət ri'pēd·ər }

intermediate result [COMPUT SCI] A quantity or value derived from an operation performed in the course of a program or subroutine which is itself used as an operand in further operations. { ‚in·tər'mēd·ē·ət ri'zəlt }

intermediate state [CRYO] A state of partial superconductivity that occurs when a magnetic field slightly less than the critical field is applied to a superconducting material below its critical temperature. [QUANT MECH] A state through which a system may pass during transition from an initial state to a final state. { ‚in·tər'mēd·ē·ət ¦stāt }

intermediate storage [COMPUT SCI] The portion of the computer storage facilities that usually stores information in the processing stage. { ‚in·tər'mēd·ē·ət 'stȯr·ij }

intermediate total [COMPUT SCI] A sum that is produced when there is a change in the value of control data at a level that is neither the most nor the least significant. { ‚in·tər'mēd·ē·ət ¦tōd·əl }

intermediate trunk distributing frame [ELEC] A frame which mounts terminal blocks for connecting linefinders and first selectors. { ‚in·tər'mēd·ē·ət 'trəŋk di'strib·yəd·iŋ frām }

intermediate value theorem [MATH] If $f(x)$ is a continuous real-valued function on the closed interval from a to b, then, for any y between the least upper bound and the greatest lower bound of the values of f, there is an x between a and b with $f(x) = y$. { ‚in·tər'mēd·ē·ət ¦val·yü 'thir·əm }

intermediate vector boson [PARTIC PHYS] One of the three fundamental particles that transmits the weak nuclear force in the same manner that the photon transmits the electromagnetic force. { ‚in·tər'mēd·ē·ət ¦vek·tər 'bō‚sän }

intermediate vertex [MATH] A vertex in an s-t network that is neither the source nor the terminal. { ‚in·tər'mēd·ē·ət 'vər‚teks }

intermedin [BIOCHEM] A hormonal substance produced by the intermediate portion of the hypophysis of certain animal species which influences pigmentation; similar to melanocyte-stimulating hormone in humans. { ‚in·tər'mēd·ən }

intermembranous ossification [HISTOL] Ossification within connective tissue with no prior formation of cartilage. { ¦in·tər'mem·brə·nəs ‚äs·ə·fə'kā·shən }

intermeningeal [ANAT] Between any two of the three meninges covering the brain and spinal cord. { ‚in·tər·men·ən'jē·əl }

intermenstrual [PHYSIO] Between periods of menstruation. { ‚in·tər'men·strəl }

intermenstrual flow See metrorrhagia. { ‚in·tər'men·strəl 'flō }

intermesh See interfit. { ¦in tər'mesh }

intermetallic alloys [MET] Ordered alloys having a superlattice crystal structure. Unlike conventional alloys, they have a strong chemical arrangement that reduces the mobility of atoms and results in good structural stability, higher melting temperatures, and lower densities. However, most are brittle due to their complex crystal structure, resulting in poor fracture resistance. { ‚in·tər·mə‚tal·ik a‚lȯis }

intermetallic compound See electron compound. { ¦in·tər·me'tal·ik 'käm‚paúnd }

intermetameric [ANAT] Between adjacent metameres. { ¦in·tər¦med·ə¦mer·ik }

intermetatarsal [ANAT] Between adjacent bones of the metatarsus. { ‚in·tər‚med·ə'tär·səl }

intermitotic [CELL MOL] Of or pertaining to a stage of the cell cycle between two successive mitoses. { ‚in·tər‚mī'täd·ik }

intermittency [PHYS] The alternation in time of a dynamical system between nearly periodic and chaotic behavior. { ‚in·tər'mit·ən·sē }

intermittency effect [GRAPHICS] A reduction in the density of a photographic film when the exposing light is interrupted at a very high frequency, even though the total light exposure is held constant. { ‚in·tər'mit·ən·sē i‚fekt }

intermittent claudication [MED] Cramping pain or weakness in the lower extremities during exercise, caused by occlusion of the arteries. { ‚in·tər¦mit·ənt klȯ·də'kā·shən }

intermittent current [ELEC] A unidirectional current that flows and ceases to flow at irregular or regular intervals. [OCEANOGR] A unidirectional current interrupted at intervals. { ¦in·tər¦mit·ənt 'kə·rənt }

intermittent defect [ENG] A defect that is not continuously present. { ¦in·tər¦mit·ənt 'dē‚fek· }

intermittent-duty rating [ENG] An output rating based on operation of a device for specified intervals of time rather than continuous duty. Also known as intermittent rating. { ¦in·tər¦mit·ənt ¦düd·ē 'rād·iŋ }

intermittent firing [MECH ENG] Cyclic firing whereby fuel and air are burned in a furnace for frequent short time periods. { ¦in·tər¦mit·ənt 'fīr·iŋ }

intermittent gas lift [PETRO ENG] A gas-drive oil reservoir that is valved and timed for intermittent activity. { ¦in·tər¦mit·ənt 'gas ˌlift }

intermittent light [NAV] In marine operations, a light having equal periods of light and darkness. { ¦in·tər¦mit·ənt 'līt }

intermittent operation [ENG] Condition in which a device operates normally for a time, then becomes defective for a time, with the process repeating itself at regular or irregular intervals. { ¦in·tər¦mit·ənt ˌäp·ə·'rā·shən }

intermittent quick-flashing light [NAV] In marine operations, a light showing flashes for several seconds, followed by an equal period of darkness. { ¦in·tər¦mit·ənt ¦kwik ¦flash·iŋ 'līt }

intermittent rating See intermittent-duty rating. { ¦in·tər¦mit·ənt 'rād·iŋ }

intermittent scanning [ELECTR] Scans of an antenna beam at irregular intervals to increase difficulty of detection by intercept receivers. { ¦in·tər¦mit·ənt 'skan·iŋ }

intermittent spring [HYD] A spring that ceases flow after a long dry spell but flows again after heavy rains. { ¦in·tər¦mit·ənt 'spriŋ }

intermittent stream [HYD] A stream which carries water a considerable portion of the time, but which ceases to flow occasionally or seasonally because bed seepage and evapotranspiration exceed the available water supply. { ¦in·tər¦mit·ənt 'strēm }

intermittent weld [MET] A weld in which the continuity is broken by recurring unwelded spaces. { ¦in·tər¦mit·ənt 'weld }

intermittent work [IND ENG] A type of task requiring moderate to highly demanding physical effort that is interrupted by short periods of rest or light work lasting a few seconds to a few minutes. { ¦in·tər¦mit·ənt 'wərk }

intermitter [PETRO ENG] A device installed in a well that can be regulated to allow wide-open flow for short periods several times a day and then be shut off. { ˌin·tər'mid·ər }

intermodulation [ELECTR] Modulation of the components of a complex wave by each other, producing new waves whose frequencies are equal to the sums and differences of integral multiples of the component frequencies of the original complex wave. { ˌin·tər,maj·ə'lā·shən }

intermodulation distortion [ELECTR] Nonlinear distortion characterized by the appearance of output frequencies equal to the sums and differences of integral multiples of the input frequency components; harmonic components also present in the output are usually not included as part of the intermodulation distortion. { ˌin·tər,maj·ə'lā·shən di,stór·shən }

intermodulation interference [ELECTR] Interference that occurs when the signals from two undesired stations differ by exactly the intermediate-frequency value of a superheterodyne receiver, and both signals are able to pass through the preselector due to poor selectivity. { ˌin·tər,maj·ə'lā·shən ˌin·tər'fir·əns }

intermolecular force [PHYS CHEM] The force between two molecules; it is that negative gradient of the potential energy between the interacting molecules, if energy is a function of the distance between the centers of the molecules. { ˌin·tər·mə'lek·yə·lər 'fōrs }

intermontane [GEOL] Located between or surrounded by mountains. { ¦in·tər¦män,tān }

intermontane glacier [GEOL] A glacier that is formed by the confluence of several valley glaciers and occupies a trough between separate ranges of mountains. { ¦in·tər¦män,tān 'glā·shər }

intermontane trough [GEOL] **1.** A subsiding area in an island arc of the ocean, lying between the stable elements of a region. **2.** A basinlike area between mountains. { ¦in·tər¦män,tān 'tróf }

intermural [ANAT] Between the walls of an organ. { ¦in·tər'myūr·əl }

intermuscular [ANAT] Between muscles. { ¦in·tər'məs·kyə·lər }

intermuscular hernia See interstitial hernia. { ¦in·tər'məs·kyə·lər 'hər·nē·ə }

intermuscular septum [ANAT] A connective-tissue partition between muscles. { ¦in·tər'məs·kyə·lər 'sep·təm }

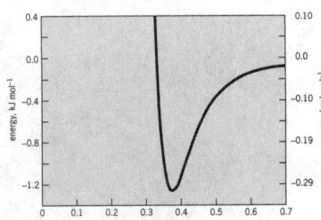

INTERMOLECULAR FORCE

Intermolecular potential energy of two argon atoms, illustrating the general form of the interaction energy between two molecules or atoms. (*After R. A. Aziz and H. H. Chen, An accurate intermolecular potential for Ar, J. Chem. Phys., 67:5719-5726, 1977*)

---

internal absorptance [ELECTROMAG] The value of absorptance, corrected to eliminate the effects of scattering and of reflection from the surfaces of the substance; that is, the ratio of the radiant power absorbed between the entry and exit surfaces of the substance to the radiant power leaving the entry surface. { in'tərn·əl əb'sórp·təns }

internal absorption See Auger effect. { in'tərn·əl əp'sórp·shən }

internal acoustic meatus [ANAT] An opening in the hard portion of the temporal bone for passage of the facial and acoustic nerves and internal auditory vessels. { in'tərn·əl ə'kü·stik mē'ād·əs }

internal ankle height [ANTHRO] The measure of the vertical distance from the lower end of the tibia to the floor. { in'tərn·əl 'aŋ·kəl ˌhīt }

internal arithmetic [COMPUT SCI] Arithmetic operations carried out in a computer's arithmetic unit within the central processing unit. { in'tərn·əl ə'rith·mə,tik }

internal biomechanical environment [IND ENG] A concept that is used in ergonomic design and considers that muscles, bones, and tissues are subject to the same Newtonian mechanical forces as are objects external to the body. { in¦tərn·əl ¦bī·ō·mi¦kan·ə·kəl in'vī·ərn·mənt }

internal brake [MECH ENG] A friction brake in which an internal shoe follows the inner surface of the rotating brake drum, wedging itself between the drum and the point at which it is anchored; used in motor vehicles. { in'tərn·əl 'brāk }

internal bremmsstrahlung See inner bremsstrahlung. { in'tərn·əl 'brem,shträ·lùŋ }

internal broaching [MECH ENG] The removal of material on internal surfaces, by means of a tool with teeth of progressively increasing size moving in a straight line or other prescribed path over the surface, other than for the origination of a hole. { in'tərn·əl 'brōch·iŋ }

internal buffer [COMPUT SCI] A portion of a computer's main storage used to temporarily hold data that is being transferred into and out of main storage. { in'tərn·əl 'bəf·ər }

internal cache See primary cache. { in'tərn·əl 'kash }

internal capsule [NEUROSCI] A layer of nerve fibers on the outer side of the thalamus and caudate nucleus, which it separates from the lenticular nucleus. { in'tərn·əl 'kap·səl }

internal carotid [ANAT] A main division of the common carotid artery, distributing blood through three sets of branches to the cerebrum, eye, forehead, nose, internal ear, trigeminal nerve, dura mater, and hypophysis. Also known as internal carotid artery. { in'tərn·əl kə'räd·əd }

internal carotid artery See internal carotid. { in'tərn·əl kə'räd·əd 'ärd·ə·rē }

internal carotid nerve [NEUROSCI] A sympathetic nerve which forms networks of branches around the internal carotid artery and its branches. { in'tərn·əl kə'räd·əd 'nərv }

internal cast See steinkern. { in'tərn·əl 'kast }

internal clocking [COMPUT SCI] Synchronization of the electronic circuitry of a device by a timing clock within the device itself. { in'tərn·əl 'kläk·iŋ }

internal combustion engine [MECH ENG] A prime mover in which the fuel is burned within the engine and the products of combustion serve as the thermodynamic fluid, as with gasoline and diesel engines. { in'tərn·əl kəm'bəs·chən ,en·jən }

internal conversion [NUC PHYS] A nuclear de-excitation process in which energy is transmitted directly from an excited nucleus to an orbital electron, causing ejection of that electron from the atom. { in'tərn·əl kən'vər·zhən }

internal conversion coefficient See conversion coefficient. { in'tərn·əl kən'vər·zhən ˌkō·i'fish·ənt }

internal cycle time [COMPUT SCI] The time required to change the information in a single register of a computer, usually a fraction of the cycle time of the main memory. Also known as clock time. { in'tərn·əl 'sī·kəl ,tīm }

internal data transfer [COMPUT SCI] The movement of data between registers in a computer's central processing unit or between a register and main storage. { in'tərn·əl 'dad·ə ,tranz·fər }

internal dielectric field See dielectric field. { in'tərn·əl ,dī·ə'lek·trik 'fēld }

internal diffusion [CHEM ENG] The diffusion of liquid or gaseous reactants to the innermost pore depths of an adsorbent-base catalyst, necessary for full catalytic effect. { in'tərn·əl di'fyü·zhən }

internal drift current [OCEANOGR] Motion in an underlying layer of water caused by shearing stresses and friction created by current in a top layer that has different density. { in'tərn·əl 'drift ,kə·rənt }

internal-drum laser scanning [GRAPHICS] A method for the direct exposure of printing plates in which a laser beam is swept across a plate that is placed on the inside of a stationary drum by a rotating mirror inclined at an angle of 45° to the axis of the drum; the beam is modulated by video signals from a similar reading system or digital signals from a computer. { in'tərn·əl 'drəm 'lā·zər ,skan·iŋ }

internal ear See inner ear. { in'tərn·əl 'ir }

internal elastic membrane [HISTOL] A sheet of elastin found between the tunica intima and the tunica media in medium- and small-caliber arteries. { in'tərn·əl i'las·tik 'mem,brān }

internal energy [THERMO] A characteristic property of the state of a thermodynamic system, introduced in the first law of thermodynamics; it includes intrinsic energies of individual molecules, kinetic energies of internal motions, and contributions from interactions between molecules, but excludes the potential or kinetic energy of the system as a whole; it is sometimes erroneously referred to as heat energy. { in'tərn·əl 'en·ər·jē }

internal erosion [GEOL] Erosion effected within a compacting sediment by movement of water through the larger pores. { in'tərn·əl i'rō·zhən }

internal fertilization [PHYSIO] Fertilization of the egg within the body of the female. { in'tərn·əl ,fərd·əl·ə'zā·shən }

internal fistula [ANAT] A fistula which has no opening through the skin. { in'tərn·əl 'fis·chə·lə }

internal floating-head exchanger [MECH ENG] Tube-and-shell heat exchanger in which the tube sheet (support for tubes) at one end of the tube bundle is free to move. { in'tərn·əl 'flōd·iŋ ,hed iks'chānj·ər }

internal font [GRAPHICS] A typeface or set of typefaces that is stored in the permanent memory of a printer. Also known as built-in font. { in'tərn·əl 'fänt }

internal force [MECH] A force exerted by one part of a system on another. { in'tərn·əl 'fōrs }

internal friction [FL MECH] See viscosity. [MECH] **1.** Conversion of mechanical strain energy to heat within a material subjected to fluctuating stress. **2.** In a powder, the friction that is developed by the particles sliding over each other; it is greater than the friction of the mass of solid that comprises the individual particles. { in'tərn·əl 'frik·shən }

internal furnace [MECH ENG] A boiler furnace having a firebox within a water-cooled heating surface. { in'tərn·əl 'fər·nəs }

internal gas drive [PETRO ENG] A primary oil recovery process in which oil is displaced from the reservoir by the expansion of the gas originally dissolved in the liquid. Also known as dissolved-gas drive; gas depletion drive; solution gas drive. { in'tərn·əl 'gas ,drīv }

internal gear [DES ENG] An annular gear having teeth on the inner surface of its rim. { in'tərn·əl 'gir }

internal granular layer [HISTOL] The fourth layer of the cerebral cortex. { in'tərn·əl 'gran·yə·lər ,lā·ər }

internal grinder [MECH ENG] A machine designed for grinding the surfaces of holes. { in'tərn·əl 'grīn·dər }

internal hemorrhage [COMPUT SCI] A condition in which a computer program continues to run following an error but produces dubious results and may adversely affect other programs or the performance of the entire system. [MED] Bleeding within a body cavity or organ that is concealed from an observer. { in'tərn·əl 'hem·rij }

internal hernia [MED] A hernia of intraabdominal contents occurring within the abdominal cavity. { in'tərn·əl 'hər·nē·ə }

internal hydrocephaly See obstructive hydrocephaly. { in'tərn·əl ¦hī·drə'sef·ə·lē }

internal iliac artery [ANAT] The medial terminal division of the common iliac artery. { in'tərn·əl ¦il·ē,ak 'ärd·ə·rē }

internal interrupt [COMPUT SCI] A signal for attention sent to a computer's central processing unit by another component of the computer. { in'tərn·əl 'int·ə,rəpt }

internalization [PSYCH] A mental mechanism operating outside of and beyond conscious awareness by which certain external attributes, attitudes, or standards are taken within oneself. { in,tərn·əl·ə'zā·shən }

internal label [COMPUT SCI] An identifier providing a name for data that is recorded with the data in a storage medium. { in'tərn·əl 'lā·bəl }

internal line [QUANT MECH] A component of a Feynman graph (in the diagrammatic presentation of perturbative quantum field theory) describing the propagation of a virtual particle whose momentum is integrated over all possible values. { in'tərn·əl 'līn }

internal loss See loss. { in'tərn·əl 'lȯs }

internally fired boiler [MECH ENG] A fire-tube boiler containing an internal furnace which is water-cooled. { in'tərn·əl·ē ¦fīrd 'bȯil·ər }

internally stable set See independent set. { in,tərn·əl·ē ,stā·bəl 'set }

internally stored program [COMPUT SCI] A sequence of instructions, stored inside the computer in the same storage facilities as the computer data, as opposed to external storage on tape, disk, drum, or cards. { in'tərn·əl·ē ,stȯrd 'prō·grəm }

internally tangent circles [MATH] Two circles, one of which is inside the other, that have a single point in common. { in,tərn·əl·ē ,tan·jənt 'sər·kəlz }

internal mechanical environment [IND ENG] A concept that considers parts of the human body, such as muscles, bones, and tissues, in terms of how they are subject to Newtonian mechanics in their interaction with the external environment. { in'tərn·əl ¦mi¦kan·ə·kəl in'vī·rən·mənt }

internal memory See internal storage. { in'tərn·əl 'mem·rē }

internal mix atomizer [MECH ENG] A type of pneumatic atomizer in which gas and liquid are mixed prior to the gas expansion through the nozzle. { in'tərn·əl ¦miks 'ad·ə,mīz·ər }

internal operation [MATH] For a set S, a function whose domain is a set of members of S or a set of ordered sequences of members of S, and whose range is a subset of S. { in,tərn·əl ,äp·ə'rā·shən }

internal oxidation [MET] The subsurface oxidation of components of an alloy due to oxygen diffusion into the metal. { in'tərn·əl ,äk·sə'dā·shən }

internal phase See disperse phase. { in'tərn·əl 'fāz }

internal photoelectric effect [SOLID STATE] A process in which the absorption of a photon in a semiconductor results in the excitation of an electron from the valence band to the conduction band. { in'tərn·ə ¦fōd·ō·ə'lek·trik i,fekt }

internal photoionization See Auger effect. { in'tərn·əl ,fōd·ō,ī·ə·nə'zā·shən }

internal pressure See intrinsic pressure. { in'tərn·əl 'presh·ər }

internal reader [COMPUT SCI] A device that reads jobs and data into a computer from on-line storage, emulating a card reader. { in'tərn·əl 'rēd·ər }

internal reflectance spectroscopy See attenuated total reflectance. { in'tərn·əl ri¦flek·təns spek'träs·kə·pē }

internal reflection [OPTICS] The reflection of electromagnetic radiation in a given medium from the boundary with a less dense medium. { in,tərn·əl ri'flek·shən }

internal resistance [ELEC] The resistance within a voltage source, such as an electric cell or generator. { in'tərn·əl ri'zis·təns }

internal respiration [PHYSIO] The gas exchange which occurs between the blood and tissues of an organism. { in'tərn·əl ,res·pə'rā·shən }

internal schema [COMPUT SCI] The physical configuration of data in a data base. { in'tərn·əl 'skē·mə }

internal secretion [PHYSIO] A secreted substance that is absorbed directly into the blood. { in'tərn·əl si'krē·shən }

internal sedimentation [GEOL] Accumulation of clastic or chemical sediments derived from the surface of, or within, a more or less consolidated carbonate sediment (mud or silt); deposited in secondary cavities formed in the host rock (after its deposition) by bending of laminae or by internal erosion or solution. { in'tərn·əl ,sed·ə·mən'tā·shən }

internal sorting [COMPUT SCI] The sorting of a list of items by a computer in which the entire list can be brought into the main computer memory and sorted in memory. { in'tərn·əl 'sȯrd·iŋ }

internal spring safety relief valve [ENG] A spring-loaded

INTERNAL GEAR

A helical internal gear and pinion. (*Fellows Corp.*)

INTERNAL VIBRATOR

Appearance of concrete after internal vibrator has consolidated it; vibrator appears on the right.

INTERNATIONAL DATE LINE

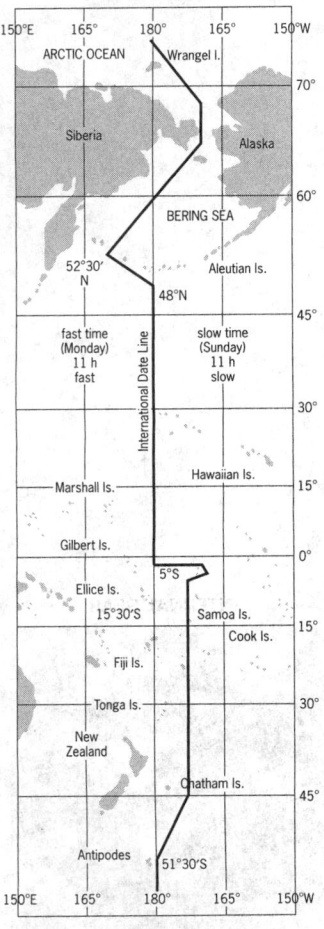

Map showing the international date line.

valve with a portion of the operating mechanism located inside the pressure vessel. { in'tərn·əl ¦spriŋ 'sáf·tē ri'lēf ,valv }

internal standard [SPECT] The principal line in spectrum analysis by the logarithmic sector method, a quantitative spectroscopy procedure. { in'tərn·əl 'stan·dərd }

internal storage [COMPUT SCI] The total memory or storage that is accessible automatically to a computer without human intervention. Also known as internal memory. { in'tərn·əl 'stór·ij }

internal storage capacity [COMPUT SCI] The quantity of data that can be retained simultaneously in internal storage. { in'tərn·əl 'stór·ij kə,pas·əd·ē }

internal stress [MECH] A stress system within a solid that is not dependent on external forces. Also known as residual stress. { in'tərn·əl 'stres }

internal table [COMPUT SCI] A table or array that is coded directly into a computer program and is compiled along with the rest of the program. { in'tərn·əl 'tā·bəl }

internal tangent [MATH] For two circles, each exterior to the other, a line that is tangent to both circles and that separates them. { in,tərn·əl 'tan·jənt }

internal thread [DES ENG] A screw thread cut on the inner surface of a hollow cylinder. { in'tərn·əl 'thred }

internal transmittance [ELECTROMAG] The value of transmittance, corrected to eliminate the effects of scattering and of reflection from the surfaces of the substance; that is, the ratio of the radiant power reaching the exit surface of the substance to the radiant power leaving the entry surface. { in'tərn·əl 'tranz'mit·əns }

internal vibrator [MECH ENG] A vibrating device which is drawn vertically through placed concrete to achieve proper consolidation. { in'tərn·əl 'vī,brād·ər }

internal wave [FL MECH] A wave motion of a stably stratified fluid in which the maximum vertical motion takes place below the surface of the fluid. { in'tərn·əl 'wāv }

internal work [IND ENG] Manual work done by a machine operator while the machine is automatically operating. Also known as fill-up work; inside work. [THERMO] The work done in separating the particles composing a system against their forces of mutual attraction. { in'tərn·əl 'wərk }

international ampere [ELEC] The current that, when flowing through a solution of silver nitrate in water, deposits silver at a rate of 0.001118 gram per second; it has been superseded by the ampere as a unit of current, and is equal to approximately 0.999850 ampere. { ¦in·tər¦nash·ən·əl 'am,pir }

international analysis code [METEOROL] An internationally recognized code for communicating details of synoptic chart analyses. Abbreviated IAC. { ¦in·tər¦nash·ən·əl ə'nal·ə·səs ,kōd }

international angstrom [PHYS] A unit of length, equal to 1/6438.4696 of the wavelength of the red cadmium line in dry air at standard atmospheric pressure, at a temperature of 15°C containing 0.03% by volume of carbon dioxide; equal to 1.0000002 angstroms. Abbreviated IA. { ¦in·tər¦nash·ən·əl 'aŋ·strəm }

international atomic time [HOROL] Time based on atomic clocks operating in conformity with the definition of the second as the International System unit of time. Abbreviated IAT. { ¦in·tər¦nash·ən·əl ə¦täm·ik 'tīm }

international broadcasting [COMMUN] Radio broadcasting for public entertainment between different countries, on frequency bands between 5950 and 21,750 kilohertz, assigned by international agreement. { ¦in·tər¦nash·ən·əl 'bród,kast·iŋ }

international cable code See Morse cable code. { ¦in·tər¦nash·ən·əl 'kā·bəl ,kōd }

international call sign [COMMUN] Call sign assigned according to the provisions of the International Telecommunication Union to identify a radio station; the nationality of the radio station is identified by the first or the first two characters. { ¦in·tər¦nash·ən·əl 'kól ,sīn }

international candle [OPTICS] A unit of luminous intensity, now replaced by the candela; as defined in the United States, it was a specified fraction of the average luminous intensity radiated in a horizontal direction by a group of 45 carbon-filament lamps preserved at the National Bureau of Standards when the lamps were operated at a specified voltage. Also known as standard candle. { ¦in·tər¦nash·ən·əl 'kand·əl }

International Celestial Reference Frame [ASTRON] A celestial reference frame made up from the positions of approximately 400 extragalactic radio sources observed with very long baseline interferometry. { ¦in·tər¦nash·ən·əl si¦les·chəl 'ref·rəns ,frām }

International Celestial Reference System [ASTRON] A system that is realized by the International Celestial Reference Frame, made up from the positions of extragalactic radio sources, and that encompasses the standard reference frames, the transformations between them, and all the constants and motions involved, as well as the time scales specified for the reference frames and origins involved. It was adopted by the International Astronomical Union in 1994 as the fundamental reference system. { ¦in·tər¦nash·ən·əl si¦les·chəl 'ref·rəns ,sis·təm }

international code signal [COMMUN] Code adopted by many nations for international communications; it uses combinations of letters in lieu of words, phrases, and sentences; the letters are transmitted by the hoisting of international alphabet flags or by transmitting their dot and dash equivalents in the international Morse code. Also known as international signal code. { ¦in·tər¦nash·ən·əl 'kōd ,sig·nəl }

international control frequency bands [COMMUN] Radio-frequency bands assigned in the United States to links between stations used for international communication and their associated control centers. { ¦in·tər¦nash·ən·əl kən¦trōl 'frē·kwən·sē ,banz }

international control station [COMMUN] Fixed station in the fixed public control service associated directly with the international fixed public radio communications service. { ¦in·tər¦nash·ən·əl kən¦trōl ,stā·shən }

International Critical Tables [PHYS] A seven-volume series of tables of numerical data in physics, chemistry, and technology, published in 1926–1930, prepared by experts who gave the "best" value which could be derived from all the data available at the time. Abbreviated ICT. { ¦in·tər¦nash·ən·əl 'krid·ə·kəl ,tā·bəlz }

international date line [ASTRON] A jagged arbitrary line, roughly equal to the 180° meridian, where a date change occurs: if the line is crossed from east to west a day is skipped, if from west to east the same day is repeated. { ¦in·tər¦nash·ən·əl 'dāt ,līn }

international ellipsoid of reference [GEOD] The reference ellipsoid, based upon the Hayford spheroid, the semimajor axis of which is 6,378,388 meters; the flattening or ellipticity equals 1/297; by computation the semiminor axis is 6,356,911.946 meters. Also known as international spheroid. { ¦in·tər¦nash·ən·əl ə'lip,sóid əv 'ref·rəns }

international fixed public radio communications service [COMMUN] Fixed service, the stations of which are open to public correspondence; this service is intended to provide radio communications between the United States and its territories and foreign or overseas points. { ¦in·tər¦nash·ən·əl ¦fixt ¦pəb·lik ¦rād·ē·ō kə,myü·nə'kā·shən ,sər·vəs }

International Geophysical Year [GEOPHYS] An internationally accepted period, extending from July 1957 through December 1958, for concentrated and coordinated geophysical exploration, primarily of the solar and terrestrial atmospheres. Abbreviated IGY. { ¦in·tər¦nash·ən·əl jē·ō'fiz·ə·kəl ,yir }

international gravity formula [GEOD] A formula for the acceleration of gravity at the earth's surface, stating that the acceleration of gravity is equal to $9.780327[1 + 5.3024 \times 10^{-3} \sin^2 \phi - 5.8 \times 10^{-6} \sin^2 2\phi]$ m/s^2, where ϕ is the latitude. { ¦in·tər¦nash·ən·əl 'grav·əd·ē ,fór·myə·lə }

international henry [ELECTROMAG] A unit of electrical inductance which has been superseded by the henry, and is equal to 1.00049 henry. Also known as quadrant; secohm. { ¦in·tər¦nash·ən·əl 'hen·rē }

International Ice Patrol [OCEANOGR] An organization established in 1914 to protect shipping by providing iceberg warnings. { ¦in·tər¦nash·ən·əl 'īs pə,trōl }

international index numbers [METEOROL] A system of designating meteorological observing stations by number, established and administered by the World Meteorological Organization; under this scheme, specified areas of the world are divided into blocks, each bearing a two-number designator; stations within each block have an additional unique three-number designator, the numbers generally increasing from east to west and from south to north. { ¦in·tər¦nash·ən·əl 'in,deks ,nəm·bərz }

international Morse code *See* continental code. { ¦in·tər¦nash·ən·əl ¦mors 'kōd }

international nautical mile [NAV] A unit of length equal to 1852 meters. { ¦in·tər¦nash·ən·əl ¦nȯd·ə·kəl 'mīl }

international ohm [ELEC] A unit of resistance, equal to that of a column of mercury of uniform cross section that has a length of 160.3 centimeters and a mass of 14.4521 grams at the temperature of melting ice; it has been superseded by the ohm, and is equal to 1.00049 ohms. { ¦in·tər¦nash·ən·əl 'ōm }

International Phonetic Alphabet [LING] A phonetic transcription system the purpose of which is to represent with graphic symbols the distinctive sounds of human speech, regardless of language. { ¦in·tər¦nash·ən·əl fə¦ned·ik 'al·fə·bet }

International Polar Year [METEOROL] The years 1882 and 1932, during which participating nations undertook increased observations of geophysical phenomena in polar (mostly arctic) regions; the observations were largely meteorological, but included such as auroral and magnetic studies. { ¦in·tər¦nash·ən·əl 'pō·lər ¦yir }

international practical temperature scale [THERMO] Temperature scale based on six points: the water triple point, the boiling points of oxygen, water, sulfur, and the solidification points of silver and gold; designated as °C, degrees Celsius, or t_{int}; replaced in 1990 by the international temperature scale. { ¦in·tər¦nash·ən·əl ¦prak·tə·kəl 'tem·prə·chər ¦skāl }

International Quiet Sun Year [GEOPHYS] An international cooperative effort, similar to the International Geophysical Year and extending through 1964 and 1965, to study the sun and its terrestrial and planetary effects during the minimum of the 11-year cycle of solar activity. Abbreviated IQSY. Also known as the International Year of the Quiet Sun. { ¦in·tər¦nash·ən·əl ¦kwī·ət 'sən ¦yir }

international radio silence [COMMUN] Three-minute periods of radio silence, on the frequency of 500 kilohertz only, commencing 15 and 45 minutes after each hour, during which all marine radio stations must listen on that frequency for distress signals of ships and aircraft. { ¦in·tər¦nash·ən·əl ¦rād·ē·ō 'sī·ləns }

international rules of the road [NAV] In marine operations, international regulations for preventing collisions at sea. { ¦in·tər¦nash·ən·əl ¦rülz əv thə 'rōd }

international signal code *See* international code signal. { ¦in·tər¦nash·ən·əl 'sig·nəl ¦kōd }

international spheroid *See* international ellipsoid of reference. { ¦in·tər¦nash·ən·əl 'sfir¸ȯid }

international standard annealed copper [MET] An annealed pure copper having a resistivity of 1.7241 microhm-centimeter at 20°C, which is taken as 100% conductivity. { ¦in·tər¦nash·ən·əl ¦stan·dərd ə¦nēld 'käp·ər }

international synoptic code [METEOROL] A synoptic code approved by the World Meteorological Organization in which the observable meteorological elements are encoded and transmitted in words of five numerical digits length. { ¦in·tər¦nash·ən·əl sə¦näp·tik ¦kōd }

international system of electrical units [ELEC] System of electrical units based on agreed fundamental units for the ohm, ampere, centimeter, and second, in use between 1893 and 1947, inclusive; in 1948, the Giorgi, or meter-kilogram-second-absolute system, was adopted for international use. { ¦in·tər¦nash·ən·əl ¦sistəm əv i¦lek·trə·kəl 'yü·nəts }

International System of Units [PHYS] A system of physical units in which the fundamental quantities are length, time, mass, electric current, temperature, luminous intensity, and amount of substance, and the corresponding units are the meter, second, kilogram, ampere, kelvin, candela, and mole; it has been given official status and recommended for universal use by the General Conference on Weights and Measures. Also known (in French) as Système International d'Unités. Abbreviated SI (in all languages). { ¦in·tər¦nash·ən·əl ¦sis·təm əv 'yü·nəts }

international table British thermal unit *See* British thermal unit. { ¦in·tər¦nash·ən·əl ¦tā·bəl ¦brid·ish 'thər·məl ¸yü·nət }

international table calorie *See* calorie. { ¦in·tər¦nash·ən·əl ¦tā·bəl 'kal·ə·rē }

international telecommunications service [COMMUN] Telecommunications service between offices or stations of different states, or between mobile stations which are not in the same state, or are subject to different states. { ¦in·tər¦nash·ən·əl ¸tel·ə·kə¸myü·nə¦kā·shənz ¸sər·vəs }

international telegraph alphabet *See* CCIT 2 code. { ¦in·tər¦nash·ən·əl 'tel·ə¸graf al·fə¸bet }

International Telegraphic Consultative Committee code 2 *See* CCIT 2 code. { ¦in·tər¦nash·ən·əl ¦tel·ə¦graf·ik kən¦səl·tə·div kə¦mid·ē ¦kōd 'tü }

international temperature scale [THERMO] A standard temperature scale, adopted in 1990, that approximates the thermodynamic scale, based on assigned temperature values of 17 thermodynamic equilibrium fixed points and prescribed thermometers for interpolation between them. Abbreviated ITS-90. { ¦in·tər¦nash·ən·əl tem·prə·chər 'skāl }

international thread [DES ENG] A standardized metric system in which the pitch and diameter of the thread are related, with the thread having a rounded root and flat crest. { ¦in·tər¦nash·ən·əl 'thred }

International Union of Pure and Applied Chemistry [CHEM] An international scientific (nongovernmental) organization, recognized as the world authority on chemical nomenclature, terminology, standardized methods for measurement, atomic weights, and many other critically evaluated data. Abbreviated IUPAC. { ¦in·tər¦nash·ən·əl ¸yün·yən əv ¦pyür ən ə¦plīd 'kem·ə·strē }

international unit [BIOL] A quantity of a vitamin, hormone, antibiotic, or other biological that produces a specific internationally accepted biological effect. { ¦in·tər¦nash·ən·əl 'yü·nət }

international volt [ELEC] A unit of potential difference or electromotive force, equal to 1/1.01858 of the electromotive force of a Weston cell at 20°C; it has been superseded by the volt, and is equal to 1.00034 volts. { ¦in·tər¦nash·ən·əl 'vōlt }

International Year of the Quiet Sun *See* International Quiet Sun Year. { ¦in·tər¦nash·ən·əl 'yir əv thə ¦kwī·ət 'sən }

internet [COMMUN] A system of local area networks that are joined together by a common communications protocol. { 'in·tər¸net }

Internet [COMPUT SCI] A worldwide system of interconnected computer networks, communicating by means of TCP/IP and associated protocols. { 'in·tər¸net }

Internet Mail Access Protocol [COMPUT SCI] An Internet standard for directly reading and manipulating e-mail messages stored on remote servers. { ¦in·tər¸net ¸māl 'ak¸ses ¸prōd·ə¸kȯl }

Internet protocol [COMMUN] The set of standards responsible for ensuring that data packets transmitted over the Internet are routed to their intended destinations. Abbreviated IP. { 'in·tər¸net 'prōd·ə¸kȯl }

Internet telephony [COMMUN] Phone calls routed over the Internet by analog-to-digital conversion of speech signals. { ¸in·tər¸net tə'lef·ə·nē }

internetting [COMPUT SCI] Connections and communications paths between separate data communications networks that allow transfer of messages. { ¦in·tər¦ned·iŋ }

internides [GEOL] The internal part of an orogenic belt, farthest away from the craton, which is commonly the site of a eugeosyncline during its early phases and is later subjected to plastic folding and plutonism. Also known as primary arc. { in'tər nə¸dēz }

internode [BIOL] The interval between two nodes, as on a stem or along a nerve fiber. { 'in·tər¸nōd }

internuclear [SCI TECH] Located between nuclei. { ¦in·tər¦nü·klē·ər }

internuclear distance [PHYS CHEM] The distance between two nuclei in a molecule. { ¦in·tər¦nü·klē·ər ¸dis·təns }

internuncial neuron [NEUROSCI] A neuron located in the spinal cord which connects motor and sensory neurons. { ¦in·tər¦nən·chəl 'nür¸än }

interoceptor [PHYSIO] A sense receptor located in visceral organs and yielding diffuse sensations. { ¸in·tə·rō'sep·tər }

interocular diameter [ANTHRO] The measure of the distance between the internal canthi. { ¦in·tər¦äk·yə·lər dī'am·əd·ər }

interocular distance [ANAT] The distance between the centers of rotation of the human eyes. { ¦in·tər¦äk·yə·lər 'dis·təns }

interoffice trunk [COMMUN] A direct trunk between local central offices in the same exchange. { ¦in·tər¦ȯf·əs 'trəŋk }

interorbital [ANAT] Between the orbits of the eyes. { ¦in·tər'ȯr·bəd·əl }

interparietal [ANAT] Between the parietal bones. { ¦in·tər·pə'rī·əd·əl }

interparietal hernia *See* interstitial hernia. { ¦in·tər·pə'rī·əd·əl 'hər·nē·ə }

interpass temperature [MET] In a multiple-pass weld, the lowest temperature of the deposited weld metal before the next run is started. { 'in·tər,pas ,tem·prə·chər }

interpenetrating polymer network [ORG CHEM] Two or more polymer components, each of which is a crosslinked three-dimensional network, one of which is formed (crosslinked) in the presence of the other. The polymer networks are physically entangled with, but not covalently bonded to, each other. Characteristically, these networks do not dissolve in solvent or flow when heated. Abbreviated IPN. { ¦in·tər,pen·ə,trād·iŋ ,päl·ə·mər 'net,wərk }

interpenetration twin [CRYSTAL] Two or more individual crystals so twinned that they appear to have grown through one another. Also known as penetration twin. { ¦in·tər,pen·ə'trā·shən ¦twin }

interpersonal therapy [PSYCH] A form of psychotherapy in which emphasis is placed on enhancing the patient's ability to cope with stresses, improving interpersonal communications, increasing morale, and helping the patient deal with the effects of the depressive disorder. { ,in·tər,pər·sən·əl 'ther·ə·pē }

interphase [CELL MOL] Also known as interkinesis. **1.** The period between succeeding mitotic divisions. **2.** The period between the first and second meiotic divisions in those organisms where nuclei are reconstituted at the end of the first division. [CHEM] A region between the two phases of a newly created interface that contains particles of both phases. { 'in·tər,fāz }

interphase reactor [ELEC] A type of current-equalizing reactor that is connected between two parallel silicon controlled rectifier converters and provides balanced system operation when both converters are conducting by acting as an inductive voltage divider. { 'in·tər,fāz rē,ak·tər }

interphase transformer [ELECTR] Autotransformer or a set of mutually coupled reactors used in conjunction with three-phase rectifier transformers to modify current relations in the rectifier system to increase the number of anodes of different phase relations which carry current at any instant. { 'in·tər,fāz tranz,fòr·mər }

interphone [COMMUN] An intercommunication system using headphones and microphones for communication between adjoining or nearby studios or offices, or between crew locations on an aircraft, vessel, or tank or other vehicle. Also known as talk-back circuit. { 'in·tər,fōn }

interplanar spacing [CRYSTAL] The perpendicular distance between successive parallel planes of atoms in a crystal. { ¦in·tər,plā·nər 'spās·iŋ }

interplanetary dust [ASTRON] Dust particles between the planets. { ¦in·tər'plan·ə,ter·ē 'dəst }

interplanetary flight [AERO ENG] Flight through the region of space between the planets, under the primary gravitational influence of the sun. { ¦in·tər'plan·ə,ter·ē 'flīt }

interplanetary magnetic field [ASTROPHYS] The magnetic field between the planets. { ¦in·tər'plan·ə,ter·ē mag¦ned·ik 'fēld }

interplanetary medium [ASTRON] That part of space containing electromagnetic radiation, dust, gas, and plasma between the planets. { ¦in·tər'plan·ə,ter·ē 'mēd·ē·əm }

interplanetary probe [AERO ENG] An instrumented spacecraft that flies through the region of space between the planets. { ¦in·tər'plan·ə,ter·ē 'prōb }

interplanetary space [ASTRON] The region that extends beyond near-space away from earth to the other planets in the solar system. { ¦in·tər'plan·ə,ter·ē 'spās }

interplanetary spacecraft [AERO ENG] A spacecraft designed for interplanetary flight. { ¦in·tər'plan·ə,ter·ē 'spās,kraft }

interplanetary transfer orbit [AERO ENG] An elliptical trajectory tangent to the orbits of both the departure planet and the target planet. { ¦in·tər'plan·ə,ter·ē 'tranz·fər ,òr·bət }

interpluvial [GEOL] Pertaining to an episode or period of geologic time that was dryer than the pluvial period occurring before or after it. { ,in·tər'plü·vē·əl }

interpolation [MATH] A process used to estimate an intermediate value of one (dependent) variable which is a function of a second (independent) variable when values of the dependent variable corresponding to several discrete values of the independent variable are known. { in,tər·pə'lā·shən }

interpole *See* commutating pole. { 'in·tər,pōl }

interpolymer [ORG CHEM] A mixed polymer made from two or more starting materials. { ¦in·tər'päl·ə·mər }

interposition trunk [COMMUN] Trunk which connects two positions of a large switchboard so that a line on one position can be connected to a line on another position. { ,in·tər·pə'zish·ən ,trəŋk }

interpret [COMPUT SCI] To print on a punched card the information punched in that card. { in'tər·prət }

interpreter [COMPUT SCI] **1.** A program that translates and executes each source program statement before proceeding to the next one. Also known as interpretive routine. **2.** A machine that senses a punched card and prints the punched information on that card. Also known as punched-card interpreter. **3.** *See* conversational compiler. { in'tər·prəd·ər }

interpretive code *See* interpretive language. { in'tər·prəd·iv ,kōd }

interpretive language [COMPUT SCI] A computer programming language in which each instruction is immediately translated and acted upon by the computer, as opposed to a compiler which decodes a whole program before a single instruction can be executed. Also known as interpretive code. { in'tər·prəd·iv 'laŋ·gwij }

interpretive programming [COMPUT SCI] The writing of computer programs in an interpretive language, which generally uses mnemonic symbols to represent operations and operands and must be translated into machine language by the computer at the time the instructions are to be executed. { in'tər·prəd·iv 'prō,gram·iŋ }

interpretive trace program [COMPUT SCI] An interpretive routine that provides a record of the machine code into which the source program is translated and of the result of each step, or of selected steps, of the program. { in'tər·prəd·iv 'trās ,prō·grəm }

interprocedure metric [COMPUT SCI] A software metric that estimates the complexity of a module or computer program based on the way that the data are used, organized, and allocated in relationship with some other modules. { ,in·tər·prə¦sē·jər 'me·trik }

interprocess communication [COMPUT SCI] The communication between computer programs running concurrently under the control of the same operating system. { ¦in·tər,prä·səs kə,myü·ə'kā·shən }

interproglottid gland [INV ZOO] Any of a number of cell clusters or glands arranged transversely along the posterior margin of the proglottids of certain tapeworms. { ¦in·tər,prō'g·läd·əd ,gland }

interpterygoid [ZOO] A space between palatal plates in certain chordates. { ¦in·tər'ter·ə,gòid }

interpulmonary [ANAT] Located between the lungs. { ¦in·tər'púl·mə,ner·ē }

interpulse time [MET] In resistance welding, the time between successive pulses of an impulse. { 'in·tər,pəls ,tīm }

interpupillary diameter [ANTHRO] A measure of the distance between the centers of the pupils, as the subject looks straight ahead. { ¦in·tər'pyü·pə,ler·ē dī'am·əd·ər }

interquartile range [STAT] The distance between the top of the lower quartile and the bottom of the upper quartile of a distribution. { ¦in·tər'kwòr,tīl ,rānj }

interradial canal [INV ZOO] Any of the radially arranged gastrovascular canals in certain jellyfishes and ctenophores. { ¦in·tər'rād·ē·əl kə,nal }

interradius [INV ZOO] The area between two adjacent arms in echinoderms. { ,in·tər'rād·ē·əs }

interray [INV ZOO] A division of the radiate body of echinoderms. { ¦in·tər'rā }

interrecord gap *See* record gap. { ¦in·tər'rek·ərd ,gap }

interrenal [ANAT] Located between the kidneys. { ¦in·tər'rēn·əl }

interrogation [COMMUN] The transmission of a radio-frequency pulse, or combination of pulses, intended to trigger a transponder or group of transponders, a racon system, or an IFF system, in order to elicit an electromagnetic reply. Also known as challenging signal. { in,ter·ə'gā·shən }

interrogation suppressed time delay [COMMUN] Overall fixed time delay between transmission of an interrogation and reception of the reply to this interrogation at zero distance. { in,ter·ə'gā·shən sə¦prest 'tīm di,lā }

interrogator [ELECTR] **1.** A radar transmitter which sends out a pulse that triggers a transponder; usually combined in a single unit with a responsor, which receives the reply from a transponder and produces an output suitable for actuating a display of some navigational parameter. Also known as challenger; interrogator-transmitter. **2.** *See* interrogator-responsor. { in'ter·ə,gād·ər }

interrogator-responsor [ELECTR] A transmitter and receiver combined, used for sending out pulses to interrogate a radar beacon and for receiving and displaying the resulting replies. Also known as interrogator. { in'ter·ə,gād·ər ri'spän·sər }

interrogator-transmitter *See* interrogator. { in'ter·ə,gād·ər tranz'mid·ər }

interrupt [COMPUT SCI] **1.** To stop a running program in such a way that it can be resumed at a later time, and in the meanwhile permit some other action to be performed. **2.** The action of such a stoppage. { 'int·ə,rəpt }

interrupt-driven system [COMPUT SCI] An operating system in which the interrupt system is the mechanism for reporting all changes in the states of hardware and software resources, and such changes are the events that induce new assignments of these resorces to meet work-load demands. { 'int·ə,rəpt ,driv·ən ,sis·təm }

interrupted aging [MET] A technique for aging material in several steps; the material is brought to room temperature after each step. { 'int·ə,rəp·təd 'āj·iŋ }

interrupted continuous wave [COMMUN] A continuous wave that is interrupted at a constant audio-frequency rate high enough to give several interruptions for each keyed code dot. Abbreviated ICW. { int·ə,rəp·təd kən¦tin·yə·wəs 'wāv }

interrupted current [ELEC] A current produced by opening and closing at regular intervals a circuit that would otherwise carry a steady current or one that varied continuously with time. { int·ə,rəp·təd 'kə·rənt }

interrupted dc tachometer [ENG] A type of impulse tachometer in which the frequency of pulses generated by the interrupted direct current of an ignition-circuit primary of an internal combustion engine is used to measure the speed of the engine. { 'int·ə,rəp·təd ¦dē¦sē tə'käm·əd·ər }

interrupted fire [ORD] Automatic fire delivered in short series of bursts. { 'int·ə,rəp·təd 'fīr }

interrupted gene *See* split gene. { in'tə,rəp·təd 'jēn }

interrupted mating [GEN] A technique for mapping bacterial genes by determining the sequence of gene transfer between conjugating bacteria. { in'tə,rəp·təd 'mād·iŋ }

interrupted quenching [MET] Quenching in which a material is intermittently removed from the quenching medium while it is still at a higher temperature than the medium. { 'int·ə,rəp·təd 'kwench·iŋ }

interrupted quick flashing light [NAV] In marine operations, a light showing quick flashes for several seconds, then a period of darkness. { 'int·ə,rəp·təd ¦kwik ¦flash·iŋ 'līt }

interrupted screw [DES ENG] A screw with longitudinal grooves cut into the thread, and which locks quickly when inserted into a similar mating part. { 'int·ə,rəp·təd 'skrü }

interrupter [ELEC] An electric, electronic, or mechanical device that periodically interrupts the flow of a direct current so as to produce pulses. [ORD] A barrier in a fuse which prevents transmission of an explosive effect to some element beyond the interrupter. { 'int·ə,rəp·tər }

interrupter gear [ORD] A synchronizing gear for machine guns, so called because it interrupts the firing mechanism of the gun or guns to allow a propeller blade to pass the muzzle. { 'int·ə,rəp·tər ,gir }

interrupter vibrator [ELEC] A mechanical device used to change direct current to alternating current. { 'int·ə,rəp·tər 'vī,brād·ər }

interrupt handler [COMPUT SCI] A section of a computer program or of the operating system that takes control when an interrupt is received and performs the operations required to service the interrupt. { 'int·ə,rəpt ,hand·lər }

interrupting capacity [ELEC] Maximum power in the arc that a circuit breaker or fuse can successfully interrupt without restrike or violent failure; rated in volt-amperes for alternating-current circuits and watts for direct-current circuits. { ,int·ə'rəp·tiŋ kə,pas·əd·ē }

interrupt mask [COMPUT SCI] A technique of suppressing certain interrupts and allowing the control program to handle these masked interrupts at a later time. { 'int·ə,rəpt ,mask }

interrupt mode *See* hold mode. { 'int·ə,rəpt ,mōd }

interrupt priorities [COMPUT SCI] The sequence of importance assigned to attending to the various interrupts that can occur in a computer system. { 'in·tə,rəpt prī,är·ə·dēz }

interrupt routine [COMPUT SCI] A program that responds to an interrupt by carrying out prescribed actions. { 'int·ə,rəpt rü,tēn }

interrupt signal [COMPUT SCI] A control signal which requests the immediate attention of the central processing unit. { 'int·ə,rəpt ,sig·nəl }

interrupt system [COMPUT SCI] The means of interrupting a program and proceeding with it at a later time; this is accomplished by saving the contents of the program counter and other specific registers, storing them in reserved areas, starting the new instruction sequence, and upon completion, reloading the program counter and registers to return to the original program, and reenabling the interrupt. { 'int·ə,rəpt ,sis·təm }

interrupt trap [COMPUT SCI] A program-controlled technique which either recognizes or ignores an interrupt, depending upon a switch setting. { 'int·ə,rəpt ,trap }

interrupt vector [COMPUT SCI] A list comprising the locations of various interrupt handlers. { 'int·ə,rəpt ,vek·tər }

interscapular [ANAT] Between the shoulders or shoulder blades. { ¦in·tər¦skap·yə·lər }

intersect [ENG] To find a position by the triangulation method. [SCI TECH] To pass through or across. { ,in·tər'sekt }

Intersecting Storage Rings [NUCLEO] Proton storage rings, located at Geneva, Switzerland, in which counterrotating protons with energies of up to 31 gigaelectronvolts injected from a proton synchrotron are made to undergo nearly head-on collisions. Abbreviated ISR. { ¦in·tər¦sek·tiŋ 'stōr·ij ,riŋz }

intersection [CIV ENG] **1.** A point of junction or crossing of two or more roadways. **2.** A surveying method in which a plane table is used alternately at each end of a measured baseline. [MATH] **1.** The point, or set of points, that is common to two or more geometric configurations. **2.** For two sets, the set consisting of all elements common to both of the sets. Also known as meet. **3.** For two fuzzy sets A and B, the fuzzy set whose membership function has a value at any element x that is the minimum of the values of the membership functions of A and B at x. **4.** The intersection of two Boolean matrices A and B, with the same number of rows and columns, is the Boolean matrix whose element c_{ij} in row i and column j is the intersection of corresponding elements a_{ij} in A and b_{ij} in B. { ,in·tər'sek·shən }

intersection angle [CIV ENG] The angle of deflection at the intersection point between the straights of a railway or highway curve. { ,in·tər'sek·shən ,aŋ·gəl }

intersection data [COMPUT SCI] Data which are meaningful only when associated with the concatenation of two segments. { ,in·tər'sek·shən ,dad·ə }

intersection point [CIV ENG] That point where two straights or tangents to a railway or road curve would meet if extended. { ,in·tər'sek·shən ,pȯint }

intersegmental [BIOL] Situated between or involving segments. [EMBRYO] Situated between the primordial segments of the embryo. { ¦in·tər·seg'ment·əl }

intersegmental reflex [NEUROSCI] An unconditioned reflex arc connecting input and output by means of afferent pathways in the dorsal spinal roots and efferent pathways in the ventral spinal roots. { ¦in·tər·seg'ment·əl 'rē,fleks }

intersertal [PETR] Referring to the texture of a porphyritic igneous rock in which the groundmass forms a small proportion of the rock, filling the interstices between unoriented feldspar laths. { ¦in·tər¦sərd·əl }

intersex [PHYSIO] An individual who is intermediate in sexual constitution between male and female. { 'in·tər,seks }

interspace [ANAT] An interval between the ribs or the fibers or lobules of a tissue or organ. [BUILD] An air space. [PHYS] An interval of space or time. { 'in·tər,spās }

interspersed repeats [CELL MOL] Repetitive deoxyribonucleic acid sequences. { 'in·tər'spərst ri'pēs }

interspersion [CELL MOL] A regular pattern of alternating sequences of repetitious and nonrepetitious deoxyribonucleic acid in the genome of eukaryotes. [ECOL] **1.** An intermingling of different organisms within a community. **2.** The level or degree of intermingling of one kind of organism with others in the community. { ‚in·tər‚spər·zhən }

interspinal [ANAT] Situated between or connecting spinous processes; interspinous. { ¦in·tər¦spīn·əl }

interstadial [GEOL] Pertaining to a period during a glacial stage in which the ice retreated temporarily. { ‚in·tər'städ·ē·əl }

interstage transformer [ELECTR] A transformer used to provide coupling between two stages. { 'in·tər‚stāj tranz‚fòr·mər }

interstation noise suppression See intercarrier noise suppression. { ¦in·tər¦stā·shən 'nòiz sə‚presh·ən }

interstellar [ASTRON] Between the stars. { ¦in·tər¦stel·ər }

interstellar extinction [ASTRON] The dimming of light from stars due to its absorption and scattering by dust grains in the interstellar medium. { ¦in·tər¦stel·ər ik'stiŋk·shən }

interstellar lines [ASTRON] Dark, narrow lines in the spectra of stars, caused by absorption of radiation by a gaseous medium in space. { ¦in·tər¦stel·ər 'līnz }

interstellar matter [ASTRON] The gaseous and dust material between the stars. { ¦in·tər¦stel·ər 'mad·ər }

interstellar probe [AERO ENG] An instrumentated spacecraft propelled beyond the solar system to obtain specific information about interstellar environment. { ¦in·tər¦stel·ər 'prōb }

interstellar space [ASTRON] The space between the stars. { ¦in·tər¦stel·ər 'spās }

interstellar travel [AERO ENG] Space flight between stars. { ¦in·tər¦stel·ər 'trav·əl }

intersternite [INV ZOO] An intersegmental plate on the ventral surface of the abdomen in insects. { ¦in·tər'stər‚nīt }

interstice [GEOL] A pore space within a rock or soil. [SOLID STATE] A space or volume between atoms of a lattice, or between groups of atoms or grains of a solid structure. { in'tərs·təs }

interstitial [CRYSTAL] A crystal defect in which an atom occupies a position between the regular lattice positions of a crystal. [SCI TECH] Of, pertaining to, or situated in a space between two things. { ¦in·tər¦stish·əl }

interstitial atom [CRYSTAL] A displaced atom which is forced into a nonequilibrium site within a crystal lattice. { ¦in·tər¦stish·əl 'ad·əm }

interstitial cell [HISTOL] A cell that is not peculiar to or characteristic of a particular organ or tissue but which comprises fibrous tissue binding other cells and tissue elements; examples are neuroglial cells and Leydig cells. { ¦in·tər¦stish·əl 'sel }

interstitial-cell-stimulating hormone See luteinizing hormone. { ¦in·tər¦stish·əl 'sel 'stim·yə‚lād·iŋ ‚hòr‚mōn }

interstitial-cell tumor [MED] A benign tumor of the testes composed of Leydig cells. Also known as interstitioma; Leydig-cell tumor. { ¦in·tər¦stish·əl 'sel 'tü·mər }

interstitial compound [CHEM] A compound of a transition metal and hydrogen, boron, carbon, or nitrogen whose crystals have a close-packed structure of the metal ions, with the nonmetal atoms being located in the interstices. [SOLID STATE] A binary compound in which atoms of one element (usually a light, nonmetallic element) occupy spaces between atoms of the crystal lattice formed by the other element (usually a heavy, metallic element). { ¦in·tər¦stish·əl 'käm‚paùnd }

interstitial emphysema [MED] Escape of air from the alveoli into the interstices of the lung, commonly due to trauma or violent cough. { ¦in·tər¦stish·əl ‚em·fə'sē·mə }

interstitial endometriosis [MED] The presence of endometrial tissue in the form of the stroma throughout the myometrium. Also known as endolymphatic stromomyosis; fibromyosis; parathelioma; stromal endometriosis; stromal myosis; stromatosis. { ¦in·tər¦stish·əl ‚en·dō·mə'trē'ō·səs }

interstitial-free steel [MET] An aluminum-killed steel with an extra-low carbon content, nominally 0.005%, in which the residual carbon is combined with niobium (columbium), titanium, or some similar element with a strong affinity for carbon. { ‚int·ər'stish·əl 'frē 'stēl }

interstitial gland [HISTOL] **1.** Groups of Leydig cells which secrete angiogens. **2.** Groups of epithelioid cells in the ovarian medulla of some lower animals. { ¦in·tər¦stish·əl 'gland }

interstitial hepatitis [MED] Pathologic deterioration and death of parenchymal cells in the liver associated with infiltration of lymphocytes and monocytes in the portal canals. Also

known as acute nonsuppurative hepatitis; nonspecific hepatitis. { ¦in·tər¦stish·əl ‚hep·ə'tīd·əs }

interstitial hernia [MED] Protrusion of the intestine between the muscular planes of the abdominal wall. Also known as intermuscular hernia; interparietal hernia. { ¦in·tər¦stish·əl 'hər‚nē·ə }

interstitial implants [MED] Solid or encapsulated radiation sources, made in the form of seeds, wires, or other shapes, to be inserted directly into tissue that is to be irradiated. { ¦in·tər¦stish·əl 'im‚plans }

interstitial impurity [SOLID STATE] An atom which is not normally found in a solid, and which is located at a position in the lattice structure where atoms or ions normally do not exist. { ¦in·tər¦stish·əl im'pyùr·əd·ē }

interstitial inflammation [MED] Inflammation of the interstitial tissues of an organ. { ¦in·tər¦stish·əl ‚in·flə'mā·shən }

interstitial keratitis [MED] Inflammation of the cornea in which the iris is almost completely hidden by the diffuse haziness of the corneal tissue. { ¦in·tər¦stish·əl ‚ker·ə'tīd·əs }

interstitial lamella [HISTOL] Any of the layers of bone between adjacent Haversian systems. { ¦in·tər¦stish·əl lə'mel·ə }

interstitial myocarditis [MED] Inflammation of the myocardium accompanied by cellular infiltration of interstitial tissues. Also known as Fiedler's myocarditis. { ¦in·tər¦stish·əl ‚mī·ə‚kär'dīd·əs }

interstitial nephritis See pyelonephritis. { ¦in·tər¦stish·əl nə'frīd·əs }

interstitial plasma-cell pneumonia See Pneumocystis carinii pneumonia. { ¦in·tər¦stish·əl 'plaz·mə ‚sel nə'mō·nē·ə }

interstitial pneumonia [MED] Inflammation of the lungs, particularly the stroma, including peribronchial tissue and interalveolar septa. { ¦in·tər¦stish·əl nə'mō·nē·ə }

interstitial radiation [MED] Radiation of tissues by implantation of a radioactive source material. { ¦in·tər¦stish·əl 'rād·ē'ā·shən }

interstitial water [HYD] Subsurface water contained in pore spaces between the grains of rock and sediments. { ¦in·tər¦stish·əl 'wòd·ər }

interstitial water saturation [HYD] The water content of a subterranean reservoir formation. { ¦in·ter¦stish·əl ¦wòd·ər 'sach·ə'rā·shən }

interstitioma See interstitial-cell tumor. { 'in·tər'stich·ē'ō·mə }

intersymbol interference [COMMUN] In a transmission system, extraneous energy from the signal in one or more keying intervals which tends to interfere with the reception of the signal in another keying interval, or the disturbance which results. { ¦in·tər'sim·bəl ‚in·tər'fir·əns }

intersystem communications [COMPUT SCI] The ability of two or more computer systems to share input, output, and storage devices, and to send messages to each other by means of shared input and output channels or by channels that directly connect central processors. { ¦in·tər'sis·təm kə‚myü·nə'kā·shənz }

intertergite [INV ZOO] One of the small plates between the tergites of certain insects. { ¦in·tər'tər‚jīt }

interterminal switching [CIV ENG] The movement of railroad cars from one line to another within a switching area. { ¦in·tər'tər·mən·əl 'swich·iŋ }

intertidal zone [OCEANOGR] The part of the littoral zone above low-tide mark. { ¦in·tər'tīd·əl ‚zōn }

intertoll trunk [COMMUN] A trunk between toll offices in different telephone exchanges. { 'in·tər‚tōl 'trəŋk }

intertropical convergence zone [METEOROL] The axis, or a portion thereof, of the broad trade-wind current of the tropics; this axis is the dividing line between the southeast trades and the northeast trades (of the Southern and Northern hemispheres, respectively). Also known as equatorial convergence zone; meteorological equator. { ¦in·tər'träp·ə·kəl kən'vər·jəns ‚zōn }

intertropical front [METEOROL] The interface or transition zone occurring within the equatorial trough between the Northern and Southern hemispheres. Also known as equatorial front; tropical front. { ¦in·tər'träp·ə·kəl 'frənt }

intertube burner [MECH ENG] A burner which utilizes a nozzle that discharges between adjacent tubes. { 'in·tər‚tüb ‚bər·nər }

intertubercular sulcus [ANAT] A deep groove on the anterior surface of the upper end of the humerus, separating the greater and lesser tubercles; contains the tendon of the long head of the biceps brachii muscle. Also known as bicipital groove. { ¦in·tər·tə¦bər·kyə·lər 'səl·kəs }

interurban [GEOGR] Connecting or extending between urban areas. { ¦in·tər¦ər·bən }

interval [ACOUS] The spacing in pitch or frequency between two sounds; the frequency interval is the ratio of the frequencies or the logarithm of this ratio. [MATH] A set of numbers which consists of those numbers that are greater than one fixed number and less than another, and that may also include one or both of the end numbers. [PHYS] The time separating two events, or the distance between two objects. [RELAT] **1.** In special relativity, the Lorentz invariant quantity $c^2(\Delta t)^2 - (\Delta x)^2 - (\Delta y)^2 - (\Delta z)^2$, where c is the speed of light, Δt is the difference in the time coordinates of two specified events, and Δx, Δy, and Δz are differences in their x, y, and z coordinates, respectively. **2.** In general relativity, a generalization of this concept, namely the sum over the indices μ and ν of $g_{\mu\nu}dx^\mu dx^\nu$, where dx^μ and dx^ν are the differences in the x^μ and x^ν coordinates of two specified neighboring events, and $g_{\mu\nu}$ is an element of the metric tensor. { 'in·tər·vəl }

interval arithmetic [COMPUT SCI] A method of numeric computation in which each variable is specified as lying within some closed interval, and each arithmetic operation computes an interval containing all values that can result from operating on any numbers selected from the intervals associated with the operands. Also known as range arithmetic. { 'in·tər·vəl ə'rith·mə·tik }

interval estimate [STAT] An estimate which specifies a range of values for a population parameter. { 'in·tər·vəl ,es·tə·mət }

interval estimation [STAT] A technique that expresses uncertainty about an estimate by defining an interval, or range of values, and indicates the certain degree of confidence with which the population parameter will fall within the interval. { 'in·tər·vəl ,es·tə,mā·shən }

intervallum [INV ZOO] The space between the walls of pleosponges. { ¦in·tər¦val·əm }

interval measurement [STAT] A method of measuring quantifiable data that assumes an exact knowledge of the quantitative difference between the objects being scaled. Also known as cardinal measurement. { 'in·tər·vəl ,mezh·ər·mənt }

interval of convergence [MATH] The interval consisting of the real numbers for which a specified power series possesses a limit. { 'in·tər·vəl əv kən'vər·jəns }

interval of Sturm *See* astigmatic interval. { ¦in·tər·vəl əv 'stərm }

intervalometer [ORD] An electrical device used in bombing, by which data is preset in order to drop a desired number of bombs at a constant predetermined interval. { ,in·tər·və'läm·əd·ər }

interval scale [STAT] A rule or system for assigning numbers to objects in such a way that the difference between any two objects is reflected in the difference in the numbers assigned to them; used in interval measurement. { 'in·tər·vəl ,skāl }

interval timer [ENG] A device which operates a set of contacts during a preset time interval and, at the end of the interval, returns the contacts to their normal positions. Also known as timer. { 'in·tər·vəl ,tīm·ər }

intervascular [ANAT] Located between or surrounded by blood vessels. { ¦in·tər'vas·kyə·lər }

intervening sequence [GEN] The one or more segments of a split gene that are transcribed but not included in the final messenger ribonucleic acid; each is flanked by two exons. Also known as intron. { ¦in·tər¦vēn·iŋ 'sē·kwəns }

interventricular foramen [ANAT] Either one of the two foramens that connect the third ventricle of the brain with each lateral ventricle. Also known as foramen of Monro. { ¦in·tər·ven'trik·yə·lər fə'rā·mən }

interventricular septal defect [MED] A congenital malformation of the septum between the ventricles of the heart. { ¦in·tər·ven'trik·yə·lər ¦sep·təl 'dē,fekt }

interventricular septum [ANAT] The muscular wall between the heart ventricles. Also known as ventricular septum. { ¦in·tər·ven'trik·yə·lər 'sep·təm }

intervertebral [ANAT] Being or located between the vertebrae. { ¦in·tər'vərd·ə·bəl }

intervillous spaces [HISTOL] Spaces in the placenta which communicate with the maternal blood vessels. { ¦in·tər'vil·əs 'spās·əz }

intestinal crura [INV ZOO] The main intestinal branches in certain trematodes. { in'tes·tən·əl 'krūr·ə }

intestinal digestion [PHYSIO] Conversion of food to an assimilable form by the action of intestinal juices. { in'tes·tən·əl di'jes·chən }

intestinal dyspepsia [MED] Disturbed digestion due to diminished secretion of intestinal juices or lack of tonus in the wall of the intestine. { in'tes·tən·əl dis'pep·sē·ə }

intestinal hormone [BIOCHEM] Either of two hormones, secretin and cholecystokinin, secreted by the intestine. { in'tes·tən·əl 'hȯr,mōn }

intestinal juice [PHYSIO] An alkaline fluid composed of the combined secretions of all intestinal glands. { in'tes·tən·əl 'jüs }

intestinal lipodystrophy *See* Whipple's disease. { in'tes·tən·əl ¦lip·ə'dis·trə·fē }

intestinal villi [ANAT] Fingerlike projections of the small intestine, composed of a core of vascular tissue covered by epithelium and containing smooth muscle cells and an efferent lacteal end capillary. { in'tes·tən·əl 'vil,ī }

intestine [ANAT] The tubular portion of the vertebrate digestive tract, usually between the stomach and the cloaca or anus. { in'tes·tən }

in-the-seam mining [MIN ENG] The usual method of mining characterized by the driving of development shafts into the coal seam. { in thə ¦sēm 'mīn·iŋ }

in the wind [NAV] In the direction from which the wind is blowing; windward; used particularly in reference to a heading or to the position of an object. { ,in thə 'wind }

intima [HISTOL] The innermost coat of a blood vessel. Also known as tunica intima. { 'in·tə·mə }

intimate blend [TEXT] A combination of two or more staple fibers in a spun yarn that has been blended so that the individual fibers do not retain their individual characteristics. { 'in·tə·mət 'blend }

intimate ion pair *See* contact ion pair. { 'in·tə·mət 'ī,än ,per }

intorsion [BIOL] Inward rotation of a structure about a fixed point or axis. { in'tȯr·shən }

into the solid [MIN ENG] Of a shot, going into the coal beyond the point to which the coal can be broken by the blast. Also known as on the solid. { ,in·tə thə 'säl·əd }

intoxication [MED] **1.** Poisoning. **2.** The state produced by overindulgence in alcohol. { in,täk·sə'kā·shən }

intraabdominal [ANAT] Occurring or being within the cavity of the abdomen. { ¦in·trə·ab'däm·ən·əl }

intraatrial heart block [MED] A type of heart block which shows a broad, notched P wave of longer than normal duration on the electrocardiographic record. { ¦in·trə'ā·trē·əl 'härt ,bläk }

intrabeam viewing [OPTICS] The viewing condition in which the eye is exposed to all or part of a laser beam. { 'in·trə,bēm 'vyü·iŋ }

intracartilaginous ossification *See* endochondral ossification. { ¦in·trə,kärd·əl'aj·ə·nəs ,äs·ə·fə'kā·shən }

intracavernous aneurysm [MED] A dilation of the wall of the internal carotid artery within the cavernous sinus. { ¦in·tər'kav·ər·nəs 'an·yə,riz·əm }

intracavity absorption spectroscopy [SPECT] A highly sensitive technique in which an absorbing sample is placed inside the resonator of a broad-band dye laser, and absorption lines are detected as dips in the laser emission spectrum. { ¦in·trə'kav·əd·ē əb'sȯrp·shən spek'träs·kə·pē }

intracellular [CELL MOL] Within a cell. { ¦in·trə'sel·yə·lər }

intracellular canaliculi [CELL MOL] A system of minute canals within certain gland cells which are thought to drain the glandular secretions. { ¦in·trə'sel·yə·lər ,kan·əl'ik·yə,lī }

intracellular digestion [PHYSIO] Digestion which takes place within the cytoplasm of the organism, as in many unicellular protozoans. { ¦in·trə'sel·yə·lər di'jes·chən }

intracellular enzyme [BIOCHEM] An enzyme that remains active only within the cell in which it is formed. Also known as organized ferment. { ¦in·trə'sel·yə·lər en,zīm }

intracellular signaling protein [CELL MOL] One of a series of intracellular proteins that mediates a cell's response when an extracellular signal molecule binds to a receptor protein on the cell's surface, either by changing the cell's metabolism or shape, or movement, or gene expression. { ˌin·trəˈselˈyə·lər ˈsigˈnəl·iŋ ˌprōˌtēn }

intracellular symbiosis [CELL MOL] Existence of a self-duplicating unit within the cytoplasm of a cell, such as a kappa particle in *Paramecium*, which seems to be an infectious agent and may influence cell metabolism. { ˈinˈtrəˈselˈyə·lər ˌsimˈbēˈō·səs }

intracervical [ANAT] Located within the cervix of the uterus. { ˈinˈtrəˈser·və·kəl }

intracistron complementation [GEN] The process whereby two different mutant alleles, each of which determines in homozygotes an inactive enzyme, determine the formation of the active enzyme when present in the same nucleus. { ˈinˈtrəˈsis·trən ˌkäm·plə·mənˈtā·shən }

intraclast [GEOL] A fragment of limestone formed by erosion within a basin of deposition and redeposited there to form a new sediment. { ˈinˈtrəˌklast }

intracloud discharge See cloud discharge. { ˈinˈtrəˌklaùd ˈdisˌchärj }

intracluster medium [ASTRON] A hot, tenuous gas that fills the space between the members of a cluster of galaxies and emits x-rays. { ˌinˈtrəˈkləsˈtər ˌmēd·ē·əm }

Intracoastal Waterway [NAV] An inside protected route extending through New Jersey; from Norfolk, Virginia, to Key West, Florida; across Florida from St. Lucie Inlet to Fort Myers, Charlotte Harbor, Tampa Bay, and Tarpon Springs; and from Carabelle, Florida, to Brownsville, Texas. { ˈinˈtrəˈkōst·əl ˈwòd·ər,wā }

Intracoastal Waterway charts [NAV] Charts of scale 1:40,000, for navigating the Intracoastal Waterway, and designed especially for small-boat and yacht operators. { ˈinˈtrəˈkōst·əl ˈwòd·ər,wā ˌchärts }

intracortical [ANAT] Occurring or located within the cortex. { ˈinˈtrəˈkórd·ə·kəl }

intracranial [ANAT] Within the cranium. { ˈinˈtrəˈkrā·nē·əl }

intracranial aneurysm [MED] Dilation of a cerebral artery. { ˈinˈtrəˈkrā·nē·əl ˈan·yə,riz·əm }

intracranial angiography [MED] Roentgenography of the blood vessels within the cranial cavity following the intravascular injection of a radiopaque material. { ˈinˈtrəˈkrā·nē·əl ˌan·jēˈäg·rə·fē }

intracratonic basin See autogeosyncline. { ˈinˈtrəˈkrəˈtän·ik ˈbā·sən }

intracrystalline See transcrystalline. { ˈinˈtrəˈkristˈəl·ən }

intracytoplasmic [CELL MOL] Being or occurring within the cytoplasm of a cell. { ˈinˈtrəˌsīd·əˈplaz·mik }

intradermal [ANAT] Within the skin. { ˈinˈtrəˈdər·məl }

intradermal nevus [MED] A lesion containing melanocytes and located principally or completely within the derma. { ˈinˈtrəˈdər·məl ˈnē·vəs }

intradermopalpebral [ANAT] Within the skin of the eyelid. { ˌinˈtrəˌdərm·ō·palˈpeb·rəl }

intrados [ARCH] The inner curved surface of an arch. { ˈinˈtrəˌdäs }

intraductal [ANAT] Within a duct. { ˈinˈtrəˈdək·təl }

intradural [ANAT] Within the dura mater. { ˈinˈtrəˈdùr·əl }

intraembryonic [EMBRYO] Within the embryo. { ˈinˈtrəˌemˈbrē·än·ik }

intraepidermal [ANAT] Within the epidermis. { ˈinˈtrəˌepˈəˈdər·məl }

intraepidermal epithelioma [MED] Carcinoma in situ, of either the squamous-cell or basal-cell type. { ˈinˈtrəˌepˈəˈdər·məl ˌepˈə,thēˈlēˈō·mə }

intraepithelial [ANAT] Within the epithelium. { ˈinˈtrəˌepˈəˈthēˈlē·əl }

intraesophageal [ANAT] Within the esophagus. { ˈinˈtrəˌəˌsäfˈəˈjēˈəl }

intrafascicular [BOT] Located or occurring within a vascular bundle. { ˈinˈtrəˈfəˈsikˈyə·lər }

intraformational breccia [PETR] A rock resulting from cracking and desiccation-shrinking of a mud after withdrawal of water followed by almost contemporaneous sedimentation. { ˈinˈtrəˌfòrˈmāshˈən·əl ˈbrech·ə }

intraformational conglomerate [GEOL] **1.** A conglomerate in which clasts and the matrix are contemporaneous in origin.

2. A conglomerate formed in the midst of a geologic formation. { ˈinˈtrəˌfòrˈmāshˈən·əl kənˈgläm·ə·rət }

intraformational fold [GEOL] A minor fold confined to a sedimentary layer lying between undeformed beds. { inˈtrəˌfòrˈmāshˈən·əl ˈfōld }

intrafusal fiber [HISTOL] Any of the striated muscle fibers contained in a muscle spindle. { ˈinˈtrəˈfyüz·əl ˈfī·bər }

intragenic [MOL BIO] Within a gene, in referring to certain events. { ˈinˈtrəˈjen·ik }

intragenic recombination [CELL MOL] Recombination occurring between the mutons of one cistron. { ˈinˈtrəˈjen·ik rē,käm·bəˈnā·shən }

intragenic suppression [GEN] The restoration of a suppressed function or character as a consequence of a second mutation located within the same gene as the original or first mutation. { ˈinˈtrəˈjen·ik səˈpresh·ən }

intrahepatic [ANAT] Within the liver. { ˈinˈtrə·heˈpad·ik }

intrajugular process [ANAT] **1.** A small, curved process on some occipital bones which partially or completely divides the jugular foramen into lateral and medial parts. **2.** A small process on the hard portion of the temporal bone which completely or partly separates the jugular foramen into medial and lateral parts. { ˈinˈtrəˈjəg·yə·lər ˌpräˈsəs }

intralaminar nuclei [NEUROSCI] A diffuse group of nuclei located in the internal medullary lamina of the thalamus. { ˌinˈtrəˈla·mən·ər ˈnü·klē,ī }

intraline distance [ENG] The minimum distance permitted between any two buildings within an explosives operating line; to protect buildings from propagation of explosions due to blast effect. { ˈinˈtrəˌlīn ˈdis·təns }

intraluminal [ANAT] Within the lumen of a structure. { ˌinˈtrəˈlü·mən·əl }

intramarginal [BIOL] Within a margin. { ˌinˈtrəˈmär·jən·əl }

intramedullary [ANAT] Within the bone marrow. [NEUROSCI] **1.** Within the tissues of the spinal cord or medulla oblongata. **2.** Within the adrenal medulla. { ˌinˈtrə·məˈdəl·ə·rē }

intramembranous [HISTOL] Formed or occurring within a membrane. { ˌinˈtrəˈmem·brə·nəs }

intramembranous ossification [HISTOL] Formation of bone tissue directly from connective tissue without a preliminary cartilage stage. { ˌinˈtrəˈmem·brə·nəs ˌäsˈəˈfəˈkā·shən }

intramural [ANAT] Within the substance of the walls of an organ. { ˈinˈtrəˈmyùr·əl }

intramuscular [ANAT] Lying within or going into the substance of a muscle. { ˈinˈtrəˈməsˈkyə·lər }

intranet [COMPUT SCI] A private network, based on Internet protocols, that is accessible only within an organization. Intranets are set up for many purposes, including e-mail, access to corporate databases and documents, and videoconferencing, as well as buying and selling goods and services. { ˈinˈtrəˌnet }

intranuclear cascade model [NUC PHYS] A model of nuclear collisions that assumes a series of independent nucleon-nucleon collisions between particles that act like billiard balls. { ˈinˈtrəˈnü·klē·ər kasˈkād ˌmäd·əl }

intraocular [ANAT] Occurring within the globe of the eye. { ˌinˈtrəˈäkˈyə·lər }

intraocular pressure [PHYSIO] The hydrostatic pressure within the eyeball. { ˌinˈtrəˈäkˈyə·lər ˈpresh·ər }

intraoptical light sighting system [OPTICS] A target sighting system used with infrared pyrometers in which a visible, pulsating light is directed through the infrared optics, and gives an indication of the exact field of view, not just the centerpoint. { ˌinˈtrəˈäpˈtə·kəl ˈlīt ˌsīd·iŋ ˌsis·təm }

intraparietal [ANAT] **1.** Within the wall of an organ or cavity. **2.** Within the parietal region of the cerebrum. **3.** Within the body wall. { ˌinˈtrə·pəˈrī·əd·əl }

intrapartum [MED] Occurring during parturition. { ˈinˈtrəˈpär·dəm }

intraperitoneal [ANAT] **1.** Within the peritoneum. **2.** Within the peritoneal cavity. { ˈinˈtrəˌper·ə·təˈnē·əl }

intrapetiolar [BOT] **1.** Enclosed by the base of the petiole. **2.** Between the petiole and the stem. { ˈinˈtrəˈpedˈēˈäl·yə·lər }

intraprocedure metric [COMPUT SCI] A software metric that determines the complexity of a computer program as a function of the relationships of the different modules constituting the program, generally by constructing a flow graph and deriving the complexity from this graph.   { ˌin·trə·prə¦sē·jər 'me·trik }

intrapulmonic [ANAT] Being or occurring within the lungs. { ¦in·trə·pủl'män·ik }

intraspecific [BIOL] Being within or occurring among the members of the same species. { ¦in·trə·spə·'sif·ik }

intraspinal block [MED] Anesthesia of the spinal column by injection of an anesthetic into the spinal canal. { ¦in·trə¦spīn·əl 'bläk }

intrastratal solution [GEOCHEM] A chemical attrition of the constituents of a rock after deposition. { ¦in·trə¦strad·əl sə'lü·shən }

intratelluric [GEOL] **1.** Pertaining to a phenocryst that is formed earlier than its matrix. **2.** Pertaining to a period in which igneous rocks crystallized prior to their eruption. **3.** Located, formed, or originating at great depths within the earth. { ¦in·trə·tə'lyur·ik }

intrathecal [ANAT] Within the subarachnoid space. { ¦in·trə'thē·kəl }

intrathoracic [ANAT] Within the thoracic cavity. { ¦in·trə·thə'ras·ik }

intratracheal [ANAT] Being or occurring within the trachea. { ¦in·trə'trā·kē·əl }

intrauterine [ANAT] Being or occurring within the uterus. { ¦in·trə'yüd·ə·rən }

intravaginal [ANAT] **1.** Being or occurring within the vagina. **2.** Located within a tendon sheath. [BOT] Located within a sheath, referring to branches of grass. { ¦in·trə'vaj·ə·nəl }

intravascular [ANAT] Within blood vessels or within a blood vessel. { ˌin·trə'vas·kyə·lər }

intravenous [ANAT] Located within, or going into, the veins. { ˌin·trə'vē·nəs }

intraventricular heart block [MED] Prolongation of the process of ventricular excitation, measured by the QRS complex of the electrocardiogram. Also known as arborization block; parietal block; peri-infarction block. { ˌin·trə·ven'trik·yə·lər 'härt ˌbläk }

intravesical [ANAT] Within the urinary bladder. { ˌin·trə'ves·ə·kəl }

intravital [BIOL] Occurring while the cell or organism is alive. { ˌin·trə'vīd·əl }

intravital stain [BIOL] A nontoxic dye injected into the body to selectively stain certain cells or tissues. { ˌin·trə'vīd·əl 'stān }

intrazonal soil [GEOL] A group of soils with well-developed characteristics that reflect the dominant influence of some local factor of relief, parent material, or age over the usual effect of vegetation and climate. { ˌin·trə'zōn·əl 'sȯil }

intrinsic asthma [MED] Asthma caused by a respiratory tract infection. { in'trin·sik 'az·mə }

intrinsic-barrier diode [ELECTR] A *pin* diode, in which a thin region of intrinsic material separates the *p*-type region and the *n*-type region. { in'trin·sik ¦bar·ē·ər 'dī·ōd }

intrinsic-barrier transistor [ELECTR] A *pnip* or *npin* transistor, in which a thin region of intrinsic material separates the base and collector. { in'trin·sik ¦bar·ē·ər tran'zis·tər }

intrinsic conductivity [SOLID STATE] The conductivity of a semiconductor or metal in which impurities and structural defects are absent or have a very low concentration. { in'trin·sik ˌkän·dək'tiv·əd·ē }

intrinsic contact potential difference [ELEC] True potential difference between two perfectly clean metals in contact. { in'trin·sik ¦kän,takt pə¦ten·chəl 'dif·ərns }

intrinsic detector [ENG] A semiconductor detector of electromagnetic radiation that utilizes the generation of electron-hole pairs across the semiconductor band gap. { in'trin·sik di'tek·tər }

intrinsic electric strength [ELEC] The extremely high dielectric strength displayed by a substance at low temperatures. { in¦trin·sik i¦lek·trik ¦streŋkth }

intrinsic equations of a curve [MATH] The equations describing the radius of curvature and torsion of a curve as a function of arc length; these equations determine the curve up to its position in space. Also known as natural equations of a curve. { in'trin·sik i¦kwā·zhənz əv ə 'kərv }

intrinsic factor [BIOCHEM] A substance, produced by the stomach, which combines with the extrinsic factor (vitamin B$_{12}$) in food to yield an antianemic principle; lack of the intrinsic

factor is believed to be a cause of pernicious anemia. Also known as Castle's intrinsic factor. { in'trin·sik 'fak·tər }

intrinsic flux density *See* intrinsic induction. { in'trin·sik 'fləks ˌden·səd·ē }

intrinsic geometry of a surface [MATH] The description of the intrinsic properties of a surface. { in'trin·sik jē'äm·ə·trē əv ə 'sərfəs }

intrinsic induction [ELECTROMAG] The vector difference between the magnetic flux density at a given point and the magnetic flux density which would exist there, for the same magnetic field strength, if the point were in a vacuum. Symbol Bi. Also known as intensity of magnetization; intrinsic flux density; magnetic polarization. { in'trin·sik in'dək·shən }

intrinsic layer [ELECTR] A layer of semiconductor material whose properties are essentially those of the pure undoped material. { in'trin·sik 'lā·ər }

intrinsic luminosity [ASTROPHYS] The total amount of radiation emitted by a star over a specified range of wavelengths. { in'trin·sik ˌlü·mə'näs·əd·ē }

intrinsic mobility [SOLID STATE] The mobility of the electrons in an intrinsic semiconductor. { in'trin·sik mō'bil·əd·ē }

intrinsic nerve supply [NEUROSCI] The nerves contained entirely within an organ or structure. { in'trin·sik 'nərv sə,plī }

intrinsic parity [PARTIC PHYS] A quantum number, equal to +1 or -1, which is assigned to particles so that the product of the intrinsic parities of the particles composing a system times the parity of the system's wave function yields the total parity. { in'trin·sik 'par·əd·ē }

intrinsic photoconductivity [SOLID STATE] Photoconductivity associated with excitation of charge carriers across the band gap of a material. { in'trin·sik ¦fōd·ō,kän,dək'tiv·əd·ē }

intrinsic photoemission [SOLID STATE] Photoemission which can occur in an ideally pure and perfect crystal, in contrast to other types of photoemission which are associated with crystal defects. { in'trin·sik ¦fōd·ō·i'mish·ən }

intrinsic pressure [PHYS] Pressure in a fluid resulting from inward forces on molecules near the fluid surface, caused by attraction between molecules. Also known as internal pressure. { in'trin·sik 'presh·ər }

intrinsic procedure *See* built-in function. { in'trin·sik prə'sē·jər }

intrinsic property [MATH] **1.** For a curve, a property that can be stated without reference to the coordinate system. **2.** For a surface, a property that can be stated without reference to the surrounding space. [SOLID STATE] A property of a substance that is not seriously affected by impurities or imperfections in the crystal structure. { in'trin·sik 'präp·ərd·ē }

intrinsic protein *See* integral membrane protein. { in¦trin·zik 'prō,tēn }

intrinsic semiconductor [SOLID STATE] A semiconductor in which the concentration of charge carriers is characteristic of the material itself rather than of the content of impurities and structural defects of the crystal. Also known as *i*-type semiconductor. { in'trin·sik ¦sem·i·kən,dək·tər }

intrinsic temperature range [SOLID STATE] In a semiconductor, the temperature range in which its electrical properties are essentially not modified by impurities or imperfections within the crystal. { in'trin·sik 'tem·prə·chər ˌrānj }

intrinsic tracer [NUC PHYS] An isotope that is present naturally in a form suitable for tracing a given element through chemical and physical processes. { in'trin·sik 'trā·sər }

intrinsic variable star [ASTRON] A star that is variable not because of an eclipse. { in'trin·sik ¦ver·ē·ə·bəl 'stär }

intrinsic viscosity [PHYS CHEM] The ratio of a solution's specific viscosity to the concentration of the solute, extrapolated to zero concentration. Also known as limiting viscosity number. { in'trin·sik vi'skäs·əd·ē }

introductory column [MIN ENG] The highest and first column that is inserted in casing a borehole. { ¦in·trə¦dək·tə·rē 'käl·əm }

introfaction [CHEM] Change in fluidity and specific wetting properties (for impregnation acceleration) of an impregnating compound, caused by an introfier (impregnation accelerator). { ¦in·trə¦fak·shən }

introgressive hybridization [GEN] The spreading of genes of a species into the gene complex of another due to hybridization between numerically dissimilar populations and extensive backcrossing. { ¦in·trə¦gres·iv ˌhī·brəd·ə'zā·shən }

introitus [ANAT] An opening or entryway, especially the opening into the vagina. { in'trō·ə·dəs }

introjection [PSYCH] The symbolic absorption into and toward oneself of concepts and feelings generated toward another person or object; motivates irrational behavior toward oneself. { 'in·trə'jek·shən }

intromission [ZOO] The act or process of inserting one body into another, specifically, of the penis into the vagina. { ,in·trə'mish·ən }

intromittent [ZOO] Adapted for intromission; applied to a copulatory organ. { ¦in·trə¦mit·ənt }

intron *See* intervening sequence. { 'in,trän }

introrse [BIOL] Turned inward or toward the axis. { 'in,trórs }

introspective diplopia *See* physiologic diplopia. { ¦in·trə¦spek·tiv də'plō·pē·ə }

introversion [MED] The act or process of turning in upon itself, as a hollow organ. [PSYCH] Preoccupation with the self associated with diminished interest in external events. { ¦in·trə¦vər·zhən }

introvert [PSYCH] An individual whose interests are self-directed, and not directed toward the outside world. [ZOO] **1.** A structure capable of introversion. **2.** To turn inward. { 'in·trə,vərt }

intrusion [GEOL] **1.** The process of emplacement of magma in preexisting rock. Also known as injection; invasion; irruption. **2.** A large-scale sedimentary injection. Also known as sedimentary intrusion. **3.** Any rock mass formed by an intrusive process. { in'trü·zhən }

intrusion grouting [ENG] A method of placing concrete by intruding the mortar component in position and then converting it into concrete as it is introduced into voids. { in'trü·zhən ,graud·iŋ }

intrusive [PETR] Pertaining to material forced while still in a fluid state into cracks or between layers of rock. { in'trü·siv }

intubation [MED] The introduction of a tube into a hollow organ to keep it open, especially into the larynx to ensure the passage of air. { ,in,tü'bā·shən }

intumescence [MATER] The property of a material to swell when heated; intumescent materials in bulk and sheet form are used as fireproofing agents. { ,in·tü'mes·əns }

intussusception [MED] Passing of a portion of a structure into another part of the same structure, such as the invagination of a part of the intestine. { ,in·tə·sə'sep·shən }

inulase [BIOCHEM] An enzyme produced by certain molds that catalyzes the conversion of inulin to levulose. { 'in·yə,lās }

inulin [BIOCHEM] A polysaccharide made up of polymerized fructofuranose units. { 'in·yə·lən }

inunction [MED] Act of applying an oil or fatty material, especially rubbing an ointment into the skin as a therapeutic measure. { i'nəŋk·shən }

inundation [HYD] Flooding, by the rise and spread of water, of a land surface that is not normally submerged. { ,i·nən'dā·shən }

inundative control [AGR] The mass production and periodic release of large numbers of biocontrol agents to achieve controlling densities. { ,i·nən¦dād·iv kən'trōl }

in utero [EMBRYO] Within the uterus, referring to the fetus. { in 'yüd·ə,rō }

in vacuo [PHYS] In a vacuum. { in 'vak·yə·wō }

invaded zone [PETRO ENG] Transitional downhole area between the area invaded completely by drilling mud and uncontaminated bulk of the reservoir. { in'vād·əd ,zōn }

invagination [EMBRYO] The enfolding of a part of the wall of the blastula to form a gastrula. [PHYSIO] **1.** The act of ensheathing or becoming ensheathed. **2.** The process of burrowing or enfolding to form a hollow space within a previously solid structure, as the invagination of the nasal mucosa within a bone of the skull to form a paranasal sinus. { in,vaj·ə'nā·shən }

invar [MATER] An alloy (64% iron–36% nickel) that exhibits almost no thermal expansion over the temperature range of −50 to 150°C (−58 to 302°F). { 'in,vär }

invariable line [MECH] A line which is parallel to the angular momentum vector of a body executing Poinsot motion, and which passes through the fixed point in the body about which there is no torque. { in'ver·ē·ə·bəl 'līn }

invariable plane [MECH] A plane which is perpendicular to the angular momentum vector of a rotating rigid body not subject to external torque, and which is always tangent to its inertia ellipsoid. { in'ver·ē·ə·bəl 'plān }

invariance [MATH] *See* invariant property. [OPTICS] Any property of a light beam that remains constant when the light is reflected or refracted at one or more surfaces. [PHYS] The property of a physical quantity or physical law of being unchanged by certain transformations or operations, such as reflection of spatial coordinates, time reversal, charge conjugation, rotations, or Lorentz transformations. Also known as symmetry. { in'ver·ē·əns }

invariance principle [PHYS] Any principle which states that a physical quantity or physical law possesses invariance under certain transformations. Also known as symmetry law; symmetry principle. [RELAT] In general relativity, the principle that the laws of motion are the same in all frames of reference, whether accelerated or not. { in'ver·ē·əns ,prin·sə·pəl }

invariant [MATH] **1.** An element x of a set E is said to be invariant with respect to a group G of mappings acting on E if $g(x) = x$ for all g in G. **2.** A subset F of a set E is said to be invariant with respect to a group G of mappings acting on E if $g(x)$ is in F for all x in F and all g in G. **3.** For an algebraic equation, an expression involving the coefficients that remains unchanged under a rotation or translation of the coordinate axes in the cartesian space whose coordinates are the unknown quantities. { in'ver·ē·ənt }

invariant function [MATH] A function f on a set S is said to be invariant under a transformation T of S into itself if $f(Tx) = f(x)$ for all x in S. { in'ver·ē·ənt 'fəŋk·shən }

invariant measure [MATH] A Borel measure m on a topological space X is invariant for a transformation group (G,X,π) if for all Borel sets A in X and all elements g in G, $m(A_g) = m(A)$, where A_g is the set of elements equal to $\pi(g,x)$ for some x in A. { in'ver·ē·ənt 'mezh·ər }

invariant plane [ASTRON] The plane that is perpendicular to the total angular momentum of the solar system and passes through its center of mass. [ATOM PHYS] The plane perpendicular to the total angular momentum (orbital plus spin) of an atom. { in'ver·ē·ənt 'plān }

invariant property [MATH] A mathematical property of some space unchanged after the application of any member from some given family of transformations. Also known as invariance. { in'ver·ē·ənt 'präp·ərd·ē }

invariant subgroup *See* normal subgroup. { in'ver·ē·ənt 'səb,grup }

invariant subspace [MATH] For a bounded operator on a Banach space, a closed linear subspace of the Banach space such that the operator takes any point in the subspace to another point in the subspace. { in,ver·ē·ənt 'səb,spās }

invasion [GEOL] **1.** The movement of one material into a porous reservoir area that has been occupied by another material. **2.** *See* intrusion; transgression. [MED] **1.** The phase of an infectious disease during which the pathogen multiplies and is distributed; precedes signs and symptoms. **2.** The process by which microorganisms enter the body. { in'vā·zhən }

invasion efficiency [PETRO ENG] Completeness of invasion of a reservoir formation by a fluid. { in'vā·zhən i,fish·ən·sē }

inventory [ENG] The amount of plastic in the heating cylinder or barrel in injection molding or extrusion. { 'in·vən,tòr·ē }

inventory control [IND ENG] Systematic management of the balance on hand of inventory items, involving the supply, storage, distribution, and recording of items. { 'in·vən,tòr·ē kən,trōl }

inverna [METEOROL] A southeast wind of Lake Maggiore, Italy. { in'vər·nä }

inverse [MATH] **1.** The additive inverse of a real or complex number a is the number which when added to a gives 0; the multiplicative inverse of a is the number which when multiplied with a gives 1. **2.** The inverse of a fractional ideal I of an integral domain R is the set of all elements x in the quotient field K of R such that xy is in I for all y in I. **3.** For a set S with a binary operation $x \circ y$ that has an identity element e, the inverse of a member, x, of S is another member, $\bar{x}$, of S for which $x \circ \bar{x} = \bar{x} \circ x = e$. { 'in,vərs }

inverse beta decay [NUC PHYS] A reaction providing evidence for the existence of the neutrino, in which an antineutrino (or neutrino) collides with a proton (or neutron) to produce a neutron (or proton) and a positron (or electron). { 'in,vərs 'bäd·ə di,kā }

inverse bremsstrahlung [ATOM PHYS] The absorption by an electron of a photon in a strong electric field such as that surrounding an atomic nucleus. Also known as free-free absorption. { 'in,vərs 'brem,shträ·lùŋ }

inverse cam [MECH ENG] A cam that acts as a follower instead of a driver. { 'in,vərs 'kam }

inverse Compton effect [QUANT MECH] A process in which relativistic particles give up some of their energy to long-wavelength radiation, converting it to shorter-wavelength radiation. { 'in,vərs 'kam·tən i,fekt }

inverse cosecant See arc cosecant. { 'in,vərs kō'sē,kant }

inverse cosine See arc cosine. { 'in,vərs 'kō,sīn }

inverse cotangent See arc cotangent. { 'in,vərs kō'tan·jənt }

inverse current [ELECTR] The current resulting from an inverse voltage in a contact rectifier. { 'in,vərs 'kə·rənt }

inverse curves [MATH] A pair of curves such that every point on one curve is the inverse point of some point on the other curve, with respect to a fixed circle. { 'in,vərs 'kərvz }

inverse cylindrical orthomorphic chart See transverse Mercator chart. { 'in,vərs sə¦lin·drə·kəl ¦òr·thə'mòr·fik ,chärt }

inverse cylindrical orthomorphic projection See transverse Mercator projection. { 'in,vərs sə¦lin·drəkəl ¦ò·thə'mòr·fik prə'jek·shən }

inverse direction [ELECTR] The direction in which the electron flow encounters greater resistance in a rectifier, going from the positive to the negative electrode; the opposite of the conducting direction. Also known as reverse direction. { 'in,vərs də'rek·shən }

inverse electrode current [ELECTR] Current flowing through an electrode in the direction opposite to that for which the tube is designed. { 'in,vərs i'lek,trōd ,kə·rənt }

inverse element [MATH] In a group G the inverse of an element g is the unique element g^{-1} such that $g \cdot g^{-1} = g^{-1} \cdot g = e$, where $\cdot$ denotes the group operation and e is the identity element. { 'in,vərs 'el·ə·mənt }

inverse feedback See negative feedback. { 'in,vərs 'fēd,bak }

inverse function [MATH] An inverse function for a function f is a function g whose domain is the range of f and whose range is the domain of f with the property that both f composed with g and g composed with f give the identity function. { 'in,vərs 'fəŋk·shən }

inverse function theorem [MATH] If f is a continuously differentiable function of euclidean n-space to itself and at a point x_0 the matrix with the entry $(\partial f_i/\partial x_j)_{x_0}$ in the ith row and jth column is nonsingular, then there is a continuously differentiable function $g(y)$ defined in a neighborhood of $f(x_0)$ which is an inverse function for $f(x)$ at all points near x_0. { 'in,vərs 'fəŋk·shən ,thir·əm }

inverse hour See inhour. { 'in,vərs 'aùr }

inverse hyperbolic cosecant See arc-hyperbolic cosecant. { ¦in,vərs ¦hī·pər,bäl·ik kō'sē,kant }

inverse hyperbolic cosine See arc-hyperbolic cosine. { ¦in,vərs ¦hī·pər,bäl·ik 'kō,sīn }

inverse hyperbolic cotangent See arc-hyperbolic cotangent. { ¦in,vərs ¦hī·pər,bäl·ik kō'tan·jənt }

inverse hyperbolic function [MATH] An inverse function of a hyperbolic function; that is, an arc-hyperbolic sine, arc-hyperbolic cosine, arc-hyperbolic tangent, arc-hyperbolic cotangent, arc-hyperbolic secant, or arc-hyperbolic cosecant. Also known as antihyperbolic function; arc-hyperbolic function. { ¦in,vərs ¦hī·pər,bäl·ik 'fəŋk·shən }

inverse hyperbolic secant See arc-hyperbolic secant. { ¦in,vərs ¦hī·pər,bäl·ik 'sē,kant }

inverse hyperbolic sine See arc-hyperbolic sine. { ¦in,vərs ¦hī·pər,bäl·ik 'sīn }

inverse hyperbolic tangent See arc-hyperbolic tangent. { ¦in,vərs ¦hī·pər,bäl·ik 'tan·jənt }

inverse image See pre-image. { ¦in,vərs 'im·ij }

inverse implication [MATH] The implication that results from replacing both the antecedent and the consequent of a given implication with their negations. { 'in,vərs ,im·plə'kā·shən }

inverse latitude See transverse latitude. { 'in,vərs 'lad·ə,tüd }

inverse limiter [ELECTR] A transducer, the output of which is constant for input of instantaneous values within a specified range and a linear or other prescribed function of the input for inputs above and below that range. { 'in,vərs 'lim·əd·ər }

inverse logarithm See antilogarithm. { 'in,vərs 'läg·ə,rith·əm }

inversely proportional quantities [MATH] Two variable quantities whose product remains constant. { in¦vərs·lē prə¦pór·shən·əl 'kwän·əd·ēz }

inverse magnetostriction See magnetic tension effect. { 'in,vərs mag,ned·ō'strik·shən }

inverse-mapping theorem [MATH] The theorem that the inverse of a linear, one-to-one, continuous mapping between two Banach spaces or two Fréchet spaces is also continuous. { ¦in,vərs 'map·iŋ ,thir·əm }

inverse matrix [MATH] The inverse of a nonsingular matrix A is the matrix A^{-1} where $A \cdot A^{-1} = A^{-1} \cdot A = I$, the identity matrix. Also known as reciprocal matrix. { 'in,vərs 'mā·triks }

inverse Mercator chart See transverse Mercator chart. { 'in,vərs mər'kād·ər ,chärt }

inverse Mercator projection See transverse Mercator projection. { 'in,vərs mər'kād·ər prə'jek·shən }

inverse micelle See inverted micelle. { 'in,vərs mī'sel }

inverse network [ELEC] Two two-terminal networks are said to be inverse when the product of their impedances is independent of frequency within the range of interest. { 'in,vərs 'net,wərk }

inverse neutral telegraph transmission [COMMUN] Form of transmission in which marking signals are zero current intervals and spacing signals are current pulses of either polarity. { 'in,vərs 'nü·trəl 'tel·ə,graf tranz,mish·ən }

inverse operator [MATH] The inverse of an operator L is the operator which is the inverse function of L. { 'in,vərs 'äp·ə,rād·ər }

inverse parallel See transverse parallel. { 'in,vərs 'par·ə,lel }

inverse peak voltage [ELECTR] **1.** The peak value of the voltage that exists across a rectifier tube or x-ray tube during the half cycle in which current does not flow. **2.** The maximum instantaneous voltage value that a rectifier tube or x-ray tube can withstand in the inverse direction (with anode negative) without breaking down and becoming conductive. { 'in,vərs ¦pēk 'vōl·tij }

inverse piezoelectric effect [SOLID STATE] The contraction or expansion of a piezoelectric crystal under the influence of an electric field, as in crystal headphones; also occurs at pn junctions in some semiconductor materials. { 'in,vərs pē¦ā·zō·i¦lek·trik i,fekt }

inverse points [MATH] A pair of points lying on a diameter of a circle or sphere such that the product of the distances of the points from the center equals the square of the radius. { 'in,vərs 'póins }

inverse probability principle See Bayes' theorem. { 'in,vərs ,präb·ə'bil·əd·ē ,prin·sə·pəl }

inverse problem [CONT SYS] The problem of determining, for a given feedback control law, the performance criteria for which it is optimal. { 'in,vərs 'präb·ləm }

inverse ranks [STAT] Ranking responses to treatments from largest response to smallest response. { 'in,vərs 'raŋks }

inverse ratio [MATH] The reciprocal of the ratio of two quantities. Also known as reciprocal ratio. { ¦in,vərs 'rā·shō }

inverse relation [MATH] For a relation R, the inverse relation R^{-1} is the relation such that the ordered pair (x, y) belongs to R^{-1} if and only if (y, x) belongs to R. { 'in,vərs 'ri'lā·shən }

inverse rhumb line See transverse rhumb line. { 'in,vərs 'rəm ,līn }

inverse scattering theory [PHYS] The discipline that determines the nature of the scattering object, or an interaction potential energy, in a scattering process or collision, from knowledge of the amplitudes of the scattered fields. { 'in,vərs 'skad·ə·riŋ ,thē·ə·rē }

inverse secant See arc secant. { 'in,vərs 'sē,kant }

inverse segregation [MET] Segregation in a cast metal in which an excess of lower-melting metal occurs in the earlier-freezing portions because liquid metal enters cavities developed in the earlier-solidified metal. { in,vərs ,seg·rə'gā·shən }

inverse sine See arc sine. { 'in,vərs 'sīn }

INVERSE-SQUARE LAW

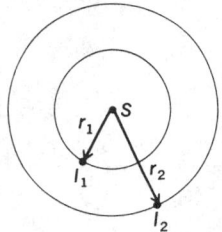

Point source S emitting energy of intensity I. The inverse-square law states that $I_1/I_2 = r_2^2/r_1^2$.

inverse-square law [PHYS] Any law in which a physical quantity varies with distance from a source inversely as the square of that distance. { 'in,vərs ¦skwer ,ló }

inverse Stark effect [SPECT] The Stark effect as observed with absorption lines, in contrast to emission lines. { 'in,vərs 'stärk i,fekt }

inverse substitution [MATH] A substitution that precisely nullifies the effect of a given substitution. { ¦in,vərs ,səb·stə'tü·shən }

inverse tangent See arc tangent. { ¦in,vərs 'tan·jənt }

inverse trigonometric function [MATH] An inverse function of a trigonometric function; that is, an arc sine, arc cosine, arc tangent, arc cotangent, arc secant, or arc cosecant. Also known as antitrigonometric function. { ¦in,vərs ,trig·ə·nə,me·trik 'fəŋk·shən }

inverse variation [MATH] **1.** A relationship between two variables wherein their product is equal to a constant. **2.** An equation or function expressing such a relationship. { 'in,vərs ,ver·e'ā·shən }

inverse video See reverse video. { 'in,vərs ¦vid·ē·ō }

inverse voltage [ELECTR] The voltage that exists across a rectifier tube or x-ray tube during the half cycle in which the anode is negative and current does not normally flow. { 'in,vərs 'vōl·tij }

inverse Zeeman effect [SPECT] A splitting of the absorption lines of atoms or molecules in a static magnetic field; it is the Zeeman effect observed with absorption lines. { 'in,vərs 'zē·mən i,fekt }

inversion [CHEM] Change of a compound into an isomeric form. [COMMUN] The process of scrambling speech for secrecy by beating the voice signal with a fixed, higher audio frequency and using only the difference frequencies. [CRYSTAL] A change from one crystal polymorph to another. Also known as transformation. [ELEC] The solution of certain problems in electrostatics through the use of the transformation in Kelvin's inversion theorem. [GEN] A type of chromosomal rearrangement in which two breaks take place in a chromosome and the orientation of the fragment between breaks is reversed in rejoining. [GEOL] **1.** Development of inverted relief through which anticlines are transformed into valleys and synclines are changed into mountains. **2.** The occupancy by a lava flow of a ravine or valley that occurred in the side of a volcano. **3.** A diagenetic process in which unstable minerals are converted to a more stable form without a change in chemical composition. [MATH] **1.** Given a point O lying in a plane or in space, a mapping of the plane or of space, excluding the point O, into itself in which every point is mapped into its inverse point with respect to a circle or sphere centered at O. **2.** The interchange of two adjacent members of a sequence. [MECH ENG] The conversion of basic four-bar linkages to special motion linkages, such as parallelogram linkage, slider-crank mechanism, and slow-motion mechanism by successively holding fast, as ground link, members of a specific linkage (as drag link). [MED] The act or process of turning inward or upside down. [METEOROL] A departure from the usual decrease or increase with altitude of the value of an atmospheric property, most commonly temperature. [OPTICS] The formation of an inverted image by an optical system. [PHYS] The simultaneous reflection of all three directions in space, so that each coordinate is replaced by the negative of itself. Also known as space inversion. [SOLID STATE] The production of a layer at the surface of a semiconductor which is of opposite type from that of the bulk of the semiconductor, usually as the result of an applied electric field. [THERMO] A reversal of the usual direction of a variation or process, such as the change in sign of the expansion coefficient of water at 4°C, or a change in sign in the Joule-Thomson coefficient at a certain temperature. { in'vər·zhən }

inversion axis See rotation-inversion axis. { in'vər·zhən ,ak·səs }

inversion heterozygote [GEN] A diploid organism in which one member of a pair of homologous chromosomes has an inverted gene sequence and the other has the normal gene sequence. { in'vər·zhən ,hed·ə·rə'zī,gōt }

inversion layer [METEOROL] The atmosphere layer through which an inversion occurs. { in'vər·zhən ,lā·ər }

inversion ratio [PHYS] The negative of the ratio in a maser medium of the difference in populations between two nondegenerate energy states under a condition of population inversion, to the population difference at equilibrium. { in'vər·zhən ,rā·shō }

inversion spectrum [SPECT] Lines in the microwave spectra of certain molecules (such as ammonia) which result from the quantum-mechanical analog of an oscillation of the molecule between two configurations which are mirror images of each other. { in'vər·zhən ,spek·trəm }

inversion symmetry [PHYS] The principle that the laws of physics are unchanged by the operation of inversion; it is violated by the weak interactions. { in'vər·zhən 'sim·ə·trē }

inversion temperature [ENG] The temperature to which one junction of a thermocouple must be raised in order to make the thermoelectric electromotive force in the circuit equal to zero, when the other junction of the thermocouple is held at a constant low temperature. [THERMO] The temperature at which the Joule-Thomson effect of a gas changes sign. { in'vər·zhən ,tem·prə·chər }

invert [CIV ENG] The floor or bottom of a conduit. { 'in,vərt }

invertase See saccharase. { in'vər,tās }

Invertebrata [INV ZOO] A division of the animal kingdom including all animals which lack a spinal column; has no taxonomic status. { in,vərd·ə'bräd·ə }

invertebrate [INV ZOO] An animal lacking a backbone and internal skeleton. { in'vərd·ə,brət }

invertebrate pathology [INV ZOO] All studies having to do with the principles of pathology as applied to invertebrates. { in'vərd·ə,brət pə'thäl·ə·jē }

invertebrate zoology [ZOO] A branch of biology that deals with the study of Invertebrata. { in'vərd·ə,brət zō'äl·ə·jē }

inverted See overturned. { in'vərd·əd }

inverted amplifier [ELECTR] A two-tube amplifier in which the control grids are grounded and the input signal is applied between the cathodes; the grid then serves as a shield between the input and output circuits. { in'vərd·əd 'am·plə,fī·ər }

inverted arch [CIV ENG] An arch with the crown downward, below the line of the springings; commonly used in tunnels and foundations. Also known as inflected arch. { in'vərd·əd 'ärch }

inverted compass [NAV] A marine magnetic compass designed and installed for observation from below the compass card; frequently used as a telltale compass. Also known as hanging compass; overhead compass. { in'vərd·əd 'käm·pəs }

inverted engine [MECH ENG] An engine in which the cylinders are below the crankshaft. { in'vərd·əd 'en·jən }

inverted file [COMPUT SCI] **1.** A file, or method of file organization, in which labels indicating the locations of all documents of a given type are placed in a single record. **2.** A file whose usual order has been inverted. { in'vərd·əd 'fīl }

inverted heading and bench method See heading-overhand bench method. { in'vərd·əd ¦hed·iŋ ən 'bench ,meth·əd }

inverted image [OPTICS] An image in which up and down, as well as left and right, are interchanged; that is, an image that results from rotating the object 180° about a line from the object to the observer; such images are formed by most astronomical telescopes. Also known as reversed image. { in'vərd·əd 'im·ij }

inverted L antenna [ELECTROMAG] An antenna consisting of one or more horizontal wires to which a connection is made by means of a vertical wire at one end. { in'vərd·əd ¦el an,ten·ə }

inverted micelle [PHYS CHEM] An aggregate of colloidal dimension in which the polar groups are concentrated in the interior and the lipophilic groups extend outward into the solvent. Also known as inverse micelle. { in¦vərd·əd mī'sel }

inverted microscope [OPTICS] A microscope in which the body of the microscope, including the objective and the ocular, are below the stage, the illumination for transmitted light is above the stage, and with opaque materials, the vertical illuminator is used under the stage near the objective. { in'vərd·əd 'mī·krə,skōp }

inverted plunge [GEOL] A plunge of a fold whose inclination has been carried past the vertical, so that the plunge is less than 90° in the direction opposite from the original attitude; younger rocks plunge beneath the older rocks. { in'vərd·əd 'plənj }

inverted pseudoplastic fluid *See* dilatant fluid. { inˈvərd·əd ˌsüd·ōˈplas·tik ˈflü·əd }

inverted relief [GEOL] A topographic configuration that is opposite to that of the geologic structure, for example, where a valley occupies the site of an anticline. { inˈvərd·əd riˈlēf }

inverted repeats [GEN] Two copies of the same nucleotide sequence oriented in opposite directions on the same molecule. Also known as IR sequences. { inˈvərd·əd riˈpēts }

inverted siphon [CIV ENG] A pressure pipeline crossing a depression or passing under a highway; sometimes called a sag line from its U-shape. { inˈvərd·əd ˈsī·fən }

inverted terminal repeats [CELL MOL] Related or identical sequences of deoxyribonucleic acid in inverted form occurring at opposite ends of some transposons. { inˈvərd·əd ˌtər·mə·nəl riˈpēts }

inverted vee [ELECTROMAG] **1.** A directional antenna consisting of a conductor which has the form of an inverted V, and which is fed at one end and connected to ground through an appropriate termination at the other. **2.** A center-fed horizontal dipole antenna whose arms have ends bent downward 45°. { inˈvərd·əd ˈvē }

invert-emulsion mud [PETRO ENG] A drilling mud whose dispersed phase is fresh or salt water and whose continuous phase is some type of oil. { ˈinˌvərt iˈməl·shən ˌməd }

inverter [ELEC] A device for converting direct current into alternating current; it may be electromechanical, as in a vibrator or synchronous inverter, or electronic, as in a thyratron inverter circuit. Also known as dc-to-ac converter; dc-to-ac inverter. [ELECTR] *See* phase inverter. { inˈvərd·ər }

inverter circuit *See* NOT circuit. { inˈvərd·ər ˌsər·kət }

invertible element [MATH] An element x of a groupoid with a unit element e for which there is an element $\bar{x}$ such that $x \circ \bar{x} = \bar{x} \circ x = e$. { inˈvərd·ə·bəl ˈel·ə·mənt }

invertible ideal [MATH] A fractional ideal I of an integral domain R such that R is equal to the set of elements of the form xy, where x is in I and y is in the inverse of I. { inˈvərd·ə·bəl iˈdēl }

invertin *See* saccharase. { inˈvərt·ən }

inverting amplifier [ELECTR] Amplifier whose output polarity is reversed as compared to its input; such an amplifier obtains its negative feedback by a connection from output to input, and with high gain is widely used as an operational amplifier. { inˈvərd·iŋ ˈam·pləˌfī·ər }

inverting function [ELECTR] A logic device that inverts the input signal, so that the output is out of phase with the input. { inˈvərd·iŋ ˌfəŋk·shən }

inverting parametric device [ELECTR] Parametric device whose operation depends essentially upon three frequencies, a harmonic of the pump frequency and two signal frequencies, of which the higher signal frequency is the difference between the pump harmonic and the lower signal frequency. { inˈvərd·iŋ ˌpar·əˌme·trik diˈvīs }

inverting prism *See* erecting prism. { inˈvərd·iŋ ˈpriz·əm }

inverting telescope [OPTICS] A telescope that inverts the usual telescopic image, allowing the object to be seen right side up. { inˈvərd·iŋ ˈtel·əˌskōp }

inverting terminal [ELECTR] The negative input terminal of an operational amplifier; a positive-going voltage at the inverting terminal gives a negative-going output voltage. { inˈvərd·iŋ ˈtər·mən·əl }

invert level [ENG] The level of the lowest portion at any given section of a liquid-carrying conduit, such as a drain or a sewer, and which determines the hydraulic gradient available for moving the contained liquid. { ˈinˌvərt ˌlev·əl }

invert oil mud [PETRO ENG] A drilling mud constituting an emulsion in which water is the dispersed phase. { ˈinˌvərt ˈoil ˌməd }

invert sugar [FOOD ENG] A mixture consisting of 50% glucose and 50% fructose, obtained by hydrolysis of sucrose, which absorbs water readily; used in the food industry. { ˈinˌvərt ˌshüg·ər }

investing bone *See* dermal bone. { inˈvest·iŋ ˌbōn }

investment casting [MET] A casting method designed to achieve high dimensional accuracy for small castings by making a mold of refractory slurry, which sets at room temperature, surrounding a wax pattern which is then melted out to leave a mold without joints. { inˈves·mənt ˌkast·iŋ }

investment compound [MATER] A mixture containing a refractory filler, a binder, and a liquid vehicle which is used to make molds for investment casting. { inˈves·mənt ˌkämˌpaund }

investment process *See* lost-wax process. { inˈves·mənt ˌpräs·əs }

inviscid flow [FL MECH] Flow of an inviscid fluid. Also known as frictionless flow; ideal flow; nonviscous flow. { inˈvis·əd ˌflō }

inviscid fluid [FL MECH] A fluid which has no viscosity; it therefore can support no shearing stress, and flows without energy dissipation. Also known as ideal fluid; nonviscous fluid; perfect fluid. { inˈvis·əd ˈflü·əd }

invisible hinge [DES ENG] A door hinge whose parts are not exposed when the door is closed. { inˈviz·ə·bəl ˈhinj }

invisible image [PHYS] An image having a form which cannot be perceived by unaided vision, such as a latent image on a photographic emulsion. { inˈviz·ə·bəl ˈim·ij }

invisible writing ink [GRAPHICS] Ink that remains invisible until the color is brought out by the application of heat or chemical; made with sal ammoniac or salts of metals. { inˈviz·ə·bəl ˈwrīd·iŋ ˌiŋk }

in vitro [BIOL] Pertaining to a biological reaction taking place in an artificial apparatus. { ˈvē·trō }

in vivo [BIOL] Pertaining to a biological reaction taking place in a living cell or organism. { ˈvē·vō }

involucrate [BOT] Having an involucre. { ˌinˈvəˈlü·krət }

involucre [BOT] Bracts forming one or more whorls at the base of an inflorescence or fruit in certain plants. { ˈin·vəˌlü·kər }

involucrum [ANAT] **1.** The covering of a part. **2.** New bone laid down by periosteum around a sequestrum in osteomyelitis. { ˌin·vəˈlü·krəm }

involuntary fiber *See* smooth muscle fiber. { inˈväl·ənˌter·ē ˈfī·bər }

involuntary muscle [PHYSIO] Muscle not under the control of the will; usually consists of smooth muscle tissue and lies within a vescus, or is associated with skin. { inˈväl·ənˌter·ē ˈməs·əl }

involute [BIOL] Being coiled, curled, or rolled in at the edge. [MATH] **1.** A curve produced by any point of a perfectly flexible inextensible thread that is kept taut as it is wound upon or unwound from another curve. **2.** A curve that lies on the tangent surface of a given space curve and is orthogonal to the tangents to the given curve. **3.** A surface for which a given surface is one of the two surfaces of center. { ˈinˈvəˈlüt }

involute gear tooth [DES ENG] A gear tooth whose profile is established by an involute curve outward from the base circle. { ˈinˈvəˈlüt ˈgir ˌtüth }

involute spline [DES ENG] A spline having the same general form as involute gear teeth, except that the teeth are one-half the depth and the pressure angle is 30°. { ˈinˈvəˈlüt ˈsplīn }

involute spline broach [MECH ENG] A broach that cuts multiple keys in the form of internal or external involute gear teeth. { ˈinˈvəˈlüt ˈsplīn ˌbrōch }

involution [BIOL] A turning or rolling in. [EMBRYO] Gastrulation by ingrowth of blastomeres around the dorsal lip. [MATH] **1.** Any transformation that is its own inverse. **2.** In particular, a correspondence between the points on a line that is its own inverse, given algebraically by $x' = (ax + b)/(cx - a)$, where $a^2 + bc \neq 0$. **3.** A correspondence between the lines passing through a given point on a plane such that corresponding lines pass through corresponding points of an involution of points on a line. [MED] **1.** The retrogressive change to their normal condition that organs undergo after fulfilling their functional purposes, as the uterus after pregnancy. **2.** The period of regression or the processes of decline or decay which occur in the human constitution after middle life. { ˌin·vəˈlü·shən }

involutional melancholia *See* involutional psychosis. { ˌin·vəˈlü·shən·əl ˌmel·ən·kōˈlē·ə }

involutional psychosis [PSYCH] A prolonged psychotic reaction occurring in late middle life, characterized by depression and paranoid ideas. Also known as involutional melancholia. { ˌin·vəˈlü·shən·əl sīˈkō·səs }

involution form [CYTOL] A cell with a bizarre configuration caused by abnormal culture conditions. { ˌin·vəˈlü·shən ˌform }

involvucel [BOT] A secondary involucre. { inˈväl·və·səl }

involvucellate [BOT] Possessing an involvucel. { inˈväl·vəˌselˌāt }

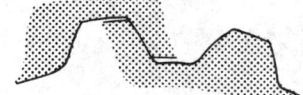

INVOLUTE SPLINE

Diagram of an involute spline profile.

inward bound [NAV] Heading toward the land or up a harbor, away from the open sea. { 'in·wərd 'baund }

inward-outward dialing system [COMMUN] Dialing system whereby calls within the local exchange area may be dialed directly to and from base private branch exchange telephone stations without the assistance of the base private branch exchange operator; CENTREX, a service offered by some telephone companies, is a form of inward-outward dialing. { ¦in·wərd ¦aut,wərd 'dī·liŋ ,sis·təm }

inyoite [MINERAL] Ca₂B₆O₁₁·13H₂O A colorless, monoclinic mineral consisting of a hydrous calcium borate; hardness is 2 on Mohs scale, and specific gravity is 2. { 'in·yō,īt }

Io [ASTRON] A satellite of Jupiter; its diameter is 2300 miles (3700 kilometers). Also known as Jupiter I. [NUC PHYS] See ionium. { 'ī,ō }

I/O See input/output.

IOCS See input/output control system.

iodargyrite [MINERAL] AgI A yellowish or greenish hexagonal mineral composed of native silver iodide, usually occurring in thin plates. Also known as iodyrite. { ,ī·ə'där·jə,rīt }

iodate [INORG CHEM] A salt of iodic acid containing the IO₃⁻ radical; sodium and potassium iodates are the most important salts and are used in medicine. { 'ī·ə,dāt }

iodcyanin See cyanine dye. { 'ī·od,sī·ə·nən }

iodic acid [INORG CHEM] HIO₃ Water-soluble, moderately strong acid; colorless or white powder or crystals; decomposes at 110°C; used in analytical chemistry and medicine. { ī'äd·ik 'as·əd }

iodic acid anhydride See iodine pentoxide. { ī'äd·ik 'as·əd an'hī,drīd }

iodide [CHEM] **1.** A compound which contains the iodine atom in the −1 oxidation state and which may be considered to be derived from hydriodic acid (HI); examples are KI and NaI. **2.** A compound of iodine, such as CH₃CH₂I, in which the iodine has combined with a more electropositive group. { 'ī·ə,dīd }

iodide process [MET] A refining process in which a metal, such as titanium or zirconium, is combined with iodine vapor and then the iodide volatilized and decomposed at high temperatures to yield a pure solid metal. { 'ī·ə,dīd 'prä·səs }

iodine [CHEM] A nonmetallic halogen element, symbol I, atomic number 53, atomic weight 126.9045; melts at 114°C, boils at 184°C; the poisonous, corrosive, dark plates or granules are readily sublimed; insoluble in water, soluble in common solvents; used as germicide and antiseptic, in dyes, tinctures, and pharmaceuticals, in engraving lithography, and as a catalyst and analytical reagent. { 'ī·ə,dīn }

iodine-131 [NUC PHYS] A radioactive, artificial isotope of iodine, mass number 131; its half-life is 8 days with beta and gamma radiation; used in medical and industrial radioactive tracer work; moderately radiotoxic. { 'ī·ə,dīn ¦wən¦thərd·ē'wən }

iodine bisulfide See sulfur iodine. { 'ī·ə,dīn bī'səl,fīd }

iodine cyanide [INORG CHEM] ICN Poisonous, colorless needles with pungent aroma and acrid taste; melts at 147°C; soluble in water, alcohol, and ether; used in taxidermy as a preservative. Also known as cyanogen iodide. { 'ī·ə,dīn 'sī·ə,nīd }

iodine disulfide See sulfur iodine. { 'ī·ə,dīn dī'səl,fīd }

iodine number [ANALY CHEM] A measure of the iodine absorbed in a given time by a chemically unsaturated material, such as a vegetable oil or a rubber; used to measure the unsaturation of a compound or mixture. Also known as iodine value. { 'ī·ə,dīn ,nəm·bər }

iodine pentoxide [INORG CHEM] I₂O₅ White crystals, decomposing at 275°C, very soluble in water, insoluble in absolute alcohol, ether, and chloroform; used as an oxidizing agent to oxidize carbon monoxide to dioxide at ordinary temperatures, and in organic synthesis. Also known as iodic acid anhydride. { 'ī·ə,dīn pen'täk,sīd }

iodine solution See tincture of iodine. { 'ī·ə,dīn sə,lü·shən }

iodine test [ANALY CHEM] Placing a few drops of potassium iodide solution on a sample to detect the presence of starch; test is positive if sample turns blue. { 'ī·ə,dīn ,test }

iodine tincture See tincture of iodine. { 'ī·ə,dīn ,tiŋk·chər }

iodine value See iodine number. { 'ī·ə,dīn ,val·yü }

iodized oil [MATER] A thick, viscous, oily liquid with a garliclike odor that is an iodine addition product of vegetable oil, containing about 40% organically combined iodine; used in medicine as a radiopaque medium for radiography. { ī·ə,dīzd 'oil }

iodized salt [FOOD ENG] Common table salt which has been treated with iodide to provide iodine as a nutritional supplement. { 'ī·ə,dīzd 'solt }

iodoacetic acid [ORG CHEM] CH₂ICOOH White or colorless crystals that are soluble in water and alcohol, and melt at 82–83°C; used in biological research for its inhibitive effect on enzymes. { ¦ī·ō·dō·ə¦sēd·ik 'as·əd }

iodoalkane [ORG CHEM] An alkane hydrocarbon in which an iodine atom replaces one or more hydrogen atoms in the molecule; an example is iodomethane, CH₃I, better known as methyl iodide. { ī¦ō·dō·al'kān }

iodobromite [MINERAL] Ag(Br,Cl,I) An isometric mineral composed of chloride, iodide, and bromide of silver; it is isomorphous with cerargyrite and bromyrite. { ī¦ō·də'brō,mīt }

5-iodo-2′-deoxyuridine [PHARM] C₉H₁₁IN₂O₅ A crystalline substance, used as an antiviral agent for the eye. Abbreviated IDU; IDUR; IUDR. { ¦fīv ī¦ō·dō ¦tü¦prīm dē,äk·sē'yür·ə,dēn }

iodoeasin See easin. { ī¦ō·dō'ē·ə·sən }

iodoethane See ethyl iodide. { ī¦ō·dō'eth,ān }

iodoethylene See tetraiodoethylene. { ī¦ō·dō'eth·ə,lēn }

iodoform [ORG CHEM] CHI₃ A yellow, hexagonal solid; melting point 119°C; soluble in chloroform, ether, and water; has weak bactericidal qualities and is used in ointments for minor skin diseases. Also known as triiodomethane. { ī'ō·də,form }

iodohydrocarbon [ORG CHEM] A hydrocarbon in which an iodine atom replaces one or more hydrogen atoms in the molecule, as in an alkane, aromatic, or olefin. { ī¦ō·də,hī·drə'kär·bən }

iodomethane See methyl iodide. { ī¦ō·də'meth,ān }

iodometry [ANALY CHEM] An application of iodine chemistry to oxidation-reduction titrations for the quantitative analysis in certain chemical compounds, in which iodine is used as a reductant and the iodine freed in the associated reaction is titrated, usually in neutral or slightly acid mediums with a standard solution of a reductant such as sodium thiosulfate or sodium arsenite; examples of chemicals analyzed are copper(III), gold(VI), arsenic(V), antimony(V), chlorine, and bromine. { ,ī·ə'däm·ə·trē }

iodonium [INORG CHEM] A halonium ion such as H₂I⁺ or R₂I⁺; it may be open-chain or cyclic. { ī·ə'dōn·ē·əm }

iodophor [CHEM] Any compound that is a carrier of iodine. { i'äd·ə,for }

iodopsin [BIOCHEM] The visual pigment found in the retinal cones, consisting of retinene, combined with photopsin. { ,ī·ə'däp·sən }

iodosobenzene [ORG CHEM] C₆H₅IO A yellowish-white amorphous solid that explodes at 200°C, soluble in hot water and alcohol; a strong oxidizing agent. { ,ī·ə¦dō·sō'ben,zēn }

iodoxybenzene [ORG CHEM] C₆H₅IO₂ Clear white crystals that explode at 227–228°C, slightly soluble in water, insoluble in chloroform, acetone, and benzene; a strong oxidizing agent. { ¦ī·ə¦däk·sē'ben,zēn }

iodyrite See iodargyrite. { ī'äd·ə,rīt }

Ioffe bars [PL PHYS] Heavy current-carrying bars that are used to increase plasma stability in some types of controlled fusion reactor. { 'yäf·ē ,bärz }

Ioffe effect [SOLID STATE] An effect in which the simultaneous exposure of an ionic crystal such as rock salt to a mechanical stress and a solvent results in an increase in its plasticity. { 'yäf·ē i,fekt }

IOGEN See input/output generation. { 'ī,ō,jen }

ion [CHEM] An isolated electron or positron or an atom or molecule which by loss or gain of one or more electrons has acquired a net electric charge. { 'ī,än }

ion accelerator [NUCLEO] A linear accelerator in which ions are accelerated by an electric field in a standing-wave pattern that is set up in a resonant cavity by external oscillators or amplifiers. { 'ī,än ak'sel·ə,rād·ər }

ion-acoustic wave [PL PHYS] A longitudinal compression wave in the ion density of a plasma which can occur at high electron temperatures and low frequencies, caused by a combination of ion inertia and electron pressure. { 'ī,än ə¦kü·stik 'wāv }

ion atmosphere See ion cloud. { 'ī,än 'at·mə,sfir }

ion backscattering [SOLID STATE] Large-angle elastic scattering of monoenergetic ions in a beam directed at a metallized film on silicon or some other thin multilayer system. { 'ī,än 'bak,skad·ə·riŋ }

ion-beam mixing [ENG] A process in which bombardment of a solid with a beam of energetic ions causes the intermixing of atoms of two separate phases originally present in the near-surface region. { 'ī,än ¦bēm ,miks·iŋ }

ion-beam scanning [ELECTR] The process of analyzing the mass spectrum of an ion beam in a mass spectrometer either by changing the electric or magnetic fields of the mass spectrometer or by moving a probe. { 'ī,än ,bēm ,skan·iŋ }

ion-beam thinning See ion machining. { 'ī,än ,bēm ¦thin·iŋ }

ion burn See ion spot. { 'ī,än ,bərn }

ion chamber See ionization chamber. { 'ī,än ,chām·bər }

ion channel See ionic channel. { 'ī,än ,chan·əl }

ion cloud [GEOPHYS] An inhomogeneity or patch of unusually great ion density in one of the regular regions of the ionosphere; such patches occur quite often in the E region. [PHYS CHEM] A slight preponderance of negative ions around a positive ion in an electrolyte, and vice versa, according to the Debye-Hückel theory. Also known as ion atmosphere. { 'ī,än ,klaůd }

ion column [GEOPHYS] The trail of ionized gases in the trajectory of a meteoroid entering the upper atmosphere; a part of the composite phenomenon known as a meteor. Also known as meteor trail. { 'ī,än ,käl·əm }

ion concentration See ion density. { 'ī,än ,kän·sən'trā·shən }

ion counter See ionization counter. { 'ī,än ,kaůnt·ər }

ion crystal See Coulomb crystal. { 'ī,än ,krist·əl }

ion current [PHYS] The electric current resulting from motion of ions. { 'ī,än ,kə·rənt }

ion cyclotron frequency [ELECTROMAG] The angular frequency of the motion of an ion in a uniform magnetic field in a plane perpendicular to the field. { 'ī,än 'sī·klə,trän ,frē·kwən·sē }

ion-cyclotron-resonance mass spectrometer [SPECT] A device for detecting and measuring the mass distribution of ions orbiting in an applied magnetic field, either by applying a constant radio-frequency signal and varying the magnetic field to bring ion frequencies equal to the applied radio frequency sequentially into resonance, or by rapidly varying the radio frequency and applying Fourier transform techniques. { 'ī,än 'sī·klə,trän 'rez·ən·əns 'mas spek'träm·əd·ər }

ion density [PHYS] The number of ions per unit volume. Also known as ion concentration. { 'ī,än 'den·səd·ē }

ion detector [ANALY CHEM] Device for detection of presence or concentration of liquid solution ions, such as with a pH meter or by conductimetric techniques. { 'ī,än di,tek·tər }

ion emission [PHYS] The ejection of ions from the surface of a substance into the surrounding space. { 'ī,än i,mish·ən }

ion engine [AERO ENG] An engine which provides thrust by expelling accelerated or high-velocity ions; ion engines using energy provided by nuclear reactors are proposed for space vehicles. { 'ī,än ,en·jən }

ion exchange [PHYS CHEM] A chemical reaction in which mobile hydrated ions of a solid are exchanged, equivalent for equivalent, for ions of like charge in solution; the solid has an open, fishnetlike structure, and the mobile ions neutralize the charged, or potentially charged, groups attached to the solid matrix; the solid matrix is termed the ion exchanger. { 'ī,än iks,chānj }

ion-exchange chromatography [ANALY CHEM] A chromatographic procedure in which the stationary phase consists of ion-exchange resins which may be acidic or basic. { 'ī,än iks,chānj krō·mə'täg·rə·fē }

ion-exchange electrolyte cell [ELEC] Fuel cell which operates on hydrogen and oxygen in the air, similar to the standard hydrogen-oxygen fuel cell with the exception that the liquid electrolyte is replaced by an ion-exchange membrane; operation is at atmospheric pressure and room temperature. { 'ī,än iks,chānj i'lek·trə,līt ,sel }

ion exchanger [PHYS CHEM] A solid or liquid material containing ions that are exchangeable with other ions with a like charge that are present in a solution in which the material is insoluble. { 'ī,än iks,chānj·ər }

ion-exchange resin [MATER] A synthetic resin that can combine or exchange ions with a solution; such a resin produces the exchange of sodium for calcium ions in the softening of hard water. { 'ī,än iks chānj ,rez·ən }

ion exclusion [CHEM] Ion-exchange resin system in which the mobile ions in the resin-gel phase electrically neutralize the immobilized charged functional groups attached to the resin, thus preventing penetration of solvent electrolyte into the resin-gel phase; used in separations where electrolyte is to be excluded from the resin, but not nonpolar materials, as the separation of salt from nonpolar glycerin. { 'ī,än iks,klü zhən }

ion-exclusion chromatography [ANALY CHEM] Chromatography in which the adsorbent material is saturated with the same mobile ions (cationic or anionic) as are present in the sample-carrying eluent (solvent), thus repelling the similar sample ions. { 'ī,än iks,klü zhən ,krō·mə'täg·rə·fē }

ion fractionation [CHEM ENG] Separation of cations or anions from an ionic solution by use of a membrane permeable to the desired ion; equipment includes electrodialyzers and ion-fractionation stills. { 'ī,än ,frak·shə'nā·shən }

ion gage See ionization gage. { 'ī,än ,gāj }

ion gun See ion source. { 'ī,än ,gən }

ionic bond [PHYS CHEM] A type of chemical bonding in which one or more electrons are transferred completely from one atom to another, thus converting the neutral atoms into electrically charged ions; these ions are approximately spherical and attract one another because of their opposite charge. Also known as electrovalent bond. { ī'än·ik 'bänd }

ionic channel [CELL MOL] A transmembrane protein structure that forms an aqueous pore that allows only certain ion species to pass through the membrane. Also known as ion channel. { ī,än·ik 'chan·əl }

ionic charge [PHYS] 1. The total charge of an ion. 2. The charge of an electron; the charge of any ion is equal to this electron charge in magnitude, or is an integral multiple of it. { ī'än·ik 'chärj }

ionic column [ARCH] A column characterized by the spiral ornament of its head. { ī'än·ik 'käl·əm }

ionic conductance [PHYS CHEM] The contribution of a given type of ion to the total equivalent conductance in the limit of infinite dilution. { ī'än·ik kən'dək·təns }

ionic conduction [SOLID STATE] Electrical conduction of a solid due to the displacement of ions within the crystal lattice. { ī'än·ik kən'dək·shən }

ionic conductivity [SOLID STATE] The portion of the electrical conductivity of a solid that results from ionic conduction. { ī'än·ik ,kän,dək'tiv·əd·ē }

ionic crystal [CRYSTAL] A crystal in which the lattice-site occupants are charged ions held together primarily by their electrostatic interaction. { ī'än·ik 'krist·əl }

ionic dissociation [PHYS CHEM] Dissociation that results in the production of ions. { ī'än·ik di,sō·sē'ā·shən }

ionic equilibrium [PHYS CHEM] The condition in which the rate of dissociation of nonionized molecules is equal to the rate of combination of the ions. { ī'än·ik ,ē·kwə'lib·rē·əm }

ionic equivalent conductance [PHYS CHEM] The contribution made by each ion species of a salt toward an electrolyte's equivconductance. { ī'än·ik i¦kwiv·ə·lənt kən'dək·təns }

ionic focusing See gas focusing. { ī'än·ik 'fō·kəs·iŋ }

ionic gel [CHEM] A gel with ionic groups attached to the structure of the gel; the groups cannot diffuse out into the surrounding solution. { ī'än·ik 'jel }

ionic heated cathode [ELECTR] Hot cathode heated primarily by ionic bombardment of the emitting surface. { ī'än·ik ,hēd·əd 'kath,ōd }

ionicity [CHEM] The ionic character of a solid. { ,ī·ə'nis·əd·ē }

ionic lattice [CRYSTAL] The lattice of an ionic crystal. { ī'än·ik 'lad·əs }

ionic membrane [CHEM ENG] Semipermeable membrane that conducts electricity; the application of an electric field to the membrane achieves an electrophoretic movement of ions through the membrane; used in electrodialysis. { ī'än·ik 'mem brān }

ionic mobility [PHYS] The ratio of the average drift velocity of an ion in a liquid or gas to the electric field. { ī'än·ik mō'bil·əd·ē }

ionic polymerization [ORG CHEM] Polymerization that proceeds via ionic intermediates (carbonium ions or carbanions) than through neutral species (olefins or acetylenes). { ī'än·ik pə‚lim·ə·rə'zā·shən }

ionic radii [PHYS CHEM] Radii which can be assigned to ions because the rapid variation of their repulsive interaction with distance makes them repel like hard spheres; these radii determine the dimensions of ionic crystals. { ī'än·ik 'rād·ē‚ī }

ionic ratio [OCEANOGR] The ratio by weight of a major constituent of seawater to the chloride ion content; for example, SO_4/Cl = 0.1396, Ca/Cl = 0.02150, Mg/Cl = 0.06694. { ī'än·ik 'rā·shō }

ionic semiconductor [SOLID STATE] A solid whose electrical conductivity is due primarily to the movement of ions rather than that of electrons and holes. { ī'än·ik ‚sem·i·kən‚dək·tər }

ionic solid [SOLID STATE] A solid made up of ions held together primarily by their electrostatic interaction. { i'än·ik 'säl·əd }

ionic strength [PHYS CHEM] A measure of the average electrostatic interactions among ions in an electrolyte; it is equal to one-half the sum of the terms obtained by multiplying the molality of each ion by its valence squared. { ī'än·ik 'streŋkth }

ion implantation [ENG] A process of introducing impurities into the near-surface regions of solids by directing a beam of ions at the solid. { 'ī‚än ‚im‚plan'tā·shən }

ion irradiation [PHYS] Bombardment of a substance by high-velocity ions. { 'ī‚ə i‚rād·ē'ā·shən }

ionite See anauxite. { 'ī·ə‚nīt }

ionium [NUC PHYS] A naturally occurring radioisotope, symbol Io, of thorium, atomic weight 230. { ī'ō·nē·əm }

ionization [CHEM] A process by which a neutral atom or molecule loses or gains electrons, thereby acquiring a net charge and becoming an ion; occurs as the result of the dissociation of the atoms of a molecule in solution ($NaCl \rightarrow Na^+ + Cl^-$) or of a gas in an electric field ($H_2 \rightarrow 2H^+$). { ‚ī·ə·nə'zā·shən }

ionization arc-over [ELEC] **1.** Arcing across terminals or contacts due to ionization of the adjacent air or gas. **2.** Arcing across satellite antenna terminals as the satellite passes through the ionized regions of the ionosphere. { ‚ī·ə·nə'zā·shən 'är‚kō·vər }

ionization chamber [NUCLEO] A particle detector which measures the ionization produced in the gas filling the chamber by the fast-moving charged particles as they pass through. Also known as ion chamber. { ‚ī·ə·nə'zā·shən ‚chām·bər }

ionization constant [PHYS CHEM] Analog of the dissociation constant, where k = [H^+][A^-]/[HA]; used for the application of the law of mass action to ionization; in the equation HA represents the acid, such as acetic acid. { ‚ī·ə·nə'zā·shən ‚kän·stənt }

ionization counter [NUCLEO] An ionization chamber in which there is no internal amplification by gas multiplication; used for counting ionizing particles. Also known as ion counter. { ‚ī·ə·nə'zā·shən ‚kaunt·ər }

ionization cross section [PHYS] The cross section for a particle or photon to undergo a collision with an atom, thus removing or adding one or more electrons to the atom. { ‚ī·ə·nə'zā·shən 'krós ‚sek·shən }

ionization current See gas current. { ‚ī·ə·nə'zā·shən ‚kə·rənt }

ionization degree [PHYS CHEM] The proportion of potential ionization that has taken place for an ionizable material in a solution or reaction mixture. { ‚ī·ə·nə'zā·shən di‚grē }

ionization density [ELECTR] The density of ions in a gas. { ‚ī·ə·nə'zā·shən ‚den·səd·ē }

ionization energy [ATOM PHYS] The amount of energy needed to remove an electron from a given kind of atom or molecule to an infinite distance; usually expressed in electron volts, and numerically equal to the ionization potential in volts. { ‚ī·ə·nə'zā·shən ‚en·ər·jē }

ionization front [ASTRON] A transition region that separates interstellar gas in which a given atomic species (usually hydrogen) is mostly ionized from interstellar gas in which it is mostly neutral. { ‚ī·ə·nə'zā·shən ‚frənt }

ionization front accelerator [NUCLEO] A particle accelerator in which the negative charge at the front of an intense relativistic electron beam is used to trap, focus, and accelerate a bunch of ions to high energies. { ‚ī·ə·nə'zā·shən ‚frənt ik'sel·ə‚rād·ər }

ionization gage [ELECTR] An instrument for measuring low gas densities by ionizing the gas and measuring the ion current.

Also known as ion gage; ionization vacuum gage. { ‚ī·ə·nə'zā·shən ‚gāj }

ionization isomer [CHEM] One of two or more compounds that have identical molecular formulas but different ionic forms. { ‚ī·ə·nə'zā·shən 'ī·sə·mər }

ionization potential [ATOM PHYS] The energy per unit charge needed to remove an electron from a given kind of atom or molecule to an infinite distance; usually expressed in volts. Also known as ion potential. { ‚ī·ə·nə'zā·shən pə'ten·chəl }

ionization radiation See ionizing radiation. { ‚ī·ə·nə'zā·shən ‚rād·ē'ā·shən }

ionization source See ion source. { ‚ī·ə·nə'zā·shən ‚sórs }

ionization spectrometer See Bragg spectrometer. { ‚ī·ə·nə'zā·shən spek'träm·əd·ər }

ionization temperature [STAT MECH] The temperature at which the average kinetic energy of gas molecules having a Maxwell distribution equals the ionization energy. { ‚ī·ə·nə'zā·shən ‚tem·prə·chər }

ionization time [ELECTR] Of a gas tube, the time interval between the initiation of conditions for and the establishment of conduction at some stated value of tube voltage drop. { ‚ī·ə·nə'zā·shən ‚tīm }

ionization vacuum gage See ionization gage. { ‚ī·ə·nə'zā·shən 'vak·yəm ‚gāj }

ionized atom [CHEM] An atom with an excess or deficiency of electrons, so that it has a net charge. { 'ī·ə‚nīzd 'ad·əm }

ionized gas [PHYS] A gas, some of whose atoms or molecules have undergone ionization. { 'ī·ə‚nīzd 'gas }

ionized layers [GEOPHYS] Layers of increased ionization within the ionosphere produced by cosmic radiation; responsible for absorption and reflection of radio waves and important in connection with communications and tracking of satellites and other space vehicles. { 'ī·ə‚nīzd 'lā·ərz }

ionizing event [PHYS] Any occurrence in which an ion or group of ions is produced; for example, by passage of charged particles through matter. { 'ī·ə‚niz·iŋ i‚vent }

ionizing radiation [NUCLEO] **1.** Particles or photons that have sufficient energy to produce ionization directly in their passage through a substance. Also known as ionization radiation. **2.** Particles that are capable of nuclear interactions in which sufficient energy is released to produce ionization. { 'ī·ə‚niz·iŋ ‚rād·ē'ā·shən }

ion kinetic energy spectrometry [SPECT] A spectrometric technique that uses a beam of ions of high kinetic energy passing through a field-free reaction chamber from which ionic products are collected and energy analyzed; it is a generalization of metastable ion studies in which both unimolecular and bimolecular reactions are considered. { 'ī‚än ki‚ned·ik 'en·ər·jē spek'träm·ə·trē }

ion laser [OPTICS] A gas laser in which stimulated emission takes place between two energy levels of an ion; gases used include argon, krypton, neon, and xenon; examples include helium-cadmium lasers and metal vapor lasers. { 'ī‚än ‚lā·zər }

ion machining [ENG] Use of a high-velocity ion beam to remove material from a surface. Also known as ion beam thinning; ion milling. { 'ī‚än mə'shēn·iŋ }

ion mean life [PHYS CHEM] The average time between the ionization of an atom or molecule and its recombination with one or more electrons, or its loss of excess electrons. { 'ī‚än ‚mēn 'līf }

ion microprobe See secondary ion mass spectrometer. { 'ī‚än 'mī·krə‚prōb }

ion microprobe mass spectrometer [ENG] A type of secondary ion mass spectrometer in which primary ions are focused on a spot 1–2 micrometers in diameter, mass-charge separation of secondary ions is carried out by a double focusing mass spectrometer or spectrograph, and a magnified image of elemental or isotopic distributions on the sample surface is produced using synchronous scanning of the primary ion beam and an oscilloscope. { 'ī‚än 'mī·krə‚prōb ‚mas spek'träm·əd·ər }

ion microscope See field-ion microscope. { 'ī‚än 'mī·krə‚skōp }

ion migration [ELEC] Movement of ions produced in an electrolyte, semiconductor, and so on, by the application of an electric potential between electrodes. { 'ī‚än mī'grā·shən }

ion milling See ion machining. { 'ī‚än ‚mil·iŋ }

ionogenic group [PHYS CHEM] A fixed group of atoms in an ion exchanger that is either ionized or capable of dissociation into fixed ions and mobile counterions. { ¦ī·ə·nə¦jen·ik ¦grüp }

ionogram [ENG] A record produced by an ionosonde, that is, a graph of the virtual height of the ionosphere plotted against frequency. { ī'än·ə¸gram }

ionography [ANALY CHEM] A type of electrochromatography involving migration of ions. { ī·ə'näg·rə·fē }

ionomer [ORG CHEM] Polymer with covalent bonds between the elements of the chain, and ionic bonds between the chains. { ī'än·ə·mər }

ionomer resin [ORG CHEM] A polymer which has ethylene as the major component, but which contains both covalent and ionic bonds. { ī'än·ə·mər 'rez·ən }

ionone [ORG CHEM] $C_{13}H_{20}O$ A colorless to light yellow liquid with a boiling point of 126–128°C at 12 mmHg (1600 pascals); soluble in alcohol, ether, and mineral oil; used in perfumery, flavoring, and vitamin A production. Also known as irisone. { ī'ə¸nōn }

ionophone [ENG ACOUS] A high-frequency loudspeaker in which the audio-frequency signal modulates the radio-frequency supply to an arc maintained in a quartz tube, and the resulting modulated wave acts directly on ionized air to create sound waves. { ī'än·ə¸fōn }

ionophore [BIOCHEM] Any of a class of compounds, generally cyclic, having the ability to carry ions across lipid barriers due to the property of cation selectivity; examples are valinomycin and nonactin. { ī'än·ə¸fōr }

ionosonde [ENG] A radar system for determining the vertical height at which the ionosphere reflects signals back to earth at various frequencies; a pulsed vertical beam is swept periodically through a frequency range from 0.5 to 20 megahertz, and the variation of echo return time with frequency is photographically recorded. { ī'än·ə¸sänd }

ionosphere [GEOPHYS] That part of the earth's upper atmosphere which is sufficiently ionized by solar ultraviolet radiation so that the concentration of free electrons affects the propagation of radio waves; its base is at about 40 or 50 miles (70 or 80 kilometers) and it extends to an indefinite height. { ī'än·ə¸sfir }

ionospheric disturbance [GEOPHYS] A temporal variation in electron concentration in the ionosphere that is caused by solar activity and that makes the heights of the ionosphere layers go beyond the normal limits for a location, date, and time of day. { ¸ī¸än·ə'sfir·ik di'stər·bəns }

ionospheric D scatter meteor burst [GEOPHYS] Phenomenon affecting ionospheric scatter communications resulting from the penetration of meteors through the D region of the ionospheric layer. { ¸ī¸än·ə'sfir·ik 'dē ¸skad·ər 'mēd·ē·ər ¸bərst }

ionospheric error [COMMUN] Variation in the character of the ionospheric transmission path or paths used by the radio waves of electronic navigation systems which, if not compensated, will produce an error in the information generated by the system. { ¸ī¸än·ə'sfir·ik 'er·ər }

ionospheric propagation [COMMUN] Propagation of radio waves over long distances by reflection from the ionosphere, useful at frequencies up to about 25 megahertz. { ¸ī¸än·ə'sfir·ik ¸präp·ə'gā·shən }

ionospheric recorder [ELECTR] A radio device for determining the distribution of virtual height with frequency, and the critical frequencies of the various layers of the ionosphere. { ¸ī¸än·ə'sfir·ik ri'kórd·ər }

ionospheric scatter [COMMUN] A form of scatter propagation in which radio waves are scattered by the lower E layer of the ionosphere to permit communication over distances from 600 to 1400 miles (1000 to 2250 kilometers) when using the frequency range of about 25 to 100 megahertz. { ¸ī¸än·ə'sfir·ik 'skad·ər }

ionospheric storm [GEOPHYS] A turbulence in the F region of the ionosphere, usually due to a sudden burst of radiation from the sun; it is accompanied by a decrease in the density of ionization and an increase in the virtual height of the region. { ¸ī¸än·ə'sfir·ik 'stórm }

ionospheric wave See sky wave. { ¸ī¸än·ə'sfir·ik 'wāv }

ion pair [NUCLEO] A positive ion and an equal-charge negative ion, usually an electron, that are produced by the action of radiation on a neutral atom or molecule. { 'ī¸än ¸per }

ion-permeable membrane [MATER] A film or sheet of a substance which is preferentially permeable to some species or types of ions. { ¸ī¸än ¦pər·mē·ə·bəl 'mem¸brān }

ion potential See ionization potential. { 'ī¸än pə¸ten·chəl }

ion probe See secondary ion mass spectrometer. { 'ī¸än ¸prōb }

ion propulsion [AERO ENG] Vehicular motion caused by reaction from the high-speed discharge of a beam of electrically charged minute particles usually positive ions, that are accelerated in an electrostatic field and ejected behind the vehicle. { 'ī¸än prə'pəl·shən }

ion pump [ELECTR] A vacuum pump in which gas molecules are first ionized by electrons that have been generated by a high voltage and are spiraling in a high-intensity magnetic field, and the molecules are then attracted to a cathode, or propelled by electrodes into an auxiliary pump or an ion trap. { 'ī¸än ¸pəmp }

ion retardation [CHEM ENG] Sorbent extraction of strong electrolytes with an anion-exchange resin in which a cationic monomer has been polymerized, or vice versa. { 'ī¸än ¸rē¸tär'dā·shən }

ion scattering spectroscopy [SPECT] A spectroscopic technique in which a low-energy (about 1000 electronvolts) beam of inert-gas ions is directed at a surface, and the energies and scattering angles of the scattered ions are used to identify surface atoms. Abbreviated ISS. { 'ī¸än ¸skad·ə·riŋ spek'träs·kə·pē }

ion-selective field-effect transistor [ELECTR] A field-effect transistor whose gate electrode is sensitive to certain ions in an electrolyte, so that the gain of the transistor depends on the concentration of these ions. Abbreviated ISFET. { 'ī¸än si¸lek·tiv 'fēld i¸fekt tran'zis·tər }

ion-solid interaction [SOLID STATE] An atomic process that occurs as a result of the collision of energetic ions, atoms, or molecules with condensed matter. { 'ī¸än ¦säl·əd ¸in·tər'ak·shən }

ion source [ELECTR] A device in which gas ions are produced, focused, accelerated, and emitted as a narrow beam. Also known as ion gun; ionization source. { 'ī¸än ¸sórs }

ion spot [ELECTR] Of a cathode-ray tube screen, an area of localized deterioration of luminescence caused by bombardment with negative ions. Also known as ion burn. { 'ī¸än ¸spät }

iontophoresis [MED] A medical treatment used to drive positive or negative ions into a tissue, in which two electrodes are placed in contact with tissue, one of the electrodes being a pad of absorbent material soaked with a solution of the material to be administered, and a voltage is applied between the electrodes. { ¸ī¸än·tə·fə'rē·səs }

ion trap [ELECTR] **1.** An arrangement whereby ions in the electron beam of a cathode-ray tube are prevented from bombarding the screen and producing an ion spot, usually employing a magnet to bend the electron beam so that it passes through the tiny aperture of the electron gun, while the heavier ions are less affected by the magnetic field and are trapped inside the gun. **2.** A metal electrode, usually of titanium, into which ions in an ion pump are absorbed. { 'ī¸än ¸trap }

iophobia [PSYCH] An abnormal fear of poison. { ¸ī·ə'fō·bē·ə }

Iospilidae [INV ZOO] A small family of pelagic polychaetes assigned to the Errantia. { ¸ī·ə'spil·ə¸dē }

Io torus [ASTRON] A doughnut-shaped region of dense plasma that orbits Jupiter at the radial distance of the satellite Io and results from ionization by solar ultraviolet radiation of gases emitted from Io in volcanic eruptions. { ¸ī¸ō 'tór·əs }

Iowan glaciation [GEOL] The earliest substage of the Wisconsin glacial stage; occurred more than 30,000 years ago. { 'ī·ə·wən ¸glā·sē'ā·shən }

ioxynil [ORG CHEM] $C_7H_3I_2NO$ A colorless solid with a melting point of 212–213°C; used for postemergence control of seedling weeds in cereals and sports turf. { ī'äk·sə¸nil }

ioxynil octanoate [ORG CHEM] $C_{15}H_{17}I_2NO_2$ A waxy solid with a melting point of 59–60°C; insoluble in water; used as an insecticide for cereals and sugarcane. { ī'äk·sə¸nil ¸äk·tə'nō·ət }

IPA See International Phonetic Alphabet. { ¸ī¸pē'ā }

IP address [COMPUT SCI] A computer's numeric address, such as 128.201.86.290, by which it can be located within a network. { ¸ī¸pē ə¸dres }

IPC See propham.

ipecac [BOT] Any of several low, perennial, tropical South American shrubs or half shrubs in the genus *Cephaelis* of the family Rubiaceae; the dried rhizome and root, containing emetine, cephaeline, and other alkaloids, is used as an emetic and expectorant. { 'ip·ə,kak }

Ipidae [INV ZOO] The equivalent name for Scolytidae. { 'ip·ə,dē }

IPL [COMPUT SCI] **1.** Collective term for a series of list-processing languages developed principally by A. Newell, H. A. Simon, and J. C. Shaw. Derived from Information Processing Language. **2.** *See* initial program load.

IPL button *See* bootstrap button. { 'ī,pē'el 'bət·ən }

IPN *See* interpenetrating polymer network.

ipsilateral *See* homolateral. { ¦ip·sə¦lad·ə·rəl }

ipsonite [GEOL] The final stage of weathered asphalt; a black, infusible substance, only slightly soluble in carbon disulfide, containing 50–80% fixed carbon and very little oxygen. { 'ip·sə,nīt }

IQ *See* intelligence quotient.

IQSY *See* International Quiet Sun Year.

Ir *See* iridium.

IRBM *See* intermediate-range ballistic missile.

irdome [OPTICS] A dome used to protect an infrared detector and its optical elements, generally made from quartz, silicon, germanium, sapphire, calcium aluminate, or other material having high transparency to infrared radiation. Derived from infrared dome. { 'ī'är,dōm }

IR drop *See* resistance drop. { 'ī¦är 'dräp }

IRG *See* record gap.

IR gene *See* immune response gene. { 'ī'är ¦jēn }

Iridaceae [BOT] A family of monocotyledonous herbs in the order Liliales distinguished by three stamens and an inferior ovary. { ,ir·ə'dās·ē,ē }

iridectomy [MED] Surgical removal of part of the iris. { ,ir·ə'dek·tə·mē }

iridescence [OPTICS] A rainbow color effect exhibited in various bodies as a result of interference in a thin film (as of soap bubbles or mother of pearl) or of diffraction of light reflected from a ribbed surface (as of the plumage of some birds). { ,ir·ə'des·əns }

iridescent cloud [METEOROL] An ice-crystal cloud which exhibits brilliant spots or borders of colors, usually red and green, observed up to about 30° from the sun. { ,ir·ə'des·ənt 'klaud }

iridescent layer *See* schiller layer. { ,i·ri'des·ənt ,lā·ər }

iridic chloride [INORG CHEM] IrCl₄ A hygroscopic brownish-black mass, soluble in water and alcohol; used to analyze for nitric acid, HNO₃, and in analytical microscopic work. Also known as iridium chloride; iridium tetrachloride. { i'rid·ik 'klȯr,īd }

iridium [CHEM] A metallic element, symbol Ir, atomic number 77, atomic weight 192.2, in the platinum group; insoluble in acids, melting at 2454°C. [MET] A silver-white, brittle, hard metal used in jewelry, electric contacts, electrodes, resistance wires, and pen tips. { i'rid·ē·əm }

iridium-192 [NUC PHYS] Radioactive isotope of iridium with a 75-day half-life; β and γ radiation; used in cancer treatment and for radiography of light metal castings. { i'rid·ē·əm ¦wən¦nīn·tē'tü }

iridium chloride *See* iridic chloride. { i'rid·ē·əm 'klȯr,īd }

iridium tetrachloride *See* iridic chloride. { i'rid·ē·əm ,te·trə'klȯr,īd }

iridocele [MED] A rupture of the cornea through which a portion of the iris protrudes. { i'rid·ə,sēl }

iridocyclitis [MED] Inflammation of the iris and the ciliary body. { ¦ir·ə·dō·sī'klīd·əs }

iridocyclochoroiditis [MED] Inflammation of the iris, the ciliary body, and the choroid. Also known as uveitis. { ¦ir·ə·dō¦sī·klō,kȯ·rȯi'dīd·əs }

iridocyte [HISTOL] A specialized cell in the integument of certain animal species which is filled with iridescent crystals of guanine and a variety of lipophores. Also known as guanophore; iridophore. { i'rid·ə,sīt }

iridophore *See* iridocyte. { i'rid·ə,fȯr }

iridosmine [MET] A natural iridium-osmium alloy composed of 10–77% iridium, 17–80% osmium, 0–10% platinum, 0–17% rhodium, 0–9% ruthenium, 0–2% iron, and 0–1% copper; used for surgical needles and compass bearings and for hardening platinum. { ,i·rə'däz,mēn }

Iridoviridae [VIROL] A family of double-stranded deoxyribonucleic acid-containing animal viruses that infects invertebrates and is characterized by an icosahedral virion that has a yellow-green glow in centrifuged pellets. { i,rid·ə'vir·ə,dē }

Iridovirus [VIROL] A genus of the animal-virus family Iridoviridae; characterized by the blue-to-purple iridescence emitted by purified viral particles and infected larvae. { i'rid·ə,vī·rəs }

iris [ANAT] A pigmented diaphragm perforated centrally by an adjustable pupil which regulates the amount of light reaching the retina in vertebrate eyes. [BOT] Any plant of the genus *Iris*, the type genus of the family Iridaceae, characterized by linear or sword-shaped leaves, erect stalks, and bright-colored flowers with the three inner perianth segments erect and the outer perianth segments drooping. [ELECTROMAG] A conducting plate mounted across a waveguide to introduce impedance; when only a single mode can be supported, an iris acts substantially as a shunt admittance and may be used for matching the waveguide impedance to that of a load. Also known as diaphragm; waveguide window. [OPTICS] A circular mechanical device, whose diameter can be varied continuously, which controls the amount of light reaching the film of a camera. Also known as iris diaphragm. { 'ī·rəs }

irisation [METEOROL] The coloration exhibited by iridescent clouds and at times along the borders of lenticular clouds. { ,ī·rə'sā·shən }

iris diaphragm *See* iris. { 'ī·rəs 'dī·ə,fram }

Irish linen [TEXT] Thin linen fabric woven of Irish flax. { 'ī·rish 'lin·ən }

Irish moss *See* carrageen. { 'ī,rish 'mȯs }

Irish potato *See* potato. { 'ī·rish pə'tād·ō }

Irish Sea [GEOGR] A marginal sea of the Atlantic Ocean between Ireland and England, approximately 53°N latitude and 5°W longitude. { 'ī·rish 'sē }

irisone *See* ionone. { 'ī·rə,sōn }

iritis [MED] Inflammation of the iris. { ə'rīd·əs }

I²R loss *See* copper loss. { 'ī,skwerd'är ,lȯs }

IRM *See* isothermal remanent magnetization.

Irminger Current [OCEANOGR] An ocean current that is one of the terminal branches of the Gulf Stream system, flowing west off the southern coast of Iceland. { 'ər·miŋ·ər 'kə·rənt }

iron [CHEM] A silvery-white metallic element, symbol Fe, atomic number 26, atomic weight 55.847, melting at 1530°C. [MET] A heavy, magnetic, malleable and ductile metal occurring in meteorites and combined in a wide range of ores and most igneous rocks; it is one of the most widely used metals, and plays a role in biological processes. { 'ī·ərn }

iron-55 [NUC PHYS] Radioactive isotope of iron, symbol ⁵⁵Fe, with a 2.91-year half-life; highly toxic. { 'ī·ərn ¦fif·tē'fīv }

iron-59 [NUC PHYS] Radioactive isotope of iron, symbol ⁵⁹Fe, 46.3-day half-life; β and γ radiation; highly toxic; used to study metallic welds, corrosion mechanisms, engine wear, and bodily functions. { 'ī·ərn ¦fif·tē'nīn }

iron acetate *See* ferrous acetate. { 'ī·ərn 'as·ə,tāt }

iron acetate liquor [MATER] Black liquor containing 5–5.5% iron and sometimes copperas or tannin; results from pyroligneous acid attack on iron filings; used for dyeing and printing with logwood, and as a mordant for alizarine and nitroso dyes. Also known as black liquor; black mordant; iron liquor. { 'ī·ərn 'as·ə,tāt 'lik·ər }

Iron Age [ARCHEO] Period characterized by production and widespread use of iron, starting approximately 1000 B.C. { 'ī·ərn ¦āj }

iron alloy [MET] An alloy having iron as the principal component. { 'ī·ərn 'al,ȯi }

iron alum *See* halotrichite. { 'ī·ərn 'al·əm }

iron ammonium sulfate *See* ferric ammonium sulfate; ferrous ammonium sulfate. { 'ī·ərn ə'mō·nē·əm 'səl,fāt }

iron and steel sheet piling [MIN ENG] A technique that uses iron and steel piling instead of wood to drive a shaft through loose, wet ground near the surface. { 'ī·ərn ən ¦stēl ¦shēt 'pīl·iŋ }

Ironarc process [MET] An ultra-high-temperature smelting process employing plasma chemistry to process refractory materials, such as zirconia. { 'ī·ərn,ärk ,prä·səs }

iron arsenate *See* ferrous arsenate. { 'ī·ərn 'ärs·ən,āt }

iron bacteria [MICROBIO] The common name for bacteria

capable of oxidizing ferrous iron to the ferric state. { 'ī·ərn bak'tir·ē·ə }

iron-binding protein [BIOCHEM] A serum protein, such as hemoglobin, for the transport of iron ions. { 'ī·ərn ˌbīnd·iŋ 'prō,tēn }

iron black [CHEM] Fine black antimony powder used to give a polished-steel look to papier-maché and plaster of paris; made by reaction of zinc with acid solution of an antimony salt and precipitation of black antimony powder. { 'ī·ərn 'blak }

iron blue [INORG CHEM] Ferric ferrocyanide used as blue pigment by the paint industry for permanent body and trim paints; also used in blue ink, in paper dyeing, and as a fertilizer ingredient. { 'ī·ərn 'blü }

iron bromide See ferric bromide. { 'ī·ərn 'brō,mīd }

iron carbide See cementite. { 'ī·ərn 'kär,bīd }

iron carbonyl See iron pentacarbonyl. { 'ī·ərn 'kär·bə,nil }

iron castings [MET] Shapes cast in molds from iron. { 'ī·ərn 'kast·iŋz }

iron cement [MATER] A mixture of small iron pieces with ammonium chloride, used to join iron or steel surfaces. { 'ī·ərn si'ment }

iron chloride See ferric chloride; ferrous chloride. { 'ī·ərn 'klȯr,īd }

iron citrate See ferric citrate. { 'ī·ərn 'sī,trāt }

iron-Constantan [MET] Dual-metal combination for thermocouple junctions, used for temperature measurement in oxidizing or reducing atmospheres. { 'ī·ərn kän'stant·ən }

iron cordierite See sekaninaite. { 'ī·ərn 'kȯr·dē·ə,rīt }

iron core [ELECTROMAG] A core made of solid or laminated iron, or some other magnetic material which may contain very little iron. { 'ī·ərn 'kȯr }

iron-core choke See iron-core coil. { 'ī·ərn 'kȯr 'chōk }

iron-core coil [ELECTROMAG] A coil in which solid or laminated iron or other magnetic material forms part or all of the magnetic circuit linking its winding. Also known as iron-core choke; magnet coil. { 'ī·ərn 'kȯr 'kȯil }

iron-core transformer [ELECTROMAG] A transformer in which laminations of iron or other magnetic material make up part or all of the path for magnetic lines of force that link the transformer windings. { 'ī·ərn 'kȯr tranz'fȯr·mər }

iron count [CHEM ENG] An analytic determination of the iron compounds in a product stream; reflects the occurrence and the extent of corrosion. { 'ī·ərn ,kau̇nt }

iron-deficiency anemia [MED] Hypochromic microcytic anemia due to excessive loss, deficient intake, or poor absorption of iron. Also known as nutritional hypochromic anemia. { 'ī·ərn də,fish·ən·sē ə'nē·mē·ə }

iron dichloride See ferrous chloride. { 'ī·ərn dī'klȯr,īd }

iron-dust core [ELECTROMAG] A core made by mixing finely powdered magnetic material with an insulating binder and molding under pressure to form a rod-shaped core that can be moved into or out of a coil or transformer to vary the inductance or degree of coupling for tuning purposes. { 'ī·ərn ,dəst ,kȯr }

irone [ORG CHEM] $C_{14}H_{22}O$ A colorless liquid terpene; a component of essential oil from the orrisroot; used in perfumes. { 'ī,rōn }

iron ferrocyanide See ferric ferrocyanide. { 'ī·ərn ,fer·ə'sī·ə,nīd }

iron fluoride See ferric fluoride. { 'ī·ərn 'flu̇r,īd }

iron formation [GEOL] Sedimentary, low-grade iron ore bodies consisting mainly of chert or fine-grained quartz and ferric oxide segregated in bands or sheets irregularly mingled. { 'ī·ərn fȯr'mā·shən }

iron foundry [MET] A building in which iron castings are made. { 'ī·ərn 'fau̇n·drē }

iron glance See specularite. { 'ī·ərn ,glans }

iron hat See gossan. { 'ī·ərn 'hat }

iron hydroxide See ferric hydroxide. { 'ī·ərn hī'dräk,sīd }

ironing [MET] Reducing the wall thickness of a deep-drawn object by reducing the clearance between punch and die. { 'ī·ərn·iŋ }

iron liquor See iron acetate liquor. { 'ī·ərn 'lik·ər }

iron loss See core loss. { 'ī·ərn ,lȯs }

iron metabolism [BIOCHEM] The chemical and physiological processes involved in absorption of iron from the intestine and in its role in erythrocytes. { 'ī·ərn mə'tab·ə,liz·əm }

iron metavanadate See ferric vanadate. { 'ī·ərn ,med·ə'van·ə,dāt }

iron meteorite [ASTRON] A type of meteorite that consists mainly of iron and nickel and is several times heavier than any ordinary rock. { 'ī·ərn 'mēd·ē·ə,rīt }

iron mica See lepidomelane. { 'ī·ərn 'mī·kə }

iron monoxide See ferrous oxide. { 'ī·ərn mə'näk,sīd }

iron-nickel alloy [MET] An iron alloy containing 20–80% nickel; it has high permeability and low hysteresis losses at low flux densities and is more readily rolled into thin laminations than silicon steels. { 'ī·ərn 'nik·əl 'al,ȯi }

iron nitrate See ferric nitrate. { 'ī·ərn 'nī,trāt }

iron nonacarbonyl [INORG CHEM] $Fe_2(CO)_9$ Orange-yellow crystals that break down at 100°C to yield tetracarbonyl, slightly soluble in alcohol and acetone, almost insoluble in water, ether, and benzene. { 'ī·ərn ,nō·nə'kär·bə,nil }

Ironoidea [INV ZOO] A superfamily of presumably carnivorous, fresh-water and soil-dwelling nematodes in the order Enoplida, having two circlets of cephalic sensory organs, a cylindrical, elongate stoma armed either anteriorly with three large teeth or posteriorly with small teeth, and cuticularized stomatal walls. { ,ī·rə'nȯid·ē·ə }

iron ore [GEOL] Rocks or deposits containing compounds from which iron can be extracted. { 'ī·ərn 'ȯr }

iron-ore cement [MATER] A cement in which iron ore is used instead of clay or shale. { 'ī·ərn ,ȯr si'ment }

iron oxalate See ferrous oxalate. { 'ī·ərn 'äk·sə,lāt }

iron oxide [INORG CHEM] Any of the hydrated, synthetic, or natural oxides of iron: ferrous oxide, ferric oxide, ferriferous oxide. { 'ī·ərn 'äk,sīd }

iron oxide process [CHEM ENG] A process by which a gas is passed through iron oxide and wood shavings to remove sulfides. { 'ī·ərn 'äk,sīd prä·səs }

iron pentacarbonyl [INORG CHEM] $Fe(CO)_5$ An oily liquid that decomposes upon exposure to light, soluble in most organic solvents; used as a source of a pure iron catalyst and for magnet cores. Also known as iron carbonyl. { 'ī·ərn ,pen·tə'kär·bə,nil }

iron phosphate See ferric phosphate. { 'ī·ərn 'fäs,fāt }

iron-porphyrin protein [BIOCHEM] Any protein containing iron and porphyrin; examples are hemoglobin, the cytochromes, and catalase. { 'ī·ərn'fȯr·fə·rən 'prō,tēn }

iron pyrites See pyrite. { 'ī·ərn 'pī,rīts }

iron red [MATER] Any of the pigments made from red varieties of ferric oxide. { 'ī·ərn 'red }

iron resinate See ferric resinate. { 'ī·ərn 'rez·ən,āt }

iron-retention agent [MATER] Complexing agent that ties dissolved iron into complex ions to prevent reprecipitation and wellbore plugging after well acidizing. { 'ī·ərn ri'ten·chən ,ā·ənt }

iron saffron See Indian red. { 'ī·ərn 'saf·rən }

iron scurf [MATER] Glazing material used to color blue bricks; a mixture of stone and iron particles from grinding of gun barrels. { 'ī·ərn 'skərf }

ironshot [MINERAL] Pertaining to a mineral with streaks or spots of iron or iron ore. { 'ī·ərn,shät }

iron sight [ORD] Any metallic gunsight, such as a blade sight, as distinguished from an optical or computing sight. { 'ī·ərn 'sīt }

iron soldering [MET] Soldering in which a soldering iron provides the required heat. { 'ī·ərn 'säd·ə·riŋ }

iron spar See siderite. { 'ī·ərn 'spär }

iron spinel See hercynite. { 'ī·ərn spə'nel }

iron stearate See ferric stearate. { 'ī·ərn 'stir,āt }

ironstone [PETR] An iron-rich sedimentary rock, either deposited directly as a ferruginous sediment or resulting from chemical replacement. { 'ī·ərn,stōn }

iron-stony meteorite See stony-iron meteorite. { 'ī·ərn ,stō·nē 'mēd·ē·ə,rīt }

iron sulfate See ferric sulfate; ferrous sulfate. { 'ī·ərn 'səl,fāt }

iron sulfide See ferrous sulfide. { 'ī·ərn 'səl,fīd }

iron tetracarbonyl [INORG CHEM] $Fe_3(CO)_{12}$ Dark-green lustrous crystals that break down at 140–50°C; soluble in organic solvents. Also known as tri-iron dodecacarbonyl. { 'ī·ərn ,te·trə'kär·bə,nil }

iron whiskers [MET] Single-crystal pure iron filaments or fibers. { 'ī·ərn 'wis·kərz }

iron winds [METEOROL] Northeasterly winds of Central America, prevalent during February and March, and blowing steadily for several days at a time. { 'ī·ərn 'winz }

ironwood [BOT] Any of a number of hardwood trees in the

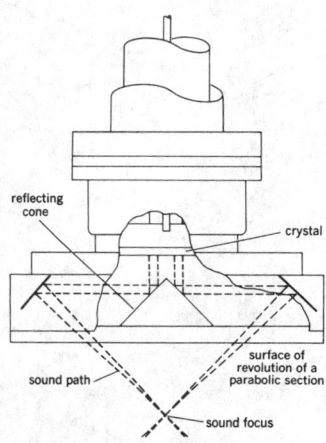

IRRADIATION

reflecting cone

crystal

sound path

surface of revolution of a parabolic section

sound focus

Reflector-type focusing irradiator, a device for producing irradiation by sound waves in the megahertz range. The sound waves emanating from the circular crystal plate impinge on a 90° included angle cone, from which surface they are directed to a surface of revolution generated by a section of a parabola. Focusing is obtained by reflection of the waves from this surface.

IRREGULAR CLEAVAGE

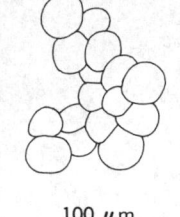

100 μm

Pattern of cells in irregular cleavage.

United States, including the American hornbeam, the buckwheat, and the eastern hophornbeam. { 'ī·ərn ˌwüd }

irradiance See radiant flux density. { i'rād·ē·əns }

irradiation [BIOPHYS] Subjection of a biological system to sound waves of sufficient intensity to modify their structure or function. [ENG] The exposure of a material, object, or patient to x-rays, gamma rays, ultraviolet rays, or other ionizing radiation. [OPTICS] An optical illusion which makes bright objects appear larger than they really are. { i,rād·ē'ā·shən }

irradiation cataract [MED] A cataract that develops slowly following prolonged or intense irradiation, as by radium or roentgen rays. Also known as cyclotron cataract; radiation cataract. { i,rād·ē'ā·shən 'kad·ə,rakt }

irradiation correction [NAV] A correction to the readings of sextant altitudes made necessary by irradiation that is caused by the apparent enlargement of the bright surface of a celestial body against the darker background of the sky. { i,rād·ē'ā·shən kə'rek·shən }

irradiation cystitis [MED] Inflammation of the urinary bladder following radiation therapy of pelvic organs. { i,rād·ē'ā·shən sis'tīd·əs }

irrational algebraic expression [MATH] An algebraic expression that cannot be written as a quotient of polynomials. { i¦rash·ən·əl ˌal·jə¦brā·ik ik'spresh·ən }

irrational equation [MATH] An equation having an unknown raised to some fractional power. Also known as radical equation. { i'rash·ən·əl i'kwā·zhən }

irrationality of dispersion [OPTICS] The effect whereby spectra produced by prisms of different types of glass are not geometrically similar. { i,rash·ə¦nal·əd·ē əv di'spər·zhən }

irrational number [MATH] A number which is not the quotient of two integers. { i'rash·ən·əl 'nəm·bər }

irrational radical [MATH] A radical that is not equivalent to a rational number or expression. { i'rash·ən·əl 'rad·ə·kəl }

irreducible element [MATH] An element x of a ring which is not a unit and such that every divisor of x is improper. { ir·ə'düs·ə·bəl 'el·ə·mənt }

irreducible equation [MATH] An equation that is equivalent to one formed by setting an irreducible polynomial equal to zero. { ir·ə'dü·sə·bəl i'kwā·zhən }

irreducible function See irreducible polynomial. { ir·ə'dü·sə·bəl 'fəŋk·shən }

irreducible lambda expression [MATH] A lambda expression that cannot be converted to a reduced form by a sequence of applications of the renaming and reduction rules. { ir·ə'dü·sə·bəl 'lam·də ik,spresh·ən }

irreducible module [MATH] A module whose only submodules are the module itself and the module that consists of the element 0. { ir·i¦dü·sə·bəl 'mäj·əl }

irreducible polynomial [MATH] A polynomial is irreducible over a field K if it cannot be written as the product of two polynomials of lesser degree whose coefficients come from K. Also known as irreducible function. { ir·ə'dü·sə·bəl ˌpäl·ə'nō·mē·əl }

irreducible representation of a group [MATH] A representation of a group as a family of linear operators of a vector space V where there is no proper closed subspace of V invariant under these operators. { ir·ə'dü·sə·bəl ˌrep·rə·zən'tā·shən əv ə 'grüp }

irreducible saturation [PETRO ENG] In a permeable reservoir, that condition in which the nonwetting phase saturation is so large that the wetting phase can be reduced no more. { ir·ə'dü·sə·bəl ˌsach·ə'rā·shən }

irreducible tensor [MATH] A tensor that cannot be written as the inner product of two tensors of lower degree. { ir·ə'dü·sə·bəl 'ten·sər }

irregular [BOT] Lacking symmetry, as of a flower having petals unlike in size or shape. { i'reg·yə·lər }

irregular cleavage [EMBRYO] Division of a zygote into random masses of cells, as in certain cnidarians. { i'reg·yə·lər 'klē·vij }

irregular cluster [ASTRON] A type of galaxy cluster that has an overall amorphous appearance, usually showing little overall symmetry or central concentration and often composed of several distinct clumps of galaxies. { i'reg·yə·lər 'kləs·tər }

irregular connective tissue [HISTOL] A loose or dense connective tissue with fibers irregularly woven and irregularly distributed; collagen is the dominant fiber type. { i'reg·yə·lər kə'nek·tiv ˌtish·ü }

irregular crystal [METEOROL] A snow particle, sometimes covered by a coating of rime, composed of small crystals randomly grown together; generally, component crystals are so small that the crystalline form of the particle can be seen only through a magnifying glass or microscope. { i'reg·yə·lər 'krist·əl }

irregular element [IND ENG] An element whose frequency of occurrence is irregular but predictable. Also known as incidental element. { i'reg·yə·lər 'el·ə·mənt }

irregular galaxy [ASTRON] A galaxy which shows no definite order or shape, except that of a general flattened appearance. { i'reg·yə·lər 'gal·ik·sē }

Irregularia [INV ZOO] An artificial assemblage of echinoderms in which the anus and periproct lie outside the apical system, the ambulacral plates remain simple, the primary radioles are hollow, and the rigid test shows some degree of bilateral symmetry. { i,reg·yə'lar·ē·ə }

irregular iceberg See pinnacled iceberg. { i'reg·yə·lər 'īs·bərg }

irregular polymer [CHEM] A polymer whose molecular structure does not consist of only one species of constitutional unit in a single sequential arrangement. { i¦reg·yə·lər 'päl·i·mər }

irregular variable star [ASTRON] A star with no fixed period. { i'reg·yə·lər ¦ver·ē·ə·bəl 'stär }

irrespirable atmosphere [MIN ENG] Atmosphere in a coal mine requiring workers to wear breathing apparatus because of poisonous gas or insufficient oxygen as a result of an explosion from firedamp or coal dust, or mine fires. { i'res·pə·rə·bəl 'at·mə,sfir }

irreversible energy loss [THERMO] Energy transformation process in which the resultant condition lacks the driving potential needed to reverse the process; the measure of this loss is expressed by the entropy increase of the system. { i·ri'vər·sə·bəl 'en·ər·je ˌlös }

irreversible process [THERMO] A process which cannot be reversed by an infinitesimal change in external conditions. { i·ri'vər·sə·bəl 'prä·səs }

irreversible thermodynamics See nonequilibrium thermodynamics. { i·ri'vər·sə·bəl ¦thər·mə·dī'nam·iks }

irrigation [CIV ENG] Artificial application of water to arable land for agricultural use. [MED] Therapeutic washing out by means of a continuous stream of water. { ir·ə'gā·shən }

irrigation canal [CIV ENG] An artificial open channel for transporting water for crop irrigation. { ir·ə'gā·shən kə,nal }

irrigation pipe [CIV ENG] A conduit of connected pipes for transporting water for crop irrigation. { ir·ə'gā·shən ,pīp }

irritability [PHYSIO] **1.** A condition or quality of being excitable; the power of responding to a stimulus. **2.** A condition of abnormal excitability of an organism, organ, or part, when it reacts excessively to a slight stimulation. { ir·əd·ə'bil·əd·ē }

irritable bowel syndrome [MED] A gastrointestinal disorder of unknown cause that is characterized by increased intestinal motility, causing recurrent abdominal pain, constipation or diarrhea that may alternate, and sensation of gaseousness and bloating. { ir·id·ə·bəl 'baul ,sin,drōm }

irritable colon [MED] Any of several disturbed colonic functions associated with anxiety or emotional stress. Also known as adaptive colitis; mucous colitis; spastic colon; unstable colon. { 'ir·əd·ə·bəl 'kō·lən }

irritant gas [MATER] A nonlethal gas, causing irritation of the skin and flow of tears; any one of the family of tear gases used for training and riot control. { 'ir·ət·ənt ,gas }

irrotational flow [FL MECH] Fluid flow in which the curl of the velocity function is zero everywhere, so that the circulation of the velocity about any closed curve vanishes. Also known as acyclic motion; irrotational motion. { ¦ir·ə'tā·shən·əl 'flō }

irrotational motion See irrotational flow. { ¦ir·ə'tā·shən·əl 'mō·shən }

irrotational strain [GEOL] Strain in which the orientation of the axes of strain does not change. Also known as nonrotational strain. { ¦ir·ə'tā·shən·əl 'strān }

irrotational vector field [MATH] A vector field whose curl is identically zero; every such field is the gradient of a scalar function. Also known as lamellar vector field. { ¦ir·ə'tā·shən·əl ,fēld }

irrotational wave See compressional wave. { ¦ir·ə'tā·shən·əl 'wāv }

irruption See intrusion. { i'rəp·shən }

IR sequences See inverted repeats. { 'ī'är ˌsē·kwən·səs }

Irvingtonian [GEOL] A stage of geologic time in southern California, in the lower Pleistocene, below the Rancholabrean. { ˌər·viŋ'tō·nē·ən }

isallobar [METEOROL] A line of equal change in atmospheric pressure during a specified time interval; an isopleth of pressure tendency; a common form is drawn for the three-hourly local pressure tendencies on a synoptic surface chart. { ī'sal·əˌbär }

isallobaric [METEOROL] Of equal or constant pressure change; this may refer either to the distribution of equal pressure tendency in space or to the constancy of pressure tendency with time. { ī¦sal·ə¦bar·ik }

isallobaric high See pressure-rise center. { ī¦sal·ə¦bar·ik 'hī }

isallobaric low See pressure-fall center. { ī¦sal·ə¦bar·ik 'lō }

isallobaric maximum See pressure-rise center. { ī¦sal·ə¦bar·ik 'mak·sə·məm }

isallobaric minimum See pressure-fall center. { ī¦sal·ə¦bar·ik 'min·ə·məm }

isallobaric wind [METEOROL] The wind velocity whose Coriolis force exactly balances a locally accelerating geostrophic wind. Also known as Brunt-Douglas isallobaric wind. { ī¦sal·ə¦bar·ik 'wind }

isallohypse See height-change line. { ī'sal·ə‚hips }

isallohypsic wind [METEOROL] An isallobaric wind, using height tendency in a constant-pressure surface instead of pressure tendency in a constant-height surface. { ī'sal·ə¦hip·sik 'wind }

isallotherm [METEOROL] A line connecting points of equal change in temperature within a given time period. { ī'sal·ə‚thərm }

isanabat [METEOROL] A line drawn through points of equal vertical component of wind velocity; positive values indicate upward motion, negative values indicate downward motion. { ī'san·ə‚bat }

isanakatabar [METEOROL] A line on a chart of equal atmospheric-pressure range during a specified time interval. { ī¦san·ə¦kad·ə‚bär }

isanomal See isanomalous line. { ‚ī·sə'näm·əl }

isanomalous line [METEOROL] A line drawn through geographical points having equal anomaly of some meteorological quantity. Also known as isanomal. { ‚ī·sə'näm·ə·ləs ¦līn }

isano oil [MATER] A pale-yellow, viscous drying oil obtained from the nut of an African tree; used as a varnish oil. { i'sä·nō ‚óil }

isanthous [BOT] Having regular flowers. { ī'san·thəs }

isarithm See isopleth. { 'ī·sə‚rith·əm }

isatin [ORG CHEM] $C_6H_5NO_2$ An indole substituted with oxygen at carbon position 2 and 3; crystallizes as red needles that are soluble in hot water; used in dye manufacture. { 'ī·sə·tən }

isaurore See isochasm. { 'ī·sə‚rór }

ISB modulation See independent-sideband modulation. { 'ī¦es¦bē ‚mäj·ə'lā·shən }

I scan See I scope. { 'ī ‚skan }

ischemia [MED] Localized tissue anemia as a result of obstruction of the blood supply or to vasoconstriction. { i'skē·mē·ə }

ischemic necrosis [MED] Local tissue death due to impaired blood supply. { i'skē·mik ne'krō·səs }

ischemic neuropathy [MED] Nerve lesions characterized by numbness, tingling, and pain with loss of sensory and motor functions of the parts involved, due to obstruction of the blood supply to the nerves. { i'skē·mik nu'räp·ə·thē }

ischemic paralysis [MED] Impaired motor function due to obstructed circulation to the area. { i'skē·mik pə'ral·ə·səs }

ischemic tubulorrhexis See lower nephron nephrosis. { i'skē·mik ‚tü·byə·lə'rek·səs }

ischiopodite [INV ZOO] The segment nearest the basipodite of walking legs in certain crustaceans. Also known as ischium. { ‚is·kē'äp·ə‚dīt }

ischiorectal region [ANAT] The region between the ischium and the rectum. { ¦is·kē·ə'rek·təl ‚rē·jən }

ischium [ANAT] Either of a pair of bones forming the dorso-posterior portion of the vertebrate pelvis; the inferior part of the human pelvis upon which the body rests in sitting. [INV ZOO] See ischiopodite. { 'is·kē·əm }

ischium-pubis index [ANAT] The ratio (length of pubis ×

100/length of ischium) by which the sex of an adult pelvis may usually be determined; the index is greater than 90 in females, and less than 90 in males. { 'is·kē·əm 'pyü·bəs ‚in‚deks }

Ischnacanthidae [PALEON] The single family of the acanthodian order Ischnacanthiformes. { ‚isk·nə'kan·thə‚dē }

Ischnacanthiformes [PALEON] A monofamilial order of extinct fishes of the order Acanthodii; members were slender, lightly armored predators with sharp teeth, deeply inserted fin spines, and two dorsal fins. { ‚isk·nə‚kan·thə'fór‚mēz }

I scope [ELECTR] A cathode-ray scope on which a single signal appears as a circular segment whose radius is proportional to the range and whose circular length is inversely proportional to the error of aiming the antenna, true aim resulting in a complete circle; the position of the arc, relative to the center, indicates the position of the target relative to the beam axis. Also known as broken circle indicator; I indicator; I scan. { 'ī ‚skōp }

ISDN See integrated services digital network.

ISDN modem [ELECTR] A device that converts signals used in a computer to signals that can be transmitted over the integrated services digital network, and vice versa. { ‚ī‚es‚dē‚en 'mō‚dem }

Isectolophidae [PALEON] A family of extinct ceratomorph mammals in the superfamily Tapiroidea. { ī‚sek·tə'läf·ə‚dē }

isenergic flow [THERMO] Fluid flow in which the sum of the kinetic energy, potential energy, and enthalpy of any part of the fluid does not change as that part is carried along with the fluid. { ī·sə‚nər·jik 'flō }

isenthalpic expansion [THERMO] Expansion which takes place without any change in enthalpy. { ¦īs·ən¦thal·mik ik 'span·chən }

isenthalpic process [THERMO] A process that is carried out at constant enthalpy. { ¦ī·sən¦thal·pik 'prä‚ses }

isentrope [THERMO] A line of equal or constant entropy. { 'īs·ən‚trōp }

isentropic [THERMO] Having constant entropy; at constant entropy. { ¦īs·ən'träp·ik }

isentropic chart [METEOROL] A constant-entropy chart; a synoptic chart presenting the distribution of meteorological elements in the atmosphere on a surface of constant potential temperature (equivalent to an isentropic surface); it usually contains the plotted data and analysis of such elements as pressure (or height), wind, temperature, and moisture at that surface. { ¦īs·ən'träp·ik 'chärt }

isentropic compression [THERMO] Compression which occurs without any change in entropy. { ¦īs·ən'träp·ik kəm'presh·ən }

isentropic condensation level See lifting condensation level. { ¦īs·ən'träp·ik ‚känd·ən'sā·shən ‚lev·əl }

isentropic expansion [THERMO] Expansion which occurs without any change in entropy. { ¦īs·ən'träp·ik ik'span·chən }

isentropic flow [THERMO] Fluid flow in which the entropy of any part of the fluid does not change as that part is carried along with the fluid. { ¦īs·ən'träp·ik 'flō }

isentropic map [GEOL] A map indicating constant entropy function for facies. { ¦īs·ən'träp·ik 'map }

isentropic mixing [METEOROL] Any atmospheric mixing process which occurs within an isentropic surface; the fact that many atmospheric motions are reversible adiabatic processes renders this type of mixing important, and exchange coefficients have been computed therefor. { ¦īs·ən'träp·ik 'mik·siŋ }

isentropic process [THERMO] A change that takes place without any increase or decrease in entropy, such as a process which is both reversible and adiabatic. { ¦īs·ən'träp·ik 'prä‚ses }

isentropic surface [METEOROL] A surface in space in which potential temperature is everywhere equal. { ¦īs·ən'träp·ik 'sər·fəs }

isentropic thickness chart [METEOROL] A thickness chart of an atmospheric layer bounded by two selected isentropic surfaces (surfaces of constant potential temperature); the thickness of such a layer is directly proportional to the static instability of that layer; hence, these charts have been called instability charts. Also known as thick-thin chart. { ¦īs·ən'träp·ik 'thik·nəs ‚chärt }

isentropic weight chart [METEOROL] A chart of atmospheric pressure difference between two selected isentropic surfaces (surfaces of constant potential temperature); the greater the pressure difference the greater the weight of the air column separating the two surfaces. { ¦īs·ən'träp·ik 'wāt ‚chärt }

isethionic acid [ORG CHEM] $CH_2OH·CH·SO_2OH$ A water-soluble liquid, boiling at 100°C; used in the manufacture of detergents. { ¦īs·ə·thī'än·ik 'as·əd }

ISFET See ion-selective field-effect transistor. { 'is‚fet }

ishikawaite [MINERAL] A black, orthorhombic mineral consisting essentially of uranium, iron, rare earth, and columbium oxide. { ‚ish·ē'kä·wə‚īt }

ishkyldite [MINERAL] $Mg_{15}Si_{11}O_{27}(OH)_{20}$ A mineral composed of a basic silicate of magnesium. { 'ish·kəl‚dīt }

I signal [ELECTR] The in-phase component of the chrominance signal in color television, having a bandwidth of 0 to 1.5 megahertz, and consisting of $+0.74(R − Y)$ and $−0.27(B − Y)$, where Y is the luminance signal, R is the red camera signal, and B is the blue camera signal. { 'ī ‚sig·nəl }

Ising coupling [SOLID STATE] A model of coupling between two atoms in a lattice, used to study ferromagnetism, in which the spin component of each atom along some axis is taken to be +1 or −1, and the energy of interaction is proportional to the negative of the product of the spin components along this axis. { 'ī·ziŋ ‚kəp·liŋ }

isinglass [MATER] A gelatin made from the dried swim bladders of sturgeon and other fishes; used in glues, cements, and printing inks. Also known as fish gelatin; ichthyocolla. [MINERAL] Sheet mica, usually in the form of single cleavage plates; used in furnace and stove doors. { 'īz·ən‚glas }

Ising model [SOLID STATE] A crude model of a ferromagnetic material or an analogous system, used to study phase transitions, in which atoms in a one-, two-, or three-dimensional lattice interact via Ising coupling between nearest neighbors, and the spin components of the atoms are coupled to a uniform magnetic field. { 'ī·ziŋ ‚mäd·əl }

island [GEOGR] A tract of land smaller than a continent and surrounded by water; normally in an ocean, sea, lake, or stream. { 'ī·lənd }

island arc [GEOGR] A group of volcanic islands, usually situated in a curving arch-like pattern that is convex toward the open ocean, having a deep trench or trough on the convex side and usually enclosing a deep basin on the concave side; formed by volcanic activity associated with oceanic plate subduction at convergent plate margins. Also known as volcanic arc. { 'ī·lənd ‚ärk }

island mountain See inselberg. { 'ī·lənd ¦maùnt·ən }

island of automation [IND ENG] A single robotic system or other automatically operating machine that functions independently of any other machine or process. { 'ī·lənd əv ‚öd·ə'mā·shən }

island of Langerhans See islet of Langerhans. { 'ī·lənd əv 'läŋ·gər‚hänz }

island of Reil [ANAT] The insula of the cerebral hemisphere. { 'ī·lənd əv 'rīl }

islet-cell carcinoma [MED] A metastatic tumor of pancreatic cells of the islet of Langerhans in the pancreas. { 'ī·lət ¦sel ‚kärs·ən'ō·mə }

islet-cell tumor [MED] A benign tumor of the pancreatic islet cells. Also known as insulinoma; insuloma; Langerhansian adenoma. { 'ī·lət ¦sel 'tü·mər }

islet of Langerhans [HISTOL] A mass of cell cords in the pancreas that is of an endocrine nature, secreting insulin and a minor hormone like lipocaic. Also known as island of Langerhans; islet of the pancreas. { 'ī·lət əv 'läŋ·gər‚hänz }

islet of the pancreas See islet of Langerhans. { 'ī·lət əv thə 'pan·krē·əs }

iso- [CHEM] A prefix indicating an isomer of an element in which there is a difference in the nucleus when compared to the most prevalent form of the element. [ORG CHEM] A prefix indicating a single branching at the end of the carbon chain. { 'ī·sō }

isoacceptor [MOL BIO] Any of several species of transfer ribonucleic acid that can accept the same amino acid. { 'ī·sō·ak'sep·tər }

isoactyl thioglycolate [ORG CHEM] $HSCH_2COOCH_2 C_7H_{15}$ A colorless liquid with a slight fruity odor and a boiling point of 125°C; used in antioxidants, insecticides, oil additives, and plasticizers. { 'ī·sō¦akt·əl ‚thī·ə'glī·kə‚lāt }

isoagglutinin [IMMUNOL] An agglutinin which acts upon the red blood cells of members of the same species. Also known as isohemagglutinin. { 'ī·sō·ə'glut·ən·ən }

isoalkane [ORG CHEM] An alkane with a branched chain whose next-to-last carbon atom is bonded to a single methyl group. { 'ī·sō'al‚kān }

isoalkyl group [ORG CHEM] A group of atoms resulting from the removal of a hydrogen atom from a methyl group situated at the end of the straight-chain segment of an isoalkane. { 'ī·sō'al·kəl ‚grüp }

isoallele [GEN] An allele whose phenotype is indistinguishable from that of a different mutant allele at the same locus. { 'ī·sō·ə'lēl }

isoalloxazine mononucleotide See riboflavin 5′-phosphate. { ‚ī·sō·ə'läk·sə‚zēn ‚mänō'nü·klē·ə‚tīd }

isoamyl acetate See amyl acetate. { 'ī·sō¦am·əl 'as·ə‚tāt }

isoamyl alcohol See isobutyl carbinol. { 'ī·sō¦am·əl 'al·kə‚hól }

isoamyl benzoate [ORG CHEM] $C_6H_5COOC_5H_{11}$ Colorless liquid with fruity aroma; boils at 260°C; soluble in alcohol, insoluble in water; used in flavors and perfumes. Also known as amyl benzoate. { 'ī·sō¦am·əl 'ben·zə‚wāt }

isoamyl bromide [ORG CHEM] $(CH_3)_2CHCH_2CH_2Br$ A colorless liquid with a boiling point of 120–121°C; miscible with alcohol and with ether; used in organic synthesis. { 'ī·sō¦am·əl 'brō‚mīd }

isoamyl butyrate [ORG CHEM] $C_5H_{11}COOC_3H_7$ A water-white liquid boiling at 150–180°C; soluble in alcohol and ether; used as a solvent and plasticizer for cellulose acetate and in flavor extracts. { 'ī·sō¦am·əl 'byüd·ə‚rāt }

isoamyl chloride [ORG CHEM] $C_5H_{11}Cl$ Water-insoluble, colorless liquid boiling at 100°C; it can be any one of several compounds, such as 1-chloro-3-methylbutane, $(CH_3)_2CH-(CH_2)_2Cl$, or mixtures thereof; used as a solvent, in inks, for soil fumigation, and as a chemical intermediate. { 'ī·sō¦am·əl 'klór‚īd }

isoamyl nitrite See amyl nitrite. { 'ī·sō¦am·əl 'nī‚trīt }

isoamyl salicylate See amyl salicylate. { 'ī·sō¦am·əl sə'lis·ə‚lāt }

isoamyl valerate [ORG CHEM] $C_4H_9CO_2C_5H_{11}$ Clear liquid with apple aroma; boils at 204°C; soluble in alcohol and ether, insoluble in water; used in medicine and fruit flavors. { 'ī·sō¦am·əl 'val·ə‚rāt }

isoantibody [IMMUNOL] An antibody formed in response to immunization with tissue constituents derived from an individual of the same species. { 'ī‚sō'ant·i‚bäd·ē }

isoantigen [IMMUNOL] An antigen in an individual capable of stimulating production of a specific antibody in another member of the same species. Also known as alloantigen. { 'ī·sō'ant·ə·jən }

isobar [METEOROL] A line drawn through all points of equal atmospheric pressure along a given reference surface, such as a constant-height surface (notably mean sea level on surface charts), an isentropic surface, or the vertical plan of a synoptic cross section. [NUC PHYS] One of two or more nuclides having the same number of nucleons in their nuclei but differing in their atomic numbers and chemical properties. [PHYS] **1.** A line connecting points of equal pressure along a given surface in a physical system. **2.** A line connecting points of equal pressure on a graph plotting thermodynamic variables. { 'ī·sə‚bär }

isobaric [THERMO] Of equal or constant pressure, with respect to either space or time. { ¦ī·sə'bär·ik }

isobaric analog states See analog states. { ¦ī·sə'bär·ik 'an·ə‚läg ‚stāts }

isobaric chart See constant-pressure chart. { ¦ī·sə'bär·ik 'chärt }

isobaric contour chart See constant-pressure chart. { ¦ī·sə'bär·ik 'kän·túr ‚chärt }

isobaric divergence [METEOROL] The horizontal divergence in a constant-pressure surface; expressed in a system of coordinates with pressure as an independent variable. { ¦ī·sə'bär·ik də'vər·jəns }

isobaric equivalent temperature See equivalent temperature. { ¦ī·sə'bär·ik i¦kwiv·ə·lənt 'tem·prə·chər }

isobaric map [METEOROL] A map depicting points in the atmosphere of equal barometric pressure. { ¦ī·sə'bär·ik 'map }

isobaric process [THERMO] A thermodynamic process of a gas in which the heat transfer to or from the gaseous system causes a volume change at constant pressure. { ¦ī·sə'bär·ik 'prä·səs }

isobaric spin See isotopic spin. { ¦ī·sə'bär·ik 'spin }

isobaric surface [METEOROL] A surface on which the pressure is uniform. Also known as constant-pressure surface. { ¦ī·sə¦bär·ik 'sər·fəs }

isobaric topography See height pattern. { ¦ī·sə¦bär·ik tə'päg·rə·fē }

isobaric vorticity [METEOROL] Relative vorticity in a constant-pressure surface, that is, expressed in a system of coordinates with pressure as an independent variable. { ¦ī·sə¦bär·ik vör'tis·əd·ē }

isobath [OCEANOGR] A contour line connecting points of equal water depths on a chart. Also known as depth contour; depth curve; fathom curve. { 'ī·sə,bath }

isobathytherm [OCEANOGR] A line or surface showing the depth in oceans or lakes at which points have the same temperatures. { ¦ī·sə'bath·ə,thərm }

isobiochore [ECOL] A boundary line on a map connecting world environments that have similar floral and faunal constituents. { ¦ī·sə'bī·ə,kór }

isobits [COMPUT SCI] Binary digits having the same value. { 'ī·sə,bits }

isobornyl acetate [ORG CHEM] $C_{10}H_{17}OOCCH_3$ A colorless liquid with an odor of pine needles and a boiling point of 220–224°C; soluble in fixed oils and mineral oil; used in toiletries and soaps and antiseptics, and as a flavoring agent. { ¦ī·sə'bórn·əl 'as·ə,tāt }

isobornyl thiocyanoacetate [ORG CHEM] $C_{10}H_{17}OOCCH_2SCN$ An oily, yellow liquid; soluble in alcohol, benzene, chloroform, and ether; used in medicine and as an insecticide. { ¦ī·sə'bórn·əl ¦thī·ə¦sī·ə·nō'as·ə,tāt }

isobront [METEOROL] A line drawn through geographical points at which a given phase of thunderstorm activity occurred simultaneously. Also known as homobront. { 'ī·sə,bränt }

Isobryales [BOT] An order of mosses in which the plants are slender to robust and up to 36 inches (90 centimeters) in length. { ¦ī·sō·brī'ā·lēz }

isobutane [ORG CHEM] $(CH_3)_2CHCH_3$ A colorless, stable gas, noncorrosive to metals, nonreactive with water; boils at −11.7°C; used as a chemical intermediate, refrigerant, and fuel. { ¦ī·sō'byü,tān }

isobutanol See isobutyl alcohol. { ¦ī·sō'byüt·ən,ól }

isobutene See isobutylene. { ¦ī·sō'byü,tēn }

isobutyl [ORG CHEM] The radical $(CH_3)_2CHCH_2-$, occurring, for example, in isobutanol (isobutyl alcohol), $(CH_3)_2CHCH_2OH$. { ¦ī·sō'byüd·əl }

isobutyl acetate [ORG CHEM] $C_4H_9OOCCH_3$ Colorless liquid with fruitlike aroma; soluble in alcohols, ether, and hydrocarbons, insoluble in water; boils at 116°C; used as a solvent for lacquer and nitrocellulose. { ¦ī·sō'byüd·əl 'as·ə,tāt }

isobutyl alcohol [ORG CHEM] $(CH_3)_2CHCH_2OH$ A colorless liquid that is a by-product of the synthetic production of methanol, boils at 107°C; soluble in water, ether, and alcohol; used as a solvent in paints and lacquers, in organic synthesis, and in resin coatings. Also known as isobutanol; isopropylcarbinol; 2-methyl-1-propanol. { ¦ī·sō'byüd·əl 'al·kə,hól }

isobutyl aldehyde [ORG CHEM] $(CH_3)_2CHCHO$ Colorless, transparent liquid with pungent aroma; soluble in alcohol, insoluble in water; boils at 64°C; used as a chemical intermediate. Also known as isobutyraldehyde. { ¦ī·sō'byüd·əl 'al·də,hīd }

isobutyl carbinol [ORG CHEM] $(CH_3)_2CH(CH_2)_2OH$ Colorless liquid with pungent taste and disagreeable aroma; soluble in alcohol and ether, slightly soluble in water; boils at 132°C; used as a chemical intermediate and solvent, and in pharmaceutical products and medicines. Also known as isoamyl alcohol. { ¦ī·sō'byüd·əl 'kär·bə,nól }

isobutylene [ORG CHEM] $(CH_3)_2CCH_2$ Flammable, colorless, volatile liquid boiling at −7°C; easily polymerized; used in gasolines, as a chemical intermediate, and to make butyl rubber. Also known as isobutene. { ¦ī·sō'byüd·əl,ēn }

isobutyl isobutyrate [ORG CHEM] $(CH_3)_2CHCOOCH_2CH(CH_3)_2$ A colorless liquid with a fruity odor and a boiling point of 148.7°C; soluble in alcohol and ether; used for flavoring and as an insect repellent. Abbreviated IBIB. { ¦ī·sō'byüd·əl ¦ī·sō'byüd·ə,rāt }

isobutyraldehyde See isobutyl aldehyde. { ¦ī·sō,byüd·ə'ral·də,hīd }

isobutyric acid [ORG CHEM] $(CH_3)_2CHCOOH$ Colorless liquid boiling at 154°C; soluble in water, alcohol, and ether; used as a chemical intermediate and disinfectant, in flavor and perfume bases, and for leather treating. { ¦ī·sō·byü'tir·ik 'as·əd }

isobutyryl [ORG CHEM] $(CH_3)_2C·CHO$ The radical group from isobutyric acid, $(CH_3)_2CHCOOH$. { ¦ī·sō'byüd·ə·rəl }

isocandle diagram [OPTICS] A diagram showing the distribution of light from a lighting system in various directions by means of contours connecting directions of equal luminous intensity, projected in a suitable manner. { ¦ī·sə¦kan·dəl 'dī·ə,gram }

isocarb [GEOCHEM] A line on a map that connects points of equal content of fixed carbon in coal. { 'ī·sə,kärb }

isocarpic [BOT] Having the same number of carpels and perianth divisions. { ,ī·sə'kär·pik }

isocenter [GRAPHICS] The unique point common to the principal plane of a tilted photograph and the plane of an assumed truly vertical photograph taken from the same camera station and having an equal principal distance. { 'ī·sə,sen·tər }

isoceraunic [METEOROL] Indicating or having equal frequency or intensity of thunderstorm activity. Also spelled isokeraunic. { ¦ī·sə·sə'rón·ik }

isoceraunic line [METEOROL] A line drawn through geographical points at which some phenomenon connected with thunderstorms has the same frequency or intensity; used for lines of equal frequency of lightning discharges. { ¦ī·sō·sə'rón·ik 'līn }

isocercal [VERT ZOO] Of the tail fin of a fish, having the upper and lower lobes symmetrical and the vertebral column gradually tapering. { ¦ī·sə'sər·kəl }

isocetyl laurate [ORG CHEM] $C_{11}H_{23}COOC_{16}H_{33}$ An oily, combustible liquid, soluble in most organic solvents; used in cosmetics and pharmaceuticals and as a plasticizer and textile softener. { ¦ī·sə¦sēd·əl 'ló,āt }

isochasm [GEOPHYS] A line connecting points on the earth's surface at which the aurora is observed with equal frequency. Also known as isaurore. { 'ī·sə,kaz·əm }

isochela [INV ZOO] **1.** A chela having two equally developed parts. **2.** A chelate spicule with both ends identical. { ¦ī·sə'kē·lə }

isochemical metamorphism [PETR] Theoretically, a metamorphism involving no great change in its chemical composition. Also known as treptomorphism. { ¦ī·sō'kem·ə·kəl ,med·ə'mór,fiz·əm }

isochemical series [PETR] A series of rocks with identical chemical compositions. { ¦ī·sō'kem·ə·kəl 'sir·ēz }

isochor See isochore. { 'ī·sə,kór }

isochore [GEN] In mammals and birds, an organizational unit of chromosome DNA that is usually characterized by a fairly constant proportion of guanine-cytosine base pairs throughout its length (usually 200–1000 kilobase pairs long); genes are concentrated in the G-C rich isochores. [PHYS] A graph that shows the variation of one quantity with another; for example, the variation of pressure with temperature, when the volume of the substance is held constant. Also known as isochor; isometric. { 'ī·sə,kór }

isochoric [PHYS] Taking place without change in volume. Also known as isovolumic. { ¦ī·sə'kór·ik }

isochromatic [OPTICS] **1.** Pertaining to a variation of certain quantities related to light (such as density of the medium through which the light is passing, index of refraction), in which the color or wavelength of the light is held constant. **2.** Pertaining to lines connecting points of the same color. { ¦ī·sō·krə'mad·ik }

isochromatic fringe pattern [OPTICS] A pattern of bands, each of uniform color, observed when a plate is placed in a polariscope and subjected to stress, making it birefringent. { ¦ī·sō·krə'mad·ik 'frinj ,pad·ərn }

isochromosome [CELL MOL] An abnormal chromosome with a medial centromere and identical arms formed as a result of transverse, rather than longitudinal, splitting of the centromere. { ¦ī·sō'krō·mə,sōm }

isochronal test [PETRO ENG] Short-time back-pressure test for low-permeability reservoirs that otherwise require excessively long times for pressure stabilization when wells are shut in. { ī'sä·krən·əl 'test }

isochron [GEOCHEM] A line on a graph defined by data for rocks of the same age with the same initial lead isotopic composition, the slope of which is proportional to the age. Also known as geochron. { 'ī·sə krän }

isochrone [MATH] *See* semicubical parabola. [PHYS] A line on a chart connecting all points having the same time of occurrence of particular phenomena or of a particular value of a quantity. { 'ī·sə,krōn }

isochronism [MECH] The property of having a uniform rate of operation or periodicity, for example, of a pendulum or watch balance. { ī'sä·krə,niz·əm }

isochronon [HOROL] A clock designed to keep very accurate time. { ī·sə'krō,nän }

isochronous [PHYS] Having a fixed frequency or period. { ī'sä·krə·nəs }

isochronous circuits [ELEC] Circuits having the same resonant frequency. { ī'sä·krə·nəs 'sər·kəts }

isochronous communications [COMMUN] Synchronization of a data communications network from timing signals provided by the network itself. { ī'sä·krə·nəs kə,myü·nə'kā·shənz }

isochronous curve [MATH] A curve with the property that the time for a particle to reach a lowest point on the curve if it starts from rest and slides without friction does not depend on the particle's starting point. { ī'sä·krə·nəs 'kərv }

isochronous governor [MECH ENG] A governor that keeps the speed of a prime mover constant at all loads. Also known as astatic governor. { ī'sä·krə·nəs 'gəv·ər·nər }

isocirculator [ELECTROMAG] A circulator that has an absorber in one of its terminals and thereby acts as an isolator. { ī·sō'sər·kyə,lād·ər }

isocitric acid [BIOCHEM] HOOCCH₂CH(COOH)CH(OH)COOH An isomer of citric acid that is involved in the Krebs tricarboxylic acid cycle in bacteria and plants. { ī·sə'si·trik 'as·əd }

isoclasite [MINERAL] Ca₂(PO₄)(OH)·2H₂O A white mineral composed of a basic hydrous calcium phosphate; occurring in small crystals or columnar forms. { ī·sə'klā,sīt }

isoclinal *See* isoclinic line. { ī·sə'klīn·əl }

isoclinal chart [GEOPHYS] A chart showing isoclinic lines. Also known as isoclinic chart. { ī·sə'klīn·əl 'chärt }

isocline [GEOL] A fold of strata so tightly compressed that parts on each side dip in the same direction. { 'ī·sə,klīn }

isoclinic chart *See* isoclinal chart. { ī·sə'klin·ik 'chärt }

isoclinic line [GEOPHYS] A line connecting points on the earth's surface which have the same magnetic dip. Also known as isoclinal. [SOLID STATE] A line joining points in a plate at which the principal stresses have parallel directions. { ī·sə'klin·ik 'līn }

isocoding mutation [GEN] A point mutation that changes a codon's nucleotide sequence to one that codes for the same amino acid specified by the initial codon. { ī·sə,kōd·iŋ myü'tā·shən }

isoconcentration [CHEM ENG] Constant concentration values. { ī·sō,käns·ən'trā·shən }

isoconcentration map [CHEM ENG] Map or diagram of a liquid or gas system's concentration with respect to a single component of the system, shown by constant-concentration contour lines. { ī·sō,käns·ən'trā·shən ,map }

isocrackate [MATER] The liquid products of the process of isocracking. { ī·sō'kra,kāt }

isocracking [CHEM ENG] A hydrocracking process for conversion of hydrocarbons into more valuable, lower-boiling products; operates at relatively low temperatures and pressures in the presence of hydrogen and a catalyst. { ī·sō'krak·iŋ }

isocrinida [INV ZOO] An order of stalked articulate echinoderms with nodal rings of cirri. { ¦ī·sō'krī·nə·də }

isocyanate [ORG CHEM] 1. One of a group of neutral derivatives of primary amines; its formula is R−N=C=O, where R may be an alkyl or aryl group; an example is 2,4-toluene diisocyanate. 2. Any compound containing the isocyanato functional group. { ī·sō'sī·ə,nāt }

isocyanate resin [ORG CHEM] A linear alkyd resin lengthened by reaction with isocyanates, then treated with a glycol or diamine to cross-link the molecular chain; the product has good abrasion resistance. { ī·sō'sī·ə,nāt 'rez·ən }

isocyanato group [ORG CHEM] A functional group (−N=C−O) which forms isocyanates by replacing the hydrogen atom of a hydrocarbon. { ī·sō·sī'an·ə,tō ,grüp }

isocyanic acid [ORG CHEM] HN=C=O One of two forms of cyanic acid; a gas used as an intermediate in the preparation of polyurethane and other resins. { ī·sō·sī'an·ik 'as·əd }

isocyanide [ORG CHEM] A compound with the general formula RN≡C in which the hydrogen of a hydrocarbon has been replaced by the −N≡C group. { ī·sō'sī·ə,nīd }

isocyanine [ORG CHEM] Any one of a series of dyes whose structure has two heterocyclic or quinoline rings connected by an odd number chain of carbon atoms containing conjugated double bonds; for example, cyanine blue. { ī·sō'sī·ə,nēn }

isocyanuric acid *See* fulminuric acid. { ī·sō,sī·ə¦nür·ik 'as·əd }

isocyclic compound [ORG CHEM] A compound in which the ring structure is made up of one kind of atom. { ¦ī·sō'sī·klik 'käm,paùnd }

isodecyl chloride [ORG CHEM] C₁₀H₂₁Cl A colorless liquid with a boiling point of 210.6°C; used as a solvent and in extractants, cleaning compounds, pharmaceuticals, insecticides, and plasticizers. { ¦ī·sə·əl 'klȯr,īd }

isodesmic structure [SOLID STATE] An ionic crystal structure in which all bonds are of the same strength, so that no distinct groups of atoms are formed. { ¦ī·sə¦dez·mik 'strək·chər }

isodiametric [BIOL] Having equal diameters or dimensions. { ¦ī·sō,dī·ə'me·trik }

isodiapheres [NUC PHYS] Nuclides which have the same difference in the number of neutrons and protons. { ¦ī·sə'dī·ə,firz }

isodisomy [GEN] A type of uniparental disomy in which two copies of the same chromosome are inherited from one parent, with resultant homozygosity at all gene loci on the chromosome. { ¦ī·sə,dī'sō·mē }

isodisperse [CHEM] 1. Having dispersed particles, of colloidal dimensions, that are all of the same size. 2. Dispersible in solutions with the same pH value. { ,īs·ə·di'spərs }

isodont [VERT ZOO] 1. Having all teeth alike. 2. Of a snake, having the maxillary teeth of equal length. { 'ī·sə,dänt }

isodose curve [NUCLEO] A curve, drawn on a chart of an object, connecting points receiving equal doses of radiation. { 'ī·sə,dōs ,kərv }

isodrosotherm [METEOROL] An isogram of dew-point temperature. { ,ī·sə'dräs·ə,thərm }

isodulcitol *See* rhamnose. { ,ī·sə'dəl·sə,tȯl }

isodynamic [MECH] Pertaining to equality of two or more forces or to constancy of a force. { ¦ī·sō'dī'nam·ik }

isodynamic line [GEOPHYS] One of the lines on a map of a magnetic field that connect points having equal strengths of the earth's field. { ¦ī·sō·dī'nam·ik 'līn }

isoelectric [ELEC] Pertaining to a constant electric potential. { ¦ī·sō·i'lek·trik }

isoelectric focusing [PHYS CHEM] Protein separation technique in which a mixture of protein molecules is resolved into its components by subjecting the mixture to an electric field in a supporting gel having a previously established pH gradient. Also known as electrofocusing. { ¦ī·sō·i'lek·trik 'fō·kəs·iŋ }

isoelectric point [PHYS CHEM] The pH value of the dispersion medium of a colloidal suspension at which the colloidal particles do not move in an electric field. { ¦ī·sō·i'lek·trik 'pȯint }

isoelectric precipitation [CHEM] Precipitation of materials at the isoelectric point (the pH at which the net charge on a molecule in solution is zero); proteins coagulate best at this point. { ¦ī·sō·i'lek·trik prə,sip·ə'tā·shən }

isoelectronic [ATOM PHYS] Pertaining to atoms having the same number of electrons outside the nucleus of the atom. { ¦ī·sō·i,lek'trän·ik }

isoelectronic principle [CHEM] The concept that molecules having the same number of electrons and the same number of atoms whose atomic masses are greater than that of hydrogen (heavy atoms) tend to have similar electronic structures, similar chemical properties, and heavy-atom geometries. { ¦ī·sō·i,lek'trän·ik 'prin·sə·pəl }

isoelectronic sequence [SPECT] A set of spectra produced by different chemical elements ionized so that their atoms or ions contain the same number of electrons. { ¦ī·sō·i,lek'trän·ik 'sē·kwəns }

isoenzyme [BIOCHEM] Any of the electrophoretically distinct forms of an enzyme, representing different polymeric states but having the same function. Also known as isozyme. { ¦ī·sō'en,zīm }

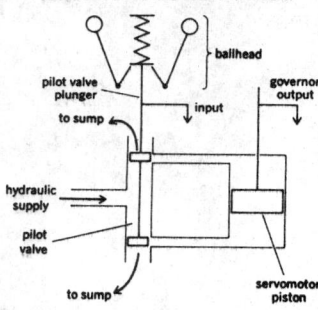

ISOCHRONOUS GOVERNOR

ballhead
pilot valve plunger
governor output
input
to sump
hydraulic supply
pilot valve
to sump
servomotor piston

Diagram of isochronous governor.

ISOCITRIC ACID

CH₂COOH
|
CHCOOH
|
CHOHCOOH

Structural formula of isocitric acid.

ISOELECTRIC POINT

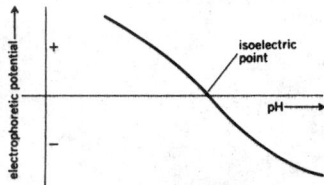

electrophoretic potential
isoelectric point
pH

Graph showing the isoelectric point where particles are electrophoretically inert.

Isoetaceae [BOT] The single family assigned to the order Isoetales in some systems of classification. { ‚ī·sō·ə'tās·ē‚ē }

Isoetales [BOT] A monotypic order of the class Isoetopsida containing the single genus *Isoetes*, characterized by long, narrow leaves with a spoonlike base, spirally arranged on an underground cormlike structure. { ‚ī·sō·ə'tā·lēz }

Isoetatae *See* Isoetopsida. { ‚ī·sō'ed·ə‚tē }

Isoetopsida [BOT] A class of the division Lycopodiophyta; members are heterosporous and have a distinctive appendage, the ligule, on the upper side of the leaf near the base. { ‚ī·sō·ə'täp·sə·də }

isoeugenol [ORG CHEM] $C_{10}H_{12}O_2$ An oily liquid prepared from eugenol by heating, slightly soluble in water; used in the manufacture of vanillin. { ‚ī·sō'yü·jə‚nȯl }

isofacies map [GEOL] A stratigraphic map showing the distribution of one or more facies within a particular stratigraphic unit. { 'ī·sə‚fā·shēz ‚map }

isoflavone [BIOCHEM] $C_{15}H_{10}O_2$ A colorless, crystalline ketone, occurring in many plants, generally in the form of a hydroxy derivative. { 'ī·sō'fla‚vōn }

isoflurophate [PHARM] $[(CH_3)_2CHO]_2P(O)F$ A liquid that forms hydrofluoric acid in the presence of moisture: used as a cholinergic drug in eye diseases in humans and as a miotic drug in animals. { ‚ī·sə'flùr·ə‚fāt }

isofootcandle *See* isolux. { 'ī·sō'fùt‚kand·əl }

isoforming [CHEM ENG] A petroleum refinery process in which olefinic naphtha is contacted with an alumina catalyst at high temperature and low pressure to produce isomers of higher octane number. { 'ī·sə‚fȯr·miŋ }

isofronts-preiso code [METEOROL] A code in which data on isobars and fronts at sea level (or earth's surface) are encoded and transmitted; a modified form of the international analysis code. { 'ī·sə‚frəns ‚prē'ī·sō ‚kōd }

isogal [GEOPHYS] A contour line on a map connecting points of equal gravity values on the earth's surface. { 'ī·sə‚gal }

isogam [GEOPHYS] A line joining points on the earth's surface having the same value of the acceleration of gravity. { 'ī·sə‚gam }

isogamete [BIOL] A reproductive cell that is morphologically similar in both male and female and cannot be distinguished on form alone. { 'ī·sō'ga‚mēt }

isogamy [BIOL] Sexual reproduction by union of gametes or individuals of similar form or structure. { ī'säg·ə·mē }

isogenic [GEN] Having the same genotype, as all organisms of an inbred strain. [IMMUNOL] Referring to cells, tissues, or organs used in transplantation that originate in identical species. { 'ī·sə·jə'nē·ik }

isogeneratae [BOT] A class of brown algae distinguished by having an isomorphic alternation of generations. { ‚ī·sō‚jen·ə'rä‚tē }

isogeotherm *See* geoisotherm. { 'ī·sō'jē·ə‚thərm }

isogonal conjugates *See* isogonal lines. { ī'säg·ən·əl 'kän·jə·gəts }

isogonal lines [MATH] Lines that pass through the vertex of an angle and make equal angles with the bisector of the angle. Also known as isogonal conjugates. { ī'säg·ən·əl 'līnz }

isogonal transformation [MATH] A mapping of the plane into itself which leaves the magnitudes of angles between intersecting lines unchanged but may reverse their sense. { ī'säg·ən·əl ‚tranz·fər'mā·shən }

isogonic line [GEOPHYS] **1.** Any of the lines on a chart or map showing the same direction of the wind vector. **2.** Any of the lines on a chart or map connecting points of equal magnetic variation. { 'ī·sə‚gän·ik ‚līn }

isogony [BIOL] Growth of parts at such a rate as to maintain relative size differences. { ī'säg·ə·nē }

isogor [PETRO ENG] Constant gas-oil ratio. { 'ī·sə‚gȯr }

isogor map [PETRO ENG] Oil reservoir contour-line map that shows constant gas-oil ratios. { 'ī·sə‚gȯr ‚map }

isograd [GEOL] A line on a map joining those rocks comprising the same metamorphic grade. { 'ī·sə‚grad }

isogradient [METEOROL] A line connecting points having the same horizontal gradient of atmospheric pressure, temperature, and so on. { 'ī·sə'grād·ē·ənt }

isograft [BIOL] A tissue transplant from one organism to another organism which is genetically identical. { 'ī·sə‚graft }

isogram *See* isopleth. { 'ī·sə‚gram }

isograph [ELECTR] An electronic calculator that ascertains

both real and imaginary roots for algebraic equations. [GRAPHICS] An instrument combining the functions of a protractor and a set square and comprising two short straightedges connected by a large circular joint with an angular-degree scale. { 'ī·sə‚graph }

isogriv [NAV] A line drawn on a map or chart joining points of equal grivation. { 'ī·sə‚griv }

isogriv chart [NAV] A chart showing isogrivs. { 'ī·sə‚griv ‚chärt }

isogyre [OPTICS] A dark band in an interference figure located at those points that correspond to directions of transmission through the crystal plate in which the polarization of the incident light is not affected by passing through the plate. { 'ī·sə'jīr }

isohaline [OCEANOGR] **1.** Of equal or constant salinity. **2.** A line on a chart connecting all points of equal salinity. { ‚ī·sō'hā‚lēn }

isoheight *See* contour line. { 'ī·sə‚hīt }

isohel [METEOROL] A line drawn through geographical points having the same duration of sunshine (or other function of solar radiation) during any specified time period. { 'ī·sō‚hel }

isohemagglutinin *See* isoagglutinin. { 'ī·sō‚hē·mə'glüt·ən·ən }

isohemolysin [IMMUNOL] A hemolysin produced by an individual injected with erythrocytes from another individual of the same species. { 'ī·sō·hə'mäl·ə·sən }

isohemolysis [IMMUNOL] Hemolysis induced by the action of an isohemolysin. { 'ī·sō·hə'mäl·ə·səs }

isohexane [ORG CHEM] C_6H_{14} A liquid mixture of isomeric hydrocarbons, flammable and explosive, insoluble in water, soluble in most organic solvents, boils at 54–61°C; used as a solvent, freezing-point depressant, and chemical intermediate. { 'ī·sō'hek‚sān }

isohume [GEOL] A line of a map or chart connecting points of equal moisture content in a coal bed. [METEOROL] A line drawn through points of equal humidity on a given surface; an isopleth of humidity; the humidity measures used may be the relative humidity or the actual moisture content (specific humidity or mixing ratio). { 'ī·sə‚hyüm }

isohydric [CHEM] Referring to a set of solutions with the same hydrogen ion concentration and not affecting the conductivity of each of the various solutions on mixing. { 'ī·sə'hī·drik }

isohyet [METEOROL] A line drawn through geographic points recording equal amounts of precipitation for a specified period or for a particular storm. { 'ī·sə'hī·ət }

isohypse *See* contour line. { 'ī·sə‚hips }

isohypsic chart *See* constant-height chart. { ‚ī·sə'hip·sik ‚chärt }

isohypsic surface *See* constant-height surface. { ‚ī·sə'hip·sik 'sər·fəs }

isoimmunization [IMMUNOL] Immunization of an individual by the introduction of antigens from another individual of the same species. { 'ī·sō‚im·yə·nə'zā·shən }

isoinertial [BIOPHYS] Pertaining to the force of a human muscle that is applied to a constant mass in motion. { 'ī·sō·i'nərsh·əl }

isokeraunic *See* isoceraunic. { 'ī·sō·kə'rȯn·ik }

isokinetic [BIOPHYS] Pertaining to the force of a human muscle that is applied during constant velocity of motion. { ‚ī·sə·ki'ned·ik }

isokinetic relationship [PHYS CHEM] A linear relationship that exists between the enthalpies and entropies of activation of a series of related reactions. { ‚ī·sə·ki'ned·ik ri'lā·shən‚ship }

isokinetic sampling [ENG] Any technique for collecting airborne particulate matter in which the collector is so designed that the airstream entering it has a velocity equal to that of the air passing around and outside the collector. { ‚ī·sə·ki'ned·ik 'sam·pliŋ }

isokinetic temperature [PHYS CHEM] The actual or virtual temperature at which rates of all members of a series of related reactions are equal. { ‚ī·sə·ki'ned·ik 'tem·prə·chər }

Isolaimoidea [INV ZOO] A superfamily of rather large, free-living soil nematodes in the order Isolaimida, characterized by six hollow tubes around the oral opening, two circlets of six circumoral sensilla, the absence of amphids, and an elongated triradiate stoma with thickened anterior walls. { ‚ī·sō·lə'mȯid·ē·ə }

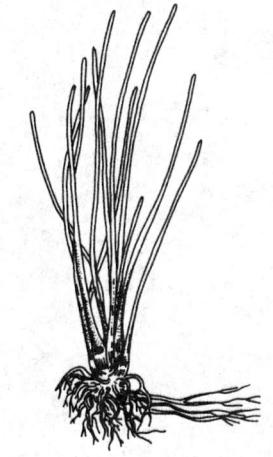

ISOETALES

Isoetes, the entire plant. *(From A. J. Eames, in E. W. Sinnott and K. S. Wilson, Botany: Principles and Problems, 6th ed., McGraw-Hill, 1963)*

isolate [CHEM ENG] To separate two portions of a process system by means of valving or line blanks; used as safety measure during maintenance or repair, or to redirect process flows. [ELEC] To disconnect a circuit or piece of equipment from an electric supply system. [GEN] A population so cut off from others that mating occurs only within the group. { 'ī·sə‚lāt }

isolated camera [ELECTR] **1.** A television camera that views a particular portion of a scene of action and produces a tape which can then be used either immediately for instant replay or for video replay at a later time. **2.** The technique of video replay involving such a camera. { 'ī·sə‚lād·əd 'kam·rə }

isolated footing [CIV ENG] A concrete slab or block under an individual load or column. { 'ī·sə‚lād·əd 'füd·iŋ }

isolated location [COMPUT SCI] A location in a computer memory which is protected by some hardware device so that it cannot be addressed by a computer program and its contents cannot be accidentally altered. { 'ī·sə‚lād·əd lō'kā·shən }

isolated point [MATH] **1.** A point p in a topological space is an isolated point of a set if p is in the set and there is a neighborhood of p which contains no other points of the set. **2.** A point that satisfies the equation for a plane curve C but has a neighborhood that includes no other point of C. Also known as acnode; hermit point. { 'ī·sə‚lād·əd 'point }

isolated set [MATH] A set consisting entirely of isolated points. { 'ī·sə‚lād·əd 'set }

isolated subgroup [MATH] An isolated subgroup of a totally ordered Abelian group G is a subgroup of G which is also a segment of G. { ‚īs·ə‚lād·əd 'səb‚grüp }

isolated system See closed system. { 'ī·sə‚lād·əd 'sis·təm }

isolated vertex [MATH] A vertex of a graph that has no edges incident to it. { 'ī·sə‚lād·əd 'vər‚teks }

isolating mechanism [GEN] A geographic barrier or biological difference that prevents mating or genetic exchange between individuals of different populations or species. { 'ī·sə‚lād·iŋ mek·ə‚niz·əm }

isolating switch [ELEC] A switch intended for isolating an electric circuit from the source of power; it has no interrupting rating and is intended to be operated only after the circuit has been opened by some other means. { 'ī·sə‚lād·iŋ ‚swich }

isolation [CHEM] Separation of a pure chemical substance from a compound or mixture; as in distillation, precipitation, or absorption. [COMPUT SCI] The ability of a logic circuit having more than one input to ensure that each input signal is not affected by any of the others. [EVOL] The restriction or limitation of gene flow between distinct populations due to barriers to interbreeding. [MED] Separation of an individual with a communicable disease from other, healthy individuals. [MICROBIO] Separation of an individual or strain from a natural, mixed population. [PHYSIO] Separation of a tissue, organ, system, or other part of the body for purposes of study. [PSYCH] Dissociation of a memory or thought from the emotions or feelings associated with it. { ‚ī·sə'lā·shən }

isolation amplifier [ELECTR] An amplifier used to minimize the effects of a following circuit on the preceding circuit. { ‚ī·sə'lā·shən 'am·plə‚fī·ər }

isolation diode [ELECTR] A diode used in a circuit to allow signals to pass in only one direction. { ‚ī·sə'lā·shən 'dī‚ōd }

isolation network [ELEC] A network inserted in a circuit or transmission line to prevent interaction between circuits on each side of the insertion point. { ‚ī·sə'lā·shən 'net‚wərk }

isolation test [ENG] A leak detection method which isolates the evacuated system from the pump, followed by observation of the rate of pressure rise. { ‚ī·sə'lā·shən ‚test }

isolation transformer [ELEC] A transformer inserted in a system to separate one section of the system from undesired influences of other sections. { ‚ī·sə'lā·shən tranz'fór·mər }

isolator [ELECTR] A passive attenuator in which the loss in one direction is much greater than that in the opposite direction; a ferrite isolator for waveguides is an example. [ENG] Any device that absorbs vibration or noise, or prevents its transmission. { 'ī·sə‚lād·ər }

isolead curve [ORD] A curved line, on a chart or diagram, used to show how far ahead of a moving target a weapon must be aimed to allow for the time the projectile takes to reach the target; the isolead curve connects points of equal lead on the chart or diagram. { 'ī·sə‚lēd ‚kərv }

isolecithal See homolecithal. { ‚ī·sō'les·ə·thəl }

isoleucine [BIOCHEM] $C_6H_{13}O_2$ An essential monocarboxylic amino acid occurring in most dietary proteins. { ‚ī·sō'lü‚sēn }

isolith [ELECTR] Integrated circuit of components formed on a single silicon slice, but with the various components interconnected by beam leads and with circuit parts isolated by removal of the silicon between them. [GEOL] A line on a contour-type map that denotes the aggregate thickness of a single lithology in a stratigraphic succession composed of one or more lithologies. { 'ī·sə‚lith }

isolith map [GEOL] A contour-line map depicting the thickness of an exclusive lithology. { 'ī·sə‚lith ‚map }

isolux [OPTICS] A curve or surface connecting points at which light intensity is the same. Also known as isofootcandle; isophot. { 'ī·sə‚ləks }

isomagnetic [GEOPHYS] Of or pertaining to lines connecting points of equality in some magnetic element. { ‚ī·sō·mag'ned·ik }

isomer [CHEM] One of two or more chemical substances having the same elementary percentage composition and molecular weight but differing in structure, and therefore in properties; there are many ways in which such structural differences occur; one example is provided by the compounds n-butane, $CH_3(CH_2)_2CH_3$, and isobutane, $CH_3CH(CH_3)_2$. [NUC PHYS] One of two or more nuclides having the same mass number and atomic number, but existing for measurable times in different quantum states with different energies and radioactive properties. Also known as nuclear isomer. { 'ī·sə·mər }

isomerase [BIOCHEM] An enzyme that catalyzes isomerization reactions. { ī'säm·ə‚rās }

isomeric shift [PHYS CHEM] Shift in the Mössbauer resonance caused by the effect of the valence of the atom on the interaction of the electron density at the nucleus with the nuclear charge. Also known as chemical shift. { ‚ī·sə‚mer·ik 'shift }

isomeric transition [NUC PHYS] A radioactive transition from one nuclear isomer to another of lower energy. { ‚ī·sə‚mer·ik tran'zish·ən }

isomerism [BIOL] The condition of having two or more comparable parts made up of identical numbers of similar segments. [CHEM] The phenomenon whereby certain chemical compounds have structures that are different although the compounds possess the same elemental composition. [NUC PHYS] The occurrence of nuclear isomers. { ī'säm·ə‚riz·əm }

isomerization [CHEM] A process whereby a compound is changed into an isomer; for example, conversion of butane into isobutane. { ī‚säm·ə·rə'zā·shən }

isomerous [BIOL] Characterized by isomerism. { ī'säm·ə·rəs }

Isometopidae [INV ZOO] A family of hemipteran insects in the superfamily Cimicimorpha. { ‚ī·sō·mə'täp·ə‚dē }

isometric See isochore. { ‚ī·sə'me·trik }

isometric contraction [PHYSIO] A contraction in which muscle tension is increased, but the muscle is not shortened because the resistance cannot be overcome. Also known as static contraction. { ‚ī·sə'me·trik kən'trak·shən }

isometric drawing [GRAPHICS] A method of nonperspective pictorial drawing in which the object being drawn is turned so that three mutually perpendicular edges are equally foreshortened. { ‚ī·sə'me·trik 'dró·iŋ }

isometric forms [MATH] Two bilinear forms f and g on vector spaces E and F for which there exists a linear isomorphism of E onto F such that $f(x,y)=g(\sigma x,\sigma y)$ for all x and y in E. { ‚ī·sə'me·trik 'fórmz }

isometric particle [VIROL] A plant virus particle that appears at first sight to be spherical when viewed in the electron microscope, but which is actually an icosahedron, possessing 20 sides. { ‚ī·sə'me·trik 'pärd·ə·kəl }

isometric process [THERMO] A constant-volume, frictionless thermodynamic process in which the system is confined by mechanically rigid boundaries. { ‚ī·sə'me·trik 'prä·səs }

isometric projection See axonometric projection. { ‚ī·sə'me·trik prə'jek·shən }

isometric spaces [MATH] Two spaces between which an isometry exists. { ‚ī·sə'me·trik 'spā·səs }

isometric system [CRYSTAL] The crystal system in which the forms are referred to three equal, mutually perpendicular axes. Also known as cubic system. { ‚ī·sə'me·trik 'sis·təm }

isometric work [BIOPHYS] Physiologic work that is performed by muscles in terms of energy utilization and heat production and involves muscular contraction that is not accompanied by movement. Also known as static work. { 'ī·sə'me·trik 'wərk }

isometry [MATH] **1.** A mapping f from a metric space X to a metric space Y where the distance between any two points of X equals the distance between their images under f in Y. **2.** A linear isomorphism σ of a vector space E onto itself such that, for a given bilinear form g, $g(\sigma x, \sigma y) = g(x,y)$ for all x and y in E. { ī'säm·ə·trē }

isometry class [MATH] A set consisting of all bilinear forms (on vector spaces over a given field) which are isometric to a given form. { ī'säm·ə·trē ˌklas }

isomolecule See nonlinear molecule. { ˈī·sō'mäl·ə·kyül }

isomorph See isomorphic mineral. { 'ī·sə·mȯrf }

isomorphic mineral [MINERAL] Any two or more crystalline mineral compounds having different chemical composition but identical structure, such as the garnet series or the feldspar group. Also known as isomorph. { ˈī·sə'mȯr·fik 'min·rəl }

isomorphic systems [MATH] Two algebraic structures between which an isomorphism exists. { ˌī·sə'mȯr·fik 'sis·təmz }

isomorphism [MATH] A one to one function of an algebraic structure (for example, group, ring, module, vector space) onto another of the same type, preserving all algebraic relations; its inverse function behaves likewise. [PHYS CHEM] A condition present when an ion at high dilution is incorporated by mixed crystal formation into a precipitate, even though such formation would not be predicted on the basis of crystallographic and ionic radii; an example is coprecipitation of lead with potassium chloride. [SCI TECH] The quality or state of being identical or similar in form, shape, or structure, such as between organisms resulting from evolutionary convergence, or crystalline forms of similar composition. { ˈī·sə'mȯr·fiz·əm }

isomorphism problem [MATH] For two simple graphs with the same numbers of vertices and edges, the problem of determining whether there exist correspondences between these vertices and edges such that there is an edge between two vertices in one graph if and only if there is an edge between the corresponding vertices in the other. { ˌī·sə'mȯr·fiz·əm ˌpräb·ləm }

isoneph [METEOROL] A line drawn through all points on a map having the same amount of cloudiness. { 'ī·sə·nef }

isoniazid [PHARM] $C_6H_7N_3O$ A drug used as a tuberculostatic. Also known as isonicotinic acid hydrazide. { ˌī·sə'nī·ə·zəd }

isonicotinic acid [ORG CHEM] $C_6H_5NO_2$ White platelets or powder, slightly soluble in water, sublimes at 260°C; used in the manufacture of isonicotinic acid hydrazide, an antitubercular agent. { ˈī·sə·nik·ə'tin·ik 'as·əd }

isonicotinic acid hydrazide See isoniazid. { ˈī·sə·nik·ə'tin·ik ˌas·əd 'hī·drə·zīd }

isonitrosoacetophenone [ORG CHEM] $C_8H_7NO_2$ Platelike crystals with a melting point of 126–128°C; soluble in alkalies and alkali carbonates; used to detect ferrous ions and palladium. { ˈī·sə·nī'trō·sō·as·ə'täf·ə·nōn }

isonymous substitution [GEN] Any deoxyribonucleic acid base-pair substitution within an exon that does not result in change in the amino acid sequence for which it codes. Also known as synonymous substitution. { ī'sän·ə·məs ˌsəb·stə'tü·shən }

isooctane [ORG CHEM] $(CH_3)_2CHCH_2C(CH_3)_3$ Flammable, colorless liquid boiling at 99°C; slightly soluble in alcohol and ether, insoluble in water; used in motor fuels and as a chemical intermediate. { ˈī·sō'äk·tān }

isooctyl alcohol [ORG CHEM] $C_7H_{15}CH_2OH$ Mixture of isomers from oxo-process synthesis; boils at 182–195°C; used as a chemical intermediate, resin solvent, emulsifier, and antifoaming agent. { ˈī·sō'äkt·əl 'al·kə·hȯl }

isopach map [GEOL] Map of the areal extent and thickness variation of a stratigraphic unit; used in geological exploration for oil and for underground structural analysis. { 'ī·sə·pak ˌmap }

isopachous line [GEOL] One of the lines drawn on a map to indicate equal thickness. { ˈī·sō'pak·əs ˌlīn }

isoparaffin [ORG CHEM] A branched-chain version of a straight-chain (normal) saturated hydrocarbon; for example, isooctane, or 2,2,4-trimethyl pentane, $(CH_3)_3C_5H_9$, is the branched-chain version of n-octane, $CH_3(CH_2)_6CH_3$. { ˈī·sō'par·ə·fən }

isopathic principle [PSYCH] The apparently paradoxical rule according to which the cause cures the effect, as when a feeling of guilt is relieved by an exhibition of guilt, namely hate. { ˌī·sə'path·ik 'prin·sə·pəl }

isopectic [CLIMATOL] A line on a map connecting points at which ice begins to form at the same time of winter. { ˌī·sə'pek·tik }

isopentane [ORG CHEM] $CH_3CHCH_3CH_2CH_3$ Flammable, colorless liquid with pleasant aroma; boils at 28°C; soluble in oils, ether, and hydrocarbons, insoluble in water; used as a solvent and chemical intermediate. Also known as 2-methylbutane. { ˈī·sō'pen·tān }

isopentanoic acid [ORG CHEM] C_4H_9COOH A colorless, combustible liquid with a boiling point of 183.2°C; used for manufacture of plasticizers, pharmaceuticals, and synthetic lubricants. { ˈī·sə·pen·tə'nō·ik 'as·əd }

isopentyl unit See isoprene unit. { ˌī·sə'pent·əl ˌyü·nət }

isoperimetric figures [MATH] Figures whose perimeters are equal. { ˈī·sō·per·ə'me·trik 'fig·yərz }

isoperimetric inequality [MATH] The statement that the area enclosed by a plane curve is equal to or less than the square of its perimeter divided by 4π. { ˈī·sə·per·ə'me·trik ˌin·i'kwäl·əd·ē }

isoperimetric line [MAP] A line on a map projection that indicates no variation from exact scale. { ˈī·sō·per·ə'me·trik 'līn }

isoperimetric problem [MATH] In the calculus of variations this problem deals with finding a closed curve in the plane which encloses the greatest area given its length as fixed. { ˈī·sō·per·ə'me·trik 'präb·ləm }

isoperm [PETRO ENG] One of the lines of equal (constant) permeability plotted on a reservoir map. { 'ī·sə·pərm }

isophene [BIOL] A line on a chart connecting those places within a given region where a particular biological phenomenon (as the flowering of a certain plant) occurs at the same time. { 'ī·sə·fēn }

isophorone [ORG CHEM] $COCHC(CH_3)CH_2C(CH_3)_2CH_2$ A water-white liquid boiling at 215°C; used as a solvent for lacquers and polyvinyl and nitrocellulose resins. { ˈī·sə'fȯ·rōn }

isophot See isolux. { 'ī·sə·fät }

isophotometer [OPTICS] A direct-recording photometer that automatically scans and measures optical density of all points in a film transparency or plate, and plots the measured density values in a quantitative two-dimensional isodensity tracing of the scanned areas. { ˈī·sō·fə'täm·əd·ər }

isophthalic acid [ORG CHEM] $C_5H_4(COOH)_2$ Colorless crystals subliming at 345°C; slightly soluble in water, soluble in alcohol and acetic acid, and insoluble in benzene; used as an intermediate for polyester and polyurethane resins, and as a plasticizer. Also known as meta-phthalic acid. { ˈī·sō'thal·ik 'as·əd }

isophyllous [BOT] Having foliage leaves of similar form on a plant or stem. { ˈī·sə'fil·əs }

isopiestic [PHYS] Denoting equal or constant pressure. { ˈī·sə'pī·es·tik }

isopiestic line [HYD] A line indicating on a map the piezometric surface of an aquifer. { ˈī·sə'pī·es·tik ˌlīn }

isopleth [MATH] The straight line which cuts the three scales of a nomograph at values satisfying some equation. Also known as index line. [METEOROL] **1.** A line of equal or constant value of a given quantity with respect to either space or time. Also known as isogram. **2.** More specifically, a line drawn through points on a graph at which a given quantity has the same numerical value (or occurs with the same frequency) as a function of the two coordinate variables. Also known as isarithm. { 'ī·sə·pleth }

isopluvial [METEOROL] A line on a map drawn through geographical points having the same pluvial index. { ˈī·sō'plü·vē·əl }

Isopoda [INV ZOO] An order of malacostracan crustaceans characterized by a cephalon bearing one pair of maxillipeds in addition to the antennae, mandibles, and maxillae. { ī'säp·ə·də }

isopolymolybdate [INORG CHEM] A class of compounds formed by the acidification of a molybdate solution, or in some cases by heating normal molybdates. { ˈī·sō·päl·i·mə'lib·dāt }

isopolytungstate [INORG CHEM] A compound formed by the condensation of tungstate compounds, usually classified into metatungstates, such as $Na_6W_{12}O_{40} \cdot xH_2O$, and paratungstates, such as $Na_{10}W_{12}O_{41} \cdot xH_2O$. { ¦ī·sō¸päl·i'təŋ, stāt }

isopor [GEOPHYS] An imaginary line connecting points on the earth's surface having the same annual change in a magnetic element. { 'ī·sə¸pȯr }

isopotential level See potentiometric surface. { ¦ī·sō·pə'ten·chəl 'lev·əl }

isopotential map [PETRO ENG] A contour-line map to show the initial or calculated daily rate of oil well production in a multiwell field. { ¦ī·sō·pə'ten·chəl 'map }

isoprecipitin [IMMUNOL] A precipitin effective only against the serum of individuals of the same species from which it is derived. { ¦ī·sō·prə'sip·ə·tən }

isoprene [ORG CHEM] C_5H_8 A conjugated diolefin; a mobile, colorless liquid having a boiling point of 34.1°C; insoluble in water, soluble in alcohol and ether; polymerizes readily to form dimers and high-molecular-weight elastomer resins. { 'ī·sə¸prēn }

isoprene unit [ORG CHEM] The five-carbon structural unit characteristic of terpenes. Also known as isopentyl unit. { 'ī·sə¸prēn ¸yü·nət }

isoprenoid See terpene. { ¸ī·sə·'prē¸nȯid }

isopropaline [ORG CHEM] $C_{15}H_{23}N_3O_4$ An orange liquid with limited solubility in water; used as a preemergence herbicide for control of grass and broadleaf weeds on tobacco. { ¦ī·sə'prō·pə¸lēn }

isopropanol See isopropyl alcohol. { ¦ī·sə'prō·pə¸nȯl }

isopropanolamine [ORG CHEM] $CH_3CH(OH)CH_2NH_2$ A combustible liquid with a faint ammonia odor and a boiling point of 159.9°C; soluble in water; used as an emulsifying agent and for dry-cleaning soaps, wax removers, cosmetics, plasticizers, and insecticides. { ¦ī·sə¸prō·pə'nal·ə¸mēn }

isopropenyl acetate [ORG CHEM] $CH_3CO_2C(CH_3)=CH_2$ A liquid with a boiling point of 97°C; used for acylation of potential enols. { ¸ī·sə'prō·pə·nəl 'as·ə¸tāt }

2-isopropoxyphenyl *N*-methylcarbamate [ORG CHEM] $C_{11}H_{15}O_3N$ A colorless solid with a melting point of 91°C; used as an insecticide for cockroaches, flies, mosquitoes, and lawn insects. { ¦tü ¦ī·sō·prə¦päk·sē'fen·əl ¦en ¸meth·əl'kär·bə¸māt }

isopropyl [ORG CHEM] The radical $(CH_3)_2CH$, from isopropane; an example of its occurrence is in isopropyl alcohol, $(CH_3)_2CHOH$. { ¦ī·sə'prō·pəl }

isopropyl acetate [ORG CHEM] $CH_3COOCH(CH_3)_2$ A colorless, aromatic liquid with a boiling point of 89.4°C; used as a solvent and for paints and printing inks. { ¦ī·sə'prō·pəl 'as·ə¸tāt }

isopropyl alcohol [ORG CHEM] $(CH_3)_2CHOH$ A colorless liquid that boils at 82.4°C; soluble in water, ether, and ethanol; used in manufacturing of acetone and its derivatives, of glycerol, and as a solvent. Also known as isopropanol; 2-propanol; *sec*-propyl alcohol. { ¦ī·sə'prō·pəl 'al·kə¸hȯl }

isopropylamine [ORG CHEM] $(CH_3)_2CHNH_2$ A volatile, colorless liquid with a boiling point of 32.4°C; used as a solvent and in the manufacture of pharmaceuticals, dyes, insecticides, and bactericides. Also known as 2-aminopropane. { ¦ī·sə·prō'pil·ə¸mēn }

isopropyl - 2 - (*N*-benzoyl-3-chloro-4-fluoroanilino)propionate [ORG CHEM] $C_{19}H_{19}O_3NClF$ Off-white crystals with a melting point of 56–57°C; used as a postemergence herbicide for wild oats and barley. { ¦ī·sə'prō·pəl 'tü ¦en ¦ben·zə·wəl ¦thrē ¦klȯr·ō ¦fȯr ¸flùr·ō¦an·ə·lō 'prō·pē·ə¸nāt }

isopropyl 4,4′-dibromobenzilate [ORG CHEM] $C_{17}H_{16}O_3Br_2$ A brownish solid with a melting point of 77°C; solubility in water is less than 0.5 part per million at 20°C; used as a miticide for deciduous fruit and citrus. { ¦ī·sə'prō·pəl ¦fȯr ¦fȯr¸prīm dī¸brō·mō'ben·zə¸lāt }

isopropyl 4,4′-dichlorobenzilate [ORG CHEM] $C_{17}H_{16}O_3Cl_2$ A white powder with a melting point of 70–72°C; solubility in water is less than 10 parts per million at 20°C; used as a miticide for spider mites on apple and pear trees. { ¦ī·sə'prō·pəl ¦fȯr ¦fȯr¸prīm dī¸klȯr·ō'ben·zə¸lāt }

isopropyl ether [ORG CHEM] $(CH_3)_2CHOCH(CH_3)_2$ Water-soluble, flammable, colorless liquid with etherlike aroma; boils at 68°C; used as a solvent and extractant, in paint and varnish removers, and in spotting formulas. Also known as diisopropyl ether. { ¦ī·sə'prō·pəl 'ē·thər }

N-4-isopropylphenyl-*N*′,*N*′-dimethylurea [ORG CHEM] $(CH_3)_2CHC_6H_4NHCON(CH_3)_2$ A crystalline solid with a

melting point of 151–153°C; solubility in water is 170 parts per million; used as an herbicide for wheat, barley, and rye. { ¦en ¦fȯr ¦ī·sə¦prō·pəl¦fen·əl ¸en¸prīm ¸en¸prīm dī¸meth·əl·yù'rē·ə }

***ortho*-isopropylphenyl-methylcarbamate** [ORG CHEM] $C_{11}H_{15}O_2N$ A white, crystalline compound with a melting point of 88–89°C; used as an insecticide for rice and cacao crops. Also known as MIPC. { ¦ȯr·thō ¦ī·sə¦prō·pəl¦fen·əl ¸meth·əl'kär·bə¸māt }

isoprostanes [BIOCHEM] A class of natural products that are isomeric with prostaglandins but are formed in vivo by the nonenzymatic, free-radical oxygenation of arachidonic acid. { ¸ī·sō'prä¸stānz }

isoproterenol [PHARM] $C_{11}H_{17}NO_3$ A sympathomimetic amine used as a bronchodilator. { ¸ī·sə¸prōd·ə'rē¸nȯl }

Isoptera [INV ZOO] An order of Insecta containing morphologically primitive forms characterized by gradual metamorphosis, lack of true larval and pupal stages, biting and prognathous mouthparts, two pairs of subequal wings, and the abdomen joined broadly to the thorax. { ī'säp·tə·rə }

isoptic [MATH] The locus of the intersection of tangents to a given curve that meet at a specified constant angle. { ī'säp·tik }

isopulegol [ORG CHEM] $C_{10}H_{17}OH$ An alcohol derived from terpene as a water-white liquid that has a mintlike odor; used in making perfumes. { ¸ī·sō'pyü·lə¸gȯl }

isopulse system [COMMUN] In adaptive communications, a pulse coding system wherein the number of information pulses transmitted is indicated by special inserted pulses. { 'ī·sə¸pəls ¸sis·təm }

isopycnic [METEOROL] A line on a chart connecting all points of equal or constant density. [PHYS] Of equal or constant density, with respect to either space or time. { ¦ī·sō¦pik·nik }

isopycnic level [METEOROL] Specifically, a level surface in the atmosphere, at about 5 miles (8 kilometers) altitude, where the air density is approximately constant in space and time; this level corresponds to the maximum upper-tropospheric interdiurnal pressure variation. { ¦ī·sō¦pik·nik ¸lev·əl }

isopygous [INV ZOO] Having a pygidium and a cephalon of equal size, as in certain trilobites. { ī'säp·ə·gəs }

isoquinoline [ORG CHEM] $C_6H_4CHNCHCH$ Colorless liquid boiling at 243°C; soluble in most organic solvents and dilute mineral acids, insoluble in water; derived from coal tar or made synthetically; used to make dyes, insecticides, pharmaceuticals, and rubber accelerators, and as a chemical intermediate. { ¦ī·sə'kwin·ə¸lēn }

isosafrole [ORG CHEM] $C_{10}H_{10}O_2$ A liquid with the odor of anise that is obtained from safrole, and that boils at 253°C; used to make perfumes and flavors. { ¦ī·sō'sa¸frōl }

isosbestic point [PHYS CHEM] During a chemical reaction, a point in the absorption spectrum (that is, a wavelength) where at least two chemical species (for example, reactant and product) have identical molar absorption coefficients, which remain constant as the reaction proceeds. A stable isosbestic point is evidence that a reaction is proceeding without forming an intermediate or multiple products. { ¸ī·səs¸bes·tik 'point }

isosceles spherical triangle [MATH] A spherical triangle that has two equal sides. { ī'säs·ə¸lēz ¦sfer·ə·kəl 'trī¸aŋ·gəl }

isosceles triangle [MATH] A triangle with two sides of equal length. { ī'säs·ə¸lēz 'trī¸aŋ·gəl }

isoschizomer [BIOCHEM] One of two or more restriction endonucleases that cleave a deoxyribonucleic acid molecule at the same site. { ¸ī·sə'siz·ə·mər }

isoseismal [GEOPHYS] Pertaining to points having equal intensity of earthquake shock, or to a line on a map of the earth's surface connecting such points. { ¦ī·sə'sīz·məl }

isoshear [METEOROL] A line on a chart of equal magnitude of vertical wind shear. { 'ī·sə¸shir }

isospin See isotopic spin. { 'ī·sə¸spin }

isospin multiplet [PARTIC PHYS] A collection of elementary particles which have approximately the same mass and the same quantum numbers except for charge, but have a sequence of charge values, $(Y/2) - I$, $(Y/2) - I + 1, \ldots, (Y/2) + I$ times the proton charge, where Y is an integer known as the hypercharge, and I is an integer or half-integer known as the isospin; examples are the pions ($Y = 0, I = 1$) and the nucleons ($Y = 1, I = 1/2$). Also known as charge multiplet; particle multiplet. { 'ī·sə¸spin 'məl·tə·plət }

ISOPRENE

Structural formula of isoprene.

ISOPRENE UNIT

Structure of the isoprene unit.

Isospondyli [VERT ZOO] A former equivalent name for Clupeiformes. { ¦ī·sə'spän·də¸lī }

isospore [BIOL] A spore that does not display sexual dimorphism. { 'ī·sə¸spór }

isosporiasis [MED] Infection with coccidia of the genus *Isospora*. { ¸ī·sə·spə'rī·ə·səs }

isostasy [GEOPHYS] A theory of the condition of approximate equilibrium in the outer part of the earth, such that the gravitational effect of masses extending above the surface of the geoid in continental areas is approximately counterbalanced by a deficiency of density in the material beneath those masses, while deficiency of density in ocean waters is counterbalanced by an excess in density of the material under the oceans. { ī'säs·tə¸sē }

isostatic adjustment *See* isostatic compensation. { ¦ī·sə'stad·ik ə'jəs·mənt }

isostatic anomaly [GEOPHYS] A gravity anomaly based on a generalized hypothesis that the gravitational effect of masses above sea level is approximately compensated by a density deficiency of the subsurface materials. { ¦ī·sə'stad·ik ə'näm·ə·lē }

isostatic compacting [MET] In powder metallurgy, a process in which pressure from a gas or liquid is applied uniformly to a metal powder contained in a flexible mold. { ¸ī·sō¦stad·ik 'käm¸pak·tiŋ }

isostatic compensation [GEOL] The process in which lateral transport at the surface of the earth by erosion or deposition is compensated by lateral movements in a subcrustal layer. Also known as isostatic adjustment; isostatic correction. { ¦ī·sə'stad·ik ¸käm·pən'sā·shən }

isostatic correction *See* isostatic compensation. { ¦ī·sə'stad·ik kə'rek·shən }

isostatics [MECH] In photoelasticity studies of stress analyses, those curves, the tangents to which represent the progressive change in principal-plane directions. Also known as stress trajectories. Also known as stress lines. { ¦ī·sə'stad·iks }

isostatic surface [MECH] A surface in a three-dimensional elastic body such that at each point of the surface one of the principal planes of stress at that point is tangent to the surface. { ¦ī·sə'stad·ik 'sər¸fəs }

isostemonous [BOT] Having the number of stamens of a flower equal to the number of perianth divisions. { ¦ī·sə¦stē·mə·nəs }

isosteric [CHEM] Referring to similar electronic arrangements in chemical compounds. [PHYS] Of equal or constant specific volume with respect to either time or space. { ¦ī·sə¦ster·ik }

isosterism [PHYS CHEM] A similarity in the physical properties of ions, compounds, or elements, as a result of electron arrangements that are identical or similar. { ī'säs·tə¸riz·əm }

isostructural [CRYSTAL] Pertaining to crystalline materials that have corresponding atomic positions, and have a considerable tendency for ionic substitution. { ¦ī·sō'strək·chə·rəl }

isosulfur map [PETRO ENG] A contour-line map to show the percentage of sulfur in underground crude oil. { ¦ī·sō'səl·fər ¸map }

isosynthesis [ORG CHEM] A process in which mixtures of hydrogen and carbon monoxide are reacted over a thorium oxide catalyst (sometimes mixed with additional substances) to produce branched hydrocarbons. { ¦ī·sō'sin·thə·səs }

isotach [METEOROL] A line in a given surface connecting points with equal wind speed. Also known as isokinetic; isovel. { 'ī·sə¸tak }

isotach chart [METEOROL] A synoptic chart showing the distribution of wind by means of isotachs. { 'ī·sə¸tak ¸chärt }

isotachophoresis [PHYS CHEM] A variant of electrophoresis in which ionic species move with equal velocity in the presence of an electric field. { ¦ī·sə¸tak·ə·fə'rē·səs }

isotactic [ORG CHEM] Designating crystalline polymers in which substituents in the asymmetric carbon atoms have the same (rather than random) configuration in relation to the main chain. { ¦ī·sə¦tak·tik }

isoteniscope [ENG] An instrument for measuring the vapor pressure of a liquid, consisting of a U tube containing the liquid, one arm of which connects with a closed vessel containing the same liquid, while the other connects with a pressure gage where the pressure is adjusted until the levels in the arms of the U tube are equal. { ¸ī·sə'ten·ə¸skōp }

isothere [CLIMATOL] A line on a map connecting points having the same mean summer temperature. { 'ī·sə¸thir }

isotherm [GEOPHYS] A line on a chart connecting all points of equal or constant temperature. [THERMO] A curve or formula showing the relationship between two variables, such as pressure and volume, when the temperature is held constant. Also known as isothermal. { 'ī·sə¸thərm }

isothermal [THERMO] **1.** Having constant temperature; at constant temperature. **2.** *See* isotherm. { ¦ī·sə¦thər·məl }

isothermal annealing [MET] Transformation of an austenitic steel to ferrite and pearlite at constant temperature. Also known as isothermal transformation. { ¦ī·sə¦thər·məl ə'nēl·iŋ }

isothermal atmosphere [METEOROL] An atmosphere in hydrostatic equilibrium, in which the temperature is constant with height and the pressure decreases exponentially upward. Also known as exponential atmosphere. { ¦ī·sə¦thər·məl 'at·mə¸sfir }

isothermal calorimeter [THERMO] A calorimeter in which the heat received by a reservoir, containing a liquid in equilibrium with its solid at the melting point or with its vapor at the boiling point, is determined by the change in volume of the liquid. { ¦ī·sə¦thər·məl ¸kal·ə'rim·əd·ər }

isothermal chart [GEOPHYS] A map showing the distribution of air temperature (or sometimes sea-surface or soil temperature) over a portion of the earth or at some level in the atmosphere; places of equal temperature are connected by lines called isotherms. { ¦ī·sə¦thər·məl 'chärt }

isothermal compression [THERMO] Compression at constant temperature. { ¦ī·sə¦thər·məl kəm'presh·ən }

isothermal equilibrium [METEOROL] The state of an atmosphere at rest, uninfluenced by any external agency, in which the conduction of heat from one part to another has produced, after a sufficient length of time, a uniform temperature throughout its entire mass. Also known as conductive equilibrium. [THERMO] The condition in which two or more systems are at the same temperature, so that no heat flows between them. { ¦ī·sə¦thər·məl ¸ē·kwə'lib·rē·əm }

isothermal expansion [THERMO] Expansion of a substance while its temperature is held constant. { ¦ī·sə¦thər·məl ik'span·chən }

isothermal flow [THERMO] Flow of a gas in which its temperature does not change. { ¦ī·sə¦thər·məl 'flō }

isothermal layer [METEOROL] The approximately isothermal region of the atmosphere immediately above the tropopause. [THERMO] A layer of fluid, all points of which have the same temperature. { ¦ī·sə¦thər·məl 'lā·ər }

isothermal magnetization [THERMO] Magnetization of a substance held at constant temperature; used in combination with adiabatic demagnetization to produce temperatures close to absolute zero. { ¦ī·sə¦thər·məl ¸mag·nə·tə'zā·shən }

isothermal process [THERMO] Any constant-temperature process, such as expansion or compression of a gas, accompanied by heat addition or removal from the system at a rate just adequate to maintain the constant temperature. { ¦ī·sə¦thər·məl 'prä·səs }

isothermal remanent magnetization [GEOPHYS] A spurious magnetization induced by lightning strikes that produce large surface electrical currents. Abbreviated IRM. { ¦ī·sə¦thər·məl ¦rem·ə·nənt ¸mag·nə·tə'zā·shən }

isothermal titration calorimeter [BIOPHYS] An instrument that directly measures the energetics (through heat effects) associated with biochemical reactions or processes occurring at constant temperatures. { ¦ī·sə¦thər·məl tī¸trā·shən ¸kal·ə'rim·əd·ər }

isothermal transformation [MET] *See* isothermal annealing. [THERMO] Any transformation of a substance which takes place at a constant temperature. { ¦ī·sə¦thər·məl ¸tranz·fər'mā·shən }

isothermal treatment [MET] Heat treatment of metals at constant temperature. { ¦ī·sə¦thər·məl 'trēt·mənt }

isothermobath [OCEANOGR] A line connecting points having the same temperature in a diagram of a vertical section of the ocean. { ¸ī·sə¦thər·mə¸bath }

isotherm ribbon [METEOROL] A zone of crowded isotherms on a synoptic upper-level chart; the temperature gradient is many times greater than normally encountered in the atmosphere. { 'ī·sə¸thərm ¸rib·ən }

isothiocyanate [ORG CHEM] A compound of the type

$R-N=C=S$, where R may be an alkyl or aryl group; an example is mustard oil. Also known as sulfocarbimide. { ˈɪ·sə,thī·ō'sī·ə,nāt }

isotimic [METEOROL] Pertaining to a quantity which has equal value in space at a particular time. { ˈɪ·səˈtim·ik }

isotimic line [METEOROL] On a given reference surface in space, a line connecting points of equal value of some quantity; most of the lines drawn in the analysis of synoptic charts are isotimic lines. { ˈɪ·səˈtim·ik 'līn }

isotimic surface [METEOROL] A surface in space on which the value of a given quantity is everywhere equal; isotimic surfaces are the common reference surfaces for synoptic charts, principally constant-pressure surfaces and constant-height surfaces. { ˈɪ·səˈtim·ik 'sər·fəs }

isotone [NUC PHYS] One of several nuclides having the same number of neutrons in their nuclei but differing in the number of protons. { 'ɪ·sə,tōn }

isotonic [PHYSIO] 1. Having uniform tension, as the fibers of a contracted muscle. 2. Of a solution, having the same osmotic pressure as the fluid phase of a cell or tissue. { ˈɪ·səˈtän·ik }

isotonic sodium chloride solution See normal saline. { ˈɪ·səˈtän·ik ˈsōd·ē·əm 'klór,īd sə,lü·shən }

isotope [NUC PHYS] One of two or more atoms having the same atomic number but different mass number. { 'ɪ·sə,tōp }

isotope abundance [NUC PHYS] The ratio of the number of atoms of a particular isotope in a sample of an element to the number of atoms of a specified isotope, or to the total number of atoms of the element. { 'ɪ·sə,tōp əˈbən·dəns }

isotope dilution [NUCLEO] The introduction of a radioisotope into stable isotopes of an element in order to make volume, mass, and age measurements of the element. { 'ɪ·sə,tōp də,lü·shən }

isotope-dilution analysis [ANALY CHEM] Variation on paper-chromatography analysis; a labeled radioisotope of the same type as the one being quantitated is added to the solution, then quantitatively analyzed afterward via radioactivity measurement. { 'ɪ·sə,tōp də,lü·shən ə,nal·ə·səs }

isotope effect [PHYS CHEM] The effect of difference of mass between isotopes of the same element on nonnuclear physical and chemical properties, such as the rate of reaction or position of equilibrium, of chemical reactions involving the isotopes. [SOLID STATE] Variation of the transition temperatures of the isotopes of a superconducting element in inverse proportion to the square root of the atomic mass. { 'ɪ·sə,tōp i,fekt }

isotope exchange [NUCLEO] 1. Exchange of places by two atoms, but different isotopes, of the same element in two different molecules, or in different locations of the same molecule. 2. The transfer of isotopically tagged atoms from one chemical form or valence state to another, without net chemical reaction. { 'ɪ·sə,tōp iks'chānj }

isotope-exchange reaction [CHEM] A chemical reaction in which interchange of the atoms of a given element between two or more chemical forms of the element occurs, the atoms in one form being isotopically labeled so as to distinguish them from atoms in the other form. { 'ɪ·sə,tōp iks'chānj rē,ak·shən }

isotope farm [BOT] A carbon-14 (^{14}C) growth chamber, or greenhouse, arranged as a closed system in which plants can be grown in an atmosphere of carbon dioxide (CO_2) containing ^{14}C and thus become labeled with ^{14}C; isotope farms also can be used with other materials, such as heavy water (D_2O), phosphorus-35 (^{35}P), and so forth, to produce biochemically labeled compounds. { 'ɪ·sə,tōp ,färm }

isotope fractionation [NUCLEO] Natural or artificial alteration of the isotopic composition of an element via processes of diffusion, evaporation, and chemical exchange, utilizing small differences in physical and chemical properties of isotopes. { 'ɪ·sə,tōp ,frak·shə,nā·shən }

isotope lamp [ELECTR] A discharge lamp containing gas of a single isotope and thus producing highly monochromatic light. { 'ɪ·sə,tōp ,lamp }

isotope separation [NUCLEO] The physical separation of different stable isotopes of an element from one another. { 'ɪ·sə,tōp ,sep·ə,rā·shən }

isotope shift [SPECT] A displacement in the spectral lines due to the different isotopes of an element. { 'ɪ·sə,tōp ,shift }

isotopic age determination See radiometric dating. { ˈɪ·səˈtäp·ik 'āj di,tər·mə,nā·shən }

isotopic carrier [CHEM] A carrier that differs from the trace it is carrying only in isotopic composition. { ˈɪ·səˈtäp·ik 'kar·ē·ər }

isotopic chronometer [NUCLEO] A method of determining the age of geological, archeological, or other samples by measuring the amount of a particular radioisotope and of its daughter isotope in a sample. { ˈɪ·səˈtäp·ik krə'näm·əd·ər }

isotopic element [NUC PHYS] An element which has more than one naturally occurring isotope. { ˈɪ·səˈtäp·ik 'el·ə·mənt }

isotopic enrichment [NUCLEO] The process by which the relative abundances of the isotopes of a given element are altered in a batch, thus producing a form of the element enriched in a particular isotope. { ˈɪ·səˈtäp·ik in'rich·mənt }

isotopic exchange [PHYS CHEM] A process in which two atoms belonging to different isotopes of the same element exchange valency states or locations in the same molecule or different molecules. { ˈɪ·səˈtäp·ik iks'chānj }

isotopic incoherence [PHYS] Incoherence in the scattering of neutrons from a crystal lattice due to differences in scattering lengths of different isotopes of the same element. { ˈɪ·səˈtäp·ik ,iŋ·kō'hir·əns }

isotopic indicator See isotopic tracer. { ˈɪ·səˈtäp·ik 'in·də,kād·ər }

isotopic irradiation [NUCLEO] The subjection of a material to radiation from radioactive isotopes for therapeutic or other purposes. { ˈɪ·səˈtäp·ik i,rād·ē'ā·shən }

isotopic label See isotopic tracer. { ˈɪ·səˈtäp·ik 'lā·bəl }

isotopic molecule [NUCLEO] A molecule in which the nucleus of one of the atoms is a special isotope. { ˈɪ·səˈtäp·ik 'mäl·ə,kyül }

isotopic number See neutron excess. { ˈɪ·səˈtäp·ik 'nəm·bər }

isotopic parity See G parity. { ˈɪ·səˈtäp·ik 'par·əd·ē }

isotopic spin [NUC PHYS] A quantum-mechanical variable, resembling the angular momentum vector in algebraic structure whose third component distinguished between members of groups of elementary particles, such as the nucleons, which apparently behave in the same way with respect to strong nuclear forces, but have different charges. Also known as isobaric spin; isospin; i-spin. { ˈɪ·səˈtäp·ik 'spin }

isotopic tracer [CHEM] An isotope of an element, either radioactive or stable, a small amount of which may be incorporated into a sample material (the carrier) in order to follow the course of that element through a chemical, biological, or physical process, and also follow the larger sample. Also known as isotopic indicator; isotopic label; label; tag. { ˈɪ·səˈtäp·ik 'trā·sər }

isotron [NUCLEO] A device for sorting isotopes of an element in which ions are accelerated to a fixed energy in a strong electric field, and a radio-frequency field then selects ions according to their velocity, which is inversely proportional to the square root of their mass. { 'ɪ·sə,trän }

isotropic [BIOL] Having a tendency for equal growth in all directions. [CYTOL] An ovum lacking any predetermined axis. [PHYS] Having identical properties in all directions. { ˈɪ·səˈträ·pik }

isotropic antenna See unipole. { ˈɪ·səˈträ·pik an'ten·ə }

isotropic dielectric [ELEC] A dielectric whose polarization always has a direction that is parallel to the applied electric field, and a magnitude which does not depend on the direction of the electric field. { ˈɪ·səˈträ·pik ,dī·ə'lek·trik }

isotropic fabric [PETR] A random orientation in space of the elements that compose a rock. { ˈɪ·səˈträ·pik 'fab·rik }

isotropic fluid [FL MECH] A fluid whose properties are not dependent on the direction along which they are measured. { ˈɪ·səˈträ·pik 'flü·əd }

isotropic flux [PHYS] Radiation, or a flow of particles or matter, which reaches a location from all directions with equal intensity. { ˈɪ·səˈträ·pik 'fləks }

isotropic gain of an antenna See absolute gain of an antenna. { ˈɪ·səˈträ·pik 'gān əv ən an'ten·ə }

isotropic material [PHYS] A material whose properties are not dependent on the direction along which they are measured. { ˈɪ·səˈträ·pik mə'tir·ē·əl }

isotropic noise [ELECTROMAG] Random noise radiation which reaches a location from all directions with equal intensity. { ˈɪ·səˈträ·pik 'nóiz }

isotropic plasma [PL PHYS] A plasma whose properties,

such as pressure, are not dependent on the direction along which they are measured. { 'ī·sə¦trä·pik 'plaz·mə }

isotropic radiation [ELECTROMAG] Radiation which is emitted by a source in all directions with equal intensity, or which reaches a location from all directions with equal intensity. { 'ī·sə¦trä·pik ˌrād·ē'ā·shən }

isotropic radiator [PHYS] An energy source that radiates uniformly in all directions. { 'ī·sə¦trä·pik 'rād·ē·ˌad·ər }

isotropic turbulence [FL MECH] Turbulence whose properties, especially statistical correlations, do not depend on direction. { 'ī·sə¦trä·pik 'tər·byə·ləns }

isotropic universe [ASTRON] A universe postulated to have the same properties when viewed from all directions. { 'ī·sə¦trä·pik 'yü·nə·ˌvərs }

isotropy [PHYS] The quality of a property which does not depend on the direction along which it is measured, or of a medium or entity whose properties do not depend on the direction along which they are measured. { ī'sä·trə·pē }

isotropy group [MATH] For an operation of a group G on a set S, the isotropy group of an element s of S is the set of elements g in G such that $gs = s$. { 'ī·sə·ˌtrō·pē ˌgrüp }

isotypes [IMMUNOL] **1.** A series of antigens, for example, blood types, common to all members of a species but differentiating classes and subclasses within the species. **2.** Different classes of immunoglobulins that have the same antigenic specificity. { 'ī·sə·ˌtīps }

isotypic [CRYSTAL] Pertaining to a crystalline substance whose chemical formula is analogous to, and whose structure is like, that of another specified compound. { 'ī·sə¦tip·ik }

isovalent conjugation [PHYS CHEM] An arrangement of bonds in a conjugated molecule such that alternative structures with an equal number of bonds can be written; an example occurs in benzene. { 'ī·sə¦vā·lənt kən'jəŋk·shən }

isovalent hyperconjugation [PHYS CHEM] An arrangement of bonds in a hyperconjugated molecule such that the number of bonds is the same in the two resonance structures but the second structure is energetically less favorable than the first structure; examples are $H_3\equiv C-C^+H_2$ and $H_3\equiv C-CH_2$. { 'ī·sə¦vā·lənt ˌhī·pər·ˌkän·jə'gā·shən }

isovaleral *See* isovaleraldehyde. { 'ī·sō¦val·ə·rəl }

isovaleraldehyde [ORG CHEM] $(CH_3)_2CHCH_2CHO$ A colorless liquid with an applelike odor and a boiling point of 92°C; soluble in alcohol and ether; used in perfumes and pharmaceuticals and for flavoring. { 'ī·sō¦val·ər'al·də·ˌhīd }

isovaleric acid [ORG CHEM] $(CH_3)_2CHCH_2COOH$ Colorless liquid with disagreeable taste and aroma; boils at 176°C; soluble in alcohol and ether; found in valeriana, hop, tobacco, and other plants; used in flavors, perfumes, and medicines. { 'ī·sō·və'ler·ik 'as·əd }

2-isovaleryl-1,3-indandione [ORG CHEM] $C_{14}H_{14}O_3$ A yellow, crystalline compound with a melting point of 67–68°C; insoluble in water; used as a rodenticide. { 'tü 'ī·sō'val·ə·ˌril ¦wən ¦thrē ˌin·dən'dī·ˌōn }

isovel *See* isotach. { 'ī·sə·ˌvel }

isovolumic *See* isochoric. { 'ī·sō¦väl·yə·mik }

isozyme *See* isoenzyme. { 'ī·sə·ˌzīm }

ISP *See* imperial smelting process; Internet Service Provider.

i-spin *See* isotopic spin. { 'ī ˌspin }

ISR *See* Intersecting Storage Rings.

Israel's theorem [RELAT] A theorem of general relativity essentially proving that the Schwarzschild solution is the unique solution of Einstein's equations describing nonrotating black holes in empty space and that the Reissner-Nordstrom solution is the unique solution describing nonrotating charged black holes. { 'iz·rē·əlz ˌthir·əm }

ISS *See* ion scattering spectroscopy.

isthmus [BIOL] A passage or constricted part connecting two parts of an organ. [GEOGR] A narrow strip of land having water on both sides and connecting two large land masses. [MATH] *See* bridge. { 'is·məs }

Istiophoridae [VERT ZOO] The billfishes, a family of oceanic perciform fishes in the suborder Scombroidei. { ˌis·tē·ə'fòr·ə·ˌdē }

ISTS *See* impulsive stimulated thermal scattering.

Isuridae [VERT ZOO] The mackerel sharks, a family of pelagic, predacious galeoids distinguished by a heavy body, nearly symmetrical tail, and sharp, awllike teeth. { i'sùr·ə·ˌdē }

itabirite [GEOL] A laminated, metamorphosed, oxide-facies iron formation in which the original chert or jasper bands have been recrystallized into megascopically distinguished grains of quartz and in which the iron is present as thin layers of hematite, magnetite, or martite. { ˌēd·ə'bi·ˌrīt }

itacolumite [PETR] A fine-grained, thin-bedded sandstone or a schistose quartzite that contains mica, chlorite, and talc and that exhibits flexibility when split into slabs. Also known as articulite. { ˌid·ə'käl·ə·ˌmīt }

itaconic acid [ORG CHEM] $CH_2:C(COOH)CH_2COOH$ A colorless crystalline compound that decomposes at 165°C, prepared by fermentation with *Aspergillus terreus;* used as an intermediate in organic synthesis and in resins and plasticizers. { ¦id·ə¦kän·ik 'as·əd }

itatartaric acid [ORG CHEM] $C_5H_8O_6$ A compound produced experimentally by fermentation; formed as a minor product, 5.8% of total acidity produced, of an itaconic-acid producing strain of *Aspergillus niger* { ¦id·ə¦tär·də·rik 'as·əd }

IT calorie *See* calorie. { 'ī·tē ˌkal·ə·rē }

itch [PHYSIO] An irritating cutaneous sensation allied to pain. { ich }

item [COMPUT SCI] A set of adjacent digits, bits, or characters which is treated as a unit and conveys a single unit of information. { 'īd·əm }

item advance [COMPUT SCI] A technique of efficiently grouping records to optimize the overlap of read, write, and compute times. { 'īd·əm əd·ˌvans }

item design [COMPUT SCI] The specification of what fields make up an item, the order in which the fields are to be recorded, and the number of characters to be allocated to each field. { 'īd·əm di·ˌzīn }

item size [COMPUT SCI] The length of an item expressed in characters, words, or blocks. { 'īd·əm ˌsīz }

iterated integral [MATH] An integral over an area or volume designated to be performed by successive integrals over line segments. { 'īd·ə·ˌrād·əd 'ir·t·ə·grəl }

iteration *See* iterative method. { ˌīd·ə'rā·shən }

iteration process [COMPUT SCI] The process of repeating a sequence of instructions with minor modifications between successive repetitions. { ˌīd·ə'rā·shən ˌprä·səs }

iterations per second [COMPUT SCI] In computers, the number of approximations per second in iterative division; the number of times an operational cycle can be repeated in 1 second. { ˌīd·ə'rā·shənz pər 'sek·ənd }

iterative array [COMPUT SCI] In a computer, an array of a large number of interconnected identical processing modules, used with appropriate driver and control circuits to permit a large number of simultaneous parallel operations. { 'īd·ə·ˌrād·iv ə'rā }

iterative division [COMPUT SCI] In computers, a method of dividing by use of the operations of addition, subtraction, and multiplication; a quotient of specified precision is obtained by a series of successively better approximations. { 'īd·ə·ˌrād·iv di'vizh·ən }

iterative filter [ELECTR] Four-terminal filter that provides iterative impedance. { 'īd·ə·ˌrād·iv 'fil·tər }

iterative impedance [ELECTR] Impedance that, when connected to one pair of terminals of a four-terminal transducer, will cause the same impedance to appear between the other two terminals. { 'īd·ə·ˌrād·iv im'pēd·əns }

iterative method [MATH] Any process of successive approximation used in such problems as numerical solution of algebraic equations, differential equations, or the interpolation of the values of a function. Also known as iteration. { 'īd·ə·ˌrād·iv 'meth·əd }

iterative process [MATH] A process for calculating a desired result by means of a repeated cycle of operations, which comes closer and closer to the desired result; for example, the arithmetical square root of N may be approximated by an iterative process using additions, subtractions, and divisions only. { 'īd·ə·ˌrād·iv 'prä·səs }

iterative routine [COMPUT SCI] A computer program that obtains a result by carrying out a series of operations repetitiously until some specified condition is met. { 'īd·ə·ˌrād·iv rü'tēn }

iteroparity [BIOL] Reproduction that occurs repeatedly over the life of the individual. { ˌīd·ə·rə'par·əd·ē }

iteroparous [ZOO] Capable of breeding or reproducing multiple times. { ˌīd·ə·rə'par·əs }

Ithomiinae [INV ZOO] The glossy-wings, a subfamily of weak-flying lepidopteran insects having on the wings broad,

---

ITACONIC ACID

Structural formula of itaconic acid.

transparent areas in which the scales are reduced to short hairs. { ,ith·ə'mī·ə,nē }

Itonididae [INV ZOO] The gall midges, a family of orthorrhaphous dipteran insects in the series Nematocera; most are plant pests. { ,id·ə'nid·ə,dē }

Itô's formula See stochastic chain rule. { 'ē,tōz ,fȯr·myə·lə }

Itô's integral See stochastic integral. { 'ē,tōz ,int·ə·grəl }

ITS See intelligent transportation system.

IVS-90 See international temperature scale.

ITV See industrial television.

I-type magma [GEOL] Magma formed from igneous source materials. { 'ī ,tīp ¦mag·mə }

i-type semiconductor See intrinsic semiconductor. { 'ī ,tīp ¦sem·i·kən'dək·tər }

ium ion [ORG CHEM] A positively charged group of atoms in which a charged nonmetallic ion other than carbon or silicon possesses a closed-shell electron configuration; often joined to a root word, as in carbonium ion. { 'ī·əm 'ī,än }

IUPAC See International Union of Pure and Applied Chemistry. { ,ī'yü,pak or 'ī,yü¦pē¦ā¦sē }

Ivanov reagent [ORG CHEM] A reagent that is similar to a Grignard reagent, and that is formed by reacting an arylacetic acid or its sodium salt with isopropyl magnesium halide. { ē·və·nȯf rē,ā·jənt }

ivory [MATER] The ivory-white material composing the tusks

and teeth of the elephant; specific gravity is 1.87; takes a high polish and is used for ornamental parts and art objects, and formerly for piano keys. { 'īv·rē }

ivory black [MATER] Animal black made by calcining ivory; used as a pigment. { 'īv·rē 'blak }

ivory board [MATER] A highly finished cardboard that is clay-coated on both sides; used for art printing and menu cards. { 'īv·rē ,bȯrd }

ivory point [ENG] A small pointer extending downward from the top of the cistern of a Fortin barometer; the level of the mercury in the cistern is adjusted so that it just comes in contact with the end of the pointer, thus setting the zero of the barometer scale. { 'īv·rē ¦pȯint }

Ixion [PHYS] An experimental magnetic-mirror device used for research on controlled fusion; involves study of plasma rotation in a magnetic-mirror confinement system using crossed electric and magnetic fields. { 'ik·sē·ən }

Ixodides [INV ZOO] The ticks, a suborder of the Acarina distinguished by spiracles behind the third or fourth pair of legs. { ,ik'säd·ə,dēz }

Izod test [MET] An impact test in which a falling pendulum strikes a fixed, usually notched specimen with 120 foot-pounds (163 joules) of energy at a velocity of 11.5 feet (3.5 meters) per second; the height of the pendulum swing after striking is a measure of the energy absorbed and thus indicates impact strength. { 'ī,zäd ,test }

J

J *See* joule.

jaagsiekte [VET MED] A contagious disease of sheep, sometimes of goats and guinea pigs, resembling the more benign and diffuse forms of bronchiolar carcinoma in humans. { 'yäg‚sēk·tə }

Jablochkoff candle [ELECTR] An early type of arc lamp in which carbons were placed side by side and separated by plaster of paris. { yə'bläch‚kóf ‚kand·əl }

Jacanidae [VERT ZOO] The jacanas or lily-trotters, constituting the single family of the avian superfamily Jacanoidea. { jə'kan·ə‚dē }

Jacanoidea [VERT ZOO] A monofamilial superfamily of colorful marshbirds distinguished by greatly elongated toes and claws, long legs, a short tail, and a straight bill. { ‚jak·ə'nóid·ē·ə }

jacaranda *See* carob wood. { ‚jak·ə'ran·də }

Jaccarino-Peter effect [SOLID STATE] The production of superconductivity in certain ferromagnetic metals through the application of an external magnetic field that compensates for the polarization of the conduction electrons. Also known as compensation effect. { ‚jak·ə‚rē·nō 'pēd·ər i‚fekt }

jacinth *See* zircon. { 'jas·ənth }

jack [ELEC] A connecting device into which a plug can be inserted to make circuit connections; may also have contacts that open or close to perform switching functions when the plug is inserted or removed. [MECH ENG] A portable device for lifting heavy loads through a short distance, operated by a lever, a screw, or a hydraulic press. *See* sphalerite. [TEXT] **1.** A frame in lace-manufacturing equipment that has horizontal bars to support fixed vertical wires, against which bobbins containing the yarn can freely revolve. **2.** An oscillating lever that raises the harness of a dobby loom. { jak }

jackal [VERT ZOO] **1.** *Canis aureus.* A wild dog found in southeastern Europe, southern Asia, and northern Africa. **2.** Any of various similar Old World wild dogs; they resemble wolves but are smaller and more yellowish. { 'jak·əl }

jackbit [DES ENG] A drilling bit used to provide the cutting end in rock drilling; the bit is detachable and either screws on or is taper-fitted to a length of drill steel. Also known as ripbit. { 'jak‚bit }

jack chain [DES ENG] **1.** A chain made of light wire, with links arranged in figure-eights with loops at right angles. **2.** A toothed endless chain for moving logs. { 'jak ‚chān }

jacket [MECH ENG] The space around an engine cylinder through which a cooling liquid circulates. [NUCLEO] A thin container for one or more fuel slugs, used to prevent the fuel from escaping into the coolant of a reactor. Also known as can; cartridge. [ORD] **1.** Cylinder of steel covering and strengthening the breech end of a gun or howitzer tube. **2.** The water jacket on some machine guns. [PETRO ENG] The support structure of a steel offshore production platform; it is fixed to the seabed by piling, and the superstructure is mounted on it. { 'jak·ət }

jacket crown [MED] An artificial crown of a tooth consisting of a covering of porcelain or resin. { 'jak·ət ‚kraún }

jacketed pipe [DES ENG] A double-walled pipe in which liquids that are too viscous for pipeline transport at normal temperatures flow through the inner pipe that is surrounded by a pipe circulating hot fluids. { 'jak·əd·əd 'pīp }

jacket face [GRAPHICS] The index-readable side of a microfilm. { 'jak·ət ‚fās }

jacket gage [GRAPHICS] Acetate thickness measurement of microfilm in thousandths of an inch. { 'jak·ət ‚gāj }

jackhammer *See* pneumatic hammer. { 'jak‚ham·ər }

jacking [TEXT] A process in spinning to provide extra twist or draft to the roving in a mule. { 'jak·iŋ }

jack ladder [ENG] A V-shaped trough holding a toothed endless chain, and used to move logs from pond to sawmill. [NAV ARCH] *See* Jacob's ladder. { 'jak ‚lad·ər }

jackleg [ENG] A supporting bar used with a jackhammer. { 'jak‚leg }

jack line [PETRO ENG] A steel cable or rod that connects the arms of the central pumping engine to the two or more wells that are being pumped. { 'ak ‚līn }

jack plane [DES ENG] A general-purpose bench plane measuring over 1 foot (30 centimeters) in length. { 'jak ‚plān }

jack post [MIN ENG] Timber used where a coal seam is separated by a rock band and one bench is loaded out before the other. { 'jak ‚pōst }

jack rafter [BUILD] A short, secondary, or simulated rafter. { 'jak ‚raf·tər }

jackscrew [MECH ENG] **1.** A jack operated by a screw mechanism. Also known as screw jack. **2.** The screw of such a jack. { 'jak‚skrü }

jackshaft [MECH ENG] A countershaft, especially when used as an auxiliary shaft between two other shafts. { 'jak‚shaft }

Jacksonian epilepsy [MED] Recurrent Jacksonian seizures. { jak'sō·nē·ən 'ep·ə‚lep·sē }

Jacksonian seizure [MED] A focal seizure originating in one part of the motor or sensorimotor cortex and manifested usually by spasmodic contractions of or crawling or burning sensations of the skin; may become generalized, leading to loss of consciousness. { jak'sō·nē·ən 'sē·zhər }

jack spool [TEXT] A large spool used to hold wool strands or yarn. { 'jak ‚spül }

jack staff [NAV ARCH] A staff or flagpole attached to the bow of a ship, or the end of a spar projecting from the bow. { 'jak ‚staf }

jackstay [NAV ARCH] A rope, rod, or pipe rove through the eyebolts fitted on a yard or mast for the purpose of attaching sails to the yard or mast. { 'jak‚stā }

jack timber [MIN ENG] A timber such as a rafter that is shorter than others with which it is used. { 'jak ‚tim·bər }

jack truss [BUILD] A minor truss in a hip roof where the roof has a reduced section. { 'jak ‚trəs }

Jacobian [MATH] The Jacobian of functions $f_i(x_1, x_2, \ldots, x_n)$, $i = 1, 2, \ldots, n$, of real variables x_i is the determinant of the matrix whose ith row lists all the first-order partial derivatives of the function $f_i(x_1, x_2, \ldots, x_n)$. Also known as Jacobian determinant. { jə'kō·bē·ən }

Jacobian determinant *See* Jacobian. { jə'kō·bē·ən di'tər·mə·nənt }

Jacobian elliptic function [MATH] For m a real number between 0 and 1, and u a real number, let ϕ be that number such that

$$\int_0^\phi d\theta/(1 - m \sin^2 \theta)^{1/2} = u;$$

the 12 Jacobian elliptic functions of u with parameter m are $\text{sn}(u|m) = \sin \phi$, $\text{cn}(u|m) = \cos \phi$, $\text{dn}(u|m) = (1 - m \sin^2 \phi)^{1/2}$, the reciprocals of these three functions, and the quotients of any two of them. { jə'kō·bē·ən ə‚lip·tik 'fəŋk·shən }

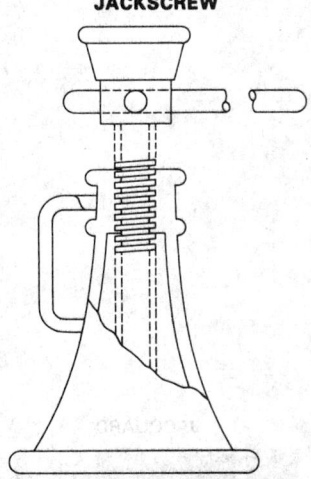

JACKSCREW

Jackscrew used to raise heavy objects.

Jacobian matrix [MATH] The matrix used to form the Jacobian. { jə'kō·bē·ən 'mā·triks }

Jacobi canonical matrix [MATH] A form to which any matrix can be reduced by a collineatory transformation, with zeros below the principal diagonal and characteristic roots as elements of the principal diagonal. { jə'kōb·ē kə'nän·ə·kəl 'mā·triks }

Jacobi condition [MATH] In the calculus of variations, a differential equation used to study the extremals in a variational problem. { jə'kō·bē kən,dish·ən }

Jacobi ellipsoid [ASTROPHYS] A triaxial ellipsoid that can be formed by the surface of a homogeneous, self-gravitating body rotating uniformly with sufficient high angular velocity. { jä'kō·bē i'lip,sóid }

Jacobi polynomials [MATH] Polynomials that are constructed from the hypergeometric function and satisfy the differential equation $(1 - x^2)y'' + [\beta - \alpha - (\alpha + \beta + 2)x]y' + n(\alpha + \beta + n + 1)y = 0$, where n is an integer and α and β are constants greater than -1; in certain cases these generate the Legendre and Chebyshev polynomials. { jə'kō·bē ,päl·ə'nō·mē·əlz }

Jacobi's method [MATH] **1.** A method of determining the eigenvalues of a Hermitian matrix. **2.** A method for finding a complete integral of the general first-order partial differential equation in two independent variables; it involves solving a set of six ordinary differential equations. { jə'kō·bēz ,meth·əd }

Jacobi's theorem [MATH] The proposition that a periodic, analytic function of a complex variable is simply periodic or doubly periodic. { jə'kō·bēz ,thir·əm }

Jacobi's transformations [MATH] Transformations of Jacobian elliptic functions to other Jacobian elliptic functions given by change of parameter and variable. { jə'kō·bēz ,tranz·fər'mā·shənz }

Jacobshavn Glacier [HYD] A glacier on the west coast of Greenland at latitude 68°N; it is the most productive glacier in the Northern Hemisphere, calving about 1400 icebergs yearly. { 'yä·kəps,häf·ən 'glä·shər }

jacobsite [MINERAL] $MnFe_3O_4$ A black magnetic mineral composed of an oxide of manganese and iron; a member of the magnetite series. { 'jä·kəb,zīt }

Jacob's ladder [NAV ARCH] A portable ladder, having rope, chain, or wire sides, and wooden or iron rungs, slung over the ship's side for temporary use. Also known as jack ladder. { 'jā·kəbz 'lad·ər }

Jacobsoniidae [INV ZOO] The false snout beetles, a small family of coleopteran insects in the superfamily Dermestoidea. { jä·kəb·sə'nī·ə,dē }

Jacobson radical See radical. { 'jā·kəb·sən ,rad·ə·kəl }

Jacobson's cartilage See vomeronasal cartilage. { 'jā·kəb·sənz ,kärt·lij }

Jacobson's organ [VERT ZOO] An olfactory canal in the nasal mucosa which ends in a blind pouch; it is highly developed in reptiles and vestigial in humans. { 'jā·kəb·sənz ,ȯr·gən }

Jacob's staff See cross-staff. { 'jā·kəbz 'staf }

Jacobs taper [DES ENG] A machine tool used for mounting drill chucks in drilling machines. { 'jā·kəbz 'tā·pər }

jacquard [TEXT] **1.** A loom or knitting machine for weaving figured fabrics and whose apparatus is controlled by punched cards. **2.** A fabric of jacquard weave. { ja,kärd }

Jacquemart's reagent [ANALY CHEM] Analytical reagent used to test for ethyl alcohol; consists of an aqueous solution of mercuric nitrate and nitric acid. { zhak'märz rē,ā·jənt }

jacupirangite [PETR] An ultramafic plutonic rock that is part of the ijolite series; composed chiefly of titanaugite and magnetite, with a smaller amount of nepheline. { jə'kü·pə·rən,jīt }

Jadassohn's nevus See nevus sebaceus. { 'yä·də,zōnz 'nē·vəs }

jade [MINERAL] A hard, compact, dark-green or greenish-white gemstone composed of either jadeite or nephrite. Also known as jadestone. { jād }

jadeite [MINERAL] $NaAl(SiO_3)_2$ A clinopyroxene mineral occurring as green, fibrous monoclinic crystals; the most valuable variety of jade. { 'jā,dīt }

jadeitite [PETR] A type of metamorphic rock composed of jadeite associated with small amounts of feldspar or feldspathoids. { 'jād·ə,tīt }

jadestone See jade. { 'jād,stōn }

Jaeger method [FL MECH] A method of determining surface tension of a liquid in which one measures the pressure required to cause air to flow from a capillary tube immersed in the liquid. { 'yä·gər ,meth·əd }

Jaeger-Steinwehr method [THERMO] A refinement of the Griffiths method for determining the mechanical equivalent of heat, in which a large mass of water, efficiently stirred, is used, the temperature rise of the water is small, and the temperature of the surroundings is carefully controlled. { 'yā·gər 'shtīn·ver ,meth·əd }

jaff [ORD] A combination of electronic and chaff jamming. { jaf }

jag bolt [DES ENG] An anchor bolt with barbs on a flaring shank. { 'jag,bȯlt }

jaguar [VERT ZOO] *Felis onca.* A large, wild cat indigenous to Central and South America; it is distinguished by a buff-colored coat with black spots, and has a relatively large head and short legs. { 'jag,wär }

Jahn-Teller effect [PHYS CHEM] The effect whereby, except for linear molecules, degenerate orbital states in molecules are unstable. { 'yän 'tel·ər i,fekt }

Jak-STAT pathway [CELL MOL] A rapid signal transduction pathway used by a variety of cytokines and growth factors to alter gene expression. Binding of a cytokine or growth factor to its receptor activates Jak (janus kinase—a cytoplasmic tyrosine kinase) and triggers it to phosphorylate and stimulate STAT (signal transducers and activators of transcription—a gene regulatory protein) to travel to the nucleus and induce transcription of a specific gene. { 'jak,stat 'path,wā }

jalap [MATER] An orange or reddish solid or a yellowish to brown powder with acrid taste and slight odor; the dried tuberous root of a Mexican plant (*Exogonium purga*), or the drug prepared from it; used as a cathartic in medicine. { 'jal·əp }

jalousie [BUILD] A window that consists of a number of long, narrow panels, each hinged at the top. { 'jal·ə·sē }

Jamaica bayberry See bayberry. { jə'mā·kə 'bā,ber·ē }

jamb [BUILD] The vertical member on the side of an opening, as a door or window. [MIN ENG] A vein or large block of rock that prevents miners from following a vein of ore. { jam }

jamb brick [MATER] A brick with one rounded corner; used to provide a rounded edge on wall openings. { 'jam ,brik }

jamb liner [BUILD] A small strip of wood applied to the edge of a window jamb to increase its width for use in thicker walls. { 'jam ,līn·ər }

James concentrator [MIN ENG] A concentration table whose deck is divided into two sections; one section contains riffles for the coarse material, and the other section is smooth to allow settling of the fine particles which will not settle on a riffled surface. { 'jāmz 'käns·ən,trād·ər }

jamesonite [MINERAL] $Pb_4FeSb_6S_{14}$ A lead-gray to gray-black mineral that crystallizes in the orthorhombic system, occurs in acicular crystals with fibrous or featherlike forms, and has a metallic luster. Also known as feather ore; gray antimony. { 'jām·sə,nīt }

Jamin effect [FL MECH] Resistance to flow of a column of liquid divided by air bubbles in a capillary tube, even when subjected to a substantial pressure difference between the ends of the tube. { jə'mēn i,fekt }

Jamin refractometer [OPTICS] An instrument for measuring the index of refraction of a gas in which two light beams from a common source are each passed through an evacuated tube and recombined, and the displacement of interference fringes is noted as gas is slowly admitted into one of the tubes. { jə'mēn ,rē,frak'täm·əd·ər }

jammer [ELECTR] A transmitter used in jamming of radio or radar transmissions. Also known as electronic jammer. { 'jam·ər }

jammer finder [ELECTR] Radar which attempts to obtain the range of the target by training a highly directional pencil beam on a jamming source. Also known as burn-through. { 'jam·ər ,fīn·dər }

jamming [ELECTR] Radiation or reradiation of electromagnetic waves so as to impair the usefulness of a specific segment of the radio spectrum that is being used by the enemy for communication or radar. Also known as active jamming; electronic jamming. { 'jam·iŋ }

jam nut See locknut. { 'jam ,nət }

Janecke coordinates [CHEM ENG] Use of a rectangular or Ponchon-type diagram to plot the solvent content of liquid-liquid equilibrium phases; used for solvent-extraction design calculations. { 'yä·nə·kē kō,ȯrd·ən·əts }

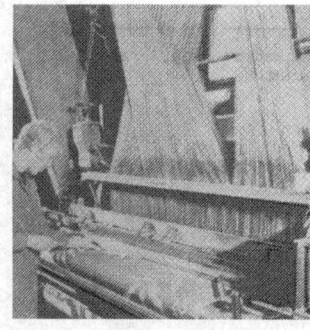

JACQUARD

Weaving linen on a jacquard damask loom.

jansky [ASTROPHYS] A unit of measurement of flux density, in units of watt · meter^{-2} · hertz^{-1}; 1 jansky is 10^{-26} W · m^{-2} · Hz^{-1}. Abbreviated Jy. { 'jans·kē }

J antenna [ELECTROMAG] Antenna having a configuration resembling a J, consisting of a half-wave antenna end-fed by a parallel-wire quarter-wave section. { 'jā ant,en·ə }

January thaw [CLIMATOL] A period of mild weather popularly supposed to recur each year in late January in New England and other parts of the northeastern United States. { 'jan·yə,wer·ē 'thȯ }

Janus [ASTRON] A satellite of Saturn which orbits at a mean distance of 151,000 kilometers (94,000 miles) and has an irregular shape with an average diameter of 190 kilometers (120 miles). { 'jā·nəs }

Janus system [NAV] A technique used in Doppler navigator design which utilizes radar or sound beams directed forward and astern for computation of ground-speed components; by heterodyning the returned pulses, the Doppler shift can be determined independently of the transmitted frequency. { 'jā·nəs ,sis·təm }

japan [MATER] A glossy, black baking paint or varnish that consists primarily of a hard asphalt base. { jə'pan }

Japan Current See Kuroshio Current. { jə'pan 'kə·rənt }

Japanese encephalitis [MED] A human viral infection epidemic in Japan, transmitted by the common house mosquito (*Culex pipiens*) and characterized by severe inflammation of the brain. { 'jap·ə¦nēz in,sef·ə'līd·əs }

Japanese peppermint oil [MATER] An oil distilled from *Mentha arvensis*, grown in Japan, Brazil, and the United States; the oil has less odor than peppermint oil; used for the production of menthol. { 'jap·ə¦nēz 'pep·ər,mint ,ȯil }

japanning [MET] The finishing of metal objects with japan. { jə'pan·iŋ }

Japan paper [MATER] A special paper made with an irregular mottled effect on the surface; used for greeting cards and other decorative applications. { jə'pan 'pā·pər }

Japan tallow See Japan wax. { jə'pan ¦tal·ō }

Japan wax [MATER] A pale-yellow wax with rancid aroma obtained from the berries of sumac; soluble in benzene and naphtha, insoluble in water; melts at 53°C; used in wax products, polishes, and soaps, and as a beeswax substitute. Also known as Japan tallow; sumac wax. { jə'pan ¦waks }

Japygidae [INV ZOO] A family of wingless insects in the order Diplura with forcepslike anal appendages; members attack and devour small soil arthropods. { jə'pij·ə,dē }

jar [ELEC] A unit of capacitance equal to 1000 statfarads, or approximately 1.11265×10^{-9} farad; it is approximately equal to the capacitance of a Leyden jar; this unit is now obsolete. { jär }

jar coupling See jars. { 'jär,kəp·liŋ }

jardiniere glaze [MATER] A type of unfritted soft and hard lead glaze; contains oxides of lead, zinc, potassium, calcium, aluminum, and silicon. { 'järd·ən¦ir ,glāz }

jarlite [MINERAL] NaSr$_3$Al$_3$F$_{16}$ A colorless to brownish mineral composed of aluminofluoride of sodium and strontium. { 'yär,līt }

jarosite [MINERAL] KFe$_3$(SO$_4$)$_2$(OH)$_6$ An ocher-yellow or brown alunite mineral having rhombohedral crystal structure. Also known as utahite. { jə'rō,sīt }

jarosite process [MET] A zinc electrometallurgical process in which ferric ions are precipitated from feed solutions in the form of jarosite, a hydrous sulfate of iron and potassium or sodium. { jə'rō,sīt ,prä·səs }

jars [PETRO ENG] A series of links in the drill string to connect drill cables to the drill bit; sets up the uneven motion on the upstroke that helps free the string of tools. Also known as jar coupling. { järz }

jasmine oil [MATER] A colorless fragrant essential oil from flowers of a jasmine, as *Jasminum officinale* or *J. grandiflorum*; the oil is extracted from the flowers by enfleurage and is used in perfumery. { 'jaz·mən ,ȯil }

jasmone [ORG CHEM] C$_{11}$H$_{16}$O A liquid ketone found in jasmine oil and other essential oils from plants. { 'jaz,mōn }

jaspagate See agate jasper. { 'jas·pə·gāt }

jaspé [TEXT] Fabric which has a series of faint stripes formed by light, medium, and dark threads of the same color. { zha'spā }

jasper [PETR] A dense, opaque to slightly translucent cryptocrystalline quartz containing iron oxide impurities; characteristically red. Also known as jasperite; jasperoid; jaspis. { 'jas·pər }

jasperite See jasper. { 'jas·pə,rīt }

jasperoid See jasper. { 'jas·pə,rȯid }

jaspilite [PETR] A compact siliceous rock resembling jasper and containing iron oxides in bands. { 'jas·pə,līt }

jaspis See jasper. { 'jas·pəs }

jaspoid See tachylite. { 'jas pȯid }

JATO engine [AERO ENG] Derived from jet-assisted-takeoff engine. **1.** An auxiliary jet-producing unit or units, usually rockets, for additional thrust. **2.** A JATO bottle or unit; the complete auxiliary power system used for assisted takeoff. { 'jā·dō ,en·jən }

jauch See jauk. { yaůk }

jauk [METEOROL] A local name for the foehn in the Klagenfurt basin of Austria; it may come from the south, but is developed as a north foehn. Also spelled jauch. { yaůk }

jaundice [INV ZOO] See grasserie. [MED] Yellow coloration of the skin, mucous membranes, and secretions resulting from hyperbile-rubinemia. Also known as icterus. { 'jȯn·dəs }

jaundice of newborn [MED] Jaundice in infants during the first few days after birth, due to various causes. { 'jȯn·dəs əv 'nü,bȯrn }

Java [COMPUT SCI] An object-oriented programming language based on C++ that was designed to run in a network such as the Internet; mostly used to write programs, called applets, that can be run on Web pages.

Java black rot [PL PATH] A fungus disease of stored sweet potatoes caused by *Diplodia tubericola*; the inside of the root becomes black and brittle. { ¦jäv·ə 'blak ,rät }

Java cotton See kapok. { ¦jäv·ə 'kat·ən }

Java man [PALEON] An overspecialized, apelike form of *Homo sapiens* from the middle Pleistocene having a small brain capacity, low cranial vault, and massive browridges. { jäv·ə 'man }

JavaScript [COMPUT SCI] A scripting language that is added to standard HTML to create interactive documents. { 'jäv·ə,skript }

Java virtual machine [COMPUT SCI] An interpreter that translates Java bytecode into actual machine instructions in real time. Abbreviated JVM. { ¦jäv·ə ,vər·chə·wəl mə'shēn }

jaw [ANAT] Either of two bones forming the skeleton of the mouth of vertebrates: the upper jaw or maxilla, and the lower jaw or mandible. [ENG] A notched part that permits a railroad-car axle box to move vertically. [GEOL] The side of a narrow passage such as a gorge. { jȯ }

jawbreaker See jaw crusher. { 'jȯ,brāk·ər }

jaw clutch [MECH ENG] A clutch that provides positive connection of one shaft with another by means of interlocking faces; may be square or spiral; the most common type of positive clutch. { 'jȯ ,kləch }

jaw crusher [MECH ENG] A machine for breaking rock between two steel jaws, one fixed and the other swinging. Also known as jawbreaker. { 'jȯ ,krəsh·ər }

jawless vertebrate [VERT ZOO] The common name for members of the Agnatha. { jȯ·ləs 'vərd·ə·brət }

J bolt [DES ENG] A J-shaped bolt, threaded on the long leg of the J. { 'jā ,bōlt }

J box See junction box. { 'jā ,bäks }

J-carrier system [COMMUN] Broad-band carrier system, providing 12 telephone channels, which uses frequencies up to about 140 kilohertz by means of effective four-wire transmission on a single open-wire pair. { 'jā ,kar·ē·ər ,sis·təm }

J display [ELECTR] A modified radarscope A display in which the time base is a circle; the target signal appears as an outward radial deflection from the time base. { 'jā di,splā }

Jeans flux [ASTRON] For a particular constituent of a planetary atmosphere, the number of atoms or molecules that escape from the atmosphere, per unit area per unit time, by virtue of their thermal motions. { 'jēnz ,fləks }

Jeans length [ASTROPHYS] A critical length such that oscillations in homogeneous, infinite media with wavelengths greater than this length are gravitationally unstable. { 'jēnz ,leŋkth }

Jeans viscosity equation [THERMO] An equation which

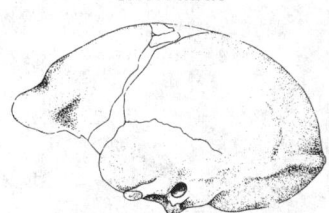

JAVA MAN

Lateral view of the cranium of *Homo erectus* II, one of the first specimens of Java man. (*Carnegie Institution of Washington, as used in M. F. Ashley Montagu, An Introduction to Physical Anthropology, 2d ed., Charles C. Thomas, 1951*)

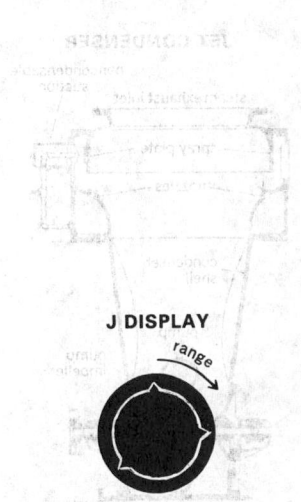

J DISPLAY

J type of radar display showing the signals as radial pips.

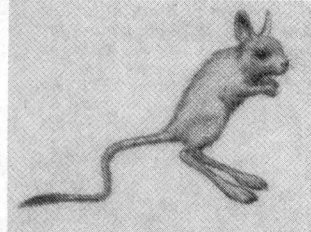

JERBOA

Jerboa, with body 3–6 inches (7–15 centimeters) long and tail up to 8 inches (20 centimeters) long.

states that the viscosity of a gas is proportional to the temperature raised to a constant power, which is different for different gases. { 'jēnz vi'skäs·əd·ē i‚kwä·zhən }

jeep [MECH ENG] A one-quarter-ton, four-wheel-drive utility vehicle in wide use in all United States military services. { jēp }

jeffersonite [MINERAL] $Ca(Mn,Zn,Fe)Si_2O_6$ A dark-green or greenish-black mineral composed of pyroxene. { 'jef·ər·sə‚nīt }

Jeffrey crusher [MIN ENG] A crusher to break soft minerals, such as limestone. Also known as whizzer mill. { 'jef·rē ‚krəsh·ər }

Jeffrey diaphragm jig [MIN ENG] A plunger-type jig with the plunger beneath the screen. { 'jef·rē 'dī·ə‚fram ‚jig }

Jeffrey molveyor [MIN ENG] A string of short conveyors on driven wheels connected together to run alongside a heading or room conveyor; used to keep a continuous miner in operation at all times. { 'jef·rē 'mäl‚vā·ər }

Jeffrey single-roll crusher [MIN ENG] A simple type of crusher for coal, with a drum to which are bolted toothed segments designed to grip the coal, thus forcing it down into the crushing opening. { 'jef·rē ‚siŋ·gəl ‚rōl ‚krəsh·ər }

Jeffrey swing-hammer crusher [MIN ENG] A crusher with swing arms on a revolving shaft for crushing coal, ore, or other material against the iron casing of the crusher; a screen at the bottom allows sufficiently fine pieces to pass through. { 'jef·rē ‚swiŋ ‚ham·ər ‚krəsh·ər }

Jeffrey-Traylor vibrating feeder [MIN ENG] A feed chute vibrated electromagnetically in a direction oblique to its surface; rate of movement of rock depends on amplitude and frequency of vibration. { 'jef·rē 'trā·lər ‚vī‚brād·iŋ 'fēd·ər }

Jeffrey-Traylor vibrating screen [MIN ENG] A vibrating screen whose action results from an oscillating armature and a stationary coil. { 'jef·rē 'trā·lər ‚vī‚brād·iŋ 'skrēn }

jejunitis [MED] Inflammation of the jejunum. { jē·jə'nīd·əs }

jejunostomy [MED] The making of an artificial opening through the abdominal wall into the jejunum. { jē·jə'näs·tə·mē }

jejunum [ANAT] The middle portion of the small intestine, extending between the duodenum and the ileum. { jə'jü·nəm }

jellied gasoline See gelatinized gasoline. { 'jel·ēd 'gas·ə‚lēn }

jellium model [PHYS CHEM] A model describing the delocalized valence electrons in a metallic atom cluster in which the positive charge is regarded as being smeared out over the entire volume of the cluster while the valence electrons are free to move within this homogeneously distributed, positively charged background. [SOLID STATE] A model of electron-electron interactions in a metal in which the positive charge associated with the ion cores immersed in the sea of conduction electrons is replaced by a uniform positive background charge terminating along a plane that represents the surface of the metal. { 'jel·ē·əm ‚mäd·əl }

jelly See ulmin. { 'jel·ē }

jellyfish [INV ZOO] Any of various free-swimming marine cnidarians belonging to the Hydrozoa or Scyphozoa and having a bell- or bowl-shaped body. Also known as medusa. { 'jel·ē‚fish }

jelly fungus [MYCOL] The common name for many members of the Heterobasidiomycetidae, especially the orders Tremallales and Dacromycetales, distinguished by a jellylike appearance or consistency. { 'jel·ē ‚fəŋ·gəs }

jelutong See pontianak gum. { 'jel·ə‚tóŋ }

Jennerian vaccine See smallpox vaccine. { jə'nir·ē·ən vak'sēn }

Jenner's stain See May-Grünwald stain. { 'jen·ərz ‚stān }

jenny [VERT ZOO] **1.** A female animal, as a jenny wren. **2.** A female donkey. { 'jen·ē }

Jensen's inequality [MATH] **1.** A general inequality satisfied by a convex function

$$f\left(\sum_{i=1}^{n} a_i x_i\right) \leq \sum_{i=1}^{n} a_i f(x_i)$$

where the x_i are any numbers in the region where f is convex and the a_i are nonnegative numbers whose sum is equal to 1.

2. If $a_1, a_2, \ldots, a_n$ are positive numbers and $s > t > 0$, then $(a_1{}^s + a_2{}^s + \cdots + a_n{}^s)^{1/s}$ is less than or equal to $(a_1{}^t + a_2{}^t + \cdots + a_n{}^t)^{1/t}$. { 'jen·sənz ‚in·i'kwäl·əd·ē }

Jensen's sarcoma [VET MED] A transmissible malignant tumor originally observed in a rat inoculated with acid-fast bacteria from a cow with pseudotuberculous enteritis. { 'jen·sənz sär'kō·mə }

Jeppel's oil See bone oil. { 'jep·əlz ‚oil }

jerboa [VERT ZOO] The common name for 25 species of rodents composing the family Dipodidae; all are adapted for jumping, having extremely long hindlegs and feet. { jər'bō·ə }

jeremejevite [MINERAL] $AlBO_3$ A colorless or yellowish mineral composed of aluminum borate that occurs in hexagonal crystals. { ‚yer·ə'mā·ə‚vīt }

Jeremiassen crystallizer [CHEM ENG] Device used to grow solid crystals in a supersaturated liquid solution and to separate them from it. { ‚yer·ə'mī·ə·sən 'krist·əl‚īz·ər }

jerk [MECH] **1.** The rate of change of acceleration; it is the third derivative of position with respect to time. **2.** A unit of rate of change of acceleration, equal to 1 foot (30.48 centimeters) per second squared per second. { jərk }

jerker line [PETRO ENG] A line that radiates from a common point of power to the jack of several wells, permitting the wells to be pumped by a single power unit. { 'jərk·ər ‚līn }

jerkinhead [ARCH] Section of a roof hipped for only part of its height, forming a truncated gable on the wall below. { 'jər·kən‚hed }

jerk pump [MECH ENG] A pump that supplies a precise amount of fuel to the fuel injection valve of an internal combustion engine at the time the valve opens; used for fuel injection. { 'jərk ‚pəmp }

jerry can [ORD] A 5-gallon (19-liter), flat-sided, narrow can adapted from a German-made can, easily stacked and transported, and adapted by special openings for discharging fuel. { 'jer·ē ‚kan }

jersey [TEXT] A knitted wool, cotton, polyester, rayon, or other fabric with a slight rib on one side. { 'jər·zē }

jet [ASTRON] A narrow, elongated feature in the radio or optical map of an active galaxy, quasar, or object in the Milky Way Galaxy, believed to represent an energetic outflow of gas from a compact astronomical object. [FL MECH] A strong, well-defined stream of compressible fluid, either gas or liquid, issuing from an orifice or nozzle or moving in a contracted duct. [PARTIC PHYS] A group of particles issuing in approximately the same direction from a high-energy collision of elementary particles, believed to consist of decay products of a member of a quark-antiquark pair created in the collision. { jet }

jet aircraft [AERO ENG] An aircraft with a jet engine or engines. { 'jet 'er‚kraft }

jet bit [DES ENG] A modification of a drag bit or a roller bit that utilizes the hydraulic jet principle to increase drilling rate. { 'jet ‚bit }

jet boat [NAV ARCH] A boat propelled by one or more engines that expel powerful jets of water. { 'jet ‚bōt }

jet coal [GEOL] A hard, lustrous, pure black variety of lignite, occurring in isolated masses in bituminous shale; thought to be derived from waterlogged driftwood. Also known as black amber. { 'jet ‚kōl }

jet compressor [MECH ENG] A device, utilizing an actuating nozzle and a combining tube, for the pumping of a compressible fluid. { 'jet kəm‚pres·ər }

jet condenser [MECH ENG] A direct-contact steam condenser utilizing the aspirating effect of a jet for the removal of noncondensables. { 'jet kən'den·sər }

jet drilling [MECH ENG] A drilling method that utilizes a chopping bit, with a water jet run on a string of hollow drill rods, to chop through soils and wash the cuttings to the surface. Also known as wash boring. { 'jet ‚dril·iŋ }

jet-effect wind [METEOROL] A wind which is increased in speed through the channeling of air by some mountainous configuration, such as a narrow mountain pass or canyon. { 'jet i‚fekt ‚wind }

jet engine [AERO ENG] An aircraft engine that derives all or most of its thrust by reaction to its ejection of combustion products (or heated air) in a jet and that obtains oxygen from the atmosphere for the combustion of its fuel. [MECH ENG] Any

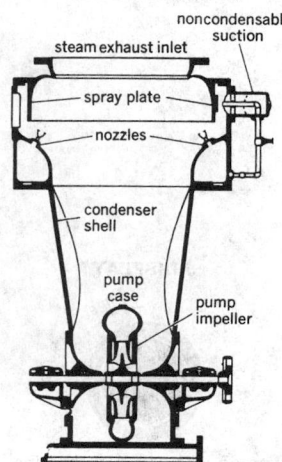

JET CONDENSER

noncondensable suction
steam exhaust inlet
spray plate
nozzles
condenser shell
pump case
pump impeller

A low-level jet condenser. (C. H. Wheeler Manufacturing Co.)

engine that ejects a jet or stream of gas or fluid, obtaining all or most of its thrust by reaction to the ejection. { ¦jet ¦en·jən }

jet-flame drill [MIN ENG] A mining drill that utilizes a high-velocity flame to spall out a hole. { 'jet ¦flām ‚dril }

jet flap [AERO ENG] A sheet of fluid discharged at high speed close to the trailing edge of a wing so as to induce lift over the whole wing. { 'jet 'flap }

jetfoil [NAV ARCH] A hydrofoil craft propelled by one or more gas turbine engines that expel powerful jets of water. { 'jet‚fȯil }

jet fuel [MATER] Special grade of kerosine with a flash point of 125°F (52°C), used for jet aircraft; may have methane or naphthene added to produce a 110°F (43°C) flash point, for military aircraft. { 'jet ¦fyül }

jet hole [ENG] A borehole drilled by use of a directed, forceful stream of fluid or air. { 'jet ‚hōl }

jet-membrane method [NUCLEO] A method of uranium isotope separation in which a rarefied vapor jet interacts with a background gas of uranium hexafluoride, causing preferential diffusion of uranium-235 into the jet; the enriched gas is extracted from the interaction region between the jet and the uranium hexafluoride by using small channels. { 'jet 'mem ‚brān ‚meth·əd }

jet mill See fluid-energy mill. { 'jet ‚mil }

jet mixer [MECH ENG] A type of flow mixer or line mixer, depending on impingement of one liquid on the other to produce mixing. { 'jet 'mik·sər }

jet molding [ENG] Molding method in which most of the heat is applied to the material to be molded as it passes through a nozzle or jet, rather than in a conventional heating cylinder. { 'jet ‚mōl·diŋ }

jet nozzle [DES ENG] A nozzle, usually specially shaped, for producing a jet, such as the exhaust nozzle on a jet or rocket engine. { 'jet 'näz·əl }

jet-piercing drill See fusion-piercing drill. { 'jet ¦pir·siŋ ‚dril }

jet propulsion [AERO ENG] The propulsion of a rocket or other craft by means of a jet engine. [ENG] Propulsion by means of a jet of fluid. { 'jet prə¦pəl·shən }

jet pump [MECH ENG] A pump in which an accelerating jet entrains a second fluid to deliver it at elevated pressure. { 'jet ‚pəmp }

jetsam [ENG] Articles that sink when thrown overboard, particularly those jettisoned for the purpose of lightening a vessel in distress. { 'jet·səm }

jet spinning [ENG] Production of plastic fibers in which a directed blast or jet of hot gas pulls the molten polymer from a die lip; similar to melt spinning. { 'jet ¦spin·iŋ }

jet streak [METEOROL] A region within the jet stream exhibiting wind speeds higher than the jet stream itself. { 'jet ‚strēk }

jet stream [AERO ENG] The stream of gas or fluid expelled by any reaction device, in particular the stream of combustion products expelled from a jet engine, rocket engine, or rocket motor. [METEOROL] A relatively narrow, fast-moving wind current flanked by more slowly moving currents; observed principally in the zone of prevailing westerlies above the lower troposphere, and in most cases reaching maximum intensity with regard to speed and concentration near the tropopause. { 'jet ‚strēm }

jetting [CIV ENG] A method of driving piles or well points into sand by using a jet of water to break the soil. [ENG] During molding of plastics, the turbulent flow of molten resin from an undersized gate or thin section into a thicker mold section, as opposed to laminar, progressive flow. { 'jed·iŋ }

jetting tool [PETRO ENG] Downhole device that jets a high-pressure, sand-laden fluid stream to clean out wellbore holes, to disintegrate perforating pipe, and to perform other operations. { 'jed·iŋ ‚tül }

jettison [ENG] The throwing overboard of objects, especially to lighten a craft in distress. { 'jed·ə·sən }

jet tones [ACOUS] Unsteady tones produced when a stream of air issues into still air from an orifice. { 'jet ‚tōnz }

jetty See groin. { 'jed·ē }

Jevons effect [METEOROL] The effect upon the measurement of rainfall caused by the presence of the rain gage; in 1861 W.S. Jevons pointed out that the rain gage causes a disturbance in airflow past it, and this carries part of the rain past the gage which would normally be captured. { 'jev·ənz i‚fekt }

jewel [ENG] **1.** A bearing usually made of synthetic corundum and used in precision timekeeping devices, gyros, and other instruments. **2.** A bearing lining of soft metal, used in railroad cars, for example. { 'jül }

jeweler's rouge See ferric oxide. { 'jü·lərz 'rüzh }

jewelry alloy [MET] Any ductile, malleable alloy, usually bronze, of good corrosion resistance, used as a base metal in jewelry. { 'jül·rē ‚al‚ȯi }

jezekite See morinite. { 'jez·ə‚kīt }

J factor [THERMO] A dimensionless equation used for the calculation of free convection heat transmission through fluid films. { 'jā ‚fak·tər }

JFET See junction field-effect transistor. { 'jā‚fet }

J function [GEOPHYS] A dimensionless mathematical relationship to correlate capillary pressure data of similar geologic formations. { 'jā ‚fəŋk·shən }

jib [NAV ARCH] A triangular sail bent to a foremast stay. { 'jib }

jib boom [MECH ENG] An extension that is hinged to the upper end of a crane boom. [NAV ARCH] A spar used as an extension of the bowsprit on sailing ships. { 'jib ‚büm }

jib crane [MECH ENG] Any of various cranes having a projecting arm (jib). { 'jib ‚krān }

jib end [MIN ENG] The delivery end in conveyor systems in which a jib is fitted to deliver the load in advance of and remote from the drive. { 'jib ‚end }

jig [ENG] A machine for dyeing piece goods by moving the cloth at full width (open width) through the dye liquor on rollers. [MECH ENG] A device used to position and hold parts for machining operations and to guide the cutting tool. [MIN ENG] A vibrating device in which coal is cleaned and ore is concentrated in water. { 'jig }

jig back [MECH ENG] An aerial ropeway with a pair of containers that move in opposite directions and are loaded or stopped alternately at opposite stations but do not pass around the terminals. Also known as reversible tramway; to-and-fro ropeway. { 'jig ‚bak }

jig borer [MECH ENG] A machine tool resembling a vertical milling machine designed for locating and drilling holes in jigs. { 'jig ‚bȯr·ər }

jigger See jigging conveyor. { 'jig·ər }

jigger boss [MIN ENG] A first-line supervisor in some western United States mines. { 'jig·ər ‚bȯs }

jiggering [ENG] A mechanization of the ceramic-forming operation consisting of molding the outside of a piece by throwing plastic clay on a plaster of paris mold, placing the mold and clay on a rotating head, and forming the inner surface by forcing a template or jigger knife against the clay; method used in mass-producing dinnerware. { 'jig·ə·riŋ }

jigging [MIN ENG] A gravity method which separates mineral from gangue particles by utilizing an effective difference in settling rate through a periodically dilated bed. { 'jig·iŋ }

jigging conveyor [MIN ENG] A series of steel troughs suspended from the roof of the stope or laid on rollers on its floor, and given reciprocating motion mechanically, to move mineral. Also known as chute conveyor; jigger; pan conveyor. { 'jig·iŋ kən‚vā·ər }

jig grinder [MECH ENG] A precision grinding machine used to locate and grind holes to size, especially in hardened steels and carbides. { 'jig ‚grīn·dər }

jigsaw [MECH ENG] A tool with a narrow blade suitable for cutting intricate curves and lines. { 'jig‚sȯ }

jig washer [MIN ENG] A coal or mineral washer for relatively coarse material; the broken ore is placed on a screen and pulsed vertically with water; the heavy portion passes through the screen and the light portion goes over the sides. { 'jig ‚wäsh·ər }

jim crow [DES ENG] A device with a heavy buttress screw thread used for bending rails by hand. { 'jim ‚krō }

jimsonweed [BOT] Datura stramonium. A tall, poisonous annual weed having large white or violet trumpet-shaped flowers and globose prickly fruits. Also known as apple of Peru. { 'jim‚sən‚wēd }

J indicator See J scope. { 'jā ‚in·də‚kād·ər }

jird [VERT ZOO] Any one of the diminutive rodents composing related species of the genus Meriones which are inhabitants of northern Africa and southwestern Asia; they serve as experimental hosts for studies of schistosomiasis. { jərd }

JIT See just-in-time.

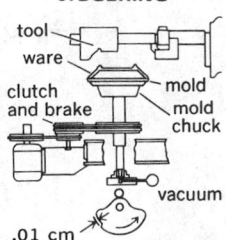

JIGGERING

tool
ware
clutch and brake
mold
mold chuck
vacuum
.01 cm

The jiggering tool and cam-raised jiggering apparatus used for plastic forming of clay. (From W. D. Kingery, ed., Ceramic Fabrication Processes, Technology Press, MIT, and Wiley, 1958)

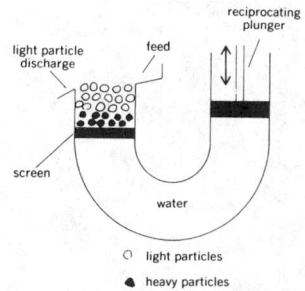

JIGGING

light particle discharge
reciprocating plunger
feed
screen
water
○ light particles
◆ heavy particles

The principle of the jigging method for beneficiation of ore.

jitter [COMMUN] In facsimile, distortion in the received copy caused by momentary errors in synchronism between the scanner and recorder mechanisms; does not include slow errors in synchronism due to instability of the frequency standards used in the facsimile transmitter and recorder. [ELECTR] Small, rapid variations in a waveform due to mechanical vibrations, fluctuations in supply voltages, control-system instability, and other causes. { 'jid·ər }

jittered pulse recurrence frequency [COMMUN] Random variation of the pulse repetition period; provides a discrimination capability against repeater-type jammers. { 'jid·ərd ¦pəls ri¦kə·rəns 'frē·kwən·sē }

j-j coupling [ATOM PHYS] A process for building up many-electron wave functions; the spin and orbital functions of each particle are combined to form eigenfunctions of the particle's total angular momentum, and then the wave functions of all the particles are combined to form eigenfunctions of the total angular momentum of the system; this coupling is used when the spin-orbit interaction is strong compared to the electrostatic interaction. { ¦jā¦jā ˌkəp·liŋ }

J-K flip-flop [ELECTR] A storage stage consisting only of transistors and resistors connected as flip-flops between input and output gates, and working with charge-storage transistors; gives a definite output even when both inputs are 1. { ¦jā¦kā 'flip·ˌfläp }

jnd *See* just-noticeable difference.

joaquinite [MINERAL] $NaBa_2Ce_2Fe(Ti,Nb)_2Si_8O_{26}(OH,F)_2$ A honey-yellow mineral composed of sodium iron titanium silicate, occurring in orthorhombic crystals. { wä'kē·ˌnīt }

job [COMPUT SCI] A unit of work to be done by the computer; it is a single entity from the standpoint of computer installation management, but may consist of one or more job steps. [IND ENG] **1.** The combination of duties, skills, knowledge, and responsibilities assigned to an individual employee. **2.** A work order. { jäb }

job analysis [IND ENG] A detailed study of the work performed, the facilities required, the working conditions, and the skills required to complete a specific job. Also known as job study. { 'jäb ə,nal·ə·səs }

jobber's reamer [DES ENG] A machine reamer that is solid with straight or helical flutes and taper shanks. { 'jäb·ərz ˌrē·mər }

job breakdown [IND ENG] Separation of an operation into elements. Also known as operation breakdown. { 'jäb 'brāk,daún }

job characteristic *See* job factor. { 'jäb ,kar·ik·tə,ris·tik }

job class [COMPUT SCI] The set of jobs on a computer system whose resource requirements (for the central processing unit, memory, and peripheral devices) fall within specified ranges. [IND ENG] A group of jobs involving a similar type of work, difficulty of performance, or range of pay. Also known as job family; job grade; labor grade. { 'jäb ,klas }

job classification [IND ENG] Designating job classes on the basis of job factors or level of pay, or on the basis of job evaluation. { 'jäb ,klas·ə·fə,kā·shən }

job control block [COMPUT SCI] A group of data containing the execution-control data and the job identification when the job is initiated as a unit of work to the operating system. { 'jäb kən,trōl ,bläk }

job control language *See* command language. { 'jäb kən,trōl ,laŋ·gwij }

job control statement [COMPUT SCI] Any of the statements used to direct an operating system in its functioning, as contrasted to data, programs, or other information needed to process a job but not intended directly for the operating system itself. Also known as control statement. { 'jäb kən,trōl ,stāt·mənt }

job description [IND ENG] A detailed description of the essential activities required to perform a task. { 'jäb di ,skrip·shən }

job design [IND ENG] The arrangement of tasks over a work shift with the goal of achieving technological and organizational requirements as well as reducing sources of fatigue and human error. Also known as work design. { 'jäb di,zīn }

job entry system [COMPUT SCI] A part of the operating system of a large computer system that accepts and schedules jobs for execution and controls the printing of output. { 'jäb ¦en·trē ,sis·təm }

job evaluation [IND ENG] Orderly qualitative appraisal of each job or position in an establishment either by a point system

for the specific job characteristics or by comparison of job factors; used for establishing a job hierarchy and wage plans. { 'jäb i,val·yə'wā·shən }

job factor [IND ENG] An essential job element which provides a basis for selecting and training employees and establishing the wage plan for the job. Also known as job characteristic. { 'jäb ,fak·tər }

job family *See* job class. { 'jäb ,fam·lē }

job flow control [COMPUT SCI] Control over the order in which jobs are handled by a computer in order to use the central processing units and the units under the computer's control as efficiently as possible. { 'jäb ,flō kən,trōl }

job grade *See* job class. { 'jäb ,grād }

job library [COMPUT SCI] A partitioned data set, or a concatenation of partitioned data sets, used as the primary source of object programs (load modules) for a particular job, and more generally, as a source of runnable programs from which all or most of the programs for a given job will be selected. { 'jäb ,li,brer·ē }

Jo block *See* Johansson block. { 'jō ,bläk }

job management program [COMPUT SCI] A control program in a computer's operating system that initials and schedules jobs. { 'jäb ¦man·ij·mənt ,prō·grəm }

job mix [COMPUT SCI] The distribution of the jobs handled by a computer system among the various job classes. { 'jäb, miks }

job-oriented terminal [COMPUT SCI] A terminal, such as a point-of-sale terminal, at which data taken directly from a source can enter a communication network directly. { ¦jäb ¦òr·ē,ent·əd 'tər·mən·əl }

job plan [IND ENG] The organized approach to production management involving formal, step-by-step procedures. { 'jäb ,plan }

job processing control [COMPUT SCI] The section of the control program responsible for initiating operations, assigning facilities, and proceeding from one job to the next. { 'jäb ¦prä·ses·iŋ kən,trōl }

job queue [COMPUT SCI] A set of computer programs that are ready to be executed in a prescribed order. { 'jäb ,kyü }

job safety analysis [IND ENG] A method of studying a job by breaking it down into its components to determine any possible hazards it may involve and the qualifications needed by those who perform it. { ¦jäb ¦sāf·tē ə,nal·ə·səs }

job schedule [CONT SYS] A control program that selects from a job queue the next job to be processed. { 'jäb ,sked·yül }

job shop [IND ENG] A manufacturing facility that generates a variety of products in relatively low numbers and in batch lots. { 'jäb ,shäp }

job stacking [COMPUT SCI] The presentation of jobs to a computer system, each job followed by another. { 'jäb ,stak·iŋ }

job step [COMPUT SCI] A unit of work in a job stream. { 'jäb ,step }

job stream [CONT SYS] A collection of jobs in a job queue. { 'jäb ,strēm }

job study *See* job analysis. { 'jäb ,stəd·ē }

job swapping [COMPUT SCI] Temporary suspension of job processing by a computer so that higher-priority jobs can be handled. { 'jäb ,swäp·iŋ }

jochwinde [METEOROL] The mountain-gap wind of the Tauern Pass in the Alps. { 'yōk,vin·də }

jodfenphos [ORG CHEM] $C_8H_8O_3Cl_2IPS$ A crystalline compound with a melting point of 76°C; slight solubility in water; used as an insecticide in homes, farm buildings, and industrial sites. { 'yòd·fən,fäs }

jog [CRYSTAL] A shift in a dislocation from one crystal plane to another. { jäg }

jogging [ELEC] Quickly repeated opening and closing of a circuit to produce small movements of the driven machine. Also known as inching. { 'jäg·iŋ }

joggle [DES ENG] **1.** A flangelike offset on a flat piece of metal. **2.** A projection or notch on a sheet of building material to prevent protrusion. **3.** A dowel for joining blocks of masonry. [MIN ENG] Trusses or sets of timbers joined for taking pressure at right angles. { 'jäg·əl }

joggled frame [NAV ARCH] A frame offset to fit the laps of the shell plating. { 'jäg·əld 'frām }

joggled plating [NAV ARCH] A plating construction used in riveted steel ships in which one or both edges of a plate are offset with respect to the edge of an adjacent one; it became essentially obsolete with the advent of all-welded ships. { 'jäg·əl 'plād·iŋ }

joggle joint [CIV ENG] In masonry or stonework, a joint between two blocks in which a projection on one fits into a recess in another. { 'jäg·əl ˌjȯint }

joggle piece *See* joggle post. { 'jäg·əl ˌpēs }

joggle post [BUILD] 1. A post constructed of two or more sections of lumber joined by joggles. 2. A king post with notches or shoulders at its lower end that provide support for the feet of the struts. Also known as joggle piece. { 'jäg·əl ˌpōst }

jog method [NAV] Following one or more doglegs to avoid a direct approach, as when it is desired to delay arrival at a destination. { 'jäg ˌmeth·əd }

Johann crystal geometry [CRYSTAL] The focusing shape of a diffracting crystal for x-ray dispersion used in electron-probe microanalysis; less stringent than Johannson crystal geometry. { 'yō¸hän ˌkrist·əl jē'äm·ə·trē }

johannite [MINERAL] $Cu(UO_2)_2(SO_4)_2 \cdot 6H_2O$ An emerald green to apple green, triclinic mineral consisting of a hydrated basic copper and uranium sulfate. { jō'ha¸nīt }

johannsenite [MINERAL] $CaMnSi_2O_6$ A clove-brown, grayish, or greenish clinopyroxene mineral composed of a silicate of calcium and manganese; a member of the pyroxene group. { jō'han·sə¸nīt }

Johannson crystal geometry [CRYSTAL] The full-focusing shape of a diffracting crystal for x-ray dispersion used in electron-probe microanalyzers; more stringent than Johann crystal geometry. { jō'han·sən ˌkrist·əl jē'äm·ə·trē }

Johansson block [DES ENG] A type of gage block ground to an accuracy of at least 1/100,000 inch (0.25 micrometer). Also known as Jo block. { jo'han·sən ¸bläk }

Johne's disease [VET MED] A chronic inflammation of the intestinal tract of sheep, cattle, and deer caused by *Mycobacterium paratuberculosis*. { 'yō·nəz di¸zēz }

Johnson and Lark-Horowitz formula [SOLID STATE] A formula according to which the resistivity of a metal or degenerate semiconductor resulting from impurities which scatter the electrons is proportional to the cube root of the density of impurities. { 'jän·sən ən ¦lärk 'här·ə¸witz ˌfȯr·myə·lə }

Johnson concentrator [MIN ENG] A device used to separate heavy particles such as metallic gold from auriferous pulp; composed of a shell in the shape of a cylinder that is lined with rubber grooves parallel to the inclined axis. { 'jän·sən 'käns·ən¸trād·ər }

Johnson-Morgan system *See* UBV system. { 'jän·sən 'mȯr·gən ¸sis·təm }

Johnson noise *See* thermal noise. { 'jän·sən ¸nȯiz }

Johnson-Rahbeck effect [PHYS] An increase in frictional force between two electrodes in contact with a semiconductor that arises when a potential difference is applied between the electrodes. { 'jän·sən 'rä¸bek i¸fekt }

johnstrupite [MINERAL] A mineral that is composed of a complex silicate of cerium and other metals, approximately $(Ca,Na)_3(Ce,Ti,Zr)(SiO_4)_2F$; occurs in prismatic crystals. { 'jän·strə¸pīt }

join [COMPUT SCI] A portion of a robotic control program that directs an activity to resume after it has been interrupted. [MATH] 1. The join of two elements of a lattice is their least upper bound. 2. *See* union. { jȯin }

joiner plans [NAV ARCH] Plans showing the arrangement of quarters and living spaces on a ship. { 'jȯin·ər ¸planz }

join-irreducible member [MATH] A member, *A*, of a lattice or ring of sets such that, if *A* is equal to the join of two other members, *B* and *C*, then *A* equals *B* or *A* equals *C*. { 'jȯin ¸ir·i¸dü·sə·bəl 'mem·bər }

joint [ANAT] A contact surface between two individual bones. Also known as articulation. [ELEC] A juncture of two wires or other conductive paths for current. [ENG] The surface at which two or more mechanical or structural components are united. [GEOL] A fracture that traverses a rock and does not show any discernible displacement of one side of the fracture relative to the other. { jȯint }

joint bar [CIV ENG] A rigid steel member used in pairs to join, hold, and align rail ends. { 'jȯint ¸bär }

joint block [GEOL] A body of rock that is bounded by joints. { 'jȯint ¸bläk }

joint body *See* joint mouse. { 'jȯint ¸bäd·ē }

joint buildup sequence [MET] The sequence in which weld beads are deposited relative to the cross section of a multiple-pass joint. { 'jȯint 'bil·dəp ¸sē·kwəns }

joint capsule [ANAT] A sheet of fibrous connective tissue enclosing a synovial joint. { 'jȯint ¸kap·səl }

joint clearance [ENG] The distance between mating surfaces of a joint. { 'jȯint ¸klir·əns }

joint compound [MATER] A material used primarily to lubricate the threads of pipe joints and secondarily to prevent joint leakage. { 'jȯint ¸käm¸paund }

joint distribution [STAT] For two random variables *Z* and *W*, the distribution which gives the probability that $Z = z$ and $W = w$ for all values *z* and *w* of *Z* and *W* respectively. { 'jȯint ¸dis·trə¸byü·shən }

joint drag *See* kink band. { 'jȯint ¸drag }

jointed-arm robot [CONT SYS] A robot whose arm is constructed of rigid members connected by rotary joints. Also known as revolute-coordinate robot. { 'jȯint·əd ¸ärm 'rō¸bät }

joint efficiency [MET] A numerical value expressed as the ratio of the strength of a riveted, welded, or brazed joint to the strength of the parent metal. { 'jȯint ə¸fish·ən·sē }

jointer [ENG] 1. Any tool used to prepare, make, or simulate joints, such as a plane for smoothing wood surfaces prior to joining them, or a hand tool for inscribing grooves in fresh cement. 2. A file for making sawteeth the same height. 3. An attachment to a plow that covers discarded material. 4. A worker who makes joints, particularly a construction worker who cuts stone to proper fit. 5. A pipe of random length made from two joined, relatively short lengths. { 'jȯint·ər }

jointer gage [DES ENG] An attachment to a bench vise that holds a board at any angle desired for planing. { 'jȯint·ər ¸gāj }

jointing [CIV ENG] Caulking of masonry joints. [GEOL] A condition of rock characterized by joints. [ENG] A basic woodworking process for trueing or smoothing one surface of a workpiece by using a single peripheral cutting head in order to prepare the workpiece for further processing. { 'jȯint·iŋ }

joint marginal distribution [STAT] The distribution obtained by summing the joint distribution of three random variables over all possible values of one of these variables. { 'jȯint ¸märj·ən·əl ¸dis·trə'byü·shən }

joint mouse [PATH] A small loose body within a synovial joint, frequently calcified, derived from synovial membrane, organized fibrin fragments of articular cartilage, or arthritic osteophytes. Also known as joint body. { 'jȯint ¸maus }

joint penetration [MET] The distance extended into a weld joint by the weld metal or fusion zone. { 'jȯint ¸pen·ə'trā·shən }

Joint Photographic Experts Group [COMPUT SCI] An international group that sets standards for continuous-tone image (still and video) coding. { 'jȯint ¸fōd·ə¸graf·iks 'ek¸sparts ¸grüp }

joint plane [GEOL] The surface of fracturing or potential fracture of a joint. { 'jȯint ¸plān }

joint pole [ELEC] Pole used in common by two or more utility companies. { 'jȯint 'pōl }

joint ring [DES ENG] A pipe-joint flange whose outside diameter is less than the diameter of the circle containing the connecting bolts and thus fits inside the bolts. { 'jȯint ¸riŋ }

joint set [GEOL] A group of parallel joints in a geologic formation. { 'jȯint ¸set }

joint space [CONT SYS] The space defined by a vector whose components are the translational and angular displacements of each joint of a robotic link. { 'jȯint ¸spās }

joint system [GEOL] Two or more joint sets. { 'jȯint ¸sis·təm }

joint variation [MATH] The relation of a variable *x* to two other variables *y* and *z* wherein *x* is proportional to the product of *y* and *z*. { 'jȯint ¸ver·ē'ā·shən }

joint vein [GEOL] A small vein in a joint. { 'jȯint ¸vān }

joist [CIV ENG] A steel or wood beam providing direct support for a floor. { jȯist }

joist anchor *See* wall anchor. { 'jȯist ¸aŋ·kər }

jojoba [AGR] *Simmondsia californica*. A shrub adapted to the arid portions of the southwestern United States and Mexico; liquid extracted from the seed may be used as a substitute for sperm oil. { hō'hō·bə }

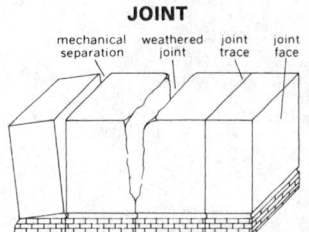

JOINT

mechanical separation   weathered joint   joint trace   joint face

Aspects of systematic joint, a type of geological joint.

Jolly balance [ENG] A spring balance used to measure specific gravity of mineral specimens by weighing a specimen when in the air and when immersed in a liquid of known density. { 'jal·ē ,bal·əns }

jolt-and-jumble tests [ORD] A standardized program of tests intended to simulate the shocks to which various components of ammunition are subjected in transportation and handling. { 'jōlt ən 'jəm·bəl ,tests }

jolt molding [ENG] A process for shaping refractory blocks in which a mold containing prepared batch is jolted mechanically to consolidate the material. { 'jōlt ¦mōl·diŋ }

Joly photometer [OPTICS] A photometer consisting of two equal paraffin wax or opal glass blocks separated by a thin opaque sheet; the positions of two light sources under comparison are adjusted until the two blocks appear equally bright. Also known as wax-block photometer. { ¦jäl·ē fō'täm·əd·ər }

Joly steam calorimeter [ENG] **1.** A calorimeter in which the mass of steam that condenses on a specimen and a pan holding it is measured, as well as the mass of steam that condenses on an empty pan. **2.** *See* differential steam calorimeter. { ¦jäl·ē ¦stēm ,kal·ə'rim·əd·ər }

Jominy end quench test *See* Jominy test. { 'jäm·ə·nē 'end ,kwench ,test }

Jominy test [MET] A hardenability test in which a steel bar is heated to the desired austenitizing temperature and quench-hardened at one end and then measured for hardness along its length, beginning at the quenched end. Also known as Jominy end quench test. { 'jäm·ə·nē ,test }

Jonathan freckle [PL PATH] A storage disease of apples characterized by small circular discolorations in the skin. { 'jän·ə·thən ,frek·əl }

Jonathan spot [PL PATH] A disease of apples characterized by circular depressed necrotic areas around the lenticels. { 'jän·ə·thən ,spät }

Jones reductor [CHEM] A device used to chemically reduce solutions, such as ferric salt solutions, consisting of a vertical tube containing granular zinc into which the solution is poured. { 'jōnz ri,dək·tər }

Jones riffle [MIN ENG] An apparatus used to reduce the size of a sample to a desired weight; consists of a hopper which passes samples to a series of open-bottom pockets, each of which divides the sample into two equal parts, and the next pass of each part gives a quarter of the original sample, and so on until the desired sample is obtained. { 'jōnz ,rif·əl }

Jones splitter [MIN ENG] A device used to reduce the volume of a sample, consisting of a belled, rectangular container, the bottom of which is fitted with a series of narrow slots or alternating chutes designed to cast material in equal quantities to opposite sides of the device. { 'jōnz ,splid·ər }

Jones sucker rod [PETRO ENG] Connecting rod between the subsurface pump and the lifting or pumping device on the surface; serves to lift oil out of the cased hole. { 'jōnz 'sək·ər ,räd }

jonquil oil [MATER] A colorless oil obtained from flowers of jonquil (*Narcissus jonquilla*); used in perfumes. Also known as narcissus oil. { 'jän·kwəl ,oil }

Joppeicidae [INV ZOO] A monospecific family of hemipteran insects included in the Pentatomorpha; found in the Mediterranean regions. { jäp·ə'īs·ə,dē }

joran *See* juran. { 'jōr·ən }

jordan [MECH ENG] A machine or engine used to refine paper pulp, consisting of a rotating cone, with cutters, that fits inside another cone, also with cutters. { 'jord·ən }

Jordan algebra [MATH] A nonassociative algebra over a field in which the products satisfy the Jordan identity $(xy)x^2 = x(yx^2)$. { zhòr'dän ,al·jə·brə }

Jordan arc *See* simple arc. { zhòr'dän ,ärk }

Jordan condition [MATH] A condition for the convergence of a Fourier series of a function f at a number x, namely, that there be a neighborhood of x on which f is of bounded variation. { zhòr'dän kən¦dish·ən }

Jordan content [MATH] For a set whose exterior Jordan content and interior Jordan content are equal, the common value of these two quantities. Also known as content. { zhòr'dän ,kän,tent }

Jordan curve [MATH] A simple closed curve in the plane, that is, a curve that is closed, connected, and does not cross itself. { zhòr'dän ,kərv }

Jordan curve theorem [MATH] The theorem that in the plane every simple closed curve separates the plane into two parts. { zhòr'dän ,kərv ,thir·əm }

Jordan form [MATH] A matrix that has been transformed into a Jordan matrix is said to be in Jordan form. { 'zhòr·dän ,fòrm }

Jordan-Hölder theorem [MATH] The theorem that for a group any two composition series have the same number of subgroups listed, and both series produce the same quotient groups. { zhòr'dän 'hùl·dər ,thir·əm }

jordanite [MINERAL] $(Pb,Tl)_{13}As_7S_{23}$ A lead-gray mineral composed of lead arsenic sulfide, occurring as monoclinic crystals. { 'jord·ən,īt }

Jordan lag [ELECTROMAG] A type of magnetic viscosity in which the angular lag of the magnetic induction behind a sinusoidally varying magnetic field strength, and also the energy loss per cycle, is independent of frequency. { 'jord·ən ,lag }

Jordan matrix [MATH] A matrix whose elements are equal and nonzero on the principal diagonal, equal to 1 on the diagonal immediately above, and equal to 0 everywhere else. { zhòr'dän ,mā·triks }

jordanon *See* microspecies. { 'jord·ən,än }

Jordan's rule [EVOL] The rule that organisms which are closely related tend to occupy adjacent rather than identical or distant ranges. [VERT ZOO] The rule that fishes in areas of low temperatures tend to have more vertebrae than those in warmer waters. { 'jord·ənz ,rül }

Jordan sunshine recorder [ENG] A sunshine recorder in which the time scale is supplied by the motion of the sun; it consists of two opaque metal semicylinders mounted with their curved surfaces facing each other; each of the semicylinders has a short narrow slit in its flat side; sunlight entering one of the slits falls on light-sensitive paper (blueprint paper) which lines the curved side of the semicylinder. { 'jord·ən 'sən,shīn ri,kord·ər }

Jordan-Wigner commutation rules [QUANT MECH] Rules obtained by replacing commutators of creation and destruction operators by anticommutators; applicable to fermion fields in a quantized field theory. { 'jord·ən 'wig·nər ,käm·yə'tā·shən rülz }

joseite [MINERAL] $Bi_3Te(Si,S)$ A mineral composed of telluride of bismuth containing sulfur and selenium. { zhə'zā,īt }

josen *See* hartite. { 'jō·sən }

josephinite [MINERAL] A mineral consisting of an alloy of iron and nickel; occurs naturally in stream gravel. { ¦jō·zə¦fē,nīt }

Josephson constant [PHYS] **1.** The quantity $K_J = 2e/h$, which appears in the equations for the alternating-current Josephson effect, where e is the magnitude of the charge of the electron and h is Planck's constant. **2.** The conventional value of this quantity adopted by international agreement on January 1, 1990, to establish a standard for the volt, $K_{J-90} = 493,597.9$ gigahertz per volt. { 'jō·sef·sən ,kän·stənt }

Josephson current [CRYO] The current across a Josephson junction in the absence of voltage across the junction, resulting from the Josephson effect. { 'jō·səf·sən ,kə·rənt }

Josephson effect [CRYO] The tunneling of electron pairs through a thin insulating barrier between two superconducting materials. Also known as Josephson tunneling. { 'jō·səf·sən i,fekt }

Josephson equation [CRYO] An equation according to which the Josephson current is a sinusoidally varying function of the applied magnetic field. { 'jō·səf·sən i,kwä·zhən }

Josephson junction [CRYO] A thin insulator separating two superconducting materials; it displays the Josephson effect. { 'jō·səf·sən ,jəŋk·shən }

Josephson penetration depth [CRYO] A measure of the distance that a magnetic field extends into a Josephson junction. { 'jō·səf·sən 'pen·ə'trā·shən ,depth }

Josephson tunneling *See* Josephson effect. { 'jō·səf·sən ,tən·əl·iŋ }

Joshi effect [ELECTR] The change in the current passing through a gas or vapor when the gas or vapor is irradiated with visible light. { 'jō·shē i,fekt }

JOSS [COMPUT SCI] A time-sharing language, designed for concurrent use by a number of people, each at his own console typewriter, and for programs of moderate size and running time. { 'jäs }

Joukowski profile [FL MECH] An airfoil profile with a cusp-shaped trailing edge, resulting from the Joukowski transformation of a circle which passes through the point $z = a$ and which is located so that the point $z = -a$ does not lie outside the circle. { yü'kȯf·skē ‚prō·fīl }

Joukowski transformation [FL MECH] A conformal mapping used to transform circles into airfoil profiles for the purpose of studying fluid flow past the airfoil profiles; it assigns to each complex number z the number $w = z + (a^2/z)$. { yü'kȯf·skē ‚tranz·fər‚mā·shən }

joule [MECH] The unit of energy or work in the meter-kilogram-second system of units, equal to the work done by a force of 1 newton magnitude when the point at which the force is applied is displaced 1 meter in the direction of the force. Symbolized J. Also known as newton-meter of energy. { jül or jaül }

Joule and Playfairs' experiment [THERMO] An experiment in which the temperature of the maximum density of water is measured by taking the mean of the temperatures of water in two columns whose densities are determined to be equal from the absence of correction currents in a connecting trough. { ¦jül and 'plā‚färz ik‚sper·ə·mənt }

Joule calorimeter [ENG] Any electrically heated calorimeter, such as that used in the Griffiths method. { ¦jül ‚kal·ə'rim· əd·ər }

Joule-Clausius velocity [STAT MECH] A quantity used in the description of the kinetic behavior of a gas, equal to the square root of the ratio of the pressure of the gas to one-third of its density. { ¦jül 'klaüz·ē·əs və‚läs·əd·ē }

Joule cycle See Brayton cycle. { 'jül ‚sī·kəl }

Joule effect [PHYS] **1.** The heating effect produced by the flow of current through a resistance. **2.** A change in the length of a ferromagnetic substance which occurs parallel to an applied magnetic field. Also known as Joule magnetostriction; longitudinal magnetostriction. { 'jül i‚fekt }

Joule equivalent [THERMO] The numerical relation between quantities of mechanical energy and heat; the present accepted value is 1 fifteen-degrees calorie equals 4.1855 ± 0.0005 joules. Also known as mechanical equivalent of heat. { 'jül i‚kwiv·ə·lənt }

Joule experiment [THERMO] **1.** An experiment to detect intermolecular forces in a gas, in which one measures the heat absorbed when gas in a small vessel is allowed to expand into a second vessel which has been evacuated. **2.** An experiment to measure the mechanical equivalent of heat, in which falling weights cause paddles to rotate in a closed container of water whose temperature rise is measured by a thermometer. { 'jül ik‚sper·ə·mənt }

Joule heat [ELEC] The heat which is evolved when current flows through a medium having electrical resistance, as given by Joule's law. { 'jül ‚hēt }

Joule-Kelvin effect See Joule-Thomson effect. { 'jül 'kel· vən i‚fekt }

Joule magnetostriction See Joule effect. { 'jül mag‚ned· ō'ri‚strik·shən }

Joule's law [ELEC] The law that when electricity flows through a substance, the rate of evolution of heat in watts equals the resistance of the substance in ohms times the square of the current in amperes. [THERMO] The law that at constant temperature the internal energy of a gas tends to a finite limit, independent of volume, as the pressure tends to zero. { 'jülz ‚lȯ }

Joule-Thomson coefficient [THERMO] The ratio of the temperature change to the pressure change of a gas undergoing isenthalpic expansion. { 'jül 'täm·sən ‚kō·ə‚fish·ənt }

Joule-Thomson effect [THERMO] A change of temperature in a gas undergoing Joule-Thomson expansion. Also known as Joule-Kelvin effect. { 'jül 'täm·sən i‚fekt }

Joule-Thomson expansion [THERMO] The adiabatic, irreversible expansion of a fluid flowing through a porous plug or partially opened valve. Also known as Joule-Thomson process. { 'jül 'täm·sən ik‚span·chən }

Joule-Thomson inversion temperature [THERMO] A temperature at which the Joule-Thomson coefficient of a given gas changes sign. { 'jül 'täm·sən in'vər·zhən ‚tem·prə·chər }

Joule-Thomson process See Joule-Thomson expansion. { 'jül 'täm·sən ‚prä·səs }

Joule-Thomson valve [CRYO] A valve through which a gas is allowed to expand adiabatically, resulting in lowering of its temperature; used in production of liquid hydrogen and helium. { 'jül 'täm·sən ‚valv }

journal [MECH ENG] That part of a shaft or crank which is supported by and turns in a bearing. { 'jərn·əl }

journal bearing [MECH ENG] A cylindrical bearing which supports a rotating cylindrical shaft. { 'jərn·əl ‚ber·iŋ }

journal box [ENG] A metal housing for a journal bearing. { 'jərn·əl ‚bäks }

journal friction [MECH ENG] Friction of the axle in a journal bearing arising mainly from viscous sliding friction between journal and lubricant. { 'jərn·əl ‚frik·shən }

journaling [COMPUT SCI] Recording processes or transactions for backup or accounting purposes. { 'jər·nəl·iŋ }

journal oil [MATER] A special grade of lubricating oil for use on journal bearings. { 'jərn·əl ‚ȯil }

JOVIAL [COMPUT SCI] A procedure-oriented language derived from ALGOL, commonly used in programming command and control procedures. { 'jō·vē·əl }

Jovian planet [ASTRON] Any of the four major planets (Jupiter, Saturn, Uranus, and Neptune) that are at a greater distance from the sun than the terrestrial planets (Mercury, Venus, Earth, and Mars). { 'jō·vē·ən 'plan·ət }

Jovian Van Allen belts [ASTROPHYS] The extended belts of high-energy charged particles that are trapped in Jupiter's magnetic field and cause the microwave nonthermal emission of radio waves observed in the band from about 3 to 70 centimeters. { 'jō·vē·ən ‚van 'al·ən ‚belts }

Joy double-ended miner [MIN ENG] A cutter-loader for continuous mining on a longwall face. { 'jȯi ¦dəb·əl ‚end·əd 'mīn·ər }

Joy extensible conveyor [MIN ENG] A type of belt conveyor consisting of a head section and a tail section, each mounted on crawler tracks and independently driven; used between a loader or continuous miner and the main transport. { 'jȯi ik¦sten·sə·bəl kən'vā·ər }

Joy extensible steel band [MIN ENG] A hydraulically driven system linking a continuous miner to the main transport; the steel band is coiled on the drivehead. { 'jȯi ik¦sten·sə·bəl ¦stēl 'band }

Joy loader [MIN ENG] A loading machine which uses mechanical arms to collect coal or ore onto an apron that is pushed onto the broken material. { 'jȯi ¦lōd·ər }

Joy longwall loading machine [MIN ENG] A modified Joy loader comprising a hydraulically elevated loading head fitted with mechanical gathering arms. { 'jȯi ¦lȯŋ‚wȯl 'lōd·iŋ mə‚shēn }

Joy microdyne [MIN ENG] A dust collector that wets and traps dust pulled into it and releases the dust as a slurry to be removed by pumps; used at the return end of tunnels or hard headings. { 'jȯi 'mī·krə‚dīn }

Joy miner [MIN ENG] A continuous miner weighing about 15 tons (13,600 kilograms) and made up of a turntable, a ripper bar, and a discharge boom conveyor; used mainly in coal headings and in extraction of coal pillars. { 'jȯi 'mīn·ər }

joystick [AERO ENG] A lever used to control the motion of an aircraft; fore-and-aft motion operates the elevators, while lateral motion operates the ailerons. [ENG] A two-axis displacement control operated by a lever or ball, for XY positioning of a device or an electron beam. { 'jȯi‚stik }

Joy-Sullivan hydrodrill rig [MIN ENG] A drill rig set on a jib or boom which can be moved to and locked in any position by hydraulic power controlled from the drill carriage. { 'jȯi 'səl·ə·vən 'hī drō‚dril ‚rig }

Joy transloader [MIN ENG] A rubber-tired, self-propelled machine for loading, transporting, and dumping. { 'jȯi 'tran‚z‚lōd·ər }

Joy walking miner [MIN ENG] A continuous miner designed to make thin seams; a walking mechanism is used instead of caterpillar tracks for moving the miner. { 'jȯi ¦wȯk·iŋ 'mīn· ər }

JP-4 [MATER] Jet engine test fuel made up of 35% light petroleum distillates and 65% gasoline distillates.

JP-5 [MATER] Military jet engine fuel made of specially refined kerosine with specified flash and freezing points.

J particle [PARTIC PHYS] A neutral meson which has a mass of 3095 megaelectronvolts, spin quantum number 1, and negative parity and charge parity; it has an anomalously long lifetime of approximately 10^{-20} second (corresponding to a width of

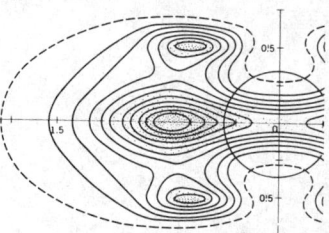

JOVIAN VAN ALLEN BELTS

Contours of radio emission from Jupiter originating from one-half of the Van Allen belt, and extending far beyond the planet's disk.

approximately 70 kiloelectronvolts). Also known as psi particle (symbolized ψ). { 'jā ‚pärd·ə·kəl }

jpd *See* just-perceptible difference.

JPEG [GRAPHICS] Graphics file format for compressed still images, particularly photographic images found on the World Wide Web; developed by the Joint Photographic Experts Group. { 'jā‚peg }

J scan *See* J scope. { 'jā ‚skan }

J scope [ELECTR] A modification of an A scope in which the trace appears as a circular range scale near the circumference of the cathode-ray tube face, the signal appearing as a radial deflection of the range scale; no bearing indication is given. Also known as J indicator; J scan. { 'jā ‚skōp }

J-shaped distribution [STAT] A frequency distribution that is extremely asymmetrical in that the initial (or final) frequency group contains the highest frequency, with succeeding frequencies becoming smaller (or larger) elsewhere; the shape of the curve roughly approximates the letter "J" lying on its side. { ¦jā ‚shāpt ‚dis·trə'byü·shən }

judgment sample [STAT] Sample selection in which personal views or opinions of the individual doing the sampling enter into the selection. { 'jəj·mənt ‚sam·pəl }

Judson powder [MATER] An explosive containing sodium nitrate, coal, sulfur, and some nitroglycerin. { 'jəds·ən ‚paùd·ər }

jugal [ANAT] Pertaining to the zygomatic bone. [VERT ZOO] In lower vertebrates, a bone lying below the orbit of the eye. { 'jü·gəl }

Jugatae [INV ZOO] The equivalent name for Homoneura. { 'jü·gə‚tē }

jugate [BIOL] Structures which are joined together. { 'jü‚gāt }

Juglandaceae [BOT] A family of dicotyledonous plants in the order Juglandales having unisexual flowers, a solitary basal ovule in a unilocular inferior ovary, and pinnately compound, exstipulate leaves. { ‚jü‚glan'dās·ē‚ē }

Juglandales [BOT] An order of dicotyledonous plants in the subclass Hamamelidae distinguished by compound leaves; includes hickory, walnut, and butternut. { ‚jü‚glan'dā·lēz }

juglone [ORG CHEM] $C_{10}H_6O_3$ A naphthoquinone derivative that occurs naturally in black walnuts and is toxic to plants. { 'jəg‚lōn }

jugular [ANAT] Pertaining to the region of the neck above the clavicle. { 'jəg·yə·lər }

jugular compression [PATH] A test for a spinal subarachnoid block by noting the rate of rise and fall of the spinal fluid pressure following compression and release of the jugular veins. { 'jəg·yə·lər kəm'presh·ən }

jugular foramen [ANAT] An opening in the cranium formed by the jugular notches of the occipital and temporal bones for passage of an internal jugular vein, the ninth, tenth, and eleventh cranial nerves, and the inferior petrosal sinus. { 'jəg·yə·lər fə'rā·mən }

jugular foramen syndrome [MED] Injury to the jugular foramen, usually due to a basilar skull fracture, with resultant paralysis of ninth, tenth, and eleventh cranial nerves. { 'jəg·yə·lər fə'rā·mən ‚sin‚drōm }

jugular process [ANAT] A rough process external to the condyle of the occipital bone. { 'jəg·yə·lər ‚prä·səs }

jugular vein [ANAT] The vein in the neck which drains the brain, face, and neck into the innominate. { 'jəg·yə·lər ‚vān }

jugum [BOT] One pair of opposite leaflets of a pinnate leaf. [INV ZOO] **1.** The most posterior and basal portion of the wing of an insect. **2.** A crossbar connecting the two arms of the brachidium in some brachiopods. { 'jü·gəm }

Julian calendar [ASTRON] A calendar (replaced by the Gregorian calendar) in which the year was 365.25 days, with the fraction allowing for an extra day every fourth year (leap year); there were 12 months, each 30 or 31 days except for February which had 28 days or in leap year 29. { 'jül·yən 'kal·ən·dər }

Julian date [ASTRON] The sum of the Julian day number and the fraction of a day elapsed since the previous noon. { 'jül·yən ¦dāt }

Julian day [ASTRON] The number of each day, as reckoned consecutively since the beginning of the present Julian period on January 1, 4713 B.C.; it is used primarily by astronomers to avoid confusion due to the use of different calendars at different times and places; the Julian day begins at noon, 12 hours later than the corresponding civil day. { 'jül·yən ¦dā }

Julian ephemeris century [ASTRON] The unit of ephemeris time (ET) in Simon Newcomb's formulas which relate the orbital position of the earth to ephemeris time; the Julian ephemeris century is subdivided into 36,525 days, and 1 ephemeris day = 86,400 ephemeris seconds. { 'jül·yən ə'fem·ə·rəs 'sen·chə‚rē }

Julia set [MATH] For a polynomial, p, with degree greater than 1, the Julia set of p is the boundary of the set of complex numbers, z, such that the sequence $p(z), p^2(z), \ldots, p_n(z), \ldots$ is bounded, where $p^2(z) = p[p(z)]$, and so forth. { 'jül·yə ‚set }

julienite [MINERAL] $Na_2Co(SCN)_4 \cdot 8H_2O$ A blue, tetragonal mineral consisting of a hydrated sodium cobalt thiocyanate. { 'jül·yə‚nīt }

Juliet [ASTRON] A satellite of Uranus orbiting at a mean distance of 39,990 miles (64,360 kilometers) with a period of 11 hours 52 minutes, and with a diameter of about 52 miles (84 kilometers). { 'jül·ē·ət }

jumbo *See* drill carriage. { 'jəm‚bō }

jump [COMPUT SCI] A transfer of control which terminates one sequence of instructions and begins another sequence at a different location. Also known as branch; transfer. { jəmp }

jump discontinuity [MATH] A point a where for a real-valued function $f(x)$ the limit on the left of $f(x)$ as x approaches a and the limit on the right both exist but are distinct. { 'jəmp dis‚känt·ən'ü·əd·ē }

jumper [ELEC] A short length of conductor used to make a connection between two points or terminals in a circuit or to provide a path around a break in a circuit. { 'jəm·pər }

jumper tube [MECH ENG] A short tube used to bypass the flow of fluid in a boiler or tubular heater. { 'jəmp ‚tüb }

jump fire [FOR] A fire carried ahead of a forest fire by windborne burning material. { 'jəmp ‚fīr }

jump function [MATH] A function used to represent a sampled data sequence arising in the numerical study of linear difference equations. { 'jəmp ‚fəŋk·shən }

jumping a claim [MIN ENG] **1.** Taking possession of a mining claim which has been abandoned. **2.** Taking possession of a mining claim that is liable to forfeiture because the requirements of the law are unfulfilled. **3.** Taking possession of a mining claim by stealth, fraud, or force. { 'jəm·piŋ ə 'klām }

jumping gene [GEN] A mobile genetic entity, such as a transposon. { 'jəmp·iŋ ¦jēn }

jumping trace routine [COMPUT SCI] A trace routine which is primarily concerned with providing a record of jump instructions in order to show the sequence of program steps that the computer followed. { 'jəm·piŋ ¦trās rü‚tēn }

jump phenomenon [CONT SYS] A phenomenon occurring in a nonlinear system subjected to a sinusoidal input at constant frequency, in which the value of the amplitude of the forced oscillation can jump upward or downward as the input amplitude is varied through either of two fixed values, and the graph of the forced amplitude versus the input amplitude follows a hysteresis loop. { 'jəmp fə‚näm·ə·nən }

jump resonance [CONT SYS] A jump discontinuity occurring in the frequency response of a nonlinear closed-loop control system with saturation in the loop. { 'jəmp ‚rez·ən·əns }

jump vector [COMPUT SCI] A list of entry-point addresses for various sections of a computer program; used by the program to branch to a section that performs a desired function. Also known as vector; vector table. { 'jəmp ‚vek·tər }

Juncaceae [BOT] A family of monocotyledonous plants in the order Juncales characterized by an inflorescence of diverse sorts, vascular bundles with abaxial phloem, and cells without silica bodies. { ‚jəŋ'kās·ē‚ē }

Juncales [BOT] An order of monocotyledonous plants in the subclass Commelinidae marked by reduced flowers and capsular fruits with one too many anatropous ovules per carpel. { ‚jəŋ'kā·lēz }

junction [CIV ENG] A point of intersection of roads or highways, especially where one terminates. [ELEC] *See* major node. [ELECTR] A region of transition between two different semiconducting regions in a semiconductor device, such as a *pn* junction, or between a metal and a semiconductor. [ELECTROMAG] A fitting used to join a branch waveguide at an angle to a main waveguide, as in a tee junction. Also known as waveguide junction. { 'jəŋk·shən }

junctional complex [CYTOL] Any specialized area of intercellular adhesion. { 'jəŋk·shən·əl ¦käm‚pleks }

junctional nevus [MED] A skin lesion containing nevus cells at the junction of the epidermis and dermis. { 'jəŋk·shən·əl 'nē·vəs }

junctional receptor [PHYSIO] An acetylcholine receptor which occurs in clusters in a muscle membrane at the nerve-muscle junction. { 'jəŋk·shən·əl ri'sep·tər }

junction battery [NUCLEO] A nuclear-type battery in which a radioactive material, such as strontium-90, irradiates a *pn* silicon junction. { 'jəŋk·shən ¦bad·ə·rē }

junction box [ENG] A protective enclosure into which wires or cables are led and connected to form joints. Also known as J box. { 'jəŋk·shən ¦bäks }

junction buoy [NAV] A buoy marking the junction of two channels or two parts of a channel when proceeding seaward. { 'jəŋk·shən ¦bȯi }

junction capacitance *See* barrier capacitance. { 'jəŋk·shən kə'pas·əd·əns }

junction capacitor [ELECTR] An integrated-circuit capacitor that uses the capacitance of a reverse-biased *pn* junction. { 'jəŋk·shən kə'pas·əd·ər }

junction detector [NUCLEO] A reverse-biased semiconductor junction functioning as a solid ionization chamber to produce an electric output pulse whose amplitude is linearly proportional to the energy deposited in the junction depletion layer by the incident ionizing radiation. { 'jəŋk·shən di¦tek·tər }

junction diode [ELECTR] A semiconductor diode in which the rectifying characteristics occur at an alloy, diffused, electrochemical, or grown junction between *n*-type and *p*-type semiconductor materials. Also known as junction rectifier. { 'jəŋk·shən ¦dī·ōd }

junction field-effect transistor [ELECTR] A field-effect transistor in which there is normally a channel of relatively low-conductivity semiconductor joining the source and drain, and this channel is reduced and eventually cut off by junction depletion regions, reducing the conductivity, when a voltage is applied between the gate electrodes. Abbreviated JFET. { 'jəŋk·shən 'fēld i¦fekt tran¦zis·tər }

junction filter [ELECTR] A combination of a high-pass and a low-pass filter that is used to separate frequency bands for transmission over separate paths. { 'jəŋk·shən ¦fil·tər }

junction isolation [ELECTR] Electrical isolation of a component on an integrated circuit by surrounding it with a region of a conductivity type that forms a junction, and reverse-biasing the junction so it has extremely high resistance. { 'jəŋk·shən ¦ī·sə'lā·shən }

junction laser [OPTICS] A laser in which a junction in a semiconductor serves as the source of the coherent laser beam. { 'jəŋk·shən ¦lā·zər }

junction loss [COMMUN] In telephone circuits, that part of the repetition equivalent assignable to interaction effects arising at trunk terminals. { 'jəŋk·shən ¦lȯs }

junction magnetoresistance *See* tunneling magnetoresistance. { 'jəŋk·shən mag¦ned·ō·ri'zis·təns }

junction phenomena [ELECTR] Phenomena which occur at the boundary between two semiconductor materials, or a semiconductor and a metal, such as the existence of an electrostatic potential in the absence of current flow, and large injection currents which may arise when external voltages are applied across the junction in one direction. { 'jəŋk·shən fə¦näm·ə·nə }

junction point *See* branch point. { 'jəŋk·shən ¦pȯint }

junction pole [ELEC] Pole at the end of a transposition section of an open-wire line or the pole common to two adjacent transposition sections. { 'jəŋk·shən ¦pōl }

junction rectifier *See* junction diode. { 'jəŋk·shən ¦rek·tə·fī·ər }

junction sequence [MOL BIO] Either of the two terminal regions of the intron in ribonucleic acid precursors. { 'jəŋk·shən ¦sē·kwəns }

junction station [ELECTR] Microwave relay station that joins a microwave radio leg or legs to the main or through route. { 'jəŋk·shən ¦stā·shən }

junction streamer [GEOPHYS] The streamer process by which negative charge centers at successively higher altitudes in a thundercloud are believed to be "tapped" for discharge by lightning. { 'jəŋk·shən ¦strēm·ər }

junction transistor [ELECTR] A transistor in which emitter and collector barriers are formed between semiconductor regions of opposite conductivity type. { 'jəŋk·shən tran¦zis·tər }

junction transposition [ELEC] Transposition located at the junction pole between two transposition sections of an open-wire line. { 'jəŋk·shən ¦tranz·pə'zish·ən }

junctor [ELEC] In crossbar systems, a circuit extending between frames of a switching unit and terminating in a switching device on each frame. { 'jəŋk·tər }

June solstice [ASTRON] Summer solstice in the Northern Hemisphere. { 'jün 'säl·stəs }

Jungermanniales [BOT] The leafy liverworts, an order of bryophytes in the class Marchantiatae characterized by chlorophyll-containing, ribbonlike or leaflike bodies and an undifferentiated thallus. { ¦jəŋ·gər‚man·ē'ā·lēz }

Jungermanniidae [BOT] A subclass of liverworts of the class Hepticopsida, division Bryophyta, distinguished by little or no tissue differentiation or organized into erect or prostrate stems with leafy appendages, leaves generally one cell in thickness and mostly arranged in three rows, with the third row of under-leaves commonly reduced or even lacking, and oil bodies usually present in all cells. { ¦jəŋ·gər'man·ə‚dē }

Jungian psychology *See* analytic psychology. { ‚yüŋ·ē·ən sī'käl·ə·jē }

jungle [ECOL] An impenetrable thicket of second-growth vegetation replacing tropical rain forest that has been disturbed; lower growth layers are dense. { 'jəŋ·gəl }

jungle yellow fever [MED] A form of yellow fever endemic in forested areas of Brazil. { 'jəŋ·gəl 'yel·ō 'fē·vər }

Jung's theorem [MATH] The theorem that a set of diameter 1 in an *n*-dimensional Euclidean space is contained in a closed ball of radius $[n/(2n + 2)]^{1/2}$.

junior [OPTICS] A 1000- or 2000-watt Fresnel spotlight. { 'jün·yər }

juniper berry oil [MATER] Colorless oil that darkens and thickens in air; bitter taste, turpentinelike aroma; derived from the dried ripe fruit of the common juniper (*Juniperus communis*), used in medicine, gins, and liqueurs, and in veterinary practice. Also known as juniper oil. { 'jü·nə·pər ‚ber·ē ‚ȯil }

juniperic acid [ORG CHEM] $C_{15}H_{32}O_3$ A crystalline hydroxy acid that melts at 95°C, obtained from waxy exudations from conifers. { ¦jü·nə¦per·ik 'as·əd }

juniper oil *See* juniper berry oil. { 'jü·nə·pər ‚ȯil }

juniper tar oil *See* cade oil. { 'jü·nə·pər ¦tär ‚ȯil }

juniper wood oil [MATER] Oil made by diluting juniper berry oil with turpentine oil or by distilling turpentine oil over juniper wood; used as an external medicine and in veterinary practice. { 'jü·nə·pər ¦wu̇d ‚ȯil }

junk deoxyribonucleic acid *See* selfish deoxyribonucleic acid. { ¦jəŋk dē¦äk·sē‚rī·bō·nü¦klē·ik ‚as·əd }

Junkers engine [MECH ENG] A double-opposed-piston, two-cycle internal combustion engine with intake and exhaust ports at opposite ends of the cylinder. { 'yu̇ŋ·kərz 'en·jən }

junk wind [METEOROL] A south or southeast monsoon wind, favorable for the sailing of junks; the wind is known in Thailand, China, and Japan. { 'jəŋk‚wind }

Juno [ASTRON] An asteroid with a diameter of about 150 miles (242 kilometers), mean distance from the sun of 2.67 astronomical units, and S-type surface composition; the third asteroid discovered, it is sometimes grouped with Ceres, Pallas, and Vesta as the Big Four, although it is about 14 in size rank. { 'jü·nō }

junta [METEOROL] A wind blowing through Andes Mountain passes, sometimes reaching hurricane force. { 'hun·tə }

Jupiter [ASTRON] The largest planet in the solar system, and the fifth in order of distance from the sun; semimajor axis = 485×10^6 miles (780×10^6 kilometers); sidereal revolution period = 11.86 years; mean orbital velocity = 8.2 miles per second (13.2 kilometers per second); inclination of orbital plane to ecliptic = 1.03; equatorial diameter = 38,700 miles (142,700 kilometers); polar diameter = 82,800 miles (133,300 kilometers); mass = about 318.4 (earth = 1). { 'jü·pəd·ər }

Jupiter I–XVII [ASTRON] The 17 known satellites of Jupiter. They are also named: I, Io; II, Europa; III, Ganymede; IV, Callisto; V, Amalthea; VI, Himalia; VII, Elara; VIII, Pasiphae; IX, Sinope; X, Lysithea; XI, Carme; XII, Ananke; XIII, Leda; XIV, Thebe; XV, Adrastea; XVI, Metis; and XVII, 1999 J1. { 'jü·pəd·ər ¦wən thru 'sev·en‚tēn }

Jupiter trojan *See* Trojan asteroid. { ¦jü·pəd·ər 'trō·jən }

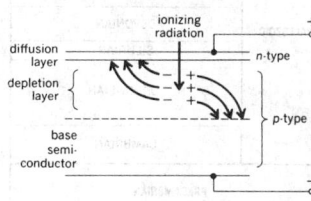

JUNCTION DETECTOR

Silicon junction detector.

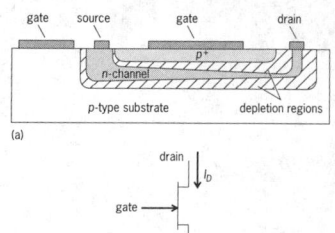

JUNCTION FIELD-EFFECT TRANSISTOR

Cross section of an *n*-channel junction field-effect transistor.

JUPITER

Appearance of Jupiter from the Hubble Space Telescope. (*Space Telescope Science Institute; Jet Propulsion Laboratory; NASA*)

JURASSIC

CENOZOIC	QUATERNARY	
	TERTIARY	
MESOZOIC	CRETACEOUS	
	JURASSIC	
	TRIASSIC	
PALEOZOIC	PERMIAN	
	CARBONIFEROUS	PENNSYLVANIAN
		MISSISSIPPIAN
	DEVONIAN	
	SILURIAN	
	ORDOVICIAN	
	CAMBRIAN	
	PRECAMBRIAN	

Chart showing the geologic periods and eras, with the Jurassic in the Mesozoic.

Jura See Jurassic. { 'jùr·ə }

juran [METEOROL] A wind blowing from the Jura Mountains in Switzerland from the northwest toward Lake Geneva; it is a cold and snowy wind and may be very turbulent, especially in spring. Also spelled joran. { 'jùr·ən }

Jurassic [GEOL] Also known as Jura. **1.** The second period of the Mesozoic era of geologic time. **2.** The corresponding system of rocks. { jə'ras·ik }

Jurin rule [FL MECH] The rule that a height to which a liquid rises in a capillary tube is twice the liquid's surface tension times the cosine of its contact angle with the capillary, divided by the product of the liquid's weight density and the internal radius of the tube. { 'jùr·ən ,rül }

jurupaite [MINERAL] $(Ca,Mg)_2(Si_2O_5)(OH)_2$ A mineral composed of hydrous calcium magnesium silicate. { hə'rüp·ə,īt }

jury rig [MIN ENG] Any makeshift or temporary device, rig, or piece of equipment. { 'jùr·ē ,rig }

jury rudder [NAV ARCH] A temporary device used to steer a boat when the rudder is out of commission. { 'jùr·ē ,rəd·ər }

justification [GRAPHICS] In type composition, the adjustment of spacing in each line of type so that all lines are filled out to the same desired length. Also known as fill justification; full justification. { ,jəs·tə·fə'kā·shən }

justify [COMPUT SCI] To shift data so that they assume a particular position relative to one or more reference points, lines, or marks in a storage medium. { 'jəs·tə,fī }

justifying space [GRAPHICS] In type composition, a variable space used between words so that each line of type will attain a desired length. { 'jəs·tə,fī·iŋ ,spās }

justifying typewriter [GRAPHICS] A typewriter that is designed to type test material with regular right-hand margins or left-hand margins or both, especially for reproduction by offset lithography; the process usually requires two typings of the text because correct word spacing cannot be determined until the first typing is done; some typewriters in combination with a computer properly programmed can justify in one typing. { 'jəs·tə,fī·iŋ 'tīp,wrīd·ər }

just-in-time [IND ENG] A systems approach to developing and operating a manufacturing system so that the least amount of resources is expended in producing the final products. Abbreviated JIT. { ¦jəst in 'tīm }

just-noticeable difference [PSYCH] The smallest difference between luminances or colors of areas, usually adjacent to each other, that can be easily discerned or is obvious from ordinary observation. Also known as difference limen; difference threshold. Abbreviated jnd. { ¦jəst ¦nōd·ə·sə·bəl 'dif·rəns }

just-perceptible difference [PSYCH] The smallest difference between luminances or colors of areas, usually adjacent to each other, that can be discerned by careful observation under the most favorable conditions. Abbreviated jpd. { 'jəst pər'sep·tə·bəl 'dif·rəns }

just scale [ACOUS] A diatonic scale rendered in the just tuning system. { 'jəst 'skāl }

just ton See ton. { 'jəst 'tən }

just tuning [ACOUS] A tuning system generated by octave rearrangements of the notes of three consecutive triads, each having the frequency ratio 4:5:6, with the highest note of one triad serving as the lowest note of the next triad. { 'jəst 'tün·iŋ }

jute [BOT] Either of two Asiatic species of tall, slender, half-shrubby annual plants, *Corchorus capsularis* and *C. olitorius*, in the family Malvaceae, useful for their fiber. { jüt }

jute board [MATER] A fiberboard made of jute fiber. { 'jüt ,bȯrd }

jute paper [MATER] A strong paper composed principally of jute fiber. { 'jüt ,pā·pər }

juvenile cell See metamyelocyte. { 'jü·vən·əl 'sel }

juvenile hormone See neotenin. { 'jü·vən·əl 'hȯr,mōn }

juvenile melanoma [MED] A benign compound nevus in young people; resembles a malignant melanoma histologically. { 'jü·vən·əl ,mel·ə'nō·mə }

juvenile-onset diabetes [MED] A form of diabetes mellitus which develops early in life and presents much more severe symptoms than the more common maturity-onset diabetes. { 'jü·vən·əl ¦ȯn,set ,dī·ə'bēd·ēz }

juvenile rift [GEOL] A stage of continental breakup before the onset of actual spreading which precedes the generation of new oceanic lithosphere. { 'jü·və·nəl 'rift }

juvenile water See magmatic water. { 'jü·vən·əl 'wȯd·ər }

juvite [PETR] A light-colored nepheline syenite in which the feldspar is exclusively or predominantly orthoclase and the potassium oxide content is higher than the sodium oxide content. { 'jü,vīt }

JVM See Java virtual machine.

Jynginae [VERT ZOO] The wrynecks, a family of Old World birds in the order Piciformes; a subfamily of the Picidae in some systems of classification. { jin'jī,nē }

K

k *See* kilo-; kilobit.

K *See* cathode; kilobyte; potassium.

kA *See* kiloampere.

K-A age [GEOL] The radioactive age of a rock determined from the ratio of potassium-40 (^{40}K) to argon-40 (^{40}A) present in the rock. { 'kā'ā ,āj }

kaavie [METEOROL] In Scotland, a heavy fall of snow. { 'kô·vē }

Ka band [COMMUN] A band of frequencies extending from 33 to 36 gigahertz, corresponding to wavelengths of 9.09 to 8.34 millimeters. { kā'ā ,band }

kachchan [METEOROL] A hot, dry west or southwest wind of foehn type in the lee of the Sri Lanka (Ceylon) hills during the southwest monsoon in June and July; it is well developed at Batticaloa on the east coast, where it is strong enough to overcome the sea breeze and bring maximum temperatures of nearly 100°F (38°C). { ,käch,chän }

K acid [ORG CHEM] $C_{10}H_4NH_2OH(SO_3H)_2$ An acid derived from naphthylamine trisulfonic acid; used in dye manufacture. { 'kā ,as·əd }

kadaya gum *See* karaya gum. { kə'dī·ə,gəm }

K-A decay [NUC PHYS] Radioactive decay of potassium-40 (^{40}K) to argon-40 (^{40}A), as the nucleus of potassium captures an orbital electron and then decays to argon-40; the ratio of ^{40}K to ^{40}A is used to determine the age of rock (K-A age). { 'kā'ā di,kā }

kaempferol [BIOCHEM] A flavonoid with a structure similar to that of quercetin but with only one hydroxyl in the B ring; acts as an enzyme cofactor and causes growth inhibition in plants. { 'kemp·fə,ról }

Kahler's disease *See* multiple myeloma. { 'kä·lərz di,zēz }

Kahn flocculation test [PATH] A macroscopic precipitin test for identification of the antibody resulting from syphilitic infection, by using an antigen prepared from normal beef heart. { 'kän ,fläk·yə'lä·shən ,test }

kainite [MINERAL] $MgSO_4 \cdot KCl \cdot 3H_2O$ A white, gray, pink, or black monoclinic mineral, occurring in irregular granular masses; used as a fertilizer and as a source of potassium and magnesium compounds. { 'kī,nīt }

kainosite [MINERAL] $Ca_2(Ce,Y)_2(SiO_4)_3CO_3 \cdot H_2O$ A yellowish-brown mineral composed of a hydrous silicate and carbonate of calcium, cerium, and yttrium. { 'kī·nə,sīt }

kairomone [PHYSIO] A chemical produced by an organism that benefits the recipient, which is an individual of a different species. { 'kī·rə,mōn }

Kaiser effect [ACOUS] An effect observed in most metals, in which acoustic emissions are not observed during the reloading of a material until the stress exceeds its previous high value. { 'kī·zər i,fekt }

Kaiserling's method [BIOL] A method for preserving organ specimens and retaining their color by fixing in a solution of formalin, water, potassium nitrate, and potassium acetate, immersing in ethyl alcohol to restore color, and preserving in a solution of glycerin, aqueous arsenious acid, water, potassium acetate, and thymol. { 'kī·zər·liŋz ,meth·əd }

Kakeya problem [MATH] The problem of finding the plane figure of least area within which a unit line segment can be moved continuously so as to return to its original position with its end points reversed; in fact, there is no such minimum area. { kä'kā·ə ,präb·ləm }

kaki [BOT] *Diospyros kaki.* The Japanese persimmon; it provides a type of ebony wood that is black with gray, yellow, and brown streaks, has a close, even grain, and is very hard. { 'kä·kē }

kakidrosis [MED] Secretion of sweat having a disagreeable odor. { ,kä·kə'drō·səs }

kakorrhaphiophobia [PSYCH] Abnormal fear of failure. { ,kä·kə,raf·ē·ə'fō·bē·ə }

kala azar [MED] Visceral leishmaniasis due to the protozoan *Leishmania donovani,* transmitted by certain sandflies (*Phlebotomus*); characterized by chronic, irregular fever, enlargement of the spleen and liver, emaciation, anemia, and leukopenia. { ,käl·ə ə'zär }

kal Baisakhi [METEOROL] A short-lived dusty squall at the onset of the southwest monsoon (April–June) in Bengal. { ,käl 'bī·sä·kē }

kale [BOT] Either of two biennial crucifers, *Brassica oleracea* var *acephala* and *B. fimbriata,* in the order Capparales, grown for the nutritious curled green leaves. { käl }

kaleidoscope [OPTICS] An optical toy consisting of a tube containing two plane mirrors placed at an angle of 60° and mounted so that a symmetrical pattern produced by multiple reflection is observed through a peephole at one end when objects (such as pieces of colored glass) at the other end are suitably illuminated. { kə'līd·ə,skōp }

kalema [OCEANOGR] A very heavy surf breaking on the Guinea coast of Africa during the winter. { kə'lä·mə }

kalfax [GRAPHICS] An emulsion sensitive to ultraviolet light coated on a Mylar base, processed by heat. { 'kal,faks }

kaliborite [MINERAL] $HKMg_2B_{12}O_{21} \cdot 9H_2O$ A colorless to white mineral composed of a hydrous borate of potassium and magnesium Also known as paternoite. { ,kal·ə'bó,rīt }

kalicinite [MINERAL] $KHCO_3$ A colorless to white or yellowish, monoclinic mineral consisting of potassium bicarbonate; occurs in crystalline aggregates. { kə'lis·ən,īt }

kalinite [MINERAL] $KAl(SO_4)_2 \cdot 11H_2O$ A birefringent mineral of the alum group composed of a hydrous sulfate of potassium and aluminum, occurring in fibrous form. Also known as potash alum. { 'kal·ə,nīt }

kaliophilite [MINERAL] $KAlSiO_4$ A rare hexagonal tectosilicate mineral found in volcanic rocks; high in potassium and low in silica, it is dimorphous with kalsilite. Also known as facellite; phacellite. { ,kal·ē'äf·ə,līt }

kalium *See* potassium. { 'käl·ēl·əm }

kalkowskite [MINERAL] $Fe_2Ti_3O_9$ A rare, brownish or black mineral composed of an oxide of iron and titanium, usually with small amounts of rare-earth elements, niobium, and tantalum. { kal'kóf,sīt }

kallidin I *See* bradykinin. { kə'līd·ən 'wən }

kallitype [GRAPHICS] An early photographic process using paper sensitized with ferrous oxalate and a silver salt, and a developer containing borax and Rochelle salt. { 'kal·ə,tīp }

Kalman filter [CONT SYS] A linear system in which the mean squared error between the desired output and the actual output is minimized when the input is a random signal generated by white noise. { 'kal·mən ,fil·tər }

Kalotermitidae [INV ZOO] A family of relatively primitive, lower termites in the order Isoptera. { ,kal·ə,tər'mid·ə,dē }

kalsilite [MINERAL] $KAlSiO_4$ A rare mineral from volcanic rocks in southwestern Uganda; the crystal system is hexagonal; kalsilite is dimorphous with kaliophilite and sometimes contains sodium. { 'kal·sə,līt }

kalsomine *See* calcimine. { 'kal·sə,mīn }

kalunite [MINERAL] The naturally occurring form of alum. { 'kal·ə‚nīt }

Kaluza theory [RELAT] An attempted unified field theory in which the four-dimensional world that one observes is taken to be a projection of a five-dimensional continuum. { kə'lü·zə ‚thē·ə·rē }

kamacite [MINERAL] A mineral composed of a nickel-iron alloy and comprising with taenite the bulk of most iron meteorites. { 'kam·ə‚sīt }

kame [GEOL] A low, long, steep-sided mound of glacial drift, commonly stratified sand and gravel, deposited as an alluvial fan or delta at the terminal margin of a melting glacier. { kām }

kame terrace [GEOL] A terracelike ridge deposited along the margins of glaciers by meltwater streams flowing adjacent to the valley walls. { kām ‚ter·əs }

Kamptozoa [INV ZOO] An equivalent name for Entoprocta. { ‚kam·tə'zō·ə }

kampyle of Eudoxus [MATH] A plane curve whose equation in cartesian coordinates x and y is $x^4 = a^2(x^2 + y^2)$, where a is a constant. { kam'pīl əv yü'däk·səs }

KAM theorem See Kolmogorov-Arnold-Moser theorem. { 'kam ‚thir·əm }

kanamycin [MICROBIO] $C_{18}H_{36}O_{11}N_4$ A water-soluble, basic antibiotic produced by strains of *Streptomyces kanamyceticus;* the sulfate salt is effective in infections caused by gram-negative bacteria. { ‚kan·ə'mīs·ən }

kanban [IND ENG] An inventory control system for tracking the flow of in-process materials through the various operations of a just-in-time production process. Kanban means "card" or "ticket" in Japanese. { 'kan¦ban }

kangaroo [VERT ZOO] Any of various Australian marsupials in the family Macropodidae generally characterized by a long, thick tail that is used as a balancing organ, short forelimbs, and enlarged hindlegs adapted for jumping. { ‚kaŋ·gə'rü }

Kanji [COMPUT SCI] A set of Chinese characters that are employed by users of the Chinese language to code information in computer programs and on visual displays. { 'kän·jē }

Kansan glaciation [GEOL] The second glaciation of the Pleistocene epoch in North America; began about 400,000 years ago, after the Aftonian and before the Yarmouth interglacials. { 'kan·zən ‚glā·sē'ā·shən }

Kansasii disease [MED] A mycobacterial tuberculosislike infection caused by *Mycobacterium kansasii,* an orange-yellow acid-fast bacterium. { kan'zas·ē‚ī di‚zēz }

kansite See mackinawite. { 'kan‚zīt }

kaolin [MINERAL] Any of a group of clay minerals, including kaolinite, nacrite, dickite, and anauxite, with a two-layer crystal in which silicon-oxygen and aluminum-hydroxyl sheets alternate; approximate composition is $Al_2O_3 \cdot 2SiO_2 \cdot 2H_2O$. [PETR] A soft, nonplastic white rock composed principally of kaolin minerals. Also known as bolus alba; white clay. { 'kā·ə·lən }

kaolinite [MINERAL] $Al_2Si_2O_5(OH)_4$ A common hydrous aluminum silicate mineral found in sediments, soils, hydrothermal deposits, and sedimentary rocks. It is a member of the kaolin group of minerals, which include dickite, halloysite, nacrite, ordered kaolinite, and disordered kaolinite. { 'kā·ə·lə‚nīt }

kaolinization [GEOL] The forming of kaolin by the weathering of aluminum silicate minerals or other clay minerals. { ‚kā·ə·lə·nə'zā·shən }

kaon See K meson. { 'kā‚än }

kaonic atom [ATOM PHYS] An atom consisting of a negatively charged kaon orbiting around an ordinary nucleus. { kā'än·ik 'ad·əm }

Kapteyn selected areas [ASTRON] Certain areas in the Milky Way Galaxy that the astronomer J.C. Kapteyn suggested be studied intensively in order to determine the structure of the galaxy. Also known as selected areas. { 'kap¦ət·ən si¦lek·təd 'er·ē·əz }

Kapteyn's star [ASTRON] A star 13.0 light-years from the solar system, absolute magnitude 11.2, spectrum type M0; has a large proper motion. { 'kap·əd·ənz ‚stär }

Kapitza balance [ENG] A magnetic balance for measuring susceptibilities of materials in large magnetic fields that are applied for brief periods. { ka'pit·sə ‚bal·əns }

Kapitza expander [CHEM ENG] Reciprocating-piston gas expander used for helium liquefaction; relies on close fit rather than packing or rings on the pistons. { 'kä·pit·sə ik¦span·dər }

Kapitza resistance [CRYO] A thermal resistance to the flow of heat across the interface between liquid helium and a solid. { 'kä·pit·sə ri¦zis·təns }

Kaplan turbine [MECH ENG] A propeller-type hydraulic turbine in which the positions of the runner blades and the wicket gates are adjustable for load change with sustained efficiency. { 'kap·lən ‚tər·bən }

kapoc oil See kapok oil. { 'kā‚päk ‚óil }

kapok [BOT] Silky fibers that surround the seeds of the kapok or ceiba tree. Also known as ceiba; Java cotton; silk cotton. { 'kā‚päk }

kapok oil [MATER] Yellow-green oil with pleasant aroma and taste; soluble in alcohol, ether, and chloroform; derived by pressing seeds of the kapok tree; used in edible oils and soap stock. Also spelled kapoc oil. { 'kā‚päk ‚óil }

kapok tree [BOT] *Ceiba pentandra.* A tree of the family Bombacaceae which produces pods containing seeds covered with silk cotton. Also known as silk cotton tree. { 'kā‚päk ‚trē }

kappa curve [MATH] A plane curve whose equation in cartesian coordinates x and y is $(x^2 + y^2) y^2 = a^2x^2$, where a is a constant. Also known as Gutschoven's curve. { 'kap·ə ‚kərv }

kappa particle [CYTOL] A self-duplicating nucleoprotein particle found in various strains of *Paramecium* and thought to function as an infectious agent; classed as an intracellular symbiont, occupying a position between the viruses and the bacteria and organelles. { 'kap·ə ‚pard·ə·kəl }

Kaposi's sarcoma [MED] A multifocal malignant or benign neoplasm that occurs in the skin and sometimes in lymph nodes or viscera and is composed of primitive tissue that is involved in blood vessel formation. { kə'pō·sēz sär'kōm·ə }

karaburan [METEOROL] A violent northeast wind of Central Asia occurring during spring and summer; it carries clouds of dust (which darken the sky) instead of snow. Also known as black buran; black storm. { ‚kä·rə·bù¦rän }

karajol [METEOROL] On the Bulgarian coast, a west wind which usually follows rain and persists 1–3 days. { ‚kä·rə'yól }

karat [MET] A unit for designating the fineness of gold in an alloy; represents a twenty-fourth part; thus, 18-karat gold is 18/24 or 75% pure. { 'kar·ət }

karaya gum [MATER] The exudation from the Indian tree (*Sterculia urens*); white to dark-colored tears; used in pharmaceuticals, textiles, and foods, and as tragacanth gum substitute. Also known as Indian tragacanth; kadaya gum. { kə'rī·ə ‚gəm }

karbutilate [ORG CHEM] $C_{14}H_{21}N_3O_3$ An off-white solid with a melting point of 176–177°C; used as a herbicide on noncroplands, railroad rights-of-way, and plant sites. { kär'byüd·əl‚āt }

karema [METEOROL] A violent east wind on Lake Tanganyika in Africa. { kə'rē·mə }

karif [METEOROL] A strong southwest wind on the southern shore of the Gulf of Aden, especially at Berbera, Somaliland, during the southwest monsoon. { kä'rēf }

Karl Fischer reagent [ANALY CHEM] A solution of 8 moles pyridine to 2 moles sulfur dioxide, with the addition of about 15 moles methanol and then 1 mole iodine; used to determine trace quantities of water by titration. { 'kärl 'fish·ər rē'ā·jənt }

Karl Fischer technique [ANALY CHEM] A method of determining trace quantities of water by titration; the Karl Fischer reagent is added in small increments to a glass flask containing the sample until the color changes from yellow to brown or a change in potential is observed at the end point. { 'kärl 'fish·ər tek'nēk }

Kármán constant [FL MECH] A dimensionless number formed from the velocity of turbulent flow parallel to a plane wall, the distance from the wall, the shear stress, and the density of the fluid; for a wide range of flow patterns it has a constant value. { 'kär‚män ‚kän·stənt }

Kármán swirling flow problem [MATH] The problem of describing fluid motion above a rotating infinite plane disk when the fluid at infinity does not rotate. { 'kär‚män ¦swir·liŋ ¦flō ‚präb·ləm }

Kármán-Tsien method [FL MECH] A method of approximating equations for two-dimensional compressible flow which yields a simple rule for estimating compressibility effects of subsonic flow. { ¦kär‚män 'tsyen ‚meth·əd }

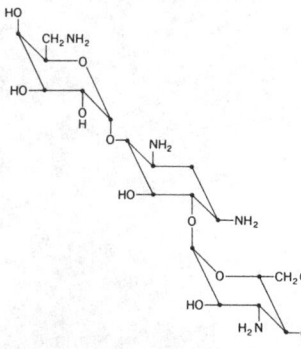

KANAMYCIN

Structural formula of kanamycin A.

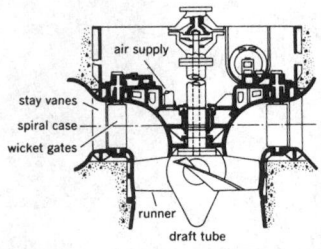

KAPLAN TURBINE

air supply
stay vanes
spiral case
wicket gates
runner
draft tube

Cross section of a Kaplan type of hydraulic turbine installation.

Kármán vortex street [FL MECH] A double row of line vortices in a fluid which, under certain conditions, is shed in the wake of cylindrical bodies when the relative fluid velocity is perpendicular to the axis of the cylinder. { 'kär,män 'vȯr,-teks ,strēt }

Karmarkar's algorithm [MATH] A method for solving linear programming problems that has a polynomial time bound and appears to be faster than the simplex method for many complex problems. { 'kär·mə,kärz 'al·gə,rith·əm }

Karnaugh map [ELECTR] A truth table that has been rearranged to show a geometrical pattern of functional relationships for gating configurations; with this map, essential gating requirements can be recognized in their simplest form. { 'kär·nō ,map }

Karnian See Carnian. { 'kär·nē·ən }

karoo See karroo. { kə'rü }

Karp circuit [ELECTR] A slow-wave circuit used at millimeter wavelengths for backward-wave oscillators. { 'kärp ,sər·kət }

karren [GEOL] Furrows or channels formed on the surface of soluble bedrock by dissolution of a portion of the rock. Also known as lapies. { kar·ən }

Karrer method [CHEM ENG] An industrial method for the chemical synthesis of riboflavin. { 'kar·ər ,meth·əd }

karroo [GEOGR] A dry, broad, level, elevated area found especially in southern Africa, often rising to considerable elevations in terrace formations; does not support vegetation in the dry season but supports grass during the wet season. Also spelled karoo. { kə'rü }

Karroo System [GEOL] Glaciated strata formed in Permian times in southern Africa. { kə'rü ,sis·təm }

karrusel [HOROL] A revolving escapement that reduces position errors. { ,kar·ə'sel }

karst [GEOL] A topography formed over limestone, dolomite, or gypsum and characterized by sinkholes, caves, and underground drainage. { kärst }

karst base level [GEOL] The level below which karstification ceases in an area of karst topography. { 'kärst ¦bās ,lev·əl }

karstbora [METEOROL] The bora of the Yugoslavian coast. { 'kärst,bȯr·ə }

karst fenster See karst window. { 'kärst ,fen·stər }

karstification [GEOL] Formation of the features of karst topography by the chemical, and sometimes mechanical, action of water in a region of limestone, dolomite, or gypsum bedrock. { ,kär·stə·fə'kā·shən }

karst plain [GEOL] A plain on which karst features are developed. { 'kärst ,plān }

karst window [GEOL] An area over a subterranean stream that is open to the surface and appears as a depression at whose bottom the stream is visible. Also known as karst fenster. { 'kärst ,win·dō }

Karumiidae [INV ZOO] The termitelike beetles, a small family of coleopteran insects in the superfamily Cantharoidea distinguished by having a tenth tergum. { ,kar·ə'mī·ə,dē }

Karush-Kuhn-Tucker conditions [MATH] A system of equations and inequalities which the solution of a nonlinear programming problem must satisfy when the objective function and the constraint functions are differentiable. { ¦kär·əsh ¦kyün 'tək·ər kən,dish·ənz }

karyocyte See normoblast. { 'kar·ē·ə,sīt }

karyodesma See nucleodesma. { ,kar·ē·ə'dez·mə }

karyogamy [CELL MOL] Fusion of gametic nuclei, as in fertilization. { ,kar·ē'äg·ə·mē }

karyokinesis [CELL MOL] Nuclear division characteristic of mitosis. { 'kar·ē·ō·kə'nē·səs }

karyolymph [CELL MOL] The clear material composing the ground substance of a cell nucleus. { 'kar·ē·ə,limf }

karyolysis [CELL MOL] Dissolution of a cell nucleus. { ,kar·ē'äl·ə·səs }

karyomastigont [INV ZOO] Pertaining to members of the protozoan order Oxymonadida; individuals can be uni- or multinucleate, and unattached forms give rise to two pairs of flagella. { ,kar·ē·ō'mas·tə,gänt }

karyoplasm See nucleoplasm. { 'kar·ē·ə,plaz·əm }

karyoplasmic ratio See nucleocytoplasmic ratio. { ,kar·ē·ə'plaz·mik 'rā·shō }

karyorrhexis [CELL MOL] Fragmentation of a nucleus with scattering of the pieces in the cytoplasm. { ,kar·ē·ə'rek·səs }

karyosphere [CELL MOL] The fraction of nuclear volume to which the chromosomes are confined in nuclei that are rich in karyolymph. { 'kar·ē·ə,sfir }

karyotype [CELL MOL] **1.** The complement of chromosomes characteristic of an individual, species, genus, or other grouping. **2.** An organized array of the chromosomes from a single cell, grouped according to size, centromere position, and banding pattern, if any. { 'kar·ē·ə,tīp }

kasha cloth [TEXT] **1.** A soft-napped fabric made from the hair fibers of Tibetan goats with a slight crosswise streaked effect. **2.** A tan, occasionally mottled, cotton flannel lining material. { 'kash·ə ,klȯth }

kasolite [MINERAL] $Pb(UO_2)SiO_4·H_2O$ Yellow-ocher mineral composed of a hydrous lead uranium silicate, occurring in monoclinic crystals. { 'kas·ə,līt }

kasugamycin [MICROBIO] $C_{14}H_{28}ClN_3O_{10}$ A white, crystalline antibiotic used as a fungicide for rice crops. Also known as kasugamycin hydrochloride; kasumin. { kə,sü·gə'mīs·ən }

kasugamycin hydrochloride See kasugamycin. { kə,sü·gə'mīs·ən ,hī·drə'klȯr,īd }

kasumin See kasugamycin. { kə'sü·mən }

katabaric See katallobaric. { kad·ə'bar·ik }

katabatic wind See gravity wind. { ¦kad·ə¦bad·ik 'wind }

katafront [METEOROL] A front (usually a cold front) at which warm air descends the frontal surface (except, presumably, in the lowest layers). { 'kad·ə,frənt }

katallobaric [METEOROL] Of or pertaining to a decrease in atmospheric pressure. Also known as katabaric. { kə,tal·ə'bar·ik }

katallobaric center See pressure-fall center. { kə,tal·ə'bar·ik 'sen·tər }

Kata thermometer [ENG] An alcohol thermometer used to measure low velocities in air circulation, by heating the large bulb of the thermometer above 100°F (38°C) and noting the time it takes to cool from 100 to 95°F (38 to 35°C) or some other interval above ambient temperature, the time interval being a measure of the air current at that location. { 'kad·ə thər'mäm·əd·ər }

katazone [GEOL] The lowest depth zone of metamorphism; features include high temperatures (500–700°C), strong hydrostatic pressure, and little or no shearing stress. { 'kad·ə,zōn }

katchung oil See peanut oil. { kə'chəŋ ,ȯil }

Kater's reversible pendulum [MECH] A gravity pendulum designed to measure the acceleration of gravity and consisting of a body with two knife-edge supports on opposite sides of the center of mass. { 'kā·dərz ri¦vər·sə·bəl 'pen·jə·ləm }

katharometer [ENG] An instrument for detecting the presence of small quantities of gases in air by measuring the resulting change in thermal conductivity of the air. Also known as thermal conductivity cell. { ,kath·ə'räm·əd·ər }

Kathlaniidae [INV ZOO] A family of nematodes assigned to the Ascaridina by some authorities and to the Oxyurina by others. { ,kath·lə'nī·ə,dē }

katoptric system [OPTICS] An optical system such that, when the object is displaced in a direction parallel to the axis, the image is displaced in the opposite direction (in contrast to a dioptric system). Also known as contracurrent system. { kə'täp·trik ,sis·təm }

katoptrite See catoptrite. { kə'täp,trīt }

Kauertz engine [MECH ENG] A type of cat-and-mouse rotary engine in which the pistons are vanes which are sections of a right circular cylinder; two pistons are attached to one rotor so that they rotate with constant angular velocity, while the other two pistons are controlled by a gear-and-crank mechanism, so that angular velocity varies. { 'kaú·ərts ,en·jən }

kauri-butanol value [ANALY CHEM] The measure of milliliters of paint or varnish petroleum thinner needed to cause cloudiness in a solution of kauri gum in butyl alcohol. { 'kaú·rē 'byüt·ən,ȯl ,val·yü }

kauri gum [MATER] Hard copal resins from kauri pine (*Agathis australis*); used in lacquers and varnishes. { 'kaú·rē ,gəm }

kaus [METEOROL] A moderate to gale-force southeasterly wind in the Persian Gulf, accompanied by gloomy weather, rain, and squalls; it is most frequent between December and April. Also known as cowshee; sharki. { kaús }

kavaburd See cavaburd. { 'kä·və,bərd }

kaver See caver. { 'kā·vər }

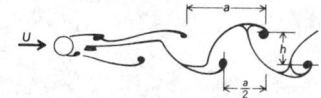

Kármán vortex street showing double row of line vortices. Here U = stream speed, h = perpendicular distance between the two lines of vortices, and a = distance between successive vortices on the same line.

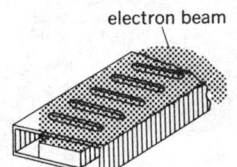

KARP CIRCUIT

electron beam

Karp circuit, used at millimeter wavelengths.

KARST

Sinkholes in Meade County, Kansas. Depth to the underground water is about 50 feet (15 meters). (*U.S. Geographical Survey*)

kay *See* key.   { kā }

kayser [SPECT] A unit of reciprocal length, especially wave number, equal to the reciprocal of 1 centimeter.   Also known as rydberg.   { 'kī·zər }

Kayser-Fleischer ring [PATH] A ring of golden-brown or brownish-green pigment behind the limbic border of the cornea, due to the deposition of copper.   { 'kī·zər 'flī·shər ˌriŋ }

Kazanian [GEOL] A European stage of geologic time: Upper Permian (above Kungurian, below Tatarian).   { kə'zä·nē·ən }

kb *See* kilobar; kilobase.

K band [COMMUN] A band of radio frequencies extending from 10,900 to 36,000 megahertz, corresponding to wavelengths of 2.75 to 0.834 centimeters.   [SOLID STATE] An optical absorption band which appears together with an *F*-band and has a lower intensity and shorter wavelength than the latter.   { 'kā ˌband }

K-band single-access service [COMMUN] A service provided by the Tracking and Data Relay Satellite System, with return-link data rates up to 300 and 800 megabits per second for the Ku and Ka bands, respectively, and forward-link data at 25 megabits per second in both bands.   Abbreviated KSA.   { ¦kā¦band ¦siŋ·gəl 'ak,ses ˌsər·vəs }

K bentonite *See* potassium bentonite.   { 'kā 'ben·tə,nīt }

kbit *See* kilobit.   { 'kā,bit }

Kbit *See* kilobit.   { 'kā,bit }

kbp *See* kilobase.

kbyte *See* kilobyte.   { 'kā,bīt }

Kbyte *See* kilobyte.   { 'kā,bīt }

kc *See* kilohertz.

kcal *See* kilocalorie.

K capture [NUC PHYS] A type of beta interaction in which a nucleus captures an electron from the *K* shell of atomic electrons (the shell nearest the nucleus) and emits a neutrino.   { 'kā ˌkap·chər }

K-carrier system [COMMUN] Carrier system providing 12 telephone channels with a bandwidth up to 60 kilohertz, either on a four-wire cable system or on microwave (line-of-sight) and trophospheric scatter radio systems.   { 'kā ˌkar·ē·ər ˌsis·təm }

K corona [ASTRON] The inner portion of the sun's corona, having a continuous spectrum caused by electron scattering.   { 'kā kə,rō·nə }

K damage [ORD] **1.** Combat damage sufficient to cause a vehicle to be destroyed.   **2.** Combat damage such that an aircraft will fall out of control immediately after the damage occurs.   { 'kā ,dam·ij }

KDD *See* knowledge discovery in databases.

K display [ELECTR] A modified radarscope A display in which a target appears as a pair of vertical deflections instead of as a single deflection; when the radar antenna is correctly pointed at the target in azimuth, the deflections are of equal height; when the antenna is not correctly pointed, the difference in pulse heights is an indication of direction and magnitude of azimuth pointing error.   { 'kā di,splā }

KdV soliton *See* Korteweg-de Vries soliton.   { 'kā·dē,vē 'säl·ə,tän }

kedge [NAV] To move, as a vessel, by carrying out an anchor, letting it go, and hauling the ship up to the anchor.   { kej }

kedge anchor [NAV ARCH] A light anchor that is used to warp or kedge a ship.   { 'kej ,aŋ·kər }

keel [NAV ARCH] A steel beam or timber, or a series of steel beams and plates or timbers joined together, extending along the center of the bottom of a ship from stem to stern and often projecting below the bottom, to which the frames and hull plating are attached.   [VERT ZOO] The median ridge on the breastbone in certain birds.   Also known as carina.   { kēl }

Keel *See* Carina.   { kēl }

keel block [CIV ENG] A docking block used to support a ship's keel.   [MET] A simple shape from which a test casting is made in the form of a large head, which is removed and discarded, with a keel on the bottom.   { 'kēl ,bläk }

keel condenser [NAV ARCH] Pipes, made of material that will not induce electrolysis with other parts of a ship, placed outside the ship near the keel for condensing steam.   { 'kēl kən,den·sər }

keel molding [ARCH] A molding with two ogees, the central fillet of which projects like a keel.   { 'kēl ,mōl·diŋ }

keelson [NAV ARCH] A structure of timbers or steel beams which are bolted to the top of a keel to increase its strength.   Also spelled kelson.   { 'kel·sən }

KEKULÉ STRUCTURE

Two resonance Kekulé structures of toluene.

Keene's cement [MATER] An anhydrous calcined gypsum mixed with an accelerator; used as a hard-finish plaster.   { 'kēnz si,ment }

keep-alive circuit [ELECTR] A circuit used with a transmit-receive (TR) tube or anti-TR tube to produce residual ionization for the purpose of reducing the initiation time of the main discharge.   { ¦kēp ə'līv ˌsər·kət }

keep-alive electrode *See* ignitor.   { ¦kēp ə'līv i'lek,trōd }

keeper [ELECTROMAG] A bar of iron or steel placed across the poles of a permanent magnet to complete the magnetic circuit when the magnet is not in use, to avoid the self-demagnetizing effect of leakage lines.   Also known as magnet keeper.   { 'kēp·ər }

keeps *See* folding boards.   { kēps }

Keesom force *See* orientation force.   { 'kā·səm ,förs }

Keesom relationship [PHYS CHEM] An equation for the potential energy associated with the interaction of the dipole moments of two polar molecules.   { 'kā·səm ri'lā·shən,ship }

Keewatin [GEOL] A division of the Archeozoic rocks of the Canadian Shield.   { kē|wät·ən }

kefir [FOOD ENG] A low-alcohol-content beverage prepared from cow's milk treated with *Lactobacillus casei* and then fermented by the action of *Saccharomyces kefir* on kefir grain.   { ke'fir }

keg buoy [NAV] A buoy consisting of a keg to which is attached a small pole with a flag, used by fishing crews to mark the position of a trawl line.   { 'keg ,böi }

Kegel karst *See* cone karst.   { 'kā·gəl ,kärst }

kehoeite [MINERAL] An amorphous mineral composed of a basic hydrous calcium aluminum zinc phosphate, occurring in massive form.   { 'kē·ō,īt }

Keilor skull [PALEON] An Australian fossil type specimen of *Homo sapiens* from the Pleistocene.   { 'kē·lər ,skəl }

kei function [MATH] A function that is expressed in terms of modified Bessel functions of the second kind in a manner similar to that in which the bei function is expressed in terms of Bessel functions.   { 'kī ,fəŋk·shən }

Kekeya needle problem [MATH] The problem of finding the smallest area of a plane region in which a line segment of unit length can be continuously moved so that it returns to its original position after turning through 360°.   { kā,kē·ə 'nēd·əl ,präb·ləm }

Kekulé structure [ORG CHEM] A molecular structure of a cyclic conjugated system that is depicted with alternating single and double bonds.   { 'kā·kə,lā ,strək·chər }

Keldysh theory [ATOM PHYS] A theory of multiphoton ionization, in which an atom is ionized by rapid absorption of a sufficient number of photons; it predicts that the ionization rate depends primarily upon the ratio of the mean binding electric field to the peak strength of the incident electromagnetic field, and upon the ratio of the binding energy to the energy of photons in the field.   { 'kel·dish ,thē·ə·rē }

K electron [ATOM PHYS] An electron in the *K* shell.   { 'kā i'lek,trän }

Kell blood group system [IMMUNOL] A family of antigens found in erythrocytes and designated K, k, Kp^a, Kp^b, and Ku; antibodies to the K antigen, which occurs in about 10% of the population of England, have been associated with hemolytic transfusion reactions and with hemolytic disease.   { 'kel 'bləd ,grüp ,sis·təm }

kellering [MECH ENG] Three-dimensional machining of a contoured surface by tracer-milling the die block or punch; the cutter path is controlled by a tracer that follows the contours on a die model.   { 'kel·ə·riŋ }

Kellner eyepiece [OPTICS] A Ramsden eyepiece with an achromatic eye lens.   { 'kel·nər 'ī,pēs }

Kellogg equation [THERMO] An equation of state for a gas, of the form

$$p = RT\rho + \sum_{n=2}^{\infty} [b_n T - a_n - (c_n/T^2)]\rho^n$$

where p is the pressure, T the absolute temperature, ρ the density, R the gas constant, and a_n, b_n, and c_n are constants.   { 'kel,äg i,kwā·zhən }

kelly [PETRO ENG] A pipe attached to the top of a drill string and turned during drilling; transmits twisting torque from the rotary machinery to the drill string and ultimately to the bit. { 'kel·ē }

Kelly ball test [ENG] A test for the consistency of concrete using the penetration of a half sphere; a 1-inch (2.5-centimeter) penetration by the Kelly ball corresponds to about 2 inches (5 centimeters) of slump. { 'kel·ē 'bȯl ,test }

kelly bushing [PETRO ENG] A device added to the rotary table through which the kelly passes, so that the torque of the table is transmitted to the kelly and the drill. { 'kel·ē ,bu̇sh·iŋ }

keloid [MED] A firm, elevated fibrous formation of tissue at the site of a scar. { 'kē,lȯid }

keloid acne [MED] Acnelike infection of hair follicles, especially on the nape of the neck, resulting in hard, white or reddish keloids and scarring. Also known as folliculitis keloidalis. { 'kē,lȯid 'ak·nē }

kelp [BOT] The common name for brown seaweed belonging to the Laminariales and Fucales. { kelp }

kelsher [METEOROL] In England, a heavy fall of rain. { 'kel·shər }

kelson See keelson. { 'kel·sən }

kelvin [ELEC] A name formerly given to the kilowatt-hour. Also known as thermal volt. [THERMO] A unit of absolute temperature equal to 1/273.16 of the absolute temperature of the triple point of water. Symbolized K. Formerly known as degree Kelvin. { 'kel·vən }

Kelvin absolute temperature scale [THERMO] A temperature scale in which the ratio of the temperatures of two reservoirs is equal to the ratio of the amount of heat absorbed from one of them by a heat engine operating in a Carnot cycle to the amount of heat rejected by this engine to the other reservoir; the temperature of the triple point of water is defined as 273.16 K. Also known as Kelvin temperature scale. { 'kel·vən ¦ab·sə,lüt 'tem·prə·chər ,skāl }

Kelvin balance [ELECTROMAG] An ammeter in which the force between two coils in series that carry the current to be measured, one coil being attached to one arm of a balance, is balanced against a known weight at the other end of the balance arm. { 'kel·vən ¦bal·əns }

Kelvin body [MECH] An ideal body whose shearing (tangential) stress is the sum of a term proportional to its deformation and a term proportional to the rate of change of its deformation with time. Also known as Voigt body. { 'kel·vən ,bäd·ē }

Kelvin bridge [ELEC] A specialized version of the Wheatstone bridge network designed to eliminate, or greatly reduce, the effect of lead and contact resistance, and thus permit accurate measurement of low resistance. Also known as double bridge; Kelvin network; Thomson bridge. { 'kel·vən ,brij }

Kelvin equation [THERMO] An equation giving the increase in vapor pressure of a substance which accompanies an increase in curvature of its surface; the equation describes the greater rate of evaporation of a small liquid droplet as compared to that of a larger one, and the greater solubility of small solid particles as compared to that of larger particles. { 'kel·vən i,kwā·zhən }

Kelvin guard-ring capacitor [ELEC] A capacitor with parallel circular plates, one of which has a guard ring separated from the plate by a narrow gap; it is used as a standard, whose capacitance can be accurately calculated from its dimensions. { 'kel·vən 'gärd ,riŋ kə,pas·əd·ər }

Kelvin-Helmholtz contraction [ASTROPHYS] A contraction of a star once it is formed and before it is hot enough to ignite its hydrogen; the contraction converts gravitational potential energy into heat, some of which is radiated, with the remainder used to raise the internal temperature of the star. { 'kel·vən 'helm,hōlts kən,trak·shən }

Kelvin-Helmholtz instability [FL MECH] An instability that occurs at the interface between two fluid layers if their relative motion is sufficiently large, and eventually results in the disruption of the interface. { ¦kel·vin 'helm,hōls ,in·stə'bil·əd·ē }

Kelvin network See Kelvin bridge. { 'kel·vən ,net,wərk }

Kelvin relations See Thomson relations. { 'kel·vən ri,lā·shənz }

Kelvin replenisher [ELEC] A simple electrostatic generator in which curved metal plates attached to an insulating arm rotate between larger curved plates, and the contacts of the smaller plates with wipers connecting them to the larger plates and to each other result in the accumulation of charge on the smaller plates, energy being supplied by the rotation of the arm. { ¦kel·vən ri'plen·əsh·ər }

Kelvin scale [THERMO] The basic scale used for temperature definition; the triple point of water (comprising ice, liquid, and vapor) is defined as 273.16 K; given two reservoirs, a reversible heat engine is built operating in a cycle between them, and the ratio of their temperatures is defined to be equal to the ratio of the heats transferred. { 'kel·vən ,skāl }

Kelvin's circulation theorem [FL MECH] The theorem that, if the external forces acting on an inviscid fluid are conservative and if the fluid density is a function of the pressure only, then the circulation along a closed curve which moves with the fluid does not change with time. { 'kel·vənz ,sər·kyə'lā·shən ,thir·əm }

Kelvin's formula See Thomson formula. { 'kel·vənz ,fȯr·myə·lə }

Kelvin's skin effect See skin effect. { 'kel·vən 'skin i,fekt }

Kelvin's minimum-energy theorem [FL MECH] The theorem that the irrotational motion of an incompressible, inviscid fluid occupying a simply connected region has less kinetic energy than any other fluid motion consistent with the boundary condition of zero relative velocity normal to the boundaries of the region. { 'kel·vənz ¦min·ə·məm 'en·ər·jē ,thir·əm }

Kelvin's statement of the second law of thermodynamics [THERMO] The statement that it is not possible that, at the end of a cycle of changes, heat has been extracted from a reservoir and an equal amount of work has been produced without producing some other effect. { 'kel·vənz 'stāt·mənt əv thə 'sek·ənd ,lȯ əv ,thər·mō·dī'nam·iks }

Kelvin temperature scale [THERMO] **1.** An International Temperature Scale which agrees with the Kelvin absolute temperature scale within the limits of experimental determination. **2.** See Kelvin absolute temperature scale. { 'kel·vən 'tem·prə·chər ,skāl }

Kelvin time scale [ASTROPHYS] The time that would be required for a star to contract gravitationally from infinity to its present radius solely through radiation of thermal energy. Also known as thermal time scale. { 'kel·vən 'tīm ,skāl }

Kelvin wave [OCEANOGR] **1.** An eastward-propagating internal gravity wave that crosses the Pacific Ocean along the equator and has no north-south velocity component. **2.** A type of wave progression in relatively confined water bodies where, because of Coriolis force, the wave is higher to the right of direction of advance (in the Northern Hemisphere). { 'kel·vən ,wāv }

kelyphite See corona. { kē·lə,fīt }

kelyphytic border See kelyphytic rime. { ¦kē·lə¦fid·ik 'bȯr·dər }

kelyphytic rime [PETR] A peripheral zone of pyroxene or amphibole developed around olivine in some igneous rocks. Also known as kelyphytic border. { ¦kē·lə¦fid·ik 'rīm }

kemp [TEXT] Short, wavy, coarse wool or hair fibers, usually white, which contain air spaces so that they resist dyeing and spinning. { kemp }

Kempe chain [MATH] A subgraph of a graph whose vertices have been colored, consisting of vertices which have been assigned a given color or colors and arcs connecting pairs of such vertices. { 'kem·pə ,chān }

kempite [MINERAL] $Mn_2(OH)_3Cl$ An emerald-green orthorhombic mineral composed of a basic manganese oxychloride, occurring in small crystals. { 'kem,pīt }

kenaf [AGR] Hibiscus cannabinus. An annual, short-day, herbaceous plant of the Malvaceae family that is cultivated for its stem fibers. Kenaf is sometimes used to refer to Hibiscus sabdariffa var. altissima. { kə'naf }

kenching [FOOD ENG] In the process of curing fish by salting, the step in which the brine is permitting to drain away. { 'ken·chiŋ }

Kendall effect [COMMUN] A spurious pattern or other distortion in a facsimile record caused by unwanted modulation products arising from the transmission of a carrier signal; occurs principally when the width of one side band is greater than half the facsimile carrier frequency. { 'kend·əl ,fekt }

Kendall's rank correlation coefficient [STAT] A statistic used as a measure of correlation in nonparametric statistics when the data are in ordinal form. Also known as Kendall's tau. { 'ken·dəlz ¦raŋk ,kä·rə'lā·shən ,kō·ə,fish·ənt }

Kendall's tau See Kendall's rank correlation coefficient. { ¦ken·dəlz 'tȯ }

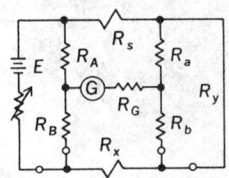

KELVIN BRIDGE

Circuit of the Kelvin bridge. E = battery; R_A, R_B = main ratio resistors; R_a, R_b = auxiliary ratio; R_x = unknown; R_s = standard; R_y = a heavy copper yoke of low resistance connected between the unknown and standard resistors; R_G = resistance in series with galvanometer G.

Kennard packet   [QUANT MECH]   A wave packet for which the product of the root-mean-square deviations of position and momentum from their respective mean values is as small as possible, being equal to Planck's constant divided by 4π. { 'ken·ərd ¡pak·ət }

Kennedy and Pancu circle   [MECH]   For a harmonic oscillator subject to hysteretic damping and subjected to a sinusoidally varying force, a plot of the in-phase and quadrature components of the displacement of the oscillator as the frequency of the applied vibration is varied. { ¦ken·ə·dē ən 'pän·chü ¡sər·kəl }

Kennedy key   [DES ENG]   A square taper key fitted into a keyway of square section and driven from opposite ends of the hub. { 'ken·ə·dē ¡kē }

Kennelly-Heaviside layer   See E layer. { 'ken·əl·ē 'hev·ē¡sīd ¡lā·ər }

kenophobia   [PSYCH]   An abnormal fear of large, empty spaces. { ¡ken·ə'fō·bē·ə }

Kenoran orogeny   See Algoman orogeny. { kə'nȯr·ən ȯ'räj·ə·nē }

kenotron   [ELEC]   A high-vacuum diode designed to serve as a rectifier in appliances requiring high voltage and low current. { 'ken·ə¡trän }

kenozooecium   [INV ZOO]   The outer, nonliving, hardened portion of a kenozooid. { ¦kēn·ə·zō'ē·sē·əm }

kenozooid   [INV ZOO]   A type of bryozoan heterozooid possessing a slender tubular or boxlike chamber, completely enclosed and lacking an aperture. { ¦kēn·ə¦zō¡ȯid }

kentrolite   [MINERAL] $Pb_2Mn_2Si_2O_9$ A dark reddish-brown mineral composed of a lead manganese silicate. { 'ken·trə¡līt }

Kentucky coffee tree   [BOT]   Gymnocladus dioica. An extremely tall, dioecious tree of the order Rosales readily recognized when in fruit by its leguminous pods containing heavy seeds, once used as a coffee substitute. { kən'tək·ē 'kȯf·ē ¡trē }

Kenyapithecus   [PALEON]   An early member of Hominidae from the Miocene. { ¡ken·yə'pith·ə·kəs }

kenyte   [MINERAL]   A variety of phonolite containing olivine in addition to anorthoclase feldspar, nepheline, acmite-augite, sodic amphibole, apatite, and opaque oxides. { 'ke¡nīt }

kep interlock   [MIN ENG]   A system designed to prevent the lowering of a shaft conveyance before all keps are fully withdrawn, and to indicate the position of the keps. { 'kep 'int·ər¡läk }

Keplerian ellipse   See Keplerian orbit. { ke'plir·ē·ən i'lips }

Keplerian motion   [ASTRON]   Orbital movement of a body about another that is not disturbed by the presence of a third celestial body. { ke'plir·ē·ən 'mō·shən }

Keplerian orbit   [ASTRON]   An elliptical orbit of a celestial body about another, the latter at a focus of the ellipse. Also known as Keplerian ellipse. { ke'plir·ē·ən 'ȯr·bət }

Keplerian telescope   [OPTICS]   A telescope that forms a real intermediate image in the focal plane and can be used for introducing a reticle or a scale into the focal plane. { ke'plir·ē·ən 'tel·ə¡skōp }

Kepler's equations   [ASTRON]   The mathematical relationship between two different systems of angular measurements of the position of a body in an ellipse. { 'kep·lərz i'kwā·zhənz }

Kepler's laws   [ASTRON]   Three laws, determined by Johannes Kepler, that describe the motions of planets in their orbits: the orbits of the planets are ellipses with the sun at a common focus; the line joining a planet and the sun sweeps over equal areas during equal intervals of time; the squares of the periods of revolution of any two planets are proportional to the cubes of their mean distances from the sun. { 'kep·lərz 'lȯz }

Kepler's supernova   [ASTRON]   A supernova that appeared in the constellation Ophiuchus in October 1604 and was visible until March 1606. { 'kep·lərz ¡sü·pər'nō·və }

keps   See folding boards. { keps }

kerabitumen   See kerogen. { ¦ker·ə·bə'tü·mən }

keratectomy   [MED]   Surgical removal of a portion of the cornea. { ¡ker·ə'tek·tə·mē }

keratin   [BIOCHEM]   Any of various albuminoids characteristic of epidermal derivatives, such as nails and feathers, which are insoluble in protein solvents, have a high sulfur content, and generally contain cystine and arginine as the predominating amino acids. { 'ker·əd·ən }

keratinized tissue   [HISTOL]   Any tissue with a high keratin

Branch of Kentucky coffee tree. (*Gymnocladus dioica*).

content, such as the epidermis or its derivatives. { 'ker·əd·ə¡nīzd 'tish·ü }

keratinocyte   [HISTOL]   A specialized epidermal cell that synthesizes keratin. { kə'rat·ən·ə¡sīt }

keratinous degeneration   [CYTOL]   The occurrence of keratin granules in the cytoplasm of a cell, other than a keratinocyte. { kə'rat·ən·əs di¡jen·ə'rā·shən }

keratitis   [MED]   Inflammation of the cornea. { ¡ker·ə'tīd·əs }

keratitis rosacea   [MED]   The occurrence of small, sterile infiltrates at the periphery of the cornea. { ¡ker·ə'īd·əs rō'zā·shə }

keratoconjunctivitis   [MED]   Concurrent inflammation of the cornea and the conjunctiva. Also known as shipyard eye. { ¦ker·əd·ō·kən¡jəŋk·tə'vīd·əs }

keratohyalin   [HISTOL]   Granules in the stratum granulosum of keratinized stratified squamous epithelium which become keratin. { ¡ker·əd·ō'hī·ə·lən }

keratoid   [MATH]   A plane curve whose equation in cartesian coordinates x and y is $y^2 = x^2 y + x^5$. { 'ker·ə¡tȯid }

keratoid cusp   [MATH]   A cusp of a curve which has one branch of the curve on each side of the common tangent. Also known as single cusp of the first kind. { 'ker·ə¡tȯid ¡kəsp }

keratomalacia   [MED]   Degeneration of the cornea characterized by infiltration and keratinization of the epithelium, eventually leading to thinning and perforation of the cornea; generally occurs in vitamin A deficiency. { ¡ker·əd·ō·mə'lā·shə }

keratophyre   [PETR]   Any dike rock or salic lava that is characterized by the presence of albite or albite oligoclase, chlorite, epidote, and calcite. { 'ker·əd·ō¡fīr·ər }

keratoplasty   [MED]   A plastic operation on the cornea, especially the transplantation of a portion of the cornea. { 'ker·əd·ō¡plas·tē }

keratosis   [MED]   Any disease of the skin characterized by an overgrowth of the cornified epithelium. { ¡ker·ə'tō·səs }

keratosis follicularis   See Darrier's disease. { ¡ker·ə¡tōs·əs fə¡lik·yə'lär·əs }

kerf   [ENG]   A cut made in wood, metal, or other material by a saw or cutting torch.   [MIN ENG]   A narrow, deep cut made in the face of coal to facilitate mining. { kərf }

ker function   [MATH]   A function that is expressed in terms of modified Bessel functions of the second kind in a manner similar to that in which the ber function is expressed in terms of Bessel functions. { 'ker ¡fəŋk·shən }

Kerguelen faunal region   [ECOL]   A marine littoral faunal region comprising a large area surrounding Kerguelen Island in the southern Indian Ocean. { 'kər·gə·lən 'fȯn·əl ¡rē·jən }

kerma   [NUCLEO]   The kinetic energy imparted to charged particles in a unit mass of material by uncharged particles such as neutrons; it may be expressed as joules per kilogram or ergs per gram. { 'kər·mə }

kermesite   [MINERAL] Sb_2S_2O A cherry-red mineral occurring as tufts of capillary crystals, and formed from an alteration of stibnite. Also known as antimony blende; purple blende; pyrostibite; red antimony. { 'kər·mə¡zīt }

Kern counter   See dust counter. { 'kərn ¡kaùn·tər }

kernel   [ATOM PHYS]   An atom that has been stripped of its valence electrons, or a positively charged nucleus lacking the outermost orbital electrons.   [BOT]   **1.** The inner portion of a seed.   **2.** A whole grain or seed of a cereal plant, such as corn or barley.   [COMPUT SCI]   **1.** A computer program that must be modified before it can be used on a particular computer. **2.** The programs that form the most essential part of a computer's operating system.   [MATH]   **1.** For any mapping f from a group A to a group B, the kernel of f, denoted ker f, is the set of all elements a of A such that $f(a)$ equals the identity element of B.   **2.** For a homomorphism h from a group G to a group H, this consists of all elements of G which h sends to the identity element of H.   **3.** For Fredholm and Volterra integral equations, this is the function $K(x,t)$.   **4.** For an integral transform, the function $K(x, t)$ in the transformation which sends the function $f(x)$ to the function $\int K(x,t)f(t)dt = F(x)$.   **5.** See null space. { 'kərn·əl }

kernel blight   [PL PATH]   Any of several fungus diseases of barley caused chiefly by *Gibberella zeae*, *Helminthosporium sativum*, and *Alternaria* species shriveling and discoloring the grain. { 'kərn·əl ¡blīt }

kernel ice   [HYD]   In aircraft icing, an extreme form of rime ice, that is, very irregular, opaque, and of low density; it forms at temperatures of $-15°C$ and lower. { 'kərn·əl ‚īs }

kernel spot [PL PATH] A fungus disease of the pecan kernel caused by *Coniothyrium caryogenum* and characterized by dull-brown roundish spots. { 'kərn·əl ‚spät }

kernicterus [PATH] Deposition of bilirubin in the gray matter of the brain and spinal cord, especially in the basal ganglia, accompanied by nerve cell degeneration. { kər'nik·tə·rəs }

Kernig's sign [MED] In meningeal irritation, with the patient lying face up and the thigh flexed at the hip, the pain and spasm of the hamstring muscles when an attempt is made to completely extend the leg at the knee. { 'kər‚nigz ‚sin }

kerning [GRAPHICS] Adjusting the spacing between certain letters during typesetting so that part of each letter overhangs adjacent letters. { 'kərn·iŋ }

kernite [MINERAL] $Na_2B_4O_7·4H_2O$ A colorless to white hydrous borate mineral crystallizing in the monoclinic system and having vitreous luster; an important source of boron. Also known as rasorite. { 'kər‚nīt }

kerogen [GEOL] The complex, fossilized organic material present in sedimentary rocks, especially in shales; converted to petroleum products by distillation. Also known as kerabitumen; petrologen. { 'ker·ə·jən }

kerogen shale *See* oil shale. { 'ker·ə·jən ‚shāl }

kerosine [MATER] A refined petroleum fraction used as a fuel for heating and cooking, jet engines, lamps, and weed burning and as a base for insecticides; specific gravity is about 0.8; components are mostly paraffinic and naphthenic hydrocarbons in the C_{10} to C_{14} range. Also known as lamp oil. { 'ker·ə‚sēn }

kerosine distillate [MATER] The distilled cut in the 150–300°C range from petroleum or shale oil; used as lamp, stove, or illuminating oil, as a solvent, and as a component of jet aircraft fuels. Also known as burning oil. { 'ker·ə‚sēn 'dist·əl·ət }

kerosine propellant [MATER] A propellant consisting of highly refined, low-aromatic liquid propellant distillate with a gravity range not exceeding three degrees American Petroleum Institute gravity at 60°F (15.6°C); may contain additives. { 'ker·ə‚sēn prə'pel·ənt }

kerosine shale *See* torbanite. { 'ker·ə‚sēn ‚shāl }

Kerr cell [OPTICS] A glass cell containing a dielectric liquid that exhibits the Kerr effect, such as nitrobenzene, in which is inserted the two plates of a capacitor, used to observe the Kerr effect on light passing through the cell. { 'kər ‚sel }

Kerr constant [OPTICS] A measure of the strength of the Kerr effect in a substance, equal to the difference between the extraordinary and ordinary indices of refraction divided by the product of the light's wavelength and the square of the electric field. { 'kər 'kän·stənt }

Kerr effect *See* electrooptical Kerr effect. { 'kər i‚fekt }

Kerr magnetooptical effect *See* magnetooptic Kerr effect. { 'kər mag‚ned·ō'äp·tə·kəl i‚fekt }

Kerr-Newman solution [RELAT] A solution to Einstein's field equations that describes a rotating, charged black hole. { ‚kər 'nü·mən sə‚lü·shən }

Kerr solution [RELAT] A solution to Einstein's field equations that describes a rotating, uncharged, axisymmetric black hole. { 'kər sə‚lü·shən }

kersantite [PETR] Dark dike rocks consisting mostly of biotite, plagioclase, and augite. { kər'zan‚tīt }

kersey *See* melton. { 'kər·zē }

ketal [ORG CHEM] 1. Former term for the $=CO$ group, as in dimethyl ketal (acetone). 2. Any of the ketone acetates from condensation of alkyl orthoformates with ketones in the presence of alcohols. { 'kē‚tal }

ketene [ORG CHEM] C_2H_2O A colorless, toxic, highly reactive gas, with disagreeable taste; boils at $-56°C$; soluble in ether and acetone, and decomposes in water and alcohol; used as an acetylating agent in organic synthesis. { 'kē‚tēn }

ketene lamp [CHEM ENG] An electrically heated Chromel filament by the means of which acetone is hydrolyzed to produce ketene. { 'kē‚tēn ‚lamp }

ketimide [ORG CHEM] A compound that is represented by $R_2:C:NX$, where X is an acyl radical. { 'ked·ə‚mīd }

ketimine [ORG CHEM] An organic compound that contains the divalent group $\rangle C=NH$; a Schiff base is an example. { 'ked·ə‚mēn }

keto- [ORG CHEM] Organic chemical prefix for the keto or carbonyl group, $C:O$, as in a ketone. { 'kēd·ō }

keto acid [ORG CHEM] A compound that is both an acid and a ketone; an example is β-acetoacetic acid. { 'kēd·ō ‚as·əd }

ketoacidosis *See* ketosis. { ‚kēd·ō‚as·ə'dō·səs }

ketoadipic acid [BIOCHEM] $C_6H_8O_5$ An intermediate product in the metabolism of lysine to glutaric acid. { ‚kēd·ō·ə‚dip·ik 'as·əd }

ketogenesis [BIOCHEM] Production of ketone bodies. { ‚kēd·ō'jen·ə·səs }

ketogenic hormone [BIOCHEM] A factor originally derived from crude anterior hypophysis extract which stimulated fatty-acid metabolism; now known as a combination of adrenocorticotropin and the growth hormone. Also known as fat-metabolizing hormone. { ‚kēd·ə‚jen·ik 'hȯr‚mōn }

ketogenic substance [BIOCHEM] Any foodstuff which provides a source of ketone bodies. { ‚kēd·ə‚jen·ik 'səb·stəns }

ketoglutarate [ORG CHEM] A salt or ester of ketoglutaric acid. { ‚kēd·ə'glüd·ə‚rāt }

ketoglutaric acid [BIOCHEM] $C_5H_6O_5$ A dibasic keto acid occurring as an intermediate product in carbohydrate and protein metabolism. { ‚kēd·ō·glü‚tar·ik 'as·əd }

ketohexose [BIOCHEM] Any monosaccharide composed of a six-carbon chain and containing one ketone group. { ‚kēd·ō'hek‚sōs }

ketolase [BIOCHEM] A type of enzyme that catalyzes cleavage of carbohydrates at the carbonyl carbon position. { 'kēd·əl‚ās }

ketolysis [BIOCHEM] Dissolution of ketone bodies. { kē'täl·ə·səs }

ketone [ORG CHEM] One of a class of chemical compounds of the general formula $RR'CO$, where R and R' are alkyl, aryl, or heterocyclic radicals; the groups R and R' may be the same or different, or incorporated into a ring; the ketones, acetone, and methyl ethyl ketone are used as solvents, and ketones in general are important intermediates in the synthesis of organic compounds. { 'kē‚tōn }

ketone body [BIOCHEM] Any of various ketones which increase in blood and urine in certain conditions, such as diabetic acidosis, starvation, and pregnancy. Also known as acetone body. { 'kē‚tōn ‚bäd·ē }

ketonemia *See* acetonemia. { ‚kēd·ə'nē·mē·ə }

ketonuria [MED] Presence of ketone bodies in the urine. { ‚kēd·ə'nür·ē·ə }

ketose [BIOCHEM] A carbohydrate that has a ketone group. { 'kē‚tōs }

ketosis [MED] Excess amounts of ketones in the body, especially associated with diabetes mellitus. Also known as ketoacidosis. { kē'tō·səs }

ketosteroid [BIOCHEM] One of a group of neutral steroids possessing keto substitution, which produces a characteristic red color with *m*-dinitrobenzene in an alkaline solution; these compounds are principally metabolites of adrenal cortical and gonadal steroids. { ‚kēd·ō'stir‚ȯid }

Ketteler formula [ELECTROMAG] The case of Sellmeier's equation where only two characteristic frequencies are involved. { 'ket·lər ‚fȯr·myə·lə }

kettle [GEOL] 1. A bowl-shaped depression with steep sides in glacial drift deposits that is formed by the melting of glacier ice left behind by the retreating glacier and buried in the drift. Also known as kettle basin; kettle hole. 2. *See* pothole. { 'ked·əl }

kettle basin *See* kettle. { 'ked·əl ‚bās·ən }

kettle hole *See* kettle. { 'ked·əl ‚hōl }

kettle reboiler [CHEM ENG] Tube-and-shell heat exchange device in which liquid is vaporized on the shell side from heat transferred from hot liquid flowing through the tubes; dome space allows liquid-vapor separation above the tube bundle. { 'ked·əl rē'bȯil·ər }

ket vector [QUANT MECH] A vector in Hilbert space specifying the state of a system (opposed to bra vector); represented by the symbol $|\rangle$, with a letter or one or more indices inserted to distinguish it from other vectors. { ‚ket‚vek·tər }

Keuper [GEOL] A European stage of geologic time, especially in Germany; Upper Triassic. { 'kȯip·ər }

keV *See* kiloelectronvolt.

Kew barometer [ENG] A type of cistern barometer; no adjustment is made for the variation of the level of mercury in the cistern as pressure changes occur; rather, a uniformly contracting scale is used to determine the effective height of the mercury column. { 'kyü bə'ram·əd·ər }

KETENE

Structural formula of ketene.

KETONE

Structural formula of a ketone.

Keweenawan [GEOL] The younger of two Precambrian time systems that constitute the Proterozoic period in Michigan and Wisconsin. { ¦kē·wē¦nō·ən }

key [BUILD] **1.** Plastering that is forced between laths to secure the rest of the plaster in place. **2.** The roughening on a surface to be glued or plastered to increase adhesiveness. [CIV ENG] A projecting portion that serves to prevent movement of parts at a construction joint. [COMPUT SCI] A data item that serves to uniquely identify a data record. [DES ENG] **1.** An instrument that is inserted into a lock to operate the bolt. **2.** A device used to move in some manner in order to secure or tighten. **3.** One of the levers of a keyboard. **4.** See machine key. [ELEC] **1.** A hand-operated switch used for transmitting code signals. Also known as signaling key. **2.** A special lever-type switch used for opening or closing a circuit only as long as the handle is depressed. Also known as switching key. [ENG] The pieces of core causing a block in a core barrel, the removal of which allows the rest of the core in the barrel to slide out. [GEOL] A cay, especially one of the islets off the south of Florida. Also spelled kay. [PETRO ENG] A hooklike wrench fitted to the square of a sucker rod to pull and run each sucker rod of a pumping oil well. [SYST] An arrangement of the distinguishing features of a taxonomic group to serve as a guide for establishing relationships and names of unidentified members of the group. { kē }

key access [COMPUT SCI] Locating data in a file by using the value of a key. { 'kē ,ak,ses }

key activity [IND ENG] An activity that possesses major significance. Also known as milestone activity. { ¦kē ak'tiv·əd·ē }

key auto-key cipher [COMMUN] A stream cipher in which the cryptographic bit stream generated at a given time is determined by the cryptographic bit stream generated at earlier times. { 'kē 'ȯd·ō ,kē ,sī·fər }

key bed [GEOL] Also known as index bed; key horizon; marker bed. **1.** A stratum or body of strata that has distinctive characteristics so that it can be easily identified. **2.** A bed whose top or bottom is employed as a datum in the drawing of structure contour maps. { 'kē ,bed }

keyboard [ENG] A set of keys or control levers having a systematic arrangement and used to operate a machine or other piece of equipment such as a typewriter, typesetter, processing unit of a computer, or piano. { 'kē,bȯrd }

keyboard enhancer [COMPUT SCI] Software that expands the functions of a computer keyboard by allowing the user to implement functions or enter predefined segments of text with a single keystroke. Also known as keyboard processor. { 'kē,bȯrd in,han·sər }

keyboard entry [COMPUT SCI] A piece of information fed manually into a computing system by means of a set of keys, such as a typewriter. { 'kē,bȯrd 'en·trē }

keyboard inquiry [COMPUT SCI] A question asked a computer concerning the status of a program being run, or concerning the value achieved by a specific variable, by means of a console typewriter. { 'kē,bȯrd ¦in·kwə·rē }

keyboardless typesetter [COMPUT SCI] An automatic typesetting machine that has no keyboard and is operated by perforated tape at a speed of 12–15 lines per minute; the text is punched on tape at separate keyboard machines. { 'kē,bȯrd·ləs 'tīp,sed·ər }

keyboard lockout [COMPUT SCI] An arrangement for preventing transmission from a particular keyboard while other transmissions are taking place on the same circuit. { 'kē,bȯrd 'läk,au̇t }

keyboard lockup [COMPUT SCI] A condition in which entries typed on a keyboard are ignored by a terminal. { 'kē,bȯrd 'läk,əp }

keyboard mapping [COMPUT SCI] The process of assigning the meaning of keys on a computer keyboard. { 'kē,bȯrd ,map·iŋ }

keyboard perforator [ENG] A typewriterlike device that prepares punched paper tape for communications or computing equipment. { 'kē,bȯrd 'pər·fə,rād·ər }

keyboard printer [COMPUT SCI] A computer input device that includes a keyboard and a printer that prints the keyed-in data and often also prints computer output information. { 'kē,bȯrd 'print·ər }

keyboard processor [COMPUT SCI] **1.** The circuitry in a computer keyboard that converts keystrokes into the appropriate character codes. **2.** See keyboard enhancer. { 'kē,bȯrd ,präs,es·ər }

keyboard send/receive [ELECTR] A manual teleprinter that can transmit or receive. Abbreviated KSR. Also known as keyboard teleprinter. { 'kē,bȯrd ¦send·ri'sēv }

keyboard teleprinter See keyboard send/receive. { 'kē,bȯrd 'tel·ə,print·ər }

keyboard template [COMPUT SCI] A card that is placed adjacent to the function keys of a computer keyboard and identifies their use for a particular software environment. { 'kē,bȯrd ,tem·plət }

key cabinet [ELECTR] A case, installed on a customer's premises, to permit different lines to the control office to be connected to various telephone stations; it has signals to indicate originating calls and busy lines. { 'kē ,kab·ə·nət }

key change [COMPUT SCI] The occurrence, in a file of records which have been sorted according to their keys and are being read into a computer, of a record whose key differs from that of its immediate predecessor. { 'kē ,chānj }

key compression [COMPUT SCI] A technique used to reduce the number of bits contained in a key. { 'kē kəm,presh·ən }

key cut [MIN ENG] In strip mine operations, the section excavated adjacent to the new highwall. { 'kē ,kət }

key day See control day. { 'kē ,dā }

key-disk machine [COMPUT SCI] A keyboard machine used to record data directly on a magnetic disk. { 'kē ,disk mə,shēn }

keyed clamp [ELECTR] Clamping circuit in which the time of clamping is determined by a control signal. { 'kēd 'klamp }

keyed clamp circuit [ELECTR] A clamp circuit in which the time of clamping is controlled by separate voltage or current sources, rather than by the signal itself. Also known as synchronous clamp circuit. { 'kēd 'klamp ,sər·kət }

keyed sequential access method [COMPUT SCI] A method for locating data in a file either directly, by using the value of a key within a particular record, or sequentially, according to the values of the keys in all the records of the file. Abbreviated KSAM. { 'kēd si'kwen·chəl 'ak,ses ,meth·əd }

key entry [COMPUT SCI] The entering of data into a computer by means of a keyboard. { 'kē ,en·trē }

keyer [ELECTR] Device which changes the output of a transmitter from one condition to another according to the intelligence to be transmitted. { 'kē·ər }

keyer adapter [ELECTR] Device which detects a modulated signal and produces the modulating frequency as a direct-current signal of varying amplitude. { 'kē·ər ə,dap·tər }

Keyes equation [THERMO] An equation of state of a gas which is designed to correct the van der Waals equation for the effect of surrounding molecules on the term representing the volume of a molecule. { 'kēz i,kwā·zhən }

Keyes process [CHEM ENG] A distillation process used to obtain absolute alcohol; benzene is added to a constant-boiling 95% alcohol-water solution, and on distillation anhydrous alcohol leaves the bottom of the column. { 'kēz ,präs·əs }

key field [COMPUT SCI] A field in a segment or record that holds the value of a key to that record. { 'kē ,fēld }

key grasp See pinch grasp. { 'kē ,grasp }

keyhole [DES ENG] A hole or a slot for receiving a key. [MET] A welding method wherein the heat source, because of its concentration, causes a hole through the surface immediately ahead of the molten weld metal in the direction of welding; the hole is filled as the welding progresses, ensuring complete joint penetration. [ORD] Of a bullet, to strike a target after tumbling in flight so that the long axis of the bullet and the line of flight are not the same; usually caused by failure of the bullet to receive sufficient spin from the rifling in the barrel. { 'kē,hōl }

keyhole saw [DES ENG] A fine compass saw with a blade 11–16 inches (28–41 centimeters) long. { 'kē,hōl ,sȯ }

keyhole specimen [MET] A metal specimen containing a keyhole shaped notch and used in certain impact tests. { 'kē,hōl ,spes·ə·mən }

key horizon See key bed. { 'kē hə,rīz·ən }

keying [CIV ENG] Establishing a mechanical bond in a construction joint. [ELEC] The forming of signals, such as for telegraph transmission, by modulating a direct-current or other carrier between discrete values of some characteristic. { 'kē·iŋ }

KEYED CLAMP CIRCUIT

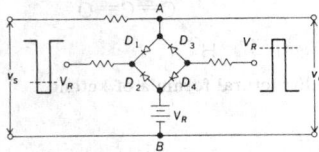

Four-diode bidirectional keyed clamp circuit. D = diode; V_R = reference voltage; v_S = voltage when switch is in series with reference voltage; A, B, = terminals.

keying error rate [COMMUN] The ratio of the number of characters incorrectly transmitted to the total number of characters in a message. { 'kē·iŋ 'er·ər ‚rāt }

keying frequency [COMMUN] In facsimile, the maximum number of times a second that a black-line signal occurs when scanning the subject copy. { 'kē·iŋ ‚frē·kwən·sē }

keying interval [COMMUN] In a periodically keyed transmission system, one of the set of intervals starting from a change in state and equal in length to the shortest time between changes of state. { 'kē·iŋ ‚int·ər·vəl }

keying sequence [COMMUN] A sequence of letters or numbers that enciphers or deciphers a polyalphabetic substitution cipher character by character. { 'kē·iŋ ‚sē·kwəns }

keying wave See marking wave. { 'kē·iŋ ‚wāv }

key job [IND ENG] A job that has been evaluated and is considered representative of similar jobs in the same labor market and is used as a benchmark to evaluate the similar jobs and to establish non-key-job wages. { 'kē ‚jäb }

key joint [CIV ENG] A mortar joint with a concave pointing. { 'kē ‚jöint }

keyless ringing [COMMUN] Form of machine ringing on a manual telephone switchboard which is started automatically by the insertion of the calling plug into the jack of the called line. { 'kē·ləs 'riŋ·iŋ }

keyline [GRAPHICS] An outline drawing on completed art for the purpose of indicating exactly the shape, position, and size of elements such as halftones or line sketches. { 'kē‚līn }

keylock switch [ELEC] A switch that can be operated only by inserting and turning a key such as that used in ordinary locks. { 'kē‚läk ‚swich }

keypad [COMPUT SCI] A cluster of special-purpose keys to one side of the regular typing keys on a terminal keyboard. { 'kē‚pad }

key plate [GRAPHICS] A plate used in printing two-color stamps that prints the central design of the stamp. { 'kē ‚plāt }

key pulse [COMMUN] System of signaling where numbered keys are depressed instead of using a dial. { 'kē ‚pəls }

key punch [COMPUT SCI] A keyboard-actuated device that punches holes in a card; it may be a hand-feed punch or an automatic feed punch. { 'kē ‚pənch }

key seat [DES ENG] See keyway. [PETRO ENG] A groove along the side of an oil well hole and parallel to its axis which results from incorrect handling of the pipe. { 'kē ‚sēt }

keyseater [MECH ENG] A machine for milling beds or grooves in mechanical parts which receive keys. { 'kē‚sēd·ər }

keyshelf [COMMUN] Horizontal shelf of a manual telephone switchboard on which are mounted the keys by which the operator switches one or more of the switchboard circuits. { 'kē‚shelf }

keystone [ARCH] Wedge-shaped stone at the crown of an arch. [MATER] Small crushed stone used as filler for the large aggregate in bituminous bound roads. { 'kē‚stōn }

keystone distortion [COMMUN] Distortion produced by scanning in a rectilinear manner, with constant-amplitude sawtooth waves, a plane target area which is not normal to the average direction of the beam. { 'kē‚stōn di'stór·shən }

keystoning [ELECTR] Producing a keystone-shaped (wider at the top than at the bottom, or vice versa) scanning pattern because the electron beam in the television camera tube is at an angle with the principal axis of the tube. { 'kē‚stōn·iŋ }

keyswitch [COMPUT SCI] A switch that is operated by depressing a key on the keyboard of a data entry terminal. { 'kē‚swich }

key telephone system [COMMUN] A telephone system consisting of phones with several keys, connecting cables, and relay switching apparatus, which does not need a special operator to handle incoming or outgoing calls and which generally permits users to select one of several possible lines and to hold calls. { 'kē 'tel·ə‚fōn ‚sis·təm }

key telephone unit [COMMUN] A small mounting plate with relays which performs pickup and hold switching functions in a key telephone system. { 'kē 'tel·ə‚fōn ‚yü·nət }

key-to-disk system [COMPUT SCI] A data-entry system in which information entered on several keyboards is collected on different sections of a magnetic disk, and the data are extracted from the disk when complete, and are copied onto a magnetic tape or another disk for further processing on the main computer. { 'kē tə 'disk ‚sis·təm }

key-to-tape system [COMPUT SCI] A data-entry system consisting of several keyboards connected to a central controlling unit, typically a minicomputer, which collects information from each keyboard and then directs it to a magnetic tape. { 'kē tə 'tāp ‚sis·təm }

key transformation [COMPUT SCI] A function that assigns integer values to keys. { 'kē ‚tranz·fər'mā·shən }

key value [COMPUT SCI] The actual characters contained in a key. { 'kē 'val·yü }

keyway [DES ENG] **1.** An opening in a lock for passage of a flat metal key. **2.** The pocket in the driven element to provide a driving surface for the key. **3.** A groove or channel for a key in any mechanical part. Also known as key seat. [ENG] An interlocking channel or groove in a cement or wood joint to provide reinforcement. { 'kē‚wā }

keyword [COMPUT SCI] A group of letters and numbers in a specific order that has special significance in a computer system. { 'kē‚wərd }

keyword-in-context index [COMPUT SCI] A computer-generated listing of titles of documents, produced on a line printer, with the keywords lined up vertically in a fixed position within the title and arranged in alphabetical order. Abbreviated KWIC index. { 'kē‚wərd in 'kän‚tekst ‚in‚deks }

keyword-out-of-context index [COMPUT SCI] A computer-generated listing of document titles with their keywords listed separately, arranged in the alphabetical order of the keywords. Abbreviated KWOC index. { 'kē‚wərd aút əv 'kän‚tekst ‚in‚deks }

keyword parameter [COMPUT SCI] A parameter whose significance is indicated by a keyword, usually with an equal sign linking the two. { 'kē‚wərd pə'ram·əd·ər }

keyword search [COMPUT SCI] A method of filing and locating information through the use of keywords that describe the content of records. { 'kē‚wərd ‚sərch }

keyword spotting [ENG ACOUS] An approach to task-oriented speech understanding through detecting a limited number of keywords that would most likely express the intent of a speaker, rather than attempting to recognize every word in an utterance. { 'kē‚wərd ‚spät·iŋ }

K factor [GRAPHICS] See base-altitude ratio. [NUCLEO] A measure of the energy of the gamma rays produced by a particular type of emitter; it is the gamma-ray dose rate in roentgens per hour at a distance of 1 centimeter from a source having a radioactive disintegration rate of 1 millicurie (3.7 × 10^7 disintegrations per second). [ORD] In artillery ground fire, a factor to be applied to the actual range to a point in order to determine the range which must be fired to hit that point; it is the result of registration and the solution of a meteorological message, and is expressed as plus or minus so many yards per thousand units of actual range. { 'kā ‚fak·tər }

K feldspar See potassium feldspar. { 'kā ‚fel‚spär }

kg See kilogram; kilogram force.

kg-cal See kilocalorie.

kgf See kilogram force.

kgf-m See meter-kilogram.

K gun [ORD] A type of U.S. Navy gun in the shape of a K for firing depth charges. { 'kā ‚gən }

kg-wt See kilogram force.

khamsin [METEOROL] A dry, dusty, and generally hot desert wind in Egypt and over the Red Sea; it is generally southerly or southeasterly, occurring in front of depressions moving eastward across North Africa or the southeastern Mediterranean. { kam'sēn }

khelin See khellin. { 'kel·ən }

khellin [PHARM] $C_{14}H_{12}O_5$ A synthetic compound that crystallizes from methanol solution, has a bitter taste, melts at 154–155°C, and is more soluble in water than in organic solvents; used in medicine as an antispasmodic, a coronary vasodilator, and a bronchodilator. Also spelled chellin; khelin. Also known as visammin. { 'kel·ən }

khibinite See mosandrite. { 'kib·ə‚nīt }

KHN filter See state-variable filter { 'kā‚äch'en fil·tər }

kHz See kilohertz.

kibble See hoppit. { 'kib·əl }

kibli See ghibli. { 'kib·lē }

kick [PETRO ENG] Entry of fluid into the wellbore when the pressure of the column of drilling fluid is insufficient to withstand the pressure of the fluids in the formation being drilled. [ORD] Violent backward movement of a gun after being fired, caused by the rearward force of the propellant gases acting on the gun. { kik }

kickback [MECH ENG] A backward thrust, such as the backward starting of an internal combustion engine as it is cranked, or the reverse push of a piece of work as it is fed to a rotary saw. { 'kik,bak }

kickdown [MECH ENG] **1.** Shifting to lower gear in an automotive vehicle. **2.** The device for shifting. { 'kik,daun }

kick over [MECH ENG] To start firing; applied to internal combustion engines. { 'kik ,ō·vər }

kickpipe [BUILD] A short pipe protecting an electrical cable at the point where it emerges from a floor. { 'kik,pīp }

kickplate [BUILD] A plate used on the bottom of doors and cabinets or on the risers of steps to protect them from shoe marks. Also known as toeplate. { 'kik,plāt }

Kick's law [ENG] The law that the energy needed to crush a solid material to a specified fraction of its original size is the same, regardless of the original size of the feed material. { 'kiks ,lò }

kick-sorter See pulse-height analyzer. { 'kik ,sórd·ər }

kick starter [MECH ENG] A mechanism for starting the operation of a motor by thrusting with the foot. { 'kik ,stärd·ər }

kick wheel [ENG] A potter's wheel worked by a foot pedal. { 'kik ,wēl }

Kidd blood group system [IMMUNOL] The erythrocyte antigens defined by reactions to anti-Jk^a antibodies, originally found in the mother (Mrs. Kidd) of the erythroblastotic infant, and to anti-Jk^b antibodies. { 'kid 'bləd ,grüp ,sis·təm }

kidney [ANAT] Either of a pair of organs involved with the elimination of water and waste products from the body of vertebrates; in humans they are bean-shaped, about 5 inches (12.7 centimeters) long, and are located in the posterior part of the abdomen behind the peritoneum. { 'kid·nē }

kidney joint [ELECTROMAG] Flexible joint, or air-gap coupling, used in the waveguide of certain radars and located near the transmitting-receiving position. { 'kid·nē ,jóint }

kidney ore [MINERAL] A form of hematite found in compact masses, concretions, or nodules that are kidney-shaped. { 'kid·nē ,ór }

kidney stone See nephrite. { 'kid·nē ,stōn }

Kienböck's disease See osteochondrosis. { kēn,beks di,zēz }

kier [TEXT] A tank in which unfinished cotton is boiled under pressure to remove foreign materials such as sizing or wax. { kir }

kier boiling [TEXT] A scouring process for cotton and linen in which the fibers are treated for removal of impurities by boiling with a 1% solution of caustic soda. { 'kir ,bóil·iŋ }

kieselguhr See diatomaceous earth. { 'kē·zəl,gùr }

kieserite [MINERAL] MgSO$_4$·H$_2$O A white mineral that crystallizes in the monoclinic system, is composed of hydrous magnesium sulfate, and occurs in saline residues. { 'kē·zə,rīt }

Kikuchi lines [CRYSTAL] A pattern consisting of pairs of white and dark parallel lines, obtained when an electron beam is scattered (diffracted) by a crystalline solid; the pattern gives information on the structure of the crystal. { kē'kü·chē ,līnz }

Kiliani reaction [ORG CHEM] A method of synthesizing a higher aldose from a lower aldose; monosaccharides, such as aldehydes and ketones, react with hydrogen cyanide to form cyanohydrins, which are hydrolyzed to hydroxy acids, converted to lactones, and reduced to aldoses with sodium amalgams. { ,kil·ē'an·ē rē,ak·shən }

Kilkenny coal See anthracite. { kil'ken·ē 'kōl }

kill [MATER] To treat in such a way as to destroy undesirable properties; for example, neutralization of an acid by the addition of an alkali. [MET] To add a strong deoxidizer, such as silicon or aluminum, to molten steel in order to stop the reaction between carbon and oxygen forming gaseous carbon monoxide and dioxide during solidification. [PETRO ENG] **1.** In drilling, to prevent well blowout by appropriate measures. **2.** In oil production, to halt well production so that reconditioning of the well may proceed. { kil }

killed spirits [MATER] An aqueous solution of zinc(II) chloride used as a flux for solder. { 'kild 'spir·əts }

killed steel [MET] Thoroughly deoxidized steel, for example, by additions of aluminum or silicon, in which the reaction between carbon and oxygen during solidification is suppressed. { 'kild 'stēl }

killed vaccine [IMMUNOL] A suspension of killed microorganisms used as antigens to produce immunity. Also known as inactivated vaccine. { 'kild vak'sēn }

killer [SOLID STATE] An impurity that inhibits luminescence in a solid. { 'kil·ər }

killer circuit [ELECTR] Vaccum tube or tubes and associated circuits in which are generated the blanking pulses used to temporarily disable a radar set. { 'kil·ər ,sər·kət }

killer pulse [ELECTR] Blanking pulse generated by a killer circuit. { 'kil·ər ,pəls }

killer stage See color killer circuit. { 'kil·ər ,stāj }

killer whale [VERT ZOO] Orcinus orca. A predatory, cosmopolitan cetacean mammal, about 30 feet (9 meters) long, found only in cold waters. { 'kil·ər ,wāl }

Killing's equations [MATH] The equations for an isometry-generating vector field in a geometry. { 'kil·iŋz i,kwā·zhənz }

Killing vector [MATH] An element of a vector field in a geometry that generates an isometry. { 'kil·iŋ ,vek·tər }

kill probability [ORD] The probability that, given a hit, a single projectile or missile will kill the target against which it is fired. { 'kil ,präb·ə,bil·əd·ē }

kill spool [PETRO ENG] A device fitted to the wellhead that connects to a high-pressure line through which heavy drilling fluid is pumped into the well to offset the natural pressure of the well. { 'kil ,spül }

kiln [ENG] A heated enclosure used for drying, burning, or firing materials such as ore or ceramics. { kil }

kilo- [SCI TECH] A prefix representing 10^3 or 1000. Abbreviated k. { 'ki·lō, 'kē·lō }

kiloampere [ELEC] A metric unit of current flow equal to 1000 amperes. Abbreviated kA. { 'ki·lō'am,pir }

kilobar [MECH] A unit of pressure equal to 1000 bars (100 megapascals). Abbreviated kb. { 'kil·ə,bär }

kilobase [GEN] Unit of length equal to 1000 base pairs in deoxyribonucleic acid or 1000 nitrogenous bases in ribonucleic acid. Abbreviated kb; kbp. { 'kil·ō,bās }

kilobit [COMPUT SCI] A unit of information content equal to 1024 bits. Abbreviated kbit; Kbit. Symbolized k. { 'kil·ə,bit }

kilobyte [COMPUT SCI] A unit of information content equal to 1024 bytes. Abbreviated kbyte; Kbyte. Symbolized K. { 'kil·ə,bīt }

kilocalorie [THERMO] A unit of heat energy equal to 1000 calories. Abbreviated kcal. Also known as kilogram-calorie (kg-cal); large calorie (Cal). { 'kil·ə,kal·ə·rē }

kilocycle See kilohertz. { 'kil·ə,sī·kəl }

kiloelectronvolt [PHYS] A nit of energy, equal to 1000 electronvolts. Abbreviated keV. { 'ki·lō·i'lek,trän,vōlt }

kilogram [MECH] **1.** The unit of mass in the meter-kilogram-second system, equal to the mass of the international prototype kilogram stored at Sèvres, France. Abbreviated kg. **2.** See kilogram force. { 'kil·ə,gram }

kilogram-calorie See kilocalorie. { 'kil·ə,gram 'kal·ə·rē }

kilogram-equivalent weight [CHEM] A unit of mass 1000 times the gram-equivalent weight. { 'kil·ə,gram i'kwiv·ə·lənt 'wāt }

kilogram force [MECH] A unit of force equal to the weight of a 1-kilogram mass at a point on the earth's surface where the acceleration of gravity is 9.80665 m/s^2. Abbreviated kgf. Also known as kilogram (kg); kilogram weight (kg-wt). { 'kil·ə,gram 'fòrs }

kilogram-meter See meter-kilogram. { 'kil·ə,gram 'mēd·ər }

kilogram weight See kilogram force. { 'kil·ə,gram 'wāt }

kilohertz [PHYS] A unit of frequency equal to 1000 hertz. Abbreviated kHz. Also known as kilocycle (kc). { 'kil·ə,hərts }

kilohm [ELEC] A unit of electrical resistance equal to 1000 ohms. Abbreviated K; kohm. { 'kil,ōm }

kilojoule [PHYS] A unit of energy or work equal to 1000 joules. Abbreviated kJ. { 'kil·ə,jül }

kiloliter [MECH] A unit of volume equal to 1000 liters or to 1 cubic meter. Abbreviated kl. { 'kil·ə,lēd·ər }

kilomega- See giga-. { 'kil·ə'meg·ə }

kilomegacycle See gigahertz. { 'kil·ə'meg·ə,sī·kəl }

kilomegahertz See gigahertz. { 'kil·ə'meg·ə,hərtz }

kilometer [MECH] A unit of length equal to 1000 meters. Abbreviated km. { 'kil·ə,mēd·ər }

kiloparsec [ASTRON] A distance of 1000 parsecs (3260 light-years). { 'kil·ə'pär,sek }

kiloton [PHYS] A unit used in specifying the yield of a fission or fusion bomb, equal to the explosive power of 1000 metric tons of trinitrotoluene (TNT). Abbreviated kt. { 'kil·ə,tän }

kilovar [ELEC] A unit equal to 1000 volt-amperes reactive. Abbreviated kvar. { 'kil·ə,vär }

kilovolt [ELEC] A unit of potential difference equal to 1000 volts. Abbreviated kV. { 'kil·ə,vōlt }

kilovolt-ampere [ELEC] A unit of apparent power in an alternating-current circuit, equal to 1000 volt-amperes. Abbreviated kVA. { 'kil·ə,vōlt 'am,pir }

kilovoltmeter [ELEC] A voltmeter which measures potential differences on the order of several kilovolts. { 'kil·ə,vōlt,mēd·ər }

kilovolts peak [ELECTR] The peak voltage applied to an x-ray tube, expressed in kilovolts. Abbreviated kVp. { 'kil·ə,vōlts 'pēk }

kilowatt [PHYS] A unit of power equal to 1000 watts. Abbreviated kW. { 'kil·ə,wät }

kilowatt-hour [ELEC] A unit of energy or work equal to 1000 watt-hours. Abbreviated kWh; kW-hr. Also known as Board of Trade Unit. { 'kil·ə,wät ,aůr }

Kimberley reefs [GEOL] Gold-bearing reefs in southern Africa that lie above the Main reef and Bird reef groups. Also known as battery reefs. { 'kim·bər·lē ,rēfs }

kimberlite [PETR] A form of mica periodite that is formed mainly of phenocrysts, olivine, phlogopite, and subordinate melilite with minor amounts of pyroxene, apatite, perovskite, and opaque oxides. { 'kim·bər,līt }

Kimmelstiel-Wilson disease See intercapillary glomerulosclerosis. { 'kim·əl,stēl 'wil·sən di,zēz }

Kimmeridgian [GEOL] A European stage of geologic time; middle Upper Jurassic, above Oxfordian, below Portlandian. { ,kim·ə'rij·ē·ən }

kimono flannel See flannelette. { kə'mō·nə ,flan·əl }

kimzeyite [MINERAL] $Ca_3(Zr,Ti)_2(Al,Si)_3O_{12}$ A mineral of the garnet group. { 'kim·zē,īt }

kinase [BIOCHEM] Any enzyme that catalyzes phosphorylation reactions. { 'kī,nās }

Kind-Chaudron process [MIN ENG] A technique used to sink a large-diameter deep shaft; a pilot shaft of smaller diameter is first dug, then enlarged until the full diameter is reached; a lining with a moss box at the bottom is forced into place when water is found. { 'kint shō'drōn prä·səs }

Kinderhookian [GEOL] Lower Mississippian geologic time, above the Chautauquan of Devonian, below Osagian. { ,kin·dər'hůk·ē·ən }

K indicator See K scope. { 'kā ,in·də,kād·ər }

kinematically admissible motion [MECH] Any motion of a mechanical system which is geometrically compatible with the constraints. { ,kin·ə'mad·ə·klē id'mis·ə·bəl 'mō·shən }

kinematic boundary condition [FL MECH] The condition that the component of fluid velocity perpendicular to a solid boundary must vanish on the boundary itself; when the boundary is a fluid surface, the condition applies to the vector difference of velocities across the interface. { kin·ə'mad·ik 'baůn·drē kən,dish·ən }

kinematic chain [ANAT] A group of body segments that are connected by joints so that the segments operate together to provide a wide range of motion for a limb. { 'kin·ə,mad·ik 'chān }

kinematic fluidity [FL MECH] The reciprocal of the kinematic viscosity. { kin·ə'mad·ik flü'id·əd·ē }

kinematics [MECH] The study of the motion of a system of material particles without reference to the forces which act on the system. { 'kin·ə'mad·iks }

kinematic similarity [FL MECH] A relationship between fluid-flow systems in which corresponding fluid velocities and velocity gradients are in the same ratios at corresponding locations. { 'kin·ə'mad·ik ,sim·ə'lar·əd·ē }

kinematic viscosity [FL MECH] The absolute viscosity of a fluid divided by its density. Also known as coefficient of kinematic viscosity. { 'kin·ə'mad·ik vi'skäs·əd·ē }

kineplasty [MED] An amputation of a limb in which tendons are arranged in the stump to permit their use in moving parts of the prosthetic appliance. Also spelled cineplasty. { 'kin·ə,plas·tē }

kinescope See picture tube. { 'kin·ə,skōp }

kinescope recording [COMMUN] A motion picture film made by photographing images on the face of the picture tube in a television monitor or receiver, to permit repeating the same television program later and at different stations. Also known as television recording. { 'kin·ə,skōp ri,kȯrd·iŋ }

kinesiatrics [MED] The treatment of disease by systematic active or passive movements. Also known as kinesitherapy; kinetotherapy. { kə,nēz·ē'a·triks }

kinesin [CELL MOL] An enzyme that hydrolyzes adenosine triphosphate to provide energy to power anterograde [from (−) to (+)] movement along microtubules. { kī'nēs·ən }

kinesiology [PHYSIO] The study of human motion through anatomical and mechanical principles. { kə,nēz·ē'äl·ə·jē }

kinesis [PHYSIO] The general term for physical movement, including that induced by stimulation, for example, light. { ki'nē·səs }

kinesitherapy See kinesiatrics. { kə,nēz·ə'ther·ə·pē }

kinesthesis [PHYSIO] The system of sensitivity present in the muscles and their attachments. { ,kin·əs'thē·səs }

kinesthetic apraxia See motor apraxia. { ,kin·əs'thed·ik ā'prak·sē·ə }

kinetic [SCI TECH] Pertaining to or producing motion. { kə'ned·ik }

kinetic art [GRAPHICS] The use of material objects in motion to produce an artistic effect. { kə'ned·ik 'ärt }

kinetic energy [MECH] The energy which a body possesses because of its motion; in classical mechanics, equal to one-half of the body's mass times the square of its speed. { kə'ned·ik 'en·ər·jē }

kinetic energy ammunition [ORD] Ammunition designed to inflict damage to fortifications, armored vehicles, or ships by reason of the kinetic energy of the missile upon impact; the damage may consist of shattering, spalling, or piercing; the missile may be solid, or may contain an explosive charge, intended to function after penetration. { kə'ned·ik 'en·ər·jē ,am·yə'nish·ən }

kinetic equilibrium See dynamic equilibrium. { kə'ned·ik ,ē·kwə'lib·rē·əm }

kinetic friction [MECH] The friction between two surfaces which are sliding over each other. { kə'ned·ik 'frik·shən }

kinetic intermediate [CELL MOL] A structural form that occurs transiently during protein folding. { kə'ned·ik ,in·tər'mēd·ē·ət }

kinetic lead [ORD] The correction or allowance made for the relative motion of a target when computing the lead angle in gunnery. { kə'ned·ik 'lēd }

kinetic momentum [MECH] The momentum which a particle possesses because of its motion; in classical mechanics, equal to the particle's mass times its velocity. { kə'ned·ik mə'men·təm }

kinetic potential See Lagrangian. { kə'ned·ik pə'ten·chəl }

kinetic pressure [FL MECH] The kinetic energy per unit volume of a fluid, equal to one-half the product of its density and the square of its velocity. { kə'ned·ik 'presh·ər }

kinetic reaction [MECH] The negative of the mass of a body multiplied by its acceleration. { kə'ned·ik rē'ak·shən }

kinetics [MECH] The dynamics of material bodies. { kə'ned·iks }

kinetic stress [STAT MECH] A stress which arises, in a theory taking the motions of individual molecules into account, from the existence of a velocity distribution of molecules, an example is the pressure of an ideal gas. { kə'ned·ik 'stres }

kinetic theory [STAT MECH] A theory which attempts to explain the behavior of physical systems on the assumption that they are composed of large numbers of atoms or molecules in vigorous motion; it is further assumed that energy and momentum are conserved in collisions of these particles, and that statistical methods can be applied to deduce the particles' average behavior. Also known as molecular theory. { kə'ned·ik 'thē·ə·rē }

kinetid [CELL MOL] In eukaryotic cells, any locomotory structure, that is, a cilium or flagellum. { ki'ned·əd }

kinetin [BIOCHEM] $C_{10}H_9ON_5$ A cytokinin formed in many plants which has a stimulating effect on cell division. { 'kin·ə·tən }

kinetochore [CELL MOL] Within the centromere, the granule upon which the spindle fibers attach. { kə'ned·ə,kȯr }

kinetoplast [CELL MOL] A genetically autonomous, membrane-bound organelle associated with the basal body at the base of flagella in certain flagellates, such as the trypanosomes. Also known as parabasal body. { kə'ned·ə,plast }

KINETIN

Structural formula for kinetin.

Kinetoplastida [INV ZOO] An order of colorless protozoans in the class Zoomastigophorea having pliable bodies and possessing one or two flagella in some stage of their life. { kə‚ned·ə'plas·tə·də }

kinetosome See basal body. { kə'ned·ə‚sōm }

kinetotherapy See kinesiatrics. { kə‚ned·ə'ther·ə·pē }

king closer [CIV ENG] In masonry work, a rectangular brick having one corner cut diagonally to half the end of the brick and used to fill an opening in a course larger than half a brick. Also known as beveled closer. { ¦kiŋ ¦klōz·ər }

kingdom [SYST] One of the primary divisions that include all living organisms: most authorities recognize two, the animal kingdom and the plant kingdom, while others recognize three or more, such as Protista, Plantae, Animalia, and Mycota. { 'kiŋ·dəm }

Kingdon trap [ELEC] A thin charged wire for confining charged particles; ions are attracted toward the wire, but their angular momentum causes them to spiral around the wire in trajectories that have a low probability of hitting the wire. { 'kin·dən trap }

kingfisher [VERT ZOO] The common name for members of the avian family Alcedinidae; most are tropical Old World species characterized by short legs, long bills, bright plumage, and short wings. { 'kiŋ‚fish·ər }

kingpin [MECH ENG] The pin for articulation between an automobile stub axle and an axle-beam or steering head. Also known as swivel pin. { 'kiŋ‚pin }

king post [BUILD] In a roof truss, the central vertical member against which the rafters abut and which supports the tie beam. [NAV ARCH] A short, strong post for supporting a cargo boom on a cargo ship. Also known as derrick post; samson post. { 'kiŋ ‚pōst }

king post truss [BUILD] A wooden roof truss having two principal rafters held by a horizontal tie beam, a king post upright between tie beam and ridge, and usually two struts to the rafters from a thickening at the king post foot. { 'kiŋ ‚pōst ‚trəs }

king's blue See cobalt blue. { 'kiŋz 'blü }

kingston valve [NAV ARCH] A sea valve so arranged that the pressure of the sea forces the valve on its seat or closes it, thus differing from most valves which are so arranged that the pressure is in the direction of opening of the valve. { 'kiŋ·stən ‚valv }

kinic acid See quinic acid. { 'kin·ik 'as·əd }

kinin [PHARM] Any of several pharmacologically active polypeptides that act as hypotensives, contracting isolated smooth muscles and increasing capillary permeability; an example is bradykinin. { 'kī·nən }

kink [ENG] A tightened loop in a wire rope resulting in permanent deformation and damage to the wire. [MOL BIO] A bend between two helical segments of deoxyribonucleic acid achieved by unstacking one base pair and twisting the polynucleotide backbone. { kiŋk }

kink band [GEOL] A deformation band in a single crystal or in foliated rocks in which the orientation is changed due to slipping on several parallel slip planes. Also known as joint drag; knick band; knick zone. { 'kiŋk ‚band }

kink instability [PL PHYS] A type of hydromagnetic instability in which the ionized gas and its magnetic confining field tend to form a loop or kink, which then grows steadily larger. Also known as sausage instability. { 'kiŋk ‚in·stə'bil·əd·ē }

kinocilium [CELL MOL] A type of cilium containing one central pair of microfibrils and nine peripheral pairs; they extend from the apex of hair cells in all vertebrate ears except mammals. { ¦kin·ə'sil·ē·əm }

kinomere See centromere. { 'kin·ə‚mir }

kinoplasm [CELL MOL] The substance of the protoplasm that is thought to form astral rays and spindle fibers. { 'kin·ə‚plaz·əm }

Kinorhyncha [INV ZOO] A class of the phylum Aschelminthes consisting of superficially segmented microscopic marine animals lacking external ciliation. { ‚kin·ə'riŋ·kē·ə }

Kinosternidae [VERT ZOO] The mud and musk turtles, a family of chelonian reptiles in the suborder Cryptodira found in North, Central, and South America. { ‚kin·ə'stər·nə‚dē }

kintal See hundredweight. { 'kint·əl }

kinzigite [PETR] A coarse-grained metamorphic rock that is formed principally of garnet and biotite, with K feldspar, quartz, mica, cordierite, and sillimanite. { 'kin·zə‚gīt }

kip [MECH] A 1000-pound (453.6-kilogram) load. { kip }

Kirchhoff formula [THERMO] A formula for the dependence of vapor pressure p on temperature T, valid over limited temperature ranges; it may be written $\log p = A - (B/T) - C \log T$, where A, B, and C are constants. { 'kərk‚hōf ‚fȯr·mya·lə }

Kirchhoff's current law [ELEC] The law that at any given instant the sum of the instantaneous values of all the currents flowing toward a point is equal to the sum of instantaneous values of all the currents flowing away from the point. Also known as Kirchhoff's first law. { 'kərk‚hōfs 'kə·rənt ‚lȯ }

Kirchhoff's equations [THERMO] Equations which state that the partial derivative of the change of enthalpy (or of internal energy) during a reaction, with respect to temperature, at constant pressure (or volume) equals the change in heat capacity at constant pressure (or volume). { 'kərk‚hōfs i‚kwā·zhənz }

Kirchhoff's first law See Kirchhoff's current law. { 'kərk‚hōfs 'fərst ‚lȯ }

Kirchhoff's law [ELEC] Either of the two fundamental laws dealing with the relation of currents at a junction and voltages around closed loops in an electric network; they are known as Kirchhoff's current law and Kirchhoff's voltage law. [THERMO] The law that the ratio of the emissivity of a heat radiator to the absorptivity of the same radiator is the same for all bodies, depending on frequency and temperature alone, and is equal to the emissivity of a blackbody. Also known as Kirchhoff's principle. { 'kərk‚hōfs ‚lȯ }

Kirchhoff's principle See Kirchhoff's law. { 'kərk‚hōfs 'prin·sə·pəl }

Kirchhoff's second law See Kirchhoff's voltage law. { 'kərk‚hōfs 'sek·ənd ‚lȯ }

Kirchhoff's voltage law [ELEC] The law that at each instant of time the algebraic sum of the voltage rises around a closed loop in a network is equal to the algebraic sum of the voltage drops, both being taken in the same direction around the loop. Also known as Kirchhoff's second law. { 'kərk‚hōfs 'vōl·tij ‚lȯ }

Kirchhoff theory [OPTICS] A theory of diffraction of light which gives a mathematical formulation of Huygens' principle, based on the wave equation and Green's theorem, and enables quantitative determination of the amplitude and phase at any point to a very close approximation. { 'kərk‚hōf ‚thē·ə·rē }

Kirchhoff vapor pressure formula [THERMO] An approximate formula for the variation of vapor pressure p with temperature T, valid over a limited temperature range; it is $\ln p = A - B/T - C \ln T$, where A, B, and C are constants. { ¦kirch‚hōf 'vā·pər ‚pre·shər ‚fȯr·mya·lə }

Kirkbyacea [PALEON] A monomorphic superfamily of extinct ostracods in the suborder Beyrichicopina, all of which are reticulate. { ‚kərk·bē'ās·ē·ə }

Kirkbyidae [PALEON] A family of extinct ostracods in the superfamily Kirkbyacea in which the pit is reduced and lies below the middle of the valve. { kərk'bē·ə‚dē }

Kirkendall effect [MET] The phenomenon whereby a marker placed at the interface between an alloy and a metal moves toward the alloy region when the temperature of the system is raised to the point where diffusion can occur. { 'kərk·ən‚dȯl i‚fekt }

Kirkman triple system [MATH] A resolvable balanced incomplete block design with block size k equal to 3. { ¦kərk·mən ‚trip·əl 'sis·təm }

Kirkwood-Brinkely's theory [MECH] In terminal ballistics, a theory formulating the scaling laws from which the effect of blast at high altitudes may be inferred, based upon observed results at ground level. { 'kərk‚wùd 'briŋk·lēz ‚thē·ə·rē }

Kirkwood gaps [ASTRON] Regions in the main zone of asteroids where almost no asteroids are found. { 'kərk‚wùd ‚gaps }

kirovite [MINERAL] $(Fe,Mg)SO_4 \cdot 7H_2O$ A mineral composed of a hydrous sulfate of iron and magnesium; it is isomorphous with malanterite and pisanite. { 'kir·ə‚vīt }

Kirsten propeller [NAV ARCH] A vertical-axis propeller whose vertical blades are interlocked by gears so that each one makes half a revolution about its axis for each revolution of the whole propeller. { 'kər·stən prə‚pel·ər }

Kiruna method [MIN ENG] A borehole-inclination survey method whereby the electrolytic deposition of copper from a solution is used to make a mark on the inside of a metal container. { kə'rü·nə ‚meth·əd }

KINGFISHER

The belted kingfisher with white underparts, a blue-gray back, and a band across the chest.

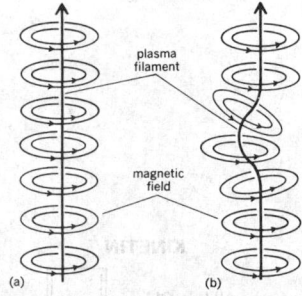

KINK INSTABILITY

Plasma-magnetic-field configuration for cylindrical pinch. *(a)* Equilibrium configuration of plasma filament and magnetic field generated by axial current flow through plasma. *(b)* Onset of kink instability.

kirwanite [MINERAL] A type of anthracite coal with a metallic luster. { 'kər·wə,nīt }

kish [MET] Free graphite that floats to the surface of molten hypereutectic cast iron as it cools. { kish }

kissing disease See infectious mononucleosis. { 'kis·iŋ di,zēz }

kiss-roll coating [ENG] Procedure for coating a substrate web in which the coating roll carries a metered film of coating material; part of the film transfers to the web, part remains on the roll. { 'kis ,rōl ,kōd·iŋ }

Kistiakowsky-Fishtine equation [PHYS CHEM] An equation to calculate latent heats of vaporization of pure compounds; useful when vapor pressure and critical data are not available. { ,kis·tē·ə'kȯf·skē fə'shtīn i,kwā·zhən }

Kitasatoa [MICROBIO] A genus of bacteria in the family Actinoplanaceae; club-shaped sporangia, each containing a single chain of diplococcuslike uniflagellate spores. { kə|ta·sə 'tō·ə }

kite observation [METEOROL] An atmospheric sounding by means of instruments carried aloft by a kite. { 'kīt ,äb·sər'vā·shən }

kitol [ORG CHEM] $C_{40}H_{60}O_2$ One of the provitamins of vitamin A derived from whale liver oil; crystallizes from methanol solution. { 'kē,tȯl }

Kiviat graph [COMPUT SCI] A circular diagram used in computer performance evaluation, in which variables are plotted on axes of the circle with 0% at the center of the circle and 100% at the circumference, and variables which are "good" and "bad" as they approach 100% are plotted on alternate axes. { 'kiv·ē·ət ,graf }

kiwi [VERT ZOO] The common name for three species of nocturnal ratites of New Zealand composing the family Apterygidae; all have small eyes, vestigial wings, a long slender bill, and short powerful legs. { 'kē,wē }

kJ See kilojoule.

Kjeldahl method [ANALY CHEM] Quantitative analysis of organic compounds to determine nitrogen content by interaction with concentrated sulfuric acid; ammonia is distilled from the NH_4SO_4 formed. { 'kel,däl ,meth·əd }

KK damage [ORD] Combat damage of such extent or nature that an aircraft will disintegrate immediately after the damage occurs. { 'kā'kā ,dam·ij }

kl See kiloliter.

klapperstein See rattle stone. { 'kläp·ər,shtīn }

klaprothite [MINERAL] $Cu_6Bi_4S_9$ A gray mineral composed of copper bismuth sulfide. { 'klap·rə,thīt }

klaxon [ENG ACOUS] A diaphragm horn sometimes operated by hand. { 'klak·sən }

klebelsbergite [MINERAL] A mineral composed of basic antimony sulfate, occurring between crystals of stibnite. { 'klā·bəlz,bər,gīt }

Klebsiella [MICROBIO] A genus of bacteria in the family Enterobacteriaceae; nonmotile, encapsulated rods arranged singly, in pairs, or in chains; some species are human pathogens. { ,kleb·zē'el·ə }

Klebsiella pneumoniae [MICROBIO] An encapsulated pathogenic bacterium that causes severe pneumonitis in humans. Formerly known as Friedlander's bacillus; pneumobacillus. { ,kleb·zē'el·ə nə'mō·nē,ī }

Klebsiella rhinoscleromatis [MICROBIO] A gram-negative, nonmotile, pathogenic species of bacteria that causes the upper respiratory disease rhinoscleroma. { ,kleb·zē,el·ə ,rī·nō ,skler·ə'mäd·əs }

Klebs-Loeffler bacillus See Corynebacterium diphtheriae. { 'kläps 'lef·lər bə,sil·əs }

Klein bottle [MATH] The nonorientable surface having only one side with no inside or outside; it resembles a bottle pulled into itself. { 'klīn ,bäd·əl }

Klein-Gordon equation [QUANT MECH] A wave equation describing a spinless particle which is consistent with the special theory of relativity. Also known as Schrödinger-Klein-Gordon equation. { 'klīn 'gȯrd·ən i,kwä·zhən }

Kleinian group [MATH] A group of conformal mappings of a Riemann surface onto itself which is discontinuous at one or more points and is not discontinuous at more than two points. { 'klī·nē·ən ,grüp }

kleinite [MINERAL] A yellow to orange mineral composed of a basic oxide, sulfate, and chloride of mercury and ammonium. { 'klī,nīt }

Kleinmann-Low Nebula [ASTRON] A cool, extended source of infrared radiation in the Orion Nebula, probably a collapsing cloud of gas containing embedded protostars. Abbreviated KL Nebula. { 'klīn,män 'lō 'neb·yə·lə }

Klein-Nishina formula [QUANT MECH] A formula, based on the Dirac electron theory without radiative correction, for the differential cross section for scattering of a photon by an unbound electron. { 'klīn ni'shē·nə ,fȯr·mya·lə }

Klein paradox [QUANT MECH] The paradox whereby, according to the Dirac electron theory, an electron can penetrate into a potential barrier which is greater than twice the rest energy of the electron (about 1 MeV) by making a transition from a positive energy state to a negative energy state, provided the potential change occurs over a distance on the order of a Compton wavelength or less. { 'klīn ,par·ə,däks }

Klein-Rydberg method [PHYS CHEM] A method for determining the potential energy function of the distance between the nuclei of a diatomic molecule from the molecule's vibrational and rotational levels. { 'klīn 'rid,berg ,meth·əd }

Kleinschmidt spread [CELL MOL] A visualization technique for electron microscopy in which molecules are mounted in a positively charged protein monolayer, which is spread on the surface of water, and then are transferred to a hydrophobic grid. { 'klīn,shmit ,spred }

Klein's four-group [MATH] The noncyclic group of order four. { 'klīnz fȯr ,grüp }

Klein's hypothesis [ASTRON] A theory of the overall structure of the universe that regards the visible universe as part of a large but finite astronomical system called a metagalaxy, which may itself belong to a much larger bounded system. { 'klīnz hī,päth·ə·səs }

Klein's reagent [CHEM] Saturated solution of borotungstate; used to separate minerals by specific gravity. { 'klīnz rē,ā·jənt }

klendusity [BOT] The tendency of a plant to resist disease due to a protective covering, such as a thick cuticle, that prevents inoculation. { klen'dü·səd·ē }

kleptomania [PSYCH] An obsessive desire to steal; stolen objects are usually petty and useless, being of symbolic value only. { ,klep·tə'mā·nē·ə }

kleptophobia [PSYCH] **1.** An abnormal fear of thieves or of being robbed. **2.** An abnormal fear of becoming a kleptomaniac. { ,klep·tə'fō·bē·ə }

Klinefelter's syndrome [MED] A complex of symptoms associated with hypogonadism in males as an accompaniment of an anomaly of the sex chromosomes; somatic cells are found to have a Y chromosome and more than one X chromosome. { 'klīn,fel·tərz ,sin,drōm }

Kline flocculation test [PATH] A microscopic precipitin test for identification of the antibody resulting from syphilitic infection. { 'klīn ,fläk·yə'lā·shən ,test }

Klinkenberg correction [PETRO ENG] Mathematical conversion of laboratory air-permeability measurements (made on formation material) into equivalent liquid-permeability values. { 'kliŋk·ən,bərg kə,rek·shən }

klinorhynchy [BIOL] The property of a downwardly bent face. { 'klīn·ə,riŋ·kē }

klinotaxis [BIOL] Positive orientation movement of a motile organism induced by a stimulus. { klī·nə'tak·səs }

klint [GEOL] An exhumed coral reef or bioherm that is more resistant to the processes of erosion than the rocks that enclose it so that the core remains in relief as hills and ridges. { klint }

klintite [GEOL] The dense, hard dolomite composing a klint; gives to the core a strength and resistance to erosion. { 'klin,tīt }

klippe [GEOL] A block of rock that is separated from underlying rocks by a fault that usually has a gentle dip. { klip }

KL Nebula See Kleinmann-Low Nebula. { 'kā 'el 'neb·yə·lə }

klockmannite [MINERAL] CuSe A slate gray mineral consisting of copper selenide; occurs in granular aggregates. { 'kläk·mə,nīt }

Kloedenellacea [PALEON] A dimorphic superfamily of extinct ostracods in the suborder Kloedenellocopina having the posterior part of one dimorph longer and more inflated than the other dimorph. { ,klēd·ən·ə'lās·ē·ə }

Kloedenellocopina [PALEON] A suborder of extinct ostracods in the order Paleocopa characterized by a relatively straight dorsal border with a gently curved or nearly straight ventral border. { ,klēd·ən,el·ə'käp·ə·nə }

kloof wind [METEOROL] A cold southwest wind of Simons Bay, South Africa. { 'klüf ,wind }

K/L ratio [NUC PHYS] The ratio of the number of internal conversion electrons emitted from the *K* shell of an atom during de-excitation of a nucleus to the number of such electrons emitted from the *L* shell. { 'kā'el ,rā·shō }

kludge [COMPUT SCI] A poorly designed data-processing system composed of ill-fitting mismatched components. { klüj }

klydonograph [ENG] A device attached to electric power lines for estimating certain electrical characteristics of lightning by means of the figures produced on photographic film by the lightning-produced surge carried over the lines; the size of the figure is a function of the potential and polarity of the lightning discharge. { klī'dän·ə,graf }

klystron [ELECTR] An evacuated electron-beam tube in which an initial velocity modulation imparted to electrons in the beam results subsequently in density modulation of the beam; used as an amplifier in the microwave region or as an oscillator. { 'klī,strän }

klystron generator [ELECTR] Klystron tube used as a generator, with its second cavity or catcher directly feeding waves into a waveguide. { 'klī,strän ¦jen·ə,rād·ər }

klystron oscillator See velocity-modulated oscillator. { 'klī,strän ¦äs·ə,lād·ər }

klystron repeater [ELECTR] Klystron tube operated as an amplifier and inserted directly in a waveguide in such a way that incoming waves velocity-modulate the electron stream emitted from a heated cathode; a second cavity converts the energy of the electron clusters into waves of the original type but of greatly increased amplitude and feeds them into the outgoing guide. { 'klī,strän ri¦pēd·ər }

km See kilometer.

K meson [PART PHYS] **1.** Collective name for four pseudoscalar mesons, having masses of about 495 MeV (megaelectronvolts) and decaying via weak interactions: K^+, K^-, K^0_S and K^0_L; they consist of two isotopic spin doublets, the (K^+, K^0) doublet and its antiparticle doublet $(K^-, \overline{K}^0)$, having hypercharge or strangeness of $+1$ and -1 respectively, where K^0 and $\overline{K}^0$ are certain combinations of K^0_L and K^0_S states. Also known as kaon. **2.** Collective name for any meson resonance belonging to an isotopic doublet with hypercharge $+1$ or -1, denoted $K_{JP}(m)$ or $\overline{K}_{JP}(m)$ respectively, where m is the mass in megaelectronvolts, and J and P are the spin and parity. { 'kā ¦mā,sän }

K Monel [MET] A nonmagnetic, age-hardenable alloy of nickel (28–34%) and copper and 2.75% aluminum that can be heat-treated after finishing to produce a material that is both corrosion-resistant and extra strong. { 'kā mō,nel }

knap [MIN ENG] **1.** To break rock. **2.** To improve the grade of ore by removing low-grade material manually by using a hammer. { nap }

knapping hammer [ENG] A steel hammer used for breaking and shaping stone. { 'nap·iŋ ,ham·ər }

knapsack problem [MATH] The problem, given a set of integers $\{A_1, A_2, \ldots, A_n\}$ and a target integer B, of determining whether a subset of the A_i can be selected without repetition so that their sum is the target B. { 'nap,sak ,präb·ləm }

kneaded eraser [MATER] An eraser made of unvulcanized rubber whose shape can be readily changed by the user for removing pencil marks from paper. { 'nēd·əd i'rā·sər }

kneader [FOOD ENG] Mixer for doughy materials in which the dough is repeatedly pulled out, folded back on itself, and pushed down to join the separate layers into a homogeneous mixture. { 'nēd·ər }

knebelite [MINERAL] $(Fe,Mn)_2SiO_4$ A mineral composed of an iron manganese silicate. { 'nā·bəlīt }

knee [ANAT] **1.** The articulation between the femur and the tibia in humans. Also known as genu. **2.** The corresponding articulation in the hindlimb of a quadrupedal vertebrate. [MECH ENG] In a knee-and-column type of milling machine, the part which supports the saddle and table and which can move vertically on the column. [MET] The lower supporting structure for an arm in a resistance welding machine. { nē }

knee brace [BUILD] A stiffener between a column and a supported truss or beam to provide greater rigidity in a building frame under transverse loads. { 'nē ,brās }

kneecap See patella. { 'nē,kap }

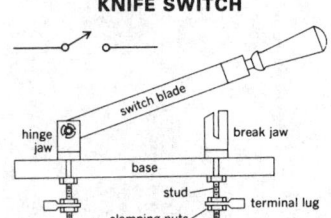

KNIFE SWITCH

Knife switch with schematic symbol of single switch shown at top.

kneecap height [ANTHRO] A measure of the vertical distance taken from the floor at the base of the heel to the top of the muscle mass near the end of the thigh bone as the subject sits with both feet on the floor. { 'nē,kap ,hīt }

knee frequency See break frequency. { 'nē ,frē·kwən·sē }

kneeler [CIV ENG] In masonry, a stone cut to provide a break in the horizontal-vertical pattern to begin the curve or angle of an arch or vault. { 'nēl·ər }

knee pad [ENG] A protective cushion, usually made of sponge rubber, that can be strapped to a worker's knee. { 'nē ,pad }

knee rafter [BUILD] A brace placed diagonally between a principal rafter and a tie beam. { 'nē ,raf·tər }

knee switch [ENG] A control mechanism operated with knee movements by a seated worker. { 'nē ,swich }

knee tool [MECH ENG] A tool holder with a shape resembling a knee, such as the holder for simultaneous cutting and interval operations on a screw machine or turret lathe. { 'nē ,tül }

knee wall [BUILD] A partition that forms a side wall or supports roof rafters under a pitched roof. { 'nē ,wòl }

Kneriidae [VERT ZOO] A small family of tropical African fresh-water fishes in the order Gonorynchiformes. { nə'rī·ə,dē }

knick See knickpoint. { nik }

knick band See kink band. { 'nik ,band }

knickpoint [GEOL] A point of sharp change of slope, especially in the longitudinal profile of a stream or of its valley. Also known as break; knick; nick; nickpoint; rejuvenation head; rock step. { 'nik ,pòint }

knick zone See kink band. { 'nik ,zōn }

knife [DES ENG] A sharp-edged blade for cutting. { nīf }

knife coating [ENG] Procedure for coating a continuous-web substrate in which coating thickness is controlled by the distance between the substrate and a movable knife or bar. { 'nīf ,kōd·iŋ }

knife-edge [DES ENG] A sharp narrow edge resembling that of a knife, such as the fulcrum for a lever arm in a measuring instrument. { 'nīf ,ej }

knife-edge bearing [MECH ENG] A balance beam or lever arm fulcrum in the form of a hardened steel wedge; used to minimize friction. { 'nīf ,ej ,ber·iŋ }

knife-edge cam follower [DES ENG] A cam follower having a sharp narrow edge or point like that of a knife; useful in developing cam profile relationships. { 'nīf ,ej 'kam ,fäl·ə·wòr }

knife-edge refraction [ELECTROMAG] Radio propagation effect in which the atmospheric attenuation of a signal is reduced when the signal passes over and is diffracted by a sharp obstacle such as a mountain ridge. { 'nīf ,ej ri'frak·shən }

knife file [DES ENG] A tapered file with a thin triangular cross section resembling that of a knife. { 'nīf ,fīl }

knife harrow [AGR] A type of harrow that consists of a frame holding a number of knives which scrape and partly invert the soil surface to smooth it and destroy small weeds. { 'nīf,har·ō }

knife-line attach [MET] Intergranular corrosion of an alloy adjacent to a weldment after heating the joint above the sensitization temperature. { 'nīf ¦līn ə,tach }

knife switch [ELEC] An electric switch consisting of a metal blade hinged at one end to a stationary jaw, so that the blade can be pushed over to make contact between spring clips. { 'nīf ,swich }

Knight shift [PHYS] A shift of the nuclear magnetic resonance frequency in a metal to higher values than that of the same nucleus in a diamagnetic compound. { 'nīt ,shift }

knik wind [METEOROL] Local name for a strong southeast wind in the vicinity of Palmer in the Matanuska Valley of Alaska; it blows most frequently in the winter, although it may occur at any time of year. { kə'nik ,wind }

Knill-Laflamme bound [COMPUT SCI] In the theory of quantum computation, a necessary condition for an algorithm that encodes N logical qubits into N' carrier qubits (with N' larger than N) to correct any error on any M carrier qubits; namely, that N' be equal to or larger than $4M + N$. { kə¦nil lə¦fläm ¦baùnd }

knitting [TEXT] Making a fabric by interlocking loops of yarn by means of needles or wires. { 'nid·iŋ }

knob [DES ENG] A component that is placed on a control shaft to facilitate manual rotation of the shaft; sometimes has a pointer or markings to indicate shaft position. [GEOL] **1.**

A rounded eminence, such as a knoll, hillock, or small hill or mountain, and especially a prominent or isolated hill with steep sides. **2.** A peak or other projection at the top of a hill or mountain. { näb }

knob-and-tube wiring [ELEC] An electric wiring method used for light and power circuits that uses open insulated wiring on solid insulators; now obsolete and illegal in most countries. { ¦näb ən 'tüb ¸wir·iŋ }

knocker See shell knocker. { 'näk·ər }

knock intensity [ENG] The intensity of knock (detonation) recorded when testing a motor gasoline for octane or knock rating. { 'näk in¸ten·səd·ē }

knockmeter [ENG] A fuels-testing device used to measure the output of the detonation meter used in American Society for Testing and Materials knock-test ratings of motor fuels. { 'näk ¸mēd·ər }

knock-off [MECH ENG] **1.** The automatic stopping of a machine when it is operating improperly. **2.** The device that causes automatic stopping. { 'näk¸óf }

knock-off bit See detachable bit. { 'näk¸óf ¸bit }

knock-on atom [SOLID STATE] An atom which is knocked out of its equilibrium position in a crystal lattice by an energetic bombarding particle, and is displaced many atomic distances away into an interstitial position, leaving behind a vacant lattice site. { 'näk¸ón ¸ad·əm }

knockout [ENG] A partially cutout piece in metal or plastic that can be forced out when a hole is needed. [GEN] Inactivation of a specific gene in a laboratory organism in order to study gene function. { 'näk¸aút }

knockout pin See ejector pin. { 'näk¸aút ¸pin }

knockout vessel [CHEM ENG] A vessel, drum, or trap used to remove fluid droplets from flowing gases. { 'näk¸aút ¸ves·əl }

knock rating [ENG] Rating of gasolines according to knocking tendency. { 'näk¸aút ¸rād·iŋ }

knock suppressor [MATER] A material added to motor fuel to retard or prevent detonation and resultant knock in reciprocating internal combustion engines; an example is tetraethyllead. { 'näk¸aút sə¸pres·ər }

Knoevenagel reaction [ORG CHEM] The condensation of aldehydes with compounds containing an activated methylene (=CH₂) group. { kə'nē·və¸näg·əl rē¸ak·shən }

knoll [GEOL] A mound rising less than 3300 feet (1000 meters) from the sea floor. Also known as sea knoll. { ¸nōl }

Knoop hardness [MET] The relative microhardness of a material, such as metal, determined by the Knoop indentation test. { 'nüp ¸härd·nəs }

Knoop indentation test [MET] A diamond pyramid hardness test employing the Knoop indenter; hardness is determined by the depth to which the Knoop indenter penetrates. { 'nüp ¸in¸den'tā·shən ¸test }

Knoop indenter [MET] A diamond indenter which has a rhombic base with diagonals in a 1:7 ratio and included apical angles of 130° and 172°30′; used in the Knoop indentation test. { 'nüp in'den·tər }

knop [TEXT] A yarn with pronounced knots of different color or of different material appearing at intervals. Also known as knotted yarn; nub yarn. { näp }

knopite [MINERAL] A cerium-bearing variety of perovskite. { 'nä¸pīt }

Knorr synthesis [ORG CHEM] A condensation reaction carried out in either glacial acetic acid or an aqueous alkali in which an α-aminoketone combines with an α-carbonyl compound to form a pyrrole; possibly the most versatile pyrrole synthesis. { 'nór ¸sin·thə·səs }

knot [COMPUT SCI] See deadlock. [MATER] A scar on lumber marking a place where a branch grew out of the tree trunk. [MATH] In the general case, a knot consists of an embedding of an n-dimensional sphere in an $(n + 2)$-dimensional sphere; classically, it is an interlaced closed curve, homeomorphic to a circle. [ORG CHEM] A chiral structure in which rings containing 50 or more members have a knotlike configuration. [PHYS] A speed unit of 1 nautical mile (1.852 kilometers) per hour, equal to approximately 0.51444 meters per second. { nät }

knotted yarn See knop. { 'näd·əd 'yärn }

knot theory [MATH] The topological and algebraic study of knots emphasizing their classification and how one may be continuously deformed into another. { 'nät ¸thē·ə·rē }

knowbot [COMPUT SCI] A program which, when given a request, searches and retrieves information on the Internet. Also known as intelligent agent, knowledge robot. { 'nō¸bät }

knowledge base [COMPUT SCI] A collection of facts, assumptions, beliefs, and heuristics that are used in combination with a database to achieve desired results, such as a diagnosis, an interpretation, or a solution to a problem. { 'näl·ij ¸bās }

knowledge-based system [COMPUT SCI] A computer system whose usefulness derives primarily from a data base containing human knowledge in a computerized format. { 'näl·ij ¸bāst ¸sis·təm }

knowledge discovery in databases [COMPUT SCI] The process of identifying valid, novel, potentially useful, and ultimately understandable structure in data. Abbreviated KDD. { ¦näl·ij di¦skəv·ə·rē in 'dad·ə¸bās·əs }

knowledge engineer [COMPUT SCI] An individual who constructs the knowledge base of an expert system. { 'näl·ij ¸en·jə¸nir }

knowledge robot See knowbot. { 'näl·ij ¸rō¸bät }

known-good die [ELECTR] An unpackaged, fully tested integrated circuit chip. { ¸nōn ¦gùd 'dī }

Knox and Oxborne furnace [MIN ENG] A continuously working shaft furnace for roasting quicksilver ores, having the fireplace built in the masonry at one side; the fuel is wood. { 'näks ən 'äks·bərn ¸fər·nəs }

knoxvillite See copiapite. { 'näks·vi¸līt }

knuckle [MIN ENG] The place on an incline where there is a sudden change in grade. { 'nək·əl }

knuckle joint [DES ENG] A hinge joint between two rods in which an eye on one piece fits between two flat projections with eyes on the other piece and is retained by a round pin. { 'nək·əl ¸jóint }

knuckle joint press [MECH ENG] A short-stroke press in which the slide is actuated by a crank attached to a knuckle joint hinge. { 'nək·əl ¸jóint ¸pres }

knuckle man [MIN ENG] A worker who connects mine cars to and disconnects them from cables and also couples cars into trains. { 'nək·əl ¸man }

knuckle pin [DES ENG] The pin of a knuckle joint. { 'nək·əl ¸pin }

knuckle post [MECH ENG] A post which acts as the pivot for the steering knuckle in an automobile. { 'nək·əl ¸pōst }

Knudsen cell [PHYS CHEM] A vessel used to measure very low vapor pressures by measuring the mass of vapor which escapes when the vessel contains a liquid in equilibrium with its vapor. { kə'nüd·sən ¸cel }

Knudsen cosine law [PHYS] A law which states that the probability of a gas molecule leaving a solid surface in a given direction within a solid angle $d\omega$ is proportional to cos θ $d\omega$, where θ is the angle between the direction and the normal to the surface. { kə'nüd·sən 'kō¸sīn ¸ló }

Knudsen flow See free molecule flow. { kə'nüd·sən ¸flō }

Knudsen gage [ENG] An instrument for measuring very low pressures, which measures the force of a gas on a cold plate beside which there is an electrically heated plate. { kə'nüd·sən ¸gāj }

Knudsen-Langmuir equation [CHEM ENG] Relationship of molecular distillation rate to vapor or saturation pressure, solution temperature, and molecular weight during evaporation and no-recycle condensation. { kə'nüd·sən 'laŋ¸myùr i¸kwā·zhən }

Knudsen number [FL MECH] The ratio of the mean free path length of the molecules of a fluid to a characteristic length; used to describe the flow of low-density gases. { kə'nüd·sən ¸nəm·bər }

Knudsen reversing water bottle [ENG] A type of frameless reversing bottle for collecting water samples; carries reversing thermometers. { kə 'nüd·sən ri¦vərs·iŋ 'wód·ər ¸bäd·əl }

Knudsen's equation [PHYS] An equation for the amount of gas that flows through a tube in free molecule flow, $q\sqrt{2\pi\Delta_p d^3/6l}\sqrt{\rho}$, where q is the volume of gas measured at unit pressure that flows through the tube per second, Δp is the difference between the pressures at the ends of the tube, d is the inside diameter of the tube, l is the length of the tube, and ρ is the density of the gas at unit pressure. { kə'nüd·sənz i¸kwā·shən }

Knudsen's tables [OCEANOGR] Hydrographical tables published by Martin Knudsen in 1901 to facilitate the computation

KNORR SYNTHESIS

An α-aminoketone combines with an α-methylene carbonyl compound to form a pyrrole.

of results of seawater chlorinity titrations and hydrometer temperature readings, and their conversion to salinity and density. { kə'nüd·sən ˌtā·bəlz }

Knudsen vacuum gage [ENG] Device to measure negative gas pressures; a rotatable vane is moved by the pressure of heated molecules, proportionately to the concentration of molecules in the system. { kə'nüd·sən 'vak·yəm ˌgāj }

knurl [ENG] To provide a surface, usually a metal, with small ridges or knobs to ensure a firm grip or as a decorative feature. { nərl }

koala [VERT ZOO] *Phascolarctos cinereus.* An arboreal marsupial mammal of the family Phalangeridae having large hairy ears, gray fir, and two clawed toes opposing three others on each limb. { kō'äl·ə }

Kobayashi potential [MATH] A solution of Laplace's equation in three dimensions constructed by superposition of the solutions obtained by separation of variables in cylindrical coordinates. { ˌkō·bī'yä·shē pə,ten·chəl }

kobellite [MINERAL] $Pb_2(Bi,Sb)_2S_5$ A blackish-gray mineral composed of antimony bismuth lead sulfide. { 'kō·bəlīt }

Kochab [ASTRON] The brighter of the two stars called the Guardian of the Pole in the constellation Ursa Minor. { 'kä·ˌkäb }

Koch curve [MATH] A fractal which can be constructed by a recursive procedure; at each step of this procedure every straight segment of the curve is divided into three equal parts and the central piece is then replaced by two similar pieces. { 'kōk ˌkərv }

Koch freezing process [MIN ENG] A process used to sink a shaft through a formation such as clay that will not sustain a shaft; magnesium chloride cooled to about $-30°C$ is circulated through pipes sunk in the ground until the ground is frozen. { 'kōk 'frēz·iŋ ˌprä·səs }

Koch's postulates [MICROBIO] A set of laws elucidated by Robert Koch: the microorganism identified as the etiologic agent must be present in every case of the disease; the etiologic agent must be isolated and cultivated in pure culture; the organism must produce the disease when inoculated in pure culture into susceptible animals; a microorganism must be observed in and recovered from the experimentally diseased animal. Also known as law of specificity of bacteria. { 'kōks 'päs·chə·ləts }

Koebe function [MATH] The analytic function $k(z) = z(1 - z)^{-2} = z + 2z^2 + 3z^3 + \cdots$, that maps the unit disk onto the entire complex plane minus the part of the negative real axis to the left of $-^1/_4$. { 'kā·bē ˌfəŋk·shən }

koechlinite [MINERAL] Bi_2MoO_6 A greenish-yellow orthorhombic mineral composed of a bismuth molybdate. { 'kek·lə,nīt }

Koehler lamp [MIN ENG] A naphtha-burning flame safety lamp for use in gaseous mines. { 'kā·lər ˌlamp }

koembang [METEOROL] A dry foehnlike wind from the southeast or south in Cheribon and Tegal in Java, caused by the east monsoon which develops a jet effect in passing through the gaps in the mountain ranges and descends on the leeward side. { 'küm,baŋ }

koenenite [MINERAL] $Mg_5Al_2(OH)_{12}Cl_4$ A very soft mineral composed of a basic magnesium aluminum chloride. { 'kō·nə,nit }

Koepe hoist *See* Koepe winder. { 'kep·ə ˌhȯist }

Koepe shear [MIN ENG] A wheel used in place of a winding drum in the Koepe winder; made up of a cast steel hub with steel arms and a welded rim. { 'kep·ə ˌsher }

Koepe winder [MIN ENG] A hoisting system in which the winding drum is replaced by large wheels or sheaves over which passes an endless rope. Also known as Koepe hoist. { 'kep·ə ˌwīn·dər }

Koepe winder brake [MIN ENG] A device that works directly on the Koepe shear to slow or stop the hoist; can be applied by the engineman's brake lever or by safety devices. { 'kep·ə ˌwīn·dər ˌbrāk }

koettigite [MINERAL] $Zn_3(AsO_4)_2 \cdot 8H_2O$ A carmine mineral composed of a hydrated zinc arsenate. { 'ked·i,gīt }

Kohler illumination [OPTICS] A method of illumination for the optical microscope used with coiled filaments or other sources of irregular form or brightness; an image of the filament large enough to fill the iris opening is focused on the condenser which is focused so that the image of the iris diaphragm on the lamp is in focus with the specimen, and the lamp iris is opened only enough to fill the field of view; the iris of the

KOALA

The koala, a slow-moving aboreal animal.

microscope is opened only enough to illuminate the back aperture of the objective; no ground glass is used. { 'kō·lər iˌlü·mə'nā·shən }

Kohler's disease *See* osteochondrosis. { 'kō·lərz di,zēz }

kohlrabi [BOT] A biennial crucifer, designated *Brassica caulorapa* and *B. oleracea* var. *caulo-rapa,* of the order Capparales grown for its edible turniplike, enlarged stem. { kōl'rä·bē }

Kohlrausch law [PHYS CHEM] **1.** The law that every ion contributes a definite amount to the equivalent conductance of an electrolyte in the limit of infinite dilution, regardless of the presence of other ions. **2.** The law that the equivalent conductance of a very dilute solution of a strong electrolyte is a linear function of the concentration. { 'kōl,raůsh ˌlō }

Kohlrausch method [PHYS CHEM] A method of measuring the electrolytic conductance of a solution using a Wheatstone bridge. { 'kōl,raůsh ˌmeth·əd }

kohm *See* kilohm. { ˌkä,ōm }

Kohn effect [SOLID STATE] A sharp change in the phonon dispersion curve of a material when the wave-number vector of the phonons corresponds to the diameter of the Fermi sphere, because of the production of standing waves. { 'kōn i,fekt }

Kohoutek's comet [ASTRON] A comet that was discovered on March 7, 1973, at a distance of 4 astronomical units from the sun, and reached a perihelion of less than 0.1 astronomical unit at the end of 1973. Also known as Comet Kohoutek. { kə'hō·teks ˌkäm·ət }

koilonychia [MED] Spoon-shaped deformity of the fingernails, which may be familial or associated with a disease, such as iron-deficiency anemia. Also known as spoon nail. { ˌkȯil·ō'nik·ē·ə }

koilorachic [MED] Having the lumbar spinal region concave ventrally. { ˌkȯil·ə,rak·ik }

Kojic acid [ORG CHEM] $C_6H_6O_4$ A crystalline antibiotic with a melting point of 152–154°C; soluble in water, acetone, and alcohol; used in insecticides and as an antifungal and antimicrobial agent. { 'kō·jik ,as·əd }

koktaite [MINERAL] $(NH_4)_2Ca(SO_4)_2 \cdot H_2O$ A mineral composed of a hydrous calcium ammonium sulfate. { 'käk·tə,īt }

kolbeckite [MINERAL] A blue to gray mineral composed of a hydrous beryllium aluminum calcium silicate and phosphate. Also known as sterrettite. { 'kōl,be,kīt }

Kolbe hydrocarbon synthesis [ORG CHEM] The production of an alkane by the electrolysis of a water-soluble salt of a carboxylic acid. { 'kōl·bə ,hī·drə'kär·bən ,sin·thə·səs }

Kolbe-Schmitt synthesis [ORG CHEM] The reaction of carbon dioxide with sodium phenoxide at 125°C to give salicyclic acid. { 'kōl·bə 'shmit ,sin·thə·səs }

Kollsman window [AERO ENG] A small window on the dial face of an aircraft pressure altimeter in which the altimeter setting in inches of mercury is indicated. { 'kōls·mən ,win·dō }

Kolmer test [PATH] A complement-fixation test for syphilis and other diseases. { 'kōl·mər ,test }

Kolmogorov-Arnold-Moser theorem [PHYS] A theorem that oscillatory motions in conservative dynamical systems persist when small perturbations are added to the system. Abbreviated KAM theorem. { ˌkȯl·mə'gȯ·rȯf 'ar·nəld 'mō·zər ,thir·əm }

Kolmogorov consistency conditions [MATH] For each finite subset F of the real numbers or integers, let P_F denote a probability measure defined on the Borel subsets of the cartesian product of $k(F)$ copies of the real line indexed by elements in F, where $k(F)$ denotes the number of elements in F; the family $\{P_F\}$ of measures satisfy the Kolmogorov consistency conditions if given any two finite sets F_1 and F_2 with F_1 contained in F_2, the restriction of P_{F_2} to those sets which are independent of the coordinates in F_F which are not in F_1 coincides with P_{F_1}. { ,kȯl·mə'gȯ·rȯf kən'sis·tən·sē kən,dish·ənz }

Kolmogorov inequalities [MATH] For each integer K let X_k be a random variable with finite variance σ_k and suppose $\{X_k\}$ is an independent sequence which is uniformly bounded by some constant c; then for every $\varepsilon > 0$, and integer n,

$$1 - (\varepsilon + 2c)^2 \Big/ \sum_{k=1}^{n} \sigma_k^2 \leq \text{Prob} \left\{ \max_{k \leq n} |S_k + ES_k| \geq \varepsilon \right\}$$

and

$$\frac{1}{\varepsilon^2} \sum_{k=1}^{n} \sigma_k^2 \geq \text{Prob} \{\max_{k \leq n} |S_k + ES_k| \geq \varepsilon\};$$

here

$$S_k = \sum_{i=1}^{k} X_i$$

and ES_k denotes the expected value of S_k. { ˌkȯlˈmə'gȯ·rȯf inˈi'kwäl·əd·ēz }

Kolmogorov inertial subrange [FL MECH] The middle portion of the turbulence spectrum, between the low-wave-number (long-wave) part and the high-wave-number (short-wave) part. { ˌkȯlˈmə'gȯ·rȯf i'nər·shəl 'səb,ranj }

Kolmogorov-Sinai invariant [MATH] An isomorphism invariant of measure-preserving transformations; if T is a measure-preserving transformation on a probability space, the Kolmogorov-Sinai invariant is the least upper bound of the set of entropies of T given each finite partition of the probability space. Also known as entropy of a transformation. { ˌkȯlˈmə'gȯ·rȯf 'sī,nī inˌver·ē·ənt }

Kolmogorov-Smirnov test [STAT] A procedure used to measure goodness of fit of sample data to a specified population; critical values exist to test goodness of fit. { ˌkȯlˈməˈgȯr·əf 'smir,nȯf ,test }

Kolmogorov space See T0 space. { ˌkȯlˈmə'gȯr·əf ,spās }

Kolosov-Muskhelishvili formulas [MECH] Formulas which express plane strain and plane stress in terms of two holomorphic functions of the complex variable $z = x + iy$, where x and y are plane coordinates. { ˌkȯlˈə,sȯf ˌmùsh'kel·ish,vil·ē ˌfȯr·myə·ləz }

komatiite [PETR] A mantle-derived igneous rock with a high content of magnesium, particularly magnesium oxide. { kō'mäd·ē,īt }

Komodo dragon [VERT ZOO] *Varanus komodoensis.* A predatory reptile of the family Varanidae found only on the island of Komodo; it is the largest living lizard and may grow to 10 feet (3 meters). { kə'mō·dō 'drag·ən }

kona [METEOROL] A stormy, rain-bringing wind from the southwest or south-southwest in Hawaii; it blows about five times a year on the southwest slopes, which are in the lee of the prevailing northeast trade winds. { 'kō·nə }

kona cyclone [METEOROL] A slow-moving extensive cyclone which forms in subtropical latitudes during the winter season. Also known as kona storm. { 'kō·nə 'sī,klōn }

kona storm See kona cyclone. { 'kō·nə 'stȯrm }

Kondo alloy [MET] A dilute alloy of a magnetic material in a nonmagnetic host which exhibits the Kondo effect. { 'kän·dō ,al,ȯi }

Kondo effect [MET] The large anomalous increase in the resistance of certain dilute alloys of magnetic materials in nonmagnetic hosts as the temperature is lowered. { 'kän·dō i,fekt }

Kondo resonance See Abrikosov-Suhl resonance. { 'kän·dō ,rez·ən·əns }

Kondo temperature [MET] The temperature below which the Kondo effect predominates for a specified magnetic impurity and host material. { 'kän·dō ,tem·prə·chər }

kongsbergite [MINERAL] A silver-rich variety of a native amalgam composed of silver (95) and mercury (5). { 'kaŋz,bər,gīt }

konig [OPTICS] The X tristimulus value. { 'kō·nig }

König-Egerváry theorem [MATH] The theorem that, for a matrix in which each entry is either 0 or 1, the largest number of 1's that can be chosen so that no two selected 1's lie in the same row or column equals the smallest number of rows and columns that must be deleted to eliminate all the 1's. { 'kərn·ik 'e·ger,vär·yi ,thir·əm }

Königsberg bridge problem [MATH] The problem of walking across seven bridges connecting four landmasses in a specified manner exactly once and returning to the starting point; this is the original problem which gave rise to graph theory. { 'kərn·iks,bərg 'brij ,präb·ləm }

König's theorem [MATH] The theorem that the largest possible number of edges in a matching of a bipartite graph equals the smallest possible number of edges in an edge cover of that graph. { 'kər·nigz ,thir·əm }

konimeter [ENG] An air-sampling device used to measure dust as in a cement mill or a mine; a measured volume of air drawn through a jet impacts on a glycerin-jelly-coated glass surface; the particles are counted with a microscope. { kō'nim·əd·ər }

koninckite [MINERAL] $FePO_4 \cdot 3H_2O$ A yellow mineral composed of a hydrous ferric phosphate. { 'kō·niŋ,kīt }

koniscope [ENG] An instrument which indicates the presence of dust particles in the atmosphere. Also spelled coniscope. { 'kän·ə,skōp }

Konowaloff rule [PHYS CHEM] An empirical rule which states that in the vapor over a liquid mixture there is a higher proportion of that component which, when added to the liquid, raises its vapor pressure, than of other components. { ˌkȯ·nə'vä·lȯf ,rül }

Koonungidae [INV ZOO] A family of Australian crustaceans in the order Anaspidacea with sessile eyes and the first thoracic limb modified for digging. { kü'nən·jə,dē }

kopfring [ORD] A metal ring which is attached to the nose of a bomb to reduce its penetration in earth or water. { 'käpf,riŋ }

Koplik's sign [MED] Small red spots surrounded by white areas seen in the mucous membrane of the mouth in the prodromal stage of measles. Also known as Koplik's spots. { 'käp·liks ,sīn }

Koplik's spots See Koplik's sign. { 'käp·liks ,späts }

kopophobia [PSYCH] An abnormal fear of fatigue or exhaustion. { ˌkap·ə'fō·bē·ə }

Köppen-Supan line [METEOROL] The isotherm connecting places which have a mean temperature of 10°C (50°F) for the warmest month of the year. { 'kep·ən sü'pän ,līn }

koppite [MINERAL] Mineral composed of a form of pyrochlore containing cerium, iron, and potassium. { 'kä,pīt }

Kopp's law [PHYS CHEM] The law that for solids the molal heat capacity of a compound at room temperature and pressure approximately equals the sum of heat capacities of the elements in the compound. { 'käps ,lȯ }

Korfmann arch saver [MIN ENG] A machine that uses a controlled hydraulic system to withdraw steel arches. { 'kȯrf·mən 'ärch ,sāv·ər }

Korfmann power loader [MIN ENG] A cutter-loader that is able to cart and load in both directions; its components are four drilling heads and one cutter chain surrounding them. { 'kȯrf·mən 'paù·ər ,lōd·ər }

kornelite [MINERAL] $Fe_2(SO_4)_3 \cdot 7H_2O$ A colorless to brown mineral composed of hydrous ferric sulfate. { 'kȯrn·ə,lī }

Korner's method [ORG CHEM] A method for determining the absolute position of substituents for positional isomers in benzene by the experimental production of positional isomers from a given disubstituted benzene. { 'kȯr·nərz ,meth·əd }

kornerupine [MINERAL] $(Mg,Fe,Al)_{20}(Si,B)_9O_{43}$ A colorless, yellow, brown, or sea-green mineral composed of magnesium iron borosilicate. { ˌkȯr·nə'rü,pēn }

Korn shell [COMPUT SCI] An enhanced version of the Bourne shell used in Unix systems. { 'kȯrn ,shel }

Korsakoff's neurosis See Korsakoff's syndrome. { 'kȯr·sə,kȯfs nə,rō·səs }

Korsakoff's psychosis See Korsakoff's syndrome. { 'kȯr·sə,kȯfs sī,kō·səs }

Korsakoff's syndrome [PSYCH] A form of amnesic-confabulatory syndrome characterized by confusion, loss of memory, retrograde amnesia with compensatory confabulation, and polyneuritis; seen in chronic alcoholism and other cases of vitamin B deficiency. Also known as Korsakoff's neurosis; Korsakoff's psychosis. { 'kȯr·sə,kȯfs ,sin,drōm }

Korshun method [ANALY CHEM] Microdetermination of carbon and hydrogen in organic compounds; the sample is prepyrolyzed (cracked) in a shortage of oxygen, then oxidized in an excess of oxygen. { 'kȯr·shən ,meth·əd }

Korteweg-de Vries soliton [PHYS] A soliton composed of a single oscillation, in contrast to an envelope soliton. Abbreviated KdV soliton. { 'kȯrd·ə,vek də'vrēz 'säl·ə,tän }

Kort nozzle [NAV ARCH] A cylindrical tube enclosing a screw propeller on a ship to decrease the tip losses and increase the efficiency; the walls of the tube are streamlined where clear of the ship's hull; adopted in many river towboats and recently in large tankers. { 'kȯrt ,näz·əl }

kosmochlor See ureyite. { 'käz·mə,klȯr }

kossava [METEOROL] A cold, very squally wind descending from the east or southeast in the region of the Danube "Iron

Gate" through the Carpathians, continuing westward over Belgrade, then spreading northward to the Rumanian and Hungarian borderlands and southward as far as Nish.   { 'kȯ·sə,vä }

Kossel effect   [PHYS]   The production of a series of cones of reflected x-rays by characteristic x-rays generated by atoms in a single crystal.   { 'kȯs·əl i,fekt }

Kossel lines   [PHYS]   Conic sections recorded on a flat film from the cones generated in the Kossel effect.   { 'kȯs·əl ,linz }

Kossel-Sommerfeld law   [SPECT]   The law that the arc spectra of the atom and ions belonging to an isoelectronic sequence resemble each other, especially in their multiplet structure.   { 'käs·əl 'zȯm·ər,felt ,lȯ }

Kosterlitz-Thouless transition   [CRYO]   The transition from the superfluid to the normal state in thin films of helium-3, which proceeds through the unbinding of vortices in the phase of the order parameter having opposite directions of rotation.   { 'käs·tər,lits tü'les tran,zish·ən }

kotoite   [MINERAL] $Mg_3(BO_3)_2$ An orthorhombic borate mineral; it is isostructural with jimboite.   { 'kōd·ə,wīt }

Kourbatoff's reagents   [MET]   Etching agents used for microanalysis of carbon steels; there are four different formulations, three with nitric acid, one with hydrochloric acid.   { 'kůr·bə,tȯfs rē,ā·jəns }

kovar   [MET]   An alloy (54% iron–29% nickel–17% cobalt) that exhibits low thermal expansion; widely used for electronic devices with glass-to-metal seals because its expansion is low enough to match that of glass.   { 'kō,vär }

Kovat's retention indexes   [ANALY CHEM]   Procedure to identify compounds in gas chromatography; the behavior of a compound is indicated by its position on a scale of normal alkane values (for example, methane = 100, ethane = 200).   { 'kō·vats ri'ten·chən ,in,dek·səs }

Kozak sequence   [MOL BIO]   A nucleotide sequence in the 5' untranslated messenger ribonucleic acid region that allows ribosomes to recognize the initiator codon.   { 'kō,zak ,sē·kwəns }

Kozeny-Carmen equation   [FL MECH]   Equation for streamline flow of fluids through a powdered bed.   { ,kō,zā·nē 'kär·mən i,kwā·zhən }

Kozeny's equation   [PETRO ENG]   Mathematical relationship of flow network permeability to capillary pore dimensions; used for reservoir calculations.   { ,kō,zā·nēz i,kwā·zhən }

Kr   See krypton.

kraft paper   [MATER]   A strong paper or cardboard made from sulfate-process wood pulp; unbleached varieties are used for wrapping paper and shipping cartons.   { 'kraft ,pā·pər }

kraft process   See sulfate pulping.   { 'kraft ,prä·səs }

kraft pulping   See sulfate pulping.   { 'kraft ,pəlp·iŋ }

Krakatao winds   [METEOROL]   A layer of easterly winds over the tropics at an altitude of about 11 to 14.5 miles (18 to 24 kilometers), which tops the mid-tropospheric westerlies (the antitrades), is at least 3.5 miles (6 kilometers) deep, and is based at about 1.2 miles (2 kilometers) above the tropopause.   { 'krak·ə,taů ,winz }

Kramers-Kronig relation   [OPTICS]   A relation between the real and imaginary parts of the index of refraction of a substance, based on the causality principle and Cauchy's theorem.   { 'krä·mərz 'krō·nig ri,lā·shən }

Kramer's theorem   [SOLID STATE]   The theorem that the states of a system consisting of an odd number of electrons in an external electrostatic field are at least twofold degenerate.   { 'krä·mərz ,thir·əm }

Krassowski ellipsoid of 1938   [GEOD]   The reference ellipsoid of which the semimajor axis is 6,378,245 meters and the flattening or ellipticity equals 1/298.3.   { kra'sȯv·skē ə'lip,-sȯid əv 'nīn,tēn ,thər·dē'āt }

K ratio   [AERO ENG]   The ratio of propellant surface to nozzle throat area.   { 'ka ,rā·shō }

kraurosis   [MED]   A progressive, sclerosing, shriveling process of the skin, due to glandular atrophy.   { krȯ'rō·səs }

Krause rolling mill   [MET]   A type of rolling mill in which the rolls translate as well as rotate, accomplishing a high reduction of thickness for the single passage of a metal sheet.   { 'kraůs 'rōl·iŋ ,mil }

Krause's corpuscle   [NEUROSCI]   One of the spheroid nerve-end organs resembling lamellar corpuscles, but having a more delicate capsule; found especially in the conjunctiva, the mucosa of the tongue, and the external genitalia; they are believed to be cold receptors.   Also known as end bulb of Krause.   { 'kraůs·əz ,kȯr·pə·səl }

krausite   [MINERAL] $KFe(SO_4)_2 \cdot H_2O$ A yellowish-green mineral composed of hydrous potassium iron sulfate.   { 'kraů,sīt }

Krawtchouk polynomials   [MATH]   Families of polynomials which are orthogonal with respect to binomial distributions.   { ¦kräv,chək ,päl·ə'nō·mē·əlz }

Krebs cycle   [BIOCHEM]   A sequence of enzymatic reactions involving oxidation of a two-carbon acetyl unit to carbon dioxide and water to provide energy for storage in the form of high-energy phosphate bonds.   Also known as citric acid cycle; tricarboxylic acid cycle.   { 'krebz ,sī·kəl }

Krebs-Henseleit cycle   [BIOCHEM]   A cyclic reaction pathway involving the breakdown of arginine to urea in the presence of arginase.   { 'krebz 'hen·sə,līt ,sī·kəl }

Krebspest   [INV ZOO]   A fatal fungus disease of crayfish caused by *Aphanomyces mystaci*.   { 'krebs,pest }

Krein-Milman property   [MATH]   The property of some topological vector spaces that any bounded closed convex subset is the closure of the convex span of its extreme points.   { ¦krīn 'mil·mən ,präp·ərd·ē }

Krein-Milman theorem   [MATH]   The theorem that in a locally convex topological vector space, any compact convex set K is identical with the intersection of all convex sets containing the extreme points of K.   { 'krīn 'mil·mən ,thir·əm }

kremastic water   See vadose water.   { krə'mas·tik ¦wȯd·ər }

kremersite   [MINERAL] $[(NH_4),K]_2FeCl_5 \cdot H_2O$ A red mineral composed of hydrous potassium ammonium iron chloride, occurring in octahedral crystals.   { 'krem·ər,zīt }

Kremser formula   [CHEM ENG]   Equation for calculating distillation-column material balances and equilibrium, assuming the ideal distribution law, that is, the concentrations in the two phases (vapor and liquid) are proportional to each other.   { 'krem·zər ,fȯr·myə·lə }

krennerite   [MINERAL] $AuTe_2$ A silver-white to pale-yellow mineral composed of gold telluride and often containing silver.   Also known as white tellurium.   { 'kren·ə,rīt }

kribergite   [MINERAL] $Al_5(PO_4)_3(SO_4)(OH)_4 \cdot 2H_2O$ White, chalklike mineral composed of hydrous basic aluminum sulfate and phosphate.   { 'krib·ər,gīt }

Krigar-Menzel law   [MECH]   A generalization of the second Young-Helmholtz law which states that when a string is bowed at a point which is at a distance of p/q times the string's length from one of the ends, where p and q are relative primes, then the string moves back and forth with two constant velocities, one of which is $q - 1$ times as large as the other.   { ¦krē·gər 'menz·əl ,lȯ }

kriging   [MIN ENG]   A geostatistical method of evaluating mine reserves based on a mathematical function known as a semivariogram.   { 'krīj·iŋ }

krill   [INV ZOO]   A name applied to planktonic crustaceans that constitute the diet of many whales, particularly whalebone whales.   { kril }

kringle domain   See kringle region.   { 'kriŋ·gəl dō,mān }

kringle region   [BIOCHEM]   A unique protein structural configuration composed of three disulfide bonds.   Also known as kringle domain.   { 'kriŋ·gəl ,rē·jən }

krohnkite   [MINERAL] $Na_2Cu(SO_4)_2 \cdot 2H_2O$ An azure-blue monoclinic mineral composed of hydrous copper sodium sulfate, occurring in massive form.   { 'kreŋ,kīt }

krokidolite   See crocidolite.   { krə'kid·əl,īt }

Kroll process   [MET]   A reduction process for the production principally of titanium metal sponge from titanium tetrachloride.   { 'krȯl ,prä·səs }

Kronecker delta   [MATH]   The function or symbol δ_{ij} dependent upon the subscripts i and j which are usually integers; its value is 1 if $i = j$ and 0 if $i \neq j$.   { 'krō·nek·ər ,del·tə }

Kronecker product   [MATH]   Given two different representations of the same group, their Kronecker product is a representation of the group constructed by taking direct products of matrices from the respective representations.   { 'krō·nek·ər ,präd·əkt }

Kronig-Penney model [SOLID STATE] An idealized one-dimensional model of a crystal in which the potential energy of an electron is an infinite sequence of periodically spaced square wells. { 'krō·nig 'pen·ē ,mäd·əl }

Krukenberg's tumor [MED] Bilateral carcinoma of the ovaries; originally described as primary, but now denoting a metastatic form usually of gastric origin. { 'krü·kən,bərgz ,tü·mər }

krummholz [ECOL] Stunted alpine forest vegetation. Also known as elfinwood. { 'krùm,hōlts }

Krupp ball mill [MIN ENG] An ore pulverizer in which the grinding is done by chilled iron or steel balls of various sizes moving against each other and the die ring, composed of five perforated spiral plates, each of which overlaps the next; material is discharged through a cylindrical screen. { 'krùp 'bòl ,mil }

Kruskal coordinates [RELAT] Coordinate system used in general relativity to describe in a nonsingular manner the geometry of a nonrotating black hole in empty space. { 'krús·kəl kō'órd·ən·əts }

Kruskal diagram [RELAT] A space-time diagram displaying the Schwarzschild metric in a form that eliminates the formal singularity that appears at the Schwarzschild radius in the form in which the metric is usually written. { 'krús·kəl ,dī·ə,gram }

kryolithionite [MINERAL] $Na_3Li_3(AlF_6)_2$ Variety of spodumene found in Greenland; has a crystal structure resembling that of garnet. { ,krī·ə'lith·ē·ə,nīt }

kryptoclimate See cryptoclimate. { ,krip·tō'klī·mət }

kryptoclimatology See cryptoclimatology. { ,krip·tō,klī·mə'täl·ə·jē }

krypton [CHEM] A colorless, inert gaseous element, symbol Kr, atomic number 36, atomic weight 83.80; it is odorless and tasteless; used to fill luminescent electric tubes. { 'krip·tän }

krypton-86 [NUC PHYS] An isotope of krypton, atomic mass 86; used in measurement of the standard meter. { 'krip·tän ,ād·ē'siks }

krypton lamp [ELEC] An arc lamp filled with krypton; one type pierces fog for 1000 feet (300 meters) or more and is used to light airplane runways at night. { 'krip·tän ,lamp }

KSAM See keyed sequential access method. { 'kā,sam }

KSA service See K-band single-access service. { 'kā,es'ā ,sər·vəs }

K scan See K scope. { 'kā ,skan }

K scope [ELECTR] A modified form of A scope on which one signal appears as two pips, the relative amplitudes of which indicate the error of aiming the antenna. Also known as K indicator; K scan. { 'kā ,skōp }

K selection [ECOL] Selection favoring species that reproduce slowly where a resource is constant but available in limited quantities; population is maintained at or near the carrying capacity (K) of the habitat. { 'kā si,lek·shən }

K shell [ATOM PHYS] The innermost shell of electrons surrounding the atomic nucleus, having electrons characterized by the principal quantum number 1. { 'kā ,shel }

K-14 sight [ORD] A gyroscopic computing gunsight employing a mechanical range-control system. { 'kā 'fòr,tēn ,sīt }

K-18 sight [ORD] A gyroscopic computing gunsight employing an electrical range-control system. { 'kā 'ā,tēn ,sīt }

k-space See wave-vector space. { 'kā ,spās }

KSR See keyboard send/receive.

K star [ASTRON] A star of spectral type K, a cool orange to red star with a surface temperature of about 3600–5000 K (6000–8500°F), and a spectrum resembling that of sunspots in which hydrogen lines have been greatly weakened. { 'kā ,stär }

kt See kiloton.

K theory [MATH] The study of the mathematical structure resulting from associating an abelian group $K(X)$ with every compact topological space X in a geometrically natural way, with the aid of complex vector bundles over X. Also known as topological K theory. { 'kā ,thē·ə·rē }

K transfer [ORD] In artillery ground fire, the shift of fire from one point to another in the transfer limits of the piece, the actual range being corrected by application of the K. { 'kā ,tranz·fər }

K truss [BUILD] A building truss in the form of a K due to

the orientation of the vertical member and two oblique members in each panel. { 'kā ,trəs }

Ku See kurchatovium.

Ku band [COMMUN] A band of frequencies extending from 15.35 to 17.25 gigahertz, corresponding to wavelengths of 1.95 to 1.74 centimeters. { 'kyü ,band or 'kā,yü ,band }

Ku-band fixed satellite service [COMMUN] Satellite communication at and near the Ku band, with the uplink frequency in bands from 12.75 to 13.25 gigahertz and 14.0 to 14.5 gigahertz and the downlink frequency in a band from 10.7 to 11.7 gigahertz. { 'kyü band ,fikst 'sad·əl,īt sər·vis }

Kubelka-Munk model [OPTICS] A widely used theoretical model of reflectance; the model supposes that some light passing through a homogeneous sample is scattered and absorbed so that the light is attenuated in both directions. { kü'bel·kə 'məŋk ,mäd·əl }

kudzu [BOT] Any of various perennial vine legumes of the genus *Pueraria* in the order Rosales cultivated principally as a forage crop. { 'kùd,zü }

Kuehneosauridae [PALEON] The gliding lizards, a family of Upper Triassic reptiles in the order Squamata including the earliest known aerial vertebrates. { ,kyün,nē·ō'sòr·ə,dē }

Kuiper Belt [ASTRON] A vast reservoir of icy bodies in the region of the solar system beyond the orbit of the planet Neptune, with diameters estimated in the 100–1200-kilometer (60–750-mile) range. Also known as Edgeworth-Kuiper belt. { 'kī·pər ,belt }

Kuiper Belt dust [ASTRON] A disk-shaped cloud of dust that is believed to be produced by collisions between members of the Kuiper Belt. { 'kī·pər ,belt ,dəst }

kukersite [GEOL] An organic sediment rich in remains of the alga *Gloexapsamorphe prisca*; found in the Ordovician of Estonia. { 'kü·kər,sīt }

Kullenberg piston corer [MECH ENG] A piston-operated coring device used to obtain 2-inch-diameter (5-centimeter) core samples. { 'kəl·ən,bərg 'pis·tən ,kór·ər }

Kumakhov optics [OPTICS] Systems composed of bundles of glass fibers that behave as waveguides to propagate x-rays or neutrons by multiple internal refflections, and that can be bent or tapered to concentrate, collimate, or focus the radiation. Also known as polycapillary optics. { kü'mä,kòf ,äp·tiks }

kumiss [FOOD ENG] A low-alcohol-content beverage made of fermented mare's or cow's milk by the action of *Lactobacillus casei, Streptococcus lactis,* and yeasts. { 'kü·məs }

kumquat [BOT] A citrus shrub or tree of the genus *Fortunella* in the order Sapindales grown for its small, flame- to orange-colored edible fruit having three to five locules filled with an acid pulp, and a sweet, pulpy rind. { 'kəm,kwät }

Kundt effect [OPTICS] **1.** The occurrence of a very large magnetic rotation when polarized light passes through very thin films of pure ferromagnetic materials. **2.** See Faraday effect. { 'kùnt i,fekt }

Kundt rule [SPECT] The rule that the optical absorption bands of a solution are displaced toward the red when its refractive index increases because of changes in composition or other causes. { 'kùnt ,rül }

Kundt's constant [OPTICS] A measure of the strength of the Faraday effect in a material, equal to the ratio of the Faraday rotation to the product of the path length and the magnetization of the material; it depends only on the temperature for any magnetic material. { 'kùns ,kän·stənt }

Kundt tube [ACOUS] A tube used to measure the speed of sound; it is filled with air or other gas and contains a light powder which becomes lumped at nodes, giving the length of standing waves generated in the tube. { 'kùnt ,tüb }

Kungurian [GEOL] A European stage of geologic time; Middle Permian, above Artinskian, below Kazanian. { kùn'gùr·ē·ən }

kunzite [MINERAL] A pinkish gem variety of spodumene. { 'kùnt,sīt }

Kupffer cell [HISTOL] One of the fixed macrophages lining the hepatic sinusoids. { 'kùp·fər ,sel }

Kuratowski closure-complementation problem [MATH] The problem of showing that at most 14 distinct sets can be obtained from a subset of a topological space by repeated operations of closure and complementation. { ,kùr·ə'tòf·skē ,klō·zhər ,käm·plə·men'tā·shən ,präb·ləm }

Kuratowski graphs [MATH] Two graphs which appear in

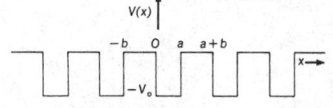
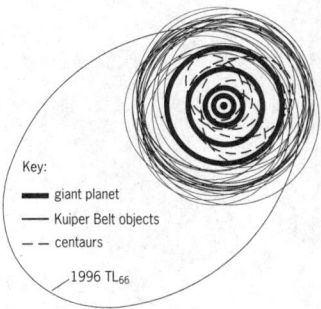
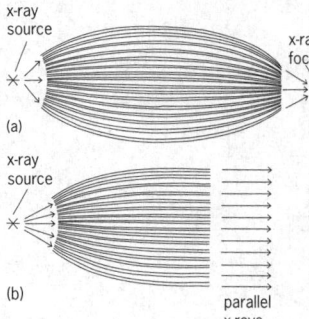

Kuratowski's theorem, the complete graph K_5 with five vertices and the bipartite graph $K_{3,3}$. { ˌkur·ə'tȯv‚skē ˌgrafs }

Kuratowski's lemma [MATH] Each linearly ordered subset of a partially ordered set is contained in a maximal linearly ordered subset. { kur·ə'tȯv‚skēz 'lem·ə }

Kuratowski's theorem [MATH] The proposition that a graph is nonplanar if and only if it has a subgraph which is either a Kuratowski graph or a subdivision of a Kuratowski graph. { ˌkur·ə'tȯv‚skēz ‚thir·əm }

kurchatovium [CHEM] The name suggested by workers in the Soviet Union for element 104. Symbolized Ku. { ˌkər·chə'tō·vē·əm }

Kureppa number [MATH] A number of the form $!n = 0! + 1! + \cdots + (n - 1)!$, where n is a positive integer. { ku'rep·ə ‚nəm·bər }

Kurie plot [NUC PHYS] Graph used in studying beta decay, in which the square root of the number of beta particles whose momenta (or energy) lie within a certain narrow range, divided by a function worked out by Fermi, is plotted against beta-particle energy; it is a straight line for allowed transitions and some forbidden transitions, in accord with the Fermi beta-decay theory. Also known as Fermi plot. { 'kyur·ē ‚plät }

kurnakovite [MINERAL] $Mg_2B_6O_{11} \cdot 13H_2O$ A white mineral composed of hydrous magnesium borate. { kur'näk·ə‚vīt }

Kuroshio [OCEANOGR] A fast ocean current originating off the southeast coast of Luzon, Philippines, and flowing northeastward off the coasts of China and Japan into the upper waters of the north Pacific Ocean. It carries large quantities of warm water from the tropics into the midlatitude regions, and is an important agent in redistributing global heat. { ˌku·rə'shē·ō }

Kuroshio Countercurrent [OCEANOGR] A component of the Kuroshio system flowing south and southwest between latitudes 155° and 160°E about 44 miles (70 kilometers) from the coast of Japan on the right-hand side of the Kuroshio Current. { ˌku·rə'shē·ō 'kaunt·ər‚kə·rənt }

Kuroshio extension [OCEANOGR] A general term for the warm, eastward-transitional flow that connects the Kuroshio and the North Pacific currents. { ˌku·rə'shē·ō ik'sten·shən }

Kuroshio system [OCEANOGR] A system of ocean currents which includes part of the North Equatorial Current, the Tsushima Current, the Kuroshio Current, and the Kuroshio extension. { ˌku·rə'shē·ō ‚sis·təm }

Kurthia [MICROBIO] A genus of gram-positive, aerobic, rod-shaped to coccoid bacteria in the coryneform group; metabolism is respiratory. { 'kər·thē·ə }

Kurtoidei [VERT ZOO] A monogeneric suborder of perciform fishes having a unique ossification that encloses the upper part of the swim bladder, and an occipital hook in the male for holding eggs during brooding. { kər'tȯid·ē‚ī }

kurtorachic [MED] Having the lumbar spinal region concave dorsally. { ¦kərd·ə¦rak·ik }

kurtosis [STAT] The extent to which a frequency distribution is concentrated about the mean or peaked; it is sometimes defined as the ratio of the fourth moment of the distribution to the square of the second moment. { kər'tō·səs }

Kusnezovia [MICROBIO] A genus of bacteria of uncertain affiliation; coccoid, nonmotile cells attach to substrate and reproduce by budding. { ‚küz·nə'zō·vē·ə }

kutnahorite [MINERAL] $Ca(Mn,Mg,Fe)(CO_3)_2$ A rare carbonate of calcium and manganese, found with some magnesium and iron substituting for manganese; forms rhombohedral crystals and is isomorphous with dolomite. { ‚kət·nə'hȯr‚īt }

Kutorginida [PALEON] An order of extinct brachiopod mollusks that is unplaced taxonomically. { ‚küd·ər'jīn·ə·də }

Kutta-Joukowski airfoil [FL MECH] A class of airfoils that may be produced by mapping circles with the complex variable transform $w = z + (c^2/z)$. { 'küd·ə jü'kȯv‚skē 'er‚fȯil }

Kutta-Joukowski condition [FL MECH] A boundary condition or fluid flow about an airfoil which requires that the circulation of the flow be such that a streamline leaves the trailing edge of the airfoil smoothly, or, equivalently, that the fluid velocity at the trailing edge be finite. { ¦küd·ə jü'kȯv‚skē kən‚dish·ən }

Kutta-Joukowski equation [FL MECH] An equation which states that the lift force exerted on a body by an ideal fluid, per unit length of body perpendicular to the flow, is equal to the product of the mass density of the fluid, the linear velocity of the fluid relative to the body, and the fluid circulation. Also known as Kutta-Joukowski theorem. { 'küd·ə jü'kȯv‚skēi‚kwā·zhən }

Kutta-Joukowski theorem See Kutta-Joukowski equation. { 'küd·ə jü'kȯv‚skē ‚thir·əm }

kV See kilovolt.

kVA See kilovolt-ampere.

kvar See kilovar. { 'kā‚vär }

K virus [VIROL] A group 2 papovavirus affecting rats and mice. { 'kā ¦vī·rəs }

kVp See kilovolts peak.

kW See kilowatt.

kwashiorkor [MED] A nutritional deficiency disease in infants and young children, mainly in the tropics, caused primarily by a diet low in proteins and rich in carbohydrates. Also known as nutritional dystrophy. { ¦kwä·shē¦ȯr·kər }

kWh See kilowatt-hour.

kW-hr See kilowatt-hour.

KWIC index See keyword-in-context index. { 'kwik ‚in‚deks }

KWOC index See keyword-out-of-context index. { 'kwäk ‚in‚deks }

kyanite [MINERAL] Al_2SiO_5 A blue or light-green neosilicate mineral; crystallizes in the triclinic system, and luster is vitreous to pearly; occurs in long, thin bladed crystals and crystalline aggregates. Also known as cyanite; disthene; sappare. { 'kī·ə‚nīt }

kyanize [CHEM ENG] To saturate wood with mercuric chloride as a decay preventive. { 'kī·ə‚nīz }

Kyasanur Forest virus [VIROL] A group B arbovirus recognized as an agent that causes hemorrhagic fever. { kī'az·ə‚nur ¦fär·əst 'vī·rəs }

kymograph [IND ENG] A device used to measure extremely short work time intervals by using a system of transducers that are activated by an operator performing a job, with the impulses recorded as a function of time. [MED] A device for recording internal body movements by making tracings with a stylus on a revolving smoked drum. { 'kī·mə‚graf }

kymography [MED] Recording of the movements of internal organs by a kymograph. { kī'mäg·rə·fē }

kynurenic acid [BIOCHEM] $C_{10}H_7O_3N$ A product of tryptophan metabolism found in the urine of mammals. { ¦kin·yə¦ren·ik 'as·əd }

kynurenine [BIOCHEM] $C_{10}H_{12}O_3N_2$ An intermediate product of tryptophan metabolism occurring in the urine of mammals. { ¦kin·yə're‚nēn }

kyphos [MED] The anteroposterior hump in the spine occurring in kyphoscoliosis. { 'kī‚fōs }

kyphoscoliosis [MED] Lateral curvature of the spine accompanied by rotation of the vertebrae. { ¦kī·fō‚skō·lē'ō·səs }

kyphosis [MED] Angular curvature of the spine, usually in the thoracic region. Also known as humpback; hunchback. { kī'fō·səs }

kyrohydratic point [OCEANOGR] The temperature at which a particular salt crystallizes in brine which is trapped by frozen seawater, the eutectic temperature of that salt. { ¦kī·rō·hī'drad·ik ‚pȯint }

kytoon [AERO ENG] A captive balloon used to maintain meteorological equipment aloft at approximately a constant height; it is streamlined, and combines the aerodynamic properties of a balloon and a kite. { 'kī‚tün }

L

l *See* liter.

L *See* lambert; liter.

La *See* lanthanum.

LAAS *See* Local-Area Augmentation System.

Labarraque's solution [MATER] Aqueous solution of 4–6% sodium hypochlorite and 4–6% sodium chloride with sodium hydroxide or sodium carbonate stabilizer; used as disinfectant. { 'la·bə¦raks sə¦lü·shən }

labbé [METEOROL] An infrequent, moderate to strong southwest wind that occurs only in March in Provence (southeastern France), bringing mild, humid, and very cloudy or rainy weather, while on the coast it raises a rough sea. { la'bā }

labdanum oil [MATER] An essential oil, golden yellow with ambergris aroma; soluble in alcohol, chloroform, and ether; derived from gum resin of various rockroses, such as *Cistus ladaniferus;* used in perfumes. Also known as ladanum oil. { 'lab·də·nəm ,oil }

label [ARCH] *See* hoodmold. [COMPUT SCI] A data item that serves to identify a data record (much in the same way as a key is used), or a symbolic name used in a program to mark the location of a particular instruction or routine. [NUCLEO] *See* isotopic tracer. { 'lā·bəl }

label alignment [COMPUT SCI] The manner in which text is aligned in the cells of a particular spreadsheet. { 'lā·bəl ə,līn·mənt }

label constant *See* location constant. { 'lā·bəl ,kän·stənt }

label data type [COMPUT SCI] A scalar data type that refers to locations in the computer program. { 'lā·bəl 'dad·ə ,tīp }

labeled cargo [IND ENG] Cargo of a dangerous nature, such as explosives and flammable or corrosive liquids, which is designated by different-colored labels to indicate the requirements for special handling and storage. { 'lā·bəld ,kär·gō }

labeled graph [MATH] A graph whose vertices are distinguished by names. { 'lā·bəld ,graf }

labeling index [CELL MOL] A measure of the mitotic activity of a cell population, defined as the number of cells in the S phase of the growth cycle divided by the total cells in the population. [NUCLEO] In autoradiography, the proportion of cells that are labeled by a radioisotope. { 'lā·bəl·iŋ ,in,deks }

labellate [BIOL] Having a labellum. { lə'bel,lāt }

labellum [BOT] The median membrane of the corolla of an orchid often differing in size and morphology from the other two petals. [INV ZOO] **1.** A prolongation of the labrum in certain beetles and true bugs. **2.** In Diptera, either of a pair of sensitive fleshy lobes consisting of the expanded end of the labium. { lə'bel·əm }

label paper [GRAPHICS] Any of the various papers used for can and bottle labels; made for both offset and letterpress printing, they have a smooth side for printing and a rough side to receive an adhesive. { 'lā·bəl ,pā·pər }

label record [COMPUT SCI] A tape record containing information concerning the file on that tape, such as format, record length, and block size. { 'lā·bəl ,re·kərd }

labial gland [ANAT] Any of the small, tubular mucous and serous glands underneath the mucous membrane of mammalian lips. [INV ZOO] A salivary gland, or modification thereof, opening at the base of the labium in certain insects. { 'lā·bē·əl ¦gland }

labial palp [INV ZOO] **1.** Either of a pair of fleshy appendages on either side of the mouth of certain bivalve mollusks. **2.** A jointed appendage attached to the labium of certain insects. { 'lā·bē·əl ¦palp }

labial papilla [INV ZOO] Any of the sensory bristles around the mouth of many nematodes; they are jointed projections of the cuticle. { 'lā·bē·əl pə pil·ə }

labia majora [ANAT] A pair of outer skin folds forming the lateral border of the vulva in the female. { ,lā·bē·ə mə'jór·ə }

labia minora [ANAT] A pair of inner skin folds, at the inner surfaces of the labia majora, surrounding the vulva in the female. { ,lā·bē·ə mə'nor·ə }

Labiatae [BOT] A large family of dicotyledonous plants in the order Lamiales; members are typically aromatic and usually herbaceous or merely shrubby. { ,lā·bē'ā,tē }

labiate [ANAT] Having liplike margins that are thick and fleshy. [BIOL] Having lips. [BOT] Having the limb of a tubular calyx or corolla divided into two unequal overlapping parts. { 'lā·bē·ət }

labile [PSYCH] Unstable in mood. [SCI TECH] Also known as metastable. **1.** Readily changed, as by heat, oxidation, or other processes. **2.** Moving from place to place. { 'lā,bīl }

labile factor *See* proaccelerin. { 'lā,bīl ,fak·tər }

labile oscillator [ELECTR] An oscillator whose frequency is controlled from a remote location by wire or radio. { 'lā,bīl 'äs·ə,lād·ər }

lability [PSYCH] Very rapid fluctuations in intensity and modality of emotions; seen in the affective reaction or in certain organic brain disorders. { lā'bil·əd·ē }

labite [MINERAL] $MgSi_3O_6(OH)_2·H_2O$ A mineral composed of hydrous basic silicate of magnesium. { 'lā,bīt }

labium [BIOL] **1.** A liplike structure. **2.** The lower lip, as of a labiate corolla or of an insect. { 'lā·bē·əm }

labium majus [ANAT] Either of the two outerfolds of skin that forms the lateral border of the vulva. { 'lā·bē·əm 'mä·jəs }

labium minus [ANAT] Either of the two inner folds, at the inner surfaces of the labia majora, surrounding the vulva in the female. { 'lā·bē·əm 'mē·nəs }

laboratory [SCI TECH] A place for experimental study. { 'lab·rə,tór·ē }

laboratory coordinate system [MECH] A reference frame attached to the laboratory of the observer, in contrast to the center-of-mass system. { 'lab·rə,tór·ē kō'órd·ən,ət ,sis·təm }

laboratory pack film [GRAPHICS] Unexposed film which is wound on a plain core (no flange), usually in 100-foot (30-meter) lengths. { 'lab·rə,tór·ē ,pak ,film }

laboratory sample [ANALY CHEM] A sample of a material to be tested or analyzed that is prepared from a gross sample and retains the latter's composition. { ¦lab·rə,tór·ē ¦sam·pəl }

labor cost [IND ENG] That part of the cost of goods and services attributable to wages, especially for direct labor. { 'lā·bər ,kòst }

labor factor [IND ENG] The ratio of the number of hours required to perform a task under project conditions to the number of hours required to perform an identical task under standard conditions of work measurement. { 'lā·bər ,fak·tər }

labor grade *See* job class. { 'lā·bər ,grād }

labor relations [IND ENG] The management function that deals with a company's work force; usually the term is restricted to relations with organized labor. { 'lā·bər ri,lā·shənz }

Laboulbeniales [MYCOL] An order of ascomycetous fungi made up of species that live primarily on the external surfaces of insects. { lə,bül·ben·ē'ā·lēz }

LABIATAE

Monarda fistulosa, and eastern North American species of wild bergamot. (*Courtesy of A. W. Ambler, National Audubon Society*)

La Bour centrifugal pump [MIN ENG] A self-priming centrifugal pump with a trap which ensures sufficient water for the pump to function, and a separator to remove the entrained air in the water. { lə'bŭr sen¦trif·ə·gəl 'pəmp }

Labrador Current [OCEANOGR] A current that flows southward from Baffin Bay, through the Davis Strait, and southwestward along the Labrador and Newfoundland coasts. { 'lab·rə¦dȯr ¦kə·rənt }

labradorite [MINERAL] A gray, blue, green, or brown plagioclase feldspar with composition ranging from $Ab_{50}An_{50}$ to $Ab_{30}An_{70}$, where $Ab = NaAlSi_3O_8$ and $An = CaAl_2Si_2O_8$; in the course of formation when the natural material cools, the feldspar sometimes exhibits a variously colored luster. Also known as Labrador spar. { 'lab·rə¦dȯ¦rīt }

Labrador spar See labradorite. { 'lab·rə¦dȯr ¦spär }

Labridae [VERT ZOO] The wrasses, a family of perciform fishes in the suborder Percoidei. { 'lab·rə¦dē }

labrum [INV ZOO] **1.** The upper lip of certain arthropods, lying in front of or above the mandibles. **2.** The outer edge of a gastropod shell. { 'lā·brəm }

labyrinth [ANAT] **1.** Any body structure full of intricate cavities and canals. **2.** The inner ear. [ENG ACOUS] A loudspeaker enclosure having air chambers at the rear that absorb rearward-radiated acoustic energy, to prevent it from interfering with the desired forward-radiated energy. { 'lab·ə¦rinth }

labyrinthine placenta See hemochorial placenta. { ¦lab·ə'rin¦thēn plə'sen·tə }

labyrinthine reflex [PHYSIO] The involuntary response to stimulation of the vestibular apparatus in the inner ear. { ¦lab·ə'rin¦thēn 'rē¦fleks }

labyrinthine syndrome See Ménière's syndrome. { ¦lab·ə'rin¦thēn 'sin¦drōm }

labyrinthitis [MED] Inflammation of the labyrinth of the inner ear. { ¦lab·ə·rən'thīd·əs }

Labyrinthodontia [PALEON] A subclass of fossil amphibians descended from crossopterygian fishes, ancestral to reptiles, and antecedent to at least part of other amphibian types. { ¦lab·ə¦rin·thə'dän·chə }

labyrinth seal [ENG] A minimum-leakage seal that offers resistance to fluid flow while providing radial or axial clearance; a labyrinth of circumferential knives or touch points provides for successive expansion of the fluid being piped; used for gas pipes, steam engines, and turbines. { 'lab·ə¦rinth ¦sēl }

Labyrinthulia [INV ZOO] A subclass of the protozoan class Rhizopoda containing mostly marine, ovoid to spindle-shaped, uninucleate organisms that secrete a network of filaments (slime tubes) along which they glide. { ¦lab·ə·rən'thül·ē·ə }

Labyrinthulida [INV ZOO] The single order of the protozoan subclass Labyrinthulia. { ¦lab·ə·rən'thül·ə·də }

lac [MATER] A resinous material secreted by some insects that live on the sap of certain trees, principally in India; used in the manufacture of shellac. { lak }

Lacaille's constellations [ASTRON] The 14 southern constellations identified by N. L. de Lacaille in 1763: Antlia, Callum, Circinus, Crux, Fornax, Horologium, Mensa, Microscopium, Norma, Octans, Pictor, Reticulum, Sculptor, and Telescopium. { lə'käz ¦kän·stə'lā·shənz }

laccal [BIOCHEM] $C_{17}H_{31}C_6H_3(OH)_2$ A phenol compound which is found in the sap of lacquer trees, and which can be isolated in crystalline form. { 'la¦kȯl }

laccase [BIOCHEM] Any of a class of plant oxidases which catalyze the oxidation of phenols. { 'la¦kās }

laccate [BIOL] Having a lacquered appearance. { 'la¦kāt }

Lacciferinae [INV ZOO] A subfamily of scale insects in the superfamily Coccoidea in which the male lacks compound eyes, the abdomen is without spiracles in all stages, and the apical abdominal segments of nymphs and females do not form a pygidium. { lak·sə'fer·ə¦nē }

laccolith [GEOL] A body of igneous rock intruding into sedimentary rocks so that the overlying strata have been notably lifted by the force of intrusion. { 'lak·ə¦lith }

lace [TEXT] A patterned, openwork fabric made by hand with needles or hooks, or by machinery. { lās }

lacerate [MED] To inflict a wound by tearing. { 'las·ə¦rāt }

lacerated [BIOL] Having a deeply and irregularly incised margin or apex. { 'las·ə¦rād·əd }

laceration [MED] A wound made by tearing. { ¦las·ə'rā·shən }

Lacerta [ASTRON] A small northern constellation lying

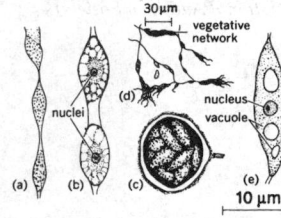

LABYRINTHULIA

Labyrinthulia. *Labyrinthula zopfi:* (a) portion of living network, (b) two organisms stained, (c) encysted stage *(after Valkoanov).* L. *macrocystis:* (d) vegetative network, (e) single organism stained *(after Cienkowski).* (*From R. P. Hall, Protozoology, Prentice-Hall, 1953*)

between Cygnus and Andromeda, and adjoining the northern boundary of Pegasus. Also known as Lizard. { lə'sərd·ə }

Lacertidae [VERT ZOO] A family of reptiles in the suborder Sauria, including all typical lizards, characterized by movable eyelids, a fused lower jaw, homodont dentition, and epidermal scales. { lə'sərd·ə¦dē }

lachesne [ORG CHEM] $C_{20}H_{26}ClNO_3$ A compound that crystallizes from a solution of ethanol and acetone, and whose melting point is 213°C; used in ophthalmology. Also known as chloride benzilate. { lə'shēn }

Lachnospira [MICROBIO] A genus of weakly gram-positive, anaerobic bacteria of uncertain affiliation; motile, curved rods that ferment glucose and are found in the rumen of bovine animals. { lak'näs·pə·rə }

lacing [CIV ENG] **1.** A lightweight metallic piece that is fixed diagonally to two channels or four angle sections, forming a composite strut. **2.** A course of brick, stone, or tiles in a wall of rubble to give strength. **3.** A course of upright bricks forming a bond between two or more arch rings. **4.** Distribution steel in a slab of reinforced concrete. **5.** A light timber fastened to pairs of struts or walings in the timbering of excavations (including mines). [ELEC] Tying insulated wires together to support each other and form a single neat cable, with separately laced branches. { 'lās·iŋ }

laciniate [BIOL] **1.** Having a fringed border. **2.** Narrowly and deeply incised to form irregular lobes, which may be pointed. { lə'sin·ē¦āt }

lacmus See litmus. { 'lak·məs }

lac operon [GEN] Three adjacent linked genes which code for the enzymes that act sequentially in lactose utilization in many bacteria. { 'lak 'äp·ə¦rän }

lacquer [MATER] A material which contains a substantial quantity of a cellulose derivative, most commonly nitrocellulose but sometimes a cellulose ester, such as cellulose acetate or cellulose butyrate, or a cellulose ether such as ethyl cellulose; used to give a glossy finish, especially on brass and other bright metals. { 'lak·ər }

lacquer diluent [MATER] An organic liquid with no solvent power added to lacquer formulations to reduce viscosity and to adjust flow or other properties. { 'lak·ər 'dil·yə·wənt }

lacquer print [TEXT] A textile design in which lustrous lacquer is used for coloring. { 'lak·ər ¦print }

lacquer tree See varnish tree. { 'lak·ər ¦trē }

lacrimal [ANAT] Pertaining to tears, tear ducts, or tear-secreting organs. { 'lak·rə·məl }

lacrimal apparatus [ANAT] The functional and structural mechanisms for secreting and draining tears; includes the lacrimal gland, lake, puncta, canaliculi, sac, and nasolacrimal duct. { 'lak·rə·məl ¦ap·ə'rad·əs }

lacrimal bone [ANAT] A small bone located in the anterior medial wall of the orbit, articulating with the frontal, ethmoid, maxilla, and inferior nasal concha. { 'lak·rə·məl ¦bōn }

lacrimal canal See nasolacrimal canal. { 'lak·rə·məl kə¦nal }

lacrimal canaliculus [ANAT] A small tube lined with stratified squamous epithelium which runs vertically a short distance from the punctum of each eyelid and then turns horizontally in the lacrimal part of the lid margin to the lacrimal sac. Also known as lacrimal duct. { 'lak·rə·məl ¦kan·ə'lik·yə·ləs }

lacrimal duct See lacrimal canaliculus. { 'lak·rə·məl ¦dəkt }

lacrimal gland [ANAT] A compound tubuloalveolar gland that secretes tears. Also known as tear gland. { 'lak·rə·məl ¦gland }

lacrimal sac [ANAT] The dilation at the upper end of the nasolacrimal duct within the medial canthus of the eye. Also known as dacryocyst. { 'lak·rə·məl ¦sak }

lacrimation [PHYSIO] **1.** Normal secretion of tears. **2.** Excessive secretion of tears, as in weeping. { ¦lak·rə'mā·shən }

lacrimator See tear gas. { 'lak·rə¦mād·ər }

lacroixite [MINERAL] A pale yellowish-green mineral composed of basic phosphate of aluminum, calcium, manganese, and sodium (often with fluorine), occurring as crystals. { lə'krwä¦zīt }

LACT See lease automatic custody transfer.

lactalbumin [BIOCHEM] A simple protein contained in milk which resembles serum albumin and is of high nutritional quality. { ¦lak¦tal'byü·mən }

lactam [ORG CHEM] An internal (cyclic) amide formed by heating gamma (γ) and delta (δ) amino acids; thus γ-aminobutyric acid readily forms γ-butyrolactam (pyrrolidone); many lactams have physiological activity. { 'lak,tam }

lactase [BIOCHEM] An enzyme that catalyzes the hydrolysis of lactose to dextrose and galactose. { 'lak,tās }

lactase deficiency syndrome [MED] Diarrhea induced by ingestion of a lactose-containing food such as milk, secondary to a congenital or acquired deficiency of lactase. { 'lak,tās də,fish·ən·sē ,sin,drōm }

lactate [ORG CHEM] A salt or ester of lactic acid in which the acidic hydrogen of the carboxyl group has been replaced by a metal or an organic radical. [PHYSIO] To secrete milk. { 'lak,tāt }

lactate dehydrogenase [BIOCHEM] A zinc-containing enzyme which catalyzes the oxidation of several α-hydroxy acids to corresponding α-keto acids. { 'lak,tāt dē'hī·drə·jə,nās }

lactation [PHYSIO] Secretion of milk by the mammary glands. { lak'tā·shən }

lacteal [ANAT] One of the intestinal lymphatics that absorb chyle. [PHYSIO] Pertaining to or resembling milk. { 'lak·tē·əl }

lactescent [BIOL] Having a milky appearance. [PHYSIO] Secreting milk or a milklike substance. { lak'tes·ənt }

lactic acid [BIOCHEM] $C_3H_6O_3$ A hygroscopic α-hydroxy acid, occurring in three optically isomeric forms: L form, in blood and muscle tissue as a product of glucose and glycogen metabolism; D form, obtained by fermentation of sucrose; and DL form, a racemic mixture present in foods prepared by bacterial fermentation, and also made synthetically. { 'lak·tik 'as·əd }

lactic dehydrogenase [BIOCHEM] An enzyme that catalyzes the dehydrogenation of L-lactic acid to pyruvic acid. Abbreviated LDH. { 'lak·tik dē'hī·drə·jə,nās }

lactic dehydrogenase virus [VIROL] A virus of the rubella group which infects mice. { 'lak·tik dē'hī·drə·jə,nās ,vī·rəs }

lactide [ORG CHEM] A cyclic, intermolecular, double ester formed from α-hydroxy acids; most lactides are relatively low melting solids and are easily hydrolyzed by base to form salts of the parent acid, such as sodium lactate. { 'lak,tīd }

lactiferous duct [BOT] A tubular channel consisting of latex vessels or latex cells; carries the latex produced by the plant. { lak'tif·ə·rəs 'dəkt }

lactim [ORG CHEM] A tautomeric enol form of a lactam with which it forms an equilibrium whenever the lactam nitrogen carries a free hydrogen. { 'lak·təm }

lactin See lactose. { 'lak·tən }

lactivorous [ZOO] Feeding on milk. { lak'tiv·ə·rəs }

Lactobacillaceae [MICROBIO] The single family of gram-positive, asporogenous, rod-shaped bacteria; they are saccharoclastic, and produce lactate from carbohydrate metabolism. { ,lak·tō·bə,sil·ē'ās·ē,ē }

Lactobacilleae [MICROBIO] Formerly a tribe of rod-shaped bacteria in the family Lactobacillaceae. { ,lak·tō·bə'sil·ē,ē }

Lactobacillus [MICROBIO] Lactic acid bacteria, the single genus of the family Lactobacillaceae; found in dairy products, meat products, fruits, beer, wine, and other food products. { ,lak·tō·bə'sil·əs }

lactoferrin [BIOCHEM] An iron-binding protein found in milk, saliva, tears, and intestinal and respiratory secretions that interferes with the iron metabolism of bacteria; in conjunction with antibodies, it plays an important role in resistance to certain infectious diseases. { ,lak·tō'fer·ən }

lactoflavin See riboflavin. { ,lak·tō'flav·ən }

lactogenic hormone See prolactin. { ,lak·tə,jen·ik 'hȯr,mōn }

lactoglobulin [BIOCHEM] A crystalline protein fraction of milk, which is soluble in half-saturated ammonium sulfate solution and insoluble in pure water. { ,lak·tō'gläb·yə·lən }

lactometer [ENG] A hydrometer used to measure the specific gravity of milk. { lak'täm·əd·ər }

lactonase [BIOCHEM] The enzyme that catalyzes the hydrolysis of 6-phosphoglucono-Δ-lactone to 6-phosphogluconic acid in the pentose phosphate pathway. { 'lak·tə,nās }

lactone [ORG CHEM] An internal cyclic mono ester formed by gamma (γ) or delta (δ) hydroxy acids spontaneously; thus γ-hydroxybutyric acid forms γ-butyrolactone. { 'lak,tōn }

lactonitrile [ORG CHEM] $CH_3CHOHCN$ A straw-colored liquid boiling at 183°C; soluble in water, insoluble in carbon disulfide and petroleum ether; used as a solvent, and as a chemical intermediate in making esters of lactic acid. Also known as acetaldehyde cyanohydrin. { ,lak,tō'nī,tril }

lactonization [ORG CHEM] The process in which a lactone is formed by intramolecular attack of a hydroxyl group on an activated carbonyl group. { ,lak·tə·nə'zā·shən }

lactophenene See para-lactophenetide. { lak'täf·ə,nēn }

para-lactophenetide [PHARM] $C_{11}H_{15}NO_3$ A water-soluble compound, crystallizing from ethyl acetate and hexane solution, and melting at 117–118°C; used in medicine as an analgesic and antipyretic. { ,par·ə ,lak·tō'fen·ə,tīd }

lactoprene [MATER] Any of several synthetic rubbers that have good resistance to hydrocarbon oil, ozone, oxygen, and other weather elements excepting cold, and that are polymers or copolymers of an acrylic acid ester. { 'lak·tə,prēn }

lactose [BIOCHEM] $C_{12}H_{22}O_{11}$ A disaccharide composed of D-glucose and D-galactose which occurs in milk. Also known as lactin; milk sugar. { 'lak,tōs }

lactosuria [MED] Presence of lactose in the urine. { ,lak·tōs'yur·ē·ə }

lactosyl ceramide [BIOCHEM] A neutral glycosphingolipid that is abundant in leukocyte membranes. { ,lak·tə,sil 'ser·ə,mīd }

lactron [TEXT] A thread made from natural rubber latex by extrusion. { 'lak·trən }

lacuna [BIOL] A small space or depression. [HISTOL] A cavity in the matrix of bone or cartilage which is occupied by the cell body. { lə'kü·nə }

lacunaris See lacunosus. { ,lak·yə'nar·əs }

lacunar system [INV ZOO] A series of intercommunicating spaces branching from two longitudinal vessels in the hypodermis of many acanthocephalans. { lə'kü·nər ,sis·təm }

lacunary space [MATH] A region in the complex plane that lies entirely outside the domain of a particular monogenic analytic function. { lə,kü·nə·rē 'spās }

lacunosus [METEOROL] A cloud variety characterized more by the appearance of the spaces between the cloud elements than by the elements themselves, the gaps being generally rounded, often with fringed edges, and the overall appearance being that of a honeycomb or net; it is the negative of clouds composed of separate rounded elements. Formerly known as lacunaris. { ,lak·yə'nō·səs }

lacustrine [GEOL] Belonging to or produced by lakes. { lə'kəs·trən }

lacustrine sediments [GEOL] Sediments that are deposited in lakes. { lə'kəs·trən 'sed·ə·məns }

lacustrine soil [GEOL] Soil that is uniform in texture but variable in chemical composition and that has been formed by deposits in lakes which have become extinct. { lə'kəs·trən 'sȯil }

Lacydonidae [INV ZOO] A benthic family of pelagic errantian polychaetes. { ,las·ə'dän·ə,dē }

ladanum oil See labdanum oil. { 'lad·ən·əm ,ȯil }

ladar [OPTICS] A missile-tracking system that uses a visible light beam in place of a microwave radar beam to obtain measurements of speed, altitude, direction, and range of missiles. Derived from laser detecting and ranging. Also known as colidar; laser radar. { 'lā,där }

ladder [ENG] A structure, often portable, for climbing up and down; consists of two parallel sides joined by a series of crosspieces that serve as footrests. { 'lad·ər }

ladder attenuator [ELECTR] A type of ladder network designed to introduce a desired, adjustable loss when working between two resistive impedances, one of which has a shunt arm that may be connected to any of various switch points along the ladder. { 'lad·ər ə'ten·yə,wād·ər }

ladder-bucket dredge See bucket-ladder dredge. { 'lad·ər ,bək·ət ,drej }

ladder diagram [CONT SYS] A diagram used to program a programmable controller, in which power flows through a network of relay contacts arranged in horizontal rows called rungs between two vertical rails on the side of the diagram containing the symbolic power. { 'lad·ər ,dī·ə,gram }

ladder ditcher See ladder trencher. { 'lad·ər 'dich·ər }

ladder dredge See bucket-ladder dredge. { 'lad·ər 'drej }

ladder drilling [MECH ENG] An arrangement of retractable drills with pneumatic powered legs mounted on banks of steel ladders connected to a holding frame; used in large-scale rock

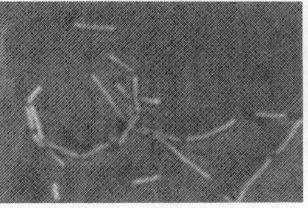

Conversion of γ-aminobutyric acid to γ-butyrolactam.

LACTOBACILLEAE

20 μm

Photomicrograph showing morphology of *Lactobacillus brevis*.

LACUNAR SYSTEM

Lacunar system of *Moniliformis*, dorsal view, showing regular circular branches. (*After Meyer in L. H. Hyman, The Invertebrates, vol. 3, McGraw-Hill, 1951*)

tunneling, with the advantage that many drills can be worked at the same time by a small labor force. { 'lad·ər ,dril·iŋ }

ladder fire [ORD] A method of establishing a gun range to a target by firing a succession of salvos with established differences in elevation. { 'lad·ər ,fīr }

ladder jack [ENG] A scaffold support which hooks onto a ladder. { 'lad·ər ,jak }

ladder network [ELECTR] A network composed of a sequence of H, L, T, or pi networks connected in tandem; chiefly used as an electric filter. Also known as series-shunt network. { 'lad·ər 'net,wərk }

ladder road See ladderway. { 'lad·ər ,rōd }

ladder track [CIV ENG] A main track that joins successive body tracks in a railroad yard. { 'lad·ər ,trak }

ladder trencher [MECH ENG] A machine that digs trenches by means of a bucket-ladder excavator. Also known as ladder ditcher. { 'lad·ər ¦trench·ər }

ladderway [MIN ENG] Also known as ladder road; manway. **1.** Mine shaft between two main levels, equipped with ladders. **2.** The particular shaft, or compartment of a shaft, containing ladders. { 'lad·ər,wā }

laddic [ELECTR] Multiaperture magnetic structure resembling a ladder, used to perform logic functions; operation is based on a flux change in the shortest available path when adjacent rungs of the ladder are initially magnetized with opposite polarity. { 'lad·ik }

Ladenburg f value See oscillator strength. { 'läd·ən,bərg 'ef ,val·yü }

Ladinian [GEOL] A European stage of geologic time: upper Middle Triassic (above Anisian, below Carnian). { lə'din·ē·ən }

ladle [DES ENG] A deep-bowled spoon with a long handle for dipping up, transporting, and pouring liquids. [MET] A receptacle for transporting and pouring molten metal. { 'lād·əl }

LADR See linear accelerator-driven reactor.

Laemobothridae [INV ZOO] A family of lice in the order Mallophaga including parasites of aquatic birds, especially geese and coots. { ,lē·mə'bäth·rə,dē }

Laennec's cirrhosis See portal cirrhosis. { 'lā'neks sə,rō·səs }

Lafarge cement [MATER] A cement made of plaster of paris, lime, and marble powder; used in mortar for marble and limestone pieces because it is nonstaining. { lə'färzh si,ment }

Lafond's Tables [METEOROL] A set of tables and associated information for correcting reversing thermometers and computing dynamic height anomalies, compiled by E. C. Lafond and published by the U.S. Navy Hydrographic Office. { lə'fänz ,tā·bəlz }

lag [CIV ENG] A flat piece of material, usually wood, used to wedge timber or steel supports against the ground and to make secure the space between supports. [ELECTR] A persistence of the electric charge image in a camera tube for a small number of frames. [PHYS] **1.** The difference in time between two events or values considered together. **2.** See lag angle. { lag }

lagan [ENG] A heavy object thrown overboard and buoyed to mark its location for future recovery. { 'lag·ən }

lag angle [PHYS] The negative of phase difference between a sinusoidally varying quantity and a reference quantity which varies sinusoidally at the same frequency, when this phase difference is negative. Also known as angle of lag; lag. { 'lag ,aŋ·gəl }

lag bolt See coach screw. { 'lag ,bōlt }

lag coefficient See time constant. { 'lag ,kō·i,fish·ənt }

lag correlation [STAT] The strength of the relationship between two elements in an ordered series, usually a time series, where one element lags a specific number of places behind the other elements. { 'lag ,kä·rə,lā·shən }

lag deposit [GEOL] Residual accumulation of coarse, unconsolidated rock and mineral debris left behind by the winnowing of finer material. { 'lag di,päz·ət }

Lagenidiales [MYCOL] An order of aquatic fungi belonging to the class Phycomycetales characterized by a saclike to limited hyphal thallus and zoospores having two flagella. { ,la·jə,nid·ē'ā·lēz }

lageniform [BIOL] Flask-shaped. { lə'jen·ə,fòrm }

lager beer [FOOD ENG] A variety of beer produced by bottom fermentation and allowed to age from 6 weeks to 6 months;

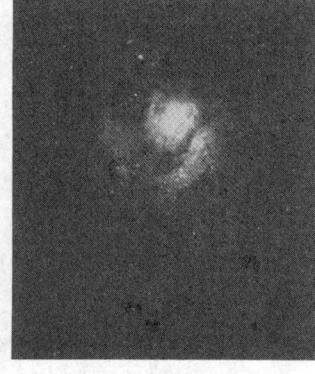

Lagoon Nebula NGC 6523 and star cluster NGC 6530. (*Curtis Schmidt Telescope, University of Michigan*)

it is usually dry, light in color, and well carbonated. { 'lag·ər ¦bir }

lag fault [GEOL] A minor low-angle thrust fault occurring within an overthrust; it develops when one part of the mass is thrust farther than an adjacent higher or lower part. { 'lag ,fòlt }

lagging [CIV ENG] **1.** Horizontal wooden strips fastened across an arch under construction to transfer weight to the centering form. **2.** Wooden members positioned vertically to prevent cave-ins in earthworking. [CYTOL] Pertaining to chromosomes that show little or no movement during metaphase and anaphase of meiosis or mitosis. [MATER] Asbestos and magnesia plaster that is used as a thermal insulation on process equipment and piping. { 'lag·iŋ }

lagging coil [ELEC] A small coil used to compensate for the lagging current in the voltage coil of an alternating-current watthour meter. { 'lag·iŋ ,kòil }

lagging current [ELEC] An alternating current that reaches its maximum value up to 90° behind the voltage that produces it. { 'lag·iŋ ,kə·rənt }

lagging load See inductive load. { 'lag·iŋ ,lōd }

lagging network See integral network. { 'lag·iŋ ,net,wərk }

lagging strand [CELL MOL] In deoxyribonucleic acid (DNA) replication, the 3′ to 5′ DNA strand that is discontinuously synthesized as a series of Okazaki fragments in the 5′ to 3′ direction. { 'lag·iŋ ,strand }

lag gravel [GEOL] Residual accumulations of particles that are coarser than the material that has blown away. { 'lag ,grav·əl }

lag-lead network See lead-lag network. { 'lag 'lēd ,net,wərk }

lag network See integral network. { 'lag ,net,wərk }

Lagomorpha [VERT ZOO] The order of mammals including rabbits, hares, and pikas; differentiated from rodents by two pairs of upper incisors covered by enamel, vertical or transverse jaw motion, three upper and two lower premolars, fused tibia and fibula, and a spiral valve in the cecum. { ,lag·ə'mòr·fə }

lagoon [GEOGR] **1.** A shallow sound, pond, or lake generally near but separated from or communicating with the open sea. **2.** A shallow fresh-water pond or lake generally near or communicating with a larger body of fresh water. { lə'gün }

Lagoon Nebula [ASTRON] A patchy, luminous gaseous nebula that appears to be surrounded by a much larger region of cold, neutral hydrogen. { lə'gün 'neb·yə·lə }

lag phase [MICROBIO] The period of physiological activity and diminished cell division following the addition of inoculum of bacteria to a new culture medium. { 'lag ,fāz }

Lagrange bracket [MECH] Given two functions of coordinates and momenta in a system, their Lagrange bracket is an expression measuring how coordinates and momenta change jointly with respect to the two functions. { lə'gränj ,brak·ət }

Lagrange function See Lagrangian. { lə'gränj ,fəŋk·shən }

Lagrange-Hamilton theory [MECH] The formalized study of continuous systems in terms of field variables where a Lagrangian density function and Hamiltonian density function are introduced to produce equations of motion. { lə'gränj 'ham·əl·tən ,thē·ə·rē }

Lagrange-Helmholtz equation See Helmholtz equation. { lə'gränj 'helm,hōlts i,kwā·zhən }

Lagrange's equations [MECH] Equations of motion of a mechanical system for which a classical (non-quantum-mechanical) description is suitable, and which relate the kinetic energy of the system to the generalized coordinates, the generalized forces, and the time. Also known as Lagrangian equations of motion. { lə'grän·jəz i,kwā·zhənz }

Lagrange's formula See mean value theorem. { lə'grän·jəz ,fòr·myə·lə }

Lagrange's theorem [MATH] In a group of finite order, the order of any subgroup must divide the order of the entire group. { lə'grän·jəz ,thir·əm }

Lagrange stream function [FL MECH] A scalar function of position used to describe steady, incompressible two-dimensional flow; constant values of this function give the streamlines, and the rate of flow between a pair of streamlines is equal to the difference between the values of this function on the streamlines. Also known as current function; stream function. { lə'gränj 'strēm ,fəŋk·shən }

Lagrangian [MECH] **1.** The difference between the kinetic energy and the potential energy of a system of particles, expressed as a function of generalized coordinates and velocities from which Lagrange's equations can be derived. Also known as kinetic potential; Lagrange function. **2.** For a dynamical system of fields, a function which plays the same role as the Lagrangian of a system of particles; its integral over a time interval is a maximum or a minimum with respect to infinitesimal variations of the fields, provided the initial and final fields are held fixed. { lə'grän·jē·ən }

Lagrangian coordinates *See* generalized coordinates. { lə'grän·jē·ən ko'ord·ən·əts }

Lagrangian current measurement [OCEANOGR] Observation of the speed direction of an ocean current by means of a device, such as a parachute drogue, which follows the water movement. { lə'grän·jē·ən 'kə·rənt ,mezh·ər·mənt }

Lagrangian density [MECH] For a dynamical system of fields or continuous media, a function of the fields, of their time and space derivatives, and the coordinates and time, whose integral over space is the Lagrangian. { lə'grän·jē·ən 'den·səd·ē }

Lagrangian description *See* Lagrangian method. { lə 'grän·jē·ən di'skrip·shən }

Lagrangian equations of motion *See* Lagrange's equations. { lə'grän·jē·ən i¦kwā·zhənz əv 'mō·shən }

Lagrangian function [MECH] The function which measures the difference between the kinetic and potential energy of a dynamical system. { lə'grän·jē·ən ,fəŋk·shən }

Lagrangian generalized velocity *See* generalized velocity. { lə'grän·jē·ən ¦jen·rə,līzd və'läs·əd·ē }

Lagrangian method [FL MECH] A method of studying fluid motion and the mechanics of deformable bodies in which one considers volume elements which are carried along with the fluid or body, and across whose boundaries material does not flow; in contrast to Euler method. Also known as Lagrangian description. { lə'grän·jē·ən ,meth·əd }

Lagrangian multipliers [MATH] A technique whereby potential extrema of functions of several variables are obtained. Also known as undetermined multipliers. { lə'grän·jē·ən 'məl·tə,plī·ərz }

Lagrangian points [ASTRON] Five points in the orbital plane of two massive objects orbiting about a common center of gravity at which a third object of negligible mass can remain in equilibrium; three points of instable equilibrium are located on the line passing through the centers of mass of the two bodies, and two points of stable equilibrium are located in the orbit of the less massive body, 60° ahead of or behind it. { lə'grän·jē·ən ,poins }

Lagriidae [INV ZOO] The long-jointed bark beetles, a family of coleopteran insects in the superfamily Tenebrionoidea. { lə'grī·ə,dē }

lag screw *See* coach screw. { 'lag ,skrü }

lag time [ELEC] The time between the application of current and rupture of the circuit within the detonator. { 'lag ,tīm }

Laguerre polynomials [MATH] A sequence of orthogonal polynomials which solve Laguerre's differential equation for positive integral values of the parameter. { lə'ger ,päl·ə'nō·mē·əlz }

Laguerre's differential equation [MATH] The equation $xy'' + (1 - x)y' + \alpha y = 0$, where α is a constant. { lə'gerz ,dif·ə¦ren·chəl i'kwä·zhən }

Lagynacea [INV ZOO] A superfamily of foraminiferan protozoans in the suborder Allogromiina having a free or attached test that has a membranous to tectinous wall and a single, ovoid, tubular, or irregular chamber. { ,lag·ə'nās·ē·ə }

lahar [GEOL] **1.** A mudflow or landslide of pyroclastic material occurring on the flank of a volcano. **2.** The deposit of mud or land so formed. { 'lä,här }

laid fabric [TEXT] Fabric in which the warp threads are bonded in a material, such as latex, there being no weft threads. { 'lād ,fab·rik }

laid paper [MATER] A paper with a pattern of parallel lines spaced so as to give a ribbed effect. { 'lād ,pā·pər }

Lainer effect [GRAPHICS] The effect of speeding the process of photographic development when potassium iodide is added to the developer. { 'lān·ər i,fekt }

laitance [MATER] Weak material, consisting principally of lime, that is formed on the surface of concrete, especially when excess water is mixed with the cement. { 'lat·əns }

lake [HYD] An inland body of water, small to moderately large, with its surface water exposed to the atmosphere.

[MATER] Any of a large group of dyes that have been combined with or adsorbed by salts of calcium, barium, chromium, aluminum, phosphotungstic acid, or phosphomolybdic acid; used for textile dyeing. Also known as color lake. { lāk }

lake ball [ECOL] A spherical mass of tangled, waterlogged fibers and other filamentous material of living or dead vegetation, produced mechanically along a lake bottom by wave action, and usually impregnated with sand and fine-grained mineral fragments. Also known as burr ball; hair ball. { 'lāk ,bòl }

lake breeze [METEOROL] A wind, similar in origin to the sea breeze but generally weaker, blowing from the surface of a large lake onto the shores during the afternoon; it is caused by the difference in surface temperature of land and water, as in the land and sea breeze system. { 'lāk ,brēz }

lake copper [MET] A pure type of copper produced from ores taken from the Lake Superior region; has high conductivity. { 'lāk ,käp·ər }

lake effect [METEOROL] Generally, the effect of any lake in modifying the weather about its shore and for some distance downwind; in the United States, this term is applied specifically to the region about the Great Lakes. { 'lāk i,fekt }

lake effect storm [METEOROL] A severe snowstorm over a lake caused by the interaction between the warmer water and unstable air above it. { 'lāk i,fekt ,storm }

lake ore *See* bog iron ore. { 'lāk ,òr }

lake peat [GEOL] A sedimentary peat formed near lakes. { 'lāk ,pēt }

lake plain [GEOL] One of the surfaces of the earth that represent former lake bottoms; these featureless surfaces are formed by deposition of sediments carried into the lake by streams. { 'lāk ,plān }

Lalande cell [ELEC] A type of wet cell that uses a zinc anode and cupric oxide cathode cast as flat plates or hollow cylinders, and an electrolyte of sodium hydroxide in aqueous solution (caustic soda). { lə'länd ,sel }

laliophobia [PSYCH] An abnormal fear of talking or stuttering. { ,lal·ē·ə'fō·bē·ə }

lally column [CIV ENG] A hollow and nearly circular steel column that supports girders or beams. { 'läl·ē ,käl·əm }

lalopathy [MED] Any disorder of speech or disturbance of language. { la'läp·ə·thē }

laloplegia [MED] Inability to speak, caused by paralysis of the muscles concerned in speech, except those of the tongue. { ,lal·ə'plē·jə }

Lamarckism [EVOL] The theory that organic evolution takes place through the inheritance of modifications caused by the environment, and by the effects of use and disuse of organs. { lə'mär,kiz·əm }

lamb [VERT ZOO] A young sheep. { lam }

lamb-blasts *See* lambing storm. { 'lam,blasts }

lambda [MECH] A unit of volume equal to 10^{-6} liter or 10^{-9} cubic meter. { 'lam·də }

lambda bacteriophage [MICROBIO] A temperate phage that infects *Escherichia coli* and then undergoes one of two life cycles: (1) lytic, in which it infects the host cell, replicates, and causes the cell to lyse (burst) as new phage progeny emerge; or (2) lysogenic, in which it infects the bacterial host cell, integrates its deoxyribonucleic acid into the host's genome, and goes into a dormant phase during which it replicates along with the host chromosome until it is induced to undergo lytic growth. { 'lam·də bak¦tir·ē·ə,fāj }

lambda calculus [MATH] A mathematical formalism to model the mathematical notion of substitution of values for bound variables. { 'lam·də ,kal·kyə·ləs }

lambda chain [IMMUNOL] A 22-kilodalton protein that is one of the two forms of smaller polypeptide chains (known as light chains) that occur in immunoglobulins. { 'lam·də ,chān }

lambda dispatch [IND ENG] The solution of the problem of finding the most economical use of generators to supply a given quantity of electric power, using the method of Lagrange multipliers, which are symbolized λ. { 'lam·də di,spach }

lambda expression [MATH] An expression used to define a function in the lambda calculus; for example, the function $f(x) = x + 1$ is defined by the expression $\lambda x(x + 1)$. { 'lam·də ik,spresh·ən }

lambda hyperon [PARTIC PHYS] **1.** A quasi-stable baryon,

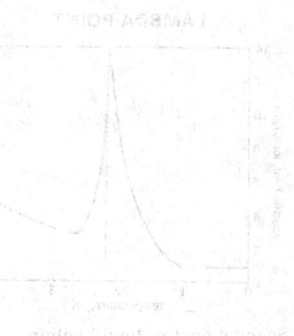

forming an isotopic singlet, having zero charge and hyper-charge, a spin of $1/2$, positive parity, and mass of approximately 1115.7 megaelectronvolts. Designated Λ. Also known as lambda particle. **2.** Any baryon resonance having zero hyper-charge and total isotopic spin; designated $\Lambda_{JP}(m)$, where m is the mass of the baryon in megaelectronvolts, and J and P are its spin and parity (if known). { 'lam·də 'hī·pə,rän }

lambda leak [CRYO] A leak of liquid helium II through small holes where normal liquids cannot pass. Also known as superleak. { 'lam·də ,lēk }

lambda particle *See* lambda hyperon. { 'lam·də ,pard·ə·kəl }

lambda point [CRYO] The temperature (2.1780 K), at atmospheric pressure, at which the transformation between the liquids helium I and helium II takes place; a special case of the thermodynamics definition. [THERMO] A temperature at which the specific heat of a substance has a sharply peaked maximum, observed in many second-order transitions. { 'lam·də ,pȯint }

lambda sulfur [CHEM] One of the two components of plastic (or gamma) sulfur; soluble in carbon disulfide. { 'lam·də ,səl·fər }

lamb dysentery [VET MED] A bacterial infection and inflammation of the intestinal tract of lambs caused by *Clostridium perfringens*, chiefly along the English-Scottish border. { 'lam 'dis·ən,ter·ē }

lambert [OPTICS] A unit of luminance (photometric brightness) that is equal to $1/\pi$ candela per square centimeter, or to the uniform luminance of a perfectly diffusing surface emitting or reflecting light at the rate of 1 lumen per square centimeter. Abbreviated L. { 'lam·bərt }

Lambert bearing [NAV] A bearing as measured on a Lambert conformal chart or plotting sheet; approximates a great-circle bearing. { 'lam·bərt ,ber·iŋ }

Lambert-Beer law *See* Bouguer-Lambert-Beer law. { 'lam·bərt 'bir ,lȯ }

Lambert conformal chart [MAP] A chart on the Lambert conformal projection. { 'lam·bərt kən'fȯr·məl ,chärt }

Lambert conformal projection [MAP] A conformal conic projection with two standard parallels, or a conformal conic map projection in which the surface of a sphere or spheroid, such as the earth, is conceived as developed on a cone which intersects the sphere or spheroid at two standard parallels; the cone is then spread out to form a plane which is the map. { 'lam·bərt kən¦fȯr·məl prə'jek·shən }

Lambert course [NAV] Course as measured on a Lambert conformal chart or plotting sheet. { 'lam·bərt ,kȯrs }

Lambert's law [OPTICS] **1.** The law that the illumination of a surface by a light ray varies as the cosine of the angle of incidence between the normal to the surface and the incident ray. **2.** The law that the luminous intensity in a given direction radiated or reflected by a perfectly diffusing plane surface varies as the cosine of the angle between that direction and the normal to the surface. **3.** *See* Bouguer-Lambert law. { 'lam·bərts ,lȯ }

Lambert surface [THERMO] An ideal, perfectly diffusing surface for which the intensity of reflected radiation is independent of direction. { 'lam·bərt ,sər·fəs }

lambing storm [METEOROL] A slight fall of snow in the spring in England. Also known as lamb-blasts; lamb-showers; lamb-storm. { 'lam·iŋ ,stȯrm }

Lamb shift [ATOM PHYS] A small shift in the energy levels of a hydrogen atom, and of hydrogenlike ions, from those predicted by the Dirac electron theory, in accord with principles of quantum electrodynamics. { 'lam ¦shift }

Lamb-shift source [NUCLEO] A device for producing a beam of polarized ions from one-electron atoms in their $2s$ excited state; the $2s$ and $2p$ levels, initially separated by the Lamb shift, are mixed by magnetic and electric fields, and selected nuclear magnetic substrates are then depopulated, leaving the remaining atoms in a state with large nuclear polarization. { 'lam ,shift ,sȯrs }

lamb-showers *See* lambing storm. { 'lam ,shau·ərz }

lamb-storm *See* lambing storm. { 'lam,stȯrm }

lamb's wool [VERT ZOO] The first fleece taken from a sheep up to 7 months old, having natural tapered fiber tip and spinning qualities superior to those of wool taken from previously shorn sheep. { 'lamz ,wul }

Lamb wave [ACOUS] *See* plate wave. [ELECTROMAG] Electromagnetic wave propagated over the surface of a solid whose thickness is comparable to the wavelength of the wave. { 'lam ,wāv }

lamé [TEXT] A fabric, usually of silk, rayon, or polyester, ornamented with flat metal threads. { la'mā }

Lamé constants [MECH] Two constants which relate stress to strain in an isotropic, elastic material. { lä'mā ,kän·stəns }

Lamé functions [MATH] Functions that arise when Laplace's equation is separated in ellipsoidal coordinates. { lä'mā ,fəŋk·shənz }

lamella [ANAT] A thin scale or plate. [CIV ENG] A thin member made of reinforced concrete, metal, or wood that is joined with similar members in an overlapping pattern to form an arch or a vault. { lə'mel·ə }

lamella arch [CIV ENG] An arch consisting basically of a series of intersecting skewed arches made up of relatively short straight members; two members are bolted, riveted, or welded to a third piece at its center. { lə'mel·ə ,ärch }

lamellar bone [HISTOL] Any bone with a microscopic structure consisting of thin layers or plates. { lə'mel·ər ¦bōn }

lamellar chloroplast [CELL MOL] A type of chloroplast in which the layered structure extends more or less uniformly through the whole chloroplast body. { lə'mel·ər 'klȯr·ə,plast }

lamellar crystal [CRYSTAL] A polycrystalline substance whose grains are in the form of thin sheets. { lə'mel·ər 'krist·əl }

lamella roof [BUILD] A large span vault built of members connected in a diamond pattern. { lə'mel·ə ,rüf }

lamellar vector field *See* irrotational vector field. { lə'mel·ər 'vek·tər ,fēld }

Lamellibranchiata [INV ZOO] An equivalent name for Bivalvia. { lə,mel·ə,braŋ·kē'äd·ə }

Lamellisabellidae [INV ZOO] A family of marine animals in the order Thecanephria. { lə,mel·ə·sə'bel·ə,dē }

Lamé polynomials [MATH] Polynomials which result when certain parameters of Lamé functions assume integral values, and which are used to express physical solutions of Laplace's equation in ellipsoidal coordinates. { lä'mā ,päl·ə'nō·mē·əlz }

Lamé's equations [MATH] A general collection of second-order differential equations which have five regular singularities. { lä'māz i,kwā·zhənz }

Lamé's relations [MATH] Six independent relations which when satisfied by the covariant metric tensor of a three-dimensional space provide necessary and sufficient conditions for the space to be euclidean. { lä'māz ri,lā·shənz }

Lamé wave functions [MATH] Functions which arise when the wave equation is separated in ellipsoidal coordinates. Also known as ellipsoidal wave functions. { lä'mā 'wāv ,fəŋk·shənz }

Lamiaceae [BOT] An equivalent name for Labiatae. { ,lā·mē'ās·ē,ē }

Lamiales [BOT] An order of dicotyledonous plants in the subclass Asteridae marked by its characteristic gynoecium, consisting of usually two biovulate carpels, with each carpel divided between the ovules by a false partition, or with the two halves of the carpel seemingly wholly separate. { ,lā·mē'ā·lēz }

lamina [BOT] *See* blade. [ANAT] A thin sheet or layer of tissue; a scalelike structure. [GEOL] A thin, clearly differentiated layer of sedimentary rock or sediment, usually less than 1 centimeter thick. [MATER] A flat or curved arrangement of unidirectional or woven fibers in a matrix. { 'lam·ə·nə }

lamina cribrosa [ANAT] **1.** The portion of the sclera which is perforated for the passage of the optic nerve. **2.** The fascia covering the saphenous opening in the thigh. **3.** The anterior or posterior perforated space of the brain. **4.** The perforated plates of bone through which pass branches of the cochlear part of the vestibulocochlear nerve. { 'lam·ə·nə krə'brō·sə }

laminal placentation [BOT] Condition in which the ovules occur on the inner surface of the carpels. { 'lam·ən·əl ,plas·ən'tā·shən }

laminar [SCI TECH] **1.** Arranged in thin layers. **2.** Pertaining to viscous streamline flow without turbulence. { 'lam·ə·nər }

laminar boundary layer [FL MECH] A thin layer over the surface of a body immersed in a fluid, in which the fluid velocity relative to the surface increases rapidly with distance

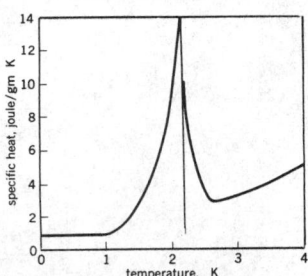

LAMBDA POINT

Specific heat of liquid helium computed as a function of temperature. Helium I is the high-temperature liquid; helium II occurs below the lambda point.

from the surface and the flow is laminar. { 'lam·ə·nər 'baùn·drē ˌlā·ər }

laminar composite [MATER] A composite material that consists of two or more layers of different materials that are bonded together. { 'lam·ə·nər kəm'päz·ət }

laminar flow [FL MECH] Streamline flow of an incompressible, viscous Newtonian fluid; all particles of the fluid move in distinct and separate lines. { 'lam·ə·nər 'flō }

laminar flow control [AERO ENG] The removal of a small amount of boundary-layer air from the surface of an aircraft wing with the result that the airflow is laminar rather than turbulent; frictional drag is greatly reduced. { 'lam·ə·nər 'flō kən,trōl }

Laminariales [BOT] An order of brown, large, structurally complicated, often highly differentiated members, commonly called kelps, of the algal class Phaeophyceae; distinctive features include a life history in which microscopic, filamentous, dioecious gametophytes alternate with a massive, parenchymatous sporophyte, and a mature sporophyte typically consisting of a holdfast, stipe, and one or more blades. { ˌlam·ə,nar·ē'ā·lēz }

Laminariophyceae [BOT] A class of algae belonging to the division Phaeophyta. { ˌlam·i,nar·ē·ō'fīs·ē,ē }

laminar sublayer [FL MECH] The laminar boundary layer underlying a turbulent boundary layer. { 'lam·ə·nər 'səb,lā·ər }

laminar wing [AERO ENG] A low-drag wing in which the distribution of thickness along the chord is so selected as to maintain laminar flow over as much of the wing surface as possible. { 'lam·ə·nər 'wiŋ }

laminate [MATER] A sheet of material made of several different bonded layers. { 'lam·ə,nāt }

laminated composite [MATER] A composite material consisting of layers of various materials. { 'lam·ə,nād·əd kəm'päz·ət }

laminated contact [ELEC] Switch contact made up of a number of laminations, each making individual contact with the opposite conducting surface. { 'lam·ə,nād·əd 'kän,takt }

laminated core [ELECTROMAG] An iron core for a coil transformer, armature, or other electromagnetic device, built up from laminations stamped from sheet iron or steel and more or less insulated from each other by surface oxides and sometimes also by application of varnish. { 'lam·ə,nād·əd 'kòr }

laminated glass See nonshattering glass. { 'lam·ə,nād·əd 'glas }

laminated metal [MET] A sheet or bar of composite metal composed of two or more bonded layers. { 'lam·ə,nād·əd 'med·əl }

laminated plastic [MATER] A thin sheet made of superposed layers of plastic bonded or impregnated with resin or compressed under heat. { 'lam·ə,nād·əd 'plas·tik }

laminated spring [DES ENG] A flat or curved spring made of thin superimposed plates and forming a cantilever or beam of uniform strength. { 'lam·ə,nād·əd 'spriŋ }

laminated wood [MATER] Board or timber composed of layers of wood glued together with the grains parallel. { 'lam·ə,nād·əd 'wùd }

lamina terminalis [ANAT] The layer of gray matter in the brain connecting the optic chiasma and the anterior commissure where the latter becomes continuous with the rostral lamina. { 'lam·ə·nə ˌtər·mə'nāl·is }

lamination [GRAPHICS] A plastic protective film on a printed sheet that has been bonded by heat and pressure. [MATER] One of the thin punchings of iron or steel used in building up a laminated core for a magnetic circuit. [MED] An operation in embryotomy in which the skull is cut in slices. [SCI TECH] Arrangement in layers. { 'lam·ə,nā·shən }

laminectomy [MED] Surgical removal of the lateral portion of the neural arch from one or more vertebrae. { ˌlam·ə'nek·tə·mē }

laminite [GEOL] Any sedimentary rock composed of millimeter- or finer-scale layers. { 'lam·ə,nīt }

laminography See sectional radiography. { ˌlam·ə'näg·rə·fē }

Lami's theorem [MECH] When three forces act on a particle in equilibrium, the magnitude of each is proportional to the sine of the angle between the other two. { la'mēz ˌthir·əm }

lamp [ENG] A device that produces light, such as an electric lamp. { lamp }

lampadite [MINERAL] A mineral composed chiefly of hydrous manganese oxide with as much as 18% copper oxide and often cobalt oxide. { 'lam·pə,dīt }

lamp bank [ELEC] A number of incandescent lamps connected in parallel or series to serve as a resistance load for full-load tests of electric equipment. { 'lamp ,baŋk }

lampblack [MATER] A grayish-black amorphous, practically pure form of carbon made by burning oil, coal tar, resin, or other carbonaceous substance in an insufficient supply of air; used in making paints, lead pencils, metal polishes, electric brush carbons, crayons, and carbon papers. { 'lamp,blak }

lampbrush chromosome [CELL MOL] An exceptionally large chromosome characterized by fine lateral projections which are associated with active ribonucleic acid and protein synthesis. { 'lamp ,brəsh 'krō·mə,sōm }

lamp cabin See lamp room. { 'lamp ,kab·ən }

lamp-charging rack [MIN ENG] Mine-lamp-charging racks which allow miners to store lamp units for recharging after daily use. { 'lamp ,chär·iŋ ,rak }

lamp cord [ELEC] Two twisted or parallel insulated wires, usually no. 18 or no. 20, used chiefly for connecting electric equipment to wall outlets. { 'lamp ,kòrd }

lamp depreciation [ELEC] The decrease in amount of light emitted by a lamp during its operating life. { 'lamp di,prē·shē,ā·shən }

lampholder [ELEC] A device designed to connect an electric lamp to a circuit and to support it mechanically. { 'lamp,hōld·ər }

lamphouse [ENG] **1.** The light housing in a motion picture projector, located behind the projector head ordinarily consisting of a carbon arc lamp operating on direct current at about 60 volts, a concave reflector behind the arc which collects the light and concentrates it on the film, and cooling devices. **2.** A box with a small hole containing an electric lamp and a concave mirror behind it, used as a concentrated source of light in a microscope, photographic enlarger, or other instrument. { 'lamp,haùs }

lamping [MIN ENG] Use of a portable ultraviolet lamp to reveal fluorescent minerals in prospecting. { 'lam·piŋ }

lamp inrush current [ELEC] The surge of current that occurs when an incandescent lamp is turned on. { 'lamp 'in,rəsh ,kər·ənt }

lampman [MIN ENG] A person responsible for maintaining and servicing miners' lamps. { 'lamp·mən }

lamp oil See kerosine. { 'lamp ,òil }

lamprey [VERT ZOO] The common name for all members of the order Petromyzonida. { 'lam·prē }

Lampridiformes [VERT ZOO] An order of teleost fishes characterized by a compressed, often ribbonlike body, fins composed of soft rays, a ductless swim bladder, and protractile maxillae among other distinguishing features. { ˌlam·prid·ə'fòr,mēz }

lamprobolite See basaltic hornblende. { ˌlam·prə'bō,līt }

Lamprocystis [MICROBIO] A genus of bacteria in the family Chromatiaceae; cells are spherical and motile, have gas vacuoles, and contain bacteriochlorophyll *a* on vesicular photosynthetic membranes. { ˌlam·prə'sis·təs }

lamp room [MIN ENG] A room or building at the surface of a mine for charging, servicing, and issuing all cap, hand, and flame safety lamps. Also known as lamp cabin; lamp station. { 'lamp ,rüm }

Lampropedia [MICROBIO] A genus of gram-negative, obligately anaerobic cocci of uncertain affiliation; cells form pairs, tetrads, or flat squared tablets. { ˌlam·prə'pēd·ē·ə }

lamprophyllite [MINERAL] Na₂SrTiSi₂O₈ A mineral composed of titanium strontium sodium silicate. { ˌlam·prə'fi,līt }

lamprophyre [PETR] Any of a group of igneous rocks characterized by a porphyritic texture in which abundant, large crystals of dark-colored minerals appear set in a not visibly crystalline matrix. { 'lam·prə,fīr }

lampshade paper [MATER] Paper that is translucent and either flame-resistant or flame-retardant; often made of wood pulp, vegetable parchment, or laminated glassine. { 'lamp,shād ,pā·pər }

lamp station See lamp room. { 'lamp ,stā·shən }

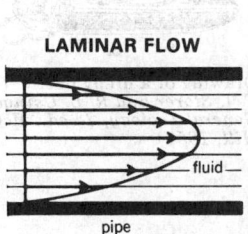

LAMINAR FLOW

Laminar flow in a circular pipe. In this case the velocity adjacent to the wall is zero and increases to a maximum in the center of the pipe.

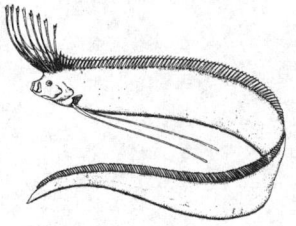

LAMPRIDIFORMES

Oarfish (*Regalecus glesne*); length to over 20 feet (6 meters).

LAMPYRIDAE

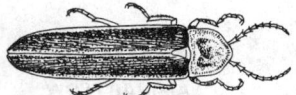

Drawing of a firefly beetle. (*From T. I. Storer and R. L. Usinger, General Zoology, 3d ed., McGraw-Hill, 1957*)

Lampyridae [INV ZOO] The firefly beetles, a large cosmopolitan family of coleopteran insects in the superfamily Cantharoidea. { lam'pir·ə,dē }

LAN *See* local-area network. { lan }

lanac [NAV] A proprietary navigation system that depends on secondary radar to aid airplanes to avoid midair collisions and maintain a specified altitude prior to landing. Derived from laminar, air navigation, and collision. { 'la,nak }

Lanarkian [GEOL] A European stage of geologic time forming part of the lower Upper Carboniferous, above Lancastrian and below Yorkian, equivalent to lowermost Westphalian. { lə'när·kē·ən }

lanarkite [MINERAL] Pb_2OSO_4 A white, greenish, or gray monoclinic mineral consisting of basic lead sulfate, with specific gravity of 6.92; formed by action of heat and air on galena. { 'lan·ər,kīt }

lanatoside [BIOCHEM] Any of three natural glycosides from the leaves of *Digitalis lanata*; on hydrolysis with acid, it yields one molecule of D-glucose, three molecules of digitoxose, and one molecule of acetic acid; all three glycosides are cardioactive. { lə'nad·ə,sīd }

Lancashire boiler [MECH ENG] A cylindrical steam boiler consisting of two longitudinal furnace tubes which have internal grates at the front. { 'laŋ·kə·shir ,bȯil·ər }

Lancastrian [GEOL] A European stage of geologic time forming part of the lower Upper Carboniferous, above Viséan and below Lanarkian. { laŋ'kas·trē·ən }

lance [MED] To cut or open, as with a lancet. [MET] To cut into but not through the piece of work. { lans }

Lance [ORD] A surface-to-surface missile that has inertial guidance and a liquid propulsion system, to provide artillery support for infantry, armored, mechanized, and airborne divisions. { lans }

lance door [MECH ENG] The door to a boiler furnace through which a hand lance is inserted. { 'lans ,dȯr }

Lancefield groups [MICROBIO] Antigenically determined categories for classification of β-hemolytic streptococci. { 'lans,fēld ,grüps }

lancelet [ZOO] The common name for members of the subphylum Cephalochordata. { 'lans·lət }

lanceolate [BIOL] Shaped like the head of a lance. { 'lan·sē·ə,lāt }

Lanceolidae [INV ZOO] A family of bathypelagic amphipod crustaceans in the suborder Hyperiidea. { ,lan·sē'äl·ə,dē }

lancet [MED] A sharp-pointed, double-edged cutting instrument used to make small incisions. { 'lan·sət }

lancet arch *See* acute arch. { 'lan·sət ,ärch }

lancet window [ARCH] A narrow window with a sharply pointed top. { 'lan·sət ,win·dō }

Lanchester balancer [MECH ENG] A device for balancing four-cylinder engines; consists of two meshed gears with eccentric masses, driven by the crankshaft. { 'lan·chə·stər 'bal·ən·sər }

Lanchester's rule [MECH] The rule that a torque applied to a rotating body along an axis perpendicular to the rotation axis will produce precession in a direction such that, if the body is viewed along a line of sight coincident with the torque axis, then a point on the body's circumference, which initially crosses the line of sight, will appear to describe an ellipse whose sense is that of the torque. { 'lan,ches·tərz ,rülz }

Lanczos's method [MATH] A transformation method for diagonalizing a matrix in which the matrix used to transform the original matrix to triple-diagonal form is formed from a set of column vectors that are determined by a recursive process. { 'län,chȯz·əs ,meth·əd }

land [AERO ENG] Of an aircraft, to alight on land or a ship deck. [DES ENG] The top surface of the tooth of a cutting tool, behind the cutting edge. [ELECTR] **1.** One of the regions between pits on a track on an optical disk. **2.** *See* terminal area. [ENG] **1.** In plastics molding equipment, the horizontal bearing surface of a semipositive or flash mold to allow excess material to escape; or the bearing surface along the top of the screw flight in a screw extruder; or the surface of an extrusion die that is parallel to the direction of melt flow. **2.** The surface between successive grooves of a diffraction grating or phonograph record. [GEOGR] The portion of the earth's surface that stands above sea level. [MET] In the preparation of a pipe length for welding, the edge of the tube wall that remains perpendicular to the bore after the pipe end has been beveled.

[ORD] One of the raised ridges in the bore of a rifled gun barrel. { land }

land accretion [CIV ENG] Gaining land in a wet area, such as a marsh or by the sea, by planting maritime plants to encourage silt deposition or by dumping dredged materials in the area. Also known as land reclamation. { 'land ə,krē·shən }

land and sea breeze [METEOROL] The complete cycle of diurnal local winds occurring on seacoasts due to differences in surface temperature of land and sea; the land breeze component of the system blows from land to sea, and the sea breeze blows from sea to land. { 'land ən 'sē ,brēz }

Landau damping [PL PHYS] Damping of a plasma oscillation wave which occurs in situations where the particles of the plasma are able to increase their average energy at the expense of the wave, and thus to damp it out, even in cases where the dissipative effects of collisions are unimportant. { 'lan,daů ,dam·piŋ }

Landau fluctuations [NUCLEO] Variations in the losses of energy of different particles in a thin detector, resulting from random variations in the number of collisions and in the energies lost in each collision of the particle. { 'lan,daů ,flək·chə'wā·shənz }

Landau-Ginzburg theory *See* Ginzburg-Landau theory. { 'lan,daů 'ginz·bərg ,thē·ə·rē }

Landau levels [SOLID STATE] Energy levels of conduction electrons which occur in a metal subjected to a magnetic field at very low temperatures and which are quantized because of the quantization of the electron motion perpendicular to the field. { 'lan,daů ,lev·əlz }

Landau-Levich-Derjaguin picture [FL MECH] A theory of fluid coating at low velocities, according to which the thickness of the film that forms when a solid is drawn out of a bath results from a balance between (1) the effects of viscosity, which causes a macroscopic entrainment of liquid by the solid, and (2) surface tension, which resists the film entrainment, so that the film thickness is proportional to the capillary number raised to the 2/3 power. { ¦lan,daů ¦lev·ich 'der·zhə,gēn ,pik·chər }

landblink [METEOROL] A yellowish glow observed over snow-covered land in the polar regions. { 'land,bliŋk }

land breeze [METEOROL] A coastal breeze blowing from land to sea, caused by the temperature difference when the sea surface is warmer than the adjacent land; therefore, the land breeze usually blows by night and alternates with a sea breeze which blows in the opposite direction by day. { 'land ,brēz }

land bridge [GEOGR] A strip of land linking two landmasses, often subject to temporary submergence, but permitting intermittent migration of organisms. { 'land ,brij }

land drainage [CIV ENG] The removal of water from land to improve the soil as a medium for plant growth and a surface for land management operations. { 'land ,drān·ij }

land-earth station [COMMUN] A facility that routes calls from mobile stations via satellite to and from terrestrial telephone networks. Abbreviated LES. { ¦land ¦ərth ,stā·shən }

land effect *See* coastal refraction. { 'land i,fekt }

Landé Γ-permanence rule [ATOM PHYS] The rule that the sum of the shifts of energy levels produced by the spin-orbit interaction, over a series of states having the same spin and orbital angular momentum quantum numbers (or the same total angular momentum quantum numbers for individual electrons) but different total angular momenta, and having the same total magnetic quantum number, is independent of the strength of an applied magnetic field. { län'dā ¦gam·ə ¦pər·mə·nəns ,rül }

Landé g factor [ATOM PHYS] Also known as g factor. **1.** The negative ratio of the magnetic moment of an electron or atom, in units of the Bohr magneton, to its angular momentum, in units of Planck's constant divided by 2π. **2.** The ratio of the difference in energy between two energy levels which differ only in magnetic quantum number to the product of the Bohr magneton, the applied magnetic field, and the difference between the magnetic quantum numbers of the levels; identical to the first definition for free atoms. Also known as Landé splitting factor; spectroscopic splitting factor. [NUC PHYS] The ratio of the magnetic moment of a nucleon, in units of the nuclear magneton, to its angular momentum in units of Planck's constant divided by 2π. { län'dā 'jē ,fak·tər }

Landé interval rule [ATOM PHYS] The rule that when the spin-orbit interaction is weak enough to be treated as a perturbation, an energy level having definite spin angular momentum and orbital angular momentum is split into levels of differing total angular momentum, so that the interval between successive levels is proportional to the larger of their total angular momentum values. { län′dā ′int·ər·vəl ‚rül }

Landenian [GEOL] A European stage of geologic time: upper Paleocene (above Montian, below Ypresian of Eocene). { lan′den·ē·ən }

lander [AERO ENG] A spacecraft that is designed to land on a celestial body. [MIN ENG] **1.** A worker at one of the levels of a mine shaft to unload rock and to load drilling and blasting supplies to be lowered. **2.** In the quarry industry, one who guides, steadies, and loads trucks or railroad cars with the blocks of stone hoisted from the quarry floor. Also known as top hooker. **3.** In metal mining, one who cleans skips by directing a blast of compressed air into them through a hose, records number of loaded skips hoisted to surface, and loads railroad cars with ore from bins. **4.** In coal mining, one who works with shaft sinking crew at the top of the shaft or at a level immediately above shaft bottom, dumping rock into mine cars from a bucket in which it is raised. Also known as bucket dumper; landing tender; top lander. **5.** The worker who receives the loaded bucket or tub at the mouth of the shaft. Also known as banksman. { ′lan·dər }

landesite [MINERAL] A brown mineral consisting of a hydrated phosphate of iron and manganese. { ′lan·də‚sīt }

Landé splitting factor *See* Landé g factor. { län′dā ′splid·iŋ ‚fak·tər }

landfall [NAV] **1.** The first sighting of land when approaching from seaward; by extension, the term is sometimes used to refer to the first contact with land by any means, as by radar. **2.** A navigational procedure in which an aircraft, after a relatively long overwater flight, turns on to a line of position passing through its destination and follows the line to the destination. { ′lan‚fȯl }

landfast ice *See* fast ice. { ′lan‚fast ′īs }

landfill [CIV ENG] Disposal of solid waste by burying in layers of earth in low ground. { ′lan‚fil }

landform [GEOGR] All the physical, recognizable, naturally formed features of land, having a characteristic shape; includes major forms such as a plain, mountain, or plateau, and minor forms such as a hill, valley, or alluvial fan. { ′lan‚fȯrm }

landform map *See* physiographic diagram. { ′lan‚fȯrm ‚map }

land hemisphere [GEOGR] The half of the globe, with its pole located at 47.25°N 2.5°W, in which most of the earth's land area is concentrated. { ′land ‚hem·ə‚sfir }

Landholt fringe [OPTICS] A black fringe that crosses the darkened field which is produced when a brilliant source of light is viewed through two Nicol prisms oriented with their principal axes at right angles to one another. { ′land‚hōlt ‚frinj }

land ice [HYD] Any part of the earth's seasonal or perennial ice cover which has formed over land as the result, principally, of the freezing of precipitation. { ′land ‚īs }

landing [CIV ENG] A place where boats receive or discharge passengers, freight, and so on. [MIN ENG] **1.** Level stage in a shaft at which cages are loaded and discharged. **2.** The top or bottom of a slope, shaft, or inclined plane. [NAV] The termination of an aircraft's flight or of a ship's voyage. { ′land·iŋ }

landing aid [NAV] A lamp, searchlight, radio beacon, radar device, communicating device, or any system of such devices for aiding aircraft in an approach and landing. Also known as landing system. { ′land·iŋ ‚ād }

landing area [AERO ENG] An area intended primarily for landing and takeoff of aircraft. { ′land·iŋ ‚er·ē·ə }

landing beacon [NAV] A beacon transmitting a beam to guide aircraft in making a landing. { ′land·iŋ ‚bē·kən }

landing chart [NAV] An aeronautical chart showing obstructions in the immediate vicinity of an aerodrome and the layout of the runways or landing area for use in landing and taxiing. { ′land·iŋ ‚chärt }

landing circle [AERO ENG] The approximately circular path flown by an airplane to get into the landing pattern; used particularly with naval aircraft landing on an aircraft carrier. { ′land·iŋ ‚sər·kəl }

landing compass [NAV] A compass taken ashore so as to be unaffected by deviation; if reciprocal bearings of the landing compass and the magnetic compass on board a ship are observed, the deviation of the magnetic compass can be determined. { ′land·iŋ ‚kam·pəs }

landing craft [NAV ARCH] A combat vessel employed in amphibious warfare to transport mobile equipment, amphibious vehicles, tanks, general cargo, and personnel, and to discharge such directly on to the beach; the vessel ranges from 36 to 135 feet (11 to 41 meters) in length and differs from landing ships in that it is not designed for long transoceanic voyages. { ′land·iŋ ‚kraft }

landing direction indicator [NAV] A device which indicates visually to pilots of aircraft the direction designated for landing or takeoff. { ′land·iŋ də‚rek·shən ′in·də‚kād·ər }

landing flap [AERO ENG] A movable airfoil-shaped structure located aft of the rear beam or spar of the wing; extends about two-thirds of the span of the wing and functions to substantially increase the lift, permitting lower takeoff and landing speeds. { ′land·iŋ ‚flap }

landing flare [NAV] A pyrotechnic flare sometimes dropped from an aircraft for illumination during a night landing. { ′land·iŋ ‚fler }

landing gear [AERO ENG] Those components of an aircraft or spacecraft that support and provide mobility for the craft on land, water, or other surface. [MECH ENG] A pair of small wheels at the forward end of a semitrailer to support the vehicle when it is detached from the tractor. { ′land·iŋ ‚gir }

landing light [AERO ENG] One of the floodlights mounted on the leading edge of the wing and below the nose of the fuselage to enable an airplane to land at night. { ′land·iŋ ‚līt }

landing load [AERO ENG] The load on an aircraft's wings produced during landing; depends on descent velocity and landing attitude. { ′land·iŋ ‚lōd }

landing ship [NAV ARCH] Large-type assault ship generally over 200 feet (60 meters) long which is designed for long sea voyages and for rapid unloading over or onto a beach. { ′land·iŋ ‚ship }

landing stage [CIV ENG] A platform, usually floating and attached to the shore, for the discharge and embarkation of passengers, freight, and so on. { ′land·iŋ ‚stāj }

landing strip [AERO ENG] A portion of the landing area prepared for the landing and takeoff of aircraft in a particular direction; it may include one or more runways. Also known as air strip. { ′land·iŋ ‚strip }

landing system *See* landing aid. { ′land·iŋ ‚sis·təm }

landing tee *See* wind tee. { ′land·iŋ ‚tee }

landing tender *See* lander. { ′land·iŋ ‚ten·dər }

landing zone [COMPUT SCI] The data-free area on the surface of a hard disk over which the read-write head comes to rest when the computer is shut off and the disk stops rotating. { ′land·iŋ ‚zōn }

landline [ELEC] A communications cable on or under the earth's surface, in contrast to a submarine cable. { ′lan‚līn }

landlocked [GEOGR] Pertaining to a harbor which is surrounded or almost completely surrounded by land. { ′land‚läkt }

landmark [CELL MOL] Any distinctive feature that can be used to identify a chromosome. [ENG] Any fixed natural or artificial monument or object used to designate a land boundary. [NAV] A conspicuous natural or artificial object near or on land, other than an established aid to navigation, and observable by eye or radar; used in the piloting type of navigation. { ′lan‚märk }

landmark beacon [NAV] A light beacon, flashing white and red, marking the location of a natural or constructed landmark or a point on the Federal airways. { ′lan‚märk ‚bē·kən }

land measure [MECH] **1.** Units of area used in measuring land. **2.** Any system for measuring land. { ′land ‚mezh·ər }

land mile *See* mile. { ′lan ‚mīl }

land mine [ORD] A container filled with high explosives or chemicals, placed on the ground or lightly covered, and fitted with a fuse or a firing device or both. { ′lan ‚mīn }

land mobile satellite [AERO ENG] A geosynchronous satellite employed in the land mobile satellite service. { ′land ‚mō·bəl ′sad·əl‚īt }

land mobile-satellite service [COMMUN] A mobile-satellite service in which the mobile earth stations are located on land. Abbreviated LMSS. { ′land ‚mō·bəl ′sad·əl‚īt ‚sər·vəs }

land mobile service [COMMUN] Mobile service between

LANDING CRAFT

The LCU-1613 used to discharge vehicles, mobile equipment, general cargo, and personnel directly onto a beach. (*Official U. S. Navy photograph*)

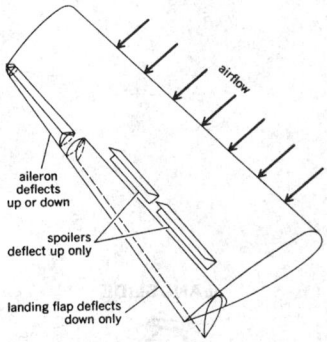

LANDING FLAP

Drawing of a section of an airplane wing showing the position and movement of the landing flap.

base stations and mobile stations, or between land mobile stations. { 'land ¦mō·bəl ¦sər·vəs }

land mobile station [COMMUN] Mobile station in the land mobile service, capable of surface movement within the geographical limits of a country or continent. { 'land ¦mō·bəl ¦stā·shən }

land navigation [NAV] Navigation of vehicles across land or ice, generally used in connection with the crossing of a region devoid of roads or landmarks, so that methods similar to those employed in air or marine navigation must be used. { 'land ¸nav·ə'gā·shən }

land pebble See land pebble phosphate. { 'land ¸peb·əl }

land pebble phosphate [GEOL] A pebble phosphate in a clay or sand bed below the ground surface; a small amount of uranium is often present and is recovered as a by-product; used as a source of phosphate fertilizer. Also known as land pebble; land rock; matrix rock. { 'land ¦peb·əl 'fäs¸fāt }

land plaster [MATER] Finely ground gypsum, used as a fertilizer and as a corrective for soil with excess sodium and potassium carbonates. { 'land ¸plas·tər }

land reclamation [CIV ENG] See land accretion. [MIN ENG] The process by which seriously disturbed land surfaces are stabilized against the hazards of water and wind erosion. { 'land ¸rek·lə'mā·shən }

land return See ground clutter. { 'land ri¸tərn }

land rock See land pebble phosphate. { 'land ¸räk }

Landry-Guillain-Barré syndrome [MED] A diffuse motor-neuron paresis, rapid in onset, and usually ascending and symmetrical in distribution, with proximal involvement greater than distal, and motor deficits greater than sensory. Also known as Guillain-Barré syndrome; Landry's paralysis. { lan'drē ¦gē¸yan bə'rā ¸sin¸drōm }

Landry's paralysis See Landry-Guillain-Barré syndrome. { lan'drēz pə'ral·ə·səs }

landscape [GEOGR] The distinct association of landforms that can be seen in a single view. [GRAPHICS] A printing orientation in which the printed lines are parallel to the wide side of the paper. { 'lan¸skāp }

landscape agate [MINERAL] A type of chalcedony that is translucent and contains inclusions which give it an appearance reminiscent of familiar natural scenes. Also known as fortification agate. { 'lan¸skāp ¸ag·ət }

landscape architecture [CIV ENG] The art of arranging and fitting land for human use and enjoyment. { 'lan¸skāp 'är·kə¸tek·chər }

landscape ecology [ECOL] The study of the ecological effects of spatial patterning of ecosystems. { 'lan¸skāp ē'käl·ə·jē }

landscape engineer [CIV ENG] A person who applies engineering principles and methods to planning, design, and construction of natural scenery arrangements on a tract of land. { 'lan¸skāp ¸en·jə'nīr }

land sky [METEOROL] The relatively dark appearance of the underside of a cloud layer when it is over land that is not snow-covered, used largely in polar regions with reference to the sky map; it is brighter than water sky, but much darker than iceblink or snowblink. { 'land ¸skī }

landslide [GEOL] The perceptible downward sliding or falling of a relatively dry mass of earth, rock, or combination of the two under the influence of gravity. Also known as landslip. { 'lan¸slīd }

landslide track [GEOL] An exposed path in rock or earth created as the result of a landslide. { 'lan¸slīd ¸trak }

landslip See landslide. { 'lan¸slip }

land station [COMMUN] Station in the mobile service not intended for operation while in motion. { 'land ¸stā·shən }

land surveyor [CIV ENG] A specialist who measures land and its natural features and any constructed features such as buildings or roads for drawing to scale as plans or maps. { 'land sər¸vā·ər }

land tie [CIV ENG] A rod or chain connecting an outside structure such as a retaining wall to a buried anchor plate. { 'land ¸tī }

land transportation frequency bands [COMMUN] A group of radio-frequency bands between 25 megahertz and 30,000 megahertz allocated for use by taxicabs, railroads, buses, and trucks. { 'land ¸tranz·pər¦tā·shən 'frē·kwən·sē ¸banz }

land transportation radio services [COMMUN] Any service of radio communications operated by and for the sole use

of certain land transportation carriers, the radio transmitting facilities of which are defined as fixed, land, or mobile stations. { 'land ¸tranz·pər¦tā·shən 'rād·ē·ō ¸sər·vəs·əz }

land-use classes [CIV ENG] Categories into which land areas can be grouped according to present or potential economic use. { 'land ¸yüs ¸klas·əz }

land-use map [MAP] A map showing land-use classes as well as other earth surface features such as roads, manufacturing plants, and harbors. { 'land ¸yüs ¸map }

lane [CIV ENG] An established route, as an air lane, shipping lane, or highway traffic lane. [NAV] One of the sections of the coverage area of a pair of Decca stations in which any phase relationship may be measured. { lān }

Lane-Emden equation See Emden equation. { 'lān 'em·dən i¸kwā·zhən }

Lane-Emden function See Emden function. { 'lān 'em·dən ¸fəŋk·shən }

Lane's law [ASTROPHYS] For the contraction of a star that is assumed to be a sphere of perfect gas, the law that the temperature of the perfect-gas sphere is inversely proportional to its radius. { 'lānz ¸lò }

langbanite [MINERAL] An iron-black hexagonal mineral composed of silicate and oxide of manganese, iron, and antimony, occurring in prismatic crystals. { 'läŋ·bə¸nīt }

langbeinite [MINERAL] $K_2Mg_2(SO_4)_3$ Colorless, yellowish, reddish, or greenish hexagonal mineral with vitreous luster, found in salt deposits; used in the fertilizer industry as a source of potassium sulfate. { 'läŋ¸bī¸nīt }

Langelier index [CHEM] A measure, based on pH, of the degree of calcium carbonate saturation in water, where negative values indicate that corrosion may result (pH below 7, dissolves calcium carbonate), and positive values indicate that scale deposition may result (pH above 7, precipitates calcium carbonate). { ¸länzh·əl¸yā 'in¸deks }

Langerhans cell [HISTOL] **1.** A type of cytotrophoblast in the human chorionic vesicle which is thought to secrete chorionic gonadotropin. **2.** A highly branched dendritic cell of the mammalian epidermis showing a lobulated nucleus and a diagnostic organelle resembling a tennis racket. { 'läŋ·ər¸hänz ¸sel }

Langerhansian adenoma See islet-cell tumor. { ¦läŋ·ər¦han¸zē·ən ¸ad·ən'ō·mə }

Lange's nerve [INV ZOO] One of the paired cords of nervous tissue lying in the wall of the radial perihemal canal of asteroids. { 'läŋ·əz ¸nərv }

Langevin-Debye formula [STAT MECH] A formula for the polarizability of a dielectric material or the paramagnetic susceptibility of a magnetic material, in which these quantities are the sum of a temperature-independent contribution and a contribution arising from the partial orientation of permanent electric or magnetic dipole moments which varies inversely with the temperature. Also known as Langevin-Debye law. { länzh·van də'bī ¸fòr·myə·lə }

Langevin-Debye law See Langevin-Debye formula. { länzh·van də'bī ¸lò }

Langevin function [ELECTROMAG] A mathematical function, $L(x)$, which occurs in the expressions for the paramagnetic susceptibility of a classical (non-quantum-mechanical) collection of magnetic dipoles, and for the polarizability of molecules having a permanent electric dipole moment; given by $L(x) = \coth x - 1/x$. { länzh·van ¸fəŋk·shən }

Langevin ion See large ion. { länzh·van ¸ī¸än }

Langevin ion-mobility theories [ELECTR] Two theories developed to calculate the mobility of ions in gases; the first assumes that atoms and ions interact through a hard-sphere collision and have a constant mean free path, while the second assumes that there is an attraction between atoms and ions arising from the polarization of the atom in the ion's field, in addition to hard-sphere repulsion for close distances of approach. { länzh·van ¦ī¸än mō'bil·əd·ē ¸thē·ə·rēz }

Langevin ion-recombination theory [ELECTR] A theory predicting the rate of recombination of negative with positive ions in an ionized gas on the assumption that ions of opposite sign approach one another under the influence of mutual attraction, and that their relative velocities are determined by ion mobilities; applicable at high pressures, above 1 or 2 atmospheres. { länzh·van ¦ī¸än rē¸käm·bə'nā·shən ¸thē·ə·rē }

Langevin radiation pressure [ACOUS] A measure of acoustic radiation pressure, equal to the difference between the mean

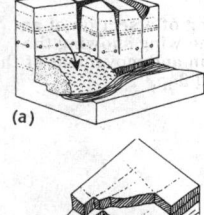

LANDSLIDE

Two principal types of landslide: *(a)* debris fall and *(b)* rockslide.

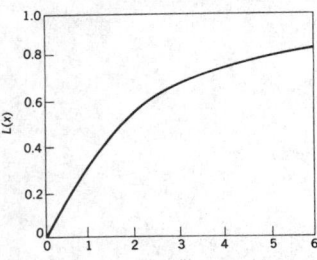

LANGEVIN FUNCTION

Plot of Langevin function.

pressure on an absorbing or reflecting wall and that in the same acoustic medium, at rest, behind the wall. { länzh·van ˌrād·ē'ā·shən ˌpresh·ər }

Langevin theory of diamagnetism [ELECTROMAG] A theory based on the idea that diamagnetism results from electronic currents caused by Larmor precession of electrons inside atoms. { länzh·van ˌthē·ə·rē əv ˌdī·ə'mag·nəˌtiz·əm }

Langevin theory of paramagnetism [ELECTROMAG] A theory which treats a substance as a classical (non-quantum-mechanical) collection of permanent magnetic dipoles with no interactions between them, having a Boltzmann distribution with respect to energy of interaction with an applied field. { länzh·van ˌthē·ə·rē əv ˌpar·ə'mag·nəˌtiz·əm }

langite [MINERAL] A blue to green mineral composed of basic hydrous copper sulfate. { 'laŋˌīt }

lang lay [DES ENG] A wire rope lay in which the wires of each strand are twisted in the same direction as the strands. { 'laŋ ˌlā }

langley [PHYS] A unit of energy per unit area commonly employed in radiation theory; equal to 1 gram-calorie per square centimeter. { 'laŋ·lē }

Langmuir-Blodgett film [PHYS CHEM] A highly ordered monomolecular film that results from compressing a surface layer of amphiphilic molecules into a floating monolayer and transferring it to a substrate by dipping. { 'laŋˌmyür 'bläj·ət ˌfilm }

Langmuir-Child equation See Child's law. { 'laŋˌmyür 'chīld iˌkwā·zhən }

Langmuir circulation [OCEANOGR] A form of motion found in the near-surface layer of lakes and oceans under windy conditions, and observed as streaks of bubbles, seaweed, or flotsam forming into lines running roughly parallel to the wind, called windrows. { 'laŋˌmyür ˌsər·kyəˌlā·shən }

Langmuir dark space [ELECTR] A nonluminous region surrounding a negatively charged probe inserted in the positive column of a glow discharge. { 'laŋˌmyür 'därk ˌspās }

Langmuir diffusion pump [ENG] A type of diffusion pump in which the mercury vapor emerges from a nozzle, giving it motion in a direction away from the high-vacuum side of the pump. { 'laŋˌmyür di'fyü·zhən ˌpəmp }

Langmuir effect [SOLID STATE] The ionization of atoms of low ionization potential that come into contact with a hot metal with a high work function. { 'laŋˌmyür iˌfekt }

Langmuir isotherm equation [PHYS CHEM] An equation, useful chiefly for gaseous systems, for the amount of material adsorbed on a surface as a function of pressure, while the temperature is held constant, assuming that a single layer of molecules is adsorbed; it is $f = ap/(1 + ap)$, where f is the fraction of surface covered, p is the pressure, and a is a constant. { 'laŋˌmyür 'īs·əˌthərm iˌkwā·zhən }

Langmuir plasma frequency [PL PHYS] The frequency of nonpropagating oscillations in a plasma; in rationalized mks units, it is $(ne^2/\epsilon_0 m)^{1/2}$, where e and m are the charge and mass of the oscillating electrons or ions, n is their number density, and ϵ_0 is the permittivity of empty space. Also known as plasma frequency. { 'laŋˌmyür 'plaz·mə ˌfrē·kwən·sē }

Langmuir probe [PL PHYS] A device for measuring the temperature and electron density of a plasma, consisting of an electrode in contact with the plasma whose potential is varied while the resulting collection currents are measured. { 'laŋˌmyür ˌprōb }

Langmuir wave [PL PHYS] A longitudinal, electrostatic wave that propagates in a plasma, because of variations in the plasma's electron density. { 'laŋˌmyür ˌwāv }

Lang's vesicle [INV ZOO] A seminal bursa in many polyclad flatworms. { 'laŋz ˌves·əˌkəl }

language [COMPUT SCI] The set of words and rules used to construct sentences with which to express and process information for handling by computers and associated equipment. [LING] The system of phonetic communication used by humans; worldwide, there are approximately 6000 distinct languages in current use. { 'laŋˌgwij }

language converter [COMPUT SCI] A device which translates a form of data (such as that on microfilm) into another form of data (such as that on magnetic tape). { 'laŋˌgwij kənˌvərd·ər }

language subset [COMPUT SCI] A portion of a programming language that can be used alone; usually applied to small

computers that do not have the capability of handling the complete language. { 'laŋˌgwij 'səbˌset }

language theory [MATH] A branch of automata theory which attempts to formulate the grammar of a language in mathematical terms; it has been applied to automatic language translation and to the construction of higher-level programming languages and systems such as the propositional calculus, nerve networks, sequential machines, and programming schemes. { 'laŋˌgwij ˌthē·ə·rē }

language translator [COMPUT SCI] **1.** Any assembler or compiler that accepts human-readable statements and produces equivalent outputs in a form closer to machine language. **2.** A program designed to convert one computer language to equivalent statements in another computer language, perhaps to be executed on a different computer. **3.** A routine that performs or assists in the performance of natural language translations, such as Russian to English, or Chinese to Russian. { 'laŋˌgwij ˌtranzˌlād·ər }

Languriidae [INV ZOO] The lizard beetles, a cosmopolitan family of coleopteran insects in the superfamily Cucujoidea. { ˌlaŋ·gə'rī·əˌdē }

Langweiler charge See traveling charge. { 'laŋˌwīl·ər ˌchärj }

Laniatores [INV ZOO] A suborder of arachnids in the order Phalangida having flattened, often colorful bodies and found chiefly in tropical areas. { ˌlan·ē·ə'tór·ēz }

lanolin [MATER] The hydrous sheep's-wool wax (primarily cholesterol esters of higher fatty acids) derived as a by-product from the preparation of raw wool for the spinner; used as a base for emollients in cosmetics and shampoos. { 'lan·ə·lən }

lanosterol [BIOCHEM] $C_{30}H_{50}O$ An unsaturated sterol occurring in wool fat and yeast. { lə'näs·təˌról }

lansan [METEOROL] A strong southeast trade wind of the New Hebrides and East Indies. { ˌlän̩sän }

lansfordite [MINERAL] $MgCO_3 \cdot 5H_2O$ A mineral composed of hydrous basic carbonate of magnesium when extracted from the earth, changing to nesquehonite after exposure to the air. { lanz·fərˌdīt }

L antenna [ELECTROMAG] An antenna that consists of an elevated horizontal wire having a vertical down-lead connected at one end. { 'el anˌten·ə }

lantern [ENG] A portable lamp. { 'lan·tərn }

lantern clock [HOROL] **1.** A wall-mounted clock with weights and pendulum outside the case. **2.** A 17th-century clock characterized by a dome and open fretwork. { 'lan·tərn ˌkläk }

lantern fish [VERT ZOO] The common name for the deep-sea teleost fishes composing the family Myctophidae and distinguished by luminous glands that are widely distributed upon the body surface. { 'lan·tərn ˌfish }

lantern pinion [DES ENG] A pinion with bars (between parallel disks) instead of teeth. { 'lan·tərn ˌpin·yən }

lantern ring [DES ENG] A ring or sleeve around a rotating shaft; an opening in the ring provides for forced feeding of oil or grease to bearing surfaces; particularly effective for pumps handling liquids. { 'lan·tərn ˌriŋ }

lantern slide See projection slide. { 'lan·tərn ˌslīd }

lanthana See lanthanum oxide. { 'lar·thə·nə }

lanthanide contraction [ATOM PHYS] A phenomenon encountered in the rare-earth elements; the radii of the atoms of the members of the series decrease slightly as the atomic numbers increase; starting with element 58 in the periodic table, the balancing electron fills in an inner incomplete $4f$ shell as the charge on the nucleus increases. { 'lan·thəˌnīd kənˌtrak·shən }

lanthanide series [CHEM] Rare-earth elements of atomic numbers 57 through 71; their chemical properties are similar to those of lanthanum, atomic number 57. { 'lan·thəˌnīd ˌsir·ēz }

lanthanite [MINERAL] $(La,Ce)_2(CO_3)_3 \cdot 8H_2O$ A colorless, white, pink, or yellow mineral composed of hydrous lanthanum carbonate, occurring in crystals or in earthy form. { 'lan·thəˌnīt }

lanthanum [CHEM] A chemical element, symbol La, atomic number 57, atomic weight 138.91; it is the second most abundant element in the rare-earth group. [MET] A white, soft, malleable metal; tarnishes in moist air; a major component of misch metal. { 'lan·thə·nəm }

LANG LAY

Drawing of lang lay method of winding wire rope.

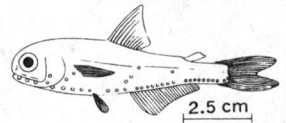

LANTERN FISH

2.5 cm

Drawing of the lantern fish (*Myctophym punctatum*); luminous glands are shown as small circles.

lanthanum-doped lead zirconate-lead titanate See lead lanthanum zirconate titanate. { 'lan·thə·nəm ¦dōpt 'led 'zərk·ən‚āt ¦led 'tīt·ən‚āt }

lanthanum nitrate [INORG CHEM] $La(NO_3)_3 \cdot 6H_2O$ Hygroscopic white crystals melting at 40°C; soluble in alcohol and water; used as an antiseptic and in gas mantles. { 'lan·thə·nəm 'nī‚trāt }

lanthanum oxide [INORG CHEM] La_2O_3 A white powder melting at about 2000°C; soluble in acid, insoluble in water; used to replace lime in calcium lights and in optical glass. Also known as lanthana; lanthanum sesquioxide; lanthanum trioxide. { 'lan·thə·nəm 'äk‚sīd }

lanthanum sesquioxide See lanthanum oxide. { 'lan·thə·nəm ‚ses·kwē'äk‚sīd }

lanthanum sulfate [INORG CHEM] $La_2(SO_4)_3 \cdot 9H_2O$ White crystals; slightly soluble in water, soluble in alcohol; used for atomic weight determinations for lanthanum. { 'lan·thə·nəm 'səl‚fāt }

lanthanum trioxide See lanthanum oxide. { 'lan·thə·nəm trī'äk‚sīd }

Lanthonotidae [VERT ZOO] A family of lizards (Sauria) belonging to the Anguimorpha line; restricted to North Borneo. { lan·thə'näd·ə‚dē }

lanugo [ANAT] A downy covering of hair, especially that seen on the fetus or persisting on the adult body. { lə'nü·gō }

lanyard [NAV ARCH] A small rope or line used to fasten something in ships, especially one passing through deadeyes and used to extend shrouds or stays. { 'lan·yərd }

lap [CIV ENG] The length by which a reinforcing bar must overlap the bar it will replace. [MATER] An abrasive material used for lapping. [MET] A defect caused by folding and then rolling or forging a hot metal fin or corner onto a surface without welding. Also known as fold. { lap }

lapachoic acid See lapachol. { lə'päch·ə·wik 'as·əd }

lapachol [BIOCHEM] $C_{15}H_{14}O_3$ A yellow crystalline compound obtained from lapacho, a hardwood in Argentina and Paraguay. Also known as lapachoic acid; targusic acid. { lə'pä‚chōl }

laparoscopy [MED] A method of visually examining the peritoneal cavity by means of a long slender endoscope equipped with sheath, obturator, biopsy forceps, a sphygmomanometer bulb and tubing, scissors, and a syringe; the endoscope is introduced into the peritoneal cavity through a small incision in the abdominal wall. Also known as peritoneoscopy. { ‚lap·ə'räs·kə·pē }

laparotomy [MED] A surgical incision through the abdominal wall into the abdominal cavity. { ‚lap·ə'räd·ə·mē }

lap dissolve [ELECTR] Changeover from one television scene to another so that the new picture appears gradually as the previous picture simultaneously disappears. [GRAPHICS] See dissolve. { 'lap di‚zälv }

lapel microphone [ENG ACOUS] A small microphone that can be attached to a lapel or pocket on the clothing of the user, to permit free movement while speaking. { lə'pel ¦mī·krə‚fōn }

lapidary [SCI TECH] 1. Of or pertaining to precious stones. 2. The art of cutting precious stones. 3. A person skilled in such art. { 'lap·ə‚der·ē }

lapidicolous [ECOL] Living under a stone. { ¦lap·ə¦dik·ə·ləs }

lapies See karren. { lə'pēz }

lapilli [GEOL] Pyroclasts that range from 0.04 to 2.6 inches (1 to 64 millimeters) in diameter. { lə'pi‚lī }

lapilli-tuff [GEOL] A pyroclastic deposit that is indurated and consists of lapilli in a fine tuff matrix. { lə'pi‚lī ‚təf }

lapis lazuli [PETR] An azure-blue, violet-blue, or greenish-blue, translucent to opaque crystalline rock used as a semiprecious stone; composed chiefly of lazurite and calcite with some haüyne, sodalite, and other minerals. Also known as lazuli. { ¦lap·is 'laz·ə·lē }

lap joint [ENG] A simple joint between two members made by overlapping the ends and fastening them together with bolts, rivets, or welding. { 'lap ‚jóint }

Laplace irrotational motion [FL MECH] Irrotational flow of an inviscid, incompressible fluid. { lə'pläs ‚ir·ō'tā·shən·əl ‚mō·shən }

Laplace law See Ampère law. { lə'pläs ‚lò }

Laplace operator [MATH] The linear operator defined on differentiable functions which gives for each function the sum of all its nonmixed second partial derivatives. Also known as Laplacian. { lə'pläs ‚äp·ə‚rād·ər }

Laplace's equation [ACOUS] An equation for the speed c of sound in a gas; it may be written $c = \sqrt{\gamma p/\rho}$, where p is the pressure, ρ is the density, and γ is the ratio of specific heats. [MATH] The partial differential equation which states that the sum of all the nonmixed second partial derivatives equals 0; the potential functions of many physical systems satisfy this equation. { lə'pläs·əz i‚kwā·zhən }

Laplace's expansion [MATH] An expansion by means of which the determinant of a matrix may be computed in terms of the determinants of all possible smaller square matrices contained in the original. { lə'pläs·əz ik‚span·chən }

Laplace's measure of dispersion [STAT] The expected value of the absolute value of the difference between a random variable and its mean. { lə'pläs·əz ¦mezh·ər əv di'spər·zhən }

Laplace transform [MATH] For a function $f(x)$ its Laplace transform is the function $F(y)$ defined as the integral over x from 0 to ∞ of the function $e^{-yx}f(x)$. { lə'pläs 'trans‚fòrm }

Laplacian See Laplace operator. { lə'pläs·ē·ən }

Laplacian speed of sound [FL MECH] The phase speed of a sound wave in a compressible fluid under the assumption that the expansions and compressions are adiabatic. { lə'pläs·ē·ən ‚spēd əv 'saùnd }

Laporte selection rule [ATOM PHYS] The rule that an electric dipole transition can occur only between states of opposite parity. { lə'pòrt si'lek·shən ‚rül }

lappet [TEXT] 1. An attachment for a loom, used to insert floating warp threads into a fabric. 2. A lightweight patterned fabric such as dotted swiss made with such an attachment. [ZOO] A lobe or flaplike projection, such as on the margin of a jellyfish or the wattle of a bird. { 'lap·ət }

lapping [ELECTR] Moving a quartz, semiconductor, or other crystal slab over a flat plate on which a liquid abrasive has been poured, to obtain a flat polished surface or to reduce the thickness a carefully controlled amount. [MET] Polishing with a material such as cloth, lead, plastic, wood, iron, or copper having fine abrasive particles incorporated or rubbed into the surface. { 'lap·iŋ }

lap-rivet [MET] To rivet a lap joint. { 'lap ‚riv·ət }

lapse line [METEOROL] A curve showing the variation of temperature with height in the free air. { 'laps ‚līn }

lapse rate [METEOROL] 1. The rate of decrease of temperature in the atmosphere with height. 2. Sometimes, the rate of change of any meteorological element with height. { 'laps ‚rāt }

lap siding [BUILD] Beveled boards used for siding that are similar to clapboards but longer and wider. [CIV ENG] Two railroad sidings, the turnout of one overlapping that of the other. { 'lap ‚sīd·iŋ }

lapstrake [NAV ARCH] A method of hull construction where each continuous band of hull planking (strake) is lapped on the outside of the one beneath. { 'lap‚strāk }

laptop computer See notebook computer. { 'lap‚täp kəm‚pyüd·ər }

lap weld [MET] A welded lap joint. { 'lap ‚weld }

lap winding [ELEC] A two-layer winding in which each coil is connected in series to the adjacent coil. { 'lap ‚wīn·diŋ }

Laramic orogeny See Laramidian orogeny. { 'lar·ə·mik ò'räj·ə·nē }

Laramide orogeny See Laramidian orogeny. { 'lar·ə·məd ò'räj·ə·nē }

Laramide revolution See Laramidian orogeny. { 'lar·ə·məd ‚rev·ə'lü·shən }

Laramidian orogeny [GEOL] An orogenic era typically developed in the eastern Rocky Mountains; phases extended from Late Cretaceous until the end of the Paleocene. Also known as Laramic orogeny; Laramide orogeny; Laramide revolution. { ‚lar·ə'mid·ē·ən ò'räj·ə·nē }

Laray viscometer [ENG] An instrument designed to measure viscosity and other properties of ink. { lə'rā vi'skäm·əd·ər }

larch [BOT] The common name for members of the genus *Larix* of the pine family, having deciduous needles and short, spurlike branches which annually bear a crown of needles. { lärch }

larch canker [PL PATH] A destructive fungus disease of larch and sometimes pine and fir caused by *Dasyscypha willkommii*

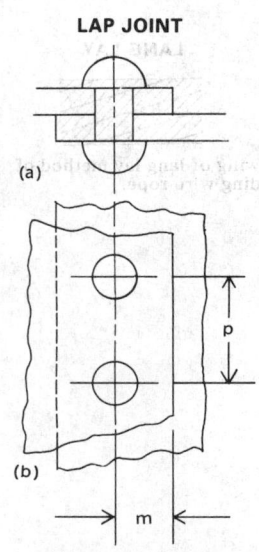

LAP JOINT

An example of a lap joint involving plates and rivets. *(a)* Side view, *(b)* top view.

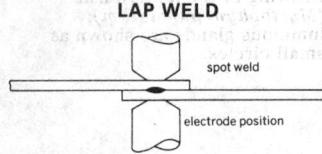

LAP WELD

Drawing of a lap weld showing how the edge of one plate is welded to the surface of the other plate.

and characterized by flat, depressed cankers on the twigs and branches. { 'lärch ¦kaŋ·kər }

lard [FOOD ENG] A solid fat prepared by rendering the fatty tissue from hogs. { lärd }

larderillite [MINERAL] $(NH_4)B_5O_8·2H_2O$ A white mineral composed of hydrous ammonium borate, occurring as a crystalline powder. { 'lär·də're¸līt }

lard ice See grease ice. { 'lärd ¸īs }

lardite See agalmatolite. { 'lär¸dīt }

lard oil [MATER] Yellowish to colorless oil with characteristic aroma and bland taste; melts at $-2°C$; soluble in carbon disulfide, ether, benzene, and chloroform; main components are olein and glycerides of solid fatty acids; used as a lubricant, wool oil, and illuminant, and in soap manufacture. { lärd ¸oil }

large calorie See kilocalorie. { 'lärj 'kal·ə·rē }

large dyne See newton. { 'lärj 'dīn }

large-eddy simulation [FL MECH] A technique for the prediction of complex turbulent flows in which the contribution of the large, energy-containing scales of motion is computed directly, and only the effect of the smallest scales of turbulence is modeled. { ¦lärj ¦ed·ē ¸sim·yə'lā·shən }

large intestine See colon. { 'lärj in'tes·tən }

large ion [METEOROL] An ion created by a small ion attaching to an Aitken nucleus; it is characterized by relatively large mass and low mobility. Also known as heavy ion; Langevin ion; slow ion. { 'lärj 'ī¸än }

Large Magellanic Cloud [ASTRON] An irregular cloud of stars in the constellation Doradus; it is 160,000 light-years away and nearly 30,000 light-years in diameter. Abbreviated LMC. Also known as Nubecula Major. { 'lärj ¦maj·ə¦lan·ik 'klaůd }

large nuclei [OCEANOGR] Particles of concentrated seawater or crystalline salt in the marine atmosphere having radii larger than 10^{-5} centimeter. { 'lärj 'nü·klē¸ī }

large number hypothesis [PHYS] The hypothesis that there is a physical basis for the approximate equality of two numbers on the order of 10^{40}: the ratio of the electrostatic to the gravitational force between a proton and an electron in a hydrogen atom, and the ratio of the age of the universe to the time required for light to cross an elementary particle diameter. { 'lärj ¸nəm·bər hī'päth·ə·səs }

large polaron [SOLID STATE] An electron in a crystal lattice together with the surrounding lattice deformation, for the case in which the deformation extends over many lattice sites so that the lattice can be treated as a continuum. { 'lärj 'pō·lə¸rän }

large scale [MAP] A scale of sufficient size to permit the plotting of much detail with exactness. [METEOROL] A scale such that the curvature of the earth may not be considered negligible; this scale is applicable to the high tropospheric long-wave patterns, with four or five waves around the hemisphere in the middle latitudes. { 'lärj 'skāl }

large-scale convection [METEOROL] Organized vertical motion on a larger scale than atmospheric free convection associated with cumulus clouds; the patterns of vertical motion in hurricanes or in migratory cyclones are examples of such convection. { 'lärj ¦skāl kən'vek·shən }

large-scale integrated circuit [ELECTR] A very complex integrated circuit, which contains well over 100 interconnected individual devices, such as basic logic gates and transistors, placed on a single semiconductor chip. Abbreviated LSI circuit. Also known as chip circuit; multiple-function chip. { 'lärj ¦skāl ¸int·ə¸grād·əd 'sər·kət }

large-scale integrated memory See semiconductor memory. { 'lärj ¦skāl ¸int·ə¸grād·əd 'mem·rē }

large-systems control theory [CONT SYS] A branch of the theory of control systems concerned with the special problems that arise in the design of control algorithms (that is, control policies and strategies) for complex systems. { 'lärj ¸sis·təmz kən'trōl ¸thē·ə·rē }

Largidae [INV ZOO] A family of hemipteran insects in the superfamily Pentatomorpha. { 'lär·jə¸dē }

Laridae [VERT ZOO] A family of birds in the order Charadriiformes composed of the gulls and terns. { 'lar·ə¸dē }

Larinae [VERT ZOO] A subfamily of birds in the family Laridae containing the gulls and characterized by a thick, slightly hooked beak, a square tail, and a stout white body, with shades of gray on the back and the upper wing surface. { 'lar·ə¸nē }

Larissa [ASTRON] A satellite of Neptune orbiting at a mean distance of 45,700 miles (73,600 kilometers) with a period of 13.3 hours, and with a diameter of about 120 miles (190 kilometers). { lə'ris·ə }

larixinic acid See maltol. { ¦lar·ik¦sin·ik 'as·əd }

Larmor formula [ELECTROMAG] The rate at which energy is radiated by a nonrelativistic, accelerated charge is $2q^2a^2/3c^3$, where q is the particle's charge in esu (electrostatic units), a is its acceleration, and c is the speed of light. { 'lär·mór ¸fór·myə·lə }

Larmor frequency [ELECTROMAG] The angular frequency of the Larmor precession, equal in esu (electrostatic units) to the negative of a particle's charge times the magnetic induction divided by the product of twice the particle's mass and the speed of light. { 'lär·mór 'frē·kwən·sē }

Larmor orbit [ELECTROMAG] The motion of a charged particle in a uniform magnetic field, which is a superposition of uniform circular motion in a plane perpendicular to the field, and uniform motion parallel to the field. { 'lär·mór ¸ór·bət }

Larmor precession [ELECTROMAG] A common rotation superposed upon the motion of a system of charged particles, all having the same ration of charge to mass, by a magnetic field. { 'lär·mór prē¸sesh·ən }

Larmor radius [ELECTROMAG] For a charged particle moving transversely in a uniform magnetic field, the radius of curvature of the projection of its path on a plane perpendicular to the field. Also known as gyromagnetic radius. { 'lär·mór ¸rād·ē·əs }

Larmor's theorem [ELECTROMAG] The theorem that for a system of charged particles, all having the same ratio of charge to mass, moving in a central field of force, the motion in a uniform magnetic induction B is, to first order in B, the same as a possible motion in the absence of B except for the superposition of a common precession of angular frequency equal to the Larmor frequency. { 'lär·mórz ¸thir·əm }

larnite [MINERAL] $β-Ca_2SiO_4$ A gray mineral that is a metastable monoclinic phase of calcium orthosilicate, stable from 520 to 670°C. Also known as belite. { 'lär¸nīt }

LARR See linear accelerator-regenerator reactor.

larry [MIN ENG] **1.** A car with a hopper bottom and adjustable chutes for feeding coke ovens. Also known as lorry. **2.** See barney. { 'lar·ē }

larsenite [MINERAL] $PbZnSiO_4$ A colorless or white mineral composed of lead zinc silicate, occurring in orthorhombic crystals. { 'lars·ən¸īt }

Larsen's pile [MIN ENG] A collection of hollow cylinders that increases resistance against bending and crumpling; useful for sinking a shaft in sand or gravel. { 'lärs·ənz ¦pīl }

Larsen's spiles [MIN ENG] Steel sheet made in various forms to resist bending; used in place of wooden spiles in timbering a weak roof. { 'lärs·ənz ¦spīlz }

Larson-Miller parameter [MECH] The effects of time and temperature on creep, being defined empirically as $P = T(C + \log t) × 10^{-3}$, where $T =$ test temperature in degrees Rankine (degrees Fahrenheit + 460) and $t =$ test time in hours; the constant C depends upon the material but is frequently taken to be 20. { 'lärs·ən 'mil·ər pə'ram·əd·ər }

larva [INV ZOO] An independent, immature, often vermiform stage that develops from the fertilized egg and must usually undergo a series of form and size changes before assuming characteristic features of the parent. { 'lär·və }

Larvacea [INV ZOO] A class of the subphylum Tunicata consisting of minute planktonic animals in which the tail, with dorsal nerve cord and notochord, persists throughout life. { lär'vā·shē·ə }

Larvaevoridae [INV ZOO] The tachina flies, a large family of dipteran insects in the suborder Cyclorrhapha distinguished by a thick covering of bristles on the body; most are parasites of arthropods. { ¸lär·və'vór·ə¸dē }

larva migrans [INV ZOO] Fly larva, *Hypoderma* or *Gastrophilus*, that produces a creeping eruption in the dermis. [MED] Infestation of the dermis by varicus burrowing nematode larvae, producing a creeping eruption that may become contaminated with bacteria. { 'lär·və 'm¯¸granz }

larvicide [MATER] A pesticide used to kill larvae. { 'lär·və¸sīd }

larvikite [PETR] An alkali syenite consisting of cryptoperthite or anorthoclase in rhombic crystals; used as an ornamental building material. { 'lär·vi¸kīt }

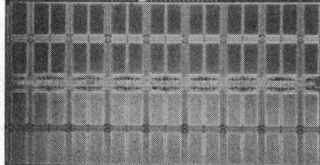

LARGE-SCALE INTEGRATED CIRCUIT

Photomicrograph of an MOS large-scale integrated circuit used as 4,194,304-bit random-access memory with an area of 0.12 in.^{2} (78 mm^2). (*IBM Corp.*)

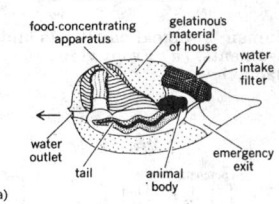

LARVACEA

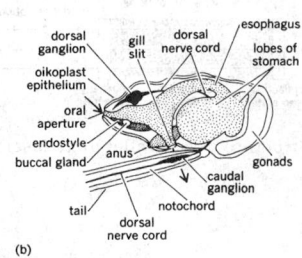

Oikopleura dioica, a larvacean. (*a*) In its house, a gelatinous tunic with strainers to filter food. (*b*) Body and anterior tail region. Arrows indicate the direction of water flow.

larviporous [INV ZOO] Feeding on larva, referring especially to insects. { lär′vip·ə·rəs }

laryngeal pouch [VERT ZOO] A lateral saclike expansion of the cavity of the larynx that is greatly developed in certain monkeys. { lə′rin·jē·əl ′pau̇ch }

laryngectomy [MED] Surgical removal of all or part of the larynx. { ‚lar·ən′jek·tə·mē }

laryngismus stridulus [MED] **1.** Spasmodic croup. **2.** The laryngeal spasm sometimes seen in hypocalcemic states. { ‚lar·ən′jiz·məs ′strij·ə·ləs }

laryngitis [MED] Inflammation of the larynx. { ‚lar·ən′jīd·əs }

laryngology [MED] The science of anatomy, physiology, and diseases of the larynx. { ‚lar·ən′gäl·ə·jē }

laryngopharynx [ANAT] The lower portion of the pharynx, lying adjacent to the larynx. Also known as hypopharynx. { lə′riŋ·gō′far·iŋks }

laryngophone [ENG ACOUS] A microphone designed to be placed against the throat of a speaker, to pick up voice vibrations directly without responding to background noise. { lə′riŋ·gə‚fōn }

laryngoscope [MED] A tubular instrument, combining a light system and a telescopic system, used in the visualization of the interior larynx and adaptable for diagnostic, therapeutic, and surgical procedures. { lə′riŋ·gə‚skōp }

laryngospasm [MED] Sudden and uncontrollable closure of the larynx; often seen in anaphylactic reactions. { lə′riŋ·gə‚spaz·əm }

laryngotracheal groove [EMBRYO] A channel in the floor of the pharynx serving as the anlage of the respiratory system. { lə′riŋ·gō′trā·kē·əl ‚grüv }

laryngotracheal diphtheria [MED] A type of diphtheria that causes hoarseness, croupy cough, difficulty in breathing, and pallor. { lə‚riŋ·gō‚trā·kē·əl dif′thir·ē·ə }

laryngotracheitis [MED] Inflammation of the larynx and trachea. { lə′riŋ·gō‚trā·kē′īd·əs }

laryngotracheobronchitis [MED] Acute inflammation of the mucosa of the larynx, trachea, and bronchi. { lə′riŋ·gō′trā·kē·ō‚braŋ′kīd· əs }

laryngotracheobronchitis virus See croup-associated virus. { lə′riŋ·gō′trā·kē·ō‚braŋ′kīd·əs ′vī·rəs }

larynx [ANAT] The complex of cartilages and related structures at the opening of the trachea into the pharynx in vertebrates; functions in protecting the entrance of the trachea, and in phonation in higher forms. { ′lar‚iŋks }

Lasater's bubble-point pressure correction [PETRO ENG] Relation of the gas-oil ratio in a high-pressure oil reservoir to the bubble-point-pressure factor. { ′las·əd·ərz ′bəb·əl‚point ′presh·ər kə‚rek·shən }

LASCR See light-activated silicon controlled rectifier.

LASCS See light-activated silicon controlled switch.

laser [OPTICS] An active electron device that converts input power into a very narrow, intense beam of coherent visible or infrared light; the input power excites the atoms of an optical resonator to a higher energy level, and the resonator forces the excited atoms to radiate in phase. Derived from light amplification by stimulated emission of radiation. { ′lā·zər }

laser altimeter [NAV] An altimeter in which a laser beam, modulated by radio frequencies, is directed downward from an aircraft, and the laser light reflected from the terrain is picked up by a telescope system, sensed by a photomultiplier, and phase-compared with the transmitted signal to obtain the round-trip propagation time; the height above ground equals one-half the product of the propagation time and the speed of light. { ′lā·zər al′tim·əd·ər }

laser amplifier [ELECTR] A laser which is used to increase the output of another laser. Also known as light amplifier. { ′lā·zər ′am·plə‚fī·ər }

laser anemometer [ENG] An anemometer in which the wind being measured passes through two perpendicular laser beams, and the resulting change in velocity of one or both beams is measured. { ′lā·zər an·ə′mäm·əd·ər }

laser beam [OPTICS] A narrow beam of coherent, powerful, and nearly monochromatic electromagnetic radiation emitted by a laser. { ′lā·zər ‚bēm }

laser-beam cutting See laser cutting. { ′la·zər ‚bēm ‚kəd·iŋ }

laser-beam printer [GRAPHICS] A nonimpact printer that operates at well over 10,000 lines per minute, using a low-power laser to produce image-forming charges a line at a time

on the photoconductive surface of a drum; dry powder that adheres only to charged areas is applied to the drum, transferred to plain paper, and fused by heat. { ′lā·zər ‚bēm ‚print·ər }

laser camera [OPTICS] An airborne camera system for night photography in which a laser beam is split into two beams; one beam, which is almost invisible, scans the ground, while the second beam is modulated by a detector of light reflected from the ground area being scanned, and is in turn swept back and forth over a moving film by the same scanner. { ′lā·zər ‚kam·rə }

laser ceilometer [ENG] A ceilometer in which the time taken by a light pulse from a ground laser to travel straight up to a cloud ceiling and be reflected to a receiving photomultiplier is measured and converted into a cathode-ray display that indicates cloud-base height. { ′lā·zər sē′läm·əd·ər }

laser communication [COMMUN] Optical communication in which the light source is a laser whose beam is modulated for voice, video, or data communication over wide information bandwidths, typically 1 gigahertz or more. { ′lā·zər kə‚myü·nə′kā·shən }

laser cooling [ATOM PHYS] A method of slowing atoms in an atomic beam to very low velocities, by directing a beam of properly tuned laser light opposite to the atomic beam, and compensating for the Doppler shift of the slowing atoms by varying the laser frequency or Zeeman-shifting the atomic levels with a varying magnetic field. { ′lā·zər ‚kül·iŋ }

laser cutting [MET] A process by which a laser beam impinges on the workpiece in order to heat and sever the piece. Also known as laser-beam cutting. { ′lā·zər ‚kəd·iŋ }

laser deposition [MET] A vapor deposition process in which films are grown on substrates from a material that is evaporated by laser radiation. { ′lā·zər dep·ə′zish·ən }

laser detecting and ranging See ladar. { ′lā·zər di′tek·tiŋ ən ′rān·jiŋ }

laser diode See semiconductor laser. { ′lā·zər ′dī‚ōd }

laser disk storage See optical disk storage. { ′lā·zər ′disk ‚stȯr·ij }

laser Doppler velocimeter [OPTICS] A type of laser velocimeter used for determining the velocity of a fluid flow from the Doppler shift in the frequency of laser light scattered from particles in the fluid. { ′lā·zər ′däp·lər ‚vel·ə′sim·əd·ər }

laser drill [OPTICS] A drill in which concentrated light from a ruby laser generates intense heat for drilling holes as small as 0.0001 inch (2.5 micrometers) in diameter in tungsten, gemstones, and other hard materials. { ′lā·zər ‚dril }

laser earthquake alarm [ENG] An early-warning system proposed for earthquakes, involving the use of two lasers with beams at right angles, positioned across a known geologic fault for continuous monitoring of distance across the fault. { ′lā·zər ′ərth‚kwāk ə‚lärm }

laser extensometer [OPTICS] A device which uses interference of laser beams to measure small changes in distance; it can operate between points as much as 0.6 mile (1 kilometer) apart, and has been used to measure effects produced by earth tides. { ′lā·zər ‚ek‚sten′säm·əd·ər }

laser/fiber-optic gyroscope See fiber-optic gyroscope. { ′lā·zər ′fī·bər ‚äp·tik ′jī·rə‚skōp }

laser flash tube [ELECTR] A high-power, air-cooled or water-cooled xenon flash tube designed to produce high-intensity flashes for pumping applications. { ′lā·zər ′flash ‚tüb }

laser fusion [NUCLEO] The use of an intense beam of laser light to heat a small pellet of deuterium and tritium to a temperature of about 100,000,000°C, as required for initiating a fusion reaction. Also known as laser-induced fusion. { ′lā·zər ′fyü·zhən }

laser glazing [MET] A surface alloying process in which a continuous high-energy carbon dioxide laser traverses the surface of a metal part, creating a thin layer of melt. { ′lā·zər ′glāz·iŋ }

laser guidance [NAV] Guidance in which the target is continuously illuminated by a laser beam from an aircraft or other location so that missiles, bombs, or projectiles equipped with suitable seeker heads can home in on the laser energy reflected by the target. { ′lā·zər ′gīd·əns }

laser gyro [ENG] A gyro in which two laser beams travel in opposite directions over a ring-shaped path formed by three or more mirrors; rotation is thus measured without the use of a spinning mass. Also known as ring laser. { ′lā·zər ′jī‚rō }

laser heterodyne spectroscopy [SPECT] A high-resolution

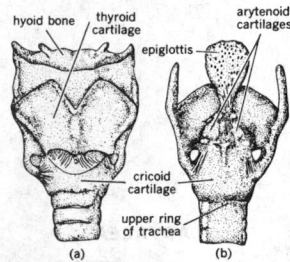

LARYNX

hyoid bone, thyroid cartilage, epiglottis, arytenoid cartilages, cricoid cartilage, upper ring of trachea

(a)   (b)

Human laryngeal cartilages and ligaments. (a) Front view. (b) Back view.

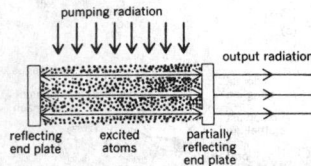

LASER

pumping radiation

output radiation

reflecting end plate, excited atoms, partially reflecting end plate

Structure of a parallel-plate laser.

spectroscopic technique, used in astronomical and atmospheric observations, in which the signal to be measured is mixed with a laser signal in a solid-state diode, producing a difference-frequency signal in the radio-frequency range. { 'lā·zər ¦hed·ə·rə‚dīn spek'träs·kə·pē }

laser-holography storage [COMPUT SCI] A computer storage technology in which information is stored in microscopic spots burned in a holographic substrate by a laser beam, and is read by sensing a lower-energy laser beam that is transmitted through these spots. { 'lā·zər hō¦läg·rə·fē ‚stȯr·ij }

laser-induced fluorescence imaging [FL MECH] A flow visualization method in which fluorescent tracers are excited by a laser beam. { ¦lā·zər in‚düst flu̇r¦es·ənt 'im·ij·iŋ }

laser-induced fusion *See* laser fusion. { 'lā·zər in‚düst 'fyü·zhən }

laser-induced nuclear polarization [NUC PHYS] A technique for making the spin vectors of an ensemble of nuclei point preferentially in one direction by means of an optical pumping process using either circularly or linearly polarized laser light. Abbreviated LINUP. { 'lā·zər in‚düst 'nü·klē·ər ‚pō·lə·rə'zā·shən }

laser infrared radar *See* lidar. { 'lā·zər ¦in·frə¦red 'rā‚där }

laser interferometer [OPTICS] An interferometer which uses a laser as a light source; because of the monochromaticity and high intrinsic brilliance of laser light, it can operate with path differences in the interfering beams of hundreds of meters, in contrast to a maximum of about 20 centimeters (8 inches) for classical interferometers. { 'lā·zər ‚in·tər·fə'räm·əd·ər }

laser intrusion detector [ENG] A photoelectric intrusion detector in which a laser is a light source that produces an extremely narrow and essentially invisible beam around the perimeter of the area being guarded. { 'lā·zər in'trü·zhən di‚tek·tər }

laser jamming [ORD] An electronic countermeasure in which a continuous-wave laser directs jamming energy back to a hostile laser receiver to prevent it from interfering with the use of laser rangefinders, radars, and tracking equipment during an attack. { 'lā·zər ¦jam·iŋ }

laser memory [COMPUT SCI] A computer memory in which a controlled laser beam acts on individual and extremely small areas of a photosensitive or other type of surface, for storage and subsequent readout of digital data or other types of information. { 'lā·zər ‚mem·rē }

laser photobiology [BIOPHYS] The interaction of laser light with biological molecules, and the applications to biology and medicine. { ¦lā·zər ‚fōd·ō·bī'äl·ə·jē }

laser photocoagulator [MED] A laser combined with an ophthalmoscope for directing bursts of coherent light through a human eye to burn selected points on a detached retina; subsequent healing of the burns causes scars that weld the retina back into position. { 'lā·zər ‚fōd·ō·kə'wag·yə‚lād·ər }

laser printer [GRAPHICS] An extremely fast printer that uses a laser to form areas of static electric charge which attract a metallic powder to the paper. { 'lā·zər ‚print·ər }

laser radar *See* ladar. { 'lā·zər ‚rā‚där }

laser radiation detector [ELECTR] A photodetector that responds primarily to the coherent visible, infrared, or ultraviolet light of a laser beam. { 'lā·zər ‚rād·ē'ā·shən di‚tek·tər }

laser rangefinder [OPTICS] A portable rangefinder using a battery-powered ruby laser in combination with an optical telescope to aim a laser beam and a photomultiplier for picking up the laser beam reflected from the target. { 'lā·zər 'rānj‚find·ər }

laser ranging [ENG] A technique for determining the distance to a target by precise measurement of the time required for a laser pulse to travel from a transmitter to a reflector on the target and return to a detector. { 'lā·zər ‚rānj·iŋ }

laser recorder [COMMUN] An image reproducer that resembles a facsimile system, in which a laser beam is initially modulated by the video signal and swept over photographic film or paper to reproduce an image received over wire or radio communication systems. { 'lā·zər ri'kȯrd·ər }

laserscope [ENG] A pulsed high-power laser used with appropriate scanning and imaging devices to sense objects over the sea at night or in fog and provide three-dimensional images on a viewing screen. { 'lā·zər‚skōp }

laser scriber [ENG] A laser-cutting setup used in place of a diamond scriber for dicing thin slabs of silicon, gallium

arsenide, and other semiconductor materials used in the production of semiconductor diodes, transistors, and integrated circuits; also used for scribing sapphire and ceramic substrates. { 'lā·zər ¦skrī·ər }

laser seismometer [ENG] A laser interferometer system that detects seismic strains in the earth by measuring changes in distance between two granite piers located at opposite ends of an evacuated pipe through which a helium-neon or other laser beam makes a round trip; movements as small as 80 nanometers (one-eighth the wavelength of the 632.8-nanometer helium-neon laser radiation) can be detected. { 'lā·zər sīz'mäm·əd·ər }

laser-solid interaction [SOLID STATE] Interaction of laser light with a solid, especially the thermal effects of absorption of a high-intensity laser beam. { 'lā·zər 'säl·əd ‚in·tər'ak·shən }

laser spectroscopy [SPECT] A branch of spectroscopy in which a laser is used as an intense, monochromatic light source; in particular, it includes saturation spectroscopy, as well as the application of laser sources to Raman spectroscopy and other techniques. { 'lā·zər spek'träs·kə‚pē }

laser spectrum [PHYS] The spectrum that includes all optical wavelengths, ranging from infrared through visible light to ultraviolet, in which coherent radiation can be produced by various types of lasers. { 'lā·zər ‚spek·trəm }

laser threshold [ELECTR] The minimum pumping energy required to initiate lasing action in a laser. { 'lā·zər ¦thresh‚hōld }

laser tracking [ENG] Determination of the range and direction of a target by echoed coherent light. { 'lā·zər 'trak·iŋ }

laser transit [ENG] A transit in which a laser is mounted over the sighting telescope to project a clearly visible narrow beam onto a small target at the survey site. { 'lā·zər ¦tranz·ət }

laser trap [OPTICS] A device for confining atoms, molecules, and larger neutral particles up to 10 micrometers in diameter, consisting of a focused laser beam tuned to a frequency below an atomic resonance, which attracts the particles toward regions of high laser intensity. { 'lā·zər ‚trap }

laser-triggered switch [ELEC] A high-voltage high-power switch that consists of a spark gap triggered into conduction by a laser beam. { 'lā·zər ‚trig·ərd 'swich }

laser tweezers [OPTICS] A laser trap used to hold microscopic organisms and their organelles and move them through the objective of an optical microscope without apparent damage. Also known as optical tweezers. { 'lā·zər ‚twē·zərz }

laser velocimeter [OPTICS] Any velocity measuring instrument that makes use of a laser. { 'lā·zər ‚vel·ə'sim·əd·ər }

laser welding [MET] Micro-spot welding with a laser beam. { 'lā·zər ¦wel·diŋ }

lashing [ENG] A rope, chain, or wire used for binding, fastening, or wrapping. [MIN ENG] Planks nailed inside of frames or sets in a shaft to keep them in place. Also known as listing. { 'lash·iŋ }

lashing chain [MIN ENG] A short chain to attach tubs to an overrope in endless rope haulage by wrapping the chain around the rope. { 'lash·iŋ ‚chān }

lash-up [ENG] A model or test sample of equipment required in the testing of a new concept or idea which is in the embryo stage. { 'lash‚əp }

lasing [OPTICS] Generation of visible or infrared light waves having very nearly a single frequency by pumping or exciting electrons into high-energy states in a laser. { 'lāz·iŋ }

Lasiocampidae [INV ZOO] The tent caterpillars and lappet moths, a family of cosmopolitan (except New Zealand) lepidopteran insects in the suborder Heteroneura. { ‚las·ē·ō'kam·pə‚dē }

Laspeyre's index [STAT] A weighted aggregate price index with base-year quantity weights. Also known as base-year method. { lä'perz ‚in‚deks }

Lassa fever [MED] An acute, highly communicable exotic infection that is endemic in western Africa. Caused by an arenavirus (the Lassa virus), it is characterized by high fever, weakness, headaches, mouth ulcers, hemorrhages under the skin, heart and kidney failure, and a high mortality rate. { 'läs·ə ‚fē·vər }

lasso cell *See* adhesive cell. { 'las·ō ‚sel }

last in, first out [IND ENG] A method of determining the inventory costs by transferring the costs of material to the product in reverse chronological order. Abbreviated LIFO. { ‚last 'in ‚first 'au̇t }

last-mask read-only memory [COMPUT SCI] A read-only memory in which the final mask used in the fabrication process determines the connections to the internal transistors, and these connections in turn determine the data pattern that will be read out when the cell is accessed. Also known as contact-mask read-only memory. { 'last ¦mask ¦rēd ¦ōn·lē 'mem·rē }

last quarter [ASTRON] The phase of the moon at western quadrature, half of the illuminated hemisphere being visible from the earth; has the characteristic half-moon shape. { 'last 'kwȯrd·ər }

latch [ELECTR] An electronic circuit that reverses and maintains its state each time that power is applied. [ENG] **1.** Any of various closing devices on a door that fit into a hook, notch, or cavity in the frame. **2.** In plastics fabrication, a device used to hold together the two members of a mold. [MIN ENG] To make an underground survey with a dial and chain, or to mark out upon the surface, with the same instruments, the position of the workings underneath. { lach }

latch bolt [DES ENG] A self-acting spring bolt with a beveled head. { 'lach ‚bōlt }

latch-in relay [ELEC] A relay that maintains its contacts in the last position assumed, even without coil energization. { 'lach‚in 'rē‚lā }

latch needle [TEXT] A fine steel needle for machine knitting which has at its ends a butt and a short hook with a latch. { 'lach ‚nēd·əl }

latch-up phenomenon [ELECTR] In a bipolar or MOS integrated circuit, the generation of photocurrents by ionizing radiation which can provide a trigger signal for a parasitic *pnpn* circuit and possibly result in permanent damage or operational failure if the circuit remains in this state. { 'lach ‚əp fə‚näm·ə‚nän }

late binding [COMPUT SCI] The assignment of data types (such as integer or string) to variables at the time of execution of a computer program, rather than during the compilation phase. { ¦lāt 'bīnd·iŋ }

late blight [PL PATH] A fungus blight disease in which symptoms do not appear until late in the growing season and vary for different species. { 'lāt ¦blīt }

late gene [GEN] Any gene expressed after the onset of deoxyribonucleic acid replication in viruses or eukaryotes. { 'lāt ¦jēn }

latency [COMPUT SCI] The waiting time between the order to read/write some information from/to a specified place and the beginning of the data-read/write operation. [MED] The stage of an infectious disease, other than the incubation period, in which there are neither clinical signs nor symptoms. [PHYSIO] The period between the introduction of and the response to a stimulus. [PSYCH] The phase between the Oedipal period and adolescence, characterized by an apparent cessation of psychosexual development. { 'lat·ən·sē }

latent bud [BOT] An axillary bud whose development is inhibited, sometimes for many years, due to the influence of apical and other buds. Also known as dormant bud. { 'lāt·ənt 'bəd }

latent defect [IND ENG] A flaw or other imperfection in any article which is discovered after delivery; usually, latent defects are inherent weaknesses which normally are not detected by examination or routine tests, but which are present at time of manufacture and are aggravated by use. { 'lāt·ənt 'dē‚fekt }

latent failure [NAV] An undetected degradation in the operation of a navigational aid. { ‚lāt·ənt 'fāl·yər }

latent heat [THERMO] The amount of heat absorbed or evolved by 1 mole, or a unit mass, of a substance during a change of state (such as fusion, sublimation or vaporization) at constant temperature and pressure. { 'lāt·ənt 'hēt }

latent heat of fusion *See* heat of fusion. { 'lāt·ənt ¦hēt əv 'fyü·zhən }

latent heat of sublimation *See* heat of sublimation. { 'lāt·ənt ¦hēt əv ‚səb·lə'mā·shən }

latent heat of vaporization *See* heat of vaporization. { 'lāt·ənt ¦hēt əv ‚vā·pə·rə'zā·shən }

latent Herschel effect *See* Herschel effect. { 'lāt·ənt 'hər·shəl i‚fekt }

latent homosexuality [PSYCH] Homosexual tendencies present in the unconscious but not felt or expressed overtly. { 'lāt·ənt ‚hō·mə‚sek·shə'wal·əd·ē }

latent image [GRAPHICS] An invisible image produced by the physical or chemical effects of light on the individual crystals (usually silver halide) of photographic emulsions; the development process makes the image visible, in the negative. { 'lāt·ənt 'im·ij }

latent instability [METEOROL] The state of that portion of a conditionally unstable air column lying above the level of free convection; latent instability is released only if an initial impulse on a parcel gives it sufficient kinetic energy to carry it through the layer below the level of free convection, within which the environment is warmer than the parcel. { 'lāt·ənt ‚in·stə'bil·ad·ē }

latent load [MECH ENG] Cooling required to remove unwanted moisture from an air-conditioned space. { 'lāt·ənt ‚lōd }

latent period [MED] Any stage of an infectious disease in which there are no clinical signs of symptoms of the infection. [PHYSIO] The period between the introduction of a stimulus and the response to it. [VIROL] The initial period of phage growth after infection during which time virus nucleic acid is manufactured by the host cell. { 'lāt·ənt ¦pir·ē·əd }

latent root *See* eigenvalue. { 'lāt·ənt 'rüt }

latent virus [VIROL] A virus that remains dormant within body cells but can be reactivated by conditions such as reduced host defenses, toxins, or irradiation, to cause disease. { ¦lāt·ənt 'vī·rəs }

latent-virus infection [MED] A chronic, inapparent virus infection in which a virus-host equilibrium is established. [VIROL] A persistent viral infection in which there is little or no demonstrable presence of the virus and disease symptoms for a long time between episodes of recurrent outbreaks. { 'lāt·ənt ¦vī·rəs in‚fek·shən }

laterad [ANAT] Toward the lateral aspect. { 'lad·ə‚rad }

lateral [ANAT] At, pertaining to, or in the direction of the side; on either side of the medial vertical plane. [ENG] In a gas distribution or transmission system, a pipe branching away from the central, primary part of the system. [MIN ENG] **1.** In horizon mining, a hard heading branching off a horizon along the strike of the seams. **2.** A horizontal mine working. { 'lad·ə·rəl }

lateral aberration [OPTICS] **1.** The distance from the axis of an optical system at which a ray intersects a plane perpendicular to the axis through the focus of paraxial rays. **2.** The difference between the reciprocals of the image distances for paraxial and rim rays. **3.** For chromatic aberration, the difference in sizes of the images of an object for two different colors. { 'lad·ə·rəl ‚ab·ə'rā·shən }

lateral acceleration [AERO ENG] The component of the linear acceleration of an aircraft or missile along its lateral, or Y, axis. { 'lad·ə·rəl ak‚sel·ə'rā·shən }

lateral accretion [GEOL] The digging away of material at the outer bank of a meandering stream and the simultaneous building up to the water level by deposition of material brought there by pushing and rolling along the stream bottom. { 'lad·ə·rəl ə'krē·shən }

lateral area [MATH] The area of a surface with the bases (if any) excluded. { 'lad·ə·rəl 'er·ē·ə }

lateral axis [NAV ARCH] The athwartship line approximately through the center of gravity of a craft, around which it pitches. { 'lad·ə·rəl 'ak·səs }

lateral bud [BOT] Any bud that develops on the side of a stem. { 'lad·ə·rəl 'bəd }

lateral chromatic aberration *See* chromatic difference of magnification. { 'lad·ə·rəl krō¦mad·ik ‚ab·ə'rā·shən }

lateral compliance [ENG ACOUS] That characteristic of a stylus based on the force required to move it from side to side as it follows the grooves of a phonograph record. { 'lad·ə·rəl kəm'plī·əns }

lateral cone *See* adventive cone. { 'lad·ə·rəl kōn }

lateral controller [AERO ENG] A primary flight control mechanism, generally a part of the longitudinal controller, which controls the ailerons; often resembles an automobile steering wheel but may be a control column. { 'lad·ə·rəl kən'trōl·ər }

lateral deflection angle [ORD] The horizontal angle representing the difference between the azimuth of the target at the instant of firing and the azimuth at which the gun must be pointed in order to hit the target; it is the algebraic sum of the principal lateral deflection angle and the lateral pointing correction. { 'lad·ə·rəl di'flek·shən ‚aŋ·gəl }

lateral deflection setting [ORD] The setting on the lateral deflection scale of the sighting mechanism of the gun, corresponding to the lateral deflection angle. { 'lad·ə·rəl di'flek·shən ,sed·iŋ }

lateral deviation [ORD] Horizontal distance between the point of impact or burst and the gun-target line. { 'lad·ə·rəl ,dē·vē'ā·shən }

lateral displacement [NAV] The number of nautical miles an aircraft has been displaced, perpendicular to its effective air path, determined by a pressure pattern formula. Also known as lateral drift; pressure pattern displacement. { 'lad·ə·rəl di'splās·mənt }

lateral drift *See* lateral displacement. { 'lad·ə·rəl 'drift }

lateral erosion [GEOL] The action of a stream in undermining a bank on one side of its channel so that material falls into the stream and disintegrates; simultaneously, the stream shifts toward the bank that is being undercut. { 'lad·ə·rəl i'rō·zhən }

lateral extensometer [ENG] An instrument used in photoelastic studies of the stresses on a plate; it measures the change in the thickness of the plate resulting from the stress at various points. { 'lad·ə·rəl ,ek,sten'säm·əd·ər }

lateral face [MATH] The lateral face for a prism or pyramid is any edge or face which is not part of a base. { 'lad·ə·rəl 'fās }

lateral fault [GEOL] A fault along which there has been strike separation. Also known as strike-separation fault. { 'lad·ə·rəl 'fȯlt }

lateral flow spillway *See* side-channel spillway. { 'lad·ə·rəl ,flō 'spil,wā }

lateral hermaphroditism [MED] A form of human hermaphroditism in which there is an ovary on one side and a testis on the other. { 'lad·ə·rəl hər'maf·rə,dīd,iz·əm }

lateralia [INV ZOO] Paired sensory bristles on the lateral aspect of the head of gnathostomulids. { ,lad·ə'rāl·yə }

lateral inversion [OPTICS] The effect produced by a mirror in reversing images from left to right. Also known as perversion. { 'lad·ə·rəl in'vər·zhən }

lateral jump [ORD] The horizontal angle between the plane of fire and the plane of departure. { 'lad·ə·rəl 'jəmp }

lateral lemniscus [ANAT] The secondary auditory pathway arising in the cochlear nuclei and terminating in the inferior colliculus and medial geniculate body. { 'lad·ə·rəl lem'nis·kəs }

lateral line [INV ZOO] A longitudinal lateral line along the sides of certain oligochaetes consisting of cell bodies of the layer of circular muscle. [VERT ZOO] A line along the sides of the body of most fishes, often distinguished by differently colored scales, which marks the lateral line organ. { 'lad·ə·rəl 'līn }

lateral-line organ [VERT ZOO] A small, pear-shaped sense organ in the skin of many fishes and amphibians that is sensitive to pressure changes in the surrounding water. { 'lad·ə·rəl ,līn ,ȯr·gən }

lateral-line system [VERT ZOO] The complex of lateral-line end organs and nerves in skin on the sides of many fishes and amphibians. { 'lad·ə·rəl ,līn ,sis·təm }

lateral magnification [OPTICS] The ratio of some linear dimension, perpendicular to the optical axis, of an image formed by an optical system, to the corresponding linear dimension of the object. Also known as magnification. { 'lad·ə·rəl ,mag·nə·fə'kā·shən }

lateral meristem [BOT] Strips or cylinders of dividing cells located parallel to the long axis of the organ in which they occur; the lateral meristem functions to increase the diameter of the organ. { 'lad·ə·rəl 'mer·ə,stem }

lateral mirage [OPTICS] A very rare type of mirage in which the apparent position of an object appears displaced to one side of its true position. { 'lad·ə·rəl mə'räzh }

lateral moraine [GEOL] Drift material, usually thin, that was deposited by a glacier in a valley after the glacier melted. { 'lad·ə·rəl mə'rān }

lateral observation [ORD] Observation of gunfire from a point considerably to the right or left of the line of fire. { 'lad·ə·rəl ,äb·zər'vā·shən }

lateral parity check [COMPUT SCI] The number of one bits counted across the width of the magnetic tape; this number plus a one or a zero must always be odd (or even), depending upon the manufacturer. { 'lad·ə·rəl 'par·əd·ē ,chek }

lateral planation [GEOL] Reduction in land in interstream areas in a plane parallel to the stream profile; the reduction is caused by lateral movement of the stream against its banks. { 'lad·ə·rəl plā'nā·shən }

lateral pointing correction [ORD] That part of the lateral deflection angle due to causes other than the travel of the target, such as wind, drift, and lateral adjustment correction. { 'lad·ə·rəl 'pȯint·iŋ kə,rek·shən }

lateral quadrupole [ACOUS] A sound source resulting from a variation of a component of the velocity of matter in a direction perpendicular to the velocity component. [ELECTROMAG] An electric or magnetic quadrupole which produces a field equivalent to that of two equal and opposite electric or magnetic dipoles separated by a small distance perpendicular to the direction of the dipoles. { 'lad·ə·rəl 'kwäd·rə,pōl }

lateral recording [ENG ACOUS] A type of disk recording in which the groove modulation is parallel to the surface of the recording medium so that the cutting stylus moves from side to side during recording. { 'lad·ə·rəl ri'kȯrd·iŋ }

lateral root [BOT] A root branch arising from the main axis. { 'lad·ə·rəl 'rüt }

lateral sclerosis *See* amyotrophic lateral sclerosis; primary lateral sclerosis. { 'lad·ə·rəl sklə'rō·səs }

lateral search *See* profiling. { 'lad·ə·rəl 'sərch }

lateral secretion [GEOL] A supposed phenomenon whereby a lode's or vein's mineral content is derived from the adjacent wall rock. { 'lad·ə·rəl si'krē·shən }

lateral separation [NAV] In air-traffic control, the separation between aircraft on parallel course lines. { 'lad·ə·rəl ,sep·ə'rā·shən }

lateral sewer [CIV ENG] A sewer discharging into a branch or other sewer and having no tributary sewer. { 'lad·ə·rəl 'sü·ər }

lateral shear interferometer [OPTICS] An interferometer in which a wavefront is interfered with a shifted version of itself, resulting in fringes along which the slope or derivative of the wavefront is constant. Also known as differential interferometer. { 'lad·ə·rəl ,shir ,in·ter·fə'räm·əd·ər }

lateral support [CIV ENG] Horizontal propping applied to a column, wall, or pier across its smallest dimension. { 'lad·ə·rəl sə'pȯrt }

lateral system [NAV] In marine operations, a system of buoyage in which the shape, color, and number distinction are assigned in accordance with the buoy's location in respect to navigable waters; when used to mark a channel, the buoys are assigned colors to indicate the side they mark and numbers to indicate their sequence along the channel; the lateral system is used in the United States. { 'lad·ə·rəl 'sis·təm }

lateral ventricle [ANAT] The cavity of a cerebral hemisphere; communicates with the third ventricle by way of the interventricular foramen. { 'lad·ə·rəl 'ven·trə·kəl }

laterite [GEOL] Weathered material composed principally of the oxides of iron, aluminum, titanium, and manganese; laterite ranges from soft, earthy, porous soil to hard, dense rock. { 'lad·ə,rīt }

lateritic soil [GEOL] **1.** Soil containing laterite. **2.** Any reddish soil developed from weathering. Also known as latosol. { ,lad·ə¦rid·ik 'sȯil }

laterization [GEOL] Those conditions of weathering that lead to removal of silica and alkalies, resulting in a soil or rock with high concentrations of iron and aluminum oxides (laterite). { ,lad·ə·rə'zā·shən }

laterolog [ENG] A downhole resistivity measurement method wherein electric current is forced to flow radially through the formation in a sheet of predetermined thickness; used to measure the resistivity in hard-rock reservoirs as a method of determining subterranean structural features. { 'lad·ər,läg }

late-type star [ASTROPHYS] A star with relatively low surface temperature, in spectral class K or M. { 'lāt ,tīp ,stär }

latewood [BOT] The portion of the annual ring that is formed after formation of earlywood has ceased. { 'lāt,wůd }

latex [MATER] **1.** Milky colloid in which natural or synthetic rubber or plastic is suspended in water. **2.** An elastomer product made from latex. { 'lā,teks }

latex cell [BIOL] A coenocytic cell of a lactiferous duct in a latex-producing plant. { 'lā,teks ,sel }

latex cement [MATER] A highly adhesive solvent solution of latex. { 'lā,teks si'ment }

latex paint [MATER] A paint consisting of a water suspension or emulsion of latex combined with pigments and additives

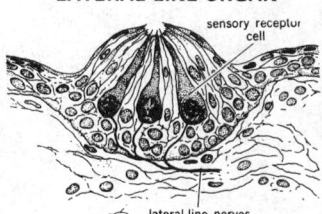

LATERAL-LINE ORGAN

sensory receptor cell

lateral-line nerves

Schematic drawing of a lateral-line sense organ in skin of adult salamander, the common aquatic vermilion spotted newt.

such as binders and suspending agents. Also known as latex water paint. { 'lā,teks 'pānt }

latex vessel [BOT] An elongated cell joined end to end with other like cells to form a type of lactiferous duct. { 'lā,teks ,ves·əl }

latex water paint See latex paint. { 'lā,teks 'wȯd·ər ,pānt }

lath [CIV ENG] **1.** A narrow strip of wood used in making a level base, as for plaster or tiles, or in constructing a light framework, as a trellis. **2.** A sheet of material used as a base for plaster. { lath }

lath brick [MATER] A long, narrow brick. { 'lath ,brik }

lath crib See lath frame. { 'lath ,krib }

lath door-set See lath frame. { 'lath 'dȯr,set }

lathe [MECH ENG] A machine for shaping a workpiece by gripping it in a holding device and rotating it under power against a suitable cutting tool for turning, boring, facing, or threading. { lāth }

lath frame [MIN ENG] A weak construction of laths, surrounding a main crib, the space between being for the insertion of piles. Also known as lath crib; lath door-set. { 'lath ,frām }

lathing board See backup strip. { 'lath·iŋ ,bȯrd }

Lathridiidae [INV ZOO] The minute brown scavenger beetles, a large cosmopolitan family of coleopteran insects in the superfamily Cucujoidea. { ,lath·rə'dī·ə,dē }

lathyrism [MED] Poisoning produced by ingestion of vetch (*Lathyrus*) and characterized by spastic paraplegia and decreased connective-tissue tensile strength. Also known as neurolathyrism. { 'lath·ə,riz·əm }

laticifer [BOT] A latex duct found in the mid-cortex of certain plants. { lā'tis·ə·fər }

latiferous [BOT] Containing or secreting latex. { lā'tif·ə·rəs }

Latimeridae [VERT ZOO] A family of deep-sea lobefin fishes (Coelacanthiformes) known from a single living species, *Latimeria chalumnae*. { ,lad·ə'mer·ə,dē }

Latin rectangle [MATH] An $r \times n$ matrix, with n equal to or greater than r in which each row is a permutation of the numbers 1, 2, . . . , n, and no number appears in a column more than once. { 'lat·ən 'rek,taŋ·gəl }

Latin square [MATH] An $n \times n$ square array of n different symbols, each symbol appearing once in each row and once in each column; these symbols prove useful in ordering the observations of an experiment. { 'lat·ən 'skwer }

latite [PETR] A not visibly crystalline rock of volcanic origin composed chiefly of sodic plagioclase and alkali feldspar with subordinate quantities of dark-colored minerals in a finely crystalline to glassy groundmass. { 'lā,tīt }

latitude [GEOD] Angular distance from a primary great circle or plane, as on the celestial sphere or the earth. { 'lad·ə,tüd }

latitude effect [GEOPHYS] The variation of a quantity with latitude; applied particularly to the increase in cosmic-ray intensity with increasing magnetic latitude. { 'lad·ə,tüd i,fect }

latitude factor [NAV] The change in latitude along a celestial line of position per 1-minute change in longitude. { 'lad·ə,tüd ,fak·tər }

latitude line [NAV] A line of position extending in an approximate east-west direction. { 'lad·ə,tüd ,līn }

latitude variation [GEOPHYS] A periodic change in the latitude of any position on the earth's surface, caused by the polar variation. { 'lad·ə,tüd ,ver·ē'ā·shən }

latosol See lateritic soil. { 'lad·ə,sȯl }

latrappite [MINERAL] $(Ca,Na)(Nb,Ti,Fe)O_3$ A variety of the mineral perovskite. { 'la·trə,pīt }

latrine [ENG] A toilet facility, either fixed or of a portable nature, such as is maintained underground for use by miners. { lə,trēn }

latrine cleaner [MIN ENG] A laborer who brings toilet cars in a mine to the surface on a cage and flushes the contents into a sewer. Also called sanitary nipper. { lə,trēn ,klēn·ər }

latten [MET] A thin metal sheet, particularly of brass or similar alloy, hot-rolled steel, or tin-covered iron, used for ornamental purposes. { 'lat·ən }

lattice [CIV ENG] A network of crisscrossed strips of metal or wood. [CRYSTAL] A regular periodic arrangement of points in three-dimensional space; it consists of all those points P for which the vector from a given fixed point to P has the form $n_1\mathbf{a} + n_2\mathbf{b} + n_3\mathbf{c}$, where n_1, n_2, and n_3 are integers, and $\mathbf{a}$, $\mathbf{b}$, and $\mathbf{c}$ are fixed, linearly independent vectors. Also

known as periodic lattice; space lattice. [MATH] A partially ordered set in which each pair of elements has both a greatest lower bound and least upper bound. [NAV] A pattern formed by two or more families of intersecting lines of position, such as the hyperbolic lines of position from two or more loran stations. [NUCLEO] An orderly array or pattern of nuclear fuel elements and moderator in a reactor or critical assembly. { 'lad·əs }

lattice Boltzmann method [STAT MECH] A numerical method of solving the Boltzmann transport equation, using a finite difference approximation on a discrete phase space in which both discrete particle velocities and discrete spatial locations are assumed. Abbreviated LBM. { ¦lad·əs 'bōlts,män ,meth·əd }

lattice constant [CRYSTAL] A parameter defining the unit cell of a crystal lattice, that is, the length of one of the edges of the cell or an angle between edges. Also known as lattice parameter. { 'lad·əs ,kän·stənt }

lattice defect See crystal defect. { 'lad·əs ,dē,fekt }

lattice drainage pattern See rectangular drainage pattern. { 'lad·əs 'drān·ij ,pad·ərn }

lattice dynamics [SOLID STATE] The study of the thermal vibrations of a crystal lattice. Also known as crystal dynamics. { 'lad·əs dī,nam·iks }

lattice energy [SOLID STATE] The energy required to separate ions in an ionic crystal an infinite distance from each other. { 'lad·əs ,en·ər·jē }

lattice field theory See lattice-gauge theory. { 'lad·əs ,fēld ,thē·ə·rē }

lattice filter [ELECTR] An electric filter consisting of a lattice network whose branches have L-C parallel-resonant circuits shunted by quartz crystals. { 'lad·əs ,fil·tər }

lattice-gauge theory [PARTIC PHYS] A formulation of the theory of hadron structure in which the continuum of spacetime is replaced by a discrete set of points or sites, and quarks move through this structure by sequential hops between neighboring sites, interacting with gauge gluons which are represented by fields located on bonds connecting the lattice sites. Also known as lattice field theory. { 'lad·əs ,gāj ,thē·ə·rē }

lattice girder [CIV ENG] An open girder, beam, or column built from members joined and braced by intersecting diagonal bars. Also known as open-web girder. { 'lad·əs ,gərd·ər }

lattice network [ELEC] A network that is composed of four branches connected in series to form a mesh; two nonadjacent junction points serve as input terminals, and the remaining two junction points serve as output terminals. { 'lad·əs 'net,wərk }

lattice parameter See lattice constant. { 'lad·əs pə,ram·əd·ər }

lattice polarization [SOLID STATE] Electric polarization of a solid due to displacement of ions from equilibrium positions in the lattice. { 'lad·əs pō·lə·rə'zā·shən }

lattice reactor [NUCLEO] A heterogeneous nuclear reactor in which both fuel and moderator are in the form of long rods. { 'lad·əs rē¦ak·tər }

lattice scattering [SOLID STATE] Scattering of electrons by collisions with vibrating atoms in a crystal lattice, reducing the mobility of charge carriers in the crystal and thereby affecting its conductivity. { 'lad·əs ,skad·ə·riŋ }

lattice truss [CIV ENG] A truss that resembles latticework because of diagonal placement of members connecting the upper and lower chords. { 'lad·əs ,trəs }

lattice vibration [SOLID STATE] A periodic oscillation of the atoms in a crystal lattice about their equilibrium positions. { 'lad·əs vī'brā·shən }

lattice wave [SOLID STATE] A disturbance propagated through a crystal lattice in which atoms oscillate about their equilibrium positions. { 'lad·əs ,wāv }

lattice winding [ELEC] A winding made of lattice coils and used for electric machines. { 'lad·əs ,wīn·diŋ }

lattice-wound coil See honeycomb coil. { 'lad·əs ,waùnd ,kȯil }

lattissimus dorsi [ANAT] The widest muscle of the back; a broad, flat muscle of the lower back that adducts and extends the humerus, is used to pull the body upward in climbing, and is an accessory muscle of respiration. { lə'tis·ə·məs 'dȯr·sē }

Lattorfian See Tongrian. { lə'tȯr·fē·ən }

latus rectum [MATH] The length of a chord through the

LATIMERIDAE

Living coelacanth, *Latimeria chalumnae*; 5 feet (1.5 meters). (*After P. P. Grasse, ed., Traite de Zoologie, tome 13, fasc. 3, 1958*)

focus and perpendicular to the axis of symmetry in a conic section. { ¦lad·əs 'rek·təm }

Iauan [MATER] Wood from any of several genera of trees of the Philippines, Malaya, and Sarawak; resembles mahogany but shrinks and swells more with changes in moisture. { lə'wän }

laubmannite [MINERAL] $Fe_3Fe_6(PO_4)_4(OH)_2$ Mineral composed of basic ferrous iron phosphate and ferric iron phosphate. { 'laüb·mə,nīt }

laudanidine [ORG CHEM] $C_{20}H_{25}NO_4$ An optically active alkaloid found in opium that crystallizes as prisms from an alcohol solution, and melts at 185°C. Also known as *l*-laudanine; tritopine. { lö'dan·ə,dēn }

laudanine [ORG CHEM] $C_{20}H_{25}NO_4$ An optically inactive alkaloid derived from alkaline mother liquors from morphine extraction; it crystallizes in orthorhombic prisms from alcohol and chloroform; the prisms melt at 167°C, and are soluble in hot alcohol, benzene, and chloroform. Also known as *dl*-laudanidine. { lö'dan·ə,nēn }

laudanosine [ORG CHEM] $C_{21}H_{27}NO_4$ An alkaloid that is the methyl ether of laudanine; the optically active form crystallizes from dilute alcohol and melts at about 115°C; the levorotatory active form crystallizes from light petroleum solution and melts at 89°C. { lö'dan·ə,sēn }

Laue camera [CRYSTAL] The apparatus used in the Laue method; the x-ray beam usually enters through a hole in the x-ray film, which records beams bent through an angle of nearly 180° by the crystal; less commonly, the film is placed beyond the crystal. { 'laü·ə ,kam·rə }

Laue condition [CRYSTAL] **1.** The condition for a vector to lie in a Laue plane: its scalar product with a specified vector in the reciprocal lattice must be one-half of the scalar product of the latter vector with itself. **2.** *See* Laue equations. { 'laü·ə kən,dish·ən }

Laue equations [CRYSTAL] Three equations which must be satisfied for an x-ray beam of specified wavelength to be diffracted through a specified angle by a crystal; they state that the scaler products of each of the crystallographic axial vectors with the difference between unit vectors in the directions of the incident and scattered beams, are integral multiples of the wavelength. Also known as Laue condition. { 'laü·ə i,kwā·zhənz }

Laue method [CRYSTAL] A method of studying crystalline structures by x-ray diffraction, in which a finely collimated beam of polychromatic x-rays falls on a single crystal whose orientation can be set as desired, and diffracted beams are recorded on a photographic film. { 'laü·ə ,meth·əd }

Laue pattern [CRYSTAL] The characteristic photographic record obtained in the Laue method. { 'laü·ə ,pad·ərn }

Laue plane [CRYSTAL] A plane which is the perpendicular bisector of a vector in the reciprocal lattice; such planes form the boundaries of Brillouin zones. { 'laü·ə ,plān }

Laue theory [CRYSTAL] A theory of diffraction of x-rays by crystals, based on the Laue equations. { 'laü·ə ,thē·ə·rē }

laughing gas *See* nitrous oxide. { 'laf·iŋ ,gas }

Laughlin state [CRYO] The simplest type of quantum Hall state, which contains only one component of incompressible fluid and has a filling factor equal to $1/m$, where m is an integer. { 'läk·lin ,stāt }

Laugiidae [PALEON] A family of Mesozoic fishes in the order Coelacanthiformes. { laü'jī·ə,dē }

laumonite *See* laumontite. { lö'mä,nīt }

laumontite [MINERAL] $CaAl_2Si_4O_{12}·4H_2O$ A white zeolite mineral crystallizing in the monoclinic system; loses water on exposure to air, eventually becoming opaque and crumbling. Also known as laumonite; lomonite; lomontite. { lö'män,tīt }

launch [AERO ENG] **1.** To send off a rocket vehicle under its own rocket power, as in the case of guided aircraft rockets, artillery rockets, and space vehicles. **2.** To send off a missile or aircraft by means of a catapult or by means of inertial force, as in the release of a bomb from a flying aircraft. **3.** To give a space probe an added boost for flight into space just before separation from its launch vehicle. { lönch }

launch complex [AERO ENG] The composite of facilities and support equipment needed to assemble, check out, and launch a rocket vehicle. { 'lönch ,käm,pleks }

launcher [ORD] **1.** A device designed to support and launch a rocket or rocket shell. **2.** A device on a rifle for firing a grenade. { 'lön·chər }

launching [CIV ENG] The act or process of floating a ship after only hull construction is completed; in some cases ships are not launched until after all construction is completed. [ELECTROMAG] The process of transferring energy from a coaxial cable or transmission line to a waveguide. { 'lön·chiŋ }

launching angle [AERO ENG] The angle between the horizontal plane and the longitudinal axis of a rocket or missile at the time of launching. { 'lön·chiŋ ,aŋ·gəl }

launching cradle [CIV ENG] A framework made of wood to support a vessel during launching from sliding ways. { 'lön·chiŋ ,krād·əl }

launching pad *See* launch pad. { 'lönch·iŋ ,pad }

launching ramp [AERO ENG] A ramp used for launching an aircraft or missile into the air. { 'lön·chiŋ ,ramp }

launching site [AERO ENG] **1.** A site from which launching is done. **2.** The platform, ramp, rack, or other installation at such a site. { 'lön·chiŋ ,sīt }

launching tube [ORD] A tube used to guide a rocket missile or other projectile during launching. { 'lön·chiŋ ,tüb }

launching ways [CIV ENG] Two (or more) sets of long, heavy timbers arranged longitudinally under the bottom of a ship during building and launching, with one set on each side, and sloping toward the water; the lower set, or ground ways, remain stationary and support the upper set, or sliding ways, which carry the weight of the ship after the shores and keel blocks are removed. { 'lön·chiŋ ,wāz }

launch pad [AERO ENG] The load-bearing base or platform from which a rocket vehicle is launched. Also known as launching pad; pad. { 'lönch ,pad }

launch vehicle [AERO ENG] A rocket or other vehicle used to launch a probe, satellite, or the like. Also known as booster. { 'lönch ,ve·ə·kəl }

launch window [AERO ENG] The time period during which a spacecraft or missile must be launched in order to achieve a desired encounter, rendezvous, or impact. { 'lönch ,win·dō }

launder [ENG] An inclined channel or trough for the conveyance of a liquid, such as for water in mining and construction engineering or for molten metal. { 'lön·dər }

launder screen [MIN ENG] A screen used in a launder for the sizing and dewatering of small sizes of anthracite. { 'lön·dər ,skrēn }

launder separation process [MIN ENG] A hydraulic process for separating heavy gravity product from the lighter product that flows above it; a stream of fluid carries material down a channel that has draws to separate the heavy material from the light material. { 'lön·dər ,sep·ə'rā·shən ,prä·səs }

launder washer [MIN ENG] A type of coal washer in which the coal is separated from the refuse by stratification due to hindered settling while being carried in aqueous suspension through the launder. { 'lön·dər ,wäsh·ər }

laundry blue [MATER] Dye-containing solution used to give a blue tint to laundry-yellowed white cottons and linens; usually contains Prussian blue. { 'lön·drē ,blü }

LAUNS *See* local-area underwater navigation system. { lönz }

lauoho o pele *See* Pele's hair. { ,lä·ü'ō,hō ō 'pe,lē }

Lauraceae [BOT] The laurel family of the order Magnoliales distinguished by definite stamens in series of three, a single pistil, and the lack of petals. { lö'rās·ē,ē }

Laurales [BOT] An order of dicotyledonous flowering, mostly woody plants of the class Magnoliopsida, division Magnoliophyta; commonly have scattered spherical cells containing volatile oils, leaves usually simple and mostly entire, and flowers often pollinated by beetles. { lo'rā·lēz }

Laurasia [GEOL] A continent theorized to have existed in the Northern Hemisphere; supposedly it broke up to form the present northern continents about the end of the Pennsylvanian period. { lö'rā·zhə }

Lauratonematidae [INV ZOO] A family of marine nematodes of the superfamily Enoploidea; many females possess a cloaca. { ,lör·ə·tō·nē'mad·ə,dē }

laurel forest *See* temperate rainforest. { 'lör·əl ,fär·əst }

laurel wax *See* bayberry wax. { 'lör·əl ,waks }

laurence [OPTICS] A shimmering seen over a hot surface on a calm, cloudless day, caused by the unequal refraction of light by innumerable convective air columns of different temperatures and densities. { 'lör·əns }

Laurence-Moon-Biedl syndrome [MED] A hereditary endocrine disorder of the pituitary or other hypothalamic structures transmitted as a dominant mutant gene and characterized

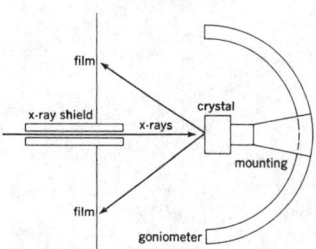

LAUE CAMERA

Schematic diagram of back-reflection Laue camera. Orientation of single crystal can be set as desired by system of goniometer circles. Diffracted beams, bent through angles of nearly 180°, are recorded on film.

principally by girdle-type obesity, mental retardation, and hypogenitalism. { 'lȯr·əns 'mün 'bēld·əl ˌsin̩ˌdrōm }

Laurent expansion [MATH] An infinite series in which an analytic function $f(z)$ defined on an annulus about the point z_0 may be expanded, with nth term $a_n(z - z_0)^n$, n ranging from $-\infty$ to ∞, and $a_n = 1/(2\pi i)$ times the integral of $f(t)/(t - z_0)^{n+1}$ along a simple closed curve interior to the annulus. Also known as Laurent series. { lȯ'rän ik,span·chən }

Laurent half-shade plate [OPTICS] A device used to determine the direction of polarization of plane polarized light; it consists of a quartz plate of special thickness that covers half of the plane polarized beam, followed by a plane polarization analyzer. { lȯ'rän 'haf ˌshäd ˌplāt }

Laurentian Plateau See Laurentian Shield. { lȯ'ren·chən pla'tō }

Laurentian Shield [GEOL] A Precambrian plateau extending over half of Canada from Labrador southwest along Hudson Bay and northwest to the Arctic Ocean. Also known as Canadian Shield; Laurentian Plateau. { lȯ'ren·chən 'shēld }

Laurentide ice sheet [HYD] A major recurring glacier that at its maximum completely covered North America east of the Rockies from the Arctic Ocean to a line passing through the vicinity of New York, Cincinnati, St. Louis, Kansas City, and the Dakotas. { lȯr·ənˌtīd 'īs ˌshēt }

Laurent series See Laurent expansion. { lȯ'ränz ˌsir·ēz }

Laurer's canal [INV ZOO] In certain flukes, a canal which passes from the oviduct to the ventral surface of the body. { 'lȯr·ərz kəˌnal }

lauric acid [ORG CHEM] $CH_3(CH_2)_{10}COOH$ A fatty acid melting at 44°C, boiling at 225°C (100 mmHg; 13,332 pascals); colorless needles soluble in alcohol and ether, insoluble in water; found as the glyceride in vegetable fats, such as coconut and laurel oils; used for wetting agents, in cosmetics, soaps, resins, and insecticides, and as a chemical intermediate. { 'lȯr ik 'as·əd }

laurionite [MINERAL] $Pb(OH)Cl$ A colorless mineral composed of basic lead chloride, occurring in prismatic crystals; it is dimorphous with paralaurionite. { 'lȯr·ē·əˌnīt }

laurisilva See temperate rainforest. { ˌlȯr·əˌsil·və }

laurite [MINERAL] RuS_2 A black mineral composed of ruthenium sulfide (often with osmium), occurring as small crystals or grains. { 'lȯˌrīt }

Lauritsen electroscope [ELEC] A rugged and sensitive electroscope in which a metallized quartz fiber is the sensitive element. { 'laȯ·rət·sən i'lek·trəˌskōp }

lauryl alcohol [ORG CHEM] $CH_3(CH_2)_{11}OH$ A colorless solid which is obtained from coconut oil fatty acids, has a floral odor, and boils at 259°C; used in detergents, lubricating oils, and pharmaceuticals. { 'lȯr·əl 'al·kəˌhȯl }

lauryl aldehyde [ORG CHEM] $CH_3(CH_2)_{10}CHO$ A constituent of an essential oil from the silver fir; a colorless solid or a liquid, with a floral odor, that is soluble in 90% alcohol; used in perfumes. { 'lȯr·əl 'al·dəˌhīd }

lauryl mercaptan [ORG CHEM] $C_{12}H_{25}SH$ Pale-yellow or water-white liquid with mild odor; insoluble in water, soluble in organic solvents; used to manufacture plastics, pharmaceuticals, insecticides, fungicides, and elastomers. { 'lȯr·əl mər'kapˌtan }

lausenite [MINERAL] $Fe_2(SO_4)_3 \cdot 6H_2O$ A white, monoclinic mineral consisting of hydrated ferric sulfate; occurs in lumpy aggregates of fibers. { 'lȯs·ənˌīt }

Lauson engine [ENG] Single-cylinder engine used in screening tests prior to the L-series lube oil tests (such as L-1 or L-2 tests). { 'laȯz·ən ˌen·jən }

lautal [MET] A hard aluminum alloy with small percentages of copper and silicon and traces of iron, manganese, or magnesium. { 'laȯˌtal }

lautarite [MINERAL] $Ca(IO_3)_2$ A monoclinic mineral composed of calcium iodate that occurs in prismatic crystals. { 'laȯd·əˌrīt }

lauter tub [FOOD ENG] A tank with a perforated false bottom in which the extract (wort) is separated from the insoluble grain residue during mashing of barley malt. { 'laȯd·ər ˌtəb }

lautite [MINERAL] $CuAsS$ A mineral composed of copper sulfide and copper arsenide. { 'laȯˌtīt }

Lauxaniidae [INV ZOO] A family of myodarian cyclorrhaphous dipteran insects in the subsection Acalypteratae; larvae are leaf miners. { lȯk'sə'nī·əˌdē }

LAV See lymphadenopathy-associated virus.

lava [GEOL] **1.** Molten extrusive material that reaches the earth's surface through volcanic vents and fissures. **2.** The rock mass formed by consolidation of molten rock issuing from volcanic vents and fissures, consisting chiefly of magnesium silicate; used for insulators. { 'lä·və }

lava blisters [GEOL] Small, steep-sided swellings that are hollow and raised on the surfaces of some basaltic lava flows; formed by gas bubbles pushing up the lava's viscous surface. { 'lä·və ˌblis·tərz }

lava cone [GEOL] A volcanic cone that was formed of lava flows. { 'lä·və ˌkōn }

lava dome See shield volcano. { 'lä·və ˌdōm }

lava field [GEOL] A wide area of lava flow; it is commonly several square kilometers in area and forms along the base of a large compound volcano or on the flanks of shield volcanoes. { 'lä·və ˌfēld }

lava flow [GEOL] **1.** A lateral, surficial stream of molten lava issuing from a volcanic cone or from a fissure. **2.** The solidified mass of rock formed when a lava stream congeals. { 'lä·və ˌflō }

lava fountain [GEOL] A jetlike eruption of lava that issues vertically from a volcanic vent or fissure. Also known as fire fountain. { 'lä·və ˌfaȯnt·ən }

lavage [MED] The therapeutic washing out of an organ. { lə'väzh }

lava lake [GEOL] A lake of lava that is molten and fluid; usually contained within a summit volcanic crater or in a pit crater on the flanks of a shield volcano. { 'lä·və ˌlāk }

Laval nozzle See de Laval nozzle. { lə'väl ˌnäz·əl }

lava plateau [GEOL] An elevated tableland or flat-topped highland that is several hundreds to several thousands of square kilometers in area; underlain by a thick succession of lava flows. { 'lä·və pla'tō }

lava tube [GEOL] A long, tubular opening under the crust of solidified lava. { 'lä·və ˌtüb }

lavender oil [MATER] Colorless to yellow or green-yellow essential oil with sweet aroma and bitter taste; distilled from fresh flowers of several species of lavender (*Lavandula*); main components are linalool, linalyl acetate, geraniol, cumarin, furfurol, and borneol; used in perfumery, and in medicine as a stimulant. { 'lav·ən·dər ˌȯil }

lavender spike oil See spike oil. { 'lav·ən·dər 'spīk ˌȯil }

lavenite [MINERAL] $(Na,Ca)_3Zr(Si_2O_7)(O,OH,F)_2$ A mineral composed of complex silicate, occurring in prismatic crystals. { 'lä·vəˌnīt }

Laves phases [MET] Alloy phases which have the general formula AB_2, and the crystal structures of either $MgCu_2$ (cubic) or $MgZn_2$ or $MgNi_2$ (both hexagonal). { 'lä·vəz ˌfāz·əz }

law [SCI TECH] A regularity which applies to all members of a broad class of phenomena. { lȯ }

lawn [TEXT] A sheer cotton or cotton and polyester fabric made of combed or carded yarn. { lȯn }

lawnmower [ELECTR] Type of radio-frequency preamplifier used with radar receivers. [ENG] A helix-type recorder mechanism. [MECH ENG] A machine for cutting grass on lawns. { 'lȯnˌmō·ər }

law of action and reaction See Newton's third law. { 'lȯ əv 'ak·shən ən 'rē,ak·shən }

law of closure [PSYCH] In Gestalt psychology, a law of organization that assumes an innate tendency to perceive incomplete objects as complete, to close up or fill in gaps in sensory inputs, and to view asymmetric and unbalanced stimuli as symmetric and balanced. { 'lȯ əv 'klō·zhər }

law of constant angles [CRYSTAL] The law that the angles between the faces of a crystal remain constant as the crystal grows. { 'lȯ əv ˈkän·stənt 'aŋ·gəlz }

law of constant heat summation See Hess's law. { 'lȯ əv ˈkän·stənt 'hēt səˌmā·shən }

law of contradiction [MATH] A principle of logic whereby a proposition cannot be both true and false. { ˌlȯ əv ˌkän·trə'dik·shən }

law of corresponding states [CHEM] The law that when, for two substances, any two ratios of pressure, temperature, or volume to their respective critical properties are equal, the third ratio must equal the other two. { 'lȯ əv 'kär·əˌspän·diŋ 'stāts }

law of corresponding times [MECH] The principle that the times for corresponding motions of dynamically similar systems are proportional to L/V and also to $\sqrt{L/G}$, where L is a typical dimension of the system, V a typical velocity, and G a typical force per unit mass. { ¦lȯ ǝv ¦kär·ǝ¦spänd·iŋ 'tīmz }

law of cosines [MATH] Given a triangle with angles A, B, and C and sides a, b, c opposite these angles respectively: $a^2 = b^2 + c^2 - 2bc \cos A$. { 'lȯ ǝv 'kō,sīnz }

law of definite composition See law of definite proportion. { 'lȯ ǝv ¦def·ǝ·nǝt ,käm·pǝ'zish·ǝn }

law of definite proportion [CHEM] The law that a given chemical compound always contains the same elements in the same fixed proportion by weight. Also known as law of definite composition. { 'lȯ ǝv ¦def·ǝ·nǝt prǝ'pȯr·shǝn }

law of electric charges [ELEC] The law that like charges repel, and unlike charges attract. { 'lȯ ǝv i¦lek·trik 'chärj·ǝz }

law of electromagnetic induction See Faraday's law of electromagnetic induction. { 'lȯ ǝv i¦lek·trō·mag¦ned·ik in'dǝk·shǝn }

law of electrostatic attraction See Coulomb's law. { 'lȯ ǝv i¦lek·trǝ¦stad·ik ǝ'trak·shǝn }

law of equal areas [ASTRON] The second of Kepler's laws. { 'lȯ ǝv ¦ē·kwǝl 'er·ē·ǝz }

law of exponents [MATH] Any of the laws $a^m a^n = a^{m+n}$, $a^m/a^n = a^{m-n}$, $(a^m)^n = a^{mn}$, $(ab)^n = a^n b^n$, $(a/b)^n = a^n/b^n$; these laws are valid when m and n are any integers, or when a and b are positive and m and n are any real numbers. Also known as exponential law. { 'lȯ ǝv ik'spō·nǝns }

law of flotation [FL MECH] The principle that an object floating in a fluid displaces its own weight of fluid. { 'lȯ ǝv flō'tā·shǝn }

law of gravitation See Newton's law of gravitation. { 'lȯ ǝv ,grav·ǝ'tā·shǝn }

law of large numbers [STAT] The law that if, in a collection of independent identical experiments, $N(B)$ represents the number of occurrences of an event B in n trials, and p is the probability that B occurs at any given trial, then for large enough n it is unlikely that $N(B)/n$ differs from p by very much. Also known as Bernoulli theorem. { 'lȯ ǝv ¦lärj 'nǝm·bǝrz }

law of magnetism [ELECTROMAG] The law that like poles repel, and unlike poles attract. { 'lȯ ǝv 'mag·nǝ,tiz·ǝm }

law of mass action [CHEM] The law stating that the rate at which a chemical reaction proceeds is directly proportional to the molecular concentrations of the reacting compounds. { 'lȯ ǝv ¦mas 'ak·shǝn }

law of minimum [BIOL] The law that those essential elements for which the ratio of supply to demand (A/N) reaches a minimum will be the first to be removed from the environment by life processes; it was proposed by J. von Liebig, who recognized phosphorus, nitrogen, and potassium as minimum in the soil; in the ocean the corresponding elements are phosphorus, nitrogen, and silicon. Also known as Liebig's law of minimum. { 'lȯ ǝv 'min·ǝ·mǝm }

law of parallel solenoids [PHYS] The law that under stationary conditions isopycnals and isobars must be parallel at all levels, and isobars and isopycnals at one level must be parallel to those at all other levels. { 'lȯ ǝv ¦par·ǝ,lel 'sō·lǝ,nȯidz }

law of partial pressures See Dalton's law. { 'lȯ ǝv ¦pär·shǝl 'presh·ǝrz }

law of quadrants [MATH] **1.** The law that any angle of a right spherical triangle (except a right angle) and the side opposite it are in the same quadrant. **2.** The law that when two sides of a right spherical triangle are in the same quadrant the third side is in the first quadrant, and when two sides are in different quadrants the third side is in the second quadrant. { 'lȯ ǝv 'kwäd·rǝns }

law of rational intercepts See Miller law. { 'lȯ ǝv ¦rash·ǝn·ǝl 'int·ǝr,seps }

law of reciprocity [GRAPHICS] The rule that the exposure time required to produce a specified density in a photographic emulsion is proportional to the reciprocal of the light intensity; it fails for low light intensities. { ,lȯ ǝv ,rēs·ǝ'präs·ǝd·ē }

law of reflection See reflection law. { 'lȯ ǝv ri'flek·shǝn }

law of signs [MATH] The product or quotient of two numbers is positive if the numbers have the same sign, negative if they have opposite signs. { 'lȯ ǝv 'sīnz }

law of sines [MATH] Given a triangle with angles A, B, and C and sides a, b, c opposite these angles respectively: $\sin A/a = \sin B/b = \sin C/c$. { 'lȯ ǝv 'sīnz }

law of species [MATH] The law that one-half the sum of two angles in a spherical triangle and one-half the sum of the

two opposite sides are of the same species, in that they are both acute or both obtuse angles. { 'lȯ ǝv 'spē·shēz }

law of specificity of bacteria See Koch's postulates. { 'lȯ ǝv ¦spes·ǝ¦fis·ǝd·ē ǝv bak'tir·ē·ǝ }

law of storms [METEOROL] Historically, the general statement of the manner in which the winds of a cyclone rotate about the cyclone's center, and the way that the entire disturbance moves over the earth's surface. { 'lȯ ǝv 'stȯrmz }

law of superposition [GEOL] The law that strata underlying other strata must be the older if there has been neither overthrust nor inversion. { 'lȯ ǝv ,sü·pǝr·pǝ'zish·ǝn }

law of the excluded middle [MATH] A principle of logic whereby a proposition is either true or false but cannot be both true and false. Also known as principle of dichotomy. { ¦lȯ ǝv thē ik,sklüd·ǝd 'mid·ǝl }

law of tangents [MATH] Given a triangle with angles A, B, and C and sides a, b, c opposite these angles respectively: $(a - b)/(a + b) = [\tan 1/2(A - B)]/[\tan 1/2(A + B)]$. { 'lȯ ǝv 'tan·jǝns }

law of the mean See mean value theorem. { ¦lȯ ǝv thǝ 'mēn }

Lawrence tube See chromatron. { 'lär·ǝns ,tüb }

lawrencite [MINERAL] (Fe,Ni)Cl₂ A brown or green mineral composed of ferrous chloride and found as an abundant accessory mineral in iron meteorites. { 'lär·ǝn sīt }

lawrencium [CHEM] A chemical element, symbol Lr, atomic number 103; isotopes with mass numbers 251–263 have been discovered, all unstable; mass number 262 has the longest half-life (3.6 hours). { 'lȯ'ren·sē·ǝm }

laws of refraction See Snell laws of refraction. { 'lȯz ǝv ri frak·shǝn }

Lawson criterion [PL PHYS] The requirement for the energy produced by fusion in a plasma to exceed that required to produce the confined plasma; it states that for a mixture of deuterium and tritium in the temperature range from 1×10^8 to 5×10^8 degrees Celsius, the product of the ionic density and the confinement time must be about 10^{14} seconds per cubic centimeter. { 'lȯs·ǝn krī,tir·ē·ǝn }

lawsonite [MINERAL] $CaAl_2(Si_2O_7)(OH)_2 \cdot H_2O$ A colorless or grayish-blue mineral crystallizing in the orthorhombic system; found in gneisses and schists. { 'lȯs·ǝn,īt }

laxative [PHARM] An agent that stimulates bowel movement and relieves constipation. { 'lak·sǝd·iv }

lay [DES ENG] The direction, length, or angle of twist of the strands in a rope or cable. [MET] The direction of the prevailing surface pattern on a piece of metal after grinding, cutting, lapping, or other processing. [MIN ENG] A share of profit. { lā }

lay-by [MIN ENG] Siding in single-track underground tramming road. { 'lā,bī }

layer [COMPUT SCI] One of the divisions within which components or functions are isolated in a computer system with layered architecture or a communications system with layered protocols. [GEOL] A tabular body of rock, ice, sediment, or soil lying parallel to the supporting surface and distinctly limited above and below. [GEOPHYS] One of several strata of ionized air, some of which exist only during the daytime, occurring at altitudes between 30 and 250 miles (50 and 400 kilometers); the layers reflect radio waves at certain frequencies and partially absorb others. [MET] The stratum of weld metal consisting of one or more passes and lying parallel to the welding surface. { 'lā·ǝr }

layer capacitance See cathode interface capacitance. { 'lā·ǝr kǝ'pas·ǝd·ǝns }

layer depth [OCEANOGR] **1.** The thickness of the mixed layer in an ocean. **2.** The depth to the top of the thermocline. { 'lā·ǝr ,depth }

layer depth effect [GEOPHYS] The weakening of a sound beam or seismic pulse because of abnormal spreading in passing from a positive gradient layer to an underlying negative layer. { 'lā·ǝr ¦depth i,fekt }

layered architecture [COMPUT SCI] A technique used in designing computer software, hardware, and communications in which system or network components are isolated in layers so that changes can be made in one layer without affecting the others. { 'lā·ǝrd 'är·kǝ,tek·chǝr }

layered complex [GEOL] An igneous rock body of large dimensions, 5–300 miles (8–480 kilometers) across and as much as 23,000 feet (7000 meters) thick, within which distinct subhorizontal stratification, or layering, is apparent and may be continuous over great distances, in some cases more than 60 miles (100 kilometers). { ¦lā·ərd ′käm‚pleks }

layered-protocols technique [COMMUN] A technique for isolating the functions required in a data communications network so that these functions can be set up in a modular fashion and changes can be made in one area without affecting the others. { ′lā·ərd ′prōd·ə‚kȯlz tek′nēk }

layer impedance See cathode interface impedance. { ′lā·ər im′pēd·əns }

layering [BOT] A propagation method by which root formation is induced on a branch or a shoot attached to the parent stem by covering the part with soil. [ECOL] A stratum of plant forms in a community, such as mosses, shrubs, or trees in a bog area. { ′lā·ə·riŋ }

layering of firedamp [MIN ENG] The formation of a layer of firedamp at the roof of a mine working and above the ventilating air current. { ′lā·ə·riŋ əv ′fīr‚damp }

layer lattice See layer structure. { ′lā·ər ‚lad·əs }

layer loading [MIN ENG] A procedure whereby the coal is placed in the railroad cars in horizontal layers. { ′lā·ər ‚lōd·iŋ }

layer of no motion [OCEANOGR] A layer, assumed to be at rest, at some depth in the ocean. { ′lā·ər əv ¦nō ′mō·shən }

layer silicate See phyllosilicate. { ′lā·ər ‚sil·ə·kət }

layer structure [CRYSTAL] A crystalline structure found in substances such as graphites and clays, in which the atoms are largely concentrated in a set of parallel planes, with the regions between the planes comparatively vacant. Also known as layer lattice. { ′lā·ər ‚strək·chər }

layer winding [ELEC] Coil-winding method in which adjacent turns are laid evenly and side by side along the length of the coil form; any number of additional layers may be wound over the first, usually with sheets of insulating material between the layers. { ′lā·ər ‚wīn·diŋ }

lay off [ENG] The process of fairing a ship's lines or an airplane's in a mold loft in order to make molds and templates for structural units. [NAV] The act of steering a ship away from the shore, a pier, or another ship. { ′lā ‚ȯf }

layout [GRAPHICS] A design drawing or graphical statement of the overall form of a component, system, or device, which is usually prepared during innovative stages of a design. { ′lā‚aut }

layout character [COMPUT SCI] A control character that determines the form in which the output data generated by a printer or display device are arranged. Also known as format effector. { ′lā‚aut ‚kar·ik·tər }

lay ratio [ELEC] The ratio of the axial length of one complete turn of the helix formed by the core of a cable or the wire of a stranded conductor, to the mean diameter of the cable. { ′lā ‚rā·shō }

lay-up [ENG] Production of reinforced plastics by positioning the reinforcing material (such as glass fabric) in the mold prior to impregnation with resin. { ′lā‚əp }

lazaret [NAV ARCH] A space between decks above the afterpeak. { ′laz·ə‚ret }

lazuli See lapis lazuli. { ′laz·ə·lē }

lazulite [MINERAL] $(Mg,Fe)Al_2(OH)_2(PO_4)_2$ A violet-blue or azure-blue mineral with vitreous luster; composed of basic aluminum phosphate and occurring in small masses or monoclinic crystals; hardness is 5–6 on Mohs scale, and specific gravity is 3.06–3.12. Also known as berkeyite; blue spar; false lapis. { ′laz·ə‚līt }

lazurite [MINERAL] $(Na,Ca)_8(Al,Si)O_{24}(S,SO_4)$ A blue or violet-blue feldspathoid mineral crystallizing in the isometric system; the chief mineral constituent of lapis lazuli. { ′laz·ə‚rīt }

lazy evaluation See demand-driven execution. { ′lā·zē i‚val·yə′wā·shən }

lazy H antenna [ELECTROMAG] An antenna array in which two or more dipoles are stacked one above the other to obtain greater directivity. { ′lā·zē ′āch an‚ten·ə }

lazy jack [ENG] A device that accommodates changes in length of a pipeline or similar structure through the motion of two linked bell cranks. { ′lā·zē ′jak }

lb See pound.

L band [COMMUN] A band of radio frequencies between 1 and 2 gigahertz. { ′el ‚band }

lb ap See pound.

lb apoth See pound.

lbf See pound.

lbf-ft See foot-pound.

lb t See pound.

lb tr See pound.

LBM See lattice Boltzmann method.

LC₅₀ See lethal concentration 50. { ¦el¦sē ′fif·tē }

LCA See life-cycle assessment.

LCAO See linear combination of atomic orbitals.

L capture [NUC PHYS] A type of generalized beta interaction in which a nucleus captures an electron from the L shell of atomic electrons (the shell second closest to the nucleus). { ′el ‚kap·chər }

L carrier system [COMMUN] Telephone carrier system used on coaxial cable systems, and microwave (line of sight), and tropospheric scatter radio systems; it occupies a frequency band from 68 kilohertz to over 8 megahertz. { ′el ′kar·ē·ər ‚sis·təm }

LCD See liquid crystal display.

LC filter See inductive filter. { ¦el¦sē ‚fil·tər }

L chain See light chain. { ′el ‚chān }

LCL See less-than-carload; lifting condensation level.

lcm See least common multiple.

LC ratio [ELEC] The inductance of a circuit in henrys divided by capacitance in farads. { ¦el¦sē ‚rā·shō }

LD₅₀ See lethal dose 50.

LDA localizer [NAV] A facility of comparable utility and accuracy to the standard ILS (instrument landing system) localizer, but which is not part of a full ILS and may not be in line with the runway. { ¦el¦dē′ā ′lō·kə‚liz·ər }

LDH See lactic dehydrogenase.

L display [ELECTR] A radarscope display in which the target appears as two horizontal pulses or blips, one extending to the right and one to the left from a central vertical time base; when the radar antenna is correctly aimed in azimuth at the target, both blips are of equal amplitude; when not correctly aimed, the relative blip amplitudes indicate the pointing error; the position of the signal along the baseline indicates target distance; the display may be rotated 90° when used for elevation aiming instead of azimuth aiming. { ′el di‚splā }

LDL See low-density lipoprotein.

LDM See limited-distance modem.

LDPE See low-density polyethylene.

L/D ratio [ENG] Length to diameter ratio, a frequently used engineering relationship. { ¦el′dē ‚rā·shō }

leachate [CHEM] A solution formed by leaching. [GEOCHEM] A liquid that has percolated through soil and dissolved some soil materials in the process. { ′lē‚chāt }

leaching [CHEM ENG] The dissolving, by a liquid solvent, of soluble material from its mixture with an insoluble solid; leaching is an industrial separation operation based on mass transfer; examples are the washing of a soluble salt from the surface of an insoluble precipitate, and the extraction of sugar from sugarbeets. [GEOCHEM] The separation or dissolving out of soluble constituents from a rock or ore body by percolation of water. [MIN ENG] Dissolving soluble minerals or metals out of the ore, as by the use of percolating solutions such as cyanide or chlorine solutions, acids, or water. Also known as lixiviation. { ′lēch·iŋ }

leach ion-exchange flotation process [MET] A method of extraction developed for treatment of copper ores not amenable to direct flotation; the metal is dissolved by leaching, for example, with sulfuric acid, in the presence of an ion-exchange resin; the resin recaptures the dissolved metal and is then recovered in a mineralized froth by the flotation process. { ′lēch ¦ī‚än iks¦chānj flō′tā·shən ‚prä·səs }

leach material [MIN ENG] Material sufficiently mineralized to be economically recoverable by leaching. { ′lēch mə‚tir·ē·əl }

leach pile [MIN ENG] Mineralized materials stacked so as to permit wanted minerals to be effectively and selectively dissolved by leaching. { ′lēch ‚pīl }

lead [CHEM] A chemical element, symbol Pb, atomic number 82, atomic weight 207.19. [DES ENG] The distance that a screw will advance or move into a nut in one complete turn. [ELEC] A wire used to connect two points in a circuit. [ENG] A mass of lead attached to a line, as used for sounding at sea. [GEOL] A small, narrow passage in a cave. [GRAPHICS] A thin strip of metal used during the composition process to space lines of type. [MET] A soft, heavy metal with a silvery-bluish color; when freshly cut it is malleable and ductile; occurs naturally, mostly in combination; used principally in alloys in pipes, cable sheaths, type metal, and shields against radioactivity. [ORD] **1.** The action of aiming ahead of a moving target with a gun, bomb, rocket, or torpedo so as to hit the target, including whatever action is necessary to correct for deflection. **2.** The distance between the moving target and the point at which the gun or missile is aimed. **3.** The number of diameters for one complete turn of the rifling. **4.** An explosive train component which consists of a column of high explosive, usually small in diameter, used to transmit detonation from one detonating component to a succeeding high explosive component; it is generally used to transmit the detonation from a detonator to a booster charge. [PHYS] *See* lead angle. { led }

lead acetate [ORG CHEM] $Pb(C_2H_3O_2)_2 \cdot 3H_2O$ Poisonous, water-soluble white crystals decomposing at 280°C; loses water at 75°C; used in hair dyes, medicines, and textile mordants, for waterproofing, for manufacture of varnishes and pigments, and as an analytical reagent. Also known as sugar of lead. { 'led 'as·ə₁tāt }

lead-acid battery [ELEC] A storage battery in which the electrodes are grids of lead containing lead oxides that change in composition during charging and discharging, and the electrolyte is dilute sulfuric acid. { 'led ₁as·əd 'bad·ə·rē }

lead angle [DES ENG] The angle that the tangent to a helix makes with the plane normal to the axis of the helix. [MET] The angle at the point of welding between an electrode and a line perpendicular to the weld axis. [ORD] **1.** The angle between the line of sight to a moving target and the line of sight to a point ahead of the target. **2.** A dropping angle. [PHYS] The phase difference between a sinusoidally varying quantity and a reference quantity which varies sinusoidally at the same frequency, when this phase difference is positive. Also known as angle of lead; lead; phase lead. { 'lēd 'aŋ·gəl }

lead antimonite [INORG CHEM] $Pb_3(SbO_4)_2$ Poisonous, water-insoluble orange-yellow powder; used as a paint pigment and to stain glass and ceramics. Also known as antimony yellow; Naples yellow. { 'led an'tim·ə₁nīt }

lead arsenate [INORG CHEM] $Pb_3(AsO_4)_2$ Poisonous, water-insoluble white crystals; soluble in nitric acid; used as an insecticide. { 'led 'ärs·ən₁āt }

lead azide [INORG CHEM] $Pb(N_3)_2$ Unstable, colorless needles that explode at 350°C; lead azide is shipped submerged in water to reduce sensitivity; used as a detonator for high explosives. { 'led 'ā₁zīd }

lead-base babbitt [MET] Alloy of 10–15% antimony, 2–10% tin, up to 0.2% copper, sometimes with arsenic, the remainder being lead; used as a bearing metal; a variation used in diesel engine and railway bearings contains alkaline-earth metals. Also known as white-metal bearing alloy. { 'led ₁bās 'bab·ət }

lead-base grease [MATER] A mixture of soap and mineral oil, usually prepared by the reaction of lead oxide and fatty acid; holds up to extreme pressure and is useful as a gear lubricant. { 'led ₁bās 'grēs }

lead-bismuth-cooled reactor [NUCLEO] A nuclear reactor that uses a molten lead-bismuth alloy as the coolant. { 'led ₁biz·məth ₁küld rē₁ak·tər }

lead borate [INORG CHEM] $Pb(BO_2)_2 \cdot H_2O$ Poisonous, water-insoluble white powder; soluble in dilute nitric acid; used as varnish and paint drier, for galvanoplastic work, in lead glass, and in waterproofing paints. { 'led 'bȯr₁āt }

lead bromide [INORG CHEM] $PbBr_2$ An alcohol-insoluble white powder melting at 373°C, boiling at 916°C; slightly soluble in hot water. { 'led 'brō₁mīd }

lead bronze [MET] An alloy of 60–70% copper, up to 2% nickel, and up to 15% tin with the balance lead; used as a bearing metal. { 'led 'bränz }

lead bullet [ORD] In small-arms ammunition, a bullet composed of lead or of a composition with a high percentage of lead. { 'led 'bu̇l·ət }

lead burning *See* lead welding. { 'led ₁bər·niŋ }

lead carbonate [INORG CHEM] $PbCO_3$ Poisonous, acid-soluble white crystals decomposing at 315°C; insoluble in alcohol and water; used as a paint pigment. { 'led 'kär·bə₁nāt }

lead-chamber process [CHEM ENG] A process for the preparation of impure or dilute (60–78) sulfuric acid; sulfur dioxide is oxidized by moist air with nitrogen oxide catalysts in a series of lead-lined chambers, the Gay-Lussac tower and the Glover tower; used primarily in the manufacture of fertilizer. { 'led ₁chām·bər ₁prä·səs }

lead chart [ORD] A chart or table giving the leads necessary to strike a moving target under various conditions, such as the distance to the target and its speed and direction of travel. { 'lēd ₁chärt }

lead chloride [INORG CHEM] $PbCl_2$ Poisonous white crystals melting at 498°C, boiling at 950°C; slightly soluble in hot water, insoluble in alcohol and cold water; used to make lead salts and lead chromate pigments and as an analytical reagent. { 'led 'klȯr₁īd }

lead chromate [INORG CHEM] $PbCrO_4$ Poisonous, water-insoluble yellow crystals melting at 844°C; soluble in acids; used as a paint pigment. { 'led 'krō₁māt }

lead compensation [CONT SYS] A type of feedback compensation primarily employed for stabilization or for improving a system's transient response; it is generally characterized by a series compensation transfer function of the type

$$G_c(s) = K \frac{(s - z)}{(s - p)}$$

where $z < p$ and K is a constant. { 'lēd ₁käm·pən'sā·shən }

lead computer [ORD] A device for calculating leads to hit a moving target, often placed in the sights or mount of a machine gun that automatically gives the proper lead when the target is tracked by the gun. { 'lēd kəm₁pyüd·ər }

lead-cooled reactor [NUCLEO] A nuclear reactor that uses molten lead as the coolant, to transport the energy released in the fission process away from the fuel rods in the reactor core and to keep the fuel rods and their clad from overheating. { 'led ₁küld rē₁ak·tər }

lead-covered cable [ELEC] A cable whose conductors are protected from moisture and mechanical damage by a lead sheath. { 'led ₁kəv·ərd 'kā·bəl }

lead curve [CIV ENG] The curve in a railroad turnout between the switch and the frog. { 'lēd ₁kərv }

lead cyanide [INORG CHEM] $Pb(CN)_2$ Poisonous white to yellow powder; slightly soluble in water, decomposed by acids; used in metallurgy. { 'led 'sī·ə₁nīd }

lead dioxide [INORG CHEM] PbO_2 Poisonous brown crystals that decompose when heated; insoluble in water and alcohol, soluble in glacial acetic acid; used as an oxidizing agent, in electrodes, batteries, matches, and explosives, as a textile mordant, in dye manufacture, and as an analytical reagent. Also known as anhydrous plumbic acid; brown lead oxide; lead peroxide. { 'led dī'äk₁sīd }

leaded alloy [MET] An alloy, especially of brass, bronze, or steel, to which lead is added to improve machinability and mechanical properties. { 'led·əd 'al₁ȯi }

leaded gasoline [MATER] Motor gasoline into which a small amount of TEL (tetraethyllead) has been added to increase octane number or rating. { 'led·əd 'gas·ə₁lēn }

leaded zinc oxide [MATER] A mixture of zinc oxide and basic lead sulfate; used in paints. { 'led·əd 'ziŋk 'äk₁sīd }

lead encephalopathy [MED] Degeneration of the neurons of the brain accompanied by cerebral edema, due to lead poisoning. { 'led ₁en·sef·ə'läp·ə·thē }

lead equivalent [NUCLEO] The thickness of lead that gives the same reduction in radiation dose rate as the material in question. { 'led i₁kwiv·ə·lənt }

leader [BUILD] *See* downspout. [COMPUT SCI] A record which precedes a group of detail records, giving information about the group not present in the detail records; for example, "beginning of batch 17." [ENG] The unrecorded length of magnetic tape that enables the operator to thread the tape through the drive and onto the take-up reel without losing data or recorded music, speech, or such. [GEOPHYS] The streamer which initiates the first phase of each stroke of a lightning discharge; it is a channel of very high ion density which propagates through the air by the continual establishment of an electron avalanche ahead of its tip. Also known as leader streamer. [GRAPHICS] **1.** A short piece of blank film at each end of a strip of photographic film or reel of motion picture film, used to thread the film through the mechanism of a camera or projector. **2.** Lines or rows of dashes or dots used to guide the reader's eye across a printed page. [MECH ENG]

LEAD-CHAMBER PROCESS

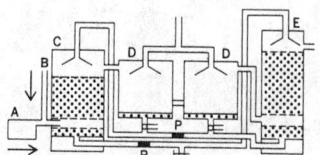

Schematic drawing of the lead-chamber process for manufacture of sulfuric acid, with concentration of 60–78%. A, sulfur or pyrite burners; B, inlet for nitrogen oxides; C, Glover tower; D, lead chamber; E, Gay-Lussac tower; P, pumps.

In a hot-air heating system, a duct that conducts heated air to an outlet. { 'lēd·ər }

leader cable [NAV] A cable used as a navigation aid, in which the path to be followed is defined by the magnetic field produced by current flowing through the cable. { 'lēd·ər ,kā·bəl }

leader label [COMPUT SCI] A record appearing at the beginning of a magnetic tape to uniquely identify the tape as one required by the system. { 'lēd·ər ,lā·bəl }

leader streamer See leader. { 'lēd·ər ,strēm·ər }

leader stroke [GEOPHYS] The entire set of events associated with the propagation of any leader between cloud and ground in a lightning discharge. { 'lēd·ər ,strōk }

lead fluoride [INORG CHEM] PbF_2 A crystalline solid with a melting point of 824°C; used for laser crystals and electronic and optical applications. { 'led 'flur,īd }

lead foil [MET] A foil of lead or of lead alloy containing, for example, 10–12% tin and 1% copper. { 'led ¦fȯil }

lead formate [ORG CHEM] $Pb(CHO_2)_2$ Poisonous, water-soluble brownish-white crystals that decompose at 190°C; used as an analytical reagent. { 'led 'fȯr,māt }

lead glance See galena. { 'led ¦glans }

lead glass [MATER] Glass into which lead oxide is incorporated to give high refractive index, optical dispersion, and surface brilliance; used in optical glass. { 'led ¦glas }

lead halide [INORG CHEM] PbX_2, where X is a halogen (such as F, Br, Cl, or I). { 'led 'ha,līd }

lead hexafluorosilicate [INORG CHEM] $PbSiF_6 \cdot 2H_2O$ Poisonous, colorless, water-soluble crystals; used in the electrolytic method for refining lead. { 'led ¦hek·sə,flur·ə'sil·ə,kāt }

leadhillite [MINERAL] $Pb_4(SO_4)(CO_3)_2(OH)_2$ A yellowish or greenish- or grayish-white monoclinic mineral consisting of basic sulfate and carbonate of lead; dimorphous with susanite. { 'led,hi,līt }

lead-I-lead junction [SOLID STATE] A Josephson junction consisting of two pieces of lead separated by a thin insulating barrier of lead oxide. Abbreviated Pb-I-Pb junction. { 'led ¦ī 'led ,jəŋk·shən }

lead-in [ELEC] A single wire used to connect a single-terminal outdoor antenna to a receiver or transmitter. Also known as down-lead. { 'lēd,in }

lead-in-air indicator [MIN ENG] An instrument that utilizes reagents to measure the concentration of lead in the air. { 'lēd,in ¦er 'in·də,kād·ər }

leading [GRAPHICS] Space inserted between lines of type to open them up vertically. { 'led·iŋ }

leading character elimination [COMPUT SCI] A method of data compression used for dictionaries that are stored in alphabetical order, in which the coding for each word has two parts: the number of characters in common with the previous word, and the unique suffix. { ¦lēd·iŋ 'kar·ik·tər i,lim·ə'nā·shən }

leading current [ELEC] An alternating current that reaches its maximum value up to 90° ahead of the voltage that produces it. { 'lēd·iŋ ¦kə·rənt }

leading edge [AERO ENG] The front edge of an airfoil or wing. [DES ENG] The surfaces or inset cutting points on a bit that face in the same direction as the rotation of the bit. [PHYS] The major portion of the rise of a pulse. { 'lēd·iŋ 'ej }

leading edge slat [AERO ENG] A small airfoil attached to the leading edge of a wing of an aircraft that automatically improves airflow at large angles of attack. { 'lēd·iŋ ¦ej ,slat }

leading fire [ORD] Fire delivered ahead of a moving target to allow for its motion. { 'lēd·iŋ 'fīr }

leading heading [MIN ENG] **1.** The heading of a pair of parallel headings that is a short distance in front of the other; used to drain a mining area. **2.** A heading dug into the solid coal before the advance of the general face. { 'lēd·iŋ 'hed·iŋ }

leading light [NAV] A light or lights arranged to indicate the path to be followed. { 'lēd·iŋ 'līt }

leading load [ELEC] Load that is predominately capacitive, so that its current leads the voltage applied to the load. { 'lēd·iŋ ¦lōd }

leading pad [COMPUT SCI] Characters that fill unused space at the left end of a data field. { 'lēd·iŋ 'pad }

leading phase [ELEC] In three-phase power measurement, the phase whose voltage is leading upon that of one of the other phases by 120°. { 'lēd·iŋ ¦fāz }

lead-in groove [DES ENG] A blank spiral groove at the outside edge of a disk recording, generally of a pitch much greater than that of the recorded grooves, provided to bring the pickup stylus quickly to the first recorded groove. Also known as lead-in spiral. { 'lēd,in ,grüv }

leading stone See lodestone. { 'lēd·iŋ ,stōn }

leading strand [CELL MOL] In deoxyribonucleic acid (DNA) replication, the 5′ to 3′ DNA strand that is synthesized with few or no interruptions. { 'lēd·iŋ ,strand }

leading truck [MECH ENG] A swiveling frame with wheels under the front end of a locomotive. { 'lēd·iŋ 'trək }

leading zeros [MATH] Zeros preceding the first nonzero integer of a number. { 'lēd·iŋ 'zir·ōz }

lead-in insulator [ELEC] A tubular insulator inserted in a hole drilled through a wall, through which the lead-in wire can be brought into a building. { 'lēd,in 'in·sə,lād·ər }

lead-in spiral See lead-in groove. { 'lēd,in 'spī·rəl }

lead iodide [INORG CHEM] PbI_2 Poisonous, water- and alcohol-insoluble golden-yellow crystals melting at 402°C, boiling at 954°C; used in photography, medicine, printing, mosaic gold, and bronzing. { 'led 'ī·ə,dīd }

lead joint [ENG] A pipe joint made by caulking with lead wool or molten lead. { 'led ,jȯint }

lead-lag ballast [ELEC] A ballast for a pair of fluorescent lamps, one operating on leading current and the other on lagging current, to diminish the stroboscopic effect. { ¦led 'lag ,bal·əst }

lead-lag network [CONT SYS] Compensating network which combines the characteristics of the lag and lead networks, and in which the phase of a sinusoidal response lags a sinusoidal input at low frequencies and leads it at high frequencies. Also known as lag-lead network. { ¦led 'lag 'net,wərk }

lead lanthanum zirconate titanate [MATER] A ferroelectric, ceramic, electrooptical material whose optical properties can be changed by an electric field or by being placed in tension or compression; used in optoelectronic storage and display devices. Abbreviated PLZT. Also known as lanthanum-doped lead zirconate-lead titanate. { 'led 'lan·thə·nəm 'zərk·ən,āt 'tīt·ən,āt }

lead line See sounding line. { 'led ,līn }

lead lining [ENG] Lead sheeting used to line the inside surfaces of liquid-storage vessels and process equipment to prevent corrosion. { 'led 'līn·iŋ }

lead marcasite See sphalerite. { 'led 'mär·kə,zīt }

lead metallurgy [MET] The science and technology of lead. { 'led 'med·əl,ər·jē }

lead metasilicate See lead silicate. { 'led ,med·ə'sil·ə,kāt }

lead molybdate [INORG CHEM] $PbMoO_4$ Poisonous, acid-soluble yellow powder; insoluble in water and alcohol; used in pigments and as an analytical reagent. { 'led mə'lib,dāt }

lead monoxide [INORG CHEM] PbO Yellow, tetragonal crystals that melt at 888°C and are soluble in alkalies and acids; used in storage batteries, ceramics, pigments, and paints. Also known as litharge; plumbous oxide; yellow lead oxide. { 'led mə'näk,sīd }

lead naphthenate [MATER] Soft, combustible, alcohol-soluble, transparent, resinous material, melting about 100°C; made from addition of lead salt to solution of sodium naphthenate; used as a paint and varnish drier, wood preservative, catalyst, insecticide, and lubricating oil additive. { 'led 'naf·thə,nāt }

lead network See derivative network. { 'led ,net,wərk }

lead nitrate [INORG CHEM] $Pb(NO_3)_2$ Strongly oxidizing, poisonous, water- and alcohol-soluble white crystals that decompose at 205–223°C; used as a textile mordant, paint pigment, and photographic sensitizer and in medicines, matches, explosives, tanning, and engraving. { 'led 'nī,trāt }

lead ocher See massicot. { 'led 'ō·kər }

lead oleate [ORG CHEM] $Pb(C_{18}H_{33}O_2)_2$ Poisonous, water-insoluble, white, ointmentlike material; soluble in alcohol, benzene, and ether; used in varnishes, lacquers, and high-pressure lubricants, and as a paint drier. { 'led 'ō·lē,āt }

lead orthoplumbate See lead tetroxide. { 'led ,ȯr·thō 'pləm,bāt }

lead-out groove [DES ENG] A blank spiral groove at the end of a disk recording, generally of a pitch much greater than that of the recorded grooves, connected to either the locked or eccentric groove. Also known as throw-out spiral. { 'lēd,aut ,grüv }

lead-over groove [DES ENG] A groove cut between separate selections or sections on a disk recording to transfer the pickup stylus from one cut to the next. Also known as cross-over spiral. { 'lēd,ō·vər ,grüv }

lead oxide red *See* lead tetroxide. { 'led ¦äk,sīd 'red }

lead palsy *See* lead polyneuropathy. { 'led ¦pól·zē }

lead peroxide *See* lead dioxide. { 'led pə'räk,sīd }

lead phosphate [INORG CHEM] Pb_3PO_2 A poisonous, white powder that melts at 1014°C; soluble in nitric acid and in fixed alkali hydroxide; used as a stabilizer in plastics. { 'led 'fäs,fāt }

lead pigments [CHEM] Chemical compounds of lead used in paints to give color; examples are white lead; basic lead carbonate; lead carbonate; lead thiosulfate; lead sulfide; basic lead sulfate (sublimed white lead); silicate white lead; basic lead silicate; lead chromate; basic lead chromate; lead oxychloride; and lead oxide (monoxide and dioxide). { 'led 'pig·məns }

lead poisoning [MED] Poisoning due to ingestion or absorption of lead over a prolonged period of time; characterized by colic, brain disease, anemia, and inflammation of peripheral nerves. { 'led 'póiz·ən·iŋ }

lead polyneuropathy [MED] A distal polyneuropathy, affecting mainly the wrist and hand, seen principally in adults with chronic lead poisoning; characterized by weakness, paresthesias, pain, and glove-and-stocking anesthesia. Also known as lead palsy. { 'led 'päl·ē·nü'räp·ə·thē }

lead rail [CIV ENG] In an ordinary rail switch, the turnout rail lying between the rails of the main track. { 'lēd ,rāl }

lead resinate [ORG CHEM] $Pb(C_{28}H_{29}O_2)_2$ Poisonous, insoluble, brown, lustrous, translucent lumps; used as a paint and varnish drier and for textile waterproofing. { 'led 'rez·ən,āt }

lead screw [MECH ENG] A threaded shaft used to convert rotation to longitudinal motion; in a lathe it moves the tool carriage when cutting threads; in a disk recorder it guides the cutter at a desired rate across the surface of an ungrooved disk. { 'lēd ,skrü }

lead silicate [INORG CHEM] $PbSiO_3$ Toxic, insoluble white crystals; used in ceramics, paints and enamels, and to fireproof fabrics. Also known as lead metasilicate. { 'led 'sil·ə,kāt }

lead-silver babbitt [MET] Alloy of lead with 10–15% antimony, 2.5–5.1% silicon, up to 5% tin, and up to 0.2% copper; used as a bearing metal. { 'led 'sil·vər 'bab·ət }

lead-soap lubricant [MATER] Hard, high-melting-point, extreme-pressure lubricant made of lead salts saponified with fats. { 'led ,sōp 'lü·brə·kənt }

lead sodium hyposulfate *See* lead sodium thiosulfate. { 'led 'sōd·ē·əm ,hī·pō'səl,fāt }

lead sodium thiosulfate [INORG CHEM] $Na_4Pb(S_2O_3)_3$ Poisonous, small, white, heavy crystals that are soluble in thiosulfate solutions; used in the manufacture of matches. Also known as lead sodium hyposulfate; sodium lead hyposulfate; sodium lead thiosulfate. { 'led 'sōd·ē·əm ,thī·ə'səl,fāt }

lead solder [MET] Solder composed of a lead alloy. { 'led 'säd·ər }

lead spar *See* anglesite. { 'led 'spär }

lead stearate [ORG CHEM] $Pb(C_{18}H_{35}O_2)_2$ Poisonous white powder; soluble in alcohol and ether, insoluble in water; used as a lacquer and varnish drier and in high-pressure lubricants. { 'led 'stir,āt }

lead sulfate [INORG CHEM] $PbSO_4$ Poisonous white crystals melting at 1170°C; slightly soluble in hot water, insoluble in alcohol; used in storage batteries and as a paint pigment. { 'led 'səl,fāt }

lead sulfide [INORG CHEM] PbS Blue, metallic, cubic crystals that melt at 1120°C, derived from the mineral galena or by reacting hydrogen sulfide gas with a solution of lead nitrate; used in semiconductors and ceramics. Also known as plumbous sulfide. { 'led 'səl,fīd }

lead sulfide cell [ELECTR] A cell used to detect infrared radiation; either its generated voltage or its change of resistance may be used as a measure of the intensity of the radiation. { 'led 'səl,fīd 'sel }

lead susceptibility [CHEM ENG] The increase in octane number of gasoline imparted by the addition of a specified amount of TEL (tetraethyllead). { 'led sə,sep·tə'bil·əd·ē }

lead telluride [INORG CHEM] PbTe A crystalline solid that is very toxic if inhaled or ingested; melts at 902°C; used as a

semiconductor and photoconductor in the form of single crystals. { 'led 'tel·yə,rīd }

lead tetraacetate [ORG CHEM] $Pb(CH_3COO)_4$ Crystals that are faintly pink or colorless; melts at 175°C; used as an oxidizing agent in organic chemistry, cleaving 1,2-diols to form aldehydes or ketones. { 'led ,te·trə'as·ə,tāt }

lead tetroxide [INORG CHEM] Pb_3O_4 A poisonous, bright-red powder, soluble in excess glacial acetic acid and dilute hydrochloric acid; used in medicine, in cement for special applications, in manufacture of colorless glass, and in ship paint. Also known as lead orthoplumbate; lead oxide red; red lead. { 'led ,te'träk,sīd }

lead thiocyanate [INORG CHEM] $Pb(SCN)_2$ Yellow, monoclinic crystals, soluble in potassium thiocyanate and slightly soluble in water; used in the powder mixture that primes small arm cartridges, in dyes, and in safety matches. { 'led ,thī·ō'sī·ə,nāt }

lead time [IND ENG] The time allowed or required to initiate and develop a piece of equipment that must be ready for use at a given time. { 'lēd ,tīm }

lead titanate [INORG CHEM] $PbTiO_3$ A water-insoluble, pale-yellow solid; used as coloring matter in paints. { 'led 'tīt·ən,āt }

lead track [CIV ENG] A distance measured along a straight railroad track from a switch to a frog. { 'lēd ,trak }

lead tungstate [INORG CHEM] $PbWO_4$ A yellowish powder, melting at 1130°C; insoluble in water, soluble in acid; used as a pigment. Also known as lead wolframate. { 'led 'təŋ·stə,nāt }

lead-208 [NUC PHYS] Lead isotope, atomic mass number of 208, which is formed by the radioactive decay of thorium. { 'led 'tü¦ō'āt }

lead vanadate [INORG CHEM] $Pb(VO_3)_2$ A water-insoluble, yellow powder; used as a pigment and for the preparation of other vanadium compounds. { 'led 'van·ə,dāt }

lead vitriol *See* anglesite. { 'led 'vit·rē,ōl }

lead welding [MET] Welding of lead by fusion. Incorrectly known as lead burning. { 'led ,wel·diŋ }

lead wire [ENG] One of the heavy wires connecting a firing switch with the cap wires. { 'lēd ,wīr }

lead wolframate *See* lead tungstate. { 'led 'wul·frə,māt }

lead wool [MATER] A coarse lead fiber used to caulk pipe joints. { 'led 'wul }

lead zirconate titanate [MATER] A ferroelectric, ceramic, electrooptic material that has lower optical transparency than lead lanthanum zirconate titanate but similar other properties. Abbreviated PZT. { 'led 'zərk·ən,āt 'tīt·ən,āt }

leaf [BOT] A modified aerial appendage which develops from a plant stem at a node, usually contains chlorophyll, and is the principal organ in which photosynthesis and transpiration occur. [BUILD] **1.** A separately movable division of a folding or sliding door. **2.** One of a pair of doors or windows. **3.** One of the two halves of a cavity wall. [COMPUT SCI] *See* terminal vertex. { lēf }

leaf blight [PL PATH] Any of various blight diseases which cause browning, death, and falling of the leaves. { 'lēf ,blīt }

leaf blotch [PL PATH] A plant disease characterized by discolored areas in the leaves with indistinct or diffuse margins. { 'lēf ,bläch }

leaf bud [BOT] A bud that produces a leafy shoot. { 'lēf bəd }

leaf cast [PL PATH] Any of several diseases of conifers characterized by falling of the needles. { 'lēf ,kast }

leaf curl [PL PATH] A fungus or viral disease of plants marked by the curling of leaves. { 'lēf 'kərl }

leaf cushion [BOT] The small part of the thickened leaf base that remains after abscission in various conifers, and also in some extinct plants. { 'lēf ,kush·ən }

leaf drop [PL PATH] Premature falling of leaves, associated with disease. { 'lēf ,dräp }

leaf fiber [BOT] A long, multiple-celled fiber extracted from the leaves of many plants that is used for cordage, such as sisal for binder, and abaca for manila hemp. { 'lēf ,fī·bər }

leaf gap [BOT] The place where the vascular bundle of the stem interrupts above a leaf trace as a result of the diversion of vascular tissue from the stem into a leaf, occurring in many vascular plants. { 'lēf ,gap }

leafhopper [INV ZOO] The common name for members of the homopteran family Cicadellidae. { 'lēf,häp·ər }

leaflet [BOT] **1.** A division of a compound leaf. **2.** A small or young foliage leaf. { 'lēf·lət }

leaflet projectile [ORD] A projectile designed for or adapted to usage as a carrier for leaflets. { 'lēf·lət prə‚jek·təl }

leaf miner [INV ZOO] Any of the larvae of various insects which burrow into and eat the parenchyma of leaves. { 'lēf ‚mīn·ər }

leaf mold [GEOL] A soil layer or compost consisting principally of decayed vegetable matter. { 'lēf ‚mōld }

leaf mottle [PL PATH] A fungus disease characterized by chlorotic mottling of the leaves; for example, caused by *Verticillium dahliae* in sunflower. { 'lēf ‚mäd·əl }

leaf-nosed [VERT ZOO] Having a leaflike membrane on the nose, as certain bats. { 'lēf ‚nōzd }

leaf of Descartes *See* folium of Descartes. { 'lēf əv dā'kärt }

leaf primordium [BOT] An immature leaf that arises as an emergence on the flanks of the apical meristem of the shoot tip. { 'lēf prī'mór·dē·əm }

leaf roll [PL PATH] Any of several virus diseases characterized by upward or inward rolling of the leaf margins. { 'lēf ‚rōl }

leaf rot [PL PATH] Any plant disease characterized by breakdown of leaf tissues; for example, caused by *Pellicularia koleroga* in coffee. { 'lēf ‚rät }

leaf rust [PL PATH] Any rust disease that primarily affects leaves; common in coffee, alfalfa, and wheat, barley, and other cereals. { 'lēf ‚rəst }

leaf scald [PL PATH] A bacterial disease of sugarcane caused by *Bacterium albilineans* which invades the vascular tissues, causing creamy or grayish streaking and withering of the leaves. { 'lēf ‚skóld }

leaf scar [BOT] A mark on a stem, formed by secretion of suberin and a gumlike substance, showing where a leaf has abscised. { 'lēf ‚skär }

leaf scorch [BOT] Any of several disorders and fungus diseases marked by a burned appearance of the leaves; for example, caused by the fungus *Diplocarpon earliana* in strawberry. { 'lēf ‚skórch }

leaf sight [ORD] Rear sight for small arms, hinged so that it can be raised for aiming or lowered to keep it from being broken when not in use, and containing a peep sight that can be moved up and down to make adjustments for range. { 'lēf ‚sīt }

leaf spot [PL PATH] Any of various diseases or disorders characterized by the appearance of well-defined discolored spots on the leaves. { 'lēf ‚spät }

leaf spring [DES ENG] A beam of cantilever design, firmly anchored at one end and with a large deflection under a load. Also known as flat spring. { 'lēf ‚spriŋ }

leaf stripe [PL PATH] Any of various plant diseases characterized by striped discolorations on the foliage. { 'lēf ‚strīp }

leaf trace [BOT] A section of the vascular bundle that leads from the stele to the base of the leaf. { 'lēf ‚trās }

league [MECH] A unit of length equal to 3 miles or 4828.032 meters. { lēg }

leak [PL PATH] A watery rot of fruits and vegetables caused by various fungi, such as *Rhizopus nigricans* in strawberry. { lēk }

leakage [ENG] Undesired and gradual escape or entry of a quantity, such as loss of neutrons by diffusion from the core of a nuclear reactor, escape of electromagnetic radiation through joints in shielding, flow of electricity over or through an insulating material, and flow of magnetic lines of force beyond the working region. [MIN ENG] An unintentional diversion of ventilation air from its designed path. [PHYS CHEM] A phenomenon occurring in an ion-exchange process in which some influent ions are not absorbed by the ion-exchange bed and appear in the effluent. { 'lēk·ij }

leakage coefficient *See* leakage factor. { 'lēk·ij ‚kō·ə'fish·ənt }

leakage conductance [ELEC] The conductance of the path over which leakage current flows; it is normally a low value. { 'lēk·ij kən'dək·təns }

leakage current [ELEC] **1.** Undesirable flow of current through or over the surface of an insulating material or insulator. **2.** The flow of direct current through a poor dielectric in a capacitor. [ELECTR] The alternating current that passes through a rectifier without being rectified. { 'lēk·ij ‚kə·rənt }

leakage factor [ELECTROMAG] The total magnetic flux in an electric rotating machine or transformer divided by the useful flux that passes through the armature or secondary winding. Also known as leakage coefficient. { 'lēk·ij ‚fak·tər }

leakage flux [ELECTROMAG] Magnetic lines of force that go beyond their intended path and do not serve their intended purpose. [NUCLEO] The number of neutrons which pass outward through a unit area at the surface of a reactor core in a unit time, and are not reflected back into the core. { 'lēk·ij ‚fləks }

leakage halo [GEOCHEM] The dispersion of elements along channels and paths followed by mineralizing solutions leading into and away from the central focus of mineralization. { 'lēk·ij ‚hā·lō }

leakage indicator [ELEC] An instrument used to measure or detect current leakage from an electric system to earth. Also known as earth detector. { 'lēk·ij ‚in·də‚kād·ər }

leakage inductance [ELECTROMAG] Self-inductance due to leakage flux in a transformer. { 'lēk·ij in'dək·təns }

leakage intake system [MIN ENG] A ventilation circuit with two adjacent intake roadways leading to the coal face. { 'lēk·ij 'in‚tāk ‚sis·təm }

leakage radiation [ELECTROMAG] In a radio transmitting system, radiation from anything other than the intended radiating system. { 'lēk·ij ‚rād·ē'ā·shən }

leakage rate [ENG] Flow rate of all leaks from an evacuated vessel. { 'lēk·ij ‚rāt }

leakage reactance [ELECTROMAG] Inductive reactance due to leakage flux that links only the primary winding of a transformer. { 'lēk·ij rē'ak·təns }

leakage resistance [ELEC] The resistance of the path over which leakage current flows; it is normally high. { 'lēk·ij ri'zis·təns }

leak detector [ENG] An instrument used for finding small holes or cracks in the walls of a vessel; the helium mass spectrometer is an example. { 'lēk di‚tek·tər }

leaking mode [GEOPHYS] A surface seismic wave which is imperfectly trapped, so that its energy leaks or escapes across a layer boundary, causing some attenuation. Also known as leaky wave. { 'lēk·iŋ ‚mōd }

leak-off rate [PETRO ENG] The rate at which a fracturing fluid flows from the fracture into the surrounding formation. { 'lēk ‚óf ‚rāt }

leak test pressure [MECH ENG] The inlet pressure used for a standard quantitative seat leakage test. { 'lēk ‚test ‚presh·ər }

leaky [ELEC] Pertaining to a condition in which the leakage resistance has dropped so much below its normal value that excessive leakage current flows; usually applied to a capacitor. [GEN] Pertaining to a genetic block that is incomplete. [CELL MOL] Pertaining to a protein coded for by a mutant gene that shows subnormal activity. { 'lēk·ē }

leaky mutant gene [GEN] An allele with reduced activity relative to that of the normal allele. { 'lēk·ē 'myüt·ənt 'jēn }

leaky wave *See* leaking mode. { 'lēk·ē 'wāv }

leaky-wave antenna [ELECTROMAG] A wide-band microwave antenna that radiates a narrow beam whose direction varies with frequency; it is fundamentally a perforated waveguide, thin enough to permit flush mounting for aircraft and missile radar applications. { 'lēk·ē 'wāv an‚ten·ə }

lean [MATER] **1.** Of concrete or mortar, containing little or insufficient cement. **2.** Of clay, deficient in plasticity. **3.** Of coal, having little or no volatile matter. **4.** Of lime, containing impurities. **5.** Of fuel mixture, expecially for internal combustion engines, being low in combustible component. **6.** Of ore, being low-grade. { lēn }

lean fuel mixture *See* lean mixture. { 'lēn 'fyül ‚miks·chər }

lean gas [MATER] **1.** In natural-gas absorption processes (such as natural-gasoline recovery from natural gas), gas from which desired liquid components have been removed. **2.** Natural gas poor in butane-and-heavier liquids. { 'lēn 'gas }

leaning wheel grader [CIV ENG] A grader with skewed wheels to help cut or spread the soil. { 'lēn·iŋ 'wēl 'grād·ər }

lean manufacturing [IND ENG] A production system consisting of manufacturing cells linked together with a functionally integrated system for inventory and production control that uses less of the key resources needed to make goods. { 'lēn ‚man·ə'fak·chər·iŋ }

lean manufacturing cells [IND ENG] Typically U-shaped manufacturing cells in which workers, cross-trained on all the

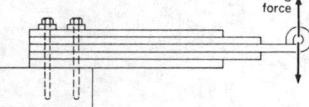

LEAF SPRING

Drawing of a leaf spring showing response to perpendicular force in either direction.

related processes, move from machine to machine in counterclockwise loops. { |lēn ¦man·ə'fak·chər·iŋ selz }

lean mixture [MECH ENG] A fuel-air mixture containing a low percentage of fuel and a high percentage of air, as compared with a normal or rich mixture. Also known as lean fuel mixture. { 'lēn ¦miks·chər }

lean oil [MATER] Absorbent oil from which absorbed gas has been stripped; an example is absorber oil in a natural-gasoline plant from which absorbed liquids (ethane, propane, butane) have been removed. { 'lēn ¦oil }

lean-to [BUILD] A single-pitched roof whose summit is supported by the wall of a higher structure. { 'lēn₁tü }

lean-to roof See shed roof. { 'lēn₁tü ₁rüf }

leapfrog system [MIN ENG] A system used in mining coal on a longwall face; self-advancing supports are used, with alternate supports advancing as each web of coal is removed. { 'lēp₁fräg ₁sis·təm }

leapfrog test [COMPUT SCI] A computer test using a special program that performs a series of arithmetical or logical operations on one group of storage locations, transfers itself to another group, checks the correctness of the transfer, then begins the series of operations again; eventually, all storage positions will have been tested. { 'lēp₁fräg ₁test }

leap year [ASTRON] A year with 366, and not 365, days. { 'lēp ₁yir }

lear See lehr. { lir }

learning [PSYCH] The gathering, processing, storage, and recall of information received through the senses. { 'lər·niŋ }

learning control [CONT SYS] A type of automatic control in which the nature of control parameters and algorithms is modified by the actual experience of the system. { 'lərn·iŋ kən₁trōl }

learning curve [PSYCH] Graphical representation of the relationship between acquisition of knowledge or skill, and the amount of practice or trials. { 'lər·niŋ ₁kərv }

learning machine [COMPUT SCI] A machine that is capable of improving its future actions as a result of analysis and appraisal of past actions. { 'lər·niŋ mə₁shēn }

lease [IND ENG] **1.** Contract between landowner and another granting the latter the right to use the land, usually upon payment of an agreed rental, bonus, or royalty. **2.** A piece of land that is leased. [TEXT] A means of keeping warp threads in position during weaving and beaming by passing them alternately over and under a set of rods. { lēs }

lease automatic custody transfer [PETRO ENG] Automatic unattended system to receive and record oil produced from a drilling lease, then to transfer the contents to a pipeline. Abbreviated LACT. { 'lēs ¦öd·ə¦mad·ik 'kəs·tə·dē ₁trans·fər }

leased facility [COMMUN] A collection of communication lines dedicated to a particular service; sometimes the lines have a predetermined path through system switching equipment. { 'lēst fə'sil·əd·ē }

lease-distribution system [PETRO ENG] Any of the electrical distribution systems serving oil-field pump motors and other electrical equipment. { 'lēs dis·trə'byü·shən ₁sis·təm }

lease rod [TEXT] A rod used to position warp threads during weaving. { 'lēs ₁räd }

lease tank [PETRO ENG] An oil-field storage tank that stores oil flowing from designated wells. Also known as production tank. { 'lēs ₁taŋk }

least-action principle See principle of least action. { ¦lēst 'ak·shən ₁prin·sə·pəl }

least common denominator [MATH] The least common multiple of the denominators of a collection of fractions. { 'lēst 'käm·ən di'näm·ə₁näd·ər }

least common multiple [MATH] The least common multiple of a set of quantities (for example, numbers or polynomials) is the smallest quantity divisible by each of them. Abbreviated lcm. { 'lēst 'käm·ən 'məl·tə·pəl }

least-energy principle [MECH] The principle that the potential energy of a system in stable equilibrium is a minimum relative to that of nearby configurations. { ¦lēst 'en·ər·jē ₁prin·sə·pəl }

least frequently used [COMPUT SCI] A technique for using main storage efficiently, in which new data replace data in storage locations that have been used least often, as determined by an algorithm. { 'lēst ¦frē·kwənt·lē 'yüzd }

least recently used [COMPUT SCI] A technique for using main storage efficiently, in which new data replace data in

storage locations that have not been accessed for the longest period, as determined by an algorithm. { 'lēst ¦rē·sənt·lē ₁yüzed }

least significant bit [COMPUT SCI] The bit that carries the lowest value or weight in binary notation for a numeral; for example, when 13 is represented by binary 1101, the 1 at the right is the least significant bit. Abbreviated LSB. { 'lēst sig¦nif·i·kənt 'bit }

least significant character [COMPUT SCI] The character in the rightmost position in a number or word. { ¦lēst sig¦nif·i·kənt 'kar·ik·tər }

least-squares estimate [STAT] An estimate obtained by the least-squares method. { 'lēst 'skwerz ₁es·tə·mət }

least-squares method [STAT] A technique of fitting a curve close to some given points which minimizes the sum of the squares of the deviations of the given points from the curve. { ¦lēst 'skwerz ₁meth·əd }

least-time principle See Fermat's principle. { ¦lēst 'tīm ₁prin·sə·pəl }

least upper bound [MATH] The least upper bound of a subset A of a set S with ordering < is the smallest element of S which is greater than or equal to every element of A. Abbreviated lub. Also known as supremum (sup). { ¦lēst ¦əp·ər 'baùnd }

least-upper-bound axiom [MATH] The statement that any set of real numbers that has an upper bound also has a least upper bound. { ¦lēst ₁əp·ər 'baùnd 'ak·sē·əm }

least-work theory [MECH] A theory of statically indeterminate structures based on the fact that when a stress is applied to such a structure the individual parts of it are deflected so that the energy stored in the elastic members is minimized. { ¦lēst 'wərk ₁thē·ə·rē }

leather [MATER] Dressed hide or skin of an animal. { 'leth·ər }

leather rot [PL PATH] A hard rot of strawberry caused by the fungus *Phytophthora cactorum*. { 'leth·ər ₁rät }

leaven [FOOD ENG] **1.** A substance (such as yeast) used to produce fermentation. **2.** A substance such as yeast or baking powder used to produce a gas to lighten dough. { 'lev·ən; }

leaving group [ORG CHEM] The group of charged or uncharged atoms that departs during a substitution or displacement reaction. Also known as nucleofuge. { 'lēv·iŋ ₁grüp }

Lebesgue exterior measure [MATH] A measure whose value on a set S is the greatest lower bound of the Lebesgue measures of open sets that contain S. Also known as exterior measure; outer measure. { lə'beg ik¦stir·ē·ər 'mezh·ər }

Lebesgue integral [MATH] The generalization of Riemann integration of real valued functions, which allows for integration over more complicated sets, existence of the integral even though the function has many points of discontinuity, and convergence properties which are not valid for Riemann integrals. { lə'beg ₁int·ə·grəl }

Lebesgue interior measure [MATH] A measure whose value on a set S is the least upper bound of the Lebesgue measures of the closed sets contained in S. Also known as inner measure; interior measure. { lə'beg in¦tir·ē·ər 'mezh·ər }

Lebesgue measure [MATH] A measure defined on subsets of euclidean space which expresses how one may approximate a set by coverings consisting of intervals { lə'beg ₁mezh·ər }

Lebesgue number [MATH] The Lebesgue number of an open cover of a compact metric space X is a positive real number so that any subset of X whose diameter is less than this number must be completely contained in a member of the cover. { lə'beg ₁nəm·bər }

Lebesgue-Stieltjes integral [MATH] A Lebesgue integral of the form

$$\int_b^a f(x)\, d\phi(x)$$

where ϕ is of bounded variation; if $\phi(x) = x$, it reduces to the Lebesgue integral of $f(x)$; if $\phi(x)$ is differentiable, it reduces to the Lebesgue integral of $f(x)\phi'(x)$. { lə'beg 'stēlt·yəs ₁int·ə·grəl }

Lecanicephaloidea [INV ZOO] An order of tapeworms of the subclass Cestoda distinguished by having the scolex divided into two portions; all species are intestinal parasites of elasmobranch fishes. { le₁kan·ə₁sef·ə'lói·d ē·ə }

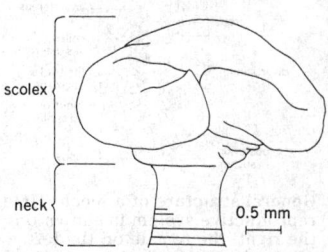

LECANICEPHALOIDEA

scolex

neck

0.5 mm

Anterior end of a lecanicephaloid tapeworm showing scolex and neck.

LECLANCHÉ CELL

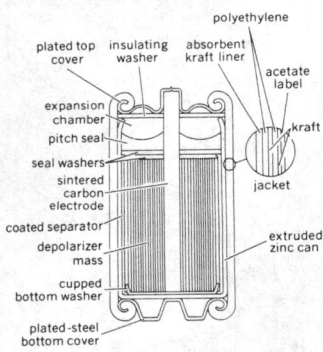

Modern Leclanché dry cell.
(Bright Star Industries Inc.)

LEDE ROOM

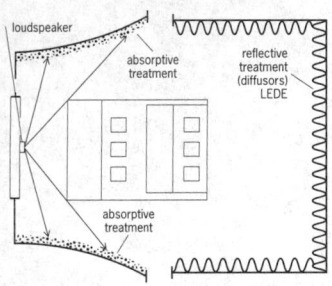

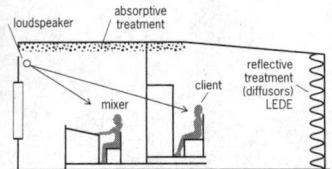

LEDE room with a live rear wall and a sound-absorbent environment at the frontal sidewalls.

LEECH

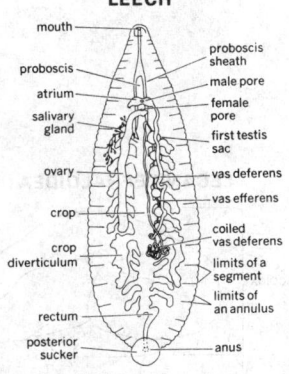

General structure of a leech. Male reproductive system is shown on the right, the female on the left.
(From K. H. Mann, A key to the British freshwater leeches, Freshwater Biol. Assoc. Sci. Publ., 14:3–21, 1954)

Lecanoraceae [BOT] A temperate and boreal family of lichens in the order Lecanorales characterized by a crustose thallus and a distinct thalloid rim on the apothecia. { le·kə·nō'rās·ē,ē }

Lecanorales [BOT] An order of the Ascolichenes having open, discoid apothecia with a typical hymenium and hypothecium. { le·kə·nō'rā·lēz }

lechatelierite [MINERAL] A natural silica glass, occurring in fulgurites and impact craters and formed by the melting of quartz sand at high temperatures generated by lightning or by the impact of a meteorite. { le,shäd·əl'ī,rīt }

Le Chatelier's principle [PHYS] The principle that when an external force is applied to a system at equilibrium, the system adjusts so as to minimize the effect of the applied force. { lə'shäd·əl,yāz ,prin·sə·pəl }

Lecher line See Lecher wires. { 'lek·ər ,līn }

Lecher wires [ELECTROMAG] Two parallel wires that are several wavelengths long and a small fraction of a wavelength apart, used to measure the wavelength of a microwave source that is connected to one end of the wires; a shorting bar which slides along the wires is used to determine the position of standing-wave nodes. Also known as Lecher line; Lecher wire wavemeter. { 'lek·ər ,wīrz }

Lecher wire wavemeter See Lecher wires. { 'lek·ər ¦wīr 'wāv,mēd·ər }

Lecideaceae [BOT] A temperate and boreal family of lichens in the order Lecanorales; members lack a thalloid rim around the apothecia. { lə,sid·ē'ās·ē,ē }

lecithin [BIOCHEM] Any of a group of phospholipids having the general composition $CH_2OR_1 \cdot CHOR_2 \cdot CH_2OPO_2OHR_3$, in which R_1 and R_2 are fatty acids and R_3 is choline, and with emulsifying, wetting, and antioxidant properties. [MATER] **1.** A mixture of phosphatides and oil obtained by drying the separate gums from the degumming of soybean oil; consists of the phosphatides (lecithin), cephalin, other fatlike phosphorus–containing compounds, and 30–35% entrained soybean oil; may be treated to produce more refined grades; used in foods, cosmetics, and paints. Also known as commercial lecithin; crude lecithin; soybean lecithin; soy lecithin. **2.** A waxy mixture of phosphatides obtained by refining commercial lecithin to remove the soybean oil and other materials; used in pharmaceuticals. Also known as refined lecithin. { 'les·ə·thən }

lecithinase [BIOCHEM] An enzyme that catalyzes the breakdown of a lecithin into its constituents. { 'les·ə·thə,nās }

lecithinase A [BIOCHEM] An enzyme that catalyzes the removal of only one fatty acid from lecithin, yielding lipolecithin. { 'les·ə·thə,nās 'ā }

lecithinase C [BIOCHEM] An enzyme that catalyzes the removal of the nitrogenous base of lecithin to produce the base and a phosphatidic acid. { 'les·ə·thə,nās 'sē }

lecithinase D [BIOCHEM] An enzyme that catalyzes the removal of the phosphorylated base from lecithins, producing α-β-diglyceride. { 'les·ə·thə,nās 'dē }

Leclanché cell [ELEC] The common dry cell, which is a primary cell having a carbon positive electrode and a zinc negative electrode in an electrolyte of sal ammoniac and a depolarizer. { lə¦klän¦shä ,sel }

lecontite [MINERAL] $Na(NH_4,K)SO_4 \cdot 2H_2O$ A colorless mineral composed of a hydrous sodium potassium ammonium sulfate; found in bat guano. { lə'kän,tīt }

le cri du chat syndrome [MED] A complex of congenital malformations resulting from a deletion in chromosome 4 or 5 and characterized by mental retardation and the production of a catlike cry. { lə ¦krē dü 'shä ,sin,drōm }

lectin [BIOCHEM] Any of various proteins that agglutinate erythrocytes and other types of cells and also have other properties, including mitogenesis, agglutination of tumor cells, and toxicity toward animals; found widely in plants, predominantly in legumes, and also occurring in bacteria, fish, and invertebrates. { 'lek·tən }

lectotype [SYST] A specimen selected as the type of a species or subspecies if the type was not designated by the author of the classification. { 'lek·tə,tīp }

Lecythidaceae [BOT] The single family of the order Lecythidales. { ¦les·ə·thə'dās·ē,ē }

Lecythidales [BOT] A monofamilial order of dicotyledonous tropical woody plants in the subclass Dilleniidae; distinguished by entire leaves, valvate sepals, separate petals,

numerous centrifugal stamens, and a syncarpous, inferior ovary with axile placentation. { ¦les·ə·thə'dä·lēz }

LED See light-emitting diode.

Leda [ASTRON] A small satellite of Jupiter with a diameter probably less than 5 miles (8 kilometers), orbiting at a mean distance of 6.88×10^6 miles (1.11×10^7 kilometers). Also known as Jupiter XIII. { 'lēd·ə }

ledeburite [MET] The eutectic of the iron-carbon system, the constituents being cementite and austenite at high temperatures; cooling decomposes the austenite to ferrite and cementite. { 'lā·də,bù,rīt }

Lederberg technique [MICROBIO] A method for rapid isolation of individual bacterial cells for demonstrating the spontaneous origin of bacterial mutants. { 'lā·də,berk tek,nēk }

LEDE room [ENG ACOUS] A control room in a sound-recording studio in which the rear wall is made reflective or diffusive, while the dead or sound-absorbent treatment is applied to the frontal sidewalls near the loudspeaker to prevent lateral reflections from mixing with direct signals from the loudspeaker. Derived from live-end-dead-end room. { 'lē,dē ,rüm }

ledge [BUILD] A horizontal timber on the back of a batten door or on a framed and braced door. [ENG] **1.** A raised edge or molding. **2.** A narrow shelf projecting from the side of a vertical structure. **3.** A horizontal timber that supports the put-logs of scaffolding. [GEOL] **1.** A narrow, shelflike ridge or rock protrusion, much longer than high, and usually horizontal, formed in a rock wall or on a cliff face. **2.** A ridge of rocks found underwater, especially one near a shore or connected with and bordering a shore. [MET] In-gate for a foundry mold. [MIN ENG] **1.** A mining quarry exposure. **2.** An outcrop associated with a mineral deposit. { lej }

ledged door See batten door. { 'lejd 'dór }

ledger [CIV ENG] A main horizontal member of formwork, supported on uprights and supporting the soffit of the formwork. [ENG] The horizontal support for a scaffold platform. { 'lej·ər }

ledger balance [COMMUN] Facility used with message switching equipment to check the number of addresses received with the number of addresses transmitted, to ensure that no messages are lost within the center. { 'lej·ər ,bal·əns }

ledger paper [GRAPHICS] A heavy white paper made from rag pulp, chemical wood pulp, or a mixture of both and well suited for ink and wash drawings; it is a strong, durable, and permanent paper, and the best varieties are fully opaque. { 'lej·ər ,pāpər }

Ledian [GEOL] Lower upper Eocene geologic time. Also known as Auversian. { 'lēd·ē·ən }

Ledoux bell meter [ENG] A type of manometer used to measure the difference in pressure between two points generated by any one of several types of flow measurement devices such as a pitot tube; it is equipped with a shaped plug which makes the reading of the meter directly proportional to the flow rate. { lə'dü 'bel ,mēd·ər }

Leduc current [ELEC] An asymmetrical alternating current obtained from, or similar to that obtained from, the secondary winding of an induction coil; used in electrobiology. { lə'dük ,kə·rənt }

Leduc effect See Righi-Leduc effect. { lə'dük i,fekt }

Leduc law See Amagat-Leduc rule. { lə'dük ,lò }

lee [SCI TECH] The side of an object, such as an island or a ship, away from the direction in which the wind is coming, and sheltered from wind or waves. { lē }

Leeaceae [BOT] A family of dicotyledonous plants in the order Rhamnales distinguished by solitary ovules in each locule, simple to compound leaves, a small embryo, and hypogynous flowers. { lē'ās·ē,ē }

leech [INV ZOO] The common name for members of the annelid class Hirudinea. { lēch }

LEED See low-energy electron diffraction.

lee dune [GEOL] A dune formed to the leeward of a source of sand or of an obstacle. { 'lē ,dün }

lee eddies [FL MECH] The small, irregular motions or eddies produced immediately in the rear of an obstacle in a turbulent fluid. { 'lē ,ed·ēz }

leek [BOT] *Allium porrum.* A biennial herb known only by cultivation; grown for its mildly pungent succulent leaves and thick cylindrical stalk. { lēk }

Lee-Norse miner [MIN ENG] A continuous miner for driving headings in medium or thick coal seams. { 'lē 'nórs 'mīn·ər }

leer *See* lehr. { ler }

Lee's disk [THERMO] A device for determining the thermal conductivity of poor conductors in which a thin, cylindrical slice of the substance under study is sandwiched between two copper disks, a heating coil is placed between one of these disks and a third copper disk, and the temperatures of the three copper disks are measured. { 'lēz ,disk }

Leeson disk [OPTICS] The screen sometimes used in a grease-spot photometer, in which the translucent spot at the center is star-shaped to provide a fine line of demarcation. { 'lē·sən ,disk }

lee tide *See* leeward tidal current. { 'lē ,tīd }

lee trough *See* dynamic trough. { 'lē ,tróf }

leeward [SCI TECH] **1.** Situated away from the wind. **2.** On the lee side. { 'lü·ərd, 'lē·wərd }

leeward tidal current [OCEANOGR] A tidal current setting in the same direction as that in which the wind is blowing. Also known as lee tide; leeward tide. { 'lē·wərd ¦tīd·əl 'kə·rənt }

leeward tide *See* leeward tidal current. { 'lē·wərd ,tīd }

lee wave [FL MECH] Any wave disturbance which is caused by, and is therefore stationary with respect to, some barrier in the fluid flow. { 'lē ,wāv }

leeway [NAV] The off-course, leeward motion of a vessel due to wind or current; may be expressed as distance, speed, or angular difference between course steered and course through the water. { 'lē,wā }

LE factor [PATH] A substance in body fluids, especially the blood, of patients with systemic lupus erythematosus, and sometimes of those with other diseases. Derived from lupus erythematosus factor. { 'el¦e ,fak·tər }

left bank [GEOGR] The bank of a stream or river on the left of an observer when he is facing in the direction of flow, or downstream. { 'left ¦baŋk }

left-continuous function [MATH] A function $f(x)$ of a real variable is left-continuous at a point c if $f(x)$ approaches $f(c)$ as x approaches c from the left, that is, $x < c$ only. { 'left kən¦tin·yə·wəs 'fəŋk·shən }

left coset [MATH] A left coset of a subgroup H of a group G is a subset of G consisting of all elements of the form ah, where a is a fixed element of G and h is any element of H. { ¦left 'käs·ət }

left-hand [DES ENG] Of drilling and cutting tools, screw threads, and other threaded devices, designed to rotate clockwise or cut to the left. { 'left ¦hand }

left-hand derivative [MATH] The limit of the difference quotient $[f(x) - f(c)]/[x - c]$ as x approaches c from the left, that is, $x < c$ only. { 'left ,hand də'riv·əd·iv }

left-handed [CRYSTAL] Having a crystal structure with a mirror-image relationship to a right-handed structure. [DES ENG] *See* left-laid. { 'left 'hand·əd }

left-handed coordinate system [MATH] **1.** A three-dimensional rectangular coordinate system such that when the thumb of the left hand extends in the positive direction of the first (or x) axis, the fingers fold in the direction in which the second (or y) axis could be rotated about the first axis to coincide with the third (or z) axis. **2.** A coordinate system of a Riemannian space which has negative scalar density function. { 'left ¦hand·əd kō'órd·ən·ət ,sis·təm }

left-handed curve [MATH] A space curve whose torsion is positive at a given point. Also known as sinistrorse curve; sinistrorsum. { 'left ¦hand·əd ,kərv }

left-hand limit *See* limit on the left. { 'left ¦hand 'lim·ət }

left-hand polarization [ELECTROMAG] In elementary-particle discussions, circular or elliptical polarization of an electromagnetic wave in which the electric field vector at a fixed point in space rotates in the left-hand sense about the direction of propagation; in optics, the opposite convention is used; in facing the source of the beam, the electric vector is observed to rotate counterclockwise. { 'left ¦hand ,pō·lə·rə'zā·shən }

left-hand rule [ELECTROMAG] **1.** For a current-carrying wire, the rule that if the fingers of the left hand are placed around the wire so that the thumb points in the direction of electron flow, the fingers will be pointing in the direction of the magnetic field produced by the wire. **2.** For a current-carrying wire in a magnetic field, such as a wire on the armature of a motor, the rule that if the thumb, first, and second fingers of the left hand are extended at right angles to one another, with the first finger representing the direction of magnetic lines of force and the second finger representing the direction of current flow, the thumb will be pointing in the direction of force on the wire. Also known as Fleming's rule. { 'left ¦hand ,rül }

left-hand screw [DES ENG] A screw that advances when turned counterclockwise. { 'left ¦hand 'skrü }

left-hand taper [ELEC] A taper in which there is greater resistance in the counterclockwise half of the operating range of a rheostat or potentiometer (looking from the shaft end) than in the clockwise half. { 'left ¦hand 'tā·pər }

left identity [MATH] In a set in which a binary operation ∘ is defined, an element e with the property $e ∘ a = a$ for every element a in the set. { ¦left i'den·ə·dē }

left inverse [MATH] For a set S with a binary operation $x∘y$ that has an identity element e, the left inverse of a member, x, of S is another member, $\bar{x}$, of S for which $\bar{x} ∘ x = e$. { ¦left in'vərs }

left-invertible element [MATH] An element x of a groupoid with a unit element e for which there is an element $\bar{x}$ such that $\bar{x} ∘ x = e$. { ¦left in,vərd·ə·bəl 'el·ə·mənt }

left-justify [COMPUT SCI] To shift the contents of a register so that the left, or most significant, digit is at some specified position. [GRAPHICS] To align characters horizontally so that the leftmost character of a string is in a specified position. Also known as flush left. { 'left 'jəs·tə·fī }

left-laid [DES ENG] The lay of a wire or fiber rope or cable in which the individual wires or fibers in the strands are twisted to the right and the strands to the left. Also known as left-handed; regular-lay left twist. { 'left ¦lād }

left lateral fault [GEOL] A fault in which movement is such that an observer walking toward the fault along an index plane (a bed, vein, or dike) would turn to the left to find the other part of the displaced index plane. Also known as sinistral fault. { 'left ¦lad·ə·rəl 'fólt }

left module [MATH] A module over a ring in which the product of a member x of the module and a member a of the ring is written ax. { 'left ,maj·əl }

left rudder [NAV] The operation of moving the rudder, and consequently the bow of the ship, to port. { 'left ¦rəd·ər }

left splicing junction *See* donor splicing site. { 'left ,splīs·iŋ jəŋk·shən }

left value [COMPUT SCI] The memory address of a symbolic variable in a computer program. Abbreviated lvalue. { 'left ,val·yü }

left ventricular thrust *See* apex impulse. { 'left ven¦trik·yə·lər 'thrəst }

leg [ANAT] The lower extremity of a human limb, between the knee and the ankle. [COMPUT SCI] The sequence of instructions that is followed in a computer routine from one branch point to the next. [ENG] **1.** Anything that functionally or structurally resembles an animal leg. **2.** One of the branches of a forked or jointed object. **3.** One of the main upright members of a drill derrick or tripod. [GEOPHYS] A single cycle of more or less periodic motion in a wave train on a seismogram. [MATH] Either side adjacent to the right angle of a right triangle. [MECH ENG] The case that encloses the vertical part of the belt carrying the buckets within a grain elevator. [MET] In a fillet weld, the distance between the root and the toe. [MIN ENG] **1.** In mine timbering, a prop or upright member of a set or frame. **2.** A stone that has to be wedged out from beneath a larger one. [NAV] **1.** One part of a craft's track, consisting of a single course line. **2.** A track identified by an aid to navigation. [ZOO] An appendage or limb used for support and locomotion. { leg }

legacy system [COMPUT SCI] A computer system that has been in operation for a long time, and whose functions are too essential to be disrupted by upgrading or integration with another system. { 'leg·ə·sē ,sis·təm }

legena [VERT ZOO] An appendage of the sacculus containing sensory areas in the inner ear of tetrapods; termed the cochlea in humans. { lə'jē·nə }

legend [GRAPHICS] A title or explanation on a chart, diagram, or other illustration. { 'lej·ənd }

Legendre contact transformation *See* Legendre transformation. { lə'zhän·drə 'kän,tak ,tranz·fər,mā·shən }

Legendre equation [MATH] The second-order linear homogeneous differential equation $(1 - x^2)y'' - 2xy' + v(v + 1)y = 0$, where v is real and nonnegative. { lə'zhän·drə i‚kwā·zhən }

Legendre function [MATH] Any solution of the Legendre equation. { lə'zhän·drə ‚fəŋk·shən }

Legendre polynomials [MATH] A collection of orthogonal polynomials which provide solutions to the Legendre equation for nonnegative integral values of the parameter. { lə'zhän·drə ‚päl·i'nō·mē·əlz }

Legendre's symbol [MATH] The symbol $(c|p)$, where p is an odd prime number, and $(c|p)$ is equal to 1 if c is a quadratic residue of p, and is equal to -1 if c is not a quadratic residue of p. { lə'zhän·drəz ‚sim·bəl }

Legendre transformation [FL MECH] The basis for a version of the hodograph method for compressible flow in which a replacement is made not only of the independent variables but also of the dependent variables, that is, of the velocity potential and the stream function. [MATH] A mathematical procedure in which one replaces a function of several variables with a new function which depends on partial derivatives of the original function with respect to some of the original independent variables. Also known as Legendre contact transformation. { lə'zhän·drə ‚tranz·fər'mā·shən }

Legionella pneumonia *See* Legionnaire's disease. { ‚lē·jə‚nel·ə nə'mō·nyə }

Legionnaire's disease [MED] A type of pneumonia usually caused by infection with the bacterium *Legionella pneumophila* that was first observed at an American Legion convention in Philadelphia, Pennsylvania, in 1976. Symptoms include headache, fever reaching 102–$105°F$ (32–$41°C$), muscle aches, a generalized feeling of discomfort, cough, shortness of breath, chest pains, and sometimes abdominal pain and diarrhea. Also known as Legionella pneumonia. { ‚lē·jə'nerz di‚zēz }

legrandite [MINERAL] $Zn_{14}(OH)(AsO_4)_9·12H_2O$ A yellow to nearly colorless mineral composed of basic hydrous zinc arsenate. { lə'gran‚dīt }

legume [BOT] A dry, dehiscent fruit derived from a single simple pistil; common examples are alfalfa, beans, peanuts, and vetch. { lə'gyüm }

legume forage [AGR] Any plant of the legume family (Leguminosae) used for livestock feed, grazing, hay, or silage. { lə'gyüm ¦fär·ij }

Leguminosae [BOT] The legume family of the plant order Rosales characterized by stipulate, compound leaves, 10 or more stamens, and a single carpel; many members harbor symbiotic nitrogen-fixing bacteria in their roots. { lə‚gyüm·ə'nō·sē }

leg wire [ENG] One of the two wires forming a part of an electric blasting cap or squib. { 'leg ‚wīr }

Lehigh jig [MIN ENG] A plunger-type jig in which check valves open on the upstroke of the plunger, makeup water is introduced with the feed, the screen plate is at two levels, and the bottom of the discharge end is hinged; used to wash anthracite. { 'lē‚hī jig }

lehiite [MINERAL] $(Na,K)_2Ca_5Al_8(PO_4)_8(OH)_{12}·6H_2O$ White mineral composed of hydrous basic calcium aluminum phosphate. { 'lē‚hīt }

Lehmann process [MIN ENG] A process for treating coal by disintegration and separation of the petrographic constituents. { 'lā·mən ‚prä·səs }

lehr [ENG] A long oven in which glass is cooled and annealed after being formed. Also spelled lear; leer. { ler }

Leibnitz's rule [MATH] A formula to compute the nth derivative of the product of two functions f and g:

$$d^n(f \cdot g)/dx^n = \sum_{k=0}^{n} \binom{n}{k} d^{n-k} f/dx^{n-k} \cdot d^k g/dx^k$$

where

$$\binom{n}{k} = n!/(n - k)! \, k!$$

{ 'līb‚nit·səz ‚rül }

Leibnitz's test [MATH] If the sequence of positive numbers a_n approaches zero monotonically, then the series

$$\sum_{n=1}^{\infty} (-1)^n a_n$$

is convergent. { 'līb‚nit·səz ‚test }

Leidenfrost point [THERMO] The lowest temperature at which a hot body submerged in a pool of boiling water is completely blanketed by a vapor film; there is a minimum in the heat flux from the body to the water at this temperature. { 'līd·ən‚fróst ‚póint }

Leidenfrost's phenomenon [THERMO] A phenomenon in which a liquid dropped on a surface that is above a critical temperature becomes insulated from the surface by a layer of vapor, and does not wet the surface as a result. { 'līd·ən‚frósts fə‚nam·ə‚nän }

leifite [MINERAL] $Na_2AlSi_4O_{10}F$ A colorless mineral composed of fluoride and silicate of sodium and aluminum. { 'lē‚fīt }

leightonite [MINERAL] $K_2Ca_2Cu(SO_4)_4·2H_2O$ A pale-blue mineral composed of hydrous sulfate of copper, calcium, and potassium. { 'lāt·ən‚īt }

Leiodidae [INV ZOO] The round carrion beetles, a cosmopolitan family of coleopteran insects in the superfamily Staphylinoidea; commonly found under decaying bark. { lī'äd·ə‚dē }

leiomyofibroma [MED] A benign tumor composed of smooth muscle cells and fibrocytes. { ‚lī·o‚mī·ō·fī'brō·mə }

leiomyoma [MED] A benign tumor composed of smooth muscle cells. { ‚lī·ō·mī'ō·mə }

leiomyosarcoma [MED] A malignant tumor composed of anaplastic smooth muscle cells. { ‚lī·ō‚mī·ō·sär'kō·mə }

leiosporous [MYCOL] Having smooth spores. { lī'äs·pə‚rəs }

Leishman-Donovan bodies [PATH] Small, oval protozoans lacking flagella and undulating membranes, found within macrophages of the skin, liver, and spleen in leishmanial infections such as kala-azar and mucocutaneous leishmaniasis. { 'lēsh·mən 'dän·ə·vən ‚bäd·ēz }

Leishmania [INV ZOO] A genus of flagellated protozoan parasites that are the etiologic agents of several diseases of humans, such as leishmaniasis. { lēsh'man·ē·ə }

Leishmania donovani [INV ZOO] The protozoan parasite that causes kala-azar. { lēsh'man·ē·ə ¦dan·ō¦vän·ē }

Leishmania infantum [INV ZOO] The protozoan parasite that causes infantile leishmaniasis. { lēsh'man·ē·ə in'fan‚təm }

leishmaniasis [MED] Any of several infections caused by *Leishmania* species. { ‚lēsh·mə'nī·ə·səs }

LEIT *See* light emission via inelastic tunneling.

LEIT device [ELECTR] A light source consisting of two crossed, thin metal-film strips separated by a very thin insulating layer and attached to a battery to produce light emission via inelastic tunneling (LEIT). { 'el¦ē¦ī'tē di‚vīs }

Leitneriales [BOT] A monofamilial order of flowering plants in the subclass Hamamelidae; members are simple-leaved, dioecious shrubs with flowers in catkins, and have a superior, pseudomonomerous ovary with a single ovule. { ‚līt·nir·ē'ā·lēz }

lek [ZOO] A gathering place for courtship. Also known as arena. { lek }

Lelapiidae [INV ZOO] A family of calcaronean sponges in the order Sycettida characterized by a rigid skeleton composed of tracts or bundles of modified triradiates. { le·lə'pī·ə‚dē }

L electron [ATOM PHYS] An electron in the L shell. { 'el i‚lek‚trän }

LEM *See* lunar excursion module. { lem }

lemma [BOT] Either of the pair of bracts that are borne above the glumes and enclose the flower of a grass spikelet. [MATH] A mathematical fact germane to the proof of some theorem. { 'lem·ə }

lemma of duBois-Reymond [MATH] A continuous function $f(x)$ is constant in the interval (a,b) if for certain functions g whose integral over (a,b) is zero, the integral over (a,b) of f times g is zero. { 'lem·ə əv dyüb'wä rā'mōn }

lemming [VERT ZOO] The common name for the small burrowing rodents composing the subfamily Microtinae. { 'lem·iŋ }

Lemnaceae [BOT] The duckweeds, a family of monocotyledonous plants in the order Arales; members are small, free-floating, thalloid aquatics with much reduced flowers that form a miniature spadix. { lem'nās·ē‚ē }

lemniscate of Bernoulli [MATH] The locus of points (x,y) in the plane satisfying the equation $(x^2 + y^2)^2 = a^2 (x^2 - y^2)$ or, in polar coordinates (r,θ), the equation $r^2 = a^2 \cos 2\theta$. { lem'nis·kət əv ber'nü·ē }

lemniscate of Gerono See eight curve. { lem'nis·kət əv je'rän·ō }

lemniscus [NEUROSCI] A secondary sensory pathway of the central nervous system, usually terminating in the thalamus. { lem'nis·kəs }

lemon [BOT] *Citrus limon.* A small evergreen tree belonging to the order Sapindales cultivated for its acid citrus fruit which is a modified berry called a hesperidium. { 'lem·ən }

lemongrass oil [MATER] An essential oil with the odor of lemon, distilled from either of two lemongrasses (*Cymbopogon citratus* or *C. flexuosus*) in the East Indies; contains citral, citronellol, and geraniol; used as a perfume and as a flavoring. { 'lem·ən‚gras ‚öil }

lemon oil [MATER] A yellow essential oil squeezed from lemon rind; it is high in citral, limonene, terpinol, and citronellol; used in soap, perfumes, and flavors. { 'lem·ən ‚öil }

lemur [VERT ZOO] The common name for members of the primate family Lemuridae; characterized by long tails, foxlike faces, and scent glands on the shoulder region and wrists. { 'lē·mər }

Lemuridae [VERT ZOO] A family of prosimian primates of Madagascar belonging to the Lemuroidea; all members are arboreal forest dwellers. { lə'myùr·ə‚dē }

Lemuroidea [VERT ZOO] A suborder or superfamily of Primates including the lemurs, tarsiers, and lorises, or sometimes simply the lemurs. { ‚lem·ə'ròid·ē·ə }

Lenard rays [ELECTR] Cathode rays produced in air by a Lenard tube. { 'lā‚närt ‚rāz }

Lenard's mass absorption law See mass absorption law. { 'lā·närdz ¦mas əp¦sòrp·chən ‚lò }

Lenard spiral [ENG] A type of magnetometer consisting of a spiral of bismuth wire and a Wheatstone bridge to measure changes in the resistance of the wire produced by magnetic fields and as a result of the transverse magnetoresistance of bismuth. { 'lā·närd ‚spī·rəl }

Lenard tube [ELECTR] An early experimental electron-beam tube that had a thin glass or metallic foil window at the end opposite the cathode, through which the electron beam could pass into the atmosphere. { 'lā‚närt ‚tüb }

lengenbachite [MINERAL] $Pb_6(Ag,Cu)_2As_4S_{13}$ A steel gray mineral consisting of lead, silver, and copper arsenic sulfide. { 'leŋ·ən‚bä‚kīt }

length [MECH] Extension in space. { leŋkth }

length between perpendiculars [NAV ARCH] The summer load waterline as measured from the fore side of the stern post or to the center of the rudder stock. { 'leŋkth bi‚twēn ‚pər·pən'dik·yə·lərz }

length block [COMPUT SCI] The total number of records, words, or characters contained in one block. { 'leŋkth ‚bläk }

lengthened dipole [ELECTROMAG] An antenna element with lumped inductance to compensate an end loss. { 'leŋk·thənd 'dī‚pōl }

lengthening joint [ENG] A joint between two members running in the same direction. { 'leŋk·thə‚niŋ ‚jòint }

lengthening reaction [PHYSIO] Sudden inhibition of the stretch reflex when extensor muscles are subjected to an excessive degree of stretching by forceful flexion of a limb. { 'leŋk·thə‚niŋ rē‚ak·shən }

length-height index [ANTHROPO] The ratio of the basion-bregma height of the skull to its greatest length multiplied by 100. { 'leŋkth'hīt ‚in‚deks }

length of a curve [MATH] A curve represented by $x = x(t)$, $y = y(t)$ for $t_1 \le t \le t_2$, with $x(t_1) = x_1$, $x(t_2) = x_2$, $y(t_1) = y_1$, $y(t_2) = y_2$, has length from (x_1,y_1) to (x_2,y_2) given by the integral from t_1 to t_2 of the function $\sqrt{(dx/dt)^2 + (dy/dt)^2}$. { 'leŋkth əv ə 'kərv }

length of barrel See length of bore. { 'leŋkth əv 'bar·əl }

length of bore [ORD] The length of the bore, measured from the rear face of the tube or barrel to the muzzle, expressed in inches or calibers. Also known as length of barrel; length of cannon; length of tube. { 'leŋkth əv 'bòr }

length of cannon See length of bore. { 'leŋkth əv 'kan·ən }

length of lay [DES ENG] The distance measured along a line parallel to the axis of the rope in which the strand makes one complete turn about the axis of the rope, or the wires make a complete turn about the axis of the strand. { 'leŋkth əv 'lā }

length of record [CLIMATOL] The period during which observations have been maintained at a meteorological station, and which serves as the frame of reference for climatic data at that station. { 'leŋkth əv 'rek·ərd }

length of shot [ENG] The depth of the shothole, in which powder is placed, or the size of the block of coal or rock to be loosened by a single blast, measured parallel with the hole. [MIN ENG] In open-pit mining, the distance from the first drill hole to the last drill hole along the bank. { 'leŋkth əv 'shät }

length of tube See length of bore. { 'leŋkth əv 'tüb }

length on waterline [NAV ARCH] The length along the designed waterline from the forepart of the stem to the afterpart of the stern. { 'leŋkth òn 'wòd·ər‚līn }

length overall [NAV ARCH] The total length of a ship's hull from the foremost to the aftermost points. { 'leŋkth ¦ō·vər'òl }

lenitic See lentic. { lə'nid·ik }

Lennard-Jones potential [PHYS CHEM] A semiempirical approximation to the potential of the force between two molecules, given by $v = (A/r^{12}) - (B/r^6)$, where r is the distance between the centers of the molecules, and A and B are constants. { 'len·ərd 'jōnz pə‚ten·chəl }

leno [TEXT] A mesh-weave cotton or cotton-and-polyester fabric made by twisting two warp yarns around each other. { 'lē·nō }

lens [ANAT] A transparent, encapsulated, nearly spherical structure located behind the pupil of vertebrate eyes, and in the complex eyes of many invertebrates, that focuses light rays on the retina. Also known as crystalline lens. [COMMUN] A dielectric or metallic structure that is highly transparent to radio waves and can bend them to produce a desired radiation pattern; used with antennas for radar and microwave relay systems. [ELECTR] See electron lens. [ELECTROMAG] See magnetic lens. [GEOL] **1.** A geologic deposit that is thick in the middle and converges toward the edges, resembling a convex lens. **2.** An irregularly shaped formation consisting of a porous, permeable sedimentary deposit surrounded by impermeable rock. [MATER] See acoustic lens. [OPTICS] A curved piece of ground and polished or molded material, usually glass, used for the refraction of light, its two surfaces having the same axis; or two or more such surfaces cemented together. Also known as optical lens. { lenz }

lens antenna [ELECTROMAG] A microwave antenna in which a dielectric lens is placed in front of the dipole or horn radiator to concentrate the radiated energy into a narrow beam or to focus received energy on the receiving dipole or horn. { 'lenz an‚ten·ə }

lens coating [OPTICS] A transparent substance coated on an optical surface to derive maximum light transmission. { 'lenz ‚kōd·iŋ }

lens crystallins [CELL MOL] A diverse family of water-soluble proteins that constitute up to 90% of the proteins found in the eye lens and together play a structural role by orienting themselves to facilitate refraction of light; some of the individual proteins exhibit other distinct functions when expressed elsewhere in the organism. { ¦lenz 'kris·tə‚linz }

lens element [OPTICS] Separate component lens of a multi-element lens. { 'lenz ‚el·ə·mənt }

lens equation [OPTICS] Any equation which relates the distance of a point object from some well-defined reference point in an optical system to the distance of its image from a similar point. { 'lenz i‚kwā·zhən }

Lense-Thirring effect See dragging of inertial frames. { 'len·zə 'tir·iŋ i‚fekt }

lens placode [EMBRYO] The ectodermal anlage of the lens of the eye; its formation is induced by the presence of the underlying optic vesicle. { 'lenz 'pla‚kōd }

lens shim [OPTICS] Thin piece of material used to position and focus a lens. { 'lenz ‚shim }

lens stop See diaphragm. { 'lenz stäp }

LEMMING

A lemming, a small burrowing rodent with short legs and stout claws adapted for digging.

LEMUR

Ring-tailed lemur, measuring 4 feet (1.2 meters), including tail.

LENTICEL

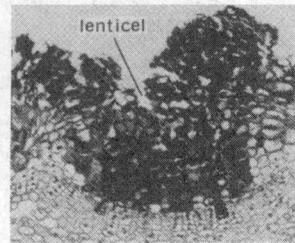

0.2 mm

A lenticel as seen in a cross section of the stem periderm of *Sambucus nigra*.

lens tissue [MATER] Specially prepared paperlike material for cleaning lenses. { 'lenz ,tish·ü }

lentic [ECOL] Of or pertaining to still waters such as lakes, reservoirs, ponds, and bogs. Also spelled lenitic. { 'len·tik }

lenticel [BOT] A loose-structured opening in the periderm beneath the stomata in the stem of many woody plants that facilitates gas transport. { 'len·tə·səl }

lenticle [GEOL] A bed or rock stratum or body that is lens-shaped. { 'len·tə·kəl }

lenticular [OPTICS] Of or pertaining to a lens. [SCI TECH] Having the shape of a lentil or double convex lens. { len'tik·yə·lər }

lenticular cloud See lenticularis. { len'tik·yə·lər 'klaud }

lenticular film process [GRAPHICS] An additive color-photography process in which the camera lens carries a three-color banded filter, the film is embossed on the back with minute lenses facing the camera lens, development by reversal produces a photograph broken up into minute elements each of which is an image of the filter, and projection through a similar filter gives a color reproduction. { len'tik·yə·lər 'film ,prä·səs }

lenticular galaxy [ASTRON] A galaxy of type S0, consisting of a nucleus surrounded by a disklike structure without arms, and containing little gas and few if any young stars. { len'tik·yə·lər 'gal·ik·sē }

lenticularis [METEOROL] A cloud species, the elements of which have the form of more or less isolated, generally smooth lenses; the outlines are sharp. Also known as lenticular cloud. { len,tik·yə'lar·əs }

lenticule [GRAPHICS] One of the minute lenses embossed on the back of a color photographic film in the lenticular film process. { 'len·tə,kyül }

lentiginose [ANAT] Of or pertaining to pigment spots in the skin; freckled. { len'tij·ə,nōs }

lentil [BOT] *Lens esculenta*. A seminivy annual legume having pinnately compound, vetchlike leaves; cultivated for its thin, lens-shaped, edible seed. [GEOL] **1.** A rock body that is lens-shaped and enclosed in a stratum of different material. **2.** A rock stratigraphic unit that is a subdivision of a formation and has limited geographic extent; it thins out in all directions. { 'lent·əl }

lentor See stoke. { 'len,tor }

Lenz's law [ELECTROMAG] The law that whenever there is an induced electromotive force (emf) in a conductor, it is always in such a direction that the current it would produce would oppose the change which causes the induced emf. { 'lenz·əz ,lo }

leo [MECH] A unit of acceleration, equal to 10 meters per second per second; it has rarely been employed. { 'lē·ō }

Leo [ASTRON] A northern constellation, right ascension 11 hours, declination 15° north. Also known as Lion. { 'lē·ō }

LEO See low-altitude earth orbit. { 'lē·ō or ¦el¦ē'ō }

Leo I system [ASTRON] A dwarf elliptical galaxy about 890,000 light-years distant, having a diameter of about 5000 light-years, and a luminosity about 5×10^6 that of the sun. { 'lē·ō 'wən ,sis·təm }

Leo II system [ASTRON] A dwarf elliptical galaxy about 700,000 light-years distant, having a diameter of about 5000 light-years and a luminosity about 600,000 times that of the sun. { 'lē·ō 'tü ,sis·təm }

Leo Minor [ASTRON] A northern constellation, right ascension 10 hours, declination 35° north. Also known as Lesser Lion. { 'lē·ō 'mīn·ər }

Leonardian [GEOL] A North American provincial series: Lower Permian (above Wolfcampian, below Guadalupian). { ¦lā·ə¦när·dē·ən }

Leon firedamp tester [MIN ENG] A device, based on a form of Wheatstone bridge, that is used to detect firedamp. { 'lē,än 'fīr,damp ,tes·tər }

Leonids [ASTRON] A meteor shower, the radiant of which lies in the constellation Leo; it is visible between November 10 and 15. { 'lē·ə·nədz }

leonite [MINERAL] $K_2Mg(SO_4)_2 \cdot 4H_2O$ A colorless, white, or yellowish mineral composed of hydrous magnesium potassium sulfate, occurring in monoclinic crystals. { 'lē·ə,nīt }

leopard [VERT ZOO] *Felis pardus*. A species of wildcat in the family Felidae found in Africa and Asia; the coat is characteristically buff-colored with black spots. { 'lep·ərd }

leopoldite See sylvite. { 'lē·ə,pōl,dīt }

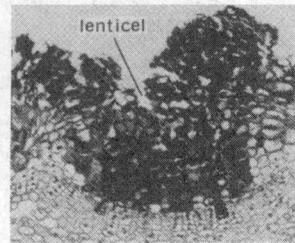

LEO

right ascension

Line pattern of the constellation Leo. The grid lines represent the coordinates of the sky. The apparent brightness, or magnitude, of the stars is shown by the sizes of the dots, which are graded by appropriate numbers.

Leotichidae [INV ZOO] A small Oriental family of hemipteran insects in the superfamily Leptopodoidea. { ,lē·ə'tik·ə,dē }

Lepadomorpha [INV ZOO] A suborder of barnacles in the order Thoracica having a peduncle and a capitulum which is usually protected by calcareous plates. { ,lep·ə·də'mor·fə }

LEPD See low-energy positron diffraction.

leper [MED] A person afflicted with Hansen's disease. { 'lep·ər }

Leperditicopida [PALEON] An order of extinct ostracods characterized by very thick, straight-backed valves which show unique muscle scars and other markings. { ,le·pər,did·ə'käp·ə·də }

Leperditillacea [PALEON] A superfamily of extinct paleocopan ostracods in the suborder Kloedenellocopina including the unisulcate, nondimorphic forms. { ,le·pər,did·ə'lās·ē·ə }

Lepiceridae [INV ZOO] Horn's beetle, a family of Central American coleopteran insects composed of two species. { ,lep·ə'ser·ə,dē }

lepidine [ORG CHEM] $C_9H_6NCH_3$ An alkaloid derived as an oily liquid from cinchona bark; boils at 266°C; soluble in ether, benzene, and alcohol; used in organic synthesis. { 'lep·ə,dēn }

lepidoblastic [PETR] Of the texture of a metamorphic rock, having a fabric of minerals characterized as flaky or scaly, such as mica. { ¦lep·ə·dō¦blas·tik }

Lepidocentroida [INV ZOO] The name applied to a polyphyletic assemblage of echinoids that are now regarded as members of the Echinocystitoida and Echinothurioida. { ,lep·ə·dō,sen'troid·ə }

lepidocrocite [MINERAL] α-FeO(OH) A ruby- or blood-red mineral crystallizing in the orthorhombic system; it is associated with limonite in iron ores and is a component of meteorites. { ,lep·ə·dō'krō,sīt }

Lepidodendrales [PALEOBOT] The giant club mosses, an order of extinct lycopods (Lycopodiopsida) consisting primarily of arborescent forms characterized by dichotomous branching, small amounts of secondary vascular tissue, and heterospory. { ,lep·ə·dō,den'drä·lēz }

lepidolite [MINERAL] $K(Li,Al)_3(Si,Al)_4O_{10}(F,OH)_2$ A rose-colored mineral of the mica group crystallizing in the monoclinic system. Also known as lithionite; lithium mica. { lə'pid·əl,īt }

lepidomelane [MINERAL] A black variety of biotite that is characterized by the presence of large amounts of ferric iron. Also known as iron mica. { ,lep·ə·dō'me,lān }

lepidophyllous [BOT] Having scaly leaves. { ¦lep·ə·dō¦fil·əs }

Lepidoptera [INV ZOO] Large order of scaly-winged insects, including the butterflies, skippers, and moths; adults are characterized by two pairs of membranous wings and sucking mouthparts, featuring a prominent, coiled proboscis. { ,lep·ə'däp·tə·rə }

Lepidorsirenidae [VERT ZOO] A family of slender, obligate air-breathing, eellike fishes in the order Dipteriformes having small thin scales, slender ribbonlike paired fins, and paired ventral lungs. { ,lep·ə·dō·sə'ren·ə,dē }

Lepidosaphinae [INV ZOO] A family of homopteran insects in the superfamily Coccoidea having dark-colored, noncircular scales. { ,lep·ə·dō'saf·ə,nē }

Lepidosauria [VERT ZOO] A subclass of reptiles in which the skull structure is characterized by two temporal openings on each side which have reduced bony arcades, and by the lack of an antorbital opening in front of the orbit. { ,lep·ə·dō'sor·ē·ə }

Lepidotrichidae [INV ZOO] A family of wingless insects in the order Thysanura. { ,lep·ə·dō'trik·ə,dē }

Lepismatidae [INV ZOO] A family of silverfish in the order Thysanura characterized by small or missing compound eyes. { ,lep·əz'mad·ə,dē }

Lepisostei [VERT ZOO] An equivalent name for Semionotiformes. { ,lep·ə'säs·tē,ī }

Lepisosteidae [VERT ZOO] A family of fishes in the order Semionotiformes. { ,lep·ə·säs'tē·ə,dē }

Lepisosteiformes [VERT ZOO] An equivalent name for Semionotiformes. { ,lep·ə,säs·tē·ə'for,mēz }

lepisphere [PETR] A microspherical aggregate of platy, blade-shaped crystals of opal-CT. { 'lep·ə,sfir }

Lepley-Hull hypothesis [PSYCH] An explanation of the

empirical phenomenon in serial learning which postulates both direct and remote associations among items, with the former excitatory and the latter inhibitory. { 'lep·lē 'həl ˌhī͞ˌpäth·ə·ses }

Leporidae [VERT ZOO] A family of mammals in the order Lagomorpha including the rabbits and hares. { lə'pór·ə͵dē }

Lepospondyli [PALEON] A subclass of extinct amphibians including all forms in which the vertebral centra are formed by ossification directly around the notochord. { ˌlep·ə'spänd·əl͵ī }

lepospondylous [VERT ZOO] Having the notochord enclosed by cylindrical vertebrae shaped like an hourglass in longitudinal section. { ˌlep·ə'spänd·əl·əs }

leproma [MED] The cutaneous nodular lesion of leprosy. { le'prō·mə }

lepromatous leprosy [MED] A severely debilitating form of Hansen's disease characterized by the presence of multiple nodular lesions (lepromata) on the skin. { lə͵prä·məd·əs 'lep·rə·sē }

lepromin [IMMUNOL] An emulsion of ground lepromata containing the leprosy bacillus; used for intradermal skin tests in Hansen's disease. { lə'prō·mən }

leprophobia [PSYCH] An abnormal fear of Hansen's disease. { ˌlep·rə'fō·bē·ə }

leprosarium [MED] An institution for the treatment of lepers. { ˌlep·rə'sar·ē·əm }

leprosy See Hansen's disease. { 'lep·rə·sē }

Leptaleinae [INV ZOO] A subfamily of the Formicidae including largely arboreal ant forms which inhabit plants in tropical and subtropical regions. { ˌlep·tə'lī͵nē }

Leptictidae [PALEON] A family of extinct North American insectivoran mammals belonging to the Proteutheria which ranged from the Cretaceous to middle Oligocene. { ˌlep'tik·tə͵dē }

leptine [ANTHRO] Having a high, a narrow, or a high and narrow forehead with an upper facial index of 55–60 on the skull and of 53–57 on the living person. { 'lep͵tīn }

Leptinidae [INV ZOO] The mammal nest beetles, a small European and North American family of coleopteran insects in the superfamily Staphylinoidea. { lep'tin·ə͵dē }

leptite [PETR] A quartz-feldspathic metamorphic rock that is fine-grained with little or no foliation; formed by regional metamorphism of the highest grade. { 'lep͵tīt }

Leptocardii [ZOO] The equivalent name for Cephalochordata. { ˌlep·tə'kärd·ē͵ī }

leptocephalous larva [VERT ZOO] The marine larva of the fresh-water European eel *Anguilla vulgaris*. { ˌlep·tə'sef·ə·ləs 'lär·və }

leptocephaly [ANTHRO] Abnormal narrowness and tallness of the head or skull. { ˌlep·tə'sef·ə·lē }

leptocercal [VERT ZOO] Of the tail of a fish, tapering to a long, slender point. { ˌlep·tə'sər·kəl }

Leptochoeridae [PALEON] An extinct family of palaeodont artiodactyl mammals in the superfamily Dichobunoidea. { ˌlep·tə'kir·ə͵dē }

Leptodactylidae [VERT ZOO] A large family of frogs in the suborder Procoela found principally in the American tropics and Australia. { ˌlep·tə͵dak'til·ə͵dē }

leptodactylous [VERT ZOO] Having slender toes, as certain birds. { ˌlep·tə'dak·tə·ləs }

Leptodiridae [INV ZOO] The small carrion beetles, a cosmopolitan family of coleopteran insects in the superfamily Staphylinoidea. { ˌlep·tə'dir·ə͵dē }

leptogeosyncline [GEOL] A deep oceanic trough that has not been filled with sedimentation and is associated with volcanism. { ˌlep·tə͵jē·ō'sin͵klīn }

leptokurtic distribution [STAT] A distribution in which the ratio of the fourth moment to the square of the second moment is greater than 3, which is the value for a normal distribution; it appears to be more heavily concentrated about the mean, or more peaked, than a normal distribution. { ˌlep·tə͵kərd·ik ˌdi·strə'byü·shən }

Leptolepidae [PALEON] An extinct family of fishes in the order Leptolepiformes representing the first teleosts as defined on the basis of the advanced structure of the caudal skeleton. { ˌlep·tə'lep·ə͵dē }

Leptolepiformes [PALEON] An extinct order of small, ray-finned teleost fishes characterized by a relatively strong, ossified axial skeleton, thin cycloid scales, and a preopercle with an elongated dorsal portion. { ˌlep·tə͵lep·ə'fōr͵mēz }

Leptomedusae [INV ZOO] A suborder of hydrozoan cnidarians in the order Hydroida characterized by the presence of a hydrotheca. { ˌlep·tō·mə'dü·sē }

leptomeninges [ANAT] The pia mater and arachnoid considered together. { ˌlep·tō·mə'nin·jēz }

leptomeningitis [MED] Inflammation of the leptomeninges of the brain or the spinal cord. { ˌlep·tō͵men·ən'jīd·əs }

Leptomitales [MYCOL] An order of aquatic Phycomycetes characterized by a hyphal thallus, or basal rhizoids and terminal hyphae, and zoospores with two flagella. { ˌlep·tō·mī'tā·lēz }

lepton [PARTIC PHYS] A fermion having a mass smaller than the proton mass; leptons interact with electromagnetic and gravitational fields, but beyond this they interact only through weak interactions. { 'lep͵tän }

lepton conservation [PARTIC PHYS] The principle that the number of electrons and *e*-neutrinos minus the number of positrons and *e*-antineutrinos is unchanged in any interaction; similarly, the number of negatively charged muons and μ-neutrinos minus the number of positively charged muons and μ-antineutrinos is unchanged. { 'lep͵tän ˌkän·sər͵vā·shən }

lepton era [ASTRON] The period in the early universe, following the hadron era, during which electrons, positrons, neutrinos, and photons were present in nearly equal numbers; roughly between 10^{-4} and 20 seconds after the big bang. { 'lep͵tän ˌir·ə }

leptonic decay [PARTIC PHYS] Decay of an elementary particle in which at least some of the products are leptons. { lep 'tän·ik di'kā }

lepton number [PARTIC PHYS] A conserved quantum number, equal to the number of leptons minus the number of antileptons in a system. { 'lep͵tän ˌnəm·bər }

leptophos [ORG CHEM] $C_{13}H_{10}BrCl_2O_2PS$ A white solid with a melting point of 70.2–70.6°C; slight solubility in water; used as an insecticide on vegetables, fruit, turf, and ornamentals. Also known as *O*-(4-bromo-2,5-dichlorophenyl) *O*-methyl phenylphosphorothioate. { 'lep·tə͵fäs }

leptophyll [ECOL] A growth-form class of plants having a leaf surface area of 0.04 square inch (25 square millimeters) or less; common in alpine and desert habitats. { 'lep·tə͵fil }

Leptopodidae [INV ZOO] A tropical and subtropical family of hemipteran insects in the superfamily Leptopodoidea distinguished by the spiny body and appendages. { ˌlep·tə'päd·ə͵dē }

Leptopodoidea [INV ZOO] A superfamily of hemipteran insects in the subdivision Geocorisae. { ˌlep·tə·pə'dóid·ē·ə }

leptoquark [PARTIC PHYS] A hypothetical elementary particle; a colored technipion with a mass of the order of 100–300 GeV, which would decay into a lepton plus a quark. { 'lep·tō͵kwärk }

leptoquark boson [PARTIC PHYS] A charged-vector gauge boson postulated in grand unified theories of color and electroweak forces, with a mass of the order of 10^{15} GeV, that can change quarks to leptons or quarks to antiquarks, and is responsible for proton decay. { 'lep·tō͵kwärk 'bō͵sän }

leptorrhine [ANTHRO] Having a long, narrow nose with a nasal index of less than 47 on the skull or of less than 70 on the living person. { 'lep·tə͵rīn }

Leptosomatidae [VERT ZOO] The cuckoo rollers, a family of Madagascan birds in the order Coraciiformes composed of a single species distinguished by the downy covering on the newly hatched young. { ˌlep·tə·sə'mad·ə͵dē }

Leptospira [MICROBIO] A genus of bacteria in the family Spirochaetaceae; thin, helical cells with bent or hooked ends. { ˌlep·tə'spī·rə }

leptospirosis [MED] Infection with spirochetes of the genus *Leptospira*. { ˌlep·tə·spə'rō·səs }

leptospirosis icterohemorrhagia See Weil's disease. { ˌlep·tə·spə'rō·səs ˌik·tə·rō͵hem·ə'raj·ē͵ə }

leptosporangium [BOT] A sporangium derived from a single actively dividing cell in a meristem. { ˌlep·tō·spə'ran·jē·əm }

leptostaphyline [ANTHRO] Having a high, narrow palate with a palatal index of less than 80 on the skull. { ˌlep·tə͵staf·ə͵līn }

Leptostraca [INV ZOO] A primitive group of crustaceans considered as one of a series of Malacostraca distinguished by

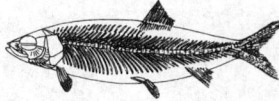

LEPTOLEPIDAE

Leptolepis dubia restoration. *(From A. S. Woodward, Catalogue of the Fossil Fishes in the British Museum (Natural History), pt. 3, 1895)*

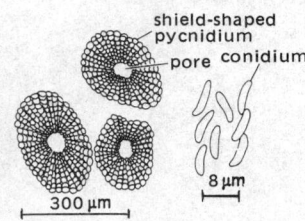

LEPTOSTROMATACEAE

shield-shaped pycnidium
pore conidium

8 μm
300 μm

Pycnidia and conidia of
Leptothyrium vulgare.

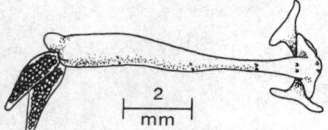

LERNAEIDAE

2 mm

Lernaea cyprinacea, the female,
showing the branched processes
of the head that anchor the
parasite deep in the flesh of the
host.

an additional abdominal somite that lacks appendages, and a telson bearing two movable articulated prongs. { lep'täs·trə·kə }

Leptostromataceae [MYCOL] A family of fungi of the order Sphaeropsidales; pycnidia are black and shield-shaped, circular or oblong, and slightly asymmetrical; included are some fruit-tree pathogens. { ˌlep·tə͵strō·mə'tās·ē͵ē }

leptotene [CYTOL] The first stage of meiotic prophase, when the chromosomes appear as thin threads having well-defined chromomeres. { 'lep·tə͵tēn }

Leptothrix [MICROBIO] A genus of sheathed bacteria; single cells are motile by means of polar or subpolar flagella, and sheaths are encrusted with iron or manganese oxides. { 'lep·tə͵thriks }

Leptotrichia [MICROBIO] A genus of bacteria in the family Bacteroidaceae; straight or slightly curved rods with pointed or rounded ends arranged in filaments. { ˌlep·tə'trik·ē·ə }

Leptotyphlopidae [VERT ZOO] A family of small, harmless, burrowing circumtropical snakes (Serpentes) in the order Squamata; teeth are present only on the lower jaw and are few in number. { ˌlep·tō·ti'fläp·ə͵dē }

Lepus [ASTRON] A southern constellation, right ascension 5.5 hours, declination 20°S. Also known as Hare. [VERT ZOO] The type genus of the family Leporidae, comprising the typical hares. { 'lē·pəs }

leren [AGR] *Calathea allouia.* A tuber food crop of the family Marautaceae cultivated in the Caribbean and in Central and South America. { 'ler·ən }

Lernaeidae [INV ZOO] A family of copepod crustaceans in the suborder Caligoida; all are fixed ectoparasites, that is, they penetrate the skin of fresh-water fish. { lər'nē·ə͵dē }

Lernaeopodidae [INV ZOO] A family of ectoparasitic crustaceans belonging to the Lernaeopodoida; individuals are attached to the walls of the fishes' gill chambers by modified second maxillae. { ˌlər·nē·ə'päd·ə͵dē }

Lernaeopodoida [INV ZOO] The fish maggots, a group of ectoparasitic crustaceans characterized by a modified postembryonic development reduced to two or three stages, a free-swimming larva, and the lack of external signs of physical maturity in adults. { ˌlər·nē·ə'póid·ē·ə }

LES See land-earth station.

lesbian [PSYCH] **1.** Pertaining to female homosexuality. **2.** A female homosexual. { 'lez·bē·ən }

Lesch-Nyhan syndrome [MED] A hereditary disease of male children, transmitted as an X-linked recessive, characterized by hyperuricemia, deficiency of hypoxanthine-guanine phosphoribosyl transferase, mental retardation, spastic cerebral palsy, choreathetosis, and self-mutilating biting. { 'lesh 'nī͵han ͵sin͵drōm }

lesion [BIOL] A structural or functional alteration due to injury or disease. [MOL BIO] A damaged site in a gene, chromosome, or protein molecule. { 'lē·zhən }

Leskeineae [BOT] A suborder of mosses in the order Hypnobryales; plants are not complanate, paraphyllia are frequently present, leaves are costate, and alar cells are not generally differentiated. { ˌles'kī·nē͵ē }

Leslie cube [THERMO] A metal box, with faces having different surface finishes, in which water is heated and next to which a thermopile is placed in order to compare the heat emission properties of different surfaces. { 'lez·lē ͵kyüb }

Leslie effect [ENG ACOUS] A dynamic timbre-changing effect created by rotating one or more directional speakers inside a cabinet such that a mixture of Doppler-shifted reflections is generated in the output of an electronic instrument. { 'lez·lē i͵fekt }

lespedeza [BOT] Any of various legumes of the genus *Lespedeza* having trifoliate leaves, small purple pea-shaped blossoms, and one seed per pod. { ˌles·pə'dē·zə }

lesser circulation See pulmonary circulation. { 'les·ər ͵sər·kyə'lā·shən }

Lesser Dog See Canis Minor. { 'les·ər ͵dóg }

lesser ebb [OCEANOGR] The weaker of two ebb currents occurring during a tidal day. { 'les·ər ͵eb }

lesser flood [OCEANOGR] The weaker of two flood currents occurring during a tidal day. { 'les·ər ͵fläd }

Lesser Lion See Leo Minor. { 'les·ər ͵lī·ən }

lesser omentum [ANAT] A fold of the peritoneum extending from the lesser curvature of the stomach to the transverse hepatic fissure. { 'les·ər ō'men·təm }

less-than-carload [IND ENG] Too light to fill a freight car and therefore not eligible for carload rate. Abbreviated LCL. { 'les thən 'kär͵lód }

leste [METEOROL] Spanish nautical term for east wind, specifically the hot, dry, dusty easterly or southeasterly wind which blows from the Atlantic coast of Morocco out to Madeira and the Canary Islands; it is a form of sirocco, occurring in front of depressions advancing eastward. { 'les·tä }

Lestidae [INV ZOO] A family of odonatan insects belonging to the Zygoptera; distinguished by the wings being held in a V position while at rest. { 'les·tə͵dē }

Lestoniidae [INV ZOO] A monospecific family of hemipteran insects in the superfamily Pentatomorpha found only in Australia. { ˌles·tə'nī·ə͵dē }

LET See stopping power.

letdown [AERO ENG] Gradual and orderly reduction in altitude, particularly in preparation for landing. { 'let͵daún }

lethal area [ORD] In terminal ballistics, a figure of merit, having the dimension of area, which permits prediction of the number of casualties a missile may be expected to produce when employed under specified conditions; an equation states the relationship between the lethal area and the numerous factors affecting its numerical value. { 'lē·thəl 'er·ē·ə }

lethal concentration 50 [PHYSIO] In a fire, the concentration of a gas or smoke that will kill 50% of the test animals within a given time of exposure. Abbreviated LC_{50}. { 'leth·əl ͵käns·ən͵trā·shən 'fif·tē }

lethal dose 50 [PHARM] The dose of a substance which is fatal to 50% of the test animals. Abbreviated LD_{50}. Also known as median lethal dose. { 'lē·thəl ͵dōs 'fif·tē }

lethal equivalent value [GEN] The product of the mean number of deleterious genes carried by each member of a population and the mean probability that each gene will cause premature death when homozygous. { ͵lē·thəl i͵kwiv·ə·lənt 'val·yü }

lethal gene [GEN] A gene mutation that causes premature death in heterozygotes if dominant, and in homozygotes if recessive. Also known as lethal mutation. { 'lē·thəl 'jēn }

lethal mutation See lethal gene. { 'lē·thəl myü'tā·shən }

lethal radius [ORD] The distance from point of burst or ground zero at which a projectile, missile, or the like will probably destroy a target or kill persons. { 'lē·thəl 'rād·ē·əs }

lethargy [MED] A morbid condition of drowsiness or stupor; mental torpor. [NUCLEO] A measure of the energy which has been lost by a neutron, equal to the natural logarithm of the ratio of the initial energy of a neutron to its energy at any given point in the slowing-down process. { 'leth·ər·jē }

Letinula edodes [MYCOL] The second most widely cultivated mushroom in the world, it is native to Asia and is touted for its medicinal properties (cholesterol reduction and antitumor and immunostimulating activities) and flavorful addition to foods. Also known as the shiitake mushroom. { lə͵tin·yə·lə ē'dō͵dēz }

letovicite [MINERAL] $(NH_3)_3H(SO_4)_2$ A mineral composed of acid ammonium sulfate. { ˌled·ə'vi͵sīt }

letter [COMMUN] A character used in an alphabet generally representing one or more sounds of a spoken language. { 'led·ər }

letter code [COMPUT SCI] A Baudot code function which cancels errors by causing the receiving terminal to print nothing. { 'led·ər ͵kōd }

Letterer-Siwe disease [MED] A fatal disease of infants and young children, of unknown etiology, characterized by hyperplasia of the reticuloendothelial system without lipid storage. { 'led·ə·rər 'zē·və di͵zēz }

letter-perfect printer See letter-quality printer. { 'led·ər ͵pər·fekt 'print·ər }

letterpress [GRAPHICS] A method of printing by impressing paper on a raised inked surface; the oldest printing process, it employs type or plates cast or engraved in relief on materials such as metal, wood, or rubber. Also known as relief press. { 'led·ər͵pres }

letter-quality printer [COMPUT SCI] A printer that produces high-quality output. Also known as correspondence printer; letter-perfect printer. { 'led·ər ͵kwäl·əd·ē 'print·ər }

letterset [GRAPHICS] A method of printing that uses a blanket for transferring an image from plate to paper; unlike offset lithography, it uses a relief plate and requires no dampening system. Also known as dry offset. { 'led·ər͵set }

letter spacing [GRAPHICS] Space inserted between letters of a word to open it up horizontally. { 'led·ər ‚spas·iŋ }

letters patent See patent. { 'led·ərz 'pat·ənt }

letters shift [COMMUN] Abbreviated LTRS. **1.** A movement of a teletypewriter carriage which permits printing of alphabetic characters in an appropriate, generally linear sequence. **2.** The control which actuates this movement. { 'led·ərz 'shift }

lettuce [BOT] *Lactuca sativa.* An annual plant of the order Asterales cultivated for its succulent leaves; common varieties are head lettuce, leaf or curled lettuce, romaine lettuce, and iceberg lettuce. { 'led·əs }

leucaenine See mimosine. { 'lü·sə‚nēn }

leucaenol See mimosine. { 'lü·sə‚nól }

Leucaltidae [INV ZOO] A family of calcinean sponges in the order Leucettida having numerous small, interstitial, flagellated chambers. { lü'kal·tə‚dē }

Leucascidae [INV ZOO] A family of calcinean sponges in the order Leucettida having a radiate arrangement of flagellated chambers. { lü'kas·ə‚dē }

leucenine See mimosine. { 'lü·sə‚nēn }

leucenol See mimosine. { 'lü·sə‚nól }

Leucettida [INV ZOO] An order of calcareous sponges in the subclass Calcinea having a leuconoid structure and a distinct dermal membrane or cortex. { lü'sed·ə·də }

leucine [BIOCHEM] $C_6H_{13}O_2N$ A monocarboxylic essential amino acid obtained by hydrolysis of protein-containing substances such as milk. { 'lü‚sēn }

leucine amino peptidase [BIOCHEM] An enzyme that acts on peptides to catalyze the release of terminal amino acids, especially leucine residues, having free α-amino groups. { 'lü‚sēn ə‚mē·nō 'pep·tə‚dās }

leucite [MINERAL] $KAlSi_2O_6$ A white or gray rock-forming mineral belonging to the feldspathoid group; at ordinary temperatures the mineral exists as aggregates of trapezohedral crystals with glassy fracture; hardness is 5.5–6.0 on Mohs scale, and specific gravity is 2.45–2.50. Also known as amphigene; grenatite; vesuvian; Vesuvian garnet; white garnet. { 'lü‚sīt }

leucite phonolite [PETR] An extrusive rock composed of alkali feldspar, mafic minerals, and leucite. { 'lü‚sīt 'fän·əl‚īt }

leucitite [PETR] A fine-grained or porphyritic extrusive rock or hypabyssal igneous rock composed mostly of pyroxene and leucite. { 'lü·sə‚tīt }

leucitohedron See trapezohedron. { ‚lü·sə·tō'hē·drən }

leuco base [ORG CHEM] Any group of colorless derivatives of triphenylmethane dyes that are produced by reducing the dye and are capable of being reconverted to the original dye by oxidation. Also known as leuco compound. { 'lü·kō ‚bās }

leucochalcite See olivenite. { ‚lü·kō'kal‚sīt }

leuco compound See leuco base. { 'lü·ko ‚käm‚paund }

leucocratic [PETR] Light-colored as applied to igneous rock containing 0–50% dark-colored minerals. { ‚lü·kə‚krad·ik }

Leucodontineae [BOT] A family of mosses in the order Isobryales with foliated branches, often bearing catkins. { ‚lü·kə·dän'tin·ē‚ē }

leucoline See quinoline. { 'lü·kō‚lēn }

leucon [INV ZOO] A type of sponge having the choanocytes restricted to flagellated chambers inserted between the incurrent and excurrent canals, and a reduced or absent paragastric cavity. { 'lü‚kän }

Leuconostoc [MICROBIO] A genus of bacteria in the family Streptococcaceae; spherical or lenticular cells occurring in pairs or chains; ferment glucose with production of levorotatory lactic acid (heterofermentative), ethanol, and carbon dioxide. { ‚lü·kə'näs‚täk }

leucophanite [MINERAL] $(Na,Ca)_2BeSi_7(O,F,OH)_7$ Greenish mineral composed of beryllium sodium calcium silicate containing fluorine and occurring in glassy, tabular crystals. { ‚lü·kō'fa‚nīt }

leucophore [HISTOL] A white reflecting chromatophore. { 'lü·kə‚fór }

leucophosphite [MINERAL] $K_2Fe_4(PO_4)_4(OH)_2 \cdot 9H_2O$ White mineral composed of hydrous basic phosphate of potassium and iron. { ‚lü·kə'fäs‚fīt }

leucoplast [BOT] A nonpigmented plastid; capable of developing into a chromoplast. { 'lü·kə‚plast }

leucopyrite See loellingite. { ‚lü·kə'pī‚rīt }

Leucosiidae [INV ZOO] The purse crabs, a family of true crabs belonging to the Oxystomata. { ‚lü·kə'sī·ə‚dē }

leucosin [BIOCHEM] A simple protein of the albumin type found in wheat and other cereals. { 'lü·kə·sən }

Leucosoleniida [INV ZOO] An order of calcareous sponges in the subclass Calcaronea characterized by an asconoid structure and the lack of a true dermal membrane or cortex. { ‚lü·kə‚sō·lə'nī·ə·də }

leucosphenite [MINERAL] $Na_4BaTi_2Si_{10}O_{27}$ A white mineral composed of sodium barium silicotitanate and occurring as wedge-shaped crystals. { ‚lü·kə'sfē‚nīt }

Leucospidae [INV ZOO] A small family of hymenopteran insects in the superfamily Chalcidoidea distinguished by a longitudinal fold in the forewings. { lü'käs·pə‚dē }

leucosporous [MYCOL] Having white spores. { lü'käs·pə·rəs }

Leucothoidae [INV ZOO] A family of amphipod crustaceans in the suborder Gammaridea including semiparasitic and commensal species. { lü'kä·thói‚dē }

Leucothrix [MICROBIO] The type genus of the family Leucotrichaceae; cells do not form sulfur deposits. { 'lü·kə‚thriks }

Leucotrichaceae [MICROBIO] A family of bacteria in the order Cytophagales; long, colorless, unbranched filaments having conspicuous cross-walls and containing cylindrical or ovoid cells; filaments attach to substrate. { ‚lü·kə·tri‚käs·ē‚ē }

leucovorin [PHARM] Folinic acid used as a calcium salt to counteract the toxic effects of folic acid antagonists and for treatment of megaloblastic anemias. { lü'käv·ə·rən }

leucoxene [MINERAL] A mineral composed of rutile with some anatase or sphene; occurs in igneous rocks, usually as an alteration product of ilmenite. { lü'käk‚sēn }

leukemia [MED] Any of several diseases of the hemopoietic system characterized by uncontrolled leukocyte proliferation. Also known as leukocythemia. { lü'kē·mē·ə }

leukemia virus [MED] See leukovirus. { lü'kē·mē·ə ‚vī·rəs }

leukemic reticuloendotheliosis [MED] A rare, usually chronic disorder characterized by proliferation of hairy cells, probably B lymphocytes, in reticuloendothelial organs and blood. Also known as hairy cell leukemia. { lü‚kēm·ik re‚tik·yə·lō‚en·dō‚thē·lē'ō·səs }

leukemoid [MED] Similar to leukemia, that is, characterized by the presence of immature leukocytes in the blood, but of a different etiology. { 'lü·kə‚móid }

leukocidin [BIOCHEM] A toxic substance released by certain bacteria which destroys leukocytes. { ‚lü·kə'sīd·ən }

leukocyte [HISTOL] A colorless, ameboid blood cell having a nucleus and granular or nongranular cytoplasm. Also known as white blood cell; white corpuscle. { 'lü·kə‚sīt }

leukocythemia See leukemia. { ‚lü·kə‚sī'thē·mē·ə }

leukocytolysin [BIOCHEM] A leukocyte-disintegrating lysin. { ‚lü·kə‚sī'täl·ə·sən }

leukocytopenia See leukopenia. { ‚lü·kə‚sīd·ə'pē·nē·ə }

leukocytopoiesis [PHYSIO] Formation of leukocytes. { ‚lü·kə‚sīd·ə‚pói'ē·səs }

leukocytosis [MED] Elevation of the leukocyte count to values above the normal limit. { ‚lü·kə‚sī'tō·səs }

leukoderma [MED] Defective skin pigmentation, especially the congenital absence of pigment in patches or bands. { ‚lü·kə'dər·mə }

leukodystrophy [MED] A condition thought to result from an inborn error of metabolism and characterized by progressive degeneration of the white matter of the cerebrum, or by defective buildup of myelin. { ‚lü·kə'dis·trə·fē }

leukoencephalitis [MED] Inflammation of the white matter of the cerebrum. { ‚lü·kō‚in‚sef·ə'līd·əs }

leukoerythroblastic anemia See myelophthisic anemia. { ‚lü·kō·ə‚rith·rə'blas·tik ə'nē·mē·ə }

leukoerythroblastosis See myelophthisic anemia. { ‚lü·kō·ə‚rith·rə‚bla'stō·səs }

leukol See quinoline. { 'lü‚kól }

leukolymphosarcoma See leukosarcoma. { ‚lü·kō‚lim·fō·sär'kō·mə }

leukoma [MED] A large and dense opacity of the cornea as a result of an ulcer, wound, or inflammation, which presents an appearance of ground glass. { lü'kō·mə }

leukonychia [MED] Whitish discoloration or spotting of the fingernails. { ‚lü·kō nik·ē·ə }

leukopenia [MED] A reduction in the leukocyte count to

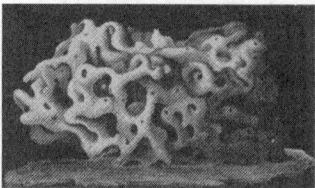

LEUCALTIDAE

Leucaltis clathria showing the small chambers.

LEUCINE

Structural formula of a leucine.

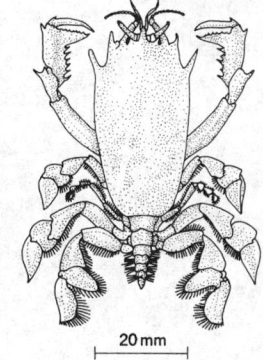

LEUCOSIIDAE

20 mm

Drawing of a purse crab, *Persephona punctata.* *(Smithsonian Institution)*

values below the normal limit. Also known as leukocytopenia. { ¦lü·kō'pē·nē·ə }

leukophoresis [MED] The selective removal of large quantities of white blood cells from a donor's blood while returning other cells to the donor. { ¦lü·kō·fə'rē·səs }

leukoplakia [MED] Formation of thickened white patches on mucous membranes, particularly of the mouth and vulva. { ¦lü·kō'plā·kē·ə }

leukorrhea [MED] A whitish, mucopurulent discharge from the female genital canal. { ¦lü·kə'rē·ə }

leukosarcoma [MED] Lymphosarcoma accompanied by a small number of anaplastic lymphoid cells in the peripheral blood. Also known as leukolymphosarcoma; sarcoleukemia. { ¦lü·kə·sär'kō·mə }

leukosis [MED] An excess of white blood cells. { lü'kō·səs }

leukotomy See lobotomy. { lü'käd·ə·mē }

leukotriene [BIOCHEM] Any of a family of oxidized metabolites of certain polyunsaturated fatty acids, predominantly arachidonic acid, that mediate responses in allergic reactions and inflammations, produced in specific cells upon stimulation. { ¦lü·kō'trī,ēn }

leukovirus [VIROL] A major group of animal viruses including those causing leukemia in birds, mice, and rats. Also known as leukemia virus. { ¦lü·kō¦vī·rəs }

leuneburgite [MINERAL] $Mg_3B_2(PO_4)_2(OH)_6 \cdot 5H_2O$ A colorless mineral consisting of a hydrous basic phosphate of magnesium and boron. { 'lü·nən,bər,gīt }

leurocristine See vincristine. { ¦lü·rə'kris,tēn }

levan [BIOCHEM] $(C_6H_{10}O_5)_n$ A polysaccharide consisting of repeating units of D-fructose and produced by a range of microorganisms, such as *Bacillus mesentericus*. { 'le,van }

levante [METEOROL] The Spanish and most widely used term for an east or northeast wind occurring along the coast and inland from southern France to the Straits of Gibraltar; it is moderate or fresh, mild, very humid, and rainy, and occurs with a depression over the western Mediterranean Sea. { lə'vän·tə }

levantera [METEOROL] A persistent east wind of the Adriatic, usually accompanied by cloudy weather. { ¦le·vən'ter·ə }

Levantine Basin [OCEANOGR] A basin in the Mediterranean Ocean between Asia Minor and Egypt. { 'le·vən,tēn 'bās·ən }

levator [MED] An instrument used for raising a depressed portion of the skull. [PHYSIO] Any muscle that raises or elevates a part. { lə'vād·ər }

leveche [METEOROL] A warm wind in Spain, either a foehn or a hot southerly wind in advance of a low-pressure area moving from the Sahara Desert. { lə'vā·chä }

levee [CIV ENG] **1.** A dike for confining a stream. **2.** A pier along a river. [GEOL] **1.** An embankment bordering one or both sides of a sea channel or the low-gradient seaward part of a canyon or valley. **2.** A low ridge sometimes deposited by a stream on its sides. { 'lev·ē }

level [CIV ENG] **1.** A surveying instrument with a telescope and bubble tube used to take level sights over various distances, commonly 100 feet (30 meters). **2.** To make the earth surface horizontal. [COMMUN] A specified position on an amplitude scale (for example, magnitude) applied to a signal waveform, such as reference white level and reference black level in a standard television signal. [COMPUT SCI] **1.** The status of a data item in COBOL language indicating whether this item includes additional items. **2.** See channel. [DES ENG] A device consisting of a bubble tube that is used to find a horizontal line or plane. Also known as spirit level. [ELEC] A single bank of contacts, as on a stepping relay. [ELECTR] **1.** The difference between a quantity and an arbitrarily specified reference quantity, usually expressed as the logarithm of the ratio of the quantities. **2.** A charge value that can be stored in a given storage element of a charge storage tube and distinguished in the output from other charge values. [MIN ENG] **1.** Mine workings that are at the same elevation. **2.** A gutter for the water to run in. [STAT] In factorial experiments, the quantitative or qualitative intensity at which a particular value of a factor is held fixed during an experiment. { 'lev·əl }

level above threshold [PHYSIO] Also known as sensation level. **1.** The pressure level of a sound in decibels above its threshold of audibility for the individual observer. **2.** In general, the level of any psychophysical stimulus, such as light,

above its threshold of perception. { 'lev·əl ə,bəv 'thresh,hōld }

level bombing [ORD] Releasing bombs in level flight. { 'lev·əl 'bäm·iŋ }

level compensator [ELECTR] **1.** Automatic transmission-regulating feature or device used to minimize the effects of variations in amplitude of the received signal. **2.** Automatic gain control device used in the receiving equipment of a telegraph circuit. { 'lev·əl 'käm·pən,sād·ər }

level converter [ELECTR] An amplifier that converts nonstandard positive or negative logic input voltages to standard DTL or other logic levels. { 'lev·əl kən,vərd·ər }

level crosscut [MIN ENG] A horizontal crosscut. { 'lev·əl 'krós,kət }

level drive [MIN ENG] A horizontal shaft which allows access to the length of a deposit and forms the basis for the splitting of the deposit into levels. { 'lev·əl 'drīv }

leveled element time See normal element time. { 'lev·əld ,el·ə,ment 'tīm }

leveled time See normal time. { 'lev·əld ,tīm }

leveler [ENG] A back scraper, drag, or other form of device for smoothing land. { 'lev·ə·lər }

level error [ASTRON] **1.** The difference between the apparent altitude of a celestial object above the apparent horizon and its true altitude above the celestial horizon. **2.** The angle between the east-west mechanical axis of a transit telescope and the horizontal plane. { 'lev·əl ,er·ər }

level fold See nonplunging fold. { 'lev·əl 'fōld }

level indicator [ENG] An instrument that indicates liquid level. [ENG ACOUS] An indicator that shows the audio voltage level at which a recording is being made; may be a volume-unit meter, neon lamp, or cathode-ray tuning indicator. { 'lev·əl 'in·də,kād·ər }

leveling [ENG] Adjusting any device, such as a launcher, gun mount, or sighting equipment, so that all horizontal or vertical angles will be measured in the true horizontal and vertical planes. [IND ENG] A method of performance rating which seeks to rate the principal factors that cause the speed of motions rather than speed itself; it considers that the level at which the operator works is influenced by effort and skill. [MET] Flattening rolled sheet by evening out irregularities, using a roller or tensile straining. [MIN ENG] Measurement of rises and falls, heights, and contour lines. Also known as gallery. { 'lev·ə·liŋ }

leveling action [MET] The property exhibited by a plating solution in making the coating smoother than the base metal. { 'lev·ə·liŋ ,ak·shən }

leveling instrument [ENG] An instrument for establishing a horizontal line of sight, usually by means of a spirit level or a pendulum device. { 'lev·ə·liŋ ,in·strə·mənt }

leveling screw [ENG] An adjusting screw used to bring an instrument into level. { 'lev·ə·liŋ ,skrü }

level measurement [MECH] The determination of the linear vertical distance between a reference point or datum plane and the surface of a liquid or the top of a pile of divided solid. { 'lev·əl 'mezh·ər·mənt }

level of burst [ORD] Point at which a projectile bursts. { 'lev·əl əv 'bərst }

level off [AERO ENG] To bring an aircraft to level flight after an ascent or descent. { 'lev·əl 'òf }

level-off position [AERO ENG] That position over which a craft ends an ascent or descent and begins relatively horizontal motion. { 'lev·əl ¦òf pə,zish·ən }

level of free convection [METEOROL] The level at which a parcel of air lifted dry and adiabatically until saturated, and lifted saturated and adiabatically thereafter, would first become warmer than its surroundings in a conditionally unstable atmosphere. Abbreviated LFC. { 'lev·əl əv ¦frē kən'vek·shən }

level of nondivergence [METEOROL] A level in the atmosphere throughout which the horizontal velocity divergence is zero; although in some meteorological situation there may be several such surfaces (not necessarily level), the level of nondivergence usually considered is that mid-tropospheric surface which separates the major regions of horizontal convergence and divergence associated with the typical vertical structure of the migratory cyclonic-scale weather systems. { 'lev·əl əv ,nän·də'vər·jəns }

level of saturation See water table. { 'lev·əl əv ,sach·ə'rā·shən }

level of significance [STAT] For a test, the probability of false rejection of the null hypothesis. Also known as significance level. { 'lev·əl əv sig'nif·i·kəns }

level 1 cache *See* primary cache. { 'lev·əl 'wən ˌkash }

level point *See* point of fall. { 'lev·əl ˌpȯint }

level rod [ENG] A straight rod or bar, with a flat face graduated in plainly visible linear units with zero at the bottom, used in measuring the vertical distance between a point of the earth's surface and the line of sight of a leveling instrument that has been adjusted to the horizontal position. { 'lev·əl ˌräd }

level scheme *See* energy-level diagram. { 'lev·əl ˌskēm }

level set [COMPUT SCI] A revision of a software package in which most or all of the executable programs are replaced with improved versions. { 'lev·əl ˌset }

level shifting [ELECTR] Changing the logic level at the interface between two different semiconductor logic systems. { 'lev·əl ˌshif·tiŋ }

level surface [ENG] A surface which is perpendicular to the plumb line at every point. [GEOPHYS] *See* geopotential surface. { 'lev·əl ˌsər·fəs }

level 2 cache *See* secondary cache. { ˌlev·əl ˌtü kash }

level valve [MECH ENG] A valve operated by a lever which travels through a maximum arc of 180°. { 'lev·əl ˌvalv }

level width [QUANT MECH] A measure of the spread in energy of an unstable state, equal to the difference between the energies at which intensity of emission or absorption of photons or particles, or the cross section for a reaction, is one-half its maximum value. { 'lev·əl 'width }

Levenstein process [CHEM ENG] A process for the manufacture of mustard gas from ethene, $CH_2 = CH_2$, and sulfur chloride, S_2Cl_2. { 'le·vənˌstīn ˌprä·səs }

lever [ENG] A rigid bar, pivoted about a fixed point (fulcrum), used to multiply force or motion; used for raising, prying, or dislodging an object. { 'lev·ər, lē·vər }

leverage [MECH] The multiplication of force or motion achieved by a lever. { 'lev·rij }

lever escapement [HOROL] A clock movement in which the balance wheel is connected to the escapement by a lever attached to a roller; the wheel swings through a much larger angle than does a pendulum. { 'lev·ər isˌkāp·mənt }

leveret [VERT ZOO] A young hare. { 'lev·rət }

Leverett function [FL MECH] A dimensionless number used in studying two-phase flow in porous mediums, written as $(\xi/e)^{1/2}(p/\sigma)$, where ξ is the permeability of a medium (as defined by Darcy's law), e is the medium's porosity, σ is the surface tension between two liquids flowing through it, and p is the capillary pressure. { 'lev·rət ˌfəŋk·shən }

lever shears [DES ENG] A shears in which the input force at the handles is related to the output force at the cutting edges by the principle of the lever. Also known as alligator shears; crocodile shears. { 'lev·ər ˌshirz }

lever switch [ELEC] A switch having a lever-shaped operating handle. { 'lev·ər ˌswich }

Levi-Civita symbol [MATH] A symbol $\epsilon_{i,j,\ldots,s}$ where $i, j, \ldots, s$ are n indices, each running from 1 to n; the symbol equals zero if any two indices are identical, and 1 or -1 otherwise, depending on whether $i, j, \ldots, s$ form an even or an odd permutation of $1, 2, \ldots, n$. { 'lā·vē chē·vē'tä ˌsim·bəl }

levigate [BOT] *See* glabrous. [CHEM] **1.** To separate a finely divided powder from a coarser material by suspending in a liquid in which both substances are insoluble. Also known as elutriation. **2.** To grind a moist solid to a fine powder. { 'lev·əˌgāt }

levigated abrasive [MATER] A fine abrasive powder for final burnishing of metals or for metallographic polishing; the powder is usually processed to make it chemically neutral. { 'lev·əˌgād·əd ə'brā·siv }

levisticum oil *See* lovage oil. { lə'vis·tə·kəm ˌȯil }

levitated vehicle [MECH ENG] A train or other vehicle which travels at high speed at some distance above an electrically conducting track by means of levitation. { 'lev·əˌtād·əd 'vē·ə·kəl }

levitation [MIN ENG] In froth flotation, raising of particles in a froth to the surface of the pulp, to facilitate separation of selected minerals in the froth. [PHYS] The use of a force that does not involve physical contact to balance gravity, such as that associated with an electric or magnetic field, or electromagnetic or acoustic radiation. { ˌlev·ə'tā·shən }

levitation heating [MET] Providing heat through high-frequency magnetic fields; employed in levitation melting. { ˌlev·ə'tā·shən ˌhēd·iŋ }

levitation melting [MET] Melting metal out of contact with a supporting material by using the induced current provided by a high-frequency surrounding magnetic field to suspend the melt. { ˌlev·ə'tā·shən ˌmel·tiŋ }

Leviviridae [VIROL] A family of single-stranded ribonucleic acid phages (including MS2 virus) characterized by icosahedral particles. { ˌlev·ē'vir·əˌdī }

Levivirus [VIROL] The only genus of the family Leviviridae. { 'lev·iˌvī·rəs }

levo *See* levorotatory enantiomer. { 'lē·vō }

levocardia [MED] Location of the heart on the left side associated with visceral situs inversus and congenital cardiac disease. { ˌlēv·ə'kär·dē·ə }

levodopa [PHARM] $C_9H_{11}NO_4$ Crystals or crystalline powder, soluble in dilute hydrochloric acid and in formic acid; used as an anticholinergic drug and in the treatment of Parkinson's disease. Abbreviated L-dopa. { ˌlev·ə'dō·pə }

levophobia [PSYCH] An abnormal fear of objects to the left of the body. { ˌlēv·ə'fō·bē·ə }

levorotation [OPTICS] Rotation of the plane of polarization of plane polarized light in a counterclockwise direction, as seen by an observer facing in the direction of light propagation. Also known as levulorotation. { ˌlē·və·rō'tā·shən }

levorotatory enantiomer [ORG CHEM] An optically active substance that rotates the plane of plane-polarized light counterclockwise. Symbolized *l*. Abbreviated levo. { ˌlē·vō'rōt·əˌtȯr·ē əˌnan·tē'ō·mər }

levotropic cleavage [EMBRYO] Spiral cleavage with the cells displaced counterclockwise. { 'lē·və'träp·ik 'klēv·ij }

levulinic acid [ORG CHEM] $CH_3COCH_2CH_2COOH$ Crystalline compound forming plates or leaflets that melt at 37°C; freely soluble in alcohol, ether, and chloroform; used in the manufacture of pharmaceuticals, plastics, rubber, and synthetic fibers. { ˌlev·yə'lin·ik 'as·əd }

levulorotation *See* levorotation. { ˌlē·vyə·lō·rō'tā·shən }

levulose [BIOCHEM] Levorotatory D-fructose. { 'lev·yəˌlōs }

levulose tolerance test [PATH] A liver function test based on the observation that blood sugar increases in cases of hepatic disease following oral administration of levulose. { 'lev·yəˌlōs 'täl·ə·rəns ˌtest }

levyine *See* levynite. { lā'vēˌīn }

levyite *See* levynite. { lā'vēˌīt }

levyne *See* levynite. { lā'vēn }

levynite [MINERAL] $NaCa_2Al_7Si_{11}O_{36} \cdot 15H_2O$ A white or light-colored mineral of the zeolite group, composed of hydrous silicate of aluminum, sodium, and calcium, and occurring in rhombohedral crystals. Also known as levyine; levyite; levyne. { lā'veˌnīt }

lewis [DES ENG] A device for hoisting heavy stones; employs a dovetailed tenon that fits into a mortise in the stone. { 'lü·əs }

Lewis acid [CHEM] A substance that can accept an electron pair from a base; thus, $AlCl_3$, BF_3, and SO_3 are acids. { 'lü·əs ˌas·əd }

Lewis base [CHEM] A substance that can donate an electron pair; examples are the hydroxide ion, OH^-, and ammonia, NH_3. { 'lü·əs ˌbās }

Lewis blood group system [IMMUNOL] An antigen, designated by Le^a, first recognized in a Mrs. Lewis, occurring in about 22% of the population, detected by anti-Le^a antibodies; primarily composed of soluble antigens of serum and body fluids like saliva, with secondary absorption by erythrocytes. { 'lü·əs 'bləd ˌgrüp ˌsis·təm }

lewis bolt [DES ENG] A bolt with an enlarged, tapered head that is inserted into masonry or stone and fixed with lead; used as a foundation bolt. { 'lü·əs ˌbōlt }

Lewis formula *See* Lewis structure. { 'lü·is ˌfȯr·myə·lə }

Lewis gun [ORD] A gas-operated machine gun with a horizontal drum magazine, manufactured in several modifications; now obsolete. { 'lü·əs ˌgən }

lewis hole [MIN ENG] A series of two or more holes drilled as closely together as possible, but then connected by knocking out the thin partition between them, thus forming one wide hole having its greatest diameter in a plane with the desired rift. { 'lü·əs ˌhōl }

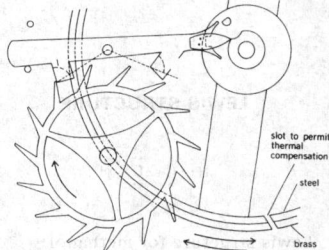

LEVER ESCAPEMENT

The lever escapement, used in good spring clocks and watches.

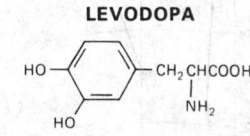

LEVODOPA

Structural formula of levodopa.

lewisite [MINERAL] $(Ca,Fe,Na)_2$ A titanian romeite mineral. [ORG CHEM] $C_2H_2AsCl_3$ An oily liquid, colorless to brown or violet; forms a toxic gas, used in World War I. { 'lü·ə,sīt }

Lewis-Matheson method [CHEM ENG] Trial-and-error calculation method for the design of multicomponent distillation columns, or for the determination of the separating ability of an existing column. { 'lü·əs 'math·ə·sən ,meth·əd }

Lewis number [PHYS] **1.** A dimensionless number used in studies of combined heat and mass transfer, equal to the thermal diffusivity divided by the diffusion coefficient. Symbolized Le; N_{Le}. **2.** Sometimes, the reciprocal of this quantity. { 'lü·əs ,nəm·bər }

lewis pin [MIN ENG] A pin used for attachment to a key block; it is placed in a shallow drill hole with a wedge at either side. { 'lü·əs ,pin }

Lewis-Rayleigh afterglow [ELECTR] A golden yellow light emitted by nitrogen gas following the passage of an electric discharge, associated with recombination of nitrogen atoms. { 'lü·əs ,rā·lē 'af·tər,glō }

Lewis structure [CHEM] A structural formula in which electrons are represented by dots; two dots between atoms represent a covalent bond. Also known as electron-dot formula; Lewis formula. { 'lü·is ,strək·chər }

lewistonite [MINERAL] $(Ca,K,Na)_5(PO_4)_3(OH)$ White mineral composed of basic calcium potassium sodium phosphate. { 'lü·ə·stə,nīt }

Lexell's Comet [ASTRON] A small comet that approached to within 2,000,000 miles (3,200,000 kilometers) of earth in 1770; it has not been seen since. { 'lek·selz ,käm·ət }

lexicographic order [MATH] Given sets A and B with a common ordering $<$, one defines an ordering between all sequences (finite or infinite) of elements of A and of elements of B by $(a_1,a_2,\ldots) < (b_1,b_2,\ldots)$ if either $a_i = b_i$ for every i, or $a_n < b_n$, where n is the first place in which they differ; this is the way words are ordered in a dictionary. { ,lek·sə·kō,graf·ik 'ör·dər }

Leyden jar [ELEC] An early type of capacitor, consisting simply of metal foil sheets on the inner and outer surfaces of a glass jar. { 'līd·ən ,jär }

Leydig cell [HISTOL] One of the interstitial cells of the testes; thought to produce androgen. { 'lī·dig ,sel }

Leydig-cell tumor See interstitial-cell tumor. { 'lī·dig ,sel ,tü·mər }

LF See low-frequency.

LFC See level of free convection.

LF loran See low-frequency loran. { ,el,ef 'lör,an }

L form [MICROBIO] A variant form of bacterial cells that has lost its cell wall because of the action of penicillin. { 'el ,förm }

LGV See lymphogranuloma venereum.

LH See luteinizing hormone.

L-head engine [MECH ENG] A type of four-stroke cycle internal combustion engine having both inlet and exhaust valves on one side of the engine block which are operated by pushrods actuated by a single camshaft. { 'el ,hed 'en·jən }

lherzolite [PETR] Peridotite composed principally of olivine with orthopyroxene and clinopyroxene. { 'lərt·sə,līt }

l'Hôpital's cubic See Tschirnhausen's cubic. { lō·pē·tälz 'kyü·bik }

l'Hôpital's rule [MATH] A rule useful in evaluating indeterminate forms: if both the functions $f(x)$ and $g(x)$ and all their derivatives up to order $(n-1)$ vanish at $x = a$, but the nth derivatives both do not vanish or both become infinite at $x = a$, then

$$\lim_{x \to a} f(x)/g(x) = f^{(n)}(a)/g^{(n)}(a),$$

$f^{(n)}$ denoting the nth derivative. { lō·pē·tälz 'rül }

LH-RH See luteinizing-hormone releasing hormone.

l'Huilier's equation [MATH] An equation used in the solution of a spherical triangle, involving tangents of various functions of its angles and sides. { lə·wē'yāz i,kwā·zhən }

Li See lithium.

liana [BOT] A woody or herbaceous climbing plant with roots in the ground. { lē'än·ə }

Liapunov function See Lyapunov function. { 'lyä·pü·nöf ,fəŋk·shən }

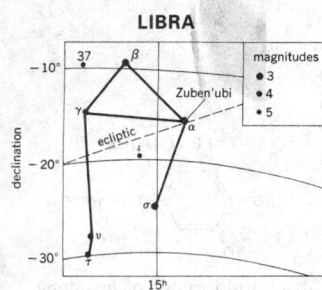
Lias See Liassic. { 'lī·as }

Liassic [GEOL] The Lower Jurassic period of geologic time. Also known as Lias. { lī'as·ik }

Libby counter See black carbon counter. { 'lib·ē ,kaúnt·ər }

Libby effect [GEOCHEM] The increase, since about 1950, in the carbon-14 content of the atmosphere, produced by the detonation of thermonuclear devices. { 'lib·ē i,fekt }

LiBeB process See l-process. { ,el,ī'bē,ē'bē ,prä·səs }

libeccio [METEOROL] Italian name for a southwest wind, used in northern Corsica for the west or southwest wind which blows throughout the year, especially in winter when it is often stormy. { li'bech·ō }

Libellulidae [INV ZOO] A large family of odonatan insects belonging to the Anisoptera and distinguished by a notch on the posterior margin of the eyes and a foot-shaped anal loop in the hindwing. { ,lī·bə'lü·lə,dē }

libethenite [MINERAL] $Cu_2(PO_4)OH$ An olive-green mineral composed of basic copper sulfate, occurring as small prismatic crystals or in masses. { lə'beth·ə,nīt }

libido [PSYCH] **1.** Sexual desire. **2.** The sum total of all instinctual forces; psychic energy or drive usually associated with the sexual instinct. { lə'bē·dō }

Libman-Sacks endocarditis [MED] Inflammation of, accompanied by the presence of watery vegetations on, the endocardium; complicates systemic lupus erythematosus. { 'lib·mən 'saks ,en·dō,kär'dīd·əs }

Libra [ASTRON] A southern constellation, right ascension 15 hours, declination 15° south. Also known as Balance. { 'lē·brə }

librarian [COMPUT SCI] The program which maintains and makes available all programs and routines composing the operating system. { lī'brer·ē·ən }

library [COMPUT SCI] **1.** A computerized facility containing a collection of organized information used for reference. **2.** An organized collection of computer programs together with the associated program listings, documentation, users' directions, decks, and tapes. { 'lī,brer·ē }

library routine [COMPUT SCI] A computer program that is part of some program library. { 'lī,brer·ē rü,tēn }

library software [COMPUT SCI] The collection of programs and routines in the library of a computer system. { 'lī,brer·ē 'söft,wer }

library tape [COMPUT SCI] A magnetic tape that is kept in a stored, indexed collection for ready use and is made generally available. { 'lī,brer·ē ,tāp }

libration [PHYS] Any oscillatory rotational motion, such as that of the moon, or of a molecule in a solid which does not have enough energy to make full rotations. { lī'brā·shən }

libration in latitude See lunar libration. { lī'brā·shən in 'lad·ə,tüd }

libriform [BOT] Elongated or thick-walled. { 'lib·rə ,förm }

Libytheidae [INV ZOO] The snout butterflies, a family of cosmopolitan lepidopteran insects in the suborder Heteroneura distinguished by long labial palps; represented in North America by a single species. { ,lib·ə'thē·ə,dē }

licanic oil [MATER] A fatty acid found in drying oils that are in paint formulations. { lə'kan·ik ,oil }

Liceaceae [MYCOL] A family of plasmodial slime molds in the order Liceales. { ,lī·sē'ās·ē,ē }

Liceales [MYCOL] An order of plasmodial slime molds in the subclass Myxogastromycetidae. { ,lī·sē'ā·lēz }

licensed material [NUCLEO] Source material, special nuclear material, or by-product material received, possessed, used, or transferred under a general or special license issued by the Nuclear Regulatory Commission. { 'lī·sənst mə'tir·ē·əl }

lichen [BOT] The common name for members of the Lichenes. { 'lī·kən }

Lichenberger figures See Lichtenberg figures. { 'lī·kən,bər·gər ,fig·yərz }

lichen blue See litmus. { 'lī·kən 'blü }

Lichenes [BOT] A group of organisms consisting of fungi and algae growing together symbiotically. { lī'kē·nēz }

Lichenes Imperfecti [BOT] A class of the Lichenes containing species with no known method of sexual reproduction. { lī'kē·nēz ,im·pər'fek·tī }

lichenification [MED] The process whereby the skin becomes leathery and hardened; often the result of chronic pruritis and the irritation produced by scratching or rubbing eruptions. { lī,ken·ə·fə'kā·shən }

lichenology [BOT] The study of lichens. { ,lī·kə'näl·ə·jē }

lichenometry [GEOL] Measurement of the diameter of lichens growing on exposed rock surfaces; used for dating geomorphic features, particularly of glacial origin. { ,lī·kə'näm·ə·trē }

lichenophagous [ZOO] Feeding on lichens. { ,lī·kə'näf·ə·gəs }

lichen planus [MED] A dermatologic disease of unknown etiology that also occurs in the mouth, on the tongue, or on the lips as smooth lacy networks of white lines or, less commonly, as white patches that may become ulcerative. { ,līk·ən 'plan·əs }

Lichnophorina [INV ZOO] A suborder of ciliophoran protozoans belonging to the Heterotrichida. { ,līk·nə'fōr·ə·nə }

Lichtenberg figures [ELEC] Patterns produced on a photographic emulsion, or in fine powder spread over the surface of a solid dielectric, by an electric discharge produced by a high transient voltage. Also known as Lichenberger figures. { 'lik·tən·bərg ,fig·yərz }

licorice [BOT] *Glycyrrhiza glabra.* A perennial herb of the legume family (Leguminosae) cultivated for its roots, which when dried provide a product used as a flavoring in medicine, candy, and tobacco and in the manufacture of shoe polish. { 'lik·rəs }

lidar [OPTICS] An instrument in which a laser generates intense infrared pulses in beam widths as small as 30 seconds of arc; beam reflections and scattering effects of clouds, smog layers, and some atmospheric discontinuities are measured by radar techniques; it can also be used for tracking weather balloons, smoke puffs, and rocket trails. Derived from laser infrared radar. { 'lī,där }

lidocaine [ORG CHEM] $C_{14}H_{22}N_2O$ A crystalline compound, used as a local anesthetic. Also known as lignocaine. { 'līd·ə,kān }

Lie algebra [MATH] The algebra of vector fields on a manifold with additive operation given by pointwise sum and multiplication by the Lie bracket. { 'lē ,al·jə·brə }

liebigite [MINERAL] $Ca_2U(CO_3)_4 \cdot 10H_2O$ An apple- or yellow-green mineral composed of hydrous uranium calcium carbonate; occurs as a coating or concretion in rock. { 'lē·bi,gīt }

Liebig's law of the minimum See law of minimum. { 'lē·bigz ,lo əv thə 'min·ə·məm }

Liebmann effect [OPTICS] The effect whereby it is more difficult to visually distinguish contrasting forms when they have the same luminance and different chromaticities than when they have different luminances and the same chromaticity. { 'lēb,mən i,fekt }

Lie bracket [MATH] Given vector fields *X,Y* on a manifold *M,* their Lie bracket is the vector field whose value is the difference between the values of *XY* and *YX.* { 'lē ,brak·ət }

lie detector [ENG] An instrument that indicates or records one or more functional variables of a person's body while the person undergoes the emotional stress associated with a lie. Also known as polygraph; psychintegroammeter. { 'lī dī,tek·tər }

Lie group [MATH] A topological group which is also a differentiable manifold in such a way that the group operations are themselves analytic functions. { 'lē ,grüp }

Liénard-Wiechert potentials [ELECTROMAG] The retarded and advanced electromagnetic scalar and vector potentials produced by a moving point charge, expressed in terms of the (retarded or advanced) position and velocity of the charge. { 'lē,närt 'vē·kərt pə,ten·chəlz }

Liesegang banding [GEOL] Colored or compositional rings or bands in a fluid-saturated rock due to rhythmic precipitation. Also known as Liesegang rings. { 'lēz·ə,gäŋ ,band·iŋ }

Liesegang rings See Liesegang banding. { 'lēz·ə,gäŋ ,riŋz }

Lieskeela [MICROBIO] A genus of sheathed bacteria; cells are nonmotile, and sheaths are not attached and may be encrusted with iron and manganese oxides. { |lēs·kē|el·ə }

lifeboat [NAV ARCH] A small boat hoisted on davits or carried on one of the upper decks of a vessel, which can be quickly lowered into the water in case of an emergency. { 'līf,bōt }

life cycle [BIOL] The functional and morphological stages through which an organism passes between two successive primary stages. { 'līf ,sī·kəl }

life-cycle assessment [SYS ENG] A methodology that identifies the environmental impacts associated with the life cycle of a material or product in a specific application, thus identifying opportunities for improvement in environmental performance. Abbreviated LCA. { 'līf ,sī·kəl ə,ses·mənt }

life-cycle cost [ENG] A measurement of the total cost of using equipment over the entire time of service of the equipment; includes initial, operating, and maintenance costs. { 'līf ,sī·kəl ,kȯst }

life expectancy [BIOL] The expected number of years that an organism will live based on statistical probability. [ENG] The predicted useful service life of an item of equipment. { 'līf ik'spek·tən·sē }

life form [ECOL] The form characteristically taken by a plant at maturity. { 'līf ,fȯrm }

life of mine [MIN ENG] The time in which, through the employment of the available capital, the ore reserves—or such reasonable extension of the ore reserves as conservative geological analysis may justify—will be extracted. { 'līf əv 'mīn }

life preserver [ENG] A buoyant device that is used to prevent drowning by supporting a person in the water. { 'līf pri,zər·vər }

life raft [NAV ARCH] A very buoyant raft designed to be used by people forced into the water; made of a metal tube covered with wood or canvas, of balsa wood, or of rubber which may be automatically inflated. { 'līf ,raft }

life-saving station See Coast Guard station. { 'līf ,sāv·iŋ ,stā·shən }

life support system [ENG] A system providing atmospheric control and monitoring, such as a breathing mixture supply system, air purification and filtering system, or carbon dioxide removal system; used in oceanographic submersibles and spacecraft. { 'līf sə,pȯrt ,sis·təm }

life test [CHEM ENG] In petroleum testing, an American Society for Testing and Materials oxidation test made on inhibited steam-turbine oils to determine their stability under oxidizing conditions. [ENG] A test in which a device is operated under conditions that simulate a normal lifetime of use, to obtain an estimate of service life. { 'līf ,test }

lifetime See mean life. { 'līf,tīm }

life zone [ECOL] A portion of the earth's land area having a generally uniform climate and soil, and a biota showing a high degree of uniformity in species composition and adaptation. { 'līf ,zōn }

LIFO See last in, first out. { 'lī,fō }

lift [FL MECH] See aerodynamic lift. [MECH ENG] See elevator. [MIN ENG] **1.** The vertical height traveled by a cage in a shaft. **2.** The distance between the first level and the surface or between any two levels. **3.** Any of the various gangways from which coal is raised at a slope colliery. { lift }

lift bridge [CIV ENG] A drawbridge whose movable spans are raised vertically. { 'lift ,brij }

lift coefficient [AERO ENG] The quantity $C_L = 2L/\rho V^2 S$, where *L* is the lift of a whole airplane wing, ρ is the mass density of the air, *V* is the free-stream velocity, and *S* is the wing area; this is also applicable to other airfoils. { 'lift ,kō·i,fish·ənt }

lift-drag ratio [AERO ENG] The lift of an aerodynamic form, such as an airplane wing, divided by the drag. { 'lift 'drag ,rā·shō }

lifter [MIN ENG] A shothole drilled near the floor when tunneling, and fired subsequent to the cut and relief holes. { 'lif·tər }

lifter case [MIN ENG] The sleeve or tubular part attached to the lower end of the inner tube of M-design core barrels and some other types of core barrels, in which is fitted a core lifter. Also known as core-catcher case; core-gripper case; core-lifter case; core-spring case; inner-tube extension; ring-lifter case; spring-lifter case. { 'lif·tər ,kās }

lifter flight [DES ENG] Spaced plates or projections on the inside surfaces of cylindrical rotating equipment (such as rotary dryers) to lift and shower the solid particles through the gas-drying stream during their passage through the dryer cylinder. { 'lif·tər ,flīt }

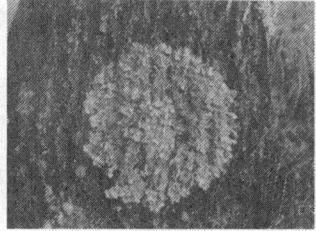

lifter roof [ENG] Gas storage tank in which the roof is raised by the incoming gas as the tank fills. { 'lif·tər ‚rüf }

lift fan [AERO ENG] A special turbofan engine used primarily for lift in VTOL/STOL aircraft and often mounted in a wing with vertical thrust axis. { 'lift ‚fan }

lift fire [ORD] **1.** Command to advance the range of fire by elevating the muzzle of a weapon. **2.** Command to cease or suspend fire. { 'lift ‚fīr }

lifting [MATH] **1.** Given a fiber bundle $(\overline{X}, B, p)$ and a continuous map of a topological space $\overline{Y}$ to B, $g: \overline{Y} \to B$, lifting entails finding a continuous map $\overline{g}: \overline{Y} \to \overline{X}$ such that the function g is the composition $p \circ \overline{g}$. **2.** *See* translation. { 'lift·iŋ }

lifting block [MECH ENG] A combination of pulleys and ropes which allows heavy weights to be lifted with least effort. { 'lift·iŋ ‚bläk }

lifting body [AERO ENG] A maneuverable, rocket-propelled, wingless craft that can travel both in the earth's atmosphere, where its lift results from its shape, and in outer space, and that can land on the ground. { 'lift·iŋ ‚bäd·ē }

lifting condensation level [METEOROL] The level at which a parcel of moist air lifted dry adiabatically would become saturated. Abbreviated LCL. Also known as isentropic condensation level (ICL). { 'lift·iŋ ‚kän‚den'sā·shən ‚lev·əl }

lifting device [ENG] A device to manually open a pressure relief valve by decreasing the spring loading in order to determine if the valve is in working order. { 'lift·iŋ di‚vīs }

lifting dog [ENG] **1.** A component part of the overshot assembly that grasps and lifts the inner tube or a wire-line core barrel. **2.** A clawlike hook for grasping cylindrical objects, such as drill rods or casing, while raising and lowering them. { 'lift·iŋ ‚dog }

lifting guard [MIN ENG] Fencing placed around the mouth of a shaft and lifted out of the way by the ascending cage. { 'lift·iŋ ‚gärd }

lifting magnet [ELECTROMAG] A type of electromagnet in which a material to be held or moved is initially placed in contact with the magnet, in contrast to a traction magnet. Also known as holding magnet. [ENG] A large circular, rectangular, or specially shaped magnet used for handling pig iron, scrap iron, castings, billets, rails, and other magnetic materials. { 'lift·iŋ ‚mag·nət }

lifting reentry [AERO ENG] A reentry into the atmosphere by a space vehicle where aerodynamic lift is used, allowing a more gradual descent, greater accuracy in landing at a predetermined spot; it can accommodate greater errors in the guidance system and greater temperature control. { 'lift·iŋ rē'en·trē }

lifting reentry vehicle [AERO ENG] A space vehicle designed to utilize aerodynamic lift upon entering the atmosphere. { 'lift·iŋ rē'en·trē ‚vē·ə·kəl }

lifting task [IND ENG] A task that involves application of a moment to the vertebral column of the worker. { 'lift·iŋ ‚task }

lifting torque [BIOPHYS] A measure of the stress arising from the performance of a task that requires lifting; it is the product of the weight of the load and the load's distance from a point within the vertebral column that serves as a fulcrum. { 'lift·iŋ ‚tork }

lift-off [AERO ENG] The action of a rocket vehicle as it leaves its launch pad in a vertical ascent. { 'lif‚tof }

lift pump [MECH ENG] A pump for lifting fluid to the pump's own level. { 'lift ‚pəmp }

lift-slab construction [CIV ENG] Pouring reinforced concrete roof and floor slabs at ground level, then lifting them into position after hardening. { 'lift ‚slab kən‚strək·shən }

lift truck [MECH ENG] A small hand- or power-operated dolly equipped with a platform or forklift. { 'lift ‚trək }

lift valve [MECH ENG] A valve that moves perpendicularly to the plane of the valve seat. { 'lift ‚valv }

ligament [ENG] The section of solid material in a tube sheet or shell between adjacent holes. [HISTOL] A flexible, dense white fibrous connective tissue joining, and sometimes encapsulating, the articular surfaces of bones. { 'lig·ə·mənt }

ligamentum nuchae *See* nuchal ligament. { ‚lig·ə'men·təm 'nü‚kē }

ligand [CHEM] The molecule, ion, or group bound to the central atom in a chelate or a coordination compound; an example is the ammonia molecules in $[Co(NH_3)_6]^{3+}$. { 'lī·gənd }

ligand membrane [CHEM] A solvent immiscible with water and a reagent and acting as an extractant and complexing agent for an ion. { 'lī·gənd 'mem‚brān }

ligase [BIOCHEM] An enzyme that catalyzes the union of two molecules, involving the participation of a nucleoside triphosphate which is converted to a nucleoside diphosphate or monophosphate. Also known as synthetase. { 'lī‚gās }

ligate tar [MATER] A soft tar produced by destructive distillation of lignite; can be used without further refining as diesel fuel or can be redistilled to yield a substance resembling coal tar distillate. { 'lī‚gāt ‚tär }

ligation [MED] Surgical tying of vessels or ducts with a ligature. [CELL MOL] **1.** The process of joining two adjacent nitrogenous bases separated by a nick in one strand of a deoxyribonucleic acid duplex or of two linear nucleic acid molecules. **2.** Formation of a phosphodiester bond to join adjacent nucleotides in the same nucleic acid chain. { lī'gā·shən }

ligature [MED] A cord or thread used for tying vessels and ducts. { 'lig·ə·chər }

light [OPTICS] **1.** Electromagnetic radiation with wavelengths capable of causing the sensation of vision, ranging approximately from 400 (extreme violet) to 770 ananometers (extreme red). Also known as light radiation; visible radiation. **2.** More generally, electromagnetic radiation of any wavelength; thus, the term is sometimes applied to infrared and ultraviolet radiation. { līt }

light absorption [OPTICS] The process in which energy of light radiation is transferred to a medium through which it is passing. { 'līt əb‚sorp·shən }

light-activated silicon controlled rectifier [ELECTR] A silicon-controlled rectifier having a glass window for incident light that takes the place of, or adds to the action of, an electric gate current in providing switching action. Abbreviated LASCR. Also known as photo-SCR; photothyristor. { 'līt ‚ak·tə‚vād·əd ‚sil·ə·kən kən‚trōld 'rek·tə‚fī·ər }

light-activated silicon controlled switch [ELECTR] A semiconductor device that has four layers of silicon alternately doped with acceptor and donor impurities, but with all four of the p and n layers made accessible by terminals; when a light beam hits the active light-sensitive surface, the photons generate electron-hole pairs that make the device turn on; removal of light does not reverse the phenomenon; the switch can be turned off only by removing or reversing its positive bias. Abbreviated LASCS. { 'līt ‚ak·tə‚vād·əd ‚sil·ə·kən kən‚trōld 'swich }

light adaptation [PHYSIO] The disappearance of dark adaptation; the chemical processes by which the eyes, after exposure to a dim environment, become accustomed to bright illumination, which initially is perceived as quite intense and uncomfortable. { 'līt ‚ad‚ap'tā·shən }

light amplification by stimulated emission of radiation *See* laser. { 'līt ‚am·plə·fə'kā·shən bī ‚stim·yə‚lād·əd i‚mish·ən əv ‚rād·ē'ā·shən }

light amplifier [ELECTR] **1.** Any electronic device which, when actuated by a light image, reproduces a similar image of enhanced brightness, and which is capable of operating at very low light levels without introducing spurious brightness variations (noise) into the reproduced image. Also known as image intensifier. **2.** *See* laser amplifier. { 'līt ‚am·plə‚fī·ər }

light antiaircraft artillery [ORD] Conventional antiaircraft artillery pieces, usually under 90-millimeter, the weight of which in a trailed mount, including on-carriage fire control, does not exceed 20,000 pounds (9000 kilograms); self-propelled versions are rated in the same category as the trailed version. { 'līt ‚ant·ē'er‚kraft är‚til·ə·rē }

light artillery [ORD] All guns and howitzers of 105-millimeter caliber (4.13-inch bore) or smaller. { 'līt är'til·ə·rē }

light beacon [NAV] A beacon from which a light is displayed for directing ships. { 'līt ‚bē·kən }

light-beam galvanometer *See* d'Arsonval galvanometer. { 'līt ‚bēm ‚gal·və'näm·əd·ər }

light-beam oscillograph [ELECTROMAG] An oscillograph in which a beam of light, focused to a point by a lens, is reflected from a tiny mirror attached to the moving coil of a galvanometer onto a photographic film moving at constant speed. { 'līt ‚bēm ä'sil·ə‚graf }

light-beam pickup [ENG ACOUS] A phonograph pickup in which a beam of light is a coupling element of the transducer. { 'līt ‚bēm 'pik‚əp }

light blasting [ENG] Loosening of shallow or small outcrops of rock and breaking boulders by explosives. { 'līt 'blast·iŋ }

light bomber [AERO ENG] Any bomber with a gross weight of less than 100,000 pounds (45,000 kilograms), including bombs; for example, the A-20 and A-26 bombers in World War II. { 'līt 'bäm·ər }

light bulb See incandescent lamp. { 'līt ,bəlb }

light carrier injection [ELECTR] A method of introducing the carrier in a facsimile system by periodic variation of the scanner light beam, the average amplitude of which is varied by the density changes of the subject copy. Also known as light modulation. { 'līt 'kar·ē·ər in,jek·shən }

light-case bomb [ORD] A type of general-purpose bomb having a thin, light metal casing and designed to accomplish damage by blast primarily. { 'līt ,kās ,bäm }

light chain [IMMUNOL] The smaller of the two types of chains found in immunoglobulin molecules, molecular weight 23,500. Also known as B chain; L chain. { 'līt 'chān }

light chopper [ELECTR] A rotating fan or other mechanical device used to interrupt a light beam that is aimed at a phototube, to permit alternating-current amplification of the phototube output and to make its output independent of strong, steady ambient illumination. { 'līt 'chäp·ər }

light climate See illumination climate. { 'līt ,klī·mət }

light cone [RELAT] The set of all points in space-time that are reached by signals traveling at the speed of light from a specified point, or from which signals traveling at the speed of light reach that point. Also known as null cone. { 'līt ,kōn }

light crude [PETRO ENG] Crude oil having a high proportion of low-viscosity, low-molecular-weight hydrocarbons. { 'līt 'krüd }

light cruiser [NAV ARCH] A naval vessel whose armaments consist primarily of 6-inch (152-millimeter) guns. { 'līt 'krü·zər }

light curve [ASTROPHYS] A graph showing the variations in brightness of a celestial object; the stellar magnitude is usually shown on the vertical axis, and time is the horizontal coordinate. { 'līt ,kərv }

light cylinder See velocity-of-light cylinder. { 'līt ,sil·ən·dər }

light-day [ASTRON] The distance traveled by light in 1 day in a vacuum. { 'līt,dā }

light-distribution photometer [OPTICS] A device which measures the luminous intensity of a light source in various directions; the light source is fixed, and a mirror system is rotated about an axis passing through the centers of the light source and a photocell so that the light emitted by the source in any direction perpendicular to this axis is reflected to the photocell. Also known as distribution photometer. { 'līt dis·trə,byü·shən fə'täm·əd·ər }

light-drawn [MET] Cold-worked very slightly; for copper or copper alloy tubing, drawing entails between 10 and 25% reduction in area. { 'līt ,drón }

light-echo model [ASTRON] An explanation of superluminal motion wherein an outburst from the center of a quasar illuminates regions at successively greater radii. { 'līt ,ek·ō ,mäd·əl }

lighted buoy [NAV] In marine operations, a buoy with a light having definite characteristics for detection and identification during darkness. { 'līd·əd 'bói }

light emission via inelastic tunneling [ELECTR] A process in which electrons tunneling through a thin insulating layer separating two metals excite surface plasmons which then scatter from surface and structural discontinuities, radiating visible light. Abbreviated LEIT. { 'līt i,mish·ən ,vē·ə 'in·ə,las·tik 'tən·əl·iŋ }

light-emitting diode [ELECTR] A rectifying semiconductor device which converts electrical energy into electromagnetic radiation. The wavelength of the emitted radiation ranges from the near-ultraviolet to the near-infrared, that is, from about 400 to over 1500 nanometers. Abbreviated LED. { 'līt i,mid·iŋ 'dī,ōd }

light-emitting polymer See polymer light-emitting diode. { 'līt ə,mid·iŋ 'pöl·ə·mər }

light ends [MATER] The lower-boiling components of a mixture of hydrocarbons, such as those evaporated or distilled off easily in comparison to the bulk of the mixture; for hydrocarbon mixtures, usually considered to be butane and lighter. { 'līt 'enz }

lightening hole [CIV ENG] An opening cut into a strengthening member that decreases its weight without significantly altering its strength. { 'līt·niŋ ,hōl }

lighter [NAV ARCH] A barge used to load and unload ships not lying at piers, or to move cargo around a harbor. { 'līd·ər }

lighterage [IND ENG] **1.** Loading or unloading ships by means of a lighter. **2.** The fee charged for this operation. { 'līd·ə·rij }

lighter-than-air craft [AERO ENG] An aircraft, such as a dirigible, that weighs less than the air it displaces. { 'līd·ər thən 'er ,kraft }

light exposure [OPTICS] A measure of the total amount of light falling on a surface; equal to the integral over time of the luminance of the surface. Also known as exposure. { 'līt ik,spō·zhər }

light filter See color filter. { 'līt ,fil·tər }

light float [NAV] A boatlike structure (smaller than a light vessel and usually unmanned) used instead of a light buoy in waters where strong streams or currents are experienced, or when a greater elevation than that of a light buoy is necessary. { 'līt ,flōt }

light fog [GRAPHICS] A graying of the image produced by exposure to white light. { 'līt ,fäg }

light freeze [METEOROL] The condition when the surface temperature of the air drops to below the freezing point of water for a short time period, so that only the tenderest plants and vines are adversely affected. { 'līt ,frēz }

light frost [HYD] A thin and more or less patchy deposit of hoarfrost on surface objects and vegetation. { 'līt ,fróst }

light-gating cathode-ray tube [ELECTR] A cathode-ray tube in which the electron beam varies the transmission or reflection properties of a screen that is positioned in the beam of an external light source. { 'līt ,gād·iŋ ,kath,ōd 'rā ,tüb }

light guide See optical fiber. { 'līt ,gīd }

light gun [ELECTR] A light pen mounted in a gun-type housing. { 'līt ,gən }

light homing [ORD] Guidance of a missile to its target by using light which the target emits. { 'līt ,hōm·iŋ }

light-hour [ASTRON] The distance traveled by light in 1 hour in a vacuum. { 'līt,aú·ər }

lighthouse [NAV] A structure equipped with a powerful light to aid in navigation. { 'līt,haús }

lighthouse model [ASTRON] An explanation of superluminal motion wherein a rotating beam from a quasar illuminates stationary clouds that reflect the radio waves to earth, analogous to a lighthouse beacon. { 'līt,haús ,mäd·əl }

lighthouse tube See disk-seal tube. { 'līt,haús ,tüb }

light howitzer [ORD] A complete projectile-firing weapon with a medium muzzle velocity and a curved trajectory, a bore diameter of over 30 millimeters, through 125 millimeters. { 'līt 'haú·ət·sər }

light hydrogen See protium. { 'līt 'hī·drə·jən }

lighting branch circuit [ELEC] A circuit that supplies power to outlets for lighting fixtures only. { 'līd·iŋ 'branch ,sər·kət }

lighting-off torch [ENG] A torch used to ignite a fuel oil burner; it consists of asbestos cloth wrapped around an iron rod and soaked with oil. { 'līd·iŋ ,óf ,tórch }

light-inspection car [MECH ENG] A railway motorcar weighing 400–600 pounds (180–270 kilograms) and having a capacity of 650–800 pounds (295–360 kilograms). { 'līt in'spek·shən ,kär }

light intensity See luminous intensity. { 'līt in,ten·səd·ē }

light ion See small ion. { 'līt 'ī än }

light list [NAV] A publication tabulating navigational lights, with their locations, candlepowers, and other characteristics, to assist in their identification, and details of any accompanying fog signals. { 'līt ,list }

light list number [NAV] In marine operations, a number assigned each aid to navigation (except unlighted buoys), published in the light list for identification purposes. { 'līt ,list ,nəm·bər }

light machine gun [ORD] **1.** Any aircraft machine gun of .30 caliber or smaller. **2.** In U.S. Army usage, any lightweight machine gun, including the .30-caliber air-cooled machine gun and excluding fully automatic rifles, submachine guns, and machine pistols. **3.** In specific nonofficial context, any fully automatic rifle or any machine gun having total weight, with

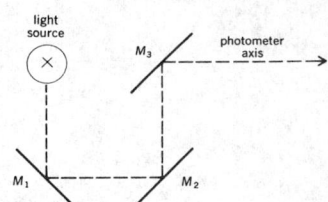

LIGHT-DISTRIBUTION PHOTOMETER

Mirror system in a light-distribution photometer. M_1, M_2, M_3 are mirrors.

mount, between 20–30 pounds (9–14 kilograms) approximately; for example, the Browning automatic rifle and the Lewis machine gun. { 'līt mə'shēn ˌgən }

light meromyosin [BIOCHEM] The smaller of two fragments obtained from the muscle protein myosin following limited proteolysis by trypsin or chymotrypsin. { 'līt ˌmer·ə'mī·ə·sən }

light metal [MET] A metal or alloy of low density, especially aluminum and magnesium alloys. { 'līt ˌmed·əl }

light meter [ENG] A small, portable device for measuring illumination; an exposure meter is a specific application, being calibrated to give photographic exposures. { 'līt ˌmēd·ər }

light microscope See optical microscope. { 'līt ˌmī·krə·skōp }

light microsecond [ELECTROMAG] Distance a light wave travels in free space in one-millionth of a second. { 'līt ˌmī·krə·sek·ənd }

light mineral [MINERAL] **1.** A rock with minerals that have a specific gravity lower than a standard, usually 2.85. **2.** A light-colored mineral. { 'līt ˌmin·rəl }

light modulation See light carrier injection. { 'līt ˌmäj·ə·lā·shən }

light modulator [ELECTR] The combination of a source of light, an appropriate optical system, and a means for varying the resulting light beam to produce an optical sound track on motion picture film. { 'līt ˌmäj·ə·lād·ər }

light-negative [ELECTR] Having negative photoconductivity, hence decreasing in conductivity (increasing in resistance) under the action of light. { 'līt ˌneg·ə·tiv }

lightness constancy [PSYCH] The tendency to perceive little variation in the lightness of objects despite enormous differences in the amount of light falling upon them and in the amount of light reaching the eye from them (luminance). { 'līt·nəs ˌkän·stən·sē }

lightning [GEOPHYS] An abrupt high-current electric discharge that occurs in the atmospheres of the earth and other planets and that has a path length ranging from hundreds of feet to tens of miles. Lightning occurs in thunderstorms because vertical air motions and interactions between cloud particles cause a separation of positive and negative charges. { 'līt·niŋ }

lightning arrester [ELEC] A protective device designed primarily for connection between a conductor of an electrical system and ground to limit the magnitude of transient overvoltages on equipment. Also known as arrester; surge arrester. { 'līt·niŋ ə,res·tər }

lightning channel [GEOPHYS] The irregular path through the air along which a lightning discharge occurs. { 'līt·niŋ ,chan·əl }

lightning conductor [ELEC] A conductor designed to carry the current of a lightning discharge from a lightning rod to ground. { 'līt·niŋ kən,dək·tər }

lightning discharge [GEOPHYS] The series of electrical processes by which charge is transferred within the atmosphere along a channel of high ion density between electric charge centers of opposite sign. { 'līt·niŋ 'dis,chärj }

lightning flash [GEOPHYS] In atmospheric electricity, the total observed luminous phenomenon accompanying a lightning discharge. { 'līt·niŋ ,flash }

lightning generator [ELEC] A high-voltage power supply used to generate surge voltages resembling lightning, for testing insulators and other high-voltage components. { 'līt·niŋ ,jen·ə,rād·ər }

lightning protection [ELEC] Means, such as lightning rods and lightning arresters, of protecting electrical systems, buildings, and other property from lightning. { 'līt·niŋ prə,tek·shən }

lightning recorder See sferics receiver. { 'līt·niŋ ri,kórd·ər }

lightning rod [ELEC] A metallic rod set up on an exposed elevation of a structure and connected to a low-resistance ground to intercept lightning discharges and to provide a direct conducting path to ground for them. { 'līt·niŋ ,räd }

lightning stroke [GEOPHYS] Any one of a series of repeated discharges comprising a single lightning discharge (or lightning flash); specifically, in the case of the cloud-to-ground discharge, a leader plus its subsequent return streamer. { 'līt·niŋ ,strōk }

lightning surge [ELEC] A transient disturbance in an electric circuit due to lightning. { 'līt·niŋ ,sərj }

lightning switch [ELEC] A manually operated switch used

LIGHT PEN

Electronic light pen is used by designer on a cathode-ray tube to instruct computer in the drafting of a schematic. (*Lockheed*)

to connect a radio antenna to ground during electrical storms, rather than to the radio receiver. { 'līt·niŋ ,swich }

light-of-the-night-sky See airglow. { 'līt əv thə ˌnīt 'skī }

light oil [MATER] **1.** A coal tar fraction obtained by distillation; boiling range, 110–210°C; used as a source of benzene, toluene, phenol, and cresols. Also known as coal tar light oil. **2.** Any oil having a boiling range of 110–210°C. { 'līt ˌóil }

light-operated switch [ELECTR] A switch that is operated by a beam or pulse of light, such as a light-activated silicon controlled rectifier. { 'līt ˌäp·ə,rād·əd 'swich }

light panel See electroluminescent panel. { 'līt ,pan·əl }

light pen [ELECTR] A tiny photocell or photomultiplier, mounted with or without fiber or plastic light pipe in a pen-shaped housing; it is held against a cathode-ray screen to make measurements from the screen or to change the nature of the display. { 'līt ,pen }

light pillar See sun pillar. { 'līt ,pil·ər }

light pipe [OPTICS] A solid, transparent plastic rod that transmits light from one end to the other even when bent. { 'līt ,pīp }

light-positive [ELECTR] Having positive photoconductivity; selenium ordinarily has this property. { 'līt ˌpäz·ə·tiv }

light projector [OPTICS] A device designed to produce controlled beams of light that can be projected over considerable distances. { 'līt prə,jek·tər }

light quantum See photon. { 'līt ,kwän·təm }

light radiation See light. { 'līt ,rād·ē,ā·shən }

light radius See velocity-of-light radius. { 'līt ,rād·ē·əs }

light ratio [ASTROPHYS] A number (2.512) that expresses the ratio of a star's light to that of another star that is one magnitude fainter or brighter. { 'līt ,rā·shō }

light ray [OPTICS] A beam of light having a small cross section. { 'līt ,rā }

light-red silver ore See proustite. { 'līt ,red 'sil·vər ,ór }

light reflex [PHYSIO] **1.** The postural orientation response of certain aquatic forms stimulated by the source of light; receptors may be on the ventral or dorsal surface. **2.** The response in which the pupil dilates when light levels are lowered, and constricts when light levels are raised. { 'līt ,rē,fleks }

light relay See photoelectric relay. { 'līt ,rē,lā }

light-ruby silver See proustite. { 'līt ,rü·bē 'sil·vər }

light scattering [OPTICS] The process in which energy is removed from a beam of light radiation and reemitted without appreciable change in wavelength. { 'līt ,skad·ə·riŋ }

light-scattering photometry [ANALY CHEM] Use of optical methods to measure the extent of scattering of light by particles suspended in fluids or by macromolecules in solution. { 'līt ,skad·ər·iŋ fə'täm·ə·trē }

light section car [MECH ENG] A railway motorcar weighing 750–900 pounds (340–408 kilograms) and propelled by 4–6-horsepower (3000–4500-watt) engines. { 'līt ,sek·shən ,kär }

light sector [NAV] **1.** A sector in which a navigational light has a distinctive color differing from that of adjoining sectors. **2.** A sector in which a navigational light is visible. { 'līt ,sek·tər }

light-sensitive [ELECTR] Having photoconductive, photoemissive, or photovoltaic characteristics. Also known as photosensitive. { 'līt 'sen·səd·iv }

light-sensitive cell See photodetector. { 'līt ˌsen·səd·iv 'sel }

light-sensitive detector See photodetector. { 'līt ˌsen·səd·iv di'tek·tər }

light-sensitive tube See phototube. { 'līt ˌsen·səd·iv 'tüb }

light sensor photodevice See photodetector. { 'līt ,sen·sər 'fōd·ō·di,vīs }

light-sheet method [FL MECH] A method of flow visualization in which a light beam is broadened by a cylindrical lens to illuminate a plane sheet of fluid; the light scattered from tracer particles in the illuminated plane exhibits the flow pattern. { 'līt ,shēt ,meth·əd }

lightship [NAV] A distinctively marked vessel anchored or moored at a charted point, to serve as an aid to navigation. { 'līt,ship }

light source [OPTICS] A lamp used to supply radiant energy, as for an optical microscope, projector, or photoelectric control system. { 'līt ,sórs }

light stability [COMPUT SCI] In optical character recognition, the ability of an image to retain its spectral appearance when exposed to radiant energy. { 'līt stə,bil·əd·ē }

light station [NAV] A group of buildings including a lighthouse and additional buildings housing personnel, fog signal, radio-beacon, and any other equipment associated with the lighthouse. { 'līt ,stā·shən }

light table [GRAPHICS] A table fitted with a transparent or translucent glass top illuminated with artificial light from below. { 'līt ,tā·bəl }

light time [ASTRON] The time required for light to travel from a distant object to the earth. { 'līt ,tīm }

light transmission [OPTICS] The process in which light travels through a medium without being absorbed or scattered. { 'līt tranz,mish·ən }

light valve [ELECTR] **1.** A device whose light transmission can be made to vary in accordance with an externally applied electrical quantity, such as volatage, current, electric field, or magnetic field, or an electron beam. **2.** Any direct-view electronic display optimized for reflecting or transmitting an image with an independent collimated light source for projection purposes. { 'līt ,valv }

light water [MATER] A water solution of perfluorocarbon compounds mixed with a polyoxyethylene thickener; used as a fire-fighting agent. [NUCLEO] Water in which both the hydrogen atoms in each molecule are of the isotope protium. { 'līt 'wód·ər }

light waterline [NAV ARCH] The waterline of a vessel without cargo. { 'līt 'wód·ər,līn }

light-water reactor [NUCLEO] A nuclear reactor that uses ordinary water as moderator, in contrast to heavy water. { 'līt ¦wód·ər rē'ak·tər }

light watt [OPTICS] A unit of luminous power equal to the luminous power of light of a single wavelength λ whose radiant power is $1/V_\lambda$ watts, where V_λ is the value of the luminosity function at λ. { 'līt ,wät }

lightweight aggregate [MATER] A lightweight inert material, such as foamed slag, vermiculite, clinker, and perlite, used in unreinforced concrete for making structures of low weight and high insulation. { 'līt,wāt 'ag·rə·gət }

lightweight concrete [MATER] A type of concrete made with lightweight aggregate. { 'līt,wāt 'kän,krēt }

light well [ARCH] A shaft within a building that is open to the outside at the top to admit daylight and fresh air through windows set into the sides of the shaft. { 'līt ,wel }

light-year [ASTROPHYS] A unit of measurement of astronomical distance; it is the distance light travels in one sidereal year and is equivalent to 9.461×10^{12} kilometers or 5.879×10^{12} miles. { 'līt ,yir }

Ligiidae [INV ZOO] A family of primitive terrestrial isopods in the suborder Oniscoidea. { lə'jī·ə,dē }

lignaloe oil See linaloe wood oil. { lī'na·lō ,oil }

ligneous [BIOL] Of, pertaining to, or resembling wood. { 'lig·nē·əs }

lignify [BOT] To convert cell wall constituents into wood or woody tissue by chemical and physical changes. { 'lig·nə,fī }

lignin [BIOCHEM] A substance that together with cellulose forms the woody cell walls of plants and cements them together. [MATER] A colorless to brown substance removed from paper-pulp sulfite liquor. { 'lig·nən }

ligninase [BIOCHEM] Any of a group of enzymes that breaks down lignin. { 'lig·nə,nās }

lignin plastic [ORG CHEM] A plastic based on resins derived from lignin; used as a binder or extender. { 'lig·nən 'plas,tik }

lignite [GEOL] Coal of relatively recent origin, intermediate between peat and bituminous coal; often contains patterns from the wood from which it formed. Also known as brown coal; earth coal. { 'lig,nīt }

lignite A See black lignite. { 'lig,nīt 'ā }

lignite B See brown lignite. { 'lig,nīt 'bē }

lignite wax See montan wax. { 'lig,nīt 'waks }

lignitious coal [MINERAL] A type of coal containing 75–84% elemental carbon. { lig'nish·əs ,kōl }

lignocaine See lidocaine. { 'lī·nə,kān }

lignocellulose [BIOCHEM] Any of a group of substances in woody plant cells consisting of cellulose and lignin. { ¦lig·nō'sel·yə,lōs }

lignosa [BOT] Woody vegetation. { lig'nō·sə }

lignosulfonate [ORG CHEM] Any of several substances manufactured from waste liquor of the sulfate pulping process of soft wood; used in the petroleum industry to reduce the viscosity of oil well muds and slurries, and as extenders in glues, synthetic resins, and cements. { lig·nō'səl·fə,nāt }

lignumvitae [BOT] *Guaiacum sanctum.* A medium-sized evergreen tree in the order Sapindales that yields a resin or gum known as gum guaiac or resin of guaiac. Also known as hollywood lignumvitae. { lig·nəm'vīd·ē }

ligroin See petroleum ether. { 'lig·rə·wən }

ligulate [BOT] **1.** Strap-shaped. **2.** Having ligules. { 'lig·yə·lət }

ligule [BOT] **1.** A small outgrowth in the axis of the leaves in Selaginellales. **2.** A thin outgrowth of a foliage leaf or leaf sheath. [INV ZOO] A small lobe on the parapodium of certain polychaetes. { 'lig·yül }

likelihood [MATH] The likelihood of a sample of independent values of $x_1, x_2, \ldots , x_n$, with $f(x)$ the probability function, is the product $f(x_1) \circ f(x_2) \circ \cdots \circ f(x_n)$. { 'līk·lē,húd }

likelihood ratio [STAT] The probability of a random drawing of a specified sample from a population, assuming a given hypothesis about the parameters of the population, divided by the probability of a random drawing of the same sample, assuming that the parameters of the population are such that this probability is maximized. { 'līk·lē,húd ,rā·shō }

likelihood ratio test [STAT] A procedure used in hypothesis testing based on the ratio of the values of two likelihood functions, one derived from the hypothesis being tested and one without the constraints of the hypothesis under test. { ¦līk·lē,húd 'rā·shō ,test }

like terms See similar terms. { ¦līk ′tərmz }

Liliaceae [BOT] A family of the order Liliales distinguished by six stamens, typically narrow, parallel-veined leaves, and a superior ovary. { ,lil·ē·'ās·ē,ē }

Liliales [BOT] An order of monocotyledonous plants in the subclass Liliidae having the typical characteristics of the subclass. { ,lil·ē'ā·lēz }

Liliatae See Liliopsida. { lə'lī·ə,tē }

Liliidae [BOT] A subclass of the Liliopsida; all plants are syncarpous and have a six-membered perianth, with all members petaloid. { lə'lī·ə,dē }

Liliopsida [BOT] The monocotyledons, making up a class of the Magnoliophyta; characterized generally by a single cotyledon, parallel-veined leaves, and stems and roots lacking a well-defined pith and cortex. { ,lil·ē·'äp·səd·ə }

lillianite [MINERAL] $Pb_3Bi_2S_6$ A steel-gray mineral composed of lead bismuth sulfide. { 'lil·ē·ə,nīt }

lilliputian star See brown dwarf. { ,lil·ə,pyü·shən 'stär }

Lilly controller [MECH ENG] A device on steam and electric winding engines that protects against overspeed, overwind, and other incidents injurious to workers and the engine. { 'lil·ē kən'trōl·ər }

lily [BOT] **1.** Any of the perennial bulbous herbs with showy unscented flowers constituting the genus *Lilium.* **2.** Any of various other plants having similar flowers. { 'lil·ē }

lily-pad ice See pancake ice. { 'lil·ē ,pad ,īs }

Limacidae [INV ZOO] A family of gastropod mollusks containing the slugs. { lī'mas·ə,dē }

limaçon [MATH] The locus of points of the plane which in polar coordinates (r,θ) satisfy the equation $r = a \cos \theta + b$. Also known as Pascal's limaçon. { 'lim·ə,sän *or* ,lim·ə'sōn }

limb [ANAT] An extremity or appendage used for locomotion or prehension, such as an arm or a leg. [ASTRON] The circular outer edge of a celestial body; the half with the greater altitude is called the upper limb, and the half with the lesser altitude, the lower limb. [BOT] A large primary tree branch. [DES ENG] **1.** The graduated margin of an arc or circle in an instrument for measuring angles, as that part of a marine sextant carrying the altitude scale. **2.** The graduated staff of a leveling rod. [GEOL] One of the two sections of an anticline or syncline on either side of the axis. Also known as flank. { limb }

limbate [BIOL] Having a part of one color bordered with a different color. { 'lim,bāt }

limb brightening [ASTRON] An observed increase in the intensity of radio, extreme ultraviolet, or x-radiation from the sun or another star from its center to its limb. { 'lim ,brīt·ən·iŋ }

limb bud [EMBRYO] A mound-shaped lateral proliferation of the embryonic trunk; the anlage of a limb. { 'lim ,bəd }

limb darkening [ASTROPHYS] An observed darkening near

LILIACEAE

Carolina lily (*Lilium michauxii*). (*Courtesy of John H. Gerard, from National Audubon Society*)

the surface of the sun's limb as compared to its brighter center. { 'lim ˌdär·kə·niŋ }

limber [ORD] A two-wheeled vehicle designed primarily to support the trail section of a gun carriage while in transit. { 'lim·bər }

limber board [NAV ARCH] One of several movable planks for covering the bilge-water passages on either side of the keelson. Also known as bilge board. { 'lim·bər ˌbȯrd }

limber hole [NAV ARCH] A drainage hole running through the bottom of the frame of ship. { 'lim·bər ˌhōl }

limbic system [ANAT] The inner edge of the cerebral cortex in the medial and temporal regions of the cerebral hemispheres. { 'lim·bik ˌsis·təm }

limb-kinetic apraxia See motor apraxia. { 'lim kə¦ned·ik ā¦prak·sē·ə }

limburgite [PETR] A dark, glass-rich igneous rock with abundant large crystals of olivine and pyroxene and with little or no feldspar. { 'lim·bər,gīt }

limbus [BIOL] A border clearly defined by its color or structure, as the margin of a bivalve shell or of the cornea of the eye. { 'lim·bəs }

lime [BOT] *Citrus aurantifolia.* A tropical tree with elliptic oblong leaves cultivated for its acid citrus fruit which is a hesperidium. { līm }

LIME

Tahiti (Persian) lime (*Citrus aurantifolia*).

lime-and-cement mortar [MATER] Mortar made of mortar cement, lime putty or hydrated lime, and sand in proportions, by volume, normally of one part cement, one or two lime, and five or six sand; suited for all kinds of masonry. { ¦līm ən si¦ment 'mȯrd·ər }

lime glass [MATER] A type of glass containing a high proportion of lime; used in many commercial glass products, such as bottles. { 'līm ˌglas }

lime grease [MATER] A type of grease that emulsifies less readily than one made with a soda base and therefore is used in an environment where water may occur. { 'līm ˌgrēs }

lime kiln [CHEM ENG] Furnace-type apparatus, usually a long, tilted cylinder that is slowly rotated, used to heat calcium carbonate, CaCO₃, above 900°C to produce lime. { 'līm ˌkil }

limelight [ENG] A light source once used in spotlights; it consisted of a block of lime heated to incandescence by means of an oxyhydrogen flame torch. { 'līm,līt }

lime liniment See carron oil. { 'līm 'lin·ə·mənt }

lime mortar [MATER] A mixture of hydrated lime, sand, and water having a compressive strength up to 400 pounds per square inch (2.8 × 10⁶ pascals); used for interior non-load-bearing walls in buildings. { 'līm ¦mȯrd·ər }

lime oil [MATER] **1.** An edible essential oil squeezed from the rind of lime and other citrus fruit, whose components are limonene and citral; used in flavorings and perfumes. **2.** The distilled essential oil from citron. { 'līm,ȯil }

lime-pan playa [GEOL] A playa with a smooth, hard surface composed of calcium carbonate. { 'līm ¦pan 'plī·ə }

lime putty [MATER] A puttylike cement made from lime slaked in water. { 'līm ¦pəd·ē }

lime sour [TEXT] The first treatment with a weak acid solution given to cotton goods in the finishing process. Also known as gray sour. { 'līm ˌsau̇r }

limestone [PETR] **1.** A sedimentary rock composed dominantly (more than 95) of calcium carbonate, principally in the form of calcite; examples include chalk and travertine. **2.** Any rock containing 80% or more of calcium carbonate or magnesium carbonate. { 'līm,stōn }

limestone log [ENG] A log that employs an electrical resistivity element in the form of four symmetrically arranged current electrodes to give accurate readings in borehole surveying of hard formations. { 'līm,stōn 'läg }

limestone pebble conglomerate [GEOL] A well-sorted conglomerate composed of limestone pebbles resulting from special conditions involving rapid mechanical erosion and short transport distances. { 'līm,stōn ¦peb·əl kən¦gläm·ə·rət }

lime water [PHARM] An alkaline aqueous solution of calcium hydroxide; used in medicine as an antacid. { 'līm ˌwȯd·ər }

limicolous [ECOL] Living in mud. { lī'mik·ə·ləs }

liminal contrast See threshold contrast. { 'lim·ə·nəl 'kän,trast }

liminal length [PHYSIO] The amount of tissue that must be raised above threshold of excitability such that an action potential will propagate. { ˌlim·ən·əl 'leŋkth }

liming [AGR] Treating the soil with lime. [CHEM ENG] Soaking hides and skins in milk of lime and causing them to swell, to facilitate the removal of hair. { 'līm·iŋ }

limit [MATH] **1.** A function *f*(*x*) has limit *L* as *x* tends to *c* if given any positive number ε (no matter how small) there is a positive number δ such that if *x* is in the domain of *f*, *x* is not *c*, and |*x* − *c*| < δ, then |*f*(*x*) − *L*| < ε; written

$$\lim_{x \to c} f(x) = L$$

2. A sequence {*aₙ*:*n* = 1, 2, . . .} has limit *L* if given a positive number ε (no matter how small) there is a positive integer *N* such that for all integers *n* greater than *N*, |*aₙ* − *L*| < ε. { 'lim·ət }

limit check [COMPUT SCI] A check to determine if a value entered into a computer system is within acceptable minimum and maximum values. { 'lim·ət ˌchek }

limit control [MECH ENG] **1.** In boiler operation, usually a device, electrically controlled, that shuts down a burner at a prescribed operating point. **2.** In machine-tool operation, a sensing device which terminates motion of the workpiece or tool at prescribed points. { 'lim·ət kən,trōl }

limit cycle [MATH] For a differential equation, a closed trajectory *C* in the plane (corresponding to a periodic solution of the equation) where every point of *C* has a neighborhood so that every trajectory through it spirals toward *C*. { 'lim·ət ˌsīk·əl }

limit dimensioning method [DES ENG] Method of dimensioning and tolerancing wherein the maximum and minimum permissible values for a dimension are stated specifically to indicate the size or location of the element in question. { 'lim·ət də,men·chən·iŋ ˌmeth·əd }

limited-access data [COMPUT SCI] Data to which only authorized users have access. { 'lim·əd·əd ¦ak·ses 'dad·ə }

limited-access highway See expressway. { 'lim·əd·əd ¦ak·ses 'hī,wā }

limited chromosome [CYTOL] A chromosome that occurs only in germ cell nuclei. { ¦lim·əd·əd ¦krō·mə,sōm }

limited-degree-of-freedom robot [CONT SYS] Robot whose end effector can be positioned and oriented in fewer than six degrees of freedom. { 'lim·əd·əd di'grē əv 'frē·dəm 'rō,bät }

limited-distance modem [COMMUN] A modem used only for communications within a building in order to improve the signal quality where a long distance exists between the terminal and the computer. [COMPUT SCI] A device designed to transmit and receive signals over relatively short distances, typically less than 5 miles (8 kilometers). Abbreviated LDM. Also known as line driver. { 'lim·əd·əd ¦dis·təns 'mō,dem }

limited-entry decision table [COMPUT SCI] A decision table in which the condition stub specifies exactly the condition or the value of the variable. { 'lim·əd·əd ¦en·trē di'sizh·ən ˌtā·bəl }

limited integrator [ELECTR] A device used in analog computers that has two input signals and one output signal whose value is proportional to the integral of one of the input signals with respect to the other as long as this output signal does not exceed specified limits. { 'lim·əd·əd 'int·ə,grād·ər }

limited-pressure cycle See mixed cycle. { 'lim·əd·əd ¦presh·ər ˌsī·kəl }

limited proportionality region [NUCLEO] The range of operating voltages of a radiation counter within which the charge collected is proportional to the charge liberated by the initial events, but saturates for larger initial events. { ¦lim·ət·əd prə,pȯr·shə'nal·əd·ē ˌrē·jən }

limited-rotation hydraulic actuator [MECH ENG] A type of hydraulic actuator that produces limited reciprocating rotary force and motion; used for lifting, lowering, opening, closing, indexing, and transferring movements; examples are the piston-rack actuator, single-vane actuator, and double-vane actuator. { 'lim·əd·əd rō¦tā·shən hī¦drȯ·lik 'ak·chə,wād·ər }

limited-sequence robot See fixed-stop robot. { 'lim·əd·əd ¦sē·kwəns 'rō,bät }

limited signal [ELECTR] Radar signal that is intentionally limited in amplitude by the dynamic range of the radar system. { 'lim·əd·əd 'sig·nəl }

limited space-charge accumulation mode [ELECTR] A mode of operation of a Gunn diode in which the frequency of operation is set by a resonant circuit to be much higher than the transit-time frequency so that domains have insufficient time to form while the field is above threshold and, as a result, the sample is maintained in the negative conductance state during a large fraction of the voltage cycle. Abbreviated LSA mode. { ˈlim·əd·əd ˌspās ˌchärj əˌkyü·myəˈlā·shən ˌmōd }

limited traverse [ORD] Restricted movement of a gun to right or left, caused by a mechanical device or by a natural obstacle. { ˈlim·əd·əd trəˈvərs }

limiter [ELECTR] An electronic circuit used to prevent the amplitude of an electronic waveform from exceeding a specified level while preserving the shape of the waveform at amplitudes less than the specified level. Also known as amplitude limiter; amplitude-limiting circuit; automatic peak limiter; clipper; clipping circuit; limiter circuit; peak limiter. { ˈlim·əd·ər }

limiter circuit See limiter. { ˈlim·əd·ər ˌsər·kət }

limit governor [MECH ENG] A mechanical governor that takes over control from the main governor to shut the machine down when speed reaches a predetermined excess above the allowable rate. Also known as topping governor. { ˈlim·ət ˌgəv·ər·nər }

limit inferior [MATH] Also known as lower limit. **1.** The limit inferior of a sequence whose nth term is a_n is the limit as N approaches infinity of the greatest lower bound of the terms a_n for which n is greater than N; denoted by

$$\lim \inf_{n\to\infty} a_n \text{ or } \underline{\lim}_{n\to\infty} a_n$$

2. The limit inferior of a function f at a point c is the limit as ϵ approaches zero of the greatest lower bound of $f(x)$ for $|x - c| < \epsilon$ and $x \neq c$; denoted by

$$\lim \inf_{x\to c} f(x) \text{ or } \underline{\lim}_{x\to c} f(x)$$

3. For a sequence of sets, the set consisting of all elements that belong to all but a finite number of the sets in the sequence. Also known as restricted limit. { ˈlim·ət inˈfir·ē·ər }

limiting [ELECTR] A desired or undesired amplitude-limiting action performed on a signal by a limiter. Also known as clipping; peak clipping. { ˈlim·əd·iŋ }

limiting current density [PHYS CHEM] The maximum current density to achieve a desired electrode reaction before hydrogen or other extraneous ions are discharged simultaneously. { ˈlim·əd·iŋ ˌkə·rənt ˌden·səd·ē }

limiting density [PHYS CHEM] The density of a gas when the ratio of density per unit pressure is extrapolated to zero pressure, the point at which a gas exhibits ideal-gas behavior. { ˈlim·əd·iŋ ˌden·səd·ē }

limiting friction See static friction. { ˈlim·əd·iŋ ˌfrik·shən }

limiting mean [ANALY CHEM] The value that the average approaches as the number of measurements made in a stable chemical measurement process increases indefinitely. { ˈlim·əd·iŋ ˈmēn }

limiting oxygen index [MATER] For a specific material, the lowest concentration of oxygen in the atmosphere, expressed in percent, that will support sustained combustion of the material. Abbreviated LOI. { ˈlim·əd·iŋ ˈäks·ə·jən ˌinˌdeks }

limiting ray [ACOUS] Any ray which is tangent to a plane at which the velocity of propagation of sound has a maximum value, either at a boundary of the medium of propagation or at a level where the velocity gradient changes sign. { ˈlim·əd·iŋ ˈrā }

limiting reagent [CHEM] In a chemical reaction, the reagent that controls the quantity of product which can be formed. { ˈlim·əd·iŋ rēˌā·jənt }

limiting viscosity number See intrinsic viscosity. { ˈlim·əd·iŋ viˈskäs·əd·ē ˌnəm·bər }

limit lines [IND ENG] Lines on a chart designating specification limits. { ˈlim·ət ˌlīnz }

limit-load design See ultimate-load design. { ˈlim·ət ˌlōd diˌzīn }

limit of detection [ANALY CHEM] The quantity or concentration that represents the smallest measure of an analyte that can be detected with reasonable certainty by a given analytical procedure. { ˈlim·ət əv diˈtek·shən }

limit of fire [ORD] **1.** Any angular limit, established for safety

purposes, for firing at a towed aerial target. **2.** The boundary of an area within which gunfire is placed. { ˈlim·ət əv ˈfīr }

limit of resolution [OPTICS] The minimum distance or angular separation between two point objects which allows them to be resolved according to the Rayleigh criterion. { ˈlim·ət əv ˌrez·əˈlü·shən }

limit of the atmosphere [GEOPHYS] The level at which the atmospheric density becomes the same as the density of interplanetary space, which is usually taken to be about one particle per cubic centimeter. { ˈlim·ət əv thə ˈat·məˌsfir }

limit on the left [MATH] The limit on the left of the function f at a point c is the limit of f at c which would be obtained if only values of x less than c were taken into account; more precisely, it is the number L which has the property that for any positive number ϵ, there is a positive number δ so that if z is the domain of f and $0 < (c - x) < \delta$ then $|f(x) - L| < \epsilon$; denoted by

$$\lim_{x\to c^-} f(x) = L \text{ or } f(c^-) = L$$

Also known as left-hand limit. { ˈlim·ət ȯn thə ˈleft }

limit on the right [MATH] The limit on the right of the function $f(x)$ at a point c is the limit of f at c which would be obtained if only values of x greater than c were taken into account; more precisely, it is the number L which has the property that for any positive number ϵ there is a positive number δ so that if x is in the domain of f and $0 < (x - c) < \delta$, then $|f(x) - L| < \epsilon$; denoted by

$$\lim_{x\to c^+} f(x) = L \text{ or } f(c^+) = L$$

Also known as right-hand limit. { ˈlim·ət ȯn thə ˈrīt }

limit point See cluster point. { ˈlim·ət ˌpȯint }

limit priority [COMPUT SCI] An upper bound to the dispatching priority that a task can assign to itself or any of its subtasks. { ˈlim·ət prīˈär·əd·ē }

limit ratio [ELECTR] Ratio of peak value to limited value, or comparison of such ratios. { ˈlim·ət ˌrā·shō }

limits [DES ENG] In dimensioning, the maximum and minimum values prescribed for a specific dimension; the limits may be of size if the dimension concerned is a size dimension, or they may be of location if the dimension concerned is a location dimension. { ˈlim·əts }

limits of integration [MATH] The end points of the interval over which a function is being integrated. { ˈlim·əts əv ˌint·əˈgrā·shən }

limit state [CIV ENG] The condition beyond which a structure or a structural member is deemed unsafe due to one or more loads or load effects. { ˈlim·ət ˌstāt }

limit stop [ORD] An arm or part used to limit angular motions, as of a gun turret. { ˈlim·ət ˌstäp }

limit superior [MATH] Also known as upper limit. **1.** The limit superior of a sequence whose nth term is a_n is the limit as N approaches infinity of the least upper bound of the terms a_n for which n is greater than N; denoted by

$$\lim \sup_{n\to\infty} a_n \text{ or } \overline{\lim}_{n\to\infty} a_n$$

2. The limit superior of a function f at a point c is the limit as ϵ approaches zero of the least upper bound of $f(x)$ for $|x - c| < \epsilon$ and $x \neq c$; denoted by

$$\lim \sup_{x\to c} f(x) \text{ or } \overline{\lim}_{x\to c} f(x)$$

3. For a sequence of sets, the set consisting of all elements that belong to infinitely many of the sets in the sequence. Also known as complete limit. { ˈlim·ət səˈpir·ē·ər }

limit switch [ELEC] A switch designed to cut off power automatically at or near the limit of travel of a moving object controlled by electrical means. { ˈlim·ət ˌswich }

limit velocity [MECH] In armor and projectile testing, the lowest possible velocity at which any one of the complete penetrations is obtained; since the limit velocity is difficult to obtain, a more easily obtainable value, designated as the ballistic limit, is usually employed. { ˈlim·ət vəˈläs·əd·ē }

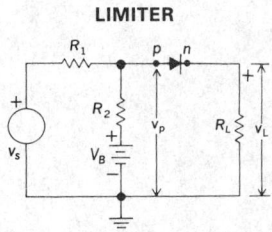

LIMITER

Circuit diagram of series diode clipping.

limivorous [ZOO] Feeding on mud, as certain annelids, for the organic matter it contains. { li'miv·ə·rəs }

Limnebiidae [INV ZOO] The minute moss beetles, a family of coleopteran insects in the superfamily Hydrophiloidea. { ˌlim·nə'bī·ə,dē }

limnetic [ECOL] Of, pertaining to, or inhabiting the pelagic region of a body of fresh water. { lim'ned·ik }

Limnichidae [INV ZOO] The minute false water beetles, a cosmopolitan family of coleopteran insects in the superfamily Dryopoidea. { lim'nik·ə,dē }

limnimeter [ENG] A type of tide gage for measuring lake level variations. { lim'nim·əd·ər }

limnite See bog iron ore. { 'lim,nīt }

Limnocharitaceae [BOT] A family of monocotyledonous plants in the order Alismatales characterized by schizogenous secretory canals, multiaperturate pollen, several or many ovules, and a horseshoe-shaped embryo. { ˌlim·nō,kar·ə'tās·ē,ē }

limnograph [ENG] A recording made on a limnimeter. { 'lim·nə,graf }

limnology [ECOL] The science of the life and conditions for life in lakes, ponds, and streams. { lim'näl·ə·jē }

Limnomedusae [INV ZOO] A suborder of hydrozoan cnidarians in the order Hydroida characterized by naked hydroids. { ˌlim·nō·mə'dü·sē }

limnoplankton [BIOL] Plankton found in fresh water, especially in lakes. { ˌlim·nō'plaŋk·tən }

Limnoriidae [INV ZOO] The gribbles, a family of isopod crustaceans in the suborder Flabellifera that burrow into submerged marine timbers. { ˌlim·nə'rī·ə,dē }

limonene [ORG CHEM] $C_{10}H_{16}$ A terpene with a lemon odor that is optically active and is found in oils from citrus fruits and in oils from peppermint and spearmint; a colorless, water-insoluble liquid that boils at 176°C. { 'lim·nə,lēn }

limoniform [BOT] Lemon-shaped. { li'män·ə,f orm }

limonite [MINERAL] A group of brown or yellowish-brown, amorphous, naturally occurring ferric oxides of variable composition; commonly formed secondary material by oxidation of iron-bearing minerals; a minor ore of iron. Also known as brown hematite; brown iron ore. { 'lī·mə,nīt }

limpet [INV ZOO] Any of several species of marine gastropod mollusks composing the families Patellidae and Acmaeidae which have a conical and tentlike shell with ridges extending from the apex to the border. { 'lim·pət }

Limulacea [INV ZOO] A group of horseshoe crabs belonging to the Limulida. { ˌlim·yə'lās·ē·ə }

Limulida [INV ZOO] A subgroup of Xiphosurida including all living members of the subclass. { lə'myül·ə·də }

Limulodidae [INV ZOO] The horseshoe crab beetles, a family of coleopteran insects in the superfamily Staphylinoidea. { ˌlim·myə'läd·ə,dē }

Limulus [INV ZOO] The horseshoe crab; the type genus of the Limulacea. { 'lim·yə·ləs }

linac See linear accelerator. { 'lin,ak }

Linaceae [BOT] A family of herbaceous or shrubby dicotyledonous plants in the order Linales characterized by mostly capsular fruit, stipulate leaves, and exappendiculate petals. { lī'nās·ē,ē }

Linales [BOT] An order of dicotyledonous plants in the subclass Orsidae containing simple-leaved herbs or woody plants with hypogynous, regular, syncarpous flowers having five to many stamens which are connate at the base. { lī'nā·lēz }

linaloe wood oil [MATER] An essential oil derived from fruit and wood of *Bursera* species; a colorless to yellow liquid, soluble in fixed oils and alcohol; used in perfumery and for flavoring. Also known as lignaloe oil; Mexican linaloe oil. { lē'näl·ō·ā ˌwúd ,óil }

linalool [ORG CHEM] $(CH_3)_2C:CH(CH_2)_2CCH_3OHCH:CH_2$ A terpene that is a colorless liquid, has a bergamot odor, boils at 195–196°C, and is found in many essential oils, particularly bergamot and rosewood; used as a flavoring agent and in perfumes. Also known as coriandrol. { lə'näl·ə,wòl }

linalyl acetate [ORG CHEM] $(CH_3)_2C:CH(CH_2)_2CCH_3(OCOCH_3)CH:CH_2$ The acetic acid ester of linalool, a colorless oily liquid with a bergamot odor that boils at 108–110°C; used in perfumes and as a flavoring agent. { 'lin·ə,lil 'as·ə,tāt }

linarite [MINERAL] $PbCu(SO_4)(OH)_2$ A deep-blue mineral composed of basic lead copper sulfate and occurring as monoclinic crystals. { 'lī·nə,rīt }

lincomycin [MICROBIO] $C_{18}H_{34}O_6N_2S·HCl$ A monobasic crystalline antibiotic, produced by *Streptomyces lincolnensis*, that is active as lincomycin hydrochloride mainly toward grampositive microorganisms. { ˌlin·kə'mīs·ən }

lindackerite [MINERAL] $Cu_6Ni_3(AsO_4)_4(SO_4)(OH)_4$ A light-green or apple-green mineral composed of hydrous basic sulfate and arsenate of nickel and copper; occurs in tabular crystals or massive. { lin'dak·ə,rīt }

lindane [INORG CHEM] The gamma isomer of 1,2,3,4,5,6-hexachlorocyclohexane, constituting a persistent, bioaccumulative pesticide and a neurotoxin. { 'lin,dān }

Lindeck potentiometer [ELEC] A potentiometer in which an unknown potential difference is balanced against a known potential difference derived from a fixed resistance carrying a variable current; the converse of most potentiometers. { 'lin,dek pə,ten·chē'äm·əd·ər }

Linde copper sweetening [CHEM ENG] A petroleum-refinery process to treat gasolines and distillates with a slurry of clay and cupric chloride to remove mercaptans. { 'lin·də 'käp·ər ,swēt·ən·iŋ }

Linde drill See fusion-piercing drill. { 'lin·də ,dril }

Lindelöf space [MATH] A topological space where if a family of open sets covers the space, then a countable number of these sets also covers the space. { 'lin·də,löf ,spās }

Lindelöf theorem [MATH] The proposition that there is a countable subcover of each open cover of a subset of a space whose topology has a countable base. { 'lin·də,lef ,thir·əm }

Lindemann electrometer [ELEC] A variant of the quadrant electrometer, designed for portability and insensitivity to changes in position, in which the quadrants are two sets of plates about 6 millimeters apart, mounted on insulating quartz pillars; a needle rotates about a taut silvered quartz suspension toward the oppositely charged plates when voltage is applied to it, and its movement is observed through a microscope. { 'lin·də·mən ē,lek'träm·əd·ər }

Lindemann glass [MATER] A lithium borate-beryllium oxide glass having no element higher in atomic number than oxygen; used as window material for low-voltage x-ray tubes because it will pass x-rays of extremely long wavelength, such as Grenz rays. { 'lin·də·mən ,glas }

Lindemann theory [SOLID STATE] A theory of the melting point of solids according to which solids melt when the amplitude of oscillation of the atoms becomes so great that neighboring atoms collide. { 'lin·də·mən ,thē·ə·rē }

linden See basswood. { 'lin·dən }

Linde process [CRYO] A cyclic process for liquefying gases in which compressed gas is cooled by Joule-Thomson expansion through a valve to a pressure of about 40 atmospheres (4 megapascals), further cools the incoming gas in a heat exchanger, and is compressed for the next cycle. { 'lin·də ,prä,ses }

Linde's rule [SOLID STATE] The rule that the increase in electrical resistivity of a monovalent metal produced by a substitutional impurity per atomic percent impurity is equal to $a + b(v - 1)^2$, where a and b are constants for a given solvent metal and a given row of the periodic table for the impurity, and v is the valence of the impurity. { 'lin·dəz ,rül }

lindgrenite [MINERAL] $Cu_3(MoO_4)_2(OH)_2$ A green mineral composed of basic copper molybdate. { 'lin·grə,nīt }

L indicator See L scope. { 'el ˌin·də,kād·ər }

lindstromite [MINERAL] $PbCuBi_3S_6$ A lead-gray to tin-white mineral composed of bismuth copper lead sulfide. { 'linz·trə,mīt }

line [BOT] A unit of length, equal to $1/12$ inch, or approximately 2.117 millimeters; it is most frequently used by botanists in describing the size of plants. [ELECTR] **1.** The path covered by the electron beam of a television picture tube in one sweep from left to right across the screen. **2.** One horizontal scanning element in a facsimile system. **3.** See trace. [MATH] The set of points $(x_1, \ldots, x_n)$ in euclidean space, each of whose coordinates is a linear function of a single parameter t; $x_i = f_i(t)$. Also known as straight line. { līn }

LINE See long interspersed nucleotide element.

linea alba [ANAT] A tendinous ridge extending in the median line of the abdomen from the pubis to the tiphoid process and formed by the blending of aponeuroses of the oblique and transverse muscles of the abdomen. { 'lin·ē·ə 'al·bə }

LIMPET

Two views of the rough keyhole limpet (*Diodora aspera*).

lineage [GEN] Individuals descended from a common progenitor. { 'lin·ē·ij }

lineage structure [CRYSTAL] An imperfection structure characterizing a crystal, parts of which have slight differences in orientation. { 'lin·ē·ij ‚strək·chər }

lineament [ASTRON] A prominent linear feature on the lunar surface. [GEOL] A straight or gently curved, lengthy topographic feature expressed as depressions or lines of depressions. Also known as linear. [GRAPHICS] A structurally controlled line on an aerial photograph; applied to lines representing beds, veins, faults, rock boundaries, and such. { 'lin·ē·ə·mənt }

line-and-staff organization [IND ENG] A form of organization structure which combines functional subunits with staff officers in line functions. { ‚līn ən ‚staf ‚òr·gə·nə‚zā·shən }

line and trunk group [COMMUN] A group consisting of four-wire circuits, incoming private automatic branch exchange trunks, and intertoll trunk groups. { ‚līn ən ‚trəŋk ‚grüp }

linear [CONT SYS] Having an output that varies in direct proportion to the input. [GEOL] *See* lineament. [SCI TECH] **1.** Of or relating to a line. **2.** Having a single dimension. { 'lin·ē·ər }

linear accelerator [NUCLEO] A particle accelerator which accelerates electrons, protons, or heavy ions in a straight line by the action of alternating voltages. Also known as linac. { 'lin·ē·ər ik'sel·ə‚rād·ər }

linear accelerator breeder [NUCLEO] A device which uses a linear accelerator to generate neutrons that produce fissile fuel for nuclear reactors. Also known as electronuclear breeder. { 'lin·ē·ər ik'sel·ə‚rād·ər 'brēd·ər }

linear accelerator-driven reactor [NUCLEO] A nuclear reactor which consists of a linear accelerator and a subcritical reactor target containing depleted, natural, or slightly enriched uranium or thorium, that produces a net amount of energy when supplied with neutrons generated by stopping a high-energy proton beam from a linear accelerator. Abbreviated LADR. { 'lin·ē·ər ik'sel·ə‚rād·ər ‚driv·ən rē'ak·tər }

linear accelerator-regenerator reactor [NUCLEO] A device in which neutrons, generated by stopping a high-energy proton beam from a linear accelerator, irradiate nuclear fuel to bring it to a level of reactivity where it can be used in place in a conventional power reactor. Abbreviated LARR. { 'lin·ē·ər ik'sel·ə‚rād·ər ri'jen·ə‚rād·ər rē'ak·tər }

linear actuator [MECH ENG] A device that converts some kind of power, such as hydraulic or electric power, into linear motion. { 'lin·ē·ər 'ak·chə‚wād·ər }

linear aerospike engine [AERO ENG] A variant of the aerospike engine with a plug nozzle that is V-shaped rather than axisymmetric. { ‚lin·ē·ər 'er·ō‚spīk ‚en·jən }

linear algebra [MATH] The study of vector spaces and linear transformations. { 'lin·ē·ər 'al·jə·brə }

linear algebraic equation [MATH] An equation in some algebraic system where the unknowns occur linearly, that is, to the first power. { 'lin·ē·ər ‚al·jə‚brā·ik i'kwā·zhən }

linear amplifier [ELECTR] An amplifier in which changes in output current are directly proportional to changes in applied input voltage. { 'lin·ē·ər 'am·plə‚fī·ər }

linear array [ELECTROMAG] An antenna array in which the dipole or other half-wave elements are arranged end to end on the same straight line. Also known as collinear array. { 'lin·ē·ər ə'rā }

linear-array camera [ELECTR] A solid-state television camera that has only a single row of light-sensitive elements or pixels. { 'lin·ē·ər ə‚rā 'kam·rə }

linear birefringence [OPTICS] Birefringence effects which are proportional to applied stresses. { 'lin·ē·ər ‚bī·ri'frin·jəns }

linear bounded automaton [COMPUT SCI] A nondeterministic, one-tape Turing machine whose read/write head is confined to move only on a restricted section of tape initially containing the input. { 'lin·ē·ər ‚baünd·əd ò'täm·ə‚tän }

linear burning rate [ORD] The distance normal to any burning surface of the propellant grain burned through in unit time. { 'lin·ē·ər 'bər·niŋ ‚rāt }

linear circuit *See* linear network. { 'lin·ē·ər 'sər·kət }

linear cleavage [GEOL] The property of metamorphic rocks of breaking into long planar fragments. { 'lin·ē·ər 'klē·vij }

linear collider [NUCLEO] A colliding-beam device consisting of two linear accelerators which fire intense bunches of charge particles at each other so that they collide. { 'lin·ē·ər kə'līd·ər }

linear collision cascade [SOLID STATE] A sputtering event in which the bombarding projectile collides directly with a small number of target atoms, which collide with others, and the sharing of energy then proceeds through many generations before one or more target atoms are ejected; the density of atoms in motion remains sufficiently small so that collisions between atoms can be ignored. { 'lin·ē·ər kə'lizh·ən ‚kas‚kād }

linear combination [MATH] A linear combination of vectors $v_1, \ldots, v_n$ in a vector space is any expression of the form $a_1 v_1 + a_2 v_2 + \cdots + a_n v_n$, where the a_i are scalars. { 'lin·ē·ər ‚käm·bə'nā·shən }

linear combination of atomic orbitals [PHYS] A method of constructing approximate wave functions for molecular orbitals or for electrons in solids, by taking sums of atomic orbitals of the component atoms, each centered on an atom in the structure and multiplied by a coefficient, and then varying these coefficients to minimize the energy of the wave function. Abbreviated LCAO. { ‚lin·ē·ər ‚käm·bə‚nā·shən əv ə‚täm·ik 'òrb·əd·əlz }

linear comparator [ELECTR] A comparator circuit which operates on continuous, or nondiscrete, waveforms. Also known as continuous comparator. { 'lin·ē·ər kəm'par·əd·ər }

linear computing element [ELEC] A linear circuit in an analog computer. { 'lin·ē·ər kəm'pyüd·iŋ ‚el·ə·mənt }

linear conductor antenna [ELECTROMAG] An antenna consisting of one or more wires which all lie along a straight line. { 'lin·ē·ər kən'dək·tər an‚ten·ə }

linear congruence [MATH] The relation between two quantities that have the same remainder on division by a given integer, where the quantities are polynomials of, at most, the first degree in the variables involved. { 'lin·ē·ər kən'grü·əns }

linear control [ELEC] Rheostat or potentiometer having uniform distribution of graduated resistance along the entire length of its resistance element. { 'lin·ē·ər kən'trōl }

linear control system [CONT SYS] A linear system whose inputs are forced to change in a desired manner as time progresses. { 'lin·ē·ər kən'trōl ‚sis·təm }

linear density [PHYS] The quantity of anything distributed along a line per unit length of line. { 'lin·ē·ər 'den·səd·ē }

linear dependence [MATH] The property of a set of vectors $v, \ldots, v_n$ in a vector space for which there exists a linear combination such that $a_1 v_1 + \cdots + a_n v_n = 0$, and at least one of the scalars a_i is not zero. { 'lin·ē·ər di'pen·dəns }

linear detection [ELECTR] Detection in which the output voltage is substantially proportional, over the useful range of the detecting device, to the voltage of the input wave. { 'lin·ē·ər di'tek·shən }

linear differential equation [MATH] A differential equation in which all derivatives occur linearly, and all coefficients are functions of the independent variable. { 'lin·ē·ər ‚dif·ren·chəl i'kwā·zhən }

linear discriminant function [STAT] A function, used in conjunction with a set of threshold values in a classification procedure, whose values are linear combinations of the values of selected variables. { ‚lin·ē·ər di‚skrim·ə·nənt ‚fəŋk·shən }

linear distortion [ELECTR] Amplitude distortion in which the output signal envelope is not proportional to the input signal envelope and no alien frequencies are involved. { 'lin·ē·ər di'stòr·shən }

linear electrical constants of a uniform line [ELEC] Series resistance, series inductance, shunt conductance, and shunt capacitance per unit length of line. { 'lin·ē·ər i‚lek·trə·kəl 'kän·stəns əv ə ‚yü·nə‚fòrm 'līn }

linear electrical parameters *See* transmission-line parameters. { 'lin·ē·ər i‚lek·trə·kəl pə'ram·əd·ərz }

linear element [MATH] On a surface determined by equations $x = f(u,v)$, $y = g(u,v)$, and $z = h(u,v)$, the element of length ds given by $ds^2 = E\,du^2 + 2F\,du\,dv + G\,dv^2$, where E, F, and G are functions of u and v. { 'lin·ē·ər 'el·ə·mənt }

linear energy transfer *See* stopping power. { 'lin·ē·ər 'en·ər·jē ‚tranz·fər }

linear equation [MATH] A linear equation in the variables $x_1, \ldots, x_n$, and y is any equation of the form $a_1 x_1 + a_2 x_2 + \cdots + a_n x_n = y$. { 'lin·ē·ər i'kwā·zhən }

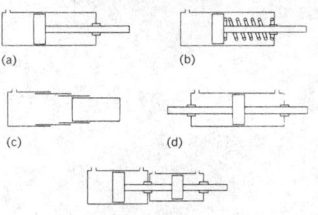

LINEAR ACTUATOR

(a) (b) (c) (d) (e)

Hydraulic cylinder types of linear actuators. (a) Single-acting, external return. (b) Single-acting spring return. (c) Telescopic. (d) Double-acting double-ended rod. (e) Tandem cylinders.

linear expansion [PHYS] Expansion of a body in one direction. { 'lin·ē·ər ik'span·chən }

linear expansity See coefficient of linear expansion. { 'lin·ē·ər ik'span·səd·ē }

linear extrapolation distance [NUCLEO] The extrapolation distance of a medium sustaining a neutron chain reaction, based on extrapolation of the neutron flux density just inside the medium by a linear function. { ¦lin·ē·ər ik'strap·ə¸lā·shən ¸dis·təns }

linear feedback control [CONT SYS] Feedback control in a linear system. { 'lin·ē·ər 'fēd¸bak kən¸trōl }

linear flow structure See platy flow structure. { 'lin·ē·ər 'flō ¸strək·chər }

linear form [MATH] A homogeneous polynomial of the first degree. { 'lin·ē·ər 'form }

linear fractional transformations See Möbius transformations. { 'lin·ē·ər ¦frak·shən·əl ¸tranz·fər'mā·shənz }

linear function See linear transformation. { 'lin·ē·ər 'fəŋk·shən }

linear functional [MATH] A linear transformation from a vector space to its scalar field. { 'lin·ē·ər 'fəŋk·shən·əl }

linear hypothesis See linear model. { 'lin·ē·ər hī'päth·ə·səs }

linear independence [MATH] The property of a set of vectors $v_1, \ldots, v_n$ in a vector space where if $a_1v_1 + a_2v_2 + \cdots + a_nv_n = 0$, then all the scalars $a_i = 0$. { 'lin·ē·ər ¸in·də'pen·dəns }

linear inequalities [MATH] A collection of relations among variables x_i, where at least one relation has the form $\Sigma_i a_i x_i \geq 0$. { 'lin·ē·ər ¸in·i'kwäl·əd·ēz }

linear integrated circuit [ELECTR] An integrated circuit that provides linear amplification of signals. { 'lin·ē·ər ¦int·ə¸grād·əd 'sər·kət }

linear interpolation [MATH] A process to find a value of a function between two known values under the assumption that the three plotted points lie on a straight line. { 'lin·ē·ər in¸tər·pə'lā·shən }

linearity [MATH] The property whereby a mathematical system is well behaved (in the context of the given system) with regard to addition and scalar multiplication. [PHYS] The relationship that exists between two quantities when a change in one of them produces a directly proportional change in the other. { ¸lin·ē'ar·əd·ē }

linearity control [ELECTR] A cathode-ray-tube control which varies the distribution of scanning speed throughout the trace interval. Also known as distribution control. { ¸lin·ē'ar·əd·ē kən¸trōl }

linearization [CONT SYS] **1.** The modification of a system so that its outputs are approximately linear functions of its inputs, in order to facilitate analysis of the system. **2.** The mathematical approximation of a nonlinear system, whose departures from linearity are small, by a linear system corresponding to small changes in the variables about their average values. [CELL MOL] Conversion of a circular deoxyribonucleic acid molecule into a linear molecule. { ¸lin·ē·ər·ə'zā·shən }

linearized theory of fluid flow [FL MECH] An approximate method for solving aerodynamic problems; it treats the flow of an inviscid gas past a body whose geometry and motion are such that the disturbance velocities caused by its introduction into some previously known flow are small compared with the speed of sound; as a result, the equations of motion can be approximated by retaining only those terms which are linear in disturbance or perturbation velocities, pressures, densities, and so forth. { ¦lin·ē·ə¸rīzd ¸thē·ə·rē əv 'flü·əd ¸flō }

linear light [NAV] In marine operations, a luminous signal having perceptible length, as contrasted with a point light, which does not have perceptible length. { 'lin·ē·ər ¸līt }

linear - logarithmic intermediate - frequency amplifier [ELECTR] Amplifier used to avoid overload or saturation as a protection against jamming in a radar receiver. { 'lin·ē·ər ¸läg·ə'rith·mik ¸in·tər¸mē·dē·ət ¦frē·kwən·sē 'am·plə¸fī·ər }

linearly dependent quantities [MATH] Quantities that satisfy a homogeneous linear equation in which at least one of the coefficients is not zero. { 'lin·ē·ər·lē di¦pen·dənt 'kwän·tə¸tēz }

linearly disjoint extensions [MATH] Two extension fields E and F of a field k contained in a common field L, such that any finite set of elements in E that is linearly independent when E is regarded as a vector space over k remains linearly independent when E is regarded as a vector space over F. { ¦lin·ē·ər·lē ¸dis¦jóint ik'sten·chənz }

linearly graded junction [ELECTR] A pn junction in which the impurity concentration does not change abruptly from donors to acceptors, but varies smoothly across the junction, and is a linear function of position. { 'lin·ē·ər·lē ¦grād·əd 'jəŋk·shən }

linearly independent quantities [MATH] Quantities which do not jointly satisfy a homogeneous linear equation unless all coefficients are zero. { 'lin·ē·ər·lē ¸in·də¦pen·dənt 'kwän·əd·ēz }

linearly ordered set [MATH] A set with an ordering $\leq$ such that for any two elements a and b either $a \leq b$ or $b \leq a$. Also known as chain; completely ordered set; serially ordered set; simply ordered set; totally ordered set. { 'lin·ē·ər·lē ¦ór·dərd 'set }

linear magnetic amplifier [ELECTR] A magnetic amplifier employing negative feedback to make its output load voltage a linear function of signal current. { 'lin·ē·ər mag¦ned·ik 'am·plə¸fī·ər }

linear manifold [MATH] A subset of a vector space which is itself a vector space with the induced operations of addition and scalar multiplication. { 'lin·ē·ər 'man·ə¸fōld }

linear meter [ENG] A meter in which the deflection of the pointer is proportional to the quantity measured. { 'lin·ē·ər 'mēd·ər }

linear model [STAT] A mathematical model in which linear equations connect the random variables and the parameters. Also known as linear hypothesis. { 'lin·ē·ər 'mäd·əl }

linear modulation [COMMUN] Modulation in which the amplitude of the modulation envelope (or the deviation from the resting frequency) is directly proportional to the amplitude of the intelligence signal at all modulation frequencies. { 'lin·ē·ər ¸mäj·ə'lā·shən }

linear molecule [PHYS CHEM] A molecule whose atoms are arranged so that the bond angle between each is 180°; an example is carbon dioxide, CO_2. { 'lin·ē·ər 'mäl·ə¸kyül }

linear momentum See momentum. { 'lin·ē·ər mə'men·təm }

linear motion See rectilinear motion. { 'lin·ē·ər 'mō·shən }

linear motor [ELEC] An electric motor that has in effect been split and unrolled into two flat sheets, so that the motion between rotor and stator is linear rather than rotary. { 'lin·ē·ər 'mōd·ər }

linear network [ELEC] A network in which the parameters of resistance, inductance, and capacitance are constant with respect to current or voltage, and in which the voltage or current of sources is independent of or directly proportional to other voltages and currents, or their derivatives, in the network. Also known as linear circuit. { 'lin·ē·ər 'net¸wərk }

linear operator See linear transformation. { 'lin·ē·ər 'äp·ə¸rād·ər }

linear order [MATH] Any order $<$ on a set S with the property that for any two elements a and b in S exactly one of the statements $a < b$, $a = b$, or $b < a$ is true. Also known as complete order; serial order; simple order; total order. { 'lin·ē·ər 'ór·dər }

linear oscillator See harmonic oscillator. { 'lin·ē·ər 'äs·ə¸läd·ər }

linear parallax See absolute stereoscopic parallax. { 'lin·ē·ər 'par·ə¸laks }

linear parallel texture [PETR] The parallel texture of a rock in which the constituents are parallel to a line, not just to a plane as in plane parallel texture. { 'lin·ē·ər ¦par·ə¸lel 'teks·chər }

linear-phase [ELECTR] Pertaining to a filter or other network whose image phase constant is a linear function of frequency. { 'lin·ē·ər ¸fāz }

linear polarization [OPTICS] Polarization of an electromagnetic wave in which the electric vector at a fixed point in space remains pointing in a fixed direction, although varying in magnitude. Also known as plane polarization. { 'lin·ē·ər ¸pō·lə·rə'zā·shən }

linear polymer [ORG CHEM] A polymer whose molecule is arranged in a chainlike fashion with few branches or bridges between the chains. { 'lin·ē·ər 'päl·ə·mər }

linear power amplifier [ELECTR] A power amplifier in which the signal output voltage is directly proportional to the signal input voltage. { 'lin·ē·ər 'paù·ər ¸am·plə¸fī·ər }

linear programming [MATH] The study of maximizing or minimizing a linear function $f(x_1, \ldots, x_n)$ subject to given constraints which are linear inequalities involving the variables x_i. { 'lin·ē·ər 'prō,gram·iŋ }

linear-quadratic-gaussian problem [CONT SYS] An optimal-state regulator problem, containing Gaussian noise in both the state and measurement equations, in which the expected value of the quadratic performance index is to be minimized. Abbreviated LQG problem. { 'lin·ē·ər kwə'drad·ik 'gaús·ē·ən ,präb·ləm }

linear rectifier [ELECTR] A rectifier, the output current of voltage of which contains a wave having a form identical with that of the envelope of an impressed signal wave. { 'lin·ē·ər 'rek·tə,fī·ər }

linear regression [STAT] The straight line running among the points of a scatter diagram about which the amount of scatter is smallest, as defined, for example, by the least squares method. { 'lin·ē·ər ri'gresh·ən }

linear regulator problem [CONT SYS] A type of optimal control problem in which the system to be controlled is described by linear differential equations and the performance index to be minimized is the integral of a quadratic function of the system state and control functions. Also known as optimal regulator problem; regulator problem { 'lin·ē·ər 'reg·yə,lād·ər ,präb·ləm }

linear repeater [ELECTR] A repeater used in communication satellites to amplify input signals a fixed amount, generally with traveling-wave tubes or solid-state devices operating in their linear region. { 'lin·ē·ər ri'pēd·ər }

linear scale See uniform scale. { 'lin·ē·ər 'skāl }

linear scanning [ENG] Radar beam which moves with constant angular velocity through the scanning sector, which may be a complete 360°. { 'lin·ē·ər 'skan·iŋ }

linear space See vector space. { 'lin·ē·ər 'spās }

linear span [MATH] See span. { ,lin·ē·ər 'span }

linear speed method [ORD] Method of calculating artillery firing data in which the future position of a moving target is determined by finding the direction of flight and the ground speed of the target; by multiplying the ground speed by the time of flight of the projectile, the future position is determined. { 'lin·ē·ər 'spēd ,meth·əd }

linear Stark effect [ATOM PHYS] A splitting of spectral lines of hydrogenlike atoms placed in an electric field; each energy level of principal quantum number n is split into $2n - 1$ equidistant levels of separation proportional to the field strength. { 'lin·ē·ər 'stärk i,fekt }

linear stopping power See stopping power. { 'lin·ē·ər 'stäp·iŋ ,paù·ər }

linear strain [MECH] The ratio of the change in the length of a body to its initial length. Also known as longitudinal strain. { 'lin·ē·ər 'strān }

linear sweep [ELECTR] A cathode-ray sweep in which the beam moves at constant velocity from one side of the screen to the other, then suddenly snaps back to the starting side. { 'lin·ē·ər 'swēp }

linear-sweep delay circuit [ELECTR] A widely used form of linear time-delay circuit in which the input signal initiates action by a linear sawtooth generator, such as the bootstrap or Miller integrator, whose output is then compared with a calibrated direct-current reference voltage level. { 'lin·ē·ər 'swēp di,lā ,sər·kət }

linear-sweep generator [ELECTR] An electronic circuit that provides a voltage or current that is a linear function of time; the waveform is usually recurrent at uniform periods of time. { 'lin·ē·ər 'swēp jen·ə,rād·ər }

linear system [CONT SYS] A system in which the outputs are components of a vector which is equal to the value of a linear operator applied to a vector whose components are the inputs. [MATH] A system where all the interrelationships among the quantities involved are expressed by linear equations which may be algebraic, differential, or integral. { 'lin·ē·ər 'sis·təm }

linear system analysis [CONT SYS] The study of a system by means of a model consisting of a linear mapping between the system inputs (causes or excitations), applied at the input terminals, and the system outputs (effects or responses), measured or observed at the output terminals. { 'lin·ē·ər 'sis·təm ə'nal·ə·səs }

linear taper [ELEC] A taper that gives the same change in resistance per degree of rotation over the entire range of a potentiometer. { 'lin·ē·ər 'tā·pər }

linear time base [ELECTR] A time base that makes the electron beam of a cathode-ray tube move at a constant speed along the horizontal time scale. { 'lin·ē·ər 'tīm ,bās }

linear topological space See topological vector space. { 'lin·ē·ər ,täp·ə,läj·ə·kəl 'spās }

linear transducer [ELECTR] A transducer for which the pertinent measures of all the waves concerned are linearly related. { 'lin·ē·ər tranz'dü·sər }

linear transformation [MATH] A function T defined in a vector space E and having its values in another vector space over the same field, such that if f and g are vectors in E, and c is a scalar, then $T(f + g) = Tf + Tg$ and $T(cf) = c(Tf)$. Also known as homogeneous transformation; linear function; linear operator. { 'lin·ē·ər ,tranz·fər'mā·shən }

linear trend [STAT] A first step in analyzing a time series, to determine whether a linear relationship provides a good approximation to the long-term movement of the series; computed by the method of semiaverages or by the method of least squares. { 'lin·ē·ər 'trend }

linear unit [ELECTR] An electronic device used in analog computers in which the change in output, due to any change in one of two or more input signals, is proportional to the change in that input and does not depend upon the values of the other inputs. { 'lin·ē·ər 'yü·nət }

linear variable-differential transformer [ELECTR] A transformer in which a diaphragm or other transducer sensing element moves an armature linearly inside the coils of a differential transformer, to change the output voltage by changing the inductances of the coils in equal but opposite amounts. Abbreviated LVDT. { 'lin·ē·ər 'ver·ē·ə·bəl 'dif·ə,ren·chəl tranz'fôr·mər }

linear velocity See velocity. { 'lin·ē·ər və'läs·əd·ē }

line at infinity See ideal line. { 'līn at in'fin·əd·ē }

lineation [GEOL] Any linear structure on or within a rock; examples are ripple marks and flow lines. { ,lin·ē'ā·shən }

line balance [ELEC] **1.** Degree of electrical similarity of the two conductors of a transmission line. **2.** Matching impedance, equaling the impedance of the line at all frequencies, that is used to terminate a two-wire line. { 'līn ,bal·əns }

line-balance converter See balun. { 'līn ,bal·əns kən,vərd·ər }

line block See line cut. { 'līn ,bläk }

line blow [METEOROL] A strong wind on the equator side of an anticyclone, probably so called because there is little shifting of wind direction during the blow, as contrasted with the marked shifting which occurs with a cyclonic windstorm. { 'līn ,blō }

line brattice [MIN ENG] A partition in an opening to divide it into intake and return airways. { 'līn 'brad·əs }

line broadening [SPECT] An increase in the range of wavelengths over which the characteristic absorption or emission of a spectral line takes place, due to a number of causes such as collision broadening and Doppler broadening. { 'līn ,brȯd·ən·iŋ }

line-building-out network See impedance-matching network. { 'līn 'bild·iŋ ,aut ,net,wərk }

linecasting machine [GRAPHICS] A composing machine which assembles a line of type, casts it, and distributes the matrices to a magazine. { 'līn,kast·iŋ mə,shēn }

line characteristic distortion [COMMUN] A kind of teletypewriter transmission distortion caused when the lengths of the received signal impulses are affected by the presence of changing current transitions in wire circuits. { 'līn ,kar·ik·tə'ris·tik di,stȯr·shən }

line circuit [ELEC] **1.** Relay equipment associated with each station connected to a dial or manual switchboard. **2.** A circuit to interconnect an individual telephone and a channel terminal. { 'līn ,sər·kət }

line clinometer [ENG] A clinometer designed to be inserted between rods at any point in a string of drill rods. { 'līn klī'näm·əd·ər }

line code [COMPUT SCI] The single instruction required to solve a specific type of problem on a special-purpose computer. { 'līn ,kōd }

line conditioning [COMMUN] The addition of compensating reactances to a data transmission line to reduce amplitude and phase delays over certain frequency bands. { 'līn kən,dish·ə,niŋ }

line conductor [ELEC] A metal used as a conductor in a

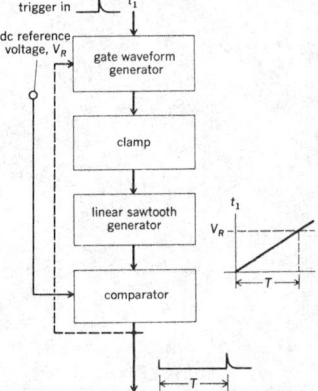

**LINEAR-SWEEP
DELAY CIRCUIT**

Elements of linear-sweep delay circuit. T = delay time; V_R = reference voltage; t_1 = time.

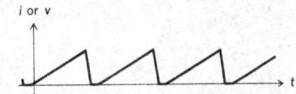

LINEAR-SWEEP GENERATOR

Sawtooth waveform of a linear-sweep generator. Current i or voltage v is plotted against time t.

power line; the most frequently used conductors are copper and aluminum. { 'līn kən,dək·tər }

line-controlled blocking oscillator [ELECTR] A circuit formed by combining a monostable blocking oscillator with an open-circuit transmission line in the regenerative circuit; it is capable of generating pulses with large amounts of power. { 'līn kən,trōld ¦bläk·iŋ 'äs·ə,lād·ər }

line conversion [GRAPHICS] A picture obtained by photographing continuous-tone pictures with line film, and no screen, thus reducing all the values in the original picture to black or white. { 'līn kən,vər·zhən }

line copy [GRAPHICS] Copy that can be reproduced directly without using a halftone screen. { 'līn ,käp·ē }

line cord [ELEC] A two-wire cord terminating in a two-prong plug at one end and connected permanently to a radio receiver or other appliance at the other end; used to make connections to a source of power. Also known as power cord. { 'līn ,kórd }

line-cord resistor [ELEC] An asbestos-enclosed wire-wound resistor incorporated in a line cord along with the two regular wires. { 'līn ,kórd ri,zis·tər }

line cut [GRAPHICS] A relief printing plate made by photographing a design and then transferring the negative onto a zinc or copper plate that is then developed, with the lines that will form the printing surface being protected and the rest of the plate etched down; used exclusively for the reproduction of materials executed in black (or a color) and white, with no intermediate shades of gray (or tones). Also known as line block; line engraving; line etching; line plate. { 'līn ,kət }

line defect *See* dislocation. { 'līn di,fekt }

line discipline [COMPUT SCI] The rules that govern exactly how data are transferred between locations in a communications network. { 'līn ,dis·ə·plən }

line displacement [ASTROPHYS] Widening or shifting of spectral lines of celestial objects arising from several causes, such as gas under high pressure. { 'līn di,splās·mənt }

line dot matrix [COMPUT SCI] A line printer that uses the dot matrix printing technique. Also known as parallel dot character printer. { 'līn ¦dät 'mā,triks }

line drilling [MIN ENG] The combined methods of drilling and broaching for the primary cut in quarrying; deep, closely spaced holes are drilled in a straight line by means of a reciprocating drill, and webs between holes are removed by a drill or a flat broaching tool. { 'līn ,dril·iŋ }

line driver [COMPUT SCI] *See* limited-distance modem. [ELECTR] An integrated circuit that acts as the interface between logic circuits and a two-wire transmission line. { 'līn ,drīv·ər }

line drop [ELEC] The voltage drop existing between two points on a power line or transmission line, due to the impedance of the line. { 'līn ,dräp }

line-drop compensator [ELEC] A device that restores the voltage lost when electricity is transmitted along a wire. { 'līn ¦dräp 'käm·pən,sād·ər }

line-drop signal [COMMUN] Signal associated with a subscriber line on a manual switchboard. { 'līn ¦dräp ,sig·nəl }

line editor [COMPUT SCI] A text-editing system that stores a file of discrete lines of text to be printed out on the console (or displayed) and manipulated on a line-by-line basis, so that editing operations are limited and are specified for lines identified by a specific number. { 'līn ,ed·əd·ər }

line engraving *See* line cut. { 'līn in,grāv·iŋ }

line equalizer [ELEC] An equalizer containing inductance or capacitance, inserted in a transmission line to modify the frequency response of the line. { 'līn 'ē·kwə,līz·ər }

line etching *See* line cut. { 'līn ,ech·iŋ }

line facility [COMMUN] A transmission line in a communication system, together with amplifiers spaced at regular intervals to offset attenuation in the line. { 'līn fə,sil·əd·ē }

line fault [ELEC] A defect, such as an open circuit, short circuit, or ground, in an electric line for transmission or distribution of power or of speech, music, or other content. { 'līn ,fólt }

line feed [COMPUT SCI] **1.** Signal that causes a printer to feed the paper up a discrete number of lines. **2.** Rate at which paper is fed through a printer. { 'līn ,fēd }

line fill [COMMUN] Ratio of the number of connected main telephone stations on a line to the nominal main station capacity of that line. { 'līn ,fil }

line filter [ELEC] **1.** A filter inserted between a power line

and a receiver, transmitter, or other unit of electric equipment to prevent passage of noise signals through the power line in either direction. Also known as power-line filter. **2.** A filter inserted in a transmission line or high-voltage power line for carrier communication purposes. { 'līn ,fil·tər }

line filter balance [COMMUN] Network designed to maintain phantom group balance when one side of the group is equipped with a carrier system. { 'līn ,fil·tər ,bal·əns }

line finder [COMMUN] A switching device that automatically locates an idle telephone or telegraph circuit going to the desired destination. [COMPUT SCI] A device that automatically advances the platen of a line printer or typewriter. { 'līn ,fīn·dər }

line-finder shelf [COMMUN] Usually, 20 line finders with the equipment required for connecting any of the associated calling telephones to a selector or connector which will receive the dial pulses from the calling telephone. { 'līn ,fīn·dər ,shelf }

line-finder switch [COMMUN] In telephony, an automatic switch for seizing selector apparatus which provides dial tone to the calling party. { 'līn ,fīn·dər ,swich }

line flux [ELECTROMAG] A local inductive field of a telephone or power line. { 'līn ,fləks }

line-formula method [ORG CHEM] A system of notation for hydrocarbons showing the chemical elements, functional groups, and ring systems in linear form; an example is acetone, CH_3COCH_3. { 'līn ,fór·myə·lə ,meth·əd }

line frequency [ELECTR] The number of times per second that the scanning spot sweeps across the screen in a horizontal direction in a television system. Also known as horizontal frequency; horizontal line frequency. { 'līn ,frē·kwən·sē }

line-frequency blanking pulse *See* horizontal blanking pulse. { 'līn ,frē·kwən·sē 'blaŋk·iŋ ,pəls }

line functions [IND ENG] Organizational functions having direct authority and responsibility. { 'līn ,fəŋk·shənz }

line gage *See* type gage. { 'līn ,gāj }

line gale *See* equinoctial storm. { 'līn ,gāl }

line graph [MATH] A graph in which successive points representing the value of a variable at selected values of the independent variable are connected by straight lines. { 'līn ,graf }

line hydrophone [ENG ACOUS] A directional hydrophone consisting of one straight-line element, an array of suitably phased elements mounted in line, or the acoustic equivalent of such an array. { 'līn 'hī·drə,fōn }

Lineidae [INV ZOO] A family of the Heteronemertini. { li'nē·ə,dē }

line impedance [ELECTROMAG] The impedance measured across the terminals of a transmission line. { 'līn im,pēd·əns }

line influence [ELECTROMAG] The effect of a local inductive field around a telephone line. { 'līn ,in·flü·əns }

line integral [MATH] **1.** For a curve in a vector space defined by $x = x(t)$, and a vector function V defined on this curve, the line integral of V along the curve is the integral over t of the scalar product of $V[x(t)]$ and dx/dt; this is written $\int V \cdot dx$. **2.** For a curve which is defined by $x = x(t)$, $y = y(t)$, and a scalar function f depending on x and y, the line integral of f along the curve is the integral over t of $f[x(t,y(t)] \cdot \sqrt{(dx/dt)^2 + (dy/dt)^2}$; this is written $\int f ds$, where $ds = \sqrt{(dx)^2 + (dy)^2}$ is an infinitesimal element of length along the curve. **3.** For a curve in the complex plane defined by $z = z(t)$, and a function f depending on z, the line integral of f along the curve is the integral over t of $f[z(t)] (dz/dt)$; this is written $\int f dz$. { 'līn ¦int·ə·grəl }

line interlace *See* interlaced scanning. { 'līn 'in·tər,lās }

line item [COMPUT SCI] Any data that is considered to be of equal importance to other data in the same file. { 'līn ,īd·əm }

line lengthener [ELECTROMAG] Device for altering the electrical length of a waveguide or transmission line without altering other electrical characteristics, or the physical length. { 'līn ,leŋk·thə·nər }

line level [COMMUN] Signal level in decibels at a particular position on a transmission line. [ENG] A small spirit level fitted with hooks at each end so that it can be hung on a horizontally stretched line. { 'līn ,lev·əl }

line link [COMMUN] A frame on which several hundred telephone lines appear in a crossbar switching system. Abbreviated LL. { 'līn ,liŋk }

linellae [INV ZOO] Thin organic threads in the tests of some xenophyophores. { lə'nel·ē }

line location [ELEC] The location of power and communications lines when two or more such lines run along the same route; they should either be used jointly, or located with respect to each other so as to avoid unnecessary crossings, conflicts, and inductive exposures. { 'līn lō,kā·shən }

line loop [COMMUN] Portion of a telephone circuit that includes a user's telephone set and the pair of wires that connect it with the distributing frame of a central office. { 'līn ,lüp }

line-loop resistance [ELEC] Metallic resistance of the line wires that extend from an individual telephone set to the dial central office. { 'līn ,lüp ri,zis·təns }

line loss [ELEC] Total of the various energy losses occurring in a transmission line. [ENG] The quantity of gas that is lost in a distribution system or pipeline. { 'līn ,lòs }

line lubricator See line oiler. { 'līn ,lü·brə,kād·ər }

line map See planimetric map. { 'līn ,map }

line microphone [ENG ACOUS] A highly directional microphone consisting of a single straight-line element or an array of small parallel tubes of different lengths, with one end of each abutting a microphone element. Also known as machine-gun microphone. { 'līn ,mī·krə,fōn }

line misregistration [COMPUT SCI] In character recognition, the improper appearance of a line of characters, on site in a character reader, with respect to a real or imaginary horizontal line. { 'līn ,mis,rej·ə'strā·shən }

line mixer See flow mixer. { 'līn ,mik·sər }

linen [TEXT] A cloth made from flax fibers, noted for its strength, weavability, durability, and minimum discharge of lint. { 'lin·ən }

line noise [COMMUN] Noise originating in a transmission line from such causes as poor joints and inductive interference from power lines. { 'līn ,nòiz }

line number [COMPUT SCI] A number at the beginning or end of each line of a computer program that specifies its position in a sequence. { 'līn ,nəm·bər }

line of action [MECH ENG] The locus of contact points as gear teeth profiles go through mesh. { 'līn əv 'ak·shən }

line of aim [ORD] A line from a person's eye, as that of a gunner or bombardier, through a sight, along which aim is taken. { 'līn əv 'ām }

line of apsides [ASTRON] 1. The line connecting the two points of an orbit that are nearest and farthest from the center of attraction, as the perigee and apogee of the moon or the perihelion and aphelion of a planet. 2. The length of this line. { 'līn əv 'ap·sə,dēz }

line of balance [IND ENG] A production planning system that schedules key events leading to completion of an assembly on the basis of the delivery date for the completed system. Abbreviated LOB. { 'līn əv 'bal·əns }

line of code [COMPUT SCI] A single statement in a programming language. { 'līn əv 'kōd }

line of collimation [OPTICS] In a surveying telescope, the imaginary line through the optical center of the object glass and the cross-hair intersection in the diaphragm. { 'līn əv ,käl·ə'mā·shən }

line of curvature [MATH] A curve on a surface whose tangent lies along a principal direction at each point. { 'līn əv 'kər·və·chər }

line of departure [NAV] The initial position of a scouting line, from which scouts proceed on their prescribed courses for search. [ORD] 1. The direction of a projectile at the instant it clears the muzzle of the gun. 2. The direction of a bomb or rocket at the instant of launching. { 'līn əv di'pär·chər }

line of electrostatic induction [ELEC] A unit of electric flux equal to the electric flux associated with a charge of 1 statcoulomb. { 'līn əv i,lek·trə,stad·ik in'dək·shən }

line of elevation [ORD] The prolongation of the bore of a gun when the piece is set to fire. { 'līn əv ,el·ə'vā·shən }

line of fall [MECH] The line tangent to the ballistic trajectory at the level point. { 'līn əv 'fòl }

line of flight [MECH] The line of movement, or the intended line of movement, of an aircraft, guided missile, or projectile in the air. { 'līn əv 'flīt }

line of flux See line of force. { 'līn əv ,fləks }

line of force [PHYS] An imaginary line in a field of force (such as an electric, magnetic, or gravitational field) whose tangent at any point gives the direction of the field at that point; the lines are spaced so that the number through a unit area perpendicular to the field represents the intensity of the field. Also known as flux line; line of flux. { 'līn əv ,fòrs }

line of impact [MECH] A line tangent to the trajectory of a missile at the point of impact. { 'līn əv 'im,pakt }

line of magnetic induction See maxwell. { 'līn əv mag,ned·ik in'dək·shən }

line of observation [ORD] A line from a position finder to a target at the exact time of a recorded observation. { 'līn əv ,äb·zər'vā·shən }

line of position [NAV] A line indicating a series of possible positions of a craft, determined by observation or measurement. Also known as position line. { 'līn əv pə'zish·ən }

line of retirement [NAV] The position of a scouting line when it has reached its outer limit, and a search to the rear is initiated. { 'līn əv ri'tīr·mənt }

line of return [NAV] The final position of the scouting line, where individual scouts leave their stations and return to their bases. { 'līn əv ri'tərn }

line of sight [ELECTROMAG] The straight line for a transmitting radar antenna in the direction of the beam. [SCI TECH] A straight, unobstructed path or line between two points, as between an observer's eye and a target. { 'līn əv 'sīt }

line-of-sight velocity See radial velocity. { 'līn əv 'sīt və'läs·əd·ē }

line of soundings [NAV] A series of soundings obtained by a vessel when it is under way. { 'līn əv 'saund·iŋz }

line of striction [MATH] The locus of the central points of the rulings of a given ruled surface. { 'līn əv 'strik·shən }

line of strike See strike. { 'līn əv 'strīk }

line of support [MATH] Relative to a convex region in a plane, a line that contains at least one point of the region but is such that a half-plane on one side of the line contains no points of the region. { 'līn əv sə'pòrt }

line of thrust [MECH] Locus of the points through which the resultant forces pass in an arch or retaining wall. { 'līn əv 'thrəst }

line of tunnel [ENG] The width marked by the exterior lines or sides of a tunnel. { 'līn əv 'tən·əl }

line oiler [MECH ENG] An apparatus inserted in a line conducting air or steam to an air- or steam-activated machine that feeds small controllable amounts of lubricating oil into the air or steam. Also known as air-line lubricator; line lubricator. { 'līn ,òi·lər }

Lineolaceae [MICROBIO] A family of bacteria in some systems of classification that includes coenocytic members (*Lineola*) of the Caryophanales. { ,lin·ē·ə'lās·ē,ē }

lineolate [BIOL] Marked with fine lines. { 'lin·ē·ə,lāt }

line pack [ENG] The actual amount of gas in a pipeline or distribution system. { 'līn ,pak }

line pad [ELECTR] Pad inserted between a program amplifier and a transmission line, to isolate the amplifier from impedance variations of the line. { 'līn ,pad }

line pair [SPECT] In spectrographic analysis, a particular spectral line and the internal standard line with which it is compared to determine the concentration of a substance. { 'līn ,per }

line parameters See transmission-line parameters. { 'līn pə,ram·əd·ərz }

line plate See line cut. { 'līn ,plāt }

line printer [COMPUT SCI] A device that prints an entire line in a single operation, without necessarily printing one character at a time. { 'līn ,print·ər }

line printing [COMPUT SCI] The printing of an entire line of characters as a unit. { 'līn ,print·iŋ }

line profile [ASTROPHYS] A curve that indicates the internal variation in intensity of a spectral line of a celestial body. { 'līn ,prō,fīl }

line pulsing [ELECTR] Method of pulsing a transmitter in which an artificial line is charged over a relatively long period of time and then discharged through the transmitter tubes in a short interval determined by the line characteristic. { 'līn ,pəls·iŋ }

liner [DES ENG] A replaceable tubular sleeve inside a hydraulic or pump-pressure cylinder in which the piston travels. [ENG] A string of casing in a borehole. [MET] 1. The cylindrical chamber that holds the billet for extrusion. 2. The slab

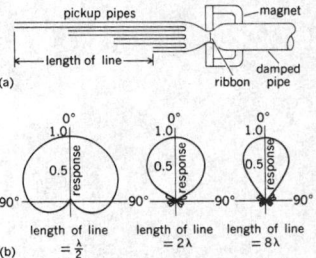

LINE MICROPHONE

Line microphone. *(a)* Sectional view showing open-ended pipes. *(b)* Directivity patterns. The maximum voltage response is arbitrarily chosen as unity.

of coating metal that is placed on the core alloy and is subsequently rolled down to form a clad composite. [MIN ENG] **1.** A foot piece for uprights in timber sets. **2.** Timber supports erected to reinforce existing sets which are beginning to collapse due to heavy strata pressure. **3.** A bar put up between two other bars to assist in carrying the roof. **4.** Replaceable facings inside a grinding mill. [NAV ARCH] A merchant vessel engaged in regular, usually high-speed service. { 'līn·ər }

line radiation [ELECTROMAG] Electromagnetic radiation from a power line caused mainly by corona pulses; gives rise to radio interference. { 'līn ,rād·ē,ā·shən }

liner bushing [DES ENG] A bushing, provided with or without a head, that is permanently installed in a jig to receive the renewable wearing bushings. Also known as master bushing. { 'līn·ər ,bu̇sh·iŋ }

line reflection [COMMUN] Reflection of a signal at the end of a transmission line, at the junction of two or more lines, or at a substation. { 'līn ri,flek·shən }

line regulation [ELEC] The maximum change in the output voltage or current of a regulated power supply for a specified change in alternating-current line voltage, such as from 105 to 125 volts. { 'līn ,reg·yə'lā·shən }

line relay [ELEC] Relay which is controlled over a subscriber line or trunkline. { 'līn rē,lā }

line rod See range rod. { 'līn ,räd }

liner plate cofferdam [CIV ENG] A cofferdam made from steel plates about 16 inches (41 centimeters) high and 3 feet (91 centimeters) long, and corrugated for added stiffness. { 'līn·ər ,plāt 'kȯf·ər,dam }

lines [NAV ARCH] The outline of a ship, either as projected onto one of three perpendicular planes or as viewed visually. Also known as ship's lines. { līnz }

LINES See long interspersed elements. { līnz }

line scanner [ENG] An infrared imaging device which utilizes the motion of a moving platform, such as an aircraft or satellite, to scan infrared radiation from the terrain. Also known as thermal mapper. { 'līn ,skan·ər }

line segment [MATH] A connected piece of a line. { 'līn ,seg·mənt }

line-segment formula See bond-line formula. { 'līn ,seg·mənt ,fȯr·myə·lə }

line-sequential color television [COMMUN] A color television system in which an entire line is one color, with colors changing from line to line in a red, blue, and green sequence. { 'līn si¦kwən·chəl 'kəl·ər 'tel·ə,vizh·ən }

line shafting [MECH ENG] One or more pieces of assembled shafting to transmit power from a central source to individual machines. { 'līn ,shaft·iŋ }

line side [ELEC] Terminal connections to an external or outstation source, such as data terminal connections to a communications circuit connecting to another data terminal. { 'līn ,sīd }

line skew [COMPUT SCI] In character recognition, a form of line misregistration, when the string of characters to be recognized appears in a uniformly slanted condition with respect to a real or imaginary baseline. { 'līn ,skyü }

linesman [ENG] **1.** A worker who sets up and repairs communication and power lines. **2.** An assistant to a surveyor. { 'līnz·mən }

line source [OPTICS] An idealized source of light consisting of an infinitely long line from which light is emitted with uniform intensity. { 'līn ,sȯrs }

line space lever [MECH ENG] A lever on a typewriter used to move the carriage to a new line. { 'līn ¦spās ,lev·ər }

line spectrum [SPECT] **1.** A spectrum of radiation in which the quantity being studied, such as frequency or energy, takes on discrete values. **2.** Conventionally, the spectra of atoms, ions, and certain molecules in the gaseous phase at low pressures; distinguished from band spectra of molecules, which consist of a pattern of closely spaced spectral lines which could not be resolved by early spectroscopes. { 'līn ,spek·trəm }

line speed [COMMUN] Maximum rate at which signals may be transmitted over a given channel, usually in bauds or bits per second. { 'līn ,spēd }

lines per minute [COMPUT SCI] A measure of the speed of the printer. Abbreviated LPM. { 'līnz pər 'min·ət }

line squall [METEOROL] A squall that occurs along a squall line. { 'līn ,skwȯl }

line-stabilized oscillator [ELECTR] Oscillator in which a

section of line is used as a sharply selective circuit element for the purpose of controlling the frequency. { 'līn ,stā·bə,līzd 'äs·ə,lād·ər }

line storm See equinoctial storm. { 'līn ,stȯrm }

line strength [ATOM PHYS] The intensity of a spectrum line. { 'līn ,streŋkth }

line stretcher [ELECTROMAG] Section of waveguide or rigid coaxial line whose physical length is variable to provide impedance matching. { 'līn ,strech·ər }

line switching [COMMUN] A telephone switching system in which a switch attached to a subscriber line connects an originating call to an idle part of the switching apparatus. [ELECTR] Connecting or disconnecting the line voltage from a piece of electronic equipment. { 'līn ,swich·iŋ }

line switching concentrator [COMMUN] Switching center used between a group of users and the switching center to reduce the number of trunks and increase efficiency of switching equipment usage (sometimes referred to as statistical multiplexing). { 'līn ¦swich·iŋ 'kän·sən,trād·ər }

line synchronizing pulse See horizontal synchronizing pulse. { 'līn ,siŋ·krə,nīz·iŋ ,pəls }

line timbers [MIN ENG] Timbers placed along the sides of the track of a working place in rows according to a predetermined plan. { 'līn ,tim·bərz }

line-to-ground fault [ELEC] A defect in a power or communications line in which faulty insulation allows the conductor to make contact with the earth. { 'līn tə 'grau̇nd ,fȯlt }

line transducer [ELECTR] A special type of electret transducer consisting essentially of a coaxial cable with polarized dielectric, and with the center conductor and shield serving as electrodes; mechanical excitation resulting in a deformation of the shield at any point along the length of the cable produces an electrical output signal. { 'līn tranz,dü·sər }

line transformer [ELEC] Transformer connecting a transmission line to terminal equipment; used for isolation, line balance, impedance matching, or additional circuit connections. { 'līn tranz,fȯr·mər }

line trap [ELEC] A filter consisting of a series inductance shunted by a tuning capacitor, inserted in series with the power or telephone line for a carrier-current system to minimize the effects of variations in line attenuation and reduce carrier energy loss. { 'līn ,trap }

line tuning [ELEC] Adjustment of the frequency of carrier current of a communication system to tune out the reactance of a capacitor with suitable inductance. { 'līn ,tün·iŋ }

line-turn See maxwell-turn. { 'līn ,tərn }

line turnaround [COMMUN] The time required for a half-duplex circuit to reverse the direction of transmission. { 'līn 'tərn·ə,rau̇nd }

line unit [ELECTR] Electric control device used to send, receive, and control the impulses of a teletypewriter. { 'līn ,yü·nət }

line up [MIN ENG] **1.** A command signifying that the drill runner wants the hoisting cable attached to the drill stem, threaded through the sheave wheel, or wound on the hoist drum. **2.** To reposition a drill so that the drill stem is centered over the parallel to a newly collared drill hole. { 'līn ¦əp }

line-use ratio [COMMUN] As applied to facsimile broadcasting, the ratio of the available line to the total length of scanning line. { 'līn ¦yüs ,rā·shō }

line voltage [ELEC] The voltage provided by a power line at the point of use. { 'līn ,vōl·tij }

line-voltage regulator [ELEC] A regulator that counteracts variations in power-line voltage, so as to provide an essentially constant voltage for the connected load. { 'līn ,vōl·tij 'reg·yə,lād·ər }

line vortex [FL MECH] A type of fluid motion in which fluid flows approximately in circles about a line, at speeds inversely proportional to the distance from the line, so that there is an infinite concentration of vorticity on the line, and vorticity vanishes elsewhere. { 'līn 'vȯr,teks }

Lineweaver-Burk equation [BIOCHEM] A double reciprocal form of the Michaelis-Menten equation, written as $1/V_0 = K_m/V_{max}[S] + 1/V_{max}$, that yields a straight-line plot for reactions obeying basic Michaelis-Menten kinetics. It provides a more accurate estimate of V_{max} and is helpful in analyzing enzyme inhibition. { ¦līn,wēv·ər 'bərk i,kwā·zhən }

linewidth [ATOM PHYS] A measure of the width of the band of frequencies of radiation emitted or absorbed in an atomic or molecular transition, given by the difference between the upper and lower frequencies at which the intensity of radiation reaches half its maximum value. { 'līn,width }

lingual artery [ANAT] An artery originating in the external carotid and supplying the tongue. { 'liŋ·gwəl 'ärd·ə·rē }

lingual gland [ANAT] A serous, mucous, or mucoserous gland lying deep in the mucous membrane of the mammalian tongue. { 'liŋ·gwəl 'gland }

lingual nerve [NEUROSCI] A branch of the mandibular nerve having somatic sensory components and innervating the mucosa of the floor of the mouth and the anterior two-thirds of the tongue. { 'liŋ·gwəl 'nərv }

lingual tonsil [ANAT] An aggregation of lymphoid tissue composed of 35–100 separate tonsillar units occupying the posterior part of the tongue surface. { 'liŋ·gwəl 'tän·səl }

Linguatulida [INV ZOO] The equivalent name for Pentastomida. { ,liŋ·gwə'tül·ə·də }

Linguatuloidea [INV ZOO] A suborder of pentastomid arthropods in the order Porocephalida; characterized by an elongate, ventrally flattened, annulate, posteriorly attenuated body, simple hooks on the adult, and binate hooks in the larvae. { liŋ,gwach·ə'lóid·ē·ə }

linguistic competence [PSYCH] The knowledge of language. { liŋ'gwis·tik 'käm·pə·təns }

linguistic model [COMPUT SCI] A method of automatic pattern recognition in which a class of patterns is defined as those patterns satisfying a certain set of relations among suitably defined primitive elements. Also known as syntactic model. { liŋ'gwis·tik 'mäd·əl }

linguistic performance [PSYCH] The production and comprehension of speech. { liŋ,gwis·tik pər'fòr·məns }

linguistics [LING] The study of human speech in its various aspects, especially units of language, phonetics, syntax, semantics, and grammar. { liŋ'gwis·tiks }

lingula [ANAT] A tongue-shaped organ, structure, or part thereof. { 'liŋ·gyə·lə }

Lingulacea [INV ZOO] A superfamily of inarticulate brachiopods in the order Lingulida characterized by an elongate, biconvex calcium phosphate shell, with the majority having a pedicle. { ,liŋ·gyə'lās·ē·ə }

lingulate [BIOL] Tongue- or strap-shaped. { 'liŋ·gyə,lāt }

Lingulida [INV ZOO] An order of inarticulate brachiopods represented by two living genera, *Lingula* and *Glottidia*. { liŋ'gyül·ə·də }

linguloid ripple mark *See* linguoid ripple mark. { 'liŋ·gyə,lòid 'rip·əl ,märk }

linguoid current ripple *See* linguoid ripple mark. { 'liŋ·gwóid ,kə·rənt ,rip·əl }

linguoid ripple mark [GEOL] An aqueous current ripple mark with tonguelike projections which are formed by action of a current of water and which point into the current. Also known as cuspate ripple mark; linguloid ripple mark; linguoid current ripple. { 'liŋ·gwóid 'rip·əl ,märk }

ling-zhi *See* Ganoderma lucidum. { ¦liŋ¦tsē }

liniment [PHARM] A heat-generating liquid that is thinner than ointment and is applied to the skin with friction. { 'lin·ə·mənt }

lining [MATER] A material used to protect inner surfaces, as of tunnels, pipes, or process equipment. { 'līn·iŋ }

lining bar [DES ENG] A crowbar with a pinch, wedge, or diamond point at its working end. { 'līn·iŋ ,bär }

lining pole *See* range rod. { 'līn·iŋ ,pōl }

lining sight [MIN ENG] An instrument consisting of a plate with a slot in the middle, and the means of suspending it; used with a plumbline for directing the courses of underground drifts or headings. { 'līn·iŋ ,sīt }

linin net [CELL MOL] The reticulum composed of chromatinic or oxyphilic substances in a cell nucleus. { 'lī·nən ¦net }

linishing *See* belt grinding. { 'li·nə·shiŋ }

link [CIV ENG] A standardized part of a surveyor's chain, which is 7.92 inches (20.1168 centimeters) in the Gunter's chain and 1 foot (30.48 centimeters) in the engineer's chain. [COMMUN] General term used to indicate the existence of communications facilities between two points. [COMPUT SCI] *See* hyperlink. [DES ENG] **1.** One of the rings of a chain. **2.** A connecting piece in the moving parts of a machine. { liŋk }

linkage [COMPUT SCI] In programming, coding that connects

two separately coded routines. [ELECTROMAG] *See* flux linkage. [GEN] Failure of nonallelic genes to recombine at random in meiosis as a result of their being located within the same chromosome. [MECH ENG] A mechanism that transfers motion in a desired manner by using some combination of bar links, slides, pivots, and rotating members. { 'liŋ·kij }

linkage disequilibrium [GEN] The occurrence in a population of certain combinations of linked alleles in greater proportion than expected from the allele frequencies at the loci. { ¦liŋ·kij dis,ē·kwə'lib·rē·əm }

linkage editor [COMPUT SCI] A service routine that converts the output of assemblers and compilers into a form that can be loaded and executed. { 'liŋ·kij ,ed·əd·ər }

linkage group [GEN] The set of gene loci that show significantly less than 50% recombination with one or more other genes within the set; given enough mapped genes, it corresponds to one complete chromosome of the complement. { 'liŋ·kij ,grüp }

linkage map [GEN] A diagrammatic representation of the linear order and genetic distance between genes in a linkage group. { 'liŋ·kij ,map }

link bar *See* hinged bar. { 'liŋk ,bär }

link chute adapter [ORD] A unit attached to a gun to allow a link ejection chute to be fastened to the gun, and to lead ejected links to the chute. { 'liŋk 'shüt ə,dap·tər }

link circuit [ELECTROMAG] Closed loop used for coupling purposes; it generally consists of two coils, each having a few turns of wire, connected by a twisted pair of wires or by other means, with each coil placed over, near, or in one of the two coils that are to be coupled. { 'liŋk ,sər·kət }

link control message [COMMUN] **1.** Message sent over a link of a network to condition the link to handle transmissions in a prearranged manner. **2.** Message used only between a pair of terminals for the conditioning of the link for digital system control. { 'liŋk kən'trōl ,mes·ij }

link coupling [ELECTROMAG] Modification of inductive coupling where the two coils are connected together by a short length of transmission line, with each coil inductively coupled to the coil of a separate tuned circuit. { 'liŋk ,kəp·liŋ }

linked ammunition [ORD] Cartridges fastened side by side with metal links, forming a belt for ready feed to a machine gun. { ,liŋkt ,am·yə'nish·ən }

linked list *See* chained list. { 'liŋkt 'list }

link ejection chute [ORD] A chute or passage attached to a machine gun, through which links are thrown or conveyed to a desired point after being separated from the cartridges; it may be either fixed or flexible. { 'liŋk i'jek·shən ,shüt }

link encryption [COMMUN] The application of on-line crypto-operation to the individual links of relay systems so that all messages passing over the link are encrypted in their entirety. { 'liŋk en'krip·shən }

linker *See* linker deoxyribonucleic acid. { 'liŋ·kər }

linker-delinker [ORD] A machine designed to assemble or disassemble a metallic disintegrating linked belt for ammunition. { 'liŋ·kər 'dē,liŋ·kər }

linker deoxyribonucleic acid [CELL MOL] **1.** A short, synthetic deoxyribonucleic acid (DNA) molecule that contains the recognition site for a specific restriction endonuclease. Also known as linker. **2.** A segment of DNA to which lysine-rich histone is bound and which connects the adjacent nucleosomes of a chromosome. { ¦liŋ·kər dē,äk·sē,rī·bō·nü¦klē·lk 'as·əd }

Linke scale [METEOROL] A type of cyanometer; used to measure the blueness of the sky; it is simply a set of eight cards of different standardized shades of blue, numbered (evenly) 2 to 16; the odd numbers are used by the observer if the sky color lies between any of the given shades. Also known as blue-sky scale. { 'liŋk ,skāl }

link field [COMPUT SCI] The first word of a message buffer, used to point to the next buffer on the message queue. { 'liŋk ,fēld }

link group [COMMUN] A collection of links that employ the same multiplex terminal equipment. { 'liŋk grüp }

linking loader [COMPUT SCI] A loader which combines the functions of a relocating loader with the ability to combine a number of program segments that have been independently compiled into an executable program. { 'liŋk·iŋ 'lōd·ər }

link-loading machine [ORD] Machine that quickly loads ammunition into interlocking metal links, which in turn form

LINGULACEA

Internal molds of Cambrian lingulacean *Lingulella.* *(a)* Pedicle valve. *(b)* Brachial valve. *(From C. D. Walcott, Cambrian Brachiopoda, USGS Monogr. no. 51, 1912)*

an ammunition belt for certain types of automatic weapons. { 'liŋk,lōd·iŋ mə,shēn }

link relatives method [STAT] A method for computing indexes by dividing the value of a magnitude in one period by the value in the previous period. { ¦liŋk 'rel·ə·tivz ,meth·əd }

link stretch [ORD] The change in the center to center distance of the individual rounds of belted ammunition as the load is applied. { 'liŋk ,strech }

link V belt [DES ENG] A V belt composed of a large number of rubberized-fabric links joined by metal fasteners. { 'liŋk 'vē ,belt }

lin-log amplifier [ELECTR] Automatic gain control amplifier that operates in a linear manner for low-amplitude input signals, but responds in a logarithmic manner to high-amplitude input signals. { ¦lin ¦läg 'am·plə,fī·ər }

linnaeite [MINERAL] $(Co,Ni)_3S_4$ A steel-gray mineral with a coppery-red tarnish, occurring in isometric crystals; an ore of cobalt. Also known as cobalt pyrites; linneite. { lə'nē,īt }

linneite See linnaeite. { lə'nē,īt }

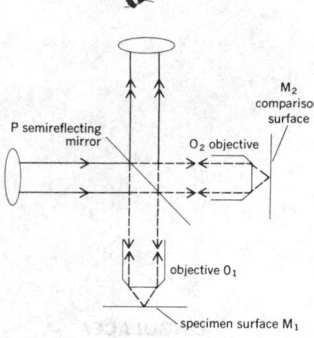

The Linnik interference microscope for reflecting specimens.

Linnik interference microscope [OPTICS] A type of interference microscope used for studying the surface structure of reflecting specimens; light from a source is divided by a semireflecting mirror into two beams, one of which is focused through an objective onto the specimen surface, the other onto a comparison surface; after reflection from the respective surfaces, the beams are reunited by the mirror. { 'lin·ik ,in·tər¦fir·əns 'mī·krə,skōp }

linoleic acid [BIOCHEM] $C_{17}H_{31}COOH$ A yellow unsaturated fatty acid, boiling at 229°C (14 mmHg), occurring as a glyceride in drying oils; obtained from linseed, safflower, and tall oils; a principal fatty acid in plants, and considered essential in animal nutrition; used in medicine, feeds, paints, and margarine. Also known as linolic acid; 9,12-octadecadienoic acid. { ¦lin·e¦lē·ik 'as·əd }

linolenate [BIOCHEM] A salt or ester of linolenic acid. { ,lin·ə'lē,nāt }

linolenic acid [BIOCHEM] $C_{17}H_{29}COOH$ One of the principal unsaturated fatty acids in plants and an essential fatty acid in animal nutrition; a colorless liquid that boils at 230°C (17 mmHg or 2266 pascals), soluble in many organic solvents; used in medicine and drying oils. Also known as 9,12,15-octadecatrienoic acid. { ¦lin·ə¦lin·ik 'as·əd }

linolenyl alcohol [ORG CHEM] $C_{18}H_{32}O$ A colorless, combustible solid used for paints, paper, leather, and flotation processes. Also known as octadecatrienol. { ¦lin·ə¦lēn·əl 'al·kə,hól }

linoleum [MATER] A floor covering made by applying a mixture of gelled linseed oil, pigments, fillers, and other materials to a burlap backing, and curing to produce a hard, resilient sheet. { lə'nō·lē·əm }

linolic acid See linoleic acid. { lə'nō·lik 'as·əd }

linseed See flaxseed. { 'lin,sēd }

linseed cake [MATER] The residue formed during pressing of commercial linseed oil; used for cattle feed and fertilizer. { 'lin,sēd ,kāk }

linseed oil [MATER] A product made from the seeds of the flax plant by crushing and pressing either with or without heat; formulated in various grades and with various drying agents and used as a vehicle in oil paints and as a component of oil varnishes. { 'lin,sēd ,óil }

lint [MATER] During the first stage of processing cotton, the fiber that is separated from the seeds in a cotton gin. { lint }

lintel [BUILD] A horizontal member over an opening, such as a door or window, usually carrying the wall load. { 'lint·əl }

linter [MECH ENG] A machine for removing fuzz linters from ginned cottonseed. { 'lin·tər }

linters [MATER] Short residual fibers that adhere to ginned cottonseed; used for making fabrics that do not require long fibers, as plastic fillers, and in the manufacture of cellulosic plastics. { 'lin·tərz }

LINUP See laser-induced nuclear polarization. { 'līn,əp }

Linux [COMPUT SCI] A freely available, open-source Unix-like operating system kernel capable of running on many different types of computer hardware; first released in 1991. { 'lin·əks }

LIOCS [COMPUT SCI] Set of routines handling buffering, blocking, label checking, and overlap of input/output with processing. Derived from logical input/output control system. { 'lī,äks }

lion [VERT ZOO] Felis leo. A large carnivorous mammal of the family Felidae distinguished by a tawny coat and blackish tufted tail, with a heavy blackish or dark-brown mane in the male. { 'lī·ən }

Lion See Leo. { 'lī·ən }

Liopteridae [INV ZOO] A small family of hymenopteran insects in the superfamily Cynipoidea. { ,lī·əp'ter·ə,dē }

Liouville equation [STAT MECH] An equation which states that the density of points representing an ensemble of systems in phase space which are in the neighborhood of some given system does not change with time. { 'lyü,vēl i,kwā·zhən }

Liouville function [MATH] A function $\lambda(n)$ on the positive integers such that $\lambda(1) = 1$, and for $n \geq 2$, $\lambda(n)$ is -1 raised to the number of prime factors of n, with repeated factors counted the number of times they appear. { 'lyü,vēl ,fəŋk·shən }

Liouville-Neumann series [MATH] An infinite series of functions constructed from the given functions in the Fredholm equation which under certain conditions provides a solution. Also known as Neumann series. { 'lyü,vēl 'nói,män ,sir·ēz }

Liouville number [MATH] An irrational number x such that for any integer n there exist integers p and q, with q greater than 1, for which the absolute value of $x - (p/q)$ is less than $1/q^n$. { 'lyü,vēl ,nəm·bər }

Liouville's theorem [MATH] Every function of a complex variable which is bounded and analytic in the entire complex plane must be constant. { 'lyü,vēlz ,thir·əm }

lip [ANAT] A fleshy fold above and below the entrance to the mouth of mammals. [CIV ENG] A parapet placed on the downstream margin of a millrace or apron in order to minimize scouring of the river bottom. [DES ENG] Cutting edge of a fluted drill formed by the intersection of the flute and the lip clearance angle, and extending from the chisel edge at the web to the circumference. [MED] The margin of an open wound. [SCI TECH] The edge of a hollow cavity or container. { lip }

Lipalian [GEOL] A hypothetical geologic period that supposedly antedated the Cambrian. { lə'pal·yən }

Liparidae [INV ZOO] The equivalent name for Lymantriidae. { lə'par·ə,dē }

lipase [BIOCHEM] An enzyme that catalyzes the hydrolysis of fats or the breakdown of lipoproteins. { 'lī,pās }

lipemia [MED] The presence of a fine emulsion of fatty substance in the blood. Also known as lipidemia; lipoidemia. { li'pē·mē·ə }

Liphistiidae [INV ZOO] A family of spiders in the suborder Liphistiomorphae in which the abdomen shows evidence of true segmentation by the presence of tergal and sternal plates. { ,lif·ə'stī·ə,dē }

Liphistiomorphae [INV ZOO] A suborder of arachnids in the order Araneida containing families with a primitively segmented abdomen. { lə,fis·tē·ə'mór·fē }

lipid [BIOCHEM] One of a class of compounds which contain long-chain aliphatic hydrocarbons and their derivatives, such as fatty acids, alcohols, amines, amino alcohols, and aldehydes; includes waxes, fats, and derived compounds. Also known as lipin; lipoid. { 'lip·əd }

lipid bilayer [CELL MOL] The foundational structure of plasma membranes; it is composed of two layers of phospholipids positioned such that their polar hydrophilic heads face outward and their nonpolar hydrophobic tails are directed inward, blocking entry of water and water-soluble material into the cell. { ¦lip·id 'bī,lā·ər }

lipidemia See lipemia. { ,lip·ə'dē·mē·ə }

lipid histiocytosis [MED] **1.** Any collection of histiocytes containing lipids. **2.** See Niemann-Pick disease. { 'lip·əd ¦his·tē·ō·sī'tō·səs }

lipid metabolism [BIOCHEM] The physiologic and metabolic processes involved in the assimilation of dietary lipids and the synthesis and degradation of lipids. { 'lip·əd me'tab·ə,liz·əm }

lipid nephrosis [MED] A chronic kidney disease of children associated with thickening of the basement membranes of glomeruli and characterized by edema, presence of protein in the urine, and abnormally high blood levels of albumin and cholesterol. { 'lip·əd ne'frō·səs }

lipidosis [MED] The generalized deposition of fat or fatty substances in reticuloendothelial cells. Also known as lipoidosis. { ,lip·ə'dō·səs }

lipid pneumonia [MED] **1.** Pneumonia resulting from aspiration of oily substances, such as nose drops. **2.** Deposition of lipids in tissues of chronically inflamed lungs. Also known as lipoid pneumonia. { 'lip·əd nü'mō·nyə }

lipid proteinosis [MED] A hereditary disorder characterized by extracellular deposits of phospholipid-protein conjugate involving various areas of the body, including the skin and air passages. { 'lip·əd ˌprō·dē·ə'nō·səs }

lipid storage disease [MED] Any of various rare diseases characterized by the accumulation of large histiocytes containing lipids throughout reticuloendothelial tissues; examples are Goucher's disease, Niemann-Pick disease, and amaurotic familial idiocy. { 'lip·əd ˌstór·ij di₁zēz }

lipin [BIOCHEM] **1.** A compound lipid, such as a cerebroside. **2.** See lipid. { 'lip·ən }

lipoblastoma See liposarcoma. { ˌlip·ə·bla'stō·mə }

lipochondrodystrophy See Hurler's syndrome. { ˌlip·ə₁kän·drō'dis·trə·fē }

lipochrome [BIOCHEM] Any of various fat-soluble pigments, such as carotenoid, occurring in natural fats. Also known as chromolipid. { 'lip·ə₁krōm }

lipodystrophy [MED] A disturbance of fat metabolism in which the subcutaneous fat disappears over some regions of the body, but is unaffected in others. { ˌlip·ə'dis·trə·fē }

lipofuscin [BIOCHEM] Any of a group of lipid pigments found in cardiac and smooth muscle cells, in macrophages, and in parenchyma and interstitial cells; differential reactions include sudanophilia, Nile blue staining, fatty acid, glycol, and ethylene. { ˌlip·ə'fyüs·ən }

lipogranuloma [MED] A small mass of fatty tissue associated with granulomatous inflammation. { ˌlip·ə₁gran·yə'lō·mə }

lipoic acid [BIOCHEM] $C_8H_{14}O_2S_2$ A compound which participates in the enzymatic oxidative decarboxylation of α-keto acids in a stage between thiamine pyrophosphate and coenzyme A. { lī'pō·ik 'as·əd }

lipoid [BIOCHEM] **1.** A fatlike substance. **2.** See lipid. { 'lī₁póid }

lipoidemia See lipemia. { ˌlī₁pói'dē·mē·ə }

lipoidosis See lipidosis. { ˌlī₁pói'dō·səs }

lipoid pneumonia See lipid pneumonia. { 'lī₁póid nü'mō·nyə }

lipolysis [PHYSIO] The release of fat from adipose tissue. { lə'päl·ə·səs }

lipoma [MED] A benign tumor composed of fat cells. { lī'pō·mə }

lipomatosis [MED] **1.** Multiple lipomas. **2.** Obesity. { ˌlī₁pō·mə'tō·səs }

lipomelanotic reticulosis [MED] A form of lymph node hyperplasia characterized by preservation of the architectural structure, inflammatory exudate, and hyperplasia of the reticulum cells which show phagocytosis of hemosiderin, melanin, and occasionally fat. Also known as dermatopathic lymphadenitis. { ˌlip·ə₁mel·ə'näd·ik ri₁tik·yə'lō·səs }

Lipomycetoideae [MICROBIO] A subfamily of oxidative yeasts in the family Saccharomycetaceae characterized by budding cells and a saclike appendage which develops into an ascus. { ˌlip·ə₁mī·sə'tóid·ē₁ē }

lipomyxoma See liposarcoma. { ˌlip·ə·mik'sō·mə }

lipophilic [CHEM] **1.** Having a strong affinity for fats. **2.** Promoting the solubilization of lipids. { ˌlip·ə'fil·ik }

lipophobic [CHEM] Lacking an affinity for, repelling, or failing to absorb or adsorb fats. { ˌlip·ə'fōb·ik }

lipophore [HISTOL] A chromatophore which contains lipochrome. { 'lip·ə₁fór }

lipophosphoglycan [BIOCHEM] A class of glycosyl phosphatidylinositol attached to a large polysaccharide structure that coats the surfaces of many parasitic protozoa, such as *Leishmania donovani.* { ˌlip·ō₁fäs·fə'glī·kən }

lipoplex [MED] A deoxyribonucleic acid-liposome complex used as a gene delivery vehicle in nonviral gene therapy. { 'lip·ə₁pleks }

lipopolysaccharide [BIOCHEM] Any of a class of conjugated polysaccharides consisting of a polysaccharide combined with a lipid. { ˌlip·ō₁päl·ē'sak·ə₁rīd }

lipoprotein [BIOCHEM] Any of a class of conjugated proteins consisting of a protein combined with a lipid. { ˌlip·ə'prō₁tēn }

liposarcoma [MED] A sarcoma originating in adipose tissue.

Also known as embryonal-cell lipoma; fetal fat-cell lipoma; infiltrating lipoma; lipoblastoma; lipomyxoma; myxolipoma; myxoma lipomatodes. { ˌlip·ə·sär'kō·mə }

liposome [CYTOL] One of the fatty droplets occurring in the cytoplasm, particularly of an egg. { 'lip·ə₁sōm }

lipostat [MED] The set point of body weight. { 'lip·ə₁stat }

Lipostraca [PALEON] An order of the subclass Branchiopoda erected to include the single fossil species *Lepidocaris rhyniensis.* { li'päs·trə·kə }

lipotropic [BIOCHEM] Having an affinity for lipid compounds. [PHARM] Having a preventive or curative effect on the deposition of excessive fat in abnormal sites. { ˌlip·ə'trä·pik }

lipotropic hormone [BIOCHEM] Any hormone having lipolytic activity on adipose tissue. { ˌlip·ə'trä·pik 'hór₁mōn }

Lipotyphla [VERT ZOO] A group of insectivoran mammals composed of insectivores which lack an intestinal cecum and in which the stapedial artery is the major blood supply to the brain. { ˌlip·ə'tī·fē·ə }

lipoxidase [BIOCHEM] An enzyme catalyzing the oxidation of the double bonds of an unsaturated fatty acid. { li'päk·sə₁dās }

lipper [OCEANOGR] **1.** Slight ruffling or roughness appearing on a water surface. **2.** Light spray originating from small waves. { 'lip·ər }

Lippich prism [OPTICS] A Nicol prism which is placed in the eyepiece of a polarimeter, covering half the field of view, to identify the character of polarized light emerging from the instrument. { 'lip·ik ₁priz·əm }

Lippmann effect [PHYS] A change in surface tension that results from a potential difference across the interface between two immiscible liquid conductors. { 'lip·mən i₁fekt }

Lippmann electrometer See capillary electrometer. { 'lip·mən ₁i₁lek'träm·əd·ər }

Lippmann fringes [OPTICS] Interference fringes in standing electromagnetic waves generated when light is reflected by a mercury coating at the back of a special fine-grained photographic emulsion; originally used in color photography. { 'lip·mən ₁frin·jəz }

Lipschitz condition [MATH] A function f satisfies such a condition at a point b if $|f(x) - f(b)| \leq K|x - b|$, with K a constant, for all x in some neighborhood of b. { 'lip₁shits kən₁dish·ən }

Lipschitz mapping [MATH] A function f from a metric space to itself for which there is a positive constant K such that, for any two elements in the space, a and b, the distance between $f(a)$ and $f(b)$ is less than or equal to K times the distance between a and b. { 'lip₁shits ₁map·iŋ }

lip-sync [COMMUN] Synchronization of sound and motion picture so that facial movements of speech coincide with the sounds. { 'lip ₁siŋk }

liptinite See exinite. { 'lip·tə₁nīt }

liq pt See pint.

liquation [MET] **1.** Separation of fusible metals from less fusible ones by applying heat. **2.** The partial melting of an alloy. { lī'kwā·shən }

liquefaction [PHYS] A change in the phase of a substance to the liquid state; usually, a change from the gaseous to the liquid state, especially of a substance which is a gas at normal pressure and temperature. { ˌlik·wə'fak·shən }

liquefied gas [MATER] A gaseous compound or mixture converted to the liquid phase by cooling or compression; examples are liquefied petroleum gas (LPG), liquefied natural gas (LNG), liquid oxygen, and liquid ammonia. { 'lik·wə₁fīd 'gas }

liquefied natural gas [MATER] A product of natural gas which consists primarily of methanes; its critical temperature is about −100°F (−73°C), and thus it must be liquefied by cooling to cryogenic temperatures and must be well insulated to be held in the liquid state; used as a domestic fuel. Abbreviated LNG. { 'lik·wə₁fīd 'nach·rəl ₁gas }

liquefied petroleum gas [MATER] A product of petroleum gases; principally propane and butane, it must be stored under pressure to keep it in a liquid state; it is often stored in metal cylinders (bottled gas) and used as fuel for tractors, trucks, and buses, and as a domestic cooking or heating fuel in rural areas. Abbreviated LPG. { 'lik·wə₁fīd pə'trō·lē·əm ₁gas }

liquefier [ENG] Equipment or system used to liquefy gases;

usually employs a combination of compression, heat exchange, and expansion operations. { 'lik·wə,fī·ər }

liqueur [FOOD ENG] An alcoholic beverage prepared by combining a spirit, usually brandy, with certain flavorings and sugar. { li'kər }

liquid [PHYS] A state of matter intermediate between that of crystalline substances and gases in which a substance has the capacity to flow under extremely small shear stresses and conforms to the shape of a confining vessel, but is relatively incompressible, lacks the capacity to expand without limit, and can possess a free surface. { 'lik·wəd }

liquid A [CRYO] A phase of superfluid helium-3 in which the helium-3 pairs only occur in those two of the three possible nuclear spin states in which the nuclear spins are parallel, and these pairs couple coherently to give macroscopic orbital and spin angular momenta and anisotropic superfluid properties. Also known as A phase. { 'lik·wəd 'ā }

liquid A₁ [CRYO] A phase of liquid helium-3 intermediate between liquid A and liquid B that appears only in the presence of a magnetic field and then only in a narrow portion of the pressure-temperature diagram, and in which only pairs of one of the three possible nuclear spin states are superfluid. Also known as A₁ phase. { 'lik·wəd 'ā 'wən }

liquid air [PHYS] Air in the liquid state obtained as a faintly bluish, transparent, mobile, intensely cold liquid by compressing purified air and cooling it to a temperature below the boiling points of its principal components, nitrogen and oxygen; used chiefly as a refrigerant. { 'lik·wəd 'er }

liquid asphalt See residual oil. { 'lik·wəd 'as,fȯlt }

liquid B [CRYO] A phase of superfluid helium-3 in which pairs of all three possible nuclear spin states are coupled to give superfluid properties that are isotropic except in the more subtle aspects of the spin configuration. Also known as B phase. { 'lik·wəd 'bē }

liquid blast cleaning [MET] Cleaning metal surfaces with a suspension of abrasive in water accelerated to high velocities by compressed air, or by a centrifugal wheel. { 'lik·wəd 'blast ,klēn·iŋ }

liquid blocking [PETRO ENG] The blocking or plugging of the sand around an injection-well borehole, usually caused by lubricant carryover from compressors. { 'lik·wəd 'bläk·iŋ }

liquid bright gold [MATER] Any of several gold compounds applied to ceramics in the form of varnish which is dried and heated to redness, decomposing the compound and leaving a thin film of gold firmly attached to the underlying ceramic; used in decorating china and for the production of printed electrical circuits on ceramics. { 'lik·wəd 'brīt ,gōld }

liquid-bubble tracer [FL MECH] A method of observing the motion of a liquid by following tiny particles of an immiscible liquid of the same density as the moving liquid. { 'lik·wəd ¦bəb·əl ¦trā·sər }

liquid carburizing [MET] Surface hardening of steel by immersion into a molten bath consisting of cyanides and other salts, for example, at 1600–1750°F (850–950°C). { 'lik·wəd 'kär·bə,rīz·iŋ }

liquid chromatography [ANALY CHEM] A form of chromatography employing a liquid as the moving phase and a solid or a liquid on a solid support as the stationary phase; techniques include column chromatography, gel permeation chromatography, and partition chromatography. { 'lik·wəd ,krō·mə'täg·rə·fē }

liquid-column gage See U-tube manometer. { 'lik·wəd ¦käl·əm ,gāj }

liquid compass [ENG] A compass in a bowl filled with liquid. { 'lik·wəd 'käm·pəs }

liquid-cooled dissipator See cold plate. { 'lik·wəd ¦küld 'dis·ə,pād·ər }

liquid-cooled engine [MECH ENG] An internal combustion engine with a jacket cooling system in which liquid, usually water, is circulated to maintain acceptable operating temperatures of machine parts. { 'lik·wəd ¦küld 'en·jən }

liquid cooling [ENG] Use of circulating liquid to cool process equipment and hermetically sealed components such as transistors. { 'lik·wəd 'kül·iŋ }

liquid crystal [PHYS CHEM] A liquid which is not isotropic; it is birefringent and exhibits interference patterns in polarized light; this behavior results from the orientation of molecules parallel to each other in large clusters. { 'lik·wəd 'krist·əl }

liquid crystal display [ELECTR] A digital display that consists of two sheets of glass separated by a sealed-in, normally transparent, liquid crystal material; the outer surface of each glass sheet has a transparent conductive coating such as tin oxide or indium oxide, with the viewing-side coating etched into character-forming segments that have leads going to the edges of the display; a voltage applied between front and back electrode coatings disrupts the orderly arrangement of the molecules, darkening the liquid enough to form visible characters even though no light is generated. Abbreviated LCD. { 'lik·wəd 'krist·əl di'splā }

liquid crystal polymers [ORG CHEM] Aromatic polyester copolymers that have characteristically high-temperature resistance, yet can be melted and molded. Upon melting, the polymer chains undergo parallel ordering in the direction of the flow, resulting in superior mechanical properties in that direction. { ¦lik·wəd ¦krist·əl ¦päl·ə·mərs }

liquid degeneracy [STAT MECH] A process in which a liquid cooled below a certain temperature loses the entropy associated with disordered motion of its molecules, without becoming a solid. { ¦lik·wəd di'jen·ə·rə·sē }

liquid-dielectric capacitor [ELEC] A capacitor in which the plate assemblies are mounted in a tank filled with a suitable oil or liquid dielectric. { 'lik·wəd ,dī·ə¦lek·trik kə'pas·əd·ər }

liquid dioxide See nitrogen dioxide. { 'lik·wəd dī'äk,sīd }

liquid-dominated hydrothermal reservoir [GEOL] Any geothermal system mainly producing superheated water (often termed brines); hot springs, fumaroles, and geysers are the surface expressions of hydrothermal reservoirs; an example is the hot-brine region in the Imperial Valley-Salton Sea area of southern California. { 'lik·wəd ¦däm·ə,nād·əd ,hī·drə¦thər·məl 'rez·əv,wär }

liquid-drop model [NUC PHYS] A model of the nucleus in which it is compared to a drop of incompressible liquid, and the nucleons are analogous to molecules in the liquid; used to study binding energies, fission, collective motion, decay, and reactions. Also known as drop model. { 'lik·wəd ¦dräp ,mäd·əl }

liquid extraction See solvent extraction. { 'lik·wəd ik'strak·shən }

liquid-filled porosity [GEOL] The condition in porous rock or sand formations in which pore spaces contain fresh or salt water, liquid petroleum, pressure-liquefied butane or propane, or tar. { 'lik·wəd ¦fild pə'räs·əd·ē }

liquid filter [CHEM ENG] A device for the removal of solids or coalesced droplets out of a liquid stream by use of a filter medium, such as a screen, cartridge, or granular bed. { 'lik·wəd 'fil·tər }

liquid flow [FL MECH] The flow or movement of materials in the liquid phase. { 'lik·wəd 'flō }

liquid fluorine [CRYO] Cold, liquefied fluorine gas; used as a cryogenic propellant. { 'lik·wəd 'flu̇r,ēn }

liquid fuel [MATER] A rocket fuel which is liquid under the conditions in which it is utilized in the rocket. Also known as liquid propellant. { 'lik·wəd 'fyül }

liquid fuse unit [ELEC] Fuse unit in which the fuse link is immersed in a liquid, or provision is made for drawing the arc into the liquid when the fuse link melts. { 'lik·wəd 'fyüz ,yü·nət }

liquid gas [PHYS] A gas in the liquid state. { 'lik·wəd 'gas }

liquid glass See sodium silicate. { 'lik·wəd 'glas }

liquid grease [MATER] Lubricating oil of light or medium grade that is thickened with calcium soap. { 'lik·wəd 'grēs }

liquid helium [CRYO] The state of helium which exists at atmospheric pressure at temperatures below -268.95°C (4.2 K), and for temperatures near absolute zero at pressures up to about 25 atmospheres (2.53 megapascals); has two phases, helium I and helium II. { 'lik·wəd 'hē·lē·əm }

liquid holdup [FL MECH] A condition in two-phase flow through a vertical pipe; when gas flows at a greater linear velocity than the liquid, slippage takes place and liquid holdup occurs. { 'lik·wəd 'hōl,dəp }

liquid honing See vapor blasting. { 'lik·wəd 'hȯn·iŋ }

liquid hydrocarbon [ORG CHEM] A hydrocarbon that has been converted from a gas to a liquid by pressure or by reduction in temperature; usually limited to butanes, propane, ethane, and methane. { 'lik·wəd 'hī·drə,kär·bən }

liquid hydrogen [CRYO] Hydrogen that exists as a liquid at

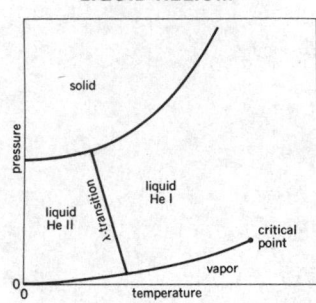

LIQUID HELIUM

Phase diagram for He⁴ (not to scale). There are two liquid phases, helium I and helium II.

atmospheric pressure, at −252.7°C (20.46 K); used for high-impulse rocket fuels. { 'lik·wəd 'hī·drə·jən }

liquid-hydrogen bubble chamber [NUCLEO] A bubble chamber in which the active liquid is hydrogen; particularly useful in research on elementary particles produced in high-energy interactions, because the hydrogen provides a dense target of protons. { 'lik·wəd ¦hī·drə·jən 'bəb·əl ‚chām·bər }

liquid-in-glass thermometer [ENG] A thermometer in which the thermally sensitive element is a liquid contained in a graduated glass envelope; the indication of such a thermometer depends upon the difference between the coefficients of thermal expansion of the liquid and the glass; mercury and alcohol are liquids commonly used in meteorological thermometers. { 'lik·wəd in ¦glas thər'mäm·əd·ər }

liquid-in-metal thermometer [ENG] A thermometer in which the thermally sensitive element is a liquid contained in a metal envelope, frequently in the form of a Bourdon tube. { 'lik·wəd in ¦med·əl thər'mäm·əd·ər }

liquid insulator [MATER] A liquid with a resistivity greater than about 10^{14} ohm-centimeters, such as a petroleum oil, silicone oil, or halogenated aromatic hydrocarbon. { 'lik·wəd 'in·sə‚lād·ər }

liquid ionization chamber [NUCLEO] A particle detector in which the gas filling a conventional ionization chamber is replaced by an exceedingly pure liquid, usually a liquefied noble gas. { 'lik·wəd ‚ī·ə·nə'zā·shən ‚chām·bər }

liquid junction emf [PHYS CHEM] The emf (electromotive force) generated at the area of contact between the salt bridge and the test solution in a pH cell electrode. { 'lik·wəd ¦jəŋk·shən ¦ē¦em'ef }

liquid junction potential See diffusion potential. { 'lik·wəd ¦jəŋk·shən pə'ten·chəl }

liquid knockout See impingement. { 'lik·wəd 'nä‚kaut }

liquid laser [OPTICS] A laser whose active material is dissolved in a liquid contained in a transparent cylindrical shell; rare-earth ions in suitable dissolved molecules and organic dye solutions are used. { 'lik·wəd 'lā·zər }

liquid level control [ENG] Regulation of the linear vertical distance between the surface of a liquid and some reference point. { 'lik·wəd 'lev·əl kən‚trōl }

liquid limit [GEOL] The moisture content boundary that exists between the plastic and semiliquid states of a sediment. { 'lik·wəd 'lim·ət }

liquid-liquid chemical reaction [CHEM] Chemical reaction in which the reactants, two or more, are liquids. { 'lik·wəd ¦lik·wəd ¦kem·ə·kəl rē'ak·shən }

liquid-liquid distribution [CHEM] The process in which a dissolved substance is transferred from one liquid phase to another, immiscible liquid phase. { 'lik·wəd 'lik·wəd ‚dis·trə'byü·shən }

liquid-liquid extraction [CHEM ENG] The removal of a soluble component from a liquid mixture by contact with a second liquid, immiscible with the carrier liquid in which the component is preferentially soluble. { 'lik·wəd 'lik·wəd ik'strak·shən }

liquid measure [MECH] A system of units used to measure the volumes of liquid substances in the United States; the units are the fluid dram, fluid ounce, gill, pint, quart, and gallon. { 'lik·wəd 'mezh·ər }

liquid-metal embrittlement [MET] The rapid loss of mechanical properties of a metal or an alloy due to contact with certain liquid metals. { 'lik·wəd ¦med·əl em'brid·əl·mənt }

liquid-metal fuel cell [ELEC] A fuel cell that uses molten potassium and bismuth as reactants and a molten salt electrolyte; has very high power output, but a relatively short life. { 'lik·wəd ¦med·əl 'fyül ‚sel }

liquid-metal MHD generator [ELEC] A system for generating electric power in which the kinetic energy of a flowing, molten metal is converted to electric energy by magnetohydrodynamic (MHD) interaction. { 'lik·wəd ¦med·əl ¦em¦äch¦dē 'jen·ə‚rād·ər }

liquid-metal nuclear fuel [NUCLEO] A nuclear fuel consisting of a solution of uranium or plutonium in a molten metal such as bismuth. { 'lik·wəd ¦med·əl ¦nü·klē·ər 'fyül }

liquid methane [CRYO] Methane that has been cooled to at least −161°C; used for cryogenic applications and for tankship transport of methane. { 'lik·wəd 'meth‚ān }

liquid nitrogen [CRYO] Nitrogen that exists as a liquid at atmospheric pressure, at −195°C (77.4 K); used in research work, cryogenics, and cryosurgery. { 'lik·wəd 'nī·trə·jən }

liquid oxygen [CRYO] Oxygen that exists as a liquid at atmospheric pressure, at −182.97°C (90.18 K); a pale-blue, transparent, mobile liquid. { 'lik·wəd 'äk·sə·jən }

liquid-oxygen explosive [MATER] Sawdust or other carbonaceous material formed into a cartridge and dipped into liquid oxygen, to use in blasting. Abbreviated LOX. { 'lik·wəd ¦äk·sə·jən ik'splō·siv }

liquid penetrant test [ENG] A penetrant method of nondestructive testing used to locate defects open to the surface of nonporous materials; penetrating liquid is applied to the surface, and after 1–30 minutes excess liquid is removed, and a developer is applied to draw the penetrant out of defects, thus showing their location, shape, and size. { 'lik·wəd 'pen·ə·trənt ‚test }

liquid petrolatum See white mineral oil. { 'lik·wəd ‚pe·trə'lād·əm }

liquid-phase epitaxy [SOLID STATE] A process for growing thin epitaxial layers on a crystalline substrate in which the substrate is sequentially brought into contact with solutions that are at the desired composition and may be supersaturated or cooled to achieve growth. Abbreviated LPE. { 'lik·wəd ¦fāz 'ep·ə‚tak·sē }

liquid-phase hydrogenation [CHEM ENG] Hydrogen reaction with liquid-phase hydrogenatable material, such as unsaturated aliphatic or aromatic hydrocarbons. { 'lik·wəd ‚fāz ‚hī·drə·jə'nā·hsən }

liquid pint See pint. { 'lik·wəd 'pīnt }

liquid piston rotary compressor [MECH ENG] A rotary compressor in which a multiblade rotor revolves in a casing partly filled with liquid, for example, water. { 'lik·wəd ¦pis·tən ¦rōd·ə·rē kəm'pres·ər }

liquid poison [NUCLEO] A neutron-absorbing liquid that can be injected quickly into the cooling system of a nuclear reactor by explosive-actuated valves; used for automatic or manual scramming to shut down a reactor. { 'lik·wəd 'pòiz·ən }

liquid propellant See liquid fuel. { 'lik·wəd prə'pel·ənt }

liquid rheostat [ELECTR] A variable-resistance type of voltage regulator in which the variable-resistance element is liquid, usually water; carbon electrodes are raised or lowered in the liquid to change resistance ratings and control voltage flow. { 'lik·wəd 'rē·ə‚stat }

liquid rosin See tall oil. { 'lik·wəd 'räz·ən }

liquid scintillation detector [NUCLEO] Scintillation counter in which the sensitive material is a liquid, such as p-terphenyl dissolved in toluene, placed in a glass or metal container. { 'lik·wəd ‚sint·əl'ā·shən di‚tek·tər }

liquid seal [CHEM ENG] **1.** The depth of liquid above an opening from which gas or vapor issues, as for a riser in a distillation-column tray. **2.** Product drawoff in which a depth of liquid prevents the outflow of gas or vapor. { 'lik·wəd 'sēl }

liquid-sealed meter [ENG] A type of positive-displacement meter for gas flows consisting of a cylindrical chamber that is more than half filled with water and divided into four rotating compartments formed by trailing vanes; gas entering through the center shaft into one compartment after another forces rotation that allows the gas then to exhaust out the top as it is displaced by the water. Also known as drum meter. { 'lik·wəd ‚sēld 'mēd·ər }

liquid semiconductor [ELECTR] An amorphous material in solid or liquid state that possesses the properties of varying resistance induced by charge carrier injection. { 'lik·wəd 'sem·i·kən‚dək·tər }

liquid-solid chemical reaction [CHEM] Chemical reaction in which at least one of the reactants is a liquid, and another of the reactants is a solid. { 'lik·wəd 'säl·əd ¦kem·ə·kəl rē'ak·shən }

liquid-solid equilibrium See solid-liquid equilibrium. { 'lik·wəd 'säl·əd ‚ē·kwə'lib·rē·əm }

liquid-sorbent dehumidifier [MECH ENG] A sorbent type of dehumidifier consisting of a main circulating fan, sorbent-air contactor, sorbent pump, and reactivator; dehumidification and reactivation are continuous operations, with a small part of the sorbent constantly bled off from the main circulating system and reactivated to the concentration required for the desired effluent dew point. { 'lik·wəd ¦sòr·bənt ‚dē·yü'mid·ə‚fī·ər }

liquid sulfur dioxide-benzene process [CHEM ENG] A petroleum-refinery process using a mixed solvent (SO_2 and

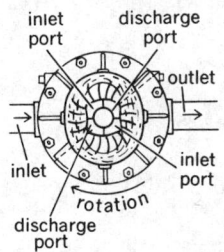

LIQUID PISTON ROTARY COMPRESSOR

Schematic of the compressor showing its components.

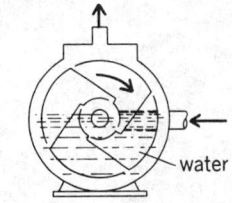

LIQUID-SEALED METER

Liquid-sealed drum-type gas flowmeter.

benzene) to dewax lubricating oils or improve their viscosity indices. { 'lik·wəd 'səl·fər dī'äk‚sīd ben'zēn ‚prä·səs }

liquidus line [THERMO] For a two-component system, a curve on a graph of temperature versus concentration which connects temperatures at which fusion is completed as the temperature is raised. { 'lik·wəd·əs ‚līn }

liquid-vapor chemical reaction [CHEM] Chemical reaction in which at least one of the reactants is a liquid, and another of the reactants is a vapor. { 'lik·wəd ¦vā·pər ¦kem·ə·kəl rē'ak·shən }

liquid-vapor equilibrium [PHYS CHEM] The equilibrium relationship between the liquid and its vapor phase for a partially vaporized compound or mixture at specified conditions of pressure and temperature; for mixtures, it is expressed by $K = x/y$, where K is the equilibrium constant, x the mole fraction of a key component in the vapor, and y the mole fraction of the same key component in the liquid. Also known as vapor-liquid equilibrium. { 'lik·wəd ¦vā·pər ‚ē·kwə'lib·rē·əm }

liquid-water content See water content. { 'lik·wəd ¦wód·ər ‚kän‚tent }

liquor [CHEM ENG] **1.** Supernatant liquid decanted from a liquid-solids mixture in which the solids have settled. **2.** Liquid overflow from a liquid-liquid extraction unit. [FOOD ENG] **1.** Sugarcane sap before it is crystallized into sugar. **2.** A strong distilled alcoholic beverage. [PHARM] A solution of a medicinal substance in water. { 'lik·ər }

liquor finish [MET] A bright, smooth finish on wet-drawn wire achieved by using fermented-grain mash liquor as a lubricant. { 'lik·ər ‚fin·ish }

lirella [BOT] A long, narrow apothecium with a medial longitudinal furrow, occurring in certain lichens. { lə'rel·ə }

Liriopeidae [INV ZOO] The phantom craneflies, a family of dipteran insects in the suborder Orthorrhapha distinguished by black and white banded legs. { ‚lir·ē·ə'pē·ə‚dē }

liroconite [MINERAL] $Cu_2Al(AsO_4)(OH)_4·4H_2O$ A light-blue or yellowish-green mineral composed of basic hydrous aluminum copper arsenate, occurring in monoclinic crystals. { lī'räk·ə‚nīt }

liskeardite [MINERAL] $(Al,Fe)_3(AsO_4)(OH)_6·5H_2O$ A soft, white mineral composed of basic hydrous aluminum iron arsenate. { li'skär‚dīt }

lisle [TEXT] Fine-quality, tightly twisted, long-staple cotton yarn with a sleek surface produced by passing it near a gas flame to remove fuzz. { līl }

LISP [COMPUT SCI] An interpretive language developed for the manipulation of symbolic strings of recursive data; can also be used to manipulate mathematical and arithmetic logic. Derived from list processing language. { lisp }

Lissajous figure [PHYS] The path of a particle moving in a plane when the components of its position along two perpendicular axes each undergo simple harmonic motions and the ratio of their frequencies is a rational number. Also known as Bowditch curve. { ¦lē·sə¦zhü ‚fig·yər }

Lissamphibia [VERT ZOO] A subclass of Amphibia including all living amphibians; distinguished by pedicellate teeth and an operculum-plectrum complex of the middle ear. { ‚li‚sam'fib·ē·ə }

list [COMPUT SCI] **1.** A last-in, first-out storage organization, usually implemented by software, but sometimes implemented by hardware. **2.** In FORTRAN, a set of data items to be read or written. [ENG] To lean to one side, or deviate from the vertical. { list }

listening station [ENG] A radio or radar receiving station that is continuously manned for various purposes, such as for radio direction finding or for gaining information about enemy electronic devices. { 'lis·ən·iŋ ‚stā·shən }

Listeria [MICROBIO] A genus of small, gram-positive, motile coccoid rods of uncertain affiliation; found in animal and human feces. { li'stir·ē·ə }

listeriosis [MED] A bacterial disease of humans and some animals caused by *Listeria monocytogenes*; occurs primarily as meningitis or granulomatosis infantiseptica in humans, and takes many forms, such as meningoencephalitis, distemperlike disease, or generalized infection, in animals. { li‚stir·ē'ō·səs }

listing See lashing. { 'list·iŋ }

Listomatic camera [GRAPHICS] A machine used in photocopying which photographs data on tabulating cards and prints it in columnar form on roll film; the film negative is then used for printing by photolithography. { ‚li·stə'mad·ik ‚kam·rə }

list processing [COMPUT SCI] A programming technique in which list structures are used to organize memory. { 'list ¦prä‚ses·iŋ }

list processing language See LISP. { 'list ¦prä‚ses·iŋ ‚laŋ·gwij }

listserv [COMPUT SCI] The software (server) used to maintain an electronic mailing list. Also known as list server. { 'list‚sərv }

list server See listserv. { 'list ‚sər·vər }

list structure [COMPUT SCI] A set of data items, connected together because each element contains the address of a successor element (and sometimes of a predecessor element). { 'list ‚strək·chər }

litchi See lychee. { 'lī‚chē }

liter [MECH] A unit of volume or capacity, equal to 1 decimeter cubed, or 0.001 cubic meter, or 1000 cubic centimeters. Abbreviated l; L. { 'lēd·ər }

literal constant [MATH] A letter denoting a constant. { 'lid·ə·rəl 'kän‚stənt }

literal expression [MATH] An expression or equation in which the constants are represented by letters. { 'lid·ə·rəl ik'spresh·ən }

literal notation [MATH] The use of letters to denote numbers, known or unknown. { 'lid·ə·rəl nō'tā·shən }

literal operand [COMPUT SCI] An operand, usually occurring in a source language instruction, whose value is specified by a constant which appears in the instruction rather than by an address where a constant is stored. { 'lid·ə·rəl ¦äp·ə‚rand }

liter-atmosphere [PHYS] A unit of energy equal to the work done on a piston by a fluid at a pressure of 1 standard atmosphere (101,325 pascals) when the piston sweeps out a volume of 1 liter; equal to 101.325 joules. { 'lēd·ə·r ¦at·mə‚sfir }

lithamide See lithium amide. { 'lith·ə‚mīd }

litharenite [PETR] A sandstone that contains more than 25% detrital rock fragments, and more rock fragments than feldspar grains. { li'thär·ə‚nīt }

litharge See lead monoxide. { 'li‚thärj }

litharge-glycerin cement [MATER] Mixture of glycerin, water, and litharge (lead monoxide) to give, when cured, an acid-resistant cement. { 'li‚thärj ¦glis·ə·rən si'ment }

lithemia See hyperuricemia. { lə'thē·mē·ə }

lithian muscovite [MINERAL] A form of the mineral lepidolite containing 3–4% lithium oxide and having a modified two-layer monoclinic muscovite structure. { 'lith·ē·ən 'məs·kə‚vīt }

lithiasis [MED] The formation of calculi in the body. { lə'thī·ə·səs }

lithic [PETR] Pertaining to stone. { 'lith·ik }

lithic graywacke [PETR] A low-grade graywacke, that is, containing an abundance of unstable materials, especially a sandstone containing less than 75% quartz and chert, 15–75% detrital clay matrix, and more rock fragments than feldspar grains. { 'lith·ik 'grā‚wak·ə }

lithic sandstone [PETR] A sandstone that contains more rock fragments than feldspar grains. { 'lith·ik 'san‚stōn }

lithic tuff [GEOL] **1.** A tuff that is mostly crystalline rock fragments. **2.** An indurated volcanic ash deposit whose fragments are composed of previously formed rocks that first solidified in the volcanic vent and were then blown out. { 'lith·ik 'təf }

lithifaction See lithification. { ‚lith·ə'fak·shən }

lithification [GEOL] **1.** Conversion of a newly deposited sediment into an indurated rock. Also known as lithifaction. **2.** Compositional change of coal to bituminous shale or other rock. { ‚lith·ə·fə'kā·shən }

lithionite See lepidolite. { 'lith·ē·ə‚nīt }

lithiophilite [MINERAL] $Li(Mn,Fe)PO_4$ A salmon-pink or clove-brown mineral crystallizing in the orthorhombic system; isomorphous with triphylite. { ‚lith·ē'äf·ə‚līt }

lithiophorite [MINERAL] $(Al,Li)MnO_2(OH)_2$ A mineral composed of basic manganese aluminum lithium oxide. { ‚lith·ē'äf·ə‚rīt }

Lithistida [PALEON] An order of fossil sponges in the class Demospongia having a reticulate skeleton composed of irregular and knobby siliceous spicules. { lə'this·tə·də }

lithium [CHEM] A chemical element, symbol Li, atomic number 3, atomic weight 6.939; an alkali metal. { 'lith·ē·əm }

lithium aluminum hydride [INORG CHEM] $LiAlH_4$ A compound made by the reaction of lithium hydride and aluminum chloride; a powerful reducing agent for specific linkages in complex molecules; used in organic synthesis. { 'lith·ē·əm ə'lü·mə·nəm 'hī,drīd }

lithium amide [INORG CHEM] $LiNH_2$ A compound crystallizing in the cubic form, and melting at 380–400°C; used in organic synthesis. Also known as lithamide. { 'lith·ē·əm 'am,īd }

lithium battery [ELEC] A solid-state battery with a lithium anode, an iodine-polyvinyl pyridine cathode, and an electrolyte consisting of a layer of lithium iodide; used in cardiac pacemakers. { 'lith·ē·əm 'bad·ə·rē }

lithium bromide [INORG CHEM] $LiBr·H_2O$ A white, deliquescent, granular powder with a bitter taste, melting at 547°C; soluble in alcohol and glycol; used to add moisture to air-conditioning systems and as a sedative and hypnotic in medicine. { 'lith·ē·əm 'brō,mīd }

lithium carbonate [INORG CHEM] Li_2CO_3 A colorless, crystalline compound that melts at 700°C and has slight solubility in water; used in ceramic industries in the manufacture of powdered glass for porcelain enamel formulation. { 'lith·ē·əm 'kär·bə,nāt }

lithium cell [CHEM] An electrolytic cell for the production of metallic lithium. [ELEC] A primary cell for producing electrical energy by using lithium metal for one electrode immersed in usually an organic electrolyte. { 'lith·ē·əm ,sel }

lithium chloride [INORG CHEM] $LiCl·2H_2O$ A colorless, water-soluble compound, forming octahedral crystals and melting at 614°C; used to form concentrated brine in commercial air-conditioning systems and as a pyrotechnic in welding and brazing fluxes. { 'lith·ē·əm 'klȯr,īd }

lithium citrate [ORG CHEM] $Li_3C_6H_5O_7·4H_2O$ White powder that decomposes when heated; slightly soluble in alcohol; soluble in water; used in beverages and pharmaceuticals. { 'lith·ē·əm 'sī,trāt }

lithium-drifted germanium crystal [ELECTR] A high-resolution junction detector, used especially for more penetrating gamma-radiation and higher-energy electrons, produced by drifting lithium ions through a germanium crystal to produce an intrinsic region where impurity-based carrier generation centers are deactivated, sandwiched between a *p* layer and an *n* layer. { 'lith·ē·əm ,drif·təd jər'mā·nē·əm 'krist·əl }

lithium-drifted silicon detector [NUCLEO] A type of junction detector fabricated by applying a reverse bias to a *pn* junction consisting of a lithium-diffused *n*-type region on *p*-type silicon, causing the lithium ions to migrate toward the negative side and to compensate negative ions, fixed in the silicon crystal lattice, to create a wide depletion layer. { 'lith·ē·əm ,drif·təd 'sil·ə·kən di'tek·tər }

lithium fluoride [INORG CHEM] LiF Poisonous, white powder melting at 870°C, boiling at 1670°C; insoluble in alcohol, slightly soluble in water, and soluble in acids; used as a heat-exchange medium, as a welding and soldering flux, in ceramics, and as crystals in infrared instruments. { 'lith·ē·əm 'flu̇r,īd }

lithium fluoride dosimetry [NUCLEO] A method of dosimetry in which the radiation dose received by a sample of the phosphor lithium fluoride is determined by measuring the thermoluminescent output of the phosphor upon heating, following the irradiation. { 'lith·ē·əm 'flu̇r,īd dō'sim·ə·trē }

lithium grease [MATER] Heat-stable, water-resistant lubricating grease with lithium salts of higher fatty acids (or lithium soaps of fatty glycerides) as a base; used for low-temperature service in aircraft. { 'lith·ē·əm 'grēs }

lithium halide [INORG CHEM] A binary compound of lithium, LiX, where X is a halide; examples are lithium chloride, LiCl, and lithium fluoride, LiF. { 'lith·ē·əm 'hal,īd }

lithium hydride [INORG CHEM] LiH Flammable, brittle, white, translucent crystals; decomposes in water; insoluble in ether, benzene, and toluene; used as a hydrogen source and desiccant, and to prepare lithium amide and double hydrides. { 'lith·ē·əm 'hī,drīd }

lithium hydroxide [INORG CHEM] $LiOH; LiOH·H_2O$ Colorless crystals; used as a storage-battery electrolyte, as a carbon dioxide absorbent, and in lubricating greases and ceramics. { 'lith·ē·əm hī'dräk,sīd }

lithium iodide [INORG CHEM] $LiI; LiI·3H_2O$ White, water- and alcohol-soluble crystals; LiI melts at 446°C; $LiI·3H_2O$ loses water at 72°C; used in medicine, photography, and mineral waters. { 'lith·ē·əm 'ī·ə,dīd }

lithium mica *See* lepidolite. { 'lith·ē·əm 'mī·kə }

lithium molybdate [INORG CHEM] Li_2MoO_4 Water-soluble white crystals melting at 705°C; used as a catalytic cracking (petroleum) catalyst and as a mill additive for steel. { 'lith·ē·əm mə'lib,dāt }

lithium nitrate [INORG CHEM] $LiNO_3$ Water- and alcohol-soluble colorless powder melting at 261°C; used as a heat-exchange medium and in ceramics, pyrotechnics, salt baths, and refrigeration systems. { 'lith·ē·əm 'nī,trāt }

lithium perchlorate [INORG CHEM] $LiClO_4·3H_2O$ A compound with high oxygen content (60% available oxygen), used as a source of oxygen in rockets and missiles. { 'lith·ē·əm pər'klȯr,āt }

lithium star [ASTRON] A peculiar giant star of spectral type G or M whose spectrum displays a high abundance of lithium. { 'lith·ē·əm ,stär }

lithium stearate [ORG CHEM] $LiC_{18}H_{35}O_2$ A white, crystalline compound with a melting point of 220°C; used in cosmetics, plastics, and greases, and as a corrosion inhibitor in petroleum. { 'lith·ē·əm 'stir,āt }

lithium-sulfur battery [ELEC] A storage battery in which the cells use a molten lithium cathode and a molten sulfur anode separated by a molten salt electrolyte that consists of lithium iodide, potassium iodide, and lithium chloride. { 'lith·ē·əm ¦səl·fər 'bad·ə·rē }

lithium tetraborate [INORG CHEM] $Li_2B_4O_7·5H_2O$ White crystals that lose water at 200°C; insoluble in alcohol, soluble in water; used in ceramics. { 'lith·ē·əm ,te·trə'bȯr,āt }

lithium titanate [INORG CHEM] Li_2TiO_3 A water-insoluble white powder with strong fluxing ability when used in titanium-containing enamels; also used as a mill additive in vitreous and semivitreous glazes. { 'lith·ē·əm 'tī·tən,āt }

Lithobiomorpha [INV ZOO] An order of chilopods in the subclass Pleurostigmophora; members are anamorphic and have 15 leg-bearing trunk segments, and when eyes are present, they are ocellar. { ,lith·ō,bī·ə'mȯr·fə }

lithocholic acid [BIOCHEM] $C_{24}H_{40}O_3$ A crystalline substance with a melting point of 184–186°C; soluble in hot alcohol; found in ox, human, and rabbit bile. { ¦lith·ə,kä·lik 'as·əd }

lithoclase [GEOL] A naturally produced rock fracture. { 'lith·ə,klās }

lithocyst [BOT] Epidermal plant cell in which cytoliths are formed. [INV ZOO] One of the minute sacs containing lithites in many invertebrates; thought to function in audition and orientation. { 'lith·ə,sist }

lithocyte [INV ZOO] A special cell in anthomedusae containing a statolith. { 'lith·ə,sīt }

Lithodidae [INV ZOO] The king crabs, a family of anomuran decapods in the superfamily Paguridea distinguished by reduced last pereiopods and by the asymmetrical arrangement of the abdominal plates in the female. { lə'thäd·ə,dē }

lithodomous [ZOO] Burrowing in rock. { lə'thäd·ə·məs }

lithofacies [GEOL] A subdivision of a specified stratigraphic unit distinguished on the basis of lithologic features. { ,lith·ə'fā·shēz }

lithofacies map [GEOL] The facies map of an area based on lithologic characters; shows areal variation in all aspects of the lithology of a stratigraphic unit. { ,lith·ə'fā·shēz ,map }

lithogenesis [PATH] The process of formation of calculi or stones. [PETR] The branch of science dealing with the formation of rocks, especially the formation of sedimentary rocks. { ,lith·ə'jen·ə·səs }

lithogeochemical survey [GEOCHEM] A geochemical survey that involves the sampling of rocks. { ,lith·ō,jē·ō'kem·ə·kəl 'sər,vā }

lithograph [GRAPHICS] Originally, a reproduction of a writing sample or a drawing made from a litho stone onto which the writing or drawing had been drawn with a greasy ink or crayon; now, a reproduction from litho metal plates produced by photolithography and run on an offset press. { 'lith·ə,graf }

lithographic film [GRAPHICS] Orthochromatic film used in the production of lithographic plates. { ,lith·ə'graf·ik 'film }

lithographic limestone [GEOL] A dense, compact, fine-grained crystalline limestone having a pale creamy-yellow or grayish color. Also known as lithographic stone; litho stone. { ,lith·ə'graf·ik 'līm,stōn }

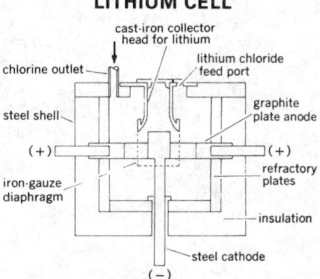

LITHIUM CELL

Cross section of lithium cell.

lithographic plate [GRAPHICS] A metal plate, usually having little porosity, on which an image is produced for lithographic printing. { ¦lith·ə¦graf·ik 'plāt }

lithographic stone *See* lithographic limestone. { ¦lith·ə'graf·ik 'stōn }

lithographic texture [GEOL] The texture of certain calcareous sedimentary rocks characterized by grain size of less than 1/256 millimeter and having a smooth appearance. { ¦lith·ə'graf·ik 'teks·chər }

lithography [ELECTR] A technique used for integrated circuit fabrication in which a silicon slice is coated uniformly with a radiation-sensitive film, the resist, and an exposing source (such as light, x-rays, or an electron beam) illuminates selected areas of the surface through an intervening master template for a particular pattern. [GRAPHICS] A printing process in which a design is sketched with an oily ink or a litho crayon on a flat, smooth stone; in printing, the entire surface of the stone is wetted, and the design areas repel the water, but accept a greasy ink; a clean impression is then made by pressing a sheet of paper against the surface of the stone and running the whole through a press. { lə'thäg·rə·fē }

lithologic map [GEOL] A kind of geologic map showing the rock types of a particular area. { ¦lith·ə¦läj·ik 'map }

lithologic unit *See* rock-stratigraphic unit. { ¦lith·ə¦läj·ik 'yü·nət }

lithology [GEOL] The description of the physical character of a rock as determined by eye or with a low-power magnifier, and based on color, structures, mineralogic components, and grain size. { lə'thäl·ə·jē }

lithol red [MATER] Any of various pigments derived from combination of β-naphthol and Tobias acid; available as sodium, barium, and calcium toners and lakes; used in outside, drum, and toy enamels. { 'li,thȯl 'red }

lithometeor [METEOROL] The general term for dry atmospheric suspensoids, including dust, haze, smoke, and sand. { ¦lith·ə'mēd·ē·ər }

lithomorphic [GEOL] Referring to a soil whose characteristics are derived from events or conditions of a former period. { ¦lith·ə¦mȯr·fik }

lithopedion [MED] A retained fetus that has become calcified. { ¦lith·ə'pē·dē,än }

lithophagous [ZOO] Feeding on stone, as certain mollusks. { lə'thäf·ə·gəs }

lithophile [GEOCHEM] **1.** Pertaining to elements that have become concentrated in the silicate phase of meteorites or the slag crust of the earth. **2.** Pertaining to elements that have a greater free energy of oxidation per gram of oxygen than iron. Also known as oxyphile. { 'lith·ə,fīl }

lithophone *See* lithopone. { 'lith·ə,fōn }

lithophysa [GEOL] A large spherulitic hollow or bubble in glassy basalts and certain rhyolites. Also known as stone bubble. { ¦lith·ə'fīs·ə }

lithophyte [ECOL] A plant that grows on rock. { 'lith·ə,fīt }

lithopone [MATER] A white pigment produced as a filtered, heated, quenched precipitate from reaction of barium sulfide and zinc sulfide; used as a pigment for paint, ink, filled leather, paper, linoleum, oilcloth, and cosmetics. Also known as Charlton white; Griffith's white; lithophone; Orr's white; zinc baryta white; zinc sulfide white. { 'lith·ə,pōn }

lithosere [ECOL] A succession of plant communities that originate on rock. { 'lith·ə,sir }

lithosiderite *See* stony-iron meteorite. { ¦lith·ə'sīd·ə,rīt }

lithosol [GEOL] A group of shallow soils lacking well-defined horizons and composed of imperfectly weathered fragments of rock. { 'lith·ə,sȯl }

lithospar [MINERAL] A combination of spodumene and feldspar which occurs naturally. { 'lith·ə,spär }

lithosphere [GEOL] **1.** The rigid outer crust of rock on the earth about 50 miles (80 kilometers) thick, above the asthenosphere. Also known as oxysphere. **2.** Since the development of plate tectonics theory, a term referring to the rigid, upper 60 miles (100 kilometers) of the crust and upper mantle, above the asthenosphere. { 'lith·ə,sfir }

lithostatic pressure *See* ground pressure. { ¦lith·ə¦stad·ik 'presh·ər }

litho stone *See* lithographic limestone. { ¦lith·ō ,stōn }

lithostratic unit *See* rock-stratigraphic unit. { ¦lith·ə¦strad·ik 'yü·nət }

lithostratigraphic unit *See* rock-stratigraphic unit. { ¦lith·ə,strad·ə'graf·ik 'yü·nət }

lithostratigraphy [GEOL] A branch of stratigraphy concerned with the description and interpretation of sedimentary successions in terms of their lithic character. { ¦lith·ō·strə'tig·rə·fē }

lithostyle [INV ZOO] A static organ in Narcomedusae. Also known as tentaculocyst. { 'lith·ə,stīl }

lithotomy [MED] Surgical removal of a calculus. { lə'thäd·ə·mē }

lithotope [GEOL] **1.** The environment under which a sediment is deposited. **2.** An area of uniform sedimentation. { 'lith·ə,tōp }

lithotripsy *See* extracorporeal shock-wave lithotripsy. { 'lith·ə,trip·sē }

lithotype [GEOL] A macroscopic band in humic coals, analyzed on the basis of physical characteristics rather than botanical origin. { 'lith·ə,tīp }

lithuria [MED] A condition marked by excess of uric (lithic) acid or its salts in the urine. { li'thyur·ē·ə }

litmus [MATER] Blue, water-soluble powder from various lichens, especially *Variolaria lecanora* and *V. rocella*; turns red in solutions at pH 4.5, and blue at pH 8.3; used as an acid-base indicator. Also known as lacmus; lichen blue. { 'lit·məs }

litmus paper [MATER] White, unsized paper saturated by litmus in water; used as a pH indicator. { 'lit·məs ,pā·pər }

Litopterna [PALEON] An order of hoofed, herbivorous mammals confined to the Cenozoic of South America; characterized by a skull without expansion of the temporal or squamosal sinuses, a postorbital bar, primitive dentition, and feet that were three-toed or reduced to a single digit. { ,lid·əp'tər·nə }

lit-par-lit [GEOL] Pertaining to the penetration of bedded, schistose, or other foliate rocks by innumerable narrow sheets and tongues of granitic rock. { 'lē,pär'lē }

Little Bear *See* Ursa Minor. { 'lid·əl 'ber }

little brother [METEOROL] A subsidiary tropical cyclone that sometimes follows a more severe disturbance. { 'lid·əl 'brəth·ər }

little cherry disease [PL PATH] A virus disease of sweet cherries characterized by small, angular pointed fruits which retain the bright red color of immaturity and never reach mature size. { 'lid·əl ¦cher·ē di,zēz }

Little Dipper *See* Ursa Minor. { 'lid·əl 'dip·ər }

little-drop technique [BIOL] A method for isolating single cells in which a drop of a cellular suspension containing a single cell, as determined by microscopic examination, is transferred with a capillary pipet to an appropriate culture medium. { 'lid·əl 'dräp ,tek,nēk }

Little Fox *See* Vulpecula. { 'lid·əl 'fäks }

little giant [MIN ENG] A jointed iron nozzle used in hydraulic mining. { 'lid·əl 'ji·ənt }

Little Horse *See* Equuleus. { 'lid·əl 'hȯrs }

Little Ice Age [GEOL] A period of expansion of mountain glaciers, marked by climatic deterioration, that began about 5500 years ago and extended to as late as A.D. 1550–1850 in some regions, as the Alps, Norway, Iceland, and Alaska. { 'lid·əl 'īs ,āj }

Little John [ORD] Name applied to a U.S. Army rocket system (318-millimeter) consisting of a surface-to-surface tactical missile, similar to but smaller than Honest John, and employing solid fuel. { 'lid·əl 'jän }

little leaf [PL PATH] Any of various plant diseases and disorders characterized by chlorotic, underdeveloped, and sometimes distorted leaves. { 'lid·əl ,lēf }

little LEO system [COMMUN] A system of small satellites in low earth orbit (LEO) that provides messaging, data, and location services but does not have the capability of voice transmission. { ¦lit·əl ¦lē·ō ,sis·təm }

little peach disease [PL PATH] A virus disease of the peach tree in which the fruit is dwarfed and delayed in ripening, the leaves yellow, and the tree dies. { 'lid·əl ¦pēch di,zēz }

Little Ruler *See* Regulus. { 'lid·əl 'rül·ər }

Little's disease [MED] Spastic diplegia of infants which is characterized by spasticity of the lower extremities; involves degenerative and atrophic cerebral changes as well as congenital malformation. { 'lid·əlz di,zēz }

Littlewood conjecture [MATH] The statement that there exists a number *C* such that, whenever $n_1, n_2, \ldots, n_N$ are *N* distinct integers, the integral over *x* from $-\pi$ to π of the absolute value of the sum from $k = 1$ to $k = N$ of the exponential functions of in_kx is greater than $2\pi\, C \log N$.   { 'lid·əl‚wùd kən‚jek·chər }

littoral current [OCEANOGR] A current, caused by wave action, that sets parallel to the shore; usually in the nearshore region within the breaker zone. Also known as alongshore current; longshore current.   { 'lit·ə·rəl 'kə·rənt }

littoral drift [GEOL] Materials moved by waves and currents of the littoral zone. Also known as longshore drift.   { 'lit·ə·rəl 'drift }

littoral sediments [GEOL] Deposits of littoral drift.   { 'lit·ə·rəl 'sed·ə·məns }

littoral transport [GEOL] The movement of littoral drift.   { 'lit·ə·rəl 'tranz‚pórt }

littoral zone [ECOL] Of or pertaining to the biogeographic zone between the high- and low-water marks.   { 'lit·ə·rəl ‚zōn }

Littorinacea [PALEON] An extinct superfamily of gastropod mollusks in the order Prosobranchia.   { ‚lid·ə·rə'nās·ē·ə }

Littorinidae [INV ZOO] The periwinkles, a family of marine gastropod mollusks in the order Pectinibranchia distinguished by their spiral, globular shells.   { ‚lid·ə'rin·ə‚dē }

Littrow grating spectrograph [SPECT] A spectrograph having a plane grating at an angle to the axis of the instrument, and a lens in front of the grating which both collimates and focuses the light.   { 'li‚trō 'grād·iŋ 'spek·trə‚graf }

Littrow mounting [SPECT] The arrangement of the grating and other components of a Littrow grating spectrograph, which is analogous to that of a Littrow quartz spectrograph.   { 'li‚trō ‚maùnt·iŋ }

Littrow prism [OPTICS] A prism having angles of 30, 60, and 90°, silvered on the side opposite the 60° angle; a lens used with it can serve both as a telescope and as a collimator.   { 'li‚trō ‚priz·əm }

Littrow quartz spectrograph [SPECT] A spectrograph in which dispersion is accomplished by a Littrow quartz prism with a rear reflecting surface that reverses the light; a lens in front of the prism acts as both collimator and focusing lens.   { 'li‚trō ‚kwórts 'spek·trə‚graf }

lituate [BOT] Having a forked member or part with the ends turned slightly outward, as in certain fungi.   { 'lich·ə·wət }

Lituolacea [INV ZOO] A superfamily of benthic marine foraminiferans in the suborder Textulariina having a multilocular, rectilinear, enrolled or uncoiled test with a simple to labyrinthic wall.   { ‚lich·ə'lās·ē·ə }

lituus [MATH] The trumpet-shaped plane curve whose points in polar coordinates (r, θ) satisfy the equation $r^2 = a/\theta$.   { 'lich·ə·wəs }

litzendraht wire See litz wire.   { 'lits·ən‚drät ‚wīr }

litz wire [ELEC] Wire consisting of a number of separately insulated strands woven together so each strand successively takes up all possible positions in the cross section of the entire conductor, to reduce skin effect and thereby reduce radio-frequency resistance. Derived from litzendraht wire.   { 'lits‚wīr }

live [COMMUN] Being broadcast directly at the time of production, instead of from recorded or filmed program material. [ELEC] See energized.   { līv }

live ammunition [ORD] Ammunition containing explosives or active chemicals, as distinguished from inert or drill ammunition.   { 'līv ‚am·yə'nish·ən }

live axle [MECH ENG] An axle to which wheels are rigidly fixed.   { 'līv 'ak·səl }

live center [MECH ENG] A lathe center that fits into the headstock spindle.   { 'līv 'sen·tər }

live chassis [ELECTR] A radio, television, or other chassis that has a direct chassis connection to one side of the alternating-current line.   { 'līv 'cha‚sē }

live data [COMPUT SCI] Actual data that are employed during the final testing of a computer system, as opposed to test data.   { 'līv 'dad·ə }

live end [ACOUS] The end of a radio studio that gives almost complete reflection of sound waves.   { 'līv ‚end }

live-end-dead-end room See LEDE room.   { 'līv‚end 'ded‚end ‚rüm }

live load [MECH] A moving load or a load of variable force acting upon a structure, in addition to its own weight.   { 'līv 'lōd }

live load allowance [ENG] The permissible load that may be

added to a completed building structure, including installations, equipment, and personnel.   { 'līv ‚lōd ə‚laù·əns }

live oil [MATER] An oil containing dissolved gas.   { 'līv ‚óil }

liver [ANAT] A large vascular gland in the body of vertebrates, consisting of a continuous parenchymal mass covered by a capsule; secretes bile, manufactures certain blood proteins and enzymes, and removes toxins from the systemic circulation. [MATER] Intermediate layer of dark-colored, oily material formed by hydrolyzation of acid sludge from sulfuric acid treatment of petroleum oil; insoluble in weak acid and oil.   { 'liv·ər }

liver failure [MED] Severe functional disability of the liver marked clinically by a variety of signs and symptoms, including jaundice, coma, and abnormal blood levels of such things as ammonia, bilirubin, and alkaline phosphatase.   { 'liv·ər ‚fāl·yər }

liver fluke [INV ZOO] Any trematode, especially *Clonorchis sinensis*, that lodges in the biliary passages within the liver.   { 'liv·ər ‚flük }

live-roller conveyor [MECH ENG] Conveying machine which moves objects over a series of rollers by the application of power to all or some of the rollers.   { 'līv ‚rōl·ər kən‚vā·ər }

live room [ACOUS] A room having a minimum of sound-absorbing material.   { 'līv ‚rüm }

liver phosphorylase [BIOCHEM] An enzyme that catalyzes the breakdown of liver glycogen to glucose-1-phosphate.   { 'liv·ər ‚fäs'fór·ə‚lās }

liverwort [BOT] The common name for members of the Marchantiatae.   { 'liv·ər‚wórt }

live steam [MECH ENG] Steam that is being delivered directly from a boiler under full pressure.   { 'līv 'stēm }

livestock [AGR] Animals, such as cattle, sheep, pigs, chickens, and horses, that are utilized for various purposes on a farm or ranch.   { 'līv‚stäk }

live system [COMPUT SCI] A computer system on which all testing has been completed so that it is fully operational and ready for production work. Also known as production system.   { 'līv 'sis·təm }

live-virus vaccine [IMMUNOL] A suspension of attenuated live viruses injected to produce immunity.   { 'līv 'vī·rəs vak'sēn }

liveware [COMPUT SCI] The people involved in the operation of a computer system, thought of as a component of the system along with hardware and software.   { 'līv‚wer }

living fossil [BIOL] A living species belonging to an ancient stock otherwise known only as fossils.   { 'liv·iŋ 'fäs·əl }

Livingstone sphere [ENG] A clay atmometer in the form of a sphere; evaporation indicated by this instrument is supposed to be somewhat representative of that from plant growth.   { 'liv·iŋ‚stən ‚sfir }

livingstonite [MINERAL] $HgSb_4S_7$ A lead-gray mineral with red streak and metallic luster; a source of mercury.   { 'liv·iŋ‚stə‚nīt }

livor mortis [PATH] The reddish-blue discoloration of the cadaver that occurs in the dependent portions of the body due to gradual gravitational flow of unclotted blood.   { 'lī‚vòr 'mórd·əs }

livre [MECH] A unit of mass, used in France, equal to 0.5 kilogram.   { 'lēv·rə }

lixiviate [CHEM ENG] To extract a soluble component from a solid mixture by washing or percolation processes.   { lik 'siv·ē‚āt }

lixuration See leaching.   { ‚lik·syü'rā·shən }

lizard [VERT ZOO] Any reptile of the suborder Sauria.   { 'liz·ərd }

Lizard See Lacerta.   { 'liz·ərd }

lizard-hipped dinosaur [PALEON] The name applied to members of the Saurichia because of the comparatively unspecialized three-pronged pelvis.   { 'liz·ərd ‚hipt 'dī·nə‚sòr }

L joint See primary flat joint.   { 'el ‚jóint }

Ljungström heater [MECH ENG] Continuous, regenerative, heat-transfer air heater (recuperator) made of slow-moving rotors packed with closely spaced metal plates or wires with a housing to confine the hot and cold gases to opposite sides.   { 'yùŋ·strəm ‚hēd·ər }

Ljungström steam turbine [MECH ENG] A radial outward-flow turbine having two opposed rotation rotors.   { 'yuŋ·strəm ‚stēm 'tər·bən }

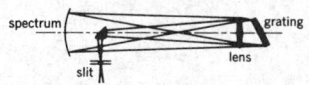

LITTROW GRATING SPECTROGRAPH

Littrow mounting of a plane grating.

LITTROW QUARTZ SPECTROGRAPH

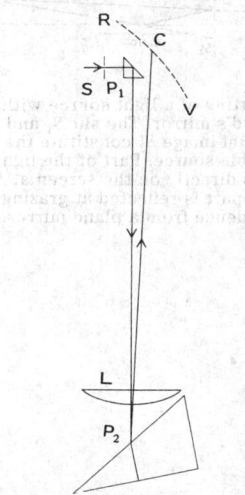

Littrow quartz spectrograph, typical arrangement for double-beam recording instrument. S, slit; P_1, totally reflecting quartz prism; L, autocollimating quartz lens; P_2, Littrow quartz prism; C, camera; RV, red to violet spectrum.

LIVERWORT

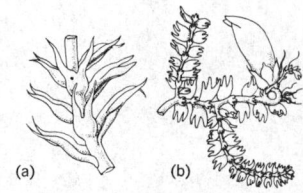

Leafy liverworts of the Jungermanniales. (a) *Herberta*, showing three ranks of equal bifid leaves. (b) *Lepidozia*, ventral aspect, showing reduced ventral leaves. (*From E. W. Sinnott and K. S. Wilson, Botany: Principles and Problems, 5th ed., McGraw-Hill, 1955*)

LK virus [VIROL] A type of equine herpesvirus. { ¦el¦kā 'vī·rəs }

LL *See* line link.

llama [VERT ZOO] Any of three species of South American artiodactyl mammals of the genus *Lama* in the camel family; differs from the camel in being smaller and lacking a hump. { 'yäm·ə }

Llandellian [GEOL] Upper Middle Ordovician geologic time. { lan'del·yən }

Llandoverian [GEOL] Lower Silurian geologic time. { ¦lan·də¦vir·ē·ən }

llano [ECOL] A savannah of Spanish America and the southwestern United States generally having few trees. { 'yä·nō }

Llanvirnian [GEOL] Lower Middle Ordovician geologic time. { lan'vir·nē·ən }

llebetjado [METEOROL] In northeastern Spain, a hot, squally wind descending from the Pyrenees and lasting for a few hours. { ¦yä·bet'hä·dō }

LLL circuit *See* low-level logic circuit. { ¦el¦el'el ˌsər·kət }

Lloyd's mirror interference [OPTICS] The interference pattern produced when part of the light from a slit falls directly on a screen, and part is reflected from a mirror whose surface makes a small angle with the incident beam. { 'loidz ¦mir·ər ˌin·tər'fir·əns }

lm *See* lumen.

LM *See* lunar excursion module.

L/M [NUC PHYS] The ratio of the number of internal conversion electrons emitted from the *L* shell in the de-excitation of a nucleus to the number of such electrons emitted from the *M* shell.

LMC *See* Large Magellanic Cloud.

lm-hr *See* lumen-hour.

lm-sec *See* lumen-second.

LMSS *See* land mobile-satellite service.

lm/w *See* lumen per watt.

LMXRB *See* low-mass x-ray binary.

L network [ELECTR] A network composed of two branches in series, with the free ends connected to one pair of terminals; the junction point and one free end are connected to another pair of terminals. { 'el ˌnet,wərk }

LNG *See* liquefied natural gas.

LNG ship [NAV ARCH] A specially designed, insulated tanker for shipping liquefied natural gas. { ¦el¦en'jē ˌship }

loach [VERT ZOO] The common name for fishes composing the family Cobitidae; most are small and many are eel-shaped. { lōch }

load [COMPUT SCI] **1.** To place data into an internal register under program control. **2.** To place a program from external storage into central memory under operator (or program) control, particularly when loading the first program into an otherwise empty computer. **3.** An instruction, or operator control button, which causes the computer to initiate the load action. **4.** The amount of work scheduled on a computer system, usually expressed in hours of work. [ELEC] **1.** A device that consumes electric power. **2.** The amount of electric power that is drawn from a power line, generator, or other power source. **3.** The material to be heated by an induction heater or dielectric heater. Also known as work. [ELECTR] The device that receives the useful signal output of an amplifier, oscillator, or other signal source. [ENG] **1.** To place ammunition in a gun, bombs on an airplane, explosives in a missile or borehole, fuel in a fuel tank, cargo or passengers into a vehicle, and the like. **2.** The quantity of gas delivered or required at any particular point on a gas supply system; develops primarily at gas-consuming equipment. [MECH] **1.** The weight that is supported by a structure. **2.** Mechanical force that is applied to a body. **3.** The burden placed on any machine, measured by units such as horsepower, kilowatts, or tons. [MIN ENG] Unit of weight of ore used in the South African diamond mines; equal to 1600 pounds (725 kilograms); the equivalent of about 16 cubic feet (0.453 cubic meter) of broken ore. { lōd }

load-and-carry equipment [MECH ENG] Earthmoving equipment designed to load and transport material. { ¦lōd ən 'kar·ē iˌkwip·mənt }

load-and-go [COMPUT SCI] An operating technique with no stops between the loading and execution phases of a program; may include assembling or compiling. { ¦lōd ən 'gō }

load-bearing tile [MATER] A tile with the capacity to support superimposed loads. { 'lōd ¦ber·iŋ ˌtīl }

load-break switch [ELEC] An electric switch in a circuit with several hundred thousand volts, designed to carry a large amount of current without overheating the open position, having enough insulation to isolate the circuit in closed position, and equipped with arc interrupters to interrupt the load current. { 'lōd ¦brāk ˌswich }

load-carrying capacity [MECH ENG] The greatest weight that the end effector of a robot can manipulate without reducing its level of performance. { 'lōd ¦kar·ē·iŋ kəˌpas·əd·ē }

load cast [GEOL] An irregularity at the base of an overlying stratum, usually sandstone, that projects into an underlying stratum, usually shale or clay. { 'lōd ¦kast }

load cell [ELEC] A device which measures large pressures by applying the pressure to a piezoelectric crystal and measuring the voltage across the crystal; the cell plus a recording mechanism constitutes a strain gage. { 'lōd ¦sel }

load characteristic [ELECTR] Relation between the instantaneous values of a pair of variables such as an electrode voltage and an electrode current, when all direct electrode supply voltages are maintained constant. Also known as dynamic characteristic. { 'lōd ¦kar·ik·tə'ris·tik }

load chart [IND ENG] A graph showing the amount of work still to be performed by a factory producing unit such as a machine or assembly group. { 'lōd ˌchärt }

load circuit [ELECTR] Complete circuit required to transform power from a source such as an electron tube to a load. { 'lōd ˌsər·kət }

load circuit efficiency [ELECTR] Ratio between useful power delivered by the load circuit to the load and the load circuit power input. { 'lōd ¦sər·kət i'fish·ən·sē }

load compensation [CONT SYS] Compensation in which the compensator acts on the output signal after it has generated feedback signals. Also known as load stabilization. { 'lōd käm·pən'sā·shən }

load controller [MIN ENG] A device to control the load and prevent spillage on a gathering conveyor receiving coal or mineral from several loading points or subsidiary conveyors; it is a simplified weightometer. { 'lōd kənˌtrōl·ər }

load curve [ELEC] A graph that plots the power supplied by an electric power system versus time. { 'lōd ˌkərv }

load deflection [MECH ENG] The change in position of a body when a load is applied to it. { 'lōd diˌflek·shən }

load diagram [CIV ENG] A diagram showing the distribution and intensity of loads on a structure. { 'lōd ˌdī·əˌgram }

load divider [ELEC] Unit for distributing power to various units. { 'lōd diˌvīd·ər }

loaded concrete [MATER] Concrete to which elements of high atomic number or capture cross section have been added to increase its effectiveness as a radiation shield in nuclear reactors. { 'lōd·əd kän'krēt }

loaded line [ELEC] Wire line in which loading coils have been inserted at regular intervals to reduce attenuation and phase lag at the frequencies within the band used. { 'lōd·əd līn }

loaded motional impedance *See* motional impedance. { 'lōd·əd ¦mō·shən·əl im'pēd·əns }

loaded Q [ELEC] The Q factor of an impedance which is connected or coupled under working conditions. Also known as working Q. [ELECTROMAG] The Q factor of a specific mode of resonance of a microwave tube or resonant cavity when there is external coupling to that mode. { 'lōd·əd kyü }

loaded wheel [ENG] A grinding wheel that is dull as a result of becoming filled with particles from the material being ground. { 'lōd·əd 'wēl }

loader [COMPUT SCI] A computer program that takes some other program from an input or storage device and places it in memory at some predetermined address. [MECH ENG] A machine such as a mechanical shovel used for loading bulk materials. [ORD] Mechanical device which loads guns with cartridges. { 'lōd·ər }

load factor [ELEC] The ratio of average electric load to peak load, usually calculated over a 1-hour period. [MECH] The ratio of load to the maximum rated load. { 'lōd ˌfak·tər }

load impedance [ELECTR] The complex impedance presented to a transducer by its load. { 'lōd imˌpēd·əns }

loading [CHEM ENG] Condition of vapor overcapacity in a liquid-vapor-contact tower, in which rising vapor lifts or holds

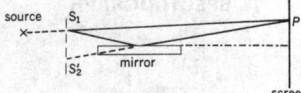

LLOYD'S MIRROR INTERFERENCE

Splitting of a light source with Lloyd's mirror. The slit S_1 and its virtual image S_2' constitute the double source. Part of the light falls directly on the screen at P, and part is reflected at grazing incidence from a plane mirror.

falling liquid. [ELEC] The addition of inductance to a transmission line to improve its transmission characteristics throughout a given frequency band. Also known as electrical loading. [ENG] **1.** Buildup on a cutting tool of the material removed in cutting. **2.** Filling the pores of a grinding wheel with material removed in the grinding process. [ENG ACOUS] Placing material at the front or rear of a loudspeaker to change its acoustic impedance and thereby alter its radiation. [FL MECH] **1.** The relative concentration of particles in a flowing fluid. **2.** In particular, the ratio of particle mass flow to fluid mass flow. [MET] Filling of a die cavity with powdered metal. [NUCLEO] Placing fuel in a nuclear reactor. { 'lōd·iŋ }

loading angle [ORD] Angle of elevation specified for loading a particular weapon with its ammunition. { 'lōd·iŋ ,aŋ·gəl }

loading board [ENG] A device that holds preforms in positions corresponding to the multiple cavities in a compression mold, thus facilitating the simultaneous insertion of the preforms. { 'lōd·iŋ ,bȯrd }

loading coil [ELECTROMAG] **1.** An iron-core coil connected into a telephone line or cable at regular intervals to lessen the effect of line capacitance and reduce distortion. Also known as Pupin coil; telephone loading coil. **2.** A coil inserted in series with a radio antenna to increase its electrical length and thereby lower the resonant frequency. { 'lōd·iŋ ,kȯil }

loading density [ENG] The number of pounds of explosive per foot length of drill hole. [ORD] A term applied specifically to explosive charges of projectiles, bombs, warheads, and so on; it is the quantity of explosive per unit volume, usually expressed as grams per cubic centimeter. { 'lōd·iŋ ,den·səd·ē }

loading device [COMPUT SCI] Equipment from which programs or other data can be transferred or copied into a computer. { 'lōd·iŋ di,vīs }

loading disk [ELECTROMAG] Circular metal piece mounted at the top of a vertical antenna to increase its natural wavelength. { 'lōd·iŋ ,disk }

loading head [MECH ENG] The part of a loader which gathers the bulk materials. { 'lōd·iŋ ,hed }

loading noise [ACOUS] The component of propeller noise that is related to the lift and drag forces acting on the propeller blade. { 'lōd·iŋ ,nȯiz }

loading pan [MIN ENG] A box or scoop into which broken rock is shoveled in a sinking shaft while the hoppit is traveling in the shaft. { 'lōd·iŋ ,pan }

loading program [COMPUT SCI] Program used to load other programs into computer memory. Also known as bootstrap program. { 'lōd·iŋ ,prō·grəm }

loading rack [ENG] The shelter and associated equipment for the withdrawal of liquid petroleum or a chemical product from a storage tank and loading it into a railroad tank car or tank truck. { 'lōd·iŋ ,rak }

loading routine See input routine. { 'lōd·iŋ rü,tēn }

loading space [ENG] Space in a compression mold for holding the plastic molding material before it is compressed. { 'lōd·iŋ ,spās }

loading station [MECH ENG] A device which receives material and puts it on a conveyor; may be one or more plates or a hopper. { 'lōd·iŋ ,stā·shən }

loading tray [ENG] A tray with a sliding bottom used to simultaneously load the plastic charge into the cavities of a multicavity mold. [ORD] **1.** Trough-shaped carrier on which heavy projectiles are placed so that they can be more easily and safely slipped into the breech of a gun. **2.** Hollowed slide which guides the projectiles into the breech of some types of automatic weapons. { 'lōd·iŋ ,trā }

loading weight [ENG] Weight of a powder put into a container. { 'lōd·iŋ ,wāt }

load isolator [ELECTROMAG] Waveguide or coaxial device that provides a good energy path from a signal source to a load, but provides a poor energy path for reflections from a mismatched load back to the signal source. { 'lōd ,ī·sə,lād·ər }

load leveling [ELEC] A method for reducing the large fluctuations that occur in electricity demand, for example by storing excess electricity during periods of low demand for use during periods of high demand. { 'lōd ,lev·ə·liŋ }

load limit [CIV ENG] The maximum weight that can be supported by a structure. [MECH ENG] The maximum recommended or permitted overall weight of a container or a cargo-carrying vehicle that is determined by combining the weight of the empty container or vehicle with the weight of the load. { 'lōd ,lim·ət }

load line [ELECTR] A straight line drawn across a series of tube or transistor characteristic curves to show how output signal current will change with input signal voltage when a specified load resistance is used. [NAV ARCH] A line, painted or cut on the outside of a ship, which marks the maximum waterline when the ship is loaded with the greatest cargo which it can carry safely. { 'lōd ,līn }

load loss [ELEC] The sum of the copper loss of a transformer, due to resistance in the windings, plus the eddy current loss in the winding, plus the stray loss. { 'lōd ,lȯs }

load metamorphism See static metamorphism. { 'lōd ,med·ə'mȯr,fiz·əm }

load module [COMPUT SCI] A program in a form suitable for loading into memory and executing. { 'lōd ,mä·jül }

load oil [PETRO ENG] Oil used as a fracturing fluid in a formation to stimulate a well to produce. { 'lōd ,ȯil }

load point [COMPUT SCI] Preset point on a magnetic tape from which reading or writing will start. { 'lōd ,pȯint }

load power [ELEC] Of an energy load, the average rate of flow of energy through the terminals of that load when connected to a specified source. { 'lōd ,pau̇·ər }

load profile [ENG] A measure of the time distribution of a building's energy requirements, including the heating, cooling, and electrical loads. { 'lōd ,prō,fīl }

load regulation [ELEC] The maximum change in the output voltage or current of a regulated power supply for a specified change in load conditions. { 'lōd ,reg·yə,lā·shən }

load shedding [ELEC] A procedure in which parts of an electric power system are disconnected in an attempt to prevent failure of the entire system due to overloading. { 'lōd ,shed·iŋ }

load shifting [ELEC] In an electric power system, the transfer of loads from times of peak demand to off-peak time periods. { 'lōd ,shift·iŋ }

load stabilization See load compensation. { 'lōd ,stā·bə·lə,zā·shən }

loadstone See lodestone. { 'lōd,stōn }

load stress [MECH] Stress that results from a pressure or gravitational load. { 'lōd ,stres }

load water [PETRO ENG] Water used to prime a well after an acidizing procedure. { 'lōd ,wȯd·ər }

load waterline [NAV ARCH] The waterline of a fully loaded vessel. { 'lōd 'wȯd·ər,līn }

load water plane [NAV ARCH] The water plane of a fully loaded vessel. { 'lōd wȯd·ər ,plān }

loam [GEOL] Soil mixture of sand, silt, clay, and humus. [MET] Molding material consisting of sand, silt, and clay used over backup material for producing massive castings, usually of iron or steel. { lōm }

loaming [GEOCHEM] In geochemical prospecting, a method in which samples of material from the surface are tested for traces of a sought-after metal; its presence on the surface presumably indicates a near-surface ore body. { 'lōm·iŋ }

LOB See line of balance.

Lobachevski geometry [MATH] A system of planar geometry in which the euclidean parallel postulate fails; any point p not on a line L has at least two lines through it parallel to L. Also known as Bolyai geometry; hyperbolic geometry. { lō·bə'chef·skē jē'äm·ə·trē }

lobar pneumonia [MED] An acute febrile disease involving one or more lobes of the lung, usually following pneumococcal infection. { 'lō,bär nu̇'mō·nyə }

lobar sclerosis [MED] Neuroglial proliferation accompanied by atrophy of a cerebral lobe leading to mental and neurological deficits; most common in infants and children who have suffered prolonged hypoxia. { 'lō,bär sklə'rō·səs }

Lobata [INV ZOO] An order of the Ctenophora in which the body is helmet-shaped. { lō'bäd·ə }

lobate [BIOL] Having lobes. [VERT ZOO] Of a fish, having the skin of the fin extend onto the bases of the fin rays. { 'lō,bāt }

lobate rill mark [GEOL] A flute cast formed by current action. { 'lō,bāt 'ril ,märk }

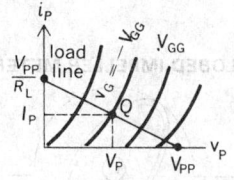

LOAD LINE

Load line drawn across characteristic curves giving plate current, i_P, as function of plate voltage, v_P, for various values of grid supply voltage, V_{GG}. V_{PP} = plate supply voltage, R_L = load resistance, v_G = grid voltage. Quiescent point, Q, determines quiescent plate current, I_P, and quiescent plate voltage, V_P.

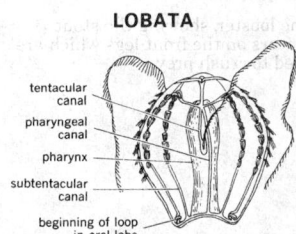

LOBATA

tentacular canal
pharyngeal canal
pharynx
subtentacular canal
beginning of loop in oral lobe

Bolinopsis mikado.

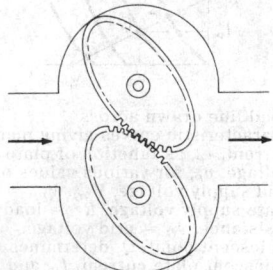

LOBED IMPELLER METER

Meshing impellers in the lobed impeller meter. Arrows show direction of fluid flow.

LOBSTER

The lobster, showing the stout pincers on the front legs which are used to crush prey.

lobe [BIOL] A rounded projection on an organ or body part. [DES ENG] A projection on a cam wheel or a noncircular gear wheel. [ELECTROMAG] A part of the radiation pattern of a directional antenna representing an area of stronger radio-signal transmission. Also known as radiation lobe. [ENG ACOUS] A portion of the directivity pattern of a transducer representing an area of increased emission or response. [HYD] A curved projection on the margin of a continental ice sheet. { lōb }

lobectomy [MED] Surgical removal of a lobe of an organ, particularly of a lung. { lō'bek·tə·mē }

lobed impeller meter [ENG] A type of positive displacement meter in which a fluid stream is separated into discrete quantities by rotating, meshing impellers driven by interlocking gears. { lōbd im'pel·ər ˌmēd·ər }

lobefin fish [VERT ZOO] The common name for members composing the subclass Crossopterygii. { 'lōbˌfin ˌfish }

lobe-half-power width [ELECTROMAG] In a plane containing the direction of the maximum energy of a lobe, the angle between the two directions in that plane about the maximum in which the radiation intensity is one-half the maximum value of the lobe. { 'lōbˌhaf ˈpau̇·ər ˌwidth }

lobeline [PHARM] $C_{22}H_{27}NO_2$ A crystalline compound isolated from the herb and seeds of Indian tobacco (*Labelia inflata*); melting point is 130–131°C; soluble in hot alcohol, chloroform, and benzene; used in medicine as a respiratory stimulant. { 'lō·bəˌlēn }

lobe penetration [ELECTROMAG] Penetration of the radar coverage of a station which is not limited by pulse repetition frequency, scope limitations, or the screening angle at the azimuth of penetration. { lōb ˌpen·ə'trā·shən }

lobe switching *See* beam switching. { lōb ˌswich·iŋ }

lobing [ELECTROMAG] Formation of maxima and minima at various angles of the vertical plane antenna pattern by the reflection of energy from the surface surrounding the radar antenna; these reflections reinforce the main beam at some angles and detract from it at other angles, producing fingers of energy. { 'lōb·iŋ }

loblolly pine [BOT] *Pinus taeda.* A hard yellow pine of the central and southeastern United States having a reddish-brown fissured bark, needles in groups of three, and a full bushy top. { 'läbˌläl·ē 'pīn }

lobopodia [INV ZOO] Broad, thick pseudopodia. { ˌlō·bə'päd·ē·ə }

Lobosia [INV ZOO] A subclass of the protozoan class Rhizopodea generally characterized by lobopodia. { lō'bō·sē·ə }

lobotomy [MED] An operative section of the fibers between the frontal lobes of the brain. Also known as leukotomy; prefrontal lobotomy. { lō'bäd·ə·mē }

Lobry de Bruyn-Ekenstein transformation [ORG CHEM] The change in which an aldose sugar treated with dilute alkali results in a mixture of an epimeric pair and 2-keto-hexose due to the production of enolic forms in the presence of hydroxyl ions, followed by a rearrangement. { lō'brē·də'brīn 'ā·kənˌshtīn ˌtrans·fərˌmā·shən }

lobster [INV ZOO] The common name for several bottom-dwelling decapod crustaceans making up the family Homaridae which are commercially important as a food item. { 'läb·stər }

lobular pneumonia *See* bronchopneumonia. { 'läb·yə·lər nu̇'mō·nyə }

lobule [BIOL] **1.** A small lobe. **2.** A division of a lobe. { 'läb·yül }

local action [ELEC] **1.** Internal losses of a battery caused by chemical reactions producing local currents between different parts of a plate. **2.** Quantitatively, the percentage loss per month in the capacity of a battery on open circuit, or the amount of current needed to keep the battery fully charged. [MET] Electrochemical corrosion resulting from the action of local cells. { 'lō·kəl 'ak·shən }

local algebra [MATH] An algebra A over a field F which is the sum of the radical of A and the subalgebra consisting of products of elements of F with the multiplicative identity of A. { ˈlō·kəl 'al·jə·brə }

local anesthetic [PHARM] A drug which induces loss of sensation only in the region to which it is applied. { 'lō·kəl ˌan·əs'thed·ik }

local angular momentum [METEOROL] Angular momentum about an arbitrarily located vertical axis which is fixed with respect to the earth. { 'lō·kəl 'aŋ·gyə·lər mə'men·təm }

local apparent noon [ASTRON] Twelve o'clock local apparent time, or the instant the apparent sun is over the upper branch of the local meridian. { 'lō·kəl ə¦par·ənt ˈnün }

local apparent time [ASTRON] The arc of the celestial equator, or the angle at the celestial pole, between the lower branch of the local celestial meridian and the hour circle of the apparent or true sun, measured westward from the lower branch of the local celestial meridian through 24 hours. { 'lō·kəl ə¦par·ənt ¦tīm }

Local-Area Augmentation System [NAV] A type of local-area DGPS, developed by the Federal Aviation Administration in the United States, which is intended for aviation users in category II and III precision approaches and which broadcasts in the 108–117.95-megahertz band presently used for very high frequency omindirectional range (VOR) systems and instrument landing systems (ILS). Abbreviated LAAS. { ˌlōk·əl ¦er·ē·ə ˌȯg·mən'tā·shən ˌsis·təm }

local-area DGPS [NAV] A version of differential GPS which provides error corrections in the vicinity of a single reference receiver based on the measurements by that receiver. { ˌlōk·əl ¦er·ē·ə ¦dē¦jē¦pē'es }

local-area network [COMPUT SCI] A communications network connecting various hardware devices together within a building by means of a continuous cable or an in-house voice-data telephone system. Also known as LAN. { 'lō·kəl¦er·ē·ə 'netˌwȯrk }

local-area underwater navigation system [NAV] A system applied chiefly in the inspection and maintenance of offshore structures, in which time differences between the reception of acoustic signals from a chain of transmitters are used to determine the positions of divers and of crewed and uncrewed underwater vehicles. Abbreviated LAUNS. { 'lō·kəl ¦er·ē·ə ¦ən·dər¦wȯd·ər ˌnav·ə'gā·shən ˌsis·təm }

local arm *See* Orion arm. { 'lō·kəl 'ärm }

local attraction *See* local magnetic disturbance. { 'lō·kəl ə'trak·shən }

local base [MATH] For a point x in a topological space, a family of neighborhoods of x such that every neighborhood of x contains a member of the family. Also known as base for the neighborhood system. { ¦lō·kəl 'bās }

local base level *See* temporary base level. { 'lō·kəl 'bās ¦lev·əl }

local battery [ELEC] Battery that actuates the telegraphic station recording instruments, as distinguished from the battery furnishing current to the line. { 'lō·kəl 'bad·ə·rē }

local-battery telephone set [ELECTR] Telephone set for which the transmitter current is supplied from a battery, or other current supply circuit, individual to the telephone set; the signaling current may be supplied from a local hand generator or from a centralized power source. { 'lō·kəl 'bad·ə·rē 'tel·əˌfōn ˌset }

local buckling [MECH] Buckling of thin elements of a column section in a series of waves or wrinkles. { 'lō·kəl 'bək·liŋ }

local cell [ELEC] A galvanic cell resulting from differences in potential between adjacent areas on the surface of a metal immersed in an electrolyte. { 'lō·kəl 'sel }

local central office [COMMUN] A telephone central office, which terminates subscriber lines and makes connections with other central offices, usually equipped to serve 10,000 main telephones of its immediate community. { 'lō·kəl ¦sen·trəl 'ȯf·əs }

local change [OCEANOGR] The time rate of change of a scalar quantity (such as temperature, salinity, pressure, or oxygen content) in a fixed locality. { 'lō·kəl 'chānj }

local circuit [COMMUN] Circuit to a main or auxiliary circuit which can be made available at any station or patched from point to point through one or more stations. { 'lō·kəl 'sər·kət }

local civil time [ASTRON] United States terminology during 1925–1952 for local mean time. { 'lō·kəl ¦siv·əl ¦tīm }

local cluster of stars *See* local star system. { 'lō·kəl ¦kləs·tər əv ¦stärz }

local coefficient [MATH] By using fiber bundles where the fiber is a group, one may generalize cohomology theory for spaces; one uses such bundles as the algebraic base for such a theory and calls the bundle a system of local coefficients. { 'lō·kəl ˌkō·i'fish·ənt }

local coefficient of heat transfer [THERMO] The heat transfer coefficient at a particular point on a surface, equal to the amount of heat transferred to an infinitesimal area of the surface at the point by a fluid passing over it, divided by the product of this area and the difference between the temperatures of the surface and the fluid. { 'lō·kəl ‚kō·i'fish·ənt əv 'hēt ‚tranz·fər }

local control [COMMUN] System or method of radio-transmitter control whereby the control functions are performed directly at the transmitter. { 'lō·kəl kən'trōl }

local controller *See* first-level controller. { 'lō·kəl kən'trōl·ər }

local coordinate system [MATH] The coordinate system about a point which is induced when the global space is locally euclidean. { 'lō·kəl kō'ȯrd·ən·ət ‚sis·təm }

local derivative [FL MECH] The rate of change of a quantity *f* with respect to time at a fixed point of a fluid, $\partial f/\partial t$; it is related to the individual derivative df/dt through the expression $\partial f/\partial t = df/dt - V \cdot \nabla f$, where *f* is a thermodynamic property $f(x,y,z,t)$ of the fluid, *V* the vector velocity of the fluid, and ∇ the del operator. { 'lō·kəl də'riv·əd·iv }

local device [COMPUT SCI] Peripheral equipment that is linked directly to a computer or other supporting equipment, without an intervening communications channel. { 'lō·kəl di'vīs }

local distortion [MATH] The absolute value of the derivative of an analytic function at a given point. { 'lō·kəl di'stȯr·shən }

local exchange *See* exchange. { 'lō·kəl iks'chānj }

local extra observation [METEOROL] An aviation weather observation taken at specified intervals, usually every 15 minutes, when there are impending aircraft operations and when weather conditions are below certain operational weather limits; the observation includes ceiling, sky condition, visibility, atmospheric phenomena, and pertinent remarks. { 'lō·kəl 'ek·strə ‚äb·zər'vā·shən }

local first selector [COMMUN] The second portion of a line that connects to a calling line, through a line primary switch, to a local second selector and special service second selector, and returns a dial tone to the calling subscribers. { 'lō·kəl ¦fərst si¦lek·tər }

local forecast [METEOROL] Generally, any weather forecast of conditions over a relatively limited area, such as a city or airport. { 'lō·kəl 'fȯr‚kast }

Local Group [ASTRON] A group of at least 20 known galaxies in the vicinity of the sun; the Andromeda Spiral is the largest of the group, and the Milky Way Galaxy is the second largest. { 'lō·kəl 'grüp }

local hidden-variable theory *See* hidden-variable theory of the second kind. { 'lō·kəl 'hid·ən 'ver·ē·ə·bəl ‚thē·ə·rē }

local hour angle [ASTRON] Angular distance west of the local celestial meridian. { 'lō·kəl 'aur ‚aŋ·gəl }

local immunity [IMMUNOL] Immunity localized in a specific tissue or region of the body. { 'lō·kəl i'myü·nəd·ē }

local inflow [HYD] The water that enters a stream between two stream-gaging stations. { 'lō·kəl 'in‚flō }

local invariance [PHYS] The property of physical laws which remain unchanged under a specified set of symmetry transformations even when these transformations are chosen independently at every point of space and time. { 'lō·kəl in'ver·ē·əns }

locality [PHYS] The condition that two events at spatially separated locations are entirely independent of each other, provided that the time interval between the events is less than that required for a light signal to travel from one location to the other. { lō'kal·əd·ē }

localization [COMPUT SCI] Imposing some physical order upon a set of objects, so that a given object has a greater probability of being in some particular regions of space than in others. { ‚lō·kə·lə'zā·shən }

localization discrimination suppression [ACOUS] The ability of human hearing to process changes in the location of a sound source without significant disruption from reflections. { ‚lōk·ə·lə¦zā·shən di‚skrim·ənā·shən sə‚presh·ən }

localization dominance [ACOUS] In human hearing, the effect whereby the perceived location of sound from a source combined with reflected sounds is dominated by the sound source. { ‚lōk·əl·ə¦zā·shən 'däm·ə·nəns }

localization principle [MATH] The principle that the convergence of the Fourier series of a function at a point depends only on the behavior of the function in some arbitrarily small neighborhood of that point. { ‚lō·kə·lə'zā·shən ‚prin·sə·pəl }

localized state [QUANT MECH] A state of motion in which an electron may be found anywhere within a region of a material of linear extent smaller than that of the material. { 'lō·kə‚līzd 'stāt }

localized vector [MECH] A vector whose line of application or point of application is prescribed, in addition to its direction. { 'lō·kə‚līzd 'vek·tər }

localized wave solution [PHYS] A solution to the multidimensional wave equation in which the energy is concentrated in certain regions of space and time. { ‚lōk·ə‚līzd 'wāv sə‚lü·shən }

localizer [NAV] A directional radio beacon to provide aircraft with signals for lateral guidance with respect to the runway centerline. { 'lō·kə‚līz·ər }

local level [NAV] The plane normal to the local vertical. { 'lō·kəl 'lev·əl }

local line *See* local loop. { 'lō·kəl 'līn }

local loop [COMMUN] A telephone line terminating at the local central office. Also known as local line. { 'lō·kəl 'lüp }

local lunar time [ASTRON] The arc of the celestial equator, or the angle at the celestial pole, between the lower branch of the local celestial meridian and the hour circle of the moon, measured westward from the lower branch of the local celestial meridian through 24 hours; local hour angle of the moon, expressed in time units, plus 12 hours; local lunar time at the Greenwich meridian is called Greenwich lunar time. { 'lō·kəl ¦lü·nər 'tīm }

locally arcwise connected topological space [MATH] A topological space in which every point has an arcwise connected neighborhood, that is, an open set any two points of which can be joined by an arc. { 'lō·kə·lē ärk‚wīz kə‚nek·təd ¦täp·ə¦läj·ə·kəl ‚spās }

locally compact topological space [MATH] A topological space in which every point lies in a compact neighborhood. { 'lō·kə·lē kəm'pakt ¦täp·ə¦läj·ə·kəl ‚spās }

locally connected topological space [MATH] A topological space in which every point has a connected neighborhood. { 'lō·kə·lē kə'nek·təd ¦täp·ə¦läj·ə·kəl ‚spās }

locally convex space [MATH] A Hausdorff topological vector space *E* such that every neighborhood of any point *x* belonging to *E* contains a convex neighborhood of *x*. { 'lō·kə·lē 'kän‚veks ‚spās }

locally euclidean topological space [MATH] A topological space in which every point has a neighborhood which is homeomorphic to a euclidean space. { 'lō·kə·lē yü'klid·ē·ən ¦täp·ə¦läj·ə·kəl ‚spās }

locally finite family of sets [MATH] A family of subsets of a topological space such that each point of the topological space has a neighborhood that intersects only a finite number of these subsets. { ¦lō·kə·lē ¦fī‚nīt ¦fam·lē əv 'sets }

locally integrable function [MATH] A function is said to be locally integrable on an open set *S* in *n*-dimensional euclidean space if it is defined almost everywhere in *S* and has a finite integral on compact subsets *S*. { ¦lō·kə·lē ¦int·ə·grə·bəl 'faŋk·shən }

locally one to one [MATH] A function is locally one to one if it is one to one in some neighborhood of each point. { 'lō·kə·lē 'wən tə 'wən }

locally trivial bundle [MATH] A bundle for which each point in the base has a neighborhood *U* whose inverse image under the projection map is isomorphic to a cartesian product of *U* with a space isomorphic to the fibers of the bundle. { ¦lō·kə·lē ¦triv·ē·əl 'bən·dəl }

local Mach number [AERO ENG] The Mach number of an isolated section of an airplane or its airframe. { 'lō·kəl 'mäk ‚nəm·bər }

local magnetic disturbance [GEOPHYS] An anomaly of the magnetic field of the earth, extending over a relatively small area, due to local magnetic influences. Also known as local attraction. { 'lō·kəl mag'ned·ik di'stər·bəns }

local maximum [MATH] A local maximum of a function *f* is a value $f(c)$ of *f* where $f(x) \leq f(c)$ for all *x* in some neighborhood of *c*; if $f(c)$ is a local maximum, *f* is said to have a local maximum at *c*. { 'lō·kəl 'mak·sə·məm }

local mean noon [ASTRON] Twelve o'clock local mean

time, or the instant the mean sun is over the upper branch of the local meridian; local mean noon at the Greenwich meridian is called Greenwich mean noon. { 'lō·kəl ¦mēn ¦nün }

local mean time [ASTRON] The arc of the celestial equator, or the angle at the celestial pole, between the lower branch of the local celestial meridian and the hour circle of the mean sun, measured westward from the lower branch of the local celestial meridian through 24 hours. { 'lō·kəl ¦mēn ¦tīm }

local meridian [ASTRON] The meridian through any particular position which serves as the reference for local time. { 'lō·kəl mə'rid·ē·ən }

local minimum [MATH] A local minimum of a function f is a value $f(c)$ of f where $f(x) \geq f(c)$ for all x in some neighborhood of c; if $f(c)$ is a local minimum, f is said to have a local minimum at c. { 'lō·kəl 'min·ə·məm }

local networking [CONT SYS] The system of communication linking together the components of a single robot. { 'lō·kəl 'net,wərk·iŋ }

local noon [ASTRON] Noon at the local meridian. { 'lō·kəl 'nün }

local oscillator [ELECTR] The oscillator in a superheterodyne receiver, whose output is mixed with the incoming modulated radio-frequency carrier signal in the mixer to give the frequency conversions needed to produce the intermediate-frequency signal. { 'lō·kəl 'äs·ə,lād·ər }

local-oscillator injection [ELECTR] Adjustment used to vary the magnitude of the local oscillator signal that is coupled into the mixer. { 'lō·kəl 'äs·ə,lād·ər in¦jek·shən }

local-oscillator radiation [ELECTR] Radiation of the fundamental or harmonics of the local oscillator of a superheterodyne receiver. { 'lō·kəl 'äs·ə,lād·ər ,räd·ē'ā·shən }

local peat [GEOL] Peat formed by groundwater. Also known as basin peat. { 'lō·kəl ,pēt }

local preheating [MET] The heating of a specific portion of a material or structure prior to the performance of a joining or fabrication process. { 'lō·kəl prē'hēd·iŋ }

local procurement [ORD] **1.** Procurement of supplies or equipment in the continental United States by other than a centralized purchasing office, such as purchase by an installation of supplies and equipment for use of that installation. **2.** Procurement of supplies or equipment for its own use in an area outside the United States by a United States military command located in that area. { 'lō·kəl prə'kyùr·mənt }

local property [MATH] A property of an object (such as a space, function, curve, or surface) whose specification is based on the behavior of the object in the neighborhoods of certain points. { 'lō·kəl ,präp·ərd·ē }

local quasi-F martingale [MATH] A stochastic process $\{X_t\}$ such that the process obtained from $\{X_t\}$ by stopping it when it reaches n or $-n$ is a quasi-F martingale for each integer n. { 'lō·kəl 'kwä·zē¦ef 'mart·ən,gāl }

local register [COMPUT SCI] One of a relatively small number (usually less than 32) of high-speed storage elements in a computer system which may be directly referred to by the instructions in the programs. Also known as general register. { 'lō·kəl 'rej·ə·stər }

local relief [GEOL] The vertical difference in elevation between the highest and lowest points of a land surface within a specified horizontal distance or in a limited area. Also known as relative relief. { 'lō·kəl ri'lēf }

local ring [MATH] A ring with only one maximal ideal. { 'lō·kəl 'riŋ }

local second selector [COMMUN] A selector that interconnects a local first selector to a connector switch which is controlled and directed by impulses received from the local first selector. { 'lō·kəl ¦sek·ənd si¦lek·tər }

local side [COMMUN] Terminal connections to an internal or in-station source such as data terminal connections to input or output devices. { 'lō·kəl ,sīd }

local sidereal noon [ASTRON] Zero hour local sidereal time, or the instant the vernal equinox is over the upper branch of the local meridian; local sidereal noon at the Greenwich meridian is called Greenwich sidereal noon. { 'lō·kəl sə¦dir·ē·əl ¦nün }

local sidereal time [ASTRON] The arc of the celestial equator, or the angle at the celestial pole which is between the upper branch of the local celestial meridian and the hour circle of the vernal equinox. { 'lō·kəl sə¦dir·ē·əl ¦tīm }

local solution [MATH] A function which solves a system of equations only in a neighborhood of some point. { 'lō·kəl sə'lü·shən }

local standard of rest [ASTRON] A frame of reference in which the velocities of neighboring stars average out to zero. { 'lō·kəl 'stan·dərd əv 'rest }

local star cloud See local star system. { 'lō·kəl 'stär ,klaùd }

local star system [ASTRON] The group of stars of which the sun is a member. Also known as local cluster of stars; local star cloud. { 'lō·kəl 'stär ,sis·təm }

local storage [COMPUT SCI] The collection of local registers in a computer system. { 'lō·kəl 'stòr·ij }

local storm [METEOROL] A storm of mesometeorological scale; thus, thunderstorms, squalls, and tornadoes are often put in this category. { 'lō·kəl 'stòrm }

local structural discontinuity [MECH] The effect of intensified stress on a small portion of a structure. { 'lō·kəl 'strək·chə·rəl dis,känt·ən'ü·əd·ē }

Local Supercluster [ASTRON] A great flattened system of groups and clusters of galaxies, about 1.5 to 2×10^8 light-years (1.4 to 1.9×10^{24} meters) across, which includes the local group of galaxies and the Virgo Cluster. Also known as Virgo Supercluster. { 'lō·kəl 'sü·pər,kləs·tər }

local time [ASTRON] **1.** Time based upon the local meridian as reference, as contrasted with that based upon a zone meridian, or the meridian of Greenwich. **2.** Any time kept locally. { 'lō·kəl 'tīm }

local trunk [COMMUN] Trunk between local and long-distance switchboards, or between local and private branch exchange switchboards. { 'lō·kəl 'trəŋk }

local variable [COMPUT SCI] A variable which can be accessed (used or changed) only in one block of a computer program. { 'lō·kəl 'ver·ē·ə·bəl }

local vertical [NAV] A reference direction determined through the use of a plumb-bob or level. { 'lō·kəl 'vərd·ə·kəl }

local winds [METEOROL] Winds which, over a small area, differ from those which would be appropriate to the general pressure distribution, or which possess some other peculiarity. { 'lō·kəl 'winz }

locant [CHEM] The portion of a chemical name, usually a number or a letter, that designates the position of an atom or group of atoms in a formula unit. { 'lō,kant }

locate [MIN ENG] To mark out the boundaries of a mining claim and establish the right of possession. { 'lō,kāt }

located vector [MATH] An ordered pair of points in n-dimensional Euclidean space. { ¦lō,kād·əd 'vek·tər }

locate mode [COMPUT SCI] A method of communicating with an input/output control system (IOCS), in which the address of the data involved, but not the data themselves, is transferred between the IOCS routine and the program. { 'lō,kāt ,mōd }

locating [MECH ENG] A function of tooling operations accomplished by designing and constructing the tooling device so as to bring together the proper contact points or surfaces between the workpiece and the tooling. { 'lō,kād·iŋ }

locating hole [MECH ENG] A hole used to position the part in relation to a cutting tool or to other parts and gage points. { 'lō,kād·iŋ ,hōl }

locating surface [MECH ENG] A surface used to position an item being manufactured in a numerical control or robotic system for clamping. { 'lō,kād·iŋ ,sər·fəs }

location [COMPUT SCI] Any place in which data may be stored; usually expressed as a number. { lō'kā·shən }

location analysis [DES ENG] An initial step in the design of a robotic system consisting of a detailed study of all aspects of the placement of components such as work stations, buffers, and materials-handling equipment, as well as accessories, tools, and workpieces within a work station. { lō'kā·shən ə,nal·ə·səs }

location constant [COMPUT SCI] A number that identifies an instruction in a computer program, written in a higher-level programming language, and used to refer to this instruction at other points in the program. Also known as label constant. { lō'kā·shən ,kän·stənt }

location counter See instruction counter. { lō'kā·shən ,kaùnt·ər }

location damages [MIN ENG] Compensation by an operator to the surface owner for injury to the surface or to growing crops, resulting from the drilling of a well. { lō'kā·shən ˌdam·ij·əz }

location dimension [DES ENG] A dimension which specifies the position or distance relationship of one feature of an object with respect to another. { lō'kā·shən də'men·chən }

location fit [DES ENG] The characteristic wherein mechanical sizes of mating parts are such that, when assembled, the parts are accurately positioned in relation to each other. { lō'kā·shən ˌfit }

location notice [MIN ENG] A written sign placed prominently on a claim, showing the locator's name and describing the claim's extent and boundaries. { lō'kā·shən ˌnōd·əs }

location plan [MIN ENG] A scale map of the projected mine development indicating, among other things, proposed shafts and works in relation to existing surface features. { lō'kā·shən ˌplan }

location principle [MATH] A principle useful in locating the roots of an equation stating that if a continuous function has opposite signs for two values of the independent variable, then it is zero for some value of the variable between these two values. { lō'kā·shən ˌprin·sə·pəl }

location work [MIN ENG] Labor required by law to be done on mining claims within 60 days of location, in order to establish ownership. { lō'kā·shən ˌwərk }

locator [ENG] A radar or other device designed to detect and locate airborne aircraft. { 'lō,kād·ər }

locellus [BOT] **1.** In some legumes, a secondary compartment of a unilocular ovary that is formed by a false partition. **2.** One of the two cavities of a pollen sac. { lō'sel·əs }

lochia [MED] The discharge from the uterus and vagina during the first few weeks after labor. { 'lō·kē·ə }

lociation [ECOL] One of the subunits of a faciation, distinguished by the relative abundance of a dominant species. { lō·sē'ā·shən }

lock [CIV ENG] A chamber with gates on both ends connecting two sections of a canal or other waterway, to raise or lower the water level in each section. [DES ENG] A fastening device in which a releasable bolt is secured. [ELECTR] To fasten onto and automatically follow a target by means of a radar beam. [ENG] *See* air lock. [MET] A condition in forging in which the flash line is in more than one plane. [ORD] **1.** Position of a safety mechanism which prevents a weapon from being fired. **2.** Fastening device used to secure against accidental movement, as on a control surface. **3.** To secure or make safe, as to set the safety on a weapon. { läk }

lockalloy [MET] A beryllium-base alloy composed of 62% beryllium and 38% aluminum; used as a structural aerospace alloy because of low density and high (47,000 pounds per square inch or 3.2×10^8 pascals) yield strength. { 'läk·ə,lȯi }

lock bolt [ENG] **1.** The bolt of a lock. **2.** A bolt equipped with a locking collar instead of a nut. **3.** A bolt for adjusting and securing parts of a machine. { 'läk ,bōlt }

lock chamber [CIV ENG] A compartment between lock gates in a canal. { 'läk ,chām·bər }

locked-coil rope [DES ENG] A completely smooth wire rope that resists wear, made of specially formed wires arranged in concentric layers about a central wire core. Also known as locked-wire rope. { 'läkt ,kȯil ,rōp }

locked groove [DES ENG] A blank and continuous groove placed at the end of the modulated grooves on a disk recording to prevent further travel of the pickup. Also known as concentric groove. { 'läkt 'grüv }

locked-in line [COMMUN] A telephone line that remains established after the caller has hung up. { 'läkt,in ,līn }

locked oscillator [ELECTR] A sine-wave oscillator whose frequency can be locked by an external signal to the control frequency divided by an integer. { 'läkt 'äs·ə,lād·ər }

locked-oscillator detector [ELECTR] A frequency-modulation detector in which a local oscillator follows, or is locked to, the input frequency; the phase difference between local oscillator and input signal is proportional to the frequency deviation, and an output voltage is generated proportional to the phase difference. { 'läkt 'äs·ə,lād·ər di,tek·tər }

locked-rotor current [ELEC] The current drawn by a stalled electric motor. { 'läkt ,rōd·ər ,kə·rənt }

locked-wire rope *See* locked-coil rope. { 'läkt ,wīr ,rōp }

lock frame [ORD] A mechanical unit in certain firearms, used to assist in unlocking the bolt from the barrel after recoil has started. { 'läk ,frām }

lock front [DES ENG] On a door lock or latch, the plate through which the latching or locking bolt (or bolts) projects. { 'läk ,frənt }

lock gate [CIV ENG] A movable barrier separating the water in an upper or lower section of waterway from that in the lock chamber. { 'läk ,gāt }

lock-in [ELECTR] Shifting and automatic holding of one or both of the frequencies of two oscillating systems which are coupled together, so that the two frequencies have the ratio of two integral numbers. { 'läk ,in }

lock-in amplifier [ELECTR] An amplifier that uses some form of automatic synchronization with an external reference signal to detect and measure very weak electromagnetic radiation at radio or optical wavelengths in the presence of very high noise levels. { 'läk,in 'am·plə,fī·ər }

locking [ELECTR] Controlling the frequency of an oscillator by means of an applied signal of constant frequency. [ENG] Automatic following of a target by a radar antenna. { 'läk·iŋ }

locking fastener [DES ENG] A fastening used to prevent loosening of a threaded fastener in service, for example, a seating lock, spring stop nut, interference wedge, blind, or quick release. { 'läk·iŋ 'fas·nər }

locking lugs [ORD] Metal projections on the bolt of a small-arms weapon which cam into recesses cut in the side of the receiver to lock the weapon prior to firing. { 'läk·iŋ ,ləgz }

lockjaw *See* tetanus. { 'läk ,jȯ }

lock joint [DES ENG] A joint made by interlocking the joined elements, with or without other fastening. { 'läk ,jȯint }

locknut [DES ENG] **1.** A nut screwed down firmly against another or against a washer to prevent loosening. Also known as jam nut. **2.** A nut that is self-locking when tightened. **3.** A nut fitted to the end of a pipe to secure it and prevent leakage. { 'läk,nət }

lock-on [ELECTR] **1.** The procedure wherein a target-seeking system (such as some types of radars) is continuously and automatically following a target in one or more coordinates (for example, range, bearing, elevation). **2.** The instant at which radar begins to track a target automatically. { 'läk ,ȯn }

lockout [COMMUN] **1.** In a telephone circuit controlled by two voice-operated devices, the inability of one or both subscribers to get through, because of either excessive local circuit noise or continuous speech from one or both subscribers. Also known as receiver lockout system. **2.** In mobile communications, an arrangement of control circuits whereby only one receiver can feed the system at one time to avoid distortion. Also known as receiver lockout system. [COMPUT SCI] **1.** In computer communications, the inability of a remote terminal to achieve entry to a computer system until project programmer number, processing authority code, and password have been validated against computer-stored lists. **2.** The precautions taken to ensure that two or more programs executing simultaneously in a computer system do not access the same data at the same time, make unauthorized changes in shared data, or otherwise interfere with each other. **3.** Preventing the central processing unit of a computer from accessing storage because input/output operations are taking place. **4.** Preventing input and output operations from taking place simultaneously. { 'läk,aut }

lockout circuit [ELECTR] A switching circuit which responds to concurrent inputs from a number of external circuits by responding to one, and only one, of these circuits at any time. Also known as finding circuit; hunting circuit. { 'läk,aut ,sər·kət }

lock primer [ORD] Primer intended to be placed by hand in the firing lock of a gun; it is used for initiation of the ignition of bag charges. { 'läk ,prīm·ər }

lock rail [BUILD] An intermediate horizontal structural member of a door, between the vertical stiles, at the height of the lock. { 'läk ,rāl }

lockset [ENG] **1.** A complete lock including the lock mechanism, keys, plates, and other parts. **2.** A jig or template for making cuts in a door for holding a lock. { 'läk,set }

lock up [GRAPHICS] To position a form in the frame in which type and plates are locked up for letterpress printing. { 'läk ,əp }

lock-up relay [ELEC] A relay that locks in its energized position either by permanent magnetic biasing which can be released only by applying a reverse magnetic pulse or by auxiliary contacts that keep its coil energized until the circuit is interrupted. { 'läk,əp rē,lā }

lock washer [DES ENG] A solid or split washer placed underneath a nut or screw that prevents loosening by exerting pressure. { 'läk ,wäsh·ər }

loco disease [VET MED] Poisoning in livestock resulting from ingestion of selenium-containing plants (loco weed); characterized by atrophy, delirium, convulsions, and stupor, often terminating in death. { 'lō·kō di,zēz }

locomotion [SCI TECH] Progressive movement, as of an animal or a vehicle. { ,lō·kə'mō·shən }

locomotive [MECH ENG] A self-propelling machine with flanged wheels, for moving loads on railroad tracks; utilizes fuel (for steam or internal combustion engines), compressed air, or electric energy. { ,lō·kə'mōd·iv }

locomotive boiler [MECH ENG] An internally fixed horizontal fire-tube boiler with integral furnace; the doubled furnace walls contain water which mixes with water in the boiler shell. { ,lō·kə'mōd·iv 'bȯil·ər }

locomotive crane [MECH ENG] A crane mounted on a railroad flatcar or a special chassis with flanged wheels. Also known as rail crane. { ,lō·kə'mōd·iv 'krān }

locomotive gradient [MIN ENG] The gradient set by law for a locomotive haulage; maximum is 1 in 15, but the limit for practical purposes is 1 in 25. { ,lō·kə'mōd·iv 'grād·ē·ənt }

locomotive haulage [MIN ENG] The use of locomotive-hauled mine cars to carry coal ore, workers, and materials in a mine. { ,lō·kə'mōd·iv 'hȯl·ij }

locomotor ataxia *See* tabes dorsalis. { ,lō·kə'mōd·ər ə'tak·sē·ə }

locomotor system [ZOO] Appendages and associated parts, such as muscles, joints, and bones, concerned with motor activities and locomotion of the animal body. { ,lō·kə'mōd·ər ,sis·təm }

loco weed [BOT] Any species of *Astragalus* containing selenium taken up from the soil. { 'lō·kō ,wēd }

loculate [BIOL] Having, or divided into, loculi. { 'läk·yə·lət }

locule [BOT] A small chamber in plant tissue within which specialized structures may develop, such as within an ovary or anther. { 'lä,kyül }

loculicidal [BOT] Pertaining to dehiscence that extends along the dorsal midline of a carpel. { 'läk·yə,lis·əd·əl }

Loculoascomycetes [MYCOL] A class in the subdivision Ascomycotina composed of organisms that form a well-developed mycelium which bears the sexual (ascus) and asexual (conidium) states, and that are distinguished from other ascomycetes by their method of ascocarp formation and their ascus structure. { ,lō·kə·lō,as·kō·mī'sēd,ēz }

loculus [BIOL] A small cavity or chamber. { 'läk·yə·ləs }

locus [GEN] The fixed position of a gene in a chromosome, occupied by an allele. [MATH] A collection of points in a Euclidean space whose coordinates satisfy one or more algebraic conditions. { 'lō·kəs }

locus control region [GEN] A segment of DNA that controls the chromatin structure and thus the potential for replication and transcription of an entire gene cluster, such as the beta-globin cluster in vertebrates. { 'lō·kəs kən'trōl ,rē·jən }

locust [BOT] Either of two species of commercially important trees, black locust (*Robinia pseudoacacia*) and honey locust (*Gladitsia triacanthos*), in the family Leguminosae. [INV ZOO] The common name for various migratory grasshoppers of the family Locustidae. { 'lō·kəst }

Locustidae [INV ZOO] A family of insects in the order Orthoptera; antennae are usually less than half the body length, hindlegs are adapted for jumping, and the ovipositor is multipartite. { lō'kəs·tə,dē }

LOD score *See* logarithm of the odds score. { ,el,ō'dē ,skȯr }

lodar [NAV] A direction finder used to determine the direction of arrival of loran signals, free of night effect, by observing the separately distinguishable ground and sky-wave loran signals on a cathode-ray oscilloscope and positioning a loop antenna to obtain a null indication of the component selected to be most suitable. Also known as lorad. { 'lō,där }

lode [GEOL] A fissure in consolidated rock filled with mineral; usually applied to metalliferous deposits. { lōd }

lode claim [MIN ENG] That portion of a vein or lode, and

of the adjoining surface, which has been acquired by a compliance with the law, both Federal and state. { 'lōd ,klām }

loden cloth [TEXT] A fleecy fabric woven from coarse wool in the Tyrolese area of Austria and Germany; it is naturally water-repellent. { 'lōd·ən ,klȯth }

lodestone [MINERAL] The naturally occurring magnetic iron oxide, or magnetite, possessing polarity, and attracting iron objects to itself. Also known as Hercules stone; leading stone; loadstone. { 'lōd,stōn }

lodicule [BOT] One of the minute membranous bodies found at the base of the carpel in most flowering grasses; usually occurs in pairs. { 'läd·ə,kyül }

lodos [METEOROL] A southerly wind on the Black Sea coast of Bulgaria. { 'lȯ·dōs }

lodranite [GEOL] A stony iron meteorite composed of bronzite and olivine within a fine network of nickel-iron. { 'lō·drə,nīt }

Loeffler's syndrome [MED] Extensive infiltration of the lung by eosinophils, and eosinophilia of the peripheral circulation. Also known as eosinophilic pneumonitis. { 'lef·lərz ,sin,drōm }

loellingite [MINERAL] FeAs₂ A silver-white to steel-gray mineral composed of iron arsenide with some cobalt, nickel, antimony, and sulfur; isomorphous with arsenopyrite; a source of arsenic. Also known as leucopyrite; lauollingite. { 'lel·iŋ,īt }

loess [GEOL] An essentially unconsolidated, unstratified calcareous silt; commonly it is homogeneous, permeable, and buff to gray in color, and contains calcareous concretions and fossils. { les }

loess kindchen [GEOL] An irregular or spheroidal nodule of calcium carbonate that is found in loess. { 'les ,kint·chən }

loeweite [MINERAL] Na₄Mg₂(SO₄)₄·5H₂O A white to pale-yellow mineral composed of hydrous sulfate of sodium and magnesium. { 'lā·və,īt }

lofar [NAV] A submarine detection system using autocorrelation techniques for long-range analysis of patterned sound picked up at the low-frequency end of the sound spectrum by underwater hydrophones of the Caesar submarine detection system. Derived from low-frequency acquisition and ranging. { 'lō,fär }

loft [BUILD] **1.** An upper part of a building. **2.** A work area in a factory or warehouse. [TEXT] **1.** The quality of resilience possessed by wool that permits it to return to its original shape after deformation. **2.** The degree of bulkiness of manufactured fibers and blends. { lȯft }

loft bombing [ORD] A method of aerial bombing in which the delivery plane approaches the target at a very low altitude, makes a definite pull-up at a given point, releases bomb at predetermined point during the pull-up, and tosses the missile on the target. { 'lȯft ,bäm·iŋ }

loft building [BUILD] A building with a large open floor area. { 'lȯft ,bild·iŋ }

log [COMMUN] A written record of radio and television station operating data, required by law. [COMPUT SCI] A record of computer operating runs, including tapes used, control settings, halts, and other pertinent data. [ENG] The record of, or the act or process of recording, events or the type and characteristics of the rock penetrated in drilling a borehole as evidenced by the cuttings, core recovered, or information obtained from electronic devices. [MATER] Unshaped timber either rough or squared. [NAV] **1.** An instrument for measuring the speed or distance or both traveled by a vessel. **2.** A written record of the movements of a craft, with regard to courses, speeds, positions, and other information of interest to navigators, and of important happenings aboard the craft. **3.** A written record of specific related information, such as that concerning performance of an instrument. { läg }

Loganiaceae [BOT] A family of mostly woody dicotyledonous plants in the order Gentianales; members lack a latex system and have fully united carpels and axile placentation. { lō,gān·ē'ā·sē,ē }

Logan slabbing machine [MIN ENG] A machine that has three cutting chains; two are horizontal—one at the base of the coal seam, the other at a distance from the floor; the third is mounted vertically and shears off the coal at the back of the cut; a short conveyor transfers the coal to the face conveyor. { 'lō·gən 'slab·iŋ mə,shēn }

logarithm [MATH] **1.** The real-valued function log *u* defined

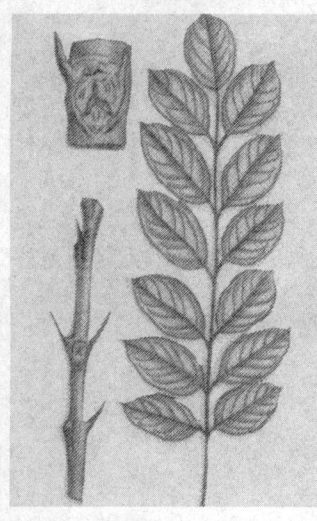
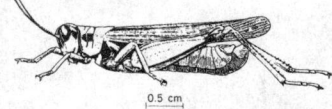

by log $u = v$ if $e^v = u$, e^v denoting the exponential function. Also known as hyperbolic logarithm; Naperian logarithm; natural logarithm. **2.** An analog in complex variables relative to the function e^z.   { 'läg·ə,ri**th**·əm }

logarithmically convex function [MATH] A function whose logarithm is a convex function.   { ,läg·ə,ri**th**·mik·lē ¦kän¦veks ,fəŋk·shən }

logarithmic amplifier [ELECTR] An amplifier whose output signal is a logarithmic function of the input signal.   { 'läg·ə,ri**th**·mik 'am·plə,fī·ər }

logarithmic coordinate paper [MATH] Paper ruled with two sets of mutually perpendicular, parallel lines spaced according to the logarithms of consecutive numbers, rather than the numbers themselves.   { 'läg·ə,ri**th**·mik kō'órd·ən·ət ,pā·pər }

logarithmic coordinates [MATH] In the plane, logarithmic coordinates are defined by two coordinate axes, each marked with a scale where the distance between two points is the difference of the logarithms of the two numbers.   { 'läg·ə,ri**th**·mik kō'órd·ən·əts }

logarithmic curve [MATH] A curve whose equation in cartesian coordinates is $y = \log ax$, where a is greater than 1.   { 'läg·ə,ri**th**·mik 'kərv }

logarithmic decrement [PHYS] The natural logarithm of the ratio of the amplitude of one oscillation to that of the next which has the same polarity, when no external forces are applied to maintain the oscillation.   { 'läg·ə,ri**th**·mik 'dek·rə·mənt }

logarithmic derivative [MATH] The logarithmic derivative of a function $f(z)$ of a real (complex) variable is the ratio $f'(z)/f(z)$, that is, the derivative of $\log f(z)$.   { 'läg·ə,ri**th**·mik də'riv·əd·iv }

logarithmic differentiation [MATH] A technique often helpful in computing the derivatives of a differentiable function $f(x)$; set $g(x) = \log f(x)$ where $f(x) \neq 0$, then $g'(x) = f'(x)/f(x)$, and if there is some other way to find $g'(x)$, then one also finds $f'(x)$.   { 'läg·ə,ri**th**·mik ,dif·ə,ren·chē'ā·shən }

logarithmic diode [ELECTR] A diode that has an accurate semilogarithmic relationship between current and voltage over wide and forward dynamic ranges.   { 'läg·ə,ri**th**·mik 'dī,ōd }

logarithmic distribution [STAT] A frequency distribution whose value at any integer $n = 1, 2, \ldots$ is $\lambda^n/(-n)\log(1-\lambda)$, where λ is fixed.   { 'läg·ə,ri**th**·mik ,dis·trə'byü·shən }

logarithmic equation [MATH] An equation which involves a logarithmic function of some variable.   { 'läg·ə,ri**th**·mik i'kwā·zhən }

logarithmic fast time constant [ELECTR] Constant false alarm rate scheme which has a logarithmic intermediate-frequency amplifier followed by a fast time constant circuit.   { 'läg·ə,ri**th**·mik 'fast ¦tīm ,kän·stənt }

logarithmic growth *See* exponential growth.   { 'läg·ə,ri**th**·mik 'grōth }

logarithmic multiplier [ELECTR] A multiplier in which each variable is applied to a logarithmic function generator, and the outputs are added together and applied to an exponential function generator, to obtain an output proportional to the product of two inputs.   { 'läg·ə,ri**th**·mik 'məl·tə,plī·ər }

logarithmic potential [PHYS] A potential function that is proportional to the logarithm of some coordinate; for example, a straight, electrically charged cylinder of circular cross section and effectively infinite length gives rise to an electrostatic potential that is the sum of a constant and a term proportional to the logarithm of the distance from the cylinder's axis.   { 'läg·ə,ri**th**·mik pə'ten·chəl }

logarithmic profile of velocity [FL MECH] The mean velocity parallel to a boundary of a fluid in turbulent motion as a function of distance from the boundary, on the assumption that the shearing stress is independent of distance from the boundary, and the mixing length is proportional either to the distance from the boundary or to the ratio of the first derivative of the profile of velocity itself to the second derivative.   { 'läg·ə,ri**th**·mik 'prō,fīl əv və'läs·əd·ē }

logarithmic scale [MATH] A scale in which the distances that numbers are at from a reference point are proportional to their logarithms.   { 'läg·ə,ri**th**·mik 'skāl }

logarithmic series [MATH] The expansion of the natural logarithm of $1 + x$ in a Maclaurin series; namely, $x - x^2/2 + x^3/3 - \cdots$.   { 'läg·ə,ri**th**·mik 'sir·ēz }

logarithmic spiral [MATH] The spiral plane curve whose

points in polar coordinates (r,θ) satisfy the equation $\log r = a\theta$. Also known as equiangular spiral.   { 'läg·ə,ri**th**·mik 'spī·rəl }

logarithmic transformation [STAT] The replacement of a variate y with a new variate $z = \log y$ or $z = \log (y + c)$, where c is a constant; this operation is often performed when the resulting distribution is normal, or if the resulting relationship with another variable is linear.   { 'läg·ə,ri**th**·mik ,tranz·fər'mā·shən }

logarithmic trigonometric function [MATH] The logarithm of the corresponding trigonometric function.   { 'läg·ə,ri**th**·mik 'trig·ə·nə,me·trik ,fəŋk·shən }

logarithmic velocity profile [METEOROL] The theoretical variation of the mean wind speed with height in the surface boundary layer under certain assumptions.   { 'läg·ə,ri**th**·mik və'läs·əd·ē ,prō,fīl }

logarithm of the odds score [GEN] A measure of the likelihood that genes are linked, expressed as the logarithm of the odds that an observed data set from the families is due to linkage at a specific map distance rather than to independent assortment on nonlinked genes. Abbreviated LOD score.   { ¦läg·ə,rith·əm əv thē 'ädz ,skór }

logbook [COMPUT SCI] A bound volume in which operating data of a computer is noted.   [NAV] A book in which all affairs and events of navigational importance of a ship are recorded, such as speed and ship's progress.   { 'läg,bùk }

loggia [ARCH] A roofed open arcade on the side of a building.   { 'lō·jē·ə }

logging [ENG] Continuous recording versus depth of some characteristic datum of the formations penetrated by a drill hole; for example, resistivity, spontaneous potential, conductivity, fluid content, radioactivity, or density.   [FOR] The cutting and removal of the woody stem portions of forest trees.   { 'läg·iŋ }

logic [ELECTR] **1.** The basic principles and applications of truth tables, interconnections of on/off circuit elements, and other factors involved in mathematical computation in a computer.   **2.** General term for the various types of gates, flip-flops, and other on/off circuits used to perform problem-solving functions in a digital computer.   [MATH] The subject that investigates, formulates, and establishes principles of valid reasoning.   { 'läj·ik }

logical addition [MATH] The additive binary operation of a Boolean algebra.   { 'läj·ə·kəl ə'dish·ən }

logical comparison [COMPUT SCI] The operation of comparing two items in a computer and producing a one output if they are equal or alike, and a zero output if not alike.   { 'läj·ə·kəl kəm'par·ə·sən }

logical connectives [MATH] Symbols which link mathematical statements; these symbols represent the terms "and," "or," "implication," and "negation."   { 'läj·ə·kəl kə'nek·tivz }

logical construction [COMPUT SCI] A simple logical property that determines the type of characters which a particular code represents; for example, the first two bits can tell whether a character is numeric or alphabetic.   { 'läj·ə·kəl kən'strək·shən }

logical data independence [COMPUT SCI] A data base structured so that changing the logical structure will not affect its accessibility by the program reading it.   { 'läj·ə·kəl ¦dad·ə ,in·də'pen·dəns }

logical data type [COMPUT SCI] A scalar data type in which a data item can have only one of two values: true or false. Also known as Boolean data type.   { 'läj·ə·kəl 'dad·ə ,tīp }

logical decision [COMPUT SCI] The ability to select one of many paths, depending upon intermediate programming data.   { 'läj·ə·kəl di'sizh·ən }

logical device table [COMPUT SCI] A table that is used to keep track of information pertaining to an input/output operation on a logical unit, and that contains such information as the symbolic name of the logical unit, the logical device type and the name of the file currently attached to it, the logical input/output request currently pending on the device, and a pointer to the buffers currently associated with the device.   { 'läj·ə·kəl di¦vīs ,tā·bəl }

logical drive [COMPUT SCI] A data storage unit, such as a subpartition of a hard drive or an array of storage units, recognized and handled according to the logic of the operating system like a single physical drive.   { 'läj·i·kəl ¦drīv }

logical expression [COMPUT SCI] Two arithmetic expressions connected by a relational operator indicating whether an expression is greater than, equal to, or less than the other, or connected by a logical variable, logical constant (true or false), or logical operator. { 'läj·ə·kəl ik'spresh·ən }

logical field [COMPUT SCI] A data field whose variables can take on only two values, which are designated yes and no, true and false, or 0 and 1. { 'läj·ə·kəl 'fēld }

logical file [COMPUT SCI] A file as seen by the program accessing it. { 'läj·ə·kəl 'fīl }

logical flow chart [COMPUT SCI] A detailed graphic solution in terms of the logical operations required to solve a problem. { 'läj·ə·kəl 'flō ,chärt }

logical function See propositional function.. { 'läj·ə·kəl 'faŋk·shən }

logical gate See switching gate. { 'läj·ə·kəl 'gāt }

logical instruction [COMPUT SCI] A digital computer instruction which forms a logical combination (on a bit-by-bit basis) of its operands and leaves the result in a known location. { 'läj·ə·kəl in'strək·shən }

logically equivalent statements [MATH] Two statements that are equivalent because of their logical form rather than their mathematical content. { ‚läj·i·klē i‚kwiv·ə·lənt 'stāt·məns }

logical multiplication [MATH] The multiplicative binary operation of a Boolean algebra. { 'läj·ə·kəl məl·tə·plə'kā·shən }

logical network [COMPUT SCI] **1.** A collection of computers that is presented as a single network to the user, although it may encompass more than one physical network. **2.** A part of a network of computers that is set up to function as a separate network. { 'läj·i·kəl ‚net‚wərk }

logical page [COMPUT SCI] A unit of computer storage consisting of a specified number of bytes. { 'läj·ə·kəl 'pāj }

logical record [COMPUT SCI] A group of adjacent, logically related data items. { 'läj·ə·kəl 'rek·ord }

logical security [COMPUT SCI] Mechanisms internal to a computing system that are used to protect against internal misuse of computing time and unauthorized access to data. { 'läj·ə·kəl sə'kyùr·əd·ē }

logical shift [COMPUT SCI] A shift operation that treats the operand as a set of bits, not as a signed numeric value or character representation. { 'läj·ə·kəl 'shift }

logical sum [COMPUT SCI] A computer addition in which the result is 1 when either one or both input variables is 1, and the result is 0 when the input variables are both 0. { 'läj·ə·kəl 'səm }

logical symbol [COMPUT SCI] A graphical symbol used to represent a logic element. { 'läj·ə·kəl 'sim·bəl }

logical unit [COMPUT SCI] An abstraction of an input/output device in the form of an additional name given to the device in a computer program. { 'läj·ə·kəl 'yü·nət }

logic-arithmetic unit See arithmetical unit. { 'läj·ik·ə'rith·mə·tik ‚yü·nət }

logic bomb [COMPUT SCI] A computer program that destroys data, generally immediately after it has been loaded. { 'läj·ik ‚bäm }

logic card [ELECTR] A small fiber chassis on which resistors, capacitors, transistors, magnetic cores, and diodes are mounted and interconnected in such a way as to perform some computer function; computers employing this type of construction may be repaired by removing the faulty card and replacing it with a new card. { 'läj·ik ‚kärd }

logic chip [COMPUT SCI] An integrated circuit that performs logic functions. { 'läj·ik ‚chip }

logic circuit [COMPUT SCI] A computer circuit that provides the action of a logic function or logic operation. Also known as logic gate. { 'läj·ik ‚sər·kət }

logic design [COMPUT SCI] The design of a computer at the level which considers the operation of each functional block and the relationships between the functional blocks. { 'läj·ik di‚zīn }

logic diagram [COMPUT SCI] A graphical representation of the logic design or a portion thereof; displays the existence of functional elements and the paths by which they interact with one another. { 'läj·ik ‚dī·ə‚gram }

logic element [COMPUT SCI] A hardware circuit that performs a simple, predefined transformation on its input and presents the resulting signal as its output. Occasionally known as functor. { 'läj·ik ‚el·ə·mənt }

logic error [COMPUT SCI] An error in programming that is caused by faulty reasoning, resulting in the program's functioning incorrectly if the instructions containing the error are encountered. { 'läj·ik ‚er·ər }

logic gate See logic circuit. { 'läj·ik ‚gāt }

logic high [ELECTR] The electronic representation of the binary digit 1 in a digital circuit or device. { 'läj·ik 'hī }

logic level [ELECTR] One of the two voltages whose values have been arbitrarily chosen to represent the binary numbers 1 and 0 in a particular data-processing system. { 'läj·ik ‚lev·əl }

logic low [ELECTR] The electronic representation of the binary digit 0 in a digital circuit or device. { 'läj·ik ‚lō }

logic operation [COMPUT SCI] A nonarithmetical operation in a computer, such as comparing, selecting, making references, matching, sorting, and merging, where logical yes-or-no quantities are involved. { 'läj·ik ‚äp·ə'rā·shən }

logic operator [COMPUT SCI] A rule which assigns, to every combination of the values "true" and "false" among one or more independent variables, the value "true" or "false" to a dependent variable. { 'läj·ik ‚äp·ə‚rād·ər }

logic section See arithmetical unit. { 'läj·ik ‚sek·shən }

logic-seeking printer [COMPUT SCI] A line printer that examines each line to be printed so that it can save time by skipping over blank spaces. { 'läj·ik ¦sēk·iŋ 'print·ər }

logic swing [ELECTR] The voltage difference between the logic levels used for 1 and 0; magnitude is chosen arbitrarily for a particular system and is usually well under 10 volts. { 'läj·ik ‚swiŋ }

logic switch [ELECTR] A diode matrix or other switching arrangement that is capable of directing an input signal to one of several outputs. { 'läj·ik ‚swich }

logic unit [COMPUT SCI] A separate unit which exists in some computer systems to carry out logic (as opposed to arithmetic) operations. { 'läj·ik ‚yü·nət }

logic word [COMPUT SCI] A machine word which represents an arbitrary set of digitally encoded symbols. { 'läj·ik ‚wərd }

logistic curve [STAT] **1.** A type of growth curve, representing the size of a population y as a function of time t: $y = k/(1 + e^{-kbt})$, where k and b are positive constants. Also known as Pearl-Reed curve. **2.** More generally, a curve representing a function of the form $y = k/(1 + e^{cf(t)})$, where c is a constant and $f(t)$ is some function of time. { lə'jis·tik 'kərv }

logistic growth [BIOL] Population growth in which the growth rate decreases with increasing number of individuals until it becomes zero when the population reaches a maximum. { lə'jis·tik 'grōth }

log line See current line. { 'läg ‚līn }

log-mean temperature difference [THERMO] The log-mean temperature difference $T_{LM} = (T_2 - T_1)/\ln T_2/T_1$, where T_2 and T_1 are the absolute (K or °R) temperatures of the two extremes being averaged; used in heat transfer calculations in which one fluid is cooled or heated by a second held separate by pipes or process vessel walls. { 'läg ¦mēn 'tem·prə·chər ‚dif·rəns }

lognormal distribution [STAT] A probability distribution in which the logarithm of the parameter has a normal distribution. { 'läg‚nor·məl ‚di·strə'byü·shən }

logo [COMPUT SCI] A high-level, interactive programming language that features a triangular shape called a turtle which can be moved about an electronic display through the use of familiar English-word commands. [GRAPHICS] See logotype. { 'lō·gō }

log-off [COMPUT SCI] The procedure for a user to disconnect from a computer system, including the release of resources that were assigned to the user. { 'läg ‚of }

logomania [MED] Logorrhea so excessive as to be a form of a manic state; new words may be invented to keep up the garrulity. { ¦lō·gə'mā·nē·ə }

log-on [COMPUT SCI] The procedure for users to identify themselves to a computer system for authorized access to their programs and information. { 'läg‚on }

logoplegia [MED] Loss of ability to articulate, usually due to paralysis of the speech organs. { ‚lō·gə'plē·jē·ə }

logorrhea [MED] Excessive, usually rapid, incoherent, and uncontrollable talkativeness. { ‚lō·gə'rē·ə }

logotype [GRAPHICS] A single slug or line of type cast in one piece, that carries one or more words such as the name of a firm or a product; a special style of type may be used, and the logotype may include a trademark or other art work with the type. Also known as logo. [SYST] The selection or designation of a genotype after the generic name was published. { 'lō·gə‚tīp }

log-periodic antenna [ELECTROMAG] A broad-band antenna which consists of a sheet of metal with two wedge-shaped cutouts, each with teeth cut into its radii along circular arcs; characteristics are repeated at a number of frequencies that are equally spaced on a logarithmic scale. { ‚läg ‚pir·ē‚ad·ik an'ten·ə }

log rule [FOR] A table showing expected log output in board feet or other units. { läg ‚rül }

log volume [FOR] The cubic volume of a log computed inside the bark as determined by any of several formulas; parameters are the cross-sectional areas of log midpoint, large end and small end of log, and log length. { läg ‚väl·yəm }

logwood See hematoxylon. { läg‚wud }

LOH See loss of heterozygosity.

LOI See limiting oxygen index.

loiasis [MED] A filariasis of tropical Africa, caused by the filaria *Loa*, and characterized by diurnal periodicity of microfilariae in the blood, and transient cutaneous swelling caused by migrating adult worms. { ‚lòi'ā·səs }

loktal base [ELECTR] A special base for small vacuum tubes, so designed that it locks the tube firmly in a corresponding special eight-pin loktal socket; the tube pins are sealed directly into the glass envelope. { 'läk·təl ‚bās }

löllingite See loellingite. { 'lel·iŋ‚īt }

lolly ice [OCEANOGR] Saltwater frazil, a heavy concentration of which is called sludge. { 'läl·ē ‚īs }

lombarde [METEOROL] An easterly wind (from Lombardy) that predominates along the French-Italian frontier, and comes from the High Alps; in winter it is violent and forms snow drifts in the mountain valleys; in the plains it is gentle and very dry. { lòm'bär·də }

Lombard effect [ACOUS] The change in a talker's articulation effort when he or she speaks in a noisy environment; for example, trying to raise the voice or to make the voice better understood by the listener. { 'lum‚bärd i‚fekt }

loment [BOT] A dry, indehiscent single-celled fruit that is formed from a single superior ovary; splits transversely in numerous segments at maturity. { 'lō‚ment }

lomonite See laumontite. { 'lō'mä‚nīt }

Lomonosov ridge [GEOGR] An undersea ridge which subdivides the Arctic Basin, extending from Ellesmere Land to the New Siberian Islands. { lō·mō'nò‚sóf ‚rij }

lomontite See laumontite. { lō'män‚tīt }

Lonchaeidae [INV ZOO] A family of minute myodarian cyclorrhaphous dipteran insects in the subsection Acalyptreatae. { län'kē·ə‚dē }

London equations [SOLID STATE] Equations for the time derivative and the curl of the current in a superconductor in terms of the electric and magnetic field vectors respectively, derived in the London superconductivity theory. { 'lən·dən i‚kwā·zhənz }

London force See dispersion force; van der Waals force. { 'lun·dən ‚fórs }

London penetration depth [SOLID STATE] A measure of the depth which electric and magnetic fields can penetrate beneath the surface of a superconductor from which they are otherwise excluded, according to the London superconductivity theory. { 'lən·dən ‚pen·ə'trā·shən ‚depth }

London shrinkage [TEXT] Method of shrinking by a cold-water bath, followed by drying and pressing. { 'lən·dən 'shriŋ·kij }

London superconductivity theory [SOLID STATE] An extension of the two-fluid model of superconductivity, in which it is assumed that superfluid electrons behave as if the only force acting on them arises from applied electric fields, and that the curl of the superfluid current vanishes in the absence of a magnetic field. { 'lən·dən ‚sü·pər‚kän‚däk'tiv·əd·ē ‚thē·ə·rē }

London superfluidity theory [CRYO] A theory, based on the fact that helium-4 obeys Bose-Einstein statistics, in which helium-4 is treated as an ideal Bose-Einstein gas, and its superfluid component is equated with the finite fraction of the atoms of such a gas which are in the ground state at very low temperatures. { 'lən·dən ‚sü·pər‚flü'id·əd·ē ‚thē·ə·rē }

lone-pair electrons [PHYS CHEM] A nonbonding pair of electrons in the valence shell of an atom. { ‚lōn ‚per i'lek‚tränz }

long-base-line system [COMMUN] System in which the distance separating ground stations approximates the distance to the target being tracked. { ‚lòŋ 'bās ‚līn ‚sis·təm }

long bone [ANAT] A bone in which the length markedly exceeds the width, as the femur or the humerus. { ‚lòŋ 'bōn }

long card [COMPUT SCI] A full-size printed circuit board that is plugged into an expansion slot in a microcomputer. { 'lòŋ ‚kärd }

long clay [MATER] A clay used in ceramics that has a high degree of plasticity. { 'lòŋ ‚klā }

long column [CIV ENG] A column so slender that bending is the primary deformation, generally having a slenderness ratio greater than 120–150. { 'lòŋ ‚käl·əm }

long-conductor antenna See long-wire antenna. { 'lòŋ kən‚däk·tər an‚ten·ə }

long-day plant [BOT] A plant that flowers in response to a long photoperiod. { 'lòŋ ‚dā ‚plant }

long-day response [PHYSIO] A photoperiodic response that is evoked by increasing day lengths and decreasing night lengths. { 'lòŋ ‚dā ri'späns }

long discharge [ELEC] **1.** A capacitor or other electrical charge accumulator which takes a long time to leak off. **2.** A gaseous electrical discharge in which the length of the discharge channel is very long compared with its diameter; lightning discharges are natural examples of long discharges. Also known as long spark. { 'lòŋ 'dis‚chärj }

long-distance loop [COMMUN] Line from a subscriber's station directly to a long-distance switchboard. { 'lòŋ ‚dis·təns 'lüp }

long-distance navigation [NAV] Navigation performed in an area where aids to navigation are spaced more than 200 miles (322 kilometers) apart, exclusive of the short-range, the approach and landing zone, and the airport zone. { 'lòŋ ‚dis·təns ‚nav·ə'gā·shən }

long-distance xerography [COMMUN] A facsimile system that uses a cathode-ray scanner at the microwave transmitting terminal; at the receiving terminal, a lens projects the received cathode-ray image onto the selenium-coated drum of a xerographic copying machine. { 'lòŋ ‚dis·təns zi'räg·rə·fē }

long division [MATH] **1.** Division of numbers in which the divisor contains more than one digit. **2.** Division of algebraic quantities in which the divisor contains more than one term. { ‚lòŋ di'vizh·ən }

longeron [AERO ENG] A principal longitudinal member of the structural framework of a fuselage, nacelle, or empennage boom. { 'län·jə‚rän }

long-haul carrier system [COMMUN] An intercity telephone communication system in which a frequency-division multiplexed signal modulates a subcarrier. { 'lòŋ ‚hól 'kar·ē·ər ‚sis·təm }

long-haul radio [COMMUN] A microwave radio system for transmitting telephone, video, data, and telegraph signals over distances on the order of 4000 miles (6500 kilometers) or more on line-of-sight paths between a series of repeaters which demodulate the signal to intermediate frequency and then remodulate. { 'lòŋ ‚hól 'rād·ē·ō }

long hole [MIN ENG] An underground borehole and blasthole exceeding 10 feet (3 meters) in depth or requiring the use of two or more lengths of drill steel or rods coupled together to attain the desired depth. { 'lòŋ ‚hól }

long-hole drill [MIN ENG] A rotary- or a percussive-type drill used to drill long holes. { 'lòŋ ‚hól ‚dril }

long-hole jetting [MIN ENG] A hydraulic mining system consisting essentially of drilling a hole down the pitch of the vein, replacing the drilling head with a jet cutting head, and then retracting the drill column with the jets in operation to remove the coal. { 'lòŋ ‚hól ‚jed·iŋ }

long hundredweight See hundredweight. { 'lòŋ 'hən·drəd‚wāt }

Longhurst-Hardy plankton sampler [ENG] A nonquantitative metal-shrouded net for trapping plankton. { 'lòŋ‚hərst 'här·dē 'plaŋk·tən ‚sam·plər }

longicollous [BIOL] Having a long beak or neck. { län·jə'käl·əs }

Longidorinae [INV ZOO] A subfamily of nematodes belonging to the Dorylaimoidea including economically important plant parasites. { ‚län·jə'dór·ə‚nē }

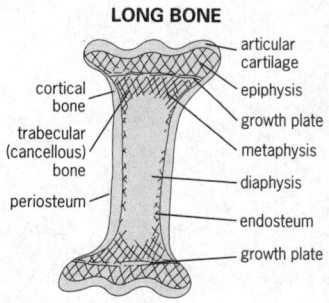

LONG BONE

articular cartilage
epiphysis
growth plate
metaphysis
diaphysis
endosteum
growth plate

cortical bone
trabecular (cancellous) bone
periosteum

Diagram of a long bone.

long ink [GRAPHICS] Printing ink that has the characteristic of good flow on the ink rollers of a printing press. { 'lȯŋ 'iŋk }

long interspersed elements [GEN] Families of deoxyribonucleic acid sequences, many truncated, that are inserted singly but in large numbers throughout the genome in many mammals, constituting about 15% of the human genome; some full-length copies are transposons. Abbreviated LINES. { ¦lȯŋ ¦in·tər,spərst 'el·ə·məns }

long interspersed nucleotide element [CELL MOL] In mammalian deoxyribonucleic acid, any of the 5–10-kilobase repeated sequences that are grouped with the nonviral retroposons. Abbreviated LINE. { ¦lȯŋ ¦in·tər,spərst 'nü·klē·ə,tīd ,el·ə·mənt }

Longipennes [VERT ZOO] An equivalent name for Charadriiformes. { ¦län·jə'pen·ēz }

longitude [GEOD] Angular distance, along the Equator, between the meridian passing through a position and, usually, the meridian of Greenwich. { 'län·jə,tüd }

longitude factor [NAV] The change in longitude along a celestial line of position per 1-minute change in latitude. { 'län·jə,tüd ,fak·tər }

longitude line [NAV] A line of position extending in a north-south direction generally. { 'län·jə,tüd ,līn }

longitude method [NAV] In celestial navigation, the establishing of a line of position from the observation of the altitude of a celestial body by assuming a latitude (or longitude), and calculating the longitude (or latitude) through which the line of position passes, and the azimuth; the line of position is drawn through the point thus found, perpendicular to the azimuth. { 'län·jə,tüd ,meth·əd }

longitude signal [NAV] A radio or telegraphic time signal transmitted by a station of known longitude and received at stations whose longitude can then be calculated from differences in local time. { 'län·jə,tüd ,sig·nəl }

longitudinal [SCI TECH] Pertaining to the lengthwise dimension. { ,län·jə'tüd·ən·əl }

longitudinal aberration [OPTICS] **1.** The distance along the optical axis from the focus of paraxial rays to the point where rays coming from the outer edges of its lens or reflecting surface intersect this axis. **2.** In chromatic aberration, the distance along the optical axis between the foci of two standard colors. { ,län·jə'tüd·ən·əl ,ab·ə'rā·shən }

longitudinal acceleration [MECH] The component of the linear acceleration of an aircraft, missile, or particle parallel to its longitudinal, or X, axis. { ,län·jə'tüd·ən·əl ak,sel·ə'rā·shən }

longitudinal baffle [CHEM ENG] Baffle sheets or plates within a process vessel (such as a heat exchanger) that are parallel to the long dimension of the vessel; used to direct fluid flow in the desired flow pattern. { ,län·jə'tüd·ən·əl 'baf·əl }

longitudinal bulkhead [NAV ARCH] A partition wall running fore and aft, made of planking or plating. { ,län·jə'tüd·ən·əl 'bəlk,hed }

longitudinal center of gravity [NAV ARCH] That point at which the combined weight of all the items that constitute a ship's weight are considered to be concentrated; usually taken as either aft or forward of the middle perpendicular or the midship frame. { ,län·jə'tüd·ən·əl 'sen·tər əv 'grav·əd·ē }

longitudinal circuit [ELEC] Circuit formed by one telephone wire (or by two or more telephone wires in parallel) with return through the earth or through any other conductors except those which are taken with the original wire or wires to form a metallic telephone circuit. { ,län·jə'tüd·ən·əl 'sər·kət }

longitudinal coefficient [NAV ARCH] The ratio of the volume of water displaced by a ship to the product of its length on the waterline and the area of its midship section below the water plane. Also known as prismatic coefficient. { ,län·jə'tüd·ən·əl ,kō·i'fish·ənt }

longitudinal controller [AERO ENG] A primary flight control mechanism which controls pitch attitude; located in the cockpit, this may be a control column or a side stick. { ,län·jə'tüd·ən·əl kən'trōl·ər }

longitudinal current [ELEC] Current which flows in the same direction in the two wires of a parallel pair using the earth or other conductors for a return path. { ,län·jə'tüd·ən·əl 'kə·rənt }

longitudinal direction [MET] The principal direction of flow in a plastically deformed metal. { ,län·jə'tüd·ən·əl di'rek·shən }

longitudinal drum boiler [MECH ENG] A boiler in which the axis of the horizontal drum is parallel to the tubes, both lying in the same plane. { ,län·jə'tüd·ən·əl 'drəm ,bȯil·ər }

longitudinal dune [GEOL] A type of linear dune ridge that extends parallel to the direction of the dominant dune-building winds. { ,län·jə'tüd·ən·əl 'dün }

longitudinal fault [GEOL] A fault parallel to the trend of the surrounding structure. { ,län·jə'tüd·ən·əl 'fȯlt }

longitudinal flow reactor [CHEM ENG] Theoretical reactor system in which there is no longitudinal mixing (back mixing) of reactants and products as they flow through the reactor, but in which there is complete radial (side-to-side) mixing. { ,län·jə'tüd·ən·əl 'flō rē,ak·tər }

longitudinal frame [NAV ARCH] Any of the frames of a ship running fore and aft. { ,län·jə'tüd·ən·əl 'frām }

longitudinal framed ship [NAV ARCH] A ship constructed of widely spaced, transverse frames, which support a number of small fore and aft frames. { ,län·jə'tüd·ən·əl ,frāmd 'ship }

longitudinal magnetoresistance [ELECTROMAG] The change of electrical resistance produced in a current-carrying metal or semiconductor upon application of a magnetic field parallel to the current flow. { ,län·jə'tüd·ən·əl mag¦ned·ō·ri'zis·təns }

longitudinal magnetostriction See Joule effect. { ,län·jə'tüd·ən·əl mag¦ned·ō·ri'strik·shən }

longitudinal mass [RELAT] The ratio of a force acting on a relativistic particle in the direction of its velocity to the resulting acceleration; equal to $m_0(1 - v^2/c^2)^{-3/2}$, where m_0 is the particle's rest mass, v is its speed, and c is the speed of light. { ,län·jə'tüd·ən·əl 'mas }

longitudinal-mode delay line [ELECTR] A magnetostrictive delay in which signals are propagated by means of longitudinal vibrations in the magnetostrictive material. { ,län·jə'tüd·ən·əl ¦mōd di'lā ,līn }

longitudinal parity [COMMUN] Parity associated with bits recorded on one track in a data block, to indicate whether the number of recorded bits in the block is even or odd. { ,län·jə'tüd·ən·əl 'par·əd·ē }

longitudinal parity check [COMMUN] The count for even or odd parity of all the bits in a message as a precaution against transmission error. Also known as horizontal parity check. { ,län·jə'tüd·ən·əl 'par·əd·ē ,chek }

longitudinal quadrupole [ACOUS] A sound source resulting from a variation of a component of the velocity of matter in a direction parallel to the velocity component. [ELECTROMAG] An electric or magnetic quadrupole which produces a field equivalent to that of two equal and opposite electric or magnetic dipoles separated by a small distance parallel to the direction of the dipoles. Also known as axial quadrupole. { ,län·jə'tüd·ən·əl 'kwäd·rə,pōl }

longitudinal redundancy check [COMMUN] A method of checking for errors, in which data are arranged in blocks according to some rule, and the correctness of each character in the block is determined according to the rule. Abbreviated LRC. { ,län·jə'tüd·ən·əl ri'dən·dən·sē ,chek }

longitudinal resistance seam welding [MET] The performance of resistance seam welding parallel to the throat depth of the welding machine. { ,län·jə'tüd·ən·əl ri¦zis·təns 'sēm ,weld·iŋ }

longitudinal section [SCI TECH] A section taken through the lengthwise dimension of a structure, organism, or other object. { ,län·jə'tüd·ən·əl 'sek·shən }

longitudinal separation [NAV] In air-traffic control, the separation of aircraft along a longitudinal line. { ,län·jə'tüd·ən·əl ,sep·ə'rā·shən }

longitudinal sequence [MET] The sequence in which successive welds are deposited along the length of a continuous weld. { ,län·jə'tüd·ən·əl 'sē·kwəns }

longitudinal stability [ENG] The ability of a ship or aircraft to recover a horizontal position after a vertical motion of its ends about a horizontal axis perpendicular to the centerline. { ,län·jə'tüd·ən·əl stə'bil·əd·ē }

longitudinal strain See linear strain. { ,län·jə'tüd·ən·əl 'strān }

longitudinal stream See subsequent stream. { ,län·jə'tüd·ən·əl 'strēm }

longitudinal study [PSYCH] The study of a group of individuals at regular intervals over a relatively long period of time. { ,län·jə'tüd·ən·əl 'stəd·ē }

longitudinal vibration [MECH] A continuing periodic change in the displacement of elements of a rod-shaped object in the direction of the long axis of the rod. { ¦län·jə'tüd·ən·əl vī'brā·shən }

longitudinal wave [PHYS] A wave in which the direction of some vector characteristic of the wave, for example, the displacement of particles of the transmitting medium, is along the direction of propagation. { ¦län·jə'tüd·ən·əl 'wāv }

long-life items [ORD] Ordnance items which have an estimated average service life of 5 years or more. { 'lȯŋ ¦līf ¦īd·əmz }

long-line current [ELEC] A current that flows through the earth from an anodic to a cathodic area and returns along an underground pipe or other metal structure, often over a considerable distance and as the result of concentration cell action. { 'lȯŋ ¦līn ¦kə·rənt }

long-line effect [ELECTR] An effect occurring when an oscillator is coupled to a transmission line with a bad mismatch; two or more frequencies may then be equally suitable for oscillation, and the oscillator jumps from one of these frequencies to another as its load changes. { 'lȯŋ ¦līn i¸fekt }

long-lines engineering [COMMUN] Engineering performed to develop, modernize, or expand long-haul, point-to-point communications facilities using radio, microwave, or wire circuits. { 'lȯŋ ¦līnz 'en·jə'nir·iŋ }

long-nose pliers [DES ENG] Small pincer with long, tapered jaws. { 'lȯŋ ¸nōz 'plī·ərz }

long oil [MATER] Varnish containing a large percentage of oil. { 'lȯŋ ¦ȯil }

long-period tide [OCEANOGR] A tide or tidal current constituent with a period which is independent of the rotation of the earth but which depends upon the orbital movement of the moon or of the earth. { 'lȯŋ ¸pir·ē·əd ¦tīd }

long-period variable [ASTRON] A variable star with a period from about 100 to more than 600 days. { 'lȯŋ ¸pir·ē·əd 'ver·ē·ə·bəl }

long-persistence screen [ELECTR] A fluorescent screen containing phosphorescent compounds that increase the decay time, so a pattern may be seen for several seconds after it is produced by the electron beam. { 'lȯŋ pər¸sis·təns 'skrēn }

long-pillar work [MIN ENG] A coal-winning technique used in underground mining in which large pillars of coal are left as the face is advanced; finally all the pillars are removed together. { 'lȯŋ 'pil·ər ¸wərk }

long-playing record [ENG ACOUS] A 10- or 12-inch (25.4- or 30.48-centimeter) phonograph record that operates at a speed of 33⅓ rpm (revolutions per minute) and has closely spaced grooves, to give playing times up to about 30 minutes for one 12-inch side. Also known as LP record; microgroove record. { 'lȯŋ ¸plā·iŋ 'rek·ərd }

long primer [ORD] Primer with relatively long body, designed to provide central ignition for a propelling charge. { 'lȯŋ 'prī·mər }

long radius [MATH] The distance from the center of a regular polygon to a vertex. { 'lȯŋ ¸rād·ē·əs }

long-range chart [NAV] An aeronautical chart of small scale, covering a large area, designed for long flights in which celestial or electronic navigation is expected to be used principally. { 'lȯŋ ¸rānj 'chärt }

long-range forecast [METEOROL] A weather forecast covering periods from 48 hours to a week in advance (medium-range forecast), and ranging to even longer forecasts over periods of a month, a season, and so on. { 'lȯŋ ¸rānj 'fȯr¸kast }

long-range materiel requirements [ORD] Ordnance items required by operational and organizational concepts established for a period 5 to 10 years hence. { 'lȯŋ ¸rānj mə¸tir·ē'el ri¸kwīr·məns }

long-range navigation zone [NAV] **1.** A zone in which navigational aids furnish service at distances greater than 200 miles (322 kilometers). **2.** See loran. { 'lȯŋ ¸rānj ¸nav·ə'gā·shən ¸zōn }

long-range order [SOLID STATE] A tendency for some property of atoms in a lattice (such as spin orientation or type of atom) to follow a pattern which is repeated every few unit cells. { 'lȯŋ ¸rānj 'ȯr·dər }

long residuum [MATER] Residue from crude-oil distillation in a petroleum refinery when a relatively small proportion of the feed is distilled overhead. { 'lȯŋ rə'zij·yə·wəm }

long run frequency [STAT] The ratio of the number of

occurrences of an event in a large number of trials to the number of trials. { 'lȯŋ ¸rən 'frē·kwən·sē }

Long's coefficient [PATH] The number 2.6, multiplied by the last two figures of the urine specific gravity, determined at 25°C, to derive the number of grams of solids per liter of urine. { 'lȯŋz ¸kō·i'fish·ənt }

longshore bar [GEOL] A ridge of sand, gravel, or mud built on the seashore by waves and currents, generally parallel to the shore and submerged by high tides. Also known as offshore bar. { 'lȯŋ¸shȯr ¸bär }

longshore current See littoral current. { 'lȯŋ¸shȯr ¸kə·rənt }

longshore drift See littoral drift. { 'lȯŋ¸shȯr ¸drift }

longshore trough [GEOL] A long, wide, shallow depression of the sea floor parallel to the shore. { 'lȯŋ¸shȯr ¸trȯf }

long span [ENG] Span of open wire exceeding 250 feet (76 meters) in length. { 'lȯŋ ¦span }

long-span steel framing [BUILD] Framing system used when there is a greater clear distance between supports than can be spanned with rolled beams; girders, simple trusses, arches, rigid frames, and cantilever suspension spans are used in this system. { 'lȯŋ ¸span ¸stēl 'frām·iŋ }

long spark See long discharge. { 'lȯŋ ¦spärk }

long-staple cotton [TEXT] A grade of cotton whose staple is 11/8 inches (29 millimeters) or over in length. { 'lȯŋ ¸stā·pəl ¸kät·ən }

long string [PETRO ENG] The last string of casing that is set in a well, through the producing zone. Also known as oil string. { 'lȯŋ ¦striŋ }

long-tail pair [ELECTR] A two-tube or transistor circuit that has a common resistor (tail resistor) which gives strong negative feedback. { 'lȯŋ ¸tāl ¸per }

long-term memory [PSYCH] The storage of information indefinitely so that it can be used again and again at a later time. { ¦lȯŋ ¸tərm 'mem·ə·rē }

long-term potentiation [NEUROSCI] A long-lasting increase in synaptic efficacy, believed to be involved in information storage in the brain. { ¦lȯŋ ¸tərm pə¸ten·chē'ā·shən }

long-term predictor [COMMUN] An electric filter that removes redundancies in a signal associated with long-term correlations so that information can be transmitted more efficiently. { ¦lȯŋ ¸tərm prə'dik·tər }

long-term repeatability [CONT SYS] The close agreement of positional movements of a robotic system repeated under identical conditions over long periods of time. { 'lȯŋ ¸tərm ri¸pēd·ə'bil·əd·ē }

long-time burning oil [MATER] Carefully refined kerosine used in railway semaphore signal lamps; heavy ends are removed so that the oil will burn without charring the wick. { 'lȯŋ ¸tīm ¦bərn·iŋ ¸ȯil }

long-time trend See secular trend. { ¦lȯŋ ¸tīm 'trend }

long tom [MIN ENG] A trough, longer than a rocker, for washing gold-bearing earth. { 'lȯŋ 'täm }

Long Tom [ORD] Popular name for 155-millimeter self-propelled gun. { 'lȯŋ 'täm }

long ton See ton. { 'lȯŋ 'tən }

long-tube vertical evaporator [CHEM ENG] A liquid evaporator in which the material is force-fed into the bottom of a bundle of long, vertical tubes; hot liquid on the outsides of the tubes transfers heat to the rising liquid feed, causing partial evaporation. { 'lȯŋ ¸tüb ¦vərd·ə·kəl i'vap·ə¸rād·ər }

longwall coal cutter [MIN ENG] Compact machine driven by electricity or compressed air which cuts into the coal on relatively long faces, with its jib at right angles to its body. { 'lȯŋ¸wȯl 'kōl ¸kəd·ər }

longwall peak stoping [MIN ENG] An underland stoping method in which the rapid advance of the face goes on, and by working the faces at a 60° angle to the strike, the peak travels down the dip at twice the face advance rate. { 'lȯŋ¸wȯl 'pēk ¸stōp·iŋ }

longwall pillar working [MIN ENG] A technique used to extract coal pillars left behind in long-pillar working. { 'lȯŋ¸wȯl 'pil·ər ¸wȯrk·iŋ }

longwall retreating [MIN ENG] A longwall working system in which all the roadways are in the solid coal seam and the waste areas are left behind; developing headings are driven close to the boundary or limit, and the coal is taken out by the longwall retreating toward the shaft. { 'lȯŋ¸wȯl ri'trēd·iŋ }

longwall system [MIN ENG] A method of mining in which the faces are advanced from the shaft toward the boundary, and the roof is allowed to cave in behind the miners as work progresses. { 'loŋ,wȯl ,sis·təm }

long wave [COMMUN] An electromagnetic wave having a wavelength longer than the longest broadcast-band wavelength of about 545 meters, corresponding to frequencies below about 550 kilohertz. [METEOROL] With regard to atmospheric circulation, a wave in the major belt of westerlies which is characterized by large length (thousands of kilometers) and significant amplitude; the wavelength is typically longer than that of the rapidly moving individual cyclonic and anticyclonic disturbances of the lower troposphere. Also known as major wave; planetary wave. { 'loŋ ¦wāv }

long-wavelength infrared radiation [ELECTROMAG] Infrared radiation having a wavelength greater than 8 micrometers. { 'loŋ ¦wāv,leŋkth ¦in·frə¦red ,rād·ē'ā·shən }

long-wave radio [COMMUN] A radio which can receive frequencies below the lowest broadcast frequency of 550 kilohertz. { 'loŋ ,wāv 'rād·ē·ō }

long-wire antenna [ELECTROMAG] An antenna whose length is a number of times greater than its operating wavelength, so as to give a directional radiation pattern. Also known as long-conductor antenna. { 'loŋ ,wīr an'ten·ə }

long-wool [TEXT] A straight wool with a shine or gloss and a staple length of 7 to 20 inches (18 to 51 centimeters). { 'loŋ ,wul }

lonsdaleite [MINERAL] A mineral composed of a form of carbon; found in meteorites. { 'länz,dā,līt }

loo [METEOROL] A hot wind from the west in India. { lü }

lookahead [COMPUT SCI] A procedure in which a processor is preparing one instruction in a computer program while executing its predecessor. { 'luk·ə,hed }

lookahead tree See game tree. { 'luk·ə,hed ,trē }

look angle [AERO ENG] The elevation and azimuth at which a particular satellite is predicted to be found at a specified time. [ENG] The solid angle in which an instrument operates effectively, generally used to describe radars, optical instruments, and space radiation detectors. { 'luk ,aŋ·gəl }

look-back time [ASTRON] The time in the past at which the light now being received from a distant object was emitted. { 'luk ,bak ,tīm }

look box [CHEM ENG] Box with glass windows built into distillation-column rundown lines (or other flow lines) so that the stream of condensate from the condenser can be watched. { 'luk ,bäks }

lookout [BUILD] A horizontal wood framing member that extends out from the studs to the end of rafters and overhangs a part of a roof, such as a gable. { 'luk,aut }

lookout station [ENG] A structure or place on shore at which personnel keep watch of events at sea or along the shore. { 'luk,aut ,stā·shən }

lookout tower [ENG] In marine operations, any tower surmounted by a small house in which a watch is habitually kept, as distinguished from an observation tower in which no watch is kept. { 'luk,aut ,tau·ər }

look-through [ELECTR] **1.** When jamming, a technique whereby the jamming emission is interrupted irregularly for extremely short periods to allow monitoring of the victim signal during jamming operations. **2.** When being jammed, the technique of observing or monitoring a desired signal during interruptions in the jamming signal. { 'luk ,thrü }

look-up [COMPUT SCI] An operation or process in which a table of stored values is scanned (or searched) until a value equal to (or sometimes, greater than) a specified value is found. { 'luk,əp }

look-up table [COMPUT SCI] A stored matrix of data for reference purpose. { 'luk,əp ,tā·bəl }

loom [METEOROL] The glow of light below the horizon produced by greater-than-normal refraction in the lower atmosphere; it occurs when the air density decreases more rapidly with height than in the normal atmosphere. [TEXT] A machine on which fabrics are woven. { lüm }

looming [OPTICS] A form of mirage in which images of objects normally hidden below the horizon are seen in the sky, sometimes upside down; a common occurrence in the Far North. { 'lüm·iŋ }

Loomis-Wood diagram [SPECT] A graph used to assign lines in a molecular spectrum to the various branches of rotational bands when these branches overlap, in which the difference between observed wave numbers and wave numbers extrapolated from a few lines that apparently belong to one branch are plotted against arbitrary running numbers for that branch. { 'lü·məs 'wud ,dī·ə,gram }

loon [VERT ZOO] The common name for birds composing the family Gaviidae, all of which are fish-eating diving birds. { lün }

loop [AERO ENG] A flight maneuver in which an airplane flies a circular path in an approximately vertical plane, with the lateral axis of the airplane remaining horizontal, that is, an inside loop. [COMPUT SCI] A sequence of computer instructions which are executed repeatedly, but usually with address modifications changing the operands of each iteration, until a terminating condition is satisfied. [ELEC] **1.** A closed path or circuit over which a signal can circulate, as in a feedback control system. **2.** Commercially, the portion of a connection from central office to subscriber in a telephone system. **3.** See mesh. [ELECTROMAG] See coupling loop; loop antenna. [ENG] **1.** A reel of motion picture film or magnetic tape whose ends are spliced together, so that it may be played repeatedly without interruption. **2.** A closed circuit of pipe in which materials and components may be placed to test them under different conditions of temperature, irradiation, and so forth. [MATH] A line which begins and ends at the same point of the graph. [PHYS] **1.** A closed curve on a graph, such as a hysteresis loop. **2.** That part of a standing wave where the vertical motion is greatest and the horizontal velocities are least. **3.** [ASTRON] See antinode. { lüp }

loop antenna [ELECTROMAG] A directional-type antenna consisting of one or more complete turns of a conductor, usually tuned to resonance by a variable capacitor connected to the terminals of the loop. Also known as loop. { 'lüp an,ten·ə }

loopback check See echo check. { 'lüp,bak ,chek }

loopback switch [ELECTR] A switch at the end of a telephone line that is used to test the line and, when closed, reflects received signals to the sender. { 'lüp,bak ,swich }

loop body [COMPUT SCI] The set of statements to be performed iteratively with the range of a loop. { 'lüp ,bäd·ē }

loop check See echo check. { 'lüp ,chek }

loop checking [COMMUN] Sending signals from the central office to test the integrity of local loops. { 'lüp ,chek·iŋ }

loop circuit [COMMUN] Common communications circuit shared by more than two parties; when applied to a teletypewriter operation, all machines print all data entered on the loop. { 'lüp ,sər·kət }

loop control See photoelectric loop control. { 'lüp kən,trōl }

loop coupling [ELECTROMAG] A method of transferring energy between a waveguide and an external circuit, by inserting a conducting loop into the waveguide, oriented so that electric lines of flux pass through it. { 'lüp ,kəp·liŋ }

loop dialing [COMMUN] Return-path method of dialing in which the dial pulses are sent out over one side of the interconnecting line or trunk and are returned over the other side; limited to short-haul traffic. { 'lüp ,dī·liŋ }

looped pipeline [PETRO ENG] A pipeline that is paralleled (looped) by a second pipeline, both of which serve the same liquid or gas source and destination. { 'lüpt 'pīp,līn }

loop filter [ELECTR] A low-pass filter, which may be a simple RC filter or may include an amplifier, and which passes the original modulating frequencies but removes the carrier-frequency components and harmonics from a frequency-modulated signal in a locked-oscillator detector. { 'lüp ,fil·tər }

loop flow See parallel flow. { 'lüp ,flō }

loop gain [CONT SYS] The ratio of the magnitude of the primary feedback signal in a feedback control system to the magnitude of the actuating signal. [ELECTR] Total usable power gain of a carrier terminal or two-wire repeater; maximum usable gain is determined by, and may not exceed, the losses in the closed path. { 'lüp ,gān }

loop head [COMPUT SCI] The first instruction of a loop, which contains the mode of execution, induction variable, and indexing parameters. { 'lüp ,hed }

looping [ENG] Laying a parallel pipeline along another, or along just a section of it, to increase capacity. { 'lüp·iŋ }

looping mill [MET] An arrangement of mills such that a hot bar discharged from one mill is fed into a second mill in the opposite direction. { 'lüp·iŋ ,mil }

LOOP ANTENNA

Shape of loop antenna.

loop lake *See* oxbow lake. { 'lüp ,lāk }

loop-mile [ELEC] Length of wire in a mile of two-wire line. { 'lüp ¦mīl }

Loop Nebula [ASTRON] A large, bright gaseous nebula in the Large Magellanic Cloud; its diameter is about 260 light-years. Also known as 30 Doradus; Tarantula. { 'lüp 'neb·yə·lə }

loop network *See* ring network. { 'lüp 'net,wərk }

loop of Henle [ANAT] The U-shaped portion of a renal tubule formed by a descending and an ascending tubule. Also known as Henle's loop. { 'lüp əv 'hen·lē }

loop of retrogression [ASTRON] A loop in the apparent path of a planet, relative to the stars, that is described when the planet undergoes retrograde motion. { 'lüp əv ,re·trə'gresh·ən }

loop pattern [FOREN SCI] A type of fingerprint pattern in which one or more of the ridges enter on either side of the impression, recurve, touch, or pass an imaginary line from the delta to the core and terminate on the entering side. { 'lüp ,pad·ərn }

loop pulsing [COMMUN] Regular, momentary interruptions of the direct-current path at the sending end of a transmission line. Also known as dial pulsing. { 'lüp ,pəls·iŋ }

loop rating [HYD] A rating curve that has higher values of discharge for a certain stage when the river is rising than it does when the river is falling; thus, the curve (stage versus discharge) describes a loop with each rise and fall of the river. { 'lüp ,rād·iŋ }

loop ratio *See* loop transfer function. { 'lüp ,rā·shō }

loop seal [CHEM ENG] Antivapor seal for liquid drawoffs from process or storage vessels; liquid drawoff is made to flow through an immersed loop or beneath an obstruction, thus sealing off vapor flow. { 'lüp ,sēl }

loopstick antenna *See* ferrite-rod antenna. { 'lüp,stik an ,ten·ə }

loop stop [COMPUT SCI] A small closed loop that is entered to stop the progress of a computer program, usually when some condition occurs that requires intervention by the operator or that should be brought to the operator's attention. Also known as stop loop. { 'lüp ,stäp }

loop strength *See* loop tenacity. { 'lüp ,streŋkth }

loop tenacity [ENG] A measure of the strength of a fibrous material determined by a test in which two linked loops of the material are pulled against each other to determine if the material will cut or crush itself. Also known a loop strength. { 'lüp tə,nas·əd·ē }

loop test [ELEC] A telephone or telegraph line test that is made by connecting a faulty line to good lines in such a way as to form a loop in which measurements can be made to determine the position of the fault. { 'lüp ,test }

loop transfer function [CONT SYS] For a feedback control system, the ratio of the Laplace transform of the primary feedback signal to the Laplace transform of the actuating signal. Also known as loop ratio. { 'lüp 'tranz·fər ,fəŋk·shən }

loop transmission [COMMUN] A series connection of subscribers in a communications system, as in the Advanced Research Project Agency computer network or McCullough burglar alarm loops. { 'lüp tranz,mish·ən }

loop transmittance [CONT SYS] **1.** The transmittance between the source and sink created by the splitting of a specified node in a signal flow graph. **2.** The transmittance between the source and sink created by the splitting of a node which has been inserted in a specified branch of a signal flow graph in such a way that the transmittance of the branch is unchanged. { 'lüp tranz,mit·əns }

loop tunnel [ENG] A tunnel which is looped or folded back on itself to gain grade in a tunnel location. { 'lüp ,tən·əl }

loop-type pit bottom [MIN ENG] An arrangement at the pit bottom in which loaded cars are fed to the cage on one side only, and the empties are returned by a loop roadway to the same place. { 'lüp ,tīp 'pit ,bäd·əm }

loop-type radio range [NAV] An A–N radio range employing two loop antennas placed at right angles to one another. { 'lüp ,tīp 'rād·ē·ō ,rānj }

loopway [MIN ENG] A double-track loop in a main single-track haulage plane at which mine cars may pass. { 'lüp,wā }

loose connective tissue [HISTOL] A type of irregularly arranged connective tissue having a relatively large amount of ground substance. { 'lüs kə'nek·tiv 'tish·ü }

loose coupling [ELEC] Coupling of a degree less than the critical coupling. { 'lüs 'kəp·liŋ }

loose-detail mold [ENG] A plastics mold with parts that come out with the molded piece. { 'lüs ¦dē,tāl ,mōld }

loose fit [DES ENG] A fit with enough clearance to allow free play of the joined members. { 'lüs ,fit }

loose ground [MIN ENG] Broken, fragmented, or loosely cemented bedrock material that tends to slough from sidewalls into a borehole and must be supported, as with timber sets. Also known as broken ground. { 'lüs ,graúnd }

loose-joint butt [DES ENG] A knuckle hinge in which the pin on one half slides easily into a slot on the other half. { 'lüs jóint 'bət }

loose kernel smut [PL PATH] A type of kernel smut disease distinguished by the ruptured spore-containing gall. { 'lüs ¦kərn·əl 'smət }

loose list [COMPUT SCI] A list, some of whose cells are empty and thus do not contain records of the file. Also known as thin list. { 'lüs 'list }

loosely coupled computer [COMPUT SCI] A computer that can function by itself and can also be connected to other computers to exchange data when necessary. { 'lüs·lē ¦kəp·əld kəm'pyüd·ər }

loose pulley [MECH ENG] In belt-driven machinery, a pulley which turns freely on a shaft so that the belt can be shifted from the driving pulley to the loose pulley, thereby causing the machine to stop. { 'lüs 'púl·ē }

Looser-Milkman syndrome *See* Milkman's syndrome. { 'lü·sər 'milk·mən ,sin,drōm }

loose round [ORD] Defective cartridge in which the bullet is loose in the cartridge case. { 'lüs ¦raund }

Lopadorrhynchidae [INV ZOO] A small family of pelagic polychaete annelids belonging to the Errantia. { ,lō·pə·dō'riŋk·ə,dē }

loparite [MINERAL] $(Ce,Na,Ca)_2(Ti,Nb)_2O_6$ A brown to black mineral; a variety of perovskite containing alkalies and cerium. { 'lō·pə,rīt }

lopezite [MINERAL] $K_2Cr_2O_7$ An orange-red mineral composed of potassium dichromate. { 'lä·pə,zīt }

Lophialetidae [PALEON] A family of extinct perissodactyl mammals in the superfamily Tapiroidea. { ,lä·fē·ə'led·ə,dē }

Lophiiformes [VERT ZOO] A modified order of actinopterygian fishes distinguished by the reduction of the first dorsal fin to a few flexible rays, the first of which is on the head and bears a terminal bulb; includes anglerfish and allies. { ,lä·fē·ə'fór,mēz }

lophine [ORG CHEM] $C_{21}H_{16}O_2$ A colorless, crystalline, water-insoluble compound that melts at 275°C; used as an indicator in fluorescent neutralization tests. { 'lō,fēn }

Lophiodontidae [PALEON] An extinct family of perissodactyl mammals in the superfamily Tapiroidea. { ,lä·fē·ə'dänt·ə,dē }

lophocercous [VERT ZOO] Having a ridgelike caudal fin that lacks rays. { ,lä·fə'sər·kəs }

lophocyte [INV ZOO] A specialized cell of uncertain function beneath the dermal membrane of certain Demospongiae which bears a process terminating in a tuft of fibrils. { 'läf·ə,sīt }

lophodont [VERT ZOO] Having molar teeth whose grinding surfaces have transverse ridges. { 'läf·ə,dänt }

Lophogastrida [INV ZOO] A suborder of free-swimming marine crustaceans in the order Mysidacea characterized by imperfect fusion of the sixth and seventh abdominal somites, seven pairs of gills and brood lamellae, and natatory, biramous pleopods. { ,läf·ə'gas·trə·də }

lophophore [INV ZOO] A food-gathering organ consisting of a fleshy basal ridge or lobe, from which numerous ciliated tentacles arise; found in Bryozoa, Phoronida, and Brachiopoda. { 'läf·ə,fór }

lophotrichous [CELL MOL] Having a polar tuft of flagella. { lə'fä·trə·kəs }

lopolith [GEOL] A large, floored intrusive body that is sunken centrally into the shape of a basin due to sagging of the underlying country rock. { 'läp·ə,lith }

lopping shears [DES ENG] Long-handled shears used for pruning branches. { 'läp·iŋ ,shirz }

lorac [NAV] A radio navigation system which utilizes phase comparison techniques to provide hyperbolic lines of position; the system permits an unlimited number of users. Derived from long-range accuracy. { 'lór,ak }

Loop fingerprint patterns.
(a) Twelve-count ulna loop.
(b) Radial loop. (Federal Bureau of Investigation)

LOPADORRHYNCHIDAE

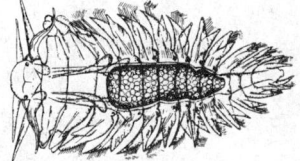

Lopadorrhynchus, dorsal view.

LOPHOCYTE

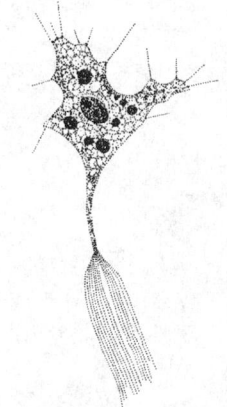

Lophocyte of a freshwater sponge showing terminal tuft of fibrils.

LOPHOGASTRIDA

Lophogaster typicus, adult male.

lorad *See* lodar.   { 'lȯr,ad }

loran [NAV] The designation of a family of radio navigation systems by which hyperbolic lines of position are determined by measuring the difference in the times of reception of synchronized pulse signals from two or more fixed transmitters. Derived from long-range navigation. { 'lȯr,an }

loran A [NAV] A medium-frequency radio navigation system by which hyperbolic lines of position are determined by measuring the difference in the times of reception of synchronized pulse signals from two fixed transmitters. Also known as standard loran. { 'lȯr,an 'ā }

loran C [NAV] A low-frequency radio navigation system by which hyperbolic lines of position are determined by measuring the difference in the times of reception of synchronized pulse signals from two fixed transmitters; as compared to loran A, time difference measurements are increased in accuracy through utilizing phase comparison techniques in addition to relatively coarse matches of pulse envelopes of received signals within the loran C receiver. { 'lȯr,an 'sē }

loran chain [NAV] **1.** A system or combination of three or more loran A stations, forming two or more pairs of stations for loran A navigation. **2.** A system or combination of a master loran C station and two or more loran C slave stations, forming two or more pairs of stations for loran C navigation. { 'lȯr,an ,chān }

loran chart [NAV] A chart showing loran lines of position along with a limited amount of topographic detail. { 'lȯr,an ,chärt }

loran D [NAV] Tactical loran system that uses the coordinate converter of low-frequency loran C and can operate in conjunction with inertial systems on aircraft, independently of ground facilities and without radiating radio-frequency energy that could reveal the aircraft's location. { 'lȯr,an 'dē }

lorandite [MINERAL] $TlAsS_2$ A cochineal- to carmine-red or dark lead-gray mineral composed of thallium sulfarsenide, occurring in monoclinic form. { 'lä·rən,dīt }

loran fix [NAV] A fix established by means of loran lines of position. { 'lȯr,an ,fiks }

loran line [NAV] A line of position on a loran chart; each line is the locus of points whose distances from two fixed stations differ by a constant amount. { 'lȯr,an ,līn }

loran rate [NAV] **1.** The frequency channel and pulse repetition rate by which a pair of loran stations is identified. **2.** By extension, the term refers to a pair of transmitting stations, their signals, and resulting lines of position. { 'lȯr,an,rāt }

loranskite [MINERAL] $(Y,Ce,Ca,Zr)TaO_4$ A black mineral composed of an oxide of yttrium, cerium, calcium, tantalum, and zirconium. { lə'ran,skīt }

loran tables [NAV] A series of tables containing tabular data for constructing loran hyperbolic lines of position. { 'lȯr,an ,tā·bəlz }

Loranthaceae [BOT] A family of dicotyledonous plants in the order Santalales in which the ovules have no integument and are embedded in a large, central placenta. { ,lȯ,ran 'thās·ē,ē }

lorate [BIOL] Strap-shaped. { 'lȯr,āt }

lordosis [MED] Exaggerated forward curvature of the lumbar spine. { lȯr,dō·səs }

Lorentz-Boltzmann equation [STAT MECH] An approximation to the Boltzmann transport equation for states that are near equilibrium, which shows that the Maxwell-Boltzmann distribution applies at equilibrium. { 'lȯr,ens ,bōlts·mən i,kwā·zhən }

Lorentz conductivity theory *See* classical conductivity theory. { 'lȯr,ens ,kän,dək'tiv·əd·ē ,thē·ə·rē }

Lorentz contraction *See* FitzGerald-Lorentz contraction. { 'lȯr,ens kən'trak·shən }

Lorentz electron [ELECTROMAG] A model of the electron as a damped harmonic oscillator; used to explain the variation of the real and imaginary parts of the index of refraction of a substance with frequency. { 'lȯr,ens i'lek,trän }

Lorentz equation [ELECTROMAG] The equation of motion for a charged particle, which sets the rate of change of its momentum equal to the Lorentz force. { 'lȯr,ens i'kwā·zhən }

Lorentz factor [RELAT] An important parameter in special relativity, equal to $1/\sqrt{1-(v/c)^2}$, where c is the speed of light and v is the constant relative velocity of two frames of reference. { 'lȯr,ens ,fak·tər }

Lorentz-FitzGerald contraction *See* FitzGerald-Lorentz contraction. { 'lȯr,ens fits'jer·əld kən,trak·shən }

Lorentz force [ELECTROMAG] The force on a charged particle moving in electric and magnetic fields, equal to the particle's charge times the sum of the electric field and the cross product of the particle's velocity with the magnetic flux density. { 'lȯr,ens ,fȯrs }

Lorentz-force density [ELECTROMAG] The force per unit volume on a charge density and current density, assuming that these densities arise from large numbers of charged particles experiencing a Lorentz force. { 'lȯr,ens ,fȯrs ,den·səd·ē }

Lorentz four-vector *See* four-vector. { 'lȯr,ens 'fȯr ,vek·tər }

Lorentz frame [RELAT] Any of the family of inertial coordinate systems, with three space coordinates and one time coordinate, used in the special theory of relativity; each frame is in uniform motion with respect to all the other Lorentz frames, and the interval between any two events is the same in all frames. { 'lȯr,ens ,frām }

Lorentz gage [ELECTROMAG] Any gage in which the sum of the divergence of the vector potential and the partial derivative of the scalar potential divided by the speed of light (in Gaussian units) vanishes identically; it is always possible to find a gage satisfying this condition. { 'lȯr,ens ,gāj }

Lorentz gas [ELECTR] A model of completely ionized gas in which ions are assumed to be stationary and interactions between electrons are neglected. { 'lȯr,ens ,gas }

Lorentz group [MATH] The group of all Lorentz transformations of euclidean four-space with composition as the operation. { 'lȯr,ens ,grüp }

Lorentz-Heaviside system *See* Heaviside-Lorentz system. { 'lȯr,ens 'hev·ē,sīd ,sis·təm }

Lorentz invariance [RELAT] The property, possessed by the laws of physics and of certain physical quantities, of being the same in any Lorentz frame, and thus unchanged by a Lorentz transformation. { 'lȯr,ens in,ver·ē·əns }

Lorentz line-splitting theory [ATOM PHYS] A theory predicting that when a light source is placed in a strong magnetic field, its spectral lines are each split into three components, one of them retaining the zero-field frequency, and the other two shifted upward and downward in frequency by the Larmor frequency (the normal Zeeman effect). { 'lȯr,ens 'līn ,splid·iŋ ,thē·ə·rē }

Lorentz local field [ELEC] In a theory of electric polarization, the average electric field due to the polarization at a molecular site that is calculated under the assumption that the field due to polarization by molecules inside a small sphere centered at the site may be neglected. Also known as Mossotti field. { 'lȯr,ens ,lō·kəl ,fēld }

Lorentz-Lorenz equation [OPTICS] The equation that results from replacing the relative dielectric constant with the square of the index of refraction in the Clausius-Mossotti equation. { 'lȯr,ens 'lȯr,ens i,kwä·shən }

Lorentz-Lorenz molar refraction *See* molar refraction. { 'lȯr,ens 'lȯr,ens ,mō·lər ri'frak·shən }

Lorentz matrix [RELAT] A matrix whose product with a vector whose components are the space and time coordinates of an event yields a vector whose components are new coordinates derived from the original ones by a Lorentz transformation. { 'lȯr,ens ,mā·triks }

Lorentz number [PL PHYS] The ratio of the velocity of a fluid to the velocity of light. Symbolized N_{Lo}. [SOLID STATE] The thermal conductivity of a metal divided by the product of its temperature and its electrical conductivity, according to the Wiedemann-Franz law. { 'lȯr,ens ,nəm·bər }

Lorentz polarization factor [OPTICS] A geometric factor in the equation for the intensity of x-rays or other radiation diffracted through a given angle by a crystalline substance. { 'lȯr,ens ,pō·lə·rə'zā·shən ,fak·tər }

Lorentz relation *See* Wiedemann-Franz law. { 'lȯr,ens ri,lā·shən }

Lorentz theory of light sources [ATOM PHYS] A theory according to which light is emitted by vibrations of electrons, which are damped harmonic oscillators attached to atoms. { 'lȯr,ens 'thē·ə·rē əv 'līt ,sȯrs·əz }

Lorentz transformation [MATH] Any linear transformation of euclidean four space which preserves the quadratic form $q(x,y,z,t) = t^2 - x^2 - y^2 - z^2$. [RELAT] Any of the family of mathematical transformations used in the special theory of relativity to relate the space and time variables of different Lorentz frames. { 'lȯr,ens ,tranz·fər,mā·shən }

Lorentz unit [SPECT] A unit of reciprocal length used to measure the difference, in wave numbers, between a (zero field) spectrum line and its Zeeman components; equal to $eH/4\pi mc^2$, where H is the magnetic field strength, c is the speed of light, and e and m are the charge and mass of the electron respectively (gaussian units). { 'lȯr,ens ,yü·nət }

Lorenz attractor [PHYS] The strange attractor for the solution of a system of three coupled, nonlinear, first-order differential equations that are encountered in the study of Rayleigh-Bénard convection; it is highly layered and has a fractal dimension of 2.06. Also know as Lorenz butterfly. { 'lȯr,ens ə,trak·tər }

Lorenz butterfly See Lorenz attractor. { ,lȯr·ənz 'bəd·ər,flī }

Lorenz curve [STAT] A graph for showing the concentration of ownership of economic quantities such as wealth and income; it is formed by plotting the cumulative distribution of the amount of the variable concerned against the cumulative frequency distribution of the individuals possessing the amount. { 'lȯr,ens ,kərv }

lorettoite [MINERAL] $Pb_7O_6Cl_2$ A honey-yellow to brownish-yellow mineral composed of lead oxychloride. { lə'red·ə,wīt }

L organisms [MICROBIO] Pleomorphic forms of bacteria occurring spontaneously, or favored by agents such as penicillin, which lack cell walls and grow in minute colonies; transition may be reversible under certain conditions. { 'el ,ȯr·gə,niz·əmz }

lorhumb line [NAV] A line along which the rates of change of the values of two families of hyperbolas are constants. { 'lȯr·əm ,līn }

lorica [INV ZOO] A hard shell or case in certain invertebrates, as in many rotifers and protozoans; functions as an exoskeleton. { lə'rī·kə }

Loricaridae [VERT ZOO] A family of catfishes in the suborder Siluroidei found in the Andes. { ,lȯr·ə·kə'rī·ə,dē }

Loricata [INV ZOO] An equivalent name for Polyplacophora. { ,lȯr·ə'käd·ə }

loricate [INV ZOO] Of, pertaining to, or having a lorica. { 'lȯr·ə,kāt }

loris [VERT ZOO] Either of two slow-moving, nocturnal, arboreal primates included in the family Lorisidae: the slender loris (*Loris tardigradus*) and slow loris (*Nycticebus coucang*). { 'lȯr·əs }

Lorisidae [VERT ZOO] A family of prosimian primates comprising the lorises of Asia and the galagos of Africa. { lə'ris·ə,dē }

lorry See larry. { 'lȯr·ē }

Loschmidt number [PHYS] The number of molecules in 1 cubic centimeter of an ideal gas at 1 atmosphere pressure and 0°C, equal to approximately 2.687×10^{19}. Symbolized n_0. { 'lō,shmit ,nəm·bər }

lose returns See lost circulation.

lose water See lost circulation. { 'lüz 'wȯd·ər }

loseyite [MINERAL] $(Mn,Zn)_7(CO_3)_2(OH)_{10}$ A bluish-white or brownish, monoclinic mineral consisting of a basic carbonate of manganese and zinc. { 'lō·zē,īt }

losing stream See influent stream. { 'lüs·iŋ ,strēm }

loss [COMMUN] See transmission loss. [ENG] Power that is dissipated in a device or system without doing useful work. Also known as internal loss. { lȯs }

loss angle [ELECTROMAG] A measure of the power loss in an inductor or a capacitor, equal to the amount by which the angle between the phasors denoting voltage and current across the inductor or capacitor differs from 90°. { 'lȯs ,aŋ·gəl }

loss cone [PL PHYS] A cone in the velocity space of particles in a plasma confined by magnetic mirrors; particles with velocities in the cone are not trapped by the mirrors and are lost out of the system. { 'lȯs ,kōn }

loss-cone instability [PL PHYS] An instability in a plasma confined between magnetic mirrors. { 'lȯs ,kōn ,in·stə'bil·əd·ē }

loss current [ELEC] The current which passes through a capacitor as a result of the conductivity of the dielectric and results in power loss in the capacitor. [ELECTROMAG] The component of the current across an inductor which is in phase with the voltage (in phasor notation) and is associated with power losses in the inductor. { 'lȯs ,kə·rənt }

losser circuit [ELEC] Resonant circuit having sufficient high-frequency resistance to prevent sustained oscillation at the resonant frequency. { 'lȯs·ər ,sər·kət }

loss evaluation [ELEC] A method of achieving an economic balance between buyer and seller in adding material to a transformer design to get lower losses, in which one calculates a value in dollars per kilowatt for load loss and for no-load loss. { 'lȯs i,val·yə'wā·shən }

Lossev effect See injection electroluminescence. { ,lȯ,sef i,fekt }

loss factor [ELEC] The power factor of a material multiplied by its dielectric constant; determines the amount of heat generated in a material. { 'lȯs ,fak·tər }

loss function [MATH] In decision theory, the function, dependent upon the decision and the true underlying distributions, which expresses the loss produced in taking the decision. { 'lȯs ,fəŋk·shən }

loss-in-weight feeder [MECH ENG] A device to apportion the output of granulated or powdered solids at a constant rate from a feed hopper; weight-measured decrease in hopper content actuates further opening of the discharge chute to compensate for flow loss as the hopper overburden decreases; used in the chemical, fertilizer, and plastics industries. { 'lȯs in 'wāt ,fēd·ər }

lossless data compression [COMMUN] Data compression in which the recovered data are assured to be identical to the source. { 'lȯs,les 'dad·ə kəm,presh·ən }

lossless junction [ELECTROMAG] A waveguide junction in which all the power incident on the junction is reflected from it. { 'lȯs·ləs ,jəŋk·shən }

lossless material [PHYS] An ideal material that dissipates none of the energy of electromagnetic or acoustic waves passing through it. { 'lȯs·ləs mə'tir·ē·əl }

loss modulation See absorption modulation. { 'lȯs ,mäj·ə,lā·shən }

loss of head [FL MECH] Energy decrease between two points in a hydraulic system due to such causes as friction, bends, obstructions, or expansions. { 'lȯs əv 'hed }

loss-of-head gage [ENG] A gage on a rapid sand filter, which indicates loss of head for a filtering operation. { 'lȯs əv 'hed ,gāj }

loss of heterozygosity [GEN] In a heterozygote, the loss of one of the two alleles at one or more loci in a cell lineage or cancer cell population due to chromosome loss, deletion, or mitotic crossing-over. Abbreviated LOH. { 'lȯs əv ,hed·e·rō·zī'gäs·əd·ē }

loss of information See walk down. { 'lȯs əv ,in·fər,mā·shən }

lossy attenuator [ELECTROMAG] In waveguide technique, a length of waveguide deliberately introducing a transmission loss by the use of some dissipative material. { 'lȯs·ē ə'ten·yə,wād·ər }

lossy data compression [COMMUN] Data compression in which controlled degradation of the data is allowed. { 'lȯs·ē 'dad·ə kəm,presh·ən }

lossy line [ELEC] 1. Cable used in test measurements which has a large attenuation per unit length. 2. Transmission line designed to have a high degree of attenuation. { 'lȯs·ē 'līn }

lossy material [PHYS] A material that dissipates energy of electromagnetic or acoustic energy passing through it. { 'lȯs·ē mə'tir·ē·əl }

lost circulation [PETRO ENG] A condition that occurs when the drilling fluid escapes into crevices or porous sidewalls of a borehole and does not return to the collar of the drill hole. Also known as lose returns; lose water; lost returns; lost water. { 'lȯst ,sər·kyə'lā·shən }

lost circulation material See bridging material. { 'lȯst ,sər·kyə'lā·shən mə,tir·ē·əl }

lost cluster [COMPUT SCI] Disk records that are not associated with a file name in a disk directory. { 'lȯst 'kləs·tər }

lost hole [PETRO ENG] A well that cannot be worked further because of a serious problem, such as a blowout. { 'lȯst 'hōl }

lost motion [MECH ENG] The delay between the movement of a driver and the movement of a follower. { 'lȯst 'mō·shən }

lost returns See lost circulation. { 'lȯst ri'tərnz }

lost stream [HYD] 1. A stream that disappears from the surface into an underground channel without reappearing in the same or even a neighboring drainage basin. 2. An evaporated stream in a desertlike region. { 'lȯst 'strēm }

LOUSE

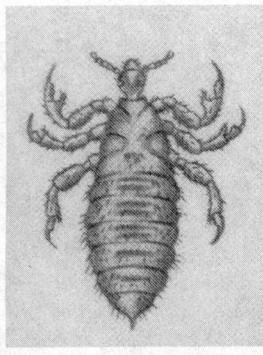

The head louse of humans.

lost time [ENG ACOUS] The period in a frequency-modulation sonar, just after flyback, during which the sound field must be reestablished; its duration equals travel time of the signal to and from the target. { 'lȯst ,tīm }

lost water *See* lost circulation. { 'lȯst 'wȯd·ər }

lost-wax process [MET] A method used in investment casting in which a wax pattern between a two-layered mold is removed by melting and replaced with molten metal; used for casting bronze statues and in jewelry casting. Also known as investment process. { 'lȯst¦waks ,prä·səs }

lot [ANALY CHEM] A specimen of bulk material that is to undergo chemical analysis. [CIV ENG] A piece of land with fixed boundaries. [IND ENG] A quantity of material, such as propellant, the units of which were manufactured under identical conditions. Also known as lot batch. { lät }

lot batch *See* lot. { 'lät ,bach }

lotic [ECOL] Of or pertaining to swiftly moving waters. { 'lōd·ik }

lot line [CIV ENG] The legal boundary line of a piece of property. { 'lät ,līn }

lot number [IND ENG] Identification number assigned to a particular quantity or lot of material from a single manufacturer. { 'lät ,nəm·bər }

lot plot method [IND ENG] A variables acceptance sampling plan based on the frequency plot of a random sample of 50 items taken from a lot. { 'lät ¦plät ,meth·əd }

lotrite *See* pumpellyite. { 'lō,trīt }

lot sample *See* gross sample. { 'lät ,sam·pəl }

lot tolerance percent defective [IND ENG] The percent of defectives in a lot which is considered bad and should be rejected for some specified fraction, usually 90, of the time. { 'lät ¦täl·ə·rəns pər¦sent di'fek·tiv }

loturine *See* harman. { 'lä·chə,rēn }

loudness [ACOUS] The magnitude of the physiological sensation produced by a sound, which varies directly with the physical intensity of sound but also depends on frequency of sound and waveform. { 'lau̇d·nəs }

loudness analyzer [ELECTR] An instrument that produces a cathode-ray display which shows the loudness of airborne sounds at a number of subdivisions of part or all of the audio spectrum. { 'lau̇d·nəs ,an·ə,līz·ər }

loudness control [ENG ACOUS] A combination volume and tone control that boosts bass frequencies when the control is set for low volume, to compensate automatically for the reduced response of the ear to low frequencies at low volume levels. Also known as compensated volume control. { 'lau̇d·nəs kən,trōl }

loudness level [ACOUS] The level of a sound, in phons, equal to the sound pressure level in decibels, relative to 0.0002 microbar, of a pure 1000-hertz tone that is judged to be equally loud by listeners. { 'lau̇d·nəs ,lev·əl }

loudness recruitment [ACOUS] An abnormal increase in perceived loudness as a sound is intensified. { 'lau̇d·nəs ri,krüt·mənt }

loudness unit [ACOUS] A unit of loudness equal to the loudness of a sound having a loudness level of 0 phon; the loudness unit has been replaced by the sone. { 'lau̇d·nəs ,yü·nət }

loudspeaker [ENG ACOUS] A device that converts electrical signal energy into acoustical energy, which it radiates into a bounded space, such as a room, or into outdoor space. Also known as speaker. { 'lau̇d,spēk·ər }

loudspeaker dividing network *See* crossover network. { 'lau̇d,spēk·ər di'vīd·iŋ ,net,wərk }

loudspeaker voice coil *See* voice coil. { 'lau̇d,spēk·ər 'vȯis ,kȯil }

loughlinite [MINERAL] $Na_2Mg_3Si_6O_{16}\cdot 8H_2O$ A pearly-white mineral that resembles asbestos, consisting of a hydrous silicate of sodium and magnesium. { 'lȯf·lə,nīt }

louping ill [VET MED] A virus disease of sheep, similar to encephalomyelitis, transmitted by the tick *Ixodes racinus*. Also known as ovine encephalomyelitis; trembling ill. { 'lüp·iŋ ,il }

louping-ill virus [VIROL] A group B arbovirus that is infectious in sheep, monkeys, mice, horses, and cattle. { 'lüp·iŋ ,il ,vī·rəs }

louse [INV ZOO] The common name for the apterous ectoparasites composing the orders Anoplura and Mallophaga. { lau̇s }

louver [BUILD] An opening in a wall or ceiling with slanted or sloping slats to allow sunlight and ventilation and exclude rain; may be fixed or adjustable, and may be at the opening of a ventilating duct. Also known as outlet ventilator. [ENG] Any arrangement of fixed or adjustable slatlike openings to provide ventilation. [ENG ACOUS] An arrangement of concentric or parallel slats or equivalent grille members used to conceal and protect a loudspeaker while allowing sound waves to pass. { 'lü·vər }

lovage oil [MATER] A yellow-brown essential oil, soluble in alcohol and fixed oils, obtained from the root and fruit of *Levisticum officinale*; used for flavors and perfumes. Also known as levisticum oil. { 'ləv·ij ,ȯil }

lovchorrite *See* mosandrite. { 'ləv·kȯ,rīt }

Love wave [GEOPHYS] A horizontal dispersive surface wave, multireflected between internal boundaries of an elastic body, applied chiefly in the study of seismic waves in the earth's crust. { 'ləv ,wāv }

Lovibond tintometer [OPTICS] A colorimeter which compares a solution or object under examination with a series of slides of each of three colors. { 'lō·və,bänd tin'täm·əd·ər }

lovozerite [MINERAL] $(Na,K)_2(Mn,Ca)ZrSi_6O_{16}\cdot 3H_2O$ Mineral composed of hydrous silicate of sodium, potassium, manganese, calcium, and zirconium. { lō'vä·zə,rīt }

low *See* depression. { lō }

low-alloy steel [MET] A hardenable carbon steel generally containing not more than about 1% carbon and one or more of the following alloyed components: < (less than) 2% manganese, < 4% nickel, < 2% chromium, < 0.6% molybdenum, and < 0.2% vanadium. { ¦lō ¦al,ȯi 'stēl }

low aloft *See* upper-level cyclone. { ¦lō ə'lȯft }

low-altitude bombing [ORD] Horizontal bombing at altitudes between about 900 and 8000 feet (0.27 and 2.4 kilometers). Also known as low-level bombing. { 'lō ,al·tə,tüd ,bäm·iŋ }

low-altitude earth orbit [AERO ENG] An artificial satellite orbit whose altitude is less than about 1000 miles (1600 kilometers) above the earth's surface. Abbreviated LEO. { ¦lō ¦al·tə,tüd 'ərth ,ȯr·bət }

low-angle bombing [ORD] Bombing from an airplane at a slight dive angle. { 'lō ,aŋ·gəl ,bäm·iŋ }

low-angle fault [GEOL] A fault that dips at an angle less than 45°. { 'lō ,aŋ·gəl ¦fȯlt }

low-angle fire [ORD] Gunfire delivered at angles of elevation below the elevation that corresponds to the maximum range of the piece, so that ranges increase with increases in angles of elevation. { 'lō ,aŋ·gəl ¦fīr }

low-angle scattering *See* small-angle scattering. { 'lō ,aŋ·gəl 'skad·ər·iŋ }

low-angle thrust *See* overthrust. { 'lō ,aŋ·gəl ¦thrəst }

low-approach system [NAV] A means for furnishing guidance in the vertical and horizontal planes to aircraft during descent from an initial approach altitude to a point near the ground. { 'lō ə'prōch ,sis·təm }

low boiler [MATER] A fast-evaporating solvent used in lacquer thinner to give a rapid initial set; boiling point is 70–100°C. { 'lō 'bȯil·ər }

low-boiling butene-2 *See* butene-2. { 'lō ,bȯil·iŋ ¦byü,tēn 'tü }

lowboy [MECH ENG] A trailer with low ground clearance for hauling construction equipment. { 'lō,bȯi }

low brass [MET] Brass containing 20% zinc, 80% copper. { 'lō ¦bras }

low-carbon steel [MET] Steel containing 0.15% or less of carbon. { 'lō ,kär·bən 'stēl }

low clouds [METEOROL] Types of clouds, the mean level of which is between the surface and 6500 feet (1980 meters); the principal clouds in this group are stratocumulus, stratus, and nimbostratus. { 'lō 'klau̇dz }

low core [COMPUT SCI] The locations with the lower addresses in a computer's main storage, usually used to store control values needed to run the system and other critical information and instructions. { 'lō ,kȯr }

low-definition television [COMMUN] Television that involves less than about 200 scanning lines per complete image. { 'lō ,def·ə,nish·ən 'tel·ə,vizh·ən }

low-density dynamite [MATER] Any dynamite containing up to 80% ammonium nitrate as the principal explosive ingredient. { 'lō ,den·səd·ē 'dī·nə,mīt }

low-density lipoprotein [BIOCHEM] A lipoprotein containing more lipids than protein that transports cholesterol from the liver to various tissues throughout the body. Abbreviated LDL. { 'lō ,den·səd·ē ,lī·pō'prō,tēn }

low-density polyethylene [ORG CHEM] A thermoplastic polymer with a density of 0.910–0.940 gram per cubic centimeter (0.526–0.543 ounce per cubic inch). Abbreviated LDPE. { 'lō ,den·səd·ē ,päl·ē'eth·ə,lēn }

low-energy electron diffraction [SOLID STATE] A technique for studying the atomic structure of single crystal surfaces, in which electrons of uniform energy in the approximate range 5–500 electronvolts are scattered from a surface, and those scattered electrons that have lost no energy are selected and accelerated to a fluorescent screen where the diffraction pattern from the surface can be observed. Abbreviated LEED. { 'lō ,en·ər·jē i,lek,trän di'frak·shən }

low-energy environment [GEOL] An aqueous sedimentary environment in which there is standing water with a general lack of wave or current action, permitting accumulation of very fine-grained sediments. { 'lō ,en·ər·jē in'vī·ərn·mənt }

low-energy physics [PHYS] That part of physics which studies microscopic phenomena involving energies of several million electronvolts or less, such as the arrangement of electrons in an atom or a solid, and the arrangement of protons and neutrons within the atomic nucleus, and the nature of forces between these particles. { 'lō ,en·ər·jē ,fiz·iks }

low-energy positron diffraction [SOLID STATE] A technique for studying the atomic structure of solid surfaces in which a narrow beam of low-energy monoenergetic positrons is made to strike a solid surface, and the diffracted beams in certain directions that are permitted by the regular array of surface atoms are observed. Abbreviated LEPD. { 'lō ,en·ər·jē 'päz·ə,trän di,frak·shən }

Lowenhertz thread [DES ENG] A screw thread that differs from U.S. Standard form in that the angle between the flanks measured on an axial plane is 53°8′; height equals 0.75 times the pitch, and width of flats at top and bottom equals 0.125 times the pitch. { 'lō·ən,hərts ,thred }

lower atmosphere [METEOROL] That part of the atmosphere in which most weather phenomena occur (that is, the troposphere and lower stratosphere); in other contexts, the term implies the lower troposphere. { 'lō·ər 'at·mə,sfir }

Lower Austral life zone [ECOL] A term used by C.H. Merriam to describe the southern portion of the Austral life zone, characterized by accumulated temperatures of 18,000°F (10,000°C). { 'lō·ər 'ós·trəl 'līf ,zōn }

lower bound [MATH] **1.** A lower bound of a subset *A* of a set *S* is a point of *S* which is smaller than every element of *A*. **2.** A lower bound on a function *f* with values in a partially ordered set *S* is an element of *S* which is smaller than every element in the range of *f*. { 'lō·ər 'baúnd }

lower branch [ASTRON] That half of a meridian or celestial meridian from pole to pole which passes through the antipode or nadir of a place. { 'lō·ər 'branch }

Lower Cambrian [GEOL] The earliest epoch of the Cambrian period of geologic time, ending about 540,000,000 years ago. { 'lō·ər 'kam·brē·ən }

lower chord [CIV ENG] The bottom member of a truss. { 'lō·ər 'kórd }

lower control limit [IND ENG] The horizontal line drawn on a control chart at a specified distance below the central line; points plotted below the lower control limit indicate that the process may be out of control. { 'lō·ər kən'trōl ,lim·ət }

Lower Cretaceous [GEOL] The earliest epoch of the Cretaceous period of geologic time, extending from about 140- to 120,000,000 years ago. { 'lō·ər krə'tā·shəs }

lower critical field [SOLID STATE] The magnetic field strength below which magnetic flux is completely excluded from type II superconductor and above which it penetrates the superconductor as microscopic filaments called fluxoids. Symbolized H_{c1}. { 'lō·ər 'krid·i·kəl 'feld }

lower culmination *See* lower transit. { 'lō·ər ,kəl·mə'nā·shən }

lower curtate [COMPUT SCI] The lower or bottom part of a punch card; on a standard punch card, it contains the punch positions designated 1 through 9. { 'lō·ər 'kûr,tāt }

Lower Devonian [GEOL] The earliest epoch of the Devonian period of geologic time, extending from about 400- to 385,000,000 years ago. { 'lō·ər də'vō·nē·ən }

lower esophageal ring *See* Schatzki's ring. { ,lō·ər i,saf·ə,jē·əl 'riŋ }

lower half-power frequency [ELECTR] The frequency on an amplifier response curve which is smaller than the frequency for peak response and at which the output voltage is $1/\sqrt{2}$ of its midband or other reference value. { 'lō·ər 'haf ,paú·ər 'frē·kwən·sē }

lower heating value *See* low heat value. { 'lō·ər 'hēd·iŋ ,val·yü }

lower high water [OCEANOGR] The lower of two high tides occurring during a tidal day. { 'lō·ər 'hī ,wód·ər }

Lower Jurassic [GEOL] The earliest epoch of the Jurassic period of geologic time, extending from about 185- to 170,000,000 years ago. { 'lō·ər jü'ras·ik }

lower limb [ASTRON] That half of the outer edge of a celestial body having the least altitude. { 'lō·ər 'lim }

lower limit *See* limit inferior. { 'lō·ər 'lim·ət }

lower low water [OCEANOGR] The lower of two low tides occurring during a tidal day. { 'lō·ər 'lō ,wód·ər }

lower mantle [GEOL] The portion of the mantle below a depth of about 600 miles (1000 kilometers). Also known as inner mantle; mesosphere; pallasite shell. { 'lō·ər mant·əl }

Lower Mississippian [GEOL] The earliest epoch of the Mississippian period of geologic time, beginning about 350,000,000 years ago. { 'lō·ər ,mis·ə'sip·ē·ən }

lower motor neuron [NEUROSCI] An efferent neuron which has its body located in the anterior gray column of the spinal cord or in the brainstem nuclei, and its axon passing by way of a peripheral nerve to skeletal muscle. Also known as final common pathway. { 'lō·ər 'mōd·ər ,nü,rän }

lower motor neuron disease [MED] An injury to any part of a lower motor neuron, characterized by flaccid paralysis of the muscle, diminished or absent reflexes, and progressive atrophy of the muscle. { 'lō·ər 'mōd·ər 'nü,rän di,zēz }

lower nephron nephrosis [MED] Retrogressive kidney changes following traumatic injury and other conditions producing shock; sometimes accompanied by distal and collecting tubule necrosis. Also known as acute tubular necrosis; crush kidney; hemoglobinuric nephrosis; ischemic tubulorrhexis. { 'lō·ər 'ne,frän nə'frō·səs }

Lower Ordovician [GEOL] The earliest epoch of the Ordovician period of geologic time, extending from about 490- to 460,000,000 years ago. { 'lō·ər ,ór·də'vish·ən }

lower pair [MECH ENG] A link in a mechanism in which the mating parts have surface (instead of line or point) contact. { 'lō·ər 'pər }

Lower Pennsylvanian [GEOL] The earliest epoch of the Pennsylvanian period of geologic time, beginning about 310,000,000 years ago. { 'lō·ər ,pen·səl'vā·nyən }

Lower Permian [GEOL] The earliest epoch of the Permian period of geologic time, extending from about 275- to 260,000,000 years ago. { 'lō·ər 'pər·mē·ən }

lower pitch limit [ACOUS] Minimum frequency, for a sinusoidal sound wave, that will produce a pitch sensation. { 'lō·ər 'pich ,lim·ət }

lower plate *See* footwall. { 'lō·ər ,plāt }

lower punch [MET] In powder metallurgy, the portion of the die forming the bottom of the die cavity. { 'lō·ər 'pənch }

lower rib height [ANTHRO] The measure of the vertical distance taken from the lower edge of the last front-attached rib to the floor as the subject stands. { 'lō·ər 'rib ,hīt }

lower semicontinuous function [MATH] A real-valued function $f(x)$ is lower semicontinuous at a point x_0 if, for any small positive number ϵ, $f(x)$ is always greater that $f(x_0) - \epsilon$ for all *x* in some neighborhood of x_0. { 'lō·ər ,sem·ē·kən'tin·yə·wəs ,fəŋk·shən }

lower sideband [COMMUN] The sideband containing all frequencies below the carrier-frequency value that are produced by an amplitude-modulation process. { 'lō·ər 'sīd,band }

lower-sideband upconverter [ELECTR] Parametric amplifier in which the frequency, power, impedance, and gain considerations are the same as for the nondegenerate amplifier; here, however, the output is taken at the difference frequency, or the lower sideband, rather than the signal-input frequency. { 'lō·ər 'sīd,band 'əp·kən,vərd·ər }

Lower Silurian [GEOL] The earliest epoch of the Silurian

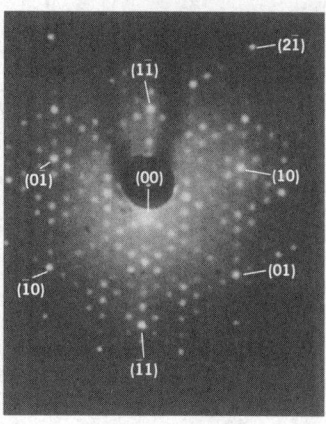

Low-energy electron diffraction pattern from a clean silicon single-crystal (111) surface (7 × 7 structure). Electron energy = 85 electronvolts. Numbers are indices (h, k) labeling diffraction spots of the surface.

period of geologic time, beginning about 420,000,000 years ago. { 'lō·ər sə'lùr·ē·ən }

Lower Sonoran life zone [ECOL] A term used by C.H. Merriam to describe the climate and biotic communities of subtropical deserts and thorn savannas in the southwestern United States. { 'lō·ər sə'nòr·ən 'lif ,zōn }

lower transit [ASTRON] Transit across the lower branch of the celestial meridian. Also known as lower culmination. { 'lō·ər 'trans·ət }

Lower Triassic [GEOL] The earliest epoch of the Triassic period of geologic time, extending from about 230- to 215,000,000 years ago. { 'lō·ər trī'as·ik }

lower yield point [MET] In annealed carbon steels, the lowest value of stress after the initial dropoff and before the load begins to rise continuously. { 'lō·ər 'yēld ,pöint }

lowest required radiating power [COMMUN] The smallest power output of an antenna which will suffice to maintain a specified grade of broadcast service. Abbreviated LRRP. { 'lō·əst ri¦kwird 'rād·ē,ād·iŋ ,paù·ər }

lowest safe waterline [MECH ENG] The lowest water level in a boiler drum at which the burner may safely operate. { 'lō·əst ¦sāf 'wòd·ər,līn }

lowest unoccupied molecular orbital [PHYS CHEM] The lowest-energy molecular orbital that is occupied by electrons. Abbreviated LUMO. { ¦lō·əst ən¦äk·yə,pīd mə¦lek·yə·lər 'òr·bəd·əl }

lowest useful high frequency [COMMUN] The lowest high frequency that is effective at a specified time for ionospheric propagation of radio waves between two specified points. Abbreviated LUHF. { 'lō·əst ¦yüs·fəl 'hī ,frē·kwən·sē }

low-expansion alloy [MET] An alloy whose dimensions do not vary appreciably with temperature. { 'lō ik,span·chən 'al,òi }

low explosive [MATER] An explosive which when used in its normal manner deflagrates or burns rather than detonates; used when shattering must be prevented. { 'lō ik,splō·siv }

low-freezing dynamite [MATER] A dynamite designed to work under freezing conditions; part of the nitroglycerin of the straight dynamite is replaced by nitrated sugar, nitrotoluene, nitrated polymerized glycerin, or ethylene glycol dinitrate. { 'lō ,frēz·iŋ 'dī·nə,mīt }

low-frequency [COMMUN] A Federal Communications Commission designation for the band from 30 to 300 kilohertz in the radio spectrum. Abbreviated LF. { 'lō ,frē·kwən·sē }

low-frequency acquisition and ranging See lofar. { 'lō ,frē·kwən·sē ,ak·wə'zish·ən ən 'rānj·iŋ }

low-frequency antenna [ELECTROMAG] An antenna designed to transmit or receive radiation at frequencies of less than about 300 kilohertz. { 'lō ,frē·kwən·sē an'ten·ə }

low-frequency compensation [ELECTR] Compensation that serves to extend the frequency range of a broad-band amplifier to lower frequencies. { 'lō ,frē·kwən·sē ,käm·pə'sā·shən }

low-frequency current [ELEC] An alternating current having a frequency of less than about 300 kilohertz. { 'lō ,frē·kwən·sē 'kə·rənt }

low-frequency cutoff [ELECTR] A frequency below which the gain of a system or device decreases rapidly. { 'lō ,frē·kwən·sē 'kə,dòf }

low-frequency cycle [MET] In resistance welding, one positive- and one negative-current pulse within a heat time at a lower frequency than the electrical power source. { 'lō ,frē·kwən·sē 'sī·kəl }

low-frequency gain [ELECTR] The gain of the voltage amplifier at frequencies less than those frequencies at which this gain is close to its maximum value. { 'lō ,frē·kwən·sē 'gān }

low-frequency impedance corrector [ELEC] Electric network designed to be connected to a basic network, or to a basic network and a building-out network, so that the combination will simulate, at low frequencies, the sending-end impedance, including dissipation, of a line. { 'lō ,frē·kwən·sē im'pēd·əns kə,rek·tər }

low-frequency induction furnace [ENG] An induction furnace in which current flow at the commercial power-line frequency is induced in the charge to be heated. { 'lō ,frē·kwən·sē in¦dək·shən ,fər·nəs }

low-frequency loran [NAV] A modification of standard

loran, which operates in the low-frequency range of approximately 100 to 200 kilohertz to increase range over land and during daytime, and which matches cycles rather than envelopes of pulses to obtain a more accurate fix. Abbreviated LF loran. Also known as cycle-matching loran. { 'lō ,frē·kwən·sē 'lòr,an }

low-frequency padder [ELECTR] In a superheterodyne receiver, a small adjustable capacitor connected in series with the oscillator tuning coil and adjusted during alignment to obtain correct calibration of the circuit at the low-frequency end of the tuning range. { 'lō ,frē·kwən·sē 'pad·ər }

low-frequency propagation [ELECTROMAG] Propagation of radio waves at frequencies between 30 and 300 kilohertz. { 'lō ,frē·kwən·sē ,präp·ə'gā·shən }

low-frequency spectrum [SPECT] Spectrum of atoms and molecules in the microwave region, arising from such causes as the coupling of electronic and nuclear angular momenta, and the Lamb shift. { 'lō ,frē·kwən·sē 'spek·trəm }

low-frequency transconductance [ELECTR] The change in the plate current of a vacuum tube divided by the change in the control-grid voltage that produces it, at frequencies small enough for these two quantities to be considered in phase. { 'lō ,frē·kwən·sē 'tranz·kən'dək·təns }

low-frequency tube [ELECTR] An electron tube operated at frequencies small enough so that the transit time of an electron between electrodes is much smaller than the period of oscillation of the voltage. { 'lō ,frē·kwən·sē 'tüb }

low-grade [MATER] An arbitrary designation of dynamites of less strength than 40. [MIN ENG] Pertaining to ores that have a relatively low content of minerals. Also known as coarse; lean. [SCI TECH] Being of inferior quality. { 'lō ,grād }

low-heat cement [MATER] A chemically altered portland cement with a low initial heat liberation. { 'lō ,hēt si'ment }

low heat value [THERMO] The heat value of a combustion process assuming that none of the water vapor resulting from the process is condensed out, so that its latent heat is not available. Also known as lower heating value; net heating value. { 'lō 'hēt ,val·yü }

low-helix drill [DES ENG] A two-flute twist drill with a lower helix angle than a conventional drill. Also known as slow-spiral drill. { 'lō ,hē·liks ,dril }

low-hydrogen electrode [MET] A covered electrode used in arc welding that provides an atmosphere low in hydrogen. { 'lō ,hī·drə·jən i'lek,trōd }

low-impedance measurement [ELECTR] The measurement of an impedance which is small enough to necessitate use of indirect methods. { 'lō im,pēd·əns 'mezh·ər·mənt }

low-impedance switching tube [ELECTR] A gas tube which has a static impedance on the order of 10,000 ohms, but zero or negative dynamic impedance, and therefore can be used as a relay and transmits information with negligible loss as well. { 'lō im,pēd·əns 'swich·iŋ ,tüb }

low index [METEOROL] A relatively low value of the zonal index which, in middle latitudes, indicates a relatively weak westerly component of wind flow (usually implying stronger north-south motion), and the characteristic weather attending such motion; a circulation pattern of this type is commonly called a low-index situation. { 'lō ¦in,deks }

low-intensity atomizer [MECH ENG] A type of electrostatic atomizer operating on the principle that atomization is the result of Rayleigh instability, in which the presence of charge in the surface counteracts surface tension. { 'lō in,ten·səd·ē 'ad·ə,mīz·ər }

low-intensity reciprocity failure [GRAPHICS] The loss of sensitivity of a photographic plate at low light levels and long exposure times, so that the exposing intensity required to produce a given photographic density is no longer proportional to the reciprocal of the exposure time. { 'lō in,ten·səd·ē ,res·ə'präs·əd·ē ,fāl·yər }

low-key photograph [GRAPHICS] A photograph which has mostly dark tones and very few light or white areas. { 'lō ,kē 'fōd·ə,graf }

low level [ELECTR] The less positive of the two logic levels or states in a digital logic system. { 'lō ,lev·əl }

low-level bombing See low-altitude bombing. { 'lō ,lev·əl 'bäm·iŋ }

low-level condenser [MECH ENG] A direct-contact water-cooled steam condenser that uses a pump to remove liquid from a vacuum space. { 'lō ‚lev·əl kən'den·sər }

low-level counting [NUCLEO] The measurement of very small amounts of radioactivity, such as that generated by long-lived natural radioactive isotopes, and isotopes produced by cosmic rays and nuclear explosions. { 'lō ‚lev·əl 'kaŭnt·iŋ }

low-level language [COMPUT SCI] A computer language consisting of mnemonics that directly correspond to machine language instructions; for example, an assembler that converts the interpreted code of a higher-level language to machine language. { 'lō ‚lev·əl 'laŋ·gwij }

low-level logic circuit [ELECTR] A modification of a diode-transistor logic circuit in which a resistor and capacitor in parallel are replaced by a diode, with the result that a relatively small voltage swing is required at the base of the transistor to switch it on or off. Abbreviated LLL circuit. { 'lō ‚lev·əl 'läj·ik ‚sər·kət }

low-level modulation [ELECTR] Modulation produced at a point in a system where the power level is low compared with the power level at the output of the system. { 'lō ‚lev·əl ‚maj·ə'lā·shən }

low-lift truck [MECH ENG] A hand or powered lift truck that raises the load sufficiently to make it mobile. { 'lō ‚lift ‚trək }

low-loss [ELEC] Having a small dissipation of electric or electromagnetic power. { 'lō ‚lòs }

low-mass x-ray binary [ASTRON] A binary system consisting of a low-mass (typically less than 1 solar mass), late-type star and a neutron star or black hole that accretes material through Roche-lobe overflow, resulting in the emission of relatively soft x-rays. Abbreviated LMXRB. { ‚lō ‚mas ‚eks‚rā 'bī‚ner·ē }

low-melting glass [MATER] Glass to which selenium, thallium, arsenic, or sulfur is added to give melting points of 260–660°F (127–349°C). { 'lō ‚mel·tiŋ 'glas }

low-melting solder See soft solder. { 'lō ‚mel·tiŋ 'säd·ər }

low-moor bog [GEOL] A bog that is at or slightly below the ground water table. { 'lō ‚mür 'bäg }

low-moor peat [GEOL] Peat found in low-moor bogs or swamps and containing little or no sphagnum. Also known as fen peat. { 'lō ‚mür 'pēt }

low-noise amplifier [ELECTR] An amplifier having very low background noise when the desired signal is weak or absent; field-effect transistors are used in audio preamplifiers for this purpose. { 'lō ‚nòiz 'am·plə‚fī·ər }

low-noise preamplifier [ELECTR] A low-noise amplifier placed in a system prior to the main amplifier, sometimes close to the source; used to establish a satisfactory noise figure at an early point in the system. { 'lō ‚nòiz prē'am·plə‚fī·ər }

low-order [COMPUT SCI] Pertaining to the digit which contributes the smallest amount to the value of a numeral, or to its position, or to the rightmost position of a word. { 'lō ‚òr·dər }

low-order burst [ORD] Functioning of a projectile or bomb in which the explosive fails to attain a high-order detonation. { 'lō ‚òr·dər 'bərst }

low-pass band-pass transformation See frequency transformation. { 'lō ‚pas 'band ‚pas ‚tranz·fər‚mā·shən }

low-pass filter [ELEC] A filter that transmits alternating currents below a given cutoff frequency and substantially attenuates all other currents. { 'lō ‚pas 'fil·tər }

low-population zone [ENG] An area of low population density sometimes required around a nuclear installation; the number and density of residents is of concern in providing, with reasonable probability, that effective protection measures can be taken if a serious accident should occur. { 'lō ‚päp·ə'lā·shən ‚zōn }

low-power television station [COMMUN] A television broadcasting facility limited in transmitter output so as to provide reception in only a local area, with a typical service area radius of 3–16 miles (5–26 kilometers). Abbreviated LPTV station. { 'lō ‚paù·ər 'tel·ə‚vish·ən ‚stā·shən }

low-pressure area [MECH ENG] The point in a bearing where the pressure is the least and the area or space for a lubricant is the greatest. { 'lō ‚presh·ər 'er·ē·ə }

low-pressure fluid flow [FL MECH] Flow of fluids below atmospheric pressures, particularly gases and vapors following ideal gas laws, in pipes, fittings, and other common configurations. { 'lō ‚presh·ər 'flü·əd ‚flō }

low-pressure laminate [MATER] A plastic laminate molded and cured at pressures in general of 400 pounds per square inch (approximately 27 atmospheres or 2.8×10^6 pascals). { 'lō ‚presh·ər 'lam·ə‚nat }

low-pressure torch [ENG] A type of torch in which acetylene enters a mixing chamber, where it meets a jet of high-pressure oxygen; the amount of acetylene drawn into the flame is controlled by the velocity of this oxygen jet. Also known as injector torch. { 'lō ‚presh·ər 'tórch }

low-pressure well [PETRO ENG] An oil or gas well with a shut-in wellhead pressure of less than 2000 pounds per square inch absolute (1.38×10^7 pascals). { 'lō ‚presh·ər 'wel }

low-Q filter [ELECTR] A filter in which the energy dissipated in each cycle is a fairly large fraction of the energy stored in the filter. { 'lō ‚kyü 'fil·tər }

low quartz [MINERAL] Quartz that has been formed below 573°C; the tetrahedral crystal structure is less symmetrically arranged than a quartz formed at a higher temperature. { 'lō 'kwórtz }

low-rank graywacke [PETR] A graywacke that is nonfeldspathic. { 'lō ‚raŋk 'grā‚wak·ə }

low-rank metamorphism [GEOL] A metamorphic process that occurs under conditions of low to moderate pressure and temperature. { 'lō ‚raŋk ‚med·ə'mór·fiz·əm }

low-reactance grounding [ELEC] Use of grounding connections with a moderate amount of inductance to effect a moderate reduction in the short-circuit current created by a line-to-ground fault. { 'lō rē‚ak·təns 'graùnd·iŋ }

low-reflection film [OPTICS] A transparent film covering a glass surface, designed so that a small proportion of the light incident will be reflected and a correspondingly large proportion transmitted into the glass. { 'lō ri‚flek·shən 'film }

low relief [GRAPHICS] Sculpture having only a slight projection. Also known as bas-relief. { 'lō ri'lēf }

low-residual-phosphorus copper [MET] Deoxidized copper with a 0.004–0.012% residual phosphorus content. { 'lō ri'zij·ə·wəl ‚fäs·fə·rəs ‚käp·ər }

Lowry process [ENG] A system for wood preservation which uses atmospheric pressure at the start and then introduces preservative into the wood in a vacuum. { 'laù·rē 'prä·səs }

low-shaft furnace [MET] A blast furnace having a short shaft; used to produce pig iron, ferroalloys, alumina, and other products from low-grade ores using low-grade fuel. { 'lō ‚shaft 'fər·nəs }

low side [COMPUT SCI] The part of a controller or other remote device that communicates with terminals or other remote devices, rather than with the host computer. { 'lō ‚sīd }

low-speed wind tunnel [ENG] A wind tunnel that has a speed up to 300 miles (480 kilometers) per hour and the essential features of most wind tunnels. { 'lō ‚spēd 'win ‚tən·əl }

low-surface-brightness galaxy [ASTRON] A galaxy whose spatial density of stars is so low that it is almost invisible. { 'lō 'sər·fəs ‚brīt·rəs ‚gal·ik·sē }

low-technology robot [CONT SYS] The simplest type of robot, with only two or three degrees of freedom, and only the end points of motion specified, using fixed and adjustable stops. { 'lō tek'näl·ə·jē 'rō‚bät }

low-temperature carbonization [CHEM ENG] Low-temperature destructive distillation of coal to produce liquid products. { 'lō ‚tem·prə·chər ‚kär·bə·nə'zā·shən }

low-temperature coke [MATER] Coke produced at temperatures of 500–750°C, used chiefly for house heating, particularly in England. Also known as char. { 'lō ‚tem·prə·chər 'kōk }

low-temperature hygrometry [ENG] The study that deals with the measurement of water vapor at low temperatures; the techniques used differ from those of conventional hygrometry because of the extremely small amounts of moisture present at low temperatures and the difficulties imposed by the increase of the time constants of the standard instruments when operated at these temperatures. { 'lō ‚tem·prə·chər hī'gräm·ə·trē }

low-temperature physics [CRYO] A study of the properties of gross matter at low temperatures, especially at temperatures so low that the quantum character of the substance becomes observable in effects such as superconductivity, superfluid liquid helium, magnetic cooling, and nuclear orientation. { 'lō ‚tem·prə·chər 'fiz·iks }

low-temperature production [CRYO] Production of temperatures from about 80 K down to about 10^{-6} K by techniques such as isentropic expansion of gases, refrigeration cycles, and

LOW-LEVEL LOGIC CIRCUIT

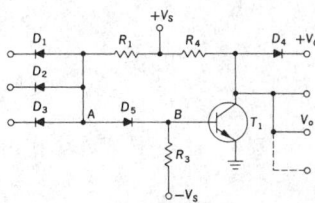

Circuit diagram of low-level logic circuit. D_1–D_5 = diodes; R_1–R_4 = resistors; T = transistor; $+V_s$ = positive supply voltage; B = common node.

adiabatic demagnetization. { 'lō ,tem·prə·chər prə'dək·shən }

low-temperature separation [CHEM ENG] Liquid condensate recovery from wet gases at temperatures of 20 to −20°F (−6.7 to −28.9°C), the temperature range at which the gas-oil separator operates. { 'lō ,tem·prə·chər ,sep·ə'rā·shən }

low-temperature thermometry [CRYO] The assignment of numbers on the Kelvin absolute temperature scale to achievable and reproducible low-temperature states, and the choice and calibration of suitable instruments for the practical measurement of low temperatures, such as thermocouples, and resistance, vapor-pressure, gas, and magnetic thermometers. { 'lō ,tem·prə·chər thər'mäm·ə·trē }

low tide *See* low water. { 'lō 'tīd }

low-tide terrace [GEOL] A flat area of a beach adjacent to the low-water line. { 'lō ¦tīd 'ter·əs }

low-tier system [COMMUN] A wireless telephone system that provides high-quality and low-delay voice and data capabilities but has small cells. { 'lō 'tēr ,sis·təm }

low velocity [MECH] Muzzle velocity of an artillery projectile of 2499 feet (762 meters) per second or less. { 'lō və'läs·əd·ē }

low-velocity drop [ORD] The act or process of delivering personnel, supplies, or equipment from aircraft in flight, utilizing sufficient parachute retardation to prevent injury or damage upon ground impact. { 'lō və¦läs·əd·ē ,dräp }

low-velocity layer [GEOPHYS] A layer in the solid earth in which seismic wave velocity is lower than the layers immediately below or above. { 'lō və¦läs·əd·ē ,lā·ər }

low-velocity star [ASTRON] One of the Population I stars in the spiral arms of a galaxy which participate in the galactic rotation, thus exhibiting low velocity with respect to the sun and high velocity with respect to the galactic center. { 'lō və¦läs·əd·ē 'stär }

low-volatile coal [GEOL] A coal that is nonagglomerating, has 78% to less than 86% fixed carbon, and 14% to less than 22% volatile matter. { 'lō ¦väl·ət·əl 'kōl }

low voltage [ELEC] **1.** Voltage which is small enough to be regarded as safe for indoor use, usually 120 volts in the United States. **2.** Voltage which is less than that needed for normal operation; a result of low voltage may be burnout of electric motors due to loss of electromotive force. { 'lō 'vōl·tij }

low-voltage relay [COMMUN] A relay that responds to the drop in voltage (increase in current) when a telephone line becomes active; used to activate interception and eavesdropping equipment. { 'lō ¦vōl·tij 'rē,lā }

low-voltage winding [ELECTROMAG] The coil of wire wound around the core of a power transformer which has the smaller number of turns, and therefore the lower voltage. { 'lō ¦vōl·tij 'wīnd·iŋ }

low water [OCEANOGR] The lowest limit of the surface water level reached by the lowering tide. Also known as low tide. { 'lō 'wȯd·ər }

low-water fuel cutoff [MECH ENG] A float device which shuts off fuel supply and burner when boiler water level drops below the lowest safe waterline. { 'lō ,wȯd·ər ,fyül ,kə,dȯf }

low-water inequality [OCEANOGR] The difference between the heights of two successive low tides. { 'lō ,wȯd·ər ,in·i'kwäl·əd·ē }

low-water interval *See* low-water lunitidal interval. { 'lō ,wȯd·ər 'in·tər·vəl }

low-water lunitidal interval [GEOPHYS] For a specific location, the interval of time between the transit (upper or lower) of the moon and the next low water. Also known as low-water interval. { 'lō ,wȯd·ər ¦lün·ə¦tīd·əl 'in·tər·vəl }

low-water neaps *See* mean low-water neaps. { 'lō ,wȯd·ər 'nēps }

low-water springs *See* mean low-water springs. { 'lō ,wȯd·ər 'spriŋz }

low-zone tolerance [IMMUNOL] Immunologic tolerance induced by repeated administration of very low doses of a protein antigen. { 'lō ¦zōn 'täl·ə·rəns }

LOX *See* liquid-oxygen explosive. { läks }

loxodont [VERT ZOO] Having molar teeth with shallow hollows between the ridges. { 'läk·sə,dänt }

loxodrome *See* rhumb line. { 'läk·sə,drōm }

loxodromic spiral [MATH] A curve on a surface of revolution which cuts the meridians at a constant angle other than 90°. { ¦lak·sə¦dräm·ik 'spī·rəl }

loxolophodont [VERT ZOO] Having crests on the molar teeth that connect three of the tubercles and with the fourth or posterior inner tubercle being rudimentary or absent. { ¦läk·sə¦läf·ə,dänt }

Loxonematacea [PALEON] An extinct superfamily of gastropod mollusks in the order Prosobranchia. { ,läk·sə,ne·mə'tās·ē·ə }

lozenge file [DES ENG] A small file with four sides and a lozenge-shaped cross section; used in forming dies. { 'läz·ənj ,fīl }

L pad [ENG ACOUS] A volume control having essentially the same impedance at all settings. { 'el ,pad }

LPE *See* liquid-phase epitaxy.

LPF process [MIN ENG] Recovery of metals from tailings by a sequence of leaching, precipitation, and flotation. { ¦el¦pē'ef ,präs·əs }

LPG *See* liquefied petroleum gas.

LPM *See* lines per minute.

LP record *See* long-playing record. { ¦el¦pē 'rek·ərd }

l-process [NUC PHYS] The synthesis of certain light nuclides through the breakup of heavier nuclides, probably by cosmic-ray bombardment of the interstellar medium. Also known as LiBeB process. { 'el ,prä·səs }

LPTV station *See* low-power television station. { ¦el¦pē¦tē've ,stā·shən }

LQG problem *See* linear-quadratic-gaussian problem. { ¦el¦kyü¦jē ,präb·ləm }

Lr *See* lawrencium.

LRC *See* longitudinal redundancy check.

LRRP *See* lowest required radiating power.

LSA diode [ELECTR] A microwave diode in which a space charge is developed in the semiconductor by the applied electric field and is dissipated during each cycle before it builds up appreciably, thereby limiting transit time and increasing the maximum frequency of oscillation. Derived from limited space-charge accumulation diode. { ¦el¦es'ā 'dī,ōd }

LSA mode *See* limited space-charge accumulation mode. { ¦el¦es'ā ,mōd }

LSB *See* least significant bit.

L scan *See* L scope. { 'el ,skan }

L scope [ELECTR] A cathode-ray scope on which a trace appears as a vertical or horizontal range scale, the signals appearing as left and right horizontal (or up and down vertical) deflections as echoes are received by two antennas, the left and right (or up and down) deflections being proportional to the strength of the echoes received by the two antennas. Also known as L indicator; L scan. { 'el ,skōp }

LS coupling *See* Russell-Saunders coupling. { ¦el¦es ,kəp·liŋ }

LSD *See* dock landing ship; lysergic acid diethylamide.

LSD-25 *See* lysergic acid diethylamide.

L shell [ATOM PHYS] The second shell of electrons surrounding the nucleus of an atom, having electrons whose principal quantum number is 2. { 'el ,shel }

LSI circuit *See* large-scale integrated circuit. { ¦el¦es'ī ,sər·kət }

L-1 test [ENG] A 480-hour engine test in a single-cylinder Caterpillar diesel engine to determine the detergency of heavy-duty lubricating oils. { ¦el 'wən ,test }

L-2 test [ENG] An engine test made in a single-cylinder Caterpillar diesel engine to determine the oiliness of an engine oil. Also known as scoring test. { ¦el 'tü ,test }

L-3 test [ENG] An engine test in a four-cylinder Caterpillar engine to determine stability of crankcase oil at high temperatures and under severe operating conditions. { ¦el 'thrē ,test }

L-4 test [ENG] An engine test in a six-cylinder spark-ignition Chevrolet engine to evaluate crankcase oil oxidation stability, bearing corrosion, and engine deposits. { ¦el 'fȯr ,test }

L-5 test [ENG] An engine test in a General Motors diesel engine to determine detergency, corrosiveness, ring sticking, and oxidation stability properties of lubricating oils. { ¦el 'fīv ,test }

LTPD *See* lot tolerance percent defective.

LTRS *See* letters shift.

Lu *See* lutetium.

lub *See* least upper bound.

lubber line *See* lubber's line. { 'ləb·ər ,līn }

lubber's line [NAV] A reference line on any direction-indicating instrument, marking the reading which coincides with the heading. Also known as lubber line; lubber's point. { 'ləb·ərz ,līn }

lubber's-line error [NAV] In a magnetic compass, the angular difference between the heading as indicated by a lubber's line, and the actual heading; this error is caused by faulty calibration. { 'ləb·ərz ,līn ,er·ər }

lubber's point See lubber's line. { 'ləb·ərz ,póint }

lube cut [MATER] The distilled fraction of crude oil with suitable boiling range and viscosity to yield a lubricating oil when it is completely refined. Also known as lube-oil distillate; lube stock. { 'lüb ,kət }

lube oil See lubricating oil. { 'lüb ,óil }

lube-oil distillate See lube cut. { 'lüb ,óil 'dis·tə,lāt }

lube stock See lube cut. { 'lüb ,stäk }

lubricant [MATER] A substance used to reduce friction between parts or objects in relative motion. { 'lü·brə·kənt }

lubricant additive [MATER] Any material added to lubricants (greases or oils) to give the product special properties, such as resistance to extremes of pressure, cold, or heat, improved viscosity, and detergency. { 'lü·brə·kənt ,ad·əd·iv }

lubricated gasoline [MATER] A motor gasoline into which a lubricant has been added. { 'lü·brə,kād·əd 'gas·ə,lēn }

lubricating film [MATER] A thin layer of oil or grease applied between rubbing surfaces. { 'lü·brə,kād·iŋ ,film }

lubricating grease [MATER] A solid or semisolid lubricant consisting of a thickening agent (soap or other additives) in a fluid lubricant (usually petroleum lubricating oil). { 'lü·brə,kād·iŋ ,grēs }

lubricating oil [MATER] Selected fractions of refined petroleum or other oils (with or without additives) used to lessen friction between moving surfaces. Also known as lube oil. { 'lü·brə,kād·iŋ ,óil }

lubrication action [MATER] The ability of the lubricant to maintain a fluid film between solid surfaces and to prevent their physical contact. { ,lü·brə'kā·shən ,ak·shən }

lubricator [ENG] A device for applying a lubricant. { 'lü·brə,kād·ər }

lubricity [MATER] The ability of a material to lubricate. { lü'bris·əd·ē }

Lucanidae [INV ZOO] The stag beetles, a cosmopolitan family of coleopteran insects in the superfamily Scarabaeoidea. { lü'kan·ə,dē }

Lucas numbers [MATH] The terms of the Fibonacci sequence whose first two terms are 1 and 3. { 'lü·kəs ,nəm·bərz }

lucca oil See olive oil. { 'lü·kə ,óil }

lucerne See alfalfa. { lü'sərn }

Lucibacterium [MICROBIO] A genus of light-emitting bacteria in the family Vibrionaceae; motile, asporogenous rods with peritrichous flagella. { ,lü·si,bak'tir·ē·əm }

luciferase [BIOCHEM] An enzyme that catalyzes the oxidation of luciferin. { lü'sif·ə,rās }

luciferin [BIOCHEM] A species-specific pigment in many luminous organisms that emits heatless light when combined with oxygen. { lü'sif·ə·rən }

Luciocephalidae [VERT ZOO] A family of fresh-water fishes in the suborder Anabantoidei. { ,lü·sē·ō·sə'fal·ə,dē }

Luckiesh-Moss visibility meter [ENG] A type of photometer that consists of two variable-density filters (one for each eye) that are adjusted so that an object seen through them is just barely discernible; the reduction in visibility produced by the filters is read on a scale of relative visibility related to a standard task. { lü'kēsh 'mòs ,viz·ə'bil·əd·ē ,mēd·ər }

Lüders' lines [MET] Surface markings on a metal caused by flow of the material strained beyond its elastic limit. Also known as deformation bands; Hartmann lines; Lüders' bands; Piobert lines; stretcher strains. { 'lüd·ərz ,līnz }

Ludian [GEOL] A European stage of geologic time in the uppermost Eocene, above the Bartonian and below the Tongrian of the Oligocene. { 'lü·dē·ən }

ludlamite [MINERAL] $(Fe,Mg,Mn)_3(PO_4)_2 \cdot 4H_2O$ A green mineral crystallizing in the monoclinic system and occurring in small, transparent crystals. { 'ləd·lə,mīt }

Ludlovian [GEOL] A European stage of geologic time; Upper Silurian, below Gedinnian of Devonian, above Wenlockian. { ləd'lō·vē·ən }

ludwigite [MINERAL] $(Mg,Fe)_2FeBO_5$ A blackish-green mineral that crystallizes in the monoclinic system and occurs in fibrous masses; isomorphous with ronsenite. { 'ləd,wi,gīt }

Ludwig's angina [MED] Acute streptococcal cellulitis of the floor of the mouth. { 'ləd,wigz 'an·jə·nə }

Ludwig-Soret effect [THERMO] A phenomenon in which a temperature gradient in a mixture of substances gives rise to a concentration gradient. { 'lüd,vik sə'rā i fekt }

Luenberger observer [CONT SYS] A compensator driven by both the inputs and measurable outputs of a control system. { 'lün,bərg·ər əb'zər·vər }

lueneburgite [MINERAL] $Mg_3B_2(OH)_6(PO_4)_2 \cdot 6H_2O$ A colorless mineral composed of hydrous basic phosphate of magnesium and boron. { 'lü·nə·bər,gīt }

lueshite [MINERAL] $NaNbO_3$ An orthorhombic mineral having perovskite-type structure; it is dimorphous with natroniobite. { 'lü·əs,hīt }

lug [DES ENG] A projection or head on a metal part to serve as a cap, handle, support, or fitting connection. { ləg }

luganot [METEOROL] A strong south or south-southeast wind of Lake Garda, Italy. { lü'gä,nót }

lug bolt [DES ENG] **1.** A bolt with a flat extension or hook instead of a head. **2.** A bolt designed for securing a lug. { 'ləg ,bōlt }

lug brick [MATER] A brick with lugs for spacing adjacent bricks. { 'ləg ,brik }

Luggin probe [PHYS CHEM] A device which transmits a significant current density on the surface of an electrode to measure its potential. { 'ləg·ən ,prōb }

Lugol solution [CHEM] A solution of 5 grams of iodine and 10 grams of potassium iodide per 100 milliliters of water; used in medicine. { 'lü,gól sə'lü·shən }

LUHF See lowest useful high frequency.

Luidiidae [INV ZOO] A family of echinoderms in the suborder Paxillosina. { lü·ə'dī·ə,dē }

Luisian [GEOL] A North American stage of geologic time: Miocene (above Relizian, below Mohnian). { lü'ē·shən }

Lukasiewicz notation See Polish notation. { lü,kä·shē'ā,vits nō,tā·shən }

lum See trolley. { ləm }

lumbago [MED] Backache in the lumbar or lumbosacral region. { ,ləm'bā,gō }

lumbang oil [MATER] Colorless or yellow liquid with pleasant aroma and bland taste; soluble in alcohol, ether, chloroform, and carbon disulfide; expressed from candlenut; used as an illuminant and wood preservative, and in paints, calking, and soap manufacture. Also known as candlenut oil. { 'ləm'bäŋ ,óil }

lumbar artery [ANAT] Any of the four or five pairs of branches of the abdominal aorta opposite the lumbar region of the spine; supplies blood to loin muscles, skin on the sides of the abdomen, and the spinal cord. { 'ləm,bär 'ärd·ə·rē }

lumbar nerve [NEUROSCI] Any of five pairs of nerves arising from lumbar segments of the spinal cord; characterized by motor, visceral sensory, somatic sensory, and sympathetic components; they innervate the skin and deep muscles of the lower back and the lumbar plexus. { 'ləm,bär 'nərv }

lumbar vertebrae [ANAT] Those vertebrae located between the lowest ribs and the pelvic girdle in all vertebrates. { 'ləm,bär 'vərd·ə,brā }

lumber [MATER] Logs that have been sawed and prepared for market. { 'ləm·bər }

lumberg [OPTICS] A unit of luminous energy equal to the luminous energy corresponding to a radiant energy of $1/K$ ergs, where K is the luminous efficiency in lumens per watt. Formerly known as lumerg. { 'ləm,bərg }

lumbodorsal fascia [ANAT] The sheath of the erector spinae muscle alone, or the sheaths of the erector spinae and the quadratus lumborum muscles. { ,ləm·bō'dor·səl 'fā·shə }

lumbosacral plexus [NEUROSCI] A network formed by the anterior branches of lumbar, sacral, and coccygeal nerves which for descriptive purposes are divided into the lumbar, sacral, and pudendal plexuses. { ,ləm·bō'sak·rəl 'plek·səs }

Lumbricidae [INV ZOO] A family of annelid worms in the order Oligochaeta; includes the earthworm. { ləm'bris·ə,dē }

Lumbriclymeninae [INV ZOO] A subfamily of mud-swallowing sedentary worms in the family Maldanidae. { ,ləm·bri·klī'men·ə,nē }

Lumbriculidae [INV ZOO] A family of aquatic annelids in the order Oligochaeta. { ,ləm·bri'kyül·ə,dē }

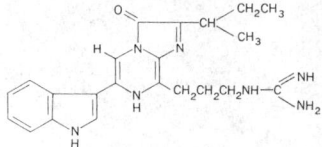

LUCIFERIN

Structural formula of *Cypridina* luciferin.

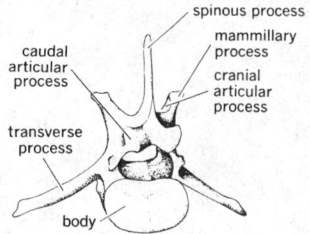

LUMBAR VERTEBRAE

Fifth lumbar vertebra of the dog, the caudal lateral aspect showing the body or centrum. (*From M. E. Miller, G. C. Christensen, and H. E. Evans, Anatomy of the Dog, Saunders, 1964*)

Lumbricus [INV ZOO] A genus of earthworms recognized as the type genus of the family Lumbricidae. { 'ləm·brə·kəs }

Lumbrineridae [INV ZOO] A family of errantian polychaetes in the superfamily Eunicea. { ,ləm·bri'ner·ə,dē }

lumen [ANAT] The interior space within a tubular structure, such as within a blood vessel, a duct, or the intestine. [OPTICS] The unit of luminous flux, equal to the luminous flux emitted within a unit solid angle (1 steradian) from a point source having a uniform intensity of 1 candela, or to the luminous flux received on a unit surface, all points of which are at a unit distance from such a source. Symbolized lm. [SCI TECH] The space within a tube. { 'lü·mən }

lumen-hour [OPTICS] A unit of quantity of light (luminous energy), equal to the quantity of light radiated or received for a period of 1 hour by a flux of 1 lumen. Abbreviated lm-hr. { 'lü·mən ¦au̇r }

lumen per watt [OPTICS] The unit of luminosity factor and of luminous efficacy. Abbreviated lm/w. { 'lü·mən pər 'wät }

lumen-second [OPTICS] A unit of quantity of light (luminous energy), equal to the quantity of light radiated or received for a period of 1 second by a flux of 1 lumen. Abbreviated lm-sec. { 'lü·mən ¦sek·ənd }

lumerg *See* lumberg. { 'lü,mərg }

luminaire [ELEC] An electric lighting fixture, wall bracket, portable lamp, or other complete lighting unit designed to contain one or more electric lighting sources and associated reflectors, refractors, housing, and such support for those items as necessary. { 'lü·mə¦ner }

luminance [OPTICS] The ratio of the luminous intensity in a given direction of an infinitesimal element of a surface containing the point under consideration, to the orthogonally projected area of the element on a plane perpendicular to the given direction. Formerly known as brightness. { 'lü·mə·nəns }

luminance carrier *See* picture carrier. { 'lü·mə·nəns ,kar·ē·ər }

luminance channel [COMMUN] A path intended primarily for the luminance signal in a color television system. { 'lü·mə·nəns ,chan·əl }

luminance factor [OPTICS] The ratio of the luminance of a body when illuminated and observed under certain conditions to that of a perfect diffuser under the same conditions. { 'lü·mə·nəns ,fak·tər }

luminance primary [COMMUN] One of the three transmission primaries whose amount determines the luminance of a color in a color television system. { 'lü·mə·nəns 'prī,mer·ē }

luminance signal [COMMUN] The color television signal that is intended to have exclusive control of the luminance of the picture. Also known as Y signal. { 'lü·mə·nəns ,sig·nəl }

luminescence [PHYS] Light emission that cannot be attributed merely to the temperature of the emitting body, but results from such causes as chemical reactions at ordinary temperatures, electron bombardment, electromagnetic radiation, and electric fields. { ,lü·mə'nes·əns }

luminescent [PHYS] Capable of exhibiting luminescence. { ,lü·mə'nes·ənt }

luminescent cell *See* electroluminescent panel. { ,lü·mə'nes·ənt 'sel }

luminescent center [SOLID STATE] A point-lattice defect in a transparent crystal that exhibits luminescence. { ,lü·mə'nes·ənt 'sen·tər }

luminescent dye [MATER] A dye that is made luminous by excitation with an outside energy source; used in luminous paint. { ,lü·mə'nes·ənt 'dī }

luminescent screen [ELECTR] The screen in a cathode-ray tube, which becomes luminous when bombarded by an electron beam and maintains its luminosity for an appreciable time. { ,lü·mə'nes·ənt 'skrēn }

luminol [ORG CHEM] $C_8H_7N_3O_2$ A white, water-soluble, crystalline compound that melts at 320°C; used in an alkaline solution for analytical testing in chemistry. Also known as 3-aminophthalic hydrazide. { 'lü·mə,nȯl }

luminophor [PHYS] A luminescent material that converts part of the absorbed primary energy into emitted luminescent radiation. Also known as fluophor; fluor; phosphor. { lü'min·ə,fȯr }

luminosity [NUCLEO] A measure of the performance of a colliding-beam system, equal to the reaction rate or number or interactions per second divided by the interaction cross section. [OPTICS] *See* luminosity factor. { ,lü·mə'näs·əd·ē }

luminosity classes [ASTRON] A classification of stars in an orderly sequence according to their absolute brightness. { ,lü·mə'näs·əd·ē ,klas·əz }

luminosity curve *See* luminosity function. { ,lü·mə'näs·əd·ē ,kərv }

luminosity factor [OPTICS] The ratio of luminous flux in lumens emitted by a source at a particular wavelength to the corresponding radiant flux in watts at the same wavelength; thus this is a measure of the visual sensitivity of the eye. Also known as luminosity. { ,lü·mə'näs·əd·ē ,fak·tər }

luminosity function [ASTRON] The functional relationship between stellar magnitude and the number and distribution of stars of each magnitude interval. Also known as relative luminosity factor. [OPTICS] A standard measure of the response of an eye to monochromatic light at various wavelengths; the function is normalized to unity at its maximum value. Also known as luminosity curve; spectral luminous efficiency; visibility function. { ,lü·mə'näs·əd·ē ,fəŋk·shən }

luminosity monitor [NUCLEO] A device, located on the inside of the detector of a colliding-beam accelerator near the two entering beams, that gives a signal proportional to the total number of collisions that occur at the interaction point. { ,lü·mə'näs·əd·ē ,män·əd·ər }

luminous blue variable [ASTRON] Any of a small group of high-luminosity, unstable, hot supergiant stars that have irregular eruptions or ejections with greatly enhanced mass outflow (10^{-5} to 10^{-4} solar mass per year). { 'lüm·ən·əs ¦blü 'ver·ē·ə·bəl }

luminous cloud *See* sheet lightning. { 'lü·mə·nəs 'klau̇d }

luminous coefficient [OPTICS] A measure of the fraction of the radiant power of a light source which contributes to its luminous properties, equal to the average of the luminosity function at various wavelengths, weighted according to the spectral intensity of the source. Also known as luminous efficiency. { 'lü·mə·nəs ,kō·i'fish·ənt }

luminous efficacy [OPTICS] **1.** The ratio of the total luminous flux in lumens emitted by a light source over all wavelengths to the total radiant flux in watts. Formerly known as luminous efficiency. **2.** The ratio of the total luminous flux emitted by a light source to the power input of the source; expressed in lumens per watt. { 'lü·mə·nəs ,ef·ə·kə·sē }

luminous efficiency *See* luminous coefficient; luminous efficacy. { 'lü·mə·nəs i'fish·ən·sē }

luminous emittance [OPTICS] The emittance of visible radiation weighted to take into account the different response of the human eye to different wavelengths of light; in photometry, luminous emittance is always used as a property of a self-luminous source, and therefore should be distinguished from luminance. Also known as luminous exitance. { 'lü·mə·nəs i'mit·əns }

luminous energy [OPTICS] The total radiant energy emitted by a source, evaluated according to its capacity to produce visual sensation; measured in lumen-hours or lumen-seconds. { 'lü·mə·nəs 'en·ər·jē }

luminous exitance *See* luminous emittance. { 'lü·mə·nəs 'ek·səd·əns }

luminous flux [OPTICS] The time rate of flow of radiant energy, evaluated according to its capacity to produce visual sensations; measured in lumens. { 'lü·mə·nəs 'fləks }

luminous flux density *See* illuminance. { 'lü·mə·nəs 'fləks 'den·səd·ē }

luminous intensity [OPTICS] The luminous flux incident on a small surface which lies in a specified direction from a light source and is normal to this direction, divided by the solid angle (in steradians) which the surface subtends at the source of light. Also known as light intensity. { 'lü·mə·nəs in'ten·səd·ē }

luminous mass [ASTRON] The mass of a celestial object inferred from its luminosity or the luminosities of its components. { 'lü·mə·nəs 'mas }

luminous meteor [METEOROL] According to United States weather observing practice, any one of a number of atmospheric phenomena which appear as luminous patterns in the sky, including halos, coronas, rainbows, aurorae, and their many variations, but excluding lightning (an igneous meteor or electrometeor). { 'lü·mə·nəs 'mēd·ē·ər }

luminous nebula [ASTRON] A nebula made bright by radiation from stars in the vicinity. { 'lü·mə·nəs 'neb·yə·lə }

luminous paint [MATER] A type of paint in which luminous pigments are used. { 'lü·mə·nəs 'pānt }

luminous pigment [MATER] A pigment that absorbs light energy and radiates visible light when exposed to ultraviolet light; made of phosphors such as strontium, zinc, and cadmium sulfides. { 'lü·mə·nəs 'pig·mənt }

luminous quantities [OPTICS] Physical quantities used in photometry, such as luminous intensity and luminance, which are based on the response of the human eye, and are thus weighted to take into account the difference in response at different wavelengths of light. { 'lü·mə·nəs 'kwän·əd·ēz }

luminous range [NAV] The distance at which a marine light may be seen in clear weather, expressed in nautical miles. { 'lü·mə·nəs 'rānj }

luminous sensitivity [ELECTR] For a phototube, the quotient of the anode current by the incident luminous flux. { 'lü·mə·nəs ,sen·sə'tiv·əd·ē }

luminous time ratio [NAV] Of a navigational light, the ratio of the length of a flash to the period of rotation. { 'lü·mə·nəs 'tīm ,rā·shō }

luminous visibility diagram [NAV] A diagram by which the luminous ranges, as given in light lists, may be adjusted to various conditions of visibility. { 'lü·mə·nəs ,viz·ə'bil·əd·ē ,dī·ə,gram }

Lummer-Brodhun sight box [OPTICS] A device, having a series of prisms, for viewing simultaneously the two sides of a white diffuse plaster screen illuminated by light sources whose luminous intensities are being compared. { 'lüm·ər 'bröd,hün 'sīt ,bäks }

Lummer-Gehrcke plate [OPTICS] An interferometer consisting of a glass or quartz plate with parallel surfaces and sizable thickness in which multiple reflections take place. { 'lüm·ər 'ger·kə ,plāt }

LUMO See lowest unoccupied molecular orbital. { ¦lü,mō or ¦el¦yü¦em'ō }

lump coal [MIN ENG] Bituminous coal that passes through a 6-inch (15-centimeter) round mesh in initial screening. { 'ləmp ,kōl }

lumpectomy [MED] Surgical removal of a tumor in the breast along with a small amount of surrounding tissue. { ləm'pek·tə·mē }

lumped constant [ELEC] A single constant that is electrically equivalent to the total of that type of distributed constant existing in a coil or circuit. Also known as lumped parameter. { 'ləmpt 'kän·stənt }

lumped-constant network [ELEC] An analytical tool in which distributed constants (inductance, capacitance, and resistance) are represented as hypothetical components. { 'ləmpt ¦kän·stənt 'net,wərk }

lumped discontinuity [ELECTROMAG] An analytical tool in the study of microwave circuits in which the effective values of inductance, capacitance, and resistance representing a discontinuity in a waveguide are shown as discrete components of equivalent value. { 'ləmpt ,dis,känt·ən'ü·əd·ē }

lumped element [ELECTROMAG] A section of a transmission line designed so that electric or magnetic energy is concentrated in it at specified frequencies, and inductance or capacitance may therefore be regarded as concentrated in it, rather than distributed over the length of the line. { 'ləmpt 'el·ə·mənt }

lumped impedance [ELECTROMAG] An impedance concentrated in a single component rather than distributed throughout the length of a transmission line. { 'ləmpt im'pēd·əns }

lumped parameter See lumped constant. { 'ləmpt pə'ram·əd·ər }

lumper [SYST] A taxonomist who tends to recognize large taxa. { 'ləm·pər }

lumpy jaw See actinomycosis. { 'ləm·pē ,jö }

Luna [ASTRON] A name for the moon. { 'lü·nə }

lunabase [ASTRON] The basic rocks that make up the dark portions of the lunar surface. Also known as marebase; marial rocks. { 'lü·nə,bās }

Luna program [AERO ENG] A series of Soviet space probes launched for flight missions to the moon. { 'lü·nə ,prō·grəm }

lunar appulse [ASTRON] An eclipse of the moon in which the penumbral shadow of the earth falls on the moon. Also known as penumbral eclipse. { 'lü·nər 'a,pəls }

lunar atmosphere [ASTROPHYS] The volatile elements postulated to have been present on the moon's surface at one time. { 'lü·nər 'at·mə,sfir }

lunar atmospheric tide [METEOROL] An atmospheric tide due to the gravitational attraction of the moon; the only detectable components are the 12-lunar-hour or semidiurnal component, as in the oceanic tides, and two others of very nearly the same period; the amplitude of this atmospheric tide is so small that it is detected only by careful statistical analysis of a long record. { 'lü·nər ,at·mə,sfir·ik 'tīd }

lunar caustic [MATER] A form of toughened silver nitrate consisting of 97–98% silver nitrate and 2–3% silver chloride. Also known as fused silver nitrate; molded silver nitrate. { 'lün·ər 'kòs·tik }

lunar crater [ASTRON] A crater on the moon's surface. { 'lü·nər 'krād·ər }

lunar crust [ASTRON] The outer layer of the moon. { 'lü·nər 'krəst }

lunar day [ASTRON] The time interval between two successive crossings of the meridian by the moon. { 'lü·nər 'dā }

lunar dust [ASTRON] Small particles adhering to the moon's surface. { 'lü·nər 'dəst }

lunar eclipse [ASTRON] Obscuration of the full moon when it passes through the shadow of the earth. { 'lü·nər i'klips }

lunar ephemeris [ASTRON] A computed list of positions the moon will occupy in the sky on certain dates. { 'lü·nər i'fem·ə·rəs }

lunar excursion module [AERO ENG] A manned spacecraft designed to be carried on top of the Apollo service module and having its own power plant for making a manned landing on the moon and a return from the moon to the orbiting Apollo spacecraft. Abbreviated LEM. Also known as lunar module (LM). { 'lü·nər ik¦skər·zhən ¦maj·ül }

lunar flight [AERO ENG] Flight by a spacecraft to the moon. { 'lü·nər 'flīt }

lunar geology See selenology. { 'lü·nər jē'äl·ə·jē }

lunar inequality [ASTRON] Variation in the moon's motion in its orbit, due to attraction by other bodies of the solar system. [GEOPHYS] A minute fluctuation of a magnetic needle from its mean position, caused by the moon. { 'lü·nər in·i'kwäl·əd·ē }

lunar interval [ASTRON] The difference in time between the transit of the moon over the Greenwich meridian and a local meridian; the lunar interval equals the difference between the Greenwich and local intervals of a tide or current phase. { 'lü·nər 'in·tər·vəl }

lunarite [ASTRON] The rocks that make up the bright portions of the lunar surface. { 'lü·nə,rīt }

lunar libration [ASTRON] **1.** The effect wherein the face of the moon appears to swing east and west about 8° from its central position each month. Also known as apparent libration in longitude. **2.** The state wherein the inclination of the moon's polar axis allows an observer on earth to see about 59% of the moon's surface. Also known as libration in latitude. **3.** The small oscillation with which the moon rocks back and forth about its mean rotation rate. Also known as physical libration of the moon. { 'lü·nər lī'brā·shən }

lunar magnetic field [ASTROPHYS] The magnetic field of the moon. { 'lü·nər mag¦ned·ik 'fēld }

lunar mass [ASTROPHYS] The mass of the moon. { 'lü·nər 'mas }

lunar meteoroid [ASTRON] A meteoric particle before it strikes the moon. { 'lü·nər 'mēd·ē·ə,ròid }

lunar module See lunar excursion module. { 'lü·nər 'maj·ül }

lunar month [ASTRON] The period of revolution of the moon about the earth, especially a synodical month. { 'lü·nər 'mənth }

lunar mountain [ASTRON] A mountain on the moon. { 'lü·nər 'maùnt·ən }

lunar node [ASTRON] A node of the moon's orbit. { 'lü·nər 'nōd }

lunar nodule [ASTRON] A rock nodule found on the moon. { 'lü·nər 'näj·ül }

lunar noon [ASTRON] The instant at which the sun is over the upper branch of any meridian of the moon. { 'lü·nər 'nün }

lunar nutation [ASTRON] A nodding motion of the earth's axis caused by the inclination of the moon's orbit to the ecliptic; it can displace the celestial pole by 9 seconds of arc from

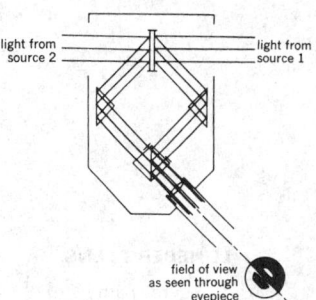

LUMMER-BRODHUN SIGHT BOX

Lummer-Brodhun contrast sight box.

its mean position and has a period of 18.6 years. { 'lü·nər nü'tā·shən }

lunar orbit [AERO ENG] Orbit of a spacecraft around the moon. { 'lü·nər 'ör·bət }

lunar polarization [ASTROPHYS] Polarization of light by the moon's surface. { 'lü·nər ‚pō·lə·rə'zā·shən }

lunar pole [ASTRON] A pole of the moon. { 'lü·nər 'pōl }

lunar probe [AERO ENG] Any space probe launched for flight missions to the moon. { 'lü·nər 'prōb }

lunar rainbow See moonbow. { 'lü·nər 'rān‚bō }

lunar rock [ASTRON] Rock found on the moon. { 'lü·nər 'räk }

lunar satellite [AERO ENG] A satellite making one or more revolutions about the moon. { 'lü·nər 'sad·əl‚īt }

lunar spacecraft [AERO ENG] A spacecraft designed for flight to the moon. { 'lü·nər 'spās‚kraft }

lunar tide [OCEANOGR] The portion of a tide produced by forces of the moon. { 'lü·nər 'tīd }

lunar time [ASTRON] **1.** Time based upon the rotation of the earth relative to the moon; it may be designated as local or Greenwich, as the local or Greenwich meridian is used as the reference. **2.** Time on the moon. { 'lü·nər 'tīm }

lunar topology [ASTRON] Topology of the moon. { 'lü·nər tə'päl·ə·jē }

lunar year [ASTRON] A time interval comprising 12 lunar (synodic) months. { 'lü·nər 'yir }

lunate [BIOL] Crescent-shaped { 'lü‚nāt }

lunate bar [GEOL] A crescent-shaped bar of sand that is frequently found off the entrance to a harbor. { 'lü‚nāt 'bär }

lunation [ASTRON] The time period between two successive new moons. { lü'nā·shən }

Lundegardh vaporizer [ANALY CHEM] A device used for emission flame photometry in which a compressed air aspirator vaporizes the solution within a chamber; smaller droplets are carried into the fuel-gas stream and to the burner orifice where the solvent is evaporated, dissociated, and optically excited. { 'lün·də‚gard 'vā·pə‚rīz·ər }

lune [MATH] A section of a plane bounded by two circular arcs, or of a sphere bounded by two great circles. { lün }

Luneberg lens [ELECTROMAG] A type of antenna consisting of a dielectric sphere whose index of refraction varies with distance from the center of the sphere so that a beam of parallel rays falling on the lens is focused at a point on the lens surface diametrically opposite from the direction of incidence, and, conversely, energy emanating from a point on the surface is focused into a plane wave. Accurately spelled Luneburg lens. { 'lü·nə‚bərg ‚lenz }

Luneburg lens See Luneberg lens. { 'lü·nə‚bərg ‚lenz }

lune of Hippocrates [MATH] **1.** A section of the plane, bounded by two circular arcs, whose area equals that of a polygon used in constructing the circles. **2.** One of a small finite number of sections of the plane, each bounded by two circular arcs, such that the sum of their areas equals that of a polygon used in constructing the circles. { ‚lün əv hip'äk·rə‚tēz }

lunette [GEOL] A broad, low crescentic mound of wind-blown fine silt and clay. [ORD] Towing ring in the trial plate or tongue of a towed vehicle, such as a gun carriage or trailer, used for attaching the towed vehicle to the prime mover or towing vehicle. { lü'net }

lung [ANAT] Either of the paired air-filled sacs, usually in the anterior or anteroventral part of the trunk of most tetrapods, which function as organs of respiration. { ləŋ }

lung bud [EMBRYO] A primary outgrowth of the embryonic trachea; the anlage of a primary bronchus and all its branches. { 'ləŋ ‚bəd }

lungfish [VERT ZOO] The common name for members of the Dipnoi; all have lungs that arise from a ventral connection with the gut. { 'ləŋ‚fish }

lung-governed breathing apparatus [ENG] A breathing apparatus in which the oxygen that is supplied to the wearer is governed by the wearer's demand. { 'ləŋ ‚gəv·ərnd 'brēth·iŋ ‚ap·ə‚rad·əs }

lungworm [INV ZOO] Any of the nematodes that are parasites of terrestrial and marine nematodes, most commonly found in the respiratory tract, characterized by a reduced or absent stoma capsule, and an oral opening surrounded by six well-developed lips. { 'ləŋ‚wərm }

lunisolar precession [ASTROPHYS] Precession of the earth's equinox caused by the gravitational attraction of the sun and moon. { ‚lü·nə'sō·lər prē'sesh·ən }

lunisolar tides [OCEANOGR] Harmonic tidal constituents attributable partly to the development of both the lunar tide and the solar tide and partly to the lunisolar synodic fortnightly constituent. { ‚lü·nə'sō·lər 'tīdz }

lunitidal interval [OCEANOGR] The period between the moon's upper or lower transit over a specified meridian and a specified phase of the tidal current following the transit. { ‚lü·nə'tīd·əl 'in·tər·vəl }

lunk [COMMUN] Access line that terminates in an automatic dial exchange where it functions as an access line for subscribers, but functions as a trunk for the automatic dial exchange equipment. { ləŋk }

lunule [BIOL] A crescent-shaped organ, structure, or mark. { 'lün‚yül }

lupine [BOT] A leguminous plant of the genus *Lupinus* with an upright stem, leaves divided into several digitate leaflets, and terminal racemes of pea-shaped blossoms. { 'lü‚pən }

lupinidine See sparteine. { lü'pin·ə‚dēn }

Lupus [ASTRON] A southern constellation lying between Centaurus and Scorpius. Also known as Wolf. { 'lü·pəs }

lupus erythematosus [MED] An acute or subacute febrile collagen disease characterized by a butterfly-shaped rash over the cheeks and perilingual erythema. { 'lü·pəs ‚er·ə‚thē·mə'tō·səs }

lupus erythematosus factor See LE factor. { 'lü·pəs ‚er·ə‚thē·mə'tō·səs 'fak·tər }

Lupus Loop [ASTRON] A very old supernova remnant, about 400–600 parsecs distant, that forms an extended source of radio waves and soft x-rays. { 'lü·pəs ‚lüp }

lupus vulgaris [MED] True tuberculosis of the skin; a slow-developing, scarring, and deforming disease, often asymptomatic, frequently involving the face, and occurring in a wide variety of appearances. { 'lü·pəs vəl'gar·əs }

lusec [PHYS] A unit used for the measurement of power of evacuation of a vacuum pump, equal to the power associated with a leak rate of 1 liter per second at a pressure of 1 millitorr, or to approximately 1.33322×10^{-4} watt. { 'lü‚sek }

Lusitanian [GEOL] Lower Jurassic geologic time. { ‚lü·sə'tan·ē·ən }

luster [OPTICS] The appearance of a surface dependent on reflected light; types include metallic, vitreous, resinous, adamantine, silky, pearly, greasy, dull, and earthy; applied to minerals, textiles, and many other materials. { 'ləs·tər }

lustering [TEXT] A series of processes used to treat cottons to improve their sales appeal; the processes are mercerizing, frictioning, and filling. { 'ləs·tə·riŋ }

lusterless paint [MATER] Paint which absorbs light rays so that no shine or polish appears on its surface; used extensively on U.S. Army vehicles. { 'ləs·tər·ləs 'pānt }

luster mottlings [GEOL] The spotted, shimmering appearance of certain rocks caused by reflection of light from cleavage faces of crystals that contain small inclusions of other minerals. { 'ləs·tər ‚mät·liŋz }

lutaceous [GEOL] Claylike. { lü'tā·shəs }

lute [MATER] A substance, such as cement or clay, for packing a joint or coating a porous surface to produce imperviousness to gas or liquid. { lüt }

lutecite [GEOL] A fibrous, chalcedony-like quartz with optical anomalies that have led to its being considered a distinct species. { 'lüd·ə‚sīt }

lutein [BIOCHEM] **1.** A dried, powdered preparation of corpus luteum. **2.** See xanthophyll. { 'lüd·ē·ən }

luteinization [PHYSIO] Acquisition of characteristics of cells of the corpus luteum by ovarian follicle cells following ovulation. { ‚lüd·ē·ə·nə'zā·shən }

luteinizing hormone [BIOCHEM] A glycoprotein hormone secreted by the adenohypophysis of vertebrates which stimulates hormone production by interstitial cells of gonads. Abbreviated LH. Also known as interstitial-cell-stimulating hormone (ICSH). { 'lüd·ē·ə‚nīz·iŋ 'hòr‚mōn }

luteinizing-hormone releasing hormone [BIOCHEM] A small peptide hormone released from the hypothalamus which acts on the pituitary gland to cause release of luteinizing hormone. Abbreviated LH-RH. { 'lüd·ē·ə‚nīz·iŋ ‚hòr‚mōn ri'lēs·iŋ ‚hòr‚mōn }

luteoma [MED] A tumor of the ovary composed of cells resembling those of the corpus luteum. { ‚lüd·ē'ō·mə }

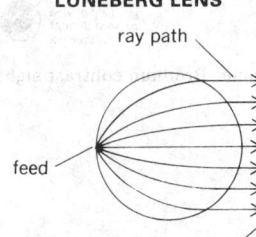

LUNEBERG LENS

ray path

feed

plane-phase wavefront

Luneberg lens with dielectric sphere between feed and plane-phase wavefront.

luteotropic hormone *See* prolactin. { ¦lüd·ē·ə¦träp·ik 'hȯr‚mōn }

Luteoviridae [VIROL] A family of plant viruses containing positive-sense, single-stranded ribonucleic acid; includes the genera *Luteovirus*, *Polerovirus*, and *Enamovirus*. { ‚lüd·ē· o'vir·ə‚dī }

Luteovirus [VIROL] A genus of plant viruses in the family Luteoviridae that is characterized by icosahedral particles containing one molecule of linear, positive-sense, single-stranded ribonucleic acid; barley yellow dwarf virus is the type species. Also known as barley yellow dwarf virus group. { 'lüd·ē· ō‚vī·rəs }

lutetium [CHEM] A chemical element, symbol Lu, atomic number 71, atomic weight 174.967; a very rare metal and the heaviest member of the rare-earth group. { lü'tē·shəm }

Lutheran blood group [IMMUNOL] The erythrocyte antigens defined by reactions with an antibody designated anti-Lu^a, initially detected in the serum of a multiply transfused patient with lupus erythematosus, who developed antibodies against erythrocytes of a donor named Lutheran, and by anti-Lu^b. { 'lüth·rən 'bləd ‚grüp }

lutite [GEOL] A consolidated rock or sediment formed principally of clay or clay-sized particles. { 'lü‚tīt }

Lutjanidae [VERT ZOO] The snappers, a family of perciform fishes in the suborder Percoidei. { lü'chan·ə‚dē }

lux [OPTICS] A unit of illumination, equal to the illumination on a surface 1 square meter in area on which there is a luminous flux of 1 lumen uniformly distributed, or the illumination on a surface all points of which are at a distance of 1 meter from a uniform point source of 1 candela. Symbolized lx. Also known as meter-candle. { ləks }

Luxemburg effect [COMMUN] Cross modulation between two radio signals during their passage through the ionosphere, due to the nonlinearity of the propagation characteristics of free charges in space. { 'lùk·səm‚bərg i‚fekt }

luxon *See* troland. { 'lək‚sän }

Luzin's theorem [MATH] Given a measurable function f which is finite almost everywhere in a euclidean space, then for every number $\epsilon > 0$ there is a continuous function g which agrees with f, except on a set of measure less than ϵ. Also spelled Lusin's theorem. { ‚lü·zēnz ‚thir·əm }

luzonite *See* enargite. { 'lü·zə‚nīt }

lvalue *See* left value. { 'el‚val·yü }

LVDT *See* linear variable-differential transformer.

L wave [GEOPHYS] A phase designation for an earthquake wave that is a surface wave, without respect to type. { 'el ‚wāv }

lx *See* lux.

Lyapunov exponent [PHYS] One of a number of coefficients that describe the rates at which nearby trajectories in phase space converge or diverge, and that provide estimates of how long the behavior of a mechanical system is predictable before chaotic behavior sets in. { li·pù'nȯf ik‚spō·nənt }

Lyapunov function [MATH] A function of a vector and of time which is positive-definite and has a negative-definite derivative with respect to time for nonzero vectors, is identically zero for the zero vector, and approaches infinity as the norm of the vector approaches infinity; used in determining the stability of control systems. Also spelled Liapunov function. { lē'ap·ə‚nȯf ‚fəŋk·shən }

Lyapunov stability criterion [CONT SYS] A method of determining the stability of systems (usually nonlinear) by examining the sign-definitive properties of an associated Lyapunov function. { lē'ap·ə‚nȯf stə'bil·əd·ē krī‚tir·ē·ən }

lyase [BIOCHEM] An enzyme that catalyzes the nonhydrolytic cleavage of its substrate with the formation of a double bond; examples are decarboxylases. { 'lī‚ās }

lyate ion [CHEM] The anion that is produced when a solvent molecule loses a proton (hydrogen nucleus), for example, the hydroxide ion is the lyate ion of water. { 'lī‚āt ‚ī·ən }

Lycaenidae [INV ZOO] A family of heteroneuran lepidopteran insects in the superfamily Papilionoidea including blue, gossamer, hairstreak, copper, and metalmark butterflies. { lī'sēn·ə‚dē }

Lycaeninae [INV ZOO] A subfamily of the Lycaenidae distinguished by functional prothoracic legs in the male. { lī'sēn·ə‚nē }

lychee [BOT] A tree of the genus *Litchi* in the family Sapindaceae, especially *L. chinensis* which is cultivated for its edible fruit, a one-seeded berry distinguished by the thin, leathery, rough pericarp that is red in most varieties. Also spelled litchi. { 'lī‚chē }

lychnisc [INV ZOO] A hexactin in which the central part of the spicule is surrounded by a system of 12 struts. { 'lik·nisk }

Lychniscosa [INV ZOO] An order of sponges in the subclass Hexasterophora in which parenchymal megascleres form a rigid framework and are all or in part lychniscs. { ‚lik·ni'skō·sə }

lycine *See* betaine. { 'lī‚sēn }

lycopene [BIOCHEM] $C_{40}H_{5C}$ A red, crystalline hydrocarbon that is the coloring matter of certain fruits, as tomatoes; it is isomeric with carotene. { 'lī·kə‚pēn }

lycoperdonosis [MED] A respiratory disease caused by inhalation of spores from the puffball mushroom, *Lycoperdon*. { ‚lī·kō‚pər·də'nō·səs }

lycophore larva [INV ZOO] A larva of certain cestodes characterized by cilia, large frontal glands, and 10 hooks. Also known as decanth larva. { 'lī·kə‚fȯr ‚lär·və }

Lycopodiales [BOT] The type order of Lycopodiopsida. { ‚lī·kə·pō·dī'ā·lēz }

Lycopodiatae *See* Lycopodiopsida. { ‚lī·kə·pō'dī·ə‚tē }

Lycopodineae [BOT] The equivalent name for Lycopodiopsida. { ‚lī·kə·pō'din·ē‚ē }

Lycopodiophyta [BOT] A division of the subkingdom Embryobionta characterized by a dominant independent sporophyte, dichotomously branching roots and stems, a single vascular bundle, and small, simple, spirally arranged leaves. { ‚lī· kə·pō·dī'äf·əd·ə }

Lycopodiopsida [BOT] The lycopods, the type class of Lycophodiophyta. { ‚lī·kə·pō·dī'äp·səd·ə }

lycopodium [MATER] A yellow powder prepared from the spores of *Lycopodium clavatum*; used as a desiccant and absorbent. { ‚lī·kə'pōd·ē·əm }

Lycopsida [BOT] Former subphylum of the Embryophyta now designated as the division Lycopodiophyta. { lī'käp· sə·də }

Lycoriidae [INV ZOO] A family of small, dark-winged dipteran insects in the suborder Orthorrhapha. { ‚lī·kə'rī·ə‚dē }

Lycosidae [INV ZOO] A family of hunting spiders in the suborder Dipneumonomorphae that actively pursue their prey. { lī'käs·ə‚dē }

Lycoteuthidae [INV ZOO] A family of squids. { ‚lī·kə'tü· thə‚dē }

Lyctidae [INV ZOO] The large-winged beetles, a large cosmopolitan family of coleopteran insects in the superfamily Bostrichoidea. { 'lik·tə‚de }

Lyddane-Sachs-Teller relation [SOLID STATE] For an infinite ionic crystal, the relation $\epsilon(0)/\epsilon(\infty) = \omega_L{}^2/\omega_T{}^2$, where $\epsilon(0)$ is the crystal's static dielectric constant, $\epsilon(\infty)$ is the dielectric constant at a frequency at which electronic polarizability is effective but ionic polarizability is not, ω_L is the frequency of longitudinal optical phonons with zero wave vectors, and ω_T is the frequency of transverse optical phonons with large wave vector. { lə'dän 'saks 'tel·ər ri‚lā·shən }

lyddite [MATER] An explosive composed chiefly of picric acid. { 'li‚dīt }

Lydian stone *See* basanite. { 'lid·ē·ən 'stōn }

lydite *See* basanite. { 'li‚dīt }

lye [INORG CHEM] **1.** A solution of potassium hydroxide or sodium hydroxide used as a strong alkaline solution in industry. **2.** The alkaline solution that is obtained from the leaching of wood ashes. { lī }

Lygaeidae [INV ZOO] The lygaeid bugs, a large family of phytophagous hemipteran insects in the superfamily Lygaeoidea. { lī'jē·ə‚dē }

Lygaeoidea [INV ZOO] A superfamily of pentatomorphan insects having four-segmented antennae and ocelli. { ‚lī· jē'ȯid·ē·ə }

Lyginopteridaceae [PALEOBOT] An extinct family of the Lyginopteridales including monostelic pteridosperms having one or two vascular traces entering the base of the petiole. { ‚lī·jə·näp‚ter·ə'dās·ē‚ē }

Lyginopteridales [PALEOBOT] An order of the Pteridospermae. { ‚lī·jə·näp‚ter·ə'dā·lēz }

Lyginopteridatae [PALEOBOT] The equivalent name for Pteridospermae. { ‚lī·jə·näp·fə'rid·əd‚ē }

Lyman-alpha radiation [SPECT] Radiation emitted by hydrogen associated with the spectral line in the Lyman series whose wavelength is 121.5 nanometers. { 'lī·mən 'al·fə ,rād· ē'ā·shən }

Lyman band [SPECT] A band in the ultraviolet spectrum of molecular hydrogen, extending from 125 to 161 nanometers. { 'lī·mən ,band }

Lyman continuum [SPECT] A continuous range of wavelengths (or wave numbers or frequencies) in the spectrum of hydrogen at wavelengths less than the Lyman limit, resulting from transitions between the ground state of hydrogen and states in which the single electron is freed from the atom. { 'lī·mən kən'tin·yə·wəm }

Lyman ghost [SPECT] A false line observed in a spectroscope as a result of a combination of periodicities in the ruling. { 'lī·mən ,gōst }

Lyman limit [SPECT] The lower limit of wavelengths of spectral lines in the Lyman series (912 angstrom units), or the corresponding upper limit in frequency, energy of quanta, or wave number (equal to the Rydberg constant for hydrogen). { 'lī·mən ,lim·ət }

Lyman series [SPECT] A group of lines in the ultraviolet spectrum of hydrogen covering the wavelengths of 121.5–91.2 nanometers. { 'lī·mən ,sir·ēz }

Lymantriidae [INV ZOO] The tussock moths, a family of heteroneuran lepidopteran insects in the superfamily Noctuoidea; the antennae of males is broadly pectinate and there is a tuft of hairs on the end of the female abdomen. { ,lī·mən'trī·ə,dē }

Lyme borreliosis See Lyme disease. { ,līm bə,rel·ē'ō·səs }

Lyme disease [MED] A complex multisystem human illness caused by the tick-borne spirochete *Borrelia burgdorferi*. Also known as Lyme borreliosis. { 'līm di,zēz }

Lymexylonidae [INV ZOO] The ship timber beetles composing the single family of the coleopteran superfamily Lymexylonoidea. { lə,mek·sə'län·ə,dē }

Lymexylonoidea [INV ZOO] A monofamilial superfamily of wood-boring coleopteran insects in the suborder Polyphaga characterized by a short neck and serrate antennae. { lə,mek·sə·lə'nȯid·ē·ə }

lymph [HISTOL] The colorless fluid which circulates through the vessels of the lymphatic system. { limf }

lymphadenitis [MED] Inflammation of lymph nodes. { ,lim,fad·ən'īd·əs }

lymphadenoid goiter See struma lymphomatosa. { ,lim'fad· ən,ȯid 'gȯid·ər }

lymphadenopathy [MED] Enlargement or disease of lymph nodes. { ,lim,fad·ən'äp·ə·thē }

lymphadenopathy-associated virus [VIROL] A former designation for the human immunodeficiency virus. Abbreviated LAV. { ,lim,fad·ən'äp·ə·thē ə¦sō·shē,ād·əd 'vī·rəs }

lymphadenosis [MED] Neoplasia or hyperplasia of lymph nodes. { ,lim,fad·ən'ō·səs }

lymphagogue [PHARM] An agent that stimulates lymph flow. { 'lim·fə,gäg }

lymphangiectasis [MED] Dilation in the wall of a lymphatic vessel. { ,lim,fan·jē'ek·tə·səs }

lymphangiectomy [MED] Surgical removal of a pathologic lymphatic channel, as for cancer. { ,lim,fan·jē'ek·tə·mē }

lymphangioendothelial sarcoma See lymphangiosarcoma. { ,lim,fan·jē·ō,en·dō'thē·lē·əl sär'kō·mə }

lymphangioendothelioma [MED] A tumor composed of aggregations of lymphatic vessels, between which are large mononuclear cells presumed to be endothelial cells. { ,lim ,fan·jē·ō,en,dō,thē·lē'ō·mə }

lymphangiofibroma [MED] A benign tumor composed of lymphangiomatous and fibromatous elements. { ,limi,fan·jē· ō,fī'brō·mə }

lymphangioma [MED] An abnormal mass of lymphatic vessels. { ,lim,fan·jē'ō·mə }

lymphangiosarcoma [MED] A sarcoma whose parenchymal cells form vascular channels resembling lymphatics. Also known as lymphangioendothelial sarcoma. { ,lim,fan·jē·ō· sär'kō·mə }

lymphangitis [MED] Inflammation of lymphatic vessels. { ,lim,fan'jīd·əs }

lymphatic See lymph vessel. { lim'fad·ik }

lymphatic system [ANAT] A system of vessels and nodes conveying lymph in the vertebrate body, beginning with capillaries in tissue spaces and eventually forming the thoracic ducts which empty into the subclavian veins. { lim'fad·ik ,sis·təm }

lymphatic tissue [HISTOL] Tissue consisting of networks of lymphocytes and reticular and collagenous fibers. Also known as lymphoid tissue. { lim'fad·ik ,tish·ü }

lymphedema [MED] Edema resulting from lymph vessel obstruction. { ,lim·fə¦dē·mə }

lymph gland See lymph node. { 'limf ,gland }

lymph heart [VERT ZOO] A muscular expansion of a lymphatic vessel which contracts, driving lymph to the veins, as in amphibians. { 'limf ,härt }

lymph node [ANAT] An aggregation of lymphoid tissue surrounded by a fibrous capsule; found along the course of lymphatic vessels. Also known as lymph gland. { 'limf ,nōd }

lymphoblast [HISTOL] Precursor of a lymphocyte. { 'lim·fə,blast }

lymphoblastic leukemia See acute lymphocytic leukemia. { ,lim·fə¦blas·tik lü'kē·mē·ə }

lymphoblastosis [MED] An excessive number of lymphoblasts in peripheral blood; occasionally found in tissues. { ,lim·fō,bla'stō·səs }

lymphocyte [HISTOL] An agranular leukocyte formed primarily in lymphoid tissue; occurs as the principal cell type of lymph and composes 20–30% of the blood leukocytes. { 'lim·fə,sīt }

lymphocyte transformation See transformation. { ¦lim·fə,sīt ,tranz·fər'mā·shən }

lymphocytic angina See infectious mononucleosis. { ¦lim· fə¦sid·ik 'an·jə·nə }

lymphocytic choriomeningitis [MED] An acute viral meningitis caused by a specific virus endemic in mice; characterized clinically by rapid onset of symptoms of meningeal irritation, pleocytosis and often a rise in protein in the cerebrospinal fluid, and a short, benign course with recovery. { ¦lim·fə¦sid·ik ,kȯr· ē·ō,men·ən'jīd·əs }

lymphocytic leukemia [MED] A type of leukemia in which lymphocytic cells predominate. { ¦lim·fə¦sid·ik lü'kē·mē·ə }

lymphocytic lymphoma [MED] A malignant neoplasm of lymphoid tissue composed predominantly of lymphocytic cells. { ¦lim·fə¦sid·ik lim'fō·mə }

lymphocytic sarcoma See lymphosarcoma. { ¦lim·fə¦sid·ik sär'kō·mə }

lymphocytopenia [MED] Reduction of the absolute number of lymphocytes per unit volume of peripheral blood. Also known as lymphopenia. { ,lim·fə,sīd·ə'pē·nē·ə }

lymphocytosis [MED] An abnormally high lymphocyte count in the blood. { ,lim·fə,sī'tō·səs }

lymphocytotropic [IMMUNOL] Having an affinity for lymphocytes. { ,lim·fō,sīd·ə'träp·ik }

lymphoepithelioma [MED] A squamous-cell carcinoma of the nasopharynx whose parenchymal cells resemble elements of the reticuloendothelial system. { ,lim·fō,ep·ə,thē·lē'ō· mə }

lymphogranuloma inguinale See lymphogranuloma venereum. { ,lim·fə,gran·yə'lō·mə ,iŋ·gwə'nä·lē }

lymphogranulomatosis See Hodgkin's disease. { ,lim· fə,gran·yə,lō·mə'tō·səs }

lymphogranuloma venereum [MED] A systemic infectious venereal disease caused by a microorganism belonging to the PLT-Bedsonia group, characterized by enlargement of inguinal lymph nodes and genital ulceration. Abbreviated LGV. Also known as lymphogranuloma inguinale; lymphopathia venereum; venereal bubo. { ,lim·fə,gran·yə'lō·mə və'nir·ē·əm }

lymphoid cell [HISTOL] A mononucleocyte that resembles a leukocyte. { 'lim,fȯid ,sel }

lymphoid hemoblast of Pappenheim See pronormoblast. { 'lim,fȯid 'hē·mə,blast əv 'päp·ən,hīm }

lymphoid organ [ANAT] An organ that produces lymphocytes or is associated with lymphocyte function, for example, the lymph nodes, spleen, and thymus. { 'lim,fȯid ,ȯr·gən }

lymphoid tissue See lymphatic tissue. { 'lim,fȯid ,tish·ü }

lymphokine [IMMUNOL] A cytokine released from T lymphocytes after contact with an antigen. { 'lim·fə,kīn }

lymphokine-activated killer cell [IMMUNOL] A cytotoxic cell that is able to lyse certain cell lines resistant to natural killer cells. { ¦lim·fə,kīn ¦ak·tə,vād·əd 'kil·ər ,sel }

lymphoma [MED] Any neoplasm, usually malignant, of the lymphoid tissues. { lim'fō·mə }

lymphopathia venereum See lymphogranuloma venereum. { ,lim·fə'path·ē·ə və'nir·ē·əm }

lymphopenia See lymphocytopenia. { ,lim·fə'pē·nē·ə }

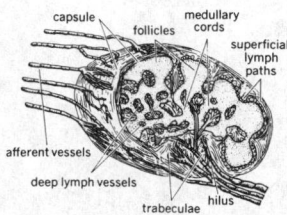

LYMPH NODE

capsule
medullary cords
follicles
superficial lymph paths
afferent vessels
deep lymph vessels
trabeculae
hilus

Diagram of a lymph node.

lymphopoiesis [PHYSIO] The production of lymph. { ˌlim·fəˌpȯi'ē·səs }

lymphoproliferative virus group *See* Gammaherpesvirinae. { ˈlim·fōˈprōˈlif·rəd·iv 'vī·rəs ˌgrüp }

lymph sinus [ANAT] One of the tracts of diffuse lymphatic tissue between the cords and nodules, and between the septa and capsule of a lymph node. { 'limf ˌsī·nəs }

lymph vessel [ANAT] A tubular passage for conveying lymph. Also known as lymphatic. { 'limf ˌves·əl }

lyngbyatoxin A [BIOCHEM] An indole alkaloid toxin produced by *Lyngbya majuscula.* { 'liŋ·bē·əˌtäk·sən 'ā }

lynx [VERT ZOO] Any of several wildcats of the genus *Lynx* having long legs, short tails, and usually tufted ears; differs from other felids in having 28 instead of 30 teeth. { liŋks }

Lyomeri [VERT ZOO] The equivalent name for Saccopharyngiformes. { lī'äm·əˌrī }

Lyon hypothesis [GEN] The concept that mammalian females are X-chromosome mosaics as a result of the random inactivation of one X chromosome in some embryonic cells and their descendants and of the other X chromosome in the rest. { 'lī·ən hīˌpäth·ə·səs }

lyonium ion [CHEM] The cation that is produced when a solvent molecule is protonated. { lī'än·ē·əm ˌī·ən }

lyons velvet [TEXT] A stiff crisp velvet with a short dense erect pile. { lē'oŋ ˈvel·vət }

lyophilic [CHEM] Referring to a substance which will readily go into colloidal suspension in a liquid. { ˈlī·əˈfil·ik }

lyophilization [CHEM ENG] Rapid freezing of a material, especially biological specimens for preservation, at a very low temperature followed by rapid dehydration by sublimation in a high vacuum. { līˌäf·ə·lə'zā·shən }

lyophobic [CHEM] Referring to a substance in a colloidal state that has a tendency to repel liquids. { ˈlī·əˈfō·bik }

Lyopomi [VERT ZOO] An equivalent name for Notacanthiformes. { lī'äp·əˌmī }

Lyot division [ASTRON] The gap between rings B and C of Saturn. Also known as French division. { 'lyō di,vizh·ən }

Lyot filter *See* birefringent filter. { 'lyō ˌfil·tər }

lyotopic series *See* Hofmeister series. { ˈlī·əˈtäp·ik 'sir·ēz }

lyotropic liquid crystal [PHYS CHEM] A liquid crystal prepared by mixing two or more components, one of which is polar in character (for example, water). { ˈlī·əˈträp·ik ˈlik·wəd 'krist·əl }

Lyra [ASTRON] A northern constellation; right ascension 19 hours, declination 40° north; its first-magnitude star, Vega, is a navigational star and the most brilliant star in this part of the sky. { 'lī·rə }

α Lyrae *See* Vega. { ˈal·fə 'lī·rə }

Lyrids [ASTRON] An important meteor shower occurring about April 22; it is regular and predictable, but not heavy, the hourly rate usually being about 7–10. { 'lī·rədz }

Lysaretidae [INV ZOO] A family of errantian polychaete worms in the superfamily Eunicea. { ˌlī·sə'red·əˌdē }

lyse [CYTOL] To undergo lysis. { līz }

Lysenkoism [BIOL] A pseudoscientific theory that flourished in the Soviet Union from the early 1930s to the mid-1960s; advocated by T. D. Lysenko, who called it agrobiology, it was claimed to be a revolutionary fusion of agronomy and biological science, and it opposed traditional biology and the gene concept but supported the inheritance of acquired characteristics. { lī'seŋ·kōˌiz·əm }

lysergic acid [ORG CHEM] $C_{16}H_{16}N_2O_2$ A compound that crystallizes in the form of hexagonal plates that melt and decompose at 240°C; derived from ergot alkaloids; used as a psychotomimetic agent. { lə'sər·jik 'as·əd }

lysergic acid diethylamide [ORG CHEM] $C_{15}H_{15}N_2CON$-$(C_2H_5)_2$ A psychotomimetic drug synthesized from compounds derived from ergot. Abbreviated LSD; LSD-25. { lə'sərjik ˌas·əd dīˌeth·əl'am·əd }

Lysianassidae [INV ZOO] A family of pelagic amphipod crustaceans in the suborder Gammaridea. { ˌlī·sē·ə'nas·əˌdē }

lysigenous [BIOL] Of or pertaining to the space formed following lysis of cells. { lī'sij·ə·nəs }

lysimeter [ENG] An instrument for measuring the water percolating through soils and determining the materials dissolved by the water. { lī'sim·əd·ər }

lysin [IMMUNOL] A substance, particularly antibodies, capable of lysing a cell. { 'līs·ən }

lysine [BIOCHEM] $C_6H_{14}O_2N_2$ An essential, basic amino acid obtained from many proteins by hydrolysis. { 'lī,sēn }

Lysiosquillidae [INV ZOO] A family of crustaceans in the order Stomatopoda. { ˌlī·sē·ō'skwil·əˌdē }

lysis [CELL MOL] Dissolution of a cell or tissue by the action of a lysin. [MED] **1.** Gradual decline in the manifestations of a disease, especially an infectious disease. Also known as defervescence. **2.** Gradual fall of fever. { 'lī·səs }

Lysithea [ASTRON] A small satellite of Jupiter with a diameter of about 15 miles (24 kilometers), orbiting at a mean distance of about 7.30×10^6 miles (11.75×10^6 kilometers). Also known as Jupiter X. { lī'sith·ē·ə }

lysocline [OCEANOGR] The level or ocean depth at which the rate of solution of calcium carbonate increases significantly. { 'lī·səˌklīn }

lysogen [MICROBIO] A bacterial or prokaryotic cell whose genome contains integrated viral deoxyribonucleic acid. { 'līs·əˌjen }

lysogeny [MICROBIO] Lysis of bacteria, with the liberation of bacteriophage particles. { lī'säj·ə·nē }

lysosome [CELL MOL] A specialized cell organelle surrounded by a single membrane and containing a mixture of hydrolytic (digestive) enzymes. { 'lī·səˌsōm }

lysosomotropic drug [PHARM] A drug which can enter selectively the lysosomes of certain cell types; may be useful in chemotherapy for destruction of tumor cells. { ˌlī·səˌsōmə'träp·ik 'drəg }

lysozyme [BIOCHEM] An enzyme present in certain body secretions, principally tears, which acts to hydrolyze certain bacterial cell walls. { 'lī·səˌzīm }

lyssacine [INV ZOO] An early stage of the skeletal network in hexactinellid sponges. { 'lī·səˌsēn }

Lyssacinosa [INV ZOO] An order of sponges in the subclass Hexasterophora in which parenchymal megascleres are typically free and unconnected but are sometimes secondarily united. { ˈlī·sə·səˈnō·sə }

Lyssavirus [VIROL] A genus of the viral family Rhabdoviridae that is characterized by a bullet-shaped enveloped virion covered with projections that contains one molecule of linear, negative-sense, single-stranded ribonucleic acid, the causative agent of rabies. { 'līs·əˌvī·rəs }

lyssophobia [PSYCH] **1.** Abnormal fear of hydrophobia. **2.** Abnormal fear of becoming insane. { ˌlī·sə'fō·bē·ə }

lysyl oxidase [BIOCHEM] Any of a group of enzymes found in bone and connective tissue which oxidize terminal amino groups of lysine residues in tropocollagen molecules to aldehyde residues. { 'lī·səl 'äk·səˌdās }

lytic infection [MICROBIO] Penetration of a host cell by lytic phage. { 'lid·ik in'fek·shən }

lytic phage [VIROL] Any phage that causes host cells to lyse. { 'lid·ik 'fāj }

lytic reaction [CELL MOL] A reaction that leads to lysis of a cell. { 'lid·ik rē'ak·shən }

LYSARETIDAE

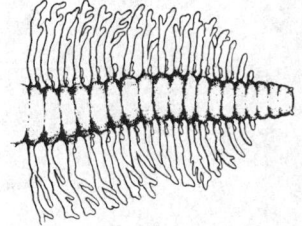

Anterior end of *Iphitime* in dorsal view.

LYSERGIC ACID DIETHYLAMIDE

Structural formula of lysergic acid diethylamide.

LYSINE

Structural formula of lysine.

m *See* meter; milli-.

M *See* mega-; megabyte; molarity.

M$_f$ [MET] In a carbon steel, the temperature at which martensite formation finishes during cooling of austenite.

M$_s$ [MET] In a carbon steel, the temperature at which martensite formation begins during cooling of austenite.

mA *See* milliampere.

MAA *See* methanearsonic acid.

maar [GEOL] A volcanic crater that was created by violent explosion but not accompanied by igneous extrusion; frequently, it is filled by a small circular lake. { mär }

maarad [TEXT] A type of Egyptian cotton cultivated from American pima seeds. { 'mä,rad }

MAC *See* message authentication code.

macadam [CIV ENG] Uniformly graded stones consolidated by rolling to form a road surface; may be bound with water or cement, or coated with tar or bitumen. { mə'kad·əm }

macadamia nut [BOT] The hard-shelled seed obtained from the fruit of a tropical evergreen tree, *Macadamia ternifolia*. { ,mak·ə'dā·mē·ə ,nət }

macaluba *See* mud volcano. { ,mä·kə'lü·bə }

Macaluso-Corbino effect [PHYS] Anomalously large Faraday rotation exhibited by a medium at wavelengths in the neighborhood of an absorption line. { ,mäk·ə,lü·sō kȯr'bē·nō i,fekt }

macanilla oil [MATER] An oil obtained from the nuts of the palm *Guilielma garipaes* of Venezuela and Central America. { 'mak·ə,nil·ə ,ȯil }

macaque [VERT ZOO] The common name for 12 species of Old World monkeys composing the genus *Macaca*, including the Barbary ape and the rhesus monkey. { mə'kak }

Macarthur and Forest cyanide process [MET] A process for recovering gold by leaching the pulped gold ore with a solution of 0.2–0.8% potassium cyanide and next with water; the gold is then obtained by precipitation on zinc or aluminum or by electrolysis. { mə'kär·thər ən 'fär·əst 'sī·ə,nīd ,präs·əs }

Macassar oil [MATER] **1.** A fat obtained from seeds of kusam, used in hair dressing, cooking, and illumination. **2.** Any of several fatty oils or oily preparations that have similar properties and are used in hair preparations. { mə'kas·ər ,ȯil }

macaw [VERT ZOO] The common name for large South and Central American parrots of the genus *Ara* and related genera; individuals are brilliantly colored with a long tail, a hooked bill, and a naked area around the eyes. { mə'kȯ }

Macbeth illuminometer [OPTICS] A type of portable visual photometer in which the light to be measured is balanced by a Lummer-Brodhun sight box against a comparison lamp, whose apparent brightness can be varied by moving it along a tube; a control box supplies a calibrated current to the comparison lamp, and calibrated optical filters can be placed in the light paths to correct for color differences in the comparison and measured sources and to extend the range of the instrument. { mək'beth ə,lüm·ə'näm·əd·ər }

Macchiavello stain [MICROBIO] A differential staining procedure for rickettsiae, in which the fixed tissue smear is stained with basic fuchsin, differentiated with a 0.5% solution of citric acid, and counterstained with a 1% solution of methylene blue; rickettsiae stain red, and tissue cells stain blue. { ,mäk·ē·ə'vel·ō ,stān }

MacCullagh's formula [PHYS] A formula for the potential due to a distribution of mass or charge at an external point: the potential V at a point P resulting from a distribution of mass or positive charge centered about a point O is $V = (kM/r) + (k/2r^3)(A + B + C - 3I) + O(1/r^4)$, where r is the distance from O to P, k is the gravitational or electrostatic constant, M is the total mass or charge, A, B, and C are the principal moments about O, I is the moment about OP, and $O(1/r^4)$ is a quantity that falls off at least as rapidly as $1/r^4$. { mə'kəl·əz ,fȯr·myə·lə }

MacDonald functions *See* modified Hankel functions. { mək'dän·əld ,fəŋk·shənz }

mace [FOOD ENG] Spice made from the covering of the nutmeg. { mās }

macedonite [MINERAL] PbTiO$_3$ A mineral composed of an oxide of lead and titanium. [PETR] A basaltic rock that contains orthoclase, sodic plagioclase, biotite, olivine, and rare pyriboles. { ,mas·ə'dä,nīt }

mace oil [MATER] An essential oil obtained by distillation from mace and containing pinene and dipentene; used in flavoring. { 'mās ,ȯil }

maceral [GEOL] The microscopic organic constituents found in coal. { 'mas·ə,ral }

maceration [CHEM ENG] The process of extracting fragrant oils from flower petals by immersing them in hot molten fat. [SCI TECH] The process of softening or wearing away a material by wetting it or steeping it in a liquid. { ,mas·ə'rā·shən }

macgovernite [MINERAL] Mn$_5$(AsO$_3$)SiO$_3$(OH)$_2$ A mineral composed of basic manganese arsenite and silicate. Also spelled mcgovernite. { mə'gəv·ər,nīt }

Machaeridea [INV ZOO] A class of homolozoan echinoderms in older systems of classification. { ,mak·ə'rid·ē·ə }

Mach angle [FL MECH] The vertex half angle of the Mach cone generated by a body in supersonic flight. { 'mäk ,aŋ·gəl }

Mach cone [FL MECH] **1.** The cone-shaped shock wave theoretically emanating from an infinitesimally small particle moving at supersonic speed through a fluid medium; it is the locus of the Mach lines. **2.** The cone-shaped shock wave generated by a sharp-pointed body, as at the nose of a high-speed aircraft. { 'mäk ,kōn }

Mach disk [FL MECH] A structure visible on a schlieren photograph of a supersonic air jet exhausting from a nozzle at low pressure into higher-pressure air at rest; it is formed by the focusing and strengthening of oblique shock waves emanating from the edges of the nozzle as they approach the jet axis. { 'mäk ,disk }

machete [DES ENG] A knife with a broad blade 2 to 3 feet (60 to 90 centimeters) long. { mə'shed·ē *or* mə'ched·ē }

Mach front *See* Mach stem. { 'mäk ,frənt }

Machilidae [INV ZOO] A family of insects belonging to the Thysanura having large compound eyes and ocelli and a monocondylous mandible of the scraping type. { mə'kil·ə,dē }

machinability [MET] **1.** The ability of a metal to be machined. **2.** The difficulty or ease with which a metal can be machined. { mə,shēn·ə'bil·əd·ē }

machinability index [MET] A numerical value that designates the degree of difficulty or ease with which a particular material can be machined; originally based on turning B1112 steel at 180 feet per minute (0.9144 meter per second) with a high-speed tool for an index of 100; with replacement of high-speed steels with carbides in turning operations, it has been found that the machinability index of a given material changes with the type of operation and the tool material. { mə,shēn·ə'bil·əd·ē ,in,deks }

MACADAMIA NUT

Macadamia integrifolia. (*a*) Mature nuts. (*b*) Nuts without husks. (*c*) Nuts in husk showing method of dehiscence. (*From R. A. Jaynes, ed., Handbook of North American Nut Trees, Humphrey Press, 1969*)

MACH CONE

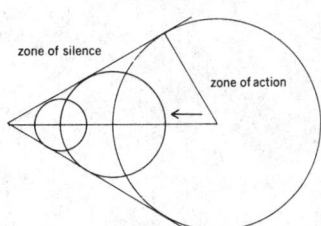

Generation of Mach cone (def. 1) by particle moving through fluid at supersonic speed. Arrow indicates direction of particle's motion. Cone is tangent to spherical surfaces of disturbances generated by particle at successive times, and separates zones of action and silence.

machinable *See* machine-sensible. { məˈshēn·ə·bəl }

machinable carbide [MET] Titanium carbide in a matrix of Ferro-Tic C tool steel. { məˈshēn·ə·bəl ˈkär,bīd }

Mach indicator *See* Machmeter. { ˈmäk ,in·də,kād·ər }

machine [COMPUT SCI] **1.** A mechanical, electric, or electronic device, such as a computer, tabulator, sorter, or collator. **2.** A simplified, abstract model of an internally programmed computer, such as a Turing machine. [MECH ENG] A combination of rigid or resistant bodies having definite motions and capable of performing useful work. { məˈshēn }

machine address [COMPUT SCI] The actual and unique internal designation of the location at which an instruction or datum is to be stored or from which it is to be retrieved. { məˈshēn əˈdres }

machine attention time [IND ENG] Time during which a machine operator must observe the machine's functioning and be available for immediate servicing, while not actually operating or servicing the machine. Also known as service time. { məˈshēn əˈten·chən ,tīm }

machine available time [COMPUT SCI] The time during which a computer has its power turned on, is not undergoing maintenance, and is thought to be operating properly. { məˈshēn əˈvāl·ə·bəl ,tīm }

machine bolt [DES ENG] A heavy-weight bolt with a square, hexagonal, or flat head used in the automotive, aircraft, and machinery fields. { məˈshēn ,bōlt }

machine capability [IND ENG] A qualitative or quantitative statement of the performance potential of a specific item of power equipment. { məˈshēn ,kā·pəˈbil·əd·ē }

machine check [COMPUT SCI] A check that tests whether the parts of equipment are functioning properly. Also known as hardware check. { məˈshēn ,chek }

machine-check indicator [COMPUT SCI] A protective device which turns on when certain conditions arise within the computer; the computer can be programmed to stop or to run a separate correction routine or to ignore the condition. { məˈshēn ,chek ,in·də,kād·ər }

machine code [COMPUT SCI] **1.** A computer representation of a character, digit, or action command in internal form. **2.** A computer instruction in internal format, or that part of the instruction which identifies the action to be performed. **3.** The set of all instruction types that a particular computer can execute. { məˈshēn ,kōd }

machine conditions [COMPUT SCI] A component of a task descriptor that specifies the contents of all programmable registers in the processor, such as arithmetic and index registers. { məˈshēn kənˈdish·ənz }

machine controlled time [IND ENG] The time necessary for a machine to complete the automatic portion of a work cycle. Also known as independent machine time; machine element; machine time. { məˈshēn kənˈtrōld ˈtīm }

machine cut [MIN ENG] A groove or slot made horizontally or vertically in a coal seam by a coal cutter, as a step to shot firing. { məˈshēn ,kət }

machine cycle [COMPUT SCI] **1.** The shortest period of time at the end of which a series of events in the operation of a computer is repeated. **2.** The series of events itself. { məˈshēn ,sī·kəl }

machine-dependent [COMPUT SCI] Referring to programming languages, programs, systems, and procedures that can be used only on a particular computer or on a line of computers manufactured by a single company. { məˈshēn diˌpen·dənt }

machine design [DES ENG] Application of science and invention to the development, specification, and construction of machines. { məˈshēn diˌzīn }

machine drill [MECH ENG] Any mechanically driven diamond, rotary, or percussive drill. { məˈshēn ,dril }

machine element [DES ENG] **1.** Any of the elementary mechanical parts, such as gears, bearings, fasteners, screws, pipes, springs, and bolts used as essentially standardized components for most devices, apparatus, and machinery. **2.** *See* machine controlled time. { məˈshēn ,el·ə·mənt }

machine error [COMPUT SCI] A deviation from correctness in computer-processed data, caused by equipment failure. { məˈshēn ,er·ər }

machine file [DES ENG] A file that can be clamped in the chuck of a power-driven machine. { məˈshēn ,fīl }

machine forging [MET] Forging operations performed in and by certain machines. { məˈshēn ,fȯrj·iŋ }

MACHINE BOLT

A hexagon-head machine bolt with nut. *(Reynolds Metals Co.)*

machine gun [ORD] **1.** A weapon that automatically fires small-arms ammunition, caliber .60 or 15.24 millimeters or under, and is capable of sustained rapid fire. **2.** To riddle a target with machine gun fire. { məˈshēn ,gən }

machine-gun microphone *See* line microphone. { məˈshēn ,gən ˈmī·krə,fōn }

machine-hour [IND ENG] A unit representing the operation of one machine for 1 hour; used in the determination of costs and economics. { məˈshēn ,au̇r }

machine idle time [IND ENG] Time during a work cycle when a machine is idle, awaiting completion of manual work. { məˈshēn ˈīd·əl ,tīm }

machine-independent [COMPUT SCI] Referring to programs and procedures which function in essentially the same manner regardless of the machine on which they are carried out. { məˈshēn ,in·dəˈpen·dənt }

machine instruction [COMPUT SCI] A set of digits, binary bits, or characters that a computer can recognize and act upon, and that, when interpreted or decoded, indicates the action to be performed and which operand is to be involved in the action. { məˈshēn inˌstrək·shən }

machine instruction statement [COMPUT SCI] A statement consisting usually of a tag, an operating code, and one or more addresses. { məˈshēn inˌstrək·shən ,stāt·mənt }

machine interference [IND ENG] A situation in which two or more units of equipment simultaneously require service. { məˈshēn ,in·tərˈfir·əns }

machine interruption [COMPUT SCI] A halt in computer operations followed by the beginning of a diagnosis procedure, as a result of an error detection. { məˈshēn ,int·əˈrəp·shən }

machine key [DES ENG] A piece inserted between a shaft and a hub to prevent relative rotation. Also known as key. { məˈshēn ,kē }

machine language [COMPUT SCI] The set of instructions available to a particular digital computer, and by extension the format of a computer program in its final form, capable of being executed by a computer. { məˈshēn ,laŋ·gwij }

machine language code [COMPUT SCI] A set of instructions appearing as combinations of binary digits. { məˈshēn ˌlaŋ·gwij ˈkōd }

machine learning [COMPUT SCI] The process or technique by which a device modifies its own behavior as the result of its past experience and performance. { məˈshēn ,lərn·iŋ }

machine loading [IND ENG] **1.** Feeding work into a machine. **2.** Planning the amount of use of a unit of equipment during a given time period. { məˈshēn ,lōd·iŋ }

machine logic [COMPUT SCI] The structure of a computer, the operation it performs, and the type and form of data used internally. { məˈshēn ,läj·ik }

machine mining [MIN ENG] The use of power machines and equipment in the excavation and extraction of ore or coal. { məˈshēn ˌmīn·iŋ }

machine oil [MATER] Medium-density lubricating oil used for machine parts. { məˈshēn ,ȯil }

machine operator [COMPUT SCI] The person who manipulates the computer controls, brings up and closes down the computer, and can override a number of computer decisions. { məˈshēn ,äp·ə,rād·ər }

machine-oriented language *See* computer-oriented language. { məˈshēn ,ȯr·ē|en·təd ˈlaŋ·gwij }

machine-oriented programming system [COMPUT SCI] A system written in assembly language (or macro code) directly oriented toward the computer's internal language. { məˈshēn ˌȯr·ē,ent·əd ˈprō,gram·iŋ ,sis·təm }

machine-paced operation [IND ENG] The proportion of an operation cycle during which the machine controls the speed of work progress. { məˈshēn ˌpāst ,äp·əˈrā·shən }

machine pistol [ORD] A pistol capable of fully automatic fire. { məˈshēn ,pis·təl }

machine processible form [COMPUT SCI] Any input medium such as a punch card, paper tape, or magnetic tape. { məˈshēn ,präs,es·ə·bəl ˈfȯrm }

machine rating [MECH ENG] The power that a machine can draw or deliver without overheating. { məˈshēn ,rād·iŋ }

machine-readable *See* machine-sensible. { məˈshēn ˈrēd·ə·bəl }

machine-recognizable *See* machine-sensible. { məˈshēn ,rek·igˈnīz·ə·bəl }

machine ringing [COMMUN] In a telephone system, ringing

which is started either mechanically or by an operator, after which it continues automatically until the call is answered or abandoned. { mə'shēn ,riŋ·iŋ }

machine run See run. { mə'shēn ,rən }

machinery [MECH ENG] A group of parts or machines arranged to perform a useful function. { mə'shēn·rē }

machine screw [DES ENG] A blunt-ended screw with a standardized thread and a head that may be flat, round, fillister, or oval, and may be slotted, or constructed for wrenching; used to fasten machine parts together. { mə'shēn ,skrü }

machine script [COMPUT SCI] Any data written in a form that can immediately be used by a computer. { mə'shēn ,skript }

machine-sensible [COMPUT SCI] Capable of being read or sensed by a device, usually by one designed and built specifically for this task. Also known as machinable; machine-readable; machine-recognizable; mechanized. { mə'shēn ¦sen·sə·bəl }

machine-sensible information [COMPUT SCI] Information in a form which can be read by a specified machine. { mə'shēn ¦sen·sə·bəl ,in·fər'mā·shən }

machine setting See mechanical setting. { mə'shēn ,sed·iŋ }

machine shop [ENG] A workshop in which work, metal or other material, is machined to specified size and assembled. { mə'shēn ,shäp }

machine shot capacity [ENG] In injection molding, the maximum weight of a given thermoplastic resin which can be displaced by a single stroke of the injection ram. { mə'shēn 'shät kə,pas·əd·ē }

machine-spoiled time [COMPUT SCI] Computer time wasted on production runs that cannot be completed or whose results are made worthless by a computer malfunction, plus extensions of running time on runs that are hampered by a malfunction. { mə'shēn ¦spóild ,tīm }

machine steel [MET] Plain carbon steel with a 0.2–0.3% carbon content. { mə'shēn ,stēl }

machine switching system See automatic exchange. { mə'shēn 'swich·iŋ ,sis·təm }

machine taper [MECH ENG] A taper that provides a connection between a tool, arbor, or center and its mating part to ensure and maintain accurate alignment between the parts; permits easy separation of parts. { mə'shēn ,tā·pər }

machine-tight [ENG] The extent of the tightening of a screwed fitting that can be accomplished without damaging or stripping the thread. { mə'shēn ,tīt }

machine time See machine controlled time. { mə'shēn ,tīm }

machine tool [MECH ENG] A stationary power-driven machine for the shaping, cutting, turning, boring, drilling, grinding, or polishing of solid parts, especially metals. { mə'shēn ,tül }

machine-tool control [COMPUT SCI] The computer control of a machine tool for a specific job by means of a special programming language. { mə'shēn ,tül kən,trōl }

machine translation See mechanical translation. { mə'shēn tranz'lā·shən }

machine utilization [ENG] The percentage of time that a machine is actually in use. { mə'shēn ,yüd·əl·ə'zā·shən }

machine wall [MIN ENG] In a coal mine, the face at which the mining machine is working. { mə'shēn ,wól }

machine welding [MET] Welding with a machine under the control and observation of an operator; may be loaded and unloaded either manually or mechanically. { mə'shēn 'weld·iŋ }

machine word [COMPUT SCI] The fundamental unit of information in a word-organized digital computer, consisting of a fixed number of binary bits, decimal digits, characters, or bytes. { mə'shēn ,wərd }

machining [MECH ENG] Performing various cutting or grinding operations on a piece of work. { mə'shēn·iŋ }

machining center [MECH ENG] Manufacturing equipment that removes metal under computer numerical control by making use of several axes and a variety of tools and operations. { mə'shēn·iŋ ,sen·tər }

machining stress [MET] Residual stress in the work caused by machining. { mə'shēn·iŋ ,stress }

machinist's file [DES ENG] A type of double-cut file that removes metal fast and is used for rough metal filing. { mə'shē·nəsts ,fil }

Machin's formula [MATH] The formula

$$\pi/4 = 4 \arctan(1/5) - \arctan(1/239)$$

which has been used to compute the value of π. { 'mä·chənz ,fór·myə·lə }

Mach line [FL MECH] **1.** A line representing a Mach wave. **2.** See Mach wave. { 'mäk ,līn }

Machmeter [ENG] An instrument that measures and indicates speed relative to the speed of sound, that is, indicates the Mach number. Also known as Mach indicator. { 'mäk ,mēd·ər }

Mach number [FL MECH] The ratio of the speed of a body or of a point on a body with respect to the surrounding air or other fluid, or the ratio of the speed of a fluid, to the speed of sound in the medium. Symbolized N_{Ma}. Also known as relative Mach number. { 'mäk ,nəm·bər }

macho [ASTRON] A massive object in the halo of the Milky Way Galaxy, detected by the effects of gravitational lensing; such objects are hypothesized to account for the dark matter in the universe. Derived from massive compact halo object. { 'mä·chō }

machopolyp See machozooid. { ¦ma·kō'päl·əp }

machozooid [INV ZOO] A defensive polyp equipped with stinging organs in certain hydroid colonies. Also known as machopolyp. { ¦ma·kō'zō,óid }

Mach principle [RELAT] The principle that the motion of a particle is only meaningful when referred to the rest of the matter in the universe; this motion is thus determined by the distribution of this matter and is not an intrinsic property of an absolute space. { 'mäk ,prin·sə·pəl }

Mach reflection [FL MECH] The reflection of a shock wave from a rigid wall in which the shock strength of the reflected wave and the angle of reflection both have the smaller of the two values which are theoretically possible. { 'mäk ri,flek·shən }

Mach refractometer See Mach-Zehnder interferometer. { 'mäk ,rē,frak'täm·əd·ər }

Mach stem [FL MECH] A shock wave or front formed above the surface of the earth by the fusion of direct and reflected shock waves resulting from an airburst bomb. Also known as Mach front. { 'mäk ,stem }

Mach wave [FL MECH] Also known as Mach line. **1.** A shock wave theoretically occurring along a common line of intersection of all the pressure disturbances emanating from an infinitesimally small particle moving at supersonic speed through a fluid medium, with such wave considered to exert no changes in the condition of the fluid passing through it. **2.** A very weak shock wave appearing, for example, at the nose of a very sharp body, where the fluid undergoes no substantial change in direction. { 'mäk ,wav }

Mach-Zehnder interferometer [OPTICS] A variation of the Michelson interferometer used mainly in measuring the spatial variation of the index of refraction of a gas; the device has two semitransparent mirrors and two wholly reflecting mirrors at alternate corners of a rectangle, and half the beam travels along each side of the rectangle. Also known as Mach refractometer. { 'mäk 'tsän·dər ,in·tər·fə'räm·əd·ər }

M acid [ORG CHEM] $NH_2C_{10}H_5(OH)SO_3H$ A sulfonic acid formed by alkaline fusion of a disulfonic acid of α-naphthylamine; used as a dye intermediate. { 'em ,as·əd }

mackayite [MINERAL] $FeTe_2O_5(OH)$ A green mineral composed of basic iron tellurite. { 'mak·ē,īt }

mackerel [VERT ZOO] The common name for perciform fishes composing the subfamily Scombroidei of the family Scombridae, characterized by a long slender body, pointed head, and large mouth. { 'mak·rəl }

mackerel shark [VERT ZOO] The common name for isurid galeoid elasmobranchs making up the family Isuridae; heavy-bodied fish with sharp-edged, awllike teeth and a nearly symmetrical tail. { 'mak·rəl 'shärk }

mackerel sky [METEOROL] A sky with considerable cirrocumulus or small-element altocumulus clouds, resembling the scales on a mackerel. { 'mak·rəl 'skī }

mackinawite [MINERAL] $(Fe,Ni)S$ A tetragonal mineral occurring as a corrosion product in iron pipes. Also known as kansite. { mə'kin·ə,wīt }

Mack's cement [MATER] Cement made of dehydrated gypsum with a small amount of calcined sodium sulfate and potassium sulfate. { 'maks si'ment }

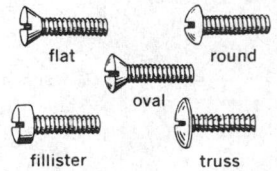
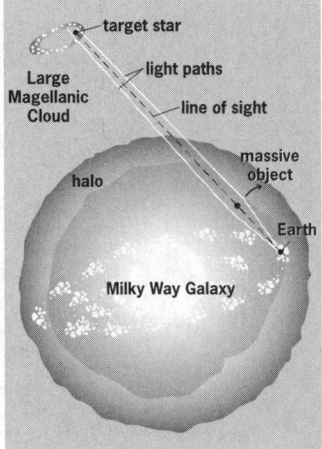
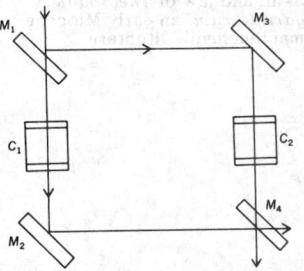
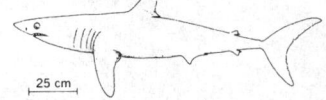

Macky effect [METEOROL] The reduction of the effective dielectric strength of air when waterdrops are present. { 'mak·ē i,fekt }

Mac-Lane system [MIN ENG] A means of conveying dirt to the top of a heap; an inclined rail track goes from the loading station at the bottom of the heap to an extending frame at the top; the rope that hauls the tubs with dirt up the rail track passes around a return sheave in the extending frame; a tub is tipped over at the top by a gear and the dirt discharged. { 'mak 'lān ,sis·təm }

Maclaurin-Cauchy test See Cauchy's test for convergence. { mə'klȯr·ən kō'shē ,test }

Maclaurin expansion [MATH] The power series representation of a function arising from Maclaurin's theorem. { mə'klȯr·ən ik,span·chən }

Maclaurin series [MATH] The power series in the Maclaurin expansion. { mə'klȯr·ən ,sir·ēz }

Maclaurin spheroid [ASTRON] A spheroid formed by the surface of a homogeneous, self-gravitating mass in uniform rotation. { mə'klȯr·ən 'sfir,ȯid }

Maclaurin's theorem [MATH] The theorem giving conditions when a function, which is infinitely differentiable, may be represented in a neighborhood of the origin as an infinite series with nth term $(1/n!) \cdot f^{(n)}(0) \cdot x^n$, where f^n denotes the nth derivative. { mə'klȯr·ənz ,thir·əm }

macle [CRYSTAL] A twinned crystal. [LAP] A twin structure in a diamond. [MINERAL] **1.** A dark or discolored spot in a mineral specimen. **2.** See chiastolite. { 'mak·əl }

Macleod equation [FL MECH] An equation which states that the fourth root of the surface tension of a liquid is proportional to the difference between the densities of the liquid and of its vapor. { mə'klaȯd i,kwā·zhən }

MacMichael degree [FL MECH] An arbitrary unit used in measuring viscosity with a type of Couette viscometer; its size depends on the stiffness of the suspension of the inner cylinder of the viscometer. { mik'mī·kəl di,grē }

Macquer's salt See potassium arsenate. { mə'kerz ,sȯlt }

macrandrous [BOT] Having both antheridia and oogonia on the same plant; used especially for certain green algae. { ma'kran·drəs }

Macraucheniidae [PALEON] A family of extinct herbivorous mammals in the order Litopterna; members were proportioned much as camels are, and eventually lost the vertebral arterial canal of the cervical vertebrae. { ,ma,krȯ·kə'nī·ə,dē }

macrencephaly [MED] The condition of having an abnormally large brain. { ,ma·kren'sef·ə·lē }

Macristiidae [VERT ZOO] A family of oceanic teleostean fishes assigned by some zoologists to the order Ctenothrissiformes. { ,mak·rə'stī·ə,dē }

macro See macroinstruction. { 'mak·rō }

macro- [SCI TECH] Prefix meaning large. { 'mak·rō or ,mak·rə }

macroanalysis [ANALY CHEM] Qualitative or quantitative analysis of chemicals that are in quantities of the order of grams. { 'mak·rō·ə'nal·ə·səs }

macroanalytical balance [ENG] A relatively large type of analytical balance that can weigh loads of up to 200 grams to the nearest 0.1 milligram. { 'mak·rō,an·ə'lid·ə·kəl 'bal·əns }

macroassembler [COMPUT SCI] A program made up of one or more sequences of assembly language statements, each sequence represented by a symbolic name. { ¦mak·rō·ə'sem·blər }

macroblast of Naegeli See pronormoblast. { 'mak·rə,blast əv 'neg·ə·lē }

macroblepharia [MED] The condition of having abnormally large eyelids. { ,mak·rə·blə'far·ē·ə }

macrobrachia [MED] The condition of having excessive arm development. { ,mak·rə'brāk·ē·ə }

macrocephalus [MED] An individual with an abnormally large head. { ,mak·rə 'sef·ə·ləs }

macrocephaly [MED] The condition of having an abnormally large head. { 'mak·rə'sef·ə·lē }

macrocheiria [MED] The condition of having abnormal hand enlargement. { 'mak·rō 'kī·rə }

macroclastic [PETR] Rock that is composed of fragments that are visible without magnification. { 'mak·rə'klas·tik }

macroclimate [CLIMATOL] The climate of a large geographic region. { ¦mak·rō'klī·mət }

macrocode [COMPUT SCI] A coding and programming language that assembles groups of computer instructions into single instructions. { 'mak·rə,kōd }

macroconsumer [ECOL] A large consumer which ingests other organisms or particulate organic matter. Also known as biophage. { 'mak·rō·kən'sü·mər }

macrocrania [MED] The condition of having abnormally large skull size compared with face size. { 'mak·rō'krā·nē·ə }

macrocrystalline [PETR] **1.** Pertaining to the texture of holocrystalline rock in which the constituents are visible without magnification. **2.** Pertaining to the texture of a rock with grains or crystals greater than 0.75 millimeter in diameter in recrystallized sediment. { 'mak·rō'krist·əl·ən }

macrocycle See macrocyclic compound. { 'mak·rō,sī·kəl }

macrocyclic [MYCOL] Of a rust fungus, having binuclear spores as well as teliospores and sporidia, or having a life cycle that is long or complex. { 'mak·rō'sī·klik }

macrocyclic compound [ORG CHEM] An organic compound containing a large ring, that is, a closed chain of 12 or more carbon atoms; examples include crown ethers, cryptands, spherands, carcerands, cyclodextrins, cyclophanes, and calixarenes. Also known as macrocycle. { ¦mak·rō,sī·klik 'kām ,paȯnd }

Macrocypracea [INV ZOO] A superfamily of marine ostracodes in the suborder Podocopa having all thoracic legs different from each other, greatly reduced furcae, and long, thin Zenker's organs. { 'mak·rō·sə'prās·ē·ə }

macrocyst [MED] A large cyst, visible to the naked eye. { 'mak·rə,sist }

macrocyte [HISTOL] An erythrocyte whose diameter or mean corpuscular volume exceed that of the mean normal by more than two standard deviations. Also known as macronormocyte. { 'mak·rə,sīt }

macrocytic anemia [MED] A form of anemia characterized by the presence of macrocytes in the blood. { ¦mak·rə¦sid·ik ə'nē·mē·ə }

macrocytosis [MED] Presence of macrocytes in the blood. { ,mak·rə,sī'tō·səs }

macrodactyly [MED] The condition of having abnormally large fingers or toes. { ¦mak·rō'dak·tə·lē }

Macrodasyoidea [INV ZOO] An order of wormlike invertebrates of the class Gastrotricha characterized by distinctive, cylindrical adhesive tubes in the cuticle which are moved by delicate muscle strands. { ,mak·rō,da·sē'ȯid·ē·ə }

macrodefinition [COMPUT SCI] A statement that defines a macroinstruction and the set of ordinary instructions which it replaces. { ¦mak·rō,def·ə'nish·ən }

macrodome [CRYSTAL] Dome of a crystal in which planes are parallel to the longer lateral axis. { 'mak·rə,dōm }

macrodontia [MED] The condition of having abnormally large teeth. { ,mak·rə'dän·chə }

macroelement [IND ENG] An element of a work cycle whose time span is long enough to be observed and measured with a stopwatch. { ¦mak·rō'el·ə·mənt }

macroencapsulation [CELL MOL] The envelopment of a large mass of xenotransplanted cells or tissue in planar membranes, hollow fibers, or diffusion chambers to isolate the cells from the body, thereby avoiding the immune responses that the foreign cells could initiate, and also to allow the desired metabolites (such as insulin and glucose for pancreatic islet cells) to diffuse in and out of the membrane. { ,mak·rō·in,kap·sə'lā·shən }

macroetch See deep-etch. { 'mak·rō,ech·iŋ }

macroevolution [EVOL] The larger course of evolution by which the categories of animal and plant classification above the species level have been evolved from each other and have differentiated into the forms within each. { ¦mak·rō,ev·ə'lü·shən }

macroexpansion [COMPUT SCI] Instructions generated by a macroinstruction and inserted into an assembly language program. { ¦mak·rō·ik'span·chən }

macrofacies [GEOL] A collection of sedimentary facies that are related genetically. { ¦mak·rō¦fā·shēz }

macrofauna [ECOL] **1.** Widely distributed fauna. **2.** Fauna of a macrohabitat. [ZOO] Animals visible to the naked eye. { ¦mak·rō¦fȯn·ə }

macroflora [BOT] Plants which are visible to the naked eye. [ECOL] **1.** Widely distributed flora. **2.** Flora of a macrohabitat. { ¦mak·rō¦flȯr·ə }

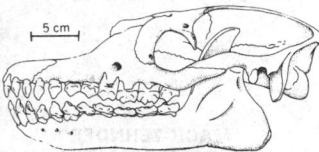

MACRAUCHENIIDAE

|← 5 cm →|

Skull and jaw of *Theosodon garrettorum*, an early Miocene macraucheniid litoptern.

macro flow chart [COMPUT SCI] A graphical representation of the overall logic of a computer program in which entire segments or subroutines of the program are represented by single blocks and no attempt is made to specify the detailed operation of the program. { 'mak·rō 'flō ,chärt }

macrofollicular adenoma [MED] **1.** A benign tumor of the thyroid with enlarged follicles. **2.** A type of malignant lymphoma. { ¦mak·rō·fə'lik·yə·lər ,ad·ən'ō·mə }

macrofossil [PALEON] A fossil large enough to be observed with the naked eye. { ¦mak·rō'fäs·əl }

macrogamete [BIOL] The larger, usually female gamete produced by a heterogamous organism. { ¦mak·rō'ga,mēt }

macrogeneration [COMPUT SCI] The creation of many machine instructions from one macroword. { mak·rō¦jen·ə¦rā·shən }

macrogenerator See macroprocessor. { ¦mak·rō'jen·ə,rād·ər }

macroglia [NEUROSCI] That portion of the neuroglia composed of astrocytes. { mə'kräg·lē·ə }

macroglobulin [BIOCHEM] Any gamma globulin with a sedimentation constant of 195. { ¦mak·rə'gläb·yə·lən }

macroglobulinemia [MED] **1.** Abnormal increase in macroglobulins in the blood. **2.** A disease characterized by proliferation of lymphocytes and plasmocytes and abnormally high macroglobulin blood levels. { ,mak·rə,gläb·yə·lə'nē·mē·ə }

macroglossia [MED] Enlargement of the tongue. { ¦mak·rō'gläs·ē·ə }

macrograph [GRAPHICS] A photograph or representation of an object that may be about 10 times natural size, as large as the natural object, or slightly smaller. { 'mak·rə,graf }

macrogyria [MED] The condition of having congenitally enlarged brain convolutions. { ,mak·rə'jī·rē·ə }

macrohabitat [ECOL] An extensive habitat presenting considerable variation of the environment, containing a variety of ecological niches, and supporting a large number and variety of complex flora and fauna. { ¦mak·rō'hab·ə,tat }

macroinstruction [COMPUT SCI] An instruction in a higher-level language which is equivalent to a specific set of one or more ordinary instructions in the same language. Also known as macro. { ¦mak·rō·in'strək·shən }

macrolanguage [COMPUT SCI] A computer language that manipulates stored strings in which particular sites of the string are marked so that other strings can be inserted in these sites when the stored string is brought forth. { 'mak·rō,laŋ·gwij }

macro lens [OPTICS] A camera lens designed to focus at very short distances and to form an image as large as the subject. { 'ma·krō ,lenz }

Macrolepidoptera [INV ZOO] A former division of Lepidoptera that included the larger moths and butterflies. { ¦mak·rō,lep·ə'däp·trə }

macrolibrary [COMPUT SCI] A collection of prewritten specialized but unparticularized routines (or sets of statements) which reside in mass storage. { 'mak·rō,lī,brer·ē }

macrolide [ORG CHEM] A large ring molecule with many functional groups bonded to it. { 'mak·rə,līd }

macrolide antibiotic [MICROBIO] A basic antibiotic characterized by a macrocyclic ring structure. { 'mak·rə,līd ,ant·i,bī'äd·ik }

macrolymphocyte [HISTOL] A large lymphocyte. { ¦mak·rō'lim·fə,sīt }

macromastia [MED] The condition of having abnormally enlarged breasts. { ,mak·rō'mas·tē·ə }

macromechanics See composite macromechanics. { ¦mak·rō·mə'kan·iks }

macromelia [MED] The condition of having abnormally large arms or legs. { ¦mak·rō'mēl·yə }

macromere [EMBRYO] Any of the large blastomeres composing the vegetative hemisphere of telolecithal morulas and blastulas. { 'mak·rō,mir }

macrometeorology [METEOROL] The study of the largest-scale aspects of the atmosphere, such as the general circulation, and weather types. { ¦mak·rō,mēd·ē·ə'räl·ə·jē }

macrometer [OPTICS] Instrument that has two mirrors and a focusing telescope with which the ranges of distant objects can be found. { ma'kräm·əd·ər }

macromolecular [ORG CHEM] Composed of or characterized by large molecules. { ¦mak·rō·mə'lek·yə·lər }

macromolecule [ORG CHEM] A large molecule in which there is a large number of one or several relatively simple structural units, each consisting of several atoms bonded together. { ¦mak·rō'mäl·ə,kyül }

Macromonas [MICROBIO] A genus of gram-negative, chemolithotrophic bacteria; large, motile, cylindrical to bean-shaped cells containing sulfur granules and calcium carbonate inclusions. { ¦mak·rə'mō·nəs }

macromutation [GEN] Any genetic change that leads to a pronounced phenotypic alteration. { ,mak·rə·myü'tā·shən }

macronormocyte See macrocyte. { ,mak·rə'nór·mə,sīt }

macronucleus [INV ZOO] A large, densely staining nucleus of most ciliated protozoans, believed to influence nutritional activities of the cell. { ¦mak·rō'nü·klē·əs }

macronutrient [BIOCHEM] An element required by animals or plants in large amounts. { ¦mak·rō'nü·trē·ənt }

macroparameter [COMPUT SCI] The character in a macro operand which will complete an open subroutine created by the macroinstruction. { ,mak·rō·pə'ram·əd·ər }

macrophage [HISTOL] A large phagocyte of the reticuloendothelial system. Also known as a histiocyte. { 'mak·rə,fāj }

macrophage inflammatory protein-1 [IMMUNOL] A protein produced by macrophages which has inflammatory and chemoattractant properties; it exists in two forms, MIP-1α and MIP-1β. { 'mak·rə,fāj in¦flam·ə,tór·ē 'prō,tēn ¦wən }

macrophagy [BIOL] Feeding on large particulate matter. { 'mak·rə,fā·jē }

macrophotography [GRAPHICS] The photography of a subject so that the final image is either unmagnified or magnified to no more than 10 times the object. { ¦mak·rō·fə'täg·rə·fē }

macrophreate [INV ZOO] A comatulid with a large, deep cavity in the calyx. { ¦mak·rō'frē,āt }

macrophyllous [BOT] Having large or long leaves. { ¦mak·rō'fil·əs }

macrophyte [ECOL] A macroscopic plant, especially one in an aquatic habitat. { 'mak·rə,fīt }

macropinacoid See front pinacoid. { ¦mak·rō'pin·ə,kóid }

macropinocytosis [CELL MOL] A mechanism of endocytosis in which large droplets of fluid are trapped underneath extensions (ruffles) of the cell surface. { ,mak·rō,pin·ə,sī'tō·səs }

Macropodidae [VERT ZOO] The kangaroos, a family of Australian herbivorous mammals in the order Marsupialia. { ,mak·rə'päd·ə,dē }

macropodous [BOT] **1.** Having a large or long hypocotyl. **2.** Having a long stem or stalk. { ma'kräp·ə·dəs }

macropore [CHEM] A pore in a catalytic material whose width is greater than 0.05 micrometer. [GEOL] A pore in soil of a large enough size so that water is not held in it by capillary attraction. { 'mak·rə,pór }

macroporous resin [ORG CHEM] A member of a class of very small, highly cross-linked polymer particles penetrated by channels through which solutions can flow; used as ion exchanger. Also known as macroreticular resin. { ¦mak·rə,pór·əs 'rez·ən }

macroprocessor [COMPUT SCI] A piece of software which replaces each macroinstruction in a computer program by the set of ordinary instructions which it stands for. Also known as macrogenerator. { ¦mak·rə'präs,es·ər }

macroprogramming [COMPUT SCI] The process of writing machine procedure statements in terms of macroinstructions. { ¦mak·rō'prō,gram·iŋ }

macroprosopus [MED] An individual with an abnormally large face. { ¦mak·rō·prə'sō·pəs }

macropsia [MED] A disturbance of vision in which objects seem larger than they are. Also known as megalopia. { ma'kräp·sē·ə }

macropterous [ZOO] Having large or long wings or fins. { ma'kräp·tə·rəs }

macroreticular resin See macroporous resin. { ¦mak·rō·rə'tik·yə·lər 'rez·ən }

macrorheology [MECH] A branch of rheology in which materials are treated as homogeneous or quasi-homogeneous, and processes are treated as isothermal. { ¦mak·rō·rē'äl·ə·jē }

Macroscelidea [VERT ZOO] A monofamilial order of mammals containing the elephant shrews and their allies. { ,mak·rō·sə'lid·ē·ə }

Macroscelididae [VERT ZOO] The single, African family of the mammalian order Macroscelidea. { ,mak·rō·sə'lid·ə,dē }

macrosclereid [BOT] A sclereid cell that is rod-like and formed from the embryonic epidermis of certain seed coats. { ˌmak·rō 'skler·ē·əd }

macroscopic [SCI TECH] Large enough to be observed by the naked eye. { ¦mak·rə¦skäp·ik }

macroscopic anisotropy [ENG] Phenomenon in electrical downhole logging wherein electric current flows more easily along sedimentary strata beds than perpendicular to them. { ¦mak·rə¦skäp·ik ˌan·ə'sä·trə,pē }

macroscopic cross section [PHYS] The sum of the cross sections of an atom in a substance. { ¦ma·krə¦skäp·ik 'krös ˌsek·shən }

macroscopic property [NUCLEO] A nuclear reactor property that can be treated independently of other factors. [THERMO] *See* thermodynamic property. { ¦mak·rə¦skäp·ik 'präp·ərd·ē }

macroscopic state [STAT MECH] Any state of a system as described by actual or hypothetical observations of its macroscopic statistical properties. Also known as macrostate. { ¦mak·rə¦skäp·ik 'stāt }

macroscopic stress [MET] Residual stress in a metal in a distance comparable to the gage length of strain measurement specimens and therefore detectable by x-ray or dissection techniques. Also known as macrostress. { ¦mak·rə¦skäp·ik 'stres }

macroscopic theory [PHYS] A theory concerning only phenomena observable with the naked eye or with an ordinary light microscope, and not with the behavior of atoms, molecules, or their constituents which may underlie these phenomena. { ¦mak·rə¦skäp·ik 'thē·ə·rē }

macroscopy [SCI TECH] The study or observation of objects visible to the unaided eye. { ma'kräs·kə·pē }

macroseptum [INV ZOO] A primary septum in certain anthozoans. { ¦mak·rō¦sep·təm }

macroskeleton [COMPUT SCI] A definition of a macroinstruction in a precise but content-free way, which can be particularized by a processor as directed by macroinstruction parameters. Also known as model. { ¦mak·rō¦skel·ə·tən }

macrosonics [ACOUS] The technology of sound at signal amplitudes so large that linear approximations are not valid, as in the use of ultrasonics for cleaning or drilling. { ¦mak·rō¦sän·iks }

macrosporangium [BOT] A spore case in which macrospores are produced. Also known as megasporangium. { ¦mak·rə·spə'ran·jē·əm }

macrospore [BOT] The larger of two spore types produced by heterosporous plants; the female gamete. Also known as megaspore. [INV ZOO] The larger gamete produced by certain radiolarians; the female gamete. { 'mak·rə,spór }

macrosporogenesis [BOT] In angiosperms, the formation of macrospores and the production of the embryo sac from one or occasionally several cells of the subepidermal cell layer within the ovule of a closed ovary. Also known as megasporogenesis. { ¦mak·rō¦spór·ō'jen·ə·səs }

macrostate *See* macroscopic state. { 'mak·rō,stāt }

Macrostomida [INV ZOO] An order of rhabdocoels having a simple pharynx, paired protonephridia, and a single pair of longitudinal nerves. { ˌmak·rə'stäm·ə·də }

Macrostomidae [INV ZOO] A family of rhabdocoels in the order Macrostomida; members are broad and flattened in shape and have paired sex organs. { ˌmak·rə'stäm·ə,dē }

macrostress *See* macroscopic stress. { 'mak·rō,stres }

macrostructure [MET] Structure of an etched metal visible to the naked eye or at magnifications up to 10 diameters. [SCI TECH] The external forms of a structure that can be observed without magnification, for example, the cubic crystal form of sodium chloride. { ¦mak·rō¦strək·chər }

macrostylous [BOT] **1.** Having long styles. **2.** Having long styles and short stamens. { ¦mak·rō¦stī·ləs }

macrosystem [COMPUT SCI] A language in which words represent a number of machine instructions. { 'mak·rō,sis·təm }

Macrotermitinae [INV ZOO] A subfamily of termites in the family Termitidae. { ˌmak·rō,tər'mid·ə,nē }

macrothermophyte *See* megathermophyte. { ˌmak·rə'thər·mə,fīt }

macrotome [ENG] A device for making large anatomical sections. { 'mak·rə,tōm }

Macrouridae [VERT ZOO] The grenadiers, a family of actinopterygian fishes in the order Gadiformes in which the body

tapers to a point, and the dorsal, caudal, and anal fins are continuous. { mə'krúr·ə,dē }

Macroveliidae [INV ZOO] A family of hemipteran insects in the subdivision Amphibicorisae. { ˌmak·rō·və'lī·ə,dē }

macro virus [COMPUT SCI] A virus that hides inside document and spreadsheet files used by popular word processing and spreadsheet applications. { ¦mak·rō 'vī·rəs }

Macrura [INV ZOO] A group of decapod crustaceans in the suborder Reptantia including eryonids, spiny lobsters, and true lobsters; the abdomen is extended and bears a well-developed tail fan. { mə'krúr·ə }

macrurous [ZOO] Having a long tail. { mə'krúr·əs }

macula [ANAT] Any anatomical structure having the form of a spot or stain. { 'mak·yə·lə }

macula densa [MED] A thickening of the epithelium of the ascending limb of the loop of Henle. { 'mak·yə·lə 'den·sə }

macula lutea [ANAT] A yellow spot on the retina; the area of maximum visual acuity, being made up almost entirely of retinal cones. { 'mak·yə·lə 'lüd·ē·ə }

maculate [BOT] Marked with speckles or spots. { 'mak·yə·lət }

maculopapule [MED] A small, circumscribed, discolored elevation of the skin; a macule and papule combined. { ¦mak·yə·lō¦pap·yül }

maculose [GEOL] Of a group of contact-metamorphosed rocks or their structures, having spotted or knotted character. { 'mak·yə,lōs }

MAD *See* magnetic anomaly detector; multiwavelength anomalous dispersion. { ¦em¦ā'dē *or* mad }

madarosis [MED] Loss of the eyelashes or eyebrows. { ˌmad·ə'rō·səs }

madder [MATER] The root of the madder plant (*Rubia tinctorium*), pulverized and used as source of glucosides to produce alizarin by fermentation. Also known as gamene. { 'mad·ər }

madder lake [MATER] Bluish-red, transparent pigment produced from alizarin red; used to make stains and inks, and as a component of artists' oil colors. { 'mad·ər 'lāk }

Madelung constant [SOLID STATE] A dimensionless constant which determines the electrostatic energy of a three-dimensional periodic crystal lattice consisting of a large number of positive and negative point charges when the number and magnitude of the charges and the nearest-neighbor distance between them is specified. { 'mä·də,lùŋ ,kän·stənt }

madia oil [MATER] An oil that is made by crushing the seeds of the melosa plant; used as a substitute for olive oil. { 'mäd·ē·ə ,óil }

madistor [ELECTR] A cryogenic semiconductor device in which injection plasma can be steered or controlled by transverse magnetic fields, to give the action of a switch. { ma'dis·tər }

Madreporaria [INV ZOO] The equivalent name for Scleractinia. { ˌmad·rə·pə'rar·ē·ə }

madreporite [INV ZOO] A delicately perforated sieve plate at the distal end of the stone canal in echinoderms. { 'mad·rə,pór,īt }

Madsen impedance meter [ENG] An instrument for measuring the acoustic impedance of normal and deaf ears, based on the principle of the Wheatstone bridge. { 'mad·zən im'pēd·əns ,mēd·ər }

MADT *See* microalloy diffused transistor.

madura foot *See* mycetoma. { 'maj·ə·rə ,fút }

maduromycosis *See* mycetoma. { ¦maj·ə·rō,mī'kō·səs }

maelstrom [OCEANOGR] A powerful and often destructive water current caused by the combined effects of high, wind-generated waves and a strong, opposing tidal current. { 'māl·strəm }

Maestrichtian [GEOL] A European stage of geologic time: Upper Cretaceous (above Menevian, below Fastiniogian). { ma'strik·tē·ən }

maestro [METEOROL] A northwesterly wind with fine weather which blows, especially in summer, in the Adriatic, most frequently on the western shore; it is also found on the coasts of Corsica and Sardinia. { 'mī·strō }

Maffei system [ASTRON] A cluster of external galaxies that lie very close to the galactic plane in the constellation Perseus and are heavily obscured by galactic dust. { mä'fā·ē ,sis·təm }

mafic mineral [MINERAL] **1.** A mineral that is composed predominantly of the ferromagnesian rock-forming silicates. **2.** In general, any dark mineral.   { 'maf·ik 'min·rəl }

mafura tallow [MATER] A bitter-tasting, heavy vegetable fat obtained from the nuts of the mafura tree (*Trichilia emetica*); used for soaps, candles, and ointments.   { mə'fə·rə ¦tal·ō }

MAG *See* maximum available gain.

magamp *See* magnetic amplifier.   { 'mag,amp }

magazine [COMPUT SCI] **1.** A holder of microfilm or magnetic recording media strips. **2.** *See* input magazine. [ENG] **1.** A storage area for explosives. **2.** A building, compartment, or structure constructed and located for the storage of explosives or ammunition. [ORD] That part of a gun or firearm that holds ammunition ready for chambering.   { ¦mag·ə¦zēn }

magazine filler [ORD] An item which attaches to the rear top portion of a rifle magazine to retain and position a cartridge clip to facilitate loading.   { ¦mag·ə¦zēn 'fil·ər }

magazine safety [ORD] A mechanism for an automatic pistol, to make firing impossible unless the weapon contains the magazine.   { ¦mag·ə¦zēn 'sāf·tē }

Magellanic Clouds [ASTRON] Two irregular clouds of stars that are the nearest galaxies to the galactic system; both the Large and Small Magellanic Clouds are identified as Irregular in the classification of E.P. Hubble. Also known as Nubeculae.   { ¦maj·ə¦lan·ik 'klaùdz }

Magellanic Stream [ASTRON] A long, thin inhomogeneous filament of gas which extends 120° from the region between the Magellanic Clouds to a point near the south galactic pole.   { ¦maj·ə¦lan·ik 'strēm }

Magellanic System [ASTRON] An envelope of neutral hydrogen that includes both the Magellanic Clouds.   { ¦maj·ə¦lan·ik 'sis·təm }

Magelonidae [INV ZOO] A monogeneric family of spioniform annelid worms belonging to Sedentaria.   { ,maj·ə'län·ə,dē }

magenstrasse [ANAT] Gastric canal.   { 'mäg·ən,shträ·sə }

magenta *See* fuchsin.   { mə'jen·tə }

Maggi-Righi-Leduc effect [PHYS] A phenomenon in which the thermal conductivity of a conductor changes when it is placed in a magnetic field.   { 'mä·jē 'rē·gē lə'dùk i,fekt }

maggot [INV ZOO] Larva of a dipterous insect.   { 'mag·ət }

maggot therapy [MED] Implantation of sterile cultivated maggots of the bluebottle fly into wounds in the treatment of chronic soft tissue infections and chronic osteomyelitis.   { 'mag·ət ,ther·ə·pē }

maghemite [MINERAL] γ-Fe$_2$O$_3$ A mineral form of iron oxide that is strongly magnetic and a member of the magnetite series.   { mag'he,mīt }

magic acid [INORG CHEM] A superacid consisting of equal molar quantities of fluorosulfonic acid (HSO$_3$F) and antimony pentafluoride (SbF$_5$).   { 'maj·ik 'as·əd }

magic eye *See* cathode-ray tuning indicator.   { 'maj·ik ¦ī }

magic numbers [NUC PHYS] The integers 8, 20, 28, 50, 82, 126; nuclei in which the number of protons, neutrons, or both is magic have a stability and binding energy which is greater than average, and have other special properties. [PHYS CHEM] Numbers of atoms or molecules for which certain atom or molecular clusters have an unusually high abundance.   { 'maj·ik 'nəm·bərz }

magic square [MATH] **1.** A square array of integers where the sum of the entries of each row, each column, and each diagonal is the same. **2.** A square array of integers where the sum of the entries in each row and each column (but not necessarily each diagonal) is the same. Also known as semimagic square.   { 'maj·ik 'skwer }

magic tee *See* hybrid tee.   { 'maj·ik 'tē }

magister of sulfur [CHEM] Amorphous sulfur produced by acid precipitation from solutions of hyposulfites or polysulfides.   { mə'jis·tər əv 'səl·fər }

maglev *See* magnetic levitation.   { 'mag,lev }

magma [GEOL] The molten rock material from which igneous rocks are formed.   { 'mag·mə }

magma chamber [GEOL] A larger reservoir in the crust of the earth that is occupied by a body of magma.   { 'mag·mə ,chām·bər }

magma geothermal system [GEOL] A geothermal system in which the dominant source of heat is a large reservoir of igneous magma within an intrusive chamber or lava pool; an

example is the Yellowstone Park area of Wyoming.   { 'mag·mə ¦jē·ō'thər·məl ,sis·təm }

magma province *See* petrographic province.   { 'mag·mə ,präv·əns }

magmatic differentiation [PETR] **1.** The process by which the different types of igneous rocks are derived from a single parent magma. **2.** The process by which ores are formed by solidification from magma. Also known as magmatic segregation.   { mag'mad·ik ,dif·ə,ren·chē'ā·shən }

magmatic rock [PETR] A rock derived from magma.   { mag'mad·ik 'räk }

magmatic segregation *See* magmatic differentiation.   { mag'mad·ik ,seg·rə'gā·shən }

magmatic stoping [GEOL] A process of igneous intrusion in which magma gradually works its way upward by breaking off and engulfing blocks of the country rock. Also known as stoping.   { mag'mad·ik 'stōp·iŋ }

magmatic water [HYD] Water derived from or existing in molten igneous rock or magma. Also known as juvenile water.   { mag'mad·ik 'wòd·ər }

magmatism [PETR] The formation of igneous rock from magma.   { 'mag·mə,tiz·əm }

magmosphere *See* pyrosphere.   { 'mag·mə,sfir }

magn [ELECTROMAG] A unit of absolute permeability equal to 1 henry per meter; it was proposed by the former Soviet Union, but has not won general acceptance.   { 'mäg·ən }

magnafacies [GEOL] A major, continuous belt of deposits that is homogeneous in lithologic and paleontologic characteristics and that extends obliquely across time planes or through several time-stratigraphic units.   { ¦mag·nə'fā·shēz }

magnesia [INORG CHEM] Magnesium oxide that is processed for a particular purpose.   { mag'nē·zhə }

magnesia brick [MATER] A type of refractory brick composed of magnesium oxide with about 15% of other oxides. Also known as magnesite brick.   { mag'nē·zhə ,brik }

magnesia cement *See* magnesium oxychloride cement.   { mag'nē·zhə si'ment }

magnesia magma *See* milk of magnesia.   { mag'nē·zhə 'mag·mə }

magnesia mica *See* biotite.   { mag'nē·zhə 'mī·kə }

magnesia mixture [ANALY CHEM] Reagent used to analyze for phosphorus; consists of the filtered liquor from an aqueous mixture of ammonium chloride, magnesium sulfate, and ammonia.   { mag'nē·zhə ,miks·chər }

magnesian calcite [MINERAL] (Ca,Mg)CO$_3$ A variety of calcite consisting of randomly substituted magnesium carbonate in a disordered calcite lattice. Also known as magnesium calcite.   { mag'nē·zhən ,kal,sīt }

magnesian limestone [PETR] Limestone with at least 90% calcite, a maximum of 10% dolomite, an approximate magnesium oxide equivalent of 1.1–2.1, and an approximate magnesium carbonate equivalent of 2.3–4.4.   { mag'nē·zhən 'līm,stōn }

magnesian marble [PETR] A type of magnesian limestone that has been metamorphosed; contains some dolomite. Also known as dolomitic marble.   { mag'nē·zhən 'mär·bəl }

magnesia refractory [MATER] Heat-and corrosion-resistant material made of magnesium oxide; used in cement or brick form to line high-temperature process vessels or furnaces.   { mag'nē·zhə ri'frak·trē }

magnesiochromite [MINERAL] MgCr$_2$O$_4$ A mineral of the spinel group composed of magnesium chromium oxide; it is isomorphous with chromite. Also known as magnochromite.   { mag,nē·zhō·'krō,mīt }

magnesiocopiapite [MINERAL] MgFe$_4$(SO$_4$)$_6$(OH)$_2$·20H$_2$O A mineral of the copiapite group composed of hydrous basic magnesium and iron sulfate; it is isomorphous with copiapite and cuprocopiapite.   { mag,nē·zhō·'kō·pē·ə,pīt }

magnesioferrite [MINERAL] (Mg,Fe)Fe$_2$O$_4$ A black, strongly magnetic mineral of the magnetite series in the spinel group. Also known as magnoferrite.   { mag,nē·zhō'fe,rīt }

magnesite [MINERAL] MgCO$_3$ The mineral form of magnesium carbonate, usually massive and white, with hexagonal symmetry; specific gravity is 3, and hardness is 4 on Mohs scale. Also known as giobertite.   { 'mag·nə,sīt }

magnesite brick *See* magnesia brick.   { 'mag·nə,sīt ,brik }

magnesite wheel [ENG] A grinding wheel made with magnesium oxychloride as the bonding agent.   { 'mag·nə,sīt ,wēl }

magnesium [CHEM] A metallic element, symbol Mg,

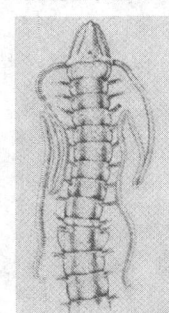

atomic number 12, atomic weight 24.305. [MET] A silvery-white, lightweight, malleable, ductile metal, used in metallurgical and chemical processes, photography, pyrotechny, and light alloys. { mag'nē·zē·əm }

magnesium acetate [ORG CHEM] $Mg(OOCCH_3)_2 \cdot 4H_2O$ or $Mg(OOCCH_3)_2$ A compound forming colorless crystals that are soluble in water and melt at 80°C; used in textile printing, in medicine as an antiseptic, and as a deodorant. { mag'nē·zē·əm 'as·ə,tāt }

magnesium anode [ELEC] Bar of magnesium buried in the earth, connected to an underground cable to prevent cable corrosion due to electrolysis. { mag'nē·zē·əm 'an,ōd }

magnesium arsenate [INORG CHEM] $Mg_3(AsO_4)_2 \cdot xH_2O$ A white, poisonous, water-insoluble powder used as an insecticide. { mag'nē·zē·əm 'ärs·ən,āt }

magnesium benzoate [ORG CHEM] $Mg(C_7H_5O_2)_2 \cdot 3H_2O$ A crystalline white powder melting at 200°C; soluble in alcohol and hot water; used in medicine. { mag'nē·zē·əm 'ben·zə,wāt }

magnesium bomb [ORD] **1.** An incendiary bomb in which the burning agent is magnesium. **2.** A magnesium flare for use from aircraft. { mag'nē·zē·əm 'bäm }

magnesium borate [INORG CHEM] $3MgO \cdot B_2O_3$ Crystals that are white or colorless and transparent; soluble in alcohol and acids, slightly soluble in water; used as a fungicide, antiseptic, and preservative. { mag'nē·zē·əm 'bȯr,āt }

magnesium boride See magnesium diboride. { mag'nē·zē·əm 'bȯr,īd }

magnesium bromate [INORG CHEM] $Mg(BrO_3)_2 \cdot 6H_2O$ A white crystalline compound, insoluble in alcohol, soluble in water; a fire hazard; used as an analytical reagent. { mag'nē·zē·əm 'brō,māt }

magnesium bromide [INORG CHEM] $MgBr_2 \cdot 6H_2O$ Deliquescent, colorless, bitter-tasting crystals, melting at 172°C; soluble in water, slightly soluble in alcohol; used in medicine and in the synthesis of organic chemicals. { mag'nē·zē·əm 'brō,mīd }

magnesium calcite See magnesian calcite. { mag'nē·zē·əm 'kal,sīt }

magnesium carbonate [INORG CHEM] $MgCO_3$ A water-insoluble, white powder, decomposing at about 350°C; used as a refractory material. { mag'nē·zē·əm 'kär·bə,nāt }

magnesium cell [ELEC] A primary cell in which the negative electrode is made of magnesium or one of its alloys. { mag'nē·zē·əm ¦sel }

magnesium chlorate [INORG CHEM] $Mg(ClO_3)_2 \cdot 6H_2O$ A white powder, bitter-tasting and hygroscopic; slightly soluble in alcohols, soluble in water; used in medicine. { mag'nē·zē·əm 'klȯr,āt }

magnesium chloride [INORG CHEM] $MgCl_2 \cdot 6H_2O$ Deliquescent white crystals; soluble in water and alcohol; used in disinfectants and fire extinguishers, and in ceramics, textiles, and paper manufacture. { mag'nē·zē·əm 'klȯr,īd }

magnesium-copper sulfide rectifier [ELECTR] Dry-disk rectifier consisting of magnesium in contact with copper sulfide. { mag'nē·zē·əm 'käp·ər ¦səl,fīd 'rek·tə,fī·ər }

magnesium diboride [INORG CHEM] MgB_2 A crystalline intermetallic compound, produced as a black powder, that becomes superconducting at the unusually high temperature of 39 K (−389°F; −234°C); melts at 800°C. Also known as magnesium boride. { mag'nē·zē·əm dī'bȯr,īd }

magnesium dust [MET] Magnesium metal powder; flammable; used in photographic flash lights and pyrotechnics. { mag'nē·zē·əm 'dəst }

magnesium flare [ORD] A flare using magnesium as the illuminating agent. { mag'nē·zē·əm 'fler }

magnesium fluoride [INORG CHEM] MgF_2 White, fluorescent crystals; insoluble in water and alcohol, soluble in nitric acid; melts at 1263°C; used in ceramics and glass. Also known as magnesium flux. { mag'nē·zē·əm 'flu̇r,īd }

magnesium fluosilicate [INORG CHEM] $MgSiF_6 \cdot 6H_2O$ Water-soluble, efflorescent white crystals; used in ceramics, in mothproofing and waterproofing, and as a concrete hardener. Also known as magnesium silicofluoride. { mag'nē·zē·əm 'flu̇·ə'sil·ə,kāt }

magnesium flux See magnesium fluoride. { mag'nē·zē·əm 'fləks }

magnesium formate [ORG CHEM] $Mg(CHO_2)_2 \cdot 2H_2O$ Colorless, water-soluble crystals; insoluble in alcohol and ether;

used in analytical chemistry and medicine. { mag'nē·zē·əm 'fȯr,māt }

magnesium gluconate [ORG CHEM] $Mg(C_6H_{11}O_7)_2 \cdot 2H_2O$ An odorless, tasteless, water-soluble powder; used in medicine. { mag'nē·zē·əm 'glü·kə,nāt }

magnesium halide [INORG CHEM] A compound formed from the metal magnesium and any of the halide elements; an example is magnesium bromide. { mag'nē·zē·əm 'ha,līd }

magnesium hydrate See magnesium hydroxide. { mag'nē·zē·əm 'hī,drāt }

magnesium hydride [INORG CHEM] MgH_2 A hydride compound formed from the metal magnesium; it decomposes violently in water, and in a vacuum at about 280°C. { mag'nē·zē·əm 'hī,drīd }

magnesium hydroxide [INORG CHEM] $Mg(OH)_2$ A white powder, very slightly soluble in water, decomposing at 350°C; used as an intermediate in extraction of magnesium metal, and as a reagent in the sulfite wood pulp process. Also known as magnesium hydrate. { mag'nē·zē·əm hī'dräk,sīd }

magnesium hyposulfite See magnesium thiosulfate. { mag'nē·zē·əm ,hī·pō'səl,fīt }

magnesium iodide [INORG CHEM] $MgI_2 \cdot 8H_2O$ Crystalline powder, white and deliquescent, discoloring in air; soluble in water, alcohol, and ether; used in medicine. { mag'nē·zē·əm 'ī·ə,dīd }

magnesium-iron mica See biotite. { mag'nē·zē·əm 'ī·ərn 'mī·ka }

magnesium lactate [ORG CHEM] $Mg(C_3H_5O_3)_2 \cdot 3H_2O$ Bitter-tasting, water-soluble white crystals; slightly soluble in alcohol; used in medicine. { mag'nē·zē·əm 'lak,tāt }

magnesium lime [MATER] Lime containing more than 20% magnesium oxide; slakes more slowly, evolves less heat, expands less, sets more rapidly, and produces higher-strength mortars than does high-calcium quicklime. { mag'nē·zē·əm 'līm }

magnesium-manganese dioxide cell [ELEC] Type of electrochemical (dry) cell battery in which the active elements are magnesium and manganese dioxide. { mag'nē·zē·əm 'maŋ·gə,nēs dī'äk,sīd ,sel }

magnesium methoxide [ORG CHEM] $(CH_3O)_2Mg$ Colorless crystals that decompose when heated; used as a catalyst, dielectric coating, and cross-linking agent, and to form gels. Also known as magnesium methylate. { mag'nē·zē·əm me'thäk,sīd }

magnesium methylate See magnesium methoxide. { mag'nē·zē·əm 'meth·ə,lāt }

magnesium nitrate [INORG CHEM] $Mg(NO_3)_2 \cdot 6H_2O$ Deliquescent white crystals; soluble in alcohol and water; a fire hazard; used as an oxidizing material in pyrotechnics. { mag'nē·zē·əm 'nī,trāt }

magnesium oleate [ORG CHEM] $Mg(C_{18}H_{33}O_2)_2$ Water-insoluble, yellowish mass; soluble in hydrocarbons, alcohol, and ether; used as a plasticizer lubricant and emulsifying agent, and in varnish driers and dry-cleaning solutions. { mag'nē·zē·əm 'ō·lē,āt }

magnesium oxide [INORG CHEM] MgO A white powder that (depending on the method of preparation) may be light and fluffy, or dense; melting point 2800°C; insoluble in acids, slightly soluble in water; used in making refractories, and in cosmetics, pharmaceuticals, insulation, and medicine. { mag'nē·zē·əm 'äk,sīd }

magnesium oxychloride cement [MATER] Cement made by adding a magnesium chloride solution to magnesia; used for interior flooring. Also known as magnesia cement. { mag'nē·zē·əm ,äk·sə'klȯr,īd si'ment }

magnesium perchlorate [INORG CHEM] $Mg(ClO_4)_2 \cdot 6H_2O$ White, deliquescent crystals; soluble in water and alcohol; explosive when in contact with reducing materials; used as a drying agent for gases. { mag'nē·zē·əm pər'klȯr,āt }

magnesium peroxide [INORG CHEM] MgO_2 A tasteless, odorless white powder; soluble in dilute acids, insoluble in water; a fire hazard; used as a bleaching and oxidizing agent, and in medicine. { mag'nē·zē·əm pə'räk,sīd }

magnesium phosphate [INORG CHEM] A compound with three forms: monobasic, $MgH_4(PO_4)_2 \cdot 2H_2O$, used in medicine and wood fireproofing; dibasic, $MgHPO_4 \cdot 3H_2O$, used in medicine and as a plastics stabilizer; tribasic, $Mg_3(PO_4)_2 \cdot 8H_2O$, used in dentifrices, as an adsorbent, and in pharmaceuticals. { mag'nē·zē·əm 'fäs,fāt }

magnesium salicylate [ORG CHEM] $Mg(C_7H_5O)_3 \cdot 4H_2O$ Efflorescent colorless crystals; soluble in water and alcohol; used in medicine. { mag'nē·zē·əm sə'lis·ə,lāt }

magnesium silicate [INORG CHEM] $3MgSiO_3 \cdot 5H_2O$ White, water-insoluble powder, containing variable proportions of water of hydration; used as a filler for rubber and in medicine. { mag'nē·zē·əm 'sil·ə,kāt }

magnesium silicofluoride See magnesium fluosilicate. { mag'nē·zē·əm ,sil·ə·kō'flur,īd }

magnesium-silver chloride cell [ELEC] A reserve primary cell that is activated by adding water; active elements are magnesium and silver chloride. { mag'nē·zē·əm 'sil·vər 'klòr,īd ,sel }

magnesium stearate [ORG CHEM] $Mg(C_{18}H_{35}O_2)_2$ Tasteless, odorless white powder; soluble in hot alcohol, insoluble in water; melts at 89°C; used in paints and medicine, and as a plastics stabilizer and lubricant. Also known as dolomol. { mag'nē·zē·əm 'stir,āt }

magnesium sulfate [INORG CHEM] $MgSO_4$ Colorless crystals with a bitter, saline taste; soluble in glycerol; used in fireproofing, textile processes, ceramics, cosmetics, and fertilizers. { mag'nē·zē·əm 'səl,fāt }

magnesium sulfite [INORG CHEM] $MgSO_3 \cdot 6H_2O$ A white, crystalline powder; insoluble in alcohol, slightly soluble in water; used in medicine and paper pulp. { mag'nē·zē·əm 'səl,fīt }

magnesium thiosulfate [INORG CHEM] $MgS_2O_3 \cdot 6H_2O$ Colorless crystals that lose water at 170°C; used in medicine. Also known as magnesium hyposulfite. { mag'nē·zē·əm ,thī·ə'səl,fāt }

magnesium trisilicate [INORG CHEM] $Mg_2Si_3O_8 \cdot 5H_2O$ A white, odorless, tasteless powder; insoluble in water and alcohol; used as an industrial odor absorbent and in medicine. { mag'nē·zē·əm ,trī'sil·ə,kāt }

magnesium tungstate [INORG CHEM] $MgWoO_4$ White crystals, insoluble in alcohol and water, soluble in acid; used in luminescent paint and for fluorescent x-ray screens. { mag'nē·zē·əm 'təŋ,stāt }

magneson [ORG CHEM] $C_{12}H_9N_3O_4$ A brownish-red powder, soluble in dilute aqueous sodium hydroxide; used in the detection of magnesium and molybdenum. { mag·nə,sän }

magnesyn [ELEC] A portion of a repeater unit; a two-pole permanently magnetized rotor within a three-phase two-pole delta-connected stator which carries the indicating pointer and is free to rotate in any direction. { mag·nə,sin }

magnet [ELECTROMAG] A piece of ferromagnetic or ferrimagnetic material whose domains are sufficiently aligned so that it produces a net magnetic field outside itself and can experience a net torque when placed in an external magnetic field. { 'mag·nət }

magnet alloy [MET] An alloy such as Alnico or Alcomax having strong magnetic properties; used in making permanent magnets. { 'mag·nət ¦al,öi }

magnet coil See iron-core coil. { 'mag·nət ,köil }

magnet grate [MIN ENG] A series of magnetized bars used to trap and remove tramp iron from a flow of pulverized or granulated dry solids passing over the grate; used to protect crushing or grinding equipment. { 'mag·nət ,grāt }

magnetic [ELECTROMAG] Pertaining to magnetism or a magnet. { mag'ned·ik }

magnetically focused tube [ELECTR] An image tube in which electrons from the photocathode are accelerated by electric fields and forced into tight spiral paths as they are further accelerated by a uniform magnetic field down the center of the tube. { mag'ned·ə·klē 'fō·kəst 'tüb }

magnetically hard alloy [MET] A ferromagnetic alloy that can be permanently magnetized after the removal of an externally applied magnetic field. { mag'ned·ə·klē ¦härd 'al,öi }

magnetically soft alloy [MET] A ferromagnetic alloy which is capable of being magnetized upon application of an external magnetic field, but which returns to a nonmagnetic condition when the field is removed. { mag'ned·ə·klē ¦sòft 'al,öi }

magnetic amplifier [ELECTR] A device that employs saturable reactors to modulate the flow of alternating-current electric power to a load in response to a lower-energy-level direct-current input signal. Abbreviated magamp. Also known as transductor. { mag'ned·ik 'am·plə,fī·ər }

magnetic amplitude [NAV] Angular distance of a celestial body north or south of the prime vertical circle and relative to magnetic east or west. { mag'ned·ik 'am·plə,tüd }

magnetic analysis inspection [MET] A nondestructive inspection method to determine the presence of variations in magnetic flux in ferromagnetic materials of constant cross section caused by defects, variations in hardness, discontinuities, or other irregularities. { mag'ned·ik ə'nal·ə·səs in,spek·shən }

magnetic anisotropy [ELECTROMAG] The dependence of the magnetic properties of some materials on direction. { mag'ned·ik ,an·ə'sä·trə,pē }

magnetic annealing [MET] Annealing and cooling in a strong magnetic field. { mag'ned·ik ə'nēl·iŋ }

magnetic annual change [GEOPHYS] The amount of secular change in the earth's magnetic field which occurs in 1 year. Also known as annual magnetic change. { mag'ned·ik 'an·yə·wəl 'chärj }

magnetic annual variation [GEOPHYS] The small, systematic temporal variation in the earth's magnetic field which occurs after the trend for secular change has been removed from the average monthly values. Also known as annual magnetic variation. { mag'ned·ik 'an·yə·wəl ,ver·ē'ā·shən }

magnetic anomaly detector [ELECTROMAG] A sensitive magnetometer carried at the end of a boom projecting from the tail of a patrol plane which can detect very small changes in the earth's magnetic field caused by a ferrous object, such as a submerged submarine; used to pinpoint a submarine's location, for effective deployment of suitable weapons. Abbreviated MAD. { mag'ned·ik ə'näm·ə·lē di,tek·tər }

magnetic axis [ELECTROMAG] A line through the center of a magnet such that the torque exerted on the magnet by a magnetic field in the direction of this line equals 0. [PL PHYS] The single line of force that closes on itself after one revolution in a magnetic field with a rotational transform. { mag'ned·ik 'ak·səs }

magnetic azimuth [NAV] Azimuth relative to magnetic north. { mag'ned·ik 'az·ə·məth }

magnetic balance [ENG] **1.** A device for determining the repulsion or attraction between magnetic poles, in which one magnet is suspended and the forces needed to cancel the effects of bringing a pole of another magnet close to one end are measured. **2.** Any device for measuring the small forces involved in determining paramagnetic or diamagnetic susceptibility. { mag'ned·ik 'bal·əns }

magnetic bay [GEOPHYS] A small magnetic disturbance whose magnetograph resembles an indentation of a coastline; on earth, magnetic bays occur mainly in the polar regions and have a duration of a few hours. { mag'ned·ik 'bā }

magnetic bearing [MECH ENG] A device incorporating magnetic forces to cause a shaft to levitate and float in a magnetic field without any contact between the rotating and stationary elements. [NAV] Bearing relative to magnetic north, with the compass bearing corrected for deviation. { mag'ned·ik 'ber·iŋ }

magnetic bias [ELECTROMAG] A steady magnetic field applied to the magnetic circuit of a relay or other magnetic device. { mag'ned·ik 'bī·əs }

magnetic blowout [ELECTROMAG] **1.** A permanent magnet or electromagnet used to produce a magnetic field that lengthens the arc between opening contacts of a switch or circuit breaker, thereby helping to extinguish the arc. **2.** See blowout. { mag'ned·ik 'blō,aùt }

magnetic bottle [PL PHYS] A magnetic field used to confine or contain a plasma in controlled fusion experiments. { mag'ned·ik 'bäd·əl }

magnetic brake [MECH ENG] A friction brake under the control of an electromagnet. { mag'ned·ik 'brāk }

magnetic bubble [SOLID STATE] A cylindrical stable (nonvolatile) region of magnetization produced in a thin-film magnetic material by an external magnetic field; direction of magnetization is perpendicular to the plane of the material. Also known as bubble. { mag'ned·ik 'bəb·əl }

magnetic bubble electromagnetic pulse [ELECTROMAG] A relatively slow pulse of electromagnetic radiation generated by the expanding conducting plasma shell of bomb material and other nearby matter from a nuclear explosion. { mag'ned·ik 'bəb·əl i¦lek·trō·mag'ned·ik 'pəls }

magnetic bubble memory See bubble memory. { mag'ned·ik ¦bəb·əl ,mem·rē }

MAGNET

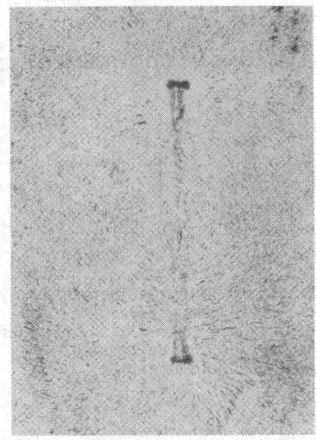

Photograph of the iron-filing map of the magnetic field of a permanent bar magnet. Note that the magnetic flux lines can be traced by the lines of iron filings, which act like tiny compass needles.

magnetic card [COMPUT SCI] A card with a magnetic surface on which data can be stored by selective magnetization. { mag'ned·ik 'kärd }

magnetic card file [COMPUT SCI] A direct-access storage device in which units of data are stored on magnetic cards contained in one or more magazines from which they are withdrawn, when addressed, to be carried at high speed past a read/write head. { mag'ned·ik 'kärd ‚fīl }

magnetic cell [ELECTR] One unit of a magnetic memory, capable of storing one bit of information as a zero state or a one state. { mag'ned·ik 'sel }

magnetic character [COMPUT SCI] A character printed with magnetic ink, as on bank checks, for reading by machines as well as by humans. { mag'ned·ik 'kar·ik·tər }

magnetic character figure See C index. { mag'ned·ik ‚kar·ik·tər ‚fig·yər }

magnetic character reader [COMPUT SCI] A character reader that reads special type fonts printed in magnetic ink, such as those used on bank checks, and feeds the character data directly to a computer for processing. { mag'ned·ik ‚kar·ik·tər ‚rēd·ər }

magnetic character sorter [COMPUT SCI] A device that reads documents printed with magnetic ink; all data read are stored, and records are sorted on any required field. Also known as magnetic document sorter-reader. { mag'ned·ik ‚kar·ik·tər ‚sórd·ər }

magnetic charge See magnetic grenade. { mag'ned·ik 'chärj }

magnetic chart [NAV] A chart showing the distribution of one of the magnetic elements, as by isogonic lines, or of its secular change. { mag'ned·ik 'chärt }

magnetic chuck [MECH ENG] A chuck in which the workpiece is held by magnetic force. { mag'ned·ik 'chək }

magnetic circuit [ELECTROMAG] A group of magnetic flux lines each forming a closed path, especially when this circuit is regarded as analogous to an electric circuit because of the similarity of its magnetic field equations to direct-current circuit equations. { mag'ned·ik 'sər·kət }

magnetic clutch See magnetic fluid clutch; magnetic friction clutch. { mag'ned·ik 'kləch }

magnetic coercive force See coercive force. { mag'ned·ik kō'ər·siv 'fòrs }

magnetic compass [NAV] A compass depending for its directive force upon the attraction of the horizontal component of the earth's magnetic field for a magnetized needle or sensing element free to turn with a minimum of friction in any horizontal direction. { mag'ned·ik 'käm·pəs }

magnetic compass table See deviation table. { mag'ned·ik 'käm·pəs ‚tā·bəl }

magnetic confinement [PL PHYS] The containment of a plasma within a region of space by the forces of magnetic fields on the charged particles in the gas. { mag'ned·ik kən'fīn·mənt }

magnetic constant [ELECTROMAG] The absolute permeability of empty space, equal to 1 electromagnetic unit in the centimeter-gram-second system, and to $4\pi \times 10^{-7}$ henry per meter or, numerically, to 1.25664×10^{-6} henry per meter in the International System of units. Symbolized μ_0. { mag'ned·ik 'kän·stənt }

magnetic cooling See adiabatic demagnetization. { mag'ned·ik 'kül·iŋ }

magnetic core Also known as core. [ELECTR] A configuration of magnetic material, usually a mixture of iron oxide or ferrite particles mixed with a binding agent and formed into a tiny doughnutlike shape, that is placed in a spatial relationship to current-carrying conductors, and is used to maintain a magnetic polarization for the purpose of storing data, or for its nonlinear properties as a logic element. Also known as memory core. [ELECTROMAG] A quantity of ferrous material placed in a coil or transformer to provide a better path than air for magnetic flux, thereby increasing the inductance of the coil and increasing the coupling between the windings of a transformer. { mag'ned·ik 'kòr }

magnetic core multiplexer [COMPUT SCI] A device which channels many bit inputs into a single output. { mag'ned·ik ‚kòr 'məl·tə‚plek·sər }

magnetic core storage [COMPUT SCI] A computer storage system in which each of thousands of magnetic cores stores one bit of information; current pulses are sent through wires threading through the cores to record or read out data. Also known as core memory; core storage. { mag'ned·ik ‚kòr 'stòr·ij }

magnetic coupling [ELECTROMAG] For a pair of particles or systems, the effect of the magnetic field created by one system on the magnetic moment or angular momentum of the other. { mag'ned·ik 'kəp·liŋ }

magnetic course [NAV] Course relative to magnetic north, with the compass course corrected for deviation. { mag'ned·ik 'kòrs }

magnetic crack detection See magnetic-particle test. { mag'ned·ik 'krak di‚tek·shən }

magnetic cumulative generator See flux-compression generator. { mag'ned·ik ‚kyü·myə‚lād·iv 'jen·ə‚rād·ər }

magnetic Curie temperature [SOLID STATE] The temperature below which a magnetic material exhibits ferromagnetism, and above which ferromagnetism is destroyed and the material is paramagnetic. { mag'ned·ik 'kyùr·ē ‚tem·prə·chər }

magnetic cutter [ENG ACOUS] A cutter in which the mechanical displacements of the recording stylus are produced by the action of magnetic fields. { mag'ned·ik 'kəd·ər }

magnetic daily variation See magnetic diurnal variation. { mag'ned·ik ‚dā·lē ver·ē'ā·shən }

magnetic damping [ELECTROMAG] Damping of a mechanical motion by means of the reaction between a magnetic field and the current generated by the motion of a coil through the magnetic field. { mag'ned·ik 'dam·piŋ }

magnetic declination See declination. { mag'ned·ik ‚dek·lə'nā·shən }

magnetic deflection [ELECTR] Deflection of an electron beam by the action of a magnetic field, as in a television picture tube. { mag'ned·ik di'flek·shən }

magnetic delay line [ELECTR] Delay line, used for the storage of data in a computer, consisting essentially of a metallic medium along which the velocity of the propagation of magnetic energy is small compared to the speed of light; storage is accomplished by the recirculation of wave patterns containing information, usually in binary form. { mag'ned·ik di'lā ‚līn }

magnetic deviation [NAV] The angle between the magnetic meridian and the axis of a compass card, expressed in degrees east or west to indicate in which direction the end of the compass card is offset from magnetic north. { mag'ned·ik ‚dē·vē'ā·shən }

magnetic diffusivity [ELECTROMAG] A measure of the tendency of a magnetic field to diffuse through a conducting medium at rest; it is equal to the partial derivative of the magnetic field strength with respect to time divided by the Laplacian of the magnetic field, or to the reciprocal of $4\pi\mu\sigma$, where μ is the magnetic permeability and σ is the conductivity in electromagnetic units. { mag'ned·ik ‚di‚fyü'siv·əd·ē }

magnetic dip See inclination. { mag'ned·ik 'dip }

magnetic dipole [ELECTROMAG] An object, such as a permanent magnet, current loop, or particle with angular momentum, which experiences a torque in a magnetic field, and itself gives rise to a magnetic field, as if it consisted of two magnetic poles of opposite sign separated by a small distance. { mag'ned·ik 'dī‚pōl }

magnetic dipole antenna [ELECTROMAG] Simple loop antenna capable of radiating an electromagnetic wave in response to a circulation of electric current in the loop. { mag'ned·ik 'dī‚pōl an‚ten·ə }

magnetic dipole density See magnetization. { mag'ned·ik 'dī‚pōl ‚den·səd·ē }

magnetic dipole moment [ELECTROMAG] A vector associated with a magnet, current loop, particle, or such, whose cross product with the magnetic induction (or alternatively, the magnetic field strength) of a magnetic field is equal to the torque exerted on the system by the field. Also known as dipole moment; magnetic moment. { mag'ned·ik 'dī‚pōl ‚mō·mənt }

magnetic direction [GEOD] Horizontal direction expressed as angular distance from magnetic north. { mag'ned·ik də'rek·shən }

magnetic disk See disk. { mag'ned·ik 'disk }

magnetic displacement See magnetic flux density. { mag'ned·ik di'splās·mənt }

magnetic diurnal variation [GEOPHYS] Oscillations of the earth's magnetic field which have a periodicity of about a day and which depend to a close approximation only on local time

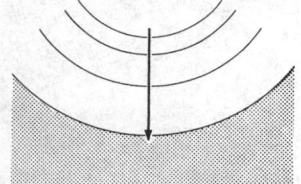

MAGNETIC CONFINEMENT

Stable magnetic confinement by a convex magnetic field. The lines of force are convex to the plasma; arrow indicates decrease of magnetic field strength.

and geographic latitude. Also known as magnetic daily variation. { mag'ned·ik dī'ərn·əl ,ver·ē'ā·shən }

magnetic document sorter-reader *See* magnetic character sorter. { mag'ned·ik ¦däk·yə·mənt 'sord·ər ¦rēd·ər }

magnetic domain *See* ferromagnetic domain. { mag'ned·ik də'mān }

magnetic domain memory *See* domain-tip memory. { mag'ned·ik də'mān ,mem·rē }

magnetic double refraction [OPTICS] The double refraction of light passing through certain substances when the substance is placed in a transverse magnetic field. { mag'ned·ik 'dəbəl ri,frak·shən }

magnetic drag dynamometer *See* eddy-current brake. { mag'ned·ik ¦drag ,dī·nə'mäm·əd·ər }

magnetic drum *See* drum. { mag'ned·ik 'drəm }

magnetic drum receiving equipment [ELECTR] Radar developed for detection of targets beyond line of sight using ionospheric reflection and very low power. { mag'ned·ik 'drəm ri'sēv·iŋ i,kwip·mənt }

magnetic drum storage *See* drum. { mag'ned·ik ¦drəm 'stór·ij }

magnetic earphone [ENG ACOUS] An earphone in which variations in electric current produce variations in a magnetic field, causing motion of a diaphragm. { mag'ned·ik 'ir,fōn }

magnetic element [ENG] That part of an instrument producing or influenced by magnetism. [GEOPHYS] Magnetic declination, dip, or intensity at any location on the surface of the earth. { mag'ned·ik 'el·ə·mənt }

magnetic energy [ELECTROMAG] The energy required to set up a magnetic field. { mag'ned·ik 'en·ər·jē }

magnetic equator [GEOPHYS] That line on the surface of the earth connecting all points at which the magnetic dip is zero. Also known as aclinic line. { mag'ned·ik i'kwād·ər }

magnetic ferroelectric [SOLID STATE] A substance which possesses both magnetic ordering and spontaneous electric polarization. { mag'ned·ik ¦fer·ō·i'lek·trik }

magnetic field [ELECTROMAG] **1.** One of the elementary fields in nature; it is found in the vicinity of a magnetic body or current-carrying medium and, along with electric field, in a light wave; charges moving through a magnetic field experience the Lorentz force. **2.** *See* magnetic field strength. { mag'ned·ik 'fēld }

magnetic field intensity *See* magnetic field strength. { mag'ned·ik 'fēld in,ten·səd·ē }

magnetic field sensor [ENG] A proximity sensor that uses a combination of a reed switch and a magnet to detect the presence of a magnetic field. { mag'ned·ik fēld ,sen·sər }

magnetic field strength [ELECTROMAG] An auxiliary vector field, used in describing magnetic phenomena, whose curl, in the case of static charges and currents, equals (in meter-kilogram-second units) the free current density vector, independent of the magnetic permeability of the material. Also known as magnetic field; magnetic field intensity; magnetic force; magnetic intensity; magnetizing force. { mag'ned·ik 'fēld ,streŋkth }

magnetic film *See* magnetic thin film. { mag'ned·ik 'film }

magnetic filter [CHEM ENG] Filtration device in which the filter screen is magnetized to trap and remove fine iron from liquids or liquid suspensions being filtered. { mag'ned·ik 'fil·tər }

magnetic firing circuit [ELECTR] A type of firing circuit in which the capacitor is discharged through the igniter by saturating a reactor, which is connected in series with the capacitor; often used in ignitron rectifiers to obtain longer life and greater reliability than is possible with thyratron firing tubes. { mag'ned·ik 'fīr·iŋ ,sər·kət }

magnetic flaw detector [ELECTROMAG] A flaw detector in which a ferrous object is magnetized with an electromagnet or permanent magnet and sprayed with magnetic particles or a solution containing fine suspended magnetic particles which outline surface or near-surface flaws. { mag'ned·ik 'flȯ di,tek·tər }

magnetic fluid [MATER] A mixture of iron particles in oil or other liquid; viscosity increases sharply in a strong magnetic field. { mag'ned·ik 'flü·əd }

magnetic fluid clutch [MECH ENG] A friction clutch that is engaged by magnetizing a liquid suspension of powdered iron located between pole pieces mounted on the input and output

shafts. Also known as magnetic clutch. { mag'ned·ik ¦flü·əd 'kləch }

magnetic flux [ELECTROMAG] **1.** The integral over a specified surface of the component of magnetic induction perpendicular to the surface. **2.** *See* magnetic lines of force. { mag'ned·ik 'fləks }

magnetic flux density [ELECTROMAG] A vector quantity that is used as a quantitative measure of magnetic field; the force on a charged particle moving in the field is equal to the particle's charge times the cross product of the particle's velocity with the magnetic flux density (SI units). Also known as magnetic displacement; magnetic induction; magnetic vector. { mag'ned·ik 'fləks ,den·səd·ē }

magnetic flux quantum [ELEC] A fundamental unit of magnetic flux, the total magnetic flux in a fluxoid in a type II superconductor, equal to $h/(2e)$, where h is Planck's constant and e is the magnitude of the electron charge, or approximately 2.07×10^{-15} weber. { mag,ned·ik 'fləks ,kwän·təm }

magnetic focusing [ELECTROMAG] Focusing a beam of electrons or other charged particles by using the action of a magnetic field. { mag'ned·ik 'fō·kə·siŋ }

magnetic force *See* magnetic field strength. { mag'ned·ik 'fȯrs }

magnetic force microscopy [ENG] The use of an atomic force microscope to measure the gradient of a magnetic field acting on a tip made of a magnetic material, by monitoring the shift of the natural frequency of the cantilever due to the magnetic force as the tip is scanned over the sample. { mag¦ned·ik ¦fȯrs mī'krä·skə·pē }

magnetic force parameter [PL PHYS] A dimensionless number used in magnetofluid dynamics, equal to the product of the square of the magnetic permeability, the square of the magnetic field strength, the electrical conductivity, and a characteristic length, divided by the product of the mass density and the fluid velocity. Symbolized N. { mag'ned·ik ¦fȯrs pə'ram·əd·ər }

magnetic force welding [MET] A welding process in which the mechanical force is exerted by a magnetic field. { mag'ned·ik ¦fȯrs 'weld·iŋ }

magnetic forming [MET] The forming of metal into desired shapes by using strong magnetic fields, produced by charging a large capacitor bank and then discharging it into an induction coil in less than 10^{-6} second, to push the metal against a forming die. { mag'ned·ik 'fȯr·miŋ }

magnetic friction clutch [MECH ENG] A friction clutch in which the pressure between the friction surfaces is produced by magnetic attraction. Also known as magnetic clutch. { mag'ned·ik 'frik·shən ,kləch }

magnetic gap [ELECTROMAG] The space between a magnet's pole faces. { mag'ned·ik 'gap }

magnetic grenade [ORD] Small explosive charge with attached magnets, designed to be thrown or placed against tanks; the magnets hold the explosive in place until detonated by a time fuse. Also known as magnetic charge. { mag'ned·ik grə'nād }

magnetic groups *See* Shubnikov groups. { mag'ned·ik 'grüps }

magnetic hardness comparator [ENG] A device for checking the hardness of steel parts by placing a unit of known proper hardness within an induction coil; the unit to be tested is then placed within a similar induction coil, and the behavior of the induction coils compared; if the standard and test units have the same magnetic properties, the hardness of the two units is considered to be the same. { mag'ned·ik 'härd·nəs kəm,par·əd·ər }

magnetic head [ELECTR] The electromagnet used for reading, recording, or erasing signals on a magnetic disk, drum, or tape. Also known as magnetic read/write head. { mag'ned·ik 'hed }

magnetic heading [NAV] Heading relative to magnetic north, with the compass heading corrected for deviation. { mag'ned·ik 'hed·iŋ }

magnetic hysteresis [ELECTROMAG] Lagging of changes in the magnetization of a substance behind changes in the magnetic field as the magnetic field is varied. Also known as hysteresis. { mag'ned·ik ,his·tə'rē·səs }

magnetic induction *See* magnetic flux density. { mag'ned·ik in'dək·shən }

MAGNETIC-INK CHARACTER RECOGNITION

E-13B font in magnetic ink on the bottom of a typical bank check to permit electronic processing.

MAGNETIC MICROPHONE

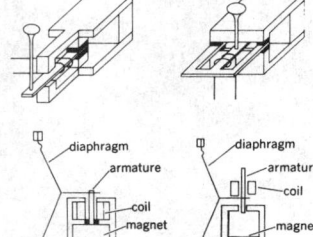

Two magnetic microphones of the balanced-armature type, perspective and sectional views. *(From H. F. Olson, Acoustical Engineering, Van Nostrand, 1957)*

magnetic induction gyroscope [ENG] A gyroscope without moving parts, in which alternating- and direct-current magnetic fields act on water doped with salts which exhibit nuclear paramagnetism. { mag′ned·ik in¦dək·shən ′jī·rə,skōp }

magnetic ink [MATER] Ink containing magnetic particles to permit reading of printed characters by a magnetic character reader as well as by humans. { mag′ned·ik ′iŋk }

magnetic-ink character recognition [COMPUT SCI] That branch of character recognition which involves the sensing of magnetic-ink characters for the purpose of determining the character's most probable identity. Abbreviated MICR. { mag′ned·ik ¦iŋk ′kar·ik·tər ,rek·ig,nish·ən }

magnetic inspection oil [MET] A light petroleum oil, such as kerosine or naphtha, to which has been added fine ferromagnetic particles (usually colored black or red for contrast) to form an inspection penetrant; when the penetrant is applied to a metal surface being inspected, the ferrous particles accumulate in any surface cracks by magnetic attraction, thereby permitting the cracks to be discernible. { mag′ned·ik in′spek·shən ,oil }

magnetic inspection paste [MET] A paste containing ferromagnetic particles designed to be added to a light distilled petroleum oil, such as kerosine or naphtha, to form an inspection penetrant; when the inspection penetrant is applied to a metal, the ferrous particles accumulate in any surface cracks by magnetic attraction, thereby permitting the cracks to be discerned. { mag′ned·ik in′spek·shən ,pāst }

magnetic inspection powder [MET] A dry powder containing ferromagnetic particles colored gray, black, or red for contrast, designed to be dusted on metal parts being inspected by a magnetic inspection machine; the ferrous powder accumulates in any surface cracks (flaws) by magnetic attraction, thereby permitting the cracks to be readily discerned; if the ferrous particles are fluorescent, surface cracks will be brilliantly illuminated under black light. { mag′ned·ik in′spek·shən ,paůd·ər }

magnetic intensity See magnetic field strength. { mag′ned·ik in′ten·səd·ē }

magnetic iron ore See magnetite. { mag′ned·ik ′ī·ərn ′ȯr }

magnetic latitude [GEOPHYS] Angular distance north or south of the magnetic equator. { mag′ned·ik ′lad·ə,tüd }

magnetic leakage [ELECTROMAG] Passage of magnetic flux outside the path along which it can do useful work. { mag′ned·ik ′lēk·ij }

magnetic lens [ELECTROMAG] A magnetic field with axial symmetry, capable of converging beams of charged particles of uniform velocity and of forming images of objects placed in the path of such beams; the field may be produced by solenoids, electromagnets, or permanent magnets. Also known as lens. { mag′ned·ik ′lenz }

magnetic levitation [ELECTROMAG] Contactless, frictionless support of objects through the controlled use of magnetic forces to balance gravitational forces. Abbreviated maglev. { mag′ned·ik ,lev·ə′tā·shən }

magnetic lines of flux See magnetic lines of force. { mag′ned·ik ′līnz əv ′fləks }

magnetic lines of force [ELECTROMAG] Lines used to represent the magnetic induction in a magnetic field, selected so that they are parallel to the magnetic induction at each point, and so that the number of lines per unit area of a surface perpendicular to the induction is equal to the induction. Also known as magnetic flux; magnetic lines of flux. { mag′ned·ik ′līnz əv ′fȯrs }

magnetic local anomaly [GEOPHYS] A localized departure of the geomagnetic field from its average over the surrounding area. { mag′ned·ik ,lō·kəl ə′näm·ə·lē }

magnetic loudspeaker [ENG ACOUS] Loudspeaker in which acoustic waves are produced by mechanical forces resulting from magnetic reactions. Also known as magnetic speaker. { mag′ned·ik ′laůd,spēk·ər }

magnetic Mach number [PL PHYS] A dimensionless number equal to the ratio of the velocity of a fluid to the velocity of Alfvén waves in the fluid. Symbolized M_{Ma}. { mag′ned·ik ′mäk ,nəm·bər }

magnetic material [ELECTROMAG] A material exhibiting ferromagnetism. { mag′ned·ik mə′tir·ē·əl }

magnetic memory See magnetic storage. { mag′ned·ik ′mem·rē }

magnetic memory plate [ELECTR] Magnetic memory consisting of a ferrite plate having a grid of small holes through which the read-in and read-out wires are threaded; printed wiring may be applied directly to the plate in place of conventionally threaded wires, permitting mass production of plates having a high storage capacity. { mag′ned·ik ¦mem·rē ,plāt }

magnetic merging See field line reconnection. { mag′ned·ik ′mər·jiŋ }

magnetic meridian [GEOPHYS] A line which is at any point in the direction of horizontal magnetic force of the earth; a compass needle without deviation lies in the magnetic meridian. { mag′ned·ik mə′rid·ē·ən }

magnetic microphone [ENG ACOUS] A microphone consisting of a diaphragm acted upon by sound waves and connected to an armature which varies the reluctance in a magnetic field surrounded by a coil. Also known as reluctance microphone; variable-reluctance microphone. { mag′ned·ik ′mī·krə,fōn }

magnetic mine [ORD] An underwater mine intended to be detonated when the hull of a passing vessel causes a change in the magnetic field at the mine. { mag′ned·ik ′mīn }

magnetic mine detector [ORD] An electrical device for the detection and indication of metallic mines buried in the earth or on the surface of land or water. { mag′ned·ik ′mīn di,tek·tər }

magnetic mirror [PL PHYS] A magnetic field used in controlled-fusion experiments to reflect charged particles into the central region of a magnetic bottle; reflection occurs in the region where the magnetic field increases abruptly in strength. { mag′ned·ik ′mir·ər }

magnetic modulator [ELECTR] A modulator in which a magnetic amplifier serves as the modulating element for impressing an intelligence signal on a carrier. { mag′ned·ik ′mäj·ə,lād·ər }

magnetic moment See magnetic dipole moment. { mag′ned·ik ′mō·mənt }

magnetic monopole [ELECTROMAG] A hypothetical particle carrying magnetic charge; it would be a source for a magnetic field in the same way that a charged particle is a source for an electric field. Also known as monopole. { mag′ned·ik ′män·ə,pōl }

magnetic multipole [ELECTROMAG] One of a series of types of static or oscillating distributions of magnetization, which is a magnetic multipole of order 2; the electric and magnetic fields produced by a magnetic multipole of order 2^n are equivalent to those of two magnetic multipoles of order 2^{n-1} of equal strength but opposite sign, separated from each other by a short distance. { mag′ned·ik ′məl·tə,pōl }

magnetic multipole field [ELECTROMAG] The electric and magnetic fields generated by a static or oscillating magnetic multipole. { mag′ned·ik ¦məl·tə,pōl ,fēld }

magnetic needle [ELECTROMAG] **1.** A bar magnet or collection of bar magnets which is hung so as to show the direction of the magnetic field. **2.** In particular, a slender bar magnet, pointed at both ends, that is pivoted or freely suspended in a magnetic compass. { mag′ned·ik ′nēd·əl }

magnetic north [GEOPHYS] At any point on the earth's surface, the horizontal direction of the earth's magnetic lines of force (direction of a magnetic meridian) toward the north magnetic pole; a particular direction indicated by the needle of a magnetic compass. { mag′ned·ik ′nȯrth }

magnetic nuclear resonance See nuclear magnetic resonance. { mag′ned·ik ¦nü·klē·ər ′rez·ən·əns }

magnetic number [PL PHYS] A dimensionless number used in magnetofluid dynamics, equal to the square root of the magnetic force parameter. Symbolized R_M. { mag′ned·ik ′nəm·bər }

magnetic observatory [GEOPHYS] A geophysical measuring station employing some form of magnetometer to measure the intensity of the earth's magnetic field. { mag′ned·ik əb′zər·və,tȯr·ē }

magnetic octupole moment [ELECTROMAG] A quantity characterizing a distribution of magnetization; obtained by integrating the product of the divergence of the magnetization, the third power of the distance from the origin, and a spherical harmonic $Y*3_m$ over the magnetization distribution. { mag′ned·ik ¦äk·tə,pōl ,mō·mənt }

magnetic oscillograph [ELECTROMAG] An instrument that records a trace measuring one component of the earth's magnetic field. { mag′ned·ik ä′sil·ə,graf }

magnetic Oseen number [PL PHYS] A dimensionless number used in magnetofluid dynamics, equal to $\frac{1}{2}(1 - N_{AL}^2)R_M$, where N_{AL} is the Alfvén number, and R_M is the magnetic number. Symbolized k. { mag′ned·ik ü′sän ‚nəm·bər }

magnetic particle test [MET] A nondestructive test to determine the existence and extent of macrodefects such as cracks in ferromagnetic materials; discontinuities in the material create variations of magnetic field which are outlined by fine magnetic particles. Also known as magnetic crack detection. { mag′ned·ik ′pärd·ə·kəl ‚test }

magnetic pendulum [ELECTROMAG] A bar magnet which is hung by a thread or balanced on a pivot so that it oscillates in a horizontal plane when disturbed and released in a magnetic field having a horizontal component. { mag′ned·ik ′pen·jə·ləm }

magnetic permeability See permeability. { mag′ned·ik ‚pər·mē·ə′bil·əd·ē }

magnetic pickup See variable-reluctance pickup. { mag′ned·ik ′pi‚kəp }

magnetic pinch See pinch effect. { mag′ned·ik ′pinch }

magnetic polarization See intrinsic induction. { mag′ned·ik ‚pō·lə·rə′zā·shən }

magnetic pole [ELECTROMAG] **1.** One of two regions located at the ends of a magnet that generate and respond to magnetic fields in much the same way that electric charges generate and respond to electric fields. **2.** A particle which generates and responds to magnetic fields in exactly the same way that electric charges generate and respond to electric fields; the particle probably does not have physical reality, but it is often convenient to imagine that a magnetic dipole consists of two magnetic poles of opposite sign, separated by a small distance. [GEOPHYS] In geomagnetism, either of the two points on the earth's surface where the magnetic meridians converge, that is, where the magnetic field is vertical. Also known as dip pole. { mag′ned·ik ′pōl }

magnetic pole strength [ELECTROMAG] The magnitude of a (fictional) magnetic pole, equal to the force exerted on the pole divided by the magnetic induction (or, alternatively, by the magnetic field strength). Also known as pole strength. { mag′ned·ik ′pōl ‚streŋkth }

magnetic potential See magnetic scalar potential. { mag′ned·ik pə′ten·chəl }

magnetic potentiometer [ENG] Instrument that measures magnetic potential differences. { mag′ned·ik pə‚ten·chē′äm·əd·ər }

magnetic pressure [PL PHYS] A function, proportional to the square of the magnetic induction, such that the force exerted by a magnetic field on an electrically conducting fluid (excluding the force associated with curvature of magnetic flux lines) is the same as the force that would be exerted by a hydrostatic pressure equal to this function. { mag′ned·ik ′presh·ər }

magnetic pressure transducer [ENG] A type of pressure transducer in which a change of pressure is converted into a change of magnetic reluctance or inductance when one part of a magnetic circuit is moved by a pressure-sensitive element, such as a bourdon tube, bellows, or diaphragm. { mag′ned·ik ′presh·ər tranz‚dü·sər }

magnetic prime vertical [GEOPHYS] The vertical circle through the magnetic east and west points of the horizon. { mag′ned·ik ′prīm ′vərd·ə·kəl }

magnetic printing [ELECTR] The permanent and usually undesired transfer of a recorded signal from one section of a magnetic recording medium to another when these sections are brought together, as on a reel of tape. Also known as crosstalk; magnetic transfer. { mag′ned·ik ′print·iŋ }

magnetic probe [ELECTROMAG] A small coil inserted in a magnetic field to measure changes in field strength. { mag′ned·ik ′prōb }

magnetic profile [GEOPHYS] A profile of a geologic structure showing magnetic anomalies. { mag′ned·ik ′prō‚fīl }

magnetic prospecting [ENG] Carrying out airborne or ground surveys of variations in the earth's magnetic field, using a magnetometer or other equipment, to locate magnetic deposits of iron, nickel, or titanium, or nonmagnetic deposits which either contain magnetic gangue minerals or are associated with magnetic structures. { mag′ned·ik ′prä‚spek·tiŋ }

magnetic pulley [ENG] Magnetized pulley device for a conveyor belt; removes tramp iron from dry products being moved by the belt. { mag′ned·ik ′púl·ē }

magnetic pumping [ELECTROMAG] A method of moving a conducting liquid by applying a magnetic field which varies with time. [PL PHYS] A method of heating a plasma to a high ion temperature by applying an oscillating electromagnetic field. { mag′nec·ik ′pəmp·iŋ }

magnetic quadrupole lens [ELECTROMAG] A magnetic field generated by four magnetic poles of alternating sign arranged in a circle; used to focus beams of charged particles in devices such as electron microscopes and particle accelerators. { mag′ned·ik ′kwä·drə‚pōl ‚lenz }

magnetic quantum number [ATOM PHYS] The eigenvalue of the component of an angular momentum operator in a specified direction, such as that of an applied magnetic field, in units of Planck's constant divided by 2π. { mag′ned·ik ‚kwän·təm ‚nəm·bər }

magnetic random access memory [COMPUT SCI] A nonvolatile memory in which submicrometer-sized magnetic structures store digital information in their magnetic orientation. Abbreviated MRAM. { mag′ned·ik ‚ran·dəm ′ak‚ses ‚mem·rē }

magnetic read/write head See magnetic head. { mag′ned·ik ′rēd ′rīt ‚hed }

magnetic recorder [ELECTR] An instrument that records information, generally in the form of audio-frequency or digital signals, on magnetic tape or magnetic wire as magnetic variations in the medium. { mag′ned·ik ri′kórd·ər }

magnetic recording [ELECTR] Recording by means of a signal-controlled magnetic field. { mag′ned·ik ri′kórd·iŋ }

magnetic recording paper [MATER] A particle-oriented paper in which both machine-readable and visible traces can be produced by a magnetic recording head; reusable because the trace can be erased by a combination of alternating-current and direct-current magnetic fields. { mag′ned·ik ri′kórd·iŋ ‚pā·pər }

magnetic reed switch See reed switch. { mag′ned·ik ′rēd ‚swich }

magnetic refrigerator [CRYO] A device for keeping substances cooled to about 0.2 K, in which a working substance consisting of a paramagnetic salt undergoes a cycle of processes which approximates a Carnot cycle between a high-temperature reservoir consisting of a liquid-helium bath at 1.2 K and a low-temperature reservoir consisting of the substance to be cooled, and isentropic cooling of the working substance is accomplished by demagnetization. { mag′ned·ik ri′frij·ə‚rād·ər }

magnetic relaxation [PHYS] The approach of a magnetic system to an equilibrium or steady-state condition, over a period of time. { mag′ned·ik ‚rē‚lak′sā·shən }

magnetic reluctance See reluctance. { mag′ned·ik ri′lək·təns }

magnetic reluctivity See reluctivity. { mag′ned·ik ‚rē‚lək′tiv·əd·ē }

magnetic reproducer [ELECTR] An instrument which moves a magnetic recording medium, such as a tape, wire, or disk, past an electromagnetic transducer that converts magnetic signals on the medium into electric signals. { mag′ned·ik ‚rē·prə′dü·sər }

magnetic reproducing [ELECTR] The conversion of information on magnetic tape or magnetic wire, which was originally produced by electric signals, back into electric signals. { mag′ned·ik ‚rē·prə′dü·siŋ }

magnetic resonance [PHYS] A phenomenon exhibited by the magnetic spin systems of certain atoms whereby the spin systems absorb energy at specific (resonant) frequencies when subjected to magnetic fields alternating at frequencies which are in synchronism with natural frequencies of the system. Also known as spin resonance. { mag′ned·ik ′rez·ən·əns }

magnetic resonance imaging [ENG] A technique in which an object placed in a spatially varying magnetic field is subjected to a pulse of radio-frequency radiation, and the resulting nuclear magnetic resonance spectra are combined to give cross-sectional images. Abbreviated MRI. { mag′ned·ik ′rez·ən·əns ′im·ij·iŋ }

magnetic reversal [GEOPHYS] A reversal of the polarity of the earth's magnetic field that has occurred at about one-million-year intervals. { mag′ned·ik ri′vər·səl }

magnetic Reynolds number [PL PHYS] A dimensionless number used to compare the transport of magnetic lines of force in a conducting fluid to the leakage of such lines from the fluid, equal to a characteristic length of the fluid times the fluid velocity, divided by the magnetic diffusivity. Symbolized R_M. { mag′ned·ik ′ren·əlz ‚nəm·bər }

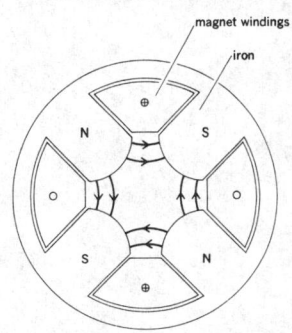

MAGNETIC QUADRUPOLE LENS

magnet windings
iron

Section normal to the lens axis of magnetic quadrupole lens.

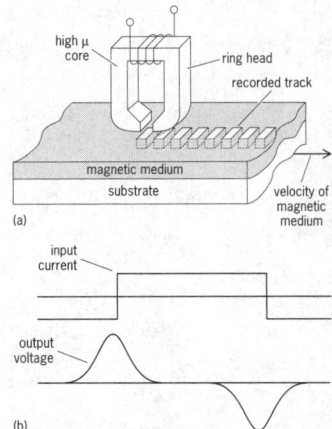

MAGNETIC RECORDING

high μ core
ring head
recorded track
magnetic medium
substrate
velocity of magnetic medium
(a)
input current
output voltage
(b)

Writing and reading process. (*a*) Motion of the magnetic medium past the electromagnet in the form of a ring head. (*b*) Variation with time of the input current and the output voltage.

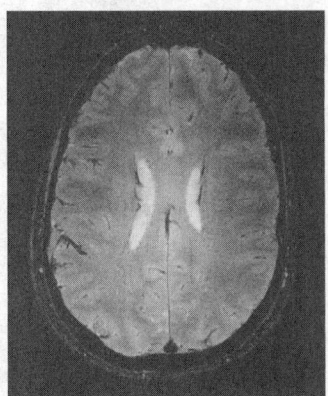

MAGNETIC RESONANCE IMAGING

Axial image of a human brain by an ultrahigh-frequency magnetic resonance imaging scanner at a magnetic field strength of 8 tesla and frequency of 340 megahertz.

magnetic rigidity [ELECTROMAG] A measure of the momentum of a particle moving perpendicular to a magnetic field, equal to the magnetic induction times the particle's radius of curvature. [PL PHYS] The existence of restoring forces which resist displacements of a conducting fluid when a magnetic field is present. { mag′ned·ik ri′jid·əd·ē }

magnetic rotation [OPTICS] **1.** In a weak magnetic field, the rotation, of the plane of polarization of fluorescent light emitted perpendicular to the field and perpendicular to the propagation direction of the incident light. **2.** *See* Faraday effect. { mag′ned·ik rō′tā·shən }

magnetic rubber [MATER] Synthetic rubber to which magnetic metal powder is added; produced in sheets or strips. { mag′ned·ik ′rəb·ər }

magnetics [ELECTROMAG] The study of magnetic phenomena, comprising magnetostatics and electromagnetism. { mag′ned·iks }

magnetic saturation [ELECTROMAG] The condition in which, after a magnetic field strength becomes sufficiently large, further increase in the magnetic field strength produces no additional magnetization in a magnetic material. Also known as saturation. { mag′ned·ik ˌsach·ə′rā·shən }

magnetic scalar potential [ELECTROMAG] The work which must be done against a magnetic field to bring a magnetic pole of unit strength from a reference point (usually at infinity) to the point in question. Also known as magnetic potential. { mag′ned·ik ¦skāl·ər pə′ten·chəl }

magnetic scanning [SPECT] The magnetic field sorting of ions into their respective spectrums for analysis by mass spectroscopy; accomplished by varying the magnetic field strength while the electrostatic field is held constant. { mag′ned·ik ′skan·iŋ }

magnetic scattering [PHYS] Scattering of neutrons as a result of the interaction of the magnetic moment of the neutron with the magnetic moments of atoms or other particles. { mag′ned·ik ′skad·ə·riŋ }

magnetic secular change [GEOPHYS] The gradual variation in the value of a magnetic element which occurs over a period of years. { mag′ned·ik ¦sek·yə·lər ′chānj }

magnetic separator [ENG] A machine for separating magnetic from less magnetic or nonmagnetic materials by using strong magnetic fields; used for example, in tramp iron removal, or concentration and purification. { mag′ned·ik ′sep·ə‚rād·ər }

magnetic separatrix [ELECTROMAG] A surface that forms the boundary between an internal region of closed magnetic surfaces and an external region of open field lines. { mag′ned·ik ′sep·rə‚triks }

magnetic shell [ELECTROMAG] Two layers of magnetic charge of opposite sign, separated by an infinitesimal distance. { mag′ned·ik ′shel }

magnetic shielding *See* magnetostatic shielding. { mag′ned·ik ′shēld·iŋ }

magnetic shift register [COMPUT SCI] A shift register in which the pattern of settings of a row of magnetic cores is shifted one step along the row by each new input pulse; diodes in the coupling loops between cores prevent backward flow of information. { mag′ned·ik ′shift ‚rej·ə·stər }

magnetic shunt [ELECTROMAG] Piece of iron, usually adjustable as to position, used to divert a portion of the magnetic lines of force passing through an air gap in an instrument or other device. { mag′ned·ik ′shənt }

magnetic sound track [ENG ACOUS] A magnetic tape, attached to a motion picture film, on which a sound recording is made. { mag′ned·ik ′saún ‚trak }

magnetic source imaging [ENG] A method of mapping electric currents within an object, particularly currents associated with biological activity, by using an array of SQUID magnetometers to detect the resulting magnetic fields surrounding the object. Abbreviated MSI. { mag‚ned·ik ′sórs ‚im·ij·iŋ }

magnetic spark chamber [NUCLEO] A spark chamber in a magnetic field up to 20,000 gauss (2 tesla), in which the sign of the charge and the momentum of charged particles can be measured by measuring the curvature of their trajectories. { mag′ned·ik ′spärk ‚chām·bər }

magnetic speaker *See* magnetic loudspeaker. { mag′ned·ik ′spēk·ər }

magnetic spectrograph [NUCLEO] A magnetic spectrometer that provides a permanent record of the distribution of intensity versus momentum of a beam of charged particles. { mag′ned·ik ′spek·trə‚graf }

magnetic spectrometer [NUCLEO] A device for measuring the momentum of charged particles, or their distribution of intensity versus momentum, by passing the particles through a magnetic field which bends their paths in proportion to their momentum. { mag′ned·ik spek′träm·əd·ər }

magnetic spin transistor *See* magnetic switch. { mag‚ned·ik ‚spin tran′zis·tər }

magnetic star [ASTRON] A star with an unusually strong magnetic field. { mag′ned·ik ′stär }

magnetic station [GEOPHYS] A facility equipped with instruments for measuring local variations in the earth's magnetic field. { mag′ned·ik ′stā·shən }

magnetic stepping motor *See* stepper motor. { mag′ned·ik ′step·iŋ ‚mōd·ər }

magnetic storage [COMPUT SCI] A device utilizing magnetic properties of materials to store data; may be roughly divided into two categories, moving (drum, disk, tape) and static (core, thin film). Also known as magnetic memory. { mag′ned·ik ′stòr·ij }

magnetic storm [GEOPHYS] A worldwide disturbance of the earth's magnetic field; frequently characterized by a sudden onset, in which the magnetic field undergoes marked changes in the course of an hour or less, followed by a very gradual return to normalcy, which may take several days. Also known as geomagnetic storm. { mag′ned·ik ′stòrm }

magnetic strain energy [SOLID STATE] The potential energy of a magnetic domain, subject to both a tensile stress and a magnetic field, associated with the domain's magnetostriction expansion. { mag′ned·ik ′strān ‚en·ər·jē }

magnetic stratigraphy *See* paleomagnetic stratigraphy. { mag′ned·ik strə′tig·rə·fē }

magnetic stress [PL PHYS] The force which acts across a surface in a conducting fluid because of curving or stretching of magnetic flux lines. { mag′ned·ik ′stres }

magnetic stress tensor [PL PHYS] A second-rank tensor, proportional to the dyad product of the magnetic induction with itself, whose divergence gives that part of the force of a magnetic field on a unit volume of conducting fluid which is due to curvature or stretching of magnetic flux lines. { mag′ned·ik ′stres ‚ten·sər }

magnetic stripe [COMPUT SCI] A small length of magnetic tape on a card or badge, containing data that is machine-readable. { mag′ned·ik ′strīp }

magnetic striped ledger [COMPUT SCI] A ledger sheet used on a special typing device which stores the coded data on a magnetic strip on the sheet while typing out the data on the sheet; the magnetic strip can be read directly by a special reader linked to a computer. { mag′ned·ik ¦strīpt ′lej·ər }

magnetic superconductor [SOLID STATE] A superconductor which is not magnetic in the ordinary sense, but which contains elements with large magnetic moments or large spin. { mag′ned·ik ′sü·pər·kən‚dək·tər }

magnetic survey [GEOPHYS] **1.** Magnetometer map of variations in the earth's total magnetic field; used in petroleum exploration to determine basement-rock depths and geologic anomalies. **2.** Measurement of a component of the geomagnetic field at different locations. { mag′ned·ik ′sər‚vā }

magnetic susceptibility [ELECTROMAG] The ratio of the magnetization of a material to the magnetic field strength; it is a tensor when these two quantities are not parallel; otherwise it is a simple number. Also known as susceptibility. { mag′ned·ik sə‚sep·tə′bil·əd·ē }

magnetic switch [ELECTR] A switching device consisting of three metallic layers (a paramagnetic layer between two ferromagnetic layers), whose action is based on electron spin and is controlled by a small magnetic field. Also known as bipolar spin device; bipolar spin switch; magnetic spin transistor; spin transistor; spin valve. { mag′ned·ik ′swich }

magnetic tape [ELECTR] A plastic, paper, or metal tape that is coated or impregnated with magnetizable iron oxide particles; used in magnetic recording. { mag′ned·ik ′tāp }

magnetic-tape core [ELECTR] Toroidal core formed by winding a strip of thin magnetic core material around a form. { mag′ned·ik ¦tāp ′kór }

magnetic tape file operation [COMPUT SCI] All the jobs

MAGNETIC SWITCH

Schematic of a magnetic switch in off state. F_1, F_2 = ferromagnetic layers; P = paramagnetic layer; N = nonmagnetic counterelectrode; current I = 0.1–10 milliampere.

related to creating, sorting, inputting, and maintenance of magnetic tapes in a magnetic tape environment. { mag′ned·ik ′tāp ′fīl ‚äp·ə‚rā·shən }

magnetic tape group [COMPUT SCI] A cabinet containing two or more magnetic tape units, each of which can operate independently, but which sometimes share one or more channels with which they communicate with a central processor. Also known as tape cluster; tape group. { mag′ned·ik ‚tāp ‚grüp }

magnetic tape librarian [COMPUT SCI] Routine which provides a computer the means to automatically run a sequence of programs. { mag′ned·ik ‚tāp lī‚brer·ē·ən }

magnetic tape master file [COMPUT SCI] A magnetic tape consisting of a set of related elements such as is found in a payroll, an inventory, or an accounts receivable; a master file is, as a rule, periodically updated. { mag′ned·ik ‚tāp ′mas·tər ′fīl }

magnetic tape parity [COMPUT SCI] A check performed on the data bits on a tape; usually an odd (or even) condition is expected and the occurrence of the wrong parity indicates the presence of an error. { mag′ned·ik ‚tāp ′par·əd·ē }

magnetic tape reader [ELECTR] A computer device that is capable of reading information recorded on magnetic tape by transforming this information into electric pulses. { mag′ned·ik ‚tāp ‚rēd·ər }

magnetic tape station [COMPUT SCI] On-line device that provides write, read, and erase data on magnetic tape to permit high-speed storage of data. { mag′ned·ik ‚tāp ‚stā·shən }

magnetic tape storage [COMPUT SCI] Storage of binary information on magnetic tape, generally on 5 to 10 tracks, with up to several thousand bits per inch (more than a thousand bits per centimeter) on each track. { mag′ned·ik ‚tāp ‚stör·ij }

magnetic tape switching unit [COMPUT SCI] A device which permits the computer operator to bring into play any number of tape drives as required by the system. { mag′ned·ik ‚tāp ′swich·iŋ ‚yü·nət }

magnetic tape terminal [COMPUT SCI] Device which converts pulses in series to pulses in parallel while checking for bit parity prior to the entry in buffer storage. { mag′ned·ik ‚tāp ′tər·mən·əl }

magnetic tape unit [COMPUT SCI] A computer unit that usually consists of a tape transport, reading and recording heads, and associated electric and electronic equipment. { mag′ned·ik ‚tāp ‚yü·nət }

magnetic temporal variation [GEOPHYS] Any change in the earth's magnetic field which is a function of time. { mag′ned·ik ′tem·pə·rəl ‚ver·ē′ā·shən }

magnetic tension effect [ELECTROMAG] The ability of stresses on a ferromagnetic material to alter its remanence. Also known as inverse magnetostriction. { mag′ned·ik ′ten·shən i‚fekt }

magnetic test coil See exploring coil. { mag′ned·ik ′test ‚kȯil }

magnetic thermometer [SOLID STATE] A sample of a paramagnetic salt whose magnetic susceptibility is measured and whose temperature is then calculated from the inverse relationship between the two quantities; useful at temperatures below about 1 K. { mag′ned·ik thər′mäm·əd·ər }

magnetic thin film [SOLID STATE] A sheet or cylinder of magnetic material less than 5 micrometers thick, usually possessing uniaxial magnetic anisotropy; used mainly in computer storage and logic elements. Also known as ferromagnetic film; magnetic film. { mag′ned·ik ′thin ‚film }

magnetic track [NAV] The direction of the track relative to magnetic north. { mag′ned·ik ′trak }

magnetic transducer [ELECTROMAG] A device for transforming mechanical into electrical energy, which consists of a magnetic field including a variable-reluctance path and a coil surrounding all or a part of this path, so that variation in reluctance leads to a variation in the magnetic flux through the coil and a corresponding induced emf (electromotive force). { mag′ned·ik tranz′dü·sər }

magnetic transfer See magnetic printing. { mag′ned·ik ′tranz·fər }

magnetic tunnel junction [ELECTR] A magnetic storage and switching device in which two magnetic layers are separated by an insulating barrier, typically aluminum oxide, that is only 1–2 nanometers thick, allowing an electronic current whose magnitude depends on the orientation of both magnetic layers to tunnel through the barrier when it is subject to a small electric bias. { mag′ned·ik ′tən·əl ‚jəŋk·shən }

magnetic variation [GEOPHYS] Small changes in the earth's magnetic field in time and space. { mag′ned·ik ‚ver·ē′ā·shən }

magnetic vector See magnetic flux density. { mag′ned·ik ′vek·tər }

magnetic vector potential See vector potential. { mag′ned·ik ‚vek·tər pə‚ten·chəl }

magnetic viscosity [ELECTROMAG] The existence of a time delay between a change in the magnetic field applied to a ferromagnetic material and the resulting change in magnetic induction which is too great to be explained by the existence of eddy currents. [PL PHYS] The effect, possessed by a magnetic field in the absence of sizable mechanical forces or electric fields, of damping motions of a conducting fluid perpendicular to the field similar to ordinary viscosity. { mag′ned·ik vis′käs·əd·ē }

magnetic wave [SOLID STATE] The spread of magnetization from a small portion of a substance where an abrupt change in the magnetic field has taken place. { mag′ned·ik ′wāv }

magnetic wave device [ELECTROMAG] A device that depends on magnetoelastic or magnetostatic wave propagation through or on the surface of a magnetic or dielectric material. { mag′ned·ik ′wāv di‚vīs }

magnetic well [PL PHYS] A configuration of magnetic fields used to contain a plasma in controlled fusion experiments, in which the plasma is confined in a central region surrounded by fields which keep it from escaping in any direction. { mag′ned·ik ′wel }

magnetic wind direction [METEOROL] The direction, with respect to magnetic north, from which the wind is blowing; distinguished from true wind direction. { mag′ned·ik ′wind də‚rek·shən }

magnetic wire [MATER] A wire made from magnetic material suitable for magnetic recording. { mag′ned·ik ′wīr }

magnetic x-ray scattering [ELECTROMAG] A process in which the electric and magnetic fields of incident x-rays interact with electronic magnetic moments, giving rise to magnetic reradiation. { mag′ned·ik ′eks‚rā ‚skad·ə·riŋ }

magnetism [PHYS] Phenomena involving magnetic fields and their effects upon materials. { ′mag·nə‚tiz·əm }

magnetite [MINERAL] An opaque iron-black and streak-black isometric mineral and member of the spinel structure type, usually occurring in octahedrals or in granular to massive form; hardness is 6 on Mohs scale, and specific gravity is 5.20. Also known as magnetic iron ore; octahedral iron ore. { ′mag·nə‚tīt }

magnetization [ELECTROMAG] **1.** The property and in particular, the extent of being magnetized; quantitatively, the magnetic moment per unit volume of a substance. Also known as magnetic dipole density; magnetization intensity. **2.** The process of magnetizing a magnetic material. { ‚mag·nəd·ə′zā·shən }

magnetization curve See B-H curve; normal magnetization curve. { ‚mag·nəd·ə′zā·shən ‚kərv }

magnetization intensity See magnetization. { ‚mag·nəd·ə′zā·shən in′ten·səd·ē }

magnetizing current [ELEC] The current that flows through the primary winding of a power transformer when no loads are connected to the secondary winding; this current establishes the magnetic field in the core and furnishes energy for the no-load power losses in the core. Also known as exciting current. { ′mag·nə‚tiz·iŋ ‚kə·rənt }

magnetizing force See magnetic field strength. { ′mag·nə‚tīz·iŋ ‚fȯrs }

magnet keeper See keeper. { ′mag·nət ‚kēp·ər }

magneto [ELEC] An alternating-current generator that uses one or more permanent magnets to produce its magnetic field; frequently used as a source of ignition energy on tractor, marine, industrial, and aviation engines. Also known as magnetoelectric generator. { mag′nēd·ō }

magnetoacoustics [PHYS] The study of the effects of magnetic fields on acoustical phenomena, such as various oscillations in the attenuation of ultrasonic sound waves by a crystal placed in a magnetic field at a very low temperature, as the magnetic field strength or sound frequency is varied. { mag‚nēd·ō·ə′kü·stiks }

magnetoaerodynamics [PL PHYS] Study of the properties

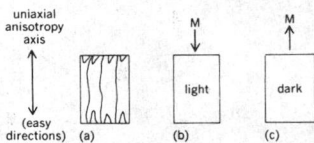

MAGNETIC THIN FILM

Typical domain patterns in a thin film viewed by use of the Kerr magnetooptic technique. *(a)* Magnetic film demagnetized. *(b)* Magnetization pointing downward. *(c)* Magnetization pointing upward.

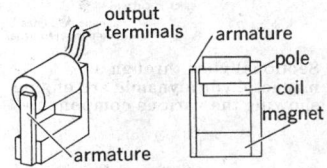

MAGNETIC TRANSDUCER

Perspective and sectional view of a single-pole and armature magnetic transducer.

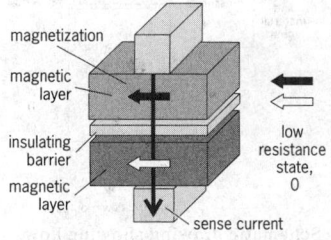

MAGNETIC TUNNEL JUNCTION

Magnetic components used in the magnetic tunnel junction.

and characteristics of, and the forces exerted by, highly ionized air and other gases; applied principally to study of reentering ballistic missiles and spacecraft. { mag¦nēd·ō¦er·ō·dī'nam·iks }

magneto anemometer [ENG] A cup anemometer with its shaft mechanically coupled to a magnet; both the frequency and amplitude of the voltage generated are proportional to the wind speed, and may be indicated or recorded by suitable electrical instruments. { mag¦nēd·ō ,an·ə'mäm·əd·ər }

magnetocaloric effect [THERMO] The reversible change of temperature accompanying the change of magnetization of a ferromagnetic material. { mag¦nēd·ō·kə'lȯr·ik i,fekt }

magnetocardiograph [MED] An instrument that records the intensity of the magnetic field of the heart as produced by currents sent through the body by measuring the voltages associated with each heart beat. Abbreviated MCG. { mag¦nēd·ō'kär·dē·ə,graf }

magnetochemistry [PHYS CHEM] A branch of chemistry which studies the interrelationship between the bulk magnetic properties of a substance and its atomic and molecular structure. { mag¦nēd·ō'kem·ə·strē }

magnetoelastic coupling [SOLID STATE] The interaction between the magnetization and the strain of a magnetic material. { mag¦nēd·ō·i'las·tik 'kəp·liŋ }

magnetoelasticity [SOLID STATE] Phenomenon in which an elastic strain alters the magnetization of a ferromagnetic material. { mag¦nēd·ō,i,las'tis·əd·ē }

magnetoelectric effect [SOLID STATE] A linear coupling between magnetization and polarization found in certain magnetic ferroelectrics, such as $BaMnF_4$ at low temperatures. { mag¦nēd·ō·i'lek·trik i,fekt }

magnetoelectric generator See magneto. { mag¦nēd·ō·i'lek·trik 'jen·ə,rād·ər }

magnetoelectricity [ELECTROMAG] Magnetic techniques for generating voltages, such as in an ordinary generator. [SOLID STATE] The appearance of an electric field in certain substances, such as chromic oxide (Cr_2O_3), when they are subjected to a static magnetic field. { mag¦nēd·ō,i,lek'tris·əd·ē }

magnetoelectronics [ELECTR] The use of electron spin (as opposed to charge) in electronic devices. Also known as spin electronics; spintronics. { mag,ned·ō·i·lek'trän·iks }

magnetoencephalogram [BIOPHYS] A measurement of the brain's magnetic field. Abbreviated MEG. { mag,ned·ō·in'sef·ə·lə,gram }

magnetoencephalography [MED] A method to detect the brain's electrical activity with an array of SQUID magnetometers positioned over the head. { mag,ned·ō·in,sef·ə'läg·rə·fē }

magnetofluid [PHYS CHEM] A Newtonian or shear-thinning fluid whose flow properties become viscoplastic when it is modulated by a magnetic field. { ¦mag·nəd·ō'flü·əd }

magnetofluid dynamics [PHYS] **1.** The study of the motion of an electrically conducting metal, such as mercury, in the presence of electric and magnetic fields. **2.** See magnetohydrodynamics. { ¦mag·nəd·ō'flü·əd dī'nam·iks }

magnetogalvanic effect See galvanomagnetic effect. { mag,nēd·ō·gal'van·ik i,fekt }

magnetogas dynamics [PL PHYS] The science of motion in a plasma under the influence of mechanical, electric, and magnetic forces. { mag¦nēd·ō,gas dī'nam·iks }

magnetogram [ASTRON] An image of the sun showing magnetic fields, obtained using a spectroheliograph that has been modified to use the Zeeman effect. { mag'ned·ə,gram }

magnetograph [ELECTROMAG] A set of three variometers attached to a suitable recording unit, which records the components of the magnetic field vector in each of three perpendicular directions. { mag'ned·ə,graf }

magnetohydrodynamic arcjet [AERO ENG] An electromagnetic propulsion system utilizing a plasma that is heated in an electric arc and then adiabatically expanded through a nozzle and further accelerated by a crossed electric and magnetic field. { mag¦nēd·ō,hī·drə·dī'nam·ik 'ärk,jet }

magnetohydrodynamic electromagnetic pulse [ELECTROMAG] A relatively slow pulse of electromagnetic radiation generated by a nuclear explosion, and comprising magnetic bubble electromagnetic pulse and atmospheric heave electromagnetic pulse. { mag¦nēd·ō,hī·drō·dī'nam·ik i¦lek·trō·mag,ned·ik 'pəls }

magnetohydrodynamic generator [ELEC] A system for generating electric power in which the kinetic energy of a flowing conducting fluid is converted to electric energy by a magnetohydrodynamic interaction. Abbreviated MHD generator. { mag¦nēd·ō,hī·drə·dī'nam·ik 'jen·ə,rād·ər }

magnetohydrodynamic instability [PL PHYS] An instability of a plasma in which the plasma expands while moving into a region of weaker magnetic field, until it is expelled from the field. Also known as hydromagnetic instability. { mag¦nēd·ō,hī·drə·dī'nam·ik ,in·stə'bil·əd·ē }

magnetohydrodynamics [PHYS] The study of the dynamics or motion of an electrically conducting fluid, such as an ionized gas or liquid metal, interacting with a magnetic field. Abbreviated MHD. Also known as hydromagnetics; magnetofluid dynamics. { mag¦nēd·ō,hī·drə·dī'nam·iks }

magnetohydrodynamic stability [PL PHYS] The condition of a plasma in which fluctuations in density, pressure, velocity, or the distribution of particles in phase space, die out rather than increase. Also known as hydromagnetic stability. { mag¦nēd·ō,hī·drə·dī'nam·ik stə'bil·əd·ē }

magnetohydrodynamic turbulence [PL PHYS] Motion of a plasma in which velocities and pressures fluctuate irregularly. { mag¦nēd·ō,hī·drə·dī'nam·ik 'tər·byə·ləns }

magnetohydrodynamic wave [PHYS] Wave motion in an electrically conducting fluid, such as plasma or liquid metal, in a strong magnetic field at a frequency much less than that of the ion cyclotron frequency. Also known as hydromagnetic wave. { mag¦nēd·ō,hī·drə·dī'nam·ik 'wāv }

magneto ignition system [ELECTROMAG] An ignition system in which the voltage required to cause a flow of current in the primary winding of the ignition coil is generated by a set of permanent magnets, instead of being supplied by a battery. { mag¦nēd·ō ig'nish·ən ,sis·təm }

magnetoionic duct [GEOPHYS] Duct along the geomagnetic lines of force which exhibits waveguide characteristics for radio-wave propagation between conjugate points on the earth's surface. { mag¦nēd·ō·ī'än·ik 'dəkt }

magnetoionic theory [GEOPHYS] The theory of the combined effect of the earth's magnetic field and atmospheric ionization on the propagation of electromagnetic waves. { mag¦nēd·ō·ī'än·ik 'thē·ə·rē }

magnetoionic wave component [GEOPHYS] Either of the two elliptically polarized wave components into which a linearly polarized wave incident on the ionosphere is separated because of the earth's magnetic field. { mag¦nēd·ō·ī'än·ik 'wāv kəm,pō·nənt }

magnetomechanical factor [PHYS] The gyromagnetic ratio of an atom or substance (magnetic dipole moment divided by angular momentum) divided by the quantity $e/2mc$, where e and m are the charge (in esu, or electrostatic units) and mass of the electron respectively, and c is the speed of light. Also known as g factor. { mag¦nēd·ō·mi'kan·ə·kəl 'fak·tər }

magnetomechanics [PHYS] The study of the effects which the magnetization of a material and its strain have on each other. { mag¦nēd·ō·mi'kan·iks }

magnetometer [ENG] An instrument for measuring the magnitude and sometimes also the direction of a magnetic field, such as the earth's magnetic field. { ,mag·nə'täm·əd·ər }

magnetomotive force [ELECTROMAG] The work that would be required to carry a magnetic pole of unit strength once around a magnetic circuit. Abbreviated mmf. { mag¦nēd·ō'mōd·iv 'fȯrs }

magneton [PHYS] A unit of magnetic moment used for atomic, molecular, or nuclear magnets, such as the Bohr magneton, Weiss magneton, or nuclear magneton. { 'mag·nə,tän }

magneton number [PHYS] The ratio of the magnetic moment per atom, ion, or molecule of a paramagnetic or ferromagnetic material to the Bohr magneton. { 'mag·nə,tän ,nəm·bər }

magnetooptical modulator [ELECTROMAG] An arrangement for modulating a beam of light by passing it through a single crystal of yttrium iron garnet, which provides intensity modulation by using a magnetic field to produce optical rotation. { mag¦nēd·ō¦äp·tə·kəl 'mäj·ə,lād·ər }

magnetooptical shutter [OPTICS] A device in which light passes through crossed Nicol prisms and a glass cell containing a liquid displaying the Faraday effect between the prisms; light can pass through the system only when a magnetic field is applied to the cell at an angle of 45° to the polarization planes of both prisms. { mag¦nēd·ō¦äp·tə·kəl 'shəd·ər }

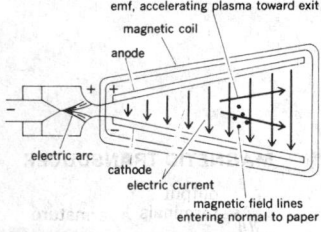

MAGNETOHYDRODYNAMIC ARCJET

emf, accelerating plasma toward exit
magnetic coil
anode
electric arc
cathode
electric current
magnetic field lines entering normal to paper

Sectional view through a magnetohydrodynamic arc engine showing the various components.

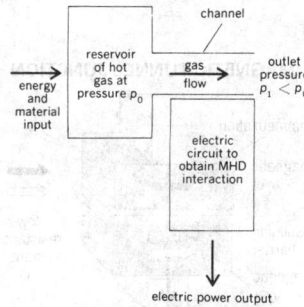

MAGNETOHYDRODYNAMIC GENERATOR

channel
reservoir of hot gas at pressure p_0
gas flow
outlet pressure $p_1 < p_0$
energy and material input
electric circuit to obtain MHD interaction
electric power output

Schematic drawing showing how the generator produces electric power.

magnetooptical switch [COMPUT SCI] A thin-film modulator which acts on a laser beam by polarization, causing the beam to emerge from the output prism at a different angle. { mag¦nēd·ō¦äp·tə·kəl ′swich }

magnetooptic disk [COMMUN] A data storage device in which information is stored in small magnetic marks along tracks on a rotating disk; the information is read by sensing the change in polarization of reflected focused light and can be altered by using a higher-power focused light spot to locally heat the medium and, with the application of an external magnetic field, switch the magnetic domains of the material. { mag‚ned·ō‚äp·tik ′disk }

magnetooptic Kerr effect [OPTICS] Changes produced in the optical properties of a reflecting surface of a ferromagnetic substance when the substance is magnetized; this applies especially to the elliptical polarization of reflected light, when the ordinary rules of metallic reflection would give only plane polarized light. Also known as Kerr magnetooptical effect. { mag¦nēd·ō¦äp·tik ′kər i‚fekt }

magnetooptic material [OPTICS] A material whose optical properties are changed by an applied magnetic field. { mag¦nēd·ō¦äp·tik mə′tir·ē·əl }

magnetooptic recording [ENG] An erasable data storage technology in which data are stored on a rotating disk in a thin magnetic layer that may be switched between two magnetization states by the combination of a magnetic field and a pulse of light from a diode laser. { mag‚ned·ō‚äp·tik ri′kȯrd·iŋ }

magnetooptics [OPTICS] The study of the effect of a magnetic field on light passing through a substance in the field. { mag¦nēd·ō¦äp·tiks }

magnetopause [GEOPHYS] A boundary that marks the transition from the earth's magnetosphere to the interplanetary medium. { mag′nēd·ə‚pȯz }

magnetoplasmadynamics [ELECTROMAG] The generation of electric current by shooting a beam of ionized gas through a magnetic field, to give the same effect as moving copper bars near a magnet. { mag¦nēd·ə‚plaz·mə·dī′nam·iks }

magnetoplumbite [MINERAL] $(Pb,Mn)_2Fe_6O_{11}$ Black mineral consisting of a ferric oxide of plumbite and manganese, and occurring in acute metallic hexagonal crystals. { mag‚nēd·ə′pləm‚bīt }

magnetoresistance [ELECTR] The change in the electrical resistance of a material when it is subjected to an applied magnetic field; this property has widespread application in sensors and magnetic read heads. [ELECTROMAG] The change in electrical resistance produced in a current-carrying conductor or semiconductor on application of a magnetic field. { mag¦nēd·ō·ri′zis·təns }

magnetoresistive memory [ELECTR] A random-access memory that uses the magnetic state of small ferromagnetic regions to store data, plus magnetoresistive devices to read the data, all integrated with silicon integrated-circuit electronics. { mag‚ned·ō·ri‚zis·tiv ′mem·rē }

magnetoresistivity [ELECTROMAG] The change in resistivity produced in a current-carrying conductor or semiconductor on application of a magnetic field. { mag‚nēd·ō·ri‚zis′tiv·əd·ē }

magnetoresistor [ELECTR] Magnetic field-controlled variable resistor. { mag¦nēd·ō·ri′zis·tər }

magnetorheological fluid [MATER] A low-viscosity fluid containing a suspension of micrometer-size magnetic particles that increases in viscosity proportionally to the strength of an applied magnetic field; it is used as an adaptive shock absorber for actively controlled viscous damping. { mag‚ned·ō‚rē·ə‚läj·ə·kəl ′flü·əd }

magnetosheath [GEOPHYS] The relatively thin region between the earth's magnetopause and the shock front in the solar wind. { mag′nēd·ə‚shēth }

magnetosomes [MICROBIO] Intracellular, membrane-bound iron mineral crystals, often magnetite, that enable magnetotactic bacteria to orient and move in the direction of the earth's magnetic field; in marine environments, magnetosomes may contain the iron sulfide mineral greigite. { mag′ned·ə‚sōmz }

magnetosphere [GEOPHYS] The region of the earth in which the geomagnetic field plays a dominant part in controlling the physical processes that take place; it is usually considered to begin at an altitude of about 60 miles (100 kilometers)

and to extend outward to a distant boundary that marks the beginning of interplanetary space. { mag‚nēd·ə‚sfir }

magnetospheric plasma [GEOPHYS] A low-energy plasma with particle energies less than a few electronvolts that permeates the entire region of the earth's magnetosphere. { mag′nēd·ə‚sfir·ik ′plaz·mə }

magnetospheric ring current [GEOPHYS] A belt of charged particles around the earth whose perturbations give rise to ionospheric storms. { mag′nēd·ə‚sfir·ik ′riŋ ‚kə·rənt }

magnetospheric substorm [GEOPHYS] A disturbance of particles and magnetic fields in the magnetosphere; occurs intermittently, lasts 1 to 3 hours, and is accompanied by various phenomena sensible from the earth's surface, such as intense auroral displays and magnetic disturbances, particularly in the nightside polar regions. { mag′nēd·ə‚sfir·ik ′səb‚stȯrm }

magnetostatic [ELECTROMAG] Pertaining to magnetic properties that do not depend upon the motion of magnetic fields. { mag¦nēd·ə¦stad·ik }

magnetostatic mode [SOLID STATE] A spin wave in a magnetic material whose wavelength is greater than about one-tenth the size of the sample. { mag¦nēd·ə¦stad·ik ′mōd }

magnetostatics [ELECTROMAG] The study of magnetic fields that remain constant with time. Also known as static magnetism. { mag¦nēd·ō¦stad·iks }

magnetostatic shielding [ELECTROMAG] The use of an enclosure made of a high-permeability magnetic material to prevent a static magnetic field outside the enclosure from reaching objects inside it, or to confine a magnetic field within the enclosure. Also known as magnetic shielding. { mag¦nēd·ō¦stad·ik ′shēld·iŋ }

magnetostratigraphy [GEOL] A branch of stratigraphy in which sedimentary successions are described and interpreted in terms of remanent magnetization. { mag‚ned·ō·strə′tig·rə·fē }

magnetostriction [ELECTROMAG] The dependence of the state of strain (dimensions) of a ferromagnetic sample on the direction and extent of its magnetization. { mag‚ned·ō′strik·shən }

magnetostriction transducer [ELECTROMAG] A transducer used with sonar equipment to change an alternating current to sound energy at the same frequency and to form the sound energy into a beam; its operation depends on the interaction between the magnetization and the deformation of a material having magnetostrictive properties. { mag‚ned·ō′strik·shən tranz′dü·sər }

magnetostrictive delay line [ELECTROMAG] A delay line made of nickel or other magnetostrictive material, in which the amount of delay is determined by a shock wave traveling through the length of the line at the speed of sound. { mag¦nēd·ō¦strik·tiv di′lā ‚līn }

magnetostrictive filter [ELECTR] Filter network which uses the magnetostrictive phenomena to form high-pass, low-pass, band-pass, or band-elimination filters; the impedance characteristic is the inverse of that of a crystal. { mag¦nēd·ō¦strik·tiv ′fil·tər }

magnetostrictive loudspeaker [ENG ACOUS] Loudspeaker in which the mechanical forces result from the deformation of a material having magnetostrictive properties. { mag¦nēd·ō¦strik·tiv ′laud‚spēk·ər }

magnetostrictive microphone [ENG ACOUS] Microphone which depends for its operation on the generation of an electromotive force by the deformation of a material having magnetostrictive properties. { mag¦nēd·ō¦strik·tiv ′mī·krə‚fōn }

magnetostrictive oscillator [ELECTR] An oscillator whose frequency is controlled by a magnetostrictive element. { mag¦nēd·ō¦strik·tiv ′äs·ə‚lād·ər }

magnetostrictive resonator [SOLID STATE] Ferromagnetic rod so designed that it can be excited magnetically into resonant vibration at one or more definite and known frequencies. { mag¦nēd·ō¦strik·tiv ′rez·ən‚ād·ər }

magnetostrictor [ELECTROMAG] A device for converting electric oscillations to mechanical oscillations by employing the property of magnetostriction. { mag¦nēd·ō¦strik·tər }

magnetoswitchboard exchange [COMMUN] A manual exchange at which the subscribers and operators call and clear by means of magnetoelectric generators. { mag‚nēd·ō′swich‚bȯrd iks‚chānj }

magnetotactic bacteria [MICROBIO] A group of bacteria containing iron mineral crystals in intracellular structures,

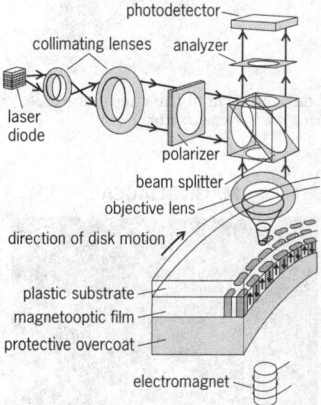

Simplified diagram of magnetooptic recording device.

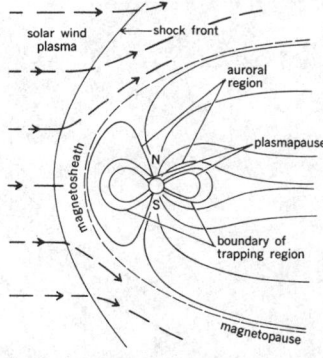

Configuration of the magnetosphere in the plane containing the sun-earth line and the geomagnetic axis; position of magnetosheath is shown.

called magnetosomes, which enable the bacteria to orient and migrate along magnetic field lines. { mag,ned·ə,tak·tik bak'tir·ē·ə }

magnetotail [GEOPHYS] The portion of the magnetosphere extending from earth in the direction away from the sun for a variable distance of the order of 1000 earth radii. { mag'ned·ō,tāl }

magneto telephone set [ELEC] Local battery telephone set in which current for signaling by the telephone station is supplied from a local hand generator, usually a magneto. { mag'nēd·ō 'tel·ə,fōn ,set }

magnetotellurics [GEOPHYS] A geophysical exploration technique that measures natural electromagnetic fields to image subsurface electrical resistivity, providing information about the earth's interior composition and structure since naturally occurring rocks and minerals exhibit a broad range of electrical resistivities. { mag,ned·ō·tə'lür·iks }

magnetovision [ENG] A method of measuring and displaying magnetic field distributions in which scanning results from a thin-film Permalloy magnetoresistive sensor are processed numerically and presented in the form of a color map on a video display unit. { mag'ned·ə,vizh·ən }

magnet power [ELECTROMAG] The electric power supplied to the coils of an electromagnet. { 'mag·nət ,paů·ər }

magnetron [ELECTR] One of a family of crossed-field microwave tubes, wherein electrons, generated from a heated cathode, move under the combined force of a radial electric field and an axial magnetic field in such a way as to produce microwave radiation in the frequency range 1–40 gigahertz; a pulsed microwave radiation source for radar, and continuous source for microwave cooking. { 'mag·nə,trän }

magnetron oscillator [ELECTR] Oscillator circuit employing a magnetron tube. { 'mag·nə,trän 'äs·ə,lād·ər }

magnetron pulling [ELECTR] Frequency shift of a magnetron caused by factors which vary the standing waves or the standing-wave ratio on the radio-frequency lines. { 'mag·nə,trän 'půl·iŋ }

magnetron pushing [ELECTR] Frequency shift of a magnetron caused by faulty operation of the modulator. { 'mag·nə,trän 'push·iŋ }

magnetron vacuum gage [ELECTR] A vacuum gage that is essentially a magnetron operated beyond cutoff in the vacuum being measured. { 'mag·nə,trän 'vak·yəm ,gāj }

magnet wire [ELEC] The insulated copper or aluminum wire used in the coils of all types of electromagnetic machines and devices. { 'mag·nət ,wīr }

magnification [OPTICS] **1.** A measure of the effectiveness of an optical system in enlarging or reducing an image; the magnification may be lateral, longitudinal, or angular. **2.** See lateral magnification. { ,mag·nə·fə'kā·shən }

magnifier See simple microscope. { 'mag·nə,fī·ər }

magnifying glass [OPTICS] **1.** Any device that uses a simple lens which enlarges the object being viewed. **2.** See simple microscope. { 'mag·nə,fī·iŋ ,glas }

magnifying power [OPTICS] The ratio of the tangent of the angle subtended at the eye by an image formed by an optical system, to the tangent of the angle subtended at the eye by the corresponding object at a distance for convenient viewing. { 'mag·nə,fī·iŋ ,paů·ər }

magnistor [ELECTR] A device that utilizes the effects of magnetic fields on injection plasmas in semiconductors such as indium antimonide. { 'mag'nis·tər }

magnitude [ASTRON] The relative luminance of a celestial body; the smaller (algebraically) the number indicating magnitude, the more luminous the body. Also known as stellar magnitude. [GEOPHYS] A measure of the amount of energy released by an earthquake. [MATH] See absolute value. { 'mag·nə,tüd }

magnitude method [ORD] Method of adjusting gunfire for range when the amount and direction of the deviation are known. { 'mag·nə,tüd ,meth·əd }

magnitude ratio [ASTRON] The ratio (2.512) of relative brightness of two celestial bodies differing in magnitude by 1.0. { 'mag·nə,tüd 'rā·shō }

magnitude system [ASTRON] A system for designating the relative brightness of stars when photography is used; emulsions of different color sensitivities, used with color filters, permit measurements of starlight of different wavelengths with

corresponding determination of magnitude at these wavelengths. { 'mag·nə,tüd ,sis·təm }

magno [MET] An alloy of 95.5% nickel and 4.5% manganese, used in the manufacture of incandescent lamps and radio tubes. { 'mag,nō }

magnocellular [CELL MOL] Having large cell bodies; said of various nuclei of the central nervous system. { ,mag·nə'sel·yə·lər }

magnochromite See magnesiochromite. { ,mag·nə'krō,mīt }

magnoferrite See magnesioferrite. { ,mag·nə'fe,rīt }

Magnolia [BOT] A genus of trees, the type genus of the Magnoliaceae, with large, chiefly white, yellow, or pinkish flowers, and simple, entire, usually large evergreen or deciduous alternate leaves. { mag'nōl·yə }

Magnolia acuminata See cucumber tree. { mag·nōl·yə ə,kü·mi'näd·ə }

Magnoliaceae [BOT] A family of dicotyledonous plants of the order Magnoliales characterized by hypogynous flowers with few to numerous stamens, stipulate leaves, and uniaperturate pollen. { mag,nō·lē'ās·ē,ē }

Magnoliales [BOT] The type order of the subclass Magnoliidae; members are woody plants distinguished by the presence of spherical ethereal oil cells and by a well-developed perianth of separate tepals. { mag,nō·lē'ā·lēz }

Magnoliatae See Magnoliopsida. { ,mag·nō'lī·ə,tē }

Magnoliidae [BOT] A primitive subclass of flowering plants in the class Magnoliopsida generally having a well-developed perianth, numerous centripetal stamens, and bitegmic, crassinucellate ovules. { ,mag·nō'lī·ə,dē }

Magnoliophyta [BOT] The angiosperms, a division of vascular seed plants having the ovules enclosed in an ovary and well-developed vessels in the xylem. { mag,nō·lē'äf·əd·ə }

Magnoliopsida [BOT] The dicotyledons, a class of flowering plants in the division Magnoliophyta generally characterized by having two cotyledons and net-veined leaves, with vascular bundles borne in a ring enclosing a pith. { mag,nō·lē'äp·sə·də }

magnon [SOLID STATE] A quasi-particle which is introduced to describe small departures from complete ordering of electronic spins in ferro-, ferri-, antiferro-, and helimagnetic arrays. Also known as quantized spin wave. { 'mag,nän }

magnophorite [MINERAL] $NaKCaMg_5Si_8O_{23}OH$ A monoclinic mineral composed of a basic silicate of sodium, potassium, calcium, and magnesium; member of the amphibole group. { ,mag·nə'fòr,īt }

magnum [ANAT] Large, as in foramen magnum. { 'mag·nəm }

Magnus effect [FL MECH] A force on a rotating cylinder in a fluid flowing perpendicular to the axis of the cylinder; the force is perpendicular to both flow direction and cylinder axis. Also known as Magnus force. { 'mäg·nəs i,fekt }

Magnus force See Magnus effect. { 'mäg·nəs ,fòrs }

Magnus moment [FL MECH] A torque associated with the Magnus effect, such as moments about the pitch and yaw axes of a missile or aircraft due to rotation about the roll axis. { 'mäg·nəs ,mō·mənt }

mag-slip See synchro. { 'mag,slip }

maguey [MATER] A fiber obtained from the agave (Agave cantala); maguey fibers are white, stiff, brilliant, and light in weight, and are used chiefly for binder twine. { mə'gā }

mahlstick [GRAPHICS] A stick held in the palette hand of a painter and used as a support for his painting hand; it is a light rod of wood often with a leather-covered ball at one end. Also known as maulstick; rest stick. { 'mòl,stik }

mahogany [BOT] Any of several tropical trees in the family Meliaceae of the Geraniales. [MATER] The hard wood of these trees, especially the red or yellow-brown wood of the West Indies mahogany tree (Swietenia mahagoni). { mə'häg·ə·nē }

mahogany acid [MATER] A dark-colored mixture of sulfonic acid derived from petroleum; the salts are used as emulsifying agents and in lubricants. { mə'häg·ə·nē 'as·əd }

mahogany soap [MATER] The sodium salt of crude or refined petroleum sulfonic acids, used as flotation agents and to increase the oil absorption of mineral pigments in paint. { mə'häg·ə·nē 'sōp }

mahubarana fat [MATER] A pale–yellow solid oil, melting point 40–44°C, obtained from seeds of trees of the genus Boldoa; used for soaps and candles. { ,mä·hə·bä'rä·nə ,fat }

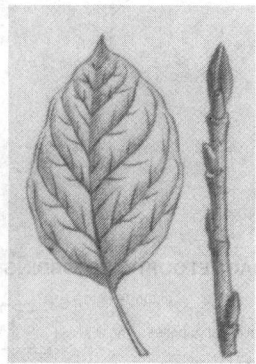

MAGNOLIA

Leaf and twig of the cucumber tree, *Magnolia acuminata*.

MAGNOLIACEAE

Flower of the tulip tree (*Liriodendron tulipifera*). (*Photograph by F. W. Westlake, from National Audubon Society*)

maidenhead *See* hymen. { 'mād·ən,hed }

maieusiophobia [PSYCH] An abnormal fear of childbirth. { mā,yü·sē·ə'fō·bē·ə }

mail box [COMPUT SCI] **1.** A portion of a computer's main storage that can be used to hold information about other devices. **2.** Computer storage facilities designed to hold electronic mail. { 'māl ,bäks }

mailbox name [COMPUT SCI] The first part of an electronic mail address, which identifies the storage space that has been set aside in a computer to receive a user's electronic mail messages. Also known as username. { 'māl ,bäks ,nām }

mailing list [COMMUN] A list of users of the Internet or another computer network who all receive copies of electronic mail messages. { 'māl·iŋ ,list }

Maillard reaction [BIOCHEM] A reaction in which the amino group in an amino acid tends to form condensation products with aldehydes; believed to cause the Browning reaction when an amino acid and a sugar coexist, evolving a characteristic flavor useful in food preparations. { mī'yär rē,ak·shən }

mail merge [COMPUT SCI] The process of combining a form letter with a list of names and addresses to produce individualized letters. { 'māl ,mərj }

main [ELEC] **1.** One of the conductors extending from the service switch, generator bus, or converter bus to the main distribution center in interior wiring. **2.** *See* power transmission line. [ENG] A duct or pipe that supplies or drains ancillary branches. { mān }

main-and-tail haulage [MIN ENG] A single-track haulage system that is operated by a haulage engine with two drums, each with a separate rope. { 'mān ən 'tāl 'hȯl·ij }

main-and-transfer bus [ELEC] A substation switching arrangement similar to a single bus but with an additional transfer bus provided. { 'mān ən 'tranz·fər ,bəs }

main-band deoxyribonucleic acid [MOL BIO] A peak band of deoxyribonucleic acid obtained by density gradient centrifugation. { ¦mān ¦band ,dē,äk·sē,rī·bō·nü'klē·ik 'as·əd }

main bang [ELECTR] Transmitted pulse, within a radar system. { 'mān 'baŋ }

main bearing [MECH ENG] One of the bearings that support the crankshaft in an internal combustion engine. { ¦mān 'ber·iŋ }

main clock *See* master clock. { 'mān 'kläk }

main controller [COMPUT SCI] A control unit assigned to direct the other control units in a computer system. { 'mān kən'trō·lər }

main crosscut [MIN ENG] A crosscut that traverses the entire mining field and penetrates all deposits. { 'mān 'krȯs,kət }

main deck [NAV ARCH] **1.** The uppermost complete deck of a naval vessel extending over its entire length and width. **2.** The uppermost deck of a merchant ship for which it is possible to close all openings securely. { 'mān 'dek }

main diagonal *See* principal diagonal. { ¦mān dī'ag·ən·əl }

main distributing frame [ELEC] Frame which terminates the permanent outside lines entering the central office building on one side and the subscriber-line multiple cabling, trunk multiple cabling, and so on, used for associating an outside line with any desired terminal on the other side; it usually carries the control-office protective devices, and functions as a test point between line and office. Also known as main frame. { 'mān di'strib·yəd·iŋ ,frām }

Maindroniidae [INV ZOO] A family of wingless insects belonging to the Thysanura proper. { ,mān·drō'nī·ə,dē }

main effect [STAT] The effect of the change in level of one factor in a factorial experiment measured independently of other variables. { ¦mān i'fekt }

main entry [MIN ENG] The principal entry or set of entries driven through the coalbed from which cross entries, room entries, or rooms are turned. { 'mān 'en·trē }

main exciter [ELEC] Exciter which supplies energy for the field excitation of a principal electric machine. { 'mān ik'sīd·ər }

main fans [MIN ENG] Fans that produce the general ventilating current of the mine, being of large capacity and permanently installed. { 'mān 'fanz }

main firing [ENG] The firing of a round of shots by means of current supplied by a transformer fed from a main power supply. { 'mān 'fīr·iŋ }

main frame [COMPUT SCI] **1.** A large computer. **2.** The part of a computer that contains the central processing unit, main

storage, and associated control circuitry. Also known as frame. [ELEC] *See* main distributing frame. { 'mān ,frām }

main haulage [MIN ENG] The section of the haulage system which moves the coal from the secondary or intermediate haulage system to the shaft or mine opening. { 'mān 'hȯl·ij }

main haulage conveyor [MIN ENG] A conveyor used to transport material in the main haulage section of the mine, between the intermediate haulage conveyor and a car-loading point or the outside. { 'mān 'hȯl·ij kən,vā·ər }

main instruction buffer [COMPUT SCI] A section of storage in the instruction unit, 16 bytes in length, used to hold prefetched instructions. { 'mān in'strək·shən ,bəf·ər }

main joint *See* master joint. { 'mān 'jȯint }

mainland [GEOGR] A continuous body of land that constitutes the main part of a country or continent. { 'mān·lənd }

main-line locomotive [MIN ENG] A large, high-powered locomotive which hauls trains of cars over the main haulage system. { 'mān ,līn ,lō·kə'mōd·iv }

main lobe *See* major lobe. { 'mān 'lōb }

main loop [COMPUT SCI] A set of instructions that constitute the primary structure of a repetitive computer program. { 'mān ,lüp }

mainmast [NAV ARCH] The principal mast of a sailing ship. { 'mān·məst }

main memory *See* main storage. { 'mān 'mem·rē }

main path [COMPUT SCI] The principal branch of a routine followed by a computer in the process of carrying out the routine. { 'mān 'path }

main program [COMPUT SCI] **1.** The central part of a computer program, from which control may be transferred to various subroutines and to which control is eventually returned. Also known as main routine. **2.** *See* executive routine. { 'mān 'prō·grəm }

main return [MIN ENG] The main return airway of a mine. { 'mān ri'tərn }

main rope *See* pull rope. { 'mān ,rōp }

main-rope haulage system [MIN ENG] A system of haulage for hauling loaded trains of tubs or cars up, or lowering them down, a comparatively steep gradient which is not steep enough, in the latter case, for a self-acting incline. { 'mān ,rōp 'hȯl·ij ,sis·təm }

main routine *See* executive routine; main program. { 'mān rü'tēn }

mainsail [NAV ARCH] The principal sail of a sailing vessel, carried on the mainmast. { 'mān·səl }

main sequence [ASTRON] The band in the spectrum luminosity diagram which has the great majority of stars; their energy derives from core burning of hydrogen into helium. { 'mān 'sē·kwəns }

main sequence star [ASTRON] **1.** Any of those stars in the smooth curve termed the main sequence in a Hertzsprung-Russell diagram. **2.** *See* dwarf star. { 'mān ¦sē·kwəns 'stär }

main shaft [MECH ENG] The line of shafting receiving its power from the engine or motor and transmitting power to other parts. { 'mān 'shaft }

main stage [AERO ENG] **1.** In a multistage rocket, the stage that develops the greatest amount of the thrust, with or without booster engines. **2.** In a single-stage rocket vehicle powered by one or more engines, the period when full thrust (at or above 90%) is attained. **3.** A sustainer engine, considered as a stage after booster engines have fallen away. { 'mān 'stāj }

main station [COMMUN] Telephone station with a distinct call number designation, directly connected to a central office. { 'mān 'stā·shən }

main storage [COMPUT SCI] A digital computer's principal working storage, from which instructions can be executed or operands fetched for data manipulation. Also known as main memory. { 'mān 'stȯr·ij }

main stream [HYD] The principal or largest stream of a given area or drainage system. Also known as master stream; trunk stream. { 'mān 'strēm }

main stroke *See* return streamer. { 'mān 'strōk }

main sweep [ELECTR] On certain fire-control radar, the longest range scale available. { 'mān 'swēp }

maintainability [ENG] **1.** The ability of equipment to meet operational objectives with a minimum expenditure of maintenance effort under operational environmental conditions in which scheduled and unscheduled maintenance is performed.

2. Quantitatively, the probability that an item will be restored to specified conditions within a given period of time when maintenance action is performed in accordance with prescribed procedures and resources. { män,tā·nə'bil·əd·ē }

maintenance [IND ENG] The upkeep of industrial facilities and equipment. { 'mānt·ən·əns }

maintenance engineering [IND ENG] The function of providing policy guidance for maintenance activities, and of exercising technical and management review of maintenance programs. { 'mānt·ən·əns ,en·jə'nir·iŋ }

maintenance kit [ENG] A collection of items not all having the same basic name, which are of a supplementary nature to a major component or equipment; the items within the collection may provide replacement parts and facilitate such functions as inspection, test repair, or preventive types of maintenance, for the specific purpose of restoring and improving the operational status of a component or equipment comparable to its original capacity and efficiency. { 'mānt·ən·əns ,kit }

maintenance pack [COMPUT SCI] A disk drive that is used to store copies of computer programs for the purpose of applying and testing changes made in the course of software maintenance. { 'mānt·ən·əns ,pak }

maintenance routine [COMPUT SCI] A computer program designed to detect conditions which may give rise to a computer malfunction in order to assist a service engineer in performing routine preventive maintenance. { 'mānt·ən·əns rü,tēn }

maintenance time [COMPUT SCI] The time required for both corrective and preventive maintenance of a computer or other components of a computer system. { 'mānt·ən·əns ,tīm }

maintenance vehicle [ENG] Vehicle used for carrying parts, equipment, and personnel for maintenance or evacuation of vehicles. { 'mānt·ən·əns ,vē·ə·kəl }

main thermocline [OCEANOGR] A thermocline that is deep enough in the ocean to be unaffected by seasonal temperature changes in the atmosphere. Also known as permanent thermocline. { 'mān 'thər·mə,klīn }

maitake mushroom See Grifola frondosa. { ,mä·i,tä·ke 'məsh,rüm }

maize [BOT] Zea mays. Indian corn, a tall cereal grass characterized by large ears. { māz }

Majac mill [MIN ENG] A mill for dry-grinding mica by means of fluid energy; consists of a chamber which contains two horizontal, directly opposing jets and into which mica is fed continuously from a screw conveyor. { 'mī·äk ,mil }

Majidae [INV ZOO] The spider, or decorator, crabs, a family of decapod crustaceans included in the Brachyura; members are slow-moving animals that often conceal themselves by attaching seaweed and sessile animals to their carapace. { 'maj·ə,dē }

Majorana effect [OPTICS] The effect in which a transverse magnetic field acting on a colloidal solution, such as a sol of iron oxide, produces optical anisotropy, resulting in magnetic birefringence. { mä·jə'ran·ə i,fekt }

Majorana force [NUC PHYS] A force between two nucleons postulated to explain various phenomena, which can be derived from a potential containing an operator which exchanges the nucleons' positions but not their spins. { ,mä·jə'ran·ə ,fòrs }

Majorana neutrino [PARTIC PHYS] A particle described by a wave function that satisfies the Dirac equation with mass equal to zero, and that is self-charge-conjugate. { ,mä·jə'ran·ə nü'trē·nō }

major arc [MATH] The longer of the two arcs produced by a secant of a circle. { 'mā·jər 'ärk }

major assembly [ENG] A self-contained unit of individual identity; a completed assembly of component parts ready for operation, but utilized as a portion of, and intended for further installation in, an end item or major item. { 'mā·jər ə'sem·blē }

major axis [MATH] The longer of the two axes with respect to which an ellipse is symmetric. { 'mā·jər 'ak·səs }

major combination [ORD] A single composite unit of mechanical equipment inherently complete for independent use and consisting of one or more major items; as issued, it is complete in respect to both equipment and spare parts, including items furnished by services other than the issuing service; for example, a tank, complete with armament, equipment, and spare parts. { 'mā·jər ,käm·bə'nā·shən }

MAIN THERMOCLINE

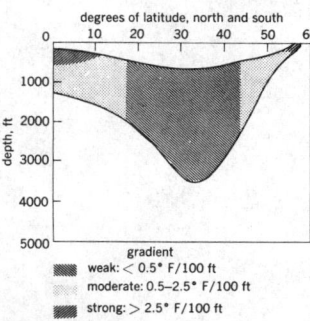

Diagram showing variations with latitude in average depth, thickness and temperature gradient of the main thermocline.

---

major cycle [COMPUT SCI] The time interval between successive appearances of a given storage position in a serial-access computer storage. { 'mā·jər 'sī·kəl }

major defect [IND ENG] Defect which causes serious malfunctioning of a product. { 'mā·jər 'dē,fekt }

major depression [PSYCH] A type of affective disorder characterized by major episodes of depression without intervening manic episodes. { 'mā·jər di'presh·ən }

major diameter [DES ENG] The largest diameter of a screw thread, measured at the crest for an external (male) thread and at the root for an internal (female) thread. { 'mā·jər dī'am·əd·ər }

major diatonic scale [ACOUS] A diatonic scale in which the relative sizes of the sequence of intervals are approximately 2,2,1,2,2,2,1. { 'mā·jər 'dī·ə'tan·ik 'skāl }

major fog signal [NAV] A sound signal which has a normal range of reception in excess of 2 miles (3.2 kilometers), to aid watercraft in avoiding obstacles. { 'mā·jər 'fäg ,sig·nəl }

major fold [GEOL] A large-scale fold with which minor folds are usually associated. { 'mā·jər 'fōld }

major gene [GEN] Any gene individually associated with pronounced phenotypic effects. { 'mā·jər jēn }

major histocompatibility complex [IMMUNOL] In vertebrates, a family of genes that encode cell surface glycoproteins that regulate interactions among cells of the immune system, some components of the complement system, and perhaps other related functions connected with intercell recognition. Abbreviated MHC. { 'mā·jər 'hi·stō·kəm'pad·ə'bil·əd·ē 'käm,pleks }

major histocompatibility molecule [IMMUNOL] Any of two classes of immunoregulatory cell-surface glycoproteins that are encoded by the major histocompatibility gene; class I molecules are found on almost every nucleated body cell and present antigens from the cytoplasm to cytotoxic T cells, whereas class II molecules are found only on macrophages, B cells, and CD4+ T cells and present antigens from outside the cell to helper T cells. { 'mā·jər ,his·tō·kəm,pad·ə'bil·əd·ē 'mäl·ə,kyül }

major hysteria [PSYCH] A conversion reaction manifested in movements that suggest a generalized convulsion. { 'mā·jər hi'ster·ē·ə }

major immunogene complex [IMMUNOL] A genetic region containing loci that code for lymphocyte surface antigens, histocompatibility antigens, immune response gene products, and proteins of the complement system. Abbreviated MIC. { 'mā·jar ,ə'myü·nə·jēn 'käm,pleks }

majorite [MINERAL] $Mg_3(Fe,Al,Si)_2(SiO_4)_3$ A garnet mineral that forms in the deep upper mantle in response to the gradual dissolution of pyroxene due to increasing pressure, first identified as an inclusion in diamond. { 'mā·jə,rīt }

major item [ORD] An end item, a group of end items individually classified by the responsible technical service, or an assembled group of items as procured or issued for a specific tactical role, excluding combinations required to complete the assigned tactical mission. { 'mā·jər 'īd·əm }

majority [MATH] A logic operator having the property that if P, Q, R are statements, then the function (P, Q, R, . . .) is true if more than half the statements are true, or false if half or less are true. { mə'jär·əd·ē }

majority carrier [ELECTR] The type of carrier, that is, electron or hole, that constitutes more than half the carriers in a semiconductor. { mə'jär·əd·ē 'kar·ē·ər }

majority element See majority gate. { mə'jär·əd·ē 'el·ə·mənt }

majority emitter [ELECTR] Of a transistor, an electrode from which a flow of minority carriers enters the interelectrode region. { mə'jär·əd·ē i'mid·ər }

majority gate [COMPUT SCI] A logic circuit which has one output and several inputs, and whose output is energized only if a majority of its inputs are energized. Also known as majority element; majority logic. { mə'jär·əd·ē 'gāt }

majority logic See majority gate. { mə'jär·əd·ē 'läj·ik }

major joint See master joint. { 'mā·jər 'jòint }

major key [COMPUT SCI] The primary key for identifying a record. { 'mā·jər ,kē }

major light [NAV] A light of high candlepower and reliability exhibited from a fixed structure ashore or on a marine site (except range lights). { 'mā·jər 'līt }

major lobe [ELECTROMAG] Antenna lobe indicating the

direction of maximum radiation or reception.　Also known as main lobe.　{ 'mā·jər 'lōb }

major node [ELEC] A point in an electrical network at which three or more elements are connected together.　Also known as junction.　{ 'mā·jər 'nōd }

major operation [MED] An extensive, difficult, and potentially dangerous surgical procedure, usually requiring general anesthesia.　{ 'mā·jər ‚äp·ə'rā·shən }

major planet [ASTRON] Any of the four planets that are larger than earth: Jupiter, Saturn, Neptune, and Uranus.　{ 'mā·jər 'plan·ət }

major relay station [ELECTR] Tape relay station which has two or more trunk circuits connected thereto to provide an alternate route or to meet command requirements.　{ 'mā·jər 'rē‚lā ‚stā·shən }

major repair [ENG] Repair work on items of material or equipment that need complete overhaul or substantial replacement of parts, or that require special tools.　{ 'mā·jər ri per }

major trough [METEOROL] A long-wave trough in the large-scale pressure pattern of the upper troposphere.　{ 'mā·jər 'tróf }

major wave See long wave.　{ 'mā·jər 'wāv }

make [ELEC] Closing of relay, key, or other contact.　{ 'māk }

make-and-break circuit [ELEC] A circuit that is alternately opened and closed.　{ 'māk ən 'brāk ‚sər·kət }

make-break operation [COMMUN] A circuit operation in which there is a cessation of current flow as a pulse transmission occurs.　{ 'māk 'brāk ‚äp·ə‚rā·shən }

make-busy [COMMUN] A switch whose activation makes a dial telephone line or group of telephone lines appear to be busy and thereby prevents completion of incoming calls.　{ 'māk 'biz·ē }

make contact [ELEC] Contact of a device which closes a circuit upon the operation of the device (normally open).　{ 'māk ‚kän‚takt }

makeready [GRAPHICS] **1.** The careful leveling of relief printing plates on the bed of the press so that they yield the best possible impression.　**2.** Final preparations and adjustments that must be made preliminary to printing, especially those that are required to compensate for irregularities in type or plates.　{ 'māk‚red·ē }

make the land [NAV] To sight and approach or reach land from seaward.　{ 'māk thə 'land }

makeup [GRAPHICS] The arrangement of lines of type and art into pages or sections of suitable length.　{ 'mā‚kəp }

makeup air [ENG] The volume of air required to replace air exhausted from a given space.　{ 'māk‚əp ‚er }

makeup gas [PETRO ENG] Gas injected into a reservoir to maintain a constant pressure, thus preventing retrograde condensation.　{ 'māk‚əp ‚gas }

makeup time [COMPUT SCI] The time required to rerun programs on a computer because of operator errors and other problems.　{ 'māk‚əp ‚tīm }

makeup water [CHEM ENG] Water feed needed to replace that which is lost by evaporation or leakage in a closed-circuit, recycle operation.　{ 'mā‚kəp ‚wód·ər }

make way [NAV] To progress through the water.　{ 'māk 'wā }

making current [ELEC] The peak value attained by the current during the first cycle after a switch, circuit breaker, or similar apparatus is closed.　{ 'māk·iŋ ‚kə·rənt }

making hole [PETRO ENG] During a drilling operation, deepening the well bore, with reference to progress at a given time.　{ 'māk·iŋ 'hōl }

makroskelic [ANTHRO] Being long-legged relative to the trunk length, with a skelic index of 95–100.　{ ‚mak·rə‚skel·ik }

Maksutov-Schmidt telescope See meniscus-Schmidt telescope.　{ mak'sü‚tóf 'shmit ‚tel·ə‚skōp }

Maksutov system [OPTICS] A catadioptric telescope optical system capable of covering a large field (60° and more); used to survey large areas of the sky.　{ mak'sü‚tóf ‚sis·təm }

mal- [SCI TECH] A combining form meaning bad, wrong, irregular, abnormal, inadequate.　{ mal }

Malachiidae [INV ZOO] An equivalent name for Malyridae.　{ ‚mal·ə'kī·ə‚dē }

malachite [MINERAL] $Cu_2CO_3(OH)_2$ A bright-green monoclinic mineral consisting of a basic carbonate of copper and

usually occurring in massive forms or in bundles of radiating fibers; specific gravity is 4.05, and hardness is 3.5–4 on Mohs scale.　{ 'mal·ə‚kīt }

malacia [MED] Abnormal softening of tissues of an organ or other body structure.　{ mə'lāsh·ə }

Malacobothridia [INV ZOO] A subclass of worms in the class Trematoda; members typically have one or two soft, flexible suckers and are endoparasitic in vertebrates and invertebrates.　{ ‚mal·ə·kō·bə'thrid·ē·ə }

Malacocotylea [INV ZOO] The equivalent name for Digenea.　{ ‚mal·ə·kō‚käd·əl'ē·ə }

malacolite See diopside.　{ 'mal·ə·kə‚līt }

malacology [INV ZOO] The study of mollusks.　{ ‚mal·ə'kál·ə·jē }

malacoplakia [MED] The accumulation of modified histiocytes (malacoplakia cells) to produce soft, pale, elevated plaques, usually in the urinary bladder of middle-aged women.　{ ‚mal·ə·kə'plā·kē·ə }

malacoplakia cell [PATH] A large histiocyte, occasionally multinucleate and containing Michaelis-Gutmann calcospherules in the cytoplasm; seen in malacoplakia.　{ ‚mal·ə·kə'plā·kē·ə ‚sel }

Malacopoda [INV ZOO] A subphylum of invertebrates in the phylum Oncopoda.　{ ‚mal·ə'käp·ə·də }

Malacopterygii [VERT ZOO] An equivalent name for Clupeiformes in older classifications.　{ ‚mal·ə·kō·tə‚rij·ē‚ī }

Malacostraca [INV ZOO] A large, diversified subclass of Crustacea including shrimps, lobsters, crabs, sow bugs, and their allies; generally characterized by having a maximum of 19 pairs of appendages and trunk limbs which are sharply differentiated into thoracic and abdominal series.　{ ‚mal·ə'kä·strə·kə }

maladie du sommel See African sleeping sickness.　{ ‚mäl·ə'dē dyü sò'mä }

maladjustment [PSYCH] Failure to conform or inadequate conformity due to the inability or a lack of motivation to change one's feelings or attitudes to adjust to the demands of the environment.　{ ‚mal·ə'jəs·mənt }

malady [MED] A disorder, disease, or illness.　{ 'mal·əd·ē }

malaise [MED] A general state of ill-being or the feeling of poor health.　{ mə'lāz }

Malapteruridae [VERT ZOO] A family of African catfishes in the suborder Siluroidei.　{ mə‚lap·tə'rür·ə‚dē }

malar [ANAT] Of or pertaining to the cheek or to the zygomatic bone.　{ 'mā·lər }

malar bone See zygomatic bone.　{ 'mā·lər ‚bōn }

malaria [MED] A group of human febrile diseases with a chronic relapsing course caused by hemosporidian blood parasites of the genus *Plasmodium,* transmitted by the bite of the *Anopheles* mosquito.　{ mə'ler·ē·ə }

malaria pigment [PATH] Dark-brown, amorphous, microcrystalline and birefringent pigment found in parasitized erythrocytes, especially with malarial parasites, and in littoral phagocytes of spleen, liver, and bone marrow.　{ mə'ler·ē·ə ‚pig·mənt }

malar stripe [VERT ZOO] **1.** The area extending from the corner of the mouth backward and down in birds.　**2.** Area on side of throat below the base of the lower mandible.　{ 'mā·lər ‚strīp }

malassimilation [MED] Faulty or inadequate assimilation.　{ ‚mal·ə‚sim·ə'lā·shən }

malate dehydrogenase See malic enzyme.　{ 'ma‚lāt dē'hī·drə·jə‚nās }

malathion [ORG CHEM] $C_{10}H_{19}O_6PS_2$ A yellow liquid, slightly soluble in water; malathion is the generic name for *S*-1,2-bis(ethoxycarbonyl)ethyl *O,O*-dimethylphosphorodithioate; used as an insecticide.　{ ‚mal·ə'thī‚än }

Malayan filariasis [MED] Filariasis of humans caused by *Brugia malayi,* transmitted by *Mansonia* and *Anopheles* mosquitoes.　{ mə'lā·ən ‚fil·ə'rī·ə·səs }

malchite [PETR] A fine-grained lamprophyre with small, rare phenocrysts or hornblende, labradorite, and sometimes biotite embedded in a matrix of hornblende, andesine, and some quartz.　{ 'mal·kīt }

Malcidae [INV ZOO] A small family of Ethiopian and Oriental hemipternan insects in the superfamily Pentatomorpha.　{ 'mal·sə‚dē }

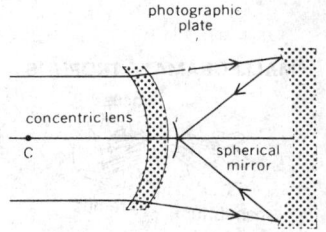

MAKSUTOV SYSTEM

Optics of Maksutov system; *C* is the center of curvature of the mirror.

Malcodermata [INV ZOO] The equivalent name for Cantharoidea. { ‚mal·kō'dər·məd·ə }

Maldacena duality [PARTIC PHYS] A form of weak-strong duality between black holes in superstring theory and ordinary quantum field theory with many fields. { ¦mäl·də‚sän·ə dü'al·əd·ē }

Maldanidae [INV ZOO] The bamboo worms, a family of mud-swallowing annelids belonging to the Sedentaria. { mal'dan·ə‚dē }

Maldaninae [INV ZOO] A subfamily of the Maldanidae distinguished by cephalic and anal plaques with the anal aperture located dorsally. { mal'dan·ə‚nē }

mal de pinto *See* pinta. { ‚mal de 'pin·tō }

maldonite [MINERAL] Au₂Bi A pinkish silver-white mineral consisting of gold and bismuth; occurs in massive granular form. { 'mal·də‚nīt }

male [BOT] A flower lacking pistils. [ZOO] **1.** Of or pertaining to the sex that produces spermatozoa. **2.** An individual of this sex. { māl }

maleate [ORG CHEM] An ester or salt of maleic acid. { mə'lē‚āt }

male climacteric [PHYSIO] A condition presumably due to loss of testicular function, associated with an elevated urinary excretion of gonadotropins and symptoms of loss of sexual desire and potency, hot flashes, and vasomotor instability. { 'māl klə'mak·tə·rik }

male connector [ELEC] An electrical connector with protruding contacts for joining with a female connector. { 'māl kə'nek·tər }

male heterogamety [GEN] The production by males of some species, such as humans and *Drosophila*, of two types of gametes that differ in their sex chromosome content. { 'māl ¦hed·ə·rō¦gam·əd·ē }

male homogamety [GEN] The production by males of some species, as in birds, of a single type of gamete with respect to sex chromosome content. { 'māl ¦hō·mō¦gam·əd·ē }

maleic acid [ORG CHEM] HOOCCH:CHCOOH A colorless, crystalline dibasic acid; soluble in water, acetone, and alcohol; melting point 130–131°C; used in textile processing, and as an oil and fat preservative. { mə'lā·ik 'as·əd }

maleic anhydride [ORG CHEM] C₄H₂O₃ Colorless crystals, soluble in acetone, hydrolyzing in water; used to form polyester resins. Also known as 2,5-furandione. { mə'lā·ik an'hī‚drīd }

maleic hydrazide [ORG CHEM] C₄N₂H₄O₂ Solid material, decomposing at 260°C; slightly soluble in alcohol and water; used as a weed killer and growth inhibitor. { mə'lā·ik 'hī·drə‚zīd }

malenclave [HYD] A body of contaminated groundwater surrounded by uncontaminated water. { ‚mal'än‚klāv }

male pseudohermaphroditism *See* androgyny. { māl ‚süd·ō·hər'maf·rə‚did‚iz·əm }

male sterility [PHYSIO] The condition in which male gametes are absent, deficient in number, or nonfunctional. { māl stə'ril·əd·ē }

male Turner's syndrome *See* Ullrich-Turner syndrome. { māl 'tər·nərz ‚sin‚drōm }

malezal swamp [ECOL] A swamp resulting from drainage of water over an extensive plain with a slight, almost imperceptible slope. { mə'lēz·əl ‚swämp }

malformation [MED] A deformity of a part of the body resulting from abnormal development. { ‚mal·fər'mā·shən }

malfunction [SCI TECH] Failure to function normally. { mal'fəŋk·shən }

malfunction routine [COMPUT SCI] A program used in troubleshooting. { mal'fəŋk·shən rü‚tēn }

malic acid [BIOCHEM] COOH·CH₂·CHOH·COOH Hydroxysuccinic acid: a dibasic hydroxy acid existing in two optically active isomers and a racemate form; found in apples and many other fruits. { 'mal·ik 'as·əd }

malic dehydrogenase [BIOCHEM] An enzyme in the Krebs cycle that catalyzes the conversion of L-malic acid to oxaloacetic acid. { 'mal·ik dē'hī·drə·jə‚nās }

malic enzyme [BIOCHEM] An enzyme which utilizes nicotinamide-adenine dinucleotide phosphate (NADP) to catalyze the oxidative decarboxylation of malic acid to pyruvic acid and carbon dioxide. Also known as malate dehydrogenase. { 'mal·ik 'en‚zīm }

malicious code [COMPUT SCI] Programming code that is capable of causing harm to availability, integrity of code or data, or confidentiality in a computing system; encompasses Trojan horses, viruses, worms, and trapdoors. { mə¦lish·əs 'kōd }

malignant [CYTOL] Pertaining to cells that have undergone phenotypic transformation by oncogenes or protooncogenes. [MED] **1.** Endangering the life or health of an individual. **2.** Of or pertaining to the growth and proliferation of certain neoplasms which terminate in death if not checked by treatment. { mə'līg·nənt }

malignant catarrh [VET MED] A catarrhal fever of cattle caused by a virus and characterized by acute inflammation and edema of the respiratory and digestive systems. { mə'līg·nənt kə'tär }

malignant disease [MED] Any disease that endangers the life of an individual over a short period of time, especially cancer. { mə'līg·nənt di'zēz }

malignant edema [VET MED] Inflammatory edema associated with certain infections, especially an acute wound infection in wild and domestic animals. { mə'līg·nənt i'dē·mə }

malignant embolus [MED] A blood-borne mass of malignant cells which have become dislodged from the parent neoplasm. { mə'līg·nənt 'em·bə·ləs }

malignant glaucoma [MED] A form of glaucoma associated with severe pain and rapidly leading to blindness. { mə'līg·nənt glau'kō·mə }

malignant hypertension [MED] A severe form of hypertension with a rapid course leading to progressive cardiac and renal vascular disease. Also known as accelerated hypertension. { mə'līg·nənt ‚hī·pər'ten·chən }

malignant malaria *See* falciparum malaria. { mə'līg·nənt mə'ler·ē·ə }

malignant pustule [MED] The commonest form of anthrax in humans, resulting from contamination of the skin; characterized by a necrotic pustule surrounded by an area of edema and vesicles containing yellow fluid. Also known as cutaneous anthrax. { mə'līg·nənt 'pəs·chül }

malignant rhabdomyoma *See* rhabdomyosarcoma. { mə'līg·nənt ‚rab·də·mī'ō·mə }

malines [TEXT] **1.** Bobbin lace which has a design outlined by a lustrous thread on a fine six-sided mesh ground. **2.** A sheer, hexagonal-mesh fabric. Also known as tulle. { mə'lēn }

malinger [MED] To pretend or exaggerate illness in order to avoid responsibilities. { mə'liŋ·gər }

malladrite [MINERAL] Na₂SiF₆ A hexagonal mineral composed of sodium fluosilicate, occurring as small crystals in volcanic holes in Vesuvius. { mə'lä‚drīt }

mallardite [MINERAL] MnSO₄·7H₂O A pale-rose, monoclinic mineral composed of hydrous manganese sulfate. { mə'lär‚dīt }

malleable [MET] Capable of undergoing plastic deformation without rupture; a property characteristic of metals. { 'mal·yə·bəl }

malleable brass *See* Muntz metal. { 'mal·yə·bəl 'bras }

malleable iron [MET] White cast iron which has been rendered malleable by heat treatment. { 'mal·yə·bəl 'ī·ərn }

malleableize [MET] To render a material malleable, such as by heat-treating white cast iron. { 'mal·yə·bə‚līz }

malleable pig iron [MET] A grade of pig iron suitable for production of white cast iron from which malleable iron is made. { 'mal·yə·bəl 'pig ‚ī·ərn }

malleate trophus [INV ZOO] A type of crushing masticatory apparatus in rotifers that are incidentally predatory, such as brachionids. { 'mal·ē‚āt ‚trō·fəs }

mallee *See* tropical scrub. { 'mä·lē }

malleolus [ANAT] A projection on the distal end of the tibia and fibula at the ankle. { mə'lē·ə·ləs }

malleoramate trophus [INV ZOO] An intermediate type of rotiferan masticatory apparatus having a looped manubrium and teeth on the incus (comprising the fulcrum and rami); developed for grinding. { ‚mal·ē·ə‚ra‚mät ‚trō·fəs }

mallet [DES ENG] An implement with a barrel-shaped head made of wood, rubber, or other soft material; used for driving another tool, such as a chisel, or for striking a surface without causing damage. { 'mal·ət }

malleus [ANAT] The outermost, hammer-shaped ossicle of the middle ear; attaches to the tympanic membrane and articulates with the incus. { 'mal·ē·əs }

MALLEORAMATE TROPHUS

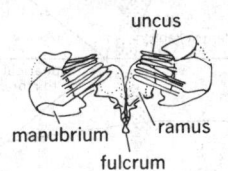

uncus

manubrium   ramus

fulcrum

Malleoramate trophus of a bdelloid rotifer, ventral view.

Mallophaga [INV ZOO] Biting lice, a comparatively small order of wingless insects characterized by five-segmented antennae, distinctly developed mandibles, one or two terminal claws on each leg, and a prothorax developed as a distinct segment. { məˈläf·ə·gə }

Mallory bodies [PATH] Oval acidophilic hyalin inclusion bodies seen in the cytoplasm of hepatic cells in Laennec's cirrhosis. Also known as alcoholic hyaline. { ˈmal·ə·rē ˌbäd·ēz }

Mallory bonding [DES ENG] Hermetically sealing polished silicon chips to polished glass plates by placing the two pieces together, heating them to about 350°C (662°F), and applying approximately 8000 volts across the assembly. { ˈmal·ə·rē ˌbänd·iŋ }

Mallory-Weiss syndrome [MED] Painless vomiting of blood secondary to lacerations of the distal esophagus and esophagogastric junction, usually a result of prolonged violent vomiting, coughing, or hiccuping. { ˈmla·ə·rē ˈwīs ˌsin̩ˌdrōm }

malloseismic [GEOPHYS] Referring to an area that is likely to experience destructive earthquakes several times in a century. { ˌmal·ə·ˈsīz·mik }

malm *See* marl. { mäm }

Malm [GEOL] The Upper Jurassic geologic series, above Dogger and below Cretaceous. { mäm }

malm brick [MATER] A brick made of natural malm or an artificial mixture consisting of pulverized chalk, sand, and bits of coke or furnace clinker. { ˈmäm ˌbrik }

malm rubber [MATER] A comparatively soft malm brick that can be worked to a desired shape by rubbing. { ˈmäm ˌrəb·ər }

malmstone [MATER] A name applied to chert when it is used in building and paving. { ˈmäm̩ˌstōn }

malnutrition [MED] Defective nutrition due to inadequate intake of nutrients or to their faulty digestion, assimilation, or metabolism. { ˌmal·nü'trish·ən }

malocclusion [MED] Faulty occlusion of the teeth. { ˌmal·ə'klü·zhən }

malodor [PHYSIO] A bad or foul odor. { malˈōd·ər }

malodorant *See* odorant. { malˈōd·ə·rənt }

malonamide nitrile *See* cyanoacetamide. { mə'län·ə·məd 'nī·trəl }

malonic acid [ORG CHEM] $CH_2(COOH)_2$ A white, crystalline dicarboxylic acid, melting at 132–134°C; used to manufacture pharmaceuticals. { mə'län·ik 'as·əd }

malonic ester *See* ethyl malonate. { mə'län·ik 'es·tər }

malonic mononitrile *See* cyanoacetic acid. { mə'län·ik 'män·ō'nī·trəl }

malonyl [ORG CHEM] $CH_2(COO)_2$ A bivalent functional group formed from malonic acid. { 'mal·ə,nil }

Malotte's metal [MET] A fusible alloy composed of 46% bismuth, 20% lead, and 34% tin; melts at 96–123°C. { mə'läts ˌmed·əl }

Malpighiaceae [BOT] A family of dicotyledonous plants in the order Polygalales distinguished by having three carpels, several fertile stamens, five petals that are commonly fringed or toothed, and indehiscent fruit. { mal,pig·ē'ās·ē,ē }

Malpighian corpuscle [ANAT] **1.** A lymph nodule of the spleen. **2.** *See* renal corpuscle. { mal'pig·ē·ən 'kȯr·pə·səl }

Malpighian layer [HISTOL] The germinative layer of the epidermis. { mal'pig·ē·ən ˌlā·ər }

Malpighian pyramid *See* renal pyramid. { mal'pig·ē·ən 'pir·ə·məd }

Malpighian tubule [INV ZOO] Any of the blind tubes that open into the posterior portion of the gut in most insects and certain other arthropods and excrete matter or secrete substances such as silk. { mal'pig·ē·ən 'tü,byül }

malposition [MED] Abnormal position of an organ or other body structure, or of the fetus. { ˌmal·pə'zish·ən }

malpractice [MED] Improper or injurious medical or surgical treatment, through carelessness, ignorance, or intent. { mal'prak·təs }

malpresentation [MED] Abnormal position of the child at birth, making normal delivery difficult or impossible. { ˌmal,prē·zən'tā·shən }

malt [FOOD ENG] A nutrient material made from grain, commonly barley, which has been soaked, allowed to germinate, and dried. { mȯlt }

Malta fever *See* brucellosis. { 'mȯl·tə 'fē·vər }

maltase [BIOCHEM] An enzyme that catalyzes the conversion of maltose to dextrose. { 'mȯl,tās }

malt beverage [FOOD ENG] Any of various fermented alcoholic beverages, including beer, ale, stout, and porter; barley malt is the principal ingredient. { 'mȯlt 'bev·rij }

Malter effect [SOLID STATE] A phenomenon in which a metal with a nonconducting surface film has a large coefficient of secondary electron emission; this is particularly notable in aluminum whose surface has been oxidized and then coated with cesium oxide. { 'mȯl·tər i,fekt }

Malthusianism [BIOL] The theory that population increases more rapidly than the food supply unless held in check by epidemics, wars, or similar phenomena. { mal'thü·zhə,niz·əm }

malting [FOOD ENG] Converting the starches of a distillery mash to fermentable sugar by the enzymes of the malt. { 'mȯlt·iŋ }

maltobiose *See* maltose. { 'mȯl·tō'bī,ōs }

maltol [ORG CHEM] $C_6H_6O_3$ Crystalline substance with a melting point of 161–162°C and a fragrant caramellike odor; used as a flavoring agent in bread and cakes. Also known as larixinic acid. { 'mȯl,tȯl }

maltose [BIOCHEM] $C_{12}H_{22}O_1$ A crystalline disaccharide that is a product of the enzymatic hydrolysis of starch, dextrin, and glycogen; does not appear to exist free in nature. Also known as maltobiose; malt sugar. { 'mȯl,tōs }

maltose phosphorylase [BIOCHEM] An enzyme which reacts maltose with inorganic phosphate to yield glucose and glucose-1-phosphate. { 'mȯl,tōs ,fäs'fȯr·ə,lās }

maltosuria [MED] Presence of maltose in the urine. { ,mȯl·tə'syùr·ē·ə }

malt sugar *See* maltose. { 'mȯlt 'shù̇g·ər }

malt whiskey [FOOD ENG] Whiskey produced in a pot still from malted barley. { 'mȯlt 'wis·kē }

malunion [MED] Faulty union of the pieces of a fractured bone. { mal'yün·yən }

Malus *See* Pyxis. { mä·ləs }

Malus cosine-squared law [OPTICS] The law that if a beam of plane polarized light passes through a Nicol prism, the intensity of light emerging from the prism is proportional to the square of the cosine between the plane of polarization of the incident light and the plane of polarization of the prism. { 'mä·ləs ˌkō sīn 'skwerd ,lȯ }

Malus' law of rays [OPTICS] The law that an orthotomic system of rays is still orthotomic after the rays have been reflected and refracted any number of times. { 'mä·ləs ˌlȯ əv 'rāz }

Malvaceae [BOT] A family of herbaceous dicotyledons in the order Malvales characterized by imbricate or contorted petals, mostly unilocular anthers, and minutely spiny, multiporate pollen. { mal'vās·ē,ē }

Malvales [BOT] An order of flowering plants in the subclass Dilleniidae having hypogynous flowers with valvate calyx, mostly separate petals, numerous centrifugal stamens, and a syncarpous pistil. { mal'vā·lēz }

malysite [MINERAL] $FeCl_3$ A halogen mineral deposited by sublimation; found most commonly at Mount Vesuvius, Italy. { 'mal·ə,sīt }

mamarron oil [MATER] A cream-colored fat high in lauric acid and similar to coconut oil in characteristics and odor; obtained from a species of *Attalea* palm. { mə'mar·ən ,ȯil }

mamelon [BIOL] Any dome-shaped protrusion or elevation. [GEOL] A small, rounded volcano which forms over a vent as a result of the slow extrusion of viscous, silicic lava. { 'mam·ə·lən }

mamm-, mammo- [ANAT] A combining form meaning breast. { 'mam·ō }

mamma [ANAT] A milk-secreting organ characterizing all mammals. { 'mam·ə }

mamma aberrans [MED] A supernumerary breast. { mam·ə ,ab·ə'ränz }

mammal [VERT ZOO] A member of Mammalia. { 'mam·əl }

Mammalia [VERT ZOO] A large class of warm-blooded vertebrates containing animals characterized by mammary glands, a body covering of hair, three ossicles in the middle ear, a muscular diaphragm separating the thoracic and abdominal cavities, red blood cells without nuclei, and embryonic development in the allantois and amnion. { mə'māl·yə }

MALLOPHAGA

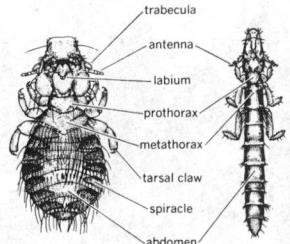

Morphology of two typical mallophagans.

MALTOSE

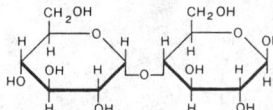

Formula for maltose.

MALVACEAE

The swamp rose mallow (*Hibiscus moscheutos*). (*Courtesy of A. W. Ambler, from National Audubon Society*)

mammary [ANAT] Of or pertaining to the mamma, or breast. { 'mam·ə·rē }

mammary gland [PHYSIO] A highly modified sebaceous gland that secretes milk; a unique anatomical feature of mammals. { 'mam·ə·rē ˌgland }

mammary lymphatic plexus [ANAT] A network of anastomosing lymphatic vessels in the walls of the ducts and between the lobules of the mamma; also functions to drain skin, areola, and nipple. { 'mam·ə·rē lim'fad·ik 'plek·səs }

mammary region [ANAT] The space on the anterior surface of the chest between a line drawn through the lower border of the third rib and one drawn through the upper border of the xiphoid cartilage. { 'mam·ə·rē ˌrē·jən }

mammary ridge [EMBRYO] An ectodermal thickening forming a longitudinal elevation on the chest between the limbs from which the mammary glands develop. { 'mam·ə·rē ˌrij }

mammary-stimulating hormone [BIOCHEM] **1.** Estrogen and progesterone considered together as the hormones which induce proliferation of the mammary ductile and acinous elements respectively. **2.** *See* prolactin. { 'mam·ə·rē ˌstim·yəˌlād·iŋ 'hȯr·mōn }

mammary tumor agent [VIROL] A milk-borne virus that produces mammary cancer in mice with the appropriate genotype. { 'mam·ə·rē 'tü·mər ˌāj·ənt }

mammectomy *See* mastectomy. { ma'mek·tə·mē }

mammillary [ANAT] **1.** Of or pertaining to the nipple. **2.** Breast- or nipple-shaped. [MINERAL] Of or pertaining to an aggregate of crystals in the form of a rounded mass. { 'ma·mə,ler·ē }

mammillary body [NEUROSCI] Either of two small, spherical masses of gray matter at the base of the brain in the space between the hypophysis and oculomotor nerve, which receive and relay olfactory impulses. { 'ma·mə,ler·ē 'bäd·ē }

mammillary line [ANAT] A vertical line passing through the center of the nipple. { 'ma·mə,ler·ē 'līn }

mammillary process [ANAT] One of the tubercles on the posterior part of the superior articular processes of the lumbar vertebrae. { 'ma·mə,ler·ē 'präs·əs }

mammillary structure *See* pillow structure. { 'ma·mə,ler·ē 'strək·chər }

mammillitis [MED] Inflammation of the nipple. { ˌmam·ə'līd·əs }

mammogen *See* prolactin. { ˌmam·ə'jen }

mammogenic hormone [BIOCHEM] **1.** Any hormone that stimulates or induces development of the mammary gland. **2.** *See* prolactin. { ˌmam·ə'jen·ik 'hȯr·mōn }

mammography [MED] Radiographic examination of the breast, performed with or without injection of the glandular ducts with a contrast medium. { ma'mäg·rə·fē }

mammoplasia [PHYSIO] Development of breast tissue. { ˌmam·ə'plā·zhə }

mammoplasty [MED] Plastic surgery performed to alter the shape of the breast. { 'mam·ə,plas·tē }

mammoth [PALEON] Any of various large Pleistocene elephants having long, upcurved tusks and a heavy coat of hair. { 'mam·əth }

mammotropin *See* prolactin. { ˌmam·ə'trä·pən }

Mammutinae [PALEON] A subfamily of extinct proboscidean mammals in the family Mastodontidae. { mə'myüt·ən,ē }

man [ARCH] **1.** *Homo sapiens.* A member of the human race. **2.** An adult human male. { man }

management control system [IND ENG] Any one of the various systems used by a contractor to plan, control the cost, and schedule the work required to undertake and complete a project. { 'man·ij·mənt ˌkən'trōl ˌsis·təm }

management engineering *See* industrial engineering. { 'man·ij·mənt ˌen·jə'nir·iŋ }

management game [IND ENG] A training exercise in which prospective decision makers act out managerial decision-making roles in a simulated environment. Also known as business game; operational game. { 'man·ij·mənt ˌgām }

management information system [COMMUN] A communication system in which data are recorded and processed to form the basis for decisions by top management of an organization. Abbreviated MIS. { 'man·ij·mənt ˌin·fər'mā·shən ˌsis·təm }

manandonite [MINERAL] $Li_4Al_{14}B_4Si_6O_{29}(OH)_{24}$ A white

mineral composed of basic borosilicate of lithium and aluminum. { mə'nan·də,nīt }

manasseite [MINERAL] $Mg_6Al_2(OH)_{16}(CO_3)\cdot4H_2O$ A hexagonal mineral composed of basic hydrous magnesium and aluminum carbonate; it is dimorphous with hydrotalcite. { mə'nas·ē,īt }

MANAV *See* Marine Integrated Navigation.

Manchester coding *See* phase encoding. { 'man·chə·stər ˌkōd·iŋ }

Manchester plate [ELEC] A storage battery consisting of a heavy alloy grid with circular openings into which are pressed pure lead buttons that are made from lead tape by crimping and rolling to develop a large surface area and are coated with lead peroxide, PbO_2. { 'man·chə·stər ˌplāt }

Mancini method *See* single radial immunodiffusion. { man'sē·nē ˌmeth·əd }

mandarin [BOT] A large and variable group of citrus fruits in the species *Citrus reticulata* and some of its hybrids; many varieties of the trees are compact with willowy twigs and small, narrow, pointed leaves; includes tangerines, King oranges, Temple oranges, and tangelos. { 'man·də·rən }

mandarin oil [MATER] Golden-yellow or olive-green essential oil with refreshing aroma, obtained from the peel of mandarin oranges; chief constituents are limonene and esters; used in medicine and as flavoring. Also known as tangerine oil. { 'man·də·rən ˌȯil }

mandatory layer [METEOROL] A layer of the atmosphere between two consecutive (or any two) specified mandatory levels. { 'man·də,tȯr·ē 'lā·ər }

mandatory level [METEOROL] One of several constant-pressure levels in the atmosphere for which a complete evaluation of data from upper-air observations is required. Also known as mandatory surface. { 'man·də,tȯr·ē 'lev·əl }

mandatory surface *See* mandatory level. { 'man·də,tȯr·ē 'sər·fəs }

Mandelbrot dimensionality *See* fractal dimensionality. { 'män·dəl,brōt di,men·shə'nal·əd·ē }

Mandelbrot set [MATH] The set of complex numbers, c, for which the sequence $s_0, s_1, \ldots$ is bounded, where $s_0 = 0$, and $s_{n+1} = s_n^2 + c$. { 'män·dəl,brōt ˌset }

mandelic acid [ORG CHEM] $C_6H_5CHOHCOOH$ A white, crystalline compound, melting at 117–119°C, darkening upon exposure to light; used in organic synthesis. { man'del·ik 'as·əd }

mandelic acid nitrile *See* mandelonitrile. { man'del·ik 'as·əd 'nī·trəl }

mandelonitrile [ORG CHEM] $C_6H_5CH(OH)CN$ A liquid used to prepare bitter almond water. Also known as mandelic acid nitrile. { man'del·ō'nī·trəl }

Mandelstam plane [PARTIC PHYS] A method of plotting energy versus scattering angle of three reactions, each having two particles both before and after scattering, which can be derived from each other by the crossing principle; the three reactions are on an equal footing, and poles in the scattering amplitude representing exchanged particles lie along straight lines. { 'mänd·əl,shtäm ˌplän }

Mandelstam representation [PARTIC PHYS] For a reaction in which there are two particles both before and after scattering: an expression, containing several integrals, for a function related to the scattering amplitude; the arguments of the function are the center-of-mass energy and scattering angle, extended to complex values; the function is conjectured to be analytic in these variables except for certain cuts and to have values along these cuts which give the scattering amplitude of the reaction, and of the two reactions derivable from it by the crossing principle. { 'mänd·əl,shtäm ˌrep·ri,zən'tā·shən }

mandible [ANAT] **1.** The bone of the lower jaw. **2.** The lower jaw. [INV ZOO] Any of various mouthparts in many invertebrates designed to hold or bite into food. { 'man·də·bəl }

mandibular arch [EMBRYO] The first visceral arch in vertebrates. { man'dib·yə·lər 'ärch }

mandibular cartilage [EMBRYO] The bar of cartilage supporting the mandibular arch. { man'dib·yə·lər 'kärt·lij }

mandibular fossa [ANAT] The depression in the temporal bone that articulates with the mandibular condyle. { man'dib·yə·lər 'fäs·ə }

mandibular gland *See* submandibular gland. { man'dib·yə·lər 'gland }

MAMMOGRAPHY

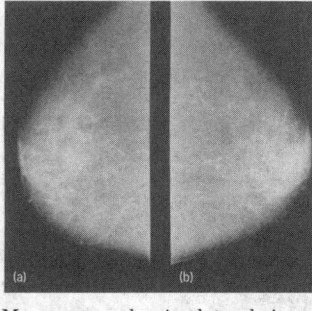

Mammogram showing lateral view of *(a)* left breast and *(b)* right breast. *(American College of Radiology)*

MANCHESTER PLATE

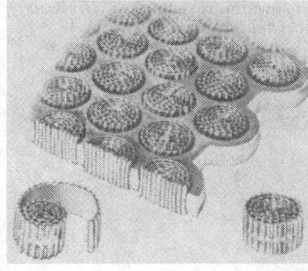

Section of Manchester plate with detail of lead button. *(ESB Inc.)*

mandibular nerve [NEUROSCI] A mixed nerve branch of the trigeminal nerve; innervates various structures of the lower jaw and face. { man'dib·yə·lər 'nərv }

Mandibulata [INV ZOO] A subphylum of Arthropoda; members possess a pair of mandibles which characterize the group. { man,dib·yə'läd·ə }

mandrel [ENG] The core around which continuous strands of impregnated reinforcement materials are wound to fabricate hollow objects made of composite materials. [MECH ENG] A shaft inserted through a hole in a component to support the work during machining. [MET] A metal bar serving as a core around which other metals are cast, forged, or extruded, forming a true central hole. { 'man·drəl }

mandrel forging See saddling. { 'man·drəl ,fórj·iŋ }

mandrel hanger [PETRO ENG] A device used to provide a liquid- or gas-tight seal (blowout preventer) between oil-well tubing and the tubing head. { 'man·drəl ,haŋ·ər }

mandrel press [MECH ENG] A press for driving mandrels into holes. { 'man·drəl ,pres }

mandrill [VERT ZOO] *Mandrillus sphinx.* An Old World cercopithecoid monkey found in west-central Africa and characterized by large red callosities near the ischium and by blue ridges on each side of the nose in males. { 'man·drəl }

maneb [ORG CHEM] Mn[SSCH(CH$_2$)NHCSS] A generic term for manganese ethylene-1,2-bisdithiocarbamate; irritating to eyes, nose, skin, and throat; used as a fungicide. { 'ma,neb }

maneuverability [NAV] The rate at which the orientation of a craft can be changed. { mə,nüv·rə'bil·əd·ē }

maneuvering board [NAV] A polar coordinate plotting sheet devised to facilitate solution of problems involving relative movement of naval vessels. { mə'nüv·riŋ ,bórd }

maneuvering craft [NAV] A craft whose movements are defined relative to a given craft called the reference craft; it may be a maneuvering ship, maneuvering aircraft, or maneuvering vehicle. { mə'nüv·riŋ ,kraft }

manganate [INORG CHEM] **1.** Salts that have manganese in the anion. **2.** In particular, a salt of manganic acid formed by fusion of manganese dioxide with an alkali. { 'maŋ·gə,nāt }

manganese [CHEM] A metallic element, symbol Mn, atomic weight 54.938, atomic number 25; a transition element whose properties fall between those of chromium and iron. [MET] A hard, brittle, grayish-white metal used chiefly in making steel. { 'maŋ·gə,nēs }

manganese acetate [ORG CHEM] Mn(C$_2$H$_3$O$_2$)$_2$·4H$_2$O A pale-red crystalline compound melting at 80°C; soluble in water and alcohol; used in textile dyeing, as a catalyst, and for leather tanning. { 'maŋ·gə,nēs 'as·ə,tāt }

manganese-aluminum [MET] A hardener alloy employed for making additions of manganese to aluminum alloys such as Duralumin; typical composition is 25% manganese, 75% aluminum. { 'maŋ·gə,nēs ə'lü·mə·nəm }

manganese binoxide See manganese dioxide. { 'maŋ·gə,nēs bi'näk,sīd }

manganese black See manganese dioxide. { 'maŋ·gə,nēs 'blak }

manganese borate [INORG CHEM] MnB$_4$O$_7$ Water-insoluble, reddish-white powder; used as a varnish and oil drier. { 'maŋ·gə,nēs 'bór,āt }

manganese-boron [MET] Manganese alloyed with boron; used as an ingredient for hardening and deoxidizing bronze. { 'maŋ·gə,nēs 'bór,än }

manganese brass [MET] A brass containing about 70% copper, 29% zinc, and 1.3% manganese. { 'maŋ·gə,nēs 'bras }

manganese bromide See manganous bromide. { 'maŋ·gə,nēs 'brō,mīd }

manganese bronze [MET] A type of brass or bronze containing about 59% copper, 39% zinc, 1.5% iron, 1% tin, and 0.1% manganese; another composition by the same name contains about 66% copper, 23% zinc, 3% iron, 4.5% aluminum, and 3.7% manganese. { 'maŋ·gə,nēs 'bränz }

manganese carbonate [INORG CHEM] MnCO$_3$ Rose-colored crystals found in nature as rhodocrosite; soluble in dilute acids, insoluble in water; used in medicine, in fertilizer, and as a paint pigment. { 'maŋ·gə,nēs 'kär·bə,nāt }

manganese citrate [ORG CHEM] Mn$_3$(C$_6$H$_5$O$_7$)$_2$ A white powder, water-insoluble in the presence of sodium citrate; used in medicine. { 'maŋ·gə,nēs 'sī,trāt }

manganese dioxide [INORG CHEM] MnO$_2$ A black, crystalline, water-insoluble compound, decomposing to manganese sesquioxide, Mn$_2$O$_3$, and oxygen when heated to 535°C; used as a depolarizer in certain dry-cell batteries, as a catalyst, and in dyeing of textiles. Also known as battery manganese; manganese binoxide; manganese black; manganese peroxide. { 'maŋ·gə,nēs dī'äk,sīd }

manganese epidote See piemontite. { maŋ·gə,nēs 'ep·ə,dōt }

manganese fluoride See manganous fluoride. { 'maŋ·gə,nēs 'flur,īd }

manganese gluconate [ORG CHEM] Mn(C$_6$H$_{11}$O$_7$)$_2$·2H$_2$O A pinkish powder, insoluble in benzene and alcohol, soluble in water; used in medicine, in vitamin tablets, and as a feed additive and dietary supplement. { 'maŋ·gə,nēs 'glü·kə,nāt }

manganese green See barium manganate. { 'maŋ·gə,nēs 'grēn }

manganese halide [INORG CHEM] Compound of manganese with a halide, such as chlorine, bromine, fluorine, or iodine. { maŋ·gə,nēs 'ha,līd }

manganese heptoxide [INORG CHEM] Mn$_2$O$_7$ A compound formed as an explosive dark-green oil by the action of concentrated sulfuric acid on permanganate compounds. { 'maŋ·gə,nēs hep'täk,sīd }

manganese hydroxide See manganous hydroxide. { 'maŋ·gə,nēs hī'dräk,sīd }

manganese hypophosphite [INORG CHEM] Mn(H$_2$PO$_2$)$_2$·H$_2$O Odorless, tasteless pink crystals which explode if heated with oxidants; used in medicine. { 'maŋ·gə,nēs ,hī·pō'fäs,fīt }

manganese iodide See manganous iodide. { 'maŋ·gə,nēs 'ī·ə,dīd }

manganese lactate [ORG CHEM] Mn(C$_3$H$_5$O$_3$)$_2$·3H$_2$O Pale-red crystals; insoluble in water and alcohol; used in medicine. { 'maŋ·gə,nēs 'lak,tāt }

manganese linoleate [ORG CHEM] Mn(C$_{18}$H$_{31}$O$_2$)$_2$ A dark-brown mass, soluble in linseed oil; used in pharmaceutical preparations and as a varnish and paint drier. { 'maŋ·gə,nēs lə'nō·lē,āt }

manganese monoxide See manganese oxide. { 'maŋ·gə,nēs mə'näk,sīd }

manganese naphthenate [ORG CHEM] Hard brown resinous mass, soluble in mineral spirits; melts at 135°C; contains 6% manganese in commercial masses; used as a paint and varnish drier. { 'maŋ·gə,nēs 'naf·thə,nāt }

manganese nodule [GEOL] Small, irregular black to brown concretions consisting chiefly of manganese salts and manganese oxide minerals; formed in oceans as a result of pelagic sedimentation or precipitation. { 'maŋ·gə,nēs 'naj·ül }

manganese oleate [ORG CHEM] Mn(C$_{18}$H$_{33}$O$_2$)$_2$ Granular brown mass, soluble in oleic acid and ether, insoluble in water; used in medicine and as a varnish drier. { 'maŋ·gə,nēs 'ō·lē,āt }

manganese oxalate [ORG CHEM] MnC$_2$O$_4$·2H$_2$O A white crystalline compound, soluble in dilute acids, only slightly soluble in water; used as a paint and varnish drier. { 'maŋ·gə,nēs 'äk·sə,lāt }

manganese oxide [INORG CHEM] MnO Green powder, soluble in acids, insoluble in water; melts at 1650°C; used in medicine, in textile printing, as a catalyst, in ceramics, and in dry batteries. Also known as manganese monoxide; manganous oxide. { 'maŋ·gə,nēs 'äk,sīd }

manganese peroxide See manganese dioxide. { 'maŋ·gə,nēs pə'räk,sīd }

manganese resinate [ORG CHEM] Mn(C$_{20}$H$_{25}$O$_2$)$_2$ Water-insoluble mass, flesh-colored or brownish black; used as a varnish and oil drier. { 'maŋ·gə,nēs 'rez·ən,āt }

manganese silicate See manganous silicate. { 'maŋ·gə,nēs 'sil·ə,kāt }

manganese-silicon [MET] An alloy that contains 73–78% silicon, 20–25% manganese, a maximum of 1.5% iron, and a maximum of 0.25% carbon; used for adding manganese and silicon to metals. { 'maŋ·gə,nēs 'sil·ə·kən }

manganese star [ASTRON] A star that has an anomalously high ratio of manganese to iron. { 'maŋ·gə,nēs ,stär }

manganese steel See Hadfield manganese steel. { 'maŋ·gə,nēs 'stēl }

manganese sulfate See manganous sulfate. { 'maŋ·gə,nēs 'səl,fāt }

MANDRILL

The mandrill (*Mandrillus sphinx*).

manganese sulfide *See* manganous sulfide. { 'maŋ·gə‚nēs 'səl‚fīd }

manganese-titanium [MET] An alloy usually composed of 38% manganese, 29% titanium, 8% aluminum, 3% silicon, 22% iron, and no carbon; used as a deoxidizer for high-grade steels and for nonferrous alloys. { 'maŋ·gə‚nēs tī'tān·ē·əm }

manganic fluoride [INORG CHEM] MnF_3 Poisonous red crystals, decomposed by heat and water; used as a fluorinating agent. { man'gan·ik 'flùr‚īd }

manganic hydroxide [INORG CHEM] $Mn(OH)_3$ A brown powder that rapidly loses water to form $MnO(OH)$; used in ceramics and as a fabric pigment. Also known as hydrated manganic hydroxide. { man'gan·ik hī'dräk‚sīd }

manganic oxide [INORG CHEM] Mn_2O_3 Hard black powder, insoluble in water, soluble in cold hydrochloric acid, hot nitric acid, and sulfuric acid; occurs in nature as manganite. Also known as manganese sesquioxide. { man'gan·ik 'äk‚sīd }

manganite [MINERAL] $MnO(OH)$ A brilliant steel-gray or black polymorphous mineral; crystallizes in the orthorhombic system. Also known as gray manganese ore. { 'maŋ·gə‚nīt }

manganolangbeinite [MINERAL] $K_2Mn_2(SO_4)_3$ A rose-red, isometric mineral composed of potassium manganese sulfate; occurs in lava on Vesuvius. { ‚maŋ·gə·nō'laŋ‚bī‚nīt }

manganosite [MINERAL] MnO An emerald-green isometric mineral occurring in small octahedrons that blacken on exposure; hardness is 5–6 on Mohs scale, and specific gravity is 5.18. { ‚maŋ·gə'nō‚sīt }

manganous bromide [INORG CHEM] $MnBr_2·4H_2O$ Water-soluble, deliquescent red crystals. Also known as manganese bromide. { 'maŋ·gə·nəs 'brō‚mīd }

manganous chloride [INORG CHEM] $MnCl_2·4H_2O$ Water-soluble, deliquescent rose-colored crystals melting at 88°C; used as a catalyst and in paints, dyeing, and pharmaceutical preparations. { 'maŋ·gə·nəs 'klór‚īd }

manganous fluoride [INORG CHEM] MnF_2 Reddish powder, insoluble in water, soluble in acid. Also known as manganese fluoride. { 'maŋ·gə·nəs 'flùr‚īd }

manganous hydroxide [INORG CHEM] $Mn(OH)_2$ Heat-decomposable white-pink crystals; insoluble in water and alkali, soluble in acids; occurs in nature as pyrochroite. Also known as manganese hydroxide. { 'maŋ·gə·nəs hī'dräk‚sīd }

manganous iodide [INORG CHEM] $MnI_2·4H_2O$ Water-soluble, deliquescent yellowish-brown crystals. Also known as manganese iodide. { 'maŋ·gə·nəs 'ī·ə‚dīd }

manganous silicate [INORG CHEM] $MnSiO_3$ Water-insoluble red crystals or yellowish-red powder; occurs in nature as rhodonite. Also known as manganese silicate. { 'maŋ·gə·nəs 'sil·ə‚kāt }

manganous sulfate [INORG CHEM] $MnSO_4·4H_2O$ Water-soluble, translucent, efflorescent rose-red prisms; melts at 30°C; used in medicine, textile printing, and ceramics, as a fungicide and fertilizer, and in paint manufacture. Also known as manganese sulfate. { 'maŋ·gə·nəs 'səl‚fāt }

manganous sulfide [INORG CHEM] MnS An almost water-insoluble powder that decomposes on heating; used as a pigment and as an additive in making steel. Also known as manganese sulfide. { 'maŋ·gə·nəs 'səl‚fīd }

manganous sulfite [INORG CHEM] $MnSO_3$ Grayish-black or brownish-red powder, soluble in sulfur dioxide, insoluble in water. { 'maŋ·gə·nəs 'səl‚fīt }

mange [VET MED] Infestation of the skin of mammals by certain mites (Sarcoptoidea) which burrow into the epidermis; characterized by multiple lesions accompanied by severe itching. { mānj }

Manger *See* Praesepe. { 'mān·jər }

Mangin mirror [OPTICS] A negative meniscus lens whose shallower surface is silvered to act as a spherical mirror while the other surface corrects for spherical aberration of the reflecting surface; used in searchlights and aircraft gunsights. { män‚zhan 'mir·ər }

mangle gearing [MECH ENG] Gearing for producing reciprocating motion; a pinion rotating in a single direction drives a rack with teeth at the ends and on both sides. { 'maŋ·gəl ‚gir·iŋ }

mango [BOT] *Mangifera indica.* A large evergreen tree of the sumac family (Anacardiaceae), native to southeastern Asia, but now cultivated in Africa, tropical America, Florida, and California for its edible fruit, a thick-skinned, yellowish-red, fleshy drupe. { 'maŋ·gō }

mangrove [BOT] A tropical tree or shrub of the genus *Rhizophora* characterized by an extensive, impenetrable system of prop roots which contribute to land building. [MATER] Liquid derived from the mangrove tree *Rhizophora mucronata* and used in the leather tanning industry. { 'maŋ‚grōv }

mangrove swamp [ECOL] A tropical or subtropical marine swamp distinguished by the abundance of low to tall trees, especially mangrove trees. { 'maŋ‚grōv ‚swämp }

Manhattan Project [ENG] A United States project lasting from August 1942 to August 1946, which developed the atomic energy program, with special reference to the atomic bomb. { man'hat·ən ‚prä‚jekt }

manhead *See* manhole. { 'man‚hed }

manhole [ENG] An opening to provide access to a tank or boiler, to underground passages, or in a deck or bulkhead of a ship; usually covered with a cast iron or steel plate. Also known as access hole; manhead. { 'man‚hōl }

manhole coaming [NAV ARCH] The raised framework or stiffening around the edge of a manhole in the deck of a ship to strengthen the opening and provide a support for the cover. { 'man‚hōl ‚kōm·iŋ }

man-hour [IND ENG] A unit of measure representing one person working for one hour. { 'man‚aùr }

mania [PSYCH] Excessive enthusiasm or excitement; a violent desire or passion; manifestation of a psychotic disorder. { 'mān·yə }

-mania [PSYCH] A combining form denoting obsession, abnormal preoccupation, or compulsion. { 'mān·yə }

manic-depressive illness *See* bipolar disorder. { ‚man·ik di‚pres·iv 'il·nəs }

Manidae [VERT ZOO] The pangolins, a family of mammals comprising the order Pholidota. { 'man·ə‚dē }

manifest constant [COMPUT SCI] A value that is assigned to a symbolic name at the beginning of a computer program and is not subject to change during execution. { 'man·ə‚fest 'kän·stənt }

manifest content [PSYCH] Any idea, feeling, or action considered to be the conscious expression of repressed motives or desires, particularly the remembered content of a dream or fantasy which conceals and distorts the unconscious meaning. { 'man·ə‚fest 'kän‚tent }

manifest covariance [RELAT] Property of an expression composed of Lorentz invariant numbers and operators, four-vectors, and tensors in such a way that its Lorentz covariance is immediately obvious. { 'man·ə‚fest kō'ver·ē·əns }

manifest stimulus [PSYCH] The obvious or external basis for anxiety, fear, or dread, such as the immediate or apparent cause for a phobic reaction. { 'man·ə‚fest 'stim·yə·ləs }

manifold [ENG] The branch pipe arrangement which connects the valve parts of a multicylinder engine to a single carburetor or to a muffler. [MATH] A topological space which is locally euclidean; there are four types: topological, piecewise linear, differentiable, and complex, depending on whether the local coordinate systems are obtained from continuous, piecewise linear, differentiable, or complex analytic functions of those in euclidean space; intuitively, a surface. { 'man·ə‚fōld }

manifolding [ENG] The gathering of multiple-line fluid inputs into a single intake chamber (intake manifold), or the division of a single fluid supply into several outlet streams (distribution manifold). { 'man·ə‚fōld·iŋ }

manifold of states [ATOM PHYS] A set of states sufficient to form a representation of an operator or a Lie group of operators. { 'man·ə‚fōld əv 'stāts }

manifold paper [MATER] An extremely thin paper used for making duplicate copies, such as onionskin paper. { 'man·ə‚fōld ‚pā·pər }

manifold pressure [MECH ENG] The pressure in the intake manifold of an internal combustion engine. { 'man·ə‚fōld ‚presh·ər }

manihot *See* cassava. { 'man·ə‚hät }

manikin [ENG] A correctly proportioned doll-like figure that is jointed and will assume any human position and hold it; useful in art to draw a human figure in action, or in medicine to show the relations of organs by means of movable parts. [MED] **1.** A model of a term fetus; used in the teaching of obstetrics. **2.** A model of an adult human used to teach first aid and basic nursing skills. { 'man·ə·kən }

Manila copal *See* Manila resin. { mə'nil·ə 'kō·pəl }

Manila hemp *See* abaca. { mə'nil·ə 'hemp }

Manila maguey *See* cantala. { mə'nil·ə mə'gā }

Manila paper [MATER] Yellowish paper or Bristol board; the term originally referred to paper manufactured from Manila hemp. { mə'nil·ə ,pā·pər }

Manila resin [MATER] A type of resin extracted from trees of the genus *Agathis* in the Philippines that is soluble in ethyl and methyl alcohol, insoluble in water; used in printing ink, varnishes, paints, and linoleum. Also known as Manila copal. { mə'nil·ə 'rez·ən }

manioc *See* cassava. { man·ē,äk }

manipulated variable [COMPUT SCI] Variable whose value is being altered to bring a change in some condition. { mə'nip·yə,lād·əd 'ver·ē·ə·bəl }

manipulation [MED] Skillful use of the hands in moving body parts, as reducing a dislocation, or changing the position of a fetus. [SCI TECH] Use of the hands in the performance of a task. { mə,nip·yə'lā·shən }

manipulative grasp *See* tripodal grasp. { mə'nip·yə·ləd·iv 'grasp }

manipulative skill [IND ENG] The ability of a worker to handle an object with the appropriate control and speed of movement required by a task. { mə'nip·yə·ləd·iv 'skil }

manipulators [CONT SYS] An armlike mechanism on a robotic system that consists of a series of segments, usually sliding or jointed which grasp and move objects with a number of degrees of freedom, under automatic control. [ENG] *See* remote manipulator. { mə'nip·yə,lād·ərz }

manjak [MATER] A variety of grahamite, manjak is the blackest of the asphalts; used for insulation and varnishes. { 'man,jak }

mankato stone [PETR] A variety of limestone containing more than 49% calcium carbonate, with about 4.5% alumina and some silica. { man'kād·ō ,stōn }

manketti oil [MATER] A light-yellow, viscous varnish oil, obtained from the nuts of the African tree *Ricinodendron rautonemii*. { man'ked·ē ,oil }

man-machine chart *See* human-machine chart. { 'man mə'shēn 'chärt }

man-machine system *See* human-machine system. { 'man mə'shēn 'sis·təm }

manna [MATER] The concrete, yellowish, saccharine exudation of the flowering ash (*Fraxinus ornus*); contains mannitol, sugar, mucilage, and resin and has been used as a mild laxative. { 'man·ə }

mannan [BIOCHEM] Any of a group of polysaccharides composed chiefly or entirely of D-mannose units. { ma,nan }

manna sugar *See* mannitol. { 'man·ə ,shüg·ər }

manned orbiting laboratory [AERO ENG] An earth-orbiting satellite which contains instrumentation and personnel for continuous measurement and surveillance of the earth, its atmosphere, and space. Abbreviated MOL. { 'mand 'òr·bəd·iŋ 'lab·rə,tòr·ē }

manned spacecraft [AERO ENG] A vehicle capable of sustaining a person above the terrestrial atmosphere. { 'mand 'spās,kraft }

Mannesmann mill [MET] A mill consisting of two rolls mounted with their axes slightly inclined. { 'män·əs,män ,mil }

Mannesmann process [MET] A process for making seamless tubing by forcing a billet between the rolls of a Mannesmann mill so as to pierce the center, and then forcing the metal over a mandrel to form the central bore. { 'män·əs,män ,präs·əs }

Mannich condensation reaction *See* Mannich reaction. { 'män·ik ,kän·dən'sā·shən rē,ak·shən }

Mannich reaction [ORG CHEM] Condensation of a primary or secondary amine or ammonia (usually as the hydrochloride) with formaldehyde and a compound containing at least one reactive hydrogen atom, for example, acetophenone. Also known as Mannich condensation reaction. { 'män·ik rē,ak·shən }

Manning equation [FL MECH] An equation used to compute the velocity of uniform flow in an open channel. { 'man·iŋ i,kwā·zhən }

mannite *See* mannitol. { 'ma,nīt }

mannitol [ORG CHEM] $C_6H_8(OH)_6$ A straight-chain alcohol with six hydroxyl groups; a white, water-soluble, crystalline powder; used in medicine and as a dietary supplement. Also known as manna sugar; mannite. { 'man·ə,tól }

mannitol hexanitrate [ORG CHEM] $C_6H_8(ONO_2)_6$ Explosive colorless crystals; soluble in alcohol, acetone, and ether, insoluble in water, melts at 112°C; used in explosives and medicine. { 'man·ə,tól ,hek·sə'nī,trāt }

mannose [BIOCHEM] $C_6H_{12}O_6$ A fermentable monosaccharide obtained from manna. { 'ma,nōs }

Mann-Whitney test [STAT] A procedure used in nonparametric statistics to determine whether the means of two populations are equal. { ,man 'wit·nē ,test }

manocryometer [THERMO] An instrument for measuring the change of a substance's melting point with change in pressure; the height of a mercury column in a U-shaped capillary supported by an equilibrium between liquid and solid in an adjoining bulb is measured, and the whole apparatus is in a thermostat. { ,man·ō,krī'äm·əd·ər }

manometer [ENG] A double-leg liquid-column gage used to measure the difference between two fluid pressures. { mə'näm·əd·ər }

manometric capsule [ACOUS] A device for studying air vibrations in a pipe or resonator, consisting of a rubber membrane which is stretched over a hole in the pipe, or over the end of a flange attached to such a hole, and apparatus for measuring vibrations of the membrane. { ,man·ə'me·trik 'kap·səl }

manometry [ENG] The use of manometers to measure gas and vapor pressures. { mə'näm·ə·trē }

manostat [ENG] Fluid-filled, upside-down manometer-type device used to control pressures within an enclosure, as for laboratory analytical distillation systems. { 'man·ə,stat }

mansard roof [ARCH] A roof with two slopes on all sides, the lower slope being steeper than the upper one. { 'man,särd 'rüf }

M-A-N scavenging system [MECH ENG] A system for removing used oil and waste gases from a cylinder of an internal combustion engine in which the exhaust ports are located above the intake ports on the same side of the cylinder, so that gases circulate in a loop, leaving a dead spot in the center of the loop. { 'em,ā,en 'skav·ənj·iŋ ,sis·təm }

mansfieldite [MINERAL] $Al(AsO_4)\cdot 2H_2O$ A white to pale-gray orthorhombic mineral composed of hydrous aluminum arsenate; it is isomorphous with scorodite. { 'manz,fēl,dīt }

mansonelliasis [MED] A parasitic infection of humans by the filarioid nematode *Mansonella ozzardi*. { ,man·sən·ə'lī·ə·səs }

Mantidae [INV ZOO] A family of predacious orthopteran insects characterized by a long, slender prothorax bearing a pair of large, grasping legs, and a freely moving head with large eyes. { 'man·tə,dē }

mantis [INV ZOO] The common name for insects composing the family Mantidae. { 'man·təs }

mantissa [COMPUT SCI] A fixed point number composed of the most significant digits of a given floating-point number. Also known as fixed-point part; floating-point coefficient. [MATH] The positive decimal part of a common logarithm. { man'tis·ə }

mantle [ANAT] Collectively, the convolutions, corpus callosum, and fornix of the brain. [BIOL] An enveloping layer, as the external body wall lining the shell of many invertebrates, or the external meristematic layers in a stem apex. [ENG] A lacelike hood or envelope (sack) of refractory material which, when positioned over a flame and heated to incandescence, gives light. [GEOL] The intermediate shell zone of the earth below the crust and above the core (to a depth of 2160 miles or 3480 kilometers). [MET] That part of the outer wall and casing of a blast furnace located above the hearth. [VERT ZOO] The back and wing plumage of a bird if distinguished from the rest of the plumage by a uniform color. { 'mant·əl }

mantle cavity [INV ZOO] The space between mantle and body proper in bivalve mollusks. { 'mant·əl ,kav·əd·ē }

mantled gneiss dome [GEOL] A dome in metamorphic terrains that has a remobilized core of gneiss surrounded by a concordant sheath of the basal part of the overlying metamorphic sequence. { 'mant·əld 'nīs ,dōm }

mantle lobe [INV ZOO] Either of the flaps on the dorsal

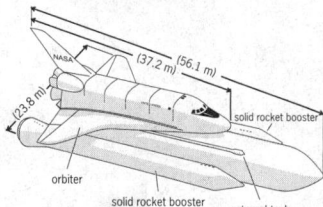

Space shuttle vehicle. 1 meter = 3.3 feet (*NASA*)

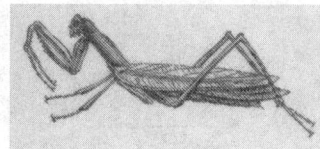

The praying mantis (*Stagomantis carolina*).

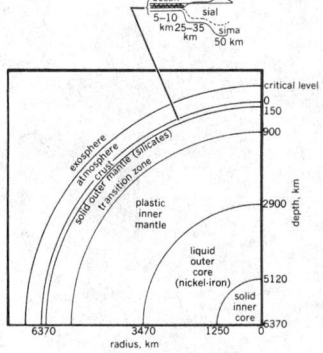

The principal layers of the earth.

and ventral sides of the mantle in bivalve mollusks. { 'mant· əl ˌlōb }

mantle rock *See* regolith. { 'mant·əl ˌräk }

mantlet [ORD] Protective shield or armor, as in front of a gun, or attached to the front of a tank. { 'mant·lət }

manto [GEOL] A sedimentary or igneous ore body occurring in flat-lying depositional layers. { 'man ˌtō }

Mantodea [INV ZOO] An order equivalent to the family Mantidae in some systems of classification. { man'tō·dē·ə }

Mantoux test [IMMUNOL] An intradermal test for tuberculin sensitivity, that is, for past or present infection with tubercle bacilli. { man'tü ˌtest }

manual casing hanger [PETRO ENG] A device in the bowl of the lowermost (or intermediate) casing head to suspend the next smaller casing string and to provide a seal between the suspended casing and the casing-head bowl. { 'man·yə·wəl 'kās·iŋ ˌhaŋ·ər }

manual central office [COMMUN] Central office of a manual telephone system. { 'man·yə·wəl ¦sen·trəl 'óf·əs }

manual control unit [CONT SYS] A portable, hand-held device that allows an operator to program and store instructions related to robot motions and positions. Also known as programming unit. { 'man·yə·wəl kən'trōl ˌyü·nət }

manual direction finder *See* manual radio direction finder. { 'man·yə·wəl di'rek·shən ˌfīn·dər }

manual element [IND ENG] A specific measurable subdivision of a work cycle or operation that is completed entirely by hand or with the use of tools. { 'man·yə·wəl 'el·ə·mənt }

manual exchange [COMMUN] Any exchange where calls are completed by an operator. { 'man·yə·wəl iks'chānj }

manual input [COMPUT SCI] The entry of data by hand into a device at the time of processing. { 'man·yə·wəl 'in,pùt }

manually controlled work *See* effort-controlled cycle. { 'man·yə·lē kən¦trōld 'wərk }

manual number generator *See* manual word generator. { 'man·yə·wəl 'nəm·bər ˌjen·ə,rād·ər }

manual operation [COMPUT SCI] Any processing operation performed by hand. { 'man·yə·wəl ˌäp·ə'rā·shən }

manual radio direction finder [NAV] A radio direction finder which requires manual rotation of an antenna or goniometer, in contrast with an automatic radio direction finder, which indicates automatically and continuously the great-circle direction of the transmitter to which it is tuned. Also known as manual direction finder. { 'man·yə·wəl 'rād·ē·ō di'rek·shən ˌfīn·dər }

manual rate-aided tracking [ELECTR] Radar circuit which tracks individual targets by computing the velocity from position fixes inserted manually into the circuitry. { 'man·yə·wəl 'rāt ¦ād·əd 'trak·iŋ }

manual ringing [COMMUN] Ringing which is started by the manual operation of a key and continues only while the key is held in operation. { 'man·yə·wəl 'riŋ·iŋ }

manual spinning [MET] A sheet-metal forming process that forms the material over a rotating mandrel with little or no change in the thickness of the original blank. Also known as conventional spinning. { 'man·yə·wəl 'spin·iŋ }

manual switchboard [ELEC] Telephone switchboard in which the connections are made manually, by plugs and jacks, or by keys. { 'man·yə·wəl 'swich,bórd }

manual switching [ELECTR] Method by which manual connection is made between two or more teletypewriter circuits. { 'man·yə·wəl 'swich·iŋ }

manual system [COMPUT SCI] A system involving data processing which does not make use of stored-program computing equipment; by this somewhat arbitrary definition, systems using other types of tabulating equipment, such as the card-programmed calculator, are considered to be manual. { 'man·yə·wəl 'sis·təm }

manual telephone set [ELECTR] Telephone set not equipped with a dial. { 'man·yə·wəl 'tel·ə,fōn ˌset }

manual telephone system [COMMUN] A telephone system in which connections between customers are ordinarily established manually by telephone operators in accordance with orders given verbally by calling parties. { 'man·yə·wəl 'tel·ə,fōn ˌsis·təm }

manual time *See* hand time. { 'man·yə·wəl 'tīm }

manual tracking [ENG] System of tracking a target in which all the power required is supplied manually through the tracking handwheels. { 'man·yə·wəl 'trak·iŋ }

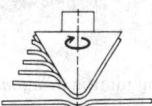

MANUAL SPINNING

Drawing of manual-spinning method of sheet-metal forming, showing the rotating mandrel.

manual welding [MET] A welding method in which the operator manually guides an electrode, clamped in a hand-held electrode holder. { 'man·yə·wəl 'weld·iŋ }

manual word generator [COMPUT SCI] A device into which an operator can enter a computer word by hand, either for direct insertion into memory or to be held until it is read during the execution of a program. Also known as manual number generator. { 'man·yə·wəl 'wərd ˌjen·ə,rād·ər }

manubrium [ANAT] **1.** The triangular cephalic portion of the sternum in humans and certain other mammals. **2.** The median anterior portion of the sternum in birds. **3.** The process of the malleus. [BOT] A cylindrical cell that projects inward from the middle of each shield composing the antheridium in stoneworts. [INV ZOO] The elevation bearing the mouth in hydrozoan polyps. { mə'nü·brē·əm }

manufactured gas [MATER] A gaseous fuel that is manufactured from soft coal or from various petroleum products; the gas mixture is composed of producer gas or carbureted water gas. { ˌman·ə'fak·chərd 'gas }

manufacturer's part number [IND ENG] Identification number of symbol assigned by the manufacturer to a part, subassembly, or assembly. { ˌman·ə'fak·chər·ərz 'pärt ˌnəm·bər }

manure [MATER] Animal excreta collected from stables and barnyards with or without litter; used to enrich the soil. { mə'núr }

manure salts [INORG CHEM] Potash salts that have a high proportion of chloride and 20–30% potash; used in fertilizers. { mə'núr ˌsólts }

manus [ANAT] The hand of a human or the forefoot of a quadruped. [INV ZOO] The proximal enlargement of the propodus of the chela of arthropods. { 'mä·nəs }

manus valga [MED] Clubhand with deviation of the ulna. { 'mä·nəs 'väl·gə }

manway *See* ladderway. { 'man,wā }

many-body force [PHYS] A force exerted on a particle, in the presence of two or more other particles, which differs from the vector sum of the forces which would be exerted on it if each of the other particles were present alone. { 'men·ē 'bäd· ē ˌfórs }

many-body problem [MECH] The problem of predicting the motions of three or more objects obeying Newton's laws of motion and attracting each other according to Newton's law of gravitation. Also known as *n*-body problem. { 'men·ē 'bäd· ē ˌpräb·ləm }

many-body theory [PHYS] A scheme for calculating physical quantities for systems with large numbers of particles, without finding details of each particle's motion, often at temperatures close to absolute zero. { 'men·ē 'bäd·ē ˌthē·ə· rē }

manyplies *See* omasum. { 'men·ē,plīz }

many-to-many correspondence [COMPUT SCI] A structure that establishes relationships between items in a data base, such that one unit of data can relate to many units, and many units can relate back to one unit and to other units as well. { 'men· ē tə 'men·ē ˌkär·ə'spän·dəns }

many-worlds interpretation *See* Everett-Wheeler interpretation. { ¦men·ē 'wərlz ˌin·tər·prə,tā·shən }

map [COMPUT SCI] **1.** An output produced by an assembler, compiler, linkage editor, or relocatable loader which indicates the (absolute or relocatable) locations of such elements as programs, subroutines, variables, or arrays. **2.** By extension, an index of the storage allocation on a magnetic disk or drum. [GRAPHICS] A representation, usually on a plane surface, of all or part of the surface of the earth, celestial sphere, or other area shows relative size and position, according to a given projection, of the physical features represented and such other information as may be applicable to the purpose intended. *See* mapping. { map }

MAP *See* microtubule-associated protein.

map chart [MAP] A representation of a land-sea area, using the characteristics of a map to represent the land area and the characteristics of a chart to represent the sea area, with such special characteristics as to make the map chart most useful in military operations, particularly amphibious operations. { map ˌchärt }

map design [MAP] The systematic process of arranging and assigning meaning to elements on a map for the purpose of communicating geographic knowledge in a pleasing format. { 'map di,zīn }

map distance [GEN] The frequency of meiotic recombination between linked genes, usually expressed in map units. { 'map ,dis·təns }

maple [BOT] Any of various broad-leaved, deciduous trees of the genus *Acer* in the order Sapindales characterized by simple, opposite, usually palmately lobed leaves and a fruit consisting of two long-winged samaras. [MATER] The hard, light-colored, close-grained wood, especially from sugar maple (*A. saccharum*). { 'mā·pəl }

maple sugar [FOOD ENG] Sugar obtained from boiled maple syrup. { 'mā·pəl 'shůg·ər }

maple syrup [FOOD ENG] Syrup made from the sap of certain maple trees, particularly the sugar maple. { 'mā·pəl 'sir·əp }

maple syrup urine disease [MED] A hereditary metabolic disorder caused by deficiency of branched-chain keto acid decarboxylase; characterized by the maple-syrup-like odor of urine. Also known as branched-chain ketoaciduria. { 'mā·pəl ˌsir·əp 'yůr·ən di,zēz }

map parallel See axis of homology. { 'map 'par·ə,lel }

mapping [GRAPHICS] Preparation of a map or engaging in a mapping operation. [MATH] **1.** Any function or multiple-valued relation. Also known as map. **2.** In topology, a continuous function. { 'map·iŋ }

mapping radar [NAV] Radar carried on an aircraft as an aid to navigation, which displays on a cathode-ray tube imagery of the ground in the vicinity of the aircraft. { 'map·iŋ ,rā,där }

map plotting [METEOROL] The process of transcribing weather information onto maps, diagrams, and so on; it usually refers specifically to decoding synoptic reports and entering those data in conventional station-model form on synoptic charts. Also known as map spotting. { 'map ,pläd·iŋ }

map projection See projection. { 'map prə,jek·shən }

map range [ORD] The range from an artillery piece to any point as scaled or computed from a map. { 'map ,rānj }

map reading [MAP] Interpretation of the symbols, lines, abbreviations, and terms appearing on maps. [NAV] The determination of position by identification of landmarks with their representations on a map or chart. { 'map ,rēd·iŋ }

map scale [MAP] The ratio between a distance on a map and the corresponding distance on the earth, often represented as 1:80,000 (natural scale) or 30 miles (48.27 kilometers) to an inch. { 'map ,skāl }

map spotting See map plotting. { 'map ,späd·iŋ }

map symbol [MAP] A character, letter, or similar graphic representation used on a map to indicate some object, characteristic, and so on. { 'map ,sim·bəl }

map unit [GEN] A measure of genetic distance corresponding to a recombination frequency of 1% or 1 centiMorgan (cM). { 'map ,yü·nət }

map vertical See geographic vertical. { 'map 'vərd·ə·kəl }

maquis [ECOL] A type of vegetation composed of shrubs, or scrub, usually not exceeding 10 feet (3 meters) in height, the majority having small, hard, leathery, often spiny or needlelike drought-resistant leaves and occurring in areas with a Mediterranean climate. { mä'kē }

maraging steel [MET] High-strength, low-carbon iron-nickel alloy in which a martensitic structure is formed on cooling; contains 7–6% nickel, 0–11% cobalt, 0–5% molybdenum, and small percentages of titanium, aluminum, and columbium; hardening is accomplished by heating the quenched alloy at 400–500°C. { mär'ā·jiŋ ,stēl }

Marangoni effect [CHEM ENG] The effect that a disturbance of the liquid-liquid interface (due to interfacial tension) has on mass transfer in a liquid-liquid extraction system. { ,mär·äŋ'gō·nē i,fekt }

Marantaceae [BOT] A family of monocotyledonous plants in the order Zingiberales characterized by one functional stamen with a single functional pollen sac, solitary ovules in each locule, and mostly arillate seeds. { ,mar·ən'tās·ē,ē }

marantic [MED] **1.** Of or pertaining to marasmus. **2.** Of or pertaining to slowed circulation. { mə'ran·tik }

marantic endocarditis [MED] Nonbacterial thrombic endocarditis, usually associated with neoplasm or other debilitating disease. { mə'ran·tik ,en·dō,kär'dīd·əs }

marasmus [MED] Chronic severe wasting of the tissues of the body, particularly in children, due to malnutrition. { mə'raz·məs }

Marattiaceae [BOT] A family of ferns coextensive with the order Marattiales. { mə,rad·ē'ās·ē,ē }

Marattiales [BOT] An ancient order of ferns having massive eusporangiate sporangia in sori on the lower side of the circinate leaves. { mə,rad·ē'ā·lēz }

marble [PETR] **1.** Metamorphic rock composed of recrystallized calcite or dolomite. **2.** Commercially, any limestone or dolomite taking polish. { 'mär·bəl }

marble bone disease See osteopetrosis. { 'mär·bəl ˌbōn di,zēz }

marble dust See marble flour. { 'mär·bəl ,dəst }

marble flour [MATER] Finely divided marble chips; used as a filler or abrasive in hand soaps and for casting. Also known as marble dust. { 'mär·bəl ,flaů·ər }

marble paper [GRAPHICS] A decorated paper, with a coloration that resembles marble, used as end leaves in blank books and often in printed books. { 'mär·bəl ,pā·pər }

marble shot [PETRO ENG] An explosive shot in open-hole well completions in which glass marbles are packed around the explosive in the wellbore; the marbles become projectiles that help break up the formation. { 'mär·bəl ,shät }

marbling [ENG] The use of antiquing techniques to achieve the appearance of marble in a paint film. { 'mär·bliŋ }

Marburg virus [VIROL] A large virus transmitted to humans by the grivet monkey (*Cercopithecus aethiops*). { 'mär,bůrg ,vī·rəs }

marcasite [MINERAL] FeS_2 A pale bronze-yellow to nearly white mineral, crystallizing in the orthorhombic system; hardness is 6–6.5 on Mohs scale, and specific gravity is 4.89. { 'mär·kə,sīt }

marcescent [BOT] Withering without falling off. { mär'ses·ənt }

Marcgraviaceae [BOT] A family of dicotyledonous shrubs or vines in the order Theales having exstipulate leaves with scanty or no endosperm, two integuments, and highly modified bracts. { märk,grā·vē'ās·ē,ē }

march [METEOROL] The variation of any meteorological element throughout a specific unit of time, such as a day, month, or year; as the daily march of temperature, the complete cycle of temperature during 24 hours. { märch }

Marchantiales [BOT] The thallose liverworts, an order of the class Marchantiopsida having a flat body composed of several distinct tissue layers, smooth-walled and tuberculate-walled rhizoids, and male and female sex organs borne on stalks on separate plants. { mär,shan·tē'ā·lēz }

Marchantiatae See Marchantiopsida. { mär,shan·tē,ā·tē }

Marchantiidae [BOT] A subclass of liverworts of the class Hepaticopsida, having gametophytes with ribbonlike or rosette-shaped thalli and generally reduced sporophytes. { ,mär·shən'tī·ə,dē }

Marchantiopsida [BOT] The liverworts, a class of lower green plants; the plant body is usually a thin, prostrate thallus with rhizoids on the lower surface. { mär,shan·tē'äp·sə·də }

March equinox See vernal equinox. { 'märch 'ē·kwə,näks }

Marconaflo slurry transport [MIN ENG] A system which recovers solids from ore or tailings piles and mixes them with water to produce a transportable slurry in pipelines. { 'mär·kän·ə,flō 'slər·ē ,trans,pórt }

Marconi antenna [ELECTROMAG] Antenna system of which the ground is an essential part, as distinguished from a Hertz antenna. { mär'kō·nē an'ten·ə }

Marcq-St.-Hilaire method [NAV] A mathematical process of deriving from an observation the information needed for establishing a line of position. { mark san·tē,ler 'meth·əd }

Marcy mill [MIN ENG] A ball mill with a vertical grate diaphragm placed near the discharge end; screens for sizing the material are located between the diaphragm and the end of the tube. { 'mär·sē ,mil }

mare [ASTRON] **1.** One of the large, dark, flat areas on the lunar surface. **2.** One of the less well-defined areas on Mars. [VERT ZOO] A mature female horse or other equine. { 'mär·ā, mer }

marebase See lunabase. { 'mär·ā,bās }

marekanite [GEOL] Rounded to subangular obsidian bodies that occur in masses of perlite. { ˌmär·ə,ka,nīt }

mare's tail See precipitation trajectory. { 'merz ,tāl }

Marfan's syndrome [MED] A hereditary connective-tissue

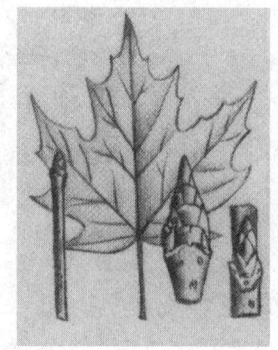

MAPLE

Twig, leaf, and terminal and axillary buds of the sugar maple (*Acer saccharum*).

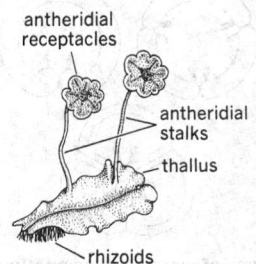

MARCHANTIALES

antheridial receptacles

antheridial stalks

thallus

rhizoids

Marchantia, a genus in Marchantiales; example of male gametophyte (antheridial plant). (*From H. J. Fuller and O. Tippo, College Botany, Holt, 1949*)

disorder transmitted as an autosomal dominant; manifested by skeletal and ocular changes and by congenital heart disease. { 'mär,fanz ,sin,drōm }

margaric acid *See* n-heptadecanoic acid. { mär'gär·ik 'as·əd }

margarine [FOOD ENG] An emulsified food fat product composed of processed vegetable oils or animal fats or both, cultured milk, salt, and emulsifiers. { 'mär·jə·rən *or* 'mär·gə·rən }

margarine oil [FOOD ENG] Edible vegetable oil used to make oleomargarine. { 'mär·jə·rən ,oil }

margarite [GEOL] A string of beadlike globulites; commonly found in glassy igneous rocks. [MINERAL] $Ca-Al_2(Al_2Si_2)-O_{10}(OH)_2$ A pink, reddish, or yellow, brittle mica mineral. { 'mär·gə,rīt }

Margaritiferidae [INV ZOO] A family of gastropod mollusks with nacreous shells that provide an important source of commercial pearls. { mär¦gär·ə·də'fer·ə,dē }

Margarodinae [INV ZOO] A subfamily of homopteran insects in the superfamily Coccoidea in which abdominal spiracles are present in all stages of development. { ,mär·gə'räd·ən,ē }

margarosanite [MINERAL] $PbCa_2(SiO_3)_3$ A colorless or snow-white triclinic mineral composed of lead calcium silicate, occurring in lamellar masses. { ,mär·gə'rōs·ən,īt }

margin [GEOGR] The boundary around a body of water. [GRAPHICS] The blank area at the vertical and horizontal edges of a printed page. [SCI TECH] An outside limit. { 'mär·jən }

marginal blight [PL PATH] A bacterial disease of lettuce caused by *Pseudomonas marginalis*, characterized by brownish marginal discoloration of the foliage. { 'mär·jən·əl 'blīt }

marginal checking [ELECTR] A preventive-maintenance procedure in which certain operating conditions, such as supply voltage or frequency, are varied about their normal values in order to detect and locate incipient defective units. { 'mär·jən·əl 'chek·iŋ }

marginal chlorosis [PL PATH] A virus disease characterized by yellowing or blanching of leaf margins; common disease of peanut plants. { 'mär·jən·əl klə'rō·səs }

marginal cost [IND ENG] The extra cost incurred for an extra unit of output. { 'mär·jən·əl 'kȯst }

marginal dimensionality [STAT MECH] The largest number of spatial dimensions for which nonlinear effects are important in calculating the behavior of a substance near a critical point. { 'mär·jən·əl di,men·shə'nal·əd·ē }

marginal escarpment [GEOL] A seaward slope of a marginal plateau with a gradient of 1:10 or more. { 'mär·jən·əl e'skärp·mənt }

marginal fissure [GEOL] A magma-filled fracture bordering an igneous intrusion. { 'mär·jən·əl 'fish·ər }

marginally outer trapped surface [RELAT] A spacelike, two-dimensional surface in a space-time such that outgoing null rays perpendicular to the surface are neither diverging nor converging. { ¦mär·jen·əl·ē ¦aud·ər ¦trapt 'sər·fəs }

marginal moraine *See* terminal moraine. { 'mär·jən·əl mə'rān }

marginal placentation [BOT] Arrangement of ovules near the margins of carpels. { 'mär·jən·əl ,plas·ən'tā·shən }

marginal plain *See* outwash plain. { 'mär·jən·əl 'plān }

marginal plateau [GEOL] A relatively flat shelf adjacent to a continent and similar topographically to, but deeper than, a continental shelf. { 'mär·jən·əl pla'tō }

marginal probability [STAT] Probability expressed by the two conditional probability distributions which arise from the joint distribution of two random variables. { 'mär·jən·əl ,präb·ə'bil·əd·ē }

marginal product [IND ENG] The extra unit of output obtained by one extra unit of some factor, all other factors being held constant. { 'mär·jən·əl 'präd·əkt }

marginal relay [ELEC] Relay with a small margin between its nonoperative current value (maximum current applicable without operation) and its operative value (minimum current that operates the relay). { 'mär·jən·əl 'rē,lā }

marginal revenue [IND ENG] The extra revenue achieved by selling an extra unit of output. { 'mär·jən·əl 'rev·ə,nü }

marginal salt pan [GEOL] A natural, coastal salt pan. { 'mär·jən·əl 'sȯlt ,pan }

marginal sea [GEOGR] A semiclosed sea adjacent to a continent and connected with the ocean at the water surface. { 'mär·jən·əl 'sē }

marginal sinus [ANAT] **1.** One of the small, bilateral sinuses of the dura mater which skirt the edge of the foramen magnum, usually uniting posteriorly to form the occipital sinus. **2.** *See* terminal sinus. [EMBRYO] An enlarged venous sinus incompletely encircling the margin of the placenta. { 'mär·jən·əl 'sī·nəs }

marginal test [ELECTR] A test of electronic equipment in which conditions are varied until failures occur or faults can be detected, allowing measurement of permissible operating margins. { 'mär·jən·əl 'test }

marginal thrust [GEOL] One of a series of faults bordering an igneous intrusion and crossing both the intrusion and the wall rock. Also known as marginal upthrust. { 'mär·jən·əl 'thrəst }

marginal ulcer [MED] A peptic ulcer of the jejunum on the efferent margin of a gastrojejunostomy. { 'mär·jən·əl 'əl·sər }

marginal upthrust *See* marginal thrust. { 'mär·jən·əl 'əp,thrəst }

marginate [BOT] Having a distinct margin or border. { 'mär·jə,nāt }

margin line [NAV ARCH] A line drawn parallel to the bulkhead deck of a ship at the side and 3 inches (76 millimeters) below the upper surface of that deck. { 'mär·jən ,līn }

margin-notched card *See* edge-notched card. { 'mär·jən ¦nächt ,kärd }

margin of safety [DES ENG] A design criterion, usually the ratio between the load that would cause failure of a member or structure and the load that is imposed upon it in service. { 'mär·jən əv 'sāf·tē }

margin plate [NAV ARCH] The plate forming the sides of the inner bottom tank of a ship. { 'mär·jən ,plāt }

margosa oil *See* neem oil. { mär'gōs·ə ,oil }

Margoulis number *See* Stanton number. { mär'gü·ləs ,nəm·bər }

Margules equation *See* Witte-Margules equation. { mär'gü·ləs i,kwā·zhən }

marialite [MINERAL] $3NaAlSi_3O_8·NaCl$ A scapolite mineral that is isomorphous with meronite. { mə'rē·ə,līt }

marial rocks *See* lunabase. { 'mär·ē·əl 'räks }

maricolous [ECOL] Living in the sea. { mə'rik·ə·ləs }

mariculture [AGR] The cultivation of marine organisms, plant and animal, for purposes of human consumption. { 'mar·ə,kəl·chər }

Marie's ataxia [MED] A hereditary ataxia combining features of cerebellar, posterior column, and pyramidal tract lesions, with onset after age 20, normal or exaggerated deep tendon reflexes, and frequently optic atrophy and oculomotor palsies, but no clubfeet or scoliosis. { mə'rēz ā'tak·sē·ə }

Marie's disease [MED] Rheumatic spondylitis involving the spine only, or invading the shoulders and hips. { mə'rēz di,zēz }

Marie-Strümpell disease *See* rheumatoid spondylitis. { mə'rē 'strim·pəl di,zēz }

marigram [OCEANOGR] A graphic record of the rising and falling movements of the tide expressed as a curve. { 'mar·ə,gram }

marigraph [ENG] A self-registering gage that records the heights of the tides. { 'mar·ə,graf }

marihuana *See* marijuana. { ,mar·ə'wän·ə }

marijuana [BOT] The Spanish name for the dried leaves and flowering tops of the hemp plant (*Cannabis sativa*), which have narcotic ingredients and are smoked in cigarettes. Also spelled marihuana. { ,mar·ə'wän·ə }

marina [CIV ENG] A harbor facility for small boats, yachts, and so on, where supplies, repairs, and various services are available. { mə'rē·nə }

marine [OCEANOGR] Pertaining to the sea. { mə'rēn }

marine abrasion [GEOL] Erosion of the ocean floor by sediment moved by ocean waves. Also known as wave erosion. { mə'rēn ə'brā·zhən }

marine alidade [NAV] An instrument used aboard ship in conjunction with, but not including, a pair of binoculars to indicate the relative bearing of a target or object; it is usually fastened to a solid mounting with the zero degree position parallel to the centerline of the ship. { mə'rēn 'al·ə,dād }

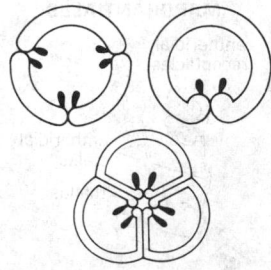

MARGINAL PLACENTATION

Types of marginal placentation; ovules in black.

MARIJUANA

Branch of the marijuana (*Cannabis sativa*) plant.

marine arch *See* sea arch. { mə'rēn 'ärch }

marine biocycle [ECOL] A major division of the biosphere composed of all biochores of the sea. { mə'rēn 'bī·ō,sī·kəl }

marine biology [BIOL] A branch of biology that deals with those living organisms which inhabit the sea. { mə'rēn bī'äl·ə·jē }

marine boiler [NAV ARCH] A steam boiler designed to meet requirements of ship operation in propulsion, running of auxiliary machinery, generation of electricity, and heating. { mə'rēn 'bȯil·ər }

marine bridge *See* sea arch. { mə'rēn 'brij }

marine cave *See* sea cave. { mə'rēn 'kāv }

marine chart *See* nautical chart. { mə'rēn 'chärt }

marine climate [CLIMATOL] A regional climate which is under the predominant influence of the sea, that is, a climate characterized by oceanity; the antithesis of a continental climate. Also known as maritime climate; oceanic climate { mə'rēn 'klī·mət }

marine-cut terrace [GEOL] A terrace or platform cut by wave erosion of marine origin. Also known as wave-cut terrace. { mə'rēn ¦kət 'ter·əs }

marine distiller [NAV ARCH] A machine that distills fresh water from sea water on a ship; evaporation must be carried out at temperatures below 60°C to prevent the sea water from forming a hard, insoluble scale, and the boiling liquid must therefore be kept below atmospheric pressure. Also known as marine evaporator. { mə'rēn di'stil·ər }

marine ecology [ECOL] An integrative science that studies the basic structural and functional relationships within and among living populations and their physical-chemical environments in marine ecosystems. { mə,rēn ē'käl·ə·jē }

marine engine [NAV ARCH] An engine that propels a waterborne vessel. { mə'rēn 'en·jən }

marine engineering [ENG] The design, construction, installation, operation, and maintenance of main power plants, as well as the associated auxiliary machinery and equipment, for the propulsion of ships. { mə'rēn ,en·jə'nir·iŋ }

marine evaporator *See* marine distiller. { mə'rēn i'vap·ə,rād·ər }

marine forecast [METEOROL] A forecast, for a specified oceanic or coastal area, of weather elements of particular interest to maritime transportation, including wind, visibility, the general state of the weather, and storm warnings. { mə'rēn 'fȯr,kast }

marine forensics [NAV ARCH] A branch of naval architecture and marine engineering that uses comprehensive analysis and reverse engineering to determine how and why a ship sank. { mə,rēn fə'ren·siks }

marine geology *See* geological oceanography. { mə'rēn jē'äl·ə·jē }

marine glue [MATER] An adhesive that is insoluble in water; usually made of rubber and shellac, sometimes with resins. { mə'rēn 'glü }

Marine Integrated Navigation [NAV] A navigation system, designed chiefly for large tankers, that routes and integrates information from remote sensors and presents anticollision and ship-handling information through visual display units cited on a central compact console. Abbreviated MANAV. { mə'rēn ¦int·ə,grād·əd ,nav·ə'gā·shən }

marine jet *See* waterjet. { mə'rēn 'jet }

marine light [NAV] A luminous or lighted aid to navigation intended primarily for marine navigation. { mə'rēn 'līt }

marine littoral faunal region [ECOL] A geographically determined division of that portion of the zoosphere composed of marine animals. { mə'rēn 'lit·ə·rəl 'fȯn·əl ,rē·jən }

marine marsh [ECOL] A flat, savannalike land expanse at the edge of a sea; usually covered by water during high tide. { mə'rēn 'märsh }

marine meteorology [METEOROL] That part of meteorology which deals mainly with the study of oceanic areas, including island and coastal regions; in particular, it serves the practical needs of surface and air navigation over the oceans. { mə'rēn ,mēd·ē·ə'räl·ə·jē }

marine microbiology [MICROBIO] The study of the microorganisms living in the sea. { mə'rēn ,mī·krō·bī'äl·ə·jē }

marine navigation [NAV] The process of directing the movements of watercraft from one point to another; the process, always present in some form when a vessel is under way and not drifting, varies with the type of craft, its mission, and its area of operation. { mə'rēn ,nav·ə'gā·shən }

marine propeller [NAV ARCH] A component of a ship-propulsion power plant which converts engine torque force into propulsive force or thrust, thus overcoming the hull resistance of a moving ship by creating a sternward accelerated column of water. { mə'rēn prə'pel·ər }

marine radio beacon [NAV] A radio-beacon station which produces service primarily for the guidance of ships. { mə'rēn 'rād·ē·ō ,bē·kən }

marine railway [CIV ENG] A type of dry dock consisting of a cradle of wood or steel with rollers on which the ship may be hauled out of the water along a fixed inclined track leading up the bank of a waterway. { mə'rēn 'rāl,wā }

marine rainbow [OPTICS] A rainbow seen in ocean spray. Also known as sea rainbow. { mə'rēn 'rān·bō }

marine salina [GEOGR] A body of salt water found along an arid coast and separated from the sea by a sand or gravel barrier. { mə'rēn səl'lēn·ə }

Marinesco-Sjögren-Garland syndrome [MED] A hereditary, congenital form of cerebral ataxia transmitted as an autosomal recessive; characterized by mental retardation, cataracts, minor skeletal anomalies, and hypertension. { mar·ə'nes·kō 'shō·grän 'gär·lənd ,sin,drōm }

marine sextant [NAV] A sextant designed primarily for ship navigation. { mə'rēn 'sek·stənt }

Marinesian *See* Bartonian. { mar·ə'nē·zhē·ən }

marine snow [OCEANOGR] A concentration of living and dead organic material and inorganic debris of the sea suspended at density boundaries such as the thermocline. { mə'rēn 'snō }

marine stack *See* stack. { mə'rēn 'stak }

marine swamp [GEOGR] An area of low, salty, or brackish water found along the shore and characterized by abundant grasses, mangrove trees, and similar vegetation. Also known as paralic swamp. { mə'rēn 'swämp }

marine terminal [CIV ENG] That part of a port or harbor with facilities for docking, cargo-handling, and storage. { mə'rēn 'term·ən·əl }

marine terrace [GEOL] A seacoast terrace formed by the merging of a wave-built terrace and a wave-cut platform. Also known as sea terrace; shore terrace. { mə'rēn 'ter·əs }

marine traffic [NAV] Traffic on the waterways. { mə'rēn 'traf·ik }

marine transgression *See* transgression. { mə'rēn tranz'gresh·ən }

marine weather observation [METEOROL] The weather as observed from a ship at sea, usually taken in accordance with procedures specified by the World Meteorological Organization. { mə'rēn 'weth·ər ,äb·zər,vā·shən }

Mariotte's law *See* Boyle's law. { ¦mar·ē¦äts 'lō }

marita [INV ZOO] An adult trematode. { mə'rīd·ə }

maritime air [METEOROL] A type of air whose characteristics are developed over an extensive water surface and which, therefore, has the basic maritime quality of high moisture content in at least its lower levels. { 'mar·ə,tīm 'er }

maritime climate *See* marine climate. { 'mar·ə,tīm 'klī·mət }

maritime DGPS [NAV] A type of local-area DGPS that makes use of existing radio beacons which broadcast error corrections over a limited range to marine and land users. { ¦mar·ə,tīm dē¦jē¦pē¦es }

maritime frequency bands [COMMUN] In the United States, a collection of radio frequencies allocated for communication between coast stations and ships or between ships. { 'mar·ə,tīm 'frē·kwən·sē ,banz }

maritime law [NAV] Law that concerns navigation and commerce on the oceans and other navigable bodies of water. { 'mar·ə,tīm 'lō }

maritime mobile satellite service [COMMUN] A mobile satellite service in which the mobile earth stations are located on board ships. Abbreviated MMSS. { ,mar·ə,tīm 'mō·bəl 'sad·əlīt ,sər·vəs }

maritime mobile service [COMMUN] A mobile service between coast stations and ship stations, or between ship stations, in which survival craft stations may also participate. { 'mar·ə,tīm 'mō·bəl 'sər·vəs }

maritime polar air [METEOROL] Polar air initially possessing similar properties to those of continental polar air, but in passing over warmer water it becomes unstable with a higher moisture content. { 'mar·ə,tīm 'pō·lər 'er }

MARINE RAILWAY

Bow-end view of a marine railway which has been pulled out of water.

maritime position [NAV] The location of a seaport or other point along a coast. { 'mar·ə,tīm pə'zish·ən }

maritime satellite *See* MARISAT. { 'mar·ə,tīm 'sad·əl,īt }

maritime tropical air [METEOROL] The principal type of tropical air, produced over the tropical and subtropical seas; it is very warm and humid, and is frequently carried poleward on the western flanks of the subtropical highs. { 'mar·ə,tīm 'träp·ə·kəl ¦er }

marjoram [BOT] Any of several perennial plants of the genera *Origanum* and *Majorana* in the mint family, Labiatae; the leaves are used as a food seasoning. { 'mär·jə·rəm }

marjoram oil [MATER] A colorless essential liquid whose chief components are terpenes, obtained from marjoram plants of the genus *Origanum*; used as a perfume in soaps, and in flavorings. { 'mär·jə·rəm ,óil }

mark [COMMUN] The closed-circuit condition in telegraphic communication, during which the signal actuates the printer; the opposite of space. [COMPUT SCI] A distinguishing feature used to signal some particular location or condition. [NAV] **1.** A charted conspicuous object, structure, or light serving as an indicator for guidance or warning to craft; a beacon; it may be a day-beacon or sea-mark depending upon its location, or a day-mark or lighted beacon depending upon its period of usefulness. **2.** Fathoms marked on a lead line. [ORD] A designation followed by a serial number, used to identify models of military equipment. [STAT] The name or value given to a class interval; frequently, the value of the midpoint or the integer nearest the midpoint. { märk }

mark detection [COMPUT SCI] That class of character recognition systems which employs coded documents, in the form of boxes or windows, in order to convey intended information by means of pencil or ink marks made in specific boxes. { 'märk di,tek·shən }

marker [IMMUNOL] Any antigen that serves to distinguish cell types. [ORD] A sign or signal for marking a location on land or water; frequently contains pyrotechnics. { 'märk·ər }

marker beacon [NAV] A low-power radio beacon transmitting a signal to designate a small area, as an aid to navigation. { 'märk·ər ,bē·kən }

marker bed [GEOL] **1.** A stratified unit with distinctive characteristics making it an easily recognized geologic horizon. **2.** A rock layer which accounts for a characteristic portion of a seismic refraction time-distance curve. **3.** *See* key bed. { 'märk·ər ,bed }

marker buoy [NAV] **1.** A temporary buoy used in surveying to mark a location of particular interest, such as a shoal or reef. **2.** *See* station buoy. { 'märk·ər ,bói }

marker gene [GEN] A gene with a known location on a chromosome and a clear-cut phenotype. { 'märk·ər ,jēn }

market analysis [IND ENG] The collection and evaluation of data concerned with the past, present, or future attributes of potential consumers for a product or service. { 'mar·kət ə,nal·ə·səs }

mark-hold [COMMUN] The transmission of a steady mark to indicate that there is no traffic over a telegraph channel; the upper marking frequency of a duplex channel (2225 hertz) is used to disable echo suppressors which may interfere with data communications. { ¦märk ¦hóld }

Mark-Houwink equation [PHYS CHEM] The relationship between intrinsic viscosity and molecular weight for homogeneous linear polymers. { 'märk 'haú,wink i,kwā·zhən }

marking and spacing intervals [COMMUN] Intervals of closed and open conditions in transmission circuits. { ¦märk·iŋ ən ¦spās·iŋ 'in·tər·vəlz }

marking bias [COMMUN] Bias distortion that lengthens the marking impulse. { 'märk·iŋ ,bī·əs }

marking current [ELEC] Magnitude and polarity of current in the line when the receiving mechanism is in the operating position. { 'märk·iŋ ,kə·rənt }

marking-end distortion [COMMUN] End distortion that lengthens the marking impulse. { 'märk·iŋ ¦end di,stór·shən }

marking pulse [ELEC] In a teletypewriter, the signal interval during which time the teletypewriter selector unit is operated. { 'märk·iŋ ,pəls }

marking wave [ELEC] In telegraphic communications, that portion of the emission during which the active portions of the code character are being transmitted. Also known as keying wave. { 'märk·iŋ ,wāv }

Markov-based model [COMPUT SCI] A model that represents a computer system by a Markov chain, which represents the set of all possible states of the system, with the possible transitions between these states. { 'mär,kóf ,bāst ,mäd·əl }

Markov chain [MATH] A Markov process whose state space is finite or countably infinite. { 'mar,kóf ,chän }

Markov inequality [STAT] If x is a random variable with probability P and expectation E, then, for any positive number a and positive integer n, $P(|x| \geq a) \leq E(|x|^n/a^n)$. { 'mar,kóf ,in·i'kwäl·əd·ē }

Markovnikoff's rule [ORG CHEM] In an addition reaction, the additive molecule RH adds as H and R, with the R going to the carbon atom with the lesser number of hydrogen atoms bonded to it. { mär'kóv·nə,kófs ,rül }

Markov process [MATH] A stochastic process which assumes that in a series of random events the probability of an occurrence of each event depends only on the immediately preceding outcome. { 'mär,kóf prä·səs }

mark reading [COMPUT SCI] In character recognition, that form of mark detection which employs a photoelectric device to locate and convey intended information; the information appears as special marks on sites (windows) within the document coding area. { 'märk ,rēd·iŋ }

mark sensing [COMPUT SCI] In character recognition, that form of mark detection which depends on the conductivity of graphite pencil marks to locate and convey intended information; the information appears as special marks on sites (windows) within the document coding area. { 'märk ,sens·iŋ }

mark-space multiplier [ELECTR] A multiplier used in analog computers in which one input controls the mark-to-space ratio of a square wave while the other input controls the amplitude of the wave, obtained by a smoothing operation, is proportional to the average value of the signal. Also known as time-division multiplier. { ¦märk ¦spās 'məl·tə,plī·ər }

mark-space ratio *See* mark-to-space ratio. { ¦märk ¦spās 'rā·shō }

mark-to-space ratio [ELECTR] The ratio of the duration of the positive-amplitude part of a square wave to that of the negative-amplitude part. Also known as mark-space ratio. { ¦märk ¦tə ¦spās 'rā·shō }

mark-to-space transition [COMMUN] The process of switching from a mark to a space. { ¦märk ¦tə ¦spās tran'zish·ən }

markup [COMPUT SCI] The process of adding information (tags) to an electronic document that are not part of the content but describe its structure or elements. { 'märk,əp }

markup language [COMPUT SCI] A set of rules and procedures for markup. { 'märk,əp ,laŋ·gwij }

marl [GEOL] A deposit of crumbling earthy material composed principally of clay with magnesium and calcium carbonate; used as a fertilizer for lime-deficient soils. Also known as malm. [TEXT] Two yarns of different colors or kinds twisted around each other. { märl }

marline [NAV ARCH] A tarred two-stranded, left-handed hemp about 1/8 inch (3 millimeters) in diameter; used for neat seizings. { 'mär·lən }

marline spike [NAV ARCH] A tapered metal tool used to separate the strands of rope in splicing, and as a lever in marling and seizing. { 'mär·lən ,spīk }

marlite *See* marlstone. { 'mär,līt }

marlstone [PETR] **1.** A consolidated rock that has about the same composition as marl; considered to be an earthy or impure argillaceous limestone. Also known as marlite. **2.** A hard ferruginous rock of the Middle Lias in England. { 'märl,stōn }

marly [GEOL] Pertaining to, containing, or resembling marl. { 'mär·lē }

marmatite [MINERAL] A dark-brown to black mineral composed of iron-bearing sphalerite. Also known as christophite. { 'mär·mə,tīt }

marmolite [MINERAL] A pale-green serpentine mineral, occurring in thin laminations; a variety of chrysotile. { 'mär·mə,līt }

marmon clampband [DES ENG] A metal band that wraps around the circumference of a special cylindrical joint between two structures, holding the structures together. { 'mär·mən 'klamp,ban }

Marmor [GEOL] A North American stage of Middle Ordovician geologic time, forming the lower subdivision of Chazyan, above Whiterock and below Ashby. { 'mär͵mȯr }

marmoset [VERT ZOO] Any of 10 species of South American primates belonging to the family Callithricidae; individuals are primitive in that they have claws rather than nails and a nonprehensile tail. { 'mär·mə͵set }

marmot [VERT ZOO] Any of several species of stout-bodied, short-legged burrowing rodents of the genus *Marmota* in the squirrel family Sciuridae. { 'mär·mət }

marocain crepe [TEXT] A heavy crepe produced from hard-twisted wool, silk, or manufactured fibers with coarse filling giving a cross-ribbed effect. { ͵mar·ə͵kān 'krāp }

marquenching *See* martempering. { 'mär͵kwench·iŋ }

marriage theorem [MATH] The proposition that a family of *n* subsets of a set *S* with *n* elements is a system of distinct representatives for *S* if any *k* of the subsets, $k = 1, 2, \ldots, n$, together contain at least *k* distinct elements. Also known as Hall's theorem. { 'mar·ij ͵thir·əm }

marrite [MINERAL] $PbAgAsS_3$ A monoclinic mineral, occurring as small crystals in Valais, Switzerland. { 'mä͵rīt }

marrubium [BOT] *Marrubium vulgari.* An aromatic plant of the mint family, Labiatae; leaves have a bitter taste and are used as a tonic and anthelmintic. Also known as hoarhound; horehound. { mə'rü·bē·əm }

Mars [ASTRON] The planet fourth in distance from the sun; it is visible to the naked eye as a bright red star, except for short periods when it is near its conjunction with the sun; its diameter is about 4150 miles (6700 kilometers). { märz }

Marsden chart [METEOROL] A system for showing the distribution of meteorological data on a chart, especially over the oceans; using a Mercator map projection, the world between 80°N and 70°S latitudes is divided into Marsden "squares," each of 10° latitude by 10° longitude and systematically numbered to indicate position; each square may be divided into quarter squares, or into 100 one-degree subsquares numbered from 00 to 99 to give the position to the nearest degree. { 'märz·dən ͵chärt }

Marseilles fever *See* fièvre boutonneuse. { mär'sā ͵fē·vər }

Marseilles soap [MATER] A castile soap to which is added olive oil and soda. { mär'sā ͵sōp }

marsh [ECOL] A transitional land-water area, covered at least part of the time by estuarine or coastal waters, and characterized by aquatic and grasslike vegetation, especially without peatlike accumulation. { märsh }

Marsh-Berzelius test *See* Marsh test. { 'märsh ber'zā·lē·əs ͵test }

marsh gas [GEOCHEM] Combustible gas, consisting chiefly of methane, produced as a result of decay of vegetation in stagnant water. { 'märsh ͵gas }

marshite [MINERAL] CuI A reddish, oil-brown isometric mineral composed of cuprous iodide and occurring as crystals; hardness is 2.5 on Mohs scale, and specific gravity is 5.6. { 'mär͵shīt }

marsh ore *See* bog iron ore. { 'märsh ͵ȯr }

Marsh test [ANALY CHEM] A test for the presence of arsenic in a compound; the substance to be tested is mixed with granular zinc, and dilute hydrochloric acid is added to the mixture; gaseous arsine forms, which decomposes to a black deposit of arsenic, when the gas is passed through a heated glass tube. Also known as Marsh-Berzelius test. { 'märsh ͵test }

Marsileales [BOT] A small monofamilial order of heterosporous, leptosporangiate ferns (Polypodiophyta); leaves arise on long stalks from the rhizome, and sporangia are enclosed in modified folded leaves or leaf segments called sporocarps. { mär͵sil·ē'ā·lēz }

Marsipobranchii [VERT ZOO] An equivalent name for Cyclostomata. { ͵mär·sə·pō'braŋ·kē͵ī }

Mars pigments [INORG CHEM] A group of five pigments produced when milk of lime is added to a ferrous sulfate solution, and the precipitate is calcined; color is controlled by calcination temperature to give yellow, orange, brown, red, or violet. { 'märz ͵pig·məns }

Mars probe [AERO ENG] A United States uncrewed spacecraft intended to be sent to the vicinity of the planet Mars, such as in the Mariner or Viking programs. { 'märz ͵prōb }

marsupial [VERT ZOO] **1.** A member of the Marsupialia. **2.** Having a marsupium. **3.** Of, pertaining to, or constituting a marsupium. { mär'sü·pē·əl }

marsupial anteater [VERT ZOO] *Myrmecobius fasciatus.* An anteater belonging to the Marsupialia. Also known as banded anteater. { mär'sü·pē·əl 'ant͵ēd·ər }

marsupial frog [VERT ZOO] Any of various South American tree frogs which carry the eggs in a pouch on the back. { mär'sü·pē·əl 'fräg }

Marsupialia [VERT ZOO] The single order of the mammalian infraclass Metatheria, characterized by the presence of a marsupium in the female. { mär͵sü·pē'ā·lē·ə }

marsupialization [MED] Surgical evacuation of pancreatic, hydatid, and other cysts when complete removal and closure are not possible. { mär͵sü·pē·ə·lə'zā·shən }

marsupial mole [VERT ZOO] *Notoryctes typhlops.* A marsupial of Australia that resembles the euterian mole. { mär͵sü·pē·əl 'mōl }

Marsupicarnivora [VERT ZOO] An order proposed to include the polydactylous and polyprotodont carnivorous superfamilies of Marsupialia. { mär͵sü·pə·kär'niv·ə·rə }

marsupium [VERT ZOO] A fold of skin that forms a pouch enclosing the mammary glands on the abdomen of most marsupials. { mär'sü·pē·əm }

martempering [MET] Quenching austenitized steel to a temperature just above, or in the upper part of, the martensite range, holding it at this point until the temperature is equalized throughout, and then cooling in air to room temperature. Also known as marquenching. { 'mär͵tem·pə·riŋ }

marten [VERT ZOO] Any of seven species of carnivores of the genus *Martes* in the family Mustelidae which resemble the weasel but are larger and of a semiarboreal habit. { 'märt·ən }

martensite [MET] A metastable transitional structure formed by a shear process during a phase transformation, characterized by an acicular or needlelike pattern; in carbon steel it is a hard, supersaturated solid solution of carbon in a body-centered tetragonal lattice of iron. { 'mär͵ten͵zīt }

martensite range [MET] The temperature interval between the temperature (M_s) at which formation of martensite initiates and the temperature (M_f) at which the formation is complete. { 'mär͵ten͵zīt ͵rānj }

martensitic stainless steel [MET] A hard, quenched magnetic martensitic steel containing principally 11–18% chromium and 0.1–1.2% carbon. { ͵mär͵ten'zid·ik 'stān·ləs ͵stēl }

martensitic steel [MET] Quenched carbon steel composed chiefly of martensite. { ͵mär͵ten'zid·ik 'stēl }

martensitic structure [MET] Of, pertaining to, or having the structure of martensite, that is, an interstitial, supersaturated solid solution of carbon in iron having a body-centered tetragonal lattice; the microstructure is characterized by an acicular or needlelike pattern. { ͵mär͵ten'zid·ik 'strək·chər }

martensitic transformation [MET] A phase transformation which occurs in some metals, resulting in the formation of martensite. Also known as shear transformation. { ͵mär͵ten'zid·ik ͵tranz·fər'mā·shən }

Martens wedge [OPTICS] A wedge-shaped piece of quartz used to rotate the plane of polarization of linearly polarized light. { 'märt·ənz ͵wej }

martingale [STAT] A sequence of random variables $x_1, x_2, \ldots,$ where the conditional expected value of x_{n+1} given $x_1, x_2, \ldots, x_n$, equals x_n. { 'märt·ən͵gāl }

Martin's cement [MATER] A gypsum cement made with potassium carbonate instead of alum. { 'märt·ənz si͵ment }

martite [MINERAL] Hematite occurring in iron-black octahedral crystals pseudomorphous after magnetite. { 'mär͵tīt }

martonite [MATER] A poison gas composed of 20% chloroacetone and 80% bromoacetone; acts as a powerful lacrimator. { 'märt·ən͵īt }

Marvin sunshine recorder [ENG] A sunshine recorder in which the time scale is supplied by a chronograph, and consisting of two bulbs (one of which is blackened) that communicate through a glass tube of small diameter, which is partially filled with mercury and contains two electrical contacts; when the instrument is exposed to sunshine, the air in the blackened bulb is warmed more than that in the clear bulb; the warmed air expands and forces the mercury through the connecting tube to a point where the electrical contacts are shorted by the mercury; this completes the electrical circuit to the pen on the chronograph. { 'mär·vən 'sən͵shīn ri͵kȯrd·ər }

Marx circuit [ELEC] An electric circuit used in an impulse generator in which capacitors are charged in parallel through charging resistors, and then connected in series and discharged

MARMOSET

The common marmoset.

MARMOT

Though short-legged and heavy, the marmot runs, jumps, and climbs with speed and agility.

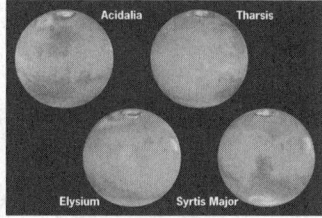

MARS

Acidalia　Tharsis

Elysium　Syrtis Major

Four images showing Mars in 90°-longitude steps through one complete rotation. The Hubble Space Telescope obtained these views in 1999. Note the bright and dark albedo markings and bright polar caps. (*NASA/Space Telescope Science Institute*)

MARTEN

Though scarce, the American marten is important as predator on mice and other injurious rodents.

through the test piece by the simultaneous sparkover of spark gaps. { 'märks ˌsər·kət }

Marx effect [SOLID STATE] The effect wherein the energy of photoelectrons emitted from an illuminated surface is decreased when the surface is simultaneously illuminated by light of lower frequency than that causing the emission. { 'märks iˌfekt }

mA·s See milliampere-second.

mascagnite [MINERAL] $(NH_4)_2SO_4$ A yellowish-gray mineral found in guano, near burning coal beds, or as lava incrustation; specific gravity is 1.77; hardness is 2–2.5 on Mohs scale. { ma'skanˌyīt }

mascaret See bore. { ¦mas·kə¦ret }

Mascheroni's constant See Euler's constant. { ˌmäsk·ə'rōˌnēz 'kän·stənt }

mascon [GEOL] A large, high-density mass concentration below a ringed mare on the surface of the moon. { 'masˌkän }

masculine [BIOL] Having an appearance or qualities distinctive for a male. { 'mas·kyə·lən }

masculine pelvis [ANAT] A female pelvis similar to the normal male pelvis in having a deeper cavity and more conical shape. Also known as android pelvis. { 'mas·kyə·lən 'pel·vəs }

masculine protest [PSYCH] The struggle to dominate, exhibited primarily by women but to some extent also by men, with the desire to escape identification with the feminine role. { 'mas·kyə·lən 'prōˌtest }

masculinize [PHYSIO] To cause a female or a sexually immature animal to take on male secondary sex characteristics. { 'mas·kyə·ləˌnīz }

masculinoma [MED] Adrenocorticoid adenoma of the ovary. { ˌmas·kyə·lə'nō·mə }

maser [PHYS] A device for coherent amplification or generation of electromagnetic waves in which an ensemble of atoms or molecules, raised to an unstable energy state, is stimulated by an electromagnetic wave to radiate excess energy at the same frequency and phase as the stimulating wave. Derived from microwave amplification by stimulated emission of radiation. Also known as paramagnetic amplifier. { 'mā·zər }

maser amplifier [ELECTR] A maser which is used to increase the power produced by another maser. { 'mā·zər 'am·pləˌfīər }

MA service See multiple-access service. { ¦em'ā ˌsər·vəs }

mash [FOOD ENG] **1.** Mixture of grain and other ingredients fermented to produce whiskey. **2.** Malted barley or other grain mixed with water to prepare wort for brewing operations. { mash }

mash seam weld [MET] A seam weld at a lap joint in which the overall lap thickness is reduced plastically to the approximate thickness of one of the lapped parts. { 'mash 'sēm ˌweld }

mask [COMPUT SCI] A pattern of characters used to control the retention or elimination of portions of another pattern of characters. Also known as extractor. [DES ENG] A frame used in front of a television picture tube to conceal the rounded edges of the screen. [ELECTR] A thin sheet of metal or other material containing an open pattern, used to shield selected portions of a semiconductor or other surface during a deposition process. [ENG] A protective covering for the face or head in the form of a wire screen, a metal shield, or a respirator. [GRAPHICS] **1.** In color separation photography, an intermediate negative or positive that is used to correct color. **2.** In offset lithography, opaque material that protectively covers open or selected areas of a printing plate during the exposure process. [MET] A protective device in thermal spraying against blasting or coating effects which are reflected from the substrate surface. { mask }

maskable interrupt [COMPUT SCI] An interrupt that can be allowed to occur or prevented from occurring by software. { ¦mas·kə·bəl 'int·əˌrəpt }

masked messenger ribonucleic acid See maternal messenger ribonucleic acid. { ¦maskd 'mes·ən·jer ¦rī·bo·nü'klē·ik 'as·əd }

mask face [MED] An expressionless face seen in certain degenerative and inflammatory diseases of the basal ganglia and the extrapyramidal system; voluntary movements are near normal while involuntary movements are infrequent. { 'mask ˌfās }

masking [ACOUS] The amount by which the threshold of audibility of a sound is raised by the presence of another sound;

the unit customarily used is the decibel. Also known as audio masking; aural masking. [COMPUT SCI] **1.** Replacing specific characters in one register by corresponding characters in another register. **2.** Extracting certain characters from a string of characters. [ELECTR] **1.** Using a covering or coating on a semiconductor surface to provide a masked area for selective deposition or etching. **2.** A programmed procedure for eliminating radar coverage in areas where such transmissions may be of use to the enemy for navigation purposes, by weakening the beam in appropriate directions or by use of additional transmitters on the same frequency at suitable sites to interfere with homing; also used to suppress the beam in areas where it would interfere with television reception. [ENG] Preventing entrance of a tracer gas into a vessel by covering the leaks. { 'mask·iŋ }

masking agent See masking reagent. { 'mask·iŋ ˌā·jənt }

masking reagent [ANALY CHEM] A substance that decreases the concentration of a free metal ion or ligand by conversion into an essentially unreactive form, thus preventing undesirable chemical reactions that would interfere with the determination. Also known as masking agent. { 'mask·iŋ rēˌā·jənt }

mask matching [COMPUT SCI] In character recognition, a method employed in character property detection in which a correlation or match is attempted between a specimen character and each of a set of masks representing the characters to be recognized. { 'mask ˌmach·iŋ }

mask register [COMPUT SCI] Filter which determines the parts of a word which are to be tested. { 'mask ˌrej·ə·stər }

mask word [COMPUT SCI] A word modifier used in a logical AND operation. { 'mask ˌwərd }

masochism [PSYCH] Pleasure derived from experiencing physical or psychological pain. { 'mas·əˌkiz·əm }

masonry [CIV ENG] A construction of stone or similar materials such as concrete or brick. { 'mās·ən·rē }

masonry cement [MATER] A blended cement, made by combining either natural or portland cements with fattening materials such as hydrated lime and, sometimes, with air-entraining mixtures; used in the mortar of brick and block masonry. { 'mās·ən·rē siˌment }

masonry dam [CIV ENG] A dam constructed of stone or concrete blocks set in mortar. { 'mās·ən·rē ˌdam }

masonry drill [DES ENG] A drill tipped with cemented carbide for drilling in concrete or masonry. { 'mās·ən·rē ˌdril }

masonry nail [DES ENG] Spiral-fluted nail designed to be driven into mortar joints in masonry. { 'mās·ən·rē ˌnāl }

mason's hydrated lime [MATER] Any hydrated lime suitable for use in mortars, base-coat plasters, and concrete. { 'mās·ənz ¦hī·drād·əd 'līm }

Mason's theorem [CONT SYS] A formula for the overall transmittance of a signal flow graph in terms of transmittances of various paths in the graph. { 'mās·ənz ˌthir·əm }

mass [MECH] A quantitative measure of a body's resistance to being accelerated; equal to the inverse of the ratio of the body's acceleration to the acceleration of a standard mass under otherwise identical conditions. { mas }

mass absorption coefficient [PHYS] The linear absorption coefficient divided by the density of the medium. { 'mas əb'sórp·shən ˌkō·iˌfish·ənt }

mass absorption law [NUCLEO] The law that the absorption of electrons with speeds greater than one-fifth that of light depends only on the mass of absorbing matter in the electron's path and not on its chemical composition. Also known as Lenard's mass absorption law. { 'mas əb'sórp·shən ˌló }

mass action law [PHYS CHEM] The law that the rate of a chemical reaction for a uniform system at constant temperature is proportional to the concentrations of the substances reacting. Also known as Guldberg and Waage law. { 'mas ¦ak·shən ˌló }

massage [COMPUT SCI] To process data, primarily to convert it into a more useful form or into a form that will simplify processing. [MED] The act of rubbing, kneading, or stroking the superficial parts of the body with the hand or with an instrument, for therapeutic purposes. { mə'säzh }

mass-analyzed ion kinetic energy spectrometry [SPECT] A type of ion kinetic energy spectrometry in which the ionic products undergo mass analysis followed by energy analysis. Abbreviated MIKES. { 'mas ¦an·əˌlīzd 'ī·ən kə¦ned·ik 'en·ər·jē spek'träm·ə·trē }

mass attraction vertical [GEOPHYS] The vertical which is a function only of the distribution of mass and is unaffected by forces resulting from the motions of the earth. { 'mas ə¦trak·shən ˌverd·ə·kəl }

mass bleaching [INV ZOO] A disease affecting coral reefs in which a reduction in the number of zooxanthellae (symbiotic plants) causes corals to lose their characteristic brown color over a period of several weeks and take on a brilliant white appearance. { mas 'blēch·iŋ }

mass burning rate [CHEM ENG] The loss in mass per unit time by materials burning under specified conditions. [ORD] Rate of consumption of propellant charge, usually expressed in pounds per second. { 'mas 'bərn·iŋ ˌrāt }

mass color See masstone. { 'mas ˌkəl·ər }

mass communication [COMMUN] Communication which is directed to or reaches an appreciable fraction of the population. { 'mas kəˌmyü·nəˈkā·shən }

mass concrete [CIV ENG] Concrete set without structural reinforcement. { 'mas ˌkän,krēt }

mass conversion [COMPUT SCI] The transfer of data from one computer system to another, in which all the data is converted in a single operation, rather than in gradual increments. { 'mas kənˌvər·zhən }

mass data multiprocessing [COMPUT SCI] The basic concept of time sharing, with many inquiry stations to a central location capable of on-line data retrieval. { 'mas ¦dad·ə ˌməl·tiˈprä,ses·iŋ }

mass defect [NUC PHYS] The difference between the mass of an atom and the sum of the masses of its individual components in the free (unbound) state. { 'mas 'dē,fekt }

mass-distance [ENG] The mass carried by a vehicle multiplied by the distance it travels. { 'mas ¦dis·təns }

mass divergence [FL MECH] The divergence of the momentum field, a measure of the rate of net flux of mass out of a unit volume of a system; in symbols, $\nabla \cdot \rho V$, where ρ is the fluid density, V the velocity vector, and ∇ the del operator. { 'mas dəˈvər·jəns }

massecuite [FOOD ENG] Sugar industry term for sugar-molasses mixture prior to the removal of the molasses. { maˈskwēt }

mass-energy conservation [RELAT] The principle that energy cannot be created or destroyed; however, one form of energy is that which a particle has because of its rest mass, equal to this mass times the square of the speed of light. { 'mas 'en·ər·jē ˌkän·sər,vā·shən }

mass-energy relation [RELAT] The relation whereby the total energy content of a body is equal to its inertial mass times the square of the speed of light. { 'mas 'en·ər·jē ri,lā·shən }

Massenfilter See quadrupole spectrometer. { 'mäs·ən,fil·tər }

mass erosion [GEOL] A process in which the direct application of gravitational body stresses causes earth and rocks to fall and be carried downslope. Also known as gravity erosion. { 'mas iˈrō·zhən }

masseter [ANAT] The masticatory muscle, arising from the zygomatic arch and inserted into the lower jaw. { məˈsēd·ər }

mass extinction See faunal extinction. { 'mas ikˈstiŋk·shən }

Massey formula [ATOM PHYS] A formula for the probability that an excited atom approaching the surface of a metal will emit secondary electrons. { 'mas·ē ˌfȯr·myə·lə }

mass flow [ENG] A pattern of powder flow occurring in hoppers that is characterized by the powder flowing at every point, including points adjacent to the hopper wall. [FL MECH] The mass of a fluid in motion which crosses a given area in a unit time. { 'mas 'flō }

mass-flow bin [ENG] A bin whose hopper walls are sufficiently steep and smooth to cause flow of all the solid, without stagnant regions, whenever any solid is withdrawn. { 'mas ¦flō ,bin }

mass flowmeter [ENG] An instrument that measures the mass of fluid that flows through a pipe, duct, or open channel in a unit time. { 'mas ¦flō,mēd·ər }

mass formula [NUC PHYS] An equation giving the atomic mass of a nuclide as a function of its atomic number and mass number. { 'mas ,fȯr·myə·lə }

mass-haul curve [CIV ENG] A curve showing the quantity of excavation in a cutting which is available for fill. { 'mas ¦hȯl ,kərv }

mass heaving [GEOL] A comprehensive expansion of the ground due to freezing. { 'mas 'hēv·iŋ }

massicot [MINERAL] PbO A yellow, orthorhombic mineral consisting of lead monoxide; found in the western and southern United States. Also known as lead ocher. { 'mas·əˌkät }

Massieu function [THERMO] The negative of the Helmholtz free energy divided by the temperature. { maˈsyü ,fəŋk·shən }

massif [GEOL] A massive block of rock within an erogenic belt, generally more rigid than the surrounding rocks, and commonly composed of crystalline basement or younger plutons. { maˈsēf }

massive [GEOL] Of a mineral deposit, having a large concentration of ore in one place. [MINERAL] Of a mineral, lacking an internal structure. [PALEON] Of corallum, composed of closely packed corallites. [PETR] **1.** Of a competent rock, being homogeneous, isotropic, and elastically perfect. **2.** Of a metamorphic rock, having constituents which do not show parallel orientation and are not arranged in layers. **3.** Of igneous rocks, being homogeneous over wide areas and lacking layering, foliation, cleavage, or similar features. { 'mas·iv }

massive compact halo object See macho. { ¦mas·iv ¦käm ,pakt 'hā·lō ,äb,jekt }

mass law of sound insulation [CIV ENG] The rule stating that sound insulation for a single wall is determined almost wholly by its weight per unit area; doubling the weight of the partition increases the insulation by 5 decibels. { 'mas ¦lȯ əv saund ,in·sə,lā·shən }

mass-luminosity relation [ASTROPHYS] A relation between stellar magnitudes and mass of the stars; when the absolute magnitudes of stars are plotted versus the logarithms of their masses, the points fall closely along a smooth curve. { 'mas ,lü·mə'näs·əd·ē ri,lā·shən }

mass-memory unit [COMPUT SCI] Drum or disk memory that provides rapid access bulk storage for messages that are awaiting availability of outgoing channels. { 'mas 'mem·rē ,yü·nət }

mass movement [GEOL] Movement of a portion of the land surface as a unit. { 'mas 'müv·mənt }

mass number [NUC PHYS] The sum of the numbers of protons and neutrons in the nucleus of an atom or nuclide. Also known as nuclear number; nucleon number. { 'mas ,nəm·bər }

mass operator [QUANT MECH] An operator which is added to the Lagrangian in a quantized field theory in order to eliminate certain infinite quantities, and whose sum with the mechanical mass gives the observed mass. { 'mas 'äp·əˌrād·ər }

mass ratio [AERO ENG] The ratio of the mass of the propellant charge of the rocket to the total mass of the rocket when charged with the propellant. { 'mas 'rā·shō }

mass reactance See acoustic mass reactance. { 'mas rēˈak·təns }

mass reflex [NEUROSCI] A spread of reflexes suggesting lack of control by higher cortical centers; seen in normal newborns, in persons under the influence of drugs or in severe emotional states, and in encephalopathy or high spinal cord transections. { 'mas 'rē,fleks }

mass renormalization [QUANT MECH] The mathematical operation of adding the mass which a particle possesses because of its self interaction, to its mechanical mass in order to obtain its measured mass. { 'mas rē,nȯr·mə·lə'zā·shən }

mass resistivity [ELEC] The product of the electrical resistance of a conductor and its mass, divided by the square of its length; the product of the electrical resistivity and the density. { 'mas ,rē,zis'tiv·əd·ē }

mass shift [NUC PHYS] The portion of the isotope shift which results from the difference between the nuclear masses of different isotopes. { 'mas 'shift }

mass spectrograph [ENG] A mass spectroscope in which the ions fall on a photographic plate which after development shows the distribution of particle masses. { 'mas 'spek·trəˌgraf }

mass spectrometer [ENG] A mass spectroscope in which a slit moves across the paths of particles with various masses, and an electrical detector behind it records the intensity distribution of masses. { 'mas spek'träm·əd·ər }

mass spectrometry [ANALY CHEM] An analytical technique for identification of chemical structures, determination of mixtures, and quantitative elemental analysis, based on application of the mass spectrometer. { 'mas spek'träm·ə·trē }

mass spectroscope [ENG] An instrument used for

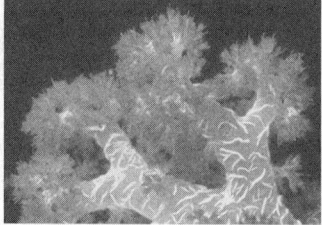

MASS BLEACHING

White-striped red soft coral due to mass bleaching.

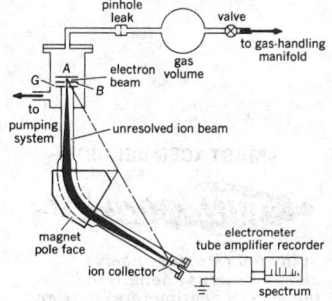

MASS SPECTROMETER

Schematic diagram of mass spectrometer tube. Electric field caused by potential difference of several volts between plates *A* and *B* draws ions through slit in *B*. Ions are further accelerated by potential difference of hundreds or thousands of volts between *B* and *G*.

determining the masses of atoms or molecules, in which a beam of ions is sent through a combination of electric and magnetic fields so arranged that the ions are deflected according to their masses. { 'mas 'spek·trə,skōp }

mass spectrum [PARTIC PHYS] A plot of masses of elementary particles, including unstable states. Also known as particle spectrum. [PHYS] A display, record, or plot of the distribution in mass, or in mass-to-charge ratio, of ionized atoms, molecules, or molecular fragments. { 'mas 'spek·trəm }

mass stopping power [NUCLEO] The decrease, per unit surface density traversed, in kinetic energy of an ionizing particle passing through matter; equal to the linear energy transfer (energy loss per unit path length) divided by the density of the material. { 'mas 'stäp·iŋ ,pau̇·ər }

mass storage [COMPUT SCI] A computer storage with large capacity, especially one whose contents are directly accessible to a computer's central processing unit. { 'mas 'stȯr·ij }

mass-storage executive capability [COMPUT SCI] The ability of the executive to relieve operators from handling cards, tapes, and the like and achieve a more efficient operation. { 'mas ¦stȯr·ij ig¦zek·yəd·iv ,kā·pə'bil·əd·ē }

mass-storage system [COMPUT SCI] A computer system containing a large number of storage devices, with one of these devices containing the master file of the operating system, routines, and library routines. { 'mas ¦stȯr·ij ,sis·təm }

mass susceptibility [PHYS CHEM] Magnetic susceptibility of a compound per gram. Also known as specific susceptibility. { 'mas sə,sep·tə'bil·əd·ē }

mass-to-charge ratio [ANALY CHEM] In analysis by mass spectroscopy, the measurement of the sample mass as a ratio to its ionic charge. { ¦mas tə 'chärj ,rā·shō }

masstone [MATER] The undiluted color of a pigment or a pigmented paint coating. Also known as mass color. { 'ma,stōn }

mass-transfer rate [PHYS] The measurement of the movement of matter as a function of time. { ¦mas 'tranz·fər ,rāt }

mass transport [FL MECH] **1.** Carrying of loose materials in a moving medium such as water or air. **2.** The movement of fluid, especially water, from one place to another. { 'mas 'tranz,pȯrt }

mass units [MECH] Units of measurement having to do with masses of materials, such as pounds or grams. { 'mas ,yü·nəts }

mass velocity [FL MECH] The weight flow rate of a fluid divided by the cross-sectional area of the enclosing chamber or conduit; for example, lb/(h·ft²). { 'mas və,läs·əd·ē }

mass wasting [GEOL] Dislodgement and downslope transport of loose rock and soil material under the direct influence of gravitational body stresses. { 'mas ,wāst·iŋ }

mast [ENG] **1.** A vertical metal pole serving as an antenna or antenna support. **2.** A slender vertical pole which must be held in position by guy lines. **3.** A drill, derrick, or tripod mounted on a drill unit, which can be raised to operating position by mechanical means. **4.** A single pole, used as a drill derrick, supported in its upright or operating position by guys. [MECH ENG] A support member on certain industrial trucks, such as a forklift, that provides guideways for the vertical movement of the carriage. [NAV ARCH] A long wooden or metal pole or spar, usually vertical, on the deck or keel of a ship, to support other spars which in turn support or are attached to sails, as well as derricks. { mast }

mast-, masto- [ANAT] A combining form denoting breast; denoting mastoid. { mast, 'mas·dō }

Mastacembeloidei [VERT ZOO] The spiny eels, a suborder of perciform fishes that are eellike in shape and have the pectoral girdle suspended from the vertebral column. { ,mas·tə,sem·bə'lȯid·ē,ī }

mastalgia [MED] Pain in the breast. { ma'stal·jə }

mastatrophy [MED] Atrophy of the breast. { mast'a·trə·fē }

mastax [INV ZOO] The muscular pharynx in rotifers. { 'ma,staks }

mast cell [HISTOL] A connective-tissue cell with numerous large, basophilic, metachromatic granules in the cytoplasm. { 'mast ,sel }

mast-cell disease See mastocytosis. { 'mast ,sel di,zēz }

mastectomy [MED] Surgical removal of the breast. Also known as mammectomy. { ma'stek·tə·mē }

MASTACEMBELOIDEI

Spiny eel (*Mastacembelus circumcinctus*); length to 7 inches (18 centimeters). (*After H. M. Smith, The Fresh-Water Fishes of Siam or Thailand, U.S. Nat. Mus. Bull. no. 188, 1945*)

master [ENG] **1.** A device which controls subsidiary devices. **2.** A precise workpiece through which duplicates are made. [ENG ACOUS] See master phonograph record. [NAV] See master station. { 'mas·tər }

master alloy [MET] An alloy of selected elements that can be added to a charge of molten metal to provide a desired composition or texture or to deoxidize the material. Also known as foundry alloy. { 'mas·tər 'al,ȯi }

master antenna television system [COMMUN] A network that distributes television signals from a common antenna to apartments or dwellings under collective ownership. Abbreviated MATV system. { 'mas·tər an¦ten·ə 'tel·ə,vizh·ən ,sis·təm }

master arm [ENG] A component of a remote manipulator whose motions are automatically duplicated by a slave arm, sometimes with changes of scale in displacement or force. { 'mas·tər 'ärm }

masterbatch [MATER] A plastic, rubber, or elastomer mixture in which there is a high additives concentration, such as rubber with carbon black, or plastic with color pigment; used to proportion additives accurately into large bulks of plastic, rubber, or elastomer. { 'mas·tər,bach }

master bushing See liner bushing. { 'mas·tər 'bu̇sh·iŋ }

master clock [COMPUT SCI] The electronic or electric source of standard timing signals, often called clock pulses, required for sequencing the operation of a computer. Also known as main clock; master synchronizer; master timer. { 'mas·tər 'kläk }

master compass [NAV] That part of a remote-indicating compass system which determines direction for transmission to various repeaters. { 'mas·tər 'käm·pəs }

master console See console. { 'mas·tər 'kän,sōl }

master control [COMMUN] The control console that contains the main program controls for a radio or television transmitter or network. [COMPUT SCI] A computer program, oriented toward applications, which carries out the highest level of control in a hierarchy of programs, routines, and subroutines. { 'mas·tər kən¦trōl }

master control interrupt [COMPUT SCI] A signal which causes the master control program to take over control of a computer system. { 'mas·tər kən¦trōl 'in·tə,rəpt }

master cylinder [MECH ENG] The container for the fluid and the piston, forming part of a device such as a hydraulic brake or clutch. { 'mas·tər 'sil·ən·dər }

master data [COMPUT SCI] A set of data which are rarely changed, or changed in a known and constant manner. { 'mas·tər 'dad·ə }

master equation [ATOM PHYS] An equation which determines the rate of change of the population of an energy level in terms of the populations of other levels and transition probabilities. { 'mas·tər i'kwā·zhən }

master file [COMPUT SCI] **1.** A computer file containing relatively permanent information, usually updated periodically, such as subscriber records or payroll data other than time worked. **2.** A computer file that is used as an authoritative source of data in carrying out a particular job on the computer. { 'mas·tər 'fīl }

master frequency meter See integrating frequency meter. { 'mas·tər 'frē·kwən·sē ,mēd·ər }

master gage [DES ENG] A locating device with fixed hole locations or part positions; locates in three dimensions and generally occupies the same space as the part it represents. { 'mas·tər 'gāj }

master gain [ELECTR] Control of overall gain of an amplifying system as opposed to varying the gain of several individual inputs. { 'mas·tər 'gān }

master group [COMMUN] In carrier telephony, ten supergroups (600 voice channels) multiplexed together and treated as a unit. { 'mas·tər ,grüp }

master gyro compass [NAV] A gyro compass for controlling one or more remote indicators, called gyro repeaters. { 'mas·tər 'jī·rō ,käm·pəs }

master instruction tape [COMPUT SCI] A computer magnetic tape on which all programs for a system of runs are recorded. { 'mas·tər in'strək·shən ,tāp }

master joint [GEOL] A persistent joint plane of greater than average extent, generally constituting the dominant jointing of an area. Also known as main joint; major joint. { 'mas·tər 'jȯint }

master layout [DES ENG] A permanent template record laid out in reference planes and used as a standard of reference in the development and coordination of other templates. { 'mas·tər 'lā,aut }

master mechanic [ENG] The supervisor, as at the mine, in charge of the maintenance and installation of equipment. { 'mas·tər mə'kan·ik }

master mode [COMPUT SCI] The mode of operation of a computer system exercised by the operating system or executive system, in which a privileged class of instructions, which user programs cannot execute, is permitted. Also known as monitor mode; privileged mode. { 'mas·tər ,mōd }

master multivibrator [ELECTR] Master oscillator using a multivibrator unit. { 'mas·tər ,məl·ti'vī,brād·ər }

master oscillator [ELECTR] An oscillator that establishes the carrier frequency of the output of an amplifier or transmitter. { 'mas·tər 'äs·ə,lād·ər }

master-oscillator power amplifier [ELECTR] Transmitter using an oscillator followed by one or more stages of radio-frequency amplification. { 'mas·tər 'äs·ə,lād·ər 'paú·ər ,am·plə,fī·ər }

master phonograph record [ENG ACOUS] The negative metal counterpart of a disk recording, produced by electroforming as one step in the production of phonograph records. Also known as master. { 'mas·tər 'fō·nə,graf ,rek·ord }

master plan position indicator [ELECTR] In a radar system, a plan position indicator which controls remote indicators or repeaters. { 'mas·tər 'plan pə¦zish·ən 'in·də,kād·ər }

master program file [COMPUT SCI] The tape record of all programs for a system of runs. { 'mas·tər 'prō·grəm ,fīl }

master record [COMPUT SCI] The basic updated record which will be used for the next run. { 'mas·tər 'rek·ord }

master routine See executive routine. { 'mas·tər rü'tēn }

master scheduler [COMPUT SCI] A program in a job entry system that assigns priorities to jobs submitted for execution. { 'mas·tər 'sked·yə·lər }

master/slave manipulator [ENG] A mechanical, electromechanical, or hydromechanical device which reproduces the hand or arm motions of an operator, enabling the operator to perform manual motions while separated from the site of the work. { 'mas·tər 'slāv mə'nip·yə,lād·ər }

master/slave mode [COMPUT SCI] The feature ensuring the protection of each program when more than one program resides in memory. { 'mas·tər 'slāv ,mōd }

master/slave system [COMPUT SCI] A system of interlinked computers under the control of one computer (master computer). { 'mas·tər 'slāv ,sis·təm }

master station [NAV] In a radio navigation system such as loran, the controlling station of two or more synchronized transmitting stations. Also known as master. { 'mas·tər 'stā·shən }

master stream See main stream. { 'mas·tər ,strēm }

Master's two-step test See two-step test. { 'mas·tərz 'tü ,step ,test }

master switch [ELEC] **1.** Switch that dominates the operation contactors, relays, or other magnetically operated devices. **2.** Switch electrically ahead of a number of individual switches. { 'mas·tər ,swich }

master synchronization pulse [COMMUN] In telemetry, a pulse distinguished from other telemetering pulses by amplitude and duration, used to indicate the end of a sequence of pulses. { 'mas·tər ,siŋ·krə·nə'zā·shən ,pəls }

master synchronizer See master clock. { 'mas·tər 'siŋ·krə,nīz·ər }

master system tape [COMPUT SCI] A monitor program centralizing the control of program operation by loading and executing any program on a system tape. { 'mas·tər 'sis·təm ,tāp }

master tape [COMPUT SCI] A magnetic tape that contains data which must not be overwritten, such as an executive routine or master file; updating a master tape means generating a new master tape onto which supplementary data have been added. { 'mas·tər 'tāp }

master terminal [COMPUT SCI] A computer terminal that is used to monitor and control a computer system. { 'mas·tər 'tər·mən·əl }

master timer See master clock. { 'mas·tər 'tīm·ər }

masthead bombing [ORD] Very low bombing, for example, against shipping. { 'mast,hed ,bäm·iŋ }

mastic [MATER] **1.** A glasslike, brittle, yellow to greenish yellow resinous exudation of the mastic tree (*Pistacia lentiscus*); used in medicine, condiments, adhesive, incense, and lacquer. Also known as mastiche; mastix; pistachia galls. **2.** Mixture of finely powdered rock and asphaltic material used for highway construction. { 'mas·tik }

mastic asphalt See asphalt mastic. { 'mas·tik 'as,fölt }

masticate [CHEM ENG] To process rubber on a machine to make it softer and more pliable before mixing with other substances. [PHYSIO] To chew. { 'mas·tə,kāt }

masticatory [PHARM] A medicine to be chewed but not swallowed. { 'mas·tə·kə,tòr·ē }

mastiche See mastic. { 'mas·tik }

mastic oil [MATER] Colorless essential oil with balsamic odor; the chief constituents are pinenes; used in medicine. { 'mas·tik ,cil }

Mastigamoebidae [INV ZOO] A family of ameboid protozoans possessing one or two flagella, belonging to the order Rhizomastigida. { ,mas·tə·gə'mē·bə,dē }

mastigoneme [CYTOL] Any of the fine hairlike appendages that extend from the shaft of the flagellum in certain motile cells. { mas·tə'gō,nēm }

Mastigophora [INV ZOO] A superclass of the Protozoa characterized by possession of flagella. { ,mas·tə'gäf·ə·rə }

mastitis [MED] Inflammation of the breast. { ma'stīd·əs }

mastix See mastic. { 'mas·tiks }

mastocytoma [VET MED] A local proliferation of mast cells forming a tumorous nodule; seen most frequently in dogs, but occasionally noted in humans. { ¦mas·tə,sī'tō·mə }

mastocytosis [MED] Excessive mast cell proliferation. Also known as mast-cell disease. { ¦mas·tə,sī'tō·səs }

mastodon [PALEON] A member of the Mastodontidae, especially the genus *Mammut*. { 'mas·tə,dän }

Mastodontidae [PALEON] An extinct family of elephantoid proboscideans that had low-crowned teeth with simple ridges and without cement. { ,mas·tə'dän·tə,dē }

mastodynia [MED] A type of cystic hyperplasia of the breast marked by an increase of connective tissue in the breast, without a proportionate increase in glandular epithelium. { ,mas·tə'din·ē·ə }

mastoid [ANAT] **1.** Breast-shaped. **2.** The portion of the temporal bone where the mastoid process is located. { 'ma,stóid }

mastoid air cell See mastoid cell. { 'ma,stóid 'er ,sel }

mastoid antrum [ANAT] An air-filled space between the upper portion of the middle ear and the mastoid cells. { 'ma,stóid 'an·trəm }

mastoid canaliculus [ANAT] A small canal opening just above the stylomastoid foramen; gives passage to the auricular branch of the vagus nerve. { 'ma,stóid ,kan·ə'lik·yə·ləs }

mastoid cell [ANAT] One of the compartments in the mastoid portion of the temporal bone, connected with the mastoid antrum and lined with a mucous membrane. Also known as mastoid air cell; mastoid sinus. { 'ma,stóid ,sel }

mastoid foramen [ANAT] A small opening behind the mastoid process. { 'ma,stóid fə'rā·mən }

mastoid fossa [ANAT] The depression behind the suprameatal spine on the lateral surface of the temporal bone. { 'ma,stóid 'fäs·ə }

mastoiditis [MED] Inflammation of the mastoid cells. { ,ma,stói'dīd·əs }

mastoid process [ANAT] A nipple-shaped, inferior projection of the mastoid portion of the temporal bone. { 'ma,stóid 'prä·səs }

mastoid sinus See mastoid cell. { 'ma,stóid 'sī·nəs }

mastopathy [MED] Any disease or pain of the mammary gland. Also known as mazopathy. { ma'stäp·ə·thē }

mastopexy [MED] Surgical fixation of a pendulous breast. Also known as mazopexy. { 'mas·tə,pek·sē }

mastoplasia [MED] Hypertrophy of breast tissue. Also known as mastoplastia. { ,mas·tə'plā·zhə }

mastoplastia See mastoplasia. { ,mas·tə'plas·tē·ə }

mastoplasty [MED] Plastic surgery on the breast. { 'mas·tə,plas·tē }

mastorrhagia [MED] Hemorrhage from the breast. { ,mas·tə'ra·jē·ə }

Mastotermitidae [INV ZOO] A family of lower termites in the order Isoptera with a single living species, in Australia. { ,mas·tō·tər'mid·ə,dē }

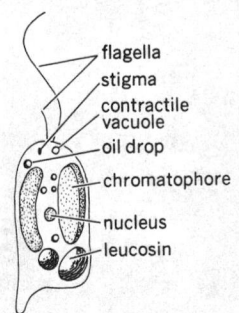

MASTIGOPHORA

flagella
stigma
contractile vacuole
oil drop
chromatophore
nucleus
leucosin

Ochromonas ludibunda, a holophytic and holozoic mastigophoran.

mastotomy [MED] Incision of the breast. { ma'städ·ə·mē }

mast step [NAV ARCH] A wooden or steel foundation supporting a mast. { 'mast ˌstep }

mat [CIV ENG] **1.** A steel or concrete footing under a post. **2.** Mesh reinforcement in a concrete slab. **3.** A heavy steel-mesh blanket used to suppress rock fragments during blasting. [MATER] Randomly distributed felt or glass fibers used in reinforced-plastics lay-up molding. [MIN ENG] An accumulation of broken mine timbers, rock, earth, and other debris coincident with the caving system of mining. { mat }

Matanuska wind [METEOROL] A strong, gusty, northeast wind which occasionally occurs during the winter in the vicinity of Palmer, Alaska. { ˌmad·ə'nüs·kə 'wind }

match [COMPUT SCI] A data-processing operation similar to a merge, except that instead of producing a sequence of items made up from the input sequences, the sequences are matched against each other on the basis of some key. [ENG] **1.** A charge of gunpowder put in a paper several inches long and used for igniting explosives. **2.** A short flammable piece of wood, paper, or other material tipped with a combustible mixture that bursts into flame through friction. [IMMUNOL] To select blood donors whose erythrocytes are compatible with those of the recipient. [MATH] *See* biconditional operation. { mach }

matched edges [ENG] Die face edges machined at right angles to each other to provide for alignment of the dies in machining equipment. { 'macht 'ej·əz }

matched-field processing [ACOUS] A signal-processing technique that has a variety of applications in underwater acoustics and is based on the comparison of measured data with predictions for the data that are calculated from a model of underwater sound propagation. { ¦macht ˌfēld 'prä·ses·iŋ }

matched filter [COMPUT SCI] In character recognition, a method employed in character property detection in which a vertical projection of the input character produces an analog waveform which is then compared to a set of stored waveforms for the purpose of determining the character's identity. [ELECTR] A filter with the property that, when the input consists of noise in addition to a specified desired signal, the signal-to-noise ratio is the maximum which can be obtained in any linear filter. { 'macht 'fil·tər }

matched groups [STAT] Groups of individuals or objects chosen so that the mean values (or some other characteristic) of some variable are the same for all the groups, in order to minimize the variation due to this variable. { ¦macht 'grüps }

matched impedance [ELEC] An impedance of a load which is equal to the impedance of a generator, so that maximum power is delivered to the load. { 'macht im¦pēd·əns }

matched load [ELECTR] A load having the impedance value that results in maximum absorption of energy from the signal source. { 'macht 'lōd }

matched-metal molding [ENG] Forming of reinforced-plastic articles between two close-fitting metal molds mounted in a hydraulic press. { 'macht ¦med·əl ˌmōld·iŋ }

matched pairs [STAT] The design of an experiment for paired comparison in which the assignment of subjects to treatment or control is not completely at random, but the randomization is restricted to occur separately within each pair. { 'macht 'perz }

matched pulse intercepting [COMMUN] System for intercepting calls on party lines in a terminal-per-line office; operates on a ground pulse which is matched in time with the intercepted station's particular ringing frequency. { 'macht ¦pəls ˌin·tər'sep·tiŋ }

matched terrace *See* paired terrace. { 'macht 'ter·əs }

matched transmission line [ELEC] Transmission line terminated with a load equivalent to its characteristic impedance. { 'macht tranz'mish·ən ˌlīn }

match gate *See* equivalence gate. { 'mach ˌgāt }

matching [COMPUT SCI] A computer problem-solving method in which the current situation is represented as a schema to be mapped into the desired situation by putting the two in correspondence. [ELEC] Connecting two circuits or parts together with a coupling device in such a way that the maximum transfer of energy occurs between the two circuits, and the impedance of either circuit will be terminated in its image. [MATH] A set of edges in a graph, no two of which have a vertex in common. Also known as independent edge set.

[NAV] The bringing of two or more signals or indications into suitable position or condition preliminary to making a measurement, as on a loran indicator or a sky compass. { 'mach·iŋ }

matching diaphragm [ELECTROMAG] Diaphragm consisting of a slit in a thin sheet of metal, placed transversely across a waveguide for matching purposes; the orientation of the slit with respect to the long dimension of the waveguide determines whether the diaphragm acts as a capacitive or inductive reactance. { 'mach·iŋ 'dī·ə,fram }

matching distribution [STAT] The distribution of number of matches obtained if N tickets labeled 1 to N are drawn at random one at a time and laid in a row, and a match is counted when a ticket's label matches its position. { 'mach·iŋ ˌdi·strə'byü·shən }

matching impedance [ELEC] Impedance value that must be connected to the terminals of a signal-voltage source for proper matching. { 'mach·iŋ im¦pēd·əns }

matching section [ELECTROMAG] A section of transmission line, a quarter or half wavelength long, inserted between a transmission line and a load to obtain impedance matching. { 'mach·iŋ 'sek·shən }

matching stub [ELECTROMAG] Device placed on a radio-frequency transmission line which varies the impedance of the line; the impedance of the line can be adjusted in this manner. { 'mach·iŋ ˌstəb }

match plate [MET] A plate on which metal-casting patterns are mounted or formed as an integral part, to facilitate the molding operation. { 'mach ˌplāt }

match processing [COMPUT SCI] The checking of two or more units of data for common characteristics. { 'mach 'prä,ses·iŋ }

mate [BIOL] **1.** To pair for breeding. **2.** To copulate. { māt }

mate killers [VIROL] Paramecia that contain the mu phage and cause their sensitive partners to die after conjugation. { 'māt ˌkil·ərz }

material balance [CHEM ENG] A calculation to inventory material inputs versus outputs in a process system. { mə'tir·ē·əl 'bal·əns }

material implication *See* implication. { mə¦tir·ē·əl ˌim·plə'kā·shən }

materialization [PHYS] The direct conversion of energy into mass, as in pair production. { mə,tir·ē·ə·lə'zā·shən }

material particle [MECH] An object which has rest-mass and an observable position in space, but has no geometrical extension, being confined to a single point. Also known as particle. { mə'tir·ē·əl 'pärd·ə·kəl }

material requirements planning [IND ENG] A formal computerized approach to inventory planning, manufacturing scheduling, supplier scheduling, and overall corporate planning. Abbreviated MRP. { mə'tir·ē·əl ri'kwīr·məns ˌplan·iŋ }

materials control [IND ENG] Inventory control of materials involved in manufacturing or assembly. { mə'tir·ē·əlz kən,trōl }

materials handling [ENG] The loading, moving, and unloading of materials. { mə'tir·ē·əlz ,hand·liŋ }

materials science [ENG] The study of the nature, behavior, and use of materials applied to science and technology. { mə'tir·ē·əlz ,sī·əns }

materials testing reactor [NUCLEO] A nuclear reactor designed mainly for studying the behavior of materials and equipment subjected to large fluxes of neutrons and other radiation. { mə'tir·ē·əlz ,test·iŋ rē,ak·tər }

material unit [GEOL] A stratigraphic unit based on rocks and their fossil content without time implication. { mə'tir·ē·əl ,yü·nət }

material well [CHEM ENG] In a plastics process, the space provided in a compression or transfer mold to allow for the bulk factor. { mə'tir·ē·əl ,wel }

materia medica [PHARM] **1.** The science that treats of the sources, properties, and preparation of medicinal substances. **2.** The materials from which medicinal substances are prepared. **3.** A treatise on the subject. { mə'tir·ē·ə 'med·ə·kə }

materiel [ORD] **1.** Things of all kinds required for the equipment, maintenance, operation, and support of military activities, both combat and noncombat. **2.** In a restricted sense, those things used in combat or logistic support operations, such as weapons, motor vehicles, or special-purpose clothing, as distinguished from items of ordinary use, such as uniforms, food, or medicine. { mə¦tir·ē'el }

maternal [BIOL] Of, pertaining to, or related to a mother. { mə'tərn·əl }

maternal behavior [PSYCH] The pattern of care given an offspring by its mother. { mə‚tərn·əl bi'hāv·yər }

maternal effect [GEN] Determination of characters of the progeny by the maternal parent; mediated by the genetic constitution of the mother. { mə'tərn·əl i‚fekt }

maternal impressions [PSYCH] The congenital developmental effects formerly thought to be produced upon the fetus in the uterus by mental impressions of a vivid character received by the mother during pregnancy. { mə'tərn·əl im'presh·ənz }

maternal inheritance [GEN] The acquisition of characters transmitted through the cytoplasm of the egg. { mə'tərn·əl in'her·əd·əns }

maternal messenger ribonucleic acid [CELL MOL] In certain oocytes, messenger ribonucleic acid that is stored during oogenesis for translation during early embryogenesis. Also known as masked messenger ribonucleic acid. { mə'tərn·əl ¦mes·ən·jər ‚rī·bō·nü¦klē·ik 'as·əd }

maternal mortality rate [MED] The number of deaths reported as due to puerperal causes in a calendar year per 100 live births reported in the same year and place. { mə'tərn·əl mór'tal·əd·ē ‚rāt }

maternal placenta [EMBRYO] The outer placental layer, developed from the decidua basalis. { mə'tərn·əl plə'sen·tə }

maternity [BIOL] **1.** Motherhood. **2.** The state of being pregnant. { mə'tərn·əd·ē }

mat foundation [CIV ENG] A large, thick, usually reinforced concrete mat which transfers loads from a number of columns. or columns and walls, to the underlying rock or soil. Also known as raft foundation. { 'mat faùn'dā·shən }

math coprocessor See numeric processor extension. { ¦math 'kō‚prä‚ses·ər }

mathematical analysis See analysis. { ¦math·ə¦mad·ə·kəl ə'nal·ə·səs }

mathematical biology [BIOL] A discipline that encompasses all applications of mathematics, computer technology, and quantitative theorizing to biological systems, and the underlying processes within the systems. { ¦math·ə¦mad·ə·kəl bī'äl·ə·jē }

mathematical biophysics [BIOL] A discipline which attempts to utilize mathematics to explain biophysical processes. { ¦math·ə¦mad·ə·kəl ‚bī·ō'fiz·iks }

mathematical check [COMPUT SCI] A programmed computer check of a sequence of operations, using the mathematical properties of that sequence. { ¦math·ə¦mad·ə·kəl 'chek }

mathematical climate [CLIMATOL] An elementary generalization of the earth's climatic pattern, based entirely on the annual cycle of the sun's inclination; this early climatic classification recognized three basic latitudinal zones (the summerless, intermediate, and winterless), which are now known as the Frigid, Temperate, and Torrid Zones, and which are bounded by the Arctic and Antarctic Circles and the Tropics of Cancer and Capricorn. { ¦math·ə¦mad·ə·kəl 'klī·mət }

mathematical ecology [ECOL] The application of mathematical theory and technique to ecology. { ¦math·ə¦mad·ə·kəl ē'käl·ə·jē }

mathematical forecasting See numerical forecasting. { ¦math·ə¦mad·ə·kəl 'fòr‚kast·iŋ }

mathematical function program [COMPUT SCI] A set of routinely used mathematical functions, such as square root, which are efficiently coded and called for by special symbols. { ¦math·ə¦mad·ə·kəl 'fəŋk·shən ‚prō·grəm }

mathematical geography [GEOGR] The branch of geography that deals with the features and processes of the earth, and their representations on maps and charts. { ¦math·ə¦mad·ə·kəl jē'äg·rə·fē }

mathematical geology [GEOL] The branch of geology concerned with the study of probability distributions of values of random variables involved in geologic processes. { ¦math·ə¦mad·ə·kəl jē'äl·ə·jē }

mathematical induction [MATH] A general method of proving statements concerning a positive integral variable: if a statement is proven true for $x = 1$, and if it is proven that, if the statement is true for $x = 1, \ldots, n$, then it is true for $x = n + 1$, it follows that the statement is true for any integer. Also known as complete induction; method of infinite descent; proof by descent. { ¦math·ə¦mad·ə·kəl in'dək·shən }

mathematical logic [MATH] The study of mathematical theories from the viewpoint of model theory, recursive function theory, proof theory, and set theory. { ¦math·ə¦mad·ə·kəl 'läj·ik }

mathematical model [MATH] **1.** A mathematical representation of a process, device, or concept by means of a number of variables which are defined to represent the inputs, outputs, and internal states of the device or process, and a set of equations and inequalities describing the interaction of these variables. **2.** A mathematical theory or system together with its axioms. { ¦math·ə¦mad·ə·kəl 'mäd·əl }

mathematical physics [PHYS] The study of the mathematical systems which represent physical phenomena; particular areas are, for example, quantum and statistical mechanics and field theory. { ¦math·ə¦mad·ə·kəl 'fiz·iks }

mathematical probability [MATH] The ratio of the number of mutually exclusive, equally likely outcomes of interest to the total number of such outcomes when the total is exhaustive. Also known as a priori probability. { ¦math·ə¦mad·ə·kəl ‚präb·ə'bil·əd·ē }

mathematical programming See optimization theory. { ¦math·ə¦mad·ə·kəl 'prō‚gram·iŋ }

mathematical software [COMPUT SCI] The set of algorithms used in a computer system to solve general mathematical problems. { ¦math·ə¦mad·ə·kəl 'sóft‚wer }

mathematical subroutine [COMPUT SCI] A computer subroutine in which a well-defined mathematical function, such as exponential, logarithm, or sine, relates the output to the input. { ¦math·ə¦mad·ə·kəl 'səb·rü‚tēn }

mathematical system [MATH] A structure formed from one or more sets of undefined objects, various concepts which may or may not be defined, and a set of axioms relating these objects and concepts. { ¦math·ə¦mad·ə·kəl 'sis·təm }

mathematical table [MATH] A listing of the values of a function of one or several variables at a series of values of the arguments, usually equally spaced. { ¦math·ə¦mad·ə·kəl 'tā·bəl }

mathematics [SCI TECH] The deductive study of shape, quantity, and dependence; the two main areas are applied mathematics and pure mathematics, the former arising from the study of physical phenomena, the latter the intrinsic study of mathematical structures. { ¦math·ə¦mad·iks }

Matheson joint [DES ENG] A wrought-pipe joint made by enlarging the end of one pipe length to receive the male end of the next length. { 'math·ə·sən ‚jóint }

Mathieu equation [MATH] A differential equation of the form $y'' + (a + b \cos 2x)y = 0$, whose solution depends on periodic functions. { ma'tyü i ‚kwā·zhən }

Mathieu functions [MATH] Any solution of the Mathieu equation which is periodic and an even or odd function. { ma'tyü ‚fəŋk·shənz }

matico [BOT] *Piper angustifolium.* An aromatic wild pepper found in tropical America whose leaves are rich in volatile oil, gums, and tannins; leaves were used medicinally as a stimulant and hemostatic. { mə'tē·kō }

matico oil [MATER] An alcohol-soluble, yellowish-brown volatile oil distilled from the leaves or flowers of matico; chief constituents are asarone and methyl eugenol. { mə'tē·kō ‚óil }

matildite [MINERAL] $AgBiS_2$ An iron black to gray, orthorhombic mineral consisting of silver bismuth sulfide; occurrence is massive or granular. { mə'til‚dīt }

matinal [METEOROL] The morning winds, that is, an east wind. { 'mat·ən·əl }

mating [BIOL] The meeting of individuals for sexual reproduction. { 'mād·iŋ }

mating type [MICROBIO] The genetically determined mating behavior characteristic of certain species of microorganisms; only different mating types can conjugate. { 'mād·iŋ ‚tīp }

matlockite [MINERAL] $PbFCl$ A mineral consisting of lead chloride and fluoride. { 'mat·lə‚kīt }

m-atm See meter-atmosphere.

mat packs [MIN ENG] Timbers laid side by side into a solid mass 2 by $2^{1}/_{2}$ feet (61 by 76 centimeters) square and 4–6 inches (10–15 centimeters) thick, kept together by wires through holes drilled through the edges of the mass, carried into a mine and built up to make effective roof supports. { 'mat ‚paks }

Matricaria [BOT] A genus of weedy herbs having a strong

odor and white and yellow disk flowers; chamomile oil is obtained from certain species. { ‚ma·trə'kar·ē·ə }

matric forces [GEOL] Forces acting on soil water that are independent of gravity but exist due to the attraction of solid surfaces for water, the attraction of water molecules for each other, and a force in the air-water interface due to the polar nature of water. { 'mā·trik ‚fȯrs·əz }

matrix [ANALY CHEM] The analyte as considered in terms of its being an assemblage of constituents, each with its own properties. [COMPUT SCI] A latticework of input and output leads with logic elements connected at some of their intersections. [ELECTR] **1.** The section of a color television transmitter that transforms the red, green, and blue camera signals into color-difference signals and combines them with the chrominance subcarrier. Also known as color coder; color encoder; encoder. **2.** The section of a color television receiver that transforms the color-difference signals into the red, green, and blue signals needed to drive the color picture tube. Also known as color decoder; decoder. [ENG] A recessed mold in which something is formed or cast. [GRAPHICS] **1.** In a type-casting machine, the portion of the mold that forms the letter face. **2.** A heavy, unsized, unfinished paper that is used for molds for stereotype plates. **3.** A master negative from which characters are projected in a photocomposition process. [HISTOL] **1.** The intercellular substance of a tissue. Also known as ground substance. **2.** The epithelial tissue from which a toenail or fingernail develops. [MATER] A binding agent used to make an agglomerate mass. [MATH] A rectangular array of numbers or scalars from a vector space. [MET] **1.** The principal component of an alloy. **2.** The precisely shaped form used as the cathode in electroforming. [MYCOL] The substrate on or in which fungus grows. [PETR] The continuous, fine-grained material in which large grains of a sediment or sedimentary rock are embedded. Also known as groundmass. { 'mā·triks }

matrix algebra [MATH] An algebra whose elements are matrices and whose operations are addition and multiplication of matrices. { 'mā·triks 'al·jə·brə }

matrix algebra tableau [COMPUT SCI] The current matrix at the end of an iteration while running a linear program. { 'mā·triks 'al·jə·brə ta'blō }

matrix-array camera [ELECTR] A solid-state television camera that has a rectangular array of light-sensitive elements or pixels. { 'mā·triks ə'rā ‚kam·rə }

matrix association region [CELL MOL] In eukaryotic interphase nuclei, any of the specific sites to which the chromatin loop domains are anchored. { ¦ma·triks ə‚sō·sē'ā·shən ‚rē·jən }

matrix calculus [MATH] The treatment of matrices whose entries are functions as functions in their own right with a corresponding theory of differentiation; this has application to the study of multidimensional derivatives of functions of several variables. { 'mā·triks 'kal·kyə·ləs }

matrix effects [ANALY CHEM] **1.** The enhancement or suppression of minor element spectral lines from metallic oxides during emission spectroscopy by the matrix element (such as graphite) used to hold the sample. **2.** The combined effect exerted by the various constituents of the matrix on the measurements of the analysis. { 'mā·triks i‚feks }

matrix element [MATH] One of the set of numbers which form a matrix. [QUANT MECH] The scalar product of a member of a complete, orthogonal set of vectors, representing states, with a vector which results from applying a specified operator to another member of this set. { 'mā·triks ‚el·ə·mənt }

matrix fiber See biconstituent fiber. { 'mā·triks ‚fī·bər }

matrix game [MATH] A game involving two persons, which gives rise to a matrix representing the amount received by the two players. Also known as rectangular game. { 'mā·triks ‚gām }

matrix isolation [SPECT] A spectroscopic technique in which reactive species can be characterized by maintaining them in a very cold, inert environment while they are examined by an absorption, electron-spin resonance, or laser excitation spectroscope. { 'mā·triks ‚i·sə'lā·shən }

matrix mechanics [QUANT MECH] The theory of quantum mechanics developed by using the Heisenberg picture and representing operators by their matrix elements between eigenfunctions of the Hamiltonian operator; Heisenberg's original formulation of quantum mechanics. { 'mā·triks mə'kan·iks }

matrix metalloproteinase [BIOCHEM] Any member of a family of at least 19 structurally related zinc-dependent neutral endopeptidases collectively capable of degrading essentially all components of extracellular matrix. { ¦mā·triks mə‚tal·ō'prō·tē·ə‚nās }

matrix of a linear transformation [MATH] A unique matrix A, such that for a specified linear transformation L from one vector space to another, and for specified finite bases in each space, L applied to a vector is equal to A times that vector. { 'mā·triks əv ə 'lin·ē·ər ‚tranz·fər'mā·shən }

matrix porosity [GEOL] Core-sample porosity determined from a small sample of the core, in contrast to total porosity, where the whole core is used. { 'mā·triks pə'räs·əd·ē }

matrix printing [COMPUT SCI] High-speed printing in which characterlike configurations of dots are printed through the proper selection of wire ends from a matrix of wire ends. Also known as stylus printing; wire printing. { 'mā·triks 'print·iŋ }

matrix rock See land pebble phosphate. { 'mā·triks ‚räk }

matrix sound system [ENG ACOUS] A quadraphonic sound system in which the four input channels are combined into two channels by a coding process for recording or for stereo frequency-modulation broadcasting and decoded back into four channels for playback of recordings or for quadraphonic stereo reception. { 'mā·triks 'saund ‚sis·təm }

matrix spectrophotometry [SPECT] Spectrophotometric analysis in which the specimen is irradiated in sequence at more than one wavelength, with the visible spectrum evaluated for the energy leaving for each wavelength of irradiation. { 'mā·triks ‚spek·trō·fə'täm·ə·trē }

matrix storage [COMPUT SCI] A computer storage in which coordinates are used to address the locations or circuit elements. Also known as coordinate storage. { 'mā·triks ‚stȯr·ij }

matrix theory [MATH] The algebraic study of matrices and their use in evaluating linear processes. { 'mā·triks ‚thē·ə·rē }

matrix velocity [GEOPHYS] The velocity of sound through a formation's rock matrix during an acoustic-velocity log. { 'mā·triks və'läs·əd·ē }

matroclinous inheritance [GEN] Inheritance in which the offspring more closely resemble the female parent than the male parent. { ‚ma·trə‚klī·nəs in'her·əd·əns }

matromycin See oleandomycin. { ‚ma·trə‚mīs·ən }

matte [MATER] Dull, as applied to appearance of a surface. [MET] An impure metallic sulfide mixture produced by smelting the sulfide ores of such metals as copper, lead, or nickel. { mat }

matte dip [MET] An etching solution which reacts with the surface of a metal, giving a dull finish. { 'mat ‚dip }

matte feeder [IND ENG] A heavy-duty apron feeder composed of thick steel flights attached to a solid chain-link mat supported by closely spaced rollers. { 'mat ‚fēd·ər }

matter [PHYS] The substance composing bodies perceptible to the senses; includes any entity possessing mass when at rest. { 'mad·ər }

matter era [ASTRON] The period in the evolution of the universe, beginning roughly 10^5 years after the big bang, when the universe had cooled to the point at which electrons and protons were able to form neutral hydrogen atoms, and continuing to the present time, during which matter, in the form of atoms, is dominant over radiation. { 'mad·ər ‚ir·ə }

matter wave See de Broglie wave. { 'mad·ər ‚wāv }

matte smelting [MET] Smelting of copper-bearing materials in a reverberatory furnace. { 'mat ‚smelt·iŋ }

Matteuci effect [PHYS] A phenomenon in which an electric potential difference appears between the ends of a ferromagnet that is twisted in a magnetic field. { mad·ə'ü·chē i‚fekt }

Matthias' rules [SOLID STATE] Several empirical rules giving the dependence of the transition temperatures of superconducting metals and alloys on the position of the metals in the periodic table and in the composition of the alloys. { mə'thī·əs ‚rülz }

Matthiessen sinker method [THERMO] A method of determining the thermal expansion coefficient of a liquid, in which the apparent weight of a sinker when immersed in the liquid is measured for two different temperatures of the liquid. { ¦math·ə·sən 'siŋ·kər ‚meth·əd }

Matthiessen's rule [SOLID STATE] An empirical rule which states that the total resistivity of a crystalline metallic specimen is the sum of the resistivity due to thermal agitation of the metal ions of the lattice and the resistivity due to imperfections in the crystal. { 'math·ə·sənz ‚rül }

mattock [DES ENG] A tool with the combined features of an adz, an ax, and a pick. { 'mad·ək }

mattress [CIV ENG] A woven mat, often of wire and cement blocks, used to prevent erosion of dikes, jetties, or river banks. { 'ma·trəs }

mattress array See billboard array. { 'ma·trəs ə'rā }

maturase [MOL BIO] Any enzyme encoded by self-splicing introns that catalyzes excision of the intron from its own primary transcript. { 'mach·ə‚rās }

maturation [BIOL] **1.** The process of coming to full development. **2.** The final series of changes in the growth and formation of germ cells. [VIROL] The process that leads to incorporation of viral genomes into capsids and complete virions. { ‚mach·ə'rā·shən }

mature [BIOL] **1.** Being fully grown and developed. **2.** Ripe. [FOOD ENG] Having attained the final state of processing, as certain wines. [GEOL] **1.** Pertaining to a topography or region, and to its landforms, having undergone maximum development and accentuation of form. **2.** Pertaining to the third stage of textural maturity of a clastic sediment. [PSYCH] Having the emotional qualities of a well-adjusted adult. { mə'chùr }

matureland [GEOL] The land surface which is characteristic of the mature stage in the erosion cycle. { mə'chùr‚land }

mature soil See zonal soil. { mə'chùr 'sóil }

maturity [GEOL] **1.** The second stage of the erosion cycle in the topographic development of a landscape or region characterized by numerous and closely spaced mature streams, reduction of level surfaces to slopes, large well-defined drainage systems, and the absence of swamps or lakes on the uplands. Also known as topographic maturity. **2.** A stage in the development of a shore or coast that begins with the attainment of a profile of equilibrium. **3.** The extent to which the texture and composition of a clastic sediment approach the ultimate end product. **4.** The stage of stream development at which maximum vigor and efficiency has been reached. { mə'chùr·əd·ē }

maturity index [GEOL] A measure of the progress of a clastic sediment in the direction of chemical or mineralogic stability; for example, a high ratio of quartz + cherts to feldspar + rock fragments indicates a highly mature sediment. { mə'chùr·əd·ē ‚in‚deks }

maturity-onset diabetes [MED] A type of diabetes mellitus which develops later in life; characterized by more gradual development and less severe symptoms than juvenile-onset diabetes. { mə'chùr·əd·ē ‚ón‚set ‚dī·ə'bēd·əs }

MATV system See master antenna television system. { ‚em‚ā‚tē‚vē 'sis·təm }

maucherite [MINERAL] $Ni_{11}As_8$ A reddish silver-white mineral composed of nickel arsenide. { 'maù·chə‚rīt }

maul See rammer. { mól }

maulstick See mahlstick. { 'mól‚stik }

maunder minimum [ASTRON] A period of time from about 1650 to 1710 when the sun did not appear to have sunspots. { 'món·dər 'min·ə·məm }

Maunoir's hydrocele [MED] A congenital lymphatic cyst of the neck. { mōn'wärz 'hī·drə‚sēl }

Maupertius' principle [MECH] The principle of least action is sufficient to determine the motion of a mechanical system. { mō'pər·shəs ‚prin·sə·pəl }

Mauriac syndrome [MED] A complex of symptoms associated with diabetes mellitus in children including retarded growth, obesity, and enlargement of the liver, probably related to inadequate control of the condition. { 'mó·rē·ak ‚sin‚drōm }

Mauritius hemp [BOT] A hard fiber obtained from the leaves of the cabuya, grown on the island of Mauritius; not a true hemp. { mó'rish·əs 'hemp }

mavar See parametric amplifier. { 'mā‚vär }

max See maximum. { maks }

max-flow min-cut theorem [IND ENG] In the analysis of networks, the concept that for any network with a single source and sink, the maximum feasible flow from source to sink is equal to the minimum cut value for any of the cuts of the network. [MATH] See Ford-Fulkerson theorem. { ‚maks'flō ‚min'kət ‚thir·əm }

maxilla [ANAT] **1.** The upper jawbone. **2.** The upper jaw.

[INV ZOO] Either of the first two pairs of mouthparts posterior to the mandibles in certain arthropods. { mak'sil·ə }

maxillary air sinus See maxillary sinus. { 'mak·sə‚ler·ē ‚er ‚sī·nəs }

maxillary antrum See maxillary sinus. { 'mak·sə‚ler·ē 'an·trəm }

maxillary arch See palatomaxillary arch. { 'mak·sə‚ler·ē 'ärch }

maxillary artery [ANAT] A branch of the external carotid artery which supplies the deep structures of the face (internal maxillary) and the side of the face and nose (external maxillary). { 'mak·sə‚ler·ē 'ärd·ə·rē }

maxillary hiatus [ANAT] An opening in the maxilla connecting the nasal cavity with the maxillary sinus. { 'mak·sə‚ler·ē hī'ād·əs }

maxillary nerve [NEUROSCI] A somatic sensory branch of the trigeminal nerve; innervates the meninges, the skin of the upper portion of the face, upper teeth, and mucosa of the nose, palate, and cheeks. { 'mak·sə‚ler·ē 'nərv }

maxillary process of the embryo [EMBRYO] An outgrowth of the dorsal part of the mandibular arch that forms the lateral part of the upper lip, the upper cheek region, and the upper jaw except the premaxilla. { 'mak·sə‚ler·ē ‚prä·səs əv thē 'em·brē·ō }

maxillary sinus [ANAT] A paranasal air cavity in the body of the maxilla. Also known as maxillary air sinus; maxillary antrum. { 'mak·sə‚ler·ē 'sī·nəs }

maxilliped [INV ZOO] One of the three pairs of crustacean appendages immediately posterior to the maxillae. { mak'sil·ə‚ped }

maxillo-alveolar index [ANTHRO] Ratio of the breadth to the length of the alveolar arch, multiplied by 100. { ‚mak·sə‚lō al‚vē·ə·lər 'in‚deks }

maxillofacial [MED] Pertaining to the jaws and face. { ‚mak·sə·lō 'fā·shəl }

maximal analytic extension [RELAT] An extension, in a real analytic manner, past all coordinate singularities of a solution to Einstein's equations of general relativity. { ‚mak·sə·məl ‚an·ə‚lid·ik ik'sten·chən }

maximal breathing capacity [PHYSIO] The greatest respiratory minute volume which an individual can produce during a given period of extremely forceful breathing. { ‚mak·sə·məl 'brēth·iŋ kə‚pas·əd·ē }

maximal chain [MATH] A sequence of $n + 1$ subsets of a set of n elements, such that the first member of the sequence is the empty set and each member of the sequence is a proper subset of the next one. { ‚mak·sə·məl 'chān }

maximal element See maximal member. { ‚mak·sə·məl 'el·ə·mənt }

maximal flow [IND ENG] Maximum total flow from the source to the sink in a connected network. { ‚mak·sə·məl 'flō }

maximal ideal [MATH] An ideal I in a ring R which is not equal to R, and such that there is no ideal containing I and not equal to I or R. { ‚mak·sə·məl ī'dēl }

maximal independent set [MATH] An independent set of vertices of a graph which is not a proper subset of another independent set. { ‚mak·sə·məl ‚in·də‚pen·dənt 'set }

maximal member [MATH] In a partially ordered set a maximal member is one for which no other element follows it in the ordering. Also known as maximal element. { ‚mak·sə·məl 'mem·bər }

maximal planar graph [MATH] A planar graph to which no new arcs can be added without forcing crossings and hence violating planarity. { ‚mak·sə·məl ‚plān·ər ‚graf }

maximax criterion [MATH] In decision theory, one of several possible prescriptions for making a decision under conditions of uncertainty; it prescribes the strategy which will maximize the maximum possible profit. { ‚mak·sə‚maks krī‚tir·ē·ən }

maxim criterion [MATH] One of several prescriptions for making a decision under conditions of uncertainty; it prescribes the strategy which will maximize the minimum profit. Also known as maximin criterion. { ‚mak·səm krī‚tir·ē·ən }

maximin [MATH] **1.** The maximum of a set of minima. **2.** In the theory of games, the largest of a set of minimum possible gains, each representing the least advantageous outcome of a particular strategy. { 'mak·sə‚min }

maximin criterion See maxim criterion. { 'mak·sə‚min krī‚tir·ē·ən }

maximizing a function [MATH] Finding the largest value assumed by a function. { 'mak·sə,mīz·iŋ ə 'faŋk·shən }

maximum [MATH] The maximum of a real-valued function is the greatest value it assumes. Abbreviated max. { 'mak·sə·məm }

maximum allowable rate of hazardously misleading information [NAV] The maximum allowable number of integrity failures by a navigational system per unit time. { ¦mak·sə·məm ə¦laü·ə·bəl ¦rāt əv ¦haz·ərd·əs·lē mis¦lēd·iŋ ,in·fər'mā·shən }

maximum allowable working pressure [MECH ENG] The maximum gage pressure in a pressure vessel at a designated temperature, used for the determination of the set pressure for relief valves. { 'mak·sə·məm ə¦laü·ə·bəl 'wərk·iŋ ,presh·ər }

maximum-and-minimum thermometer [ENG] A thermometer that automatically registers both the maximum and the minimum temperatures attained during an interval of time. { 'mak·sə·məm ən 'min·ə·məm thər'mäm·əd·ər }

maximum angle of inclination [MECH ENG] The maximum angle at which a conveyor may be inclined and still deliver an amount of bulk material within a given time. { 'mak·sə·məm 'aŋ·gəl əv ,in·klə'nā·shən }

maximum available gain [ELECTR] The theoretical maximum power gain available in a transistor stage; it is seldom achieved in practical circuits because it can be approached only when feedback is negligible. Abbreviated MAG. { 'mak·sə·məm ə¦vāl·ə·bəl 'gān }

maximum average power output [ELECTR] In television, the maximum of radio-frequency output power that can occur under any combination of signals transmitted, averaged over the longest repetitive modulation cycle. { 'mak·sə·məm ¦av·rəj ¦paú·ər 'aút,pút }

maximum belt slope [MECH ENG] A slope beyond which the material on the belt of a conveyor tends to roll downhill. { 'mak·sə·məm 'belt ,slōp }

maximum belt tension [MECH ENG] The total of the starting and operating tensions in a conveyor. { 'mak·sə·məm 'belt ,ten·chən }

maximum breathing capacity [PHYSIO] The greatest volume of air an individual can breathe voluntarily in 10–30 seconds; expressed as liters per minute. Abbreviated MBC. { 'mak·sə·məm 'breth·iŋ kə,pas·əd·ē }

maximum cardinality matching See maximum matching. { ,mak·sə·məm ,kärd·ən'al·əd·ē ,mach·iŋ }

maximum continuous load [MECH ENG] The maximum load that a boiler can maintain for a designated length of time. { 'mak·sə·məm kən¦tin·yə·wəs 'lōd }

maximum credible accident [NUCLEO] The most serious nuclear reactor accident that can be hypothesized from an adverse combination of equipment malfunction, operating errors, and other reasonable foreseen causes. { 'mak·sə·məm ¦kred·ə·bəl 'ak·sə·dənt }

maximum demand [ELEC] The greatest average value of the power, apparent power, or current consumed by a customer of an electric power system, the averages being taken over successive time periods, usually 15 or 30 minutes in length. { 'mak·sə·məm di'mand }

maximum depression [ORD] The maximum vertical angle below the horizontal plane at which a piece of artillery can be laid and still deliver effective fire. { 'mak·sə·məm di'presh·ən }

maximum ebb [OCEANOGR] The greatest speed of an ebb current. { 'mak·sə·məm 'eb }

maximum effective range [ORD] The greatest distance at which a weapon may be expected to fire accurately to inflict casualties or damage. { 'mak·sə·məm i¦fek·tiv 'rānj }

maximum elevation [ORD] The greatest vertical angle at which a gun or launcher can be laid; usually limited by the mechanical structure of the piece. { 'mak·sə·məm ,el·ə'vā·shən }

maximum flood [OCEANOGR] The greatest speed of a flood current. { 'mak·sə·məm 'fləd }

maximum flow problem [MATH] The problem of finding a feasible flow in an *s-t* network with the largest possible flow value for a given weight function. { ¦mak·sə·məm 'flō ,präb·ləm }

maximum gradability [MECH ENG] Steepest slope a vehicle can negotiate in low gear; usually expressed in percentage of slope, namely, the ratio between the vertical rise and the horizontal distance traveled; sometimes expressed by the angle between the slope and the horizontal. { 'mak·sə·məm ,grād·ə'bil·əd·ē }

maximum independent set [MATH] An incident set of vertices of a graph such that there is no other independent set with more vertices. { ¦mak·sə·məm ,in·də,pen·dənt 'set }

maximum keying frequency [ELECTR] In facsimile, the frequency in hertz that is numerically equal to the spot speed divided by twice the horizontal dimension of the spot. { 'mak·sə·məm 'kē·iŋ ,frē·kwən·sē }

maximum likelihood method [STAT] A technique in statistics where the likelihood distribution is so maximized as to produce an estimate to the random variables involved. { 'mak·sə·məm 'līk·lē·,hùd ,meth·əd }

maximum matching [MATH] A matching of edges in a graph such that no other matching has a greater number of edges. Also known as maximum cardinality matching. { ¦mak·sə·məm 'mach·iŋ }

maximum-minimum principle See min-max theorem. { 'mak·sə·məm 'min·ə·məm ,prin·sə·pəl }

maximum modulating frequency [ELECTR] Highest picture frequency required for a facsimile transmission system; the maximum modulating frequency and the maximum keying frequency are not necessarily equal. { 'mak·sə·məm 'mäj·ə,lād·iŋ ,frē·kwən·sē }

maximum-modulus theorem [MATH] For a complex analytic function in a closed bounded simply connected region its modulus assumes its maximum value on the boundary of the region. { 'mak·sə·məm 'mäj·ə·ləs ,thir·əm }

maximum operating frequency [COMPUT SCI] The highest rate at which the modules perform iteratively and reliably. { 'mak·sə·məm 'äp·ə,rād·iŋ ,frē·kwən·sē }

maximum ordinate [MECH] Difference in altitude between the origin and highest point of the trajectory of a projectile. { 'mak·sə·məm 'òrd·ən·ət }

maximum permissible concentration [MED] The maximum quantity/unit volume of radioactive material in air, water, and foodstuffs that is not considered an undue risk to human health. { 'mak·sə·məm pər'mis·ə·bəl ,kän·sən'trā·shən }

maximum permissible dose [MED] The dose of ionizing radiation that a person may receive in his lifetime without appreciable bodily injury. { 'mak·sə·məm pər'mis·ə·bəl 'dōs }

maximum producible oil index [PETRO ENG] An approximation of the maximum amount of oil per bulk formation volume that is producible with water drive. { 'mak·sə·məm prə¦düs·ə·bəl 'oil ,in,deks }

maximum production life [MECH ENG] The length of time that a cutting tool performs at cutting conditions of maximum tool efficiency. { 'mak·sə·məm prə'dək·shən ,līf }

maximum retention time [ELECTR] Maximum time between writing into and reading an acceptable output from a storage element of a charge storage tube. { 'mak·sə·məm ri'ten·chən ,tīm }

maximum signal level [ELECTR] In an amplitude-modulated facsimile system, the level corresponding to copy black or copy white, whichever has the highest amplitude. { 'mak·sə·məm 'sig·nəl ,lev·əl }

maximum sound pressure [ACOUS] For any given cycle of a periodic wave, the maximum absolute value of the instantaneous sound pressure occurring during that cycle. { 'mak·sə·məm 'saúnd ,presh·ər }

maximum subsidence [GEOL] The maximum amount of subsidence in a basin. { 'mak·sə·məm səb'sīd·əns }

maximum sustainable yield [OCEANOGR] **1.** In fishery management, the highest average fishing level over time that does not reduce a stock's abundance in balance with the stock's reproductive and growth capacities under a given set of environmental conditions. **2.** A level of fishing that, if approached, should signal caution rather than increased fishing. { ¦mak·sə·məm sə¦stān·ə·bul 'yēld }

maximum thermometer [ENG] A thermometer that registers the maximum temperature attained during an interval of time. { 'mak·sə·məm thər'mäm·əd·ər }

maximum thrust [ORD] The highest thrust recorded on the thrust-time trace in missile testing. { 'mak·sə·məm 'thrəst }

maximum unambiguous range [ELECTROMAG] The range beyond which the echo from a pulsed radar signal returns after generation of the next pulse, and can thus be mistaken as a short-range echo of the next cycle. { 'mak·sə·məm ˌən·amˌbig·yə·wəs 'rānj }

maximum undistorted power output [ELECTR] Of a transducer, the maximum power delivered under specified conditions with a total harmonic output not exceeding a specified percentage. { 'mak·sə·məm ˌən·diˌstȯrd·əd 'paů·ər ˌaůt̯půt }

maximum usable frequency [COMMUN] The upper limit of the frequencies that can be used at a specified time for point-to-point radio transmission involving propagation by reflection from the regular ionized layers of the ionosphere. Abbreviated MUF. { 'mak·sə·məm ˌyü·zə·bəl 'frē·kwən·sē }

maximum-value theorem [MATH] The theorem that there is a point in the domain of a real-valued function at which the function has its greatest value if this domain is compact. { ˌmak·sə·məm 'val·yü ˌthir·əm }

maximum-wind and shear chart [METEOROL] A synoptic chart on which are plotted the altitudes of the maximum wind speed, the maximum wind velocity (wind direction optional), plus the velocity of the wind at mandatory levels both above and below the level of maximum wind. Also known as max-wind and shear chart. { 'mak·sə·məm 'wind ən 'shir ˌchärt }

maximum-wind level [METEOROL] The height at which the maximum wind speed occurs, determined in a winds-aloft observation. Also known as max-wind level. { 'mak·sə·məm 'wind ˌlev·əl }

maximum-wind topography [METEOROL] The topography of the surface of maximum wind speed. Also known as max-wind topography. { 'mak·sə·məm 'wind tə̇'päg·rə·fē }

maximum working area [IND ENG] That portion of the working area that is readily accessible to the hands of a worker when in his normal operating position. { 'mak·sə·məm 'wərk·iŋ ˌer·ē·ə }

maximum zonal westerlies [METEOROL] The average west-to-east component of wind over the continuous 20° belt of latitude in which this average is a maximum; it is usually found, in the winter season, in the vicinity of 40–60° north latitude. { 'mak·sə·məm ˌzōn·əl 'wes·tərˌlēz }

maxoplasia [MED] Degenerative disease of the breast. { ˌmak·sō'plā·zhə }

maxwell [ELECTROMAG] A centimeter-gram-second electromagnetic unit of magnetic flux, equal to the magnetic flux which produces an electromotive force of 1 abvolt in a circuit of one turn linking the flux, as the flux is reduced to zero in 1 second at a uniform rate; equal to 10^{-8} weber. Abbreviated Mx. Also known as abweber (abWb); line of magnetic induction. { 'mak·swel }

Maxwell body See Maxwell liquid. { 'mak·swel ˌbäd·ē }

Maxwell-Boltzmann density function See Maxwell-Boltzmann distribution. { 'mak·swel 'bōlts·mən 'den·səd·ē ˌfəŋk·shən }

Maxwell-Boltzmann distribution [STAT MECH] Any function giving the probability (or some function proportional to it) that a molecule of a gas in thermal equilibrium will have values of certain variables within given infinitesimal ranges, assuming that the gas molecules obey classical mechanics, and possibly making other assumptions; examples are the Maxwell distribution and the Boltzmann distribution. Also known as Maxwell-Boltzmann density function. { 'mak·swel 'bōlts·mən ˌdi·strə,byü·shən }

Maxwell-Boltzmann equation See Boltzmann transport equation. { 'mak·swel 'bōlts·mən iˌkwā·zhən }

Maxwell-Boltzmann statistics [STAT MECH] The classical statistics of identical particles, as opposed to the Bose-Einstein or Fermi-Dirac statistics. Also known as Boltzmann statistics. { 'mak·swel 'bōlts·mən stə'tis·tiks }

Maxwell bridge [ELEC] A four-arm alternating-current bridge used to measure inductance (or capacitance) in terms of resistance and capacitance (or inductance); bridge balance is independent of frequency. Also known as Maxwell-Wien bridge; Wien-Maxwell bridge. { 'mak·swel 'brij }

Maxwell distribution [STAT MECH] A function giving the number of molecules of a gas in thermal equilibrium whose velocities lie within a given, infinitesimal range of values, assuming that the molecules obey classical mechanics, and do not interact. Also known as Maxwellian distribution. { 'mak·swel ˌdi·strə,byü·shən }

Maxwell effect [OPTICS] Double refraction of a viscous liquid having anisotropic molecules, which results from components of the velocity gradient perpendicular to the fluid velocity itself. { 'mak·swel iˌfekt }

Maxwell equal-area rule [THERMO] At temperatures for which the theoretical isothermal of a substance, on a graph of pressure against volume, has a portion with positive slope (as occurs in a substance with liquid and gas phases obeying the van der Waals equation), a horizontal line drawn at the equilibrium vapor pressure and connecting two parts of the isothermal with negative slope has the property that the area between the horizontal and the part of the isothermal above it is equal to the area between the horizontal and the part of the isothermal below it. { 'mak·swel ˌē·kwəl 'er·ē·ə ˌrül }

Maxwell field equations [ELECTROMAG] Four differential equations which relate the electric and magnetic fields to electric charges and currents, and form the basis of the theory of electromagnetic waves. Also known as electromagnetic field equations; Maxwell equations. { 'mak,swel 'fēld iˌkwā·zhənz }

Maxwellian distribution See Maxwell distribution. { mak ˌswel·ē·ən ˌdi·strə byü·shən }

Maxwellian distribution law [STAT MECH] Equation relating the statistical distribution of speeds and energies of molecules of a pure gas at a uniform temperature where there are no convection currents. { mak,swel·ē·ən ˌdi·strə'byü·shən ˌlȯ }

Maxwellian equilibrium [STAT MECH] Thermal equilibrium of a gas, or of some group of particles, in which the velocity distribution of the particles is the Maxwell distribution corresponding to the temperature of the object with which they are in equilibrium. { mak,swel·ē·ən ē·kwə'lib·rē·əm }

Maxwellian gas [STAT MECH] A gas whose molecules have the Maxwell distribution of velocities. { mak,swel·ē·ən 'gas }

Maxwellian view [OPTICS] A method of using an optical instrument in which a real image of a light source is focused on the pupil of the eye, instead of using an eyepiece. { mak ˌswel·ē·ən 'vyü }

Maxwell liquid [FL MECH] A liquid whose rate of deformation is the sum of a term proportional to the shearing stress acting on it and a term proportional to the rate of change of this stress. Also known as Maxwell body. { 'mak,swel ˌlik·wəd }

Maxwell primaries [OPTICS] The primary colors in a system of colorimetry devised by J.C. Maxwell; they are cyan, green, and magenta. { 'mak,swel 'prī,mer·ēz }

Maxwell relation [ELECTROMAG] According to Maxwell's electromagnetic theory, that relation wherein the dielectric constant of a substance equals the square of its index of refraction. [THERMO] One of four equations for a system in thermal equilibrium, each of which equates two partial derivatives, involving the pressure, volume, temperature, and entropy of the system. { 'mak,swel ri'lā·shən }

Maxwell's coefficient of diffusion [FL MECH] A number in an equation for the difference between mean velocities of two gases which are allowed to mix, which determines the contribution to this quantity of the concentration gradient. { 'mak ˌswelz ˌkō·i·fish·ənt əv di'fyü·zhən }

Maxwell's cyclic currents See mesh currents. { 'mak,swelz 'sī·klik 'kə·rəns }

Maxwell's demon See demon of Maxwell. { 'mak,swelz 'dē·mən }

Maxwell's displacement current See displacement current. { 'mak,swelz di'splās·mənt ˌkə·rənt }

Maxwell's electromagnetic theory [ELECTROMAG] A mathematical theory of electric and magnetic fields which predicts the propagation of electromagnetic radiation, and is valid for electromagnetic phenomena where effects on an atomic scale can be neglected. { 'mak,swelz iˌlek·trō·mag'ned·ik 'thē·ə·rē }

Maxwell's law [ELECTROMAG] A movable portion of a circuit will always move in such a direction as to give maximum magnetic flux linkages through the circuit. { 'mak,swelz 'lȯ }

Maxwell's stress functions [MECH] Three functions of position, ϕ_1, ϕ_2, and ϕ_3, in terms of which the elements of the stress tensor σ of a body may be expressed, if the body is in equilibrium and is not subjected to body forces; the elements of the stress tensor are given by $\sigma_{11} = \partial^2\phi_2/\partial x_3^2 + \partial^2\phi_3/\partial x_2^2$, $\sigma_{23} = -\partial^2\phi_1/\partial x_2\partial x_3$, and cyclic permutations of these equations. { 'mak,swelz 'stres ˌfəŋk·shənz }

Maxwell's stress tensor [ELECTROMAG] A second-rank tensor whose product with a unit vector normal to a surface gives the force per unit area transmitted across the surface by an electromagnetic field. { 'mak,swelz 'stres ,ten·sər }

Maxwell's theorem [MECH] If a load applied at one point *A* of an elastic structure results in a given deflection at another point *B*, then the same load applied at *B* will result in the same deflection at *A*. { 'mak,swelz 'thir·əm }

Maxwell's theory of light [OPTICS] An application of Maxwell's electromagnetic theory in which light is treated as a propagating electromagnetic wave. { 'mak,swelz ,thē·ə·rē əv 'līt }

Maxwell triangle [OPTICS] Color-matching chromaticity values plotted on an *x,y* diagram. Also known as *x,y* chromaticity diagram. { 'mak,swel 'trī,aŋ·gəl }

maxwell-turn [ELECTROMAG] A centimeter-gram-second electromagnetic unit of flux linkage, equal to the flux linkage of a coil consisting of one complete loop of wire through which passes a magnetic flux of one maxwell. Also known as line-turn. { 'mak,swel ,tərn }

Maxwell-Wagner mechanism [ELEC] A capacitor consisting of two parallel metal plates with two layers of material between them, one with vanishing conductivity, the other with finite conductivity and vanishing electric susceptibility. { 'mak,swel 'wag·nər 'mek·ə,niz·əm }

Maxwell-Wien bridge *See* Maxwell bridge. { 'mak,swel 'wēn 'brij }

max-wind and shear chart *See* maximum-wind and shear chart. { 'maks 'wind ən 'shir ,chärt }

max-wind level *See* maximum-wind level. { 'maks 'wind ,lev·əl }

max-wind topography *See* maximum-wind topography. { 'maks 'wind tə'päg·rə·fē }

Mayall's object [ASTRON] A peculiar object that consists of a ring, a cigar-shaped galaxy, and a bridge that appears to connect them. { 'mā·ôlz ,äb·jikt }

mayer [THERMO] A unit of heat capacity equal to the heat capacity of a substance whose temperature is raised 1° Celsius by 1 joule. { 'mī·ər }

Mayer condensation theory [STAT MECH] A theory of the condensation and critical state of a system of chemically saturated molecules, in which the system is assumed to consist of independent clusters of molecules. { 'mī·ər ,kän·dən'sā·shən ,thē·ə·rē }

Mayer's formula [THERMO] A formula which states that the difference between the specific heat of a gas at constant pressure and its specific heat at constant volume is equal to the gas constant divided by the molecular weight of the gas. { 'mī·ərz ,fôr·myə·lə }

mayfly [INV ZOO] The common name for insects composing the order Ephemeroptera. { 'mā,flī }

May-Grünwald stain [MATER] A saturated solution of methylene blue eosinate in methyl alcohol; used to stain blood. Also known as Jenner's stain. { 'mī 'grün,vält ,stān }

mayonnaise [FOOD ENG] A semisolid emulsion prepared by whipping a mixture of raw eggs or egg yolks, vegetable oil, vinegar or lemon juice, salt, and condiments; used as a dressing. { 'mā·ə,nāz }

maz-, mazo- [EMBRYO] A combining form denoting placenta. { māz,'mā·zō }

mazaedium [BOT] The fruiting body of certain lichens, with the spores lying in a powdery mass in the capitulum. [MYCOL] A slimy layer on the hymenial surface of some ascomycetous fungi. { mə'zē·dē·əm }

maze [PSYCH] A network of paths, blind alleys, and compartments; used in intelligence tests and in experimental psychology for developing learning curves. { māz }

mazopathy [MED] **1.** Disease of the placenta. **2.** *See* mastopathy. { mā'zäp·ə·thē }

mazopexy *See* mastopexy. { 'mā·zə,pek·sē }

Mazzoni's corpuscle [NEUROSCI] A specialized encapsulated sensory nerve end organ on a tendon in series with muscle fibers. { mät'zō·nēz ,kòr·pə·səl }

mb *See* millibar; millibarn.

MBC *See* maximum breathing capacity.

MBE *See* molecular-beam epitaxy.

Mbit *See* megabit. { 'em,bit }

MBT *See* mercaptobenzothiazole.

Mbyte *See* megabyte. { 'em,bīt }

mc *See* millihertz.

Mc *See* megahertz.

McArdle's syndrome [MED] A hereditary metabolic disorder caused by deficiency of muscle phosphorylase, with abnormal glycogen deposition in skeletal muscle leading to muscle fiber destruction. Also known as myophosphorylase deficiency glycogenosis. { mə'kärd·əlz ,sin,drōm }

M-C asphalt *See* medium-curing asphalt. { ¦em¦sē 'as,fôlt }

McBurney's incision [MED] A short diagonal incision in the lower right quadrant in which the muscle fibers are separated rather than cut; used for appendectomy. { mək'bər·nēz in,sizh·ən }

McBurney's point [ANAT] A point halfway between the umbilicus and the anterior superior iliac spine; a point of extreme tenderness in appendicitis. { mək'bər·nēz ,pòint }

McCabe's cyclomatic number [COMPUT SCI] The total number of decision statements in a computer program plus one; a measure of the complexity of the program. { mə,kābz ,sī·klə,mad·ik 'nəm·bər }

McCabe-Thiele diagram [CHEM ENG] Graphical method for calculation of the number of theoretical plates or contacting stages required for a given binary distillation operation. { mə'kāb 'tēl·ə ,dī·ə,gram }

mcd *See* millicurie-destroyed.

M center [SOLID STATE] A color center consisting of an *F* center combined with two ion vacancies. { 'em ,sen·tər }

MCG *See* magnetocardiograph.

McGinty [MIN ENG] Three sheaves over which a rope is passed so as to take a course somewhat like that of the letter M. { mə'gin·tē }

mcgovernite *See* macgovernite. { mə'gəv·ər,nīt }

McGurk effect [PHYSIO] An auditory illusion discovered by H. McGurk that demonstrates the important contribution made by visible face movements to normal speech perception. { mə'gərk i,fekt }

MCHF approximation *See* multiconfiguration Hartree-Fock approximation. { ¦em¦sē¦āch¦ef ə,präk·sə'mā·shən }

McLeod gage [FL MECH] A type of instrument used to measure vacuum by measuring the height of a column of mercury supported by the gas whose pressure is to be measured, when this gas is trapped and compressed into a capillary tube. { mə'klaùd ,gāj }

McLuckie gas detector [MIN ENG] A portable nonautomatic means of analyzing air that can be used underground or aboveground if the air sample is brought out of the mine in small rubber bladders. { mə'klək·ē 'gas də,tek·tər }

McMath telescope [OPTICS] A unique 60-inch (1.5-meter) solar telescope at Kitt Peak, Arizona, that has an unconventional configuration; the sun's light is reflected from an 80-inch (2.0-meter) mirror into a long, fixed tube. { mik'math 'tel·ə,skōp }

McNally-Carpenter centrifuge [MIN ENG] A machine that removes water from fine coal; it has a cone-shaped rotating vertical element, into the top of which the wet feed is put by gravity; a distributing disk forces the material onto the screen or basket lining the cone; as the material approaches the cone bottom, the drying action increases until the dry coal is discharged from the bottom. { mik'nal·ē 'kär·pən·tər 'sen·trə,fyüj }

McNally-Norton jig [MIN ENG] A device used to clean raw coal by carrying it to a wash box and using water whose level rises and falls as a result of air pulsations; the incoming coal is suspended by the pulsating water while the heavier refuse sinks to the bottom; the coal then spills over into the next wash box where the process is repeated, and the clean coal is discharged by the dewatering screens. { mik'nal·ē 'nòrt·ən 'jig }

McNally tube [ELECTR] Reflex klystron tube, the frequency of which may be electrically controlled over a wide range; used as a local oscillator. { mik'nal·ē ,tüb }

McNally-Vissac dryer [MIN ENG] A coal dryer that operates on the convection from the heavy forced draft from a coal-fired furnace; it consists of an inclined reciprocating screen over which coal moves; moisture is removed by the hot air from the furnace passing down through the bed of coal. { mik'nal·ē ,vi,sak ,drī·ər }

M-component hypergammaglobulinemia [MED] A form of hypergammaglobulinemia characterized by a single, prominent, more or less narrow band occurring from the slow gamma

---

MAYFLY

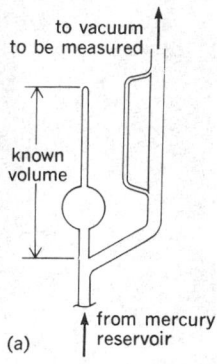

|← 35 mm →|

Mayfly, adult male, *Hexagenia* species, in usual resting position. *(From A. H. Morgan, Field Book of Ponds and Streams, copyright 1930 by G. P. Putnam's Sons)*

McLEOD GAGE

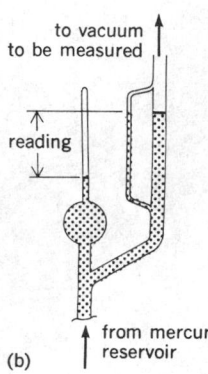

McLeod gage in *(a)* filling (charging) position, and *(b)* measuring position.

to the fast alpha-1 region of the electrophoretic strip. { 'em kəm,pō·nənt ¦hī·pər,gam·ə,gläb·yə·lə'nē·mē·ə }

M contour [CONT SYS] A line on a Nyquist diagram connecting points having the same magnitude of the primary feedback ratio. { 'em ,kän·tùr }

McQuaid-Ehn test [MET] A test for determining the grain-size characteristics of a steel in which a sample is carburized for 8 hours at 1700°F (927°C) and cooled slowly; the high-carbon case on slow cooling will reject cementite at the austenite grain boundaries and, by polishing and etching, the grains will clearly be seen under a microscope. { mə'kwād 'ān ,test }

MCT See MOS-controlled thyristor.

M-cyclin [CELL MOL] A regulatory protein that binds with M-kinase, a cyclin-dependent kinase, activating the cell's entry into mitosis from the G2 stage of the cell cycle. { 'em¦sīk·lən }

md See millidarcy.

Md See mendelevium.

M damage [ORD] Damage causing immobilization of a combat vehicle. { 'em,dam·ij }

M-derived filter [ELECTR] A filter consisting of a series of T or pi sections whose impedances are matched at all frequencies, even though the sections may have different resonant frequencies. { 'em di,rīvd 'fil·tər }

M-design bit [DES ENG] A long-shank, box-threaded core bit made to fit M-design core barrels. { 'em di,zīn ,bit }

M-design core barrel [DES ENG] A double-tube core barrel in which a $2\frac{1}{2}$°-taper core lifter is carried inside a short tubular sleeve coupled to the bottom end of the inner tube, and the sleeve extends downward inside the bit shank to within a very short distance behind the face of the core bit. { 'em di,zīn 'kòr ,bar·əl }

MDF bearing indicator [NAV] A manual direction finder (MDF) indicator operated manually and used with a radio direction finder to indicate the relative, magnetic, or true bearing (or reciprocal) of a transmitter. { 'em¦dē'ef 'ber·iŋ ,in·də,kād·ər }

M display [ELECTR] A modified radarscope A display in which target distance is determined by moving an adjustable pedestal signal along the baseline until it coincides with the horizontal position of the target deflection. { 'em di,splā }

MEA See 2-aminoethanethiol; minimum enroute instrument altitude.

meaconing [ELECTROMAG] A system for receiving electromagnetic signals and rebroadcasting them with the same frequency so as, for instance, to confuse navigation; a confusion reflector, such as chaff, is an example. { 'mē·kə·niŋ }

meadow [ECOL] A vegetation zone which is a low grassland, dense and continuous, variously interspersed with forbs but few if any shrubs. Also known as pelouse; Wiesen. [ENG] Range of air-fuel ratio within which smooth combustion may be had. { 'med·ō }

meadow ore See bog iron ore. { 'med·ō ,òr }

meager set [MATH] A set that is a countable union of nowhere-dense sets. Also known as set of first category. { ¦mē·gər 'set }

mealybug [INV ZOO] Any of various scale insects of the family Pseudococcidae which have a powdery substance covering the dorsal surface; all are serious plant pests. { 'mē·lē,bəg }

Mealy machine [COMPUT SCI] A sequential machine in which the output depends on both the current state of the machine and the input. { 'mē·lē mə,shēn }

mean [MATH] A single number that typifies a set of numbers, such as the arithmetic mean, the geometric mean, or the expected value. Also known as mean value. { mēn }

mean-average boiling point [CHEM ENG] Pseudo boiling point for a hydrocarbon mixture; calculated from the American Society for Testing and Materials distillation curve's volumetric average boiling point. { ¦mēn ¦av·rij 'bòil·iŋ ,pòint }

mean British thermal unit See British thermal unit. { 'mēn ¦brid·ish 'thər·məl ,yü·nət }

mean calorie [THERMO] One-hundredth of the heat needed to raise 1 gram of water from 0 to 100°C. { mēn 'kal·ə,rē }

mean camber line [AERO ENG] A line on a cross section of a wing of an aircraft which is equidistant from the upper and lower surfaces of the wing. { 'mēn 'kam·bər ,līn }

mean carrier frequency [ELECTR] Average carrier frequency of a transmitter corresponding to the resting frequency

in a frequency-modulated system. { 'mēn 'kar·ē·ər ,frē·kwən·sē }

mean chart [METEOROL] Any chart on which isopleths of the mean value of a given meteorological element are drawn. Also known as mean map. { 'mēn ,chärt }

mean chord [AERO ENG] That chord of an airfoil that is equal to the sum of all the airfoil's chord lengths divided by the number of chord lengths; equivalently, that chord whose length is equal to the area of the airfoil section divided by the span. { 'mēn 'kòrd }

mean curvature [MATH] Half the sum of the principal curvatures at a point on a surface. { ¦mēn 'kər·və·chər }

mean depth [HYD] Average water depth in a stream channel or conduit computed by dividing the cross-sectional area by the surface width. { 'mēn 'depth }

meander [HYD] A sharp, sinuous loop or curve in a stream, usually part of a series. [OCEANOGR] A deviation of the flow pattern of a current. { mē'an·dər }

meander bar See point bar. { mē'an·dər ,bär }

meander belt [GEOL] The zone along the floor of a valley across which a meandering stream periodically shifts its channel. { mē'an·dər ,belt }

meander core [GEOL] A hill encircled by a stream meander. Also known as rock island. { mē'an·dər ,kòr }

meandering stream [HYD] A stream having a pattern of successive meanders. Also known as snaking stream. { mē'an·də·riŋ 'strēm }

meander niche [GEOL] A conical or crescentic opening in the wall of a cave formed by downward and lateral stream erosion. { mē'an·dər ,nich }

meander plain [GEOL] A plain built by the meandering process, or a plain of lateral accretion. { mē'an·dər ,plān }

meander scar [GEOL] A crescentic, concave mark on the face of a bluff or valley wall formed by a meandering stream. { mē'an·dər ,skär }

meander spur [GEOL] An undercut projection of high land that extends into the concave part of, and is enclosed by, a meander. { mē'ar·dər ,spər }

mean deviation See average deviation. { 'mēn ,dē·vē'ā·shən }

mean difference [STAT] The average of the absolute values of the $n(n-1)/2$ differences between pairs of elements in a statistical distribution that has n elements. { 'mēn 'dif·rəns }

mean diurnal high-water inequality [OCEANOGR] The average difference between the heights of the two high waters of each tidal day over a 19-year period; it is obtained by subtracting the mean of all high waters from the mean of the higher high waters. { 'mēn dī'ərn·əl 'hī ,wòd·ər ,in·i'kwäl·əd·ē }

mean diurnal low-water inequality [OCEANOGR] Half the average difference between the heights of the two low waters of each tidal day over a 19-year period; it is obtained by subtracting the mean of all lower low waters from the mean of the low waters. { 'mēn dī'ərn·əl 'lō ,wòd·ər ,in·i'kwäl·əd·ē }

mean draft [NAV ARCH] The mean of the drafts at the bow and the stern; for a vessel with a straight keel, the draft at the midpoint of the waterline length. { 'mēn 'draft }

mean effective pressure [MECH ENG] A term commonly used in the evaluation for positive displacement machinery performance which expresses the average net pressure difference in pounds per square inch on the two sides of the piston in engines, pumps, and compressors. Abbreviated mep; mp. Also known as mean pressure. { 'mēn i¦fek·tiv 'presh·ər }

mean evolute [MATH] The envelope of the planes that are orthogonal to the normals of a given surface and cut the normals halfway between the centers of principal curvature of the surface. { ¦mēn 'ev·ə,lüt }

mean free path [ACOUS] For sound waves in an enclosure, the average distance sound travels between successive reflections in the enclosure. [PHYS] The average distance traveled between two similar events, such as elastic collisions of molecules in a gas, of electrons or phonons in a crystal, or of neutrons in a moderator. { 'mēn ¦frē 'path }

mean height of burst [ORD] Average of the heights of bursts of a group of shots fired with the same firing data. { 'mēn 'hīt əv 'bərst }

mean higher high water [OCEANOGR] The average height

of higher high waters at a place over a 19-year period. { 'mēn ‖hī·ər 'hī 'wȯd·ər }

mean high water [OCEANOGR] The average height of all high waters recorded at a given place over a 19-year period. { 'mēn 'hī 'wȯd·ər }

mean high-water lunitidal interval [OCEANOGR] The average interval of time between the transit (upper or lower) of the moon and the next high water at a place. Also known as corrected establishment. { 'mēn ‖hī ‖wȯd·ər 'lü·nə‖tīd·əl 'in·tər·vəl }

mean high-water neaps [OCEANOGR] The average height of the high waters of neap tides. Also known as neap high water. { 'mēn ‖hī ‖wȯd·ər 'nēps }

mean high-water springs [OCEANOGR] The average height of the high waters of spring tides. Also known as high-water springs; spring high water. { 'mēn ‖hī ‖wȯd·ər 'spriŋz }

mean latitude [GEOD] Half the arithmetical sum of the latitudes of two places on the same side of the equator; mean latitude is labeled N or S to indicate whether it is north or south of the equator. { 'mēn 'lad·ə,tüd }

mean life [PHYS] The average time during which a system, such as an atom, nucleus, or elementary particle, exists in a specified form; for a radionuclide or an excited state of an atom or nucleus, it is the reciprocal of the decay constant. Also known as average life; lifetime. { 'mēn 'līf }

mean lower low water [OCEANOGR] The average height of the lower low waters at a place over a 19-year period. { 'mēn ‖lō·ər 'lō 'wȯd·ər }

mean lower low-water springs [OCEANOGR] The average height of lower low-water springs at a place. { 'mēn ‖lō·ər ‖lō‖wȯd·ər 'spriŋz }

mean low water [OCEANOGR] The average height of all low waters recorded at a given place over a 19-year period. { 'mēn 'lō 'wȯd·ər }

mean low-water lunitidal interval [OCEANOGR] The average interval of time between the transit (upper or lower) of the moon and the next low water at a place. { 'mēn ‖lō 'wȯd·ər ‖lü·nə‖tīd·əl 'in·tər·vəl }

mean low-water neaps [OCEANOGR] The average height of the low water at neap tides. Also known as low-water neaps; neap low water. { 'mēn ‖lō 'wȯd·ər 'nēps }

mean low-water springs [OCEANOGR] The average height of the low waters of spring tides; this level is used as a tidal datum in some areas. Also known as low-water springs; spring low water. { 'mēn ‖lō ‖wȯd·ər 'spriŋz }

mean map *See* mean chart. { 'mēn ,map }

mean motion [ASTRON] The speed which a planet or its satellite would have if it were moving in a circular orbit with radius equal to its distance from the sun or a central planet with a period equal to its actual period. { 'mēn 'mō·shən }

mean neap range *See* neap range. { 'mēn 'nēp ,rānj }

mean neap rise [OCEANOGR] The height of mean high-water neaps above the chart datum. { 'mēn 'nēp ,rīz }

mean noon [ASTRON] Twelve o'clock mean time, or the instant the mean sun is over the upper branch of the meridian; it may be either local or Greenwich, depending upon the reference meridian. { 'mēn 'nün }

mean normal stress [MECH] In a system stressed multiaxially, the algebraic mean of the three principal stresses. { 'mēn ‖nȯrm·əl 'stres }

mean place [ASTRON] The position of a star on the celestial sphere as it would be observed from the center of the sun, referred to the mean celestial equator and celestial equinox for the beginning of the year of observation. { 'mēn 'plās }

mean point of impact [ORD] The point which is at the geometrical center of all the points of impact of the several shots of a salvo, excluding wild shots; when firing time-fused projectiles, the point of detonation of such a projectile is considered to be the point of impact. { 'mēn 'pȯint əv 'im,pakt }

mean point of impact error [ORD] Distance between the mean point of impact and the target. { 'mēn 'pȯint əv 'im,pakt ,er·ər }

mean power [ELECTR] For a radio transmitter, the power supplied to the antenna transmission line by a transmitter during normal operation, averaged over a time sufficiently long compared with the period of the lowest frequency encountered in the modulation; a time of 1/10 second during which the mean power is greatest will be selected normally. { 'mēn 'pau̇·ər }

mean pressure *See* mean effective pressure. { 'mēn 'presh·ər }

mean profile [ASTRON] The waveform of a pulsar's periodic emission, averaged synchronously over several hundred pulses or more. Also known as integrated profile; pulse window. { 'mēn 'prō,fīl }

mean proportional [MATH] For two numbers a and b, a number x, such that $x/a = b/x$. { 'mēn prə'pȯr·shən·əl }

mean range [OCEANOGR] The difference in the height between mean high water and mean low water. [ORD] Average distance reached by a group of shots fired with the same firing data. [SCI TECH] The average difference in the extreme values of a variable quantity. { 'mēn 'rānj }

mean rank method [STAT] A method of handling data which has the same observed frequency occurring at two or more consecutive ranks; it consists of assigning the average of the ranks as the rank for the common frequency. { 'mēn 'raŋk ,meth·əd }

mean rise interval [OCEANOGR] The average interval of time between the transit (upper or lower) of the moon and the middle of the period of rise of the tide at a place; it may be either local or Greenwich, depending on the transit to which it is referred, but the local interval is assumed unless otherwise specified. { 'mēn 'rīz ,in·tər·vəl }

mean rise of tide [OCEANOGR] The height of mean high water above the chart datum. { 'mēn 'rīz əv 'tīd }

mean river level [HYD] The average height of the surface of a river at any point for all stages of the tide over a 19-year period. { 'mēn 'riv·ər ,lev·əl }

mean sea level [OCEANOGR] The average sea surface level for all stages of the tide over a 19-year period, usually determined from hourly height readings from a fixed reference level. { 'mēn 'sē ,lev·əl }

means-ends analysis [COMPUT SCI] A method of problem solving in which the difference between the form of the data in the present and desired situations is determined, and an operator is then found to transform from one into the other, or, if this is not possible, objects between the present and desired objects are created, and the same procedure is then repeated on each of the gaps between them. { 'mēnz 'enz ə,nal·ə·səs }

mean sidereal time [ASTRON] Sidereal time adjusted for nutation, to eliminate slight irregularities in the rate. { 'mēn sī'dir·ē·əl 'tīm }

mean solar day [ASTRON] The duration of one rotation of the earth on its axis, with respect to the mean sun; the length of the mean solar day is 24 hours of mean solar time or 24^h 03^m 56.555^s of mean sidereal time. { 'mēn 'sō·lər 'dā }

mean solar second [ASTRON] A unit equal to 1/86,400 of a mean solar day. { 'mēn 'sō·lər 'sek·ənd }

mean solar time [ASTRON] Time that has the mean solar second as its unit, and is based on the mean sun's motion. { 'mēn 'sō·lər 'tīm }

mean specific heat [THERMO] The average over a specified range of temperature of the specific heat of a substance. { 'mēn spə'sif·ik 'hēt }

mean spherical intensity [OPTICS] The luminous intensity of a light source averaged over all directions. { 'mēn 'sfer·ə·kəl in'ten·səd·ē }

mean spring range *See* spring range. { 'mēn 'spriŋ ,rānj }

mean spring rise [OCEANOGR] The height of mean high-water springs above the chart datum. { 'mēn 'spriŋ ,rīz }

mean square [STAT] The arithmetic mean of the squares of the differences of a set of values from some given value. { 'mēn 'skwer }

mean-square deviation [STAT] A measure of the extent to which a collection $v_1, v_2, \ldots, v_n$ of numbers is unequal; it is given by the expression $(1/n)[(v_1 - \bar{v})^2 + \cdots + (v_n - \bar{v})^2]$, where $\bar{v}$ is the mean of the numbers. { 'mēn 'skwer dē·vē'ā·shən }

mean-square error [STAT] The residual or error sum of squares divided by the number of degrees of freedom of the sum; gives an estimate of the error or residual variance. { ‖mēn ‖skwer 'er·ər }

mean-square-error criterion [CONT SYS] Evaluation of the performance of a control system by calculating the square root of the average over time of the square of the difference between the actual output and the output that is desired. { 'mēn 'skwer 'er·ər krī,tir·ē·ən }

mean-square velocity [PHYS] The average value of the

square of the velocities of a group of particles, such as the molecules of a gas.   { ¦mēn ¦skwər və'läs·əd·ē }

mean stress [MECH] **1.** The algebraic mean of the maximum and minimum values of a periodically varying stress.   **2.** *See* octahedral normal stress.   { 'mēn 'stres }

mean sun [ASTRON] A fictitious sun conceived to move eastward along the celestial equator at a rate that provides a uniform measure of time equal to the average apparent time; used as a reference for reckoning time, such as mean time or zone time.   { 'mēn 'sən }

mean temperature [METEOROL] The average temperature of the air as indicated by a properly exposed thermometer during a given time period, usually a day, month, or year.   { 'mēn 'tem·prə·chər }

mean temperature difference [CHEM ENG] In heat exchange calculations, a pseudo average temperature difference between the warmer and colder fluids at inlet and outlet conditions.   { 'mēn 'tem·prə·chər ¦dif·rəns }

mean terms [MATH] The second and third terms of a proportion.   { 'mēn 'tərmz }

Meantes [VERT ZOO] The mud eels, a small suborder of the Urodela including three species of aquatic eellike salamanders with only anterior limbs.   { mē'an¦tēz }

mean tide *See* half tide.   { 'mēn 'tīd }

mean tide level [OCEANOGR] The tide level halfway between mean high water and mean low water.   { 'mēn 'tīd ¦lev·əl }

mean time [ASTRON] Time based on the rotation of the earth relative to the mean sun.   { 'mēn 'tīm }

mean time between failures [COMPUT SCI] A measure of the reliability of a computer system, equal to average operating time of equipment between failures, as calculated on a statistical basis from the known failure rates of various components of the system.   Abbreviated MTBF.   { 'mēn 'tīm bi¦twēn 'fāl·yərz }

mean-time screw [HOROL] A screw in a watch balance, used to regulate precisely the speed of oscillation.   { 'mēn ¦tīm ¦skrü }

mean time to failure [ENG] A measure of reliability of a piece of equipment, giving the average time before the first failure.   { 'mēn 'tīm tə 'fāl·yər }

mean time to repair [ENG] A measure of reliability of a piece of repairable equipment, giving the average time between repairs.   { 'mēn 'tīm tə ri'per }

mean-tone scale [ACOUS] A musical scale formed by giving the major third a ratio of exactly 5:4 and adapting other intervals to equalize them.   { 'mēn ¦tōn ¦skāl }

mean trajectory [MECH] The trajectory of a missile that passes through the center of impact or center of burst.   { 'mēn trə'jek·trē }

mean value [MATH] **1.** For a function $f(x)$ defined on an interval (a,b), the integral from a to b of $f(x)\,dx$ divided by $b - a$.   **2.** *See* mean.   { 'mēn 'val·yü }

mean value theorem [MATH] The proposition that, if a function $f(x)$ is continuous on the closed interval $[a,b]$ and differentiable on the open interval (a,b), then there exists x_0, $a < x_0 < b$, such that $f(b) - f(a) = (b - a)f'(x_0)$.   Also known as first law of the mean; Lagrange's formula; law of the mean.   { 'mēn 'val·yü ¦thir·əm }

mean velocity [PHYS] The average value of the velocities of a group of particles, such as the molecules of a gas.   { 'mēn və'läs·əd·ē }

mean water level [OCEANOGR] The average surface level of a body of water.   { 'mēn 'wȯd·ər ¦lev·əl }

measles [MED] An acute, highly infectious viral disease with cough, fever, and maculopapular rash; the appearance of Koplik spots on the oral mucous membranes marks the onset. Also known as rubeola.   { 'mē·zəlz }

measles encephalitis [MED] Acute disseminated encephalitis following measles.   { 'mē·zəlz in¦sef·ə'līd·əs }

measles immune globulin [IMMUNOL] Sterile human globulin used to provide passive immunization against measles.   { 'mē·zəlz i'myün 'gläb·yə·lən }

measles virus vaccine [IMMUNOL] A suspension of live attenuated or inactivated measles virus used for active immunization against measles.   { 'mē·zəlz 'vī·rəs vak¦sēn }

measly [FOOD ENG] Of meat, containing larval tapeworms. [MED] Infected with measles.   { 'mēz·lē }

measurable function [MATH] **1.** A real valued function f defined on a measurable space X, where for every real number

a all those points *x* in *X* for which $f(x) \geq a$ form a measurable set.   **2.** A function on a measurable space to a measurable space such that the inverse image of a measurable set is a measurable set.   { 'mezh·rə·bəl 'fəŋk·shən }

measurable set [MATH] A member of the sigma-algebra of subsets of a measurable space.   { 'mezh·rə·bəl 'set }

measurable space [MATH] A set together with a sigma-algebra of subsets of this set.   { 'mezh·rə·bəl 'spās }

measurand transmitter [COMMUN] A telemetry transmitter that transmits a signal modulated according to the values of the quantity being measured.   { 'mezh·ə¦rand tranz¦mid·ər }

measure [MATH] A nonnegative real valued function m defined on a sigma-algebra of subsets of a set S whose value is zero on the empty set, and whose value on a countable union of disjoint sets is the sum of its values on each set.   { 'mezh·ər }

measured daywork [IND ENG] Work done for an hourly wage on which specific productivity levels have been determined but which provides no incentive pay.   { ¦mezh·ərd 'dā¦wȯrk }

measured drawing [ARCH] An architectural representation drawn to the scale of an existing building.   { ¦mezh·ərd 'drȯ·iŋ }

measured drilling depth [ENG] The apparent depth of a borehole as measured along its longitudinal axis.   { 'mezh·ərd 'dril·iŋ ¦depth }

measured mile [CIV ENG] The distance of 1 mile (1609.344 meters), the units of which have been accurately measured and marked.   [NAV] A length of 1 nautical mile (1852 meters), the limits of which have been accurately measured and are indicated by ranges ashore; used by vessels to calibrate logs, engine revolution counters, and such, and to determine speed.   { 'mezh·ərd 'mīl }

measured ore *See* developed reserves.   { 'mezh·ərd 'ȯr }

measured relieving capacity [DES ENG] The measured amounts of fluid which can be exhausted through a relief device at its rated operating pressure.   { 'mezh·ərd ri'lēv·iŋ kə,pas·əd·ē }

measured service [COMMUN] Telephone service for which charge is made according to the measured amount of usage.   { 'mezh·ərd 'sər·vəs }

measured spectrum *See* spectrogram.   { 'mezh·ərd 'spek·trəm }

measured work [IND ENG] Work, operations, or cycles for which a standard has been set.   { ¦mezh·ərd 'wərk }

measurement [SCI TECH] The process of determining the value of some quantity in terms of a standard unit.   { 'mezh·ər·mənt }

measurement ton *See* ton.   { 'mezh·ər·mənt ¦tən }

measure of location [STAT] A statistic, such as the mean, median, quartile, or mode; it has the property for the mean that if a constant is added to each value the same constant must also be added to the location measure.   { ¦mezh·ər əv lō'kā·shən }

measure-preserving transformation [MATH] A transformation T of a measure space S into itself such that if E is a measurable subset of S then so is $T^{-1}E$ (the set of points mapped into E by T) and the measure of $T^{-1}E$ is then equal to that of E.   { ¦mezh·ər pri¦zərv·iŋ ¦tranz·fər mā·shən }

measure space [MATH] A set together with a sigma-algebra of subsets of the set and a measure defined on this sigma-algebra.   { 'mezh·ər ¦spās }

measure theory [MATH] The study of measures and their applications, particularly the integration of mathematical functions.   { 'mezh·ər ¦thē·ə·rē }

measure zero [MATH] **1.** A set has measure zero if it is measurable and the measure of it is zero.   **2.** A subset of Euclidean n-dimensional space which has the property that for any positive number ϵ there is a covering of the set by n-dimensional rectangles such that the sum of the volumes of the rectangles is less than ϵ.   { 'mezh·ər ¦zir·ō }

measuring chute [MIN ENG] An ore bin or coal bin installed adjacent to the shaft bottom in skip winding and having a capacity equal to that of the skip used; ensures rapid, correct loading of skips without spillage.   Also known as measuring pocket.   { 'mezh·ə·riŋ ¦shüt }

measuring day [MIN ENG] The day when work is measured and recorded for assessing the wages.   { 'mezh·ə·riŋ ¦dā }

measuring machine [ENG] A device in which an astronomical photographic plate is viewed through a fixed low-power microscope with cross-hairs and which is mounted on a carriage that is moved by micrometer screws equipped with scales, in order to measure the relative positions of images on the plate. { 'mezh·ə·riŋ mə,shēn }

measuring pocket See measuring chute. { 'mezh·ə·riŋ ,päk·ət }

measuring tank [ENG] A tank that has been calibrated and fitted with devices to measure a volume of liquid and then release it. Also known as dump tank; metering tank. { 'mezh·ə·riŋ ,taŋk }

meatal plate [EMBRYO] A mass of ectodermal cells on the bottom of the branchial groove in a 2-month embryo. { mē'ād·əl 'plāt }

meatotomy [MED] Incision into and enlargement of a meatus. { ,mē·ə'täd·ə·mē }

meatus [ANAT] A natural opening or passage in the body. { mē'ād·əs }

mechanical [ENG] Of, pertaining to, or concerned with machinery or tools. [GRAPHICS] A finished copy that usually contains hand lettering, type proofs, and art especially positioned and mounted so that a photochemical reproduction can be made on a letterpress, offset, or other printing plate. Also known as keyline layout; paste-up. { mi'kan·ə·kəl }

mechanical advantage [MECH ENG] The ratio of the force produced by a machine such as a lever or pulley to the force applied to it. Also known as force ratio. { mi'kan·ə·kəl əd'van·tij }

mechanical alloying [MET] A materials-processing method for assembling metal constituents with a controlled microstructure by repeated welding, fracturing, and rewelding of a mixture of powder particles, generally in a high-energy ball mill. { mi'kan·i·kəl ə'lói·iŋ }

mechanical analog [IND ENG] A mechanical model of a nonmechanical system that responds to an input with an output corresponding to the response of the real system. { mi'kan·i·kəl 'an·ə,läg }

mechanical analog computer [COMPUT SCI] A machine aid to computation in which variables are represented as continuously variable displacements or motions of mechanical elements, such as gears and shafts. { mi'kan·ə·kəl 'an·ə,läg kəm'pyüd·ər }

mechanical analysis [MECH ENG] Mechanical separation of soil, sediment, or rock by sieving, screening, or other means to determine particle-size distribution. { mi'kan·ə·kəl ə'nal·ə·səs }

mechanical area [BUILD] The areas in a building that include equipment rooms, shafts, stacks, tunnels, and closets used for heating, ventilating, air conditioning, piping, communication, hoisting, conveying, and electrical services. { mi'kan·ə·kəl 'er·ē·ə }

mechanical bearing cursor See bearing cursor. { mi'kan·ə·kəl 'ber·iŋ ,kər·sər }

mechanical birefringence [OPTICS] A change in the double refraction of a solid material when it is subjected to stress. Also known as stress birefringence. { mi'kan·ə·kəl ,bī·ri'frin·jəns }

mechanical classification [MECH ENG] A sorting operation in which mixtures of particles of mixed sizes, and often of different specific gravities, are separated into fractions by the action of a stream of fluid, usually water. { mi'kan·ə·kəl ,klas·ə·fə'kā·shən }

mechanical classifier [MECH ENG] Any of various machines that are commonly used to classify mixtures of particles of different sizes, and sometimes of different specific gravities; the Dorr classifier is an example. { mi'kan·ə·kəl 'klas·ə,fī·ər }

mechanical comparator [ENG] A contact comparator in which movement is amplified usually by a rack, pinion, and pointer or by a parallelogram arrangement. { mi'kan·ə·kəl kəm'par·əd·ər }

mechanical computer [COMPUT SCI] A machine such as a Charles Babbage's analytical engine. { mi'kan·ə·kəl kəm'pyüd·ər }

mechanical damping [ENG ACOUS] Mechanical resistance which is generally associated with the moving parts of an electromechanically transducer such as a cutter or a reproducer. { mi'kan·ə·kəl 'damp·iŋ }

mechanical dewaxing [PETRO ENG] A dewaxing procedure

in which cooled oil is forced through wax presses, separating the solid wax particles from the oil. { mi'kan·ə·kəl dē'wak·siŋ }

mechanical dialer See automatic dialer. { mi'kan·ə·kəl 'dī·lər }

mechanical differential analyzer [COMPUT SCI] An analog computer using interconnected surfaces to solve differential equations, such as the device developed by Vannevar Bush at Massachusetts Institute of Technology. { mi'kan·ə·kəl dif·ə'ren·chəl 'an·ə,līz·ər }

mechanical draft [MECH ENG] A draft that depends upon the use of fans or other mechanical devices; may be induced or forced. { mi'kan·ə·kəl 'draft }

mechanical-draft cooling tower [MECH ENG] Cooling tower that depends upon fans for introduction and circulation of its air supply. { mi'kan·ə·kəl ¦draft kül·iŋ ,taů·ər }

mechanical drawing [GRAPHICS] Drawing with the aid of instruments. { mi'kan·ə·kəl 'dró·iŋ }

mechanical dysmenorrhea [MED] Painful menstruation due to mechanical obstruction of the discharge of menstrual fluids. Also known as obstructive dysmenorrhea. { mi'kan·ə·kəl di,smen·ə'rē·ə }

mechanical efficiency [MECH ENG] In an engine, the ratio of brake horsepower to indicated horsepower. { mi'kan·ə·kəl i'fish·ən·sē }

mechanical engineering [MECH ENG] The branch of engineering concerned with energy conversion, mechanics, and mechanisms and devices for diverse applications, ranging from automotive parts through nanomachines. { mi'kan·ə·kəl ,en·jə'nir·iŋ }

mechanical equation of state [MET] An equation that expresses the relation of stress, strain, strain rate, and temperature for a metal. { mi'kan·ə·kəl i'kwā·zhən əv 'stāt }

mechanical equivalent of heat [THERMO] The amount of mechanical energy equivalent to a unit of heat. { mi'kan·ə·kəl i'kwiv·ə·lənt əv 'hēt }

mechanical equivalent of light [OPTICS] The ratio of the radiant power emitted by a monochromatic light source whose wavelength is that at which the sensitivity of phototopic vision is greatest (about 555 nanometers), to its luminous flux measured in lumens. { mi'kan·ə·kəl i'kwiv·ə·lənt əv 'līt }

mechanical erosion See corrasion. { mi'kan·ə·kəl i'rō·zhən }

mechanical expression See expression. { mi'kan·ə·kəl ik'spresh·ən }

mechanical filter [ELECTR] Filter, used in intermediate-frequency amplifiers of highly selective superheterodyne receivers, consisting of shaped metal bars, rods, or disks that act as coupled mechanical resonators when used with piezoelectric or magnetostrictive input and output transducers and coupled by small-diameter wires. Also known as mechanical wave filter. [PETRO ENG] Granule-packed steel shell used to filter suspended floc or undissolved solids out of treated waterflood water; granules can be graded sand and gravel, anthracite coal, graphitic ore, or aluminum-oxide plates with granular filter medium. { mi'kan·ə·kəl 'fil·tər }

mechanical flotation cell [MIN ENG] A device that separates minerals from ore water pulp; it consists of a cell in which the pulp is kept mixed and moving by an impeller at the bottom of the cell; the impeller pulls air down the standpipe and disperses it as bubbles through the pulp; the floatable minerals concentrate in the froth above, and the pulp is removed by a scraper. { mi'kan·ə·kəl flō'tā·shən ,sel }

mechanical gripper [MECH ENG] A robot component that uses movable, fingerlike levers to grasp objects. { mi'kan·ə·kəl 'grip·ər }

mechanical hygrometer [ENG] A hygrometer in which an organic material, most commonly a bundle of human hair, which expands and contracts with changes in the moisture in the surrounding air or gas is held under slight tension by a spring, and a mechanical linkage actuates a pointer. { mi'kan·ə·kəl hī'gräm·əd·ər }

mechanical hysteresis [MECH] The dependence of the strain of a material not only on the instantaneous value of the stress but also on the previous history of the stress; for example, the elongation is less at a given value of tension when the tension is increasing than when it is decreasing. { mi'kan·ə·kəl ,his·tə'rē·səs }

mechanical impedance [MECH] The complex ratio of a phasor representing a sinusoidally varying force applied to a

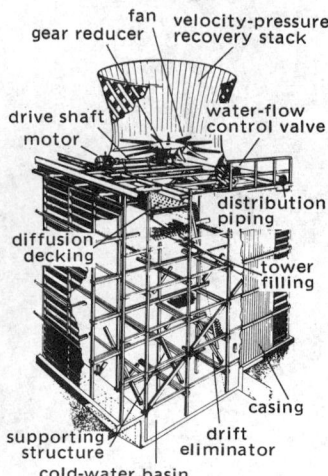

MECHANICAL-DRAFT COOLING TOWER

An induced-draft cross-flow cooling tower showing component parts. *(N. P. Green and Associates)*

system to a phasor representing the velocity of a point in the system. { mi'kan·ə·kəl im'pēd·əns }

mechanical instability *See* absolute instability. { mi'kan·ə·kəl ,in·stə'bil·əd·ē }

mechanical integrator [COMPUT SCI] A mechanical device which draws the graph of the integral of a function when a tracing point is passed over a graph of the function. { mi'kan·ə·kəl 'int·ə,grād·ər }

mechanical jamming *See* passive jamming. { mi'kan·ə·kəl 'jam·iŋ }

mechanical lift dock [CIV ENG] A type of dry dock or marine elevator in which a vessel, after being placed on the keel and bilge blocks in the dock, is bodily lifted clear of the water so that work may be performed on the underwater body. { mi'kan·ə·kəl 'lift 'däk }

mechanical linkage [MECH ENG] A set of rigid bodies, called links, joined together at pivots by means of pins or equivalent devices. { mi'kan·ə·kəl 'liŋ·kij }

mechanical loader [MECH ENG] A power machine for loading mineral, coal, or dirt. { mi'kan·ə·kəl 'lōd·ər }

mechanically foamed plastic [MATER] A foamed plastic having its cellular structure produced by gases that are physically incorporated. { mi'kan·ə·klē ¦fōmd 'plas·tik }

mechanical mass [QUANT MECH] The part of a particle's mass which is supposed to exist in the absence of any interaction of the particle with itself through a field. { mi'kan·ə·kəl 'mas }

mechanical metallurgy [MET] The science and technology of the behavior of metals relating to mechanical forces imposed on them; includes rolling, extruding, deep drawing, bending, and other processes. { mi'kan·ə·kəl 'med·əl,ər·jē }

mechanical modulator [ELEC] A device that varies a carrier wave by moving some part of a circuit element. { mi'kan·ə·kəl 'mäj·ə,lād·ər }

mechanical mucking [ENG] Loading of dirt or stone in tunnels or mines by machines. { mi'kan·ə·kəl 'mək·iŋ }

mechanical mule [ORD] Popular name for a lightweight, low-silhouette United States infantry light weapons carrier, powered by an opposed-cylinder engine. { mi'kan·ə·kəl 'myül }

mechanical ohm [MECH] A unit of mechanical resistance, reactance, and impedance, equal to a force of 1 dyne divided by a velocity of 1 centimeter per second. { mi'kan·ə·kəl 'ōm }

mechanical oil valve [PETRO ENG] A float-operated liquid-level control valve used to control liquid flow out of oil-lease gas-oil separator tank systems. { mi'kan·ə·kəl 'óil ,valv }

mechanical oscillograph *See* direct-writing recorder. { mi'kan·ə·kəl ä'sil·ə,graf }

mechanical patent [ENG] A patent granted for an inventive improvement in a process, manufacture, or machine. { mi'kan·ə·kəl 'pat·ənt }

mechanical plating [MET] Deposition of one metal on another by a cold-peening process, such as tumbling. { mi'kan·ə·kəl 'plād·iŋ }

mechanical plotting board *See* coordinate plotter. { mi'kan·ə·kəl 'plād·iŋ bòrd }

mechanical press [MECH ENG] A press whose slide is operated by mechanical means. { mi'kan·ə·kəl 'pres }

mechanical property [MECH] A property that involves a relationship between stress and strain or a reaction to an applied force. { mi'kan·ə·kəl 'präp·ərd·ē }

mechanical puddling *See* vibration puddling. { mi'kan·ə·kəl 'pəd·liŋ }

mechanical pulp [MATER] Wood pulp produced by grinding and soaking the wood fibers. Also known as groundwood pulp. { mi'kan·ə·kəl 'pəlp }

mechanical pulping [MECH ENG] Mechanical, rather than chemical, recovery of cellulose fibers from wood; unpurified, finely ground wood is made into newsprint, cheap Manila papers, and tissues. { mi'kan·ə·kəl 'pəlp·iŋ }

mechanical pump [MECH ENG] A pump through which fluid is conveyed by direct contact with a moving part of the pumping machinery. { mi'kan·ə·kəl 'pəmp }

mechanical reactance [MECH] The imaginary part of mechanical impedance. { mi'kan·ə·kəl rē'ak·təns }

mechanical rectifier [ELEC] A rectifier in which rectification is accomplished by mechanical action, as in a synchronous vibrator. { mi'kan·ə·kəl 'rek·tə,fī·ər }

mechanical refrigeration [MECH ENG] The removal of heat by utilizing a refrigerant subjected to cycles of refrigerating thermodynamics and employing a mechanical compressor. { mi'kan·ə·kəl ri,frij·ə'rā·shən }

mechanical replacement [COMPUT SCI] The replacement of one piece of hardware by another piece of hardware at the instigation of the manufacturer. { mi'kan·ə·kəl ri'plās·mənt }

mechanical resistance *See* resistance. { mi'kan·ə·kəl ri'zis·təns }

mechanical rotational impedance *See* rotational impedance. { mi'kan·ə·kəl rō'tā·shən·əl im'pēd·əns }

mechanical rotational reactance *See* rotational reactance. { mi'kan·ə·kəl rō'tā·shən·əl rē'ak·təns }

mechanical rotational resistance *See* rotational resistance. { mi'kan·ə·kəl rō'tā·shən·əl ri zis·təns }

mechanical scale [ENG] A weighing device that incorporates a number of levers with precisely located fulcrums to permit heavy objects to be balanced with counterweights or counterpoises. { mi'kan·ə·kəl 'skāl }

mechanical scanner [COMPUT SCI] In optical character recognition, a device that projects an input character into a rotating disk, on the periphery of which is a series of small, uniformly spaced apertures; as the disk rotates, a photocell collects the light passing through the apertures. { mi'kan·ə·kəl 'skan·ər }

mechanical seal [MECH ENG] Mechanical assembly that forms a leakproof seal between flat, rotating surfaces to prevent high-pressure leakage. { mi'kan·ə·kəl 'sēl }

mechanical sediment *See* clastic sediment. { mi'kan·ə·kəl 'sed·ə·mənt }

mechanical separation [MECH ENG] A group of industrial operations by means of which particles of solid or drops of liquid are removed from a gas or liquid, or are separated into individual fractions, or both, by gravity separation (settling), centrifugal action, and filtration. { mi'kan·ə·kəl ,sep·ə'rā·shən }

mechanical setting [MECH ENG] Producing bits by setting diamonds in a bit mold into which a cast or powder metal is placed, thus embedding the diamonds and forming the bit crown; opposed to hand setting. Also known as cast setting; machine setting; sinter setting. { mi'kan·ə·kəl 'sed·iŋ }

mechanical shovel [MECH ENG] A loader limited to level or slightly graded drivages; when full, the shovel is swung over the machine, and the load is discharged into containers or vehicles behind. { mi'kan·ə·kəl 'shəv·əl }

mechanical splice [ENG] A splice made to terminate wire rope by pressing one or more metal sleeves over the rope junction. { mi'kan·ə·kəl 'splīs }

mechanical spring *See* spring. { mi'kan·ə·kəl 'spriŋ }

mechanical stage [ENG] A stage on a microscope provided with a mechanical device for positioning or changing the position of a slide. { mi'kan·ə·kəl 'stāj }

mechanical stepping motor [ELEC] A device in which a voltage pulse through a solenoid coil causes reciprocating motion by a solenoid plunger, and this is transformed into rotary motion through a definite angle by ratchet-and-pawl mechanisms or other mechanical linkages. { mi'kan·ə·kəl 'step·iŋ ,mōd·ər }

mechanical stoker *See* automatic stoker. { mi'kan·ə·kəl 'stōk·ər }

mechanical telemetry [COMMUN] Telemetry by mechanical linkages through shafts and gear trains over distances of a few feet, or by propagation of pressure or acoustic waves through fluid media over distances of several hundred feet. { mi'kan·ə·kəl tə'lem·ə·trē }

mechanical tilt [ELECTR] 1. Vertical tilt of the mechanical axis of a radar antenna. 2. The angle indicated by the tilt indicator dial. { mi'kan·ə·kəl 'tilt }

mechanical torque converter [MECH ENG] A torque converter, such as a pair of gears, that transmits power with only incidental losses. { mi'kan·ə·kəl 'tork kən,vərd·ər }

mechanical translation [COMPUT SCI] Automatic translation of one language into another by means of a computer or other machine that contains a dictionary look-up in its memory, along with the programs needed to make logical choices from synonyms, supply missing words, and rearrange word order as required for the new language. Also known as machine translation. { mi'kan·ə·kəl tranz'lā·shən }

mechanical turbulence [METEOROL] Irregular air movement in the lower atmosphere resulting from obstructions, for example, tall buildings. { mi kan·ə·kəl 'tər·byə·ləns }

MECHANICAL LIFT DOCK

An electromechanical platform-lift dry dock, capacity 4800 tons (4350 metric tons). A submarine is on the multiwheel cradle. (*Pearlson Engineering Co.*)

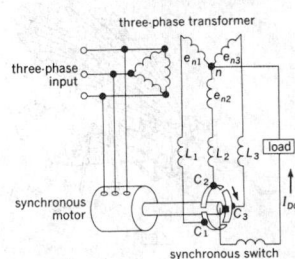

MECHANICAL RECTIFIER

Circuit of mechanical rectifier used on three-phase system; e_{n1}, e_{n2}, and e_{n3} are transformers. L_1, L_2, and L_3 are nonlinear reactors which provide good commutation at contacts C_1, C_2, and C_3, and limit short circuit current.

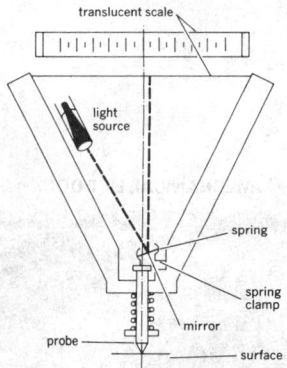

MECHANOOPTICAL VIBROMETER

translucent scale

light source

spring

spring clamp

mirror

probe

surface

Diagram of mechanooptical vibrometer. *(General Electric Co.)*

MECHATRONICS

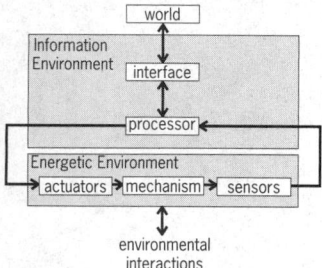

world

Information Environment

interface

processor

Energetic Environment

actuators → mechanism → sensors

environmental interactions

Generalized mechatronic system.

MECOPTERA

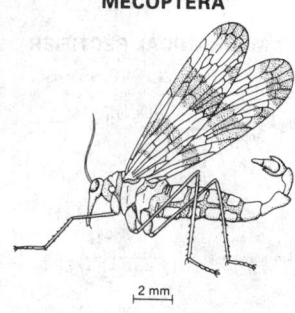

2 mm

Male scorpion fly *(Panorpa).*

mechanical twin [MET] A twin formed in a metal crystal by plastic deformation, involving shear of the lattice. { mi'kan·ə·kəl 'twin }

mechanical units [MECH] Units of length, time, and mass, and of physical quantities derivable from them. { mi'kan·ə·kəl ,yü·nəts }

mechanical vapor diffusion [MET] A process in which a metal and a metal surface to which the first metal is to be added are vaporized by a pulsating arc and become an integral part of the total mass. Abbreviated MVD. { mi'kan·ə·kəl 'vāp·ər də'fyü·shen }

mechanical vibration [MECH] The continuing motion, often repetitive and periodic, of parts of machines and structures. { mi'kan·ə·kəl vī'brā·shən }

mechanical wave filter *See* mechanical filter. { mi'kan·ə·kəl 'wāv ,fil·tər }

mechanical weathering [GEOL] The process of weathering by which physical forces break down or reduce a rock to smaller and smaller fragments, involving no chemical change. Also known as physical weathering. { mi'kan·ə·kəl 'weth·ə·riŋ }

mechanical working [MET] Formation of a desired shape and physical properties of a metal by subjecting it to pressure by rolls, presses, or hammers. { mi'kan·ə·kəl 'wərk·iŋ }

mechanical yielding prop [MIN ENG] A steel prop in which yield is controlled by friction between two sliding surfaces or telescopic tubes. Also known as friction yielding prop. { mi'kan·ə·kəl 'yēld·iŋ ,präp }

mechanics [PHYS] **1.** In the original sense, the study of the behavior of physical systems under the action of forces. **2.** More broadly, the branch of physics which seeks to formulate general rules for predicting the behavior of a physical system under the influence of any type of interaction with its environment. { mi'kan·iks }

mechanic's rule [MATH] A rule for estimating the square root of a number x whereby an estimate e is made of $\sqrt{x}$, a new estimate is made by taking the quantity $e' = (1/2)[e + (x/e)]$, and this procedure is repeated as many times as required to achieve the desired accuracy. { mi'kan·iks ,rül }

mechanism [CHEM] A detailed description of the course of a chemical reaction as it proceeds from the reactants to the products, with as complete a characterization as possible of the reaction steps and intermediate species. Also known as reaction path. [MECH ENG] That part of a machine which contains two or more pieces so arranged that the motion of one compels the motion of the others. { mek·ə,niz·əm }

mechanize [MECH ENG] **1.** To substitute machinery for human or animal labor. **2.** To produce or reproduce by machine. [ORD] To equip a military force with armed and armored vehicles. { mek·ə,nīz }

mechanized *See* machine-sensible. { mek·ə,nīzd }

mechanized dew-point meter *See* dew-point recorder. { 'mek·ə,nīzd 'dü ,point ,mēd·ər }

mechanized gun [ORD] Gun mounted on, and fired from, a motor vehicle; usually an artillery gun mounted on an armored, wheeled or track-laying vehicle. { 'mek·ə,nīzd ,gən }

mechanocaloric effect [CRYO] An effect resulting from the fact that a temperature gradient in helium II is invariably accompanied by a pressure gradient, and conversely; examples are the fountain effect, and the heating of liquid helium left behind in a container when part of it leaks out through a small orifice. { ,mek·ə·nō·kə'lor·ik i,fekt }

mechanochemical effect [PHYS CHEM] Changes in the dimensions of certain polymers, particularly photoelectrolytic gels and crystalline polymers, in response to changes in their chemical environment. { ,mek·ə·nō'kem·ə·kəl i,fekt }

mechanochemistry [PHYS CHEM] The study of the conversion of mechanical energy into chemical energy in polymers. { ,mek·ə·nō'kem·ə·strē }

mechanomotive force [MECH] The root-mean-square value of a periodically varying force. { ,mek·ə·nō,mōd·iv ,fors }

mechanooptical vibrometer [ENG] A vibrometer in which the motion given to a probe by a surface whose vibration amplitude is to be measured is used to rock a mirror; a light beam reflected from the mirror and focused onto a scale provides an indication of the vibration amplitude. { ,mek·ə·nō'äp·tə·kəl vī'bräm·əd·ər }

mechanophotochemistry [PHYS CHEM] The study of changes in the dimensions of certain photoresponsive polymers upon exposure to light. { ,mek·ə·nō,fō·dō'kem·ə·strē }

mechanoreceptor [PHYSIO] A receptor that provides the organism with information about such mechanical changes in the environment as movement, tension, and pressure. { ,mek·ə·nō·ri'sep·tər }

mechatronics [ENG] A branch of engineering that incorporates the ideas of mechanical and electronic engineering into a whole, and, in particular, covers those areas of engineering concerned with the increasing integration of mechanical, electronic, and software engineering into a production process. { ,mek·ə'trän·iks }

Meckel's cartilage [EMBRYO] The cartilaginous axis of the mandibular arch in the embryo and fetus. { 'mek·əlz ,kärt·lij }

Meckel's diverticulum [EMBRYO] The persistent blind end of the yolk stalk forming a tube connected with the lower ileum. { 'mek·əlz ,dī·vər'tik·yə·ləm }

meconin [ORG CHEM] $C_{10}H_{10}O_4$ A neutral principle of opium; white crystals, soluble in hot water and alcohol and melting at 102–103°C. Also known as opianyl. { 'mek·ə·nən }

meconium [EMBRYO] A greenish mass of mucous, desquamated epithelial cells, lanugo, and vernix caseosa that collects in the fetal intestine, becoming the first fecal discharge of the newborn. { mə'kō·nē·əm }

meconium ileus [MED] Intestinal obstruction in the newborn with cystic fibrosis due to trypsin deficiency. { mə'kō·nē·əm 'il·ē·əs }

Mecoptera [INV ZOO] The scorpion flies, a small order of insects; adults are distinguished by the peculiar prolongation of the head into a beak, which bears chewing mouthparts. { me'käp·tə·rə }

mecystasis [PHYSIO] Increase in muscle length with maintenance of the original degree of tension. { me'sis·tə·səs }

media [HISTOL] The middle, muscular layer in the wall of a vein, artery, or lymph vessel. { 'mē·dē·ə }

media conversion [COMPUT SCI] The transfer of data from one storage type (such as punched cards) to another storage type (such as magnetic tape). { 'mē·dē·ə kən,vər·zhən }

media conversion buffer [COMPUT SCI] Large storage area, such as a drum, on which data may be stored at low speed during nonexecution time, to be later transferred at high speed into core memory during execution time. { 'mē·dē·ə kən,vər·zhən ,bəf·ər }

mediad [ANAT] Toward the median line or plane of the body or of a part of the body. { 'mē·dē,ad }

medial [ANAT] **1.** Being internal as opposed to external (lateral). **2.** Toward the midline of the body. [SCI TECH] Located in the middle. { 'mē·dē·əl }

medial arteriosclerosis [MED] Calcification of the tunica media of small and medium-sized muscular arteries. Also known as medial calcinosis; Mönckeberg's arteriosclerosis. { 'mē·dē·əl ,är·tir·ē·ō·sklə'rō·səs }

medial calcinosis *See* medial arteriosclerosis. { 'mē·dē·əl ,kal·sə'nō·səs }

medial lemniscus [ANAT] A lemniscus arising in the nucleus gracilis and nucleus cuneatus of the brain, crossing immediately as internal arcuate fibers, and terminating in the posterolateral ventral nucleus of the thalamus. { 'mē·dē·əl lem'nis·kəs }

medial moraine [GEOL] **1.** An elongate moraine carried in or upon the middle of a glacier and parallel to its sides. **2.** A moraine formed by glacial abrasion of a rocky protuberance near the middle of a glacier. { 'mē·dē·əl mə'rān }

medial necrosis [MED] Death of cells in the tunica media of arteries. Also known as medionecrosis. { 'mē·dē·əl ne'krō·səs }

medial sclerosis [MED] An uncommon degenerative arterial disease that is characterized by ring-like calcifications that occur within the media (middle layer) of muscular arteries of the arms, legs, or genital tract; it rarely occurs in men or women under the age of 50, and its cause is obscure. { ,mēd·ē·əl sklə'rō·səs }

media migration [CHEM ENG] Carryover of fibers or other filter material by liquid effluent from a filter unit. { 'mē·dē·ə mī'grā·shən }

media mill *See* shot mill. { 'mēd·ē·ə ,mil }

median [MATH] **1.** Any line in a triangle which joins a vertex to the midpoint of the opposite side. **2.** The line that joins the midpoints of the nonparallel sides of a trapezoid. Also known as midline. [SCI TECH] Located in the middle.

[STAT] An average of a series of quantities or values; specifically, the quantity or value of that item which is so positioned in the series, when arranged in order of numerical quantity or value, that there are an equal number of items of greater magnitude and lesser magnitude. { 'mē·dē·ən }

median effective dose *See* effective dose 50. { 'mē·dē·ən i'fek·tiv 'dōs }

median infective dose *See* infective dose 50. { 'mē·dē·ən in'fek·tiv 'dōs }

median lethal dose *See* lethal dose 50. { 'mē·dē·ən 'lēth·əl 'dōs }

median lethal time [MICROBIO] The period of time required for 50% of a large group of organisms to die following a specific dose of an injurious agent, such as a drug or radiation. { 'mē·dē·ən 'lēth·əl ˌtīm }

median mass [GEOL] A less disturbed structural block in the middle of an orogenic belt, bordered on both sides by orogenic structure, thrust away from it. Also known as betwixt mountains; Zwischengebirge. { 'mē·dē·ən 'mas }

median maxillary cyst [MED] Cystic dilation of embryonal inclusions in the incisive fossa or between the roots of the central incisors. Also known as nasopalatine cyst. { 'mē·dē·ən 'mak·səˌler·ē ˌsist }

median nasal process [EMBRYO] The region below the frontonasal sulcus between the olfactory sacs; forms the bridge and mobile septum of the nose and various parts of the upper jaw and lip. { 'mē·dē·ən 'nāz·əl ˌprä·səs }

median nerve test [MED] A test for loss of function of the median nerve by having the patient abduct the thumb at right angles to the palm with fingertips in contact and forming a pyramid. { 'mē·dē·ən 'nərv ˌtest }

median particle diameter [GEOL] The middlemost particle diameter of a rock or sediment, larger than 50% of the diameter in the distribution and smaller than the other 50%. { 'mē·dē·ən 'pärd·ə·kəl dīˌam·əd·ər }

median point [MATH] The point at which all three medians of a triangle intersect. { 'med·ē·ən ˌpoint }

median strip [CIV ENG] A paved or planted section dividing a highway into lanes according to direction of travel. { 'mē·dē·ən 'strip }

mediastinitis [MED] Inflammation of the mediastinum. { ˌmē·dēˌas·təˈnīd·əs }

mediastinum [ANAT] **1.** A partition separating adjacent parts. **2.** The space in the middle of the chest between the two pleurae. { ˌmē·dē·əˈstī·nəm }

medical bacteriology [MED] A branch of medical microbiology that deals with the study of bacteria which affect human health, especially those which produce disease. { 'med·ə·kəl bakˌtir·ēˈäl·ə·jē }

medical chemical engineering [CHEM ENG] The application of chemical engineering to medicine, frequently involving mass transport and separation processes, especially at the molecular level. { 'med·ə·kəl 'kem·ə·kəl ˌen·jəˈnir·iŋ }

medical climatology [MED] The study of the relation between climate and disease. { 'med·ə·kəl ˌklī·məˈtäl·ə·jē }

medical control systems [MED] Physiological and artificial systems that control one or more physiological variables or functions of the human body. { ˌmed·ə·kəl kənˈtrōl ˌsis·təmz }

medical electronics [ELECTR] A branch of electronics in which electronic instruments and equipment are used for such medical applications as diagnosis, therapy, research, anesthesia control, cardiac control, and surgery. { 'med·ə·kəl iˌlekˈträn·iks }

medical entomology [MED] The study of insects that are vectors for diseases and parasitic infestations in humans and domestic animals. { 'med·ə·kəl ˌen·təˈmäl·ə·jē }

medical ethics [MED] Principles and moral values of proper medical conduct. { 'med·ə·kəl 'eth·iks }

medical examiner [FOREN SCI] A professionally qualified physician duly authorized and charged by a governmental unit to determine facts concerning causes of death, particularly deaths not occurring under natural circumstances, and to testify thereto in courts of law. { 'med·ə·kəl igˈzam·ə·nər }

medical frequency bands [COMMUN] A collection of radio frequency bands allocated to medical equipment in the United States. { 'med·ə·kəl 'frē·kwən·sē ˌbanz }

medical geography [MED] The study of the relation between geographic factors and disease. { 'med·ə·kəl jēˈäg·rə·fē }

medical history [MED] An account of a patient's past and present state of health obtained from the patient or relatives. { 'med·ə·kəl 'his·trē }

medical imaging [MED] The production of visual representations of body parts, tissues, or organs, for use in clinical diagnosis; encompasses x-ray methods, magnetic resonance imaging, single-photon-emission and positron-emission tomography, and ultrasound. { 'med·ə·kəl 'im·ij·iŋ }

medical information systems [MED] Standardized methods of collection, evaluation or verification, storage, and retrieval of data about a patient. { ˌmed·ə·kəl ˌin·fərˈmā·shən ˌsis·təmz }

medical microbiology [MED] The study of microorganisms which affect human health. { 'med·ə·kəl ˌmī·krō·bīˈäl·ə·jē }

medical mycology [MED] A branch of medical microbiology that deals with fungi that are pathogenic to humans. { 'med·ə·kəl mīˈkäl·ə·jē }

medical parasitology [MED] A branch of medical microbiology which deals with the relationship between humans and those animals which live in or on them. { 'med·ə·kəl ˌpar·əsiˈtäl·ə·jē }

medical protozoology [MED] A branch of medical microbiology that deals with the study of Protozoa which are parasites of humans. { 'med·ə·kəl ˌprō·dō·zōˈäl·ə·jē }

medical radiography [MED] The use of x-rays to produce photographic images for visualizing internal anatomy as an aid in diagnosis. { 'med·ə·kəl ˌrād·ēˈäg·rə·fē }

medication [MED] **1.** A medicinal substance. **2.** Treatment by or administration of a medicine. { ˌmed·əˈkā·shən }

medicinal [MED] Of, pertaining to, or having the nature of medicine. { məˈdis·ən·əl }

medicinal oil [MATER] A highly refined, colorless, tasteless and odorless petroleum oil used medicinally as an internal lubricant and for the manufacture of salves and ointments. Also known as mineral oil. { məˈdis·ən·əl ˈoil }

medicine [MED] **1.** Any agent administered for the treatment of disease. **2.** The science and art of treating and healing. { 'med·ə·sən }

medina quartzite [MINERAL] A variety of quartz containing 97.8% silica; melting point is about 1700°C. { məˈdē·nə ˈkwȯrtˌsīt }

medionecrosis [MED] Necrosis occurring in the tunica media of an artery. { ˌmē·dē·ō·nəˈkrō·səs }

mediterranean *See* mesogeosyncline. { ˌmed·ə·təˈrā·nē·ən }

Mediterranean anemia *See* thalassemia. { ˌmed·ə·təˈrā·nē·ən əˈnē·mē·ə }

Mediterranean climate [CLIMATOL] A type of climate characterized by hot, dry, sunny summers and a winter rainy season; basically, this is the opposite of a monsoon climate. Also known as etesian climate. { ˌmed·ə·təˈrā·nē·ən ˈklī·mət }

Mediterranean faunal region [ECOL] A marine littoral faunal region including that offshore portion of the Atlantic Ocean from northern France to near the Equator. { ˌmed·ə·təˈrā·nē·ən ˈfȯn·əl ˌrē·jən }

Mediterranean fever *See* brucellosis. { ˌmed·ə·təˈrā·nē·ən ˈfē·vər }

mediterranean sea [GEOGR] A deep epicontinental sea that is connected with the ocean by a narrow channel. { ˌmed·ə·təˈrā·nē·ən ˈsē }

Mediterranean Sea [GEOGR] A sea that lies between Europe, Asia Minor, and Africa and is completely landlocked except for the Strait of Gibraltar, the Bosporus, and the Suez Canal; total water area is 965,000 square miles (2,501,000 square kilometers). { ˌmed·ə·təˈrā·nē·ən ˈsē }

medium [CHEM ENG] **1.** The carrier in which a chemical reaction takes place. **2.** Material of controlled pore size used to remove foreign particles or liquid droplets from fluid carriers. [COMPUT SCI] The material, or configuration thereof, on which data are recorded; usually not applied to disk, drum, or core, but to storable, removable media, such as paper tape, cards, and magnetic tape. [PHYS] That entity in which objects exist and phenomena take place; examples are free space and various fluids and solids. { 'mē·dē·əm }

medium-altitude bombing [ORD] Horizontal bombing with the height of release at an altitude between 8000 and 15,000 feet (2400 and 4600 meters). { 'mē·dē·əm ˌal·təˌtüd ˈbäm·iŋ }

medium-altitude earth orbit [AERO ENG] An artificial-satellite orbit whose altitude is between about 1000 and 8000 miles (1600 and 14,500 kilometers) above the earth's surface. Abbreviated MEO. { ‚mēd·ē·əm ¦al·tə‚tüd 'ərth ‚ȯr·bət }

medium antiaircraft artillery [ORD] Conventional antiaircraft artillery pieces, 90 millimeters or larger, the weight of which in a trailed mount, excluding on-carriage fire control, does not exceed 40,000 pounds (18,000 kilograms). { 'mē·dē·əm ¦an·tē'er‚kraft är‚til·ə·rē }

medium artillery [ORD] Artillery which includes guns of caliber greater than 105 millimeters but less than 155 millimeters, and howitzers of caliber greater than 105 millimeters but not greater than 155 millimeters. { 'mē·dē·əm är'til·ə·rē }

medium boiler [MATER] A solvent, intermediate in volatility between high and low boilers, which is added to lacquer thinner and influences flow and freedom from an orange peel condition during drying; boiling point is 115–145°C. { 'mē·dē·əm 'bȯil·ər }

medium carbon steel [MET] Steel containing 0.15–0.30% carbon. { 'mē·dē·əm ¦kär·bən ‚stēl }

medium-curing asphalt [MATER] Liquid product composed of asphalt cement and a kerosine-type diluent. Also known as M-C asphalt. { 'mē·dē·əm ¦kyùr·iŋ 'as‚fȯlt }

medium frequency [COMMUN] A Federal Communications Commission designation for the band from 300 to 3000 kilohertz in the radio spectrum. Abbreviated mf. { 'mē·dē·əm 'frē·kwən·sē }

medium-frequency propagation [COMMUN] Radio propagation at broadcast frequencies where skip is not an important factor. { 'mē·dē·əm ¦frē·kwən·sē ‚präp·ə'gā·shən }

medium-frequency tube [ELECTR] An electron tube operated at frequencies between 300 and 3000 kilohertz, at which the transit time of an electron between electrodes is much smaller than the period of oscillation of the voltage. { 'mē·dē·əm ¦frē·kwən·sē ‚tüb }

medium howitzer [ORD] A complete projectile-firing weapon, with a medium muzzle velocity and a curved trajectory; the bore diameter is from 126 through 200 millimeters. { 'mē·dē·əm 'haù·ət·sər }

medium-range forecast [METEOROL] A forecast of weather conditions for a period of 48 hours to a week in advance. Also known as extended-range forecast. { 'mē·dē·əm ¦rānj 'fȯr‚kast }

medium-scale integration [ELECTR] Fabrication of solid-state integrated circuits having more than about 12 gate-equivalent circuits. Abbreviated MSI. { 'mē·dē·əm ¦skāl ‚int·ə'grā·shən }

medium-technology robot [CONT SYS] An automatically controlled machine that employs servomechanisms and microprocessor control units. { 'mē·dē·əm tek¦näl·ə·jē 'rō‚bät }

medium-volatile bituminous coal [GEOL] Bituminous coal consisting of 23–31% volatile matter. { 'mē·dē·əm ‚väl·ə·təl bə'tü·mə·nəs 'kōl }

medius [ANAT] The middle finger. { 'mē·dē·əs }

medulla [ANAT] **1.** The central part of certain organs and structures such as the adrenal glands and hair. **2.** Marrow, such as of bone or the spinal cord. [BOT] **1.** Pith. **2.** The central spongy portion of some fungi. [NEUROSCI] See medulla oblongata. { mə'dəl·ə }

medulla oblongata [NEUROSCI] The somewhat pyramidal, caudal portion of the vertebrate brain which extends from the pons to the spinal cord. Also known as medulla. { mə'dəl·ə ‚äb‚lȯŋ'gäd·ə }

medullary carcinoma [MED] A form of poorly differentiated adenocarcinoma, usually of the breast, grossly well circumscribed, gray-pink, and firm. Also known as encephaloid carcinoma. { mə'dəl·ə·rē ‚kärs·ən'ō·mə }

medullary cord [ANAT] Dense lymphatic tissue separated by sinuses in the medulla of a lymph node. [EMBRYO] A primary invagination of the germinal epithelium of the embryonic gonad that differentiates into rete testis and seminiferous tubules or into rete ovarii. { mə'dəl·ə·rē 'kȯrd }

medullary ray [BOT] An extension of pith between vascular bundles in the plant stem. Also known as pith ray. { me'dəl·ə·rē ‚rā }

medulloblastoma [MED] A malignant neoplasm of the brain with a tendency to metastasize in the meninges. { mə‚dəl·ə·bla'stō·mə }

medulloepithelioma [MED] **1.** A locally invasive tumor of

MEGACHILIDAE

Drawing of a leaf-cutting bee. (From T. I. Storer and R. L. Usinger, General Zoology, 3d ed., McGraw-Hill, 1957)

the eye arising from the ciliary epithelium of the iris. **2.** See ependymoma. { mə‚dəl·ō‚ep·ə‚thē'lē·ō·mə }

Medullosaceae [PALEOBOT] A family of seed ferns; these extinct plants all have large spirally arranged petioles with numerous vascular bundles. { mə'dəl·ō'sās·ē‚ē }

medusa See jellyfish. { mə'düs·ə }

meerschaum See sepiolite. { 'mir‚shóm }

meet [MATH] **1.** The meet of two elements of a lattice is their greatest lower bound. **2.** See intersection. { mēt }

meet-irreducible member [MATH] A member, A, of a lattice or ring of sets such that, if A is equal to the meet of two other members, B and C, then A equals B or A equals C. { ¦mēt ‚ir·i'dü·sə·bəl ‚mem·bər }

MEG See magnetoencephalogram.

mega- [SCI TECH] A prefix representing 10^6, or one million. Abbreviated M. { 'meg·ə }

megabit [COMPUT SCI] A unit of information content equal to 1,048,576 (1024 × 1024) bits. Abbreviated Mbit. { 'meg·ə‚bit }

megabyte [COMPUT SCI] A unit of information content equal to 1,048,576 (1024 × 1024) bytes. Abbreviated Mbyte. Symbolized M. { 'meg·ə‚bīt }

megacanthopore [INV ZOO] A large prominent tube, commonly projecting as a spine in a mature region of a bryozoan colony. { ‚meg·ə'kan·thə‚pór }

Megachilidae [INV ZOO] The leaf-cutting bees, a family of hymenopteran insects in the superfamily Apoidea. { ‚meg·ə'kil·ə‚dē }

Megachiroptera [VERT ZOO] The fruit bats, a group of Chiroptera restricted to the Old World; most species lack a tail, but when present it is free of the interfemoral membrane. { ¦meg·ə·kī'räp·tə·rə }

megacine [MICROBIO] The bacteriocidin produced by Bacillus megaterium. { 'meg·ə·sən }

megacolon [MED] Hypertrophy and dilation of the colon associated with prolonged constipation. { 'meg·ə‚kō·lən }

megacryst [PETR] Any crystal or grain in an igneous or metamorphic rock that is significantly larger than the surrounding matrix. { 'meg·ə‚krist }

megacycle See megahertz. { 'meg·ə‚sī·kəl }

megacyclothem [GEOL] A cycle of or combination of related cyclothems. { ‚meg·ə'sī·klə‚them }

Megadermatidae [VERT ZOO] The false vampires, a family of tailless bats with large ears and a nose leaf; found in Africa, Australia, and the Malay Archipelago. { ‚meg·ə·dər'mad·ə‚dē }

megaelectronvolt [PHYS] A unit of energy commonly used in nuclear and particle physics, equal to the energy acquired by an electron in falling through a potential of 1,000,000 volts. Abbreviated MeV. { ¦meg·ə·i'lek‚trän‚vōlt }

megaelectronvolt-curie [NUCLEO] A unit of radioactive power, equal to the power generated by 1 curie emitting an average energy of 10^6 electronvolts per disintegration; equal to approximately 5.92805 × 10^{-3} watt. Abbreviated MeV Ci. { ¦meg·ə·i'lek‚trän‚vōlt 'kyúr·ē }

megaflops [COMPUT SCI] A unit of computer speed, equal to 10^6 flops. { 'meg·ə‚fläps }

megagametophyte [BOT] The female gametophyte in plants having two types of spores. { ¦meg·ə·gə'mēd·ə‚fīt }

megagauss [ELECTROMAG] A unit of magnetic induction equal to 10^6 gauss or 100 tesla. { 'meg·ə‚gaùs }

megagauss physics [PHYS] The production, measurement, and application of megagauss fields, as produced by discharge of capacitor banks or explosive flux-compression techniques. { 'meg·ə‚gaùs 'fiz·iks }

megahertz [PHYS] Unit of frequency, equal to 1,000,000 hertz. Abbreviated MHz. Also known as megacycle (Mc). { 'meg·ə‚hərts }

megakaryocyte [HISTOL] A giant bone-marrow cell characterized by a large, irregularly lobulated nucleus; precursor to blood platelets. { ¦meg·ə'kar·ē·ə‚sīt }

megakaryocytopenia See megakaryophthisis. { ¦meg·ə‚kar·ē·ə‚sīd·ə'pē·nē·ə }

megakaryophthisis [MED] A scarcity of megakaryocytes in the bone marrow. Also known as megakaryocytopenia. { ¦meg·ə‚kar·ē'äf·thə·səs }

megalith [ARCHEO] One of a group of large stones arranged in some pattern in a prehistoric monument. { 'meg·ə‚lith }

megaloblast [PATH] A large nucleated erythroblast

appearing in bone marrow in vitamin B_{12} or folic acid deficiency. { 'meg·ə·lō,blast }

megaloblastic anemia [MED] Anemia characterized by the occurrence of megaloblasts in the bone marrow and blood. { ¦meg·ə·lō¦blas·tik ə'nē·mē·ə }

megaloblast of Sabin See pronormoblast. { 'meg·ə·lō,blast əv 'sā·bən }

megalocardia [MED] Abnormal enlargement of the heart. { ¦meg·ə·lō'kär·dē·ə }

megalocephaly [MED] The condition of having a head whose maximum fronto-occipital circumference is greater than two standard deviations above the mean for age and sex. { meg·ə·lō'sef·ə·lē }

Megalodontoidea [INV ZOO] A superfamily of hymenopteran insects in the suborder Symphyta. { ¦meg·ə·lō,dän'tóid·ē·ə }

megalomania [PSYCH] The delusion of greatness and omnipotence characterizing certain psychotic reactions. { ,meg·ə·lō'mā·nē·ə }

Megalomycteroidei [VERT ZOO] The mosaic-scaled fishes, a monofamilial suborder of the Cetomimiformes; members are rare species of small, elongate deep-sea fishes with degenerate eyes and irregularly disposed scales. { ,meg·ə·lō,mik·tə'róid·ē,ī }

megalopia See macropsia. { ,meg·ə·lä·pē·ə }

megalops larva [INV ZOO] A preimago stage of certain crabs having prominent eyes and chelae. { 'meg·ə,läps ,lär·və }

Megaloptera [INV ZOO] A suborder included in the order Neuroptera by some authorities. { ,meg·ə'läp·tə·rə }

Megalopygidae [INV ZOO] The flannel moths, a small family of lepidopteran insects in the suborder Heteroneura. { ¦meg·ə·lō'pij·ə,dē }

megalosphere [INV ZOO] The initial, large-chambered shell of sexual individuals of certain dimorphic species of Foraminifera. { 'meg·ə·lō,sfir }

megaloureter [MED] Abnormal enlargement of a ureter. { ,meg·ə·lō'yùr·əd·ər }

-megaly [MED] A combining form denoting abnormal enlargement. { 'meg·ə·lē }

Megamerinidae [INV ZOO] A family of myodarian cyclorrhaphous dipteran insects in the subsection Acalypteratae. { ,meg·ə·mə'rin·ə,dē }

megaparsec [ASTRON] A unit equal to 1,000,000 parsecs. { ,meg·ə'pär,sek }

megapel display [COMPUT SCI] A computer graphics display that handles 10^6 or more pixels (pels). { 'meg·ə,pel di,splā }

megaphenic [GEN] Pertaining to genetic or environmental factors that are individually of large effect relative to the phenotypic standard deviation. { ,meg·ə'fē·nik }

megaphone [ACOUS] A conical or rectangular horn used to amplify or direct the sound of a speaker's voice. { 'meg·ə,fōn }

megaphyllous [BOT] Having large leaves or leaflike extensions. { ¦meg·ə'fil·əs }

Megapodiidae [VERT ZOO] The mound birds and brush turkeys, a family of birds in the order Galliformes; distinguished by their method of incubating eggs in mounds of dirt or in decomposing vegetation. { ,meg·ə·pə'dī·ə,dē }

megarectum [MED] Abnormal enlargement of the rectum. { ¦meg·ə¦rek·təm }

megaripple [GEOL] A large sand wave. { 'meg·ə,rīp·əl }

megasclere [INV ZOO] A large sclerite. { 'meg·ə,sklir }

megasecond [MECH] A unit of time, equal to 1,000,000 seconds. Abbreviated Ms; Msec. { 'meg·ə,sek·ənd }

Megasphaera [MICROBIO] A genus of bacteria in the family Veillonellaceae; relatively large cells occurring in pairs arranged in chains. { mə'gas·fə·rə }

megasporangium See macrosporangium. { ¦meg·ə·spə'ran·jē·əm }

megaspore See macrospore. { 'meg·ə,spór }

megaspore mother cell See megasporocyte. { 'meg·ə,spór 'məth·ər ,sel }

megasporocyte [BOT] A diploid cell from which four megaspores are produced by meiosis. Also known as megaspore mother cell. { ¦meg·ə·spór·ə,sīt }

megasporogenesis See macrosporogenesis. { ,meg·ə,spór·ə'jen·ə·səs }

megasporophyll [BOT] A leaf bearing megasporangia. { ,meg·ə'spór·ə,fil }

megass See bagasse. { 'meg,as }

megatectonics [GEOL] The tectonics of the very large structural features of the earth. { ,meg·ə,tek'tän·iks }

megathermophyte [ECOL] A plant that requires great heat and abundant moisture for normal growth. Also known as macrothermophyte. { ,meg·ə'thər·mə,fīt }

Megathymiinae [INV ZOO] The giant skippers, a subfamily of lepidopteran insects in the family Hesperiidae. { ,meg·ə·thə'mī·ə,nē }

megaton [PHYS] The energy released by 1,000,000 metric tons of chemical high explosive calculated at a rate of 1000 calories per gram, or a total of 4.18×10^{15} joules; used principally in expressing the energy released by a nuclear bomb. Abbreviated MT. { 'meg·ə,tən }

megaton weapon [ORD] A nuclear fission or fusion bomb capable of exploding with megaton energy. { 'meg·ə,tən 'wep·ən }

megatron See disk-seal tube. { 'meg·ə,trän }

megavolt [ELEC] A unit of potential difference or emf (electromotive force), equal to 1,000,000 volts. Abbreviated MV. { 'meg·ə,vōlt }

megawatt [MECH] A unit of power, equal to 1,000,000 watts. Abbreviated MW. { 'meg·ə,wät }

megawatt-day per ton [NUCLEO] A unit used for expressing the burnup of fuel in a reactor; specifically, the number of megawatt-days of heat output per metric ton of fuel in the reactor. { 'meg·ə,wät ¦dā pər 'tən }

megawatt electric [NUCLEO] Unit of the electric power of a nuclear reactor, as opposed to thermal power. Abbreviated MW(E). { 'meg·ə,wät i'lek·trik }

megawatt thermal [NUCLEO] Unit of the thermal power of a nuclear reactor, as opposed to electric power. Abbreviated MW(Th). { 'meg·ə,wät 'thər·məl }

megawatt year of electricity [ELEC] A unit of electric energy, equal to the energy from a power of 1,000,000 watts over a period of 1 tropical year, or to 3.1557×10^{13} joules. Abbreviated MWYE. { 'meg·ə,wät ¦yir əv i,lek'tris·əd·ē }

meglumine [PHARM] See N-methyl glucamine. { 'me·glə,mīn }

megohm [ELEC] A unit of resistance, equal to 1,000,000 ohms. { 'me,gōm }

megohmmeter [ELEC] An instrument which is used for measuring the high resistance of electrical materials of the order of 20,000 megohms at 1000 volts; one direct-reading type employs a permanent magnet and a moving coil. { 'me,gōm,mēd·ər }

Mehlis' gland [INV ZOO] One of the large unicellular glands around the ootype of flatworms. { 'mā·ləs ,gland }

Meibomian cyst See chalazion. { mī'bō·mē·ən 'sist }

Meibomian gland See tarsal gland. { mī'bō·mē·ən 'gland }

meibomianitis [MED] Inflammation of the tarsal glands. { mī,bō·mē·ə'nīd·əs }

Meig's syndrome [MED] A complex of symptoms associated with ovarian fibroma including abnormal accumulation of serous fluid in the pleural and peritoneal cavities. { 'megz ,sin,drōm }

Meijer transform [MATH] The Meijer transform of a function $f(x)$ is the function $F(y)$ defined as the integral from 0 to ∞ of $\sqrt{xy}K_n(xy)f(x)dx$, where K_n is a modified Bessel function. { 'mā·ər ,tranz,fórm }

Meinertellidae [INV ZOO] A family of wingless insects belonging to the Microcoryphia. { ,mī·nər'tel·ə,dē }

Meinzer unit See permeability coefficient. { 'mīnt·sər ,yü·nət }

meiocyte [CELL MOL] A cell undergoing meiotic division. { 'mī·ə,sīt }

meiofauna [ECOL] Small benthic animals ranging in size between macrofauna and microfauna; includes interstitial animals. { ¦mī·ə'fón·ə }

meioflora [ECOL] Small benthic plants ranging in size between macroflora and microflora; includes interstitial plants. { ¦mī·ə'flór·ə }

meionite [MINERAL] $3CaAl_2Si_2O_8 \cdot CaCO_3$ A scapolite mineral composed of calcium aluminosilicate and calcium carbonate; it is isomorphous with marialite. { 'mī·ə,nīt }

meiosis [CELL MOL] A type of cell division occurring in

diploid or polyploid tissues that results in a reduction in chromosome number, usually by half. { mī'ō·səs }

meiospore [BIOL] A spore produced as the result of meiosis. { 'mī·ə,spȯr }

meiotic drive [GEN] Preferential meiotic segregation favoring one chromosome over its homologue. { mē¦äd·ik 'drīv }

Meissner effect [SOLID STATE] The expulsion of magnetic flux from the interior of a piece of superconducting material as the material undergoes the transition to the superconducting phase. Also known as flux jumping; Meissner-Ochsenfeld effect. { 'mīs·nər i‚fekt }

Meissner-Ochsenfeld effect *See* Meissner effect. { 'mīs·nər 'äk·sən‚feld i‚fekt }

Meissner oscillator [ELECTR] An electron-tube oscillator in which the grid and plate circuits are inductively coupled through an independent tank circuit which determines the frequency. { 'mīs·nər ‚äs·ə‚läd·ər }

Meissner's corpuscle [NEUROSCI] An ovoid, encapsulated cutaneous sense organ presumed to function in touch sensation in hairless portions of the skin. { 'mīs·nərz 'kȯr·pə·səl }

Meissner's plexus *See* submucous plexus. { 'mīs·nərz 'plek·səs }

meitnerium [CHEM] A chemical element, symbolized Mt, atomic number 109, a synthetic element; the seventeenth transuranium element. { mīt'nir·ē·əm }

MEK *See* methyl ethyl ketone.

mel [ACOUS] A unit of pitch, equal to one-thousandth of the pitch of a simple tone whose frequency is 1000 hertz and whose loudness is 40 decibels above a listener's threshold. { mel }

mel-, melo- [SCI TECH] A combining form denoting dark or black; denoting or pertaining to melanin. { mel, 'mel·ō }

melamine [ORG CHEM] $C_3H_6N_4$ A white crystalline compound that is slightly soluble in water, melts at 354°C and is a cyclic trimer of cyanamide; used to make melamine resins and in tanning of leather. { 'mel·ə‚mēn }

melamine resin [MATER] An amino resin made from formaldehyde and melamine; it is used as a molding compound with fillers added to it; it may also be used for laminating. { 'mel·ə‚mēn 'rez·ən }

Melamphaidae [VERT ZOO] A family of bathypelagic fishes in the order Beryciformes. { ‚mel·əm'fā·ə‚dē }

Melampsoraceae [MYCOL] A family of parasitic fungi in the order Uredinales in which the teleutospores are laterally united to form crusts or columns. { ‚mel·əm·sə'rās·ē‚ē }

melancholia [PSYCH] A disordered mental condition of psychotic proportion characterized by severe depression. { ‚mel·ən'kō·lē·ə }

Melanconiaceae [MYCOL] The single family of the order Melanconiales. { ‚mel·ən‚kō·nē'ās·ē‚ē }

Melanconiales [MYCOL] An order of the class Fungi Imperfecti including many plant pathogens commonly causing anthracnose; characterized by densely aggregated cnidophores on an acervulus. { ‚mel·ən‚kō·nē'ā·lēz }

Melandryidae [INV ZOO] The false darkling beetles, a family of coleopteran insects in the superfamily Tenebrionoidea. { ‚mel·ən'drī·ə‚dē }

Melanesia [GEOGR] A group of islands in the Pacific Ocean northeast of Australia. { mel·ə'nē·zhə }

mélange [GEOL] A heterogeneous medley or mixture of rock materials; specifically, a mappable body of deformed rocks consisting of a pervasively sheared, fine-grained, commonly pelitic matrix, thoroughly mixed with angular and poorly sorted inclusions of native and exotic tectonic fragments, blocks, or slabs, of diverse origins and geologic ages, that may be as much as several kilometers in length. Also known as block clay. { mā'länzh }

melangeur [FOOD ENG] A machine used in chocolate manufacture for mixing chocolate liquor with sugar and cocoa butter. Also known as paste mixer. { ‚mā‚län'jər }

melanic *See* melanocratic. { me'lan·ik }

melaniline *See* diphenylguanidine. { mel'an·ə·lən }

melanin [BIOCHEM] Any of a group of brown or black pigments occurring in plants and animals. { 'mel·ə·nən }

melanoblast [HISTOL] **1.** Precursor cell of melanocytes and melanophores. **2.** An immature pigment cell in certain vertebrates. **3.** A mature cell that elaborates melanin. { mə'lan·ə‚blast }

melanoblastoma [MED] A malignant tumor composed principally of melanoblasts. { ¦mel·ə·nō·bla'stō·mə }

melanocarcinoma [MED] A malignant melanoma derived from epithelial tissue. { ¦mel·ə·nō‚kärs·ən'ō‚mə }

melanocerite [MINERAL] $(Ca,Ce,Y)_8(BO_3)(SiO_4)_4(F,OH)_4$ A brown or black rhombohedral mineral composed of complex silicate, borate, fluoride, tantalate, or other anion of cerium, yttrium, calcium, and other metals; occurs as crystals. { ¦mel·ə·nō'se‚rīt }

melanocratic [GEOL] Dark-colored, referring to igneous rock containing at least 50–60% mafic minerals. Also known as chromocratic; melanic. { ¦mel·ə·nō¦krad·ik }

melanocyte [HISTOL] A cell containing dark pigments. { mə'lan·ə‚sīt }

melanocyte-stimulating hormone [BIOCHEM] A protein substance secreted by the intermediate lobe of the pituitary of humans which causes dispersion of pigment granules in the skin; similar to intermedins in other vertebrates. Abbreviated MSH. Also known as melanophore-dilating principle; melanophore hormone. { mə'lan·ə‚sīt 'stim·yə‚lād·iŋ 'hȯr‚mōn }

melanocytoma [MED] A benign tumor composed principally of melanocytes. { mə‚lan·ə‚sī'tō·mə }

melanocytosis [MED] An excessive number of melanocytes. { mə‚lan·ə‚sī'tō·səs }

melanoderma [MED] Abnormal darkening of the skin. { ¦mel·ə·nō¦dər·mə }

melanogen [BIOCHEM] A colorless precursor of melanin. { mə'lan·ə·jən }

melanogenesis [BIOCHEM] The formation of melanin. { ‚mel·ə·nō'jen·ə·səs }

melanoglosia [MED] Any blackening of the tongue associated with certain disorders and diseases in animals and humans. { ‚mel·ə·nō'gläs·ē·ə }

melanoid [MED] Dark-colored; resembling melanin. { 'mel·ə‚nȯid }

melanoma [MED] **1.** A malignant tumor composed of anaplastic melanocytes. **2.** A benign or malignant tumor composed of melanocytes. { mel·ə'nō·mə }

melanomatosis [MED] **1.** Widespread distribution of melanoma. **2.** Diffuse melanotic pigmentation of the meninges. { ‚mel·ə‚nō·mə'tō·səs }

melanophage [HISTOL] A phagocytic cell which engulfs and contains melanin. { mə'lan·ə‚fāj }

melanophlogite [MINERAL] A mineral composed chiefly of silicon dioxide and containing some carbon and sulfur. { ¦mel·ə·nō'flō‚jīt }

melanophore [HISTOL] A type of chromatophore containing melanin. { mə'lan·ə‚fȯr }

melanophore-dilating principle *See* melanocyte-stimulating hormone. { mə'lan·ə‚fȯr 'dī‚lād·iŋ ‚prin·sə·pəl }

melanophore hormone *See* melanocyte-stimulating hormone. { mə'lan·ə‚fȯr 'hȯr‚mōn }

melanoprotein [BIOCHEM] A conjugated protein in which melanin is the associated chromagen. { ‚mel·ə·nō'prō‚tēn }

melanosarcoma [MED] (Obsolete) A malignant melanoma. { ¦mel·ə·nō·sär'kō·mə }

melanose [PL PATH] **1.** A fungus disease of grapevine caused by *Septoria ampelina*; leaves are infected and fall off. **2.** A fungus disease of citrus trees and fruits caused by *Diaporthe citri*, characterized by hard, brown, usually gummy elevations on the rind, twigs, and leaves. { 'mel·ə‚nōs }

melanosis coli [PATH] Melanotic pigmentation of the mucosa of the colon in large numbers of minute foci. { ‚mel·ə'nō·səs 'kō‚lī }

melanosis iridis [PATH] Abnormal melanotic pigmentation of the iris. { ‚mel·ə'nō·səs i'rīd·əs }

melanosome [CYTOL] An organelle which contains melanin and in which tyrosinase activity is not demonstrable. { mə'lan·ə‚sōm }

melanostibian [MINERAL] $Mn(Sb,Fe)O_3$ A black mineral consisting of iron and manganese antimonite; occurs as foliated masses and as striated crystals. { ‚mel·ə·nō'stib·ē·ən }

melanotekite [MINERAL] $Pb_2Fe_2Si_2O_9$ A black or dark-gray mineral composed of lead iron silicate. { ‚mel·ə·nō'tek‚īt }

melanotic freckle [MED] An unevenly pigmented macule that sometimes develops on the skin, usually on the face, of an individual beyond middle life and enlarges progressively.

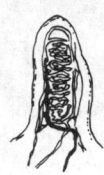

MEISSNER'S CORPUSCLE

Drawing of Meissner's corpuscle. *(From F. A. Geldard, Human Senses, 2d ed., copyright © 1972 by John Wiley & Sons, Inc.; reprinted by permission)*

Also known as Hutchinson's freckle.　{ ,mel·ə'näd·ik frek· əl }

melanovanadite　[MINERAL]　$Ca_2V_{10}O_{25}$　A black mineral composed of a complex oxide of calcium and vanadium.　{ ¦mel·ə·nō'van·ə,dīt }

melanterite　[MINERAL]　$FeSO_4 \cdot 7H_2O$　A green mineral occurring mainly in fibrous or concretionary masses, or in short, monoclinic, prismatic crystals; hardness is 2 on Mohs scale, and specific gravity is 1.90.　{ mə'lan·tə,rīt }

melanuria　[MED]　The presence of black pigment in the urine.　{ ,mel·ə'nur·ē·ə }

melaphyre　[PETR]　Altered basalt, especially of Carboniferous and Permian age.　{ 'mel·ə,fīr }

Melasidae　[INV ZOO]　The equivalent name for Eucnemidae.　{ mə'las·ə,dē }

Melastomataceae　[BOT]　A large family of dicotyledonous plants in the order Myrtales characterized by an inferior ovary, axile placentation, up to twice as many stamens as petals (or sepals), anthers opening by terminal pores, and leaves with prominent, subparallel longitudinal ribs.　{ ,mel·ə,stō·mə'tās·ē,ē }

melatonin　[BIOCHEM]　A hormone secreted by the pineal gland that acts on melanophores in the skins of amphibians and reptiles to concentrate the melanin in the center of the cells, lightening the body surface; in higher vertebrates it conveys information about time that influences reproduction and circadian physiology.　{ ,mel·ə'tōn·ən }

Melde's experiment　[MECH]　An experiment to study transverse vibrations in a long, horizontal thread when one end of the thread is attached to a prong of a vibrating tuning fork, while the other passes over a pulley and has weights suspended from it to control the tension in the thread.　{ 'mel·dēz ik,sper· ə·mənt }

Meleagrididae　[VERT ZOO]　The turkeys, a family of birds in the order Galliformes characterized by a bare head and neck.　{ ,mel·ē·ə'grid·ə,dē }

M electron　[ATOM PHYS]　An electron whose principal quantum number is 3.　{ 'em i,lek,trän }

melena　[MED]　The discharge of stools colored black by altered blood.　{ mə'lē·nə }

Meliaceae　[BOT]　A family of dicotyledonous plants in the order Sapindales characterized by mostly exstipulate, alternate leaves, stamens mostly connate by their filaments, and syncarpous flowers.　{ ,mel·ē'ās·ē,ē }

melilite　[MINERAL]　A sorosilicate mineral group of complex composition $[(Na,Ca)_2(Mg,Al)(Si,Al)_2O_7]$ crystallizing in the tetragonal system; luster is vitreous to resinous, and color is white, yellow, greenish, reddish, or brown; hardness is 5 on Mohs scale, and specific gravity varies from 2.95 to 3.04.　{ 'mel·ə,līt }

melilitite　[PETR]　An extrusive rock that is generally olivine-free and composed of more than 90% mafic mineral such as melilite and augite, with minor amounts of feldspathoids and sometimes plagioclase.　{ mə'lil·ə,tīt }

Melinae　[VERT ZOO]　The badgers, a subfamily of carnivorous mammals in the family Mustelidae.　{ 'mel·ə,nē }

Melinninae　[INV ZOO]　A subfamily of sedentary annelids belonging to the Ampharetidae which have a conspicuous dorsal membrane, with or without dorsal spines.　{ mə'lin·ə,nē }

melioidosis　[VET MED]　An endemic bacterial disease, primarily of rodents but occasionally communicable to humans, caused by *Pseudomonas pseudomallei* and characterized by infectious granulomas.　{ ,mel·ē,oi'dō·səs }

Meliolaceae　[MYCOL]　The sooty molds, a family of ascomycetous fungi in the order Erysiphales, with dark mycelia and conidia.　{ ,mel·ē·ə'lās·ē,ē }

meliphane　*See* meliphanite.　{ 'mel·ə,fān }

meliphanite　[MINERAL]　$(Ca,Na)_2Be(Si,Al)_2(O,OH,F)_7$　A yellow, red, or black mineral composed of sodium calcium beryllium fluosilicate. Also known as meliphane.　{ mə'lif·ə,nīt }

melissic acid　[ORG CHEM]　$CH_3(CH_2)_{28}COOH$　Fatty acid found in beeswax; soluble in benzene and hot alcohol; melts at 90°C; used in biochemical research.　{ mə'lis·ik 'as·əd }

melissopalynology　[PALEOBOT]　A branch of palynology that deals with the analysis of bee pollen loads (pollen collected from flowers and then carried back to the hive on the bee's hindlegs) and the pollen component within honeys.　{ mə,lis· ə,pal·ə'näl·ə·jē }

melissophobia　[PSYCH]　An abnormal fear of bees.　{ mə ,lis·ə'fō·bē·ə }

melitose　*See* raffinose.　{ 'mel·ə,tōs }

melitriose　*See* raffinose.　{ mə'lī·trē,ōs }

Melittidae　[INV ZOO]　A family of hymenopteran insects in the superfamily Apoidea.　{ mə'lid·ə,dē }

melituria　[MED]　The presence of sugar in the urine.　{ ,mel· ə'tur·ē·ə }

Mellin transform　[MATH]　The transform $F(s)$ of a function $f(t)$ defined as the integral over t from 0 to ∞ of $f(t)t^{s-1}$.　{ me'lën ,tranz,form }

mellitate　[ORG CHEM]　An ester or salt of mellitic acid.　{ 'mel·ə,tāt }

mellite　[MINERAL]　$Al_2[C_6(COO)_6] \cdot 18H_2O$　A honey-colored mineral with resinous luster composed of the hydrous aluminum salt of mellitic acid, occurring as nodules in brown coal; it is in part a product of vegetable decomposition.　{ 'me,līt }

mellitic acid　[ORG CHEM]　$C_6(COOH)_6$　A water-soluble compound forming colorless needles that melt at 287°C.　{ mə'lid·ik 'as·əd }

melodeon　[ELECTR]　Broadband panoramic receiver used for countermeasures reception; all types of received electromagnetic radiation are presented as vertical pips on a frequency-calibrated cathode-ray indicator screen.　{ mə'lōd·ē·ən }

Meloidae　[INV ZOO]　The blister beetles, a large cosmopolitan family of coleopteran insects in the superfamily Meloidea; characterized by soft, flexible elytra and the strongly vesicant properties of the body fluids.　{ mə'lō·ə,dē }

Meloidea　[INV ZOO]　A superfamily of coleopteran insects in the suborder Polyphaga.　{ mə'lòid·ē·ə }

melon　[BOT]　Either of two soft-fleshed edible fruits, muskmelon or watermelon, or varieties of these. [VERT ZOO] A round mass of fat on the forehead of some cetaceans between the blowhole and nose.　{ 'mel·ən }

melonite　[MINERAL]　$NiTe_2$　A reddish-white mineral composed of nickel telluride.　{ 'mel·ə,nīt }

melt　[CHEM]　**1.** To change a solid to a liquid by the application of heat.　**2.** A melted material.　[MET]　A charge of molten metal.　{ melt }

meltback transistor　[ELECTR]　A junction transistor in which the junction is made by melting a properly doped semiconductor and allowing it to solidify again.　{ 'melt'back tran'- zist·ər }

meltdown　[NUCLEO]　An accident in a nuclear reactor in which melting of the fuel core occurs.　{ 'melt,daùn }

melter　[ENG]　A chamber used for melting.　{ 'melt·ər }

melt extraction　[MET]　A rapid quenching process in which the molten metal is brought into contact with the periphery of a rotating heat-extracting disk; quench rates exceed 1,000,000 K per second.　{ 'melt ik,strak·shən }

melt extractor　[ENG]　A device used to feed an injection mold, separating molten feed material from partially molten pellets.　{ 'melt ik,strak·tər }

melt-fabricable　[MATER]　Referring to a plastic material that can be shaped as a melt without decomposing, and is capable of being extruded.　{ 'melt ¦fab·rə·kə·bəl }

melt fracture　[MECH]　Melt flow instability through a die during plastics molding, leading to helical, rippled surface irregularities on the finished product.　{ 'melt ,frak·chər }

melt index　[ENG]　Number of grams of thermoplastic resin at 190°C that can be forced through a 0.0825-inch (2.0955-millimeter) orifice in 10 minutes by a 2160-gram force.　{ 'melt ,in,deks }

melting　*See* fusion.　{ 'melt·iŋ }

melting furnace　[ENG]　A furnace in which the frit for glass is melted.　{ 'melt·iŋ fər·nəs }

melting level　[METEOROL]　The altitude at which ice crystals and snowflakes melt as they descend through the atmosphere.　{ 'melt·iŋ ,lev·əl }

melting loss　[MET]　Weight loss due to volatilization or oxidation during metal melting in a foundry.　{ 'melt·iŋ ,lòs }

melting point　[THERMO]　**1.** The temperature at which a solid of a pure substance changes to a liquid. Abbreviated mp.　**2.** For a solution of two or more components, the temperature at which the first trace of liquid appears as the solution is heated.　{ 'melt·iŋ ,pòint }

MELINAE

The American badger (*Taxidea taxus*).

MELOIDAE

Blister beetle. (*From T. I. Storer and R. L. Usinger, General Zoology, 3d ed., McGraw-Hill, 1957*)

melting profile [BIOCHEM] A plot of the degree of denaturation of the strands in a nucleic acid duplex in a specified time as a function of temperature. { 'melt·iŋ ˌprō,fīl }

melting rate [MET] In electric arc welding, the weight or length of electrode melted in a specified unit of time. Also known as burn-off rate; melt-off rate. { 'melt·iŋ ˌrāt }

melting ratio [MET] The ratio of metal weight to fuel weight in a melting process. { 'melt·iŋ ˌrā·shō }

melting temperature [BIOCHEM] The temperature at which denaturing occurs for half of the double helices of deoxyribonucleic acid. { 'melt·iŋ ˌtem·prə·chər }

melt instability [MECH] Instability of the plastic melt flow through a die. { 'melt ˌin·stə'bil·əd·ē }

melt loading [ORD] Process of melting solid explosive by heat and pouring into bombs, projectiles, and the like to solidify. Also known as cast loading. { 'melt ˌlōd·iŋ }

melt-off rate See melting rate. { 'melt,óf ˌrāt }

melton [TEXT] A fabric with all-wool or cotton warp and woolen weft; the face is napped carefully to raise the nap straight up, showing the weave clearly. Also known as beaver cloth; kersey. { 'mel·tən }

melt spinning [TEXT] A process by which nylon, polyester, or glass is melted to allow it to be extruded into fibers through a spinneret. { 'melt,spin·iŋ }

melt strength [MECH] Strength of a molten plastic. { 'melt ˌstreŋkth }

melt-through [NUCLEO] An accident in a nuclear reactor in which melting of the fuel core (meltdown) leads to runaway melting of nuclear fuel out of the bottom of the reactor, down through the concrete mat below, and into the earth. Also known as China syndrome. { 'melt ˌthrü }

meltwater [HYD] Water derived from melting ice or snow, especially glacier ice. { 'melt,wȯd·ər }

Melusinidae [INV ZOO] A family of orthorrhaphous dipteran insects in the series Nematocera. { ˌmel·ə'sin·ə,dē }

Melyridae [INV ZOO] The soft-winged flower beetles, a large family of cosmopolitan coleopteran insects in the superfamily Cleroidea. { mə'lir·ə,dē }

member [CIV ENG] A structural unit such as a wall, column, beam, or tie, or a combination of any of these. [GEOL] A rock stratigraphic unit of subordinate rank comprising a specially developed part of a varied formation. [MATH] **1.** An individual object that belongs to a set. Also known as element. **2.** For an equation, the expression on either side of the equality sign. { 'mem·bər }

membership function [MATH] The characteristic function of a fuzzy set, which assigns to each element in a universal set a value between 0 and 1. { 'mem·bər,ship ,fəŋk·shən }

Membracidae [INV ZOO] The treehoppers, a family of homopteran insects included in the series Auchenorrhyncha having a pronotum that extends backward over the abdomen, and a vertical upper portion of the head. { mem'bras·əd·ē }

membrane [BUILD] In built-up roofing, a weather-resistant (flexible or semiflexible) covering consisting of alternate layers of felt and bitumen, fabricated in a continuous covering and surfaced with aggregate or asphaltic material. [CHEM ENG] **1.** The medium through which the fluid stream is passed for purposes of filtration. **2.** The ion-exchange medium used in dialysis, diffusion, osmosis and reverse osmosis, and electrophoresis. [HISTOL] A thin layer of tissue surrounding a part of the body, separating adjacent cavities, lining cavities, or connecting adjacent structures. { 'mem,brān }

membrane analogy [MECH] A formal identity between the differential equation and boundary conditions for a stress function for torsion of an elastic prismatic bar, and those for the deflection of a uniformly stretched membrane with the same boundary as the cross section of the bar, subjected to a uniform pressure. { 'mem,brān ə,nal·ə·jē }

membrane bone See dermal bone. { 'mem,brān ˌbōn }

membrane carrier [CELL MOL] Any protein that facilitates the movement of small molecules across cell membranes. { 'mem,brān ,kar·ē·ər }

membrane curing See membrane waterproofing. { 'mem,brān ,kyúr·iŋ }

membrane distillation [CHEM ENG] A separation method that uses a nonwetting, microporous membrane, with a liquid feed phase on one side and a condensing permeate phase on

the other. Also known as membrane evaporation; thermopervaporation; transmembrane distillation. { 'mem,brān ,distə'lā·shən }

membrane evaporation See membrane distillation. { 'mem ,brān i,vap·ə'rā·shən }

membrane keyboard [COMPUT SCI] A flat keyboard, used with microcomputers and hand-held calculators, that consists of two closely spaced membranes separated by a flat sheet called a spacer with holes corresponding to the keys. { 'mem ,brān 'kē,bȯrd }

membrane mimetic chemistry [ORG CHEM] The study of processes and reactions that have been developed by using information obtained from biological membrane systems. { ˌmem,brān mi,med·ik 'kem·ə·strē }

membrane potential [PHYSIO] A potential difference across a living cell membrane. { 'mem,brān pə,ten·chəl }

membrane separation [CHEM ENG] The use of thin barriers (membranes) between miscible fluids for separating a mixture; a suitable driving force across the membrane, for example concentration or pressure differential, leads to preferential transport of one or more feed components. { 'mem,brān ,sep·ə'rā·shən }

membrane stress [MECH] Stress which is equivalent to the average stress across the cross section involved and normal to the reference plane. { 'mem,brān ,stres }

membrane waterproofing [CIV ENG] Curing concrete, especially in pavements, by spraying a liquid material over the surface to form a solid, impervious layer which holds the mixing water in the concrete. Also known as membrane curing. { 'mem,brān 'wȯd·ər,prüf·iŋ }

membranous glomerulonephritis [MED] A type of glomerulonephritis characterized by thickening of the basement membrane due to deposition of electron-dense material. { 'mem·brə·nəs glaˌmer·yə·lō·ne'frīd·əs }

membranous labyrinth [ANAT] The membranous portion of the inner ear of vertebrates. { 'mem·brə·nəs 'lab·ə,rinth }

membranous pregnancy [MED] Gestation in which there has been a rupture of the amniotic sac and the fetus is in direct contact with the wall of the uterus. { 'mem·brə·nəs 'preg·nən·sē }

membranous urethra [ANAT] The part of the urethra between the two facial layers of the urogenital diaphragm. { 'mem,brə·nəs yü'rē·thrə }

MEMC See methoxyethylmercury chloride.

memex [COMPUT SCI] A hypothetical machine described by Vannevar Bush, which would store written records so that they would be available almost instantly by merely pushing the right button for the information desired. { 'me,meks }

memistor [ELEC] Nonmagnetic memory device consisting of a resistive substrate in an electrolyte; when used in an adaptive system, a direct-current signal removes copper from an anode and deposits it on the substrate, thus lowering the resistance of the substrate; reversal of the current reverses the process, raising the resistance of the substrate. { me'mis·tər }

memomotion study [IND ENG] A technique of work measurement and methods analysis using a motion picture camera operated at less than normal camera speed. Also known as camera study; micromotion study. { ˌmem·ōˌmō·shən ,stəd·ē }

memory [COMPUT SCI] Any apparatus in which data may be stored and from which the same data may be retrieved; especially, the internal, high-speed, large-capacity working storage of a computer, as opposed to external devices. Also known as computer memory. [PSYCH] The recollection of past events or sensations, or the performance of previously learned skills without practice. { 'mem·rē }

memory address register [COMPUT SCI] A special register containing the address of a word currently required. { 'mem·rē 'ad,res ,rej·ə·stər }

memory bank [COMPUT SCI] A physical section of a computer memory, which may be designed to handle information transfers independently of other such transfers in other such sections. { 'mem·rē ,baŋk }

memory buffer register [COMPUT SCI] A special register in which a word is stored as it is read from memory or just prior to being written into memory. { 'mem·rē 'bəf·ər ,rej·ə·stər }

memory capacity See storage capacity. { 'mem·rē kə'pas·əd·ē }

memory card [COMPUT SCI] A small card, typically with

MEMBRACIDAE

Membracids on stems. (a) Adult. (b) Nymphs. (Courtesy of C. H. Hanson)

dimensions of about 2 × 3 inches (5 × 8 centimeters), that can store information, usually in integrated circuits or magnetic strips. { 'mem·rē ˌkärd }

memory cell [COMPUT SCI] A single storage element of a memory, together with associated circuits for storing and reading out one bit of information. { 'mem·rē ˌsel }

memory chip See semiconductor memory. { 'mem·rē ˌchip }

memory contention [COMPUT SCI] A situation in which two different programs, or two parts of a program, try to read items in the same block of memory at the same time. { 'mem·rē kən'ten·chən }

memory core See magnetic core. { 'mem·rē ˌkȯr }

memory cycle See cycle time. { 'mem·rē ˌsī·kəl }

memory dump See storage dump. { 'mem·rē ˌdəmp }

memory dump routine [COMPUT SCI] A debugging routine which produces a listing of a consecutive section of memory, either numbers or instructions, at selected points in a program. { 'mem·rē ˌdəmp rü̇ˌtēn }

memory element [COMPUT SCI] Any component part of core memory. { 'mem·rē ˌel·ə·mənt }

memory expansion card [COMPUT SCI] A printed circuit board that contains additional storage and can be plugged into a computer to increase its storage capacity. { 'mem·rē ik'span·chən ˌkärd }

memory fill See storage fill. { 'mem·rē ˌfil }

memory gap [COMPUT SCI] A gulf in access time, capacity, and cost of computer storage technologies between fast, expensive, main-storage devices and slow, high-capacity, inexpensive secondary-storage devices. Also known as access gap. { 'mem·rē ˌgap }

memory guard [COMPUT SCI] Built-in safety devices which prevent a program or a programmer from accessing certain memory areas reserved for the central processor. Also known as memory protect. { 'mem·rē ˌgärd }

memory hierarchy [COMPUT SCI] A ranking of computer memory devices, with devices having the fastest access time at the top of the hierarchy, and devices with slower access times but larger capacity and lower cost at lower levels. { 'mem·rē 'hī·ər·ˌär·kē }

memory lockout register [COMPUT SCI] A special register containing the limiting addresses of an area in memory which may not be accessed by the program. { 'mem·rē 'läk·ˌau̇t ˌrej·ə·stər }

memory management [COMPUT SCI] **1.** The allocation of computer storage in a multiprogramming system so as to maximize processing efficiency. **2.** The collection of routines for placing, fetching, and removing pages or segments into or out of the main memory of a computer system. { 'mem·rē ˌman·ij·mənt }

memory map [COMPUT SCI] The list of variables, constants, identifiers, and their memory locations when a FORTRAN program is being run. Also known as memory map list. { 'mem·rē ˌmap }

memory map list See memory map. { 'mem·rē ˌmap ˌlist }

memory mapping [COMPUT SCI] The method by which a computer translates between its logical address space and its physical address space. { 'mem·rē ˌmap·iŋ }

memory overlay [COMPUT SCI] The efficient use of memory space by allowing for repeated use of the same areas of internal storage during the different stages of a program; for instance, when a subroutine is no longer required, another routine can replace all or part of it. { 'mem·rē 'ō·vər·ˌlā }

memory port [COMPUT SCI] A logical connection through which data are transferred in or out of main memory under control of the central processing unit. { 'mem·rē ˌpȯrt }

memory power [COMPUT SCI] A relative characteristic pertaining to differences in access time speeds in different parts of memory; for instance, access time from the buffer may be a tenth of the access time from core. { 'mem·rē ˌpau̇·ər }

memory print See storage dump. { 'mem·rē ˌprint }

memory printout [COMPUT SCI] A listing of the contents of memory. { 'mem·rē ˌprint·ˌau̇t }

memory protect See memory guard. { 'mem·rē prə'tekt }

memory protection See storage protection. { 'mem·rē prə'tek·shən }

memory-reference instruction [COMPUT SCI] A type of instruction usually requiring two machine cycles, one to fetch the instruction, the other to fetch the data at an address (part of the instruction itself) and to execute the instruction. { 'mem·rē ˌref·rəns inˌstrək·shən }

memory register See storage register. { 'mem·rē ˌrej·ə·stər }

memory search routine [COMPUT SCI] A debugging routine which has as an essential feature the scanning of memory in order to locate specified instructions. { 'mem·rē 'sərch rü̇ˌtēn }

memory-segmentation control [COMPUT SCI] Address-computing logic to address words in memory with dynamic allocation and protection of memory segments assigned to different users. { 'mem·rē ˌseg·mən'tā·shən kənˌtrōl }

memory sniffer [COMPUT SCI] A diagnostic routine that continually tests the computer memory while the machine is in operation. { 'mem·rē ˌsnif·ər }

memory storage [COMPUT SCI] The sum total of the computer's storage facilities, that is, core, drum, disk, cards, and paper tape. { 'mem·rē ˌstȯr·ij }

memory switch See ovonic memory switch. { 'mem·rē ˌswich }

memory trace [PHYSIO] See engram. [PSYCH] An experience intentionally forgotten but not fully repressed, which may result in the development of a neurotic conflict. { 'mem·rē 'trās }

memory tube See storage tube. { 'mem·rē ˌtüb }

memory twist See false twist. { 'mem·rē ˌtwist }

memory typewriter See electronic typewriter. { 'mem·rē ˌtīp·ˌrīd·ər }

memotron [ELECTR] An electrical-visual storage tube which is capable of bistable visual-signal display, controllable in duration from a few milliseconds to infinity, and which is suited to specialized oscillography. { 'mem·ə·ˌträn }

MEMS See micro-electro-mechanical system. { memz or ˌemˌēˌem'es }

MEMS microphone [ENG ACOUS] A very small microphone, generally less than 1 millimeter, that can be incorporated directly onto an electronic chip and commonly uses a small thin membrane fabricated on the chip to detect sound. { ˌmemz or ˌemˌēˌem'es 'mī·krəˌfōn }

men-, meno- [PHYSIO] A combining form denoting menses. { men, 'men·ō }

menacme [PHYSIO] The period of a woman's life during which menstruation persists. { mə'nak·mē }

menadione [ORG CHEM] $C_1H_8O_2$ (2-methyl-1,4-naphthoquinone) A compound used as a vitamin K supplement; important in blood clotting. { ˌmen·ə'dī·ōn }

ménage number [MATH] One of the numbers M_n that count the number of ways, once n wives are seated in alternate seats about a circular table, that their husbands can be seated in the seats between them so that no husband sits next to his wife. { mā'näzh ˌnəm·bər }

ménage problem See problème des ménages. { mā'näzh ˌoräb·ləm }

menarche [PHYSIO] The onset of menstruation. { mə'när·kē }

mendelevium [CHEM] Synthetic radioactive element, symbol Md, with atomic number 101; made by bombarding lighter elements with light nuclei accelerated in cyclotrons. { ˌmen·də'lē·vē·əm }

Mendelian genetics [GEN] Scientific study of the role of the nuclear genome in heredity, as opposed to cytoplasmic inheritance. { men'dēl·yən jə'ned·iks }

Mendelian population [GEN] A group of interbreeding individuals; the total allelic gene content of the group is called their gene pool. { men'dēl·yən ˌpäp·yə'lā·shən }

Mendelian ratio [GEN] The ratio of occurrence of various phenotypes in F_1 and F_2 generations in any cross involving characters controlled by nuclear genes. { men'dēl·yən 'rā·shō }

Mendelism [GEN] The basic laws of inheritance as formulated by Mendel. { 'men·də'līz·əm }

Mendel's laws [GEN] Two basic principles of genetics formulated by Mendel: the law of segregation of alleles of a unit factor (gene), and the law of independent assortment of alleles of different unit factors. { 'men·dəlz ˌlȯz }

mendip [GEOL] **1.** A buried hill that is exposed as an inlier. **2.** A coastal-plain hill that was originally an offshore island. { 'men·ˌdip }

mendipite [MINERAL] $Pb_3Cl_2O_2$ A white orthorhombic

mineral consisting of an oxide and chloride of lead. { 'men·də‚pīt }

mendozite [MINERAL] NaAl(SO₄)₂·11H₂O A monoclinic mineral of the alum group composed of hydrous sodium aluminum sulfate. { 'men·də‚zīt }

meneghinite [MINERAL] CuPb₁₃Sb₇S₂₄ A blackish lead gray mineral consisting of lead antimony sulfide. { 'men·ə'gē‚nīt }

Menelaus' theorem [MATH] If *ABC* is a triangle and *PQR* is a straight line that cuts *AB*, *CA*, and the extension of *BC* at *P*, *Q*, and *R* respectively, then (*AP/PB*)(*CQ/QA*)(*BR/CR*) = 1. { ¦men·ə¦lā·əs ‚thir·əm }

Menetrier's disease [MED] Benign, diffuse hypertrophic gastritis; symptoms include vomiting, diarrhea, weight loss, and excessive secretion of mucus { 'men·ə‚trirz di‚zēz }

Menger's theorem [MATH] A theorem in graph theory which states that if *G* is a connected graph and *A* and *B* are disjoint sets of points of *G*, then the minimum number of points whose deletion separates *A* and *B* is equal to the maximum number of disjoint paths between *A* and *B* { 'meŋ·ərz ‚thir·əm }

menhaden oil [MATER] A combustible drying oil that is soluble in benzene or ether, and is derived by cooking or pressing the menhaden fish { men'hād·ən ‚oil }

Ménière's syndrome [MED] A disease of the inner ear characterized by deafness, vertigo, and tinnitus; possibly an allergic process. Also known as labyrinthine syndrome { 'mā·nē¦erz ‚sin‚drōm }

meninges [ANAT] The membranes that cover the brain and spinal cord; there are three in mammals and one or two in submammalian forms { mə'nin·jēz }

meninginococcemia [MED] **1.** The presence of meningococci in the blood **2.** A clinical disorder consisting of fever, skin hemorrhages, varying degrees of shock, and meningococci in the blood { mə‚niŋ·gō‚käk'sē·mē·ə }

meningioma [MED] A localized tumor composed of meningeal cells, involving the meninges and other central nervous system structures. Also known as meningothelioma. { mə‚nin·jē'ō·mə }

meningism [MED] A condition in which signs and symptoms suggest meningitis, but clinical evidence for the disease is absent. Also known as meningismus. { mə'nin‚jiz·əm }

meningismus *See* meningism. { ‚men·ən'jiz·məs }

meningitis [MED] Inflammation of the meninges of the brain and spinal cord, caused by viral, bacterial, and protozoan agents. { ‚men·ən'jīd·əs }

meningitophobia [PSYCH] An abnormal fear of meningitis. { ‚men·ən‚jīd·ə'fō·bē·ə }

meningocele [MED] Hernia of the meninges through a defect in the skull or vertebral column, forming a cyst filled with cerebrospinal fluid. { mə'niŋ·gə‚sēl }

meningococcal meningitis [MED] Inflammation of the meninges caused by the bacterium *Neisseria meningitidis* (meningococcus). { mə¦niŋ·gə¦käk·əl ‚men·ən'jīd·əs }

meningococcus [MICROBIO] Common name for *Neisseria meningitidis*. { mə¦niŋ·gə¦käk·əs }

meningoencephalitis [MED] Inflammation of the brain and its meninges. { mə¦niŋ·gō·in‚sef·ə'līd·əs }

meningoencephalocele [MED] A protrusion of the brain and its membranes through a defect in the skull. { mə¦niŋ·gō·in‚sef·ə·lō‚sēl }

meningoencephalomyelitis [MED] Combined inflammation of the meninges, brain, and spinal cord. { mə¦niŋ·gō·in‚sef·ə·lō‚mī·ə'līd·əs }

meningoencephalopathy [MED] Disease of the brain and its meninges. { mə¦niŋ·gō·in‚sef·ə'läp·ə‚thē }

meningomyelitis [MED] Inflammation of the spinal cord and its membranes. { mə‚niŋ·gō‚mī·ə'līd·əs }

meningomyocele [MED] Hernia of the spinal cord and its meninges through a defect in the vertebral column. { mə¦niŋ·gō'mī·ə‚sēl }

meningothelioma *See* meningioma. { mə'niŋ·gə‚thē·lē'ō·mə }

meningothelium [HISTOL] Epithelium of the arachnoid which envelops the brain. { mə‚niŋ·gə'thē·lē·əm }

meningovascular [MED] Involving both the meninges and the cerebral blood vessels. { mə‚niŋ·gə'vas·kyə·lər }

meningovascular syphilis [MED] Syphilis of the central nervous system involving the formation of gummas of the leptomeninges and endarteritis of cerebral vessels. { mə‚niŋ·gə'vas·kyə·lər 'sif·ə·ləs }

meninx [ANAT] Any one of the three membranes covering the brain and spinal cord. { 'me‚niŋks }

meninx primitiva [VERT ZOO] The single membrane covering the brain and spinal cord of certain submammalian vertebrates. { 'me‚niŋks ‚prim·ə'tī·və }

menisc-, menisco- [SCI TECH] A combining form denoting crescentic, sickle-shaped, semilunar; denoting meniscus, semilunar cartilage. { mə'nis·kō }

meniscectomy [MED] Surgical removal of a meniscus or semilunar cartilage. { ‚men·ə'sek·tə‚mē }

meniscitis [MED] Inflammation of the semilunar cartilages. { ‚men·ə'sīd·əs }

Meniscotheriidae [PALEON] A family of extinct mammals of the order Condylarthra possessing selenodont teeth and molarized premolars. { mə‚nis·kō·thə'rī·ə‚dē }

meniscus [ANAT] A crescent-shaped body, especially an interarticular cartilage. [FL MECH] The free surface of a liquid which is near the walls of a vessel and which is curved because of surface tension. [MET] In reference to a solder joint, the minimum angle at which the solder tapers from the joint to the flat area. { mə'nis·kəs }

meniscus lens [OPTICS] A lens with one convex surface and one concave surface. { mə'nis·kəs 'lenz }

meniscus-Schmidt telescope [OPTICS] A variant of the Schmidt system in which the corrector plate is replaced by a weaker corrector plate followed by a meniscus lens. Also known as Maksutov-Schmidt telescope; Schmidt-Maksutov telescope. { mə'nis·kəs 'shmit 'tel·ə‚skōp }

Menispermaceae [BOT] A family of dicotyledonous woody vines in the order Ranunculales distinguished by mostly alternate, simple leaves, unisexual flowers, and a dioecious habit. { ‚men·ə·spər'mās·ē‚ē }

menometrorrhagia [MED] Excessive uterine bleeding during menstruation, plus irregular uterine bleeding at other times. { ‚men·ə‚me·trə'rā·jē·ə }

menopause [PHYSIO] The natural physiologic cessation of menstruation, usually occurring in the last half of the fourth decade. Also known as climacteric. { 'men·ə‚pòz }

menoplania [MED] Bleeding during menstruation from a part of the body other than the uterus. { ‚men·ə'plā·nē·ə }

Menoponidae [INV ZOO] A family of biting lice (Mallophaga) adapted to life only upon domestic and sea birds. { ‚men·ə'pän·ə‚dē }

menorrhagia [MED] Excessive bleeding during menstruation. Also known as hypermenorrhea. { ‚men·ə'rā·jē·ə }

menorrhalgia [MED] Pelvic pain occurring at the menstrual period. { ‚men·ə'ral·jē·ə }

menorrhea [PHYSIO] The normal flow of the menses. [MED] Excessive menstruation. { ‚men·ə'rē·ə }

menostasis [MED] Suppression of the menstrual flow. { mə'näs·tə·səs }

menostaxis [MED] Prolonged menstruation. { ‚men·ə'stak·səs }

menses *See* menstruation. { 'men‚sēz }

menstrual age [EMBRYO] The age of an embryo or fetus calculated from the first day of the mother's last normal menstruation preceding pregnancy. { 'men·strə·wəl 'āj }

menstrual cycle [PHYSIO] The periodic series of changes associated with menstruation and the intermenstrual cycle; menstrual bleeding indicates onset of the cycle. { 'men·strə·wəl 'sī·kəl }

menstrual period [PHYSIO] The time of menstruation. { 'men·strə·wəl 'pir·ē·əd }

menstruate [PHYSIO] To discharge the products of menstruation. { 'men·strə‚wāt }

menstruation [PHYSIO] The periodic discharge of sanguineous fluid and sloughing of the uterine lining in women from puberty to the menopause. Also known as menses. { ‚men·strə'wā·shən }

menstruum [MATER] A solvent, commonly one that extracts certain principles from entire plant or animal tissues. { 'men·strə·wəm }

mensuration [MATH] The measurement of geometric quantities; for example, length, area, and volume. [SCI TECH] The act or process of measuring. { ‚men·sə'rā·shən }

mental [ANAT] Pertaining to the chin. Also known as genial. [PSYCH] **1.** Pertaining to the mind, psyche, or inner

MENISCUS LENS

Two types of miniscus lens. (*a*) Positive. (*b*) Negative.

self.   **2.** Pertaining to the intellectual or cognitive functions.   **3.** Imaginary or unreal, as when a pain is said to be purely mental.   { 'men·təl }

mental aberration   [PSYCH]   A departure from normal mental function.   { 'men·təl ,ab·ə'rā·shən }

mental adjustment   [PSYCH]   The act or process by which an individual adapts his attitudes, traits, or feelings to the social environment.   { 'men·təl ə'jəs·mənt }

mental age   [PSYCH]   The degree of mental development of an individual in terms of the chronological age of the average individual of equivalent mental ability; specifically, a score derived from intelligence tests.   { 'men·təl 'āj }

mental deficiency   [PSYCH]   A condition characterized by intellectual retardation, social inadequacy, and persistent dependency.   { 'men·təl di'fish·ən·sē }

mental health   [PSYCH]   A relatively enduring state of being in which an individual has effected an integration of his instinctual drives in a way that is reasonably satisfying to himself as reflected in his zest for living and his feeling of self-realization.   { 'men·təl 'helth }

mental hygiene   [PSYCH]   That branch of hygiene dealing with the preservation of mental and emotional health.   { 'men·təl 'hī,jēn }

mental illness   [PSYCH]   Any form of mental aberration; usually refers to a chronic or prolonged disorder in which there are wide deviations from the normal.   { 'men·təl 'il·nəs }

mental retardation   [PSYCH]   An abnormal slowness of mental function and behavior patterns relative to age and development.   { 'men·təl ,rē,tär'dā·shən }

mental telepathy   [PSYCH]   A form of extrasensory perception in which one person is aware of an external event through direct sensory perception, and another person, not in the same place, also becomes aware of the event but not through direct sensory perception.   { 'men·təl tə'lep·ə·thē }

mentation   [PSYCH]   Mental activity.   { men'tā·shən }

Menthaceae   [BOT]   An equivalent name for Labiatae.   { men'thās·ē,ē }

menthane   [ORG CHEM] $C_{10}H_{20}$ A colorless, water-insoluble liquid hydrocarbon; used in organic synthesis.   { 'men,thān }

menthene   [ORG CHEM] $C_{10}H_{18}$ A colorless, water-insoluble, liquid hydrocarbon; used in organic synthesis.   { 'men,thēn }

menthol   [ORG CHEM] $CH_3C_6H_9(C_3H_7)OH$ An alcohol-soluble, white crystalline compound that may exist in levo form or a mixture of dextro and levo isomers; used in medicines and perfumes, and as a flavoring agent.   Also known as peppermint camphor.   { 'men,thȯl }

menthone   [ORG CHEM] $C_{10}H_{18}O$ Oily, colorless ketonic liquid with slight peppermint odor; slightly soluble in water, soluble in organic solvents.   { 'men,thōn }

menthyl   [ORG CHEM] $C_{10}H_{19}$ A univalent radical that is derived from menthol by removal of the hydroxyl group.   { 'men·thəl }

menton-philtrum   [ANTHRO]   A measure of the distance from the midpoint of the lower edge of the chin to the midpoint of the philtrum or vertical groove of the upper lip.   { 'men,tän 'fil·trəm }

menton-supramentale   [ANTHRO]   The measurement of the distance taken from the midpoint of the lower edge of the chin to the supramentale, or the angle between the chin and the lower lip.   { 'men,tän ,sü·prə'men,tāl }

mentum   [ANAT]   The chin.   [BOT]   A projection formed by union of the sepals at the base of the column in some orchids.   [INV ZOO]   **1.** A projection between the mouth and foot in certain gastropods.   **2.** The median or basal portion of the labium in insects.   { 'men·təm }

menu   [COMPUT SCI]   A list of computer functions appearing on a video display terminal which indicates the possible operations that a computer can perform next, only one of which can be selected by the operator.   { 'men·yü }

menu bar   [COMPUT SCI]   **1.** In a graphical user interface, a horizontal strip near the top of the screen or a window, containing the titles of available pull-down menus.   **2.** A horizontal or vertical strip containing the names of currently available commands.   { 'men·yü ,bär }

menu-driven system   [COMPUT SCI]   An interactive computer system in which the operator requests the processing to be performed by making selections from a series of menus.   { 'men·yü ¦driv·ən 'sis·təm }

Menurae   [VERT ZOO]   A small suborder of suboscine perching birds restricted to Australia, including the lyrebirds and scrubbirds.   { mə'nyūr·ē }

Menuridae   [VERT ZOO]   The lyrebirds, a family of birds in the suborder Menurae notable for their vocal mimicry.   { mə'nyūr·ə dē }

MEO   See medium-altitude earth orbit.

mep   See mean effective pressure.

meperidine hydrochloride   [PHARM] $C_{15}H_{21}NO_2$ A narcotic compound that is used medicinally as an analgesic and sedative.   { mə¦pir·ə,dēn ,hī·drə'klȯr·īd }

méplat   [ORD]   The flat nose formed by truncation of the ogival portion of a projectile or point fuse.   { ,mā,plä }

meprobromate   [PHARM] $C_9H_{18}N_2O_4$ The compound 2-methyl-2-n-propyl-1,3-propanediol dicarbamate, a tranquilizer with anticonvulsant, muscle relaxant, and sedative actions.   { ¦me·prō'brō,māt }

meq   See milliequivalent.   { mek }

mer-, mero-   [SCI TECH]   A combining form meaning part or partial.   { 'mər·ō }

-mer   [ORG CHEM]   A combining form denoting the repeating structure unit of any high polymer.   { mər }

meralluride   [PHARM] $C_9H_{16}HgN_2O_6$ A diuretic consisting of succinamic acid and theophylline, in approximately molecular proportions, administered as the sodium derivative.   { mə'ral·yə,rīd }

Meramecian   [GEOL]   A North American provincial series of geologic time: Upper Mississippian (above Osagian, below Chesterian).   { ,mer·ə'mē·shən }

meraspis   [PALEON]   Advanced larva of a trilobite; stage in which the pygidium begins to form.   { mə'rap·səs }

merbromin   [ORG CHEM] $C_{20}H_8O_6Na_2Br_2Hg$ A green crystalline powder that gives a deep-red solution in water; used as an antiseptic.   { mər'brō·mən }

Mercalli scale   [GEOPHYS]   A 12-point scale for classifying the magnitude of an earthquake.   { mer'käl·ē ,skāl }

mercallite   [MINERAL] $KHSO_4$ A colorless or sky blue, orthorhombic mineral consisting of potassium acid sulfate; occurs as stalactites composed of minute crystals.   { mər'kal,īt }

mercapt-, mercapto-   [CHEM]   A combining form denoting the presence of the thiol (SH) group.   { mər'kap·tō }

mercaptal   [ORG CHEM]   A group of organosulfur compounds that contain the group $=C(SR)_2$.   { mər'kap,tal }

mercaptan   [ORG CHEM]   A group of organosulfur compounds that are derivatives of hydrogen sulfide in the same way that alcohols are derivatives of water; have a characteristically disagreeable odor, and are found with other sulfur compounds in crude petroleum; an example is methyl mercaptan.   Also known as thiol.   { mər'kap,tan }

mercaptide   [ORG CHEM]   A compound consisting of a metal and a mercaptan.   { mər'kap,tīd }

mercaptoacetic acid   See thioglycolic acid.   { mər¦kap·tō·ə¦sēd·ik 'as·əd }

2-mercaptobenzoic acid   See thiosalicylic acid.   { ¦tü mər¦kap·tō·ben'zō·ik 'as·əd }

mercaptobenzothiazole   [ORG CHEM] C_7H_5NS A yellow powder, melting at 164–174°C; used in rubber as a vulcanization accelerator with stearic acid.   Abbreviated MBT.   { mər¦kap·tō,ben·zō'thī·ə,zōl }

mercapto compound   See sulfhydryl compound.   { mər'kap·tō ,käm,paund }

mercaptoethanol   [ORG CHEM] $HSCH_2CH_2OH$ Mobile liquid, water-white; soluble in water, benzene, ether, and most organic solvents; boils at 157°C; used as a solvent, chemical intermediate, and reducing agent.   { mər,kap·tō'eth·ə,nȯl }

mercaptol   [ORG CHEM]   A compound formed by combining a mercaptal and a ketone.   { mər'kap,tȯl }

mercaptosuccinic acid   See thiomalic acid.   { mər,kap·tō·sək'sin·ik 'as·əd }

Mercator bearing   See rhumb bearing.   { mər'kād·ər ,ber·iŋ }

Mercator chart   [MAP]   A chart on the Mercator projection, commonly used for marine navigation.   Also known as equatorial cylindrical orthomorphic chart.   { mər'kād·ər ,chärt }

Mercator course   See rhumb-line course.   { mər'kād·ər ,kȯrs }

Mercator direction   [NAV]   Horizontal direction of a rhumb line, expressed as angular distance from a reference direction. Also known as rhumb direction. { mər'kād·ər di'rek·shən }

Mercator projection [MAP] A conformal cylindrical map projection in which the surface of a sphere or spheroid, such as the earth, is conceived as developed on a cylinder tangent along the Equator; meridians appear as equally spaced vertical lines, and parallels as horizontal lines drawn farther apart as the latitude increases, such that the correct relationship between latitude and longitude scales at any point is maintained. { mər'kād·ər prə‚jek·shən }

Mercator sailing [NAV] A method of solving the various problems involving course, distance, difference of latitude, difference of longitude, and departure by considering them in the relation in which they are plotted on a Mercator chart. { mər'kād·ər ‚sāl·iŋ }

Mercer engine [MECH ENG] A revolving-block engine in which two opposing pistons operate in a single cylinder with two rollers attached to each piston; intake ports are uncovered when the pistons are closest together, and exhaust ports are uncovered when they are farthest apart. { 'mər·sər ‚en·jən }

mercerization [TEXT] A technique used to increase luster, dye absorptivity, and strength in cotton and linen goods; the cloth is put into a heated solution of caustic soda at a controlled temperature, then washed, neutralized, and rinsed. { ‚mər·sə·ri'zā·shən }

mercerizing assistant [MATER] A wetting agent, such as cresylic acid and derivatives or oils, used to increase the penetration of textile mercerization baths. { 'mər·sə‚riz·iŋ ə·‚sis·tənt }

merchantable tree height [FOR] The usable portion of the tree stem; for single-stemmed trees this is the length from an assumed stump height to an arbitrary upper-stem diameter. { ¦mər·chənt·ə·bəl 'trē ‚hīt }

merchant mill [MET] A mill, consisting of a group of stands of three rolls each, used to roll rounds, flats, or squares of smaller dimensions than could be rolled on a bar mill. { 'mər·chənt ‚mil }

merchant ship [NAV ARCH] A power-driven ship employed in commercial transport on the oceans and large inland bodies of water such as the Great Lakes. { 'mər·chənt ‚ship }

mercurial horn ore See calomel. { mər'kyùr·ē·əl 'hòrn ‚òr }

mercurialism [MED] Chronic type of mercury poisoning. Also known as hydrargyrism. { mər'kyùr·ē·ə‚liz·əm }

mercurial nephrosis [MED] Nephrosis caused by poisoning with mercury bichloride. { mər'kyùr·ē·əl ne'frō·səs }

mercurial tremor [MED] A fine muscular tremor observed in persons with mercurialism or poisoning by other heavy metals. { mər'kyùr·ē·əl 'trem·ər }

mercuric [INORG CHEM] The mercury ion with a 2+ oxidation state, for example $Hg(NO_3)_2$. { mər'kyùr·ik }

mercuric acetate [ORG CHEM] $Hg(C_2H_3O_2)_2$ Poisonous, light-sensitive white crystals; soluble in alcohol and water; used in medicine and as a catalyst in organic synthesis. Also known as mercury acetate. { mər'kyùr·ik 'as·ə‚tāt }

mercuric arsenate [INORG CHEM] $HgHAsO_4$ A poisonous yellow powder; soluble in hydrochloric acid, insoluble in water; used in antifouling and waterproof paints and in medicine. Also known as mercury arsenate; mercury arseniate. { mər'kyùr·ik 'ärs·ən‚āt }

mercuric barium iodide [INORG CHEM] $HgI_2·BaI_2·5H_2O$ Crystals that are yellow or reddish and deliquescent; soluble in alcohol and water; used in aqueous solution as Rohrbach's solution for mineral separation on the basis of density. Also known as barium mercury iodide; mercury barium iodide. { mər'kyùr·ik 'bar·ē·əm 'ī·ə‚dīd }

mercuric benzoate [ORG CHEM] $Hg(C_7H_5O_2)_2·H_2O$ Poisonous white crystals, sensitive to light, melting at 165°C; slightly soluble in alcohol and water; used in medicine. Also known as mercury benzoate. { mər'kyùr·ik 'ben·zə‚wāt }

mercuric bromide [INORG CHEM] $HgBr_2$ Poisonous white crystals, sensitive to light, melting at 235°C; soluble in alcohol and ether; used in medicine. Also known as mercury bromide. { mər'kyùr·ik 'brō‚mīd }

mercuric chloride [INORG CHEM] $HgCl_2$ An extremely toxic compound that forms white, rhombic crystals which sublime at 300°C and are soluble in alcohol or benzene; used for the manufacture of other mercuric compounds, as a fungicide, and in medicine and photography. Also known as bichloride of mercury; corrosive sublimate. { mər'kyùr·ik 'klòr‚īd }

mercuric cyanate See mercury fulminate. { mər'kyùr·ik 'sī·ə‚nāt }

mercuric cyanide [INORG CHEM] $Hg(CN)_2$ Poisonous, colorless, transparent crystals that darken in light, decompose when heated; soluble in water and alcohol; used in photography, medicine, and germicidal soaps. Also known as mercury cyanide. { mər'kyùr·ik 'sī·ə‚nīd }

mercuric fluoride [INORG CHEM] HgF_2 Poisonous, transparent crystals that decompose when heated; moderately soluble in alcohol and water; used to synthesize organic fluorides. { mər'kyùr·ik 'flùr‚īd }

mercuric iodide [INORG CHEM] HgI_2 Poisonous red crystals that turn yellow when heated to 150°C; soluble in boiling alcohol; used in medicine and in Nessler's and Mayer's reagents. { mər'kyùr·ik 'ī·ə‚dīd }

mercuric lactate [ORG CHEM] $Hg(C_3H_5O_3)_2$ A poisonous white powder that decomposes when heated; soluble in water; used in medicine. { mər'kyùr·ik 'lak‚tāt }

mercuric nitrate [INORG CHEM] $Hg(NO_3)_2·H_2O$ Poisonous, colorless crystals that decompose when heated; soluble in water and nitric acid, insoluble in alcohol; a fire hazard; used in medicine, in nitrating organic aromatics, and in felt manufacture. Also known as mercury nitrate; mercury pernitrate. { mər'kyùr·ik 'nī‚trāt }

mercuric oleate [ORG CHEM] $Hg(C_{18}H_{33}O_2)_2$ A poisonous yellowish-to-red liquid or solid mass; insoluble in water; used in medicine and antifouling paints, and as an antiseptic. Also known as mercury oleate. { mər'kyùr·ik 'ōl·ē‚āt }

mercuric oxide [INORG CHEM] HgO A compound of mercury that exists in two forms, red mercuric oxide and yellow mercuric oxide; the red form decomposes upon heating, is insoluble in water, and is used in pigments and paints, and in ceramics; the yellow form is insoluble in water, decomposes upon heating, and is used in medicine. Also known as mercury oxide; red precipitate; yellow precipitate. { mər'kyùr·ik 'äk‚sīd }

mercuric phosphate [INORG CHEM] $Hg_3(PO_4)_2$ Poisonous yellowish or white powder; insoluble in alcohol and water, soluble in acids; used in medicine. Also known as mercury phosphate; trimercuric orthophosphate. { mər'kyùr·ik 'fäs‚fāt }

mercuric salicylate [ORG CHEM] $Hg(C_7H_5O_3)_2$ Poisonous, white powder; odorless and tasteless; almost insoluble in water and alcohol; variable composition; used in medicine. Also known as salicylated mercury. { mər'kyùr·ik sə'lis·ə‚lāt }

mercuric stearate [ORG CHEM] $Hg(C_{17}H_{35}CO_2)_2$ Poisonous yellow powder; soluble in fatty acids, slightly soluble in alcohol; used as a germicide and in medicine. Also known as mercury stearate. { mər'kyùr·ik 'stir‚āt }

mercuric sulfate [INORG CHEM] $HgSO_4$ A toxic, white, crystalline powder, soluble in acids; used in medicine, as a catalyst, and for galvanic batteries. Also known as mercury persulfate; mercury sulfate. { mər'kyùr·ik 'səl‚fāt }

mercuric sulfide [INORG CHEM] HgS **1.** The black variety is a poisonous powder; insoluble in water, alcohol, and nitric acid, soluble in sodium sulfide solution; sublimes at 583°C; used as a pigment. Also known as black mercury sulfide; ethiops mineral. **2.** The red variety is a poisonous powder; insoluble in water and alcohol; sublimes at 446°C; used as a medicine and pigment. Also known as Chinese vermilion; quicksilver vermilion; red mercury sulfide; vermilion. { mər'kyùr·ik 'səl‚fīd }

mercuric thiocyanate [INORG CHEM] $Hg(SCN)_2$ Poisonous white powder; soluble in alcohol, slightly soluble in water; decomposes when heated; used in photography. Also known as mercury thiocyanate. { mər'kyùr·ik ‚thī·ə'sī·ə‚nāt }

mercurous [INORG CHEM] Referring to mercury with a valence of 1; for example, mercurous chloride, Hg_2Cl_2, where the mercury is covalently bonded, as $Cl—Hg—Hg—Cl$. { mər'kyùr·əs }

mercurous acetate [ORG CHEM] $HgC_2H_3O_2$ Poisonous colorless plates or scales; decomposed by boiling water and by light; soluble in dilute nitric acids, slightly soluble in water. Also known as mercury acetate; mercury protoacetate. { mər'kyùr·əs 'as·ə‚tāt }

mercurous bromide [INORG CHEM] $HgBr$ Poisonous white powder, crystals, or fibrous mass; odorless and tasteless; darkens in light; soluble in hot sulfuric acid and fuming nitric

MERCER ENGINE

(a)
pistons　track
intake
intake　block
rollers

(b)
compression　combustion
spark plug　compression

(c)
exhaust　transfer
transfer

Mercer engine. (*a*) Pistons are closest together and intake ports are open admitting fresh charge. (*b*) Pistons separate as combustion takes place, compressing gases behind pistons, forcing roller to move outward, and rotating the entire engine block at the same time. (*c*) Pistons are farthest apart, exhaust ports are opened, gases are purged, and compressed fresh charge is transferred to region between pistons.

acid, insoluble in alcohol and ether; used in medicine. Also known as mercury bromide. { mər'kyür·əs 'brō,mīd }

mercurous chlorate [INORG CHEM] $Hg_2(ClO_3)_2$ Poisonous white crystals that decompose at 250°C; soluble in alcohol and water; explodes in contact with combustible substances. Also known as mercury chlorate. { mər'kyür·əs 'klór,āt }

mercurous chloride [INORG CHEM] Hg_2Cl_2 Odorless, nonpoisonous white crystals that darken in light; insoluble in water, alcohol, and ether; melts at 302°C; used in medicine and pyrotechnics. Also known as mercury monochloride; mercury protochloride; mild mercury chloride. { mər'kyür·əs 'klór,īd }

mercurous chromate [INORG CHEM] Hg_2CrO_4 Red powder with variable composition; decomposes when heated; soluble in nitric acid, insoluble in water and alcohol; used to color ceramics green. Also known as mercury chromate. { mər'kyür·əs 'krō,māt }

mercurous iodide [INORG CHEM] Hg_2I_2 Odorless, tasteless, poisonous yellow powder; darkens when heated; insoluble in water, alcohol, and ether; sublimes at 140°C; used as external medicine. Also known as mercury protoiodide. { mər'kyür·əs 'ī·ə,dīd }

mercurous oxide [INORG CHEM] Hg_2O A poisonous black powder; insoluble in water, soluble in acids; decomposes at 100°C. { mər'kyür·əs 'äk,sīd }

mercurous phosphate [INORG CHEM] Hg_3PO_4 Light-sensitive white powder with variable composition; insoluble in alcohol and water, soluble in nitric acids; used in medicine. Also known as mercury phosphate; trimercurous orthophosphate. { mər'kyür·əs 'fäs,fāt }

mercurous sulfate [INORG CHEM] Hg_2SO_4 Poisonous yellow-to-white powder; soluble in hot sulfuric acid or dilute nitric acid, insoluble in water; used as a catalyst and in laboratory batteries. { mər'kyür·əs 'səl,fāt }

mercury [CHEM] A metallic element, symbol Hg, atomic number 80, atomic weight 200.59, existing at room temperature as a silvery, heavy liquid. Also known as quicksilver. { 'mər·kyə·rē }

Mercury [ASTRON] The planet nearest to the sun; it is visible to the naked eye shortly after sunset or before sunrise when it is nearest to its greatest angular distance from the sun. { 'mər·kyə·rē }

mercury acetate See mercuric acetate; mercurous acetate. { 'mər·kyə·rē 'as·ə,tāt }

mercury arc [ELECTR] An electric discharge through ionized mercury vapor, giving off a brilliant bluish-green light containing strong ultraviolet radiation. { 'mər·kyə·rē 'ärk }

mercury-arc rectifier [ELECTR] A gas-filled rectifier tube in which the gas is mercury vapor; small sizes use a heated cathode, while larger sizes rated up to 8000 kilowatts and higher use a mercury-pool cathode. Also known as mercury rectifier; mercury-vapor rectifier. { 'mər·kyə·rē 'ärk 'rek·tə,fī·ər }

mercury arsenate See mercuric arsenate. { 'mər·kyə·rē 'ärs·ən,āt }

mercury arseniate See mercuric arsenate. { 'mər·kyə·rē är'sē·nē,āt }

mercury ballistic [NAV] A system of reservoirs and connecting tubes containing mercury used with a type of nonpendulous gyro compass; the action of gravity on this system provides the torques and resultant precessions required to convert the gyroscope into a compass. { 'mər·kyə·rē bə'lis·tik }

mercury barium iodide See mercuric barium iodide. { 'mər·kyə·rē ba·rē·əm 'ī·ō·dīd }

mercury barometer [ENG] An instrument which determines atmospheric pressure by measuring the height of a column of mercury which the atmosphere will support; the mercury is in a glass tube closed at one end and placed, open end down, in a well of mercury. Also known as Torricellian barometer. { 'mər·kyə·rē bə'räm·əd·ər }

mercury benzoate See mercuric benzoate. { 'mər·kyə·rē 'ben·zə,wāt }

mercury bromide See mercuric bromide; mercurous bromide. { 'mər·kyə·rē 'brō,mīd }

mercury-cathode cell [CHEM ENG] Electrolytic cell used to manufacture chlorine and caustic soda from sodium chloride brine; includes Castner and DeNora cells. { 'mər·kyə·rē 'kath,ōd ,sel }

mercury cell [ELEC] A primary dry cell that delivers an essentially constant output voltage throughout its useful life by

means of a chemical reaction between zinc and mercury oxide; widely used in hearing aids. Also known as mercury oxide cell. { 'mər·kyə·rē ,sel }

mercury chlorate See mercurous chlorate. { 'mər·kyə·rē 'klór,āt }

mercury chromate See mercurous chromate. { 'mər·kyə·rē 'krō,māt }

mercury cyanide See mercuric cyanide. { 'mər·kyə·rē 'sī·ə,nīd }

mercury delay line [ELECTR] An acoustic delay line in which mercury is the medium for sound transmission. Also known as mercury memory; mercury storage. { 'mər·kyə·rē di'lā ,līn }

mercury fulminate [ORG CHEM] $Hg(CNO)_2$ A gray, crystalline powder; explodes at the melting point; soluble in alcohol, ammonium hydroxide, and hot water; used for explosive caps and detonators. Also known as mercuric cyanate. { 'mər·kyə·rē 'fül·mə,nāt }

mercury jet magnetometer [ENG] A type of magnetometer in which the magnetic field strength is determined by measuring the electromotive force between electrodes at opposite ends of a narrow pipe made of insulating material, through which mercury is forced to flow. { ¦mər·kyə·rē ¦jet ,mag·nə'täm·əd·ər }

mercury lamp See mercury-vapor lamp. { 'mər·kyə·rē ,lamp }

mercury-manganese star [ASTRON] A star of spectral type B8 or B9 that has a variable spectrum displaying excesses of phosphorus, manganese, gallium, strontium, yttrium, zirconium, platinum, and mercury, and lacks a strong global magnetic field. { 'mər·kyə·rē 'maŋ·gə,nēs 'stär }

mercury manometer [ENG] A manometer in which the instrument fluid is mercury; used to record or control difference of pressure or fluid flow. { 'mər·kyə·rē mə'näm·əd·ər }

mercury memory See mercury delay line. { 'mər·kyə·rē 'mem·rē }

mercury monochloride See mercurous chloride. { 'mər·kyə·rē ,män·ə'klór,īd }

mercury naphthenate [ORG CHEM] Poisonous dark-amber liquid; soluble in mineral oils; used in gasoline antiknock compounds and as a paint antimildew promoter. { 'mər·kyə·rē 'naf·thə,nāt }

mercury nitrate See mercuric nitrate. { 'mər·kyə·rē 'nī,trāt }

mercury oleate See mercuric oleate. { 'mər·kyə·rē 'ōl·ē,āt }

mercury oxide See mercuric oxide. { 'mər·kyə·rē 'äk,sīd }

mercury oxide cell See mercury cell. { 'mər·kyə·rē ¦äk,sīd ,sel }

mercury pernitrate See mercuric nitrate. { 'mər·kyə·rē pər'nī,trāt }

mercury persulfate See mercuric sulfate. { 'mər·kyə·rē pər'səl,fāt }

mercury phosphate See mercuric phosphate; mercurous phosphate. { 'mər·kyə·rē 'fäs,fāt }

mercury-pool cathode [ELECTR] A cathode of a gas tube consisting of a pool of mercury; an arc spot on the pool emits electrons. { 'mər·kyə·rē ,pül 'kath,ōd }

mercury-pool rectifier See pool-cathode mercury-arc rectifier. { 'mər·kyə·rē ,pül rek·tə,fī·ər }

mercury protoacetate See mercurous acetate. { 'mər·kyə·rē ,prō·dō'as·ə,tāt }

mercury protochloride See mercurous chloride. { 'mər·kyə·rē ,prō·dō'klór,īd }

mercury protoiodide See mercurous iodide. { 'mər·kyə·rē ,prō·dō'ī·ə,dīd }

mercury stearate See mercuric stearate. { 'mər·kyə·rē 'stir,āt }

mercury storage See mercury delay line. { 'mər·kyə·rē 'stór·ij }

mercury sulfate See mercuric sulfate. { 'mər·kyə·rē 'səl,fāt }

mercury switch [ELEC] A switch that is closed by making a large globule of mercury move up to the contacts and bridge them; the mercury is usually moved by tilting the entire switch. { 'mər·kyə·rē ,swich }

mercury tank [ELECTR] A container of mercury, with pairs of transducers at opposite ends, used in a mercury delay line. { 'mər·kyə·rē ,taŋk }

mercury thermometer [ENG] A liquid-in-glass thermometer or a liquid-in-metal thermometer using mercury as the liquid. { 'mər·kyə·rē thər'mäm·əd·ər }

MERCURY-VAPOR LAMP

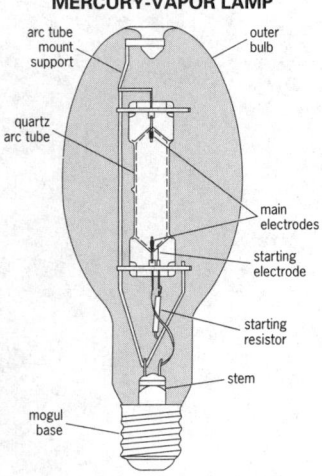

arc tube
mount
support

outer
bulb

quartz
arc tube

main
electrodes

starting
electrode

starting
resistor

stem

mogul
base

High-pressure mercury-vapor lamp.

mercury thiocyanate *See* mercuric thiocyanate. { 'mər·kyə·rē ˌthī·ə'sī·əˌnāt }

mercury tube *See* mercury-vapor tube; pool tube. { 'mər·kyə·rē ˌtüb }

mercury-vapor lamp [ELECTR] A lamp in which light is produced by an electric arc between two electrodes in an ionized mercury-vapor atmosphere; it gives off a bluish-green light rich in ultraviolet radiation. Also known as mercury lamp. { 'mər·kyə·rē ˈvā·pər ˌlamp }

mercury-vapor rectifier *See* mercury-arc rectifier. { 'mər·kyə·rē ˈvā·pər 'rek·təˌfī·ər }

mercury-vapor tube [ELECTR] A gas tube in which the active gas is mercury vapor. Also known as mercury tube. { 'mər·kyə·rē ˈvā·pər ˌtüb }

mercury-wetted reed switch [ELEC] A reed switch containing a pool of mercury at one end and normally operated vertically; the contacts on the reeds are covered with a mercury film by capillary action; each operation of the switch renews this mercury film contact, thereby increasing the operating life of the switch many times. { 'mər·kyə·rē ˈwed·əd 'rēd ˌswich }

mere [HYD] A large pond or a shallow lake. { mīr }

merganser [VERT ZOO] Any of several species of diving water fowl composing a distinct subfamily of Anatidae and characterized by a serrate bill adapted for catching fish. { mər'gan·sər }

merge [COMPUT SCI] To create an ordered set of data by combining properly the contents of two or more sets of data, each originally ordered in the same manner as the output data set. Also known as mesh. { mərj }

merged-transistor logic *See* integrated injection logic. { ¦mərjd tran¦zis·tər 'läj·ik }

merge search [COMPUT SCI] A procedure for searching a table in which both the table and file records must first be ordered in the same sequence on the key involved, and the table is searched sequentially until a table-record key equal to or greater than the file-record key is found, upon which the file record is processed if its key is equal, and the process is repeated with the next file record, starting at the table position where the previous search terminated. { 'mərj ˌsərch }

merge sort [COMPUT SCI] To produce a single sequence of items ordered according to some rule, from two or more previously ordered or unordered sequences, without changing the items in size, structure, or total number; although more than one pass may be required for a complete sort, items are selected during each pass on the basis of the entire key. { 'mərj ˌsort }

merging routine [COMPUT SCI] A program that creates a single sequence of items, ordered according to some rule, out of two or more sequences of items, each sequence ordered according to the same rule. { 'mərj·iŋ rüˌtēn }

Merian's formula [OCEANOGR] A formula for the period of a seiche, $T = (1/n)(2L/\sqrt{gd})$, where n is the number of nodes, L is the horizontal dimension of the basin measured in the direction of wave motion, g is the acceleration of gravity, and d is the depth of the water. { 'mer·ē·ənz ˌför·myə·lə }

mericarp [BOT] An individual, one-seeded carpel of a schizocarp. { 'mer·əˌkärp }

mericlinal chimera [BIOL] An organism or organ composed of two genetically different tissues, one of which partly surrounds the other. { ¦mer·əˌklīn·əl 'kim·ə·rə }

meridian [ASTRON] **1.** A great circle passing through the poles of the axis of rotation of a planet or satellite. **2.** *See* celestial meridian. [GEOD] A north-south reference line, particularly a great circle through the geographical poles of the earth. { mə'rid·ē·ən }

meridian altitude [ASTRON] The altitude of a celestial body when it is on the celestial meridian of the observer, bearing 000° or 180° true. { mə'rid·ē·ən ˌal·təˌtüd }

meridian angle [ASTRON] Angular distance east or west of the local celestial meridian; the arc of the celestial equator, or the angle at the celestial pole, between the upper branch of the local celestial meridian and the hour circle of a celestial body, measured eastward or westward from the local celestial meridian through 180°, and labeled E or W to indicate the direction of measurement. { mə'rid·ē·ən ˌaŋ·gəl }

meridian angle difference [ASTRON] The difference between two meridian angles, particularly between the meridian angle of a celestial body and the value used as an argument for entering into a table. Also called hour angle difference. { mə'rid·ē·ən ¦aŋ·gəl 'dif·rəns }

meridian circle *See* transit circle. { mə'rid·ē·ən ˌsər·kəl }

meridian observation [ASTRON] Measurement of the altitude of a celestial body on the celestial meridian of the observer, or the altitude so measured. { mə'rid·ē·ən ˌäb·sər'vā·shən }

meridian passage [ASTRON] The passage of a celestial body across an observer's meridian. { mə'rid·ē·ən ˌpas·ij }

meridian photometer [ASTRON] An instrument in which mirrors are used to bring the light from two stars which are at or near the celestial meridian simultaneously, but at different altitudes, to a common focus, to compare their brightness. { mə'rid·ē·ən fə'täm·əd·ər }

meridian sailing [NAV] Following a true course of 000° or 180°, or sailing along a meridian; under these conditions the dead-reckoning latitude is assumed to change 1 minute for each mile run, and the dead-reckoning longitude is assumed to remain unchanged. { mə'rid·ē·ən ˌsāl·iŋ }

meridian section [MATH] The intersection of a surface of revolution with a plane that contains the axis of revolution. { mə'rid·ē·ən ˌsek·shən }

meridian telescope [OPTICS] Any telescope used to make observations in the plane of the meridian, such as a transit telescope or zenith telescope. { mə'rid·ē·ən 'tel·əˌskōp }

meridian transit *See* transit; transit circle. { mə'rid·ē·ən ˌtran·zət }

meridional [GEOL] Pertaining to longitudinal movements or directions, that is, northerly or southerly. { mə'rid·ē·ən·əl }

meridional cell [GEOPHYS] A very large-scale convection circulation in the atmosphere or ocean which takes place in a meridional plane, with northward and southward currents in opposite branches of the cell, and upward and downward motion in the equatorward and poleward ends of the cell. { mə'rid·ē·ən·əl 'sel }

meridional circulation [METEOROL] An atmospheric circulation in a vertical plane oriented along a meridian; it consists, therefore, of the vertical and the meridional (north or south) components of motion only. [OCEANOGR] The exchange of water masses between northern and southern oceanic regions. { mə'rid·ē·ən·əl ˌsər·kyə'lā·shən }

meridional difference [MAP] The difference between the meridional parts of any two given parallels of latitude; this difference is found by subtraction if the two parallels are on the same side of the equator, and by addition if on opposite sides. Also known as difference of meridional parts. { mə'rid·ē·ən·əl 'dif·rəns }

meridional flow [METEOROL] A type of atmospheric circulation pattern in which the meridional (north and south) component of motion is unusually pronounced; the accompanying zonal component is usually weaker than normal. [OCEANOGR] Current moving along a meridian. { mə'rid·ē·ən·əl 'flō }

meridional focus *See* primary focus. { mə'rid·ē·ən·əl 'fō·kəs }

meridional front [METEOROL] A front in the South Pacific separating successive migratory subtropical anticyclones; such fronts are essentially in the form of great arcs with meridians of longitudes as chords; they have the character of cold fronts. { mə'rid·ē·ən·əl 'frənt }

meridional index [METEOROL] A measure of the component of air motion along meridians, averaged, without regard to sign, around a given latitude circle. { mə'rid·ē·ən·əl 'in·deks }

meridional parts [MAP] The length of the arc of a meridian between the equator and a given parallel on a mercator chart, expressed in units of 1 minute of longitude at the equator. { mə'rid·ē·ən·əl 'pärts }

meridional plane [OPTICS] A plane containing the axis of an optical system. Also known as tangential plane. { mə'rid·ē·ən·əl 'plān }

meridional ray [OPTICS] A ray that lies within a plane which also contains the axis of an optical system. { mə'rid·ē·ən·əl 'rā }

meridional wind [METEOROL] The wind or wind component along the local meridian, as distinguished from the zonal wind. { mə'rid·ē·ən·əl 'wind }

Meridosternata [INV ZOO] A suborder of echinoderms including various deep-sea forms of sea urchins. { ¦mer·ə·dō·stər'näd·ə }

merino [TEXT] A fine, soft fabric made of wool from the Merino sheep. Also known as botany. { mə'rē·nō }

merinthophobia [PSYCH] An abnormal fear of being tied up. { məˌrin·thə'fō·bē·ə }

merismite [PETR] A type of chorismite in which penetration of the diverse units is irregular. { mə'riz‚mīt }

meristem [BOT] Formative plant tissue composed of undifferentiated cells capable of dividing and giving rise to other meristematic cells as well as to specialized cell types; found in growth areas. { 'mer·ə‚stem }

meristic [BIOL] Pertaining to a change in number or in geometric relation of parts of an organism. [ZOO] Of, pertaining to, or divided into segments. { mə'ris·tik }

merit [ELECTR] A performance rating that governs the choice of a device for a particular application; it must be qualified to indicate type of rating, as in gain-bandwidth merit or signal-to-noise merit. { 'mer·ət }

merit pay plan [IND ENG] Work performed for a set hourly wage that varies from one pay period to another as a function of the worker's productivity, but never declines below a guaranteed minimum wage. { 'mer·ət ‚pā ‚plan }

Merkel's corpuscles [NEUROSCI] Touch receptors consisting of flattened platelets at the tips of certain cutaneous nerves. { 'mər·kəlz ‚kər·pə·səlz }

Mermithidae [INV ZOO] A family of filiform nematodes in the superfamily Mermithoidea; only juveniles are parasitic. { mər'mith·ə‚dē }

Mermithoidea [INV ZOO] A superfamily of nematodes composed of two families, both of which are invertebrate parasites. { mər·mə'thóid·ē·ə }

meroblastic [EMBRYO] Of or pertaining to an ovum that undergoes incomplete cleavage due to large amounts of yolk. { ¦mer·ə¦blas·tik }

merocrine [PHYSIO] Pertaining to glands in which the secretory cells undergo cytological changes without loss of cytoplasm during secretion. { 'mer·ə·krən }

merocrystalline See hypocrystalline. { ¦mer·ə'krist·əl·ən }

merogony [EMBRYO] The normal or abnormal development of a part of an egg following cutting, shaking, or centrifugation of the egg before or after fertilization. { mə'räj·ə·nē }

merohedral [CRYSTAL] Of a crystal class in a system, having a general form with only one-half, one-fourth, or one-eighth the number of equivalent faces of the corresponding form in the holohedral class of the same system. Also known as merosymmetric. { ¦mer·ə¦hē·drəl }

meromictic [HYD] Of or pertaining to a lake whose water is permanently stratified and therefore does not circulate completely throughout the basin at any time during the year. { ¦mer·ə¦mik·tik }

meromixis [GEN] Genetic exchange in bacteria involving a unidirectional transfer of a partial genome. { ‚mer·ə'mik·səs }

meromorphic function [MATH] A function of complex variables which is analytic in its domain of definition save at a finite number of points which are poles. { ¦mer·ə¦mor·fik 'fəŋk·shən }

meromyarian [INV ZOO] Having few muscle cells in each quadrant as seen in cross section; applied especially to nematodes. { ¦mer·ə·mī¦ar·ē·ən }

meromyosin [BIOCHEM] Protein fragments of a myosin molecule, produced by limited proteolysis. { ‚mer·ə'mī·ə·sən }

Meropidae [VERT ZOO] The bee-eaters, a family of brightly colored Old World birds in the order Coraciiformes. { mə'räp·ə‚dē }

meroplankton [BIOL] Plankton composed of floating developmental stages (that is, eggs and larvae) of the benthos and nekton organisms. Also known as temporary plankton. { ‚mer·ə'plaŋk·tən }

merospermy [CELL MOL] Fusion of an egg with an anucleate sperm. { mi'räs·pər·mē }

merosporangium [MYCOL] A cylindrical sporangium containing sporangiospores in a row or chain-like formation. { ‚mer·ə·spə'ran·jē·əm }

Merostomata [INV ZOO] A class of primitive arthropods of the subphylum Chelicerata distinguished by their aquatic mode of life and the possession of abdominal appendages which bear respiratory organs; only three living species are known. { ‚mer·ə'stō·mə·də }

merosymmetric See merohedral. { ¦mer·ə·si¦me·trik }

-merous [BIOL] Combining form that denotes having such parts or so many parts. { mər·əs }

Merozoa [INV ZOO] The equivalent name for Cestoda. { ‚mer·ə'zō·ə }

merozoite [INV ZOO] An ameboid trophozoite in some sporozoans produced from a schizont by schizogony. { ‚mer·ə'zō‚īt }

merozygote [MICROBIO] In bacteria, a zygote that has some diploid and some haploid genetic material because a chromosomal fragment was transferred by the F+ mate. { ‚mir·ə'zī‚gōt }

merrihueite [MINERAL] $(K,Na)_2(Fe,Mg)_5Si_{12}O_{30}$ A silicate mineral found only in meteorites. { ‚mer·ə'hwā‚īt }

Merrill-Crowe process [MET] Removal of gold from cyanide solution by deoxygenation followed by precipitation on zinc dust, the work being completed by filtration to give the resultant auriferous gold slimes. { 'mer·əl 'krō ‚prä·səs }

merrillite [MINERAL] $Ca_3(PO_4)_2$ Colorless phosphate mineral found only in meteorites. { 'mer·ə‚līt }

Merrington effect [FL MECH] The pronounced expansion of a non-Newtonian fluid when it emerges from a nozzle so that the diameter of the emerging stream can be several times the nozzle diameter. { 'mer·iŋ·tən i‚fekt }

Mersenne number [MATH] A number of the form $2^p - 1$, where p is a prime number. { mər'sen ¦nəm·bər }

Mersenne prime [MATH] A Mersenne number that is also a prime number. { mər'sen ¦prīm }

Mersenne's law [MECH] The fundamental frequency of a vibrating string is proportional to the square root of the tension and inversely proportional both to the length and the square root of the mass per unit length. { mər'senz ‚lo }

Mersey yellow coal See tasmanite. { 'mər·zē ‚yel·ō ‚kōl }

merthiolate See thimerosal. { mər'thī·ə‚lāt }

Merton grating [OPTICS] A type of diffraction grating which is produced by a process in which a helical thread is cut on a cylinder, and errors are smoothed by cutting a second thread further along the same cylinder with a Merton nut. { 'mərt·ən ‚grād·iŋ }

Merton nut [DES ENG] A nut whose threads are made of an elastic material such as cork, and are formed by compressing the material into a screw. { 'mərt·ən ‚nət }

merwinite [MINERAL] $Ca_3MgSi_2O_8$ A rare colorless or pale-green neosilicate mineral crystallizing in the monoclinic system; occurs in granular aggregates showing polysynthetic twinning; hardness is 6 on Mohs scale, and specific gravity is 3.15 { 'mər·wə‚nīt }

merycism See rumination. { 'mer·ə‚siz·əm }

Merycoidodontidae [PALEON] A family of extinct tylopod ruminants in the superfamily Merycoidodontoidea. { ‚mer·ə‚kóid·ə'dän·tə‚dē }

Merycoidodontoidea [PALEON] A superfamily of extinct ruminant mammals in the infraorder Tylopoda which were exceptionally successful in North America. { mer·ə‚kóid·ə‚dän'tóid·ē·ə }

merzlota See frozen ground. { ‚merz'lō·tə }

mes-, meso- [SCI TECH] A combining form denoting mid-, middle, medial; medium, moderate, intermediate; mesentery; mesodermal. { 'mes·ō or 'mez·ō }

mesa [GEOGR] A broad, isolated, flat-topped hill bounded by a steep cliff or slope on at least one side; represents an erosion remnant. { 'mē·sə }

mesa-butte [GEOGR] A butte formed as the result of erosion and reduction of a mesa. { 'mā·sə ‚byüt }

Mesacanthidae [PALEON] An extinct family of primitive acanthodian fishes in the order Acanthodiformes distinguished by a pair of small intermediate spines, large scales, superficially placed fin spines, and a short branchial region. { ‚mes·ə'kan·thə‚dē }

mesaconic acid [ORG CHEM] $C_5H_6O_4$ An unsaturated dibasic acid, an isomer of citraconic acid, that melts at 202°C. Also known as methyl fumaric acid. { ‚mes·ə'kän·ik 'as·əd }

mesa device [ELECTR] Any device produced by diffusing the surface of a germanium or silicon wafer and then etching down all but selected areas, which then appear as physical plateaus or mesas. { 'mā·sə di‚vīs }

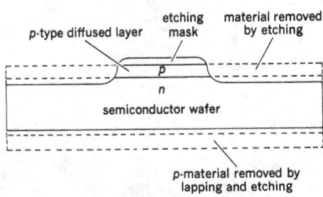

MESA DIODE

Mesa structure of a high-speed diffused silicon diode.

MESCALINE

Structural formula of mescaline.

MESOBILIVERDIN

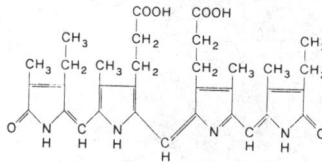

Structural formula of mesobiliverdin.

mesa diode [ELECTR] A diode produced by diffusing the entire surface of a large germanium or silicon wafer and then delineating the individual diode areas by a photoresist-controlled etch that removes the entire diffused area except the island or mesa at each junction site. { 'mā·sə ‚dī‚ōd }

mesa plain [GEOGR] A flat-topped summit of a hilly mountain. { 'mā·sə ‚plān }

mesappendix [ANAT] The mesentery of the vermiform appendix. { ‚mes·ə'pen·diks }

mesarch [BOT] Having metaxylem on both sides of the protoxylem in a siphonostele. [ECOL] Originating in a mesic environment. { 'me‚zärk }

mesarteritis [MED] Inflammation of the tunica media of an artery. { ¦mez‚ärd·ə'rīd·əs }

mesa transistor [ELECTR] A transistor in which a germanium or silicon wafer is etched down in steps so the base and emitter regions appear as physical plateaus above the collector region. { 'mā·sə tran'zis·tər }

mescal buttons [BOT] The dried tops from the cactus *Lophophora williamsii*; capable of producing inebriation and hallucinations. { me'skal ‚bət·ənz }

mescaline [ORG CHEM] $C_{11}H_{17}NO_3$ The alkaloid 3,4,5-trimethoxyphenethylamine, found in mescal buttons; produces unusual psychic effects and visual hallucinations. { 'mes·kə‚lēn }

mesectoderm [EMBRYO] The portion of the mesenchyme arising from ectoderm. { mə'zek·tə‚dərm }

mesencephalon [EMBRYO] The middle portion of the embryonic vertebrate brain; gives rise to the cerebral peduncles and the tectum. Also known as midbrain. { ¦mez·ən'sef·ə‚län }

mesenchymal cell [HISTOL] An undifferentiated cell found in mesenchyme and capable of differentiating into various specialized connective tissues. { ¦mez·ən¦kī·məl ‚sel }

mesenchymal epithelium [HISTOL] A layer of squamous epithelial cells lining subdural, subarachnoid, and perilymphatic spaces, and the chambers of the eyeball. { ¦mez·ən¦kī·məl ‚ep·ə'thē·lē·əm }

mesenchymal hyalin [PATH] A form of hyalin which results from degeneration or necrosis of nonepithelial tissue, usually of muscle, as in Zenker's hyaline necrosis, or of blood vessels. { ¦mez·ən¦kī·məl 'hī·ə·lən }

mesenchymal tissue [EMBRYO] Undifferentiated tissue composed of branching cells embedded in a coagulable fluid matrix. { ¦mez·ən¦kī·məl 'tish·ü }

mesenchyme [EMBRYO] That part of the mesoderm from which all connective tissues, blood vessels, blood, lymphatic system proper, and the heart are derived. { 'mez·ən‚kīm }

mesenchymoma [MED] A tumor composed of cells resembling those of embryonic mesenchyme, or of mesenchyme with its derivatives. { ¦mez·ən·kī'mō·mə }

mesendoderm [EMBRYO] Embryonic tissue which differentiates into mesoderm and endoderm. { mə'zen·də‚dərm }

mesenteric [ANAT] Of or pertaining to the mesentery. { ¦mez·ən¦ter·ik }

mesenteric artery [ANAT] Either of two main arterial branches arising from the abdominal aorta: the inferior, supplying the descending colon and the rectum, and the superior, supplying the small intestine, the cecum, and the ascending and transverse colon. { ¦mez·ən¦ter·ik 'ärd·ə·rē }

mesenteric lymphadenitis [MED] Inflammation of the lymph nodes in the mesentery. { ¦mez·ən¦ter·ik lim‚fad·ən'īd·əs }

mesenteron [EMBRYO] *See* midgut. [INV ZOO] Central gastric cavity in an actinozoan. { me'zen·tə‚rän }

mesentery [ANAT] A fold of the peritoneum that connects the intestine with the posterior abdominal wall. { 'mez·ən‚ter·ē }

mesentoderm [EMBRYO] **1.** The entodermal portion of the mesoderm. **2.** Undifferentiated tissue from which entoderm and mesoderm are derived. **3.** That part of the mesoderm which gives rise to certain structures of the digestive tract. { mə'zen·tə‚dərm }

mesethmoid [ANAT] A bone or cartilage in the center of the ethmoid region of the vertebrate skull; usually constitutes the greater portion of the nasal septum. { me'zeth‚mȯid }

MESFET *See* metal semiconductor field-effect transistor. { 'mes‚fet }

mesh [COMPUT SCI] *See* merge. [DES ENG] A size of

screen or of particles passed by it in terms of the number of openings occurring per linear inch in each direction. Also known as mesh size. [ELEC] A set of branches forming a closed path in a network so that if any one branch is omitted from the set, the remaining branches of the set do not form a closed path. Also known as loop. [MATH] *See* fineness. [MECH ENG] Engagement or working contact of teeth of gears or of a gear and a rack. [MIN ENG] **1.** A closed path traversed through the network in ventilation surveys. **2.** The size of diamonds as determined by sieves. [TEXT] Any fabric, knitted or woven, with an open, fine or coarse texture. { mesh }

mesh analysis [ELEC] A method of electrical circuit analysis in which the mesh currents are taken as independent variables and the potential differences around a mesh are equated to 0. { 'mesh ə'nal·ə·səs }

mesh connection *See* delta connection. { 'mesh kə‚nek·shən }

mesh currents [ELEC] The currents which are considered to circulate around the meshes of an electric network, so that the current in any branch of the network is the algebraic sum of the mesh currents of the meshes to which that branch belongs. Also known as cyclic currents; Maxwell's cyclic currents. { 'mesh kə·rəns }

mesh impedance [ELEC] The ratio of the voltage to the current in a mesh when all other meshes are open. Also known as self-impedance. { 'mesh im'pēd·əns }

mesh network [COMMUN] A communications network in which each node has at least two links to other nodes. { 'mesh ‚net‚wərk }

mesh size *See* mesh. { 'mesh ‚sīz }

mesh texture *See* reticulate. { 'mesh ‚teks·chər }

mesh weld [MET] A seam weld in which the finished weld is only slightly thicker than the sheets, and the lap disappears. { 'mesh ‚weld }

mesic [ECOL] **1.** Of or pertaining to a habitat characterized by a moderate amount of water. **2.** Of or pertaining to a mesophyte. { 'me·zik }

mesic atom [PARTIC PHYS] An atom in which one of the electrons is replaced by a negative muon or meson orbiting close to or within the nucleus. Also known as mesonic atom. { 'me·zik 'ad·əm }

mesic molecule [PARTIC PHYS] A molecule in which one of the electrons is replaced by a negative muon or meson orbiting close to or within one of the nuclei. Also known as mesonic molecule. { 'me·zik 'mäl·ə‚kyül }

mesitylene [ORG CHEM] C_9H_{12} A colorless fragrant liquid that boils at 164.7°C (328.6°F); it is an aromatic hydrocarbon that is part of the benzene series, and occurs naturally in coal tar or is synthesized from acetone. { mə'sid·əl‚ēn }

mesityl oxide [ORG CHEM] $(CH_3)_2C=CHCOCH_3$ A colorless, oily liquid with a honeylike odor; solidifies at −41.5°C; used as a solvent for resins, particularly vinyl resins, many gums, and nitrocellulose; also used in lacquers, paints, and varnishes. { 'mez·ə‚til 'äk‚sīd }

mesmerism [PSYCH] Hypnotism induced by animal magnetism, a supposed force passing from operator to subject. { 'mez·mə‚riz·əm }

meso- [CHEM] A prefix meaning intermediate or middle, as in denoting inactive optical isomers, the form of intermediate inorganic acid, the middle position in cyclic organic compounds, or a ring system with middle ring positions. { 'me·zō }

mesoappendix [ANAT] The mesentery of the vermiform appendix. { ¦me·zō·ə'pen·diks }

mesobenthos [OCEANOGR] The sea bottom at depths of 100–500 fathoms (180–900 meters). { ¦me·zō¦ben‚thäs }

mesobilirubin [BIOCHEM] $C_{33}H_{40}O_6N_4$ Yellow, crystalline by-product of bilirubin reduction. { ¦me·zō‚bil·i'rü·bən }

mesobilirubinogen [BIOCHEM] $C_{33}H_{44}O_6N_4$ Colorless, crystalline by-product of bilirubin reduction; may be converted to urobilin, stercobilinogen, or stercobilin. { ¦me·zō‚bil·i·rə'bin·ə·jən }

mesobiliverdin [BIOCHEM] $C_{28}H_{38}O_6N_4$ A structural isomer of phycoerythrin and phycocyanobilin released by certain biliprotein by treatment with alkali. { ¦me·zō‚bil·i'vərd·ən }

mesoblast [EMBRYO] Undifferentiated mesoderm of the middle layer of the embryo. { 'me·zō‚blast }

mesoblastema *See* mesoderm. { ‚me·zō·bla'stē·mə }

mesocardium [ANAT] Epicardium covering the blood vessels which enter and leave the heart. [EMBRYO] The mesentery supporting the embryonic heart. { ,me·zō'kärd·ē·əm }

mesocarp [BOT] The middle layer of the pericarp. { 'mez·ə,kärp }

mesocercaria [INV ZOO] The developmental stage in the second intermediate host of *Alaria*, a digenetic trematode. { ,me·zō·sər'kar·ē·ə }

mesoclimate [CLIMATOL] **1.** The climate of small areas of the earth's surface which may not be representative of the general climate of the district. **2.** A climate characterized by moderate temperatures, that is, in the range 20–30°C. Also known as mesothermal climate. { ¦me·zō¦klī·mət }

mesoclimatology [CLIMATOL] The study of mesoclimates. { ,me·zō,klī·mə'täl·ə·jē }

mesocolon [ANAT] The part of the mesentery that is attached to the colon. { ¦me·zō'kō·lən }

mesoconch [ANTHRO] Having moderately rounded orbits with an orbital index of 83 to 89. { 'mez·ə,kaŋk }

mesocranial [ANTHRO] Having a medium-sized skull with a cranial index of 75–80. { ¦me·zō¦krā·nē·əl }

mesocratic [PETR] Of igneous rock, being intermediate in color between leucocratic and melanocratic due to equal amounts of light and dark constituents. { ,mez·ə'krad·ik }

mesocrystalline [PETR] Of a crystalline rock, containing crystals whose diameters are intermediate between microcrystalline and macrocrystalline rock. { ,mez·ə'krist·əl·ən }

mesocyclone [METEOROL] A cyclonic circulation interior to a convective storm. { ¦me·zō¦sī,klōn }

mesoderm [EMBRYO] The third germ layer, lying between the ectoderm and endoderm; gives rise to the connective tissues, muscles, urogenital system, vascular system, and the epithelial lining of the coelom. Also known as mesoblastema. { 'mez·ə,dərm }

mesodermal tumor [MED] A tumor composed of cells normally derived from the mesoderm. { ¦mez·ə¦dər·məl 'tü·mər }

mesogaster [ANAT] The mesentery of the stomach. { 'mez·ə,gas·tər }

Mesogastropoda [INV ZOO] The equivalent name for Pectinibranchia. { ,mez·ə·ga'sträp·ə·də }

mesogen *See* mesogenic unit. { 'mez·ə,jen }

mesogenic unit [PHYS CHEM] A component of a molecule that induces a mesomorphic or liquid crystalline phase. Also known as mesagen. { ¦mez·ə¦jen·ik 'yü·nət }

mesogeosyncline [GEOL] A geosyncline between two continents. Also known as mediterranean. { ¦me·zō,jē·ō'sin,klīn }

mesoglea [INV ZOO] The gelatinous layer between the ectoderm and endoderm in cnidarians and certain sponges. { ¦me·zō'glē·ə }

mesognathic [ANTHRO] Designating a condition of the upper jaw in which it has a mild degree of anterior projection with respect to the profile of the facial skeleton, when the skull is oriented on the Frankfort horizontal plane; having a gnathic index of 98.0 to 102.9. { ¦mez·ə¦nath·ik }

mesogranulation [ASTRON] An intermediate scale of convection on the sun, giving rise to a system of convective cells whose size (2500–3700 miles or 4000–6000 kilometers) and lifetime (3–10 hours) lie between those of granulation and supergranulation. { ,mez·ō,gran·yə'lā·shən }

mesogranule [ASTRON] One of the convective cells in the mesogranulation observed on the sun. { ,mez·ə'gran·yül }

mesohaline [OCEANOGR] **1.** Referring to estuarine water with salinity ranging 5–18 parts per thousand. **2.** Referring to moderately brackish water. { ,me·sō'ha·lēn }

Mesohippus [PALEON] An early ancestor of the modern horse; occurred during the Oligocene. { ¦me·zō'hip·əs }

mesoinositol *See* myoinositol. { ¦me·zō·i'näs·ə,tól }

meso-ionic compound [ORG CHEM] Any of a class of five-membered ring heterocycles and their benzo derivatives which possess a sextet of pi electrons in association with the atoms composing the ring but which cannot be represented satisfactorily by any one covalent or polar structure. { ¦mez·ō·ī¦än·ik 'käm,paùnd }

mesokaryotic [INV ZOO] Pertaining to an organism that shares characteristics of both prokaryotic and eukaryotic organisms. { ,mez·ə,kar·ē'äd·ik }

mesokurtic distribution [STAT] A distribution in which the ratio of the fourth moment to the square of the second moment equals 3, which is the value for a normal distribution. { ,mes·ə¦kərd·ik ,di·strə'byü·shən }

mesolamella [INV ZOO] A thin gelatinous membrane between the epidermis and gastrodermis in hydrozoans. { ¦me·zō·lə'mel·ə }

mesolite [MINERAL] $Na_2Ca_2Al_6Si_9O_{30}\cdot 8H_2O$ Zeolite mineral composed of hydrous sodium calcium aluminosilicate, usually found in white or colorless tufts of acicular crystals; used as cation exchangers or molecular sieves. { 'mez·ə,līt }

mesomere [EMBRYO] The muscle-plate region between the epimere and hypomere in vertebrates. { 'me·zō,mir }

mesomerism *See* resonance. { mə'säm·ə,riz·əm }

mesometeorology [METEOROL] That portion of the science of meteorology concerned with the study of atmospheric phenomena on a scale larger than that of micrometeorology, but smaller than the cyclonic scale. { ,me·zō,mē·dē·ə'räl·ə·jē }

mesometrium [ANAT] The part of the broad ligament attached directly to the uterus. { ,mez·ə'mē·trē·əm }

mesomorph [PSYCH] A somatotype characterized by an athletic physique. { 'mez·ə,mòrf }

mesomorphism [PHYS CHEM] A state of matter intermediate between a crystalline solid and a normal isotropic liquid, in which long rod-shaped organic molecules contain dipolar and polarizable groups. { ,mez·ə¦mòr,fiz·əm }

meson [PARTIC PHYS] Any elementary (noncomposite) particle with strong nuclear interactions and baryon number equal to zero. { 'me,sän }

meson capture [PARTIC PHYS] Process in which an atomic nucleus acquires a negative muon or meson which circles it in a tightly bound orbit until it decays. { 'me,sän ,kap·chər }

mesonephric duct [EMBRYO] The efferent duct of the mesonephros. Also known as Wolffian duct. { ¦me·zō¦nef·rik 'dəkt }

mesonephric fold *See* mesonephric ridge. { ¦me·zō¦nef·rik 'fōld }

mesonephric ridge [EMBRYO] A fold of the dorsal wall of the coelom lateral to the mesentery formed by development of the mesonephros. Also known as mesonephric fold. { ¦me·zō¦nef·rik 'rij }

mesonephros [EMBRYO] One of the middle of three pairs of embryonic renal structures in vertebrates; persists in adult fish and is replaced by the metanephros in higher forms. { ¦mez·ə'ne,frōs }

mesonic atom *See* mesic atom. { me'zän·ik 'ad·əm }

mesonic molecule *See* mesic molecule. { me'zän·ik 'mäl·ə,kyül }

mesonic x-ray [PARTIC PHYS] An x-ray emitted by a mesic atom when the muon or meson makes a transition from one bound state to another. { me'zän·ik 'eks,rā }

meson resonance [PARTIC PHYS] Any elementary particle with a baryon number of zero which decays through strong interactions, and therefore has an extremely short lifetime on the order of 10^{-23} second. { 'me,zän 'rez·ən·əns }

Mesonychidae [PALEON] A family of extinct mammals of the order Condylarthra. { ,me,zän'kid·ə,dē }

mesopause [METEOROL] The top of the mesosphere; corresponds to the level of minimum temperature at 50 to 60 miles (80 to 95 kilometers). { 'mez·ə,póz; }

mesopeak [METEOROL] The temperature maximum at about 30 miles (50 kilometers) in the mesosphere. { 'me·zō,pēk }

mesophile [BIOL] An organism, as certain bacteria, that grows at moderate temperature. { 'mez·ə,fīl }

mesophily [ECOL] Physiological response of organisms living in environments with moderate temperatures and a fairly high, constant amount of moisture. { 'mez·ə,fil·ē }

mesophyll [BOT] Parenchymatous tissue between the upper and lower epidermal layers in foliage leaves. { 'mez·ə,fil }

mesophyte [ECOL] A plant requiring moderate amounts of moisture for optimum growth. { 'mez·ə,fīt }

mesopore [CHEM] A pore in a catalytic material whose width ranges from 2 nanometers to 0.05 micrometer. [PALEON] A tube paralleling the autopore or chamber in fossil bryozoans. { 'mez·ə,pór }

mesopterygium [VERT ZOO] The middle one of three basal cartilages in the pectoral fin of sharks and rays. { ,me,zäp·tə'rij·ē·əm }

mesoptic vision [PHYSIO] Vision in which the human eye's

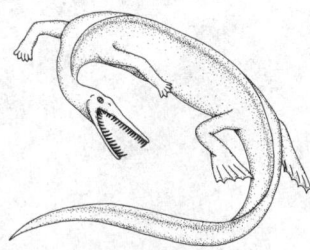

MESOSAURIA

Restoration of *Mesosaurus*.

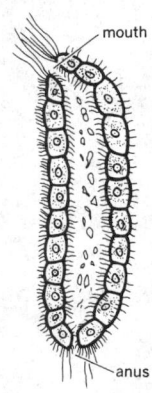

MESOZOA

mouth

anus

Salinella, in longitudinal section.

MESOZOIC

PRECAMBRIAN		
CAMBRIAN		
ORDOVICIAN		
SILURIAN	PALEOZOIC	
DEVONIAN		
Mississippian	CARBON-IFEROUS	
Pennsylvanian		
PERMIAN		
TRIASSIC		
JURASSIC	MESOZOIC	
CRETACEOUS		
TERTIARY	CENOZOIC	
QUATERNARY		

Chart showing the position of the Mesozoic era in relation to the other eras and to the periods of geologic time.

spectral sensitivity is changing from the photoptic state to the scotoptic state. { me'zäp·tik 'vizh·ən }

mesorchium [EMBRYO] The mesentery that supports the embryonic testis in vertebrates. { me'zór·kē·əm }

mesorrhine [ANTHRO] Having a nose of moderate size: nasal index is 47–51 on the skull and 70–85 on the living person. { 'mez·ə‚rīn }

mesosalpinx [ANAT] The portion of the broad ligament forming the mesentery of the uterine tube. { ¦me·zō'sal‚piŋks }

Mesosauria [PALEON] An order of extinct aquatic reptiles which is known from a single genus, *Mesosaurus*, characterized by a long snout, numerous slender teeth, small forelimbs, and webbed hindfeet. { ‚me·zō'sór·ē·ə }

mesoscale eddies *See* mode eddies. { 'me·zō‚skāl 'ed·ēz }

mesoscale motion [GEOPHYS] Motion of winds and ocean currents over regional areas with sizes of 6–60 miles (10–100 kilometers). { ‚mes·ō‚skāl 'mō·shən }

mesoscopic [PHYS] Pertaining to a size regime, intermediate between the microscopic and the macroscopic, that is characteristic of a region where a large number of particles can interact in a quantum-mechanically correlated fashion. { ¦mez·ə¦skäp·ik }

mesoscopic physics [PHYS] A subdiscipline of condensed-matter physics that focuses on the properties of solids in a size range intermediate between bulk matter and individual atoms or molecules. { ‚mez·ə‚skäp·ik 'fiz·iks }

mesosere [ECOL] A sere originating in a mesic habitat and characterized by mesophytes. { 'me·zō‚sir }

mesosiderite [GEOL] A stony-iron meteorite containing about equal amounts of silicates and nickel-iron, with considerable troilite. Also known as grahamite. { ¦me·zō'sīd·ə‚rīt }

mesosoma [INV ZOO] **1.** The anterior portion of the abdomen in certain arthropods. **2.** The middle of the body of some invertebrates, especially when the phylogenetic segmentation pattern cannot be determined. { ¦me·zō'sō·mə }

mesosome [MICROBIO] An extension of the cell membrane within a bacterial cell; possibly involved in cross-wall formation, cell division, and the attachment of daughter chromosomes following deoxyribonucleic acid replication. { 'mez·ə‚sōm }

mesosphere [GEOL] *See* lower mantle. [METEOROL] The atmospheric shell between about 28–35 and 50–60 miles (45–55 and 80–95 kilometers), extending from the top of the stratosphere to the mesopause; characterized by a temperature that generally decreases with altitude. { 'mez·ə‚sfir }

mesostasis [GEOL] The last-formed interstitial material, either glassy or aphanitic, of an igneous rock. { ¦me·zō'stā·səs }

mesosternum [ANAT] The middle portion of the sternum in vertebrates. Also known as gladiolus. [INV ZOO] The ventral portion of the mesothorax in insects. { ¦me·zō'stər·nəm }

Mesostigmata [INV ZOO] The mites, a suborder of the Acarina characterized by a single pair of breathing pores (stigmata) that are located laterally in the middle of the idiosoma between the second and third, or third and fourth, legs. { ¦me·zō‚stig'mäd·ə }

Mesosuchia [PALEON] A suborder of extinct crocodiles of the Late Jurassic and Early Cretaceous. { ‚me·zō'sü·kē·ə }

Mesotaeniaceae [BOT] The saccoderm desmids, a family of fresh-water algae in the order Conjugales; cells are oval, cylindrical, or rectangular and have simple, undecorated walls in one piece. { ‚me·zō‚tē·nē'ās·ē‚ē }

Mesotardigrada [INV ZOO] An order of tardigrades which combines certain echiniscoidean features with eutardigradan characters. { ¦me·zō‚tär'dig·rə·də }

mesotheca [INV ZOO] The middle lamina of bifoliate bryozoan colonies. { ¦me·zō'thē·kə }

mesothelioma [MED] A primary benign or malignant tumor composed of cells resembling the mesothelium. { ‚me·zō‚thē·lē'ō·mə }

mesothelium [ANAT] The simple squamous-cell epithelium lining the pleural, pericardial, peritoneal, and scrotal cavities. [EMBRYO] The lining of the wall of the primitive body cavity situated between the somatopleure and splanchnopleure. { ‚me·zō'thē·lē·əm }

mesotherm [ECOL] A plant that grows successfully at moderate temperatures. { 'mez·ə‚thərm }

mesothermal [MINERAL] Of a hydrothermal mineral deposit, formed at great depth at temperatures of 200–300°C. { ¦mez·ə¦thər·məl }

mesothermal climate *See* mesoclimate. { ¦mez·ə¦thər·məl 'klī·mət }

mesothorax [INV ZOO] The middle of three somites composing the thorax in insects. { ¦me·zō'thó‚raks }

mesotil [GEOL] A semiplastic or semifriable derivative of chemically weathered till; forms beneath a partially drained area. { 'mez·ə‚til }

mesovarium [ANAT] A fold of the peritoneum that connects the ovary with the broad ligament. { ‚me·zō'var·ē·əm }

Mesoveliidae [INV ZOO] The water treaders, a small family of hemipteran insects in the subdivision Amphibicorisae having well-developed ocelli. { ‚me·zō·və'lī·ə‚dē }

mesoxalyurea *See* alloxan. { mə‚zäk·sē·al'yur·ē·ə }

Mesozoa [INV ZOO] A division of the animal kingdom sometimes ranked intermediate between the Protozoa and the Metazoa; composed of two orders of small parasitic, wormlike organisms. { ‚mez·ə'zō·ə }

Mesozoic [GEOL] A geologic era from the end of the Paleozoic to the beginning of the Cenozoic; commonly referred to as the Age of Reptiles. { ¦mez·ə'zō·ik }

mesozone [PETR] The intermediate depth zone of metamorphism in metamorphic rock characterized by moderate temperatures (300–500°C), hydrostatic pressure, and shearing stress. { 'mez·ə‚zōn }

mesozooid [INV ZOO] A type of bryozoan heterozooid that produces slender tubes (mesozooecia or mesopores), internally subdivided by many closely spaced diaphragms, that open as tiny polygonal apertures. { ¦mez·ə¦zō‚óid }

mesquite [BOT] Any plant of the genus *Prosopis*, especially *P. juliflora*, a spiny tree or shrub bearing sugar-rich pods; an important livestock feed. { mə'skēt }

message [COMMUN] A series of words or symbols, transmitted with the intention of conveying information. [COMPUT SCI] An arbitrary amount of information with beginning and end defined or implied: usually, it originates in one place and is intended to be transmitted to another place. { 'mes·ij }

message accounting [COMMUN] Use of equipment to make records of telephone calls for billing purposes. { 'mes·ij ə‚kaúnt·iŋ }

message authentication [COMMUN] Security measure designed to establish the authenticity of a message by means of an authenticator within the transmission derived from certain predetermined elements of the message itself. { 'mes·ij ó‚then·tə'kā·shən }

message authentication code [COMPUT SCI] The encrypted personal identification code appended to the message transmitted to a computer; the message is accepted only if the decrypted code is recognized as valid by the computer. Abbreviated MAC. { 'mes·ij ó‚then·tə'kā·shən ‚kōd }

message blocking [COMMUN] The division of messages into blocks having a fixed number of bytes in order to provide consistent work units and thereby simplify the design of data communications networks. { 'mes·ij ‚bläk·iŋ }

message buffer [COMPUT SCI] One of a number of sections of computer memory, which contains a message that can be transmitted between tasks in the computer system to request service and receive replies from tasks, and which is stored in a system buffer area, outside the address spaces of tasks. { 'mes·ij ‚bəf·ər }

message center [COMMUN] A communications facility charged with the responsibility for acceptance, preparation for transmission, transmission, receipt and delivery of messages. { 'mes·ij ‚sen·tər }

message display console [COMPUT SCI] A cathode-ray tube on which is displayed information requested by the user. { 'mes·ij di'splā ‚kän‚sōl }

message exchange [COMPUT SCI] A device which acts as a buffer between a communication line and a computer and carries out communication functions. { 'mes·ij iks‚chānj }

message indicator [COMMUN] Element placed within a message to serve as a guide to the selection or derivation and application of the correct key to facilitate the prompt decryption of the message. { 'mes·ij ‚in·də'kād·ər }

message interpolation [COMMUN] Data message insertion during intersyllable periods or speech pauses on a busy voice channel without breaking down the voice connection or noticeably affecting the voice transmission. { 'mes·ij in,tər·pə'lā·shən }

message keying element [COMMUN] That part of the key which changes with every message. { 'mes·ij 'kē·iŋ ,el·ə·mənt }

message-oriented applications [COMMUN] Applications of data communications that involve medium-size data transfers in the range of hundreds to a few thousand bytes or characters, and are usually unidirectional information flows from source to destination. { 'mes·ij ˌȯr·ē,ent·əd ˌap·lə'kā·shənz }

message queuing [COMPUT SCI] The stacking of messages according to some priority rule as the messages await processing. { 'mes·ij ,kyü·iŋ }

message reference block [COMMUN] A set of signals denoting the beginning or end of a message. { 'mes·ij 'ref·rəns ,bläk }

message registration [COMMUN] A method for counting the number of completed charged calls which originate from a particular telephone line, making one scoring for each local call and more than one scoring for calls between zones. { 'mes·ij ,rej·ə,strā·shən }

message routing [COMMUN] Selection of the communication path over which a message is sent. { 'mes·ij ,rüd·iŋ }

message switching [COMMUN] A system in which data transmitted between stations on different circuits within a network are routed through central points. { 'mes·ij ,swich·iŋ }

message trailer [COMMUN] The last part of a data communications message that signals the end of the message and may also contain control information such as a check character. { 'mes·ij ,trā·lər }

messaging [COMMUN] Electronic communication in which a message is sent directly to its destination without being stored en route. { 'mes·ij·iŋ }

messenger [ENG] A small, cylindrical metal weight that is attached around an oceanographic wire and sent down to activate the tripping mechanism on various oceanographic devices. [NAV ARCH] A light line used to haul in a larger line or hawser. { 'mes·ən·jər }

messenger cable [COMMUN] A cable made of stranded steel which supports aerial cables between poles. { 'mes·ən·jər ,kā·bəl }

messenger ribonucleic acid [NEUROSCI] A linear sequence of nucleotides which is transcribed from and complementary to a single strand of deoxyribonucleic acid and which carries the information for protein synthesis to the ribosomes. Abbreviated mRNA. { 'mes·ən·jər ¦rī·bō·nü'klē·ik 'as·əd }

Messier number [ASTRON] A number by which star clusters and nebulae are listed in Messier's catalog; for example, the Andromeda Galaxy is M 31. { me'syā ,nəm·bər }

Messier's catalog [ASTRON] A listing of 103 star clusters and nebulae compiled in 1784. { me'syāz 'kad·əl,äg }

mesurand [SCI TECH] A quantity that is to be measured. { 'mezh·ə,rand }

mesyl *See* methylsulfonyl. { 'mes·əl }

meta- [ORG CHEM] A prefix for benzene-ring compounds when two side chains are connected to carbon atoms with an unsubstituted carbon atom between them. { 'med·ə }

metaanthracite [GEOL] Anthracite coal containing at least 98% fixed carbon. { ¦med·ə'an·thrə,sīt }

metabentonite [GEOL] Altered bentonite, formed by compaction or metamorphism; it swells very little and lacks the usual high colloidal properties of bentonite. { ¦med·ə'bent·ən,īt }

metabiosis [ECOL] An ecological association in which one organism precedes and prepares a suitable environment for a second organism. { ,med·ə·bī'ō·səs }

metabolic [PHYSIO] Of or pertaining to metabolism. { ,med·ə'bäl·ik }

metabolic block [BIOCHEM] A nonfunctional reaction in a metabolic pathway due to a defective enzyme whose normal counterpart catalyzes the reaction. { ,med·ə¦bäl·ik 'bläk }

metabolic cost [IND ENG] The amount of energy consumed as the result of performing a given work task; usually expressed in calories. { ,med·ə,bäl·ik 'kȯst }

metabolic disorder [MED] Any disorder that involves an alteration in the normal metabolism of carbohydrates, lipids, proteins, water, and nucleic acids; evidenced by various syndromes and diseases. { ,med·ə'bäl·ik dis'ȯrd·ər }

metabolism [PHYSIO] The physical and chemical processes by which foodstuffs are synthesized into complex elements (assimilation, anabolism), complex substances are transformed into simple ones (disassimilation, catabolism), and energy is made available for use by an organism. { mə'tab·ə,liz·əm }

metabolite [BIOCHEM] A product of intermediary metabolism. { mə'tab·ə,līt }

metabolize [PHYSIO] To transform by metabolism; to subject to metabolism. { mə'tab·ə,līz }

metacarpus [ANAT] The portion of a hand or forefoot between the carpus and the phalanges. { ¦med·ə'kär·pəs }

metacenter [FL MECH] The intersection of a vertical line through the center of buoyancy of a floating body, slightly displaced from its equilibrium position, with a line connecting the center of gravity and the equilibrium center of buoyancy; the floating body is stable if the metacenter lies above the center of gravity. { 'med·ə,sen·tər }

metacentric [CELL MOL] Having the centromere near the middle of the chromosome. { ¦med·ə'sen·trik }

metacentric diagram [NAV ARCH] A curve indicating the height of metacenter (generally above base) for all drafts to which the vessel may be loaded. { med·ə'sen·trik 'dī·ə,gram }

metacentric height [NAV ARCH] The vertical distance between a ship's center of gravity and its metacenter, for transverse or longitudinal inclinations, as specified. { med·ə'sen·trik 'hīt }

metacercaria [INV ZOO] Encysted cercaria of digenetic trematodes; the infective form. { ¦med·ə·sər'kar·ē·ə }

metacestode [INV ZOO] Encysted larva of a tapeworm; occurs in the intermediate host. { ¦med·ə'ses,tōd }

meta character [COMPUT SCI] A character in a computer programming language system that has some controlling role with respect to other characters with which it may be associated. { 'med·ə ,kar·ik·tər }

Metachlamydeae [BOT] An artificial group of flowering plants, division Magnoliophyta, recognized in the Englerian system of classification; consists of families of dicotyledons in which petals are characteristically fused, forming a sympetalous corolla. { ,med·ə·klə'mid·ē,ē }

metachromasia [CHEM] **1.** The property exhibited by certain pure dyestuffs, chiefly basic dyes, of coloring certain tissue elements in a different color, usually of a shorter wavelength absorption maximum, than most other tissue elements. **2.** The assumption of different colors or shades by different substances when stained by the same dye. Also known as metachromatism. { ,med·ə·krō'mē·zhə }

metachromatic granules [CELL MOL] Granules which assume a color different from that of the dye used to stain them. { ¦med·ə·krō'mad·ik 'gran·yülz }

metachromatic leukodystrophy [MED] A hereditary degenerative disease transmitted as an autosomal recessive, due to sulfatase A deficiency, with excess accumulation of sulfated lipids responsible for metachromasia in various tissues. Abbreviated MLD. Also known as sulfatide lipidosis. { ¦med·ə·krō'mad·ik ,lü·kə'dis·trə·fē }

metachromatic stain [MATER] A stain which changes apparent color when absorbed by certain cell constituents. { ¦med·ə·krō'mad·ik 'stān }

metachromatism *See* metachromasia. { ¦med·ə'krō·mə ,tiz·əm }

metachrome yellow [ORG CHEM] $C_{13}H_8N_3NaO_5$ A yellow dye that is slightly soluble in water. { ¦med·ə,krōm 'yel·ō }

metachrosis [VERT ZOO] The ability of some animals to change color by the expansion and contraction of chromatophores. { ,med·ə'krō·səs }

metacinnabar [MINERAL] HgS A black isometric mineral that represents an ore of mercury. Also known as metacinnabarite. { ,med·ə'sin·ə,bär }

metacinnabarite *See* metacinnabar. { ,med·ə'sin·ə·bə,rīt }

metacneme [INV ZOO] A secondary mesentery in many zoantharians. { 'me,tak,nēm }

metacompact space [MATH] A topological space with the property that every open covering F is associated with a point-finite open covering G, such that every element of G is a subset of an element of F. { ,med·ə¦käm,pakt 'spās }

metacompiler [COMPUT SCI] A compiler that is used chiefly to construct compilers for other programming languages. { 'med·ə·kəm,pī·lər }

metacone [VERT ZOO] **1.** The posterior of three cusps of primitive upper molars. **2.** The posteroexternal cusp of an upper molar in higher vertebrates, especially mammals. { 'med·ə‚kōn }

metaconid [VERT ZOO] The posteroexternal cusp of a lower molar in mammals; corresponds with the metacone. { ‚med·ə'kō·nəd }

Metacopina [PALEON] An extinct suborder of ostracods in the order Podocopida. { ‚med·ə'käp·ə·nə }

metacryst [PETR] A large crystal, such as garnet, formed in metamorphic rock by recrystallization. Also known as metacrystal. { 'med·ə‚krist }

metacrystal *See* metacryst. { ‚med·ə'krist·əl }

metadata [COMPUT SCI] A description of the data in a source, distinct from the actual data; for example, the currency by which prices are measured in a data source for purchasing goods. { 'med·ə‚dad·ə }

metadyne [ELECTR] A type of rotating magnetic amplifier having more than one brush per pole, used for voltage regulation or transformation. { 'med·ə‚dīn }

metagalaxy [ASTRON] The total assemblage of recognized galaxies; essentially this represents the entire material universe. { ‚med·ə'gal·ik·sē }

metagenesis [BIOL] The phenomenon in which one generation of certain plants and animals reproduces asexually, followed by a sexually reproducing generation. Also known as alternation of generations. { ‚med·ə'jen·ə·səs }

metagranulocyte *See* metamyelocyte. { ‚med·ə'gran·yə‚līt }

metahalloysite [GEOL] A term used in Europe for the less hydrous form of halloysite. Also known as halloysite in the United States. { ‚med·ə·hə'lȯi‚sīt }

metaharmosis *See* metharmosis. { ‚med·ə·här'mō·səs }

metahewettite [MINERAL] $CaV_6O_{16} \cdot 9H_2O$ A deep red, probably orthorhombic mineral consisting of hydrated calcium vanadate; occurs as pulverulent masses. { ‚med·ə'hyü·ə‚tīt }

metahohmannite [MINERAL] $Fe_2(SO_4)_2(OH)_2 \cdot 3H_2O$ An orange mineral consisting of a hydrated basic iron sulfate; occurs as pulverulent masses. { ‚med·ə'hō·mə‚nīt }

metahydrate sodium carbonate [INORG CHEM] $Na_2CO_3 \cdot H_2O$ Water-soluble, white crystals with an alkaline taste, loses water at 109°C, melts at 851°C; used in medicine, photography, and water pH control, and as a food additive. Also known as crystal carbonate; soda crystals. { ‚med·ə'hī‚drāt 'sōd·ē·əm 'kär·bə‚nāt }

metaigneous [PETR] Pertaining to metamorphic rock formed from igneous rock. { ‚med·ə'ig·nē·əs }

metakaryocyte *See* normoblast. { ‚med·ə'kar·ē·ə‚sīt }

metal [ASTRON] In stellar spectroscopy, any element heavier than helium. [MATER] An opaque crystalline material usually of high strength with good electrical and thermal conductivities, ductility, and reflectivity; properties are related to the structure, in which the positively charged ions are bonded through a field of free electrons which surrounds them forming a close-packed structure. { 'med·əl }

metal ague *See* metal fume fever. { 'med·əl ‚āg }

metal-air battery *See* air-depolarized battery. { 'med·əl ‚er 'bad·ə·rē }

metal alkyl [ORG CHEM] One of the family of organometallic compounds, a combination of an alkyl organic radical with a metal atom or atoms. { 'med·əl 'al·kəl }

metalanguage [COMPUT SCI] A programming language that uses symbols to represent the syntax of other programming languages, and is used chiefly to write compilers for those languages. { 'med·ə‚laŋ·gwij }

metal antenna [ELECTROMAG] An antenna which has a relatively small metal surface, in contrast to a slot antenna. { 'med·əl an'ten·ə }

metal arc cutting [MET] Cutting metal with the heat of an arc between a metal electrode and the base metal. { 'med·əl 'ärk ‚kəd·iŋ }

metal arc welding [MET] Arc welding using covered metal electrodes. { 'med·əl 'ärk ‚weld·iŋ }

metal casting [MET] A metal-forming process whereby molten metal is poured into a cavity or mold and, when cooled, solidifies, taking on the characteristic shape of the mold. { 'med·əl ‚kast·iŋ }

metal ceramic *See* cermet. { 'med·əl sə'ram·ik }

metal-clad substation [ELEC] An electric power substation housed in a metal cabinet, either indoors or outdoors. { 'med·əl ‚klad 'səb‚stā·shən }

metal cluster compound [CHEM] A compound in which two or more metal atoms aggregate so as to be within bonding distance of one another and each metal atom is bonded to at least two other metal atoms; some nonmetal atoms may be associated with the cluster. { 'med·əl ‚kləs·tər 'käm‚paund }

metal coating [MET] A thin film of metal bonded to a base material in order to add specific surface properties, such as corrosion or oxidation resistance, color, wear resistance, or optical characteristics. { 'med·əl 'kōd·iŋ }

metaldehyde [ORG CHEM] $(CH_3CHO)_n$ White acetaldehyde-polymer prisms; soluble in organic solvents, insoluble in water; used as a pesticide or fuel. { me'tal·də‚hīd }

metal detector [ELECTR] An electronic device for detecting concealed metal objects, such as guns, knives, or buried pipelines, generally by radiating a high-frequency electromagnetic field and detecting the change produced in that field by the ferrous or nonferrous metal object being sought. Also known as electronic locator; metal locator; radio metal locator. { 'med·əl di‚tek·tər }

metal distribution ratio [MET] In plating operations, the ratio of the thickness of metal deposited on a near portion of a cathode to that deposited on a far portion of the cathode. { 'med·əl ‚dis·trə'byü·shən ‚rā·shō }

metal dye [MATER] Any of the special dyes, such as alizarin cyanin RR or alizarin green S, used to color oxided surfaces of aluminum or steel. { 'med·əl ‚dī }

metal-enhanced star formation [ASTRON] The hypothesis that stars form preferentially from regions with higher-than-average atomic number in a chemically inhomogeneous interstellar medium. { 'med·əl in‚hanst 'stär fȯr‚mā·shən }

metal-film resistor [ELEC] A resistor in which the resistive element is a thin film of metal or alloy, deposited on an insulating substrate of an integrated circuit. { 'med·əl ‚film ri'zis·tər }

metal-foil paper [MATER] Paper backed with metal foils, manufactured in a number of vivid colors. { 'med·əl ‚fȯil 'pā·pər }

metal forming [MET] Any manufacturing process by which parts of components are fabricated by shaping or molding a piece of metal stock. { 'med·əl ‚fȯrm·iŋ }

metal fouling [ORD] Deposits of metal that collect in the bore of a gun, coming from the jackets or rotating bands of projectiles. { 'med·əl ‚faul·iŋ }

metal fume fever [MED] A febrile influenzalike occupational disorder following the inhalation of finely divided particles and fumes of metallic oxides. Also known as brass chills; brass founder's ague; galvo; Monday fever; metal ague; polymer fume fever; spelter shakes; teflon shakes; zinc chills. { 'med·əl ‚fyüm 'fē·vər }

metal halide lamp [ELECTR] A discharge lamp in which metal halide salts are added to the contents of a discharge tube in which there is a high-pressure arc in mercury vapor; the added metals generate different wavelengths, to give substantially white light at an efficiency approximating that of high-pressure sodium lamps. { 'med·əl 'ha‚līd ‚lamp }

metaliding [MET] A process of depositing a metal as an alloy on a substrate from a fused complex metal salt. { 'med·əl‚īd·iŋ }

metalimnion *See* thermocline. { ‚med·ə'lim·nē‚än }

metal inert-gas welding [MET] A welding procedure in which an electric current heats one metal which then joins to a second metal, with an inert gas preventing oxidation. { 'med·əl i‚nərt ‚gas 'weld·iŋ }

metal-in-gap head [ELECTR] A ring head in which the gap in the ring is lined with a metallic material having a higher saturation magnetization in order to extend the maximum field of the head. Abbreviated MIG head. { ‚med·əl in ‚gap 'hed }

metal-insulator semiconductor [SOLID STATE] Semiconductor construction in which an insulating layer, generally a fraction of a micrometer thick, is deposited on the semiconducting substrate before the pattern of metal contacts is applied. Abbreviated MIS. { 'med·əl ‚in·sə‚lād·ər 'sem·i·kən'dək·tər }

metal-insulator transition [SOLID STATE] The change of certain low-dimensional conductors from metals to insulators as the temperature is lowered through a certain value, due to the lattice distortion and band gap accompanying the onset of a charge-density wave. { 'med·əl ˌin·sə͵läd·ər tran'zish·ən }

metal ion indicator [ANALY CHEM] A substance, usually a dyestuff, that changes color after forming a metal ion complex with a color different from that of the uncomplexed indicator. Also known as complexation indicator. { ˌmed·əl ˈīˌän ˌin-də͵kād·ər }

metal lath [ENG] A mesh of metal used to provide a base for plaster. { 'med·əl 'lath }

metallation [ORG CHEM] The direct replacement of a hydrogen atom by a metal atom in an organic molecule to form a carbon-metal bond. { ˌmed·ə'lā·shən }

metal leaf [MET] Metal sheet, thinner than foil, formed by beating. { 'med·əl 'lef }

metallic [OPTICS] Having a brilliant mineral luster characteristic of metals. [SCI TECH] Pertaining to metals. { mə'tal·ik }

metallic bond [PHYS CHEM] The type of chemical bond that is present in all metals, and may be thought of as resulting from a sea of valence electrons which are free to move throughout the metal lattice. { mə'tal·ik 'bänd }

metallic circuit [ELEC] Wire circuit of which the ground or earth forms no part. { mə'tal·ik 'sər·kət }

metallic corrosion [MET] Destruction of a metal by dissolution, oxidation, or other chemical reaction of the metal with its environment. { mə'tal·ik kə'rō·zhən }

metallic disk rectifier See metallic rectifier. { mə'tal·ik ˌdisk 'rek·tə͵fī·ər }

metallic electrode arc lamp [ELEC] A type of arc lamp in which light is produced by luminescent vapor introduced into the arc by evaporation from the cathode; the anode is solid copper, and the cathode is formed of magnetic iron oxide with titanium as the light-producing element and other chemicals to control steadiness and vaporization. { mə'tal·ik i͵lek͵trōd ärk ˌlamp }

metallic element [CHEM] An element generally distinguished (from a nonmetallic one) by its luster, electrical conductivity, malleability, and ability to form positive ions. { mə'tal·ik 'el·ə·mənt }

metallic glass See glassy alloy. { mə'tal·ik 'glas }

metallic hydrogen [PHYS CHEM] **1.** A phase of hydrogen believed to occur at extremely high pressures, in which the material transforms to a conducting molecular solid. **2.** A phase of hydrogen believed to occur at still higher pressures, in which the molecular bonds that exist at lower pressures are broken and an atomic solid with the structure of an alkali metal is formed. { mə'tal·ik 'hī·drə·jən }

metallic insulator [ELECTROMAG] Section of transmission line used as a mechanical support device; the section is an odd number of quarter-wavelengths long at the frequency of interest, and the input impedance becomes high enough so that the section effectively acts as an insulator. { mə'tal·ik 'in·sə͵läd·ər }

metallic mortar [MATER] Mortar made with ceramic oxide binders and containing a high percentage of lead powder; mixed with water to form plasters or for casting sections and blocks; used for x-ray and nuclear installation shielding. { mə'tal·ik 'mȯrd·ər }

metallic nuclear fuel [NUCLEO] A fissionable isotope of a metallic element, or an alloy containing such an isotope, used as the energy source for a nuclear reactor. { mə'tal·ik 'nü·klē·ər 'fyül }

metallic paint [MATER] **1.** Paint used for covering metal surfaces; the pigment is commonly iron oxide. **2.** Paint with a metal pigment. { mə'tal·ik 'pānt }

metallic paper [MATER] **1.** A paper coated with zinc white, or with clay and other materials; the surface can be marked on with metal points (of silver, aluminum, or gold, for example), but it cannot be erased. **2.** Paper coated with finely flaked metal. { mə'tal·ik 'pā·pər }

metallic pigment [MATER] Thin, opaque aluminum or copper alloy flakes that are incorporated into plastic masses to produce metallike effects. { mə'tal·ik 'pig·mənt }

metallic rectifier [ELECTR] A rectifier consisting of one or more disks of metal under pressure-contact with semiconductor coatings or layers, such as a copper oxide, selenium, or silicon rectifier. Also known as contact rectifier; dry-disk rectifier; dry-plate rectifier; metallic-disk rectifier; semiconductor rectifier. { mə'tal·ik 'rek·tə͵fī·ər }

metallic soap [ORG CHEM] A salt of stearic, oleic, palmitic, lauric, or erucic acid with a heavy metal such as cobalt or copper; used as a drier in paints and inks, in fungicides, decolorizing varnish, and waterproofing. { mə'tal·ik 'sōp }

metalliferous [MINERAL] Pertaining to mineral deposits from which metals can be extracted. { ˌmed·əl'if·ə·rəs }

metallize [ENG] To coat or impregnate a metal or nonmetal surface with a metal, as by metal spraying or by vacuum evaporation. { 'med·əl͵īz }

metallized capacitor [ELEC] A capacitor in which a film of metal is deposited directly on the dielectric to serve in place of a separate foil strip; has self-healing characteristics. { 'med·əl͵īzd kə'pas·əd·ər }

metallized-paper capacitor [ELEC] A modification of a paper capacitor in which metal foils are replaced by extremely thin films of metal deposited on the paper; if a breakdown occurs, these films burn away in the area of the breakdown. { 'med·əl͵īzd ˌpāp·ər kə'pas·əd·ər }

metallized resistor [ELEC] A resistor made by depositing a thin film of high-resistance metal on the surface of a glass or ceramic rod or tube. { 'med·əl͵īzd ri'zis·tər }

metallized slurry blasting [ENG] The breaking of rocks by using slurried explosive medium containing a powdered metal, such as powdered aluminum. { 'med·əl͵īzd ˌslər·ē 'blast·iŋ }

metallized wood [MATER] Wood impregnated with molten metal, filling the cells in the wood to increase hardness, compressive strength, and flexural strength; the wood becomes an electrical conductor lengthwise of the grain. { 'med·əl͵īzd 'wùd }

metallobiochemistry See bioinorganic chemistry. { mə͵tal·ō͵bī·ō'kem·ə·strē }

metallo-carbohedrene [CHEM] A member of a class of molecular clusters in which atoms of an early transition metal (scandium through nickel in the third period of the periodic table) are bonded with carbon atoms in a cagelike network. { mə͵tal·ō ˌkär·bə'hed͵rēn }

metal locator See metal detector. { 'med·əl ˌlō͵kād·ər }

metallocene [ORG CHEM] Organometallic coordination compound which is obtained as a cyclopentadienyl derivative of a transition metal or a metal halide. { mə'tal·ə͵sēn }

metallocene catalyst [ORG CHEM] A molecular structure with a well-defined single catalytic site, consisting of an organometallic coordination compound in which one or two cyclopentadienyl rings (with or without substituents) are bonded to a central transition-metal atom; used to produce uniform polyolefins with unique structures and physical properties. { mə'tal·ə͵sēn 'kad·ə͵list }

metallochaperones [BIOCHEM] **1.** Metalloproteins that aid in the insertion of the appropriate metal ion into a metalloenzyme. **2.** A family of proteins that shuttles metal ions to specific intracellular locations where metalloenzymes bind to the metal ions and use them as cofactors to carry out essential biochemical reactions. { mə͵tal·ō'shap·ə͵rōnz }

metallocycle [ORG CHEM] A compound whose structure consists of a cyclic array of atoms of which one is a metal atom; frequently the ring contains three or four carbon atoms and one transition-metal atom. { mə'tal·ə͵sī·kəl }

metalloenzymes [BIOCHEM] Metalloproteins that catalyze important cellular reactions. { mə͵tal·ō en͵zīmz }

metallogenic province [GEOL] A region characterized by a particular mineral assemblage, or by one or more specific types of mineralization. Also known as metallographic province. { mə͵tal·ə'jen·ik 'präv·əns }

Metallogenium [MICROBIO] A genus of bacteria of uncertain affiliation; coccoid cells that attach to substrate; they germinate directly or form groups of elementary bodies by budding, and filaments form from these bodies. { mə͵tal·ə'jē·nē·əm }

metallograph [OPTICS] An optical microscope equipped with a camera for both visual observation and photography of the structure and constitution of a metal or alloy. { mə'tal·ə͵graf }

metallographic province See metallogenic province. { mə'tal·ə͵graf·ik 'präv·əns }

metallographic test [MET] A test to determine the structural composition of a metal as shown at low and high magnification and by x-ray diffraction methods; tests include macroexamination, microexamination, and x-ray diffraction studies. { mə'tal·ə͵graf·ik 'test }

metallography [MET] The study of the structure of metals

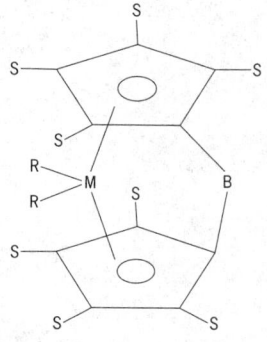

METALLOCENE CATALYST

Generic structure of a metallocene catalyst. S represents ring substituents, M the type of transition metal and its substituents R, and B the bridge, if present.

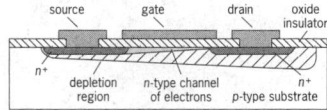
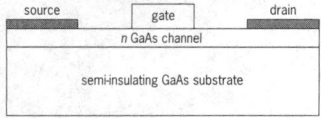
and alloys by various methods, especially by the optical and the electron microscope, and by x-ray diffraction. { ˌmed·əl'äg·rə·fē }

metalloid [CHEM] An element whose properties are intermediate between those of metals and nonmetals. Also known as semimetal. { 'med·ə,lȯid }

metallophobia [PSYCH] An abnormal fear of metal or metallic objects. { me,tal·ə'fō·bē·ə }

metalloporphyrin [BIOCHEM] A compound, such as heme, consisting of a porphyrin combined with a metal such as iron, copper, silver, zinc, or magnesium. { mə,tal·ō'pȯr·fə·rən }

metalloprotein [BIOCHEM] A protein enzyme containing a metallic atom as an inherent portion of its molecule. { mə,tal·ō'prō,tēn }

metallostatic pressure [MET] Pressure developed within a volume of molten metal. { mə¦tal·ə,stad·ik 'presh·ər }

metallothionein [BIOCHEM] A group of vertebrate and invertebrate proteins that bind heavy metals; it may be involved in zinc homeostasis and resistance to heavy-metal toxicity. { mə,tal·ō'thī·ə,nēn }

metallurgical balance sheet [MET] Material balance of a metallurgical process. { ˌmed·əl'ər·jə·kəl 'bal·əns ,shēt }

metallurgical coke [MATER] Coke resulting from high-temperature retorting of suitable coal; a dense, crush-resistant fuel for use in shaft furnaces. { ˌmed·əl'ər·jə·kəl 'kōk }

metallurgical dust [MET] A mixture of particles of elements and nonmetallic and metallic compounds. { ˌmed·əl'ər·jə·kəl 'dəst }

metallurgical engineer [ENG] A person who specializes in metallurgical engineering. { ˌmed·əl'ər·jə·kəl ,en·jə'nir }

metallurgical engineering [ENG] Application of the principles of metallurgy to the engineering sciences. { ˌmed·əl'ər·jə·kəl ,en·jə'nir·iŋ }

metallurgical fume [MET] A mixture of fine particles of elements and metallic and nonmetallic compounds either sublimed or condensed from the vapor state. { ˌmed·əl'ər·jə·kəl 'fyüm }

metallurgical microscope [ENG] A microscope used in the study of metals, usually optical. { ˌmed·əl'ər·jə·kəl 'mī·krə,skōp }

metallurgy [SCI TECH] The science and technology of metals and alloys. { ˌmed·əl'ər·jē }

metal matrix composite [MATER] A material in which a continuous metallic phase (the matrix) is combined with another phase (the reinforcement) to strengthen the metal and increase high-temperature stability. The reinforcement is typically a ceramic in the form of particulates, platelets, whiskers, or fibers. The metals are typically alloys of aluminum, magnesium, or titanium. { ¦med·əl ¦mā·triks kəm'päz·ət }

metal mining [MIN ENG] The industry that supplies the various metals and associated products. { 'med·əl ,mīn·iŋ }

metal-nitride-oxide semiconductor [SOLID STATE] A semiconductor structure that has a double insulating layer; typically, a layer of silicon dioxide (SiO_2) is nearest the silicon substrate, with a layer of silicon nitride (Si_3N_4) over it. Abbreviated MNOS. { 'med·əl ¦nī,trīd ¦äk,sīd 'sem·i·kən,dək·tər }

metal-organic chemical vapor deposition [SOLID STATE] A technique for growing thin layers of compound semiconductors in which metal organic compounds, having the formula MR_x, where M is a group III metal and R is an organic radical, are decomposed near the surface of a heated substrate wafer, in the presence of a hydride of a group V element. Abbreviated MOCVD. { 'med·əl ȯr'gan·ik 'kem·ə·kəl 'vā·pər ,dep·ə'zish·ən }

metal oxide resistor [ELEC] A metal-film resistor in which an oxide of a metal such as tin is deposited as a film onto an insulating substrate. { 'med·əl ¦äk,sīd ri'zis·tər }

metal oxide semiconductor [SOLID STATE] A metal insulator semiconductor structure in which the insulating layer is an oxide of the substrate material; for a silicon substrate, the insulating layer is silicon dioxide (SiO_2). Abbreviated MOS. { 'med·əl ¦äk,sīd 'sem·i·kən,dək·tər }

metal oxide semiconductor field-effect transistor [ELECTR] A field-effect transistor having a gate that is insulated from the semiconductor substrate by a thin layer of silicon dioxide. Abbreviated MOSFET; MOST; MOS transistor. Formerly known as insulated-gate field-effect transistor (IGFET). { 'med·əl ¦äk,sīd 'sem·i·kən,dək,tər 'fēld i,fekt tran'zis·tər }

metal oxide semiconductor integrated circuit [ELECTR] An integrated circuit using metal oxide semiconductor transistors; it can have a higher density of equivalent parts than a bipolar integrated circuit. { 'med·əl ¦äk,sīd 'sem·i·kən,dək·tər 'int·ə,grād·əd 'sər·kət }

metal-petal basket See cementing basket. { 'med·əl ¦ped·əl ,bas·kət }

metal plating See plating. { 'med·əl 'plād·iŋ }

metal pointing See pointing. { 'med·əl 'pȯint·iŋ }

metal powder [MET] A finely divided metal or alloy. { 'med·əl ,paủd·ər }

metal replacement See immersion plating. { 'med·əl ri'plās·mənt }

metal-rich star [ASTRON] A star in which the ratio of metals (elements heavier than helium) to hydrogen is greater than that of the Hyades. { 'med·əl ¦rich 'stär }

metal rolling See rolling. { 'med·əl ,rōl·iŋ }

metal screen [GRAPHICS] An intensifying screen consisting of a metal which emits secondary electrons and x-rays when bombarded by x-rays. { 'med·əl 'skrēn }

metal semiconductor field-effect transistor [ELECTR] A field-effect transistor that uses a thin film of gallium arsenide, with a Schottky barrier gate formed by depositing a layer of metal directly onto the surface of the film. Abbreviated MESFET. { 'med·əl 'sem·i·kən,dək·tər 'fēld i,fekt tran'zis·tər }

metal-slitting saw [MECH ENG] A milling cutter similar to a circular saw blade but sometimes with side teeth as well as teeth around the circumference; used for deep slotting and sinking in cuts. { 'med·əl 'slid·iŋ 'sȯ }

metal spinning See spinning. { 'med·əl ,spin·iŋ }

metal spraying [ENG] Coating a surface with droplets of molten metal or alloy by using a compressed gas stream. { 'med·əl 'sprā·iŋ }

metal-to-metal tap [COMMUN] A tapping procedure in which actual contact is made with the target pair. { 'med·əl tə 'med·əl 'tap }

metal vapor laser [OPTICS] An ion laser based on vaporization of a solid or liquid metal, such as cadmium, calcium, copper, lead, manganese, selenium, strontium, and tin, vaporized with a buffer gas such as helium. { 'med·əl ,vā·pər 'lā·zər }

metamathematics [MATH] The study of the principles of deductive logic as they are used in mathematical logic. { ¦med·ə,math·ə'mad·iks }

metamer [ORG CHEM] One of two or more chemical compounds that exhibits isomerism with the others. { 'med·ə·mər }

metamere [ZOO] One of the linearly arranged similar segments of the body of metameric animals. Also known as somite. { 'med·ə,mir }

metamerism [ZOO] The condition of an animal body characterized by the repetition of similar segments (metameres), exhibited especially by arthropods, annelids, and vertebrates in early embryonic stages and in certain specialized adult structures. Also known as segmentation. { mə'tam·ə,riz·əm }

metamict [MINERAL] Of a radioactive mineral, exhibiting lattice disruption due to radiation damage while the original external morphology is retained. { 'med·ə,mikt }

metamorphic aureole See aureole. { ¦med·ə¦mȯr·fik 'ȯr·ē,ōl }

metamorphic breccia [PETR] Breccia formed by metamorphism. { ¦med·ə¦mȯr·fik 'brech·ə }

metamorphic differentiation [PETR] Processes by which different mineral assemblages develop in some sequence from an initially uniform parent rock. { ¦med·ə,mȯr·fik ,dif·ə,ren·chē'ā·shen }

metamorphic facies [PETR] All rocks of any composition that have reached chemical equilibrium with respect to certain ranges of pressure and temperature during metamorphism, characterized by the stability of specific index minerals. Also known as densofacies. { ¦med·ə¦mȯr·fik 'fā·shēz }

metamorphic facies series [PETR] A group of metamorphic facies characteristic of an individual area, represented in a pressure-temperature diagram by a curve or group of curves illustrating the range of the different types of metamorphism and metamorphic facies. { ¦med·ə¦mȯr·fik 'fā·shēz ,sir·ēz }

metamorphic overprint See overprint. { ¦med·ə¦mȯr·fik 'ō·vər,print }

metamorphic rock [PETR] A rock formed from preexisting solid rocks by mineralogical, structural, and chemical changes, in response to extreme changes in temperature, pressure, and shearing stress. 　{ ¦med·ə¦mȯr·fik 'rak }

metamorphic rock reservoir [GEOL] Uncommon type of formation for oil reservoir; developed when secondary porosity results from fracturing or weathering. 　{ ¦med·ə¦mȯr·fik ¦rak 'rez·əv,wär }

metamorphic zone See aureole. 　{ ¦med·ə¦mȯr·fik 'zōn }

metamorphism [PETR] The mineralogical and structural changes of solid rock in response to environmental conditions at depth in the earth's crust. 　{ ¦med·ə¦mȯr,fiz·əm }

metamorphosis [BIOL] **1.** A structural transformation. **2.** A marked structural change in an animal during postembryonic development. [MED] A degenerative change in tissue or organ structure. 　{ ¦med·ə¦mȯr·fə·səs }

metamyelocyte [HISTOL] A granulocytic cell intermediate in development between the myelocyte and granular leukocyte; characterized by a full complement of cytoplasmic granules and a bean-shaped nucleus. Also known as juvenile cell; metagranulocyte. 　{ ¦med·ə¦mī·ə·lō,sīt }

metanauplius [INV ZOO] A primitive larval stage of certain decapod crustaceans characterized by seven pairs of appendages; follows the nauplius stage. 　{ ¦med·ə'nȯ·plē·əs }

metanephridium [INV ZOO] A type of nephridium consisting of a tubular structure lined with cilia which opens into the coelomic cavity. 　{ ¦med·ə·ne'frid·ē·əm }

metanephrine [BIOCHEM] An inactive metabolite of epinephrine (3-O-methylepinephrine) that is excreted in the urine; it is recovered and measured as a test for pheochromocytoma. 　{ ,med·ə'ne,frən }

metanephros [EMBRYO] One of the posterior of three pairs of vertebrate renal structures; persists as the definitive or permanent kidney in adult reptiles, birds, and mammals. 　{ ,med·ə'ne,fräs }

metanillic acid [ORG CHEM] $C_6H_4(NH_2)SO_3H$ A water-soluble, crystalline compound with sulfanilic acid; used in medicines and dyes. 　{ ¦med·ə¦nil·ik 'as·əd }

metanitricyte See normoblast. 　{ ,med·ə'nī·trə,sīt }

metaphase [CELL MOL] **1.** The phase of mitosis during which centromeres are arranged on the equator of the spindle. **2.** The phase of the first meiotic division when centromeric regions of homologous chromosomes come to lie equidistant on either side of the equator. 　{ 'med·ə,fāz }

metaphase plate [CELL MOL] At metaphase, the imaginary plane located half way between the spindle poles along which the kinetochore microtubules align their chromosomes. 　{ 'med·ə,fāz ,plāt }

metaphloem [BOT] The primary phloem that forms after differentiation of the protophloem. 　{ ¦med·ə'flō·əm }

metaphyseal aclasis See multiple hereditary exostoses. 　{ mə¦taf·ə¦sē·əl a'klas·əs }

metaphysis See epiphyseal plate. 　{ mə'ta·fə·səs }

Metaphyta [BIOL] A kingdom set up to include mosses, ferns, and other plants in some systems of classification. 　{ mə'taf·əd·ə }

metaplasia [PATH] Transformation of one form of tissue to another. 　{ ¦med·ə'plā·zhə }

metaplasm [CELL MOL] The ergastic substance of protoplasm. 　{ 'med·ə,plaz·əm }

metapodium [ANAT] **1.** The metatarsus in bipeds. **2.** The metatarsus and metacarpus in quadrupeds. [INV ZOO] Posterior portion of the foot of a mollusk. 　{ ¦med·ə'pō·dē·əm }

metapodosoma [INV ZOO] Portion of the body bearing the third and fourth pairs of legs in Acarina. 　{ ,med·ə,päd·ə'sō·mə }

metapterygium [VERT ZOO] The posterior one of three basal cartilages in the pectoral fin of sharks and rays. 　{ mə¦tap·tə¦rij·ē·əm }

metaquartzite [MINERAL] A quartzite formed by metamorphic recrystallization. 　{ ¦med·ə'kwȯrt,zīt }

metarheology [MECH] A branch of rheology whose approach is intermediate between those of macrorheology and microrheology; certain processes that are not isothermal are taken into consideration, such as kinetic elasticity, surface tension, and rate processes. 　{ ,med·ə·rē'äl·ə·jē }

metaripple [GEOL] An asymmetrical sand ripple. 　{ 'med·ə,rip·əl }

metarossite [MINERAL] $CaV_2O_6·2H_2O$ A light yellow mineral consisting of hydrated calcium vanadate; occurs as masses and veinlets. 　{ ¦med·ə'rȯ,sīt }

metarubricyte See normoblast. 　{ ,med·ə'rü·brə,sīt }

metascope [ELECTR] An infrared receiver used for converting pulsed invisible infrared rays into visible signals for communication purposes; also used with an infrared source for reading maps in darkness. 　{ 'med·ə,skōp }

metascutellum [INV ZOO] The scutellum of the metathorax in insects. 　{ ¦med·ə·skü'tel·əm }

metasediment [GEOL] A sediment or sedimentary rock which shows evidence of metamorphism. [PETR] Metamorphic rock formed from sedimentary rock. 　{ ¦med·ə'sed·ə·mənt }

metasicula [INV ZOO] The succeeding part of the sicula or colonial tube of graptolites. 　{ ¦med·ə'sik·yə·lə }

metasideronatrite [MINERAL] $Na_4Fe_2(SO_4)_4(OH)_2·3H_2O$ A yellow mineral composed of basic hydrous iron sodium sulfate. 　{ ¦med·ə,sid·ə·rə'nā,trīt }

metasilicate [MINERAL] A salt of the hypothetical metasilicic acid H_2SiO_3. Also known as bisilicate. 　{ ¦med·ə'sil·ə,kāt }

metasoma [INV ZOO] The posterior region of the body of certain invertebrates, a term used especially when the phylogenetic segmentation pattern cannot be identified. 　{ ,med·ə'sō·mə }

metasomatic [PETR] Pertaining to the process or the result of metasomatism. 　{ ¦med·ə·sō'mad·ik }

metasomatism [PETR] A variety of metamorphism in which one mineral or a mineral assemblage is replaced by another of different composition without melting. 　{ ,med·ə'sō·mə, tiz·əm }

metastable See labile. 　{ ¦med·ə'stā·bəl }

metastable equilibrium [PHYS] A condition in which a system returns to equilibrium after small (but not large) displacements; it may be represented by a ball resting in a small depression on top of a hill. [PHYS CHEM] A state of pseudo-equilibrium having higher free energy than the true equilibrium state. 　{ ¦med·ə'stā·bəl ,ē·kwə'lib·rē·əm }

metastable ion [ANALY CHEM] In mass spectroscopy, an ion formed by a secondary dissociation process in the analyzer tube (formed after the parent or initial ion has passed through the accelerating field). 　{ ¦med·ə'stā·bəl 'ī,än }

metastable phase [PHYS CHEM] Existence of a substance as either a liquid, solid, or vapor under conditions in which it is normally unstable in that state. 　{ ¦med·ə'stā·bəl ¦fāz }

metastable state [QUANT MECH] An excited stationary energy state whose lifetime is unusually long. 　{ ¦med·ə'stā·bəl ¦stāt }

metastasis [MED] Transfer of the causal agent (cell or microorganism) of a disease from a primary focus to a distant one through the blood or lymphatic vessels. [PHYS] A transition of an electron or nucleon from one bound state to another in an atom or molecule, or the capture of an electron by a nucleus. 　{ mə'tas·tə·səs }

metastasize [MED] To be transferred by metastasis. 　{ mə'tas·tə,sīz }

metastatic anemia See myelophthisic anemia. 　{ ,med·ə'stad·ik ə'nē·mē·ə }

metastatic capacity [MED] The malignancy of a tumor. 　{ ,med·ə,stad·ik kə'pas·əd·ē }

metasternum [INV ZOO] The ventral portion of the metathorax in insects. 　{ ¦med·ə'stər·nəm }

metastoma [INV ZOO] Median plate posterior to the mouth in certain crustaceans and related arthropods. 　{ mə'tas·tə·mə }

Metastrongylidae [INV ZOO] A family of roundworms belonging to the Strongyloidea; species are parasitic in sheep, cattle, horses, dogs, and other domestic animals. 　{ ¦med·ə,strän'jil·ə,dē }

Metastrongyloidea [INV ZOO] A superfamily of parasitic nematodes, characterized by a reduced or absent stoma capsule and an oral opening surrounded by six well-developed lips. 　{ ¦med·ə,strän·jə'lȯid·ē·ə }

metatarsal [ANAT] Of or pertaining to the metatarsus. 　{ ¦med·ə'tär·səl }

metatarsalgia [MED] Tenderness and burning pain in the metatarsal region. 　{ ,med·ə,tär'sal·jē·ə }

metatarsus [ANAT] The part of a foot or hindfoot between the tarsus and the phalanges. 　{ ¦med·ə'tär·səs }

Metatheria [VERT ZOO] An infraclass of therian mammals including a single order, the Marsupialia; distinguished by a

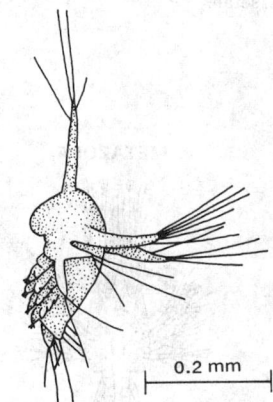

METANAUPLIUS

Lateral view of metanauplius larva of penaeid shrimp. *(Smithsonian Institution)*

0.2 mm

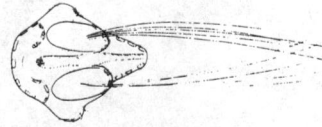

METATROCH

Metatroch of planktonic sabellarian larva, showing early segmentation and long swimming setae.

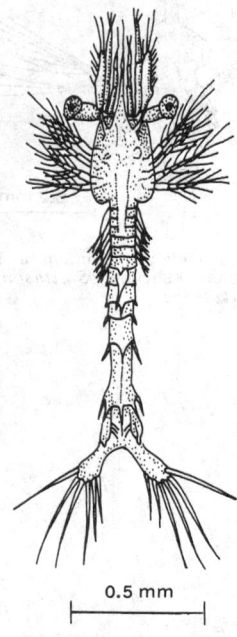

METAZOEA

0.5 mm

Metazoea larva of a penaeid shrimp.

METEORITE CRATER

Canyon Diablo, Arizona, meteorite crater. The rocks immediately surrounding this impact crater are turned upward.

small braincase, a total of 50 teeth, the inflected angular process of the mandible, and a pair of marsupial bones articulating with the pelvis. { ¦med·ə'thir·ē·ə }

metathesis [CHEM] A reaction involving the exchange of elements or groups as in the general equation AX + BY → AY + BX. { mə'tath·ə·səs }

metathetical salts [CHEM] Salts that form a four-component, ternary equilibrium system in which there are four possible binary systems, resulting in two quadruple points. { ¦med·ə¦thed·ə·kəl 'sȯls }

metathorax [INV ZOO] Posterior segment of the thorax in insects. { ¦med·ə'thȯr‚aks }

metatitanic acid See titanic acid. { ¦med·ə·tī'tan·ik 'as·əd }

metatorbernite [MINERAL] Cu(UO₂)₂(PO₄)₂·8H₂O A green secondary mineral composed of hydrous copper uranium phosphate; similar to torbernite, but with less water content. { ¦med·ə'tȯr·bər‚nīt }

metatroch [INV ZOO] A segmented larval form following the trochophore in annelids. { 'med·ə‚träk }

metavariable [COMPUT SCI] One of the elements of a formal language, corresponding to the parts of speech of a natural language. Also known as component name; phrase name. { ¦med·ə'ver·ē·ə·bəl }

metavariscite [MINERAL] AlPO₆·2H₂O A green monoclinic mineral composed of hydrous aluminum phosphate; it is isomorphous with phosphosiderite. { ¦med·ə'var·ə‚sīt }

metavauxite [MINERAL] FeAl₂(PO₄)₂(OH)₂·8H₂O A colorless mineral composed of hydrous basic phosphate of iron and aluminum; similar to vauxite, but with more water. { ¦med·ə'vȯk‚sīt }

metavoltine [MINERAL] A yellowish-brown or orange-brown to greenish-brown, hexagonal mineral consisting of a hydrated basic sulfate of iron and potassium; occurs in tabular form or as aggregates. { ¦med·ə'vȯl‚tēn }

metaxylem [BOT] Primary xylem differentiated after and distinguished from protoxylem by thicker tracheids and vessels with pitted or reticulated walls. { ¦med·ə'zī·ləm }

metazeunerite [MINERAL] Cu(UO₂)₂(AsO₄)₂·8H₂O A grass to emerald green, tetragonal mineral consisting of a hydrated arsenate of copper and uranium; occurs in tabular form. { ¦med·ə'zȯi·nə‚rīt }

Metazoa [ZOO] The multicellular animals that make up the major portion of the animal kingdom; cells are organized in layers or groups as specialized tissues or organ systems. { ‚med·ə'zō·ə }

metazoea [INV ZOO] The last zoea of certain decapod crustaceans; metamorphoses into a megalopa. { ‚med·ə·zō'ē·ə }

metencephalon [EMBRYO] The cephalic portion of the rhombencephalon; gives rise to the cerebellum and pons. { ‚med·in'sef·ə‚län }

meteor [ASTRON] The phenomena which accompany a body from space (a meteoroid) in its passage through the atmosphere, including the flash and streak of light and the ionized trail. { 'mēd·ē·ər }

meteor bumper [AERO ENG] A thin shield around a space vehicle designed to dissipate the energy of impacting meteoric particles. { 'mēd·ē·ər ‚bəm·pər }

Meteoriaceae [BOT] A family of mosses in the order Isobryales in which the calyptra is frequently hairy. { ‚mēd·ē‚ȯr·ē'ās·ē‚ē }

meteoric ionization [ASTROPHYS] Ionization resulting from collisional interactions of a meteoroid and its vaporization products with the air. { ‚mēd·ē'ȯr·ik ‚ī·ə·nə'zā·shən }

meteoric scatter [COMMUN] A form of scatter propagation in which meteor trails serve to scatter radio waves back to earth. { ‚mēd·ē'ȯr·ik 'skad·ər }

meteoric stone See stony meteorite. { ‚mēd·ē'ȯr·ik 'stōn }

meteoric theory See impact theory. { ‚mēd·ē'ȯr·ik 'thē·ə·rē }

meteoric water [HYD] Groundwater which originates in the atmosphere and reaches the zone of saturation by infiltration and percolation. { ‚mēd·ē'ȯr·ik 'wȯd·ər }

meteorism [MED] Presence of abdominal gas causing severe distention. { 'mēd·ē·ə‚riz·əm }

meteorite [GEOL] Any meteoroid that has fallen to the earth's surface. { 'mēd·ē·ə‚rīt }

meteorite crater [GEOL] An impact crater on the surface of the earth or of a celestial body caused by a meteorite; a characteristic feature on the earth is the upturned rim, which

formed as the rocks rebounded following the impact. { 'mēd·ē·ə‚rīt ‚krād·ər }

meteoritic theory See impact theory. { ‚mēd·ē·ə'rid·ik 'thē·ə·rē }

meteorogram [ENG] A record obtained from a meteorograph. [METEOROL] A chart in which meteorological variables are plotted against time. { ‚med·ē'ȯr·ə‚gram }

meteorograph [ENG] An instrument that measures and records meteorological data such as air pressure, temperature, and humidity. { ‚med·ē'ȯr·ə‚graf }

meteoroid [ASTRON] Any solid object moving in interplanetary space that is smaller than a planet or asteroid but larger than a molecule. { 'mēd·ē·ə‚rȯid }

meteorolite See stony meteorite. { ‚mēd·ē'ȯr·ə‚līt }

meteorological [METEOROL] Of or pertaining to meteorology or weather. { ‚mēd·ē·ə·rə'läj·ə·kəl }

meteorological balloon [ENG] A balloon, usually of high-quality neoprene, polyethylene, or Mylar, used to lift radiosondes to high altitudes. { ‚mēd·ē·ə·rə'läj·ə·kəl bə'lün }

meteorological chart [METEOROL] A weather map showing the spatial distribution, at an instant of time, of atmospheric highs and lows, rain clouds, and other phenomena. { ‚mēd·ē·ə·rə'läj·ə·kəl 'chärt }

meteorological check point in gunnery [ORD] Arbitrarily selected point for which meteorological corrections are determined as a time-saving expedient; these corrections are applied to any target located within transfer limits of the meteorological check point. { ‚mēd·ē·ə·rə'läj·ə·kəl 'chek ‚pȯint in 'gən·ə·rē }

meteorological correction [ORD] Adjustment made in the firing data of a gun or other weapon to allow for the effect of wind, air pressure, and so forth, on the flight of a projectile. { ‚mēd·ē·ə·rə'läj·ə·kəl kə'rek·shən }

meteorological data [METEOROL] Facts pertaining to the atmosphere, especially wind, temperature, and air density. { ‚mēd·ē·ə·rə'läj·ə·kəl 'dad·ə }

meteorological datum plane [ORD] Reference plane assumed as a basis or starting point for atmospheric data furnished to artillery; its altitude is that of the meteorological station. { ‚mēd·ē·ə·rə'läj·ə·kəl 'dad·əm ‚plān }

meteorological equator [METEOROL] **1.** The parallel of latitude 5° north, so called because it is the annual mean latitude of the equatorial trough. **2.** See equatorial trough; intertropical convergence zone. { ‚mēd·ē·ə·rə'läj·ə·kəl i'kwād·ər }

meteorological frequency bands [COMMUN] A collection of radio and microwave frequency bands allocated for use by radiosondes and ground-based radars used in weather forecasting in the United States. { ‚mēd·ē·ə·rə'läj·ə·kəl 'frē·kwən·sē ‚banz }

meteorological instrumentation [ENG] Apparatus and equipment used to obtain quantitative information about the weather. { ‚mēd·ē·ə·rə'läj·ə·kəl ‚in·strə·mən'tā·shən }

meteorological minima [METEOROL] Minimum values of meteorological elements prescribed for specific types of flight operation. { ‚mēd·ē·ə·rə'läj·ə·kəl 'min·ə·mə }

meteorological optics [OPTICS] A branch of atmospheric physics or physical meteorology in which optical phenomena occurring in the atmosphere are described and explained. Also known as atmospheric optics. { ‚mēd·ē·ə·rə'läj·ə·kəl 'äp·tiks }

meteorological radar [METEOROL] A remote sensing device that transmits and receives microwave radiation for the purpose of detecting and measuring weather phenomena; includes Doppler radar, which is used to determine air motions (to detect tornadoes), and multiparameter radar, which provides information on the phase (ice or liquid), shapes, and sizes of hydrometeors. { ‚mēd·ē·ə·rə'läj·ə·kəl 'rā‚där }

meteorological range [METEOROL] An empirically consistent measure of the visual range of a target; a concept developed to eliminate from consideration the threshold contrast and adaptation luminance, both of which vary from observer to observer. Also known as standard visibility; standard visual range. { ‚mēd·ē·ə·rə'läj·ə·kəl 'ränj }

meteorological rocket [ENG] Small rocket system used to extend observation of atmospheric character above feasible limits for balloon-borne observing and telemetering instruments. Also known as rocketsonde. { ‚mēd·ē·ə·rə'läj·ə·kəl 'räk·ət }

meteorological satellite [AERO ENG] Earth-orbiting spacecraft carrying a variety of instruments for measuring visible and invisible radiations from the earth and its atmosphere. { ˌmed·ē·ə·rə'läj·ə·kəl 'sad·əl₁īt }

meteorological solenoid [METEOROL] A hypothetical tube formed in space by the intersection of a set of surfaces of constant pressure and a set of surfaces of constant specific volume of air. Also known as solenoid. { ˌmed·ē·ə·rə'läj·ə·kəl 'sō·lə₁nȯid }

meteorological tide [OCEANOGR] A change in water level caused by local meteorological conditions, in contrast to an astronomical tide, caused by the attractions of the sun and moon. { ˌmed·ē·ə·rə'läj·ə·kəl 'tīd }

meteorology [SCI TECH] The science concerned with the atmosphere and its phenomena; the meteorologist observes the atmosphere's temperature, density, winds, clouds, precipitation, and other characteristics and aims to account for its observed structure and evolution (weather, in part) in terms of external influence and the basic laws of physics. { ˌmed·ē·ə'räl·ə·jē }

meteor shower [ASTRON] A number of meteors with approximately parallel trajectories. { 'mēd·ē·ər ˌshaù·ər }

meteor stream [ASTRON] A group of meteoric bodies with nearly identical orbits. { 'mēd·ē·ər ˌstrēm }

meteor trail *See* ion column. { 'mēd·ē·ər ˌtrāl }

meter [MECH] The international standard unit of length, equal to the length of the path traveled by light in vacuum during a time interval of 1/299,792,458 of a second. Abbreviated m. [ENG] A device for measuring the value of a quantity under observation; the term is usually applied to an indicating instrument alone. { 'mēd·ər }

meter-atmosphere [PHYS] The depth of an equivalent atmosphere of a given gas, in meter-atmospheres, is equal to the depth in meters that the atmosphere would have if it were composed entirely of the gas in question and in the same amount as exists in the actual atmosphere, and had a uniform temperature and pressure of 0°C and 1 standard atmosphere. Abbreviated m-atm. Also known as atmo-meter. { 'mēd·ər 'at·mə₁sfir }

meter bar [ENG] A metal bar for mounting a gas meter, having fittings at the ends for the inlet and outlet connections of the meter. { 'mēd·ər ˌbär }

meter bridge [ELEC] A uniform resistance wire 1 meter in length, mounted above a scale marked in millimeters, with terminals added to make the device usable as either part of a Wheatstone bridge or of a potentiometer. { 'mēd·ər ˌbrij }

meter-candle *See* lux. { 'mēd·ər 'kan·dəl }

meter density [ENG] In an energy distribution system, the number of meters per unit area or per unit length. { 'mēd·ər ˌden·səd·ē }

meter factor [ENG] A factor used with a meter to correct for ambient conditions, for example, the factor for a fluid-flow meter to compensate for such conditions as liquid temperature change and pressure shrinkage. { 'mēd·ər ˌfak·tər }

metering installation [PETRO ENG] Oil-production receiving system that includes with the tank battery a metering separator, metering treater, or other type of meter used in conjunction with test separators or emulsion treaters. { 'mēd·ə·riŋ ˌin·stə'lā·shən }

metering pin *See* metering rod. { 'mēd·ə·riŋ ˌpin }

metering pump [CHEM ENG] Plunger-type pump designed to control accurately small-scale fluid-flow rates; used to inject small quantities of materials into continuous-flow liquid streams. Also known as proportioning pump. { 'mēd·ə·riŋ ˌpəmp }

metering rod [ENG] A device consisting of a long metallic pin of graduated diameters fitted to the main nozzle of a carburetor (on an internal combustion engine) or passage leading thereto in such a way that it measures or meters the amount of gasoline permitted to flow by it at various speeds. Also known as metering pin. { 'mēd·ə·riŋ ˌräd }

metering screw [MECH ENG] An extrusion-type screw feeder or conveyor section used to feed pulverized or doughy material at a constant rate. { 'mēd·ə·riŋ ˌskrü }

metering separator [PETRO ENG] Oil-field process vessel that performs the dual functions of gas-oil separation and liquids metering. { 'mēd·ə·riŋ ˌsep·ə₁rād·ər }

metering tank *See* measuring tank. { 'mēd·ə·riŋ ˌtaŋk }

metering valve [MECH ENG] In an automotive hydraulic braking system, a valve that momentarily delays application of the front disk brakes until the rear drum brakes begin to act. { 'mēd·ə·riŋ ˌvalv }

meter-kilogram [MECH] **1.** A unit of energy or work in a meter-kilogram-second gravitational system, equal to the work done by a kilogram-force when the point at which the force is applied is displaced 1 meter in the direction of the force; equal to 9.80665 joules. Abbreviated m-kgf. Also known as meter kilogram-force. **2.** A unit of torque, equal to the torque produced by a kilogram-force acting at a perpendicular distance of 1 meter from the axis of rotation. Also known as kilogram-meter (kgf-m). { 'mēd·ər 'kil·ə₁gram }

meter kilogram-force *See* meter-kilogram. { 'mēd·ər 'kil·ə₁gram 'fȯrs }

meter-kilogram-second-ampere system [PHYS] A system of electrical and mechanical units in which length, mass, time, and electric current are the fundamental quantities, and the units of these quantities are the meter, the kilogram, the second, and the ampere respectively. Abbreviated mksa system. Also known as Giorgi system; practical system. { 'mēd·ər 'kil·ə₁gram 'sek·ənd 'am₁pir ˌsis·təm }

meter-kilogram-second system [MECH] A metric system of units in which length, mass, and time are fundamental quantities, and the units of these quantities are the meter, the kilogram, and the second respectively. Abbreviated mks system. { 'mēd·ər 'kil·ə₁gram 'sek·ənd ˌsis·təm }

meter oil [MATER] High-purity grade of oil used to lubricate the moving elements of meters. { 'mēd·ər ˌȯil }

meter prover [ENG] A device that determines the accuracy of a gas meter; a quantity of air is collected over water or oil in a calibrated cylindrical bell, and then the bell is allowed to sink into the liquid, forcing the air through the meter; the calibrated measurement is then compared with the reading on the meter dial. { 'mēd·ər ˌprü·vər }

meter-proving tank *See* calibrating tank. { 'mēd·ər ˌprü·viŋ ˌtaŋk }

meter run [ENG] The length of straight, unobstructed fluid-flow conduit preceding an orifice or venturi meter. { 'mēd·ər ˌrən }

meter sensitivity [ENG] The accuracy with which a meter can measure a voltage, current, resistance, or other quantity. { 'mēd·ər ˌsen·sə'tiv·əd·ē }

meter sizing factor [FL MECH] A dimensionless number used in calculating the rate of flow of fluid through a pipe from the readings of a flowmeter that measures the drop in pressure when the fluid is forced to flow through a circular orifice; it is equal to $K(d/D)^2$, where K is the flow coefficient, d is the orifice bore diameter, and D is the internal diameter of the pipe. { 'mēd·ər 'sīz·iŋ ˌfak·tər }

meter stop [MECH ENG] A valve installed in a water service pipe for control of the flow of water to a building. { 'mēd·ər ˌstäp }

meter-ton-second system [MECH] A modification of the meter-kilogram-second system in which the metric ton (1000 kilograms) replaces the kilogram as the unit of mass. { 'mēd·ər 'tən 'sek·ənd ˌsis·təm }

meter-type relay [ELEC] A relay that uses a meter movement having a contact-bearing pointer which moves toward or away from a fixed contact mounted on the meter scale. { 'mēd·ər ˌtīp ˌrē·lā }

meter wheel [ENG] A special block used to support the oceanographic wire paid out over the side of a ship; attached directly or connected by means of a speedometer cable to a gearbox which measures the length of wire. { 'mēd·ər ˌwēl }

metestrus [PHYSIO] The beginning of the luteal phase following estrus. { med'es·trəs }

methacrolein [ORG CHEM] $CH_2C(CH_3)CHO$ Liquid with 68°C boiling point; slightly soluble in water; used to make resins and copolymers { mə'thak·rə·lən }

methacrylate ester [ORG CHEM] $CH_2{:}C(CH_3)COOR$ Methacrylic acid ester in which R can be methyl, ethyl, isobutyl, or 50-50 *n*-butyl-isobutyl groups; used to make thermoplastic polymers or copolymers. { meth'ak·rə₁lāt 'es·tər }

methacrylic acid [ORG CHEM] $CH_2C(CH_3)COOH$ Easily polymerized, colorless liquid melting at 15–16°C; soluble in water and most organic solvents; used to make water-soluble polymers and as a chemical intermediate. { ˌmeth·ə'kril·ik 'as·əd }

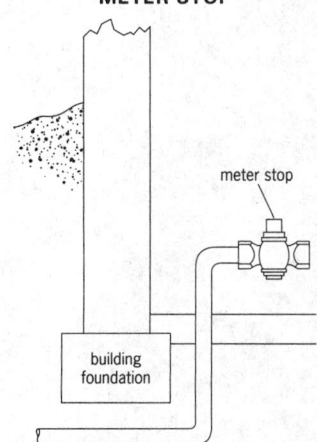

METER STOP

Diagram of a meter stop on a water service pipe.

METHIONINE

CH₃
|
S
|
CH₂
|
CH₂
|
C
/ \
H₂N   COOH
|
H

Structural formula of methionine.

methacrylic polymer [ORG CHEM] A polymer whose monomer is a methacrylic ester with the general formula $H_2C=C(CH_3)COOR$. { ¦meth·ə¦kril·ik 'pal·ə·mər }

methacrylonitrile [ORG CHEM] $CH_2:C(CH_3)CN$ Clear, colorless liquid boiling at 90°C; used to make solvent-resistant thermoplastic polymers and copolymers. { ¦meth·ə·krə'län·ə,tril }

methadone [PHARM] $C_{21}H_{27}NO$ The compound 6-(dimethylamino)-4,4-diphenyl-3-heptanone, a narcotic analgesic, administered in the hydrochloride form for maintenance treatment of heroin addiction. { 'meth·ə,dän }

methallyl alcohol [ORG CHEM] $H_2C:C(CH_3)CH_2OH$ Flammable, toxic, water-soluble, colorless liquid boiling at 115°C; has pungent aroma; soluble in most organic solvents; used as a chemical intermediate. Also spelled methyl allyl alcohol. { meth'al·əl 'al·kə,höl }

methanal See formaldehyde. { 'meth·ə,nal }

methanamide See formamide. { meth'an·ə·mid }

methanation [CHEM ENG] In coal gasification, the catalytic conversion of hydrogen and carbon monoxide to methane. { ¦meth·ə'nā·shən }

methane [ORG CHEM] CH_4 A colorless, odorless, and tasteless gas, lighter than air and reacting violently with chlorine and bromine in sunlight, a chief component of natural gas; used as a source of methanol, acetylene, and carbon monoxide. Also known as methyl hydride. { 'meth,ān }

methanearsonic acid [ORG CHEM] $CH_3AsO(OH)_2$ A white solid with a melting point of 161°C; very soluble in water; used as an herbicide for cotton crops and for noncrop areas. Abbreviated MAA. { ¦meth,ān·är'sän·ik 'as·əd }

methane drainage See firedamp drainage. { 'meth,ān ,drān·ij }

methane hydrate [CHEM] Methane gas trapped or dissolved in ice formed in deep-sea sediments. { ¦mēth,ān 'hī,drāt }

methane indicator [MIN ENG] A portable analytical instrument that can determine the methane content in the mine air at the place where the sample is taken; air is brought into the instrument through an aspirator bulb and passed through a cartridge filter to remove moisture. { 'meth,ān ,in·də,kād·ər }

methane monitoring system [MIN ENG] A system that samples methane content in mine air continuously and feeds this information into an electrical device that cuts off power in each mining machine when the methane content rises above a predetermined level. { 'meth,ān 'män·ə·triŋ ,sis·təm }

methane monooxygenase [BIOCHEM] A nonheme iron enzyme that catalyzes the oxidation of methane to methanol. { ,meth,ān ,män·ō'ak·sə·jə,nās }

methane-oxidizing bacteria [MICROBIO] Bacteria that derive energy from oxidation of methane. { 'meth,ān ¦äk·sə,dīz·iŋ bak'tir·ē·ə }

methanesulfonic acid [ORG CHEM] CH_3SO_2OH A solid with a melting point of 20°C; used as a catalyst in polymerization, esterification, and alkylation reactions, and as a solvent. Also known as methysulfonic acid. { ¦meth,ān·səl'fän·ik 'as·əd }

methanethiol See methyl mercaptan. { mə,than·ə'thī,ól }

Methanobacteriaceae [MICROBIO] The single family of methane-producing bacteria; anaerobes which obtain energy via formation of methane. { ¦meth·ə·nō,bak·tir·ē'ās·ē,ē }

methanogen [BIOL] An organism carrying out methanogenesis, requiring completely anaerobic conditions for growth; considered by some authorities to be distinct from bacteria. { mə'than·ə·jən }

methanogenesis [BIOCHEM] The biosynthesis of the hydrocarbon methane; common in certain bacteria. Also known as bacterial methanogenesis. { ¦meth·ə·nō'jen·ə·səs }

methanoic acid See formic acid. { ,meth·ə'nō·ik 'as·əd }

methanol See methyl alcohol. { 'meth·ə,nól }

Methanomonadaceae [MICROBIO] Formerly a family of bacteria in the suborder Pseudomonadineae; members identified as gram-negative rods are able to use carbon monoxide (*Carboxydomonas*), methane (*Methanomonas*), and hydrogen (*Hydrogenomonas*) as their sole source of energy for growth. { ¦meth·ə·nō,män·ə'dās·ē,ē }

methanotroph [MICROBIO] A bacterial organism that can use methane as its only source of carbon and energy. { mə'than·ə,träf }

metharmosis [GEOL] Changes that occur in a buried sediment after uplift or consolidation but before the onset of weathering. Also spelled metaharmosis. { mə'thär·mə'səs }

methemoglobin See ferrihemoglobin. { met,hē·mə'glō·bən }

methemoglobinemia [MED] The presence of methemoglobin in the blood. { ¦met,hē·mə·glō·bə'nē·mē·ə }

methemoglobinuria [MED] The presence of methemoglobin in the urine. { ¦met,he·mə·glō·bə'nyùr·ē·ə }

methenyl See methine group. { 'meth·ə,nil }

methidathion [ORG CHEM] $C_4H_{11}O_4N_2PS_3$ A colorless, crystalline compound with a melting point of 39–40°C; used as an insecticide and miticide for pests on alfalfa, citrus, and cotton. { mə,thid·ə'thī,än }

methide [ORG CHEM] A binary compound consisting of methyl and, most commonly, a metal, such as sodium (sodium methide, $NaCH_3$). { 'me,thīd }

methine group [ORG CHEM] $HC\equiv$ A radical consisting of a single carbon and a single hydrogen. Also known as methenyl; methylidyne. { 'me,thēn 'grüp }

methionic acid [ORG CHEM] $CH_2(SO_3H)_2$ An acid that exists as hygroscopic crystals; used in organic synthesis. { ¦meth·ē¦än·ik 'as·əd }

methionine [BIOCHEM] $C_5H_{11}O_2NS$ An essential amino acid; furnishes both labile methyl groups and sulfur necessary for normal metabolism. { mə'thī·ə,nēn }

method of bisectors [NAV] As applied to celestial lines of position, the movement of each of three or four intersecting lines of position in equal amounts, in the same direction, toward or away from the celestial bodies, so as to bring them as nearly as possible to a common intersection; when there are more than four lines of position, the lines of position in the same general direction are combined to reduce the data to not more than four lines of position. { 'meth·əd əv 'bī,sek·tərz }

method of exhaustion [MATH] A method of finding areas and volumes by finding an increasing or decreasing sequence of sets whose areas or volumes are known and less than or greater than the desired area or volume, and then showing that the area or volume between the boundaries of the approximating sets and the boundary of the set to be measured approaches zero (is exhausted). { ,meth·əd əv ig'zòs·chən }

method of images [ELEC] In electrostatics, a method of determining the electric fields and potentials set up by charges in the vicinity of a conductor, in which the conductor and its induced surface charges are replaced by one or more fictitious charges. [PETRO ENG] Method of calculating the interference between reservoirs by assuming a mirror image of one reservoir on the far side of a geologic fault. [PHYS] Any method of solving magnetostatic, hydrodynamic, and other problems involving boundary conditions at the interface between two media, in which fictitious objects, such as magnetic dipoles and sources and sinks of fluid, are introduced to satisfy the boundary conditions; these methods are generalizations of the method in electrostatics. { 'meth·əd əv 'im·ij·əz }

method of infinite descent See mathematical induction. { ¦meth·əd əv ¦in·fə·nət di'sent }

method of joints [ENG] Determination of stresses for joints at which there are not more than two unknown forces by the methods of the stress polygon, resolution, or moments. { 'meth·əd əv 'jòins }

method of mixtures [THERMO] A method of determining the heat of fusion of a substance whose specific heat is known, in which a known amount of the solid is combined with a known amount of the liquid in a calorimeter, and the decrease in the liquid temperature during melting of the solid is measured. { 'meth·əd əv 'miks·chərz }

method of moments [STAT] A method of estimating the parameters of a frequency distribution by first computing as many moments of the distribution as there are parameters to be estimated and then using a function that relates the parameters to moments. { ¦meth·əd əv 'mō·məns }

method of moving averages [STAT] A series of averages where each average is the mean value of the time series over a fixed interval of time, and where all possible averages of the length are included in the analysis; used to smooth data in a time series. { ¦meth·əd əv ¦müv·iŋ 'av·rij·əz }

method of semiaverages [STAT] A method for providing a quick estimate of a linear regression line, in which data are divided into two equal sets and the means of the two sets or two other points representative of each set are determined and a straight line drawn through them. { ¦meth·əd əv 'sem·ē‚av·rij·əz }

methods design [IND ENG] Design for a new, more efficient method of job performance. { 'meth·ədz di‚zīn }

methods engineering [IND ENG] A technique used by management to improve working methods and reduce labor costs in all areas where human effort is required. { 'meth·ədz ‚en·jə'nir·iŋ }

methods study [IND ENG] An analysis of the methods in use, of the means and potentials for their improvement, and of reducing costs. { 'meth·ədz ‚stəd·ē }

methotrexate *See* amethopterin. { ‚meth·ə'trek‚sāt }

methoxide [ORG CHEM] A compound formed from a metal and the methoxy radical; an example is sodium methoxide. Also known as methylate. { mə'thäk‚sīd }

methoxy- [ORG CHEM] OCH_3- A combining form indicating the oxygen-containing methane radical, found in many organic solvents, insecticides, and plasticizer intermediates. { mə'thäk·sē }

methoxychlor [ORG CHEM] $Cl_3CCH(C_6H_4OCH_3)_2$ White, water-insoluble crystals melting at 89°C; used as an insecticide. Also known as DMDT; methoxy DDT. { me'thäk·si‚klȯr }

methoxy DDT *See* methoxychlor. { me'thäk·sē ¦de¦dē'tē }

2-methoxyethanol [ORG CHEM] $CH_3OCH_2CH_2OH$ A poisonous liquid, used as a solvent for low-viscosity cellulose acetate, natural and some synthetic resins, and alcohol-soluble dyes, and also used in dyeing leather. { ¦tü mə‚thäk·sē'eth·ə‚nȯl }

methoxyethylmercury chloride [ORG CHEM] $CH_3OCH_2CH_2HgCl$ A white, crystalline compound with a melting point of 65°C; used as a fungicide in diseases of sugarcane, pineapples, seed potatoes, and flower bulbs, and as seed dressings for cereals, legumes, and root crops. Abbreviated MEMC. { mə¦thäk·sē¦eth·əl¦mər·kyə·rē 'klȯr‚īd }

4-methoxy-2-hydroxybenzophenone *See* oxybenzone. { ¦fȯr mə¦thäk·sē ¦tü ‚hī¦dräk·sē·ben'zä·fə‚nōn }

methoxyl [ORG CHEM] CH_3O- A functional group which is univalent. { mə'thäk·səl }

methyl [ORG CHEM] The alkyl group derived from methane and usually written CH_3-. Also known as carbinyl. { 'meth·əl }

methyl abietate [ORG CHEM] $C_{19}H_{29}COOCH_3$ Colorless to yellow liquid boiling at 365°C; miscible with most organic solvents; used as a solvent and plasticizer for lacquers, varnishes, and coatings. { 'meth·əl 'ab·ē·ə‚tāt }

methyl acetate [ORG CHEM] $CH_3CO_2CH_3I$ Flammable, colorless liquid with fragrant odor; boils at 54°C; partially soluble in water, miscible with hydrocarbon solvents; used as a solvent and extractant. { 'meth·əl 'as·ə‚tāt }

methylacetic acid *See* propionic acid. { ¦meth·əl·ə¦sēd·ik ‚as·əd }

methyl acetoacetate [ORG CHEM] $CH_3COCH_2CO_2CH_3$ Alcohol-soluble, colorless liquid boiling at 172°C; used as a chemical intermediate and as a solvent for cellulosics. { 'meth·əl ¦as·əd·ō¦as·ə‚tāt }

methyl acetone [MATER] Flammable, water-white liquid, a mixture of acetone, methanol, and methyl acetate in various proportions; miscible with water, oils, and hydrocarbons; used as a solvent. { 'meth·əl 'as·ə‚tōn }

methyl acetophenone [ORG CHEM] $CH_3C_6H_4COCH_3$ Fragrant (coumarin aroma), colorless or pale-yellow liquid, soluble in alcohol; used in perfumery. { 'meth·əl ‚as·ə'täf·ə‚nōn }

methyl acrylate [ORG CHEM] $CH_2:CHCOOCH_3$ A readily polymerized, volatile, colorless liquid boiling at 80°C; slightly soluble in water; used as a chemical intermediate and in making polymers. { 'meth·əl 'ak·rə‚lāt }

methylal [ORG CHEM] $CH_3OCH_2OCH_3$ Flammable, volatile, colorless liquid boiling at 42°C; soluble in ether, hydrocarbons, and alcohol, partially soluble in water; used as a solvent and chemical intermediate, and in perfumes, adhesives, coatings. Also known as formal. { 'meth·ə‚lal }

methyl alcohol [ORG CHEM] CH_3OH A colorless, toxic, flammable liquid, boiling at 64.5°C; miscible with water, ether, alcohol; used in manufacture of formaldehyde, chemical synthesis, antifreeze for autos, and as a solvent. Also known as methanol; wood alcohol. { 'meth·əl 'al·kə‚hȯl }

methyl allyl alcohol *See* methallyl alcohol. { 'meth·əl 'al·əl 'al·kə‚hȯl }

methyl allyl chloride [ORG CHEM] $CH_2:C(CH_3)CH_2Cl$ Volatile, flammable, colorless liquid boiling at 72°C; has disagreeable odor; used as an insecticide and fumigant, and for chemical synthesis. { 'meth·əl 'al·əl 'klȯr‚īd }

methylamine [ORG CHEM] CH_3NH_2 A colorless gas that is highly toxic and flammable; used to prepare dyes, and as a chemical intermediate. { ¦meth·ə·lə¦mēn }

N-methyl-*para*-aminophenol [ORG CHEM] $CH_3NHC_6H_4OH$ Colorless, combustible needles with a melting point of 87°C; soluble in water, alcohol, and ether; used as a photographic developer. { ¦en ¦meth·əl ¦par·ə ¦am·ə·nō'fē‚nȯl }

methyl amyl acetate [ORG CHEM] $CH_3COOCH(CH_3)CH_2CH(CH_3)_2$ Toxic, flammable, colorless liquid with mild, agreeable odor; boils at 146°C; used as nitrocellulose lacquer solvent. Also known as methyl isobutyl carbinol acetate. { 'meth·əl 'am·əl 'as·ə‚tāt }

methyl amyl alcohol [ORG CHEM] $(CH_3)_2CHCH_2CHOHCH_3$ Toxic, flammable, colorless liquid; boils at 132°C; miscible with water and most organic solvents; used as a solvent and as a chemical intermediate. Also known as methyl isobutyl carbinol (MIBC). { 'meth·əl 'am·əl 'al·kə ‚hȯl }

methyl amyl carbinol [ORG CHEM] $CH_3(CH_2)_4CHOHCH_3$ Colorless liquid with mild aroma; boils at 160°C; miscible with most organic liquids; used as an ore-flotation frothing agent and as a synthetic-resin solvent. { 'meth·əl 'am·əl 'kär·bə‚nȯl }

methyl-*n*-amyl ketone [ORG CHEM] $CH_3(CH_2)_4COCH_3$ Stable, water-white liquid; miscible with organic lacquer solvents, slightly soluble in water; used as an inert reaction medium and as a solvent for nitrocellulose lacquers. Also known as 2-heptanone. { 'meth·əl ¦en ¦am·əl 'kē‚tōn }

N-methylaniline [ORG CHEM] $C_6H_5NH(CH_3)$ Oily liquid, colorless to reddish-brown; soluble in water and organic solvents; boils at 190°C; used as an acid acceptor, solvent, and chemical intermediate. { ¦en ¦meth·əl'an·ə·lən }

α-methylanisalacetone [ORG CHEM] $CH_3OC_6H_4CH:CHCOCH_2CH_3$ A white to pale yellow, combustible solid with a melting point of 60°C; used as a flavoring. { ¦al·fə ¦meth·əl‚an·ə·sə'las·ə‚tōn }

methyl anisole *See* methyl *para*-cresol. { 'meth·əl 'an·ə‚sōl }

methyl anthranilate [ORG CHEM] $H_2NC_6H_4CO_2CH_3$ A yellowish to colorless liquid, slightly soluble in water; used in flavoring and in perfumery. Also known as artificial neroli oil. { 'meth·əl ən'thran·ə‚lāt }

2-methyl anthraquinone *See* tectoquinone. { ¦tü ¦meth·əl ‚an·thrə‚kwē'nōn }

methyl arachidate [ORG CHEM] $CH_3(CH_2)_{18}COOCH_3$ A waxlike solid with a melting point of 45.8°C; soluble in alcohol and ether; used in medical research and as a reference standard for gas chromatography. Also known as methyl eicosanoate. { 'meth·əl ə'rak·ə‚dāt }

methylarsinic sulfide [ORG CHEM] CH_3AsS A colorless compound whose flakes melt at 110°C; insoluble in water; used as a fungicide in treating cotton seeds. Also known as rhizoctol. { ¦meth·əl·är¦sin·ik 'səl‚fīd }

methylate *See* methoxide. { 'meth·ə‚lāt }

methylated cap [MOL BIO] A modified guanine nucleotide that terminates a messenger ribonucleic acid molecule. { ¦meth·ə‚lād·əd 'kap }

methylated spirits [MATER] Denatured ethanol produced by addition of about 9.5% methanol, 0.5% pyridine, and a blue dye. { 'meth·ə‚lād·əd 'spir·əts }

methylation [ORG CHEM] A chemical process for introducing a methyl group (CH_3-) into an organic compound. { ‚meth·ə'lā·shən }

methyl behenate [ORG CHEM] $CH_3(CH_2)_{20}COOCH_3$ A combustible, waxlike solid with a melting point of 53.2°C; soluble in alcohol and ether; used in medical and biochemical research and as a reference standard for gas chromatography. Also known as methyl docosanoate. { 'meth·əl bə'he‚nāt }

methylbenzene *See* toluene. { ¦meth·əl¦ben‚zēn }

methylbenzethonium chloride [ORG CHEM] $C_{27}H_{44}O_2Cl·H_2O$ Colorless crystals with a melting point of 161–163°C; soluble in alcohol, hot benzene, chloroform, and water; used as a bactericide. { ‚meth·əl‚ben·zə'thō·nē·əm 'klȯr‚īd }

methyl benzoate [ORG CHEM] $C_6H_5CO_2CH_3$ Colorless, fragrant liquid boiling at 199°C; slightly soluble in alcohol and water, soluble in ether; used in perfumery and as a solvent. Also known as niobe oil. { ¦meth·əl 'ben·zə‚wāt }

methyl *ortho*-benzoylbenzoate [ORG CHEM] C_6H_5CO-$C_6H_4COOCH_3$ A colorless, combustible liquid with a boiling point of 351°C; slightly soluble in water; used as a plasticizer. { 'meth·əl ¦or·thō₁ben·zə₁wil'ben·zə₁wāt }

α-methylbenzyl acetate [ORG CHEM] $C_6H_5CH(CH_3)$-$OOCCH_3$ A colorless, combustible liquid with a strong floral odor; soluble in glycerin, mineral oil, and 70% alcohol; used in perfumes and as a flavoring. { ¦al·fə ¦meth·əl¦ben·zəl 'as·ə₁tāt }

α-methylbenzyl alcohol [ORG CHEM] $C_6H_5CH(CH_3)OH$ A colorless, combustible liquid with a mild floral odor and a boiling point of 204°C; soluble in water; used in perfumes and dyes and as a flavoring agent. { ¦al·fə ¦meth·əl¦ben·zəl 'al·kə₁hȯl }

α-methylbenzylamine [ORG CHEM] $C_6H_5CH(CH_3)NH_2$ A colorless, combustible liquid with a boiling point of 188.5°C; soluble in most organic solvents; used as an emulsifying agent. { ¦al·fə ₁meth·əl·ben'zal·ə₁mēn }

α-methylbenzyl ether [ORG CHEM] $C_6H_5CH(CH_3)OCH$-$(CH_3)C_6H_5$ A straw-colored, combustible liquid with a boiling point of 286.3°C; at 760 mmHg (101,325 pascals); slightly soluble in water; used as a solvent and as a synthetic rubber softener. { ¦al·fə ¦meth·əl¦ben·zəl 'ē·thər }

methyl blue [ORG CHEM] Dark-blue powder or dye; sodium triphenyl *para*-rosaniline sulfonate; used as a biological and bacteriological stain and as an antiseptic. { 'meth·əl 'blü }

methyl borate See trimethyl borate. { 'meth·əl 'bȯr₁āt }

methyl bromide [ORG CHEM] CH_3Br A toxic, colorless gas that forms a crystalline hydrate with cold water; used in synthesis of organic compounds, and as a fumigant. { 'meth·əl 'brō₁mīd }

2-methyl-1,3-butadiene See isoprene. { ¦tü 'meth·əl ¦wən ¦thrē ₁byüd·ə'dī₁ēn }

2-methylbutanal See 2-methylbutyraldehyde. { ¦tü ₁meth·əl'byüt·ən₁al }

2-methylbutane See isopentane. { ¦tü ₁meth·əl'byü₁tān }

2-methyl-1-butanol [ORG CHEM] $C_5H_{12}O$ A liquid with a boiling point of 128°C, miscible with alcohol and with ether, slightly soluble in water; used as a solvent, in organic synthesis, and as an additive in oils and paints. { ¦tü ¦meth·əl ¦wən 'byüt·ən₁ȯl }

methyl butene [ORG CHEM] C_5H_{10} Either of two colorless, flammable, volatile liquid isomers; soluble in alcohol, insoluble in water: 3-methyl-1-butene boils at 20°C, is used as a chemical intermediate and in the manufacture of high-octane fuel, and is also known as isopropylethylene; 3-methyl-2-butene boils at 38°C, is used as an anesthetic and high-octane fuel and as a chemical intermediate, and is also known as trimethylethylene. { 'meth·əl 'byü₁tēn }

2-methyl-2-butene See amylene. { ¦tü 'meth·əl ¦tü 'byü₁tēn }

methyl butyl ketone [ORG CHEM] $CH_3COC_4H_9$ A liquid boiling at 127°C; soluble in water, alcohol, and ether; used as a solvent. Also known as propylacetone. { 'meth·əl 'byüd·əl 'kē₁tōn }

methylbutynol [ORG CHEM] $HC:CCOH(CH_3)_2$ Water-miscible, colorless liquid boiling at 104°C; soluble in most organic solvents; used as a stabilizer for chlorinated organic compounds, as a solvent, and as a chemical intermediate. { ₁meth·əl'byüt·ən₁ȯl }

2-methylbutyraldehyde [ORG CHEM] $CH_3CH_2CH(CH_3)$-CHO A combustible liquid with a boiling point of 92.93°C; soluble in alcohol and ether; used as a brightener in electroplating. Also known as 2-methylbutanal. { ¦tü ₁meth·əl₁byüd·ə'ral·də₁hīd }

methyl butyrate [ORG CHEM] $CH_3CH_2CH_2COOCH_3$ Liquid boiling at 102°C; used as a solvent for cellulosic materials. { 'meth·əl 'byüd·ə₁rāt }

methyl caprate [ORG CHEM] $CH_3(CH_2)_8COOCH_3$ A colorless, combustible liquid with a boiling point of 244°C; soluble in alcohol and ether; used in the manufacture of detergents, stabilizers, plasticizers, textiles, and lubricants. Also known as methyl decanoate. { 'meth·əl 'ka₁prāt }

methyl caproate [ORG CHEM] $CH_3(CH_2)_4COOCH_3$ Colorless liquid boiling at 150°C; soluble in alcohol and ether, insoluble in water; used as an intermediate to make caproic acid. Also known as methyl hexanoate. { 'meth·əl 'kap·rə₁wāt }

methyl caprylate [ORG CHEM] $CH_3(CH_2)_6COOCH_3$ Colorless liquid boiling at 193°C; soluble in ether and alcohol, insoluble in water; used as an intermediate to make caprylic acid. { 'meth·əl 'kap·rə₁lāt }

methyl carbonate [ORG CHEM] $CO(OCH_3)_2$ Water-insoluble, colorless liquid boiling at 91°C; has pleasant odor; miscible with acids and alkalies; used as a chemical intermediate. { 'meth·əl 'kär·bə₁nāt }

methylcellulose [ORG CHEM] A grayish-white powder derived from cellulose; swells in water to a colloidal solution; soluble in glacial acetic acid; used in water-based paints and ceramic glazes, for leather tanning, and as a thickening and sizing agent, adhesive, and food additive. Also known as cellulose methyl ether. { 'meth·əl'sel·yə₁lōs }

methyl chloride See chloromethane. { 'meth·əl 'klȯr₁īd }

methyl chloroacetate [ORG CHEM] $ClHC_2COOCH_3$ Colorless liquid boiling at 131°C; miscible with ether and alcohol, slightly soluble in water; used as a solvent. { 'meth·əl ₁klȯr·ō'as·ə₁tāt }

methyl chlorocarbonate See methyl chloroformate. { 'meth·əl ₁klȯr·ō'kär·bə₁nāt }

methyl chloroform See trichloroethane. { 'meth·əl 'klȯr·ə₁fȯrm }

methyl chloroformate [ORG CHEM] $ClCOOCH_3$ A toxic, corrosive, colorless liquid with a boiling point of 71.4°C; soluble in benzene, ether, and methanol; used as a lacrimator in military poison gas and for insecticides. Also known as methyl chlorocarbonate. { 'meth·əl 'klȯrə¦fȯr₁māt }

methyl cinnamate [ORG CHEM] $C_6H_5CH:CHCO_2CH_3$ A white crystalline compound with strawberry aroma; soluble in ether and alcohol, insoluble in water; boils at 260°C; used to flavor confectioneries and in perfumes. { 'meth·əl 'sin·ə₁māt }

methyl *para*-cresol [ORG CHEM] $CH_3C_6H_4OCH_3$ Colorless liquid with floral aroma; used in perfumery. Also known as methyl anisole. { 'meth·əl 'par·ə 'krē₁sȯl }

methyl cyanoacetate [ORG CHEM] $CNCH_2COOCH_3$ A toxic, combustible, colorless liquid with a boiling point of 203°C; soluble in water, ether, and alcohol; used in pharmaceuticals and dyes. { 'meth·əl ¦sī·ə₁nō'as·ə₁tāt }

methyl cyclohexane [ORG CHEM] C_7H_{14} Colorless liquid boiling at 101°C; used as a cellulosic solvent and as a chemical intermediate. Also known as hexahydrotoluene. { 'meth·əl ¦sī·klō'hek₁sān }

methyl cyclohexanol [ORG CHEM] $CH_3C_6H_{10}OH$ A toxic, colorless liquid with menthol aroma; a mixture of three isomers; used as a solvent for lacquer and cellulosics, as a lubricant antioxidant, and in detergents and textile soaps. { 'meth·əl ¦sī·klō'hek·sə₁nȯl }

methyl cyclohexanone [ORG CHEM] $CH_3C_5H_9CO$ A toxic, clear to pale-yellow liquid with acetonelike aroma; a mixture of cyclic ketones; used as a solvent and in lacquers. { 'meth·əl ¦sī·klō'hek·sə₁nōn }

methylcyclopentadiene dimer [ORG CHEM] $C_{12}H_{16}$ A flammable, colorless liquid with a boiling range of 78-183°C; soluble in alcohol, benzene, and ether; used in high-energy fuels, plasticizers, dyes, and pharmaceuticals. { ¦meth·əl₁sī₁klō₁pen·tə'dī₁ēn 'dī·mər }

methyl cyclopentane [ORG CHEM] $C_5H_9CH_3$ Flammable, colorless liquid boiling at 72°C; used as a chemical intermediate. { 'meth·əl ¦sī·klō'pen₁tān }

methyl decanoate See methyl caprate. { 'meth·əl də'kan·ə₁wāt }

methyl-*N*-(3,4-dichlorophenyl)carbamate See swep. { 'meth·əl ¦en ¦thrē ¦fȯr dī₁klȯr·ō'fen·əl 'kär·bə₁māt }

methyldichlorosilane [ORG CHEM] CH_3SiHCl_2 A colorless liquid with a melting point of −91°C (−130°F) and boiling point of 41°C (106°F). Also known as dichloromethylsilane. { 'meth·əl·dī₁klȯr·ō 'sī₁lān }

methyl diethanolamine [ORG CHEM] $CH_3N(C_2H_4OH)_2$ A colorless liquid miscible with water and benzene; has amine aroma; boils at 247°C; used as a chemical intermediate and as an acid-gas absorbent. { 'meth·əl dī₁eth·ə'näl·ə₁mēn }

methyl dioxolane [ORG CHEM] $C_4H_7O_2$ Water-soluble, clear liquid boiling at 81°C; used as a solvent and extractant. { 'meth·əl dī'äk·sə₁lān }

methyl docosanoate See methyl behenate. { 'meth·əl ₁dō·kə'san·ə₁wāt }

methyl eicosanoate See methyl arachidate. { 'meth·əl ₁ī·kə'san·ə₁wāt }

methylene [ORG CHEM]　$-CH_2-$　A radical that contains a bivalent carbon.　{ 'meth·ə,lēn }

methylene blue [ORG CHEM]　Dark green crystals or powder; soluble in water (deep blue solution), alcohol, and chloroform. $C_{16}H_{18}N_3SCl·3H_2O$ used in medicine; $(C_{16}H_{18}N_3SCl)_2·ZnCl_2·H_2O$ used as a textile dye, biological stain, and indicator.　Also known as methylthionine chloride.　{ 'meth·ə,lēn 'blü }

methylene bromide [ORG CHEM]　CH_2Br_2　Colorless, clear liquid boiling at 97°C; miscible with organic solvents, slightly soluble in water; used as a solvent and chemical intermediate. Also known as dibromomethane.　{ 'meth·ə,lēn 'brō,mīd }

methylene chloride [ORG CHEM]　CH_2Cl_2　A colorless liquid, practically nonflammable and nonexplosive; used as a refrigerant in centrifugal compressors, a solvent for organic materials, and a component in nonflammable paint-remover mixtures.　{ 'meth·ə,lēn 'klór,īd }

methylene iodide [ORG CHEM]　CH_2I_2　Yellow liquid boiling at 180°C; soluble in ether and alcohol, insoluble in water; used as a chemical intermediate and to separate mineral mixtures. Also known as diiodomethane.　{ 'meth·ə,lēn 'ī·ə,dīd }

methylene oxide See formaldehyde.　{ 'meth·ə,lēn 'äk,sīd }

methylene succinic acid See itaconic acid.　{ 'meth·ə,lēn sək¦sin·ik 'as·əd }

methyl ester [ORG CHEM]　An ester that forms methanol when hydrolyzed.　{ 'meth·əl 'es·tər }

methyl ether See dimethyl ether.　{ 'meth·əl 'ē·thər }

methylethylcellulose [ORG CHEM]　A combustible, white to cream-colored, fibrous solid or powder; disperses in cold water, forming solutions which undergo reversible transformation from sol to gel; used as an emulsifier and foaming agent.　{ ¦meth·əl¦eth·əl'sel·yə,lōs }

methyl ethylene See propylene.　{ 'meth·əl 'eth·ə,lēn }

methyl ethyl ketone [ORG CHEM]　$CH_3COC_2H_5$　A water-soluble, colorless liquid that is miscible in oil; used as a solvent in vinyl films and nitrocellulose coatings, and as a reagent in organic synthesis.　Also known as ethyl methyl ketone; MEK.　{ 'meth·əl ¦eth·əl ¦kē,tōn }

methyl formate [ORG CHEM]　$HCOOCH_3$　A flammable, colorless liquid with a boiling point of 31.8°C; soluble in ether, water, and alcohol; used in military poison gases and larvicides, and as a fumigant.　{ 'meth·əl 'fór,māt }

methyl fumaric acid See mesaconic acid.　{ 'meth·əl fyü'mar·ik 'as·əd }

2-methylfuran [ORG CHEM]　$C_4H_3OCH_3$　A colorless liquid with ether flike aroma; boils at 64°C; used as a chemical intermediate.　{ ¦tü ,meth·ə'fyür,än }

methyl furoate [ORG CHEM]　$C_4H_3OCO_2CH_3$　Colorless liquid that turns yellow in light; soluble in ether and alcohol, insoluble in water; used as a solvent and chemical intermediate.　{ 'meth·əl 'fyür·ə,wāt }

N-methyl glucamine [PHARM]　$C_7H_{17}NO_5$　A crystalline compound with a melting point of 128-129°C; used in medicine as a drug for leishmaniasis.　Also known as meglumine.　{ ¦en ¦meth·əl 'glü·kə,mēn }

methyl glucoside [ORG CHEM]　$C_7H_{14}O_6$　Odorless, water-soluble white crystals; used to make resins, drying oils, plasticizers, and surfactants.　{ 'meth·əl 'glü·kə,sīd }

methyl glycocoll See sarcosine.　{ 'meth·əl 'glī·kə,kól }

methyl heptane [ORG CHEM]　C_8H_{18}　Either of two colorless, water-insoluble liquids, soluble in alcohol and ether, used as chemical intermediates: 2-methylheptane boils at 118°C, is flammable; 4-methylheptane boils at 122°C.　{ ,meth·əl 'hep,tān }

methylheptenone [ORG CHEM]　$(CH_3)_2C:CH(CH_2)_2COCH_3$　A combustible, colorless liquid with a boiling point of 173–174°C; a constituent of many essential oils; used in perfumes and for flavoring.　{ ,meth·əl'hep·tə,nōn }

2-methylhexane [ORG CHEM]　C_7H_{16}　Colorless liquid boiling at 90°C; insoluble in alcohol and water; used as a chemical intermediate.　Also known as ethyl isobutylmethane.　{ ¦tü ¦meth·əl'hek,sān }

methyl hexanoate See methyl caproate.　{ 'meth·əl ,hek'san·ə,wāt }

methyl hexyl ketone [ORG CHEM]　$CH_3COC_6H_{13}$　A combustible, colorless liquid with a boiling point of 173.5°C; soluble in alcohol, hydrocarbons, ether, and esters; used in perfumes and as a flavoring and odorant.　{ ,meth·əl 'hek·səl 'kē,tōn }

methyl hydride See methane.　{ 'meth·əl 'hī,drīd }

methyl hydroxystearate [ORG CHEM]　$C_{19}H_{38}O_3$　A white, waxy material; slightly soluble in organic solvents, insoluble in water; used in cosmetics, inks, and adhesives.　{ 'meth·əl hī,dräk·sē'stir,āt }

methylidyne See methine group.　{ me'thil·ə,dīn }

3-methylindole See skatole.　{ ¦thrē meth·ə'lin,dōl }

methyl iodide [ORG CHEM]　CH_3I　Flammable colorless liquid that turns brown in light; boils at 42°C; soluble in ether and alcohol, insoluble in water; used as a chemical intermediate, in medicine, and in analytical chemistry.　Also known as iodomethane.　{ 'meth·əl 'ī·ə,dīd }

methyl isobutyl carbinol See methyl amyl alcohol.　{ 'meth·əl ,ī·sō'byüd·əl 'kär·bə,nól }

methyl isobutyl carbinol acetate See methyl amyl acetate.　{ 'meth·əl ,ī·sō'byüd·əl 'kär·bə,nól 'as·ə,tāt }

methyl isobutyl ketone [ORG CHEM]　$(CH_3)_2CHCH_2COCH_3$　Flammable colorless liquid with pleasant aroma; boils at 116°C, miscible with most organic solvents; used as a solvent, extractant, and chemical intermediate.　Also known as hexone.　{ 'meth·əl ,ī·sō'byüd·əl 'kē,tōn }

methylisothiocyanate [ORG CHEM]　C_2H_3NS　A crystalline compound, with a melting point of 35–36°C; soluble in alcohol and ether; used as a pesticide and in amino acid sequence analysis.　Also known as methyl mustard oil.　{ ¦meth·əl¦ī·sō,thī·ə sī·ə,nāt }

methyl jasmonate [BOT]　An enzymatic product of lipid metabolism produced via activation of lipoxygenase enzymes in most plant species when tissues are damaged or infected by pathogens; it acts as an external signal effecting communication among plants.　{ ,meth·əl 'jaz·mə,nāt }

methyl lactate [ORG CHEM]　$CH_3CHCHCOOCH_3$　Liquid boiling at 145°C; miscible with water and most organic liquids; used as a solvent for lacquers, stains, and cellulosic materials.　{ 'meth·əl 'lak,tāt }

methyl laurate [ORG CHEM]　$CH_3(CH_2)_{10}COOCH_3$　Water-insoluble, clear, colorless liquid boiling at 262°C; used as a chemical intermediate to make rust removers, and for leather treatment.　{ 'meth·əl 'ló,rāt }

methyl linoleate [ORG CHEM]　$C_{19}H_{34}O_2$　A combustible, colorless liquid with a boiling point of 212°C; soluble in alcohol and ether; used in the manufacture of detergents, emulsifiers, lubricants, and textiles, and in medical research.　{ 'meth·əl lə'nō·lē,āt }

methyl mercaptan [ORG CHEM]　CH_3SH　Colorless, toxic, flammable gas with unpleasant odor; boils at 6.2°C; insoluble in water, soluble in organic solvents; used as a chemical intermediate.　Also known as methanethiol.　{ 'meth·əl mər'kap,tan }

methylmercury compound [ORG CHEM]　Any member of a class of toxic compounds containing the methyl-mercury group, CH_3Hg.　{ ¦meth·əl'mər·kyə·rē ,käm,paúnd }

methylmercury cyanide See methylmercury nitrile.　{ ¦meth·əl'mər·kyə·rē 'sī·ə,nīd }

methylmercury nitrile [ORG CHEM]　CH_3HgCN　A crystalline solid with a melting point of 95°C; soluble in water; used as a fungicide to treat seeds of cereals, flax, and cotton.　Also known as methylmercury cyanide.　{ ¦meth·əl'mər·kyə·rē 'nī·trəl }

methyl methacrylate [ORG CHEM]　$CH_2C(CH_3)COOCH_3$　A flammable, colorless liquid, soluble in most organic solvents but insoluble in water; used as a monomer for polymethacrylate resins.　{ ¦meth·əl mə'thak·rə,lāt }

methyl mustard oil See methylisothiocyanate.　{ 'meth·əl 'məs·tərd ,óil }

methyl myristate [ORG CHEM]　$CH_3(CH_2)_{12}COOCH_3$　A colorless liquid with a boiling point of 186.8°C; used in the manufacture of detergents, plasticizers, resins, textiles, and animal feeds, and as a flavoring.　Also known as methyl tetradecanoate.　{ 'meth·əl mə'ri,stāt }

methylnaphthalene [ORG CHEM]　$C_{10}H_7CH_3$　A solid melting at 34°C; used in insecticides and organic synthesis.　{ ,meth·əl'naf·thə lēn }

methyl nitrate [ORG CHEM]　CH_3NO_3　Explosive liquid boiling at 60°C; slightly soluble in water, soluble in ether and alcohol; used as a rocket propellant.　{ 'meth·əl 'nī,trāt }

methyl nonanoate [ORG CHEM]　$CH_3(CH_2)_7COOCH_3$　A colorless liquid with a fruity odor and a boiling point of 213.5°C; soluble in alcohol and ether; used in perfumes and flavors, and for medical research.   Also known as methyl pelargonate.   { 'meth·əl nə'nan·ə,wāt }

methyl nonyl ketone   [ORG CHEM] $CH_3COC_9H_{19}$ An oily liquid with a boiling point of 225°C; soluble in two parts of 70% alcohol; used in perfumes and flavoring.   Also known as 2-undecanone.   { ¦meth·əl ¦nō·nəl 'kē,tōn }

methyl oleate   [ORG CHEM] $C_{17}H_{33}COOCH_3$ Amber liquid with faint fatty odor; soluble in organic liquids, mineral spirits, and vegetable oil, insoluble in water; used as a plasticizer and softener.   { 'meth·əl 'ōl·ē,āt }

methylol riboflavin   [ORG CHEM]   An orange to yellow powder, soluble in water; used as a nutrient and in medicine.   { 'meth·ə,lól 'rī·bə,flā·vən }

methylol urea   [ORG CHEM] $H_2NCONHCH_2OH$ Water-soluble, colorless crystals melting at 111°C; used to treat textiles and wood, and in the manufacture of resins and adhesives.   { 'meth·ə,lól yù'rē·ə }

Methylomonadaceae   [MICROBIO]   A family of gram-negative, aerobic bacteria which utilize only one-carbon compounds as a source of carbon.   { ,meth·ə·lō,män·ə'dās·ē,ē }

methylotrophic bacteria   [MICROBIO]   Bacteria that are capable of growing on methane derivatives as their sole source of carbon and metabolic energy.   { ¦meth·ə·lə¦trä·fik bak'tir·ē·ə }

methyl palmitate   [ORG CHEM] $CH_3(CH_2)_{14}COOCH_3$ A colorless liquid with a boiling point of 211.5°C; soluble in alcohol and ether; used in the manufacture of detergents, resins, plasticizers, lubricants, and animal feed.   { 'meth·əl 'pal·mə,tāt }

α-methyl-*para*-tyrosine   [ORG CHEM] $C_{10}H_{13}NO_3$ A crystalline compound which acts as the inhibitor of the first and rate-limiting reaction in the biosynthesis of catecholamine; used as an inhibitor of tyrosine hydroxylase.   { ¦al·fə 'meth·əl ¦par·ə 'tī·rə,sēn }

methyl pelargonate   See methyl nonanoate.   { 'meth·əl pə'lär·gə,nāt }

3-methylpentane   [ORG CHEM] C_6H_{14} Flammable, colorless liquid; insoluble in water, soluble in alcohol; boils at 64°C; used as a chemical intermediate.   { ¦thrē ¦meth·əl'pen,tān }

2-methylpentanoic acid   [ORG CHEM] $(CH_3)_2CH(CH_2)_2-COOH$ A colorless liquid with a boiling point of 197°C; soluble in alcohol, benzene, and acetone; used for plasticizers, vinyl stabilizers, and metallic salts.   { ¦tü ¦meth·əl,pen·tə¦nō·ik 'as·əd }

methylpentene polymer   [ORG CHEM]   Thermoplastic material based on 4-methylpentene-1; has low gravity, excellent electrical properties, and 90% optical transmission.   { ¦meth·əl'pen,tēn 'päl·ə·mər }

methyl pentose   [ORG CHEM]   **1.** Any compound that is a methyl derivative of a five carbon sugar.   **2.** In particular, the compound $CH_3(CHOH)_4CHO$.   { 'meth·əl 'pen,tōs }

methyl phenyl acetate   [ORG CHEM] $C_6H_5CH_2COOCH_3$ A colorless liquid with honey odor; used to flavor tobacco and in perfumery.   { ¦meth·əl ¦fen·əl 'as·ə,tāt }

methylphosphoric acid   [ORG CHEM] $CH_3H_2PO_4$ A straw-colored liquid used for textile- and paper-processing compounds, as a rust remover, and in soldering flux.   { ,meth·əl,fäs'fòr·ik 'as·əd }

methyl propionate   [ORG CHEM] $CH_3CH_2COOCH_3$ A flammable, colorless liquid with a boiling range of 78.0–79.5°C; soluble in most organic solvents; used as a solvent for cellulose nitrate, in lacquers, varnishes, and paints, and for flavoring.   { 'meth·əl 'prō·pē·ə,nāt }

methyl propyl carbinol   [ORG CHEM] $CH_3CHOHC_3H_7$ Colorless liquid boiling at 119°C; miscible with ether and alcohol, slightly soluble in water; used as a pharmaceuticals intermediate and as a paint and lacquer solvent.   Also known as *sec-n*-amyl alcohol; 2-pentanol.   { ¦meth·əl ¦prō·pəl 'kär·bə,nól }

N-methyl-2-pyrrolidone   [ORG CHEM] C_5H_9NO A liquid boiling at 202°C; miscible with water, castor oil, and organic solvents; used as a chemical intermediate and as a solvent for petroleum and resins, and in PVC spinning.   { ¦en 'meth·əl ¦tü pə'räl·ə,dōn }

methyl red   [ORG CHEM] $(CH_3)_2NC_6H_4NNC_6H_4COOH$ A dark red powder or violet crystals with a melting point of 180°C; soluble in alcohol, ether, and glacial acetic acid; used as an acid-base indicator (pH 4.2–6.2).   { 'meth·əl 'red }

methyl red test   [MICROBIO]   A cultural test for the ability of bacteria to ferment carbohydrate to form acid; uses methyl red as the indicator.   { 'meth·əl ¦red ,test }

methyl ricinoleate   [ORG CHEM] $C_{19}H_{36}O_3$ Clear, low-viscosity fluid used as a wetting agent, cutting oil additive, lubricant, and plasticizer.   { 'meth·əl ,ris·ən'ōl·ē,āt }

methyl salicylate   [ORG CHEM] $C_6H_4OHCOOCH_3$ A colorless, yellow, or reddish liquid, slightly soluble in water, boiling at 222.2°C, with an odor of wintergreen; used in medicine and perfumery, and as a solvent for cellulose derivatives.   Also known as betula oil; gaultheria oil; wintergreen oil.   { 'meth·əl sə'lis·ə,lāt }

3-methylsalicylic acid   [ORG CHEM] $C_8H_8O_3$ A white to reddish, crystalline compound with a melting point of 165–166°C; soluble in chloroform, alcohol, ether, and alkali hydroxides; used to make dyes.   { ¦thrē ¦meth·əl¦sal·ə¦sil·ik 'as·əd }

methyl silicone   [ORG CHEM] $[(CH_3)_2SiO]_x$, $[C(CH_3)_2Si_2O_3]_y$, etc.   The common varieties of silicones with properties of oil, resin, or rubber, depending on molecular size and arrangement.   { 'meth·əl 'sil·ə,kōn }

methyl stearate   [ORG CHEM] $C_{17}H_{35}COOCH_3$ Colorless crystals melting at 39°C; soluble in alcohol and ether, insoluble in water; used as an intermediate for stearic acid manufacture.   { 'meth·əl 'stir,āt }

methyl styrene   See vinyltoluene.   { 'meth·əl 'stī,rēn }

α-methyl styrene   [ORG CHEM] $C_6H_5C(CH_3):CH_2$ Colorless, toxic, polymerizable liquid boiling at 165°C; used to produce polystyrene resins.   { ¦al·fə ¦meth·əl 'stī,rēn }

methyl sulfate   See dimethyl sulfate.   { 'meth·əl 'səl,fāt }

methyl sulfide   [ORG CHEM] $(CH_3)_2S$ Flammable, colorless liquid with disagreeable aroma; soluble in ether and alcohol, insoluble in water; boils at 38°C; used as a chemical intermediate.   Also known as dimethyl sulfide.   { 'meth·əl 'səl,fīd }

methylsulfonic acid   See methanesulfonic acid.   { ¦meth·əl·səl¦fän·ik 'as·əd }

methylsulfonyl   [ORG CHEM]   A functional group with the formula CH_3SO_2-.   Also known as mesyl.   { ¦meth·əl 'səl·fə,nil }

methyl tertiary butyl ether   [ORG CHEM] $CH_3OC(CH_3)_3$ A volatile, flammable, colorless liquid, with a boiling point of 55°C (131°F) and a terpene-like odor; originally used in gasoline as an octane enhancer and lead substitute; more recently used to reduce engine exhaust emissions.   Abbreviated MTBE.   { ¦meth·əl ¦tər·shē,er·ē ¦byüd·əl 'ē·thər }

methyl tetradecanoate   See methyl myristate.   { 'meth·əl ,te·tra·də'kan·ə,wāt }

4-methyl-5-thiazole ethanol   [ORG CHEM] C_6H_9NOS A viscous, oily liquid; soluble in alcohol, ether, benzene, chloroform, and water; used as an intermediate in the synthesis of vitamin B_1 and as a sedative and hypnotic.   { ¦fòr 'meth·əl ¦fīv 'thī·ə·zōl 'eth·ə,nòl }

methylthionine chloride   See methylene blue.   { ¦meth·əl'thī·ə,nēn 'klòr,īd }

methylthiouracil   [PHARM] $C_5H_6N_2OS$ A crystalline compound; used as a medicine for overactivity of the thyroid.   Also known as 6-methyl-2-thiouracil.   { ¦meth·əl¦thī·ō'yùr·ə·səl }

6-methyl-2-thiouracil   See methylthiouracil.   { ¦siks 'meth·əl ¦tü ¦thī·ō'yùr·ə·səl }

methyl transferase   [BIOCHEM]   Any of a group of enzymes which catalyze the reaction of S-adenosyl methionine with a suitable acceptor to yield the methylated acceptor molecule and S-adenosyl homocysteine.   { 'meth·əl 'tranz·fə,rās }

methyltrichlorosilane   [ORG CHEM] CH_3SiCl_3 A colorless liquid with a pungent odor, boiling point of 66°C (150.8°F), and melting point of −90°C (−130°F).   Also known as trichloromethylsilane.   { ¦meth·əl ,trī,klòr·ō'sī,lān }

methyl violet   [ORG CHEM]   A derivative of pararosaniline, used as an antiallergen and bactericide, acid-base indicator, biological stain, and textile dye.   Also known as crystal violet; gentian violet.   { 'meth·əl 'vī·lət }

Metis   [ASTRON]   The innermost known satellite of Jupiter, having an orbital radius of 79,510 miles (127,960 kilometers) and dimensions of 37 × 21 miles (60 × 34 kilometers).   Also known as Jupiter XVI.   { 'mēd·əs }

metonic cycle   [ASTRON]   A time period of 235 lunar months, or 19 years; after this period the phases of the moon occur on the same days of the same months.   { me'tän·ik 'sī·kəl }

metopan hydrochloride   [PHARM] $C_{18}H_{21}O_3N·HCl$ A morphine derivative; a white, crystalline powder soluble in water; used as a sedative. { 'med·ə‚pan ‚hī·drə'klȯr‚īd }

metraterm [INV ZOO] The distal portion of the uterus in trematodes. { 'mē·trə‚tərm }

metria [MED] **1.** Any pathologic condition of the uterus. **2.** Any uterine inflammation occurring within the 6 weeks following childbirth. { 'mē·trē·ə }

metric [MATH] A real valued "distance" function on a topological space *X* satisfying four rules: for *x*, *y*, and *z* in *X*, the distance from *x* to itself is zero; the distance from *x* to *y* is positive if *x* and *y* are different; the distance from *x* to *y* is the same as the distance from *y* to *x*; and the distance from *x* to *y* is less than or equal to the distance from *x* to *z* plus the distance from *z* to *y* (triangle inequality). { 'me·trik }

metricate [SCI TECH] To use the metric system in expressing all physical quantities. { 'me·trə‚kāt }

metric carat *See* carat. { 'me·trik ‚kar·ət }

metric centner [MECH] **1.** A unit of mass equal to 50 kilograms. **2.** A unit of mass equal to 100 kilograms. Also known as quintal. { 'me·trik 'sent·nər }

metric grain [MECH] A unit of mass, equal to 50 milligrams; used in commercial transactions in precious stones. { 'me·trik 'grān }

metric horsepower [PHYS] A unit of power, equal to 75 meter kilograms-force per second; equal to 735.49875 watts. { 'me·trik ‚hȯrs‚pau̇·ər }

metric line *See* millimeter. { 'me·trik 'līn }

metric ounce *See* mounce. { 'me·trik 'au̇ns }

metric slug *See* metric-technical unit of mass. { 'me·trik 'sləg }

metric space [MATH] Any topological space which has a metric defined on it. { 'me·trik 'spās }

metric system [MECH] A system of units used in scientific work throughout the world and employed in general commercial transactions and engineering applications; its units of length, time, and mass are the meter, second, and kilogram respectively, or decimal multiples and submultiples thereof. { 'me·trik ‚sis·təm }

metric-technical unit of mass [MECH] A unit of mass, equal to the mass which is accelerated by 1 meter per second per second by a force of 1 kilogram-force; it is equal to 9.80665 kilograms. Abbreviated TME. Also known as hyl; metric slug. { 'me·trik ‚tek·ni·kəl 'yü·nət əv 'mas }

metric tensor [MATH] A second rank tensor of a Riemannian space whose components are functions which help define magnitude and direction of vectors about a point. Also known as fundamental tensor. { 'me·trik 'ten·sər }

metric thread gearing [DES ENG] Gears that may be interchanged in change-gear systems to provide feeds suitable for cutting metric and module threads. { 'me·trik 'thred ‚gir·iŋ }

metric ton *See* tonne. { 'me·trik 'tən }

metric waves [ELECTROMAG] Radio waves having wavelengths between 1 and 10 meters, corresponding to frequencies between 30 and 300 megahertz (the very-high-frequency band). { 'me·trik 'wāvz }

Metridiidae [INV ZOO] A family of zoantharian cnidarians in the order Actiniaria. { ‚me·trə'dī·ə‚dē }

metriocranic [ANTHRO] Having a skull that is moderately high compared with its width, with a breadth-height index of 92 to 98. { ‚me·trē·ō‚krā·nik }

metritis [MED] Inflammation of the uterus, usually involving both the endometrium and myometrium. { mə'trīd·əs }

metrizable space [MATH] A topological space on which can be defined a metric whose topological structure is equivalent to the original one. { mə'trīz·ə·bəl 'spās }

metrology [PHYS] The science of measurement. { mə'träl·ə‚jē }

metrorrhagia [MED] Uterine bleeding during the intermenstrual cycle. Also known as intermenstrual flow; polymenorrhea. { ‚mē·trə'rā‚jē·ə }

metrorrhea [MED] Any pathologic discharge from the uterus. { ‚mē·trə'rē·ə }

metrorrhexis [MED] Rupture of the uterine wall. { ‚mē·trə‚rek·səs }

metrosalpingitis [MED] Inflammation of the uterus and oviducts. { ‚mē·trə‚sal·pən'jīd·əs }

metrostaxis [MED] Slight, chronic bleeding from the uterus. { ‚mē·trə'stak·səs }

Metzgeriales [BOT] An order of liverworts in the subclass Jungermanniidae, class Hepaticopsida, distinguished by archegonia produced behind a growing apex, a flat elongated gametophyte with no tissue differentiation or surface pores and, less commonly, a stem with two rows of leaves. { ‚mets·gə‚rē'ā·lēz }

Meusnier's theorem [MATH] A theorem stating that the curvature of a surface curve equals the curvature of the normal section through the tangent to the curve divided by the cosine of the angle between the plane of this normal section and the osculating plane of the curve. { mən'yāz ‚thir·əm }

MeV *See* megaelectronvolt.

mevalonic acid [ORG CHEM] $HO_2C_5H_9COOH$ A dihydroxy acid used in organic synthesis. { ¦mev·ə¦lan·ik 'as·əd }

MeV Ci *See* megaelectronvolt-curie.

mexacarbate [ORG CHEM] $C_{12}H_{18}N_2O_2$ A tan solid with a melting point of 85°C; used to control insect pests of trees, flowers, and shrubs. { ¦mek·sə'kär‚bāt }

Mexican linaloe oil *See* linaloe wood oil. { 'mek·si·kən lē'nä‚lō·ə ‚oil }

Mexican onyx *See* onyx marble. { 'mek·si·kən 'än·iks }

Meyer atomic volume curve [ATOM PHYS] A graph of the atomic volumes of the elements versus their atomic numbers; it reveals a periodicity, with peaks at the alkali elements and valleys at the transition elements. { 'mī·ər ə¦täm·ik ‚väl·yəm 'kərv }

meyerhofferite [MINERAL] $Ca_2B_6O_{11}\cdot 7H_2O$ A colorless, hydrated borate mineral that crystallizes in the triclinic system. { 'mī·ər‚häf·ə‚rīt }

Meyliidae [INV ZOO] A family of free-living nematodes in the superfamily Desmoscolecoidea. { ma'lī·ə‚dē }

Meziridae [INV ZOO] A family of hemipteran insects in the superfamily Aradoidea. { mə'zir·ə‚dē }

mezzograph [GRAPHICS] A halftone that has a grained surface rather than crossline screen dots. { 'met·sə‚graf }

mezzotint [GRAPHICS] An engraving process in which a copper or steel plate is first entirely roughened by rubbing Carborundum between it and another plate and by using a steel chisel with an edge set with minute teeth which rock into the plate at various angles; the rough-grain plate is burnished with a steel instrument to produce appropriate white areas of the design. { 'met·sə‚tint }

mf *See* medium frequency.

mF *See* millifarad.

Mflop *See* million floating-point operations per second.

MFSK *See* multiple-frequency-shift keying.

mg *See* milligram.

mG *See* milligauss.

Mg *See* magnesium.

mGal *See* milligal.

M82 galaxy [ASTRON] An active, variable spiral galaxy that exhibits strong emission from its center in the radio band, 10^4 to 10^6 times greater than that of normal spirals, and ejection of gases at speeds up to 620 miles (1000 kilometers) per second. { ¦em ¦ād·ē¦tü 'gal·ik·sē }

mg h *See* milligram-hour.

M glass [MATER] A glass with a high content of beryllium oxide and a high modulus of elasticity. { 'em ‚glas }

mH *See* millihenry.

MHC *See* major histocompatibility complex.

MHD *See* magnetohydrodynamics.

MHD generator *See* magnetohydrodynamic generator. { ¦em¦āch'dē 'jen·ə‚rād·ər }

mho *See* siemens. { mō }

mHz *See* millihertz.

MHz *See* megahertz.

mi *See* mile.

Miacidae [PALEON] The single, extinct family of the carnivoran superfamily Miacoidea. { mī'as·ə‚dē }

Miacoidea [PALEON] A monofamilial superfamily of extinct carnivoran mammals; a stem group thought to represent the progenitors of the earliest member of modern carnivoran families. { ‚mī·ə'kȯid·ē·ə }

miagite *See* corsite. { 'mī·ə‚jīt }

miargyrite [MINERAL] $AgSbS_2$ An iron-black to steel-gray mineral that crystallizes in the monoclinic system. { mī'är‚jə‚rīt }

miarolithite [PETR] A chorismite type of igneous rock having miarolitic cavities or vestiges thereof. { ‚mē·ə'rō·lə‚thīt }

miarolitic [PETR] Of igneous rock, characterized by small

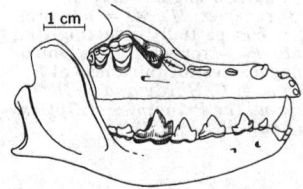

MIACOIDEA

Right side of *Miacis* jaw shown in ventral (upper) and lateral (lower) views.

irregular cavities into which well-formed crystals of the rock-forming mineral protrude. { ¦mē·ə·rō¦lid·ik }

MIBC *See* methyl amyl alcohol.

MIC *See* major immunogene complex. *See* microwave integrated circuit.

mica [MINERAL] A group of phyllosilicate minerals (with sheetlike structures) of general formula $(K,Na,Ca)(Mg,Fe,Li,Al)_{2-3}(Al,Si)_4O_{10}(OH,F)_2$ characterized by low hardness $(2-2^1/_2)$ and perfect basal cleavage. { 'mī·kə }

mica book [MINERAL] A crystal of mica, usually large and irregular, whose cleavage plates resemble the leaves of a book. Also known as book. { 'mī·kə ¸bùk }

mica capacitor [ELEC] A capacitor whose dielectric consists of thin rectangular sheets of mica and whose electrodes are either thin sheets of metal foil stacked alternately with mica sheets, or thin deposits of silver applied to one surface of each mica sheet. { 'mī·kə kə'pas·əd·ər }

micaceous [GEOL] Pertaining to or resembling mica. { mī'kā·shəs }

micaceous arkose [PETR] A sandstone containing 25–90% feldspars and feldspathic crystalline rock fragments, 10–50% micas and micaceous metamorphic rock fragments, and 0–65% quartz, chert, and metamorphic quartzite. { mī'kā·shəs 'är¸kōs }

mica schist [PETR] A schist which is composed essentially of mica and quartz and whose characteristic foliation is mainly due to the parallel orientation of the mica flakes. { 'mī·kə ¸shist }

micellar catalysis [CHEM] Enhancement of the rate of a chemical reaction in solution by the addition of a surfactant, so that the reaction proceeds in the environment of surfactant aggregates. { mī¦sel·ər kə'tal·ə·səs }

micellar flooding [PETRO ENG] A two-step enhanced oil recovery process in which a surfactant slug is injected into the well followed by a larger slug of water containing a high-molecular-weight polymer which pushes the chemicals through the field and improves mobility and sweep efficiency. Also known as microemulsion flooding; surfactant flooding. { mī¦sel·ər 'fləd·iŋ }

micelle [CELL MOL] A submicroscopic structural unit of protoplasm built up from polymeric molecules. [PHYS CHEM] A colloidal aggregate of a unique number (between 50 and 100) of amphipathic molecules, which occurs at a well-defined concentration known as the critical micelle concentration. { mī'sel }

Michaelis constant [BIOCHEM] A constant K_m such that the initial rate of reaction V, produced by an enzyme when the substrate concentration is high enough to saturate the enzyme, is related to the rate of reaction v at a lower substrate concentration c by the formula $V = v(1 + K_m/c)$. { mi'kā·ləs ¸kän·stənt }

Michaelis-Menten equation [BIOCHEM] A mathematical equation expressing the hyperbolic relationship between the initial velocity, V_o, and the substrate concentration, $[S]$, in a number of enzyme-catalyzed reactions such that $V_o = V_{max}[S]/K_m + [S]$, where V_{max} is the maximum velocity and K_m is the Michaelis constant. { ¸mik·ä¦ə·ləs 'men·tən i¸kwā·zhən }

Michaelson actinograph [ENG] A pyrheliometer of the bimetallic type used to measure the intensity of direct solar radiation; the radiation is measured in terms of the angular deflection of a blackened bimetallic strip which is exposed to the direct solar beams. { 'mī·kəl·sən ak'tin·ə¸graf }

Michel parameter [PARTIC PHYS] A number appearing in an equation for the momentum spectrum of muon decay, which depends on the nature of the weak interactions; the number is equal to 3/4 in any two-component neutrino theory before radiative corrections are taken into account. { mi'shel pə¸ram·əd·ər }

Michelson interferometer [OPTICS] An interferometer in which light strikes a partially reflecting plate at an angle of 45°, the light beams reflected and transmitted by the plate are both reflected back to the plate by mirrors, and the beams are recombined at the plate, interfering constructively or destructively depending on the distances from the plate to the two mirrors. { 'mī·kəl·sən ¸in·tər·fə'räm·əd·ər }

Michelson-Morley experiment [OPTICS] An experiment which uses a Michelson interferometer to determine the difference between the speeds of light in two perpendicular directions. { 'mī·kəl·sən 'mòr·lē ik¸sper·ə·mənt }

Michelson stellar interferometer [OPTICS] An instrument for measuring angular diameters of astronomical objects, in which a system of mirrors directs two parallel beams of light into a telescope, and angular diameter is determined from the maximum distance between the beams at which interference fringes are observable. { 'mī·kəl·sən ¦stel·ər ¸in·tər·fə'räm·əd·ər }

michenerite [MINERAL] A silver-white mineral (PdBiTe) that is a major source of palladium. { 'mich·ə·nə¸rīt *or* 'mich·nə¸rīt }

Michigan cut [MIN ENG] A technique used to break off ore at a heading; a large hole or series of small holes at the center of the heading are drilled parallel to the tunnel direction but not charged with explosive; other holes are drilled in the heading and charged so that upon detonation they break out toward the uncharged holes. { 'mich·ə·gən ¦kət }

Michigan tripod [MIN ENG] A support for a drilling outfit; consists of three debarked pine or fir timber poles about 25 feet (7.6 meters) long whose butt ends are about 12 inches (30 centimeters) in diameter; a sheave suspended from a clevis at the top of the tripod is aligned over the hoisting drum and the borehole; there is a minimum of 22 feet (6.7 meters) of headroom above the drill floor. { 'mich·ə·gən 'trī¸päd }

Michler's ketone *See* tetramethyldiaminobenzophenone. { 'mik·lərz 'kē¸tōn }

mickey-mouse [COMPUT SCI] To play with something new, such as hardware, software, or a system, until a feel is gotten for it and the proper operating procedure is discovered, understood, and mastered. { ¦mik·ē 'maùs }

MICR *See* magnetic-ink character recognition.

micracanthopore [INV ZOO] Small, minute tubes projecting from the surface of bryozoan colonies. { ¸mī·krə'kan·thrə¸pòr }

micrencephaly [MED] The condition of having an abnormally small brain. Also spelled microencephaly. { ¦mī·kren¦sef·ə·lē }

micril *See* gammil. { 'mī·krəl }

micrinite [PETR] An opaque granular variety of inertinite of medium hardness showing no plant-cell structure. { 'mī·krə¸nīt }

micrite [PETR] A semiopaque crystalline limestone matrix that consists of chemically precipitated calcite mud, whose crystals are generally 1–4 micrometers in diameter. { 'mī¸krīt }

micro- [MATH] A prefix representing 10^{-6}, or one-millionth. [SCI TECH] **1.** A prefix indicating smallness, as in microwave. **2.** A prefix indicating extreme sensitivity, as in microradiometer and microphone. { 'mī·krō }

microabscess [MED] A small abscess. { ¦mī·krō'ab¸ses }

microaccelerometer [ENG] A MEMS device developed for the automotive industry to control air-bag inflation. { ¸mī·krō·ik¸sel·ə·rə'täm·əd·ər }

microactuator [ENG] A very small actuator, with physical dimensions in the submicrometer to millimeter range, generally batch-fabricated from silicon wafers. { ¸mī·krō'ak·chə¸wād·ər }

microaerophilic [MICROBIO] Pertaining to those microorganisms requiring free oxygen but in very low concentration for optimum growth. { ¦mī·krō¦er·ə¦fil·ik }

micro air vehicle [ENG] A very small airborne autonomous vehicle that can operate inside a building using primarily visual and other sensory information to navigate. { ¦mī·krō 'er ¸vē·ə·kəl }

microalloy diffused transistor [ELECTR] A microalloy transistor in which the semiconductor wafer is first subjected to gaseous diffusion to produce a nonuniform base region. Abbreviated MADT. { 'mī·krō¸al¸òi di'fyüzd tran'zis·tər }

microalloy transistor [ELECTR] A transistor in which the emitter and collector electrodes are formed by etching depressions, then electroplating and alloying a thin film of the impurity metal to the semiconductor wafer, somewhat as in a surface-barrier transistor. { 'mī·krō¸al¸òi tran'zis·tər }

microammeter [ELEC] An ammeter whose scale is calibrated to indicate current values in microamperes. { ¦mī·krō'a¸mēd·ər }

microampere [ELEC] A unit of current equal to one-millionth of an ampere. Abbreviated μA. { ¦mī·krō'am¸pir }

microanalysis [ANALY CHEM] Identification and chemical

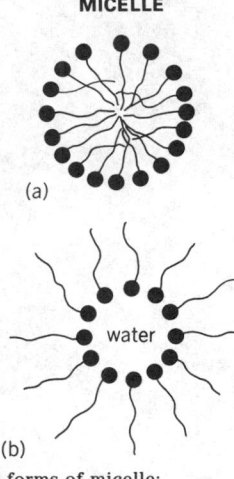

MICELLE

Two forms of micelle:
(a) spherical and *(b)* reversed.

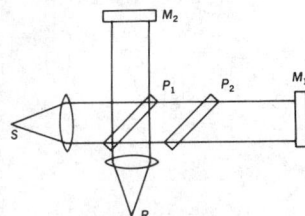

MICHELSON INTERFEROMETER

Michelson interferometer,
S = narrow angle source;
R = receiver; M_1, M_2 = mirrors;
P_1 = 50% partially reflecting
plate; P_2 = reflector plate which
compensates for thickness of P_1.
*(From A. C. Hardy and F. H.
Perrin, The Principles of Optics,
McGraw-Hill, 1932)*

analysis of material on a small scale so that specialized instruments such as the microscope are needed; the material analyzed may be on the scale of 1 microgram.　{ ¦mī·krō·ə'nal·ə·səs }

microanatomy [ANAT] Anatomical study of microscopic tissue structures.　{ ¦mī·krō·ə'nad·ə·mē }

microaneurysm [MED] Dilation of the wall of a capillary, characteristic of certain disease entities.　{ ¦mī·krō'an·yə,riz·əm }

microangiopathy [MED] The development of lesions in small blood vessels throughout the body.　{ ,mī·krō,an·jē'äp·ə·thē }

microangstrom [MECH] A unit of length equal to one-millionth of an angstrom, or 10^{-16} meter. Abbreviated μA.　{ ¦mī·krō'aŋ·strəm }

microautoradiograph [GRAPHICS] An image which is produced by placing a specimen containing radioactive material (usually a radioactive tracer) in close contact with a photographic film and optically enlarging the developed image.　{ ¦mī·krō,ȯd·ō'rād·ē·ō,graf }

microbalance [ENG] A small, light type of analytical balance that can weigh loads of up to 0.1 gram to the nearest microgram.　{ ¦mī·krō'bal·əns }

microbar See barye.　{ 'mī·krə,bär }

microbarm [GEOPHYS] That portion of the record of a microbarograph between any two or a specified small number of the successive crossings of the average pressure level in the same direction; analogous to microseism.　{ 'mī·krə,bärm }

microbarogram [ENG] The record or trace made by a microbarograph.　{ ¦mī·krō'bar·ə,gram }

microbarograph [METEOROL] A type of aneroid barograph designed to record atmospheric pressure variations of very small magnitude.　{ ¦mī·krō'bar·ə,graf }

microbarom [ACOUS] An infrasound wave that originates with surface waves on the seas or oceans, having a period of 4–7 seconds and a sound pressure level of about 85 decibels.　{ 'mī·krō,bar·əm }

microbe [MICROBIO] A microorganism, especially a bacterium of a pathogenic nature.　{ 'mī,krōb }

microbeam [ELECTROMAG] An x-ray beam with submicrometer dimensions.　{ 'mī·krō,bēm }

microbial ecology [ECOL] The study of interrelationships between microorganisms and their living and nonliving environments.　{ mī,krōb·ē·əl ē'käl·ə·jē }

microbial insecticide [MICROBIO] Species-specific bacteria which are pathogenic for and used against injurious insects.　{ mī'krō·bē·əl in'sek·tə,sīd }

microbicide [MATER] An agent that kills microbes.　{ mī'krō·bə,sīd }

microbiologist [MICROBIO] A scientist that studies a wide range of microorganisms in various subdisciplines of biology, such as bacteriology, mycology, parasitology, and virology.　{ ,mī·krō·bī'äl·ə·jist }

microbiology [MICROBIO] The science and study of microorganisms, including protozoans, algae, fungi, bacteria, viruses, and rickettsiae.　{ ¦mī·krō·bī'äl·ə·jē }

microbiophobia [PSYCH] An abnormal fear of microbes.　{ ,mī·krə'fō·bē·ə }

microbit [COMPUT SCI] A unit of information equal to one-millionth of a bit.　{ 'mī·krə,bit }

microbody [CELL MOL] Any of three distinct classes (peroxisomes, glyoxysomes, and microperoxisomes) of cytoplasmic organelles that are bounded by a single membrane and contain a variety of enzymes.　{ 'mī·krə,bäd·ē }

microbreccia [GEOL] A poorly sorted sandstone containing large, angular sand particles in a fine silty or clayey matrix.　{ 'mī·krō'brech·ə }

microbridge [CRYO] A Josephson junction formed by configuration of thin superconducting films.　{ 'mī·krō,brij }

microburst [METEOROL] A downdraft with horizontal extent of about 2.5 miles (4 kilometers) or less, associated with atmospheric convection, often a thundershower.　{ 'mī·krō,bərst }

microcaliper log [PETRO ENG] A detailed and accurate record of drill-hole diameter; used to detect caved sections and to verify the presence of mud cake.　{ ¦mī·krō'kal·ə·pər 'läg }

microcalorimeter [ENG] A calorimeter for measuring very small amounts of heat, in which the heat source and a small heating coil are placed in identical vessels and the amount of current through the coil is varied until the temperatures of the

vessels are identical, as indicated by thermocouples.　{ ,mī·krō,kal·ə'rim·əd·ər }

microcanonical ensemble [STAT MECH] A collection of systems describing a single isolated system of specified energy; its members are uniformly distributed over a part of phase space whose energies lie within an infinitesimal range.　{ ¦mī·krō·kə'nän·ə·kəl än'säm·bəl }

microcapacitor [ELECTR] Any very small capacitor used in microelectronics, usually consisting of a thin film of dielectric material sandwiched between electrodes.　{ ¦mī·krō·kə'pas·əd·ər }

microcapsule [CHEM ENG] A capsule with a plastic or wax-like coating having a diameter anywhere from well below 1 micrometer to over 2000 micrometers.　{ 'mī·krō,kap·səl }

microcard [GRAPHICS] A type of microtext, consisting of photographic prints 7.5 by 12.5 centimeters in size prepared from 16- or 35-millimeter film, commonly at a reduction of 20 diameters.　{ 'mī·krō,kärd }

microcell [CELL MOL] A micronucleus within a layer of cytoplasm and a membrane.　{ 'mī·krə,sel }

microcentrum [CELL MOL] The centrosome, or a group of centrosomes, functioning as the dynamic center of a cell.　{ 'mī·krō,sen·trəm }

microcephalus [MED] An individual with microcephaly.　{ ,mī·krō'sef·ə·ləs }

microcephaly [MED] The condition of having an abnormally small head, with a circumference less than two standard deviations below the mean.　{ ,mī·krō'sef·ə·lē }

microceratous [INV ZOO] Having short antennae.　{ ,mī·krō'ser·ə·təs }

microcercous cercaria [INV ZOO] A cercaria with a very short broad tail.　{ mi·krō¦sər·kəs sər'kar·ē·ə }

microchannel plate [ELECTR] A plate that consists of extremely small cylinder-shaped electron multipliers mounted side by side, to provide image intensification factors as high as 100,000. Also known as channel plate multiplier.　{ ¦mī·krō¦chan·əl 'plāt }

microchemistry [BIOCHEM] The chemistry of individual cells and minute organisms. [CHEM] The study of chemical reactions, using small quantities of materials, frequently less than 1 milligram or 1 milliliter, and often requiring special small apparatus and microscopical observation.　{ ¦mī·krō'kem·ə·strē }

microchip See chip.　{ 'mī·krō,chip }

Microchiroptera [VERT ZOO] A suborder of the mammalian order Chiroptera composed of the insectivorous bats.　{ ¦mī·krō·kī'räp·tə·rə }

microchronometer [HOROL] A spring-driven, fast-moving clock capable of indicating time intervals as small as 1/2000 of a minute; used as a timing device in micromotion studies.　{ ¦mī·krō·krə näm·əd·ər }

microcircuitry [ELECTR] Electronic circuit structures that are orders of magnitude smaller and lighter than circuit structures produced by the most compact combinations of discrete components. Also known as microelectronic circuitry; microminiature circuitry.　{ ¦mī·krō'sər·kə·trē }

microcirculation [PHYSIO] The flow of blood or lymph in the vessels of the microcirculatory system.　{ ¦mī·krō,sər·kyə'lā·shən }

microcirculatory system [ANAT] Those vessels of the blood and lymphatic systems which are visible only with a microscope.　{ ¦mī·krō'sər·kyə·lə,tȯr·ē ,sis·təm }

microclimate [CLIMATOL] The local, rather uniform climate of a specific place or habitat, compared with the climate of the entire area of which it is a part.　{ ¦mī·krō'klī·mət }

microclimatology [CLIMATOL] The study of a microclimate, including the study of profiles of temperature, moisture and wind in the lowest stratum of air, the effect of the vegetation and of shelterbelts, and the modifying effect of towns and buildings.　{ ¦mī·krō,klī·mə'täl·ə·jē }

microcline [MINERAL] $KAlSi_3O_8$ A triclinic potassium-rich feldspar, usually containing minor amounts of sodium; may be clear, white, pale-yellow, brick-red, or green, and is generally characterized by crosshatch twinning.　{ 'mī·krə,klīn }

microcneme [INV ZOO] Microsepta in certain anemones.　{ 'mī·krə,nēm }

Micrococcaceae [MICROBIO] A family of gram-positive

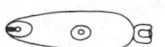
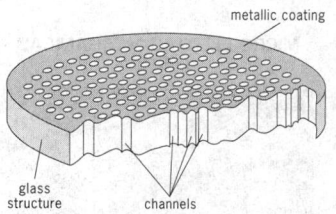

cocci; chemoorganotrophic organisms with respiratory or fermentative metabolism. { ¦mī·krō·kak'sas·ē,ē }

microcode [COMPUT SCI] A code that employs microinstructions; not ordinarily used in programming. { 'mī·krō,kōd }

microcomputer [COMPUT SCI] **1.** A digital computer whose central processing unit resides on a single semiconductor integrated circuit chip, a microprocessor. **2.** An electronic device, typically consisting of a microprocessor central processing unit, semiconductor memory (RAM), graphics display, and keyboard. Typical configurations also include a hard disk for persistent memory, a compact disk drive, a disk drive which allows removable disks to be used to move data in and out of the machine, and a pointing device. { ¦mī·krō·kəm'pyüd·ər }

microcomputer development system [COMPUT SCI] A complete microcomputer system that is used to test both the software and hardware of other microcomputer-based systems. { ¦mī·krō·kəm'pyüd·ər di'vel·əp·mənt ,sis·təm }

microconsumer See decomposer. { ¦mī·krō·kən'sü·mər }

microcontroller [ELECTR] A microcomputer, microprocessor, or other equipment used for precise process control in data handling, communication, and manufacturing. { ¦mī·krō·kən'trōl·ər }

microcopy [GRAPHICS] A photographic reproduction that is too small to be read without magnification. { 'mī·krō,käp·ē }

microcoquina [PETR] A clastic limestone composed wholly or partially of cemented sand-size particles of shell detritus. { ¦mī·krō·kə'kē·nə }

Microcotyloidea [INV ZOO] A superfamily of ectoparasitic trematodes in the subclass Monogenea. { ¦mī·krō,käd·əl'öid·ē·ə }

microcoulomb [ELEC] A unit of electric charge equal to one-millionth of a coulomb. Abbreviated μC. { ¦mī·krō'kü,läm }

microcrack See microfissure. { 'mī·krō,krak }

microcrystalline [CRYSTAL] Composed of or containing crystals that are visible only under the microscope. { ¦mī·krō'krist·əl·ən }

microcrystalline wax [MATER] A petroleum wax containing small, indistinct crystals, and having a higher molecular weight, melting point, and viscosity than paraffin wax; used in laminated paper and electrical coil coating. { ¦mī·krō'krist·əl·ən 'waks }

Microcyprini [VERT ZOO] The equivalent name for Cyprinodontiformes. { ,mī·krō·sə'prē,nē }

microcyst [MED] A very small cyst. { 'mī·krə,sist }

microcyte [MED] A red blood cell whose diameter or mean corpuscular volume or both are more than two standard deviations below the normal mean. Also known as microerythrocyte. { 'mī·krə,sīt }

microcythemia [MED] Blood characterized by the presence of small red blood cells. { ¦mī·krō·sī'thē·mē·ə }

microcytic anemia [MED] Any form of anemia in which small erythrocytes occur in the blood. { ¦mī·krə¦sid·ik ə'nē·mē·ə }

microcytosis [MED] A blood disorder characterized by a preponderance of microcytes. { ,mī·krə·sī'tō·səs }

microdactyly [MED] A condition of abnormal smallness of fingers or toes. { ¦mī·krō'dak·tə·lē }

microdensitometer [SPECT] A high-sensitivity densitometer used in spectroscopy to detect spectrum lines too faint on a negative to be seen by the human eye. { ¦mī·krō,den·sə'täm·əd·ər }

microdiagnostic program [COMPUT SCI] A microprogram that tests a specific hardware component, such as a bus or store location, for faults. { ¦mī·krō,dī·əg'näs·tik 'prō·gram }

microdialysis [ANALY CHEM] A technique for sampling biological systems in which a short length of hollow-fiber dialysis membrane is implanted into any tissue or fluid compartment, and through which compounds in the extracellular fluid are collected for subsequent analysis. { ,mī·krō·dī'al·ə·səs }

microdiffusiometer [ENG] A type of diffusiometer in which diffusion is measured over microscopic distances, greatly reducing the time required for the measurement and the effects of vibration and temperature changes. { ,mī·krō·də'fyüz·ər }

microdisk [COMPUT SCI] A small floppy disk with a diameter between 3 and 4 inches (7 and 10 centimeters). Also known as microfloppy disk. { 'mī·krō,disk }

microdisk laser [OPTICS] A very small semiconductor laser that consists of a quantum well structure formed into a disk, such that total internal reflection of photons traveling around the perimeter of the disk results in high-Q whispering-gallery resonances. { 'mī·krō,disk ,lā·zər }

microdissection [BIOL] Dissection under a microscope. { ¦mī·krō·di'sek·shən }

Microdomatacea [PALEON] An extinct superfamily of gastropod mollusks in the order Aspidobranchia. { ,mī·krə,dō·mə'tās·ē·ə }

microearthquake [GEOPHYS] An earthquake with a low intensity, usually less than 3 on the Richter scale. Also known as microquake. { ,mī·krō'ərth,kwāk }

microelectrode [ENG] **1.** In biological research, an electrode with a microscopic tip dimension that may be placed adjacent to or inside a cell for the purpose of recording the electric potentials of single cells, passing electrical currents, or injecting electrically charged substances into the cell. **2.** In physical chemistry, a minute electrode used to perform electrolysis of small quantities of material. { ,mī·krō·i'lek,trōd }

microelectrolysis [PHYS CHEM] Electrolysis of small quantities of material. { ¦mī·krō·i,lek'träl·ə·səs }

micro-electro-mechanical system [ENG] A system in which micromechanisms are coupled with microelectronics, most commonly fabricated as microsensors or microactuators. Abbreviated MEMS. Also known as microsystem. { ¦mī·krō·i,lek·trə·mə'kan·ə·kəl ,sis·təm }

microelectronic circuitry See microcircuitry. { ¦mī·krō·i,lek'trän·ik 'sər·kə·trē }

microelectronics [ELECTR] The technology of constructing circuits and devices in extremely small packages by various techniques. Also known as microminiaturization; microsystem electronics. { ¦mī·krō·i,lek'trän·iks }

microelectrophoresis [ANALY CHEM] Direct microscopic observation and measurement of the velocity of migration of ions or other charged bodies through a solution toward oppositely charged electrodes. Also known as optical cytopherometry. { ¦mī·krō·i,lek·trə·fə'rē·səs }

microelement [ELECTR] Resistor, capacitor, transistor, diode, inductor, transformer, or other electronic element or combination of elements mounted on a ceramic wafer 0.025 centimeter thick and about 0.75 centimeter square; individual microelements are stacked, interconnected, and potted to form micromodules. [IND ENG] An element of a work cycle whose time span is too short to be observed by the unaided eye. { ¦mī·krō'el·ə·mənt }

microemulsion [MATER] A thermodynamically stable dispersion of two immiscible liquids, stabilized by surfactants; it is typically clear because the dispersed droplets are less than 100 nanometers in diameter. { ¦mī·krō·i'məl·shən }

microemulsion flooding See micellar flooding. { ¦mī·krō·i'məl·shən 'fläd·iŋ }

microencapsulation [CHEM ENG] Enclosing of materials in capsules from well below 1 micrometer to over 2000 micrometers in diameter. { ¦mī·krō·in,kap·sə'lā·shən }

microencephaly See micrencephaly. { ¦mī·krō·en'sef·ə·lē }

microengineering [ENG] The design and production of small, three-dimensional objects, usually for manufacture in high volumes at low cost. { ,mī·krō,en·jə'nir·iŋ }

microenvironment [ECOL] The specific environmental factors in a microhabitat. { ¦mī·krō·in'vī·ərn·mənt }

microerythrocyte See microcyte. { ¦mī·krō·ə'rith·rə,sīt }

microevolution [EVOL] **1.** Evolutionary processes resulting from the accumulation of minor changes over a relatively short period of time; evolutionary changes due to gene mutation. **2.** Evolution of species. { ¦mī·krō,ev·ə'lü·shən }

microfabrication [ENG] The technology of fabricating microsystems from silicon wafers, using standard semiconductor process technologies in combination with specially developed processes. { ¦mī·krō,fab·rə'kā·shən }

microfacies [PETR] The composition, features, or appearance of a rock or mineral in thin section under the microscope. { ¦mī·krō'fā·shēz }

microfarad [ELEC] A unit of capacitance equal to one-millionth of a farad. Abbreviated μF. { ¦mī·krō'far·əd }

microfauna [ECOL] Microscopic animals such as protozoa and nematodes. { ¦mī·krə,fön·ə }

microfibril [CELL MOL] The submicroscopic unit of a microscopic cellular fiber. { ¦mī·krō'fī·brəl }

microfiche [GRAPHICS] A microfilm card or sheet used in

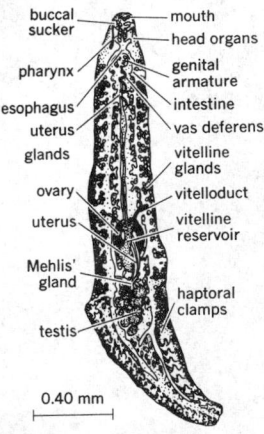

MICROCOTYLOIDEA

buccal sucker — mouth
— head organs
pharynx — genital armature
esophagus — intestine
uterus — vas deferens
glands — vitelline glands
ovary — vitelloduct
uterus — vitelline reservoir
Mehlis' gland
— haptoral clamps
testis

0.40 mm

Ventral view of *Heteraxinoides xanthophilis* (Hargis), an ectoparasite of the spot fish (*Leiostomus xanthurus*).

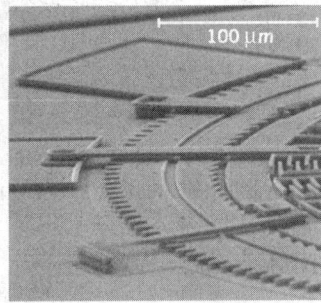

MICRO-ELECTRO-MECHANICAL SYSTEM

100 μm

Close-up of a torsional ratcheting actuator fabricated by surface micromachining. (*J. Jakubczala, Sandia National Laboratories*)

some information storage systems; consists of a film format about 4 by 6 inches (10 by 15 centimeters) containing microimages of type and other information, with a title heading large enough to be read by the unaided eye. { ¦mī·krə‚fēsh }

microfilament [CELL MOL] One of the cytoplasmic fibrous structures, about 5 nanometers in diameter, virtually identical to actin; thought to be important in the processes of phagocytosis and pinocytosis. { ¦mī·krō'fil·ə·mənt }

microfilaria [INV ZOO] Slender, motile prelarval forms of filarial nematodes measuring 150–300 micrometers in length; adult filaria are mammalian parasites. { ¦mī·krō·fə'lar·ē·ə }

microfilm [GRAPHICS] Greatly reduced film records of such things as books, newspapers, engineering drawings, reports, and manuscripts; copies are made on fine-grain film of 16, 35, 70, and 105-millimeter size, permitting easy storage and handling. { 'mī·krə‚film }

microfilm plotter [GRAPHICS] A device for plotting graphs on microfilm, generally from instructions on a magnetic tape or disk. { 'mī·krə‚film ‚pläd·ər }

microfiltration [CHEM ENG] A membrane separation process in which particles greater than about 20 nanometers in diameter are screened out of a liquid in which they are suspended. { ‚mi·krō·fil'trā·shən }

microfissure [MET] A crack of microscopic dimensions. Also known as microcrack. { ¦mī·krō'fish·ər }

microfloppy disk See microdisk. { ¦mī·krō‚fläp·ē ‚disk }

microflora [BOT] Microscopic plants. [ECOL] The flora of a microhabitat. { ¦mī·krō'flor·ə }

microfluid [FL MECH] A fluid in which the effects of local motion of contained material particles on properties and behavior of the fluid are not disregarded. { 'mī·krō‚flü·əd }

microfluoroscope [ENG] A fluoroscope in which a very fine-grained fluorescent screen is optically enlarged. { ¦mī·krō'flùr·ə‚skōp }

microforge [ENG] In micromanipulation techniques, an optical-mechanical device for controlling the position of needles or pipets in the field of a low-power microscope by a simple micromanipulator. { 'mī·krə‚fórj }

microform [GRAPHICS] A miniature replica of data, such as microfiche or microfilm. { 'mī·krə‚fòrm }

microfossil [PALEON] A small fossil which is studied and identified by means of the microscope. { ¦mī·krō'fäs·əl }

microgamete [BIOL] The smaller, or male gamete produced by heterogametic species. { ¦mī·krō'ga‚mēt }

microgametocyte [BIOL] A cell that gives rise to microgametes. { ¦mī·krō·gə'mēd·ə‚sīt }

microgametophyte [BOT] The male gametophyte in plants having two types of spores. { ¦mī·krō·gə'mēd·ə‚fīt }

microgammil See gammil. { ¦mī·krō'gam·əl }

microgamy [BIOL] Sexual reproduction by fusion of the small male and female gametes in certain protozoans and algae. { mī'kräg·ə·mē }

microgastria [MED] A condition of abnormal smallness of the stomach. { ¦mī·krō'gas·trē·ə }

microgenesis [BIOL] Abnormally small development of a part. { ¦mī·krō'jen·ə·səs }

microgenitalism [MED] Having abnormally small genitalia. { ¦mī·krō'jen·əd·əl‚iz·əm }

microgeography [GEOGR] The detailed empirical geographical study on a small scale of a specific locale. { ‚mī·krə·jē'äg·rə·fē }

microglia [NEUROSCI] Small neuroglia cells of the central nervous system having long processes and exhibiting ameboid and phagocytic activity under certain pathologic conditions. { mī'kräg·lē·ə }

microglossia [MED] A condition of abnormal smallness of the tongue. { ¦mī·krō'gläs·ē·ə }

micrognathia [MED] A condition of abnormal smallness of the jaws, particularly the mandible. { ¦mī·krō'nā·thē·ə }

microgram [MECH] A unit of mass equal to one-millionth of a gram. Abbreviated μg. { 'mī·krə‚gram }

micrograph [ENG] An instrument for making very tiny writing or engraving. [GRAPHICS] A graphic reproduction of the surface of a prepared specimen at a magnification greater than 10 diameters. { 'mī·krə‚graf }

micrographic [SCI TECH] Having graphic texture distinguishable only with the aid of a microscope. { ¦mī·krə‚graf·ik }

micrographics [GRAPHICS] The technology of reproducing

information on film in greatly reduced form, such as microfilm or microfiche, which requires enlargement to be readable. { ¦mī·krə‚graf·iks }

microgravity [MECH] A state of very weak gravity, such that the gravitational acceleration experienced by an observer inside the system in question is of the order of one-millionth of that on earth. { ‚mī·krō'grav·əd·ē }

microgroove record See long-playing record. { 'mī·krə‚grüv ‚rek·ərd }

microgyria [MED] A condition of abnormal smallness of the gyri of the brain. { ¦mī·krə‚jī·rē·ə }

microhabitat [ECOL] A small, specialized, and effectively isolated location. { ¦mī·krō'hab·ə‚tat }

microhardness [MET] Hardness of microscopic areas of a metal or alloy. { mī·krō'härd·nəs }

micro heat pipe [ENG] A very small heat pipe that has a diameter between about 100 micrometers and 2 millimeters (0.004 and 0.08 inch) and a triangular cross section or other cross section with sharp corners, and that uses the sharp corner regions instead of a wick to return the working fluid from the condenser to the evaporator; it has potential applications in the electronics (cooling circuit chips), medical, space, and aircraft industries. { ¦mī·krō 'hēt ‚pīp }

microheterogeneity [CHEM] A small variation in the chemical structure of a molecule that does not result in a significant change in properties. { ‚mī·krō‚hed·ə·rə·jə'nē·əd·ē }

microhm [ELEC] A unit of resistance, reactance, and impedance, equal to 10^{-6} ohm. { 'mī·krōm }

microholography See x-ray holography. { ¦mī·krō·hō'läg·rə·fē }

Microhylidae [VERT ZOO] A family of anuran amphibians in the suborder Diplasiocoela including many heavy-bodied forms with a pointed head and tiny mouth. { ‚mī·krō'hī·lə‚dē }

microhysteresis effect [SOLID STATE] Hysteresis that results from the motion of domain walls lagging behind an applied magnetic or elastic stress when these walls are held up by dislocations and other imperfections in the material. { ¦mī·krō‚his·tə'rē·səs i‚fekt }

microimage [COMPUT SCI] A single image stored on a microform medium. { 'mī·krō‚im·ij }

microincineration [CHEM] Reduction of small quantities of organic substances to ash by application of heat. { ¦mī·krō·in‚sin·ə'rā·shən }

microinfarct [MED] A very small infarct. { ¦mī·krō 'in‚färkt }

microinjection [CELL MOL] Injection of cells with solutions by using a micropipet. { ¦mī·krō·in'jek·shən }

microinstruction [COMPUT SCI] The portion of a microprogram that specifies the operation of individual computing elements and such related subunits as the main memory and the input/output interfaces; usually includes a next-address field that eliminates the need for a program counter. { ¦mī·krō· in'strək·shən }

microinterferometer [OPTICS] Functional combination of a microscope with an interferometer; used to study thin films, platings, or transparent coatings. { ¦mī·krō‚int·ə·fə'räm·əd· ər }

microinvasion [MED] Invasion by tumor, especially a squamous-cell carcinoma of the uterine cervix, a very short distance into the tissues beneath the point of origin. { mī·krō·in'vā· zhən }

microirradiation [BIOPHYS] A technique in which a laser beam is focused through the objective of a microscope onto a single cell to study the photosensitivity of its various parts. { ‚mī·krō·i‚rād·ē'ā·shən }

microjacket [GRAPHICS] A holder for microfilm consisting of two wavy sheets of transparent plastic that are bonded together to create tubular channels in which the strips are stored. { 'mī·krə‚jak·ət }

microlaser See single-atom laser. { 'mī·krō‚lā·zər }

microlaterolog [PETRO ENG] Modification of the downhole microlog in which extra electrodes focus electric current into a trumpet-shaped area; gives greater resistivity-measurement resolution than does the microlog. { ¦mī·krō'lad·ə·rə‚läg }

microlayer [OCEANOGR] The thin zone beneath the surface of the ocean or any free water surface within which physical processes are modified by proximity to the air-water boundary. { 'mī·krō‚lā·ər }

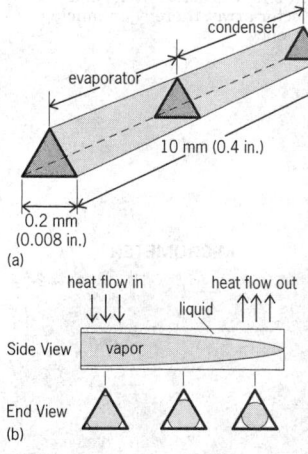

MICRO HEAT PIPE

condenser

evaporator

10 mm (0.4 in.)

0.2 mm (0.008 in.)

(a)

heat flow in　　heat flow out

liquid

Side View　vapor

End View

(b)

Diagrams of (a) concept and (b) operation.

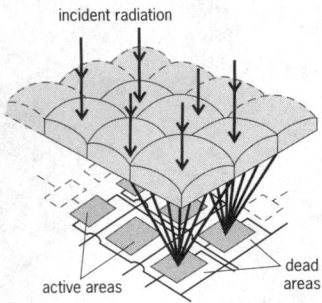

incident radiation

active areas

dead areas

The array focuses all of the incident light onto the active areas of a detector array, thereby enhancing efficiency.

MICROMETER

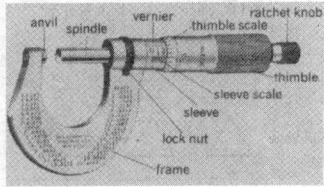

Machinist's outside caliper with micrometer reading 0.250 inch (6.35 millimeters), showing component parts. (*L. S. Starrett Co.*)

microlens array [OPTICS] An array of very small lenses with diameters between 20 micrometers and 1 millimeter; used in a variety of applications, including integral photography, photocopying, facsimile, and high-speed parallel switching networks. { 'mī·krō‚lenz ə‚rā }

microlensing [ASTRON] A phenomenon in which a foreground star acts as a gravitational lens when it happens to pass in front of a background star, causing the background starlight to brighten and bend through a ring-shaped region. { 'mī·krō‚lenz·iŋ }

Microlepidoptera [INV ZOO] A former division of Lepidoptera. { ¦mī·krō‚lep·ə'däp·tə·rə }

microlite [CRYSTAL] A microscopic crystal which polarizes light. Also known as microlith. [MINERAL] $(Na,Ca)_2$ $(Ta,Nb)_2O_6(O,OH,F)$ A pale-yellow, reddish, brown, or black isometric mineral composed of sodium calcium tantalum oxide with a small amount of fluorine; it is isomorphous with pyrochlore. Also known as djalmaite. { 'mī·krə‚līt }

microlith *See* microlite. [MED] A calculus of microscopic size. { 'mī·krə‚lith }

microlithiasis [MED] The presence of numerous microliths. { ¦mī·krō·li'thī·ə·səs }

microlithiasis alveolaris pulmonum [MED] A rare form of pulmonary calcification of unidentified etiology in which microliths, and larger osseous nodules, are found. { ¦mī·krō·li'thī·ə·səs ‚al·vē·ə'lar·əs pùl'mō·nəm }

microlithography [MATER] The transfer of a pattern or image from one medium to another, as from a mask to a wafer, with image features in the micrometer range or smaller. { ‚mī·krō·li'thäg·rə·fē }

microlithology [PETR] Microscopic study of the characteristics of rocks. { ¦mī·krō·li'thäl·ə·jē }

microlitic [PETR] Of the texture of a porphyritic igneous rock, having a groundmass composed of an aggregate of microlites in a generally glassy base. { ¦mī·krə‚lid·ik }

microlock [ELECTR] **1.** Satellite telemetry system that uses phase-lock techniques in the ground receiving equipment to achieve extreme sensitivity. **2.** A lock by a tracking station upon a minitrack radio transmitter. **3.** The system by which this lock is effected. { 'mī·krə‚läk }

microlog [PETRO ENG] A drill-hole resistivity log recorded with electrodes mounted at short distances from each other in the face of a rubber-padded microresistivity sonde. { 'mī·krə‚läg }

micromachine [MATER] A micrometer-size mechanical device; compared with an integrated circuit, it has some mechanical parts that stand above the substrate or move freely over it. { 'mī·krō·mə‚shēn }

micromachining [ENG] The use of standard semiconductor process technologies in combination with specially developed processes to fabricate miniature mechanical devices and components on silicon and other materials. { 'mī·krō·mə‚shēn·iŋ }

micromainframe [COMPUT SCI] A main frame of a computer placed on one or more integrated circuit chips. { ¦mī·krō'mān‚frām }

Micromalthidae [INV ZOO] A family of coleopteran insects in the superfamily Cantharoidea; the single species is the telephone pole beetle. { ‚mī·krō'mól·thə‚dē }

micromania [PSYCH] A delusional state in which the patient believes himself diminutive in size and mentally inferior. { ‚mī·krə'mā·nē·ə }

micromanipulation [BIOL] The techniques and practice of microdissection, microvivisection, microisolation, and microinjection. { ¦mī·krō·mə‚nip·yə'lā·shən }

micromanipulator [ENG] A device for holding and moving fine instruments for the manipulation of microscopic specimens under a microscope. { ¦mī·krō·mə'nip·yə‚lād·ər }

micromanometer [ENG] Any manometer that is designed to measure very small pressure differences. { ¦mī·krō·mə'näm· əd·ər }

micromaser *See* single-atom laser. { 'mī·krə‚māz·ər }

micromechanical display [ENG] A video display based on an array of mirrors on a silicon chip that can be deflected by electrostatic forces. Abbreviated MMD. { ‚mī·krō·mə‚kan· i·kəl di'splā }

micromechanics [ENG] **1.** The design and fabrication of micromechanisms. **2.** *See* composite micromechanics. { ¦mī·krō·mə'kan·iks }

micromechanism [ENG] A mechanical component with submillimeter dimensions and corresponding tolerances of the order of 1 micrometer or less. { ¦mī·krō'mek·ə‚niz·əm }

micromechatronics [ENG] The branch of engineering concerned with micro-electro-mechanical systems. { ¦mī·kro ‚mek·ə'trän·iks }

micromere [EMBRYO] A small blastomere of the upper or animal hemisphere in eggs that undergo uneven cleavage. { 'mī·krə‚mir }

micrometeorite [ASTRON] A very small meteorite or meteoritic particle with a diameter generally less than a millimeter. { ¦mī·krō'mē·də·ə‚rīt }

micrometeorite penetration [AERO ENG] Penetration of the thin outer shell (skin) of space vehicles by small particles traveling in space at high velocities. { ¦mī·krō'mē·də·ə‚rīt ‚pen·ə'trā·shən }

micrometeoroid [ASTRON] A very small meteoroid with diameter generally less than a millimeter. { ¦mī·krō'mē·də· ə‚róid }

micrometeorology [METEOROL] That portion of the science of meteorology that deals with the observation and explanation of the smallest-scale physical and dynamic occurrences within the atmosphere; studies are confined to the surface boundary layer of the atmosphere, that is, from the earth's surface to an altitude where the effects of the immediate underlying surface upon air motion and composition become negligible. { ¦mī· krō‚mē·dē·ə'räl·ə·jē }

micrometer [ENG] **1.** An instrument attached to a telescope or microscope for measuring small distances or angles. **2.** A caliper for making precise measurements; a spindle is moved by a screw thread so that it touches the object to be measured; the dimension can then be read on a scale. Also known as micrometer caliper. [MECH] A unit of length equal to one-millionth of a meter. Abbreviated μm. Also known as micron (μ). { mī'kräm·əd·ər }

micrometer caliper *See* micrometer. { mī'kräm·əd·ər 'kal· ə·pər }

micrometer of mercury *See* micron. { mī'kräm·əd·ər əv 'mər·kyə·rē }

micromicro- *See* pico-. { ¦mī·krō¦mī·krō }

micromicrowatt *See* picowatt. { ¦mī·krō¦mī·krō'wät }

micromini [COMPUT SCI] The central processing unit of a minicomputer placed on one of more integrated circuit chips. { ¦mī·krō'min·ē }

microminiature circuitry *See* microcircuitry. { ¦mī·krō¦min· ə·char 'sər·kə·trē }

microminiaturization *See* microelectronics. { ¦mī·krō‚min·ə· chə·rə'zā·shən }

micromodule [ELECTR] Cube-shaped, plug-in, miniature circuit composed of potted microelements; each microelement can consist of a resistor, capacitor, transistor, or other element, or a combination of elements.

micromolding [ENG] An alternative technique to micromachining for fabricating microsystems, in which a sacrificial material serves as a mold to which a deposited material conforms. { 'mī·krō‚mōld·iŋ }

Micromonospora [MICROBIO] A genus of bacteria in the family Micromonosporaceae; the mycelium is well developed, branched, and septate; single spores are formed on hyphae.

Micromonosporaceae [MICROBIO] A family of bacteria in the order Actinomycetales; aerial hyphae are formed in all genera except *Micromonospora;* saprophytic soil organisms.

Micromonospora purpurea [MICROBIO] The bacterium that produces the antibiotic gentamycin.

micromotion film [IND ENG] A record of a specific task made with motion picture film or video tape in which each component of the activity is recorded in an individual frame. { 'mī·krō‚mō·shən ‚film }

micromotion study *See* memomotion study. { ¦mī·krō¦mō· shən 'stəd·ē }

micromotor [MATER] A micromachine and forerunner of micro-electro-mechanical systems. { 'mī·krə‚mōd·ər }

micron [MECH] **1.** A unit of pressure equal to the pressure exerted by a column of mercury 1 micrometer high, having a density of 13.5951 grams per cubic centimeter, under the standard acceleration of gravity; equal to 0.133322387415 pascal; it differs from the millitorr by less than one part in seven million. Also known as micrometer of mercury. **2.** *See* micrometer. { 'mī‚krän }

micronekton [ECOL] Active pelagic crustaceans and other forms intermediate between thrusting nekton and feebler-swimming plankton. { ˌmī·krə'nek,tän }

micronized clay [MATER] A pure kaolin pulverized to a fineness of 400 to 800 mesh; used as a filler material in rubber. { 'mī·krə,nīzd 'klā }

micronized mica [MATER] Powdered mica of a fineness of 400 to 1000 mesh; used as a filler. { 'mī·krə,nīzd 'mī·kə }

micronucleus [INV ZOO] The smaller, reproductive nucleus in multinucleate protozoans. { ¦mī·krō'nü·klē·əs }

micronutrient [BIOCHEM] An element required by animals or plants in small amounts. { ¦mī·krō'nü·trē·ənt }

microoperation [COMPUT SCI] Any clock-timed step of an operation. { ¦mī·krō,äp·ə'rā·shən }

microoptics [OPTICS] A technology that utilizes optical elements that range in diameter from 20 micrometers to 1 millimeter. { ¦mī·krō'äp·tiks }

micro-opto-electro-mechanical system [ENG] A microsystem that combines the functions of optical, mechanical, and electronic components in a single, very small package or assembly. Abbreviated MOEMS. { ¦mī·krō¦äp·tō i¦lek·trō mə¦kan·ə·kəl 'sis·təm }

micro-opto-mechanical system [ENG] A microsystem that combines optical and mechanical functions without the use of electronic devices or signals. Abbreviated MOMS. { ¦mī·krō ¦op·to·mə¦kan·ə·kəl ,sis·təm }

microorganism [MICROBIO] A microscopic organism, including bacteria, protozoans, yeast, viruses, and algae. { ¦mī·krō'or·gə,niz·əm }

micropaleontology [PALEON] A branch of paleontology that deals with the study of microfossils. { ¦mī·krō,pā·lē·ən'täl·ə·jē }

micropane [GRAPHICS] A microphotograph on glass. { 'mī·krə,pān }

micropegmatite [PETR] Microcrystalline graphic granite. { ¦mī·krō'peg·mə,tīt }

microperf [COMPUT SCI] A type of continuous-feed computer paper having extremely small perforations along the separations and edges which give separated pages the appearance of standard typewriter paper. { 'mī·krə,pərf }

microperthite [MINERAL] Perthite in which the lamellae are visible only under the microscope. { ¦mī·krō'pər,thīt }

Micropezidae [INV ZOO] A family of myodarian cyclorrhaphous dipteran insects in the subsection Acalypteratae. { ¦mī·krō'pez·ə,dē }

microphage [HISTOL] A small phagocyte, especially a neutrophil. { 'mī·krə,fāj }

microphagy [BIOL] Feeding on minute organisms or particles. { mī'kräf·ə·jē }

microphakia [MED] Congenital condition of abnormal smallness of the crystalline lens. { ¦mī·krō'fā·kē·ə }

microphenic [GEN] Pertaining to genetic or environmental factors that are numerous but individually of small effect relative to the phenotypic standard deviation. { ¦mī·krə'fen·ik }

microphone [ENG ACOUS] An electroacoustic device containing a transducer which is actuated by sound waves and delivers essentially equivalent electric waves. { 'mī·krə,fōn }

microphone transducer [ENG ACOUS] A device which converts variation in the position or velocity of some body into corresponding variations of some electrical quantity, in a microphone. { 'mī·krə,fōn tranz'dü·sər }

microphonics [ELECTR] Noise caused by mechanical vibration of the elements of an electron tube, component, or system. Also known as microphonism. { ¦mī·krə'fän·iks }

microphonism See microphonics. { mī'krä·fə,niz·əm }

microphotograph [GRAPHICS] A microscopically small photograph, requiring magnification to be readable. { ¦mī·krə'fōd·ə,graf }

microphotometer [ENG] A photometer that provides highly accurate illumination measurements; in one form, the changes in illumination are picked up by a phototube and converted into current variations that are amplified by vacuum tubes. { ¦mī·krō·fə'täm·əd·ər }

microphthalmus [MED] A condition characterized by an abnormally small eyeball. Also known as nanophthalmus. { ¦mī,kräf'thal·məs }

microphyllous [BOT] 1. Having small leaves. 2. Having leaves with a single, unbranched vein. { ¦mī·krō¦fil·əs }

microphyric [PETR] Of the texture of an igneous rock, containing microscopic phenocrysts (longest dimension 0.2 millimeter). Also known as microporphyritic. { ¦mī·krō¦fir·ik }

Microphysidae [INV ZOO] A palearctic family of hemipteran insects in the subfamily Cimicimorpha. { ¦mī·krə'fīs·ə,dē }

microphyte [ECOL] 1. A microscopic plant. 2. A dwarfed plant due to unfavorable environmental conditions. { 'mī·krə,fīt }

micropinocytosis [CELL MOL] A mechanism of endocytosis in which fluid droplets are internalized by indentations (caveolae) on the surface membrane which pinch off as tiny internal vesicles (micropinosomes). { ¦mī·krō,pin·ə·sī'tō·səs }

micropinosome [CELL MOL] A very tiny vesicle that is pinched off from the plasma membrane of a cell during micropinocytosis. { ¦mī·krō'pin·ə,sōm }

micropipet [ENG] 1. A pipet with capacity of 0.5 milliliter or less, to measure small volumes of liquids with a high degree of accuracy; types include lambda, straight-bore, and Lang-Levy. 2. A fine-pointed pipette used for microinjection. { ¦mī·krō·pī'pet }

micropipette [MED] An extremely fine glass capillary tube with a narrow tip on the order of micrometers that is used to extract or deliver minute quantities of fluid. { ¦mī·krō·pī'pet }

microplankton [ECOL] Zooplankton between 20 and 200 micrometers in size. { 'mī·krə,plaŋk·tən }

micropoikilitic [PETR] Of the texture of an igneous rock, having poikilitic character visible only under the microscope. { ¦mī·krō,pȯi·kə'lid·ik }

micropore [CHEM] A pore in a catalytic material whose diameter is less than 2 nanometers. [GEOL] A pore small enough to hold water against the pull of gravity and to retard water flow. { 'mī·krə,pȯr }

microporosity [MET] Extremely fine porosity, visible only with the aid of a microscope. { ¦mī·krō·pə'räs·əd·ē }

microporous barrier [CHEM ENG] A metallic or plastic membrane with micrometer-sized pores used for dialysis and other membrane-separation processes. { ¦mī·krō'pȯr·əs 'bar·ē·ər }

microporphyritic See microphyric. { ¦mī·krō pȯr·fə'rid·ik }

microprism [OPTICS] A usually circular area in the focusing screen of a camera viewfinder that is made up of tiny prisms and causes the image in the viewfinder to blur if the subject is out of focus. { 'mī·krə,priz·əm }

microprobe [SPECT] An instrument for chemical microanalysis of a sample, in which a beam of electrons is focused on an area less than a micrometer in diameter, and the characteristic x-rays emitted as a result are dispersed and analyzed in a crystal spectrometer to provide a qualitative and quantitative evaluation of chemical composition. Also known as x-ray microprobe. { 'mī·krə,prōb }

microprobe spectrometry [SPECT] Microanalysis of a sample, using a microprobe. { 'mī·krə,prōb spek'träm·ə·trē }

microprocessing unit [ELECTR] A microprocessor with its external memory, input/output interface devices, and buffer, clock, and driver circuits. Abbreviated MPU. { ¦mī·krō'prä,ses·iŋ ,yü·nət }

microprocessor [ELECTR] A single silicon chip on which the arithmetic and logic functions of a computer are placed. { ¦mī·krō'prä,ses·ər }

microprocessor intertie and communication system [COMMUN] A data communications system which provides the communication network with its own dedicated processing resources and reduces in-terminal response time, compensating for the capacity used up by communications terminals. Abbreviated MICS. { mī·krō'prä,ses·ər 'in·tər,tī ən kə,myü·nə'kā·shən ,sis·təm }

microprogram [COMPUT SCI] A computer program that consists only of basic elemental commands which directly control the operation of each functional element in a microprocessor. { ¦mī·krō'prō·grəm }

microprogrammable instruction [COMPUT SCI] An instruction that does not refer to a core memory address and that can be microprogrammed, thus specifying various commands within one instruction. { ¦mī·krō·prə'gram·ə·bəl in'strək·shən }

microprogramming [COMPUT SCI] Transformation of a computer instruction into a sequence of elementary steps

MICRO-OPTO-ELECTRO-MECHANICAL SYSTEM

Scanning electron microscope image of a simple MOEMS tilting mirror, actuated by a micro-electro-mechanical rack-and-pinion drive that pushes the hinged mirror up from the right side. (*Sandia National Laboratories*)

(microinstructions) by which the computer hardware carries out the instruction. { ¦mī·krō'prō,gram·iŋ }

micropsia [MED] A visual disturbance in which objects appear undersized. { mī'kräp·sē·ə }

Micropterygidae [INV ZOO] The single family of the lepidopteran superfamily Micropterygoidea; members are minute moths possessing toothed, functional mandibles and lacking a proboscis. { mī,kräp·tə'rij·ə,dē }

Micropterygoidea [INV ZOO] A monofamilial superfamily of lepidopteran insects in the suborder Homoneura. { mī,kräp·tə·rə'góid·ē·ə }

micropulsation [GEOPHYS] A short-period geomagnetic variation in the range of about 0.2–600 seconds, typically exhibiting an oscillatory waveform. { ¦mī·krō·pəl'sā·shən }

micropump See electroosmotic driver. { 'mī·krə,pəmp }

micropycnometer [ENG] A small-volume pycnometer with a capacity from 0.25 to 1.6 milliliters; weighing precision is 1 part in 10,000, or better. { ,mī·krō·pik'näm·əd·ər }

Micropygidae [INV ZOO] A family of echinoderms in the order Diadematoida that includes only one genus, *Micropyga*, which has noncrenulate tubercles and umbrellalike outer tube feet. { mī·krə'pij·ə,dē }

micropyle [BOT] A minute opening in the integument at the tip of an ovule through which the pollen tube commonly enters; persists in the seed as an opening or a scar between the hilum and point of radicle. { 'mī·krə,pīl }

microquake See microearthquake.

microquasar [ASTRON] An object in the Milky Way Galaxy that contains a black hole or neutron star and emits jets of material that exhibit superluminal motion and resemble those emitted from quasars. { ¦mī·krō'kwā,zär }

microradiogram [PHYS] A two-dimensional x-ray image of a sample, produced by one type of x-ray microscope used in microradiography; all levels of the sample object are imaged into essentially a single focal plane for subsequent microphotographic enlargement. { ¦mī·krō'rād·ē·ə,gram }

microradiograph [GRAPHICS] An enlarged radiographic image on photographic film produced either by increasing the distance from specimen to photographic plate to secure inherent enlargement of divergent x-ray beams, or by optical enlargement of a developed image. { ¦mī·krō'rād·ē·ə,graf }

microradiography [ANALY CHEM] Technique for the study of surfaces of solids by monochromatic-radiation (such as x-ray) contrast effects shown via projection or enlargement of a contact radiograph. [GRAPHICS] The radiography of small objects having details too fine to be seen by the unaided eye, with optical enlargement of the resulting negative. { ¦mī·krō,rād·ē'äg·rə·fē }

microradiometer [ELECTR] A radiometer used for measuring weak radiant power, in which a thermopile is supported on and connected directly to the moving coil of a galvanometer. Also known as radiomicrometer. { ¦mīkrō,rād·ē'äm·əd·ər }

microreactor [CHEM ENG] A microsystem for chemical and biochemical reactions, including separation, fluid handling, and unit operations of chemical engineering, as well as analytical systems. Its small reaction volumes and high heat and mass transfer rates allow for precise adjustment of process conditions, short response times, and defined residence times, resulting in greater process control and higher yields and selectivity. { ¦mī·krō·rē'ak·tər }

micro-reciprocal-degree See mired. { ¦mī·krō ri'sip·rə·kəl di'grē }

microrefractometry [OPTICS] The measurement of refractive indices of microscopic objects; this is often done by immersing an object in a series of mediums of graded refractive index until one is found that makes the object invisible in a phase-contrast microscope. { ¦mī·krō,rē,frak'täm·ə·trē }

microrelief [GEOGR] Irregularities of the land surface causing variations in elevation amounting to no more than a few feet. { 'mī·krō·ri,lēf }

microresistivity survey [PETRO ENG] General term for downhole resistivity surveys of oil-bearing formations; includes microlog and microlaterolog surveys. { ¦mī·krō,rē,zis'tiv·əd·ē 'sər,vā }

microrheology [MECH] A branch of rheology in which the heterogeneous nature of dispersed systems is taken into account. { ¦mī·krō·rē'äl·ə,jē }

microsatellite deoxyribonucleic acid [CELL MOL] Any of a series of repeated motifs of a two to six base-pair sequence

that are scattered throughout the nuclear genome and used as landmarks during physical mapping. Abbreviated msDNA. { ,mī·krō¦sad·əl,īt dē,äk·sē,rī·bō·nü,klē·ik 'as·əd }

Microsauria [PALEON] An order of Carboniferous and early Permian lepospondylous amphibians. { ,mī·krō'sór·ē·ə }

microscale motion [GEOPHYS] Local motion of winds and ocean currents over areas with sizes of 300 feet (100 meters) or less. { 'mī·krə,skāl ,mō·shən }

microsclere [INV ZOO] A minute sclerite in Porifera. { 'mī·krə,sklir }

microscope [OPTICS] An instrument through which minute objects are enlarged by means of a lens or lens system; principal types include optical, electron, and x-ray. { 'mī·krə,skōp }

microscope stage [OPTICS] The platform on which specimens are placed for microscopic examination. { 'mī·krə,skōp ,stāj }

microscopic [OPTICS] See microscopical. [SCI TECH] Of extremely small size. { ¦mī·krə¦skäp·ik }

microscopical [OPTICS] Also known as microscopic. **1.** Of or pertaining to the microscope. **2.** Visible only under a microscope. { ,mi·krə¦skäp·ə·kəl }

microscopical diagnosis [PATH] Identification of a disease by microscopic examination of specimens taken from the patient. { ,mi·krə¦skäp·ə·kəl ,dī·əg'nō·səs }

microscopic anisotropy [PETRO ENG] Phenomenon in electrical downhole logging wherein electric current flows most easily along the water-filled interstices, usually parallel to sedimentary bed strata. { ¦mī·krə¦skäp·ik ,an·ə'sä·trə·pē }

microscopic reversibility [STAT MECH] A principle which requires that in a system at equilibrium any molecular process and its reverse take place at the same average rate. Also known as reversibility principle. { ¦mī·krə¦skäp·ik ri,vər·sə'bil·əd·ē }

microscopic state [STAT MECH] The state of a system as specified by the actual properties of each individual, elemental component, in the ultimate detail permitted by the uncertainty principle. Also known as microstate. { ¦mī·krə¦skäp·ik 'stāt }

microscopic stress [MET] Residual stress ranging from compression to tension in a metal within a distance often comparable to the grain size. Also known as microstress. { ¦mī·krə¦skäp·ik 'stres }

microscopic theory [PHYS] A theory concerned with the interactions of atoms, molecules, or their constituents, involving distances on the order of 10^{-10} meter or less, which underlie observable phenomena. { ¦mī·krə¦skäp·ik 'thē·ə·rē }

microscopist [SCI TECH] An individual skilled in the use of the microscope. { mī'kräs·kə,pist }

microscopy [OPTICS] The interpretive application of microscope magnification to the study of materials that cannot be properly seen by the unaided eye. { mī'kräs·kə·pē }

microsecond [MECH] A unit of time equal to one-millionth of a second. Abbreviated μs. { ¦mī·krə,sek·ənd }

microsegregation [MET] Segregation within a grain, crystal, or particle of microscopic size. { ¦mī·krō,seg·rə'gā·shən }

microseism [GEOPHYS] A weak, continuous, oscillatory motion in the earth having a period of 1–9 seconds and caused by a variety of agents, especially atmospheric agents; not related to an earthquake. { 'mī·krə,sīz·əm }

microseismic instrument [MIN ENG] An instrument for the study of roof strata and supports; it is inserted in holes, drilled at selected points, for listening to subaudible vibrations that precede rock failure. { ¦mī·krə¦sīz·mik 'in·strə·mənt }

microsensor [ENG] A submicrometer- to millimeter-size device that converts a nonelectrical physical or chemical quantity, such as pressure, acceleration, temperature, or gas concentration, into an electrical signal; it is generally able to offer better sensitivity, accuracy, dynamic range, and reliability, as well as lower power consumption, compared to larger counterparts. { 'mī·krō,sen·sər }

microseptum [INV ZOO] An incomplete or imperfect mesentery in zoantharians. { ¦mī·krō'sep·təm }

microshrinkage [MET] A casting defect consisting of interdendritic voids, visible only at magnifications over 10 diameters. { ¦mī·krō'shriŋ·kij }

microsilica See silica fume. { ,mī·krō'sil·ə·kə }

microsome [CYTOL] **1.** A fragment of the endoplasmic reticulum. **2.** A minute granule of protoplasm. { 'mī·krə,sōm }

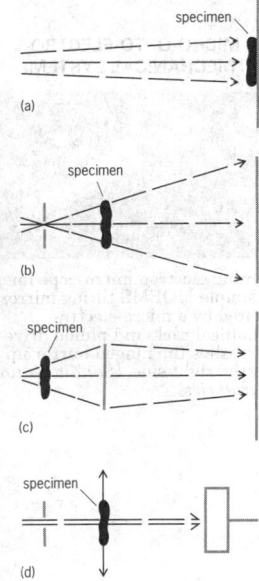

MICRORADIOGRAPHY

specimen

(a)

specimen

(b)

specimen

(c)

specimen

(d)

Schematics of techniques. X-ray paths are indicated by broken lines. (*a*) Contact. (*b*) Projection. (*c*) Imaging. (*d*) Scanning.

MICROSAURIA

2.5 cm

Microbrachis, a microsaur from the Late Pennsylvanian epoch.

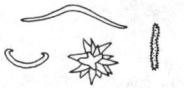

MICROSCLERE

Shapes of various microscleres.

microspec function [COMPUT SCI] The set of microinstructions which performs a specific operation in one or more machine cycles. { 'mī·krə,spek ,fəŋk·shən }

microspecies [ECOL] A small, localized species population that is clearly differentiated from related forms. Also known as jordanon. { |mī·krō'spē·shēz }

microspectrograph [SPECT] A microspectroscope provided with a photographic camera or other device for recording the spectrum. { |mī·krō'spek·trə,graf }

microspectrophotometer [SPECT] A split-beam or double-beam spectrophotometer including a microscope for the localization of the object under study, and capable of carrying out spectral analyses within the dimensions of a single cell. { |mī·kro|spek·trə·fə'täm·əd·ər }

microspectroscope [SPECT] An instrument for analyzing the spectra of microscopic objects, such as living cells, in which light passing through the sample is focused by a compound microscope system, and both this light and the light which has passed through a reference sample are dispersed by a prism spectroscope, so that the spectra of both can be viewed simultaneously. { |mī·krō'spek·trə,skōp }

microsphere [MATER] A sphere sized from about 0.5 to 100 micrometers and made of any material. { 'mī·krə,sfir }

microspherulitic [PETR] Of the texture of an igneous rock, having spherulitic character visible only under the microscope. { |mī·krō,sfer·ə'lüd·ik }

microspike [CELL MOL] Any of the narrow cytoplasmic projections that extend or retract from the surface of a cell and may have a sensory function. { 'mi·krə,spīk }

Microsporaceae [BOT] A monogeneric family of green algae in the suborder Ulotrichineae; the chloroplast is a parietal network. { ,mī·krō·spə'rās·ē,ē }

microsporangium [BOT] A sporangium bearing microspores. { |mī·krō·spə'ran·jē·əm }

microspore [BOT] The smaller spore of heterosporous plants; gives rise to the male gametophyte. { 'mī·krə,spór }

microspore mother cell See microsporocyte. { 'mī·krə,spór 'məth·ər ,sel }

Microsporida [INV ZOO] The single order of the class Microsporidea. { ,mī·krə'spór·ə·də }

Microsporidae [INV ZOO] The equivalent name for Sphaeriidae. { ,mī·krə'spór·ə,dē }

Microsporidea [INV ZOO] A class of Cnidospora characterized by the production of minute spores with a single intrasporal filament or one or two intracapsular filaments and a single sporoplasm; mainly intracellular parasites of arthropods and fishes. { ,mī·krə·spə'rid·ē·ə }

microsporidiosis [VET MED] Infection with microsporidians. { ,mī·krō·spə,rid·ē'ō·səs }

microsporocyte [BOT] A diploid cell from which four microspores are produced by meiosis. Also known as microspore mother cell. { 'mī·krō·spór·ə,sīt }

microsporogenesis [BOT] In angiosperms, formation of microspores and production of the male gametophyte. { ,mī·krə,spór·ə'jen·ə·səs }

microsporophyll [BOT] A sporophyll bearing microsporangia. { 'mī·krō'spór·ə,fil }

microspot See redox blemish. { 'mī·krō,spät }

microstate See microscopic state. { 'mī·krō,stāt }

microstress See microscopic stress. { 'mī·krə,stres }

microstrip [ELECTROMAG] A strip transmission line that consists basically of a thin-film strip in intimate contact with one side of a flat dielectric substrate, with a similar thin-film ground-plane conductor on the other side of the substrate. { 'mī·krə,strip }

microstructure [SCI TECH] The structure of an object, organism, or material as revealed by a microscope of a magnification over 10 times. { 'mī·krō'strək·chər }

microstylolite [PETR] A stylolite in which the surface relief is less than 1 millimeter. { ,mī·krə'stīl·ə,līt }

microsurgery [BIOL] Surgery on single cells by micromanipulation. { |mī·krō'sər·jə·rē }

microsystem See micro-electro-mechanical system. { 'mī·krō,sis·təm }

microsystem electronics See microelectronics. { 'mī·krə,sis·təm i,lek'trän·iks }

Microtatobiotes [BIOL] An artificial taxonomic category, comprising two unrelated groups of biological entities, the rickettsiae and the viruses. { mī|krád·ə·dō,bī'ōd·ēz }

microtectonics See structural petrology. { |mī·krō,tek'tän·iks }

microtektite [GEOL] An extremely small tektite, 1 millimeter or less in diameter. { ,mī·krə'tek,tīt }

microtherm [ECOL] A plant requiring a mean annual temperature range of 0–14°C for optimum growth. { 'mī·krə,thərm }

microthermal climate [CLIMATOL] A temperature province in both of C.W. Thornthwaite's climatic classifications, generally described as a "cool" or "cold winter" climate. { 'mī·krə'thər·məl 'klī·mət }

microthrowing power [PHYS CHEM] Relative ability of an electroplating solution to deposit metal in a small, shallow aperture or crevice not exceeding a few thousandths of an inch in dimensions. { 'mī·krō'thrō·iŋ ,pau·ər }

Microtinae [VERT ZOO] A subfamily of rodents in the family Muridae that includes lemmings and muskrats. { mī'krät·ən,ē }

microtome [ENG] An instrument for cutting thin sections of tissues or other materials for microscopical examination. { 'mī·krə,tōm }

microtomy [BIOL] Cutting of thin sections of specimens with a microtome. { mī'kräd·ə·mē }

microtrabecular lattice [CELL MOL] A network of thin filaments that interconnect the cytoplasmic filaments. { ,mī·krə·trə|bek·yə·lər 'lad·əs }

Microtragulidae [PALEON] A group of saltatorial caenolistoid marsupials that appeared late in the Cenozoic and paralleled the small kangaroos of Australia. { ,mī·krō·trə'gyül·ə,dē }

microtrichia [INV ZOO] Small hairs on the integument of various insects, especially on the wings. { ,mī·krō'trik·ē·ə }

microtron [NUCLEO] A type of circular particle accelerator for accelerating electrons to energies of several megaelectron-volts, in which the time of successive revolutions of the particles increases by exactly one cycle of the accelerating radio-frequency voltage, so that synchronism is maintained. { 'mī·krə,trän }

microtubule [CELL MOL] One of the hollow tubelike filaments found in certain cell components, such as cilia and the mitotic spindle, and composed of repeating subunits of the protein tubulin. { 'mī·krō'tüb·yül }

microtubule-associated protein [CELL MOL] A protein that enhances the rate and extent of the polymerization of intracellular microtubules or that modifies their properties once formed. Abbreviated MAP. { 'mī·krō,tüb·yül a|sō·sē,ād·ed 'prō,tēn }

microvillus [CELL MOL] One of the filiform processes that form a brush border on the surfaces of certain specialized cells, such as intestinal epithelium. { 'mī·krō'vil·əs }

Microviridae [VIROL] A family of nontailed bacterial viruses (bacteriophages) that is characterized by icosahedral particles containing a single-stranded circular deoxyribonucleic acid genome; members infect enterobacteria. { ,mī·krō'vir·ə,dī }

Microvirus [VIROL] The only genus of the family Microviridae. { 'mī·krō,vī·rəs }

microvitrain [GEOL] A coal lithotype; fine vitrain-like lenses or laminae in clarain. { 'mī·krō'vi,trān }

microvolt [ELEC] A unit of potential difference equal to one-millionth of a volt. Abbreviated μV. { 'mī·krə,vōlt }

microvoltmeter [ELECTR] A voltmeter whose scale is calibrated to indicate voltage values in microvolts. { |mī·krō'vōlt,mēd·ər }

microvolts per meter [ELECTROMAG] Field strength of antenna which is the ratio of the antenna voltage in microvolts to the antenna length in meters, as measured at a given point. { 'mī·krə,vols pər 'mēd·ər }

microwatt [MECH] A unit of power equal to one-millionth of a watt. Abbreviated μW. { 'mī·krə,wät }

microwave [ELECTROMAG] An electromagnetic wave which has a wavelength between about 0.3 and 30 centimeters, corresponding to frequencies of 1–100 gigahertz; however, there are no sharp boundaries distinguishing microwaves from infrared and radio waves. { 'mī·krə,wāv }

microwave acoustics [ACOUS] The production and study of elastic vibrations in materials at microwave frequencies, on the order of 10^5 to 10^{11} hertz, such as in single-crystal delay lines used in radar systems. { 'mī·krə,wāv ə'küs·tiks }

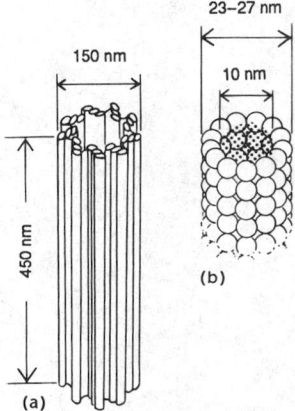

MICROTUBULE

23–27 nm

150 nm

10 nm

450 nm

(a)   (b)

Diagram of mature centriole showing (a) arrangement of the microtubules and (b) structure of a microtubule.

microwave amplification by stimulated emission of radiation *See* maser. { 'mī·krə,wāv ,am·plə·fə'kā·shən bī 'stim·yə,lād·əd i'mish·ən əv ,rād·ē'ā·shən }

microwave amplifier [ELECTR] A device which increases the power of microwave radiation. { 'mī·krə,wāv 'am·plə,fī·ər }

microwave antenna [ELECTROMAG] A combination of an open-end waveguide and a parabolic reflector or horn, used for receiving and transmitting microwave signal beams at microwave repeater stations. { 'mī·krə,wāv an'ten·ə }

microwave attenuator [ELECTROMAG] A device that causes the field intensity of microwaves in a waveguide to decrease by absorbing part of the incident power; usually consists of a piece of lossy material in the waveguide along the direction of the electric field vector. { 'mī·krə,wāv ə'ten·yə,wād·ər }

microwave background *See* cosmic microwave radiation. { 'mī·krə,wāv 'bak,graúnd }

microwave bridge [ELECTROMAG] A microwave circuit equivalent to an ordinary electrical bridge and used to measure impedance; consists of six waveguide sections arranged to form a multiple junction. { 'mī·krə,wāv ,brij }

microwave cavity *See* cavity resonator. { 'mī·krə,wāv ,kav·ə·dē }

microwave circuit [ELECTROMAG] Any particular grouping of physical elements, including waveguides, attenuators, phase changers, detectors, wavemeters, and various types of junctions, which are arranged or connected together to produce certain desired effects on the behavior of microwaves. { 'mī·krə,wāv ,sər·kət }

microwave circulator *See* circulator. { 'mī·krə,wāv 'sər·kyə,lād·ər }

microwave communication [COMMUN] Transmission of messages using highly directional microwave beams, which are generally relayed by a series of microwave repeaters spaced up to 50 miles (80 kilometers) apart. { 'mī·krə,wāv kə,myü·nə'kā·shən }

microwave detector [ELECTR] A device that can demonstrate the presence of a microwave by a specific effect that the wave produces, such as a bolometer, or a semiconductor crystal making a pinpoint contact with a tungsten wire. { 'mī·krə,wāv di,tek·tər }

microwave device [ELECTR] Any device capable of generating, amplifying, modifying, detecting, or measuring microwaves, or voltages having microwave frequencies. { 'mī·krə,wāv di,vīs }

microwave early warning [ENG] High-power, long-range radar with a number of indicators, giving high resolution, and with a large traffic-handling capacity; used for early warning of missiles. { 'mī·krə,wāv ,ər·lē 'wor·niŋ }

microwave filter [ELECTROMAG] A device which passes microwaves of certain frequencies in a transmission line or waveguide while rejecting or absorbing other frequencies; consists of resonant cavity sections or other elements. { 'mī·krə,wāv ,fil·tər }

microwave frequency [PHYS] A frequency on the order of 10^9–10^{11} hertz. { 'mī·krə,wāv ,frē·kwən·sē }

microwave generator *See* microwave oscillator. { 'mī·krə,wāv 'jen·ə,rād·ər }

microwave gyrator *See* gyrator. { 'mī·krə,wāv 'jī,rād·ər }

microwave heating [ELECTROMAG] Heating of food by means of electromagnetic energy in or just below the microwave spectrum for cooking, dehydration, sterilization, thawing, and other purposes. { 'mī·krə,wāv 'hēd·iŋ }

microwave hop [COMMUN] A microwave communications channel between two stations with directive antennas that are aimed at each other. { 'mī·krə,wāv 'häp }

microwave impedance measurement [ENG] The determination of parameters, associated with microwave propagation in transmission lines or waveguides, which are generalizations of the impedance concept at lower frequencies and are derived from ratios of electric- or magnetic-field amplitudes. { 'mī·krə,wāv im'pēd·əns ,mezh·ər·mənt }

microwave integrated circuit [ELECTR] A microwave circuit that uses integrated-circuit production techniques involving such features as thin or thick films, substrates, dielectrics, conductors, resistors, and microstrip lines, to build passive assemblies on a dielectric. Abbreviated MIC. { 'mī·krə,wāv 'int·ə,grād·əd 'sər·kət }

microwave landing system [NAV] A system of ground equipment which generates guidance beams at microwave frequencies for guiding aircraft to landings; it is intended to replace the present lower-frequency instrument landing system. Abbreviated MLS. { 'mī·krə,wāv 'land·iŋ ,sis·təm }

microwave link *See* microwave repeater. { 'mī·krə,wāv ,liŋk }

microwave maser [PHYS] A maser which emits microwave radiation. { 'mī·krə,wāv 'mā·zər }

microwave network [COMMUN] A series of microwave repeaters, spaced up to 50 miles (80 kilometers) apart, which relay messages over long distances using highly directional microwave beams. { 'mī·krə,wāv 'net,wərk }

microwave noise standard [ENG] An electrical noise generator of calculable intensity that is used to calibrate other noise sources by using comparison methods. { 'mī·krə,wāv 'noiz ,stan·dərd }

microwave optics [ELECTROMAG] The study of those properties of microwaves which are analogous to the properties of light waves in optics. { 'mī·krə,wāv 'äp,tiks }

microwave oscillator [ELECTR] A type of electron tube or semiconductor device used for generating microwave radiation or voltage waveforms with microwave frequencies. Also known as microwave generator. { 'mī·krə,wāv 'äs·ə,lād·ər }

microwave oven [ENG] An oven that uses microwave heating for fast cooking of meat and other foods. { 'mī·krə,wāv 'əv·ən }

microwave position-fixing [NAV] A navigation system using three shore-based unattended transmitters and a ship-based rotary radar-type antenna that scans the signals from the three shore stations; equipment aboard ship permits a three-point-fix type of plot. { 'mī·krə,wāv pə'zish·ən,fik·siŋ }

microwave pumping [ELECTROMAG] The use of microwaves to produce large departures from thermal equilibrium in the relative populations of selected quantized states of different energy in atomic, molecular, or nuclear systems. { 'mī·krə,wāv 'pəmp·iŋ }

microwave radiometer *See* radiometer. { 'mī·krə,wāv ,rād·ē'äm·əd·ər }

microwave receiver [ELECTR] Complete equipment that is needed to convert modulated microwaves into useful information. { 'mī·krə,wāv ri'sē·vər }

microwave reflectometer [ELECTROMAG] A pair of single-detector couplers on opposite sides of a waveguide, one of which is positioned to monitor transmitted power, and the other to measure power reflected from a single discontinuity in the line. { 'mī·krə,wāv ,rē,flek'täm·əd·ər }

microwave refractometer [ELECTROMAG] An instrument that measures the index of refraction of the atmosphere by measuring the travel time of microwave signals through each of two precision microwave transmission cavities, one of which is hermetically sealed to serve as a reference. { 'mī·krə,wāv ,rē,frak'täm·əd·ər }

microwave relay *See* microwave repeater. { 'mī·krə,wāv 'rē,lā }

microwave repeater [COMMUN] A tower equipped with a receiver and transmitter for picking up, amplifying, and passing on in either direction the signals sent over a microwave network by highly directional microwave beams. Also known as microwave link; microwave relay. { 'mī·krə,wāv ri'pēd·ər }

microwave resonance cavity *See* cavity resonator. { 'mī·krə,wāv 'rez·ən·əns ,kav·əd·ē }

microwave solid-state device [ELECTR] A semiconductor device for the generation or amplification of electromagnetic energy at microwave frequencies. { 'mī·krə,wāv ¦säl·əd ¦stāt di'vīs }

microwave spectrometer [SPECT] An instrument which makes a graphical record of the intensity of microwave radiation emitted or absorbed by a substance as a function of frequency, wavelength, or some related variable. { 'mī·krə,wāv spek'träm·əd·ər }

microwave spectroscope [SPECT] An instrument used to observe the intensity of microwave radiation emitted or absorbed by a substance as a function of frequency, wavelength, or some related variable. { 'mī·krə,wāv 'spek·trə,skōp }

microwave spectroscopy [SPECT] The methods and techniques of observing and the theory for interpreting the selective absorption and emission of microwaves at various frequencies by solids, liquids, and gases. { 'mī·krə,wāv spek'träs·kə·pē }

microwave spectrum [ELECTROMAG] The range of wavelengths or frequencies of electromagnetic radiation that are designated microwaves. [SPECT] A display, photograph, or plot of the intensity of microwave radiation emitted or absorbed by a substance as a function of frequency, wavelength, or some related variable. { 'mī·krə‚wāv 'spek·trəm }

microwave thermography [MED] A method of measuring temperature through the detection of microwave radiation emitted from heated tissue during a therapeutic procedure. { 'mī·krə‚wāv thər'mäg·rə·fē }

microwave transmission line [ELECTROMAG] A material structure forming a continuous path from one place to another and capable of directing the transmission of electromagnetic energy along this path. { 'mī·krə‚wāv tranz'mish·ən ‚līn }

microwave tube [ELECTR] A high-vacuum tube designed for operation in the frequency region from approximately 3000 to 300,000 megahertz. { 'mī·krə‚wāv ‚tüb }

microwave waveguide See waveguide. { 'mī·krə‚wāv 'wāv‚gīd }

microwave wavemeter [ELECTROMAG] Any device for measuring the free-space wavelengths (or frequencies) of microwaves; usually made of a cavity resonator whose dimensions can be varied until resonance with the microwaves is achieved. { 'mī·krə‚wāv 'wāv‚mēd·ər }

microzoospermia [MED] A condition of abnormal smallness of sperm in the semen. { ‚mī·krō‚zō·ō'spər·mē·ə }

micrurgy [SCI TECH] The art and science of using minute tools in a magnified field. { 'mī·krər·jē }

MICS See microprocessor intertie and communication system.

mictic [BIOL] **1.** Requiring or produced by sexual reproduction. **2.** Of or pertaining to eggs which without fertilization develop into males and with fertilization develop into amictic females, as occurs in rotifers. { 'mik·tik }

micturition See urination. { ‚mik·chə'ri·shən }

Midas [AERO ENG] A two-object trajectory-measuring system whereby two complete cotar antenna systems and two sets of receivers at each station, with the multiplexing done after phase comparison, are utilized in tracking more than one object at a time. { 'mīd·əs }

mid-Atlantic ridge [GEOL] The mid-oceanic ridge in the Atlantic. { ‚mid·ət'lan·tik 'rij }

midaxillary line [ANAT] A perpendicular line drawn downward from the apex of the axilla. { ‚mid'ak·sə‚ler·ē 'līn }

midbrain [ANAT] Those portions of the adult brain derived from the embryonic midbrain. [EMBRYO] The middle portion of the embryonic vertebrate brain. Also known as mesencephalon. { 'mid‚brān }

midclavicular line [ANAT] A vertical line parallel to and midway between the midsternal line and a vertical line drawn downward through the outer end of the clavicle. { 'mid·klə‚vik·yə·lər 'līn }

midcourse correction [AERO ENG] A change in the course of a spacecraft some time between the end of the launching phase and some arbitrary point when terminal guidance begins. { 'mid‚kȯrs kə'rek·shən }

middle body [NAV ARCH] That portion of the ship adjacent to the midship section. { 'mid·əl 'bäd·ē }

Middle Cambrian [GEOL] The geologic epoch occurring between Upper and Lower Cambrian, beginning approximately 540,000,000 years ago. { 'mid·əl 'kam·brē·ən }

middle clouds [METEOROL] Types of clouds the mean level of which is between 6500 and 20,000 feet (1980 and 6100 meters); the principal clouds in this group are altocumulus and altostratus. { 'mid·əl 'klaúdz }

middle core [COMPUT SCI] The locations with medium addresses in a computer's main storage; usually assigned to workspace for application programs. { 'mid·əl 'kȯr }

Middle Cretaceous [GEOL] The geologic epoch between the Upper and Lower Cretaceous, beginning approximately 120,000,000 years ago. { 'mid·əl krə'tā·shəs }

Middle Devonian [GEOL] The geologic epoch occurring between the Upper and Lower Devonian, beginning approximately 385,000,000 years ago. { 'mid·əl di'vō·nē·ən }

middle ear [ANAT] The middle portion of the ear in higher vertebrates; in mammals it contains three ossicles and is separated from the external ear by the tympanic membrane and from the inner ear by the oval and round windows. { 'mid·əl 'ir }

middle ground [NAV] A shoal with a channel on either side of it in a navigable part of a river, port, or harbor. { 'mid·əl 'graúnd }

middle ground buoy [NAV] In maritime operations, one of the buoys placed at each end of a middle ground, that is, a shoal with channels on both sides. { 'mid·əl 'graúnd 'bȯi }

Middle Jurassic [GEOL] The geologic epoch occurring between the Upper and Lower Jurassic, beginning approximately 170,000,000 years ago. { 'mid·əl jə'ras·ik }

middle lamella [CELL MOL] The layer of a cell wall that is derived from the phragmoplast. { 'mid·əl lə'mel·ə }

middle latitude Also known as mid-latitude. [GEOGR] A point of latitude that is midway on a north-and-south line between two parallels. [NAV] The latitude at which the arc length of the parallel separating the meridians passing through two specific points is exactly equal to the departure in proceeding from one point to the other by middle-latitude sailing. { 'mid·əl 'lad·ə‚tüd }

middle-latitude sailing [NAV] A method of converting departure into difference of longitude, or vice versa, when the true course is not 090° or 270° by assuming that such a course is steered at the middle latitude. { 'mid·əl 'lad·ə‚tüd ‚sāl·iŋ }

middle-latitude westerlies See westerlies. { 'mid·əl 'lad·ə‚tüd 'wes·tər‚lēz }

middle lobe syndrome [MED] A complex of symptoms due to enlarged lymph nodes compressing the bronchus of the right middle lobe, causing atelectasis, bronchiectasis, or chronic pneumonitis of the lobe. { 'mid·əl 'lōb 'sin‚drōm }

middle marker [NAV] That instrument-landing-system marker between the outer and inner markers. { 'mid·əl 'märk·ər }

Middle Mississippian [GEOL] The geologic epoch between the Upper and Lower Mississippian. { 'mid·əl ‚mis·ə'sip·ē·ən }

Middle Ordovician [GEOL] The geologic epoch occurring between the Upper and Lower Ordovician, beginning approximately 460,000,000 years ago. { 'mid·əl ‚ȯr·də'vish·ən }

Middle Pennsylvanian [GEOL] The geologic epoch between the Upper and Lower Pennsylvanian. { 'mid·əl ‚pen·səl'vā·nyə }

Middle Permian [GEOL] The geologic epoch occurring between the Upper and Lower Permian, beginning approximately 260,000,000 years ago. { 'mid·əl 'pər·mē·ən }

Middle Silurian [GEOL] The geologic epoch between the Upper and Lower Silurian. { 'mid·əl si'lür·ē·ən }

middle-third rule [CIV ENG] The rule that no tension is developed in a wall or foundation if the resultant force lies within the middle third of the structure. { 'mid·əl 'thərd ‚rül }

middle tones [GRAPHICS] In a black-and-white photograph or reproduction, the range of tones between highlights and shadows. { 'mid·əl 'tōnz }

Middle Triassic [GEOL] The geologic epoch occurring between the Upper and Lower Triassic, beginning approximately 215,000,000 years ago. { 'mid·əl trī'as·ik }

middle-ultraviolet lamp [ELECTR] A mercury-vapor lamp designed to produce radiation in the wavelength band from 2800 to 3200 angstrom units (280 to 320 nanometers) such as sunlamps and photochemical lamps. { 'mid·əl 'əl·trə‚vī·lət 'lamp }

middleware [MATH] Software that allows different computer programs used in a corporate network to work together. { 'mid·əl‚wer }

middling [MIN ENG] An ore product intermediate in mineral content between a concentrate and a tailing. { 'mid·liŋ }

middlings See sharps. { 'mid·liŋz }

mid-extreme tide [OCEANOGR] A level midway between the extreme high water and extreme low water occurring at a place. { 'mid ik‚strēm 'tīd }

midfan [GEOL] The portion of an alluvial fan between the fanhead and the outer, lower margins. { 'mid‚fan }

mid-frequency gain [ELECTR] The maximum gain of an amplifier, when this gain depends on the frequency; for an RC-coupled voltage amplifier the gain is essentially equal to this value over a large range of frequencies. { 'mid‚frē·kwən·sē ‚gān }

midge [INV ZOO] Any of various dipteran insects, principally of the families Ceratopogonidae, Cecidomyiidae, and Chironomidae; many are biting forms and are vectors of parasites of man and other vertebrates. { mij }

midget [MED] An individual who is abnormally small, but otherwise normal. { 'mij·ət }

midget impinger [MIN ENG] A dust-sampling impinger requiring only a 12-inch (30-centimeter) head of water for its operation. { 'mij·ət im'pin·jər }

midgut [EMBRYO] The middle portion of the digestive tube in vertebrate embryos. Also known as mesenteron. [INV ZOO] The mesodermal intermediate part of an invertebrate intestine. { 'mid,gət }

MIDI See Musical Instrument Digital Interface. { 'mid·ē }

midicomputer [COMPUT SCI] A computer having greater performance and capacity than a minicomputer and less than that of a mainframe. { 'mid·ē·kəm,pyüd·ər }

mid-infrared radiation See intermediate-infrared radiation. { ,mid¦in·frə,red ,rad·ē'ā·shən }

mid-latitude See middle latitude. { 'mid¦lad·ə,tüd }

mid-latitude westerlies See westerlies. { 'mid¦lad·ə,tüd 'wes·tər,lēz }

midline See median. { 'mid,līn }

midnight sun [ASTRON] The sun when it is visible at midnight; occurs during the summer in high latitudes, poleward of the circle at which the latitude is approximately equal to the polar distance of the sun. { 'mid,nīt 'sən }

mid-ocean canyon See deep-sea channel. { 'mid¦ō·shən 'kan·yən }

mid-oceanic ridge [GEOL] A continuous, median, seismic mountain range on the floor of the ocean, extending through the North and South Atlantic oceans, the Indian Ocean, and the South Pacific Ocean; the topography is rugged, elevation is 0.6–1.8 miles (1–3 kilometers), width is about 900 miles (1500 kilometers), and length is over 52,000 miles (84,000 kilometers). Also known as mid-ocean ridge; mid-ocean rise; oceanic ridge. { 'mid,ō·shē¦an·ik 'rij }

mid-ocean ridge See mid-oceanic ridge. { 'mid¦ō·shən 'rij }

mid-ocean rift See rift valley. { 'mid¦ō·shən 'rift }

mid-ocean rise See mid-oceanic ridge. { 'mid¦ō·shən 'rīs }

midpoint [MATH] The midpoint of a line segment is the point which separates the segment into two equal parts. { 'mid,pȯint }

midrange [ENG ACOUS] A loudspeaker designed to reproduce medium audio frequencies, generally used in conjunction with a crossover network, a tweeter, and a woofer. Also known as squawker. { 'mid,rānj }

mid-range materiel requirements [ORD] Items required as soon as possible by current operational and organizational concepts, or required by new concepts to be implemented within the next 5 years. { 'mid,rānj mə,tir·ē'el ri,kwīr·məns }

midrib [BOT] The large central vein of a leaf. { 'mid,rib }

midships [NAV ARCH] **1.** Halfway between a boat's or ship's stem and stern, at the design waterline. **2.** Halfway between the side of a boat's or ship's hull. { 'mid,ships }

midship section [NAV ARCH] A drawing of the cross section of a ship halfway between the intersections of its stem and stern with the design waterline. { 'mid,ship 'sek·shən }

midship section coefficient [NAV ARCH] The ratio of the area of the midship section below the water plane to the product of its beam and its draft. { 'mid,ship ,sek·shən ,kō·i'fish·ənt }

mid-square generator [COMPUT SCI] A procedure for generating a sequence of random numbers, in which a member of a sequence is squared and the middle digits of the resulting number form the next member of the sequence. { 'mid¦skwər 'jen·ə,rād·ər }

Mie-Grüneisen equation [THERMO] An equation of state particularly useful at high pressure, which states that the volume of a system times the difference between the pressure and the pressure at absolute zero equals the product of a number which depends only on the volume times the difference between the internal energy and the internal energy at absolute zero. { 'mē 'grü,nīz·ən i,kwā·zhən }

miersite [MINERAL] (Cu,Ag)I A canary yellow, isometric mineral consisting of copper and silver iodide. { 'mir,zīt }

Mie scattering [OPTICS] The scattering of light by a sphere of dielectric material. { 'mē ,skad·ə·riŋ }

Mie's double plate [ELEC] A device consisting of two small metal disks with insulating handles; they are held in contact in an electric field and then separated, and the charge on one of the disks is then measured to determine the electric displacement. { 'mēz ¦dəb·əl 'plāt }

MIG head See metal-in-gap head. { ¦em¦ī'jē ,hed or 'mig ,hed }

migma [GEOL] A mixture of solid rock materials and rock melt with mobility or potential mobility. { 'mig·mə }

migma plasma [PL PHYS] A hybrid physical state between a colliding beam and a plasma, which is generated by accelerating ions to energies of several megaelectronvolts and causing them to travel in self-colliding orbits in the presence of thermal, ambient electrons. { 'mig·mə ,plaz·mə }

migmatite [PETR] A mixed rock exhibiting crystalline textures in which a truly metamorphic component is streaked and mixed with obviously once-molten material of a more or less granitic character. { 'mig·mə,tīt }

migmatization [PETR] Formation of migmatite; involves either injection or in-place melting. { ,mig·mə·də'zā·shən }

migraine [MED] Recurrent paroxysmal vascular headache, commonly having unilateral onset and often associated with nausea and vomiting. { 'mī,grān }

migrant [ZOO] An animal that moves from one habitat to another. { 'mī·grənt }

migration [CHEM] The movement of an atom or group of atoms to new positions during the course of a molecular rearrangement. [CHEM ENG] See bleeding. [COMPUT SCI] Movement of frequently used data items to more accessible storage locations, and of infrequently used data items to less accessible locations. [GEN] The transfer of genetic information among populations by the movement of individuals or groups of individuals from one population into another. [GEOL] **1.** Movement of a topographic feature from one place to another, especially movement of a dune by wind action. **2.** Movement of liquid or gaseous hydrocarbons from their source into reservoir rocks. [HYD] Slow, downstream movement of a system of meanders. [MET] The uncontrolled movement of certain metals, particularly silver, from one location to another, usually with associated undesirable effects such as oxidation or corrosion. [SOLID STATE] **1.** The movement of charges through a semiconductor material by diffusion or drift of charge carriers or ionized atoms. **2.** The movement of crystal defects through a semiconductor crystal under the influence of high temperature, strain, or a continuously applied electric field. [VERT ZOO] Periodic movement of animals to new areas or habitats. { mī'grā·shən }

migration area [NUCLEO] One-sixth the mean square distance that a neutron travels in a medium from its birth in fission until its absorption. { mī'grā·shən ,er·ē·ə }

migration current [PHYS CHEM] Additional current produced by electrostatic attraction of cations to the surface of a dropping electrode; an unpredictable and undesirable effect to be avoided during analytical voltammetry. { mī'grā·shən ,kə·rənt }

migration length [NUCLEO] The square root of the migration area. { mī'grā·shən ,leŋkth }

migratory [METEOROL] Commonly applied to pressure systems embedded in the westerlies and, therefore, moving in a general west-to-east direction. { 'mī·grə,tȯr·ē }

migratory dune See wandering dune. { 'mī·grə,tȯr·ē 'dün }

MIKES See mass-analyzed ion kinetic energy spectrometry. { mīks }

Mikheyeve-Smirnov-Wolfenstein effect See MSW effect. { mi·¦khā·yev ¦smēr,nȯf 'vùlf·ən,shtīn i,fekt }

Mikulicz's disease [MED] Enlargement of salivary and lacrimal glands from any of various causes. { 'mi·kü,lich·əz dī,zēz }

mil [MATH] A unit of angular measure which, due to nonuniformity of usage, may have any one of three values: 0.001 radian or approximately 0.0572958°; 1/6400 of a full revolution or 0.05625°; 1/1000 of a right angle or 0.09°. [MECH] **1.** A unit of length, equal to 0.001 inch, or to 2.54×10^{-5} meter. Also known as milli-inch; thou. **2.** See milliliter. { mil }

milan lace [TEXT] A bobbin lace of the Belgian type with a ground with designs formed of tape or braid, in scroll or floral motifs. { mə'lan 'lās }

Milankovitch cycles [GEOPHYS] Periodic variations in the earth's position relative to the sun as the earth orbits, affecting the distribution of the solar radiation reaching the earth and causing climatic changes that have profound impacts on the abundance and distribution of organisms, best seen in the fossil record of the Quaternary Period (the last 1.6 million years). { mē·lən'kō·vich ,sīk·əlz }

milarite [MINERAL] $K_2Ca_4Be_4Al_2Si_{24}O_{62} \cdot H_2O$ A colorless to greenish, glassy, hexagonal mineral composed of a hydrous silicate of potassium, calcium, beryllium, and aluminum, occurring in crystals. { 'mē,lä,rīt }

mild abrasive [MATER] An abrasive material, such as chalk or talc, having a hardness of 1–2 on Mohs scale; used in silver polishes and window cleaners. { 'mīld ə'brā·siv }

mildew [MYCOL] **1.** A whitish growth on plants, organic matter, and other materials caused by a parasitic fungus. **2.** Any fungus producing such growth. { 'mil,dü }

mild mental retardation [PSYCH] Subnormal general intellectual functioning in which the intelligence quotient is approximately 52–67. { 'mīld 'ment·əl ,rē,tär'dā·shən }

mild mercury chloride See mercurous chloride. { 'mīld 'mər·kyə·rē 'klór,īd }

mild steel [MET] Carbon steel containing 0.05–0.25% carbon. { 'mīld 'stēl }

mile [MECH] A unit of length in common use in the United States, equal to 5280 feet, or 1609.344 meters. Abbreviated mi. Also known as land mile; statute mile. { mīl }

mileage chart [MAP] A chart showing distances between various points. { 'mīl·ij ,chärt }

mileage number [NAV] For the Mississippi River system, a number assigned to aids to navigation which gives the distance in sailing miles along the river from a reference point to the aid; the number serves as a light list number for identification purposes. { 'mīl·ij ,nəm·bər }

milepost [CIV ENG] **1.** A post placed a mile away from a similar post. **2.** A post indicating mileage from a given point. { 'mīl,pōst }

miles of relative movement [NAV] The distance, in miles, traveled relative to a reference point which is usually in motion. { 'mīlz əv 'rel·ə·tiv 'müv·mənt }

miles on course [NAV] The actual or predicted distance, in miles, traveled on any given course. { 'mīlz ȯn 'kȯrs }

milestone activity See key activity. { 'mīl,stōn ,ak'tiv·əd·ē }

mil formula [ORD] Mil relation used in gunnery; expressed by $n = W/R$, where n is the angular measurement in mils between two points, W is the lateral distance in yards between the points, and R is the mean distance to the points in thousands of yards; the mil relation is approximately true for angles less than 400 mils. { 'mil ,fȯr·myə·lə }

miliaria [MED] An acute inflammatory skin disease, the lesions consisting of vesicles and papules, which may be accompanied by a prickling or tingling sensation. Also known as heat rash; prickly heat. { ,mil·ē'ar·ē·ə }

Milichiidae [INV ZOO] A family of myodarian cyclorrhaphous dipteran insects in the subsection Acalypteratae. { ,mil·ə'kī·ə,dē }

milieu interieur [PHYSIO] The fundamental concept that the living organism exists in an aqueous internal environment which bathes all tissues and provides a medium for the elementary exchange of nutrients and waste. { mēl'yü in,tir·ē·ər }

milieu therapy [PSYCH] The treatment of mental disorder or maladjustment by making substantial changes in a patient's immediate life circumstances and environment in a way that will enhance the effectiveness of other forms of therapy. Also known as situation therapy. { mēl'yü 'ther·ə·pē }

Miliolacea [INV ZOO] A superfamily of marine or brackish foraminiferans in the suborder Miliolina characterized by an imperforate test wall of tiny, disordered calcite rhombs. { ,mil·ē·ə'lās·ē·ə }

Miliolidae [INV ZOO] A family of foraminiferans in the superfamily Miliolacea. { ,mil·ē'äl·ə,dē }

Miliolina [INV ZOO] A suborder of the Foraminiferida characterized by a porcelaneous, imperforate calcite wall. { ,mil·ē'äl·ə·nə }

military aircraft [AERO ENG] Aircraft that are designed or modified for highly specialized use by the armed services of a nation. { 'mil·i,ter·ē 'er,kraft }

military characteristics [ORD] Those characteristics of equipment found desirable or necessary to the performance of a military mission, either combat or noncombat. { 'mil·i,ter·ē ,kar·ik·tə'ris·tiks }

military department [ORD] In the United States, the Department of the Army, the Department of the Navy, or the Department of the Air Force. { 'mil·i,ter·ē di'pärt·mənt }

military engineering [ENG] Science, art, and practice involved in design and construction of defensive and offensive military works as well as construction and maintenance of transportation systems. { 'mil·i,ter·ē ,en·jə'nir·iŋ }

military geology [ENG] The application of the earth sciences to such military concerns as terrain analysis, water supply, foundations, and construction of roads and airfields. { 'mil·i,ter·ē jē'äl·ə·jē }

military grid [MAP] Two sets of parallel lines intersecting at right angles and forming squares; superimposed on maps or charts in an accurate and consistent manner to permit identification of ground locations with respect to other locations and the computation of direction and distance to other points. { 'mil·i,ter·ē 'grid }

military grid reference system [MAP] A system which uses a standard-scaled grid square, based on a point of origin on a map projection of the earth's surface in an accurate and consistent manner to permit either position referencing, or the computation of direction and distance between grid positions. { 'mil·i,ter·ē 'grid 'ref·rəns ,sis·təm }

military map [MAP] A map containing a military grid. { 'mil·i,ter·ē 'map }

military motor vehicle [ORD] Wheeled or track-laying vehicle or combined wheeled and track-laying vehicle, designed for transporting personnel or cargo for military purposes; it may be either propelled by a self-contained power unit or drawn by a vehicle containing a self-contained power unit. { 'mil·i,ter·ē 'mōd·ər ,vē·ə·kəl }

military requirement [ORD] The statement of a recognized and approved need for a new item, a weapons system, or an assemblage for military use. { 'mil·i,ter·ē ri kwīr·mənt }

military research [ORD] Investigation or experimentation to discover and interpret new facts or to analyze existing facts with the aim of applying new or revised conclusions, theories, or laws to areas of interest to the military services. { 'mil·i,ter·ē ri'sərch }

military satellite [AERO ENG] An artificial earth satellite used for military purposes; the six mission categories are communication, navigation, geodesy, nuclear test detection, surveillance, and research and technology. { 'mil·i,ter·ē 'sad·əl,īt }

military specification [ORD] A procurement specification promulgated by the military agencies and used for the procurement of military supplies and equipment. { 'mil·i,ter·ē ,spes·ə·fə'kā·shən }

military standard test [ORD] Specifications approved by the United States Department of Defense, to ensure uniformity of test conditions. { 'mil·i,ter·ē 'stan·dərd 'test }

military target [ORD] **1.** Any industrial plant, city, or other object, or any person, group of persons, or force marked as a target for destruction, damage, injury, or capture because of its direct or indirect use in the conduct or support of an enemy's military endeavor. **2.** In restricted usage, a military person, force, installation, or area marked as a target because of its use, or potential use, in direct military operations. { 'mil·i,ter·ē 'tär·gət }

military technology [ENG] The technology needed to develop and support the armament used by the military. { 'mil·i,ter·ē tek'näl·ə·jē }

military transport vehicle [ORD] Wheeled or tracked vehicle specifically designed for military purposes other than employment as a fighting vehicle, intended for transporting cargo, personnel, and equipment or for towing other vehicles over land and roads in close support of fighting vehicles and troops. { 'mil·i ter·ē ,tranz,pȯrt ,vē·ə·kəl }

military-type vehicle [ORD] Vehicle designed primarily for military purposes to meet field requirements in connection with combat and tactical operations. { 'mil·i,ter·ē ,tīp 'vē·ə·kəl }

military vessel [NAV ARCH] A ship or boat designed primarily for use in war. { 'mil·i,ter·ē 'ves·əl }

milk [CHEM] A suspension of certain metallic oxides, as milk of magnesia, iron, or bismuth. [FOOD ENG] A product derived from the lacteal secretion, practically free from colostrum, of dairy cows; contains not less than 8.25% milk solids and not less than 3.25% milk fat. [PHYSIO] **1.** The whitish fluid secreted by the mammary gland for the nourishment of the young; composed of carbohydrates, proteins, fats, mineral salts, vitamins, and antibodies. **2.** Any whitish fluid in nature resembling milk, as coconut milk. { milk }

milk-alkali syndrome [MED] A complex of symptoms associated with prolonged excessive intake of milk and soluble alkali, including hypercalcemia, renal insufficiency, milk alkalosis, conjunctivitis, and calcinosis. Also known as Burnett's

MILIOLIDAE

Test representative of Miliolidae. *(From L. H. Hyman, The Invertebrates, vol. I, McGraw-Hill, 1940)*

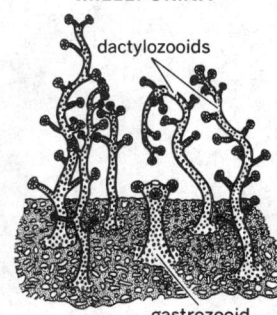

MILLEPORINA

dactylozooids

gastrozooid

Polyps of Milleporina. *(From L. H. Hyman, The Invertebrates, vol. 1, McGraw-Hill, 1940)*

MILLER INDICES

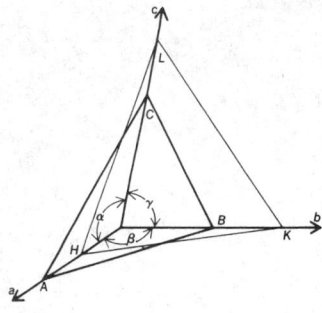

The reference system used to define Miller indices. *a, b,* and *c* are chosen as crystallographic axes, and *ABC* as unit plane, defined by angles α, β, and γ, and ratio *OA:OB:OC*. Miller indices of another plane, *HKL*, are integers proportional to *OA/OH, OB/OK,* and *OC/OL*.

syndrome; milk-drinker's syndrome. { 'milk 'al·kə·lī ˌsin·drōm }

milk-drinker's syndrome *See* milk-alkali syndrome. { 'milk ˌdriŋ·kərz 'sin·drōm }

milk factor [BIOCHEM] A filtrable, noncellular agent in the milk and tissues of certain strains of inbred mice; transmitted from the mother to the offspring by nursing. Also known as Bittner milk factor. { 'milk ˌfak·tər }

milk fever [MED] A fever occurring during the first six weeks after childbirth, believed to be caused by puerperal infection. { 'milk ˌfē·vər }

milk glass [MATER] A white, and sometimes colored, opaque glass made by adding calcium fluoride and alumina to soda-lime glass. { 'milk ˌglas }

milk intolerance [MED] Extreme sensitivity to milk due to allergy to milk protein or lactose deficiency; characterized by diarrhea, abdominal cramps, and vomiting. { 'milk in'täl·ə·rəns }

milk leg *See* phlegmasia alba dolens. { 'milk ˌleg }

milk line *See* mammary ridge. { 'milk ˌlīn }

Milkman's syndrome [MED] Decreased tubular reabsorption of phosphate, resulting in osteomalacia which gives a peculiar striped appearance (multiple pseudofractures) to the bones in roentgenograms. Also known as Looser-Milkman syndrome. { 'milkˌmanz ˌsin·drōm }

milk of magnesia [PHARM] A white suspension of magnesium hydroxide in water; used as a cathartic. Also known as magnesia magma. { 'milk əv mag'nē·zhə }

milk sugar *See* lactose. { 'milk ˌshùg·ər }

milk teeth *See* deciduous teeth. { 'milk ˌtēth }

milkweed [BOT] Any of several latex-secreting plants of the genus *Asclepias* in the family Asclepiadaceae. { 'milkˌwēd }

milky disease [INV ZOO] A bacterial disease of Japanese beetle larvae or related grubs caused by *Bacillus papilliae* and *B. lentimorbus* that penetrate the intestine and sporulate in the body cavity; blood of the grub eventually turns milky white. { 'mil·kē di,zēz }

milky quartz [MINERAL] An opaque, milk-white variety of crystalline quartz, often with a greasy luster; milkiness is due to the presence of air-filled cavities. Also known as greasy quartz. { 'mil·kē 'kwòrts }

Milky Way [ASTRON] The faint band of light which encircles the sky and results from the combined light of the many stars near the plane of our galaxy. { 'mil·kē 'wā }

Milky Way Galaxy [ASTRON] The large aggregation of stars and interstellar gas and dust of which the sun is a member. Also known as Galaxy. { 'mil·kē 'wā 'gal·ik·sē }

milky weather *See* whiteout. { 'mil·kē 'weth·ər }

mill [FOOD ENG] **1.** A machine for grinding grain into flour. **2.** A building that houses milling machines. [IND ENG] **1.** A machine that manufactures paper, textiles, or other products by the continuous repetition of some simple process or action. **2.** A building that houses machinery for manufacturing processes. [MIN ENG] **1.** An excavation made in the country rock, by a crosscut from the workings on a vein, to obtain waste for filling; it is left without timber so that the roof can fall in and furnish the required rock. **2.** A passage connecting a stope or upper level with a lower level intended to be filled with broken ore that can then be drawn out at the bottom as desired for further transportation. { mil }

millboard [MATER] Hard, strong paperboard; used for furniture panels. { 'milˌbòrd }

mill building [CIV ENG] A steel-frame building in which roof trusses span columns in the outside wall; originally, this type of building housed milling machinery, as for wood or metal, hence the name. { 'mil ˌbild·iŋ }

milled soap [MATER] Soap, such as toilet soap, made by adding color and perfume to soap chips and then passing the mixture through milling rollers and pressing in molds. { mild 'sōp }

Milleporina [INV ZOO] An order of the class Hydrozoa known as the stinging corals; they resemble true corals because of a calcareous exoskeleton. { ˌmil·ə·pə'rī·nə }

miller *See* milling machine. { 'mil·ər }

Miller-Abbott tube [MED] A double-lumen rubber tube having a balloon at the end, inserted through the nasal passage and passed through the pylorus to locate and treat intestinal obstructions. { 'mil·ər 'ab·ət ˌtüb }

Miller bridge [ELECTR] Type of bridge circuit for measuring amplification factors of vacuum tubes. { 'mil·ər ˌbrij }

Miller code [COMPUT SCI] A code used internally in some computers, in which a binary 1 is represented by a transition in the middle of a bit (either up or down), and a binary 0 is represented by no transition following a binary 1; a transition between bits represents successive 0's; in this code, the longest period possible without a transition is two bit times. { 'mil·ər ˌkōd }

Miller effect [ELECTR] The increase in the effective grid-cathode capacitance of a vacuum tube due to the charge induced electrostatically on the grid by the anode through the grid-anode capacitance. { 'mil·ər iˌfekt }

Miller generator *See* bootstrap integrator. { 'mil·ər 'jen·ə ˌrād·ər }

Miller indices [CRYSTAL] Three integers identifying a type of crystal plane; the intercepts of a plane on the three crystallographic axes are expressed as fractions of the crystal parameters; the reciprocals of these fractions, reduced to integral proportions, are the Miller indices. Also known as crystal indices. { 'mil·ər 'in·dəˌsēz }

Miller integrator [ELECTR] A resistor-capacitor charging network having a high-gain amplifier paralleling the capacitor; used to produce a linear time-base voltage. Also known as Miller time-base. { 'mil·ər 'int·əˌgrād·ər }

millerite [MINERAL] NiS A brass to bronze-yellow mineral that crystallizes in the hexagonal system and usually contains trace amounts of cobalt, copper, and iron; hardness is 3–3.5 on Mohs scale, and specific gravity is 5.5; it generally occurs in fine crystals, chiefly as nodules in clay ironstone. Also known as capillary pyrites; hair pyrites; nickel pyrites. { 'mil·əˌrīt }

Miller law [AGR] A law administered by the Food and Drug Administration that regulates the production and use of agricultural fungicides in the United States, and will not allow materials to leave poisonous residues on edible crops. [CRYSTAL] If the edges formed by the intersections of three faces of a crystal are taken as the three reference axes, then the three quantities formed by dividing the intercept of a fourth face with one of these axes by the intercept of a fifth face with the same axis are proportional to small whole numbers, rarely exceeding 6. Also known as law of rational intercepts. { 'mil·ər ˌlò }

Miller time-base *See* Miller integrator. { 'mil·ər 'tīm ˌbās }

millet [BOT] A common name applied to at least five related members of the grass family grown for their edible seeds. { 'mil·ət }

mill finish [MET] The characteristic surface finish on a rolled metal product. [TEXT] The slightly napped finish of a worsted fabric. { 'mil ˌfin·ish }

mill-head ore *See* run-of-mill. { 'mil ˌhed ˌòr }

milli- [MATH] A prefix representing 10^{-3}, or one-thousandth. Abbreviated m. { 'mil·ē }

milliammeter [ELEC] An ammeter whose scale is calibrated to indicate current values in milliamperes. { ˌmil·ē'amˌēd·ər }

milliampere [ELEC] A unit of current equal to one-thousandth of an ampere. Abbreviated mA. { ˌmil·ē'amˌpir }

milliampere-second [NUCLEO] A unit of radiation dose resulting from exposure to x-rays, equal to the dose produced by an electron beam, carrying a current of 1 milliampere, bombarding the target of an x-ray tube for 1 second. Abbreviated mA · s. { ˌmil·ē'amˌpir 'sek·ənd }

millibar [MECH] A unit of pressure equal to one-thousandth of a bar. Abbreviated mb. Also known as vac. { 'mil·əˌbär }

millibarn [NUC PHYS] A unit of cross section equal to one-thousandth of a barn. Abbreviated mb. { 'mil·əˌbärn }

millicurie-destroyed [NUCLEO] A unit of radiation dose, equal to the radiation emitted by a sample over a period during which its activity decreases by 1 millicurie; for radon-222 (for which this unit is most often used), it is approximately 133 milligram-hours. Abbreviated mcd. { ˌmil·əˌkyur·ē di 'stròid }

millicurie-of-intensity-hour *See* sievert. { ˌmil·əˌkyur·ē əv in'ten·səd·ē ˌaùr }

millicycle *See* millihertz. { 'mil·əˌsī·kəl }

millidarcy [PHYS] A unit of fluid permeability equal to one-thousandth of a darcy. Abbreviated md. { 'mil·əˌdär·sē }

milliequivalent [CHEM] One-thousandth of a compound's or an element's equivalent weight. Abbreviated meq. { ¦mil·ē·ə¦kwiv·ə·lənt }

millier *See* tonne. { mil′yā }

millifarad [ELEC] A unit of capacitance equal to one-thousandth of a farad. Abbreviated mF. { ¦mil·ē′far·əd }

milligal [MECH] A unit of acceleration commonly used in geodetic measurements, equal to 10^{-3} galileo, or 10^{-5} meter per second per second. Abbreviated mGal. { ′mil·ə,gal }

milligauss [ELECTROMAG] A unit of magnetic flux density equal to one-thousandth of a gauss. Abbreviated mG. { ′mil·ə,gaús }

milligram [MECH] A unit of mass equal to one-thousandth of a gram. Abbreviated mg. { ′mil·ə,gram }

milligram-hour [NUCLEO] A unit of radiation dose, equal to the radiation emitted by a source with an equivalent radium content of 1 milligram for a period of 1 hour. Abbreviated mgh. { ′mil·ə,gram,aúr }

millihenry [ELECTROMAG] A unit of inductance equal to one-thousandth of a henry. Abbreviated mH. { ′mil·ə,hen·rē }

millihertz [PHYS] A unit of frequency equal to one-thousandth of a hertz. Abbreviated mHz. Also known as millicycle (mc). { ′mil·ə,hərts }

millihg *See* millimeter of mercury.

milli-inch *See* mil. { ′mil·ē,inch }

milli-k [NUCLEO] A unit of reactivity; the reactivity of a reactor in milli-k is equal to $1000(k − 1)$, where k is the effective multiplication factor. { ′mil·ē,kā }

Millikan meter [ELECTR] An integrating ionization chamber in which a gold-leaf electroscope is charged a known amount and ionizing events reduce this charge, so that the resulting angle through which the gold leaf is repelled at any given time indicates the number of ionizing events that have occurred. { ′mil·ə·kən ,mēd·ər }

Millikan oil-drop experiment [ATOM PHYS] A method of determining the charge on an electron, in which one measures the terminal velocities of rise and fall of oil droplets in an electric field after the droplets have picked up charge from ionization in the surrounding gas produced by an x-ray beam. { ′mil·ə·kən ′oil,dräp ik,sper·əmənt }

milliliter [MECH] A unit of volume equal to 10^{-3} liter or 10^{-6} cubic meter. Abbreviated ml. Also known as mil. { ′mil·ə,lēd·ər }

milli-mass-unit [PHYS] One-thousandth of an atomic mass unit. Abbreviated mmu. { ¦mil·ə ¦mas ′yü·nət }

millimeter [MECH] A unit of length equal to one-thousandth of a meter. Abbreviated mm. Also known as metric line; strich. { ′mil·ə,mēd·ər }

millimeter of mercury [MECH] A unit of pressure, equal to the pressure exerted by a column of mercury 1 millimeter high with a density of 13.5951 grams per cubic centimeter under the standard acceleration of gravity; equal to 133.322387415 pascals; it differs from the torr by less than 1 part in 7,000,000. Abbreviated mmHg. Also known as millihg. { ′mil·ə,mēd·ər əv ′mər·kyə·rē }

millimeter of water [MECH] A unit of pressure, equal to the pressure exerted by a column of water 1 millimeter high with a density of 1 gram per cubic centimeter under the standard acceleration of gravity; equal to 9.80665 pascals. Abbreviated mmH₂O. { ′mil·ə,mēd·ər əv ′wódər }

millimeter wave [ELECTROMAG] An electromagnetic wave having a wavelength between 1 millimeter and 1 centimeter, corresponding to frequencies between 30 and 300 gigahertz. Also known as millimetric wave. { ′mil·ə,mēd·ər ′wāv }

millimetric wave *See* millimeter wave. { ¦mil·ə¦me·trik ′wāv }

milli-micro- *See* nano-. { ¦mil·ə¦mī′krō }

millimicron *See* nanometer. { ′mil·ə,mī′krón }

milling [MECH ENG] Mechanical treatment of materials to produce a powder, to change the size or shape of metal powder particles, or to coat one powder mixture with another. [MIN ENG] A combination of open-cut and underground mining, wherein the ore is mined in open cut and handled underground. { ′mil·iŋ }

milling cutter [DES ENG] A rotary tool-steel cutting tool with peripheral teeth, used in a milling machine to remove material from the workpiece through the relative motion of workpiece and cutter. { ′mil·iŋ ,kəd·ər }

milling machine [MECH ENG] A machine for the removal of metal by feeding a workpiece through the periphery of a rotating circular cutter. Also known as miller. { ′mil·iŋ mə,shēn }

milling ore *See* second-class ore. { ′mil·iŋ ,ór }

milling planer [MECH ENG] A planer that uses a rotary cutter rather than single-point tools. { ′mil·iŋ ,plān·ər }

milling system *See* chute system. { ′mil·iŋ ,sis·təm }

Millington reverberation formula [ACOUS] A formula that states that the reverberation time of a chamber in seconds is 0.05 times its volume in cubic feet, divided by the sum over the surfaces of the chamber of the product of the surface's area in square feet by the natural logarithm of 1 minus its absorption coefficient. { ′mil·iŋ·tən ri,vər·bə′rā·shən ,fór·myə·lə }

milling width [MIN ENG] Width of lode designated for treatment in the mill, as calculated with regard to daily tonnage. { ′mil·iŋ ,width }

million [MATH] The number 10^6, or 1,000,000. { ′mil·yən }

million electronvolts *See* megaelectronvolt. { ′mil·yən i′lek,trän,vōlts }

million floating-point operations per second [COMPUT SCI] A unit used to measure the processing speed or throughput of supercomputers or array processors. Abbreviated Mflop. { ′mil·yən ¦flōd·iŋ ¦póint ,äp·ə′rā·shənz pər ′sek·ənd }

million instructions per second [COMPUT SCI] A unit used to measure the speed at which a computer's central processing unit can process instructions. Abbreviated MIPS. { ′mil·yən in′strək·shənz pər ′sek·ənd }

millipede [INV ZOO] The common name for members of the arthropod class Diplopoda. { ′mil·ə,pēd }

Millipore filter [MICROBIO] A filter capable of ultrafine separation, used for purification and analyses of fluids, among other applications. { ′mil·ə,pór ,fil·tər }

millirad [NUCLEO] A unit of absorbed ionizing radiation dose equal to one-thousandth of a rad. Abbreviated mrad. { ′mil·ə,rad }

milliroentgen [NUCLEO] A unit of radioactive dose of electromagnetic radiation equal to one-thousandth of a roentgen. Abbreviated mr. { ′mil·ē′rent·gən }

millisecond [MECH] A unit of time equal to one-thousandth of a second. Abbreviated ms; msec. { ′mil·ə,sek·ənd }

millisecond delay cap [ENG] A delay cap with an extremely short (20–500 thousandths of a second) interval between passing of current and explosion. Also known as short-delay detonator. { ′mil·ə,sek·ənd di,lā ,kap }

millisecond pulsar *See* fast pulsar. { ′mil·ə,sek·ənd ′pəl·sär }

millisite [MINERAL] $(Na,K)CaAl_6(PO_4)_4(OH)_9 \cdot 3H_2O$ White mineral composed of a basic hydrous phosphate of sodium, potassium, calcium, and aluminum. { ′mil·ə,sīt }

millivolt [ELEC] A unit of potential difference or emf equal to one-thousandth of a volt. Abbreviated mV. { ′mil·ə,vōlt }

millivoltmeter [ELEC] A voltmeter whose scale is calibrated to indicate voltage values in millivolts. { ,mil·ə′vōlt,mēdər }

milliwatt [MECH] A unit of power equal to one-thousandth of a watt. Abbreviated mW. { ′mil·ə,wät }

mill length *See* random length. { ′mil ,leŋkth }

Millon's reagent [CHEM] Reagent used to test for proteins; made by dissolving mercury in nitric acid, diluting, then decanting the liquid from the precipitate. { mē′lónz re,ā·jənt }

mill ore [MIN ENG] An ore that must be given some preliminary treatment before a marketable grade or a grade suitable for further treatment can be obtained. { ′mil ,ór }

millrace [CIV ENG] A canal filled with water that flows to and from a waterwheel acting as the power supply for a mill. { ′mil,rās }

mill run [MIN ENG] **1.** A given quantity of ore tested for its quality by actual milling. **2.** The yield of such a test. { ′mil ,rən }

mill scale [MET] A surface layer of ferric oxide (Fe_3O_4) that forms on steel or iron during hot rolling. { ′mil ,skāl }

Mills cross [ELECTROMAG] An antenna array that consists of two antennas oriented perpendicular to each other and that produces a narrow pencil beam. { ′milz ′krós }

Mills-Crowe process [MIN ENG] Method of regeneration of

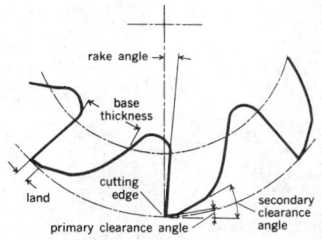

MILLING CUTTER

Typical milling cutter teeth.

cyanide liquor from the gold leaching process; the barren solution is acidified, liberating hydrocyanic acid which is separated and reabsorbed in alkaline solution. { 'milz 'krō ˌprä·səs }

millsite [MIN ENG] A plot of ground suitable for the erection of a mill, or reduction works, to be used in connection with mining operations. { 'milˌsīt }

millstone See buhrstone. { 'milˌstōn }

millwork [MATER] Ready-made products which are manufactured at a wood-planing mill or woodworking plant, such as moldings, doors, doorframes, window sashes, stairwork, and cabinets. { 'milˌwərk }

millwright [ENG] **1.** A person who plans, builds, or sets up the machinery for a mill. **2.** A person who repairs milling machines. { 'milˌrīt }

Milne method [MATH] A technique which provides numerical solutions to ordinary differential equations. { 'miln ˌmeth·əd }

Milroy's disease [MED] Familial chronic lymphedema of the lower extremities. { 'milˌroiz di zēz }

Mima mound [GEOGR] A circular or oval domelike structure composed of loose silt and soil that is believed to be generated by a combination of geomorphic processes and burrowing by animals; found in northwest North America, Africa, and southern South America. { 'mē·mə ˌmaund }

Mimas [ASTRON] A satellite of Saturn orbiting at a mean distance of 115,300 miles (186,000 kilometers). { 'mē·məs }

MIMD [COMPUT SCI] A type of multiprocessor architecture in which several instruction cycles may be active at any given time, each independently fetching instructions and operands into multiple processing units and operating on them in a concurrent fashion. Acronym for multiple-instruction-stream, multiple-data-stream. { ˌem¦ī¦em¦dē }

MIME [COMPUT SCI] The Multimedia Internet Mail Enhancements standard, describing a way of encoding binary files, such as pictures, videos, sounds, and executable files, within a normal text message in an operating-system-independent manner. { mīm }

mimeograph [GRAPHICS] A duplicating device for making copies by means of a stretched stencil and ink roller. { 'mim·ē·əˌgraf }

mimetene See mimetite. { 'mim·əˌtēn }

mimetesite See mimetite. { mə'med·əˌzīt }

mimetic [CRYSTAL] Pertaining to a crystal that is twinned or malformed but whose crystal symmetry appears to be of a higher grade than it actually is. [PETR] Of a tectonite, having a deformation fabric, formed by mimetic crystallization, that reflects and is influenced by preexisting anisotropic structure. [ZOO] Pertaining to or exhibiting mimicry. { mə'med·ik }

mimetic camouflage [ECOL] Protective coloration that is achieved by a resemblance to some other existing object, which is recognized by the predator but not associated in its mind with feeding. { məˌmed·ik 'kam·əˌfläzh }

mimetic crystallization [PETR] Recrystallization or neomineralization in metamorphism which reproduces preexistent structures. { mə'med·ik ˌkrist·əl·ə'zā·shən }

mimetite [MINERAL] Pb₅(AsO₄)₃Cl A yellow to yellowish-brown mineral of the apatite group, commonly containing calcium or phosphate; a minor ore of lead. Also known as mimetene; mimetesite. { 'mim·əˌtīt }

mimicry [ZOO] Assumption of color, form, or behavior patterns by one species of another species, for camouflage and protection. { 'mim·ə·krē }

Mimidae [VERT ZOO] The mockingbirds, a family of the Oscines in the order Passeriformes. { 'mim·əˌdē }

Mimosa See β Crucis. { mə'mō·sə }

mimosine [ORG CHEM] C₈H₁₀N₂O₄ A crystalline compound with a melting point of 235–236°C; soluble in dilute acids or bases; used as a depilatory agent. Also known as leucaenine; leucaenol; leucenine; leucenol. { mə'mōˌsēn }

Mimosoideae [BOT] A subfamily of the legume family, Leguminosae; members are largely woody and tropical or subtropical with regular flowers and usually numerous stamens. { ˌmim·ə'soidˌē·ē }

min See minim. { min }

minable [MIN ENG] Material that can be mined under present-day mining technology and economics. { 'mīn·ə·bəl }

minasragrite [MINERAL] (VO)₂H₂(SO₄)₃·15H₂O A blue, monoclinic mineral consisting of hydrated acid vanadyl sulfate;

occurs in efflorescences and as aggregates or masses. { ˌmē·näs'räˌgrīt }

mind [PSYCH] **1.** The sum total of the neural processes which receive, code, and interpret sensations, recall and correlate stored information, and act on it. **2.** The state of consciousness. **3.** The understanding, reasoning, and intellectual faculties and processes considered as a whole. **4.** The psyche, or the conscious, subconscious, and unconscious considered together. { mīnd }

Mindel glaciation [GEOL] The second glacial stage of the Pleistocene in the Alps. { 'min·dəl ˌglā·sē'ā·shən }

Mindel-Riss interglacial [GEOL] The second interglacial stage of the Pleistocene in the Alps; follows the Mindel glaciation. { 'min·dəl 'ris ˌin·tər'glā·shəl }

M indicator See M scope. { 'em ˌin·də·kād·ər }

mine [MIN ENG] An opening or excavation in the earth for extracting minerals. [ORD] An encased explosive or chemical charge designed to be positioned so that it detonates when the target touches or moves near it or when it is fired by remote control; general types are land mines and underwater mines. { mīn }

mine captain [MIN ENG] **1.** The director of work in a mine, with or without superior officials or subordinates. **2.** See mine superintendent. { 'mīn ˌkap·tən }

mine car [MECH ENG] An industrial car, usually of the four-wheel type, with a low body; the door is at one end, pivoted at the top with a latch at the bottom used for hauling bulk materials. { 'mīn ˌkär }

mine characteristic [MIN ENG] The relation between pressure p and volume Q in the ventilation of a mine of resistance R, expressed as $p = RQ^2$. { 'mīn ˌkar·ik·tə,ris·tik }

mine characteristic curve [MIN ENG] A graph derived by plotting the static or total mine head, or both, against the quantity of air; used to solve problems in mine ventilation. { 'mīn ˌkar·ik·təˌris·tik ˌkərv }

mine countermeasures [ORD] All methods for preventing or reducing damage or danger to ships, personnel, aircraft, and vehicles from mines. { 'mīn 'kaun·tərˌmezh·ərz }

mine development [MIN ENG] The operations involved in preparing a mine for ore extraction, including tunneling, sinking, crosscutting, drifting, and raising. { 'mīn di,vel·əp·mənt }

mine dust [MIN ENG] Dust from drilling, blasting, or handling rock. { 'mīn ˌdəst }

mined volume [MIN ENG] A statistic used in mine subsidence, computed by multiplying the mined area by the mean thickness of the bed or of that part of the bed which has been dug out. { 'mīnd 'väl·yəm }

mine examiner See fire boss. { 'mīn igˌzam·ə·nər }

mine fan signal system [MIN ENG] A system which indicates by electric light or electric audible signal, or both, the slowing down or stopping of a mine ventilating fan. { 'mīn ¦fan 'sig·nəl ˌsis·təm }

mine field [ORD] An area in which mines have been placed. { 'mīn ˌfēld }

mine field gap [ORD] A portion of a mine field in which no mines have been laid, of specified width to enable a friendly force to pass through the mine field in tactical formation; seldom less than 100 yards (91 meters) wide. { 'mīn ¦fēld ˌgap }

mine field lane [ORD] An unmined or demined route through any mine field, normally 8 yards (7.3 meters) wide and suitably marked; however, for lanes through enemy mine fields, the width depends on the method of breaching and the intended purpose. { 'mīn ¦fēld ˌlān }

mine fire truck [MIN ENG] A low-slung railcar designed to fight fires in mines; has a water supply and a pump to supply high pressure to the fire hoses. { 'mīn 'fīr ˌtrək }

mine foreman [MIN ENG] The worker charged with the general supervision of the underground workings of a mine and the persons employed therein. { 'mīn 'fòr·mən }

mine hoist [MIN ENG] A device for raising and lowering ore, rock, or coal in a mine, and workers and supplies. { 'mīn ˌhoist }

mine inspector [MIN ENG] Generally, the state mine inspector, as contrasted to the Federal mine inspector; inspects mines to find fire and dust hazards and inspects the safety of working areas, electric circuits, and mine equipment. { 'mīn in'spek·tər }

mine jeep [MIN ENG] An electrically driven car for underground transportation of officials, inspectors, rescue workers, and repair, maintenance, and surveying crews. { 'mīn ˌjēp }

minelite [MATER] An explosive made by mixing a chlorate compound with paraffin wax. { 'mīn·əˌlīt }

mine locomotive [MIN ENG] A low, heavy haulage engine designed for underground operation; usually propelled by electricity, gasoline, or compressed air. { 'mīn ˌlō·kəˌmōd·iv }

mine-planting equipment [ORD] A group of items for uncrating, fusing, and magazine loading of antitank mines, and the mechanical emplacement of barrier-type mine fields. { 'mīn ˌplant·iŋ iˌkwip·mənt }

mine props [MIN ENG] Sections of wood used for holding up pieces of rock in the roof of mines. { 'mīn ˌpräps }

miner [MIN ENG] **1.** A person engaged in removing ore, coal, or minerals from the earth. **2.** A machine designed for automatic mining of an ore. { 'mīn·ər }

mine radio telephone system [MIN ENG] A communication system between the dispatcher and locomotive operators; radio impulses travel along the trolley wire and down the trolley pole to the radio telephone. { 'mīn ˌrād·ē·ō 'tel·ə·fōn ˌsis·təm }

mineragraphy *See* ore microscopy. { ˌmin·ə'räg·rə·fē }

mineral [GEOL] A naturally occurring substance with a characteristic chemical composition expressed by a chemical formula; may occur as individual crystals or may be disseminated in some other mineral or rock; most mineralogists include the requirements of inorganic origin and internal crystalline structure. { 'min·rəl }

mineral acid [INORG CHEM] Any one of the major inorganic acids, such as sulfuric, nitric, or hydrochloric acids. { 'min·rəl ˌas·əd }

mineral additive [MATER] A mineral-derived substance added during grease manufacture, particularly to heavy-duty grease. { 'min·rəl 'ad·ə·tiv }

mineral black [MATER] Black pigment made from ground slate, shale, coke, coal, or slaty coal; used in inks, plastics, and coatings. { 'min·rəl ˌblak }

mineral caoutchouc *See* elaterite. { 'min·rəl 'kaů·chůk }

mineral charcoal *See* fusain. { 'min·rəl 'chär·kōl }

mineral cotton *See* mineral wool. { 'min·rəl 'kät·ən }

mineral deposit [GEOL] A mass of naturally occurring mineral material, usually of economic value. { 'min·rəl di päz·ət }

mineral dressing *See* beneficiation. { 'min·rəl ˌdres·iŋ }

mineral dye [MATER] A natural dyestuff made from minerals; examples are ochre, chrome yellow, and Prussian blue. { 'min·rəl ˌdī }

mineral economics [MIN ENG] Study and application of the processes used in management and finance connected with the discovery, exploitation, and marketing of minerals. { 'min·rəl ˌek·ə'näm·iks }

mineral engineering *See* mining engineering. { 'min·rəl ˌen·jə'nir·iŋ }

mineral facies [PETR] Rocks of any origin whose components have been formed within certain temperature-pressure limits characterized by the stability of certain index minerals. { 'min·rəl 'fā·shēz }

mineral fuel [MATER] A carbonaceous fuel mined or stripped from the earth, such as petroleum, coal, peat, shale oil, or tar sands. { 'min·rəl ˌfyül }

mineral green *See* copper carbonate. { 'min·rəl ˌgrēn }

mineralization [GEOL] **1.** The process of fossilization whereby inorganic materials replace the organic constituents of an organism. **2.** The introduction of minerals into a rock, resulting in a mineral deposit. { ˌmin·rə·lə'zā·shən }

mineralize [CHEM] To convert organic compounds to simpler inorganic compounds, namely, carbon dioxide and water (and halogen acids, if the organic substances are halogenated). [GEOL] To convert to, or impregnate with, mineral material; applied to processes of ore vein deposition and of fossilization. { 'min·rəˌlīz }

mineralizer [GEOL] A gas or fluid dissolved in a magma that aids in the concentration and crystallization of ore minerals. { 'min·rəˌlīz·ər }

mineral jelly [MATER] A viscous, thick material derived from petroleum; used as a stabilizer in explosives. { 'min·rəl ˌjel·ē }

mineral land [MIN ENG] Land which is worth more for mining than for agriculture or other use. { 'min·rəl ˌland }

mineral lard oil [MATER] A mixture of refined mineral oil with lard oil, having a fatty content of 25–30, and a flash point about 300°F (149°C). { 'min·rəl 'lärd ˌoil }

mineral lease *See* mining lease. { 'min·rəl ˌlēs }

mineral monument [MIN ENG] A permanent monument established in a mining district to provide for an accurate description of mining claims and their location. { 'min·rəl 'män·yə·mənt }

mineralocorticoic [BIOCHEM] A steroid hormone secreted by the adrenal cortex that regulates mineral metabolism and, secondarily, fluid balance. { ˌmin·rə·lō'kórd·ə·kóid }

mineralogenetic epoch [GEOL] A geologic time period during which mineral deposits formed. { ˌmin·rə·lō·jə'ned·ik 'ep·ək }

mineralogenetic province [GEOL] Geographic region where conditions were favorable for the concentration of useful minerals. { ˌmin·rə·lō·jə'ned·ik prä·vəns }

mineralogical phase rule [MINERAL] Any of several variations of the Gibbs phase rule, taking into account the number of degrees of freedom consumed by the fixing of physical-chemical variables in the natural environment; it assumes that temperature and pressure are fixed externally and that consequently the number of phases (minerals) in a system (rock) will not usually exceed the number of components. { ˌmin·rə'läj·ə·kəl 'fāz ˌrül }

mineralogist [MINERAL] A person who studies the occurrence, description, mode of formation, and uses of minerals. { ˌmin·ə'rāl·ə·jəst }

mineralography *See* ore microscopy. { ˌmin·rə'läg·rə·fē }

mineralogy [INORG CHEM] The science which concerns the study of natural inorganic substances called minerals. { ˌmin·ə'räl·ə·jē }

mineraloid [MINERAL] A naturally occurring, inorganic material that is amorphous and is therefore not considered to be a mineral. Also known as gel mineral. { 'min·rəˌloid }

mineral oil *See* medicinal oil. { 'min·rəl ˌoil }

mineral processing [MIN ENG] Procedures, such as dry and wet crushing and grinding of ore or other products containing minerals, to raise the concentration of the substance being mined. { 'min·rəl ˌprä·ses·iŋ }

mineral resources [GEOL] Valuable mineral deposits of an area that are presently recoverable and may be so in the future; includes known ore bodies and potential ore. { 'min·rəl ri'sórs·əz }

mineral right *See* mining right. { 'min·rəl ˌrīt }

mineral rubber [MATER] Asphaltine minerals (such as gilsonite and grahamite) and blown asphalts; used in compounding of rubber, coatings, and paints. { 'min·rəl 'rəb·ər }

minerals beneficiation *See* beneficiation. { 'min·rəlz ˌben·ə·fish·ē'ā·shən }

mineral seal oil [MATER] A petroleum distillate with a boiling point higher than that of kerosine; used as a solvent oil and illuminant. { 'min·rəl 'sēl ˌoil }

mineral sequence *See* paragenesis. { 'min·rəl 'sē·kwəns }

mineral soil [GEOL] Soil composed of mineral or rock derivatives with little organic matter. { 'min·rəl ˌsoil }

mineral spring [HYD] A spring whose water has a definite taste due to a high mineral content. { 'min·rəl ˌspriŋ }

mineral suite [MINERAL] **1.** A group of associated minerals in one deposit. **2.** A representative group of minerals from a certain locality. **3.** A group of specimens showing variations, as in color or form, in a single mineral species. { 'min·rəl 'swēt }

mineral tallow *See* hatchettite. { 'min·rəl 'tal·ō }

mineral tanker [NAV ARCH] A ship used for transporting dry bulk commodities in slurry form, as well as crude petroleum products. Also known as slurry carrier. { 'min·rəl 'taŋ·kər }

mineral water [HYD] Water containing naturally or artificially supplied minerals or gases. { 'min·rəl ˌwód·ər }

mineral wax *See* ozocerite. { 'min·rəl ˌwaks }

mineral wool [MATER] A fibrous substance, technically a glass, made from molten slag, rock, glass, or a selected combination of these ingredients; produced by blowing, drawing, or other means of fabricating into fine fibers; used for insulation, fireproofing, and as a filter medium. Also known as mineral cotton; rock wool; silicate cotton; slag wool. { 'min·rəl ˌwůl }

mine rescue apparatus [MIN ENG] Certain types of apparatus worn by workers and permitting them to perform in noxious or irrespirable atmospheres, such as during mine fires or following mine explosions. { ˈmīn ˈres·kyü ˌap·ə,rad·əs }

mine rescue crew [MIN ENG] A crew consisting usually of five to eight persons who are thoroughly trained in the use of mine rescue apparatus, which they wear in rescue or recovery work in a mine following an explosion or during a fire. { ˈmīn ˈres·kyü ˌkrü }

mine rescue lamp [MIN ENG] A particular type of electric safety hand lamp used in rescue operations; equipped with a lens for concentrating or diffusing the light. { ˈmīn ˈres·kyü ˌlamp }

mine resistance [MIN ENG] Resistance by a mine to the passage of an air current. { ˈmīn ri,zis·təns }

miner's friend [MIN ENG] **1.** The Davy safety lamp. **2.** A steam engine once used to pump water from underground. { ˈmīn·ərz ˈfrend }

miner's hammer [MIN ENG] A hammer used to break ore. { ˈmīn·ərz ,ham·ər }

miner's hand lamp [MIN ENG] A self-contained mine lamp with handle for carrying. { ˈmīn·ərz ˈhand ˌlamp }

miner's helmet [MIN ENG] A hat designed for miners to provide head protection and to hold the cap lamp. { ˈmīn·ərz ,hel·mət }

miner's horn [MIN ENG] A metal spoon or horn to collect ore particles in gold washing. { ˈmīn·ərz ,hȯrn }

miner's inch [MIN ENG] The quantity of water that will escape from an aperture 1 inch (2.54 centimeter) square through a 2-inch-thick (5.08-centimeter) plank, with a steady flow of water standing 6 inches (15.24 centimeters) above the top of the escape aperture, the quantity so discharged amounting to 2274 cubic feet (64.39 cubic meters) in 24 hours. { ˈmīn·ərz ˈinch }

miner's lamp [MIN ENG] Any one of a variety of lamps used by a miner to furnish light, such as oil lamps, carbide lamps, flame safety lamps, and cap lamps. { ˈmīn·ərz ,lamp }

miner's right [MIN ENG] An annual permit from the government to occupy and work mineral land. { ˈmīn·ərz ,rīt }

miners' rules [MIN ENG] Rules and regulations proclaimed by the miners of any district, relating to the location of, recording of, and the work necessary to hold possession of a mining claim. { ˈmīn·ərz ,rülz }

miner's self-rescuer [MIN ENG] A pocket gas mask effective against carbon monoxide; air passes through a cannister containing fused calcium chloride before entering the mouth. { ˈmīn·ərz ¦self ˈres·kyü·ər }

mine run [MIN ENG] The unscreened output of a mine. Also known as run-of-mine. { ˈmīn ,rən }

mine sample [MIN ENG] Coal or mineral extracted at underground exposures for analysis. { ˈmīn ,sam·pəl }

mine signal system [MIN ENG] Signal lights installed at individual track switches immediately indicating to the motorman whether or not it is safe to proceed. { ˈmīn ˈsig·nəl ,sis·təm }

mine skips [MIN ENG] Skips used to bring mined ore to the surface of a shaft. { ˈmīn ,skips }

mine superintendent [MIN ENG] A mine manager or group manager. Also known as mine captain. { ˈmīn ,sü·prən,ten·dənt }

mine surveyor [MIN ENG] The official who periodically surveys the mine workings and prepares plans for the manager. { ˈmīn sər,vā·ər }

minesweeper [NAV ARCH] A ship specially designed and equipped to remove or destroy underwater mines. [ORD] Heavy road roller pushed in front of a tank, to destroy land mines by exploding them. { ˈmīn,swēp·ər }

minesweeping [ORD] Process of detecting and removing land mines or underwater mines. { ˈmīn,swēp·iŋ }

mine track devices [MIN ENG] Track devices to provide maximum safety for haulage trains in mines. { ˈmīn ˈtrak di,vīs·əz }

mine tractor [MIN ENG] A trackless, self-propelled vehicle used to transport equipment and supplies. { ˈmīn ,trak·tər }

minette [PETR] A syenitic variety of lamprophyre composed principally of biotite phenocrysts in a matrix of orthoclase and biotite. { mə′net }

mine valuation [MIN ENG] Properly weighing the financial considerations to place a present value on mineral reserves. { ˈmīn ,val·yə′wā·shən }

mine ventilating fan [MIN ENG] A motor-driven disk, propeller, or wheel for blowing or exhausting air to provide ventilation of a mine; large units are used for stationary systems, while small portable types provide fresh air in inaccessible locations, such as dead ends. { ˈmīn ˈvent·əl,ād·iŋ ,fan }

mine ventilation system [MIN ENG] A combination of connecting airways with air pressure sources and governing devices that are instrumental in making and controlling airflow. { ˈmīn ,vent·əl′ā·shən ,sis·təm }

mine water [MIN ENG] Water pumped from mines. { ˈmīn ,wȯd·ər }

miniature electron tube [ELECTR] A small electron tube having no base, with tube electrode leads projecting through the glass bottom in positions corresponding to those of pins for either a seven-pin or nine-pin tube base. { ˈmin·ə·chər i′lek,trän ,tüb }

miniaturization [ELECTR] Reduction in the size and weight of a system, package, or component by using small parts arranged for maximum utilization of space. { ,min·ə·chə·rə′zā·shən }

minicartridge [COMPUT SCI] A self-contained package of reel-to-reel magnetic tape that resembles a cassette or cartridge but is slightly different in design and dimensions. { ˈmin·ē,kär·trij }

minicell [MICROBIO] A small anucleate bacterial cell produced by abnormal and unequal division of a parent cell. { ˈmin·ē,sel }

minichromosome [CELL MOL] A eukaryotic chromosome reduced in size by deletion of a segment of deoxyribonucleic acid. [VIROL] Viral deoxyribonucleic acid combined with histone to form a chromatin-like structure. { ,min·ē′krō·mə,sōm }

minicomputer [COMPUT SCI] A relatively small general-purpose digital computer, intermediate in size between a microcomputer and a main frame. { ˈmin·ē·kəm,pyüd·ər }

minidisk [COMPUT SCI] A floppy disk that has a diameter of 5.25 inches (approximately 13 centimeters). Also known as minifloppy disk. { ˈmin·ē,disk }

minifloppy disk *See* minidisk. { ˈmin·ē,fläp·ē ,disk }

minim [MECH] A unit of volume in the apothecaries' measure; equals $\frac{1}{60}$ fluidram (approximately 0.061612 cubic centimeter) or about 1 drop (of water). Abbreviated min. { ˈmin·əm }

minimal brain dysfunction syndrome [MED] A complex of learning and behavioral disabilities seen primarily in children of near-average or above-average intelligence exhibiting also deviations of function of the central nervous system. { ˈmin·ə·məl ¦brān dis′fəŋk·shən ,sin,drōm }

minimal element *See* minimal member. { ˈmin·ə·məl ¦el·ə·mənt }

minimal equation [MATH] **1.** An algebraic equation whose zeros define a minimal surface. **2.** *See* reduced characteristic equation. { ˈmin·ə·məl i′kwā·zhən }

minimal flight path [NAV] The flight path between two points that provides the shortest possible time enroute; it may be planned to take advantage of winds and pressure systems, and is thus not necessarily the shortest air distance between the two points. { ˈmin·ə·məl ˈflīt ,path }

minimal-latency coding *See* minimum-access coding. { ˈmin·ə·məl ¦lāt·ən·sē ,kōd·iŋ }

minimal member [MATH] In a partially ordered set, a minimal member is one for which no other element precedes it in the ordering. Also known as minimal element. { ˈmin·ə·məl ¦mem·bər }

minimal polynomial [MATH] The polynomial of least degree which both divides the characteristic polynomial of a matrix and has the same roots. { ˈmin·ə·məl ¦päl·ə′nō·mē·əl }

minimal realization [CONT SYS] In linear system theory, a set of differential equations, of the smallest possible dimension, which have an input/output transfer function matrix equal to a given matrix function $G(s)$. { ˈmin·ə·məl ,rē·ə·lə′zā·shən }

minimal recognition length [CELL MOL] The shortest length of base pairs that will form a stable deoxyribonucleic acid duplex in genetic recombination. { ¦min·ə·məl ,rek ig′nish·ən ,leŋkth }

minimal surface [MATH] A surface that has assumed a geometric configuration of least area among those into which it can readily deform. { 'min·ə·məl 'sər·fəs }

minimal transformation group [MATH] A transformation group such that every orbit is dense in the phase space. { ¦min·ə·məl ¦tranz·fər'mā·shən ¦grüp }

Minimata disease [MED] A disorder resulting from methyl mercury poisoning, which occurred in epidemic proportions in 1956 in Minimata Bay, a Japanese coastal town, where the inhabitants ate fish contaminated by industrial pollution; the most obvious symptoms are tremors and involuntary movements. { ¦min·ē'mäd·ə di¸zēz }

minimax [MATH] **1.** The minimum of a set of maxima. **2.** In the theory of games, the smallest of a set of maximum possible losses, each representing the most unfavorable outcome of a particular strategy. { 'min·ə¸maks }

minimax criterion [STAT] A concept in game theory and decision theory which requires that losses or expected losses associated with a variable that can be controlled be minimized, and thus maximizes the losses or expected losses associated with the variable that cannot be controlled. { 'min·ə¸maks krī¸tir·ē·ən }

minimax estimator [STAT] A random variable obtained by applying the minimax criterion to a risk function associated with a loss function. { 'min·ə¸maks ¸es·tə¸mād·ər }

mini-maxi regret [CONT SYS] In decision theory, a criterion which selects that strategy which has the smallest maximum difference between its payoff and that of the best hindsight choice. { ¦min·ē ¦mak·sē ri'gret }

minimax technique See min-max technique. { 'min·ē¸maks tek¸nēk }

minimax theorem [MATH] A theorem of games that the lowest maximum expected loss in a two-person zero-sum game equals the highest minimum expected gain. { 'min·ə¸maks ¸thir·əm }

minimization [MATH] The determination of the simplest expression of a Boolean function equivalent to a given one. { ¸min·ə·mə'zā·shən }

minimization principle [PHYS] A principle requiring that the final state of a system is determined by the attainment of the minimum possible value of a certain quantity. { ¸min·ə·mə'zā·shən ¸prin·sə·pəl }

minimize [COMMUN] Condition when normal message and telephone traffic is drastically reduced so messages connected with an actual or simulated emergency will not be delayed. [COMPUT SCI] In a graphical user interface environment, to reduce a window to an icon that represents the application running in the window. { 'min·ə¸mīz }

minimum [MATH] The least value that a real valued function assumes. { 'min·ə·məm }

minimum-access coding [COMPUT SCI] Coding in such a way that a minimum time is required to transfer words to and from storage, for a computer in which this time depends on the location in storage. Also known as minimal-latency coding; minimum-delay coding; minimum-latency coding. { 'min·ə·məm ¦ak¸ses ¸kōd·iŋ }

minimum-access programming [COMPUT SCI] The programming of a digital computer in such a way that minimum waiting time is required to obtain information out of the memory. Also known as forced programming; minimum-latency programming. { 'min·ə·məm ¦ak¸ses 'prō¸gram·iŋ }

minimum-access routine See minimum-latency routine. { 'min·ə·məm ¦ak¸ses rü¸tēn }

minimum-altitude bombing [ORD] Horizontal or glide bombing with the height of release at an altitude under 900 feet (274 meters); it includes masthead bombing (bombing at masthead level), which is sometimes erroneously referred to as skip bombing. { 'min·ə·məm ¦al·tə¸tüd 'bäm·iŋ }

minimum bend radius [MET] The minimum radius through which a piece of metal can be bent to form a given angle without fracturing. { 'min·ə·məm 'bend 'rād·ē·əs }

minimum configuration [COMPUT SCI] **1.** A computer system that has only essential hardware components. **2.** The smallest assortment of hardware and software components required to carry out a particular data-processing function. { 'min·ə·məm kən¸fig·yə'rā·shən }

minimum cut [MATH] For an *s-t* network, an *s-t* cut whose weight has the minimum possible value. { 'min·ə·məm ¦kət }

minimum-delay coding See minimum-access coding. { 'min·ə·məm di¸lā 'kōd·iŋ }

minimum detectable signal See threshold signal. { 'min·ə·məm di'tek·tə·bəl 'sig·nəl }

minimum deviation [OPTICS] For a prism, the smallest possible angle between the incident and refracted rays; this angle is realized when refraction is symmetrical. { 'min·ə·məm ¸dē¸vē'ā·shən }

minimum discernible signal [ELECTR] Receiver input power level that is just sufficient to produce a discernible signal in the receiver output; a receiver sensitivity test. { 'min·ə·məm di'sər·nə·bəl 'sig·nəl }

minimum distance [NAV] The shortest distance at which a navigational instrument or system will function satisfactorily. { 'min·ə·məm 'dis·təns }

minimum-distance code [COMMUN] A binary code in which the signal distance does not fall below a specified minimum value. { 'min·ə·məm ¦dis·təns 'kōd }

minimum dominating vertex set [MATH] A dominating vertex set such that there is no other dominating vertex set with fewer vertices. { 'min·ə·məm ¦däm·ə¸nād·iŋ 'vər¸teks ¸set }

minimum ebb [OCEANOGR] The least speed of a current that runs continuously ebb. { 'min·ə·məm 'eb }

minimum edge cover [MATH] An edge cover of a graph such that there is no other edge cover with fewer vertices. { 'min·ə·məm 'ej ¸kəv·ər }

minimum elevation [ORD] Lowest elevation of a weapon at which the projectile will safely clear an obstacle between the weapon and the target. { 'min·ə·məm ¸el·ə vā·shən }

minimum enroute instrument altitude [NAV] In air operations, the altitude which is attained as the result of the height of an artificial or natural object or the limitation in the performance of navigation and communication facilities. Abbreviated MEA. { 'min·ə·məm än'rüt 'in·strə·mənt ¸al·tə¸tüd }

minimum firing current [ELEC] The limit below which firing will not occur in electric blasting caps. { 'min·ə·məm 'fīr·iŋ ¸kə·rənt }

minimum flight altitude [AERO ENG] The lowest altitude at which aircraft may safely operate. { 'min·ə·məm 'flīt ¸al·tə¸tüd }

minimum flood [OCEANOGR] The least speed of a current that runs continuously flood. { 'min·ə·məm 'fləd }

minimum frontal diameter [ANTHRO] The smallest distance measured between the temporal crests, with only moderate pressure. { 'min·ə·məm 'frənt·əl dī'am·əd·ər }

minimum ionizing speed [ATOM PHYS] The smallest speed at which a charged particle passing through a gas can ionize an atom or molecule. { 'min·ə·məm 'ī·ə¸niz·iŋ ¸spēd }

minimum-latency coding See minimum-access coding. { 'min·ə·məm ¦lat·ən·sē ¸kōd·iŋ }

minimum-latency programming See minimum-access programming. { 'min·ə·məm ¦lat·ən·sē 'prō¸gram·iŋ }

minimum-latency routine [COMPUT SCI] A computer routine that is constructed so that the latency in serial-access storage is less than the random latency that would be expected if storage locations were chosen without regard for latency. Also known as minimum-access routine. { 'min·ə·məm ¦lat·ən·sē rü¸tēn }

minimum lethal dose [PHARM] The amount of an injurious agent which is the average of the smallest dose that kills and the largest dose that fails to kill. { 'min·ə·məm ¦lēth·əl 'dōs }

minimum-loss attenuator [ELECTR] A section linking two unequal resistive impedances which is designed to introduce the smallest attenuation possible. Also known as minimum-loss pad. { 'min·ə·məm ¦lòs ə'ten·yə¸wād·ər }

minimum-loss matching [ELECTR] Design of a network linking two resistive impedances so that it introduces a loss which is as small as possible. { 'min·ə·məm ¦lòs ¸mach·iŋ }

minimum-loss pad See minimum-loss attenuator. { 'min·ə·məm ¦lòs ¸pad }

minimum metal condition [DES ENG] The condition corresponding to the removal of the greatest amount of material permissible in a machined part. { 'min·ə·məm 'med·əl kən¸dish·ən }

minimum-modulus theorem [MATH] The theorem that a nonvanishing, complex analytic function in a closed, bounded, simply connected region assumes its minimum absolute value on the boundary of the region. { ¦min·ə·məm 'mäj·ləs ¸thir·əm }

minimum-phase system [CONT SYS] A linear system for which the poles and zeros of the transfer function all have negative or zero real parts. { 'min·ə·məm 'fāz ,sis·təm }

minimum range [ORD] **1.** Least range setting of a gun at which the projectile will clear an obstacle or friendly troops between the gun and the target. **2.** Shortest distance to which a gun can fire from a given position. { 'min·ə·məm 'rānj }

minimum reflux ratio [CHEM ENG] The smallest reflux ratio in a two-component liquid distillation system that will produce the desired overhead and bottom compositions. { 'min·ə·məm 'rē,fləks ,rā·shō }

minimum rent [MIN ENG] The right to work coal acquired by a mine owner by the payment of an annual rent and a royalty to the landowner (the coal or mineral owner). Also known as fixed rent. { 'min·ə·məm 'rent }

minimum resolvable temperature difference [THERMO] The change in equivalent blackbody temperature that corresponds to a change in radiance which will produce a just barely resolvable change in the output of an infrared imaging device, taking into account the characteristics of the device, the display, and the observer. Abbreviated MRTD. { 'min·ə·məm ri'zäl·və·bəl 'tem·prə·chər ,dif·rəns }

minimum safe distance [ORD] In an atomic explosion, the total distance from desired ground zero (DGZ) to friendly positions required to ensure troops safety. { 'min·ə·məm ¦sāf 'dis·təns }

minimum-shock boom [ACOUS] A sonic boom whose pressure signature is shaped to reduce the perceived amplitude of the shocks by rounding the shape of the signature near the maximum. { ¦min·ə·məm 'shäk ,büm }

minimum signal level [ELECTR] In facsimile, level corresponding to the copy white or copy black signal, whichever is the lower. { 'min·ə·məm 'sig·nəl ,lev·əl }

minimum thermometer [ENG] A thermometer that automatically registers the lowest temperature attained during an interval of time. { 'min·ə·məm thər'mäm·əd·ər }

minimum turning circle [ENG] The diameter of the circle described by the outermost projection of a vehicle when the vehicle is making its shortest possible turn. { 'min·ə·məm 'tərn·iŋ ,sər·kəl }

minimum-variance estimator [STAT] An estimator that possesses the least variance among the members of a defined class of estimators. { ¦min·i·məm 'ver·ē·əns ,es·tə,mād·ər }

minimum vertex cover [MATH] A vertex cover in a graph such that there is no other vertex cover with fewer vertices. { ,min·ə·məm 'vər,teks ,kəv·ər }

minimum wetting rate [CHEM ENG] The smallest liquid-flow rate through a packed column that will thoroughly wet the column packing. { 'min·ə·məm 'wed·iŋ ,rāt }

mining [MIN ENG] The technique and business of mineral discovery and exploitation. { 'mīn·iŋ }

mining camp [MIN ENG] A term loosely applied to any mining town. { 'mīn·iŋ ,kamp }

mining claim [MIN ENG] That portion of the public mineral lands which a miner, for mining purposes, takes and holds in accordance with mining laws. Also known as claim. { 'mīn·iŋ ,klām }

mining effect [ORD] Violent upheaval or movement of earth and the destruction or damage resulting therefrom, generally caused by an explosion below the surface of the earth; mining effect may be contrasted with the blast effect produced by an explosion on or above the surface of the earth. { 'mīn·iŋ i,fekt }

mining engineer [MIN ENG] One qualified by education, training, and experience in mining engineering. { 'mīn·iŋ ,en·jə'nir }

mining engineering [ENG] Engineering concerned with the discovery, development, and exploitation of coal, ores, and minerals, as well as the cleaning, sizing, and dressing of the product. Also known as mineral engineering. { 'mīn·iŋ ,en·jə'nir·iŋ }

mining geology [MIN ENG] The study of the structure and occurrence of mineral deposits and the geologic aspects of mine planning. { 'mīn·iŋ jē'äl·ə·jē }

mining ground [MIN ENG] Land from which a mineral substance is extracted by the process of mining. { 'mīn·iŋ ,graůnd }

mining hazard [MIN ENG] Any of the dangers unique to the winning and working of minerals and coal. { 'mīn·iŋ ,haz·ərd }

mining lease [MIN ENG] A contract to work a mine and extract mineral or other deposits from it under specified conditions. Also known as mineral lease. { 'mīn·iŋ ,lēs }

mining machine truck [MIN ENG] A truck used to transport shortwall mining machines. { 'mīn·iŋ mə,shēn ,trək }

mining on a shoestring *See* grass-roots mining. { 'mīn·iŋ ȯn ə 'shü,striŋ }

mining partnership [MIN ENG] The arrangement whereby two or more persons acquire a mining claim and actually engage in working it. { 'mīn·iŋ 'part·nər,ship }

mining property [MIN ENG] Property valued for its mining possibilities. { 'mīn·iŋ ,präp·ərd·ē }

mining retreating [MIN ENG] A process of mining by which the ore or coal is untouched until after all the gangways and such are driven, at which time the work of extraction begins at the boundary and progresses toward the shaft. { 'mīn·iŋ rē'trēd·iŋ }

mining right [MIN ENG] A right to enter upon and occupy a specific piece of ground for the purpose of working it, either by underground excavations or open workings, to obtain the mineral ores which may be deposited therein. Also known as mineral right. { 'mīn·iŋ ,rīt }

mining shield [MIN ENG] A canopy or cover for the protection of workers and machines at the face of a mechanized coal heading. { 'mīn·iŋ ,shēld }

mining title [MIN ENG] A claim, exclusive prospecting license, right, concession, or lease. { 'mīn·iŋ ,tīd·əl }

mining town [MIN ENG] A town that has arisen next to a mine or mines. { 'mīn·iŋ ,taůn }

mining width [MIN ENG] The minimum width needed to extract ore regardless of the actual width of ore-bearing rock. { 'mīn·iŋ ,width }

miniplasmid [MOL BIO] Any plasmid that has been reduced in size by means of recombinant deoxyribonucleic acid technology. { ,min·ē'plaz·mid }

miniprinter [GRAPHICS] A printer without a keyboard, driven by an external digital source such as a computer, communication line, or tape handler to print lines of alphameric data on paper; generally applied to printers that have less than 80 characters per line or lower printing speeds than 120 characters per second. { 'min·ē,print·ər }

minisolar nebula [ASTRON] A rotating flattened cloud of gas and dust from which a giant planet and its satellites formed in a manner similar to the formation of the solar system from the solar nebula. { ,min·ē,sō·lər 'neb·yə·lə }

mini-supercomputer [COMPUT SCI] A supercomputer that is about a quarter to a half as fast in vector processing as the most powerful supercomputers. { ,min·ē'sü·pər·kəm,püd·ər }

minitrack [AERO ENG] A satellite tracking system consisting of a field of separate antennas and associated receiving equipment interconnected so as to form interferometers which track a transmitting beacon in the payload itself. [ELECTR] A subminiature radio transmitter capable of sending data over 4000 miles (6500 kilometers) on extremely low power. { 'min·ē,trak }

minium [MINERAL] Pb_3O_4 A scarlet or orange-red mineral consisting of an oxide of lead; found in Wisconsin and the western United States. Also known as red lead. { 'min·ē·əm }

mink [VERT ZOO] Any of three species of slender-bodied aquatic carnivorous mammals in the genus *Mustela* of the family Mustelidae. { miŋk }

Minkowski distance function [MATH] Relative to a convex body with the origin O in its interior, the function whose value at a point P is the distance ratio OP/OQ, where Q is the point of the convex body on the ray OP that is furthest from O. { miŋ'kȯf·skē *or* min'kaů·skē 'dis·təns ,fəŋk·shən }

Minkowski electrodynamics [ELECTROMAG] An electromagnetic theory, compatible with the special theory of relativity, which takes into account the presence of matter with electric and magnetic polarization. { miŋ'kȯf·skē i¦lek·trō·dī¦nam·iks }

Minkowski metric [RELAT] The metric tensor of the Minkowski space-time used in special relativity; it is a 4 × 4 matrix whose nonzero entries lie on the diagonal, with one entry (corresponding to the time coordinate) equal to 1, and three

MINK

The American mink (*Mustela vison*), the largest mink.

entries (corresponding to space coordinates) equal to -1; sometimes, the negative of this matrix is used. { miŋ'kȯf·skē 'me·trik }

Minkowski's inequality [MATH] **1.** An inequality involving powers of sums of sequences of real or complex numbers, a_k and b_k:

$$\left[\sum_{k=1}^{\infty}|a_k+b_k|^s\right]^{1/s} \le \left[\sum_{k=1}^{\infty}|a_k|^s\right]^{1/s} + \left[\sum_{k=1}^{\infty}|b_k|^s\right]^{1/s}$$

provided $s \ge 1$. **2.** An inequality involving powers of integrals of real or complex functions, f and g, over an interval or region R:

$$\left[\int_R |f(x)+g(x)|^s\,dx\right]^{1/s}$$

$$\le \left[\int_R |f(x)|^s\,dx\right]^{1/s} + \left[\int_R |g(x)|^s\,dx\right]^{1/s}$$

provided $s \ge 1$ and the integrals involved exist. { miŋ'kȯf·skēz ,in·i'kwäl·əd·ē }

Minkowski space-time [RELAT] The space-time of special relativity; it is completely flat and contains no gravitating matter. Also known as Minkowski universe. { miŋ'kȯf·skē 'spās 'tīm }

Minkowski universe *See* Minkowski space-time. { miŋ'kȯf·skē 'yü·nə,vərs }

min-max technique [MATH] A method of approximation of a function f by a function g from some class where the maximum of the modulus of $f - g$ is minimized over this class. Also known as Chebyshev approximation; minimax technique. { 'min 'maks tek,nēk }

min-max theorem [MATH] The theorem that provides information concerning the nth eigenvalue of a symmetric operator on an inner product space without necessitating knowledge of the other eigenvalues. Also known as maximum-minimum principle. { 'min 'maks ,thir·əm }

Minnesota multiphasic personality inventory [PSYCH] An empirical scale of an individual's personality based mainly on the person's yes-or-no responses to a questionnaire of 550 items; included are special validating scales which measure the individual's test-taking attitude and degree of frankness. Abbreviated MMPI. Also known as multiphasic personality inventory (MPI). { ,min·ə'sōd·ə ,məl·tə,fāz·ik ,pər·sə'nal·əd·ē ,in·vən,tȯr·ē }

Minnesota preschool scale [PSYCH] A verbal and nonverbal test designed to measure the learning ability of children from 18 months to 6 years of age. { ,min·ə'sōd·ə 'prē,skül ,skāl }

minnow [VERT ZOO] The common name for any fresh-water fish composing the family Cyprinidae, order Cypriniformes. { 'min·ō }

minor [MATH] The minor of an entry of a matrix is the determinant of the matrix obtained by removing the row and column containing the entry. Also known as cofactor; complementary minor. { 'mīn·ər }

minor arc [MATH] The smaller of the two arcs on a circle produced by a secant. { 'mīn·ər 'ärk }

minor axis [MATH] The smaller of the two axes of an ellipse. { 'mīn·ər 'ak·səs }

minor bend [ELECTROMAG] Rectangular waveguide bent so that throughout the length of a bend a longitudinal axis of the guide lies in one plane which is parallel to the narrow side of the waveguide. { 'mīn·ər 'bend }

minor control change [COMPUT SCI] A change of function that is of relatively small magnitude or importance, resulting from a difference in minor control data between one card and the next. { 'mīn·ər kən'trōl ,chānj }

minor control data [COMPUT SCI] Control data which are at the least significant level used, or which are used to sort records into the smallest groups used; for example, if control data are used to specify state, town, and street, then the data specifying street would be minor control data. { 'mīn·ər kən'trōl ,dad·ə }

minor cycle [COMPUT SCI] The time required for the transmission or transfer of one machine word, including the space

between words, in a digital computer using serial transmission. Also known as word time. { 'mīn·ər ,sī·kəl }

minor defect [IND ENG] A defect which reduces the effectiveness of the product, without causing serious malfunctioning. { 'mīn·ər di'fekt }

minor diameter [DES ENG] The diameter of a cylinder bounding the root of an external thread or the crest of an internal thread. { 'mīn·ər dī'am·əd·ər }

minor diatonic scale [ACOUS] A diatonic scale in which the relative sizes of the sequence of intervals are approximately 2,1,2,2,2,2,1. { 'mīn·ər ¦dī·ə¦tän·ik 'skāl }

minor fog signal [NAV] A sound signal of limited power generally having a normal range of 2 miles (3.2 kilometers) or less. { 'mīn·ər 'fäg ,sig·nəl }

minority carrier [SOLID STATE] The type of carrier, electron, or hole that constitutes less than half the total number of carriers in a semiconductor. { mə'när·əd·ē 'kar·ē·ər }

minority emitter [ELECTR] Of a transistor, an electrode from which a flow of minority carriers enters the interelectrode region. { mə'när·əd·ē i'mid·ər }

minor key [COMPUT SCI] A secondary key for identifying a record. { 'mīn·ər ,kē }

minor light [NAV] An automatic unwatched light on a fixed structure usually showing low to moderate candlepower. { 'mīn·ər ¦līt }

minor lobe [ELECTROMAG] Any lobe except the major lobe of an antenna radiation pattern. Also known as secondary lobe; side lobe. { 'mīn·ər ¦lōb }

minor loop [CONT SYS] A portion of a feedback control system that consists of a continuous network containing both forward elements and feedback elements. { 'mīn·ər ¦lüp }

minor planet [ASTRON] **1.** Those planets smaller than the earth, specifically Mercury, Venus, Mars, and Pluto. **2.** *See* asteroid. { 'mīn·ər 'plan·ət }

minor relay station [ELECTR] A tape relay station which has tape relay responsibility but does not provide an alternate route. { 'mīn·ər rē,lā 'stā·shən }

minor surgery [MED] Any superficial surgical procedure involving little hazard to the life of the patient and not requiring general anesthesia. { 'mīn·ər 'sər·jə·rē }

minor switch [ELEC] Single-motion stepping switch mounted atop the telephone connectors and most commonly used for the party-line selection. { 'mīn·ər ¦swich }

minor trough [METEOROL] A pressure trough of smaller scale than a long-wave trough; it ordinarily moves rapidly and is associated with a migratory cyclonic disturbance in the lower troposphere. { 'mīn·ər ¦trȯf }

minuano [METEOROL] A cold southwesterly wind of southern Brazil, occurring during the Southern Hemisphere winter (June to September). { ,min·yə'wä·nō }

minuend [MATH] The quantity from which another quantity is to be subtracted. { 'min·yə,wend }

minus [MATH] A minus B means that the quantity B is to be subtracted from the quantity A. { 'mī·nəs }

minus angle *See* angle of depression. { 'mī·nəs 'aŋ·gəl }

minus-cement porosity [GEOL] The porosity that would characterize a sedimentary material if it contained no chemical cement. { 'mī·nəs si¦ment pə'räs·əd·ē }

minus sieve [MET] In powder metallurgy, the portion of a powder sample that passes through a specified standard sieve. { 'mī·nəs ,siv }

minus sight *See* foresight. { 'mī·nəs ,sīt }

minus sign *See* subtraction sign. { 'mī·nəs ,sīn }

minus strand [CELL MOL] A polynucleotide strand that is complementary to, and formed by transcription from, another specific polynucleotide (plus) strand, with which it produces the double-stranded (double-helix) ribonucleic acid. { 'mī·nəs ,strand }

minus zone [COMPUT SCI] The bit positions in a computer code that represent the algebraic minus sign. { 'mī·nəs ,zōn }

minute [MATH] A unit of measurement of angle that is equal to $^1/_{60}$ of a degree. Symbolized '. Also known as arcmin. [MECH] A unit of time, equal to 60 seconds. { 'min·ət }

minute hand [HOROL] The long hand on a timepiece, indicating minutes. { 'min·ət ,hand }

minute pressure [PETRO ENG] A measurement of gas well capacity, determined by shutting the gate valve and recording the gage pressure each minute, with the pressure at the end of

MINNOW

The common shiner (*Notropis cornutus*).

the first minute providing an estimate of the gas volume. { 'min·ət ˌpresh·ər }

minutes on leg [NAV] The interval of time, estimated or actual, required for an aircraft to fly a given distance at a given ground speed, usually along a single course line. { 'min·əts òn 'leg }

minute wheel [HOROL] A wheel in a dial train that is driven by the cannon pinion, and drives the hour wheel by means of a pinion. { 'min·ət ˌwēl }

minyulite [MINERAL] $KAl_2(PO_4)_2(OH,F)\cdot 4H_2O$ A white mineral composed of hydrous basic potassium aluminum phosphate. { 'min·yəˌlīt }

Miocene [GEOL] A geologic epoch of the Tertiary period, extending from the end of the Oligocene to the beginning of the Pliocene. { 'mī·əˌsēn }

miocrystalline See hypocrystalline. { ¦mī·ō'krist·əl·ən }

miogeocline [GEOL] A nonvolcanic (nonmagmatic) continental margin, characterized by carbonate, shale, and sandstone sediments. { ˌmī·ō'jē·əˌklīn }

miogeosyncline [GEOL] The nonvolcanic portion of an orthogeosyncline, located adjacent to the craton. { ¦mī·ō'jē·ō'sinˌklīn }

Miosireninae [PALEON] A subfamily of extinct sirenian mammals in the family Dugongidae. { 'mī·ō·sə'ren·əˌnē }

miosis [PHYSIO] Contraction of the pupil of the eye. { mī'ō·səs }

miotic [PHARM] **1.** Causing miosis. **2.** Any agent that causes miosis. [PHYSIO] Of or pertaining to miosis. { mī'äd·ik }

MIPC See ortho-isopropylphenyl-methylcarbamate.

MIPS See million instructions per second.

Mira [ASTRON] The first star recognized to be a periodic variable; has a period of 332 ± 9 days and its spectrum changes from M5e at maximum to M9e at minimum; it is the prototype of long-period variable stars. { 'mir·ə }

mirabilite [MINERAL] $Na_2SO_4\cdot 10H_2O$ A yellow or white monoclinic mineral consisting of hydrous sodium sulfate, occurring as a deposit from saline lakes, playas, and springs, and as an efflorescence; the pure crystals are known as Glauber's salt. { mə'rab·əˌlīt }

miracidium [INV ZOO] The ciliated first larva of a digenetic trematode; forms a sporocyst after penetrating intermediate host tissues. { ˌmī·rə'sid·ē·əm }

mirage [OPTICS] Any one of a variety of unusual images of distant objects seen as a result of the bending of light rays in the atmosphere during abnormal vertical distribution of air density. { mə'räzh }

mirage effect [COMMUN] Reception of radio waves at distances far beyond the normally expected range due to abnormal refraction caused by meteorological conditions such as abnormal vertical water-vapor and temperature gradients. { mə'räzh iˌfekt }

Miranda [ASTRON] A satellite of Uranus orbiting at a mean distance of 76,880 miles (124,000 kilometers). { mə'ran·də }

Mirapinnatoidei [VERT ZOO] A suborder of tiny oceanic fishes in the order Cetomimiformes. { ˌmir·ə'pin·ə'tóid·ē·ī }

Mira-type variable [ASTRON] A member of a class of long-period variable stars whose prototype is the star Mira and that exhibit a periodic change in brightness with a time interval between 100 and 1000 days and a range variation of 2.5 magnitudes or more, but that have some cycles that may be much brighter or fainter than others. { 'mir·əˌtīp 'ver·ē·ə·bəl }

mire [GEOL] Wet spongy earth, as of a marsh, swamp, or bog. { mīr }

mired [THERMO] A unit used to measure the reciprocal of color temperature, equal to the reciprocal of a color temperature of 10^6 kelvins. Derived from micro-reciprocal-degree. { mīrd }

Miridae [INV ZOO] The largest family of the Hemiptera; included in the Cimicomorpha, it contains herbivorous and predacious plant bugs which lack ocelli and have a cuneus and four-segmented antennae. { 'mir·əˌdē }

Miripinnati [VERT ZOO] The equivalent name for Marapinnatoidei. { ˌmir·ə·pə'näd·ē }

mirror [OPTICS] A surface which specularly reflects a large fraction of incident light. { 'mir·ər }

mirror coating [OPTICS] A thin film of highly reflective material spread over a correctly shaped glass surface to produce

a mirror; aluminum is usually used in the visible region. Also known as reflective coating. { 'mir·ər ˌkōd·iŋ }

mirror galvanometer [ELEC] A galvanometer having a small mirror attached to the moving element, to permit use of a beam of light as an indicating pointer. Also known as reflecting galvanometer. { 'mir·ər ˌgal·və'näm·əd·ər }

mirror glance See wehrlite. { 'mir·ər ˌglans }

mirror image [OPTICS] A form that is identical to another except that it is reversed, as if viewed in a mirror. { 'mir·ər 'im·ij }

mirror-image programming [CONT SYS] Programming of a robot in which the x and y axes are reversed in all instructions, in order to create mirror images of workpieces. { 'mir·ər ¦imij 'prōˌgram·iŋ }

mirror interference [OPTICS] Interference occurring between two beams, one or both of which are reflected from a mirror at a small angle. { 'mir·ər ˌin·tər'fir·əns }

mirror interferometer [ENG] An interferometer used in radio astronomy, in which the sea surface acts as a mirror to reflect radio waves up to a single antenna, where the reflected waves interfere with the waves arriving directly from the source. [OPTICS] Any interferometer which makes use of mirror interference. { 'mir·ər ˌin·tər·fə'räm·əd·ər }

mirror machine [PL PHYS] A device which confines plasma in a tube with magnetic mirrors at each end to prevent it from escaping. { 'mir·ər mə,shēn }

mirror nephoscope [ENG] A nephoscope in which the motion of a cloud is observed by its reflection in a mirror. Also known as cloud mirror; reflecting nephoscope. { 'mir·ər 'nef·əˌskōp }

mirror nuclei [NUC PHYS] A pair of atomic nuclei, each of which would be transformed into the other by changing all its neutrons into protons, and vice versa. { 'mir·ər 'nü·klē·ī }

mirror optics [OPTICS] The science and technology of mirrors which, by means of reflecting rays of light, either revert optical bundles or focus them to form images. { 'mir·ər 'äp·tiks }

mirror plane of symmetry See plane of mirror symmetry. { 'mir·ər 'plān əv 'sim·ə·trē }

mirror reflection See specular reflection. { 'mir·ər ri'flek·shən }

mirror scale [ENG] A scale with a mirror used to align the eye perpendicular to the scale and pointer when taking a reading; improves accuracy by eliminating parallax. { 'mir·ər ˌskāl }

mirror stone See muscovite. { 'mir·ər ˌstōn }

mirror transit circle [ENG] A development of the conventional transit circle in which light from a star is reflected into fixed horizontal telescopes pointing due north and south by a plane mirror that is mounted on a horizontal east-west axis and attached to a large circle with accurately calibrated markings to determine the mirror's position. { 'mir·ər 'tran·zit ˌsər·kəl }

Mirsky's theorem [MATH] The theorem that, in a finite partially ordered set, the maximum cardinality of a chain is equal to the minimum number of disjoint antichains into which the partially ordered set can be partitioned. { 'mir·skēz ˌthir·əm }

MIR technique See multiple isomorphous replacement technique. { ¦em¦ī'är tekˌnēk }

MIRV See multiple independently targeted reentry vehicle. { mərv }

MIS See management information system; metal-insulator semiconductor.

miscarriage See spontaneous abortion. { 'mis,kar·ij }

miscegenation [ANTHRO] Intermarriage between different races. { miˌsej·ə'nā·shən }

misch metal [MET] An alloy consisting of a crude mixture of cerium, lanthanum, and other rare-earth metals obtained by electrolysis of the mixed chlorides of the metals dissolved in fused sodium chloride; used in making aluminum alloys, in some steels and irons, and in coating the cathodes of glow-type voltage regulator tubes. { 'mish ,med·əl }

miscibility [CHEM] The tendency or capacity of two or more liquids to form a uniform blend, that is, to dissolve in each other; degrees are total miscibility, partial miscibility, and immiscibility. { ˌmis·ə'bil·əd·ē }

miscible-phase displacement [PETRO ENG] Method of increasing reservoir oil recovery by displacement with an oil-miscible driving fluid, such as gas or liquefied petroleum gas. { 'mis·ə·bəl ¦fāz diˌsplās·mənt }

MIOCENE

PRECAMBRIAN	PALEOZOIC					CARBON-IFEROUS							MESOZOIC	CENOZOIC
	CAMBRIAN	ORDOVICIAN	SILURIAN	DEVONIAN	Mississippian	Pennsylvanian	PERMIAN	TRIASSIC	JURASSIC	CRETACEOUS	TERTIARY	QUATERNARY		

TERTIARY					QUATERNARY	
Paleocene	Eocene	Oligocene	Miocene	Pliocene	Pleistocene	Recent

Position of the Miocene epoch in relation to other epochs and to the periods and eras of geologic time.

MIRROR MACHINE

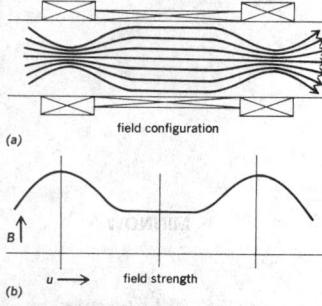

Mirror machine. (a) Longitudinal section. Lines represent magnetic lines of force. (b) Curve represents magnetic field strength B along path u in the tube. Regions where B has a large value are magnetic mirrors produced by coils and located at constrictions of field lines at each end; they confine plasma to central part.

miscible-slug process [PETRO ENG] A miscible-phase displacement in which reservoir oil recovery is increased by displacement with liquefied petroleum gas as the driving fluid. { 'mis·ə·bəl ¦sləg ¦prä·səs }

misenite [MINERAL] $K_8H_6(SO_4)_7$ A white mineral composed of native acid potassium sulfate. { mə'ze‚nīt }

miser [PETRO ENG] A well-boring bit that is tubular with a valve at the bottom, and has a screw for forcing the earth upward. Also spelled mizer. { 'mī·zər }

misfeed [ORD] Failure to supply ammunition properly, especially to a magazine-fed or belt-fed automatic gun. { 'mis‚fēd }

misfire [CHEM] Failure of fuel or an explosive charge to ignite properly. [ELECTR] Failure to establish an arc between the main anode and the cathode of an ignitron or other mercury-arc rectifier during a scheduled conducting period. { 'mis‚fīr }

misfit stream [HYD] A stream whose meanders are either too large or too small to have eroded the valley in which it flows. { 'mis‚fit ¦strēm }

mismatch [ELEC] The condition in which the impedance of a source does not match or equal the impedance of the connected load or transmission line. [MET] Failure to match forged surfaces formed in opposite dies. { 'mis‚mach }

mismatch factor See reflection factor. { 'mis‚mach ‚fak·tər }

mismatch loss [ELECTR] Loss of power delivered to a load as a result of failure to make an impedance match of a transmission line with its load or with its source. { 'mis‚mach ‚lòs }

mismatch slotted line [ELECTROMAG] A slotted line linking two waveguides which is not properly designed to minimize the power reflected or transmitted by it. { 'mis‚mach 'släd·əd 'līn }

misogamy [PSYCH] A feeling of revulsion and repugnance toward marriage. { mi'säg·ə·mē }

misogyny [PSYCH] Hatred of women. { mi'säj·ə·nē }

misology [PSYCH] Unreasoning aversion to intellectual or literary matters, or to argument or speaking. { mi'säl·ə·jē }

misoneism [PSYCH] Hatred of new circumstances or things, or change. { ‚mis·ə'nē‚iz·əm }

misopedia [PSYCH] Morbid hatred of all children, but especially of one's own. { ‚mis·ə'pē·dē·ə }

mispairing [CELL MOL] Pairing of a nucleotide in one chain of a deoxyribonucleic acid molecule that is not complementary to the nucleotide occupying the corresponding position in the other chain. { mis'per·iŋ }

mispickel See arsenopyrite. { 'mi‚spik·əl }

misregistration [COMPUT SCI] In character recognition, the improper state of appearance of a character, line, or document, on site in a character reader, with respect to a real or imaginary horizontal baseline. [GRAPHICS] Improper alignment of register marks in matching overlays on a page or in matching the front of a type page to the back. { ‚mis‚rej·ə'strā·shən }

misrepair [CELL MOL] Repair of deoxyribonucleic acid that gives rise to gene mutations or changes in chromosome structure. { 'mis‚ri‚per }

misrun [MET] Incompletely formed casting due to premature solidification of metal before the mold is filled. { mis'rən }

missed abortion [MED] A condition in which a fetus weighing less than 500 grams dies and remains in the uterus for an extended period of time. { 'mist ə'bòr·shən }

missed hole See failed hole. { 'mist 'hōl }

missed round [ENG] A round in which all or part of the explosive has failed to detonate. { 'mist 'raúnd }

missense codon [GEN] A mutant codon that directs the incorporation of a different amino acid and results in the synthesis of a protein with a sequence in which one amino acid has been replaced by a different one; in some cases the mutant protein may be unstable or less active. { 'mis‚ens 'kō‚dän }

missense mutation [CELL MOL] A mutation that converts a codon coding for one amino acid to a codon coding for another amino acid. { 'mis·əns myü'tā·shən }

missense suppressor [CELL MOL] A suppressor that incorporates the correct amino acid at the site of a codon that has been altered because of a missense mutation. { 'mis·əns sə‚pres·ər }

missile [ORD] Any object that is, or is designed to be, thrown, dropped, projected, or propelled, for the purpose of making it strike a target; examples are guided missile and ballistic missile. { 'mis·əl }

missile attitude [MECH] The position of a missile as determined by the inclination of its axes (roll, pitch, and yaw) in relation to another object, as to the earth. { 'mis·əl ‚ad·ə‚tüd }

missile checkout [ORD] Performance of procedures to determine whether all parts of the missile are apparently capable of performing their prescribed functions. { 'mis·əl 'chek‚aút }

missile countermeasure [ORD] Any device or technique intended to impair the effectiveness of an enemy missile. { 'mis·əl 'kaúnt·ər‚mezh·ər }

missile decoy [ORD] A vehicle used to simulate a missile in flight, attracting enemy radar and drawing fire so as to increase the odds for penetration by manned bombers or other weapon systems. Also known as diversionary missile. { 'mis·əl 'dē kói }

missile destruct system [ORD] That portion of a missile which, upon command or due to a failure, is capable of destroying the missile, generally for safety reasons. { 'mis·əl di‚strəkt ‚sis·təm }

missile guidance central [ORD] A collection of items specifically designed to provide target ranging facilities, and guidance and control equipment for directing the terminal flight phase of a guided missile. { 'mis·əl gīd·əns 'sen·trəl }

missile impact predictor group [ORD] A group that provides facilities for determining the point of impact of a missile. { 'mis·əl im‚pakt prə‚dik·tər ‚grüp }

missile plume [ELECTROMAG] The region of electromagnetic and other disturbances that follow a missile during reentry and make the missile more readily detectable. { 'mis·əl ‚plüm }

missile position tracking group [ORD] A group that provides facilities for determining the in-flight position of a missile. { 'mis·əl pə‚zish·ən 'trak·iŋ ‚grüp }

missile ship See guided-missile ship. { 'mis·əl ‚ship }

missile site radar [ENG] Phased array radar located at a missile launch area to provide a guidance link with interceptor missiles enroute to their targets. { 'mis·əl 'sīt 'rā‚där }

missile test [ORD] The test that includes missile checkout plus actual firing to determine whether the missile is capable of performing its operational function. { 'mis·əl ‚test }

missing error [COMPUT SCI] The result of calling for a subroutine not available in the library. { ‚mis·iŋ 'er·ər }

missing mass See dark matter. { ‚mis·iŋ 'mas }

missing mass spectrometer [PARTIC PHYS] An apparatus which measures the momentum of the recoil protons in a reaction such as $\pi^- + p \to p + (MM)^-$, in order to determine the distribution of masses of the MM system, without any detailed observations of this system. { ‚mis·iŋ ‚mas spek'träm·əd·ər }

mission adaptive wing [AERO ENG] An advanced supercritical wing design that uses smooth variable wing camber to change the shape of the wing, thereby providing high levels of aerodynamic efficiency over a large range of subsonic, transonic, and supersonic flight conditions. { 'mish·ə‚dap·tiv ¦wiŋ }

Mississippian [GEOL] The fifth period of the Paleozoic Era beginning about 350 million years ago and ending about 320 million years ago. The Mississippian System (referring to rocks) or Period (referring to the time during which these rocks were deposited) is employed in North America as the lower (or older) subdivision of the Carboniferous, as used in Europe and on other continents. { ¦mis·ə‚sip·ē·ən }

Mississippi River-type buoy [NAV] A lightweight unlighted buoy of unique design developed for use on rivers, and now used in many inland areas. { ¦mis·ə‚sip·ē 'riv·ər ¦tüp ‚bói }

Missnay-Schardin effect [ORD] The acceleration of a solid end plate (usually metal) from the face of an explosive charge under detonation, such that the end plate remains a solid and is usable as a missile. { 'mis‚nā 'shärd·ən i‚fekt }

Missourian [GEOL] A North American provincial series of geologic time: lower Upper Pennsylvanian (above Desmoinesian, below Virgilian). { mə'zür·ē·ən }

mist [FL MECH] Fine liquid droplets suspended in or falling through a moving or stationary gas atmosphere. [METEOROL] A hydrometeor consisting of an aggregate of microscopic and more or less hygroscopic water droplets suspended in the atmosphere; it produces, generally, a thin, grayish veil over the landscape; it reduces visibility to a lesser extent than fog; the relative humidity with mist is often less than 95. { mist }

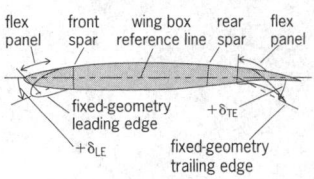

MISSION ADAPTIVE WING

flex panel   front spar   wing box reference line   rear spar   flex panel

fixed-geometry leading edge $+\delta_{TE}$

$+\delta_{LE}$ fixed-geometry trailing edge

Smooth variable-camber flap shape of mission-adaptive wing. δ_{LE} and δ_{TE} represent the deflections of the leading and trailing edges respectively.

MISSISSIPPIAN

CENOZOIC	QUATERNARY	
	TERTIARY	
MESOZOIC	CRETACEOUS	
	JURASSIC	
	TRIASSIC	
PALEOZOIC	PERMIAN	
	CARBONIFEROUS	PENNSYLVANIAN
		MISSISSIPPIAN
	DEVONIAN	
	SILURIAN	
	ORDOVICIAN	
	CAMBRIAN	
PRECAMBRIAN		

Chart showing position of the Mississippian in geologic time.

mistake [COMPUT SCI] A human action producing an unintended result, in contrast to an error in a computer operation. { mə'stāk }

mistbow See fogbow. { 'mist,bō }

mist drilling See foam drilling. { 'mist ,dril·iŋ }

mist droplet [METEOROL] A particle of mist, intermediate between a haze droplet and a fog drop. { 'mist ,dräp·lət }

mist extractor [ENG] A device that removes liquid mist or droplets from a gas stream via impingement, flow-direction change, velocity change, centrifugal force, filters, or coalescing packs. { 'mist ik,strak·tər }

mistletoe [BOT] **1.** *Viscum album.* The true, Old World mistletoe having dichotomously branching stems, thick leathery leaves, and waxy-white berries. **2.** Any of several species of green hemiparasitic plants of the family Loranthaceae. { 'mis·əl,tō }

mist projector [MIN ENG] An appliance to form a mist spray to allay dust and fume during blasting operations in a tunnel. { 'mist prə,jek·tər }

mistral [METEOROL] A north wind which blows down the Rhone Valley south of Valence, France, and into the Gulf of Lions. Strong, squally, cold, and dry, it is the combined result of the basic circulation, a fall wind, and jet-effect wind. { mə'sträl }

mistranslation [CELL MOL] Incorporation of the wrong amino acid into a protein due to misreading of a codon. { ,mis·tranz'lā·shon }

mistuning [MECH] The difference between the square of the natural frequency of vibration of a vibrating system, without the effect of damping, and the square of the frequency of an external, oscillating force. { mis'tün·iŋ }

misuse detection [COMPUT SCI] The technology that seeks to identify an attack on a computer system by its attempted effect on sensitive resources. { mis'yüs di,tek·shən }

MIT bag model [PARTIC PHYS] A model describing quark confinement in hadrons in which a hadron is viewed as a bubble of gas in a uniform, isotropic, perfect fluid, with the thermodynamic pressure of the gas replaced by the quantum pressure of quarks. Derived from Massachusetts Institute of Technology bag model. { ¦em¦ī¦tē 'bag ,mäd·əl }

Mitchell's disease See erythromelalgia. { 'mich·əlz di,zēz }

mite [INV ZOO] The common name for the acarine arthropods composing the diverse suborders Onychopalpida, Mesostigmata, Trombidiformes, and Sarcoptiformes. { mīt }

miter bend [DES ENG] A pipe bend made by mitering (angle cutting) and joining pipe ends. { 'mīd·ər ,bend }

miter box [ENG] A troughlike device of metal or wood with vertical slots set at various angles in the upright sides, for guiding a handsaw in making a miter joint. { 'mīd·ər ,bäks }

miter gate [CIV ENG] Either of a pair of canal lock gates that swing out from the side walls and meet at an angle pointing toward the upper level. { 'mīd·ər ,gāt }

miter gear [DES ENG] A bevel gear whose bevels are in 1:1 ratio. { 'mīd·ər ,gir }

miter joint [DES ENG] A joint, usually perpendicular, in which the mating ends are beveled. { 'mīd·ər ,jóint }

miter saw [DES ENG] A hollow-ground saw in diameters from 6 to 16 inches (15.24 to 40.64 centimeters), used for cutting off and mitering on light stock such as moldings and cabinet work. { 'mīd·ər ,só }

miter valve [DES ENG] A valve in which a disk fits in a seat making a 45° angle with the axis of the valve. { 'mīd·ər ,valv }

miticide [MATER] An agent that kills mites. Also known as acaricide. { 'mīd·ə,sīd }

mitochondria [CELL MOL] Minute cytoplasmic organelles in the form of spherical granules, short rods, or long filaments found in almost all living cells; submicroscopic structure consists of an external membrane system. { ,mīd·ə'kän·drē·ə }

mitochondrial crest [CELL MOL] Any of the infoldings of the mitochondrial inner membrane that extend into the matrix. { ,mīd·ə¦kän·drē·əl 'kres }

mitochondrial deoxyribonucleic acid [BIOCHEM] The circular deoxyribonucleic acid duplex, generally 5 to 10 copies, contained within a mitochondrion and maternally inherited since only the egg cell contributes significant numbers of mitochondria to the zygote. Abbreviated mtDNA. Also known as mitochondrial genome. { ,mīd·ə¦kän·drē·əl dē¦äk·sē,rī·bō·nü'klē·ik 'as·əd }

mitochondrial genome See mitochondrial deoxyribonucleic acid. { ,mīd·ə¦kän·drē·əl 'je,nōm }

mitochondrial plasmid [CELL MOL] Plasmid-like deoxyribonucleic acid molecules in mitochondria of certain higher plants and some fungi. { ,mīd·ə¦kän·drē·əl 'plaz·mid }

mitogen [CELL MOL] A compound that stimulates cells to undergo mitosis. { 'mīd·ə,jen }

mitogen-activated protein kinases [CELL MOL] A large kinase network in which upstream kinases activate downstream kinases that, in response to phosphorylation, translocate to the nucleus and activate transcription factors. Abbreviated MAPKs. { ,mīd·ə·jən ¦ak·tə,vād·əd ¦prō,tēn 'kī,nās·əs }

mitogenesis [CELL MOL] **1.** Induction of mitosis. **2.** Formation as a result of mitosis. { ,mīd·ə'jen·ə·səs }

mitomycin [MICROBIO] A complex of three antibiotics (mitomycin A, B, and C) produced by *Streptomyces caespitosus.* { ¦mīd·ə¦mīs·ən }

mitoplast [CELL MOL] **1.** A mitochondrion that has had its outer membrane removed. **2.** The cytoplast of a mitotic cell after the chromosomes are extruded. { 'mīd·ə,plast }

mitosis [CELL MOL] Nuclear division involving exact duplication and separation of the chromosome threads so that each of the two daughter nuclei carries a chromosome complement identical to that of the parent nucleus. { mī'tō·səs }

mitotic apparatus [CELL MOL] A transitory organelle-like formation that is seen during mitosis and meiosis and consists of the asters, the spindle, and the traction fibers. { mī¦täd·ik ,ap·ə'rad·əs }

mitotic center [CELL MOL] A structure that defines the poles toward which chromosomes move during mitosis and meiosis. { mī¦täd·ik 'sen·tər }

mitotic chromosomes [CELL MOL] Condensed, replicated chromosomes held together as sister chromatids by the centromere as part of the process ensuring accurate segregation and transmission of genetic material to the daughter cells during cell division. { mī¦täd·ik 'krō·mə,sōmz }

mitotic index [CELL MOL] The number of cells undergoing mitosis per thousand cells. { mī'täd·ik 'in,deks }

mitotic inhibitor [CELL MOL] A compound that inhibits mitosis. { mī'täd·ik in'hib·əd·ər }

mitotic poison [CELL MOL] A compound that prevents or affects the completion of mitosis. { mī'täd·ik 'póiz·ən }

mitral commissurotomy [MED] Any of several surgical procedures performed to relieve mitral stenosis. { 'mī·trəl ,käm·ə·sə'räd·ə·mē }

mitral stenosis [MED] Obstruction of the mitral valve, usually due to narrowing of the orifice. { 'mī·trəl stə'nō·səs }

mitral valve [ANAT] The atrioventricular valve on the left side of the heart. { 'mī·trəl 'valv }

mitriform [BIOL] Shaped like a miter. { 'mī·trə,fórm }

mitscherlichite [MINERAL] $K_2CuCl_4 \cdot 2H_2O$ A greenish-blue, tetragonal mineral consisting of potassium copper chloride dihydrate. { 'mich·ər·lə,kīt }

Mitscherlich law of isomorphism [CHEM] Substances which have similar chemical properties and crystalline forms usually have similar chemical formulas. { 'mich·ər,lik ¦lō əv ¦ī·sō'mór,fiz·əm }

Mittag-Leffler's theorem [MATH] A theorem that enables one to explicitly write down a formula for a meromorphic complex function with given poles; for a function $f(z)$ with poles at $z = z_i$, having order m_i and principal parts

$$\sum_{j=1}^{m_i} a_{ij}(z - z_i)^{-j},$$

the formula is

$$f(z) = \sum_i \left[\sum_{j=1}^{m_i} a_{ij}(z - z_i)^{-j} + p_i(z) \right] + g(z)$$

where the $p_i(z)$ are polynomials, $g(z)$ is an entire function, and the series converges uniformly in every bounded region where $f(z)$ is analytic. { 'mi,täk 'lef·lərz ,thir·əm }

mittelschmerz [MED] Pain or discomfort in the lower abdomen in women occurring midway in the intermenstrual interval, thought to be secondary to the irritation of the pelvic peritoneum

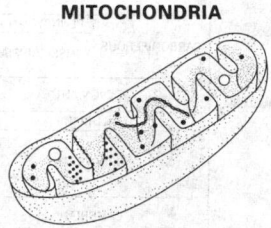

MITOCHONDRIA

Three-dimensional cut-away view of a mitochondrion.

by fluid or blood escaping from the point of ovulation in the ovary. { 'mid·əl,shmerts }

Mitteniales [BOT] An order of true mosses, class Bryopsida, characterized by branches of protonema consisting of spherical cells which reflect light from a backing of chloroplasts, thus providing a glow. { ,mit·ən'ā·lēz }

mix crystal *See* mixed crystal. { 'miks ,krist·əl }

mixed acid *See* nitrating acid. { 'mikst 'as·əd }

mixed aniline point [PHYS CHEM] The minimum temperature at which a mixture of aniline, heptane, and hydrocarbon will form a solution; related to the aromatic character of the hydrocarbon. { 'mikst 'an·ə·lən ,pȯint }

mixed aphasia [PSYCH] A combination of two or more forms of aphasia with impairment or loss of language function. { 'mikst ə'fā·zhə }

mixed-base notation [MATH] A computer number system in which a single base, such as 10 in the decimal system, is replaced by two number bases used alternately, such as 2 and 5. { 'mikst ¦bās nō'tā·shən }

mixed-base number [MATH] A number in mixed-base notation. Also known as mixed-radix number. { 'mikst ¦bās 'nəm·bər }

mixed-base oil [MATER] A crude oil that has both paraffin and asphaltum as predominating solid residuals. { 'mikst ¦bās 'ȯil }

mixed bud [BOT] A bud that contains both rudimentary leaves and rudimentary flowers. { 'mikst 'bəd }

mixed cloud [METEOROL] A cloud containing both water drops and ice crystals, hence a cloud whose composition is intermediate between that of a water cloud and that of an ice-crystal cloud. { 'mikst 'klau̇d }

mixed congruential generator [COMPUT SCI] A congruential generator in which the constant *b* in the generating formula is not equal to zero. { 'mikst ,kän·grü'en·chəl 'jen·ə,rād·ər }

mixed cryoglobulin [BIOCHEM] A cryoglobulin with a monoclonal component made of immunoglobulin belonging to two different classes, one of which is monoclonal. { 'mikst ,krī·ō'gläb·yə·lən }

mixed crystal [CRYSTAL] A crystal whose lattice sites are occupied at random by different ions or molecules of two different compounds. Also known as mix crystal. { 'mikst 'krist·əl }

mixed current [OCEANOGR] A type of tidal current characterized by a conspicuous difference in speed between the two flood currents or two ebb currents usually occurring each tidal day. { 'mikst 'kə·rənt }

mixed cycle [MECH ENG] An internal combustion engine cycle which combines the Otto cycle constant-volume combustion and the Diesel cycle constant-pressure combustion in high-speed compression-ignition engines. Also known as combination cycle; commercial Diesel cycle; limited-pressure cycle. { 'mikst 'sī·kəl }

mixed deafness [MED] Combined conductive and sensorineural impairment. { 'mikst 'def·nəs }

mixed decimal [MATH] Any decimal plus an integer. { 'mikst 'des·məl }

mixed-entry decision table [COMPUT SCI] A decision table in which the action entries may be either sequenced or unsequenced. { 'mikst ¦en·trē di'sizh·ən ,tā·bəl }

mixed flow [CHEM ENG] Flow stream existing in two or more phases, such as gas, hydrocarbon, and water. Also known as mixed-phase flow. { 'mikst 'flō }

mixed-flow fan [MIN ENG] A mine fan in which the flow is both radial and axial. { 'mikst ¦flō 'fan }

mixed-flow impeller [MECH ENG] An impeller for a pump or compressor which combines radial- and axial-flow principles. { 'mikst ¦flō im'pel·ər }

mixed forest [FOR] A forest consisting of two or more types of trees, with no more than 80% of the most common tree. { 'mikst 'fär·əst }

mixed gland [PHYSIO] A gland that secretes more than one substance, especially a gland containing both mucous and serous components. { 'mikst 'gland }

mixed graph [MATH] A graph in which directions are associated with some arcs but not with others. { 'mikst 'graf }

mixed highs [COMMUN] In color television, a method of

reproducing very fine picture detail by transmitting high-frequency components as part of luminance signals for achromatic reproduction in color pictures. { 'mikst 'hīz }

mixed indicator [ANALY CHEM] Color-change indicator for acid-base titration end points in which a mixture of two indicator substances is used to give sharper end-point color changes. { 'mikst 'in·də,kād·ər }

mixed laterality [PSYCH] The tendency, when there is a choice, to prefer to use parts of one side of the body for certain tasks and parts of the opposite side for others. { 'mikst ,lad·ə'ral·əd·ē }

mixed layer [OCEANOGR] The layer of water which is mixed through wave action or thermohaline convection. Also known as surface water. { 'mikst 'lā·ər }

mixed-layer mineral [MINERAL] A mineral having an interstratified structure consisting of alternating layers of two different clays or of a clay and some other mineral. { 'mikst ¦lā·ər 'min·rəl }

mixed-mode expression [COMPUT SCI] An expression involving operands of more than one data type. { 'mikst ¦mōd ik'spresh·ən }

mixed model [STAT] **1.** A model having both determinate and stochastic elements in its equations. **2.** A model having both difference and differential equations. **3.** A model containing both endogenous and exogenous elements. **4.** In analysis of variance for a two-way layout, the combined rows and columns. { 'mikst ,mäd·əl }

mixed nerve [NEUROSCI] A nerve containing both sensory and motor components. { 'mikst 'nərv }

mixed nucleus [METEOROL] A condensation nucleus of intermediate efficacy which, as a result of particle coagulation, contains both soluble hygroscopic matter and insoluble but wettable matter. { 'mikst 'nü·klē·əs }

mixed number [MATH] The sum of an integer and a fraction. { 'mikst 'nəm·bər }

mixed oil *See* nitrating acid. { 'mikst 'ȯil }

mixed ore [GEOL] Any ore with both oxidized and unoxidized minerals. { 'mikst 'ȯr }

mixed partial derivative [MATH] A partial derivative whose differentiations are with respect to two or more different variables. { 'mikst ¦pär·shəl də'riv·əd·iv }

mixed-phase flow *See* mixed flow. { 'mikst ¦fāz 'flō }

mixed potential [PHYS CHEM] The electrode potential of a material while more than one electrochemical reaction is occurring simultaneously. { 'mikst pə'ten·chəl }

mixed radix [MATH] Pertaining to a numeration system using more than one radix, such as the biquinary system. { 'mikst 'rā·diks }

mixed-radix number *See* mixed-base number. { 'mikst ¦rā·diks 'nəm·bər }

mixed reflection *See* spread reflection. { 'mikst ri'flek·shən }

mixed salvo [ORD] Series of shots in which some fall short of the target and some beyond it. { 'mikst 'sal·vō }

mixed sampling [STAT] The use of two or more methods of sampling; for example, in multistage sampling, if samples are drawn at random at one stage and drawn by a systematic method at another. { 'mikst 'sam·pliŋ }

mixed state [SOLID STATE] The state of a superconductor at magnetic field strengths between the lower critical field and the upper critical field, at which the magnetic field is partially excluded from the superconductor but penetrates it as microscopic filaments called fluxoids. { ¦mikst 'stāt }

mixed strategy [MATH] A method of playing a matrix game in which the player attaches a probability weight to each of the possible options, the probability weights being nonnegative numbers whose sum is unity, and then operates a chance device that chooses among the options with probabilities equal to the corresponding weights. [STAT] A concept in game theory which allows a player more than one choice of action which is determined by a chance mechanism. { ¦mikst 'strad·ə·jē }

mixed surd [MATH] A surd containing a rational factor or term, as well as irrational numbers. { ¦mikst 'sərd }

mixed tensor [MATH] A tensor with both contravariant and covariant indices. { ¦mikst 'ten·sər }

mixed tide [OCEANOGR] A tide in which the presence of a diurnal wave is conspicuous by a large inequality in the heights of either the two high tides or the two low tides usually occurring each tidal day. { 'mikst 'tīd }

mixer [ELECTR] **1.** A device having two or more inputs, usually adjustable, and a common output; used to combine separate audio or video signals linearly in desired proportions to produce an output signal. **2.** The stage in a superheterodyne receiver in which the incoming modulated radio-frequency signal is combined with the signal of a local r-f oscillator to produce a modulated intermediate-frequency signal. Also known as first detector; heterodyne modulator; mixer-first detector. [OPTICS] A nonlinear device in which two light beams are combined to form new beams having frequencies equal to the sum or the difference of the input wavelengths. { 'mik·sər }

mixer-first detector *See* mixer. { 'mik·sər ¦fərst di'tek·tər }

mixer-settler [CHEM ENG] Solvent-extraction system with alternating or combined arrangement of mixers and settlers; used for chemicals extraction, lubricating-oil refining, and uranium oxide recovery. Also known as mixer-settler extractor. { 'mik·sər 'set·lər }

mixer-settler extractor *See* mixer-settler. { 'mik·sər ¦set·lər ik'strak·tər }

mixer tube [ELECTR] A multigrid electron tube, used in a superheterodyne receiver, in which control voltages of different frequencies are impressed upon different control grids, and the nonlinear properties of the tube cause the generation of new frequencies equal to the sum and difference of the impressed frequencies. { 'mik·sər ¦tüb }

mixing [CHEM ENG] The intermingling of different materials (liquid, gas, solid) to produce a homogeneous mixture. [ELECTR] Combining two or more signals, such as the outputs of several microphones. [SCI TECH] The thorough intermingling of two or more different materials. { 'mik·siŋ }

mixing chamber [ENG] The space in a welding torch in which the gases are mixed. { 'mik·siŋ ¦chām·bər }

mixing length [PHYS] A mean length of travel, characteristic of a particular motion, over which an eddy maintains its identity; it is analogous to the mean free path of a molecule; physically, the idea implies that mixing occurs by discontinuous steps, that fluctuations which arise as eddies with different characteristics wander about, and that the mixing is done almost entirely by the small eddies. { 'mik·siŋ ¦leŋkth }

mixing ratio [METEOROL] In a system of moist air, the dimensionless ratio of the mass of water vapor to the mass of dry air; for many purposes, the mixing ratio may be approximated by the specific humidity. { 'mik·siŋ ¦rā·shō }

mixing transformation [MATH] A function of a measure space which moves the measurable sets in such a manner that, asymptotically as regards measure, any measurable set is distributed uniformly throughout the space. { 'mik·siŋ ¦tranz·fər'mā·shən }

mixing valve [ENG] Multi-inlet valve used to mix two or more fluid intakes to give a mixed product of desired composition. { 'mik·siŋ ¦valv }

mixite [MINERAL] $Cu_{11}Bi(AsO_4)_5(OH)_{10}·6H_2O$ A green to whitish mineral composed of a hydrous basic arsenate of copper and bismuth. { 'mik¦sīt }

Mixodectidae [PALEON] A family of extinct insectivores assigned to the Proteutheria; a superficially rodentlike group confined to the Paleocene of North America. { ¦mik·sə'dek·tə¦dē }

mixogram [FOOD ENG] A graphic-type record relating flour mixes from various wheats to the qualities of the resulting dough mixes. { 'mik·sə¦gram }

mixolimnion [HYD] The upper layer of a meromictic lake, characterized by low density and free circulation; this layer is mixed by the wind. { ¦mik·sō'lim·nē¦än }

mixoploidy [CELL MOL] The presence of cells having different chromosome numbers in the same cell population. { ¦mik·sə¦plóid·ē }

mixotrophic [BIOL] Obtaining nutrition by combining autotrophic and heterotrophic mechanisms. { ¦mik·sə¦träf·ik }

mixtite *See* diamictite. { 'miks¦tīt }

mixture [PHARM] A liquid medicine prepared by adding insoluble substances to a liquid medium, usually with a suspending agent. [SCI TECH] The product of mixing; components are not in a fixed proportion to each other. { 'miks·chər }

mixture ratio [AERO ENG] The ratio of the weight of oxidizer used per unit of time to the weight of fuel used per unit of time. { 'miks·chər ¦rā·shō }

mizer *See* miser. { 'mī·zər }

mizzenmast [NAV ARCH] The third mast or the mast aft of the mainmast in a sailing ship. { 'miz·ən¦mast }

mizzonite [MINERAL] A mineral of the scapolite group, composed of 54 to 57% silica. Also known as dipyre. { 'miz·ə¦nīt }

MJD *See* modified Julian date.

m-kgf *See* meter-kilogram.

MKK system *See* Morgan-Keenan-Kellman system. { ¦em¦kā'kā ¦sis·təm }

mksa system *See* meter-kilogram-second-ampere system. { ¦em¦kā¦es'ā ¦sis·təm }

mks system *See* meter-kilogram-second system. { ¦em¦kā¦es ¦sis·təm }

MK system [ASTRON] A system of classifying stars in which suffixes are added to the designations of the Harvard-Draper sequence to indicate luminosity, ranging from I for supergiants to VI for subdwarfs and white dwarfs. Also known as MKK system; spectral/luminosity classification; Yerkes system. { ¦em¦kā ¦sis·təm }

ml *See* milliliter.

MLD *See* metachromatic leukodystrophy.

MLS *See* microwave landing system.

mm *See* millimeter.

MMD *See* micromechanical display.

M meter [ENG] A class of instruments which measure the liquid water content of the atmosphere. { 'em ¦mēd·ər }

mmf *See* magnetomotive force.

mmHg *See* millimeter of mercury.

mmH₂O *See* millimeter of water.

M mode [ACOUS] A modification of the A mode of ultrasonic medical tomography used to display the movement of time-varying echo-producing structures by intensity-modulating the trace as it is swept slowly across the oscilloscope screen in a direction at right angles to the fast time-base sweep. { 'em ¦mōd }

MMPI *See* Minnesota multiphasic personality inventory.

MMSCFD [CHEM ENG] Abbreviation for million standard cubic feet per day; usually refers to gas flow.

MMSCFH [CHEM ENG] Abbreviation for million standard cubic feet per hour; usually refers to gas flow.

MMSCFM [CHEM ENG] Abbreviation for million standard cubic feet per minute; usually refers to gas flow.

MMSS *See* maritime mobile satellite service.

MMT *See* multi-mirror telescope.

mmu *See* milli-mass-unit.

Mn *See* manganese.

mnemonic [PSYCH] **1.** Aiding or pertaining to memory. **2.** A device, such as combinations of letters, pictures, or words, to stimulate recall of the facts they represent. { nə'män·ik }

mnemonic code [COMPUT SCI] A programming code that is easy to remember because the codes resemble the original words, such as MPY for multiply and ACC for accumulator. { nə'män·ik ¦kōd }

mnemonics [PSYCH] The science of the cultivation of memory functions using systematic methods. { nə'män·iks }

Mnesarchaeidae [INV ZOO] A family of lepidopteran insects in the suborder Homoneura; members are confined to New Zealand. { ¦nē·sär'kē·ə¦dē }

MNOS *See* metal-nitride-oxide semiconductor. { 'em¦nós }

Mo *See* molybdenum.

moat [GEOL] **1.** A ringlike depression around the base of a seamount. **2.** A valleylike depression around the inner side of a volcanic cone, between the rim and the lava dome. [HYD] **1.** A glacial channel in the form of a deep, wide trench. **2.** *See* oxbow lake. { mōt }

mobile [GRAPHICS] A decorative three-dimensional art object constructed of metal, glass, wood, plastic, or other materials; it is mounted in a hanging position and is free to move in any of its planes. { 'mō¦bēl }

mobile artillery [ORD] Artillery weapons designed for movement and ready conversion from traveling position to firing position; wheels or other suspension devices are not ordinarily removed in the firing position. { 'mō·bəl är'til·ə·rē }

mobile belt [GEOL] A long, relatively narrow crustal region of tectonic activity. { 'mō·bəl ¦belt }

mobile code [COMPUT SCI] Code that can be transmitted across, and executed at the other end of, a network, and is capable of running on multiple platforms, for example, Java. { ˌmōˑbəl ˈkōd }

mobile crane [MECH ENG] **1.** A cable-controlled crane mounted on crawlers or rubber-tired carriers. **2.** A hydraulic-powered crane with a telescoping boom mounted on truck-type carriers or as self-propelled models. { 'mōˑbəl 'krān }

mobile digital computer [COMPUT SCI] Large, mobile, fixed-point operation, one-address, parallel-mode type digital computer. { 'mōˑbəl ˈdijˑədˑəl kəm'pyüdˑər }

mobile drill [MIN ENG] A drill unit mounted on wheels or crawl-type tracks to facilitate moving. { 'mōˑbəl 'dril }

mobile earth station [COMMUN] An earth station intended to be used while in motion or at halts at unspecified points. { ˌmōˑbəl 'ərth ˌstāˑshən }

mobile earth terminal [COMMUN] An antenna small enough to fit in a briefcase or suitcase, used for satellite communications, especially by news-service reporters at locations that cannot be accessed by conventional transportable satellite news-gathering terminals. Abbreviated MET. { ˌmōˑbəl 'ərth ˌtərˑmənˑəl }

mobile electron [PHYS CHEM] An electron that can move readily from one atom to another within a chemical structure in response to changes in the external chemical environment. { ˌmōˑbəl ə'lekˌträn }

mobile filling [MIN ENG] Filling which is supplemented only from above, and which sinks, filling the mined-out rooms. { 'mōˑbəl 'filˑiŋ }

mobile hoist [MECH ENG] A platform hoist mounted on a pair of pneumatic-tired road wheels, so it can be towed from one site to another. { 'mōˑbəl 'hóist }

mobile loader [MECH ENG] A self-propelling power machine for loading coal, mineral, or dirt. { 'mōˑbəl 'lōdˑər }

mobile mount [ORD] Any weapons mount which is mobile, as contrasted to one which is fixed; as a mount for mobile artillery. { 'mōˑbəl 'maúnt }

mobile phase [ANALY CHEM] **1.** In liquid chromatography, the phase that is moving in the bed, including the fraction of the sample held by this phase. **2.** The carrier gas in a gas chromatography procedure. { 'mōˑbəl ˌfāz }

mobile radio [COMMUN] Radio communication in which the transmitter is installed in a vessel, vehicle, or airplane and can be operated while in motion. { 'mōˑbəl 'rādˑēˑō }

mobile-relay station [COMMUN] Base station in which the base receiver automatically tunes on the base station transmitter and which retransmits all signals received by the base station receiver; used to extend the range of mobile units, and requires two frequencies for operation. { 'mōˑbəl 'rēˌlā ˌstāˑshən }

mobile robot [CONT SYS] A robot mounted on a movable platform that transports it to the area where it carries out tasks. { 'mōˑbəl 'rōˌbät }

mobile service [COMMUN] A radiocommunication service between mobile and land stations or between mobile stations. { ˌmōˑbəl 'sərˑvəs }

mobile station [COMMUN] **1.** Station in the mobile service intended to be used while in motion or during halts at unspecified points. **2.** One or more transmitters that are capable of transmission while in motion. { 'mōˑbəl 'stāˑshən }

mobile satellite service [COMMUN] A radiocommunication service between mobile earth stations by means of one or more space stations. Abbreviated MSS. { ˌmōˑbəl 'sadˑəlˌīt ˌsərˑvəs }

mobile systems equipment [COMPUT SCI] Computers located on planes, ships, or vans. { 'mōˑbəl 'sisˑtəmz iˌkwipˑmənt }

Mobilina [INV ZOO] A suborder of ciliophoran protozoans in the order Peritrichida. { ˌmōˑbə'līˑnə }

mobility [ENG] The ability of an analytical balance to react to small load changes; affected by friction and degree of looseness in the balance components. [FL MECH] The reciprocal of the plastic viscosity of a Bingham plastic. [PHYS] Freedom of particles to move, either in random motion or under the influence of fields or forces. [SOLID STATE] See drift mobility. { mō'bilˑədˑē }

mobility coefficient [PHYS CHEM] The average speed of motion of molecules in a solution in the direction of the concentration gradient, at unit concentration and unit osmotic pressure gradient. { mō'bilˑədˑē ˌkōˑəˌfishˑənt }

mobility ratio [PETRO ENG] The ratio of the mobility of the driving fluid (such as water) to that of the driven fluid (such as gas) in a petroleum reservoir. { mō'bilˑədˑē ˌrāˑshō }

mobility tensor [PL PHYS] A second-rank tensor whose product with the electric field vector for a plane wave in a plasma gives a vector equal to the average velocity of electrons or ions; components of both vectors are in phasor notation. { mō'bilˑədˑē ˌtenˑsər }

mobility threshhold [ENG] On an analytical balance, the smallest load change that will cause a noticeable change in the weight measurement. { mō'bilˑədˑē ˌthreshˌhōld }

mobilization [GEOL] Any process by which solid rock becomes sufficiently soft and plastic to permit it to flow or to permit geochemical migration of the mobile components. { ˌmōˑbəˑlə'zāˑshən }

Möbius band [MATH] The nonorientable surface obtained from a rectangular strip by twisting it once and then gluing the two ends. Also known as Möbius strip. { 'mərˑbēˑəs ˌband }

Möbius function [MATH] The function μ of the positive integers where μ(1) = 1, μ(n) = (−1)^r if n factors into r distinct primes, and μ(n) = 0 otherwise; also, μ(n) is the sum of the primitive nth roots of unity. { 'mərˑbēˑəs ˌfəŋkˑshən }

Möbius resistor [ELEC] A nonreactive resistor made by placing strips of aluminum or other metallic tape on opposite sides of a length of dielectric ribbon, twisting the strip assembly half a turn, joining the ends of the metallic tape, then soldering leads to opposite surfaces of the resulting loop. { 'mərˑbēˑəs riˌzisˑtər }

Möbius strip See Möbius band. { 'mərˑbēˑəs ˌstrip }

Möbius transformations [MATH] These are the most commonly used conformal mappings of the complex plane; their form is $f(z) = (az + b)/(cz + d)$ where the real numbers a, b, c, and d satisfy $ad − bc ≠ 0$. Also known as linear fractional transformations. Also known as bilinear transformations; homographic transformations. { 'mərˑbēˑəs ˌtranzˑfər'māˑshənz }

Mobulidae [VERT ZOO] The devil rays, a family of batoids that are surface feeders and live mostly on plankton. { mə'byülˑəˌdē }

mocaya oil [MATER] Oil from kernels of the Paraguayan palm (*Acrocomia sclerocarpa*), found in South America and the West Indies; used in manufacture of tinplate, soaps, candles, and margarine. { mə'kīˑyə ˌóil }

mock fog [METEOROL] A simulation of true fog by atmospheric refraction. { 'mäk ˌfäg }

mock lead See sphalerite. { 'mäk 'led }

mock moon See paraselene. { 'mäk 'mün }

mock ore See sphalerite. { 'mäk 'ór }

mock silver [MET] **1.** An aluminum alloy containing 5% copper and 10% tin, or 5% copper and 5% silver. **2.** A white brass containing 55% zinc and 45% copper. { 'mäk 'silˑvər }

mock sun See paranthelion; parhelion. { 'mäk 'sən }

mock sun ring See parhelic circle. { 'mäk 'sən ˌriŋ }

mockup [ENG] A model, often full-sized, of a piece of equipment, or installation, so devised as to expose its parts for study, training, or testing. { 'mäkˌəp }

MOCVD See metal-organic chemical vapor deposition.

modacrylic [TEXT] Of a synthetic fiber, composed of less than 85% and more than 35% by weight of acrylonitrile units. { ˌmädˑə'krilˑik }

modal class [STAT] The class that contains more individuals than any other class in a statistical distribution. { 'mōdˑəl ˌklas }

modal distortion See modal noise. { ˌmōdˑəl di'stórˑshən }

modal noise [COMMUN] Interference of a multimode optical communications fiber with a laser light source when a speckle pattern in the light intensity in the fiber alters because of motion of the fiber or changes in the laser spectrum. Also known as modal distortion. { ˌmōdˑəl 'nóiz }

modal number [GEN] **1.** The typical chromosome number of a taxonomic group. **2.** The most common chromosome number of a tumor cell population. { 'mōdˑəl ˌnəmˑbər }

mode [COMMUN] Form of the information in a communication such as literal language, digital data, and video. [COMPUT SCI] One of several alternative conditions or methods of operation of a device. [ELECTROMAG] A form of propagation of guided waves that is characterized by a particular field pattern in a plane transverse to the direction of propagation. Also known as transmission mode. [PETR] The mineral composition of a rock, usually expressed as percentages of total weight or volume. [PHYS] A state of an oscillating system that corresponds to a particular field pattern and one of the possible resonant frequencies of the system. [STAT] The most frequently occurring member of a set of numbers. { mōd }

mode converter *See* mode transducer. { 'mōd kən,vərd·ər }

mode eddies [OCEANOGR] Densely packed, irregularly oval high- and low-pressure centers roughly 240 miles (400 kilometers) in diameter in which current intensities are typically tenfold greater than the local means. Also known as mesoscale eddies. { 'mōd ,ed·ēz }

mode filter [ELECTROMAG] A waveguide filter designed to separate waves of the same frequency but of different transmission modes. { 'mōd ,fil·tər }

mode jump [ELECTR] Change in mode of magnetron operation from one pulse to the next; each mode represents a different frequency and power level. { 'mōd ,jəmp }

model [COMPUT SCI] *See* macroskeleton. [SCI TECH] A mathematical or physical system, obeying certain specified conditions, whose behavior is used to understand a physical, biological, or social system to which it is analogous in some way. { 'mäd·əl }

model atmosphere [METEOROL] Any theoretical representation of the atmosphere, particularly of vertical temperature distribution. { 'mäd·əl 'at·mə,sfir }

model-based expert system [COMPUT SCI] An expert system that is based on knowledge of the structure and function of the object for which the system is designed. { ¦mäd·əl ,bāst 'ek·spərt ,sis·təm }

model basin [ENG] A large basin or tank of water where scale models of ships can be tested. Also known as model tank; towing tank. { 'mäd·əl 'bās·ən }

model-following problem [CONT SYS] The problem of determining a control that causes the response of a given system to be as close as possible to the response of a model system, given the same input. { 'mäd·əl ¦fäl·ə·wiŋ ,präb·ləm }

mode-locked laser [OPTICS] A laser designed so that several modes of oscillation with closely spaced wavelengths, in which the laser would normally oscillate, are synchronized so that a pulse of light, lasting for as little as a picosecond, is generated. { 'mōd ,läkt 'lā·zər }

model reduction [CONT SYS] The process of discarding certain modes of motion while retaining others in the model used by an active control system, in order that the control system can compute control commands with sufficient rapidity. { 'mäd·əl ri'dək·shən }

model reference system [CONT SYS] An ideal system whose response is agreed to be optimum; computer simulation in which both the model system and the actual system are subjected to the same stimulus is carried out, and parameters of the actual system are adjusted to minimize the difference in the outputs of the model and the actual system. { 'mäd·əl 'ref·rəns ,sis·təm }

model symbol [COMPUT SCI] The standard usage of geometrical figures, such as squares, circles, or triangles, to help illustrate the various working parts of a model: each symbol must, nevertheless, be footnoted for complete clarification. { 'mäd·əl ,sim·bəl }

model tank *See* model basin. { 'mäd·əl ,taŋk }

model theory [MATH] The general qualitative study of the structure of a mathematical theory. { 'mäd·əl ,thē·ə·rē }

modem [ELECTR] A combination modulator and demodulator at each end of a telephone line to convert binary digital information to audio tone signals suitable for transmission over the line, and vice versa. Also known as dataset. Derived from modulator-demodulator. { 'mō,dem }

modem eliminator [COMPUT SCI] A device that is used to connect two computers in proximity and that mimics the action of two modems and a telephone line. { 'mō,dem ə'lim·ə,nād·ər }

mode number [ELECTR] 1. The number of complete cycles during which an electron of average speed is in the drift space of a reflex klystron. 2. The number of radians of phase in the microwave field of a magnetron divided by 2π as one goes once around the anode. { 'mōd ,nəm·bər }

mode of oscillation *See* mode of vibration. { 'mōd əv ,äs·ə'lā·shən }

mode of vibration [MECH] A characteristic manner in which

a system which does not dissipate energy and whose motions are restricted by boundary conditions can oscillate, having a characteristic pattern of motion and one of a discrete set of frequencies. Also known as mode of oscillation. { 'mōd əv vī'brā·shən }

moder [GEOL] Humus consisting of plant material that is undergoing alteration from the living to the decayed state and is intermediate in acidity between mor and mull. { 'mōd·ər }

moderate breeze [METEOROL] In the Beaufort wind scale, a wind whose speed is from 11 to 16 knots (13 to 18 miles per hour or 20 to 30 kilometers per hour). { 'mäd·ə·rət 'brēz }

moderate gale [METEOROL] In the Beaufort wind scale, a wind whose speed is from 28 to 33 knots (32 to 38 miles per hour or 52 to 61 kilometers per hour). { 'mäd·ə·rət 'gāl }

moderator [NUCLEO] The material used in a nuclear reactor to moderate or slow down neutrons from the high velocities at which they are created in the fission process. { 'mäd·ə,rād·ər }

modern algebra [MATH] The study of algebraic systems such as groups, rings, modules, and fields. { 'mäd·ərn 'al·jə·brə }

modern control [CONT SYS] A control system that takes account of the dynamics of the processes involved and the limitations on measuring them, with the aim of approaching the condition of optimal control. { 'mäd·ərn kən'trōl }

Mode S [NAV] An augmentation of the Air Traffic Control Radar Beacon System in which each aircraft is equipped with a transponder that replies when interrogated with a discrete identity code. Also known as ADSEL (in Britain); discrete address beacon system or DABS (in the United States). { 'mōd 'es }

mode shift [ELECTR] Change in mode of magnetron operation during a pulse. { 'mōd ,shift }

mode skip [ELECTR] Failure of a magnetron to fire on each successive pulse. { 'mōd ,skip }

mode switch [COMPUT SCI] A preset control which affects the normal response of various components of a mechanical desk calculator. [ELECTR] A microwave control device, often consisting of a waveguide section of special cross section, which is used to change the mode of microwave power transmission in the waveguide. { 'mōd ,swich }

mode transducer [ELECTR] Device for transforming an electromagnetic wave from one mode of propagation to another. Also known as mode converter; mode transformer. { 'mōd tranz,dü·sər }

mode transformer *See* mode transducer. { 'mōd tranz,fór·mər }

MODFET *See* high-electron-mobility transistor. { 'mäd,fet }

modification [CELL MOL] In nucleic acid metabolism, any changes made to deoxyribonucleic acid or ribonucleic acid after their original incorporation into a polynucleotide chain. [ENG] A major or minor change in the design of an item, effected in order to correct a deficiency, to facilitate production, or to improve operational effectiveness. [MET] Treatment of molten aluminum alloys containing 8–13% silicon with small amounts of a sodium fluoride or sodium chloride mixture; improves mechanical properties. [SCI TECH] Any change brought about by external or internal factors. { ,mäd·ə·fə'kā·shən }

modification kit [ENG] A collection of items not all having the same basic name which are employed individually or conjunctively to alter the design of a component or equipment. { ,mäd·ə·fə'kā·shən ,kit }

modified asphalt [MATER] Asphalt modified by addition of a rosin ester or synthetic resin. { 'mäd·ə,fīd 'as,fólt }

modified base [CELL MOL] A nucleotide that is an altered form of the usual four nucleic acid bases. { 'mäd·ə,fīd 'bās }

modified Bessel equation [MATH] The differential equation $z^2 f''(z) + z f'(z) - (z^2 + n^2)f(z) = 0$, where z is a variable that can have real or complex values and n is a real or complex number. { ¦mäd·ə,fīd 'bes·əl i,kwā·zhən }

modified Bessel function of the first kind *See* modified Bessel function. { ¦mäd·ə,fīd ¦bes·əl ,fəŋk·shən əv thə 'fərst ,kīnd }

modified Bessel function of the second kind *See* modified Hankel function. { ¦mäd·ə,fīd ¦bes·əl ,fəŋk·shən əv thə 'sek·ənd ,kīnd }

modified Bessel functions [MATH] The functions defined by $I_v(x) = \exp(-iv\pi/2) J_v(ix)$, where J_v is the Bessel function of order v, and x is real and positive. Also known as modified

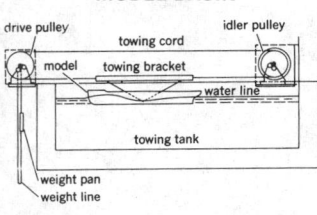

MODEL BASIN

Model basin with model towed by falling weight.

Bessel function of the first kind. { 'mäd·ə‚fīd 'bes·əl ‚fəŋk· shənz }

modified betatron [NUCLEO] A betatron in which the orbital stability properties of the beam are improved by adding a toroidal magnetic field. { 'mäd·ə‚fīd 'bād·ə‚trän }

modified constant-voltage charge [ELEC] Charging of a storage battery in which the voltage of the charging circuit is held substantially constant, but a fixed resistance is inserted in the battery circuit producing a rising voltage characteristic at the battery terminals as the charge progresses. { 'mäd·ə‚fīd ‚kän·stənt ‚vōl·tij 'chärj }

modified exponential curve [STAT] The equation resulting when a constant is added to the exponential curve equation; used to estimate trend in a nonlinear time series. { 'mäd·ə‚fīd ‚ek·spə‚nen·chəl 'kərv }

modified gunmetal [MET] Gunmetal containing about 2.5% lead; used for gears and bearings. { 'mäd·ə‚fīd 'gən‚med·əl }

modified Hankel functions [MATH] The functions defined by $K_v(x) = (i\pi/2) \exp(iv\pi/2) H_v^{(1)}(ix)$, where $H_v^{(1)}$ is the first Hankel function of order v, and x is real and positive. Also known as modified Bessel function of the second kind. { 'mäd·ə‚fīd 'häŋk·əl ‚fəŋk·shənz }

modified index of refraction [METEOROL] An atmospheric index of refraction mathematically modified so that when its gradient is applied to energy propagation over a hypothetical flat earth, it is substantially equivalent to propagation over the true curved earth with the actual index of refraction. Also known as modified refractive index; refractive modulus. { 'mäd·ə‚fīd 'in‚deks əv ri'frak·shən }

modified Julian date [ASTRON] The Julian date minus 2,400,000.5. Abbreviated MJD. { 'mäd·ə‚fīd 'jül·yən 'dāt }

modified Lambert conformal chart [MAP] A chart on the modified Lambert conformal projection. Also known as Ney's chart. { 'mäd·ə‚fīd 'lam·bərt kən‚fȯr·məl 'chärt }

modified Lambert conformal projection [MAP] A modification of the Lambert conformal projection for use in polar regions, one of the standard parallels being at latitude 89°59′58″ and the other at latitude 71° or 74°, and the parallels being expanded slightly to form complete concentric circles. Also known as Ney's projection. { 'mäd·ə‚fīd 'lam·bərt kən‚fȯr· məl prə'jek·shən }

modified Lewis acid [PHYS CHEM] An acid that is a halide ion acceptor. { 'mäd·ə‚fīd 'lü·əs ‚as·əd }

modified mean [STAT] A mean computed after elimination of observations judged to be atypical. { ‚mäd·ə‚fīd 'mēn }

modified precision approach radar [NAV] A special precision radar approach landing procedure for high performance aircraft; radar guidance is provided to a landing flare point instead of a runway touchdown point. { 'mäd·ə‚fīd prə'sizh· ən ə‚prōch 'rā‚där }

modified rayon [TEXT] A woollike rayon fiber made with additives in the spinning solution. { 'mäd·ə‚fīd 'rā‚än }

modified refractive index See modified index of refraction. { 'mäd·ə‚fīd ri‚frak·tiv 'in‚deks }

modifier [COMPUT SCI] A quantity used to alter the address of an operand in a computer, such as the cycle index. Also known as index word. [MATER] In flotation, any of the chemicals which increase the specific attraction between collector agents and particle surfaces or which increase the wettability of those surfaces. { 'mäd·ə‚fī·ər }

modifier gene [GEN] A gene that alters the phenotypic expression of a nonallelic gene. { 'mäd·ə‚fī·ər ‚jēn }

modifier register See index register. { 'mäd·ə‚fī·ər ‚rej·ə‚ stər }

modify [COMPUT SCI] **1.** To alter a portion of an instruction so its interpretation and execution will be other than normal; the modification may permanently change the instruction or leave it unchanged and affect only the current execution; the most frequent modification is that of the effective address through the use of index registers. **2.** To alter a subroutine according to a defined parameter. { 'mäd·ə‚fī }

modify structure [COMPUT SCI] A statement in a database language that allows changes to be made in the structure of the records in a file. { 'mäd·ə‚fī ‚strək·chər }

modillon [ARCH] A horizontal bracket, usually in the form of a scroll with acanthus, supporting the corona under a cornice. { mō'dil·yən }

moding [ELECTR] Defect of magnetron oscillation in which it oscillates in one or more undesired modes. { 'mōd·iŋ }

modiolus [ANAT] The central axis of the cochlea. { mə'dī· ə·ləs }

MOD room [ENG ACOUS] A control room in a sound-recording studio in which the acoustic treatment comprises a uniform disposition of the sound-absorbent material all about the room. { 'mäd ‚rüm }

modula-2 [COMPUT SCI] A general-purpose programming language that allows a computer program to be written as separate modules which can be compiled separately but can share a common code. { 'mäj·ə·lə ‚tü }

modular circuit [ELECTR] Any type of circuit assembled to form rectangular or cubical blocks that perform one or more complete circuit functions. { 'mäj·ə·lər 'sər·kət }

modular compilation [COMPUT SCI] The separate translation into machine language of the individual parts of a computer program, which are then combined into a single program by a linkage editor. { 'mäj·ə·lər ‚käm·pə'lā·shən }

modularity [COMPUT SCI] The property of functional flexibility built into a computer system by assembling discrete units which can be easily joined to or arranged with other parts or units. { ‚mäj·ə'lar·əd·ē }

modular lattice [MATH] A lattice with the property that, if x is equal to or greater than z, then for any element y, the greatest lower bound of x and v equals the least upper bound of w and z, where v is the least upper bound of y and z, and w is the greatest lower bound of x and y. { ‚mäj·ə·lər 'lad·əs }

modular programming [COMPUT SCI] The construction of a computer program from a collection of modules, each of workable size, whose interactions are rigidly restricted. { 'mäj·ə·lər 'prō‚gram·iŋ }

modular structure [BUILD] A building that is constructed of preassembled or presized units of standard sizes; uses a 4-inch (10.16-centimeter) cubical module as a reference. [ELECTR] **1.** An assembly involving the use of integral multiples of a given length for the dimensions of electronic components and electronic equipment, as well as for spacings of holes in a chassis or printed wiring board. **2.** An assembly made from modules. { 'mäj·ə·lər 'strək·chər }

modulate [ELECTR] To vary the amplitude, frequency, or phase of a wave, or vary the velocity of the electrons in an electron beam in some characteristic manner. { 'mäj·ə‚lāt }

modulated amplifier [ELECTR] Amplifier stage in a transmitter in which the modulating signal is introduced and modulates the carrier. { 'mäj·ə‚lād·əd 'am·plə‚fī·ər }

modulated carrier [COMMUN] Radio-frequency carrier wave whose amplitude, phase, or frequency has been varied according to the intelligence to be conveyed. { 'mäj·ə‚lād·əd 'kar·ē·ər }

modulated continuous wave [COMMUN] Wave in which the carrier is modulated by a constant audio-frequency tone. { 'mäj·ə‚lād·əd kən‚tin·yə·wəs 'wāv }

modulated Raman scattering [SPECT] Application of modulation spectroscopy to the study of Raman scattering; in particular, use of external perturbations to lower the symmetry of certain crystals and permit symmetry-forbidden modes, and the use of wavelength modulation to analyze second-order Raman spectra. { 'mäj·ə‚lād·əd 'rä·mən ‚skad·ə·riŋ }

modulated stage [ELECTR] Radio-frequency stage to which the modulator is coupled and in which the continuous wave (carrier wave) is modulated according to the system of modulation and the characteristics of the modulating wave. { 'mäj· ə‚lād·əd 'stāj }

modulating codon [MOL BIO] A codon that controls the frequency of transcription of a cistron. { 'mäj·ə‚lād·iŋ 'kō‚dän }

modulating electrode [ELECTR] Electrode to which a potential is applied to control the magnitude of the beam current. { 'mäj·ə‚lād·iŋ i'lek‚trōd }

modulating signal [COMMUN] Signal which causes a variation of some characteristics of a carrier. { 'mäj·ə‚lād·iŋ 'sig·nəl }

modulation [COMMUN] The process or the result of the process by which some parameter of one wave is varied in accordance with some parameter of another wave. [MECH ENG] Regulation of the fuel-air mixture to a burner in response to fluctuations of load on a boiler. { ‚mäj·ə'lā·shən }

modulation capability [ELECTR] Of an aural transmitter, the maximum percentage modulation that can be obtained without exceeding a given distortion figure. { ‚mäj·ə'lā·shən ‚kā·pə· 'bil·ə·dē }

modulation code [COMMUN] A code used to cause variations in a signal in accordance with a predetermined scheme; normally used to alter or modulate a carrier wave to transmit data. { ‚mäj·ə'lā·shən ‚kōd }

modulation crest [COMMUN] The peak amplitude of an amplitude-modulated wave. { ‚mäj·ə'lā·shən 'krest }

modulation-doped field-effect transistor *See* high-electron-mobility transistor. { ‚mäj·ə'lā·shən ¦dōpt 'fēld i¦fekt tran'zis·tər }

modulation-doped structure [SOLID STATE] An epitaxially grown crystal structure in which successive semiconductor layers contain different types of electrical dopants. { ‚mäj·ə'lā·shən ¦dōpt 'strək·chər }

modulation envelope [COMMUN] The peaks of the waveform of a modulated signal. { ‚mäj·ə'lā·shən 'en·və‚lōp }

modulation factor [COMMUN] **1.** In general, the ratio of the peak variation in the modulation actually used in a transmitter to the maximum variation for which the transmitter was designed. **2.** In an amplitude-modulated wave, the ratio (usually expressed in percent) of the peak variation of the envelope from its reference value to the reference value. Also known as index of modulation. **3.** In a frequency-modulated wave, the ratio of the actual frequency swing to the frequency swing required for 100% modulation. { ‚mäj·ə'lā·shən ‚fak·tər }

modulation index [COMMUN] The ratio of the frequency deviation to the frequency of the modulating wave in a frequency-modulation system when using a sinusoidal modulating wave. Also known as ratio deviation. { ‚mäj·ə'lā·shən ‚in‚deks }

modulation meter [ENG] Instrument for measuring the degree of modulation (modulation factor) of a modulated wave train, usually expressed in percent. { ‚mäj·ə'lā·shən ‚mēd·ər }

modulation rise [ELECTR] Increase of the modulation percentage caused by nonlinearity of any tuned amplifier, usually the last intermediate-frequency stage of a receiver. { ‚mäj·ə'lā·shən ‚rīz }

modulation spectroscopy [SPECT] A branch of spectroscopy concerned with the measurement and interpretation of changes in transmission or reflection spectra induced (usually) by externally applied perturbation, such as temperature or pressure change, or an electric or magnetic field. { ‚mäj·ə'lā·shən spek'träs·kə·pē }

modulation transformer [ENG ACOUS] An audio-frequency transformer which matches impedances and transmits audio frequencies between one or more plates of an audio output stage and the grid or plate of a modulated amplifier. { ‚mäj·ə'lā·shən tranz‚fòr·mər }

modulator [ELECTR] **1.** The transmitter stage that supplies the modulating signal to the modulated amplifier stage or that triggers the modulated amplifier stage to produce pulses at desired instants as in radar. **2.** A device that produces modulation by any means, such as by virtue of a nonlinear characteristic or by controlling some circuit quantity in accordance with the waveform of a modulating signal. **3.** One of the electrodes of a spacistor. { 'mäj·ə‚lād·ər }

modulator crystal [OPTICS] Crystal which is used to modulate a polarized light beam by the use of the Pockel's effect; useful as a modulator in laser systems. { 'mäj·ə‚lād·ər ‚krist·əl }

modulator-demodulator *See* modem. { 'mäj·ə‚lād·ər dē 'mäj·ə‚lād·ər }

modulator glow tube [ELECTR] Cold cathode recorder tube that is used for facsimile and sound-on-film recording; provides a modulated high-intensity point source of light. { 'mäj·ə‚lād·ər 'glō ‚tüb }

module [AERO ENG] A self-contained unit which serves as a building block for the overall structure in space technology; usually designated by its primary function, such as command module or lunar landing module. [COMPUT SCI] **1.** A distinct and identifiable unit of computer program for such purposes as compiling, loading, and linkage editing. **2.** One memory bank and associated electronics in a computer. [ELECTR] A packaged assembly of wired components, built in a standardized size and having standardized plug-in or solderable terminations. [ENG] A unit of size used as a basic component for standardizing the design and construction of buildings, building parts, and furniture. [MATH] A vector space in which the scalars are a ring rather than a field. { 'mäj·ül }

modulo [MATH] **1.** A group *G* modulo a subgroup *H* is the quotient group *G/H* of cosets of *H* in *G*. **2.** A technique of identifying elements in an algebraic structure in such a manner that the resulting collection of identified objects is the same type of structure. { 'mäj·ə‚lō }

modulo N [MATH] Two integers are said to be congruent modulo *N* (where *N* is some integer) if they have the same remainder when divided by *N*. { 'mäj·ə‚lō 'en }

modulo N arithmetic [MATH] Calculations in which all integers are replaced by their remainders after division by *N* (where *N* is some fixed integer.) { 'mäj·ə‚lō 'en ə'rith·mə·tik }

modulo N check [COMPUT SCI] A procedure for verification of the accuracy of a computation by repeating the steps in modulo *N* arithmetic and comparing the result with the original result (modulo *N*). Also known as residue check. { 'mäj·ə‚lō 'en 'chek }

modulo-two adder [COMPUT SCI] A logical circuit for adding one-digit binary numbers. { 'mäj·ə‚lō 'tü 'ad·ər }

modulus [MATH] **1.** The modulus of a logarithm with a given base is the factor by which a logarithm with a second base must be multiplied to give the first logarithm. **2.** *See* absolute value. { 'mäj·ə·ləs }

modulus of a congruence [MATH] A number *a*, such that two specified numbers *b* and *c* give the same remainder when divided by *a*; *b* and *c* are then said to be congruent, modulus *a* (or congruent, modulo *a*). { 'mäj·ə·ləs əv ə kən'grü·əns }

modulus of compression *See* bulk modulus of elasticity. { 'mäj·ə·ləs əv kəm'presh·ən }

modulus of continuity [MATH] For a real valued continuous function *f*, this is the function whose value at a real number *r* is the maximum of the modulus of $f(x) - f(y)$ where the modulus of $x - y$ is less than *r*; this function is useful in approximation theory. { 'mäj·ə·ləs əv ‚känt·ən'ü·əd·ē }

modulus of decay [MECH] The time required for the amplitude of oscillation of an underdamped harmonic oscillator to drop to 1/*e* of its initial value; the reciprocal of the damping factor. { ¦mäj·ə·ləs əv di'kā }

modulus of deformation [MECH] The modulus of elasticity of a material that deforms other than according to Hooke's law. { 'mäj·ə·ləs əv ‚dē‚fòr'mā·shən }

modulus of distance [ASTRON] The quantity $m - M$, where *M* is the absolute magnitude of a given star and *m* is its apparent magnitude. Also known as distance modulus. { 'mäj·ə·ləs əv 'dis·təns }

modulus of elasticity [MECH] The ratio of the increment of some specified form of stress to the increment of some specified form of strain, such as Young's modulus, the bulk modulus, or the shear modulus. Also known as coefficient of elasticity; elasticity modulus; elastic modulus. { 'mäj·ə·ləs əv i‚las'tis·əd·ē }

modulus of elasticity in shear [MECH] A measure of a material's resistance to shearing stress, equal to the shearing stress divided by the resultant angle of deformation expressed in radians. Also known as coefficient of rigidity; modulus of rigidity; rigidity modulus; shear modulus. { 'mäj·ə·ləs əv i‚las'tis·əd·ē in 'shir }

modulus of resilience [MECH] The maximum mechanical energy stored per unit volume of material when it is stressed to its elastic limit. { 'mäj·ə·ləs əv ri'zil·yəns }

modulus of rigidity *See* modulus of elasticity in shear. { 'mäj·ə·ləs əv ri'jid·əd·ē }

modulus of rupture in bending [MECH] The maximum stress per unit area that a specimen can withstand without breaking when it is bent, as calculated from the breaking load under the assumption that the specimen is elastic until rupture takes place. { 'mäj·ə·ləs əv 'rəp·chər in 'bend·iŋ }

modulus of rupture in torsion [MECH] The maximum stress per unit area that a specimen can withstand without breaking when its ends are twisted, as calculated from the breaking load under the assumption that the specimen is elastic until rupture takes place. { 'mäj·ə·ləs əv 'rəp·chər in 'tòr·shən }

modulus of simple longitudinal extension *See* axial modulus. { ¦mäj·ə·ləs əv ¦sim·pəl ‚län·jə¦tüd·ən·əl ik'sten·chən }

modulus of strain hardening *See* rate of strain hardening. { 'mäj·ə·ləs əv 'strān ‚härd·ən·iŋ }

modulus of torsion *See* torsional modulus. { 'mäj·ə·ləs əv 'tòr·shən }

modulus of volume elasticity *See* bulk modulus of elasticity. { 'mäj·ə·ləs əv 'väl·yəm i‚las'tis·əd·ē }

Moeller-Barlow disease *See* infantile scurvy. { 'məl·ər 'bär·lō di‚zēz }

moellon *See* degras. { mwe'lòn }

MOEMS *See* micro-opto-electro-mechanical system. { 'mō‚emz }

Moeritheriidae [PALEON] The single family of the extinct order Moeritherioidea. { ‚mir·ə·thə'rī·ə‚dē }

Moeritherioidea [PALEON] A suborder of extinct sirenian mammals considered as primitive proboscideans by some authorities and as a sirenian offshoot by others. { ‚mir·ə‚thir·ē'òid·ē·ə }

mofette [GEOL] A small opening emitting carbon dioxide in an area of late-stage volcanic activity. { mō'fet }

mohair [TEXT] A sleek, lustrous fabric made of angora goat fibers. { 'mō‚her }

Mohaupt effect [ORD] The effect of a metal liner introduced in a shaped charge to increase penetration. { 'mō‚haùpt i‚fekt }

mohavite *See* tincalconite. { mō'hä‚vīt }

Mohawkian [GEOL] A North American stage of middle Ordovician geologic time, above Chazyan and below Edenian. { mō'hòk·ē·ən }

mohm [MECH] A unit of mechanical mobility, equal to the reciprocal of 1 mechanical ohm. { mōm }

Mohnian [GEOL] A North American stage of geologic time: Miocene (above Luisian, below Delmontian). { 'mō·nē·ən }

Moho *See* Mohorovičić discontinuity. { 'mō·hō }

Mohole drilling [GEOL] Drilling aimed at penetration of the earth's crust, through the Mohorovičić discontinuity, to sample the mantle. { 'mō‚hōl ‚dril·iŋ }

Mohorovičić discontinuity [GEOPHYS] A seismic discontinuity that separates the earth's crust from the subjacent mantle, inferred from travel time curves indicating that seismic waves undergo a sudden increase in velocity. Also known as Moho. { ‚mō·hō'rō·və‚chich dis‚känt·ən'ü·əd·ē }

Mohr cubic centimeter [CHEM ENG] A unit of volume used in saccharimetry, equal to the volume of 1 gram of water at a specified temperature, usually 17.5°C, in which case, it is equal to 1.00238 cubic centimeters. { 'mòr 'kyü·bik 'sent·ə‚mēd·ər }

Mohr liter [CHEM ENG] A unit of volume, equal to 1000 Mohr cubic centimeters. { 'mòr 'lēd·ər }

Mohr's circle [MECH] A graphical construction making it possible to determine the stresses in a cross section if the principal stresses are known. { 'mòrz 'sər·kəl }

Mohr's salt *See* ferrous ammonium sulfate. { 'mòrz ‚sòlt }

Mohr titration [ANALY CHEM] Titration with silver nitrate to determine the concentration of chlorides in a solution; silver chromate precipitation is the end-point indicator. { 'mòr tī'trā·shən }

mohsite *See* ilmenite. { 'mō‚sīt }

Mohs scale [MINERAL] An empirical scale consisting of 10 minerals with reference to which the hardness of all other minerals is measured; it includes, from softest (designated 1) to hardest (10): talc, gypsum, calcite, fluorite, apatite, orthoclase, quartz, topaz, corundum, and diamond. { 'mōz ‚skāl }

moiety [CHEM] A part or portion of a molecule, generally complex, having a characteristic chemical or pharmacological property. { 'mòi·əd·ē }

moil [MIN ENG] A long steel wedge with a rounded point used for breaking up rocks in a mine. { mòil }

moiré [COMMUN] In television, the spurious pattern in the reproduced picture resulting from interference beats between two sets of periodic structures in the image. [GRAPHICS] Undesirable patterns that occur when a halftone is made from a previously printed halftone or steel engraving; they are caused by the conflict between the dot arrangement produced by the halftone screen and the dots or lines of the original halftone or engraving; careful rotation of the halftone screen by the photographer or engraver may minimize moiré. [TEXT] A fabric finish in which the warp has yarn of harder twist than the filling, with a surface pattern resembling water ripples that is produced by engraved rollers, heat, pressure, steam, and chemicals. { mò'rā }

moiré effect [OPTICS] The effect whereby, when one family of curves is superposed on another family of curves so that the curves cross at angles of less than about 45°, a new family of curves appears which pass through intersections of the original curves. { mò'rā i‚fekt }

moiré fringes [OPTICS] The bands which appear in the moiré effect. { mò'rā 'frin·jəz }

moiré interferometry [ENG] An optical technique that measures the components of deformation of a specimen surface in the plane of the surface by superposing a reference grating and a diffraction grating that is applied to, and deforms with, the surface. { mò'rā in·tər·fə'räm·ə·trē }

moissanite [MINERAL] SiC A carbide mineral found in meteorites; identical with artificial carborundum. { 'mòis·ən‚īt }

moist adiabat *See* saturation adiabat. { 'mòist 'ad·ē·ə‚bat }

moist-adiabatic lapse rate *See* saturation-adiabatic lapse rate. { 'mòist ‚ad·ē·ə‚bad·ik 'laps ‚rāt }

moist air [METEOROL] **1.** In atmospheric thermodynamics, air that is a mixture of dry air and any amount of water vapor. **2.** Generally, air with a high relative humidity. { 'mòist 'er }

moist climate [CLIMATOL] In C.W. Thornthwaite's climatic classification, any type of climate in which the seasonal water surplus counteracts seasonal water deficiency; thus it has a moisture index greater than zero. { 'mòist 'klīm·ət }

moist-heat sterilization [ENG] Sterilization with steam under pressure, as in an autoclave, pressure cooker, or retort; most bacteriological media are sterilized by autoclaving at 121°C, with 15 pounds (103 kilopascals) of pressure, for 20 minutes or more. { 'mòist ‚hēt ‚ster·ə·lə'zā·shən }

moist room [ENG] An enclosed space that is maintained at a specified temperature, usually 73°F (23°C), with the humidity maintained at 98% or above and that is used to cure and store test specimens of cementitious material. { 'mòist ‚rüm }

moisture [CLIMATOL] The quantity of precipitation or the precipitation effectiveness. [METEOROL] The water vapor content of the atmosphere, or the total water substance (gaseous, liquid, and solid) present in a given volume of air. [PHYS CHEM] Water that is dispersed through a gas in the form of water vapor or small droplets, dispersed through a solid, or condensed on the surface of a solid. { 'mòis·chər }

moisture adjustment [METEOROL] The adjustment of observed precipitation in a storm by the ratio of the estimated probable maximum precipitable water over the basin under study to the actual precipitable water calculated for the particular storm. { 'mòis·chər ə‚jəs·mənt }

moisture barrier [MATER] A material that retards the passage of moisture into walls. { 'mòis·chər ‚bar·ē·ər }

moisture content [MECH] The quantity of water in a mass of soil, sewage, sludge, or screenings; expressed in percentage by weight of water in the mass. { 'mòis·chər ‚kän·tent }

moisture factor [METEOROL] One of the simplest measures of precipitation effectiveness: moisture factor = P/T, where P is precipitation in centimeters and T is temperature degrees centigrade for the period in question. { 'mòis·chər ‚fak·tər }

moisture film cohesion *See* apparent cohesion. { 'mòis·chər ‚film kō‚hē·zhən }

moisture flux *See* eddy flux. { 'mòis·chər ‚fləks }

moisture gradient [ENG] The difference in moisture content between the surface and the inner portion of a section of wood. { 'mòis·chər ‚grād·ē·ənt }

moisture inversion [METEOROL] An increase with height of the moisture content of the air; specifically, the layer through which this increase occurs, or the altitude at which the increase begins. { 'mòis·chər in‚vər·zhən }

moisture loss [MECH ENG] The difference in heat content between the moisture in the boiler exit gases and that of moisture at ambient air temperature. { 'mòis·chər ‚lòs }

moisture-vapor transmission [FL MECH] The rate at which water vapor permeates a porous film (such as plastic or paper) or a wall. { 'mòis·chər ‚vā·pər tranz‚mish·ən }

Moko disease [PL PATH] Bacterial wilt of banana caused by *Pseudomonas solanacearum*. { 'mō·kō di‚zēz }

mcl *See* mole. { mōl }

MOL *See* manned orbiting laboratory.

molal average boiling point [PHYS CHEM] A pseudo boiling point for a mixture calculated as the summation of individual mole fraction-boiling point (in degrees Rankine) products. { 'mō·ləl 'av·rij 'bòil·iŋ ‚pòint }

MOIRÉ EFFECT

Two simple gratings crossed at a small angle.

molal elevation of the boiling point *See* ebullioscopic constant. { ¦mō·ləl ‚el·ə¦vā·shən əv th̲ə 'bȯil·iŋ ‚pȯint }

molal heat capacity *See* molar heat capacity. { 'mō·ləl 'hēt kə‚pas·əd·ē }

molality [CHEM] Concentration given as moles per 1000 grams of solvent. { mō'lal·əd·ē }

molal quantity [CHEM] The number of moles (gram-molecular weights) present, expressed with weight in pounds, grams, or such units, numerically equal to the molecular weight; for example, pound-mole, gram-mole. { 'mō·ləl 'kwän·əd·ē }

molal solution [CHEM] Concentration of a solution expressed in moles of solute divided by 1000 grams of solvent. { 'mō·ləl sə‚lü·shən }

molal specific heat *See* molar specific heat. { 'mō·ləl spə¦sif·ik 'hēt }

molal volume *See* molar volume. { 'mō·ləl 'väl·yəm }

molar [ANAT] **1.** A tooth adapted for grinding. **2.** Any of the three pairs of cheek teeth behind the premolars on each side of the jaws in humans. [PHYS CHEM] Denoting a physical quantity divided by the amount of substance expressed in moles. { 'mō·lər }

molar conductivity [PHYS CHEM] The ratio of the conductivity of an electrolytic solution to the concentration of electrolyte in moles per unit volume. { 'mō·lər ‚kän‚dək'tiv·əd·ē }

molar dispersion [OPTICS] In refractometry, the difference in molar refraction (refractive index) of a compound at two different light-beam wavelengths. { 'mō·lər di'spər·shən }

molar heat capacity [PHYS CHEM] The amount of heat required to raise 1 mole of a substance 1° in temperature. Also known as molal heat capacity; molecular heat capacity. { 'mō·lər 'hēt kə‚pas·əd·ē }

molarity [CHEM] Measure of the number of gram-molecular weights of a compound present (dissolved) in 1 liter of solution; it is indicated by M, preceded by a number to show solute concentration. { mō'lar·əd·ē }

molar magnetic rotation [OPTICS] A measure of the strength of the Faraday effect in a substance, equal to $M\alpha\rho'/(M'\alpha'\rho)$, where α is the angle of rotation, M is the molecular weight of the substance, ρ is its density, and α', M', and ρ' are corresponding quantities for water. { ¦mō·lər mag¦ned·ik rō'tā·shən }

molar refraction [OPTICS] Equation for the refractive index of a compound modified by the compound's molecular weight and density. Also known as the Lorentz-Lorenz molar refraction. { 'mō·lər ri'frak·shən }

molar solution [CHEM] Aqueous solution that contains 1 mole (gram-molecular weight) of solute in 1 liter of the solution. { 'mō·lər sə‚lü·shən }

molar specific heat [PHYS CHEM] The ratio of the amount of heat required to raise the temperature of 1 mole of a compound 1°, to the amount of heat required to raise the temperature of 1 mole of a reference substance, such as water, 1° at a specified temperature. Also known as molal specific heat; molecular specific heat. { 'mō·lər spə‚sif·ik ‚hēt }

molar susceptibility [PHYS CHEM] Magnetic susceptibility of a compound per gram-mole of that compound. { 'mō·lər sə‚sep·tə'bil·əd·ē }

molar volume [PHYS CHEM] The volume occupied by one mole of a substance in the form of a solid, liquid, or gas. Also known as molal volume; mole volume. { 'mō·lər 'väl·yəm }

molasse [GEOL] A paralic sedimentary facies consisting mainly of shale, subgraywacke sandstone, and conglomerate; it is more clastic and less rhythmic than the preceding flysch and is generally postorogenic. { mə'läs }

molasses [FOOD ENG] A brown viscid syrup prepared from raw sugar during sugar manufacturing processes. { mə'las·əs }

molasses/A.N. explosive [MATER] An explosive mixture consisting of ammonium nitrate mixed with molasses and water. { mə'las·əs ¦ā¦en ik'splō·siv }

mold [ENG] **1.** A pattern or template used as a guide in construction. **2.** A cavity which imparts its form to a fluid or malleable substance. [ENG ACOUS] The metal part derived from the master by electroforming in reproducing disk recordings; has grooves similar to those of the recording. [GEOL] Soft, crumbling friable earth. [GRAPHICS] **1.** To form a plastic substance by placing it in a matrix or form. **2.** The form or matrix for shaping a plastic substance. [MYCOL] Any of various woolly fungus growths. [PALEON] An

impression made in rock or earth material by an inner or outer surface of a fossil shell or other organic structure; a complete mold would be the hollow space. { mōld }

moldability [MATER] The capability of being molded. { ‚mōl·də'bil·əd·ē }

moldauite *See* moldavite. { mōl'daù‚īt }

moldavite [GEOL] A translucent, olive-to brownish-green or pale-green tektite from western Czechoslovakia, characterized by surface sculpturing due to solution etching. Also known as moldauite; pseudochrysolite; vitavite. [MINERAL] A variety of ozocerite from Moldavia. { mōl'dä‚vīt }

mold base [ENG] The assembly of all parts of an injection mold except the cavity, cores, and pins. { 'mōld ‚bās }

moldboard plow [AGR] A plow equipped with a curved iron plate (moldboard) that lifts and turns the soil. Also known as turnplow. { 'mōld‚bȯrd ‚plaù }

molded breadth [NAV ARCH] The maximum breadth of a ship measured to the outside of the frame bar usually at the midship section. { 'mōl·dəd 'bredth }

molded capacitor [ELEC] Capacitor, usually mica, that has been encased in a molded plastic insulating material. { 'mōl·dəd kə'pas·əd·ər }

molded coal [MATER] A synthetic solid fuel produced by molding charcoal refuse and coal tar into cylinders. { 'mōl·dəd 'kōl }

molded depth [NAV ARCH] The vertical distance from the base line to the molded line of the main deck at side measured at the midship section. { 'mōl·dəd 'depth }

molded-fabric bearing [DES ENG] A bearing composed of laminations of cotton or other fabric impregnated with a phenolic resin and molded under heat and pressure. { 'mōl·dəd ¦fab·rik 'ber·iŋ }

molded lines [ENG] Full-size lines of a ship or airplane which are laid out in a mold loft. [NAV ARCH] Line drawings of a ship which represent the shape inside the plating or outside the planking on a wooden vessel. { 'mōl·dəd 'līnz }

molded silver nitrate *See* lunar caustic. { 'mōl·dəd 'sil·vər 'nī‚trāt }

mold efficiency [ENG] In a multimold blow-molding system, the percentage of the total turn-around time actually required for the forming, cooling, and ejection of the formed objects. { 'mōld i‚fish·ən·sē }

molding [ARCH] A continuous contour surface of rectangular or curved profile, used on a plane surface such as a wall to effect a transition or create a decorative effect by the play of shadow and light on it. { 'mōl·diŋ }

molding board [MET] In sand-mold metal casting, a component that fits the lower half of the form or flask used for the molding operation to hold the casting pattern. { 'mōl·diŋ ‚bȯrd }

molding cycle [ENG] **1.** The time required for a complete sequence of molding operations. **2.** The combined operations required to produce a set of moldings. { 'mōl·diŋ ‚sī·kəl }

molding machine [MET] A machine that compacts sand around a pattern to form a mold. { 'mōl·diŋ mə‚shēn }

molding powder [MATER] Powdered plastic-material ingredients (such as resin, filler, pigments, and plasticizers) ready for compression in molding. { 'mōl·diŋ ‚paùd·ər }

molding press [MET] A press used to form compacts in powder metallurgy. { 'mōl·diŋ ‚pres }

molding pressure [ENG] Pressure needed to force softened plastic to fill a mold cavity. { 'mōl·diŋ ‚presh·ər }

molding sand *See* foundry sand. { 'mōld·iŋ ‚sand }

molding shrinkage [ENG] Difference in dimensions between the molding and the mold cavity, measured at normal room temperature. { 'mōl·diŋ ‚shriŋk·ij }

molding surface [MATH] A surface generated by a plane curve as its plane rolls without slipping over a cylinder. { 'mōld·iŋ ‚sər·fas }

molding time *See* curing time. { 'mōl·diŋ ‚tīm }

mold loft [ENG] A large building with a smooth wooden floor where full-size lines of a ship or airplane are laid down and templates are constructed from them to lay off the steel for cutting. { 'mōld ‚lȯft }

mold release *See* release agent. { 'mōld ri‚lēs }

mold seam *See* seam. { 'mōld ‚sēm }

mold shift [MET] The mismatch of mold halves at the parting line, resulting in a casting defect. { 'mōld ‚shift }

mold steel [MET] Steel of tool-steel quality used to make

MOLD LOFT

Interior of a mold lift. (*Newport News Shipbuilding photograph*)

molds for plastics; properties include uniform texture, good machinability with die-sinking tools, and lack of microscopic porosity. { 'mōld ,stēl }

mold wash [MET] An aqueous or alcoholic suspension or emulsion used to coat the surfaces of a mold cavity. { 'mōld ,wåsh }

mole [CHEM] An amount of substance of a system which contains as many elementary units as there are atoms of carbon in 0.012 kilogram of the pure nuclide carbon-12; the elementary unit must be specified and may be an atom, molecule, ion, electron, photon, or even a specified group of such units. Symbolized mol. [CIV ENG] A breakwater or berthing facility, extending from shore to deep water, with a core of stone or earth. [MECH ENG] A mechanical tunnel excavator. [MED] **1.** A mass formed in the uterus by the maldevelopment of all or part of the embryo or of the placenta and membranes. **2.** A fleshy, pigmented nevus. [VERT ZOO] Any of 19 species of insectivorous mammals composing the family Talpidae; the body is stout and cylindrical, with a short neck, small or vestigial eyes and ears, a long naked muzzle, and forelimbs adapted for digging. { mōl }

molecular adhesion [PHYS CHEM] A particular manifestation of intermolecular forces which causes solids or liquids to adhere to each other; usually used with reference to adhesion of two different materials, in contrast to cohesion. { mə'lek·yə·lər ad'hē·zhən }

molecular amplitude [ANALY CHEM] The difference between the molecular rotation at the extreme (peak or trough) value caused by the longer light wavelength and the molecular rotation at the extreme value caused by the shorter wavelength. { mə'lek·yə·lər 'am·plə,tüd }

molecular association [PHYS CHEM] The formation of double molecules or polymolecules from a single species as a result of specific and moderately strong intermolecular forces. { mə'lek·yə·lər ə,sō·sē'ā·shən }

molecular asymmetry See asymmetry. { mə'lek·yə·lər ,ā'sim·ə·trē }

molecular attraction [PHYS CHEM] A force which pulls molecules toward each other. { mə'lek·yə·lər ə'trak·shən }

molecular beam [PHYS] A beam of neutral molecules whose directions of motion lie within a very small solid angle. { mə'lek·yə·lər 'bēm }

molecular-beam apparatus [PHYS] A device in which a molecular beam in a vacuum is subjected to magnetic fields, oscillating fields, or other influences, and a detector measures the resulting intensity of the beam at some location; used primarily in radio-frequency spectroscopy. { mə'lek·yə·lər¦bēm ,ap·ə'rad·əs }

molecular-beam epitaxy [SOLID STATE] A technique of growing single crystals in which beams of atoms or molecules are made to strike a single-crystalline substrate in a vacuum, giving rise to crystals whose crystallographic orientation is related to that of the substrate. Abbreviated MBE. { mə'lek·yə·lər¦bēm 'ep·ə,tak·sē }

molecular binding [SOLID STATE] The force which holds a molecule at some site on the surface of a crystal. { mə'lek·yə·lər 'bind·iŋ }

molecular biology [BIOL] That part of biology which attempts to interpret biological events in terms of the physicochemical properties of molecules in a cell. { mə'lek·yə·lərbī'äl·ə·jē }

molecular biomarker [MED] A biological indicator that signals a changed physiological state, stress, or injury due to disease or the environment; for example, an elevated serum level of prostate-specific antigen is a biomarker for prostate cancer. { me¦lek·yə·lər 'bī·ō,märk·ər }

molecular biophysics [BIOPHYS] The study of the physical properties and interactions of large molecules and particles of comparable size which play important roles in biology. { mə,lek·yə·lər ,bi·ō'fiz·iks }

molecular chaperone [CELL MOL] Any of a class of cellular proteins involved in correct folding of certain polypeptide chains and their assembage into an oligomer. { mə'lek·yə·lər 'shap·ə,rän }

molecular circuit [ELECTR] A circuit in which the individual components are physically indistinguishable from each other. { mə'lek·yə·lər 'sər·kət }

molecular cloud [ASTRON] A dense cloud of interstellar gas in which molecules have formed in appreciable abundance. { mə'lek·yə·lər 'klaůd }

molecular cluster [PHYS CHEM] An assembly of molecules that are weakly bound together and display properties intermediate between those of isolated gas-phase molecules and bulk condensed media. { mə'lek·yə·lər 'kləs·tər }

molecular conductivity [PHYS CHEM] The conductivity of a volume of electrolyte containing 1 mole of dissolved substance. { mə'lek·yə·lər ,kän,dək'tiv·əd·ē }

molecular crystal [CRYSTAL] A solid consisting of a lattice array of molecules such as hydrogen, methane, or more complex organic compounds, bound by weak van der Waals forces, and therefore retaining much of their individuality. { mə'lek·yə·lər 'krist·əl }

molecular device [CHEM] An assemblage of a discrete number of molecular components (that is, a supramolecular structure) designed to achieve a specific function. { mə'lek·yə·lər di'vīs }

molecular diamagnetism [PHYS CHEM] Diamagnetism of compounds, especially organic compounds whose susceptibilities can often be calculated from the atoms and chemical bonds of which they are composed. { mə'lek·yə·lər ,dī·ə'mag·nə,tiz·əm }

molecular diameter [PHYS CHEM] The diameter of a molecule, assuming it to be spherical; has a numerical value of 10^{-8} centimeter multiplied by a factor dependent on the compound or element. { mə'lek·yə·lər dī'am·əd·ər }

molecular diffusion [FL MECH] The transfer of mass between adjacent layers of fluid in laminar flow. { mə'lek·yə·lər di'fyü·zhən }

molecular dipole [PHYS CHEM] A molecule having an electric dipole moment, whether it is permanent or produced by an external field. { mə'lek·yə·lər 'dī,pōl }

molecular distillation [CHEM] A process by which substances are distilled in high vacuum at the lowest possible temperature and with least damage to their composition. { mə'lek·yə·lər ,dis·tə'lā·shən }

molecular drag pump [ENG] A vacuum pump in which pumping is accomplished by imparting a high momentum to the gas molecules by impingement of a body rotating at very high speeds, as much as 16,000 revolutions per minute; such pumps achieve a vacuum as high as 10^{-6} torr. { mə'lek·yə·lər 'drag ,pəmp }

molecular dynamics [PHYS CHEM] A branch of physical chemistry concerned with molecular mechanisms of the elementary physical and chemical processes that control rates of reaction. { mə'lek·yə·lər di'nam·iks }

molecular effusion [FL MECH] Mass-transfer flow mechanism of free-molecule transfer through pores or orifices. { mə'lek·yə·lər i'fyü·zhən }

molecular electronics [COMPUT SCI] The use of biological or organic molecules for fabricating electronic materials with novel electronic, optical, or magnetic properties; applications include polymer light-emitting diodes, conductive-polymer sensors, pyroelectric plastics, and, potentially, molecular computational devices. { mə'lek·yə·lər i,lek'trän·iks }

molecular energy level [PHYS CHEM] One of the states of motion of nuclei and electrons in a molecule, having a definite energy, which is allowed by quantum mechanics. { mə'lek·yə·lər 'en·ər·jē ,lev·əl }

molecular engineering [ELECTR] The use of solid-state techniques to build, in extremely small volumes, the components necessary to provide the functional requirements of overall equipments, which when handled in more conventional ways are vastly bulkier. { mə'lek·yə·lər ,en·jə'nir·iŋ }

molecular entity [CHEM] A chemically or isotopically distinct atom, molecule, ion, complex, free radical, or similar unit that can be distinguished from other kinds of units. { mə'lek·yə·lər 'en·təd·ē }

molecular exclusion chromatography See gel filtration. { mə'lek·yə·lər ik¦sklü·zhən ,krō·mə'täg·rə·fē }

molecular field theory See Weiss theory. { mə'lek·yə·lər 'ēld ,thē·ə·rē }

molecular flow [FL MECH] Gas-flow phenomenon at low pressures or in small channels when the mean free path is of the same order of magnitude as the channel diameter; a gas molecule thus migrates along the channel independent of other gas molecules present. { mə'lek·yə·lər 'flō }

MOLE

The European common mole.

MOLECULAR BINDING

A model of a crystal surface showing different types of molecular binding sites. Molecule A resting on completed plane of molecules is more weakly bound than molecule B at ledge formed by incomplete plane. (General Electric Co.)

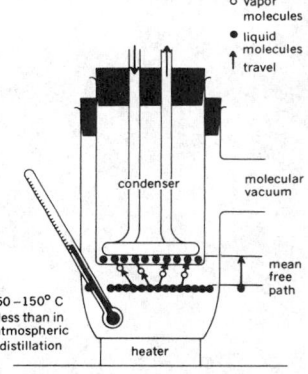

MOLECULAR DISTILLATION

○ vapor molecules
● liquid molecules
↑ travel

condenser

molecular vacuum

mean free path

50–150° C less than in atmospheric distillation

heater

Schematic diagram of process.

molecular formula [CHEM] A chemical formula that indicates the actual numbers and kinds of atoms in a molecule, but not the chemical structure. { mə'lek·yə·lər 'fòr·myə·lə }

molecular fossils See biomarkers. { mə'lek·yə·lər 'fäs·əlz }

molecular gage [ENG] Any instrument, such as a rotating viscometer gage or a decrement gage, that uses the dependence of the viscosity of a gas on its pressure to measure pressures on the order of 1 pascal or less. Also known as viscosity gage; viscosity manometer. { mə'lek·yə·lər 'gāj }

molecular gas [CHEM] A gas composed of a single species, such as oxygen, chlorine, or neon. { mə'lek·yə·lər 'gas }

molecular gas laser [OPTICS] Any gas laser in which the gas consists of molecules rather than atoms; such a laser can be operated on a large number of rotational-vibrational lines, and, at a sufficiently high pressure, these lines overlap and a wide gain region is obtained. Also known as molecular laser. { mə'lek·yə·lər ¦gas 'lā·zər }

molecular genetics [MOL BIO] The approach which deals with the physics and chemistry of the processes of inheritance. { mə'lek·yə·lər jə'ned·iks }

molecular graphics [PHYS CHEM] The use of computer graphics to display and manipulate chemical structures with sufficient accuracy that bond distances and angles may be displayed and reported and it is possible to dock or fit together two or more molecules. Also known as graphics-based molecular modeling. { mə'lek·yə·lər 'graf·iks }

molecular heat [THERMO] The heat capacity per mole of a substance. { mə'lek·yə·lər 'hēt }

molecular heat capacity See molar heat capacity. { mə'lek·yə·lər ¦hēt kə,pas·əd·ē }

molecular heat diffusion [THERMO] Transfer of heat through the motion of molecules. { mə'lek·yə·lər ¦hēt di,fyü·shən }

molecular imprinting [PHYS CHEM] A technique for creating receptor structures on a polymer surface that can selectively bind to molecules of interest; molecularly imprinted polymers are used for separations, as catalysts, and in biosensors. { mə'lek·yə·lər 'im,print·iŋ }

molecular ion [ORG CHEM] An ion that results from the loss of an electron by an organic molecule following bombardment with high-energy electrons during mass spectrometry. { mə'lek·yə·lər 'ī,än }

molecularity [PHYS CHEM] In a chemical reaction, the number of molecules which come together and form the activated complex. { mə,lek·yə'lar·əd·ē }

molecular laser See molecular gas laser. { mə'lek·yə·lər 'lā·zər }

molecular machine [CHEM] A molecular device in which the component parts can display changes (reversible movement) in their relative positions as a result of some external stimulus (such as light, electrical energy, or chemical energy), resulting in a signal (a change in a chemical or physical property of the supramolecular system) that can be used to monitor the operation of the device. { mə'lek·yə·lər mə'shēn }

molecular magnet [PHYS CHEM] A molecule having a nonvanishing magnetic dipole moment, whether it is permanent or produced by an external field. { mə'lek·yə·lər 'mag·nət }

molecular mechanics [PHYS CHEM] An empirical method of calculating the dynamics of molecules, in which bonds between atoms are represented by springs obeying Hooke's law, and additional terms representing bond angle bending, torsional interactions, and van der Waals-type interactions are included. Also known as force-field method. { mə'lek·yə·lər mi'kan·iks }

molecular mimicry [IMMUNOL] The sharing, by two organisms closely related ecologically but not phylogenetically, of common macromolecular structures that are not attributable to evolutionary conservation of these structures. { mə'lek·yə·lər 'mim·i,krē }

molecular modeling [CHEM] The use of computers for the simulation of chemical entities and processes. { mə'lek·yə·lər 'mäd·liŋ }

molecular optics [OPTICS] The study of the propagation of light and associated phenomena, such as refraction, absorption, and scattering, through collections of molecules in gases, liquids, and solids. { mə'lek·yə·lər 'äp·tiks }

molecular orbital [PHYS CHEM] A wave function describing an electron in a molecule. { mə'lek·yə·lər 'òr·bəd·əl }

molecular paramagnetism [PHYS CHEM] Paramagnetism of molecules, such as oxygen, some other molecules, and a large number of organic compounds. { mə'lek·yə·lər ¦par·ə'mag·nə,tiz·əm }

molecular pathology [PATH] The study of the bases and mechanisms of disease on a molecular or chemical level. { mə'lek·yə·lər pə'thäl·ə·jē }

molecular physics [PHYS] The study of the behavior and structure of molecules, including the quantum-mechanical explanation of several kinds of chemical binding between atoms in a molecule, directed valence, the polarizability of molecules, the quantization of vibrational, rotational, and electronic motions of molecules, and the phenomena arising from intermolecular forces. { mə'lek·yə·lər 'fiz·iks }

molecular polarizability [PHYS CHEM] The electric dipole moment induced in a molecule by an external electric field, divided by the magnitude of the field. { mə'lek·yə·lər ,pō·lə,rīz·ə'bil·əd·ē }

molecular pump [MECH ENG] A vacuum pump in which the molecules of the gas to be exhausted are carried away by the friction between them and a rapidly revolving disk or drum. { mə'lek·yə·lər 'pəmp }

molecular rearrangement See rearrangement reaction. { mə'lek·yə·lər ,rē·ə'rānj·mənt }

molecular receptor [ORG CHEM] A species that can select one of many possible binding partners and form a complex that is stabilized by interactions such as hydrogen bonding or changes in solvation. { mə'lek·yər·lər ri'sep·tər }

molecular recognition [CELL MOL] The ability of biological and chemical systems to distinguish between molecules and regulate behavior accordingly. [CHEM] The (molecular) storage and the (supramolecular) retrieval and processing of molecular structural information and interactions. { mə'lek·yə·lər ,rek·ig'nish·ən }

molecular relaxation [PHYS CHEM] Transition of a molecule from an excited energy level to another excited level of lower energy or to the ground state. { mə'lek·yə·lər ,rē,lak'sā·shən }

molecular rotation [OPTICS] In a solution of an optically active compound, the specific rotation (angular rotation of polarized light) multiplied by the compound's molecular weight. { mə'lek·yə·lər rō'tā·shən }

molecular self-assembly [ORG CHEM] The spontaneous aggregation of molecules into well-defined, stable, noncovalently bonded assemblies that are held together by intermolecular forces. { mə'lek·yə·lər ,self ə'sem·blē }

molecular sieve [CHEM] A naturally occurring or synthetic zeolite characterized by the ability to undergo dehydration with little or no change in crystal structure, thereby offering a very high surface area for adsorption of foreign molecules. { mə'lek·yə·lər 'siv }

molecular-sieve chromatography See gel filtration. { mə'lek·yə·lər 'siv ,krō·mə'täg·rə·fē }

molecular simulation [CHEM] Computational techniques for predicting many useful functional properties of chemicals and materials, including thermodynamic properties, thermochemical properties, spectroscopic properties, mechanical properties, transport properties, and morphological information. { mə'lek·yə·lər ,sim·yə'lā·shən }

molecular specific heat See molar specific heat. { mə'lek·yə·lər spə¦sif·ik 'hēt }

molecular spectroscopy [SPECT] The production, measurement, and interpretation of molecular spectra. { mə'lek·yə·lər spek'träs·kə·pē }

molecular spectrum [SPECT] The intensity of electromagnetic radiation emitted or absorbed by a collection of molecules as a function of frequency, wave number, or some related quantity. { mə'lek·yə·lər 'spek·trəm }

molecular still [CHEM] An apparatus used to conduct molecular distillation. { mə'lek·yə·lər 'stil }

molecular stopping power [NUCLEO] For an ionizing particle passing through a compound, the particle's energy loss per molecule within a unit area normal to the particle's path; equal to the linear energy transfer (energy loss per unit path length) divided by the number of molecules per unit volume. { mə'lek·yə·lər 'stäp·iŋ ,paú·ər }

molecular structure [PHYS CHEM] The manner in which electrons and nuclei interact to form a molecule, as elucidated by quantum mechanics and a study of molecular spectra. { mə'lek·yə·lər 'strək·chər }

molecular theory *See* kinetic theory.　{ mə'lek·yə·lər 'thē·ə·rē }

molecular velocity [PHYS CHEM]　The velocity of an individual molecule in a given sample of gas; the vector quantity is symbolized u, and the magnitude is symbolized u.　{ mə'lek·yə·lər və'läs·əd·ē }

molecular vibration [PHYS CHEM]　The theory that all atoms within a molecule are in continuous motion, vibrating at definite frequencies specific to the molecular structure as a whole as well as to groups of atoms within the molecule; the basis of spectroscopic analysis.　{ mə'lek·yə·lər vī'brā·shən }

molecular volume [CHEM]　The volume that is occupied by 1 mole (gram-molecular weight) of an element or compound; equals the molecular weight divided by the density.　{ mə'lek·yə·lər 'väl·yəm }

molecular weight [CHEM]　The sum of the atomic weights of all the atoms in a molecule.　Also known as relative molecular mass.　{ mə'lek·yə·lər 'wāt }

molecular-weight distribution [ORG CHEM]　Frequency of occurrence of the different molecular-weight chains in a homologous polymeric system.　{ mə'lek·yə·lər ¦wāt ˌdi·strə'byü·shən }

molecule [CHEM]　A group of atoms held together by chemical forces; the atoms in the molecule may be identical as in H_2, S_2, and S_8, or different as in H_2O and CO_2; a molecule is the smallest unit of matter which can exist by itself and retain all its chemical properties.　{ 'mäl·ə,kyül }

mole drain [CIV ENG]　A subsurface channel for water drainage; formed by pulling a solid object, usually a solid cylinder having a wedge-shaped point at one end, through the soil at the proper slope and depth.　{ 'mōl ,drān }

mole fraction [CHEM]　The ratio of the number of moles of a substance in a mixture or solution to the total number of moles of all the components in the mixture or solution.　{ 'mōl ,frak·shən }

mole mining [MIN ENG]　A method of working coal seams about 30 inches (75 centimeters) thick; a small continuous-miner type of machine is used, which is remote-controlled from the roadway, without associated supports.　{ 'mōl ,mīn·iŋ }

Molenbroeck-Chaplygin transformation [FL MECH]　A version of the hodograph method for compressible flow in which only the independent variables are replaced and no change is made in the dependent variables, that is, the velocity potential and stream function.　{ ¦mō·lən·brük chap'lē·gən ,tranz·fər,mā·shən }

mole percent [CHEM]　Percentage calculation expressed in terms of moles rather than weight.　{ 'mōl pər,sent }

moleskin [TEXT]　Rugged satin-weave cotton with a soft napped back having a one-warp, two-filling construction.　{ 'mōl,skin }

mole volume *See* molar volume.　{ 'mōl 'väl·yəm }

Molidae [VERT ZOO]　A family of marine fishes, including some species of sunfishes, in the order Perciformes.　{ 'mäl·ə,dē }

molinate [ORG CHEM]　$C_9H_{17}NOS$　A light yellow liquid with limited solubility in water; used as a herbicide to control watergrass in rice.　{ 'mäl·ə,nāt }

Molisch test [BIOCHEM]　A test used for the general detection of carbohydrates; a purple color is produced when the solution containing carbohydrate is treated with strong sulfuric acid in the presence of α-naphthol.　{ 'mōl·ish ,test }

Moller scattering [QUANT MECH]　Scattering of electrons by electrons.　{ 'mȯl·ər ,skad·ə·riŋ }

Mollicutes [MICROBIO]　The mycoplasmas, a class of prokaryotic organisms lacking a true cell wall; cells are very small to submicroscopic.　{ mə'lik·yə,dēz }

Mollier diagram [THERMO]　Graph of enthalpy versus entropy of a vapor on which isobars, isothermals, and lines of equal dryness are plotted.　{ 'mȯl·yā ,dī·ə,gram }

Mollisol [GEOL]　An order of soils having dark or very dark, friable, thick A horizons high in humus and bases such as calcium and magnesium; most have lighter-colored or browner B horizons that are less friable and about as thick as the A horizons; all but a few have paler C horizons, many of which are calcareous.　{ 'mäl·ə,sȯl }

Moll thermopile [ENG]　A thermopile used in some types of radiation instruments; alternate junctions of series-connected

manganan-constantan molybdenum, added as ferromolybdenum or calcium molybdenum; increases strength, toughness, and wear resistance.　{ 'mȯl 'thər·mə,pīl }

Mollusca [INV ZOO]　One of the divisions of phyla of the animal kingdom containing snails, slugs, octopuses, squids, clams, mussels, and oysters; characterized by a shell-secreting organ, the mantle, and a radula, a food-rasping organ located in the forward area of the mouth.　{ mə'ləs·kə }

molluscicide [MATER]　An agent that kills mollusks.　{ mə'ləs·ə,sīd }

molluscum contagiosum [MED]　A viral disease of the skin, characterized by one or more discrete, waxy, dome-shaped nodules with frequent umbilication.　{ mə'ləs·kəm kən,tä·jē'ō·səm }

mollusk [INV ZOO]　Any member of the Mollusca.　{ 'mäl·əsk }

Molniya orbit [AERO ENG]　An earth satellite orbit designed for communications satellite service coverage at high latitudes, with an orbital period of slightly less than 12 hours (semisynchronous orbit), inclination of 63.4°, and high eccentricity (0.722), so that the apogee (where the satellite lingers over the service coverage area) is at 25,000 miles (40,000 kilometers) and the perigee is at only 300 miles (500 kilometers).　{ 'mȯl·nē·ə ,ȯr·bət }

Molossidae [VERT ZOO]　The free-tailed bats, a family of tropical and subtropical insectivorous mammals in the order Chiroptera.　{ mə'läs·ə,dē }

Molpadida [INV ZOO]　An order of sea cucumbers belonging to the Apodacea and characterized by a short, plump body bearing a taillike prolongation.　{ mäl'pā·də·də }

Molpadidae [INV ZOO]　The single family of the echinoderm order Molpadida.　{ mäl'pā·də,dē }

molt [PHYSIO]　To shed an outer covering as part of a periodic process of growth.　{ mōlt }

molten-salt reactor [NUCLEO]　A nuclear reactor in which the fissile and fertile material, in the form of fluoride salts, is dissolved in the coolant, which is a molten mixture of salts such as lithium fluoride and beryllium fluoride.　Abbreviated MSR.　Also known as fused-salt reactor.　{ ¦mōlt·ən ¦sȯlt rē'ak·tər }

molting hormone [BIOCHEM]　Any of several hormones which activate molting in arthropods.　{ 'mōl·tiŋ 'hȯr,mōn }

molybdate [INORG CHEM]　A salt derived from a molybdic acid.　{ mə'lib,dāt }

molybdate orange [MATER]　Pigment that is a solid solution of lead chromate, molybdate, and sulfate; used in paints, inks, and plastics.　{ mə'lib,dāt 'är·ənj }

molybdenite [MINERAL]　MoS_2　A metallic, lead-gray mineral that crystallizes in the hexagonal system and is commonly found in scales or foliated masses; hardness is 1.5 on Mohs scale, and specific gravity is 4.7; it is chief ore of molybdenum.　{ mə'lib·də,nīt }

molybdenum [CHEM]　A chemical element, symbol Mo, atomic number 42, and atomic weight 95.94.　[MET]　A silvery-gray metal used in iron-based alloys.　{ mə'lib·də·nəm }

molybdenum cast iron [MET]　Cast iron containing small amounts of molybdenum, added as ferromolybdenum or calcium molybdenum; increases strength, toughness, and wear resistance.　{ mə'lib·də·nəm 'kast ¦ī·ərn }

molybdenum dioxide [INORG CHEM]　MoO_2　Lead-gray powder; insoluble in hydrochloric and hydrofluoric acids; used in pigment for textiles.　{ mə'lib·də·nəm dī'äk,sīd }

molybdenum disilicide [INORG CHEM]　$MoSi_2$　A dark gray, crystalline powder with a melting range of 1870–2030°C; soluble in hydrofluoric and nitric acids; used in electrical resistors and for protective coatings for high-temperature conditions.　{ mə'lib·də·nəm dī'sil·ə,sīd }

molybdenum disulfide [INORG CHEM]　MoS_2　A black lustrous powder, melting at 1185°C, insoluble in water, soluble in aqua regia and concentrated sulfuric acid; used as a dry lubricant and an additive for greases and oils.　Also known as molybdenum sulfide; molybdic sulfide.　{ mə'lib·də·nəm dī'səl,fīd }

molybdenum pentachloride [INORG CHEM]　$MoCl_5$　Hygroscopic gray-black needles melting at 194°C; reacts with water and air; soluble in anhydrous organic solvents; used as a catalyst and as raw material to make molybdenum hexacarbonyl.　{ mə'lib·də·nəm ,pen·tə'klȯr,īd }

molybdenum sesquioxide [INORG CHEM] MoO_3 Water-insoluble, gray-black powder with slight solubility in acids; used as a catalyst and as a coating for metal articles. { mə'lib·də·nəm ˌses·kwē'äkˌsīd }

molybdenum silicide [MET] A mixture of molybdenum, silicon, and iron in the proportion 60:30:10; used to introduce molybdenum into steel melts. { mə'lib·də·nəm 'sil·əˌsīd }

molybdenum steel [MET] **1.** A carbon steel containing usually less than 0.5% molybdenum to aid hardenability. **2.** A tool steel containing up to 10% molybdenum, up to 1.5% carbon, and varying amounts of chromium, vanadium tungsten, and sometimes cobalt. { mə'lib·də·nəm 'stel }

molybdenum sulfide See molybdenum disulfide. { mə'lib·də·nəm 'səlˌfīd }

molybdenum trioxide [INORG CHEM] MoO_3 A white solid at room temperature, with a melting point of 795°C; soluble in concentrated mixtures of nitric and sulfuric acids and nitric and hydrochloric acids; used as a corrosion inhibitor, in enamels and ceramic glazes, in medicine and agriculture, and as a catalyst in the petroleum industry. { mə'lib·də·nəm trī'äkˌsīd }

molybdic acid [INORG CHEM] Any acid derived from molybdenum trioxide, especially the simplest acid H_2MoO_4, obtained as white crystals. { mə'lib·dik 'as·əd }

molybdic ocher See molybdite. { mə'lib·dik 'ō·kər }

molybdic sulfide See molybdenum disulfide. { mə'lib·dik 'səlˌfīd }

molybdine See molybdite. { mə'libˌdēn }

molybdite [MINERAL] MoO_3 A mineral, much of which is actually ferrimolybdite. Also known as molybdic ocher; molybdine. { mə'libˌdīt }

molybdophyllite [MINERAL] $(Pb,Mg)_2SiO_4 \cdot H_2O$ A colorless, white, or pale-green mineral composed of a silicate of lead and magnesium. { mə'lib·dō'fiˌlīt }

moly-blacks [MATER] Lustrous, black, molybdenum-containing decorative coatings used to blacken zinc or zinc-base alloys. { 'mäl·ēˌblaks }

molysite [MINERAL] $FeCl_3$ A brownish-red or yellow mineral composed of native ferric chloride, occurring in lava at Vesuvius. { 'mäl·əˌsīt }

molysmophobia [PSYCH] Abnormal fear of infection or contamination. { məˌliz·mə'fō·bē·ə }

moment [MECH] Static moment of some quantity, except in the term "moment of inertia." [STAT] The nth moment of a distribution $f(x)$ about a point x_0 is the expected value of $(x - x_0)^n$, that is, the integral of $(x - x_0)^n df(x)$, where $df(x)$ is the probability of some quantity's occurrence; the first moment is the mean of the distribution, while the variance can be found in terms of the first and second moments. { 'mō·mənt }

momental ellipsoid [MECH] An inertia ellipsoid whose size is specified to be such that the tip of the angular velocity vector of a freely rotating object, with origin at the center of the ellipsoid, always lies on the ellipsoid's surface. Also known as energy ellipsoid. { mō'ment·əl ə'lipˌsoid }

moment coefficient [AERO ENG] The coefficients used for moment are similar to coefficients of lift, drag, and thrust, and are likewise dimensionless; however, these must include a characteristic length, in addition to the area; the span is used for rolling or yawing moment, and the chord is used for pitching moment. { 'mō·məntˌkō·i'fish·ənt }

moment diagram [MECH] A graph of the bending moment at a section of a beam versus the distance of the section along the beam. { 'mō·məntˌdī·əˌgram }

moment generating function [STAT] For a frequency function $f(x)$, a function $\phi(t)$ that is defined as the integral from $-\infty$ to ∞ of $\exp(tx) f(x)dx$, and whose derivatives evaluated at $t = 0$ give the moments of f. { 'mō·mənt ˈjen·əˌrād·iŋ 'fəŋk·shən }

moment of force See torque. { 'mō·mənt əv 'fors }

moment of inertia [MECH] The sum of the products formed by multiplying the mass (or sometimes, the area) of each element of a figure by the square of its distance from a specified line. Also known as rotational inertia. { 'mō·mənt əv i'nər·shə }

moment of momentum See angular momentum. { 'mō·mənt əv mō'ment·əm }

moment problem [STAT] The problem of finding a distribution whose moments have specified values, or of determining whether such a distribution exists. { 'mō·mənt ˌpräb·ləm }

moment sensor [ENG] A device that measures the force applied at a remote point in a robotic system. { 'mō·mənt ˌsen·sər }

momentum [MECH] **1.** Also known as linear momentum; vector momentum. **2.** For a single nonrelativistic particle, the product of the mass and the velocity of a particle. **3.** For a single relativistic particle, $m\mathbf{v}/(1 - v^2/c^2)^{1/2}$, where m is the rest-mass, $\mathbf{v}$ the velocity, and c the speed of light. For a system of particles, the vector sum of the momenta (as in the first or second definition) of the particles. { mō'ment·əm }

momentum conservation See conservation of momentum. { mōm'ment·əm ˌkän·sər'vā·shən }

momentum density [PHYS] The momentum per unit volume of any given field. { mō'ment·əm 'den·sədˌē }

momentum-transport hypothesis [FL MECH] The hypothesis that the principle of conservation of momentum is valid in turbulent eddy transfer. { mō'ment·əm ˈtranzˌport hīˌpäth·ə·səs }

momentum wave function [QUANT MECH] A function of the momenta of a system of particles and of time which results from taking Fourier transforms, over the coordinates of all the particles, of the Schrödinger wave function; the absolute value squared is proportional to the probability that the particles will have given momenta at a given time. { mō'ment·əm 'wāv ˌfəŋk·shən }

Momertz-Lentz system [MIN ENG] Placement of two winding engines alongside the top of the mine shaft using the shaft collar as a common foundation; results in practically vertical ropes and less rope oscillation. { 'mōˌmerts 'lents ˌsis·təm }

Momotidae [VERT ZOO] The motmots, a family of colorful New World birds in the order Coraciiformes. { mə'mädəˌdē }

MOMS See micro-opto-mechanical system. { mämz or ˈemˌōˈem'es }

monactine [INV ZOO] A single-rayed spicule in the sponges. { ˈmä'nak·tən }

monadelphous [BOT] Having the filaments of the stamens united into one set. { ˈmän·əˈdel·fəs }

monadic operation [COMPUT SCI] An operation on one operand, such as a negation. { mō'nad·ik ˌäp·ə'rā·shən }

Monadidae [INV ZOO] A family of flagellated protozoans in the order Kinetoplastida having two flagella of uneven length. { mə'nad·əˌdē }

monadnock [GEOL] A remnant hill of resistant rock rising abruptly from the level of a peneplain; commonly represents an outcrop of rock that has withstood erosion. Also known as torso mountain. { mə'nadˌnäk }

monalbite [MINERAL] A modification of albite with monoclinic symmetry that is stable under equilibrium conditions at temperatures (about 1000°C) near the melting point. { ˌmō'nalˌbīt }

monandrous [BOT] Having one stamen. { mə'nan·drəs }

monatomic [CHEM] Composed of one atom. { ˈmänə'täm·ik }

monatomic gas [CHEM] A gas whose molecules have only one atom; the inert gases are examples. { ˈmän·ə'täm·ik 'gas }

monaural sound [ENG ACOUS] Sound produced by a system in which one or more microphones are connected to a single transducing channel which is coupled to one or two earphones worn by the listener. { män'or·əl 'saund }

monaxon [INV ZOO] A spicule formed by growth along a single axis. { ˈmä'nakˌsän }

monazite [MINERAL] A yellow or brown rare-earth phosphate monoclinic mineral with appreciable substitution of thorium for rare-earths and silicon for phosphorus; the principal ore of the rare earths and of thorium. Also known as cryptolite. { 'män·əˌzīt }

monchiquite [PETR] A lamprophyre composed of olivine, pyroxene, and usually mica or amphibole phenocrysts embedded in a glass or analcime groundmass. { 'man·chəˌkwīt }

Mönckeberg's arteriosclerosis See medial arteriosclerosis. { 'məŋ·kəˌbərgz ärˈtir·ē·ō·sklə'rō·səs }

Monday fever See metal fume fever. { 'mənˌdā ˌfē·vər }

Mond process [MET] A process for extracting and purifying nickel whereby nickel carbonyl is first formed by reaction of the reduced metal with carbon monoxide, and then the nickel carbonyl is decomposed thermally, resulting in deposition of nickel. { 'mänd ˌprä·səs }

monellin [BIOCHEM] A sweet protein obtained from the African plant *Dioscorephyllum cumminisii*; in its natural form, it consists of two polypeptide chains and is 3000 times sweeter than sucrose on a weight basis. { mə'nel·ən }

Monera [BIOL] A kingdom that includes the bacteria and blue-green algae in some classification schemes. { mə'nir·ə }

monestrous [PHYSIO] Having a single estrous cycle per year. { män'es·trəs }

monetite [MINERAL] CaHPO₄ A yellowish-white mineral consisting of an acid calcium hydrogen phosphate, occurring in crystals. { 'män·ə‚tīt }

Monge form [MATH] An equation of a surface of the form $z = f(x,y)$, where x, y, and z are Cartesian coordinates. { 'mónzh ‚fòrm }

Monge's disease See mountain sickness. { 'mōnzh·əz di‚zēz }

Monge's theorem [MATH] For three coplanar circles, and for radii of these circles which are parallel to each other, the three outer centers of similitude of the circles taken in pairs lie on a single straight line, and any two inner centers of similitude lie on a straight line with one of the outer centers. { 'mōnzh·əz ‚thir·əm }

Mongolian spot [MED] A focal bluish-gray discoloration of the skin of the lower back, also aberrantly on the face, present at birth and fading gradually. { mäŋ'gō·lē·ən 'spät }

mongolism See Down's syndrome. { 'mäŋ·gə‚liz·əm }

mongoloid [ANTHRO] Characteristic of a Far Eastern racial stock and often American Indians; features yellow complexion, a broad flat face with small nose and prominent cheekbones, and eyes that have an epicanthal fold. [MED] Having physical characteristics associated with Down's syndrome. { 'mäŋ· gə‚lóid }

mongoose [VERT ZOO] The common name for 39 species of carnivorous mammals which are members of the family Viveridae; they are plantigrade animals and have a long slender body, short legs, nonretractile claws, and scent glands. { 'mäŋ‚güs }

Monhysterida [INV ZOO] An order of aquatic nematodes in the subclass Chromadoria. { ‚män·hi'ster·ə·də }

Monhysteroidea [INV ZOO] A superfamily of free-living nematodes in the order Monhysterida characterized by single or paired outstretched ovaries, circular to cryptospiral amphids, and a stoma which is usually shallow and unarmed. { ‚män ‚hi·stə'róid·ē·ə }

monic equation [MATH] A polynomial equation with integer coefficients, where the coefficient of the term of highest degree is +1. { 'mō·nik i'kwā·zhən }

monic polynomial [MATH] A polynomial in which the coefficient of the term of highest degree is +1 and the coefficients of the other terms are integers. { 'mō·nik ‚päl·ə'nō·mē·əl }

Moniliaceae [MYCOL] A family of fungi in the order Moniliales; sporophores are usually lacking, but when present they are aggregated into fascicles, and hyphae and spores are hyaline or brightly colored. { mə‚nil·ē'ās·ē‚ē }

Moniliales [MYCOL] An order of fungi of the Fungi Imperfecti containing many plant pathogens; asexual spores are always formed free on the surface of the material on which the organism is living, and never occur in either pycnidia or acervuli. { mə‚nil·ē'ā·lēz }

monilial vaginitis [MED] Inflammation of the vagina caused by a fungus of the genus *Monilia*. { mə'nil·ē·əl ‚vaj·ə'nīd·əs }

moniliasis See candidiasis. { mə‚nil·ē'ī·ə·səs }

moniliform [BIOL] Constructed with contractions and expansions at regular alternating intervals, giving the appearance of a string of beads. { mə'nil·ə‚fòrm }

Monilinia fructicola [MYCOL] A fungal pathogen in the class Discomycetes that causes brown rot of stone fruits. { ‚mō·nə‚lin·ē·ə ‚frük·ti'kō·lə }

monimolimnion [HYD] The dense bottom stratum of a meromictic lake; it is stagnant and does not mix with the water above. { ‚män·ə·mō'lim·nē‚än }

monimolite [MINERAL] (Pb,Ca)₃Sb₂O₈ Yellowish to brownish or greenish mineral composed of lead calcium antimony oxide; it may contain ferrous iron. { mə'nim·ə‚līt }

monitor [COMPUT SCI] To supervise a program, and check that it is operating correctly during its execution, usually by means of a diagnostic routine. [ELECTR] See video monitor. [ENG] **1.** An instrument used to measure continuously or at intervals a condition that must be kept within prescribed limits, such as radioactivity at some point in a nuclear reactor, a

variable quantity in an automatic process control system, the transmissions in a communication channel or bank, or the position of an aircraft in flight. **2.** To use meters or special techniques to measure such a condition. **3.** A person who watches a monitor. [MIN ENG] See hydraulic monitor. [VERT ZOO] Any of 27 carnivorous, voracious species of the reptilian family Varanidae characterized by a long, slender forked tongue and a dorsal covering of small, rounded scales containing pointed granules. { 'män·əd·ər }

monitor board [COMMUN] A console at which a supervising telephone operator sits and from which she or he can intercept calls being handled by other operators. { 'män·əd·ər ‚bòrd }

monitor control dump [COMPUT SCI] A memory dump routinely carried out by the system once a program has been run. { 'män·əd·ər kən'trōl ‚dəmp }

monitor display [COMPUT SCI] The facility of stopping the central processing unit and displaying information of main storage and internal registers; after manual intervention, normal instruction execution can be initiated. { 'män·əd·ər di‚splā }

monitoring amplifier [ELECTR] A power amplifier used primarily for evaluation and supervision of a program. { 'män· ə·triŋ ‚am·plə‚fī·ər }

monitoring key [ELECTR] Key which, when operated, makes it possible for an attendant or operator to listen on a telephone circuit without appreciably impairing transmission on the circuit. { 'män·ə·triŋ ‚kē }

monitor mode See master mode. { 'män·əd·ər ‚mōd }

monitor operating system [COMPUT SCI] The control of the routines which achieves efficient use of all the hardware components. { 'män·əd·ər 'äp·ə‚rād·iŋ ‚sis·təm }

monitor printer [COMMUN] A teleprinter used in a technical control facility or communications center for checking incoming teletypewriter signals. [COMPUT SCI] Input-output device, capable of receiving coded signals from the computer, which automatically operates the keyboard to print a hard copy and, when desired, to punch paper tape. { 'män·əd·ər ‚print·ər }

monitor routine See executive routine. { 'män·əd·ər rü‚tēn }

monkey [MIN ENG] **1.** An appliance for mechanically gripping or releasing the rope in rope haulage. **2.** An airway in an anthracite mine. [VERT ZOO] Any of several species of frugivorous and carrivorous primates which compose the families Cercopithecidae and Cebidae in the suborder Anthropoidea; the face is typically flattened and hairless, all species are pentadactyl, and the mammary glands are always in the pectoral region. { 'məŋ·kē }

monkeyboard [PETRO ENG] A working platform on a derrick that provides the operator a position suitable for handling the top end of the pipe. { 'məŋ·kē‚bòrd }

monkey drift [MIN ENG] A small drift driven in for prospecting purposes, or a crosscut driven to an airway above the gangway. { 'məŋ·kē ‚drift }

monkey heading [MIN ENG] A narrow, low passage in the coal, providing refuge for miners while coal is blasted. { 'məŋ·kē ‚hed·iŋ }

monkey ladder [MIN ENG] A ladder of saplings; the widely separated steps rest in the coal. { 'məŋ·kē ‚lad·ər }

monkeypox [VET MED] An animal virus that causes a smallpox-like eruption but only rarely infects humans and has little potential for interhuman spread. { 'məŋ·kē‚päks }

monkey winch [MIN ENG] A device for exerting a strong pull; consists of a framework containing a hand-operated drum, around which a steel rope 50 feet (15 meters) long is wound. { 'məŋ·kē ‚winch }

monkey wrench [DES ENG] A wrench having one jaw fixed and the other adjustable, both of which are perpendicular to a straight handle. { 'məŋ·kē ‚rench }

monk's cloth [TEXT] A heavy cotton fabric with a basketweave. { 'məŋks ‚klòth }

monkshood See aconite. { 'məŋks‚hùd }

mono- [CHEM] A prefix for chemical compounds to show a single radical; for example, monoglyceride, a glycol ester on which a single acid group is attached to the glycerol group. { 'män·ō }

monoacetate [ORG CHEM] A compound such as a salt or ester that contains one acetate group. { ‚män·ō'as·ə‚tāt }

monoacid [CHEM] **1.** An acid that has only one replaceable hydrogen. **2.** A base or an alcohol that has a single hydroxyl

MONGOOSE

The golden-brown mongoose.

MONKEY

The spider monkey (*Ateles*).

(—OH) group which can be replaced by an atom or a functional group to form a salt or ester. { ¦män·ō'as·əd }

monoamine [ORG CHEM] An amine compound that has only one amino group. { ¦män·ō'am,ēn }

monoamine oxidase [BIOCHEM] A mitochondrial enzyme which oxidatively deaminates intraneuronal biogenic amines, some of which are important neurotransmitters in the peripheral and central nervous system. { män·ō'am,ēn 'äk·sə,dās }

monoamine oxidase inhibitor [PHARM] Any drug, such as isocarboxazid and tranylcypromine, that inhibits monoamine oxidase and thereby leads to an accumulation of the amines on which the enzyme normally acts. { män·ō'am,ēn 'äk·sə,dās in'hib·əd·ər }

monoammonium tartrate See ammonium bitartrate. { ¦män·ō·ə'mō·nē·əm 'tär,trāt }

monobasic [CHEM] Pertaining to an acid with one displaceable hydrogen atom, such as hydrochloric acid, HCl. { ¦män·ō'bās·ik }

monobasic calcium phosphate See calcium phosphate. { män·ō'bās·ik 'kal·sē·əm 'fäs,fāt }

monobasic sodium phosphate [INORG CHEM] NaH_2PO_4 White crystals that are slightly hygroscopic, soluble in water, insoluble in alcohol; used in baking powders and acid cleansers, and as a cattle-food supplement. { män·ō'bās·ik 'sōd·ē·əm 'fäs,fāt }

monoblast [HISTOL] A motile cell of the spleen and bone marrow from which monocytes are derived. { 'män·ō,blast }

monoblastic leukemia See acute monocytic leukemia. { ¦män·ō'blas·tik lü'kē·mē·ə }

Monoblepharidales [MYCOL] An order of aquatic fungi in the class Phycomycetes; distinguished by a mostly hyphal thallus and zoospores with one posterior flagellum. { ¦män·ō,blef·ə·rə'dā·lēz }

monobloc projectile [ORD] Armor-piercing projectile which consists of one piece of steel, suitably heat-treated; may be provided with a false ogive to decrease air resistance. { 'män·ō,bläk prə'jek·təl }

Monobothrida [PALEON] An extinct order of monocyclic camerate crinoids. { ¦män·ō'bäth·rə·də }

monobrid circuit [ELECTR] Integrated circuit using a combination of monolithic and multichip techniques by means of which a number of monolithic circuits, or a monolithic device in combination with separate diffused or thin-film components, are interconnected in a single package. { 'män·ə,brid ,sər·kət }

monocable [MECH ENG] An aerial ropeway that uses one rope to both support and haul a load. { 'män·ō,kā·bəl }

monocalcium phosphate See calcium phosphate. { ¦män·ō'kal·sē·əm 'fäs,fāt }

monocardiogram See vectorcardiogram. { ¦män·ō'kär·dē·ə,gram }

monocarpic [BOT] Bearing fruit once and then dying. { ¦män·ō'kär·pik }

monocarpous [BOT] Having a single ovary. { ¦män·ō'kär·pəs }

Monoceros [ASTRON] A constellation, right ascension 7 hours, declination 5° south; it has mostly faint stars. { mə'näs·ə·rəs }

Monoceros Loop [ASTRON] A filamentary loop nebula about 1000 parsecs (2 × 10¹⁶ miles or 3 × 10¹⁶ kilometers) distant, which is the remnant of a supernova that took place 50,000–100,000 years ago. { mə'näs·ə·rəs 'lüp }

Monoceros R2 molecular cloud [ASTRON] A massive rotating gas cloud with an abundance of various molecules that contains a compact H II region and is a region of active star formation. { mə'näs·ə·rəs ¦är¦tü mə'lek·yə·lər klaúd }

monocharge electret [ELECTR] A type of foil electret that carries electrical charge of the same sign on both surfaces. { 'män·ō,chärj i'lek·trət }

monochlamydous [BOT] Referring to flowers having only one set of floral envelopes, that is, either a calyx or a corolla. { ¦män·ō'klam·ə·dəs }

Monochoidea [INV ZOO] A superfamily of free-living, nonparasitic nematodes in the order Mononchida, characterized by angular, distinct lips bearing papilliform cephalic sensilla, an expanded lip region that is flattened anteriorly, and a heavily cuticularized, barrel or globular stoma, with one or more teeth or denticles. { ¦män·ə'kóid·ē·ə }

monochromasia [MED] Complete color blindness in which all colors appear as shades of gray. Also known as monochromatism. { ¦män·ə·krə'mā·zhə }

monochromat [MED] An individual who suffers from total color blindness even at high light levels; such persons are typically deficient or lacking in cone receptors, so that their form vision is also poor. { ¦män·ə¦krō,mat }

monochromatic [OPTICS] Pertaining to the color of a surface which radiates light having an extremely small range of wavelengths. [PHYS] Consisting of electromagnetic radiation having an extremely small range of wavelengths, or particles having an extremely small range of energies. { ¦män·ə·krə'mad·ik }

monochromatic emissivity [THERMO] The ratio of the energy radiated by a body in a very narrow band of wavelengths to the energy radiated by a blackbody in the same band at the same temperature. Also known as color emissivity. { ,män·ə·krə'mad·ik ,ē·mi'siv·əd·ē }

monochromatic filter See birefringent filter. { män·ə·krə'mad·ik 'fil·tər }

monochromatic interference [OPTICS] Interference between beams coming from a source of monochromatic light. { män·ə·krə'mad·ik ,in·tər'fir·əns }

monochromatic light [OPTICS] Light of one color, having wavelengths confined to an extremely narrow range. { män·ə·krə'mad·ik 'līt }

monochromatic neutron beam [NUCLEO] A beam of neutrons whose energies are confined to an extremely narrow range of values. { män·ə·krə'mad·ik 'nü,trän ,bēm }

monochromatic radiation [ELECTROMAG] Electromagnetic radiation having wavelengths confined to an extremely narrow range. { ,män·ə·krə'mad·ik ,rād·ē'ā·shən }

monochromatic temperature scale [THERMO] A temperature scale based upon the amount of power radiated from a blackbody at a single wavelength. { män·ə·krə'mad·ik 'tem·prə·chər ,skāl }

monochromatism See monochromasia. { ¦män·ə¦krō·mə,tiz·əm }

monochromator [SPECT] A spectrograph in which a detector is replaced by a second slit, placed in the focal plane, to isolate a particular narrow band of wavelengths for refocusing on a detector or experimental object. { ¦män·ə¦krō,mād·ər }

monochrome [OPTICS] Having only one chromaticity. { 'män·ə,krōm }

monochrome channel [ELECTR] In a color television system, any path which is intended to carry the monochrome signal; the monochrome channel may also carry other signals. { 'män·ə,krōm ¦chan·əl }

monochrome signal [ELECTR] **1.** A signal wave used for controlling luminance values in monochrome television. **2.** The portion of a signal wave that has major control of the luminance values in a color television system, regardless of whether the picture is displayed in color or in monochrome. Also known as M signal. { 'män·ə,krōm ,sig·nəl }

monochrome television [COMMUN] Television in which the final reproduced picture is monochrome, having only shades of gray between black and white. Also known as black-and-white television. { 'män·ə,krōm 'tel·ə,vizh·ən }

Monochuloidea [INV ZOO] A superfamily of nonparasitic nematodes in the order Mononchida, distinguished by papilliform cephalic sense organs, an inconspicuous, small, slitlike amphid aperture, and a stoma with a thick-walled, slightly tapered anterior and an elongate, thin-walled posterior. { ,män·ə·kə'lóid·ē·ə }

monocistronic messenger [BIOCHEM] A messenger ribonucleic acid molecule that contains the amino acid sequence for a single polypeptide chain. { ,män·ə·sis¦trän·ik 'mes·ən·jər }

Monocleales [BOT] An order of liverworts of the subclass Marchantiidae consisting of a single genus, *Monoclea*, which has the largest gametophyte of all liverworts, and lobed spore mother cells. { ¦män·ō·klē'ā·lēz }

monoclimax [ECOL] A climax community controlled primarily by one factor, as climate. { ¦män·ō'klī,maks }

monocline [GEOL] A stratigraphic unit that dips from the horizontal in one direction only, not as part of an anticline or syncline. { 'män·ə,klīn }

monoclinic [BOT] Having both stamens and pistils in the same flower. { ¦män·ə'klin·ik }

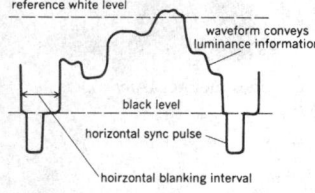

MONOCHROME SIGNAL

reference white level
waveform conveys luminance information
black level
horizontal sync pulse
hoirzontal blanking interval

Waveform sketch of a normal monochrome signal, one of the major components of the color television signal.

monoclinic system [CRYSTAL] One of the six crystal systems characterized by a single, two-fold symmetry axis or a single symmetry plane. { ¦män·ə¦klin·ik ¸sis·təm }

monoclonal antibody [IMMUNOL] A highly specific antibody produced by hybridoma cells; the antibody binds with a single antigenic determinant. { ¦män·ə¦klō·nəl 'ant·i¸bäd·ē }

monoclonal cryoglobulin [BIOCHEM] A cryoglobulin composed of immunoglobin with only one class or subclass of heavy and light chain. { ¦män·ə¦klō·nəl ¦krī·ō'gläb·yə·lən }

monocolpate pollen [BOT] Pollen grains having a single furrow. { ¦män·ə'kōl¸pāt ¸päl·ən }

monocoque [AERO ENG] A type of construction, as of a rocket body, in which all or most of the stresses are carried by the skin. { 'män·ə¸käk }

monocord switchboard [ELEC] Local battery switchboard in which each telephone line terminates in a single jack and plug. { 'män·ə¸kórd 'swich¸bórd }

monocotyledon [BOT] Any plant of the class Liliopsida; all have a single cotyledon. { ¸män·ə¸käd·əl'ēd·ən }

Monocotyledoneae [BOT] The equivalent name for Liliopsida. { ¸män·ə¸käd·əl·ə'dō·nē¸ē }

monocrepid [INV ZOO] A desma formed by secondary deposits of silica on a monaxon. { ¦män·ə¦krep·əd }

monocular vision [MED] Sight with one eye. { mə'näk·yə·lər 'vizh·ən }

Monocyathea [PALEON] A class of extinct parazoans in the phylum Archaeocyatha containing single-walled forms. { ¸män·ō·sī'ā·thē·ə }

monocyclic [BOT] Referring to flower parts arranged in a single whorl. { ¸män·ə'sī·klik }

monocyte [HISTOL] A large (about 12 micrometers), agranular leukocyte with a relatively small, eccentric, oval or kidney-shaped nucleus. { 'män·ə¸sīt }

monocytic angina *See* infectious mononucleosis. { ¦män·ə¦sid·ik 'an·jə·nə }

monocytic leukemia [MED] A form of leukemia in which monocytic cells are predominant in the blood. Also known as myelomonocytic leukemia. { ¦män·ə¦sid·ik lü'kē·mē·ə }

monocytoma [MED] A neoplasm composed principally of monocytes, usually anaplastic. { ¦män·ō¸sī'tō·mə }

monocytopenia [MED] Reduction in the number of circulating monocytes per unit volume of blood to below the minimum normal levels. { ¸män·ō¸sīd·ə'pē·nē·ə }

monocytosis [MED] Increase in the number of circulating monocytes per unit volume of blood to above the maximum normal levels. { ¸män·ō¸sī'tō·səs }

monodactylous [ZOO] Having a single digit or claw. { ¦män·ə'dak·tə·ləs }

Monodellidae [INV ZOO] A monogeneric family of crustaceans in the order Thermosbaenacea distinguished by seven pairs of biramous pereiopods on thoracomeres 2–8, and by not having the telson united to the last pleonite. { ¦män·ə'del·ə¸dē }

monodelphic [VERT ZOO] **1.** Having a single genital tract, in the female. **2.** Having a single uterus. { ¦män·ō¦del·fik }

monodisperse colloidal system [CHEM] A colloidal system in which the suspended particles have identical size, shape, and interaction. { ¦män·ō·di¦spərs kə'lóid·əl 'sis·təm }

monodispersity [ORG CHEM] Polymer system that is homogeneous in molecular weight, that is, it does not have a distribution of different molecular-weight chains within the total mass. { ¦män·ō·di'spər·səd·ē }

monodromy theorem [MATH] If a complex function is analytic at a point of a bounded simply connected domain and can be continued analytically along every curve from the point, then it represents a single-valued analytic function in the domain. { ¦män·ə¸drō·mē ¸thir·əm }

monoecious [BOT] **1.** Having both staminate and pistillate flowers on the same plant. **2.** Having archegonia and antheridia on different branches. [ZOO] Having male and female reproductive organs in the same individual. Also known as hermaphroditic. { mə'nē·shəs }

Monoedidae [INV ZOO] An equivalent name for Colydiidae. { mə'nē·də¸dē }

monoelectron oscillator *See* geonium. { ¦män·ō·i'lek¸trän 'äs·ə¸lād·ər }

monoenergetic gamma rays [PHYS] A beam of gamma rays whose energies are confined to an extremely narrow range. { ¦män·ō¸en·ər'jed·ik 'gam·ə ¸rāz }

monoenergetic radiation [PHYS] Radiation consisting of photons or particles whose energies are confined to an extremely narrow range. { ¸män·ō¸en·ər¸jed·ik ¸rād·ē'ā·shən }

monoester [ORG CHEM] An ester that has only one ester group. { ¦män·ō'es·tər }

monofier [ELECTR] Complete master oscillator and power amplifier system in a single evacuated tube envelope; electrically, it is equivalent to a stable low-noise oscillator, an isolator, and a two- or three-cavity klystron amplifier. { 'män·ə¸fī·ər }

monofilament [TEXT] A single, large, continuous filament (single-strand thread) of a natural or synthetic fiber. { ¦män·ə'fil·ə·mənt }

monofuel propulsion [AERO ENG] Propulsion system which obtains its power from a single fuel; in rocket units, the fuel furnishes both oxygen supply and the hydrocarbon for combustion. { 'män·ō¸fyül prə'pəl·shən }

monofunctional compound [ORG CHEM] An organic compound whose chemical structure possesses a single highly reactive site. { ¸män·ō¦fəŋk·shən·əl 'käm¸paúnd }

monogamous bivalent [IMMUNOL] Antigen-antibody complex in which each bivalent antibody combines with two determinant groups on a single antigen molecule. { mə¦näg·ə·məs bī'vā·lənt }

monogamy [ANTHRO] Marriage to only one person at a time. { mə'näg·ə·mē }

monogastric [VERT ZOO] Having only one digestive cavity. { ¦män·ō'gas·trik }

Monogenea [INV ZOO] A diverse subclass of the Trematoda which are principally ectoparasites of fishes; individuals have enlarged anterior and posterior holdfasts with paired suckers anteriorly and opisthaptors posteriorly. { ¸män·ə'je·nē·ə }

monogenic [GEN] Relating to or controlled by one gene. { ¸män·ə'jen·ik }

monogenic analytic function [MATH] An analytic function whose domain of definition has been extended directly or indirectly by analytic continuation as far as theoretically possible. { ¦män·ə¸jen·ik ¸an·ə¦lid·ik 'fəŋk·shən }

Monogenoidea [INV ZOO] A class of the Trematoda in some systems of classification; equivalent to the Monogenea of other systems. { ¸män·ə·jə'nóid·ē·ə }

monogeosyncline [GEOL] A primary geosyncline that is long, narrow, and deeply subsided; composed of the sediments of shallow water and situated along the inner margin of the borderlands. { ¦män·ō¸jē·ō'sin¸klīn }

monoglyceride [ORG CHEM] Any of the fatty-acid glycerol esters where only one acid group is attached to the glycerol group, for example, RCOOCH$_2$CHOHCH$_2$OH; examples are glycerol monostearate and monolaurate; used as emulsifiers in cosmetics and lubricants. { ¦män·ō'glis·ə¸rīd }

Monogonota [INV ZOO] An order of the class Rotifera, characterized by the presence of a single gonad in both males and females. { ¸män·ō·gō'näd·ə }

monogony [BIOL] Asexual reproduction. { mə'näg·ə·nē }

monogynous [BOT] Having only one pistil. [VERT ZOO] **1.** Having only one female in a colony. **2.** Consorting with only one female. { mə'näj·ə·nəs }

monohull boat [NAV ARCH] A boat with a single hull. { ¦män·ō¸həl 'bōt }

monohybrid [GEN] An individual heterozygous for alleles at one gene locus. { ¦män·ō'hī·brəd }

monoid [MATH] A semigroup which has an identity element. { 'mä¸nóid }

monoideism [PSYCH] A mental condition marked by the domination of a single idea; persistent and thorough preoccupation with one idea, but seldom an idea that is complete. { ¦män·ō'īd·ē¸iz·əm }

monokine [BIOCHEM] A cytokine released from macrophages. { 'män·ə¸kīn }

monolayer *See* monomolecular film. { 'män·ō¸lā·ər }

monolayer capacity [CHEM] **1.** In chemisorption, the amount of adsorbate required to occupy all adsorption sites on the solid surface. **2.** In physisorption, the amount of material required to cover the solid surface with a complete monolayer of the adsorbate in a close-packed array. { 'män·ə¸lā·ər kə¸pas·əd·ē }

monolith [MATER] A large concrete block. { 'män·ə¸lith }

monolithic [CIV ENG] Pertaining to concrete construction

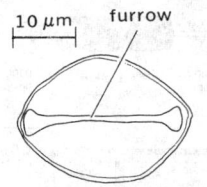

MONOCOLPATE POLLEN

10 μm   furrow

Monocolpate pollen of *Zamia floridana*.

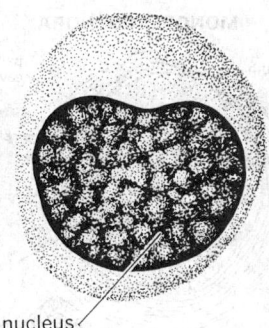

MONOCYTE

nucleus

Diagrammatic representation of a monocyte.

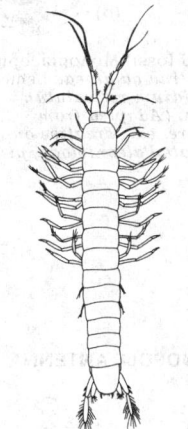

MONODELLIDAE

Drawing of a male *Monodella halophila*.

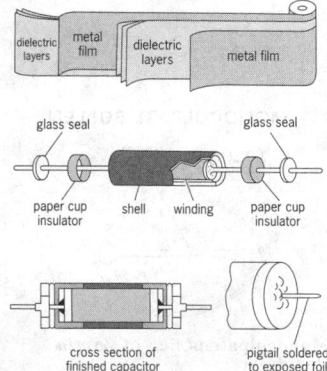

MONOLITHIC CERAMIC CAPACITOR

Ceramic capacitor constructed in chip form, showing cutaway of finished chip.

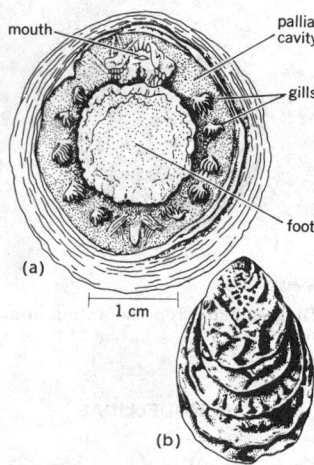

MONOPLACOPHORA

Living and fossil Monoplacophora. (a) *Neopilina galatheae* Lemche. (b) *Tryblidium reticulatum* Lindström. (*Adapted from R. C. Moore, ed., Treatise on Invertebrate Paleontology, pt. 1, 1957*)

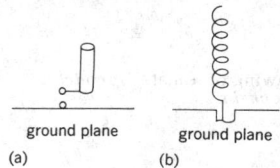

MONOPOLE ANTENNA

Types of monopole antenna with horizontal and vertical patterns. (*a*) Vertical tube. (*b*) Helical.

---

which is cast in one jointless piece. [SCI TECH] Constructed from a single crystal or other single piece of material. { ,män·ə'lith·ik }

monolithic ceramic capacitor [ELECTR] A capacitor that consists of thin dielectric layers interleaved with staggered metal-film electrodes; after leads are connected to alternate projecting ends of the electrodes, the assembly is compressed and sintered to form a solid monolithic block. { ,män·ə'lith·ik sə'ram·ik kə'pas·əd·ər }

monolithic filter [COMMUN] A device used to separate telephone communications sent simultaneously over the transmission line, consisting of a series of electrodes vacuum-deposited on a crystal plate so that the plated sections are resonant with ultrasonic sound waves, and the effect of the device is similar to that of an electric filter. { ,män·ə'lith·ik 'fil·tər }

monolithic integrated circuit [ELECTR] An integrated circuit having elements formed in place on or within a semiconductor substrate, with at least one element being formed within the substrate. { ,män·ə'lith·ik 'int·ə,grād·əd 'sər·kət }

monomer [ORG CHEM] A molecule which is capable of combining with like or unlike molecules to form a polymer; it is a repeating structure unit within a polymer. Also known as repeating unit. { 'män·ə·mər }

monomeric [CELL MOL] Having a single polypeptide chain. { män·ə'mer·ik }

monomeric unit *See* repeating unit. { ,män·ə¦mer·ik 'yü·nət }

monomethylhydrazine [CHEM] $CH_3N_2H_3$ A volatile toxic liquid that will react with carbon dioxide and oxygen. { ,mä·nō,meth·əl 'hī·drə,zēn }

monomial [MATH] A polynomial of degree one. { mə'nō·mē·əl }

monomial factor [MATH] A single factor that can be divided out of every term in a given expression. { mə'nō·mē·əl ,fak·tər }

monomineralic [PETR] Of a rock, composed entirely or principally of a single mineral. { ,män·ō,min·ə¦ral·ik }

monomino *See* square. { mə'näm·ə·nō }

Monommidae [INV ZOO] A family of coleopteran insects in the superfamily Tenebrionoidea. { mə'nam·ə,dē }

monomolecular film [PHYS CHEM] A film one molecule thick. Also known as monolayer. { ¦män·ō·mə¦lek·yə·lər 'film }

monomorphic [BIOL] Having or exhibiting only a single form. { ¦män·ə¦mȯr·fik }

mononuclear [CELL MOL] Having only one nucleus. { ¦män·ō¦nü·klē·ər }

mononucleosis [MED] Any of various conditions marked by an abnormal increase in monocytes in the peripheral blood. { ,män·ə,nü·klē'ō·səs }

monophagous [ZOO] Subsisting on a single kind of food. Also known as monotrophic. { mə'näf·ə·gəs }

Monophisthocotylea [INV ZOO] An order of the Monogenea in which the posthaptor is without discrete multiple suckers or clamps. { ,män·ə,fis·thə,käd·əl'ē·ə }

Monophlebinae [INV ZOO] A subfamily of the homopteran superfamily Coccoidea distinguished by a dorsal anus. { ,män·ə'fleb·ə,nē }

monophonic sound [ENG ACOUS] Sound produced by a system in which one or more microphones feed a single transducing channel which is coupled to one or more loudspeakers. { ¦män·ə¦fän·ik ,saùnd }

monophyletic [EVOL] Pertaining to any form evolved from a single interbreeding population. { ,män·ə·fə'led·ik }

monophyodont [VERT ZOO] Having only one set of teeth throughout life. { ¦män·ō'fī·ə,dänt }

monopinch [ELECTR] Antijam application of the monopulse technique where the error signal is used to provide discrimination against jamming signals. { 'män·ō,pinch }

Monopisthocotylea [INV ZOO] An order of trematode worms in the subclass Pectobothridia. { ,män·ə,fis·thə,käd·əl'ē·ə }

Monoplacophora [INV ZOO] A group of shell-bearing mollusks represented by few living forms; considered to be a sixth class of mollusks. { ,män·ō·plə'käf·ə·rə }

monoplegia [MED] Paralysis involving a single limb, muscle, or group of muscles. { ,män·ə'plē·jē·ə }

monoploid [GEN] 1. Having only one set of chromosomes. 2. Having the lowest haploid number of chromosomes in a polyploid series, in plants. { 'män·ə,plȯid }

---

monopodial [BOT] Stem branching in which there are lateral shoots on a primary axis. { ,män·ə'pōd·ē·əl }

monopodium [BOT] A primary axis that continues to grow while giving off successive lateral branches. { ,män·ə'pōd·ē·əm }

monopole *See* magnetic monopole. { 'män·ə,pōl }

monopole antenna [ELECTROMAG] An antenna, usually in the form of a vertical tube or helical whip, on which the current distribution forms a standing wave, and which acts as one part of a dipole whose other part is formed by its electrical image in the ground or in an effective ground plane. Also known as spike antenna. { 'män·ə,pōl an'ten·ə }

Monoposthioidea [INV ZOO] A superfamily of chiefly marine nematodes in the order Desmodorida, represented by the single family Monoposthiidae; distingushed by an annulate cuticle with spikelike ornamentation and a stoma that may or may not possess a well-developed tooth opposed by small subventral teeth. { ,män·ō,päs·thē'ȯid·ē·ə }

monopotassium L-glutamate *See* potassium glutamate. { ¦män·ō·pə'tas·ē·əm ¦el'glüd·ə,māt }

monopropellant [MATER] A rocket propellant consisting of a single substance, especially a liquid, capable of creating rocket thrust without the addition of a second substance. { ¦män·ō·prə'pel·ənt }

monoprotic acid [CHEM] An acid that has only one ionizable hydrogen atom in each molecule. { ,män·ə¦präd·ik 'as·əd }

monopulse radar [ENG] Radar in which directional information is obtained with high precision by using a receiving antenna system having two or more partially overlapping lobes in the radiation patterns. { 'män·ə,pəls 'rā,där }

Monopylina [INV ZOO] A suborder of radiolarian protozoans in the order Oculosida in which pores lie at one pole of a single-layered capsule. { ,män·ō·pə'lī·nə }

monopyroxene clinoaugite *See* clinopyroxene. { ¦män·ō·pə'räk,sēn ¦klī·nō'ȯ,gāt }

monorail [CIV ENG] A single rail used as a track; usually elevated, with cars straddling or hanging from it. { 'män·ə,rāl }

monorchid [ANAT] 1. Having one testis. 2. Having one testis descended into the scrotum. { mä'nȯr·kəd }

Monorhina [VERT ZOO] The subclass of Agnatha that includes the jawless vertebrates with a single median nostril. { ,män·ə'rī·nə }

monorhinal [ANAT] Having only one nostril. { ¦män·ə¦rī·nəl }

monosaccharide [BIOCHEM] A carbohydrate which cannot be hydrolyzed to a simpler carbohydrate; a polyhedric alcohol having reducing properties associated with an actual or potential aldehyde or ketone group; classified on the basis of the number of carbon atoms, as triose (3C), tetrose (4C), pentose (5C), and so on. { ¦män·ō¦sak·ə,rīd }

monoscope [ELECTR] A signal-generating electron-beam tube in which a picture signal is produced by scanning an electrode that has a predetermined pattern of secondary-emission response over its surface. Also known as monotron; phasmajector. { 'män·ə,skōp }

Monosigales [BOT] A botanical order equivalent to the Choanoflagellida in some systems of classification. { ,män·ō·si'gā·lēz }

monosiphonous [BIOL] Having a single central tube, as in the thallus of certain filamentous algae or the hydrocaulus of some hydrozoans. { ¦män·ō'sī·fə·nəs }

monosodium acid methanearsonate [ORG CHEM] CH_4AsNaO_3 A white, crystalline solid; melting point is 132–139°C; soluble in water; used as an herbicide for grassy weeds on rights-of-way, storage areas, and noncrop areas, and as preplant treatment for cotton, citrus trees, and turf. Abbreviated MSMA. { ,män·ə'sōd·ē·əm ¦as·əd ¦meth,än'ärs·ən,āt }

monosodium glutamate *See* sodium glutamate. { ¦män·ə'sōd·ē·əm 'glüd·ə,māt }

monosome [CELL MOL] 1. A single ribosome attached to messenger ribonucleic acid. 2. A chromosome in the diploid chromosome complement that lacks a homolog. { 'män·ə,sōm }

monosomy [GEN] The condition in which one chromosome of a pair is missing in a diploid organism. { 'män·ə,sōm·ē }

monospacing [GRAPHICS] Type spacing in which every

character occupies the same horizontal space. { 'män·ō,spās·iŋ }

monospermous [BOT] Having or producing one seed. { ¦män·ō¦spər·məs }

monosporangium [BOT] A sporangium producing monospores. { ¦män·ō·spə'ran·jē·əm }

monospore [BOT] A simple or undivided nonmotile asexual spore; produced by the diploid generation of some algae. { 'män·ə,spór }

monostable [ELECTR] Having only one stable state. { ¦män·ō¦stā·bəl }

monostable blocking oscillator [ELECTR] A blocking oscillator in which the electron tube or other active device carries no current unless positive voltage is applied to the grid. Also known as driven blocking oscillator. { ¦män·ō¦stā·bəl 'bläk·iŋ ¦äs·ə,lād·ər }

monostable circuit [ELECTR] A circuit having only one stable condition, to which it returns in a predetermined time interval after being triggered. { ¦män·ō¦stā·bəl 'sər·kət }

monostable multivibrator [ELECTR] A multivibrator with one stable state and one unstable state; a trigger signal is required to drive the unit into the unstable state, where it remains for a predetermined time before returning to the stable state. Also known as one-shot multivibrator; single-shot multivibrator; start-stop multivibrator; univibrator. { ¦män·ō¦stā·bəl ,məl·tə'vī,brād·ər }

monostat [ENG] Fluid-filled, upside-down manometer-type device used to control pressures within an enclosure, as for laboratory analytical distillation systems. { 'män·ə,stat }

monostatic radar [ENG] Conventional radar, in which the transmitter and receiver are at the same location and share the same antenna; in contrast to bistatic radar. { ¦män·ə¦stad·ik 'rā,där }

monostome [INV ZOO] A cercaria having only one mouth or sucker. { 'män·ə,stōm }

Monostylifera [INV ZOO] A suborder of the Hoplonemertini characterized by a single stylet. { ,män·ō·stī'lif·ə·rə }

monosubstituted alkene [ORG CHEM] An alkene with the general formula $RHC=CH_2$, where R is any organic group; only one carbon atom is bonded directly to one of the carbons of the carbon-to-carbon double bond. { ,män·ō¦səb·stə,tüd·əd 'al,kēn }

monoterpene [ORG CHEM] **1.** A class of terpenes with molecular formula $C_{10}H_{16}$; the members of the class contain two isoprene units. **2.** A derivative of a member of such a class. { ¦män·ō'tər,pēn }

Monotomidae [INV ZOO] The equivalent name for Rhizophagidae. { ,män·ə'täm·ə,dē }

monotone [SCI TECH] A quantity which never increases (or which never decreases) as a function of some other quantity. Also known as monotonic. { 'män·ə,tōn }

monotone convergence theorem [MATH] The integral of the limit of a monotone increasing sequence of nonnegative measurable functions is equal to the limit of the integrals of the functions in the sequence. { 'män·ə,tōn kən'vər·jəns ,thir·əm }

monotone decreasing function See monotone nonincreasing function. { ¦män·ə,tōn di¦krēs·iŋ 'fəŋk·shən }

monotone decreasing sequence [MATH] A sequence of real numbers in which each term is equal to or less than the preceding term. { ¦män·ə,tōn di¦krēs·iŋ 'sē·kwəns }

monotone function [MATH] A function which is either monotone nondecreasing or monotone nonincreasing. Also known as monotonic function. { ,män·ə,tōn ,fəŋk·shən }

monotone increasing function See monotone nondecreasing function. { ¦män·ə,tōn in¦krēs·iŋ 'fəŋk·shən }

monotone increasing sequence [MATH] A sequence of real numbers in which each term is equal to or greater than the preceding term. { ,män·ə,tōn in'krēs·iŋ ,sē·kwəns }

monotone nondecreasing function [MATH] A function which never decreases, that is, if $x \leq y$ then $f(x) \leq f(y)$. Also known as monotone increasing function; monotonically nondecreasing function. { ,män·ə,tōn ¦nän·di'krēs·iŋ ,fəŋk·shən }

monotone nondecreasing sequence [MATH] **1.** A sequence, $\{S_n\}$, of real numbers that never decreases; that is, $S_{n+1} \geq S_n$ for all n. **2.** A sequence of real-valued functions, $\{f_n\}$, defined on the same domain, D, that never decreases; that is, $f_{n+1}(x) \geq f_n(x)$ for all n and for all x in D.

monotone nonincreasing function [MATH] A function

which never increases, that is, if $x \leq y$ then $f(x) \geq f(y)$. Also known as monotone decreasing function; monotonically nonincreasing function. { ,män·ə,tōn ¦nän·in'krēs·iŋ ,fəŋk·shən }

monotone nonincreasing sequence [MATH] **1.** A sequence, $\{S_n\}$, of real numbers that never increases; that is, $S_{n+1} \leq S_n$ for all n. **2.** A sequence of real-valued functions, $\{f_n\}$, defined on the same domain, D, that never increases; that is, $f_{n+1}(x) \leq f_n(x)$ for all n and for all x in D.

monotone sequence [MATH] **1.** A sequence of real numbers that is monotone-nondecreasing or monotone-nonincreasing. **2.** A sequence of real-valued functions, defined on the same domain, that is either monotone-nondecreasing or monotone-nonincreasing. { 'män·ə,tōn 'sē·kwəns }

monotonic See monotone. { ¦män·ə¦tän·ik }

monotonically nondecreasing function See monotone nondecreasing function. { ¦män·ə¦tän·ik·lē ¦nän·di'krēs·iŋ ,fəŋk·shən }

monotonically nonincreasing function See monotone nonincreasing function. { ¦män·ə¦tän·ik·lē ¦nän·in'krēs·iŋ ,fəŋk·shən }

monotonic decreasing function See monotone nonincreasing function. { ,män·ə¦tän·ik di¦krēs·iŋ 'fəŋk·shən }

monotonic function See monotone function. { ¦män·ə¦tän·ik 'fəŋk·shən }

monotonic increasing function See monotonic nondecreasing function. { ,män·ə¦tän·ik in¦krēs·ing 'fəŋk·shən }

monotonicity [ELECTR] In an analog-to-digital converter, the condition wherein there is an increasing output for every increasing value of input voltage over the full operating range. { ,män·ə·tə nis·əd·ē }

monotonic system of sets See nested sets. { ¦män·ə¦tän·ik ¦sis·təm əv 'sets }

Monotremata [VERT ZOO] The single order of the mammalian subclass Prototheria containing unusual mammallike reptiles, or quasi-mammals. { ¦män·ō·trə'mad·ə }

monotrichous [MICROBIO] Of bacteria, having an individual flagellum at one pole. { mə'nä·trə·kəs }

monotron [ELECTR] See monoscope. [MET] A machine for determining indentation hardness by measuring the load required to dent a specimen to a constant depth with a diamond 0.625 millimeter in diameter. { 'män·ə,trän }

Monotropaceae [BOT] A family of dicotyledonous herbs or half shrubs in the order Ericales distinguished by a small, scarcely differentiated embryo without cotyledons, lack of chlorophyll, leaves reduced to scales, and anthers opening by longitudinal slits. { ,män·ō·trə'pās·ē,ē }

monotrophic [CRYSTAL] Of crystal pairs, having one of the pair always metastable with respect to the other. [ZOO] See monophagous. { ¦män·ə¦träf·ik }

monotropic [PHYS] Pertaining to an element which may exist in two or more forms, but in which one form is the stable modification at all temperatures and pressures. { ¦män·ə¦träp·ik }

monotropy coefficient [FL MECH] A coefficient ν related to the ratio of velocity coefficients, A_y/A_x, in an equation developed by P. Raethjen for the velocity profile in a fluid. { mə'nä·trə·pē ,kō·i,fish·ənt }

monotype [BIOL] A single type of organism that constitutes a species or genus. [GRAPHICS] A printing technique in which a picture is painted on a sheet of glass or metal; the picture is transferred to a sheet of paper by pressure; additional copies require that the subject be repainted on the plate. { 'män·ə,tīp }

monotype metal [MET] A type metal typically composed of 76% lead, 16% antimony, and 8% tin, with good wear resistance and compressive strength. { 'män·ə,tīp ,med·əl }

monotypic [SYST] Pertaining to a taxon that contains only one immediately subordinate taxon. { ¦män·ə¦tip·ik }

monovalent [CHEM] A radical or atom whose valency is 1. { ¦män·ō'vā·lənt }

monoxide [CHEM] A compound that contains a single oxygen atom, such as carbon monoxide, CO. { mə'näk,sīd }

monozygotic twins [BIOL] Twins which develop from a single fertilized ovum. Also known as identical twins. { ¦män·ō,zī'gäd·ik 'twinz }

mons [ANAT] An eminence. { mänz }

monsoon [METEOROL] A large-scale wind system which predominates or strongly influences the climate of large regions,

MONOSTOME

Drawing of monostome type of cercaria showing the one sucker. *(From R. M. Cable, An Illustrated Laboratory Manual of Parasitology, Burgess, 1958)*

MONOTROPACEAE

The Indian pipe (*Monotropa uniflora*). *(Photograph by Arthur Cronquist)*

and in which the direction of the wind flow reverses from winter to summer; an example is the wind system over the Asian continent. { 'män'sün }

monsoon climate [CLIMATOL] The type of climate which is found in regions subject to monsoons. { 'män·sün ‚klī·mət }

monsoon current [OCEANOGR] A seasonal wind-driven current occurring in the northern part of the Indian Ocean. { 'män'sün ‚kə·rənt }

monsoon fog [METEOROL] An advection type of fog occurring along a coast where monsoon winds are blowing, when the air has a high specific humidity and there is a large difference in the temperature of adjacent land and sea. { 'män'sün ‚fäg }

monsoon forest [ECOL] A tropical forest occurring in regions where a marked dry season is followed by torrential rain; characterized by vegetation adapted to withstand drought. { ¦män·sün 'fär·əst }

monsoon low [METEOROL] A seasonal low found over a continent in the summer and over the adjacent sea in the winter. { 'män'sün ‚lō }

mons pubis [ANAT] The eminence of the lower anterior abdominal wall above the superior rami of the pubic bones. { 'mänz 'pyü·bəs }

monster [MED] A congenitally malformed fetus which is incapable of properly performing the vital functions, or which exhibits marked structural differences from the normal. { 'män·stər }

Monstrilloida [INV ZOO] A suborder or order of microscopic crustaceans in the subclass Copepoda; adults lack a second antenna and mouthparts, and the digestive tract is vestigial. { ‚män·strə'lóid·ə }

mons veneris [ANAT] The mons pubis of the female. { 'mänz 'ven·ə·rəs }

montage [GRAPHICS] In photography, a collection of photos pasted on a common background, or a composite picture made by printing two or more negatives on a single sheet of photographic paper. { män'tazh }

montane [ECOL] Of, pertaining to, or being the biogeographic zone composed of moist, cool slopes below the timberline and having evergreen trees as the dominant life-form. { män'tān }

montanite [MINERAL] $Bi_2O_3·TeO_3·2H_2O$ A yellowish mineral consisting of a hydrated tellurate of bismuth; occurs in soft and earthy to compact form. { män'ta‚nīt }

montan wax [MATER] A hard mineral wax with a melting point of 80–90°C; white after purification and brown in the crude form; soluble in carbon tetrachloride, benzene, and chloroform; used for shoe and furniture polishes, adhesive pastes, roofing paints, and phonograph records. Also known as lignite wax. { 'män‚tan ‚waks }

montebrasite [MINERAL] $LiAlPO_4(OH)$ A mineral composed of basic lithium aluminum phosphate; it is isomorphous with amblygonite and natromontebrasite. { ‚män·tə'brä‚zīt }

Monte Carlo method [STAT] A technique which obtains a probabilistic approximation to the solution of a problem by using statistical sampling techniques. { ‚män·tē 'kär·lō ‚meth·əd }

montgomeryite [MINERAL] $Ca_2Al_2(PO_4)_3(OH)·7H_2O$ A green to colorless mineral composed of hydrous basic calcium aluminum phosphate. { mənt'gəm·rē‚īt }

Montgomery's tubercles [ANAT] Elevations in the areola of the nipple due to apocrine sweat glands; most prominent during pregnancy and lactation. { mənt'gəm·rēz 'tü·bər‚kəlz }

month [ASTRON] **1.** The period of the revolution of the moon around the earth (sidereal month). **2.** The period of the phases of the moon (synodic month). **3.** The month of the calendar (calendar month). { mənth }

Montian [GEOL] A European stage of geologic time: Paleocene (above Danian, below Thanetian). { 'män·chən }

Monticellidae [INV ZOO] A family of tapeworms in the order Proteocephaloidea, in which some or all of the organs are in the cortical mesenchyme; catfish parasites. { ‚män·tə'sel·ə‚dē }

monticellite [MINERAL] $CaMgSiO_4$ A colorless or gray mineral of the olivine structure type; isomorphous with kirsch steinite. { ‚män·tə'se‚līt }

monticulus [ANAT] The median dorsal portion of the cerebellum. { män'tik·yə·ləs }

montmorillonite [MINERAL] **1.** A group name for all clay

minerals with an expanding structure, except vermiculite. **2.** The high-alumina end member of the montmorillonite group; it is grayish, pale red, or blue and has some replacement of aluminum ion by magnesium ion. **3.** Any mineral of the montmorillonite group. { ‚mänt·mə'ril·ə‚nīt }

Montonen-Olive conjecture [PARTIC PHYS] The conjecture that unified theories of elementary particle interactions with supersymmetry have the property that, when the coupling constant between electric charges and electric fields becomes large, the electric charges become fuzzy, heavy, and strongly coupled while the magnetic charges become point-like, light, and weakly coupled, which is the reverse of the situation when the coupling constant is small. { ‚män·tə‚nən 'äl·əv kən‚jek·chər }

montroydite [MINERAL] HgO Natural mercury oxide mineral from Texas. { män'trói‚dīt }

monument [ENG] A natural or artificial (but permanent) structure that marks the location on the ground of a corner or other survey point. { 'män·yə·mənt }

monzonite [PETR] A phaneritic (visibly crystalline) plutonic rock composed chiefly of sodic plagioclase and alkali feldspar, with subordinate amounts of dark-colored minerals, intermediate between syenite and diorite. { 'män·zə‚nīt }

mood disorder See affective disorder. { 'müd dis‚órd·ər }

Moody formula [MECH ENG] A formula giving the efficiency e' of a field turbine, whose runner has diameter D', in terms of the efficiency e of a model turbine, whose runner has diameter D; $e' = 1 - (1 - e)(D/D')^{1/5}$. { 'müd·ē ‚fór·myə·lə }

Moody friction factor [FL MECH] Modification of the friction factor-Reynolds number-fluid flow relationship into which a roughness factor has been incorporated. { 'müd·ē 'frik·shən ‚fak·tər }

moon [ASTRON] **1.** The natural satellite of the earth. **2.** A natural satellite of any planet. { mün }

moonbow [OPTICS] A rainbow formed by light from the moon; the colors in a moonbow are usually very difficult to detect. Also known as lunar rainbow. { 'mün‚bō }

Mooney unit [CHEM ENG] An arbitrary unit used to measure the plasticity of raw, or unvulcanized rubber; the plasticity in Mooney units is equal to the torque, measured on an arbitrary scale, on a disk in a vessel that contains rubber at a temperature of 100°C and rotates at two revolutions per minute. { 'mün·ē ‚yü·nət }

moon illusion [OPTICS] An optical illusion whereby the moon appears larger when it is close to the horizon than when it is higher up. { 'mün i‚lü·zhən }

moon pillar [METEOROL] A halo consisting of a vertical shaft of light through the moon. { 'mün ‚pil·ər }

moon pool [PETRO ENG] In offshore drilling, an opening in the ship hull that provides access for the well drilling and completion machinery. { 'mün ‚pül }

moonrise [ASTRON] The crossing of the visible horizon by the upper limb of the ascending moon. { 'mün‚rīz }

moonset [ASTRON] The crossing of the visible horizon by the upper limb of the descending moon. { 'mün‚set }

moon shot [AERO ENG] The launching of a rocket intended to travel to the vicinity of the moon. { 'mün ‚shät }

moonstone [MINERAL] An alkali feldspar or cryptoperthite that is semitransparent to translucent and exhibits a bluish to milky-white, pearly, or opaline luster; used as a gemstone if flawless. Also known as hecatolite. { 'mün‚stōn }

moor [ECOL] See bog. [ENG] Securing a ship or aircraft by attaching it to a fixed object or a mooring buoy with chains or lines, or with anchors or other devices. { mur }

moor coal [GEOL] A friable lignite or brown coal. { 'mur ‚kōl }

Moore code [COMMUN] A binary teleprinter code with seven binary digits for each letter. { 'mur ‚kōd }

moored mine [ORD] An underwater mine with positively buoyant mine case, held at a predetermined depth beneath the surface by a cable or chain mooring attached to an anchor that rests on the bottom. { 'murd ¦mīn }

mooreite [MINERAL] $(Mg,Zn,Mn)_8(SO_4)_4(OH)_{14}·4H_2O$ A glassy white mineral composed of hydrous basic magnesium zinc manganese sulfate. { 'mur‚īt }

Moore machine [COMPUT SCI] A sequential machine in which the output depends uniquely on the current state of the machine, and not on the input. { 'mur mə‚shēn }

Moore's law [COMPUT SCI] The prediction by Gordon

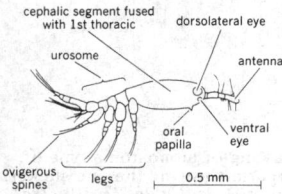

MONSTRILLOIDA

cephalic segment fused with 1st thoracic · dorsolateral eye · urosome · antenna · oral papilla · ventral eye · ovigerous spines · legs · 0.5 mm

Monstrilla reticulata, adult female, shown in lateral view.

Moore (cofounder of the Intel Corporation) that the number of transistors on a microprocessor would double periodically (approximately every 18 months). { 'mürz ‚lò }

Moore-Smith convergence [MATH] Convergence of a net to a point x in a topological space, in the sense that for each neighborhood of x there is an element a of the directed system that indexes the net such that, if b is also an element of this directed system and $b \geq a$, then x_b (the element indexed by b) is in this neighborhood. { 'mùr 'smith kən'vər·jəns }

Moore-Smith sequence See net. { ‚mür 'smith ‚sē·kwəns }

Moore-Smith set See directed set. { ‚mür 'smith ‚set }

Moore space [MATH] A topological space that has a sequence of coverings by open sets, such that each member of the sequence is a subcollection of the previous one, and such that, for any two points, x and y, of an open set S in the space, there is an open covering in the sequence such that the closure of any member of this covering that includes x is a subset of S and does not include y. { 'mür ‚spās }

mooring anchor [NAV ARCH] The second or additional anchor used to hold a ship at the mooring position. { 'mür·iŋ ‚aŋ·kər }

mooring buoy [ENG] A buoy secured to the bottom by permanent moorings and provided with means for mooring a vessel by use of its anchor chain or mooring lines; in its usual form a mooring buoy is equipped with a ring. { 'mür·iŋ ‚bòi }

mooring line [NAV ARCH] A rope or cable used for securing a ship to a pier or wharf. { 'mür·iŋ ‚līn }

mooring mast [AERO ENG] A mast or pole with fittings at the top to secure any lighter-than-air craft, such as a dirigible or blimp. { 'mür·iŋ ‚mast }

mooring winch [NAV ARCH] A hydraulic, electric, or steam machine on a ship used to haul in mooring lines when securing the ship to a pier or wharf. { 'mür·iŋ ‚winch }

moose [VERT ZOO] An even-toed ungulate of the genus *Alces* in the family Cervidae; characterized by spatulate antlers, long legs, a short tail, and a large head with prominent overhanging snout. { müs }

mor See ectohumus. { mór }

Moraceae [BOT] A family of dicotyledonous woody plants in the order Urticales characterized by two styles or style branches, anthers inflexed in the bud, and secretion of a milky juice. { mə'rās·ē‚ē }

mora fat See mowrah fat. { 'mòr·ə ‚fat }

morainal apron See outwash plain. { mə'rān·əl 'ā·prən }

morainal delta See ice-contact delta. { mə'rān·əl 'del·tə }

morainal lake [HYD] A glacial lake filling a depression resulting from irregular deposition of drift in a terminal or ground moraine of a continental glacier. { mə'rān·əl 'lāk }

morainal plain See outwash plain. { mə'rān·əl 'plān }

moraine [GEOL] An accumulation of glacial drift deposited chiefly by direct glacial action and possessing initial constructional form independent of the floor beneath it. { mə'rān }

moraine bar [GEOL] A terminal moraine serving as a bar, rising out of deep water at some distance from the shore. { mə'rān ‚bär }

moraine kame [GEOL] One of a group of kames characterized by the same topography, constitution, and position as a terminal moraine. { mə'rān ‚kām }

moraine plateau [GEOL] A relatively flat area within a hummocky moraine, generally at the same elevation as, or a little higher than, the summits of surrounding knobs. { mə'rān plə'tō }

moral idiocy [PSYCH] Inability to understand moral principles and values and to act in accordance with them, apparently without impairment of the reasoning and intellectual faculties. { 'mär·əl 'id·ē·ə·sē }

moral masochism [PSYCH] Masochism in which there is a need for punishment arising from unconscious sexual desires and reactivation of the Oedipus complex, characterized by self-destructive acts or the provocation of punishment from authority figures. { 'mär·əl 'mas·ə‚kiz·əm }

morass ore See bog iron ore. { mə'ras ‚òr }

moratronic [TEXT] A patterned fabric, usually polyester doubleknit; made on a German knitting machine which produces larger patterns than can be done on comparable jacquard knitting machines. { ‚mòr·ə'trän·ik }

moravite [MINERAL] $Fe_2(N,Fe)_4Si_7O_{20}(OH)_4$ A black mineral of the chlorite group, composed of basic iron aluminum silicate, occurring as fine scales. { mə'rä‚vīt }

Moraxella [MICROBIO] A genus of bacteria that are parasites of mucous membranes. { mə'rak·sə·lə }

morbid anatomy See pathologic anatomy. { 'mòr·bəd ə'nad·ə·mē }

morbidity [MED] **1.** The quantity or state of being diseased. **2.** The conditions inducing disease. **3.** The ratio of the number of sick individuals to the total population of a community. { mòr'bid·əd·ē }

mordant [CHEM] An agent, such as alum, phenol, or aniline, that fixes dyes to tissues, cells, textiles, and other materials by combining with the dye to form an insoluble compound. Also known as dye mordant. { 'mòrd·ənt }

mordant dye [MATER] Textile dye that requires a mordant (third substance) to bind the dye onto the fiber. { 'mòrd·ənt 'dī }

mordanting assistant [MATER] A chemical used with textile dye mordants to cause decomposition of the mordant and uniform deposition on the fibers; examples are sulfuric, oxalic, and lactic acids. { 'mòrd·ənt·iŋ ə‚sis·tənt }

mordant rouge [MATER] Aluminum acetate–acetic acid solution used in dyeing and calico printing. Also known as red acetate; red liquor. { 'mòrd·ənt 'rüzh }

Mordellidae [INV ZOO] The tumbling flower beetles, a family of coleopteran insects in the superfamily Meloidea. { mòr'del·ə‚dē }

mordenite [MINERAL] $(Ca,Na_2,K_2)_4Al_8Si_{40}O_{96}·28H_2O$ A zeolite mineral crystallizing in the orthorhombic system and found in minute crystals or fibrous concretions. Also known as arduinite; ashton tea; flokite; ptilolite. { 'mòrd·ən‚īt }

morel [MYCOL] Any fungus belonging to the genus *Morchella*, distinguished by a large, pitted, spongelike cap; it is a highly prized food, but may be poisonous when taken with alcohol. { mə'rel }

morencite See nontronite. { mə'ren‚sīt }

morenosite [MINERAL] $NiSO_4·7H_2O$ An apple-green or light-green mineral composed of hydrous nickel sulfate, occurring in crystals or fibrous crusts. Also known as nickel vitriol. { mə'ren·ə‚sīt }

Morera's stress functions [MECH] Three functions of position, ψ_1, ψ_2, and ψ_3, in terms of which the elements of the stress tensor σ of a body may be expressed, if the body is in equilibrium and is not subjected to body forces; the elements of the stress tensor are given by $\sigma_{11} = -2\partial^2\psi_1/\partial x_2\partial x_3$, $\sigma_{23} = \partial^2\psi_2/\partial x_1\partial x_2 + \partial^2\psi_3/\partial x_1\partial x_3$, and cyclic permutations of these equations. { mó'rer·əz 'stres ‚fəŋk·shənz }

Morera's theorem [MATH] If a function of a complex variable is continuous in a simply connected domain D, and if the integral of the function about every simply connected curve in D vanishes, then the function is analytic in D. { mó'rer·əz ‚thir·əm }

mores [ECOL] Groups of organisms preferring the same physical environment and having the same reproductive season. { 'mòr‚āz }

Morgan [GEN] The unit of genetic map distance (1 Morgan) between two loci that show one crossover per meiosis; 1 Morgan = 100 centimorgans. { 'mòr·gən }

Morgan equation [THERMO] A modification of the Ramsey-Shields equation, in which the expression for the molar surface energy is set equal to a quadratic function of the temperature rather than to a linear one. { 'mòr·gən i‚kwā·zhən }

morganite See vorobyevite. { 'mòr·gə‚nīt }

Morgan-Keenan-Kellman system [ASTRON] An expansion of the Harvard sequence to include luminosity, whereby a roman numeral is appended to the Harvard class to indicate position on the Hertzsprung-Russell diagram: I for supergiant, II for bright giant, III for giant, IV for subgiant, and V for dwarf or main sequence. Abbreviated MKK system. { ‚mòr·gən ‚kēn·ən 'kel·mən ‚sis·təm }

morgue [MED] A place where dead bodies are held pending identification and disposition. { mòrg }

moribund [BIOL] **1** In a dying or deathlike state. **2.** In a state of suspended life functions; dormant. { 'mòr·ə·bənd }

Moridae [VERT ZOO] A family of actinopterygian fishes in the order Gadiformes. { 'mòr·ə‚dē }

morin [ORG CHEM] $C_{15}H_{10}O_7·2H_2O$ Colorless needles soluble in boiling alcohol, slightly soluble in water; used as a mordant dye and analytical reagent. { 'mòr·ən }

Morinae [VERT ZOO] The deep-sea cods, a subfamily of the Moridae. { 'mòr·ə‚nē }

morinite [MINERAL] $Na_2Ca_3Al_3H(PO_4)_6F_6\cdot8H_2O$ A mineral composed of hydrous acid phosphate of sodium, calcium, and aluminum. Also known as jezekite. { 'mȯr·ə‚nīt }

Morisette expansion reamer [MIN ENG] A reaming device with three tapered lugs or cutters designed so that drilling pressure necessary to penetrate rock with a noncoring pilot bit forces the diamond-faced cutters of the reamer to expand outward, thereby enlarging the pilot hole sufficiently to allow the casing to follow the reamer as drilling progresses. { 'mär·ə‚set ik'span·chən ‚rē·mər }

Mormyridae [VERT ZOO] A large family of electrogenic fishes belonging to the Osteoglossiformes; African river and lake fishes characterized by small eyes, a slim caudal peduncle, and approximately equal dorsal and anal fins in most. { mȯr'mir·ə‚dē }

Mormyriformes [VERT ZOO] Formerly an order of fishes which are now assigned to the Osteoglossiformes. { ‚mȯr·mə·rə'fȯr‚mēz }

morning glory spillway See shaft spillway. { 'mȯrn·iŋ ‚glȯr·ē ‚spil‚wā }

morning gun See reveille gun. { 'mȯrn·iŋ ‚gən }

morning sickness [MED] Morning nausea associated with early pregnancy. { 'mȯrn·iŋ ‚sik·nəs }

morning star [ASTRON] A misnomer given to a planet visible to the naked eye, when it rises before the sun. { 'mȯrn·iŋ ‚stär }

morning twilight [ASTRON] The period of time between darkness and sunrise. { 'mȯrn·iŋ ‚twī‚līt }

moron [PSYCH] A mentally defective individual with a mental age between 7 and 12 years; a child with an IQ between 50 and 69. { 'mȯr‚än }

Moro reflex [PHYSIO] The startle reflex observed in normal infants from birth through the first few months, consisting of abduction and extension of all extremities, followed by flexion and abduction of the extremities. { 'mȯr·ō ‚rē‚fleks }

morph [GEN] An individual variant in a polymorphic population. { mȯrf }

morphallaxis [PHYSIO] Regeneration whereby one part is transformed into another by reorganization of tissue fragments rather than by cell proliferation. { ‚mȯr·fə'lak·səs }

Morphinae [INV ZOO] A subfamily of large tropical butterflies in the family Nymphalidae. { 'mȯr·fə‚nē }

morphine [PHARM] $C_{17}H_{19}NO_3\cdot H_2O$ A white crystalline narcotic powder, melting point 254°C, an alkaloid obtained from opium; used in medicine in the form of a hydrochloride or sulfate salt. { 'mȯr‚fēn }

***para*-morphine** See thebaine. { ¦par·ə 'mȯr‚fēn }

morphinism [MED] **1.** The condition caused by the habitual use of morphine. **2.** The morphine habit. { 'mȯr·fə‚niz·əm }

morphism [MATH] The class of elements which together with objects form a category; in most cases, morphisms are functions which preserve some structure on a set. { 'mȯr‚fiz·əm }

morphogen [BIOCHEM] Any compound that exerts a morphogenetic effect at low concentrations. { 'mȯr·fə·jən }

morphogene [GEN] Any gene involved directly or indirectly in the control of growth and morphogenesis. { 'mȯr·fə‚jēn }

morphogenesis [EMBRYO] The transformation involved in the growth and differentiation of cells and tissue. Also known as topogenesis. { ‚mȯr·fə'jen·ə·səs }

morphogenetic movement [EMBRYO] Any movement of or within a cell that changes the shape of differentiating cells or tissues. { ‚mȯr·fə·jə¦ned·ik müv·mənt }

morphogenetic region [GEOL] A region in which, under certain climatic conditions, the predominant geomorphic processes will contribute regional characteristics to the landscape that contrast with those of other regions formed under different climatic conditions. { ‚mȯr·fə·jə¦ned·ik 'rē·jən }

morphogenetic stimulus [EMBRYO] A stimulus exerted by one part of the developing embryo on another, leading to morphogenesis in the reacting part. { ‚mȯr·fə·jə¦ned·ik 'stim·yə·ləs }

morphographic map See physiographic diagram. { ¦mȯr·fə¦graf·ik 'map }

morpholine [ORG CHEM] C_4H_8ONH A hygroscopic liquid, soluble in water; used as a solvent and rubber accelerator. { 'mȯr·fə‚lēn }

morphological astronomy [ASTRON] A branch of astronomy in which the forms of celestial objects, such as galaxies, are observed, and an attempt is made to draw conclusions from these observations. { ¦mȯr·fə¦läj·ə·kəl ə'strän·ə·mē }

morphology [BIOL] A branch of biology that deals with structure and form of an organism at any stage of its life history. { mȯr'fäl·ə·jē }

morphosan [ORG CHEM] $C_{17}H_{19}NO_3\cdot CH_3Br$ A solid morphine derivative without morphine's disagreeable after effects; used in medicine. { 'mȯr·fə‚san }

morphospecies [SYST] A typological species distinguished solely on the basis of morphology. { ¦mȯr·fō'spē‚shēz }

morphotropism [CRYSTAL] Similarity of structure, axial ratios, and angles between faces of one or more zones in crystalline substances whose formulas can be derived one from another by substitution. { ¦mȯr·fō'trō‚piz·əm }

Morquio's syndrome [MED] A hereditary disease transmitted as an autosomal recessive and characterized by large quantities of keratosulfate in urine, dwarfism, and a typical facies with broad mouth, prominent maxilla, short nose, and widely spaced teeth. Also known as Brailsford-Morquio syndrome; familial osteochondrodystrophy; mucopolysaccharidosis IV. { 'mȯr·kwē‚ōz ‚sin‚drōm }

Morse cable code [COMMUN] A code used chiefly in submarine cable telegraphy, in which positive and negative current impulses of equal length represent dots and dashes, and a space is represented by the absence of current. Also known as cable code; International cable code. { 'mȯrs 'kā·bəl ‚kōd }

Morse code [COMMUN] **1.** A telegraph code for manual operating, consisting of short (dot) and long (dash) signals and various-length spaces; now used only for wire telegraphy. Also known as American Morse code. **2.** Collective term for Morse code (American Morse code) and continental code (International Morse code). { 'mȯrs 'kōd }

Morse equation [PHYS CHEM] An equation according to which the potential energy of a diatomic molecule in a given electronic state is given by a Morse potential. { 'mȯrs i‚kwā·zhən }

Morse potential [PHYS CHEM] An approximate potential associated with the distance r between the nuclei of a diatomic molecule in a given electronic state; it is $V(r) = D\{1 - \exp[-a(r - r_e)]\}^2$, where r_e is the equilibrium distance, D is the dissociation energy, and a is a constant. { 'mȯrs pə‚ten·chəl }

Morse taper reamer [DES ENG] A machine reamer with a taper shank. { 'mȯrs 'tā·pər rēm·ər }

Morse theory [MATH] The study of differentiable mappings of differentiable manifolds, which by examining critical points shows how manifolds can be constructed from one another. { 'mȯrs ‚thē·ə·rē }

Morse-Thue sequence [MATH] A sequence of binary digits defined by the number of 1's modulo 2 in successive integers when written in binary notation: 01101001.... { ¦mȯrs 'thü ‚sē·kwəns }

mortality rate [MED] For a given period of time, the ratio of the number of deaths occurring per 1000 population. Also known as death rate. { mȯr'tal·əd·ē ‚rāt }

mortar [MATER] A mixture of cement, lime, and sand used for laying bricks or masonry. [ORD] A complete projectile-firing weapon, with rifled or smooth bore, characterized by a shorter barrel, lower velocity, shorter range, and higher angle of fire than a howitzer or a gun; most present-day mortars are muzzle-loaded and of simple construction for lightness and mobility. [SCI TECH] A bowl-shaped vessel made of hard material in which solids are crushed by hand with a pestle. { 'mȯrd·ər }

mortar structure [PETR] A cataclastic structure produced by dynamic metamorphism of crystalline rocks and characterized by a mica-free aggregate of finely crushed grains of quartz and feldspar filling the interstices between or forming borders on the edges of larger, rounded relicts. Also known as cataclastic structure; murbruk structure; porphyroclastic structure. { 'mȯrd·ər ‚strək·chər }

mortise [ENG] A groove or slot in a timber for holding a tenon. { 'mȯrd·əs }

mortise and tenon [DES ENG] A type of joint, principally used for wood, in which a hole, slot, or groove (mortise) in one member is fitted with a projection (tenon) from the second member. { 'mȯrd·əs ən 'ten·ən }

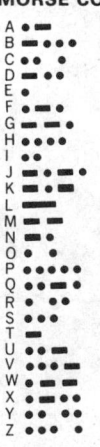

MORSE CODE

```
A  ·—
B  —···
C  —·—·
D  —··
E  ·
F  ··—·
G  ——·
H  ····
I  ··
J  ·———
K  —·—
L  ·—··
M  ——
N  —·
O  ———
P  ·——·
Q  ——·—
R  ·—·
S  ···
T  —
U  ··—
V  ···—
W  ·——
X  —··—
Y  —·——
Z  ——··
```

Morse code for the alphabet.

mortise lock [DES ENG] A lock designed to be installed in a mortise rather than on a door's surface. { 'mȯrd·əs ,läk }

mortising machine [MECH ENG] A machine employing an auger and a chisel to produce a square or rectangular mortise in wood. { 'mȯrd·ə·siŋ mə,shēn }

mortlake *See* oxbow lake. { 'mȯrt,lāk }

Morton's neuroma [MED] The thickening of the third intermetatarsal nerve over a period of years, causing a small benign fusiform tumor to eventually form in the space between the third and fourth toes; symptoms include painful burning, tingling, and numbness the third and fourth toes, often accompanied by radiating electric-like shocks. { 'mȯrt·ənz nə'rō·mə }

morula [EMBRYO] A solid mass of blastomeres formed by cleavage of the eggs of many animals; precedes the blastula or gastrula, depending on the type of egg. [INV ZOO] A cluster of immature male gametes in which differentiation occurs outside the gonad; common in certain annelids. { 'mȯr·ə·lə }

Moruloidea [INV ZOO] The only class of the phylum Mesozoa; embryonic development in the organisms proceeds as far as the morula or stereoblastula stage. { ,mȯr·ə'lȯid·ē·ə }

morvan [GEOL] The area where two peneplains intersect. Also known as skiou. { 'mȯr·vən }

MOS *See* metal oxide semiconductor.

mosaic [BIOL] An organism or part made up of tissues or cells exhibiting mosaicism. [ELECTR] A light-sensitive surface used in television camera tubes, consisting of a thin mica sheet coated on one side with a large number of tiny photosensitive silver-cesium globules, insulated from each other. [EMBRYO] An egg in which the cytoplasm of early cleavage cells is of the type which determines its later fate. [PETR] **1.** Pertaining to a granoblastic texture in a rock formed by dynamic metamorphism in which the boundaries between individual grains are straight or slightly curved. Also known as cyclopean. **2.** Pertaining to a texture in a crystalline sedimentary rock in which contacts at grain boundaries are more or less regular. [SCI TECH] A surface pattern made by the assembly and arrangement of many small pieces. { mō'zā·ik }

mosaic evolution [EVOL] The tendency of one or more characters to undergo evolutionary change at different rates than other characters in a lineage. { mō'zā·ik ,ev·ə'lü·shən }

mosaic gold *See* stannic sulfide. { mō'zā·ik ,gōld }

mosaicism [GEN] The coexistence in an individual of somatic cells of two or more genotypes or karyotypes; it is caused by gene or chromosome mutations, especially mitotic nondisjunction after fertilization. { mō'zā·ə,siz·əm }

mosaic structure [CRYSTAL] In crystals, a substructure in which neighboring regions are oriented slightly differently. { mō'zā·ik 'strək·chər }

mosandrite [MINERAL] A reddish-brown or yellowish-brown mineral composed of a silicate of sodium, calcium, titanium, zirconium, and cerium. Also known as khibinite; lovchorrite; rinkite; rinkolite. { mō'san,drīt }

mosasaur [PALEON] Any reptile of the genus *Mosasaurus*; large, aquatic, fish-eating lizards from the Cretaceous which are related to the monitors but had paddle-shaped limbs. { 'mō·sə,sȯr }

Moschcowitz's disease *See* thrombotic thrombocytopenic purpura. { 'mȯsh·kə,vit·səz di,zēz }

moschellandsbergite [MINERAL] Ag_2Hg_3 A silver-white mineral consisting of a silver and mercury compound; occurs in dodecahedral crystals and in massive and granular forms. { ,mō·shə'lanz·bər,gīt }

MOS-controlled thyristor [ELECTR] A type of thyristor in which there is a very thin metal oxide semiconductor (MOS) integrated circuit in the top surface of the high-power thyristor components, so that only a small gate current is needed to turn the entire device off or on. Abbreviated MCT. { 'em‚ō'es kən‚trōld thī'ris·tər }

moscovite *See* muscovite. { 'mäs·kə,vīt }

Moseley's law [SPECT] The law that the square-root of the frequency of an x-ray spectral line belonging to a particular series is proportional to the difference between the atomic number and a constant which depends only on the series. { 'mōz·lēz ,lȯ }

mosesite [MINERAL] $Hg_2N(SO_4,MoO_4)\cdot H_2O$ A mineral composed of a hydrous nitride of mercury and various anions. { 'mō·zə,zīt }

MOSFET *See* metal oxide semiconductor field-effect transistor. { 'mȯs,fet }

MOSFET-C filter [ELECTR] An active integrated-circuit filter in which the resistors of an active-RC filter are replaced with metal oxide semiconductor field-effect transistors (MOSFETs). { 'mȯs,fet 'sē ,fil·tər }

Mosotti field *See* Lorentz local field. { mō'säd·ē ,fēld }

mosquito [INV ZOO] Any member of the dipterous subfamily Culicinae; a slender fragile insect, with long legs, a long slender abdomen, and narrow wings. { mə'skēd·ō }

mosquito boat *See* motor torpedo boat. { mə'skēd·ō ,bōt }

moss [BOT] Any plant of the class Bryatae, occurring in nearly all damp habitats except the ocean. { mȯs }

moss agate [MINERAL] A milky or almost transparent chalcedony containing dark inclusions in a dendritic pattern. { 'mȯs 'ag·ət }

Mössbauer effect [NUC PHYS] The emission and absorption of gamma rays by certain nuclei, bound in crystals, without loss of energy through nuclear recoil, with the result that radiation emitted by one such nucleus can be absorbed by another. { 'mùs,baù·ər i,fekt }

Mössbauer spectroscopy [SPECT] The study of Mössbauer spectra, for example, for nuclear hyperfine structure, chemical shifts, and chemical analysis. { 'mùs,baù·ər spek'-träs·kə·pē }

Mössbauer spectrum [SPECT] A plot of the absorption, by nuclei bound in a crystal lattice, of gamma rays emitted by similar nuclei in a second crystal, as a function of the relative velocity of the two crystals. { 'mùs,baù·ər ,spek·trəm }

moss forest *See* temperate rainforest. { 'mȯs 'fär·əst }

mossite [MINERAL] $Fe(Nb,Ta)_2O_6$ A mineral composed of an iron tantalum oxide; it is isomorphous with tapiolite. { 'mȯ,sīt }

moss land [ECOL] An area which contains abundant moss but is not wet enough to be a bog. { 'mȯs ,land }

mossy zinc [MET] Zinc granules made by pouring molten zinc into water. { 'mȯs·ē 'ziŋk }

MOST *See* metal oxide semiconductor field-effect transistor.

most powerful test [STAT] If two tests have the same level of significance, then the test with a smaller-size type II error is the most powerful test of the two at that significance level. { 'mōst 'pau·ər·fùl 'test }

most probable position [NAV] **1.** The computed position of an aircraft determined by comparing a dead-reckoning position and a loran position determined for the same period of time, in which relative weights are given to the estimated probable errors of each. **2.** That position of a craft judged to be most accurate when an element of doubt exists as to the true position; it may be a fix, estimated position, dead-reckoning position, no-wind position, or some other position as determined from the squares of the estimated probable errors of each. { 'mōst 'präb·ə·bəl pə'zish·ən }

MOS transistor *See* metal oxide semiconductor field-effect transistor. { 'em‚ō'es tran'zis·tər }

most significant bit [COMPUT SCI] The left-most bit in a word. Abbreviated msb. { 'mōst sig'nif·i·gənt 'bit }

most significant character [COMPUT SCI] The character in the leftmost position in a number or word. { 'mōst sig'nif·i·gənt 'kar·ik·tər }

Motacillidae [VERT ZOO] The pipits, a family of passeriform birds in the suborder Oscines. { ,mōd·ə'sil·ə,dē }

moth [INV ZOO] Any of various nocturnal or crepuscular insects belonging to the lepidopteran suborder Heteroneura; typically they differ from butterflies in having the antennae feathery and rarely clubbed, a stouter body, less brilliant coloration, and proportionally smaller wings. { mȯth }

mothball [ORD] Placing military equipment into a state of long storage. { 'mȯth,bȯl }

mother [ENG ACOUS] A mold derived by electroforming from a master; used to produce the stampers from which disk records are molded in large quantities. Also known as metal positive. { 'məth·ər }

motherboard [COMPUT SCI] A common pathway over which information is transmitted between the hardware devices (the central processing unit, memory, and each of the peripheral control units) in a microcomputer. { 'məth·ər,bȯrd }

mother liquor *See* discharge liquor. { 'məth·ər ,lik·ər }

mother lode [GEOL] A main unit of mineralized matter that

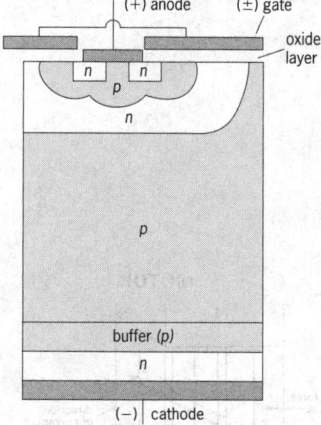

MOS-CONTROLLED THYRISTOR

Diagram of an MOS-controlled thyristor. *n* refers to the negative layer, and *p* to the positive layer.

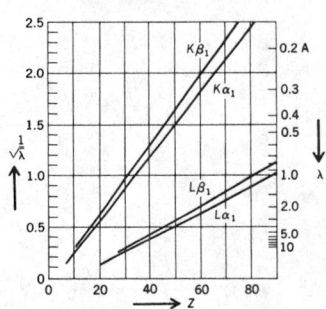

MOSELEY'S LAW

Plot of Moseley's law showing dependence of characteristic x-ray line wavelengths on atomic number.

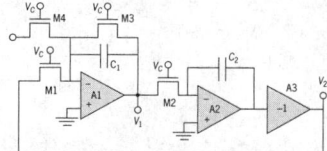

MOSFET-C FILTER

The filter consists of a loop of two integrators.

MOTH

The luna moth (*Tropaea luna*).

may not have economic value but to which workable veins are related. { 'məth·ər ‚lȯd }

mother map *See* base map. { 'məth·ər ‚map }

mother-of-coal *See* fusain. { ¦məth·ər əv 'kōl }

mother-of-emerald *See* prase. { 'məth·ər əv 'em·rəld }

mother-of-pearl [INV ZOO] The pearly iridescent internal layer of the shell of various pearl-bearing bivalve mollusks. { 'məth·ər əv 'pərl }

mother-of-pearl clouds *See* nacreous clouds. { 'məth·ər əv 'pərl ‚klau̇dz }

mother rock *See* source rock. { 'məth·ər ‚räk }

motile [BIOL] Being capable of spontaneous movement. [PSYCH] An individual characterized by motor-type mental imagery which takes the form of inner feelings of action. { mōd·əl }

motility symbiosis [ECOL] A symbiotic relationship in which motility is conferred upon an organism by its symbiont. { mō¦til·əd·ē ‚sim·bē'ō·səs }

motion [MECH] A continuous change of position of a body. { 'mō·shən }

motional electromotive force [ELECTROMAG] An electromotive force in a circuit that results from the motion of all or part of the circuit through a magnetic field. { 'mō·shən·əl i¦lek·trə¦mōd·iv 'fȯrs }

motional impedance [ELECTR] Of a transducer, the complex remainder after the blocked impedance has been subtracted from the loaded impedance. Also known as loaded motional impedance. { 'mō·shən·əl im'pēd·əns }

motional induction [ELECTROMAG] The production of an electromotive force in a circuit by motion of all or part of the circuit through a magnetic field in such a way that the circuit cuts across the magnetic flux. { 'mō·shən·əl in'dək·shən }

motion analysis [IND ENG] Detailed study of the motions used in a work task or at a given work area. { 'mō·shən ə‚nal·ə·səs }

motion-compensated coding scheme [COMMUN] A form of differential pulse-code modulation in which the motions of objects are estimated and comparisons of intensities are carried out between picture elements in successive frames spatially displaced by an amount equal to the motion of an object. { 'mō·shən ¦käm·pən‚sād·əd 'kōd·iŋ ‚skēm }

motion compensator [PETRO ENG] A device for maintaining constant weight on the bit during vertical motion of a floating offshore drilling rig. { 'mō·shən ‚käm·pən‚sād·ər }

motion cycle [IND ENG] The complete sequence of motions and activities required to complete one work cycle. { 'mō·shən ‚sī·kəl }

motion economy [IND ENG] Simplification and reduction of body motions to simplify and reduce work content. { 'mō·shən i‚kän·ə·mē }

motion picture [GRAPHICS] **1.** A sequence of filmed images viewed in rapid succession so that the illusion of continuity and motion is created. **2.** The complete thematically related content of such images. Also known as cinema; movie; moving picture. { 'mō·shən 'pik·chər }

motion picture camera [OPTICS] A camera capable of capturing action by taking a series of still pictures at regular brief intervals on a lengthy strip of film. { 'mō·shən ¦pik·chər ‚kam·rə }

motion picture pickup [ELECTR] Use of a television camera to pick up scenes directly from motion picture film. { 'mō·shən ¦pik·chər 'pik‚əp }

motion picture projector [ENG] An optical and mechanical device capable of flashing pictures taken by a motion picture camera on a viewing screen at the same frequency the action was photographed, thus producing an image that appears to move. { 'mō·shən ¦pik·chər prə‚jek·tər }

motion register [COMPUT SCI] The register which controls the go/stop, forward/reverse motion of a tape drive. { 'mō·shən ‚rej·ə·stər }

motion sickness [MED] A complex of symptoms, including nausea, vertigo, and vomiting, occurring as the result of random multidirectional accelerations of a vehicle. { 'mō·shən ‚sik·nəs }

motions pathway [IND ENG] The locus of movement of an anatomical segment in moving from one point of the workplace to another; includes the elemental increments in such motions as reaching, changing position, examining, and holding. { 'mō·shənz 'path‚wā }

motivated forgetting [PSYCH] Forgetting, such as by repression, activated by the needs of the individual. { 'mōd·ə‚vād·əd fər'ged·iŋ }

motivation [PSYCH] The comparatively spontaneous drive, force, or incentive, which partly determines the direction and strength of the response of a higher organism to a given situation; it arises out of the internal state of the organism. { ‚mōd·ə'vā·shən }

motive column [MIN ENG] The ventilating pressure in a mine in units of feet of air column; the height of a column of air whose density is the same as the air in the downcast shaft, which exerts a pressure equal to the ventilating pressure. { 'mōd·iv ‚käl·əm }

motoneuron *See* motor neuron. { ¦mōd·ə'nu̇r‚än }

motor [ELEC] A machine that converts electric energy into mechanical energy by utilizing forces produced by magnetic fields on current-carrying conductors. Also known as electric motor. [NEUROSCI] Pertaining to efferent nerves which innervate muscles and glands. [PHYSIO] That which causes action or movement. { 'mōd·ər }

motor alexia [MED] Inability to read aloud, while comprehension of the written word is preserved. { 'mōd·ər ā'lek·sē·ə }

motor aphasia [MED] A form of aphasia in which the patient knows what he wishes to say but is unable to get the words out, and is able to perceive and comprehend both spoken and written language but is unable to repeat what he sees or hears; due principally to a brain lesion. { 'mōd·ər ə'fā·zhə }

motor apraxia [MED] Inability to carry out, on command, a complex or skilled movement, though the purpose thereof is clear to the patient. Also known as kinesthetic apraxia; limb-kinetic apraxia. { 'mōd·ər ā'prak·sē·ə }

motor area [NEUROSCI] The ascending frontal gyrus containing nerve centers for voluntary movement; characterized by the presence of Betz cells. Also known as Broadman's area 4; motor cortex; pyramidal area. { 'mōd·ər ‚er·ē·ə }

motor ataxia [MED] Inability to coordinate the muscles, which becomes apparent only on body movement. { 'mōd·ər ə'tak·sē·ə }

motor benzol [MATER] Grade of benzene used in fuels for internal combustion engines. { 'mōd·ər ‚ben‚zȯl }

motorboat [NAV ARCH] A boat propelled by an internal combustion engine or an electric motor. Also known as autoboat. { 'mōd·ər‚bōt }

motorboating [ELECTR] Undesired oscillation in an amplifying system or transducer, usually of a pulse type, occurring at a subaudio or low-audio frequency. { 'mōd·ər‚bōd·iŋ }

motor branch circuit [ELEC] A branch circuit that terminates at a motor; it must have conductors with current-carrying capacity at least 125% of the motor full-load current rating, and overcurrent protection capable of carrying the starting current of the motor. { 'mōd·ər 'branch ‚sər·kət }

motor cell [BOT] *See* bulliform cell. [NEUROSCI] An efferent nerve cell in the anterior horn of the spinal cord. { 'mōd·ər ‚sel }

motor control *See* electronic motor control. { 'mōd·ər kən‚trōl }

motor-converter [ELEC] Induction motor and a synchronous converter with their rotors mounted on the same shaft and with their rotor windings connected in series; such converter operates synchronously at a speed corresponding to the sum of the numbers of poles of the two machines. { 'mōd·ər kən'vərd·ər }

motor cortex *See* motor area. { 'mōd·ər 'kȯr‚teks }

motorcycle [MECH ENG] An automotive vehicle, essentially a motorized bicycle, with two tandem and sometimes three rubber wheels. { 'mōd·ər‚sī·kəl }

motor effect [ELECTROMAG] The mutually repulsive force exerted by neighboring conductors that carry current in opposite directions. { 'mōd·ər i‚fekt }

motor element [ENG ACOUS] That portion of an electroacoustic receiver which receives energy from the electric system and converts it into mechanical energy. { 'mōd·ər ‚el·ə·mənt }

motor end plate [ANAT] A specialized area beneath the sarcolemma where functional contact is made between motor nerve fibers and muscle fibers. { 'mōd·ər 'end ‚plāt }

motor-generator set [ELEC] A motor and one or more generators that are coupled mechanically for use in changing one

---

MOTOR

Relative directions of field flux, current, and force in an electric motor.

power-source voltage to other desired voltages or frequencies. { 'mōd·ər ,jen·ə,rād·ər ,set }

motor grader See autopatrol. { 'mōd·ər ,grād·ər }

motor learning [PSYCH] In animals or humans, learning to perform some motor task in response to a given event or stimulus. { 'mōd·ər ,lərn·iŋ }

motor meter [ENG] An integrating meter which has a rotor, one or more stators, a retarding element which makes the speed of the rotor proportional to the quantity (such as power or current) whose integral over time is being measured, and a register which counts the total number of revolutions of the rotor. { 'mōd·ər ,mēd·ər }

motor nerve [NEUROSCI] A nerve composed wholly or principally of motor fibers. { 'mōd·ər ,nərv }

motor neuron [NEUROSCI] An efferent nerve cell. Also known as motoneuron. { 'mōd·ər 'nur,än }

motor noise [ACOUS] The noisy sound made by an electric motor. { 'mōd·ər ,nóiz }

motor reducer [MECH ENG] Speed-reduction power transmission equipment in which the reducing gears are integral with drive motors. { 'mōd·ər ri,dü·sər }

motorship [NAV ARCH] A ship propelled by a motor which can travel on the seas; in particular, one propelled by an internal combustion engine such as a diesel engine. { 'mōd·ər,ship }

motor speech area [ANAT] The cortical area located in the triangular and opercular portions of the inferior frontal gyrus; in right-handed people it is more developed on the left side. { 'mōd·ər 'spēch ,er·ē·ə }

motor system [PHYSIO] Any portion of the nervous system that regulates and controls the contractile activity of muscle and the secretory activity of glands. { 'mōd·ər ,sis·təm }

motor torpedo boat [NAV ARCH] A motor boat 60 to 100 feet (18 to 30 meters) long and capable of speeds of about 60 knots (110 kilometers per hour), armed with two to four torpedo tubes, machine and antiaircraft guns, and depth charges. Also known as mosquito boat; PT boat. { 'mōd·ər tór'pēd·ō ,bōt }

motortruck [MECH ENG] An automotive vehicle which is used to transport freight. { 'mōd·ər,trək }

motor unit [ANAT] The axon of an anterior horn cell, or the motor fiber of a cranial nerve, together with the striated muscle fibers innervated by its terminal branches. { 'mōd·ər ,yü·nət }

motor vehicle [MECH ENG] Any automotive vehicle that does not run on rails, and generally having rubber tires. { 'mōd·ər 'vē·ə·kəl }

mottle [MED] An effect that occurs during radiological imaging when the dose of radiation is reduced to a level where quantum effects can be observed. { 'mäd·əl }

mottled [GEOL] Of a soil, irregularly marked with spots of different colors. [GRAPHICS] Of a printed area, spotty or uneven in appearance. [PETR] Of a sedimentary rock, marked with spots of various colors. { 'mäd·əld }

mottled iron [MET] A cast iron showing gray areas that contain graphite, perlite, and sometimes ferrite, and white areas containing primarily cementite. { 'mäd·əld 'ī·ərn }

mottle-leaf [PL PATH] 1. A virus disease characterized by chlorotic mottling and wrinkling of leaves. 2. A zinc-deficiency disease characterized by partial chlorosis of the leaves and stunting of the plant. { 'mäd·əl,lēf }

mottramite [MINERAL] (Cu,Zn)Pb(VO₄)(OH) A mineral composed of a basic lead copper zinc vanadate; it is isomorphous with descloizite. Also known as cuprodescloizite; psittacinite. { 'mä·trə,mīt }

Mott scattering [QUANT MECH] 1. The scattering of identical particles due to a Coulomb force. 2. The scattering of a relativistic electron by a Coulomb field. { 'mät ,skad·ə·riŋ }

motu [GEOGR] One of a series of closely spaced coral islets separated by narrow channels; the group of islets forms a ring-shaped atoll. { 'mō·tü }

moulage [GRAPHICS] 1. The technique of making a mold. 2. A material used to make molds; it is extremely delicate and can be used on human hair or skin, or on antiques or other objects that are too fragile to be molded in rubber, gelatin, or plaster; it is applied warm with brushes or palette knives and can be reused. { mü'läzh }

moulin [HYD] A shaft or hole in the ice of a glacier which is roughly cylindrical and nearly vertical, formed by swirling meltwater pouring down from the surface. Also known as glacial mill; glacier mill; glacier pothole; glacier well; pothole. { mü'lan }

moulin pothole See giant's kettle. { mü'lan 'pät,hōl }

mounce [MECH] A unit of mass, equal to 25 grams. Also known as metric ounce. { mauns }

mound [GEOL] 1. A low, isolated, rounded natural hill, usually of earth. Also known as tuft. 2. A structure built by fossil colonial organisms. { maund }

mount [ELECTROMAG] The flange or other means by which a switching tube, or tube and cavity, is connected to a waveguide. [ENG] 1. Structure supporting any apparatus, as a gun, searchlight, telescope, or surveying instrument. 2. To fasten an apparatus in position, such as a gun on its support. [ORD] To equip; to put into operation; to go into operation, as to mount an offensive. { maunt }

mountain [GEOGR] A feature of the earth's surface that rises high above the base and has generally steep slopes and a relatively small summit area. { 'maunt·ən }

mountain and valley winds [METEOROL] A system of diurnal winds along the axis of a valley, blowing uphill and upvalley by day, and downhill and downvalley by night; they prevail mostly in calm, clear weather. { 'maunt·ən ən 'val·ē 'winz }

mountain blue [INORG CHEM] 2CuCO₃·Cu(OH)₂ Ground azurite used as a paint pigment. Also known as copper blue. { 'maunt·ən 'blü }

mountain breeze [METEOROL] A breeze that blows down a mountain slope due to the gravitational flow of cooled air. Also known as mountain wind. { 'maunt·ən 'brēz }

mountain brown ore [GEOL] Name used in Virginia for limonite or brown iron ore. { 'maunt·ən 'braun 'ór }

mountain butter See halotrichite. { 'maunt·ən 'bəd·ər }

mountain chain See mountain system. { 'maunt·ən ,chān }

mountain climate [CLIMATOL] Very generally, the climate of relatively high elevations; mountain climates are distinguished by the departure of their characteristics from those of surrounding lowlands, and the one common basis for this distinction is that of atmospheric rarefaction; aside from this, great variety is introduced by differences in latitude, elevation, and exposure to the sun; thus, there exists no single, clearly defined, mountain climate. Also known as highland climate. { 'maunt·ən ,klī·mət }

mountain cork [MINERAL] 1. A white or gray variety of asbestos composed of thick, interwoven fibers and having a corklike weight and texture. Also known as rock cork. 2. A fibrous clay mineral, such as sepiolite. { 'maunt·ən 'kórk }

mountain crystal See rock crystal. { 'maunt·ən 'krist·əl }

mountain effect [ELECTROMAG] The effect of rough terrain on radio-wave propagation, causing reflections that produce errors in radio direction-finder indications. { 'maunt·ən i,fekt }

mountain-gap wind [METEOROL] A local wind blowing through a gap between mountains. { 'maunt·ən 'gap ,wind }

mountain glacier See alpine glacier. { 'maunt·ən 'glā·shər }

mountain lion See puma. { 'maunt·ən ,lī·ən }

mountain mahogany See obsidian. { 'maunt·ən mə'häg·ə·nē }

mountain meteorology [METEOROL] The branch of meteorology that studies the effects of mountains on the atmosphere, ranging over all scales of motion. { 'maunt·ən ,mēd·ē·ə'räl·ə·jē }

mountain pediment [GEOL] A plain of combined erosion and transportation at the base of and surrounding a desert mountain range; at a distance it has the appearance of a broad triangular mass. { 'maunt·ən 'ped·ə·mənt }

mountain range [GEOGR] A succession of mountains or narrowly spaced mountain ridges closely related in position, direction, and geologic features. { 'maunt·ən ,rānj }

mountain sickness [MED] A disease occurring in persons living at high altitudes when homeostatic adjustments to the lowered atmospheric oxygen tension fail or develop disproportionately. Also known as high-altitude disease; high-altitude erythremia; Monge's disease; seroche. { 'maunt·ən ,sik·nəs }

mountain slope [GEOGR] The inclined surface that forms a mountainside. { 'maunt·ən ,slōp }

mountain soap See saponite. { 'maunt·ən ,sōp }

mountain system [GEOGR] A group of mountain ranges tied together by common geological features. Also known as mountain chain. { 'maunt·ən ,sis·təm }

mountain tallow See hatchettite. { 'maunt·ən ,tal·ō }

mountain tick fever See Colorado tick fever. { 'maunt·ən 'tik ,fē·vər }

mountain wave [METEOROL] An undulating flow of wind on the downwind, or lee, side of a mountain ridge caused by wind blowing strongly over the ridge. { 'maunt·ən ‚wāv }

mountain wind See mountain breeze. { 'maunt·ən ‚wind }

mountain wood [GEOL] **1.** A compact, fibrous, gray to brown type of asbestos which has an appearance similar to dry wood. Also known as rock wood. **2.** A fibrous clay mineral; for example, sepiolite or palygorskite. { 'maunt·ən ‚wud }

Mount Rose snow sampler [ENG] A particular pattern of snow sampler having an internal diameter of 1.485 inches (3.7719 centimeters), so that each inch of water in the sample weighs 1 ounce (28.3495 grams). { 'maunt 'rōz 'snō ‚sam·plər }

mousable interface [COMPUT SCI] A user interface that responds to input from a mouse for various functions. { ¦maus·ə·bəl 'in·tər‚fās }

mouse [COMPUT SCI] A small box-shaped device with wheels that is moved about by hand over a flat surface and generates signals to control the position of a cursor or pointer on a computer display. [VERT ZOO] Any of various rodents which are members of the families Muridae, Heteromyidae, Cricetidae, and Zapodidae; characterized by a pointed snout, short ears, and an elongated body with a long, slender, sparsely haired tail. { maus }

mousebane See aconite. { 'maus‚bān }

mouse deer See chevrotain. { 'maus ‚dir }

mouse hole [PETRO ENG] A drill hole under the derrick floor cased with a suspended drill pipe, to be connected later on the drill string. { 'maus ‚hōl }

mouse trap [ENG] A cylindrical fishing tool having the open bottom end fitted with an inward opening valve. { 'maus ‚trap }

mousse de chêne See oakmoss resin. { ‚müs də 'shen }

mousseline [TEXT] Referring to a lightweight crisp sheer fabric; usually used in connection with the name of the fiber. { ‚müs·ə'lēn }

mouth [ANAT] The oral or buccal cavity and its related structures. [ENG ACOUS] The end of a horn that has the larger cross-sectional area. [GEOGR] **1.** The place where one body of water discharges into another. Also known as influx. **2.** The entrance or exit of a geomorphic feature, such as of a cave or valley. [MIN ENG] **1.** The end of a shaft, adit, drift, entry, or tunnel emerging at the surface. **2.** The collar of a borehole. [SCI TECH] Something resembling a mouth, that is, a place where one thing enters another or an opening at the receiving end of a container or enclosure. { mauth }

mouth breadth [ANTHRO] The measure of the distance from one to the other corner of the mouth in a natural relaxed position. { 'mauth ‚bredth }

mouthfeel [FOOD ENG] An organoleptic property used to describe the overall texture of a food product. { 'mauth‚fēl }

mouth-to-mouth resuscitation [MED] A method of artificial respiration in which the rescuer's mouth is placed over the victim's mouth and air is blown forcefully into the victim's lungs every few seconds to inflate them. { 'mauth tə 'mauth ri‚səs·ə'tā·shən }

movable-active tooling [MECH ENG] Any equipment in a robotic system that is able to move and that operates under power. { 'mü·və·bəl ¦ak·tiv 'tül·iŋ }

movable bridge [CIV ENG] A bridge in which either the horizontal or vertical alignment can be readily changed to permit the passage of traffic beneath it. Often called drawbridge (an anachronism). { 'mü·və·bəl 'brij }

movable contact [ELEC] The relay contact that is mechanically displaced to engage or disengage one or more stationary contacts. Also known as armature contact. { 'mü·və·bəl 'kän‚takt }

movable-head disk drive [COMPUT SCI] A type of disk drive in which read/write heads are moved over the surface of the disk, toward and away from the center, so that they are correctly positioned to read or write the desired information. { 'mü·və·bəl ¦hed 'disk ‚drīv }

movable-passive tooling [MECH ENG] Equipment in a robotic system that moves but requires no power to operate, such as workpieces, clamps, and templates. { 'mü·və·bəl 'pas·iv 'tül·iŋ }

movable platen [ENG] The large platen at the back of an injection-molding machine to which the back half of the mold is fastened. { 'mü·və·bəl 'plat·ən }

movable-point crossing [CIV ENG] A small-angle rail crossing with two center frogs, each of which consists essentially of a knuckle rail and two opposed movable center points. { 'mü·və·bəl ¦point 'krós·iŋ }

movable propeller blade [NAV ARCH] A propeller blade cast separate from the boss or hub and attached to the boss by bolts. { 'mü·və·bəl prə'pel·ər ‚blād }

move mode [COMPUT SCI] A method of communicating between an operating program and an input/output control system in which the data records to be read or written are actually moved into and out of program-designated memory areas. { 'müv ‚mōd }

move operation [COMPUT SCI] An operation in which data is moved from one storage location to another. { 'müv ‚äp·ə‚rā·shən }

movie See motion picture. { 'mü·vē }

moving area [COMMUN] The portion of a television picture frame in which the intensity has changed since the previous frame. { 'müv·iŋ ¦er·ē·ə }

moving bed [CHEM ENG] Granulated solids in a process vessel that are circulated (moved) either mechanically or by gravity flow; used in catalytic and absorption processes. { 'müv·iŋ 'bed }

moving-bed catalytic cracking [CHEM ENG] Petroleum refining process for cracking (breaking) of long hydrocarbon molecules by use of heat, pressure, and a granular cracking catalyst that is continuously cycled between the reactor vessel and the catalyst regenerator. { 'müv·iŋ ¦bed ‚kad·əl‚id·ik 'krak·iŋ }

moving-boundary electrophoresis [ANALY CHEM] A U-tube variation of electrophoresis analysis that uses buffered solution so that all ions of a given species move at the same rate to maintain a sharp, moving front (boundary). { 'müv·iŋ ¦baun·drē i¦lek·trə·fə'rē·səs }

moving cluster [ASTRON] **1.** A star cluster with common motions. **2.** An open star cluster near the sun such that measurements may be made of the individual proper motions of the stars. { 'müv·iŋ ¦kləs·tər }

moving-coil galvanometer [ENG] Any galvanometer, such as the d'Arsonval galvanometer, in which the current to be measured is sent through a coil suspended or pivoted in a fixed magnetic field, and the current is determined by measuring the resulting motion of the coil. { 'müv·iŋ ¦koil ‚gal·və'näm·əd·ər }

moving-coil instrument [ELEC] Any instrument in which current is sent through one or more coils suspended or pivoted in a magnetic field, and the motion of the coils is used to measure either the current in the coils or the strength of the field. { 'müv·iŋ ¦koil 'in·strə·mənt }

moving-coil loudspeaker See dynamic loudspeaker. { 'müv·iŋ ¦koil 'laud‚spēk·ər }

moving-coil meter [ELEC] A meter in which a pivoted coil is the moving element. { 'müv·iŋ ¦koil 'mēd·ər }

moving-coil microphone See dynamic microphone. { 'müv·iŋ ¦koil 'mī·krə‚fōn }

moving-coil pickup See dynamic pickup. { 'müv·iŋ ¦koil 'pik‚əp }

moving-coil voltmeter [ENG] A voltmeter in which the current, produced when the voltage to be measured is applied across a known resistance, is sent through coils pivoted in the magnetic field of permanent magnets, and the resulting torque on the coils is balanced by control springs so that the deflection of a pointer attached to the coils is proportional to the current. { 'müv·iŋ ¦koil 'vōlt‚med·ər }

moving-coil wattmeter See electrodynamic wattmeter. { 'müv·iŋ ¦koil 'wät‚med·ər }

moving-conductor loudspeaker [ENG ACOUS] A loudspeaker in which the mechanical forces result from reactions between a steady magnetic field and the magnetic field produced by current flow through a moving conductor. { 'müv·iŋ kən¦dək·tər 'laud‚spēk·ər }

moving constraint [MECH] A constraint that changes with time, as in the case of a system on a moving platform. { 'müv·iŋ kən'strānt }

moving-head disk [COMPUT SCI] A disk-storage device in which one or more read-write heads are attached to a movable arm which allows each head to cover many tracks of information. { 'müv·iŋ ¦hed 'disk }

moving-iron meter [ENG] A meter that depends on current in one or more fixed coils acting on one or more pieces of soft iron, at least one of which is movable. { 'müv·iŋ |ī·ərn 'mēd·ər }

moving-iron voltmeter [ENG] A voltmeter in which a field coil is connected to the voltage to be measured through a series resistor; current in the coil causes two vanes, one fixed and one attached to the shaft carrying the pointer, to be similarly magnetized; the resulting torque on the shaft is balanced by control springs. { 'müv·iŋ |ī·ərn 'vōlt,mēd·ər }

moving load [MECH] A load that can move, such as vehicles or pedestrians. { 'müv·iŋ 'lōd }

moving-magnet voltmeter [ENG] A voltmeter in which a permanent magnet aligns itself with the resultant magnetic field produced by the current in a field coil and another permanent control magnet. { 'müv·iŋ |mag·nət 'vōlt,mēd·ər }

moving map display [NAV] An air navigation device which displays the aircraft's navigational position on moving film of an aeronautical chart; the positional information may be obtained from various systems such as Tacan, Doppler, inertial, and so forth; such display systems may indicate course and distance to destination, track over ground, and so on. { 'müv·iŋ 'map di,splā }

moving picture See motion picture. { 'müv·iŋ 'pik·chər }

moving sidewalk [CIV ENG] A sidewalk constructed on the principle of an endless belt, on which pedestrians are moved. { 'müv·iŋ 'sīd,wok }

moving submarine haven [NAV] A haven surrounding submarines in transit, extending 50 miles (80 kilometers) ahead, 100 miles (161 kilometers) astern, and 15 miles (24 kilometers) on each side of the estimated position of the submarine along the stated track. { 'müv·iŋ 'səb·mə,rēn 'hāv·ən }

moving surface ship haven [NAV] A haven which will normally be a circle with a specified radius centered on the estimated position of the ship or the guide of a group of ships. { 'müv·iŋ 'sər·fəs 'ship 'hāv·ən }

moving-target indicator [ELECTR] A device that limits the display of radar information primarily to moving targets; signals due to reflections from stationary objects are canceled by a memory circuit. Abbreviated MTI. { 'müv·iŋ 'tär·gət 'in·də,kād·ər }

moving totals [STAT] The sum of the year's figures and those of some years before and after it. { 'müv·iŋ 'tōd·əlz }

moving trihedral [MATH] For a space curve, a configuration consisting of the tangent, principal normal, and binormal of the curve at a variable point on the curve. { 'müv·iŋ trī'hē·drəl }

mowa fat See mowrah fat. { 'maú·ə ,fat }

mowrah fat [MATER] A vegetable fat extracted from the seeds of *Bassia longifola;* used to make soap. Also known as illepé fat; mora fat; mowa fat. { 'maú·rə ,fat }

Mozambique Current [OCEANOGR] The portion of the South Equatorial Current that turns and flows along the coast of Africa in the Mozambique Channel, forming one of the western boundary currents in the Indian Ocean. { ,mō·zəm'bēk 'kə·rənt }

mp See mean effective pressure; melting point.

MPEG [COMPUT SCI] The Motion Picture Experts Group video compression standards for digital video broadcasting. { 'em,peg }

MPI See Minnesota multiphasic personality inventory.

MPK See pentanone.

MPU See microprocessing unit.

MQ register [COMPUT SCI] Temporary-storage register whose contents can be transferred to or from, or swapped with, the accumulator. { 'em|kyü 'rej·ə·stər }

mr See milliroentgen.

mrad See millirad. { 'em ,rad }

MRAM See magnetic random access memory. { 'em,ram }

M region [ASTROPHYS] Any of the areas on the surface of the sun that are theoretically responsible for magnetic disturbances on the earth. { 'em ,rē·jən }

MRI See magnetic resonance imaging.

mRNA See messenger ribonucleic acid.

MRP See material requirements planning.

MRTD See minimum resolvable temperature difference.

ms See millisecond.

Ms See megasecond.

MSAS See MTSAT Satellite-Based Augmentation System. { 'em,sas or |em|es|ā'es }

msb See most significant bit.

M scan See M scope. { 'em ,skan }

MSCFD [CHEM ENG] Abbreviation for thousand standard cubic feet per day; usually refers to gas flow.

MSCFH [CHEM ENG] Abbreviation for thousand standard cubic feet per hour; usually refers to gas flow.

MSCFM [CHEM ENG] Abbreviation for thousand standard cubic feet per minute; usually refers to gas flow.

M scope [ELECTR] A modified form of A scope on which part of the time base is slightly displaced in a vertical direction by insertion of an adjustable step which serves as a range marker. Also known as M indicator; M scan. { 'em ,skōp }

msDNA See microsatellite deoxyribonucleic acid.

msec See millisecond.

Msec See megasecond.

MSG See sodium glutamate.

MSH See melanocyte-stimulating hormone.

M shell [ATOM PHYS] The third layer of electrons about the nucleus of an atom, having electrons characterized by the principal quantum number 3. { 'em ,shel }

MSI See magnetic source imaging; medium-scale integration.

M signal See monochrome signal. { 'em ,sig·nəl }

MSMA See monosodium acid methanearsonate.

MSR See molten-salt reactor.

MSS See mobile satellite service.

M star [ASTRON] A spectral classification for a star whose spectrum is characterized by the presence of titanium oxide bands; M stars have surface temperatures of 3000 K for giants and 3400 K for dwarfs. { 'em ,stär }

MSW effect [ASTROPHYS] An enhancement of neutrino oscillation by matter that is predicted to occur when electron neutrinos travel through the sun. Derived from Mikheyeve-Smirnov-Wolfenstein effect. { |em|es'dəb·ə yü i,fekt }

M synchronization [ENG] A linking arrangement between a camera lens and the flashbulb unit to allow a 15-millisecond delay of the shutter so that the bulb burns to its brightest point before the shutter opens. { 'em ,siŋ·krə·nə'zā·shən }

MT See megaton.

MTBE See methyl tertiary butyl ether.

MTBF See mean time between failures.

mtDNA See mitochondrial deoxyribonucleic acid.

M-theory [PARTIC PHYS] A highly symmetric but only partially understood theory of particles and their interactions that would be a generalization of supergravity and would be related by weak-strong duality to each of the five known superstring theories. { 'em ,thē·ə·rē }

MTI See moving-target indicator.

MTSAT Satellite-Based Augmentation System [NAV] A wide-area DGPS for air navigation whose error corrections are broadcast over geostationary satellites, developed by Japan Civil Aviation Bureau. Abbreviated MSAS. { |em|tē'sat |sad·əl,īt ,bāst ,óg·mən'tā·shən ,sis·təm }

MTTF See mean time to failure.

M-type backward-wave oscillator [ELECTR] A backward-wave oscillator in which focusing and interaction are through magnetic fields, as in a magnetron. Also known as M-type carcinotron; type-M carcinotron. { 'em ,tīp bak·wərd |wāv 'äs·ə,lād·ər }

M-type carcinotron See M-type backward-wave oscillator. { 'em ,tīp kär'sin·ə,trän }

muc-, muci-, muco- [ZOO] A combining form denoting pertaining to mucus, mucin, mucosa. { myük,myü·sē, myü·kō }

Mucedinaceae [MYCOL] The equivalent name for Moniliaceae. { myü,sed·ən'ās·ē,ē }

mucic acid [ORG CHEM] HOOC(CHOH)$_4$COOH A white, crystalline powder with a melting point of 210°C; soluble in water; used as a metal ion sequestrant and to retard concrete hardening. Also known as glactaric acid; saccharolactic acid; tetrahydroxyadipic acid. { 'myü·sik 'as·əd }

mucigel [BIOCHEM] A complex polysaccharide material that is composed of root mucilage and bacterial slime and acts to control aggregation of soil particles in the rhizosphere in the vicinity of older portions of plant roots. { 'myü·sə,jel }

mucigen [BIOCHEM] A substance from which mucin is derived; contained in mucus-secreting epithelial cells. { 'myü·sə·jən }

mucilage [MATER] **1.** A sticky material employed as an adhesive. **2.** A gummy material derived from plants. { 'myü·sə·lij }

mucin [BIOCHEM] A glycoprotein constituent of mucus and

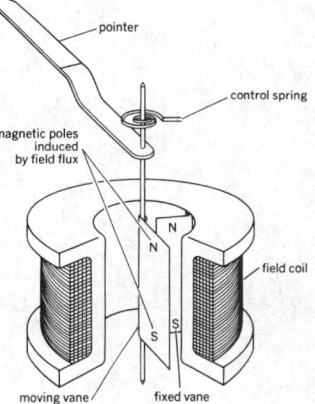

pointer

control spring

magnetic poles induced by field flux

N

N

S

field coil

moving vane

fixed vane

A representative form of the moving-iron voltmeter. *(General Electric Co.)*

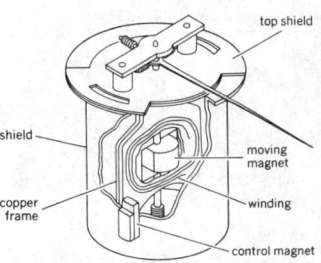

top shield

shield

moving magnet

copper frame

winding

control magnet

Moving-magnet direct-current voltmeter. *(General Electric Co.)*

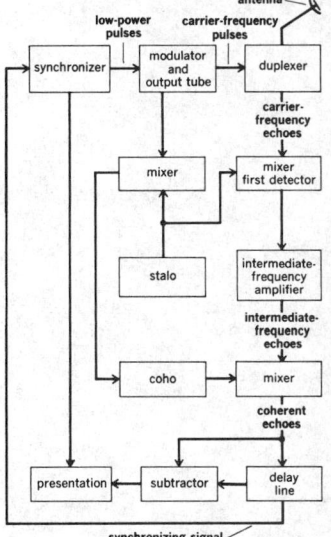

Diagram of a radar system with moving-target indicator capability. Here stalo = stable local oscillator, coho = coherent oscillator.

various other secretions of humans and lower animals. { 'myü·sən }

mucinosis [MED] Accumulations of materials containing mucin or mucinous substances in the skin; sometimes accompanied by papule and nodule formation. { ˌmyü·sə'nō·səs }

mucinous cyst See mucous cyst. { 'myüs·ən·əs 'sist }

muck [CIV ENG] Rock or earth removed during excavation. [GEOL] Dark, finely divided, well-decomposed, organic matter intermixed with a high percentage of mineral matter, usually silt, forming a surface deposit in some poorly drained areas. [MIN ENG] See waste rock. { mək }

mucker [MIN ENG] A worker who loads broken mineral into trams or pushes them from stope chute to shaft. Also known as groundman. { 'mək·ər }

mucking [ENG] Clearing and loading broken rock and other excavated materials, as in tunnels or mines. { 'mək·iŋ }

mucocele [MED] 1. Dilatation, particularly of a cavity, with mucus secretion. 2. A polypoid lesion consisting of mucus and mucus-secreting tissue. { 'myü·kə,sēl }

mucocutaneous [ANAT] Pertaining to a mucous membrane and the skin, and to the line where these join. { ˌmyü·kō·kyü'tā·nē·əs }

mucoid [BIOCHEM] 1. Any of various glycoproteins, similar to mucins but differing in solubilities and precipitation properties and found in cartilage, in the crystalline lens, and in white of egg. 2. Resembling mucus. [MICROBIO] Pertaining to large colonies of bacteria characterized by being moist and sticky. { 'myü,kȯid }

mucolytic [BIOCHEM] Effecting the dissolution, liquefaction, or dispersion of mucus and mucopolysaccharides. { ˌmyü·kə'līd·ik }

mucopolysaccharide [BIOCHEM] Any of a group of polysaccharides containing an amino sugar and uronic acid; a constituent of mucoproteins, glycoproteins, and blood-group substances. { ˌmyü·kō,päl·ē'sak·ə,rīd }

mucopolysaccharidosis [MED] Any of several inborn metabolic disorders involving mucopolysaccharides; the six types are MPS I, Hurler's syndrome; MPS II, Hunter's syndrome; MPS III, Sanfillipo's syndrome; MPS IV, Morquio's syndrome; MPS V, Scheil's syndrome; and MPS VI, Maroteaux-Lamy's syndrome. { ˌmyü·kō,päl·ē,sak·ə·rə'dō·səs }

mucoprotein [BIOCHEM] Any of a group of glycoproteins containing a sugar, usually chondroitinsulfuric or mucoitinsulfuric acid, combined with amino acids or polypeptides. { ˌmyü·kō'prō,tēn }

mucopurulent [MED] Containing mucus and pus. { ˌmyü·kō'pyur·ə·lənt }

Mucorales [MYCOL] An order of terrestrial fungi in the class Phycomycetes, characterized by a hyphal thallus and nonmotile sporangiospores, or conidiospores. { ˌmyü·kə'rā·lēz }

mucormycosis [MED] An acute, usually fulminating fungus infection of humans caused by several genera of Mucorales, including Absidia, Rhizopus, and Mucor. { ˌmyü·kȯr,mī'kō·səs }

mucosa [HISTOL] A mucous membrane. { myü'kō·sə }

mucosanguineous [MED] Containing mucus and blood. { ˌmyü·kō·saŋ'gwin·ē·əs }

mucous [PHYSIO] Of or pertaining to mucus; secreting mucus. { 'myü·kəs }

mucous cell [PHYSIO] A mucus-secreting cell. { 'myü·kəs ,sel }

mucous colitis See irritable colon. { 'myü·kəs kə'līd·əs }

mucous connective tissue [HISTOL] A type of loose connective tissue in which the ground substance is especially prominent and soft; occurs in the umbilical cord. { 'myü·kəs kə'nek·tiv 'tish·ü }

mucous cyst [MED] A retention cyst of a gland, containing a secretion rich in mucin. Also known as mucinous cyst. { 'myü·kəs ,sist }

mucous degeneration [MED] Any retrogressive change associated with abnormal production of mucus. { 'myü·kəs dē,jen·ə'rā·shən }

mucous epithelium [EMBRYO] The epidermis of an embryo, excluding the epitrichium. [HISTOL] The germinative layer of a stratified squamous epithelium. { 'myü·kəs ,ep·ə'thē·lē·əm }

mucous gland [PHYSIO] A gland that secretes mucus. { 'myü·kəs ,gland }

mucous membrane [HISTOL] The type of membrane lining cavities and canals which have communication with air; it is kept moist by glandular secretions. Also known as tunica mucosa. { 'myü·kəs 'mem,brān }

mucoviscidosis See cystic fibrosis. { ˌmyü·kō,vis·ə'dō·səs }

mucro [BIOL] An abrupt, sharp terminal tip or process. { 'myü·krō }

mucronate [BIOL] Terminated abruptly by a sharp terminal tip or process. { 'myü·krə,nāt }

mucus [PHYSIO] A viscid fluid secreted by mucous glands, consisting of mucin, water, inorganic salts, epithelial cells, and leukocytes, held in suspension. { 'myü·kəs }

mud [ENG] See slime. [GEOL] An unindurated mixture of clay and silt with water; it is slimy with a consistency varying from that of a semifluid to that of a soft and plastic sediment. [MATER] See drilling mud. [PETR] The silt plus clay portion of a sedimentary rock. { məd }

mud auger [DES ENG] A diamond-point bit with the wings of the point twisted in a shallow augerlike spiral. Also known as clay bit; diamond-point bit; mud bit. { 'məd ,ȯg·ər }

mud ball [GEOL] A rounded mass of mud or mudstone up to 8 inches (20 centimeters) in diameter in a sedimentary rock. Also known as chalazoidite; tuff ball. { 'məd ,bȯl }

mud berth [CIV ENG] A berth where a vessel rests on the bottom at low water. { 'məd ,bərth }

mud bit See mud auger. { 'məd ,bit }

mud blasting [ENG] The detonation of sticks of explosive stuck on the side of a boulder with a mud covering, so that little of the explosive energy is used in breaking the boulder. { 'məd ,blast·iŋ }

mud cake [ENG] A caked layer of clay adhering to the walls of a well or borehole, formed where the water in the drilling mud filtered into a porous formation during rotary drilling. Also known as filter cake. { 'məd ,kāk }

mud-cake resistivity [PETRO ENG] Resistivity of drilling mud cake pressed from a sample of the mud; important in mud log interpretation. { 'məd ,kāk ,rē,zis'tiv·əd·ē }

mudcap [ENG] A quantity of wet mud, wet earth, or sand used to cover a charge of dynamite or other high explosive fired in contact with the surface of a rock in mud blasting. { 'məd ,kap }

mud conditioning [PETRO ENG] In a well drilling operation, the treatment and control of drilling mud to ensure proper gel strength, viscosity, density and so on. { 'məd kən,dish·ən·iŋ }

mud cone [GEOL] A cone of sulfurous mud built around the opening of a mud volcano or mud geyser, with slopes as steep as 40° and diameters ranging upward to several hundred yards. Also known as puff cone. { 'məd ,kōn }

mud crack [GEOL] An irregular fracture formed by shrinkage of clay, silt, or mud under the drying effects of atmospheric conditions at the surface. Also known as desiccation crack; sun crack. { 'məd ,krak }

mud crack polygon See mud polygon. { 'məd ,krak 'päl·ə,gän }

mud drilling [MIN ENG] Drilling operations in which a mud-laden circulation fluid is used. Also known as mud flush drilling. { 'məd ,dril·iŋ }

mudfish See bowfin. { 'məd,fish }

mud flat [GEOL] A relatively level, sandy or muddy coastal strip along a shore or around an island; may be alternately covered and uncovered by the tide or may be covered by shallow water. Also known as flat. { 'məd ,flat }

mudflow [GEOL] A flowing mass of fine-grained earth material having a high degree of fluidity during movement. { 'məd,flō }

mud flush drilling See mud drilling. { 'məd 'fləsh ,dril·iŋ }

mud flush test [MIN ENG] A test carried out at the boring site to determine whether the mud solution is of the correct viscosity and density. { 'məd 'fləsh ,test }

mud log [PETRO ENG] A continuous record of changes in oil or gas contents of circulating drilling mud while drilling a well. { 'məd ,läg }

mudlump [GEOL] A diapiric sedimentary structure consisting of clay or silt and forming an island in deltaic areas; produced by the loading action of rapidly deposited delta front sands upon lighter-weight prodelta clays. { 'məd,ləmp }

mud pilot [NAV] A person who pilots a vessel by visually observing changes in the color of the water as the depth of the water increases or decreases. { 'məd ,pī·lət }

mud pit See slushpit. { 'məd ,pit }

mud polygon [GEOL] A nonsorted polygon whose center lacks vegetation but whose peripheral fissures contain peat and plants. Also known as mud crack polygon. { 'məd 'päl·ə‚gän }

mud pot [GEOL] A type of hot spring which contains boiling mud, typically sulfurous and often multicolored; tends to be associated with geysers and other hot springs in volcanic zones. Also known as painted pot; sulfur-mud pool. { 'məd ‚pät }

mud pulse telemetry [PETRO ENG] A technique in drilling an oil or gas well in which measurements made by downhole sensors are transmitted to the surface by encoding the data in increasing or decreasing pressure pulses. { ¦məd ‚pəls tə'lem·ə·trē }

mud puppy [VERT ZOO] Any of several American salamanders of the genera *Necturus* and *Proteus* making up the family Proteidae; distinguished by having both lungs and gills as an adult. { 'məd ‚pəp·ē }

mud-removal acid [PETRO ENG] Mixture of hydrochloric and hydrofluoric acids with inhibitors, surfactants, and demulsifiers; used to dissolve drilling-mud clays away from the drillhole face. { 'məd ri‚mü·vəl ‚as·əd }

mudsill [CIV ENG] The lowest sill of a structure, usually embedded in the earth. { 'məd‚sil }

mudslide [GEOL] A slow-moving mudflow in which movement is mainly by sliding upon a discrete boundary shear surface. { 'məd‚slīd }

mud still [ENG] An instrument used to separate oil, water, and other volatile materials in a mud sample by distillation, permitting determination of the quantities of oil, water, and total solid contents in the original sample. { 'məd ‚stil }

mudstone [GEOL] An indurated equivalent of mud in the form of a blocky or massive, fine-grained sedimentary rock containing approximately equal proportions of silt and clay; lacks the fine lamination or fissility of shale. { 'məd‚stōn }

mud sump [CHEM ENG] Upstream area in a process vessel where, because of a velocity drop, entrained solids drop out and are collected in a sump. { 'məd ‚səmp }

mud volcano [GEOL] A conical accumulation of variable admixtures of sand and rock fragments, the whole resulting from eruption of wet mud and impelled upward by fluid or gas pressure. Also known as hervidero; macaluba. { 'məd väl'kā·nō }

Mueller matrices [OPTICS] Matrix operators in a calculus used to treat polarized light; in this calculus, the light vector is split into four components one of which is the intensity of the light, and unpolarized light can be treated directly. { 'myül·ər ‚mā·trə‚sēz }

MUF *See* maximum usable frequency.

mu factor [ELECTR] Ratio of the change in one electrode voltage to the change in another electrode voltage under the conditions that a specified current remains unchanged and that all other electrode voltages are maintained constant; a measure of the relative effect of the voltages on two electrodes upon the current in the circuit of any specified electrode. { 'myü ‚fak·tər }

muffin-tin potential [SOLID STATE] A potential function used in the augmented plane-wave method and related methods of approximating the energy states of electrons in a crystal lattice, which is spherically symmetric within spheres centered at each atomic nucleus and constant in the region between these spheres. { 'məf·ən ‚tin pə‚ten·chəl }

muffle furnace [ENG] A furnace with an externally heated chamber, the walls of which radiantly heat the contents of the chamber. { 'məf·əl ‚fər·nəs }

muffler [ENG] A device to deaden the noise produced by escaping gases or vapors. { 'məf·lər }

mugearite [PETR] A dark-colored, fine-grained igneous rock in which the chief feldspar is oligoclase, plus orthoclase and olivine with some apatite and opaque oxides; originates by differentiation and volcanic crystallization of the primary magma. { myü'jē·ə‚rīt }

muggy [METEOROL] Referring to warm and especially humid weather. { 'məg·ē }

Mugilidae [VERT ZOO] The mullets, a family of perciform fishes in the suborder Mugiloidei. { myü'jil·ə‚dē }

Mugiloidei [VERT ZOO] A suborder of fishes in the order Perciformes; individuals are rather elongate, terete fishes with a short spinous dorsal fin that is well separated from the soft dorsal fin. { ‚myü·jə'lòid·ē‚ī }

mulberry [BOT] Any of various trees of the genus *Morus* (family Moraceae), characterized by milky sap and simple, often lobed alternate leaves. { 'məl‚ber·ē }

mulch [MATER] A mixture of organic material, such as straw, peat moss, or leaves, that is spread over soil to prevent evaporation, maintain an even soil temperature, prevent erosion, control weeds, and enrich soil. { məlch }

mule [MIN ENG] *See* barney. [VERT ZOO] The sterile hybrid offspring of the male ass and the mare, or female horse. { myül }

mule skinner [MIN ENG] A mule driver. { 'myül ‚skin·ər }

mull [ENG] To mix thoroughly or grind. [GEOGR] *See* headland. [GEOL] Granular forest humus that is incorporated with mineral matter. [TEXT] A thin, sheer cotton or cotton and polyester fabric. { məl }

muller [ENG] A foundry sand-mixing machine. { 'məl·ər }

Müllerian duct *See* paramesonephric duct. { mi'ler·ē·ən 'dəkt }

Müllerian duct cyst [MED] A congenital cyst arising from vestiges of the Müllerian ducts. { mi'ler·ē·ən ‚dəkt 'sist }

Müllerian mimicry [ZOO] Mimicry between two aposematic species. { mi'ler·ē·ən 'mim·ə·krē }

Muller method [MATH] A method for finding zeros of a function $f(x)$, in which one repeatedly evaluates $f(x)$ at three points, x_1, x_2, and x_3, fits a quadratic polynomial to $f(x_1)$, $f(x_2)$, and $f(x_3)$, and uses x_2, x_3, and the root of this quadratic polynomial nearest to x_3 as three new points to repeat the process. { 'məl·ər ‚meth·əd }

Müller's glass *See* hyalite. { 'mil·ərz ‚glas }

Müller's larva [INV ZOO] The ciliated larva characteristic of various members of the Polycladida; resembles a modified ctenophore. { 'mil·ərz ‚lär·və }

mulling [ENG] The combining of clay, water, and sand, prior to molding, by compressing with a roller to ensure development of optimum sand properties by the adequate distribution of ingredients. { 'məl·iŋ }

mullion [BUILD] A vertical bar separating two windows in a multiple window. [GEOL] In folded sedimentary and metamorphic rocks, a columnar structure in which the rock columns seem to intersect. { 'məl·yən }

mullite [MINERAL] $Al_6Si_2O_{13}$ An orthorhombic mineral consisting of an aluminum silicate that is resistant to corrosion and heat; used as a refractory. Also known as porcelainite. { 'mə‚līt }

mullock *See* waste rock. { 'məl·ək }

mull technique [SPECT] Method for obtaining infrared spectra of materials in the solid state; material to be scanned is first pulverized, then mulled with mineral oil. { 'məl tek‚nēk }

multi- [SCI TECH] A prefix meaning many. { 'məl·tē }

multiaccess computer [COMPUT SCI] A computer system in which computational and data resources are made available simultaneously to a number of users who access the system through terminal devices, normally on an interactive or conversational basis. { ¦məl·tē'ak‚ses kəm‚pyüd·ər }

multiaccess network *See* multiple-access network. { ¦məl·tē'ak‚ses 'net‚wərk }

multiaddress [COMPUT SCI] Referring to an instruction that has more than one address part. { ¦məl·tē'a‚dres }

multianode tube [ELECTR] Electron tube having two or more main anodes and a single cathode. { ¦məl·tē'an‚ōd ‚tüb }

multiaperture reluctance switch [ELECTR] Two-aperture ferrite storage core which may be used to provide a nondestructive readout computer memory. { ¦məl·tē'ap·ə·chər ri'lək·təns ‚swich }

multiaspect [COMPUT SCI] Pertaining to searches or systems which permit more than one aspect, or facet, of information to be used in combination, one with the other to effect identifying or selecting operations. { ¦məl·tē'as‚pekt }

multicavity klystron [ELECTR] A klystron in which there is at least one cavity between the input and output cavities, each of which remodulates the beam so that electrons are more closely bunched. { ¦məl·tē'kav·əd·ē 'klī‚strän }

multicavity magnetron [ELECTR] A magnetron in which the circuit includes a plurality of cavities, generally cut into the solid cylindrical anode so that the mouths of the cavities face the central cathode. { ¦məl·tē'kav·əd·ē 'mag·nə‚trän }

multicellular [BIOL] Consisting of many cells. { ¦məl·tē'sel·yə·lər }

MUD PUPPY

Necturus maculosus.

multicellular horn [ELECTROMAG] A cluster of horn antennas having mouths that lie in a common surface and that are fed from openings spaced one wavelength apart in one face of a common waveguide. [ENG ACOUS] A combination of individual horn loudspeakers having individual driver units or joined in groups to a common driver unit. Also known as cellular horn. { ¦məl·tē¦sel·yə·lər 'hȯrn }

multichannel analyzer See pulse-height analyzer. { ¦məl·tē'chan·əl 'an·ə,līz·ər }

multichannel communication [COMMUN] Communication in which there are two or more communication channels over the same path, such as a communication cable, or a radio transmitter which can broadcast on two different frequencies, either individually or simultaneously. { ¦məl·tē'chan·əl kə,myü·nə'kā·shən }

multichannel field-effect transistor [ELECTR] A field-effect transistor in which appropriate voltages are applied to the gate to control the space within the current flow channels. { ¦məl·tē'chan·əl 'fēld i¦fekt tran'zis·tər }

multichannel loading [COMMUN] Behavior of a multichannel communications system with all channels active. { ¦məl·tē'chan·əl 'lōd·iŋ }

multichannel telephone system [COMMUN] A telephone system in which two or more communications channels are carried over a single telephone cable or radio link. { ¦məl·tē'chan·əl 'tel·ə,fōn ,sis·təm }

multichip microcircuit [ELECTR] Microcircuit in which discrete, miniature, active electronic elements (transistor or diode chips) and thin-film or diffused passive components or component clusters are interconnected by thermocompression bonds, alloying, soldering, welding, chemical deposition, or metallization. { ¦məl·tē¦chip 'mī·krō,sər·kət }

multicipital [BIOL] Having many heads or branches arising from one point. { ,məl·tə'sip·əd·əl }

multicollector electron tube [ELECTR] An electron tube in which electrons travel to more than one electrode. { ¦məl·tē¦kə¦lek·tər i'lek,trän ,tüb }

multicollinearity [STAT] A concept in regression analysis describing the situation where, because of the high degree of correlation between two or more independent variables, it is not possible to separate accurately the effect of each individual independent variable upon the dependent variable. { ,məl·tē kō,lin·ē'ar·əd·ē }

multicompartmental genome [VIROL] In certain viruses, separation of genetic information into encapsulated nucleic acid molecules. { ,məl·tə·kəm,pärt'ment·əl 'jē,nōm }

multicompletion well See multiple-completion well. { ¦məl·tē·kəm¦plē·shən 'wel }

multicomponent distillation [CHEM ENG] The distillation separation of a single liquid feed stream containing three or more components into a single overhead product and a single bottoms product. { ¦məl·tē·kəm¦pō·nənt ,dist·əl'ā·shən }

multicomputer system [COMPUT SCI] A system consisting of more than one computer, usually under the supervision of a master computer, in which smaller computers handle input/output and routine jobs while the large computer carries out the more complex computations. { ¦məl·tē·kəm¦pyüd·ər ,sis·təm }

multiconfiguration Hartree-Fock approximation [ATOM PHYS] A natural extension of the Hartree-Fock approximation for an atom or molecule in which a number of configurations are chosen and the mixing coefficients, as well as the radial parts of the orbitals, are varied to minimize the expectation value of the energy. Abbreviated MCHF approximation. { ,məl·tē·kən,fig·yə¦rā·shən ¦här,trē 'fäk ə,prok·sə,mā·shən }

multicoupler [ELECTR] A device for connecting several receivers to one antenna and properly matching the impedances of the receivers to the antenna. { 'məl·tə,kəp·lər }

multicycle [GEOL] Pertaining to a landscape or landform produced by more than one cycle of erosion. { 'məl·tə,sī·kəl }

multicycle feeding See multiread feeding. { 'məl·tə,sī·kəl 'fēd·iŋ }

multideck cage [MIN ENG] A cage with two or more compartments or platforms to hold the mine cars. { 'məl·tə,dek 'kāj }

multideck clarifiers [ENG] Extraction units which remove pollutants from recycled plant waste water. { 'məl·tə,dek 'klar·ə,fī·ərz }

multideck screen [MIN ENG] A screen with two or more superimposed screening surfaces mounted within a common frame. { 'məl·tə,dek 'skrēn }

multideck sinking platform [MIN ENG] A sinking platform of several decks so that various shaft-sinking operations may be performed simultaneously. { 'məl·tə,dek 'siŋ·kiŋ ,plat,fȯrm }

multideck table [MIN ENG] A type of shaking table that is double-decked; while each deck is fed and discharged independently, one mechanism vibrates both. { 'məl·tə,dek 'tā·bəl }

multident See polydent. { 'məl·tə,dent }

multidentate ligand [CHEM] A ligand capable of donating two or more pairs of electrons in a complexation reaction to form coordinate bonds. { ,məl·tē¦den,tāt 'lī·gənd }

multidimensional derivative [MATH] The generalized derivative of a function of several variables which is usually represented as a matrix involving the various partial derivatives of the function. { ¦məl·tə·di¦men·shən·əl də'riv·əd·iv }

multidimensional Turing machine [COMPUT SCI] A variation of a Turing machine in which tapes are replaced by multidimensional structures. { ¦məl·tə·di¦men·shən·əl 'tùr·iŋ mə,shēn }

multidither COAT See multidither coherent adaptive optical techniques. { 'məl·tē,dith·ər ¦sē¦ō¦ā'tē or 'kō,at }

multidither coherent adaptive optical techniques [OPTICS] Adaptive optical techniques for concentrating laser radiation into as small an area as possible, involving an array of laser sources, one of which has fixed phase, while the phase of a second element is controlled by a low-amplitude phase modulation or dither, and adjusted to maximize the reflection from a target glint. Abbreviated multidither COAT. { 'məl·tē,dith·ər kō'hir·ənt ə'dap·tiv 'äp·ti·kəl tek,neks }

multidrop line [COMMUN] A telephone pair which terminates at several locations. { 'məl·tē,dräp ,līn }

multielectrode tube [ELECTR] Electron tube containing more than three electrodes associated with a single electron stream. { 'məl·tē·i'lek,trōd ,tüb }

multielement array [ELECTROMAG] An antenna array having a large number of antennas. { ¦məl·tē'el·ə·mənt ə'rā }

multielement parasitic array [ELECTROMAG] Antennas consisting of an array of driven dipoles and parasitic elements, arranged to produce a beam of high directivity. { ¦məl·tē'el·ə·mənt ,par·ə'sid·ik ə,rā }

multielement vacuum tube [ELECTR] A vacuum tube which has one or more grids in addition to the cathode and plate electrodes. { ¦məl·tē'el·ə·mənt 'vak·yəm ,tüb }

multifid [BIOL] Divided into many lobes. { 'məl·tə,fid }

multifilament [TEXT] A yarn made of two or more monofilaments, natural or synthetic. { 'məl·tē'fil·ə·mənt }

multifocal lens [OPTICS] A lens that has more than one focal length. { ¦məl·tē,fō·kəl 'lenz }

multifoil [MATH] A plane figure consisting of congruent arcs of a circle arranged around a regular polygon, with the end points of each arc located at the midpoints of adjacent sides of the polygon, and the tangents to the arcs at these points perpendicular to the sides. { 'məl·tē,fȯil }

multifuel burner [ENG] A burner which utilizes more than one fuel simultaneously for combustion. { 'məl·tē,fyül ,bər·nər }

multifunction array radar [ENG] Electronic scanning radar which will perform target detection and identification, tracking, discrimination, and some interceptor missile tracking on a large number of targets simultaneously and as a single unit. { ¦məl·tə'fəŋk·shən ə'rā 'rā,där }

multifuse igniter [ENG] A black powder cartridge that allows several fuses to be fired at the same time by lighting a single fuse. { 'məl·tə,fyüz ig'nīd·ər }

multigene family [GEN] A set of genes that arose from duplications of a single ancestral gene and variation due to independent mutations and selection acting on individual members of the duplicate genes. { 'məl·tə,jēn ,fam·lē }

multiglandular [ANAT] Of or pertaining to several glands. { ,məl·tə'glan·jə·lər }

multigraph [MATH] **1.** A graph with no loops. **2.** A graph that may have more than one edge joining a particular pair of vertices. { 'məl·tə,graf }

multigrid tube [ELECTR] An electron tube having two or more grids between cathode and anode, as a tetrode or pentode. { 'məl·tə,grid ,tüb }

multigroup diffusion [NUCLEO] Diffusion of neutrons in a material as it is regarded in the multigroup model. { 'məl·tə‚grüp di'fyü·zhən }

multigroup model [NUCLEO] A model for the behavior of neutrons in a material in which they are grouped into several energy ranges, taking into account differences in spatial behavior of neutrons in the various groups, and transfer of neutrons between groups. { 'məl·tə‚grüp 'mäd·əl }

multigun tube [ELECTR] A cathode-ray tube having more than one electron gun. { 'məl·tə‚gən ‚tüb }

multihead Turing machine [COMPUT SCI] A variation of a Turing machine in which more than one head is allowed per tape. { 'məl·tē‚hed 'tu̇r·iŋ mə‚shēn }

multihull boat [NAV ARCH] A vessel, such as a catamaran, that has two or more hulls, usually joined by a common deck. { ¦məl·tē‚həl ‚bōt }

multijob operation [COMPUT SCI] The concurrent or interleaved execution of job steps from more than one job. { 'məl·tē‚jäb ‚äp·ə'rā·shən }

multijunction solar cell [ELECTR] A solar cell made of two or more materials, each optimally efficient over a limited spectral range. Also known as multiple-junction solar cell. { ‚məl·tē¦jəŋk·shən 'sō·lər 'sel }

multilayer bit [DES ENG] A bit set with diamonds arranged in successive layers beneath the surface of the crown. { ¦məl·tē'lā·ər ‚bit }

multilayer board [ELECTR] A printed wiring board that contains circuitry on internal layers throughout the cross section of the board as well as on the external layers. { ‚məl·tē‚lā·ər 'bȯrd }

multilayer optical storage [COMPUT SCI] An extension of optical disk storage technology to the third dimension by stacking data layers one above another, with each layer separated by a spacer region. { ‚məl·tə‚lā·ər ¦äp·ti·kəl 'stȯr·ij }

multilevel address See indirect address. { ¦məl·tə'lev·əl a‚dres }

multilevel control theory [CONT SYS] An approach to the control of large-scale systems based on decomposition of the complex overall control problem into simpler and more easily managed subproblems, and coordination of the subproblems so that overall system objectives and constraints are satisfied. { ¦məl·tə'lev·əl kən'trōl 'thē·ə‚rē }

multilevel indirect addressing [COMPUT SCI] A programming device whereby the address retrieved in the memory word may itself be an indirect address that points to another memory location, which in turn may be another indirect address, and so forth. { ¦məl·tə'lev·əl ‚in·də‚rekt ə'dres·iŋ }

multilevel transmission [COMMUN] Transmission of digital information in which three or more levels of voltage are recognized as meaningful, as 0,1,2 instead of simply 0,1. { ¦məl·tə'lev·əl tranz'mish·ən }

multiline appearances [COMMUN] **1.** The ability of a telephone to receive or originate additional voice or data calls at the terminal while it is still engaged in the primary voice call. **2.** The ability to bring additional parties to a primary telephone call. { ¦məl·tē'līn ə'pir·ən·səz }

multilinear algebra [MATH] The study of functions of several variables which are linear relative to each variable. { 'məl·tē‚lin·ē·ər 'al·jə·brə }

multilinear form [MATH] A multilinear form of degree n is a polynomial expression which is linear in each of n variables. { 'məl·tə‚lin·ē·ər 'fȯrm }

multilinear function [MATH] A function of several variables that is a linear function of each variable when the other variables are given fixed values. { ‚məl·tə‚lin·ē·ər 'fəŋk·shən }

multilist organization [COMPUT SCI] A chained file organization in which each segment is indexed. { 'məl·tē‚list ‚ȯr·gə·nə'zā·shən }

Multillidae [INV ZOO] An economically important family of Hymenoptera; includes the cow killer, a parasite of bumblebee pupae. { məl'til·ə‚dē }

multilocular [BIOL] Having many small chambers or vesicles. { ‚məl·tē'äk·yə·lər }

multimedia technology [COMPUT SCI] The synergistic union of digital video, audio, computer, information, and telecommunication technologies. { ¦məl·tə‚mēd·ē·ə tek'näl·ə‚jē }

multimer [BIOCHEM] A protein molecule composed of two or more monomers. { 'məl·tə·mər }

multimeter See volt-ohm-milliammeter. { 'məl·tə‚mēd·ər or məl'tim·əd·ər }

multi-mirror telescope [OPTICS] A telescope in which light from several mirrors of similar shape is brought to a common focus by additional optical elements. Abbreviated MMT. { ‚məl·tē ‚mir·ər 'tel·ə‚skōp }

multimodal distribution [STAT] A frequency distribution that has several relative maxima. { ‚məl·tē¦mōd·əl ‚di·strə'byü·shən }

multinomial [MATH] An algebraic expression which involves the sum of at least two terms. { ¦məl·tə‚nō·mē·əl }

multinomial distribution [MATH] The joint distribution of the set of random variables which are the number of occurrences of the possible outcomes in a sequence of multinomial trials. { ¦məl·tə‚nō·mē·əl ‚di·strə'byü·shən }

multinomial theorem [MATH] The rule for expanding $(x_1 + x_2 + \cdots + x_m)^n$, where m and n are positive integers; a generalization of the binomial theorem. { ‚məl·tə‚nō·mē·əl 'thir·əm }

multinomial trials [STAT] Unrelated trials with more than two possible outcomes the probabilities of which do not change from trial to trial. { ¦məl·tə‚nō·mē·əl 'trīlz }

multipactor [ELECTR] A high-power, high-speed microwave switching device in which a thin electron cloud is driven back and forth between two parallel plane surfaces in a vacuum by a radio-frequency electric field. { 'məl·tə‚pak·tər }

multipass sort [COMPUT SCI] Computer program designed to sort more data than can be contained within the internal storage of a computer; intermediate storage, such as disk, tape, or drum, is required. { 'məl·tə‚pas 'sȯrt }

multipath See multipath transmission. { 'məl·tə‚path }

multipath cancellation [COMMUN] Occurrence of essentially complete cancellation of radio signals because of the relative amplitude and phase differences of the components arriving over separate paths. { 'məl·tə‚path ‚kan·sə'lā·shən }

multipath system [NAV] An electronic navigation system that measures differences in or otherwise compares transmission times of radio signals. { 'məl·tə‚path ‚sis·təm }

multipath transmission [ELECTROMAG] The propagation phenomenon that results in signals reaching a radio receiving antenna by two or more paths, causing distortion in radio and ghost images in television. Also known as multipath. { 'məl·tə‚path tranz'mish·ən }

multiphase flow [CHEM ENG] Mixture of two or more distinct phases (such as oil, water, and gas) flowing through a closed conduit. { 'məl·tə‚fāz ‚flō }

multiphase sampling [STAT] A sampling method in which certain items of information are drawn from the whole units of a sample and certain other items of information are taken from the subsample. { 'məl·tə‚fāz 'sam·pliŋ }

multiphasic personality inventory See Minnesota multiphasic personality inventory. { ¦məl·tə'fāz·ik ‚pər·sə'nal·əd·ē 'in·vən‚tȯr·ē }

multiphonon emission [SOLID STATE] A process of nonradiative recombination of electrons and holes in which an electron is captured into a deep level near the middle of an energy gap associated with a lattice defect, exciting lattice vibrations, and the trapped electron state captures a hole from the valence band. { ‚məl·tə'fō‚nän i'mish·ən }

multiphoton absorption [ATOM PHYS] The excitation of an atom or other microscopic system to a higher quantum state by simultaneous absorption of two or more photons which together provide the necessary energy. { ‚məl·tə'fō‚tän ab'sȯrp·shən }

multiphoton ionization [ATOM PHYS] The removal of one or more electrons from an atom or other microscopic system as the result of simultaneous absorption of two or more photons. { ‚məl·tə'fō‚tän ‚ī·ə·nə'zā·shən }

multiple [ELEC] **1.** Group of terminals arranged to make a circuit or group of circuits accessible at a number of points at any one of which connection can be made. **2.** To connect in parallel. **3.** See parallel. [MATH] The product of a number or quantity by an integer. [MET] A piece of stock cut from bar for use in a forging which provides the exact length needed for a single workpiece. { 'məl·tə·pəl }

multiple access [COMMUN] Multiplexing schemes by which multiple users who are geographically dispersed gain access to a shared telecommunications facility or channel. { ‚məl·tə·pəl 'ak‚ses }

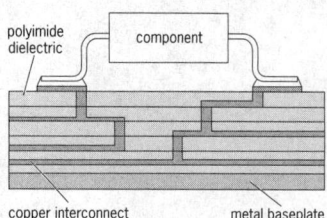

MULTILAYER BOARD

polyimide dielectric — component
copper interconnect — metal baseplate

Surface mounting of a component on a multilayer board fabricated by using additive polyimide-copper process.

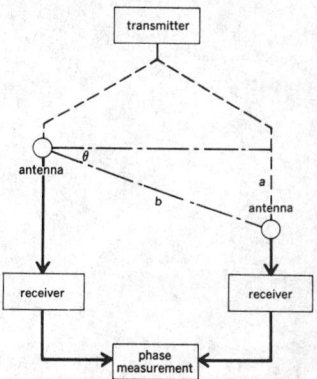

MULTIPATH SYSTEM

transmitter
antenna — θ — a — antenna
b
receiver — receiver
phase measurement

Multipath system. Transmission paths (broken lines) are assumed parallel for long distances. Here b = base line between the two antennas. Path from transmitter to one antenna is larger by length a than path to second antennna, causing phase difference $n\theta$ between signals received at two receivers.

multiple-access computer [COMPUT SCI] A computer system whose facilities can be made available to a number of users at essentially the same time, normally through terminals, which are often physically far removed from the central computer and which typically communicate with it over telephone lines. { 'məl·tə·pəl ¦ak‚ses kəm‚pyüd·ər }

multiple-access network [COMPUT SCI] A computer network that permits every computer on it to communicate with the network at any time during operation. Also known as multi-access network. { 'məl·tə·pəl ¦ak‚ses 'net‚wərk }

multiple-access service [COMMUN] One of the services of the Tracking and Data Relay Satellite System, which provides simultaneous return-link service from as many as 20 low-earth-orbiting user spacecraft, with data rates up to 3 megabits per second for each user, and a time-shared forward-link service to the user spacecraft with a maximum data rate of 300 kilobits per second, one user at a time. Abbreviated MA service. { ‚məl·tə·pəl 'ak‚ses ‚sər·vəs }

multiple accumulating registers [COMPUT SCI] Special registers capable of handling factors larger than one computer word in length. { 'məl·tə·pəl ə'kyü·myə‚lād·iŋ 'rej·ə‚stərz }

multiple-activity process chart [IND ENG] A chart showing the coordinated synchronous or simultaneous activities of a work system comprising one or more machines or individuals; separate, parallel columns indicate each machine's or person's activities as related to the other parts of the work system. { 'məl·tə·pəl ak¦tiv·əd·ē 'prä·səs ‚chärt }

multiple-address code [COMPUT SCI] A computer instruction code in which more than one address or storage location is specified; the instruction may give the locations of the operands, the destination of the result, and the location of the next instruction. { 'məl·tə·pəl ¦a‚dres ‚kōd }

multiple-address computer [COMPUT SCI] A computer whose instruction contains more than one address, for example, an operation code and three addresses A, B, C, such that the content of A is multiplied by the content of B and the product stored in location C. { 'məl·tə·pəl ¦a‚dres kəm‚pyüd·ər }

multiple-address instruction [COMPUT SCI] An instruction which has more than one address in a computer; the addresses give locations of other instructions, or of data or instructions that are to be operated upon. { 'məl·tə·pəl ¦a‚dres in‚strək‚shən }

multiple-anomaly syndrome [MED] Any syndrome associated with numerous congenital abnormalities. { 'məl·tə·pəl ə¦näm·ə·lē ‚sin‚drōm }

multiple appearance [ELEC] Jack arrangement in telephone switchboards whereby a single-line circuit appears before two or more operators. { 'məl·tə·pəl ə'pir·əns }

multiple-arch dam [CIV ENG] A dam composed of a series of arches inclined at about 45° and carried on parallel buttresses or piers. { 'məl·tə·pəl ¦ärch 'dam }

multiple-beam antenna [ELECTROMAG] An antenna or antenna array which radiates several beams in different directions. { 'məl·tə·pəl ¦bēm an'ten·ə }

multiple-beam interference [OPTICS] Interference which arises when part of a beam is reflected several times back and forth between a pair of strongly reflecting surfaces before being reflected or transmitted from the pair. { 'məl·tə·pəl ¦bēm ‚in·tər'fir·əns }

multiple-beam interferometer [OPTICS] An interferometer in which a beam is reflected several times back and forth between a pair of parallel plane surfaces; examples are the Fizeau interferometer and the Fabry-Perot interferometer. { 'məl·tə·pəl ¦bēm ‚in'tər·fə'räm·əd·ər }

multiple call transmission [COMMUN] Routing and transmitting to two or more stations are done by the operator switching the message, cross-office, to the multiple address processing unit for routing line segregation. { 'məl·tə·pəl ¦kōl tranz 'mish·ən }

multiple cancellous exostoses See multiple hereditary exostoses. { 'məl·tə·pəl 'kan·sə·ləs ‚ek·səs'tō·sēz }

multiple cartilaginous exostoses See multiple hereditary exostoses. { 'məl·tə·pəl ‚kard·əl'aj·ə·nəs ‚ek·səs'tō·sēz }

multiple cartridges [CHEM ENG] Filter medium made up of two or more filter cartridges, either fastened end to end or arranged side by side (in series or parallel flow respectively). { 'məl·tə·pəl 'kär·trə·jəz }

multiple coefficient of determination [STAT] A statistic that measures the proportion of total variation which is explained by the regression line; computed by taking the square root of the coefficient of multiple correlation. { ¦məl·tə·pəl ‚kō·ə¦fish·ənt əv di‚tər·mə'nā·shən }

multiple colloid goiter See adenomatous goiter. { ¦məl·tə·pəl 'kä‚lòid ‚gòid·ər }

multiple-completion packer [PETRO ENG] A device used to provide a seal between the outside surfaces of the two or more parallel tubing strings in a multiple-completion well, and the inside surface of the common wellbore casing. { 'məl·tə·pəl kəm¦plesh·ən ‚pak·ər }

multiple-completion well [PETRO ENG] Oil well in which there is production from more than one oil-bearing zone (different depths) with parallel tubing strings within a single wellbore casing string. Also known as multicompletion well. { 'məl·tə·pəl kəm¦plesh·ən ‚wel }

multiple computer operation [COMPUT SCI] The utilization of any one computer of a group of computers by means of linkages provided by multiplexor channels, all computers being linked through their channels or files. { 'məl·tə·pəl kəm¦pyüd·ər ‚äp·ə‚rā·shən }

multiple connector [ENG] A flow chart symbol that indicates the merging of several flow lines into one line or the dispersal of a flow line into several lines. { 'məl·tə·pəl kə'nek·tər }

multiple-contact switch See selector switch. { 'məl·tə·pəl ¦kän‚takt ‚swich }

multiple cropping [AGR] A system of growing several food crops on the same field in one year. { 'məl·tə·pəl 'kräp·iŋ }

multiple-current hypothesis [OCEANOGR] The hypothesis that the Gulf Stream, instead of being composed of a single tortuous current, actually consists of many quasipermanent currents, countercurrents, and eddies. { 'məl·tə·pəl ¦kə·rənt hī‚päth·ə·səs }

multiple decay See branching. { 'məl·tə·pəl di'kā }

multiple discharge See composite flash. { 'məl·tə·pəl 'dis‚chärj }

multiple disintegration See branching. { 'məl·tə·pəl di‚sin·tə'grā·shən }

multiple edges See parallel edges. { ‚məl·tə·pəl 'ej·əs }

multiple-effect evaporation [CHEM ENG] Series-operation energy economizer system in which heat from the steam generated (evaporated liquid) in the first stage is used to evaporate additional liquid in the second stage (by reducing system pressure), and so on, up to 10 or more effects; commonly used in the pulp and paper industry. { 'məl·tə·pəl i¦fekt i‚vap·ə'rā·shən }

multiple-effect evaporator [CHEM ENG] An evaporation system in which a series of evaporator bodies are connected so that the vapors from one body act as a heat source for the next body. { 'məl·tə·pəl i¦fkt i'vap·ə‚rād·ər }

multiple-entry system [MIN ENG] A system of access or development openings generally in bituminous coal mines involving more than one pair of parallel entries, one for haulage and fresh-air intake and the other for return air. { 'məl·tə·pəl 'en·trē ‚sis·təm }

multiple epidermis [BOT] Epidermis that is several layers thick, occurring in many species of Ficus, Begonia, and Peperomia. { 'məl·tə·pəl ‚ep·ə'dər·məs }

multiple fabric [TEXT] Material made of two or more fabrics bound together as they are woven. { 'məl·tə·pəl 'fab·rik }

multiple-factor incentive plan [IND ENG] A wage incentive plan based on productivity and other factors such as yield, material usage, and reduction of scrap. { 'məl·tə·pəl ¦fak·tər in'sen·tiv ‚plan }

multiple fault See step fault. { 'məl·tə·pəl 'fòlt }

multiple firing [ENG] Electrically firing with delay blasting caps in a number of holes at one time. { 'məl·tə·pəl 'fīr·iŋ }

multiple-frequency-shift keying [COMMUN] A modulation scheme in which a number of carrier frequencies (2, 4, 8, and so forth) are transmitted according to a group of consecutive data bits (n bits producing 2^n frequencies). Abbreviated MFSK. { 'məl·tə·pəl ¦frē·kwən·sē ¦shift ‚kē·iŋ }

multiple fruit [BOT] Any fruit derived from the ovaries and accessory structures of several flowers consolidated into one mass, such as a pineapple and mulberry. { 'məl·tə·pəl 'früt }

multiple-function chip See large-scale integrated circuit. { 'məl·tə·pəl ¦fəŋk·shən ‚chip }

multiple gun [ORD] Any group of guns mounted in one position and fired as a unit. { 'məl·tə·pəl 'gən }

multiple hereditary exostoses [MED] An inherited form

of exostosis, revealing itself at several sites in childhood or adolescence. Also known as diaphyseal aclasis; hereditary deforming chondrodysplasia; metaphyseal aclasis; multiple cancellous exostoses; multiple cartilaginous exostoses. { 'məl·tə·pəl hə'red·ə‚ter·ē ‚ek·səs'tō·sēz }

multiple hydraulic pump [PETRO ENG] Oil-well pump arrangement by which a single pump can be used in alternating operation to lift oil from two producing zones within a single multiple completion well. { 'məl·tə·pəl hī'dró·lik 'pəmp }

multiple-impulse welding [MET] Spot, upset, or projection welding in which more than one current impulse is generated during a single machine cycle. Also known as pulsation welding. { 'məl·tə·pəl ‚im‚pəls ‚weld·iŋ }

multiple independently targeted reentry vehicle [ORD] A type of intercontinental ballistics missile which carries several nuclear warheads; before the missile reenters the atmosphere, the warheads separate and follow different trajectories to various targets. Abbreviated MIRV. { 'məl·tə·pəl ‚in·də‚pen·dənt·lē 'tär·gəd·əd rē'en·trē ‚vē·ə·kəl }

multiple infarct dementia See cerebral arteriosclerosis. { 'məl·tə·pəl ‚in‚färkt di'men·chə }

multiple-instruction-stream, multiple-data-stream See MIMD. { ‚məl·tə·pəl in'strək·shən ‚strēm ‚məl·tə·pəl 'dad·ə ‚strēm }

multiple integral [MATH] An integral over a subset of n-dimensional space. { 'məl·tə·pəl 'int·ə·grəl }

multiple isomorphous replacement technique [CRYSTAL] A technique for overcoming the phase problem by growing crystals in three different isomorphic chemical forms and comparing x-ray diffraction data obtained from all three. Abbreviated MIR technique. { ‚məl·tə·pəl ‚ī·sə‚mór·fəs ri'plās·mənt tek‚nēk }

multiple jacks [ELEC] Series of jacks with tip, ring, and sleeve, respectively connected in parallel, and appearing in different panels of the face equipment of a telephone exchange. { 'məl·tə·pəl 'jaks }

multiple-junction solar cell See multijunction solar cell. { ‚məl·tə·pəl 'jəŋk·shən ‚sō·lər ‚sel }

multiple-key access [COMPUT SCI] A technique for locating stored data in a computer system by using the values contained in two or more separate key fields. { 'məl·tə·pəl 'kē 'ak‚ses }

multiple-keyboard point-of-sale system [COMPUT SCI] A point-of-sale system consisting of a group of electronic machines, without programming capability, placed at all check points and linked either to one central data collector with a magnetic tape or to a minicomputer with disk storage. { 'məl·tə·pəl ‚kē‚bórd 'póint əv 'sāl ‚sis·təm }

multiple lamp holder [ELEC] A device that can be inserted in a lamp holder to act as two or more lamp holders. Also known as current tap. { 'məl·tə·pəl 'lamp ‚hōl·dər }

multiple-length arithmetic [COMPUT SCI] Arithmetic performed by a computer in which two or more machine words are used to represent each number in the calculations, usually to achieve higher precision in the result. { 'məl·tə·pəl ‚leŋkth ə'rith·mə‚tik }

multiple-length number [COMPUT SCI] A number having two or more times as many digits as are ordinarily used in a given computer. { 'məl·tə·pəl ‚leŋkth 'nəm·bər }

multiple-length working [COMPUT SCI] Any processing of data by a computer in which two or more machine words are used to represent each data item. { 'məl·tə·pəl ‚leŋkth 'wərk·iŋ }

multiple linear correlation [STAT] An index for estimating the strength of the linear relationship between one dependent variable and two or more independent variables. { 'məl·tə·pəl ‚lin·ē·ər ‚kär·ə'lā·shən }

multiple linear regression [STAT] A technique for determining the linear relationship between one dependent variable and two or more independent variables. { 'məl·tə·pəl ‚lin·ē·ər ri'gresh·ən }

multiple-loop system [CONT SYS] A system whose block diagram has at least two closed paths, along each of which all arrows point in the same direction. { 'məl·tə·pəl 'lüp ‚sis·təm }

multiple midstop [MECH ENG] A peripheral device that allows a pick-and-place robot to swing and stop in several positions. { 'məl·tə·pəl 'mid‚stäp }

multiple-mirror telescope [OPTICS] A type of optical telescope in which images from several complete conventional telescopes that are mounted rigidly on a common frame and coaligned by an active laser and computer system are brought to a common focus by a mirror system. { 'məl·tə·pəl ‚mir·ər 'tel·ə ‚skōp }

multiple modulation [COMMUN] A succession of modulating processes in which the modulated wave from one process becomes the modulating wave for the next. Also known as compound modulation. { 'məl·tə·pəl ‚mäj·ə'lā·shən }

multiple module access [COMPUT SCI] Device which establishes priorities in storage access in a multiple computer environment. { 'məl·tə·pəl ‚mäj·yül ‚ak‚ses }

multiple myeloma [MED] A primary bone malignancy characterized by diffuse osteoporosis, anemia, hyperglobulinemia, and other clinical features. Also known as Kahler's disease. { 'məl·tə·pəl ‚mī·ə'lō·mə }

multiple neuritis See polyneuritis. { 'məl·tə·pəl nü'rīd·əs }

multiple neurofibroma See neurofibromatosis. { 'məl·tə·pəl ‚nür·ō·fī'brō·mə }

multiple neurofibromatosis See neurofibromatosis. { 'məl·tə·pəl ‚nür·ō·fī‚brō·mə'tō·səs }

multiple openings [MIN ENG] Any series of underground openings separated by rib pillars or connected at frequent intervals to form a system of rooms and pillars. { 'məl·tə·pəl 'ō·pə·niŋz }

multiple-parallel-tubing string [PETRO ENG] Two or more parallel and closely packed oil-well tubing strings used in multiple-completion wells. Also known as multiple-tubing string. { 'məl·tə·pəl 'par·ə‚lel ‚tüb·iŋ ‚striŋ }

multiple-pass weld [MET] A weld made by depositing filler metal with two or more passes in succession. { 'məl·tə·pəl ‚pas ‚weld }

multiple personality [PSYCH] A personality capable of dissociation into several or many other personalities at the same time, whereby the delusion is entertained that the one person is many separate persons; a symptom in schizophrenic patients. { 'məl·tə·pəl ‚pər·sə'nal·əd·ē }

multiple piece rate plan [IND ENG] A wage incentive plan wherein increasingly higher unit pay rates are given to the worker as his productivity increases. { 'məl·tə·pəl 'pēs ‚rāt ‚plan }

multiple point [MATH] A point of a curve through which passes more than one arc of the curve. { 'məl·tə·pəl 'póint }

multiple precision arithmetic [COMPUT SCI] Method of increasing the precision of a result by increasing the length of the number to encompass two or more computer words in length. { 'məl·tə·pəl prə'sizh·ən ə'rith·mə‚tik }

multiple pregnancy [MED] Being pregnant with more than one fetus. { 'məl·tə·pəl 'preg·nən·sē }

multiple programming [COMPUT SCI] The execution of two or more operations simultaneously. { 'məl·tə·pəl 'prō‚gram·iŋ }

multiple-purpose tester See volt-ohm-milliammeter. { 'məl·tə·pəl 'pər·pəs 'tes·tər }

multiple reflection [GEOPHYS] A seismic wave which has more than one reflection. Also known as repeated reflection; secondary reflection. [OPTICS] Reflection of light back and forth several times between a pair of strongly reflecting surfaces. { 'məl·tə·pəl ri'flek·shən }

multiple reflection echoes [ELECTROMAG] Radar echoes returned from a real target by reflection from some object in the radar beam; such echoes appear at a false bearing and false range. { 'məl·tə·pəl ri'flek·shən 'ek‚ōz }

multiple resonance [ELEC] Two or more resonances at different frequencies in a circuit consisting of two or more coupled circuits which are resonant at slightly different frequencies. [QUANT MECH] Two or more resonances at slightly different energies, resulting from the splitting of a single resonance by an interaction that is relatively weak. { 'məl·tə·pəl 'rez·ən·əns }

multiple root [MATH] A polynomial $f(x)$ has c as a multiple root if $(x - c)^n$ is a factor for some $n > 1$. Also known as repeated root. { 'məl·tə·pəl 'rüt }

multiple-row blasting [ENG] The drilling, charging, and firing of rows of vertical boreholes. { 'məl·tə·pəl 'rō 'blast·iŋ }

multiple sampling [IND ENG] A plan for quality control in which a given number of samples from a group are inspected, and the group is either accepted, resampled, or rejected, depending on the number of failures found in the samples. { 'məl·tə·pəl 'sam·pliŋ }

multiple scattering [PHYS] Process in which a particle undergoes a large number of collisions, and the total change in its momentum is the sum of the many small changes occurring during individual collisions. { 'məl·tə·pəl 'skad·ə·riŋ }

multiple sclerosis [MED] A degenerative disease of the nervous system of unknown cause in which there is demyelination followed by gliosis. { 'məl·tə·pəl sklə'rō·səs }

multiple series [ENG] A method of wiring a large group of blasting charges by connecting small groups in series and connecting these series in parallel. Also known as parallel series. { 'məl·tə·pəl 'sir·ēz }

multiple serositis See polyserositis. { 'məl·tə·pəl ,sir·ō'sīd·əs }

multiple shooting [ENG] The firing of an entire face at one time by means of connecting shot holes in a single series and shooting all holes at the same instant. { 'məl·tə·pəl 'shüd·iŋ }

multiple-shot survey [PETRO ENG] The determining and recording of drill-hole direction. { 'məl·tə·pəl ¦shät 'sər,vā }

multiple sleep latency test [MED] A test used to document pathologic sleepiness and diagnose narcolepsy in which recordings of brain waves, muscle activities, and eye movements are taken while a person spends the day in a sleep laboratory, taking naps at intervals. { ¦məl·tə·pəl 'slēp ,lāt·ən·sē ,test }

multiple-slide press [MECH ENG] A press with individual adjustable slides built into the main slide or connected independently to the main shaft. { 'məl·tə·pəl ¦slīd 'pres }

multiple spot welding [MET] Spot welding in which several spots are welded during a single machine cycle. { 'məl·tə·pəl 'spät ,weld·iŋ }

multiple-stage rocket See multistage rocket. { 'məl·tə·pəl ¦stāj 'räk·ət }

multiple-stage separator [PETRO ENG] Oilwell gas-oil separator in which wellhead pressure is reduced in several stages, with the flashing off of gas at each pressure reduction. { 'məl·tə·pəl ¦stāj 'sep·ə,rād·ər }

multiple star [ASTRON] A system of three or more stars which appear to the naked eye as a single star. { 'məl·tə·pəl 'stär }

multiple-strand conveyor [MECH ENG] A conveyor with two or more spaced strands of chain, belts, or cords as the supporting or propelling medium. { 'məl·tə·pəl ¦strand kən'vā·ər }

multiple stratification [STAT] Division of a population into two or more parts with respect to two or more variables. { 'məl·tə·pəl ,strad·ə·fə'kā·shən }

multiple switchboard [ELEC] Manual telephone switchboard in which each subscriber line is attached to two or more jacks, to be within reach of several operators. { 'məl·tə·pəl 'swich,bȯrd }

multiplet [QUANT MECH] A collection of relatively closely spaced energy levels which result from the splitting of a single energy level by an interaction which is relatively weak; examples are spin-orbit multiplets and isospin multiplets. [SPECT] A collection of relatively closely spaced spectral lines resulting from transitions to or from the members of a multiplet (as in the quantum-mechanics definition). { 'məl·tə·plət }

multiple target generator [ELECTR] An electronic countermeasures device that produces several false responses in a hostile radar set. { 'məl·tə·pəl 'tär·gət ¦jen·ə,rād·ər }

multiplet intensity rules [SPECT] Rules for the relative intensities of spectral lines in a spin-orbit multiplet, stating that the sum of the intensities of all lines which start from a common initial level, or end on a common final level, is proportional to $2J+1$, where J is the total angular momentum of the initial level or final level respectively. { 'məl·tə·plət in'ten·səd·ē ,rülz }

multiple tropopause [GEOPHYS] A frequent condition in which the tropopause appears not as a continuous single "surface" of discontinuity between the troposphere and the stratosphere, but as a series of quasi-horizontal "leaves," which are partly overlapping in steplike arrangement. { 'məl·tə·pəl 'trō·pə,pȯz }

multiple-tubing string See multiple-parallel-tubing string. { 'məl·tə·pəl 'tüb·iŋ ,striŋ }

multiple-tuned antenna [ELECTROMAG] Low-frequency antenna having a horizontal section with a multiplicity of tuned vertical sections. { 'məl·tə·pəl ¦tünd an'ten·ə }

multiple twin quad [ELEC] Quad cable in which the four conductors are arranged in two twisted pairs, and the two pairs twisted together. { 'məl·tə·pəl 'twin 'kwäd }

multiple-unit semiconductor device [ELECTR] Semiconductor device having two or more seats of electrodes associated with independent carrier streams. { 'məl·tə·pəl ¦yü·nət 'sem·i·kən,dək·tər di,vīs }

multiple-unit steerable antenna See musa. { 'məl·tə·pəl ¦yü·nət 'stir·ə·bəl an'ten·ə }

multiple-unit tube See multiunit tube. { 'məl·tə·pəl ¦yü·nət 'tüb }

multiple-valued [MATH] A relation between sets is multiple-valued if it associates to an element of one more than one element from the other; sometimes functions are allowed to be multiple-valued. { 'məl·tə·pəl 'val,yüd }

multiple-valued logic [MATH] A form of logic in which statements can have values other than the two values "true" and "false." { 'məl·tə·pəl ¦val,yüd 'läj·ik }

multiple winding [ELEC] A winding composed of several circuits connected in parallel. { 'məl·tə·pəl 'wīnd·iŋ }

multiplex [ENG] Stereoscopic device to project aerial photographs onto surfaces so that the images may be viewed in three dimensions by using anaglyphic spectacles; used to prepare topographic maps. { 'məl·tə,pleks }

multiplexer [ELECTR] A device for combining two or more signals, as for multiplex, or for creating the composite color video signal from its components in color television. Also spelled multiplexor. { 'məl·tə,plek·sər }

multiplex holography [OPTICS] A technique in which a rotating cylindrical hologram is illuminated by a tungsten-filament light bulb on the axis of the cylinder; as the cylinder rotates, or as the viewer moves around, three-dimensional images are seen inside the cylinder. { 'məl·tə,pleks hō'läg·rə·fē }

multiplexing [COMMUN] 1. A set of techniques that enable the sharing of the usable electromagnetic spectrum of a telecommunications channel (the channel pass-band) among multiple users for the transfer of individual information streams. 2. In particular, the case in which the user information streams join at a common access point to the channel. { 'məl·tə,pleks·iŋ }

multiplex mode [COMPUT SCI] The utilization of differences in operating speeds between a computer and transmission lines; the multiplexor channel scans each line in sequence, and any transmitted pulse on a line is assembled in an area reserved for this line; consequently, a number of users can be handled by the computer simultaneously. Also known as multiplexor channel operation. { 'məl·tə,pleks ,mōd }

multiplex operation [COMMUN] Simultaneous transmission of two or more messages in either or both directions over a carrier channel. { 'məl·tə,pleks ,äp·ə'rā·shən }

multiplexor See multiplexer. { 'məl·tə,plek·sər }

multiplexor channel operation See multiplex mode. { 'məl·tə,plek·sər ¦chan·əl ,äp·ə,rā·shən }

multiplexor terminal unit [COMPUT SCI] Device which permits a large number of data transmission lines to access a single computer. { 'məl·tə,plek·sər 'ter·mən·əl ,yü·nət }

multiplex transmission [COMMUN] The simultaneous transmission of two or more programs or signals over a single radio-frequency channel, such as by time division, frequency division, code division, or phase division. { 'məl·tə,pleks tranz'mish·ən }

multiple x-y recorder [ENG] Recorder that plots a number of independent charts simultaneously, each showing the relation of two variables, neither of which is time. { 'məl·tə·pəl ¦eks'wī ri,kȯrd·ər }

multiple zone [PETRO ENG] Two or more discrete oil or gas reservoirs in the same geographic area, but at different depths. { 'məl·tə·pəl ,zōn }

multiplicand [MATH] If a number x is to be multiplied by a number y, then x is called the multiplicand. { ,məl·tə·pli'kand }

multiplication [ELECTR] An increase in current flow through a semiconductor because of increased carrier activity. [MATH] Any algebraic operation analogous to multiplication of real numbers. [NUCLEO] The ratio of neutron flux in a subcritical reactor to that supplied by a neutron source; it is the factor by which, in effect, the reactor multiplies the source strength. { ,məl·tə·pli'kā·shən }

multiplication constant *See* multiplication factor. { ˌməl·tə·pliˈkā·shən ˈkän·stənt }

multiplication factor [NUCLEO] The ratio of the number of neutrons present in a reactor in any one neutron generation to that in the immediately preceding generation. Also known as multiplication constant; neutron multiplication factor. { ˌməl·tə·pliˈkā·shən ˌfak·tər }

multiplication formula [MATH] An equation that expresses a function of a multiple of a quantity in terms of functions of the quantity itself and possibly functions of other multiples of the quantity. { ˌməl·tə·pləˈkā·shən ˌfȯr·myə·lə }

multiplication on the left *See* premultiplication. { ˌməl·tə·pliˈkā·shən ȯn thə ˈleft }

multiplication on the right *See* postmultiplication. { ˌməl·tə·pliˈkā·shən ȯn thə ˈrīt }

multiplication sign [MATH] The symbol × or ·, used to indicate multiplication. Also known as times sign. { ˌməl·tə·pliˈkā·shən ˌsīn }

multiplication table [COMPUT SCI] In certain computers, a part of memory holding a table of numbers in which the computer looks up values in order to perform the multiplication operation. { ˌməl·tə·pliˈkā·shən ˌtā·bəl }

multiplication time [COMPUT SCI] The time required for a computer to perform a multiplication; for a binary number it will be equal to the total of all the addition times and all the shift times involved in the multiplication. { ˌməl·tə·pliˈkā·shən ˌtīm }

multiplicative acoustic array [ACOUS] An acoustic array of receiving elements which is divided into two parts, the signal voltages obtained from them being multiplied together. Also known as correlation array. { ˌməl·təˈplik·əd·iv əˈküs·tik əˈrā }

multiplicative congruential generator [COMPUT SCI] A congruential generator in which the constant *b* in the generating formula is equal to zero. { ˌməl·təˈplik·əd·iv ˌkän·grüˈen·chəl ˈjen·ə·rād·ər }

multiplicative identity [MATH] In a mathematical system with an operation of multiplication, denoted ×, an element 1 such that $1 \times e = e \times 1 = e$ for any element *e* in the system. { ˌməl·təˈplik·əd·iv īˈden·əd·ē }

multiplicative inverse [MATH] In a mathematical system with an operation of multiplication, denoted ×, the multiplicative inverse of an element *e* is an element $\bar{e}$ such that $e \times \bar{e} = \bar{e} \times e = 1$, where 1 is the multiplicative identity. { ˌməl·təˈplik·əd·iv ˈin·vərs }

multiplicative number-theoretic function [MATH] A number theoretic function, *f*, which has the properties that *mn* is in its range whenever *m* and *n* are, and that $f(mn) = f(m)f(n)$ whenever *m* and *n* are relatively prime. { ˌməl·təˈplik·əd·iv ˌnəm·bər ˌthē·əˈred·ik ˈfəŋk·shən }

multiplicative subset [MATH] A subset *S* of a commutative ring such that if *x* and *y* are in *S* then so is *xy*. { ˌməl·təˈplik·əd·iv ˈsəb·set }

multiplicity [MATH] **1.** A root of a polynomial $f(x)$ has multiplicity *n* if $(x − a)^n$ is a factor of $f(x)$ and *n* is the largest possible integer for which this is true. **2.** The geometric multiplicity of an eigenvalue λ of a linear transformation *T* is the dimension of the null space of the transformation $T − \lambda I$, where *I* denotes the identity transformation. **3.** The algebraic multiplicity of an eigenvalue λ of a linear transformation *T* on a finite-dimensional vector space is the multiplicity of λ as a root of the characteristic polynomial of *T*. [PHYS] In a system having Russell-Saunders coupling, the quantity $2S+1$, where *S* is the total spin quantum number. { ˌməl·təˈplis·əd·ē }

multiplier [ELEC] **1.** A resistor used in series with a voltmeter to increase the voltage range. Also known as multiplier resistor. [ELECTR] A device that has two or more inputs and an output that is a representation of the product of the quantities represented by the input signals; voltages are the quantities commonly multiplied. **2.** *See* electron multiplier. **3.** *See* frequency multiplier. [MATH] If a number *x* is to be multiplied by a number *y*, then *y* is called the multiplier. { ˈməl·təˌplī·ər }

multiplier field [COMPUT SCI] The area reserved for a multiplication, equal to the length of multiplier plus multiplicand plus one character. { ˈməl·təˌplī·ər ˌfēld }

multiplier phototube [ELECTR] A phototube with one or more dynodes between its photocathode and the output electrode; the electron stream from the photocathode is reflected off each dynode in turn, with secondary emission adding electrons to the stream at each reflection. Also known as electron-multiplier phototube; photoelectric electron-multiplier tube; photomultiplier; photomultiplier tube. { ˈməl·təˌplī·ər ˈfōd·ō̇ˌtüb }

multiplier-quotient register [COMPUT SCI] A register equal to two words in length in which the quotient is developed and in which the multiplier is entered for multiplication. { ˈməl·təˌplī·ər ˈkwō·shənt ˌrej·ə·stər }

multiplier resistor *See* multiplier. { ˈməl·təˌplī·ər riˌzis·tər }

multiplier traveling-wave photodiode [ELECTR] Photodiode in which the construction of a traveling-wave tube is combined with that of a multiplier phototube to give increased sensitivity. { ˈməl·təˌplī·ər ˈtrav·ə·liŋ ˈwāv ˌfōd·ōˈdī̇·ōd }

multiplier tube [ELECTR] Vacuum tube using secondary emission from a number of electrodes in sequence to obtain increased output current; the electron stream is reflected, in turn, from one electrode of the multiplier to the next. { ˈməl·təˌplī·ər ˌtüb }

multipling [COMMUN] Use of multidrop lines to provide for changes in telephone service patterns or requirements; unused terminals afford convenient access to wiretappers. { ˈməl·təˌpliŋ }

multiply connected region [MATH] An open set in the plane which has holes in it. { ˈməl·tə·plē kəˌnek·təd ˈrē·jən }

multiply defined symbol [COMPUT SCI] Common assembler or compiler error printout indicating that a label has been used more than once. { ˈməl·tə·plē diˌfīnd ˈsim·bəl }

multiply perfect number [MATH] An integer such that the sum of all its factors is a multiple of the integer itself. { ˈməl·tə·plē ˌpər·fikt ˈnəm·bər }

multipoint line [COMMUN] A line which is shared by two or more different tributary stations. { ˈməl·təˌpȯint ˌlīn }

multipolar [ELECTROMAG] Having more than one pair of magnetic poles. { ˈməl·təˌpō·lər }

multipolar machine [ELECTROMAG] An electric machine that has a field magnet with more than one pair of poles. { ˌməl·təˈpō·lər məˈshēn }

multipole [ELECTROMAG] One of a series of types of static or oscillating distributions of charge or magnetization; namely, an electric multipole or a magnetic multipole. { ˈməl·təˌpōl }

multipole fields [ELECTROMAG] The electric and magnetic fields generated by static or oscillating electric or magnetic multipoles. { ˈməl·təˌpōl ˌfēlz }

multipole radiation [PHYS] **1.** Electromagnetic radiation which has characteristics equivalent to those of radiation generated by an oscillating electric or magnetic multipole, and is made up of photons of well-defined angular momentum and parity. **2.** Internal conversion electrons, or positron-electron pairs having similar characteristics, emitted from an atom when the nucleus makes a transition between two energy states. { ˈməl·təˌpōl ˌrād·ē·āˈshən }

multipole transition [PHYS] A transition between two energy states of an atom or nucleus in which a quantum of multipole radiation is emitted or absorbed. { ˈməl·təˌpōl tranˈzish·ən }

multiport burner [ENG] A burner having several nozzles which discharge fuel and air. { ˈməl·təˌpȯrt ˈbər·nər }

multiport memory [COMPUT SCI] A memory shared by many processors to communicate among themselves. { ˈməl·təˌpȯrt ˈmem·rē }

multiport network analyzer [ENG] A linear, passive microwave network having five or more ports which is used for measuring power and the complex reflection coefficient in a microwave circuit. Also known as multiport reflectometer. { ˈməl·təˌpȯrt ˌnet·wərk ˈan·əˌlīz·ər }

multiport reflectometer *See* multiport network analyzer. { ˈməl·təˌpȯrt rēˌflekˈtäm·əd·ər }

multipotent cell [EMBRYO] A cell capable of giving rise to only a limited number of cell types. { ˌməl·təˌpōt·ent ˈsel }

multiprecision arithmetic [COMPUT SCI] A form of arithmetic similar to double precision arithmetic except that two or more words may be used to represent each number. { ˈməl·tə·prə sizh·ən əˈrith·mə·tik }

multiprocessing [COMPUT SCI] Carrying out of two or more sequences of instructions at the same time in a computer. { ˌməl·təˈprä·ses·iŋ }

multiprocessing system *See* multiprocessor. { ˌməl·təˈprä·ses·iŋ ˌsis·təm }

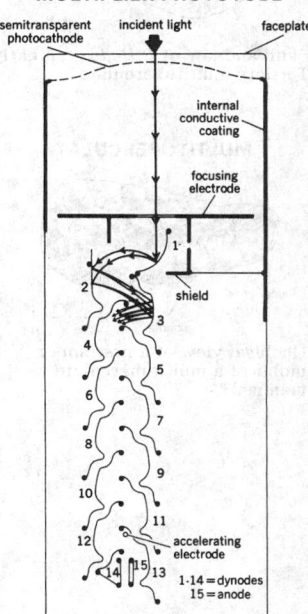

MULTIPLIER PHOTOTUBE

semitransparent photocathode — incident light — faceplate

internal conductive coating

focusing electrode

shield

accelerating electrode

1-14 = dynodes
15 = anode

Typical multiplier phototube construction.

multiprocessor [COMPUT SCI] A data-processing system that can carry out more than one program, or more than one arithmetic operation, at the same time. Also known as multiprocessing system. { ‚məl·tə'prä‚ses·ər }

multiprocessor interleaving [COMPUT SCI] Technique used to speed up processing time; by splitting banks of memory each with x microseconds access time and accessing each one in sequence 1/n-th of a cycle later, a reference to memory can be had every x/n microseconds; this speed is achieved at the cost of hardware complexity. { ‚məl·tə'prä‚ses·ər ‚in·tər'lēv·iŋ }

multiprogramming [COMPUT SCI] The interleaved execution of two or more programs by a computer, in which the central processing unit executes a few instructions from each program in succession. { ‚məl·tə'prō‚gram·iŋ }

multiprogramming executive control [COMPUT SCI] Control program structure required to handle multiprogramming with either a fixed or a variable number of tasks. { ‚məl·tə'prō‚gram·iŋ ig'zek·yəd·iv kən'trōl }

multipropellant [AERO ENG] A rocket propellant consisting of two or more substances that are fed separately to the combustion chamber. { ¦məl·tə·prə'pel·ənt }

multipurpose projectile [ORD] A projectile designed so that the type of payload can be changed; this is accomplished by using prepared loads in canister form and providing a removable base plug to permit change of canister; thus a canister containing colored smoke mixture can be replaced by, for instance, one containing leaflets. { ¦məl·tə'pər·pəs prə'jek·təl }

multiring structure [ASTRON] A formation on the moon's surface consisting of two or more craters within a larger crater. { 'məl·tə‚riŋ ‚strək·chər }

multirole programmable device [CONT SYS] A device that contains a programmable memory to store data on positioning robots and sequencing their motion. { 'məl·tə‚rōl prō¦gram·ə·bəl di'vīs }

multirope friction winder [MECH ENG] A winding system in which the drive to the winding ropes is the frictional resistance between the ropes and the driving sheaves. { 'məl·tə‚rōp 'frik·shən ‚wīn·dər }

multisegment magnetron [ELECTR] Magnetron with an anode divided into more than two segments, usually by slots parallel to its axis. { 'məl·tə‚seg·mənt 'mag·nə‚trän }

multispeed motor [ELEC] An induction motor that can rotate at any one of two or more speeds, independent of the load. { 'məl·tə‚spēd 'mōd·ər }

multistable circuit [ELECTR] A circuit having two or more stable operating conditions. { 'məl·tə‚stā·bəl 'sər·kət }

multistage [ENG] Functioning or occurring in separate steps. { 'məl·tē‚stāj }

multistage amplifier See cascade amplifier. { 'məl·tē‚stāj 'am·plə‚fī·ər }

multistage compressor [MECH ENG] A machine for compressing a gaseous fluid in a sequence of stages, with or without intercooling between stages. { 'məl·tē‚stāj kəm'pres·ər }

multistage pump [MECH ENG] A pump in which the head is developed by multiple impellers operating in series. { 'məl·tē‚stāj 'pəmp }

multistage queuing [IND ENG] A situation involving two or more sequential stages in a process, each of which involves waiting in line. { 'məl·tē‚stāj 'kyü·iŋ }

multistage rocket [AERO ENG] A vehicle having two or more rocket units, each unit firing after the one in back of it has exhausted its propellant; normally, each unit, or stage, is jettisoned after completing its firing. Also known as multiple-stage rocket; step rocket. { 'məl·tē‚stāj 'räk·ət }

multistage sampling [STAT] A sampling method in which the population is divided into a number of groups or primary stages from which samples are drawn; these are then divided into groups or secondary stages from which samples are drawn, and so on. { 'məl·tə‚stāj 'sam·pliŋ }

multistatic radar [ENG] Radar in which successive antenna lobes are sequentially engaged to provide a tracking capability without physical movement of the antenna. { 'məl·tē‚stad·ik 'rā‚där }

multistation [COMMUN] Pertaining to a network in which each station can communicate with each of the other stations. { 'məl·tē‚stā·shən }

multistator watt-hour meter [ELEC] An induction type of watt-hour meter in which several stators exert a torque on the rotor. { 'məl·tē‚stād·ər 'wat ‚aùr ‚mēd·ər }

multistrip coupler [ELECTR] A series of parallel metallic strips placed on a surface acoustic wave filter between identical apodized interdigital transducers; it converts the spatially non-uniform surface acoustic wave generated by one transducer into a spatially uniform wave received at the other transducer, and helps to reject spurious bulk acoustic modes. { 'məl·tə‚strip 'kəp·lər }

multisync monitor [COMPUT SCI] A video display monitor that automatically adjusts to the synchronization frequency of the video source from which it is receiving signals. { ¦məl·ti‚siŋk 'män·ə·tər }

multisystem coupling [COMPUT SCI] The electronic connection of two or more computers in proximity to make them act as a single logical machine. { 'məl·tə‚sis·təm 'kəp·liŋ }

multisystem network [COMPUT SCI] A data communications network that has two or more host computers with which the various terminals in the system can communicate. { 'məl·tə‚sis·təm 'net‚wərk }

multitape Turing machine [COMPUT SCI] A variation of a Turing machine in which more than one tape is permitted, each tape having its own read-write head. { 'məl·tē‚tāp 'tür·iŋ mə‚shēn }

multitasking [COMPUT SCI] The simultaneous execution of two or more programs by a single central processing unit. { ¦məl·tē'task·iŋ }

multitask operation [COMPUT SCI] A sophisticated form of multijob operation in a computer which allows a single copy of a program module to be used for more than one task. { ¦məl·tē'task ‚äp·ə'rā·shən }

multithreading [COMPUT SCI] A processing technique that allows two or more of the same type of transaction to be carried out simultaneously. { ¦məl·tə'thred·iŋ }

multitrack operation [COMPUT SCI] The selection of the next read/write head in a cylinder, usually indicated by bit zero of the operation code in the channel command word. { ¦məl·tē'trak ‚äp·ə'rā·shən }

multitrack recording system [ENG] Recording system which provides two or more recording paths on a medium, which may carry either related or unrelated recordings in common time relationship. { ¦məl·tē'trak ri'kòrd·iŋ ‚sis·təm }

Multituberculata [PALEON] The single order of the nominally mammalian suborder Allotheria; multituberculates had enlarged incisors, the coracoid bones were fused to the scapula, and the lower jaw consisted of the dentary bone alone. { ‚məl·tē·tə‚bər·kyə'läd·ə }

multituberculate [VERT ZOO] Of teeth, having several or many simple conical cusps. { ‚məl·tē·tə‚bər·kyə'lāt }

multiturn potentiometer [ELEC] A precision wire-wound potentiometer in which the resistance element is formed into a helix, generally having from 2 to 10 turns. { 'məl·tē‚tərn pə‚ten·chē'äm·əd·ər }

multiunit tube [ELECTR] Electron tube containing within one glass or metal envelope, two or more groups of electrodes, each associated with separate electron streams. Also known as multiple-unit tube. { ¦məl·tē¦yü·nət 'tüb }

multiuser system [COMPUT SCI] A computer system with multiple terminals, enabling several users, each at their own terminal, to use the computer. { ¦məl·tē¦yü·zər 'sis·təm }

multivalent See polyvalent. { ¦məl·tə'vā·lənt }

multivariable system [CONT SYS] A dynamical system in which the number of either inputs or outputs is greater than 1. { ¦məl·tē'ver·ē·ə·bəl ‚sis·təm }

multivariate analysis [STAT] The study of random variables which are multidimensional. { ¦məl·tē'ver·ē·ət ə'nal·ə·səs }

multivariate distribution [MATH] For two or more random variables, $X_1, X_2, \ldots$, and X_n, the distribution which gives the probability that $X_1 = x_1, X_2 = x_2, \ldots$, and $X_n = x_n$ for all values, $x_1, x_2, \ldots$, and x_n, of $X_1, X_2, \ldots$, and X_n respectively. { ¦məl·tē'ver·ē·ət ‚dis·trə'byü·shən }

multivator [ELEC] Type of dc-to-dc up-converter. [ELECTR] An automatic device for analyzing a number of dust samples that might be collected by spacecraft on the moon, Mars, and other planets, to detect the presence of microscopic organisms with a multiplier phototube that measures the fluorescence given off. { 'məl·tə‚vād·ər }

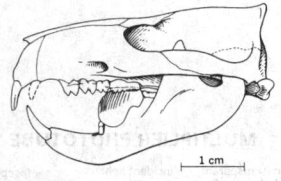

MULTITUBERCULATA

Skull and jaw of *Ptilodus*, an early Tertiary multituberculate.

MULTITUBERCULATE

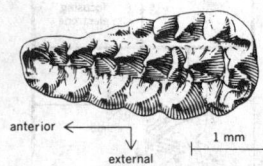

anterior
external
1 mm

Occlusal view of a first upper molar of a multituberculate mammal.

multivibrator [ELECTR] A relaxation oscillator using two tubes, transistors, or other electron devices, with the output of each coupled to the input of the other through resistance-capacitance elements or other elements to obtain in-phase feedback voltage. { ˌməl·tə'vī¦brād·ər }

multivolume file [COMPUT SCI] A file that consists of more than one physical unit of storage medium. { ¦məl·tē¦väl·yəm 'fīl }

multiwavelength anomalous dispersion [CRYSTAL] A technique for overcoming the phase problem by collecting x-ray diffraction data at several wavelengths around the absorption edge of a strongly absorbing atom. Abbreviated MAD. { ˌməl·tē¦wāv·leŋkth ə·näm·ə·ləs di'spərzh·ən }

multiway merge [COMPUT SCI] A computer operation in which three or more lists are merged into a single list. { 'məl·tē·wā 'mərj }

multiwell gas-lift system [PETRO ENG] An installation to allow gas-lift production from the various tubing strings involved in a multiple completion well. { 'məl·tē·wel 'gas ˌlift ˌsis·təm }

mu meson See muon. { 'myü ¦mā·sän }

Mumetal [MET] An alloy of high magnetic permeability, containing 14% iron, 5% copper, 1.5% chromium, and the balance nickel. { 'myü·med·əl }

mummification [MED] **1.** Drying of a part of the body into a hard mass. **2.** Dry gangrene. { ˌməm·ə·fə'kā·shən }

mumps [MED] An acute contagious viral disease characterized chiefly by painful enlargement of a parotid gland. { məmps }

mumps orchitis [MED] Inflammation of the testis due to the mumps virus. { 'məmps ȯr'kīd·əs }

Munchausen syndrome [PSYCH] A personality disorder in which the patient describes dramatic but false symptoms or simulates acute illness, happily undergoing examinations, hospitalization, and diagnostic and therapeutic manipulations, and upon discovery of the real nature of his case often leaves without notice and moves on to another hospital. { 'mùn,chauz·ən ˌsin,drōm }

mundic See pyrite. { 'mən,dik }

municipal engineering [CIV ENG] Branch of engineering dealing with the form and functions of urban areas. { myü'nis·ə·pəl ˌen·jə'nir·iŋ }

munition [ORD] **1.** In a broad sense, any and all supplies and equipment required to conduct offensive or defensive war, including war machines, ammunition, transport, fuel, food, and clothing, but excluding personnel and excluding supplies and equipment for purposes other than for direct military operations. **2.** In a restricted sense, ordnance. { myü'nish·ən }

Munsell chroma See chroma. { 'mən'sel ˌkrō·mə }

Munsell color system [OPTICS] A system for designating colors which employs three perceptually uniform scales (Munsell hue, Munsell value, Munsell chroma) defined in terms of daylight reflectance. { 'mən'sel 'kəl·ər ˌsis·təm }

Munsell hue [OPTICS] The dimension of the Munsell system of color that determines whether a color is blue, green, yellow, red, purple, or the like, without regard to its lightness or saturation. { 'mən'sel ˌhyü }

Munsell value [OPTICS] The dimension, in the Munsell system of object-color specification, that indicates the apparent luminous transmittance or reflectance of the object on a scale having approximately equal perceptual steps under the usual conditions of observation. { 'mən'sel ˌval,yü }

muntin See sash bar. { 'mənt·ən }

Muntz metal [MET] A 60/40 type of brass composed of 58–61% copper, up to 1% lead, and remainder zinc. Also known as malleable brass; yellow metal. { 'məns ˌmed·əl }

muon [PARTIC PHYS] Collective name for two semistable elementary particles with positive and negative charge, designated μ^+ and μ^- respectively, which are leptons and have a spin of $^1/_2$ and a mass of approximately 105.7 MeV. Also known as mu meson. { 'myü,än }

muon-catalyzed fusion [NUC PHYS] Nuclear fusion that occurs quickly at normal temperatures when the reacting nuclei are bound in an exotic molecule, containing in addition a positively charged muon. { ˌmyü,än ¦kad·ə¦līzd 'fyü·zhən }

muonic atom [PARTIC PHYS] An atom in which an electron is replaced by a negatively charged muon orbiting close to or within the nucleus. { myü'än·ik ¦ad·əm }

muonium [PARTIC PHYS] An atom consisting of an electron bound to a positively charged muon by their mutual Coulomb attraction, just as an electron is bound to a proton in the hydrogen atom. { myü'ō·nē·əm }

muon lepton number [PARTIC PHYS] The number of muons and muon-associated neutrinos minus the number of antimuons and muon-associated antineutrinos; it is conserved in all known interactions but may not be exactly conserved. { 'myü,än 'lep·tän ˌnəm·bər }

muon spin relaxation [PHYS] A technique for studying various phenomena in solids and liquids and chemical reactions of muonium atoms, in which a beam of polarized muons is focused on a sample and the loss of polarization of muons in the sample is monitored by observing the spatial anisotropy of electrons or positrons emitted in the muon decay. Also known as muon spin rotation; muon spin resonance. { 'myü,än ¦spin ri,lak'sā·shən }

muon spin resonance See muon spin relaxation. { 'myü,än ¦spin 'rez·ən·əns }

muon spin rotation See muon spin relaxation. { 'myü,än ¦spin rō tā·shən }

mu phage [VIROL] A temperate phage with properties similar to those of transposable genetic elements. { 'myü ˌfāj }

mural thrombus [MED] A thrombus attached to the wall of a blood vessel or mural endocardium. Also known as lateral thrombus. { 'myùr·əl 'thräm·bəs }

muramic acid [BIOCHEM] An organic acid found in the mucopeptide (murein) in the cell walls of bacteria and blue-green algae. { myù'ram·ik 'as·əd }

muramidase [BIOCHEM] Lysozyme when acting as an enzyme on the hydrolysis of the muramic acid-containing mucopeptide in the cell walls of some bacteria. { myù'ram·ə,dās }

murbruk structure See mortar structure. { 'mər,brùk ˌstrək·chər }

Murchisoniacea [PALEON] An extinct superfamily of gastropod mollusks in the order Prosobranchia. { ¦mər·chə,sən·ē'ā·shē·ə }

murein [BIOCHEM] The peptidoglycan of bacterial cell walls. { myùr·ē·ən }

muriatic acid See hydrochloric acid. { myùr·ē'ad·ik 'as·əd }

Muricacea [INV ZOO] A superfamily of gastropod mollusks in the order Prosobranchia. { ˌmyùr·ə'ka·shē·ə }

muricate [ZOO] Covered with sharp, hard points. { 'myùr·ə,kāt }

Muricidae [INV ZOO] A family of predatory gastropod mollusks in the order Neogastropoda; contains the rock snails. { ˌmyù'ris·ə,dē }

Muridae [VERT ZOO] A large diverse family of relatively small cosmopolitan rodents; distinguished from closely related forms by the absence of cheek pouches. { 'myùr·ə,dē }

muriform [BIOL] **1.** Resembling the arrangement of courses in a brick wall, especially having both horizontal and vertical septa. **2.** Pertaining to or resembling a rat or mouse. { 'myùr·ə,fȯrm }

Murinae [VERT ZOO] A subfamily of the Muridae which contains such forms as the striped mouse, house mouse, harvest mouse, and field mouse. { myù'rī,nē }

murine plague [VET MED] Infection of the rat by the bacterium *Pasteurella pestis*; transmitted from rat to rat and from rat to human by a flea. { 'myù,rīn ¦plāg }

murine typhus [MED] A relatively mild, acute, febrile illness of worldwide distribution caused by *Rickettsia mooseri*, transmitted from rats to humans by the flea and characterized by headache, macular rash, and myalgia. Also known as endemic typhus; flea-borne typhus; rat typhus; shop typhus; urban typhus. { 'myù,rīn 'tī·fəs }

murmur [MED] A blowing or roaring heart sound heard through the wall of the chest; caused by blood flow through a defective valve. { 'mər·mər }

muromontite [MINERAL] $Be_2FeY_2(SiO_4)_3$ A mineral composed of yttrium iron beryllium silicate. { ˌmyùr·ə'män,tīt }

Murphree efficiency [CHEM ENG] In a plate-distillation column, the ratio of the actual change in vapor composition when the vapor passes through the liquid on a tray (plate) to the composition change of the vapor if it were in vapor-liquid equilibrium with the tray liquid. { 'mər·frē i'fish·ən·sē }

Murray code [COMMUN] A binary code with five binary

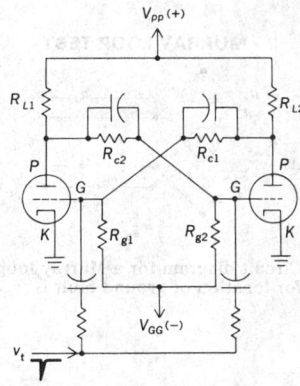

MULTIVIBRATOR

Circuit diagram of a bistable multivibrator.

MURRAY LOOP TEST

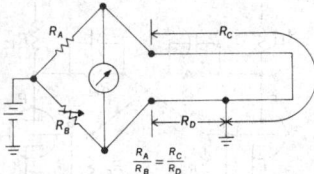

Circuit diagram for a Murray loop for location of ground fault.

MUSHROOM

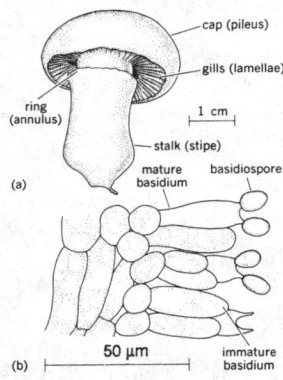

Agaricus bisporus, the common cultivated mushroom. (*a*) Basidiocarp. (*b*) Cross section of gill, showing basidia in various stages of development.

digits per letter which was developed to be used with a typewriterlike device which would punch holes in paper tape, and is now the basis of the widely used CCIT 2 code. { 'mər·ē ,kōd }

Murray loop test [ELEC] A method of localizing a fault in a cable by replacing two arms of a Wheatstone bridge with a loop formed by the cable under test and a good cable connected to the far end of the defective cable. { 'mər·ē 'lüp ,test }

Murray Valley encephalitis [MED] An acute inflammation of the brain and spinal cord caused by a virus; confined to Australia and New Guinea. Also known as Australian X disease. { 'mər·ē ,val·ē in,sef·ə'līd·əs }

murumuru oil [MATER] An oil obtained from nuts of the palm *Attalea orbignya,* containing 40% lauric acid, 35% myristic acid, and some palmitic, stearic, linoleic, and oleic acids. { mə¦rüm·ə¦rü ,ȯil }

musa [ELECTROMAG] An electrically steerable receiving antenna whose directional pattern can be rotated by varying the phases of the contributions of the individual units. Derived from multiple-unit steerable antenna. { 'myü·sə }

Musaceae [BOT] A family of monocotyledonous plants in the order Zingiberales characterized by five functional stamens, unisexual flowers, spirally arranged leaves and bracts, and fleshy, indehiscent fruit. { myü'zā·sē,ē }

Musca [ASTRON] A southern constellation, right ascension 12 hours, declination 70°S. Also known as Fly. { 'məs·kə }

muscardine diseases [INV ZOO] A group of insect diseases caused by the muscardine fungi, in which the fungal pathogen emerges from the body of the insect and covers the animal with a characteristic fungus mat. { 'məs·kər,dēn di,zēz·əs }

muscarine [ORG CHEM] $C_8H_{19}NO_3$ A quaternary ammonium compound, the toxic ingredient of certain mushrooms, as *Amanita muscaria.* Also known as hydroxycholine. { 'məs·kə,rēn }

muscarinism [MED] Poisoning due to ingestion of certain mushrooms. { məs'kar·ə,niz·əm }

Muschelkalk [GEOL] A European stage of geologic time equivalent to the Middle Triassic, above Bunter and below Keuper. { 'müsh·əl,kälk }

Musci *See* Bryopsida. { 'mə,sī }

Muscicapidae [VERT ZOO] A family of passeriform birds assigned to the Oscines; includes the Old World flycatchers or fantails. { ,məs·ə'kap·ə,dē }

Muscidae [INV ZOO] A family of myodarian cyclorrhaphous dipteran insects in the subsection Calypteratae; includes the houseflies, stable flies, and allies. { 'məs·ə,dē }

muscle [ANAT] A contractile organ composed of muscle tissue that changes in length and effects movement when stimulated. [HISTOL] A tissue composed of cells containing contractile fibers; three types are smooth, cardiac, and skeletal. { 'məs·əl }

muscle-contraction headache [MED] A type of headache characterized by dull constricting pain that can either occur intermittently or continue for days, months, or years. Also known as a tension headache. { ¦məs·əl kən,trak·shən ,hed,āk }

muscle fiber [HISTOL] The contractile cell or unit of which muscle is composed. { 'məs·əl ,fī·bər }

muscle hemoglobin *See* myoglobin. { 'məs·əl 'hē·mə,glō·bən }

muscle tone *See* tonus. { 'məs·əl ,tōn }

muscovite [MINERAL] $KAl_2(AlSi_3)O_{10}(OH)_2$ One of the mica group of minerals, occurring in some granites and abundant in pegmatites; it is colorless, whitish, or pale brown, and the crystals are tabular sheets with prominent base and hexagonal or rhomboid outline; hardness is 2–2.5 on Mohs scale, and specific gravity is 2.7–3.1. Also known as common mica; mirror stone; moscovite; Muscovy glass; potash mica; white mica. { 'məs·kə,vīt }

Muscovy glass *See* muscovite. { 'məs·kə·vē 'glas }

muscul-, musculo- [ZOO] A combining form denoting muscle, muscular. { 'məs·kə·lō }

muscular atrophy [MED] Degenerative reduction of muscle size, especially skeletal muscles, due to a lesion involving either the cell body or axon of the lower motor neuron. { 'məs·kyə·lər 'a·trə·fē }

muscular dystrophy [MED] A group of diseases characterized by degeneration of or injury to individual muscle cells, not primarily involving the nerve supply; the most common

form is Duchenne-Greisinger disease. { 'məs·kyə·lər 'dis·trə·fē }

muscularis externa [HISTOL] The layer of the digestive tube consisting of smooth muscles. { ,məs·kyə'lar·əs ek'stər·nə }

muscularis mucosae [HISTOL] Thin, deep layer of smooth muscle in some mucous membranes, as in the digestive tract. { ,məs·kyə'lar·əs myü'kō·sē }

muscular system [ANAT] The muscle cells, tissues, and organs that effect movement in all vertebrates. { 'məs·kyə·lər ,sis·təm }

musculoaponeurotic [HISTOL] Composed of muscle and of fibrous connective tissue in the form of a membrane. { ¦məs·kyə·lō¦ap·ə·nù¦räd·ik }

musculocutaneous [ANAT] Of or pertaining to muscles and skin. { ¦məs·kyə·lō·kyü'tā·nē·əs }

musculocutaneous nerve [NEUROSCI] A branch of the brachial plexus with both motor and somatic sensory components; innervates flexor muscles of the upper arm, and skin of the lateral aspect of the forearm. { ¦məs·kyə·lō·kyü'tā·nē·əs 'nərv }

musculoskeletal system [ANAT] The muscular and skeletal elements of vertebrates, considered as a functional unit. { ¦məs·kyə·lō'skel·ə·təl ,sis·təm }

musculoskeletal toxicity [MED] Adverse effects to the structure and/or function of the muscles, bones, and joints caused by exposure to a toxic chemical, such as coal dust or cadmium. Also, the bone disorders arthritis, fluorosis, and osteomalacia can result. { ,məs·kyə·lō¦skel·ət·əl ,tak'sis·əd·ē }

mushroom [MYCOL] **1.** A fungus belonging to the basidiomycetous order Agaricales. **2.** The fruiting body (basidiocarp) of such a fungus. { 'məsh,rüm }

mushroom anchor [NAV ARCH] A large anchor having the approximate shape of a mushroom, capable of grasping the ground whichever way it falls; used mainly where the bottom is sandy or muddy for permanent moorings of light ships and moorings of modern submarines. { 'məsh,rüm ,aŋ·kər }

musical acoustics [ACOUS] That part of acoustics which is relevant to the composition, performance, and appreciation of music, including the physical characteristics of sounds that may be heard as music, laws governing the action, design, and construction of musical instruments, and the effects of musical sounds upon listeners. { 'myü·zə·kəl ə'kü·stiks }

musical echo [ACOUS] A musical tone produced by the reflection of an impulsive sound from a stepped structure such as a picket fence, when reflections from successive steps reach the observer with suitable frequency. { 'myü·zə·kəl 'ek·ō }

musical instrument digital interface [COMPUT SCI] **1.** The digital standard for connecting computers, musical instruments, and synthesizers. **2.** A compression format for encoding music. Abbreviated MIDI. { ,myü·zi·kəl ¦in·strə·mənt ,dij·ə·dəl 'in·tər,fās }

musical quality *See* timbre. { 'myü·zə·kəl 'kwäl·əd·ē }

music wire [MET] High-quality, high-carbon steel wire used for making mechanical springs. { 'myü·zik ,wīr }

Musidoridae [INV ZOO] A family of orthorrhaphous dipteran insects in the series Brachycera distinguished by spear-shaped wings. { ,myü·zə'dȯr·ə,dē }

musk [PHYSIO] Any of various strong-smelling substances obtained from the musk glands of musk deer or similar animals; used in the form of a tincture as a fixative for perfume. { məsk }

musk ambrette [ORG CHEM] $C_{12}H_{16}N_2O_5$ White to yellow powder with heavy musky aroma; soluble in various oils and phthalates, insoluble in water; congeals at 83°C; used as a perfume fixative. Also known as 2,6-dinitro-3-methoxy-4-tert-butyltoluene. { 'məsk ,am,bret }

Muskat equation [PETRO ENG] Equation used to calculate oil reservoir permeability from the pressure buildup curve recorded when a producing well is shut in after a flow test. { 'məs,kat i,kwā·zhən }

musk bag *See* musk gland. { 'məsk ,bag }

muskeg [ECOL] A peat bog or tussock meadow, with variably woody vegetation. { 'mə,skeg }

musk gland [VERT ZOO] A large preputial scent gland of the musk deer and various other animals, including skunk and musk-ox. Also known as musk bag. { 'məsk ,gland }

Muskhelishvili's method [MECH] A method of solving

problems concerning the elastic deformation of a planar body that involves using methods from the theory of functions of a complex variable to calculate analytic functions which determine the plane strain of the body. { mə'skel·ish,vil·ēz ,meth·əd }

musk ketone [ORG CHEM] $C_{14}H_{18}N_2O_5$ White to yellow crystals with sweet musk aroma; soluble in various oils and phthalates, insoluble in water; used as a perfume fixative. Also known as 3,5-dinitro-2,6-dimethyl-4-tert-butylacetophenone. { 'məsk 'kē,tōn }

muskmelon [BOT] *Cucumis melo.* The edible, fleshy, globular to long-tapered fruit of a trailing annual plant of the order Violales; surface is uniform to broadly sutured to wrinkled, and smooth to heavily netted, and flesh is pale green to orange; varieties include cantaloupe, Honey Dew, Casaba, and Persian melons. { 'məsk,mel·ən }

musk-ox [VERT ZOO] *Ovibos moschatus.* An even-toed ungulate which is a member of the mammalian family Bovidae; a heavy-set animal with a shag pilage, splayed feet, and flattened horns set low on the head. { 'məs,käks }

muskrat [VERT ZOO] *Ondatra zibethica.* The largest member of the rodent subfamily Microtinae; essentially a water rat with a laterally flattened, long, naked tail, a broad blunt head with short ears, and short limbs. { 'mə,skrat }

musk xylene See musk xylol. { 'məsk 'zī,lēn }

musk xylol [ORG CHEM] $(NO_2)_3C_6(CH_3)_2C(CH_3)_3$ White to yellow crystals with powerful musk aroma; soluble in various oils and phthalates, insoluble in water; congeals at 105°C; used as a perfume fixative. Also known as musk xylene; 2,4,6-trinitro-1,3-dimethyl-5-tert-butylbenzene. { 'məsk 'zī,lōl }

muslin [TEXT] A thin to coarse-textured plain-weave cotton. { 'məz·lən }

Musophagidae [VERT ZOO] The turacos, an African family of birds of uncertain affinities usually included in the order Cuculiformes; resemble the cuckoos anatomically but have two unique pigments, turacin and turacoverdin. { ,myü·zə'faj·ə,dē }

Muspiceoidea [INV ZOO] A superfamily of parasitic nematodes in the order Dioctophymatida, distinguished by a greatly reduced neurosensory structure, the absence of (except in one species) amphids and cephalic papillae, and, in females, a reduced digestive tube; males have never been reported. { myüs,pī·sē'oid·ē·ə }

mustard [BOT] Any of several annual crucifers belonging to the genus *Brassica* of the order Capparales; leaves are lyrately lobed, flowers are yellow, and pods have linear beaks; the mustards are cultivated for their pungent seed and edible foliage, and the seeds of *B. niger* are used as a condiment, prepared as a powder, paste, or oil. { 'məs·tərd }

mustard gas [ORG CHEM] $HS(CH_2ClCH_2)_2S$ An oil with density 1.28, boiling point 215°C; used in chemical warfare. Also known as dichlorodiethylsulfide. { 'məs·tərd ,gas }

mustard oil See allyl isothiocyanate. { 'məs·tərd ,öil }

mustard-seed coal [GEOL] Anthracite that will pass through circular holes in a screen which measure 3/64 inch (1.2 millimeter) in diameter. { 'məs·tərd,sēd ,kōl }

Mustilidae [VERT ZOO] A large, diverse family of low-slung, long-bodied carnivorous mammals including minks, weasels, and badgers; distinguished by having only one molar in each upper jaw, and two at the most in the lower jaw. { mə'stil·ə,dē }

mutable gene [GEN] Any of a class of unstable genes that spontaneously mutate at a sufficiently high rate to produce mosaicism. { 'myüd·ə·bəl 'jēn }

mutagen [GEN] An agent that raises the frequency of mutation above the spontaneous or background rate. { 'myüd·ə·jən }

mutagen persistence [GEN] The stability of a mutagen in the environment or in the human body. { 'myüd·ə·jən pər'sis·təns }

mutagen specificity [GEN] The tendency of a mutagen to induce only one type of mutation. { 'myüd·ə·jən ,spes·ə'fis·əd·ē }

mutant [GEN] An individual bearing an allele that has undergone mutation and is expressed in the phenotype. { 'myüt·ənt }

mutarotation [CHEM] A change in the optical rotation of light that takes place in the solutions of freshly prepared sugars. { ,myüd·ə·rō'tā·shən }

mutase [BIOCHEM] An enzyme able to catalyze a dismutation or a molecular rearrangement. { 'myü,tās }

mutation [GEN] An abrupt change in the genotype of an organism, not resulting from recombination; genetic material may undergo qualitative or quantitative alteration, or rearrangement. { myü'tā·shən }

mutation fixation [MOL BIO] The condition of changing a premutational deoxyribonucleic acid lesion to mutation. { myü'tā·shən fik,sā·shən }

mutation hot spot [MOL BIO] Any locus in the deoxyribonucleic acid sequence or on a chromosome where mutations or aberrations occur preferentially. { myü,tā·shən 'hät ,spät }

mutator phenotype [GEN] The loss of function of one gene, such as one for the repair of damaged deoxyribonucleic acid, that greatly increases the mutation rates at other loci. { 'myü,tād·ər 'fēn·ə,tīp }

muthmannite [MINERAL] (Ag,Au)Te A bright brass yellow mineral consisting of silver-gold telluride; occurs as tabular crystals. { 'müt·mə,nīt }

Mutillidae [INV ZOO] The velvet ants, a family of hymenopteran insects in the superfamily Scolioidea. { myü'til·ə,dē }

muting circuit [ELECTR] 1. Circuit which cuts off the output of a receiver when no radio-frequency carrier greater than a predetermined intensity is reaching the first detector. 2. Circuit for making a receiver insensitive during operation of its associated transmitter. { 'myüd·iŋ ,sər·kət }

muting switch [ELEC] 1. A switch used in connection with automatic tuning systems to silence the receiver while tuning from one station to another. 2. A switch used to ground the output of a phonograph pickup automatically while a record changer is in its change cycle. { 'myüd·iŋ ,swich }

mutism [MED] Inability or refusal to speak. { 'myü,tiz·əm }

muton [MOL BIO] The smallest unit of genetic material capable of undergoing mutation. { 'myü,tän }

mutual admittance [ELEC] For two meshes of a network carrying alternating current, the ratio of the complex current in one mesh to the complex voltage in the other, when the voltage in all meshes besides these two is 0. { 'myü·chə·wəl ad'mit·əns }

mutual branch See common branch. { 'myü·chə·wəl 'branch }

mutual capacitance [ELEC] The accumulation of charge on the surfaces of conductors of each of two circuits per unit of potential difference between the circuits. { 'myü·chə·wəl kə'pas·əd·əns }

mutual conductance See transconductance. { 'myü·chə·wəl kən'dək·təns }

mutual deadlock [COMPUT SCI] A condition in which deadlocked tasks are awaiting resource assignments, and each task on a list awaits release of a resource held by the following task, with the last task awaiting release of a resource held by the first task. Also known as circular wait. { 'myü·chə·wəl 'ded,läk }

mutual exclusion rule [PHYS CHEM] The rule that if a molecule has a center of symmetry, then no transition is allowed in both its Raman scattering and infrared emission (and absorption), but only in one or the other. { 'myü·chə·wəl ik'sklü·zhən ,rül }

mutual impedance [ELEC] For two meshes of a network carrying alternating current, the ratio of the complex voltage in one mesh to the complex current in the other, when all meshes besides the latter one carry no current. { 'myü·chə·wəl im'pēd·əns }

mutual inductance [ELECTROMAG] Property of two neighboring circuits, equal to the ratio of the electromotive force induced in one circuit to the rate of change of current in the other circuit. { 'myü·chə·wəl in'dək·təns }

mutual induction [ELECTROMAG] The generation of a voltage in one circuit by a varying current in another. { 'myü·chə·wəl in'dək·shən }

mutual interference [COMMUN] Interference from two or more electrical or electronic systems which affects these systems on a reciprocal basis. { 'myü·chə·wəl ,in·tər'fir·əns }

mutualism [ECOL] Mutual interactions between two species that are beneficial to both species. { 'myü·chə·wə,liz·əm }

mutuality of phases [CHEM] The rule that if two phases, with respect to a reaction, are in equilibrium with a third phase at a certain temperature, then they are in equilibrium with

The musk-ox (*Ovibos moschatus*).

The muskrat (*Ondatra zibethica*).

Drawing of a velvet ant. *(From T. I. Storer and R. L. Usinger, General Zoology, 3d ed., McGraw-Hill, 1957)*

respect to each other at that temperature. { ‚myü·chə'wal·əd·ē əv 'făz·əz }

mutually exclusive events [STAT] Two or more events such that the occurrence of any one makes impossible the occurrence of any of the others. { ¦myü·chə·lē ik¦sklü·siv i'vens }

muzzle [ORD] The end of the barrel of a gun from which the bullet or projectile emerges. [VERT ZOO] The snout of an animal, as a dog or horse. { 'məz·əl }

muzzle energy [ORD] Kinetic energy of the projectile as it emerges from the muzzle; it is a measure of the power of the weapon. { 'məz·əl ‚en·ər·jē }

mV *See* millivolt.

MV *See* megavolt.

MVD *See* mechanical vapor diffusion.

mW *See* milliwatt.

MW *See* megawatt.

MW(E) *See* megawatt electric.

MW(Th) *See* megawatt thermal.

MWYE *See* megawatt year of electricity.

Mx *See* maxwell.

myalgia [MED] Pain in the muscles. { mī'al·jē·ə }

myasthenia [MED] Muscular weakness. { ‚mī·əs·thēn·ē·ə }

myasthenia gravis [MED] A muscle disorder of unknown etiology characterized by varying degrees of weakness and excessive fatigability of voluntary muscle. { ‚mī·əs·thēn·ē·ə 'grav·əs }

myasthenia reaction [MED] The electromyographic reaction observed in myasthenia gravis in which there is a gradual loss of intensity and duration for the tetanic contraction, and a gradual diminution in amplitude and frequency of motor unit discharges until the muscle is fatigued. { ‚mī·əs·thēn·ē·ə ri'ak·shən }

myasthenic crisis [MED] Profound myasthenia and respiratory paralysis associated with myasthenia gravis. { ¦mī·əs¦thēn·ik 'krī·səs }

myatonia [MED] Lack of muscle tone. { ‚mī·ə'tō·nē·ə }

Mycelia Sterilia [MYCOL] An order of fungi of the class Fungi Imperfecti distinguished by the lack of spores; certain members are plant pathogens. { mī'sel·yə stə'ril·yə }

mycelium [BIOL] A mass of filaments, or hyphae, composing the vegetative body of many fungi and some bacteria. { mī'sē·lē·əm }

Mycetaeidae [INV ZOO] The equivalent name for Endomychidae. { ‚mī·sə'tē·ə‚dē }

mycetocyte [INV ZOO] **1.** One of the cells clustered together to form a mycetome. **2.** An individual cell functioning like a mycetome. { mī'sēd·ə‚sīt }

mycetoma [MED] A chronic fungus or bacterial infection, usually of the feet, resulting in swelling. Also known as madura foot; maduromycosis. { ‚mī·sə'tō·mə }

mycetome [INV ZOO] One of the specialized structures in the body of certain insects for holding endosymbionts. { 'mī·sə‚tōm }

Mycetophagidae [INV ZOO] The hairy fungus beetles, a cosmopolitan family of coleopteran insects in the superfamily Cucujoidea. { mī‚sēd·ə'faj·ə‚dē }

Mycetozoa [BIOL] A zoological designation for organisms that exhibit both plant and animal characters during their life history (Myxomycetes); equivalent to the botanical Myxomycophyta. { mī‚sēd·ə'zō·ə }

Mycetozoia [INV ZOO] A subclass of the protozoan class Rhizopodea. { mī‚sēd·ə'zói·ə }

Mycobacteriaceae [MICROBIO] A family of bacteria in the order Actinomycetales; acid-fast, aerobic rods form a filamentous or myceliumlike growth. { ‚mī·kō‚bak·tir·ē'ās·ē‚ē }

mycobacterial disease [MED] Any disease caused by species of *Mycobacterium*. { ¦mī·kō·bak¦tir·ē·əl di‚zēz }

mycobactin [BIOCHEM] Any compound produced by some strains, and required for growth by other strains, of *Mycobacteria*. { ¦mī·kō¦bak·tən }

mycobiont [BOT] The fungal component of a lichen, commonly an ascomycete. { ¦mī·kə'bī‚änt }

mycology [BOT] The branch of botany that deals with the study of fungi. { mī'käl·ə·jē }

mycomycin [MICROBIO] $C_{13}H_{10}O_2$ An antibiotic produced by *Nocardia acidophilus* and a species of *Actinomyces*; characterized as a highly unsaturated aliphatic acid that shows strong

activity against *Mycobacterium tuberculosis*. { ‚mīk·ə'mīs·ən }

mycophagous [ZOO] Feeding on fungi. { mī'käf·ə·gəs }

Mycophiformes [VERT ZOO] An equivalent name for Salmoniformes. { mī‚käf·ə'fór·mēz }

Mycoplasmataceae [MICROBIO] A family of the order Mycoplasmatales; distinguished by sterol requirement for growth. { ‚mī·kō‚plaz·mə'tās·ē‚ē }

Mycoplasmatales [MICROBIO] The single order of the class Mollicutes; organisms are gram-negative, generally nonmotile, nonsporing bacteria which lack a true cell wall. { ‚mī·kō‚plaz·mə'tā·lēz }

mycorrhiza [BOT] A mutual association in which the mycelium of a fungus invades the roots of a seed plant. { ‚mīk·ə'rīz·ə }

mycorrhizal fungi [MYCOL] Fungi that form symbiotic relationships in and on the roots of host plants. { ‚mī·kə‚rīz·əl 'fən‚jī }

mycosis [MED] An infection with or a disease caused by a fungus. { mī'kō·səs }

mycosis fungoides [MED] A lymphoma of the skin, usually present in several sites when first diagnosed, that may remain confined to the skin for 10 or more years before eventually spreading to internal organs and causing death. { mī¦kōs·əs fəŋ'gói‚dēz }

Mycosphaerella sentina [MYCOL] A fungal plant pathogen that causes leaf blight of pears, a disease that destroys the leaves of pear trees. { ‚mī·kō·sfī‚rel·ə sen'tē·nə }

Mycota [MYCOL] An equivalent name for Eumycetes. { mī'käd·ə }

mycotic stomatitis *See* thrush. { mī'käd·ik ‚stō·mə'tīd·əs }

mycotoxicosis [MED] Any of a group of diseases caused by accidental or recreational ingestion of toxic fungal metabolites, such as mushroom poisoning. { ‚mī·kō‚täk·sə'kō·səs }

mycotoxin [MYCOL] A toxin produced by a fungus. { 'mī·kə‚täk·sən }

Myctophidae [VERT ZOO] The lantern fishes, a family of deep-sea forms of the suborder Myctophoidei. { mik'täf·ə‚dē }

Myctophoidei [VERT ZOO] A large suborder of marine salmoniform fishes characterized by having the upper jaw bordered only by premaxillae, and lacking a mesocoracoid arch in the pectoral girdle. { mik·tə'fóid·ē‚ī }

Mydaidae [INV ZOO] The mydas flies, a family of orthorrhaphous dipteran insects in the series Brachycera. { mī'dā·ə‚dē }

mydriasis [MED] Prolonged dilation of the pupil of the eye. { mə'drī·ə·səs }

mydriatic [PHARM] An agent which produces dilation of the pupil, such as eucatropine hydrochloride. { ¦mid·rē¦ad·ik }

myel-, myelo- [ANAT] A combining form indicating relationship to marrow, often in specific reference to the spinal cord. { 'mī·əl, 'mī·ə·lō }

myelencephalon [EMBRYO] The caudal portion of the hindbrain; gives rise to the medulla oblongata. { ‚mī·ə·lən'sef·ə‚län }

myelin [NEUROSCI] A soft, white fatty substance that forms a sheath around certain nerve fibers. { 'mī·ə·lən }

myelin sheath [NEUROSCI] An investing cover of myelin around the axis cylinder of certain nerve fibers. { 'mī·ə·lən 'shēth }

myelitis [MED] **1.** Inflammation of the spinal cord. **2.** Inflammation of the bone marrow. { ‚mī·ə'līd·əs }

myeloblast [HISTOL] The youngest precursor cell for blood granulocytes, having a nucleus with finely granular chromatin and nucleoli and intensely basophilic cytoplasm. { 'mī·ə·lō‚blast }

myeloblastemia [MED] The presence of myeloblasts in the peripheral circulation. { ‚mī·ə·lō·bla'stē·mē·ə }

myeloblastic [HISTOL] Of, pertaining to, or characterized by the presence of myeloblasts. { ‚mī·ə·lə'blas·tik }

myeloblastic leukemia *See* acute granulocytic leukemia. { ‚mī·ə·lə'blas·tik lü'kē·mē·ə }

myeloblastoma [MED] A malignant tumor composed of myeloblasts. { ‚mī·ə·lō·bla'stō·mə }

myeloblastosis [MED] Diffuse proliferation of myeloblasts, with involvement of blood, bone marrow, and other tissues and organs. { ‚mī·ə·lō·bla'stō·səs }

myelocele [ANAT] The canal of the spinal cord. [MED]

Spina bifida, with protrusion of the spinal cord. { 'mī·ə·lō,sēl }

myelocyte [HISTOL] A motile precursor cell of blood granulocytes found in bone marrow. { 'mī·ə·lə,sīt }

myelocytoma [MED] A malignant plasmacytoma. { ,mī·ə·lə·sī'tō·mə }

myelocytosis [MED] The presence of myelocytes in the blood. { ,mī·ə·lə·sī'tō·səs }

myelodysplasia [MED] Abnormal spinal cord development, especially the lumbosacral portion. { ,mī·ə·lō·dis'plā·zhə }

myeloencephalitis [MED] Inflammation of the brain and spinal cord. { ,mī·ə·lō·in,sef·ə'līd·əs }

myelofibrosis [PATH] Growth of white, fibrous connective tissue in the bone marrow. { ,mī·ə·lō·fī'brō·səs }

myelogenous leukemia See granulocytic leukemia. { ,mī·ə'läj·ə·nəs lü'kē·mē·ə }

myelogram [MED] Roentgenogram of the spinal cord, made by myelography. [PATH] Differential cell study of material extracted from bone marrow. { 'mī·ə·lə,gram }

myelography [MED] Roentgenographic visualization of the subarachnoid space, after the injection of air or an opaque medium. { ,mī·ə'läg·rə·fē }

myeloid [ANAT] **1.** Of or pertaining to bone marrow. **2.** Of or pertaining to the spinal cord. { 'mī·ə,lȯid }

myeloid cell [HISTOL] Any of the white blood cell (leukocyte) types that do not fall into the lymphocyte category. { 'mī·ə,lȯid ,sel }

myeloid leukemia See granulocytic leukemia. { 'mī·ə,lȯid lü'kē·mē·ə }

myeloid metaplasia [MED] The occurrence of hemopoietic tissue in abnormal places in the body. { 'mī·ə,lȯid ,med·ə'plā·zhə }

myeloid myeloma [MED] A malignant plasmacytoma. { 'mī·ə,lȯid ,mī·ə'lō·mə }

myeloid reaction [MED] Increased numbers of granulocytes in the bone marrow and peripheral circulation, often with the appearance of immature granulocytes in the blood. { 'mī·ə,lȯid rē,ak·shən }

myeloid tissue [HISTOL] Red bone marrow attached to argyrophile fibers which form wide meshes containing scattered fat cells, erythroblasts, myelocytes, and mature myeloid elements. { 'mī·ə,lȯid ,tish·ü }

myeloma [MED] A primary tumor of the bone marrow composed of any of the bone marrow cell types. { ,mī·ə'lō·mə }

myelomalacia [MED] Softening of the spinal cord. { ,mī·ə·lō'mā·shə }

myelomeningitis [MED] Inflammation of the spinal cord and its meninges. { ,mī·ə·lō,men·ən'jīd·əs }

myelomeningocele [MED] Spina bifida with protrusion of the spinal meninges. { ,mī·ə·lō·mə'niŋ·gō,sēl }

myelomonocyte [HISTOL] **1.** A monocyte developing in bone marrow. **2.** A blood cell intermediate between monocytes and granulocytes. { ,mī·ə·lō'män·ə,sīt }

myelomonocytic leukemia See monocytic leukemia. { ,mī·ə·lō,män·ə'sid·ik lü'kē·mē·ə }

myelopathic anemia See myelophthisic anemia. { ,mī·ə·lō,path·ik ə'nē·mē·ə }

myelophthisic anemia [MED] An anemia associated with space-occupying disorders of the bone marrow. Also known as leukoerythroblastic anemia; leukoerythroblastosis; metastatic anemia; myelopathic anemia; myelosclerotic anemia; osteosclerotic anemia. { ,mī·ə·lō'this·ik ə'nē·mē·ə }

myelophthisis [MED] **1.** Loss of bone marrow. **2.** Atrophy of the spinal cord. { ,mī·ə·lō'this·əs }

myeloplegia [MED] Spinal paralysis. { ,mī·ə·lə'plē·jē·ə }

myelopoiesis [PHYSIO] The process by which blood cells form in the bone marrow. { ,mī·ə·lō,pȯi'ē·səs }

myelosclerosis [MED] **1.** Multiple sclerosis of the spinal cord. **2.** Hardening of the bone marrow. { ,mī·ə·lō·sklə'rō·səs }

myelosclerotic anemia See myelophthisic anemia. { ,mī·ə·lō·sklə,räd·ik ə'nē·mē·ə }

myenteric [HISTOL] Of or pertaining to the muscular coat of the intestine. { ,mī·ən'ter·ik }

myenteric plexus [NEUROSCI] A network of nerves between the circular and longitudinal layers of the muscular coat of the digestive tract. Also known as Auerbach's plexus. { ,mī·ən'ter·ik 'plek·səs }

myenteron [HISTOL] The muscular coat of the intestine. { mī'ent·ə,rän }

Mygalomorphae [INV ZOO] A suborder of spiders (Araneida) including American tarantulas, trap-door spiders, and purse-web spiders; the tarantulas may attain a leg span of 10 inches (25 centimeters). { ,mig·ə·lō'mȯr,fē }

myiasis [MED] Infestation of vertebrates by the larvae, or maggots, of flies. { 'mī·ə·səs }

Myklestad method [AERO ENG] A method of determining the mode shapes and frequencies of the lateral bending modes of space vehicles, taking into account secondary effects of shear and rotary inertia, in which one imagines masses to be concentrated at a finite number of points along the beam, with elastic properties remaining constant between consecutive mass points. { 'mik·əl,stad ,meth·əd }

Mylabridae [INV ZOO] The equivalent name for Bruchidae. { mə'lab·rə,dē }

Myliogatidae [VERT ZOO] The eagle rays, a family of batoids which may reach a length of 15 feet (4.6 meters). { ,mil·ē·ə'gad·ə,dē }

mylonite [PETR] A hard, coherent, often glassy-looking rock that has suffered extreme mechanical deformation and granulation but has remained chemically unaltered; appearance is flinty, banded, or streaked, but the nature of the parent rock is easily recognized. { 'mī·lə,nīt }

mylonite gneiss [PETR] A metamorphic rock intermediate in character between mylonite and schist. { 'mī·lə,nīt 'nīs }

mylonitic structure [PETR] A structure characteristic of mylonites, produced by extreme microbrecciation and shearing which gives the appearance of a flow structure. { mī·lə,nid·ik 'strək·chər }

mylonitization [GEOL] Rock deformation produced by intense microbrecciation without appreciable chemical alteration of granulated materials. { mī,län·ə·tə'zā·shən }

Mymaridae [INV ZOO] The fairy flies, a family of hymenopteran insects in the superfamily Chalcidoidea. { mī'mar·ə,dē }

myoblast [EMBRYO] A precursor cell of a muscle fiber. { 'mī·ə,blast }

myocardial infarct [MED] An infarct in heart muscle. { ,mī·ə,kärd·ē·əl 'in,färkt }

myocardiopathy [MED] Disease of the myocardium. Also known as cardiomyopathy. { ,mī·ō,kärd·ē'äp·ə·thē }

myocarditis [MED] Inflammation of the myocardium. { ,mī·ə,kär'dīd·əs }

myocardium [HISTOL] The muscular tissue of the heart wall. { ,mī·ə'kärd·ē·əm }

myocardosis [MED] Any noninflammatory disease of the myocardium. { ,mī·ə,kär'dō·səs }

myoclonic epilepsy [MED] Recurrent irregular, arrhythmic clonic muscle spasms, usually occurring more frequently in the morning or on going to sleep and often associated with other types of seizures. { ,mī·ə,klän·ik 'ep·ə,lep·sē }

myoclonic status [MED] Continual clonic spasms lasting an hour or more. { ,mī·ə,klän·ik 'stad·əs }

myoclonus [MED] **1.** Clonic muscle spasm. **2.** Any disorder characterized by scattered, irregular, arrhythmic muscle spasms. { mī'äk·lə·nəs }

myocoel [EMBRYO] Portion of the coelom enclosed in a myotome. { 'mī·ə,sēl }

myocomma [HISTOL] A ligamentous connection between successive myomeres. Also known as myoseptum. { ,mī·ə'käm·ə }

myocyte [HISTOL] **1.** A contractile cell. **2.** A muscle cell. { 'mī·ə,sīt }

Myodaria [INV ZOO] A section of the Schizophora series of cyclorrhaphous dipterans; in this group adult antennae consist of three segments, and all families except the Conopidae have the second cubitus and the second anal veins united for almost their entire length. { ,mī·ə'dar·ē·ə }

Myodocopa [INV ZOO] A suborder of the order Myodocopida; includes exclusively marine ostracodes distinguished by possession of a heart. { ,mī·ə'däk·ə·pə }

Myodocopida [INV ZOO] An order of the subclass Ostracoda. { ,mī·ə·də,käp·ə·də }

Myodopina [INV ZOO] The equivalent name for Myodocopa. { ,mī·ə'däp·ə·nə }

myodystrophy [MED] Muscle degeneration. { ,mī·ə'dis·trə,fē }

myoelastic fiber [HISTOL] An elastic fiber associated with the smooth muscles in bronchi and bronchioles. { ¦mī·ō·i'las·tik 'fī·bər }

myoelectric potential [PHYSIO] The electrical potential created by muscle action. { ¦mī·ō·i'lek·trik pə'ten·chəl }

myoelectric prosthesis [MED] A replacement device for lost limbs that uses the electromyographic activity of a contracting muscle as a control signal; it is most commonly used for below-elbow amputees in whom elbow function is retained. { ¸mī·ō·i¸lek·trik präs'thē·səs }

myoepithelial cells [HISTOL] Contractile epithelial cells resembling smooth muscle cells that are present in glands, notably the mammary gland, and aid in secretion. { ¸mī·ō¸ep·ə'thē·lē·əl ¸sel }

myofascitis [MED] Muscular pain of obscure nature and origin in the lower back. { ¦mī·ō·fə'sīd·əs }

myofibril [CELL MOL] A contractile fibril in a muscle cell. [INV ZOO] *See* myoneme. { ¦mī·ō'fī·brəl }

myofilament [CELL MOL] The structural unit of muscle proteins in a muscle cell. { ¦mī·ō'fil·ə·mənt }

myofrisk [INV ZOO] A contractile structure surrounding the spines of certain radiolarians. { 'mī·ō¸frisk }

myoglobin [BIOCHEM] A hemoglobinlike iron-containing protein pigment occurring in muscle fibers. Also known as muscle hemoglobin; myohemoglobin. { ¦mī·ə¦glō·bən }

myoglobinuria [MED] The presence of myoglobin in the urine. { ¸mī·ə¸glō·bə'nur·ē·ə }

myohematin [BIOCHEM] A cytochrome respiratory enzyme allied to hematin. { ¦mī·ō'hē·mə·tən }

myohemoglobin *See* myoglobin. { ¦mī·ō'hē·mə¸glō·bən }

myoinositol [BIOCHEM] The commonest isomer of inositol. Also known as mesionositol. { ¦mī·ō·i'näs·ə¸tōl }

myokinase [BIOCHEM] An enzyme that catalyzes the reversible transfer of phosphate groups in adenosinediphosphate; occurs in muscle and other tissues. { ¦mī·ō'kī¸nās }

myolipoma [MED] A benign tumor composed of adipose and smooth muscle cells. { ¦mī·ō'lī¸pō·mə }

myology [MED] The study of muscles in both the normal and diseased states. { mī'äl·ə·jē }

myoma [MED] **1.** A benign uterine tumor composed principally of smooth muscle cells. **2.** Any neoplasm originating in muscle. { mī'ō·mə }

myomalacia [MED] Degeneration, with softening, of muscle tissue. { ¦mī·ō·mə'lā·shə }

myomere [EMBRYO] A muscle segment differentiated from the myotome, which divides to form the epimere and hypomere. { 'mī·ə¸mir }

myometritis [MED] Inflammation of the myometrium. { ¦mī·ō·mə'trīd·əs }

myometrium [HISTOL] The muscular tissue of the uterus. { ¸mī·ə'mē·trē·əm }

Myomorpha [VERT ZOO] A suborder of rodents recognized in some systems of classification. { ¸mī·ə'mòr·fə }

myoneme [INV ZOO] A contractile fibril in a protozoan. Also known as myofibril. { 'mī·ə¸nēm }

myoneural junction [NEUROSCI] The point of junction of a motor nerve with the muscle which it innervates. Also known as neuromuscular junction. { ¦mī·ō'nur·əl 'jəŋk·shən }

myopathia *See* myopathy. { ¦mī·ə'path·ē·ə }

myopathic facies [MED] An expressionless face with sunken cheeks and a drooping lower lip characteristic of patients with myopathies, especially myotonic dystrophy. { ¦mī·ə¦path·ik 'fā·shēz }

myopathy [MED] Any disease of the muscles. Also known as myopathia. { mī'äp·ə·thē }

myopericarditis [MED] A combination of myocarditis and pericarditis. { ¦mī·ō¸per·ə¸kär'dīd·əs }

myophagia [PATH] The invasion of degenerated muscle sarcoplasm by histiocytes. { ¦mī·ə'fā·jē·ə }

myophosphorylase deficiency glycogenosis *See* McArdle's syndrome. { ¸mī·ə¸fäs'fór·ə¸lās di'fish·ən·sē ¦glī·kō·jə'nō·səs }

myopia [MED] A condition in which the focal image is formed in front of the retina of the eye. Also known as nearsightedness. { mī'ō·pē·ə }

myoplasty [MED] Plastic surgery performed on muscle tissue. { 'mī·ə¸plas·tē }

Myopsida [INV ZOO] A natural assemblage of cephalopod mollusks considered as a suborder in the order Teuthoida

according to some systems of classification, and a group of the Decapoda according to other systems; the eye is covered by the skin of the head in all species. { mī'äp·sə·də }

myopsychopathy [MED] Any disease of the muscles associated with mental retardation or loss of intellect. { ¦mī·ō¸sī'käp·ə·thē }

myorhythmia [MED] Muscle tremor with a rate of 2–4 per second, and irregular intervals between cycles. { ¦mī·ō'rith·mē·ə }

myosarcoma [MED] A sarcoma derived from muscle. { ¦mī·ō·sär'kō·mə }

myoseptum *See* myocomma. { ¦mī·ō'sep·təm }

myosin [BIOCHEM] A muscle protein, comprising up to 50% of the total muscle proteins; combines with actin to form actomycin. { 'mī·ə·sən }

myositis [MED] Inflammation of muscle. Also known as fibromyositis. { ¸mī·ə'sīd·əs }

myositis ossificans [MED] Muscle inflammation with bone formation in muscle, tendons, or ligaments. { ¸mī·ə'sīd·əs ä'sif·ə¸kanz }

myostatic reflex *See* stretch reflex. { ¦mī·ə¦stad·ik 'rē¸fleks }

myosynovitis [MED] Inflammation of synovial membranes and surrounding musculature. { ¦mī·ə¸sīn·ə·'vīd·əs }

myotasis [PHYSIO] Stretching of a muscle. { mī'äd·ə·səs }

myotome [ANAT] A group of muscles innervated by a single spinal nerve. [EMBRYO] The muscle plate that differentiates into myomeres. [ENG] An instrument used to divide a muscle. { 'mī·ə¸tōm }

myotonia [MED] Tonic muscular spasm occurring after injury or infection. { ¸mī·ə'tō·nē·ə }

myotonia congenita intermittens *See* paramyotonia congenita. { ¸mī·ə'tō·nē·ə kən'jen·əd·ə ¸in·tər'mit·ənz }

myotonic [MED] Of, pertaining to, or characterized by myotonia. { ¦mī·ə¦tän·ik }

myotonic dystrophy [MED] A hereditary disease, transmitted as an autosomal dominant, characterized by lack of normal relaxation of muscles after contraction, slowly progressive muscular weakness and atrophy, especially of the face and neck, cataract formation, early baldness, gonadal atrophy, abnormal glucose tolerance curve, and, frequently, mental deficiency. { ¦mī·ə¦tän·ik 'dis·trə·fē }

Myoviridae [VIROL] A family of linear double-stranded deoxyribonucleic acid–containing bacterial viruses (bacteriophages) characterized by a contractile tail and an elongated (or icosahedral) head; a well-known genus is the T even phage, which contains coliphage T2. { ¸mī·ə'vir·ə¸dī }

β-myrcene [ORG CHEM] $C_{10}H_{16}$ An oily liquid with a pleasant odor; soluble in alcohol, chloroform, ether, and glacial acetic acid; used as an intermediate in the preparation of perfume chemicals. { ¦bād·ə 'mər¸sēn }

myria- [SCI TECH] A prefix representing 10^4 or 10,000. { 'mir·ē·ə }

myriametric waves [ELECTROMAG] Electromagnetic waves having wavelengths between 10 and 100 kilometers, corresponding to the very low frequency band. { ¦mir·ē·ə¦me·trik 'wāvz }

Myriangiales [MYCOL] An order of parasitic fungi of the class Ascomycetes which produce asci at various levels in uniascal locules within stromata. { ¸mir·ē¸an·jē'ā·lēz }

Myriapoda [INV ZOO] Informal designation for those mandibulate arthropods having two body tagmata, one pair of antennae, and more than three pairs of adult ambulatory appendages. { ¸mir·ē'äp·ə·də }

Myricaceae [BOT] The single family of the plant order Myricales. { ¸mir·ə'kās·ē¸ē }

Myricales [BOT] An order of dicotyledonous plants in the subclass Hamamelidae, marked by its simple, resinous-dotted, aromatic leaves, and a unilocular ovary with two styles and a single ovule. { ¸mir·ə'kā·lēz }

myricetin [ORG CHEM] $C_{15}H_{10}O_8$ A yellow, crystalline compound with a melting point of 357°C; soluble in alcohol; used as an inhibitor of adenosine triphosphatase. Also known as cannabiscetin; delphidenolon. { mə'ris·ə·tən }

Myrientomata [INV ZOO] The equivalent name for the Protura. { ¸mir·ē·ən'täm·əd·ə }

myrigoplasty [MED] A surgical procedure involving the simple repair of a persistent perforation of the tympanic membrane. { mə'rig·ə¸plas·tē }

myringitis [MED] Inflammation of the tympanic membrane. { ‚mir·ən'jīd·əs }

myriotic field [QUANT MECH] A quantized field that has creation and annihilation operators satisfying specified commutation rules, but no vacuum state. { ‚mir·ē'äd·ik 'fēld }

Myriotrochidae [INV ZOO] A family of holothurian echinoderms in the order Apodida, distinguished by eight or more spokes in each wheel-shaped spicule. { ‚mir·ē·ə'trō·kə‚dē }

myristic acid [ORG CHEM] $CH_3(CH_2)_{12}COOH$ Oily white crystals melting at 58°C; soluble in ether and alcohol, insoluble in water; used to synthesize flavor and perfume esters, and in soaps and cosmetics. { mə'ris·tik 'as·əd }

myristica oil See nutmeg oil. { mə'ris·tə·kə ‚öil }

myristyl alcohol [ORG CHEM] $C_{14}H_{29}OH$ Liquid boiling at 264°C; soluble in ether and alcohol, insoluble in water; used as a chemical intermediate, plasticizer, and perfume fixative. { mə'rist·əl 'al·kə‚hól }

myristyl mercaptan See tetradecyl mercaptan. { mə'rist·əl mər'kap‚tan }

Myrmecophagidae [VERT ZOO] A small family of arboreal anteaters in the order Edentata. { ‚mər·mə·kō'faj·ə‚dē }

myrmecophagous [ZOO] Feeding on ants. { ‚mər·mə'käf·ə·gəs }

myrmecophile [ECOL] An organism, usually a beetle, that habitually inhabits the nest of ants. { mər'mek·ə‚fīl }

myrmecophyte [ECOL] A plant that houses and benefits from the habitation of ants. { mər'mek·ə‚fīt }

myrmekite [PETR] Intergrowth of plagioclase feldspar and vermicular quartz in an igneous rock. { 'mər·mə‚kīt }

myrmekitic [PETR] **1.** Pertaining to the texture of an igneous rock marked by intergrowths of feldspar and vermicular quartz. **2.** Having characteristic properties of myrmekite. { ‚mər·mə¦kid·ik }

Myrmeleontidae [INV ZOO] The ant lions, a family of insects in the order Neuroptera; larvae are commonly known as doodlebugs. { ‚mər·mə·lē'än·tə‚dē }

Myrmicinae [INV ZOO] A large diverse subfamily of ants (Formicidae); some members are inquilines and have no worker caste. { mər'mis·ə‚nē }

myrrh [MATER] A gum resin of species of myrrh (*Commiphora*); partially soluble in water, alcohol, and ether; used in dentifrices, perfumery, and pharmaceuticals. { mər }

Myrsinaceae [BOT] A family of mostly woody dicotyledonous plants in the order Primulales characterized by flowers without staminodes, a schizogenous secretory system, and gland-dotted leaves. { ‚mər·sə'nās·ē‚ē }

Myrtaceae [BOT] A family of dicotyledonous plants in the order Myrtales characterized by an inferior ovary, numerous stamens, anthers usually opening by slits, and fruit in the form of a berry, drupe, or capsule. { mər'tās·ē‚ē }

Myrtales [BOT] An order of dicotyledonous plants in the subclass Rosidae characterized by opposite, simple, entire leaves and perigynous to epigynous flowers with a compound pistil. { mər'tā·lēz }

myrtle oil [MATER] Light-yellow liquid distilled from the flowers and leaves of the European myrtle (*Myrtus communis*); aromatic aroma; formerly used in medicine; now used for flavors and as perfume fixative. { 'mərd·əl ‚öil }

myrtle wax See bayberry wax. { 'mərd·əl ‚waks }

Mysida [INV ZOO] A suborder of the crustacean order Mysidacea characterized by fusion of the sixth and seventh abdominal somites in the adult, lack of gills, and other specializations. { 'mī·sə·də }

Mysidacea [INV ZOO] An order of free-swimming Crustacea included in the division Pericarida; adult consists of 19 somites, each bearing one pair of functionally modified, biramous appendages, and the carapace envelops most of the thorax and is fused dorsally with up to four of the anterior thoracic segments. { ‚mī·sə'dās·ē·ə }

mysis [INV ZOO] A larva of certain higher crustaceans, characterized by biramous thoracic appendages. { 'mī·səs }

mysophobia [PSYCH] An abnormal fear of contamination or of dirt. { ‚mī·sə'fō·bē·ə }

Mystacinidae [VERT ZOO] A monospecific family of insectivorous bats (Chiroptera) containing the New Zealand short-tailed bat; hindlegs and body are stout, and fur is thick. { ‚mis·tə'sin·ə‚dē }

Mystacocarida [INV ZOO] An order of primitive Crustacea; the body is wormlike and the cephalothorax bears first and second antennae, mandibles, and first and second maxillae. { ‚mis·tə·kō'kar·ə·də }

Mysticeti [VERT ZOO] The whalebone whales, a suborder of the mammalian order Cetacea, distinguished by horny filter plates of suspended from the upper jaws. { ‚mis·tə'sē‚tī }

mythophobia [PSYCH] An abnormal fear of making an incorrect statement. { ‚mith·ə'fō·bē·ə }

Mytilacea [INV ZOO] A suborder of bivalve mollusks in the order Filibranchia. { ‚mid·əl'ā·shē·ə }

Mytilidae [INV ZOO] A family of mussels in the bivalve order Anisomyaria. { mī'til·ə‚dē }

myx-, myxo- [ZOO] A combining form denoting mucus, mucous, mucin, mucinous. { 'mik·sō }

myxadenitis [MED] Inflammation of mucous glands. { ‚miks‚ad·ən'īd·əs }

myxadenoma [MED] An adenoma of a mucous gland. { ‚miks‚ad·ən'ō·mə }

myxameba [BIOL] An independent ameboid cell of the vegetative phase of Acrasiales. { ‚miks·ə'mē·bə }

myxedema [MED] A condition caused by hypothyroidism characterized by a subnormal basal metabolic rate, dry coarse hair, loss of hair, mental dullness, anemia, and slowed reflexes. { ‚miks·ə'dē·mə }

Myxicolinae [INV ZOO] A subfamily of sedentary polychaete annelids in the family Sabellidae. { ‚mik·sə'käl·ə‚nē }

Myxiniformes [VERT ZOO] The equivalent name for the Myxinoidea. { mik·sə·nə'fór·mēz }

Myxinoidea [VERT ZOO] The hagfishes, an order of eellike, jawless vertebrates (Agnatha) distinguished by having the nasal opening at the tip of the snout and leading to the pharynx, with barbels around the mouth and 6–15 pairs of gill pouches. { ‚mik·sə'nóid·ē·ə }

myxoadenoma [MED] An adenoma of a mucous gland. { ‚mik·sō‚ad·ən'ē·mə }

Myxobacterales [MICROBIO] An order of gliding bacteria; unicellular, gram-negative rods embedded in a layer of slime and capable of gliding movement; form fruiting bodies containing resting cells (myxospores) under certain environmental conditions. { ‚mik·sə‚bak·tə'rā·lēz }

myxochondrofibrosarcoma [MED] A sarcoma composed of anaplastic myxoid, chondroid, and fibrous cells. { ‚mik·sə¦kän·drō‚fī·brə‚sär'kō·mə }

Myxococcaceae [MICROBIO] A family of the order Myxobacterales; vegetative cells are straight to slightly tapered, and spherical to ovoid microcysts (myxospores) are produced. { ‚mik·sə‚käk'sās·ē‚ē }

myxofibroma of nerve sheath See neurofibroma. { ‚mik·sə‚fī'brō·mə əv 'nərv ‚shēth }

Myxogastromycetidae [MYCOL] A large subclass of plasmodial slime molds (Myxomycetes). { ¦mik·sə‚gas·trō‚mī'sed·ə‚dē }

myxolipoma See liposarcoma. { ‚mik·sō·lī'pō·mə }

myxoma [MED] A benign tumor composed of mucinous connective tissue. { mik'sō·mə }

myxoma lipomatodes See liposarcoma. { mik'sō·mə lī‚pō·mə'tō·dēz }

myxomatosis [VET MED] A virus disease of rabbits producing fever, skin lesions resembling myxomas, and mucoid swelling of mucous membranes. { ‚mik‚sō·mə'tō·səs }

Myxomycetes [BIOL] Plasmodial (acellular or true) slime molds, a class of microorganisms of the division Mycota; they are on the borderline of the plant and animal kingdoms and have a noncellular multinucleate, jellylike, creeping, assimilative stage (the plasmodium) which alternates with a myxameba stage. { ‚mik·sə‚mī'sēd·ēz }

Myxomycophyta [BOT] An order of microorganisms, equivalent to the Mycetozoia of zoological classification. { ‚mik·sə‚mī'käf·əd·ə }

Myxophaga [INV ZOO] A suborder of the Coleoptera. { mik'säf·ə·gə }

Myxophyceae [BOT] An equivalent name for the Cyanophyceae. { ‚mik·sə'fī·sē‚ē }

myxosarcoma [MED] A sarcoma whose parenchyma is composed of anaplastic myxoid cells. { ‚mik·sō·sär'kō·mə }

Myxosporida [INV ZOO] An order of the protozoan class Myxosporidea characterized by the production of spores with one or more valves and polar capsules, and by possession of a single sporoplasm. { ‚mik·sə'spór·ə·də }

50 µm

An eight-spoked spicule from the skin of an echinoderm in the Myriotrochidae.

MYSIDA

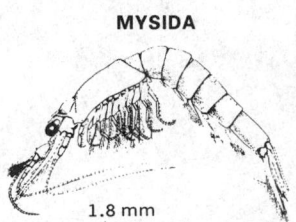

1.8 mm

Neomysis integer, adult male mysid.

MYSTACOCARIDA

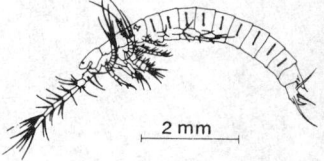

2 mm

Derocheilocaris typicus, lateral view.

Myxosporidea [INV ZOO] A class of the protozoan subphylum Cnidospora; members are parasitic in some fish, a few amphibians, and certain invertebrates. { ‚mik·sə·spə'rid·ē·ə }

myxovirus [VIROL] A group of ribonucleic-acid animal viruses characterized by hemagglutination and hemadsorption; includes influenza and fowl plague viruses and the paramyxoviruses. { 'mik·sə‚vī·rəs }

Myzopodidae [VERT ZOO] A monospecific order of insectivorous bats (Chiroptera) containing the Old World diskwinged bat of Madagascar; characterized by long ears and by a vestigial thumb with a monostalked sucking disk. { ‚mī·zə'päd·ə‚dē }

myzorhynchus [INV ZOO] An apical sucker on the scolex of certain tapeworms. { ‚mī·zə'riŋ·kəs }

Myzostomaria [INV ZOO] An aberrant group of Polychaeta; most are greatly depressed, broad, and very small, and true segmentation is delayed or absent in the adult; all are parasites of echinoderms. { ‚mī·zə·stə'mar·ē·ə }

Myzostomidae [INV ZOO] A monogeneric family of the Myzostomaria. { ‚mī·zə'stäm·ə‚dē }

n- [ORG CHEM] Chemical prefix for normal (straight-carbon-chain) hydrocarbon compounds.

N *See* newton; nitrogen; normality.

Na *See* sodium.

N.A. *See* numerical aperture.

NAA *See* naphthaleneacetic acid.

Nabidae [INV ZOO] The damsel bugs, a family of hemipteran insects in the superfamily Cimicimorpha. { 'nab·ə,dē }

nabla *See* del operator. { 'nab·lə }

Nabothian cyst [MED] Cystic distention of the Nabothian glands of the uterine cervix. { nə'bō·thē·ən 'sist }

Nabothian glands [ANAT] Mucous glands of the uterine cervix. { nə'bō·thē·ən 'glanz }

nacelle [AERO ENG] A separate streamlined enclosure on an airplane for sheltering or housing something, as the crew or an engine. [ELEC] An enclosure containing the electric generating equipment in a wind-energy conversion system. { nə'sel }

nacre [INV ZOO] An iridescent inner layer of many mollusk shells. { 'nā·kər }

nacreous [OPTICS] Having an iridescent luster resembling that of mother-of-pearl. Also known as pearly. { 'nā·krē·əs }

nacreous clouds [METEOROL] Clouds of unknown composition, whose form resembles that of cirrus or altocumulus lenticularis, and which show very strong irisation similar to that of mother-of-pearl, especially when the sun is several degrees below the horizon; they occur at heights of about 12 or 18 miles (20 or 30 kilometers). Also known as mother-of-pearl clouds. { 'nā·krē·əs 'klaùdz }

nacré velvet [TEXT] A velvet fabric with a back of one color and a filling or pile of another, giving a changeable, pearly appearance. { nə'krā 'vel·vət }

nacrite [MINERAL] $Al_2Si_2O_5(OH)_4$ A crystallized clay mineral of the kaolinite group; structurally distinct in being the most closely stacked in the *c*-axis direction. { 'nā,krīt }

NAD *See* diphosphopyridine nucleotide.

nadir [ASTRON] That point on the celestial sphere vertically below the observer, or 180° from the zenith. [OCEANOGR] The point on the sea floor that lies directly below the sonar during a survey. { 'nā·dər }

nadir point *See* photograph nadir. { 'nā·dər ,pöint }

nadorite [MINERAL] $PbSbO_2Cl$ A smoky brown or brownish-yellow to yellow, orthorhombic mineral consisting of an oxychloride of lead and antimony. { 'nad·ə,rīt }

NADP+ [CELL MOL] *See* nicotinamide adenine dinucleotide phosphate. { ,en,ā,dē,pē'pləs }

Naegeli-type leukemia [MED] A type of monocytic leukemia in which the leukocytes resemble cells of the granulocytic series. { 'nā·gə·lē ¦tīp lü'kē·mē·ə }

naftalan *See* naphthalan. { 'naf·tə,lan }

nagatelite [MINERAL] Black mineral composed of phosphosilicate of an aluminum, rare-earth elements, calcium, and iron; occurs in tabular masses. { ,nag·ə'te,līt }

nagyagite [MINERAL] $Pb_5Au(Te,Sb)_4S_{5-8}$ A lead-gray mineral consisting of a sulfide of lead, gold, tellurium, and antimony. Also known as black tellurium; tellurium glance. { 'nag·yə,jīt }

nahcolite [MINERAL] $NaHCO_3$ A white, monoclinic mineral consisting of natural sodium bicarbonate. { 'nä·kə,līt }

Naiad [ASTRON] A satellite of Neptune orbiting at a mean distance of 30,000 miles (48,000 kilometers) with a period of

7.1 hours, and a diameter of about 34 miles (54 kilometers). { 'nī,ad }

naif [MINERAL] Of a gemstone, having a true or natural luster when uncut. Also spelled naife. { nä'ēf }

naife *See* naif. { nä'ēf }

nail [ANAT] The horny epidermal derivative covering the dorsal aspect of the terminal phalanx of each finger and toe. [DES ENG] A slender, usually pointed fastener with a head, designed for insertion by impact. [ENG] To drive nails in a manner that will position and hold two or more members, usually of wood, in a desired relationship. [MED] A metallic rod with one blunt end and one sharp end, used surgically to anchor bone fragments. { nāl }

nail coat *See* devil float. { 'nāl ,kōt }

nailer [ENG] A wood strip or block which serves as a backing into which nails can be driven. { 'nāl·ər }

nailhead [DES ENG] Flat protuberance at the end of a nail opposite the point. { 'nāl,hed }

nailheaded molding [ARCH] A molding consisting of a series of short protuberances resembling nailheads. { 'nāl,hed·əd 'mōld·iŋ }

nailhead spot [PL PATH] A fungus rot of tomato caused by *Alternaric tomato* and marked by small brown to black sunken spots on the fruit. { 'nāl,hed ,spät }

nailhead striation [GEOL] A glacial striation with a definite or blunt head or point of origin, generally narrowing or tapering in the direction of ice movement and coming to an indefinite end. { 'nāl,hed strī'ā·shən }

nail set [DES ENG] A small cylindrical steel tool, usually tapered at one end, that is used to drive a nail or a brad below or flush with a wood surface. Also known as punch. { 'nāl ,set }

Nairovirus [VIROL] A genus of the viral family Bunyaviridae that causes Nairobi sheep disease. { 'nī·rə,vī·rəs }

Najadaceae { 'nī·rə,vī·rəs } [BOT] A family of monocotyledonous, submerged aquatic plants in the order Najadales distinguished by branching stems and opposite or whorled leaves. { ,näj·ə'dās·ē,ē }

Najadales [BOT] An order of aquatic and semiaquatic flowering plants in the subclass Alismatidae; the perianth, when present, is not differentiated into sepals and petals, and the flowers are usually not individually subtended by bracts. { ,näj·ə'dā·lēz }

NAK *See* negative acknowledgement. { nak *or* ,en,ā'kā }

Nakayama's lemma [MATH] The proposition that, if R is a commutative ring, I is an ideal contained in all maximal ideals of R, and M is a finitely generated module over R, and if $IM = M$, where IM denotes the set of all elements of the form am with a in I and m in M, then $M = 0$. { ,nä·kä,yä·məz 'lem·ə }

naked bud [BOT] A bud covered only by rudimentary foliage leaves. { 'nā·kəd 'bəd }

naked karst [GEOL] Karst that is developed in a region without soil cover, so that its topographic features are well exposed. { 'nā·kəd 'kärst }

naked light [MIN ENG] Open flame, such as a match or a burning cigarette, that is a fire risk in mines. { 'nā·kəd 'līt }

naked-light mine [MIN ENG] A coal mine that is nongassy, where naked lights can be used by miners. { 'nā·kəd 'līt ,mīn }

naked singularity [RELAT] A singularity that is not surrounded by an event horizon, and thus gives rise to timelike curves that violate causality. { 'nā·kəd ,siŋ·gyə'lar·əd·ē }

NANOSTRUCTURED MATERIAL

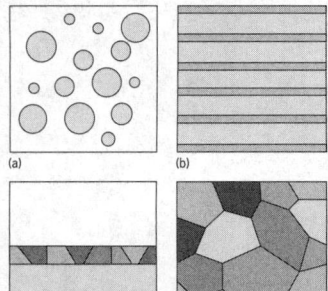

(a)   (b)

(c)   (d)

Schematic of four basic types of nanostructured materials, classified according to integral modulation dimensionality. (*a*) Dimensionality 0: clusters of any aspect ratio from 1 to infinity. (*b*) Dimensionality 1: multilayers. (*c*) Dimensionality 2: ultrafine-grained overlayers (coatings) or buried layers. (*d*) Dimensionality 3: nanophase materials. (*After R. W. Siegel, Nanostructured materials: Mind over matter, Nanostruct. Mat., 3:1–18, 1993*)

naked T Tauri star *See* weak-line T Tauri star. { ¦nāk·əd ¦tē ¦tȯr·ē ¦stär }

nakhlite [GEOL] An achondritic stony meteorite composed of an aggregate of diopside and olivine. { 'näk,līt }

naled [ORG CHEM] $C_4H_7Br_2Cl_2O_4$ A white solid with a melting point of 27°C; slight solubility in water; used as an insecticide and miticide for crops, farm buildings, and kennels, and for mosquito control. { 'nal·əd }

nalidixic acid [PHARM] $C_{12}H_{12}N_2O_3$ Pale buff, crystalline powder; melting point is 229–230°C; soluble in chloroform and in potassium hydroxide and sodium hydroxide solutions; used as an antibacterial drug in humans and animals. { ¦nal·ə¦dik·sik 'as·əd }

naltrexone [PHARM] $C_{20}H_{23}NO_4$ An opiate receptor antagonist that blocks the effects of endogenous opioids in the brain; used to treat alcoholism. { nal'trek,zōn }

Namanereinae [INV ZOO] A subfamily of largely freshwater errantian annelids in the family Nereidae. { ¦nā·mə,nə¦rē·ə,nē }

Namurian [GEOL] A European stage of geologic time; divided into a lower stage (Lower Carboniferous or Upper Mississippian) and an upper stage (Upper Carboniferous or Lower Pennsylvanian). { nə'myu̇r·ē·ən }

nancy receiver *See* infrared receiver. { 'nan·sē ri,sēv·ər }

NAND [MATH] A logic operator having the characteristic that if P, Q, R, . . . are statements, then the NAND of P, Q, R, . . . is true if at least one statement is false, false if all statements are true. Derived from NOT-AND. Also known as sheffer stroke. { nand }

NAND circuit [ELECTR] A logic circuit whose output signal is a logical 1 if any of its inputs is a logical 0, and whose output signal is a logical 0 if all of its inputs are logical 1. { 'nand ,sər,kət }

nanism [MED] Dwarfed stature due to arrested development. { 'nā,niz·əm }

nannandrous [BOT] Pertaining to species of plants in which male members are markedly smaller than females, such as in some algal species of *Oedogonium* that have antheridia produced in special dwarf filaments. { na'nan·drəs }

nannoplankton [BIOL] Minute plankton; the smallest (usually from 2 to 20 nanometers) plankton, including algae, bacteria, and protozoans. Also spelled nanoplankton. { 'nan·ō'plaŋk·tən }

nano- [BIOL] A prefix meaning dwarfed. [MATH] A prefix representing 10^{-9}, which is 0.000000001 or one-billionth of the unit adjoined. { 'nan·ō }

nanocephalus [MED] A fetus with an undersized head. { ,nan·ə'sef·ə·ləs }

nanochemistry [CHEM] The study of the synthesis and analysis of materials in the nanoscale range (1–10 nanometers), including large organic molecules, inorganic cluster compounds, and metallic or semiconductor particles. { ,nan·ō'kem·ə·strē }

nanocomposite [MATER] A material that results from the intimate mixture of two or more nanophase materials. { ,nan·ō·kəm'päz·ət }

nanocomposite material *See* nanostructured material. { ,nan·ō·kəm'päz·ət mə,tir·ē·əl }

nanoelectronics [ELECTR] The technology of electronic devices whose dimensions range from atoms up to 100 nanometers. { ,nan·ō·i,lek'trän·iks }

nanogram [MECH] One-billionth (10^{-9}) of a gram. Abbreviated ng. { 'nan·ə,gram }

nanometer [MECH] A unit of length equal to one-billionth of a meter, or 10^{-9} meter. Also known as millimicron (μm); nanon. { 'nan·ə,mēd·ər }

nanon *See* nanometer. { 'na,nän }

nanophanerophyte [ECOL] A shrub not exceeding 6.6 feet (2 meters) in height. { ¦nan·ō'fan·ə·rə,fīt }

nanophase material [MATER] **1.** A material made up of phases that have dimensions of the order of nanometers. **2.** An ultrafine single solid phase where at least one dimension is in the nanometer range, and typically dimensions are in the 1–20-nanometer range. { 'nan·ō,fāz mə,tir·ē·əl }

nanophthalmus *See* microphthalmus. { 'nan·ə,thal·məs }

nanoplankton *See* nannoplankton. { 'nan·ō'plaŋk·tən }

nanosecond [MECH] A unit of time equal to one-billionth of a second, or 10^{-9} second. { 'nan·ə,sek·ənd }

nanostructure [SOLID STATE] Something that has a physical dimension smaller than 100 nanometers, ranging from clusters of atoms to dimensional layers. { 'nan·ō,strək·chər }

nanostructured material [MATER] A material whose composition is modulated over nanometer length scales in zero, one, two, or three dimensions. Also known as nanocomposite material. { 'nan·ō,strək·chərd mə,tir·ē·əl }

nanotechnology [ENG] **1.** Systems for transforming matter, energy, and information that are based on nanometer-scale components with precisely defined molecular features. **2.** Techniques that produce or measure features less than 100 nanometers in size. { ,nan·ō·tek'näl·ə·jē }

nanozooid [INV ZOO] Dwarf zooid; bryozoan heterozooid possessing only a single tentacle. { ,nan·ō¦zō,ȯid }

Nansen bottle [ENG] A bottlelike water-sampling device with valves at both ends that is lowered into the water by wire; at the desired depth it is activated by a messenger which strikes the reversing mechanism and inverts the bottle, closing the valves and trapping the water sample inside. Also known as Petterson-Nansen water bottle; reversing water bottle. { 'nan·sən ,bäd·əl }

Nansen cast [OCEANOGR] A series of Nansen-bottle water samples and associated temperature observations resulting from one release of a messenger. { 'nan·sən ,kast }

nantokite *See* cuprous chloride. { 'nan·tə,kīt }

nap [TEXT] Fuzzy fibers on the surface of a fabric; produced by a finishing process called raising. { nap }

napalm [MATER] **1.** Aluminum soap in powder form, used to gelatinize oil or gasoline for use in napalm bombs or flame throwers. **2.** The resultant gelatinized substance. { 'nā,päm }

napalm bomb [ORD] A bomb filled with napalm; primarily an antipersonnel weapon. { 'nā,päm ,bäm }

nape [ANAT] The back of the neck. { nāp }

Naperian logarithm *See* logarithm. { nā'pir·ē·ən 'läg·ə,rith·əm }

napex [ANAT] That portion of the scalp just below the occipital protuberance. { 'nā,peks }

naphtha [MATER] **1.** Petroleum fraction with volatility between gasoline and kerosine; used as a gasoline ingredient, solvent for paints and rubber, and cleaning solvent. **2.** Aromatic solvent from coal tar, either solvent naphtha or heavy naphtha. { 'naf·thə }

naphthacene [ORG CHEM] $C_{18}H_{12}$ A hydrocarbon molecule that may be considered to be four benzene rings fused together; it is explosive when shocked; used in organic synthesis. Also known as rubene; tetracene. { 'naf·thə,sēn }

naphtha gas [MATER] Illuminating gas charged with a low-boiling-point fraction of distilled naphtha. { 'naf·thə ,gas }

naphthalan [MATER] Soft, greenish-black mass distilled from Armenian naphtha; soluble in ether and hydrocarbons, insoluble in water; melts at 70°C; used in medicine. Also known as naftalan. { 'naf·thə,lan }

naphthalene [ORG CHEM] $C_{10}H_8$ White, volatile crystals with coal tar aroma; insoluble in water, soluble in organic solvents; structurally it is represented as two benzenoid rings fused together; boiling point 218°C, melting point 80.1°C; used for moth repellents, fungicides, lubricants, and resins, and as a solvent. Also known as naphthalin; tar camphor. { 'naf·thə,lēn }

naphthaleneacetamide [ORG CHEM] $C_{12}H_{11}NO$ A colorless solid with a melting point of 183°C; used as a growth regulator for root cuttings and for thinning of apples and pears. { ¦naf·thə,lēn·ə'sed·ə·məd }

naphthaleneacetic acid [ORG CHEM] $C_{10}H_7CH_2COOH$ White, odorless crystals, melting at 132–135°C; soluble in organic solvents, slightly soluble in water; used as an agricultural spray. Abbreviated NAA. Also known as 1-naphthylacetic acid. { ¦naf·thə,lēn·ə¦sēd·ik 'as·əd }

naphthalene-1,5-disulfonic acid [ORG CHEM] $C_{10}H_6-(SO_3H)_2$ White crystals, decomposing when heated; used to make dyes. Also known as Armstrong's acid. { 'naf·thə,lēn ¦wən ¦fīv ,dī·səl'fän·ik 'as·əd }

1-naphthalenesulfonic acid [ORG CHEM] $C_{10}H_8O_3S$ A crystalline compound with a melting point of 90°C (dihydrate); soluble in water or alcohol; used to make α-naphthol. { ¦wən ¦naf·thə,lēn·səl'fän·ik 'as·əd }

naphthalic acid *See* phthalic acid. { naf'thal·ik 'as·əd }

naphthalin *See* naphthalene. { 'naf·thə·lən }

naphthene [ORG CHEM] Any of the cycloparaffin derivatives of cyclopentane (C_5H_{10}) or cyclohexane (C_6H_{12}) found in crude petroleum. { 'naf,thēn }

naphthene base [MATER] Crude oil with a high carbon and low oxygen content that leaves an asphaltic residue after refining. Also known as asphalt base. { 'naf,thēn ,bās }

naphthenic acid [ORG CHEM] Any of the derivatives of cyclopentane, cyclohexane, cycloheptane, or other naphthenic homologs derived from petroleum; molecular weights 180 to 350; soluble in organic solvents and hydrocarbons, slightly soluble in water; used as a paint drier and wood preservative, and in metals production. { naf'thēn·ik 'as·əd }

naphthenic crude [MATER] Crude petroleum containing a significant proportion of naphthenic compounds. { naf'thēn·ik 'krüd }

naphthine See hatchettite. { 'naf,thēn }

naphthionic acid [ORG CHEM] $C_{10}H_6(NH_2)SO_3H$ White powder or crystals that decompose when heated; used to manufacture dyes. { naf·thē¦än·ik 'as·əd }

α-naphthol [ORG CHEM] $C_{10}H_7OH$ Colorless to yellow powder, melting at 96°C; used to make dyes and perfumes, and in synthesis of organic molecules. { 'al·fə 'naf,thōl }

β-naphthol [ORG CHEM] $C_{10}H_7OH$ White crystals that melt at 121.6°C; insoluble in water; used to make pigments, dyes, and antioxidants. { ¦bäd·ə 'naf,thōl }

1,2-naphthoquinone [ORG CHEM] $C_{10}H_6O_2$ A golden yellow, crystalline compound that decomposes at 145–147°C; soluble in benzene and ether; used as a reagent for resorcinol and thalline. { ¦wən ¦tü ¦naf·thə·kwə'nōn }

1,4-naphthoquinone [ORG CHEM] $C_{10}H_6O_2$ Greenish-yellow powder soluble in organic solvents, slightly soluble in water; melts at 123–126°C; used as an antimycotic agent, in synthesis, and as a rubber polymerization regulator. { ¦wən ¦fȯr ¦naf·thə·kwə'nōn }

naphthoresorcinol [ORG CHEM] $C_{10}H_6(OH)_2$ Crystals with a melting point of 124–125°C; soluble in ether, alcohol, and water; used as a reagent for sugars and oils, and to determine glucuronic acid in urine. { ¦naf·thə·ri'sȯrs·ən,ȯl }

β-naphthoxyacetic acid [ORG CHEM] $C_{12}H_{10}O_3$ A crystalline compound soluble in water, with a melting point of 156°C; used as a growth regulator to set blossoms and regulate growth for pineapples, strawberries, and tomatoes. Also known as O-(2-naphthyl)glycolic acid. Abbreviated BNOA. { ¦bäd·ə naf¦thäk·sē·ə'sēd·ik 'as·əd }

2-(α-naphthoxy)-N,N-diethylpropionamide See devrinol. { ¦tü ¦al·fə naf¦thäk·sē ¦en ¦en dī,eth·əl,pro·pē¦än·ə·məd }

1-naphthylacetic acid See naphthaleneacetic acid. { ¦wən ,naf·thil·ə'sēd·ik 'as·əd }

naphthylamine [ORG CHEM] $C_{10}H_7NH_2$ White, toxic crystals, soluble in alcohol and ether; used in dyes; the two forms are α-naphthylamine, boiling at 301°C, and β-naphthylamine, boiling at 306°C. { naf'thil·ə,mēn }

2,5-naphthylamine sulfonic acid See gamma acid. { ¦tü ¦fīv naf'thil·ə,mēn səl'fän·ik 'as·əd }

β-naphthylmethyl ether [ORG CHEM] $C_{10}H_7OCH_3$ White, crystalline scales with a melting point of 72°C; soluble in alcohol and ether; used for soap perfumes. { ¦bäd·ə ¦naf,thil¦meth·əl 'ē·thər }

N-1-naphthylphthalamic acid [ORG CHEM] $C_{10}H_7NHCO$-C_6H_4COOH A crystalline solid with a melting point of 185°C; used as a preemergence herbicide. { ¦en ¦wən ¦naf·thil·thə'lam·ik 'as·əd }

α-naphthylthiocarbamide See 1-(1-naphthyl)-2-thiourea. { ¦al·fə ¦naf·thil,thī·ō'kär·bə·məd }

1-(1-naphthyl)-2-thiourea [ORG CHEM] $C_{10}H_7NHCSNH_2$ A crystalline compound with a melting point of 198°C; soluble in water, acetone, triethylene glycol, and hot alcohol; used as a poison to control the adult Norway rat. { ¦wən ¦wən ¦naf·thil ¦tü ,thī·ə·yu̇'rē·ə }

napier See neper. { 'nā·pē·ər }

Napier diagram [NAV] A diagram on which compass deviation is plotted for various headings, and the points are connected by a smooth curve, permitting deviation problems to be solved quickly without interpolation; it consists of a vertical line, usually in two parts, each part being graduated for 180° of heading, and two additional sets of lines at an angle of 60° to each other and to the vertical lines. { 'nā·pē·ər 'dī·ə,gram }

Napierian logarithm See logarithm. { nā'pir·ē·ən 'läg·ə,rith·əm }

Napier's analogies [MATH] Formulas which enable one to study the relationships between the sides and the angles of a spherical triangle. { 'nā·pē·ərz ə'nal·ə·jēz }

Napier's rules [MATH] Two rules which give the formulas necessary in the solution of right spherical triangles. { 'nā·pē·ərz ,rülz }

napiform [BOT] Turnip-shaped, referring to roots. { 'nāp·ə,fȯrm }

Naples yellow See lead antimonite. { 'nā·pəlz 'yel·ō }

NAPLPS See North American presentation-level protocol syntax. { 'nap lips }

napoleonite See corsite. { nə'pōl·yə,nīt }

Napoleonville [GEOL] A North American (Gulf Coast) stage of geologic time; a subdivision of the Miocene, above Anahuac and below Duck Lake. { nə'pōl·ē·ən,vil }

nappe [GEOL] A sheetlike, allochthonous rock unit that is formed by thrust faulting or recumbent folding or both. [MATH] One of the two parts of a conical surface defined by the vertex. { nap }

napped leather See suede. { 'napt 'leth·ər }

narbonnais [METEOROL] A wind coming from Narbonne; a north wind in the Roussillon region of southern France resembling the tramontana; if associated with an influx of arctic air, it may be very stormy with heavy falls of rain or snow. { när·bə'nā }

narceine [ORG CHEM] $C_{23}H_{27}O_8N\cdot3H_2O$ White, odorless crystals with bitter taste; soluble in alcohol and water, insoluble in ether; melts at 170°C; used in medicine. { 'när·sē,ēn }

narcissism [PSYCH] Excessive self-love. { 'när·sə,siz·əm }

narcissistic reaction [CHEM] A chemical reaction in which a reactant is converted into a product whose structure is the mirror image of the reactant molecule. { ,när·sə'sis·tik rē'ak·shən }

narcissus oil See jonquil oil. { när'sis·əs ,ȯil }

narco- Combining form meaning numbness, narcosis, or stupor. { 'när·kō }

narcoanalysis [PSYCH] Induction of a reversible sleep by intravenous injections of drugs such as amobarbital or thiopental sodium in order to elicit memories and feelings not expressed by the person in a wakeful state because of resistance. { ,när·kō·ə'nal·ə·səs }

narcolepsy [MED] A disorder of sleep mechanism characterized by two or more of four distinct symptoms: uncontrollable periods of daytime drowsiness, cataleptic attacks of muscular weakness, sleep paralysis, and vivid nocturnal or hypnogogic hallucinations. { 'när·kə,lep·sē }

narcomania [MED] Morbid physiologic or psychologic craving for narcotics to avoid painful stimuli. { ,när·kə'mā·nē·ə }

Narcomedusae [INV ZOO] A suborder of hydrozoan cnidarians in the order Trachylina; the hydroid generation is represented by an actinula larva. { ,när·kə·mə'dü,sē }

narcosis [MED] Drug-produced state of profound stupor, unconsciousness, or arrested activity. { när'kō·səs }

narcosis therapy [MED] Prolonged, drug-induced sleep as treatment for certain mental disorders. Also known as sleep therapy. { när'kō·səs ,ther·ə·pē }

narcospasm [MED] Spasm accompanied by stupor. { 'när·kō,spaz·əm }

narcosynthesis [MED] Psychotherapeutic treatment under partial anesthesia, in which abreaction is a significant factor in obtaining positive results. { ¦när·kō'sin·thə·səs }

narcotic [PHARM] A drug which in therapeutic doses diminishes awareness of sensory impulses, especially pain, by the brain; in large doses, it causes stupor, coma, or convulsions. { när käd·ik }

narcotine See noscapine. { 'när·kə,tēn }

nari See caliche. { 'när·ē }

naringin [ORG CHEM] $C_{27}H_{32}O_{14}$ A crystalline bioflavonoid with a melting point of 171°C; soluble in acetone and alcohol; used as a food supplement. Also known as aurantiin. { nə'rin·jən }

naris See nostril. { 'nar·əs }

Narizian [GEOL] A North American stage of geologic time; a subdivision of the upper Eocene, above Ulatisian and below Fresnian. { nə'rizh·ən }

narrow [GEOGR] A constricted section of a mountain pass,

1,2-NAPHTHOQUINONE

Structural formula of 1,2-naphthoquinone.

NAPHTHORESORCINOL

Structural formula of naphthoresorcinol.

valley, or cave, or a gap or narrow passage between mountains. { 'när·ō }

narrow-angle glaucoma [MED] Increased intraocular tension due to a block of the angle of the anterior chamber from contact of the iris by the trabecula. Also known as obstructive glaucoma. { 'när·ō ¦aŋ·gəl glau̇'kō·mə }

narrow-band amplifier [ELECTR] An amplifier which increases the magnitude of signals over a band of frequencies whose bandwidth is small compared to the average frequency of the band. { 'när·ō ¦band 'am·plə‚fī·ər }

narrow-band frequency modulation [COMMUN] Frequency-modulated broadcasting system used primarily for two-way voice communication, typically having a maximum deviation of 15 kilohertz or less. { 'när·ō ¦band 'frē·kwən·sē ‚mäj·ə'lā·shən }

narrow-band-pass filter [ELECTR] A band-pass filter in which the band of frequencies transmitted by the filter has a bandwidth which is small compared to the average frequency of the band. { 'när·ō ¦ban ‚pas ‚fil·tər }

narrow-band path [COMMUN] A communications path having a bandwith typically less than 20 kilohertz. { 'när·ō ¦band 'path }

narrow-band pyrometer [ENG] A pyrometer in which light from a source passes through a color filter, which passes only a limited band of wavelengths, before falling on a photoelectric detector. Also known as spectral pyrometer. { 'när·ō ¦band pī'räm·əd·ər }

narrow beam [PHYS] In measurements of the attenuation of a beam of ionizing radiation, a beam in which the scattered radiation does not reach the detector. { ¦när·ō 'bēm }

narrow-beam antenna [ELECTROMAG] An antenna which radiates most of its power in a cone having a radius of only a few degrees. { 'när·ō ¦bēm an'ten·ə }

narrow cut filter [OPTICS] An optical filter which displays an abrupt change from high transmission to complete absorption over a narrow wavelength region. { ¦när·ō ‚kət 'film }

narrow gage [CIV ENG] A railway gage narrower than the standard gage of 4 feet 8$\frac{1}{2}$ inches (143.51 centimeters). { 'när·ō ¦gāj }

narrow-gap spark chamber [NUCLEO] A type of spark chamber in which the plates are only 6 to 10 millimeters apart, so that sparks usually follow the electric field perpendicular to the plates, and coincide with the particle track at one point; the great majority of spark chambers are of this type. { 'när·ō ¦gap 'spärk ‚chäm·bər }

narrows [GEOGR] A navigable narrow part of a bay, strait, or river. { 'när·ōz }

narrow-sector recorder [ELECTR] A radio direction finder with which atmospherics are received from a limited sector related to the position of the antenna; this antenna is usually rotated continuously and the bearings of the atmospherics recorded automatically. { 'när·ō ¦sek·tər ri'kȯrd·ər }

narrow-sense heritability [GEN] The degree to which individual phenotypes are determined by the genes transmitted from the parents; expressed as the ratio of the additive genetic variance to the total phenotypic variance. { ¦när·ō ‚sens ‚her·ə·tə'bil·əd·ē }

narrow-spectrum antibiotic [MICROBIO] An antibiotic effective against a limited number of microorganisms. { 'när·ō ¦spek·trəm ‚ant·i‚bī'äd·ik }

narsarsukite [MINERAL] Na$_2$(Ti,Fe)Si$_4$(O,F) Mineral composed of sodium titanium iron fluoride and silicate. { när·sə'sə‚kīt }

narwhal [VERT ZOO] *Monodon monoceros.* An arctic whale characterized by lack of a dorsal fin, and by possession in the male of a long, twisted, pointed tusk (or rarely, two tusks) which is a source of ivory. { 'när‚wäl }

N-ary code [COMMUN] Code employing *N* distinguishable types of code elements. { 'en·ə·rē ‚kōd }

n-ary composition [MATH] A function that associates an element of a set with every sequence of *n* elements of the set. { 'en·ə·rē ‚käm·pə'zish·ən }

N-ary pulse-code modulation [COMMUN] Pulse-code modulation in which the code for each element consists of any one of *N* distinguishable types of elements. { 'en·ə·rē 'pəls ‚kōd ‚mäj·ə'lā·shən }

n-ary tree [MATH] A rooted tree in which each vertex has at most *n* successors. { 'en·ə·rē ‚trē }

nasal [ANAT] Of or pertaining to the nose. { 'nā·zəl }

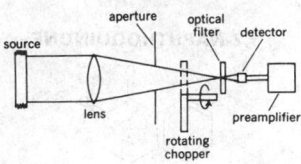

NARROW-BAND PYROMETER

source, aperture, optical filter, detector, lens, rotating chopper, preamplifier

Components of a narrow-band pyrometer.

nasal base breadth [ANTHRO] The distance measured across the alae (wings) of the nose, when they are in rest position. { 'nā·zəl ¦bās ¦bredth }

nasal bone [ANAT] Either of two rectangular bone plates forming the bridge of the nose; they articulate with the frontal, ethmoid, and maxilla bones. { 'nā·zəl ‚bōn }

nasal bridge breadth [ANTHRO] The width measured between the junctures of the cheekbone and nasal bone, just inside the internal canthi. { 'nā·zəl ¦brij ¦bredth }

nasal bridge salient [ANTHRO] The distance measured from the tip of the bony bridge in the midline of the nose to the juncture of the bony sidewall with the cheek. { 'nā·zəl ¦brij ¦sā·lē·ənt }

nasal cavity [ANAT] Either of a pair of cavities separated by a septum and located between the nasopharynx and anterior nares. { 'nā·zəl ¦kav·əd·ē }

nasal crest [ANAT] **1.** The linear prominence on the medial border of the palatal process of the maxilla. **2.** The linear prominence on the medial border of the palatine bone. **3.** The linear prominence on the internal border of the nasal bone and forming part of the nasal septum. { 'nā·zəl ¦krest }

nasal height [ANTHRO] The height of the nose measured from the nasion to the middle of the lower margin of the anterior nares. { 'nā·zəl ¦hīt }

nasal index [ANTHRO] The ratio, ×100, of the greatest width of the anterior nasal openings of the skull to the height of the nasal skeleton. { 'nā·zəl ¦in‚deks }

nasal pit *See* olfactory pit. { 'nā·zəl ‚pit }

nasal process of the frontal bone [ANAT] The downward projection of the nasal part of the frontal bone which terminates as the nasal spine. { 'nā·zəl ¦prä·səs əv thə 'frənt·əl ‚bōn }

nasal process of the maxilla [ANAT] Frontal process of the maxilla. { 'nā·zəl ¦prä·səs əv thə mak'sil·ə }

nasal root breadth [ANTHRO] The distance measured between junctures of the cheekbone and the nasal bone, just inside the internal canthi. { 'nā·zəl ¦rüt ¦bredth }

nasal root salient [ANTHRO] The distance measured between the internal canthus and the nasion. { 'nā·zəl ¦rüt 'sā·lē·ənt }

nasal septum [ANAT] The partition separating the two nasal cavities. { 'nā·zəl ¦sep·təm }

nasal tip height [ANTHRO] The distance measured between the subnasale and the pronasale. { 'nā·zəl ¦tip 'hīt }

nasal tip salient [ANTHRO] The distance measured from the nasal wing to the pronasale. { 'nā·zəl ¦tip 'sā·lē·ənt }

nascent [CHEM] Pertaining to an atom or simple compound at the moment of its liberation from chemical combination, when it may have greater activity than in its usual state. { 'nā·sənt }

nascent ribonucleic acid [MOL BIO] **1.** A ribonucleic acid (RNA) molecule in the process of being synthesized. **2.** A complete, newly synthesized RNA molecule before any alterations have been made. { ¦näs·ənt ¦rī·bō·nü‚klē·ik 'as·əd }

n'aschi [METEOROL] A northeast wind which occurs in winter on the Iranian coast of the Persian Gulf, especially near the entrance to the Gulf, and also on the Makran (West Pakistan) coast; it is probably associated with an outflow from the central Asiatic anticyclone which extends over the high land of Iran. { 'näs·chē }

Nasellina [INV ZOO] The equivalent name for Monopylina. { ‚nas·ə'lī·nə }

nasion [ANTHRO] Midpoint of the nasofrontal suture. { 'nā·zē‚än }

nasion-menton [ANTHRO] The distance measured between the nasion and the midpoint of the lower edge of the chin. { 'nā·zē‚än 'men‚tän }

Nasmyth focus [ASTRON] One of two locations in a telescope with an altitude-azimuth mounting, located on the horizontal axis on either side of the telescope structure, to which light can be reflected to come to a focus; at such a focus there is usually an observing platform to carry instrumentation. { 'nā‚smith ‚fō·kəs }

nasolacrimal canal [ANAT] The bony canal that lodges the nasolacrimal duct. Also known as lacrimal canal. { ¦nā·zō'lak·rə·məl kə'nal }

nasolacrimal duct [ANAT] The membranous duct lodged within the nasolacrimal canal; it gives passage to the tears from the lacrimal sac to the inferior meatus of the nose. { ¦nā·zō'lak·rə·məl ‚dəkt }

nasolacrimal groove [EMBRYO] The furrow, the maxillary, and the lateral nasal processes of the embryo. { ¦nā·zō'lak·rə·məl 'grüv }

nasonite [MINERAL] $Ca_4Pb_6Si_6O_{21}Cl_2$ A white mineral composed of silicate and chloride of calcium and lead and occurring in granular masses. { 'nās·ən,īt }

nasopalatine cyst See median maxillary cyst. { ¦nā·zō'pal·ə,tēn 'sist }

nasopalatine duct [EMBRYO] A canal between the oral and nasal cavities of the embryo at the point of fusion of the maxillary and palatine processes. { ¦nā·zō'pal·ə,tēn 'dəkt }

nasopharynx [ANAT] The space behind the posterior nasal orifices, above a horizontal plane through the lower margin of the palate. { ¦nā·zō'far·iŋks }

nastic movement [BOT] Movement of a flat plant part, oriented relative to the plant body and produced by diffuse stimuli causing disproportionate growth or increased turgor pressure in the tissues of one surface. { 'nas·tik 'müv·mənt }

nasturan See pitchblende. { 'nas·tə¦ran }

Nasutitermitinae [INV ZOO] A subfamily of termites in the family Termitidae, characterized by having the cephalic glands open at the tip of an elongated tube which projects anteriorly. { nə¦süd·ə·tər'mit·ən,ē }

Natalidae [VERT ZOO] The funnel-eared bats, a monogeneric family of small, tropical American insectivorous bats (Chiroptera) with large, funnellike ears. { nə'tal·ə,dē }

Natantia [INV ZOO] A suborder of decapod crustaceans comprising shrimp and related forms characterized by a long rostrum and a ventrally flexed abdomen. { nə'tan·chə }

Nathansohn's theory [OCEANOGR] The theory that nutrient salts in the lighted surface layers of the ocean are consumed by plants, accumulate in the deep ocean through sinking of dead plant and animal bodies, and eventually return to the euphotic layer through diffusion and vertical circulation of the water. { 'nā·thən·sənz ,thē·ə·rē }

Naticacea [INV ZOO] A superfamily of gastropod mollusks in the order Prosobranchia. { ,nad·ə'kā·shē·ə }

Naticidae [INV ZOO] A family of gastropod mollusks in the order Pectinibranchia comprising the moon-shell snails. { nə'tis·ə,dē }

national meridian [GEOD] A meridian chosen in a particular nation as the reference datum for determining longitude for that nation. { 'nash·ən·əl mə'rid·ē·ən }

native [BIOL] Grown, produced or originating in a specific region or country. [GEOCHEM] Pertaining to an element found in nature in a nongaseous state. { 'nād·iv }

native asphalt [GEOL] Exudations or seepages of asphalt occurring in nature in a liquid or semiliquid state. Also known as natural asphalt. { 'nād·iv 'as,fôlt }

native coal See natural coke. { 'nād·iv 'kōl }

native element [GEOL] Any of 20 elements, such as copper, gold, and silver, which occur naturally uncombined in a nongaseous state; there are three groups—metals, semimetals, and nonmetals. { 'nād·iv 'el·ə·mənt }

native language [COMPUT SCI] Machine language that is executed by the computer for which it is specifically designed, in contrast to a computer using an emulator. { 'nād·iv 'laŋ·gwij }

native metal [GEOCHEM] A metallic native element; includes silver, gold, copper, iron, mercury, iridium, lead, palladium, and platinum. { 'nād·iv 'med·əl }

native mode [COMPUT SCI] **1.** The mode of operation of a software product that is being used on a computer for which it was specifically designed, without use of an emulator. **2.** The mode of operation of a device that is carrying out the function for which it was designed and is not emulating another device. { 'nād·iv 'mōd }

native paraffin See ozocerite. { 'nād·iv 'par·ə·fən }

native state [CEL MOL] The folded protein configuration that is maintained through noncovalent interactions such as hydrophobic interactions, electrostatic interactions, and hydrogen bonds. { ¦nād·iv ¦stāt }

native uranium [GEOCHEM] Uranium as found in nature; a mixture of the fertile uranium-238 isotope (99.3%), the fissionable uranium-235 isotope (0.7%), and a minute percentage of other uranium isotopes. Also known as natural uranium; normal uranium. { 'nād·iv yə'rā·nē·əm }

natremia [MED] Excessive amounts of sodium in the blood. { nə'trē·mē·ə }

natric horizon [GEOL] A soil horizon that has the properties of an argillic horizon, but also displays a blocky, columnar, or prismatic structure and has a subhorizon with an exchangeable-sodium saturation of over 15%. { 'nā·trik hə'rīz·ən }

natrium [CHEM] Latin name for sodium; source of the symbol Na. { 'nā·trē·əm }

natriuresis [PHYSIO] Excretion of sodium in the urine. { ¦na·trē·yù'rē·səs }

natriuretic [PHARM] A medicinal agent which inhibits reabsorption of cations, particularly sodium, from urine. { ¦nā·trē·yù'red·ik }

natroalunite [MINERAL] $NaAl_3(SO_4)_2(OH)_6$ Mineral composed of basic sodium aluminum sulfate. Also known as almeriite. { ¦nā·trō'al·ə,nīt }

natrochalcite [MINERAL] $NaCu_2(SO_4)(OH)·H_2O$ An emerald-green mineral composed of hydrous basic sulfate of sodium and copper. { ¦nā·trō'kal,sīt }

natrolite [MINERAL] $Na_2Al_2Si_3O_{10}·2H_2O$ A zeolite mineral composed of hydrous silicate of sodium and aluminum; usually occurs in slender acicular or prismatic crystals. { 'nā·trə,līt }

natromontebrasite [MINERAL] $(Na,Li)Al(PO_4)(OH,F)$ A mineral composed of hydrous basic phosphate of sodium, lithium, and aluminum; it is isomorphous with montebrasite and amblygonite. Also known as fremontite. { ¦nā·trō,män·tē'brä,zīt }

natron [MINERAL] $Na_2CO_3·10H_2O$ A white, yellow, or gray mineral that crystallizes in the monoclinic system, is soluble in water, and generally occurs in solution or in saline residues. { 'nā·trən }

natron lake See soda lake. { 'nā·trən ,lāk }

natrophilite [MINERAL] $NaMn(PO_4)$ A mineral composed of sodium manganese phosphate. { nə'trä·fə,līt }

natural abundance [NUCLEO] The abundance ratio of an isotope in a naturally occurring terrestrial sample of an element. { 'nach·rəl ə'bən·dəns }

natural aging [MET] Spontaneous aging at room temperature of a supersaturated metallic solid solution. { 'nach·rəl 'āj·iŋ }

natural antenna frequency [ELECTROMAG] Lowest resonant frequency of an antenna without added inductance or capacitance. { 'nach·rəl an'ten·ə ,frē·kwən·sē }

natural arch [GEOL] **1.** A landform similar to a natural bridge but not formed by erosive agencies. **2.** See natural bridge. { 'nach·rəl 'ärch }

natural asphalt See native asphalt. { 'nach·rəl 'as,fôlt }

natural binary coded decimal system [COMPUT SCI] A particular binary coded decimal system that uses the first ten binary numbers in sequence to represent the digits 0 through 9. { 'nach·rəl 'bī·ner·ē ¦kōd·əd ¦des·məl ,sis·təm }

natural bitumen [GEOL] Native mineral pitch, tar, or asphalt. { 'nach·rəl bə'tü·mən }

natural boundary [MATH] Those points of the boundary of a region where an analytic function is defined through which the function cannot be continued analytically. { 'nach·rəl 'baùn·drē }

natural bridge [GEOL] An archlike rock formation spanning a ravine or valley and formed by erosion. Also known as natural arch. { 'nach·rəl 'brij }

natural cement [MATER] Hydraulic cement made from pulverized and heated limestone containing clay, magnesia, and iron. { 'nach·rəl si'ment }

natural circulation reactor [NUCLEO] A reactor in which the coolant (usually water) circulates without pumping, owing to the different densities in its cold and reactor-heated portions. { 'nach·rəl ,sər·kyə'lā·shən rē,ak·tər }

natural coke [GEOL] Coal that has been naturally carbonized by contact with an igneous intrusion, or by natural combustion. Also known as black coal; blind coal; carbonite; cinder coal; coke coal; cokeite; finger coal; native coal. { 'nach·rəl 'kōk }

natural convection [THERMO] Convection in which fluid motion results entirely from the presence of a hot body in the fluid, causing temperature and hence density gradients to develop, so that the fluid moves under the influence of gravity. Also known as free convection. { 'nach·rəl kən'vek·shən }

natural coordinates [FL MECH] An orthogonal, or mutually perpendicular, system of curvilinear coordinates for the description of fluid motion, consisting of an axis t tangent to the instantaneous velocity vector and an axis n normal to this velocity vector to the left in the horizontal plane, to which a vertically directed axis *z* may be added for the description of three-dimensional flow; such a coordinate system often permits a concise formulation of atmospheric dynamical problems, especially in the Lagrangian system of hydrodynamics. { 'nach·rəl kō'ȯrd·ən·əts }

natural draft [FL MECH] Unforced gas flow through a chimney or vertical duct, directly related to chimney height and the temperature difference between the ascending gases and the atmosphere, and not dependent upon the use of fans or other mechanical devices. { 'nach·rəl 'draft }

natural-draft cooling tower [MECH ENG] A cooling tower that depends upon natural convection of air flowing upward and in contact with the water to be cooled. { 'nach·rəl ¦draft 'kül·iŋ ¦taù·ər }

natural equations of a curve *See* intrinsic equations of a curve. { ¦nach·rəl i¦kwā·zhənz əv ə 'kərv }

natural fiber [TEXT] A textile fiber of mineral, plant, or animal origin. { 'nach·rəl 'fī·bər }

natural food [FOOD ENG] A type of food that is prepared with minimum processing and contains no preservatives or artificial additives. { 'nach·rəl 'füd }

natural frequency [ELECTR] The lowest resonant frequency of an antenna, circuit, or component. [PHYS] The frequency with which a system oscillates in the absence of external forces; or, for a system with more than one degree of freedom, the frequency of one of the normal modes of vibration. { 'nach·rəl 'frē·kwən·sē }

natural fuel reactor *See* natural uranium reactor. { 'nach·rəl 'fyül rē,ak·tər }

natural function [MATH] A trigonometric function, as opposed to its logarithm. { 'nach·rəl 'fəŋk·shən }

natural function generator *See* analytical function generator. { 'nach·rəl 'fəŋk·shən ¦jen·ə,rād·ər }

natural gas [MATER] A combustible, gaseous mixture of low-molecular-weight paraffin hydrocarbons, generated below the surface of the earth; contains mostly methane and ethane with small amounts of propane, butane, and higher hydrocarbons, and sometimes nitrogen, carbon dioxide, hydrogen sulfide, and helium. { 'nach·rəl 'gas }

natural gasoline [MATER] The liquid paraffin hydrocarbon contained in natural gas and recovered by compression, distillation, and absorption. { 'nach·rəl ¦gas·ə'lēn }

natural-gasoline plant [CHEM ENG] Compression, distillation, and absorption process facility used to remove natural gasoline (mostly butanes and heavier components) from natural gas. { 'nach·rəl ¦gas·ə'lēn ¦plant }

natural glass [GEOL] An amorphous, vitreous inorganic material that has solidified from magma too quickly to crystallize. { 'nach·rəl 'glas }

natural harbor [GEOGR] A harbor where the configuration of the coast provides the necessary protection. { 'nach·rəl 'här·bər }

natural immunity [IMMUNOL] Native immunity possessed by the individuals of a race, strain, or species. { 'nach·rəl i'myü·nəd·ē }

natural interference [COMMUN] Electromagnetic interference arising from natural terrestrial phenomena (called atmospheric interference), or electromagnetic interference caused by natural disturbances originating outside the atmosphere of the earth (called galactic and solar noise). { 'nach·rəl ¦in·tər'fir·əns }

naturalized [ECOL] Of a species, having become permanently established after being introduced. { 'nach·rə,līzd }

natural killer cell [HISTOL] A large, granular lymphocyte that can lyse a variety of target cells when it is activated by interferon. Abbreviated NK. { ¦nach·rəl 'kil·ər ,sel }

natural laminar flow [AERO ENG] Airflow over a portion of the wing such that local pressure decreases in the direction of flow and flow in the boundary layer is laminar rather than turbulent; frictional drag on the aircraft is greatly reduced. { ¦nach·rəl 'lam·ə·nər 'flō }

natural language [COMPUT SCI] A computer language whose rules reflect and describe current rather than prescribed usage; it is often loose and ambiguous in interpretation, meaning different things to different hearers. { 'nach·rəl 'laŋ·gwij }

natural language interaction [COMPUT SCI] The interaction of users with computer systems through the medium of natural languages. { 'nach·rəl ¦laŋ·gwij ,in·tər'ak·shən }

NATURAL-DRAFT COOLING TOWER

Photograph of natural-draft cooling tower. (*Haman, Inc.*)

natural language processing [COMPUT SCI] Computer analysis and generation of natural language text; encompasses natural language interaction and natural language text processing. { 'nach·rəl ¦laŋ·gwij 'prä,ses·iŋ }

natural language text processing [COMPUT SCI] Computer processing of natural language text into a more useful form, as in automatic text translation or text summarization. { 'nach·rəl ¦laŋ·gwij 'tekst 'prä,ses·iŋ }

natural levee [GEOL] An elongate embankment compounded of sand and silt and deposited along both banks of a river channel during times of flood. { 'nach·rəl 'lev·ē }

natural linewidth [SPECT] The part of the linewidth of an absorption or emission line that results from the finite lifetimes of one or both of the energy levels between which the transition takes place. { 'nach·rəl 'līn,width }

natural load [HYD] The quantity of sediment carried by a stable stream. { 'nach·rəl 'lōd }

natural logarithm *See* logarithm. { 'nach·rəl 'läg·ə,rith·əm }

natural number [MATH] One of the integers 1, 2, 3, ... { 'nach·rəl 'nəm·bər }

natural period [PHYS] Period of the free oscillation of a body or system; when the period varies with amplitude, the natural period is the period when the amplitude approaches zero. { 'nach·rəl 'pir·ē·əd }

natural pressure cycle [MET] A cycle in which pressure buildup conforms proportionately to the buildup of stresses due to forming. { 'nach·rəl 'presh·ər ,sī·kəl }

natural radiation *See* background radiation. { 'nach·rəl ,rād·ē'ā·shən }

natural radioactivity [NUCLEO] Radioactivity exhibited by naturally occurring radionuclides. { 'nach·rəl ,rād·ē·ō·ak'tiv·əd·ē }

natural radio-frequency interference [GEOPHYS] Natural terrestrial phenomena of an electromagnetic nature, or natural electromagnetic disturbances originating outside the atmosphere, which interfere with radio communications. { 'nach·rəl ¦rād·ē·ō ¦frē·kwən·sē ,in·tər'fir·əns }

natural red *See* purpurin. { 'nach·rəl 'red }

natural remanent magnetization [GEOPHYS] The magnetization of rock which exists in the absence of a magnetic field and has been acquired from the influence of the earth's magnetic field at the time of their formation or, in certain cases, at later times. Abbreviated NRM. { 'nach·rəl 'rem·ə·nənt ,mag·nə·tə'zā·shən }

natural resonance [PHYS] Resonance in which the period or frequency of the applied agency maintaining oscillation is the same as the natural period of oscillation of a system. { 'nach·rəl 'rez·ən·əns }

natural resource [MATER] A deposit of minerals, water, or other materials furnished by nature. { 'nach·rəl 'rē,sȯrs }

natural scale [GRAPHICS] The ratio between the linear dimensions of a chart or drawing and the actual dimensions represented, expressed as a proportion; for example, 1 inch on a chart of natural scale 1:2,000,000 represents 2,000,000 inches on the earth. Also known as representative fraction. { 'nach·rəl 'skāl }

natural science [SCI TECH] Collectively, the branches of science dealing with objectively measurable phenomena pertaining to the transformations and relationships of energy and matter; includes biology, physics, and chemistry. { 'nach·rəl 'sī·əns }

natural-seasoned lumber *See* air-dried lumber. { 'nach·rəl ¦sēz·ənd 'ləm·bər }

natural selection [EVOL] Darwin's theory of evolution, according to which organisms tend to produce progeny far above the means of subsistence; in the struggle for existence that ensues, only those progeny with favorable variations survive; the favorable variations accumulate through subsequent generations, and descendants diverge from their ancestors. { 'nach·rəl si'lek·shən }

natural splitting [MIN ENG] In mine ventilation, a flow of air dividing among the branches, of its own accord and without regulation, in inverse relation to the resistance of each airway. { 'nach·rəl 'splid·iŋ }

natural steel [MET] **1.** Steel made directly from cast iron. **2.** Steel, such as wootz, made directly from the ore. { 'nach·rəl 'stēl }

natural tunnel [GEOL] A cave that is nearly horizontal and is open at both ends. Also known as tunnel cave. { 'nach·rəl ¦tən·əl }

natural uranium *See* native uranium. { 'nach·rəl yu'rā·nē·əm }

natural uranium reactor [NUCLEO] A nuclear reactor in which natural unenriched uranium is the principal fissionable material. Also known as natural fuel reactor. { 'nach·rəl yu'rā·nē·əm rē‚ak·tər }

natural ventilation [MIN ENG] The weak and varying ventilation in a mine caused by the difference in air density between shafts. { 'nach·rəl ‚vent·əl'ā·shən }

natural ventilation pressure [MIN ENG] A pressure difference across the shaft-bottom doors caused by a lack of balance in the two vertical air columns. { 'nach·rəl ‚vent·əl'ā·shən ‚presh·ər }

natural wavelength [ELECTROMAG] Wavelength corresponding to the natural frequency of an antenna or circuit. { 'nach·rəl 'wāv‚leŋkth }

natural well [GEOL] A sinkhole or other natural opening which resembles a well extending below the water table and from which groundwater can be withdrawn. { 'nach·rəl 'wel }

natural width of energy level [PHYS] A measure of the spread in energy of an excited state of a quantized system due to spontaneous transitions to other states; quantitatively, it is the difference between the energies for which the intensity of emission from or absorption by the state, or of the scattering cross section associated with it, is one-half its maximum value, in the absence of any external influence on the system. { 'nach·rəl ¦width əv 'en·ər·jē ‚lev·əl }

Naucoridae [INV ZOO] A family of hemipteran insects in the superfamily Naucoroidea. { nȯ'kȯr·ə‚dē }

Naucoroidea [INV ZOO] The creeping water bugs, a superfamily of hemipteran insects in the subdivision Hydrocorisae; they are suboval in form, with chelate front legs. { nȯ·kə'rȯid·ē·ə }

naujaite [PETR] A coarse hypidiomorphic-granular sodalite-rich nepheline syenite that contains microcline and small amounts of albite, analcime, acmite, and sodium amphiboles and is characterized by a poikilitic texture. { 'naú·jə‚īt }

naumannite [MINERAL] Ag₂Se An iron-black mineral that crystallizes in the isometric system; consists of silver selenide, and occurs massive or in crystals; specific gravity is 8. { 'naú·mə‚nīt }

nauplius [INV ZOO] A larval stage characteristic of many groups of Crustacea; the oval, unsegmented body has three pairs of appendages: uniramous antennules, biramous antennae, and mandibles. { 'nȯ·plē·əs }

nausea [MED] Feeling of discomfort in the stomach region, accompanied by aversion to food and a tendency to vomit. { 'nȯ·zē·ə }

nautical almanac [NAV] A book published annually by the governments of the principal maritime nations which contains the astronomical data required for navigation by observations of celestial objects; an abridged version is known as the abridged nautical almanac. { 'nȯd·ə·kəl 'ȯl·mə‚nak }

nautical astronomy [NAV] The science of determining position and direction of a ship by observation of celestial objects. { 'nȯd·ə·kəl ə'strän·ə·mē }

nautical chain [MECH] A unit of length equal to 15 feet or 4.572 meters. { 'nȯd·ə·kəl 'chān }

nautical chart [NAV] A graphic representation on a plane surface of a section of the earth's sea surface constructed to include known dangers and aids to navigation. Also known as marine chart. { 'nȯd·ə·kəl 'chärt }

nautical distance [MAP] The length in nautical miles of the rhumb line joining any two places on the earth's surface. { 'nȯd·ə·kəl 'dis·təns }

nautical mile [NAV] A unit of distance used principally in navigation; for practical consideration it is usually considered the length of 1 minute of any great circle of the earth, the meridian being the great circle most commonly used; the International Hydrographic Bureau in 1929 proposed a standard length of 1852 meters, which is known as the international nautical mile. { 'nȯd·ə·kəl 'mīl }

nautical twilight [ASTRON] The interval of incomplete darkness between sunrise or sunset and the time at which the center of the sun's disk is 12° below the celestial horizon. { 'nȯd·ə·kəl 'twī‚līt }

Nautilidae [INV ZOO] A monogeneric family of cephalopod

mollusks in the order Nautiloidea; *Nautilus pompilius* is the only well-known living species. { nȯ'til·ə‚dē }

Nautiloidea [INV ZOO] A primitive order of tetrabranchiate cephalopods; shells are external and smooth, being straight or coiled and chambered with curved transverse septa. { ‚nȯd·əl'ȯid·ē·ə }

Nautilus [INV ZOO] The only living genus of the molluscan subclass Nautiloidea, containing the only living cephalopods with an external chambered shell and numerous cephalic tentacles; six species live in the western Pacific and around the East Indies. { 'nȯd·ə·ləs }

navaglobe [NAV] The portion of the navarho navigation system that provides bearing; it utilizes three antennas located at the apexes of a triangle. { 'nav·ə‚glōb }

Navaho [ORD] A U.S. Air Force surface-to-surface long-range supersonic strategic missile, powered by two ramjet engines. { 'nä·və‚hō }

Navajo sandstone [GEOL] A fossil dune formation of Jurassic age found in the Colorado Plateau of the United States. { 'nä·və‚hō 'san‚stōn }

naval architect [NAV ARCH] An individual who designs ships and makes alterations in them. { 'nā·vəl 'är·kə‚tekt }

naval architecture [ENG] The study of the physical characteristics and the design and construction of buoyant structures, such as ships, boats, barges, submarines, and floats, which operate in water; includes the construction and operation of the power plant and other mechanical equipment of these structures. { 'nā·vəl 'är·kə‚tek·chər }

naval armament [ORD] The combat equipment used in naval ships and by naval aircraft. { 'nā·vəl 'är·mə·mənt }

naval brass [MET] Brass composed of 60–62% copper, 37–39% zinc, and 0.75–1% tin; relatively resistant to corrosion by seawater. Also known as naval bronze. { 'nā·vəl 'bras }

naval bronze *See* naval brass. { 'nā·vəl 'bränz }

naval meteorology [METEOROL] The branch of meteorology which studies the interaction between the ocean and the overlying air mass, and which is concerned with atmospheric phenomena over the oceans, the effect of the ocean surface on these phenomena, and the influence of such phenomena on shallow and deep seawater. { 'nā·vəl ‚mē·dē·ə'räl·ə·jē }

naval mine [ORD] An item designed to be located under water and exploded by means of propeller vibration, magnetic attraction, contact, or remote control. { 'nā·vəl ‚mīn }

naval ship [NAV ARCH] A ship designed primarily for use in warfare, either directly in combat operations or to provide services and support such operations. { 'nā·vəl ‚ship }

naval stores [MATER] **1.** Pitch and rosin formerly used in the construction of wooden ships. **2.** All pine wood products, including rosin, turpentine, and pine oils. { 'nā·vəl ‚stȯrz }

navar [NAV] A coordinated series of radar air navigation and traffic-control aids utilizing transmissions at wavelengths of 10 and 60 centimeters to provide in an aircraft the distance and bearing from a given point, along with a display of other aircraft in the vicinity and commands from the ground; the system also provides on the ground a display of all aircraft in the vicinity, with their altitudes, identities, and means for transmitting certain commands. Derived from navigation and ranging. { 'nā‚vär }

navarho [NAV] A long-distance, low-frequency continuous-wave navigation system providing simultaneous bearing and distance information; the portion providing bearing is termed navaglobe. { 'nav·ə‚rō }

Navarroan [PALEON] A North American (Gulf Coast) stage of Upper Cretaceous geologic time, above the Tayloran and below the Midwayan of the Tertiary. { ‚nav·ə'rō·ən }

Navascreen [NAV] System for displaying and computing air-traffic control data, using information obtained from radar and other sources. { 'nav·ə‚skrēn }

NAVEAM [NAV] A British-originated radio navigational warning of dangers in the eastern Atlantic, Mediterranean Sea, and Red Sea. { 'nä‚vēm }

navel [ANAT] The umbilicus. { 'nā·vəl }

navel height [ANTHRO] The vertical distance, of a standing subject, measured from the center of the navel to the floor. { 'nā·vəl ‚hīt }

navel point *See* umbilical point. { 'nā·vəl ‚pȯint }

navicular [ANAT] A boat-shaped bone, especially the lateral bone on the radial side of the proximal row of the carpus.

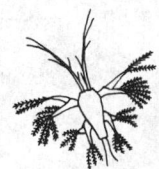

NAUPLIUS

Nauplius of the shrimp *Penaeus*. *(From T. I. Storer and R. L. Usinger, General Zoology, 4th ed., McGraw-Hill, 1965)*

NAUTILIDAE

Shell of *nautilus pompilius* which may be up to 10 inches (25 centimeters) in diameter.

[BIOL] Resembling or having the shape of a boat. { nə'vik·yə·lər }

navicular cells [PATH] Boat-shaped squamous epithelial cells filled with glycogen and prominent in the exfoliated cells of the uterine cervix of pregnant women. { nə'vik·yə·lər ‚selz }

naviculoid [BIOL] Referring to a diatom, boat-shaped. { nə'vik·yə‚lȯid }

Navier's equation [MECH] A vector partial differential equation for the displacement vector of an elastic solid in equilibrium and subjected to a body force. { nä'vyāz i‚kwā·zhən }

Navier-Stokes equations [FL MECH] The equations of motion for a viscous fluid which may be written $d\mathbf{V}/dt = -(1/\rho)\nabla p + \mathbf{F} + \nu\nabla^2\mathbf{V} + (^1/_3)\nu\nabla(\nabla\cdot\mathbf{V})$, where p is the pressure, ρ the density, F the total external force per unit mass, $\mathbf{V}$ the fluid velocity, and ν the kinematic viscosity; for an incompressible fluid, the term in $\nabla\cdot\mathbf{V}$ (divergence) vanishes, and the effects of viscosity then play a role analogous to that of temperature in thermal conduction and to that of density in simple diffusion. { nä'vyā 'stōks i‚kwā·zhənz }

navigable airspace [NAV] Airspace at and above the minimum safe flight level, including airspace needed for safe takeoff and landing. { 'nav·i·gə·bəl 'er‚spās }

navigable semicircle [METEOROL] That half of a cyclonic storm area in which the rotary and progressive motions of the storm tend to counteract each other, and the winds are in such a direction as to blow a vessel away from the storm track. { 'nav·i·gə·bəl 'sem·i‚sər·kəl }

navigating bridge See flying bridge. { 'nav·ə‚gād·iŋ ‚brij }

navigating officer [NAV] An officer serving as a navigator. { 'nav·ə‚gād·iŋ ‚ȯf·ə·sər }

navigating sextant [NAV] A sextant designed and used for observing the altitudes of celestial bodies, as contrasted with a hydrographic sextant. { 'nav·ə‚gād·iŋ ‚sek·stənt }

navigation [COMPUT SCI] In a database management system, the techniques provided for locating information within the system. [ENG] The process of directing the movement of a craft so that it will reach its intended destination; subprocesses are position fixing, dead reckoning, pilotage, and homing. { ‚nav·ə'gā·shən }

navigation accuracy measurement system [NAV] A simple height-finding radar system that employs a pulse radar with a rotating fan-beam antenna and a curve-fitting method to evaluate the accuracy of aircraft altitude-measuring equipment. { ‚nav·ə'gā·shən 'ak·yə·rə·sē 'mezh·ər·mənt ‚sis·təm }

navigational aid [NAV] An instrument, device, chart, method, or such, intended to assist in the navigation of a craft; this expression should not be confused with "aid to navigation," which refers only to devices external to a craft. { ‚nav·ə'gā·shən·əl 'ād }

navigational almanac [NAV] A publication that contains tabulated positions of astronomical objects at regular intervals to enable navigators to determine their position. { ‚nav·ə'gā·shən·əl 'ȯl·mə‚nak }

navigational planets [NAV] The four planets commonly observed for obtaining data for use in celestial navigation: Venus, Mars, Jupiter, and Saturn. { ‚nav·ə'gā·shən·əl 'plan·əts }

navigational plot [NAV] A plot of the movements of a craft. { ‚nav·ə'gā·shən·əl 'plät }

navigational satellite [AERO ENG] An artificial earth-orbiting satellite designed for use in at least four widely different navigational systems. { ‚nav·ə'gā·shən·əl 'sad·əl‚īt }

navigational triangle [NAV] In celestial navigation, the spherical triangle solved in computing altitude and azimuth and great-circle sailing problems. { ‚nav·ə'gā·shən·əl 'trī‚aŋ·gəl }

navigation computer [NAV] A computer that uses electronic or electric circuits to compute two or more navigation factors such as altitude, direction, and velocity, or to receive such data and compute course information. { ‚nav·ə'gā·shən kəm‚pyüd·ər }

navigation dam [CIV ENG] A structure designed to raise the level of a stream to increase the depth for navigation purposes. { ‚nav·ə'gā·shən ‚dam }

navigation dome See astrodome. { ‚nav·ə'gā·shən ‚dōm }

navigation head [NAV] A transshipment point on a waterway where loads are transferred between water carriers and land carriers. { ‚nav·ə'gā·shən ‚hed }

navigation lights [NAV] Statutory lights shown by aircraft and vessels during the hours between sunset and sunrise, in accordance with international agreements. { ‚nav·ə'gā·shən ‚līts }

navigation radar [NAV] A search radar used on ships primarily for navigation purposes, to provide a visual indication of bearing and distance to any object that projects above the surface of the water within the range of the radar. { ‚nav·ə'gā·shən 'rā‚där }

navigation receiver [ELECTR] An electronic device that determines a ship's position by receiving and comparing radio signals from transmitters at known locations. { ‚nav·ə'gā·shən ri‚sē·vər }

navigation system error [NAV] The difference between an aircraft's true position and the position reported by its navigation sensors. { ‚nav·ə'gā·shən ‚sis·təm ‚er·ər }

navigator [NAV] A person who navigates or is directly responsible for the navigation of a craft. { 'nav·ə‚gād·ər }

navite [MINERAL] A porphyritic basalt containing phenocrysts of altered olivine, augite, and basic plagioclase in a groundmass of labradorite and augite. { 'nā‚vīt }

NAVSTAR [NAV] A global system of up to 24 navigation satellites developed to provide instantaneous and highly accurate worldwide three-dimensional location by air, sea, and land vehicles equipped with suitable receivers. Derived from navigation system using time and ranging. { 'nav‚stär }

Navy Electronics Laboratory International Algol Compilers See NELIAC. { 'nā·vē i‚lek'trän·iks 'lab·rə‚tȯr·ē ‚in·tər'nash·ən·əl 'al‚gȯl kəm'pīl·ərz }

Navy Heavy See bunker C fuel oil. { 'nā·vē 'hev·ē }

Navy Oceanographic and Meteorological Automatic Device [OCEANOGR] A 6-meter-long, boat-shaped, moored instrumented buoy. Abbreviated NOMAD. { ‚hāv·ē ‚ō·shə·nə‚graf·ik and ‚mēd·ē·ə·rə‚läj·ə·kəl ‚ȯd·ə‚mad·ik di'vīs }

Nb See niobium.

n-body problem See many-body problem. { 'en ‚bad·ē ‚präb·ləm }

NBR See nitrile rubber.

n-cell [MATH] A set that is homeomorphic either with the set of points in n-dimensional euclidean space $(n = 1, 2, \ldots)$ whose distance from the origin is less than unity, or with the set of points whose distance from the origin is less than or equal to unity. { 'en ‚sel }

N center [SOLID STATE] A color center which arises from continued exposure to light in the F band or to x-rays and which produces a faint absorption band on the long-wavelength side of the M band. Also known as G center. { 'en ‚sen·tər }

n-channel [ELECTR] A conduction channel formed by electrons in an n-type semiconductor, as in an n-type field-effect transistor. { 'en ‚chan·əl }

n-channel metal-oxide semiconductor See NMOS. { 'en ‚chan·əl ‚med·əl ‚äk‚sīd 'sem·i·kən‚dək·tər }

n-colorable graph [MATH] A graph whose nodes can be colored using one of n colors on each node in such a way that no edge connects a pair of nodes with the same color. { ‚en ‚kəl·ə·rə·bəl 'graf }

n-component [PARTIC PHYS] Cosmic-ray particles that can take part in nuclear interactions, that is, nucleons, pions, and other baryons and mesons. { 'en kəm‚pō·nənt }

n-connected graph [MATH] A connected graph for which the removal of n points is required to disconnect the graph. { 'en kə‚nek·təd 'graf }

N curve [ELECTR] A plot of voltage against current for a negative-resistance device; its slope is negative for some values of current or voltage. { 'en ‚kərv }

Nd See neodymium.

NDGA See nordihydroguaiaretic acid.

n-dimensional space [MATH] A vector space whose basis has n vectors. { 'en di'men·shən·əl 'spās }

N display [ELECTR] Radar display in which the target appears as a pair of vertical deflections from a horizontal time base; direction is indicated by relative amplitude of the blips; target distance is determined by moving an adjustable pedestal signal along the base line until it coincides with the horizontal position of the blips; the pedestal control is calibrated in distance. { 'en di‚splā }

NDRO *See* nondestructive readout.

Ne *See* neon.

neallotype [SYST] A type specimen that, compared with the holotype, is of the opposite sex, and was collected and described later. { nē'al·ə,tīp }

NEA material *See* negative-electron affinity material. { ¦en¦ē'ā mə,tir·ē·əl }

Neanderthal man [PALEON] A type of fossil human that is a subspecies of *Homo sapiens* and is distinguished by a low broad braincase, continuous arched browridges, projecting occipital region, short limbs, and large joints. { nē'an·dər,täl 'man }

neap high water *See* mean high-water neaps. { 'nēp 'hī ,wȯd·ər }

neap low water *See* mean low-water neaps. { 'nēp 'lō ,wȯd·ər }

neap range [OCEANOGR] The mean semidiurnal range of tide when neap tides are occurring; the mean difference in height between neap high water and neap low water. Also known as mean neap range. { 'nēp ,rānj }

neap rise [OCEANOGR] The height of neap high water above the chart datum. { 'nēp ,rīz }

neaps *See* neap tide. { 'nēps }

neap tidal currents [OCEANOGR] Tidal currents of decreased speed occurring at the time of neap tides. { 'nēp 'tīd·əl ,kə·rəns }

neap tide [OCEANOGR] Tide of decreased range occurring about every 2 weeks when the moon is in quadrature, that is, during its first and last quarter. Also known as neaps. { 'nēp ,tīd }

Nearctic fauna [ECOL] The indigenous animal communities of the Nearctic zoogeographic region. { nē'ärd·ik 'fȯn·ə }

Nearctic zoogeographic region [ECOL] The zoogeographic region that includes all of North America to the edge of the Mexican Plateau. { nē'ärd·ik ¦zō·ō,jē·ə'graf·ik ,rē·jən }

near-earth object [ASTRON] An asteroid or comet whose orbit takes it within 1.3 astronomical units of the sun. { ¦nir 'ərth ,äb,jekt }

near-end crosstalk [COMMUN] A type of interference that may occur at carrier telephone repeater stations when output signals of one repeater leak into the same end of the other repeater. { 'nir ,end 'krȯs,tȯk }

nearest approach [NAV] The least distance between two objects having relative motion with respect to each other. { 'nir·əst ə'prōch }

nearest neighbors [CRYSTAL] Any pair of atoms in a crystal lattice which are as close to each other, or closer to each other, than any other pair. { 'nir·əst 'nā·bərz }

near field [ACOUS] The acoustic radiation field that is close to an acoustic source such as a loudspeaker. [ELECTROMAG] The electromagnetic field that exists within one wavelength of a source of electromagnetic radiation, such as a transmitting antenna. { 'nir ,fēld }

near-field noise *See* flow noise. { 'nir ¦fēld ,nȯiz }

near-field scanning optical microscope [OPTICS] An optical microscope in which the intensity of light focused through a pipette with an aperture at its tip is recorded as the tip is moved across the specimen in a raster pattern at a distance of much less than a wavelength. { 'nir ¦fēld 'skan·iŋ 'äp·tə·kəl 'mī·krə,skōp }

near-field scanning optical microscopy [OPTICS] A technique for making optical measurements at dimensions much smaller than the wavelength of light, by scanning a nanometric detector or radiation source in proximity to a sample surface. Also known as scanning near-field optical microscopy. { 'nir ¦fēld ,skan·iŋ ,äp·tə·kəl mī'kräs·kə·pē }

near-infrared radiation [ELECTROMAG] Infrared radiation having a relatively short wavelength, between 0.75 and about 2.5 micrometers (some scientists place the upper limit from 1.5 to 3 micrometers), at which radiation can be detected by photoelectric cells, and which corresponds in frequency range to the lower electronic energy levels of molecules and semiconductors. Also known as photoelectric infrared radiation. { 'nir ,in·frə'red ,rād·ē·ā'shən }

near-infrared spectrophotometry [ANALY CHEM] Spectrophotometry at wavelengths in the near-infrared region, generally using instruments with quartz prisms in the monochromators and lead sulfide photoconductor cells as detectors to observe absorption bands which are harmonics of bands at longer wavelengths. { 'nir ,in·frə'red ,spek·trō·fə'täm·ə·trē }

nearly free electron method [SOLID STATE] A method of approximating the energy levels of electrons in a crystal lattice by considering the potential energy resulting from atomic nuclei and from other electrons in the lattice as a perturbation on free electron states. Abbreviated NFE method. { 'nir·lē ¦frē i¦lek,trän ,meth·əd }

nearly isometric spaces [MATH] Two Banach spaces, A and B, such that for any numbers $c < 1$ and $d > 1$ there is a bijective mapping, f, from A to B such that the norm of $f(x)$ divided by the norm of x lies in the interval $[c,d]$. { ¦nir·lē ,ī·sə,me·trik 'spās·əz }

near miss [ORD] The strike of an explosive missile, especially of an aerial bomb, near but not on the object of attack, and usually close enough to cause effective damage. { 'nir 'mis }

near point [PHYSIO] The smallest distance from the eye at which a small object can be seen without blurring. { 'nir ,pȯint }

near ring [MATH] An algebraic system with two binary operations called multiplication and addition; the system is a group (not necessarily commutative) relative to addition, and multiplication is associative, and is left-distributive with respect to addition, that is, $x(y + z) = xy + xz$ for any x, y, and z in the near ring. { ¦nir ¦riŋ }

nearshore [OCEANOGR] An indefinite zone which extends from the shoreline seaward to a point beyond the breaker zone. { 'nir,shȯr }

nearshore circulation [OCEANOGR] Ocean circulation consisting of both the nearshore currents and the coastal currents. { 'nir,shȯr ,sər·kyə'lā·shən }

nearshore current system [OCEANOGR] A current system, caused mainly by wave action in and near the breaker zone, which contains four elements: the shoreward mass transport of water; longshore currents; seaward return flow, including rip currents; and the longshore movement of the expanded heads of rip currents. { 'nir,shȯr 'kə·rənt ,sis·təm }

nearsightedness *See* myopia. { 'nir¦sīd·əd·nəs }

near stars [ASTRON] Those stars in the celestial neighborhood of the sun, sometimes taken as those 22 stars within 13 light-years of the sun. { 'nir 'stärz }

nearthrosis [MED] A type of nonunion of broken ends of bones in which a cystic space resembling a joint cavity develops between poorly joined ends. { ¦nē·är'thrō·səs }

near-ultraviolet radiation [ELECTROMAG] Ultraviolet radiation having relatively long wavelength, in the approximate range from 300 to 400 nanometers. { 'nir ¦əl·trə¦vī·lət ,rād·ē'ā·shən }

near wilt [PL PATH] A fungus disease of peas caused by *Fusarium oxysporum pisi;* affects scattered plants and develops more slowly than true wilt. { 'nir ,wilt }

neat cement grout [MATER] Grout made from a mixture of cement and water. { nēt si'ment ,graut }

neat line [CIV ENG] The line defining the limits of an aspect of construction, such as an excavation or a wall. Also known as net line. [MAP] That border line which indicates the limits of an area shown on a map or chart. { 'nēt ,līn }

neat plaster [MATER] A base-coat plaster, having sand added at the job location. { 'nēt ,plas·tər }

neatsfoot oil [MATER] Pale-yellow oil with unusual odor; soluble in organic solvents and kerosine; obtained by boiling shinbones and hoofless feet of cattle; used to treat leather, as a lubricant, and to oil wool. Also known as bubulum oil; hoof oil. { 'nēts,fut ,ȯil }

neat soap [MATER] Soap in the molten state formed during manufacture, especially after fitting and settling out of nigre and lye. { 'nēt ,sōp }

Nebaliacea [INV ZOO] A small, marine order of Crustacea in the subclass Leptostraca distinguished by a large bivalve shell, without a definite hinge line, an anterior articulated rostrum, eight thoracic and seven abdominal somites, a pair of articulated furcal rami, and the telson. { nə,bā·lē'ā·shə }

Nebraskan drift [GEOL] Rock material transported during the Nebraskan glaciation; it is buried below the Kansan drift in Iowa. { nə'bras·kən 'drift }

Nebraskan glaciation [GEOL] The first glacial stage of the Pleistocene epoch in North America, beginning about 1,000,000 years ago, and preceding the Aftonian interglacial stage. { nə'bras·kən glā·sē·ā·shən }

NEAR-FIELD SCANNING OPTICAL MICROSCOPY

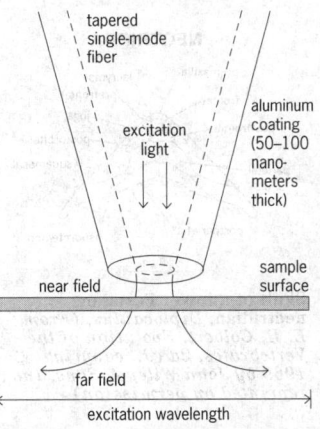

Fiber-optic probe used in near-field scanning optical microscopy and its position relative to the sample surface.

nebula [ASTRON] Interstellar clouds of gas or small particles; an example is the Horsehead Nebula in Orion. { 'neb·yə·lə }

nebular hypothesis [ASTROPHYS] A theory, proposed in 1796 by Laplace, supposing that the planets originated from the solar nebula surrounding the proto-sun; as the sun cooled, it contracted, rotated faster, and thus caused a ringlike bulging at the equator; this bulge eventually broke off and formed the planets; Laplace further theorized that the sun and other stars formed from clouds of nebulous matter; the theory in this form is not accepted. { 'neb·yə·lər hī'path·ə·səs }

nebular lines [ASTROPHYS] The spectral lines formed in the glow of bright nebulae; they arise from forbidden atomic transition which can take place because of the very low pressure in the nebula itself. { 'neb·yə·lər 'līnz }

nebular redshift [ASTROPHYS] A systematic shift observed in the spectra of all distant galaxies; the wavelength shift toward the red increases with the distance of the galaxies from the earth. { 'neb·yə·lər 'red,shift }

nebular transitions [ASTROPHYS] Those electronic transitions for doubly ionized argon and chlorine that yield the nebular lines seen in the spectra of gaseous nebulae. { 'neb·yə·lər tran'zish·ənz }

nebular variable See T Tauri star. { 'neb·yə·lər 'ver·ē·ə·bəl }

nebulite [PETR] A chorismite in which one of the textural elements occurs in nebulitic lenticular masses. { 'neb·yə,līt }

nebulitic [PETR] **1.** Having indistinct boundaries between textural elements. **2.** Of or pertaining to a nebulite. { ,neb·yə'lid·ik }

nebulium line [SPECT] An optical emission line in the spectrum of oxygen at a wavelength of 500.7 nanometers, prominent in the spectra of H II regions. { nə'būl·ē·əm ,līn }

nebulosus [METEOROL] A cloud species with the appearance of a nebulous veil, showing no distinct details; found principally in the genera cirrostratus and stratus. { 'neb·yə'lō·səs }

necessary bandwidth [COMMUN] For a given class of emission, the minimum value of the occupied bandwidth sufficient to ensure the transmission of information at the rate and with the quality required for the system employed, under specified conditions. { 'nes·ə,ser·ē 'band,width }

necessary condition [MATH] A mathematical statement that must be true if a given statement is true. { ,nes·ə,ser·ē kən'dish·ən }

neck [ANAT] The usually constricted communicating column between the head and trunk of the vertebrate body. [ENG] The part of a furnace where the flame is contracted before reaching the stack. [GEOGR] A narrow strip of land, especially one connecting two larger areas. [GEOL] See pipe. [MET] In a tensile test, that portion of the metal at which fracture is imminent during the later stages of plastic deformation in a tensile test. [OCEANOGR] The narrow band of water forming the part of a rip current where feeder currents converge and flow swiftly through the incoming breakers and out to the head. { nek }

neck breadth [ANTHRO] The diameter of the neck measured halfway between the otobasion inferior and the shoulder. { 'nek ,bredth }

neck cutoff [GEOGR] A high-angle meander cutoff formed where a stream breaks through or across a narrow meander neck, as where downstream migration of one meander has been slowed and the next meander upstream has overtaken it. { 'nek 'kət,óf }

neck depth [ANTHRO] The diameter of the neck between the tip of the thyroid cartilage and the back of the neck, measured perpendicular to the axis of the neck, with contact only. { 'nek ,depth }

neck-down [MET] **1.** A thin core used for restricting the riser neck; facilitates cutting off the riser from the casting. **2.** Localized area reduction of a test piece during plastic deformation. { 'nek ,daún }

Neckeraceae [BOT] A family of mosses in the order Isobryales distinguished by undulate leaves. { ,nek·ə'rās·ē,ē }

neck-in [ENG] When coating by extrusion, the width difference between the extruded web leaving the die and that of the coating on the surface. { 'nek,in }

necking [MET] Reducing the diameter or cross-sectional area of a tube or other piece of metal by stretching. { 'nek·iŋ }

necking down [MET] Localized reduction in cross-sectional area of a specimen during tensile deformation. { 'nek·iŋ ,daún }

neck rot [PL PATH] A fungus disease of onions caused by species of *Botrytis* and characterized by rotting of the leaves just above the bulb. { 'nek ,rät }

necr-, necro- [MED] Combining form denoting death. { ¦ne·krō }

necrobiosis [MED] Death of a cell or group of cells under either normal or pathologic conditions. { ¦ne·krō,bī'ō·səs }

Necrolestidae [PALEON] An extinct family of insectivorous marsupials. { ¦ne·krō'les·tə,dē }

necrophagous [ZOO] Feeding on dead bodies. { ne'kräf·ə·gəs }

necrophile [PSYCH] A person affected with necrophilia. { 'nek·rə,fīl }

necrophilia [PSYCH] **1.** Longing for death. **2.** See necrophilism. { ,nek·rə'fil·ē·ə }

necrophilism [PSYCH] Also known as necrophilia. **1.** Unnatural obsession with and usually erotic attraction for dead bodies. **2.** Sexual violation of a corpse. { nə'kräf·ə,liz·əm }

necrophobia [PSYCH] Abnormal dread of death and of dead bodies. { ,nek·rə'fō·bē·ə }

necropsy [MED] To perform an autopsy. { 'ne,kräp·sē }

necrosis [MED] Death of a cell or group of cells as a result of injury, disease, or other pathologic state. { nə'krō·səs }

necrotic [MED] Pertaining to, causing, or undergoing necrosis. { nə'kräd·ik }

necrotic enteritis [VET MED] A bacterial infection of young swine caused by *Salmonella suipestifer* or *S. choleraesuis* and characterized by fever and necrotic and ulcerative inflammation of the intestine. { nə'kräd·ik ,ent·ə'rīd·əs }

necrotic ring spot [PL PATH] A virus leaf spot of cherries marked by small, dark water-soaked rings which may drop out, giving the leaf a tattered appearance. { nə'kräd·ik 'riŋ ,spät }

necrotize [MED] To undergo necrosis; to become necrotic. { 'nek·rə,tīz }

necrozoospermia [MED] A condition in which spermatozoa are immobile. { ¦ne·krō,zō·ō'spər·mē·ə }

nectar [BOT] A sugar-containing liquid secretion of the nectaries of many flowers. { 'nek·tər }

nectarine [BOT] A smooth-skinned, fuzzless fruit originating as a spontaneous somatic mutation of the peach, *Prunus persica* and *P. persica* var. *nectarina*. { ¦nek·tə¦rēn }

nectary [BOT] A secretory organ or surface modification of a floral organ in many flowers, occurring on the receptacle, in and around ovaries, on stamens, or on the perianth; secretes nectar. { 'nek·tə·rē }

nectocalyx [INV ZOO] A swimming bell of a siphonophore. Also known as nectophore. { ¦nek·tō'kā·liks }

Nectonematoidea [INV ZOO] A monogeneric order of worms belonging to the class Nematomorpha, characterized by dorsal and ventral epidermal chords, a pseudocoele, and dorsal and ventral rows of bristles; adults are parasites of true crabs and hermit crabs. { ¦nek·tō,ne·mə'tóid·ē·ə }

nectophore See nectocalyx. { 'nek·tə,fór }

nectosome [INV ZOO] The part of a complex siphonophore that bears swimming bells. { 'nek·tə,sōm }

Nectridea [PALEON] An order of extinct lepospondylous amphibians characterized by vertebrae in which large fan-shaped hemal arches grow directly downward from the middle of each caudal centrum. { nek'trid·ē·ə }

Nectrioidaceae [MYCOL] The equivalent name for Zythiaceae. { ¦nek·trē,óid·ē'ās·ē,ē }

need [PSYCH] An acquired or physiological lack or deficit within the individual. { nēd }

need complementarity [PSYCH] The concept that people having different needs like each other because they provide each other with mutual satisfaction of opposed needs. { 'nēd ,käm·plə·mən'tar·əd·ē }

needle [BOT] A slender-pointed leaf, as of the firs and other evergreens. [COMPUT SCI] A slender rod or probe used to sort decks of edge-punched cards by inserting it through holes along the margin of the deck and vibrating the deck so that cards having that particular hole are retained, but those having a notch cut at that hole position drop out. [DES ENG] **1.** A device made of steel pointed at one end with a hole at the other; used for sewing. **2.** A device made of steel with a hook at one end; used for knitting. [ENG] **1.** A piece of copper or brass about ¹/₂ inch (13 millimeters) in diameter and 3 or 4

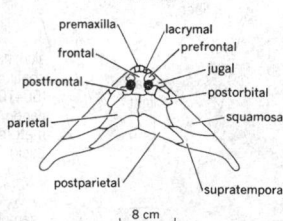

NECTRIDEA

premaxilla — lacrymal — frontal — prefrontal — jugal — postfrontal — postorbital — parietal — squamosal — postparietal — supratemporal

8 cm

Skull of a lower Permian nectridian, *Diplocaulus*. (*From E. H. Colbert, Evolution of the Vertebrates, 2d ed., copyright © 1969 by John Wiley & Sons, Inc., reprinted by permission*)

feet (90 or 120 centimeters) long, pointed at one end, thrust into a charge of blasting powder in a borehole and then withdrawn, leaving a hole for the priming, fuse, or squib. Also known as pricker. **2.** A thin pointed indicator on an instrument dial. [ENG ACOUS] *See* stylus. [GEOL] A pointed, elevated, and detached mass of rock formed by erosion, such as an aiguille. [HYD] A long, slender snow crystal that is at least five times as long as it is broad. [MINERAL] A needle-shaped or acicular mineral crystal. { 'nēd·əl }

needlebar [TEXT] A bar for mounting needles on a sewing or knitting machine. { 'nēd·əl‚bär }

needle beam [CIV ENG] A temporary member thrust under a building or a foundation for use in underpinning. { 'nēd·əl ‚bēm }

needle bearing [DES ENG] A roller-type bearing with long rollers of small diameter; the rollers are retained in a flanged cup, have no retainer, and bear directly on the shaft. { 'nēd·əl ‚ber·iŋ }

needle blow [ENG] A blow-molding technique in which air is injected into the plastic article through a hollow needle inserted in the parison. { 'nēd·əl ‚blō }

needle board [TEXT] A board that holds needles in a loom. { 'nēd·əl ‚bȯrd }

needle coal [MINERAL] Lignite containing fibrous needle-shaped masses formed from the vascular bundles of palm stems. { 'nēd·əl ‚kōl }

needle dam [CIV ENG] A barrier made of horizontal bars across a pass through a dam or of planks that can be removed in case of flooding. { 'nēd·əl ‚dam }

needle file [DES ENG] A small file with an extended tang that serves as a needle. { 'nēd·əl ‚fīl }

needle gap [ELECTR] Spark gap in which the electrodes are needle points. { 'nēd·əl ‚gap }

needle ice *See* frazil ice; pipkrake. { 'nēd·əl ‚īs }

needle nozzle [MECH ENG] A streamlined hydraulic turbine nozzle with a movable element for converting the pressure and kinetic energy in the pipe leading from the reservoir to the turbine into a smooth jet of variable diameter and discharge but practically constant velocity. { 'nēd·əl ‚näz·əl }

needle ore [MINERAL] **1.** Iron ore of very high metallic luster, found in small quantities, which may be separated into long, slender filaments resembling needles. **2.** *See* aikinite. { 'nēd·əl ‚ȯr }

needlepoint lace [TEXT] A lace which is worked over a paper pattern in a buttonhole stitch and then released from the pattern by a knife passed between the paper and the fibers. { 'nēd·əl‚pȯint 'lās }

needle scratch *See* surface noise. { 'nēd·əl ‚skrach }

needle test point [ELEC] A sharp steel probe connected to a test cord for making contact with a conductor. { 'nēd·əl 'test ‚pȯint }

needle tubing [ENG] Stainless steel tubing with outside diameters from 0.014 to 0.203 inch (0.36 to 5.16 millimeters); used for surgical instruments and radon implanters. { 'nēd·əl ‚tüb·iŋ }

needle valve [MECH ENG] A slender, pointed rod fitting in a hole or circular or conoidal seat; used in hydraulic turbines and hydroelectric systems. { 'nēd·əl ‚valv }

needle weir [CIV ENG] A type of frame weir in which the wooden barrier is constructed of vertical square-section timbers placed side by side against the iron frames. { 'nēd·əl ‚wer }

needling [CIV ENG] Underpinning the upper part of a building with horizontally placed timber or steel beams. [MIN ENG] Cutting holes or ledges in a coal bed or rock surface for receiving the ends of timber supports. { 'nēd·əl·iŋ }

need-to-know [ORD] A criterion used in security procedures in the United States that requires a person requesting classified information to establish the need to know such information in terms of the pertinent mission. { 'nēd tə 'nō }

Néel ferromagnetism *See* ferrimagnetism. { 'nā·el ‚fer·ō'mag·nə‚tiz·əm }

Néel point *See* Néel temperature. { 'nā·el ‚pȯint }

Néel's theory [SOLID STATE] A theory of the behavior of antiferromagnetic and other ferrimagnetic materials in which the crystal lattice is divided into two or more sublattices; each atom in one sublattice responds to the magnetic field generated by nearest neighbors in other sublattices, with the result that magnetic moments of all the atoms in any sublattice are parallel, but magnetic moments of two different sublattices can be different. { 'nā·elz ‚thē·ə·rē }

Néel temperature [SOLID STATE] A temperature, characteristic of certain metals, alloys, and salts, below which spontaneous nonparalleled magnetic ordering takes place so that they become antiferromagnetic, and above which they are paramagnetic. Also known as Néel point. { 'nā·el 'tem·prə·chər }

Néel wall [SOLID STATE] The boundary between two magnetic domains in a thin film in which the magnetization vector remains parallel to the faces of the film in passing through the wall. { 'nā·el ‚wȯl }

neem oil [MATER] An aromatic oil from the seeds and fruit of the neem tree (*Melia azadirachta*); contains sulfur compounds; used as an anthelmintic and as an alcohol denaturant. Also known as margosa oil; nim oil. { 'nēm ‚ȯil }

neencephalon [ANAT] The neopallium and the phylogenetically new acquisitions of the cerebellum and thalamus collectively. Also spelled neoencephalon. { ‚nē·in'sef·ə‚län }

negation [MATH] The negation of a proposition P is a proposition which is true if and only if P is false; this is often written ~P. Also known as denial. { nə'gā·shən }

negative [ELEC] Having a negative charge. [GRAPHICS] The image on film in which the dark tones of the original appear transparent, and the light tones appear black and opaque. Also known as reversed image. { 'neg·əd·iv }

negative acceleration [MECH] Acceleration in a direction opposite to the velocity, or in the direction of the negative axis of a coordinate system. { 'neg·əd·iv ik‚sel·ə'rā·shən }

negative acknowledgement [COMPUT SCI] In a data communications network, a control character returned from a receiving machine to a sending machine to indicate the presence of errors in the preceding block of data. Abbreviated NAK. { 'neg·əd·iv ik'näl·ij·mənt }

negative afterimage [PHYSIO] An afterimage that is seen on a bright background and is complementary in color to the initial stimulus. { 'neg·əd·iv 'af·tər‚im·ij }

negative angle [MATH] The angle subtended by moving a ray in the clockwise direction. { 'neg·əd·iv 'aŋ·gəl }

negative area [GEOGR] An area that is almost uncultivable or uninhabitable. [GEOL] *See* negative element. { 'neg·əd·iv 'er·ē·ə }

negative binomial distribution [STAT] The distribution of a negative binomial random variable. Also known as Pascal distribution. { 'neg·əd·iv bī'nō·mē·əl ‚di·strə'byü·shən }

negative booster [ELEC] Booster used in connection with a ground-return system to reduce the difference of potential between two points on the grounded return. { 'neg·əd·iv 'bü·stər }

negative catalysis [CHEM] A catalytic reaction such that the reaction is slowed down by the presence of the catalyst. { 'neg·əd·iv kə'tal·ə·səs }

negative charge [ELEC] The type of charge which is possessed by electrons in ordinary matter, and which may be produced in a resin object by rubbing with wool. Also known as negative electricity. { 'neg·əd·iv 'chärj }

negative conductor [ELEC] The conductor that is connected to the negative terminal of a voltage source. { 'neg·əd·iv kən'dək·tər }

negative correlation [STAT] A relation between two quantities such that when one increases the other decreases. { 'neg·əd·iv ‚kär·ə'lā·shən }

negative crystal [CRYSTAL] A crystal containing a cavity, where the form of the cavity is one of the characteristic crystal forms of the mineral in question. [OPTICS] A uniaxial crystal in which the extraordinary wave travels faster than the ordinary wave, such as calcite. { 'neg·əd·iv 'krist·əl }

negative easement [CIV ENG] An easement that can be exercised to prevent the owner of a piece of land from using it in certain ways that he or she would otherwise be entitled to. { 'neg·əd·iv 'ēz·mənt }

negative effective mass amplifiers and generators [ELECTR] Class of solid-state devices for broad-band amplification and generation of electrical waves in the microwave region; these devices use the property of the effective masses of charge carriers in semiconductors becoming negative with sufficiently high kinetic energies. { 'neg·əd·iv i‚fek·tiv 'mas 'am·plə‚fī·ərz ən 'jen·ə‚rād·ərz }

negative electricity *See* negative charge. { 'neg·əd·iv ‚i‚lek'tris·əd·ē }

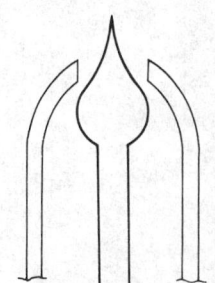
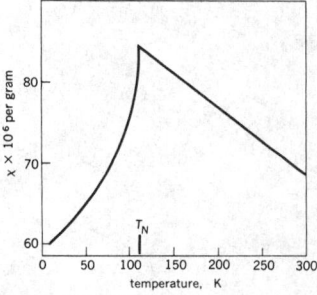

NEGATIVE MENISCUS LENS

Shape of a negative meniscus lens.

negative electrode See cathode; negative plate. { 'neg·əd·iv i'lek‚trōd }

negative electron See electron. { 'neg·əd·iv i'lek‚trän }

negative electron-affinity material [ELECTR] A material, such as gallium phosphide, whose surface has been treated with a substance, such as cesium, so that the surface barrier is reduced, band-bending occurs so that the top of the conduction band lies above the vacuum level, and the electron affinity of the substance in negative. Abbreviated NEA material. { 'neg·əd·iv i'lek‚trän ə'fin·əd·ē mə‚tir·ē·əl }

negative element [GEOL] A large structural feature or part of the earth's crust, characterized through a long geologic time period by frequent and conspicuous downward movement (subsidence) or by extensive erosion, or by an uplift that is considerably less rapid or less frequent than that of adjacent positive elements. Also known as negative area. { 'neg·əd·iv 'el·ə·mənt }

negative elongation [CRYSTAL] In a section of an anisotropic crystal, a sign of elongation that is parallel to the faster of the two plane-polarized rays. { 'neg·əd·iv ‚ē‚lóŋ'gā·shən }

negative empathy [PSYCH] Empathy which takes place against a certain resistance or unwillingness. { 'neg·əd·iv 'em·pə·thē }

negative feedback [CONT SYS] Feedback in which a portion of the output of a circuit, device, or machine is fed back 180° out of phase with the input signal, resulting in a decrease of amplification so as to stabilize the amplification with respect to time or frequency, and a reduction in distortion and noise. Also known as inverse feedback; reverse feedback; stabilized feedback. [SCI TECH] Feedback which tends to reduce the output in a system. { 'neg·əd·iv 'fēd‚bak }

negative g [MECH] In designating the direction of acceleration on a body, the opposite of positive g; for example, the effect of flying an outside loop in the upright seated position. { 'neg·əd·iv 'jē }

negative gene control [MOL BIO] Prevention of gene expression by the binding of specific repressor molecules to operator sites. { ‚neg·əd·iv 'jēn kən‚trōl }

negative glow [ELECTR] The luminous flow in a glow-discharge cold-cathode tube occurring between the cathode dark space and the Faraday dark space. { 'neg·əd·iv 'glō }

negative-grid generator [ELECTR] Conventional oscillator circuit in which oscillation is produced by feedback from the plate circuit to a grid which is normally negative with respect to the cathode, and which is designed to operate without drawing grid current at any time. { 'neg·əd·iv ‚grid 'jen·ə‚rād·ər }

negative-grid thyratron [ELECTR] A thyratron with only one grid, which serves to prevent the flow of current until its potential relative to the cathode is made less negative than a certain critical value. { 'neg·əd·iv ‚grid 'thī·rə‚trän }

negative impedance [ELECTR] An impedance such that when the current through it increases, the voltage drop across the impedance decreases. { 'neg·əd·iv im'pēd·əns }

negative-impedance repeater [ELECTR] A telephone repeater that provides an effective gain for voice-frequency signals by insertion into the line of a negative impedance that cancels out line impedances responsible for transmission losses. { 'neg·əd·iv im‚pēd·əns ri'pēd·ər }

negative indication [COMPUT SCI] A hole punched in a specified column and specified punch position on a punch card to indicate that a number represented in a particular field of the card has a negative sign. { 'neg·əd·iv ‚in·də'kā·shən }

negative integer [MATH] The additive inverse of a positive integer relative to the additive group structure of the integers. { 'neg·əd·iv 'int·ə·jər }

negative interference [GEN] A crossover exchange between homologous chromosomes which increases the likelihood of another in the same vicinity. { 'neg·əd·iv ‚in·tər'fir·əns }

negative ion [CHEM] An atom or group of atoms which by gain of one or more electrons has acquired a negative electric charge. [PHYS] An electron or negatively charged subatomic particle. { 'neg·əd·iv 'ī‚än }

negative-ion vacancy [CRYSTAL] A point defect in an ionic crystal in which a negative ion is missing from its lattice site. { 'neg·əd·iv 'ī‚än 'vā·kən·sē }

negative landform [GEOL] **1.** A relatively depressed or low-lying topographic form, such as a valley, basin, or plain. **2.**

A volcanic feature formed by a lack of material (such as a caldera). { 'neg·əd·iv 'land‚fórm }

negative lens See diverging lens. { 'neg·əd·iv 'lenz }

negative logic [ELECTR] Logic circuitry in which the more positive voltage (or current level) represents the 0 state; the less positive level represents the 1 state. { 'neg·əd·iv ‚läj·ik }

negative meniscus lens [OPTICS] A lens having one convex and one concave surface, with the radius of curvature of the convex surface greater than that of the concave surface. Also known as diverging meniscus lens. { 'neg·əd·iv mə‚nis·kəs 'lenz }

negative mirror See diverging mirror. { 'neg·əd·iv 'mir·ər }

negative modulation [ELECTR] **1.** Modulation in which an increase in brightness corresponds to a decrease in amplitude-modulated transmitter power; used in United States television transmitters and in some facsimile systems. **2.** Modulation in which an increase in brightness corresponds to a decrease in the frequency of a frequency-modulated facsimile transmitter. Also known as negative transmission. { 'neg·əd·iv ‚mäj·ə'lā·shən }

negative movement [GEOL] **1.** A downward movement of the earth's crust relative to an adjacent part of the crust, such as produced by subsidence. **2.** A relative lowering of the sea level with respect to the land, such as produced by a positive movement of the earth's crust or by a retreat of the sea. { 'neg·əd·iv 'müv·mənt }

negative nodal points [OPTICS] Two points on the axis of an optical system such that an incident ray passing through one results in an emergent ray passing through the other which makes an angle with the axis having the same magnitude but opposite sign. Also known as antinodal points. { 'neg·əd·iv 'nōd·əl ‚póins }

negative number [MATH] A real number that is less than 0. { ‚neg·əd·iv 'nəm·bər }

negative part [MATH] For a real-valued function f, this is the function, denoted f^-, for which $f^-(x) = f(x)$ if $f(x) \leq 0$ and $f^-(x) = 0$ if $f(x) > 0$. { 'neg·əd·iv 'pärt }

negative pedal [MATH] **1.** The negative pedal of a curve with respect to a point O is the envelope of the line drawn through a point P of the curve perpendicular to OP. Also known as first negative pedal. **2.** Any curve that can be derived from a given curve by repeated application of the procedure specified in the first definition. { 'neg·əd·iv 'ped·əl }

negative phase [IMMUNOL] The temporary quantitative reduction of serum antibodies immediately following a second inoculation of antigen. { 'neg·əd·iv 'fāz }

negative phase sequence [ELEC] The phase sequence that corresponds to the reverse of the normal order of phases in a polyphase system. { 'neg·əd·iv 'fāz 'sē·kwəns }

negative-phase-sequence relay [ELEC] Relay which functions in conformance with the negative-phase-sequence component of the current, voltage, or power of the circuit. { 'neg·əd·iv ‚fāz ‚sē·kwəns 'rē‚lā }

negative photokinesis [PHYSIO] The slower movement of an organism upon entering an illuminated area relative to its velocity of movement in the dark or in dim light. { ‚neg·əd·iv ‚fōd·ō·kə'nē·səs }

negative phototaxis [PHYSIO] The orientation and movement of an organism away from the source of a light stimulus. { ‚neg·əd·iv ‚fōd·ō'tak·səs }

negative picture phase [ELECTR] The video signal phase in which the signal voltage swings in a negative direction for an increase in brilliance. { 'neg·əd·iv 'pik·chər ‚fāz }

negative pion [PARTIC PHYS] A pion having a negative electric charge. { 'neg·əd·iv 'pī‚än }

negative plate [ELEC] The internal plate structure that is connected to the negative terminal of a storage battery. Also known as negative electrode. { 'neg·əd·iv 'plāt }

negative pole See south pole. { 'neg·əd·iv 'pōl }

negative potential [ELEC] An electrostatic potential which is lower than that of the ground, or of some conductor or point in space that is arbitrarily assigned to have zero potential. { 'neg·əd·iv pə'ten·chəl }

negative pressure [PHYS] A way of expressing vacuum; a pressure less than atmospheric or the standard 760 mmHg (101,325 pascals). { 'neg·əd·iv 'presh·ər }

negative principal planes [OPTICS] Two planes perpendicular to the optical axis such that objects in one plane form images in the other with a lateral magnification of −1. Also known as antiprincipal planes. { 'neg·əd·iv ¦prin·sə·pəl 'plānz }

negative principal point [OPTICS] The intersection of a negative principal plane with the optical axis. Also known as antiprincipal point. { 'neg·əd·iv ¦prin·sə·pəl 'póint }

negative rain [METEOROL] Rain which exhibits a net negative electric charge. { 'neg·əd·iv 'rān }

negative rake [MECH ENG] The orientation of a cutting tool whose cutting edge lags the surface of the tooth face. { 'neg·əd·iv 'rāk }

negative regulator [GEN] Any regulator that acts to prevent transcription or translation. { 'neg·əd·iv ˌreg·yə'lād·ər }

negative resistance [ELECTR] The resistance of a negative-resistance device. { 'neg·əd·iv ri'zis·təns }

negative-resistance device [ELECTR] A device having a range of applied voltages within which an increase in this voltage produces a decrease in the current. { 'neg·əd·iv ri¦zis·təns di'vīs }

negative-resistance oscillator [ELECTR] An oscillator in which a parallel-tuned resonant circuit is connected to a vacuum tube so that the combination acts as the negative resistance needed for continuous oscillation. { 'neg·əd·iv ri¦zis·təns 'äs·ə,lād·ər }

negative-resistance repeater [ELECTR] Repeater in which gain is provided by a series negative resistance or a shunt negative resistance, or both. { 'neg·əd·iv ri¦zis·təns ri'pēd·ər }

negative selection [IMMUNOL] The death of autoimmune lymphocytes shortly after they develop. Also known as clonal deletion. { 'neg·əd·iv si'lek·shən }

negative series [MATH] A series whose terms are all negative real numbers. { 'neg·əd·iv 'sir,ēz }

negative shoreline See shoreline of emergence. { 'neg·əd·iv 'shòr,līn }

negative sign [MATH] The symbol −, used to indicate a negative number. { 'neg·əd·iv 'sīn }

negative skewness [MATH] Skewness in which the mean is smaller than the mode. { 'neg·əd·iv 'skü·nəs }

negative staining [BIOL] A method in microscopy for demonstrating the form of cells, bacteria, and other small objects by staining the ground rather than the objects. { 'neg·əd·iv 'stān·iŋ }

negative temperature [THERMO] The property of a thermally isolated thermodynamic system whose elements are in thermodynamic equilibrium among themselves, whose allowed states have an upper limit on their possible energies, and whose high-energy states are more occupied than the low-energy ones. { 'neg·əd·iv 'tem·prə·chər }

negative temperature coefficient [PHYS] A condition wherein the resistance, length, or some other characteristic of a material decreases when temperature increases. { 'neg·əd·iv 'tem·prə·chər ˌkō·i,fish·ənt }

negative terminal [ELEC] The terminal of a battery or other voltage source that has more electrons than normal; electrons flow from the negative terminal through the external circuit to the positive terminal. { 'neg·əd·iv 'tər·mən·əl }

negative thermion See thermoelectron. { 'neg·əd·iv 'thər,mē,än }

negative-transconductance oscillator [ELECTR] Electron-tube oscillator in which the output of the tube is coupled back to the input without phase shift, the phase condition for oscillation being satisfied by the negative transconductance of the tube. { 'neg·əd·iv ˌtranz·kən'dək·təns 'äs·ə,lād·ər }

negative transfer [PSYCH] The harmful effect that occurs on learning in one situation because previous learning, in another situation, required different responses, incompatible with the new learning situation. { 'neg·əd·iv 'tranz·fər }

negative transmission See negative modulation. { 'neg·əd·iv tranz'mish·ən }

negative with respect to a measure [MATH] A set A is negative with respect to a signed measure m if, for every measurable set B, the intersection of A and B, A ∩ B, is measurable and m(A ∩ B) ≤ 0. { ¦neg·əd·iv with ri¦spekt tü ə 'mezh·ər }

negative work [IND ENG] Work that is performed with the assistance of gravity so that the muscular effort required involves only control of the load. { 'neg·əd·iv 'wərk }

negativism [PSYCH] Indifference, opposition, or resistance to suggestions, or persistent refusal to do as asked, without apparent or objective reasons. { 'neg·əd·ə,viz·əm }

negatron See dynatron; electron. { 'neg·ə,trän }

negentropy See information content. { nə'gen·trə·pē }

negotiated contract [IND ENG] A purchase or sales agreement made by a United States government agency without normally employing techniques required by formal advertising. { nə'gō·shē,ād·əc 'kän,trakt }

Negri bodies [PATH] Acidophil cytoplasmic inclusion bodies in neurons, considered diagnostic of rabies. { 'nā·grē ,bäd·ēz }

Neididae [INV ZOO] A small family of thread-legged hemipteran insects in the superfamily Lygaeoidea. { nē'id·ə,dē }

neighbor [CRYSTAL] One of a pair of atoms or ions in a crystal which are close enough to each other for their interaction to be of significance in the physical problem being studied. { 'nā·bər }

neighbor effect [GRAPHICS] An effect whereby the blackening of a photographic layer during development at a particular location depends on the blackening in neighboring areas if the developer is not agitated. { 'nā·bər i,fekt }

neighborhood of a point [MATH] A set in a topological space which contains an open set which contains the point; in Euclidean space, an example of a neighborhood of a point is an open (without boundary) ball centered at that point. { 'nā·bər,hud əv ə 'póirt }

neighboring-group participation See anchimeric assistance. { 'nā·bər·iŋ ,grüp pär,tis·ə'pā·shən }

Neil's parabola [MATH] The graph of the equation $y = ax^{3/2}$, where a is a constant. { 'nēlz pə'rab·ə·lə }

Neisseriaceae [MICROBIO] The single family of gram-negative aerobic cocci and coccobacilli; some species are human parasites and pathogens. { 'nī·sər·ē'ās·ē,ē }

Neisseria gonorrhoeae [MICROBIO] A gram-negative coccus pathogen that causes the sexually transmitted disease gonorrhea. Also known as gonococcus. { nī·sə,rē·ə ,gän·ə'rē,ī }

nektobenthos [ECOL] Those forms of marine life that exist just above the ocean bottom and occasionally rest on it. { ¦nek·tə'ben,thòs }

nekton [INV ZOO] Free-swimming aquatic animals, essentially independent of water movements. { 'nek·tən }

N electron [ATOM PHYS] An electron in the fourth (N) shell of electrons surrounding the atomic nucleus, having the principal quantum number 4. { 'en i'lek·trän }

NELIAC [COMPUT SCI] An early dialect of ALGOL, which was developed for a specific data-processing application but, unlike ALGOL, is not primarily concerned with being used for complex scientific and engineering calculations. Derived from Navy Electronics Laboratory International Algol Compilers. { 'nel·ē,ak }

Nelson diaphragm cell [CHEM ENG] Obsolete carbon-electrode type of electrolytic diaphragm cell once widely used to produce chlorine and caustic soda from brine. { 'nel·sən 'dī·ə fram ,sel }

nelsonite [PETR] A group of hypabyssal rocks composed mainly of ilmenite and apatite. { 'nel·sə,nīt }

Nelumbonaceae [BOT] A family of flowering aquatic herbs in the order Nymphaeales characterized by having roots, perfect flowers, alternate leaves, and triaperturate pollen. { nə,ləm·bə'nās·ē,ē }

nemalite [MINERAL] A fibrous brucite that contains ferrous oxide. { 'nem·ə,līt }

Nemata [INV ZOO] An equivalent name for Nematoda. { nə'mad·ə }

Nemataceae [BOT] A family of mosses in the order Hookeriales distinguished by having perichaetial leaves only. { ,nem·ə'tās·ē,ē }

nematath [GEOL] A submarine ridge across an Atlantic-type ocean basin which is not an orogenic structure, but which is composed of otherwise undeformed continental crust that has been stretched across a sphenochasm or rhombochasm. { 'nem·ə,tath }

Nemathelminthes [INV ZOO] A subdivision of the Amera which comprised the classes Rotatoria, Gastrotrichia, Kinorhyncha, Nematoda, Nematomorpha, and Acanthocephala. { ,nem·ə,thel'min·thēz }

nematicide [MATER] A chemical used to kill plant-parasitic nematodes. Also spelled nematocide. { nə'mad·ə,sīd }

nematic phase [PHYS CHEM] A phase of a liquid crystal in the mesomorphic state, in which the liquid has a single optical axis in the direction of the applied magnetic field, appears to be turbid and to have mobile threadlike structures, can flow readily, has low viscosity, and lacks a diffraction pattern. { nə'mad·ik ,fāz }

nematoblastic [PETR] Pertaining to a metamorphic rock with a homeoblastic texture due to development during recrystallization of slender prismatic crystals. { ¦nem·ə·də¦blas·tik }

Nematocera [INV ZOO] A series of dipteran insects in the suborder Orthorrhapha; adults have antennae that are usually longer than the head, and the flagellum consists of 10–65 similar segments. { ¦nem·ə'täs·ə·rə }

nematocide *See* nematicide. { nə'mad·ə,sīd }

nematocyst [INV ZOO] An intracellular effector organelle in the form of a coiled tube which may be rapidly everted in food gathering or defense by cnidarians. { nə'mad·ə,sist }

Nematoda [INV ZOO] A group of unsegmented worms which have been variously recognized as an order, class, and phylum. { nem·ə'tō·də }

nematode [INV ZOO] **1.** Any member of the Nematoda. **2.** Of or pertaining to the Nematoda. { 'nem·ə,tōd }

Nematodonteae [BOT] A group of mosses included in the subclass Eubrya in which there may be faint transverse bars on the peristome teeth. { nə¦mad·ə'dänt·ē,ē }

nematogen [INV ZOO] A reproductive phase of the Dicyemida during which vermiform larvae are formed asexually from the germ cells in the axial cells. { nə'mad·ə·jən }

nematogenic solid [PHYS CHEM] A solid which will form a nematic liquid crystal when heated. { nə'mad·ə,jen·ik 'säl·əd }

Nematognathi [VERT ZOO] The equivalent name for Siluriformes. { ¦nem·ə'täg·nə,thī }

Nematoidea [INV ZOO] An equivalent name for Nematoda. { nem·ə'tóid·ē·ə }

nematology [INV ZOO] The study of nematodes. { ,nem·ə'täl·ə·jē }

Nematomorpha [INV ZOO] A group of the Aschelminthes or a separate phylum that includes the horsehair worms. { nem·əd·ə'mór·fə }

Nematophytales [PALEOBOT] A group of fossil plants from the Silurian and Devonian periods that bear some resemblance to the brown seaweeds (Phaeophyta). { ¦nem·əd·ō·fī'tā·lēz }

Nematospora coryli [MICROBIO] A mycelial species with needle-shaped ascospores that causes yeast spot disease of various crops. { nə,mad·ə,spór·ə 'kór·ə,lē }

Nematosporoideae [BOT] A subfamily of the Saccharomycetaceae containing parasitic yeasts; two genera have been studied in culture: *Nematospora* with asci that contain eight spindle-shaped ascospores, and *Metschnikowia* whose asci contain one or two needle-shaped ascospores. { ¦nem·əd·ō·spə'róid·ē,ē }

nematozooid [INV ZOO] A zooid bearing organs of defense, in hydroids and siphonophores. { ¦nem·əd·ə'zō,óid }

nemere [METEOROL] In Hungary, a stormy, cold fall wind. { 'ne,mir·ə }

Nemertea [INV ZOO] An equivalent name for Rhynchocoela. { nə'mərd·ē·ə }

Nemertina [INV ZOO] An equivalent name for Rhynchocoela. { ,ne·mər'tī·nə }

Nemertinea [INV ZOO] An equivalent name for Rhynchocoela. { ,ne·mər'tin·ē·ə }

Nemesis [ASTRON] A hypothetical, undetected, brown-dwarf companion of the sun, in a highly elongated orbit that would cause cometary material in Oort's Cloud to fall toward the inner region of the solar system approximately once every 2.8×10^7 years. { 'nem·ə·səs }

Nemestrinidae [INV ZOO] The hairy flies, a family of dipteran insects in the series Brachycera of the suborder Orthorrhapha. { ¦nem·ə¦strin·ə,dē }

Nemichthyidae [VERT ZOO] A family of bathypelagic, eel-like amphibians in the order Apoda. { ,nem·ik'thī·ə,dē }

Nemognathinae [INV ZOO] A subfamily of the coleopteran family Meloidae; members have greatly elongate maxillae that form a poorly fitted tube. { ,nem·əg'nath·ə,nē }

nemoral [ECOL] Pertaining to or inhabiting a grove or wooded area. { 'nem·rəl }

neo-, ne- [ORG CHEM] Prefix indicating hydrocarbons where a carbon is bonded directly to at least four other carbon atoms,

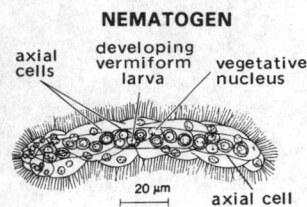

NEMATOGEN

axial cells   developing vermiform larva   vegetative nucleus

— 20 μm — axial cell

Young stem nematogen, with three axial cells, of *Dicyema schulzianum*.

such as neopentane. [SCI TECH] Prefix meaning new, or different in form; indicating a compound related to an older one, or a precursor. { 'nē·ō }

neoadjuvant chemotherapy [MED] A type of chemotherapy that is used to shrink a tumor prior to surgery or radiation. { ,nē·ō¦aj·ə·vənt ,kē·mō'ther·ə·pē }

Neoanthropinae [PALEON] A subfamily of the Hominidae in some systems of classification, set up to include *Homo sapiens* and direct ancestors of *H. sapiens*. { ¦nē·ō·an'thräp·ə,nē }

neoautochthon [GEOL] A stable basement or autochthon formed where a nappe has ceased movement and has become defunct. { ¦nē·ō·ò'täk·thən }

neoblast [INV ZOO] Any of various undifferentiated cells in annelids which migrate to and proliferate at sites of repair and regeneration. { 'nē·ə,blast }

Neocathartidae [PALEON] An extinct family of vulturelike diurnal birds of prey (Falconiformes) from the Upper Eocene. { ¦nē·ō·kə'thärd·ə,dē }

neocentric activity [CYTOL] In plants, an aberrant behavior during meiosis in which specific chromosome regions act as secondary sites of attachment for spindle fibers. { ,nē·ə'sen·trik ak¦tiv·əd·ē }

neocentromere [GEN] A functional centromere in a novel location; may lack specific classes of deoxyribonucleic acid usually present in a centromere. { ,nē·ō'sen·trə,mir }

neocerebellum [ANAT] Phylogenetically, the most recent part of the cerebellum; receives cerebral cortex impulses via the corticopontocerebellar tract. { ¦nē·ō,ser·ə'bel·əm }

Neocomian [GEOL] A European stage of Lower Cretaceous geologic time; includes Berriasian, Valanginian, Hauterivian, and Barremian. { ¦nē·ə¦kō·mē·ən }

neocortex [ANAT] Phylogenetically the most recent part of the cerebral cortex; includes all but the olfactory, hippocampal, and piriform regions of the cortex. { ¦nē·ō'kór,teks }

neocryst [GEOL] An individual crystal of a secondary mineral in an evaporite. { 'nē·ə,krist }

neodymium [CHEM] A metallic element, symbol Nd, with atomic weight 144.24, atomic number 60; a member of the rare-earth group of elements. { ,nē·ō'dim·ē·əm }

neodymium chloride [INORG CHEM] $NdCl_3 \cdot xH_2O$ Water- and acid-soluble, pink lumps; used to prepare metallic neodymium. { ,nē·ō'dim·ē·əm 'klór,īd }

neodymium glass [MATER] A glass containing small amounts of neodymium oxide; used for color television filter plates since it transmits 90% of the blue, green, and red light rays and no more than 10% of the yellow. { ,nē·ō'dim·ē·əm 'glas }

neodymium glass laser [OPTICS] An amorphous solid laser in which glass is doped with neodymium; characteristics are comparable with those of a pulsed ruby laser, but the wavelength of radiation is outside the visible range. { ,nē·ō'dim·ē·əm ¦glas 'lāz·ər }

neodymium-iron-boron magnet [MET] A permanent-magnet material that has the highest energy product known, and consists of an intermetallic phase, $Nd_2Fe_{14}B$, in a tetragonal crystal structure. { ,nē·ō'dim·ē·əm 'ī·ərn 'bó,rän ,mag·nət }

neodymium liquid laser *See* inorganic liquid laser. { ,nē·ō'dim·ē·əm ¦lik·wəd 'lāz·ər }

neodymium oxide [INORG CHEM] Nd_2O_3 A hygroscopic, blue-gray powder; insoluble in water, soluble in acids; used to color glass and in ceramic capacitors. { ,nē·ō'dim·ē·əm 'äk,sīd }

neoencephalon *See* neencephalon. { ¦nē·ō·in'sef·ə,län }

neoformation *See* neogenesis. { ¦nē·ō·fòr'mā·shən }

Neogastropoda [INV ZOO] An order of gastropods which contains the most highly developed snails; respiration is by means of ctenidia, the nervous system is concentrated, an operculum is present, and the sexes are separate. { ¦nē·ō·ga'sträp·ə·də }

Neogene [GEOL] An interval of geologic time incorporating the Miocene and Pliocene of the Tertiary period; the Upper Tertiary. { 'nē·ə,jēn }

neogenesis [GEOL] The formation of new minerals, as by diagenesis or metamorphism. Also known as neoformation. { ¦nē·ō'jen·ə·səs }

neoglaciation [GEOL] The removal of glacier ice growth in certain mountain areas during the Little Ice Age, following its shrinkage or disappearance during the Altithermal interval. { ¦nē·ō·glā·sē'ā·shən }

Neognathae [VERT ZOO] A superorder of the avian subclass Neornithes, characterized as flying birds with fully developed wings and sternum with a keel, fused caudal vertebrae, and absence of teeth. { nē'äg·nə,thē }

Neogregarinida [INV ZOO] An order of sporozoan protozoans in the subclass Gregarinia which are insect parasites. { nē·ō,greg·ə'rin·ə·də }

neohexane [ORG CHEM] C_6H_{14} Volatile, flammable, colorless liquid boiling at 50°C; used as high-octane component of motor and aviation gasolines. { nē·ō'hek,sān }

neohexane alkylation [CHEM ENG] A noncatalytic petroleum-refinery alkylation process that forms neohexane from a feed of ethylene and isobutane. { nē·ō'hek,sān ,al·kə'lā·shən }

Neolithic [ANTHRO] The period of prehistoric culture in which people made some of their stone tools by grinding the edges to create smooth surfaces; frequently defined as the time interval following the Paleolithic, beginning about 11,000–8000 years ago. { nē·ə'lith·ik }

neomagma [GEOL] Magma formed by partial or complete refusion of preexisting rocks under the conditions of plutonic metamorphism. { nē·ō'mag·mə }

neomineralization [GEOCHEM] Chemical interchange within a rock whereby its mineral constituents are converted into entirely new mineral species. { nē·ō,min·rə·lə'zā·shən }

neomorph [GEN] A mutant allele that produces an effect different from that produced by the normal allele. { 'nē·ə,mȯrf }

neomycin [MICROBIO] The collective name for several colorless antibiotics produced by a strain of *Streptomyces fradiae*; the commercial fraction $(C_{23}H_{46}N_6O_{13})$ has a broad spectrum of activity. { ¦nē·ə¦mīs·ən }

neon [CHEM] A gaseous element, symbol Ne, atomic number 10, atomic weight 20.179; a member of the family of noble gases in the zero group of the periodic table. { 'nē,än }

neonatal [MED] Pertaining to a newborn infant. { ¦nē·ə'nād·əl }

neonatal impetigo [MED] A type of impetigo occurring in the newborn, characterized by bullae and caused by staphylococci or sometimes streptococci. { nē·ə'nād·əl ,im·pə'tī,gō }

neonatal line [ANAT] A prominent incremental line formed in the neonatal period in the enamel and dentin of a deciduous tooth or of a first permanent molar. { nē·ə'nād·əl 'līn }

neonatal mortality rate [MED] The number of deaths reported among infants under 1 month of age in a calendar year per 1000 live births reported in the same year and place. { nē·ə'nād·əl mȯr'tal·əd·ē ,rāt }

neonatal myasthenia [MED] Muscle weakness and ineffective motor activities in infants born of myasthenic mothers. { nē·ə'nād·əl ,mī·əs'thē·nē·ə }

neonate [MED] A newborn infant. { 'nē·ə,nāt }

neonatology [MED] The study of the newborn up to 2 months of age. { nē·ə'näl·ə·jē }

neon glow lamp [ELECTR] A glow lamp containing neon gas, usually rated between $\frac{1}{25}$ and 3 watts, and producing a characteristic red glow; used as an indicator light and electronic circuit component. { 'nē,än 'glō ,lamp }

neon-helium laser [OPTICS] A continuous-wave gas laser using a combination of neon and helium gases to obtain a 6328-angstrom (632.8-nanometer) visible red beam. { 'nē,än ¦hē·lē·əm 'lā·zər }

neon oscillator [ELECTR] Relaxation oscillator in which a neon tube or lamp serves as the switching element. { 'nē,än 'äs·ə,lād·ər }

neon tube [ELECTR] An electron tube in which neon gas is ionized by the flow of electric current through long lengths of gas tubing, to produce a luminous red glow discharge; used chiefly in outdoor advertising signs. { 'nē,än ,tüb }

neopallium [ANAT] Phylogenetically, the new part of the cerebral cortex; formed from the region between the pyriform lobe and the hippocampus, it comprises the nonolfactory region. { nē·ō'pal·ē·əm }

neopalynology [BOT] A field of palynology concerned with extant microorganisms and disassociated microscopic parts of megaorganisms. { nē·ō,pal·ə'näl·ə·jē }

neopentane [ORG CHEM] C_5H_{12} Colorless liquid boiling at 10°C; soluble in alcohol, insoluble in water; a hydrocarbon

found as a minor component of natural gasoline. { ¦nē·ō'pen,tān }

neophane glass [MATER] A glass containing neodymium oxide to reduce glare; used for yellow sunglasses or for windshield glass. { 'nē·ə,fān ,glas }

neoplasia [MED] **1.** Formation of a neoplasm or tumor. **2.** Formation of new tissue. { ,nē·ə'plā·zhə }

neoplasm [MED] An aberrant new growth of abnormal cells or tissues; a tumor. { 'nē·ə,plaz·əm }

neoplastoid [CYTOL] Pertaining to immortal mammalian cell lines that may behave like neoplasms. { ,nē·ə'plas,tȯid }

neoprene [MATER] A synthetic rubber with outstanding resistance to ozone, weathering, various chemicals, oil, and flame, made by polymerization of chloroprene (2-chlorobutadiene-1,3); varies from amber to silver to cream in color; used in paints, putties, adhesives, shoe soles, tank linings, and rubber products. { 'nē·ə,prēn }

Neopseustidae [INV ZOO] A family of Lepidoptera in the superfamily Eriocranicidea. { ,nē·əp'sü·stə,dē }

Neoptera [INV ZOO] A section of the insect subclass Pterygota; members have a muscular and articular mechanism allowing the wings to be flexed over the abdomen when at rest. { nē'äp·tə·rə }

Neopterygii [VERT ZOO] An equivalent name for Actinopterygii. { nē,äp·tə'rij·ē,ī }

Neorhabdocoela [INV ZOO] A group of the Rhabdocoela comprising fresh-water, marine, or terrestrial forms, with a bulbous pharynx, paired protonephridia, sexual reproduction, and ventral gonopores. { ,nē·ō,rab·də'sē·lə }

Neornithes [VERT ZOO] A subclass of the class Aves containing all known birds except the fossil *Archaeopteryx*. { nē'ȯr·nə,thēz }

neosilicate [MINERAL] A structural type of silicate mineral characterized by linkage of isolated SiO_4 tetrahedra by ionic bonding only; an example is olivine. { ¦nē·ō'sil·ə·kāt }

neosome [GEOL] A geometric element of a composite rock or mineral deposit, appearing to be younger than the main rock mass. { 'nē·ə,sōm }

neossoptile [VERT ZOO] A downy feather on most newly hatched birds. { ,nē·ə'säp·təl }

neostigmine [PHARM] A quaternary ammonium cation that is used as the bromide $(C_{12}H_{19}BrN_{20}O_2)$ and methylsulfate $(C_{13}H_{22}N_2O_6S)$ salts; has anticholinesterase activity. { ¦nē·ō¦stig mēn }

neostratotype [GEOL] A stratotype established after the holostratotype has been destroyed or is otherwise not usable. { ¦nē·ō'strad·ə,tīp }

neotectonic map [GEOL] A map depicting neotectonic structures. { ¦nē·ō·tek'tän·ik 'map }

neotectonics [GEOL] The study of the most recent structures and structural history of the earth's crust, after the Miocene. { ¦nē·ō·tek'tän·iks }

neotenin [BIOCHEM] A hormone secreted by cells of the corpus allatum in arthropod larvae and nymphs; inhibits the development of adult characters. Also known as juvenile hormone. { nē'ät·ən·ər }

neoteny [VERT ZOO] A phenomenon peculiar to some salamanders, in which large larvae become sexually mature while still retaining gills and other larval features. { 'nē·ə,tē·nē *or* nē'ät·ən·ē }

Neotropical zoogeographic region [ECOL] A zoogeographic region that includes Mexico south of the Mexican Plateau, the West Indies, Central America, and South America. { ¦nē·ō träp·ə·kəl ¦zō·ō,je·ə'graf·ik 'rē·jən }

neotype [SYST] A specimen selected as type subsequent to the original description when the primary types are known to be destroyed; a nomenclatural type. { 'nē·ə,tīp }

neounitarian theory of hematopoiesis [HISTOL] A theory that under certain conditions, such as in tissue culture or in pathologic states, lymphocytes or cells resembling lymphocytes can become multipotent. { nē·ō·yü·nə'tar·ē·ən ¦thē·ə·re əv ,hem·əd·ō·pȯi'ē·səs }

neovolcanic [PETR] Referring to extrusive rocks that are of Tertiary or younger age. { nē·ō·väl'kan·ik }

nep [TEXT] Any small entanglement of textile fibers that cannot be unraveled; formed during carding or ginning. { nep }

Nepenthaceae [BOT] A family of dicotyledonous plants in

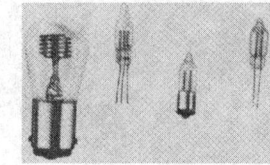

NEON GLOW LAMP

Examples of neon glow lamps.

the order Sarraceniales; includes many of the pitcher plants. { ,nep·ən'thās·ē,ē }

neper [PHYS] Abbreviated Np. Also known as napier. **1.** A unit used for expressing the ratio of two currents, voltages, or analogous quantities; the number of nepers is the natural logarithm of this ratio. **2.** A unit used for expressing the ratio of two powers (even when this ratio is not the square of the corresponding current or voltage ratio); the number of nepers is the natural logarithm of the square root of this ratio; to avoid confusion, this usage should be accompanied by a specific statement. { 'nē·pər }

nephanalysis [METEOROL] The analysis of a synoptic chart in terms of the types and amount of clouds and precipitation; cloud systems are identified both as entities and in relation to the pressure pattern, fronts, and other aspects. { nef·ə'nal·ə·səs }

nephcurve [METEOROL] In nephanalysis, a line bounding a significant portion of a cloud system, for example, a clear-sky line, precipitation line, cloud-type line, or ceiling-height line. { 'nef,kərv }

nepheline [MINERAL] A mineral of variable composition, with its purest state represented by the formula $NaAlSiO_4$; calcium, magnesium, iron (Fe^{2+} and Fe^{3+}), and titanium are usually present in only minor or trace amounts. { 'nef·ə,lēn }

nepheline basalt See olivine nephelinite. { 'nef·ə,lēn bə'sȯlt }

nepheline monzonite [PETR] A nepheline syenite in which sodic plagioclase exceeds the quantity of alkali feldspar. { 'nef·ə,lēn 'män·zə,nīt }

nepheline phonolite [PETR] The fine-grained equivalent of nepheline syenite. { 'nef·ə,lēn fän·ə,līt }

nepheline syenite [PETR] A phaneritic plutonic rock with granular texture, composed largely of alkali feldspar, nepheline, and dark-colored materials. { 'nef·ə,lēn 'sī·ə,nīt }

nephelinite [PETR] A dark-colored, aphanitic rock of volcanic origin, composed essentially of nepheline and pyroxene; texture is usually porphyritic with large crystals of augite and nepheline in a very-fine-grained matrix. { ne'fel·ə,nīt }

nephelite See nepheline. { 'nef·ə,līt }

nepheloid zone [OCEANOGR] A layer of water near the bottom of the continental rise and slope of the North Atlantic Ocean that contains suspended sediment of the clay fraction and organic matter. { 'nef·ə,lȯid ,zōn }

nephelometer [OPTICS] A type of instrument that measures, at more than one angle, the scattering function of particles suspended in a medium; information obtained may be used to determine the size of the suspended particles and the visual range through the medium. { ,nef·ə'läm·əd·ər }

nephelometry [OPTICS] **1.** The study of suspensoids using the techniques of light scattering. **2.** The study of the scattering properties of small samples of air and its suspensoids. { ,nef·ə'läm·ə·trē }

nepheloscope [ENG] An instrument for the production of clouds in the laboratory by condensation or expansion of moist air. { 'nef·ə·lə,skōp }

nephology [METEOROL] The study of clouds. { ne'fäl·ə·jē }

nephometer [ENG] A general term for instruments designed to measure the amount of cloudiness; an early type consists of a convex hemispherical mirror mapped into six parts; the amount of cloud coverage on the mirror is noted by the observer. { ne'fäm·əd·ər }

nephoscope [ENG] An instrument for determining the direction of cloud motion. { 'nef·ə,skōp }

nephr-, nephro- [ANAT] Combining form denoting kidney. { 'nef·rō }

nephrectomy [MED] Surgical removal of a kidney. { nə'frek·tə·mē }

nephric tubule See uriniferous tubule. { 'nef·rik 'tüb,yül }

nephridioblast [INV ZOO] An ecodermal precursor cell of a nephridium in certain animals. { nə'frid·ē·ə,blast }

nephridioduct [INV ZOO] The duct of a nephridium, sometimes serving as a common excretory and genital outlet. { nə'frid·ē·ə,dəkt }

nephridiopore [INV ZOO] The external opening of a nephridium. { nə'frid·ē·ə,pȯr }

nephridium [INV ZOO] Any of various paired excretory structures present in the Platyhelminthes, Rotifera, Rhynchocoela, Acanthocephala, Priapuloidea, Entoprocta, Gastrotricha,

Kinorhyncha, Cephalochorda, and some Archiannelida and Polychaeta. { nə'frid·ē·əm }

nephrite [MINERAL] An exceptionally tough, compact, fine-grained, greenish or bluish amphibole constituting the less valuable type of jade; formerly worn as a remedy for kidney diseases. Also known as greenstone; kidney stone. { ne'frīt }

nephritic [MED] **1.** Pertaining to or affected with nephritis. **2.** Pertaining to or affecting the kidney. { nə'frid·ik }

nephritis [MED] Inflammation of the kidney. { nə'frīd·əs }

nephroabdominal [ANAT] Of or pertaining to the kidneys and abdomen. { nef·rō·ab'däm·ə·nəl }

nephroblastoma See Wilms' tumor. { nef·rə·bla'stō·mə }

nephrocalcinosis [PATH] Deposition of calcium salts in the kidney tubules. { nef·rō,kal·sə'nō·səs }

nephrocoel [ANAT] The cavity of a nephrotome. { nef·rə,sēl }

nephrodystrophy See nephrosis. { nef·rə'dis·trə·fē }

nephrogenic [EMBRYO] **1.** Having the potential to develop into kidney tissue. **2.** Of renal origin. { nef·rə'jen·ik }

nephrogenic cord [EMBRYO] The longitudinal cordlike mass of mesenchyme derived from the mesomere or nephrostomal plate of the mesoderm, from which develop the functional parts of the pronephros, mesonephros, and metanephros. { nef·rə'jen·ik 'kȯrd }

nephrogenic tissue [EMBRYO] The tissue of the nephrogenic cord derived from the nephrotome plate that forms the blastema or primordium from which the embryonic and definitive kidneys develop. { nef·rə'jen·ik 'tish·ü }

nephroid [MATH] An epicycloid for which the diameter of the fixed circle is two times the diameter of the rolling circle. { 'ne,frȯid }

nephroid of Freeth See Freeth's nephroid. { 'ne,frȯid əv 'frēth }

nephrolithiasis [PATH] Formation of renal calculi. { nef·rō·li'thī·ə·səs }

nephrolithotomy [MED] Excision of renal calculi from the kidney. { nef·rō·li'thäd·ə·mē }

nephrology [MED] The study of the kidney, including diseases. { nə'fräl·ə·jē }

nephrolysin [BIOCHEM] A toxic substance capable of disintegrating kidney cells. { nə'fräl·ə·sən }

nephrolysis [MED] **1.** Dissolution of kidney tissue by the action of a nephrolysin. **2.** Surgical detachment of a kidney from surrounding adhesions. { nə'fräl·ə·səs }

nephroma [MED] A tumor of the kidney. { nə'frō·mə }

nephromegaly [MED] Enlargement of the kidney. { nef·rō'meg·ə·lē }

nephromixium [INV ZOO] A compound nephridium composed of flame cells and the coelomic funnel; functions as both an excretory organ and a genital duct. { nef·rō'mik·sē·əm }

nephron [ANAT] The functional unit of a kidney, consisting of the glomerulus with its capsule and attached uriniferous tubule. { 'nef,rän }

nephropathy [MED] **1.** Any disease of the kidney. **2.** See nephrosis. { nə'fräp·ə·thē }

nephropexy [MED] Fixation of a floating kidney by means of surgery. { 'nef·rə,pek·sē }

Nephropidae [INV ZOO] The true lobsters, a family of decapod crustaceans in the superfamily Nephropidea. { nə'fräp·ə,dē }

Nephropidea [INV ZOO] A superfamily of the decapod section Macrura including the true lobsters and crayfishes, characterized by a rostrum and by chelae on the first three pairs of pereiopods, with the first pair being noticeably larger. { nef·rə'pid·ē·ə }

nephroptosis [MED] Prolapse of the kidney. { ,ne,fräp'tō·səs }

nephrorrhaphy [MED] **1.** The stitching of a floating kidney to the posterior wall of the abdomen or to the loin. **2.** Suturing a wound in the kidney. { nə'frȯr·ə·fē }

nephros [ANAT] The kidney. { 'nef·rəs }

nephrosclerosis [MED] Sclerosis of the renal arteries and arterioles. { nef·rō·sklə'rō·səs }

nephrosis [PATH] Degenerative or retrogressive renal lesions, distinct from inflammation (nephritis) or vascular involvement (nephrosclerosis), especially as applied to tubular lesions (tubular nephritis). Also known as nephrodystrophy; nephropathy. { nə'frō·səs }

nephrostome [INV ZOO] The funnel-shaped opening of a nephridium into the coelom. { 'nef·rə,stōm }

nephrotic [MED] Pertaining to or affected by nephroses. { nə'fräd·ik }

nephrotic edema [MED] A type of edema occurring in persons with chronic lepoid nephrosis or the nephrotic stage of glomerular nephritis. { nə'fräd·ik i'dē·mə }

nephrotic syndrome [MED] A complex of symptoms, including proteinuria, hyperalbuminemia, and hyperlipemia, resulting from damage to the basement membrane of glomeruli. { nə'fräd·ik 'sin,drōm }

nephrotome [EMBRYO] The narrow mass of embryonic mesoderm connecting somites and lateral mesoderm, from which the pronephros, mesonephros, metanephros, and their ducts develop. { 'nef·rə,tōm }

nephrotomy [MED] Incision of the kidney. { nə'fräd·ə·mē }

nephsystem See cloud system. { 'nef,sis·təm }

Nephtyidae [INV ZOO] A family of errantian annelids of highly opalescent colors, distinguished by an eversible pharynx. { nef'tī·ə,dē }

Nepidae [INV ZOO] The water scorpions, a family of hemipteran insects in the superfamily Nepoidea, characterized by a long breathing tube at the tip of the abdomen, chelate front legs, and a short stout beak. { 'nep·ə,dē }

nepit See nit. { 'nep·ət }

Nepoidea [INV ZOO] A superfamily of hemipteran insects in the subdivision Hydrocorisae. { nə'pòid·ē·ə }

Nepovirus [VIROL] A genus of plant viruses in the family Comoviridae; tobacco ring spot virus is the type species. Also known as tobacco ring spot virus group. { 'nep·ə,vī·rəs }

Nepticulidae [INV ZOO] The single family of the lepidopteran superfamily Nepticuloidea. { ,nep·tə'kyül·ə,dē }

Nepticuloidea [INV ZOO] A monofamilial superfamily of heteroneuran Lepidoptera; members are tiny moths with wing spines, and the females have a single genital opening. { ,nep·tə·kyə'lòid·ē·ə }

Neptune [ASTRON] The outermost of the four giant planets, and the next to last planet, from the sun; it is 30 astronomical units from the sun, and the sidereal revolution period is 164.8 years. { 'nep·tün }

neptune powder [MATER] An explosive consisting of nitroglycerin with a more or less explosive dope. { 'nep·tün ,paùd·ər }

neptunian dike [GEOL] A sedimentary dike formed by infilling of sediment, generally sand, in an undersea fissure or hollow. { nep'tü·nē·ən 'dīk }

neptunianism See neptunism. { nep'tü·nē·ə,niz·əm }

neptunian theory See neptunism. { nep'tü·nē·ən 'thē·ə·rē }

neptunic rock [GEOL] **1.** A rock that is formed in the sea. **2.** See sedimentary rock. { nep'tün·ik 'räk }

neptunism [GEOL] The obsolete theory that all rocks of the earth's crust were deposited from or crystallized out of water. Also known as neptunianism; neptunian theory. { 'nep·tə,niz·əm }

neptunite [MINERAL] $(Na,K)_2(Fe,Mn)TiSi_4O_{12}$ Black mineral composed of silicate of sodium, potassium, iron, manganese, and titanium. { 'nep·tə,nīt }

neptunium [CHEM] A chemical element, symbol Np, atomic number 93, atomic weight 237.0482; a member of the actinide series of elements. { nep'tü·nē·əm }

neptunium decay series [CHEM] Little-known radioactive elements with short lives; produced as successive series of decreasing atomic weight when uranium-237 and plutonium-241 decay radioactively through neptunium-237 to bismuth-209. { nep'tü·nē·əm di'kā ,sir·ēz }

nepuite See garnierite. { 'nep·yə,wīt }

Nereid [ASTRON] The outermost known satellite of Neptune, orbiting at a mean distance of 3,425,900 miles (5,513,400 kilometers) with a period of 360 days, 3.1 hours, and with a diameter of about 210 miles (340 kilometers). { 'nir·ē·əd }

Nereidae [INV ZOO] A large family of mostly marine errantian annelids that have a well-defined head, elongated body with many segments, and large complex parapodia on most segments. { nə'rē·ə,dē }

Nerillidae [INV ZOO] A family of archiannelids characterized by well-developed parapodia and setae. { nə'ril·ə,dē }

Neritacea [INV ZOO] A superfamily of gastropod mollusks in the order Aspidobranchia. { ,ner·ə'tās·ē·ə }

neritic [OCEANOGR] Of or pertaining to the region of shallow water adjoining the seacoast and extending from low-tide mark to a depth of about 660 feet (200 meters). { nə'rid·ik }

Neritidae [INV ZOO] A family of primitive marine, freshwater, and terrestrial snails in the order Archaeogastropoda. { nə'rid·ə,dē }

Nernst approximation formula [THERMO] An equation for the equilibrium constant of a gas reaction based on the Nernst heat theorem and certain simplifying assumptions. { 'nernst ə,präk·sə'mā·shən ,fòr·myə·lə }

Nernst bridge [ELEC] A four-arm bridge containing capacitors instead of resistors, used for measuring capacitance values at high frequencies. { 'nernst ,brij }

Nernst effect [PHYS] The phenomenon that, when a conductor is placed in a magnetic field and an electric current flows through the conductor perpendicular to the field, a temperature gradient arises in the direction of the current. { 'nernst i,fekt }

Nernst equation [PHYS CHEM] The relationship showing that the electromotive force developed by a dry cell is determined by the activities of the reacting species, the temperature of the reaction, and the standard free-energy change of the overall reaction. { 'nernst i,kwā·zhən }

Nernst glower See Nernst lamp. { 'nernst ,glō·ər }

Nernst heat theorem [THERMO] The theorem expressing that the rate of change of free energy of a homogeneous system with temperature, and also the rate of change of enthalpy with temperature, approaches zero as the temperature approaches absolute zero. { 'nernst 'hēt ,thir·əm }

Nernst lamp [ELEC] An electric lamp consisting of a short, slender rod of zirconium oxide in open air, heated to brilliant white incandescence by current. Also known as Nernst glower. { 'nernst ,lamp }

Nernst-Lindemann calorimeter [ENG] A calorimeter for measuring specific heats at low temperatures, in which the heat reservoir consists of a metal of high thermal conductivity such as copper, to promote rapid temperature equalization; none of the material under study is more than a few millimeters from a metal surface, and the whole apparatus is placed in an evacuated vessel and heated by current through a platinum heating coil. { 'nernst 'lin·də·mən ,kal·ə'rim·əd·ər }

Nernst-Simon statement of the third law of thermodynamics [THERMO] The statement that the change in entropy which occurs when a homogeneous system undergoes an isothermal reversible process approaches zero as the temperature approaches absolute zero. { 'nernst 'sī·mən 'stāt·mənt əv thə 'thərd 'lò əv ,thər·mō·dī'nam·iks }

Nernst-Thomson rule [PHYS CHEM] The rule that in a solvent having a high dielectric constant the attraction between anions and cations is small so that dissociation is favored, while the reverse is true in solvents with a low dielectric constant. { 'nernst 'täm·sən ,rül }

Nernst zero of potential [PHYS CHEM] An electrode potential corresponding to the reversible equilibrium between hydrogen gas at a pressure of 1 standard atmosphere and hydrogen ions at unit activity. { 'nernst 'zir·ō əv pə'ten·chəl }

nerol [ORG CHEM] $C_{10}H_{17}OH$ Colorless liquid with rose-neroli odor; derived from geraniol (a trans isomer); used in perfumery. { 'ne,ròl }

nerolidol [ORG CHEM] $C_{15}H_{26}O$ A straw-colored sesquiterpene alcohol; liquid with rose and apple aroma derived from cabreuva oil, oils of orange flower, and ylang ylang; soluble in alcohol; used in perfumery. { nə'räl·ə,dòl }

neroli oil See oil of orange blossoms. { 'nər·ə·lē ,òil }

nerve [NEUROSCI] A bundle of nerve fibers or processes held together by connective tissue. { nərv }

nerve block [NEUROSCI] Interruption of impulse transmission through a nerve. { 'nərv ,bläk }

nerve cell See neuron. { 'nərv ,sel }

nerve cord [INV ZOO] Paired, ventral cords of nervous tissue in certain invertebrates, such as insects or the earthworm. [ZOO] Dorsal, hollow tubular cord of nervous tissue in chordates. { 'nərv ,kòrd }

nerve deafness [MED] Deafness due to an abnormality of the sense organs or of the nerves involved in hearing. { 'nərv ,def·nəs }

nerve ending [NEUROSCI] **1.** The structure on the distal end of an axon. **2.** The termination of a nerve. { 'nərv ,end·iŋ }

nerve fiber [NEUROSCI] The long process of a neuron, usually the axon. { 'nərv ,fī·bər }

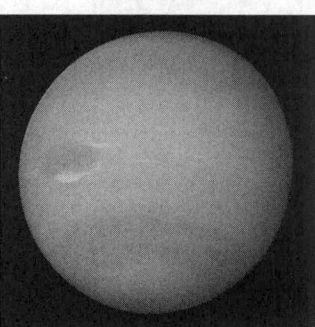

NEPTUNE

Image of Neptune from *Voyager 2*.

nerve gas [CHEM] Chemical agent which is absorbed into the body by breathing, by ingestion, or through the skin, and affects the nervous and respiratory systems and various body functions; an example is isopropylphosphonofluoridate. { 'nərv 'gas }

nerve growth factor [NEUROSCI] A multimeric protein that promotes nerve cell growth and may protect some types of nerve cells from damage, including nerve cells in the cholinergic system. { 'nərv ˌgrōth ˌfak·tər }

nerve impulse [NEUROSCI] The transient physicochemical change in the membrane of a nerve fiber which sweeps rapidly along the fiber to its termination, where it causes excitation of other nerves, muscle, or gland cells, depending on the connections and functions of the nerve. { 'nərv ˌim,pəls }

nerve net [INV ZOO] A network of continuous nerve cells characterized by diffuse spread of excitation, local and equipotential autonomy, spatial attenuation of conduction, and facilitation; occurs in cnidarians and certain other invertebrates. { 'nərv ˌnet }

nerve tracing [MED] A method used by chiropractors by which nerves are located and their pathologies are studied. { 'nərv ˌtrās·iŋ }

nerve tract [NEUROSCI] A bundle of nerve fibers having the same general origin and destination. { 'nərv ˌtrakt }

nerve trunk [NEUROSCI] Bundle of nerve secured together by connective tissue (epineurium). { 'nərv ˌtrəŋk }

nervous [NEUROSCI] **1.** Of or pertaining to nerves. **2.** Originating in or affected by nerves. **3.** Affecting or involving nerves. [PSYCH] A state or condition of nervousness. { 'nər·vəs }

nervousness [PSYCH] Hyperexcitability of the nervous system; characterized generally by restless or impulsive behavior, shaken mental poise, and an uncomfortable awareness of self. { 'nər·vəs·nəs }

nervous system [ANAT] A coordinating and integrating system which functions in the adaptation of an organism to its environment; in vertebrates, the system consists of the brain, brainstem, spinal cord, cranial and peripheral nerves, and ganglia. { 'nər·vəs ˌsis·təm }

nervous tissue [HISTOL] The nerve cells and neuroglia of the nervous system. { 'nər·vəs ˌtish·ü }

Nesiotinidae [INV ZOO] A family of bird-infesting biting lice (Mallophaga) that are restricted to penguins. { ˌnes·ē·ō'tin·ə,dē }

nesistor [ELECTR] A negative-resistance semiconductor device that is basically a bipolar field-effect transistor. { ne'zis·tər }

Nesophontidae [PALEON] An extinct family of large, shrewlike lipotyphlans from the Cenozoic found in the West Indies. { ˌnes·ə'fän·tə,dē }

nesosilicate [MINERAL] A mineral (such as olivine) composed of independent silicon-oxygen tetrahedra bonded by ionic bonds, without sharing of oxygens. { ˌnes·ō'sil·ə·kət }

nesquehonite [MINERAL] $MgCO_3 \cdot 3H_2O$ A colorless to white, orthorhombic mineral consisting of hydrated magnesium carbonate. { ˌnes·kwə'hō,nīt }

Nessler's reagent [ANALY CHEM] Mercuric iodide-potassium iodide solution, used to analyze for small amounts of ammonia. { 'nes·lərz rē,ā·jənt }

Nessler tubes [ANALY CHEM] Standardized glass tubes for filling with standard solution colors for visual color comparison with similar tubes filled with solution samples. { 'nes·lər ˌtübz }

nest [COMPUT SCI] To include data or subroutines in other items of a similar nature with a higher hierarchical level so that it is possible to access or execute various levels of data or routines recursively. [GEOL] A concentration of some relatively conspicuous element of a geologic feature, such as pebbles or inclusions, within a sand layer or igneous rock. [VERT ZOO] Abed, receptacle, or location in which the eggs of animals are laid and hatched. { nest }

nested [GEOL] **1.** Pertaining to volcanic cones, craters, or calderas that occur one within another. **2.** Pertaining to two or more calderas that intersect, having been formed at different times or by different explosions. { 'nes·təd }

nested intervals [MATH] A sequence of intervals, each of which is contained in the preceding interval. { 'nes·təd 'in·tər·vəlz }

nested sets [MATH] A family of sets where, given any two

of its sets, one is contained in the other. Also known as monotonic system of sets. { 'nes·təd 'sets }

nesting [COMPUT SCI] **1.** Inclusion of a routine wholly within another routine. **2.** Inclusion of a DO statement within a DO statement in FORTRAN. [IND ENG] A production technique in which parts with similar patterns are manufactured together. { 'nest·iŋ }

nesting storage See push-down storage. { 'nest·iŋ 'stór·ij }

Nestor [ASTRON] One of a group of asteroids whose period of revolution is approximately equal to that of Jupiter, or about 12 years (it is one of the Trojan planets). { 'nes·tər }

net [COMMUN] A number of communication stations equipped for communicating with each other, often on a definite time schedule and in a definite sequence. [ENG] **1.** Threads or cords tied together at regular intervals to form a mesh. **2.** A series of surveying or leveling stations that have been interconnected in such a manner that closed loops or circuits have been formed, or that are arranged so as to provide a check on the consistency of the measured values. Also known as network. [GEOL] **1.** In structural petrology, coordinate network of meridians and parallels, projected from a sphere at intervals of 2°; used to plot points whose spherical coordinates are known and to study the distribution and orientation of planes and points. Also known as projection net; stereographic net. **2.** A form of horizontal patterned ground whose mesh is intermediate between a circle and a polygon. [MATH] **1.** A set whose members are indexed by elements from a directed set; this is a generalization of a sequence. Also known as Moore-Smith sequence. **2.** A nondegenerate partial plane satisfying the parallel axiom. [TEXT] Any fabric made in open hexagonal mesh. { net }

Net See Reticulum. { net }

net aerial production [ECOL] The biomass or biocontent which is incorporated into the aerial parts, that is, the leaf, stem, seed, and associated organs, of a plant community. { 'net 'er·ē·əl prə'dək·shən }

net balance [HYD] The change in mass of a glacier from the time of minimum mass in one year to the time of minimum mass in the succeeding year. Also known as net budget. { 'net 'bal·əns }

net blotch [PL PATH] A fungus disease of barley caused by *Helminthosporium teres* and marked by spotting of the foliage. { 'net 'bläch }

net budget See net balance. { 'net 'bəj·ət }

net call sign [COMMUN] A call sign that represents all stations within a net. { 'net 'kól ˌsīn }

net control station [COMMUN] Communications station having the responsibility of clearing traffic and exercising circuit discipline within a net. { 'net kən'trōl ˌstā·shən }

NETD See noise equivalent temperature difference.

net density See density scale. { 'net ˌden·səd·ē }

net floor area [BUILD] Gross floor area of a building, excluding the area occupied by walls and partitions, the circulation area (where people walk), and the mechanical area (where there is mechanical equipment). { 'net 'flór ˌer·ē·ə }

net flow [MATH] The net flow at a vertex in an *s-t* network is the outflow at that vertex minus the inflow there. { 'net 'flō }

net flow area [DES ENG] The calculated net area which determines the flow after the complete bursting of a rupture disk. { 'net 'flō ˌer·ē·ə }

net head [FL MECH] The difference in elevation between the last free water surface in a power conduit above the waterwheel and the first free water surface in the conduit below the waterwheel, less the friction losses in the conduit. { 'net 'hed }

net heating value See low heat value. { 'net 'hēd·iŋ ˌval·yü }

net line See neat line. { 'net ˌlīn }

net load capacity [ENG] The weight of a material that can be handled, without failure, by a machine or process plus the weight of the container or device. { 'net ˌlōd kə'pas·əd·ē }

net loss [COMMUN] The ratio of the power at the input of a transmission system to the power at the output; expressed in nepers, it is one-half the natural logarithm of this ratio, and in decibels it is 10 times the common logarithm of the ratio. { 'net 'lós }

net plankton [ECOL] Plankton that can be removed from sea water by the process of filtration through a fine net. { 'net 'plaŋk·tən }

net positive suction head [MECH ENG] The minimum suction head required for a pump to operate; depends on liquid characteristics, total liquid head, pump speed and capacity, and impeller design.   Abbreviated NPSH.   { 'net 'päz·əd·iv ¦sək·shən ¦hed }

net power flow [ELECTROMAG] The difference between the power carried by electromagnetic waves traveling in a given direction along a waveguide and the power carried by waves traveling in the opposite direction.   { 'net 'pau̇·ər ¸flō }

net primary production [ECOL] Over a specified period of time, the biomass or biocontent which is incorporated into a plant community.   { 'net 'prīm·ə·rē prə'dək·shən }

net production rate [ECOL] The assimilation rate (gross production rate) minus the amount of matter lost through predation, respiration, and decomposition.   { 'net prə'dək·shən ¸rāt }

net radiometer [ENG] A Moll thermopile modified so that both sides are sensitive to radiation and the resulting electromotive force is proportional to the difference in intensities of radiation incident on the two sides; used to measure the difference in intensity between radiation entering and leaving the earth's surface.   { 'net ¸rād·ē'äm·əd·ər }

net slip [GEOL] On a fault, the distance between two formerly adjacent points on either side of the fault; defines direction and relative amount of displacement.   Also known as total slip.   { 'net 'slip }

net thrust [AERO ENG] The gross thrust of a jet engine minus the drag due to the momentum of the incoming air.   { 'net 'thrəst }

nettle [BOT] A prickly or stinging plant of the family Urticaceae, especially in the genus *Urtica*.   { 'ned·əl }

nettle cell See cnidoblast.   { 'ned·əl ¸sel }

net ton See ton.   { 'net 'tən }

net tonnage [NAV ARCH] The volume of the interior of a ship measured in tons (100 cubic feet or approximately 2.8317 cubic meters per ton), excluding the space occupied by fuel, the engine room, navigation machinery, and the crew's quarters.   { 'net 'tən·ij }

net-veined [BIOL] Having a network of veins, as a leaf or an insect wing.   { 'net ¸vānd }

network [COMMUN] A number of radio or television broadcast stations connected by coaxial cable, radio, or wire lines, so all stations can broadcast the same program simultaneously.   [ELEC] A collection of electric elements, such as resistors, coils, capacitors, and sources of energy, connected together to form several interrelated circuits.   Also known as electric network.   [ENG] See net.   [MATH] The name given to a graph in applications in management and the engineering sciences; to each segment linking points in the graph, there is usually associated a direction and a capacity on the flow of some quantity.   { 'net¸wərk }

network admittance [ELEC] The admittance between two terminals of a network under specified conditions.   { 'net¸wərk ad'mit·əns }

network analysis [ELEC] Derivation of the electrical properties of a network, from its configuration, element values, and driving forces.   [IND ENG] An analytic technique used during project planning to determine the sequence of activities and their interrelationship within the network of activities that will be required by the project.   Also known as network planning.   { 'net¸wərk ə'nal·ə·səs }

network analyzer [COMPUT SCI] An analog computer in which networks are used to simulate power line systems or physical systems and obtain solutions to various problems before the systems are actually built.   { 'net¸wərk 'an·ə¸līz·ər }

network architecture [COMMUN] The high-level design of a communications system, including the choice of hardware, software, and protocols.   { 'net¸wərk 'är·kə¸tek·chər }

network constant [ELEC] One of the resistance, inductance, mutual inductance, or capacitance values involved in a circuit or network; if these values are constant, the network is said to be linear.   { 'net¸wərk 'kän·stənt }

network control program [COMPUT SCI] A computer program that controls communications between multiple terminals and a mainframe.   { ¦net¸wərk kən'trōl ¸prō·grəm }

network data structure [COMPUT SCI] The arrangement of data in a computer system into interconnected groupings of information according to relationships between groupings.   { 'net¸wərk 'dad·ə ¸strək·chər }

network filter [ELEC] A combination of electrical elements (for example, interconnected resistors, coils, and capacitors) that represents relatively small attenuation to signals of a certain frequency, and great attenuation to all other frequencies.   { 'net¸wərk 'fil·tər }

network flow [ELEC] Flow of current in a network.   { 'net¸wərk 'flō }

networking [COMPUT SCI] The use of transmission lines to join geographically separated computers.   { 'net¸wərk·iŋ }

network input impedance [ELEC] The impedance between the input terminals of a network under specified conditions.   { 'net¸wərk 'in·pu̇t im¸pēd·əns }

network master relay [ELEC] Relay that performs the chief functions of closing and tripping an alternating-current low-voltage network protector.   { 'net¸wərk 'mas·tər 'rē¸lā }

network operating system [COMPUT SCI] The system software of a local-area network, which manages the network's resources, handling multiple inputs concurrently and providing necessary security.   Abbreviated NOS.   { ¸net¸wərk 'äp·ə¸rād·iŋ ¸sis·təm }

network phasing relay [ELEC] Relay which functions in conjunction with a master relay to limit closure of the network protector to a predetermined relationship between the voltage and the network voltage.   { 'net¸wərk 'fāz·iŋ 'rē¸lā }

network planning See network analysis.   { 'net¸wərk 'plan·iŋ }

network polymer [ORG CHEM] A three-dimensional material made by crosslinking.   { ¦net¸wərk 'päl·ə·mər }

network relay [ELEC] Form of voltage, power, or other type of relay used in the protection and control of alternating-current low-voltage networks.   { 'net¸wərk 'rē¸lā }

network server See file server.   { 'net¸wərk 'sər·vər }

network structure [MET] A crystal structure in a metal in which one constituent occurs primarily at grain boundaries enveloping the grains made up of other constituents.   { 'net¸wərk ¸strək·chər }

network synthesis [ELEC] Derivation of the configuration and element values of a network with given electrical properties.   { 'net¸wərk ¸sin·thə·səs }

network system [COMPUT SCI] A type of data-base management system in which data records can be related in more general structures than in a hierarchical file, permitting a given record to have more than one parent.   { 'net¸wərk ¸sis·təm }

network terminal protocol [COMMUN] A set of standards that allows the user of a computer connected to a network to log in on any other computer on the network.   Also known as TELNET.   { ¦net¸wərk ¦tərm·ən·əl ¦prōd·ə¸kȯl }

network theory [ELEC] The systematizing and generalizing of the relations between the currents, voltages, and impedances associated with the elements of an electrical network.   { 'net¸wərk ¸thē·ə·rē }

network transfer admittance [ELEC] The current that would flow through a short circuit between one pair of terminals in a network if a unit voltage were applied across the other pair.   { 'net¸wərk 'trans·fər ad¸mit·əns }

network vulnerability scan [COMPUT SCI] The process of determining the connectivity of the protected subnetwork within a security perimeter of a distributed computing system, and then testing the strength of protection at all access points to the subnetwork.   { ¦net¸wərk ¸vəl·nər·ə'bil·əd·ē ¸skan }

Neuberg blue [MATER] Pigment made up of a mixture of copper blue and iron blue.   { 'nȯi¸bərk 'blü }

Neugebauer effect [ELEC] A small change in the polarization of an optically isotropic medium in an external electric field, related to the electrooptical Kerr effect.   { 'nȯi·gə¸bau̇·ər i¸fekt }

Neumann bands See Neumann lines.   { 'nȯi¸män ¸banz }

Neumann boundary condition [MATH] The boundary condition imposed on the Neumann problem in potential theory.   { 'nȯi¸män 'bau̇n·drē kən¸dish·ən }

Neumann function [MATH] **1.** One of a class of Bessel functions arising in the study of the solutions to Bessel's differential equation.   **2.** A harmonic potential function in potential theory occurring in the study of Neumann's problem.   { 'nȯi¸män ¸fəŋk·shən }

Neumann-Kopp rule [THERMO] The rule that the heat capacity of 1 mole of a solid substance is approximately equal to the sum over the elements forming the substance of the heat capacity of a gram atom of the element times the number of atoms of the element in a molecule of the substance. { 'nȯi‚män 'kȯp ‚rül }

Neumann line [MATH] The generalization of the concept of a line occurring in Neumann's study of continuous geometry. { 'nȯi‚män ‚līn }

Neumann lines [MET] Mechanical deformation twins seen as straight, serrated narrow bands parallel to preferred planes in the crystals of an etched metal which has been strained, usually by sudden impact; most often observed along the 112 planes of body-centered-cubic ferrite. Also known as Neumann bands. { 'nȯi‚män ‚līnz }

Neumann problem [MATH] The determination of a harmonic function within a finite region of three-dimensional space enclosed by a closed surface when the normal derivatives of the function on the surface are specified. { 'nȯi‚män ‚präb·ləm }

Neumann series *See* Liouville-Neumann series. { 'nȯi‚män ‚sir·ēz }

Neumann's formula [ELECTROMAG] A formula for the mutual inductance M_{12} between two closed circuits C_1 and C_2; it is

$$M_{12} = \frac{\mu_0}{4\pi} \int_{C_1} \int_{C_2} \frac{ds_1 ds_2}{r}$$

where r is the distance between line elements ds_1 and ds_2, and μ_0 is the permeability of the empty space. { 'nȯi‚mänz ‚fȯr·myə·lə }

Neumann's principle [CRYSTAL] The principle that the symmetry elements of the point group of a crystal are included among the symmetry elements of any property of the crystal. { 'nȯi‚mänz ‚prin·sə·pəl }

Neumann's triangle [FL MECH] A triangle whose sides have lengths proportional to the surface tensions of two immiscible liquids and their interfacial tension, and directions parallel to the free surfaces of the liquids and the interface between them at a line where these three surfaces meet, when one liquid is placed on the surface of the other. { 'nȯi·mänz 'trī‚aŋ·gəl }

neural arc [NEUROSCI] A nerve circuit consisting of effector and receptor with intercalated neurons between them. { 'nùr·əl 'ärk }

neural arch *See* vertebral arch. { 'nùr·əl 'ärch }

neural canal [EMBRYO] The embryonic vertebral canal. { 'nùr·əl kə'nal }

neural cell adhesion molecule [NEUROSCI] A calcium-independent cell adhesion molecule expressed by migrating neurons that mediates intercellular binding via homophilic mechanisms (by binding to other neural cell adhesion molecules); important in neuronal aggregation. { ¦nùr·əl ‚sel ad'hē·zhen ‚mäl·ə‚kyül }

neural crest [EMBRYO] Ectoderm composing the primordium of the cranial, spinal, and autonomic ganglia and adrenal medulla, located on either side of the neural tube. { 'nùr·əl 'krest }

neural ectoderm [EMBRYO] Embryonic ectoderm which will form the neural tube and neural crest. { 'nùr·əl 'ek·tə‚dərm }

neural fold [EMBRYO] Either of a pair of dorsal longitudinal folds of the neural plate which unite along the midline, forming the neural tube. { 'nùr·əl 'fōld }

neuralgia [MED] Pain in or along the course of one or more nerves. Also known as neurodynia. { nù'ral·jē·ə }

neural groove [EMBRYO] A longitudinal groove between the neural folds of the vertebrate embryo before the neural tube is completed. { 'nùr·əl 'grüv }

neural lymphomatosis [VET MED] A form of the avian leukosis complex affecting primarily the sciatic nerve. { 'nùr·əl ‚lim·fə·mə'tō·səs }

neural network [COMPUT SCI] An information-processing device that utilizes a very large number of simple modules, and in which information is stored by components that at the same time effect connections between these modules. { 'nùr·əl 'net‚wərk }

neural plate [EMBRYO] The thickened dorsal plate of ectoderm that differentiates into the neural tube. { 'nùr·əl 'plāt }

neural prosthesis [MED] A prosthesis that bypasses a portion of the nervous system and provides electrical stimulation to existing muscles in situations where paralysis has interrupted the natural control pathways. { ‚nùr·əl präs'thē·səs }

neural spine [ANAT] The spinous process of a vertebra. { 'nùr·əl 'spīn }

neural tube [EMBRYO] The embryonic tube that differentiates into brain and spinal cord. { 'nùr·əl 'tüb }

neural tube defects [MED] Congenital defects resulting from the incomplete closure of the neural tube during embryogenesis. { ‚nùr·əl 'tüb ‚dē‚feks }

neuraminic acid [BIOCHEM] $C_9H_{17}NO_8$ An amino acid, the aldol condensation product of pyruvic acid and *N*-acetyl-D-mannosamine, regarded as the parent acid of a family of widely distributed acyl derivatives known as sialic acids. { ¦nùr·ə¦min·ik 'as·əd }

neuraminidase [MICROBIO] A bacterial enzyme that acts to split sialic acid from neuraminic acid glycosides. { ‚nùr·ə'min·ə‚dās }

neurapophysis [EMBRYO] Either of two projections on each embryonic vertebra which unite to form the neural arch. { ¦nùr·ə'päf·ə·səs }

neurapraxia [MED] Injury to a nerve in which there is localized degeneration of the myelin sheath with transient nerve block. { ¦nùr·ə'prak·sē·ə }

neurasthenia [MED] A group of symptoms, now generally subsumed in the neurasthenic neurosis, formerly ascribed to debility or exhaustion of the nerve centers. { ‚nùr·əs'thē·nē·ə }

neurasthenic neurosis [PSYCH] A neurotic disorder characterized by chronic complaints of easy fatigability, lack of energy, weakness, various aches and pains, and sometimes exhaustion. Also known as psychophysiologic nervous system reaction. { ‚nùr·əs'then·ik nù'rō·səs }

neurectomy [MED] Surgical removal of a portion of a nerve. { nù'rek·tə·mē }

neurenteric canal [EMBRYO] A temporary duct connecting the neural tube and primitive gut in certain vertebrate and tunicate embryos. { ¦nùr·ən'ter·ik kə'nal }

neurilemma [HISTOL] A thin tissue covering the axon directly, or covering the myelin sheath when present, of peripheral nerve fibers. { ‚nùr·ə'lem·ə }

neurilemmoma [MED] A solitary, encapsulated benign tumor originating in the neurilemma of peripheral, cranial, and sympathetic nerves. Also known as schwannoma. { ‚nùr·ə·lə'mō·mə }

neurine [BIOCHEM] $CH_2=CHN(CH_3)_3OH$ A very poisonous, syrupy liquid with fishy aroma; soluble in water and alcohol; a product of putrefaction of choline in brain tissue and bile, and in cadavers. Also known as trimethylvinylammonium hydroxide. { 'nü‚rēn }

neurinomatosis *See* neurofibromatosis. { ‚nùr·ə‚nō·mə'tō·səs }

neuristor [ELECTR] A device that behaves like a nerve fiber in having attenuationless propagation of signals; one goal of research is development of a complete artificial nerve cell, containing many neuristors, that could duplicate the function of the human eye and brain in recognizing characters and other visual images. { nù'ris·tər }

neurite *See* axon. { 'nù‚rīt }

neuritic plaque [PATH] Abnormal clumps of degenerating neurons surrounding a core of amyloid protein; a characteristic pathological feature found in the brains of Alzheimer's disease patients. { nù‚rid·ik 'plak }

neuritis [MED] Degenerative or inflammatory nerve lesions associated with pain, hypersensitivity, anesthesia or paresthesia, paralysis, muscular atrophy, and loss of reflexes in the innervated part of the body. { nù'rīd·əs }

neuroanatomy [ANAT] The study of the anatomy of the nervous system and nerve tissue. { ¦nùr·ō·ə'nad·ə·mē }

neuroarthropathy [MED] Joint disease associated with disease of the nervous system. { ¦nùr·ō·är'thräp·ə·thē }

neuroastrocytoma [MED] Ganglioneuroma, especially when on the floor of the third brain ventricle and in the temporal lobes and exhibiting neuronal elements within predominant astrocytic elements. { ¦nùr·ō‚as·trə‚sī'tō·mə }

neurobiotaxis [EVOL] Hypothetical migration of nerve cells and ganglia toward regions of maximum stimulation during phylogenetic development. { ¦nùr·ō‚bī·ō'tak·səs }

neuroblast [EMBRYO] Embryonic, undifferentiated neuron, derived from neural plate ectoderm. { 'nùr·ō‚blast }

neuroblastoma [MED] A malignant neoplasm composed of

anaplastic sympathicoblasts; occurs usually in the adrenal medulla of children. { ˌnu̇r·ō·bla'stō·mə }

neuroblastomatosis *See* neurofibromatosis. { ˌnu̇r·ō·bla,stō·mə'tō·səs }

neurobrucellosis [MED] Brucellosis with neurologic involvement, manifested by signs and symptoms of meningitis, encephalitis, radiculitis, or neuritis. { ˌnu̇r·ō,brü·sə'lō·səs }

neurochemistry [BIOCHEM] Chemistry of the nervous system. { ˌnu̇r·ō'kem·ə·strē }

neurochorioretinitis [MED] Chorioretinitis combined with optic neuritis. { ˌnu̇r·ō,kȯr·ē·ō,ret·ən'īd·əs }

neurochoroiditis [MED] Choroiditis combined with optic neuritis. { ˌnu̇r·ō,kȯ·rȯi'dīd·əs }

neurocirculatory asthenia [MED] A syndrome characterized by dyspnea, palpitation, chest pain, fatigue, and faintness. { ˌnu̇r·ō'sər·kyə·lə,tȯr·ē as'thē·nē·ə }

neurocoele [ANAT] The system of cavities and ventricles in the brain and spinal cord. { 'nu̇r·ō,sēl }

neurocranium [ANAT] The portion of the cranium which forms the braincase. { ˌnu̇r·ə'krā·nē·əm }

neurocutaneous [ANAT] **1.** Concerned with both the nerves and skin. **2.** Pertaining to innervation of the skin. { ˌnu̇r·ō·kyə'tā·nē·əs }

neurocyte [NEUROSCI] The body of a nerve cell. { 'nu̇r·ə,sīt }

neurodegeneration [NEUROSCI] The process in which neurons die. { ˌnu̇·rō·di,jen·ə'rā·shən }

neurodegenerative [PATH] Characterized by the gradual and progressive loss of nerve cells. { ˌnu̇·rō·di'jen·rəd·iv }

neurodermatitis [MED] A skin disorder characterized by localized, often symmetrical, patches of pruritic dermatitis with lichenification, occurring in persons of nervous temperament. { ˌnu̇r·ō,dər·mə'tīd·əs }

neurodermatosis [MED] A skin disease which is presumed to have a psychogenic component or basis. { ˌnu̇r·ō,dər·mə'tō·səs }

Neurodontiformes [PALEON] A suborder of Conodontophoridia having a lamellar internal structure. { ˌnu̇r·ō,dänt·ə'fȯr,mēz }

neurodynia *See* neuralgia. { ˌnu̇r·ō'dī·nē·ə }

neuroelectricity [PHYSIO] A current or voltage generated in the nervous system. { ˌnu̇r·ō,i,lek'tris·əd·ē }

neuroendocrine [BIOL] Pertaining to both the nervous and endocrine systems, structurally and functionally. { ˌnu̇r·ō·en·də·krən }

neuroendocrinology [BIOL] The study of the structural and functional interrelationships between the nervous and endocrine systems. { ˌnu̇r·ō,en·də·krə'näl·ə·jē }

neuroepidermal [BIOL] Pertaining to both the nerves and epidermis, structurally and functionally. { ˌnu̇r·ō,ep·ə'dər·məl }

neuroepithelioma [MED] A tumor resembling primitive medullary epithelium, containing cells of small cuboidal or columnar form with a tendency to form true rosettes, occurring in the retina, central nervous system, and occasionally in peripheral nerves. Also known as diktoma; esthesioneuroblastoma; esthesioneuroepithelioma. { ˌnu̇r·ō,ep·ə,thē·lē'ō·mə }

neuroethology [ZOO] The study of the neural basis of animal behavior. { ˌnu̇·rō·ē'thäl·ə·jē }

neurofibril [NEUROSCI] A fibril of a neuron, usually extending from the processes and traversing the cell body. { ˌnu̇r·ō'fī·brəl }

neurofibrillary tangle [PATH] The accumulation of twisted protein filaments (neurofibrils) within neurons of the cerebral cortex; a characteristic pathological feature found in the brains of Alzheimer's disease patients. { ˌnu̇·rō,fīb·rə,ler·ē 'taŋ·gəl }

neurofibroma [MED] A tumor characterized by the diffuse proliferation of peripheral nerve elements. Also known as endoneural fibroma; myxofibroma of nerve sheath; neurofibromyxoma; perineural fibroblastoma; perineural fibroma. { ˌnu̇r·ō·fī'brō·mə }

neurofibromatosis [MED] A hereditary disease characterized by the presence of neurofibromas in the skin or along the pathway of peripheral nerves. Also known as fibroma molluscum; multiple neurofibroma; multiple neurofibromatosis; neurinomatosis; neuroblastomatosis; Smith-Recklinghausen's disease. { ˌnu̇r·ō·fī,brō·mə'tō·səs }

neurofibromyxoma *See* neurofibroma. { ˌnu̇r·ō,fī·brō·mik'sō·mə }

neurofibrosarcoma [MED] A malignant tumor composed of interlacing bundles of anaplastic spindle-shaped cells which resemble those of nerve sheaths. { ˌnu̇r·ō,fī·brō·sär'kō·mə }

neurofilament [PHYSIOL] A type of intermediate filament, composed of three polypeptides (NF-L, NF-M, and NF-H), that helps support and strengthen the axons of nerve cells. { ˌnu̇·rō'fil·ə·mənt }

neurogenesis [EMBRYO] The formation of nerves. { ˌnu̇r·ō'jen·ə·səs }

neurogenic [MED] Caused or affected by a trauma, dysfunction, or disease of the nervous system. [NEUROSCI] **1.** Originating in nervous tissue. **2.** Innervated by nerves. { ˌnu̇r·ō'jen·ik }

neurogenic bladder [MED] A urinary bladder disorder due to lesions of the central or peripheral nervous system. { ˌnu̇r·ō'jen·ik 'blad·ər }

neurogenic shock [MED] Shock caused by vasodilation leading to low blood pressure and serious reduction in venous return and in cardiac output; due to such causes as injury to the central nervous system, spinal anesthesia, or reflex. { ˌnu̇r·ō'jen·ik 'shäk }

neuroglia [NEUROSCI] The nonnervous, supporting elements of the nervous system. { nu̇'räg·lē·ə }

neurohemal organ [ZOO] Any of various structures in vertebrates and some invertebrates that consist of clusters of bulbous, secretion-filled axon terminals of neurosecretory cells which function as storage-and-release centers for neurohormones. { ˌnu̇r·ō'hē·məl ,ȯr·gən }

neurohormone [NEUROSCI] A hormone produced by nervous tissue. { ˌnu̇r·ō'hȯr,mōn }

neurohumor [NEUROSCI] A hormonal transmitter substance, such as acetylcholine, released by nerve endings in the transmission of impulses. { ˌnu̇r·ō'hyü·mər }

neurohypophysis [NEUROSCI] The neural portion or posterior lobe of the hypophysis. { ˌnu̇r·ō·hī'päf·ə·səs }

neuroimmunomodulation [PHYSIOL] The influences of the nervous system upon the immune system via neural and hormonal actions. { ˌnu̇·rō,im·yə·no,mä·jə'lā·shən }

neurolathyrism *See* lathyrism. { ˌnu̇r·ō'lath·ə,riz·əm }

neuroleptic [PHARM] **1.** A drug that is useful in the treatment of mental disorders, especially psychoses. **2.** Pertaining to the actions of such a drug. { ˌnu̇r·ō'lep·tik }

neuroleptoanalgesia [MED] A state of analgesic consciousness produced by the administration of neuroleptic drugs, allowing painless surgery to be performed on a wakeful subject. { ˌnu̇r·ō,lep·tō,an·ə'jē·zhə }

neurolinguistics [LING] A branch of linguistics concerned with the biological basis of language development. { ˌnu̇r·ō·liŋ'gwis·tiks }

neurologist [MED] A person versed in neurology, usually a physician who specializes in the diagnosis and treatment of disorders of the nervous system and the study of its functioning. { nu̇'räl·ə·jəst }

neurology [MED] The study of the anatomy, physiology, and disorders of the nervous system. { nu̇'räl·ə·jē }

neuroma [MED] A tumor of the nervous system. { nu̇'rō·mə }

neuromast [VERT ZOO] A lateral-line sensory organ in fishes and other lower vertebrates consisting of a cluster of receptor cells connected with nerve fibers. { 'nu̇r·ō,mast }

neuromere [EMBRYO] An embryonic segment of the central nervous system in vertebrates. { 'nu̇r·ō,mir }

neuromodulator [NEUROSCI] A chemical agent that is released by a neurosecretory cell and acts on other neurons in a local region of the central nervous system by modulating their response to neurotransmitters. { ˌnu̇r·ō'mäj·ə,läd·ər }

neuromorphic engineering [ENG] Use of the functional principles of biological nervous systems to inspire the design and fabrication of artificial nervous systems, such as vision chips and roving robots. { nu̇·rō,mȯr·fik ,en·jə'nir·iŋ }

neuromuscular [BIOL] Pertaining to both nerves and muscles, functionally and structurally. { ˌnu̇r·ō'məs·kyə·lər }

neuromuscular junction *See* myoneural junction. { ˌnu̇r·ō'məs·kyə·lər 'jəŋk·shən }

neuromyasthenia [MED] Fatigue, headache, intense muscle pain, slight or transient muscle weakness, mental disturbances, objective signs in neurologic examination but usually normal cerebrospinal fluid findings, occurring in epidemics and thought to be viral in origin.   Also known as benign myalgic encephalomyelitis.   { ¦nur·ō¸mī·əs'thē·nē·ə }

neuromyelitis [MED]   Inflammation of the spinal cord and of nerves.   { ¦nur·ō¸mī·ə'līd·əs }

neuron [NEUROSCI]   A nerve cell, including the cell body, axon, and dendrites.   { 'nü¸rän }

neuronal interface [ENG]   An artificial synapse capable of reversible chemical-to-electrical transduction processes between neural tissue and conventional solid-state electronic devices for applications such as aural, visual, and mechanical prostheses, as well as expanding human memory and intelligence.   { nü¦rōn·əl 'in·tər¸fās }

neuron doctrine [NEUROSCI]   A doctrine that the neuron is the basic structural and functional unit of the nervous system, and that it acts upon another neuron through the synapse.   { 'nü¸rän ¸däk·trən }

neuronitis [MED]   Inflammation of a neuron; particularly, neuritis involving the cells and roots of spinal nerves.   { ¸nur·ō'nīd·əs }

neuropathy [MED]   Any   disease   affecting   neurons.   { nü'räp·ə·thē }

neuropeptide [NEUROSCI]   A polypeptide released by axons at the synapse; it may act as a neurotransmitter and have a direct effect on synapse function or as a neuromodulator, having a long-term effect on postsynaptic neurons.   { ¸nur·ō'pep¸tīd }

neuropharmacology [MED]   The science dealing with the action of drugs on the nervous system.   { ¦nur·ō¸fär·mə'käl·ə·jē }

neurophysiology [NEUROSCI]   The study of the functions of the nervous system.   { ¦nur·ō¸fiz·ē'äl·ə·jē }

neuropil [NEUROSCI]   Nervous tissue consisting of a fibrous network of nonmyelinated nerve fibers; gray matter with few nerve cell bodies; usually a region of synapses between axons and dendrites.   { 'nur·ō¸pil }

neuroplasm [CELL MOL]   Protoplasm   of   nerve   cells.   { 'nur·ō¸plaz·əm }

neuropodium [NEUROSCI]   A terminal branch of an axon.   { ¸nur·ō'pō·dē·əm }

neuropore [EMBRYO]   A terminal aperture of the neural tube before complete closure at the 20–25 somite stage.   { 'nur·ō¸pór }

neuropsychology [PSYCH]   A system of psychology based on neurology.   { ¦nur·ō·sī'käl·ə·jē }

neuropsychopathy [MED]   A mental disease based upon or manifesting itself in disorders or symptoms of the nervous system.   { ¦nur·ō·sī'käp·ə·thē }

Neuroptera [INV ZOO]   An order of delicate insects having endopterygote development, chewing mouthparts, and soft bodies.   { nü'räp·tə·rə }

neuroradiology [MED]   The roentgenology of neurologic disease.   { ¦nur·ō¸rād·ē'äl·ə·jē }

neurorrhexis [MED]   Surgical tearing away of a nerve from its origin, as in the treatment of neuralgia.   { ¦nur·ō'rek·səs }

neurosarcoma [MED]   A sarcoma composed of elements resembling those of the nervous system, or thought to be neurogenic.   { ¦nur·ō·sär'kō·mə }

neurosclerosis [MED]   Hardening   of   nervous   tissue.   { ¦nur·ō·sklə'rō·səs }

neurosecretion [NEUROSCI]   The synthesis and release of hormones by nerve cells.   { ¦nur·ō·si'krē·shən }

neurosecretory cell [NEUROSCI]   A neuron that releases one or more hormones into the circulatory system.   { ¸nur·ō·si'krēd·ə·rē ¸sel }

neurosis [PSYCH]   A category of emotional maladjustments characterized by some impairment of thinking and judgment, with anxiety as the chief symptom.   { nü'rō·səs }

neurosurgery [MED]   Surgery   of   the   nervous   system.   { ¦nur·ō'sər·jə·rē }

neurosyphilis [MED]   Syphilitic infection of the nervous system.   { ¦nur·ō'sif·ə·ləs }

neurotechnology [ENG]   The application of microfabricated devices to achieve direct contact with the electrically active cells of the nervous system (neurons).   { ¸nü·rō·tek'näl·ə·jē }

neuroticism [PSYCH]   A neurotic condition, character, or trait.   { nə'räd·ə¸siz·əm }

neurotic personality [PSYCH]   An individual who exhibits

symptoms or manifestations intermediate between normal character traits and true neurotic features.   { nə'räd·ik ¸pər·sə'nal·əd·ē }

neurotoxicity [MED]   Adverse effects on the structure or function of the central and/or peripheral nervous system caused by exposure to a toxic chemical; symptoms include muscle weakness, loss of sensation and motor control, tremors, cognitive alterations, and autonomic nervous system dysfunction.   { ¸nü·ro·täk'sis·əd·ē }

neurotoxin [BIOCHEM]   A poisonous substance in snake venom that acts as a nervous system depressant by blocking neuromuscular transmission by binding acetylcholine receptors on motor end plates, or on the innervated face of an electroplax.   { ¦nur·ō'täk·sən }

neurotransmitter [NEUROSCI]   A chemical agent that is released by a neuron at a synapse, diffuses across the synapse, and acts upon a postsynaptic neuron, a muscle, or a gland cell.   { ¸nur·ō¸tranz'mid·ər }

neurotrophic ulcer [MED]   A decubitus ulcer due to trophic disturbances following interruption or disease of afferent nerve fibers plus the factor of external trauma.   { ¦nur·ō¦träf·ik 'əl·sər }

neurotropic [BIOL]   Having an affinity for nerve tissue.   { ¦nur·ō¦träp·ik }

neurovaricosis [MED]   The formation or the presence of a varicosity on a nerve fiber.   { ¦nur·ō¸var·ə'kō·səs }

neurovascular [BIOL]   Pertaining structurally and functionally to both the nervous and vascular structures.   { ¦nur·ō'vas·kyə·lər }

neurulation [EMBRYO]   Differentiation of nerve tissue and formation of the neural tube.   { ¸nur·ə'lā·shən }

neuston [BIOL]   Minute organisms that float or swim on surface water or on a surface film of water.   { 'nü¸stän }

neutral [CHEM]   Property of a solution which is neither acidic nor basic, having the same concentration of hydrogen ions as water.   [ELEC]   Referring to the absence of a net electric charge.   [MECH ENG]   That setting in an automotive transmission in which all the gears are disengaged and the output shaft is disconnected from the drive wheels.   { 'nü·trəl }

neutral atmosphere [ENG]   An atmosphere which neither oxidizes nor reduces immersed materials.   { 'nü·trəl 'at·mə¸sfir }

neutral atom [ATOM PHYS]   An atom in which the number of electrons that surround the nucleus is equal to the number of protons in the nucleus, so that there is no net electric charge.   { 'nü·trəl 'ad·əm }

neutral axis [MECH]   In a beam bent downward, the line of zero stress below which all fibers are in tension and above which they are in compression.   { 'nü·trəl 'ak·səs }

neutral beam [PHYS]   A stream of uncharged particles.   { 'nü·trəl 'bēm }

neutral conductor [ELEC]   A conductor of a polyphase circuit or of a single-phase, three-wire circuit which is intended to have a potential such that the potential differences between it and each of the other conductors are approximately equal in magnitude and are also equally spaced in phase.   { 'nü·trəl kən'dək·tər }

neutral current interaction [PARTIC PHYS]   A weak interaction in which the charges of the interacting fermions are not changed.   { 'nü·trəl 'kə·rənt ¸in·tər¸ak·shən }

neutral-density filter [OPTICS]   An optical filter that reduces the intensity of light without appreciably changing its color; used on a camera when the lens cannot be stopped down sufficiently for use with a given film.   Also known as gray filter; neutral filter.   { 'nü·trəl ¦den·səd·ē 'fil·tər }

neutral equilibrium [PHYS]   A property of the steady state of a system which exhibits neither instability nor stability according to the particular criterion under consideration; a disturbance introduced into such an equilibrium will thus be neither amplified nor damped.   Also known as indifferent equilibrium.   { 'nü·trəl ¸ē·kwə'lib·rē·əm }

neutral estuary [GEOGR]   An estuary in which neither freshwater inflow nor evaporation dominates.   { 'nü·trəl 'es·chə¸wer·ē }

neutral fiber [MECH]   A line of zero stress in cross section of a bent beam, separating the region of compressive stress from that of tensile stress.   { 'nü·trəl 'fī·bər }

neutral filter See neutral-density filter.   { 'nü·trəl 'fil·tər }

neutral flame [CHEM]   Gas flame produced by a mixture of fuel and oxygen so as to be neither oxidizing nor reducing. { 'nü·trəl 'flām }

neutral granulation [CHEM] Propellant granulation in which the surface area of a grain remains constant during burning. { 'nü·trəl ˌgran·yə'lā·shən }

neutral ground [ELEC] Ground connected to the neutral point or points of an electric circuit, transformer, rotating machine, or system. { 'nü·trəl 'graund }

neutralism [ECOL] A neutral interaction between two species, that is, one having no evident effect on either species. { 'nü·trəˌliz·əm }

neutralization [CHEM] The process of making a solution neutral (pH = 7) by adding a base to an acid solution, or adding an acid to an alkaline (basic) solution. Also known as neutralization reaction. { ˌnü·trə·lə'zā·shən }

neutralization equivalent [CHEM] For an acid or base, the same as equivalent weight; multiplication of the neutralization equivalent by the number of acidic or basic groups in the molecule gives the molecular weight. { ˌnü·trə·lə'zā·shən iˌkwiv·ə·lənt }

neutralization fire [ORD] Fire which is delivered to cause casualties, to hamper and interrupt the firing of weapons, movement, or action, and to reduce the combat efficiency of enemy personnel. { ˌnü·trə·lə'zā·shən ˌfīr }

neutralization number [ANALY CHEM] Petroleum product test; it is the milligrams of potassium hydroxide required to neutralize the acid in 1 gram of oil; used as an indication of oil acidity. { ˌnü·trə·lə'zā·shən ˌnəm·bər }

neutralization reaction See neutralization. { ˌnü·trə·lə'zā·shən rē,ak·shən }

neutralize [CHEM] To make a solution neutral (neither acidic nor basic, pH of 7) by adding a base to an acidic solution, or an acid to a basic solution. [ELECTR] To nullify oscillation-producing voltage feedback from the output to the input of an amplifier through tube interelectrode capacitances; an external feedback path is used to produce at the input a voltage that is equal in magnitude but opposite in phase to that fed back through the interelectrode capacitance. [OPTICS] To place a lens in contact with other lenses of equal and opposite power so that the combination has zero power. [ORD] **1.** To destroy or reduce the effectiveness of enemy personnel and materiel by gunfire, bombing, or any other means. **2.** To make a toxic chemical agent harmless by chemical action. **3.** To disarm or otherwise render safe a mine, bomb, missile, or booby trap. { 'nü·trəˌlīz }

neutralized radio-frequency stage [ELECTR] Stage having an additional circuit connected to feed back, in the opposite phase, an amount of energy equivalent to what is causing the oscillation, thus neutralizing any tendency to oscillate and making the circuit function strictly as an amplifier. { 'nü·trəˌlīzd ˌrād·ē·ō 'frē·kwən·sē ˌstāj }

neutralizing antibody [IMMUNOL] An antibody that reduces or abolishes some biological activity of a soluble antigen or of a living microorganism. { 'nü·trəˌlīz·iŋ 'ant·iˌbäd·ē }

neutralizing capacitor [ELECTR] Capacitor, usually variable, employed in a radio receiving or transmitting circuit to feed a portion of the signal voltage from the plate circuit of a stage back to the grid circuit. { 'nü·trəˌlīz·iŋ kə,pas·əd·ər }

neutralizing circuit [ELECTR] Portion of an amplifier circuit which provides an intentional feedback path from plate to grid to prevent regeneration. { 'nü·trəˌlīz·iŋ ˌsər·kət }

neutralizing power of a lens [OPTICS] The power of a lens, measured by neutralizing it with trial lenses of equal and opposite power; for a spectacle lens, it may differ significantly from back vertex power. { 'nü·trəˌlīz·iŋ ˌpau·ər əv ə 'lenz }

neutralizing transformer [ELECTROMAG] A transformer installed on a communication line which produces counter electromotive forces which largely cancel out induced longitudinal voltage and allow normal operation of the line during inductive disturbances. { 'nü·trəˌlīz·iŋ tranz'för·mər }

neutralizing voltage [ELECTR] Voltage developed in the plate circuit (Hazeltine neutralization) or in the grid circuit (Rice neutralization), used to nullify or cancel the feedback through the tube. { 'nü·trəˌlīz·iŋ ˌvōl·tij }

neutral line See Busch lemniscate. { 'nü·trəl 'līn }

neutrally buoyant float See swallow float. { 'nü·trə·lē ˌbói·ənt 'flōt }

neutral molecule [PHYS CHEM] A molecule in which the number of electrons surrounding the nuclei is the same as the total number of protons in the nuclei, so that there is no net electric charge. { 'nü·trəl 'mäl·əˌkyül }

neutral mutation [GEN] A mutation that has no phenotypic effect or adaptive significance. { 'nü·trəl myü,tā·shən }

neutral oil [MATER] Light oil from dewaxed petroleum; medium or low viscosity, flash point 143–160°C. { 'nü·trəl 'òil }

neutral operation [COMMUN] System whereby marking signals are formed by current impulses of one polarity, either positive or negative, and spacing signals are formed by reducing the current to zero or nearly zero. { 'nü·trəl ˌäp·ə'rā·shən }

neutral particle [PARTIC PHYS] A particle that carries no electric charge. { 'nü·trəl 'pärd·ə·kəl }

neutral point [ELEC] Point which has the same potential as the point of junction of a group of equal nonreactive resistances connected at their free ends to the appropriate main terminals or lines of the system. [FL MECH] See hyperbolic point. [MET] In rolling mills, the point at which the speed of the work is equal to the peripheral speed of the rolls. [METEOROL] See col. [OPTICS] In atmospheric optics, one of several points in the sky for which the degree of polarization of diffuse sky radiation is zero. [PHYS] A point where two fields are equal in magnitude and opposite in direction so that the net field is zero. { 'nü·trəl ˌpoint }

neutral potassium phosphate See potassium phosphate. { 'nü·trəl pə'tas·ē·əm 'fäsˌfāt }

neutral pressure See neutral stress. { 'nü·trəl 'presh·ər }

neutral red [ORG CHEM] $(CH_3)_2NC_6H_3N_2C_6H_2CH_3NH_2 \cdot$ ClH Water- and alcohol-soluble green powder; used as pH 6.8–8.0 acid-base indicator, and as a dye to test stomach function. Also known as dimethyl diaminophenazine chloride; toluylene red. { 'rü·trəl 'red }

neutral region [ASTRON] A region on the sun's surface where the longitudinal magnetic field nearly vanishes; generally found between regions of opposite polarity. { 'nü·trəl ˌrē·jən }

neutral relay [ELEC] Relay in which the movement of the armature does not depend upon the direction of the current in the circuit controlling the armature. Also known as nonpolarized relay. { 'nü·trəl 'rēˌlā }

neutral return path [ELEC] A route from the load back to the power source, completing a circuit in an electric power distribution system, which is grounded, usually by connections to water pipes. { 'nü·trəl ri'tərn ˌpath }

neutral safety switch [ELEC] An electric switch that is connected to the ignition switch of an internal combustion engine and prevents starting the engine unless the transmission shift lever is in the neutral or park position, or the clutch pedal is depressed. { 'nü·trəl 'sāf·tē ˌswich }

neutral shoreline [GEOL] A shoreline whose essential features are independent of either the submergence of a former land surface or the emergence of a former underwater surface. { 'nü·trəl 'shòrˌlīn }

neutral species See uncharged species. { 'nü·trəl 'spē·shēz }

neutral spirits [FOOD ENG] An alcohol distillate from a fermented grain mash of 190 proof or higher that is used for producing blended whiskey. Also known as grain neutral spirits. { 'nü·trəl 'spir·əts }

neutral stability [CONT SYS] Condition in which the natural motion of a system neither grows nor decays, but remains at its initial amplitude. [METEOROL] The state of an unsaturated or saturated column of air in the atmosphere when its environmental lapse rate of temperature is equal to the dry-adiabatic lapse rate or the saturation-adiabatic lapse rate respectively; under such conditions a parcel of air displaced vertically will experience no buoyant acceleration. Also known as indifferent equilibrium; indifferent stability. { 'nü·trəl stə'bil·əd·ē }

neutral stress [HYD] The stress transmitted through the interstitial fluid of a soil or rock mass. Also known as neutral pressure; pore pressure; pore-water pressure. { 'nü·trəl 'stres }

neutral surface [MECH] A surface in a bent beam along which material is neither compressed nor extended. { 'nü·trəl 'sər·fəs }

neutral temperature [ELECTR] The temperature of the hot junction of a thermocouple at which the electromotive force of the thermocouple attains its maximum value, when the cold junction is maintained at a constant temperature of 0°C. { 'nü·trəl 'tem·prə·chər }

neutral wave [PHYS] Any wave whose amplitude does not change with time; in most contexts the wave is referred to as a stable wave, the term "neutral wave" being used when it is important to emphasize that the wave is neither damped nor amplified. { 'nü·trəl ¦wāv }

neutral zone *See* dead band. { 'nü·trəl ¦zōn }

neutrino [PHYS] A neutral particle having zero rest mass and spin $1/2$ ($h/2\pi$), where h is Planck's constant; experimentally, there are two such particles known as the e neutrino (ν_e) and the μ neutrino (ν_μ). { nü'trē·nō }

neutrino astronomy [ASTRON] The observation of neutrinos from the sun and from extrasolar astronomical sources. { nü¦trē·nō ə'strän·ə·me }

neutrino bremsstrahlung [NUC PHYS] The scattering of an electron from a nucleus with the emission of a neutrino and an antineutrino. { nü'trē·nō 'brem¦shträ·lúŋ }

neutrino oscillation [PARTIC PHYS] A phenomenon in which a neutrino in one of the three known flavor states (electron neutrino, mu neutrino, or tau neutrino) becomes a mixture of flavor states that changes back and forth periodically as the neutrino travels through space; it will occur if neutrinos have mass and if each flavor state is a mixture of different mass states. { nü¦trē·nō ¦äs·ə'lā·shən }

neutrino telescope [ASTRON] Any device for detecting and determining the directions of extraterrestrial neutrinos, such as the deep underwater muon and neutrino detector. { nü'trē·nō 'tel·ə¦skōp }

neutron [PHYS] An elementary particle which has approximately the same mass as the proton but lacks electric charge, and is a constituent of all nuclei having mass number greater than 1. { 'nü¦trän }

neutron absorber [NUCLEO] A material in which a significant number of neutrons passing through combine with nuclei and are not reemitted. { 'nü¦trän əb¦sorb·ər }

neutron absorption *See* neutron capture. { 'nü¦trän əb¦sorp·shən }

neutron activation analysis [NUCLEO] Activation analysis in which the specimen is bombarded with neutrons; identification is made by measuring the resulting radio isotopes. { 'nü¦trän ¦ak·tə'vā·shən ə¦nal·ə·səs }

neutron age *See* Fermi age. { 'nü¦trän ¦āj }

neutron albedo [NUCLEO] The probability, under specified conditions, that a neutron entering into a region through a surface will return through that surface. { 'nü¦trän al'bēd·ō }

neutron-antineutron oscillations [PARTIC PHYS] Hypothetical periodic transitions between the state of a neutron and the state of an antineutron which are predicted by certain unified gauge theories. { 'nü¦trän 'ant·i¦nü¦trän ¦äs·ə'lā·shənz }

neutron binding energy [NUC PHYS] The energy required to remove a single neutron from a nucleus. { 'nü¦trän ¦bīnd·iŋ ¦en·ər·jē }

neutron bomb *See* neutron radiation weapon. { 'nü¦trän ¦bäm }

neutron bottle [NUCLEO] A vacuum vessel for storing ultracold neutrons. { 'nü¦trän ¦bäd·əl }

neutron capture [NUC PHYS] A process in which the collision of a neutron with a nucleus results in the absorption of the neutron into the nucleus with the emission of one or more prompt gamma rays; in certain cases, beta decay or fission of the nucleus results. Also known as neutron absorption; neutron radiative capture. { 'nü¦trän ¦kap·chər }

neutron-capture cross section [NUC PHYS] The cross section for neutron capture by nuclei in a material; it is a measure of the probability that this reaction will occur. { 'nü¦trän ¦kap·chər 'kros ¦sek·shən }

neutron chopper [NUCLEO] A device that interrupts the output beam of neutrons from a nuclear reactor mechanically, to provide bursts or pulses of neutrons for research purposes. { 'nü¦trän ¦chäp·ər }

neutron counter [NUCLEO] A neutron detector which counts the number of neutrons passing through the detecting medium. { 'nü¦trän ¦kaúnt·ər }

neutron cross section [NUC PHYS] A measure of the probability that an interaction of a given kind will take place between a nucleus and an incident neutron; it is an area such that the number of interactions which occur in a sample exposed to a beam of neutrons is equal to the product of the number of

nuclei in the sample and the number of neutrons in the beam that would pass through this area if their velocities were perpendicular to it. { 'nü¦trän 'kros ¦sek·shən }

neutron cycle [NUCLEO] The life history of the neutrons in a nuclear reactor, extending from the initial fission process until all the neutrons have been absorbed or have leaked out. { 'nü¦trän ¦sī·kəl }

neutron detector [NUCLEO] Any device which detects passing neutrons, for example, by observing the charged particles or gamma rays released in nuclear reactions induced by the neutrons, or observing the recoil of charged particles caused by collisions with neutrons. { 'nü¦trän di¦tek·tər }

neutron diffraction [PHYS] The phenomenon associated with the interference processes which occur when neutrons are scattered by the atoms within solids, liquids, and gases. { 'nü¦trän di¦frak·shən }

neutron diffraction analysis [PHYS] The study of the atomic structure of solids, liquids, and gases by passing high-flux beams of thermal neutrons through them and measuring the intensity of scattered neutrons in various directions. { ¦nü¦trän di'frak·shən ə¦nal·ə·səs }

neutron diffractometer [PHYS] A diffractometer in which a beam of neutrons is used for diffraction analysis, and the intensities of the diffracted beams at different angles are measured with an ionization chamber or radiation counter. { 'nü¦trän ¦di¦frak'täm·əd·ər }

neutron drip [ASTROPHYS] The rapid increase in the abundance of free neutrons that occurs when matter becomes sufficiently dense that electrons are absorbed into nuclei. { 'nü¦trän ¦drip }

neutron drip-line [NUC PHYS] On a chart of the nuclides, plotting proton number versus neutron number, the boundary beyond which neutron-rich nuclei are unstable against neutron emission. { ¦nü¦trän 'drip ¦līn }

neutron economy [NUCLEO] The balance sheet describing the ways neutrons are used in a nuclear reactor, as in fission, leakage, and absorption. { 'nü¦trän i'kän·ə·mē }

neutron excess [NUC PHYS] The number of neutrons in a nucleus in excess of the number of protons. Also known as difference number; isotopic number. { 'nü¦trän 'ek¦ses }

neutron flux [NUCLEO] The intensity of neutron radiation, expressed as the number of neutrons passing through a unit area per unit time. Also known as neutron flux density. { 'nü¦trän ¦fləks }

neutron flux density *See* neutron flux. { 'nü¦trän 'fləks ¦den·səd·ē }

neutron-gamma well logging [ENG] Neutron well logging in which the varying intensity of gamma rays produced artificially by neutron bombardment is recorded. { 'nü¦trän ¦gam·ə ¦wel ¦läg·iŋ }

neutron hardening [NUCLEO] The increase which occurs in the average energy of thermal neutrons diffusing in a medium whose absorption cross section decreases as energy increases. { 'nü¦trän ¦härd·ən·iŋ }

neutron howitzer [NUCLEO] A device which produces a neutron beam; consists of a neutron source contained in a block of moderating material with a small hole from the source to the surface of the block, from which the beam emerges. { 'nü¦trän ¦haú·ət·sər }

neutronicism [NUC PHYS] In a nuclear reaction, the power carried by the neutrons as a fraction of the total power released in the reaction. { nü'trän·ə¦siz·əm }

neutron inelastic scattering reactions [NUCLEO] Emissions of low-energy neutrons when materials have been bombarded by fast neutrons during neutron activation analysis. Also known as n-n' reactions. { 'nü¦trän ¦in·i'las·tik ¦skad·ə·riŋ rē¦ak·shənz }

neutron irradiation [NUCLEO] The exposure of a material or object to neutrons. { 'nü¦trän i¦rād·ē'ā·shən }

neutron logging *See* neutron well logging. { 'nü¦trän ¦läg·iŋ }

neutron magnetic moment [NUC PHYS] A vector whose scalar product with the magnetic flux density gives the negative of the energy of interaction of a neutron with a magnetic field. { 'nü¦trän mag¦ned·ik ¦mō·mənt }

neutron matter [ASTROPHYS] Degenerate matter such as occurs in neutron stars in which there are 8 to 12 times as many neutrons as protons. { 'nü¦trän ¦mad·ər }

neutron multiplication factor *See* multiplication factor. { 'nü¦trän ¦məl·tə·plə'kā·shən ¦fak·tər }

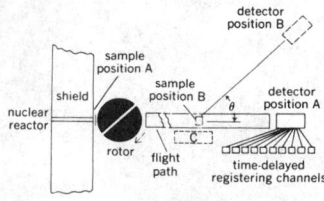

NEUTRON CHOPPER

Schematic diagram of a neutron chopper and time-of-flight apparatus. Registering channels record at successive times following passage of pulse of neutrons through time-of-flight. Sample and detector position A used to measure total cross section. Sample and detector position B used to measure differential cross section for scattering through angle θ. Detector position C used to measure absorption cross section.

neutron number [NUC PHYS] The number of neutrons in the nucleus of an atom. { 'nü‚trän ‚nəm·bər }

neutron optics [PHYS] The study of certain phenomena, for example, crystal diffraction, in which the wave character of neutrons dominates and leads to behavior similar to that of light. { 'nü‚trän 'äp·tiks }

neutron-proton scattering [NUC PHYS] A collision of a neutron with a proton, usually the nucleus of a hydrogen atom. { ¦nü‚trän 'prō‚tän ‚skad·ər·iŋ }

neutron radiation weapon [ORD] A small nuclear weapon that has a yield of 1–2 kilotons, causes extensive blast and heat damage within a radius of 430–1160 feet (130–350 meters), and emits neutron and gamma radiation lethal within a radius of 0.6–1.2 miles (1–2 kilometers). Also known as neutron bomb. { 'nü‚trän ‚rād·ē'ā·shən ‚wep·ən }

neutron radiative capture See neutron capture. { 'nü‚trän 'rād·ē‚ād·iv 'kap·chər }

neutron radiography [NUCLEO] Radiography that uses a neutron beam generated by a nuclear reactor; the neutrons are detected by placing a conventional x-ray film next to a converter screen composed of potentially radioactive materials or prompt emission materials which convert the neutron radiation to other types of radiation more easily detected by the film. { 'nü‚trän ‚rād·ē'äg·rə·fē }

neutron reflection [PHYS] Specular reflection of neutrons, either from lattice planes of crystalline substances according to the Bragg law for their de Broglie wavelength, or from highly polished surfaces of certain substances at an angle smaller than their critical angle. { 'nü‚trän ri‚flek·shən }

neutron-rich nucleus [NUC PHYS] An atomic nucleus in which the ratio of neutron number to proton number is much larger than that of nuclei found in nature. { ¦nü‚trän ‚rich 'nü·klē·əs }

neutron scattering [NUCLEO] The change in direction of neutrons caused by collision with nuclei in a material. { 'nü‚trän ‚skad·ə·riŋ }

neutron shield [ENG] A shield that protects personnel from neutron irradiation. { 'nü‚trän ‚shēld }

neutron soil-moisture meter [ENG] An instrument for measuring the water content of soil and rocks as indicated by the scattering and absorption of neutrons emitted from a source, and resulting gamma radiation received by a detector, in a probe lowered into an access hole. { 'nü‚trän 'sȯil ‚mȯis·chər ‚mēd·ər }

neutron spectrometer [NUCLEO] An instrument used to determine the energies of neutrons and the relative intensities of neutrons of different energies in a neutron beam. { 'nü‚trän spek'träm·əd·ər }

neutron spectrometry [NUC PHYS] A method of observing excited states of nuclei in which neutrons are used to bombard a target, causing nuclei to be transmuted into excited states by various nuclear reactions; the resultant excited states are determined by observing resonances in the reaction cross sections or by observing spectra of emitted particles or gamma rays. Also known as neutron spectroscopy. { 'nü‚trän spek'träm·ə·trē }

neutron spectroscopy See neutron spectrometry; slow-neutron spectroscopy. { 'nü‚trän spek'träs·kə·pē }

neutron spectrum [NUC PHYS] A plot or display of the number of neutrons at various energies, such as the neutrons emitted in a nuclear reaction, or the neutrons in a nuclear reactor. { 'nü‚trän ‚spek·trəm }

neutron star [ASTRON] A star that is supposed to occur in the final stage of stellar evolution; it consists of a superdense mass mainly of neutrons, and has a strong gravitational attraction from which only neutrinos and high-energy photons could escape so that the star is invisible. { 'nü‚trän ‚stär }

neutron therapy [MED] Medical therapy involving irradiation with neutrons. { 'nü‚trän ‚ther·ə·pē }

neutron transport theory [NUCLEO] Theory of diffusion of neutrons in a material based on the Boltzmann transport equation. { 'nü‚trän 'tranz‚pȯrt ‚thē·ə·rē }

neutron turbine [NUCLEO] A source of ultracold neutrons in which neutrons are slowed down by reflection from a set of moving curved neutron reflectors. { 'nü‚trän 'tər·bən }

neutron velocity selector [NUCLEO] An instrument that isolates and detects neutrons having a particular range of velocities. { 'nü‚trän və'läs·əd·ē si‚lek·tər }

neutron well logging [ENG] Study of formation fluid-content properties down a wellhole by neutron bombardment and detection of resultant radiation (neutrons or gamma rays). Also known as neutron logging. { 'nü‚trän ‚wel ‚läg·iŋ }

neutropenia [MED] Abnormally low number of neutrophils in the peripheral circulation. { ‚nü·trə‚pē·nē·ə }

neutrophil [HISTOL] A large granular leukocyte with a highly variable nucleus, consisting of three to five lobes, and cytoplasmic granules which stain with neutral dyes and eosin. { 'nü·trə‚fil }

neutrophilia [BIOL] Affinity for neutral dyes. [MED] An abnormal increase in leukocytes in the tissues or peripheral circulation. { ‚nü·trə¦fil·ē·ə }

neutrophilic leukemia [MED] Granulocytic leukemia in which the leukocytes resemble cells of the neutrophilic series. { ¦nü·trə¦fil·ik lü'kē·mē·ə }

neutrophilous [BIOL] Preferring an environment free of excess acid or base. { nü'träf·ə·ləs }

neutrosphere [METEOROL] The atmospheric shell from the earth's surface upward, in which the atmospheric constituents are for the most part un-ionized, that is, electrically neutral; the region of transition between the neutrosphere and the ionosphere is somewhere between 42 and 54 miles (70 and 90 kilometers), depending on latitude and season. { 'nü·trə‚sfir }

nevada [METEOROL] A cold wind descending from a mountain glacier or snowfield, for example, in the higher valleys of Ecuador. { nə'väd·ə }

Nevadan orogeny [GEOL] Orogenic episode during Jurassic and Early Cretaceous geologic time in the western part of the North American Cordillera. Also known as Nevadian orogeny; Nevadic orogeny. { nə'vad·ən ȯ'räj·ə·nē }

Nevadian orogeny See Nevadan orogeny. { nə'vad·ē·ən ȯ'räj·ə·nē }

Nevadic orogeny See Nevadan orogeny. { nə'vad·ik ȯ'räj·ə·nē }

névé [GEOGR] A geographic area of perennial snow. [HYD] An accumulation of compacted, granular snow in transition from soft snow to ice; it contains much air; the upper portions of most glaciers and ice shelves are usually composed of névé. { nā'vā }

nevus [MED] A lesion containing melanocytes. { 'nē·vəs }

nevus sebaceus [MED] A nevus formed by an aggregate of sebaceous glands. Also known as Jadassohn's nevus. { 'nē·vəs si bā·shəs }

nevyanskite [MET] A tin-white variety of iridosmine containing 35–50% osmium or more than 50% iridium; occurs in flat scales. { nev'yan‚skīt }

new achromat [OPTICS] An achromatic lens in which the component lenses are made of glasses chosen from a relatively broad selection, among which refractive index and dispersive power may vary inversely, permitting better correction of optical errors. { ¦nü 'ak·rə‚mat }

new-band service [COMMUN] A broadcasting service that is allocated a portion of the radio frequency spectrum that was not previously used. { 'nü ‚band ‚sər·vis }

newberyite [MINERAL] $MgH(PO_4) \cdot 3H_2O$ A white, orthorhombic member of the brushite mineral group; it is isostructural with gypsum. { 'nüb·rē‚īt }

new blue [MATER] Any of several iron blue types of pigments; varieties are called mendola blue and prussian blue. { 'nü 'blü }

newborn [MED] Born recently; said of human infants less than a month old, especially of those only a few days old. { 'nü‚bȯrn }

new candle See candela. { 'nü 'kand·əl }

Newcastle disease [VET MED] An acute viral disease of fowls, with respiratory, gastrointestinal, and central nervous system involvement; may be transmitted to human beings as a mild conjunctivitis. Also known as avian pneumoencephalitis; avian pseudoplague; Philippine fowl disease. { 'nü‚kas·əl di‚zēz }

Newcastle virus [VIROL] A ribonucleic acid hemagglutinating myxovirus responsible for Newcastle disease. { 'nü‚kas·əl ‚vī·rəs }

newel post [CIV ENG] **1.** A pillar at the end of an oblique retaining wall of a bridge. **2.** The post about which a circular staircase winds. **3.** A large post at the foot of a straight stairway or on a landing. { 'nü·əl ‚pōst }

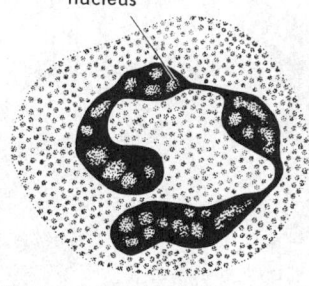

NEUTROPHIL

nucleus

Diagram of neutrophil showing multilobed nucleus.

new global tectonics [GEOL] Comprehensive theory relating the formation of mountain belts, island arcs, and ocean trenches to the relative movement of regionally extensive lithospheric plates which are delineated by the major seismic belts of the earth. { 'nü 'glō·bəl tek'tän·iks }

new inflationary cosmology [ASTRON] A modification of the original inflationary universe cosmology in which the breaking of grand unified symmetry does not involve a tunneling process and is, instead, analogous to a process in which a ball rolls down from a hill with an extremely flat top. { 'nü in'flā·shə,ner·ē käz'mäl·ə·jē }

New Jersey retort process See vertical retort process. { 'nü 'jər·zē 'rē,tȯrt ,prä·səs }

Newland's law of octaves [CHEM] An arrangement of the elements that predated Mendeleev's periodic table; Newland's arrangement was a grouping of the elements in increasing atomic weights (starting with lithium) in horizontal rows of eight elements, with each new row directly beneath the previous one. { 'nü·lənz 'lȯ əv 'äk·tivz }

newly formed ice [HYD] Ice in the first stage of formation and development. Also known as fresh ice. { 'nü·lē ,fȯrmd 'īs }

Newman-Penrose formalism [RELAT] A formalism in general relativity, based on spinor analysis, convenient for dealing with gravitational perturbations of space-time. { 'nü·mən 'pen,rōz 'fȯr·mə,liz·əm }

Newman projection for ethane.

Newman projection [ORG CHEM] A representation of the conformation of a molecule in which the viewer's eye is considered to be sighting down a carbon-carbon bond; the front carbon is represented by a point and the back carbon by a circle. { 'nü·mən prə,jek·shən }

new moon [ASTRON] The moon at conjunction, when little or none of it is visible to an observer on the earth because the illuminated side is turned away. { 'nü 'mün }

New Red Sandstone [GEOL] The red sandstone facies of the Permian and Triassic systems exposed in the British Isles. { 'nü 'red 'san,stōn }

newsgroup [COMPUT SCI] A collection of computers on a wide-area network that form a discussion group on a particular topic, such that a message generated by any computer in the group is automatically distributed over the network to all the others. Also known as forum. { 'nüz,grüp }

new snow [METEOROL] **1.** Fallen snow whose original crystalline structure has been retained and is therefore recognizable. **2.** Snow which has fallen in a single day. { 'nü 'snō }

Newson's boring method [MIN ENG] A method of boring small shafts, up to $5\frac{1}{2}$ feet (1.7 meters) in diameter using the principle of chilled-shot drilling on a large scale. { 'nü·sənz 'bȯr·iŋ ,meth·əd }

newsprint [MATER] The paper used in the publication of newspapers; an impermanent material made from mechanical wood pulp, with some chemical wood pulp. { 'nüz,print }

newt [VERT ZOO] Any of the small, semiaquatic salamanders of the genus *Triturus* in the family Salamandridae; all have an aquatic larval stage. { nüt }

newton [MECH] The unit of force in the meter-kilogram-second system, equal to the force which will impart an acceleration of 1 meter per second squared to the International Prototype Kilogram mass. Symbolized N. Formerly known as large dyne. { 'nüt·ən }

Newton-Cotes formulas [MATH] Approximation formulas for the integral of a function along a small interval in terms of the values of the function and its derivatives. { 'nüt·ən 'kōts ,fȯr·myə·ləz }

Newton formula for the stress See Newtonian friction law. { 'nüt·ən 'fȯr·myə·lə fȯr <u>thə</u> 'stres }

Newtonian attraction [MECH] The mutual attraction of any two particles in the universe, as given by Newton's law of gravitation. { nü'tō·nē·ən ə'trak·shən }

Newtonian-Cassegrain telescope [OPTICS] A modification of a Cassegrain telescope in which the light reflected from the hyperboloidal secondary mirror is again reflected from a diagonal plane mirror and focused at a point on the side of the telescope, avoiding the need to pierce the primary mirror and making the eyepiece more accessible. Also known as Cassegrain-Newtonian telescope. { nü'tō·nē·ən 'kas,gran 'tel·ə,skōp }

Newtonian flow [FL MECH] Flow system in which the fluid

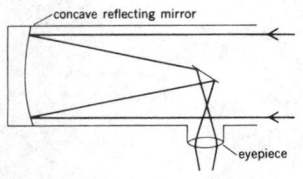

Simplified diagram of a Newtonian telescope.

performs as a Newtonian fluid, that is, shear stress is proportional to shear rate. { nü'tō·nē·ən 'flō }

Newtonian fluid [FL MECH] A simple fluid in which the state of stress at any point is proportional to the time rate of strain at that point; the proportionality factor is the viscosity coefficient. { nü'tō·nē·ən 'flü·əd }

Newtonian focus [OPTICS] The position in a Newtonian telescope at which the image is formed, located at the side of the tube near its open end. { nü'tō·nē·ən 'fō·kəs }

Newtonian friction law [FL MECH] The law that shear stress in a fluid is proportional to the shear rate; it holds only for some fluids, which are then called Newtonian. Also known as Newton formula for the stress. { nü'tō·nē·ən 'frik·shən ,lȯ }

Newtonian mechanics [MECH] The system of mechanics based upon Newton's laws of motion in which mass and energy are considered as separate, conservative, mechanical properties, in contrast to their treatment in relativistic mechanics. { nü'tō·nē·ən mi'kan·iks }

Newtonian potential [PHYS] A potential which is associated with an inverse square law of force (such as an electrostatic force), and therefore varies with distance in the same manner as a gravitational potential. { nü'tō·nē·ən pə'ten·chəl }

Newtonian reference frame [MECH] One of a set of reference frames with constant relative velocity and within which Newton's laws hold; the frames have a common time, and coordinates are related by the Galilean transformation rule. { nü'tō·nē·ən 'ref·rəns ,frām }

Newtonian speed of sound [ACOUS] An approximation to the speed of sound in a perfect gas given by the relation $c^2 = p/\rho$, where p is pressure and ρ the density. { nü'tō·nē·ən 'spēd əv 'saùnd }

Newtonian telescope [OPTICS] A reflecting telescope in which the light reflected from a concave mirror is reflected again by a plane mirror making an angle of 45° with the telescope axis, so that it passes through a hole in the side of the telescope containing the eyepiece. { nü'tō·nē·ən 'tel·ə,skōp }

Newtonian velocity [MECH] The velocity of an object in a Newtonian reference frame, S, which can be determined from the velocity of the object in any other such frame, S', by taking the vector sum of the velocity of the object in S' and the velocity of the frame S' relative to S. { nü'tō·nē·ən və'läs·əd·ē }

Newtonian viscosity [FL MECH] The viscosity of a Newtonian fluid. { nü'tō·nē·ən vi'skäs·əd·ē }

newton-meter of energy See joule. { 'nüt·ən ˌmēd·ər əv 'en·ər·jē }

newton-meter of torque [MECH] The unit of torque in the meter-kilogram-second system, equal to the torque produced by 1 newton of force acting at a perpendicular distance of 1 meter from an axis of rotation. Abbreviated N-m. { 'nüt·ən ,mēd·ər əv 'tȯrk }

Newton-Raphson formula [MATH] If c is an approximate value of a root of the equation $f(x) = 0$, then a better approximation is the number $c - [f(c)/f'(c)]$. { 'nüt·ən 'raf·sən ,fȯr·myə·lə }

Newton's alloy [MET] A fusible alloy made up of 50% bismuth, 31% lead, and 19% tin; melts at 95°C; used in applications where it is required to fall away at predetermined temperatures, as for automatic sprinkler links. Also known as Newton's metal. { 'nüt·ənz 'al,ȯi }

Newton's equations of motion [MECH] Newton's laws of motion expressed in the form of mathematical equations. { 'nüt·ənz i'kwā·zhənz əv 'mō·shən }

Newton's first law [MECH] The law that a particle not subjected to external forces remains at rest or moves with constant speed in a straight line. Also known as first law of motion; Galileo's law of inertia. { 'nüt·ənz 'fərst 'lȯ }

Newton's identity [MATH] The identity $C(n, r)C(r, k) = C(n, k)C(n - k, r - k)$, where, in general, $C(n,r)$ is the number of distinct subsets of r elements in a set of n elements (the binomial coefficient). { ˌnüt·ənz ī'den·ə,dē }

Newton's inequality [MATH] For any set of n numbers ($n = 0, 1, 2, \ldots$), the inequality $p_{r-1}p_{r+1} \leqq p_r^2$, for $1 \leqq r < n$, where p_r is the average value of the terms constituting the rth elementary symmetric function of the numbers. { ˌnüt·ənz ,in·i'kwäl·əd·ē }

Newton's law of cooling [THERMO] The law that the rate of heat flow out of an object by both natural convection and radiation is proportional to the temperature difference between the object and its environment, and to the surface area of the object.   { 'nüt·ənz 'lȯ əv 'kül·iŋ }

Newton's law of gravitation   [MECH]   The law that every two particles of matter in the universe attract each other with a force that acts along the line joining them, and has a magnitude proportional to the product of their masses and inversely proportional to the square of the distance between them.   Also known as law of gravitation.   { 'nüt·ənz 'lȯ əv ,grav·ə'tā·shən }

Newton's law of resistance   [FL MECH]   The law that the force opposing the motion of an object through a fluid at moderate velocities is proportional to the square of the velocity.   { 'nüt·ənz 'lȯ əv ri'zis·təns }

Newton's laws of motion   [MECH]   Three fundamental principles (called Newton's first, second, and third laws) which form the basis of classical, or Newtonian, mechanics, and have proved valid for all mechanical problems not involving speeds comparable with the speed of light and not involving atomic or subatomic particles.   { 'nüt·ənz 'lȯz əv 'mō·shən }

Newton's lens formula   [OPTICS]   A formula which states that the product of the distances of two conjugate points from the respective principal foci of a lens or mirror is equal to the square of the focal length.   { 'nüt·ənz 'lenz ,fȯr·myə·lə }

Newton's metal   See Newton's alloy.   { 'nüt·ənz ,med·əl }

Newton's method   [MATH]   A technique to approximate the roots of an equation by the methods of the calculus.   { 'nüt·ənz ,meth·əd }

Newton's rings   [OPTICS]   A series of circular bright and dark bands which appear about the point of contact between a glass plate and a convex lens which is pressed against it and illuminated with monochromatic light.   { 'nüt·ənz 'riŋz }

Newton's second law   [MECH]   The law that the acceleration of a particle is directly proportional to the resultant external force acting on the particle and is inversely proportional to the mass of the particle.   Also known as second law of motion.   { 'nüt·ənz 'sek·ənd 'lȯ }

Newton's square-root method   [MATH]   A technique for the estimation of the roots of an equation exhibiting faster convergence than Newton's method; this involves calculus methods and the square-root function.   { 'nüt·ənz 'skwer ,rüt ,meth·əd }

Newton's theory of lift   [FL MECH]   A theory of the forces acting on an airfoil in a fluid current in which these forces are assumed to result from the impact of particles of the fluid on the body.   { 'nüt·ənz ,thē·ə·rē əv 'lift }

Newton's theory of light   See corpuscular theory of light.   { 'nüt·ənz ,thē·ə·rē əv 'lït }

Newton's third law   [MECH]   The law that, if two particles interact, the force exerted by the first particle on the second particle (called the action force) is equal in magnitude and opposite in direction to the force exerted by the second particle on the first particle (called the reaction force).   Also known as law of action and reaction; third law of motion.   { 'nüt·ənz 'thərd 'lȯ }

1/N-expansion   [PHYS]   The grouping of scattering processes in a power series in 1/N, where N, assumed to be large, is the degeneracy of some quantum state such as the ground state of an impurity center in a solid, of the number of equivalent colors of quarks.   { ¦wən ,ō·vər ¦en ik'span·chən }

NEXRAD   See next-generation radar.   { 'neks,rad }

next-event file   [COMPUT SCI]   A portion of a computer simulation program which maintains a list of all events to be processed and updates the simulated time.   { 'nekst i'vent ,fïl }

next-generation radar   [METEOROL]   A Doppler radar, called WSR-88D, that enables forecasters to detect and give early warning for potentially severe weather.   Abbreviated NEXRAD.   { ,nekst jen·ə,rā·shən 'rā,där }

nexus   [COMMUN]   A connection or interconnection of a communications system, such as a data link or a network of branches and nodes.   [PHYSIO]   See gap junction.   { 'nek·səs }

Ney-Allen nebula   [ASTRON]   An extended source of infrared radiation in the Trapezium region of Orion which displays intense emission at a wavelength of 10 millimeters.   { 'nï 'al·ən 'neb·yə·lə }

Neyman-Pearson theory   [STAT]   A theory that determines what is the best test to use to examine a statistical hypothesis.   { 'nā·mən 'pir·sən ,thē·ə·rē }

Ney's chart   See modified Lambert conformal chart.   { 'nāz ,chärt }

Ney's projection   See modified Lambert conformal projection.   { 'nāz prə,jek·shən }

Nezelof's syndrome   [MED]   A congenital immunodeficiency disease in which T and sometimes B cells are absent, causing a lack in immune function leading to recurrent infections.   { 'nez·ə,lȯfs ,sin,drōm }

NFE method   See nearly free electron method.   { ¦en¦ef'ē ,meth·əd }

ng   See nanogram.

N galaxy   [ASTRON]   A galaxy that has the optical appearance of a strongly concentrated object with a semistellar nucleus that is surrounded by a faint halo or extension.   { 'en ,gal·ik·sē }

NGF   See nerve growth factor.

NGU   See nongonococcal urethritis.

Ni   See nickel.

niacin   See nicotinic acid.   { 'nï·ə·sən }

niacinamide   See nicotinamide.   { ,nï·ə'sin·ə·mïd }

Niagaran   [GEOL]   A North American provincial geologic series, in the Middle Silurian.   { nï'ag·rən }

nib   [ENG]   A small projecting point.   { nib }

nibble   [COMPUT SCI]   A unit of computger storage or information equal to one-half a byte.   { 'nib·əl }

nibbling   [MECH ENG]   Contour cutting of material by the action of a reciprocating punch that takes repeated small bites as the work is passed beneath it.   { 'nib·liŋ }

nicarbing   See carbonitriding.   { 'nï,kär·biŋ }

niccolite   [MINERAL]   NiAs A pale-copper-red, hexagonal mineral with metallic luster; an important ore of nickel; hardness is 5–5.5 on Mohs scale.   Also known as arsenical nickel; copper nickel; nickeline.   { 'nik·ə,lït }

niche   [ECOL]   The unique role or way of life of a plant or animal species.   [GEOL]   A shallow cave or reentrant produced by weathering and erosion near the base of a rock face or cliff or beneath a waterfall.   { nich }

niche glacier   [HYD]   A common type of small mountain glacier occupying a funnel-shaped hollow or irregular recess in a mountain slope.   { 'nich ,glā·shər }

Nichol's chart   [CONT SYS]   A plot of curves along which the magnitude M or argument α of the frequency control ratio is constant on a graph whose ordinate is the logarithm of the magnitude of the open-loop transfer function, and whose abscissa is the open-loop phase angle.   { 'nik·ȯlz ,chärt }

Nicholson's hydrometer   [ENG]   A modification of Fahrenheit's hydrometer in which the lower end of the instrument carries a scale pan to permit the determination of the relative density of a solid.   { 'nik·əl·sənz hï'dräm·əd·ər }

Nichols radiometer   [ENG]   An instrument, used to measure the pressure exerted by a beam of light, in which there are two small, silvered glass mirrors at the ends of a light rod that is suspended at the center from a fine quartz fiber within an evacuated enclosure.   { 'nik·ȯlz ,rād·ē·äm·əd·ər }

nick   [BIOCHEM]   The absence of a phosphodiester bond between adjacent nucleotides in one strand of duplex deoxyribonucleic acid.   [GEOL]   See knickpoint.   { nik }

nickase   [BIOCHEM]   An enzyme that causes single-stranded breaks in duplex deoxyribonucleic acid, allowing it to unwind.   { 'ni,kās }

nickel   [CHEM]   A chemical element, symbol Ni, atomic number 28, atomic weight 58.69.   [MET]   A silver-gray, ductile, malleable, tough metal; used in alloys, plating, coins (to replace silver), ceramics, and electronic circuits.   { 'nik·əl }

nickel-63   [NUC PHYS]   Radioactive nickel with beta radiation and 92-year half-life; derived by pile-irradiation of nickel; used in radioactive composition studies and tracer studies.   { 'nik·əl ,sik·stē'thrē }

nickel acetate   [ORG CHEM] $Ni(OOCCH_3)_2 \cdot 4H_2O$ Efflorescent green crystals that decompose upon heating; soluble in alcohol and water; used as textile dyeing mordant.   { 'nik·əl as·ə,tāt }

nickel-aluminum bronze   [MET]   An alloy that is composed of an 8–10% aluminum bronze with nickel added to increase strength, corrosion resistance, and heat resistance; used for dies, molds, cast propellers, and valve seats.   { 'nik·əl ə'lü·mə·nəm 'bränz }

nickel ammonium sulfate   [INORG CHEM] $NiSO_4 \cdot (NH_4)_2SO_4 \cdot 6H_2O$ A green, crystalline compound, soluble in water; used as a nickel electrolyte for electroplating.   Also known as ammonium nickel sulfate; double nickel salt. { 'nik·əl ə'mō·nē·əm 'səl‚fāt }

nickel-antimony glance *See* ullmannite. { 'nik·əl 'ant·ə‚mō·nē 'glans }

nickel arsenate [INORG CHEM] $Ni_3(AsO_4)_2·H_2O$ Poisonous yellow-green powder; soluble in acids, insoluble in water; used as a fat-hardening catalyst in soapmaking. { 'nik·əl 'ars·ən‚āt }

nickel bloom *See* annabergite. { 'nik·əl ‚blüm }

nickel brass *See* nickel silver. { 'nik·əl 'bras }

nickel bronze [MET] Bronze containing nickel; a common type contains 88% copper, 5% tin, 5% nickel, and 2% zinc. { 'nik·əl 'bränz }

nickel-cadmium battery [ELEC] A sealed storage battery having a nickel anode, a cadmium cathode, and an alkaline electrolyte; widely used in cordless appliances; without recharging, it can serve as a primary battery. Also known as cadmium-nickel storage cell. { 'nik·əl ‚kad·mē·əm 'bad·ə·rē }

nickel carbonate [INORG CHEM] $NiCO_3$ Light-green crystals that decompose upon heating; soluble in acid, insoluble in water; used in electroplating. { 'nik·əl 'kär·bə‚nāt }

nickel carbonyl [INORG CHEM] $Ni(CO)_4$ Colorless, flammable, poisonous liquid boiling at 43°C; soluble in alcohol and concentrated nitric acid, insoluble in water; used in gas plating (vapor decomposes at 60°C) and to produce metallic nickel. { 'nik·əl 'kär·bə‚nil }

nickel cast iron [MET] An improved-strength alloy cast iron containing a small percentage of nickel (2–5%); in larger amounts (15–36%) nickel primarily imparts corrosion resistance. { 'nik·əl 'kast 'ī·ərn }

nickel-chromium steel [MET] Steel containing nickel (0.2–3.75%) and chromium (0.3–1.5%) as alloying elements. { 'nik·əl 'krō·mē·əm 'stēl }

nickel cyanide [INORG CHEM] $Ni(CN)_2·4H_2O$ Poisonous, water-insoluble apple-green powder; melts and loses water at 200°C, decomposes at higher temperatures; used for electroplating and metallurgy. { 'nik·əl 'sī·ə‚nīd }

nickel delay line [ELECTR] An acoustic delay line in which nickel is used to transmit sound signals. { 'nik·əl di'lā ‚līn }

nickel formate [ORG CHEM] $Ni(HCOO)_2·2H_2O$ Water-soluble green crystals; used in hydrogenation catalysts. { 'nik·əl 'fòr‚māt }

nickel glance *See* gersdorffite. { 'nik·əl 'glans }

nickeline *See* niccolite. { ‚nik·ə‚lēn }

nickel iodide [INORG CHEM] NiI_2 or $NiI_2·6H_2O$ Hygroscopic black or blue-green solid; soluble in water and alcohol; sublimes when heated. { 'nik·əl 'ī·ə‚dīd }

nickel-iron battery *See* Edison battery. { 'nik·əl 'ī·ərn 'bad·ə·rē }

nickel-molybdenum iron [MET] An alloy containing 20–40% molybdenum and up to 60% nickel with some carbon added; has high acid resistance. { 'nik·əl mə'lib·də·nəm 'ī·ərn }

nickel-molybdenum steel [MET] Steel containing 0.2–0.3% molybdenum and 1.65–3.75% nickel as alloying elements. { 'nik·əl mə'lib·də·nəm 'stēl }

nickel nitrate [INORG CHEM] $Ni(NO_3)_2·6H_2O$ Fire-hazardous oxidant; deliquescent, green, water- and alcohol-soluble crystals; used for nickel plating and brown ceramic colors, and in nickel catalysts. { 'nik·əl 'nī‚trāt }

nickelocene [ORG CHEM] $(C_5H_5)_2Ni$ Dark green crystals with a melting point of 171–173°C; soluble in most organic solvents; used as an antiknock agent. { nə'kəl·ə‚sēn }

nickel ocher *See* annabergite. { 'nik·əl 'ō·kər }

nickel oxide [INORG CHEM] NiO Green powder; soluble in acids and ammonium hydroxide; insoluble in water; used to make nickel salts and for porcelain paints. Also known as green nickel oxide. { 'nik·əl 'äk‚sīd }

nickel phosphate [INORG CHEM] $Ni_3(PO_4)_2·7H_2O$ A light-green powder; soluble in acids and ammonium hydroxide, insoluble in water; used for electroplating and production of yellow nickel. { 'nik·əl 'fäs‚fāt }

nickel plating [MET] Electrolytic deposition of a metallic nickel coating. { 'nik·əl ‚plād·iŋ }

nickel pyrites *See* millerite. { 'nik·əl 'pī‚rīts }

nickel rhodium [MET] Nickel alloy with 25–80% rhodium; can also contain other metals, such as platinum or molybdenum; used for pen points, reflectors, electrodes, and chemical equipment. { 'nik·əl 'rō·dē·əm }

nickel silver [MET] A silver-white alloy composed of 52–80% copper, 10–35% zinc, and 5–35% nickel; sometimes also contains a few percent of lead and tin. Also known as German silver; nickel brass. { 'nik·əl 'sil·vər }

nickel steel [MET] Carbon steel containing up to 9% nickel as a major alloying element. { 'nik·əl 'stēl }

nickel-vanadium steel [MET] A nickel steel containing about 1.5% nickel, 1% manganese, 0.28% carbon, and 0.10% vanadium; used for high-strength cast parts. { 'nik·əl və'nā·dē·əm 'stēl }

nickel vitriol *See* morenosite. { 'nik·əl 'vit·rē‚òl }

nickpoint *See* knickpoint. { 'nik‚pòint }

Nicoletiidae [INV ZOO] A family of the insect order Thysanura proper. { ‚nik·ə·lə'tī·ə‚dē }

Nicol prism [OPTICS] A device for producing plane-polarized light, consisting of two pieces of transparent calcite (a birefringent crystal) which together form a parallelogram and are cemented together with Canada balsam. { 'nik·əl ‚priz·əm }

Nicomachinae [INV ZOO] A subfamily of the limnivorous sedentary annelids in the family Maldanidae. { ‚nik·ə'mak·ə‚nē }

nicotinamide [BIOCHEM] $C_6H_6ON_2$ Crystalline basic amide of the vitamin B complex that is interconvertible with nicotinic acid in the living organism; the amide of nicotinic acid. Also known as niacinamide. { ‚nik·ə'tin·ə‚mīd }

nicotinamide adenine dinucleotide *See* diphosphopyridine nucleotide. { ‚nik·ə'tin·ə‚mīd 'ad·ən‚ēn dī'nü·klē·ə‚tīd }

nicotinamide adenine dinucleotide oxidase [BIOCHEM] Any of a group of proteins located on the cell surface that are responsible for functions which, in concert with other membrane proteins, allow cells to enlarge following cell division. Abbreviated NOX. { ‚nik·ə‚tin·ə‚mīd ‚ad·ən·ēn di‚nü·klē·ə‚tīd 'äk·sə‚dās }

nicotinamide adenine dinucleotide phosphate [CELL MOL] A coenzyme and important component of the enzymatic systems concerned with biological oxidation-reduction systems. Abbreviated NADP+ (the reduced form is abbreviated NADPH). { ‚nik·ə¦tin·e‚mīd 'ad·ən‚ēn dī'nü·klē·ə‚tīd 'fäs‚fāt }

nicotine [ORG CHEM] $C_{10}H_{14}N_2$ A colorless liquid with a boiling point of 247.3°C; miscible with water; used as a contact insecticide fumigant in closed spaces. { 'nik·ə‚tēn }

nicotinic acid [BIOCHEM] $C_6H_5NO_2$ A component of the vitamin B complex; a white, water-soluble powder stable to heat, acid, and alkali; used for the treatment of pellagra. Also known as niacin. { ¦nik·ə¦tin·ik 'as·əd }

nictitating membrane [VERT ZOO] A membrane of the inner angle of the eye or below the eyelid in many vertebrates, and capable of extending over the eyeball. { 'nik·ə‚tād·iŋ 'mem‚brān }

nidamental gland [ZOO] Any of various structures that secrete covering material for eggs or egg masses. { ¦nīd·ə¦ment·əl 'gland }

nidicolous [ZOO] **1.** Spending a short time in the nest after hatching. **2.** Sharing the nest of another species. { nī'dik·ə·ləs }

nidus [MED] A focus of infection. [ZOO] A nest or breeding place. { 'nīd·əs }

niello [MATER] Mixture of sulfides of copper, silver, and lead, with black metallic appearance; used in ornamental inlays engraved on metals such as silver. { nē'el·ō }

Niemann-Pick disease [MED] A hereditary sphingolipidosis due to an enzyme deficiency resulting in abnormal accumulation of sphingomyelin; symptoms include anemia, enlargement of the liver, spleen, and lymph nodes, gastrointestinal disturbances, and various neurologic deficits. Also known as lipid hystiocytosis. { 'nē‚män 'pik di‚zēz }

nieve penitente [GEOL] A jagged pinnacle or spike of snow or firn, up to several meters in height. Also known as penitent. { nē'ā·vä ‚pen·ə'ten‚tā }

NIF *See* noise improvement factor.

niggliite [MINERAL] $PtSn$ or $PtTe$ A silver-white mineral consisting of a platinum telluride compound. { 'nig·lē‚īt }

night [ASTRON] The period of darkness between sunset and sunrise. { nīt }

night ape *See* bushbaby. { 'nīt ‚āp }

night blindness [MED] Reduced dark adaptation resulting from vitamin A deficiency or from retinitis pigmentosa or other

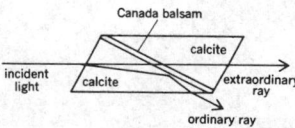

NICOL PRISM

Drawing of path of incident light through a Nicol prism. The extraordinary ray is plane-polarized.

NICOTINE

Structural formula of nicotine.

peripheral retinal disease. Also known as nyctalopia. { 'nīt ,blīnd·nəs }

night effect *See* polarization error. { 'nīt i ,fekt }

nightglow [GEOPHYS] A subdivision of airglow in which energy comes from reactions of atomic oxygen between 42 and 60 miles (70 and 100 kilometers), and from ionic recombination around 180 miles (300 kilometers). { 'nīt,glō }

night order book [NAV] A notebook in which the commanding officer of a ship writes orders with respect to courses and speeds, any special precautions concerning the speed and navigation of the ship, and all other orders for the night for the officer of the deck. { 'nīt 'ȯr·dər ,bùk }

night-sky camera [OPTICS] A simple camera mounted in a fixed direction to record trails of stars caused by the earth's rotation and the breaks in these trails caused by clouds, in order to determine the reduction, due to clouds, of the effective exposures of astronomical photographs. { 'nīt ,skī ,kam·rə }

night-sky light *See* airglow. { 'nīt ,skī 'līt }

night-sky luminescence *See* airglow. { 'nīt ,skī ,lü·mə'nes·əns }

night sweat [MED] Drenching perspiration occurring at night or during sleep in the course of certain febrile diseases. { 'nīt ,swet }

nighttime visual range *See* night visual range. { 'nī,tīm 'vizh·ə·wəl 'rānj }

night vision *See* scotopic vision. { 'nīt ,vizh·ən }

night-vision binoculars [OPTICS] Binoculars that are worn like eyeglasses but use a battery-powered television camera to pick up images; the images are viewed on tiny television picture tubes built into the binoculars. { 'nīt ,vizh·ən bə'näk·yə·lərz }

night-vision telescope [OPTICS] A telescope that has sufficient electronic amplification of images to be used at night without artificial illumination; may have television, optoelectronic, or other means of providing the necessary image amplification. { 'nīt ,vizh·ən 'tel·ə,skōp }

night visual range [OPTICS] The greatest distance at which a point source of light of a given candlepower can be perceived at night by an observer under given atmospheric conditions. Also known as nighttime visual range; penetration range; transmission range. { 'nīt 'vizh·ə·wəl 'rānj }

night wind [METEOROL] Dry squalls which occur at night in southwest Africa and the Congo; the term is loosely applied to other diurnal local winds such as mountain wind, land breeze, and midnight wind. { 'nīt ,wind }

nigre [CHEM ENG] Dark-colored layer formed between neat soap and lye during soap manufacture; contains more soap than lye, and a high concentration of salts and colored impurities. { 'nī·gər }

nigrescent [BIOL] Blackish. { nī'gres·ənt }

nigrite [MATER] A mixture of rubber with ozocerite distillation residue; used as a substitute for gutta-percha. { 'nī,grīt }

nigrosine [MATER] Any of a group of blue or black azine dyes used for coloring inks, shoe polish, leather, and wood; can be water-, alcohol-, or oil-soluble. { 'nī·grə,sēn }

nihilism [MED] Pessimism in regard to the efficacy of treatment, particularly the use of drugs. [PSYCH] The content of delusions encountered in depressed or melancholic states; the patient insists that his inner organs no longer exist, and that his relatives have passed away. { 'nī·ə,liz·əm }

Nikischov effect [PHYS] The production of electron-positron pairs in the collision of low-energy photons with high-energy photons (gamma rays). { ni'kis·chəf i,fekt }

nikkel oil [MATER] A bright-yellow liquid with a lemon and cinnamon odor obtained from the leaves and twigs of a laurel tree, *Cinnamomum zeylanicum;* contains citral and cineol; used in perfumery. { 'nik·əl ,oil }

niklesite [PETR] A pyroxenite containing the three pyroxenes: diopside, enstatite, and diallage. { 'nik·lə,sīt }

nilas [HYD] A thin elastic crust of gray-colored ice formed on a calm sea; characterized by a matte surface, and easily bent by waves and thrust into a pattern of interlocking fingers. { 'nī·ləs }

nile [NUCLEO] A unit of reactivity; the reactivity of a reactor in niles is equal to $100(1 - 1/k)$, where k is the effective multiplication factor. { nīl }

Nilionidae [INV ZOO] The false ladybird beetles, a family of coleopteran insects in the superfamily Tenebrionoidea. { ,nil·ē'än·ə,dē }

nilmanifold [MATH] The factor space of a connected nilpotent Lie group by a closed subgroup. { nil'man·ə,fōld }

nilpotent [MATH] An element of some algebraic system which vanishes when raised to a certain power. { nil'pōt·ənt }

nilradical [MATH] For an ideal, *I*, in a ring, *R*, the set of all elements, *a*, in *R* for which a^r is a member of *I* for some positive integer *n*. Also known as radical. { nil'rad·ə·kəl }

nimbostratus [METEOROL] A principal cloud type, or cloud genus, gray-colored and often dark, rendered diffuse by more or less continuously falling rain, snow, or sleet of the ordinary varieties, and not accompanied by lightning, thunder, or hail; in most cases the precipitation reaches the ground. { ¦nim·bō¦strad·əs }

nimbus [ASTRON] *See* halo. [METEOROL] A characteristic rain cloud; the term is not used in the international cloud classification except as a combining term, as cumulonimbus. { 'nim·bəs }

nim oil *See* neem oil. { 'nim ,oil }

N indicator *See* N scope. { 'en ,in·də,kād·ər }

nine-j symbol [QUANT MECH] A coefficient used in the general recoupling of four angular momenta, as in a transformation from *L-S* to *j-j* coupling in a two-electron system. Also known as X coefficient. { 'nīn 'jā ,sim·bəl }

nine-light indicator [ENG] A remote indicator for wind speed and direction used in conjunction with a contact anemometer and a wind vane; the indicator consists of a center light, connected to the contact anemometer, surrounded by eight equally spaced lights which are individually connected to a set of similarly spaced electrical contacts on the wind vane; wind speed is determined by counting the number of flashes of the center light during an interval of time; direction, indicated by the position of illuminated outer bulbs, is given to points of the compass. { 'nīn ¦līt 'in·də,kād·ər }

nine's complement [COMPUT SCI] The radix-minus-1 complement of a numeral whose radix is 10. { 'nīnz 'käm·plə·mənt }

ninhydrin [ORG CHEM] $C_9H_4O_3 \cdot H_2O$ White crystals or powder with a melting point of 240–245°C; soluble in water and alcohol; used for the detection and assay of peptides, amines, amino acids, and amino sugars. Also known as triketohydrindene hydrate. { nin'hī·drən }

niningerite [MINERAL] (Mg,Fe,Mn)S A mineral found only in meteorites. { nə'nin·jə,rīt }

ninon [TEXT] A sheer crisp fabric in plain weave of silk, nylon, or rayon. { nē'nȯn }

niobe oil *See* methyl benzoate. { 'nī·ə,bē,oil }

niobic acid [INORG CHEM] $Nb_2O_5 \cdot nH_2O$ Family of hydrates; white precipitate, soluble in inorganic acids and bases, insoluble in water; its formation is part of the analytical determination of niobium. { nī'ō·bik 'as·əd }

niobite *See* columbite. { 'nī·ə,bīt }

niobium [CHEM] A chemical element, symbol Nb, atomic number 41, atomic weight 92.9064. [MET] A platinum-gray, ductile metal with brilliant luster; used in alloys, especially stainless steels. Also known as columbium. { nī'ō·bē·əm }

niobium carbide [INORG CHEM] NbC A lavender gray powder with a melting point of 3500°C; used for carbide-tipped tools and special steels. { nī'ō·bē·əm 'kär,bīd }

nip [GEOL] **1.** A small, low cliff or break in slope which is produced by wavelets at the high-water mark. **2.** The point on the bank of a meander lake where erosion takes place due to crowding of the stream current toward the lake. **3.** Thinning of a coal seam, particularly if caused by tectonic movements. Also known as want. [MET] *See* angle of nip. [MIN ENG] *See* squeeze. { nip }

Nipher shield [ENG] A conically shaped, copper, rain-gage shield; used to prevent the formation of vertical wind eddies in the vicinity of the mouth of the gage, thereby making the rainfall catch a representative one. { 'nī·fər ,shēld }

Nipkow disk [COMPUT SCI] In optical character recognition, a disk having one or more spirals of holes around the outer edge, with successive openings positioned so that rotation of the disk provides mechanical scanning, as of a document. { 'nip·kō ,disk }

nippers [DES ENG] Small pincers or pliers for cutting or gripping. { 'nip·ərz }

nipping [NAV] The forcible closing of ice around a vessel so that it is held fast by ice under pressure. { 'nip·iŋ }

nipple [ANAT] The conical projection in the center of the mamma, containing the outlets of the milk ducts. [DES ENG] A short piece of tubing, usually with an internal or external thread at each end, used to couple pipes. Also known as bushing. { 'nip·əl }

nipple chaser [ENG] A member of a drilling crew who procures and delivers the tools and equipment necessary for an operation. { 'nip·əl ˌchā·sər }

Nippotaeniidea [INV ZOO] An order of tapeworms of the subclass Cestoda including some internal parasites of certain fresh-water fishes; the head bears a single terminal sucker. { ˌnip·ō‚tē·nē'ī·dē·ə }

Nissen stamp [MIN ENG] **1.** Machine used in crushing rock to sand sizes. **2.** An individual stamp worked in its own circular mortar box. { 'nis·ən ˌstamp }

Nissl bodies [CYTOL] Chromophil granules of nerve cells which ultrastructurally are composed of large ribosomes. { 'nis·əl ˌbäd·ēz }

nit [COMMUN] A unit of information content such that the information content of a symbol or message in nits is the negative of the natural logarithm of the probability of selecting that symbol or message from all the symbols or messages which could have been chosen. Also known as nepit. [OPTICS] A unit of luminance, equal to 1 candela per square meter. Abbreviated nt. { nit }

Nitelleae [BOT] A tribe of stoneworts, order Charales, characterized by 10 cells in two tiers of five each composing the apical crown. { ni'tel·ē‚ē }

niter See potassium nitrate. { 'nīd·ər }

niter balls [MATER] A pellet form of potassium nitrate, used as a fertilizer. Also known as sal prunella; throat balls. { 'nīd·ər ˌbólz }

niter cake See sodium bisulfate. { 'nīd·ər ˌkāk }

Nitidulidae [INV ZOO] The sap-feeding beetles, a large family of coleopteran insects in the superfamily Cucujoidea; individuals have five-jointed tarsi and antennae with a terminal three-jointed clavate expansion. { ˌnid·ə'dyül·ə‚dē }

nitrate [CHEM] **1.** A salt or ester of nitric acid. **2.** Any compound containing the ion NO_3^-. { 'nī‚trāt }

nitrate mineral [MINERAL] Any of several generally rare minerals characterized by a fundamental ionic structure of NO_3^-; examples are soda niter, niter, and nitrocalcite. { 'nī‚trāt ˌmin·rəl }

nitratine See soda niter. { 'nī‚trə‚tēn }

nitrating acid [INORG CHEM] Sulfuric-nitric acid mix used to nitrate cellulosics and aromatic chemicals. Also known as mixed acid. { 'nī‚trād·iŋ ˌas·əd }

nitration [ORG CHEM] Introduction of an NO_2^- group into an organic compound. { nī'trā·shən }

nitrene [ORG CHEM] A molecular fragment that is an uncharged, electron-deficient species containing a monocovalent nitrogen. { 'nī‚trēn }

nitric acid [INORG CHEM] HNO_3 Strong oxidant that is fire-hazardous; colorless or yellowish liquid, miscible with water; boils at 86°C; used for chemical synthesis, explosives, and fertilizer manufacture, and in metallurgy, etching, engraving, and ore flotation. Also known as aqua fortis. { 'nī‚trik 'as·əd }

nitric oxide [INORG CHEM] NO A colorless gas that, at room temperature, reacts with oxygen to form nitrogen dioxide (NO_2, a reddish-brown gas). It may be used to form other compounds. It is a crucial physiological messenger molecule thought to play a role in blood pressure regulation, control of blood clotting, immune defense, digestion, the senses of sight and smell, and possibly learning and memory. { 'nī‚trik 'äk‚sīd }

nitric oxide synthase [BIOCHEM] An enzyme that catalyzes the stepwise conversion of the amino acid L-arginine to nitric oxide and L-citrulline. There are three types: the brain (or neuronal) and epithelial nitric oxide synthases, which are always present in cells, and inducible nitric oxide synthase, which is produced as needed. { ˌnī‚trik 'äk‚sīd ˌsin‚thās }

nitride [INORG CHEM] Compound of nitrogen and a metal, such as Mg_3N_2. { 'nī‚trīd }

nitride nuclear fuel [NUCLEO] Fissionable nuclear fuel in nitride form, such as UN and UN_2. { 'nī‚trīd 'nü·klē·ər 'fyül }

nitriding [MET] Surface hardening of steel by formation of nitrides; nitrogen is introduced into the steel usually by heating in gaseous ammonia. { 'nī‚trīd·iŋ }

nitrification [MICROBIO] Formation of nitrous and nitric

acids or salts by oxidation of the nitrogen in ammonia; specifically, oxidation of ammonium salts to nitrites and oxidation of nitrites to nitrates by certain bacteria. { ˌnī‚trə·fə'kā·shən }

nitrifying bacteria [MICROBIO] Members of the family Nitrobacteraceae. { 'nī‚trə‚fī·iŋ bak'tir·ē·ə }

nitrile [ORG CHEM] RC≡N Cyanide derived by removal of water from an acid amide. { 'nī‚trīl }

nitrile-butadiene rubber See nitrile rubber. { 'nī‚trīl ˌbyüd·ə'dī‚ēn ˌrəb·ər }

nitrile resin [ORG CHEM] Any one of a family of polymers produced from acrylonitrile, various esters, butadiene, and styrene. { 'nī‚trīl ˌrez·ən }

nitrile rubber [MATER] A synthetic rubber formed by polymerization of acrylonitrile with butadiene; the structure of the polymer is $-CH_2CH=CHCH_2CH_2CH(CN)-$. Also known as acrylonitrile-butadiene rubber; acrylonitrile rubber; NBR; nitrile-butadiene rubber; NR. { 'nī‚trīl ˌrəb·ər }

nitrilotriacetic acid [ORG CHEM] $N(CH_2COOH)_3$ A white powder, melting point 240°C, with some decomposition; soluble in water; it is toxic, and birth abnormalities may result from ingestion; may be used as a chelating agent in the laboratory. Also known as NTA; TGA. { ˌnī‚trə·lō‚trī·ə'sēd·ik 'as·əd }

nitrite [CHEM] A compound containing the radical NO_2^-; can be organic or inorganic. { 'nī‚trīt }

nitro- [CHEM] Chemical prefix showing the presence of the NO_2^- radical. { 'nī·trō }

nitroalkane See nitroparaffin. { ¦nī·trō'al‚kān }

meta-nitroaniline [ORG CHEM] $NO_2C_6H_4NH_2$ Yellow crystals that melt at 112.5°C; a toxic material; used as a dye intermediate. { ¦med·ə ˌnī·trō'an·ə·lən }

ortho-nitroaniline [ORG CHEM] $NO_2C_6H_4NH_2$ Orange-red crystals that melt at 69.7°C, soluble in ethanol; a toxic material; used to manufacture dyes. { ¦ór·thō ˌnī·trō'an·ə·lən }

para-nitroaniline [ORG CHEM] $NO_2C_6H_4NH_2$ Yellow crystals that melt at 148°C; insoluble in water, soluble in ethanol; a toxic material; used to make dyes, and as a corrosion inhibitor. { ¦par·ə ˌnī·trō'an·ə·lən }

nitroaromatic [ORG CHEM] A nitrated benzene or benzene derivative, such as nitrobenzene, $C_6H_5NO_2$, or nitrobenzoic acid, $NO_2·C_6H_4·COOH$. { ¦nī·trō‚ar·ə'mad·ik }

Nitrobacteraceae [MICROBIO] The nitrifying bacteria, a family of gram-negative, chemolithotrophic bacteria; autotrophs which derive energy from nitrification of ammonia or nitrite, and obtain carbon for growth by fixation of carbon dioxide. { ¦nī·trō‚bak·tə'rās·ē‚ē }

nitrobarite See barium nitrate. { ¦nī·trō'ba‚rīt }

nitrobenzene [ORG CHEM] $C_6H_5NO_2$ Greenish crystals or a yellowish liquid, melting point 5.70°C; a toxic material; used in aniline manufacture. Also known as oil of mirbane. { ¦nī·trō'ben‚zēn }

ortho-nitrobiphenyl [ORG CHEM] $C_{12}H_9NO_2$ A crystalline compound with a sweetish odor; melting point is 36.7°C; used as a plasticizer for resins, cellulose acetate and nitrate, and polystyrenes, and as a fungicide for textiles. Abbreviated ONB. { ¦ór·thō ˌnī·trō·bī'fen·əl }

nitrobromoform See bromopicrin. { ¦nī·trō'brō·mə‚fórm }

nitrocalcite See calcium nitrate. { ¦nī·trō'kal‚sīt }

nitrocellulose See cellulose nitrate. { ¦nī·trō'sel·yə‚lōs }

nitrocellulose filter [MOL BIO] A very thin filter whose fibers selectively bind single-stranded deoxyribonucleic acid (DNA) but not double-stranded DNA or ribonucleic acid. { ¦nī·trō'sel·yə‚lōs ˌfil·tər }

nitrocellulose propellant [MATER] A single-base propellant whose main constituent is nitrocellulose, with only minor percentages of additives for stabilizing and other purposes. { ¦nī·trō'sel·yə‚lōs prə'pel·ənt }

nitrocotton See cellulose nitrate. { ¦nī·trō'kät·ən }

nitro dye [ORG CHEM] A dye with the NO_2 chromophore group in the molecules. { 'nī·trō ˌdī }

nitroethane [ORG CHEM] $CH_3CH_2NO_2$ A colorless liquid, slightly soluble in water; boils at 114°C; used as a solvent for cellulosics, resins, waxes, fats, and dyestuffs, and as a chemical intermediate. { ¦nī·trō'eth‚ān }

nitro explosive [ORG CHEM] Explosive compound containing one or more NO_2^- groups, such as nitroglycerine, $C_3H_5(ONO_2)_3$, or trinitrotoluene, $C_6H_2(CH_3)(NO_2)_3$. { 'nī·trō ik'splō·siv }

nitrogelatin See gelatin dynamite. { ¦nī·trə'jel·ət·ən }

nitrogen [CHEM] A chemical element, symbol N, atomic

number 7, atomic weight 14.0067; it is a gas, diatomic (N_2) under normal conditions; about 78% of the atmosphere is N_2; in the combined form the element is a constituent of all proteins. { 'nī·trə·jən }

nitrogen acid anhydride *See* nitrogen pentoxide. { 'nī·trə·jən ¦as·əd an'hī‚drīd }

nitrogenase [BIOCHEM] An enzyme that catalyzes a six-electron reduction of N_2 in the process of nitrogen fixation. { nī'trä·jə‚nās }

nitrogenated oil [MATER] A class of essential oils containing carbon, hydrogen, oxygen, and nitrogen; an example is oil of bitter almonds. { 'nī·trə·jə‚nād·əd ‚oil }

nitrogen balance [GEOCHEM] The net loss or gain of nitrogen in a soil. [PHYSIO] The difference between nitrogen intake (as protein) and total nitrogen excretion for an individual. { 'nī·trə·jən ‚bal·əns }

nitrogen cycle *See* carbon-nitrogen cycle. { 'nī·trə·jən ‚sī·kəl }

nitrogen dioxide [INORG CHEM] NO_2 A reddish-brown gas; it exists in varying degrees of concentration in equilibrium with other nitrogen oxides; used to produce nitric acid. Also known as dinitrogen tetroxide; liquid dioxide; nitrogen peroxide; nitrogen tetroxide. { 'nī·trə·jən dī'äk‚sīd }

nitrogen fixation [CHEM ENG] Conversion of atmospheric nitrogen into compounds such as ammonia, calcium cyanamide, or nitrogen oxides by chemical or electric-arc processes. [MICROBIO] Assimilation of atmospheric nitrogen by heterotrophic bacteria. Also known as dinitrogen fixation. { 'nī·trə·jən ‚fik¦sā'shən }

nitrogen monoxide *See* nitrous oxide. { 'nī·trə·jən mə'näk‚sīd }

nitrogen mustard [ORG CHEM] Any of the substituted mustard gases in which the sulfur is replaced by an amino nitrogen, such as for methyl bis(2-chlorethyl)amine, $(CH_2ClCH_2)_2NCH_3$; useful in cancer research. { 'nī·trə·jən 'məs·tərd }

nitrogen narcosis [MED] Narcosis caused by gaseous nitrogen at high pressure in the blood; produced in divers breathing air at depths of 100 feet (30 meters) or more. Also known as rapture of the deep. { 'nī·trə·jən när'kō·səs }

nitrogenous base [BIOCHEM] A purine or a pyrimidine derivative which is one of the three components of a nucleotide of nucleic acids. Also known as base. { nī'trä·jə·nəs 'bās }

nitrogenous fertilizer [MATER] Fertilizer materials, natural or synthesized, containing nitrogen available for fixation by vegetation, such as potassium nitrate, KNO_3, or ammonium nitrate, NH_4NO_3. { nī'trä·jə·nəs 'fərd·əl‚īz·ər }

nitrogen oxides [INORG CHEM] NO_x Chemical compounds of nitrogen and oxygen; produced primarily from the combustion of fossil fuels, they contribute to the formation of ground-level ozone. { 'nī·trə·jən 'äk‚sīdz }

nitrogen pentoxide [INORG CHEM] N_2O_5 Colorless crystals, soluble in water (forms HNO_3); decomposes at 46°C. Also known as nitrogen acid anhydride. { 'nī·trə·jən pen'täk‚sīd }

nitrogen peroxide *See* nitrogen dioxide. { 'nī·trə·jən pə'räk‚sīd }

nitrogen sequence [ASTRON] Wolf-Rayet stars in which nitrogen emission bands dominate the spectrum. { 'nī·trə·jən ‚sē·kwəns }

nitrogen solution [INORG CHEM] Mixture used to neutralize super-phosphate in fertilizer manufacture; consists of 60% ammonium nitrate, and the balance a 50% aqua ammonia solution. { 'nī·trə·jən sə‚lü·shən }

nitrogen tetroxide *See* nitrogen dioxide. { 'nī·trə·jən te'träk‚sīd }

nitrogen trifluoride [INORG CHEM] NF_3 A colorless gas that has a melting point of -206.6°C and a boiling point of -128.8°C; used as an oxidizer for high-energy fuels. { 'nī·trə·jən trī'flùr‚īd }

nitrogen trioxide [INORG CHEM] N_2O_3 Green, water-soluble liquid; boils at 3.5°C. { 'nī·trə·jən trī'äk‚sīd }

nitroglycerin [ORG CHEM] $CH_2NO_3CHNO_3CH_2NO_3$ Highly unstable, explosive, flammable pale-yellow liquid; soluble in alcohol; freezes at 13°C and explodes at 260°C; used as an explosive, to make dynamite, and in medicine. Also spelled nitroglycerine. { ‚nī·trə'glis·ə·rən }

nitroglycerine *See* nitroglycerin. { ‚nī·trə'glis·ə·rən }

nitroglycerin powder [MATER] Any explosive characterized by a low nitroglycerin content, up to 10% and a high

ammonium nitrate content of 80–85% with carbonaceous material forming the remainder. { ‚nī·trə'glis·ə·rən ‚paud·ər }

nitroguanidine [ORG CHEM] $H_2NC(NH)NHNO_2$ Explosive yellow solid, soluble in alcohol; melts at 246°C; used in explosives and smokeless powder. { ‚nī·trō'gwän·ə‚dēn }

nitromagnesite [MINERAL] $Mg(NO_3)_2 \cdot 6H_2O$ Mineral consisting of magnesium nitrate, occurring as an efflorescence in limestone caverns. { ‚nī·trō'mag·nə‚sīt }

nitrometer [ANALY CHEM] Glass apparatus used to collect and measure nitrogen and other gases evolved by a chemical reaction. Also known as azotometer. { nī'träm·əd·ər }

nitromethane [ORG CHEM] CH_3NO_2 A liquid nitroparaffin compound; oily and colorless; boils at 101°C; used as a mono-propellant for rockets, in chemical synthesis, and as an industrial solvent for cellulosics, resins, waxes, fats, and dyestuffs. { ‚nī·trō'meth‚ān }

nitron [ORG CHEM] $CN_4(C_6H_5)_3CH$ Yellow crystals, soluble in chloroform and acetone; used as reagent to detect NO_3 ion in dilute solutions. { 'nī‚trän }

nitronium [CHEM] Positively charged NO_2 ion, believed to be formed from HNO_3. Also known as nitryl ion. { nī'trō·nē·əm }

nitroparaffin [ORG CHEM] Any organic compound in which one or more hydrogens of an alkane are replaced by a nitro, or NO_2^-, group, such as nitromethane, CH_3NO_2, or nitroethane, $C_2H_5NO_2$. Also known as nitroalkane. { ‚nī·trō'par·ə·fən }

nitrophenide [PHARM] $(NO_2C_6H_4)_2S$ Yellow crystals with a melting point of 83°C; soluble in ether; used in veterinary medicine and pharmaceutical manufacture. { ‚nī·trō'fe‚nīd }

***ortho*-nitrophenol** [ORG CHEM] $C_6H_5NO_3$ A yellow, crystalline compound; melting point is 44–45°C; soluble in hot water, alcohol, benzene, ether, carbon disulfide, and alkali hydroxides; used in the commercial preparation of many compounds. { ór‚thō ‚nī·trō'fe‚nól }

***para*-nitrophenylhydrazine** [ORG CHEM] $C_6H_7N_3O_2$ An orange-red, crystalline compound with a melting point of about 157°C; soluble in hot water or hot benzene; used as a reagent for aliphatic aldehydes and ketones. { ‚par·ə ‚nī·trō'fen·əl'hī‚drə‚zēn }

nitrophosphate [MATER] A nitrogen-phosphorus fertilizer made by reacting nitric acid with phosphate rock. { ‚nī·trō'fäs‚fāt }

nitrophyte [BOT] A plant that requires nitrogen-rich soil for growth. { ‚nī·trə‚fīt }

1-nitropropane [ORG CHEM] $CH_3CH_2CH_2NO_2$ A colorless liquid with a boiling point of 132°C; used as a rocket propellant and gasoline additive. { ‚wən ‚nī·trō'prō‚pān }

2-nitropropane [ORG CHEM] $CH_3CHNO_2CH_3$ A colorless liquid with a boiling point of 120°C; used as a solvent for vinyl coatings, as a rocket propellant, and as a gasoline additive. { ‚tü ‚nī·trō'prō‚pān }

nitroso [CHEM] The radical NO^- with trivalent nitrogen. Also known as hydroximino; oximido. { nī'trō·sō }

nitrostarch [ORG CHEM] $C_{12}H_{12}(NO_2)_8O_{10}$ Orange powder, soluble in ethyl alcohol; used in explosives. Also known as starch nitrate. { 'nī·trə‚stärch }

***meta*-nitrotoluene** [ORG CHEM] $NO_2C_6H_4CH_3$ Yellow powder that melts at 15°C; insoluble in water; used in organic synthesis. { ‚med·ə ‚nī·trō'täl·yə‚wēn }

***ortho*-nitrotoluene** [ORG CHEM] $NO_2C_6H_4CH_3$ A yellow liquid boiling at 220.4°C; insoluble in water; used to produce toluidine and dyes. { ór‚thō ‚nī·trō'täl·yə‚wēn }

***para*-nitrotoluene** [ORG CHEM] $NO_2C_6H_4CH_3$ Yellow crystals that melt at 51.7°C; insoluble in water, soluble in ethanol; used to produce toluidine and to manufacture dyes. { ‚par·ə ‚nī·trō'täl·yə‚wēn }

nitrourea [ORG CHEM] $NH_2CONHNO_2$ Highly explosive white crystals, melting at 159°C; soluble in ether and alcohol, slightly soluble in water; used as a chemical intermediate. { ‚nī·trō·yú'rē·ə }

nitrous acid [INORG CHEM] HNO_2 Aqueous solution of nitrogen trioxide, N_2O_3. { 'nī·trəs 'as·əd }

nitrous oxide [INORG CHEM] N_2O Colorless, sweet-tasting gas, boiling at -90°C; slightly soluble in water, soluble in alcohol; used as a food aerosol, and as an anesthetic in dentistry and surgery. Also known as laughing gas; nitrogen monoxide. { 'nī·trəs 'äk‚sīd }

nitroxylene [ORG CHEM] $C_6H_3(CH_3)_2NO_2$ Any of three isomers occurring either as a yellow liquid or as crystalline

ORTHO-NITROPHENOL

Structural formula.

needles with a melting point of 2°C and boiling point of 246°C; soluble in alcohol and ether; used in gelatinizing accelerators for pyroxylin. { nī'träk·sə,lēn }

nitryl halide [INORG CHEM] NO_2X Compound containing a halide (X) and a nitro group (NO_2). { 'nī,tril 'ha,līd }

nitryl ion See nitronium. { 'nī,tril ,ī,än }

nival [ECOL] **1.** Characterized by or living in or under the snow. **2.** Of or pertaining to a snowy environment. { 'nī·vəl }

nival gradient [GEOL] The angle between a nival surface and the horizon. { 'nī·vəl ,grād·ē·ənt }

nival surface [GEOL] The hypothetical planar surface containing all of the different snowlines of the same geologic time period. { 'nī·vəl ,sər·fəs }

nivation [GEOL] Rock or soil erosion beneath a snowbank or snow patch, due mainly to frost action but also involving chemical weathering, solifluction, and meltwater transport of weathering products. Also known as snow patch erosion. { nī'vā·shən }

nivation cirque See nivation hollow. { nī'vā·shən ,sərk }

nivation glacier [HYD] A small, newly formed glacier; represents the initial stage of glaciation. Also known as snowbank glacier. { nī'vā·shən ,glā·shər }

nivation hollow [GEOL] A small, shallow depression formed, and occupied during part of the year, by a snow patch or snowbank that, through nivation, is thought to initiate glaciation. Also known as nivation cirque; snow niche. { nī'vā·shən,häl·ō }

nivation ridge See winter-talus ridge. { nī'vā·shən ,rij }

niveal [GEOL] Property of features and effects resulting from the action of snow and ice. { 'niv·ē·əl }

nivenite [MINERAL] UO_2 A velvet-black member of the uranite group; contains rare-earth metals cerium and yttrium; a source of uranium. { 'niv·ə,nīt }

niveoglacial [GEOL] Pertaining to the combined action of snow and ice. { ¦niv·ē·ō'glā·shəl }

niveolian [GEOL] Pertaining to simultaneous accumulation and intermixing of snow and airborne sand at the side of a gentle slope. { ¦niv·ē¦ō·lē·ən }

niveous See niveus. { 'niv·ē·əs }

niveus [BIOL] Snow-white in color. Also spelled niveous. { 'niv·ē·əs }

Nix Olympica See Olympus Mons. { 'niks ə'lim·pə·kə }

NK See natural killer cell.

n-key rollover [COMPUT SCI] The ability of a computer-terminal keyboard to remember the order in which keys were operated and pass this information to the computer even when several keys are depressed before other keys have been released. { 'en ,kē 'rōl,ō·vər }

N-level address [COMPUT SCI] A multilevel address specifying N levels of addressing. { 'en ,lev·əl 'ad,res }

N-level logic [ELECTR] An arrangement of gates in a digital computer in which not more than N gates are connected in series. { 'en ,lev·əl 'läj·ik }

NLGI number [ENG] One of a series of numbers developed by the National Lubricating Grease Institute and used to classify the consistency range of lubricating greases; NLGI numbers are based on the American Society for Testing and Materials cone penetration number. { ¦en¦el¦jē'ī ,nəm·bər }

N line [SPECT] One of the characteristic lines in an atom's x-ray spectrum, produced by excitation of an N electron. { 'en ,līn }

N-m See newton-meter of torque.

N-modular redundancy [COMPUT SCI] A generalization of triple modular redundancy in which there are N identical units, where N is any odd number. { 'en ¦mäj·ə·lər ri'dən·dən·sē }

NMOS [ELECTR] Metal-oxide semiconductors that are made on p-type substrates, and whose active carriers are electrons that migrate between n-type source and drain contacts. Derived from n-channel metal-oxide semiconductor. { 'en,mós }

NMR See nuclear magnetic resonance.

NMRR See normal-mode rejection ratio.

NMR tomography See zeugmatography. { ¦en¦em'är tə'mäg·rə·fē }

n-net [MATH] A finite net in which n lines pass through each point. { 'en ,net }

nn junction [ELECTR] In a semiconductor, a region of transition between two regions having different properties in n-type semiconducting material. { ¦en¦en ,jəŋk·shən }

n-n′ reactions See neutron inelastic scattering reactions. { ¦en ¦en,prīm rē,ak·shənz }

No See nobelium.

no-address instruction [COMPUT SCI] An instruction which a computer can carry out without using an operand from storage. { 'nō 'ad,res in,strək·shən }

no-atmospheric control [AERO ENG] Any device or system designed or set up to control a guided rocket missile, rocket craft, or the like outside the atmosphere or in regions where the atmosphere is of such tenuity that it will not affect aerodynamic controls. { 'nō ,at·mə'sfir·ik kən'trōl }

nobelium [CHEM] A chemical element, symbol No, atomic number 102; a synthetic element, in the actinium series; isotopes with mass numbers 250–260 and 262 have been produced in the laboratory, with mass number 259 having the longest known half-life, 58 minutes. { nō'bel·ē·əm }

noble gas [CHEM] A gas in group 0 of the periodic table of the elements; it is monatomic and, with limited exceptions, chemically inert. Also known as inert gas; rare gas. { 'nō·bəl 'gas }

noble-gas electron configuration [CHEM] An electron structure of an atom or ion in which the outer electron shell contains eight electrons, corresponding to the electron configuration of a noble gas, such as neon or argon. { ¦nō·bəl ¦gas i'lek,trän kən,fig·yə,rā·shən }

noble metal [MET] A metal, or alloy, such as gold, silver, or platinum having high resistance to corrosion and oxidation; used in the construction of thin-film circuits, metal-film resistors, and other metal-film devices. { 'nō·bəl 'med·əl }

noble potential [PHYS CHEM] A potential equaling or approaching that of the noble elements, such as gold, silver, or copper, of the electromotive series. { 'nō·bəl pə'ten·chəl }

no-bottom sounding [ENG] A sounding in the ocean in which the bottom is not reached. { 'nō 'bäd·əm ,saúnd·iŋ }

no-break power [ELEC] Power system designed to fulfill load requirements during the interval between the failure of the primary power and the time the auxiliary power can be made available. { 'nō 'brāk 'paú·ər }

Nocardiaceae [MICROBIO] A family of aerobic bacteria in the order Actinomycetales; mycelium and spore production is variable. { nō,kär·dē'ās·ē,ē }

nocardiosis [MED] Infection by species of the fungus Nocardia characterized by spreading granulomatous lesions. { nō,kär·dē'ō·səs }

nocerite See fluoborite. { 'nō·sə,rīt }

nociceptive reflex See flexion reflex. { ¦nō·sē·ō·ri'sep·tiv 'rē,fleks }

nociceptor [PHYSIO] A sensory nerve ending that is particularly sensitive to noxious stimuli such as chemical changes in surrounding tissue evoked by injury. { 'nō·sə,sep·tər }

noct-, nocti-, nocto-, noctu- [SCI TECH] Combining form meaning night. { näkt, 'näk·tē, 'näk·tō, 'näk·tə }

noctalbuminuria [MED] Excretion of protein in night urine only. { ,näkt·al,byü·mə'nyúr·ē·ə }

Noctilionidae [VERT ZOO] The fish-eating bats, a tropical American monogeneric family of the Chiroptera having small eyes and long, narrow wings. { näk,til·ē'än·ə,dē }

noctilucent cloud [METEOROL] A cloud of unknown composition which occurs at great heights and high altitudes; photometric measurements have located such clouds between 45 and 54 miles (75 and 90 kilometers); they resemble thin cirrus, but usually with a bluish or silverish color, although sometimes orange to red, standing out against a dark night sky. { ¦näk·tə¦lü·sənt 'klaúd }

noctiphobia [PSYCH] Abnormal fear of night or darkness. { ,näk·tə·fō'bē·ə }

Noctuidae [INV ZOO] A large family of dull-colored, medium-sized moths in the superfamily Noctuoidea; larva are mostly exposed foliage feeders, representing an important group of agricultural pests. { näk'tü·ə,dē }

Noctuoidea [INV ZOO] A large superfamily of lepidopteran insects in the suborder Heteroneura; most are moderately large moths with reduced maxillary palpi. { ,näk·tü·óid·ē·ə }

nocturia [MED] Excessive urination at night. { nak'túr·ē·ə }

nocturnal [BIOL] Active during the nighttime. [SCI TECH] Occurring during the nighttime. { näk'tərn·əl }

nocturnal emission [PHYSIO] Normal, involuntary seminal

discharge occurring during sleep in males after puberty. { näk'tərn·əl i'mish·ən }

nocturnal enuresis [MED] Involuntary nocturnal urination during sleep. { näk'tərn·əl ,en·yu̇'rē·səs }

nocturnal radiation See effective terrestrial radiation. { näk'tərn·əl ,rād·ē'ā·shən }

no-cut rounds [MIN ENG] Set of holes drilled straight into the face for blasting underground. { 'nō ,kət ,rau̇nz }

NODA See n-octyl n-decyl adipate. { 'nō·də }

nodal analysis [ELEC] A method of electrical circuit analysis in which potential differences are taken as independent variables and the sum of the currents flowing into a node is equated to 0. { 'nōd·əl ə'nal·ə·səs }

nodal line [ASTRON] The line passing through the ascending and descending nodes of the orbit of a celestial body. [PHYS] **1.** A line or curve in a two-dimensional standing-wave system, such as a vibrating diaphragm, where some specified characteristic of the wave, such as velocity of pressure, does not oscillate. **2.** A line which remains fixed during some deformation or rotation of a body or coordinate system. { 'nōd·əl ,līn }

nodal points [ELEC] Junction points in a transmission system; the automatic switches and switching centers are the nodal points in automated systems. [OCEANOGR] The no-tide points in amphidromic regions. [OPTICS] A pair of points on the axis of an optical system such that an incident ray passing through one of them results in a parallel emergent ray passing through the other. { 'nōd·əl ,pȯins }

nodal rhythm [PHYSIO] A cardiac rhythm characterized by pacemaker function originating in the atrioventricular node, with a heart rate of 40–70 per minute. { 'nōd·əl 'rith·əm }

nodal tachycardia [MED] A cardiac arrhythmia characterized by a heart rate of 140–220 per minute. { 'nōd·əl ,tak·ə'kärd·ē·ə }

nodal tissue [HISTOL] **1.** Tissue from the sinoatrial node, and the atrioventricular node and bundle and its branches, composed of a dense network of Purkinje fibers. **2.** Tissue from a lymph node. { 'nōd·əl ,tish·ü }

nodal zone [OCEANOGR] A zone in which there is a change in the prevailing direction of the littoral transport. { 'nōd·əl ,zōn }

Nodamura virus [VIROL] The type species of the *Nodavirus* genus; it infects mammals and insects. Abbreviated NOV. { 'nō·də|mu̇r·ə ,vī·rəs }

Nodaviridae [VIROL] A family of single-strand ribonucleic acid-containing viruses that infect insects. { ,nō·də'vir·ə,dī }

Nodavirus [VIROL] The genus comprising the Nodaviridae family of insect viruses. { 'nō·də,vī·rəs }

node [ANAT] **1.** A knob or protuberance. **2.** A small, rounded mass of tissue, such as a lymph node. [ASTRON] **1.** One of two points at which the orbit of a planet, planetoid, or comet crosses the plane of the ecliptic. **2.** One of two points at which a satellite crosses the equatorial plane of its primary. [BOT] A site on a plant stem at which leaves and axillary buds arise. [ELEC] See branch point. [ELECTR] A junction point within a network. [GEOL] That point along a fault at which the direction of apparent displacement changes. [IND ENG] On a graphic presentation of a project, a symbol placed at the intersection of arrows that represent activities to identify the completion or start of an activity. [MATH] See crunode; vertex. [NEUROSCI] A point of constriction along a nerve. [PHYS] A point, line, or surface in a standing-wave system where some characteristic of the wave has essentially zero amplitude. { nōd }

node cycle [ASTRON] The period of time needed for the regression of the moon's nodes to conclude a circuit of 360° of longitude; approximately equal to 18.61 Julian years. { 'nōd ,sī·kəl }

node of Ranvier [NEUROSCI] The region of a local constriction in a myelinated nerve; formed at the junction of two Schwann cells. { 'nōd əv rän'vyā }

node voltage [ELEC] The voltage at a given point in an electric network with respect to that at a node. { 'nōd ,vōl·tij }

nodical month [ASTRON] The average period of revolution of the moon about the earth with respect to the moon's ascending node, a period of 27 days 5 hours 5 minutes 35.8 seconds; also known as draconic month; draconitic month. { 'näd·ə·kəl 'mənth }

nodical year See eclipse year. { 'nōd·i·kəl ,yir }

Nodosariacea [INV ZOO] A superfamily of Foraminiferida

in the suborder Rotaliina characterized by a radial calcite test wall with monolamellar septa, and a test that is coiled, uncoiled, or spiral about the long axis. { ,nōd·ə,sar·ē'ās·ē·ə }

nodose [BIOL] Having many or noticeable protuberances; knobby. { 'nō,dōs }

no-draft forging [MET] A forging designed with little or no taper for removal from dies, and with extremely fine tolerances for closer control of grain flow during production of the final part. { 'nō ,draft ,fȯrj·iŋ }

nodular [SCI TECH] Occurring in the form of small, rounded lumps. { 'näj·ə·lər }

nodular cast iron [MET] Cast iron treated in the molten state with a master alloy containing an element such as magnesium which favors formation of spheroidal graphite. Also known as ductile iron; spheroidal graphite cast iron. { 'näj·ə·lər ¦kast 'ī·ərn }

nodular chert [GEOL] Chert occurring as nodular or concretionary segregations (chert nodules). { 'näj·ə·lər 'chərt }

nodular goiter See adenomatous goiter. { 'näj·ə·lər 'gȯid·ər }

nodular powder [MET] Irregularly shaped metal powder particles. { 'näj·ə·lər 'pau̇d·ər }

nodule [ANAT] **1.** A small node. **2.** A small aggregation of cells. [BOT] A bulbous enlargement found on roots of legumes and certain other plants, whose formation is stimulated by symbiotic, nitrogen-fixing bacteria that colonize the roots. [GEOL] A small hard mass or lump of a mineral or mineral aggregate characterized by a contrasting composition from and a greater hardness than the surrounding sediment or rock matrix in which it is embedded. [MED] A primary skin lesion, seen as a circumscribed solid elevation. { 'näj·ül }

nodules of the semilunar valves [ANAT] Small nodes in the midregion of the pulmonary and aortic semilunar valves. { 'näj·ülz əv thə ¦sem·i,lü·nər 'valvz }

nodulizing [ENG] Creation of spherical lumps from powders by working them together, coalescing them with binders, drying fluid-solid mixtures, heating, or chemical reaction. { 'näj·ə,līz·iŋ }

nodulose [BIOL] Having minute nodules or fine knobs. { 'näj·ə,lōs }

Noeggerathiales [PALEOBOT] A poorly defined group of fossil plants whose geologic range extends from Upper Carboniferous to Triassic. { ,neg·ə,rath·ē'ā·lēz }

NOEL See no observed effect level.

Noetherian module [MATH] A module in which every ascending sequence of submodules has only a finite number of distinct members. { ,nō·ə¦thir·ē·ən 'mäj·əl }

Noetherian ring [MATH] A ring is Noetherian on left ideals (or right ideals) if every ascending sequence of left ideals (or right ideals) has only a finite number of distinct members. { ,nō·ə'thir·ē·ən'riŋ }

no-fines concrete [MATER] Concrete made without sand and therefore containing a high proportion of communicating pores which provide thermal insulation and drainage. { 'nō ¦fīnz kän'krēt }

nog [MIN ENG] **1.** Roof support for stopes, formed of rectangular piles of logs squared at the ends and filled with waste rock. **2.** A wood block wedged tightly into the cut in a coal seam after the coal cutter has passed; it forms a temporary support. { näg }

nogalamycin [MICROBIO] $C_{39}H_{45}NO_{16}$ An antineoplastic antibiotic produced by *Streptomyces nogalaster*. { nō,gal·ə'mīs·ən }

no-go gage [ENG] A limit gage designed not to fit a part being tested; usually employed with a go gage to set the acceptable maximum and minimum dimension limits of the part. { 'nō ¦gō ,gāj }

no-hair theorems [RELAT] Popular name for general relativistic theorems proving that black holes are uniquely described by their mass, charge, and angular momentum. { 'nō ¦hār ¦thir·əmz }

noil [TEXT] Short-staple, wool fiber combings used in the manufacture of worsted fabrics. { nȯil }

noise [ACOUS] Sound which is unwanted, either because of its effect on humans, its effect on fatigue or malfunction of physical equipment, or its interference with the perception or detection of other sounds. [COMMUN] Unwanted electrical signal disturbances. [ELEC] Interfering and unwanted currents or voltages in an electrical device or system. [PHYS] Nonperiodic behavior of a system that results from the presence

NOISE-POWER MEASUREMENT

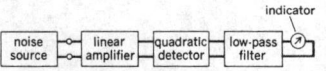

Diagram of setup for noise-power measurement.

of random driving forces, such as thermal agitation, as opposed to chaotic behavior. [SPECT] Random fluctuations of electronic signals appearing in a recorded spectrum. { nȯiz }

noise analysis [PHYS] Determination of the frequency components that make up a particular noise being studied. { ˈnȯiz əˌnal·ə·səs }

noise analyzer [ELECTR] A device used for noise analysis. { ˈnȯiz ˌan·əˌlīz·ər }

noise-canceling microphone *See* close-talking microphone. { ˈnȯiz ˌkans·liŋ ˈmī·krəˌfōn }

noise control [ACOUS] The process of obtaining an acceptable noise environment for a particular observation point or receiver, involving control of the noise source, transmission path, or receiver, or all three. { ˈnȯiz kənˌtrōl }

noise digit [COMPUT SCI] A digit, usually 0, inserted into the rightmost position of the mantissa of a floating point number during a left-shift operation associated with normalization. Also known as noisy digit. { ˈnȯiz ˌdij·ət }

noise distortion [COMMUN] Noise on a communications facility which exceeds standards governing acceptable levels and which negatively affects the signal. { ˈnȯiz diˌstȯr·shən }

noise equivalent temperature difference [THERMO] The change in equivalent blackbody temperature that corresponds to a change in radiance which will produce a signal-to-noise ratio of 1 in an infrared imaging device. Abbreviated NETD. { ˈnȯiz iˌkwiv·ə·lənt ˈtem·prə·chər ˈdif·rəns }

noise factor [ELECTR] The ratio of the total noise power per unit bandwidth at the output of a system to the portion of the noise power that is due to the input termination, at the standard noise temperature of 290 K. Also known as noise figure. { ˈnȯiz ˌfak·tər }

noise figure *See* noise factor. { ˈnȯiz ˌfig·yər }

noise filter [ELECTR] **1.** A filter that is inserted in an alternating-current power line to block noise interference that would otherwise travel through the line in either direction and affect the operation of receivers. **2.** A filter used in a radio receiver to reduce noise, usually an auxiliary low-pass filter which can be switched in or out of the audio system. { ˈnȯiz ˌfil·tər }

noise generator [ELECTR] A device which produces (usually random) electrical noise, for use in tests of the response of electrical systems to noise, and in measurements of noise intensity. Also known as noise source. { ˈnȯiz ˌjen·əˌrād·ər }

noise grade [COMMUN] Number which defines the relative noise at a particular location with respect to other locations throughout the world. { ˈnȯiz ˌgrād }

noise immission level [ACOUS] A measure of the cumulative noise energy to which an individual is exposed over time; equal to the average noise level to which the person has been exposed, in decibels, plus 10 times the logarithm of the number of years for which the individual is exposed. { ˈnȯiz iˈmish·ən ˌlev·əl }

noise improvement factor [COMMUN] In pulse modulation, the receiver output signal-to-noise ratio divided by the receiver input signal-to-noise ratio. Abbreviated NIF. Also known as improvement factor; signal-to-noise improvement factor. { ˈnȯiz imˈprüv·mənt ˌfak·tər }

noise jammer [ELECTR] **1.** An electronic jammer that emits a carrier modulated with recordings or synthetic reproductions of natural atmospheric noise; the radio-frequency carrier may be suppressed; used to discourage the enemy by simulating naturally adverse communications conditions. **2.** During World War II, a powerful transmitter modulated with white noise tuned to the approximate frequency of an enemy transmitter and used to obscure intelligible output at the receiver. { ˈnȯiz ˌjam·ər }

noise killer [ELECTR] **1.** Device installed in a circuit to reduce its interference to other circuits. **2.** *See* noise suicide circuit. { ˈnȯiz ˌkil·ər }

noiseless channel [COMMUN] In information theory, a communications channel in which the effects of random influences are negligible, and there is essentially no random error. { ˈnȯiz·ləs ˈchan·əl }

noise level [PHYS] The intensity of unwanted sound, or the magnitude of unwanted currents or voltages, averaged over a specified frequency range and time interval, and weighted with frequency in a specified manner; usually expressed in decibels relative to a specified reference. { ˈnȯiz ˌlev·əl }

noise limiter [ELECTR] A limiter circuit that cuts off all noise peaks that are stronger than the highest peak in the desired signal being received, thereby reducing the effects of atmospheric or human-produced interference. Also known as noise silencer; noise suppressor. { ˈnȯiz ˌlim·əd·ər }

noise measurement [ACOUS] The process of quantitatively determining one or more properties of acoustic noise. [ELECTR] Any of a wide range of measurements of random and nonrandom electrical noise, but usually noise-power measurement. { ˈnȯiz ˌmezh·ər·mənt }

noise-metallic [ELECTR] In telephone communications, weighted noise current in a metallic circuit at a given point when the circuit is terminated at that point in the nominal characteristic impedance of the circuit. { ˈnȯiz məˈtal·ik }

noise-modulated jamming [ELECTR] Random electronic noise that appears at the radar receiver as background noise and tends to mask the desired radar echo or radio signal. { ˈnȯiz ˌmäj·əˌlād·əd ˈjam·iŋ }

noise pollution [ACOUS] Excessive noise in the human environment. { ˈnȯiz pəˌlü·shən }

noise-power measurement [ELECTR] Measurement of the power carried by electrical noise averaged over some brief interval of time, usually by amplifying noise from the source in a linear amplifier and then using a quadratic detector followed by a low-pass filter and an indicating device. { ˈnȯiz ˌpau̇·ər ˌmezh·ər·mənt }

noise radial [ENG] The brightening of all range points on a particular plan position indicator bearing on a radar screen caused by noise reception from the indicated direction. { ˈnȯiz ˈrād·ē·əl }

noise rating number [ACOUS] The perceived noise level of the noise that can be tolerated under specified conditions; for example, the noise rating number of a bedroom is 25, that of a workshop is 65. { ˈnȯiz ˈrād·iŋ ˌnəm·bər }

noise-reducing antenna system [ELECTROMAG] Receiving antenna system so designed that only the antenna proper can pick up signals; it is placed high enough to be out of the noise-interference zone, and is connected to the receiver with a shielded cable or twisted transmission line that is incapable of picking up signals. { ˈnȯiz riˈdüs·iŋ anˈten·ə ˌsis·təm }

noise reduction [ENG ACOUS] A process whereby the average transmission of the sound track of a motion picture print, averaged across the track, is decreased for signals of low level; since background noise introduced by the sound track is less at low transmission, this process reduces noise during soft passages. { ˈnȯiz riˌdək·shən }

noise reduction coefficient [ACOUS] The average over the logarithm of frequency, in the frequency range from 256 to 2048 hertz inclusive, of the sound absorption coefficient of a material. { ˈnȯiz riˌdək·shən ˌkō·iˈfish·ənt }

noise reduction rating [ACOUS] A common method for expressing values of noise reduction or attenuation provided by different types of hearing protectors; values range from 0 to approximately 30, with higher values indicating greater amounts of noise reduction. Abbreviated NRR. { ˈnȯiz riˌdək·shən ˌrād·iŋ }

noise silencer *See* noise limiter. { ˈnȯiz ˈsī·lən·sər }

noise source *See* noise generator. { ˈnȯiz ˌsȯrs }

noise suicide circuit [ELECTR] A circuit which reduces the gain of an amplifier for a short period whenever a sufficiently large noise pulse is received. Also known as noise killer. { ˈnȯiz ˈsü·əˌsīd ˌsər·kət }

noise suppression [ELECTR] Any method of reducing or eliminating the effects of undesirable electrical disturbances, as in frequency modulation whenever the signal carrier level is greater than the noise level. { ˈnȯiz səˌpresh·ən }

noise suppressor [ELECTR] **1.** A circuit that blocks the audio-frequency amplifier of a radio receiver automatically when no carrier is being received, to eliminate background noise. Also known as squelch circuit. **2.** A circuit that reduces record surface noise when playing phonograph records, generally by means of a filter that blocks out the higher frequencies where such noise predominates. **3.** *See* noise limiter. { ˈnȯiz səˌpres·ər }

noise temperature [ELEC] The temperature at which the thermal noise power of a passive system per unit bandwidth would be equal to the actual noise at the actual terminals; the standard reference temperature for noise measurements is 290 K. { ˈnȯiz ˌtem·prə·chər }

noise testing [ELECTR] The measurement of the power dissipated in a resistance termination of given value joined to one end of a telephone or telegraph circuit when no test power is applied to the circuit. { 'nȯiz ,test·iŋ }

noise tube [ELECTR] A gas tube used as a source of white noise. { 'nȯiz ,tüb }

noise-type flowmeter [ENG] A flowmeter that measures the noise generated in a selected frequency band. { 'nȯiz ¦tīp 'flō ,mēd·ər }

noise weighting [ELECTR] Use of an electrical network to obtain a weighted average over frequency of the noise power, which is representative of the relative disturbing effects of noise in a communications system at various frequencies. { 'nȯiz ,wād·iŋ }

noisy channel [COMMUN] In information theory, a communications channel in which the effects of random influences cannot be dismissed. { 'nȯiz·ē 'chan·əl }

noisy digit See noise digit. { 'nȯiz·ē 'dij·ət }

noisy mode [COMPUT SCI] A floating-point arithmetic procedure associated with normalization in which "1" bits, rather than "0" bits, are introduced in the low-order bit position during the left shift. { 'nȯiz·ē ,mōd }

no-load current [ELEC] The current which flows in a network when the output is open-circuited. { 'nō ¦lōd 'kə·rənt }

no-load loss [ELEC] The power loss of a device that is operated at rated voltage and frequency but is not supplying power to a load. { 'nō ¦lōd 'lȯs }

no-load voltage See open-circuit voltage. { 'nō ¦lōd 'vōl·tij }

noma [MED] Spreading gangrene beginning in a mucous membrane; considered to be a malignant form of infection by fusospirochetal organisms. Also known as gangrenous stomatitis. { 'nō·mə }

NOMAD See Navy Oceanographic and Meteorological Automatic Device. { 'nō,mad }

Nomarski microscope [OPTICS] A type of interference microscope that is used to study reflecting specimens, such as metallic surfaces or metallized replicas of surfaces, and that gives a true relief image uncomplicated by variations in refractive index. { nə'mär·skē 'mī·krə,skōp }

nomenclature [SCI TECH] A systematic arrangement of the distinctive names employed in any science. { 'nō·mən,klā·chər }

nomenclature plate [ORD] Plate, usually made of metal, which is conspicuously mounted on military equipment, giving model letters, symbols, and numbers, together with other pertinent information. { 'nō·mən,klā·chər ,plāt }

nomen dubium [SYST] A proposed taxonomic name invalid because it is not accompanied by a definition or description of the taxon to which it applies. { 'nō·mən 'dü·bē·əm }

nomen nudum [SYST] A proposed taxonomic name invalid because the accompanying definition or description of the taxon cannot be interpreted satisfactorily. { 'nō·mən 'nü·dəm }

nominal band [COMMUN] Frequency band of a facsimile-signal wave equal in width to that between zero frequency and maximum modulating frequency; the frequency band occupied in the transmitting medium will, in general, be greater than the nominal band. { 'näm·ə·nəl ,band }

nominal bandwidth [COMMUN] The interval between the assigned frequency limits of a channel. [ENG] The difference between the nominal upper and lower cutoff frequencies of an acoustic or electric filter. { 'näm·ə·nəl 'band,width }

nominal decline rate [PETRO ENG] The negative slope of the curve representing the hydrocarbon production rate versus time for an oil gas reservoir. { 'näm·ə·nəl di'klīn ,rāt }

nominal diameter [GEOL] The diameter computed for a hypothetical sphere which would have the same volume as the calculated volume for a specific sedimentary particle. Also known as equivalent diameter. { 'näm·ə·nəl dī'am·əd·ər }

nominal impedance [ELEC] Impedance of a circuit under conditions at which it was designed to operate; normally specified at center of operating frequency range. { 'näm·ə·nəl im'pēd·əns }

nominal line pitch See nominal line width. { 'näm·ə·nəl 'līn ,pich }

nominal line width [COMMUN] 1. In television, the reciprocal of the number of lines per unit length in the direction of line progression. 2. In facsimile transmission, the average separation between centers of adjacent scanning or recording lines. Also known as nominal line pitch. { 'näm·ə·nəl 'līn ,width }

nominal pass-band center frequency [ENG] The geometric mean of the nominal upper and lower cutoff frequencies of an acoustic or electric filter. { 'näm·ə·nəl 'pas ,band ¦sen·tər 'frē·kwən·sē }

nominal scale measurement [STAT] A method for sorting objects into categories according to some distinguishing characteristic and attaching a name or label to each category; considered the weakest type of measurement. { 'näm·ə·nəl ¦skāl 'mezh·ər·mənt }

nominal size [DES ENG] Size used for purposes of general identification; the actual size of a part will be approximately the same as the nominal size but need not be exactly the same; for example, a rod may be referred to as $^1/_4$ inch, although the actual dimension on the drawing is 0.2495 inch, and in this case 1/4 inch is the nominal size. { 'näm·ə·nəl 'sīz }

nominal stress [MET] The stress calculated by simple elasticity theory, ignoring stress raisers and plastic flow; in tensile testing of a notched specimen, the load applied at the notch divided by the initial cross-sectional area at the notch. { 'näm·ə·nəl 'stres }

nominal value [ELEC] The value of some property (such as resistance, capacitance, or impedance) of a device at which it is supposed to operate, under normal conditions, as opposed to actual value. { 'näm·ə·nəl 'val·yü }

nomogram See nomograph. { 'näm·ə,gram }

nomograph [MATH] A chart which represents an equation containing three variables by means of three scales so that a straight line cuts the three scales in values of the three variables satisfying the equation. Also known as abac; alignment chart; nomogram. { 'näm·ə graf }

nonabelian gauge theory [PARTIC PHYS] A gauge theory in which the gauge transformations can be represented by a Lie group whose members do not commute. { ,nän·ə¦bēl·yən ¦gāj ,thē·ə·rē }

nonabelian quantum Hall state [CRYO] A quantum Hall state that is not amenable to a description in terms of electron dancing steps and that cannot be characterized by a symmetric matrix and a charge vector. { nän·ə¦bēl·yən ¦kwän·təm 'hȯl ,stāt }

nonacosane [ORG CHEM] $C_{29}H_{60}$ Colorless hydrocarbon, melting at 63°C; found in beeswax and the fat of cabbage leaves. { ,nō·nə'kō,sān }

nonacoustic coupler [ELECTR] A type of modem that is built into a microcomputer or terminal and connects it directly to a telephone line. { nän·ə'kü·stik 'kəp·lər }

nonadecane [ORG CHEM] $CH_3(CH_2)_{17}CH_3$ Flammable crystals, soluble in ether and alcohol, insoluble in water; melts at 32°C; used as a chemical intermediate. { ,nō·nə'de,kān }

nonadiabatic See diabatic. { ,nän'ad·ə·bē'ak·ik }

nonagon [MATH] A nine-sided polygon. Also known as enneagon. { nän·ə,gän }

nonahedron [MATH] A polyhedron with nine faces. { ,nō·nə'hē·drən }

nonambiguity [COMMUN] The property of a code in which any character can be recognized uniquely without reference to preceding characters or the spatial position of a character. { ¦nän,am·bə'gyü·əd·ē }

nonanal [ORG CHEM] $C_8H_{17}CHO$ A colorless liquid with an orange rose odor; used in perfumes and for flavoring. { 'nän·ə,näl }

nonane [ORG CHEM] $CH_3(CH_2)_7CH_3$ Flammable, colorless liquid, boiling at 151°C; soluble in alcohol, insoluble in water; used as a chemical intermediate. Also known as nonyl hydride. { 'nō,nän

non-A, non-B hepatitis [MED] A type of viral hepatitis that is most common among people who have received transfused blood and whose serologic tests show no evidence of hepatitis A, hepatitis B, or other types of virus such as Epstein-Barr. { ¦nän ¦ā ,nän ¦bē ,hep·ə'tīd·əs }

nonanticipatory system See causal system. { ¦nän·an'tis·ə·pə,tór·ē ,sis·təm }

nonaqueous [CHEM] Pertaining to a liquid or solution containing no water. { ¦nän'ā·kwē·əs }

nonarithmetic shift See cyclic shift. { ¦nän,a'rith'med·ik 'shift }

nonasphaltic road oil [MATER] Any of the nonhardening petroleum distillates or residual oils used to lay road dust; they

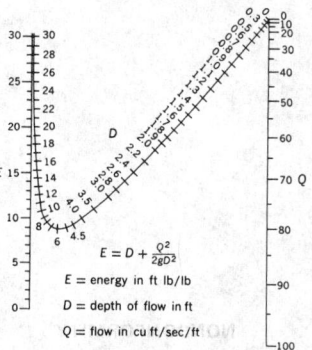

NOMOGRAPH

$$E = D + \frac{Q^2}{2gD^2}$$

E = energy in ft lb/lb
D = depth of flow in ft
Q = flow in cu ft/sec/ft

Nomograph for energy content of a rectangular channel with uniform flow.

have viscosities low enough to be applied without prior heating. { ¦nän·as'fȯl·tik 'rōd ‚ȯil }

nonassociated-gas reservoir [PETRO ENG] Formation in which gaseous hydrocarbons exist as a free phase in a reservoir that is not commercially productive of crude oil. { ¦nän·ə'sō·shē‚ād·əd ¦gas 'rez·əv‚wär }

nonassociative algebra [MATH] A generalization of the concept of an algebra; it is a nonassociative ring *R* which is a vector space over a field *F* satisfying *a*(*xy*) = (*ax*)*y* = *x*(*ay*) for all *a* in *F* and *x* and *y* in *R*. { ‚nän·ə¦sō·shəd·əv 'al·jə·brə }

nonassociative ring [MATH] A generalization of the concept of a ring; it is an algebraic system with two binary operations called addition and multiplication such that the system is a commutative group relative to addition, and multiplication is distributive with respect to addition, but multiplication is not assumed to be associative. { ‚nän·ə¦sō·shəd·əv 'riŋ }

nonatomic Boolean algebra [MATH] A Boolean algebra in which there is no element *x* with the property that if *y*·*x*=*y* for some *y*, then *y*=0. { ‚nän·ə¦täm·ik ¦bü·lē·ən 'al·jə·brə }

nonatomic measure space [MATH] A measure space in which no point has positive measure. { ¦nän·ə'täm·ik ¦mezh·ər ‚spās }

nonbanded coal [GEOL] Coal without lustrous bands, composed mainly of clarain or durain without nitrain. { 'nän‚ban·dəd 'kōl }

nonbearing wall [CIV ENG] A wall that bears no vertical weight other than its own. { 'nän‚ber·iŋ 'wȯl }

nonbenzenoid aromatic compound [ORG CHEM] A compound exhibiting aromatic character but not containing a benzene nucleus, or having one or more rings in a fused ring system that are not benzene rings. { ¦nän'ben·zə‚nȯid ¦ar·ə¦mad·ik 'käm‚paủnd }

nonblackbody [THERMO] A body that reflects some fraction of the radiation incident upon it; all real bodies are of this nature. { nän'blak‚bäd·ē }

nonblocking access [COMMUN] Connection of the incoming line or trunk made within the switching center at all times, provided that the required outgoing line or trunk is not busy. { 'nän‚bläk·iŋ 'ak‚ses }

nonbonded distance [PHYS CHEM] The distance between atoms in a molecule that are not bonded to each other. { ¦nän‚bän·dəd 'dis·təns }

non-bore-safe [ORD] Pertaining to a fuse or booster that does not include a safety device to prevent the explosion of the main charge of a projectile prematurely, while it is still in the bore of the gun. { ¦nän ¦bȯr ‚sāf }

noncaking coal [GEOL] Hard or dull coal that does not cake when heated. Also known as free-burning coal. { 'nän‚kāk·iŋ 'kōl }

Noncalcic Brown soil [GEOL] A great soil group having a slightly acidic, light-pink or reddish-brown A horizon and a light-brown or dull-red B horizon, and developed under a mixture of grass and forest vegetation in a subhumid climate. Also known as Shantung soil. { ¦nän'kal·sik 'braủn ‚sȯil }

noncapillary porosity [GEOL] The property of a volume of large interstices in a rock or soil that do not hold water by capillarity. { ¦nän'kap·ə‚ler·ē pə'räs·əd·ē }

noncarbon oil [MATER] **1.** Oil in which little or no free carbon is suspended. **2.** Oil which, upon decomposition, contributes little so-called carbon deposits. { ¦nän¦kär·bən 'ȯil }

noncentral chi-square distribution [STAT] The distribution of the sum of squares of independent normal random variables, each with unit variance and nonzero mean; used to determine the power function of the chi-square test. { ‚nän ¦sen·trəl ¦kī ‚skwer ‚dis·trə'byü·shən }

noncentral distribution [STAT] A distribution of random variables which is not normal. { 'nän‚sen·trəl ‚di·strə'byü·shən }

noncentral F distribution [STAT] The distribution of the ratio of two independent random variables, one with a noncentral chi-square distribution and one with a central chi-square distribution; used to determine the power of the *F* test in the analysis of variance. { ‚nän¦sen·trəl¦ef ‚dis·trə'byü·shən }

noncentral force [PHYS] A force between two particles that is not directed along the line connecting them; for example, a tensor force between two nucleons. { ‚nän‚sen·trəl 'fȯrs }

noncentral quadric [MATH] A quadric surface that does not have a point about which the surface is symmetrical; namely,

an elliptic or hyperbolic paraboloid, or a quadric cylinder. { 'nän‚sen·trəl 'kwä‚drik }

noncentral t distribution [STAT] A particular case of a noncentral *F* distribution; used to test the power of the *t* test. { ‚nän¦sen·trəl ¦tē ‚dis·trə'byü·shən }

nonclastic [GEOL] Of the texture of a sediment or sedimentary rock, formed chemically or organically and showing no evidence of a derivation from preexisting rock or mechanical deposition. Also known as nonmechanical. { ¦nän'kla‚stik }

noncoherent scattering [ATOM PHYS] The absorption of a photon and its reemission at a different energy (in the observer's frame of reference) by scattering atoms. { ¦nän·kō'hir·ənt 'skad·ə·riŋ }

noncohesive See cohesionless. { ‚nän·kō'hē·siv }

noncoincident demand [ELEC] The sum of the peak demands of all the utilities in a specified region, regardless of the times at which they occurred. { ¦nän·kō'in·sə·dənt di'mand }

noncombat vehicle [ORD] An unofficial term used to distinguish a vehicle not classed as a combat vehicle; that is, the vehicle lacks means for engaging in combat. { ¦nän'käm‚bat 'vē·ə·kəl }

noncommunicating hydrocephaly See obstructive hydrocephaly. { ¦nän·kə'myü·nə‚kād·iŋ ‚hī·drō'sef·ə·lē }

noncompetitive inhibition [BIOCHEM] Enzyme inhibition in which the inhibitor can combine with either the free enzyme or the enzyme-substrate complex so that the inhibitor does not compete with the substrate for the enzyme. { ¦nän·kəm'ped·əd·iv ‚in·ə'bish·ən }

noncomposite color picture signal [COMMUN] The signal in color television transmission that represents complete color picture information but excludes the line- and field-synchronizing signals. { ¦nän·kəm'päz·ət ¦kəl·ər 'pik·chər ‚sig·nəl }

noncompulsory reporting points [NAV] In air operations, those reports established by air-traffic control without the requirement for rule-making action. { ¦nän·kəm'pəl·sə·rē ri'pȯrd·iŋ ‚pȯins }

noncondensable gas [MATER] A gas from chemical or petroleum processing units (such as distillation columns or steam ejectors) that is not easily condensed by cooling; consists mostly of nitrogen, light hydrocarbons, carbon dioxide, or other gaseous materials. { ¦nän·kən'den·sə·bəl 'gas }

nonconformity [GEOL] A type of unconformity in which rocks below the surface of unconformity are either igneous or metamorphic. { ¦nän·kən'fȯr·məd·ē }

noncongression [CYTOL] The failure of pairing chromosomes, in certain stages of mitosis and meiosis, to orient in an orderly arrangement on the spindle equator. { ‚nän·kən'gre·shən }

nonconjugative plasmid [GEN] Any plasmid that prevents conjugation of its bacterial host. { nän¦kän·jə‚gād·iv 'plaz‚mid }

nonconservative element [OCEANOGR] An element in sea water which is so uncommon that a large proportion of its total composition enters and leaves the particulate phase. { ¦nän·kən'sər·vəd·iv 'el·ə·mənt }

nonconservative scattering [PHYS] Scattering that is accompanied by absorption. { ¦nän·kən'sər·vəd·iv 'skad·ə·riŋ }

nonconsumable electrode [MET] An electrode, such as of carbon or tungsten, that is not consumed during a welding or melting operation. { ¦nän·kən'sü·mə·bəl i'lek‚trōd }

noncontacting piston See choke piston. { ¦nän'kän‚tak·tiŋ 'pis·tən }

noncontacting plunger See choke piston. { ¦nän'kän‚tak·tiŋ 'plən·jər }

noncontact sensor See proximity sensor. { ¦nän'kän‚takt 'sen·sər }

noncontact thermometer See radiation pyrometer. { ¦nän 'kän‚takt thər'mäm·əd·ər }

noncontributing area [HYD] An area with closed drainage. { ¦nän·kən'trib·yəd·iŋ 'er·ē·ə }

noncoring bit [ENG] A general type of bit made in many shapes which does not produce a core and with which all the rock cut in a borehole is ejected as sludge; used mostly for blasthole drilling and in the unmineralized zones in a borehole where a core sample is not wanted. Also known as borehole bit; plug bit. { ¦nän‚kȯr·iŋ 'bit }

noncorrosive flux [MET] A soldering flux composed of

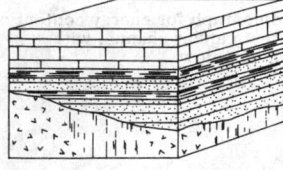

NONCONFORMITY

Drawing of a nonconformity type of unconformity showing the lack of stratification in the rocks below the break.

rosin or of rosin in a volatile solvent; the residue is nonhygroscopic, noncorrosive, and nonconducting; suitable for soldering electronic components. Also known as activated rosin flux. { ¦nän·kə¦rō·siv 'fləks }

noncovalent bonds [CHEM] Weak chemical bonds that are electrostatic and hydrophobic in nature, for example, hydrogen bonds; important in determining complex biological structures. { ¦nän·kō¦vāl·ənt 'bändz }

noncritical region [STAT] In testing hypotheses, the set of values leading to acceptance of the null hypothesis. { ‚nän ¦krit·ə·kəl 'rē·jən }

noncrossing rule [PHYS CHEM] The rule that when the potential energies of two electronic states of a diatomic molecule are plotted as a function of distance between the nuclei, the resulting curves do not cross, unless the states have different symmetry. { ¦nän'krós·iŋ 'rül }

noncyclic element [IND ENG] An element of an operation or process that does not occur in every cycle but has a frequency of occurrence that is specified by the method. { ¦nän‚sī·klik 'el·ə·mənt }

noncyclic terrace [GEOL] One of a series of terraces representing previous valley floors formed during periods when continued valley deepening accompanied lateral erosion. { ¦nän'sī·klik 'ter·əs }

nondeclarative memory See implicit memory. { ‚nän·di‚klar ·əd·iv 'mem·rē }

nondeforming steel [MET] A group of alloy steels which do not easily deform when heat-treated. Also known as nonshrinking steel. { ¦nän·di'fór·miŋ 'stēl }

nondegenerate amplifier [ELECTR] Parametric amplifier that is characterized by a pumping frequency considerably higher than twice the signal frequency; the output is taken at the signal input frequency; the amplifier exhibits negative impedance characteristics, indicative of infinite gain, and is therefore capable of oscillation. { ¦nän·di'jen·ə·rət 'am·plə‚fī·ər }

nondegenerate plane [MATH] In projective geometry, a plane in which to every line L there are at least two distinct points that do not lie on L, and to every point p there are at least two distinct lines which do not pass through p. { ¦nän· di'jen·ə·rət 'plān }

nondegenerative basic feasible solution [COMPUT SCI] In linear programming, a basic feasible solution with exactly m positive variables x_i, where m is the number of constraint equations. { ¦nän·di'jen·rəd·iv 'bā·sik 'fēz·ə·bəl sə'lü·shən }

nondelay fuse [ORD] Fuse that functions as a result of inertia of the firing pin (or primer) as the missile is retarded during penetration of the target; the inertia causes the firing pin to strike the primer (or the primer, the firing pin), initiating fuse action. { ¦nän·di‚lā 'fyüz }

nondeletable message [COMPUT SCI] A message that appears on a computer display which can be removed only by entering a specific command. { ¦nän·di'lēd·ə·bəl 'mes·ij }

nondenumerable set [MATH] A set that cannot be put into one-to-one correspondence with the positive integers or any subset of the positive integers. { ‚nän·di¦nüm·rə·bəl 'set }

nondepositional unconformity See paraconformity. { ¦nän ‚dep·ə'zish·ən·əl ‚ən·kən'fór·məd·ē }

nondestructive breakdown [ELECTR] Breakdown of the barrier between the gate and channel of a field-effect transistor without causing failure of the device; in a junction field-effect transistor, avalanche breakdown occurs at the pn junction. { ¦nän·di'strək·div 'brāk‚daún }

nondestructive evaluation [IND ENG] A technique for probing and sensing material structure and properties without causing damage (as opposed to revealing flaws and defects). { ‚nän·di‚strək·tiv i‚val·yə'wā·shən }

nondestructive read [COMPUT SCI] A reading process that does not erase the data in memory; the term sometimes includes a destructive read immediately followed by a restorative writeback. Also known as nondestructive readout (NDRO). { ¦nän·di'strək·div 'rēd }

nondestructive readout See nondestructive read. { ¦nän· di'strək·div 'rēd‚aút }

nondestructive testing [ENG] A technique for revealing flaws and defects in a material or device without damaging or destroying the test sample; includes use of x-rays, ultrasonics, radiography, and magnetic flux. { ¦nän·di'strək·div 'test·iŋ }

nondeterministic [SCI TECH] Unpredictable in terms of

observable antecedents and known laws; this is a relative term pertaining to a given state of knowledge but not necessarily implying ultimate unpredictability. { ¦nän·di‚tər·mə'nis·dik }

nondeviated absorption [PHYS] Absorption that occurs without any appreciable slowing up of waves. { ¦nän'dē· vē‚ād·əd əb'sórp·shən }

nondifferentiable programming [MATH] The branch of nonlinear programming which does not require the objective and constraint functions to be differentiable. { ¦nän‚dif·ə'ren· chə·bəl 'prō‚gram·iŋ }

nondimensional parameter See dimensionless number. { ¦nän·di'men·chən·əl pə'ram·əd·ər }

nondirectional See omnidirectional. { ¦nän·di'rek·shən·əl }

nondirectional antenna See omnidirectional antenna. { ¦nän· di'rek·shən·əl an'ten·ə }

nondirectional beacon [NAV] A beacon that provides navigational guidance over a 360° azimuth. { ¦nän·di'rek·shən·əl 'bē·kən }

nondirective therapy [PSYCH] Type of psychotherapy in which the patient is in the dominant position and is given complete freedom to express herself or himself. { ¦nän·di'rek· div 'ther·ə·pē }

nondisjunction [GEN] Failure of homologous chromosomes to separate symmetrically during cell division, with both ending up in the same daughter cell instead of one in each daughter cell. { ¦nän·dis'jəŋk·shən }

nondisjunction mosaic [GEN] A population of cells with different chromosome numbers produced when one chromosome is lost during mitosis or when both members of a pair of chromosomes are included in the same daughter nucleus; can occur during embryogenesis or adulthood. { ¦nän· dis'jəŋk·shən mō'zā·ik }

nondissipative muffler See reactive muffler. { ¦nän'dis· ə‚pād·iv 'məf·lər }

nondissipative stub [ELECTROMAG] Nondissipative length of waveguide or transmission line. { ¦nän'dis·ə‚pād·iv 'stəb }

nondivergent flow [OCEANOGR] Fluid flow in which the divergence of the ocean current field is zero. { ¦nän·di‚vər· jənt 'flō }

nondurable goods [ENG] Products that are serviceable for a comparatively short time or are consumed or destroyed in a single usage. { ‚nän¦dúr·ə·bəl 'gúdz }

nonelectromagnetic radiation [PHYS] A stream of particles other than photons, such as neutrinos, electrons, positrons, protons, neutrons, or alpha particles (all of which also have wave properties, such as diffraction), or of waves other than electromagnetic waves, such as sound waves (which also have particle properties, for example, as phonons in the case of sound). { ‚nän·i‚lek·trə·mag¦ned·ik ‚rād·ē'ā·shən }

nonene See 1-nonylene. { 'nō‚nēn }

nonequilibrium thermodynamics [THERMO] A quantitative treatment of irreversible processes and of rates at which they occur. Also known as irreversible thermodynamics. { ¦nän‚ē·kwə'lib·rē·əm ‚thər·mō·cī'nam·iks }

nonerasable storage [COMPUT SCI] A device that permits a nondestructive read, such as punched cards, electrically conductive sheets, or paper tape. { ¦nän·i'rās·ə·bəl 'stór·ij }

nonesite [PETR] A porphyritic basalt composed of enstatite, labradorite, and augite phenocrysts in a groundmass of plagioclase and augite. { 'nän·ə‚sīt }

nonessential amino acid [BIOCHEM] An amino acid which can be synthesized by an organism and thus need not be supplied in the diet. { ¦nän·i'sen·chəl ə'mē·nō 'as·əd }

noneuclidean geometry [MATH] A geometry in which one or more of the axioms of euclidean geometry are modified or discarded. { ¦nän·yü'klid·ē·ən jē äm·ə·trē }

nonexecutable statement [COMPUT SCI] A statement in a higher-level programming language which cannot be related to the instructions in the machine language program ultimately produced, but which provides the compiler with essential information from which it may determine the allocation of storage and other organizational characteristics of the final program. { ¦nän‚ek·sə'kyüd·ə·bəl 'stāt·mənt }

nonexpansive mapping [MATH] A function f from a metric space to itself such that, for any two elements in the space, a and b, the distance between $f(a)$ and $f(b)$ is not greater than the distance between a and b. { ‚nän·ik‚span·siv 'map·iŋ }

nonexpendable [ENG] Pertaining to a supply item or piece of equipment that is not consumed, and does not lose its identity, in use, as a weapon, vehicle, machine, tool, piece of furniture, or instrument.　{ ¦nän·ik'spen·də·bəl }

nonfaradaic path [PHYS CHEM] One of the two available paths for transfer of energy across an electrolyte-metal interface, in which energy is carried by capacitive transfer, that is, by charging and discharging the double-layer capacitance.　{ ¦nän,far·ə'dā·ik 'path }

nonfatal error [COMPUT SCI] An error in a computer program which does not result in termination of execution, but which causes the processor to invent an interpretation, issue a warning, and continue processing.　{ 'nän,fād·əl 'er·ər }

nonfeasible method See goal coordination method.　{ ¦nän'fē·zə·bəl 'meth·əd }

nonferrous metal [MET] Any metal other than iron and its alloys.　{ ¦nän'fer·əs 'med·əl }

nonferrous metallurgy [MET] A branch of metallurgy that deals with metals other than iron and iron-base alloys.　{ ¦nän'fer·əs 'med·əl,ər·jē }

nonflowing well [ENG] A well that yields water at the land surface only by means of a pump or other lifting device.　{ 'nän,flō·iŋ 'wel }

nonfluent aphasia [PSYCH] Aphasia characterized by effortful articulation and loss of syntax, but relatively well-preserved auditory comprehension; generally the result of injury to the speech zone anterior to the Rolandic fissure. (Broca's area). Also known as Broca's aphasia.　{ 'nän,flü·ənt ə'fā·zhə }

nonfragmenting projectile [ORD] Projectile for antiaircraft gun practice, containing a smoke-producing substance, available in various colors, which makes it possible to observe the burst without a close burst destroying the target.　{ ¦nän'frag·ment·iŋ prə'jek·təl }

nonfreezing explosive [MATER] An explosive to which 15–20% of nitroethylene glycol has been added.　{ 'nän,frēz·iŋ ik'splō·siv }

nonfrontal squall line See prefrontal squall line.　{ 'nän,frənt·əl 'skwȯl ,līn }

nonfunctional packages software [COMPUT SCI] General-purpose software which permits the user to handle her or his particular applications requirements with little or no additional program or systems design work, or to perform certain specialized computational functions.　{ nän'fəŋk·shən·əl 'pak·ij·əz 'sȯft,wer }

nongonococcal urethritis [MED] Human urethral inflammation not associated with common bacterial pathogens; thought to be caused by bacteria of the Bedsonia group. Abbreviated NGU.　{ ¦nän,gän·ə'kä·kəl ,yùr·ə'thrīd·əs }

nongraded [GEOL] Pertaining to a soil or an unconsolidated sediment consisting of particles of essentially the same size.　{ ¦nän'grād·əd }

nongranular leukocyte [HISTOL] A white blood cell, such as a lymphocyte or monocyte, with clear homogeneous cytoplasm.　{ ¦nän'gran·yə·lər 'lü·kə,sīt }

nongraphic character [COMPUT SCI] A set of signals that, when sent to a printer, results in a control action, such as carriage return, line feed, or tab, rather than the generation of a printed character.　{ ¦nän'graf·ik 'kar·ik·tər }

nonhemolysis See gamma hemolysis.　{ ,nän·hē'mäl·ə·səs }

nonhistone protein [BIOCHEM] A class of acidic proteins in the cell nucleus associated with deoxyribonucleic acid.　{ nän¦hi,stōn 'prō,tēn }

nonholonomic constraint [MATH] One of a nonintegrable set of differential equations which describe the restrictions on the motion of a system.　{ ¦nän,häl·ə'näm·ik kən'stränt }

nonholonomic system [MECH] A system of particles which is subjected to constraints of such a nature that the system cannot be described by independent coordinates; examples are a rolling hoop, or an ice skate which must point along its path.　{ ¦nän,häl·ə'näm·ik 'sis·təm }

nonhoming [CONT SYS] Not returning to the starting or home position, as when the wipers of a stepping relay remain at the last-used set of contacts instead of returning to their home position.　{ ¦nän'hōm·iŋ }

nonhoming tuning system [ELECTR] Motor-driven automatic tuning system in which the motor starts up in the direction of previous rotation; if this direction is incorrect for the new station, the motor reverses, after turning to the end of the dial, then proceeds to the desired station.　{ ¦nän'hōm·iŋ 'tün·iŋ ,sis·təm }

nonhypergolic [CHEM] Not capable of igniting spontaneously upon contact; used especially with reference to rocket fuels.　{ ,nän,hī·pər'gäl·ik }

nonideal gas [STAT MECH] A gas whose molecules have significant interaction, more than that needed to bring about the equilibrium.　{ 'nän·ī,dēl 'gas }

nonideal solution [PHYS CHEM] A solution whose behavior does not conform to that of an ideal solution; that is, the behavior is not predictable over a wide range of concentrations and temperatures by the use of Raoult's law.　{ 'nän·ī,dēl sə·lü·shən }

nonimpact printer [GRAPHICS] A line printer in which the characters are produced electrically, electronically, or optically rather than mechanically; only a single copy can be produced.　{ ¦nän'im,pakt ,print·ər }

nonimpinging injector [AERO ENG] An injector used in rocket engines which employs parallel streams of propellant usually emerging normal to the face of the injector.　{ 'nän·im,pin·jiŋ in'jek·tər }

noninductive [ELEC] Having negligible or zero inductance.　{ ,nän·in'dək·tiv }

noninductive capacitor [ELEC] A capacitor constructed so it has practically no inductance; foil layers are staggered during winding, so an entire layer of foil projects at either end for contact-making purposes; all currents then flow laterally rather than spirally around the capacitor.　{ ,nän·in'dək·tiv kə'pas·əd·ər }

noninductive resistor [ELEC] A wire-wound resistor constructed to have practically no inductance, either by using a hairpin winding or by reversing connections to adjacent sections of the winding.　{ ,nän·in'dək·tiv ri'zis·tər }

noninductive winding [ELEC] A winding constructed so that the magnetic field of one turn or section cancels the field of the next adjacent turn or section.　{ ,nän·in'dək·tiv 'wīn·diŋ }

nonine See nonyne.　{ 'nō,nīn }

nonintegrable system [MECH] A dynamical system whose motion is governed by an equation that is not an integrable differential equation.　{ ,nän¦int·i·grə·bəl ,sis·təm }

nonintelligible crosstalk [COMMUN] Crosstalk which cannot be understood regardless of its received volume, but which because of its syllabic nature is more annoying subjectively than thermal-type noise.　{ ,nän·in'tel·ə·jə·bəl 'krȯs,tȯk }

noninteracting control [CONT SYS] A feedback control in a system with more than one input and more than one output, in which feedback transfer functions are selected so that each input influences only one output.　{ ¦nän,in·tər'ak·tiŋ kən'trōl }

noninverting amplifier [ELECTR] An operational amplifier in which the input signal is applied to the ungrounded positive input terminal to give a gain greater than unity and make the output voltage change in phase with the input voltage.　{ ¦nän·in'vərd·iŋ 'am·plə,fī·ər }

noninverting parametric device [ELECTR] Parametric device whose operation depends essentially upon three frequencies, a harmonic of the pump frequency and two signal frequencies, of which one is the sum of the other plus the pump harmonic.　{ ¦nän·in'vərd·iŋ ,par·ə¦me·trik di'vīs }

Nonionacea [INV ZOO] A superfamily of Foraminiferida in the suborder Orbitoidacea, characterized by a granular calcite test wall with monolamellar septa, and a planispiral to trochospiral test.　{ ,nō·nē·ə'nās·ē·ə }

nonionic detergent [MATER] A detergent with molecules that do not ionize in aqueous solution, for example, detergents derived from condensation products of long-chain glycols and octyl or nonyl phenols.　{ ¦nän·ī'än·ik di'tər·jənt }

nonlinear [PHYS] Pertaining to a response which is other than directly or inversely proportional to a given variable.　{ 'nän,lin·ē·ər }

nonlinear acoustics [ACOUS] The study of the behavior of sufficiently large sonic and ultrasonic disturbances that nonlinear differential equations are necessary for an adequate mathematical description of the phenomena.　{ 'nän,lin·ē·ər ə'kü·stiks }

nonlinear amplifier [ELECTR] An amplifier in which a change in input does not produce a proportional change in output.　{ 'nän,lin·ē·ər 'am·plə,fī·ər }

nonlinear capacitor [ELEC] Capacitor having a mean charge characteristic or a peak charge characteristic that is not linear, or a reversible capacitance that varies with bias voltage. { 'nän‚lin·ē·ər kə'pas·əd·ər }

nonlinear circuit [ELEC] A circuit in which the current and voltage in any element that results from two sources of energy acting together is not equal to the sum of the currents or voltages that result from each of the sources acting alone. { 'nän‚lin·ē·ər 'sər·kət }

nonlinear circuit component [ELECTR] An electrical device for which a change in applied voltage does not produce a proportional change in current. Also known as nonlinear device; nonlinear element. { 'nän‚lin·ē·ər ¦sər·kət kəm'pō·nənt }

nonlinear coil [ELECTROMAG] Coil having an easily saturable core, possessing high impedance at low or zero current and low impedance when current flows and saturates the core. { 'nän‚lin·ē·ər 'kȯil }

nonlinear control system [CONT SYS] A control system that does not have the property of superposition, that is, one in which some or all of the outputs are not linear functions of the inputs. { 'nän‚lin·ē·ər kən'trōl ‚sis·təmz }

nonlinear coupler [ELECTR] A type of frequency multiplier which uses the nonlinear capacitance of a junction diode to couple energy from the input circuit, which is tuned to the fundamental, to the output circuit, which is tuned to the desired harmonic. { 'nän‚lin·ē·ər 'kəp·lər }

nonlinear crosstalk [COMMUN] Interaction between channels occupying different wavelengths in a wavelength-division-multiplexed system because of optical nonlinearities in the transmission medium. { ‚nän‚lin·ē·ər 'krȯs‚tȯk }

nonlinear crystal [SOLID STATE] A crystal in which some influence (such as stress, electric field, or magnetic field) produces a response (such as strain, electric polarization, or magnetization) which is not proportional to the influence. { 'nän‚lin·ē·ər 'krist·əl }

nonlinear damping [PHYS] Damping that is not proportional to velocity. { 'nän‚lin·ē·ər 'damp·iŋ }

nonlinear detection [ELECTR] Detection based on the curvature of a tube characteristic, such as square-law detection. { 'nän‚lin·ē·ər di'tek·shən }

nonlinear device See nonlinear circuit component. { 'nän‚lin·ē·ər di'vīs }

nonlinear dielectric [ELEC] A dielectric whose polarization is not proportional to the applied electric field. { 'nän‚lin·ē·ər ‚dī·ə'lek·trik }

nonlinear distortion [ELECTR] Distortion in which the output of a system or component does not have the desired linear relation to the input. [ENG ACOUS] The ratio of the total root-mean-square (rms) harmonic distortion output of a microphone to the rms value of the fundamental component of the output. { 'nän‚lin·ē·ər di'stȯr·shən }

nonlinear element See nonlinear circuit component. { 'nän‚lin·ē·ər 'el·ə·mənt }

nonlinear equation [MATH] An equation in variables x_1, ..., x_n, y which cannot be put into the form $a_1x_1 + \cdots + a_nx_n = y$. { 'nän‚lin·ē·ər i'kwā·zhən }

nonlinear feedback control system [CONT SYS] Feedback control system in which the relationships between the pertinent measures of the system input and output signals cannot be adequately described by linear means. { 'nän‚lin·ē·ər 'fēd‚bak kən'trōl ‚sis·təm }

nonlinear fiber amplifier [COMMUN] An optical amplifier in which nonlinear interactions (stimulated Raman and Brillouin scattering and four-wave mixing) between pump light and the signal cause transfer of power to the signal, resulting in fiber gain. { ‚nän¦lin·ē·ər ‚fī·bər 'am·plə‚fī·ər }

nonlinear inductance [ELEC] The behavior of an inductor for which the voltage drop across the inductor is not proportional to the rate of change of current, such as when the inductor has a core of magnetic material in which magnetic induction is not proportional to magnetic field strength. { 'nän‚lin·ē·ər in'dək·təns }

nonlinearity [SCI TECH] The deviation of any functional relationship from direct proportionality. { ‚nän‚lin·ē·'är·əd·ē }

nonlinear material [PHYS] A material in which some specified influence (such as stress, electric field, or magnetic field) produces a response (such as strain, electric polarization, or magnetization) which is not proportional to the influence. { 'nän‚lin·ē·ər mə'tir·ē·əl }

nonlinear molecule [ORG CHEM] A branched-chain molecule, that is, one whose atoms do not all lie along a straight line. Also known as isomolecule. { 'nän‚lin·ē·ər 'mäl·ə kyül }

nonlinear network [ELEC] A network in which the current or voltage in any element that results from two sources of energy acting together is not equal to the sum of the currents or voltages that result from each of the sources acting alone. { 'nän‚lin·ē·ər 'net‚wərk }

nonlinear optical device [OPTICS] A device based on one of a class of optical effects that result from the interaction of electromagnetic radiation from lasers with nonlinear materials. { 'nän‚lin·ē·ər 'äp·tə·kəl di‚vīs }

nonlinear optical loop mirror [OPTICS] A fiber-optic device in which a coupler splits the input light into two waves that travel in opposite directions around a fiber-optic loop and acquire different phases because of nonlinearities in the optical medium; the output power is very low for some phase differences and equals the input power for other phase differences. { ¦nän‚lin·ē·ər ‚äp·tə·kəl 'lüp 'mir·ər }

nonlinear optics [OPTICS] The study of the interaction of radiation with matter in which certain variables describing the response of the matter (such as electric polarization or power absorption) are not proportional to variables describing the radiation (such as electric field strength or energy flux). { 'nän‚lin·ē·ər 'äp‚tiks }

nonlinear oscillator [ELECTR] A radio-frequency oscillator that changes frequency in response to an audio signal; it is the basic circuit used in eavesdropping devices. { 'nän‚lin·ē·ər 'äs·ə‚läd·ər }

nonlinear physics [PHYS] The study of situations where the measure of an effect is not proportional to the measure of what is considered to be its cause. { ‚nän‚lin·ē·ər 'fiz·iks }

nonlinear programming [MATH] A branch of applied mathematics concerned with finding the maximum or minimum of a function of several variables, when the variables are constrained to yield values of other functions lying in a certain range, and either the function to be maximized or minimized, or at least one of the functions whose value is constrained, is nonlinear. { 'nän‚lin·ē·ər 'prō‚gram·iŋ }

nonlinear reactance [ELECTR] The behavior of a coil or capacitor whose voltage drop is not proportional to the rate of change of current through the coil, or the charge on the capacitor. { 'nän‚lin·ē·ər rē'ak·təns }

nonlinear refraction [OPTICS] The phenomenon whereby the refractive index of certain substances varies with light intensity. { 'nän‚lin·ē·ər ri frak·shən }

nonlinear regression See curvilinear regression. { 'nän‚lin·ē·ər ri'gresh·ən }

nonlinear resistance [ELECTR] The behavior of a substance (usually a semiconductor) which does not obey Ohm's law but has a voltage drop across it that is proportional to some power of the current. { 'nän‚lin·ē·ər ri'zis·təns }

nonlinear Schrödinger equation [OPTICS] A special form into which the Maxwell equations can be transformed in a medium with an optical nonlinearity that gives rise to self-action effects; this equation resembles the Schrödinger equation of quantum mechanics with the potential term in the latter equation replaced by a nonlinear term proportional to the local intensity of the light field, and it possesses soliton solutions. { ¦nän‚lin·ē·ər 'shrād·iŋ·ər i‚kwä·zhən }

nonlinear spectroscopy [SPECT] The study of energy levels not normally accessible with optical spectroscopy, through the use of nonlinear effects such as multiphoton absorption and ionization. { 'nän‚lin·ē·ər spek'träs·kə·pē }

nonlinear system [MATH] A system in which the interrelationships among the quantities involved are expressed by equations, some of which are not linear. [SCI TECH] A system in which outputs are not linear functions of vectors whose components represent the inputs. { 'nän‚lin·ē·ər 'sis·təm }

nonlinear taper [ELEC] Nonuniform distribution of resistance throughout the element of a potentiometer or rheostat. { 'nän lin·ē·ər 'tā·pər }

nonlinear vibration [MECH] A vibration whose amplitude is large enough so that the elastic restoring force on the vibrating object is not proportional to its displacement. { 'nän‚lin·ē·ər vī'brā·shən }

nonlinear viscoelasticity [FL MECH] The behavior of a fluid

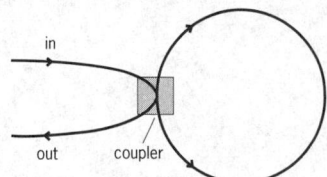

NONLINEAR OPTICAL LOOP MIRROR

The input light is split into two counterpropagating waves by the coupler.

which does not obey a first-order differential equation in stress and strain. { 'nän·lin·ē·ər ¦vis·gō·i¸las'tis·əd·ē }

non-load-bearing tile [MATER] Tile unable to carry superimposed loads. { ¦nän 'lōd ¸ber·iŋ 'tīl }

nonloaded Q [ELEC] Of an electric impedance, the Q value of the impedance without external coupling or connection. Also known as basic Q. { 'nän¸lōd·əd 'kyü }

nonlocalized bond See delocalized bond. { ¦nän'lō·kə¸līzd 'bänd }

nonlocalized electron [PHYS] An electron whose wave function is not confined to the vicinity of one or two nuclei, but is spread out over a molecule or a crystal lattice. { ¦nän'lō·kə¸līzd i'lek¸trän }

nonmagnetic [ELECTROMAG] Not magnetizable, and therefore not affected by magnetic fields. { ¦nän·mag'ned·ik }

nonmagnetic steel [MET] A steel alloy containing about 12% manganese and sometimes a small quantity of nickel; it is practically nonmagnetic at ordinary temperatures. { ¦nän·mag'ned·ik 'stēl }

nonmaintenance time [COMPUT SCI] The elapsed time during scheduled working hours between the determination of a machine failure and placement of the equipment back into operation. { ¦nän'mānt·ən·əns ¸tīm }

nonmechanical See nonclastic. { ¦nän·mi'kan·ə·kəl }

nonmetallic sheathed cable [ELEC] Assembly of two or more rubber-covered conductors in an outer sheath of nonconducting fibrous material that has been treated to make it flame-resistant and moisture-repellent. { ¦nän·mə'tal·ik 'shēthd 'kā·bəl }

nonmetric reference [ANTHRO] One of a group of features that can be counted rather than measured—for example, accessory bones in the sutures of the skull and lingual ridges in the incisor teeth. { ¦nän'me·trik 'ref·rəns }

non-minimum-phase system [CONT SYS] A linear system whose transfer function has one or more poles or zeros with positive, nonzero real parts. { ¦nän¦min·ə·məm 'fāz ¸sis·təm }

nonmultiple switchboard [ELEC] Manual telephone switchboard in which each subscriber line is attached to only one jack. { ¦nän'məl·tə·pəl 'swich¸bȯrd }

nonnegative semidefinite See positive semidefinite. { ¦nän 'neg·əd·iv ¸sem·i'def·ə·nət }

non-Newtonian fluid [FL MECH] A fluid whose flow behavior departs from that of a Newtonian fluid, so that the rate of shear is not proportional to the corresponding stress. Also known as non-Newtonian system. { ¸nän·nü'tō·nē·ən 'flü·əd }

non-Newtonian fluid flow [FL MECH] The flow behavior of non-Newtonian fluids, whose study has applications in many important problems of practical significance such as flow in tubes, extrusion, flow through dies, coating operations, rolling operations, and mixing of fluids. { ¸nän·nü'tō·nē·ən 'flü·əd ¸flō }

non-Newtonian system See non-Newtonian fluid. { ¸nän·nü'tō·nē·ən 'sis·təm }

non-Newtonian viscosity [FL MECH] The behavior of a fluid which, when subjected to a constant rate of shear, develops a stress which is not proportional to the shear. Also known as anomalous viscosity. { ¸nän·nü'tō·nē·ən vi'skäs·əd·ē }

nonnitroglycerin explosive [MATER] An explosive which contains TNT instead of nitroglycerin to sensitize ammonium nitrate, and a little aluminum powder may also be added to increase the power and sensitivity. { ¦nän¸nī·trō'glis·ə·rən ik'splō·siv }

nonnumeric character [COMPUT SCI] Any character except a digit. { ¦nän·nü'mer·ik 'kar·ik·tər }

nonnumeric programming [COMPUT SCI] Computer programming that deals with objects other than numbers. { ¦nän·nü¸mer·ik 'prō¸gram·iŋ }

nonodontogenic cyst [MED] Any oral cyst that develops from epithelium which has been sequestered in bony or soft-tissue suture lines during embryonic development. { ¦nän·ə¦dant·ə¦jen·ik 'sist }

non-ohmic [ELEC] Pertaining to a substance or circuit component that does not obey Ohm's law. { ¦nän'ō·mik }

nonoic acid [ORG CHEM] $C_8H_{17}COOH$ Any of a family of acids which are mixed isomers produced in the Fischer-Tropsch process; pelargonic acid is the straight-chain member; used as a chemical intermediate. { nō'nō·ik 'as·əd }

nonomino [MATH] One of the 1285 plane figures that can

be formed by joining nine unit squares along their sides. { nō 'näm·ə¸nō }

nonorientable surface See one-sided surface. { ¸nän¸ȯr ē¦ent·ə·bəl 'sər·fəs }

nonosteogenic fibroma [MED] A tumor of bone, usually in the shaft of long bones; characterized by whorls of spindle-shaped connective tissue cells. { ¦nän¦äs·tē·ō¦jen·ik fī'brō mə }

nonparalytic poliomyelitis [MED] Infection by poliomyelitis virus accompanied by upper respiratory or gastrointestinal symptoms, muscular pain and stiffness, and mild fever. { ¸nän¸par·ə'lid·ik ¸pō·lē·ō¸mī·ə'līd·əs }

nonparametric statistics [STAT] A class of statistical methods applicable to a large set of probability distributions used to test for correlation, location, independence, and so on. { ¦nän ¦par·ə¦me·trik stə'tis·tiks }

nonpareil [GRAPHICS] A designation to indicate a specific amount of space surrounding the type, for example, interlinear space; the space is 6 points in the American point system. { ¦nän·pə¦rel }

nonpenetrative [GEOL] Of a type of deformation, affecting only part of a rock, such as kink bands. { nän'pen·ə¸trād·iv }

nonperiodic decimal See nonrepeating decimal. { ¸nän¸pir ē¸äd·ik 'des·məl }

nonpermissive cell [VIROL] A cell that does not support replication of a virus. { ¸nän·pər¦mis·iv 'sel }

nonpersistent war gas [ORD] A chemical agent, that is, a war gas, normally effective in the open for 10 minutes or less at the point of dispersion. { ¸nän·pər'sis·tənt 'wȯr ¸gas }

nonplunging fold [GEOL] A fold with a horizontal axial surface. Also known as horizontal fold; level fold. { ¦nän¦plən·jiŋ 'fōld }

nonpoint source [CIV ENG] A dispersed source of stormwater runoff; the water comes from land dedicated to uses such as agriculture, development, forest, and land fills and enters the surface water system as sheet flow at irregular rates. { ¸nän 'pȯint ¸sȯrs }

nonpolar [CHEM] Pertaining to an element or compound which has no permanent electric dipole moment. { ¦nän'pō lər }

nonpolar bond [PHYS CHEM] A type of covalent bond in which both atoms attract the bonding electrons equally or nearly equally. { ¦nän¦pōl·ər 'bänd }

nonpolar covalent bond [PHYS CHEM] A bond in which a pair of electrons is distributed or shared equally between two atoms. { ¦nän¸pō·lər ¦kō¸vā·lənt 'bänd }

nonpolarized relay See neutral relay. { ¦nän'pō·lə¸rīzd 'rē¸lā }

nonpolar molecule [PHYS CHEM] A molecule with equal distribution of electrons among its atoms. { ¦nän¸pō·lər 'mäl·ə¸kyül }

nonpolar solvent [MATER] A solvent that does not have a permanent electric dipole moment and therefore has no tendency for intramolecular association with polar species. { ¦nän'pō·lər 'säl·vənt }

nonpositive tactile stimulus [PHYSIO] A cessation of a feedback signal from tactile sensors to the brain, such as when an object held in the hand falls. { ¦nän¦päz·əd·iv ¦tak¸tīl 'stim yə·ləs }

nonprecision [NAV] In air operations, pertaining to a navigational facility without a glide slope; does not imply an unacceptable quality of course guidance. { ¦nän·pri'sizh·ən }

nonpreemptive multitasking See cooperative multitasking. { ¸nän·prē¦em·tiv 'məl·tē¸task·iŋ }

nonprint code [COMPUT SCI] A bit combination which is interpreted as no printing, no spacing. { ¦nän¦print 'kōd }

nonpriority interrupt [COMPUT SCI] Any one of a group of interrupts which may be disregarded by the central processing unit. { ¦nän·pri'är·əd·ē 'int·ə¸rəpt }

nonprobabilistic sampling [STAT] A process in which some criterion other than the laws of probability determines the elements of the population to be included in the sample. { ¸nän¸präb·ə¦lis·tik 'sam·pliŋ }

nonprocedural language [COMPUT SCI] A programming language in which the program does not follow the actual steps a computer follows in executing a program. { ¸nän·prə'sē·jə rəl 'laŋ¸gwij }

nonproductive cough See dry cough. { ¸nän·prə¦dək·tiv 'kȯf }

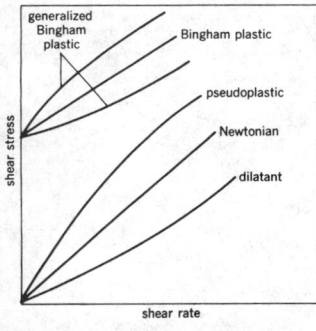

NON-NEWTONIAN FLUID

Typical flow curves, giving rate of shear as a function of shear stress, for a Newtonian fluid and for the three types of time-independent non-Newtonian fluid: Bingham plastic, pseudoplastic fluid, and dilatant fluid.

nonproductive infection *See* abortive infection. { ¦nän·prə'dək·tiv in'fek·shən }

nonpropagating soliton [FL MECH] A stable oscillation in a fluid channel that is stationary and highly localized in the length direction of the channel and can be generated when the floor of the channel is subjected to vertical, sinusoidal oscillations. { ¦nän'präp·ə,gäd·iŋ 'säl·ə,tän }

nonprotein nitrogen [BIOCHEM] The nitrogen fraction in the body tissues, excretions, and secretions, not precipitated by protein precipitants. { ¦nän¦prō,tēn 'nī·trə·jən }

nonprotic solvent [CHEM] A solvent that does not contain a hydrogen ion source. { ¦nän¦prōd·ik 'säl·vənt }

nonpsychotic organic brain syndrome [MED] Organic brain syndrome in which there is no apparent psychosis. { ¦nän·sī'käd·ik òr'gan·ik 'brān ,sin,drōm }

nonquantum mechanics [MECH] The classical mechanics of Newton and Einstein as opposed to the quantum mechanics of Heisenberg, Schrödinger, and Dirac; particles have definite position and velocity, and they move according to Newton's laws. { ¦nän¦kwän·təm mi'kan·iks }

nonradial pulsation [ASTRON] A stellar oscillation whose phase and amplitude vary with stellar latitude and longitude. { ¦nän,räd·ē·əl pəl'sā·shən }

nonradioactive reaction [NUC PHYS] A nuclear reaction whose radionuclide content is less than 0.01. { ,nän,räd·ē·ō¦ak·tiv rē'ak·shən }

nonreactive [ELEC] Pertaining to a circuit, component, or load that has no capacitance or impedance, so that an alternating current is in phase with the corresponding voltage. { ,nän·rē'ak·tiv }

nonreactive load *See* resistive load. { ,nän·rē'ak·tiv 'lōd }

nonreclosing pressure relief device [MECH ENG] A device which remains open after relieving pressure and must be reset before it can operate again. { ¦nän·rē'klōz·iŋ 'presh·ər ri,lēf di,vīs }

nonrecording rain gage [ENG] A rain gage which indicates but does not record the amount of precipitation. { ¦nän·ri'kòrd·iŋ 'rān ,gāj }

nonrecoverable error [COMPUT SCI] An error detected during computer processing that cannot be handled by the computer system and therefore causes processing to be interrupted. { ¦nän·ri'kəv·rə·bəl 'er·ər }

nonrecurring decimal *See* nonrepeating decimal. { ,nän·ri·kər·iŋ 'des·məl }

nonrecurring issue [ORD] An ordnance issue made on a one-time basis with no foreseeable subsequent demand from the consignee to whom issue was made. { ¦nän·ri'kər·iŋ 'i·shü }

nonrecursive filter [ELECTR] A digital filter that lacks feedback; that is, its output depends on present and past input values only and not on previous output values. { ,nän·ri,kər·siv 'fil·tər }

nonredundant system [COMPUT SCI] A computer system designed in such a way that only the absolute minimum amount of hardware is utilized to implement its function. { ¦nän·ri'dən·dənt 'sis·təm }

nonrelativistic approximation [PHYS] The approximation in which it is assumed that speeds of objects are small compared to the speed of light. { ¦nän,rel·ə·tə'vis·tik ə,präk·sə'mā·shən }

nonrelativistic kinematics [MECH] The study of motions of systems of objects at speeds which are small compared to the speed of light, without reference to the forces which act on the system. { ¦nän,rel·ə·tə'vis·tik ,kin·ə'mad·iks }

nonrelativistic mechanics [MECH] The study of the dynamics of systems in which all speeds are small compared to the speed of light. { ¦nän,rel·ə·tə'vis·tik mi'kan·iks }

nonrelativistic particle [RELAT] A particle whose velocity is small with respect to that of light. { ¦nän,rel·ə·tə'vis·tik 'pärd·ə·kəl }

nonrelativistic quantum mechanics [QUANT MECH] The modern theory of matter and its interaction with radiation, applicable to systems of material particles which move slowly compared to the speed of light, which are neither created nor destroyed, and whose internal structure (except for spin) either does not change or is irrelevant to the description of the system. { ¦nän,rel·ə·tə'vis·tik 'kwän·təm mi'kan·iks }

nonremovable discontinuity [MATH] A point at which a function is not continuous or is undefined, and cannot be made continuous by being given a new value at the point. { ¦nän·ri'müv·ə·bəl dis,känt·ən'ü·əd·ē }

nonrenewable fuse unit [ELEC] Fuse unit that cannot be readily restored for service after operation. { ¦nän·ri'nü·ə·bəl 'fyüz ,yü·nət }

nonrepeating decimal [MATH] An infinite decimal that fails to have any finite block of digits that eventually repeats indefinitely. Also known as nonperiodic decimal; nonrecurring decimal. { ,nän·ri,pēd·iŋ 'des·məl }

nonreproducible drawing [GRAPHICS] An opaque drawing which cannot be used to make copies by any contact (printthrough) process. { ¦nän,rē·prə'dü·sə·bəl 'dró·iŋ }

nonreproducing code [COMPUT SCI] A code which normally does not appear as such in a generated output but will result in a function such as paging or spacing. { ¦nän,rē·prə'dü·siŋ 'kōd }

nonresident routine [COMPUT SCI] Any computer routine which is not stored permanently in the memory but must be read into memory from a data carrier or external storage device. { ¦nän'rez·ə·dənt rü,tēn }

nonresidue [MATH] A nonresidue of m of order n, where m and n are integers, is an integer a such that $x^n = a + bm$, where x and b are integers, has no solution. { ,nän'rez·ə,dü }

nonresonant antenna [ELECTROMAG] A long-wire or traveling-wave antenna which does not have natural frequencies of oscillation, and responds equally well to radiation over a broad range of frequencies. { ¦nän'rez·ən·ənt an'ten·ə }

nonresonant line [ELECTROMAG] Transmission line having no reflected waves, and neither current nor voltage standing waves. { ¦nän'rez·ən·ənt 'līn }

non-return-to-zero [COMPUT SCI] A mode of recording and readout in which it is not necessary for the signal to return to zero after each item of recorded data. Abbreviated NRZ. { 'nän ri,tərn tə 'zir·ō }

nonreturn valve *See* check valve. { ¦nän·ri tərn ,valv }

nonrhegmatogenous retinal detachment [MED] Retinal detachment that is not caused by a retinal hole or tear, but occurs as a final stage of such pathologic conditions as retinopathy of prematurity or diabetic retinopathy. { ,nän,reg·mə¦täj·ən·əs ¦ret·ən·əl di'tach·mənt }

nonrigid plastic [MATER] A plastic with modulus of elasticity not greater than 50,000 pounds per square inch (3.45×10^8 pascals) at 25°C, according to standard American Society for Testing and Materials test procedures. { ¦nän¦rij·əd 'plas·tik }

nonrotating disk *See* semiconductor disk. { ¦nän'rō,tād·iŋ 'disk }

nonrotational strain *See* irrotational strain. { ,nän·rō'tā·shən·əl 'strān }

nonsaccharine sorghum *See* grain sorghum. { ¦nän'sak·ə·rən 'sòr·gəm }

nonscrollable message [COMPUT SCI] A message on a computer display that does not scroll off the top of the display as new information is written at the bottom. { ¦nän'skrō·lə·bəl 'mes·ij }

nonsegregated reservoir [PETRO ENG] Solution-gas-drive oil reservoir in which the gas does not separate from the oil as a function of height or upward movement. { ¦nän'seg·rə,gād·əd 'rez·əv,wär }

nonselective medium [MICROBIO] A culture medium that supports the growth of all genotypes. { ¦nän·si,lek·tiv 'mē·dē·əm }

nonselective mining [MIN ENG] Mining methods permitting low cost, generally by using a cheap stoping method combined with large-scale operations; can be used in deposits where the individual stringers, bands, or lenses of high-grade ore are numerous and so irregular in occurrence and separated by such thin lenses of waste that a selective method cannot be employed. { ¦nän·si,lek·tiv 'mīn·iŋ }

nonselective radiator *See* graybody. { ¦nän·si,lek·tiv 'rād·ē,ād·ər }

nonsense correlation [STAT] A correlation between two variables that is not due to any causal relationship, but to the fact that each variable is correlated with a third variable, or to random sampling fluctuations. Also known as illusory correlation. { 'nän,sens ,kär·ə,lā·shən }

nonsense mutation [MOL BIO] A mutation that changes a codon that codes for one amino acid into a codon that does not specify any amino acid (a nonsense codon). { 'nän,sens myü,tā·shən }

nonsense suppression [GEN] Suppression of the termination effect of a nonsense codon on a growing polypeptide chain, allowing return to the normal phenotype. { ¦nän¦sens sə¦presh·ən }

nonseptate *See* coencytic. { ¦nän¦sep¦tāt }

nonservo robot *See* fixed-stop robot. { ¦nän¦sər¦vō 'rō¦bät }

nonshared control unit [COMPUT SCI] A control unit relating to only one device. Also known as unipath. { ¦nän¦sherd kən¦trōl ¦yü·nət }

nonshattering glass [MATER] Two sheets of plate glass with a sheet of transparent resinoid between, the whole molded together under heat and pressure; it will crack without shattering. Also known as laminated glass; shatterproof glass. { ¦nän¦shad·ə·riŋ 'glas }

nonshorting contact switch [ELEC] Selector switch in which the width of the movable contact is less than the distance between contact clips, so that the old circuit is broken before the new circuit is completed. { ¦nän¦shórd·iŋ 'kän¦takt ¦swich }

nonshrinking steel *See* nondeforming steel. { ¦nän¦shriŋk·iŋ 'stēl }

nonsingular matrix [MATH] A matrix which has an inverse; equivalently, its determinant is not zero. { ¦nän¦siŋ·gyə·lər 'mā·triks }

nonsingular transformation [MATH] A linear transformation which has an inverse; equivalently, it has null space kernel consisting only of the zero vector. { ¦nän¦siŋ·gyə·lər ¦tranz·fər¦mā·shən }

nonsinusoidal waveform [ELEC] The representation of a wave which does not vary in a sinusoidal manner, and which therefore contains harmonics. { ¦nän¦sī·nə¦sóid·əl 'wāv ¦fòrm }

nonskid [CIV ENG] Pertaining to a surface that is roughened to reduce slipping, as a concrete floor treated with iron filings or carborundum powder, or indented while wet. { ¦nän¦skid }

nonslip concrete [MATER] Rough-surface concrete made by applying oxide grains to the mixture before it hardens; used for steps. { ¦nän¦slip 'kan¦krēt }

nonsoap grease [MATER] Mineral oil thickened with solid lubricants such as graphite, mica, talc, molybdenum sulfide, asbestos fiber, uncombined fats, or rosin oils; used as a lubricant. { ¦nän¦sōp 'grēs }

nonsorted polygon [GEOL] A form of patterned ground which has a dominantly polygonal mesh and an unsorted appearance due to the absence of border stones, and whose borders are generally marked by wedge-shaped fissures narrowing downward. { ¦nän¦sórd·əd 'päl·i¦gän }

nonspecific [MED] Not attributable to any one definite cause, as a disease not caused by one particular microorganism, or an immunity not conferred by a specific antibody. [PHARM] Of medicines or therapy, not counteracting any one causative agent. { ¦nän¦spə¦sif·ik }

nonspecific active immunotherapy [IMMUNOL] Active immunotherapy that utilizes materials that have no apparent antigenic relationship to the tumor but have modulatory effects on the immune system.

nonspecific hepatitis *See* interstitial hepatitis. { ¦nän¦spə¦sif·ik ¦hep·ə¦tīd·əs }

nonspecific immunity [IMMUNOL] Resistance attributable to factors other than specific antibodies, including genetic, age, or hormonal factors. { ¦nän¦spə¦sif·ik i¦myün·əd·ē }

nonspherical nucleus [NUC PHYS] A nucleus which appears to have a permanent ellipsoidal shape in its ground state, as suggested by a large electric quadrupole moment. { ¦nän¦sfer·ə·kəl 'nü·klē·əs }

nonsquare Banach space [MATH] A Banach space in which there are no nonzero elements, x and y, that satisfy the equation $\|x + y\| = \|x - y\| = 2\|x\| = 2\|y\|$. { ¦nän¦skwer 'bä¦näk ¦spās }

nonstandard [ORD] Differing from specifications, conditions, or procedures that have been prescribed or established; for instance, weather conditions different from those assumed in firing tables are nonstandard. { ¦nän¦stan·dərd }

nonstandard numbers [MATH] A generalization of the real numbers to include infinitesimal and infinite quantities by considering equivalence classes of infinite sequences of numbers. Also known as hyperreal numbers. { ¦nän¦stan·dərd 'nəm·bərz }

nonstop computer [COMPUT SCI] A computer system that is equipped with duplicate components or excess capacity so that a hardware or software failure will not interrupt processing. { ¦nän¦stäp kəm¦pyüd·ər }

nonstop switch [ELEC] A manual switch in an elevator car that can prevent the car from stopping at a specified floor. { ¦nän¦stäp ¦swich }

nonstorage camera tube [ELECTR] Television camera tube in which the picture signal is, at each instant, proportional to the intensity of the illumination on the corresponding area of the scene. { ¦nän¦stór·ij ¦kam·rə ¦tüb }

nonstranded rope [DES ENG] A wire rope with the wires in concentric sheaths instead of in strands, and in opposite directions in the different sheaths, giving the rope nonspinning properties. Also known as nonspinning rope. { ¦nän¦stran·dəd 'rōp }

nonstriated muscle fiber *See* smooth muscle fiber. { ¦nän¦strī¦ād·əd 'məs·əl ¦fī·bər }

nonswappable program [COMPUT SCI] A program that is given priority status so that its execution cannot be suspended to allow execution of other programs. { ¦nän¦swäp·ə·bəl 'prō·grəm }

nonsymmetrical aquifer [PETRO ENG] In an oil reservoir formation, a water-containing part that is irregular, being neither radial nor linear. { ¦nän·si¦me·trə·kəl 'ak·wə·fər }

nonsynchronous [ELEC] Not related in phase, frequency, or speed to other quantities in a device or circuit. { ¦nän¦siŋ·krə·nəs }

nonsynchronous initiation [MET] In resistance welding, random starting and stopping of transformer primary current relative to the voltage wave. { ¦nän¦siŋ·krə·nəs i¦nish·ē¦ā·shən }

nonsynchronous timer [ELECTR] A circuit at the receiving end of a communications link which restores the time relationship between pulses when no timing pulses are transmitted. { ¦nän¦siŋ·krə·nəs 'tīm·ər }

nonsynchronous transmission [ELECTR] A data transmission process in which a clock is not used to control the unit intervals within a block or a group of data signals. { ¦nän¦siŋ·krə·nəs tranz¦mish·ən }

nonsynchronous vibrator [ELECTR] Vibrator that interrupts a direct-current circuit at a frequency unrelated to the other circuit constants and does not rectify the resulting stepped-up alternating voltage. { ¦nän¦siŋ·krə·nəs 'vī¦brād·ər }

nonsyndromic hearing loss [MED] A type of hearing loss in which the individual has no other symptoms except hearing loss. { ¦hän·sin¦drō·mik 'hir·iŋ ¦lós }

nonsystematic joint [GEOL] A joint that is not part of a set. { ¦nän¦sis·tə¦mad·ik 'jóint }

nontectonite [PETR] Any rock whose fabric shows no influence of movement of adjacent grains; for example, a rock formed by mechanical settling. { ¦nän¦tek·tə¦nīt }

nonterminal vertex [MATH] A vertex in a rooted tree that has at least one successor. { ¦nän¦tər·mən·əl 'vər¦teks }

nonterminating continued fraction [MATH] A continued fraction that has an infinite number of terms. { ¦nän¦tər·mə¦nād·iŋ kən¦tin·yüd 'frak·shən }

nonterminating decimal [MATH] A decimal for which there is no digit to the right of the decimal point such that all digits farther to the right are zero. { ¦nän¦tər·mə¦nād·iŋ 'des·məl }

nonthermal decimetric emission [ELECTROMAG] A radio-wave emission above the 4-centimeter wavelength from the planet Jupiter that has a nearly constant flux between 5-centimeter and 1-meter wavelength. Also known as DIM. { ¦nän¦thər·məl ¦des·ə¦me·trik i¦mish·ən }

nonthermal radiation [PHYS] Electromagnetic radiation emitted by accelerated charged particles that are not in thermal equilibrium; aurora light and fluorescent-lamp light are examples. { ¦nän¦thər·məl ¦räd·ē¦ā·shən }

nontidal current [OCEANOGR] Any current due to causes other than tidal, as a permanent ocean current. { ¦nän¦tīd·əl 'kə·rənt }

nontransferred arc [MET] In arc welding and cutting, an arc made between the electrode and constricting nozzle, excluding the workpiece from the circuit. { ¦nän¦tranz·fərd 'ärk }

nontrivial solution [MATH] A solution of a set of homogeneous linear equations in which at least one of the variables has a value different from zero. { ¦nän¦triv·ē·əl sə¦lü·shən }

nontronite [MINERAL] $Na(Al,Fe,Si)O_{10}(OH)_2$ An iron-rich clay mineral of the montmorillonite group that represents the end member in which the replacement of aluminum by ferric ion is essentially complete. Also known as chloropal; gramenite; morencite; pinguite. { 'nän·trə,nīt }

nontropical sprue *See* gluten enteropathy. { ,nän,träp·ə·kəl 'sprü }

nonuniform flow [FL MECH] Fluid flow which does not have the same velocity at all points in a medium, at a given instant. { ¦nän'yü·nə,fȯrm 'flō }

nonuniform memory access machine [COMPUT SCI] A multiprocessor in which the memory is spread out over memory modules, which are attached to the processors, so that each processor has its own memory module. Abbreviated NUMA machine. { ,nän¦yün·i,fȯrm ¦mem·rē 'ak·ses mə,shēn }

nonvenereal syphilis [MED] Syphilis not acquired during sexual intercourse. { 'nän·və,nir·ē·əl 'sif·ə·ləs }

nonviscous flow *See* inviscid flow. { 'nän,vis·kəs 'flō }

nonviscous fluid *See* inviscid fluid. { 'nän,vis·kəs 'flü·əd }

nonviscous neutral [MATER] A neutral petroleum-derived oil with viscosity less than 135 SSU (Saybolt Seconds Universal) at 100°F (38°C). { 'nän,vis·kəs 'nü·trəl }

nonvolatile memory *See* nonvolatile storage. { 'nän'val·ə·təl 'mem·rē }

nonvolatile random-access memory [COMPUT SCI] A semiconductor storage device which has two memory cells for each bit, one of which is volatile, as in a static RAM (random-access memory), and provides unlimited read and write operations, while the other is nonvolatile, and provides the ability to retain information when power is removed. Abbreviated NV RAM. { ¦nän'val·ə·təl ¦ran·dəm 'ak,ses 'mem·rē }

nonvolatile storage [COMPUT SCI] A computer storage medium that retains information in the absence of power, such as a magnetic tape, drum, or core. Also known as nonvolatile memory. { ¦nän'val·ə·təl 'stȯr·ij }

nonwetting [MET] Of a metal or alloy, when it is molten, not adhering to or wetting the surface of a diamond which is to be set. { 'nän,wed·iŋ }

nonwetting phase [PETRO ENG] The oil phase contained in a reservoir pore structure when the reservoir fluid is two-phase (oil and water) or three-phase (oil, water, and gas). { 'nän,wed·iŋ 'fāz }

nonwetting sand [GEOL] Sand that resists infiltration of water; consists of angular particles of various sizes and occurs as a tightly packed lens. { 'nän,wed·iŋ 'sand }

nonwork unit [IND ENG] A time unit on a schedule during which work may not be performed on a given activity, for example, a weekend or a holiday. { 'nän'wərk ,yü·nət }

nonwoven fabric [TEXT] Cloth produced from a random arrangement or matting of natural or synthetic fibers held together by adhesives, heat and pressure, or needling; felt is an example. { 'nän,wōv·ən 'fab·rik }

nonyl acetate [ORG CHEM] $C_9H_{19}OOCCH_3$ Any of a family of isomers, such as *n*-nonyl acetate and diisobutyl carbinyl acetate, which are products of Fischer-Tropsch and oxo syntheses. { 'nä,nil 'as·ə,tāt }

n-nonyl acetate [ORG CHEM] $CH_3COO(CH_2)_8CH_3$ Alcohol-soluble, colorless liquid with pungent odor; boiling point 208–212°C; used in perfumery. { ¦en 'nä,nil 'as·ə,tāt }

n-nonyl alcohol [ORG CHEM] $CH_3(CH_2)_7CH_2OH$ One of a family of $C_9H_{19}OH$ isomers; a colorless liquid with rose aroma; boils at 215°C; insoluble in water, soluble in alcohol; used in perfumery and flavorings. { ¦en 'nä,nil 'al·kə,hȯl }

nonyl benzene [ORG CHEM] $C_9H_{19}C_6H_5$ Liquid boiling at 245–252°C; straw-colored with aromatic aroma; used to make surface-active agents. { 'nä,nil 'ben,zēn }

1-nonylene [ORG CHEM] C_9H_{18} Colorless liquid boiling at 150°C; soluble in alcohol, insoluble in water; used as a chemical intermediate. Also known as nonene. { ¦wən 'nän·ə,lēn }

nonyl hydride *See* nonane. { 'nän·əl 'hī,drīd }

nonyl phenol [ORG CHEM] $C_9H_{19}C_6H_4OH$ Pale-yellow liquid boiling at 283–302°C; soluble in organic solvents, insoluble in water; a mixture of monoalkyl phenol isomers, mostly para-substituted; used to make surface-active agents, resins, and plasticizers. { 'nä,nil 'fē,nȯl }

nonylphenoxyacetic acid [ORG CHEM] $C_9H_{19}C_6H_4OCH_2COOH$ A viscous, amber-colored liquid, soluble in alkali; used in turbine oils, lubricants, greases, and other materials as a corrosion inhibitor. { 'nä,nil·fe¦näk·sē·ə'sēd·ik 'as·əd }

nonyne [ORG CHEM] $CH_3(CH_2)_6≡CCH$ Water-insoluble, colorless liquid boiling at 160°C. { 'nō,nīn }

no observed effect level [MED] The highest dose at which

no effects can be observed; used as a measure of chronic toxicity. { 'nō əb¦zərvd i'fekt ,lev·əl }

noon [ASTRON] The instant at which a time reference is over the upper branch of the reference meridian. { nün }

noon constant [NAV] A predetermined value added to a meridian or ex-meridian sextant altitude to determine the latitude. { 'nün 'kän·stənt }

noon interval [ASTRON] The predicted time interval between a given instant, usually the time of a morning observation, and local apparent noon; it is used to predict the time for observing the sun on the celestial meridian. { 'nün 'in·tər·vəl }

noon sight [NAV] **1.** Measurement of the altitude of the sun at local apparent noon. **2.** The altitude so measured. { 'nün ,sīt }

NO OP [COMPUT SCI] An instruction telling the computer to do nothing, except to proceed to the next instruction in sequence. Also known as do-nothing instruction; no-operation instruction. { 'nō ,äp }

no-operation instruction *See* NO OP. { ¦nō ,äp·ə'rā·shən in'strək·shən }

noosphere *See* anthroposphere. { 'nō·ə,sfir }

nopinene *See* pinene. { 'nä·pə,nēn }

nor- [CHEM] Chemical formula prefix for normal; indicates a parent for another compound to be formed by removal of one or more carbons and associated hydrogens. { nȯr }

NOR [MATH] A logic operator having the property that if P, Q, R, . . . are statements, then the NOR of P, Q, R, . . . is true if all statements are false, false if at least one statement is true. Derived from NOT-OR. Also known as Peirce stroke relationship. { nȯr }

noradrenaline *See* norepinephrine. { ,nȯr·ə'dren·ə·lən }

noradrenergic system [NEUROSCI] A system of neurons that is responsible for the synthesis, storage, and release of the neurotransmitter norepinephrine. { nȯr¦ad·rə¦nər·jik 'sis·təm }

norbergite [MINERAL] $Mg_3SiO_4(F,OH)_2$ A yellow or pink orthorhombic mineral composed of magnesium silicate with fluoride and hydroxyl; it is a member of the humite group. { 'nȯr,bər,gīt }

NOR circuit [ELECTR] A circuit in which output voltage appears only when signal is absent from all of its input terminals. { 'nȯr ,sər·kət }

Norden bombsight [ORD] A gyroscopically stabilized synchronizing bombsight used mainly for synchronous bombing but useful for fixed-angle bombing. { 'nȯrd·ən 'bäm,sīt }

nordenskioldine [MINERAL] $CaSn(BO_3)_2$ A colorless or sulfur-, lemon-, or wine-yellow, hexagonal mineral consisting of a borate of calcium and tin; occurs in tabular form and as lenslike crystals. { 'nȯrd·ən,shēl·dən }

Nordenskjöld line [CLIMATOL] The line connecting all places at which the mean temperature of the warmest month is equal (in degrees Celsius) to $9 - 0.1k$, where k is the mean temperature of the coldest month (in degrees Fahrenheit it becomes $51.4 - 0.1k$). { 'nȯrd·ən,shēld ,līn }

Nordheim's rule [SOLID STATE] The rule that the residual resistivity of a binary alloy that contains mole fraction x of one element and $1 - x$ of the other is proportional to $x(1 - x)$. { 'nȯrd,īmz ,rül }

nordihydroguaiaretic acid [ORG CHEM] $C_{18}H_{22}O_4$ A crystalline compound with a melting point of 184–185°C; soluble in alcohols, ether, acetone, glycerol, and propyleneglycol; used as an antioxidant in fats and oils. Abbreviated NDGA. { ¦nȯr·dī¦hī·drō¦gwī·ə¦red·ik 'as·əd }

nordmarkite [PETR] A quartz-bearing alkalic syenite that has microperthite as its main component with smaller amounts of oligocase, quartz, and biotite and is characterized by granitic or trachytoid texture. { 'nȯrd,mär,kīt }

nor'easter *See* northeaster. { nȯr'ē·stər }

norepinephrine [BIOCHEM] $C_8H_{11}O_3N$ A hormone produced by chromaffin cells of the adrenal medulla; acts as a vasoconstrictor and mediates transmission of sympathetic nerve impulses. Also known as noradrenaline. { ,nȯr·ep·ə'ne·frən }

Norian [GEOL] A European stage of Upper Triassic geologic time that lies above the Carnian and below the Rhaetian. { 'nȯr·ē·ən }

norite [PETR] A coarse-grained plutonic rock composed

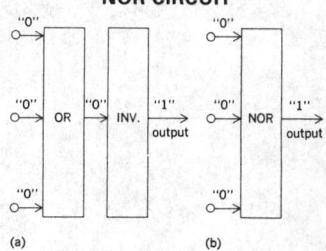

NOR CIRCUIT

NOR circuit. (*a*) OR gate followed by an inverter; if all inputs are "0", the output is "1." (*b*) OR-inverter (NOR) combination shown as a single symbol.

principally of basic plagioclase with orthopyroxene (hypersthene) as the dominant mafic material. Also known as hypersthenfels. { 'nȯ₁rīt }

norm [MATH] **1.** A scalar valued function on a vector space with properties analogous to those of the modulus of a complex number; namely: the norm of the zero vector is zero, all other vectors have positive norm, the norm of a scalar times a vector equals the absolute value of the scalar times the norm of the vector, and the norm of a sum is less than or equal to the sum of the norms. **2.** For a matrix, the square root of the sum of the squares of the moduli of the matrix entries. **3.** For a quaternion, the product of the quaternion and its conjugate. **4.** *See* absolute value. **5.** *See* fineness. [PETR] The theoretical mineral composition of a rock expressed in terms of standard mineral molecules as determined by means of chemical analyses. [QUANTMECH] **1.** The square of the modulus of a Schrödinger-Pauli wave function, integrated over the space coordinates and summed over the spin coordinates of the particles it describes. **2.** The square root of this quantity. { nȯrm }

Norma [ASTRON] A southern constellation; right ascension 16 hours, declination 50°S. Also known as Rule. { 'nȯr·mə }

normal [METEOROL] The average value of a meteorological element over any fixed period of years that is recognized as standard for the country and element concerned. { 'nȯr·məl }

normal acceleration [MECH] **1.** The component of the linear acceleration of an aircraft or missile along its normal, or Z, axis. **2.** The usual or typical acceleration. { 'nȯr·məl ak₁sel·ə'rā·shən }

normal adjustment [OPTICS] Property of an image formed by an optical system whose viewing position is similar to that of the object, such as an image at infinity formed by a telescope or an image at the viewer's near point formed by a microscope. { 'nȯr·məl ə'jəs·mənt }

normal aeration [GEOL] The complete renewal of soil air to a depth of 8 inches (20 centimeters) about once each hour. { 'nȯr·məl e'rā·shən }

normal aiming error [ORD] In bombing, an aiming error that falls within the boundary for gross errors. { 'nȯr·məl 'ām·iŋ ₁er·ər }

normal anticlinorium [GEOL] An anticlinorium in which axial surfaces of the subsidiary folds converge downward. { 'nȯr·məl₁ant·i·klə'nȯr·ē·əm }

normal axis [MECH] The vertical axis of an aircraft or missile. { 'nȯr·məl 'ak·səs }

normal barometer [ENG] A barometer of such accuracy that it can be used for the determination of pressure standards; an instrument such as a large-bore mercury barometer is usually used. { 'nȯr·məl bə'räm·əd·ər }

normal benzine [MATER] A mixture of hydrocarbons; clear, colorless, water-insoluble liquid distilled from petroleum; boils at 65–95°C; density 0.695–0.705. Also known as benzoline. { 'nȯr·məl 'ben₁zēn }

normal bonded-phase chromatography [ANALY CHEM] A technique of bonded-phase chromatography in which the stationary phase is polar and the mobile phase is nonpolar. { 'nȯr·məl ₁ban·dəd ₁fāz ₁krō·mə'täg·rə·fē }

normal bundle [MATH] If A is a manifold and B is a submanifold of A, then the normal bundle of B in A is the set of pairs (x,y), where x is in B, y is a tangent vector to A, and y is orthogonal to B. { 'nȯr·məl ₁bən·dəl }

normal chart [METEOROL] Any chart that shows the distribution of the official normal values of a meteorological element. Also known as normal map. { 'nȯr·məl 'chärt }

normal consolidation [GEOL] Consolidation of a sedimentary material in equilibrium with overburden pressure. { 'nȯr·məl kən₁säl·ə'dā·shən }

normal contour *See* accurate contour. { 'nȯr·məl 'kan₁tu̇r }

normal coordinates [MECH] A set of coordinates for a coupled system such that the equations of motion each involve only one of these coordinates. { 'nȯr·məl kō'ȯrd·ən·əts }

normal curvature [MATH] The normal curvature at a point on a surface is the curvature of the normal section to the point. { 'nȯr·məl 'kər·və·chər }

normal curve *See* Gaussian curve. { 'nȯr·məl 'kərv }

normal cycle [GEOL] A cycle of erosion whereby a region is reduced to base level by running water, especially by the action of rivers. Also known as fluvial cycle of erosion. { 'nȯr·məl 'sī·kəl }

normal density function [STAT] A normally distributed frequency distribution of a random variable x with mean e and variance σ is given by $(1/\sigma\sqrt{2\pi}) \exp [-(x - e)^2/2\sigma^2]$. { 'nȯr·məl 'den·səd·ē ₁faŋk·shən }

normal depth [FL MECH] The depth in an open channel at which a given flow has uniform velocity. { 'nȯr·məl 'depth }

normal derivative [MATH] The directional derivative of a function at a point on a given curve or surface in the direction of the normal to the curve or surface. { 'nȯr·məl di'riv·əd·iv }

normal dip *See* regional dip. { 'nȯr·məl 'dip }

normal direction flow [COMPUT SCI] The direction from left to right or top to bottom in flow charting. { 'nȯr·məl di'rek·shən ₁flō }

normal dispersion [GEOPHYS] The dispersion of seismic waves in which the recorded wave period increases with time. [OPTICS] Dispersion in which the refractive index decreases monotonically and continuously with increasing wavelength. { 'nȯr·məl di'spər·zhən }

normal displacement *See* dip slip. { 'nȯr·məl di'splās·mənt }

normal distribution [STAT] A commonly occurring probability distribution that has the form

$$(1/\sigma\sqrt{2\pi}) \int_{-\infty}^{u} \exp(-u^2/2)\,du$$

$$u = (x - e)/\sigma$$

where e is the mean and σ is the variance. Also known as Gauss' error curve; Gaussian distribution. { 'nȯr·məl ₁di·strə'byü·shən }

normal divisor *See* normal subgroup. { 'nȯr·məl di'vīz·ər }

normal effort [IND ENG] The effort expended by the average operator in performing manual work with average skill and application. { 'nȯr·məl 'ef·ərt }

normal electrode [ELEC] Standard electrode used for measuring electrode potentials. { 'nȯr·məl i'lek₁trōd }

normal element time [IND ENG] The selected or average element time adjusted to obtain the element time used by an average qualified operator. Also known as base time; leveled element time. { 'nȯr·məl ₁el·ə¦ment 'tīm }

normal equations [STAT] The set of equations arising in the least squares method whose solutions give the constants that determine the shape of the estimated function. { ¦nȯr·məl i'kwā·zhənz }

normal erosion [GEOL] Erosion effected by prevailing agencies of the natural environment, including running water, rain, wind, waves, and organic weathering. Also known as geologic erosion. { 'nȯr·məl i'rō·zhən }

normal extension [MATH] An algebraic extension of K of a field k, contained in the algebraic closure k̄ of k, such that every injective homomorphism of K into k̄, inducing the identity on k, is an automorphism of K. { 'nȯr·məl ik'sten·chən }

normal family [MATH] A family of complex functions analytic in a common domain where every sequence of these functions has a subsequence converging uniformly on compact subsets of the domain to an analytic function on the domain or to +∞. { 'nȯr·məl 'fam·lē }

normal fault [GEOL] A fault, usually of 45–90°, in which the hanging wall appears to have shifted downward in relation to the footwall. Also known as gravity fault; normal slip fault; slump fault. { 'nȯr·məl 'fȯlt }

normal fluid [CRYO] The component of liquid helium II, postulated in the two-fluid theory, that has viscosity and behaves like an ordinary fluid. { 'nȯr·məl 'flü·əd }

normal fold *See* symmetrical fold. { 'nȯr·məl 'fōld }

normal form [COMPUT SCI] The form of a floating-point number whose mantissa lies between 0.1 and 1.0. { 'nȯr·məl 'fȯrm }

normal frequencies [MECH] The frequencies of the normal modes of vibration of a system. { 'nȯr·məl 'frē·kwən₁sēz }

normal function *See* normalized function. { 'nȯr·məl 'faŋk·shən }

normal Hall effect [ELECTROMAG] The development, in a current-carrying conductor in a magnetic field, of a transverse voltage in the direction of the deflection of negative charge carriers (electrons) by the Lorentz force. { 'nȯrm·əl 'hȯl i₁fekt }

normal horizontal separation *See* offset. { 'nȯr·məl ˌhär·ə'zänt·əl ˌsep·ə'rā·shən }

normal hydrostatic pressure [HYD] In porous strata or in a well, the pressure at a given point that is approximately equal to the weight of a column of water extending from the surface to that point. { 'nȯr·məl ˌhī·drə'stad·ik 'presh·ər }

normal impact [MECH] **1.** Impact on a plane perpendicular to the trajectory. **2.** Striking of a projectile against a surface that is perpendicular to the line of flight of the projectile. { 'nȯr·məl 'im‚pakt }

normal impedance *See* free impedance. { 'nȯr·məl im'pēd·əns }

normal-incidence pyrheliometer [ENG] An instrument that measures the energy in the solar beam; it usually measures the radiation that strikes a target at the end of a tube equipped with a shutter and baffles to collimate the beam. { 'nȯr·məl ‚in·səd·əns ‚pīr‚hē·lē'äm·əd·ər }

normal incidence reflectivity [ELECTROMAG] The ratio of the energy of electromagnetic radiation reflected from the interface between two media to the energy of the incident radiation when the incident radiation travels in a direction perpendicular to the surface. { 'nȯr·məl ‚in·səd·əns ri‚flek'tiv·əd·ē }

normal induction [ELECTROMAG] Limiting induction, either positive or negative, in a magnetic material that is under the influence of a magnetizing force which varies between two specific limits. { 'nȯr·məl in'dək·shən }

normal inspection [IND ENG] The number of items inspected as specified by the sampling inspection plan at the outset; if the quality of the product improves, the number of units to be inspected is reduced; if quality deteriorates, the number of units inspected is increased. { 'nȯr·məl in'spek·shən }

normality [CHEM] Measure of the number of gram-equivalent weights of a compound per liter of solution. Abbreviated N. { nȯr'mal·əd·ē }

normalization [COMPUT SCI] Breaking down of complex data structures into flat files. { ‚nȯr·mə·lə'zā·shən }

normalize [COMPUT SCI] **1.** To adjust the representation of a quantity so that this representation lies within a prescribed range. **2.** In particular, to adjust the exponent and mantissa of a floating point number so that the mantissa falls within a prescribed range. [MATH] To multiply a quantity by a suitable constant or scalar so that it then has norm one; that is, its norm is then equal to one. [MET] To heat a ferrous alloy to some temperature above the transformation range, followed by air cooling. [QUANT MECH] To multiply a wave function by a constant so that its norm is equal to unity. [STAT] To carry out a normal transformation on a variate. { 'nȯr·mə‚līz }

normalized admittance [ELECTROMAG] The reciprocal of the normalized impedance. { 'nȯr·mə‚līzd ad'mit·əns }

normalized coupling coefficient [ELECTROMAG] Mutual inductance, expressed on a scale running from zero to one. { 'nȯr·mə‚līzd ‚kəp·liŋ ‚kō·i‚fish·ənt }

normalized current [ELECTROMAG] The current divided by the square root of the characteristic admittance of a waveguide or transmission line. { 'nȯr·mə‚līzd 'kə·rənt }

normalized function [MATH] A function with norm one; the norm is usually given by an integral $(\int |f|^p d\mu)^{1/p}$, $1 \le p < \infty$. Also known as normal function. { 'nȯr·mə‚līzd 'fəŋk·shən }

normalized impedance [ELECTROMAG] An impedance divided by the characteristic impedance of a transmission line or waveguide. { 'nȯr·mə‚līzd im'pēd·əns }

normalized Q [ELEC] The ratio of the reactive component of the impedance of a filter section to the resistive component. { 'nȯr·mə‚līzd 'kyü }

normalized standard scores [STAT] A procedure in which each set of original scores is converted to some standard scale under the assumption that the distribution of scores approximates that of a normal. { 'nȯr·mə‚līzd 'stan·dərd 'skȯrz }

normalized support function [MATH] The function that results from restricting the domain of the independent variable of the support function to the unit sphere. { 'nȯr·mə‚līzd sə'pȯrt ‚fəŋk·shən }

normalized susceptance [ELECTROMAG] The susceptance of an element of a waveguide or transmission line divided by the characteristic admittance. { 'nȯr·mə‚līzd sə'sep·təns }

normalized variate [STAT] A variate to which a normal transformation has been applied and which therefore has a normal distribution. { 'nȯr·mə‚līzd 'ver·ē·ət }

normalized voltage [ELECTROMAG] The voltage divided by the square root of the characteristic impedance of a waveguide or transmission line. { 'nȯr·mə‚līzd 'vōl·tij }

normalizer [MATH] The normalizer of a subset S of a group G is the subgroup of G consisting of all elements x such that xsx^{-1} is in S whenever s is in S. { 'nȯr·mə‚līz·ər }

normalizing section [GEN] Removal of alleles that produce divergence from the average phenotype in a population by selecting against deviant individuals. { 'nȯr·mə‚līz·iŋ si‚lek·shən }

normally dispersive waves [PHYS] Waves which move more rapidly as their wavelengths increase. { 'nȯr·mə·lē di'spər·siv 'wāvz }

normally distributed observations [STAT] Any set of observations whose histogram looks like the normal curve. { 'nȯr·mə·lē di'strib·yəd·əd ‚äb·zər'vā·shənz }

normal magnetization curve [ELECTROMAG] Curve traced on a graph of magnetic induction versus magnetic field strength in an originally unmagnetized specimen, as the magnetic field strength is increased from zero. Also known as magnetization curve. { 'nȯr·məl ‚mag·nə·tə'zā·shən ‚kərv }

normal map [MATH] A planar map in which no more than three regions meet any one point and no region completely encloses another. Also known as regular map. [METEOROL] *See* normal chart. { 'nȯr·məl 'map }

normal mass shift [NUC PHYS] The portion of the mass shift that corresponds to the variation of reduced mass, and is thus easily calculated for all transitions. { 'nȯr·məl 'mas ‚shift }

normal matrix [MATH] A matrix is normal if multiplying it on the right by its adjoint is the same as multiplying it on the left. { 'nȯr·məl 'mā‚triks }

normal mode [COMPUT SCI] Operation of a computer in which it executes its own instructions rather than those of a different computer. { 'nȯr·məl ‚mōd }

normal-mode helix [ELECTROMAG] A type of helical antenna whose diameter and electrical length are considerably less than a wavelength, and which has a radiation pattern with greatest intensity normal to the helix axis. { 'nȯr·məl ‚mōd 'hē‚liks }

normal mode of vibration [MECH] Vibration of a coupled system in which the value of one of the normal coordinates oscillates and the values of all the other coordinates remain stationary. { 'nȯr·məl ‚mōd əv vī'brā·shən }

normal-mode rejection ratio [ELECTR] The ability of an amplifier to reject spurious signals at the power-line frequency or at harmonics of the line frequency. Abbreviated NMRR. { 'nȯr·məl ‚mōd ri'jek·shən ‚rā·shō }

normal moveout [GEOPHYS] In seismic prospecting, the increase in stepout time that results from an increase in distance from source to detector when there is no dip. { 'nȯr·məl 'müv‚aút }

normal number [MATH] A number whose expansion with respect to a given base (not necessarily 10) is such that all the digits occur with equal frequency, and all blocks of digits of the same length occur equally often. { 'nȯr·məl 'nəm·bər }

normal operation [MECH ENG] The operation of a boiler or pressure vessel at or below the conditions of coincident pressure and temperature for which the vessel has been designed. { 'nȯr·məl ‚äp·ə'rā·shən }

normal operator [MATH] A linear operator where composing it with its adjoint operator in either order gives the same result. Also known as normal transformation. { 'nȯr·məl 'äp·ə‚rād·ər }

normal orientation [COMPUT SCI] In optical character recognition, that determinate position which indicates that the line elements of an inputted source document appear parallel with the document's leading edge. { 'nȯr·məl ‚ȯr·ē·ən'tā·shən }

normal pace [IND ENG] The manual pace achieved by normal effort. { 'nȯr·məl 'pās }

normal pedal curve [MATH] The normal pedal curve of a given curve C with respect to a fixed point P is the locus of the foot of the perpendicular from P to the normal to C. { 'nȯr·məl 'ped·əl ‚kərv }

normal permeability [ELECTROMAG] The permeability of a specimen whose magnetic induction and magnetic field strength lie on the normal magnetization curve. Also known as cyclic permeability. { 'nȯr·məl ‚pər·mē·ə'bil·əd·ē }

normal pitch [MECH ENG] The distance between working

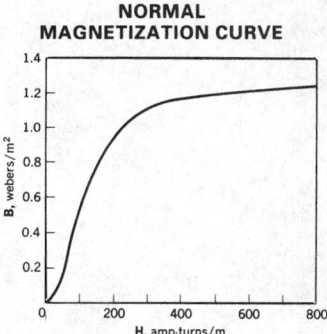

NORMAL MAGNETIZATION CURVE

Normal magnetization curve.

faces of two adjacent gear teeth, measured between the intersections of the line of action with the faces. { 'nȯr·məl 'pich }

normal plane [MATH] For a point P on a curve in space, the plane passing through P which is perpendicular to the tangent to the curve at P. { 'nȯr·məl 'plān }

normal-plate anemometer [ENG] A type of pressure-plate anemometer in which the plate, restrained by a stiff spring, is held perpendicular to the wind; the wind-activated motion of the plate is measured electrically; the natural frequency of this system can be made high enough so that resonance magnification does not occur. { 'nȯr·məl ¦plāt ‚an·ə'mäm·əd·ər }

normal polarity [GEOPHYS] Natural remanent magnetism nearly identical to the present ambient field. { 'nȯr·məl pə'lar·əd·ē }

normal potassium pyrophosphate *See* potassium pyrophosphate. { 'nȯr·məl pə'tas·ē·əm ‚pī·rō'fäs‚fāt }

normal pressure *See* standard pressure. { 'nȯr·məl 'presh·ər }

normal probability paper [STAT] Graph paper with the abscissa ruled in uniform increments and the ordinate ruled in such a way that the plot of a cumulative normal distribution is a straight line. { 'nȯr·məl ‚präb·ə'bil·əd·ē ‚pā·pər }

normal projection [MAP] A projection in which a three-dimensional object is projected onto two mutually perpendicular planes. { 'nȯr·məl prə'jek·shən }

normal range [COMPUT SCI] An interval within which results are expected to fall during normal operations. { 'nȯr·məl 'rānj }

normal ray [OPTICS] A ray that is incident perpendicularly on a surface. { 'nȯr·məl 'rā }

normal reaction [MECH] The force exerted by a surface on an object in contact with it which prevents the object from passing through the surface; the force is perpendicular to the surface, and is the only force that the surface exerts on the object in the absence of frictional forces. { 'nȯr·məl rē'ak·shən }

normal ripple mark [GEOL] An aqueous current ripple mark consisting of a simple asymmetrical ridge that may have various configurations. { 'nȯr·məl 'rip·əl ‚märk }

normal saline [PHYSIO] U.S. Pharmacopoeia title for a sterile solution of sodium chloride in purified water, containing 0.9 gram of sodium chloride in 100 milliliters; isotonic with body fluids. Also known as isotonic sodium chloride solution; normal salt solution; physiological saline; physiological salt solution; physiological sodium chloride solution; sodium chloride solution. { 'nȯr·məl 'sā‚lēn }

normal salt [CHEM] A salt in which all of the acid hydrogen atoms have been replaced by a metal, or the hydroxide radicals of a base are replaced by an acid radical; for example, Na_2CO_3. { 'nȯr·məl 'sȯlt }

normal salt solution *See* normal saline. { 'nȯr·məl 'sȯlt sə‚lü·shən }

normal section [MATH] Relative to a surface, this is a planar section produced by a plane containing the normal to a point. { 'nȯr·məl 'sek·shən }

normal segregation [MET] Segregation of the lower-melting-point constituents of an alloy primarily near the center of a casting (last portion to solidify). { 'nȯr·məl ‚seg·rə'gā·shən }

normal series [MATH] A normal series of a group G is a normal tower of subgroups of G, $G_0, G_1, \ldots, G_n$, in which $G_0 = G$ and G_n is the trivial group containing only the identity element. { 'nȯr·məl 'sir‚ēz }

normal silver sulfate *See* silver sulfate. { 'nȯr·məl 'sil·vər 'səl‚fāt }

normal slip fault *See* normal fault. { 'nȯr·məl 'slip ‚fȯlt }

normal soil [GEOL] A soil having a profile that is more or less in equilibrium with the environment. { 'nȯr·məl 'sȯil }

normal solution [CHEM] An aqueous solution containing one equivalent of the active reagent in grams in 1 liter of solution. { 'nȯr·məl sə'lü·shən }

normal space [MATH] A topological space in which any two disjoint closed sets may be covered respectively by two disjoint open sets. { 'nȯr·məl 'spās }

normal spiral galaxy [ASTRON] A galaxy that has a lens-shaped central portion with two arms that begin to coil in the same plane and in the same fashion immediately upon emerging from opposite sides of it. { 'nȯr·məl 'spī·rəl 'gal·ik‚sē }

normal state [NUC PHYS] A term sometimes used for ground state. { 'nȯr·məl 'stāt }

normal stress [MECH] The stress component at a point in a structure which is perpendicular to the reference plane. { 'nȯr·məl 'stres }

normal subgroup [MATH] A subgroup N of a group G where every expression $g^{-1}ng$ is in N for every g in G and every n in N. Also known as invariant subgroup; normal divisor. { 'nȯr·məl 'səb‚grüp }

normal surface [OPTICS] The surface that is generated by taking, at each point of the ray surface, the intersection of the tangent plane to the ray surface at that point with the perpendicular from the origin to this plane. { 'nȯr·məl 'sər·fəs }

normal synclinorium [GEOL] A synclinorium in which the axial surfaces of the subsidiary folds converge upward. { 'nȯr·məl ‚sin·klə'nȯr·ē·əm }

normal temperature and pressure *See* standard conditions. { 'nȯr·məl 'tem·prə·chər ən 'presh·ər }

normal thorium sulfate *See* thorium sulfate. { 'nȯr·məl 'thȯr·ē·əm 'səl‚fāt }

normal time [IND ENG] **1.** The time required by a trained worker to perform a task at a normal pace. **2.** The total of all the normal elemental times constituting a cycle or operation. Also known as base time; leveled time. { 'nȯr·məl 'tīm }

normal to a curve [MATH] The normal to a curve at a point is the line perpendicular to the tangent line at the point. { 'nȯr·məl tü ə 'kərv }

normal to a surface [MATH] The normal to a surface at a point is the line perpendicular to the tangent plane at that point. { 'nȯr·məl tü ə 'sər·fəs }

normal tower [MATH] A tower of subgroups, $G_0, G_1, \ldots, G_n$, such that each G_{i+1} is normal in G_i, $i = 1, 2, \ldots, n - 1$. { 'nȯr·məl 'taü·ər }

normal transformation [MATH] *See* normal operator. [STAT] A transformation on a variate that converts it into a variate which has a normal distribution. { 'nȯr·məl ‚tranz·fər'mā·shən }

normal twin [CRYSTAL] A twin crystal whose twin axis is perpendicular to the composition surface. { 'nȯr·məl 'twin }

normal uranium *See* native uranium. { 'nȯr·məl yü'rā·nē·əm }

normal volume *See* standard volume. { 'nȯr·məl 'väl·yəm }

normal water [OCEANOGR] Water whose chlorinity lies between 19.30 and 19.50 parts per thousand and has been determined to within ±0.001 per thousand. Also known as Copenhagen water; standard seawater. { 'nȯr·məl 'wȯd·ər }

normative mineral *See* standard mineral. { 'nȯr·məd·iv 'min·rəl }

normed linear space [MATH] A vector space which has a norm defined on it. Also known as normed vector space. { 'nȯrmd 'lin·ē·ər 'spās }

normed vector space *See* normed linear space. { 'nȯrmd 'vek·tər 'spās }

normoblast [HISTOL] The smallest of the nucleated precursors of the erythrocyte; slightly larger than a mature adult erythrocyte. Also known as acidophilic erythroblast; arthochromatic erythroblast; eosinophilic erythroblast; karyocyte; metakaryocyte; metanitricyte; metarubricyte. { 'nȯr·mə‚blast }

normochromatic [CELL MOL] Pertaining to cells of the erythrocytic series which have a normal staining color; attributed to the presence of a full complement of hemoglobin. { ¦nȯr·mə·krə'mad·ik }

normochromic [CELL MOL] Pertaining to erythrocytes which have a mean corpuscular hemoglobin (MCH), or color, index and a mean corpuscular hemoglobin concentration (MCHC), or saturation, index within two standard deviations above or below the mean normal. { ¦nȯr·mə'krō·mik }

normocyte [HISTOL] An erythrocyte having both a diameter and a mean corpuscular volume (MCV) within two standard deviations above or below the mean normal. { 'nȯr·mə‚sīt }

normothermia [PHYSIO] A state of normal body temperature. { ‚nȯr·mə'thər·mē·ə }

norphytane *See* pristane. { 'nȯr'fī‚tān }

Norris-Eyring reverberation formula [ACOUS] The reverberation time of a chamber, in seconds, is equal to 0.05 times its volume, in cubic feet, divided by the product of its surface area, in square feet, and the negative of the natural logarithm of 1 minus the absorption coefficient averaged over the surface. { 'när·əs 'ī·riŋ ri‚vər·bə'rā·shən ‚fȯr·myə·lə }

norte [METEOROL] **1.** The winter north wind in Spain.

2. A strong, cold northeasterly wind which blows in Mexico and on the shores of the Gulf of Mexico, and results from an outbreak of cold air from the north; actually, the Mexican extension of a norther. { 'norˈtä }

north [GEOD] The direction of the north terrestrial pole; the primary reference direction on the earth; the direction indicated by 000° in any system other than relative. { north }

North America [GEOGR] The northern of the two continents of the New World or Western Hemisphere, extending from narrow parts in the tropics to progressively broadened portions in middle latitudes and Arctic polar margins. { 'north ə'mer·i·kə }

North American anticyclone See North American high. { 'north ə'mer·i·kən ‚ant·i'sī‚klōn }

North American blastomycosis [MED] A type of blastomycosis caused by the diphasic fungus Blastomyces dermatitidis; two recognized forms are cutaneous and systemic. { 'north ə'mer·i·kən ‚bla·stō‚mī'kō·səs }

North American high [METEOROL] The relatively weak general area of high pressure which, as shown on mean charts of sea-level pressure, covers most of North America during winter. Also known as North American anticyclone. { 'north ə'mer·i·kən 'hī }

North American Nebula [ASTRON] A cloud of dust and gas in the constellation Cygnus; the density of this gas and dust is possibly a thousand times greater than the average density of interstellar gas; a much denser cloud of dust between the nebula and earth obscures portions of the emission nebula to create the appearance of the "Gulf of Mexico" and the "Atlantic Ocean." { 'north ə'mer·i·kən 'neb·yə·lə }

North American presentation-level protocol syntax [COMMUN] A format for transmitting text and graphics that allows the transmission of large amounts of information over narrow-bandwidth transmission lines. Abbreviated NAPLPS. { ‚north ə'mer·ə·kən ‚prē·zən‚tā·shən ‚lev·əl ‚prōd·ə‚kól 'sin‚taks }

North Atlantic Current [OCEANOGR] A wide, slow-moving continuation of the Gulf Stream originating in the region east of the Grand Banks of Newfoundland. { 'north at'lan·tik 'kə·rənt }

northbound node See ascending node. { 'north‚baund 'nōd }

North Cape Current [OCEANOGR] A warm current flowing northeastward and eastward around northern Norway, and curving into the Barents Sea. { 'north 'kāp ‚kə·rənt }

Northeast Drift Current [OCEANOGR] A North Atlantic Ocean current flowing northeastward toward the Norwegian Sea, gradually widening and, south of Iceland, branching and continuing as the Irminger Current and the Norwegian Current; it is the northern branch of the North Atlantic Current. { 'north‚ēst ‚drift 'kə·rənt }

northeaster [METEOROL] A northeast wind, particularly a strong wind or gale. Also spelled nor'easter. { nor'thē·stər or nó'rē·stər }

northeast storm [METEOROL] A cyclonic storm of the east coast of North America, so called because the winds over the coastal area are from the northeast; they may occur at any time of year but are most frequent and most violent between September and April. { 'north'ēst 'storm }

northeast trades [METEOROL] The trade winds of the Northern Hemisphere. { 'north'ēst 'trādz }

North Equatorial Current [OCEANOGR] Westward ocean currents driven by the northeast trade winds blowing over tropical oceans of the Northern Hemisphere. Also known as Equatorial Current. { 'north ‚ē·kwə'tór·ē·əl 'kə·rənt }

norther [METEOROL] A northerly wind. { 'nór·thər }

northerly turning error [AERO ENG] An acceleration error in the magnetic compass of an aircraft in a banked attitude during a turn, so called because it was first noted and is most pronounced during turns made from initial north-south courses; during a turn the magnetic needle is tilted from the horizontal, due to acceleration and the banking of the aircraft; in this position the compass needle will be acted upon by the vertical as well as the horizontal component of the earth's magnetic field; in addition, the compass needle is mechanically restricted in movement, due to tilt. Also known as turning error. { 'nór·thər·lē ‚tərn·iŋ ‚er·ər }

northern anthracnose [PL PATH] A fungus disease of red and crimson clovers in North America, Asia, and Europe caused by Kabatiella caulivora; depressed, linear brown lesions form on the stems and petioles. { 'nór·thərn an'thrak‚nōs }

Northern blotting [GEN] Method of ribonucleic acid (RNA) detection and identification in which the intact RNA is separated by size via gel electrophoresis, transferred (blotted) onto nitrocellulose or nylon paper, and then hybridized with labeled DNA probes. { 'nór·thərn 'bläd·iŋ }

Northern Cross See Cygnus. { 'nór·thərn 'krós }

Northern Crown See Corona Borealis. { 'nór·thərn 'kraún }

Northern Hemisphere [GEOGR] The half of the earth north of the Equator. { 'nór·thərn 'hem·i‚sfir }

Northern Hemisphere annular mode See Arctic Oscillation. { 'nór·thərn 'hem·ə‚sfir ‚an·yə·lər 'mōd }

northern lights See aurora borealis. { 'nór·thərn 'līts }

north foehn [METEOROL] A foehn condition sustained by wind flow across the Alps from north to south. { 'north 'fān }

north frigid zone [GEOGR] That part of the earth north of the Arctic Circle. { 'north 'frij·əd ‚zōn }

north geographic pole See North Pole. { 'north jē·ə'graf·ik 'pōl }

north geomagnetic pole See north pole. { 'north 'jē·ō‚mag‚ned·ik 'pōl }

Northill anchor [NAV ARCH] A lightweight kedge anchor with large flukes and a stock added to the crown perpendicular to them; used in sizes up to 100 pounds (45.4 kilograms). { 'nór‚thil 'aŋ·kər }

northing [NAV] The difference in latitude to the north between the present reckoning and the previous reckoning. { 'nór·thiŋ }

north magnetic pole See north pole. { 'north mag'ned·ik 'pōl }

North Pacific Current [OCEANOGR] The warm branch of the Kuroshio Extension flowing eastward across the Pacific Ocean. { 'north pə'sif·ik 'kə·rənt }

north point [ASTRON] The point on the celestial sphere, due north of the observer, at which the celestial meridian intersects the celestial horizon. { 'north 'póint }

north polar distance [ASTRON] The angular distance between a celestial object and the north celestial pole. { 'north 'pō·lər 'dis·təns }

north polar sequence [ASTRON] A list of stars in the vicinity of the north celestial pole whose photographic magnitudes have been measured as accurately as possible, and which are used as a basis for determining the magnitudes of other stars. { 'north 'pō·lər 'sē·kwəns }

North Polar Spur [ASTRON] A region of radio and soft x-ray emission, having a continuous spectral distribution, that extends from the galactic plane to the north galactic pole; believed to be the remnant of an old supernova. { 'north 'pō·lər spər }

north pole [ASTRON] The north celestial pole that indicates the zenith of the heavens when viewed from the north geographic pole. [ELECTROMAG] The pole of a magnet at which magnetic lines of force are considered as leaving the magnet; the lines enter the south pole; if the magnet is freely suspended, its north pole points toward the north geomagnetic pole. Also known as positive pole. [GEOPHYS] The geomagnetic pole in the Northern Hemisphere, at approximately latitude 78.5°N, longitude 69°W. Also known as north magnetic pole; north geomagnetic pole. { 'north 'pōl }

North Pole [GEOGR] The geographic pole located at latitude 90°N in the Northern Hemisphere of the earth; it is the northernmost point of the earth, and the northern extremity of the earth's axis of rotation. Also known as north geographic pole. { 'north 'pōl }

north-south effect [GEOPHYS] An effect whereby the intensity of cosmic radiation incident from the north is somewhat greater than that from the south in the Northern Hemisphere, and vice versa. { 'north 'saúth i‚fekt }

north-stabilized plan-position indicator [ENG] A heading-upward plan-position indicator; this term is deprecated because it may be confused with azimuth-stabilized plan-position indicator, a north-upward plan-position indicator. { 'north 'stā·bə‚līzd 'plan pə‚zish·ən 'in·də‚kād·ər }

North Star See Polaris. { 'north 'stär }

north temperate zone [CLIMATOL] That part of the earth between the Tropic of Cancer and the Arctic Circle. { 'north 'tem·prət ‚zōn }

northupite [MINERAL] $Na_3MgCl(CO_3)_2$ A white, yellow, gray, or colorless isometric mineral composed of magnesium sodium carbonate; occurs in octahedral crystals. { 'nȯr·thə,pīt }

north-upward plan position indicator [ENG] A plan position indicator on which north is maintained at the top of the indicator, regardless of the heading of the craft. { 'nȯrth 'əp·wərd ¦plan pə¦zish·ən 'in·də,kād·ər }

north-upward presentation [NAV] An indicator presentation, such as that of a plan position indicator or a radio direction finder, on which north is at the top of the indicator regardless of the heading of the craft. { 'nȯrth 'əp·wərd ,prez·ən'tā·shən }

northwester [METEOROL] A northwest wind. Also spelled nor'wester. { nȯrth'wes·tər or nȯr'wes·tər }

Norton equivalent circuit [ELEC] An equivalent circuit that consists of a parallel connection of a current source and a two-terminal circuit, where the current source is usually dependent on the electric signals applied to the input terminals. { ¦nȯrt·ən i'kwiv·ə·lənt ,sər·kət }

Norton's theorem [ELEC] The theorem that the voltage across an element that is connected to two terminals of a linear network is equal to the short-circuit current between these terminals in the absence of the element, divided by the sum of the admittances between the terminals associated with the element and the network respectively. { 'nȯrt·ənz ,thir·əm }

Norway Current [OCEANOGR] A continuation of the North Atlantic Current, which flows northward along the coast of Norway. Also known as Norwegian Current. { 'nȯr,wā ,kə·rənt }

Norwegian Current See Norway Current. { nȯr'wē·jən ,kə·rənt }

nor'wester See northwester. { nȯr'wes·tər }

NOS See network operating system. { ¦en¦ō'es or 'näs }

noscapine [PHARM] $C_{22}H_{23}NO_7$ A white, colorless, tasteless alkaloid obtained from opium; used as a nonaddicting antitussive. Formerly known as narcotine. { 'näs·kə,pēn }

nose [ANAT] The nasal cavities and the structures surrounding and associated with them in all vertebrates. [ENG] The foremost point or section of a bomb, missile, or something similar. [FL MECH] The dense, forward part of a turbidity current. [GEOL] **1.** A plunging anticline that is short and without closure. **2.** A projecting and generally overhanging buttress of rock. **3.** The projecting end of a hill, spur, ridge, or mountain. **4.** The central forward part of a parabolic dune. { nōz }

nosean See noselite. { 'nō·zē·ən }

nose cone [AERO ENG] A protective cone-shaped case for the nose section of a missile or rocket; may include the warhead, fusing system, stabilization system, heat shield, and supporting structure and equipment. { 'nōz ,kōn }

nose fuse [ORD] A fuse for use in the forward end (nose) of a bomb or other missile. { 'nōz ,fyüz }

nose-heavy [AERO ENG] Pertaining to an airframe in which the nose tends to sink when the longitudinal control is released in any attitude of normal flight. { 'nōz ,hev·ē }

nose height [ANTHRO] The distance measured between the nasion and the subnasale. { 'nōz ,hīt }

nose leaf [VERT ZOO] A leaflike expansion of skin on the nose of certain bats; believed to have a tactile function. { 'nōz ,lēf }

nose length [ANTHRO] The distance measured between the nasion and the pronasale. { 'nōz ,leŋkth }

noselite [MINERAL] $Na_4Al_3Si_3O_{12}\cdot SO_4$ A gray, blue or brown mineral of the sodalite group; similar to haüynite; hardness is 5.5 on Mohs scale. Also known as nosean. { 'nōz·ə,līt }

nose radius [MECH ENG] The radius measured in the back rake or top rake plane of a cutting tool. { 'nōz ,rād·ē·əs }

nose sill [ENG] A short timber located under the end of the main sill of a standard rig front of a well. { 'nōz ,sil }

nose spray [ORD] Fragments of a bursting projectile that are thrown forward in the line of flight, in contrast to base spray and side spray. { 'nōz ,sprā }

nosing [BUILD] Projection of a tread of a stair beyond the riser below it. [CIV ENG] A transverse, horizontal motion of a locomotive that exerts a lateral force on the track. { 'nōz·iŋ }

No. 6 fuel See bunker C fuel oil. { ¦nəm·bər ¦siks 'fyül }

nosocomial [MED] **1.** Pertaining to a hospital. **2.** Of disease, caused or aggravated by hospital life. { ¦näz·ə¦kō·mē·əl }

Nosodendridae [INV ZOO] The wounded-tree beetles, a small family of coleopteran insects in the superfamily Dermestoidea. { ,näz·ə'den·drə,dē }

nosology [MED] The science of classification of diseases. { nō'säl·ə·jē }

nosomania [PSYCH] An unfounded, morbid belief that one is suffering from a disease. { ,näs·ə'mā·nyə }

nosophilia [PSYCH] A morbid desire to be sick. { ,näs·ə'fil·ē·ə }

nosophobia [PSYCH] An irrational fear of acquiring a disease. { ,näz·ə'fō·bē·ə }

nostopathy [PSYCH] Pathogenic homecoming, that is, stress-precipitated illness as observed in individuals who have spent a considerable length of time in institutions such as hospitals or prisons. { näs'tap·ə·thē }

nostril [ANAT] One of the external orifices of the nose. Also known as naris. { 'näs·trəl }

nostrum [PHARM] A quack medicine. { 'näs·trəm }

Notacanthidae [VERT ZOO] A family of benthic, deep-sea teleosts in the order Notacanthiformes, including the spiny eel. { ,nōd·ə'kan·thə,dē }

Notacanthiformes [VERT ZOO] An order of actinopterygian fishes whose body is elongated, tapers posteriorly, and has no caudal fin. { ,nōd·ə,kan·thə'fȯr·mēz }

NOTAM [COMMUN] The aviation communications code word for "notice to airmen"; used to provide timely information or conditions which are essential to flight operations; these notices are controlled by the National Flight Data Center. { 'nō,tam }

notancephalia [MED] A congenital anomaly characterized by a deficiency in the occipital region of the skull. { ,nō·tan·sə'fal·ē·ə }

NOT-AND See NAND. { 'nät 'and }

notanencephalia [MED] A congenital anomaly marked by defective development or absence of the cerebellum. { ,nō·tan,en·sə'fal·ē·ə }

Notarys-Mercereau microbridge See proximity-effect microbridge. { nō'tar·əs 'mer·sə,rō 'mī·krō,brij }

notation [COMPUT SCI] See positional notation. [MATH] **1.** The use of symbols to denote quantities or operations. **2.** See positional notation. { nō'tā·shən }

notch [ELECTR] Rectangular depression extending below the sweep line of the radar indicator in some types of equipment. [ENG] A V-shaped indentation or cut in a surface or edge. [GEOL] A deep, narrow cut near the high-water mark at the base of a sea cliff. [GEOGR] A narrow passage between mountains or through a ridge, hill, or mountain. { näch }

notch acuity [MET] The severity of the stress concentration produced by a given notch in a structure; it is expressed as the ratio of the notch depth to the notch radius (depth is small compared to width, or diameter, of the narrowest cross section). { 'näch ə'kyü·əd·ē }

notch antenna [ELECTROMAG] Microwave antenna in which the radiation pattern is determined by the size and shape of a notch or slot in a radiating surface. { 'näch an,ten·ə }

notch brittleness [MET] Susceptibility of a material to brittle fracture at areas of stress concentration; in notch tensile testing, a material has notch brittleness if the notch strength lies below the tensile strength. { 'näch 'brid·əl·nəs }

notch depth [MET] The distance from the surface of a metal specimen to the bottom of the notch. { 'näch ,depth }

notch ductility [MET] Percentage reduction in area of the specimen after failure in a notched tensile test. { 'näch dək'til·əd·ē }

notched-bar test [MET] Test in which a notched metal specimen is bent with the notch in tension. { 'nächt ¦bär ,test }

notch filter [ELECTR] A band-rejection filter that produces a sharp notch in the frequency response curve of a system; used in television transmitters to provide attenuation at the low-frequency end of the channel, to prevent possible interference with the sound carrier of the next lower channel. { 'näch ,fil·tər }

notch graft [BOT] A plant graft in which the scion is inserted in a narrow slit in the stock. { 'näch ,graft }

notching [ELEC] Term indicating that a predetermined number of separate impulses are required to complete operation of

---

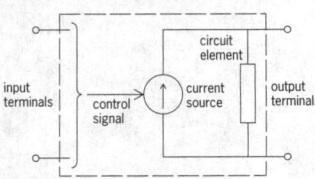

The circuit consists of the parallel connection of a current source and a two-terminal circuit.

NOTACANTHIFORMES

Spiny eel (*Notacanthus nasus*).

a relay. [MECH ENG] Cutting out various shapes from the ends or edges of a workpiece. { 'näch·iŋ }

notching press [MECH ENG] A mechanical press for notching straight or rounded edges. { 'näch·iŋ ‚pres }

notch sensitivity [MET] A measure of the reduction in strength of a metal caused by the presence of a notch. { 'näch ‚sen·sə'tiv·əd·ē }

Notch signaling [EMBRYO] An evolutionarily conserved developmental pathway utilized during the differentiation of a plethora of tissue types, in organisms as diverse as nematodes and humans. { 'näch ‚sig·nəl·iŋ }

notch strength [MET] The ratio of maximum tensional load required to fracture a notched specimen to the original minimum cross-sectional area. { 'näch ‚streŋkth }

notch test [MET] A tensile or creep test of a metal to determine the effect of a surface notch. { 'näch ‚test }

NOT circuit [ELECTR] A logic circuit with one input and one output that inverts the input signal at the output; that is, the output signal is a logical 1 if the input signal is a logical 0, and vice versa. Also known as inverter circuit. { 'nät ‚sər·kət }

note [ACOUS] **1.** A conventional sign indicating the pitch of a musical sound by its position on a staff, and the duration of the sound by its shape. **2.** The sound indicated by this sign. { nōt }

notebook computer [COMPUT SCI] A portable computer typically weighing less than 6 pounds (3 kilograms) that has a flat-panel display and miniature hard disk drives, and is powered by rechargeable batteries. Also known as laptop computer. { 'nōt‚bük kəm‚pyüd·ər }

Noteridae [INV ZOO] The burrowing water beetles, a small family of coleopteran insects in the suborder Adephaga. { nō'ter·ə‚dē }

NOT function [MATH] A logical operator having the property that if P is a statement, then the NOT of P is true if P is false, and false if P is true. { 'nät ‚faŋk·shən }

Nothosauria [PALEON] A suborder of chiefly marine Triassic reptiles in the order Sauropterygia. { ‚näth·ə'sȯr·ē·ə }

Notidanoidea [VERT ZOO] A suborder of rare sharks in the order Selachii; all retain the primitive jaw suspension of the order. { ‚nōd·ə·də'nȯid·ē·ə }

Notiomastodontinae [PALEON] A subfamily of extinct elephantoid proboscidean mammals in the family Gomphotheriidae. { ‚nōd·ē·ō‚mas·tə'dän·tə‚nē }

Notioprogonia [PALEON] A suborder of extinct mammals comprising a diversified archaic stock of Notoungulata. { ‚nōd·ē·ō·prə'gō·nē·ə }

notochord [VERT ZOO] An elongated dorsal cord of cells which is the primitive axial skeleton in all chordates; persists in adults in the lowest forms (*Branchiostoma* and lampreys) and as the nuclei pulposi of the intervertebral disks in adult vertebrates. { 'nōd·ə‚kȯrd }

notochordal canal [EMBRYO] A canal formed by a continuation of the primary pit into the head process of mammalian embryos; provides a temporary connection between the yolk sac and amnion. { ‚nōd·ə‚kȯrd·əl kə'nal }

notochordal plate [EMBRYO] A plate of cells representing the root of the head process of the embryo after the embryo becomes vesiculated. { ‚nōd·ə‚kȯrd·əl 'plāt }

Notodelphyidiformes [INV ZOO] A tribe of the Gnathostoma in some systems of classification. { ‚nōd·ə·del‚fid·ə'fȯr‚mēz }

Notodelphyoida [INV ZOO] A small group of crustaceans bearing a superficial resemblance to many insect larvae as a result of uniform segmentation, comparatively small trunk appendages, and crowding of inconspicuous oral appendages into the anterior portion of the head. { ‚nōd·ə‚del·fē'ȯid·ə }

Notodontidae [INV ZOO] The puss moths, a family of lepidopteran insects in the superfamily Noctuoidea, distinguished by the apparently three-branched cubitus. { ‚nōd·ə'dän·tə‚dē }

Notogaean [ECOL] Pertaining to or being a biogeographic region including Australia, New Zealand, and the southwestern Pacific islands. { ‚nōd·ə'jē·ən }

Notommatidae [INV ZOO] A family of rotifers in the order Monogonota including forms with a cylindrical body covered by a nonchitinous cuticle and with a slender posterior foot. { ‚nōd·ə'mat·ə‚dē }

Notomyotina [INV ZOO] A suborder of echinoderms in the order Phanerozonida in which the upper marginals alternate in position with the lower marginals, and each tube foot has a terminal sucking disk. { ‚nōd·ə·mī'ät·ən·ə }

Notonectidae [INV ZOO] The backswimmers, a family of aquatic, carnivorous hemipteran insects in the superfamily Notonectoidea; individuals swim ventral side up, aided in breathing by an air bubble. { ‚nōd·ə'nek·tə‚dē }

Notonectoidea [INV ZOO] A superfamily of Hemiptera in the subdivision Hydrocorisae. { ‚nōd·ə·nek'tȯid·ē·ə }

notopodium [INV ZOO] The dorsal branch of a parapodium in certain annelids. { ‚nōd·ə'pō·dē·əm }

Notopteridae [VERT ZOO] The featherbacks, a family of actinopterygian fishes in the order Osteoglossiformes; bodies are tapered and compressed, with long anal fins that are continuous with the caudal fin. { ‚nō‚täp'ter·ə‚dē }

NOT-OR *See* NOR. { 'nät 'ȯr }

Notoryctidae [PALEON] An extinct family of Australian insectivorous mammals in the order Marsupialia. { ‚nōd·ə'rik·tə‚dē }

Notostigmata [INV ZOO] The single suborder of the Opilioacriformes, an order of mites. { ‚nōd·ə·stig'mäd·ə }

Notostigmophora [INV ZOO] A subclass or suborder of the Chilopoda, including those centipedes embodying primitive as well as highly advanced characters, distinguished by dorsal respiratory openings. { ‚nōd·ə·stig'mäf·ə·rə }

Notostraca [INV ZOO] The tadpole shrimps, an order of crustaceans generally referred to the Branchiopoda, having a cylindrical trunk that consists of 25–44 segments, a dorsoventrally flattened dorsal shield, and two narrow, cylindrical cercopods on the telson. { nə'täs·trə·kə }

Nototheniidae [VERT ZOO] A family of perciform fishes in the suborder Blennioidei, including most of the fishes of the permanently frigid waters surrounding Antarctica. { ‚nōd·ə·thə'nī·ə‚dē }

Notoungulata [PALEON] An extinct order of hoofed herbivorous mammals, characterized by a skull with an expanded temporal region, primitive dentition, and primitive feet with five toes, the weight borne mainly by the third digit. { ‚nōd·ō‚əŋ·gyə'läd·ə }

nought state *See* zero condition. { 'nȯt ‚stāt }

noumeite *See* garnierite. { 'nü·mē‚īt }

nourishment [GEOL] The replenishment of a beach, either naturally (such as by littoral transport) or artificially (such as by deposition of dredged materials). [HYD] *See* accumulation. { 'nər·ish·mənt }

NOV *See* Nodamura virus.

nova [ASTRON] A star that suddenly becomes explosively bright, the term is a misnomer because it does not denote a new star but the brightening of an existing faint star. { 'nō·və }

novaculite [GEOL] A siliceous sedimentary rock that is dense, hard, even-textured, light-colored, and characterized by dominance of microcrystalline quartz over chalcedony. Also known as razor stone. { nə'vak·yə‚līt }

novaculitic chert [GEOL] A gray chert that fragments into slightly rough, splintery pieces. { nə‚vak·yə‚lid·ik 'chərt }

novalike symbiotic [ASTRON] A symbiotic star consisting of a red giant combined with a white dwarf, in which thermonuclear reactions of hydrogen and helium from the red giant accreted in the surface layers of the white dwarf are believed to produce outbursts of energy similar to those observed in a nova. Also known as symbiotic nova. { 'nō·və‚līk ‚sim·bē'äd·ik }

nova-like variable [ASTRON] A binary star in which one component is a white dwarf and the other is a main-sequence star that overflows its Roche lobe and feeds an accretion disk around the white dwarf, resulting in irregular variability. { 'nō·və‚līk 'ver·ē·ə·bəl }

novar [ELECTR] Beam-power tube having a nine-pin base. { 'nō‚vär }

novobiocin [MICROBIO] $C_{30}H_{36}O_{11}N_2$ A moderately broad-spectrum antibiotic produced by strains of *Streptomyces niveus* and *S. spheroides*; it is a dibasic acid and is converted either to the monosodium salt or to the calcium acid salt for pharmaceutical use. { ‚nō·və'bī·ə·sən }

novolac resin [MATER] Any of the thermoplastic phenol-formaldehyde resins made with an excess of phenol in the reaction; used in varnishes. { 'nō·və‚lak 'rez·ən }

nowcasting [METEOROL] **1.** The detailed description of the

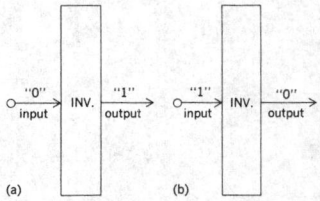

NOT CIRCUIT

NOT circuit, also known as inverter circuit. *(a)* A "0" input produces a "1" output. *(b)* A "1" input produces a "0" output.

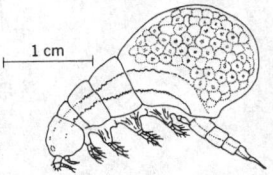

NOTODELPHYOIDA

1 cm

Doropygus psyllus, female; eggs are incubated in swollen part of thorax.

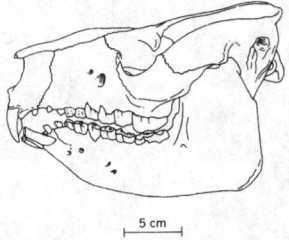

NOTOUNGULATA

5 cm

Skull and jaw of *Adinotherium ovinum*, an early Miocene toxodontid notoungulate.

current weather along with forecasts obtained by extrapolation up to about 2 hours ahead. **2.** Any area-specific forecast for the period up to 12 hours ahead that is based on very detailed observational data. { 'naȯ,kast·iŋ }

nowhere dense set [MATH] A set in a topological space whose closure has empty interior. Also known as rare set. { 'nō,wer 'dens 'set }

no-wind heading [NAV] A heading determined without allowance for wind, and hence equal to the course. { 'nō ,wind 'hed·iŋ }

no-wind position See air position. { 'nō ,wind pə'zish·ən }

nox [OPTICS] A unit of illumination, used in measuring low-level illumination, equal to 10^{-3} lux. { näks }

NOX See nicotinamide adenine dinucleotide oxidase. { näks or ¦en¦ō'eks }

noy [ACOUS] A unit of perceived noisiness equal to the perceived noisiness of random noise occupying the frequency band 910–1090 hertz at a sound pressure level of 40 decibels above 0.0002 microbar; a sound that is n times as noisy as this sound has a perceived noisiness of n noys, under the assumption that the perceived noisiness of a sound increases with physical intensity at the same rate as the loudness. { nȯi }

nozzle [DES ENG] A tubelike device, usually streamlined, for accelerating and directing a fluid, whose pressure decreases as it leaves the nozzle. { 'näz·əl }

nozzle blade [AERO ENG] Any one of the blades or vanes in a nozzle diaphragm. Also known as nozzle vane. { 'näz·əl ,blād }

nozzle-contraction-area ratio [DES ENG] Ratio of the cross-sectional area for gas flow at the nozzle inlet to that at the throat. { 'näz·əl kən'trak·shən ¦er·ē·ə ,rā·shō }

nozzle-divergence loss factor [FL MECH] The ratio between the momentum of the gases in a nozzle and the momentum of an ideal nozzle. { 'näz·əl də,vər·jəns 'lȯs ,fak·tər }

nozzle efficiency [MECH ENG] The efficiency with which a nozzle converts potential energy into kinetic energy, commonly expressed as the ratio of the actual change in kinetic energy to the ideal change at the given pressure ratio. { 'näz·əl i,fish·ən·sē }

nozzle exit area [DES ENG] The cross-sectional area of a nozzle available for gas flow measured at the nozzle exit. { 'näz·əl ¦eg·zət ,er·ē·ə }

nozzle-expansion ratio [DES ENG] Ratio of the cross-sectional area for gas flow at the exit of a nozzle to the cross-sectional area available for gas flow at the throat. { 'näz·əl ik'pan·shən ,rā·shō }

nozzle-mix gas burner [ENG] A burner in which injection nozzles mix air and fuel gas at the burner tile. { 'näz·əl ,miks 'gas ,bər·nər }

nozzle process [NUCLEO] A method of separating isotopes by allowing a gaseous compound to exhaust through a properly shaped nozzle. { 'näz·əl ,prä·səs }

nozzle throat [DES ENG] The portion of a nozzle with the smallest cross section. { 'näz·əl ,thrōt }

nozzle throat area [DES ENG] The area of the minimum cross section of a nozzle. { 'näz·əl ¦thrōt ,er·ē·ə }

nozzle thrust coefficient [AERO ENG] A measure of the amplification of thrust due to gas expansion in a particular nozzle as compared with the thrust that would be exerted if the chamber pressure acted only over the throat area. Also known as thrust coefficient. { 'näz·əl ¦thrəst ,kō·i,fish·ənt }

nozzle vane See nozzle blade. { 'näz·əl ,vān }

Np See neptunium.

NP [COMPUT SCI] The class of decision problems for which solutions can be checked in polynomial time.

NP-complete problem [COMPUT SCI] One of the hardest problems in class NP, such that, if there are any problems in class NP but not in class P, this is one of them. { ¦en¦pē kəm'plēt ,präb·ləm }

NP-hard [COMPUT SCI] Referring to problems at least as hard as or harder than any problem in NP. Given a method for solving an NP-hard problem, any problem in NP can be solved with only polynomially more work. { ¦en¦pē 'härd }

npin transistor [ELECTR] An *npn* transistor which has a layer of high-purity germanium between the base and collector to extend the frequency range. { 'en,pin tran'zis·tər }

N-P-K [CHEM ENG] The code identifying the components in a fertilizer mixture: nitrogen (N), phosphorus pentoxide (P),

and potassium oxide (K). Fertilizers are graded in the order N-P-K, with the numbers indicating the percentage of the total weight of each component. For example, 5-10-10 represents a mixture containing by weight 5% nitrogen, 10% phosphorus pentoxide, and 10% potassium oxide.

N-plus-one address instruction [COMPUT SCI] An instruction with N + 1 address parts, one of which gives the location of the next instruction to be carried out. { 'en pləs 'wən 'ad,res in,strək·shən }

npnp diode See pnpn diode. { ¦en,pē¦en,pē 'dī,ōd }

npnp transistor [ELECTR] An *npn*-junction transistor having a transition or floating layer between *p* and *n* regions, to which no ohmic connection is made. Also known as *pnpn* transistor. { ¦en,pē¦en,pē tran'zis·tər }

npn semiconductor [ELECTR] Double junction formed by sandwiching a thin layer of *p*-type material between two layers of *n*-type material of a semiconductor. { 'en,pē'en 'sem·i·kən,dək·tər }

npn transistor [ELECTR] A junction transistor having a *p*-type base between an *n*-type emitter and an *n*-type collector; the emitter should then be negative with respect to the base, and the collector should be positive with respect to the base. { 'en,pē'en tran'zis·tər }

NPO-body [ELEC] Referring to a series of temperature-compensating capacitors that have an invariant dielectric constant over a specified temperature range. { ¦en¦pē¦ō 'bäd·ē }

np semiconductor [ELECTR] Region of transition between *n*- and *p*-type material. { ¦en¦pē 'sem·i·kən,dək·tər }

NPSH See net positive suction head.

NPV See nuclear polyhedrosis virus.

NPX See numeric processor extension.

NR See nitrile rubber.

NRM See natural remanent magnetization.

NRM wind scale [METEOROL] A wind scale adapted by the United States Forest Service for use in the forested areas of the Northern Rocky Mountains (NRM); it is an adaptation of the Beaufort wind scale; the difference between these two scales lies in the specification of the visual effects of the wind; the force numbers and the corresponding wind speeds are the same in both. { ¦en¦är'em 'wind ,skāl }

N rod bit [DES ENG] A Canadian standard noncoring bit having a set diameter of 2.940 inches (74.676 millimeters). { 'en 'räd ,bit }

NRR See noise reduction rating.

NRS See nuclear reaction spectrometry.

NRZ See non-return-to-zero.

N scan See N scope. { 'en ,skan }

N scope [ELECTR] A cathode-ray scope combining the features of K and M scopes. Also known as N indicator; N scan. { 'en ,skōp }

N shell [ATOM PHYS] The fourth layer of electrons about the nucleus of an atom, having electrons characterized by the principal quantum number 4. { 'en ,shel }

NSOM See near-field scanning optical microscopy. { 'en,säm or ¦en¦es¦ō'em }

n space [MATH] A vector space over the real numbers whose basis has n vectors. { 'en ,spās }

n-sphere [MATH] The set of all points in $(n+1)$-dimensional euclidean space whose distance from the origin is unity, where n is a positive integer. { 'en ,sfir }

N star [ASTRON] An obsolete classification for a star in the carbon sequence; has about the same temperature as an M star in the Draper catalog. { 'en ,stär }

nt See nit.

NTA See nitrilotriacetic acid.

NTP See standard conditions.

n-type conduction [ELECTR] The electrical conduction associated with electrons, as opposed to holes, in a semiconductor. { 'en ,tīp kən,dək·shən }

N-type crystal rectifier [ELECTR] Crystal rectifier in which forward current flows when the semiconductor is negative with respect to the metal. { 'en ,tīp ¦krist·əl 'rek·tə,fī·ər }

n-type germanium [ELECTR] Germanium to which more impurity atoms of donor type (with valence 5, such as antimony) than of acceptor type (with valence 3, such as indium) have been added, with the result that the conduction electron density exceeds the hole density. { 'en ,tīp jər'mā·nē·əm }

n-type semiconductor [ELECTR] An extrinsic semiconductor in which the conduction electron density exceeds the hole density. { 'en ‚tīp 'sem·i·kən‚dək·tər }

nub [TEXT] A planned knot or tangle in yarn to give a desired irregular texture to a fabric. { nəb }

nubbin [GEOL] **1.** One of the isolated bedrock knobs or small hills forming the last remnants of a mountain crest or mountain range that has succumbed to desert erosion. **2.** A residual boulder, commonly granitic, occurring on a desert dome or broad pediment. { 'nəb·ən }

Nubeculae See Magellanic Clouds. { nü'bek·yə‚lē }

Nubecula Major See Large Magellanic Cloud. { nü'bek·yə lə 'mā·jər }

Nubecula Minor See Small Magellanic Cloud. { nü'bek·yə lə 'mīn·ər }

nub yarns See knop. { nəb ‚yärnz }

nucellus [BOT] The oval central mass of tissue in the ovule; contains the embryo sac. { 'nü'sel·əs }

nucha [ANAT] The nape of the neck. { 'nü·kə }

nuchal ligament [ANAT] An elastic ligament extending from the external occipital protuberance and middle nuchal line to the spinous process of the seventh cervical vertebra. Also known as ligamentum nuchae. { 'nü·kəl 'lig·ə·mənt }

nuchal organ [INV ZOO] Any of various sense organs on the prostomium of many annelids, which are sensitive to changes in the immediate environment of the individual. { 'nü·kəl ‚ȯr·gən }

nuchal rigidity [MED] Stiffness in the nape of the neck, often accompanied by pain and spasm on attempts to move the head; the most common sign of meningitis. { 'nü·kəl ri'jid·əd·ē }

nuchal tentacle [INV ZOO] Any of various filiform or thick, fleshy tactoreceptors on anterior segments of many annelids. { 'nü·kəl ‚ten·tə·kəl }

nuclear [CHEM] Pertaining to a group of atoms joined directly to the central group of atoms or central ring of a molecule. [NUCLEO] Pertaining to nuclear energy. [NUC PHYS] Pertaining to the atomic nucleus. { 'nü·klē·ər }

nuclear absorption [NUC PHYS] Absorption of energy by the nucleus of an atom. { 'nü·klē·ər əb'sȯrp·shən }

nuclear adiabatic demagnetization [CRYO] A technique for cooling substances, in which the sample is first cooled to temperatures on the order of 10^{-2} K in an extremely intense magnetic field and is then thermally isolated and removed from the field to reach temperatures on the order of 10^{-6} K. { 'nü·klē·ər ‚ad·ē·ə'bad·ik dē‚mag·nə·tə'zā·shən }

nuclear age determination See radiometric dating. { 'nü·klē·ər ‚āj di‚tər·mə'nā·shən }

nuclear angular momentum See nuclear spin. { 'nü·klē·ər 'aŋ·gyə·lər mə'men·təm }

nuclear atom [CHEM] An atomic structure consisting of dense, positively charged nucleus (neutrons and protons) surrounded by a corresponding set of negatively charged electrons. { 'nü·klē·ər 'ad·əm }

nuclear battery [NUCLEO] A primary battery in which the energy of radioactive material is converted into electric energy by solar cells or other energy converters. Also known as atomic battery; radioisotope battery; radioisotopic generator. { 'nü·klē·ər 'bad·ə·rē }

nuclear binding energy [NUC PHYS] The energy required to separate an atom into its constituent protons, neutrons, and electrons. { 'nü·klē·ər 'bīnd·iŋ ‚en·ər·jē }

nuclear boiler [NUCLEO] A nuclear reactor in which water is the primary coolant and is converted to steam, such as a pressurized-water reactor or a boiling-water reactor. { 'nü·klē·ər 'bȯil·ər }

nuclear bomb See atomic bomb. { 'nü·klē·ər 'bäm }

nuclear breeder [NUCLEO] A nuclear reactor in which more fissionable material is formed in each generation than is used up in fission. { 'nü·klē·ər 'brēd·ər }

nuclear capture [NUC PHYS] Any process in which a particle, such as a neutron, proton, electron, muon, or alpha particle, combines with a nucleus. { 'nü·klē·ər 'kap·chər }

nuclear chain reaction [NUCLEO] A succession of generation after generation of acts of nuclear division such that the neutrons set free in the nuclear disruptions of the nth generation split the fissile nuclei (^{233}U, ^{235}U, ^{239}Pu) of the ($n + 1$)st generation. Also known as chain reaction. { 'nü·klē·ər 'chān rē‚ak·shən }

nuclear chemical engineering [CHEM ENG] The branch of chemical engineering that deals with the production and use of radioisotopes. { 'nü·klē·ər ¦kem·ə·kəl ‚en·jə'nir·iŋ }

nuclear chemistry [ATOM PHYS] Study of the atomic nucleus, including fission and fusion reactions and their products. { 'nü·klē·ər 'kem·ə·strē }

nuclear cloud [NUCLEO] The cloud of dust and gas formed from the debris of a nuclear explosion. { 'nü·klē·ər 'klaůd }

nuclear collision [NUC PHYS] A collision between an atomic nucleus and another nucleus or particle. { 'nü·klē·ər kə'lizh·ən }

nuclear converter See converter. { 'nü·klē·ər kən'vərd·ər }

nuclear cross section [NUC PHYS] A measure of the probability for a reaction to occur between a nucleus and a particle; it is an area such that the number of reactions which occur in a sample exposed to a beam of particles equals the product of the number of nuclei in the sample and the number of incident particles which would pass through this area if their velocities were perpendicular to it. { 'nü·klē·ər 'krȯs ‚sek·shən }

nuclear decay mode [NUC PHYS] One of the ways in which a nucleus can undergo radioactive decay, distinguished from other decay modes by the resulting isotope and the particles emitted. { 'nü·klē·ər di'kā ‚mōd }

nuclear density [NUC PHYS] The mass per unit volume of a nucleus as a function of distance from the center of the nucleus, as determined by a number of different types of experiments which are in reasonably good agreement. { 'nü·klē·ər 'den·səd·ē }

nuclear device [NUCLEO] A nuclear explosive used for peaceful purposes, tests, or experiments. { 'nü·klē·ər di'vīs }

nuclear dimorphism [INV ZOO] In ciliated protozoas, the occurrence of two types of nuclei in a cell, each with different genetic functions. { 'nü·klē·ər dī mȯr‚fiz·əm }

nuclear electric power generation [ELEC] Large-scale generation of electric power in which the source of energy is nuclear fission, generally in a nuclear reactor, or nuclear fusion. { 'nü·klē·ər i¦lek·trik 'paů·ər ‚jen·ə‚rā·shən }

nuclear-electric propulsion [AERO ENG] A system of propulsion utilizing a nuclear reactor to generate electricity which is then used in an electric propulsion system or as a heat source for the working fluid. { 'nü·klē·ər i¦lek·trik prə‚pəl·shən }

nuclear-electric rocket engine [AERO ENG] A rocket engine in which a nuclear reactor is used to generate electricity that is used in an electric propulsion system or as a heat source for the working fluid. { 'nü·klē·ər i¦lek·trik 'räk·ət ‚en·jən }

nuclear emulsion [NUCLEO] A photographic emulsion specially designed to register individual tracks of ionizing particles. { 'nü·klē·ər i'məl·shən }

nuclear emulsion counter [NUCLEO] A device used to measure the intensity of ionizing radiation by counting the number of tracks in an emulsion which has been exposed to the radiation. { 'nü·klē·ər i'məl·shən ‚kaůnt·ər }

nuclear energy [NUCLEO] Energy released by nuclear fission or nuclear fusion. Also known as atomic energy. { 'nü·klē·ər 'en·ər·jē }

nuclear engine [NUCLEO] A type of thermal engine utilizing nuclear fission or fusion reactions to heat a working fluid for propulsive purposes. { 'nü·klē·ər 'en·jən }

nuclear engineering [NUCLEO] The branch of technology that deals with the utilization of the nuclear fission process, and is concerned with the design and construction of nuclear reactors and auxiliary facilities, the development and fabrication of special materials, and the handling and processing of reactor products. { 'nü·klē·ər ‚en·jə'nir·iŋ }

nuclear envelope [CELL MOL] A structure consisting of two membranes that surrounds the nucleus; the outermost membrane is continuous with the rough endoplasmic reticulum. { 'nü·klē·ər 'en·və‚lōp }

nuclear equation of state [NUC PHYS] Any equation that relates pressure, density, and temperature for nuclear matter. { ¦nü·klē·ər i·kwā·zhən əv 'stāt }

nuclear excavation [ENG] The use of nuclear explosions to remove earth for constructing harbors, canals, and other facilities. { 'nü·klē·ər ‚ek·skə'vā·shən }

nuclear explosion [NUCLEO] An explosion for which the energy is produced by a nuclear transformation, either fission or fusion. { 'nü·klē·ər ik'splō·zhən }

nuclear fission See fission. { 'nü·klē·ər 'fish·ən }

nuclear fluid dynamic model [NUC PHYS] A model of high-energy nuclear collisions in which nuclear matter is assumed

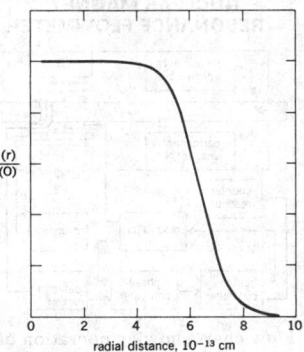

NUCLEAR DENSITY

Distribution of charge in a nucleus of gold; this is believed to be approximately proportional to the nuclear mass density, on the assumption that the distributions of protons and neutrons are approximately the same; $\rho(r)$ = density at distance r from the center; $\rho(O)$ = density at the center.

NUCLEAR FORCE

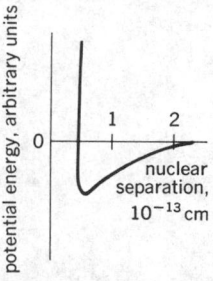

The potential operating between two nucleons due to the nuclear forces.

NUCLEAR MAGNETIC RESONANCE FLOWMETER

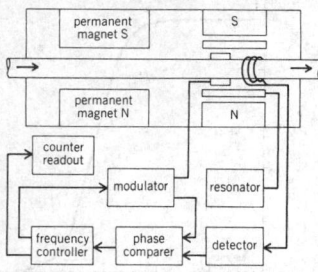

Flow chart showing operation of a nuclear magnetic resonance flowmeter.

NUCLEAR MAGNETIC RESONANCE GYROSCOPE

Experimental model of a nuclear magnetic resonance gyroscope. (*Singer Kearfott Division*)

to behave like a compressible fluid whose pressure, density, and temperature are related by an equation of state. Also known as nuclear hydrodynamic model. { 'nü·klē·ər 'flü·əd dī,nam·ik 'mäd·əl }

nuclear force [NUC PHYS] That part of the force between nucleons which is not electromagnetic; it is much stronger than electromagnetic forces, but drops off very rapidly at distances greater than about 10^{-13} centimeter; it is responsible for holding the nucleus together. { 'nü·klē·ər 'fòrs }

nuclear fuel [NUCLEO] A fissionable or fertile isotope with a reasonably long half-life, used as a source of energy in a nuclear reactor. Also known as fission fuel; reactor fuel. { 'nü·klē·ər ¦fyül }

nuclear fuel cycle See reactor fuel cycle. { 'nü·klē·ər ¦fyül ¸sī·kəl }

nuclear fuel element [NUCLEO] A piece of nuclear fuel which has been formed and coated, and is ready to be placed in a reactor fuel assembly. Also known as reactor fuel element. { 'nü·klē·ər ¦fyül ¸el·ə·mənt }

nuclear fuel form [NUCLEO] Any chemical form of fissionable nuclear fuel, such as UO_2. { 'nü·klē·ər ¦fyül ¸fòrm }

nuclear fuel pebble See nuclear fuel pellet. { 'nü·klē·ər ¦fyül ¸peb·əl }

nuclear fuel pellet [NUCLEO] A piece of nuclear fuel usually in the shape of a sphere or cylinder, used in pebble-bed reactors, inserted in graphite blocks, or used in metallic tubular fuel elements. Also known as fuel ball; nuclear fuel pebble; reactor fuel pellet. { 'nü·klē·ər ¦fyül ¸pel·ət }

nuclear fuel plate [NUCLEO] A flat or curved sandwich of metallic cladding, with nuclear fuel inside. { 'nü·klē·ər ¦fyül ¸plāt }

nuclear fuel reprocessing [NUCLEO] The periodic chemical, physical, and metallurgical treatment of materials used as fuel elements in nuclear reactors, to recover and purify the residual fissionable and fertile materials. { 'nü·klē·ər ¦fyül rē'prä,ses·iŋ }

nuclear fusion See fusion. { 'nü·klē·ər 'fyü·zhən }

nuclear ground state [NUC PHYS] The stationary state of lowest energy of an isotope. { 'nü·klē·ər 'graund ,stāt }

nuclear gyroscope [ENG] A gyroscope in which the conventional spinning mass is replaced by the spin of atomic nuclei and electrons; one version uses optically pumped mercury isotopes, and another uses nuclear magnetic resonance techniques. { 'nü·klē·ər 'jī·rə,skōp }

nuclear Hanle effect [NUC PHYS] The dependence of the linear polarization of gamma rays emitted from a nucleus on the hyperfine interaction in an external magnetic field. { ¦nü·klē·ər 'han·lē i,fekt }

nuclear heat [NUCLEO] The heat released in a nuclear reactor due to the fission process. { 'nü·klē·ər 'hēt }

nuclear hormone receptor [CELL MOL] Any of a superfamily of proteins that directly regulate transcription in response to hormones and other ligands. { ¦nü·klē·ər 'hòr,mōn ri,sep·tər }

nuclear hydrodynamic model See nuclear fluid dynamic model. { 'nü·klē·ər ¦hī·drō·dī¦nam·ik 'mäd·əl }

nuclear induction [PHYS] Magnetic induction originating in the magnetic moments of nuclei; the effect depends on the unequal population of energy states available when the material is placed in a magnetic field. { 'nü·klē·ər in'dək·shən }

nuclear isomer See isomer. { 'nü·klē·ər 'ī·sə·mər }

nuclear lamina [CELL MOL] A protein meshwork lining the inner surface of the nuclear envelope. { 'nü·klē·ər 'lam·ə·nə }

nuclear laser [OPTICS] A gas laser in which the gas molecules are excited by high-energy fission particles produced by a pulsed nuclear reactor. { 'nü·klē·ər 'lā·zər }

nuclear magnetic moment [NUC PHYS] The magnetic dipole moment of an atomic nucleus; a vector whose scalar product with the magnetic flux density gives the negative of the energy of interaction of a nucleus with a magnetic field. { 'nü·klē·ər mag'ned·ik 'mō·mənt }

nuclear magnetic resonance [PHYS] A phenomenon exhibited by a large number of atomic nuclei, in which nuclei in a static magnetic field absorb energy from a radio-frequency field at certain characteristic frequencies. Abbreviated NMR. Also known as magnetic nuclear resonance. { 'nü·klē·ər mag'ned·ik 'rez·ən·əns }

nuclear magnetic resonance flowmeter [ENG] A flowmeter in which nuclei of the flowing fluid are resonated by a

radio-frequency field superimposed on an intense permanent magnetic field, and a detector downstream measures the amount of decay of the resonance, thereby sensing fluid velocity. { 'nü·klē·ər mag'ned·ik 'rez·ən·əns 'flō,mēd·ər }

nuclear magnetic resonance gyroscope [ENG] A gyroscope that obtains information from the dynamic angular motion of atomic nuclei. { 'nü·klē·ər mag'ned·ik 'rez·ən·əns 'jī·rə,skōp }

nuclear magnetic resonance spectrometer [SPECT] A spectrometer in which nuclear magnetic resonance is used for the analysis of protons and nuclei and for the study of changes in chemical and physical quantities over wide frequency ranges. { 'nü·klē·ər mag'ned·ik 'rez·ən·əns spek'träm·əd·ər }

nuclear magnetic resonance tomography See zeugmatography. { 'nü·klē·ər mag'ned·ik 'rez·ən·əns tə'mäg·rə·fē }

nuclear magnetism [PHYS] The phenomena associated with the magnetic dipole, octupole, and higher moments of a nucleus, including the magnetic field generated by the nucleus, the force on the nucleus in an inhomogeneous magnetic field, and the splitting of nuclear energy levels in a magnetic field. { 'nü·klē·ər 'mag·nə,tiz·əm }

nuclear magnetometer [ENG] Any magnetometer which is based on the interaction of a magnetic field with nuclear magnetic moments, such as the proton magnetometer. Also known as nuclear resonance magnetometer. { 'nü·klē·ər mag·nə'täm·əd·ər }

nuclear magneton [NUC PHYS] A unit of magnetic dipole moment used to express magnetic moments of nuclei and baryons; equal to the electron charge times Planck's constant divided by the product of 4π, the proton mass, and the speed of light. { 'nü·klē·ər 'mag·nə,tän }

nuclear mass [NUC PHYS] The mass of an atomic nucleus, which is usually measured in atomic mass units; it is less than the sum of the masses of its constituent protons and neutrons by the binding energy of the nucleus divided by the square of the speed of light. { 'nü·klē·ər 'mas }

nuclear medicine [MED] A branch of medicine in which radioactive pharmaceuticals are used for imaging or other diagnostic studies. { 'nü·klē·ər 'med·ə·sən }

nuclear medicine imaging [MED] A technique for producing images of the distribution of radioactive tracers in the human body by recording with a scintillation camera the gamma rays that are emitted by the trace. { 'nü·klē·ər ¦med·ə·sən 'im·ij·iŋ }

nuclear membrane [CYTOL] The envelope surrounding the cell nucleus, separating the nucleoplasm from the cytoplasm; composed of two membranes and contains numerous pores. { 'nü·klē·ər 'mem,brān }

nuclear molecule [NUC PHYS] A quasistable entity of nuclear dimensions formed in nuclear collisions and comprising two or more nuclei that retain their identities and are bound together by strong nuclear forces. { 'nü·klē·ər 'mäl·ə,kyül }

nuclear moment [NUC PHYS] One of the various static electric or magnetic multipole moments of a nucleus. { 'nü·klē·ər 'mō·mənt }

nuclear number See mass number. { 'nü·klē·ər 'nəm·bər }

nuclear orientation [NUC PHYS] The directional ordering of an assembly of nuclear spins with respect to some axis in space. { 'nü·klē·ər ,ór·ē·ən'tā·shən }

nuclear paramagnetism [PHYS] Paramagnetism in which a substance develops a net magnetic moment because the magnetic moments of nuclei tend to point in the direction of the field. { 'nü·klē·ər ,par·ə'mag·nə,tiz·əm }

nuclear physics [PHYS] The study of the characteristics, behavior, and internal structures of the atomic nucleus. { 'nü·klē·ər 'fiz·iks }

nuclear pile See nuclear reactor. { 'nü·klē·ər 'pīl }

nuclear plaque [MYCOL] In yeast, any region on the nuclear envelope from which the spindle originates. { 'nü·klē·ər 'plak }

nuclear polarization [NUC PHYS] For a nucleus in a mixed state, with spin I and probability $p(I_z)$ that the I_z substate is populated, the polarization is the sum over allowed values of I_z of $I_z p(I_z)/I$. { 'nü·klē·ər ,pō·lə·rə'zā·shən }

nuclear polyhedrosis virus [VIROL] A Baculovirus subgroup characterized by the multiplication and formation of polyhedron-shaped inclusion bodies in the nuclei of infected host cells; used in the control of agriculture and forest insects. Abbreviated NPV. { ¦nü·klē·ər ,päl·ē·hē'drōs·əs ,vī·rəs }

nuclear pore complex [CYTOL] Any of the nonrandomly distributed, octagonal orifices in the nuclear envelope. { ¦nü·klē·ər ¦pȯr ¦käm,pleks }

nuclear potential [NUC PHYS] The potential energy of a nuclear particle as a function of its position in the field of a nucleus or of another nuclear particle. { 'nü·klē·ər pə'ten·chəl }

nuclear potential energy [NUC PHYS] The average total potential energy of all the protons and neutrons in a nucleus due to the nuclear forces between them, excluding the electrostatic potential energy. { 'nü·klē·ər pə'ten·chəl 'en·ər·jē }

nuclear potential scattering [NUC PHYS] That part of elastic scattering of particles by a nucleus which may be treated by studying the scattering of a wave which obeys the Schrödinger equation with a potential determined by the properties of the nucleus. { 'nü·klē·ər pə'ten·chəl 'skad·ə·riŋ }

nuclear power [NUCLEO] Power whose source is nuclear fission or fusion. { 'nü·klē·ər 'pau·ər }

nuclear power plant [MECH ENG] A power plant in which nuclear energy is converted into heat for use in producing steam for turbines, which in turn drive generators that produce electric power. Also known as atomic power plant. { 'nü·klē·ər 'pau·ər ,plant }

nuclear propulsion [NAV ARCH] Propulsion of a ship or submarine by an engine driven by steam generated by nuclear energy in a reactor, rather than combustion of fuel in a boiler. { 'nü·klē·ər prə'pəl·shən }

nuclear quadrupole moment [NUC PHYS] The electric quadrupole moment of an atomic nucleus. { 'nü·klē·ər 'kwäd·rə,pōl ,mō·mənt }

nuclear quadrupole resonance [PHYS] The phenomenon in which certain nuclei in a static, inhomogeneous electric field absorb energy from a radio-frequency field. { 'nü·klē·ər 'kwäd·rə,pōl ,rez·ən·əns }

nuclear radiation [NUC PHYS] A term used to denote alpha particles, neutrons, electrons, photons, and other particles which emanate from the atomic nucleus as a result of radioactive decay and nuclear reactions. { 'nü·klē·ər ,rād·ē·'ā·shən }

nuclear radiation spectroscopy [NUC PHYS] Study of the distribution of energies or momenta of particles emitted by nuclei. { 'nü·klē·ər ,rād·ē·'ā·shən spek'träs·kə·pē }

nuclear radius [NUC PHYS] The radius of a sphere within which the nuclear density is large, and at the surface of which it falls off sharply. { 'nü·klē·ər 'rād·ē·əs }

nuclear reaction [NUC PHYS] A reaction involving a change in an atomic nucleus, such as fission, fusion, neutron capture, or radioactive decay, as distinct from a chemical reaction, which is limited to changes in the electron structure surrounding the nucleus. Also known as reaction. { 'nü·klē·ər rē'ak·shən }

nuclear reaction spectrometry [SPECT] A method of determining the concentration of a given element as a function of depth beneath the surface of a sample, by measuring the yield of characteristic gamma rays from a resonance reaction occurring when the surface is bombarded by a beam of ions. Abbreviated NRS. { 'nü·klē·ər rē'ak·shən spek'träm·ə·trē }

nuclear reactor [NUCLEO] A device containing fissionable material in sufficient quantity and so arranged as to be capable of maintaining a controlled, self-sustaining nuclear fission chain reaction. Also known as atomic pile (deprecated usage); atomic reactor; fission reactor; nuclear pile (deprecated); pile (deprecated); reactor. { 'nü·klē·ər rē'ak·tər }

nuclear receptor superfamily [CELL MOL] A large family of intracellular receptor proteins that bind to hydrophobic signal molecules (such as steroid and thyroid hormones) or intracellular metabolites and are thus activated to bind to specific DNA sequences, affecting transcription. { ¦nü·klē·ər ri¦sep·tər 'sü·pər,fam·lē }

nuclear recoil [NUC PHYS] The imparting of motion to an atomic nucleus during its emission of particles in radioactive decay, or during its collision with another particle, according to the principle of conservation of momentum. { 'nü·klē·ər 'rē,kȯil }

nuclear relaxation [PHYS] The approach of a system of nuclear spins to a steady-state or equilibrium condition over a period of time, following a change in the applied magnetic field. { 'nü·klē·ər ,rē,lak'sā·shən }

nuclear resonance [NUC PHYS] **1.** An unstable excited state formed in the collision of a nucleus and a bombarding particle, and associated with a peak in a plot of cross section versus energy. **2.** The absorption of energy by nuclei from radio-frequency fields at certain frequencies when these nuclei are also subjected to certain types of static fields, as in magnetic resonance and nuclear quadrupole resonance. { 'nü·klē·ər 'rez·ən·əns }

nuclear resonance magnetometer See nuclear magnetometer. { ¦nü·klē·ər 'rez·ən·əns ,mag·nə'täm·əd·ər }

nuclear response function [NUC PHYS] The probability that a given probing particle with given initial energy is scattered from a nucleus with given energy transfer and given momentum transfer, as a function of the energy transfer and momentum transfer. { 'nü·klē·ər ri¦späns ,fəŋk·shən }

nuclear rocket See atomic rocket. { 'nü·klē·ər 'räk·ət }

nuclear scattering [NUC PHYS] The change in directions of particles as a result of collisions with nuclei. { 'nü·klē·ər 'skad·ə·riŋ }

nuclear sclerosis [MED] Hardening of the ocular lens nucleus. { 'nü·klē·ər sklə'rō·səs }

nuclear ship [NAV ARCH] A ship in which nuclear energy in a reactor, rather than combustion of fuel in a boiler, generates steam which drives the engine to propel the ship. { 'nü·klē·ər 'ship }

nuclear snow gage [ENG] Any type of gage using a radioactive source and a detector to measure, by the absorption of radiation, the water-equivalent mass of a snowpack. { 'nü·klē·ər 'snō ,gāj }

nuclear spallation See spallation. { 'nü·klē·ər spȯ'lā·shən }

nuclear species See nuclide. { 'nü·klē·ər 'spē,shēz }

nuclear spectrum [NUC PHYS] **1.** The relative number of particles emitted by atomic nuclei as a function of energy or momenta of these particles. **2.** The graphical display of data from devices used to measure these quantities. { 'nü·klē·ər 'spek·trəm }

nuclear spin [NUC PHYS] The total angular momentum of an atomic nucleus, resulting from the coupled spin and orbital angular momenta of its constituent nuclei. Also known as nuclear angular momentum. Symbolized I. { 'nü·klē·ər 'spin }

nuclear spontaneous reaction See radioactive decay. { 'nü·klē·ər spän'tā·nē·əs rē'ak·shən }

nuclear stability [NUC PHYS] The ability of an isotope to resist decay or fission. { 'nü·klē·ər stə'bil·əd·ē }

nuclear star See star. { 'nü·klē·ər 'stär }

nuclear submarine [NAV ARCH] A submarine in which nuclear energy in a reactor generates steam which drives the engine to propel the ship. { 'nü·klē·ər ,səb·mə'rēn }

nuclear superheating [NUCLEO] Superheating the steam produced in a reactor by using additional heat from a reactor; two methods may be used: recirculating the steam through the same core in which it is first produced (integral superheating) or passing the steam through a second and separate reactor. { 'nü·klē·ər ¦sü·pər¦hēd·iŋ }

nuclear teleoperator [NUCLEO] A remote manipulator that enables the operator to handle objects in a hazardous radioactive environment. { 'nü·klē·ər ,tel·ē'äp·ə,rād·ər }

nuclear thermionic converter [NUCLEO] A thermionic converter whose heat source is a nuclear reactor or radioisotope. { 'nü·klē·ər ,thər·mē'än·ik kən'vərd·ər }

nuclear time scale [ASTRON] The time it takes for a star to evolve a significant distance from the main sequence when a certain fraction of the hydrogen in its core has been converted to helium by thermonuclear reactions. { 'nü·klē·ər 'tīm ,skāl }

nuclear transfer [CYTOL] Insertion of a diploid somatic nucleus into an egg from which the nucleus has been removed. { 'nü·klē·ər 'tranz·fər }

nuclear transformation See transmutation. { 'nü·klē·ər ,tranz·fər'mā·shən }

nuclear triode detector [ELECTR] A type of junction detector that has two outputs which together determine the precise location on the detector where the ionizing radiation was incident, as well as the energy of the ionizing particle. { 'nü·klē·ər 'trī,ōd di'tek·tər }

nuclear twin-probe gage See profiling snow gage. { 'nü·klē·ər ¦twin ,prōb ,gāj }

nuclear warhead [ORD] A warhead that contains fissionable or fissionable-fusionable material. { 'nü·klē·ər 'wȯr,hed }

nuclear weapon See atomic weapon. { 'nü·klē·ər 'wep·ən }

nuclear winter [METEOROL] Predicted global-scale changes resulting from a nuclear war, in which dust raised by nuclear bursts and smoke generated in fires would cause reductions in solar energy reaching the earth's surface and reductions in surface temperatures for periods of months. { 'nü·klē·ər 'win·tər }

nuclear yield [NUCLEO] The energy released by the detonation of a nuclear weapon; measured in terms of megatons of trinitrotoluene (TNT) required to release the same energy. { 'nü·klē·ər 'yēld }

nuclear Zeeman effect [SPECT] A splitting of atomic spectral lines resulting from the interaction of the magnetic moment of the nucleus with an applied magnetic field. { 'nü·klē·ər 'zē·mən i‚fekt }

nuclease [BIOCHEM] An enzyme that catalyzes the splitting of nucleic acids to nucleotides, nucleosides, or the components of the latter. { 'nü·klē‚ās }

nucleate boiling [CHEM ENG] Boiling in which bubble formation is at the liquid-solid interface rather than from external or mechanical devices; occurs in kettle-type and natural-circulation heaters or reboilers. { 'nü·klē‚āt 'bȯil·iŋ }

nucleated glass [MATER] Glass treated with a nucleating agent to transform it into a crystalline material. { 'nü·klē‚ād·əd 'glas }

nucleation [CHEM] In crystallization processes, the formation of new crystal nuclei in supersaturated solutions. [PHYS CHEM] The formation of vapor bubbles in a superheated liquid. { ‚nü·klē'ā·shən }

nucleic acid [BIOCHEM] A large, acidic, chainlike molecule containing phosphoric acid, sugar, and purine and pyrimidine bases; two types are ribonucleic acid and deoxyribonucleic acid. { nü'klē·ik 'as·əd }

nuclein [BIOCHEM] Any of a poorly defined group of nucleic acid protein complexes occurring in cell nuclei. { 'nü·klē·ən }

nucleocapsid [VIROL] The nucleic acid of a virus and its surrounding capsid. { ‚nü·klē·ō'kap·səd }

nucleocytoplasmic ratio [CELL MOL] The ratio between the measured cross-sectional area or estimated volume of the nucleus of a cell to the volume of its cytoplasm. Also known as karyoplasmic ratio. { ‚nü·klē·ō‚sīd·ə'plaz·mik 'rā·shō }

nucleodesma [CELL MOL] A connection composed of fibrils between the nucleus and cytoplasm. Also know as karyodesma. { ‚nü·klē·ə'dez·mə }

nucleofuge See leaving group. { 'nü·klē·ə‚fyüj }

nucleogenesis [ASTROPHYS] The origin of chemical elements in the universe. { ‚nü·klē·ə'jen·ə·səs }

nucleoid [CELL MOL] A discrete region within mitochondria, chloroplasts, and prokaryotes that contain molecules of deoxyribonucleic acid. [VIROL] The ribonucleic acid (RNA) core that is enveloped by a protein capsid in RNA tumor viruses. { 'nü·klē‚ȯid }

nucleolar [CELL MOL] Of or pertaining to the nucleolus. { nü'klē·ə·lər }

nucleolus [CELL MOL] A small, spherical body composed principally of protein and located in the metabolic nucleus. Also known as plasmosome. { nü'klē·ə·ləs }

nucleon [PHYS] A collective name for a proton or a neutron; these particles are the main constituents of atomic nuclei, have approximately the same mass, have a spin of $1/2$, and can transform into each other through the process of beta decay. { 'nü·klē‚än }

nucleonics [ENG] The technology based on phenomena of the atomic nucleus such as radioactivity, fission, and fusion; includes nuclear reactors, various applications of radioisotopes and radiation, particle accelerators, and radiation-detection devices. { ‚nü·klē'än·iks }

nucleonium [ATOM PHYS] A bound state of a nucleus and an antinucleus. { ‚nü·klē·ō'nē·əm }

nucleon number See mass number. { 'nü·klē‚än ‚nəm·bər }

nucleophile [PHYS CHEM] A species possessing one or more electron-rich sites, such as an unshared pair of electrons, the negative end of a polar bond, or pi electrons. Also known as electron donor. { 'nü·klē·ə‚fīl }

nucleophilic displacement See nucleophilic substitution. { ‚nü·klē·ə|fil·ik di'splās·mənt }

nucleophilic reagent [PHYS CHEM] A reactant that gives up electrons, or a share in electrons, to other molecules or ions in the course of a chemical reaction. { ‚nü·klē·ō|fil·ik rē'ā·jənt }

nucleophilic substitution [ORG CHEM] A reaction in which a nucleophile bonds to a carbon atom in a molecule, displacing

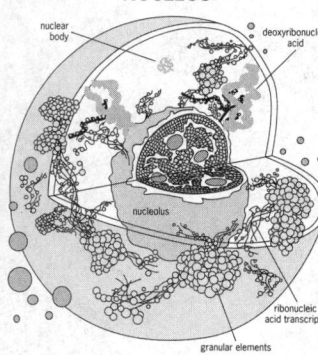

NUCLEUS

nuclear body, deoxyribonucleic acid, nucleolus, ribonucleic acid transcripts, granular elements

Three-dimensional organization of the mammalian cell nucleus.

a leaving group. Also known as nucleophilic displacement. { ‚nü·klē·ə|fil·ik ‚səb·stə'tü·shən }

nucleoplasm [CELL MOL] The protoplasm of a nucleus. Also known as karyoplasm. { 'nü·klē·ə‚plaz·əm }

nucleoprotein [BIOCHEM] Any member of a class of conjugated proteins in which molecules of nucleic acid are closely associated with molecules of protein. { ‚nü·klē·ō'prō‚tēn }

nucleor [PARTIC PHYS] A hypothetical core of a nucleon, surrounded by a hypothetical cloud of pions. { 'nü·klē‚ȯr }

nucleoreticulum [CELL MOL] Any type of network found within a nucleus. { ‚nü·klē·ō·rə'tik·yə·ləm }

nucleosidase [BIOCHEM] Any enzyme involved in splitting a nucleoside into its component base and pentose. { ‚nü·klē·ə'sī‚dās }

nucleoside [BIOCHEM] The glycoside resulting from removal of the phosphate group from a nucleotide; consists of a pentose sugar linked to a purine or pyrimidine base. { 'nü·klē·ə‚sīd }

nucleosome [CELL MOL] A morphologically repeating unit of deoxyribonucleic acid (DNA) containing 190 base pairs of DNA folded together with eight histone molecules. Also known as v-body. { 'nü·klē·ə‚sōm }

nucleospindle [CELL MOL] A mitotic spindle derived from nuclear material. { ‚nü·klē·ō'spind·əl }

nucleosynthesis [ASTROPHYS] The formation of the various nuclides present in the universe by various nuclear reactions, occurring chiefly in the early universe following the big bang, in the interiors of stars, and in supernovae. { ‚nü·klē·ō'sin·thə·səs }

nucleotidase [BIOCHEM] Any of a group of enzymes which split phosphoric acid from nucleotides, leaving nucleosides. { ‚nü·klē·ə'tī‚dās }

nucleotide [BIOCHEM] An ester of a nucleoside and phosphoric acid; the structural unit of a nucleic acid. { 'nü·klē·ə‚tīd }

nucleotide excision repair [CELL MOL] A five-step pathway that is the major repair system for removing bulky lesions in deoxyribonucleic acid. { ‚nü·klē·ə‚tīd ik'sizh·ən rē‚per }

nucleus [ASTRON] The small permanent body of a comet, believed to have a diameter between one and a few tens of kilometers, and to be composed of water and volatile hydrocarbons. [CELL MOL] A small mass of differentiated protoplasm rich in nucleoproteins and surrounded by a membrane; found in most animal and plant cells, contains chromosomes, and functions in metabolism, growth, and reproduction. [COMPUT SCI] **1.** That portion of the control program that must always be present in main storage. **2.** The main storage area used in the nucleus (first definition) and other transient control program routines. [HYD] A particle of any nature upon which, or a locus at which, molecules of water or ice accumulate as a result of a phase change to a more condensed state. [NEUROSCI] A mass of nerve cells in the central nervous system. [NUC PHYS] The central, positively charged, dense portion of an atom. Also known as atomic nucleus. [SCI TECH] A central mass about which accretion takes place. { 'nü·klē·əs }

nucleus counter [ENG] An instrument which measures the number of condensation nuclei or ice nuclei per sample volume of air. { 'nü·klē·əs ‚kau̇nt·ər }

nucleus pulposus [ANAT] The soft, fibrocartilaginous central portion of the intervertebral disk. { ‚nü·klē·əs pəl'pō·səs }

nuclide [NUC PHYS] A species of atom characterized by the number of protons, number of neutrons, and energy content in the nucleus, or alternatively by the atomic number, mass number, and atomic mass; to be regarded as a distinct nuclide, the atom must be capable of existing for a measurable lifetime, generally greater than 10^{-10} second. Also known as nuclear species; species. { 'nü‚klīd }

Nuda [INV ZOO] A class of the phylum Ctenophora distinguished by the lack of tentacles. { 'nüd·ə }

Nudechiniscidae [INV ZOO] A family of heterotardigrades in the suborder Echiniscoidea characterized by a uniform cuticle. { ‚nüd·ə·ki'nis·kə‚dē }

Nudibranchia [INV ZOO] A suborder of the Opisthobranchia containing the sea slugs; these mollusks lack a shell and a mantle cavity, and the gills are variable in size and shape. { ‚nüd·ə'braŋ·kē·ə }

nudism [PSYCH] Intolerance for wearing clothing, with a

morbid tendency for the individual to remove her or his clothing. { 'nüd,iz·əm }

nuée ardente [GEOL] A turbulent, rapidly flowing, and sometimes incandescent gaseous cloud erupted from a volcano and containing ash and other pyroclastics in its lower part. Also known as glowing cloud; Pelean cloud. { ¦nü¦ā är'dänt }

nugget [GEOL] A small mass of metal found free in nature. [MET] A weld bead. { 'nəg·ət }

nuisance parameter [STAT] A parameter to be estimated by a statistic which arises in the distribution of the statistic under some hypothesis to be tested about the parameter. { 'nü·səns pə,ram·əd·ər }

null [MATH] Indicating that an object is nonexistent or a quantity is zero. [NAV] **1.** The azimuth or elevation reading on a navigational device indicated by minimum signal output. **2.** Any of the nodal points on the radiation patterns of some antennas. { nəl }

null allele [GEN] An allele that does not produce a functional product and behaves as a recessive. { nəl ə,lēl }

nullary composition [MATH] The selection of a particular element of a set. { 'nəl·ə·rē ,käm·pə'zish·ən }

null-balance recorder [ENG] An instrument in which a motor-driven slide wire in a measuring circuit is continuously adjusted so that the voltage or current to be measured will be balanced against the voltage or current from this circuit; a pen linked to the slide wire makes a graphical record of its position as a function of time. { 'nəl 'bal·əns ri,kórd·ər }

null cell [IMMUNOL] A lymphocyte without T- or B-cell markers on its surface. { 'nəl ,sel }

null character [COMPUT SCI] A control character used as a filler in data processing; may be inserted or removed from a sequence of characters without affecting the meaning of the sequence, but may affect format or control equipment. { 'nəl ,kar·ik·tər }

null cone See light cone. { 'nəl ,kōn }

null-current circuit [ELECTR] A circuit used to measure current, in which the unknown current is opposed by a current resulting from applying a voltage controlled by a slide wire across a series resistor, and the slide wire is continuously adjusted so that the resulting current, as measured by a direct-current detector amplifier, is equal to zero. { 'nəl ¦kə·rənt 'sər·kət }

null-current measurement [ELECTR] Measurement of current using a null-current circuit. { 'nəl ¦kə·rənt 'mezh·ər·mənt }

null detection [ELEC] Altering of adjustable bridge circuit components, to obtain zero current. [NAV] Method of determining the radio direction by altering the antenna position to obtain minimum signal strength. { 'nəl di,tek·shən }

null detector See null indicator. { 'nəl di,tek·tər }

null geodesic [MATH] In a Riemannian space, a minimal geodesic curve. [RELAT] A curve in space-time which has the property that the infinitesimal interval between any two neighboring points on the curve equals zero; it represents a possible path of a light ray. Also known as zero geodesic. { 'nəl ,jē·ə'des·ik }

null hypothesis [STAT] The hypothesis that there is no validity to the specific claim that two variations (treatments) of the same thing can be distinguished by a specific procedure. { 'nəl hī'päth·ə·səs }

null indicator [ENG] A galvanometer or other device that indicates when voltage or current is zero; used chiefly to determine when a bridge circuit is in balance. Also known as null detector. { 'nəl ,in·də,kād·ər }

nulling interferometry [OPTICS] A technique in which light waves from a bright object such as a star are made to interfere and cancel each other in an optical system, allowing the observation of much fainter nearby objects that would otherwise be invisible in the glare of the bright object. { ¦nəl·iŋ ,in·tə·fə'räm·ə·trē }

nullity [MATH] The dimension of the null space of a linear transformation. { 'nəl·əd·ē }

null matrix [MATH] The matrix all of whose entries are zero. { 'nəl 'mā·triks }

null method [ENG] A method of measurement in which the measuring circuit is balanced to bring the pointer of the indicating instrument to zero, as in a Wheatstone bridge, and the settings of the balancing controls are then read. Also known as balance method; zero method. { 'nəl ,meth·əd }

null modem cable [COMPUT SCI] A cable that connects two local computers via serial ports without the use of a modem. { 'nəl 'mō,dem ,kā·bəl }

null reference system [NAV] A type of instrument landing system in which the antenna array consists of two elements, one twice as high as the other; the lower element radiates with maximum intensity along the glide path, while the upper element produces a signal null at the glide path angle. { 'nəl 'ref·rəns ,sis·təm }

null sequence [MATH] A sequence of numbers or functions which converges to the number zero or the zero function. { 'nəl 'sē·kwəns }

null set [MATH] The empty set; the set which contains no elements. { 'nəl 'set }

null space [MATH] For a linear transformation, the vector subspace of all vectors which the transformation sends to the zero vector. Also known as kernel. { 'nəl 'spās }

null surface [RELAT] A surface in space-time whose normal vector is everywhere null. { 'nəl 'sər·fəs }

null vector [MATH] A vector whose invariant length, that is, the sum over the coordinates of the vector space of the product of its covariant component and contravariant component, is equal to zero. [RELAT] In special relativity, a four vector whose spatial part in any Lorentz frame has a magnitude equal to the speed of light multiplied by its time part in that frame; a special case of the mathematics definition. { 'nəl 'vek·tər }

NUMA machine See nonuniform memory access machine. { 'rü·mə mə'shēn }

number [MATH] **1.** Any real or complex number. **2.** The number of elements in a set is the cardinality of the set. { 'nəm·bər }

number class modulo N [MATH] The class of all numbers which differ from a given number by a multiple of N. { 'nəm·bər ¦klas ¦mä·ə·lō 'en }

number cruncher [COMPUT SCI] A computer with great power to carry out computations, designed to maximize this ability rather than to process large amounts of data. { 'nəm·bər ,krən·chər }

number density [PHYS] The number of particles per unit volume. { 'nəm·bər ,den·səd·ē }

number field [MATH] Any set of real or complex numbers that includes the sum, difference, product, and quotient (except division by zero) of any two members of the set. { 'nəm·bər ,fēld }

number line See real line. { 'nəm·bər ,līn }

number record printer [COMMUN] A printer in a relay station that provides a complete automatic written record of channel numbers and the fixed routing line associated with each message that is relayed through that particular station. { 'nəm·bər ¦rek·ərd ,print·ər }

number scale [MATH] Representation of points on a line with numbers arranged in some order. { 'nəm·bər ,skāl }

Number six fuel See bunker C fuel oil. { 'nəm·bər ,siks 'fyül }

number system [MATH] **1.** A mathematical system, such as the real or complex numbers, the quaternions, or the Cayley numbers, that satisfies many of the axioms of the real number system; in general, it is a finite-dimensional vector space over the real numbers with multiplicative operation under which it is an associative or nonassociative division algebra. **2.** See numeration system. { 'nəm·bər ,sis·təm }

number-theoretic function [MATH] A function whose domain is the set of positive integers. { ¦nəm·bər ,thē·ə¦red·ik 'fəŋk·shən }

number theory [MATH] The study of integers and relations between them. { 'rəm·bər 'thē·ə·rē }

numeral [MATH] A symbol used to denote a number. { 'nüm·rəl }

numeral system See numeration system. { 'nüm·rəl ,sis·təm }

numeration [MATH] The listing of numbers in their natural order. { ,nü·mə'rā·shən }

numeration system [MATH] An orderly method of representing numbers by numerals in which each numeral is associated with a unique number. Also known as number system; numeral system. { nü·mə'rā·shən ,sis·təm }

numerator [MATH] In a fraction a/b, the numerator is the quantity a. { 'nü·mə,rād·ər }

numeric [COMPUT SCI] In computers, pertaining to data

composed wholly or partly of digits, as distinct from alphabetic. { nü'mer·ik }

numerical [MATH] Pertaining to numbers. { nü'mer·i·kəl }

numerical analysis [MATH] The study of approximation techniques using arithmetic for solutions of mathematical problems. { nü'mer·i·kəl ə'nal·ə·səs }

numerical aperture [OPTICS] A measure of the resolving power of a microscope objective, equal to the product of the refractive index of the medium in front of the objective and the sine of the angle between the outermost ray entering the objective and the optical axis. Abbreviated N.A. { nü'mer·i·kəl 'ap·ə·chər }

numerical decrement *See* decrement. { nü'mer·i·kəl 'dek·rə·mənt }

numerical display device [ELECTR] Any device for visually displaying numerical figures, such as a numerical indicator tube, a device utilizing electroluminescence, or a device in which any one of a stack of transparent plastic strips engraved with digits can be illuminated by a small light at the edge of the strip. { nü'mer·i·kəl di'splā di,vīs }

numerical equation [MATH] An equation all of whose constants and coefficients are numbers. { nü'mer·i·kəl i'kwā·zhən }

numerical forecasting [METEOROL] The forecasting of the behavior of atmospheric disturbances by the numerical solution of the governing fundamental equations of hydrodynamics, subject to observed initial conditions. Also known as dynamic forecasting; mathematical forecasting; numerical weather prediction; physical forecasting. { nü'mer·i·kəl 'fȯr,kast·iŋ }

numerical hydrodynamics [FL MECH] An amalgamation of fluid dynamics, applied mathematics, and numerical analysis, in which numerical algorithms are developed to solve equations of incompressible fluid flows about or within bodies. { nü'mer·i·kəl ,hī·drə·də'nam·iks }

numerical indicator tube [ELECTR] An electron tube capable of visually displaying numerical figures; some varieties also display alphabetical characters and commonly used symbols. { nü'mer·i·kəl 'in·də,kād·ər ,tüb }

numerical integration [MATH] The process of using a set of approximate values of a function to calculate its integral to comparable accuracy. { nü'mer·i·kəl ,int·ə'grā·shən }

numerical range [MATH] For a linear operator T of a Hilbert space into itself, the set of values assumed by the inner product of Tx with x as x ranges over the set of vectors with norm equal to 1. { nü'mer·i·kəl 'rānj }

numerical system [TEXT] A grade scale for calculating the quality of wool. Also known as count system. { nü'mer·i·kəl 'sis·təm }

numerical tape [COMPUT SCI] The tape required by a computer operating a machine tool. { nü'mer·i·kəl 'tāp }

numerical taxonomy [SYST] The numerical evaluation of the affinity or similarity between taxonomic units and the ordering of these units into taxa on the basis of their affinities. { nü'mer·i·kəl tak'sän·ə·mē }

numerical tensor [MATH] A tensor whose components are the same in all coordinate systems. { nü'mer·i·kəl 'ten·sər }

numerical value *See* absolute value. { nü'mer·i·kəl 'val·yü }

numerical weather prediction *See* numerical forecasting. { nü'mer·i·kəl 'weth·ər pri,dik·shən }

numeric character *See* digit. { nü'mer·ik 'kar·ik·tər }

numeric character set [COMPUT SCI] A character set that includes only digits and certain special characters, such as plus and minus signs and control characters. { nü'mer·ik 'kar·ik·tər ,set }

numeric coding [COMPUT SCI] Code in which only digits are used, usually binary or octal. { nü'mer·ik 'kōd·iŋ }

numeric control [COMPUT SCI] The action of programs written for specialized computers which operate machine tools. { nü'mer·ik kən'trōl }

numeric coprocessor *See* numeric processor extension. { nü'mer·ik kō'prä,ses·ər }

numeric data [COMPUT SCI] Data consisting of digits and not letters of the alphabet or special characters. { nü'mer·ik'dad·ə }

numeric format [COMPUT SCI] The manner in which numbers are displayed in the cells of a particular spreadsheet. { nü'mer·ik 'fȯr,mat }

numeric keypad [COMPUT SCI] A section of a computer keyboard that contains a group of keys, usually about 12, arranged in compact fashion for entering numeric characters efficiently. Also known as numeric pad. { nü'mer·ik 'kē,pad }

numeric pad *See* numeric keypad. { nü'mer·ik 'pad }

numeric pager [COMMUN] A receiver in a radio paging system that contains a liquid-crystal display which can show numeric messages, most commonly a telephone number. { nü'mer·ik 'pā·jər }

numeric printer [COMPUT SCI] Old type of printer which positioned its keys to print a field in one operation, rather than one digit at a time. { nü'mer·ik 'print·ər }

numeric processor extension [COMPUT SCI] A specialized integrated circuit that is added to a computer to perform high-speed floating-point mathematical calculations. Abbreviated NPX. Also known as arithmetic processor; math coprocessor; numeric coprocessor. { nü'mer·ik 'prä,ses·ər ik,sten·chən }

numeric variable [COMPUT SCI] The symbolic name of a data element whose value changes during the carrying out of a computer program. { nü'mer·ik 'ver·ē·ə·bəl }

Numididae [VERT ZOO] A family of birds in the order Galliformes commonly known as guinea fowl; there are few if any feathers on the neck or head, but there may be a crest of feathers and various fleshy appendages. { nü'mid·ə,dē }

Nummulites [INV ZOO] A genus of unicellular shelled protozoa of the order Foraminiferida (superfamily Nummulitacea, family Nummulitidae). The discoidal, lenticular, or globular test or shell can reach a diameter of about 5 inches (12 centimeters) and is composed of finely perforate calcium carbonate; it consists of planispirally enrolled whorls of many tiny undivided chambers. { 'nəm·yə,līts }

nunatak [GEOL] An isolated hill, knob, ridge, or peak of bedrock projecting prominently above the surface of a glacier and completely surrounded by glacial ice. { 'nən·ə,tak }

nun buoy [NAV] A red buoy made of two conical or truncated cone-shaped sections joined at the base; marks the starboard side when entering a channel from the sea. { 'nən ,bȯi }

N unit [OPTICS] A unit of index of refraction; a mathematical simplification designed to replace rather awkward numbers involved in the values of the index of refraction n for the atmosphere; it is defined by the relation $N = (n - 1)10^6$. { 'en ,yü·nət }

nuplex [NUCLEO] Proposed nuclear reactor which would both produce electrical power and heat water for desalination and other industrial benefits. { 'nü,pleks }

nurse cell [HISTOL] A cell type of the ovary of many animals which nourishes the developing egg cell. { 'nərs ,sel }

nurse graft [BOT] A plant graft in which the scion remains united with the stock only until roots develop on the scion. { 'nərs ,graft }

nursing [MED] The application of the principles of physical, biological, and social sciences in the physical and mental care of people. { 'nərs·iŋ }

Nusselt equation [THERMO] Dimensionless equation used to calculate convection heat transfer for heating or cooling of fluids outside a bank of 10 or more rows of tubes to which the fluid flow is normal. { 'nùs·əlt i,kwā·zhən }

Nusselt number [PHYS] A dimensionless number used in the study of mass transfer, equal to the mass-transfer coefficient times the thickness of a layer through which mass transfer is taking place divided by the molecular diffusivity. Symbolized Nu_m; N_{Nu_m}. Also known as Sherwood number (N_{Sh}). [THERMO] A dimensionless number used in the study of forced convection which gives a measure of the ratio of the total heat transfer to conductive heat transfer, and is equal to the heat-transfer coefficient times a characteristic length divided by the thermal conductivity. Symbolized N_{Nu}. { 'nùs·əlt ,nəm·bər }

nut [BOT] **1.** A fruit which has at maturity a hard, dry shell enclosing a kernel consisting of an embryo and nutritive tissue. **2.** An indehiscent, one-celled, one-seeded, hard fruit derived from a single, simple, or compound ovary. [DES ENG] An internally threaded fastener for bolts and screws. { nət }

nutating antenna [ENG] An antenna system used in conical scan radar, in which a dipole or feed horn moves in a small circular orbit about the axis of a paraboloidal reflector without changing its polarization. { 'nü,tād·iŋ an'ten·ə }

nutating-disk meter [ENG] An instrument for measuring flow of a liquid in which liquid passing through a chamber causes a disk to nutate, or roll back and forth, and the total number of rolls is mechanically counted. { 'nü₁tād·iŋ ¦disk ¦mēd·ər }

nutation [ASTRON] A slight, slow, nodding motion of the earth's axis of rotation which is superimposed on the precession of the equinoxes; it is the combination of a number of perturbations (lunar, solar, and fortnightly nutation). [BOT] The rhythmic change in the position of growing plant organs caused by variation in the growth rates on different sides of the growing apex. [MECH] A bobbing or nodding up-and-down motion of a spinning rigid body, such as a top, as it precesses about its vertical axis. { nü'tā·shən }

nutational scanner [OPTICS] An optical-mechanical system for scanning an image along a series of closely spaced parallel lines, using an oscillating plane mirror and a rotating prism. { nü'tā·shən·əl 'skan·ər }

nutator [ENG] A mechanical or electrical device used to move a radar beam in a circular, conical, spiral, or other manner periodically to obtain greater air surveillance than could be obtained with a stationary beam. { 'nü₁tād·ər }

nutgall [PL PATH] A nutlike gall. { 'nət₁gȯl }

nutmeg [BOT] *Myristica fragrans.* A dark-leafed evergreen tree of the family Myristicaceae cultivated for the golden-yellow fruits which resemble apricots; a delicately flavored spice is obtained from the kernels inside the seeds. { 'nət₁meg }

nutmeg liver [MED] Chronic passive hyperemia of the liver; the cut surface of the diseased organ resembles the cut surface of a nutmeg. { 'nət₁meg ¦liv·ər }

nutmeg oil [MATER] A pale-yellow or colorless essential oil with spicy taste and nutmeg aroma; obtained from nutmegs; soluble in alcohol, carbon bisulfide, and glacial acetic acid; chief constituents are myristin, pinene, and dipentene; used in flavors, perfumes, and medicines. Also known as myristica oil. { 'nət₁meg ₁ȯil }

nutraceutical [BIOCHEM] Any food or food ingredient that is medically beneficial to an organism, including preventing disease. { ₁nü·trə'süd·ə·kəl }

nutrient [BIOL] Providing nourishment. { 'nü·trē·ənt }

nutrient biopurification [ECOL] A process taking place within a nutrient cycle that maintains the pools of nutrient substances at optimum concentrations, to the exclusion of non-nutrient substances. { 'nü·trē·ənt ¦bī·ō₁pyùr·ə·fə'kā·shən }

nutrient foramen [ANAT] The opening into the canal which gives passage to the blood vessels of the medullary cavity of a bone. { 'nü·trē·ənt fə'rā·mən }

nutrilite [BIOCHEM] A nourishing compound. { 'nü·trə₁līt }

nutrition [BIOL] The science of nourishment, including the study of nutrients that each organism must obtain from its environment in order to maintain life and reproduce. { nü'trish·ən }

nutritional anemia [MED] Anemia resulting from certain nutritional deficiencies. { nü'trish·ən·əl ə'nē·mē·ə }

nutritional dystrophy *See* kwashiorkor. { nü'trish·ən·əl 'dis·trə·fē }

nutritional edema [MED] Edema resulting from starvation or malnutrition. { nü'trish·ən·əl i'dē·mə }

nutritional hypochromic anemia *See* iron-deficiency anemia. { nü'trish·ən·əl ¦hī·pō¦krō·mik ə'nē·mē·ə }

Nuttalliellidae [INV ZOO] A family of ticks (Ixodides) containing one rare African species, *Nuttalliella namaqua*, morphologically intermediate between the families Argasidae and Ixodidae. { nə₁tal·ē'el·ə₁dē }

nu value [OPTICS] The reciprocal of the dispersive power of a medium. Also known as constringence. { 'nü ₁val·yü }

nuvistor [ELECTR] Electron tube in which all electrodes are cylindrical, placed one inside the other with close spacing, in a ceramic envelope. { nü'vis·tər }

nux vomica [BOT] The seed of *Strychnos nux-vomica*, an Indian tree of the family Loganiaceae; contains the alkaloid strychnine, and was formerly used in medicine. { 'nəks 'väm·ə·kə }

n value *See* field index. { 'en ₁val·yü }

NV RAM *See* nonvolatile random-access memory. { ¦en¦vē 'ram }

N wave [ACOUS] The N-shaped pressure wave that is generated by passage of an aircraft at large distances and has lost all of the fine structure observed closer to the aircraft. { 'en ₁wāv }

n-way analysis [STAT] A statistical analysis in which *n* major factors are used to jointly classify the observed values, where *n* is a positive integer. { 'en ₁wā ə₁nal·ə·səs }

nybble [COMPUT SCI] A string of bits, smaller than a byte, operated on as a unit. { 'nib·əl }

Nyctaginaceae [BOT] A family of dicotyledonous plants in the order Caryophyllales characterized by an apocarpous, monocarpous, or syncarpous gynoecium, sepals joined to a tube, a single carpel, and a cymose inflorescence. { ₁nik·tə·jə'nās·ē₁ē }

nyctalopia *See* night blindness. { ₁nik·tə'lō·pē·ə }

Nycteridae [VERT ZOO] The slit-faced bats, a monogeneric family of insectivorous chiropterans having a simple, well-developed nose leaf, and large ears joined together across the forehead. { nik'ter·ə₁dē }

Nyctibiidae [VERT ZOO] A family of birds in the order Caprimulgiformes including the neotropical potoos. { ₁nik·tə'bī·ə₁dē }

nyctinasty [BOT] A nastic movement in higher plants associated with diurnal light and temperature changes. { 'nik·tə₁nas·tē }

nyctophobia [PSYCH] An abnormal fear of night and darkness. { ₁nik·tə'fō·bē·ə }

Nyctribiidae [INV ZOO] The bat tick flies, a family of myodarian cyclorrhaphous dipteran insects in the subsection Acalyptratae. { ₁nik·trə'bī·ə₁dē }

nygmata [INV ZOO] Sensory spots on wings of certain insects, such as some neuropterans. { nig'mäd·ə }

Nygolaimoidea [INV ZOO] A superfamily of predaceous nematodes of the order Dorylaimida, distinguished by an eversible stoma with a protrusible subventral mural tooth, and a bottle-shaped esophagus. { ₁nī·gō·lə'mȯid·ē·ə }

Nylander reagent [CHEM] A solution of Rochelle salt (potassium sodium tartrate), potassium or sodium hydroxide, and bismuth subnitrate in water; used to test for sugar in urine. { 'nī·lən·dər rē₁ā·jənt }

nylon [MATER] Generic name for long-chain polymeric amide molecules in which recurring amide groups are part of the main polymer chain; used to make fibers, fabrics, sheeting, and extruded forms. { 'nī₁län }

nymph [INV ZOO] Any immature larval stage of various hemimetabolic insects. { nimf }

Nymphaeaceae [BOT] A family of dicotyledonous plants in the order Nymphaeales distinguished by the presence of roots, perfect flowers, alternate leaves, and uniaperturate pollen. { ₁nim·fē'ās·ē₁ē }

Nymphaeales [BOT] An order of flowering aquatic herbs in the subclass Magnoliidae; all lack cambium and vessels and have laminar placentation. { ₁nim·fē'ā·lēz }

Nymphalidae [INV ZOO] The four-footed butterflies, a family of lepidopteran insects in the superfamily Papilionoidea; prothoracic legs are atrophied, and the well-developed patagia are heavily sclerotized. { nim'fal·ə₁dē }

Nymphalinae [INV ZOO] A subfamily of the lepidopteran family Nymphalidae. { nim'fal·ə₁nē }

nymphomania [PSYCH] Excessive sexual desire on the part of a woman. Also known as hysteromania. { ₁nim·fə'mā·nē·ə }

Nymphonidae [INV ZOO] A family of marine arthropods in the subphylum Pycnogonida; members have chelifores, five-jointed palpi, and ten-jointed ovigers. { nim'fän·ə₁dē }

Nymphulinae [INV ZOO] A subfamily of the lepidopteran family Pyralididae which is notable because some species are aquatic. { nim'fyül·ə₁nē }

Nyquist contour [CONT SYS] A directed closed path in the complex frequency plane used in constructing a Nyquist diagram, which runs upward, parallel to the whole length of the imaginary axis at an infinitesimal distance to the right of it, and returns from $+j\infty$ to $-j\infty$ along a semicircle of infinite radius in the right half-plane. { 'nī₁kwist ₁kän₁tùr }

Nyquist diagram [CONT SYS] A plot in the complex plane of the open-loop transfer function as the complex frequency is varied along the Nyquist contour; used to determine stability of a control system. { 'nī₁kwist ₁dī·ə₁gram }

Nyquist interval [COMMUN] Maximum separation in time which can be given to regularly spaced instantaneous samples

Nutmeg (*Myristica fragrans*), mature fruits. (*USDA*)

A common eastern American species of water lily (*Nymphaea odorata*). (*Photograph by Hugh Spencer, National Audubon Society*)

of a wave of specified bandwidth for complete determination of the waveform of the signal. { 'nī,kwist ,in·tər·vəl }

Nyquist rate [COMMUN] The maximum rate at which code elements can be unambiguously resolved in a communications channel with a limited range of frequencies; equal to twice the frequency range. { 'nī,kwist ,rāt }

Nyquist sampling [COMMUN] The periodic sampling of audio or video signals, in order to preserve their information content, at a rate equal to twice the highest frequency to be preserved. { 'nī,kwist ,sam·pliŋ }

Nyquist stability criterion *See* Nyquist stability theorem. { 'nī,kwist stə'bil·əd·ē krī,tir·ē·ən }

Nyquist stability theorem [CONT SYS] The theorem that the net number of counterclockwise rotations about the origin of the complex plane carried out by the value of an analytic function of a complex variable, as its argument is varied around the Nyquist contour, is equal to the number of poles of the variable in the right half-plane minus the number of zeros in the right half-plane. Also known as Nyquist stability criterion. { 'nī,kwist stə'bil·əd·ē ,thir·əm }

Nyquist's theorem [ELECTR] The mean square noise voltage across a resistance in thermal equilibrium is four times the product of the resistance, Boltzmann's constant, the absolute temperature, and the frequency range within which the voltage is measured. { 'nī,kwists ,thir·əm }

Nysa [ASTRON] An asteroid whose surface composition may resemble that of the aubrites; it has a diameter of approximately 42 miles (68 kilometers) and a mean distance from the sun of 2.42 astronomical units. { 'nī·sə }

Nysmyth's membrane [ANAT] The primary enamel cuticle which is the transitory remnants of the enamel organ and oral epithelium covering the enamel of a tooth after eruption. { 'nī,smiths ,mem,brān }

Nyssaceae [BOT] A family of dicotyledonous plants in the order Cornales characterized by perfect or unisexual flowers with imbricate petals, a solitary ovule in each locule, a unilocular ovary, and more stamens than petals. { nə'sās·ē,ē }

nystagmogram [IND ENG] A recording of saccadic eye movements, that is, quick, rhythmic, and usually involuntary oscillations of the eyes. { ni'stag·mə,gram }

nystagmus [MED] Involuntary oscillatory movement of the eyeballs. { nə'stag·məs }

nystatin [MICROBIO] $C_{46}H_{77}NO_{19}$ An antifungal antibiotic produced by *Streptomyces noursei*; used for the treatment of infections caused by *Candida* (*Monilia*) *albicans*. { 'nis·təd·ən }

nytril [TEXT] A manufactured fiber containing at least 85% of a long-chain polymer of vinylidene dinitrile, where the vinylidene dinitrile content is no less than every other unit in the polymer chain. { 'nī·trəl }

O *See* oxygen.

oak [BOT] Any tree of the genus *Quercus* in the order Fagales, characterized by simple, usually lobed leaves, scaly winter buds, a star-shaped pith, and its fruit, the acorn, which is a nut; the wood is tough, hard, and durable, generally having a distinct pattern. { ōk }

oakmoss resin [MATER] Concrete oleoresin from the oakmoss lichens *Evernia prunastri* and *E. furfuracea*; used as a perfume fixative. Also known as mousse de chêne. { 'ōk,mòs 'rez·ən }

oakum [MATER] Old hemp or jute fiber, loosely twisted and impregnated with tar or a tar derivative, used to caulk sides and decks of ships and to pack joints of pipes and caissons. { 'ōk·əm }

oak wilt [PL PATH] A fungus disease of oak trees caused by *Chalara quercina*, characterized by wilting and yellow and red discoloration of the leaves progressing from the top downward and inward. { 'ōk ,wilt }

O antigen [MICROBIO] A somatic antigen of certain flagellated microorganisms. { 'ō 'ant·i·jən }

oasis [GEOGR] An isolated fertile area, usually limited in extent and surrounded by desert, and marked by vegetation and a water supply. { ō'ā·səs }

OASIS *See* Open-Access Same-Time Information System. { ,o'ā·səs }

OASM system [PHYS] A system of electrical and mechanical units in which the fundamental quantities are electric resistance, electric current, time, and length, and the base units of these quantities are the ohm, ampere, second, and meter, respectively. { 'ō,ā,es'em ,sis·təm }

oat [BOT] Any plant of the genus *Avena* in the family Graminae, cultivated as an agricultural crop for its seed, a cereal grain, and for straw. { ōt }

oatmeal paper [MATER] A paper in which fine sawdust is added to produce a sheet with a coarse texture; it can be used as an inexpensive sketching paper for work in pastels and charcoal, or as wallpaper. { 'ōt,mēl 'pā·pər }

O attenuator [ELECTR] A dissipative attenuator in which the circuit has the form of a ladder with two rungs, and the resistances across the rungs are unequal, so that the impedances across the two pairs of terminals are unequal. { 'ō ə'ten·yə,wād·ər }

OBA *See* octave-band analyzer.

OB association [ASTRON] A grouping of very young, very hot massive stars of spectral types O and B that has not had time to disperse. { ,ō'bē ə,sō·sē'ā·shən }

obclavate [BIOL] Inversely clavate. { äb'kla,vāt }

obcordate [BOT] Referring to a leaf, heart-shaped with the notch apical. { äb'kòr,dāt }

obdiplastemonous [BOT] Having the stamens arranged in two whorls, with members of the outer whorl positioned opposite the petals. { äb,dip·lō,stē·mə·nəs }

obduction [MED] The act or instance of performing a postmortem examination. { äb'dək·shən }

obelion [ANTHRO] The point where the line which joins the parietal foramens crosses the sagittal suture. { ō'bē·lē,än }

obelisk [ARCH] A four-sided pillar, tapering toward the top. [MATH] A frustrum of a regular, rectangular pyramid. { 'äb·ə,lisk }

Obermayer's reagent [CHEM] A 0.4% solution of ferric chloride in concentrated hydrochloric acid; used to test for indican in urine, with a pale-blue or deep-violet color indicating positive. { 'ō·bər,mī·ərz rē,ā·jənt }

Oberon [ASTRON] One of the five satellites of Uranus; diameter about 870 miles (1400 kilometers). { 'ō·bə,rän }

oberwind [METEOROL] A night wind from mountains or the upper ends of lakes; a wind of Salzkammergut in Austria. { 'ō·bər,vint }

obese [ANAT] Extremely fat. { ō'bēs }

obesity [MED] An excessive accumulation of body fat which confers health risks such as diabetes, cardiovascular diseases, arthritis, and some types of tumors. { ō'bē·səd·ē }

obfuscation [PSYCH] Mental confusion. { ,äb·fə'skā·shən }

object [COMPUT SCI] **1.** Any collection of related items. **2.** The name of a single element in an object-oriented programming language. [OPTICS] A collection of points which may be regarded as a source of light rays in an optical system, whether it actually has this function (as in a real object) or does not (as in a virtual object). { 'äb·jekt }

object code [COMPUT SCI] The statements generated from source code by a compiler, constituting an intermediate step in the translation of source code into executable machine language. { 'äb·jekt ,kōd }

object computer [COMPUT SCI] The computer processing an object program; the same computer compiling the source program could, therefore, be called the source computer; such terminology is seldom used in practice. { 'äb·jekt kəm,pyüd·ər }

object contrast [OPTICS] The ratio of the difference between the brightness of an object and of the background to the brightness of the background in an image or reproduction. { 'äb·jekt 'kän,trast }

object deck [COMPUT SCI] The set of machine-readable computer instructions produced by a compiler, either in absolute format (that is, containing only fixed addresses) or, more frequently, in relocatable format. { 'äb·jekt ,dek }

object glass *See* objective. { 'äb·jekt ,glas }

objective [OPTICS] The first lens, lens system, or mirror through which light passes or from which it is reflected in an optical system; many scientists exclude mirrors from the definition. Also known as object glass. { äb'jek·tiv }

objective basic research [SCI TECH] Basic research in fields recognized as having a potential technological importance. { äb'jek·tiv 'bā·sik 'rē,sərch }

objective function [MATH] In nonlinear programming, the function, expressing given conditions for a system, which one seeks to minimize subject to given constraints. { äb'jek·tiv 'fəŋk·shən }

objective grating [OPTICS] A series of equally spaced parallel wires placed over the objective lens of a telescope; photographic magnitudes of stars are calculated from the relative brightnesses of images in the resulting diffraction pattern. { äb'jek·tiv 'grād·iŋ }

objective plane [ORD] Plane tangent to the ground or coinciding with the surface of the target, especially such a plane at the point of impact of a bomb or projectile. { äb'jek·tiv 'plān }

objective prism [OPTICS] A large prism, usually having a small angle, which is placed in front of the objective of a photographic telescope to make spectroscopic observations. { äb'jek·tiv 'priz·əm }

objective probabilities [STAT] Probabilities determined by

OAK

Terminal bud, leaf, and twig of white oak (*Quercus alba*).

the long-run relative frequency of an event. Also known as frequency probabilities. { əb'jek·tiv ,präb·ə'bil·əd·ēz }

objective sign [MED] A sign which can be detectable by someone other than the patient. { äb'jek·tiv 'sīn }

object language [COMPUT SCI] The intended and desired output language in the translation or conversion of information from one language to another. { 'äb·jekt ,laŋ·gwij }

object lens [OPTICS] The first lens through which light passes in a compound objective. { 'äb·jekt ,lenz }

object library See object program library. { 'äb·jekt 'lī ,brer·ē }

Object Management Group object model [COMPUT SCI] A model that defines common object semantics in an object-oriented computer system. Abbreviated OMG object model. { 'äb·jikt 'man·ij·mənt 'grüp 'äb·jikt ,mäd·əl }

object module [COMPUT SCI] The computer language program prepared by an assembler or a compiler after acting on a programmer-written source program. { 'äb·jekt ,mäj·ül }

object-oriented graphics See vector graphics. { 'äb,jekt ,ór·ē,en·təd 'graf·iks }

object-oriented interface [COMPUT SCI] A user interface that employs icons and a mouse. { 'äb,jekt ,ór·ē,en·təd 'in·tər,fās }

object-oriented language [COMPUT SCI] A programming language consisting of a sequence of commands directed at objects. { 'äb,jekt ,ór·ē,en·təd 'laŋ·gwij }

object-oriented programming [COMPUT SCI] A computer programming methodology that focuses on data rather than processes, with programs composed of self-sufficient modules (objects) containing all the information needed to manipulate a data structure. Abbreviated OOP. { 'äb,jekt ,ór·ē,en·təd 'prō,gram·iŋ }

object plane [OPTICS] A plane containing the real or virtual object in an optical system; usually perpendicular to the axis of the system. { 'äb·jekt ,plān }

object program [COMPUT SCI] The computer language program prepared by an assembler or a compiler after acting on a programmer-written source program. Also known as object routine; target program; target routine. { 'äb·jekt ,prō,gram }

object program library [COMPUT SCI] A collection of computer programs in the form of relocatable instructions, which reside on, and may be read from, a mass storage device. Also known as object library. { 'äb·jekt ,prō,gram ,lī,brer·ē }

object relationship [PSYCH] The attitudes and responses of one person toward another; the capacity of an individual to react appropriately to and to accept and love other people. { 'äb·jekt ri'lā·shən,ship }

object request broker [COMPUT SCI] The central component of CORBA, which passes requests from clients to the objects on which they are invoked. Abbreviated ORB. { 'äb·jikt ri'kwest ,brō·kər }

object routine See object program. { 'äb·jekt rü,tēn }

object space [OPTICS] The region of space where objects are located so that a given optical system can form images of them. { 'äb·jekt ,spās }

object time [COMPUT SCI] The time during which execution of an object program is carried out. { 'äb·jekt ,tīm }

oblanceolate [SCI TECH] Inversely lanceolate. { äb'lan·sē·ə,lāt }

oblate ellipsoid See oblate spheroid. { 'ä,blāt i'lip,sóid }

oblateness [ASTRON] The distortion from a spherical shape in which the diameter at the equator exceeds that at the poles. { ä'blāt·nəs }

oblate spheroid [MATH] The surface or ellipsoid generated by rotating an ellipse about one of its axes so that the diameter of its equatorial circle exceeds the length of the axis of revolution. Also known as oblate ellipsoid. { 'ä,blāt 'sfir,óid }

oblate spheroidal coordinate system [MATH] A three-dimensional coordinate system whose coordinate surfaces are the surfaces generated by rotating a plane containing a system of confocal ellipses and hyperbolas about the minor axis of the ellipses, together with the planes passing through the axis of rotation. { 'ō,blāt sfir,óid·əl kō'órd·ən·ət ,sis·təm }

obligate [BIOL] Restricted to a specified condition of life, as an obligate parasite. { 'äb·lə·gət }

obligate aerobe [MICROBIO] A microorganism that uses oxygen for cellular respiration and requires some free molecular oxygen in its surroundings to support growth. { 'äb·lə,gāt 'er·ōb }

obligate anaerobe [MICROBIO] A microorganism that cannot use oxygen and can grow only in the absence of free oxygen. { 'äb·li,gāt 'an·ə,rōb }

oblique [ANAT] Referring to a muscle, positioned obliquely and having one end that is not attached to bone. [BOT] Referring to a leaf, having the two sides of a blade unequal. [SCI TECH] Having a slanted direction or position. { ə'blēk }

oblique angle [MATH] An angle that is neither a right angle nor a multiple of a right angle. { ə'blēk 'aŋ·gəl }

oblique ascension [ASTRON] The arc of the celestial equator, or the angle at the celestial pole, between the hour circle of the vernal equinox and the hour circle through the intersection of the celestial equator and the eastern horizon at the instant a point on the oblique sphere rises, measured eastward from the hour circle of the vernal equinox through 24 hours. { ə'blēk ə'sen·chən }

oblique astigmatism See radial astigmatism. { ə'blēk ə'stig·mə,tiz·əm }

oblique chart [MAP] A chart on an oblique projection. { ə'blēk 'chärt }

oblique circular cone [MATH] A circular cone whose axis is not perpendicular to its base. { ə'blēk 'sər·kyə·lər 'kōn }

oblique coordinates [MATH] Magnitudes defining a point relative to two intersecting nonperpendicular lines, called axes; the magnitudes indicate the distance from each axis, measured along a parallel to the other axis; oblique coordinates are a form of cartesian coordinates. { ə'blēk kō'órd·ən·əts }

oblique cylindrical orthomorphic projection See oblique Mercator projection. { ə'blēk si'lin·drə·kəl ,ór·thə'mór·fik prə'jek·shən }

oblique equator [MAP] A great circle, the plane of which is perpendicular to the axis of an oblique projection; an oblique equator serves as the origin for measurement of oblique latitude; on an oblique Mercator projection, the oblique equator is the tangent great circle. { ə'blēk i'kwād·ər }

oblique extinction See inclined extinction. { ə'blēk ik 'stiŋk·shən }

oblique fault See diagonal fault. { ə'blēk 'fólt }

oblique fire [ORD] Fire placed on a target from a direction diagonal to the long dimension of the target, or on an enemy from a direction between the enemy's front and flank. { ə'blēk 'fīr }

oblique graticule [MAP] A fictitious graticule based upon an oblique projection. { ə'blēk 'grad·ə,kyül }

oblique-incidence reflectivity [OPTICS] The reflectivity of an interface between two media when the direction of propagation of the incident electromagnetic radiation is not perpendicular to the interface; it differs for the component whose electric vector lies in the plane containing the perpendicular to the surface and the propagation direction, and the component for which this vector is perpendicular to this plane. { ə'blēk 'in·sə·dəns ,rē,flek'tiv·əd·ē }

oblique-incidence transmission [COMMUN] Transmission of a radio wave obliquely up to the ionosphere and down again. { ə'blēk 'in·sə·dəns tranz'mish·ən }

oblique joint See diagonal joint. { ə'blēk 'jóint }

oblique lines [MATH] Lines that are neither perpendicular nor parallel. { ə'blēk 'līnz }

oblique Mercator projection [MAP] A conformal cylindrical map projection in which points on the surface of a sphere or spheroid, such as the earth, are conceived as developed by Mercator principles on a cylinder tangent along an oblique great circle. Also known as oblique cylindrical orthomorphic projection. { ə'blēk mər'kād·ər prə'jek·shən }

oblique meridian [MAP] A great circle perpendicular to an oblique equator; the reference oblique meridian is called the prime oblique meridian. { ə'blēk mə'rid·ē·ən }

oblique parallel [MAP] A circle or line parallel to an oblique equator, connecting all points of equal oblique latitude. { ə'blēk 'par·ə,lel }

oblique parallelepiped [MATH] A parallelepiped whose lateral edges are not perpendicular to its bases. { ə'blēk ,par·ə,lel·ə'pī,ped }

oblique photograph [GRAPHICS] An aerial photograph taken with the camera axis intentionally inclined between the horizontal and the vertical. { ə'blēk 'fōd·ə,graf }

oblique pole [MAP] One of the two points 90° from an oblique equator. { ə'blēk 'pōl }

oblique projection [MAP] A map projection with its axis at

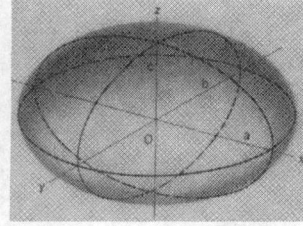

OBLATE SPHEROID

Drawing of an oblate spheroid generated by rotating an ellipse about its minor axis lying along *z* axis of coordinate system, with center of ellipse at origin of coordinates, *O*. Diameters 2*a* and 2*b* along *x* and *y* axes are equal to each other, and greater than axis of revolution 2*c*.

an oblique angle to the plane of the equator. { ə'blek prə'jek·shən }

oblique rhumb line [MAP] **1.** A line making the same oblique angle with all fictitious meridians of an oblique Mercator projection; oblique parallels and meridians may be considered special cases of the oblique rhumb line. **2.** Any rhumb line, real or fictitious, making an oblique angle with its meridians; in this sense the expression is used to distinguish such rhumb lines from parallels and meridians, real or fictitious, which may be included in the expression "rhumb line." { ə'blek 'rəm ,līn }

oblique rotator [ASTRON] A star model in which the axis of the magnetic field does not coincide with the axis of rotation. { ə'blek 'rō,tād·ər }

oblique shock *See* oblique shock wave. { ə'blek 'shäk }

oblique shock wave [FL MECH] A shock wave inclined at an oblique angle to the direction of flow in a supersonic flow field. Also known as oblique shock. { ə'blek 'shäk ,wāv }

oblique slip fault [GEOL] A fault which has slippage along both the strike and dip of the fault plane. { ə'blek 'slip ,fōlt }

oblique sphere [ASTRON] The celestial sphere as it appears to an observer between the equator and the pole, where celestial bodies appear to rise obliquely to the horizon. { ə'blek 'sfir }

oblique spherical triangle [MATH] A spherical triangle that has no right angle. { ə'blek 'sfer·ə·kəl 'trī,aŋ·gəl }

oblique strophoid [MATH] A plane curve derived from a straight line L and two points called the pole and the fixed point, where the fixed point lies on L but is not the foot of the perpendicular from the pole to the line; it consists of the locus of points on a rotating line L' passing through the pole whose distance from the intersection of L and L' is equal to the distance of this intersection from the fixed point. { ə'blek 'strō,fóid }

oblique triangle [MATH] A triangle that does not contain a right angle. { ə'blek 'trī,aŋ·gəl }

oblique valve [MECH ENG] A type of globe valve having an inclined orifice that serves to reduce the disruption of the flow pattern of the working fluid. { ə'blek 'valv }

oblique visibility *See* oblique visual range. { ə'blek ,viz·ə'bil·əd·ē }

oblique visual range [OPTICS] The greatest distance at which a specified target can be perceived when viewed along a line of sight inclined to the horizontal. Also known as oblique visibility; slant visibility. { ə'blek 'vizh·ə·wəl 'rānj }

obliquity factor [OPTICS] A function which is proportional to the amplitudes of secondary waves propagating in various directions according to Huygens' principle; it is $1 + \cos\theta$, where θ is the angle between the normal to the original wavefront and the normal to the secondary wavefront. { ə'blik·wəd·ē ,fak·tər }

obliquity of the ecliptic [ASTRON] The acute angle between the plane of the ecliptic and the plane of the celestial equator, about 23°27'. { ə'blik·wəd·ē əv thə i'klip·tik }

obliterated corner [CIV ENG] In surveying, a corner for which visible evidence of the previous surveyor's work has disappeared, but whose original position can be established from other physical evidence and testimony. { ə'blid·ə,rād·əd 'kór·nər }

obliterating endarteritis *See* endarteritis obliterans. { ə'blid·ə,rād·iŋ 'end,ärd·ə'rīd·əs }

obliteration [MED] **1.** Complete removal of an organ or other body part by disease or surgical excision. **2.** Closure of a lumen. **3.** Loss of memory or consciousness of specific events. { ə,blid·ə'rā·shən }

obliterative appendicitis [MED] Obliteration of the lumen of the appendix by fibrofatty tissue. { ə'blid·ə,rād·iv ə,pen·də'sīd·əs }

oblong mesh *See* rectangular mesh. { 'äb,lóŋ 'mesh }

O/B/O carrier *See* ore/bulk/oil carrier. { 'ō'bē'ō ,kar·ē·ər }

Oboe [NAV] An electronic navigation system utilizing a single-path round-trip system for determination of transmission times and distance; used for bombing in World War II. { 'ō,bō }

Obolellida [PALEON] A small order of Early and Middle Cambrian inarticulate brachiopods, distinguished by a shell of calcium carbonate. { ,äb·ə'lel·ə·də }

obovate [BIOL] Inversely ovate. { 'äb·ə,vāt }

obscuration [METEOROL] In United States weather observing practice, the designation for the sky cover when the sky is completely hidden by surface-based obscuring phenomena, such as fog. Also known as obscured sky cover. { ,äb·skyù'rā·shən }

obscured sky cover *See* obscuration. { əb'skyùrd 'skī ,kəv·ər }

obscure glass [MATER] Translucent glass. { əb'skyùr 'glas }

obscuring phenomenon [METEOROL] In United States weather observing practice, any atmospheric phenomenon (not including clouds) which restricts the vertical visibility or slant visibility, that is, which obscures a portion of the sky from the point of observation. { əb'skyùr·iŋ fə,näm·ə,nän }

obsequent [GEOL] Of a stream, valley, or drainage system, being in a direction opposite to that of the original consequent drainage. { 'äb·sə·kwənt }

obsequent fault-line scarp [GEOL] A fault-line scarp which faces in the direction opposite to that of the original fault scarp or in which the structurally upthrown block is topographically lower than the downthrown block. { 'äb·sə·kwənt 'fólt ,līn ,skärp }

observability [CONT SYS] Property of a system for which observation of the output variables at all times is sufficient to determine the initial values of all the state variables. { əb,zər·və'bil·əd·ē }

observable operator [QUANT MECH] A Hermitian operator with a complete, orthonormal set of eigenfunctions on the Hilbert space representing the states of a physical system; such operators are postulated to represent the observable quantities of the system. { əb'zər·və·bəl 'äp·ə,rād·ər }

observable quantity [PHYS] A measurable physical quantity. { əb'zər·və·bəl 'kwän·əd·ē }

observational day [GEOPHYS] Any 24-hour period selected as the basis for climatological or hydrological observations. { ,äb·zər'vā·shən·əl ,dā }

observation of fire [ORD] Act of watching artillery fire in order to locate the burst or impact of projectiles in respect to the target and to correct firing data; surveillance of fire; may be made from the ground or from the air. { ,äb·zər'vā·shən əv 'fīr }

observation spillover [CONT SYS] The part of the sensor output of an active control system caused by modes that have been omitted from the control algorithm in the process of model reduction. { ,äb·zər'vā·shən 'spil,ō·vər }

observation well [PETRO ENG] A special well drilled in a selected location for the purpose of observing parameters such as fluid levels and pressure changes (for example, within an oil reservoir) as production proceeds. { ,äb·zər'vā·shən ,wel }

observed fire [ORD] Fire for which the points of impact or burst can be seen by an observer on the ground, in aircraft, or on a naval vessel; it can be controlled and adjusted on the basis of the observations. { əb'zərvd 'fīr }

observed latitude [NAV] Latitude determined from observations which result in lines of position extending in a generally east-west direction. { əb'zərvd 'lad·ə,tüd }

observed longitude [NAV] Longitude determined from observations which result in lines of position extending in a generally north-south direction. { əb'zərvd 'län·jə,tüd }

observed position [ORD] Position of a moving target at the instant of observation. { əb'zərvd pə'zish·ən }

observer [CONT SYS] A linear system B driven by the inputs and outputs of another linear system A which produces an output that converges to some linear function of the state of system A. Also known as state estimator; state observer. { əb'zər·vər }

observing angle [ORD] Angle at the target between a line to the observer and a line to the gun or battery; the angular distance of an observer from the gun or battery. { əb'zər·viŋ ,aŋ·gəl }

obsession [PSYCH] Persistence of or anxious preoccupation with an idea or emotion recognized as unreasonable by the individual. { əb'sesh·ən }

obsessive-compulsive disorder [PSYCH] A type of anxiety disorder characterized by recurrent, persistent, unwanted, and unpleasant thoughts (obsessions) or repetitive, purposeful, ritualistic behaviors that the person feels driven to perform (compulsions). Abbreviated OCD. { əb'ses·iv kəm'pəl·siv dis,órd·ər }

obsessive-compulsive neurosis [PSYCH] A neurotic disorder in which anxiety relates to obsessions which the individual fights against but cannot control, and by which the person is dominated. { əb'ses·iv kəm'pəl·siv nü'rō·səs }

obsessive-compulsive personality [PSYCH] A behavioral disorder in which a person is generally characterized by chronic, excessive concern with conformity or adherence to standards, resulting in inhibited, inflexible behavior, inability to relax, and the performance of an inordinate amount of work. { əb'ses·iv kəm'pəl·siv ˌpər·sə'nal·əd·ē }

obsidian [GEOL] A jet-black volcanic glass, usually of rhyolitic composition, formed by rapid cooling of viscous lava; generally forms the upper parts of lava flows. Also known as hyalopsite; Iceland agate; mountain mahogany. { äb'sid·ē·ən }

obsidian hydration method [ARCHEO] A chemical dating method that depends on the progressive thickening of a hydrated layer on the surface of volcanic glass. { äb'sid·ē·ən hī'drā·shən ˌmeth·əd }

obsidianite *See* tektite. { äb'sid·ē·əˌnīt }

obsolescence [ENG] Decreasing value of functional and physical assets or value of a product or facility from technological changes rather than deterioration. { ˌäb·sə'les·əns }

obsolete [BIOL] A part of an organism that is imperfect or indistinct, compared with a corresponding part of similar organisms. [ENG] No longer satisfactory for the purpose for which obtained, due to improvements or revised requirements. { ˌäb·sə'lēt }

obstacle [NAV] In air operations, a natural or artificial object which by its geographical location requires that its vertical dimension be considered in establishing the permissible flight paths in an area; this leads to the concept of the fictitious height for various obstacles; for example, a moving vehicle 17 feet high is assumed to be on an interstate highway 15 feet high, and thus the area is assumed to have an obstacle 32 feet high. { 'äb·stə·kəl }

obstacle clearance [NAV] In air operations, the vertical distance between the lowest and the authorized height of operation. { 'äb·stə·kəl ˌklir·əns }

obstacle clearance box [NAV] A rectangle on an aeronautical chart which encloses a number indicating the obstacle clearance required (in feet) on an approach segment where the box appears. { 'äb·stə·kəl ˌklir·əns ˌbäks }

obstetric [MED] **1.** Of or pertaining to obstetrics. **2.** Of or pertaining to pregnancy and childbirth. { äb'ste·trik }

obstetrical analgesia [MED] Analgesia induced to diminish or obliterate the pain of childbirth. { äb'ste·trə·kəl ˌan·əl'jē·zē·ə }

obstetric forceps [MED] A perforated, double-bladed traction forceps which can be applied to the fetal head in cases of difficult labor. { äb'ste·trik 'fór·seps }

obstetrician [MED] One who practices obstetrics. { ˌäb·stə'trish·ən }

obstetrics [MED] The branch of medicine that deals with pregnancy, labor, and the puerperium. { äb'ste·triks }

obstipation [MED] Constipation that is difficult to relieve. { ˌäb·stə'pā·shən }

obstructed stream [GEOL] A stream whose valley has been blocked by a landslide, glacial moraine, sand dune, or lava flow; it frequently consists of a series of ponds or small lakes. { əb'strək·təd 'strēm }

obstruction [MED] Occlusion or stenosis of hollow viscera, ducts, and vessels. [NAV] Anything that hinders or prevents movement, particularly anything that endangers or prevents passage of a vessel or aircraft; usually refers to an isolated danger to navigation, such as a submerged rock or pinnacle in the case of marine navigation, and a tower, tall building, mountain peak, and so forth, in the case of air navigation. { əb'strək·shən }

obstruction beacon [NAV] A beacon marking an obstruction or hazard. Also known as hazard beacon. { əb'strək·shən ˌbē·kən }

obstruction buoy [NAV] A buoy marking an obstruction. { əb'strək·shən ˌbói }

obstruction light [NAV] A light indicating the presence of an obstruction. { əb'strək·shən ˌlīt }

obstruction marker [NAV] A marker indicating the presence of an obstruction. { əb'strək·shən ˌmärk·ər }

obstruction moraine [GEOL] A moraine formed where the movement of ice is obstructed, for example, by a ridge of bedrock. { əb'strək·shən mə'rān }

obstruction to vision [METEOROL] In United States weather observing practice, one of a class of atmospheric phenomena, other than the weather class of phenomena, which may reduce horizontal visibility at the earth's surface; examples are fog, smoke, and blowing snow. { əb'strək·shən tə 'vizh·ən }

obstructive apnea [MED] A pause in breathing while sleeping that lasts more than 10 seconds and is caused by a collapse of the upper airways. { əb'stək·tiv 'ap·nē·ə }

obstructive atelectasis [MED] Collapse of all or a portion of the lung due to bronchial obstruction or occlusion. Also known as absorption atelectasis. { əb'strək·tiv ˌad·əl'ek·tə·səs }

obstructive dysmenorrhea *See* mechanical dysmenorrhea. { əb'strək·tiv ˌdis·men·ə'rē·ə }

obstructive emphysema [MED] Overdistension of the lung due to partial obstruction of the air passages, which permits air to enter the alveoli but which resists expiration of the air. { əb'strək·tiv ˌem·fə'sē·mə }

obstructive glaucoma *See* narrow-angle glaucoma. { əb'strək·tiv glau'kō·mə }

obstructive hydrocephaly [MED] Accumulation of cerebrospinal fluid in the brain ventricles caused by obstruction of the passage of the fluid from the ventricles to the subarachnoid space. Also known as internal hydrocephaly; noncommunicating hydrocephaly. { əb'strək·tiv ˌhī·drə'sef·ə·lē }

obstructive jaundice [MED] Jaundice caused by mechanical obstruction of the biliary passages, preventing the outflow of bile. { əb'strək·tiv 'jón·dəs }

obtund [MED] To make dull or reduce, as to obtund sensibility. { äb'tənd }

obturating cup [ORD] An inverted cup, sealing against passage of gases of explosion, which inverts due to pressure and actuates a firing pin or other mechanism. { 'äb·təˌrād·iŋ ˌkəp }

obturation [MED] **1.** The closing of an opening or passage. **2.** A form of intestinal obstruction in which the lumen is occupied by its normal contents or by foreign bodies. [ORD] Sealing of the breech of a gun to prevent escape of propellant gases in firing. { ˌäb·tə'rā·shən }

obturator [ANAT] **1.** Pertaining to that which closes or stops up, as an obturator membrane. **2.** Either of two muscles, originating at the pubis and ischium, which rotate the femur laterally. [MED] A solid wire or rod contained within a hollow needle or cannula. [ORD] **1.** Assembly of steel spindle, mushroom head, obturator rings, and a gas-check or obturator pad of tough plastic material used as a seal to prevent the escape of propellant gases around the breechblock of guns using separate-loading ammunition, and therefore not having the obturation provided by a cartridge case. **2.** A device incorporated in a projectile to make the tube of a weapon gastight, preventing escape of gas until the projectile has left the muzzle of the weapon. { 'äb·təˌrād·ər }

obturator artery [ANAT] A branch of the internal iliac; it branches into the pubic and acetabular arteries. { 'äb·təˌrād·ər ˌärd·ə·rē }

obturator foramen [ANAT] A large opening in the pelvis, between the ischium and the pubis, that gives passage to vessels and nerves; it is partly closed by a fibrous obturator membrane. { 'äb·təˌrād·ər fə'rā·mən }

obturator membrane [ANAT] **1.** A fibrous membrane closing the obturator foramen of the pelvis. **2.** A thin membrane between the crura and foot plates of the stapes. { 'äb·təˌrād·ər 'memˌbrān }

obturator nerve [NEUROSCI] A mixed nerve arising in the lumbar plexus; innervates the adductor, gracilis, and obturator externus muscles, and the skin of the medial aspect of the thigh, hip, and knee joints. { 'äb·təˌrād·ər ˌnərv }

obturator pad [ORD] Pad of tough plastic material, forming part of an obturator. { 'äb·təˌrād·ər ˌpad }

obturator spindle [ORD] Part of the breechblock assembly of a gun which fires separate-loading ammunition; it extends through the breechblock and holds in position the various parts of the obturator, while permitting the breechblock to rotate independently about these parts. { 'äb·təˌrād·ər 'spind·əl }

obtuse [BOT] Of a leaf, having a blunt or rounded free end. { äb'tüs }

obtuse angle [MATH] An angle of more than 90° and less than 180°. { äb'tüs 'aŋ·gəl }

obtuse bisectrix [CRYSTAL] The bisectrix of the obtuse angle between the axes of a biaxial crystal. { äb'tüs bī'sek·triks }

obtuse triangle [MATH] A triangle having one obtuse angle. { äb'tüs 'trī,aŋ·gəl }

obvallate [BIOL] Surrounded by or as if by a wall. { äb 'va,lāt }

obvolute [BIOL] Overlapping. { 'äb·və,lüt }

occasional fog signal [NAV] A fog signal not sounded regularly in fog. { ə'kāzh·ən·əl 'fäg ,sig·nəl }

occasional light [NAV] A light not regularly exhibited for navigation purpose. { ə'kāzh·ən·əl ¦līt }

occidental [LAP] **1.** Property of a gemstone which is of an inferior quality (grade, luster, or value) or is an inferior variety. **2.** Property of a substitute gemstone that is misrepresented as being the genuine gem. { ,äk·sə'dent·əl }

occipital arch [INV ZOO] A part of an insect cranium lying between the occipital suture and postoccipital suture. { äk'sip·əd·əl ¦ärch }

occipital artery [ANAT] A branch of the external carotid which branches into the mastoid, auricular, sternocleidomastoid, and meningeal arteries. { äk'sip·əd·əl ¦ärd·ə·rē }

occipital bone [ANAT] The bone which forms the posterior portion of the skull, surrounding the foramen magnum. { äk'sip·əd·əl ¦bōn }

occipital condyle [ANAT] An articular surface on the occipital bone which articulates with the atlas. [INV ZOO] A projection on the posterior border of an insect head which articulates with the lateral neck plates. { äk'sip·əd·əl ¦kän,dīl }

occipital crest [ANAT] Either of two transverse ridges connecting the occipital protuberances with the foramen magnum. { äk'sip·əd·əl ¦krest }

occipital ganglion [INV ZOO] One of a pair of ganglia located just posterior to the brain in insects. { äk'sip·əd·əl ¦gaŋ·glē,än }

occipitalia [INV ZOO] An unpaired row of dorsal cilia on the head of gnathostomulids. { äk,sip·ə'tal·ē·ə }

occipital lobe [ANAT] The posterior lobe of the cerebrum having the form of a three-sided pyramid. { äk'sip·əd·əl ¦lōb }

occipital pole [ANAT] The tip of the occipital lobe of the brain. { äk'sip·əd·əl ¦pōl }

occipital protuberance [ANAT] A prominence on the surface of the occipital bone to which the ligamentum nuchae is attached. { äk'sip·əd·əl prə'tü·bə·rəns }

occipitofrontalis [ANAT] A muscle in two parts, the frontal (inserting in the skin of the forehead) and the occipital (inserting in the galea sponeurotica). { äk¦sip·əd·ō,frən'tal·is }

occiput [ZOO] The back of the head of an insert or vertebrate. { 'äk·sə,pùt }

occluded cyclone [METEOROL] Any cyclone (or low) within which there has developed an occluded front. { ə'klüd·əd 'sī,klōn }

occluded front [METEOROL] A composite of two fronts, formed as a cold front overtakes a warm front or quasi-stationary front. Also known as frontal occlusion; occlusion. { ə'klüd·əd 'frənt }

occluded gases [MIN ENG] Gases entering the mine atmosphere from feeders and blowers, and also from blasting operations. { ə'klüd·əd 'gas·əz }

occluding junction See tight junction. { ə'klüd·iŋ ,jəŋk·shən }

occlusal disharmony [MED] Increased or maldirected occlusal force on individual teeth or groups of teeth causing a malposition or functional aberration. { ə'klüs·əl dis'här·mə,nē }

occlusion [ANAT] The relationship of the masticatory surfaces of the maxillary teeth to the masticatory surfaces of the mandibular teeth when the jaws are closed. [COMPUT SCI] In computer vision, the obstruction of a view. [ENG] The retention of undissolved gas in a solid during solidification. [MED] A closing or shutting up. [METEOROL] See occluded front. [PHYS] Adhesion of gas or liquid on a solid mass, or the trapping of a gas or liquid within a mass. [PHYSIO] The deficit in muscular tension when two afferent nerves that share certain motor neurons in the central nervous system are stimulated simultaneously, as compared to the sum of tensions when the two nerves are stimulated separately. { ə'klü·zhən }

occultation [ASTRON] The disappearance of the light of a celestial body by intervention of another body of larger apparent size; especially, a lunar eclipse of a star or planet. { ,ä·kəl'tā·shən }

occult blood [PATH] Blood in body products such as feces, not detectable on gross examination. { ə'kəlt 'bləd }

occult hydrocephaly [MED] A syndrome in which the brain ventricular system is enlarged while cerebrospinal fluid pressure remains normal, causing dementia, disturbances of equilibrium, and disorders of sphincter control. { ə'kəlt ,hī·drə'sef·ə·lē }

occulting bar [ASTRON] A bar placed in the focal plane of a telescope eyepiece to cover part of the field of view, usually to cover a bright object in order to permit observation of a nearby faint object. { ə'kəlt·iŋ ,bär }

occulting disk [ASTRON] A small metal disk placed in the focal plane of the eyepiece of a telescope, usually to cover a bright object in order to permit observation of a faint one. { ə'kəlt·iŋ ,disk }

occulting light [NAV] A navigational light totally interrupted at intervals by periods of darkness; the duration of the light period is equal to or greater than that of the dark period. { ə'kəlt·iŋ ,līt }

occulting quick-flashing light [NAV] A light showing flashes for several seconds, followed by a shorter period of darkness. { ə'kə,t·iŋ ¦kwik ¦flash·iŋ ,līt }

occult mineral [MINERAL] A mineral component of rock which cannot be seen through a microscope, but whose presence can be detected by chemical analyses. { ə'kəlt 'min·rəl }

occult virus [VIROL] A virus whose presence is assumed but which cannot be recovered. { ə'kəlt 'vī·rəs }

occupational acne [MED] Acne acquired from regular exposure to acnegenic materials in certain industries; disappears when the cause is removed. { ,ä·kyə'pā·shən·əl 'ak·nē }

occupational disease [MED] A functional or organic disease caused by factors arising from the operations or materials of an individual's industry, trade, or occupation. { ,ä·kyə'pā·shən·əl di'zēz }

occupational ecology [IND ENG] A discipline concerned with the interaction of workers with the environment, and with matching humans with the environment in the most ergonomically efficient way and with minimal disturbance of the environment. { ,ä·kyə'pā shen·əl i'käl·ə·jē }

occupational medicine [MED] The branch of medicine which deals with the relationship of humans to their occupations, for the purpose of the prevention of disease and injury and the promotion of optimal health, productivity, and social adjustment. { ,ä·kyə'pā·shən·əl 'med·i·sən }

occupational neurosis [PSYCH] Any neurotic disorder manifested by the individual's inability to use those parts of the body commonly employed in the individual's occupation, such as a writer's inability to write due to a painful feeling of fatigue in the hand. { ,ä·kyə'pā·shən·əl nù'rō·səs }

occupational therapy [MED] The teaching of skills or the use of selected occupations for therapeutic or rehabilitation purposes. { ,ä·kyə'pā·shən·əl 'ther·ə·pē }

occupied bandwidth [COMMUN] Frequency bandwidth such that, below its lower and above its upper frequency limits, the mean powers radiated are each equal to 0.5% of the total mean power radiated by a given emission. { 'äk·yə,pīd 'band,width }

occupy [ENG] To set a surveying instrument over a point for the purpose of making observations or measurements. { 'äk·yə,pī }

OC curve See operating characteristic curve. { ¦ō'sē ,kərv }

ocean [OCEANOGR] The interconnected body of salt water that occupies almost three-quarters of the earth's surface. { 'ō·shən }

ocean basin [GEOL] The great depression occupied by the ocean on the surface of the lithosphere. { 'ō·shən 'bā·sən }

ocean circulation [OCEANOGR] **1.** Water current flow in a closed circular pattern within an ocean. **2.** Large-scale horizontal water motion within an ocean. { 'ō·shən ,sər·kyə'lā·shən }

ocean current [OCEANOGR] A net transport of ocean water along a definable path. { 'ō·shən 'kə·rənt }

ocean engineering [ENG] A subfield of engineering

involved with the development of new equipment concepts and the methodical improvement of techniques which allow humans to operate successfully beneath the ocean surface in order to develop and utilize marine resources. { 'ō·shən ,en·jə'nir·iŋ }

ocean ferry [NAV ARCH] A ferry vessel capable of navigating any ocean or the Gulf of Mexico, the Caribbean Sea, or the Gulf of Alaska more than 20 nautical miles (37 kilometers) offshore. { 'ō·shən ,fer·ē }

ocean floor [GEOL] The near-horizontal surface of the ocean basin. { 'ō·shən 'flȯr }

ocean-floor spreading See sea-floor spreading. { 'ō·shən ¦flȯr ‚spred·iŋ }

Oceanian [ECOL] Of or pertaining to the zoogeographic region that includes the archipelagos and islands of the central and south Pacific. { ‚ō·shē'an·ē·ən }

oceanic anticyclone See subtropical high. { ‚ō·shē'an·ik ‚ant·i'sī‚klōn }

oceanic basalt [PETR] Rocks of the oceanic island volcanoes. { ‚ō·shē'an·ik bə'sȯlt }

oceanic climate See marine climate. { ‚ō·shē'an·ik 'klī·mət }

oceanic crust [GEOL] A thick mass of igneous rock which lies under the ocean floor. { ‚ō·shē'an·ik 'krəst }

oceanic heat flow [GEOPHYS] The amount of thermal energy escaping from the earth through the ocean floor per unit area and unit time. { ‚ō·shē'an·ik 'hēt ‚flō }

oceanic high See subtropical high. { ‚ō·shē'an·ik 'hī }

oceanic island [GEOL] Any island which rises from the deep-sea floor rather than from shallow continental shelves. { ‚ō·shē'an·ik 'ī·lənd }

oceanicity [CLIMATOL] The degree to which a point on the earth's surface is in all respects subject to the influence of the sea; it is the opposite of continentality; oceanicity usually refers to climate and its effects; one measure for this characteristic is the ratio of the frequencies of maritime to continental types of air mass. Also known as oceanity. { ‚ō·shē·ə'nis·əd·ē }

oceanic province [OCEANOGR] The water of the ocean that lies seaward of the break in the continental shelf. { ‚ō·shē'an·ik 'präv·əns }

oceanic ridge See mid-oceanic ridge. { ‚ō·shē'an·ik 'rij }

oceanic rise [GEOL] A long, broad elevation of the bottom of the ocean. { ‚ō·shē'an·ik 'rīz }

oceanic stratosphere See cold-water sphere. { ‚ō·shē'an·ik 'strad·ə‚sfir }

oceanic zone [OCEANOGR] The biogeographic area of the open sea. { ‚ō·shē'an·ik 'zōn }

oceanite [PETR] A picritic basalt in which olivine is a great deal more abundant than plagioclase. { 'ō·shə‚nīt }

oceanity See oceanicity. { ‚ō·shē'an·əd·ē }

oceanization [GEOL] Process by which continental crust (sial) is converted into oceanic crust (sima). { ‚ō·shə·nə'zā·shən }

oceanodromous [VERT ZOO] Of a fish, migratory in salt water. { ¦ō·shə¦nä·drə·məs }

oceanographic dredge [ENG] A device used aboard ship to bring up large samples of deposits and sediments from the ocean bottom. { ¦ō·shə·nə¦graf·ik 'drej }

oceanographic equator [OCEANOGR] **1.** The region of maximum temperature of the ocean surface. **2.** The region in which the temperature of the ocean surface is greater than 28°C. { ¦ō·shə·nə¦graf·ik i'kwād·ər }

oceanographic model [OCEANOGR] A theoretical representation of the marine environment which relates physical, chemical, geological, biological, and other oceanographic properties. { ¦ō·shə·nə¦graf·ik 'mäd·əl }

oceanographic platform [ENG] A construction with a flat horizontal surface higher than the water, on which oceanographic equipment is suspended or installed. { ¦ō·shə·nə¦graf·ik 'plat‚fȯrm }

oceanographic ship See research ship. { ¦ō·shə·nə¦graf·ik 'ship }

oceanographic station [OCEANOGR] A geographic location at which oceanographic observations are taken from a stationary ship. { ¦ō·shə·nə¦graf·ik 'stā·shən }

oceanographic submersible [NAV ARCH] Any small research vessel designed for undersea operations. { ¦ō·shə·nə¦graf·ik səb'mər·sə·bəl }

OCEANOGRAPHIC SUBMERSIBLE

Crewed deep submersible vehicle *Deep Rover*® with an all-acrylic pressure hull. (*Deep Ocean Engineering, Inc.*)

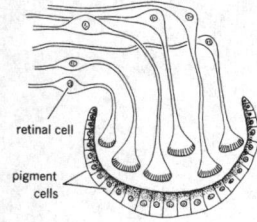

OCELLUS

retinal cell

pigment cells

Simple pigment cup ocellus of flatworm.

oceanographic survey [OCEANOGR] A study of oceanographic conditions with reference to physical, chemical, biological, geological, and other properties of the ocean. { ¦ō·shə·nə¦graf·ik 'sər‚vā }

oceanographic vessel [NAV ARCH] A research ship or other manned vehicle used in oceanography. { ¦ō·shə·nə¦graf·ik 'ves·əl }

oceanography [OCEAN] The science of the sea, including physical oceanography (the study of the physical properties of seawater and its motion in waves, tides, and currents), marine chemistry, marine geology, and marine biology. Also known as oceanology. { ‚ō·shə'näg·rə·fē }

oceanology See oceanography. { ‚ō·shə'näl·ə·jē }

ocean station vessel [NAV ARCH] A vessel assigned to patrol at a specified geographical position on the ocean; such vessels are specially equipped to make comprehensive meteorological and some oceanographic observations, and to provide meteorological observations, navigation assistance, and other services to transiting aircraft. { 'ō·shən ¦stā·shən ‚ves·əl }

ocean thermal-energy conversion [MECH ENG] The conversion of energy arising from the temperature difference between warm surface water of oceans and cold deep-ocean current into electrical energy or other useful forms of energy. Abbreviated OTEC. { 'ō·shən 'thər·məl 'en·ər·jē kən‚vər·zhən }

ocean tomography [ACOUS] A form of acoustic tomography in which an array of acoustic sources and receivers transmits and detects a pulse; the pulse travel times are used to determine temperature distributions in the ocean. { 'ō·shən tō'mäg·rə·fē }

ocean weather station [METEOROL] As defined by the World Meteorological Organization, a specific maritime location occupied by a ship equipped and staffed to observe weather and sea conditions and report the observations by international exchange. { 'ō·shən 'weth·ər ‚stā·shən }

ocellar [PETR] Of the texture of an igneous rock, having crystalline aggregates of phenocrysts arranged radially or tangentially around larger euhedral crystals or which form rounded branching forms. { ō'sel·ər }

ocellus [INV ZOO] A small, simple invertebrate eye composed of photoreceptor cells and pigment cells. [PETR] A phenocryst in an ocellar rock. { ō'sel·əs }

ocelot [VERT ZOO] *Felis pardalis.* A small arboreal wild cat, of the family Felidae, characterized by a golden head and back, silvery flanks, and rows of somewhat metallic spots on the body. { 'äs·ə‚lät }

ocher [MINERAL] A yellow, brown, or red earthy iron oxide, or any similar earthy, pulverulent metallic oxides used as pigments. { 'ō·kər }

ochlophobia [PSYCH] An abnormal fear of crowds. { ‚äk·lə'fō·bē·ə }

Ochnaceae [BOT] A family of dicotyledonous plants in the order Theales, characterized by simple, stipulate leaves, a mostly gynobasic style, and anthers that generally open by terminal pores. { äk'nās·ē‚ē }

Ochoan [GEOL] A North American provincial series that is uppermost in the Permian, lying above the Guadalupian and below the lower Triassic. { ō'chō·ən }

Ochotonidae [VERT ZOO] A family of the mammalian order Lagomorpha; members are relatively small, and all four legs are about equally long. { ‚äk·ə'tän·ə‚dē }

ochratoxin A [BIOCHEM] $C_{20}H_{18}ClNO_6$ A toxic metabolite from *Aspergillus ochraceus*; a crystalline compound which exhibits green fluorescence; melting point is 169°C; inhibits phosphorylase and mitochondrial respiration in rat liver. { ¦äk·rə¦täk·sən 'ā }

ochre mutation [GEN] Alteration of a codon to *UAA*, a stop codon that results in premature termination of the polypeptide chain in prokaryotes and eukaryotes. { 'ō·kər myü'tā·shən }

Ochrept [GEOL] A suborder of the soil order Inceptisol, with horizon below the surface, lacking clay, sesquioxides, or humus; widely distributed, occurring from the margins of the tundra region through the temperate zone, but not into the tropics. { 'ō·krept }

Ochrobium [MICROBIO] A genus of bacteria in the family Siderocapsaceae; cells are ellipsoidal to rod-shaped and are surrounded by a delicate sheath, resembling a horseshoe, that is heavily embedded with iron oxides. { ō'krō·bē·əm }

ochroleucous [BIOL] Pale ocher or buff colored. { ‚ō·krə'lü·kəs }

ochronosis [MED] A blue or brownish-blue discoloration of cartilage and connective tissue, especially around joints, caused by melanotic pigment. { ‚ō·krə'nō·səs }

Ochteridae [INV ZOO] The velvety shorebugs, the single family of the hemipteran superfamily Ochteroidea. { äk'ter·ə‚dē }

Ochteroidea [INV ZOO] A monofamilial tropical and subtropical superfamily of hemipteran insects in the subdivision Hydrocorisae; individuals are black with a silky sheen, and the antennae are visible from above. { ‚äk·tə'roid·ē·ə }

Ockham's razor [SCI TECH] The doctrine that unnecessary assumptions should be avoided in formulating hypotheses. { 'äk·əmz 'rā·zər }

ocotea oil [MATER] A volatile essential oil obtained from the wood of a Brazilian tree, *Ocotea cymbarum*; a source of safrole; used to make heliotropin and other technical preparations. { ə'kō·dē·ə ‚òil }

OCR *See* optical character recognition.

ocrea [BOT] A tubular stipule or pair of coherent stipules. { 'ä·krē·ə }

n-octadecane [ORG CHEM] $C_{18}H_{38}$ Colorless liquid boiling at 318°C; soluble in alcohol, acetone, ether, and petroleum, insoluble in water; used as a solvent and chemical intermediate. { ¦en ‚äk·tə'de‚kān }

1-octadecene [ORG CHEM] $C_{18}H_{36}$ Colorless liquid boiling at 180°C; soluble in alcohol, acetone, ether, and petroleum, insoluble in water; used as a chemical intermediate. { ¦wən ‚äk·tə'de‚sēn }

octadecenoic acid *See* oleic acid. { ‚äk·tə'des·ə‚nō·ik 'as·əd }

octadecenyl aldehyde [ORG CHEM] $C_{17}H_{35}CHO$ A flammable liquid with a boiling point of 167°C; used in the manufacture of vulcanization accelerators, rubber antioxidants, and pesticides. { ‚äk·tə'des·ə‚nəl 'al·də‚hīd }

octafluorocyclobutane [ORG CHEM] C_4F_8 A colorless gas or liquid with a boiling point of −4°C and a freezing point of −41.4°C; soluble in ether; used as a dielectric, refrigerant, and aerosol propellant. { ‚äk·tə¦flur·ō‚sī·klō'byü‚tān }

octafluoropropane [ORG CHEM] C_3F_8 A colorless gas with a boiling point of −36.7°C and a freezing point of approximately −160°C; used as a refrigerant and gaseous insulator. { ‚äk·tə‚flur·ō'prō‚pān }

octagon [MATH] A polygon with eight sides. { 'äk·tə‚gän }

octahedral borax *See* tincalconite. { ‚äk·tə¦hē·drəl 'bór‚aks }

octahedral cleavage [CRYSTAL] Crystal cleavage in the four planes parallel to the face of the octahedron. { ‚äk·tə¦hē·drəl 'klē·vij }

octahedral coordination [MINERAL] An atomic structure where six cations surround every anion, and vice versa. { ‚äk·tə¦hē·drəl kō'órd·ən‚ā·shən }

octahedral copper ore *See* cuprite. { ‚äk·tə¦hē·drəl 'käp·ər ‚ór }

octahedral group [MATH] The group of motions of three-dimensional space that transform a regular octahedron into itself. { ‚äk·tə¦hē·drəl ‚grüp }

octahedral iron ore *See* magnetite. { ‚äk·tə¦hē·drəl 'ī·ərn ‚ór }

octahedral molecule [CHEM] A molecule whose structure forms an octahedron in which a central atom possesses six valence bonds that are directed to the points of the octahedron, for example, sulfur hexafluoride (SF_6). { ‚äk·tə¦hē·drəl 'mäl·ə‚kyül }

octahedral normal stress [MECH] The normal component of stress across the faces of a regular octahedron whose vertices lie on the principal axes of stress; it is equal in magnitude to the spherical stress across any surface. Also known as mean stress. { ‚äk·tə¦hē·drəl 'nór·məl ‚stres }

octahedral plane [CRYSTAL] The plane in a cubic lattice having three numerically equal Miller indices. { ‚äk·tə¦hē·drəl 'plān }

octahedral shear stress [MECH] The tangential component of stress across the faces of a regular octahedron whose vertices lie on the principal axes of stress; it is a measure of the strength of the deviatoric stress. { ‚äk·tə¦hē·drəl 'shir ‚stres }

octahedrite [GEOL] The most common iron meteorite, containing 6–18% nickel in the metal phase and having intimate intergrowths lying parallel to the octahedral planes. [MINERAL] *See* anatase. { ‚äk·tə hē‚drīt }

octahedron [MATH] A polyhedron having eight faces, each of which is an equilateral triangle. { ‚äk·tə'hē·drən }

octal [MATH] Pertaining to the octal number system. { 'äkt·əl }

octal base [ELECTR] Tube base having a central aligning key and positioned for eight equally spaced pins. { 'äkt·əl ‚bās }

octal debugger [COMPUT SCI] A simple debugging program which permits only octal (instead of symbolic) address references. { 'äkt·əl dē'bəg·ər }

octal digit [MATH] The symbol 0, 1, 2, 3, 4, 5, 6, or 7 used as a digit in the octal number system. { 'äkt·əl 'dij·ət }

octal loading program [COMPUT SCI] Computer utility program providing a method for making changes in programs and tables existing in core memory or drum storage, by reading in words coded in octal notation on punched cards or tape. { 'äkt·əl 'lōd·iŋ ‚prō‚gram }

octal number system [MATH] A number system in which a number r is written as $n_k n_{k-1} \ldots n_1$ where $r = n_1 8^0 + n_2 8^1 + \cdots + n_k 8^{k-1}$. Also known as octonary number system. { 'äkt·əl 'nəm·bər ‚sis·təm }

octamethylcyclotetrasiloxane [ORG CHEM] $C_8H_{24}O_4Si_4$ An oily colorless liquid with a boiling point of 175°C (315°F) and melting point of 17.5°C (63.5°F); a component of silicone gel. { ‚äk·tə‚meth·əl sīk·lə‚te·trə·sə'läk‚sān }

octanal *See* octyl aldehyde. { 'äk·tə‚nal }

n-octane [ORG CHEM] C_8H_{18} Colorless liquid boiling at 126°C; soluble in alcohol, acetone, and ether, insoluble in water; used as a solvent and chemical intermediate. { ¦en 'äk‚tān }

octanedioic acid *See* suberic acid. { 'äk‚tan·dī'ō·ik 'as·əd }

octane number [ENG] A rating that indicates the tendency to knock when a fuel is used in a standard internal combustion engine under standard conditions; n-heptane is 0, isooctane is 100; different test methods yield other values variously known as research octane, motor octane, and road octane. { 'äk‚tān ‚nəm·bər }

octane requirement [MECH ENG] The fuel octane number needed for efficient operation (without knocking or spark retardation) of an internal combustion engine. { 'äk‚tān ri‚kwīr·mənt }

octane scale [ENG] Series of arbitrary numbers from 0 to 120.3 used to rate the octane number of a gasoline; n-heptane is 0 octane, isooctane is 100, and isooctane + 6 milliliters TEL (tetraethyllead) is 120.3. { 'äk‚tān ‚skāl }

Octans [ASTRON] The constellation that includes the south celestial pole. Also known as Octant. { 'äk‚tanz }

octant [MATH] **1.** One of the eight regions into which three-dimensional Euclidean space is divided by the coordinate planes of a Cartesian coordinate system. **2.** A unit of plane angle equal to 45° or π/8 radians. [NAV] A double-reflecting instrument used primarily for measuring altitudes of celestial bodies from a moving ship or airplane; has a range of 90°, and while a sextant has the capability up to 120°, the common practice applies the name sextant to both types of instruments. { 'äk·tənt }

Octant *See* Octans. { 'äk·tənt }

octant error *See* sextant error. { 'äk·tənt ‚er·ər }

octaphyllite [MINERAL] **1.** A group of mica minerals that contain eight cations per ten oxygen and two hydroxyl ions. **2.** Any mineral of this group, such as biotite. { 'äk·tə·ə'fi‚līt }

octave [ACOUS] The interval in pitch between two tones such that one tone may be regarded as duplicating at the next higher pitch the basic musical import of the other tone; the sounds producing these tones then have a frequency ratio of 2 to 1. [PHYS] The interval between any two frequencies having a ratio of 2 to 1. { 'äk·tiv }

octave-band analyzer [ENG ACOUS] A portable sound analyzer which amplifies a microphone signal, feeds it into one of several band-pass filters selected by a switch, and indicates the magnitude of sound in the corresponding frequency band on a logarithmic scale; all the bands except the highest and lowest span an octave in frequency. Abbreviated OBA. { 'äk·tiv ‚band 'an·ə‚līz·ər }

octave-band filter [ENG ACOUS] A band-pass filter in which the upper cutoff frequency is twice the lower cutoff frequency. { 'äk·tiv ‚band 'fil·tər }

octave-band oscillator [ELECTR] An oscillator that can be tuned over a frequency range of 2 to 1, so that its highest frequency is twice its lowest frequency.   { 'äk·tiv ‖band 'äs·ə‚lād·ər }

octave frequency band [PHYS] A band of frequencies whose highest frequency is twice its lowest frequency.   { 'äk·tiv 'frē·kwən·sē ‚band }

1-octene [ORG CHEM] $CH_3(CH_2)_5CHCH_2$ A colorless, flammable liquid; used as a plasticizer and in synthesis of organic compounds.   Also known as 1-caprylene; 1-octylene.   { ‖wən 'äk‚tēn }

2-octene [ORG CHEM] $CH_3(CH_2)_4CHCHCH_3$ A colorless, flammable liquid, with trans and cis forms; used to manufacture lubricants and to synthesize organic materials.   { 'tü 'äk‚tēn }

octet [ATOM PHYS] A collection of eight valence electrons in an atom or ion, which form the most stable configuration of the outermost, or valence, electron shell.   [PARTIC PHYS] A multiplet of eight elementary particles, corresponding to a representation of the approximate unitary symmetry (SU_3) of the strong interactions.   { äk'tet }

octet rule [CHEM] A concept of chemical bonding theory based on the assumption that in the formation of compounds, atoms exhibit a tendency for their valence shells either to be empty or to have a full complement of eight electrons (octet); for some elements there are more than the usual eight valence electrons in some of their compounds.   { äk‚tet ‚rül }

octillion [MATH] **1.** The number 10^{27}.   **2.** In British and German usage, the number 10^{48}.   { äk'til·yən }

Octocorallia [INV ZOO] The equivalent name for Alcyonaria.   { ‚äk·tō·kə'ril·yə }

octode [ELECTR] An eight-electrode electron tube containing an anode, a cathode, a control electrode, and five additional electrodes that are ordinarily grids.   { 'äk‚tōd }

octoid [DES ENG] Pertaining to a gear tooth form used to generate the teeth in bevel gears; the octoid form closely resembles the involute form.   { 'äk‚tóid }

octomethylene See cyclooctane.   { ‚äk·tō'meth·ə‚lēn }

octomino [MATH] One of the 369 plane figures that can be formed by joining eight unit squares along their sides.   { äk'täm·ə·nō }

octonary number system See octal number system.   { äk'tän·ə·rē 'nəm·bər ‚sis·təm }

octonary signaling [COMMUN] A communications mode in which information is passed by the presence and absence of plus and minus variation of eight discrete levels of one parameter of the signaling medium.   { 'äk·tə‚ner·ē 'sig·nə·liŋ }

octonions See Cayley numbers.   { äk'tän·yənz }

octopamine [PHARM] $C_8H_{11}NO_2$ A crystalline compound, used as an adrenergic drug.   { äk'tō·pə‚mēn }

Octopoda [INV ZOO] An order of the dibranchiate cephalopods, characterized by having eight arms equipped with one to three rows of suckers.   { äk'täp·ə·də }

Octopodidae [INV ZOO] The octopuses, in family of cephalopod mollusks in the order Octopoda.   { ‚äk·tə'päd·ə‚dē }

octopus [INV ZOO] Any member of the genus *Octopus* in the family Octopodidae; the body is round with a large head and eight partially webbed arms, each bearing two rows of suckers, and there is no shell.   { 'äk·tə‚pús }

octupole [PHYS] **1.** Two electric or magnetic quadrupoles having charge distributions of opposite signs and separated from each other by a small distance.   **2.** Any device for controlling beams of electrons or other charged particles, consisting of eight electrodes or magnetic poles arranged in a circular pattern, with alternating polarities; commonly used to correct aberrations of quadrupole systems.   { 'äk·tə‚pōl }

octyl- [ORG CHEM] Prefix indicating the eight-carbon hydrocarbon radical ($C_8H_{17}-$).   { 'äk·təl }

n-octyl acetate [ORG CHEM] $CH_3COO(CH_2)_7CH_3$ A colorless liquid with a fruity odor and a boiling point of 199°C; soluble in alcohol and other organic liquids; used for perfumes and flavoring.   { ‖en 'äkt·əl 'as·ə‚tāt }

octyl alcohol See 2-ethylhexyl alcohol.   { 'äkt·əl 'al·kə‚hól }

octyl aldehyde [ORG CHEM] $C_8H_{16}O$ A liquid aldehyde boiling at 172°C; found in essential oils of many plants; used in perfume compositions.   Also known as octanal.   { 'äkt·əl 'al·də‚hīd }

n-octyl n-decyl adipate [ORG CHEM] A liquid with a boiling range of 250–254°C; used as a low-temperature plasticizer.   Abbreviated NODA.   { ‖en 'äkt·əl ‖en 'des·əl 'ad·ə‚pāt }

n-octyl n-decyl phthalate [ORG CHEM] A clear liquid with a boiling range of 232–267°C; used as a plasticizer for vinyl resins.   { ‖en 'äkt·əl ‖en 'des·əl 'tha‚lāt }

1-octylene See 1-octene.   { ‖wən 'äk·tə‚lēn }

octyl formate [ORG CHEM] $C_8H_{17}OOCH$ A colorless liquid with a fruity odor; soluble in mineral oil; used for flavoring.   { 'äkt·əl 'fór‚māt }

n-octyl mercaptan [ORG CHEM] $C_8H_{17}SH$ Clear, colorless liquid boiling at 199°C; used as a chemical intermediate and polymerization conditioner.   { ‖en 'äkt·əl mər'kap‚tan }

octyl phenol [ORG CHEM] $C_8H_{17}C_6H_4OH$ White flakes, congealing at 73°C; soluble in organic solvents, insoluble in water; used to make surfactants, plasticizers, and antioxidants.   { 'äkt·əl 'fē‚nól }

octyne [ORG CHEM] $CHC(CH_2)_5CH_3$ Colorless hydrocarbon liquid, boiling at 125°C.   { 'äk‚tīn }

ocuba wax [MATER] A waxy fat obtained from the fruit of the myrtle *Myristica ocuba;* melting point, 40°C; used in candles.   { ə'kyü·bə ‚waks }

ocular [BIOL] Of or pertaining to the eye.   [OPTICS] See eyepiece.   { 'äk·yə·lər }

ocular myasthenia [MED] A form of myasthenia gravis that is restricted to the eye muscles; symptoms may include droopy eyelids and blurred or double vision.   { ‚äk·yə·lər ‚mī·əs'thē·nē·ə }

ocular prism [OPTICS] The prism employed in a range finder to bend the line of sight through the instrument into the eyepiece.   { 'äk·yə·lər 'priz·əm }

ocular skeleton [VERT ZOO] A rigid structure in most submammalian vertebrates consisting of a cup of hyaline cartilage enclosing the posterior part of the eye, and a thin-walled ring of intramembranous bones in the edge of the sclera at its junction with the cornea.   { 'äk·yə·lər 'skel·ə·tən }

oculist See ophthalmologist.   { 'äk·yə·ləst }

oculoagravic illusion See agravic illusion.   { äk·yə·lō·ā'grav·ik i'lü·zhən }

oculoglandular tularemia [MED] Infection by *Pasteurella tularensis* which, in addition to the usual symptoms of tularemia, causes swollen eyelids, conjunctivitis, swollen lymph nodes, and ulcers on the conjunctivae.   { ‖äk·yə·lō‖glan·jə·lər ‚tü·lə'rē·mē·ə }

oculomotor [PHYSIO] Pertaining to eye movement.   { ‖äk·yə·lō'mōd·ər }

oculomotor nerve [NEUROSCI] The third cranial nerve; a paired somatic motor nerve arising in the floor of the midbrain, which innervates all extrinsic eye muscles except the lateral rectus and superior oblique, and furnishes autonomic fibers to the ciliary and pupillary sphincter muscles within the eye.   { ‖äk·yə·lō'mōd·ər 'nərv }

oculomotor nucleus [NEUROSCI] A nucleus in the floor of the midbrain that gives rise to motor fibers of the oculomotor nerve.   { ‖äk·yə·lō'mōd·ər 'nü·klē·əs }

oculomotor paralysis [MED] Paralysis of the oculomotor nerve.   { ‖äk·yə·lō'mōd·ər pə'ral·ə·səs }

Oculosida [INV ZOO] An order of the protozoan subclass Radiolaria; pores are restricted to certain areas in the central capsule, and an olive-colored material is always present near the astropyle.   { ‚äk·yə'läs·ə·də }

OD See optical density; outside diameter.

odd-even check [COMPUT SCI] A means of detecting certain kinds of errors in which an extra bit, carried along with each word, is set to zero or one so that the total number of zeros or ones in each word is always made even or always made odd.   Also known as parity check.   { 'äd 'ē·vən ‚chek }

odd-even nucleus [NUC PHYS] A nucleus which has an odd number of protons and an even number of neutrons.   { 'äd ‚ē·vən 'nü·klē·əs }

odd function [MATH] A function $f(x)$ is odd if, for every x, $f(-x) = -f(x)$.   { 'äd ‚fəŋk·shən }

odd-leg caliper [DES ENG] A caliper in which the legs bend in the same direction instead of opposite directions.   { 'äd ‚leg 'kal·ə·pər }

odd number [MATH] A natural number not divisible by 2.   { 'äd 'nəm·bər }

odd-odd nucleus [NUC PHYS] A nucleus that has an odd number of protons and an odd number of neutrons.   { 'äd 'äd 'nü·klē·əs }

odd parity [COMPUT SCI] Property of an expression in binary code which has an odd number of ones.   [QUANT MECH] Property of a system whose state vector is multiplied by -1

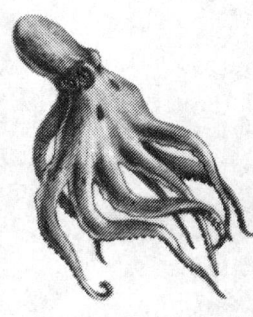

OCTOPUS

Octopus bairdi with a body length of 3 inches (7.6 centimeters) and arms up to 40 inches (1 meter) long, webbed for one-third of their length.

under the operation of space inversion, that is, the simultaneous reflection of all spatial coordinates through the origin. { 'äc 'par·əd·ē }

odd parity check [COMPUT SCI] A parity check in which the number of 0's or 1's in each word is expected to be odd; if the number is even, the check bit is 1, and if the number is odd, the check bit is 0. { 'äd 'par·əd·ē ,chek }

odd permutation [MATH] A permutation that may be represented as the result of an odd number of transpositions. { 'äd ,pər·myə'tā·shən }

odd-pinnate [BOT] Of a compound leaf, having a single leaflet at the tip of the petiole with leaflets on both sides of the petiole. Also known as imparipinnate. { 'äd 'pi,nāt }

odds ratio [STAT] The ratio of the probability of occurrence of an event to the probability of the event not occurring. { 'ädz ,rā·shō }

odd term [ATOM PHYS] A term of an atom or molecule for which the sum of the angular-momentum quantum numbers of all the electrons is odd, so that the states have odd parity; designated by a superscript o or u. { 'äd ,tərm }

odd vertex [MATH] A vertex whose degree is an odd number. { 'äd ¦vər,teks }

Odessey protractor [NAV] A device used in plotting shoran, EPI (electronic position indicator), and Raydist positions; it consists of a transparent plate on which closely spaced concentric circles are drawn; these circles represent distances from the center in terms of miles, microseconds, or lanes, constructed to fit the electronic system in use and the scale at which the survey is to be plotted. { ō'des·ē 'prō,trak·tər }

Odiniidae [INV ZOO] A family of cyclorrhaphous myodarian dipteran insects in the subsection Acalyptratae. { ,ōd·ən'ī·ə,dē }

odinite [PETR] A grayish-green lamprophyre composed of labradorite and augite or diallage; sometimes containing hornblende, phenocrysts in a groundmass of fine lath-shaped or equigranular feldspar, and a felty mesh of acicular hornblende crystals. { 'ōd·ən,īt }

ODMR See optical detection of magnetic resonance.

Odobenidae [VERT ZOO] A family of carnivorous mammals in the suborder Pinnipedia; contains a single species, the walrus (*Odobenus rosmarus*). { ,ō·dō'ben·ə,dē }

odograph [ENG] An instrument installed in a vehicle to automatically plot on a map the course and distance traveled by the vehicle. { 'ō·də,graf }

odometer [ENG] **1.** An instrument for measuring distance traversed, as of a vehicle. **2.** The indicating gage of such an instrument. **3.** A wheel pulled by surveyors to measure distance traveled. { ō'däm·əd·ər }

O'Donahue's theory [MIN ENG] A mine subsidence theory, with subsidence regarded as taking place in two stages: first, a breaking of the rocks in which the lines of fracture tend to run at right angles to the stratification; then, an inward movement from the sides, resulting in a pull or draw beyond the edges of the workings. { ō'dän·ə,hyüz ,thē·ə·rē }

Odonata [INV ZOO] The dragonflies, an order of the class Insecta, characterized by a head with large compound eyes, and wings with clear or transparent membranes traversed by networks of veins. { ,ōd·ən'ad·ə }

odont-, odonto- [VERT ZOO] A combining form meaning tooth. { ,ō,dän·tō }

odontalgia See toothache. { ,ō,dän'täl·jē·ə }

odontectomy [MED] Surgical excision of a tooth. { ,ō ,dän'tek·tə·mē }

odontexesis [MED] Removal of deposits from the surface of teeth. { ,ō,dän·tek'sē·səs }

odontoblast [HISTOL] One of the elongated, dentin-forming cells covering the dental papilla. { ō'dänt·ə,blast }

odontoblastoma See ameloblastic odontoma. { ō¦dänt·ə· bla'stō·mə }

Odontoceti [VERT ZOO] The toothed whales, a suborder of cetacean mammals distinguished by a single blowhole. { ,ō,dänt·ə'sē,tī }

odontoclast [HISTOL] A multinuclear cell concerned with resorption of the roots of milk teeth. { ō'dänt·ə,klast }

odontogenesis [EMBRYO] Formation of teeth. { ,ō,dänt· ə'jen·ə·səs }

odontogenic [HISTOL] **1.** Pertaining to the origin and development of teeth. **2.** Originating in tissues associated with teeth. { ō¦dänt·ə¦jen·ik }

odontogenic cyst [MED] A cyst originating in tissues associated with teeth. { ō¦dänt·ə¦jen·ik 'sist }

odontogenic fibroma [MED] A benign tumor originating in the mesenchymal derivatives of the tooth germ. { ō¦dänt· ə¦jen·ik fī'brō·mə }

Odontognathae [PALEON] An extinct superorder of the avian subclass Neornithes, including all large, flightless aquatic forms and other members of the single order Hesperornithiformes. { ō,dän'täg·nə,thē }

odontoid [BIOL] Toothlike. { ō'dän,tóid }

odontoid process [ANAT] A toothlike projection on the anterior surface of the axis vertebra with which the atlas articulates. { ō'dän,tóid ,prä·səs }

odontology [VERT ZOO] A branch of science that deals with the formation, development, and abnormalities of teeth. { ,ō,dän'täl·ə·jē }

odontoma [MED] A benign tumor representing a developmental excess, composed of mesodermal or octodermal tooth-forming tissue, alone or in association with the calcified derivatives of these structures. { ,ō,dän'tō·mə }

odontophobia [PSYCH] An abnormal fear of teeth. { ō,dänt·ə'fō·bē·ə }

Odontostomatida [INV ZOO] An order of the protozoan subclass Spirotrichia; individuals are compressed laterally and possess very little ciliature. { ō¦dänt·ə·stō'mad·ə·də }

odorant [MATER] Material added to odorless fuel gases to give them a distinctive odor for safety purposes; usually a sulfur- or mercaptan-containing compound. Also known as malodorant; stench; warning agent. { 'ō·də·rənt }

odoriferous homing [ELECTR] Homing on the ionized air produced by the exhaust gases of a snorkeling submarine. { ,ō·də'rif·ə·rəs 'hōm·iŋ }

odorize [CHEM ENG] To add an unpleasant odor as a safety measure to an odorless material such as fuel gas. { 'ō·də,rīz }

Oecophoridae [INV ZOO] A family of small to moderately small moths in the epidopteran superfamily Tineoidea, characterized by a comb of bristles, the pecten, on the scape of the antennae. { ,ēk·ə'fòr·ə,dē }

Oedemeridae [INV ZOO] The false blister beetles, a large family of coleopteran insects in the superfamily Tenebrionoidea. { ,ēd·ə'mer·ə,dē }

Oedipal [PSYCH] Pertaining to the Oedipus complex. { 'ēd·ə·pəl }

Oedipal phase [PSYCH] In psychoanalytic theory, the period of psychosexual development that represents a time of conflict between the child and parents, lasting approximately from 4 to 6 years of age. { 'ēd·ə·pəl ,fāz }

Oedipus complex [PSYCH] In psychoanalytic theory, the attraction and attachment of the child to the parent of the opposite sex, accompanied by feelings of envy and hostility toward the parent of the child's sex, whose displeasure and punishment the child so fears that the child represses his or her feelings toward the parent of opposite sex. { 'ēd·ə·pùs ,käm,pleks }

Oedogoniales [BOT] An order of fresh-water algae in the division Chlorophyta; characterized as branched or unbranched microscopic filaments with a basal holdfast cell. { ,ēd·ə,gō·nē'ā·lēz }

Oegophiurida [INV ZOO] An order of echinoderms in the subclass Ophiuroidea, represented by a single living genus; members have few external skeletal plates and lack genital bursae, dorsal and ventral arm plates, and certain jaw plates. { ,ēg·ə·fi'yùr·ə·də }

Oegopsida [INV ZOO] A suborder of cephalopod mollusks in the order Decapoda of one classification system, and in the order Teuthoidea of another system. { ē'gäp·sə·də }

Oehman's survey instrument [ENG] A drill-hole surveying apparatus that makes a photographic record of the compass and clinometer readings. { 'ā·mənz 'sər,vā ,in·strə·mənt }

O electron [ATOM PHYS] An electron in the fifth (O) shell of electrons surrounding the atomic nucleus, having the principal quantum number 5. { 'ō i,lek,trän }

Oepikellacea [PALEON] A dimorphic superfamily of extinct ostracods in the order Paleocopa, distinguished by convex valves and the absence of any trace of a major sulcus in the external configuration. { ē,pik·ə'läs·ē·ə }

oersted [ELECTROMAG] The unit of magnetic field strength in the centimeter-gram-second electromagnetic system of units, equal to the field strength at the center of a plane circular coil

ODONATA

Adult dragonfly showing the four membranous wings.

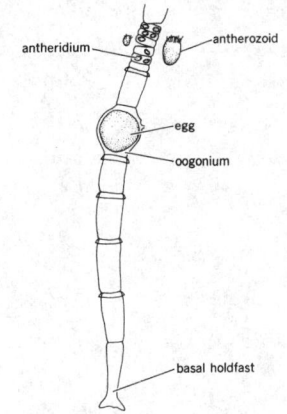

OEDOGONIALES

Oedogonium, an attached, unbranched, microscopic filament arising from a basal holdfast cell.

of one turn and 1-centimeter radius, when there is a current of $1/(2\pi)$ abamp in the coil; 1 oersted corresponds to approximately 79.577 amperes per meter. { 'ər·stəd }

Oersted experiment [ELECTROMAG] An experiment in which the deflection of a magnetic needle is observed when it is placed near a wire carrying an electric current. { 'er,sted ik,sper·ə·mənt }

Oestridae [INV ZOO] A family of cyclorrhaphous myodarian dipteran insects in the subsection Calypteratae. { 'es·trə,dē }

Oetling freezing method [MIN ENG] A method of shaft sinking by freezing the wet ground in sections as the sinking proceeds. { 'et·liŋ 'frēz·iŋ ,meth·əd }

off [ENG] Designating the inoperative state of a device, or one of two possible conditions (the other being "on") in a circuit. { of }

off-airways [AERO ENG] Pertaining to any aircraft course or track that does not lie within the bounds of prescribed airways. { 'of 'er,wāz }

off-carriage fire control [ORD] Process of controlling fire on a target with the aid of a sighting device which is not mounted directly on the weapon. { 'of ,kar·ij 'fīr kən,trōl }

off-carriage fire control equipment [ORD] Fire control items, such as directors, aiming circles, and observation instruments, which are required for directing the fire of the weapon and which are separately transported. { 'of ,kar·ij 'fīr kən,trōl i,kwip·mənt }

off-center plan position indicator [ELECTR] A plan position indicator in which the center of the display that represents the location of the radar can be moved from the center of the screen to any position on the face of the PPI. { 'of ,sent·ər 'plan pə,zish·ən 'in·də,kād·ər }

off-count mesh [DES ENG] A mesh in a wire cloth in which the count is not the same for both directions. { 'of ,kaunt 'mesh }

offense against the sine condition [OPTICS] A numerical measure of coma, equal to the sagittal coma divided by the perpendicular distance from the image point to the optical axis. { ə'fens ə'genst thə 'sīn kən,dish·ən }

offensive grenade [ORD] A hand grenade having a nonmetallic container, designed to kill or injure by blast and concussion. { ə'fen·siv grə'nād }

offhand grinding [MECH ENG] Grinding operations performed with hand-held tools. Also known as freehand grinding. { 'of,hand 'grīnd·iŋ }

off-highway truck [MIN ENG] A truck of such size, weight, or dimensions that it cannot be used on public highways. { 'of ,hī,wā 'trək }

off-highway vehicle [MECH ENG] A bulk-handling machine, such as an earthmover or dump truck, that is designed to operate on steep or rough terrain and has a height and width that may exceed highway legal limits. { 'of ,hī,wā 've·ə·kəl }

off-hook [COMMUN] The active state (closed loop) of a subscriber or PBX user loop. { 'of ,huk }

off-hook service [COMMUN] Priority telephone service for key personnel that affords a connection from caller to receiver by the simple expedient of removing the phone from its cradle or hook. { 'of ,huk ,sər·vəs }

official master drawing [GRAPHICS] That drawing which is established as the official drawing upon which any revision will originally be made and from which copies of the current drawing will originally be reproduced. { ə'fish·əl 'mas·tər 'drȯ·iŋ }

officials' inspection lamp [MIN ENG] A portable combined electric lamp and battery, fitted with a reflector to provide directional illumination. { ə'fish·əlz in'spek·shən ,lamp }

offing [NAV] That part of the visible sea a considerable distance from the shore, or that part just beyond the limits of the area in which a pilot is needed. { 'of·iŋ }

offlap [GEOL] The successive lateral contraction extent of strata (in an upward sequence) due to their deposition in a shrinking sea or on the margin of a rising landmass. Also known as regressive overlap. { 'of,lap }

off-line [COMPUT SCI] Describing equipment not connected to a computer, or temporarily disconnected from one. [ENG] **1.** A condition existing when the drive rod of the drill swivel head is not centered and parallel with the borehole being drilled. **2.** A borehole that has deviated from its intended course. **3.** A condition existing wherein any linear excavation (shaft, drift, borehole) deviates from a previously determined or intended

survey line or course. [IND ENG] State in which an equipment or subsystem is in standby, maintenance, or mode of operation other than on-line. { 'of,līn }

off-line cipher [COMMUN] Method of encrypting which is not associated with a particular transmission system and in which the resulting encrypted message can be transmitted by any means. { 'of,līn 'sī·fər }

off-line computer See course-line computer. { 'of,līn kəm'pyüd·ər }

off-line equipment [COMPUT SCI] Peripheral equipment or devices not in direct communication with the central processing unit of a computer. Also known as auxiliary equipment. { 'of,līn i'kwip·mənt }

off-line mode [COMPUT SCI] Any operation such as printing or converting which does not involve the main computer. { 'of,līn 'mōd }

off-line operation [COMPUT SCI] Operation of peripheral equipment in conjunction with, but not under the control of, the central processing unit. { 'of,līn ,äp·ə'rā·shən }

off-line processing [COMPUT SCI] Any processing which takes place independently of the central processing unit. { 'of,līn 'prä,ses·iŋ }

off-line storage [COMPUT SCI] A storage device not under control of the central processing unit. { 'of,līn 'stȯr·ij }

off-line unit [COMPUT SCI] Any operation device which is not attached to the main computer. { 'of,līn 'yü·nət }

offload [COMPUT SCI] To transfer operations from one computer to another, usually from a large computer to a smaller one. { 'of,lōd }

off-peak [SCI TECH] Not at the maximum. { 'of,pēk }

off-reef facies [GEOL] Facies of the inclined strata made up of reef detritus deposited along the seaward margin of a reef. { 'of,rēf 'fā·shēz }

off-road vehicle [MECH ENG] A conveyance designed to travel on unpaved roads, trails, beaches, or rough terrain rather than on public roads. { 'of,rōd 've·ə·kəl }

offset [BUILD] A horizontal ledge on the face of a wall or other member that is formed by diminishing the thickness of the wall at that point. Also known as setback. [COMPUT SCI] See displacement. [CONT SYS] The steady-state difference between the desired control point and that actually obtained in a process control system. [ENG] **1.** A short perpendicular distance measured to a traverse course or a surveyed line or principal line of measurement in order to locate a point with respect to a point on the course or line. **2.** In seismic prospecting, the horizontal distance between a shothole and the line of profile, measured perpendicular to the line. **3.** In seismic refraction prospecting, the horizontal displacement, measured from the detector, of a point for which a calculated depth is relevant. **4.** In seismic reflection prospecting, the correction of a reflecting element from its position on a preliminary working profile to its actual position in space. [GEOL] **1.** The movement of an upcurrent part of a shore to a more seaward position than a downcurrent part. **2.** A spur from a mountain range. **3.** A level terrace on the side of a hill. **4.** The horizontal displacement component in a fault, measured parallel to the strike of the fault. Also known as normal horizontal separation. [MAP] During construction of a map projection, the small distance added to the length of meridians on either side of the central meridian in order to determine the chart's top latitude. [MECH] The value of strain between the initial linear portion of the stress-strain curve and a parallel line that intersects the stress-strain curve of an arbitrary value of strain; used as an index of yield stress; a value of 0.2% is common. [MIN ENG] **1.** A short drift or crosscut driven from a main gangway or level. **2.** The horizontal distance between the outcrops of a dislocated bed. [NAV ARCH] One of a series of measurements of the perpendicular distance of various points on a ship's hull from the centerline and above the molded baseline; used in ship construction. [ORD] The horizontal distance of forward travel covered by the missile after it strikes the ground; this distance is measured from the center of the hole of entry to the most forward part of the missile. { 'of,set }

offset bombing [ORD] Any bombing procedure which employs a reference or aiming point other than the actual target. { 'of,set 'bäm·iŋ }

offset cab [ENG] Operator's cab positioned to one side of earthmoving equipment for greater visibility and safety. { 'of,set 'kab }

offset-center plan position indicator *See* off-center plan position indicator. { 'óf₁set ¦sen·tər 'plan pə₁zish·ən 'in·də₁kād·ər }

offset-course computer *See* course-line computer. { 'óf₁set ¦kórs kəm'pyüd·ər }

offset cylinder [MECH ENG] A reciprocating part in which the crank rotates about a center off the centerline. { 'óf₁set 'sil·ən·dər }

offset deposit [GEOL] A mineral deposit, especially of sulfides, formed partly by magmatic segregation and partly by hydrothermal solution and located near the source rock. { 'óf₁set di'päz·ət }

offset drilling [PETRO ENG] The drilling of a well on property under which oil is being drained away by a well on adjacent property to make up for the loss of oil from the first property. { 'óf₁set 'dril·iŋ }

offset duplicating [GRAPHICS] A method of duplicating in which an ink-receptive image is typed or drawn on a paper master or produced photographically on a sensitized metal plate; the master or plate is then placed on the master cylinder of an offset press; the inked image built up on the master is transferred onto a blanket cylinder; the paper picks up the inked image as it passes between the blanket cylinder and an impression cylinder. { 'óf₁set 'dü·plə₁kād·iŋ }

offset ground zero [ORD] A case where the desired ground zero in an atomic explosion is displaced some distance and direction from the target center. { 'óf₁set ₁graúnd 'zir·ō }

offset line [ENG] A secondary line established close to and roughly parallel with the primary survey line to which it is referenced by measured offsets. { 'óf₁set 'līn }

offset lithography [GRAPHICS] A system of printing that depends on the principle that the printing area accepts greasy ink while the nonprinting area is dampened with water and repels the ink; in practice, the image from a plate is offset onto the rubber blanket of an impression cylinder, and transferred to a sheet of paper. { 'óf₁set li'thäg·rə·fē }

offset paper [MATER] Paper with a certain degree of porosity as a result of coating with an alkali-swelling resin; used for offset printing. { 'óf₁set 'pā·pər }

offset plan position indicator *See* off-center plan position indicator. { 'óf₁set 'plan pə₁zish·ən 'in·də₁kād·ər }

offset plotting [ORD] Method of plotting firing data when different ranges and azimuths must be sent to each gun of a battery. { 'óf₁set 'plad·iŋ }

offset press [GRAPHICS] A printing press for offset lithography printing. { 'óf₁set ₁pres }

offset ridge [GEOL] A ridge consisting of resistant sedimentary rock that has been made discontinuous as a result of faulting. { 'óf₁set 'rij }

offset screwdriver [DES ENG] A screwdriver with the blade set perpendicular to the shank for access to screws in otherwise awkward places. { 'óf₁set 'skrü₁drīv·ər }

offset stream [GEOL] A stream displaced laterally or vertically by faulting. { 'óf₁set 'strēm }

offset voltage [ELECTR] The differential input voltage that must be applied to an operational amplifier to return the zero-frequency output voltage to zero volts, due to device mismatching at the input stage. { 'óf₁set ₁vōl·tij }

offset well [PETRO ENG] A well made by offset drilling. { 'óf₁set 'wel }

offset yield strength [MECH] That stress at which the strain surpasses by a specific amount (called the offset) an extension of the initial proportional portion of the stress-strain curve; usually expressed in pounds per square inch. { 'óf₁set 'yēld ₁streŋkth }

offshore [GEOL] The comparatively flat zone of variable width extending from the outer margin of the shoreface to the edge of the continental shelf. { 'óf¦shór }

offshore bar *See* longshore bar. { 'óf¦shór 'bär }

offshore beach *See* barrier beach. { 'óf¦shór 'bēch }

offshore current [OCEANOGR] **1.** A prevailing nontidal current usually setting parallel to the shore outside the surf zone. **2.** Any current flowing away from shore. { 'óf¦shór 'kə·rənt }

offshore drilling [PETRO ENG] The drilling of oil or gas wells into water-covered locations, usually on submerged continental shelves. { 'óf¦shór 'dril·iŋ }

offshore gas [PETRO ENG] Natural gas produced from reservoirs under the offshore continental shelves. { 'óf¦shór 'gas }

offshore mooring [CIV ENG] An anchorage serving an area for which it is not considered feasible or cost-effective to construct a dock or provide a protected harbor, and providing equipment to which ships can attach mooring lines. { 'óf¦shór 'mür·iŋ }

offshore navigation [NAV] Navigation at a distance from a coast, in contrast with coastwise navigation in the vicinity of a coast. { 'óf¦shór ₁nav·ə'gā·shən }

offshore oil [PETRO ENG] Oil produced from reservoirs under the offshore continental shelves. { 'óf¦shór 'óil }

offshore slope [GEOL] The frontal slope below the outer edge of an offshore terrace. { 'óf¦shór 'slōp }

offshore survey [PETRO ENG] Seismic geophysical survey procedures conducted over water-covered continental-shelf areas in the search for possible oil reservoirs. { 'óf¦shór 'sər₁vā }

offshore terrace [GEOL] A wave-built terrace in the offshore zone composed of gravel and coarse sand. { 'óf¦shór 'ter·əs }

offshore water [OCEANOGR] Water adjacent to land in which the physical properties are slightly influenced by continental conditions. { 'óf¦shór 'wód·ər }

offshore wind [METEOROL] Wind blowing from the land toward the sea. { 'óf¦shór 'wind }

off-site facility [CHEM ENG] In a chemical process plant, any supporting facility that is not a direct part of the reaction train, such as utilities, steam, and waste-treatment facilities. { 'óf¦sīt fə'sil·əd·ē }

off soundings [NAV] Of a vessel, navigating beyond the 100-fathom (183-meter) curve; in earlier times, the term was applied to a vessel in water deeper than could be sounded with the sounding lead. { 'óf ₁saúnd·iŋz }

off-the-road equipment [MIN ENG] Tires and earthmoving equipment designed for off-highway duty in surface mines and quarries. { 'of thə ¦rōd i'kwip·mənt }

off-the-road hauling [MIN ENG] Hauling off the public highways, and generally on the mining site or excavation site. { 'óf thə ¦rōd 'hól·iŋ }

off-the-shelf [IND ENG] Available for immediate shipment. [ORD] Referring to those items required by the military services which are generally used throughout the civilian economy and which are available through normal commercial distribution channels. { 'óf thə ¦shelf }

off time [MET] In resistance welding, usually in repetitive cycles, the time that the electrode is not in contact with the work. { 'óf ₁tīm }

ogdosymmetric class [CRYSTAL] A merohedral crystal class whose general form has one-eighth the number of equivalent faces of the corresponding holohedral form. { ¦äg·dō·si'me·trik 'klas }

ogee [ARCH] A reverse curve, shaped like an elongated letter S, as the outline of an ogee molding. { 'ō₁jē }

ogive [ARCH] **1.** An arch or rib placed diagonally across a Gothic vault. **2.** A pointed arch. [GEOL] One of a periodically repeated series of dark, curved structures occurring down a glacier that resemble a pointed arch. [ORD] The curved or tapered front of a projectile. { 'ō₁jīv }

Ohio sampler [MIN ENG] A single tube or pipe with a thread on top, and the bottom beveled and hardened for driving into the ground to obtain a soil sample. { ō'hī·ō 'sam·plər }

ohm [ELEC] The unit of electrical resistance in the rationalized meter-kilogram-second system of units, equal to the resistance through which a current of 1 ampere will flow when there is a potential difference of 1 volt across it. Symbolized Ω. { ōm }

ohmic [ELEC] Pertaining to a substance or circuit component that obeys Ohm's law. { 'ō·mik }

ohmic contact [ELEC] A region where two materials are in contact, which has the property that the current flowing through it is proportional to the potential difference across it. { 'ō·mik 'kän₁takt }

ohmic dissipation [ELECTR] Loss of electric energy when a current flows through a resistance due to conversion into heat. Also known as ohmic loss. { 'ō·mik ₁dis·ə'pā·shən }

OFFSHORE DRILLING

An offshore drilling platform of the "fixed" type, one of several types used in offshore applications. (*a*) Underwater design (*World Petroleum*). (*b*) Rig on a drilling site (*Marathon Oil Co., Findlay, Ohio*).

ohmic loss *See* ohmic dissipation. { 'ō·mik 'lós }

ohmic resistance [ELEC] Property of a substance, circuit, or device for which the current flowing through it is proportional to the potential difference across it. { 'ō·mik ri'zis·təns }

ohmmeter [ENG] An instrument for measuring electric resistance; scale may be graduated in ohms or megohms. { 'ō,mēd·ər }

Ohm's law [ELEC] The law that the direct current flowing in an electric circuit is directly proportional to the voltage applied to the circuit; it is valid for metallic circuits and many circuits containing an electrolytic resistance. { 'ōmz 'lò }

ohms per volt [ENG] Sensitivity rating for measuring instruments, obtained by dividing the resistance of the instrument in ohms at a particular range by the full-scale voltage value at that range. { 'ōmz pər 'vōlt }

OHV engine *See* overhead-valve engine. { 'ō¦äch've 'en·jən }

-oic [ORG CHEM] A suffix indicating the presence of a —COOH group, as in ethyloic (—CH₂—COOH). { 'ō·ik }

oidiophore [MYCOL] A hypha that produces oidia. { ō'id·ə,fòr }

oidium [MYCOL] One of the small, thin-walled spores with flat ends produced by autofragmentation of the vegetative hyphae in certain Eumycetes. { ō'id·ē·ə }

oikocryst [PETR] One of the enclosing crystals in a poikilitic fabric. { 'òik·ə,krist }

Oikomonadidae [INV ZOO] A family of protozoans in the order Kinetoplastida containing organisms that have a single flagellum. { ,òik·ə·mə'nad·ə,dē }

oikophobia [PSYCH] An abnormal fear of the home, or of a house. { ,òik·ə'fō·bē·ə }

oil [GEOL] *See* petroleum. [MATER] Any of various viscous, combustible, water-immiscible liquids that are soluble in certain organic solvents, as ether and naphtha; may be of animal, vegetable, mineral, or synthetic origin; examples are fixed oils, volatile or essential oils, and mineral oils. { òil }

oil accumulation *See* oil pool. { 'òil ə,kyü·mya,lā·shən }

oil ammoniac [MATER] A yellow liquid distilled from gum ammoniac; boiling point is 275°C; soluble in alcohol and benzene. { 'òil ə'mō·nē,ak }

oil asphalt [MATER] Water-insoluble, heavy black residue left after removing the tar tailings during the distillation of petroleum; used in roofing, paints, and coatings. { 'òil 'as,fólt }

oil-base mud [PETRO ENG] Drilling mud made with oil as the solvent carrier for the solids content. { 'òil ,bās ,məd }

oil bath [ENG] **1.** Oil, in a container, within which a mechanism works or into which it dips. **2.** Oil in which a piece of apparatus is submerged. **3.** Oil that is poured on a cutting tool. [MET] Oil used in tempering. { 'òil ,bath }

oil blue [INORG CHEM] Violet-blue copper sulfide pigment used in varnishes. { 'òil 'blü }

oil-break [ELEC] Property of an electrical switch, circuit breaker, or similar apparatus whose contacts separate in oil. { 'òil ,brāk }

oil buffer [ORD] A mechanism on certain types of automatic weapons, especially the caliber-.50 guns, for absorbing the shock of recoil and regulating the speed of firing. { 'òil ,bəf·ər }

oil burner [ENG] Liquid-fuel burner device using a mixture of air and vaporized or atomized oil for combustion. { 'òil ,bər·nər }

oil cake [MATER] Solid residue after removal of vegetable oils from oil-bearing seeds (such as soya beans) by expression or solvent extraction; used as fertilizer and animal feed. { 'òil ,kāk }

oil circuit breaker [ELECTR] A high-voltage circuit breaker in which the arc is drawn in oil to dissipate the heat and extinguish the arc; the intense heat of the arc decomposes the oil, generating a gas whose high pressure produces a flow of fresh fluid through the arc that furnishes the necessary insulation to prevent a restrike of the arc. { 'òil 'sər·kət ,brāk·ər }

oilcloth [TEXT] **1.** A fabric coated with a mixture of oil and clay, used as a waterproof covering. **2.** A floor covering made of a heavy fabric treated with oil paint. { 'òil,klòth }

oil column [GEOL] The difference in elevation between the highest and lowest portions of various producing zones of an oil-producing formation. { 'òil ,käl·əm }

oil cooler [MECH ENG] A small radiator used to cool the oil that lubricates an automotive engine. { 'òil ,kü·lər }

oil core [MET] A core in which sand is held together by an oil binder. { 'òil ,kór }

oil cup [ENG] A permanently mounted cup used to feed lubricant to a gear, usually with some means of regulating the flow. { 'òil ,kəp }

oil cut [PETRO ENG] A mixture of oil and drilling mud that is recovered during oil exploration. { 'òil ,kət }

oil derrick [PETRO ENG] Tower structure used during oil well drilling to aid in raising and lowering of drill and piping strings. { 'òil ,der·ik }

oil dilution valve [MECH ENG] A valve used to mix gasoline with engine oil to permit easier starting of the gasoline engine in cold weather. { 'òil di,lü·shən ,valv }

oiled silk [TEXT] Soft woven silk made waterproof by treatment with boiled linseed oil, followed by drying. { 'òild 'silk }

oil emulsion [MATER] Suspension of oil droplets in another liquid in which the oil is insoluble. { 'òil i,məl·shən }

oil-extended rubber [MATER] Synthetic rubber into which 25–50% of a petroleum oil emulsion has been incorporated to decrease cost and increase low-temperature flexibility and resilience. { 'òil ik¦stend·əd 'rəb·ər }

oil field [PETRO ENG] The surface boundaries of an area from which petroleum is obtained; may correspond to an oil pool or may be circumscribed by political or legal limits. { 'òil ,fēld }

oil-field brine [HYD] Connate waters, usually containing a high concentration of calcium and sodium salts and found during deep rock penetration by the drill. { 'òil ¦fēld ,brīn }

oil-field emulsion [PETRO ENG] A crude oil that reaches the surface as an oil-water emulsion. { 'òil ¦fēld i,məl·shən }

oil-field model [PETRO ENG] Laboratory simulation of steady-state fluid flow through porous reservoir media by electrical (Ohm's-law system), electronic (graphite-impregnated cloth), or electrolytic (gelatin, blotter, potentiometric-liquid) models. { 'òil ¦fēld ,mäd·əl }

oil-field separator *See* gas-oil separator. { 'òil ¦fēld ,sep·ə,rād·ər }

oil-filled cable [ELEC] Cable having insulation impregnated with an oil which is fluid at all operating temperatures and provided with facilities such as longitudinal ducts or channels and with reservoirs; by this means positive oil pressure can be maintained within the cable at all times, incipient voids are promptly filled during periods of expansion, and all surplus oil is adequately taken care of during periods of contraction. { 'òil ¦fild ,kā·bəl }

oil filter [ENG] Cartridge-type filter used in automotive oil-lubrication systems to remove metal particles and products of heat decomposition from the circulating oil. { 'òil ,fil·tər }

oil floor [GEOL] In a sedimentary basin, the depth below which there is no economic oil accumulation. { 'òil ,flòr }

oil fogging [ENG] Spraying a fine oil mist into the gas stream of a distribution system to alleviate the drying effects of gas on certain kinds of distribution and utilization equipment. { 'òil 'fäg·iŋ }

oil furnace [MECH ENG] A combustion chamber in which oil is the heat-producing fuel. { 'òil ,fər·nəs }

oil gas [MATER] A heating gas made by interaction of petroleum oil vapors and steam in a process similar to the water-gas reaction. { 'òil ,gas }

oil-gas process [CHEM ENG] Process to manufacture high-caloric-value fuel gas by the destructive distillation of high-boiling petroleum oils. { 'òil ,gas ,prä·səs }

oil-gas separator *See* gas-oil separator. { 'òil ,gas ,sep·ə,rād·ər }

oil-gas tar [MATER] A type of tar produced during the oil-gas process by cracking the oil vapors at high temperatures. { 'òil ,gas ,tär }

oil gland *See* uropygial gland. { 'òil ,gland }

oil groove [DES ENG] One of the grooves in a bearing which distribute and collect lubricating oil. { 'òil ,grüv }

oil hardening [MET] Quenching of carbon steel in an oil bath; the steel cools slowly, and a more uniform and desirable hardness is attained. { 'òil ,härd·ən·iŋ }

oil hole [ENG] A small hole for injecting oil for a bearing. { 'òil ,hōl }

oil-hole drill [DES ENG] A twist drill containing holes through which oil can be fed to the cutting edges. { 'òil ,hōl ,dril }

oil horizon [PETRO ENG] The upper surface of oil in a well,

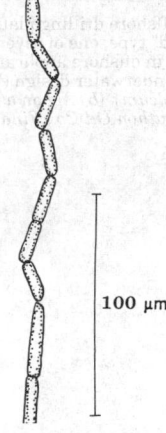

or the stratum in which the oil surface is located. { 'óil hə'rīz·ən }

oil-immersed [ELEC] Property of a transformer, reactor, regulator, or similar apparatus whose coils are immersed in an insulating liquid that is usually, but not necessarily, oil. { 'óil i,mərst }

oiliness [ENG] The effect of a lubricant to reduce friction between two solid surfaces in contact; the effect is more than can be accounted for by viscosity alone. { 'ói·lē·nəs }

oil in place [PETRO ENG] The total volume of oil estimated to be present in an oil reservoir. { 'óil in 'plās }

oil isoperms [PETRO ENG] Reservoir-map plotting areas of equal oil permeability. { 'óil 'ī·sə,pərmz }

oil length [MATER] The ratio of oil to resin in varnish; expressed as gallons of oil per 100 pounds (45.3 kilograms) of resin. { 'óil ,leŋkth }

oilless bearing [MECH ENG] A self-lubricating bearing containing solid or liquid lubricants in its material. { 'óil·les 'ber·iŋ }

oil lift [MECH ENG] Hydrostatic lubrication of a journal bearing by using oil at high pressure in the area between the bottom of the journal and the bearing itself so that the shaft is raised and supported by an oil film whether it is rotating or not. { 'óil ,lift }

oil of amber [MATER] Brown essential oil distilled from amber; miscible with alcohol; has balsamic aroma. { 'óil əv 'am·bər }

oil of bitter almond [MATER] A fatty, nondrying oil obtained from bitter almond kernels and used in cosmetics and perfumes. { 'óil əv 'bid·ər 'äm·ənd }

oil of cloves [MATER] A thin, colorless to pale-yellow liquid distilled from cloves; thickens and darkens with time; boils at 250–260°C; has aromatic aroma and pungent taste; soluble in ether and chloroform; chief component is eugenol; used in medicine, perfumery, flavoring, and soaps. Also known as caryophyllus oil; clove oil. { 'óil əv 'klōvz }

oil of coriander [MATER] A colorless or pale-yellow essential oil extracted from the dried seed of the coriander plant (*Coriandrum sativum*); used in medicines, beverages, and flavoring extracts. { 'óil əv ,kór·ē'an·dər }

oil of cubeb [MATER] A camphorous tasting, colorless, or pale green or yellow liquid with a peppery odor; distilled from cubebs; soluble in ether, alcohol, and chloroform; boils at 175–180°C; chief components are sesquiterpenes, cadinene, and dipentane; used in medicine. Also known as cubeb oil. { 'óil əv 'kyü,beb }

oil of grapefruit *See* grapefruit oil. { 'óil əv 'grāp,früt }

oil of mirbane *See* nitrobenzene. { 'óil əv 'mər,bān }

oil of orange blossoms [MATER] Bitter-tasting, fluorescent, pale-yellow essential oil with orange aroma; soluble in alcohol; main components are limonene, linalool, and geraniol; used for perfumes and flavors. Also known as neroli oil; orange flower oil. { 'óil əv 'är·inj ,bläs·əmz }

oil of peppermint [MATER] Colorless or slightly yellow essential oil with minty aroma and taste; soluble in ether, alcohol, and chloroform; derived from leaves and flowering tops of the peppermint plant (*Mentha piperita*); has high menthol content; used in medicines, flavors, perfumes, and liqueurs. Also known as peppermint oil. { 'óil əv 'pep·ər·mənt }

oil of sage [MATER] **1.** An alcohol-soluble yellow oil with sage aroma; obtained from leaves of the common sage (*Salvia officinalis*); used chiefly in flavors. Also known as dalmatian sage oil; sage oil; salvia oil. **2.** A pale-yellow oil with ambergris aroma obtained from the flowers of *Salvia sclarea*; used chiefly in perfumes. { 'óil əv 'sāj }

oil of sassafras [MATER] A pungent, aromatic, yellowish or reddish-yellow liquid obtained from the bark of American sassafras (*Sassafras albidum*); soluble in organic solvents, glacial acetic acid, and carbon disulfide; used for flavors, perfumes, and medicines. Also known as clary sage oil; sassafras oil. { 'óil əv 'sas·ə,fras }

oil of shaddock *See* grapefruit oil. { 'óil əv 'shad·ik }

oil of turpentine [MATER] Water-insoluble, colorless, volatile essential oil distilled from turpentine; contains pinene, sylvestrene, and dipentene; used as a carminative, solvent, paint vehicle, and disinfectant. { 'óil əv 'tər·pən,tīn }

oil of vitriol *See* sulfuric acid. { 'óil əv 'vit·rē,ól }

oil paint [MATER] A paint made with a vegetable oil as the filmogen. { 'óil ,pānt }

oil painting [GRAPHICS] A method of painting in which pigments are bound together, and to the canvas, by a drying oil; the oil, usually linseed, is thinned with solvents such as turpentine and mineral spirits; the pigments should dry in linseed oil to form an acceptably strong paint film. { 'óil 'pānt·iŋ }

oil pool [GEOL] An accumulation of petroleum locally confined by subsurface geologic features. Also known as oil accumulation; oil reservoir. { 'óil ,pül }

oil pump [MECH ENG] A pump of the gear, vane, or plunger type, usually an integral part of the automotive engine; it lifts oil from the sump to the upper level in the splash and circulating systems, and in forced-feed lubrication it pumps the oil to the tubes leading to the bearings and other parts. { 'óil ,pəmp }

oil reclaiming [ENG] **1.** A process in which oil is passed through a filter as it comes from equipment and then returned for reuse, in the same manner that crank case oil is cleaned by an engine filter. **2.** A method in which solids are removed from oil by treatment in settling tanks. { 'óil ri'klām·iŋ }

oil reservoir *See* oil pool. { 'óil 'rez·əv,wär }

oil-reservoir water *See* formation water. { 'óil |rez·əv,wär ,wód·ər }

oil ring [MECH ENG] **1.** A ring located at the lower part of a piston to prevent an excess amount of oil from being drawn up onto the piston during the suction stroke. **2.** A ring on a journal, dipping into an oil bath for lubrication. { 'óil ,riŋ }

oil rock [GEOL] A rock stratum containing oil. { 'óil ,räk }

oil sand [GEOL] An unconsolidated, porous sand formation or sandstone containing or impregnated with petroleum or hydrocarbons. { 'óil ,sand }

oil saturation [PETRO ENG] Measurement of the degree of saturation of reservoir pore structure by reservoir oil. { 'óil ,sach·ə,rā·shən }

oil seal [ENG] **1.** A device for preventing the entry or return of oil from a chamber. **2.** A device using oil as the sealing medium to prevent the passage of fluid from one chamber to another. { 'óil ,sēl }

oil seed [MATER] Seeds of plants from which oil can be derived by expression or solvent extraction, such as soya beans. { 'óil ,sēd }

oil seep [GEOL] The emergence of liquid petroleum at the land surface as a result of slow migration from its buried source through minute pores or fissure networks. Also known as petroleum seep. { 'óil ,sēp }

oil separator *See* gas-oil separator. { 'óil ,sep·ə,rād·ər }

oil shale [GEOL] A finely layered brown or black shale that contains kerogen and from which liquid or gaseous hydrocarbons can be distilled. Also known as kerogen shale. { 'óil ,shāl }

oilskin [TEXT] A fabric made of cotton or linen and waterproofed with linseed oil. { 'óil,skin }

oil-soluble resin [MATER] A resin that, at moderate temperatures, dissolves in, disperses in, or reacts with drying oils to produce a homogeneous film of modified characteristics. { 'óil |säl·yə·bəl 'rez·ən }

oil stain [MATER] Thin oil paint with very little pigment, used to stain wood surfaces. { 'óil ,stān }

oilstone [MATER] A whetstone used with oil. { 'óil,stōn }

oil string *See* long string. { 'óil ,striŋ }

oil switch [ELEC] A switch whose contacts are immersed in oil in order to suppress the arc and prevent the contacts from being damaged. { 'óil ,swich }

oil tanker [NAV ARCH] A very large ship which carries crude oil or other petroleum products in big tanks. { 'óil ,taŋk·ər }

oil trap [GEOL] An accumulation of petroleum which, by a combination of physical conditions, is prevented from escaping laterally or vertically. Also known as trap. { 'óil ,trap }

Oiluvium *See* Pleistocene. { ói'lü·vē·əm }

oil varnish [MATER] A varnish composed of resins dissolved in oil. { 'óil ,vär·nish }

oil-water contact *See* oil-water surface. { 'óil |wód·ər 'kän,takt }

oil-water interface *See* oil-water surface. { 'óil |wód·ər 'in·tər,fās }

oil-water surface [GEOL] The datum of a two-dimensional oil-water interface. Also known as oil-water contact; oil-water interface. { 'óil |wód·ər 'sər·fəs }

oil well [PETRO ENG] A hole drilled (usually vertically) into an oil reservoir for the purpose of recovering the oil trapped in porous formations. { 'óil ,wel }

oil-well acidizing [PETRO ENG] The use of hot hydrochloric acid (200–300°F; 93–149°C) to remove slow-dissolving and hard-to-remove wellbore scale. { 'ȯil ,wel 'as·ə,dīz·iŋ }

oil-well cement [MATER] A type of hydraulic cement which has a slow setting rate under the high temperatures obtained in oil wells; uses include support of tubing and bypassing of unwanted zones. { 'ȯil ,wel si,ment }

oil-well drive *See* reservoir drive mechanism. { 'ȯil ,wel ,drīv }

oil-well pump [PETRO ENG] Device for artificial or secondary (non-gas-lift) oil production; about 85% of the production is by ground-level sucker-rod pumps, the remainder by downhole hydraulic lift pumps. { 'ȯil ,wel ,pəmp }

oil white [MATER] Mixture of lithopone and zinc white or white lead; used as a house-paint pigment. { 'ȯil 'wīt }

oil zone [GEOL] The formation or horizon from which oil is produced, usually immediately under the gas zone and above the water zone if all three fluids are present and segregated. { 'ȯil ,zōn }

O indicator *See* O scope. { 'ō ,in·də,kād·ər }

ointment [PHARM] A semisolid preparation used for a protective and emollient effect or as a vehicle for the local or endermic administration of medicaments; ointment bases are composed of various mixtures of fats, waxes, animal and vegetable oils, and solid and liquid hydrocarbons. { 'ȯint·mənt }

oiticica oil [MATER] A light-yellowish oil obtained from seeds of the Brazilian oiticica tree (*Licania rigida*); raw oil becomes buttery unless heat-treated (semipolymerized); used principally in paint and varnish as a drying oil as a substitute for tung oil or with tung oil. { ,ȯid·ə'sē·kə ,ȯil }

okaite [PETR] An ultramafic igneous rock composed chiefly of melilite and haüyne, with accessory biotite, perovskite, apatite, calcite, and opaque oxides. { ō'kā,īt }

okapi [VERT ZOO] *Okapia johnstoni*. An artiodactylous mammal in the family Giraffidae; has a hazel coat with striped hindquarters, and the head shape, lips, and tongue are the same as those of the giraffe, but the neck is not elongate. { ō'kä·pē }

Okazaki fragment [MOL BIO] In deoxyribonucleic acid replication, a discontinuous segment in which the lagging strand is synthesized. { ,ō·kə¦zä·kē ,frag·mənt }

okenite [MINERAL] CaSi$_2$O$_4$(OH)$_2$·H$_2$O A whitish mineral consisting of calcium silicate and occurring in fibrous masses. { 'ō·kə,nīt }

okonite [MATER] Insulating material made from the vulcanization of ozokerite and resin with rubber and sulfur. { 'ō·kə,nīt }

okra [BOT] *Hibiscus esculentus*. A tall annual plant grown for its edible immature pods. Also known as gumbo. { 'ō·krə }

-ol [ORG CHEM] Chemical suffix for an −OH group in organic compounds, such as phenol (C$_6$H$_5$OH). { ȯl }

OL *See* only loadable.

Olacaceae [BOT] A family of dicotyledonous plants in the order Santalales characterized by dry or fleshy indehiscent fruit, the presence of petals, stamens, and chlorophyll, and a 2-5-celled ovary. { ,ō·lə'sās·ē,ē }

Olbers' paradox [ASTRON] If the universe were static, of infinite age, and the galaxies distributed isotropically, the distance attenuation of their light would be exactly balanced by the increase in number in successive spherical shells centered at the earth; hence the night sky would be of daylight brightness instead of dark. { 'ōl·bərz 'par·ə,däks }

old achromat [OPTICS] An achromatic lens in which the component lenses are made of glasses chosen from a limited selection, in which refractive index and dispersive power vary roughly together. { ¦ōld 'a·krə,mat }

old age [GEOL] The last stage of the erosion cycle in the development of the topography of a region in which erosion has reduced the surface almost to base level and the land forms are marked by simplicity of form and subdued relief. Also known as topographic old age. { 'ōld 'āj }

Oldham coupling *See* slider coupling. { 'ōl·dəm ,kəp·liŋ }

Oldhaminidina [PALEON] A suborder of extinct articulate brachiopods in the order Strophomenida distinguished by a highly lobate brachial valve seated within an irregular convex pedicle valve. { ¦ōl·də·mə'nī·də·nə }

oldhamite [MINERAL] CaS A pale-brown mineral known only from meteorites; unstable under earth conditions; member of the galena group with face-centered isometric structure. { 'ōl·də,mīt }

Oldham-Wheat lamp [MIN ENG] A cap lamp designed for full self-service. { 'ōl·dəm 'wēt ,lamp }

old ice [OCEANOGR] Floating sea ice that is more than 2 years old. { 'ōld 'īs }

old inflationary cosmology [ASTRON] The original version of the inflationary universe cosmology in which a quantum-mechanical tunneling process is responsible for the phase transition of the universe to a state in which grand unified symmetry is broken. { 'ōld in'flā·shə,ner·ē käz'mäl·ə·jē }

old lake [GEOL] **1.** A lake in an advanced stage of filling by sediments. **2.** A eutrophic or dystrophic lake. **3.** A lake whose shoreline exhibits an advanced stage of development. { 'ōld ¦lāk }

oldland [GEOL] **1.** An extensive area (as the Canadian Shield) of ancient crystalline rocks reduced to low relief by long, continuous erosion from which the materials of later sedimentary rocks were derived. **2.** A region of older land, projected above sea level behind a coastal plain, that supplied the material of which the coastal-plain strata were formed. { 'ōld,land }

old mountain [GEOL] A mountain that was formed before the beginning of the Tertiary Period. { 'ōld ¦maúnt·ən }

Old Red Sandstone [GEOL] A Devonian formation in Great Britain and northwestern Europe, of nonmarine, predominantly red sedimentary rocks, consisting principally of sandstone, conglomerates, and shales. { 'ōld 'red 'san,stōn }

old snow [HYD] Deposited snow in which the original crystalline forms are no longer recognizable, such as firn or spring snow. Also known as firn snow. { 'ōld 'snō }

old wives' summer [METEOROL] A period of calm, clear weather, with cold nights and misty mornings but fine warm days, which sets in over central Europe toward the end of September; comparable to Indian summer. { 'ōld ,wīvz 'səm·ər }

old workings [MIN ENG] Mines which have been abandoned, allowed to collapse, and sometimes sealed off. { 'ōld 'wərk·iŋz }

Oleaceae [BOT] A family of dicotyledonous plants in the order Scrophulariales characterized generally by perfect flowers, two stamens, axile to parietal or apical placentation, a four-lobed corolla, and two ovules in each locule. { ,ō·lē'ās·ē,ē }

oleandomycin [MICROBIO] C$_{35}$H$_{61}$O$_{12}$N A macrolide antibiotic produced by *Streptomyces antibioticus*; active mainly against gram-positive microorganisms. Also known as matromycin. { ,ō·lē,an·də'mīs·ən }

oleate [ORG CHEM] Salt made up of a metal or alkaloid with oleic acid; used for external medicines and in soaps and paints. { 'ō·lē,āt }

olecranon [ANAT] The large process at the distal end of the ulna that forms the bony prominence of the elbow and receives the insertion of the triceps muscle. { ō'lek·rə,nän }

olefiant gas *See* ethylene. { ¦ō·lə¦fī·ənt 'gas }

olefin [ORG CHEM] C$_n$H$_{2n}$ A family of unsaturated, chemically active hydrocarbons with one carbon-carbon double bond; includes ethylene and propylene. [TEXT] A manufactured fiber in which the fiber-forming substance is any long-chain synthetic polymer composed of at least 85% by weight of ethylene, propylene, or other olefin units except amorphous (noncrystalline) polyolefins qualifying as rubber. { 'ō·lə·fən }

olefin copolymer [ORG CHEM] Polymer made by the interreaction of two or more kinds of olefin monomers, such as butylene and propylene. { 'ō·lə·fən kō'päl·ə·mər }

olefin resin [ORG CHEM] Long-chain polymeric material produced by the chain reaction of olefinic monomers, such as polyethylene from ethylene, or polypropylene from propylene. { 'ō·lə·fən 'rez·ən }

oleic acid [ORG CHEM] C$_{17}$H$_{33}$COOH Yellowish, unsaturated fatty acid with lardlike aroma; soluble in organic solvents, slightly soluble in water; boils at 286°C (100 mmHg); the main component of olive and cooking oils; used in soaps, ointments, cosmetics, and ore beneficiation. Also known as octadecenoic acid; red oil. { ō'lā·ik 'as·əd }

olein [ORG CHEM] (C$_{17}$H$_{33}$COO)$_3$C$_3$H$_5$ Oleic acid triglyceride; yellow liquid melting at −5°C; slightly soluble in alcohol, soluble in chloroform, ether, and carbon tetrachloride; found in most fats and oils; used in textile lubrication. { 'ō·lē·ən }

OKRA

Okra (*Hibiscus esculentus*), branch with pods.

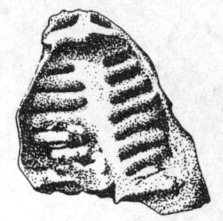

OLDHAMINIDINA

Dorsal view of the shell of *Leptodus*.

Olenellidae [PALEON] A family of extinct arthropods in the class Trilobita. { ¦ō·lə·lə'nel·ə‚dē }

oleometer [ENG] 1. A device for measuring specific gravity of oils. 2. An instrument for determining the proportion of oil in a substance. { ‚ō·lē'äm·əd·ər }

oleo oil [MATER] Yellow liquid fat used to make oleomargarine; consists of liquid olein and palmitin from cold-pressed tallow. { 'ō·lē‚ō ‚oil }

oleoresin [MATER] A resin-essential oil mixture with pungent taste; extracted from various plants; used in pharmaceutical preparations; examples are Peru, tulu, and styrax balsams. { ¦ō·lē·ō'rez·ən }

oleoresinous varnish [MATER] A varnish made by compounding the resin with oxidizable oil, such as linseed oil. { ¦ō·lē·ō'rez·ən·əs 'vär‚nish }

oleostearin [MATER] Edible solid fat from tissues of cattle (genus *Bos*); the solid remaining after oleo oil or tallow oil is removed from tallow. Also known as beef stearin. { ¦ō·lē· ō'stir·ən }

oleo strut [MECH ENG] A shock absorber consisting of a telescoping cylinder that forces oil into an air chamber, thereby compressing the air; used on aircraft landing gear. { 'ō·lē· ō ‚strət }

Olethreutidae [INV ZOO] A family of moths in the superfamily Tortricoidea whose hindwings usually have a fringe of long hairs along the basal part of the cubitus. { ‚ō·lə'thrüd·ə‚dē }

oleum [CHEM] 1. Latin name for oil. 2. *See* fuming sulfuric acid. { 'ō·lē·əm }

oleum gossypii seminis *See* cotton oil. { 'ō·lē·əm gä'sip·ē‚ī 'sem·ə·nəs }

oleum morrhuae *See* cod-liver oil. { 'ō·lē·əm mə'rü‚ī }

oleum theobromatis *See* cocoa butter. { 'ō·lē·əm ¦thē·ō· brə'mad·əs }

oleyl alcohol [ORG CHEM] $C_{18}H_{35}OH$ Clear liquid, boiling at 282–349°C; fatty alcohol derived from oleic acid; commercial grade 80–90% pure; used to make resins and surface-active agents, and as a chemical intermediate. { ō'lē·əl 'al·kə‚hól }

olfaction [PHYSIO] 1. The function of smelling. 2. The sense of smell. { äl'fak·shən }

olfactoreceptor [PHYSIO] A structure which is a receptor for the sense of smell. { äl¦fak·tō·ri‚sep·tər }

olfactory aura [MED] Prodromal disagreeable olfactory sensation preceding or characterizing an epileptic attack. { äl'fak·trē 'ór·ə }

olfactory bulb [VERT ZOO] The bulbous distal end of the olfactory tract located beneath each anterior lobe of the cerebrum; well developed in lower vertebrates. { äl'fak·trē ‚bəlb }

olfactory cell [NEUROSCI] One of the sensory nerve cells in the olfactory epithelium. { äl'fak·trē ‚sel }

olfactory foramen [ANAT] Any of the openings in the cribriform plate of the ethmoid bone through which pass the fila olfactoria of the olfactory nerves. { äl'fak·trē fə'rā·mən }

olfactory gland [PHYSIO] A type of serous gland in the nasal mucous membrane. { äl'fak·trē ‚gland }

olfactory lobe [VERT ZOO] A lobe projecting forward from the inferior surface of the frontal lobe of each cerebral hemisphere, including the olfactory bulb, tracts, and trigone; well developed in most vertebrates, but reduced in humans. { äl'fak·trē ‚lōb }

olfactory nerve [NEUROSCI] The first cranial nerve; a paired sensory nerve with its origin in the olfactory lobe and formed by processes of the olfactory cells which lie in the nasal mucosa; greatly reduced in humans. { äl'fak·trē ‚nərv }

olfactory organ [PHYSIO] Any of the small chemoreceptors in the mucous membrane lining the upper part of the nasal gravity which receive stimuli interpreted as odors. { äl'fak· trē ‚ór·gən }

olfactory pit [EMBRYO] A depression near the olfactory placode in the embryo that develops into part of the nasal cavity. Also known as nasal pit. { äl'fak·trē ‚pit }

olfactory region [ANAT] The area on and above the superior conchae and on the adjoining nasal septum where the mucous membrane has olfactory epithelium and olfactory glands. { äl'fak·trē ‚rē·jən }

olfactory stalk [ANAT] The structure that connects the olfactory bulb to the cerebrum of the vertebrate brain. { äl'fak· trē ‚stók }

olfactory tract [NEUROSCI] A narrow tract of white nerve fibers originating in the olfactory bulb and extending posteriorly to the anterior perforated substance, where it enlarges to form a lateral root (olfactory trigone). { äl'fak·trē ‚trakt }

olibanum [MATER] A gum resin distilled from the dried exudation of African and Arabian trees of the genus *Boswellia*; used as a perfume, fixative, in incense, in fumigants, and in pharmacy. Also known as frankincense; gum thus. { ō'lib· ə·nəm }

olibanum oil [MATER] Colorless oil with balsamic aroma; distilled from olibanum; soluble in ether, chloroform, and carbon disulfide; main components are pinene, phellandrene, and dipentene; used in medicine. Also known as frankincense oil. { ō'lib·ə·nəm ‚oil }

olig-, oligo- [SCI TECH] A combining form denoting few, scant, or deficiency. { 'äl·ə·gō *or* ə'lig·ə }

oligemia [MED] A state in which the total blood volume is reduced. { ‚äl·ə'gē·mē·ə }

Oligobrachiidae [INV ZOO] A monotypic family of the order Athecanephria. { ‚äl·ə·gō·brə'kī·ə‚dē }

Oligocene [GEOL] The third oldest of the seven geological epochs of the Cenozoic Era, beginning 34 million years ago and ending 24 million years ago. It corresponds to an interval of geological time (and rocks deposited during that time) from the close of the Eocene Epoch to the beginning of the Miocene Epoch. { ə'lig·ə‚sēn }

Oligochaeta [INV ZOO] A class of the phylum Annelida including worms that exhibit both external and internal segmentation, and setae which are not borne on parapodia. { ‚äl·ə· gō'kēd·ə }

oligochromemia *See* anemia. { ‚äl·ə·gō·krə'mē·mē·ə }

oligoclase [MINERAL] A plagioclase feldspar mineral with a composition ranging from $Ab_{90}An_{10}$ to $Ab_{70}An_{30}$, where $Ab = NaAlSi_3O_8$ and $An = CaAl_2O_8$. { 'äl·ə·gō‚klās }

oligoclasite [PETR] A granular plutonic rock composed almost entirely of oligoclase. Also known as oligosite. { ¦äl· ə·gō'kla‚sīt }

oligocythemia [MED] A reduction in the total number of red blood cells in the body. { ‚äl·ə·gō‚sī'thē·mē·ə }

oligodendrocyte [NEUROSCI] Glial cell responsible for elaborating myelin in the central nervous system. { äl·i· gō'den·drə‚sīt }

oligodendroglia [NEUROSCI] Small neuroglial cells with spheroidal or ovoid nuclei and fine cytoplasmic processes with secondary divisions. { ‚äl·ə·gō·den'dräg·lē·ə }

oligodendroglioma [MED] A slowly growing, large, well-defined cerebral glioma, composed of small cells with richly chromatic nuclei and scanty, poorly staining cytoplasm. { ‚äl· ə·gō·den‚dräg·ē'ō·mə }

oligodontia *See* hypodontia. { ō‚lig·ə'dän·chə }

oligodynamic action [MICROBIO] The inhibiting or killing of microorganisms by use of very small amounts of a chemical substance. { ¦äl·ə·gō‚dī nam·ik 'ak·shən }

oligogene [GEN] A gene that encodes segments of multiple structural genes and causes major phenotypic alterations when mutated.

oligomenorrhea [MED] Abnormally infrequent menstruation. { ‚äl·ə·gō‚men·ə'rē·ə }

oligomer [ORG CHEM] A molecule made up of a relatively small number of monomer units. { ə'lig·ə·mər }

Oligomera [INV ZOO] A subphylum of the phylum Vermes comprising groups with two or three coelomic divisions. { ‚äl· ə'gäm·ə·rə }

oligomeric protein [BIOCHEM] A protein composed of two or more polypeptide chains. { ‚ə‚lig·ə'mer·ik 'prō‚tēn }

oligomerous [BOT] Having one or more whorls with fewer members than other whorls of the flower. { ‚äl·ə'gäm·ə·rəs }

oligomictic [HYD] Pertaining to a lake that circulates only at rare, irregular intervals during abnormal cold spells. [PETR] Of a clastic sedimentary rock, composed of a single rock type. { ‚ə‚lig·ə'mik·tik }

oligomycin [MICROBIO] Any of a group of antifungal antibiotics produced by an actinomycete resembling *Streptomyces diastachromogenes*; the colorless, hexagonal crystals are soluble in many organic solvents. { ‚ə‚lig·ə'mīs·ən }

oligonucleotide [BIOCHEM] A polynucleotide of low molecular weight, consisting of less than 20 nucleotide polymers. { ‚äl·ə·gō'nü·klē·ə‚tīd }

oligopelic [GEOL] Property of a lake bottom deposit which contains very little clay. { ‚ə‚lig·ə'pel·ik }

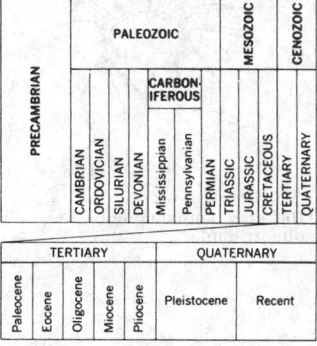

OLIGOCENE

				CARBON-IFEROUS						MESOZOIC	CENOZOIC
PRECAMBRIAN		PALEOZOIC									
	CAMBRIAN	ORDOVICIAN	SILURIAN	DEVONIAN	Mississippian	Pennsylvanian	PERMIAN	TRIASSIC	JURASSIC	CRETACEOUS	TERTIARY / QUATERNARY

TERTIARY				QUATERNARY	
Paleocene	Eocene	Oligocene	Miocene	Pliocene	Pleistocene / Recent

Position of the Oligocene epoch in the geologic time scale.

oligopeptide [ORG CHEM] A peptide composed of no more than 10 amino acids. { ‚äl·ə·gō'pep‚tīd }

oligophagous [ZOO] Eating only a limited variety of foods. { ‚äl·ə‚gäf·ə·gəs }

oligophyre [PETR] A light-colored diorite containing oligoclase phenocrysts in a groundmass of the same minerals. { ə'lig·ə‚fīr }

Oligopygidae [PALEON] An extinct family of exocyclic Euechinoidia in the order Holectypoida which were small ovoid forms of the Early Tertiary. { ‚äl·ə·gō‚pij·ə‚dē }

oligosaccharide [BIOCHEM] A sugar composed of two to eight monosaccharide units joined by glycosidic bonds. Also known as compound sugar. { ‚äl·ə·gō'sak·ə‚rīd }

oligosite See oligoclasite. { ə'lig·ə‚sīt }

oligospermia [MED] Scarcity of spermatozoa in the semen. { ‚äl·ə·gō'spər‚mē·ə }

Oligotrichida [INV ZOO] A minor order of the Spirotrichia; the body is round in cross section, and the adoral zone of membranelles is often highly developed at the oral end of the organism. { ‚äl·ə·gō'trik·ə·də }

oligotrophic [HYD] Of a lake, lacking plant nutrients and usually containing plentiful amounts of dissolved oxygen without marked stratification. { ‚äl·ə·gō‚träf·ik }

oliguria [MED] Diminished excretion of urine. { ‚äl·ə‚gyur·ē·ə }

olistolith [GEOL] An exotic block or other rock mass that has been transported by submarine gravity sliding or slumping and is included in the binder of an olistostrome. { ə'lis·tə‚lith }

olistostrome [GEOL] A sedimentary deposit composed of a chaotic mass of heterogeneous material that is intimately mixed; accumulated in the form of a semifluid body by submarine gravity sliding or slumping of unconsolidated sediments. { ə'lis·tə‚strōm }

olivaceous [BIOL] **1.** Resembling an olive. **2.** Olive colored. { ‚äl·ə'vā·shəs }

olivary nucleus [ANAT] A prominent, convoluted gray band that opens medially and occupies the upper two-thirds of the medulla oblongata. { 'äl·ə‚ver·ē 'nü‚klē·əs }

olive [BOT] Any plant of the genus *Olea* in the order Schrophulariales, especially the evergreen olive tree (*O. europea*) cultivated for its drupaceous fruit, which is eaten ripe (black olives) and unripe (green), and is of high oil content. { 'äl·əv }

olive hole [HOROL] In a watch, a jeweled bearing hole in which the sharp corners have been smoothed by grinding to reduce friction between the sides of the hole and the pivot that turns in it. { 'äl·əv ‚hōl }

olive infused oil [MATER] A synthetic olive oil made by infusing corn oil with a paste of finely ground, partly dehydrated ripe olives; contains carotene. { 'äl·əv in‚fyüzd ‚oil }

oliveiraite [MINERAL] $Zr_3Ti_2O_{10} \cdot 2H_2O$ An isotropic mineral consisting of an oxide of titanium and zirconium. { ‚äl·ə·və'rä‚īt }

olive knot [PL PATH] A bacterial disease of the olive caused by *Pseudomonas sevastonoi* and characterized by excrescences on the foliage and branches, and sometimes on the trunk. Also known as olive tubercle. { 'äl·əv ‚nät }

olivenite [MINERAL] $Cu_2(AsO_4)(OH)$ An olive-green, dull-brown, gray, or yellow mineral crystallizing in the orthorhombic system and consisting of a basic arsenate of copper. Also known as leucochalcite; wood copper. { ō'liv·ə‚nīt }

olive oil [MATER] Pale- or greenish-yellow edible oil; main components are olein and palmitin; soluble in ether, chloroform, and carbon disulfide; derived from the pulp of olive tree fruit; used in foods, ointments, linaments, and soaps, as a lubricant, and for tanning. Also known as Florence oil; lucca oil; sweet oil. { 'äl·əv ‚oil }

Oliver filter [MIN ENG] A continuous-type filter made in the form of a cylindrical drum with filter cloth stretched over the convex surface of the drum. { 'äl·ə·vər ‚fil·tər }

olivette [ELEC] Standing floodlight used in the wings for lighting stage entrances and acting areas at fairly close range; bulb wattage ranges from 500 to 1500 watts. { ‚äl·ə‚vet }

olive tubercle See olive knot. { 'äl·əv 'tü·bər·kəl }

Olividae [INV ZOO] A family of snails in the gastropod order Neogastropoda. { ō'liv·ə‚dē }

olivine [MINERAL] $(Mg,Fe_2)SiO_4$ A neosilicate group of olive-green magnesium-iron silicate minerals crystallizing in the orthorhombic system and having a vitreous luster; hardness is 6½–7 on Mohs scale; specific gravity is 3.27–3.37. { 'äl·ə‚vēn }

olivine basalt [PETR] Any of a group of olivine-bearing basalts. { 'äl·ə‚vēn bə'sólt }

olivine-bronzite chondrite [GEOL] A type of chondritic meteorite that contains about equal amounts of olivine and bronzite. { 'äl·ə‚vēn 'brän‚zīt 'kän‚drīt }

olivine diabase [PETR] An igneous rock composed principally of olivine and formed from tholeiitic magmas by differentiation in thick sills. { 'äl·ə‚vēn 'dī·ə‚bās }

olivine-hypersthene chondrite [GEOL] A type of chondritic meteorite generally containing more olivine than hypersthene; the hypersthene contains 12–20% iron, giving the meteorite a relatively dark color, and the metal grains usually contain 7–12% nickel. { 'äl·ə‚vēn 'hī·pər‚sthen 'kän‚drīt }

olivine nephelinite [PETR] An extrusive igneous rock differing in composition from nephelinite only by the presence of olivine. Also known as ankaratrite; nepheline basalt. { 'äl·ə‚vēn nə'fel·ə‚nīt }

olivine-pigeonite chondrite [GEOL] A type of chondritic meteorite in which olivine is the predominant mineral and pigeonite is secondary, and metal inclusions are usually rich in nickel. { 'äl·ə‚vēn 'pij·ə‚nīt 'kän‚drīt }

ollenite [PETR] A type of hornblende schist characterized by abundant epidote, sphene, and rutile. { 'äl·ə‚nīt }

Ollier's disease See enchondromatosis. { ōl'yäz di‚zēz }

OLRT system See on-line real-time system. { ‚ō‚el‚är'tē ‚sis·təm }

Olsen ductility test [MET] A cupping test in which a piece of sheet metal is deformed at the center by a steel ball until fracture occurs; ductility is measured by the height of the cup at the time of failure. { 'ōl·sən dək'til·ə·dē ‚test }

Olympus Mons [ASTRON] The largest volcano on Mars; it is approximately 360 miles (600 kilometers) across at its base and stands approximately 16 miles (26 kilometers) above the surrounding terrain. Also known as Nix Olympica. { ə'lim·pəs 'mänz }

omasum [VERT ZOO] The third chamber of the ruminant stomach where the contents are mixed to a more or less homogeneous state. Also known as manyplies; psalterium. { ō'mä·səm }

ombré [TEXT] Referring to colors on fabrics that are shaded or graduated from light to dark tones, or to tones that shade from one color into another. { äm'brä }

ombrometer See rain gage. { äm'bräm·əd·ər }

ombrophilous [ECOL] Able to thrive in areas of abundant rainfall. { äm'bräf·ə·ləs }

ombrophobia [PSYCH] An abnormal fear of rain. { ‚äm·brə'fō·bē·ə }

ombrophobous [ECOL] Unable to live in the presence of long, continuous rain. { äm'bräf·ə·bəs }

ombroscope [ENG] An instrument consisting of a heated, water-sensitive surface which indicates by mechanical or electrical techniques the occurrence of precipitation; the output of the instrument may be arranged to trip an alarm or to record on a time chart. { 'äm·brə‚skōp }

Omega [NAV] A worldwide radio navigation system providing navigational parameters by phase comparison of very-low-frequency (10 to 14 kilohertz), continuous-wave radio signals; terminated on September 30, 1997. { ō'meg·ə }

omega hyperon [PARTIC PHYS] A semistable baryon with a mass of approximately 1672 MeV, negative charge, spin of 3/2, and positive parity; constitutes an isotopic spin singlet. Also known as omega particle. Symbolized Ω^-. { ō'meg·ə 'hī·pə‚rän }

omega meson [PARTIC PHYS] An unstable, neutral vector meson having a mass of about 783 MeV, a width of about 8 MeV, and negative charge parity and G parity. Symbolized $\omega(783)$. { ō'meg·ə 'mä‚sän }

Omega Nebula [ASTRON] A bright H II region in the constellation Sagittarius that is both a bright far-infrared source and a double radio source. Also known as Swan Nebula. { ō'meg·ə 'neb·yə·lə }

omega particle See omega hyperon. { ō'meg·ə 'pärd·ə·kəl }

omegatron [ELECTR] A miniature mass spectrograph, about the size of a receiving tube, that can be sealed to another tube and used to identify the residual gases left after evacuation. { ō'meg·ə‚trän }

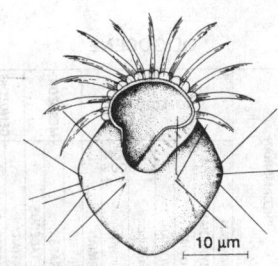

OLIGOTRICHIDA

Halteria, an example of an oligotrichid.

10 µm

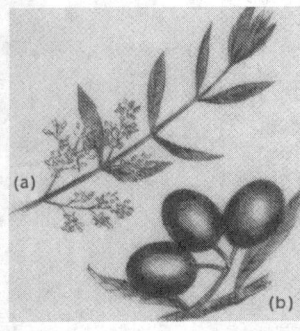

OLIVE

(a)

(b)

Olive (*Olea europeae*) branches. (*a*) Bearing small white flowers. (*b*) Bearing drupes, or fruit.

omentum [ANAT] A fold of the peritoneum connecting or supporting abdominal viscera. { ō'ment·əm }

omethioate See folimat. { ‚ō·mə'thī·ə‚wāt }

OMG object model See Object Management Group object model. { ¦ō¦em¦jē 'äb·jikt ‚mäd·əl }

omission [GEOL] The elimination or nonexposure of certain stratigraphic beds at the surface of any specified section because of disruption and displacement of the beds by faulting. { ō'mish·ən }

omission factor [COMPUT SCI] In information retrieval, the ratio obtained in dividing the number of nonretrieved relevant documents by the total number of relevant documents in the file. { ō'mish·ən ‚fak·tər }

omission solid solution [CRYSTAL] A crystal with certain atomic sites incompletely filled. { ō'mish·ən ¦säl·əd sə'lü·shən }

ommatidium [INV ZOO] The structural unit of a compound eye, composed of a cornea, a crystalline cone, and a receptor element connected to the optic nerve. { ‚äm·ə'tid·ē·əm }

ommatophore [INV ZOO] A movable peduncle that bears an eye, as in snails. { ə'mad·ə‚fór }

omnibearing [NAV] The magnetic bearing of an omnidirectional radio range. { 'äm·nə‚ber·iŋ }

omnibearing beacon [NAV] A beacon transmitter capable of providing bearing information in all horizontal directions, particularly a very-high-frequency omnidirectional radio range. { 'äm·nə‚ber·iŋ 'bē·kən }

omnibearing converter [ENG] An electromechanical device which combines an omnirange signal with heading information to furnish electrical signals for the operation of the pointer of a radio magnetic indicator. { 'äm·nə‚ber·iŋ kən'vərd·ər }

omnibearing distance navigation [NAV] Navigation based upon polar coordinates relative to a reference point. Also known as r-theta navigation; rho-theta navigation. { 'äm·nə‚ber·iŋ ¦dis·təns ‚nav·ə'gā·shən }

omnibearing distance station [NAV] A radio station having an omnidirectional radio range and a distance-measuring-equipment (DME) transponder in combination. { 'äm·nə‚ber·iŋ ¦dis·təns ‚stā·shən }

omnibearing indicator [ENG] An instrument providing automatic and continuous indication of omnibearing. { 'äm·nə‚ber·iŋ 'in·də‚kād·ər }

omnibearing line [NAV] One of an infinite number of straight lines radiating from the geographical location of a very-high-frequency omnirange. { 'äm·nə‚ber·iŋ 'līn }

omnibearing selector [ENG] A device capable of being set manually to any desired omnibearing, or its reciprocal, to control a course-line deviation indicator. Also known as radial selector. { 'äm·nə‚ber·iŋ si'lek·tər }

omnidirectional [ELECTR] Radiating or receiving equally well in all directions. Also known as nondirectional. { 'äm·nə·di'rek·shən·əl }

omnidirectional antenna [ELECTROMAG] An antenna that has an essentially circular radiation pattern in azimuth and a directional pattern in elevation. Also known as nondirectional antenna. { 'äm·nə·di'rek·shən·əl an'ten·ə }

omnidirectional hydrophone [ENG ACOUS] A hydrophone whose response is fundamentally independent of the incident sound wave's angle of arrival. { ¦äm·nə·di'rek·shən·əl 'hī·drə‚fōn }

omnidirectional range See omnirange. { ¦äm·nə·di'rek·shən·əl 'rānj }

omnidistance [NAV] The distance indicated on the meter of an aircraft's distance-measuring equipment when this equipment is interrogating and receiving replies from a transponder beacon located at an omnirange station; thus it is the slant distance. { 'äm·nə'dis·təns }

omnifocal lens [OPTICS] A bifocal eyeglass lens that is shaped to allow a smooth transition from one focus to the other. Also known as progressive lens. { 'äm·nə‚fō·kəl 'lenz }

omnigraph [ENG] An automatic acetylene cutter controlled by a mechanical pointer that traces a pattern; capable of cutting several duplicates simultaneously. { 'äm·nə‚graf }

omnimeter [ENG] A theodolite with a microscope that can be used to observe vertical angular movement of the telescope. { äm'nim·əd·ər }

omnirange [NAV] A radio aid to navigation providing direct indication of the magnetic bearing (omnibearing) of that station

from any direction. Also known as omnidirectional range. { 'äm·nə‚rānj }

omnivore [ZOO] An organism that eats both animal and vegetable matter. { 'äm·nə‚vór }

omohyoid [ANAT] **1.** Pertaining conjointly to the scapula and the hyoid bone. **2.** A muscle attached to the scapula and the hyoid bone. { ¦ō·mō'hī‚óid }

Omophronidae [INV ZOO] The savage beetles, a small family of coleopteran insects in the suborder Adephaga. { ‚o·mə'frän·ə‚dē }

omphacite [MINERAL] A grassy- to pale-green, granular or foliated, high-temperature aluminous clinopyroxene mineral with a vitreous luster that commonly occurs in the rock eclogite; a variety of augite. { 'äm·fə‚sīt }

omphalitis [MED] Inflammation of the umbilicus. { ‚äm·fə'līd·əs }

omphalomesenteric artery See vitelline artery. { ¦äm·fə·lō ‚mez·ən'ter·ik 'ärd·ə·rē }

omphalomesenteric duct See vitelline duct. { ¦äm·fə·lō ‚mez·ən'ter·ik 'dəkt }

omphalomesenteric vein See vitelline vein. { ¦äm·fə·lō ‚mez·ən'ter·ik 'vān }

omphaloproptosis [MED] Abnormal protrusion of the navel. { ¦äm·fə·lō‚präp'tō·səs }

Omphralidae [INV ZOO] A family of orthorrhaphous dipteran insects in the series Nematocera. { äm'fral·ə‚dē }

OMR See optical mark reading.

OMS See ovonic memory switch.

on [ENG] Designating the operating state of a device or one of two possible conditions (the other being "off") in a circuit. { ón }

Onagraceae [BOT] A family of dicotyledonous plants in the order Myrtales characterized generally by an inferior ovary, axile placentation, twice as many stamens as petals, a four-nucleate embryo sac, and many ovules. { ‚än·ə'grās·ē‚ē }

ONB See ortho-nitrobiphenyl.

onboard [COMPUT SCI] Referring to a computer hardware component that is built directly into the computer. { 'ón'bórd }

on-call circuit [COMMUN] A permanently designated circuit that is activated only upon request of the user; this type of circuit is usually provided when a full-period circuit cannot be justified and the duration of use cannot be anticipated; during unactivated periods, the communications facilities required for the circuit are available for other requirements. { 'ón 'kól ‚sər·kət }

on-carriage equipment [ORD] Items of supply which, although not part of the cannon carriage proper, are mounted on the carriage and remain in their respective mounts, brackets, or containers on the carriage when it is being towed. { 'ón ¦kar·ij ‚kwip·mənt }

on-carriage fire control [ORD] Process of controlling fire on a target with the aid of a sighting device mounted directly on the weapon. { 'ón ¦kar·ij 'fīr kən‚trōl }

on-carriage fire control equipment [ORD] Fire control items, such as telescope mounts, which are built into the cannon carriage or mount, or which are carried on the carriage or mount in the traveling position. { 'ón ¦kar·ij 'fīr kən‚trōl i‚kwip·mənt }

on center [BUILD] The measurement made between the centers of two adjacent members. { 'ón 'sen·tər }

once-through boiler [MECH ENG] A boiler in which water flows, without recirculation, sequentially through the economizer, furnace wall, and evaporating and superheating tubes. { 'wəns ¦thrü ‚bóil·ər }

Onchidiidae [INV ZOO] An intertidal family of sluglike pulmonate mollusks of the order Systellommatophora in which the body is oval or lengthened, with the convex dorsal integument lacking a mantle cavity or shell. { ‚äŋ·kə'dī·ə‚dē }

onchocerciasis [MED] Infection with the filaria Onchocerca volvulus; results in skin tumors, papular dermatitis, and ocular complications. { ‚äŋ·kō·sər'kī·ə·səs }

onchogryposis [MED] A thickened, ridged, and curved condition of a nail. { ‚äŋ·kō·grə'pō·səs }

Oncholaimoidea [INV ZOO] A superfamily of nematodes in the order Enoplida, characterized by a stoma armed with one dorsal tooth and two subventral teeth and sometimes with transverse rows of small denticles from its walls, and two whorls of cephalic sensilla. { ‚äŋ·kō·lə'móid·ē·ə }

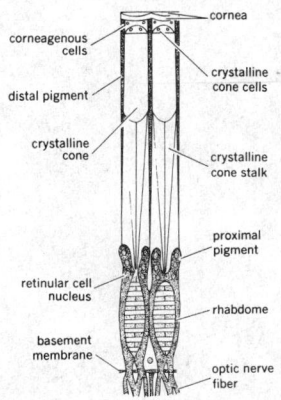

OMMATIDIUM

corneagenous cells — cornea

distal pigment — crystalline cone cells

crystalline cone — crystalline cone stalk

retinular cell nucleus — proximal pigment

basement membrane — rhabdome — optic nerve fiber

Two ommatidia from the compound eye of the crayfish *Astacus*.

oncholysis [MED] A slow process of loosening of a nail from its bed, beginning at the free edge and progressing gradually toward the root. { äŋ'käl·ə·səs }

onchomycosis [MED] Any fungus disease of the nail. { ¦äŋ·kō·mī'kō·səs }

onchosphere [INV ZOO] The hexacanth embryo identified as the earliest differentiated stage of cyclophyllidean tapeworms. { 'äŋ·kō,sfir }

oncocyte [HISTOL] A columnar-shaped cell with finely granular eosinophilic cytoplasm, found in salivary and certain endocrine glands, nasal mucosa, and other locations. { 'äŋ·kō,sīt }

oncocytoma [MED] A benign tumor composed principally of oncocytes; usually occurs in salivary glands. { ¦äŋ·kō·sī'tō·mə }

oncofetal antigen [IMMUNOL] Any of a group of antigens that are commonly present both in fetal tissue during early development of life and in adult tissue when cancer occurs. { ¦aŋk·ə,fēd·əl 'ant·i·jən }

oncogene [GEN] A gene whose mutation can lead to cancer in experimental animals and humans. { 'äŋ·kō,jēn }

oncogenesis [MED] Processes of tumor formation. { ¦äŋ·kō'jen·ə·səs }

oncogenic virus [VIROL] A virus that transforms the infected cells so that they undergo uncontrolled proliferation. { ¦äŋ·kə,jen·ik 'vī·rəs }

oncolite [GEOL] A small, variously shaped (often spheroidal), concentrically laminated, calcareous sedimentary structure resembling an oolith; formed by accretion of successive, layered masses of gelatinous sheaths of blue-green algae. { 'äŋ·kō,līt }

oncology [MED] The study of the causes, development, characteristics, and treatment of tumors. { äŋ'käl·ə·jē }

oncolytic [MED] Pertaining to destruction of cancer cells. { ,äŋ·kə'lid·ik }

oncomouse [BIOL] A laboratory mouse that carries activated human cancer genes. { 'äŋk·ə,maús }

on composition See on grade. { ¦òn ,käm·pə'zish·ən }

on-condition maintenance [IND ENG] Examination of those aspects of an installation that are predictive of pending failure, followed by performance of preventative maintenance activities before occurrence of total failure. { ¦òn kən¦dish·ən 'mānt·ən·əns }

Oncopoda [INV ZOO] A phylum of the superphylum Articulata. { äŋ'käp·ə·də }

Oncorhynchus [VERT ZOO] A genus of seven semelparous salmon species that occur naturally in the North Pacific Ocean and spawn in western North America and coastal Asia. { ,äŋ·kə'riŋ·kəs }

oncotic pressure [PHYSIO] Also known as colloidal osmotic pressure. **1.** The osmotic pressure exerted by colloids in a solution. **2.** The pressure exerted by plasma proteins. { äŋ'käd·ik 'presh·ər }

oncotomy [MED] Surgical incision of a tumor, abscess, or other swelling. { äŋ'käd·ə·mē }

Oncovirinae [VIROL] A subfamily of the Retroviridae family. { ,äŋ·kō'vir·ə,nī }

onde de choc [ACOUS] The first sound heard as the result of the passage of a high-speed projectile. { òn·də 'shòk }

ondograph [ELECTR] An instrument that draws the waveform of an alternating-current voltage step by step; a capacitor is charged momentarily to the amplitude of a point on the voltage wave, then discharged into a recording galvanometer, with the action being repeated a little further along on the waveform at intervals of about 0.01 second. { 'än·də,graf }

ondoscope [ELECTR] A glow-discharge tube used to detect high-frequency radiation, as in the vicinity of a radar transmitter; the radiation ionizes the gas in the tube and produces a visible glow. { 'än·də,skōp }

ondulé [TEXT] A wavy effect on the surface of a woven fabric that has been achieved by alternate spreading and converging of some of the warp threads. { 'än·də¦lā }

-one [ORG CHEM] Chemical suffix indicating a ketone, a substance related to starches and sugars, or an alkone. { ōn }

one-address code [COMPUT SCI] In computers, a code using one-address instructions. { 'wən ə,dres 'kōd }

one-address instruction [COMPUT SCI] A digital computer programming instruction that explicitly describes one operation and one storage location. Also known as single-address instruction. { 'wən ə,dres in'strək·shən }

one condition [COMPUT SCI] The state of a magnetic core or other computer memory element in which it represents the value 1. Also known as one state. { 'wən kən,dish·ən }

one-digit subtracter See half-subtracter. { 'wən ,dij·ət səb'trak·tər }

one-dimensional array [COMPUT SCI] A group of related data elements arranged in a single row or column. { ¦wən də¦men·chən·əl ə'rā }

one-dimensional flow [FL MECH] Fluid flow in which all flow is parallel to some straight line, and characteristics of flow do not change in moving perpendicular to this line. { 'wən di,men·chən·əl 'flō }

one-dimensional lattice [CRYSTAL] A simplified model of a crystal lattice consisting of particles lying along a straight line at either equal or periodically repeating distances. { 'wən di,men·chən·əl 'lad·əs }

one-dimensional strain [MATH] A transformation that elongates or compresses a configuration in a given direction, given by $x' = kx$, $y' = y$, $z' = z$, where k is a constant, when the direction is that of the x axis. { 'wən di,men·chən·əl 'strān }

one-ended tape Turing machine [COMPUT SCI] A variation of a Turing machine in which the tape can be extended to the right, but not to the left. { 'wən ¦end·əd ¦tāp 'tùr·iŋ mə,shēn }

one-face-centered lattice [CRYSTAL] A crystal lattice in which there are lattice points at the centers of one pair of faces in each unit cell as well as at the corners. { 'wən ¦fās ¦sen·tərd 'lad·əs }

one-group model [NUCLEO] A neutron-behavior model in which neutrons of all energies are treated as having the same characteristics. { 'wən ,grüp 'mäd·əl }

one-hundred-percent premium plan [IND ENG] A wage incentive plan wherein each unit produced by an employee in excess of standard is compensated at the same rate paid for each unit of standard production. Also known as straight piecework system; straight proportional system. { 'wən ,hən·drəd pər¦sent 'prē·mē·əm ,plan }

one-level address [COMPUT SCI] In digital computers, an address that directly indicates the location of an instruction or some data. { 'wən ,lev·əl ə'dres }

one-level code [COMPUT SCI] Any code using absolute addresses and absolute operation codes. { 'wən ,lev·əl 'kōd }

one-level subroutine [COMPUT SCI] A subroutine that does not use other subroutines during its execution. { 'wən ,lev·əl 'səb·rü,tēn }

one-line adapter [COMPUT SCI] A unit connecting central processes and permitting high-speed transfer of data under program control. { 'wən ,līn ə'dap·tər }

one-parameter semigroup [MATH] A semigroup with which there is associated a bijective mapping from the positive real numbers onto the semigroup. { ¦wən pə¦ram·əd·ər 'sem·i,grüp }

one-part code [COMMUN] Code in which the plain text elements are arranged in alphabetical or numerical order, accompanied by their code groups also arranged in alphabetical, numerical, or other systematic order. { 'wən ,pärt 'kōd }

one-particle exchange [PARTIC PHYS] A model for the interaction of two particles in which the interaction results entirely from a single virtual particle being emitted by one interacting particle and absorbed by the other. { 'wən ,pärd·ə·kəl iks'chānj }

one-pass operation [COMPUT SCI] An operating method, now standard, which produces an object program from a source program in one pass. { 'wən ,pas ,äp·ə'rā·shən }

one-piece set [MIN ENG] A single stick of timber used as a post, stull, or prop. { 'wən ,pēs 'set }

one-plus-one address instruction [COMPUT SCI] A digital computer instruction whose format contains two address parts; one address designates the operand to be involved in the operation; the other indicates the location of the next instruction to be executed. { 'wən ,pləs 'wən ə,dres in,strək·shən }

one-point compactification [MATH] The one-point compactification $\bar{X}$ of a topological space X is the union of X with a set consisting of a single element, with the topology of $\bar{X}$ consisting of the open subsets of X and all subsets of $\bar{X}$ whose complements in $\bar{X}$ are closed compact subsets in X. Also known as Alexandroff compactification. { 'wən ˌpóint kəmˌpak·tə·fə'kā·shən }

one-point perspective *See* parallel perspective. { 'wən ˌpóint pər'spek·tiv }

one-quadrant multiplier [ELECTR] Of an analog computer, a multiplier in which operation is restricted to a single sign of both input variables. { 'wən ˌkwä·drənt 'məl·tə,plī·ər }

one-sample problem [STAT] The problem of testing the hypothesis that the average of a sequence of observations or measurement of the same kind has a specified value. { 'wən ˌsam·pəl 'präb·ləm }

one's complement [COMPUT SCI] A numeral in binary notation, derived from another binary number by simply changing the sense of every digit. { 'wənz 'käm·plə·mənt }

ones-complement code [COMPUT SCI] A number coding system used in some computers, where, for any number x, $x = (1 - 2^{n-1}) \cdot a_0 + 2^{n-2} a_1 + \cdots + a_{n-1}$, where $a_i = 1$ or 0. { 'wənz 'käm·plə·mənt ˌkōd }

one-shot molding [ENG] Production of urethane-plastic foam in which the isocynate, polyol, and catalyst and other additives are mixed directly together and a foam is produced immediately. { 'wən ˌshät 'mōld·iŋ }

one-shot multivibrator *See* monostable multivibrator. { 'wən ˌshät ˌməl·tə'vī,brād·ər }

one-shot operation *See* single-step operation. { 'wən ˌshät ˌäp·ə'rā·shən }

one-sided abrupt junction [ELECTR] An abrupt junction that is realized by giving one side of the junction a high doping level compared with the other; that is, an n^+p or p^+n junction. { 'wən ˌsīd·əd ə'brəpt 'jəŋk·shən }

one-sided acceptance sampling test [IND ENG] A test against a single specification only, in which permissible values in one direction are not limited. { 'wən ˌsīd·əd ik'sep·təns 'samp·liŋ ˌtest }

one-sided limit [MATH] Either a limit on the left or a limit on the right. { 'wən ˌsīd·əd 'lim·ət }

one-sided surface [MATH] A surface such that an object resting on one side can be moved continuously over the surface to reach the other side without going around an edge; the Möbius band and the Klein bottle are examples. Also known as nonorientable surface. { 'wən ˌsīd·əd 'sər·fəs }

one-sided test [STAT] A test statistic T which rejects a hypothesis only for $T \geq d$ or $T \leq c$ but not for both (here d and c are critical values). { 'wən ˌsīd·əd 'test }

Onesquethawan [GEOL] A North American stage in the Lower and Middle Devonian, lying above the Deerparkian and below the Cazenovian. { ˌän·ə'skweth·əˌwän }

one state *See* one condition. { 'wən ˌstāt }

one-step operation *See* single-step operation. { 'wən ˌstep ˌäp·ə'rā·shən }

one-tail test *See* one-tailed test. { 'wən ˌtāl 'test }

one-tailed test [STAT] A statistical test in which the critical region consists of all values of a test statistic that are less than a given value or greater than a given value, but not both. Also known as one-tail test. { 'wən ˌtāld 'test }

one-time pad [COMMUN] A keying sequence based on random numbers that is used to code a single message and is then destroyed. { 'wən ˌtīm 'pad }

one-to-many correspondence [COMPUT SCI] A structure that establishes relationships between two types of items in a data base such that one item of the first type can relate to several items of the second type, but items of the second type can relate back to only one item of the first type. { 'wən tə ˌmen·ē ˌkär·ə'spän·dəns }

one-to-one assembler [COMPUT SCI] An assembly program which produces a single instruction in machine language for each statement in the source language. Also known as one-to-one translator. { 'wən tə 'wən ə'sem·blər }

one-to-one correspondence [MATH] A pairing between two classes of elements whereby each element of either class is made to correspond to one and only one element of the other class. { 'wən tə 'wən ˌkär·ə'spän·dəns }

one-to-one mapping *See* injection. { 'wən tə 'wən 'map·iŋ }

one-to-one translator *See* one-to-one assembler. { 'wən tə 'wən 'tranz,lād·ər }

O network [ELEC] Network composed of four impedance branches connected in series to form a closed circuit, two adjacent junction points serving as input terminals, the

remaining two junction points serving as output terminals. { 'ō ˌnet,wərk }

one-valued function *See* single-valued function. { 'wən ˌval·yüd 'fəŋk·shən }

one-way classification [STAT] The basis for the simplest case of the analysis of variance; a set of observations are categorized according to values of one variable or one characteristic. { 'wən ˌwā ˌklas·ə·fə'kā·shən }

one-way coupling [FL MECH] The property of a particle flow in which the fluid will affect the particle properties (velocity, temperature, and so forth) but the particles will not influence the fluid properties. { 'wən ˌwā 'kəp·liŋ }

one-way slab [CIV ENG] A concrete slab in which the reinforcing steel runs perpendicular to the supporting beams, that is, one way. { 'wən ˌwā 'slab }

one-way trunk [ELEC] Trunk between two central offices, used for calls that originate at one of those offices, but not for calls that originate at the other. Also known as outgoing trunk. { 'wən ˌwā 'trəŋk }

one-year ice [OCEANOGR] Sea ice formed the previous season, not yet 1 year old. { 'wən ˌyir 'īs }

on grade [CIV ENG] **1.** At ground level. **2.** Supported directly on the ground. [MET] A classification for an alloy indicating that it has a chemical composition within the range of the specified composition limits for that particular alloy. Also known as on composition. { 'ón 'grād }

on-hook [COMMUN] The idle state (open loop) of a subscriber or PBX user loop. { 'ón ˌhúk }

onion [BOT] **1.** *Allium cepa.* A biennial plant in the order Liliales cultivated for its edible bulb. **2.** Any plant of the genus *Allium.* { 'ən·yən }

onion diagram [SYS ENG] A schematic diagram of a system that is composed of concentric circles, with the innermost circle representing the core, and all the outer layers dependent on the core. { 'ən·yən ˌdī·ə,gram }

onion oil [MATER] A sharp-odored yellow liquid; derived from the bulb of the onion *Allium cepa*; soluble in ether, chloroform, and carbon disulfude; main component is allyl propyl disulfide; used in flavorings. { 'ən·yən ˌóil }

onion scab *See* onion smudge. { 'ən·yən ˌskab }

onionskin paper [MATER] A lightweight, durable bond paper; usually quite translucent, resembling the dry outer skin of an onion; used for duplicate typewriter copies and in interleaving order books. { 'ən,yən,skin 'pā·pər }

onionskin weathering [GEOL] A type of spheroidal weathering in which successive shells of decayed rock resembling the layers of an onion are produced. { 'ən·yən,skin 'weth·ə,riŋ }

onion smudge [PL PATH] A fungus disease of the onion caused by *Colletotrichum circinans* and characterized by black concentric integral rings or smutty spots on the bulb scales. Also known as onion scab. { 'ən·yən ˌsməj }

onion smut [PL PATH] A fungus disease of onion, especially seedlings, caused by *Urocystis cepulae* and characterized by elongate black blisters on the scales and foliage. { 'ən·yən ˌsmət }

onisciform [INV ZOO] Ovate and slightly flattened. { ä'nis·əˌfórm }

Oniscoidea [INV ZOO] A terrestrial suborder of the Isopoda; the body is either dorsoventrally flattened or highly vaulted, and the head, thorax, and abdomen are broadly joined. { ˌän·ə'skóid·ē·ə }

-onium [CHEM] Chemical suffix indicating a complex cation, as for hydronium, $(H_3O)^+$. { 'ō·nē·əm }

onlap [GEOL] A type of overlap characterized by regular and progressive pinching out of the strata toward the margins of a depositional basin; each unit transgresses and extends beyond the point of reference of the underlying unit. Also known as transgressive overlap. { 'ón,lap }

on-line [COMPUT SCI] Pertaining to equipment capable of interacting with a computer. [ELECTR] The state in which a piece of equipment or a subsystem is connected and powered to deliver its proper output to the system. { 'ón,līn }

on-line analyzer [MIN ENG] An instrument which monitors the content of materials at various stages in flotation or other mineral-processing flow sheets. { 'ón,līn ˌan·ə'līz·ər }

on-line central file [COMPUT SCI] An organized collection of data, such as an on-line disk file, in a storage device under direct control of a central processing unit, that serves as a

ONISCOIDEA

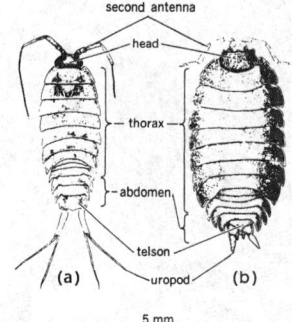

Genera of oniscoideans.
(a) Ligia, the shore slater found above tidewater mark.
(b) Porcellio, the sow bug or wood louse.

continually available source of data in applications where real-time or direct-access capabilites are required. { 'ȯn‚līn 'sen·trəl ‚fīl }

on-line cipher [COMMUN] A method of encryption directly associated with a particular transmission system, whereby messages may be encrypted and simultaneously transmitted from one station to one or more stations where reciprocal equipment is automatically operated. { 'ȯn‚līn 'sī·fər }

on-line computer system [COMPUT SCI] A computer system which is adapted to on-line operation. { 'ȯn‚līn kəm'pyüd·ər ‚sis·təm }

on-line cryptographic operation See on-line operation. { 'ȯn‚līn ‚krip·tə'graf·ik ‚äp·ə'rā·shən }

on-line data reduction [COMPUT SCI] The processing of information as rapidly as it is received by the computing system. { 'ȯn‚līn ‚dad·ə ri‚dək·shən }

on-line disk file [COMPUT SCI] A magnetic disk directly connected to the central processing unit, thereby increasing the memory capacity of the computer. { 'ȯn‚līn 'disk ‚fīl }

on-line equipment [COMPUT SCI] The equipment or devices in a system whose operation is under control of the central processing unit, and in which information reflecting current activity is introduced into the data-processing system as soon as it occurs. { 'ȯn‚līn i'kwip·mənt }

on-line inquiry [COMPUT SCI] A level of computer processing that results from adding to an expanded batch system the capability to immediately access, from any terminal, any record that is stored in the disk files attached to the computer. { 'ȯn‚līn 'in·kwə·rē }

on-line mode [COMPUT SCI] Mode of operation in which all devices are responsive to the central processor. { 'ȯn‚līn ‚mōd }

on-line operation [COMPUT SCI] Computer operation in which input data are fed into the computer directly from observing instruments or other input equipment, and computer results are obtained during the progress of the event. [COMMUN] A method of operation whereby messages are encrypted and simultaneously transmitted from one station to one or more other stations where reciprocal equipment is automatically operated to permit reception and simultaneous decryptment of the message. Also known as on-line cryptographic operation. { 'ȯn‚līn ‚äp·ə'rā·shən }

on-line real-time system [COMPUT SCI] A computer system that communicates interactively with users, and immediately returns to them the results of data processing during an interaction. Abbreviated OLRT system. { 'ȯn‚līn 'rēl ‚tīm 'sis·təm }

on-line secured communications system [COMMUN] Any combination of interconnected communications centers partially or wholly equipped for on-line cryptographic operation and capable of relaying or switching message traffic using on-line cryptographic procedures. { 'ȯn‚līn si'kyùrd kə‚myü·nə'kā·shənz ‚sis·təm }

on-line storage [COMPUT SCI] Storage controlled by the central processing unit of a computer. { 'ȯn‚līn 'stȯr·ij }

on-line tab setting [COMPUT SCI] A feature in some computer printers which allows the computer that controls the printer to issue commands to set and change the tab stops. { 'ȯn‚līn 'tab ‚sed·iŋ }

on-line typewriter [COMPUT SCI] A typewriter which transmits information into and out of a computer, and which is controlled by the central processing unit and thus by whatever program the computer is carrying out. { 'ȯn‚līn 'tīp‚rīd·ər }

only loadable [COMPUT SCI] Attribute of a load module which can be brought into main memory only by a LOAD macroinstruction given from another module. Abbreviated OL. { 'ōn·lē 'lōd·ə·bəl }

on-off control [CONT SYS] A simple control system in which the device being controlled is either full on or full off, with no intermediate operating positions. Also known as on-off system. { 'ȯn 'ȯf kən‚trōl }

on-off keying [COMMUN] Binary form of amplitude modulation in which one of the states of the modulated wave is the absence of energy in the keying interval. { 'ȯn 'ȯf ‚kē·iŋ }

on-off switch [ELEC] A switch used to turn a receiver or other equipment on or off; often combined with a volume control in radio and television receivers. { 'ȯn 'ȯf ‚swich }

on-off system See on-off control. { 'ȯn 'ȯf ‚sis·təm }

on-off tests [ELEC] Tests conducted to determine the source

of interference by switching various suspected sources on and off while observing the victim receiver. { 'ȯn 'ȯf ‚tests }

Onsager equation [PHYS CHEM] An equation which relates the measured equivalent conductance of a solution at a certain concentration to that of the pure solvent. { 'ȯn‚säg·ər i‚kwā·zhən }

Onsager reciprocal relations [THERMO] A set of conditions which state that the matrix, whose elements express various fluxes of a system (such as diffusion and heat conduction) as linear functions of the various conjugate affinities (such as mass and temperature gradients) for systems close to equilibrium, is symmetric when certain definitions are chosen for these fluxes and affinities. { 'ȯn‚säg·ər ri'sip·rə·kəl ri'lā·shənz }

Onsager theory of dielectrics [ELEC] A theory for calculating the dielectric constant of a material with polar molecules in which the local field at a molecule is calculated for an actual spherical cavity of molecular size in the dielectric using Laplace's equation, and the polarization catastrophe of the Lorentz field theory is thereby avoided. { 'ȯn‚säg·ər ‚thē·ə·rē əv ‚dī·ə'lek·triks }

onsetter [MIN ENG] The worker in charge of hoisting coal, ore, men, or materials in a mine shaft. { 'ȯn‚sed·ər }

onshore [GEOGR] Pertaining to, in the direction toward, or located on the shore. Also known as shoreside. { 'ȯn‚shȯr }

onshore wind [METEOROL] Wind blowing from the sea toward the land. { 'ȯn‚shȯr ‚wind }

on soundings [NAV] Of a vessel, navigating within the 100-fathom (183-meter) curve; in earlier times, the term was applied to a vessel in water sufficiently shallow for sounding by sounding lead. { 'ȯn ‚saùnd·iŋz }

on stream [CHEM ENG] Of a plant or process-operations unit, being in operation. { 'ȯn ‚strēm }

on-stream factor [IND ENG] The ratio of the number of operating days to the number of calendar days per year. { 'ȯn ‚strēm 'fak·tər }

on-stream time [CHEM ENG] In plant or process operations, the actual time that a unit is operating and producing product. { 'ȯn ‚strēm ‚tīm }

on the beam [ELECTR] Centered on a beam of, or on an equisignal zone of, radiant energy, as a radio range. [NAV] Bearing approximately 090° relative (on the starboard beam) or 270° relative (on the port beam); the expression is often used loosely for broad on the beam, or bearing exactly 090° or 270° relative. Also known as abeam. { 'ȯn t͟hə 'bēm }

on the bow [NAV] Bearing approximately 045° relative (on the starboard bow) or 315° relative (on the port bow); the expression is often used loosely for broad on the bow, or bearing exactly 045° or 315° relative. { 'ȯn t͟hə 'baù }

on the quarter [NAV] Bearing approximately 135° relative (on the starboard quarter) or 225° relative (on the port quarter); the expression is often used loosely for broad on the quarter, or bearing exactly 135° or 225° relative. { 'ȯn t͟hə 'kwȯrd·ər }

on the run [MIN ENG] Manner of working a seam of coal when there is sufficient inclination to cause the coal, as worked toward the rise, to fall by gravity to the gangways for loading into cars. { 'ȯn t͟hə 'rən }

on the solid [MIN ENG] **1.** Pertaining to the practice of blasting heavy charges of explosives, in lieu of undercutting or channeling. **2.** See into the solid. { 'ȯn t͟hə 'säl·əd }

on the way [ORD] An expression sent over the communication system from the firing position to the observation position when the weapon is fired, thus warning the observers to be on the alert for spotting the impact. { 'ȯn t͟hə 'wā }

ontogeny [EMBRYO] The origin and development of an organism from conception to adulthood. { än'täj·ə·nē }

on-top flight [AERO ENG] Flight above an overcast. { 'ȯn 'täp 'flīt }

Onuphidae [INV ZOO] A family of tubicolous, herbivorous, scavenging errantian annelids in the superfamily Eunicea. { än'yüf·ə‚dē }

on-vehicle materiel [ORD] Items of supply which, although not part of the vehicle proper, are issued with, and carried on, ordnance or other technical service vehicles; the items are required for vehicular maintenance, armament, fire protection, or communication, to complete the major combination; examples are guns, gun mounts, radios, flashlights, and fire extinguishers. { 'ȯn ‚vē·ə·kəl mə'tir·ē·əl }

onych-, onycho- [ZOO] A combining form denoting claw or nail. { 'än·ə‚kō }

onychia [MED] Inflammation of the nail matrix. { ō'nik·ē·ə }

Onychodontidae [PALEON] A family of Lower Devonian lobefin fishes in the order Osteolepiformes. { ¦än·ə·kō'dänt·ə,dē }

onychomycosis [MED] A fungus disease of the nails. { ¦än·ə·kō·mī'kō·səs }

Onychopalpida [INV ZOO] A suborder of mites in the order Acarina. { ¦an·ə·kō'pal·pə·də }

Onychophora [INV ZOO] A phylum of wormlike animals that combine features of both the annelids and the arthropods. { ¦än·ə'käf·ə·rə }

onychorrhexis [MED] Longitudinal striation of the nail plate, with or without the formation of fissures. { ¦än·ə·kə'rek·səs }

Onygenaceae [MYCOL] A family of ascomycetous fungi in the order Eurotiales comprising forms that inhabit various animal substrata, such as horns and hoofs. { ¦än·ə·jə'nās·ē,ē }

onyx [MINERAL] **1.** Banded chalcedonic quartz, in which the bands are straight and parallel; natural colors are usually red or brown with white, although black is occasionally encountered. **2.** *See* onyx marble. { 'än·iks }

onyx agate [MINERAL] A banded agate with straight, parallel, alternating bands of white and different tones of gray. { 'än·iks 'ag·ət }

onyx marble [MINERAL] A hard, compact, dense, generally translucent variety of calcite resembling true onyx and usually banded. Also known as alabaster; Algerian onyx; Gibraltar stone; Mexican onyx; onyx; oriental alabaster. { 'än·iks 'mär·bəl }

onyx opal [MINERAL] Common opal with straight, parallel markings. { 'än·iks 'ō·pəl }

O/O carrier *See* ore/oil carrier. { ¦ō¦ō ,kar·ē·ər }

oocyst [INV ZOO] The encysted zygote of some Sporozoa. { 'ō·ə,sist }

oocyte [HISTOL] An egg before the completion of maturation. { 'ō·ə,sīt }

oogamete [BIOL] A large, nonmotile female gamete containing reserve materials. { ¦ō·ə'ga,mēt }

oogamous [BIOL] Of sexual reproduction, characterized by fusion of a motile sperm with an oogamete. { ō'äg·ə·məs }

oogenesis [PHYSIO] Processes involved in the growth and maturation of the ovum in preparation for fertilization. { ,ō·ə'jen·ə·səs }

oogonium [BOT] The unisexual female sex organ in oogamous algae and fungi. [HISTOL] A descendant of a primary germ cell which develops into an oocyte. { ,ō·ə'gō·nē·əm }

ookinete [INV ZOO] The elongated, mobile zygote of certain Sporozoa, as that of the malaria parasite. { ,ō·ə'kī,nēt }

oolemma *See* zona pellucida. { ¦ō·ə'lem·ə }

oolicast [PETR] A small, nearly spherical feature occurring in an oolith as a result of a selective dissolution that did not destroy the matrix but left an opening that was subsequently filled. { ō'äl·ə,kast }

oolicastic porosity [PETR] The porosity produced in an oolitic rock by removal of the ooids and formation of oolicasts. { ō¦äl·ə¦kas·tik pə'räs·əd·ē }

oolite [PETR] A sedimentary rock, usually a limestone, composed principally of cemented ooliths. Also known as eggstone; roestone. { 'ō·ə,līt }

oolith [PETR] A small (0.25–2.0 millimeters), rounded accretionary body in a sedimentary rock; generally formed of calcium carbonate by inorganic precipitation or by replacement; ooliths generally exhibit concentric or radial internal structure. { 'ō·ə,lith }

oolitic chert [PETR] Chert composed chiefly of ooliths. { ,ō·ə'lid·ik 'chərt }

oolitic limestone [PETR] An even-textured limestone made up almost entirely of calcareous ooliths with essentially no matrix. { ,ō·ə'lid·ik 'līm,stōn }

oology [VERT ZOO] A branch of zoology concerned with the study of eggs, especially bird eggs. { ō'äl·ə·jē }

oomicrite [PETR] A limestone containing at least 25% ooliths and no more than 25% intraclasts in which the carbonate-mud matrix (micrite) is more abundant than the sparry-calcite cement. { ,ō·ə'mī,krīt }

oomicrudite [PETR] An oomicrite containing ooliths that are more than 1 millimeter in diameter. { ,ō·ə'mī·krə,dīt }

Oomycetes [MYCOL] A class of the Phycomycetes comprising the biflagellate water molds and downy mildews. { ,ō·ə·mī'sēd·ēz }

OOP *See* object-oriented programming.

oophagous [ZOO] Feeding or living on eggs. { ō'äf·ə·gəs }

oophoritis [MED] Inflammation of the ovaries. { ,ō·ə·fə'rīd·əs }

ooplasm [CELL MOL] Cytoplasm of an egg. { 'ō·ə,plaz·əm }

Oort dark matter [ASTRON] Matter of unknown nature that is postulated to exist in the disk of the Milky Way Galaxy in order to account for the spatial and velocity distributions of stars in the direction perpendicular to the galactic plane. { ¦ort 'därk ,mad·ər }

Oort's Cloud [ASTRON] A cloud of comets at distances from 75,000 to 150,000 astronomical units from the sun, which has been proposed as a source of comets that pass near the sun. { ¦orts ,klaüd }

oospararenite [PETR] An oosparite containing medium sand or coarse sand-sized ooliths. { ,ō·ə·spə'rar·ə,nīt }

oosparite [PETR] A limestone containing at least 25% ooliths and no more than 25% intraclasts in which the sparry-calcite cement is more abundant that the carbonate-mud matrix. { ,ō·ə'spa,rīt }

oosparrudite [PETR] An oosparite containing ooliths that are more than 1 millimeter in diameter. { ,ō·ə'spar·ə,dīt }

oospore [BOT] A spore which is produced by heterogamous fertilization and from which the sporophytic generation develops. { 'ō·ə,spor }

oostegite [INV ZOO] In many crustaceans, a platelike expansion of the basal segment of a thoracic appendage that aids in forming an egg receptacle. { ō'äs·tə,jīt }

ootype [INV ZOO] In trematodes and tapeworms, a thickening of the oviduct near the ovaries. { 'ō·ə,tīp }

oovoid [PETR] A void in the center of an incompletely replaced oolith. { 'ō·ə,void }

ooze [GEOL] **1.** A soft, muddy piece of ground, such as a bog, usually resulting from the flow of a spring or brook. **2.** A marine pelagic sediment composed of at least 30% skeletal remains of pelagic organisms, the rest being clay minerals. **3.** Soft mud or slime, typically covering the bottom of a lake or river. { üz }

opacifier [MATER] **1.** A substance, used to treat a solid rocket propellant, that absorbs light and heat and protects the propellant from deterioration until ready for use. **2.** A material used in ceramic glazes and vitreous enamels to render them nontransparent and to improve other properties. { ō'pas·ə,fī·ər }

opacite [PETR] Masses of opaque, microscopic grains in rocks, particularly in the groundmass of an igneous rock. { 'äp·ə,sīt }

opacity [OPTICS] The light flux incident upon a medium divided by the light flux transmitted by it. { ō'pas·əd·ē }

opacus [METEOROL] A variety of cloud (sheet, layer, or patch), the greater part of which is sufficiently dense to obscure the sun; found in the genera altocumulus, altostratus, stratocumulus, and stratus; cumulus and cumulonimbus clouds are inherently opaque. { ō'pā·kəs }

opal [MINERAL] A natural hydrated form of silica; it is amorphous, usually occurs in botryoidal or stalactic masses, has a hardness of 5–6 on Mohs scale, and specific gravity is 1.9–2.2. { 'ō·pəl }

opal agate [PETR] A variety of banded opal that displays different shades of color, is agatelike in structure, and consists of alternating layers of opal and chalcedony. { 'ō·pəl 'ag·ət }

opal-CT [PETR] A poorly ordered crystalline form of silica thought to be the intermediate phase in quartz chert formation. { 'ō·pəl 'sē'tē }

opalescence [OPTICS] The milky, iridescent appearance of a dense, transparent medium or colloidal system when it is illuminated by polychromatic radiation in the visible range, such as sunlight. { ,ō·pə'les·əns }

opal glass [MATER] Translucent or opaque glass, often milky white, made by adding impurities such as fluorine compounds to the melt; it appears white by reflected light but shows color images through thin sections; used for ornamental glass and as an efficient light diffuser. { 'ō·pəl ,glas }

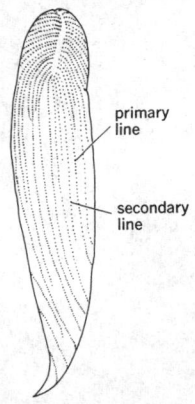

OPALINATA

An opalinid (*Protoopalina axonucleata*) showing primary and secondary lines of blepharoplasts.

Opalinata [INV ZOO] A superclass of the subphylum Sarcomastigophora containing highly specialized forms which resemble ciliates. { o̩pal·ə̩nad·ə }

opaline [MINERAL] **1.** Any of several minerals related to or resembling opal. **2.** An earthy form of gypsum. { 'ō·pə,lēn }

opalized wood *See* silicified wood. { 'ō·pə,līzd 'wùd }

opaque attritus [GEOL] Attritus that does not contain large quantities of transparent humic degradation matter. { ō'pāk ə'trīd·əs }

opaque medium [OPTICS] A medium impervious to rays of light, that is, not transparent to the human eye. [PHYS] **1.** A medium which does not transmit electromagnetic radiation of a specified type, such as that in the infrared, x-ray, ultraviolet, and microwave regions. **2.** A medium which prevents the passage of particles of a specified type. { ō'pāk ,mēd·ē·əm }

opaque projector [OPTICS] A projector designed to project the image of an opaque object, or of graphic material on an opaque support, by reflected light. { ō'pāk prə'jek·tər }

opaque sky cover [METEOROL] In United States weather observing practice, the amount (in tenths) of sky cover that completely hides all that might be above it; opposed to transparent sky cover. { ō'pāk 'skī ,kəv·ər }

opaquing [GRAPHICS] A technique in photoengraving or offset lithography in which unwanted areas are eliminated from a negative by application of an opaque substance. { ō'pāk·iŋ }

OPDAR [ENG] A laser system for measuring elevation angle, azimuth angle, and slant range of a missile during its firing period. Derived from optical direction and ranging. Also known as optical radar. { 'äp,där }

Opegraphaceae [BOT] A family of the Hysteriales characterized by elongated ascocarps; members are crustose on bark and rocks. { ,ō·pə·grə'fās·ə,ē }

open [ELEC] **1.** Condition in which conductors are separated so that current cannot pass. **2.** Break or discontinuity in a circuit which can normally pass a current. { 'ō·pən }

Open-Access Same-Time Information System [ELEC] An electronic system that uses Internet Web nodes to communicate to everyone in a fair and equitable manner information on available transmission capability and the cost of purchasing transmission services on the electric power transmission system, and allows for purchasing and reselling of transmission rights. Abbreviated OASIS. { ,ō·pən ¦ak,ses ¦sām ¦tīm ,in·fər'mā·shən ,sis·təm }

open ammunition space [ORD] Ground area prepared or improvised for storage of ammunition in open areas to supplement magazine space. { 'ō·pən ,am·yə'nish·ən ,spās }

open-angle glaucoma [MED] Bilateral, increased intraocular tension due to reduced aqueous outflow but with the angle open and the aqueous in free contact with the trabecula. { 'ō·pən ,aŋ·gəl glaù'kō·mə }

open arc [ASTRON] A crater arc in which the craters do not touch each other. { 'ō·pən 'ärk }

open-arc furnace [MET] An electrosmelting furnace in which the arc is generated above the level of the furnace feed. { 'ō·pən ¦ärk ,fər·nəs }

open architecture [COMPUT SCI] A computer architecture whose specifications are made widely available to allow third parties to develop add-on peripherals for it. { 'ō·pən 'ar·kə,tek·chər }

open association [ARCHEO] An assumed relationship between two or more objects that are found together, when it cannot be proved that they were deposited together. { 'ō·pən ə,sō·sē'ā·shən }

open ball [MATH] In a metric space, an open set about a point *x* which consists of all points that are less than a fixed distance from *x*. { 'ō·pən 'ból }

open bay [GEOGR] An indentation between two capes or headlands which is so broad and open that waves coming directly into it are nearly as high near its center as they are in adjacent parts of the open sea. { 'ō·pən 'bā }

open-belt drive [DES ENG] A belt drive having both shafts parallel and rotating in the same direction. { 'ō·pən ,belt ,drīv }

open berth [CIV ENG] An anchorage berth in an open roadstead. { 'ō·pən 'bərth }

open bundle [BOT] A vascular bundle containing cambium. { 'ō·pən 'bən·dəl }

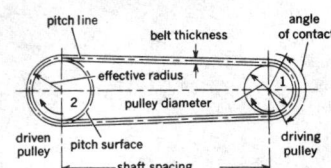

OPEN-BELT DRIVE

Diagram of open-belt drive; 1 and 2 indicate two pulleys whose shafts are parallel.

open-bus system [COMPUT SCI] A computer with an expansion bus that is designed to easily accept expansion boards. { ¦ō·pən 'bəs ,sis·təm }

open caisson [CIV ENG] A caisson in the form of a cylinder or shaft that is open at both ends; it is set in place, pumped dry, and filled with concrete. { 'ō·pən 'kā,sän }

open-cast mining *See* open-pit mining. { 'ō·pən ,kast 'mīn·iŋ }

open-cell foam [MATER] Foamed material, natural or synthetic, rigid or flexible, organic or metallic, in which there is interconnection between the cells. { 'ō·pən ,sel 'fōm }

open-center plan position indicator [ENG] A plan position indicator on which no signal is displayed within a set distance from the center. { 'ō·pən ,sen·tər 'plan pə,zish·ən 'in·də,kād·ər }

open chain [ASTRON] A crater chain in which the craters do not touch each other. { 'ō·pən 'chān }

open channel [SCI TECH] Any natural or artificial, covered or uncovered conduit in which liquid (usually water) flows with its top surface bounded by the atmosphere. { 'ō·pən 'chan·əl }

open-circle deoxyribonucleic acid *See* relaxed circular deoxyribonucleic acid. { 'ō·pən ,sər·kəl dē,äk·sē,rī·bō·nü'klē·ik 'as·əd }

open circuit [ELEC] An electric circuit that has been broken, so that there is no complete path for current flow. { 'ō·pən 'sər·kət }

open-circuited line [ELECTROMAG] A microwave discontinuity which reflects an infinite impedance. { 'ō·pən ¦sər·kəd·əd 'līn }

open-circuit grinding [MECH ENG] Grinding system in which material passes through the grinder without classification of product and without recycle of oversize lumps; in contrast to closed-circuit grinding. { 'ō·pən ¦sər·kət 'grīnd·iŋ }

open-circuit impedance [ELEC] Of a line or four-terminal network, the driving-point impedance when the far end is open. { 'ō·pən ¦sər·kət im'pēd·əns }

open-circuit jack [ELEC] Jack that normally leaves its circuit open; the circuit can be closed only by a circuit connected to the plug that is inserted in the jack. { 'ō·pən ¦sər·kət 'jak }

open-circuit potential [PHYS CHEM] Steady-state or equilibrium potential of an electrode in absence of external current flow to or from the electrode. { 'ō·pən ¦sər·kət pə'ten·chəl }

open-circuit scuba [ENG] The simplest type of scuba equipment, in which all exhaled gas is discharged directly into the water and the utilization of gas is therefore equal to the mass exhaled. { 'ō·pən ¦sər·kət 'skü·bə }

open-circuit signaling [COMMUN] Type of signaling in which no current flows while the circuit is in the idle condition. { 'ō·pən ¦sər·kət 'sig·nə·liŋ }

open-circuit voltage [ELEC] The voltage at the terminals of a source when no appreciable current is flowing. Also known as no-load voltage. { 'ō·pən ¦sər·kət 'vōl·tij }

open circular region [MATH] The interior of a circle. { ¦ō·pən 'sər·kyə·lər ,rē·jən }

open cluster [ASTRON] One of the groupings of stars that are concentrated along the central plane of the Milky Way; most have an asymmetrical appearance and are loosely assembled, and the stars are concentrated in their central region; they may contain from a dozen to many hundreds of stars. Also known as galactic cluster. { 'ō·pən 'kləs·tər }

open coast [GEOGR] A coast that is not sheltered from the sea. { 'ō·pən 'kōst }

open community [ECOL] A community which other organisms readily colonize because some niches are unoccupied. { 'ō·pən kə'myü·nəd·ē }

open covering [MATH] For a set *S* in a topological space, a collection of open sets whose union contains *S*. { 'ō·pən 'kəv·ər·iŋ }

opencut [CIV ENG] An open trench, such as across a hill. [MIN ENG] **1.** To drive headings out, or to commence working in the coal after sinking the shafts. **2.** To commence longwall working. **3.** To increase the size of a shaft when it intersects a drift. { 'ō·p·ən¦kət }

opencut mining *See* open-pit mining. { 'ō·p·ən¦kət 'mīn·iŋ }

open cycle [THERMO] A thermodynamic cycle in which new mass enters the boundaries of the system and spent exhaust leaves it; the automotive engine and the gas turbine illustrate this process. { ¦ō·pən 'sī·kəl }

open-cycle engine [MECH ENG] An engine in which the working fluid is discharged after one pass through boiler and engine. { ¦ō·pən ¦sī·kəl 'en·jən }

open-cycle gas turbine [MECH ENG] A gas turbine prime mover in which air is compressed in the compressor element, fuel is injected and burned in the combustor, and the hot products are expanded in the turbine element and exhausted to the atmosphere. { ¦ō·pən ¦sī·kəl 'gas 'tər·bən }

open-cycle reactor system [NUCLEO] A reactor system in which the coolant passes through the reactor core only once and is then discarded. { ¦ō·pən ¦sī·kəl rē'ak·tər ‚sis·təm }

open-delta connection [ELEC] An unsymmetrical transformer connection which is employed when one transformer of a bank of three single-phase delta-connected units must be cut out, because of failure. Also known as V connection. { ¦ō·pən ¦del·tə kə‚nek·shən }

open die [MET] A forming or forging die in which there is little or no restriction to the lateral flow of metal within the die set. { ¦ō·pən 'dī }

open-die forging [MET] Forging performed with open dies. { ¦ō·pən ¦dī 'fórj·iŋ }

open-ended [COMPUT SCI] Of techniques, designed to facilitate or permit expansion, extension, or increase in capability; the opposite of closed-in and artificially constrained. { ¦ō·pən ¦en·dəd }

open-ended class [STAT] The first or last class interval in a frequency distribution having no upper or lower limit. { ¦ō·pən ¦en·dəd 'klas }

open-ended system [COMPUT SCI] In character recognition, a system in which the input data to be read are derived from sources other than the computer with which the character reader is associated. { ¦ō·pən ¦en·dəd 'sis·təm }

open-end method [MIN ENG] A technique of mining pillars in which no stump is left. { ¦ō·pən ¦end 'meth·əd }

open-end wrench [DES ENG] A wrench consisting of fixed jaws at one or both ends of a handle. { ¦ō·pən ¦end 'rench }

open fault [GEOL] A fault, or section of a fault, whose two walls have become separated along the fault surface. { ¦ō·pən 'fólt }

open file [COMPUT SCI] A file that can be accessed for reading, writing, or both. { ¦ō·pən 'fīl }

open fire [MIN ENG] A fire at a roadway or at the coal face in a mine. { ¦ō·pən 'fīr }

open-flame arc [ELECTR] An electric arc which causes the anode to evaporate and be ejected as a flame. { ¦ō·pən ¦flām 'ärk }

open-flow capacity [PETRO ENG] Fluid flow rate from a gas well flowing open to the atmosphere. { ¦ō·pən ¦flō kə'pas·əd·ē }

open-flow potential [PETRO ENG] Gas flow rate in thousands of cubic feet of gas per 24 hours that would be produced by a well if the only pressure against the face of the producing formation wellbore were atmospheric pressure. { ¦ō·pən ¦flō pə'ten·chəl }

open-flow test [PETRO ENG] The flowing of wells wide open to the atmosphere with simultaneous measurement of gas-flow rate and pressure drop; used to analyze the potential of a reservoir to deliver gas to the wellbore. { ¦ō·pən ¦flō ‚test }

open-flow well [PETRO ENG] Gas well flowing open to the atmosphere. { ¦ō·pən ¦flō ‚wel }

open fold [GEOL] A fold having only moderately compressed limbs. { ¦ō·pən ¦fold }

open form [CRYSTAL] A crystal form in which the crystal faces do not entirely enclose a space. { ¦ō·pən ‚fórm }

open-fuse cutout [ELEC] Enclosed fuse cutout in which the fuse support and fuse holder are exposed. { ¦ō·pən ‚fyüz 'kə‚daut }

open half plane [MATH] A half plane that does not include any of the line that bounds it. { ¦ō·pən ¦haf 'plān }

open half space [MATH] A half space that does not include any of the plane that bounds it. { ¦ō·pən ¦haf 'spās }

open harbor [GEOGR] An unsheltered harbor exposed to the sea. { ¦ō·pən 'här·bər }

open-hearth furnace [MET] A reverberatory smelting furnace with a shallow hearth and a low roof, in which the charge is heated both by direct flame and by radiation from the roof and walls of the furnace. { ¦ōpən ¦härth 'fər·nəs }

open-hearth process [MET] A steel-making process carried out in an open-hearth furnace in which selected pig iron and malleable scrap iron are melted, with the addition of pure iron ore. { ¦ō·pən ¦härth ‚prä·səs }

open hole [ENG] **1.** A well or borehole, or a portion thereof, that has not been lined with steel tubing at the depth referred to. **2.** An unobstructed borehole. **3.** A borehole being drilled without cores. [PETRO ENG] In an oil-drilling operation, the unprotected hole which lies below the shoes of the last landed string of casing. { ¦ō·pən 'hōl }

open-hole completion [PETRO ENG] Preparation of an oil well without setting a production casing or liner opposite the producing formation so that reservoir fluids flow unrestricted into the open wellbore. Also known as barefoot completion. { ¦ō·pən ¦hōl kəm'plē·shən }

open ice [OCEANOGR] On navigable waters, ice that has broken apart sufficiently to permit passage of vessels. { ¦ō·pən 'īs }

open inflation [ASTRON] A version of the inflationary universe cosmology in which there is sufficient inflation to solve the horizon problem but not enough to result in a flat universe. { ‚ō·pən in'flā·shən }

opening [GEOGR] A break in a coastline or a passage between shoals, and so forth. [MIN ENG] **1.** A widening of a crevice, in consequence of a softening or decomposition of the adjacent rock, so as to leave a vacant space. **2.** A short heading driven between two or more parallel headings or levels for ventilation. **3.** An area in a coal mine between pillars, or between pillars and ribs. [OCEANOGR] Any break in sea ice which reveals the water. { 'ōp·ə·niŋ }

opening die [MECH ENG] A die head for cutting screws that opens automatically to release the cut thread. { 'ōp·ə·niŋ ‚dī }

opening material [MATER] A material added to plastic clay in ceramic making to speed drying and reduce shrinking. { 'ōp·ə·niŋ mə‚tir·ē·əl }

opening pressure [MECH ENG] The static inlet pressure at which discharge is initiated. { 'ōp·ə·niŋ ‚presh·ər }

open interval [MATH] An open interval of real numbers, denoted by (a,b), consists of all numbers strictly greater than a and strictly less than b. { ¦ō·pən 'in·tər·vəl }

open lake [HYD] **1.** A lake that has a stream flowing out of it. **2.** A lake whose water is free of ice or emergent vegetation. { ¦ō·pən 'lāk }

open lead [NAV] A navigable passage through ice or between rocks or shoals not covered with ice. { ¦ō·pən 'lēd }

open-link fuse [ELEC] A simple type of fuse that consists of a strip of fuse material bolted to open terminal blocks. { ¦ō·pən ¦liŋk 'fyüz }

open-loop control system [CONT SYS] A control system in which the system outputs are controlled by system inputs only, and no account is taken of actual system output. { ¦ō·pən ¦lüp kən'trōl ‚sis·təm }

open map [MATH] A function between two topological spaces which sends each open set of one to an open set of the other. { ¦ō·pən 'map }

open mapping theorem [MATH] A continuous linear function between Banach spaces which has closed range must be an open map. { ¦ō·pən 'map·iŋ ‚thir·əm }

open n-cell [MATH] A set that is homeomorphic with the set of points in n-dimensional Euclidean space (n = 1, 2, . . .) whose distance from the origin is less than unity. { ¦ō·pən en ‚sel }

open-packed structure [CRYSTAL] A crystal structure corresponding to the stacking of spheres in an orthogonal arrangement so that each sphere is in contact with six others. { ¦ō·pən ¦pakt 'strək·chər }

open pack ice [OCEANOGR] Floes of sea ice that are seldom in contact with each other, generally covering between four-tenths and six-tenths of the sea surface. { ¦ō·pən ¦pak ‚īs }

open-phase protection [ELEC] Effect of a device operating on the loss of current in one phase of a polyphase circuit to cause and maintain the interruption of power in the circuit. { ¦ō·pən ¦fāz prə'tek·shən }

open-phase relay [ELEC] Relay which functions by reason of the opening of one or more phases of a polyphase circuit, when sufficient current is flowing in the remaining phase or phases. { ¦ō·pən ¦fāz 'rē‚lā }

open-pit mining [MIN ENG] Extracting metal ores and minerals that lie near the surface by removing the overlying material and breaking and loading the ore. Also known as open-cast mining; opencut mining. { ¦ō·pən ¦pit 'mīn·iŋ }

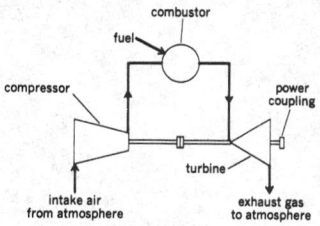

open-pit quarry [MIN ENG] A quarry in which the opening is the full size of the excavation. { ¦ō·pən ¦pit ¦kwär·ē }

open plan [BUILD] Arrangement of the interior of a building without distinct barriers such as partitions. { ¦ō·pən ¦plan }

open plug [ELEC] Plug designed to hold jack springs in their open position. { 'ō·pən ¦pləg }

open polygonal region [MATH] The interior of a polygon. { ¦ō·pən pə'lig·ən·əl ¦rē·jən }

open reading frame [MOL BIO] A stretch of triplets contained between an initator codon and a terminator codon. Abbreviated ORF. { ¦ō·pən ¦rēd·iŋ ¦frām }

open rectangular region [MATH] The interior of a rectangle. { ¦ō·pən rek'taŋ·gyə·lər ¦rē·jən }

open region *See* domain. { 'ō·pən ¦rē·jən }

open resonator *See* beam resonator. { 'ō·pən ¦rez·ən‚äd·ər }

open roadstead [NAV] An area near the shore where vessels anchor with relatively little protection from the sea. { 'ō·pən 'rōd‚sted }

open rock [GEOL] Any stratum sufficiently open or porous to contain a significant amount of water or to convey water along its bed. { 'ō·pən ¦räk }

open routine [COMPUT SCI] **1.** A routine which can be inserted directly into a larger routine without a linkage or calling sequence. **2.** A computer program that changes the state of a file from closed to open. { 'ō·pən rü‚tēn }

open sand [GEOL] A formation of sandstone that has porosity and permeability sufficient to provide good storage for oil. { 'ō·pən ¦sand }

open sea [GEOGR] **1.** That part of the ocean not enclosed by headlands, not within narrow straits, and so on. **2.** That part of the ocean outside the territorial jurisdiction of any country. { 'ō·pən 'sē }

open sentence *See* propositional function. { ¦ō·pən ¦sent·əns }

open set [MATH] A set included in a topology; equivalently, a set which is a neighborhood of each of its points; a topology on a space is determined by a collection of subsets which are called open. { 'ō·pən ¦set }

open shop [COMPUT SCI] A data-processing-center organization in which individuals from outside the data-processing community are permitted to implement their own solutions to problems. [IND ENG] A shop in which employment is not restricted to members of a labor union. { 'ō·pən ¦shäp }

open-side planer [DES ENG] A planer constructed with one upright or housing to support the crossrail and tools. { 'ō·pən ‚sīd 'plān·ər }

open-side tool block [DES ENG] A toolholder on a cutting machine consisting of a T-slot clamp, a C-shaped block, and two or more tool clamping screws. Also known as heavy-duty tool block. { 'ō·pən ‚sīd 'tül ‚bläk }

open sight [ORD] A rear gunsight having a notch. { 'ō·pən 'sīt }

open simplex [MATH] A modification of a simplex with vertices $p_0, p_1, \ldots, p_n$, in which the points of the simplex, $a_0p_0 + a_1p_1 + \cdots + a_np_n$, are restricted by the condition that each of the coefficients a_i must be greater than 0. { 'ō·pən 'sim‚pleks }

open source software [COMPUT SCI] Software that is written in such a way that others are encouraged to freely redistribute it, and all changes to the code must be made freely available. { ¦ō·pən ‚sors 'sòf‚twer }

open-space structure [GEOL] A structure in a carbonate sedimentary rock formed by a partial or complete occupation by internal sediments or cement. { 'ō·pən ‚späs 'strək·chər }

open standard [COMPUT SCI] Freely distributed. { ¦ō·pən 'stan·dərd }

open statement *See* propositional function. { ¦ō·pən 'stāt·mənt }

open stope [MIN ENG] Underground working place that is unsupported, or supported by timbers or pillars of rock. { 'ō·pən 'stōp }

open-stope method [MIN ENG] Stoping in which no regular artificial method of support is employed, although occasional props or cribs may be used to hold local patches of insecure ground. { ¦ō·pən ‚stōp 'meth·əd }

open storage [ORD] The storage of certain ordnance materiel outdoors. { 'ō·pən 'stòr·ij }

open subroutine [COMPUT SCI] A set of computer instructions that collectively perform some particular function and are inserted directly into the program each and every time that particular function is required. { 'ō·pən 'səb·rü‚tēn }

open system [COMPUT SCI] A computer system whose key software interfaces are specified, documented, and made publicly available. [HYD] A condition of freezing of the ground in which additional groundwater is available either through free percolation or through capillary movement. [THERMO] A system across whose boundaries both matter and energy may pass. { 'ō·pən 'sis·təm }

open-system architecture [COMPUT SCI] The structure of a computer network that allows different types of computers and peripheral devices from different manufacturers to be connected together. { 'ō·pən ¦sis·təm 'är·kə‚tek·chər }

open-timbered roof [BUILD] A roof in which the supporting timbers are left uncovered, forming part of the ceiling. { 'ō·pən ¦tim·bərd 'rüf }

open timbering [MIN ENG] A method of supporting the ground in a mine shaft or tunnel; supports are several feet apart, with the ground between them secured by struts. { 'ō·pən 'tim·bə·riŋ }

open traverse [ENG] A surveying traverse in which the last leg, because of error, does not terminate at the origin of the first leg. { 'ō·pən 'tra‚vərs }

open triangular region [MATH] The interior of a triangle. { ¦ō·pən trī'aŋ·gyə·lər ‚rē·jən }

open tuberculosis [MED] Tuberculosis in which tubercle bacilli are being discharged from the body; tuberculosis capable of transmission to other persons. { 'ō·pən tə‚bər·kyə'lō·səs }

open tubular column *See* capillary column. { 'ō·pən 'tü·byə·lər ‚käl·əm }

open universe [ASTRON] A cosmological model in which the volume of the universe is infinite and its expansion will continue forever. { 'ō·pən 'yü·nə‚vərs }

open valley [BUILD] A valley formed at the intersection of two roof surfaces and lined with either metal or a mineral-surfaced roofing material; the lining is exposed at the intersection. { 'ō·pən 'val·ē }

open water [ECOL] Lake water that is free from emergent vegetation, artificial obstructions, or tangled masses of underwater vegetation at very shallow depths. [HYD] Lake water that does not freeze during the winter. [OCEANOGR] Water less than one-tenth covered with floating ice. { 'ō·pən 'wòd·ər }

open-web girder *See* lattice girder. { 'ō·pən ¦web 'gərd·ər }

open well [CIV ENG] **1.** A well whose diameter is great enough (1 meter or more) for a person to descend to the water level. **2.** An artificial pond filling a large excavation in the zone of saturation up to the water table. { 'ō·pən 'wel }

open-window unit *See* sabin. { 'ō·pən ¦win·dō 'yü·nət }

open wire [ELEC] A conductor supported above the ground, separate from other conductors. { 'ō·pən 'wīr }

open-wire carrier system [COMMUN] A system for carrier telephony using an open-wire line. { 'ō·pən ¦wir 'kar·ē·ər ‚sis·təm }

open-wire feeder *See* open-wire transmission line. { 'ō·pən ¦wir 'fēd·ər }

open-wire loop [ELEC] Branch line on a main open-wire line. { 'ō·pən ¦wir 'lüp }

open-wire transmission line [ELEC] A transmission line consisting of two spaced parallel wires supported by insulators, at the proper distance to give a desired value of surge impedance. Also known as open-wire feeder. { 'ō·pən ¦wir tranz'mish·ən ‚līn }

open workings [MIN ENG] Surface workings, for example, a quarry or open-cast mine. { 'ō·pən 'wərk·iŋz }

opera glasses [OPTICS] Small binocular telescopes, usually of the Galilean type, adapted for use where magnification and field of view are secondary to compactness and cost. { 'äp·rə ‚glas·əz }

operand [COMPUT SCI] Any one of the quantities entering into or arising from an operation. { 'äp·ə‚rand }

operand-precision register [COMPUT SCI] A special register found in some minicomputers which can be programmed from 8- to 32-bit precision. { 'äp·ə‚rand pri'sizh·ən ‚rej·ə·stər }

operant conditioning [PSYCH] A form of learning in which the subject, in a given situation, tends to respond in a way that produces rewarding effects, reinforcing previous pleasurable experiences. Also known as instrumental conditioning; reinforcement conditioning. { 'äp·ə·rənt kən'dish·ə·niŋ }

operate time [COMPUT SCI] The phase of computer operation when an instruction is being carried out. [ELEC] Total elapsed time from application of energizing current to a relay coil to the time the contacts have opened or closed. { 'äp·ə,rāt ,tīm }

operating angle [ELECTR] Electrical angle of the input signal (for example, portion of a cycle) during which plate current flows in a vacuum tube amplifier. { 'äp·ə,rād·iŋ ,aŋ·gəl }

operating characteristic curve [STAT] In hypothesis testing, a plot of the probability of accepting the hypothesis against the true state of nature. Abbreviated OC curve. { 'äp·ə,rād·iŋ ,kar·ik·tə'ris·tik 'kərv }

operating delay [COMPUT SCI] Computer time lost because of mistakes or inefficiency of operating personnel or users of the system, excluding time lost because of defects in programs or data. { 'äp·ə,rād·iŋ di,lā }

operating handle [ORD] Handle or bar with which the operating lever of a gun is operated to open and close the breech. { 'äp·ə,rād·iŋ ,han·dəl }

operating instructions [COMPUT SCI] A detailed description of the actions that must be carried out by a computer operator in running a program or group of interrelated programs, usually included in the documentation of a program supplied by a programmer or systems analyst, along with the source program and flow charts. { 'äp·ə,rād·iŋ in,strək·shənz }

operating lever [ORD] Lever device on a gun with which the breech is opened and closed. { 'äp·ə,rād·iŋ ,lev·ər }

operating line [CHEM ENG] In the graphical solution of equilibrium processes (such as distillation absorption extraction), the actual liquid-vapor relationship of a key component, in contrast to a true equilibrium relationship. { 'äp·ə,rād·iŋ ,līn }

operating point [ELECTR] Point on a family of characteristic curves of a vacuum tube or transistor where the coordinates of the point represent the instantaneous values of the electrode voltages and currents for the operating conditions under study or consideration. { 'äp·ə,rād·iŋ ,point }

operating position [COMMUN] Terminal of a communications channel which is attended by an operator; usually the term refers to a single operator, such as a radio operator's position or a telephone operator's position; however, certain terminals may require more than one operating position. { 'äp·ə,rād·iŋ pə,zish·ən }

operating power [ELECTROMAG] Power that is actually supplied to a radio transmitter antenna. { 'äp·ə,rād·iŋ ,paù·ər }

operating pressure [ENG] The system pressure at which a process is operating. { 'äp·ə,rād·iŋ ,presh·ər }

operating range [ELECTR] The frequency range over which a reversible transducer is operable. [NAV] The maximum distance at which reliable service is provided by an aid to navigation. { 'äp·ə,rād·iŋ ,rānj }

operating ratio [COMPUT SCI] The time during which computer hardware operates and gives reliable results divided by the total time scheduled for computer operation. { 'äp·ə,rād·iŋ ,rā·shō }

operating slide [ORD] Mechanism in a Browning machine gun that permits opening the breech for loading, unloading, and clearing out stoppages, and closing the breech for firing. { 'äp·ə,rād·iŋ ,slīd }

operating stress [MECH] The stress to which a structural unit is subjected in service. { 'äp·ə,rād·iŋ ,stres }

operating system [COMPUT SCI] A set of programs and routines which guides a computer or network in the performance of its tasks, assists the programs (and programmers) with certain supporting functions, and increases the usefulness of the computer or network hardware. { 'äp·ə,rād·iŋ ,sis·təm }

operating system supervisor [COMPUT SCI] The control program of a set of programs which guide a computer in the performance of its tasks and which assist the program with certain supporting functions. { 'äp·ə,rād·iŋ ,sis·təm 'sü·pər,vīz·ər }

operating water level [MECH ENG] The water level in a boiler drum which is normally maintained above the lowest safe level. { 'äp·ə,rād·iŋ 'wòd·ər ,lev·əl }

operation [COMPUT SCI] **1.** A process or procedure that obtains a unique result from any permissible combination of operands. **2.** The sequence of actions resulting from the execution of one digital computer instruction. [IND ENG] A job,

usually performed in one location, and consisting of one or more work elements. [MATH] An operation of a group G on a set S is a mapping which associates to each ordered pair (g,s), where g is in G and s is in S, another element in S, denoted gs, such that, for any g,h in G and s in S, $(gh)s = g(hs)$, and $es = s$, where e is the identity element of G. { ,äp·ə'rā·shən }

operational [ENG] Of equipment such as aircraft or vehicles, being in such a state of repair as to be immediately usable. { äp·ə'rā·shən·əl }

operational advantage [NAV] In air operations, an improvement which benefits the users of an instrument procedure, such as the ability to use lower minimums. { ,äp·ə'rā·shən·əl əd'van·tij }

operational amplifier [ELECTR] An amplifier having high direct-current stability and high immunity to oscillation, generally achieved by using a large amount of negative feedback; used to perform analog-computer functions such as summing and integrating. { ,äp·ə'rā·shən·əl 'am·plə,fī·ər }

operational analysis See operational calculus. { ,äp·ə'rā·shən·əl ə'nal·ə·səs }

operational calculus [MATH] A technique by which problems in analysis, in particular differential equations, are transformed into algebraic problems, usually the problem of solving a polynomial equation. Also known as operational analysis. { ,äp·ə'rā·shən·əl 'kal·kyə·ləs }

operational game See management game. { ,äp·ə'rā·shən·əl 'gām }

operational label [COMPUT SCI] A combination of letters and digits at the beginning of the tape which uniquely identify the tape required by the system. { ,äp·ə'rā·shən·əl 'lā·bəl }

operational maintenance [ENG] The cleaning, servicing, preservation, lubrication, inspection, and adjustment of equipment; it includes that minor replacement of parts not requiring high technical skill, internal alignment, or special locative training. { ,äp·ə'rā·shən·əl 'mānt·ən·əns }

operational standby program [COMPUT SCI] The program operating in the standby computer when in the duplex mode of operation. { ,äp·ə'rā·shən·əl 'stand,bī ,prō,gram }

operational unit [GEOL] An arbitrary stratigraphic unit that is distinguished by objective criteria for some practical purpose. Also known as parastratigraphic unit. { ,äp·ə'rā·shən·əl 'yü·nət }

operational weather limits [METEOROL] The limiting values of ceiling, visibility, and wind, or runway visual range, established as safety minima for aircraft landings and takeoffs. { ,äp·ə'rā·shən·əl 'weth·ər ,lim·əts }

operation analysis [IND ENG] An analysis of all procedures concerned with the design or improvement of production, the purpose of the operation, inspection standards, materials used and the manner of handling them, the setup, tool equipment, and working conditions and methods. { ,äp·ə'rā·shən ə'nal·ə·səs }

operation analysis chart [IND ENG] A form that lists all the essential factors influencing the effectiveness of an operation. { ,äp·ə'rā·shən ə'nal·ə·səs ,chärt }

operation breakdown See job breakdown. { ,äp·ə'rā·shən 'brāk,daùn }

operation code [COMPUT SCI] A field or portion of a digital computer instruction that indicates which action is to be performed by the computer. Also known as command code. { ,äp·ə'rā·shən ,kōd }

operation cycle [COMPUT SCI] The portion of a memory cycle required to perform an operation; division and multiplication usually require more than one memory cycle to be completed. { ,äp·ə'rā·shən ,sī·kəl }

operation decoder [COMPUT SCI] A device that examines the operation contained in an instruction of a computer program and sends signals to the circuits required to carry out the operation. { ,äp·ə'rā·shən dē'kōd·ər }

operation number [COMPUT SCI] **1.** Number designating the position of an operation, or its equivalent subroutine, in the sequence of operations composing a routine. **2.** Number identifying each step in a program stated in symbolic code. { ,äp·ə'rā·shən ,nəm·bər }

operation part [COMPUT SCI] That portion of a digital computer instruction which is reserved for the operation code. { ,äp·ə'rā·shən ,pärt }

operation process chart [IND ENG] A graphic representation that gives an overall view of an entire process, including the points at which materials are introduced, the sequence of inspections, and all operations not involved in material handling. { ,äp·ə'rā·shən 'prä·səs ,chärt }

operation register [COMPUT SCI] A register used to store and decode the operation code for the next instruction to be carried out by a computer. { ,äp·ə'rā·shən ,rej·ə·stər }

operations research [MATH] The mathematical study of systems with input and output from the viewpoint of optimization subject to given constraints. [SCI TECH] The application of objective and quantitative criteria to decision making previously undertaken by empirical methods. { ,äp·ə·rā·shənz ri,sərch }

operations sequence [CONT SYS] The logical series of procedures that constitute the task for a robot. { ,äp·ə'rā·shənz ,sē·kwəns }

operation time [COMPUT SCI] The time elapsed during the interpretation and execution of an arithmetic or logic operation by a computer. { ,äp·ə'rā·shən ,tīm }

operative ankylosis See arthrodesis. { 'äp·rəd·iv ,aŋ·kə'lō·səs }

operator [COMPUT SCI] Anything that designates an action to be performed, especially the operation code of a computer instruction. [ENG] A person whose duties include the operation, adjustment, and maintenance of a piece of equipment. [GEN] A sequence at one end of an operon on which a repressor acts, thus regulating the transcription of the operon. [MATH] A function between vector spaces. { 'äp·ə,rād·ər }

operator algebra [MATH] An algebra whose elements are functions and in which the multiplication of two elements f and g is defined by composition; that is, $(fg)(x) = (f \circ g)(x) = f[g(x)]$. { 'äp·ə,rād·ər ,al·jə·brə }

operator hierarchy [COMPUT SCI] A sequence of mathematical operators which designates the order in which these operators are to be applied to any mathematical expression in a given programming language. { 'äp·ə,rād·ər 'hī·ər,är·kē }

operator interrupt [COMPUT SCI] A step whereby control is passed to the monitor, and a message, usually requiring a typed answer, is printed on the console typewriter. { 'äp·ə,rād·ər 'in·tə,rəpt }

operator process chart [IND ENG] A chart of the time relationship of the movements made by the body members of a workman performing an operation. { 'äp·ə,rād·ər 'prä·səs ,chärt }

operator productivity [IND ENG] The ratio of standard hours to actual hours for a given task. { 'äp·ə,rād·ər ,präd·ək'tiv·əd·ē }

operator's console [COMPUT SCI] Equipment which provides for manual intervention and monitoring computer operation. { 'äp·ə,rād·ərz 'kän,sōl }

operator subgoaling [COMPUT SCI] A computer problem-solving method in which the inability of the computer to take the desired next step at any point in the problem-solving process leads to a subgoal of making that step feasible. { 'äp·ə,rād·ər ,səb'gōl·iŋ }

operator theory [MATH] The general qualitative study of operators in terms of such concepts as eigenvalues, range, domain, and continuity. { 'äp·ə,rād·ər ,thē·ə·rē }

operator training [IND ENG] The process used to prepare the employee to make his expected contribution to his employer, usually involving the teaching of specialized skills. { 'äp·ə,rād·ər ,trān·iŋ }

operator utilization [IND ENG] The ratio of working time to total clock time; a ratio of 1.00 (or 100) indicates full utilization of the operator's work time. { 'äp·ə,rād·ər ,yüd·əl·ə'zā·shən }

operculum [ANAT] **1.** The soft tissue partially covering the crown of an erupting tooth. **2.** That part of the cerebrum which borders the lateral fissure. [BIOL] **1.** A lid, flap, or valve. **2.** A lidlike body process. { ō'pər·kyə·ləm }

operon [GEN] A functional unit composed of a number of adjacent cistrons on the chromosome; its transcription is regulated by a receptor sequence, the operator, and a repressor. { 'äp·ə,rän }

operon network [GEN] A group of operons and their associated regulator genes that interact such that the products of one operon activate or suppress another operon. { 'äp·ə,rän ,net,wərk }

Ophelia [ASTRON] A satellite of Uranus orbiting at a mean distance of 33,400 miles (53,760 kilometers) with a period

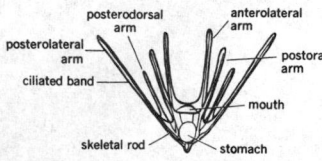

OPHIOPLUTEUS

posterodorsal arm — anterolateral arm — posterolateral arm — postoral arm — ciliated band — mouth — skeletal rod — stomach

Bilateral symmetrical larva of the brittle stars.

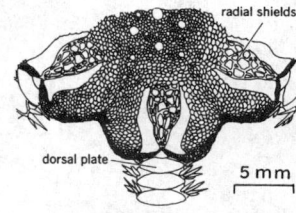

OPHIUROIDEA

radial shields — dorsal plate — 5 mm

Pectinura cylindrica, ophiuroid showing central disk and a part of one of the arms with the dorsal plate showing.

of 9 hours 3 minutes, and a diameter of about 20 miles (32 kilometers); the outer shepherding satellite for the outermost ring of Uranus. { ō'fēl·yə }

Opheliidae [INV ZOO] A family of limivorous worms belonging to the annelid group Sedentaria. { äf·ə'lī·ə,dē }

Ophiacodonta [VERT ZOO] A suborder of extinct reptiles in the order Pelycosauria, including primitive, partially aquatic carnivores. { ,af·ē·ə·kə'dän·tə }

ophicalcite [PETR] A recrystallized limestone composed of calcite and serpentine and formed by dedolomitization of a siliceous dolomite. { 'äf·ə'kal,sīt }

Ophidiidae [VERT ZOO] A family of small actinopterygian fishes in the order Gadiformes, comprising the cusk eels and brotulas. { ,äf·ə'dī·ə,dē }

ophidiophobia [PSYCH] An abnormal fear of snakes. { ä,fid·ē·ō'fō·bē·ə }

Ophiocanopidae [INV ZOO] A family of asterozoan echinoderms in the subclass Ophiuroidea. { ,äf·ē·ō·kə'näp·ə,dē }

Ophiocistioidea [PALEON] A small class of extinct Echinozoa in which the domed aboral surface of the test was roofed by polygonal plates and carried an anal pyramid. { ,äf·ē·ō,sis·tē'óid·ē·ə }

Ophioglossales [BOT] An order of ferns in the subclass Ophioglossidae. { ,äf·ē·ō·glä'sä·lēz }

Ophioglossidae [BOT] The adder's-tongue ferns, a small subclass of the class Polypodiopsida; the plants are homosporous and eusporangiate and are distinguished by the arrangement of the sporogenous tissue in the characteristic fertile spike of the sporophyte. { ,äf·ē·ō'gläs·ə,dē }

ophiolite [PETR] A distinctive assemblage of mafic plus ultramafic rocks, generally considered to be fragments of oceanic lithosphere that have been tectonically emplaced onto continental margins and island arcs. { 'äf·ē·ə,līt }

ophiolitic eclogite [PETR] Any of the eclogites which are products of early orogenic volcanism and which by later metamorphism transformed into rocks of the high-pressure facies series. { ,äf·ē·ə,lid·ik 'ek·lə,jīt }

Ophiomyxidae [INV ZOO] The single family of the echinoderm suborder Ophiomyxina distinguished by a soft, unprotected integument. { ,äf·ē·ō'mik·sə,dē }

Ophiomyxina [INV ZOO] A monofamilial suborder of ophiuroid echinoderms in the order Phrynophiurida. { ,äf·ē·ō'mik·sə·nə }

ophiopluteus [INV ZOO] The pluteus larva of brittle stars. { ,äf·ē·ō'plüd·ē·əs }

ophite [PETR] A diabase in which the ophitic structure is retained even though the pyroxene is altered to uralite. { 'ä,fīt }

ophitic [PETR] Of the holocrystalline, hypidiomorphic-granular texture of an igneous rock, exhibiting lath-shaped plagioclase crystals partly or wholly included within pyroxene crystals. { ä'fid·ik }

Ophiucus [ASTRON] A large constellation centered near right ascension 17 hours, declination 10° south; it includes a portion of the ecliptic between the constellations Scorpius and Sagittarius, although it is not considered a constellation of the zodiac. Also known as Serpent Bearer. { 'äf·i,ÿ·kəs }

Ophiurida [INV ZOO] An order of echinoderms in the subclass Ophiuroidea in which the vertebrae articulate by means of ball-and-socket joints, and the arms, which do not branch, move mainly from side to side. { ,äf·ē'yùr·ə·də }

Ophiuroidea [INV ZOO] The brittle stars, a subclass of the Asterozoa in which the arms are usually clearly demarcated from the central disk and perform whiplike locomotor movements. { äf·ē·yə'róid·ē·ə }

ophryon [ANTHRO] The point where the sagittal plane intersects an arc drawn horizontally from the frontotemporalis across the frontal bone. { 'äf·rē,än }

ophthalmectomy [MED] Excision, or enucleation, of the eye. { ,äf·thəl'mek·tə·mē }

ophthalmia [MED] Inflammation of the eye, especially involving the conjunctiva. { äf'thal·mē·ə }

ophthalmia neonatorum [MED] Inflammation of the eyes in the newborn contracted during passage through the birth canal; may be gonorrheal or purulent. { äf'thal·mē·ə ,nē·ə·nə'tór·əm }

ophthalmia nodosa [MED] Inflammation of the eye due to lodging of caterpillar hairs in the conjunctiva, cornea, or iris. { äf'thal·mē·ə nō'dō·sə }

ophthalmic [ANAT] Of or pertaining to the eye. { äf'thal·mik }

ophthalmic nerve [NEUROSCI] A sensory branch of the trigeminal nerve which supplies the lacrimal glands, upper eyelids, skin of the forehead, and anterior portion of the scalp, meninges, nasal mucosa, and frontal, ethmoid, and sphenoid air sinuses. { äf'thal·mik 'nərv }

ophthalmodynamometer [MED] An instrument which measures the pressure necessary to collapse the retinal arteries. { äf¦thal·mō,dī·nə'mäm·əd·ər }

ophthalmological [PHARM] Any drugs used in the treatment of eye disease. { äf¦thal·mə¦läj·ə·kəl }

ophthalmologist [MED] A physician who specializes in ophthalmology. Also known as oculist. { ¸äf,thal'mäl·ə·jəst }

ophthalmology [MED] The study of the anatomy, physiology, and diseases of the eye. { ¸äf,thal'mäl·ə·jē }

ophthalmomalacia [MED] Abnormal softness or subnormal tension of the eye. { äf¦thal·mə¦lā·shə }

ophthalmometer [OPTICS] **1.** An instrument for measuring refractive errors, especially astigmatism. **2.** An instrument for measuring the capacity of the chamber of the eye. **3.** An instrument for measuring the eye as a whole. { ¸äf·thal'mäm·əd·ər }

ophthalmorrhexis [MED] Rupture of the eyeball. { äf¦thal·mə¦rek·səs }

ophthalmoscope [OPTICS] An instrument, consisting essentially of a concave mirror with a hole in it and fitted with lenses of different powers, for examining the interior of the eye through the pupil. { äf'thal·mə,skōp }

opianyl *See* meconin. { 'ō·pē·ə,nil }

opiate [PHARM] **1.** A sleep-inducing drug. **2.** Any narcotic. **3.** An opium preparation. **4.** Any tranquilizing agent. { 'ō·pē·ət }

Opilioacaridae [INV ZOO] The single family of moderately large mites of the suborder Notostigmata which comprises the Opiliocariformes. { ō¦pil·ē·ō'kar·ə,dē }

Opilioacariformes [INV ZOO] A small monofamilial order of the Acari comprising large mites characterized by long legs and by the possession of a pretarsus on the pedipalp, with prominent claws. { ō¦pil·ē·ō,kar·ə'för,mēz }

opine [BIOCHEM] A type of amino acid usually not found in nature, such as that secreted by a crown gall. { 'ō,pīn }

opisometer [ENG] An instrument for measuring the length of curved lines, such as those on a map; a wheel on the instrument is traced over the line. { ¸äp·ə'säm·əd·ər }

opisthaptor [INV ZOO] A posterior adhesive organ in monogenetic trematodes. { ¸äp·əs¦thap·tər }

Opisthobranchia [INV ZOO] A subclass of the class Gastropoda containing the sea hares, sea butterflies, and sea slugs; generally characterized by having gills, a small external or internal shell, and two pairs of tentacles. { ə,pis·thə'braŋ·kē·ə }

Opisthocoela [VERT ZOO] A suborder of the order Anura; members have opisthocoelous trunk vertebrae, and the adults typically have free ribs. { ə,pis·thə'sē·lə }

opisthocoelous [ANAT] Of, related to, or being a vertebra with the centrum convex anteriorly and concave posteriorly. { ə¦pis·thə,sē·ləs }

Opisthocomidae [VERT ZOO] A family of birds in the order Galliformes, including the hoatzins. { ə,pis·thə'käm·ə,dē }

opisthocranion [ANTHRO] The point, wherever it may lie in the sagittal plane on the occipital bone, which marks the posterior extremity of the longest diameter of the skull, measured from the glabella. { ə,pis·thə'krā·nē,än }

opisthognathous [INV ZOO] Having the mouthparts ventral and posterior to the cranium. [VERT ZOO] Having retreating jaws. { ¦ä·pəs¦thäg·nə·thəs }

opisthonephros [VERT ZOO] The fundamental adult kidney in amphibians and fishes. { ə,pis·thə'ne,fräs }

Opisthopora [INV ZOO] An order of the class Oligochaeta distinguished by meganephridiostomal, male pores opening posteriorly to the last testicular segment. { ¦ä·pəs¦thäp·ə·rə }

opisthotic [ANAT] Of, relating to, or being the posterior and inferior portions of the bony elements in the inner ear capsule. { ¦ä·pəs¦thäd·ik }

opisthotonus [MED] A condition, caused by a tetanic spasm of the back muscles, in which the trunk is arched forward while the head and lower limbs are bent backward. { ¸ap·əs'thät·ən·əs }

opium [PHARM] A narcotic obtained from the unripe capsules of the opium poppy (*Papaver somniferum*); crude extract contains alkaloids such as morphine (5–15%), narcotine (2–8%), and codeine (0.1–2.5%). { 'ō·pē·əm }

opoka [PETR] A porous, flinty, and calcareous sedimentary rock, with conchoidal or irregular fracture, consisting of fine-grained opaline silica (up to 90%), and hardened by the presence of silica of organic origin. { ō'päk·ə }

Opomyzidae [INV ZOO] A family of cyclorrhaphous myodarian dipteran insects in the subsection Acalypteratae. { ¸äp·ə'mī·zə,dē }

opossum [VERT ZOO] Any member of the family Didelphidae in the order Marsupialia; these mammals are arboreal and mainly omnivorous, and have many incisors, with all teeth pointed and sharp. { ə'päs·əm }

Oppenauer oxidation [ORG CHEM] The oxidation of a primary or secondary hydroxyl compound to form the corresponding carbonyl compound; aluminum alkoxide and an excess amount of a carbonyl hydrogen acceptor, such as benzophenone or acetone, are required. { 'äp·ə,naù·ər ¸äk·sə'dā·shən }

Oppenheimer-Phillips reaction [NUC PHYS] A type of stripping reaction which can occur when a deuteron passes near a nucleus, in which the proton in the deuteron experiences Coulomb repulsion from the nucleus while the neutron is attracted to the nucleus by nuclear forces, with the result that the neutron-proton bond in the deuteron is broken, the neutron is absorbed into the nucleus, and the proton is repelled. { 'äp·ə,naù·ər 'fil·əps rē,ak·shən }

Oppenheimer-Volkoff limit [ASTRON] The upper limit on the mass of a neutron star, above which there is no stable equilibrium configuration and it is predicted that matter will collapse into a black hole. { 'äp·ən,hī·mər 'fōl,kóf ,lim·ət }

Oppenheim's disease *See* amyotonia congenita. { 'äp·ən,hīmz di,zēz }

opponent-colors theory [OPTICS] A theory of color vision according to which various processes in the visual system are capable of responding in two opposite ways; the Hering theory is an example. { ə'pō·nənt 'kəl·ərz ,thē·ə·rē }

opportunistic microorganism [MICROBIO] A normally harmless endogenous microorganism that produces disease due to fortuitous events that affect the host. { ¦äp·ər,tü¦nis·tik ¦mī·krō'ór·gə,niz·əm }

opportunistic species [ECOL] Species characterized by high reproduction rates, rapid development, early reproduction, small body size, and uncertain adult survival. { ¦äp·ər,tü¦nis·tik 'spē·shēz }

opposed engine [MECH ENG] A reciprocating engine having the pistons on opposite sides of the crankshaft, with the piston strokes on each side working in a direction opposite to the direction of the strokes on the other side. { ə'pōzd 'en·jən }

opposing wind [OCEANOGR] In wave forecasting, a wind blowing in opposition to the direction that the waves are traveling. { ə'pōz·iŋ 'wind }

opposite [BOT] **1.** Located side by side. **2.** Of leaves, being in pairs on an axis with each member separated from the other of the pair by half the circumference of the axis. { 'äp·ə·zət }

oppositely congruent figures [MATH] Two solid figures, one of which can be made to coincide with the other by a rigid motion in space combined with reflection through a plane. { 'äp·ə·zət·lē ¦kän,grü·ənt 'fig·yərz }

opposite rays [MATH] Two rays that lie on the same or parallel lines but point in opposite directions. { 'äp·ə·zət 'rāz }

opposite side [MATH] **1.** One of two sides of a polygon with an even number of sides that have the same number of sides between them along either path around the polygon from one of the sides to the other. **2.** For a given vertex of a polygon with an odd number of sides, a side of the polygon that has the same number of sides between it and the vertex along either path around the polygon. { 'äp·ə·zət 'sīd }

opposite tide [OCEANOGR] A high tide at a corresponding place on the opposite side of the earth which accompanies a direct tide. { 'äp·ə·zət 'tīd }

opposite vertices [MATH] Two vertices of a polygon with an even number of sides that have the same number of sides

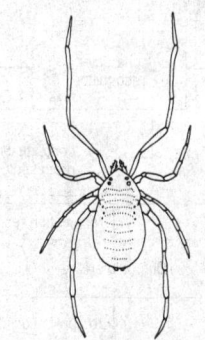

OPILIOACARIDAE

An opilioacarid mite, dorsal view.

OPOSSUM

Common opossum.

between them along either path around the polygon from one vertex to the other.　{ 'äp·ə·zət 'vərd·ə,sēz }

opposition [ASTRON] The situation of two celestial bodies having either celestial longitudes or sidereal hour angles differing by 180°; the term is usually used only in relation to the position of a superior planet or the moon with reference to the sun. [PHYS] The condition in which the phase difference between two periodic quantities having the same frequency is 180°, corresponding to one half-cycle.　{ ,äp·ə'zish·ən }

opsonic action [IMMUNOL] The effect produced upon susceptible microorganisms and other cells by opsonins, which renders them vulnerable to phagocytes.　{ äp'sän·ik 'ak·shən }

opsonic index [IMMUNOL] A numerical measure of the opsonic activity of sera, expressed as the ratio of the average number of bacteria engulfed per phagocytic cell in immune serum compared with the corresponding value for normal serum.　{ äp'sän·ik 'in,deks }

opsonin [IMMUNOL] A substance in blood serum that renders bacteria more susceptible to phagocytosis by leukocytes.　{ 'äp·sə·nən }

opsonize [IMMUNOL] To render microorganisms susceptible to phagocytosis.　{ 'äp·sə,nīz }

-opsy [MED] A combining form denoting examination, or denoting a condition of vision.　{ 'äp·sē }

optic [BIOL] Pertaining to the eye. [OPTICS] Pertaining to the lenses, prisms, and mirrors of a camera, microscope, or other conventional optical instrument.　{ 'äp·tik }

optical [OPTICS] Pertaining to or utilizing visible or near-visible light; the extreme limits of the optical spectrum are about 100 nanometers (0.1 micrometer or 3×10^{15} hertz) in the far ultraviolet and 30,000 nanometers (30 micrometers or 10^{13} hertz) in the far infrared.　{ 'äp·tə·kəl }

optical aberration [OPTICS] Deviation from perfect image formation by an optical system; examples are spherical aberration, coma, astigmatism, curvature of field, distortion, and chromatic aberration. Also known as aberration.　{ 'äp·tə·kəl ,ab·ə'rā·shən }

optical achromatism See visual achromatism.　{ 'äp·tə·kəl ā'krō·mə,tiz·əm }

optical activity [OPTICS] The behavior of substances which rotate the plane of polarization of plane-polarized light, as it passes through them. Also known as rotary polarization.　{ 'äp·tə·kəl ak'tiv·əd·ē }

optical air mass [GEOPHYS] A measure of the length of the path through the atmosphere to sea level traversed by light rays from a celestial body, expressed as a multiple of the path length for a light source at the zenith.　{ 'äp·tə·kəl 'er ,mas }

optical amplifier [ENG] An optoelectronic amplifier in which the electric input signal is converted to light, amplified as light, then converted back to an electric signal for the output.　{ 'äp·tə·kəl 'am·plə,fī·ər }

optical analysis [OPTICS] Study of properties of a substance or medium, such as its chemical composition or the size of particles suspended in it, through observation of effects on transmitted light, such as scattering, absorption, refraction, and polarization.　{ 'äp·tə·kəl ə'nal·ə·səs }

optical anisotropy [OPTICS] The behavior of a medium, or of a single molecule, whose effect on electromagnetic radiation depends on the direction of propagation of the radiation.　{ 'äp·tə·kəl ,an·ə'sä·trə·pē }

optical anomaly [PHYS CHEM] The phenomenon in which an organic compound has a molar refraction which does not agree with the value calculated from the equivalents of atoms and other structural units composing it.　{ 'äp·tə·kəl ə'näm·ə·lē }

optical antipode See enantiomorph.　{ 'äp·tə·kəl 'ant·i,pōd }

optical aspherical surface [OPTICS] An optical surface that does not form part of a sphere, such as a paraboloidal or ellipsoidal surface.　{ 'äp·tə·kəl ā'sfer·ə·kəl 'sər·fəs }

optical axis [ANAT] An imaginary straight line passing through the midpoint of the cornea (anterior pole) and the midpoint of the retina (posterior pole). [OPTICS] **1.** A line passing through a radially symmetrical optical system such that rotation of the system about this line does not alter it in any detectable way. **2.** See optic axis.　{ 'äp·tə·kəl 'ak·səs }

optical bar-code reader [COMPUT SCI] A device which uses any of various photoelectric methods to read information which

has been coded by placing marks in prescribed boxes on documents with ink, pencil, or other means.　{ 'äp·tə·kəl 'bär ,kōd ,rēd·ər }

optical bench [ENG] A rigid horizontal bar or track for holding optical devices in experiments; it allows device positions to be changed and adjusted easily.　{ 'äp·tə·kəl 'bench }

optical bistability [OPTICS] The property of a substance or device which has two stable states of transmission, high or low, for a single input light intensity.　{ 'äp·tə·kəl ,bī·stə'bil·əd·ē }

optical branch [SOLID STATE] The vibrations of an optical mode plotted on a graph of frequency versus wave number; it is separated from, and has higher frequencies than, the acoustic branch.　{ 'äp·tə·kəl 'branch }

optical calcite [MINERAL] The type of calcite used to make Nicol prisms.　{ 'äp·tə·kəl 'kal,sīt }

optical center [OPTICS] A point on the axis of a lens so that, for any ray passing through this point, the incident part and the emergent part are parallel. Also known as pole.　{ 'äp·tə·kəl 'sen·tər }

optical character recognition [COMPUT SCI] That branch of character recognition concerned with the automatic identification of handwritten or printed characters by any of various photoelectric methods. Abbreviated OCR. Also known as electrooptical character recognition.　{ 'äp·tə·kəl 'kar·ik·tər ,rek·ig,nish·ən }

optical coating [OPTICS] Either a mirror coating, or a film of the proper thickness and refractive index applied to the air-glass surface of a lens to reduce reflection.　{ 'äp·tə·kəl 'kōd·iŋ }

optical coherence microscopy [OPTICS] A variation of optical coherence tomography which uses a system of high numerical aperture to achieve resolutions comparable to that of confocal microscopy.　{ 'äp·tə·kəl kō,hir·əns mī'kräs·kə·pē }

optical coherence tomography [OPTICS] A noninvasive technique for imaging subsurface tissue structure with micrometer-scale resolution, which is based on a broadband light source and a fiber-optic Michelson interferometer.　{ 'äp·tə·kəl kō,hir·əns tə'mäg·rə·fē }

optical communication [COMMUN] The use of electromagnetic waves in the region of the spectrum near visible light for the transmission of signals representing speech, pictures, data pulses, or other information, usually in the form of a laser beam modulated by the information signal.　{ 'äp·tə·kəl kə,myü·nə'kā·shən }

optical comparator [ENG] Any comparator in which movement of a measuring plunger tilts a small mirror which reflects light in an optical system. Also known as visual comparator.　{ 'äp·tə·kəl kəm'par·əd·ər }

optical computer [COMPUT SCI] A computer that uses various combinations of holography, lasers, and mass-storage memories for such applications as ultra-high-speed signal processing, image deblurring, and character recognition.　{ 'äp·tə·kəl kəm'pyüd·ər }

optical contact [OPTICS] Contact between two surfaces in which the surfaces are separated by a distance much less than a wavelength of light, so that interference fringes are not formed.　{ 'äp·tə·kəl 'kän,takt }

optical coupler See optoisolator.　{ 'äp·tə·kəl 'kəp·lər }

optical coupling [ELECTR] Coupling between two circuits by means of a light beam or light pipe having transducers at opposite ends, to isolate the circuits electrically.　{ 'äp·tə·kəl 'kəp·liŋ }

optical crystal [CRYSTAL] Any natural or synthetic crystal, such as sodium chloride, calcium fluoride, silver chloride, potassium iodide, or stilbene, that is used in infrared and ultraviolet optics and for its piezoelectric effects.　{ 'äp·tə·kəl 'krist·əl }

optical cytophotometry See microelectrophoresis.　{ 'äp·tə·kəl ,sī·dō·fə'räm·ə·trē }

optical data storage [COMPUT SCI] The technology of placing information in a medium so that, when a light beam scans the medium, the reflected light can be used to recover the information.　{ 'äp·tə·kəl 'dad·ə ,stór·ij }

optical density [OPTICS] The degree of opacity of a translucent medium expressed by log I_0/I, where I_0 is the intensity of the incident ray, and I is the intensity of the transmitted ray. Abbreviated OD.　{ 'äp·tə·kəl 'den·səd·ē }

optical depth See optical thickness.　{ 'äp·tə·kəl 'depth }

optical detection of magnetic resonance [SPECT] A type

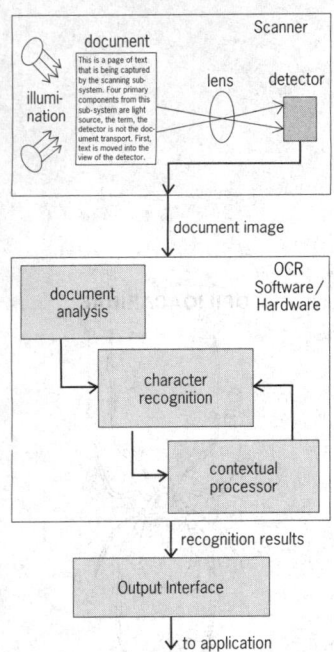

OPTICAL CHARACTER RECOGNITION

General structure of an optical character recognition system.

of electron paramagnetic resonance (EPR) spectroscopy that takes advantage of the sensitivity of electric dipole transitions, in which paramagnetic states are optically excited and the EPR signal is detected through changes in the optical absorption as the magnetic field is swept through one or more resonances. Abbreviated ODMR. { ˈäp·tə·kəl di‚tek·shən əv mag‚ned·ik ˈrez·ən·əns }

optical diffraction velocimeter See diffraction velocimeter. { ˈäp·tə·kəl diˈfrak·shən ‚vel·əˈsim·əd·ər }

optical direction and ranging See OPDAR. { ˈäp·tə·kəl diˈrek·shən ən ˈränj·iŋ }

optical disk [COMPUT SCI] A type of video disk storage device consisting of a pressed disk with a spiral groove at the bottom of which are submicrometer-sized depressions that are sensed by a laser beam. { ˈäp·tə·kəl ˈdisk }

optical disk storage [COMPUT SCI] A computer storage technology in which information is stored in submicrometer-sized holes on a rotating disk, and is recorded and read by laser beams focused on the disk. Also known as laser disk storage; video disk storage. { ˈäp·tə·kəl ˈdisk ˈstór·ij }

optical dispersion [OPTICS] Separation of different colors of light such as occurs when it passes from one medium to another or is reflected from a diffraction grating. { ˈäp·tə·kəl diˈspər·shən }

optical distance See optical path. { ˈäp·tə·kəl ˈdis·təns }

optical Doppler effect [ELECTROMAG] A change in the observed frequency of light or other electromagnetic radiation caused by relative motion of the source and observer. { ˈäp·tə·kəl ˈdäp·lər i‚fekt }

optical Doppler tomography See color Doppler optical coherence tomography. { ˈäp·tə·kəl ˈdäp·lər təˈmäg·rə·fē }

optical double star [ASTRON] Two stars not formally a physical system but that appear to be a typical double star; a false binary star whose components happen to lie nearby in the same line of sight. { ˈäp·tə·kəl ˈdəb·əl ˈstär }

optical electronic reproducer See optical sound head. { ˈäp·tə·kəl i‚lekˈträn·ik ‚rē·prəˈdü·sər }

optical element [OPTICS] A part of an optical instrument which acts upon the light passing through the instrument, such as a lens, prism, or mirror. { ˈäp·tə·kəl ˈel·ə·mənt }

optical encoder [ELECTR] An encoder that converts positional information into corresponding digital data by interrupting light beams directed on photoelectric devices. { ˈäp·tə·kəl inˈkōd·ər }

optical exaltation [PHYS CHEM] Optical anomaly in which the observed molar refraction exceeds the calculated one; most cases of optical anomaly are in this category. { ˈäp·tə·kəl ‚ek·səlˈtā·shən }

optical fiber [OPTICS] A long, thin thread of fused silica, or other transparent substance, used to transmit light. Also known as light guide. { ˈäp·tə·kəl ˈfī·bər }

optical-fiber amplifier [COMMUN] A device for amplifying signals transmitted over optical fibers, consisting of a low-loss single-mode fiber made of basic silica glass, along whose length gain is generated by coupling pump light at either or both fiber ends, or at periodic locations in between. { ‚äp·tə·kəl ‚fī·bər ˈam·plə‚fī·ər }

optical-fiber cable See optical waveguide. { ˈäp·tə·kəl ‚fī·bər ˈkā·bəl }

optical-fiber sensor [ENG] An instrument in which the physical quantity to be measured is made to modulate the intensity, spectrum, phase, or polarization of light from a light-emitting diode or laser diode traveling through an optical fiber; the modulated light is detected by a photodiode. Also known as fiber-optic sensor. { ˈäp·tə·kəl ˈfī·bər ˈsen·sər }

optical figuring [OPTICS] The final polishing or grinding process used to give glass components of optical instruments their desired shape. { ˈäp·tə·kəl ˈfig·yə·riŋ }

optical filter See filter. { ˈäp·tə·kəl ˈfil·tər }

optical flat [OPTICS] **1.** A disk of high-grade quartz glass approximately 2 centimeters thick, with a deviation in flatness usually not exceeding 0.05 micrometer all over, and a surface quality of 5 microfinish or less; used in determinations of surface contour and in comparison of lineal measurement. **2.** A plane surface, with deviations from a plane surface generally not exceeding one-tenth of a wavelength of light, used to redirect light in a telescope or other optical instrument. { ˈäp·tə·kəl ˈflat }

optical fluid-flow measurement [ENG] Any method of

measuring the varying densities of a fluid in motion, such as schlieren, interferometer, or shadowgraph, which depends on the fact that light passing through a flow field of varying density is retarded differently through the field, resulting in refraction of the rays, and in a relative phase shift among different rays. { ˈäp·tə·kəl ˈflü·əd ‚flō ‚mezh·ər·mənt }

optical frequency [PHYS] A frequency comparable to that of electromagnetic waves in the optical region, above about 3×10^{11} hertz. { ˈäp·tə·kəl ˈfrē·kwən·sē }

optical gage [ENG] A gage that measures an image of an object, and does not touch the object itself. { ˈäp·tə·kəl ˈgāj }

optical galaxy [ASTRON] One of the galaxies that appear as nearly starlike, generally having compact nuclei. { ˈäp·tə·kəl ˈgal·ik·sē }

optical glass [MATER] A type of glass which is free from imperfections, such as unmelted particles, bubbles, and chemical inhomogeneities, which would affect its transmission of light. { ˈäp·tə·kəl ˈglas }

optical guided wave [ELECTROMAG] An optical-frequency electromagnetic wave confined within an optical waveguide. { ‚äp·tə·kəl ‚gīd·əd ˈwāv }

optical harmonic [SOLID STATE] Light, generated by passing a laser beam with a power density on the order of 10^{10} watts per square centimeter or more through certain transparent materials, which has a frequency which is an integral multiple of that of the incident laser light. { ˈäp·tə·kəl härˈmän·ik }

optical haze See terrestrial scintillation. { ˈäp·tə·kəl ˈhāz }

optical horizon [GEOD] Locus of points at which a straight line from the given point becomes tangential to the earth's surface. { ˈäp·tə·kəl həˈrīz·ən }

optical indicator [ENG] An instrument which makes a plot of pressure in the cylinder of an engine as a function of piston (or volume) displacement, making use of magnification by optical systems and photographic recording; for example, the small motion of a pressure diaphragm may be transmitted to a mirror to deflect a beam of light. { ˈäp·tə·kəl ˈin·də‚kād·ər }

optical indicatrix See index ellipsoid. { ˈäp·tə·kəl inˈdik·ə‚triks }

optical information processor See optical information system. { ˈäp·tə·kəl ‚in·fərˈmā·shən ‚prä‚ses·ər }

optical information system [COMPUT SCI] A device that uses light to process information; consists of one or several light sources, a one- or two-dimensional plane of data such as a film transparency, lens, or other optical component, and a detector. Also known as optical information processor. { ˈäp·tə·kəl ‚in·fərˈmā·shən ‚sis·təm }

optical instrument [OPTICS] An optical system which acts on light in some desired way, such as to form a real or virtual image, to form an optical spectrum, or to produce light with a specified polarization or wavelength. { ˈäp·tə·kəl ˈin·strə·mənt }

optical interference [OPTICS] Interference of light waves. { ˈäp·tə·kəl ‚in·tərˈfir·əns }

optical isolator See optoisolator. { ˈäp·tə·kəl ˈī·sə‚lād·ər }

optical isomer See enantiomorph. { ˈäp·tə·kəl ˈī·sə·mər }

optical isomerism [PHYS CHEM] Existence of two forms of a molecule such that one is a mirror image of the other; the two molecules differ in that rotation of light is equal but in opposite directions. { ˈäp·tə·kəl ˈī‚säm·ə‚riz·ən }

optical Kerr effect [OPTICS] An effect in which a very strong linearly polarized light field produces anisotropy in the refractive index of an isotropic medium, usually a liquid. { ˈäp·tə·kəl ˈkər i‚fekt }

optical landing system [NAV] A shipboard gyro-stabilized device or a short-based device which indicates to the pilot his displacement from a preselected glide path. { ˈäp·tə·kəl ˈland·iŋ ‚sis·təm }

optical lantern [ENG] A device for projecting positive transparent pictures from glass or film onto a reflecting screen; it consists of a concentrated source of light, a condenser system, a holder (or changer) for the slide, a projection lens, and (usually) a blower for cooling the slide. Also known as slide projector. { ˈäp·tə·kəl ˈlan·tərn }

optical lattice [OPTICS] A regular pattern of microscopic traps for atoms, formed by the light forces in an interference pattern formed by laser beams. { ‚äp·tə·kəl ˈlad·əs }

optical length See optical path. { ˈäp·tə·kəl ˈleŋkth }

optical lens See lens. { ˈäp·tə·kəl ˈlenz }

optical lever [OPTICS] A device for measuring small angular

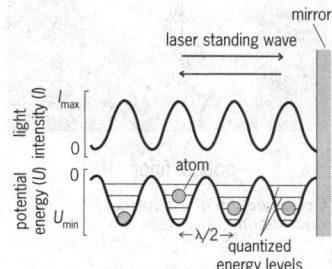

OPTICAL LATTICE

Light-induced potential formed by a laser standing wave. A series of microtraps are separated by λ/2 (one-half laser wavelength), forming an optical lattice. Atoms caught in traps occupy quantized vibrational levels.

displacements of a rotating body in which a narrow fixed beam of light is directed onto a small mirror attached to the body and the reflected beam is directed onto a screen, producing a spot of light whose position is measured. { 'äp·tə·kəl 'lev·ər }

optical lithography [ELECTR] Lithography in which an integrated circuit pattern is first created on a glass plate or mask and is then transferred to the resist by one of a number of optical techniques by using visible or ultraviolet light. { 'äp·tə·kəl li'thäg·rə·fē }

optically coupled isolator *See* optoisolator.

optically effective atmosphere [GEOPHYS] That portion of the atmosphere lying below the altitude (30–36 miles or 50–60 kilometers) from which scattered light at twilight still reaches the observer with sufficient intensity to be discerned. Also known as effective atmosphere. { 'äp·tə·klē i'fek·tiv 'at·mə,sfir }

optically pumped laser [OPTICS] A laser that uses absorption of light from an auxiliary light source to excite electrons into an upper energy state. { 'äp·tə·klē |pəmpt 'lā·zər }

optically pumped magnetometer [ENG] A type of magnetometer that measures total magnetic field intensity by observation of the precession frequency of magnetic atoms, usually gaseous rubidium, cesium, or helium, which are magnetized by irradiation with circularly polarized light of a suitable wavelength. { 'äp·tə·klē |pəmpt ,mag·nə'täm·əd·ər }

optical mark reading [COMPUT SCI] Optically sensing information encoded as a series of marks, such as lines or filled-in boxes on a test answer sheet, or some special pattern, such as the Universal Product Code. Abbreviated OMR. { 'äp·tə·kəl |märk ,rēd·iŋ }

optical maser *See* laser. { 'äp·tə·kəl 'mā·zər }

optical mask [ELECTR] A thin sheet of metal or other substance containing an open pattern, used to suitably expose to light a photoresistive substance overlaid on a semiconductor or other surface to form an integrated circuit. { 'äp·tə·kəl 'mask }

optical material [MATER] A material which is transparent to light or to infrared, ultraviolet, or x-ray radiation, such as glass and certain single crystals, polycrystalline materials (chiefly for the infrared), and plastics. { 'äp·tə·kəl mə'tir·ē·əl }

optical measurement [OPTICS] Measurement of the intensity, spectral distribution, polarization, or other characteristics of light or of infrared or ultraviolet radiation, which is emitted by or reflected from an object or passes through some medium. { 'äp·tə·kəl 'mezh·ər·mənt }

optical memory [COMPUT SCI] A computer memory that uses optical techniques which generally involve an addressable laser beam, a storage medium which responds to the beam for writing and sometimes for erasing, and a detector which reacts to the altered character of the medium when it uses the beam to read out stored data. { 'äp·tə·kəl 'mem·rē }

optical meteor [OPTICS] Any phenomenon of the atmosphere explained in terms of optical laws, such as a mirage or a halo. { 'äp·tə·kəl 'mē·dē·ər }

optical microphone [ENG ACOUS] A microphone in which the motion of a membrane is detected using a light beam reflected from it, either with the aid of an interferometer or by detecting the deflection of the beam. { 'äp·tə·kəl 'mī·krə,fōn }

optical microscope [OPTICS] An instrument used to obtain an enlarged image of a small object, utilizing visible light; in general it consists of a light source, a condenser, an objective lens, and an ocular or eyepiece, which can be replaced by a recording device. Also known as light microscope; photon microscope. { 'äp·tə·kəl 'mī·krə,skōp }

optical mode [SOLID STATE] A type of vibration of a crystal lattice whose frequency varies with wave number only over a limited range, and in which neighboring atoms or molecules in different sublattices move in opposition to each other. { 'äp·tə·kəl ,mōd }

optical model *See* cloudy-crystal-ball model. { 'äp·tə·kəl 'mäd·əl }

optical modulator [COMMUN] A device used for impressing information on a light beam. { 'äp·tə·kəl 'mäj·ə,lād·ər }

optical molasses [OPTICS] **1.** A viscous damping force exerted on neutral atoms by a pair of identical oppositely directed lasers tuned at a frequency below an atomic resonance. **2.** A large number of atoms collected and cooled in a small

volume at the intersection of the beams of three orthogonal pairs of such lasers. { 'äp·tə·kəl mə'las·əs }

optical moment [OPTICS] For a ray of light passing through an optical system, the triple product of a vector from an arbitrary origin on the optical axis to a point on the ray, a vector tangent to the ray at that point whose length equals the refractive index, and a unit vector along the optical axis; it does not depend on the point on the ray. { 'äp·tə·kəl 'mō·mənt }

optical monochromator [SPECT] A monochromator used to observe the intensity of radiation at wavelengths in the visible, infrared, or ultraviolet regions. { 'äp·tə·kəl |man·ə'kräm·əd·ər }

optical mouse [COMPUT SCI] A mouse that emits a light signal and uses its reflection from a reflective grid to determine position and movement. { 'äp·tə·kəl 'maús }

optical null method [SPECT] In infrared spectrometry, the adjustment of a reference beam's energy transmission to match that of a beam that has been passed through a sample being analyzed. { 'äp·tə·kəl |nəl ,meth·əd }

optical oceanography [OCEANOGR] That aspect of physical oceanography which deals with the optical properties of sea water and natural light in sea water. { 'äp·tə·kəl ,ō·shə'näg·rə·fē }

optical parallax [OPTICS] A fault in an optical measuring instrument in which the image being observed does not lie in the plane of the wires or marks used to make the measurement, so that motion of the observer's eye causes displacement of the image relative to these wires or marks. { 'äp·tə·kəl 'par·ə,laks }

optical parametric amplification [OPTICS] A process in which a weak signal beam is amplified at the expense of an intense pump beam simultaneously incident on a nonlinear crystal. { 'äp·tə·kəl |par·ə|me·trik ,am·plə·fə'kā·shən }

optical parametric oscillator [OPTICS] A device, employing a nonlinear dielectric, which when pumped by a laser can generate coherent light whose wavelength can be varied continuously over a wide range. { 'äp·tə·kəl |par·ə|me·trik 'äs·ə,läd·ər }

optical path [OPTICS] For a ray of light traveling along a path between two points, the optical path is the integral, over elements of length along the path, of the refractive index. Also known as optical distance; optical length. { 'äp·tə·kəl 'path }

optical-path difference *See* retardation. { 'äp·tə·kəl |path 'dif·rəns }

optical phase conjugation [OPTICS] The use of nonlinear optical effects to precisely reverse the direction of propagation of each plane wave in an arbitrary beam of light, thereby causing the return beam to exactly retrace the path of the incident beam. Also known as time-reversal reflection; wavefront reversal. { 'äp·tə·kəl |fāz ,kän·jə'gā·shən }

optical phenomena [ELECTROMAG] Phenomena associated with the generation, transmission, and detection of electromagnetic radiation in the visible, infrared, or ultraviolet regions. { 'äp·tə·kəl fə'näm·ə·nä }

optical phonon [SOLID STATE] A quantum of an optical mode of vibration of a crystal lattice. { 'äp·tə·kəl 'fō,nän }

optical plastic [MATER] A plastic which is transparent to light, occasionally used in optical systems for reasons of economy, special index-dispersion relation, light weight, and nonbrittleness. { 'äp·tə·kəl 'plas·tik }

optical printing [GRAPHICS] Any printing technique in which a process camera is used in one step to form the end item; enlargements, reductions, or same-size negatives may be formed. { 'äp·tə·kəl 'print·iŋ }

optical prism *See* prism. { 'äp·tə·kəl 'priz·əm }

optical processing [COMPUT SCI] The use of light, including visible and infrared, to handle data-processing information. { 'äp·tə·kəl 'prä,ses·iŋ }

optical projection system [OPTICS] An optical system which forms a real image of a suitably illuminated object so that it can be viewed, photographed, or otherwise observed. Also known as optical projector; projector. { 'äp·tə·kəl prə'jek·shən ,sis·təm }

optical projector *See* optical projection system. { 'äp·tə·kəl prə'jek·tər }

optical property [ELECTROMAG] One of the effects of a substance or medium on light or other electromagnetic radiation passing through it, such as absorption, scattering, refraction,

OPTICAL MICROPHONE

reflective membrane   inteferometer

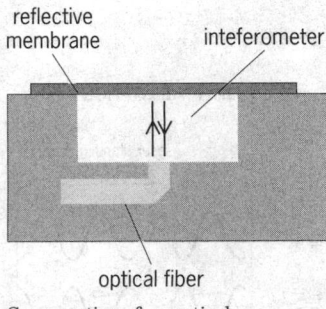

optical fiber

Cross section of an optical microphone.

and polarization. [OPTICS] Also known as reflection property. **1.** The property of an ellipse whereby rays of light emanating from one focus and reflected from a strip of polished metal at the ellipse come together at the other focus. **2.** The property of a parabola whereby rays of light emanating from the focus and reflected from a strip of polished metal at the parabola are reflected parallel to the axis of the parabola, and likewise rays parallel to the axis of the parabola are reflected and brought together at the focus. **3.** The property of a hyperbola whereby rays emanating from a focus and reflected from a strip of polished metal at the hyperbola appear to emanate from the other focus. { 'äp·tə·kəl 'präp·ərd·ē }

optical proximity sensor [ENG] A device that uses the principle of triangulation of reflected infrared or visible light to measure small distances in a robotic system. { 'äp·tə·kəl präk'sim·əd·ē ,sen·sər }

optical pulse [OPTICS] A short flash of light, used to isolate moments of time; pulses as short as 15 femtoseconds have been generated with laser and pulse compression techniques. { 'äp·tə·kəl 'pəls }

optical-pulse compression [OPTICS] The shortening of the duration of an optical pulse by techniques similar to those used in chirp radar, in which a frequency sweep is imposed on the pulse and the pulse is then compressed by using a dispersive delay line. { 'äp·tə·kəl ¦pəls kəm,presh·ən }

optical pumping [OPTICS] The process of causing strong deviations from thermal equilibrium populations of selected quantized states of different energy in atomic or molecular systems by the use of electromagnetic radiation in or near the visible region. { 'äp·tə·kəl 'pəmp·iŋ }

optical pyrometer [ENG] An instrument which determines the temperature of a very hot surface from its incandescent brightness; the image of the surface is focused in the plane of an electrically heated wire, and current through the wire is adjusted until the wire blends into the image of the surface. Also known as disappearing filament pyrometer. { 'äp·tə·kəl pī'räm·əd·ər }

optical quenching [OPTICS] Reduction in the intensity of luminescent radiation by long-wavelength, visible or infrared radiation. { 'äp·tə·kəl 'kwench·iŋ }

optical radar See OPDAR. { 'äp·tə·kəl 'rā,där }

optical rangefinder [ENG] An optical instrument for measuring distance, usually from its position to a target point, by measuring the angle between rays of light from the target, which enter the rangefinder through the windows spaced apart, the distance between the windows being termed the baselength of the rangefinder; the two types are coincidence and stereoscopic. { 'äp·tə·kəl 'rānj,fīnd·ər }

optical reader [COMPUT SCI] A computer data-entry machine that converts printed characters, bar or line codes, and pencil-shaded areas into a computer-input code format. { 'äp·tə·kəl 'rēd·ər }

optical recording [ENG] Production of a record by focusing on photographic paper a beam of light whose position on the paper depends on the quantity to be measured, as in a light-beam galvanometer. { 'äp·tə·kəl ri'kórd·iŋ }

optical rectification [OPTICS] An effect whereby one or more electromagnetic waves propagating in a nonlinear medium produce a second-order polarization that does not oscillate and, in turn, produces an electrical voltage. { 'äp·tə·kəl ,rek·tə·fə'kā·shən }

optical reflectometer [ENG] An instrument which measures on surfaces the reflectivity of electromagnetic radiation at wavelengths in or near the visible region. { 'äp·tə·kəl ,rē,flek'täm·əd·ər }

optical relay [ELECTR] An optoisolator in which the output device is a light-sensitive switch that provides the same on and off operations as the contacts of a relay. { 'äp·tə·kəl 'rē,lā }

optical-righting reflex See visual-righting reflex. { 'äp·tə·kəl ¦rīd·iŋ ,rē,fleks }

optical rotation [OPTICS] Rotation of the plane of polarization of plane-polarized light, or of the major axis of the polarization ellipse of elliptically polarized light by transmission through a substance or medium. { 'äp·tə·kəl rō'tā·shən }

optical rotatory dispersion [OPTICS] Specific rotation, considered as a function of wavelength. Abbreviated ORD. { 'äp·tə·kəl 'rōd·ə,tór·ē di'spər·zhən }

optical scanner See flying-spot scanner. { 'äp·tə·kəl 'skan·ər }

optical sight [OPTICS] A sight with lenses, prisms, or mirrors that is used in laying weapons, for aerial bombing, or for surveying. { 'äp·tə·kəl 'sīt }

optical sound head [ELECTR] The assembly in motion picture projection which reproduces photographically recorded sound; light from an incandescent lamp is focused on a slit, light from the slit is in turn focused on the optical sound track of a film, and the light passing through the film is detected by a photoelectric cell. Also known as optical electronic reproducer. { 'äp·tə·kəl 'saúnd ,hed }

optical sound recorder See photographic sound recorder. { 'äp·tə·kəl 'saúnd ri,kórd·ər }

optical sound reproducer See photographic sound reproducer. { 'äp·tə·kəl 'saúnd ,rē·prə'dü·sər }

optical spectra [SPECT] Electromagnetic spectra for wavelengths in the ultraviolet, visible and infrared regions, ranging from about 10 nanometers to 1 millimeter, associated with excitations of valence electrons of atoms and molecules, and vibrations and rotations of molecules. { 'äp·tə·kəl 'spek·trə }

optical spectrograph [SPECT] An optical spectroscope provided with a photographic camera or other device for recording the spectrum made by the spectroscope. { 'äp·tə·kəl 'spek·trə,graf }

optical spectrometer [SPECT] An optical spectroscope that is provided with a calibrated scale either for measurement of wavelength or for measurement of refractive indices of transparent prism materials. { 'äp·tə·kəl spek'träm·əd·ər }

optical spectroscope [SPECT] An optical instrument, consisting of a slit, collimator lens, prism or grating, and a telescope or objective lens, which produces an optical spectrum arising from emission or absorption of radiant energy by a substance, for visual observation. { 'äp·tə·kəl 'spek·trə,skōp }

optical spectroscopy [SPECT] The production, measurement, and interpretation of optical spectra arising from either emission or absorption of radiant energy by various substances. { 'äp·tə·kəl spek'träs·kə·pē }

optical spherical surface [OPTICS] An optical surface which forms part of a sphere. { 'äp·tə·kəl 'sfer·ə·kəl 'sər·fəs }

optical square [ENG] A surveyor's hand instrument used for laying off right angles; employs two mirrors at a 45° angle. { 'äp·tə·kəl 'skwer }

optical staining See Rheinberg illumination. { 'äp·tə·kəl 'stān·iŋ }

optical storage [COMPUT SCI] Storage of large amounts of data in permanent form on photographic film or its equivalent, for nondestructive readout by means of a light source and photodetector. { 'äp·tə·kəl 'stór·ij }

optical superposition principle [OPTICS] The principle that the optical rotation produced by a compound which is made up of two radicals of opposite optical activity is the algebraic sum of the rotations of each radical alone; not always valid. { 'äp·tə·kəl ¦sü·pər·pə¦zish·ən ,prin·sə·pəl }

optical surface [OPTICS] An interface between two media, such as between air and glass, which is used to reflect or refract light. { 'äp·tə·kəl 'sər·fəs }

optical system [OPTICS] A collection comprising mirrors, lens, prisms, and other devices, placed in some specified configuration, which reflect, refract, disperse, absorb, polarize, or otherwise act on light. { 'äp·tə·kəl ,sis·təm }

optical tape storage [COMMUN] A data storage technology in which information is stored on a tape that is wound on a spool and has a large number of parallel channels, and information is retrieved by sensing the reflected light when a light beam scans the medium. { ¦äp·tə·kəl 'tāp ,stór·ij }

optical thickness [METEOROL] **1.** In calculations of the transfer of radiant energy, the mass of a given absorbing or emitting material lying in a vertical column of unit cross-sectional area and extending between two specified levels. Also known as optical depth. **2.** Subjectively, the degree to which a cloud prevents light from passing through it; depends upon the physical constitution (crystals, drops, droplets), the form, the concentration of particles, and the vertical extent of the cloud. [OPTICS] The thickness of an optical material times its index of refraction. { 'äp·tə·kəl 'thik·nəs }

optical-to-optical interface device [OPTICS] A device that converts a noncoherently illuminated image into a coherently illuminated object, which can then be used as input to certain

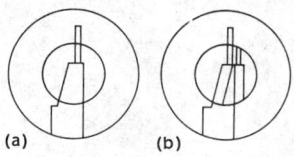

OPTICAL RANGEFINDER

View in camera coincidence rangefinder. *(a)* Images in coincidence. *(b)* Images unmatched.

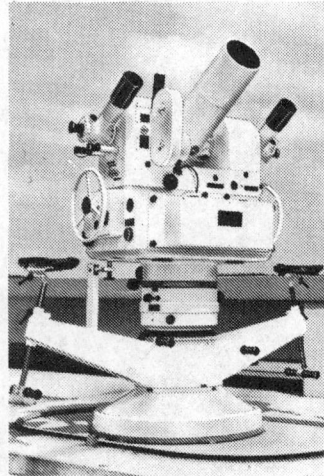

Contraves Model E cinetheodolite, a surveying theodolite having a 35-millimeter motion-picture camera with a 60- and 120-inch (1.5- and 3-meter) focal-length objectives substituted telescope. *(Fecker Systems Division, Owens-Illinois and Contraves AG)*

types of data processor. Abbreviated OTTO. { 'äp·tə·kəl tü 'äp·tə·kəl 'in·tər,fās di,vīs }

optical tracking [ENG] The determination of spatial positions of distant airplanes, missiles, and artificial satellites as a function of time, or the recording of engineering events, by precise time-correlated observations with various types of telescopes or ballistic cameras. { 'äp·tə·kəl 'trak·iŋ }

optical train [OPTICS] The series of lenses, mirrors, and prisms of an optical apparatus, such as a microscope or telescope, through which the light rays pass. { 'äp·tə·kəl 'trān }

optical transition [PHYS] A process in which an atom or molecule changes from one energy state to another and emits or absorbs electromagnetic radiation in the visible, infrared, or ultraviolet region. { 'äp·tə·kəl tran'zish·ən }

optical tweezers See laser tweezers. { 'äp·tə·kəl 'twēz·ərz }

optical twinning [CRYSTAL] Growing together of two crystals which are the same except that the structure of one is the mirror image of the structure of the other. Also known as chiral twinning. { 'äp·tə·kəl 'twin·iŋ }

optical type font [COMPUT SCI] A special type font whose characters are designed to be easily read by both people and optical character recognition machines. { 'äp·tə·kəl ¦tīp ,fänt }

optical waveguide [ELECTROMAG] A waveguide in which a light-transmitting material such as a glass or plastic fiber is used for transmitting information from point to point at wavelengths somewhere in the ultraviolet, visible-light, or infrared portions of the spectrum. Also known as fiber waveguide; optical-fiber cable. { 'äp·tə·kəl 'wāv,gīd }

optical window [OPTICS] The spectral region between 300 and 2000 nanometers (0.3 and 2 micrometers in wavelength), in which visible and near-visible radiation will pass through the earth's atmosphere. { 'äp·tə·kəl 'win·dō }

optic angle See axial angle. { 'äp·tik ,aŋ·gəl }

optic apraxia [MED] A form of apraxia in which the individual fails to represent spatial relations correctly in drawing or construction by other means. Also known as constructional apraxia. { 'äp·tik ā'prak·sē·ə }

optic-axial angle See axial angle. { 'äp·tik ¦ak·sē·əl ,aŋ·gəl }

optic axis [OPTICS] The axis in a doubly refracting medium in which the ordinary and extraordinary waves propagate with the same velocity, and double refraction vanishes. Also known as optical axis; principal axis. { 'äp·tik 'ak·səs }

optic canal [ANAT] The channel at the apex of the orbit, the anterior termination of the optic groove, just beneath the lesser wing of the sphenoid bone; it gives passage to the optic nerve and ophthalmic artery. { 'äp·tik kə'nal }

optic capsule [VERT ZOO] A cartilaginous capsule that develops around the eye in elasmobranchs and higher vertebrate embryos. { 'äp·tik 'kap·səl }

optic chiasma [NEUROSCI] The partial decussation of the optic nerves on the undersurface of the hypothalamus. { 'äp·tik kī'az·mə }

optic cup [EMBRYO] A two-layered depression formed by invagination of the optic vesicle from which the pigmented and sensory layers of the retina will develop. { 'äp·tik ,kəp }

optic disk [NEUROSCI] The circular area in the retina that is the site of the convergence of fibers from the ganglion cells of the retina to form the optic nerve. { 'äp·tik ,disk }

optic ellipse [OPTICS] Any section through the index ellipsoid. { 'äp·tik i'lips }

optic gland [INV ZOO] Either of a pair of endocrine glands in the octopus and squid which are found near the brain and produce a substance which causes gonadal maturation. { 'äp·tik ,gland }

optician [ENG] A maker of optical instruments or lenses. { äp'tish·ən }

optic lobe [INV ZOO] A lateral lobe of the forebrain in certain arthropods. [NEUROSCI] One of the anterior pair of colliculi of the mammalian corpora quadrigemina. [VERT ZOO] Either of the corpora bigemina of lower vertebrates. { 'äp·tik ,lōb }

optic nerve [NEUROSCI] The second cranial nerve; a paired sensory nerve technically consisting of three layers of special nerve cells in the retina of the eye; fibers converge to form the optic tracts. { 'äp·tik ,nərv }

optic normal [OPTICS] The axis that lies perpendicular to the optic axis. { 'äp·tik 'nór·məl }

optics [PHYS] **1.** Narrowly, the science of light and vision. **2.** Broadly, the study of the phenomena associated with the

generation, transmission, and detection of electromagnetic radiation in the spectral range extending from the long-wave edge of the x-ray region to the short-wave edge of the radio region, or in wavelength from about 1 nanometer to about 1 millimeter. { 'äp·tiks }

optic stalk [EMBRYO] The constriction of the optic vesicle which connects the embryonic eye and forebrain in vertebrates. { 'äp·tik ,stók }

optic tectum [VERT ZOO] The roof of the mesencephalon constituting a major visual center and association area of the brain of premature vertebrates. { 'äp·tik 'tek·təm }

optic tract [NEUROSCI] The band of optic nerve fibers running from the optic chiasma to the lateral geniculate body and midbrain. { 'äp·tik ,trakt }

optic vesicle [EMBRYO] An evagination of the lateral wall of the forebrain in vertebrate embryos which precedes formation of the optic cup. { 'äp·tik 'ves·ə·kəl }

optimal control theory [CONT SYS] An extension of the calculus of variations for dynamic systems with one independent variable, usually time, in which control (input) variables are determined to maximize (or minimize) some measure of the performance (output) of a system while satisfying specified constraints. { 'äp·tə·məl kən'trōl ,thē·ə·rē }

optimal feedback control [CONT SYS] A subfield of optimal control theory in which the control variables are determined as functions of the current state of the system. { 'äp·tə·məl 'fēd,bak kən,trōl }

optimal policy [MATH] In optimization problems of systems, a sequence of decisions changing the states of a system in such a manner that a given criterion function is minimized. { 'äp·tə·məl 'päl·ə·sē }

optimal programming [CONT SYS] A subfield of optimal control theory in which the control variables are determined as functions of time for a specified initial state of the system. { 'äp·tə·məl 'prō,gram·iŋ }

optimal regulator problem See linear regulator problem. { 'äp·tə·məl 'reg·yə,lād·ər ,präb·ləm }

optimal smoother [CONT SYS] An optimal filer algorithm which generates the best estimate of a dynamical variable at a certain time based on all available data, both past and future. { 'äp·tə·məl 'smüth·ər }

optimal strategy [MATH] One of the pair of mixed strategies carried out by the two players of a matrix game when each player adjusts strategy so as to minimize the maximum loss that an opponent can inflict. { 'äp·tə·məl 'strad·ə·jē }

optimal system [MATH] A system where the variables representing the various states are so determined that a given criterion function is minimized subject to given constraints. { 'äp·tə·məl 'sis·təm }

optimization [MATH] The maximizing or minimizing of a given function possibly subject to some type of constraints. [SYS ENG] **1.** Broadly, the efforts and processes of making a decision, a design, or a system as perfect, effective, or functional as possible. **2.** Narrowly, the specific methodology, techniques, and procedures used to decide on the one specific solution in a defined set of possible alternatives that will best satisfy a selected criterion. Also known as system optimization. { ,äp·tə·mə'zā·shən }

optimization theory [MATH] The specific methodology, techniques, and procedures used to decide on the one specific solution in a defined set of possible alternatives that will best satisfy a selected criterion; includes linear and nonlinear programming, stochastic programming, and control theory. Also known as mathematical programming. { ,äp·tə·mə'zā·shən ,thē·ə·rē }

optimize [COMPUT SCI] To rearrange the instructions or data in storage so that a minimum number of time-consuming jumps or transfers are required in the running of a program. { 'äp·tə,mīz }

optimized code [COMPUT SCI] A machine-language program that has been revised to remove inefficiencies and unused or unnecessary instructions so that the program is executed more quickly and occupies less storage space. { 'äp·tə,mīzd 'kōd }

optimizer [COMPUT SCI] A utility program that processes machine-language programs and generates optimized code. { 'äp·tə,mīz·ər }

optimizing control function [CONT SYS] That level in the functional decomposition of a large-scale control system which determines the necessary relationships among the variables of the system to achieve an optimal, or suboptimal, performance based on a given approximate model of the plant and its environment. { 'äp·tə,mīz·iŋ kən'trōl ,faŋk·shən }

optimum allocation [STAT] A procedure used in stratified sampling to allocate numbers of sample units to different strata to either maximize precision at a fixed cost or minimize cost for a selected level of precision. { 'äp·tə·məm ,al·ə'kā·shən }

optimum array current [ELECTROMAG] The current distribution in a broadside antenna array which is such that for a specified side-lobe level the beam width is as narrow as possible, and for a specified first null the side-lobe level is as small as possible. { 'äp·tə·məm ə'rā ,kə·rənt }

optimum bunching [ELECTR] Bunching condition required for maximum output in a velocity modulation tube. { 'äp·tə·məm 'bənch·iŋ }

optimum charge [ORD] Propelling charge, with web and propellant weight combination, which produces maximum velocity at a specified pressure. { 'äp·tə·məm 'chärj }

optimum code [COMPUT SCI] A computer code which is particularly efficient with regard to a particular aspect; for example, minimum time of execution, minimum or efficient use of storage space, and minimum coding time. { 'äp·tə·məm 'kōd }

optimum coupling See critical coupling. { 'äp·tə·məm 'kəp·liŋ }

optimum cure [CHEM ENG] The degree of vulcanization at which maximum desired property is reached. { 'äp·tə·məm 'kyùr }

optimum filter [ELECTR] An electric filter in which the mean square value of the error between a desired output and the actual output is at a minimum. { 'äp·tə·məm 'fil·tər }

optimum flight [AERO ENG] An aircraft flight so planned and navigated that it is completed under the optimum conditions of minimum time and minimum exposure to dangerous flying weather. { 'äp·tə·məm 'flīt }

optimum moisture content [GEOL] The water content at which a specified compactive force can compact a soil mass to its maximum dry unit weight. { 'äp·tə·məm 'mȯis·chər ,kän,tent }

optimum programming [COMPUT SCI] Production of computer programs that maximize efficiency with respect to some criteria such as least cost, least use of storage, least time, or least use of time-sharing peripheral equipment. { 'äp·tə·məm 'prō,gram·iŋ }

optimum reverberation time [ACOUS] The reverberation time which is most desirable for a given room size and a given use, such as speech, chamber music, or symphony orchestra. { 'äp·tə·məm ri,vər·bə'rā·shən ,tīm }

optimum separation point [PETRO ENG] In extraction of natural gasoline, the pressure and temperature conditions necessary for maximum condensation in the separators under field conditions. { 'äp·tə·məm ,sep·ə'rā·shən ,pȯint }

optimum-track ship routing [NAV] The selection of an optimum track for a transoceanic crossing by the application of long-range predictions of winds, waves, and currents to the knowledge of how the routed ship reacts to these variables. { 'äp·tə·məm ,trak 'ship ,rüd·iŋ }

optimum traffic frequency See optimum working frequency. { 'äp·tə·məm 'traf·ik ,frē·kwən·sē }

optimum working frequency [COMMUN] The most effective frequency at a specified time for ionospheric propagation of radio waves between two specified points. Also known as frequency optimum traffic; optimum traffic frequency. { 'äp·tə·məm 'wərk·iŋ ,frē·kwən·sē }

optional halt instruction [COMPUT SCI] A halt instruction that can cause a computer program to stop either before or after the instruction is obeyed if certain criteria are met. Also known as optional stop instruction. { 'äp·shən·əl 'hȯlt in,strək·shən }

optional product [COMPUT SCI] Any of various forms of documentation that may be made available with a software product, such as source code, manuals, and instructions. { 'äp·shən·əl 'präd·əkt }

optional stop instruction See optional halt instruction. { 'äp·shən·əl 'stäp in,strək·shən }

option switch [COMPUT SCI] 1. A DIP switch or jumper that activates an optional feature. 2. A software parameter that overrides a default value and thereby activates an optional feature. Also known as option toggle. { 'äp·shən ,swich }

option toggle See option switch. { 'äp·shən ,täg·əl }

optoacoustic detection method [ANALY CHEM] A method of detecting trace impurities in a gas, in which the absorption of a sample of the gas at various light frequencies is measured by directing a periodically interrupted laser beam through the sample in a spectrophone and measuring the sound generated by the optoacoustic effect at the frequency of interruption of the beam. { ¦äp·tō·ə¦küs·tik di'tek·shən ,meth·əd }

optoacoustic effect [PHYS] A phenomenon in which a periodically interrupted beam of light generates sound in a gas through which it is passing; this results from energy in the light beam being transformed first into internal motions of the gas molecules, then into random translational motions of these molecules, or heat, and finally into periodic pressure fluctuations or sound. Also known as thermoacoustic effect. { ¦äp·tō·ə¦küs·tik i,fekt }

optoacoustic modulator See acoustooptic modulator. { ¦äp·tō·ə¦küs·tik 'mäj·ə,lād·ər }

optoacoustic spectroscopy See photoacoustic spectroscopy. { ¦äp·tō·ə¦kü·stik spek'träs·kə·pē }

optocoupler See optoisolator. { ¦äp·tō'kəp·lər }

optoelectronic amplifier [ENG] An amplifier in which the input and output signals and the method of amplification may be either electronic or optical. { ¦äp·tō·i,lek'trän·ik 'am·plə,fī·ər }

optoelectronic integration [ELECTR] A technology that combines optical components with electronic components such as transistors on a single wafer to obtain highly functional circuits. { ¦äp·tō·i,lek'trän·ik ,in·tə'grā·shən }

optoelectronic isolator See optoisolator. { ¦äp·tō·i,lek'trän·ik 'ī·sə,lād·ər }

optoelectronics [ELECTR] 1. The branch of electronics that deals with solid-state and other electronic devices for generating, modulating, transmitting, and sensing electromagnetic radiation in the ultraviolet, visible-light, and infrared portions of the spectrum. 2. See photonics. { ¦äp·tō·i,lek'trän·iks }

optoelectronic scanner [ELECTR] A scanner in which lenses, mirrors, or other optical devices are used between a light source or image and a photodiode or other photoelectric device. { ¦äp·tō·i,lek'trän·ik 'skan·ər }

optoelectronic shutter [ENG] A shutter that uses a Kerr cell to modulate a beam of light. { ¦äp·tō·i,lek'trän·ik 'shəd·ər }

optogalvanic effect [PHYS] The alteration of the current through an electrical discharge by light incident on the discharge space. { ¦äp·tō·gal van·ik i,fekt }

optogalvanic spectroscopy [SPECT] A method of obtaining absorption spectra of atomic and molecular species in flames and electrical discharges by measuring voltage and current changes upon laser irradiation. { ¦äp·tō·gal¦van·ik spek'träs·kə·pē }

optoisolator [ELECTR] A coupling device in which a light-emitting diode, energized by the input signal, is optically coupled to a photodetector such as a light-sensitive output diode, transistor, or silicon controlled rectifier. Also known as optical coupler; optical isolator; optically coupled isolator; optocoupler; optoelectronic isolator; photocoupler; photoisolator. { ¦äp·tō'ī·sə,lād·ər }

optometrist [MED] One who measures the degrees of visual powers, without the aid of cycloplegic or mydriatic agents. { äp'täm·ə,trist }

optometry [MED] Measurement of visual powers. { äp'täm·ə·trē }

optophone [ENG ACOUS] A device with a photoelectric cell to convert ordinary printed letters into a series of sounds; used by the blind. { 'äp·tə,fōn }

optovoltaic effect [PHYS] The alteration of the potential difference across a discharge by light incident on the discharge space. { ¦äp·tō·vōl'tā·ik i,fekt }

OPW method See orthogonalized plane-wave method. { ¦ō¦pē'dəb·əl,yü ,meth·əd }

or [COMPUT SCI] An instruction which performs the logical operation "or" on a bit-by-bit basis for its two or more operand words, usually storing the result in one of the operand locations. Also known as OR function. [MATH] A logical operation whose result is false (or zero) only if every one of its operands is false, and true (or one) otherwise. Also known as inclusive or. { ȯr }

ora [METEOROL] A regular valley wind at Lake Garda in Italy. { 'òr·ə }

oral [ANAT] Of or pertaining to the mouth. { 'òr·əl }

oral and maxillofacial surgery [MED] A branch of dentistry that treats diseases and abnormalities of the maxillofacial region by surgical means. { 'òr·əl and ˌmak·sə·lō͝ˈfā·shəl 'sər·jə·rē }

oral arm [INV ZOO] In a jellyfish, any of the prolongations of the distal end of the manubrium. { 'òr·əl ˈärm }

oral cavity [ANAT] The cavity of the mouth. { 'òr·əl 'kav·əd·ē }

oral character [PSYCH] A Freudian term applied to persons who have undergone an unusual degree of oral stimulation during the developmental period and are characterized by an attitude of carefree indifference and by dependence on a mother figure. { 'òr·əl 'kar·ik·tər }

oral contraceptive [PHARM] Any medication taken by mouth that renders a woman nonfertile as long as the medication is continued. { 'òr·əl ˌkän·trə'sep·tiv }

oral dependence [PSYCH] In psychoanalytic theory, the desire to return to the oral phase of psychosexual development, presumably because of the sense of security and safety characterizing this stage. { 'òr·əl di'pen·dəns }

oral disc [INV ZOO] The flattened upper or free end of the body of a polyp that has the mouth in the center and tentacles around the margin. { 'òr·əl ˈdisk }

oral erotic stage [PSYCH] In psychoanalysis, the first, or receptive, part of the oral phase of psychosexual development, dominated by sucking and lasting for the first 6 to 9 months of life. { 'òr·əl ə'räd·ik ˌstāj }

oral erotism [PSYCH] The primordial pleasurable experience of nursing, reappearing in usually disguised and sublimated form in later life. { 'òr·əl 'er·ə,tiz·əm }

oral groove [INV ZOO] A depressed, groovelike peristome. { 'òr·əl ˈgrüv }

oral pathology [MED] A branch of dentistry that is concerned with the diseases of the teeth, oral cavity, and jaws, and with the oral manifestations of systemic diseases. { ˌòr·əl pə'thäl·ə·jē }

oral personality [PSYCH] An individual who is mouth-centered far beyond the age when the oral phase should have been passed, and who exhibits oral erotism and sadism in disguised and sublimated form. { 'òr·əl ˌpər·sə'nal·əd·ē }

oral phase [PSYCH] In psychoanalytic theory, the earliest of the stages of psychosexual development, lasting from birth to approximately 12 months of age, in which libidinal gratification is derived from intake of food, by sucking, and later by biting. { 'òr·əl ˈfāz }

oral primacy [PSYCH] In psychoanalytic theory, the concentration of libido upon the mouth, evidenced by using the tongue and lips to explore things and by deriving pleasure from sucking, biting, and chewing. { 'òr·əl 'prī·mə·sē }

orange [BOT] Any of various evergreen trees of the genus *Citrus*, cultivated for the edible fruit, a berry with an aromatic, leathery rind containing numerous oil glands. [OPTICS] The hue evoked in an average observer by monochromatic radiation having a wavelength in the approximate range from 597 to 622 nanometers; however, the same sensation can be produced in a variety of other ways. { 'är·inj }

orange coffee rust [PL PATH] A disease of the coffee plant caused by the rust fungus *Hemileia vastatrix*; characterized by the formation of small, powdery, pale yellow to orange spots on the lower leaf surface, followed by defoliation. { 'är·inj 'kò·fē ˌrəst }

orange flower oil *See* oil of orange blossoms. { 'är·inj ˈflaù·ər ˌoil }

orange lake [MATER] Any of various transparent orange pigments from the precipitation of an orange dyestuff on aluminum hydrate or other base; used to produce transparent coatings for metal cans and bottle caps. { 'är·inj ˈlāk }

orange mineral [MATER] A bright orange-red lead oxide pigment used in printing inks and primers. { 'är·inj 'min·rəl }

orange oil *See* bitter orange oil; sweet orange oil. { 'är·inj ˌoil }

orangeophile [HISTOL] A type of acidophile cell of the anterior lobe of the adenohypophysis, presumed to elaborate growth hormone. { ə'ran·jē·ə,fil }

orange oxide *See* uranium trioxide. { 'är·inj 'äk,sīd }

orange peel [MATER] A pebbled film surface, resembling an orange skin, on lacquer or enamel as a result of too rapid drying after spraying, or failure to exhibit the desired leveling effects. [MET] A rough, pebble-grained metal surface resulting from either plastic deformation or electropolishing. Also known as alligator effect; pebbling. { 'är·inj ˌpēl }

orange-peel bucket [DES ENG] A type of grab bucket that is multileaved and generally round in configuration. { 'är·inj ˌpēl ˌbək·ət }

orange sapphire [MINERAL] An orange variety of gem corundum (sapphire). Also known as padparadsha. { 'är·inj 'sa,fīr }

orange spectrometer [SPECT] A type of beta-ray spectrometer that consists of a number of modified double-focusing spectrometers employing a common source and a common detector, and has exceptionally high transmission. { 'är·inj spek'träm·əd·ər }

orange toner [ORG CHEM] A diazo dyestuff coupled to diacetoacetic acid anhydride; contains no sulfonic or carboxylic groups; used for printing inks. { 'är·inj 'tōn·ər }

orangutan [VERT ZOO] *Pongo pygmaeus*. The largest of the great apes, a long-armed primate distinguished by long sparse reddish-brown hair, naked face and hands and feet, and a large laryngeal cavity which appears as a pouch below the chin. { ə'raŋ·ü,tan }

O ray *See* ordinary ray. { 'ō ,rā }

ORB *See* object request broker. { òrb *or* ˌōˈär'bē }

ORB core [COMPUT SCI] The part of an object request broker that is responsible for the communication of requests. { 'òrb ˌkòr *or* ˌōˈär'bē ˌkòr }

orbicular [PETR] Of the structure of a rock, containing large quantities of orbicules. [SCI TECH] Having the form of a sphere or orb. { òr'bik·yə·lər }

orbicule [GEOL] A nearly spherical body, up to 2 centimeters (0.8 inch) or more in diameter, in which the components are arranged in concentric layers. { 'òr·bə,kyül }

Orbiniidae [INV ZOO] A family of polychaete annelids belonging to the Sedentaria; the prostomium is exposed, and the thorax and abdomen are weakly separated. { ˌòr·bə'nī·ə,dē }

Orbiniinae [INV ZOO] A subfamily of sedentary polychaete annelids in the family Orbiniidae. { òr·bə'nī·ə,nē }

orbit [ANAT] The bony cavity in the lateral front of the skull beneath the frontal bone which contains the eyeball. Also known as eye socket. [MATH] Let *G* be a group which operates on a set *S*; the orbit of an element *s* of *S* under *G* is the subset of *S* consisting of all elements *gs* where *g* is in *G*. [OCEANOGR] The path of a water particle affected by wave motion; it is almost circular in deep-water waves and almost elliptical in shallow-water waves. [PHYS] **1.** Any closed path followed by a particle or body, such as the orbit of a celestial body under the influence of gravity, the elliptical path followed by electrons in the Bohr theory, or the paths followed by particles in a circular particle accelerator. **2.** More generally, any path followed by a particle, such as helical paths of particles in a magnetic field, or the parabolic path of a comet. { 'òr·bət }

orbital [ATOM PHYS] The space-dependent part of the Schrödinger wave function of an electron in an atom or molecule in an approximation such that each electron has a definite wave function, independent of the other electrons. { 'òr·bəd·əl }

orbital angular momentum [MECH] The angular momentum associated with the motion of a particle about an origin, equal to the cross product of the position vector with the linear momentum. Also known as orbital momentum. [QUANT MECH] The angular momentum operator associated with the motion of a particle about an origin, equal to the cross product of the position vector with the linear momentum, as opposed to the intrinsic spin angular momentum. Also known as orbital moment. { 'òr·bəd·əl 'aŋ·gyə·lər mə'men·təm }

orbital current [OCEANOGR] The flow of water which follows the orbital motion of water particles in a wave. { 'òr·bəd·əl 'kə·rənt }

orbital curve [AERO ENG] One of the tracks on a primary body's surface traced by a satellite that orbits about it several times in a direction other than normal to the primary body's axis of rotation; each track is displaced in a direction opposite and by an amount equal to the degrees of rotation between each satellite orbit. { 'òr·bəd·əl 'kərv }

orbital decay [AERO ENG] The lessening of the eccentricity of the elliptical orbit of an artificial satellite. [ATOM PHYS]

ORGANGUTAN

The orangutan. (*Pongo pygmaeus*).

ORBINIIDAE

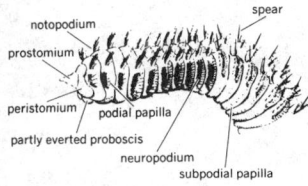

Anterior parts of the body of *Phylo*.

A change of an atom from one energy state to another of lower energy in which the orbital of one of the electrons changes { 'ȯr·bəd·əl di'kā }

orbital direction [AERO ENG] The direction that the path of an orbiting body takes; in the case of an earth satellite, this path may be defined by the angle of inclination of the path to the equator. { 'ȯr·bəd·əl di'rek·shən }

orbital electron [ATOM PHYS] An electron which has a high probability of being in the vicinity (at distances on the order of 10^{-10} meter or less) of a particular nucleus, but has only a very small probability of being within the nucleus itself. Also known as planetary electron. { 'ȯr·bəd·əl i'lek,trän }

orbital elements [PHYS] A set of seven parameters defining the orbit of a body attracted by a central, inverse-square force. { 'ȯr·bəd·əl 'el·ə·məns }

orbital fossa [INV ZOO] A depression from which the eyestalk arises on the front of the carapace of crustaceans. { 'ȯr·bəd·əl 'fäs·ə }

orbital index [ANTHRO] The ratio of the orbital height, taken at right angles to the orbital width between the upper and lower orbital margins, times 100, to the orbital width, taken between the maxillofrontale and lateral margin of the orbit in such a manner that the line of the orbital width bisects the plane of the orbital entrance. { 'ȯr·bəd·əl 'in,deks }

orbital magnetic moment [QUANT MECH] The magnetic dipole moment associated with the motion of a charged particle about an origin, rather than with its intrinsic spin. { 'ȯr·bəd·əl mag'ned·ik 'mō·mənt }

orbital moment See orbital angular momentum. { 'ȯr·bəd·əl 'mō·mənt }

orbital momentum See orbital angular momentum. { 'ȯr·bəd·əl mə'men·təm }

orbital motion [PHYS] Continuous motion of a body in a closed path, such as a circle or an ellipse, about some point. { 'ȯr·bəd·əl 'mō·shən }

orbital node [ASTRON] One of the two points at which the orbit of a planet or satellite crosses the plane of the ecliptic or equator. { 'ȯr·bəd·əl 'nōd }

orbital overlap [PHYS CHEM] The overlapping of two electron orbitals, one from each of two atoms, such that each orbital obtains a share in the electron of the other atom, forming a chemical bond. { 'ȯr·bəd·əl 'ō·vər,lap }

orbital parity [QUANT MECH] The parity associated with the wave function of a particle, or system of particles, as a function of spatial coordinates; it is opposed to intrinsic parity; if the orbital angular momentum quantum number is l, the orbital parity is $(-1)^l$. { 'ȯr·bəd·əl 'par·əd·ē }

orbital period [ASTRON] The interval between successive passages of a satellite through the same specified point in its orbit. { 'ȯr·bəd·əl 'pir·ē·əd }

orbital plane [MECH] The plane which contains the orbit of a body or particle in a central force field; it passes through the center of force. { 'ȯr·bəd·əl 'plān }

orbital rendezvous [AERO ENG] **1.** The meeting of two or more orbiting objects with zero relative velocity at a preconceived time and place. **2.** The point in space at which such an event occurs. { 'ȯr·bəd·əl 'rän·də,vü }

orbital sander [MECH ENG] An electric sander that moves the abrasive in an elliptical pattern. { 'ȯr·bəd·əl 'san·dər }

orbital symmetry [PHYS CHEM] The property of certain molecular orbitals of being carried into themselves or into the negative of themselves by certain geometrical operations, such as a rotation of 180° about an axis in the plane of the molecule, or reflection through this plane. { 'ȯr·bəd·əl 'sim·ə·trē }

orbital velocity [ASTRON] The instantaneous velocity at which an earth satellite or other orbiting body travels around the origin of its central force field. { 'ȯr·bəd·əl və'läs·əd·ē }

orbite [PETR] An igneous rock containing large phenocrysts of hornblende, or plagioclase and hornblende, in a groundmass with the composition of malachite. { 'ȯr,bīt }

orbiting collision [ATOM PHYS] An interaction between an ion and an atom in which they approach each other very closely and spend a relatively long time (several orbital periods of the atomic electrons) in proximity. { 'ȯr·bəd·iŋ kə'lizh·ən }

Orbitoidacea [INV ZOO] A superfamily of foraminiferan protozoans in the suborder Rotaliina characterized by a low trochospire or a planispiral, uncoiled or branching test composed of radial calcite with bilamellar septa. { ,ȯr·bə·tȯi'-dās·ē,ə }

orbit point [AERO ENG] A geographically defined reference point over land or water, used in stationing airborne aircraft. { 'ȯr·bət ,pȯint }

orbitron [ELECTR] A maser that uses synthetic atoms composed of free electrons orbiting long, thin, positively charged, metal wires. { 'ȯr·bə,trän }

orbit space [MATH] The orbit space of a G space X is the topological space whose points are equivalence classes obtained by identifying points in X which have the same G orbit and whose topology is the largest topology that makes the function which sends x to its orbit continuous. { 'ȯr·bət ,spās }

Orbivirus [VIROL] A genus in the family Reoviridae that is the causative agent of bluetongue. { 'ȯrb·ə,vī·rəs }

orch-, orchi-, orchid-, orchido-, orchio- [ZOO] A combining form denoting testis. { 'ȯr·kē, 'ȯr·kəd·ō, 'ȯr·kē·ō }

orchard [AGR] A group of fruit-bearing, nut-bearing, or sugar maple trees under cultivation. { 'ȯr·chərd }

orchid [BOT] Any member of the family Orchidaceae; plants have complex, specialized irregular flowers usually with only one or two stamens. { 'ȯr·kəd }

Orchidaceae [BOT] A family of monocotyledonous plants in the order Orchidales characterized by irregular flowers with only one or two stamens which are adnate to the style, and pollen grains which cohere in large masses called pollinia. { ,ȯr·kə'dās·ē,ē }

Orchidales [BOT] An order of monocotyledonous plants in the subclass Liliidae; plants are mycotropic and sometimes nongreen with numerous tiny seeds that have an undifferentiated embryo and little or no endosperm. { ,ȯr·kə'dā·lēz }

orchid mycorrhizal fungi [MYCOL] Mycorrhizal fungi that are characterized by the absence of a Hartig net and a fungal sheath and the presence of hyphal coils in the root cells. { 'ȯr·kid ,mī·kə,rīz·əl 'fən,jī }

orchiectomy [MED] Surgical removal of one or both testes. { ,ȯr·kē'ek·tə·mē }

orchil [MATER] Dark-brownish-red coloring matter derived from lichens as paste or aqueous extract; main components are orcin and orcein; used as carpet-yarn dye. Also known as orseille. { 'ȯr·chəl }

orchiopexy [MED] Surgical fixation of a testis. { 'ȯr·kē·ō,pek·sē }

orcin [ORG CHEM] $CH_3C_6H_3(OH)_2 \cdot H_2O$ White crystals with strong, sweet, unpleasant taste; soluble in water, alcohol, and ether; extracted from lichens; used in medicine and as an analytical reagent. { 'ȯr·sən }

OR circuit See OR gate. { 'ȯr ,sər·kət }

ORD See optical rotatory dispersion.

order [CHEM] A classification of chemical reactions, in which the order is described as first, second, third, or higher, according to the number of molecules (one, two, three, or more) which appear to enter into the reaction; decomposition of H_2O_2 to form water and oxygen is a first-order reaction. [MATH] **1.** A differential equation has order n if the derivatives of a function appear up to the nth derivative. **2.** The number of elements contained within a given group. **3.** A square matrix with n rows and n columns has order n. **4.** The number of poles a given elliptic function has in a parallelogram region where it repeats its values. **5.** A characteristic of infinitesimals used in their comparison. **6.** For a polynomial, the largest exponent appearing in the polynomial. **7.** The number of vertices of a graph. **8.** For a pole of an analytic function, the largest negative power in the function's Laurent expansion about the pole. **9.** For a zero point z_0 of an analytic function, the integer n such that the function near the pole has the form $g(z)(z-z_0)^n$, where $g(z)$ is analytic at z_0 and does not vanish there. **10.** For an algebraic curve or surface, the degree of its equation. **11.** For an algebra, the dimension of the underlying vector space. **12.** For a branch point of a Riemann surface, the number of sheets of the surface that join at the branch point, minus one. **13.** See ordering. [PHYS] A range of magnitudes of a quantity (and of all other quantities having the same physical dimensions) extending from some value of the quantity to some small multiple of the quantity (usually 10). Also known as order of magnitude. [SYST] A taxonomic category ranked below the class and above the family, made up either of families, subfamilies, or suborders. { 'ȯrd·ər }

order-disorder transition [SOLID STATE] The transition of an alloy or other solid solution between a state in which atoms of one element occupy certain regular positions in the lattice

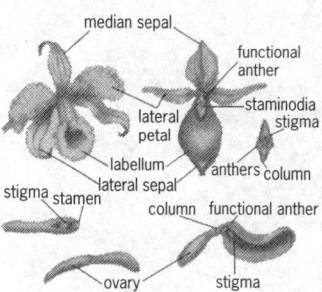

ORCHIDACEAE

median sepal
functional anther
lateral petal
staminodia stigma
labellum
anthers column
lateral sepal
stigma stamen
column functional anther
ovary
stigma

The two common floral types found in orchids. (a) *Cattleya.* (b) *Cypripedium.*

of another element, and a state in which this regularity is not present. { 'ȯrd·ər 'dis‚ȯrd·ər tran'zish·ən }

ordered array [COMPUT SCI] A set of data elements that has been arranged in rows and columns in a specified order so that each element can be individually accessed. { 'ȯrd·ərd ə'rā }

ordered field [MATH] A field with an ordering as a set analogous to the properties of less than or equal for real numbers relative to addition and multiplication. { 'ȯrd·ərd 'fēld }

ordered list [COMPUT SCI] A set of data items that has been arranged in a specified sequence to aid in processing its contents. { 'ȯrd·ərd 'list }

ordered *n*-tuple [MATH] A set of n elements, $x_1, x_2, \ldots, x_n$, written $(x_1, x_2, \ldots, x_n)$, where x_1 is distinguished as first, x_2 as second, and so on. { 'ȯrd·ərd 'en ‚təp·el }

ordered octad [MYCOL] A linear sequence of pairs of each of four haploid cells produced by a postmeiotic division within a fungal ascus. { 'ȯrd·ərd 'äk‚tad }

ordered pair [MATH] A pair of elements x and y from a set, written (x,y), where x is distinguished as first and y as second. { 'ȯrd·ərd 'per }

ordered partition [MATH] For a set A, an ordered sequence whose members are the members of a partition of A. { ‚ȯrd·ərd pär'tish·ən }

ordered quadruple [MATH] A set of four elements, distinguished as first, second, third, and fourth. { ‚ȯrd·ərd kwä'drüp·əl }

ordered rings [MATH] Rings which have an ordering on them as sets in a manner analogous to the behavior of the usual ordering of the real numbers relative to addition and multiplication. { 'ȯrd·ərd 'riŋz }

ordered tetrad [MYCOL] A linear sequence of four haploid meiotic cells within a fungal ascus. { 'ȯrd·ərd 'te‚trad }

ordered triple [MATH] A set of three elements, written (x,y,z), where x is distinguished as first, y as second, and z as third. { 'ȯrd·ərd 'trip·əl }

ordering [MATH] A binary relation, denoted $\leq$, among the elements of a set such that $a \leq b$ and $b \leq c$ implies $a \leq c$, and $a \leq b$, $b \leq a$ implies $a = b$; it need not be the case that either $a \leq b$ or $b \leq a$. Also known as order; order relation; partial ordering. [SOLID STATE] A solid-state transformation in certain solid solutions, in which a random arrangement in the lattice is transformed into a regular ordered arrangement of the atoms with respect to one another; a so-called superlattice is formed. { 'ȯrd·ə·riŋ }

orderly shutdown [COMPUT SCI] The procedures for shutting off a computer system in an organized manner, normally after all work in progress has been completed, permitting restarting of the systems without loss of transactions or data. { 'ȯrd·ər·lē 'shət‚daůn }

order of aberration [OPTICS] The sum of the powers to which the field height and the pupil coordinates are raised in describing a term in the decomposition of an aberration according to the degree of dependence on these variables. { 'ȯrd·ər əv ‚ab·ə'rā·shən }

order of degeneracy *See* degree of degeneracy. { 'ȯrd·ər əv di'jen·ə·rə·sē }

order of interference [OPTICS] The difference in the number of wavelengths along the paths of two constructively interfering rays of light. { 'ȯrd·ər əv ‚in·tər'fir·əns }

order of magnitude *See* order. { 'ȯrd·ər əv 'mag·nə‚tüd }

order of phase transition [THERMO] A phase transition in which there is a latent heat and an abrupt change in properties, such as in density, is a first-order transition; if there is not such a change, the order of the transition is one greater than the lowest derivative of such properties with respect to temperature which has a discontinuity. { 'ȯrd·ər əv 'fāz tran‚zish·ən }

order parameter [STAT MECH] A measure of the degree of ordering of a system which has value zero above the temperature of a phase transition and acquires some nonzero value below the transition temperature. { 'ȯrd·ər pə‚ram·əd·ər }

order point [IND ENG] The inventory level at which a replenishment order must be placed. { 'ȯrd·ər ‚pȯint }

order quantity [IND ENG] The number of pieces ordered to replenish the inventory. { 'ȯrd·ər ‚kwän·əd·ē }

order relation *See* ordering. { 'ȯrd·ər ri‚lā·shən }

order statistics [STAT] Variate values arranged in ascending order of magnitude; for example, first-order statistic is the smallest value of sample observations. { 'ȯr·dər stə‚tis·tiks }

order tone [COMMUN] Tone sent over a trunk to indicate that the trunk is ready to receive an order or, to the receiving operator, that an order is about to arrive. { 'ȯrd·ər ‚tōn }

ordinal number [MATH] A generalized number which expresses the size of a set, in the sense of "how many" elements. { 'ȯrd·nəl 'nəm·bər }

ordinal scale measurement [STAT] A method of measuring quantifiable data in nonparametric statistics that is considered to be stronger than nominal scale; it expresses the relationship of order by characterizing objects by relative rank. { ‚ȯrd·nəl ‚skāl 'mezh·ər·mənt }

ordinal type [COMPUT SCI] A data type whose possible values are sequential in the manner of the integers 1, 2, 3, and so forth; for example, the months January, February, and so forth. { 'ȯrd·nəl 'tīp }

ordinary cut [MIN ENG] The arrangement of drill holes in a mine in which the drill holes are symmetrical with respect to the vertical center-line of the section, extend horizontally, and make a large angle with the working face. { 'ȯrd·ən‚er·ē 'kət }

ordinary differential equation [MATH] An equation involving functions of one variable and their derivatives. { 'ȯrd·ən‚er·ē ‚dif·ə'ren·chəl i'kwā·zhən }

ordinary gear train [MECH ENG] A gear train in which all axes remain stationary relative to the frame. { 'ȯrd·ən‚er·ē 'gir ‚trān }

ordinary generating function *See* generating function. { ‚ȯrd·ən‚er·ē 'jen·ə‚rād·iŋ ‚fəŋk·shən }

ordinary index [OPTICS] The index of refraction of the ordinary ray in a crystal. { 'ȯrd·ən‚er·ē 'in‚deks }

ordinary point [MATH] A point of a curve where a curve does not cross itself and where there is a smoothly turning tangent. Also known as regular point; simple point. { 'ȯrd·ən‚er·ē 'pȯint }

ordinary ray [OPTICS] One of two rays into which a ray incident on an anisotropic uniaxial crystal is split; it obeys the ordinary laws of refraction, in contrast to the extraordinary ray. Also known as O ray. { 'ȯrd·ən‚er·ē 'rā }

ordinary sheathed explosive [MATER] A permitted explosive (one passing certain safety tests) whose safety has been further increased by a sheath of sodium bicarbonate; when the explosive is detonated, the sheath forms carbon dioxide, which tends to extinguish the flame around the detonator wave. { 'ȯrd·ən‚er·ē 'shēthd ik'splō·siv }

ordinary singular point [MATH] A singular point at which the tangents to all branches at the point are distinct. { ‚ȯrd·ən‚er·ē ‚siŋ·gyə·lər 'pȯint }

ordinary tides [OCEANOGR] Tides which have cycles of 12 to 24 hours. { 'ȯrd·ən‚er·ē 'tīdz }

ordinary-wave component [GEOPHYS] One of the two components into which an electromagnetic wave entering the ionosphere is divided under the influence of the earth's magnetic field; it has characteristics more nearly like those expected in the absence of a magnetic field. Also known as O-wave component. [OPTICS] The component of electromagnetic radiation propagating in an anisotropic uniaxial crystal whose electric displacement vector is perpendicular to the optical axis and the direction normal to the wavefront; gives rise to the ordinary ray. { 'ȯrd·ən‚er·ē ‚wāv kəm‚pō·nənt }

ordinate [MATH] The perpendicular distance of a point (x,y) of the plane from the x axis. { 'ȯrd·ən·ət }

ordnance [ENG] Military materiel, such as combat weapons of all kinds, with ammunition and equipment for their use, vehicles, and repair tools and machinery. { 'ȯrd·nəns }

ordnance service [ORD] All activities necessary to maintain in usable condition the ordnance equipment of a command and such other equipment as directed by proper authority. { 'ȯrd·nəns ‚sər·vəs }

ordnance stores [ORD] All commodities and materials used by the U.S. Army Ordnance Corps in the design, manufacture, testing, preservation, and overhaul of ordnance property or supplies. { 'ȯrd·nəns ‚stȯrz }

ordnance supplies [ORD] All military supplies assigned to the U.S. Army Ordnance Corps for storage, issue, and maintenance; ordnance supplies consist of all raw materials, completely manufactured articles, and parts of such articles assigned to the corps. { 'ȯrd·nəns sə‚plīz }

ordnance troops [ORD] Technically trained troops assigned or attached to a tactical unit to provide ordnance maintenance, supply, or technical service; they also give instruction in the use, maintenance, and adjustment of ordnance materiel. { 'ȯrd·nəns ,trüps }

Ordovician [GEOL] The second period of the Paleozoic era, above the Cambrian and below the Silurian, from approximately 500 million to 440 million years ago. { ȯrd·ə'vish·ən }

ore [GEOL] **1.** The naturally occurring material from which economically valuable minerals can be extracted. **2.** Specifically, a natural mineral compound of the elements, of which one element at least is a metal. **3.** More loosely, all metalliferous rock, though it contains the metal in a free state. **4.** Occasionally, a compound of nonmetallic substances, as sulfur ore. { ȯr }

ore bed [GEOL] An economic aggregation of minerals occurring between or in rocks of sedimentary origin. { 'ȯr ,bed }

ore bin [MIN ENG] A receptacle for ore awaiting treatment or shipment. { 'ȯr ,bin }

ore block [MIN ENG] A vein of ore which is bound above, below, and at one or both ends; it is ready for excavation. { 'ȯr ,bläk }

ore blocked out [MIN ENG] Ore exposed on three sides within a reasonable distance of each other. { 'ȯr 'bläkt ,au̇t }

orebody [GEOL] Generally, a solid and fairly continuous mass of ore, which may include low-grade ore and waste as well as pay ore, but is individualized by form or character from adjoining country rock. { 'ȯr,bäd·ē }

ore bridge [MIN ENG] A gantry crane used to load and unload stockpiles of ore. { 'ȯr ,brij }

ore/bulk/oil carrier [NAV ARCH] A ship for transporting bulk cargo such as coal and grain, and high-density cargoes such as iron ore, as well as crude petroleum products. Abbreviated O/B/O carrier. { 'ȯr ¦bəlk ¦ȯil ,kar·ē·ər }

ore car [MIN ENG] A mine car for carrying ore or waste rock. { 'ȯr ,kär }

ore carrier [NAV ARCH] A vessel designed to carry ore in bulk, and similar in construction to a collier. { 'ȯr ,kar·ē·ər }

ore chimney See pipe. { 'ȯr ,chim·nē }

ore chute [MIN ENG] An inclined passage for the transfer of ore to a lower level. { 'ȯr ,shüt }

ore cluster [GEOL] A group of interconnected ore bodies. { 'ȯr ,kləs·tər }

ore control [GEOL] A geologic feature that has influenced the ore deposition. { 'ȯr ,kən'trōl }

ore crusher [MIN ENG] A machine for breaking up masses of ore, usually prior to passing through other size-reduction equipment. { 'ȯr ,krəsh·ər }

Orectolobidae [VERT ZOO] An ancient isurid family of galeoid sharks, including the carpet and nurse sharks, which are primarily bottom feeders with small teeth and a blunt rostrum with barbels near the mouth. { ȯ,rek·tə'läb·ə,dē }

ore deposit [GEOL] Rocks containing minerals of economic value in such amount that they can be profitably exploited. { 'ȯr di,päz·ət }

ore developed [MIN ENG] Ore exposed on four sides in blocks variously prescribed. { 'ȯr di'vel·əpt }

ore district [GEOL] A combination of several ore deposits into one common whole or system. { 'ȯr ,dis,trikt }

ore dressing [MIN ENG] The cleaning of ore by the removal of certain valueless portions, as by jigging, cobbing, or vanning. { 'ȯr ,dres·iŋ }

ore expectant [MIN ENG] The whole or any part of the ore below the lowest level or beyond the range of vision. { 'ȯr ik'spek·tənt }

ore faces [MIN ENG] Those ore bodies that are exposed on one side, or show only one face. { 'ȯr ,fās·əz }

ore flotation promoter [MATER] Material that gives a water-repellent surface to mineral particles so that air bubbles will adhere and cause selective flotation. { 'ȯr flō'tā·shən prə,mōd·ər }

oregano [FOOD ENG] A spice prepared from leaves of various aromatic mints, especially wild marjoram. { ə'reg·ə,nō }

ore grader [MIN ENG] In metal mining, a person who directs the storage of iron ores in bins at shipping docks so that the various grades in each bin will contain approximate percentages of iron. { 'ȯr ,grād·ər }

oreide bronze [MET] A series of brass compositions containing 68–87% copper, 10–32% zinc, and sometimes small amounts of tin; used for hardware. { 'ō·rē,īd 'bränz }

ore in sight [MIN ENG] **1.** Ore exposed on at least three sides within reasonable distance of each other. **2.** Ore which may be reasonably assumed to exist, though not actually blocked out. **3.** See developed reserves. { 'ȯr in 'sīt }

ore intersection [MIN ENG] **1.** The point at which a borehole, crosscut, or other underground opening encounters an ore vein or deposit. **2.** The thickness of the ore-bearing deposit so traversed. { 'ȯr ,in·tər,sek·shən }

ore-lead age [GEOL] An estimate of the age of the earth made by comparing the relative progress of the two radioactive decay schemes ^{235}U-^{207}Pb and ^{238}U-^{206}Pb. { 'ȯr 'led ,āj }

ore microscopy [MINERAL] The use of a reflecting microscope to study polished sections of ore minerals. Also known as mineragraphy; mineralography. { 'ȯr mī'kräs·kə·pē }

orendite [PETR] A porphyritic extrusive rock containing phlogopite phenocrysts in a nepheline-free reddish-gray groundmass of leucite, sanidine, phlogopite, amphibole, and diopside. { 'ȯr·ən,dīt }

oreodont [PALEON] Any member of the family Merycoidodontidae. { 'ȯr·ē·ō,dänt }

ore/oil carrier [NAV ARCH] A ship capable of transporting dry, high-density cargoes, such as iron ore, as well as crude petroleum products. Abbreviated O/O carrier. { 'ȯr 'ȯil ,kar·ē·ər }

ore of sedimentation See placer. { 'ȯr əv ,sed·ə·mən'tā·shən }

ore pass [MIN ENG] A vertical or inclined passage for the downward transfer of ore. { 'ȯr ,pas }

ore pipe See pipe. { 'ȯr ,pīp }

ore pocket [MIN ENG] **1.** Excavation near the hoisting shaft into which ore from stopes is moved, preliminary to hoisting. **2.** An unusual concentration of ore in the lode. { 'ȯr ,päk·ət }

ore reduction [MIN ENG] The size reduction of solids by crushing and grinding in mineral processing plants. { 'ȯr ri,dək·shən }

ore reserve [MIN ENG] The total tonnage and average value of proved ore, plus the total tonnage and value (assumed) of the probable ore. { 'ȯr ri,zərv }

ore sampling [MIN ENG] The process in which a portion of ore is selected so that its composition will represent the average composition of the entire bulk of ore. { 'ȯr ,sam·pliŋ }

ore shoot [GEOL] **1.** A large, generally vertical, pipelike ore body that is economically valuable. Also known as shoot. **2.** A large and usually rich aggregation of mineral in a vein. { 'ȯr ,shüt }

ore/slurry/oil carrier [NAV ARCH] A ship for transporting dry bulk commodities, such as slurry, or dry iron ore products, as well as crude petroleum products. Abbreviated O/S/O carrier. { 'ȯr ¦slə·rē ¦ȯil ,kar·ē·ər }

ORF See open reading frame. { ȯrf or ¦ō¦är'ef }

OR function See or. { 'ȯr ,faŋk·shən }

organ [ANAT] A differentiated structure of an organism composed of various cells or tissues and adapted for a specific function. { 'ȯr·gən }

organdy [TEXT] A sheer, crisp fabric woven of fine combed cotton yarns. { 'ȯr·gən·dē }

organelle [CELL MOL] A specialized subcellular structure, such as a mitochondrion, having a special function; a condensed system showing a high degree of internal order and definite limits of size and shape. { 'ȯr·gə,nel }

organic [ORG CHEM] Of chemical compounds, based on carbon chains or rings and also containing hydrogen with or without oxygen, nitrogen, or other elements. { ȯr'gan·ik }

organic acid [ORG CHEM] A chemical compound with one or more carboxyl radicals (COOH) in its structure; examples are butyric acid, $CH_3(CH_2)_2COOH$, maleic acid, HOOCCHCH-COOH, and benzoic acid, C_6H_5COOH. { ȯr'gan·ik 'as·əd }

organic bonded wheel [DES ENG] A grinding wheel in which organic bonds are used to hold the abrasive grains. { ȯr'gan·ik ¦bän·dəd 'wēl }

organic brain syndrome [MED] A mental condition of multiple etiologies, resulting in diffuse impairment of brain tissue function and manifested by a complex of symptoms including impaired judgment and intellectual function, and often somatic and motor dysfunctions. { ȯr'gan·ik 'brān ,sin,drōm }

organic chelates [AGR] Chelates formed by reaction of organic compounds with the mineral end products of weathering; they enhance nutrient richness of soils by forming organo-mineral complexes that are easy for plants to absorb. { ȯr,gan·ik 'kē,lāts }

organic chemistry [CHEM] The study of the structure, preparation, properties, and reactions of carbon compounds. { ȯr′gan·ik ′kem·ə·strē }

organic coating [MATER] Material used to protect metal surfaces from chemical or atmospheric attack; includes latex paints, plastics, asphaltic materials, rubbers, and elastomers. { ȯr′gan·ik ′kōd·iŋ }

organic-cooled reactor [NUCLEO] A reactor that uses organic chemicals, such as mixtures of polyphenyls (diphenyls and terphenyls), as coolant. { ȯr′gan·ik ′küld rē′ak·tər }

organic conductor [MATER] A two-component material containing anion and cation charged species originating from a charge transfer between two inorganic molecules or between one organic molecule and one inorganic ion. { ȯr′gan·ik kən′dək·tər }

organic electrolyte cell [ELEC] A type of wet cell that is based on the use of particularly reactive metals such as lithium, calcium, or magnesium in conjunction with organic electrolytes; the best-known type is the lithium-cupric fluoride cell. { ȯr′gan·ik i′lek·trə‚līt ‚sel }

organic evolution [EVOL] The processes of change in organisms by which descendants come to differ from their ancestors, and a history of the sequence of such changes. { ȯr′gan·ik ‚ev·ə′lü·shən }

organic geochemistry [GEOCHEM] A branch of geochemistry which deals with naturally occurring carbonaceous and biologically derived substances which are of geological interest. { ȯr′gan·ik ‚jē·ō′kem·ə·strē }

organic glass [MATER] An amorphous, solid, glasslike material made of transparent plastic. { ȯr′gan·ik ′glas }

organicism See holism. { ȯr′gan·ə‚siz·əm }

organic lattice See growth lattice. { ȯr′gan·ik ′lad·əs }

organic-moderated reactor [NUCLEO] A nuclear reactor in which organic compounds are used as moderator and coolant. { ȯr′gan·ik ′mäd·ə‚rād·əd rē′ak·tər }

organic mood disorder [PSYCH] Depressions and manic episodes that occur secondary to organic illnesses, including neurologic disorders and systemic medical illnesses, and as adverse effects of drugs commonly used in the treatment of medical conditions. { ȯr′gan·ik ′müd dis‚ȯrd·ər }

organic mound See bioherm. { ȯr′gan·ik ′maund }

organic pigment [ORG CHEM] Any of the materials with organic-chemical bases used to add color to dyes, plastics, linoleum, tones, and lakes. { ȯr′gan·ik ′pig·mənt }

organic quantitative analysis [ANALY CHEM] Quantitative determination of elements, functional groups, or molecules in organic materials. { ȯr′gan·ik ′kwän·ə‚tād·iv ə‚nal·ə·səs }

organic reaction mechanism [ORG CHEM] A pathway of chemical states traversed by an organic chemical system in its passage from reactants to products. { ȯr′gan·ik rē′ak·shən ‚mek·ə‚niz·əm }

organic reef [GEOL] A sedimentary rock structure of significant dimensions erected by, and composed almost exclusively of the remains of, corals, algae, bryozoans, sponges, and other sedentary or colonial organisms. { ȯr′gan·ik ′rēf }

organic rock [PETR] A sedimentary rock composed principally of the remains of plants and animals. { ȯr′gan·ik ′räk }

organic salt [ORG CHEM] The reaction product of an organic acid and an inorganic base, for example, sodium acetate (CH_3COONa) from the reaction of acetic acid (CH_3COOH) and sodium hydroxide ($NaOH$). { ȯr′gan·ik ′sȯlt }

organic semiconductor [MATER] An organic material having unusually high conductivity, often enhanced by the presence of certain gases, and other properties commonly associated with semiconductors; an example is anthracene. { ȯr′gan·ik ′sem·i·kən‚dək·tər }

organic soil [GEOL] Any soil or soil horizon consisting chiefly of, or containing at least 30% of, organic matter; examples are peat soils and muck soils. { ȯr′gan·ik ′sȯil }

organic solvent [ORG CHEM] Liquid organic compound with the power to dissolve solids, gases, or liquids (miscibility); examples are methanol (methyl alcohol), CH_3OH, and benzene, C_6H_6. { ȯr′gan·ik ′säl·vənt }

organic texture [GEOL] A sedimentary texture resulting from the activity of organisms such as the secretion of skeletal material. { ȯr′gan·ik ′teks·chər }

organic weathering [GEOL] Biological processes and changes that contribute to the breakdown of rocks. Also known as biological weathering. { ȯr′gan·ik ′weth·ə·riŋ }

organism [BIOL] An individual constituted to carry out all life functions. { ′ȯr·gə‚niz·əm }

organismic psychology [PSYCH] A movement in psychology based on the theory that the individual is made up of elements composing a single organized system, and an element in the system cannot be evaluated independently of its position within the system. { ‚ȯr·gə′niz·mik sī′käl·ə·jē }

organizational reengineering [SYS ENG] The study, capture, and modification of the internal mechanisms or functionality of existing system-management processes and practices in an organization in order to reconstitute them in a new form and with new features, often to take advantage of newly emerged organizational competitiveness requirements, but without changing the inherent purpose of the organization itself. Also known as systems management reengineering. { ‚ȯr·gə·nə‚zā·shən·əl ‚rē‚en·jə′nir·iŋ }

organization chart [IND ENG] Graphic representation of the interrelationships within an organization, depicting lines of authority and responsibility and provisions for control. { ‚ȯr·gə·nə′zā·shən ‚chärt }

organized ferment See intracellular enzyme. { ‚ȯr·gə·nə′zā·shən ′fər·mənt }

organizer [EMBRYO] Any part of the embryo which exerts a morphogenetic stimulus on an adjacent part or parts, as in the induction of the medullary plate by the dorsal lip of the blastopore. { ′ȯr·gə‚niz·ər }

organizing pneumonia [MED] Pneumonia in which the healing process is characterized by organization and cicatrization of the exudate rather than by resolution and resorption. Also known as unresolved pneumonia. { ′ȯr·gə‚niz·iŋ nu̇′mō·nyə }

organoborane [ORG CHEM] A derivative of a borane (boron hydride) in which one or more hydrogen atoms have been replaced by functional groups. { ȯr‚gan·ə′bȯr‚ān }

organ of Corti [NEUROSCI] A specialized structure located on the basilar membrane of the mammalian cochlea, which contains rods of Corti and hair cells connected to ganglia of the cochlear nerve. Also known as spiral organ. { ′ȯr·gən əv ′kȯrd·ē }

organ of Leydig [VERT ZOO] Two large accumulations of lymphoid tissues which run longitudinally the length of the esophagus in selachian fishes. { ′ȯr·gən əv ′lī‚dig }

organogenesis [EMBRYO] The formation of an organ. { ‚ȯr‚gan·ə′jen·ə·səs }

organogenic [GEOL] Property of a rock or sediment derived from organic substances. { ȯr‚gan·ə′jen·ik }

organoleptic [PHYSIO] Having an effect or making an impression on sense organs; usually used in connection with subjective testing of food and drug products. { ȯr‚gan·ə′lep·tik }

organolite [GEOL] Any rock consisting mainly of organic material. { ȯr′gan·ə‚līt }

organometallic compound [ORG CHEM] Molecules containing carbon-metal linkage; a compound containing an alkyl or aryl radical bonded to a metal, such as tetraethyllead, $Pb(C_2H_5)_4$. { ȯr‚gan·ə·mə′tal·ik ′käm‚paund }

organophosphate [ORG CHEM] A soluble fertilizer material made up of organic phosphate esters such as glucose, glycol, or sorbitol; useful for providing phosphorus to deep-root systems. { ȯr‚gan·ə′fäs‚fāt }

organophosphorus compound [ORG CHEM] An organic compound that contains phosphorus in its chemical structure. { ȯr‚gan·ə′fäs·fə·rəs ′käm‚paund }

organoselenium compound [ORG CHEM] An organic compound that contains both selenium and carbon, and frequently other elements, such as the halogens, oxygen, sulfur, or nitrogen. { ȯr‚gan·ə·sə′lē·nē·əm ′käm‚paund }

organosilicon compound [ORG CHEM] A compound in which silicon is bonded to an organic functional group, either directly or indirectly via another atom. { ȯr‚gan·ō′sil·ə·kən ′käm‚paund }

organosol [MATER] 1. Finely divided or colloidal suspension of insoluble material in a suspending organic liquid; known as plastisol when the solid is a synthetic resin suspended in an organic liquid; used for coatings, moldings, and casting of films. 2. A dispersion of very finely divided resin particles that are suspended in an organic-liquid mixture which cannot dissolve the resin at normal temperatures. { ȯr′gan·ə‚sȯl }

organosulfur compound [ORG CHEM] One of a group of

substances which contain both carbon and sulfur. { ȯr'gar·ə̇sə̇l·fər 'käm,paủnd }

organotropic [MICROBIO] Of microorganisms, localizing in or entering the body by way of the viscera or, occasionally, somatic tissue. { ȯr¦gan·ə̇träp·ik }

organs of Zuckerkandl See aortic paraganglion. { 'ȯr·gənz əv 'tsủk·ər,känd·əl }

organza [TEXT] An organdy fabric woven from silk yarns. { ȯr'gan·zə }

organzine [TEXT] A fine silk yarn of the best quality, consisting of two or three filaments twisted together. { 'ȯr·gən,zēn }

orgasm [PHYSIO] The intense, diffuse, and subjectively pleasurable sensation experienced during sexual intercourse or genital manipulation, culminating in the male with seminal ejaculation and in the female with uterine contractions, warm suffusion, and pelvic throbbing sensations. { 'ȯr,gaz·əm }

orgasmolepsy [MED] Sudden loss of muscle tone during orgasm, accompanied by a transitory loss of consciousness. { ȯr¦gaz·mə̇lep·sē }

OR gate [ELECTR] A multiple-input gate circuit whose output is energized when any one or more of the inputs is in a prescribed state; performs the function of the logical inclusive-or; used in digital computers. Also known as OR circuit. { 'ȯr ,gāt }

Oribatei [INV ZOO] A heavily sclerotized group of free-living mites in the suborder Sarcoptiformes which serve as intermediate hosts of tapeworms. { ,ȯr·ə'bad·ē,ī }

Oribatulidae [INV ZOO] A family of oribatid mites in the suborder Sarcoptiformes. { ,ȯr·ə·bə'tül·ə,dē }

orient [COMPUT SCI] To change relative and symbolic addresses to absolute form. [ENG] **1.** To place or set a map so that the map symbols are parallel with their corresponding ground features. **2.** To turn a transit so that the direction of the 0° line of its horizontal circle is parallel to the direction it had in the preceding or initial setup, or parallel to a standard reference line. [OPTICS] The play of color upon or just below the surface of a gem-quality pearl. { 'ȯr·ē·ənt }

orientability of sound signal [ACOUS] The property of a sound signal by virtue of which a listener can estimate the direction of the location of the apparatus producing the signal. { 'ȯr·ē,ent·ə'bil·ə̇d·ē əv 'saủnd ,sig·nəl }

orientable surface [MATH] A surface for which an object resting on one side of it cannot be moved continuously over it to get to the other side without going around an edge. { ,ȯr·ē,en·tə·bəl 'sər·fəs }

oriental alabaster See onyx marble. { ,ȯr·ē'ent·əl 'al·ə,bas·tər }

oriental amethyst [MINERAL] A violet to purple variety of sapphire. { ,ȯr·ē'ent·əl 'am·ə,thist }

oriental jasper See bloodstone. { ,ȯr·ē'ent·əl 'jas·pər }

oriental linaloe [MATER] A rosewood oil distilled from highly perfumed parts of *Aquilaria agollocha* trees of Burma, eastern India, and Java. Also known as agar attar; aloe wood oil. { ,ȯr·ē'ent·əl lə'nal·ō }

oriental topaz [MINERAL] A yellow variety of corundum, used as a gem. { ,ȯr·ē'ent·əl 'tō,paz }

Oriental zoogeographic region [ECOL] A zoogeographic region which encompasses tropical Asia from the Iranian Peninsula eastward through the East Indies to, and including, Borneo and the Philippines. { ,ȯr·ē'ent·əl ,zō·ə,jē·ə'graf·ik ,rē·jən }

orientation [CRYSTAL] The directions of the axes of a crystal lattice relative to the surfaces of the crystal, to applied fields, or to some other planes or directions of interest. [ELECTROMAG] The physical positioning of a directional antenna or other device having directional characteristics. [ENG] Establishment of the correct relationship in direction with reference to the points of the compass. [MATH] **1.** A choice of sense or direction in a topological space. **2.** An ordering $p_0, p_1, \ldots, p_n$ of the vertices of a simplex, two such orderings being regarded as equivalent if they differ by an even permutation. **3.** For a simple graph, a directed graph that results from assigning a direction to each of the edges. [PHYS] **1.** The direction of some vector or set of vectors, such as the direction of the electric vector and the propagation direction of plane polarized light, or the direction of a preponderance of nuclear spins in a crystal near absolute zero, relative to some other directions of interest. **2.** Any process in which vectors associated with atoms or molecules in the substance are organized relative to

some direction, rather than pointed at random; examples include dipole moments of polar molecules in an electric field, and nuclear spins in a crystal in a magnetic field at temperatures near absolute zero. [PHYS CHEM] The arrangement of radicals in an organic compound in relation to each other and to the parent compound. [PSYCH] Determination of one's relation to the environment. { ,ȯr·ē·ən'tā·shən }

orientation diagram [GEOL] Any point or contour diagram used in structural petrology. { ,ȯr·ē·ən'tā·shən ,dī·ə,gram }

orientation effect [ELEC] Those bulk properties of a material which result from orientation polarization. [PHYS CHEM] A method of determining attractive forces among molecules, or components of these forces, from the interaction energy associated with the relative orientation of molecular dipoles. { ,ȯr·ē·ən'tā·shən i,fekt }

orientation force [PHYS CHEM] A type of van der Waals force, resulting from interaction of the dipole moments of two polar molecules. Also known as dipole-dipole force; Keesom force. { ,ȯr·ē·ən'tā·shən ,fȯrs }

orientation polarization [ELEC] Polarization arising from the orientation of molecules which have permanent dipole moments arising from an asymmetric charge distribution. Also known as dipole polarization. { ,ȯr·ē·ən'tā·shən ,pō·lə·rə,zā·shən }

orientation vector [MECH ENG] A vector whose direction indicates the orientation of a robot gripper. { ,ȯr·ē·ən'tā·shən ,vek·tər }

oriented [GEOL] Pertaining to a specimen that is so marked as to show its exact, original position in space. { 'ȯr·ē,ent·əd }

oriented core [ENG] A core that can be positioned on the surface in the same way that it was arranged in the borehole before extraction. { 'ȯr·ē,ent·əd ,kȯr }

oriented graph [MATH] A directed graph in which there is no pair of points *a* and *b* such that there is both an arc directed from *a* to *b* and an arc directed from *b* to *a*. { 'ȯr·ē,ent·əd 'graf }

oriented simplex [MATH] A simplex for which an order has been assigned to the vertices. { ,ȯr·ē,ent·əd 'sim,pleks }

oriented simplicial complex [MATH] A simplicial complex each of whose simplexes is an oriented simplex. { ,ȯr·ē,ent·əd sim¦plish·əl 'käm,pleks }

oriented-strategic research [SCI TECH] That background research considered to be required in order to achieve a specified objective. { 'ȯr·ē,ent·əd strə'tē·jik rē'sərch }

orifice [ELECTROMAG] Opening or window in a side or end wall of a waveguide or cavity resonator through which energy is transmitted. [SCI TECH] An aperture or hole. { 'ȯr·ə·fəs }

orifice gas [MET] The torch gas in a plasma arc welding or cutting process which becomes ionized in the arc to form plasma and is ejected from the orifice in a jet stream. { 'ȯr·ə·fəs 'gas }

orifice meter [ENG] An instrument that measures fluid flow by recording differential pressure across a restriction placed in the flow stream and the static or actual pressure acting on the system. { 'ȯr·ə·fəs ,mēd·ər }

orifice mixer [MECH ENG] Arrangement in which two or more liquids are pumped through an orifice constriction to cause turbulence and consequent mixing action. { 'ȯr·ə·fəs ,mik·sər }

orifice plate [DES ENG] A disk, with a hole, placed in a pipeline to measure flow. { 'ȯr·ə·fəs ,plāt }

orifice well tester [PETRO ENG] Velocity-type meter used to measure gas flow quantity from a gas well; static pressure differences before and after a sharp-edged orifice are converted to flow values. { 'ȯr·ə·fəs 'wel ,tes·tər }

origanum oil [MATER] A light-yellow essential oil obtained from herbs of the genus *Origanum*; contains carvacrol and cymene plus other components; used in flavors and pharmaceuticals. { ə'rig·ə·nəm ,ȯil }

Origem Loop [ASTRON] A loop of gas on the boundary between Orion and Gemini about 60 parsecs (1.2×10^{15} miles or 1.8×10^{15} kilometers) in radius and 1000 parsecs (2×10^{16} miles or 3×10^{16} kilometers) distant, with at least five nebulae embedded in it. { 'ȯr·ə,jem 'lüp }

origin [ANAT] The point at which the nonmoving end of a muscle is attached to a bone; it is at the proximal end of the muscle. [COMPUT SCI] Absolute storage address in relative coding to which addresses in a region are referenced. [MATH] The point of a coordinate system at which all coordinate axes meet. { 'är·ə·jən }

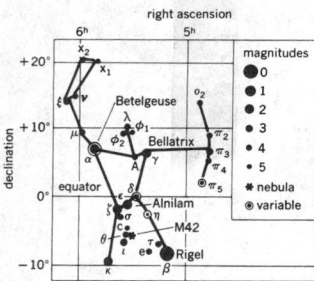

ORION

Line pattern of the constellation Orion. The grid lines represent the coordinates of the sky. The apparent brightness, or magnitude, of the stars is shown by the size of the dots, graded by appropriate numbers.

ORION NEBULA

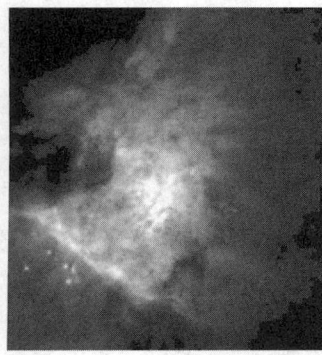

Hubble Space Telescope image of the inner portion of the Orion Nebula. (*C.R. O'Dell and S.K. Wong, Rice University; NASA*)

original dip *See* primary dip. { əˈrij·ən·əl ˈdip }

original document *See* source document. { əˈrij·ən·əl ˈdäk·yə·mənt }

original duration [IND ENG] The initial estimate of length of time required to complete a given activity. { əˈrij·ən·əl də'rā·shən }

original interstice [PETR] An interstice that formed contemporaneously with the enclosing rock. Also known as primary interstice. { əˈrij·ən·əl ˈin·tər,stīs }

original valley [GEOL] A valley formed by hypogene action or by epigene action other than that of running water. { əˈrij·ən·əl ˈval·ē }

origin of replication [CELL MOL] The nucleotide sequence from which deoxyribonucleic acid replication begins. { ˈär·ə·jən əv ,rep·lə'kā·shən }

origin of rifling [ORD] The position in a rifled gun bore at which the rifling begins; more specifically, the plane, perpendicular to the axis of the gun bore, in which the rifling starts. { ˈär·ə·jən əv ˈrīf·liŋ }

origin of the trajectory [ORD] Center of the muzzle of a gun at the instant when the projectile leaves it. { ˈär·ə·jən əv thə trə'jek·tə·rē }

O ring [DES ENG] A flat ring made from synthetic rubber, used as an airtight seal or a seal against high pressures. { ˈō ˌriŋ }

Orion [ASTRON] A northern constellation near the celestial equator, right ascension 5 hours, declination 5° north. Also known as Warrior. { əˈrī·ən }

Orion A [ASTRON] A giant molecular cloud, 100,000 times more massive than the sun, mostly made of molecular hydrogen but best traced in the 2.6-millimeter emission line of ^{13}CO, the rarer isotopic variant of carbon monoxide; the Orion Nebula is located in front of its northern part. { əˌrī·ən ˈā }

Orion arm [ASTRON] The spiral arm of the Milky Way Galaxy that has a spur in which the sun is located. Also known as local arm. { əˈrī·ən ˌärm }

Orion B [ASTRON] A giant molecular cloud located in the northern part of the constellation Orion. { əˌrī·ən ˈbē }

Orionids [ASTRON] A meteor shower seen in October in the northern hemisphere; its radiant lies in the constellation Orion. { əˈrī·əˌnidz }

Orion molecular clouds [ASTRON] Two molecular clouds in the Orion Nebula; one has about 300,000 hydrogen molecules per cubic centimeter and contains the Becklin-Nengebauer object and the Kleinmann-Low Nebula, while the other is centered on a cluster of infrared sources. { əˈrī·ən mə'lek·yə·lər ˈklaudz }

Orion Nebula [ASTRON] A luminous cloud surrounding Ori, the northern star in Orion's dagger; visible to the naked eye as a hazy object. Also known as Great Nebula of Orion. { əˈrī·ən ˈneb·yə·lə }

Orion OB association [ASTRON] A loose, gravitationally unbound grouping of hot massive stars with spectral types A, B, and O, which originated in the giant molecular clouds Orion A and Orion B, together with tens of thousands of low-mass young stars which formed from the Orion molecular clouds. { əˌrī·ən ˌōˈbē əˌsō·shē,ā·shən }

Orion spur [ASTRON] That portion of the Orion arm within which the sun is located. { əˈrī·ən ˈspər }

Orleans process [MICROBIO] An older commercial method of vinegar production in which fermentation is carried out in a large cask in which holes have been drilled to permit the introduction of air. The cask has a spigot for the withdrawal of finished vinegar, and the bung (stopper) contains a tube so that fresh wine or other substrate can be added without disturbing the film of vinegar bacteria. { ˈór·lē·ənz ˌprä·ses }

orlop deck [NAV ARCH] The lowest continuous deck of a ship having four or more decks. { ˈór,läp ˌdek }

ormer *See* abalone. { ˈór·mər }

Ormyridae [INV ZOO] A small family of hemipteran insects in the superfamily Chalcidoidea. { órˈmī·rəˌdē }

Orneodidae [INV ZOO] A small family of lepidopteran insects in the superfamily Tineoidea; adults have each wing divided into six featherlike plumes. { ˌór·nē'äd·əˌdē }

ornithine [BIOCHEM] $C_5H_{12}O_2N_2$ An amino acid occurring in the urine of some birds, but not found in native proteins. { ˈór·nəˌthēn }

ornithine cycle [BIOCHEM] A sequence of cyclic reactions in which potentially toxic products of protein catabolism are converted to nontoxic urea. { ˈór·nəˌthīn ˌsī·kəl }

Ornithischia [PALEON] An order of extinct terrestrial reptiles, popularly known as dinosaurs; distinguished by a four-pronged pelvis, and a median, toothless predentary bone at the front of the lower jaw. { ˌór·nə'this·kē·ə }

ornithology [VERT ZOO] The study of birds. { ˌór·nə'thäl·ə·jē }

Ornithomimus [PALEON] A 13-foot-long (4-meter) omnivorous theropod dinosaur from the Late Cretaceous Period that had large hips, a long tail, and strong hindlimbs, and closely resembled ostriches. { ˌór·nə·thō'mīm·əs }

Ornithopoda [PALEON] A suborder of extinct reptiles in the order Ornithischia including all bipedal forms in the order. { ˌór·nə'thäp·ə·də }

Ornithorhynchidae [VERT ZOO] A monospecific order of monotremes containing the semiaquatic platypus; characterized by a duck-billed snout, horny plates instead of teeth in the adult, and a flattened, well-developed tail. { ˌór·nə·thō'riŋ·kə,dē }

ornithosis [MED] Any form of psittacosis originating in birds other than psittacines. { ˌór·nə'thō·səs }

Ornstein-Uhlenbeck process [STAT] A stochastic process used as a theoretical model for Brownian motion. { ˈórn,stīn ˈü·lən,bek ˌprä·ses }

orocline [GEOL] An orogenic belt with a change in horizontal direction, either a horizontal curvature or a sharp bend. Also known as geoflex. { ˈór·ə,klīn }

orocratic [GEOL] Pertaining to a period of time in which there is much diastrophism. { ˌór·ə'krad·ik }

orogen *See* orogenic belt. { ˈór·ə·jən }

orogene *See* orogenic belt. { ˈór·ə,jēn }

orogenesis *See* orogeny. { ˌór·ə'jen·ə·səs }

orogenic belt [GEOL] A linear region that has undergone folding or other deformation during the orogenic cycle. Also known as fold belt; orogen; orogene. { ˌór·ə'jen·ik ˈbelt }

orogenic cycle [GEOL] A time interval during which a mobile belt evolved into an orogenic belt, passing through preorogenic, orogenic, and postorogenic stages. Also known as geotectonic cycle. { ˌór·ə'jen·ik ˈsī·kəl }

orogenic sediment [GEOL] Any sediment that is produced as the result of an orogeny or that is directly attributable to the orogenic region in which it is later found. { ˌór·ə'jen·ik ˈsed·ə·mənt }

orogenic unconformity [GEOL] An angular unconformity produced locally in a region affected by mountain-building movements. { ˌór·ə'jen·ik ˌən·kən'fór·məd·ē }

orogeny [GEOL] The process or processes of mountain formation, especially the intense deformation of rocks by folding and faulting which, in many mountainous regions, has been accompanied by metamorphism, invasion of molten rock, and volcanic eruption; in modern usage, orogeny produces the internal structure of mountains, and epeirogeny produces the mountainous topography. Also known as orogenesis; tectogenesis. { ó'räj·ə·nē }

orogeosyncline [GEOL] A geosyncline that later became an area of orogeny. { ˌór·ōˌjē·ō'sin,klīn }

orograph [ENG] A machine that records both distance and elevations as it is pushed across land surfaces; used in making topographic maps. { ˈór·ə,graf }

orographic [GEOL] Pertaining to mountains, especially in regard to their location and distribution. { ˌór·ə'graf·ik }

orographic cloud [METEOROL] A cloud whose form and extent is determined by the disturbing effects of orography upon the passing flow of air; because these clouds are linked with the form of the terrestrial relief, they generally move very slowly, if at all, although the winds at the same level may be very strong. { ˌór·ə'graf·ik ˈklaud }

orographic lifting [METEOROL] The lifting of an air current caused by its passage up and over surface elevations. { ˌór·ə'graf·ik ˈlift·iŋ }

orographic occlusion [METEOROL] An occluded front in which the occlusion process has been hastened by the retardation of the warm front along the windward slopes of a mountain range. { ˌór·ə'graf·ik ə'klü·zhən }

orographic precipitation [METEOROL] Precipitation which results from the lifting of moist air over an orographic barrier such as a mountain range; strictly, the amount so designated should not include that part of the precipitation which would be expected from the dynamics of the associated weather disturbance, if the disturbance were over flat terrain. { ¦ȯr·ə¦graf·ik prə,sip·ə'tā·shən }

orography [GEOGR] The branch of geography dealing with mountains. { ȯ'räg·rə·fē }

orohydrography [HYD] A branch of hydrography dealing with the relations of mountains to drainage. { ¦ȯr·ō·hī'dräg·rə·fē }

Oromericidae [VERT ZOO] An extinct family of camellike tylopod runinants in the superfamily Cameloidea. { ,ȯr·ə·mə'ris·ə,dē }

orometer [ENG] A barometer with a scale that indicates elevation above sea level. { ȯ'räm·əd·ər }

oropharynx [ANAT] The oral pharynx, located between the lower border of the soft palate and the larynx. { ¦ȯr·ō'far,iŋks }

orophyte [ECOL] Any plant that grows in the subalpine region. { 'ȯr·ə,fīt }

orotath [GEOL] An orogenic belt that has been stretched substantially in a lengthwise direction. { 'ȯr·ə,tath }

orotic acid [BIOCHEM] $C_4H_4O_4N_2$ A crystalline acid which is a growth factor for certain bacteria and is also a pyrimidine precursor. { ə'räd·ik 'as·əd }

Oroya fever [MED] The severe form of Carrion's disease, characterized by a sudden, severe, and rapid course, often fatal anemia, and remittent fever. { ə'rȯi·ə ,fē·vər }

orphan drug [PHARM] A pharmaceutical developed to treat a disease that afflicts relatively few people. { ¦ȯr·fən 'drəg }

orphan virus [VIROL] Any nonpathogenic virus found in the human digestive and respiratory systems. { ¦ȯr·fən 'vī·rəs }

orpiment [MINERAL] As_2S_3 A lemon-yellow mineral, crystallizing in the monoclinic system, and generally occurring in foliated or columnar masses; luster is resinous and pearly on the cleavage surface, hardness is 1.5–2 on Mohs scale, and specific gravity is 3.49. Also known as yellow arsenic. { 'ȯr·pə·mənt }

orrery [ASTRON] A model of the solar system equipped with mechanical devices to make the planets move at their correct relative velocities around the sun. { 'ȯr·ər·ē }

orris [MATER] The fragrant powder from the root of the plants *Iris florentina, I. germanica,* and *I. pallida;* used in perfume, medicine, and tooth powder. Also known as orrisroot. { 'ȯr·əs }

orris oil [MATER] A yellow, fatty, semisolid, fragrant essential oil obtained from roots of the Florentine iris; melts at 44–50°C; soluble in ether, chloroform, and alcohol; main components are myristic acid, oleic acid, and irone; used in flavoring and perfumes. { 'ȯr·əs ,ȯil }

orrisroot *See* orris. { 'ȯr·əs,rüt }

Orr's white *See* lithopone. { 'ȯrz 'wīt }

Orsat analyzer [ANALY CHEM] Gas analysis apparatus in which various gases are absorbed selectively (volumetric basis) by passing them through a series of preselected solvents. { 'ȯr,sat 'an·ə,līz·ər }

orseille *See* orchil. { ȯr'sā }

Orthacea [PALEON] An extinct group of articulate brachiopods in the suborder Orthidina in which the delthyrium is open. { ȯr'thäs·ē·ə }

Orthent [GEOL] A suborder of the soil order Entisol, well drained and of medium or fine texture, usually shallow to bedrock and lacking evidence of horizonation; occurs mostly on strong slopes. { 'ȯr·thənt }

Ortheziinae [INV ZOO] A subfamily of homopteran insects in the superfamily Coccoidea having abdominal spiracles present in all stages and a flat anal ring bearing pores and setae in immature forms and adult females. { ,ȯr·thə'zī·ə,nē }

orthicon [ELECTR] A camera tube in which a beam of low-velocity electrons scans a photoemissive mosaic that is capable of storing a pattern of electric charges; has higher sensitivity than the iconoscope. { 'ȯr·thə,kän }

Orthid [GEOL] A suborder of the soil order Aridisol, mostly well drained, gray or brownish-gray with little change from top to bottom of the soil profile; occupies younger, but not the youngest, land surfaces in deserts. { 'ȯr·thəd }

Orthida [PALEON] An order of extinct articulate brachiopods which includes the oldest known representatives of the class. { 'ȯr·thə·də }

Orthidina [PALEON] The principal suborder of the extinct Orthida, including those articulate brachiopods characterized

by biconvex, finely ribbed shells with a straight hinge line and well-developed interareas on both valves. { ȯr'thid·ən·ə }

orthite [MINERAL] Allanite in the form of slender prismatic or acicular crystals. { 'ȯr,thīt }

ortho acid [ORG CHEM] **1.** Aromatic acid with a carboxyl group in the ortho position (1,2 position). **2.** Organic acid with one added molecule of water in chemical combination; for example, $HC(OH)_3$, orthoformic acid, in contrast to HCOOH, formic acid; $H_3PO_4(P_2O_5·3H_2O)$, orthophosphoric acid, in contrast to the less hydrated form, metaphosphoric acid, $HPO_3(P_2O_5·H_2O)$. { 'ȯr·thō 'as·əd }

orthoarsenic acid *See* arsenic acid. { ¦ȯr·thō·är¦sen·ik 'as·əd }

orthoaxis [CRYSTAL] The diagonal or lateral axis perpendicular to the vertical axis in the monoclinic system. { ¦ȯr·thō'ak·səs }

orthobaric density [PHYS] The density of a liquid and of a saturated vapor with which it is at equilibrium at a given temperature. { ¦ȯr·thə¦bar·ik 'den·səd·ē }

orthobituminous coal [GEOL] Bituminous coal that contains 87–89% carbon, analyzed on a dry, ash-free basis. { ¦ȯr·thō·bə'tü·mən·əs 'kōl }

orthoboric acid *See* boric acid. { ¦ȯr·thə¦bor·ik 'as·əd }

orthocenter [MATH] The point at which the altitudes of a triangle intersect. { ¦ȯr·thō'sen·tər }

orthocephaly [ANTHRO] The condition of having the skull with a vertical index of 70.1–75. { ,ȯr·thə'sef·ə·lē }

orthoceratite [INV ZOO] Any nautiloid belonging to the genus *Orthoceras,* characterized by the presence of three longitudinal furrows on the body chamber. { ,ȯr·thə'ser·ə,tīt }

orthochem [GEOCHEM] A precipitate formed within a depositional basin or within the sediment itself by direct chemical action. { 'ȯr·thə,kem }

orthochromatic [BIOL] Having normal staining characteristics. [GRAPHICS] Pertaining to sensitized materials that can be exposed by ultraviolet, blue, and green light, but not deep orange or red. { ,ȯr·thə·krə'mad·ik }

orthochronology [GEOL] Geochronology based on a standard succession of biostratigraphically significant faunas or floras, or based on irreversible evolutionary processes. { ,ȯr·thə·krə'näl·ə·jē }

orthoclase [MINERAL] $KAlSi_3O_8$ A colorless, white, cream-yellow, flesh-reddish, or gray potassium feldspar that usually contains some sodium feldspar, either as albite or analbite or in some intermediate state; it is or appears to be monoclinic. Also known as common feldspar; orthose; pegmatolite. { 'ȯr·thə,klās }

orthoconglomerate [GEOL] A conglomerate with an intact gravel framework held together by mineral cement and deposited by ordinary water currents. { ,ȯr·thə·kən'gläm·ə·rət }

orthocumulate [PETR] A cumulate composed chiefly of one or more cumulus minerals plus the crystallization products of the intercumulus liquid. { ,ȯr·thə'kyü·myə·lət }

Orthod [GEOL] A suborder of the soil order Spodosol having accumulations of humus, aluminum, and iron; widespread in Canada and the Soviet Union. { 'ȯr,thäd }

orthodolomite [PETR] **1.** A primary dolomite, or one formed by sedimentation. **2.** A dolomite rock so well cemented that the particles interlock. { ȯr·thə'dō·lə,mīt }

orthodontics [MED] A branch of dentistry that deals with the prevention and treatment of malocclusion. { ,ȯr·thə'dän·tiks }

orthodromic projection [MAP] A map projection, which is taken from the gnomonic projection, where angles are correct at two points and all great circles are straight lines. { ¦ȯr·thə¦dräm·ik prə'jek·shən }

orthoferrosilite [MINERAL] An orthopyroxene consisting of the orthorhombic silicate $FeSiO_3$. { ¦ȯr·thō,fer·ə'sil,īt }

orthogenesis [EVOL] A unidirectional evolutionary change among a related group of animals. { ,ȯr·thə'jen·ə·səs }

orthogeosyncline [GEOL] A linear geosynclinal belt lying between continental and oceanic cratons, and having internal volcanic belts (eugeosynclinal) and external nonvolcanic belts (miogeosynclinal). Also known as geosynclinal couple; primary geosyncline. { ¦ȯr·thō,jē·ō·sin,klīn }

orthognathic [ANTHRO] Pertaining to a condition of the upper jaw in which it is in an approximately vertical relationship to the profile of the facial skeleton, when the skull is oriented

ORTHACEA
delthyrium

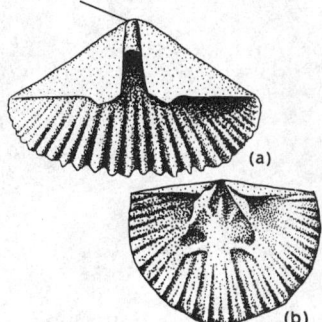

Shell of *Hesperothis.* (*a*) Pedicle valve posterior. (*b*) Brachial valve interior.

on the Frankfort horizontal plane; having a gnathic index of 97.9 or less. { ¦or·thəg¦nath·ik }

orthogneiss [GEOL] Gneiss originating from igneous rock. { 'or·thə‚nīs }

orthogonal [COMPUT SCI] **1.** An area of a computer display in which units of distance are the same horizontally and vertically so that there is no distortion. **2.** A viewing area in which positions are determined by using a cartesian coordinate system with horizontal and vertical axes. [MATH] Perpendicular, or some concept analogous to it. { or'thäg·ən·əl }

orthogonal antennas [ELECTROMAG] In radar, a pair of transmitting and receiving antennas, or a single transmitting-receiving antenna, designed for the detection of a difference in polarization between the transmitted energy and the energy returned from the target. { or'thäg·ən·əl an'ten·əz }

orthogonal basis [MATH] A basis for an inner product space consisting of mutually orthogonal vectors. { or'thäg·ən·əl 'bā·səs }

orthogonal complement [MATH] In an inner product space, the orthogonal complement of a vector **v** consists of all vectors orthogonal to **v**; the orthogonal complement of a subset S consists of all vectors orthogonal to each vector in S. { or'thäg·ən·əl 'käm·plə·mənt }

orthogonal crystal [CRYSTAL] A crystal whose axes are mutually perpendicular. { or'thäg·ən·əl 'krist·əl }

orthogonal family See orthogonal system. { or'thäg·ən·əl'fam·lē }

orthogonal functions [MATH] Two real-valued functions are orthogonal if their inner product vanishes. { or'thäg·ən·əl 'fəŋk·shənz }

orthogonal group [MATH] The group of matrices arising from the orthogonal transformations of a euclidean space. { or'thäg·ən·əl 'grüp }

orthogonality [MATH] Two geometric objects have this property if they are perpendicular. { or‚thäg·ə'nal·əd·ē }

orthogonalization [MATH] A procedure in which, given a set of linearly independent vectors in an inner product space, a set of orthogonal vectors is recursively obtained so that each set spans the same subspace. { or‚thäg·ə·nə·lə'zā·shən }

orthogonalized plane-wave method [SOLID STATE] A method of approximating the energy states of electrons in a crystal lattice: trial wave functions (the orthogonalized plane waves) are constructed which are linear combinations of plane waves and Bloch functions based on core states, and which are orthogonal to the Bloch functions, and linear combinations of these trial functions are then determined by the variational method. Abbreviated OPW method. { or'thäg·ən·əl‚īzd 'plän ‚wāv ‚meth·əd }

orthogonal Latin squares [MATH] Two Latin squares which, when superposed, have the property that the cells contain each of the possible pairs of symbols exactly once. { or'thäg·ən·əl 'lat·ən 'skwerz }

orthogonal lines [MATH] Lines which are perpendicular. { or'thäg·ən·əl 'līnz }

orthogonal matrix [MATH] A matrix whose inverse and transpose are identical. { or'thäg·ən·əl 'mā·triks }

orthogonal parity check [COMPUT SCI] A parity checking system involving both a lateral and a longitudinal parity check. { or'thäg·ən·əl 'par·əd·ē ‚chek }

orthogonal polynomial [MATH] Orthogonal polynomials are various families of polynomials, which arise as solutions to differential equations related to the hypergeometric equation, and which are mutually orthogonal as functions. { or'thäg·ən·əl päl·ə'nō·mē·əl }

orthogonal projection Also known as orthographic projection. [GRAPHICS] A two-dimensional representation formed by perpendicular intersections of lines drawn from points on the object being pictured to a plane of projection. [MATH] **1.** A continuous linear map P of a Hilbert space H onto a subspace M such that if **h** is any vector in H, $\mathbf{h} = P\mathbf{h} + \mathbf{w}$, where **w** is in the orthogonal complement of M. **2.** A mapping of a configuration into a line or plane that associates to any point of the configuration the intersection with the line or plane of the line passing through the point and perpendicular to the line or plane. { or'thäg·ən·əl prə'jek·shən }

orthogonal series [MATH] An infinite series each term of which is the product of a member of an orthogonal family of functions and a coefficient; the coefficients are usually chosen so that the series converges to a desired function. { or¦thäg·ən·əl 'sir‚ēz }

orthogonal spaces [MATH] Two subspaces F and F' of a vector space E with a scalar product g such that $g(x,x') = 0$ for any x in F and x' in F'. { or¦thäg·ən·əl 'späs·əz }

orthogonal sum [MATH] **1.** A vector space E with a scalar product is said to be the orthogonal sum of subspaces F and F' if E is the direct sum of F and F' and if F and F' are orthogonal spaces. **2.** A scalar product g on a vector space E is said to be the orthogonal sum of scalar products f and f' on subspaces F and F' if E is the orthogonal sum of F and F' (in the sense of the first definition) and if $g(x + x', y + y') = f(x, y) + f'(x', y')$ for all x,y in F and x', y' in F'. { or¦thäg·ən·əl 'səm }

orthogonal system [MATH] **1.** A system made up of n families of curves on an n-dimensional manifold in an $n + 1$ dimensional euclidean space, such that exactly one curve from each family passes through every point in the manifold, and, at each point, the tangents to the n curves that pass through that point are mutually perpendicular. **2.** A set of real-valued functions, the inner products of any two of which vanish. Also known as orthogonal family. { or'thäg·ən·əl 'sis·təm }

orthogonal trajectory [MATH] A curve that intersects all the curves of a given family at right angles. { or'thäg·ən·əl trə'jek·tə·rē }

orthogonal transformation [MATH] A linear transformation between real inner product spaces which preserves the length of vectors. { or'thäg·ən·əl ‚tranz‚fər'mā·shən }

orthogonal vectors [MATH] In an inner product space, two vectors are orthogonal if their inner product vanishes. { or'thäg·ən·əl 'vek·tərz }

orthographic chart [MAP] A chart on the orthographic projection. { or·thə¦graf·ik 'chärt }

orthographic projection [CRYSTAL] A projection for displaying the poles of a crystal in which the poles are projected from a reference sphere onto an equatorial plane by dropping perpendiculars from the poles to the plane. [GRAPHICS] See orthogonal projection. [MAP] A perspective azimuthal projection of one hemisphere produced by straight parallel lines from any point desired from an infinite distance; it is true to scale at the center only. [MATH] See orthogonal projection. { or·thə¦graf·ik prə'jek·shən }

orthohelium [ATOM PHYS] Those states of helium atoms in which the spins of the two electrons are parallel. { ¦or·thō'hē·lē·əm }

orthohexagonal axes [CRYSTAL] A set of crystallographic axes, two of which have a fixed ratio, as in hexagonal or trigonal crystals. { ¦or·thō·hek'säg·ən·əl 'ak‚sēz }

orthohydrogen [ATOM PHYS] Those states of hydrogen molecules in which the spins of the two nuclei are parallel. { ¦or·thə'hī·drə·jən }

orthohydrous coal [GEOL] Coal that contains 5–6% hydrogen, analyzed on a dry, ash-free basis. { ¦or·thə¦hī·drəs 'kōl }

orthokeratology [MED] A procedure designed to reduce or eliminate refractive anomalies and binocular dysfunctions of the eye by the programmed application of contact lenses. { ¦or·thō‚ker·ə'täl·ə·jē }

orthokinesis [BIOL] Random movement of a motile cell or organism in response to a stimulus. { ¦or·thə·ki'nē·səs }

ortholignitous coal [GEOL] Coal that contains 75–80% carbon, analyzed on a dry, ash-free basis. { ¦or·thō·lig'nīd·əs 'kōl }

orthologous locus [GEN] A gene that evolved without diverging from an ancestral locus. { or'thäl·ə·gəs ‚lō·kəs }

orthomagmatic stage [GEOL] The principal stage in the crystallization of silicates from a typical magma; up to 90% of the magma may crystallize during this stage. Also known as orthotectic stage. { ¦or·thō‚mag'mad·ik 'stäj }

orthometric correction [ENG] A systematic correction that must be applied to a measured difference in elevation since level surfaces at varying elevations are not absolutely parallel. { ¦or·thə‚me·trik kə'rek·shən }

orthometric height [ENG] The distance above sea level measured along a plumb line. { ¦or·thə‚me·trik 'hīt }

orthomimic feldspars [MINERAL] A group of feldspars that by repeated twinning simulate a higher degree of symmetry with rectangular cleavages. { ¦or·thə¦mim·ik 'fel‚spärz }

orthomorphic [MAP] Preserving the correct shape. { ¦or·thə¦mor·fik }

orthomorphic chart [MAP] A chart on which very small shapes are correctly represented. { ¦ȯr·thə¦mȯr·fik ¦chärt }

orthomorphic map projection See conformal map projection. { ¦ȯr·thə¦mȯr·fik ¦map prə¦jek·shən }

Orthomyxoviridae [VIROL] A family of negative-strand ribonucleic acid viruses characterized by enveloped, spherical, pleomorphic virions with a helical nucleocapsid containing a fragmented genome; it includes the genus *Influenzavirus* (human influenza type A). { ¦ȯr·thə‚mik·sə'vir·ə‚dī }

Orthonectida [INV ZOO] An order of Mesozoa; orthonectids parasitize various marine invertebrates as multinucleate plasmodia, and sexually mature forms are ciliated organisms. { ‚ȯr·thə'nek·tə·də }

Orthonik [MET] A magnetic alloy composed of 45–50% nickel, with the remainder iron, having a grainy structure, high permeability, and a rectangular hysteresis loop; used in magnetic cores. { 'ȯr·thə‚nik }

orthonormal coordinates [MATH] In an inner product space, the coordinates for a vector expressed relative to an orthonormal basis. { ¦ȯr·thə¦nȯr·məl kō'ȯrd·ən·əts }

orthonormal functions [MATH] Orthogonal functions f_1, f_2, ... with the additional property that the inner product of $f_n(x)$ with itself is 1. { ¦ȯr·thə¦nȯr·məl 'fəŋk·shənz }

orthonormal tetrad [RELAT] A collection of four mutually orthogonal unit vectors, three spacelike and one timelike, at a point of space-time, that specify the directions of the four axes of a locally Minkowskian coordinate system. { ¦ȯr·thə¦nȯr·məl 'te‚trad }

orthonormal vectors [MATH] A collection of mutually orthogonal vectors, each having length 1. { ¦ȯr·thə¦nȯr·məl 'vek·tȯrz }

orthopedics [MED] The branch of surgery concerned with corrective treatment of musculoskeletal deformities, diseases, and ailments by manual and instrumental measures. { ‚ȯr·thə'pēd·iks }

Orthoperidae [INV ZOO] The minute fungus beetles, a family of coleopteran insects in the superfamily Cucujoidea. { ‚ȯr·thə'per·ə‚dē }

orthophosphate [INORG CHEM] One of the possible salts of orthophosphoric acid; the general formula is M_3PO_4, where M may be potassium as in potassium orthophosphate, K_3PO_4. { ¦ȯr·thə'fäs‚fāt }

orthophosphoric acid See phosphoric acid. { ¦ȯr·thə·fäs'fȯr·ik 'as·əd }

orthophotograph [GEOL] A photographic copy, prepared from a photograph formed by a perspective projection, in which the displacements due to tilt and relief have been removed. { ‚ȯr·thə'fōd·ə‚graf }

orthophyric [PETR] Of the texture of the matrix of certain igneous rocks, having feldspar crystals with quadratic or short and stumpy rectangular cross sections. { ‚ȯr·thə'fir·ik }

orthopinacoid See front pinacoid. { ¦ȯr·thə'pin·ə‚kȯid }

orthopnea [MED] A condition in which there is difficulty in breathing except when sitting or standing upright. { ȯr'thäp·nē·ə }

orthopositronium [PARTIC PHYS] The state of positronium in which the positron and electron have parallel spins. { ¦ȯr·thō·‚päz·ə'trō·nē·əm }

Orthopsida [INV ZOO] An order of echinoderms in the subclass Euechinoidea. { ȯr'thäp·sə·də }

Orthopsidae [PALEON] A family of extinct echinoderms in the order Hemicidaroida distinguished by a camarodont lantern. { ȯr'thäp·sə‚dē }

Orthoptera [INV ZOO] A heterogeneous order of generalized insects with gradual metamorphosis, chewing mouthparts, and four wings. { ȯr'thäp·tə·rə }

orthoptic [MATH] The locus of the intersection of tangents to a given curve that meet at a right angle. { ȯr'thäp·tik }

orthopyroxene [MINERAL] A series of pyroxene minerals crystallizing in the orthorhombic system; members include enstatite, bronzite, hypersthene, ferrohypersthene, eulite, and orthoferrosilite. { ‚ȯr·thə·pə'räk‚sēn }

orthoquartzite [PETR] A clastic sedimentary rock composed almost entirely of detrital quartz grains; a quartzite of sedimentary origin. Also known as orthoquartzitic sandstone; sedimentary quartzite. { ¦ȯr·thə'kwȯrt‚sīt }

orthoquartzitic conglomerate [GEOL] A lithologically homogeneous, light-colored orthoconglomerate composed of quartzose residues that is commonly interbedded with pure quartz sandstone. Also known as quartz-pebble conglomerate. { ¦ȯr·thə·kwȯrt¦sid·ik kən'gläm·ə·rət }

orthoquartzitic sandstone See orthoquartzite. { ¦ȯr·thə·kwȯrt¦sid·ik 'san‚stȯn }

orthorhombic lattice [CRYSTAL] A crystal lattice in which the three axes of a unit cell are mutually perpendicular, and no two have the same length. Also known as rhombic lattice. { ¦ȯr·thə¦räm·bik 'lad·əs }

orthorhombic pyroxene [MINERAL] A member of the mineral series enstatite-orthoferrosilite, crystallizing in the orthorhombic system, space group *Pbca*. { ¦ȯr·thə¦räm·bik pə'räk‚sēn }

orthorhombic system [CRYSTAL] A crystal system characterized by three axes of symmetry that are mutually perpendicular and of unequal length. Also known as rhombic system. { ¦ȯr·thə¦räm·bik 'sis·təm }

Orthorrhapha [INV ZOO] A suborder of the Diptera; in this group of flies, the adult escapes from the puparium through a T-shaped opening. { ȯr'thȯr·ə·fə }

orthoschist [PETR] A schist derived from igneous rocks. { 'ȯr·thə‚shist }

orthoscope [MED] **1.** An instrument for examination of the eye through a layer of water, whereby the curvature and hence the refraction of the cornea is neutralized. **2.** An instrument used in drawing the projections of skulls. [OPTICS] A polarizing microscope in which light is transmitted by a crystal which is parallel to the microscope axis. { 'ȯr·thə‚skōp }

orthoscopic eyepiece [OPTICS] An eyepiece that consists of a single lens, made up of three cemented elements, to which a planoconvex lens is added; designed to minimize distortion and spherical aberration. { ¦ȯr·thə¦skäp·ik 'ī‚pēs }

orthoscopic system [OPTICS] An optical system that has been corrected so that distortion and spherical aberration are eliminated. Also known as rectilinear system. { ¦ȯr·thə¦skäp·ik 'sis·təm }

orthose See orthoclase. { 'ȯr‚thōs }

orthosis [MED] A device applied to a human limb to control or enhance movement or to prevent bone movement or deformity, for example, a splint or an arch support. { ȯr'thōs·əs }

orthosite [PETR] A light-colored coarse-grained igneous rock composed almost entirely of orthoclase. { 'ȯr·thə‚sīt }

orthostatic [MED] Pertaining to or caused by standing upright. { ¦ȯr·thə¦stad·ik }

orthostratigraphy [GEOL] Standard stratigraphy based on fossils which identify recognized biostratigraphic zones. { ¦ȯr·thō·strə'tig·rə·fē }

orthosymmetric crystal [CRYSTAL] A crystal that has orthorhombic symmetry. { ¦ȯr·thō·si¦me·trik 'krist·əl }

orthotectic stage See orthomagmatic stage. { ¦ȯr·thə'tek·tik ‚stāj }

orthotill [GEOL] A till formed by immediate release of material from transported ice, such as by ablation and melting. { 'ȯr·thə‚til }

orthotomic [MATH] The orthotomic of a curve with respect to a point is the envelope of the circles which pass through the point and whose centers lie on the curve. { ‚ȯr·thə'täm·ik }

orthotomic system [OPTICS] An optical system in which all the rays may be interesected at right angles by a suitably chosen surface. { ‚ȯr·thə'täm·ik ‚sis·təm }

orthotonus [MED] Tetanic muscle spasm in which the body assumes a posture of rigid straightness. { ȯr'thät·ən·əs }

Orthotrichales [BOT] An order of true mosses in the subclass Bryidae, characterized by dull, tuft- or mat-forming plants that are probably heterogeneous, making a generalized description difficult. { ‚ȯr·thō·trə'kā·lēz }

orthotronic error control [COMPUT SCI] An error check carried out to ensure correct transmission, which uses lateral and longitudinal parity checks. { ¦ȯr·thə¦trän·ik 'er·ər kən‚trōl }

orthotropic [MECH] Having elastic properties such as those of timber, that is, with considerable variations of strength in two or more directions perpendicular to one another. { ¦ȯr·thə¦trä·pik }

orthotropic deck [CIV ENG] A bridge deck constructed typically of flat steel plate and longitudinal and transverse ribs; functions in carrying traffic and acting as top flanges of floor beams. { ¦ȯr·thə¦trä·pik 'dek }

orthotropism [BOT] The tendency of a plant to grow with the longer axis oriented vertically. { ȯr'thä·trə‚piz·əm }

ORTHONECTIDA

anterior cone

testis

sperm

ovocytes

The orthonectid *Rhopalura ophiocomae*, male discharging sperm near genital pore of female.

orthotropous [BOT] Having a straight ovule with the micropyle at the end opposite the stalk. { ȯr'thä·trə·pəs }

orthotungstic acid *See* tungstic acid. { ¦ȯr·thō'təŋ·stik 'as·əd }

Orthox [GEOL] A suborder of the soil order Oxisol that is moderate to low in organic matter, well drained, and moist all or nearly all year; believed to be extensive at low altitudes in the heart of the humid tropics. { 'ȯr,thäks }

Orussidae [INV ZOO] A small family of hymenopteran insects in the superfamily Siricoidea. { ȯ'rūs,ə,dē }

orvietite [PETR] An extrusive rock composed of approximately equal amounts of plagioclase and sanidine; includes leucite, augite, minor biotite, and olivine, and accessory apatite and opaque oxides. { 'ȯr·vē·ə,tīt }

oryctocoenosis [PALEON] The part of a thanatocoenosis that has been preserved as a fossil. { ə,rik·tə·sə'nō·səs }

Os *See* osmium.

Osagean [GEOL] A provincial series of geologic time in North America; Lower Mississippian (above Kinderhookian, below Meramecian). { ō'sā·jē·ən }

osage orange [BOT] *Maclura pomifera.* A tree in the mulberry family of the Urticales characterized by yellowish bark, milky sap, simple entire leaves, strong axillary thorns, and aggregate green fruit about the size and shape of an orange. { 'ō,sāj 'är·inj }

osar *See* esker. { 'ō,sär }

osazone [BIOCHEM] Any of the compounds that contain two phenylhydrazine residues and are produced by a reaction between a reducing sugar and phenylhydrazine. { 'ō·sə,zōn }

O scan *See* O scope. { 'ō ,skan }

osciducer [ELECTR] Transducer in which information pertaining to the stimulus is provided in the form of deviation from the center frequency of an oscillator. { ¦äs·ə¦dü·sər }

oscillating conveyor [MECH ENG] A conveyor on which pulverized solids are moved by a pan or trough bed attached to a vibrator or oscillating mechanism. Also known as vibrating conveyor. { 'äs·ə,lād·iŋ kən'vā·ər }

oscillating granulator [MECH ENG] Solids size-reducer in which particles are broken by a set of oscillating bars arranged in cylindrical form over a screen of suitable mesh. { 'äs·ə,lād·iŋ 'gran·yə,lād·ər }

oscillating magnetic field [ELECTROMAG] A magnetic field which varies periodically in time. { 'äs·ə,lād·iŋ mag'ned·ik 'fēld }

oscillating screen [MECH ENG] Solids separator in which the sifting screen oscillates at 300 to 400 revolutions per minute in a plane parallel to the screen. { 'äs·ə,lād·iŋ 'skrēn }

oscillating series [MATH] A series that is divergent but not properly divergent; that is, the partial sums do not approach a limit, or become arbitrarily large or arbitrarily small. { 'äs·ə,lād·iŋ ,sir,ēz }

oscillating universe [ASTRON] An extension of the closed universe model in which the universe, after contracting toward a singularity, undergoes another big bang to begin a new cycle, and thenceforth oscillates between successive expansions and contractions, each contraction followed by a new big bang. { 'äs·ə,lād·iŋ 'yü·nə,vərs }

oscillation [CONT SYS] *See* cycling. [MATH] **1.** The oscillation of a real-valued function on an interval is the difference between its least upper bound and greatest lower bound there. **2.** The oscillation of a real-valued function at a point x is the limit of the oscillation of the function on the interval $[x - e, x + e]$ as e approaches 0. Also known as saltus. [PHYS] Any effect that varies periodically back and forth between two values. { ,äs·ə'lā·shən }

oscillation photography [SOLID STATE] A method of x-ray diffraction analysis in which a single crystal is made to oscillate through a small angle about an axis perpendicular to a beam of monochromatic x-rays or particles. { ,äs·ə¦lā·shən fə¦täg·rə·fē }

oscillation ripple *See* oscillation ripple mark. { ,äs·ə'lā·shən ,rip·əl }

oscillation ripple mark [GEOL] A symmetric ripple mark having a sharp, narrow, and relatively straight crest between broadly rounded troughs, formed by the motion of water agitated by oscillatory waves on a sandy base at a depth shallower than wave base. Also known as oscillation ripple; oscillatory ripple mark; wave ripple mark. { ,äs·ə'lā·shən 'rip·əl ,märk }

oscillator [ELECTR] **1.** An electronic circuit that converts

---

OSAGE ORANGE

Leaf and branch of *Maclura pomifera.*

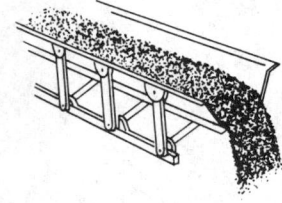

OSCILLATING CONVEYOR

Oscillating conveyor, with whole trough oscillating.

---

energy from a direct-current source to a periodically varying electric output. **2.** The stage of a superheterodyne receiver that generates a radio-frequency signal of the correct frequency to mix with the incoming signal and produce the intermediate-frequency value of the receiver. **3.** The stage of a transmitter that generates the carrier frequency of the station or some fraction of the carrier frequency. [PHYS] Any device (mechanical or electrical) which, in the absence of external forces, can have a periodic back- and-forth motion, the frequency determined by the properties of the oscillator. { 'äs·ə,lād·ər }

oscillator harmonic interference [ELECTR] Interference occurring in a superheterodyne receiver due to the interaction of incoming signals with harmonics (usually the second harmonic) of the local oscillator. { 'äs·ə,lād·ər här'män·ik ,in·tər'fir·əns }

Oscillatoriales [BOT] An order of blue-green algae (Cyanophyceae) which are filamentous and truly multicellular. { ¦äs·ə·lə,tȯr·ē'ā·lēz }

oscillator-mixer-first detector *See* converter. { 'äs·ə,lād·ər 'mik·sər ,fərst di'tek·tər }

oscillator strength [ATOM PHYS] A quantum-mechanical analog of the number of dispersion electrons having a given natural frequency in an atom, used in an equation for the absorption coefficient of a spectral line; it need not be a whole number. Also known as f value; Ladenburg f value. { 'äs·ə,lād·ər ,streŋkth }

oscillatory circuit [ELEC] Circuit containing inductance or capacitance, or both, and resistance, connected so that a voltage impulse will produce an output current which periodically reverses or oscillates. { 'äs·ə·lə,tȯr·ē 'sər·kət }

oscillatory discharge [ELEC] Alternating current of gradually decreasing amplitude which, under certain conditions, flows through a circuit containing inductance, capacitance, and resistance when a voltage is applied. { 'äs·ə·lə,tȯr·ē 'dis,chärj }

oscillatory extinction *See* undulatory extinction. { 'äs·ə·lə,tȯr·ē ik'stiŋk·shən }

oscillatory reaction [CHEM] A chemical reaction in which a variable of a chemical system exhibits regular periodic changes in time or in space. { 'äs·ə·lə,tȯr·ē rē'ak·shən }

oscillatory ripple mark *See* oscillation ripple mark. { 'äs·ə·lə,tȯr·ē 'rip·əl ,märk }

oscillatory shear [FL MECH] Application of small-amplitude oscillations to produce shear in viscoelastic fluids for the study of dynamic viscosity. { 'äs·ə·lə,tȯr·ē 'shir }

oscillatory surge [ELEC] Surge which includes both positive and negative polarity values. { 'äs·ə·lə,tȯr·ē 'sərj }

oscillatory twinning [CRYSTAL] Repeated, parallel twinning. { 'äs·ə·lə,tȯr·ē 'twin·iŋ }

oscillatory wave [PHYS] A wave composed of individual particles, each of which oscillates about a point with little, if any, permanent change in position. { 'äs·ə·lə,tȯr·ē 'wāv }

oscillistor [ELECTR] A bar of semiconductor material, such as germanium, that will oscillate much like a quartz crystal when it is placed in a magnetic field and is carrying direct current that flows parallel to the magnetic field. { ¦äs·ə'lis·tər }

oscillogram [ENG] The permanent record produced by an oscillograph, or a photograph of the trace produced by an oscilloscope. { ə'sil·ə,gram }

oscillograph [ENG] A measurement device for determining waveform by recording the instantaneous values of a quantity such as voltage as a function of time. { ə'sil·ə,graf }

oscillographic polarography [PHYS CHEM] A type of voltammetry using a dropping mercury electrode with oscillographic scanning of the applied potential; used to measure the concentration of electroactive species in solutions. { ¦äs·ə·lə¦graf·ik ,pō·lə'räg·rə·fē }

oscillograph tube [ELECTR] Cathode-ray tube used to produce a visible pattern, which is the graphical representation of electric signals, by variations of the position of the focused spot or spots according to these signals. { ə'sil·ə,graf ,tüb }

oscillometric titration [PHYS CHEM] Radio-frequency technique used for conductometric and dielectrometric titrations; the changes in conductance or dielectric properties changes the solution capacity and thus the frequency of the connected oscillator circuit. { ¦äs·ə·lō¦me·trik tī'trā·shən }

oscillometry [PHYS CHEM] Electrode measurement of oscillation-frequency changes to detect the progress of a titration of electrolytic solutions. { ˌäs·ə'läm·ə·trē }

oscilloscope See cathode-ray oscilloscope. { ə'sil·ə,skōp }

Oscillospiraceae [MICROBIO] Formerly a family of large, gram-negative, motile bacteria of the order Caryophanales which lose motility on exposure to oxygen. { ˌäs·ə·lī·spə'rās·ē,ē }

oscine See scopoline. { 'ä,sīn }

Oscines [VERT ZOO] The songbirds, a suborder of the order Passeriformes. { 'äs·ə,nēz }

O scope [ELECTR] An A scope modified by the inclusion of an adjustable notch for measuring range. Also known as O indicator; O scan. { 'ō ,skōp }

osculating circle [MATH] For a plane curve C at a point p, the limiting circle obtained by taking the circle that is tangent to C at p and passes through a variable point q on C, and then letting q approach p. { 'äs·kyə,lād·iŋ 'sər·kəl }

osculating orbit [ASTRON] The orbit which would be followed by a body such as an asteroid or comet if, at a given time, all the planets suddenly disappeared, and it then moved under the gravitational force of the sun alone. { 'äs·kyə,lād·iŋ 'ȯr·bət }

osculating plane [MATH] For a curve C at some point p this is the limiting plane obtained from taking planes through the tangent to C at p and containing some variable point p' and then letting p' approach p along C. { 'äs·kyə,lād·iŋ 'plān }

osculating sphere [MATH] For a curve C at a point p, the limiting sphere obtained by taking the sphere that passes through p and three other points on C and then letting these three points approach p independently along C. { 'äs·kyə,lād·iŋ 'sfir }

osculum [INV ZOO] An excurrent orifice in Porifera. { 'äs·kyə·ləm }

Oseen's flow [FL MECH] Fluid flow in which the velocity of flow is very small but the Reynolds number is greater than 1. { ü'sänz ,flō }

Osgood-Schlatter disease See osteochondrosis. { 'äz,gud 'shlad·ər di,zēz }

O shell [ATOM PHYS] The fifth layer of electrons about the nucleus of an atom, having electrons characterized by the principal quantum number 5. { 'ō ,shel }

Osler-Rendu-Weber disease See hereditary hemorrhagic telangiectasia. { 'ōs·lər 'rän·dü 'web·ər di,zēz }

osmate [INORG CHEM] A salt or ester of osmic acid, containing the osmate radical, $OsO_4{}^{2-}$; for example, potassium osmate (K_2OsO_4). { 'äz·mət }

osmic acid anhydride [INORG CHEM] OsO_4 Poisonous yellow crystals with disagreeable odor; melts at 40°C; soluble in water, alcohol, and ether; used in medicine, photography, and catalysis. Also known as osmium oxide; osmium tetroxide. { 'äz·mik 'as·əd an'hī,drīd }

osmium [CHEM] A chemical element, symbol Os, atomic number 76, atomic weight 190.2. [MET] A hard white metal of rare natural occurrence. { 'äz·mē·əm }

osmium oxide See osmic acid anhydride. { 'äz·mē·əm 'äk,sīd }

osmium tetroxide See osmic acid anhydride. { 'äz·mē·əm te'träk,sīd }

osmolality [CHEM] The molality of an ideal solution of a nondissociating substance that exerts the same osmotic pressure as the solution being considered. { ˌäz·mō'lal·əd·ē }

osmolarity [CHEM] The molarity of an ideal solution of a nondissociating substance that exerts the same osmotic pressure as the solution being considered. { ˌäz·mō'lar·əd·ē }

osmole [CHEM] **1.** The unit of osmolarity equal to the osmolarity of a solution that exerts an osmotic pressure equal to that of an ideal solution of a nondissociating substance that has a concentration of 1 mole of solute per liter of solution. **2.** The unit of osmolality equal to the osmolality of a solution that exerts an osmotic pressure equal to that of an ideal solution of a nondissociating substance that has a concentration of 1 mole of solute per kilogram of solvent. { 'äz,mōl }

osmometer [ANALY CHEM] A device for measuring molecular weights by measuring the osmotic pressure exerted by solvent molecules diffusing through a semipermeable membrane. { äz'mäm·əd·ər }

osmophile [MICROBIO] A microorganism adapted to media with high osmotic pressure. { 'äz·mə,fīl }

osmophobia [PSYCH] An abnormal fear of odors. { ˌäz·mə'fō·bē·ə }

osmophore [BOT] A particular flower part specialized for odor production. { 'äz·mə,for }

osmoreceptor [PHYSIO] One of a group of structures in the hypothalamus which respond to changes in osmotic pressure of the blood by regulating the secretion of the neurohypophyseal antidiuretic hormone. { 'äz·mō·ri'sep·tər }

osmoregulatory mechanism [PHYSIO] Any physiological mechanism for the maintenance of an optimal and constant level of osmotic activity of the fluid in and around the cells. { ˌäz·mō'reg·yə·lə,tȯr·ē 'mek·ə,niz·əm }

osmosis [PHYS CHEM] The transport of a solvent through a semipermeable membrane separating two solutions of different solute concentration, from the solution that is dilute in solute to the solution that is concentrated. { ä'smō·səs }

osmotic diarrhea [MED] Diarrhea resulting from a rise in the osmotic pressure of fecal contents, diminishing the absorption of water by the intestine; it abates with fasting. { äz,mäd·ik ,dī·ə'rē·ə }

osmotic fragility [PHYSIO] Susceptibility of red blood cells to lyses when placed in dilute (hypotonic) salt solutions. { äz'mäd·ik frə jil·əd·ē }

osmotic gradient See osmotic pressure. { äz'mäd·ik 'grād·ē·ənt }

osmotic pressure [PHYS CHEM] **1.** The applied pressure required to prevent the flow of a solvent across a membrane which offers no obstruction to passage of the solvent, but does not allow passage of the solute, and which separates a solution from the pure solvent. **2.** The applied pressure required to prevent passage of a solvent across a membrane which separates solutions of different concentration, and which allows passage of the solute, but may also allow limited passage of the solvent. Also known as osmotic gradient. { äz'mäd·ik 'presh·ər }

osmotic shock [PHYSIO] The bursting of cells suspended in a dilute salt solution. { äz'mäd·ik 'shäk }

osmotolerance [PHYSIO] The ability to withstand high solute concentrations. { ˌäz·mō'täl·ə·rəns }

O/S/O carrier See ore/slurry/oil carrier. { 'ō,es,ō ,kar·ē·ər }

Osos wind [METEOROL] In California, a strong northwest wind blowing from the Los Osos valley to the San Luis valley. { 'ō,sōs ,wind }

osphradium See asphradium. { äs'frad·ē·əm }

os priapi See baculum. { 'äs prī'ā·pē }

osseine [MATER] The organic residue formed when bone is dissolved in hydrochloric acid; used to make glue and gelatin. { 'äs·ē·ən }

osseous [ANAT] Bony; composed of or resembling bone. { 'äs·ē·əs }

osseous system [ANAT] The skeletal system of the body. { 'äs·ē·əs 'sis·təm }

osseous tissue HISTOL Bone tissue. { 'äs·ē·əs 'tish·ü }

ossicle [ANAT] Any of certain small bones, as those of the middle ear. [INV ZOO] Any of various calcareous bodies. { 'äs·ə·kəl }

ossify [PHYSIO] To form or turn into bone. { 'äs·ə,fī }

ossifying fibroma [MED] A benign bone tumor derived from ossiferous connective tissue. Also known as fibrous osteoma; osteogenic fibroma. { 'äs·ə,fī·iŋ fī'brō·mə }

ossipite [PETR] A coarse-grained variety of troctolite containing labradorite, olivine, magnetite, and a small amount of diallage. { 'äs·ə,pīt }

ost-, oste-, osteo- [ANAT] A combining form meaning bone. { äst, 'ä·stē 'ä·stē·ō }

O star [ASTRON] A star of spectral type O, a massive, very hot blue star with a surface temperature of at least 35,000 K (63,000°F), and a spectrum in which lines of singly ionized helium are prominent. { 'ō ,stär }

Ostariophysi [VERT ZOO] A superorder of actinopterygian fishes distinguished by the structure of the anterior four or five vertebrae which are modified as an encasement for the bony ossicles connecting the inner ear and swim bladder. { ä,stär·ē·ō'fī,sī }

Osteichthyes [VERT ZOO] The bony fishes, a class of fishlike vertebrates distinguished by having a bony skeleton, a swim bladder, a true gill cover, and mesodermal ganoid, cycloid, or ctenoid scales. { ˌä·stē'ik·thē,ēz }

osteitis [MED] Inflammation of bone. { ˌä·stē'īd·əs }

osteitis fibrosa cystica [MED] Generalized skeletal demineralization due to an increased rate of bone destruction resulting from hyperparathyroidism. Also known as Engel-Recklinghausen disease; osteitis fibrosa generalisata. { ˌä·stē'īd·əs 'fī·brō·sə 'sis·tə·kə }

osteitis fibrosa generalisata See osteitis fibrosa cystica. { ˌä·stē'īd·əs 'fī·brō·sə ˌjen·ə·rə'lis·əd·ə }

osteoarthritis See degenerative joint disease. { ˌäs·tē·ō̩är̩'thrīd·əs }

osteoarthropathy [MED] Any disease of bony articulations. { ¦äs·tē·ō̩är̩'thräp·ə·thē }

osteoblast [HISTOL] A bone-forming cell of mesenchymal origin. { 'äs·tē·ə̩blast }

osteochondritis [MED] Inflammation of both bone and cartilage. { ˌäs·tē·ō̩kän'drīd·əs }

osteochondroma [MED] A benign hamartomatous tumor originating in bone or cartilage. { ˌäs·tē·ō̩kän'drō·mə }

osteochondrosis [MED] A disease characterized by avascular necrosis of ossification centers followed by regeneration. Also known as Calvé's disease; Kienböck's disease; Köhler's disease; Osgood-Schlatter disease; Scheuermann's disease. { ˌäs·tē·ō̩kän'drō·səs }

osteoclasis [MED] Forcible fracture of a long bone without open operation, to correct a deformity. [PHYSIO] **1.** Destruction of bony tissue. **2.** Bone resorption. { ˌäs·tē'äk·lə·səs }

osteoclast [HISTOL] A large multinuclear cell associated with bone resorption. [MED] A large surgical apparatus through which leverage can be exerted to effect osteoclasis. { 'äs·tē·ə̩klast }

osteoclast differentiation factor [CELL MOL] A protein on the surface of osteoblasts that binds to receptors on osteoclast precursor cells and induces progression to osteoclasts. { ˌäs·tē·ə̩klast ˌdif·ə̩ren·chē'ā·shən ˌfak·tər }

osteocyte [HISTOL] A bone cell. { 'äs·tē·ə̩sīt }

osteodermia [MED] A condition characterized by ossification within the skin. { ¦äs·tē·ə̩'dər·mē·ə }

osteodystrophy [MED] Any defective bone formation, as in rickets or dwarfism. { ˌäs·tē·ō'di·strə·fē }

osteofibrosis [MED] Fibrosis of bone. { ¦äs·tē·ō̩fī'brō·səs }

osteogenesis [PHYSIO] Formation or histogenesis of bone. { ¦äs·tē·ō'jen·ə·səs }

osteogenesis imperfecta [MED] A disease inherited as an autosomal dominant and characterized by hypoplasia of osteoid tissue and collagen, resulting in bone fractures. { ¦äs·tē·ō'jen·ə·səs ˌim·pər'fek·tə }

osteogenesis imperfecta congenita [MED] A form of osteogenesis imperfecta in which fractures occur at or before birth. { ¦äs·tē·ō'jen·ə·səs ˌim·pər'fek·tə kən'jen·əd·ə }

osteogenic fibroma See ossifying fibroma. { ¦äs·tē·ə̩jen·ik fī'brō·mə }

osteogenic sarcoma See osteosarcoma. { ¦äs·tē·ə̩jen·ik sär'kō·mə }

Osteoglossidae [VERT ZOO] The bony tongues, a family of actinopterygian fishes in the order Osteoglossiformes. { ˌäs·tē·ō'gläs·ə̩dē }

Osteoglossiformes [VERT ZOO] An order of soft-rayed, actinopterygian fishes distinguished by paired, usually bony rods at the base of the second gill arch, a single dorsal fin, no adipose fin, and a usually abdominal pelvic fin. { ˌäs·tē·ō̩gläs·ə'fór·mēz }

osteoid [HISTOL] The young hyaline matrix of true bone in which the calcium salts are deposited. { 'äs·tē̩oid }

osteolathyrism [MED] Degeneration of bone collagen resulting from experimental administration of β-aminoproprionitrile. { ˌäs·tē·ō'lath·ə̩riz·əm }

Osteolepidae [PALEON] A family of extinct fishes in the order Osteolepiformes. { ˌäs·tē·ō'lep·ə̩dē }

Osteolepiformes [PALEON] A primitive order of fusiform lobefin fishes, subclass Crossopterygii, generally characterized by rhombic bony scales, two dorsal fins placed well back on the body, and a well-ossified head covered with large dermal plating bones. { ˌäs·tē·ō̩lep·ə'fór·mēz }

osteolith [PALEON] A fossil bone. { 'äs·tē·ə̩lith }

osteology [ANAT] The study of anatomy and structure of bone. { ˌäs·tē'äl·ə·jē }

osteolysis [MED] Degeneration of bone tissue. [PHYSIO] Resorption of bone. { ˌäs·tē'äl·ə·səs }

osteoma [MED] A benign bone tumor, especially in membrane bones of the skull. { ˌäs·tē'ō·mə }

osteomalacia [MED] Failure of bone to ossify due to a reduced amount of available calcium. Also known as adult rickets. { ¦äs·tē·ō·mə'lā·shə }

osteometry [ANAT] The study of the size and proportions of the osseous system. { ˌäs·tē'äm·ə·trē }

osteomyelitis [MED] Inflammation of bone tissue and bone marrow. { ¦äs·tē·ō̩mī·ə'līd·əs }

osteon [HISTOL] A microscopic unit of mature bone composed of layers of osteocytes and bone surrounding a central canal. Also known as Haversian system. { 'äs·tē̩än }

osteonephropathy [MED] Any syndrome involving bone changes accompanying kidney disease. { ¦äs·tē·ō·nə'fräp·ə·thē }

osteopath [MED] A physician who specializes in osteopathy. { 'äs·tē·ə̩path }

osteopathy [MED] **1.** A school of healing which teaches that the body is a vital mechanical organism whose structural and functional integrity are coordinate and interdependent, the abnormality of either constituting disease. **2.** Any disease of bone. { ˌäs·tē'äp·ə·thē }

osteopenia [MED] The reduction in bone volume and bone structural quality. { ˌäs·tē·ə'pē·nē·ə }

osteopetrosis [MED] A rare developmental error of unknown cause but of familial tendency, characterized chiefly by excessive radiographic density of most or all of the bones. Also known as marble bone disease. { ¦äs·tē·ō·pə'rō·səs }

osteophony [PHYSIO] Conduction of sound by bone. { ˌäs·tē'äf·ə·nē }

osteoplasty [MED] Plastic surgery performed on bone, particularly bone tissue replacement or reconstruction. { 'äs·tē·ə̩plas·tē }

osteopoikilosis [MED] A bone affection of unknown cause and no symptoms, characterized by ellipsoidal, dense foci in all bones. { ¦äs·tē·ō̩pói·kə'lō·səs }

osteoporosis [MED] Deossification with absolute decrease in bone tissue, resulting in enlargement of marrow and Haversian spaces, decreased thickness of cortex and trabeculae, and structural weakness. { ¦äs·tē·ō·pə'rō·səs }

osteoprotegerin [BIOCHEM] A protein that plays a central role in regulating bone mass. { ¦äs·tē·ō·prō'teg·ə·rin }

osteosarcoma [MED] A malignant tumor principally composed of anaplastic cells of mesenchymal derivation. Also known as osteogenic sarcoma. { ¦äs·tē·ō·sär'kō·mə }

osteosclereid [BOT] A sclereid cell that is rod-like with swollen ends, occurring in seed coats and in some leaves. Also known as bone cell. { ˌäs·tē·ə'skler·ē·əd }

osteosclerotic anemia See myelophthisic anemia. { ¦äs·tē·ō̩sklə'räd·ik ə'nē·mē·ə }

Osteostraci [PALEON] An order of extinct jawless vertebrates; they were mostly small, with the head and part of the body encased in a solid armor of bone, and the posterior part of the body and the tail covered with thick scales. { ˌäs·tē'äs·trə̩sī }

osteotomy [MED] **1.** Surgical division of a bone. **2.** Making a section of a bone for the purpose of correcting a deformity. { ˌäs·tē'äd·ə·mē }

ostiole [BIOL] A small orifice or pore. { 'äs·tē̩ōl }

ostium [BIOL] A mouth, entrance, or aperture. { 'äs·tē·əm }

Ostomidae [INV ZOO] The bark-gnawing beetles, a family of coleopteran insects in the superfamily Cleroidea. { ä'stäm·ə̩dē }

Ostracoda [INV ZOO] A subclass of the class Crustacea containing small, bivalved aquatic forms; the body is unsegmented and there is no true abdominal region. { ä'sträk·ə·də }

ostracoderm [PALEON] Any of various extinct jawless vertebrates covered with an external skeleton of bone which together with the Cyclostomata make up the class Agnatha. { 'ä·strə̩kō̩dərm }

Ostreidae [INV ZOO] A family of bivalve mollusks in the order Anisomyaria containing the oysters. { ä'strē·ə̩dē }

ostria [METEOROL] A warm southerly wind on the Bulgarian coast; it is considered a precursor of bad weather. Also known as auster. { 'äs·trē·ə }

ostrich [VERT ZOO] *Struthio camelus.* A large running bird with soft plumage, naked head, neck and legs, small wings,

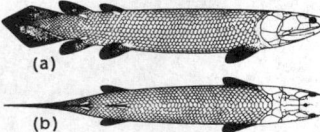

OSTEOLEPIFORMES

(a)

(b)

Devonian osteolepid *Gyroptychius agassizi*, with a length of 18 inches (46 centimeters), shown in *(a)* lateral and *(b)* dorsal aspects.

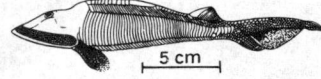

OSTRACODERM

5 cm

The ostracoderm *Hermicyclaspis*, a cephalaspid, a Lower Devonian jawless vertebrate, *(From E. H. Colbert. Evolution of the Vertebrates, 2d ed., copyright © 1969 by John Wiley & Sons, Inc.; reprinted by permission)*

and thick powerful legs with two toes on each leg; the only living species of the Struthioniformes. { 'ȯ,strich }

Ostriker-Peebles halo [ASTRON] A spherical distribution of matter of unknown nature that is postulated to exist to account for the stability of the highly flattened visible disk of the Milky Way Galaxy. { 'äs·trīk·ər 'pēb·əlz ,hā,lō }

Ostrogradski's theorem *See* Gauss' theorem. { ,ȯ·strə'gräd·skēz ,thir·əm }

Ostrya virginiana *See* American hophornbeam. { ,ȯs·trē ə vər,jin·ē'än·ə }

Ostwald coefficient [PHYS CHEM] A measure of the solubility of a gas in a liquid, equal to the volume of gas that can be dissolved in a given volume of liquid divided by the volume of liquid. { 'äst,vält ,kō·ə,fish·ənt }

Ostwald dilution law [PHYS CHEM] The law that for a sufficiently dilute solution of univalent electrolyte, the dissociation constant approximates $a^2c/(1-a)$, where c is the concentration of electrolyte and a is the degree of dissociation. { 'ȯst,vält di'lü·shən ,lȯ }

Ostwald process [CHEM ENG] An industrial preparation of nitric acid by the oxidation of ammonia; the oxidation takes place in successive stages to nitric oxide, nitrogen dioxide, and nitric acid; a catalyst of platinum gauze is used and high temperatures are needed. { 'ȯst,vält ,prä·səs }

Ostwald ripening [CHEM] Solution-crystallizer phenomenon in which small crystals, more soluble than large ones, dissolve and reprecipitate onto larger particles. { 'ȯst,vält ,rīp·ə·niŋ }

Ostwald's adsorption isotherm [THERMO] An equation stating that at a constant temperature the weight of material adsorbed on an adsorbent dispersed through a gas or solution, per unit weight of adsorbent, is proportional to the concentration of the adsorbent raised to some constant power. { 'ȯst,välts ad'sȯrp·shən 'ī·sə,thərm }

Ostwald viscometer [ENG] A viscometer in which liquid is drawn into the higher of two glass bulbs joined by a length of capillary tubing, and the time for its meniscus to fall between calibration marks above and below the upper bulb is compared with that for a liquid of known viscosity. { 'ȯst,vält vi'skäm·əd·ər }

osumilite [MINERAL] $(K,Na)(Mg,Fe^{2+})_2(Al,Fe^{3+})_3(Si,Al)_{12}O_{30}·H_2O$ A mineral that crystallizes in the hexagonal system and is commonly mistaken for cordierite. { ä'sü·mə,līt }

Oswald diagram [ANALY CHEM] Diagram used in fuel Orsat analyses by plotting percent by volume CO_2 (carbon dioxide) maximum in the fuel [ordinate] versus percent by volume O_2 (oxygen) in air [abscissa]; O_2 and CO_2 Orsat readings should fall on a line connecting these maximum values if the analysis is proceeding properly. { 'äz,wȯld 'dī·ə,gram }

ot-, oto- [ANAT] A combining form meaning ear. { ȯt, 'ō·dō }

otalgia [MED] Pain in the ear. { ō'tal·jə }

Otariidae [VERT ZOO] The sea lions, a family of carnivorous mammals in the superfamily Canoidea. { ,ōd·ə'rī·ə,dē }

otavite [MINERAL] $CdCO_3$ A mineral that crystallizes in the hexagonal system and is isostructural with calcite. { 'ōd·ə,vīt }

OTEC *See* ocean thermal energy conversion. { 'ō,tek }

Othniidae [INV ZOO] The false tiger beetles, a small family of coleopteran insects in the superfamily Tenebrionoidea. { ,ȯth'nī·ə,dē }

otic [ANAT] Of or pertaining to the ear or a part thereof. { 'ōd·ik }

-otic [SCI TECH] A suffix meaning of, pertaining to, characterized by, or causing the process. { 'äd·ik }

otic capsule [EMBRYO] A cartilaginous capsule surrounding the auditory vesicle during development, later fusing with the spheroid and occipital cartilages. { 'ōd·ik 'kap·səl }

otic ganglion [NEUROSCI] The nerve ganglion located immediately below the foramen ovale of the sphenoid bone. { 'ōd·ik 'gaŋ·glē,än }

Otitidae [INV ZOO] A family of cyclorrhaphous myodarian dipteran insects in the subsection Acalyptratae. { ō'tid·ə,dē }

otitis [MED] Inflammation of the ear. { ō'tīd·əs }

otitis externa [MED] Inflammation of the external ear. { ō'tīd·əs ek'stər·nə }

otitis media [MED] Inflammation of the middle ear. { ō'tīd·əs 'mē·dē·ə }

otocyst [EMBRYO] The auditory vesicle of vertebrate embryos. [INV ZOO] An auditory vesicle, otocell, or otidium in some invertebrates. { 'ōd·ə,sist }

otolaryngology [MED] A branch of medicine that deals with the ear, nose, and throat. Also known as otorhinolaryngology. { ,ōd·ō,lar·ən'gäl·ə·jē }

otolith [ANAT] A calcareous concretion on the end of a sensory hair cell in the vertebrate ear and in some invertebrates. { 'ōd·ə,lith }

otology [MED] A branch of medicine that deals with the ear and its diseases. { ō'täl·ə·je }

otomycosis [MED] Fungus infection of the external ear, usually caused by *Aspergillus niger* and *A. fumigatus*. { ,ōd·ə,mī'kō·səs }

Otopheidomeridae [INV ZOO] A family of parasitic mites in the suborder Mesostigmata. { ,ōd·ə,fē·dō'men·ə,dē }

otorhinolaryngology *See* otolaryngology. { ,ōd·ə,rīn·ō,lar·ən'gäl·ə·jē }

otosclerosis [MED] Sclerosis of the inner ear, causing a progressive increase in deafness. { ,ōd·ō·sklə'rō·səs }

otoscope [MED] An apparatus designed for examination of the ear and for rendering the tympanic membrane visible. { 'ōd·ə,skōp }

ototoxicity [MED] Drug- or chemical-induced damage to the ear resulting in high-frequency hearing loss and tinnitus or disequilibrium. { ,ō·tō·täk'sis·əd·ē }

OTS *See* ovonic threshold switch.

otter [ENG] *See* paravane. [VERT ZOO] Any of various members of the family Mustelidae, having a long thin body, short legs, a somewhat flattened head, webbed toes, and a broad flattened tail; all are adapted to aquatic life. { 'äd·ər }

otter trawl [NAV ARCH] A large commercial fishing trawl which uses kitelike wooden boards at the corners of the mouth of the net. { 'äd·ər ,trȯl }

Otto cycle [THERMO] A thermodynamic cycle for the conversion of heat into work, consisting of two isentropic phases interspersed between two constant-volume phases. Also known as spark-ignition combustion cycle. { 'äd·ō ,sī·kəl }

Otto engine [MECH ENG] An internal combustion engine that operates on the Otto cycle, where the phases of suction, compression, combustion, expansion, and exhaust occur sequentially in a four-stroke-cycle or two-stroke-cycle reciprocating mechanism. { 'äd·ō ,en·jən }

Otto-Lardillon method [MECH] A method of computing trajectories of missiles with low velocities (so that drag is proportional to the velocity squared) and quadrant angles of departure that may be high in which exact solutions of the equations of motion are arrived at by numerical integration and are then tabulated. { 'äd ō ,lär·dē'yȯn ,meth·əd }

ottoman [TEXT] Heavyweight fabric with pronounced crosswise rounded ribs, often padded, with heavier ribs than those of faille or bengaline. { 'äd·ō·mən }

otto of rose oil *See* rose oil. { 'äd·ō əv 'rōz ,ȯil }

ottrelite [MINERAL] A gray to black variety of chloritoid containing manganese. { 'ä·trə,līt }

O-type backward-wave oscillator [ELECTR] A backward-wave tube in which an electron gun produces an electron beam focused longitudinally throughout the length of the tube, a slow-wave circuit interacts with the beam, and at the end of the tube a collector terminates the beam. Also known as O-type carcinotron; type-O carcinotron. { 'ō ,tīp 'bak·wərd 'wāv 'äs·ə,lad·ər }

O-type carcinotron *See* O-type backward-wave oscillator. { 'ō ,tīp kär'sin·ə,trän }

O-type star [ASTRON] A spectral-type classification in the Draper catalog of stars; a star having spectral type O; a very hot, blue star in which the spectral lines of ionized helium are prominent. { 'ō ,tīp 'stär }

ouabain [ORG CHEM] $C_{29}H_{44}O_{12}·8H_2O$ White crystals that melt with decomposition at 190°C, soluble in water and ethanol; used in medicine. { wä'bī·ən }

ouachitite [PETR] A biotite monchiquite with no olivine and a glassy or analcime groundmass. { 'wä·chə,tīt }

ouari [METEOROL] A south wind of Somaliland, Africa; it is similar to the khamsin. { 'wä·rē }

Ouchterlony test [IMMUNOL] A technique used to analyze an antigen-antibody mixture, in which the components are placed in multiple wells cut into agar on a flat slide and allowed to diffuse toward one another. { 'au̇ch·tər,lȯn·ē ,test }

OTTER

North American otter (*Lutra canadensis*).

Oudeman law [PHYS CHEM] The law that the molecular rotations of the various salts of an acid or base tend toward an identical limiting value as the concentration of the solution is reduced to zero. { 'òd·ə·mən ˌlò }

Oudin test [IMMUNOL] A technique used to measure antigen concentration, in which an antigen and an antibody are held in an agar matrix in a test tube and allowed to diffuse toward one another. { 'ü·dan ˌtest }

ounce [MECH] **1.** A unit of mass in avoirdupois measure equal to 1/16 pound or to approximately 0.0283495 kilogram. Abbreviated oz. **2.** A unit of mass in either troy or apothecaries' measure equal to 480 grains or exactly 0.0311034768 kilogram. Also known as apothecaries' ounce or troy ounce (abbreviations are oz ap and oz t in the United States, and oz apoth and oz tr in the United Kingdom). { 'aùns }

ouncedal [MECH] A unit of force equal to the force which will impart an acceleration of 1 foot per second per second to a mass of 1 ounce; equal to 0.0086409346485 newton. { 'aùn·sə¦dal }

ounce metal [MET] An alloy composed of 1 ounce each of lead, tin, and zinc to 1 pound of copper. Also known as composition metal. { 'aùns ˌmed·əl }

ouricury wax [MATER] A hard brown wax obtained from leaves of the ouricury palm (*Cocos coronapa*); similar to carnauba wax in use and properties. { ¦ùr·ə·kə¦rē ˌwaks }

outage [ELEC] A failure in an electric power system. [PETRO ENG] The difference between the full or rated capacity of a barrel, tank, or tank car as compared to actual content. { 'aùd·ij }

outage method [PETRO ENG] Deduction of the liquid content of a tank by measurement of the distance from the top of the tank to the surface of the liquid; in contrast to the innage method. { 'aùd·ij ˌmeth·əd }

outboard [NAV ARCH] Toward the outside of a vessel; outside the hull. { 'aùt¦bòrd }

outboard engine [NAV ARCH] A unit assembly of engine, propeller, and vertical drive shaft used to propel a boat and usually clamped to the boat transom; power of various models ranges from 1 horsepower (approximately 750 watts) to well over 100 horsepower. Also known as outboard motor. { 'aùt¦bòrd ˈen·jən }

outboard motor See outboard engine. { 'aùt¦bòrd ˈmōd·ər }

outboard profile [NAV ARCH] A plan representing the longitudinal exterior of a vessel showing the starboard side of the shell, all deck erections, masts, yards, rigging, rails, and so on. { 'aùt¦bòrd ˈprō¦fil }

outbreed See crossbreed. { 'aùt¦brēd }

outbreeding See exogamy. { 'aùt¦brēd·iŋ }

outburst [METEOROL] Outflow from a convective event originating in cool descending air, often associated with a thunderstorm. [MIN ENG] The sudden issue of gases, chiefly methane (sometimes accompanied by coal dust), from the working face of a coal mine. { 'aùt¦bərst }

outby [MIN ENG] Toward the mine entrance or shaft and therefore away from the working face. { 'aùt¦bī }

outcrop [GEOL] Exposed stratum or body of ore at the surface of the earth. Also known as cropout. { 'aùt¦kräp }

outcrop curvature See settling. { 'aùt¦kräp ˈkər·və·chər }

outcrop map [GEOL] A type of geologic map that shows the distribution and shape of actual outcrops, leaving those areas without outcrops blank. { 'aùt¦kräp ˌmap }

outcrop water [HYD] Rain and surface water which seeps downward through outcrops of porous and fissured rock, fault planes, old shafts, or surface drifts. { 'aùt¦kräp ˌwòd·ər }

outdegree [MATH] For a vertex, *v*, in a directed graph, the number of arcs directed from *v* to other vertices. { 'aùt·di¦grē }

outer atmosphere [METEOROL] Very generally, the atmosphere at a great distance from the earth's surface; possibly best usage of the term is as an approximate synonym for exosphere. { 'aùd·ər 'at·mə,sfir }

outer automorphism [MATH] Any element of the quotient group formed from the group of automorphisms of a group and the subgroup of inner automorphisms. { 'aùd·ər ¦òd·ō'mòr,fiz·əm }

outer bar [GEOL] A bar formed at the mouth of an ebb channel of an estuary. { 'aùd·ər 'bär }

outer beach [GEOL] The part of a beach that is ordinarily dry and reached only by the waves generated by a violent storm. { 'aùd·ər 'bēch }

outer bottom [NAV ARCH] A part of a double-bottom ship, specifically, the bottom shell plating. { 'aùd·ər 'bäd·əm }

outer bremsstrahlung [PHYS] Bremsstrahlung involving the acceleration of a charged particle coming from outside the atom whose nucleus produces the acceleration, and in which the energy loss by radiation is much greater than that by ionization, usually seen in electrons with energies greater than about 50 MeV (million electronvolts). { 'aùd·ər 'brem,shträ·lùŋ }

outer core [GEOL] The outer or upper zone of the earth's core, extending to a depth of 3160 miles (5100 kilometers), and including the transition zone. { 'aùd·ər 'kòr }

outer effects [PHYS] Effects on x-ray diffraction that involve neighboring atoms or molecules. { 'aùd·ər i'feks }

outer fix [NAV] A fix in the destination terminal area, other than the approach fix, to which aircraft are normally cleared by an air route traffic-control center or a terminal area traffic-control facility, and from which aircraft are cleared to the approach fix or final approach course. { 'aùd·ər 'fiks }

outer harbor [GEOGR] The part of a harbor toward the sea, through which a vessel enters the inner harbor. { 'aùd·ər 'här·bər }

outer keel [NAV ARCH] The outer plate of a double flat-plate keel. { 'aùd·ər 'kēl }

outer mantle See upper mantle. { 'aùd·ər 'mant·əl }

outer marker [NAV] That instrument landing system (ILS) marker farthest from the approach end of the instrument runway; the marker is located 4.5 miles ± 1000 feet (7240 ± 305 meters) from the end of the runway and not more than 250 feet (75 meters) from the extended center line of the runway. { 'aùd·ər 'mär·kər }

outer measure [MATH] **1.** A function with the same properties as a measure except that it is only countably subadditive rather than countably additive; usually defined on the collection of all subsets of a given set. **2.** See Lebesgue exterior measure. { 'aùd·ər 'mezh·ər }

outer orbital complex [PHYS CHEM] A metal coordination compound in which the *d* orbital used in forming the coordinate bond is at the same energy level as the *s* and *p* orbitals. { 'aùd·ər 'òrb·əd·əl 'käm,pleks }

outer planets [ASTRON] The planets with orbits larger than that of Mars: Jupiter, Saturn, Uranus, Neptune, and Pluto. { 'aùd·ər 'plan·əts }

outer product [MATH] For any two tensors *R* and *S*, a tensor *T* each of whose indices corresponds to an index of *R* or an index of *S*, and each of whose components is the product of the component of *R* and the component of *S* with identical values of the corresponding indices. { ¦aùd·ər 'präd·əkt }

outer-shell electron See conduction electron. { 'aùd·ər ¦shel i'lek,trän }

outer space [ASTRON] A general term for any region that is beyond the earth's atmosphere. { 'aùd·ər 'spās }

outer trapped surface [RELAT] A compact, spacelike, two-dimensional surface in a space-time, such that outgoing light rays perpendicular to the surface are not diverging; whether ingoing light rays are converging or not is immaterial. { ¦aùd·ər ¦trapt 'sər·fəs }

outface See dip slope. { 'aùt,fās }

outfall [CIV ENG] The point at which a sewer or drainage channel discharges to a body of water. [HYD] The narrow part of a stream, lake, or other body of water where it drops away into a larger body. { 'aùt,fòl }

outflow [CHEM ENG] Flow of fluid product out of a process facility. [MATH] The outflow from a vertex in an *s-t* network is the sum of the flows of all the arcs that originate at that vertex. { 'aùt,flō }

outflow cave [GEOL] A cave from which a stream issues or is known to have issued. { 'aùt,flō ˌkāv }

outgassing [ASTRON] The ejection of gases trapped within a planet so that they are added to the planet's atmosphere. [ENG] The release of adsorbed or occluded gases or water vapor, usually by heating, as from a vacuum tube or other vacuum system. { 'aùt,gas·iŋ }

outgoing trunk See one-way trunk. { 'aùt,gō·iŋ 'trəŋk }

outgroup [SYST] A monophyletic taxon that is used in a phylogenetic study to resolve which of two homologous character states are apomorphic. { 'aùt,grüp }

outlet [ELEC] A power line termination from which electric power can be obtained by inserting the plug of a line cord. Also known as convenience receptacle; electric outlet; receptacle. { 'autˌlet }

outlet box [ELEC] A box at which lines in an electric wiring system terminate, so that electric appliances or fixtures may be connected. { 'autˌlet ˌbäks }

outlet glacier [HYD] A stream of ice from an ice cap to the sea. { 'autˌlet ˌglā·shər }

outlet head [HYD] The place where water leaves a lake and enters an effluent. { 'autˌlet ˌhed }

outlet ventilator *See* louver. { 'autˌlet 'vent·əlˌ ād·ər }

outlier [GEOL] A group of rocks separated from the main mass and surrounded by outcrops of older rocks. [STAT] In a set of data, a value so far removed from other values in the distribution that its presence cannot be attributed to the random combination of chance causes. { 'autˌlī·ər }

outline font *See* scalable font. { 'autˌlīn ˌfänt }

outline map [MAP] A map that presents minimal geographic information, usually only coastlines, principal streams, major civil boundaries, and large cities, leaving as much space as possible for the addition of specific data. { 'autˌlīn ˌmap }

outline processor [COMPUT SCI] A software system that organizes notes in ordinary English into an outline that serves as the basis for a document. { 'autˌlīn ˌprä,ses·ər }

out-of-line coding [COMPUT SCI] Instructions in a routine that are stored in a different part of computer storage from the rest of the instructions. { 'aut əv ¦līn ¦kōd·iŋ }

out of phase [PHYS] Having waveforms that are of the same frequency but do not pass through corresponding values at the same instant. { 'aut əv 'fāz }

out-of-service jack [ELEC] Jack associated with a test jack which removes the circuit from service when a shorted plug is inserted. { 'aut əv ¦sər·vəs ¦jak }

outpatient [MED] A patient who comes to the hospital or clinic for diagnosis and treatment but who does not occupy a bed in the institution. { 'autˌpā·shənt }

out-plant system [COMPUT SCI] A data-processing system that has one or more remote terminals from which information is transmitted to a central computer. { 'aut ˌplant ˌsis·təm }

output [COMPUT SCI] **1.** The data produced by a data-processing operation, or the information that is the objective or goal in data processing. **2.** The data actively transmitted from within the computer to an external device, or onto a permanent recording medium (paper, microfilm). **3.** The activity of transmitting the generated information. **4.** The readable storage medium upon which generated data are written, as in hard-copy output. [ELECTR] **1.** The current, voltage, power, driving force, or information which a circuit or device delivers. **2.** Terminals or other places where a circuit or device can deliver current, voltage, power, driving force, or information. [SCI TECH] The product of a system. { 'autˌput }

output area [COMPUT SCI] A part of storage that has been reserved for output data. Also known as output block. { 'aut ˌput ˌer·ē·ə }

output block [COMPUT SCI] **1.** A portion of the internal storage of a computer that is reserved for receiving, processing, and transmitting data to be transferred out. **2.** *See* output area. { 'autˌput ˌbläk }

output-bound computer [COMPUT SCI] A computer that is slowed down by its output functions. { 'autˌput ˌbaund kəmˌpyüd·ər }

output bus driver [ELECTR] A device that power-amplifies output signals from a computer to allow them to drive heavy circuit loads. { 'autˌput 'bəs ˌdrīv·ər }

output capacitance [ELECTR] Of an *n*-terminal electron tube, the short-circuit transfer capacitance between the output terminal and all other terminals, except the input terminal, connected together. { 'autˌput kə,pas·əd·əns }

output class [COMPUT SCI] An indicator of the priority of output from a computer that determines the order in which it is printed from a spool file. { 'autˌput ˌklas }

output device *See* output unit. { 'autˌput diˌvīs }

output gap [ELECTR] An interaction gap by means of which usable power can be abstracted from an electron stream in a microwave tube. { 'autˌput ˌgap }

output impedance [ELECTR] The impedance presented by a source to a load. { 'autˌput imˌpēd·əns }

output indicator [ENG] A meter or other device that is connected to a radio receiver to indicate variations in output signal

strength for alignment and other purposes, without indicating the exact value of output. { 'autˌput ˌin·də,kād·ər }

output-limited [ENG] Restricted by the need to await completion of an output operation, as in process control or data processing. { 'autˌput ˌlim·əd·əd }

output link [COMMUN] The last link in a communications chain. { 'autˌput ˌliŋk }

output meter [ENG] An alternating-current voltmeter connected to the output of a receiver or amplifier to measure output signal strength in volume units or decibels. { 'autˌput ˌmēd·ər }

output-meter adapter [ENG] Device that can be slipped over the plate prong of the output tube of a radio receiver to provide a conventional terminal to which an output meter can be connected during alignment. { 'autˌput ˌmēd·ər ə,dap·tər }

output monitor interrupt [COMPUT SCI] A data-processing step in which control is passed to the monitor to determine the precedence order for two requests having the same priority level. { 'autˌput ˌman·əd·ər 'int·ə,rəpt }

output power [ELEC] Power delivered by a system or transducer to its load. { 'autˌput ˌpau·ər }

output program *See* output routine. { 'autˌput ˌprō,gram }

output rating *See* carrier power output rating. { 'autˌput ˌrād·iŋ }

output record [COMPUT SCI] **1.** A unit of data that has been transcribed from a computer to an external medium or device. **2.** The unit of data that is currently held in the output area of a computer before being transcribed to an external medium or device. { 'autˌput ˌrek·ərd }

output resistance [ELECTR] The resistance across the output terminals of a circuit or device. { 'autˌput ri,zis·təns }

output routine [COMPUT SCI] A series of computer instructions which organizes and directs all operations associated with the transcription of data from a computer to various media and external devices by various types of output equipment. Also known as output program. { 'autˌput rü,tēn }

output shaft [MECH ENG] The shaft that transfers motion from the prime mover to the driven machines. { 'autˌput ˌshaft }

output stage [ELECTR] The final stage in any electronic equipment. { 'autˌput ˌstāj }

output standard *See* standard time. { 'autˌput ˌstan·dərd }

output transformer [ELECTR] The iron-core audio-frequency transformer used to match the output stage of a radio receiver or an amplifier to its loudspeaker or other load. { 'aut ˌput tranz,fòr·mər }

output tube [ELECTR] Power-amplifier tube designed for use in an output stage. { 'autˌput ˌtüb }

output unit [COMPUT SCI] In computers, a unit which delivers information from the computer to an external device or from internal storage to external storage. { 'autˌput ˌyü·nət }

output winding [ELECTROMAG] Of a saturable reactor, a winding, other than a feedback winding, which is associated with the load, and through which power is delivered to the load. { 'autˌput ˌwīnd·iŋ }

output word [COMPUT SCI] Any running word into which an input word is to be translated. { 'autˌput ˌwərd }

outrigger [ENG] A steel beam or lattice girder extending from a crane to provide stability by widening the base. { 'autˌrig·ər }

outside air temperature *See* indicated air temperature. { 'autˌsīd 'er ˌtem·prə·chər }

outside caliper [DES ENG] A caliper having two curved legs which point toward each other; used for measuring outside dimensions of a workpiece. { 'autˌsīd 'kal·ə·pər }

outside diameter [DES ENG] The outer diameter of a pipe, including the wall thickness; usually measured with calipers. Abbreviated OD. { 'autˌsīd dī'am·əd·ər }

outside extension [COMMUN] Telephone extension on premises separated from the main station. { 'autˌsīd ik'sten·chən }

outside fix [NAV] The fix position determined by the method of bisectors when the lines of position result from observations of objects or celestial bodies lying within a 180° arc of the horizon. { 'autˌsīd 'fiks }

outside strake [NAV ARCH] The outer strake of an in-and-out system of shell plating; a strake which laps on the inner strake and which is the thickness of the plating outside the molded frame line. { 'autˌsīd 'strāk }

OUTSIDE CALIPER

Drawing of outside caliper.

outward bound [NAV] Heading for the open sea. { 'aut·wərd 'baund }

outwash [GEOL] **1.** Sand and gravel transported away from a glacier by streams of meltwater and either deposited as a floodplain along a preexisting valley bottom or broadcast over a preexisting plain in a form similar to an alluvial fan. Also known as glacial outwash; outwash drift; overwash. **2.** Soil material washed down a hillside by rainwater and deposited on more gently sloping land. { 'aut,wäsh }

outwash apron *See* outwash plain. { 'aut,wäsh ,ā·prən }

outwash cone [GEOL] A cone-shaped deposit consisting chiefly of sand and gravel found at the edge of shrinking glaciers and ice sheets. { 'aut,wäsh ,kōn }

outwash drift *See* outwash. { 'aut,wäsh ,drift }

outwash fan [GEOL] A fan-shaped accumulation of outwash deposited by meltwater streams in front of the terminal moraine of a glacier. { 'aut,wäsh ,fan }

outwash plain [GEOL] A broad, outspread flat or gently sloping alluvial deposit of outwash in front of or beyond the terminal moraine of a glacier. Also known as apron; frontal apron; frontal plain; marginal plain; morainal apron; morainal plain; outwash apron; overwash plain; sandur; wash plain. { 'aut,wäsh ,plān }

outwash terrace [GEOL] A dissected and incised valley train or benchlike deposit extending along a valley downstream from an outwash plain or terminal moraine. { 'aut,wäsh ,ter·əs }

outwash train *See* valley train. { 'aut,wäsh ,trān }

ouvarovite *See* uvarovite. { ü'vär·ə,vīt }

oval [MATH] A curve shaped like a section of an egg. { 'ō·vəl }

ovalbumin [BIOCHEM] The major, conjugated protein of eggwhite. { ,ov·al'byü·mən }

oval of Cassini [MATH] An ovallike curve similar to a lemniscate obtained as the locus corresponding to a general type of quadratic equation in two variables x and y; it is expressed as $[(x + a)^2 + y^2] [(x - a)^2 + y^2] = k^4$, where a and k are constants. Also known as Cassinian oval. { 'ō·vəl əv kə'sē·nē }

oval window [ANAT] The membrane-covered opening into the inner ear of tetrapods, to which the ossicles of the middle ear are connected. { 'ō·vəl 'win·dō }

ovarian [ANAT] Of or pertaining to the ovaries. { ō'ver·ē·ən }

ovarian agenesis [MED] Failure of the ovaries to develop. { ō'ver·ē·ən ā'jen·ə·səs }

ovarian dysmenorrhea [MED] Dysmenorrhea caused by ovarian disease. { ō'ver·ē·ən dis,men·ə're·ə }

ovarian follicle [HISTOL] An ovum and its surrounding follicular cells, found in the ovarian cortex. { ō'ver·ē·ən 'fäl·ə·kəl }

ovarian insufficiency [MED] Deficient functioning of the ovaries, leading to amenorrhea, oligomenorrhea, or abnormal dysfunctional uterine bleeding. { ō'ver·ē·ən ,in·sə'fish·ən·sē }

ovariectomy [MED] Excision of an ovary. { ō,var·ē'ek·tə·mē }

ovariole [INV ZOO] The tubular structural unit of an insect ovary. { ō'var·ē,ōl }

ovary [ANAT] A glandular organ that produces hormones and gives rise to ova in female vertebrates. [BOT] The enlarged basal portion of a pistil that bears the ovules in angiosperms. { 'ōv·ə·rē }

oven [ENG] A heated enclosure for baking, heating, or drying. [GEOL] **1.** A rounded, saclike, chemically weathered pit or hollow in a rock (especially a granitic rock) which has an arched roof and resembles an oven. **2.** *See* spouting horn. { 'əv·ən }

over [ORD] A bomb or projectile hit beyond or past the target. { 'ō·vər }

overaging [MET] Aging at a higher temperature or for a longer time than is required to produce maximum or optimum properties. { 'ō·vər,āj·iŋ }

overall drilling time [MIN ENG] The total time for rock drilling including time for setting up, withdrawing, and moving drills, time for mechanical delays, and the time for the actual drilling. { 'ō·vər,ȯl 'dril·iŋ ,tīm }

overall efficiency [AERO ENG] The efficiency of a jet engine, rocket engine, or rocket motor in converting the total heat energy of its fuel first into available energy for the engine, then into effective driving energy. { 'ō·vər,ȯl i'fish·ən·sē }

overall plate efficiency [CHEM ENG] For a specified liquid-mixture separation in a fractionation (or distillation) tower, the ratio of actual to theoretical plates (or trays) required. { 'ō·vər,ȯl 'plāt i,fish·ən·sē }

overall response [ELECTR] The ratio between system input and output. { 'ō·vər,ȯl ri'späns }

overall stability constant [ANALY CHEM] Reaction equilibrium constant for the reaction forming soluble complexes during compleximetric titration. { 'ō·vər,ȯl stə'bil·əd·ē ,kän·stənt }

over a map [MATH] A map f from a set A to a set L is said to be over a map g from a set B to L if B is a subset of A and the restriction of f to B equals g. { 'ō·vər ə 'map }

overarching weight [MIN ENG] The pressure of the rocks over the active mine workings. { 'ō·vər,ärch·iŋ 'wāt }

overarm [MECH ENG] One of the adjustable supports for the end of a milling-cutter arbor farthest from the machine spindle. { 'ō·vər,ärm }

over a set [MATH] A map f from a set A to a set L is said to be over a set B if B is a subset of both A and L and if the restriction of F to B is the identity map on B. { 'ō·vər ə 'set }

overbank deposit [GEOL] Fine-grained sediment (silt and clay) deposited from suspension on a floodplain by floodwaters from a stream channel. { 'ō·vər,baŋk di,päz·ət }

overbanking [HOROL] A malfunction in which the escape wheel unlocks prematurely, without the fork contacting the roller jewel. { 'ō·vər,baŋk·iŋ }

overbank stage [HYDR] The height of the surface of a river as the river floods over its banks. { 'ō·vər,baŋk ,stāj }

overbending [MET] Compensation for springback in metalforming by bending the material through a greater arc than that required for the finished part. { 'ō·vər,bend·iŋ }

overbreak [CIV ENG] Rock excavated in excess of the neat lines of a tunnel or cutting. Also known as backbreak. { 'ō·vər,brāk }

overbunching [ELECTR] In velocity-modulated streams of electrons, the bunching condition produced by the continuation of the bunching process beyond the optimum condition. { 'ō·vər,bənch·iŋ }

overburden [GEOL] **1.** Rock material overlying a mineral deposit or coal seam. Also known as baring; top. **2.** Material of any nature, consolidated or unconsolidated, that overlies a deposit of useful materials, ores, or coal, especially those deposits that are mined from the surface by open cuts. **3.** Loose soil, sand, or gravel that lies above the bedrock. [MIN ENG] To charge in a furnace too much ore and flux in proportion to the amount of fuel. { 'ō·vər,bȯrd·ən }

overburdened stream *See* overloaded stream. { 'ō·vər,bȯrd·ən 'strēm }

overcast [METEOROL] **1.** Pertaining to a sky cover of 1.0 (95% or more) when at least a portion of this amount is attributable to clouds or obscuring phenomena aloft, that is, when the total sky cover is not due entirely to surface-based obscuring phenomena. **2.** Cloud layer that covers most or all of the sky; generally, a widespread layer of clouds such as that which is considered typical of a warm front. [MIN ENG] **1.** An enclosed airway to permit one air current to pass over another without interruption. **2.** To move overburden removed from coal mined from surface mines to an area from which the coal has been mined. { 'ō·vər,kast }

overcast bombing [ORD] The bombing of a target through an overcast above the target, using radar or other equipment to aid in sighting through the overcast. { 'ō·vər,kast ,bäm·iŋ }

overcheck [TEXT] A fabric pattern in which one check of different size or color is superimposed upon another. { 'ō·vər,chek }

overcoating [ENG] Extruding a plastic web beyond the edge of the substrate web in extrusion coating. { 'ō·vər,kōd·iŋ }

overcompound [ELEC] To use sufficiently many series turns in a compound-wound generator so that the terminal voltage at rated load is greater than at no load, usually to compensate for increased line drop. { 'ō·vər,käm,paund }

overconsolidation [GEOL] Consolidation of sedimentary material exceeding that which is normal for the existing overburden. { 'ō·vər,kən,säl·ə'dā·shən }

overcoupled circuits [ELECTR] Two resonant circuits which are tuned to the same frequency but coupled so closely that two response peaks are obtained; used to attain broad-band response with substantially uniform impedance. { 'ō·vər‚kap· əld 'sər·kəts }

overcritical binding [ATOM PHYS] A binding energy for electrons in atoms which is so large that a vacancy in the bound state results in the spontaneous formation of an electron-positron pair; predicted to occur when the atomic number exceeds 173. { 'ō·vər‚krid·ə·kəl 'bīnd·iŋ }

overcritical electric field [ATOM PHYS] An electric field so strong that an electron-positron pair is created spontaneously; quantum electrodynamics predicts that this will happen near a nucleus having more than approximately 173 protons. { 'ō· vər‚krid·ə·kəl i'lek·trik 'fēld }

overcuring [CHEM ENG] A condition resulting from vulcanizing longer than necessary to achieve full development of physical strength; causes softness or brittleness and impaired age-resisting quality of the material. { 'ō·vər‚kyur·iŋ }

overcurrent [ELECTR] An abnormally high current, usually resulting from a short circuit. { ‚ō·vər‚kə·rənt }

overcurrent protection See overload protection. { ‚ō·vər‚kə·rənt prə'tek·shən }

overcut [MIN ENG] A machine cut made along the top or near the top of a coal seam; sometimes used in a thick seam or a seam with sticky coal. { 'ō·vər‚kət }

overcutting machine [MIN ENG] A coal-cutting machine designed to make the cut at a desired place in the coal seam some distance above the floor. { 'ō·vər‚kəd·iŋ mə‚shēn }

overdamping [PHYS] Damping greater than that required for critical damping. { 'ō·vər‚damp·iŋ }

overdeepening [GEOL] The erosive process by which a glacier deepens and widens an inherited preglacial valley to below the level of the subglacial surface. { ‚ō·vər'dēp·ə·niŋ }

overdevelop [GRAPHICS] To process a photographic film or plate for too long a time or in a solution that is too concentrated. { ‚ō·vər·di'vel·əp }

overdominance [GEN] Monohybrid heterosis, that is, the phenotype is more pronounced in the heterozygote than in either homozygote with respect to a specified pair of alleles. { ‚ō·vər'däm·ə·nəns }

overdose [MED] An excessive dose of medicine. { 'ō· vər‚dōs }

overdraft [MET] Upward curving of a piece of metal after leaving the rolls during forming, due to higher speed of the lower roll. { 'ō·vər‚draft }

overdrilling [ENG] The act or process of drilling a run or length of borehole greater than the core-capacity length of the core barrel, resulting in loss of the core. { 'ō·vər'dril·iŋ }

overdrive [MECH ENG] An automobile engine device that lowers the gear ratio, thereby reducing fuel consumption. { 'ō·vər‚drīv }

overdriven amplifier [ELECTR] Amplifier stage which is designed to distort the input-signal waveform by permitting the grid signal to drive the stage beyond cutoff or plate-current saturation. { ‚ō·vər'driv·ən 'am·plə‚fī·ər }

overexpose [GRAPHICS] To expose a photographic film or plate with too much light or for too long a time. { ‚ō·vər· ik'spōz }

overfall dam See overflow dam. { 'ō·vər‚fol ‚dam }

overfalls [OCEANOGR] Short, breaking waves occurring when a strong current passes over a shoal or other submarine obstruction or meets a contrary current or wind. { 'ō· vər‚folz }

overfeed [MIN ENG] To attempt to make a diamond- or rock-drill bit penetrate rock at a rate in excess of that at which the optimum economical performance of the bit is attained, needlessly damaging the bit and shortening its life. { 'ō· vər‚fēd }

overfire draft [MECH ENG] The air pressure in a boiler furnace during occurrence of the main flame. { 'ō·vər‚fīr 'draft }

overflow [CIV ENG] Any device or structure that conducts excess water or sewage from a conduit or container. [COMPUT SCI] **1.** The condition that arises when the result of an arithmetic operation exeeds the storage capacity of the indicated result-holding storage. **2.** That part of the result which exceeds the storage capacity. [SCI TECH] Excess liquid which overflows its given limits. { 'ō·vər‚flō }

overflow bucket [COMPUT SCI] A unit of storage in a direct-access storage device used to hold an overflow record. { 'ō· vər‚flō ‚bək·ət }

overflow capacity [ENG] Capacity of a container measured

to its top, or to the point of overflow. { 'ō·vər‚flō kə‚pas· əd·ē }

overflow channel [CIV ENG] An artificial waterway for conducting water away from an overflowing structure such as a reservoir or canal. [GEOL] A channel or notch cut by the overflow waters of a lake, especially the channel draining meltwater from a glacially dammed lake. { 'ō·vər‚flō ‚chan·əl }

overflow check indicator See overflow indicator. { 'ō·vər‚flō 'chek ‚in·də‚kād·ər }

overflow dam [CIV ENG] A dam built with a crest to allow the overflow of water. Also known as overfall dam; spillway dam. { 'ō·vər‚flō ‚dam }

overflow error [COMPUT SCI] The condition in which the numerical result of an operation exceeds the capacity of the register. { 'ō·vər‚flō 'er·ər }

overflow groove [ENG] Small groove on a plastics mold that allows material to flow freely, to prevent weld lines and low density in the finished product and to dispose of excess material. { 'ō·vər‚flō ‚grüv }

overflow ice [HYD] Ice formed during high spring tides by water rising through cracks in the surface ice and then freezing. { 'ō·vər‚flō ‚īs }

overflow indicator [COMPUT SCI] A bistable device which changes state when an overflow occurs in the register associated with it, and which is designed so that its condition can be determined, and its original condition restored. Also known as overflow check indicator. { 'ō·vər‚flō ‚in·də‚kād·ər }

overflow pipe [ENG] Open pipe protruding above the surface of a liquid in a container, such as a distillation or absorption column or a toilet tank, to control the height of the liquid; excess liquid enters the pipe's open end and drains away. { 'ō· vər‚flō ‚pīp }

overflow record [COMPUT SCI] A unit of data whose length is too great for it to be stored in an assigned section of a direct-access storage, and which must be stored in another area from which it may be retrieved by means of a reference stored in the original assigned area in place of the record. { 'ō·vər‚flō ‚rek·ərd }

overflow spring [HYD] A type of contact spring that develops where a permeable deposit dips beneath an impermeable mantle. { 'ō·vər‚flō ‚spriŋ }

overflow storage [COMMUN] Additional storage provided in a store-and-forward-switching center to prevent the loss of messages (or parts of messages) offered to a completely filled line store. { 'ō·vər‚flō ‚stor·ij }

overflow stream [HYD] **1.** A stream containing water that has overflowed the banks of a river or another stream. Also known as spill stream. **2.** An effluent from a lake, carrying water to a stream, a sea, or another lake. { 'ō·vər‚flō ‚strēm }

overfold [GEOL] A fold that is overturned. { 'ō·vər‚fōld }

overgear [MECH ENG] A gear train in which the angular velocity ratio of the driven shaft to driving shaft is greater than unity, as when the propelling shaft of an automobile revolves faster than the engine shaft. { 'ō·vər‚gir }

overglaze color [MATER] Any of the mixtures of ground pigment and low-melting glass melting at 704–816°C; used for decorative designs fired onto china and ceramics. { 'ō· vər glāz ‚kəl·ər }

overgrinding [MIN ENG] Grinding an ore to a smaller particle size than that necessary to free the desired mineral from other materials. { 'ō·vər‚grīnd·iŋ }

overgrowth [CRYSTAL] A crystal growth in optical and crystallographic continuity around another crystal of different composition. [MINERAL] A mineral deposited on and growing in oriented, crystallographic directions on the surface of another mineral. { 'ō·vər‚grōth }

overhand cut and fill [MIN ENG] A method of mining ore in which material removed from the roof of a drive drops through chutes to a lower drive, from which the material is removed. { 'ō·vər‚hand ‚kət ən 'fil }

overhand stope [MIN ENG] A stope in which the ore above the point of entry is attacked, so that severed ore tends to gravitate toward discharge chutes, and the stope is self-draining. { 'ō·vər‚hand 'stōp }

overhand stoping [MIN ENG] A method of mining in which the ore is blasted from a series of ascending stepped benches; both horizontal and vertical holes may be employed. { 'ō· vər‚hand 'stōp·iŋ }

overhang [BUILD] The distance measured horizontally that a roof projects beyond a wall. [GEOL] The part of a salt plug that projects from the top. { 'ō·vər‚haŋ }

overhaul [ENG] A maintenance procedure for machinery involving disassembly, the inspecting, refinishing, adjusting, and replacing of parts, and reassembly and testing. { 'ō·vər‚hȯl }

overhauling [MET] Removing scale and surface defects from metal castings or slabs by cutting away surface layers. { 'ō·vər‚hȯl·iŋ }

Overhauser effect [ATOM PHYS] The effect whereby, if a radio frequency field is applied to a substance in an external magnetic field, whose nuclei have spin $\frac{1}{2}$ and which has unpaired electrons, at the electron spin resonance frequency, the resulting polarization of the nuclei is as great as if the nuclei had the much larger electron magnetic moment. { 'ō·vər‚haůz·ər i‚fekt }

overhead [CHEM ENG] Pertaining to fluid (gas or liquid) effluent from the top of a process vessel, such as a distillation column. [COMPUT SCI] The time a computer system spends doing computations that do not contribute directly to the progress of any user tasks in the system, such as allocation of resources, responding to exceptional conditions, providing protection and reliability, and accounting. [IND ENG] *See* fixed cost. { 'ō·vər‚hed }

overhead cableway [MIN ENG] A type of equipment for the removal of soil or rock, consisting of a strong overhead cable which is usually attached to towers at either end, and on which a car or traveler may run back and forth; from this car a pan or bucket may be lowered to the surface, then raised and locked to the car and transported to any position on the cable where it is desired to dump. { 'ō·vər‚hed 'kā·bəl‚wā }

overhead camshaft [MECH ENG] A camshaft mounted above the cylinder head. { 'ō·vər‚hed 'kam‚shaft }

overhead compass *See* inverted compass. { 'ō·vər‚hed 'käm·pəs }

overhead cost *See* fixed cost. { 'ō·vər‚hed ‚kȯst }

overhead fire [ORD] Fire that is delivered over the heads of friendly troops. { 'ō·vər‚hed ‚fīr }

overhead position [MET] In welding, the position by which the deposit is made from the underside of the joint. { 'ō·vər‚hed pə‚zish·ən }

overhead shovel [MECH ENG] A tractor which digs with a shovel at its front end, swings the shovel rearward overhead, and dumps the shovel at its rear end. { 'ō·vər‚hed 'shəv·əl }

overhead traveling crane [MECH ENG] A hoisting machine with a bridgelike structure moved on wheels along overhead trackage which is usually fixed to the building structure. { 'ō·vər‚hed ‚trav·ə·liŋ 'krān }

overhead-valve engine [MECH ENG] A four-stroke-cycle internal combustion engine having its valves located in the cylinder head, operated by pushrods that actuate rocker arms. Abbreviated OHV engine. Also known as valve-in-head engine. { 'ō·vər‚hed ‚valv 'en·jən }

overheat [MET] To heat a metal or alloy to such high temperatures that its physical properties are impaired. { 'ō·vər‚hēt }

overheating effect [SOLID STATE] The effect whereby, under certain conditions, a superconductor can be heated above its critical temperature without losing superconductivity. { ‚ō·vər'hēd·iŋ i‚fekt }

overhit [ORD] To hit a target with more destructive force than necessary to accomplish the desired amount of damage. { ‚ō·vər‚hit }

overite [MINERAL] $Ca_3Al_8(PO_4)_8(OH)_6\cdot15H_2O$ A mineral composed of hydrous basic calcium aluminum phosphate. { 'ō·və‚rīt }

overland flow [HYD] Water flowing over the ground surface toward a channel; upon reaching the channel, it is called surface runoff. Also known as surface flow. { 'ō·vər·lənd 'flō }

overlap [COMMUN] **1.** In teletypewriter practice, the selecting of another code group while the printing of a previously selected code group is taking place. **2.** Amount by which the effective height of the scanning facsimile spot exceeds the nominal width of the scanning line. [COMPUT SCI] To perform some or all of an operation concurrently with one or more other operations. [GEOL] **1.** Movement of an upcurrent part of a shore to a position extending seaward beyond a downcurrent part. **2.** Extension of strata over or beyond older underlying rocks. **3.** The horizontal component of separation measured parallel to the strike of a fault. [MET] **1.** Projection

of the weld metal beyond the bond at the toe of the weld. **2.** Extension of one sheet over another in spot, seam, or projection welding. { 'ō·vər‚lap }

overlap fault [GEOL] A fault structure in which the displaced strata are doubled back upon themselves. { 'ō·vər‚lap ‚fȯlt }

overlap integral [QUANT MECH] The integral over space of the product of the wave function of a particle and the complex conjugate of the wave function of another particle. { 'ō·vər‚lap 'int·ə·grəl }

overlapped memories [COMPUT SCI] An arrangement of computer memory banks in which, to cut down access time, successive words are taken from different memory banks, rewriting in one bank being overlapped by logic operations in another bank, with memory access in still another bank. { 'ō·vər‚lapt 'mem·rēz }

overlapping [COMPUT SCI] An operation whereby, if the processor determines that the current instruction and the next instruction lie in different storage modules, the two words may be retrieved in parallel. { ¦ō·vər¦lap·iŋ }

overlapping genes [GEN] Genes having nucleotide sequences that may overlap in a way that involves control genes or structural genes. { ¦ō·vər¦lap·iŋ 'jēnz }

overlapping input/output [COMPUT SCI] A procedure in which a computer system works on several programs, suspending work on a program and moving to another when it encounters an instruction for input/output operation, which is then executed when input/output operations from other programs have been carried out. { ¦ō·vər¦lap·iŋ 'in‚pût 'aůt‚pût }

overlapping orbitals [ATOM PHYS] Two orbitals (usually of electrons associated with different atoms in a molecule) for which there is a region of space where both are of appreciable magnitude. { ¦ō·vər¦lap·iŋ 'ȯrb·əd·əlz }

overlapping pair [GRAPHICS] Two aerial photographs taken at different camera locations so that part of one photograph shows terrain which is also shown on part of the other photograph. { ¦ō·vər¦lap·iŋ 'per }

overlap radar [ENG] Radar located in one sector whose area of useful radar coverage includes a portion of another sector. { 'ō·vər‚lap 'rā‚där }

overlay [CIV ENG] A repair topping of asphalt or concrete placed on a worn roadway. [COMPUT SCI] A technique for bringing routines into high-speed storage from some other form of storage during processing, so that several routines will occupy the same storage locations at different times; overlay is used when the total storage requirements for instructions exceed the available main storage. [ENG] **1.** Nonwoven fibrous mat (glass or other fiber) used as the top layer in a cloth or mat lay-up to give smooth finish to plastic products or to minimize the fibrous pattern on the surface. Also known as surfacing mat. **2.** An ornamental covering, as of wood or metal. [GRAPHICS] **1.** A sheet attached to copy or artwork and containing special instructions about reproduction or arrangement. **2.** A transparent or translucent film attached to artwork and carrying additional detail to be reproduced; in multicolor printing these overlays may represent the separation of the colors to be printed. { 'ō·vər‚lā }

overlay plywood [MATER] Plywood with a resin-treated fiber surface. { 'ō·vər‚lā 'plī‚wůd }

overlay transistor [ELECTR] Transistor containing a large number of emitters connected in parallel to provide maximum power amplification at extremely high frequencies. { 'ō·vər‚lā tran'zis·tər }

overline [COMMUN] In teletypewriter practice, the printing of one group of characters over another. { 'ō·vər‚līn }

overload [CIV ENG] A load on a structure that is greater than that for which the structure was designed. [ELECTR] A load greater than that which a device is designed to handle; may cause overheating of power-handling components and distortion in signal circuits. [GEOL] The amount of sediment that exceeds the ability of a stream to transport it and is therefore deposited. { 'ō·vər‚lōd }

overload capacity [ELEC] Current, voltage, or power level beyond which permanent damage occurs to the device considered. { 'ō·vər‚lōd kə‚pas·əd·ē }

overload current [ELECTR] A current greater than that which a circuit is designed to carry; may melt wires or damage elements of the circuit. { 'ō·vər‚lōd ‚kə·rənt }

overloaded stream [HYD] A stream so heavily loaded with sediment that its velocity is lessened and it is forced to deposit part of its load. Also known as overburdened stream. { ¦ō·vər¦lōd·əd 'strēm }

overloader [MIN ENG] A loading machine which digs with a bucket, raises the bucket, and swings it in a wide horizontal arc to the dumping point. { 'ō·vər,lōd·ər }

overloading [COMPUT SCI] The use, in some advanced programming languages, of two or more variables or subroutines with the same name; the compiler determines by inference which entity is referred to each time the name occurs. { ¦ō·vər¦lōd·iŋ }

overload level [ELEC] Level above which operation ceases to be satisfactory as a result of signal distortion, overheating, damage, and so forth. { 'ō·vər,lōd ,lev·əl }

overload protection [ELEC] Effect of a device operative on excessive current, but not necessarily on short circuit, to cause and maintain the interruption of current flow to the device governed. [MECH ENG] A safeguard against the application of excessive force against the wrist socket or end effector of a robot. Also known as overcurrent protection. { 'ō·vər,lōd prə,tek·shən }

overload relay [ELEC] A relay that opens a circuit when the load in the circuit exceeds a preset value, in order to provide overload protection; usually responds to excessive current, but may respond to excessive values of power, temperature, or other quantities. Also known as overload release. { 'ō·vər,lōd ,rē,lā }

overload release See overload relay. { 'ō·vər,lōd ri,lēs }

overmatching plate [ORD] Armor plate whose thickness exceeds the diameter of the projectile. { ¦ō·vər¦mach·iŋ 'plāt }

overmatching projectile [ORD] A projectile whose diameter exceeds the thickness of the armor plate. { ¦ō·vər¦mach·iŋ prə'jek·təl }

overmodulation [COMMUN] Amplitude modulation greater than 100%, causing distortion because the carrier voltage is reduced to zero during portions of each cycle. { ¦ō·vər,mäj·ə'lā·shən }

overnight polysomnography [MED] The overnight sleep recording of brain waves, muscle activities, eye movements, heart activity, airflow at the nose and mouth, respiratory effort, and oxygen saturation that is used to diagnose abnormalities in sleep and/or wakefulness. { ,ō·vər¦nīt ,päl·ē·səm'näg·rə·fē }

overoccult [ENG] The action of a coronagraph that occults a region whose diameter is significantly greater than that of the photosphere and thereby cuts off the inner corona from observation, as may be necessary for a coronagraph aboard a spacecraft due to limitations on spacecraft control. { ,ō·vər·ə'kəlt }

overpass [CIV ENG] **1.** A grade separation in which traffic at the higher level is raised, and traffic at the lower level moves at approximately its original level. **2.** The upper level at such a grade separation. { 'ō·vər,pas }

overplaid [TEXT] A fabric design consisting of a double plaid in which weave or the color effect is in blocks of the same or different sizes, arranged one over the other. { 'ō·vər,plad }

overpoint [CHEM] The initial boiling point in a distillation process; specifically, the temperature at which the first drop falls from the tip of the condenser into the condensate flask. { 'ō·vər,pȯint }

overpotential See overvoltage. { ¦ō·vər·pə'ten·chəl }

overpressure [FL MECH] The transient pressure, usually expressed in pounds per square inch, exceeding existing atmospheric pressure and manifested in the blast wave from an explosion. { 'ō·vər,presh·ər }

overprint [GEOCHEM] A complete or partial disturbance of an isolated radioactive system by thermal, igneous, or tectonic activities which results in loss or gain of radioactive or radiogenic isotopes and, hence, a change in the radiometric age that will be given the disturbed system. [GEOL] The development or superposition of metamorphic structures on original structures. Also known as imprint; metamorphic overprint; superprint. [GRAPHICS] **1.** To imprint over something that has been printed. **2.** To apply a varnish or lacquer to printed matter from a type or litho process, by means of a brush, spray, or roller coating. **3.** To print in a primary color over an existing color print to obtain a compound shade. { 'ō·vər,print }

override [CONT SYS] To cancel the influence of an automatic control by means of a manual control. { 'ō·və,rīd }

overriding process control [CONT SYS] Process control in which any one of several controllers associated with one control valve can be made to override another in accordance with a priority requirement of the process. { 'ō·və,rīd·iŋ 'präs·əs kən,trōl }

overrun [CIV ENG] A cleared area extending beyond the end of a runway. [COMPUT SCI] The arrival of an amount of data greater than the space allocated to it. { 'ō·və,rən }

overrunning [METEOROL] A condition existing when an air mass is in motion aloft above another air mass of greater density at the surface; this term usually is applied in the case of warm air ascending the surface of a warm front or quasi-stationary front. { 'ō·və·rən·iŋ }

overrunning clutch [MECH ENG] A clutch that allows the driven shaft to turn freely only under certain conditions; for example, a clutch in an engine starter that allows the crank to turn freely when the engine attempts to run. { 'ō·və,rən·iŋ 'kləch }

oversail [BUILD] To project beyond the general face of a structure. { 'ō·vər,sāl }

oversaturated See silicic. { ¦ō·vər'sach·ə,rād·əd }

overseeding [METEOROL] Cloud seeding in which an excess of nucleating material is released; as the term is normally used, the excess is relative to that amount of nucleating material which would, theoretically, maximize the precipitation received at the ground. { ¦ō·vər¦sēd·iŋ }

overshoot [ELECTROMAG] The reception of microwave signals where they were not intended, due to an unusual atmospheric condition that sets up variations in the index of refraction. [ENG] **1.** An initial transient response to a unidirectional change in input which exceeds the steady-state response. **2.** The maximum amount by which this transient response exceeds the steady-state response. { 'ō·vər,shüt }

overshot [ENG] **1.** A fishing tool for recovering lost drill pipe or casing. **2.** See bullet. { 'ō·vər,shät }

overshot wheel [MECH ENG] A horizontal-shaft waterwheel with buckets around the circumference; the weight of water pouring into the buckets from the top rotates the wheel. { 'ō·vər,shät ,wēl }

oversite concrete [BUILD] A layer of concrete that is installed below a slab or other type of floor surface. { ¦ō·vər,sīt 'kän,krēt }

oversize control screen [MIN ENG] A screen used to prevent the entry into a machine of coarse particles which might interfere with its operation. Also known as check screen; guard screen. { 'ō·vər,sīz kən'trōl ,skrēn }

oversize powder [MET] A metal powder having coarser particles than the maximum permitted. { 'ō·vər,sīz ,paüd·ər }

oversize rod See guide rod. { 'ō·vər,sīz ,räd }

overspeed governor [MECH ENG] A governor that stops the prime mover when speed is excessive. { 'ō·vər,spēd ,gəv·ə·nər }

overspin [MECH] In a spin-stabilized projectile, the overstability that results when the rate of spin is too great for the particular design of projectile, so that its nose does not turn downward as it passes the summit of the trajectory and follows the descending branch. Also known as overstabilization. { 'ō·vər,spin }

oversquare engine [MECH ENG] An engine with bore diameter greater than the stroke length. { 'ō·vər,skwer 'en·jən }

overstability [PL PHYS] Condition in which the restoring forces acting on an oscillation of a plasma or other conducting fluid drive the fluid back to its equilibrium state at a speed greater than its original outward speed, resulting in continually greater oscillation. { ¦ō·vər·stə'bil·əd·ē }

overstabilization See overspin. { ¦ō·vər,stā·bə·lə'zā·shən }

oversteepening [GEOL] The process by which an eroding alpine glacier steepens the sides of an inherited preglacial valley. { ¦ō·vər'stēp·ə·niŋ }

oversteer [MECH ENG] The tendency of an automotive vehicle to steer into a turn to a sharper degree than was intended by the driver; sometimes causes the vehicle's rear end to swing out. { 'ō·vər,stir }

overstep [GEOL] **1.** An overlap characterized by the regular truncation of older units of a complete sedimentary sequence by one or more later units of the sequence. **2.** A stratum deposited on the upturned edges of underlying strata. { 'ō·vər,step }

overstressed area [MIN ENG] In strata control, an area

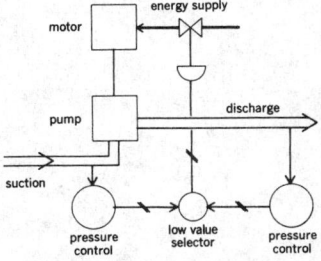

**OVERRIDING
PROCESS CONTROL**

Diagram of overriding process control in an oil pipeline. The motor driving the pipeline is controlled by suction pressure when this pressure approaches the atmospheric value, and by the discharge pressure when this is close to the maximum allowable value. The low-value selector passes the lower of its inputs, in order to prevent the noncritical controlled value from increasing the speed of the pump.

where the force is concentrated on pillars. { ¦ō·vər¦strest 'er·ē·ə }

overstressing [ENG] Cyclically stressing a material at a level higher than that used at the end of a fatigue test. { ¦ō·vər¦stres·iŋ }

over-the-horizon propagation See scatter propagation. { 'ō·vər t͟hə hə'rīz·ən ˌpräp·ə'gā·shən }

over-the-horizon radar [ELECTROMAG] Long-range radar in which the transmitted and reflected beams are bounced off the ionosphere layers to achieve ranges far beyond line of sight. { 'ō·vər t͟hə hə'rīz·ən 'rā,där }

overthrow distortion [COMMUN] Distortion caused when the maximum amplitude of the signal wavefront exceeds the steady state of amplitude of the signal wave. { 'ō·vər,thrō di,stór·shən }

overthrust [GEOL] **1.** A thrust fault that has a low dip or a net slip that is large. Also known as low-angle thrust; overthrust fault. **2.** A thrust fault with the active element being the hanging wall. { 'ō·vər,thrəst }

overthrust black See overthrust nappe. { 'ō·vər,thrəst ,blak }

overthrust fault See overthrust. { 'ō·vər,thrəst ,fólt }

overthrust nappe [GEOL] The body of rock making up the hanging wall of a large-scale overthrust. Also known as overthrust block; overthrust sheet; overthrust slice. { 'ō·vər,thrəst ,nap }

overthrust sheet See overthrust nappe. { 'ō·vər,thrəst ,shēt }

overthrust slice See overthrust nappe. { 'ō·vər,thrəst ,slīs }

overtide [OCEANOGR] A harmonic tidal component which has a speed that is an exact multiple of the speed of one development of the tide-producing force. { 'ō·vər,tīd }

overt infection [MED] A host-parasite interaction that results in some injury to the tissues of the host. { ,ō·vərt in'fek·shən }

overtone [ACOUS] **1.** A component of a complex sound whose frequency is an integral multiple, greater than 1, of the fundamental frequency. **2.** A component of a complex tone having a pitch higher than that of the fundamental pitch. [MECH] One of the normal modes of vibration of a vibrating system whose frequency is greater than that of the fundamental mode. [PHYS] A harmonic other than the fundamental component. { 'ō·vər,tōn }

overtone band [SPECT] The spectral band associated with transitions of a molecule in which the vibrational quantum number changes by 2 or more. { 'ō·vər,tōn ,band }

overtone crystal [ELECTR] Quartz crystal cut in such a manner that it will operate at a higher order than its fundamental frequency, or operate at two frequencies simultaneously as in a synthesizer. { 'ō·vər,tōn ,krist·əl }

overtopping [CIV ENG] The flow of water over a dam or embankment. { 'ō·vər,täp·iŋ }

overtravel [ORD] In machine guns, the distance the firing notch overrides the sear notch in cocking, to ensure positive engagement of the two notches. { ¦ō·vər¦trav·əl }

overtub system [MIN ENG] An endless-rope system in which the rope runs over the tubs or cars in the center of the rails. { 'ō·vər,təb ,sis·təm }

overturn [HYD] Renewal of bottom water in lakes and ponds in regions where winter temperatures are cold; in the fall, cooled surface waters become denser and sink, until the whole body of water is at 4°C; in the spring, the surface is warmed back to 4°C, and the lake is homothermous. Also known as convective overturn. { 'ō·vər,tərn }

overturned [GEOL] Of a fold or the side of a fold, tilted beyond the perpendicular. Also known as inverted; reversed. { 'ō·vər,tərnd }

overturning [CIV ENG] Failure of a retaining wall caused by the soil pressure overcoming the stability of the structure. { ¦ō·vər¦tərn·iŋ }

overvoltage [ELEC] A voltage greater than that at which a device or circuit is designed to operate. Also known as overpotential. [ELECTR] The amount by which the applied voltage exceeds the Geiger threshold in a radiation counter tube. [PHYS CHEM] The difference between electrode potential under electrolysis conditions and the thermodynamic value of the electrode potential in the absence of electrolysis for the same experimental conditions. Also known as overpotential. { ¦ō·vər¦vōl·tij }

overvoltage crowbar [ELEC] A circuit that monitors the output of a power supply and prevents the output voltage from exceeding a preset voltage, under any failure condition, by having a low resistance (crowbar) placed across the output terminals when an overvoltage occurs. { ¦ō·vər¦vōl·tij 'krō,bär }

overwash [GEOL] **1.** A mass of water representing the part of the wave advancing up a beach that runs over the highest part of the berm (or other structure) and that does not flow directly back to the sea or lake. **2.** See outwash. { 'ō·vər,wäsh }

overwash mark [GEOL] A narrow, tonguelike ridge of sand formed by overwash on the landward side of a berm. { 'ō·vər,wäsh ,märk }

overwash plain See outwash plain. { 'ō·vər,wäsh ,plān }

overwash pool [OCEANOGR] A tidal pool between a berm and a beach scarp which water enters only at high tide. { 'ō·vər,wäsh pül }

overwind [ENG] To wind a spring, rope, or cable too tightly or too far. { ¦ō·vər¦wīnd }

overwinding [MOL BIO] Supercoiling of a deoxyribonucleic acid molecule in the same direction as that of the winding of the double helix, resulting in increased tension in the two strands of the molecule. { ¦ō·vər¦wīn·diŋ }

overwrite [COMPUT SCI] To enter information into a storage location and destroy the information previously held there. { ¦ō·vər¦rīt }

ovex [ORG CHEM] ClC₆H₄OSO₂C₆H₄Cl A white, crystalline solid with a melting point of 86.5°C; soluble in acetone and aromatic solvents; used as an insecticide and acaricide. { 'ō,veks }

ovicell [INV ZOO] A broad chamber in certain bryozoans. { 'ō·və,sel }

ovicyst [INV ZOO] The pouch of a tunicate in which the eggs develop. { 'ō·və,sist }

oviduct [ANAT] A tube that serves to conduct ova from the ovary to the exterior or to an intermediate organ such as the uterus. Also known in mammals as Fallopian tube; uterine tube. { 'ō·və,dəkt }

oviger [INV ZOO] A modified leg used for carrying eggs in some pycnogonids. { 'ō·və·jər }

ovine encephalomyelitis See louping ill. { 'ō,vīn in¦sef·ə·lō,mī·ə'līd·əs }

oviparous [VERT ZOO] Producing eggs that develop and hatch externally. { ō'vip·ə·rəs }

oviposit [ZOO] To lay or deposit eggs, especially by means of a specialized organ, as found in certain insects and fishes. { 'ō·və,päz·ət }

ovipositor [INV ZOO] A specialized structure in many insects for depositing eggs. [VERT ZOO] A tubular extension of the genital orifice in most fishes. { 'ō·və,päz·əd·ər }

ovography [GRAPHICS] A printing method based upon the use of a drum coated with ovonic memory material, in which a permanent but alterable image can be "written" onto the drum by a computer-controlled energy beam, such as a laser, and printout can be achieved by electrostatic printing techniques. { ō'väg·rə·fē }

ovonic device See glass switch. { ō'vän·ik di,vīs }

ovonic memory switch [ELECTR] A glass switch which, after being brought from the highly resistive state to the conducting state, remains in the conducting state until a current pulse returns it to its highly resistive state. Abbreviated OMS. Also known as memory switch. { ō'vän·ik 'mem·rē ,swich }

ovonic threshold switch [ELECTR] A glass switch which, after being brought from the highly resistive state to the conducting state, returns to the highly resistive state when the current falls below a holding current value. Abbreviated OTS. { ō'vän·ik 'thresh,hōld ,swich }

ovotesticular hermaphroditism [MED] A rare form of hermaphroditism in which an ovotestis is present on one or both sides. { ,ō·vō,tes'tik·yə·lər hər'maf·rə·də,diz·əm }

ovoviviparous [VERT ZOO] Producing eggs that develop internally and hatch before or soon after extrusion. { ¦ō·vō,vī'vip·ə·rəs }

Ovshinsky effect [ELECTR] The characteristic of a special thin-film solid-state switch that responds identically to both positive and negative polarities so that current can be made to flow in both directions equally. { ōv'shin·skē i,fekt }

ovulation [PHYSIO] Discharge of an ovum or ovule from the ovary. { ,äv·yə'lā·shən }

ovule [BOT] A structure in the ovary of a seed plant that develops into a seed following fertilization. { 'äv‚yül }

ovum [CELL MOL] A female gamete. Also known as egg. { 'ō·vəm }

O-wave component *See* ordinary-wave component. { 'ō ‚wāv kəm‚pō·nənt }

Owen bridge [ELECTR] A four-arm alternating-current bridge used to measure self-inductance in terms of capacitance and resistance; bridge balance is independent of frequency. { 'ō·wən ‚brij }

Oweniidae [INV ZOO] A family of limivorous polychaete annelids of the Sedentaria. { ‚ō·wə'nī·ə‚dē }

owl [VERT ZOO] Any of a number of diurnal and nocturnal birds of prey composing the order Strigiformes; characterized by a large head, more or less forward-directed large eyes, a short hooked bill, and strong talons. { aul }

Owl Nebula [ASTRON] A large planetary nebula in Ursa Major which has two large, circular darker areas in an otherwise opaque spherical shell. { aul 'neb·yə·lə }

own coding [COMPUT SCI] A series of instructions added to a standard software routine to change or extend the routine so that it can carry out special tasks. { 'ōn 'kōd·iŋ }

owned program *See* proprietary program. { 'ōnd 'prō‚gram }

OW unit *See* sabin. { 'ō'dəb·əl‚yü ‚yü·nət }

oxadiazon [ORG CHEM] $C_{13}H_{18}Cl_2N_2O_3$ A white solid with a melting point of 88–90°C; slight solubility in water; used as a pre- and postemergence herbicide to control weeds in rice, turf, soybeans, peanuts, and orchards. { ‚äk·sə'dī·ə‚zän }

oxalate [ORG CHEM] Salt of oxalic acid; contains the $(COO)_2$ radical; examples are sodium oxalate, $Na_2C_2O_4$, ammonium oxalate, $(NH_4)_2C_2O_4·H_2O$, and ethyl oxalate, $C_2H_5(C_2O_4)·C_2H_5$. { 'äk·sə‚lāt }

oxalic acid [ORG CHEM] $HOOCCOOH·2H_2O$ Poisonous, transparent, colorless crystals melting at 187°C; soluble in water, alcohol, and ether; used as a chemical intermediate and a bleach, and in polishes and rust removers. { äk'sal·ik 'as·əd }

Oxalidaceae [BOT] A family of dicotyledonous plants in the order Geraniales, generally characterized by regular flowers, two or three times as many stamens as sepals or petals, a style which is not gynobasic, and the fruit which is a beakless, loculicidal capsule. { ‚äk‚sal·ə'dās·ē‚ē }

oxalite *See* humboldtine. { 'äk·sə‚līt }

oxalosis [MED] A rare hereditary metabolic disorder, inherited as an autosomal recessive, in which glyoxylic acid metabolism is impaired, resulting in overproduction of oxalic acid and deposition of calcium oxalate in body tissues. { ‚äk·sə'lō·səs }

oxaluria [MED] The presence of oxalic acid or oxalates in the urine. { ‚äk·səl'yùr·ē·ə }

oxalyl chloride [INORG CHEM] $(COCl)_2$ Toxic, colorless liquid boiling at 64°C; soluble in ether, benzene, and chloroform; used as a chlorinating agent and for military poison gas. { 'äk·sə‚lil 'klór‚īd }

oxamide [ORG CHEM] $NH_2COCONH_2$ Water-insoluble white powder, melting at 419°C; used as a stabilizer for nitrocellulose products. { 'äk·sə‚məd }

oxammite [MINERAL] $(NH_4)_2C_2O_4·H_2O$ A yellowish-white, orthorhombic mineral consisting of ammonium oxalate monohydrate; occurs as lamellar masses. { 'äk·sə‚mīt }

oxamyl [ORG CHEM] $C_7H_{13}N_3O_3S$ A white, crystalline compound with a melting point of 100–102°C; used to control pests of tobacco, ornamentals, fruits, and crops. { 'äk·sə‚mil }

oxazole [ORG CHEM] C_3H_3ON A structure that consists of a five-membered ring containing oxygen and nitrogen in the 1 and 3 position; a colorless liquid (boiling point 69–70°C) that is miscible with organic solvents and water; used to prepare other organic compounds. { 'äk·sə‚zōl }

oxbow [GEOL] The abandoned, horseshoe-shaped channel of a former stream meander after the stream formed a neck cutoff. Also known as abandoned channel. [HYD] **1.** A closely looping, U-shaped stream meander whose curvature is so extreme that only a neck of land remains between the two parts of the stream. Also known as horseshoe bend. **2.** *See* oxbow lake. { 'äks‚bō }

oxbow lake [HYD] The crescent-shaped body of water located alongside a stream in an abandoned oxbow after a neck cutoff is formed and the ends of the original bends are silted up. Also known as crescentic lake; cutoff lake; horseshoe lake; loop lake; moat; mortlake; oxbow. { 'äks‚bō ‚lāk }

oxford cloth [TEXT] A strong fabric, usually cotton or cotton blended with a synthetic fiber, finished with a silk luster. { 'äks·fərd ‚klöth }

Oxfordian [GEOL] A European stage of geologic time, in the Upper Jurassic (above Callovian, below Kimmeridgean). Also known as Divesian. { äks'fór·dē·ən }

oxidant *See* oxidizing agent. { 'äk·səd·ənt }

oxidase [BIOCHEM] An enzyme that catalyzes oxidation reactions by the utilization of molecular oxygen as an electron acceptor. { 'äk·sə‚dās }

oxidate [GEOL] A sediment made up of iron and manganese oxides and hydroxides crystallized from aqueous solution. { 'äk·sə‚dāt }

oxidation [CHEM] **1.** A chemical reaction that increases the oxygen content of a compound. **2.** A chemical reaction in which a compound or radical loses electrons, that is in which the positive valence is increased. { ‚äk·sə'dā·shən }

oxidation number [CHEM] **1.** Numerical charge on the ions of an element. **2.** *See* oxidation state. { ‚äk·sə'dā·shən ‚nəm·bər }

oxidation pond [CIV ENG] A shallow lagoon or basin in which wastewater is purified by sedimentation and aerobic and anaerobic treatment. { ‚äk·sə'dā·shən ‚pänd }

oxidation potential [PHYS CHEM] The difference in potential between an atom or ion and the state in which an electron has been removed to an infinite distance from this atom or ion. { ‚äk·sə'dā·shən pə‚ten·chəl }

oxidation-reduction indicator [ANALY CHEM] A compound whose color in the oxidized state differs from that in the reduced state. { ‚äk·sə'dā·shən ri'dək·shən ‚in·də‚kād·ər }

oxidation-reduction potential *See* redox potential. { ‚äk·sə'dā·shən ri'dək·shən pə‚ten·chəl }

oxidation-reduction reaction [CHEM] An oxidizing chemical change, where an element's positive valence is increased (electron loss), accompanied by a simultaneous reduction of an associated element (electron gain). { ‚äk·sə'dā·shən ri'dək·shən rē‚ak·shən }

oxidation state [CHEM] The number of electrons to be added (or subtracted) from an atom in a combined state to convert it to elemental form. Also known as oxidation number. { ‚äk·sə'dā·shən ‚stāt }

oxidative phosphorylation [BIOCHEM] Conversion of inorganic phosphate to the energy-rich phosphate of adenosinetriphosphatase by reactions associated with the electron transfer system. { ‚äk·sə‚dād·iv ‚fäs·fə·rə'lā·shən }

oxide [CHEM] Binary chemical compound in which oxygen is combined with a metal (such as Na_2O; basic) or nonmetal (such as NO_2; acidic). { 'äk‚sīd }

oxide-coated cathode [ELECTR] A cathode that has been coated with oxides of alkaline-earth metals to improve electron emission at moderate temperatures. Also known as Wehnelt cathode. { 'äk‚sīd ‚kōd·əd 'kath‚ōd }

oxide fuel reactor [NUCLEO] A nuclear fission reactor with fuel in the form UO_2 or PuO_2. { 'äk‚sīd ‚fyül rē'ak·tər }

oxide isolation [ELECTR] Isolation of the elements of an integrated circuit by forming a layer of silicon oxide around each element. { 'äk‚sīd ‚ī·sə'lā·shən }

oxide mineral [MINERAL] A naturally occurring material in oxide form such as silicon dioxide, SiO_2, magnetite, Fe_3O_4, or lime, CaO. { 'äk‚sīd 'min·rəl }

oxide nuclear fuel [NUCLEO] The fissionable nuclear fuel UO_2 or PuO_2. { 'äk‚sīd 'nü·klē·ər ‚fyül }

oxide passivation [ELECTR] Passivation of a semiconductor surface by producing a layer of an insulating oxide on the surface. { 'äk‚sīd ‚pas·ə·vā·shən }

oxidite *See* shale ball. { 'äk·sə‚dīt }

oxidized cellulose *See* oxycellulose. { 'äk·sə‚dīzd 'sel·yə‚lōs }

oxidized microcrystalline wax [MATER] Refined, oxidized wax from bottoms of storage tanks for solvent-extracted petroleum; used in floor polishes. { 'äk·sə‚dīzd ‚mī·krō‚krist·əl·ən 'waks }

oxidized shale *See* burnt shale. { 'äk·sə‚dīzd 'shāl }

oxidized zone [GEOL] A region of mineral deposits which has been altered by oxidizing surface waters. { 'äk·sə‚dīzd ‚zōn }

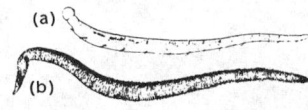

OWENIIDAE

Myriochele. (a) Entire ovigerous individual. (b) The tapering tube in which the mud-swallowing worm is securely contained.

OWL

Great horned owl (*Bubo virginianus*), a nocturnal bird of prey.

oxidizer [AERO ENG] A substance, not necessarily containing oxygen, that supports the combustion of a fuel or propellant. { 'äk·sə‚dīz·ər }

oxidizing agent [CHEM] Compound that gives up oxygen easily, removes hydrogen from another compound, or attracts negative electrons. Also known as oxidant. { 'äk·sə‚dīz·iŋ ‚ā·jənt }

oxidizing atmosphere [CHEM] Gaseous atmosphere in which an oxidation reaction occurs; usually refers to the oxidation of solids. { 'äk·sə‚dīz·iŋ 'at·mə‚sfir }

oxidizing flame [CHEM] A flame, or the portion of it, that contains an excess of oxygen. { 'äk·sə‚dīz·iŋ ‚flām }

oxidoreductase [BIOCHEM] An enzyme catalyzing a reaction in which two molecules of a compound interact so that one molecule is oxidized and the other reduced, with a molecule of water entering the reaction. { ‚äk·sə·dō·ri'dək‚tās }

oxime [ORG CHEM] Compound containing the CH(:NOH) radical; condensation product of hydroxylamine with aldehydes or ketones. { 'äk‚sēm }

oximeter [MED] A photoelectric photometer used to measure the oxygenated fraction of the hemoglobin in blood which is either circulating in a particular tissue of an intact animal or human being, or during, or shortly after, its withdrawal from the vascular system, by observation of the absorption of light transmitted through or reflected from the blood. { äk'sim·əd·ər }

oximetry [PHYSIO] Optical measurement of the degree of oxygen saturation of the blood hemoglobin by determining the variation in the color of the blood. { äk'sim·ə·trē }

oximido See nitroso. { äk'sim·ə·dō }

oxine [ORG CHEM] C_9H_6NOH White powder that darkens when exposed to light; slightly soluble in water, dissolves in ethanol, acetone, and benzene; used to prepare fungicides and to separate metals by precipitation. Also known as 8-hydroxyquinoline; oxyquinoline; 8-quinolinol. { 'äk‚sēn }

oxirane See epoxide; ethylene oxide. { 'äk·sə‚rān }

Oxisol [GEOL] A soil order characterized by residual accumulations of inactive clays, free oxides, kaolin, and quartz; mostly tropical. { 'äk·sə‚sòl }

oxo- [ORG CHEM] Chemical prefix designating the keto group, C:O. { 'äk·sō }

oxoferrite [GEOL] A variety of naturally occurring iron with some ferrous oxide in solid solution. { ‚äk·sō'fe‚rīt }

***para*-oxon** [ORG CHEM] $(C_2H_5O)_2P(O)C_6H_4NO_2$ A reddish-yellow oil with a boiling point of 148–151°C; soluble in most organic solvents; used as an insecticide. Also known as diethyl *para*-nitrophenyl phosphate. { ‚par·ə 'äk‚sän }

oxonium ion [CHEM] R_3O^+ A cation in which an oxygen atom is covalently bound to three atoms or groups of atoms. { äk'sō·nē·əm 'ī‚än }

oxo process [CHEM ENG] Catalytic process for production of alcohols, aldehydes, and other oxygenated organic compounds by reaction of olefin vapors with carbon monoxide and hydrogen. { 'äk·sō ‚prä·səs }

oxosilane See siloxane. { ‚äk·sō'si‚lān }

oxoxanthone See genicide. { ‚äk·sō'zan‚thōn }

oxyacanthine [ORG CHEM] $C_{37}H_{40}N_2O_6$ An alkaloid obtained from the root of *Berberis vulgaris*; a white, crystalline powder with a melting point of 202–214°C; soluble in water, chloroform, benzene, alcohol, and ether; used in medicine. Also known as vinetine. { ‚äk·sē·ə'kan‚thēn }

oxyacetylene cutting [ENG] The flame cutting of ferrous metals in which the preheating of the metal is accomplished with a flame produced by an oxyacetylene torch. Also known as acetylene cutting. { ‚äk·sē·ə'sed·əl‚ēn ‚kəd·iŋ }

oxyacetylene torch [ENG] A torch that mixes acetylene and oxygen to produce a hot flame for the welding or cutting of metal. Also known as acetylene torch. { ‚äk·sē·ə'sed·əl‚ēn ‚tòrch }

oxyacetylene welding [MET] A welding process in which the heat is supplied by an oxyacetylene torch. Also known as acetylene welding. { ‚äk·sē·ə'sed·əl‚ēn ‚weld·iŋ }

Oxyaenidae [PALEON] An extinct family of mammals in the order Deltatheridea; members were short-faced carnivores with powerful jaws. { ‚äk·sē'en·ə‚dē }

oxyamination See ammoxidation. { ‚äk·sē‚am·ə'nā·shən }

oxybenzone [ORG CHEM] $C_{14}H_{12}O_3$ A crystalline substance with a melting point of 66°C; used as a sunscreen agent.

Also known as 4-methoxy-2-hydroxybenzophenone. { ‚äk·sē'ben‚zōn }

oxybiotite [MINERAL] Phenocrystic biotite with increased amounts of Fe(III). { ‚äk·sē'bī·ə‚tīt }

oxycarboxin [ORG CHEM] $C_{12}H_{13}NO_4S$ An off-white, crystalline compound with a melting point of 127.5–130°C; used to control rust disease in greenhouse carnations. Also known as 5,6-dihydro-2-methyl-1,4-oxathiin-3-carboxanilide-4,4-dioxide. { ‚äk·sē·kär'bäk·sən }

oxycellulose [MATER] Cellulose mixed with reaction products from oxidation of cellulose in the presence of steam or alkalies or by strong sunlight. Also known as oxidized cellulose. { ‚äk·sē'sel·yə‚lōs }

oxycephaly [MED] A condition in which the head assumes a roughly conical shape due to premature closure of the coronal or lambdoid sutures, or to artificial pressure on the frontal and occipital regions of the infant's head. Also known as acrocephaly. { ‚äk·sē'sef·ə·lē }

oxychloride cement [MATER] A strong, hard cement composed of magnesium chloride and calcined magnesia; used for floors and stucco. Also known as Sorel cement. { ‚äk·sē'klòr‚īd si'ment }

oxy compound [CHEM] A compound containing two or more oxygen atoms that are not joined to each other but are covalently bound to other atoms in the structure. { 'äk·sē ‚käm‚paùnd }

oxygen [CHEM] A gaseous chemical element, symbol O, atomic number 8, and atomic weight 15.9994; an essential element in cellular respiration and in combustion processes; the most abundant element in the earth's crust, and about 20% of the air by volume. { 'äk·sə·jən }

oxygen absorbent [CHEM] Any material that will absorb (dissolve) oxygen into its body without reacting with it. { 'äk·sə·jən əb'sòr·bənt }

oxygenase [BIOCHEM] An oxidoreductase that catalyzes the direct incorporation of oxygen into its substrate. { 'äk·sə·jə‚nās }

oxygenate [CHEM] To treat, infuse, or combine with oxygen. [MATER] An oxygen-containing compound, such as an alcohol or an ether, used as an additive to gasoline to improve octane rating or antiknock characteristics. { 'äk·sə·jə‚nāt }

oxygenated oil [MATER] A class of essential oils containing carbon, hydrogen, and oxygen; an example is oil of cassia. { 'äk·sə·jə‚nād·əd 'òil }

oxygenator [MED] An apparatus for introducing oxygen into the blood during extracorporal circulation. { 'äk·sə·jə‚nād·ər }

oxygen bomb calorimeter [ENG] Device to measure heat of combustion; the sample is burned with oxygen in a closed vessel, and the temperature rise is noted. { 'äk·sə·jən 'bäm ‚kal·ə'rim·əd·ər }

oxygen burning [NUC PHYS] The synthesis of nuclei in stars through reactions involving the fusion of two oxygen-16 nuclei at temperatures of about 10^9 K. { 'äk·sə·jən ‚bərn·iŋ }

oxygen cell See aeration cell. { 'äk·sə·jən ‚sel }

oxygen concentration cell See differential aeration cell. { 'äk·sə·jən ‚kän·sən'trā·shən ‚sel }

oxygen corrosion [MET] The reaction of oxygen with metallic surfaces to form an oxide of the metal or alloy. { 'äk·sə·jən kə‚rōzh·ən }

oxygen cutting [ENG] Any of several types of cutting processes in which metal is removed with or without a flux by a chemical reaction of the base metal with oxygen at high temperatures. { 'äk·sə·jən ‚kəd·iŋ }

oxygen debt [PHYSIO] **1.** A bodily condition in which oxygen demand is greater than oxygen supply. **2.** The amount of oxygen needed to restore the body to a steady state after a muscular exertion. { 'äk·sə·jən ‚det }

oxygen deficit [GEOCHEM] The difference between the actual amount of dissolved oxygen in lake or sea water and the saturation concentration at the temperature of the water mass sampled. { 'äk·sə·jən ‚def·ə·sət }

oxygen distribution [OCEANOGR] The concentration of dissolved oxygen in ocean water as a function of depth, ranging from as much as 5 milliliters of oxygen per liter at the surface to a fraction of that value at great depths. { 'äk·sə·jən ‚dis·trə'byü·shən }

oxygen-18 [NUC PHYS] Oxygen isotope with atomic weight 18; found 8 parts to 10,000 of oxygen-16 in water, air, and rocks;

used in tracer experiments. Also known as heavy oxygen. { 'äk·sə·jən ā'tēn }

oxygen-flask method [ANALY CHEM] Technique to determine the presence of combustible elements; the sample is burned with oxygen in a closed flask, and combustion products are absorbed in water of dilute alkali with subsequent analysis of the solution. { 'äk·sə·jən ¦flask ¦meth·əd }

oxygen-free copper [MET] Pure copper having a conductivity greater than that of copper containing impurities such as cuprous oxide; used for the construction of high-power electron tubes because it does not release appreciable gas when hot. { 'äk·sə·jən ¦frē 'käp·ər }

oxygen furnace steel [MET] Steel made by a process in which oxygen under pressure is directed onto or into the molten metal. { 'äk·sə·jən 'fər·nəs ¦stēl }

oxygen gouging See flame gouging. { 'äk·sə·jən ¦gaúj·iŋ }

oxygen-hydrocarbon engine [AERO ENG] A rocket engine that uses a propellant consisting of liquid oxygen as the oxidizer and a hydrocarbon, such as a petroleum derivative, as the fuel. { 'äk·sə·jən ¦hī·drə'kär·bən ¦en·jən }

oxygen isotope fractionation [GEOCHEM] The use of temperature-dependent variations of the oxygen-18/oxygen-16 ratio in the carbonate shells of marine organisms, to measure water temperature at the time of deposition. { 'äk·sə·jən ¦īs·ə¦tōp ¦frak·shə'nā·shən }

oxygen-kerosine burner [ENG] Liquid-fuel device using a mixture of oxygen and vaporized or atomized kerosine for combustion. { 'äk·sə·jən 'ker·ə¦sēn ¦bərn·ər }

oxygen lance [MET] A pipe used to direct oxygen under pressure into a bath of molten steel. { 'äk·sə·jən ¦lans }

oxygen mask [ENG] A mask that covers the nose and mouth and is used to administer oxygen. { 'äk·sə·jən ¦mask }

oxygen minimum layer [HYD] A subsurface layer in which the content of dissolved oxygen is very low (or absent), lower than in the layers above and below. { 'äk·sə·jən 'min·ə·məm ¦lā·ər }

oxygen point [THERMO] The temperature at which liquid oxygen and its vapor are in equilibrium, that is, the boiling point of oxygen, at standard atmospheric pressure; it is taken as a fixed point on the International Practical Temperature Scale of 1968, at $-182.962°C$. { 'äk·sə·jən ¦point }

oxygen propellant [MATER] A propellant having a minimum assay by volume of 99.9% oxygen when gasified. { 'äk·sə·jən prə'pel·ənt }

oxygen ratio See acidity coefficient. { 'äk·sə·jən ¦rā·shō }

oxygen steelmaking [MET] The manufacture of steel from molten pig iron and steel scrap by methods which employ pure oxygen gas (99+%) and suitable fluxes to remove carbon and phosphorus (and in part, sulfur) without introducing nitrogen or hydrogen. { 'äk·sə·jən 'stēl¦māk·iŋ }

oxygen toxicity [PHYSIO] **1.** Harmful effects of breathing oxygen at pressures greater than atmospheric. **2.** A toxic effect in a living organism caused by a species of oxygen-containing reactive intermediate produced during the reduction of dioxygen. { 'äk·sə·jən täk'sis·əd·ē }

oxyheeite [MINERAL] $Pb_5Ag_2Sb_6S_{15}$ A light steel gray to silver white mineral consisting of lead and silver antimony sulfide; occurs as acicular needles or in massive form. { ¦äk·sē'hē¦īt }

oxyhemoglobin [BIOCHEM] The red crystalline pigment formed in blood by the combination of oxygen and hemoglobin, without the oxidation of iron. { ¦äk·sē'hē·mə¦glō·bən }

oxyhornblende See basaltic hornblende. { ¦äk·sē'hórn¦blend }

oxyhydrogen flame [CHEM] A flame obtained from the combustion of a mixture of oxygen and hydrogen. { ¦äk·sē'hī·drə·jən ¦flām }

oxyhydrogen welding [MET] Welding with an oxyhydrogen flame. { ¦äk·sē'hī·drə·jən ¦weld·iŋ }

oxylophyte [ECOL] A plant that thrives in or is restricted to acid soil. { 'äk'sil·ə¦fīt }

oxyl process [CHEM ENG] Modified Fischer-Tropsch process used to make alcohols, other oxygenated compounds, paraffins, and olefin hydrocarbons from carbon monoxide and hydrogen. { 'äk·səl ¦prä·səs }

Oxymonadida [INV ZOO] An order of xylophagous protozoans in the class Zoomastigophorea; colorless flagellate symbionts in the digestive tract of the roach *Cryptocercus* and of certain termites. { ¦äk·sē·mə'näd·ə·də }

oxyneurine See betaine. { ¦äk·sē'nú¦rēn }

oxyntic cell See parietal cell. { äk'sin·tik 'sel }

oxypetalous [BOT] Having sharp-pointed petals. { ¦äk·sē¦ped·əl·əs }

oxyphile See lithophile. { 'äk·sə¦fīl }

oxyphilia See eosinophilia. { ¦äk·sə'fil·ē·ə }

oxyphytia [ECOL] Discordant habitat control due to an excessively acidic substratum. { ¦äk·sə'fīd·ē·ə }

oxyquinoline See oxine. { ¦äk·sə'kwin·ə·lən }

oxyreductase [BIOCHEM] Any of a class of enzymes that catalyze electron-transfer reactions. { ¦äk·sē·ri'dək¦tās }

oxysphere See lithosphere. { 'äk·sə¦sfir }

oxysterol [BIOCHEM] Oxidized derivative of cholesterol. { ¦äk·sē'stir¦ól }

Oxystomata [INV ZOO] A subsection of the Brachyura, including those true crabs in which the first pair of pereiopods is chelate, and the mouth frame is triangular and forward. { ¦äk·sē'stō·mə·də }

Oxystomatidae [INV ZOO] A family of free-living marine nematodes in the superfamily Enoploidea, distinguished by amphids that are elongated longitudinally. { ¦äk·sē·stō'mad·ə¦dē }

oxytetracycline [MICROBIO] $C_{22}H_{24}O_9N_2$ A crystalline, amphoteric, broad-spectrum antibiotic produced by *Streptomyces rimosus*; produced commercially by fermentation. { ¦äk·sē¦te·trə'sī¦klēn }

oxytocic [MED] Hastening parturition. [PHARM] A drug that hastens parturition. { ¦äk·sē¦tō·sik }

oxytocin [BIOCHEM] $C_{43}H_{66}O_{12}N_{12}S_2$ A polypeptide hormone secreted by the neurohypophysis that stimulates contraction of the uterine muscles. { ¦äk·sē¦tō·sən }

Oxyurata [INV ZOO] The equivalent name for Oxyurina. { ¦äk·sē·yú'räd·ə }

Oxyuridae [INV ZOO] A family of the nematode superfamily Oxyuroidea. { ¦äk·sē'yúr·ə¦dē }

Oxyurina [INV ZOO] A suborder of nematodes in the order Ascaridida. { ¦äk·sē·yú'rī·nə }

Oxyuroidea [INV ZOO] A superfamily of marine nematodes in the order Enoplida; contains species that maintain the most ancestral characters known in the phylum, such as a stoma composed entirely of esophastome, which is the ancestral primary blastocoel invagination. { ¦äk·sē·yú'róid·ē·ə }

Oyashio [OCEANOGR] A cold current flowing from the Bering Sea southwest along the coast of Kamchatka, past the Kuril Islands, continuing close to the northeast coast of Japan, and reaching nearly 35°N. { ō'yä·shē·ō }

oyster [INV ZOO] Any of various bivalve mollusks of the family Ostreidae; the irregular shell is closed by a single adductor muscle, the foot is small or absent, and there is no siphon. { 'ói·stər }

oz See ounce.

ozalid [GRAPHICS] A print on light-sensitized material produced directly from a positive transparency and developed via a dry process with ammonia vapor. { 'äz·ə·ləd }

oz ap See ounce.

oz apoth See ounce.

Ozawainellidae [PALEON] A family of extinct protozoans in the superfamily Fusulinacea. { ō¦zä·wə·i'nel·ə¦dē }

ozocerite [GEOL] A natural, brown to jet black paraffin wax occurring in irregular veins; consists principally of hydrocarbons, is soluble in water, and has a variable melting point. Also known as ader wax; earth wax; fossil wax; mineral wax; native paraffin; ozokerite. [MATER] See ceresin. { ō'zäs·ə¦rīt }

ozokerite See ozocerite. { ō'zäk·ə¦rīt }

ozone [CHEM] O_3 Unstable blue gas with pungent odor; an allotropic form of oxygen; a powerful oxidant boiling at $-112°C$; used as an oxidant, bleach, and water purifier, and to treat industrial wastes. { 'ō¦zōn }

ozone generator [ENG] Apparatus that converts oxygen, O_2, into ozone, O_3, by subjecting the oxygen to an electric-brush discharge. Also known as ozonizer. { 'ō¦zōn ¦jen·ə¦rād·ər }

ozone hole See Antarctic ozone hole. { 'ō¦zōn ¦hōl }

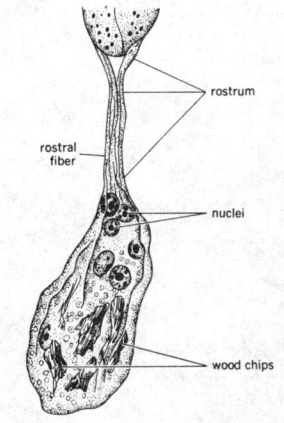

An oxymonad, *Microrhopalodina inflata*.

ozone layer *See* stratospheric ozone. { 'ō,zōn ,lā·ər }

ozonesonde [METEOROL] A balloon-borne instrument for measuring the ozone concentration at various altitudes and transmitting the data by radio. { 'ō,zōn,sänd }

ozonide [ORG CHEM] Any of the oily, thick, unstable compounds formed by reaction of ozone with unsaturated compounds; an example is oleic ozonide from the reaction of oleic acid and ozone. { 'äz·ə,nīd }

ozonization [CHEM] The process of treating, impregnating, or combining with ozone. { ,ō,zō·nə'zā·shən }

ozonizer *See* ozone generator. { 'ō,zō,nīz·ər }

ozonolysis [ANALY CHEM] The use of ozone to locate double bonds. [ORG CHEM] Oxidation of an organic substance by means of ozone. { ,ō·zə'näl·ə·səs }

ozonosphere [METEOROL] The general stratum of the upper atmosphere in which there is an appreciable ozone concentration and in which ozone plays an important part in the radiative balance of the atmosphere; lies roughly between 6 and 30 miles (10 and 50 kilometers), with maximum ozone concentration at about 12 to 15 miles (20 to 25 kilometers). Also known as ozone layer. { ō'zō·nə,sfir }

oz t *See* ounce.

oz tr *See* ounce.

P

p- *See* para-; pico-.

P *See* peta-; phosphorus; poise.

P₁ [GEN] The parental generation; parents of the F_1 generation.

pA *See* picoampere.

Pa *See* pascal; protactinium.

Paal-Knorr synthesis [ORG CHEM] A method of converting a 1,4-dicarbonyl compound by cyclization with ammonia or a primary amine to a pyrrole. { 'pȯl kə'nȯr 'sin·thə·səs }

paar [GEOL] A depression produced by the moving apart of crustal blocks rather than by subsidence within a crustal block. { pär }

Paar turbidimeter [ANALY CHEM] A visual-extinction device for measurement of solution turbidity; the length of the column of liquid suspension is adjusted until the light filament can no longer be seen. { pär ‚tər·bə'dim·əd·ər }

Paasche's index [STAT] A weighted aggregate price index with given-year quantity weights. Also known as given-year method. { 'päs·kəz 'in‚deks }

PABA *See* para-aminobenzoic acid.

PABA sodium *See* sodium para-aminobenzoate. { 'pēˌaˈbēˌa ‚sōd·ē·əm }

PABX *See* private automatic branch exchange.

paca [VERT ZOO] Any of several rodents of the genus *Cuniculus*, especially *C. paca*, with a white-spotted brown coat, found in South and Central America. { 'päk·ə }

Pacchionian bodies *See* arachnoidal granulations. { pak·ē'ō·nē·ən 'bäd·ēz }

pacemaker [MED] A pulsed battery-operated oscillator implanted in the body to deliver electric impulses to the muscles of the lower heart, either at a fixed rate or in response to a sensor that detects when the patient's pulse rate slows or ceases. Also known as cardiac pacemaker; heart pacer. { 'pās‚māk·ər }

pace rating *See* effort rating. { 'pās ‚rād·iŋ }

pachimeter [ENG] An instrument for measuring the limit beyond which shear of a solid ceases to be elastic. { pə'kim·əd·ər }

pachnolite [MINERAL] NaCaAlF₆·H₂O Colorless to white mineral composed of hydrous sodium calcium aluminum fluoride, occurring in monoclinic crystals. { 'pak·nə‚līt }

pachoidal structure *See* flaser structure. { pə'kȯid·əl ‚strək·chər }

Pachuca tank [CHEM ENG] Air-agitated, solid-liquid mixing vessel in which the air is injected into the bottom of a center draft tube; air and solids rise through the tube, with solids exiting the top of the tube and falling through the bulk of the liquid. { pə'chü·kə ‚taŋk }

pachycephalosaur [PALEON] A bone-headed dinosaur, composing the family Pachycephalosauridae. { ‚pak·ə'sef·ə·lə‚sȯr }

Pachycephalosauridae [PALEON] A family of ornithischian dinosaurs characterized by a skull with a solid rounded mass of bone 4 inches (10 centimeters) thick above the minute brain cavity. { ‚pak·ə‚sef·ə·lə'sȯr·ə‚dē }

pachyderm [VERT ZOO] Any of various nonruminant hooved mammals characterized by thick skin, including the elephants, hippopotamuses, rhinoceroses, and others. { 'pak·ə‚dərm }

pachydermatous [MED] Abnormally thick-skinned. { ‚pak·ə'dər·məd·əs }

pachydermia [MED] Abnormal thickening of the skin. { ‚pak·ə'dər·mē·ə }

pachyglossal [VERT ZOO] Of lizards, having a thick tongue. { ‚pak·ə'gläs·əl }

pachyglossia [MED] Abnormal thickness of the tongue. { 'pak·ə‚glä·sē·ə }

pachygyria [MED] A malformation of the brain characterized by its being too broad in form. { ‚pak·ə'jī·rē·ə }

pachymeningitis [MED] Inflammation of the dura mater. { ‚pak·ē‚men·ən'jīd·əs }

pachymeter [ENG] An instrument used to measure the thickness of a material, for example, a sheet of paper. { pə'kim·əd·ər }

pachynema *See* pachytene. { pə'kin·ə·mə }

pachyostosis [VERT ZOO] A bony thickening of vertebrae and ribs. { ‚pak·ē·ə'stō·səs }

Pachyrhachis problematicus [PALEON] A fossil snake showing features that are transitional between snakes and lizards. { ‚pak·ē‚rak·əs ‚präb·lə'mad·ə·kəs }

pachytene [CYTOL] The third stage of meiotic prophase during which paired chromosomes thicken, each chromosome splits into chromatids, and breakage and crossing over between nonsister chromatids occur. Also known as pachynema. { 'pak·ə‚tēn }

Pacific anticyclone *See* Pacific high. { pə'sif·ik ‚ant·i'sī‚klōn }

Pacific Equatorial Countercurrent [OCEANOGR] The Equatorial Countercurrent flowing east across the Pacific Ocean between 3° and 10°N. { pə'sif·ik ‚ek·wə'tȯr·ē·əl 'kaunt·ər‚kə·rənt }

Pacific faunal region [ECOL] A marine littoral faunal region including offshore waters west of Central America, running from the coast of South America at about 5° south latitude to the southern tip of California. { pə'sif·ik 'fȯn·əl ‚rē·jən }

Pacific high [METEORCL] The nearly permanent subtropical high of the North Pacific Ocean, centered, in the mean, at 30–40°N and 140–150°W. Also known as Pacific anticyclone. { pə'sif·ik 'hī }

Pacific North Equatorial Current [OCEANOGR] The North Equatorial Current which flows westward between 10° and 20°N in the Pacific Ocean. { pə'sif·ik 'nȯrth ‚ek·wə'tȯr·ē·əl 'kə·rənt }

Pacific Ocean [GEOGR] The largest division of the hydrosphere, having an area of 63,690 square miles (165,000,000 square kilometers) and covering 46% of the surface of the total extent of the oceans and seas; it is bounded by Asia and Australia on the west and North and South America on the east. { pə'sif·ik 'ō·shən }

Pacific South Equatorial Current [OCEANOGR] The South Equatorial Current flowing westward between 3°N and 10°S in the Pacific Ocean. { pə'sif·ik 'sauth ‚ek·wə'tȯr·ē·əl 'kə·rənt }

Pacific Standard Time *See* Pacific time. { pə'sif·ik 'stan·dərd 'tīm }

Pacific suite [PETR] A large group of igneous rocks characterized by calcic and calc-alkalic rocks, especially in the region of the circum-Pacific orogenic belt. Also known as anapeirean; circum-Pacific province. { pə'sif·ik 'swēt }

Pacific temperate faunal region [ECOL] A marine littoral faunal region including a narrow zone in the North Pacific Ocean, from Indochina to Alaska and along the west coast of the United States to about 40° north latitude. { pə'sif·ik 'tem·prət 'fȯn·əl ‚rē·jən }

PACHYCEPHALOSAURIDAE

Head of *Stegocerus*, a member of the Pachycephalosauridae. Late Cretaceous, Canada. Dome above level of eyes is solid bone.

Pacific time [ASTRON] The time for a given time zone that is based on the 120th meridian and is the eighth zone west of Greenwich. Also known as Pacific Standard Time. { pə'sif· ik 'tīm }

Pacific-type continental margin [GEOL] A continental margin typified by that of the western Pacific where oceanic lithosphere descends beneath an adjacent continent and produces an intervening island arc system. { pə'sif·ik ,tīp ,känt· ən'ent·əl 'mär·jən }

Pacinian corpuscle [NEUROSCI] An encapsulated lamellar sensory nerve ending that functions as a kinesthetic receptor. { pə'chin·ē·ən 'kȯr·pə·səl }

pack [COMPUT SCI] To reduce the amount of storage required to hold information by changing the method of encoding the data. [IND ENG] To provide protection for an article or group of articles against physical damage during shipment; packing is accomplished by placing articles in a shipping container, and blocking, bracing, and cushioning them when necessary, or by strapping the articles or containers on a pallet or skid. [MIN ENG] **1.** A pillar built in the waste area or roadside within a mine to support the mine roof; constructed from loose stones and dirt. **2.** Waste rock or timber used to support the roof or underground workings or used to fill excavations. Also known as fill. [OCEANOGR] See pack ice. [ORD] Part of a parachute assembly in which the canopy and shroud lines are folded and carried. Also known as pack assembly. { pak }

package [COMPUT SCI] A program that is written for a general and widely used application in such a way that its usefulness is not impaired by the problems of data or organization of a particular user. { 'pak·ij }

packaged circuit See rescap. { 'pak·ijd ¦sər·kət }

packaged magnetron [ELECTR] Integral structure comprising a magnetron, its magnetic circuit, and its output matching device. { 'pak·ijd 'mag·nə,trän }

package freight [IND ENG] Freight shipped in lots insufficient to fill a complete car; billed by the unit instead of by the carload. { 'pak·ij ,frāt }

package power reactor [NUC PHYS] A small nuclear power plant designed to be crated in packages small enough for transportation to remote locations. { 'pak·ij 'pau̇·ər rē,ak·tər }

packaging [ELEC] The process of physically locating, connecting, and protecting devices or components. { 'pak·ə·jiŋ }

packaging density [ELECTR] The number of components per unit volume in a working system or subsystem. { 'pak· ə·jiŋ ,den·səd·ē }

pack artillery [ORD] Artillery weapons designed for transport in sections by animals or delivery by parachute; the weapon and carriage are partially disassembled for transport and reassembled for firing from ground positions. { 'pak är'til·ə·rē }

pack assembly See pack. { 'pak ə,sem·blē }

pack builder [MIN ENG] **1.** One who builds packs or pack walls. **2.** In anthracite and bituminous coal mining, one who fills worked-out rooms, from which coal has been mined, with rock, slate, or other waste to prevent caving of walls and roofs, or who builds rough walls and columns of loose stone, heavy boards, timber, or coal along haulageways and passageways and in rooms where coal is being mined to prevent caving of roof or walls during mining operations. Also known as packer; pillar man; timber packer; waller. { 'pak ,bild·ər }

pack carburizing [MET] A method of surface hardening of steel in which parts are packed in a steel box with the carburizing compound and heated to elevated temperatures. { 'pak 'kär·bə,rīz·iŋ }

packed bed [CHEM ENG] A fixed layer of small particles or objects arranged in a vessel to promote intimate contact between gases, vapors, liquids, solids, or various combinations thereof; used in catalysis, ion exchange, sand filtration, distillation, absorption, and mixing. { 'pakt 'bed }

packed decimal [COMPUT SCI] A means of representing two digits per character, to reduce space and increase transmission speed. { 'pakt 'des·məl }

packed file [COMPUT SCI] A file that has been encoded so that it takes up less space in storage. Also known as compressed file. { ¦pakt 'fīl }

packed tower [CHEM ENG] A fractionating or absorber tower filled with small objects (packing) to bring about intimate contact between rising fluid (vapor or liquid) and falling liquid. { 'pakt 'tau̇·ər }

packed tube [CHEM ENG] A pipe or tube filled with high-heat-capacity granular material; used to heat gases when tubes are externally heated. { 'pakt 'tüb }

packer [ENG] A device that is inserted into a hole being grouted to prevent return of the grout around the injection pipe. [MIN ENG] See pack builder. [PETRO ENG] See production packer. { 'pak·ər }

packer fluid [PETRO ENG] Fluid inserted in the annulus between the tubing and casing above a packer in order to reduce pressure differentials between the formation and the inside of the casing and across the packer. { 'pak·ər ,flü·əd }

packer test [PETRO ENG] A pressure test of a sealed zone in a well. { 'pak·ər ,test }

packet [BIOL] A cluster of organisms in the form of a cube resulting from cell division in three planes. [COMMUN] A short section of data of fixed length that is transmitted as a unit. [PHYS] See wave packet. { 'pak·ət }

packet gland [INV ZOO] A cluster of gland cells opening through the epidermis of nemertines. { 'pak·ət ,gland }

packet switching See packet transmission. { 'pak·ət ,swich·iŋ }

packet transmission [COMMUN] Transmission of standardized packets of data over transmission lines rapidly by networks of high-speed switching computers that have the message packets stored in fast-access core memory. Also known as packet switching. { 'pak·ət tranz,mish·ən }

pack hardening [MET] A process of heat treating in which the workpiece is packed in a metal box together with carbonaceous material; carbon penetration is proportional to the length of heating; after treatment the workpiece is reheated and quenched. { 'pak ,härd·ən·iŋ }

pack ice [OCEANOGR] Any area of sea ice, except fast ice, composed of a heterogeneous mixture of ice of varying ages and sizes, and formed by the packing together of pieces of floating ice. Also known as ice canopy; ice pack; pack. { 'pak ,īs }

packing [CRYSTAL] Arrangement of atoms or ions in a crystal lattice. [ENG] See stuffing. [ENG ACOUS] Excessive crowding of carbon particles in a carbon microphone, produced by excessive pressure or by fusion particles due to excessive current, and causing lowered resistance and sensitivity. [GEOL] The arrangement of solid particles in a sediment or in sedimentary rock. [GRAPHICS] Paper used as a layer under the image or impression cylinder in letterpress printing or under the plate or blanket in lithographic printing in order to produce suitable pressure. [MET] In powder metallurgy, a material in which compacts are embedded during presintering or sintering operations. { 'pak·iŋ }

packing density [COMPUT SCI] The amount of information per unit of storage medium, as characters per inch on tape, bits per inch or drum, or bits per square inch in photographic storage. [ELECTR] The number of devices or gates per unit area of an integrated circuit. [GEOL] A measure of the extent to which the grains of a sedimentary rock occupy the gross volume of the rock in contrast to the spaces between the grains; equal to the cumulative grain-intercept length along a traverse in a thin section. { 'pak·iŋ ,den·səd·ē }

packing fraction [NUC PHYS] The quantity $(M - A)/A$, where M is the mass of an atom in atomic mass units and A is its atomic number. { 'pak·iŋ ,frak·shən }

packing house [FOOD ENG] **1.** A food processing plant generally requiring the use of refrigeration. **2.** A building in which livestock are slaughtered and processed, and the meat products and by-products are packed. { 'pak·iŋ ,hau̇s }

packing house pitch [MATER] Dark-brown to black by-product residue from manufacturing soap and candle stock or from refining vegetable oils, refuse, or wool grease; soluble in naphtha and carbon disulfide; used to make paints, varnishes, and tar paper, and in marine caulking and waterproofing. Also known as fatty-acid pitch. { 'pak·iŋ ,hau̇s ,pich }

packing index [CRYSTAL] The volume of ion divided by the volume of the unit cell in a crystal. { 'pak·iŋ ,in,deks }

packing proximity [GEOL] In a sedimentary rock, an estimate of the number of grains that are in contact with adjacent grains; equal to the total percentage of grain-to-grain contacts along a traverse measured on a thin section. { 'pak·iŋ präk,-sim·əd·ē }

packing radius [CRYSTAL] One-half the smallest approach distance of atoms or ions. { 'pak·iŋ ,rād·ē·əs }

packing ring See piston ring. { 'pak·iŋ ,riŋ }

packing routine [COMPUT SCI] A subprogram which compresses data so as to eliminate blanks and reduce the storage needed for a file. { 'pak·iŋ rü,tēn }

pack rolling [MET] Hot rolling of two or more sheets of metal packed together; a thin surface oxide film prevents their welding. { 'pak ,rōl·iŋ }

packsand [PETR] A very fine-grained sandstone that is so loosely consolidated by a slight calcareous cement that it can be readily cut by a spade. { 'pak,sand }

packstone [PETR] A sedimentary carbonate rock whose granular material is arranged in a self-supporting framework, yet also contains some matrix of calcareous mud. { 'pak,stōn }

pack unit [COMMUN] A compact, combination radio transmitter-receiver that can be carried or strapped on the back; some pack units are popularly known as walkie-talkies. { 'pak ,yü·nət }

pack wall [MIN ENG] A wall of dry stone built along the side of a roadway, or in the waste area, of a coal or metal mine to help support the roof and to retain the packing material and prevent its spreading into the roadway. { 'pak ,wȯl }

pactamycin [MICROBIO] An antitumor and antibacterial antibiotic produced by *Streptomyces pactum* var. *pactum*. { ,pak·tə'mīs·ən }

pad [AERO ENG] See launch pad. [ANAT] A small circumscribed mass of fatty tissue, as in terminal phalanges of the fingers or the underside of the toes of an animal, such as a dog. [ELECTR] **1.** An arrangement of fixed resistors used to reduce the strength of a radio-frequency or audio-frequency signal by a desired fixed amount without introducing appreciable distortion. Also known as fixed attenuator. **2.** See terminal area. [ENG] **1.** A layer of material used as a cushion or for protection. **2.** A projection of excess metal on a casting forging, or welded part. **3.** An area within an airstrip or airway that is used for warming up the motors of an airplane before takeoff. **4.** A block of stone or masonry set on a wall to distribute a load that is concentrated at that portion of the wall. Also known as padstone. **5.** That portion of an airstrip or airway from which an airplane leaves the ground on takeoff or first touches the ground on landing. **6.** See helipad. [MET] The brickwork that is beneath the molten iron at the base of a blast furnace. { pad }

pad deluge [AERO ENG] Water sprayed on certain launch pads during rocket launching in order to reduce the temperatures of critical parts of the pad or the rocket. { 'pad 'del,yüj }

padder [ELECTR] A trimmer capacitor inserted in series with the oscillator tuning circuit of a superheterodyne receiver to control calibration at the low-frequency end of a tuning range. { 'pad·ər }

padding [COMPUT SCI] The adding of meaningless data (usually blanks) to a unit of data to bring it up to some fixed size. { 'pad·iŋ }

paddle [AERO ENG] A large, flat, paddle-shaped support for solar cells, used on some satellites. [DES ENG] Any of various implements consisting of a shaft with a broad, flat blade or bladelike part at one or both ends. { 'pad·əl }

paddle wheel [MECH ENG] **1.** A device used to propel shallow-draft vessels, consisting of a wheel with paddles or floats on its circumference, the wheel rotating in a plane parallel to the ship's length. **2.** A wheel with paddles used to move leather in a processing vat. { 'pad·əl ,wēl }

paddle-wheel steamer [NAV ARCH] A steamer propelled by a wheel or wheels having long paddles, some of which are curved and feathering; the wheel revolves in a vertical plane parallel to the length of the ship; one type of craft has a wide stern wheel, and the other type has two narrow side wheels. { 'pad·əl ,wēl ,stēm·ər }

pad dyeing [TEXT] A process in which fabrics are dyed by passing them between rollers. { 'pad ,dī·iŋ }

Pade table [MATH] A table associated to a power series having in its pth row and qth column the ratio of a polynomial of degree q by one of degree p so that this fraction expanded into a power series agrees with the original up to the $p + q$ term. { 'päd·ə ,tā·bəl }

p-adic field [MATH] For a fixed prime number, p, the set of all p-adic numbers, with addition and multiplication defined in a natural way. { ,pē 'ad·ik 'fēld }

p-adic integer [MATH] For a fixed prime number p, a sequence of integers, $x_0, x_1, \ldots$, such that $x_n - x_{n-1}$ is divisible by p^n for all $n \geq 0$; two such sequences, x_n and y_n, are considered equal if $x_n - y_n$ is divisible by p^{n+1} for all $n \geq 0$, and the sum and product of two such sequences is defined by term-by-term addition and multiplication. { ,pē 'ad·ik 'int·i·jər }

p-adic number [MATH] For a fixed prime number p, a fraction of the form a/p^k, where a is a p-adic integer and k is a nonnegative integer; two such fractions, a/p^k and b/p^m, are considered equal if ap^m and bp^k are the same p-adic integer. { ,pē 'ad·ik 'nəm·bər }

padlock [DES ENG] An unmounted lock with a shackle that can be opened and closed; the shackle is usually passed through an eye, then closed to secure a hasp. { 'pad,läk }

padparadsha See orange sapphire. { pad'par·əd,shä }

padstone See pad. { 'pad,stōn }

paedogamy [INV ZOO] A type of autogamy in certain protozoans whereby there is mutual fertilization of gametes derived from a single cell. { pē'däg·ə·mē }

paedomorphosis [EVOL] Phylogenetic change in which adults retain juvenile characters, accompanied by an increased capacity for further change; indicates potential for further evolution. { ,pēd·ə'mȯr·fə·səs }

Paenungulata [VERT ZOO] A superorder of mammals, including proboscideans, xenungulates, and others. { pēn,əŋ·gyə'läd·ə }

Paeoniaceae [BOT] A monogeneric family of dicotyledonous plants in the order Dilleniales; members are mesophyllic shrubs characterized by cleft leaves, flowers with an intrastaminal disk, and seeds having copious endosperm. { ,pē·ə·nē'ās·ē,ē }

paesa [METEOROL] A violent north-northeast wind of Lake Garda in Italy. { pī'ä·zə }

paesano [METEOROL] A northerly night breeze, blowing down from the mountains, of Lake Garda in Italy. { pī'zä·nō }

PAF See platelet-activating factor. { pē¦ā'ef }

page [COMPUT SCI] **1.** A standard quantity of main-memory capacity, usually 512 to 4096 bytes or words, used for memory allocation and for partitioning programs into control sections. **2.** A standard quantity of source program coding, usually 8 to 64 lines, used for displaying the coding on a cathode-ray tube. { pāj }

pageable memory [COMPUT SCI] The part of a computer's main storage that is subject to paging in a virtual storage system. { 'pāj·ə·bəl 'mem·rē }

page boundary [COMPUT SCI] The address of the first (lowest) word or byte within a page of memory. { 'pāj ,baùn·drē }

page data set [COMPUT SCI] A file for storing images of pages in a virtual storage system, so that they can be returned to main storage for further processing when needed. { 'pāj 'dad·ə ,set }

page description language [COMPUT SCI] A high-level language that specifies the format of a page generated by a printer; it is translated into specific codes by any printer that supports the language. Abbreviated PDL. { 'pāj di,skrip·shən ,laŋ·gwij }

page fault [COMPUT SCI] An interruption that occurs while a page which is referred to by the program is being read into memory. { 'pāj ,fȯlt }

page printer [COMPUT SCI] A computer output device which composes a full page of characters before printing the page. { 'pāj ,print·ər }

page proof [GRAPHICS] A proof received from a compositor after the galley, and having the form of the final page, usually including any illustrations. { 'pāj ,prüf }

pager [COMMUN] A receiver in a radio paging system. { 'pāj·ər }

page reader [COMPUT SCI] In character recognition, a character reader capable of processing cut-form documents of varying sizes; sometimes capable of reading information in reel forms. { 'pāj ,rēd·ər }

page skip [COMPUT SCI] A control character that causes a printer to skip over the remainder of the current page and move to the beginning of the following page. { 'pāj ,skip }

page table [COMPUT SCI] A key element in the virtual memory technique; a table of addresses where entries are adjusted for easy relocation of pages. { 'pāj ,tā·bəl }

Paget's cells [PATH] Large, epithelial cells with clear cytoplasm found in certain breast and skin cancers. { 'paj·əts ,selz }

PAEONIACEAE

The garden peony (*Paeonia lactiflora*). (Courtesy of F. E. Westlake, from National Audubon Society)

Paget's disease [MED] **1.** A type of carcinoma of the breast that involves the nipple or areola and the larger ducts, characterized by the presence of Paget's cells. **2.** Osseous hyperplasia simultaneous with accelerated deossification. **3.** An apocrine gland skin cancer, composed principally of Paget's cells. { 'paj·əts di‚zēz }

page turning [COMPUT SCI] **1.** The process of moving entire pages of information between main memory and auxiliary storage, usually to allow several concurrently executing programs to share a main memory of inadequate capacity. **2.** In conversational time-sharing systems, the moving of programs in and out of memory on a round-robin, cyclic schedule so that each program may use its allotted share of computer time. { 'paj ‚tərn·iŋ }

pagination [GRAPHICS] The art of planning page format to allow sequence page numbering. { ‚paj·ə'nā·shən }

paging [COMPUT SCI] The scheme used to locate pages, to move them between main storage and auxiliary storage, or to exchange them with pages of the same or other computer programs; used in computers with virtual memories. { 'paj· iŋ }

paging rate [COMPUT SCI] The number of pages per second moved by virtual storage between main storage and the page data set. { 'pāj·iŋ ‚rāt }

paging system [COMMUN] A system which gives an indication to a particular individual that he is wanted at the telephone, such as by sounding a number on musical gongs, calling by name over a loudspeaker, or producing an audible signal in a radio receiver carried in the individual's pocket. { 'pāj·iŋ ‚sis·təm }

pagoda stone [GEOL] **1.** A Chinese limestone showing in section fossil orthoceratites arranged in pagodalike designs. **2.** An agate whose markings resemble pagodas. { pə'gōd· ə‚stōn }

pagodite *See* agalmatolite. { 'pag·ə‚dīt }

Paguridae [INV ZOO] The hermit crabs, a family of decapod crustaceans belonging to the Paguridea. { pə'gyùr·ə‚dē }

Paguridea [INV ZOO] A group of anomuran decapod crustaceans in which the abdomen is nearly always asymmetrical, being either soft and twisted or bent under the thorax. { ‚pag· yə'rid·ē·ə }

paha [GEOL] A low, elongated, rounded glacial ridge or hill which consists mainly of drift, rock, or windblown sand, silt, or clay but is capped with a thick cover of loess. { pä'hä }

pahoehoe [GEOL] A type of lava flow whose surface is glassy, smooth, and undulating; the lava is basaltic, glassy, and porous. Also known as ropy lava. { pə'hō·ē‚hō·ē }

paigeite [MINERAL] (Fe,Mg)FeBO₅ A black mineral composed of iron magnesium borate, occurring as fibrous aggregates. { 'pā‚jīt }

pail [DES ENG] A cylindrical or slightly tapered container. { pāl }

pain [PHYSIO] Patterns of somesthetic sensation, generally unpleasant, or causing suffering or distress. { pān }

pain spot [PHYSIO] Any of the small areas of skin overlying the endings of either very small myelinated (delta) or unmyelinated (C) nerve fibers whose stimulation, depending on the intensity and duration, results in the sensation of either pain or itching. { 'pān ‚spät }

paint [COMPUT SCI] To fill an area of a display screen or printed output with a color, shade of gray, or image. [ELECTR] Vernacular for a target image on a radarscope. [MATER] A mixture of a pigment and a vehicle, such as oil or water, that together form a liquid or paste that can be applied to a surface to provide an adherent coating that imparts color to and often protects the surface. { pānt }

paint base [MATER] A vehicle into which pigment is mixed to form a paint. { 'pānt ‚bās }

paint clay [MATER] A light-yellow to dark-reddish-brown iron- or manganese-bearing clay that mixes easily with linseed oil. { 'pānt ‚klā }

painted pot *See* mud pot. { ‚pānt·əd 'pät }

painter [METEOROL] A fog frequently experienced on the coast of Peru; the brownish deposit which it often leaves upon exposed surfaces is sometimes called Peruvian paint. Also known as Callao painter. { 'pānt·ər }

pain threshold [PHYSIO] The lowest limit for the perception of pain sensations. { 'pān ‚thresh‚hōld }

PALAEACANTHOCEPHALA

Corynosoma reductum. (From H. J. Van Cleave, Acanthocephala of North American Mammals, University of Illinois Press, 1953)

paint pot [GEOL] A mud pot containing multicolored mud. { 'pānt ‚pät }

paint program [COMPUT SCI] A graphics program that maintains images in raster format, allowing the user to simulate painting with the aid of a mouse or a graphics tablet. { 'pānt ‚prō·grəm }

paint remover [MATER] Liquid or paste formulation used to remove dried paint, varnish, enamel, or lacquer; contains solvents such as methanol, ethyl alcohol, acetone, toluene, benzene, and ethyl acetate. { 'pānt ri‚müv·ər }

paint vehicle [MATER] The liquid constituent of paint; consists of volatile solvent or thinner and a film-forming component. { 'pānt ‚vē·ə·kəl }

pair [ELEC] Two like conductors employed to form an electric circuit. [MECH ENG] Two parts in a kinematic mechanism that mutually constrain relative motion; for example, a sliding pair composed of a piston and cylinder. [SCI TECH] A set of two things that are identical or nearly so, or are designed to function as a unit. { per }

paired cable [ELEC] Cable in which the single conductors are twisted together in groups of two, none of which is arranged with others to form quads. { ¦perd 'kā·bəl }

paired comparison [STAT] A method used where order relations are more easily determined than measurements, such as studying taste preferences; in the comparison of a group of objects, each pair of objects is tested with either one or the other or neither preferred. { ¦perd kəm'par·ə·sən }

paired electron [PHYS CHEM] One of two electrons that form a valence bond between two atoms. { ¦perd i'lek‚trän }

paired terrace [GEOL] One of two stream terraces that face each other at the same elevation from opposite sides of the stream valley and represent the remnants of the same floodplain or valley floor. Also known as matched terrace. { ¦perd 'ter·əs }

pairing [ELECTR] In television, imperfect interlace of lines composing the two fields of one frame of the picture; instead of having the proper equal spacing, the lines appear in groups of two. { 'per·iŋ }

pairing element [MECH ENG] Either of two machine parts connected to permit motion. { 'per·iŋ ‚el·ə·mənt }

pairing energy [NUC PHYS] An energy associated with extra stability of pairs of nucleons of the same kind, which results in nuclei with odd numbers of neutrons and protons having a lower binding energy and being less stable than nuclei with even numbers of neutrons and protons. { 'per·iŋ ‚en·ər·jē }

pairing isomer [NUC PHYS] An excited nuclear state which has an unusually long lifetime because the microscopic motions of its constituent nucleons differ sharply from those of states of lower energy into which it is permitted to decay. { 'per· iŋ ‚ī·sə·mər }

pair production [PHYS] The conversion of a photon into an electron and a positron when the photon traverses a strong electric field, such as that surrounding a nucleus or an electron. { 'per prə‚dək·shən }

pairwise disjoint [MATH] The property of a collection of sets such that no two members of the collection have any elements in common. { 'per‚wīz dis'jóint }

paisanite *See* ailsyte. { 'pīs·ən‚īt }

Palaeacanthaspidoidei [PALEON] A suborder of extinct, placoderm fishes in the order Rhenanida; members were primitive, arthrodire-like species. { ‚pāl·ē·ə‚kan·thə·spi'dóid·ē‚ī }

Palaeacanthocephala [INV ZOO] An order of the Acanthocephala including parasitic worms characterized by fragmented nuclei in the hypodermis, lateral placement of the chief lacunar vessels, and proboscis hooks arranged in long rows. { ‚pāl· ē·ə‚kan·thə'sef·ə·lə }

Palaechinoida [PALEON] An extinct order of echinoderms in the subclass Perischoechinoidea with a rigid test in which the ambulacra bevel over the adjoining interambulacra. { ‚pāl· ē·kī'nóid·ē·ə }

Palaemonidae [INV ZOO] A family of decapod crustaceans in the group Caridea. { ‚pāl·ē'män·ə‚dē }

Palaeocaridacea [INV ZOO] An order of crustaceans in the superorder Syncarida. { ‚pāl·ē·ō‚kar·ə'dās·ē·ə }

Palaeocaridae [INV ZOO] A family of the crustacean order Palaeocaridacea. { ‚pāl·ē·ō'kar·ə‚dē }

Palaeoconcha [PALEON] An extinct order of simple, smooth-hinged bivalve mollusks. { ‚pāl·ē·ō'kaŋ·kə }

Palaeocopida [PALEON] An extinct order of crustaceans in the subclass Ostracoda characterized by a straight hinge and by the anterior location for greatest height of the valve. { ¦pal·ē·ō'käp·ə·də }

Palaeodonta [VERT ZOO] A suborder of artiodactylous mammals including piglike forms such as the extinct "giant pigs" and the hippopotami. { ¦pal·ē·ə'dän·tə }

Palaeognathae [VERT ZOO] The ratites, making up a superorder of birds in the subclass Neornithes; merged with the Neognathae in some systems of classification. { ¦pal·ē'äg·nə,thē }

Palaeoisopus [PALEON] A singular, monospecific, extinct arthropod genus related to the pycnogonica, but distinguished by flattened anterior appendages. { ¦pal·ē·ō'ī·sə·pəs }

Palaeomastodontinae [PALEON] An extinct subfamily of elaphantoid proboscidean mammals in the family Mastodontidae. { ¦pal·ē·ō,mas·tə'dänt·ən,ē }

Palaeomerycidae [PALEON] An extinct family of pecoran ruminants in the superfamily Cervoidea. { ¦pal·ē·ō·mə'ris·ə,dē }

Palaeonemertini [INV ZOO] A family of the class Anopla distinguished by the two- or three-layered nature of the body-wall musculature. { ¦pal·ē·ō,ne·mər'tī,nī }

Palaeonisciformes [PALEON] A large extinct order of chondrostean fishes including the earliest known and most primitive ray-finned forms. { ¦pal·ē·ō,nis·ə'fór·mēz }

Palaeoniscoidei [PALEON] A suborder of extinct fusiform fishes in the order Palaeonisciformes with a heavily ossified exoskeleton and thick rhombic scales on the body surface. { ¦pal·ē·ō·nis'kóid·ē,ī }

Palaeopantopoda [PALEON] A monogeneric order of extinct marine arthropods in the subphylum Pycnogonida. { ¦pal·ē·ō·pan'täp·ə·də }

Palaeopneustidae [INV ZOO] A family of deep-sea echinoderms in the order Spatangoida characterized by an oval test, long spines, and weakly developed fascioles and petals. { ¦pal·ē·ō·nü'stə,dē }

Palaeopterygii [VERT ZOO] An equivalent name for the Actinopterygii. { ¦pal·ē,äp·tə'rij·ē,ī }

Palaeoryctidae [PALEON] A family of extinct insectivorous mammals in the order Deltatheridia. { ¦pal·ē·ō'rik·tə,dē }

Palaeospondyloidea [PALEON] An ordinal name assigned to the single, tiny fish *Palaeospondylus,* known only from Middle Devonian shales in Carthness, Scotland. { ¦pal·ē·ō,spän·də'lóid·ē·ə }

Palaeotheriidae [PALEON] An extinct family of perissodactylous mammals in the superfamily Equoidea. { ¦pal·ē·ō·thə'rī·ə,dē }

palaeotheriodont [VERT ZOO] Being or having lophodont teeth with longitudinal external tubercles that are connected with inner tubercles by transverse oblique crests. { ¦pal·ē·ō¦ther·ē·ə,dänt }

palaeotropical *See* paleotropical. { ¦pal·ē·ō'träp·ə·kəl }

palagonite [GEOL] A brown to yellow altered basaltic glass found as interstitial material or amygdules in pillow lavas. { pə'lag·ə,nīt }

palagonite tuff [PETR] A pyroclastic rock composed of angular fragments of palagonite. { pə'lag·ə,nīt ¦təf }

palama [VERT ZOO] The membranous web on the feet of aquatic birds. { 'pal·ə·mə }

palasite [GEOL] The most abundant of the intermediate types of meteorites, consisting of olivine enclosed in a nickel-iron matrix. { 'pal·ə,sīt }

palatal index [ANTHRO] The ratio, multiplied by 100, of the length to the breadth of the hard palate. { 'pal·əd·əl ,in,deks }

palate [ANAT] The roof of the mouth. { 'pal·ət }

palatine bone [ANAT] Either of a pair of irregularly L-shaped bones forming portions of the hard palate, orbits, and nasal cavities. { 'pal·ə,tīn 'bōn }

palatine canal [ANAT] One of the canals in the palatine bone, giving passage to branches of the descending palatine nerve and artery. { 'pal·ə,tīn kə'nal }

palatine gland [ANAT] Any of numerous small oral glands on the palate of mammals. { 'pal·ə,tīn 'gland }

palatine process [ANAT] A thick process that projects horizontally mediad from the medial aspect of the maxilla. [EMBRYO] An outgrowth on the ventromedial aspect of the maxillary process that develops into the definite palate. { 'pal·ə,tīn 'präs·əs }

palatine suture [ANAT] The median suture joining the bones of the palate. { 'pal·ə,tīn 'sü·chər }

palatine tonsil [ANAT] Either of a pair of almond-shaped aggregations of lymphoid tissue embedded between folds of tissue connecting the pharynx and posterior part of the tongue with the soft palate. Also known as faucial tonsil; tonsil. { 'pal·ə,tīn 'tän·səl }

palatomaxillary arch [ANAT] An arch formed by the palatine, maxillary, and premaxillary bones. Also known as maxillary arch. { ¦pal·ə·dō'mak·sə,ler·ē 'ärch }

palatomaxillary index [ANTHRO] An index denoting the form of the dental arch and palate, expressed by the formula: palatomaxillary width multiplied by 100, divided by the palatomaxillary length. { ¦pal·ə·dō'mak·sə,ler·ē 'in,deks }

palatoquadrate [VERT ZOO] A series of bones or a cartilaginous rod constituting part of the roof of the mouth or upper jaw of most nonmammalian vertebrates. { ¦pal·ə·dō 'kwä,drāt }

palau [MET] A palladium-gold alloy; used as a platinum substitute in analytical chemistry. { pə'laù }

palba wax [MATER] A grayish-yellow wax from older green leaves of the palm tree *Copernicia cerifera.* { 'pal·bə ,waks }

palea [BOT] **1.** The upper, enclosing bract of a grass flower. **2.** A chaffy scale found on the receptacle of the disk flowers of some composite plants. [INV ZOO] One of the enlarged flattened setae forming the operculum of the tube of certain polychaete worms. { 'pā·lē·ə }

Palearctic [ECCL] Pertaining to a biogeographic region including Europe, northern Asia and Arabia, and Africa north of the Sahara. { ¦pal·ē'ärd·ik }

paleate [BOT] Having a covering of chaffy scales, as some rhizomes. { 'pā·lē,āt }

pale catechu *See* gambir. { 'pal 'kad·ə,chü }

paleic surface [GEOL] A smooth, preglacial erosion surface. { pə'lē·ik 'sər·fəs }

paleoagrostology [PALEOBOT] The study of fossil grasses. { ¦pal·ē·ō,ag·rə'stäl·ə·jē }

paleoalgology [PALEOBOT] The study of fossil algae. Also known as paleophycology. { ¦pal·ē·ō·al'gäl·ə·jē }

paleoanthropology [ANTHRO] A branch of anthropology concerned with the study of fossil humans. { ¦pal·ē·ō,an·thrə'päl·ə·jē }

paleobiochemistry [PALEON] The study of chemical processes used by organisms that lived in the geologic past. { ¦pal·ē·ō,bī·ō'kem·ə·strē }

paleobioclimatology [PALEON] The study of climatological events affecting living organisms for millennia or longer. { ¦pal·ē·ō,bī·ō,klī·mə'täl·ə·jē }

paleobiocoenosis [PALEON] An assemblage of organisms that lived together in the geologic past as an interrelated community. Also known as paleocoenosis. { ¦pal·ē·ō,bī·ō·sə'nō·səs }

paleobiology [PALEON] The branch of paleontology concerned with the biologic aspects of the history of life. { ¦pā·lē·ō·bī'äl·ə·jē }

paleobotanic province [GEOL] A large region defined by similar fossil floras. { ¦pal·ē·ō·bə'tan·ik 'präv·əns }

paleobotany [PALEON] The branch of paleontology concerned with the study of ancient and fossil plants and vegetation of the geologic past. { ¦pal·ē·ō'bät·ən·ē }

paleoceanography [OCEANOGR] The study of the history of the circulation, chemistry, biogeography, fertility, and sedimentation of the oceans. { ¦pal·ē·ō·shə'näg·rə·fē }

Paleocene [GEOL] The oldest of the seven geological epochs of the Cenozoic Era, spanning 65 million to 55 million years ago. Comprising the Tertiary and Quaternary periods in modern usage, it is also the oldest of the five epochs constituting the Tertiary Period. It represents an interval of geological time (and rocks deposited during that time) extending from the termination of the Cretaceous Period of the Mesozoic Era to the dawn of the Eocene Epoch. { 'pal·ē·ə,sēn }

paleochannel [GEOL] A remnant of a stream channel cut in older rock and filled by the sediments of younger overlying rock. { ¦pal·ē·ō¦chan·əl }

Paleocharaceae [PALEOBOT] An extinct group of fossil plants belonging to the Charophyta distinguished by sinistrally spiraled gyrogonites. { ¦pal·ē·ō·kə'rās·ē,ē }

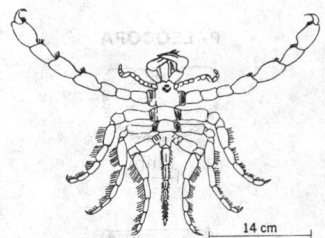

PALAEOISOPUS

Palaeoisopus problematicus.

PALAEONISCOIDEI

Elonichthys robisoni, a Lower Carboniferous palaeoniscoid from Scotland, which attained a length of 12 inches (30 centimeters).

PALEOCENE

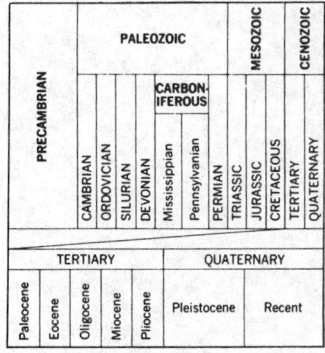

PRECAMBRIAN	PALEOZOIC									MESOZOIC			CENOZOIC
	CAMBRIAN	ORDOVICIAN	SILURIAN	DEVONIAN	CARBON-IFEROUS		PERMIAN	TRIASSIC	JURASSIC	CRETACEOUS	TERTIARY	QUATERNARY	
					Mississippian	Pennsylvanian							

TERTIARY					QUATERNARY	
Paleocene	Eocene	Oligocene	Miocene	Pliocene	Pleistocene	Recent

Chart showing the relationship of the Paleocene to the eras and periods of geologic time.

PALEOCOPA

1 mm

Exterior and interior views of right valve of *Eurychilina subradiata*.

paleoclimate [GEOL] The climate of a given period of geologic time. Also known as geologic climate. { ¦pāl·ē·ō¦klī·mət }

paleoclimatic sequence [GEOL] The sequence of climatic changes in geologic time; it shows a succession of oscillations between warm periods and ice ages, but superimposed on this are numerous shorter oscillations. { ¦pāl·ē·ō·klə¦mad·ik 'sē·kwəns }

paleoclimatology [GEOL] The study of climates in the geologic past, involving the interpretation of glacial deposits, fossils, and paleogeographic, isotopic, and sedimentologic data. { ¦pāl·ē·ō,klī·mə'täl·ə·jē }

paleocoenosis See paleobiocoenosis. { ¦pāl·ē·ō·sə'nō·səs }

Paleocopa [PALEON] An order of extinct ostracodes distinguished by a long, straight hinge. { ¸pāl·ē'äk·ə·pə }

paleocrystic ice [HYD] Sea ice generally considered to be at least 10 years old, especially well-weathered polar ice. { ¦pāl·ē·ō¦kris·tik 'īs }

paleocurrent [GEOL] Ancient fluid current flow whose orientation can be inferred by primary sedimentary structures and textures. { ¦pāl·ē·ō'kə·rənt }

paleodepth [PALEON] The water level at which an ancient organism or group of organisms flourished. { ¦pāl·ē·ō,depth }

paleoecology [PALEON] The ecology of the geologic past. { ¦pāl·ē·ō·i'käl·ə·jē }

paleoequator [GEOL] The position of the earth's equator in the geologic past as defined for a specific geologic period and based on geologic evidence. { ¦pāl·ē·ō·i'kwäd·ər }

paleofluminology [GEOL] The study of ancient stream systems. { ¦pāl·ē·ō,flü·mə'näl·ə·jē }

Paleogene [GEOL] A geologic time interval comprising the Oligocene, Eocene, and Paleocene of the lower Tertiary period. Also known as Eogene. { 'pāl·ē·ō,jēn }

paleogeographic event See palevent. { ¦pāl·ē·ō¦jē·ə'graf·ik i'vent }

paleogeographic stage See palstage. { ¦pāl·ē·ō¦jē·ə'graf·ik 'stāj }

paleogeography [GEOL] The geography of the geologic past; concerns all physical aspects of an area that can be determined from the study of the rocks. Paleogeography is used to describe the changing positions of the continents and the ancient extent of land, mountains, shallow sea, and deep ocean basins. { ¦pāl·ē·ō·jē'äg·rə·fē }

paleogeologic map [GEOL] An areal map of the geology of an ancient surface immediately below a buried unconformity, showing the geology as it appeared at some time in the geologic past at the time the surface of unconformity was completed and before the overlapping strata were deposited. { ¦pāl·ē·ō,jē·ə'läj·ik 'map }

paleogeology [GEOL] The geology of the past, applied particularly to the interpretation of the rocks at a surface of unconformity. { ¦pāl·ē·ō·jē'äl·ə·jē }

paleogeomorphology [GEOL] A branch of geomorphology concerned with the recognition of ancient erosion surfaces and the study of ancient topographies and topographic features that are now concealed beneath the surface and have been removed by erosion. Also known as paleophysiography. { ¦pāl·ē·ō,jē·ō·mȯr'fäl·ə·jē }

paleography [ANTHRO] The study of ancient modes of writing, for example, deciphering ancient writing and identifying the source or dating the text. { ¸pā·lē'äg·rə·fē }

paleoherpetology [PALEON] The study of fossil reptiles. { ¦pāl·ē·ō,hər·pə'täl·ə·jē }

paleohydrology [GEOL] The study of ancient hydrologic features preserved in rock. { ¦pāl·ē·ō·hī'dräl·ə·jē }

paleoichnology [PALEON] The study of trace fossils in the fossil state. Also spelled palichnology. { ¦pāl·ē·ō·ik'näl·ə·jē }

pale oil [MATER] A petroleum lubricating or process oil refined until its color (measured by transmitted light) is straw to pale yellow. { 'pāl ,ȯil }

paleoisotherm [GEOL] The locus of points of equal temperature for some former period of geologic time. { ¦pāl·ē·ō'ī·sə,thərm }

paleokarst [GEOL] A rock or area that has undergone the karst process and subsequently been buried under sediments. { ¦pāl·ē·ō,kärst }

paleolatitude [GEOL] The latitude of a specific area on the earth's surface in the geologic past. { ¦pāl·ē·ō'lad·ə,tüd }

paleolimnology [GEOL] **1.** The study of the past conditions and processes of ancient lakes. **2.** The study of the sediments and history of existing lakes. { ¦pāl·ē·ō·lim'näl·ə·jē }

Paleolithic [ANTHRO] The prehistoric period when people made stone tools exclusively by chipping or flaking; frequently defined by the time range from about 2,500,000 years ago to around the final retreat of the ice sheets about 10,300 years ago. { ¸pā·lē·ō'lith·ik }

paleolithologic map [GEOL] A paleogeologic map indicating lithologic variations at a buried horizon or within a restricted zone at a specific time in the geologic past. { ¦pāl·ē·ō,lith·ə'läj·ik 'map }

paleomagnetics [GEOPHYS] The study of the direction and intensity of the earth's magnetic field throughout geologic time. { ¦pāl·ē·ō·mag'ned·iks }

paleomagnetic stratigraphy [GEOPHYS] The use of natural remanent magnetization in the identification of stratigraphic units. Also known as magnetic stratigraphy. { ¦pāl·ē·ō·mag¦ned·ik strə'tig·rə·fē }

paleomalacology [PALEON] A branch of paleontology concerned with the study of mollusks. { ¦pāl·ē·ō,mal·ə'käl·ə·jē }

paleometeoritics [GEOL] The study of variation of extraterrestrial debris as a function of time over extended parts of the geologic record, especially in deep-sea sediments and possibly in sedimentary rocks, and, for more recent periods, in ice. { ¦pāl·ē·ō,mēd·ē'ȯr·iks }

paleomorphology [PALEON] The study of the form and structure of fossil remains in order to describe the original anatomy of an organism. { ¦pāl·ē·ō·mȯr'fäl·ə·jē }

paleomycology [PALEOBOT] The study of fossil fungi. { ¦pāl·ē·ō·mī'käl·ə·jē }

Paleonthropinae [PALEON] A former subfamily of fossil man in the family Hominidae; set up to include the Neanderthalers together with Rhodesian man. { ¸pāl·ē·ən'thräp·ə,nē }

paleontology [BIOL] The study of life of the past as recorded by fossil remains. { ¸pāl·ē·ən'täl·ə·jē }

paleopalynology [PALEON] A field of palynology concerned with fossils of microorganisms and of dissociated microscopic parts of megaorganisms. { ¦pāl·ē·ō,pal·ə'näl·ə·jē }

Paleoparadoxidae [PALEON] A family of extinct hippopotamuslike animals in the order Desmostylia. { ¦pāl·ē·ō,par·ə'däk·sə,dē }

paleopedology [GEOL] The study of soils of past geologic ages, including determination of their ages. { ¦pāl·ē·ō·pə'däl·ə·jē }

paleophycology See paleoalgology. { ¦pāl·ē·ō·fī'käl·ə·jē }

paleophysiography See paleogeomorphology. { ¦pāl·ē·ō,fiz·ē'äg·rə·fē }

Paleophytic [PALEOBOT] A paleobotanic division of geologic time, signifying that period during which the pteridophytes flourished, sometime between the evolution of the algae and the appearance of the first gymnosperms. Also known as Pteridophytic. { ¦pāl·ē·ə¦fid·ik }

paleoplain [GEOL] An ancient degradational plain that is buried beneath later deposits. { 'pāl·ē·ə,plān }

paleopole [GEOL] A pole of the earth, either magnetic or geographic, in past geologic time. { 'pāl·ē·ə,pōl }

Paleoptera [INV ZOO] A section of the insect subclass Pterygota including primitive forms that are unable to flex their wings over the abdomen when at rest. { ¸pāl·ē'äp·tə·rə }

paleosalinity [GEOL] The salinity of a body of water in the geologic past, as evaluated on the basis of chemical analyses of sediment or formation water. { ¦pāl·ē·ō·sə'lin·əd·ē }

paleoseismology [PALEON] The study of geological evidence for past earthquakes. { ¸pā·lē·ō·sīz'mäl·ə·jē }

paleosere [ECOL] A series of ecologic communities that have led to a climax community. { 'pāl·ē·ə,sir }

paleoslope [GEOL] The direction of initial dip of a former land surface, such as an ancient continental slope. { 'pāl·ē·ə,slōp }

paleosol [GEOL] A soil horizon that formed on the surface during the geologic past, that is, an ancient soil. Also known as buried soil; fossil soil. { 'pāl·ē·ə,sȯl }

paleosome [GEOL] A geometric element of a composite rock or mineral deposit which appears to be older than an associated younger rock element. { 'pāl·ē·ə,sōm }

paleospecies [PALEON] The species that are given ancestor and descendant status in a phyletic lineage, depending on the geological strata in which they are found. { ¦pē·lē·ō ¦spē,shēz }

paleostructure [GEOL] The geologic structure of a region or sequence of rocks in the geologic past. { ¦pal·ē·ō¦strək·chər }

paleotectonic map [GEOL] Regional map that shows the structural patterns that existed during a particular period of geologic time, for example, the Lower Cretaceous in western Canada. { ¦pal·ē·ō·tek¦tän·ik ¦map }

paleotemperature [GEOL] **1.** The temperature at which a geologic process took place in ancient past. **2.** The mean climatic temperature at a given time or place in the geologic past. { ¦pal·ē·ō¦tem·prə·chər }

paleothermal [GEOL] Pertaining to warm climates of the geologic past. { ¦pal·ē·ō¦thər·məl }

paleothermometry [GEOL] Measurement or estimation of past temperatures. { ¦pal·ē·ō·thər¦mäm·ə·trē }

paleotopography [GEOL] The topography of a given area in the geologic past. { ¦pal·ē·ō·tə¦päg·rə·fē }

paleotropical [ECOL] Of or pertaining to a biogeographic region that includes the Oriental and Ethiopian regions. Also spelled palaeotropical. { ¦pal·ē·ō¦träp·ə·kəl }

Paleozoic [GEOL] The era of geologic time from the end of the Precambrian (600 million years before present) until the beginning of the Mesozoic era (225 million years before present). { ¦pal·ē·ə¦zō·ik }

paleozoology [PALEON] The branch of paleontology concerned with the study of ancient animals as recorded by fossil remains. { ¦pal·ē·ō·zō¦äl·ə·jē }

palette [COMPUT SCI] In computer graphics, the set of colors that can be shown on a display monitor. [GEOL] A broad sheet of calcite representing a solutional remnant in a cave. Also known as shield. { ¦pal·ət }

palevent [GEOL] A relatively sudden and short-lived paleogeographic happening, such as the short, static existence of a particular depositional environment, or a rapid geographic change separating two palstages. Also known as paleogeographic event. { ¦pal·ə·vənt }

palichnology See paleoichnology. { ,pal·ik¦näl·ə·jē }

palilalia [MED] Pathologic repetition of words or phrases. { ¦pal·ə¦lal·yə }

palimpsest [GEOL] **1.** Referring to a kind of drainage in which a modern, anomalous drainage pattern is superimposed upon an older one, clearly indicating different topographic and possibly structural conditions at the time of development. **2.** In sedimentology, autochthonous sediment deposits which exhibit some of the attributes of the source sediment. [PETR] Of a metamorphic rock, having remnants of the original structure or texture preserved. { pə¦lim·səst }

palindrome [GEN] A nucleic acid sequence that is self-complementary. { ¦pal·ən,drōm }

palingenesis [EMBRYO] Unaltered recapitulation of ancestral features by the developing stages of an organism. [PETR] In-place formation of new magma by the melting of preexisting rock material. { ,pal·ən¦jen·ə·səs }

palingenetic See resurrected. { ¦pal·ən·jə¦ned·ik }

palinopsia [PSYCH] A form of pathological afterimagery that is characterized by the hallucinatory persistence of an object after the viewer has turned away. { ,pal·ə¦näp·sē·ə }

palinspastic map [GEOL] A paleogeographic or paleotectonic map showing restoration of the features to their original geographic positions, before thrusting or folding of the crustal rocks. { ¦pal·ən¦spas·tik ¦map }

Palinuridae [INV ZOO] The spiny lobsters or langoustes, a family of macruran decapod crustaceans belonging to the Scyllaridea. { ,pal·ə¦nyūr·ə,dē }

palisade cell [BOT] One of the columnar cells of the palisade mesophyll which contain numerous chloroplasts. { ,pal·ə'sād ,sel }

Palisade disturbance [GEOL] Appalachian orogenic episode occurring during Triassic time which produced a series of faultlike basins. { ,pal·ə'sād di'stər·bəns }

palisade mesophyll [BOT] A tissue system of the chlorenchyma in well-differentiated broad leaves composed of closely spaced palisade cells oriented parallel to one another, but with their long axes perpendicular to the surface of the blade. { ,pal·ə'sād ¦mez·ə,fil }

palisades [GEOL] A series of sharp cliffs. { ,pal·ə'sādz }

Palladian window [ARCH] An arched central window with a narrower, square-headed window on each side. { pə'lād·ē·ən 'win·dō }

palladium [CHEM] A chemical element, symbol Pd, atomic number 46, atomic weight 106.42. [MET] A white, ductile malleable metal that resembles platinum and follows it in abundance and importance of applications; does not tarnish at normal temperatures. { pə'lād·ē·əm }

palladium amalgam See potarite. { pə'lād·ē·əm ə'mal·gəm }

palladium barrier leak detector [ENG] A type of leak detector in which hydrogen is diffused through a barrier of hot palladium into an evacuated vacuum gage. { pə'lād·ē·əm 'bar·ē·ər 'lēk di,tek·tər }

palladium chloride [INORG CHEM] $PdCl_2$ or $PdCl_2\cdot 2H_2O$ Dark-brown, deliquescent powder that decomposes at 501°C; soluble in water, alcohol, acetone, and hydrochloric acid; used in medicine, analytical chemisty, photographic chemicals, and indelible inks. { pə'lād·ē·əm 'klȯr,īd }

palladium gold See porpezite. { pə'lād·ē·əm 'gōld }

palladium iodide [INORG CHEM] PdI_2 Black powder that decomposes above 100°C; soluble in potassium iodide solution, insoluble in water and alcohol. { pə'lād·ē·əm 'ī·ə,dīd }

palladium nitrate [INORG CHEM] $Pd(NO_3)_2$ Brown, water-soluble, deliquescent salt; used as an analytical reagent. { pə'lād·ē·əm 'nī,trāt }

palladium oxide [INORG CHEM] PdO Amber or black-green powder that decomposes at 750°C; soluble in dilute acids; used in chemical synthesis as a reduction catalyst. { pə'lād·ē·əm 'äk,sīd }

pallanesthesia [MED] Absence of pallesthesia, or vibration sense. { ,pal·lan·əs¦thē·zhə }

Pallas [ASTRON] The second-largest asteroid, with a diameter of about 343 miles (552 kilometers), mean distance from the sun of 2.77 astronomical units, and B-type (C-like) surface composition. { 'pal·əs }

pallasite [GEOL] **1.** A stony-iron meteorite composed essentially of large single glassy crystals of olivine embedded in a network of nickel-iron. **2.** An ultramafic rock, of either meteoric or terrestrial origin, which contains more than 60% iron in the former, or more iron oxides than silica in the latter. { 'pal·ə,sīt }

pallasite shell See lower mantle. { 'pal·ə,sīt ,shel }

pallet [BUILD] A flat piece of wood laid in a wall to which woodwork may be securely fastened. [ENG] **1.** A lever that regulates or drives a ratchet wheel. **2.** A hinged valve on a pipe organ. **3.** A tray or platform used in conjunction with a fork lift for lifting and moving materials. [GRAPHICS] An instrument consisting of a flat blade with a handle, used in clay work. [INV ZOO] One of a pair of plates on the siphon tubes of certain Bivalvia. [MECH ENG] One of the disks or pistons in a chain pump. { 'pal·ət }

palletize [IND ENG] To package material for convenient handling on a pallet or lift truck. { 'pal·ə,tīz }

palletized ship [NAV ARCH] A ship designed to carry palletized cargo. { 'pal·ə,tīzd 'ship }

pallet stone [HOROL] A stone or jewel face on a pallet; reduces friction and wear. { 'pal·ət ,stōn }

pallial artery [INV ZOO] The artery that supplies blood to the mantle of a mollusk. { 'pal·ē·əl ,ärd·ə·rē }

pallial chamber [INV ZOO] The mantle cavity in mollusks. { 'pal·ē·əl ,chām·bər }

pallial line [INV ZOO] A mark on the inner surface of a bivalve shell caused by attachment of the mantle. { 'pal·ē·əl ,līn }

pallial nerve [INV ZOO] One of the pair of dorsal nerves that innervate the mantle in mollusks. { 'pal·ē·əl ,nərv }

pallial sinus [INV ZOO] An inward bend in the posterior portion of the pallial line in bivalve mollusks. { 'pal·ē·əl ,sī·nəs }

palliative [PHARM] **1.** Having a soothing or relieving quality. **2.** A drug that soothes or relieves symptoms of a disease. { 'pal·ē·əd·iv }

pallium [ANAT] The cerebral cortex. [INV ZOO] The mantle of a mollusk or brachiopod. { 'pal·ē·əm }

Pallopteridae [INV ZOO] A family of myodarian cyclorrhaphous dipteran insects in the subsection Acalypteratae. { ,pal·äp¦ter·ə,dē }

pallor [MED] Paleness, especially of the skin and mucous membranes. { 'pal·ər }

PALEOZOIC

CENOZOIC	QUATERNARY	
	TERTIARY	
MESOZOIC	CRETACEOUS	
	JURASSIC	
	TRIASSIC	
PALEOZOIC	PERMIAN	
	CARBONIFEROUS	PENNSYLVANIAN
		MISSISSIPPIAN
	DEVONIAN	
	SILURIAN	
	ORDOVICIAN	
	CAMBRIAN	
	PRECAMBRIAN	

Chart showing the relationship of the Paleozoic to the other eras and to the periods of geologic time.

pall ring [CHEM ENG] A specially shaped steel ring used as packing for distillation columns. { 'pȯl ˌriŋ }

palm [ANAT] The flexor or volar surface of the hand. [BOT] Any member of the monocotyledonous family Arecaceae; most are trees with a slender, unbranched trunk and a terminal crown of large leaves that are folded between the veins. { päm }

Palmales [BOT] An equivalent name for Arecales. { pä'mā·lēz }

palmar [ANAT] Of or pertaining to the palm of the hand. { 'päm·ər }

palmar aponeurosis [ANAT] Bundles of fibrous connective tissue which radiate from the tendons of the deep fascia of the forearm toward the proximal ends of the fingers. { 'päm·ər ˌap·ə·nə'rōs·əs }

palmar arch See deep palmar arch; superficial palmar arch. { 'päm·ər 'ärch }

palmarosa oil [MATER] Colorless to light-yellow, volatile essential oil with roselike aroma obtained from a rosha grass (*Cymbopogon martinii* var. *motia*); soluble in alcohol and mineral oils; main component is geraniol; used in soaps and perfumes. Also known as East Indian geranium oil; Indian grass oil; Rusa oil; Turkish geranium oil. { pä·mə'rōs·ə ˌȯil }

palmar reflex [PHYSIO] Flexion of the fingers when the palm of the hand is irritated. { 'päm·ər 'rē·fleks }

palmate [BOT] Having lobes, such as on leaves, that radiate from a common point. [VERT ZOO] Having webbed toes. [ZOO] Having the distal portion broad and lobed, resembling a hand with the fingers spread. { 'päˌmāt }

palmately compound leaf [BOT] A leaf with leaflets that originate from a common point at the end of the petiole. { 'päˌmāt·lē ˌkäm·pau̇nd 'lēf }

palm butter [MATER] A reddish-yellow edible fatty oil expressed from putrid or fermented fruit pulp of the African oil palm (*Elaesis guineensis*); soluble in alcohol, ether, carbon disulfide, and chloroform; main components are palmitic acid, stearic acid, and glycerides of palmitic and oleic acids; melts at 27–42°C; used to make soaps and candles, as a lubricant, a color for butter substitutes, and an emollient. Also known as palm grease; palm oil. { 'päm ˌbəd·ər }

palmella stage [BOT] A stage in the life history of some unicellular flagellate algae in which the cells lose their flagella and form a gelatinous aggregation. { päl'mel·ə ˌstāj }

palmelloid [BOT] Pertaining to a colony of cells that aggregates in a gelatinous matrix, as is characteristic of blue-green algae. { päl'me·lȯid }

Palmer scan [ELECTR] Combination of circular or raster and conical radar scans; the beam is swung around the horizon, and at the same time a conical scan is performed. { 'päm·ər ˌskan }

palmetto fiber [BOT] Brush or broom fiber obtained from young leafstalks of the cabbage palm tree (*Sabal palmetto*). { päl'med·ō ˌfī·bər }

palm grease See palm butter. { 'päm ˌgrēs }

palmierite [MINERAL] (K,Na)₂Pb(SO₄)₂ A white hexagonal mineral that is composed of potassium sodium lead sulfate. { pä'mi·rīt }

palmitate [ORG CHEM] A derivative ester or salt of palmitic acid. { 'päm·əˌtāt }

palmitic acid [ORG CHEM] C₁₅H₃₁COOH A fatty acid; white crystals, soluble in alcohol and ether, insoluble in water; melts at 63.4°C, boils at 271.5°C (100 mmHg); derived from spermaceti; used to make metallic palmitates and in soaps, waterproofing, and lubricating oils. { päl'mid·ik 'as·əd }

palmitoleic acid [ORG CHEM] C₁₆H₃₀O₂ An unsaturated fatty acid, found in marine animal oils; it is a clear liquid used as a standard in chromatography. { ˌpäl·məd·ō'lē·ik 'as·əd }

palm kernel oil See palm nut oil. { 'päm ˌkər·nəl ˌȯil }

palm nut [BOT] The edible seed of the African oil palm (*Elaeis guineensis*). { 'päm ˌnət }

palm nut oil [MATER] Yellowish fatty oil; soluble in alcohol, ether, carbon disulfide, and chloroform; main components are triolein and triglycerides of stearic, palmitic, myristic, lauric, and other fatty acids; used to make soap, chocolate products, margarine, cosmetics, and candles, and as an illuminant and a color for butter substitutes. Also known as palm kernel oil. { 'päm ˌnət ˌȯil }

palm oil See palm butter. { 'päm ˌȯil }

palmtop See handheld computer. { 'pämˌtäp }

palm wax [MATER] A yellow wax from the Ecuadoran palm (*Ceroxylon andicola*); used as a beeswax substitute. { 'päm ˌwaks }

Palmyridae [INV ZOO] A mongeneric family of errantian polychaete annelids. { pal'mir·əˌdē }

palouser [METEOROL] A dust storm of northwestern Labrador. { pə'lüz·ər }

palp [INV ZOO] Any of various sensory, usually fleshy appendages near the oral aperture of certain invertebrates. { palp }

palpable [MED] **1.** Capable of being felt or touched. **2.** Evident. { 'pal·pə·bəl }

palpable coordinate [MECH] A generalized coordinate that appears explicitly in the Lagrangian of a system. { 'pal·pə·bəl kō'ȯrd·ən·ət }

palpal organ [INV ZOO] An organ on the terminal joint of each pedipalp of a male spider which functions to convey sperm to the female genital orifice. { 'pal·pəl ˌȯr·gən }

palpation [MED] Diagnostic examination by touch. { pal'pā·shən }

Palpatores [INV ZOO] A suborder of long-legged arachnids in the order Phalangida. { ˌpal·pə'tȯr·ēz }

palpebra [ANAT] The eyelid. { 'pal·pə·brə }

palpebral disk [VERT ZOO] A scale, often transparent, covering the eyelid of certain lizards. { 'pal·pə·brəl 'disk }

palpebral fissure [ANAT] The opening between the eyelids. { 'pal·pə·brəl 'fish·ər }

palpebral fold [ANAT] A fold formed by the reflection of the conjunctiva from the eyelid onto the eye. { 'pal·pə·brəl 'fōld }

Palpicornia [INV ZOO] The equivalent name for Hydrophiloidea. { ˌpal·pə'kȯr·nē·ə }

palpiger [INV ZOO] The palpi-bearing portion of an insect labium. { 'pal·pə·jər }

Palpigradida [INV ZOO] An order of rare tropical and warm-temperate arachnids; all are minute, whitish, eyeless animals with an elongate body that terminates in a slender, multisegmented flagellum set with setae. { ˌpal·pə'grad·əd·ə }

palpitate [MED] To flutter, or beat abnormally fast; applied especially to the rate of the heartbeat. { 'pal·pəˌtāt }

palpocil [INV ZOO] A fine, filamentous tactile hair. { 'pal·pəˌsil }

palpus [INV ZOO] **1.** A process on a mouthpart of an arthropod that has a tactile or gustatory function. **2.** Any similar process on other invertebrates. { 'pal·pəs }

palstage [GEOL] A period of time when paleogeographic conditions were relatively static or were changing gradually and progressively with relation to such factors as sea level, surface relief, or the distance of the shoreline from the region in question. Also known as paleogeographic stage. { 'palˌstāj }

palsy [MED] Any of various special types of paralysis, such as cerebral palsy. { 'pȯl·zē }

PAL system See phase-alternation line system. { 'pal ˌsistəm }

paludal [ECOL] Relating to swamps or marshes and to material that is deposited in a swamp environment. { pə'lüd·əl }

paludification [ECOL] Bog expansion resulting from the gradual rising of the water table as accumulation of peat impedes water drainage. { pəˌlüd·ə·fə'kā·shən }

palustrine [ECOL] Being, living, or thriving in a marsh. { pə'ləs·trən }

palygorskite [MINERAL] **1.** A chain-structure type of clay mineral. **2.** A group of lightweight, tough, fibrous clay minerals showing extensive substitution of aluminum for magnesium. { ˌpal·ə'gȯrˌskīt }

palynofacies [PALEON] An assemblage of palynomorphs in a portion of a sediment, representing local environmental conditions, but not representing the regional palynoflora. { ˌpal·ə·nō'fā·shēz }

palynology [PALEON] The study of spores, pollen, microorganisms, and microscopic fragments of megaorganisms that occur in sediments. { ˌpal·ə'näl·ə·jē }

palynomorph [PALEON] A microscopic feature such as a spore or pollen that is of interest in palynological studies. { pə'lin·əˌmȯrf }

palynostratigraphy [PALEON] The stratigraphic application of palynologic methods. { ˌpal·ə·nō·strə'tig·rə·fē }

palytoxin [BIOCHEM] A water-soluble toxin produced by

several species of *Palythoa;* considered to be one of the most poisonous substances known. { 'pal·ə,täk·sən }

PAM *See* pulse-amplitude modulation.

pamabrom [PHARM] $C_{11}H_{18}BrN_5O_3$ A water-soluble, fine white powder, decomposing at 300°C; used in medicine as a diuretic. { 'pam·ə,brōm }

pamaquine naphthoate [PHARM] $C_{42}H_{45}N_3O_7$ A yellow to orange-yellow powder, soluble in alcohol and acetone; used as an antimalarial drug. { 'pam·ə,kwēn 'naf·thə,wāt }

pampa [ECOL] An extensive plain in South America, usually covered with grass. { 'päm·pə }

pampero [METEOROL] A wind of gale force blowing from the southwest across the pampas of Argentina and Uruguay, often accompanied by squalls, thundershowers, and a sudden drop of temperature; it is comparable to the norther of the plains of the United States. { päm'per·ō }

Pamphillidae [INV ZOO] The web-spinning sawflies, a family of hymenopteran insects in the superfamily Megalodontoidea. { ,pam·fə'lī·ə,dē }

pampiniform [ANAT] Of the network of veins in the spermatic cord and in the broad ligament, having the form of a tendril. { pam'pin·ə,fórm }

pampiniform plexus [ANAT] A venous network in the spermatic cord in the male, and in the broad ligament in the female. { pam'pin·ə,fórm 'plek·səs }

pamprodactylous [VERT ZOO] Having the toes turned forward, as of certain birds. { ,pam·prə'dak·təl·əs }

pan [COMMUN] To tilt or otherwise move a television or movie camera vertically and horizontally to keep it trained on a moving object or to secure a panoramic effect. [GEOL] **1.** A shallow, natural depression or basin containing a body of standing water. **2.** A hard, cementlike layer, crust, or horizon of soil within or just beneath the surface; may be compacted, indurated, or very high in clay content. [MIN ENG] **1.** A shallow, circular, concave steel or porcelain dish in which drillers or samplers wash the drill sludge to gravity-separate the particles of heavy, dense minerals from the lighter rock powder as a quick visual means of ascertaining if the rocks traversed by the borehole contain minerals of value. **2.** The act or process of performing the above operation. [OCEANOGR] *See* pancake ice. { pan }

panabase *See* tetrahedrite. { 'pan·ə,bās }

panacinar emphysema [MED] Emphysema characterized by diffuse destruction of one lung. { ,pan·ə'sin·ər ,em·fə'sē·mə }

panadapter *See* panoramic adapter. { 'pan·ə,dap·tər }

panagglutinin [IMMUNOL] An agglutinin lacking specificity, which agglutinates erythrocytes of various types. { ,pan·ə'glüt·ən·ən }

Panagrolaimoidea [INV ZOO] A superfamily of free-living nematodes in the order Rhabditida, characterized by a broad, open, thick-walled stoma that forms a chamber as long as its breadth. { pə,na·grō·lə'móid·ē·ə }

panama [TEXT] A plain woven hopsacking of coarse-yarn basket weave, plain or in two colors, producing an effect similar to the texture of panama hats. { 'pan·ə,mä }

Panama disease [PL PATH] A fungus disease of banana caused by invasion of the vascular system by *Fusarium oxysporum cubense,* resulting in yellowing and wilting of the foliage and ultimate death of the shoots. { 'pan·ə,mä di,zēz }

pan-amalgamation process [MIN ENG] A process for extracting gold or silver from their ores; the ore is crushed and mixed with salt, copper sulfate, and mercury, and the gold or silver amalgamize with the mercury. { ¦pan·ə,mal·gə'mā·shən ,prä·səs }

Pan-American jig [MIN ENG] Mineral jig developed to treat alluvial sands; the jig cell is pulsated vertically on a flexible diaphragm seated above the stationary hutch. { ¦pan ə¦mer·ə·kən 'jig }

panarteritis [MED] **1.** Arteritis involving all the coats of an artery. **2.** *See* polyarteritis. { pan,ärd·ə'rīd·əs }

panarthritis [MED] Inflammation of several joints. { ¦pan·är'thrīd·əs }

panas oetara [METEOROL] A strong, warm, dry north wind in February in Indonesia. { pə'näs ,ō·ə'tär·ə }

panautomorphic rock *See* panidiomorphic rock. { ¦pan,ód·ə'mór·fik 'räk }

pan bolt [DES ENG] A bolt with a head resembling an upside-down pan. { 'pan ,bōlt }

pancake [MIN ENG] A concrete disk employed in stope support. [OCEANOGR] *See* pancake ice. { 'pan,kāk }

pancake auger [DES ENG] An auger having one spiral web, 12 to 15 inches (30 to 38 centimeters) in diameter, attached to the bottom end of a slender central shaft; used as removable deadman to which a drill rig or guy line is anchored. { 'pan ,kāk ¦óg·ər }

pancake coil [ELEC] A coil having the shape of a pancake, usually with the turns arranged in the form of a flat spiral. { 'pan,kāk ¦kóil }

pancake engine [MECH ENG] A compact engine with cylinders arranged radially. { 'pan,kāk ¦en·jən }

pancake forging [MET] A rough, flat, forged shape made quickly with a minimum of tooling. { 'pan,kāk ¦fórj·iŋ }

pancake ice [OCEANOGR] One or more small, newly formed pieces of sea ice, generally circular with slightly raised edges and about 1 to 10 feet (0.3 to 3 meters) across. Also known as lily-pad ice; pan; pancake; pan ice; plate ice. { 'pan,kāk ¦īs }

pancake landing [AERO ENG] Landing of an aircraft at a low forward speed and at a very high rate of descent. { 'pan ,kāk ¦land·iŋ }

pancarditis [MED] Carditis involving the endocardium, myocardium, and pericardium. { ¦pan·kär'dīd·əs }

Pancarida [INV ZOO] A superorder of the subclass Malacostraca; the cylindrical, cruciform body lacks an external division between the thorax and pleon and has the cephalon united with the first thoracomere. { pan'kar·ə·də }

panchratic eyepiece [OPTICS] A telescope eyepiece whose magnifying power can be varied by moving the erecting lens while keeping the focus at infinity.

panchromatic [GRAPHICS] Of a photographic emulsion, film, or plate, sensitive to all wavelengths within the visible spectrum, though not uniformly so. { ¦pan·krə'mad·ik }

panclimax [ECOL] Two or more related climax communities or formations having similar climate, life forms, and genera or dominants. Also known as panformation. { pan'klī,maks }

pan coefficient [METEOROL] The ratio of the amount of evaporation from a large body of water to that measured in an evaporation pan. { 'pan ,kō·i,fish·ənt }

pan conveyor [MECH ENG] A conveyor consisting of a series of pans. [MIN ENG] *See* jigging conveyor. { 'pan kən,vā·ər }

pancreas [ANAT] A composite gland in most vertebrates that produces and secretes digestive enzymes, as well as at least two hormones, insulin and glucagon. { 'pan·krē·əs }

pancreatectomy [MED] Surgical removal of the pancreas. { pan·krē·ə'tek·tə·mē }

pancreatic diarrhea [MED] Diarrhea due to deficiency of pancreatic digestive enzymes, characterized by the passage of large, greasy stools having a high fat and nitrogen content. { ¦pan·krē¦ad·ik ,dī·ə'rē·ə }

pancreatic diverticulum [EMBRYO] One of two diverticula (dorsal and ventral) from the embryonic duodenum or hepatic diverticulum that form the pancreas or its ducts. { ¦pan·krē¦ad·ik ,dī·vər tik·yə·ləm }

pancreatic duct [ANAT] The main duct of the pancreas formed from the dorsal and ventral pancreatic ducts of the embryo. { ¦pan·krē¦ad·ik ,dəkt }

pancreatic juice [PHYSIO] The thick, transparent, colorless secretion of the pancreas. { ¦pan·krē¦ad·ik ¦jüs }

pancreatic lipase *See* steapsin. { ¦pan·krē¦ad·ik 'lī,pās }

pancreatin [BIOCHEM] A cream-colored, amorphous powder obtained from the fresh pancreas of a hog; contains amylopsin, trypsin, steapsin, and other enzymes. { pan'krē·əd·ən }

pancreatitis [MED] Inflammation of the pancreas. { ,pan·krē·ə'tīd·əs }

pancreozymin [BIOCHEM] A crude extract of the intestinal mucosa that stimulates secretion of pancreatic juice. { ,pan·krē·ō'zī·mən }

pan crusher [MECH ENG] Solids-reduction device in which one or more grinding wheels or mullers revolve in a pan containing the material to be pulverized. { 'pan ,krəsh·ər }

pancytopenia [MED] Abnormally low numbers of all formed elements in the blood. { ,pan·sīd·ə'pē·nē·ə }

panda [VERT ZOO] Either of two Asian species of carnivores in the family Procyonidae; the red panda (*Ailurus fulgens*) has long, thick, red fur, with black legs; the giant panda (*Ailuropoda melanoleuca*) is white, with black legs and black patches around the eyes. { 'pan·də }

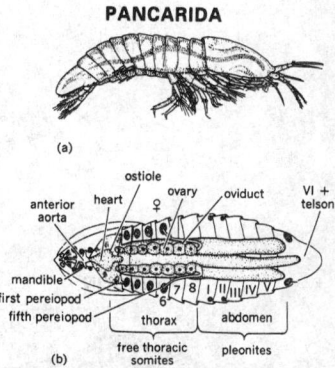

PANCARIDA

Female *Thermosbaena mirabilis.* (a) Lateral view. (b) Ventral section.

Pandanaceae [BOT] The single, pantropical family of the plant order Pandanales. { ,pan·də'nās·ē,ē }

Pandanales [BOT] A monofamilial order of monocotyledonous plants; members are more or less arborescent and sparingly branched, with numerous long, firm, narrow, parallel-veined leaves that usually have spiny margins. { ,pan·də'nā·lēz }

pandanus [BOT] Any tree of the genus *Pandanus*, which contains more than 500 species. It is a characteristic component of the vegetation in the tropics of the Old World, especially on the Pacific islands and along continental coasts. Also known as screw pine. { pan'dan·əs }

Pandaridae [INV ZOO] A family of dimorphic crustaceans in the suborder Caligoida; members are external parasites of sharks. { pan'dar·ə,dē }

pandemic [MED] Epidemic occurring over a widespread geographic area. { pan'dem·ik }

pandermite *See* priceite. { 'pan·dər,mīt }

Pandionidae [VERT ZOO] A monospecific family of birds in the order Falconiformes; includes the osprey (*Pandion haliaetus*), characterized by a reversible hindtoe, well-developed claws, and spicules on the scales of the feet. { pan·dē'än·ə,dē }

Pandora [ASTRON] A satellite of Saturn which orbits at a mean distance of 88,000 miles (142,000 kilometers), just outside the F ring; together with Prometheus, it holds this ring in place. { pan'dór·ə }

pandurate [BOT] Of a leaf, having the outline of a fiddle. { 'pan·dyùr·ət }

pane [BUILD] A sheet of glass in a window or door. [DES ENG] One of the sides on a nut or on the head of a bolt. { pān }

panel [CIV ENG] **1.** One of the divisions of a lattice girder. **2.** A sheet of material held in a frame. **3.** A distinct, usually rectangular, raised or sunken part of a construction surface or a material. [COMPUT SCI] The face of the console, which is normally equipped with lights, switches, and buttons to control the machine, correct errors, determine the status of the various CPU (central processing unit) parts, and determine and revise the contents of various locations. Also known as control panel; patch panel. [DES ENG] *See* frog. [ENG] A metallic or nonmetallic sheet on which operating controls and dials of an electronic unit or other equipment are mounted. [MIN ENG] **1.** A system of coal extraction in which the ground is laid off in separate districts or panels, pillars of extra-large size being left between. **2.** A large rectangular block or pillar of coal. { 'pan·əl }

panel board [ELECTR] *See* control board. [ENG] A drawing board with an adjustable outer frame that is forced over the drawing paper to hold and strain it. [MATER] A rigid paperboard used for paneling in buildings and automobile bodies. { 'pan·əl ,bórd }

panel code [COMMUN] Prearranged code designed for visual communications between ground units and friendly aircraft. { 'pan·əl ,kōd }

panel coil *See* plate coil. { 'pan·əl ,kóil }

panel cooling [CIV ENG] A system in which the heat-absorbing units are in the ceiling, floor, or wall panels of the space which is to be cooled. { 'pan·əl ,kül·iŋ }

panel display [ELECTR] An electronic display in which a large orthogonal array of display devices, such as electroluminescent devices or light-emitting diodes, form a flat screen. Also known as flat-panel display. { 'pan·əl di,splā }

panel heating [CIV ENG] A system in which the heat-emitting units are in the ceiling, floor, or wall panels of the space which is to be heated. { 'pan·əl ,hēd·iŋ }

panel length [CIV ENG] The distance between adjacent joints on a truss, measured along the upper or lower chord. { 'pan·əl ,leŋkth }

panel methods [FL MECH] Methods used in ship design to simplify the computation of the flow of water around a ship by assuming the flow to be frictionless and by taking advantage of the linearity of the Laplace equation, which governs the velocity potential in frictionless flow, to superpose elementary solutions to the problem on panels on the hull and (in most methods) on the free surface of the water. { 'pan·əl ,meth·ədz }

panel point [CIV ENG] The point in a framed structure where a vertical or diagonal member and a chord intersect. { 'pan·əl ,póint }

panel system [BUILD] A wall composed of factory-assembled units connected to the building frame and to each other by means of anchors. { 'pan·əl ,sis·təm }

panel wall [BUILD] A nonbearing partition between columns or piers. { 'pan·əl ,wól }

panendoscope [MED] A modification of the cystoscope, utilizing a Foroblique lens system, permitting adequate visualization of both the urinary bladder and the urethra. { pan'en·də,skōp }

Paneth cells *See* cells of Paneth. { 'pan·əth ,selz }

panethite [MINERAL] A phosphate mineral known only in meteorites; contains sodium, potassium, magnesium, calcium, iron, and manganese. { 'pan·ə,thīt }

Paneth's adsorption rule [PHYS CHEM] The rule that an element is strongly absorbed on a precipitate which has a surface charge opposite in sign to that carried by the element, provided that the resulting adsorbed compound is very sparingly soluble in the solvent. { 'pan·əths ad'sórp·shən ,rül }

panfan *See* pediplain. { 'pan,fan }

panformation *See* panclimax. { 'pan·fər,mā·shən }

Pangaea [GEOL] A postulated former supercontinent supposedly composed of all the continental crust of the earth, and later fragmented by drift into Laurasia and Gondwana. Also spelled Pangea. { pan'jē·ə }

Pangea *See* Pangaea. { pan'jē·ə }

pangene [CELL MOL] A hypothetical heredity-controlling protoplasmic particle proposed by Darwin. { 'pan,jēn }

pangenesis [BIOL] Darwin's comprehensive theory of heredity and development, according to which all parts of the body give off gemmules which aggregate in the germ cells; during development, they are sorted out from one another and give rise to parts similar to those of their origin. { pan'jen·ə·səs }

pangolin [VERT ZOO] Any of seven species composing the mammalian family Manidae; the entire dorsal surface of the body is covered with broad, horny scales, the small head is elongate, and the mouth is terminal in the snout. { 'paŋ·gə·lən }

pan head [DES ENG] The head of a screw or rivet in the shape of a truncated cone. { 'pan ,hed }

panhypopituitarism *See* hypopituitarism. { pan,hī·pō·pə'tü·ə·tə,riz·əm }

panic attack [PSYCH] Unexpected, paroxysmal episodes of anxiety and accompanying physical sensations (for example, racing heart, shortness of breath) that can occur at any time in susceptible individuals. { 'pan·ik ə,tak }

panic disorder [PSYCH] A severe and disabling psychiatric condition characterized by spontaneous attacks of sudden and intense fear. { 'pan·ik dis,órd·ər }

pan ice *See* pancake ice. { 'pan ,īs }

panic exit device [ENG] A locking device installed on an exit door to release the latch when the crash bar is pushed. Also known as fire-exit bolt; panic hardware. { ¦pan·ik ¦eg·zit di,vīs }

panic hardware *See* panic exit device. { 'pan·ik ,härd,wer }

panicle [BOT] A branched or compound raceme in which the secondary branches are often racemose as well. { 'pan·ə·kəl }

panidiomorphic rock [GEOL] An igneous rock that is completely or predominantly idiomorphic. Also known as panautomorphic rock. { ,pan¦id·ē·ō¦mór·fik 'räk }

panmixis [BIOL] Random mating within a breeding population; in a closed population this results in a high degree of uniformity. { pan'mik·səs }

Panmycin [MICROBIO] A trade name for tetracycline. { pan'mīs·ən }

panniculitis [MED] Inflammation of the layer of subcutaneous fat, especially in the abdomen. { pə,nik·yə'līd·əs }

panniculus [ANAT] A membrane or layer. { pə'nik·yə·ləs }

pannier *See* gabion. { 'pan·yər }

Pannonian [GEOL] A European stage of geologic time comprising the lower Pliocene. { pə'nō·nē·ən }

pannose [BIOL] Having a felty or woolly texture. { 'pa,nōs }

pannus [MED] **1.** Vascularization accompanied by deposition of connective tissue beneath the cornal epithelium. **2.**

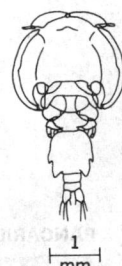

PANDARIDAE

Pandarus satyrus, male.

PANGOLIN

Giant pangolin (*Manis gigantea*).

PANICLE

Drawing of a panicle showing the branching of this type of inflorescence.

Overgrowth of connective tissue on the articular surface of a diarthrodial joint. [METEOROL] Numerous cloud shreds below the main cloud; may constitute a layer separated from the main part of the cloud or attached to it. { 'pan·əs }

panophthalmitis [MED] Inflammation of all the tissues of the eyeball. { pan,äf,thal'mīd·əs }

panoramic [OPTICS] Pertaining to a lens or optical instrument that has a wide field of view. { ¦pan·ə¦ram·ik }

panoramic adapter [ELECTR] A device designed to operate with a search receiver to provide a visual presentation on an oscilloscope screen of a band of frequencies extending above and below the center frequency to which the search receiver is tuned. Also known as panadapter. { ¦pan·ə¦ram·ik ə'dap·tər }

panoramic display [ELECTR] A display that simultaneously shows the relative amplitudes of all signals received at different frequencies. { ¦pan·ə¦ram·ik di'splā }

panoramic radar [ENG] Nonscanning radar which transmits signals over a wide beam in the direction of interest. { ¦pan·ə¦ram·ik 'rā,där }

panoramic receiver [ELECTR] Radio receiver that permits continuous observation on a cathode-ray-tube screen of the presence and relative strength of all signals within a wide frequency range. { ¦pan·ə¦ram·ik ri'sē·vər }

pan out [MIN ENG] To give a result, especially as compared with expectations; for example, in mining, the gravel may be said to pan out. { 'pan ,aut }

panplain [GEOL] A broad, level plain formed by coalescence of several adjacent flood plains. Also spelled panplane. { 'pan,plān }

panplanation [GEOL] The action or process of formation or development of a panplain. { ,pan·plə'nā·shən }

panplane See panplain. { 'pan,plān }

pan-range [ELECTR] Intensity-modulated, A-type radar indication with a slow vertical sweep applied to video; stationary targets give solid vertical deflection, and moving targets give broken vertical deflection. { 'pan ,rānj }

panspermia [BIOL] The theoretical ability of life to travel from body to body within the solar system. { pan'spər·mē·ə }

pansporoblast [INV ZOO] A sporont of cnidosporan protozoans that contains two sporoblasts. { pan'spór·ə,blast }

pan tank See rundown tank. { 'pan ,taŋk }

pantellerite [PETR] A green to black extrusive rock characterized by acmite-augite or diopside, anorthoclase, and cossyrite phenocrysts in an acmite or feldspar matrix that is either pumiceous, partly glassy, fine-grained holocrystalline trachytic, or microlitic. { pan'tel·ə,rīt }

Panthalassa [GEOL] The hypothetical proto-ocean surrounding Pangea, supposed by some geologists to have combined all the oceans or areas of oceanic crust of the earth at an early time in the geologic past. { ,pan·thə'las·ə }

panting [NAV ARCH] A series of pulsations resulting from repeated minor explosions in the furnace of a ship's boiler or from vibration of a ship's plating due to sea loads. { 'pant·iŋ }

panting beam [NAV ARCH] A beam fitted athwartship in the bow or stern of a vessel to prevent panting of the sides. { 'pant·iŋ ,bēm }

Pantodonta [PALEON] An extinct order of mammals which included the first large land animals of the Tertiary. { ,pan·tə'dän·tə }

Pantodontidae [VERT ZOO] A family of fishes in the order Osteoglossiformes; the single, small species is known as African butterflyfish because of its expansive pectoral fins. { ,pan·tə'dän·tə,dē }

pantograph [ENG] A device that sits on the top of an electric locomotive or cars in an electric train and picks up electricity from overhead wires to run the train. [GRAPHICS] A drawing instrument used for copying and consisting of four rigid bars linked together in a parallelogram form; one arm, equipped with a pencil, is connected through the bars to a pointer that is used to trace the original drawing. { 'pan·tə,graf }

pantography [ENG] System for transmitting and automatically recording radar data from an indicator to a remote point. { pan'täg·rə·fē }

Pantolambdidae [PALEON] A family of middle to late Paleocene mammals of North America in the superfamily Pantolambdoidea. { ,pan·tə'lam·də,dē }

Pantolambdodontidae [PALEON] A family of late Eocene mammals of Asia in the superfamily Pantolambdoidea. { ,pan·tə,lam·də'dän·tə,dē }

Pantolambdoidea [PALEON] A superfamily of extinct mammals in the order Pantodonta. { ,pan·tə·lam'dóid·ē·ə }

Pantolestidae [PALEON] An extinct family of large aquatic insectivores referred to the Proteutheria. { ,pan·tə'les·tə,dē }

pantometer [ENG] An instrument that measures all the angles necessary for determining distances and elevations. { pan'täm·əd·ər }

pantophagous [ZOO] Feeding on a variety of foods. { pan'täf·ə·gəs }

pantophobia [PSYCH] An abnormal fear of everything. { ,pan·tə'fō·bē·ə }

Pantophthalmidae [INV ZOO] The wood-boring flies, a family of orthorrhaphous dipteran insects in the series Brachycera. { ,pan,täf'thal·mə,dē }

Pantopoda [INV ZOO] The equivalent name for Pycnogonida. { pan'täp·ə·də }

pantothenate [BIOCHEM] A salt or ester of pantothenic acid. { ,pan·tə'the,nāt }

pantothenic acid [BIOCHEM] $C_9H_{17}O_5N$ A member of the vitamin B complex that is essential for nutrition of some animal species. Also known as vitamin B_3. { ¦pan·tə¦then·ik 'as·əd }

Pantotheria [PALEON] An infraclass of carnivorous and insectivorous Jurassic mammals; early members retained many reptilian features of the jaws. { ,pan·tə'thir·ē·ə }

pan-type car [MIN ENG] A vehicle for removing material from quarries; it is doorless, is reversible in direction, and can be dumped from either side. { 'pan ,tīp ,kär }

panuveitis [MED] Inflammation of the entire uveal tract. { pan,yü·vē'īd·əs }

Panzer-Forderer snaking conveyor [MIN ENG] An armored conveyor that is moved forward behind the coal plough by means of a traveling wedge pulled along by the plough or by means of jacks or compressed-air-operated rams attached at intervals to the conveyor structure. { 'pant·sər 'fór·də·rər 'snāk·iŋ kən'vā·ər }

panzootic [VET MED] Affecting many animals of different species. { ,pan·zō'äd·ik }

papagayo [METEOROL] A violent, northeasterly fall wind on the Pacific coast of Nicaragua and Guatemala; it consists of the cold air mass of a norte which has overridden the mountains of Central America and, being a descending wind, it brings fine, clear weather. { ,pä·pə'gī·yō }

papain [BIOCHEM] An enzyme preparation obtained from the juice of the fruit and leaves of the papaya (Carica papaya); contains proteolytic enzymes. { pə'pī·ən }

Papanicolaou's stains [CHEM] A group of stains used on exfoliated cells, particularly those from the vagina, for examination and diagnosis. { ,pä·pə'nēk·ə,lauz ,stān }

Papanicolaou test [PATH] A technique for the detection of precancerous and early noninvasive cancer by the staining and examination of exfoliated cells; used especially in the diagnosis of uterine cervical and endometrial cancer. Also known as Pap test. { ,pä·pə'nēk·ə,lau ,test }

Papaveraceae [BOT] A family of dicotyledonous plants in the order Papaverales, with regular flowers, numerous stamens, and a well-developed latex system. { pə,pav·ə'rās·ē,ē }

Papaverales [BOT] An order of dicotyledonous plants in the subclass Magnoliidae, marked by a syncarpous gynoecium, parietal placentation, and only two sepals. { pə,pav·ə'rā·lēz }

papaverine [ORG CHEM] $C_{20}H_{21}O_4N$ A white, crystalline alkaloid, melting at 147°C; soluble in acetone and chloroform, insoluble in water; used as a smooth muscle relaxant and weak analgesic, usually as the water-soluble hydrochloride salt. { pə'pav·ə,rēn }

Papaver somniferum See poppy. { ,päp·ə,ver säm'nif·ə·rəm }

paper [MATER] Felted or matted sheets of cellulose fibers, formed on a fine-wire screen from a dilute water suspension, and bonded together as the water is removed and the sheet is dried. { 'pā·pər }

paperboard [MATER] A composition board available in varying thicknesses and degrees of rigidity. { 'pā·pər,bórd }

paper capacitor [ELEC] A capacitor whose dielectric material consists of oiled paper sandwiched between two layers of metallic foil. { 'pā·pər kə'pas·əd·ər }

PAPAVERACEAE

Oriental poppy (Papaver orientale). (Photograph by John H. Gerard, from National Audubon Society)

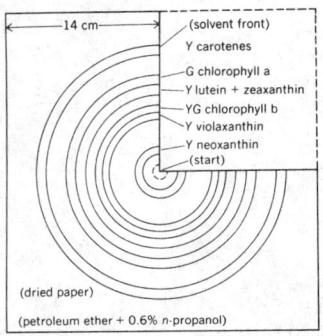

Paper chromatogram of chlorophyll showing ring separation. *G* signifies green; *Y*, yellow.

paper chromatography [ANALY CHEM] Procedure for analysis of complex chemical mixtures by the progressive absorption of the components of the unknown sample (in a solvent) on a special grade of paper. { 'pā·pər ¦krō·mə'täg·rə·fē }

paper clay [MATER] A special-grade clay that is mixed with paper pulp to add body, weight, and finish to paper products. { 'pā·pər ¦klā }

paper coating [MATER] Surface coating for paper; made from suspension of clays, starches, casein, rosin, polymers, wax, or various combinations; used to give strength and special surface qualities. { 'pā·pər ¦kōd·iŋ }

paper cutter [DES ENG] A hand-operated device to cut and trim paper, consisting of a cutting blade bolted at one end to a ruled board; when the blade is drawn flush with the board, which has a metal strip at the cutting edge, a shearing action takes place which cuts the paper cleanly and evenly. { 'pā·pər ¦kəd·ər }

paper electrochromatography [ANALY CHEM] Variation of paper electrophoresis in which the electrolyte-impregnated absorbent paper is suspended vertically and the electrodes are connected to the sides of the paper, producing a current at right angles to the downward movement of the unknown sample. { 'pā·pər i¦lek·trō¦krō·mə'täg·rə·fē }

paper electrophoresis [ANALY CHEM] A variation of paper chromatography in which an electric current is applied to the ends of the electrolyte-impregnated absorbent paper, thus moving chargeable molecules of the unknown sample toward the appropriate electrode. { 'pā·pər i¦lek·trə·fə'rē·səs }

paper insulation [MATER] Electrical insulation made of paper, chiefly from coniferous woods but also from rags, rope, and other materials, which are chemically treated, beaten into a dispersed pulp, formed into a loose sheet by filtering on a moving wire screen, and compacted into paper by calendering with heated rolls. { 'pā·pər ¦in·sə'lā·shən }

paper machine [MECH ENG] A synchronized series of mechanical devices for transforming a dilute suspension of cellulose fibers into a dry sheet of paper. { 'pā·pər mə¦shēn }

paper master [GRAPHICS] A paper printing plate that is used on an offset duplicator. { 'pā·pər 'mas·tər }

paper mill [IND ENG] A building or complex of buildings housing paper machines. { 'pā·pər ¦mil }

paperoid [MATER] A heavy composition board, generally made of rope pulp and having a reddish color; used for large expandible filing envelopes. { 'pā·pə¦rȯid }

paper shale [GEOL] A shale that easily separates on weathering into very thin, tough, uniform, and somewhat flexible layers or laminae suggesting sheets of paper. { 'pā·pər ¦shāl }

paper spar [GEOL] A crystallized variety of calcite occurring in thin lamellae or paperlike plates. { 'pā·pər ¦spär }

paper tape [COMPUT SCI] A paper ribbon in which data may be represented by means of partially or completely punched holes. { 'pā·pər ¦tāp }

paper-tape chemical analyzer [ANALY CHEM] Chemically treated paper tape that is continuously unreeled, exposed to the sample, and viewed by a phototube to measure the color change that is empirically related to changes in the sample's chemical composition. { 'pā·pər ¦tāp ¦kem·ə·kəl 'an·ə¦līz·ər }

paper-tape code [COMPUT SCI] The system by which data are represented by means of holes punched on a paper tape. { 'pā·pər ¦tāp ¦kōd }

paper-tape punch [COMPUT SCI] Device which places binary characters on a paper tape by punching holes in appropriate channels on the tape; a binary one is placed on the tape by punching a hole; a zero is indicated by the absence of a punched hole. { 'pā·pər ¦tāp ¦pənch }

paper-tape Turing machine [COMPUT SCI] A variation of a Turing machine in which a blank square can have a nonblank symbol written on it, but this symbol cannot be changed thereafter. { 'pā·pər ¦tāp 'tûr·iŋ mə¦shēn }

paper throw [COMPUT SCI] The movement of paper through a computer printer for a purpose other than printing, in which the distance traveled, and usually the speed, is greater than that of a single line spacing. { 'pā·pər ¦thrō }

papier maché [MATER] A lightweight molding material made from paper pulped with glue and other additives; dries to a hard finish that can be drilled, sanded, or painted. { ¦pā·pə·mə'shā }

Papilionidae [INV ZOO] A family of lepidopteran insects in the superfamily Papilionoidea; members are the only butterflies with fully developed forelegs bearing an epiphysis. { pə¦pil·ē'än·ə¸dē }

Papilionoidea [INV ZOO] A superfamily of diurnal butterflies (Lepidoptera) with clubbed antennae, which are rounded at the tip, and forewings that always have two or more veins. { pə¸pil·ē·ə'nȯid·ē·ə }

Papilionoideae [BOT] A subfamily of the family Leguminosae with characteristic irregular flowers that have a banner, two wing petals, and two lower petals united to form a boat-shaped keel. { pə¸pil·ē·ə'nȯid·ē¸ē }

papilla [BIOL] A small, nipplelike eminence. { pə'pil·ə }

papilla of Vater *See* ampulla of Vater. { pə'pil·ə əv 'fät·ər }

papillary adenoma of ovary *See* serous cystadenoma. { 'pap·ə¸ler·ē ¸ad·ən'ō·mə əv 'ō·və·rē }

papillary carcinoma [MED] A carcinoma characterized by fingerlike outgrowths. { 'pap·ə¸ler·ē ¸kärs·ən'ō·mə }

papillary muscle [ANAT] Any of the muscular eminences in the ventricles of the heart from which the chordae tendineae arise. { 'pap·ə¸ler·ē 'məs·əl }

papillate [BIOL] **1.** Having or covered with papillae. **2.** Resembling a papilla. Also known as papillose. { 'pap·ə¸lāt }

papilledema [MED] Edema of the optic disk. Also known as choked disk. { ¸pap·əl·ə'dē·mə }

papillocystoma *See* serous cystadenoma. { ¸pap·ə·lō·si'stō·mə }

papilloma [MED] A growth pattern of epithelial tumors in which the proliferating epithelial cells grow outward from a surface, accompanied by vascularized cores of connective tissue, to form a branching structure. { ¸pap·ə'lō·mə }

papillomatosis [MED] Widespread formation of papillomas. { ¸pap·ə¸lō·mə'tō·səs }

papillomatous [MED] Characterized by or pertaining to a papilloma. { ¸pap·ə¦läm·əd·əs }

papillose *See* papillate. { 'pap·ə¸lōs }

papite [MATER] A poison gas composed of acrolein with stannic chloride. { 'pa¸pīt }

Papovaviridae [VIROL] A family of deoxyribonucleic acid (DNA)-containing viruses characterized by a nonenveloped icosahedral virion containing double-stranded circular DNA that is complexed inside the nucleocapsid to histone proteins of host cell origin. { ¸pā·pə·və'vir·ə¸dī }

papovavirus [VIROL] A deoxyribonucleic acid-containing group of animal viruses, including papilloma and vacuolating viruses. { ¸pap·ə·və'vī·rəs }

pappataci fever *See* phlebotomus fever. { ¦päp·ə¦tä·chē ¸fē·vər }

Pappian plane [MATH] Any projective plane in which points and lines satisfy Pappus' theorem (third definition). { ¦pap·ē·ən 'plān }

Pappotheriidae [PALEON] A family of primitive, tenreclike Cretaceous insectivores assigned to the Proteutheria. { ¸pap·ə·thə'rī·ə¸dē }

pappus [BOT] An appendage or group of appendages consisting of a modified perianth on the ovary or fruit of various seed plants; adapted to dispersal by wind and other means. { 'pap·əs }

Pappus' theorem [MATH] **1.** The proposition that the area of a surface of revolution generated by rotating a plane curve about an axis in its own plane which does not intersect it is equal to the length of the curve multiplied by the length of the path of its centroid. **2.** The proposition that the volume of a solid of revolution generated by rotating a plane area about an axis in its own plane which does not intersect it is equal to the area multiplied by the length of the path of its centroid. **3.** A theorem of projective geometry which states that if A, B, and C are collinear points and A', B' and C' are also collinear points, then the intersection of AB' with $A'B$, the intersection of AC' with $A'C$, and the intersection of BC' with $B'C$ are collinear. **4.** A theorem of projective geometry which states that if A, B, C, and D are fixed points on a conic and P is a variable point on the same conic, then the product of the perpendiculars from P to AB and CD divided by the product of the perpendiculars from P to AD and BC is constant. { 'pap·əs ¸thir·əm }

paprika [BOT] *Capsicum annuum.* A type of pepper with nonpungent flesh, grown for its long red fruit from which a dried, ground condiment is prepared. { pə'prē·kə }

Pap test *See* Papanicolaou test. { 'pap ¸test }

papula [BIOL] A small papilla. { 'pap·yə·lə }

papule [MED] A solid circumscribed elevation of the skin varying from less than 0.1 to 1 centimeter in diameter. { 'pap·yül }

papulonecrotic [MED] Pertaining to papule formation with a tendency to central necrosis; applied especially to a variety of skin tuberculosis. { ,pap·yə·lō·ne'kräd·ik }

papyrus [MATER] A paperlike material made by pressing the pith of the papyrus plant in water. { pə'pī·rəs }

PAR *See* precision approach radar.

para- [ORG CHEM] Chemical prefix designating the positions of substituting radicals on the opposite ends of a benzene nucleus, for example, paraxylene, $CH_3C_6H_4CH_3$. Abbreviated p-. { 'par·ə }

paraaortic body [ANAT] One of the small masses of chromaffin tissue lying along the abdominal aorta. Also known as glomus aorticum. { ¦par·ə·ā¦ôrd·ik 'bäd·ē }

paraballoon [ELECTROMAG] Air-inflated radar antenna. { ¦par·ə·bə'lün }

parabanic acid [ORG CHEM] $C_3H_2O_3N_2$ A water-soluble cyclic compound that decomposes when heated to about 227°C; used in organic synthesis. { ¦par·ə¦ban·ik 'as·əd }

parabasal body *See* kinetoplast. { ¦par·ə¦bā·səl 'bäd·ē }

parabiosis [BIOL] Experimental joining of two individuals to study the effects of one partner upon the other. { ¦par·ə·bī'ō·səs }

parabiotic [MED] Physiologically and anatomically associated. { ,par·ə·bī'äd·ik }

parabituminous coal [GEOL] Bituminous coal that contains 84–87% carbon, analyzed on a dry, ash-free basis. { ¦par·ə·bə'tüm·ə·nəs 'kōl }

parabola [MATH] The plane curve given by an equation of the form $y = ax^2 + bx + c$. { pə'rab·ə·lə }

parabolic antenna [ELECTROMAG] Antenna with a radiating element and a parabolic reflector that concentrates the radiated power into a beam. { ¦par·ə¦bäl·ik an'ten·ə }

parabolic coordinate system [MATH] **1.** A two-dimensional coordinate system determined by a system of confocal parabolas. **2.** A three-dimensional coordinate system whose coordinate surfaces are the surfaces generated by rotating a plane containing a system of confocal parabolas about the axis of symmetry of the parabolas, together with the planes passing through the axis of rotation. { ¦par·ə¦bäl·ik kō'ôrd·ən·ət ,sis·təm }

parabolic cylinder [MATH] A cylinder whose directrix is a parabola. { ¦par·ə¦bäl·ik 'sil·ən·dər }

parabolic cylinder functions [MATH] Solutions to the Weber differential equation, which results from separation of variables of the Laplace equation in parabolic cylindrical coordinates. { ¦par·ə¦bäl·ik 'sil·ən·dər ,fəŋk·shənz }

parabolic cylindrical coordinate system [MATH] A three-dimensional coordinate system in which two of the coordinates depend on the x and y coordinates in the same manner as parabolic coordinates and are independent of the z coordinate, while the third coordinate is directly proportional to the z coordinate. { ¦par·ə¦bäl·ik si¦lin·drə·kəl kō'ôrd·ən·ət ,sis·təm }

parabolic differential equation [MATH] A general type of second-order partial differential equation which includes the heat equation and has the form

$$\sum_{i,j=1}^{n} A_{ij}(\partial^2 u/\partial x_i \partial x_j) + \sum_{i=1}^{n} B_i(\partial u/\partial x_i) + Cu + F = 0$$

where the A_{ij}, B_i, C, and F are suitably differentiable real functions of $x_1, x_2, \ldots, x_n$, and there exists at each point $(x_1, \ldots, x_n)$ a real linear transformation on the x_i which reduces the quadratic form

$$\sum_{i,j=1}^{n} A_{ij} x_i x_j$$

to a sum of fewer than n squares, not necessarily all of the same sign, while the same transformation does not reduce the B_i to 0. Also known as parabolic partial differential equation. { ¦par·ə¦bäl·ik ,dif·ə'ren·chəl i'kwä·zhən }

parabolic dune [GEOL] A long, scoop-shaped sand dune having a ground plan approximating the form of a parabola, with the horns pointing windward (upward). Also known as blowout dune. { ¦par·ə¦bäl·ik dün }

parabolic flight [AERO ENG] A space flight occurring in a parabolic orbit. { ¦par·ə¦bäl·ik 'flīt }

parabolic microphone [ENG ACOUS] A microphone used at the focal point of a parabolic sound reflector to give improved sensitivity and directivity, as required for picking up a band marching down a football field. { ¦par·ə¦bäl·ik 'mī·krə,fōn }

parabolic orbit [ASTRON] An orbit whose overall shape is like a parabola; the orbit represents the least eccentricity for escape from an attracting body. { ¦par·ə¦bäl·ik 'ôr·bət }

parabolic partial differential equation *See* parabolic differential equation. { ¦par·ə¦bäl·ik ¦pär·shəl ,dif·ə'ren·chəl i,kwä·zhən }

parabolic point [MATH] A point on a surface where the total curvature vanishes. { ¦par·ə¦bäl·ik 'pöint }

parabolic reflector [ELECTROMAG] An antenna having a concave surface which is generated either by translating a parabola perpendicular to the plane in which it lies (in a cylindrical parabolic reflector), or rotating it about its axis of symmetry (in a paraboloidal reflector). Also known as dish. [OPTICS] *See* paraboloidal reflector. { ¦par·ə¦bäl·ik ri'flek·tər }

parabolic Riemann surface *See* parabolic type. { ,par·ə,bäl·ik 'rē,män ,sər·fəs }

parabolic rule *See* Simpson's rule. { ¦par·ə¦bäl·ik 'rül }

parabolic segment [MATH] The line segment given by a chord perpendicular to the axis of a parabola. { ¦par·ə¦bäl·ik 'seg·mənt }

parabolic spiral [MATH] The curve whose equation in polar coordinates is $r^2 = a\theta$. { ¦par·ə¦bäl·ik 'spī·rəl }

parabolic type [MATH] A type of simply connected Riemann surface that can be mapped conformally onto the complex plane, excluding the origin and the point at infinity. Also known as Riemann surface. { ¦par·ə¦bäl·ik ,tīp }

parabolic velocity [ASTRON] The velocity attained by a celestial body in a parabolic orbit. { ¦par·ə¦bäl·ik və'läs·əd·ē }

paraboloid [ENG] A reflecting surface which is a paraboloid of revolution and is used as a reflector for sound waves and microwave radiation. [MATH] A surface where sections through one of its axes are ellipses or hyperbolas, and sections through the other are parabolas. { pə'rab·ə,löid }

paraboloidal antenna *See* paraboloidal reflector. { pə'rab·ə¦löid·əl an'ten·ə }

paraboloidal coordinate system [MATH] A three-dimensional coordinate system in which the coordinate surfaces form families of confocal elliptic and hyperbolic paraboloids. { pə¦rab·ə¦löid·əl kō'ôrd·ən·ət ,sis·təm }

paraboloidal reflector [ELECTROMAG] An antenna having a concave surface which is a paraboloid of revolution; it concentrates radiation from a source at its focal point into a beam. Also known as paraboloidal antenna. [OPTICS] A concave mirror which is a paraboloid of revolution and produces parallel rays of light from a source located at the focus of the parabola. Also known as parabolic reflector. { pə¦rab·ə¦löid·əl ri'flek·tər }

paraboloid of revolution [MATH] The surface obtained by rotating a parabola about its axis. { pə'rab·ə,löid əv ,rev·ə'lü·shən }

parabomb [ENG] An equipment container with a parachute which is capable of opening automatically after a delayed drop. { 'par·ə,bäm }

parabutlerite *See* butlerite. { ,par·ə'bət·lə,rīt }

paracaisson [ORD] Small, two-wheeled, hand-drawn vehicle whose body forms an aerial delivery container for artillery ammunition and which, upon being assembled, becomes a utility cart. { ¦par·ə'kā,sän }

Paracanthopterygii [VERT ZOO] A superorder of teleost fishes, including the codfishes and allied groups. { ,par·ə,kan,thäp·tə'rij·ē,ī }

paracentesis [MED] Puncture of the wall of a fluid-filled cavity by means of a hollow needle to draw off the contents. { ,par·ə·sen'tē·səs }

paracentric [DES ENG] Pertaining to a key and keyway with longitudinal ribs and grooves that project beyond the center, as used in pin-tumbler cylinder locks to deter lockpicking. { ¦par·ə¦sen·trik }

PARABOLOIDAL REFLECTOR

Parallel rays of light produced from paraboloidal reflector; dark circle is the light source.

PARACENTRIC INVERSION

Schematic drawing of the change taking place in paracentric inversion.

paracentric inversion [GEN] A type of chromosomal alteration that occurs within one arm of a chromosome and does not span the centromere. { ¦par·ə¦sen·trik in′vər·zhən }

parachor [PHYS] The molecular weight of a liquid times the fourth root of its surface tension, divided by the difference between the density of the liquid and the density of the vapor in equilibrium with it; essentially constant over wide ranges of temperature. { ′par·ə‚kȯr }

parachronology [GEOL] **1.** Practical dating and correlation of stratigraphic units. **2.** Geochronology based on fossils that supplement, or replace, biostratigraphically significant fossils. { ‚par·ə·krə′näl·ə·jē }

parachute [AERO ENG] **1.** A contrivance that opens out somewhat like an umbrella and catches the air so as to retard the movement of a body attached to it. **2.** The canopy of this contrivance. [MIN ENG] A kind of safety catch for mine shaft cages. { ′par·ə‚shüt }

parachute flare [ENG] Pyrotechnic device attached to a parachute and designed to provide intense illumination for a short period; it may be discharged from aircraft or from the surface. { ′par·ə‚shüt ‚fler }

parachute fragmentation bomb [ORD] A fragmentation bomb adapted for drop by parachute, used in low-level bombing to give the bombing plane time to escape. { ′par·ə‚shüt ‚frag·mən′tā·shən ‚bäm }

parachute-opening shock [AERO ENG] The shock or jolt exerted on a suspended parachute load when the parachute fully catches the air. { ′par·ə‚shüt ‚ōp·ə·niŋ ‚shäk }

parachute radiosonde *See* dropsonde. { ¦par·ə‚shüt ′rād·ē·ō‚sänd }

parachute troops [ORD] Troops organized and trained to be carried into battle by transport aircraft and dropped by parachute, as distinguished especially from airborne infantry. Also known as paratroops. { ′par·ə‚shüt ‚trüps }

parachute weather buoy [ENG] A general-purpose automatic weather station which can be air-dropped; it is 10 feet (3 meters) long and 22 inches (56 centimeters) in diameter, and is designed to operate for 2 months on a 6-hourly schedule, transmitting station identification, wind speed, wind direction, barometric pressure, air temperature, and sea-water temperature. { ′par·ə‚shüt ′weth·ər ‚bȯi }

paraclinal [GEOL] Referring to a stream or valley that is oriented in a direction parallel to the fold axes of a region. { ¦par·ə¦klīn·əl }

paracoccidioidomycosis *See* South American blastomycosis. { ¦par·ə·käk‚sid·ē¦ȯid·ō·mī‚kō·səs }

paracolon bacteria [MICROBIO] A group of bacteria intermediate between the *Escherichia-Aerobacter* genera and the *Salmonella-Shigella* group. { ¦par·ə′kō·lən bak′tir·ē·ə }

paracompact space [MATH] A topological space with the property that every open covering *F* is associated with a locally finite open covering *G*, such that every element of *G* is a subset of an element *F*. { ¦par·ə¦käm‚pakt ‚spās }

paracondyloid [VERT ZOO] A process on the outer side of each condyle of the occipital bone in the skull of certain mammals. { ¦par·ə′känd·əl‚ȯid }

paracone [VERT ZOO] **1.** The anterior cusp of a primitive tricuspid upper molar. **2.** The principal anterior, external cusp of an upper molar in higher forms. { ′par·ə‚kōn }

paraconformity [GEOL] A type of unconformity in which strata are parallel; there is little apparent erosion and the unconformity surface resembles a simple bedding plane. Also known as nondepositional unconformity; pseudoconformity. { ¦par·ə·kən′fȯr·məd·ē }

paraconglomerate [GEOL] A conglomerate that is not a product of normal aqueous flow but is deposited by such modes of mass transport as subaqueous turbidity currents and glacier ice; characterized by a disrupted gravel framework, often unstratified, and notable for a matrix of greater than gravel-sized fragments. { ¦par·ə·kən′gläm·ə·rət }

paraconid [VERT ZOO] **1.** The cusp of a primitive lower molar corresponding to the paracone. **2.** The anterior, internal cusp of a lower molar in higher forms. { ‚par·ə′kän·əd }

paracoquimbite [MINERAL] $Fe_2(SO_4)_3 \cdot 9H_2O$ A pale-violet rhombohedral mineral composed of hydrous ferric iron sulfate; it is dimorphous with coquimbite. { ¦par·ə·kə′kim‚bīt }

paracrate [ENG] Rigid equipment container for dropping equipment from an airplane by parachute. { ′par·ə‚krāt }

paracrine signaling [PHYSIO] Signaling in which the target cell is close to the signaling cell and the signal molecule affects only adjacent target cells. { ′par·ə‚krēn ‚sig·nəl·iŋ }

Paracrinoidea [PALEON] A class of extinct Crinozoa characterized by the numerous, irregularly arranged plates, uniserial armlike appendages, and no clear distinction between adoral and aboral surfaces. { ¦par·ə·krə′nȯid·ē·ə }

Paracucumidae [INV ZOO] A family of holothurian echinoderms in the order Dendrochirotida; the body is invested with plates and has a simplified calcareous ring. { ¦par·ə·kə′kyüm·ə‚dē }

paracyanogen [INORG CHEM] $(CN)_x$ A white solid produced by polymerization of cyanogen gas when heated to 400°C. { ¦par·ə·sī′an·ə·jən }

paracystitis [MED] Inflammation of the connective tissue surrounding the urinary bladder. { ¦par·ə·si′stīd·əs }

paradidymis [ANAT] Atrophic remains of the paragenital tubules of the mesonephros, occurring near the convolutions of the epididymal duct. { ¦par·ə′did·ə·məs }

paradox [SCI TECH] An argument which gives a contradictory conclusion. { ′par·ə‚däks }

paradoxical cold [PHYSIO] The arousal of a cold sensation by application of a hot probe to a cold point (a skin receptor that normally responds only to cold). { ¦par·ə‚däk·sə·kəl ′kōld }

paradoxical embolus [MED] An embolus which is transported to the circulation in peripheral arteries through septal defect in the heart, usually a patent foramen ovale. { ‚par·ə′däk·sə·kəl em′bə·ləs }

paradoxical warmth [PHYSIO] The arousal of a warm sensation by application of a cold probe to a warm point (a skin receptor that normally responds only to warmth). { ‚par·ə‚däk·sə·kəl ′wȯrmth }

paraesophageal cyst [MED] A bronchogenic cyst intimately connected with the esophageal wall, containing cartilage, and usually filled with a mucoid material and desquamated epithelial cells. { ¦par·ə·i‚säf·ə′jē·əl ′sist }

paraffin *See* alkane; paraffin wax. { ′par·ə·fən }

paraffin-base crude [MATER] Crude petroleum oil that contains predominately alkanes as contrasted with asphaltic- or naphthenic-base crudes; used as a source of fuels and high-grade lubricating oils. { ′par·ə·fən ′bās ′krüd }

paraffin coal [GEOL] A type of light-colored bituminous coal from which oil and paraffin are produced. { ′par·ə·fən ‚kōl }

paraffin dirt [GEOL] A clay soil appearing rubbery or curdy and occurring in the upper several inches of a soil profile near gas seeps; probably formed by biodegradation of natural gas. { ′par·ə·fən ‚dərt }

paraffin distillate [MATER] In a petroleum refinery, the distillate oils ready for pressing to produce crystalline paraffin wax and paraffin oil. { ′par·ə·fən ′dis·tə·lət }

paraffinic hydrocarbon *See* alkane. { ‚par·ə¦fin·ik ′hī·drə‚kär·bən }

paraffinicity [ORG CHEM] The paraffinic nature or composition of crude petroleum or its products. { ‚par·ə·fə′nis·əd·ē }

paraffin jelly [MATER] A light, amber-colored petrolatum; used for medicinal purposes. { ′par·ə·fən ‚jel·ē }

paraffin oil [MATER] A viscous, pale to yellow oil made from petroleum; used as a lubricant, medicine, and leather dressing. { ′par·ə·fən ‚ȯil }

paraffin press [ENG] A filter press used during petroleum refining for the separation of paraffin oil and crystallizable paraffin wax from distillates. { ′par·ə·fən ‚pres }

paraffin scale [MATER] Unrefined paraffin wax remaining in the chamber after oil has been removed from a mixture of oil and paraffin by sweating. { ′par·ə·fən ¦skāl }

paraffinum liquidum *See* white mineral oil. { ‚par·ə′fin·əm ′lik·wə·dəm }

paraffin wax [MATER] A solid, crystalline hydrocarbon mixture derived from the paraffin distillate portion of crude petroleum; used in paper coating, candles, creams, emollients, and lipsticks. Also known as ceresin wax; paraffin. { ′par·ə·fən ‚waks }

paraformaldehyde [ORG CHEM] $(HCHO)_n$ Polymer of formaldehyde where *n* is greater than 6; white, alkali-soluble solid, insoluble in alcohol, ether, and water; used as a disinfectant, fumigant, and fungicide, and to make resins. { ¦par·ə·fȯr′mal·də‚hīd }

paraganglion [NEUROSCI] Small structure associated with the sympathetic nervus system containing chromaffin tissue (hormonally active tissue related to the adrenal medulla). Also known as chromaffin body. { ¦par·ə¦gaŋ·glē‚än }

paragastric [ANAT] Located near the stomach. [INV ZOO] A cavity in Porifera into which radial canals open, and which opens to the outside through the cloaca. { ¦par·ə¦gas·trik }

paragenesis [MINERAL] **1.** The association and order of crystallization of minerals in a rock or vein. **2.** The effect of one mineral on the development of another. Also known as mineral sequence; paragenetic sequence. { ‚par·ə¦jen·ə·səs }

paragenetic [GEN] Pertaining to chromosome changes that alter gene expression but not makeup. { ¦par·ə·jə¦ned·ik }

paragenetic mineralogy [MINERAL] The study of mineral paragenesis, usually accompanying the analysis of the general geologic structures within and around the ore body. { ¦par·ə·jə¦ned·ik ‚min·ə¦räl·ə·jē }

paragenetic sequence *See* paragenesis. { ¦par·ə·jə¦ned·ik ¦sē·kwəns }

parageosyncline [GEOL] An epeirogenic geosynclinal basin located within a craton or stable area. { ¦par·ə‚jē·ō'sin‚klīn }

paragglutination [IMMUNOL] Agglutination of colon bacteria with the serum of a patient infected, or recovering from an infection, with dysentery bacilli. { ¦par·ə‚glüt·ən'ā·shən }

paraglider [AERO ENG] A triangular device on a rocket or spacecraft that consists of two flexible sections and resembles a kite; deployed to assist in guiding or landing a spacecraft or in recovering a launching rocket. { ¦par·ə‚glīd·ər }

paraglomerate [GEOL] A conglomerate which contains more matrix than gravel-sized fragments and was deposited by subaqueous turbidity flows and glacier ice rather than normal aqueous flow. Also known as conglomeratic mudstone. { ‚par·ə¦gläm·ə·rət }

paragnath [INV ZOO] **1.** One of the paired leaflike lobes of the metastoma situated behind the mandibles in most crustaceans. **2.** One of the paired lobes of the hypopharynx in certain insects. **3.** One of the small, sharp and hard jaws of certain annelids. { ¦par·əg‚nath }

paragneiss [GEOL] A gneiss showing a sedimentary parentage. { ¦par·ə‚nīs }

paragon [LAP] A perfect diamond whose mass is equal to or greater than 100 carats (20 grams). { 'par·ə‚gän }

paragonimiasis [MED] Presence of the fluke *Paragonimus westermani* in the lungs or other tissues of humans. { ‚par·ə‚gän·ə¦mī·ə·səs }

paragonite [MINERAL] $NaAl_2(AlSi_3)O_{10}(OH)_2$ A yellowish or greenish monoclinic mica species that contains sodium and usually occurs in metamorphic rock. Also known as soda mica. { pə¦rag·ə‚nīt }

paragraph [COMPUT SCI] A complete, logical sequence of instructions in the COBOL programming language, required to carry out a definable program or task. { 'par·ə‚graf }

paragraphia [PSYCH] The erroneous production of unintended words in writing that is a feature in some forms of aphasia. { ‚par·ə¦graf·ē·ə }

parahelium [ATOM PHYS] Those states of helium in which the spins of the two electrons are antiparallel, in contrast to orthohelium. Also spelled parhelium. { ‚par·ə'hē·lē·əm }

parahilgardite [MINERAL] $Ca_8(B_6O_1)_3Cl·4H_2O$ A triclinic mineral composed of hydrous borate and chloride of calcium; it is dimorphous with hilgardite. { ¦par·ə'hil·gär‚dīt }

parahopeite [MINERAL] $Zn_3(PO_4)_2·4H_2O$ A colorless mineral composed of hydrous phosphate of zinc, occurring in tabular triclinic crystals; it is dimorphous with hopeite. { ¦par·ə'hō‚pīt }

parahydrogen [ATOM PHYS] Those states of hydrogen molecules in which the spins of the two nuclei are antiparallel; known as spin isomers. { ¦par·ə'hī·drə·jən }

parainfluenza [MED] A viral condition similar to or resulting from influenza. [MICROBIO] An organism exhibiting growth characteristics of *Hemophilus influenzae*. { ¦par·ə‚in·flü'en·zə }

parakeet [VERT ZOO] Any of various small, slender species of parrots with long tails in the family Psittacidae. { 'par·ə‚kēt }

parakeratosis [PATH] Incomplete keratinization of epidermal cells characterized by retention of nuclei of cells attaining the level of the stratum corneum. { ¦par·ə‚ker·ə'tō·səs }

paralalia [MED] Disturbance of the faculty of speech, characterized by distortion of sounds, or the habitual substitution of one sound for another. { ‚par·ə'lāl·ē·ə }

paralaurionite [MINERAL] PbCl(OH) A white mineral composed of basic lead chloride; it is dimorphous with laurionite. { ‚par·ə'lòr·ē·ə‚nīt }

paraldehyde [ORG CHEM] $C_6H_{12}O_3$ Acetaldehyde polymer; colorless, flammable, toxic liquid, miscible with most organic solvents, soluble in water; melts at 12.6°C, boils at 124.5°C; used as a chemical intermediate, in medicine, and as a solvent. { pə'ral·də‚hīd }

paraldol [ORG CHEM] $(CH_3CHOHCH_2CHO)_2$ Water-soluble, white crystals, boiling at 90–100°C; used as a chemical intermediate, to make resins, and in cadmium plating baths. { par'al‚dòl }

paralexia [MED] Transposition or substitution of words or syllables in reading. { ‚par·ə'lek·sē·ə }

paralgesia [MED] **1.** Paresthesia characterized by pain. **2.** Any perverted and disagreeable cutaneous sensation, as of formication, cold, or burning. { ‚par·əl'jē·zē·ə }

paraliageosyncline [GEOL] A geosyncline developing along a present-day continental margin, such as the Gulf Coast geosyncline. { pə‚ral·yə‚jē·ō'sin‚klīn }

paralic [GEOL] Pertaining to deposits laid down on the landward side of a coast. { pə'ral·ik }

paralic coal basin [GEOL] Coal deposits formed along the margin of the sea. { pə'ral·ik 'kōl ‚bas·ən }

paralic swamp *See* marine swamp. { pə'ral·ik 'swämp }

paralimnion [HYD] The littoral part of a lake, extending from the margin to the deepest limit of rooted vegetation. { ‚par·ə'lim·nē‚än }

parallactic angle *See* position angle. { ¦par·ə¦lak·tik 'aŋ·gəl }

parallactic displacement [ASTRON] The apparent changes in the position of a star due to changes in the position of the earth as it moves around the sun. Also known as parallactic shift. { ¦par·ə¦lak·tik di'splās·mənt }

parallactic ellipse [ASTRON] An annual apparent elliptical course of a celestial body on the celestial sphere about its mean position; caused by the elliptical orbital motion of the earth. { ¦par·ə¦lak·tik i'lips }

parallactic equation [ASTRON] An inequality in the moon's motion caused by the sun's perturbing effect on the moon being greater in that half of the moon's apparent orbit around the earth when at new moon rather than at full moon. Also known as parallactic inequality. { ¦par·ə¦lak·tik i'kwā·zhən }

parallactic inequality *See* parallactic equation. { ¦par·ə¦lak·tik ‚in·i'kwäl·əd·ē }

parallactic motion [ASTRON] An apparent motion of stars away from the point in the celestial sphere toward which the sun is moving. { ¦par·ə¦lak·tik 'mō·shən }

parallactic orbit [ASTRON] The apparent orbit of a star as it appears to move once around in the sky each year; the motion is caused by the earth's orbital motion around the sun. { ¦par·ə¦lak·tik 'òr·bət }

parallactic shift *See* parallactic displacement. { ¦par·ə¦lak·tik 'shift }

parallax [OPTICS] The change in the apparent relative orientations of objects when viewed from different positions. { 'par·ə‚laks }

parallax age *See* age of parallax inequality. { 'par·ə‚laks ‚āj }

parallax correction [NAV] In celestial navigation, that sextant altitude correction made necessary because of the difference between the apparent direction from a point on the surface of the earth to a celestial body and the apparent direction from the center of the earth to the same body. { 'par·ə‚laks kə‚rek·shən }

parallax error [OPTICS] Error in reading an instrument employing a scale and pointer because the observer's eye and pointer are not in a line perpendicular to the plane of the scale. { 'par·ə‚laks ‚er·ər }

parallax in altitude [NAV] Geocentric parallax of a celestial body at any altitude; the term is used to distinguish the parallax at the given altitude from the horizontal parallax, when the body is in the horizon. { 'par·ə‚laks in 'al·tə‚tüd }

parallax inequality [OCEANOGR] The variation in the range of tide or in the speed of tidal currents due to the continual change in the distance of the moon from the earth. { 'par·ə‚laks ‚in·i'kwäl·əd·ē }

parallax-second *See* parsec. { 'par·ə‚laks ¦sek·ənd }

parallel [COMPUT SCI] Simultaneous transmission of, storage of, or logical operations on the parts of a word, character, or other subdivision of a word in a computer, using separate facilities for the various parts. [ELEC] Connected to the same pair of terminals. Also known as multiple; shunt. [GEOD] A circle on the surface of the earth, parallel to the plane of the equator and connecting all points of equal latitude. Also known as circle of longitude; parallel of latitude. [MATH] **1.** Lines are parallel in a euclidean space if they lie in a common plane and do not intersect. **2.** Planes are parallel in a Euclidean three-dimensional space if they do not intersect. **3.** A circle parallel to the primary great circle of a sphere or spheroid. **4.** A curve is parallel to a given curve C if it consists of points that are a fixed distance from C along lines perpendicular to C. [PHYS] Of two or more displacements or other vectors, having the same direction. { 'par·ə‚lel }

parallel access [COMPUT SCI] Transferral of information to or from a storage device in which all elements in a unit of information are transferred simultaneously. Also known as simultaneous access. { 'par·ə‚lel ‚ak‚ses }

parallel addition [COMPUT SCI] A method of addition by a computer in which all the corresponding pairs of digits of the addends are processed at the same time during one cycle, and one or more subsequent cycles are used for propagation and adjustment of any carries that may have been generated. { 'par·ə‚lel ə'dish·ən }

parallel algorithm [COMPUT SCI] An algorithm in which several computations are carried on simultaneously. { 'par·ə‚lel 'al·gə‚rith·əm }

parallel axiom [MATH] The axiom of an affine plane which states that if p and L are a point and line in the plane such that p is not on L, then there exists exactly one line that passes through p and does not intersect L. { ‚par·ə‚lel 'ak·sē·əm }

parallel axis theorem [MECH] A theorem which states that the moment of inertia of a body about any given axis is the moment of inertia about a parallel axis through the center of mass, plus the moment of inertia that the body would have about the given axis if all the mass of the body were located at the center of mass. Also known as Steiner's theorem. { 'par·ə‚lel ‚ak·səs ‚thir·əm }

parallel baffle muffler [DES ENG] A muffler constructed of a series of ducts placed side by side in which the duct cross section is a narrow but long rectangle. { 'par·ə‚lel ‚baf·əl 'məf·lər }

parallel buffer [ELECTR] Electronic device (magnetic core or flip-flop) used to temporarily store digital data in parallel, as opposed to series storage. { 'par·ə‚lel 'bəf·ər }

parallel by character [COMPUT SCI] The handling of all the characters of a machine word simultaneously in separate lines, channels, or storage cells. { 'par·ə‚lel bī 'kar·ik·tər }

parallel circuit [ELEC] An electric circuit in which the elements, branches (having elements in series), or components are connected between two points, with one of the two ends of each component connected to each point. { 'par·ə‚lel 'sər‚kət }

parallel communications [COMMUN] The simultaneous transmission of data over two or more communications channels. { 'par·ə‚lel kə‚myü·nə'kā·shənz }

parallel compensation See feedback compensation. { 'par·ə‚lel ‚käm·pən'sā·shən }

parallel computation [COMPUT SCI] The simultaneous computation of several parts of a problem. { 'par·ə‚lel ‚käm·pyü'tā·shən }

parallel computer [COMPUT SCI] **1.** A computer that can carry out more than one logic or arithmetic operation at one time. **2.** See parallel digital computer. { 'par·ə‚lel kəm'pyüd·ər }

parallel conversion [COMPUT SCI] The process of transferring operations from one computer system to another, during which both systems are run together for a period of time to ensure that they are producing identical results. { 'par·ə‚lel kən'vər·zhən }

parallel course computer See course-line computer. { 'par·ə‚lel ‚kȯrs kəm‚pyüd·ər }

parallel curves [MATH] Two curves such that one curve is the locus of points on the normals to the other curve at a fixed distance along the normals. { 'par·ə‚lel 'kərvz }

parallel cut [ENG] A group of parallel holes, not all charged with explosive, to create the initial cavity to which the loaded

holes break in blasting a development round. Also known as burn cut. { 'par·ə‚lel 'kət }

parallel digital computer [COMPUT SCI] Computer in which the digits are handled in parallel; mixed serial and parallel machines are frequently called serial or parallel, according to the way arithmetic processes are performed; an example of a parallel digital computer is one which handles decimal digits in parallel, although it might handle the bits constituting a digit either serially or in parallel. { 'par·ə‚lel 'dij·əd·əl kəm'pyüd·ər }

parallel displacement [MATH] A vector A at a point P of an affine space is said to be obtained from a vector B at a point Q of the space by a parallel displacement with respect to a curve connecting A and B if a vector $V(X)$ can be associated with each point X on the curve in such a manner that $A = V(P)$, $B = V(Q)$, and the values of V at neighboring points of the curve are parallel as specified by the affine connection. { 'par·ə‚lel di'splās·mənt }

parallel dot character printer See line dot matrix. { 'par·ə‚lel ‚dät 'kar·ik·tər ‚print·ər }

parallel drainage pattern [HYD] A drainage pattern characterized by regularly spaced streams flowing parallel to one another over a large area. { 'par·ə‚lel 'drān·ij ‚pad·ərn }

parallel drum [DES ENG] A cylindrical form of drum on which the haulage or winding rope is coiled. { 'par·ə‚lel 'drəm }

parallel edges [MATH] Two or more edges that join the same pair of vertices in a graph. Also known as multiple edges. { ‚par·ə‚lel 'ej·əz }

parallel element-processing ensemble [COMPUT SCI] A powerful electronic computer used by the U.S. Army to simulate tracking and discrimination of reentry vehicles as part of the ballistic missile defense research program. Abbreviated PEPE. { 'par·ə‚lel 'el·ə·mənt ‚prä‚ses·iŋ än‚säm·bəl }

parallel entry [MIN ENG] An intake airway parallel to the haulageway. { 'par·ə‚lel 'en·trē }

parallelepiped [MATH] A polyhedron all of whose faces are parallelograms. { ‚par·ə‚lel·ə'pī·pəd }

parallel evolution [EVOL] Evolution of similar characteristics in different groups of organisms. { 'par·ə‚lel ‚ev·ə'lü·shən }

parallel extinction [OPTICS] Nearly total absorption of light that is propagating in an anisotropic crystal in a direction parallel to crystal outlines or traces of cleavage planes. { 'par·ə‚lel ik'stiŋk·shən }

parallel feed [COMPUT SCI] See sideways feed. [ELECTR] Application of a direct-current voltage to the plate or grid of a tube in parallel with an alternating-current circuit, so that the direct-current and the alternating-current components flow in separate paths. Also known as shunt feed. { 'par·ə‚lel 'fēd }

parallel firing [ENG] A method of connecting together a number of detonators which are to be fired electrically in one blast. { 'par·ə‚lel 'fīr·iŋ }

parallel flow [ELEC] Also known as loop flow. **1.** The flow of electric current from one point to another in an electric network over multiple paths, in accordance with Kirchhoff's laws. **2.** In particular, the flow of electric current through electric power systems over paths other than the contractual path. { 'par·ə‚lel 'flō }

parallel fold See concentric fold. { 'par·ə‚lel 'fōld }

parallel gripper [CONT SYS] A robot end effector made up of two jawlike components that grasp objects. { 'par·ə‚lel 'grip·ər }

parallel growth See parallel intergrowth. { 'par·ə‚lel 'grōth }

parallel impedance [ELEC] One of two or more impedances that are connected to the same pair of terminals. { 'par·ə‚lel im'pēd·əns }

paralleling reactor [ELECTROMAG] Reactor for correcting the division of load between parallel-connected transformers which have unequal impedance voltages. { 'par·ə‚lel·iŋ rē'ak·tər }

parallel input/output [COMPUT SCI] Data that are transmitted into and out of a computer over several conductors simultaneously. { 'par·ə‚lel 'in‚pút 'aút‚pút }

parallel interface [ELECTR] A link between two devices in which all the information transferred between them is transmitted simultaneously over separate conductors. Also known as parallel port. { 'par·ə‚lel 'in·tər‚fās }

parallel intergrowth [CRYSTAL] Intergrowth of two or more

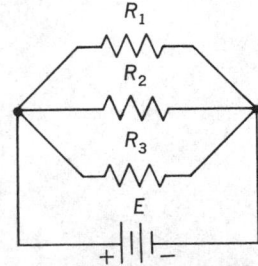

crystals in such a way that one or more axes in each crystal are approximately parallel. Also known as parallel growth. { 'par·ə,lel 'in·tər,grōth }

parallel laminate [MATER] A laminate in which all the layers of material are set approximately parallel with respect to a particular characteristic, such as the grain or the direction of tension. { 'par·ə,lel 'lam·ə·nət }

parallel linkage [MECH ENG] An automotive steering system that has a short idler arm mounted parallel to the pitman arm. { 'par·ə,lel 'liŋ·kij }

parallel middle body See dead flat. { 'par·ə,lel 'mid·əl 'bäd·ē }

parallel motion protractor [GRAPHICS] An instrument consisting essentially of a protractor and one or more arms attached to a parallel motion device, so that the movement of the arms is everywhere parallel; the protractor can be rotated and set at any position so that it can be oriented to a chart. Also known as drafting machine. { 'par·ə,lel 'mō·shən 'prō,trak·tər }

parallel muscle [ANAT] Any muscle having the long fibers arranged parallel to each other. { 'par·ə,lel 'məs·əl }

parallel of altitude [ASTRON] A circle on the celestial sphere parallel to the horizon connecting all points of equal altitude. Also known as almucantar; altitude circle. { 'par·ə,lel əv 'al·tə,tüd }

parallel of declination [ASTRON] A small circle of the celestial sphere parallel to the celestial equator. Also known as celestial parallel; circle of equal declination. { 'par·ə,lel əv ,dek·lə'nā·shən }

parallel of latitude See circle of longitude; parallel. { 'par·ə,lel əv 'lad·ə,tüd }

parallelogram [MATH] A four-sided polygon with each pair of opposite sides parallel. { ,par·ə'lel·ə,gram }

parallelogram law [MATH] The rule that the sum of two vectors is the diagonal of a parallelogram whose sides are the vectors to be added. { ,par·ə'lel·ə,gram ,lo }

parallelogram of vectors [MATH] A parallelogram whose sides form two vectors to be added and whose diagonal is the sum of the two vectors. { ,par·ə'lel·ə,gram əv 'vek·tərz }

parallel operation [COMPUT SCI] Performance of several actions, usually of a similar nature, by a computer system simultaneously through provision of individual similar or identical devices. [ELECTR] The connecting together of the outputs of two or more batteries or other power supplies so that the sum of their output currents flows to a common load. { 'par·ə,lel ,äp·ə'rā·shən }

parallelotope [MATH] A parallelepiped with sides in proportion of 1, 1/2, and 1/4. { ,par·ə'lel·ə,tōp }

parallel padding [ELEC] Method of parallel operation for two or more power supplies in which their current limiting or automatic crossover output characteristic is employed so that each supply regulates a portion of the total current, each parallel supply adding to the total and padding the output only when the load current demand exceeds the capability, or limit setting, of the first supply. { 'par·ə,lel 'pad·iŋ }

parallel perspective [GRAPHICS] A perspective such that the faces and edges of represented objects are either perpendicular or parallel to the plane of the picture. Also known as one-point perspective. { 'par·ə,lel pər'spek·tiv }

parallel-plate capacitor [ELEC] A capacitor consisting of two parallel metal plates, with a dielectric filling the space between them. { 'par·ə,lel 'plāt kə'pas·əd·ər }

parallel-plate laser [OPTICS] A laser which has two small parallel plates facing each other at a distance which is large compared with their diameters; one of them reflects light and the other is partially reflecting, so that light can bounce back and forth between the plates enough to build up a strong pulse. { 'par·ə,lel 'plāt 'lā·zər }

parallel-plate reactor [ENG] A type of plasma reactor in which a process gas is introduced into the space between two closely spaced parallel plane electrodes, and a plasma, generated by a radio-frequency excitation applied to the electrodes, acts directly on substrates placed on either electrode. { 'par·ə,lel 'plāt rē,ak·tər }

parallel-plate waveguide [ELECTROMAG] Pair of parallel conducting planes used for propagating uniform circularly cylindrical waves having their axes normal to the plane. { 'par·ə,lel 'plāt 'wāv,gīd }

parallel port See parallel interface. { 'par·ə,lel ,pórt }

parallel printer [GRAPHICS] A printer that is designed to

be connected to a parallel port of a computer. { 'par·ə,lel 'print·ər }

parallel processing [PSYCH] The processing of several pieces of information at the same time. { 'par·ə,lel 'präs·ə·siŋ }

parallel processor See multiprocessor. { 'par·ə,lel 'prä,ses·ər }

parallel programming [COMPUT SCI] A method for performing simultaneously the normally sequential steps of a computer program, using two or more processors. [ELECTR] Method of parallel operation for two or more power supplies in which their feedback terminals (voltage control terminals) are also paralleled; these terminals are often connected to a separate programming source. { 'par·ə,lel 'prō,gram·iŋ }

parallel projection [MATH] A central projection in which the center of projection is the point at infinity, so that the projectors are parallel; equivalent to an orthogonal projection. { 'par·ə,lel prə'jek·shən }

parallel radio tap [COMMUN] A telephone tapping procedure in which a battery-powered miniature radio transmitter is bridged across the target pair. { 'par·ə,lel 'rād·ē·ō ,tap }

parallel rays [MATH] **1.** Two rays lying on the same line or on parallel lines. **2.** Two rays that lie on the same line or on parallel lines, and point in the same direction. { 'par·ə,lel 'rāz }

parallel rectifier [ELECTR] One of two or more rectifiers that are connected to the same pair of terminals, generally in series with small resistors or inductors, when greater current is desired than can be obtained with a single rectifier. { 'par·ə,lel 'rek·tə,fī·ər }

parallel reliability [SYS ENG] Property of a system composed of functionally parallel elements in such a way that if one of the elements fails, the parallel units will continue to carry out the system function. { 'par·ə,lel ri'lī·ə,bil·əd·ē }

parallel representation [COMPUT SCI] The simultaneous appearance of the different bits of a digital variable on parallel bus lines. { 'par·ə,lel ,rep·ri,zen'tā·shən }

parallel resonance [ELEC] Also known as antiresonance. **1.** The frequency at which the inductive and capacitive reactances of a parallel resonant circuit are equal. **2.** The frequency at which the parallel impedance of a parallel resonant circuit is a maximum. **3.** The frequency at which the parallel impedance of a parallel resonant circuit has a power factor of unity. { 'par·ə,lel 'rez·ən·əns }

parallel resonant circuit [ELEC] A circuit in which an alternating-current voltage is applied across a capacitor and a coil in parallel. Also known as antiresonant circuit. { 'par·ə,lel 'rez·ən·ənt ,sər·kət }

parallel resonant interstage [ELECTR] A coupling between two amplifier stages achieved by means of a parallel-tuned LC circuit. { 'par·ə,lel 'rez·ən·ənt 'in·tər,stāj }

parallel ripple mark [GEOL] A ripple mark characterized by a relatively straight crest and an asymmetric profile. { 'par·ə,lel 'rip·əl ,märk }

parallel roads [GEOL] A series of horizontal beaches or wave-cut terraces occurring parallel to each other at different levels on each side of a glacial valley. { 'par·ə,lel 'rōdz }

parallel-rod oscillator [ELECTR] Ultra-high-frequency oscillator circuit in which parallel rods or wires of required length and dimensions form the tank circuits. { 'par·ə,lel 'räd 'äs·ə,lād·ər }

parallel rulers [GRAPHICS] An instrument for transferring a line parallel to itself; in its most common form it consists essentially of two parallel bars or rulers connected in such manner that when one is held in place, the other may be moved, remaining parallel to its original position. { 'par·ə,lel 'rü·lərz }

parallel running [COMPUT SCI] **1.** The running of a newly developed system in a data-processing area in conjunction with the continued operation of the current system. **2.** The final step in the debugging of a system; this step follows a system test. { 'par·ə,lel 'rən·iŋ }

parallels [ENG] **1.** Spacers located between steam plate and press platen of the mold to prevent bending of the middle section. **2.** Spacers or pressure pods located between steam plates of a mold to regulate height and prevent crushing of mold parts. { 'par·ə lelz }

parallel sailing [NAV] A method of converting departure

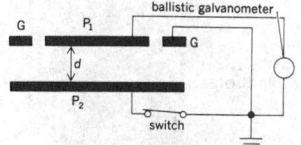

Cross section of parallel-plate capacitor with all parts of capacitor at ground potential. P_1 and P_2 are the parallel plates; G is the guard ring, used to reduce edge effects; d is the distance between plates. Distance between guard ring and plate P_1 is exaggerated.

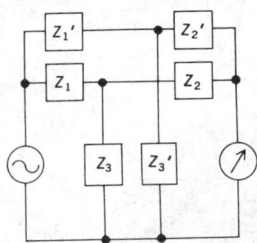

Schematic drawing of the network showing the two sets of three impedances: Z_1, Z_2, Z_3, and Z_1', Z_2', Z_3'.

PARAMAGNETIC FARADAY EFFECT

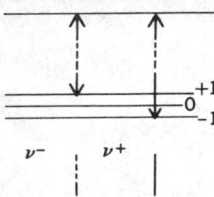

Schematic energy level diagram of ions of a paramagnetic salt. The lengths of the arrows are proportional to the frequencies of absorption lines for right-handed ($\nu+$) and left-handed ($\nu-$) circularly polarized light.

PARAMECIUM

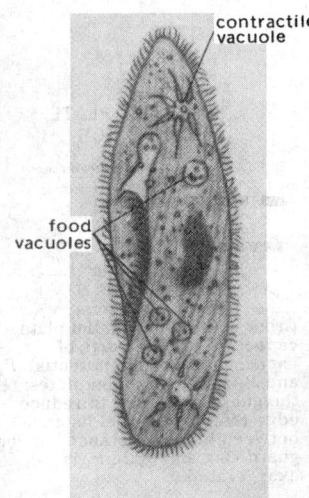

contractile vacuole

food vacuoles

A paramecium. Two contractile vacuoles and numerous food vacuoles are shown.

into difference of longitude, or vice versa, when the true course is 090° or 270°. { 'par·ə,lel 'säl·iŋ }

parallel search storage [COMPUT SCI]   A device for very rapid search of a volume of stored data to permit finding a specific item. { 'par·ə,lel 'sərch ,stòr·ij }

parallel series [ELEC]   Circuit in which two or more parts are connected together in parallel to form parallel circuits, and in which these circuits are then connected together in series so that both methods of connection appear. [ENG]   *See* multiple series. { 'par·ə,lel 'sir·ēz }

parallel shot [ENG]   In seismic prospecting, a test shot which is made with all the amplifiers connected in parallel and activated by a single geophone so that lead, lag, polarity, and phasing in the amplifier-to-oscillograph circuits can be checked. { 'par·ə,lel 'shät }

parallel-slit interferometer [OPTICS]   A type of stellar interferometer consisting of a screen with two narrow, parallel slits whose separation is adjustable, placed over the objective of a refracting telescope. { 'par·ə,lel ,slit ,in·tər·fə'räm·əd·ər }

parallel sphere [ASTRON]   The celestial sphere as it appears to an observer at the pole, where celestial bodies appear to move parallel to the horizon. { 'par·ə,lel 'sfir }

parallel storage [COMPUT SCI]   A storage device in which words (or characters or digits) can be read in or out simultaneously. { 'par·ə,lel 'stòr·ij }

parallel surfaces [MATH]   Two surfaces such that one surface is the locus of points on the normals to the other curve at a fixed distance along the normals. { 'par·ə,lel 'sər·fəs·əz }

parallel texture [PETR]   A rock texture characterized by tabular-to-prismatic crystals oriented parallel to a plane or line. { 'par·ə,lel 'teks·chər }

parallel-T network [ELEC]   A network used in capacitance measurements at radio frequencies, having two sets of three impedances, each in the form of the letter T, with the arms of the two T's joined to common terminals, and the source and detector each connected between two of these terminals.   Also known as twin-T network. { 'par·ə,lel ¦tē 'net,wərk }

parallel transfer [COMPUT SCI]   Simultaneous transfer of all bits in a storage location constituting a character or word. { 'par·ə,lel 'tranz·fər }

parallel transmission [COMPUT SCI]   The transmission of characters of a word over different lines, usually simultaneously; opposed to serial transmission. { 'par·ə,lel tranz'mish·ən }

parallel-tuned circuit [ELEC]   A circuit with two parallel branches, one having an inductance and a resistance in series, the other a capacitance and a resistance in series. { 'par·ə,lel ,tünd 'sər·kət }

parallel twin [CRYSTAL]   A twinned crystal whose twin axis is parallel to the composition surface. { 'par·ə,lel 'twin }

parallel vectors [MATH]   **1.** Two nonzero vectors such that one vector equals the product of the other vector and a nonzero scalar.   **2.** Two nonzero vectors in a vector space over the real numbers such that one vector equals the product of the other vector and a positive number. { 'par·ə,lel 'vek·tərz }

parallel-veined [BOT]   Of a leaf, having the veins parallel, or nearly parallel, to each other. { 'par·ə,lel ¦vānd }

parallel venation [BOT]   A vascular arrangement in leaves characterized by the longitudinal (or nearly so) orientation of veins of relatively uniform size. { ,par·ə,lel və'nā·shən }

parallel wire method [MIN ENG]   An electrical prospecting method employing equipotential lines or curves in searching for ore bodies. { 'par·ə,lel 'wīr ,meth·əd }

parallel wires [ELEC]   Two conductors which are parallel to each other; often used in transmission lines. { 'par·ə,lel 'wīrz }

parallochthon [GEOL]   Rocks that were brought from intermediate distances and deposited near an allochthonous mass during transit. { ,par·ə'läk,thän }

paralogous locus [GEN]   A gene that arose by duplication and later diverged in sequence or location from the parent gene. { pə'ral·ə·gəs ,lō·kəs }

paralutein cells [HISTOL]   Epithelioid cells of the corpus luteum. { ¦par·ə'lüd·ē·ən ,selz }

paralysis [MED]   Complete or partial loss of motor or sensory function. { pə'ral·ə·səs }

paralysis agitans *See* parkinsonism. { pə'ral·ə·səs 'aj·ə,tanz }

paralytic secretion [PHYSIO]   Glandular secretion occurring in a denervated gland. { ¦par·ə¦lid·ik si'krē·shən }

paralytic spinal poliomyelitis [MED]   An acute inflammatory virus disease chiefly involving the anterior horns of the gray matter of the spinal cord. { ¦par·ə¦lid·ik 'spīn·əl ¦pō·lē·ō,mī·ə'līd·əs }

paramagnetic [ELECTROMAG]   Exhibiting paramagnetism. { ¦par·ə·mag'ned·ik }

paramagnetic alloy [MET]   An alloy whose permeability is slightly greater than that of vacuum and is independent of the magnetic field strength, such as intermetallic compounds of nickel and titanium. { ¦par·ə·mag'ned·ik 'al,ói }

paramagnetic amplifier *See* maser. { ¦par·ə·mag'ned·ik 'am·plə,fī·ər }

paramagnetic analytical method [ANALY CHEM]   A method for analyzing fluid mixtures by measurement of the paramagnetic (versus diamagnetic) susceptibilities of materials when exposed to a magnetic field. { ¦par·ə·mag'ned·ik ,an·ə'lid·ə·kəl 'meth·əd }

paramagnetic cooling *See* adiabatic demagnetization. { ¦par·ə·mag'ned·ik 'kül·iŋ }

paramagnetic crystal [ELECTROMAG]   A crystal whose permeability is slightly greater than that of vacuum and is independent of the magnetic field strength. { ¦par·ə·mag'ned·ik 'krist·əl }

paramagnetic Faraday effect [OPTICS]   The Faraday effect observed in paramagnetic salts at frequencies near an absorption line of the salt which is split due to splitting of the lower energy level.   Also known as Becquerel effect. { ¦par·ə·mag'ned·ik 'far·ə,dā i,fekt }

paramagnetic iron [MET]   Iron which has been transformed from a ferromagnetic to a paramagnetic substance by application of a pressure somewhat greater than 10^5 bars (10^{10} pascals). { ¦par·ə·mag'ned·ik 'ī·ərn }

paramagnetic material [ELECTROMAG]   A material within which an applied magnetic field is increased by the alignment of electron orbits. { ¦par·ə·mag'ned·ik mə'tir·ē·əl }

paramagnetic relaxation [ELECTROMAG]   The approach of a system, which displays paramagnetism because of electronic magnetic moments of atoms or ions, to an equilibrium or steady-state condition over a period of time, following a change in the magnetic field. { ¦par·ə·mag'ned·ik ,rē,lak'sā·shən }

paramagnetic resonance *See* electron paramagnetic resonance. { ¦par·ə·mag'ned·ik 'rez·ən·əns }

paramagnetic salt [ELECTROMAG]   A salt whose permeability is slightly greater than that of vacuum and is independent of magnetic field strength; used in adiabatic demagnetization. { ¦par·ə·mag'ned·ik 'sólt }

paramagnetic spectra [SPECT]   Spectra associated with the coupling of the electronic magnetic moments of atoms or ions in paramagnetic substances, or in paramagnetic centers of diamagnetic substances, to the surrounding liquid or crystal environment, generally at microwave frequencies. { ¦par·ə·mag'ned·ik 'spek·trə }

paramagnetic susceptibility [ELECTROMAG]   The susceptibility of a paramagnetic substance, which is a positive number and is, in general, much smaller than unity. { ¦par·ə·mag'ned·ik sə,sep·tə'bil·əd·ē }

paramagnetism [ELECTROMAG]   A property exhibited by substances which, when placed in a magnetic field, are magnetized parallel to the field to an extent proportional to the field (except at very low temperatures or in extremely large magnetic fields). { ¦par·ə'mag·nə,tiz·əm }

Parameciidae [INV ZOO]   A family of ciliated protozoans in the order Holotrichia; the body has differentiated anterior and posterior ends and is bounded by a hard but elastic pellicle. { ,par·ə·mə'sī·ə,dē }

paramecium [INV ZOO]   A single-celled protozoan belonging to the family Parameciidae. { ,par·ə'mē·sē·əm }

Paramecium [INV ZOO]   The genus of protozoans composing the family Parameciidae. { ,par·ə'mē·sē·əm }

paramedical [MED]   Having a supplementary or secondary relation to medicine. { ¦par·ə'med·ə·kəl }

paramelaconite [MINERAL]   A black tetragonal mineral composed of cupric and cuprous oxides, occurring in pyramidal crystals. { ¦par·ə·mə'lak·ə,nīt }

Paramelina [VERT ZOO]   An order of marsupials that includes the bandicoots in some systems of classification. { ,par·ə'mel·ə·nə }

paramere [BIOL] One half of a bilaterally symmetrical animal or somite. [INV ZOO] Any of several paired structures of an insect, especially those on the ninth abdominal segment. { 'par·ə,mir }

paramesonephric duct [EMBRYO] An embryonic genital duct; in the female, it is the anlage of the oviducts, uterus, and vagina; in the male, it degenerates, leaving the appendix testes. Also known as Müllerian duct. { ¦par·ə·me·zə¦nef·rik 'dəkt }

parameter [CRYSTAL] Any of the axial lengths or interaxial angles that define a unit cell. [ELEC] **1.** The resistance, capacitance, inductance, or impedance of a circuit element. **2.** The value of a transistor or tube characteristic. [MATH] An arbitrary constant or variable so appearing in a mathematical expression that changing it gives various cases of the phenomenon represented. [PHYS] A quantity which is constant under a given set of conditions, but may be different under other conditions. { pə'ram·əd·ər }

parameter-driven system [COMPUT SCI] A software system whose functions and operations are controlled mainly by parameters. { pə'ram·əd·ər ¦driv·ən 'sis·təm }

parameter identification [SYS ENG] The problem of estimating the values of the parameters that govern a dynamical system from data on the observed behavior of the system. { pə'ram·əd·ər ī,dent·ə·fə'kā·shən }

parameterization [SCI TECH] The representation, in a dynamic model, of physical effects in terms of admittedly oversimplified parameters, rather than realistically requiring such effects to be consequences of the dynamics of the system. { pə,ram·əd·ər·ə'zā·shən }

parameter of distribution [MATH] For a fixed line on a ruled surface, a quantity whose magnitude is the limit, as a variable line on the surface approaches the fixed line, of the ratio of the minimum distance between the two lines to the angle between them; and whose sign is positive or negative according to whether the motion of the tangent plane to the surface is left- or right-handed as the point of tangency moves along the fixed line in a positive direction. { pə'ram·əd·ər əv ,dis·trə'byü·shən }

parameter tags [COMPUT SCI] Constants that are used by several computer programs. { pə'ram·əd·ər ,tagz }

parameter word [COMPUT SCI] A word in a computer storage containing one or more parameters that specify the action of a routine or subroutine. { pə'ram·əd·ər ,wərd }

paramethadione [PHARM] $C_7H_{11}NO_3$ An anticonvulsant primarily useful in the treatment of petit mal epilepsy. { ¦par·ə¦meth·ə'dī,ōn }

parametric acoustic array [ACOUS] A device for generating very sharp beams of sound devoid of side lobes, consisting of a source of well-collimated high-frequency sound modulated at the frequency of the sound which is to be generated. { ¦par·ə¦me·trik ə'küs·tik ə'rā }

parametric amplifier [ELECTR] A highly sensitive ultrahigh-frequency or microwave amplifier having as its basic element an electron tube or solid-state device whose reactance can be varied periodically by an alternating-current voltage at a pumping frequency. Also known as mavar; paramp; reactance amplifier. [OPTICS] A device consisting of an optically nonlinear crystal in which an optical or infrared beam draws power from a laser beam at a higher frequency and is amplified. { ¦par·ə¦me·trik 'am·plə,fī·ər }

parametric converter [ELECTR] Inverting or noninverting parametric device used to convert an input signal at one frequency into an output signal at a different frequency. { ¦par·ə¦me·trik kən'vərd·ər }

parametric curves [MATH] On a surface determined by equations $x = f(u,v)$, $y = g(u,v)$, and $z = h(u,v)$, these are families of curves obtained by setting the parameters u and v equal to various constants. { ¦par·ə¦me·trik 'kərvz }

parametric device [ELECTR] Electronic device whose operation depends essentially upon the time variation of a characteristic parameter usually understood to be a reactance. { ¦par·ə¦me·trik di'vīs }

parametric down-converter [ELECTR] Parametric converter in which the output signal is at a lower frequency than the input signal. { ¦par·ə¦me·trik 'daün,vərd·ər }

parametric equalizer [ENG ACOUS] A device that allows control over the center frequencies, bandwidths, and amplitudes (parameters) of band-pass filters that determine the frequency response of audio equipment. { ¦par·ə¦me·trik ,ē·kwə'līz·ər }

parametric equation [MATH] An equation where coordinates of points appear dependent on parameters such as the parametric equation of a curve or a surface. { ¦par·ə¦me·trik i'kwā·zhən }

parametric excitation [ENG] The method of exciting and maintaining oscillations in either an electrical or mechanical dynamic system, in which excitation results from a periodic variation in an energy storage element in a system such as a capacitor, inductor, or spring constant. { ¦par·ə¦me·trik ,ek·si'tā·shən }

parametric generation [OPTICS] A process in which a single electromagnetic wave propagating in a nonlinear medium is converted to two lower-frequency waves, the sum of whose frequencies equals the frequency of the original wave. { ¦par·ə¦me·trik ,jen·ə'rā·shən }

parametric hydrology [HYD] That branch of hydrology dealing with the development and analysis of relationships among the physical parameters involved in hydrologic events and the use of these relationships to generate, or synthesize, hydrologic events. { ¦par·ə¦me·trik hī'dräl·ə·jē }

parametric latitude *See* reduced latitude. { ¦par·ə¦me·trik 'lad·ə,tüd }

parametric mixing [OPTICS] In a medium possessing optical nonlinearities, the mixing of electromagnetic waves to form waves with frequencies linearly related to the frequency of incident radiation. { ¦par·ə¦me·trik 'mik,siŋ }

parametric oscillator [ELECTR] An oscillator in which the reactance parameter of an energy-storage device is varied to obtain oscillation. [OPTICS] A device consisting of an optically nonlinear crystal surrounded by a pair of mirrors to which is applied a relatively high-frequency laser beam and a relatively low-frequency signal, resulting in a low-frequency output whose frequency can be varied, usually by varying the indices of refraction. { ¦par·ə¦me·trik 'äs·ə,lād·ər }

parametric phase-locked oscillator *See* parametron. { ¦par·ə¦me·trik 'fāz ,läkt 'äs·ə,lād·ər }

parametric programming [COMPUT SCI] A programming approach in which data are stored in external tables or files, rather than within the program itself, and accessed by the program when needed, so that the values of these data can be changed with relative ease. { ¦par·ə¦me·trik 'prō,gram·iŋ }

parametric up-converter [ELECTR] Parametric converter in which the output signal is at a higher frequency than the input signal. { ¦par·ə¦me·trik 'əp kən,vərd·ər }

parametrized voice response system [ENG ACOUS] A voice response system which first extracts informative parameters from human speech, such as natural resonant frequencies (formants) of the speaker's vocal tract and the fundamental frequency (pitch) of the voice, and which later reconstructs speech from such stored parameters. { pə'ram·ə,trīzd 'vois ri,späns ,sis·təm }

parametron [ELECTR] A resonant circuit in which either the inductance or capacitance is made to vary periodically at one-half the driving frequency; used as a digital computer element, in which the oscillation represents a binary digit. Also known as parametric phase-locked oscillator; phase-locked oscillator; phase-locked subharmonic oscillator. { pə'ram·ə,trän }

paramo [ECOL] A biological community, essentially a grassland, covering extensive high areas in equatorial mountains of the Western Hemisphere. { 'pär·ə,mō }

paramorph [MINERAL] A mineral exhibiting paramorphism. { 'par·ə,morf }

paramorphism [MINERAL] The property of a mineral whose internal structure has changed without change in composition or external form. Formerly known as allomorphism. { ¦par·ə'mor,fiz·əm }

paramp *See* parametric amplifier. { 'par,amp }

paramutation [GEN] A mutation in which one member of a heterozygous pair of alleles permanently changes its partner allele. { ¦par·ə·myü'tā·shən }

paramylum [BIOCHEM] A reserve, starchlike carbohydrate of various protozoans and algae. { pə'ram·ə·ləm }

paramyosin [BIOCHEM] A type of fibrous protein found in the adductor muscles of bivalves and thought to form the core of a filament with myosin molecules at the surface. { ¦par·ə'mī·ə·sən }

paramyotonia congenita [MED] A heredofamilial condition characterized by recurrent muscular stiffness and weakness (myotonia) on exposure to cold, as well as on mechanical

---

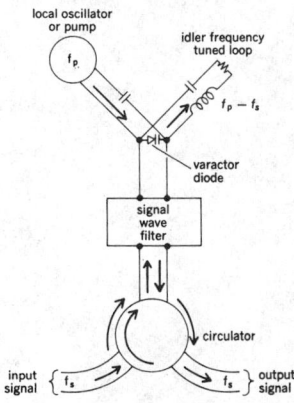

Schematic of negative-resistance-type parametric amplifier that used a varactor diode as the variable reactor. f_p = pumping frequency. Input signal, with frequency f_s, is reflected back down transmission line with increased power. Wave at idler frequency $f_p - f_s$ is not utilized outside the amplifier.

PARAPHASE AMPLIFIER

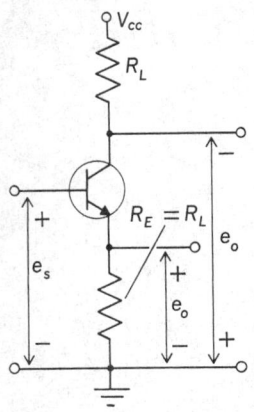

Circuit diagram of single-transistor-inverter, the simplest paraphase amplifier.
V_{cc} = collector supply voltage,
R_E = emitter resistance,
R_L = collector load,
e_s = input signal voltage,
e_o = output voltages, equal in magnitude and 180° out of phase.

irritation; transmitted as an autosomal dominant and considered to be a variety of the hyperkalemic form of periodic paralysis. Also known as Eulenburg's disease; myotonia congenita intermittens. { ‚par·ə‚mī·ə'tō·nē·ə kən'jen·əd·ə }

Paramyxoviridae [VIROL] A family of negative-strand ribonucleic acid (RNA) viruses characterized by an enveloped spherical virion containing a single-stranded, nonfragmented molecule of RNA; contains the genera *Paramyxovirus* (sendai, mumps), *Morbillivirus* (measles), and *Pneumovirus* (respiratory syncytial virus). { ‚par·ə‚mik·sə'vir·ə‚dī }

paramyxovirus [VIROL] A subgroup of myxoviruses, including the viruses of mumps, measles, parainfluenza, and Newcastle disease; all are ribonucleic acid-containing viruses and possess an ether-sensitive lipoprotein envelope. { ¦par·ə‚mik·sō'vī·rəs }

paranasal sinus [ANAT] Any of the paired sinus cavities of the human face; includes the frontal, ethmoid, and sphenoid sinuses. { ¦par·ə¦nā·zəl 'sī·nəs }

paranephritis [MED] **1.** Inflammation of the adrenal gland. **2.** Inflammation of the connective tissue adjacent to the kidney. { ¦par·ə·nə'frīd·əs }

paranesthesia [MED] Anesthesia of the body below the waist. { pə¦ran·əs'thē·zhə }

paranitraniline red *See* para red. { ¦par·ə·nə'tran·ə·lən 'red }

paranoia [PSYCH] A rare form of paranoid psychosis characterized by the slow development of a complex, internally logical system of persecutory or grandiose delusions. { ‚par·ə'nói·ə }

paranoid personality [PSYCH] An individual characterized by the tendency to be hypersensitive, rigid, extremely self-important, and jealous, and to project hostile feelings so that he or she easily becomes suspicious of others and is quick to blame them or attribute evil motives to them. { 'par·ə‚nóid ‚pər·sə'nal·əd·ē }

paranoid schizophrenia [PSYCH] A form of schizophrenia in which delusions of persecution or grandeur (or both), hallucinations, and ideas of reference predominate and sometimes are systematized. { 'par·ə‚nóid ‚skit·sə'frē·nē·ə }

paranoid state [PSYCH] Any of various psychotic disorders in which the patient exhibits persistent delusions, usually of persecution or grandeur. { 'par·ə‚nóid ‚stāt }

paranosmia [MED] A deviation in odor sensitivity involving change in odor quality. { ‚par·ə'näz·mē·ə }

paranthelion [ASTRON] A refraction phenomenon similar to a parhelion, but occurring generally at a distance of 120° (occasionally 90° and 140°) from the sun, on the parhelic circle. Also known as mock sun. { ‚par·ən'thē·lē‚än }

paranthropophytia [ECOL] Discrepant control of regions or areas due to immediate and continuous or periodic interference, as by certain cultivation practices. { ‚par·an‚thräp·ə'ff·shə }

Paranyrocidae [PALEON] An extinct family of birds in the order Anseriformes, restricted to the Miocene of South Dakota. { pə‚ran·ə'räs·ə‚dē }

Paraonidae [INV ZOO] A family of small, slender polychaete annelids belonging to the Sedenteria. { ¦par·ə'än·ə‚dē }

parapack [ENG] A package or bundle with a parachute attached for dropping from an aircraft. { 'par·ə‚pak }

Paraparchitacea [PALEON] A superfamily of extinct ostracods in the suborder Kloedenellocopina including nonsulcate, nondimorphic forms. { ¦par·ə‚pär·kə'tās·ē·ə }

paraparesis [MED] Partial paralysis of the lower extremities. { ¦par·ə·pə'rē·səs }

parapatric [ECOL] Referring to populations or species that occupy nonoverlapping but adjacent geographical areas without interbreeding. { ¦par·ə¦pa·trik }

parapatric speciation [EVOL] Gradual speciation whereby new species are created from populations that maintain overlapping geographic zones of genetic contact. { ¦par·ə¦pa·trik ‚spē·shē'ā·shən }

parapertussis [MED] An acute bacterial respiratory infection similar to mild pertussis and caused by *Bordetella pertussis*. { ¦par·ə·pər'təs·əs }

parapet [ARCH] A low retaining wall at the edge of a roof, bridge, porch, or other structure. [ORD] An elevation of earth or material which is thrown up in front of a trench or emplacement to protect the occupants from fire and observation, and over which fire may be delivered. { 'par·ə·pət }

paraphase amplifier [ELECTR] An amplifier that provides two equal output signals 180° out of phase. { 'par·ə‚fāz 'am·plə‚fī·ər }

paraphasia [PSYCH] The erroneous production of unintended words in speech that is a feature in some forms of aphasia. { ‚par·ə'fā·zhə }

paraphimosis [MED] **1.** Retraction and constriction, especially of the prepuce behind the glans penis. Also known as Spanish collar. **2.** Retraction of the eyelid behind the eyeball. { ‚par·ə·fə'mō·səs }

paraphyletic [SYST] Describing a taxonomic group that does not contain all of the descendants of its most recent common ancestor. { ‚par·ə·fī'led·ik }

paraphysis [BOT] A sterile filament borne among the sporogenous or gametogenous organs in many cryptogams. [VERT ZOO] A median evagination of the roof of the telencephalon of some lower vertebrates. { pə'raf·ə·səs }

paraplasm *See* hyaloplasm. { 'par·ə‚plaz·əm }

paraplegia [MED] Paralysis of the lower limbs. { ‚par·ə'plē·jə }

parapodium [INV ZOO] **1.** One of the short, paired processes on the sides of the body segments in certain annelids. **2.** A lateral expansion of the foot in gastropod mollusks. { ‚par·ə'pōd·ē·əm }

parapolar cell [INV ZOO] Either of the first two trunk cells in the development of certain Mesozoa. { ‚par·ə¦pō·lər 'sel }

parapositronium [PARTIC PHYS] The state of positronium in which the positron and electron have antiparallel spins. { ‚par·ə‚päz·ə'trō·nē·əm }

parapsychology [PSYCH] The study of the phenomena of extrasensory perception and psychokinesis. { ¦par·ə·sī'käl·ə‚jē }

paraquat [ORG CHEM] [CH₃(C₅H₄N)₂CH₃]·2CH₃SO₄ A yellow, water-soluble solid, used as a herbicide. { 'par·ə‚kwät }

pararammelsbergite [MINERAL] NiAs₂ A tin white, orthorhombic or pseudoorthorhombic mineral consisting of nickel diarsenide; occurrence is usually in massive form. { ¦par·ə'ram·əlz‚bər‚gīt }

para red [ORG CHEM] $C_{10}H_6(OH)NNC_6H_4NO_2$ Red pigment derived from the coupling of β-naphthol with diazotized paranitroaniline. Also known as paranitraniline red. { 'par·ə 'red }

pararipple [GEOL] A large, symmetric ripple whose surface slopes gently and which shows no assortment of grains. { 'par·ə‚rip·əl }

pararosaniline [ORG CHEM] $HOC(C_6H_4NH_2)_3$ Red to colorless crystals, melting at 205°C; soluble in ethanol, in the hydrochloride salt; used as a dye. { ¦par·ə·rō'san·ə·lən }

Parasaleniidae [INV ZOO] A family of echinacean echinoderms in the order Echinoida composed of oblong forms with trigeminate ambulacral plates. { ¦par·ə‚sal·ə'nī·ə‚dē }

paraschist [PETR] A schist derived from sedimentary rocks. { 'par·ə‚shist }

paraselene [ASTRON] A weakly colored lunar halo identical in form and optical origin to the solar parhelion; paraselenae are observed less frequently than are parhelia, because of the moon's comparatively weak luminosity. Also known as mock moon. { ¦par·ə·sə'lēn }

paraselenic circle [ASTRON] A halo phenomenon consisting of a horizontal circle passing through the moon, corresponding to the parhelic circle through the sun, and produced by reflection of moonlight from ice crystals. { ¦par·ə·sə'len·ik 'sər·kəl }

Paraselloidea [INV ZOO] A group of the Asellota that contains forms in which the first pleopods of the male are coupled along the midline, and are lacking in the female. { ¦par·ə·sə'lóid·ē·ə }

Paraseminotidae [PALEON] A family of Lower Triassic fishes in the order Palaeonisciformes. { ‚par·ə‚sem·ə'näd·ə‚dē }

parasexual cycle [GEN] A series of events leading to genetic recombination in vegetative or somatic cells even in mammals; it was first described in filamentous fungi; there are three essential steps: heterokaryosis; fusion of unlike haploid nuclei in the heterokaryon to yield heterozygous diploid nuclei; and recombination and segregation at mitosis by two independent processes, mitotic crossing-over and loss of chromosomes. { ¦par·ə'sek·shə·wəl 'sī·kəl }

parasheet [AERO ENG] A simple form of parachute in which the canopy is a single piece of material or two or more pieces sewed together; it may have any geometrical form, such as square or hexagonal, and the hem may be gathered to assist in the development of a crown when the parasheet is opened. { 'par·ə,shēt }

parasite [BIOL] An organism that lives in or on another organism of different species from which it derives nutrients and shelter. [ELEC] Current in a circuit, due to some unintentional cause, such as inequalities of temperature or of composition; particularly troublesome in electrical measurements. { 'par·ə,sīt }

parasite drag [FL MECH] The portion of the total drag of an aircraft exclusive of the induced drag of the wings. { 'par·ə,sīt ,drag }

parasitemia [MED] The presence of parasites in the blood. { ,par·ə·si'tē·mē·ə }

parasitic [ELECTR] An undesired and energy-wasting signal current, capacitance, or other parameter of an electronic circuit. { ,par·ə'sid·ik }

Parasitica [INV ZOO] A group of hymenopteran insects that includes four superfamilies of the Apocrita: Ichneumonoidea, Chalcidoidea, Cynipoidea, and Proctotrupoidea; some are phytophagous, while others are parasites of other insects. { ,par·ə'sid·ə·kə }

parasitic absorption See parasitic capture. { ,par·ə'sid·ik əb'sórp·shən }

parasitic antenna See parasitic element. { ,par·ə'sid·ik an'ten·ə }

parasitic capture [NUCLEO] Any absorption of a neutron that does not result in a fission or the production of a desired element. Also known as parasitic absorption. { ,par·ə'sid·ik 'kap·chər }

parasitic castration [BIOL] Destruction of the reproductive organs by parasites. { ,par·ə'sid·ik ka'strā·shən }

parasitic cone See adventive cone. { ,par·ə'sik·ik 'kōn }

parasitic current [ELEC] An eddy current in a piece of electrical machinery; gives rise to energy losses. { ,par·ə'sid·ik 'kə·rənt }

parasitic element [ELECTROMAG] An antenna element that serves as part of a directional antenna array but has no direct connection to the receiver or transmitter and reflects or reradiates the energy that reaches it, in a phase relationship such as to give the desired radiation pattern. Also known as parasitic antenna; parasitic reflector; passive element. { ,par·ə'sid·ik 'el·ə·mənt }

parasitic oscillation [ELECTR] An undesired self-sustaining oscillation or a self-generated transient impulse in an oscillator or amplifier circuit, generally at a frequency above or below the correct operating frequency. { ,par·ə'sid·ik ,äs·ə'lā·shən }

parasitic reflector See parasitic element. { ,par·ə'sid·ik ri'flek·tər }

parasitic stomatitis See thrush. { ,par·ə'sid·ik ,stō·mə'tīd·əs }

parasitic suppressor [ELECTR] A suppressor, usually in the form of a coil and resistor in parallel, inserted in a circuit to suppress parasitic high-frequency oscillations. { ,par·ə'sid·ik sə'pres·ər }

parasitism [ECOL] A symbiotic relationship in which the host is harmed, but not killed immediately, and the species feeding on it is benefited. { 'par·ə·sə,tiz·əm }

parasitoidism [BIOL] Systematic feeding by an insect larva on living host tissues so that the host will live until completion of larval development. { 'par·ə·sə'tóid,iz·əm }

parasitology [BIOL] A branch of biology which deals with those organisms, plant or animal, which have become dependent on other living creatures. { ,par·ə·sə'tāl·ə·jē }

parasitophobia [PSYCH] An abnormal fear of parasites. { ,par·ə,sid·ə'fō·bē·ə }

parasomnia [MED] A group of disorders characterized by abnormal movements or behavior intruding into sleep (for example, sleep walking, sleep terrors, sleep talking, nightmares, sleep paralysis, tooth grinding, and bed wetting). { ,par·ə'säm·nē·ə }

parasphenoid [VERT ZOO] A bone in the base of the skull of many vertebrates. { ,par·ə'sfē,nóid }

parastacidae [INV ZOO] A family of crayfishes assigned to the Nephropoidea. { ,par·ə'stäs·ə,dē }

parastate [ATOM PHYS] A state of a diatomic molecule in which the spins of the nuclei are antiparallel. { 'par·ə,stāt }

parastratigraphic unit See operational unit. { 'par·ə,strad·ə'graf·ik 'yü·nət }

parastratigraphy [GEOL] 1. Supplemental stratigraphy based on fossils other than those governing the prevalent orthostratigraphy. 2. Stratigraphy based on operational units. { ,par·ə·strə'tig·rə·fē }

parastratotype [GEOL] Another section in the original locality where a stratotype was defined. { ,par·ə'strad·ə,tīp }

parastyle [VERT ZOO] A small cusp anterior to the paracone of an upper molar. { 'par·ə,stīl }

Parasuchia [PALEON] The equivalent name for Phytosauria. { ,par·ə'sü·kē·ə }

parasympathetic [NEUROSCI] Of or pertaining to the craniosacral division of the autonomic nervous system. { ,par·ə,sim·pə'thed·ik }

parasympathetic nervous system [NEUROSCI] The craniosacral portion of the autonomic nervous system, consisting of preganglionic nerve fibers in certain sacral and cranial nerves, outlying ganglia, and postganglionic fibers. { ,par·ə,sim·pə'thed·ik 'nər·vəs ,sis·təm }

parasympatholytic [PHARM] Blocking the action of parasympathetic nerve fibers. { ,par·ə·sim,path·ə'lid·ik }

parasympathomimetic [PHARM] Of drugs, having an effect similar to that produced when the parasympathetic nerves are stimulated. { ,par·ə,sim·pə·thō·mi'med·ik }

paratacamite [MINERAL] $Cu_2(OH)_3Cl$ Rhombohedral mineral composed of basic copper chloride; it is dimorphous with tacamite. { par,ad·ə'ka,mīt }

parataxic distortion [PSYCH] A perceptual or judgmental distortion of interpersonal relations resulting from the observer's need to pattern his responses on previous experiences and thus defend himself against anxiety. { ,par·ə,tak·sik di'stór·shən }

parathelioma See interstitial endometriosis. { ,par·ə,thē·lē'ō·mə }

parathene [MATER] Any of a group of high-grade hydrocarbons that are extracted from lubricating oil stocks by the solvent process or refining. { 'par·ə,thēn }

parathormone [BIOCHEM] A polypeptide hormone that functions in regulating calcium and phosphate metabolism. Also known as parathyroid hormone. { ,par·ə'thór,mōn }

Parathuramminacea [PALEON] An extinct superfamily of foraminiferans in the suborder Fusulinina, with a test having a globular or tubular chamber and a simple, undifferentiated wall. { ,par·ə·thə,ram·ə'nās·ē·ə }

parathyroidectomy [MED] Excision of a parathyroid gland. { ,par·ə,thī,rói'dek·tə·mē }

parathyroid gland [ANAT] A paired endocrine organ located within, on, or near the thyroid gland in the neck region of all vertebrates except fishes. { 'par·ə'thī,róid ,gland }

parathyroid hormone See parathormone. { ,par·ə'thī,róid 'hór,mōn }

paratill [GEOL] A till formed by ice-rafting in a marine or lacustrine environment; includes deposits from ice floes and icebergs. { 'par·ə,til }

para toner [MATER] Water-insoluble red pigment made from 3-naphthol and paranitroaniline; used in paints, in printing, and to make para lakes. { 'par·ə,tōn·ər }

paratonic movement [BOT] The movement of the whole or parts of a plant due to the influence of an external stimulus, such as gravity, chemicals, heat, light, or electricity. { ,par·ə'tän·ik 'müv·mənt }

paratrachoma See inclusion conjunctivitis. { ,par·ə·tra'kō·mə }

paratrichosis [MED] A condition in which the hair is either imperfect in growth or develops in abnormal places. { ,par·ə·tri'kō·səs }

paratroch [INV ZOO] The ciliated band encircling the anus in certain trichophore larvae. { 'par·ə,trōk }

paratroops See parachute troops. { 'par·ə,trüps }

paratuberculosis See Johne's disease. { ,par·ə·tə,bər·kyə'lō·səs }

paratype [SYST] A specimen other than the holotype which is before the author at the time of original description and which is designated as such or is clearly indicated as being one of the specimens upon which the original description was based. { 'par·ə,tīp }

paratyphoid fever [MED] A bacterial disease of humans resembling typhoid fever and caused by *Salmonella paratyphi*. { ¦par'tī¸fȯid 'fē·vər }

parauterine organ [INV ZOO] In certain tapeworms, a pouchlike sac which receives and retains the embryos. { ¦par·ə'yüd·ə·rən ¸ȯr·gən }

parautochthonous [GEOL] Pertaining to a mobilized part of an autochthonous granite moved higher in the crust or into a tectonic area of lower pressure and characterized by variable and diffuse contacts with country rocks. [PETR] Pertaining to a rock that is intermediate in tectonic character between autochthonous and allochthonous. { ¦par·ə·ȯ'täk·thə·nəs }

paravane [ENG] A torpedo-shaped device with sawlike teeth along its forward end, towed with a wire rope underwater from either side of the bow of a ship to cut the cables of anchored mines. Also known as otter. { 'par·ə¸vān }

paravauxite [MINERAL] $FeAl_2(PO_4)_2(OH)_2 \cdot 8H_2O$ A colorless mineral composed of hydrous basic iron aluminum phosphate; contains more water than vauxite. { ¦par·ə'vȯk¸sīt }

parawollastonite [MINERAL] $CaSiO_3$ A monoclinic mineral composed of silicate of calcium; it is dimorphous with wollastonite. { ¸par·ə'wȯl·ə·stə¸nīt }

paraxial [SCI TECH] Lying near the axis. { par'ak·sē·əl }

paraxial rays [OPTICS] Rays which are close enough to the opical axis of a system, and thus whose directions are sufficiently close to being parallel to it, so that sines of angles between the rays and the optical axis may be replaced by the angles themselves in calculations. { par'ak·sē·əl 'rāz }

paraxial trajectory [ELEC] A trajectory of a charged particle in an axially symmetric electric or magnetic field in which both the distance of the particle from the axis of symmetry and the angle between this axis and the tangent to the trajectory are small for all points on the trajectory. { par'ak·sē·əl trə'jek·trē }

paraxonic [VERT ZOO] Pertaining to a state or condition wherein the axis of the foot lies between the third and fourth digits. { ¦par·ak'sän·ik }

Parazoa [INV ZOO] A name proposed for a subkingdom of animals which includes the sponges (Porifera). { 'par·ə 'zō·ə }

parcel method [METEOROL] A method of testing for instability in which a displacement is made from a steady state under the assumption that only the parcel or parcels displaced are affected, the environment remaining unchanged. { 'pär· səl ¸meth·əd }

parchment [MATER] **1.** The skin of a goat or sheep that has been treated so that it can be used to write upon. **2.** A drawing or written text on this material. { 'pärch·mənt }

parchment paper [MATER] Paper that has been manufactured so that its appearance resembles parchment. { 'pärch· mənt ¸pā·pər }

paregoric [PHARM] Camphorated opium tincture, a preparation of opium, camphor, benzoic acid, anise oil, glycerin, and diluted alcohol; used mainly as an antiperistaltic. { ¦par· ə¦gȯr·ik }

Pareiasauridae [PALEON] A family of large, heavy-boned terrestrial reptiles of the late Permian, assigned to the order Cotylosauria. { pə'rī·ə'sȯr·ə'dē }

parenchyma [BOT] A tissue of higher plants consisting of living cells with thin walls that are agents of photosynthesis and storage; abundant in leaves, roots, and the pulp of fruit, and found also in leaves and stems. [HISTOL] The specialized epithelial portion of an organ, as contrasted with the supporting connective tissue and nutritive framework. { pə'reŋ·kə·mə }

parenchymal jaundice [MED] Any of various forms of jaundice in which the disease is due in part to damaged liver cells. { pə'reŋ·kə·məl 'jȯn·dəs }

parenchymella *See* diploblastula. { pə¦reŋ·kə'mel·ə }

parenchymula [INV ZOO] The flagellate larva of calcinean sponges in which there is a cavity filled with gelatinous connective tissue. { ¸par·əŋ'kim·yə·lə }

parent [COMPUT SCI] An element that precedes a given element in a data structure. [NUC PHYS] A radionuclide that upon disintegration yields a specified nuclide, the daughter, either directly, or indirectly as a later member of a radioactive series. [QUANT MECH] If an *n*-electron state is written as a sum of products of 1-electron states and (*n* − 1)-electron states, the (*n* − 1)-electron states are called the parents of the *n*-electron states. { 'per·ənt }

parental imprinting [GEN] The condition whereby the extent of gene expression depends upon the sex of the parent that transmits the gene. Also known as genomic imprinting. { pə¦ren·təl 'im¸print·iŋ }

parental magma [GEOL] The naturally occurring mobile rock material from which a particular igneous rock solidified or from which another magma was derived. { pə'rent·əl 'mag·mə }

parent compound [CHEM] A chemical compound that is the basis for one or more derivatives; for example, ethane is the parent compound for ethyl alcohol and ethyl acetate. { 'per·ənt ¸käm¸paȯnd }

parenteral [MED] Outside the intestine; not via the alimentary tract. { pər'ent·ə·rəl }

parent figure [PSYCH] A person who represents essential but not necessarily ideal attributes of a father or mother and who is the object of the attitudes and responses of an individual in a parent-child relationship. { 'per·ənt ¸fig·yər }

parenthesis-free notation *See* Polish notation. { pə'ren·thə· səs ¦frē nō'tā·shən }

parent material [GEOL] The unconsolidated mineral or organic material from which the true soil develops. { 'per· ənt mə¸tir·ē·əl }

parent metal *See* base metal. { 'per·ənt ¸med·əl }

parent name [CHEM] That part of a chemical compound's name from which the name of a derivative comes; for example, ethane is the parent name for ethanol. { 'per·ənt ¸nām }

parent rock [GEOL] **1.** The rock mass from which parent material is derived. **2.** *See* source rock. { 'per·ənt ¸räk }

paresis [MED] **1.** A slight paralysis. **2.** Incomplete loss of muscular power. **3.** Weakness of a limb. { pə'rē·səs }

paresthesia [MED] Tingling, crawling, or burning sensation of the skin. { ¸par·əs'thē·zhə }

Pareto diagram [IND ENG] A histogram of defects or quality problems, classified by type and sorted in the order of descending frequency, that is used to focus on the major sources of problems. { pä're·tō ¸dī·ə¸gram }

Pareto's law [IND ENG] The principle that in most activities a small fraction (around 20%) of the total activity accounts for a large fraction (around 80%) of the result. Also known as rule of 80–20. { pə'rēd·ōz ¸lȯ }

Pareulepidae [INV ZOO] A monogeneric family of errantian polychaete annelids. { ¸par·yü'lep·ə¸dē }

parfocal eyepieces [OPTICS] Eyepieces whose lower focal points lie in the same plane, so that they can be interchanged without changing the focus of the instrument with which they are used. { pär'fō·kəl 'ī¸pēs·əz }

parging [CIV ENG] A thin coating of mortar or plaster on a brick or stone surface. { 'pärj·iŋ }

parhelic circle [ASTRON] A halo consisting of a faint white circle passing through the sun and running parallel to the horizon for as much as 360° of azimuth. Also known as mock sun ring. { pär'hē·lik }

parhelion [ASTRON] Either of two colored luminous spots that appear at points 22° (or somewhat more) on both sides of the sun and at the same elevation as the sun; the solar counterpart of the lunar paraselene. Also known as mock sun; sun dog. { pär'hēl·yən }

parhelium *See* parahelium. { pär'hē·lē·əm }

Parian cement [MATER] Gypsum plaster containing borax, which dries to a hard finish. { 'par·ē·ən si¸ment }

parietal [ANAT] Of or situated on the wall of an organ or other body structure. [BOT] Of a plant part, having a peripheral location or orientation; in particular, attached to the main wall of an ovary. { pə'rī·əd·əl }

parietal block *See* intraventricular heart block. { pə'rī·əd·əl ¸bläk }

parietal bone [ANAT] The bone that forms the side and roof of the cranium. { pə'rī·əd·əl ¸bōn }

parietal cell [HISTOL] One of the peripheral, hydrochloric acid-secreting cells in the gastric fundic glands. Also known as acid cell; delomorphous cell; oxyntic cell. { pə'rī·əd·əl ¸sel }

Parietales [BOT] An order of plants in the Englerian system; families are placed in the order Violales in other systems. { pə¸rī·ə'tā·lēz }

parietal lobe [ANAT] The cerebral lobe of the brain above the lateral cerebral sulcus and behind the central sulcus. { pə'rī·əd·əl ¸lōb }

parietal peritoneum [ANAT] The portion of the peritoneum lining the interior of the body wall. { pə'rī·əd·əl ,per·ə·tə'nē·əm }

parietal pleura [ANAT] The pleura lining the inner surface of the thoracic cavity. { pə'rī·əd·əl 'plùr·ə }

Parinaud's oculoglandular syndrome [MED] Enlargement of the lymph node around the eye and conjunctivitis. { 'pär·ə,nōz ,äk·yə·lō,glan·jə·lər 'sin,drōm }

paring [MECH ENG] A method of wood turning in which the piece is trimmed or reduced in size by cutting or shaving thin sections from the surface. { 'per·iŋ }

paring chisel [DES ENG] A long-handled chisel used to pare wood manually. { 'per·iŋ ,chiz·əl }

paring gouge [DES ENG] A long, thin concave woodworker's gouge with the cutting edge beveled on the inside of the blade. { 'per·iŋ ,gaùj }

parisite [MINERAL] $(Ce,La)_2Ca(CO_3)_3F_2$ A brownish-yellow secondary mineral composed of a carbonate and a fluoride of calcium, cerium, and lanthanum. { 'par·ə,sīt }

parison [ENG] A hollow plastic tube from which a bottle or other hollow object is blow-molded. { 'par·ə·sən }

parison swell [ENG] In blow molding, the ratio of the cross-sectional area of the parison to that of the die opening. { 'par·ə·sən ,swel }

parity [COMPUT SCI] The use of a self-checking code in a computer employing binary digits in which the total number of 1's or 0's in each permissible code expression is always even or always odd. [MATH] Two integers have the same parity if they are both even or both odd. [QUANT MECH] A physical property of a wave function which specifies its behavior under an inversion, that is, under simultaneous reflection of all three spatial coordinates through the origin; if the wave function is unchanged by inversion, its parity is 1 (or even); if the function is changed only in sign, its parity is −1 (or odd). Also known as space reflection symmetry. { 'par·əd·ē }

parity bit [COMMUN] An additional nondata bit that is attached to a set of data bits to check their validity; it is set so that the sum of one-bits in the augmented set is always odd or always even. { 'par·əd·ē ,bit }

parity check See odd-even check. { 'par·əd·ē ,chek }

parity conservation See conservation of parity. { 'par·əd·ē ,kän·sər'vā·shən }

parity error [COMPUT SCI] A machine error in which an odd number of bits are accidentally changed, so that the error can be detected by a parity check. { 'par·əd·ē ,er·ər }

parity selection rules [QUANT MECH] Rules which specify whether or not a change in parity occurs during a given type of transition of an atom, molecule, or nucleus; for example, the Laporte selection rule, or the rule that there is no parity change in an allowed β-decay transition of a nucleus. { 'par·əd·ē si'lek·shən ,rülz }

parity transformation [COMMUN] A change in value of a transmitted character denoting the number of one-bits. [PHYS] See inversion. { 'par·əd·ē ,tranz·fər'mä·shən }

parker See rep. { 'pär·kər }

Parker bound [ASTROPHYS] An upper bound on the density of magnetic monopoles that is obtained from arguments based on the existence of a galactic magnetic field. { 'pär·kər ,baùnd }

parkerite [MINERAL] $Ni_3(Bi,Pb)_2S_2$ A bright-bronze mineral composed of nickel bismuth lead sulfide. { 'pär·kə,rīt }

parkerizing [MET] Trade name for a process for the production of phosphate coating on steel articles by immersion in an aqueous solution of manganese or zinc acid with phosphate. { 'pär·kə,rīz·iŋ }

Parker model [ASTRON] A model of the solar wind that assumes the solar wind is driven by the thermal pressure of the hot coronal gas. { 'pär·kər ,mäd·əl }

Parker-Washburn boundary [SOLID STATE] A surface which separates two regions in a solid in which the crystal axes point in different directions, and which is made up of a single array of dislocations. { 'pär·kər 'wäsh·bərn baùn·drē }

Parkes process [MET] A process for recovering precious metals from lead by stirring about 2% zinc into the melt to form zinc compounds with gold and silver which can then be skimmed off the surface. { 'pärks ,prä·səs }

parking apron [CIV ENG] A hard-surfaced area used for parking aircraft. { 'pärk·iŋ ,ā·prən }

parking brake [MECH ENG] In an automotive vehicle, a brake that functions independently of the service brake and is set after the vehicle has been brought to a stop. { 'pärk·iŋ ,brāk }

parking lot [CIV ENG] An outdoor lot for parking automobiles. { 'pärk·iŋ ,lät }

parking orbit [AERO ENG] A temporary earth orbit during which the space vehicle is checked out and its trajectory carefully measured to determine the amount and time of increase in velocity required to send it into a final orbit or into space in the desired direction. { 'pärk·iŋ ,ȯr·bət }

Parkinje effect See Purkinje effect. { pər'kin·jē i,fekt }

parkinsonism [MED] A clinical state characterized by tremor at a rate of three to eight tremors per second, with "pill-rolling" movements of the thumb common, muscular rigidity, dyskinesia, hypokinesia, and reduction in number of spontaneous and autonomic movements; produces a masked facies, disturbances of posture, gait, balance, speech, swallowing, and muscular strength. Also known as paralysis agitans; Parkinson's disease. { 'pär·kən·sə,niz·əm }

Parkinson's disease See parkinsonism. { 'pär·kən·sənz di,zēz }

parkland See temperate woodland; tropical woodland. { 'pärk,land }

parkway [CIV ENG] A broad landscaped expressway which is not open to commercial vehicles. { 'pärk,wā }

Parmeliaceae [BOT] The foliose shield lichens, a family of the order Lecanorales. { ,pär·mel·ē'ās·ē,ē }

Parnidae [INV ZOO] The equivalent name for Dryopidae. { 'pär·nə,dē }

parogenetic [GEOL] Formed previous to the enclosing rock; especially said of a concretion formed in a different (older) rock from its present (younger) host. { ,par·ə'jen·ik }

paromomycin [MICROBIO] A broad-spectrum antibiotic produced by *Streptomyces rimosus forma paromomycinus;* it is effective in the treatment of intestinal amebiasis in humans. { ,par·ə·me'mīs·ən }

paronychia [MED] A suppurative inflammation about the margin of a nail. { ,par·ə'nik·ē·ə }

paronychium See perionychium. { ,par·ə'nik·ē·əm }

parotid duct [ANAT] The duct of the parotid gland. Also known as Stensen's duct. { pə'räd·əd 'dəkt }

parotidectomy [MED] Excision of a parotid gland. { pə,räd·ə'dek·tə·mē }

parotid gland [ANAT] The salivary gland in front of and below the external ear; the largest salivary gland in humans; a compound racemose serous gland that communicates with the mouth by Steno's duct. { pə'räd·əd 'gland }

parotitis [MED] Inflammation of the parotid glands. { ,par·ə'tīd·əs }

parous [MED] Pertaining to an organism that has produced offspring. { 'par·əs }

parovarian cyst [MED] A cyst of mesonephric origin arising between the layers of the mesosalpinx, adjacent to the ovary. { ,par·ō'ver·ē·ən 'sist }

parovarium See epoophoron. { ,par·ə'var·ē·əm }

paroxysm [MED] **1.** A sudden attack, or the periodic crisis in the progress of a disease. **2.** A spasm, convulsion, or seizure. **3.** A burst of electrical activity during electroencephalography in the form of spikes, or spikes and waves, which indicates cerebral dysrhythmia or epileptic discharges. [PSYCH] A sudden, uncontrollable emotional outburst. { 'par·ək,siz·əm }

paroxysmal eruption See Vulcanian eruption. { ,par·ək,siz·məl i'rəp·shən }

parquet flooring [BUILD] Wood flooring made of strips laid in a pattern to form designs. { pär'kā 'flȯr·iŋ }

parrot [VERT ZOO] Any member of the avian family Psittacidae, distinguished by the short, stout, strongly hooked beak. { 'par·ət }

Parry arcs [OPTICS] A class of halos appearing as faintly colored arcs above and below the sun; these refraction phenomena are produced by ice crystals which exhibit a preferred orientation, and are correspondingly more unusual than those associated with randomly oriented crystals. { 'par·ē ,ärks }

pars anterior [ANAT] The major secretory portion of the anterior lobe of the adenohypophysis. Also known as pars distalis. { 'pärz ,an'tir·ē·ər }

pars distalis See pars anterior. { 'pärz 'dis·tə·ləs }

PARROT

A parrot, with short hooked beak.

parsec [ASTRON] The distance at which a star would have a parallax equal to 1 second of arc; 1 parsec equals 3.258 light-years or 3.08572 × 10^{13} kilometers. Derived from parallax-second. { 'pär,sek }

parser [COMPUT SCI] The portion of a computer program that carries out parsing operations. { 'pär·sər }

parsettensite [MINERAL] $Mn_5Si_6O_{13}(OH)_8$ A copper-red mineral composed of hydrous silicate of manganese. { pär'set·ən,zīt }

Parseval's equation [MATH] The equation which states that the square of the length of a vector in an inner product space is equal to the sum of the squares of the inner products of the vector with each member of a complete orthonormal base for the space. Also known as Parseval's identity; Parseval's relation. { 'pär·sə·vəlz i,kwā·zhən }

Parseval's identity See Parseval's equation. { 'pär·sə·vəlz ī'den·əd·ē }

Parseval's relation See Parseval's equation. { 'pär·sə·vəlz re'lā·shən }

Parseval's theorem [MATH] A theorem that gives the integral of a product of two functions, $f(x)$ and $F(x)$, in terms of their respective Fourier coefficients; if the coefficients are defined by

$$a_n = (1/\pi)\int_0^{2\pi} f(x)\cos nx\,dx$$

$$b_n = (1/\pi)\int_0^{2\pi} f(x)\sin nx\,dx$$

and similarly for $F(x)$, the relationship is

$$\int_0^{2\pi} f(x)F(x)\,dx = \pi\left[{}^1\!/_2 a_0 A_0 + \sum_{n=1}^{\infty}(a_n A_n + b_n B_n)\right]$$

{ 'pär·sə·vəlz 'thir·əm }

Parshall flume [ENG] A calibrated device for measuring the flow of liquids in open conduits by measuring the upper and lower beads at a specified distance from an obstructing sill. { 'pär·shəl ,flüm }

parsimony [SCI TECH] The principle that the simplest scientific explanation is best. { 'pär·sə,mō·nē }

parsing [COMPUT SCI] A process whereby phrases in a string of characters in a computer language are associated with the component names of the grammar that generated the string. { 'pärs·iŋ }

pars intermedia [ANAT] The intermediate lobe of the adenohypophysis. { ¦pärz ,in·tər'mē·dē·ə }

parsley [BOT] *Petroselinum crispum.* A biennial herb of European origin belonging to the order Umbellales; grown for its edible foliage. { 'pär·slē }

parsley oil [MATER] Colorless or pale-greenish-yellow liquid with parsley aroma; soluble in alcohol, ether, and chloroform; distilled from parsley seeds; used in medicine. { 'pär·slē ,óil }

pars nervosa [ANAT] The inferior subdivision of the neurohypophysis. Also known as pars neuralis. { ¦pärz nər'vō·sə }

pars neuralis See pars nervosa. { ¦pärz nù'räl·əs }

parsnip [BOT] *Pastinaca sativa.* A biennial herb of Mediterranean origin belonging to the order Umbellales; grown for its edible thickened taproot. { 'pär·snəp }

parsonsite [MINERAL] $Pb_2(UO_2)(PO_4)_2 \cdot 2H_2O$ A pale-yellow to brownish mineral composed of hydrous lead uranyl phosphate, occurring as a powder. { 'pär·sən,zīt }

Parsons-stage steam turbine [MECH ENG] A steam turbine having a reaction-type stage in which the pressure drop occurs partially across the stationary nozzles and partly across the rotating blades. { 'pär·sənz ¦stāj 'stēm 'tər·bən }

pars tuberalis [ANAT] A pair of processes that grow forward or upward along the stalk of the adenohypophysis. { ¦pärz ,tü·bə'ral·əs }

part [ENG] An element of a subassembly, not normally useful by itself and not amenable to further disassembly for maintenance purposes. { pärt }

part classification [IND ENG] A coding scheme employed in automated manfacturing processes that uses four or more

digits to assign discrete products to families of parts. { 'pärt ,klas·ə·fə,kā·shən }

part detection [IND ENG] The recognition of parts and workpieces by a robot or a computer vision system. { 'pärt di,tek·shən }

part family [IND ENG] In the group technology concept, a set of related parts that can be produced by the same sequence of machining operations because of similarity in shape and geometry or similarity in production operation processes. { 'pärt ,fam·lē }

parthenita [INV ZOO] A stage, such as the sporocyst, redia, or cercaria, in the development of a fluke which reproduces parthenogenetically. { pär'then·əd·ə }

parthenocarpy [BOT] Production of fruit without fertilization. { 'pär·thə·nō,kär·pē }

parthenogenesis [INV ZOO] A special type of sexual reproduction in which an egg develops without entrance of a sperm; common among rotifers, aphids, thrips, ants, bees, and wasps. { ¦pär·thə·nō'jen·ə·səs }

parthenomerogony [EMBRYO] Development of a nucleated fragment of an unfertilized egg following parthenogenetic stimulation. { ¦pär·thə·nō·mə'räg·ə·nē }

parthenospore See azygospore. { 'pär·thə·nə,spór }

partial [ACOUS] Also known as partial tone. **1.** A simple sinusoidal physical component of a complex tone. **2.** A sound sensation component that is distinguishable as a simple tone, cannot be further analyzed by the ear, and contributes to the character of the complex sound; the frequency of a partial may be higher or lower than the basic frequency and may be an integral multiple or submultiple of the basic frequency. { 'pär·shəl }

partial bulkhead [NAV ARCH] A partition wall that does not extend across a compartment; used to strengthen the structure. { 'pär·shəl 'bəlk,hed }

partial carry [COMPUT SCI] A word composed of the carries generated at each position when adding many digits in parallel. { 'pär·shəl 'kar·ē }

partial Cauchy surface [RELAT] A spacelike surface S which is intersected only once by each timelike or null curve; "partial" means that only a portion of the future history of the space-time can be predicted from S, that is, there exists a Cauchy horizon. { 'pär·shəl 'kō·shē ,sər·fəs }

partial cleavage [EMBRYO] Cleavage in which only part of the egg divides into blastomeres. { 'pär·shəl 'klē·vij }

partial coherence [PHYS] Property of two waves whose relative phase undergoes random fluctuations which are not, however, sufficient to make the wave completely incoherent. { 'pär·shəl kō'hir·əns }

partial common battery [COMMUN] Type of telephone system in which the talking battery is supplied by each individual telephone, and the signaling and supervisory battery is supplied by the switchboard. { 'pär·shəl 'käm·ən 'bad·ə·rē }

partial condensation [CHEM ENG] The cooling (or pressurization) of a saturated vapor until a part of it is condensed out as liquid. { 'pär·shəl ,känd·ən'sā·shən }

partial correlation [STAT] The strength of the linear relationship between two random variables where the effect of other variables is held constant. { 'pär·shəl ,kär·ə'lā·shən }

partial correlation analysis [STAT] A technique used to measure the strength of the relationship between the dependent variable and one independent variable in such a way that variations in other independent variables are taken into account. { 'pär·shəl ,kär·ə'lā·shən ə,nal·ə·səs }

partial correlation coefficient [STAT] A measure of the strength of association between a dependent variable and one independent variable when the effect of all other independent variables is removed; equal to the square root of the partial coefficient of determination. { 'pär·shəl ,kä·rə'lā·shən ,kō·i,fish·ənt }

partial derivative [MATH] A derivative of a function of several variables taken with respect to one variable while holding the others fixed. { 'pär·shəl də'riv·əd·iv }

partial differential equation [MATH] An equation that involves more than one independent variable and partial derivatives with respect to those variables. { 'pär·shəl ,dif·ə'ren·chəl i,kwā·zhən }

partial dislocation [CRYSTAL] The line at the edge of an extended dislocation where a slip through a fraction of a lattice constant has occurred. { 'pär·shəl ,dis·lō'kā·shən }

partial-duration series [GEOPHYS] A series composed of all events during the period of record which exceed some set criterion; for example, all floods above a selected base, or all daily rainfalls greater than a specified amount. { 'pär·shəl dù'ra·shən ,sir·ēz }

partial eclipse [ASTRON] An eclipse in which only part of the source of light is obscured. { 'pär·shəl i'klips }

partial fractions [MATH] A collection of fractions which when added are a given fraction whose numerator and denominator are usually polynomials; the partial fractions are usually constants or linear polynomials divided by factors of the denominator of the given fraction. { 'pär·shəl 'frak·shənz }

partial function [COMPUT SCI] A partial function from a set *A* to a set *B* is a correspondence between some subset of *A* and *B* which associates with each element of the subset of *A* a unique element of *B*. { 'pär·shəl 'fəŋk·shən }

partially balanced incomplete block design [STAT] An experimental design in which, while all treatments are not represented in each block, each treatment is tested the same number of times and certain aspects of the design satisfy conditions which simplify the least squares analysis. { ¦pär·shə·lē ¦bal·ənst ,iŋ·kəm¦plēt 'bläk di,zīn }

partially ionic bond [PHYS CHEM] A chemical bond that is neither wholly ionic nor wholly covalent in character. { 'pär·shə·lē ī¦än·ik 'bänd }

partially ordered set [MATH] A set on which a partial order is defined. Also known as poset. { 'pär·shə·lē ¦ór·dərd 'set }

partially populated board [COMPUT SCI] A printed circuit board on which some but not all of the possible electronic components are mounted, leaving room for additional components. { 'pär·shə·lē ¦päp·yə,lād·əd 'bórd }

partial molal quantity [PHYS CHEM] The molal concentration of one component of a mixture of components as related to total molal concentration for all components in the mixture. { 'pär·shəl 'mō·ləl 'kwän·əd·ē }

partial molar volume [PHYS CHEM] That portion of the volume of a solution or mixture related to the molar content of one of the components within the solution or mixture. { 'pär·shəl 'mō·lər 'väl·yəm }

partial node [PHYS] That part (a point, line, or surface) of a standing wave where some characteristic of the wave field has a minimum amplitude other than zero. { 'pär·shəl 'nōd }

partial nucleate boiling [PHYS CHEM] A stage in the boiling process in which isolated vapor bubbles are released from randomly located active sites on the heater surface. { 'pär·shəl ¦nü·klē,āt 'bóil·iŋ }

partial obscuration [METEOROL] In United States weather observing practice, the designation for sky cover when part (0.1 to 0.9) of the sky is completely hidden by surface-based obscuring phenomena. { 'pär·shəl äb·skyù'rā·shən }

partial order *See* ordering. { 'par·shəl ¦órd·ər }

partial ordering *See* ordering. { 'pär·shəl 'ór·də·riŋ }

partial pediment [GEOL] **1.** A broadly planate, gravel-capped, interstream bench or terrace. **2.** A broad, planate erosion surface which is formed by the coalescence of contemporaneous, valley-restricted benches developed at the same elevation in proximate valleys, and which would produce a pediment if uninterrupted planation were to continue at this level. { 'pär·shəl 'ped·ə·mənt }

partial penetration [ORD] Penetration obtained when a projectile fails to pass through the target far enough for either the projectile itself or light from its penetration to be seen from the back of the target. { 'pär·shəl ,pen·ə'trā·shən }

partial plane [MATH] In projective geometry, a plane in which at most one line passes through any two points. { 'pär·shəl 'plān }

partial pluton [GEOL] That part of a composite intrusion representing a single intrusive episode. { 'pär·shəl 'plü,tän }

partial potential temperature [METEOROL] The temperature that the dry-air component of an air parcel would attain if its actual partial pressure were changed to 1000 millibars (10^5 pascals). { 'pär·shəl pə'ten·chəl 'tem·prə·chər }

partial pressure [PHYS] The pressure that would be exerted by one component of a mixture of gases if it were present alone in a container. { 'pär·shəl 'presh·ər }

partial-pressure maintenance [PETRO ENG] The partial replacement of produced gas in an oil reservoir by gas injection to maintain a portion of the initial reservoir pressure. { 'pär·shəl 'presh·ər ,mānt·ən·əns }

partial pressure suit [AERO ENG] A skintight suit which does not completely enclose the body but which is capable of exerting pressure on the major portion of the body in order to counteract an increased oxygen pressure in the lungs. { 'pär·shəl 'presh·ər ,süt }

partial product [MATH] The product of a multiplicand and one digit of a multiplier that contains more than one digit. { 'pär·shəl 'präd·əkt }

partial-read pulse [ELECTR] Current pulse that is applied to a magnetic memory to select a specific magnetic cell for reading. { 'pär·shəl 'rēd 'pəls }

partial recursive function [MATH] A function that can be computed by using a Turing machine for some inputs but not necessarily for all inputs. { 'pär·shəl rē,kər·siv 'fəŋk·shən }

partial regression coefficient [STAT] Statistics in the population multiple linear regression equation that indicate the effect of each independent variable on the dependent variable with the influence of all the remaining variables held constant; each coefficient is the slope between the dependent variable and each of the independent variables. { 'pär·shəl ri'gresh·ən ,kō·ə,fish·ənt }

partial-response maximum-likelihood technique [COMMUN] A method of constructing a digital data stream from an analog signal by using information acquired by sampling the analog waveform at selected instants of time rather than using the entire waveform, and then applying the Viterbi algorithm to find the most likely sequence of bits. Abbreviated PRML technique. { ¦pär·shəl ri,späns 'mak·sə·məm ¦līk·lē,hùd tek,nēk }

partial-select output [ELECTR] The voltage response produced by applying partial-read or partial-write pulses to an unselected magnetic cell. { 'pär·shəl si¦lekt 'aùt,pùt }

partial sum [MATH] A partial sum of an infinite series is the sum of its first *n* terms for some *n*. { 'pär·shəl 'səm }

partial thermoremanent magnetization [GEOPHYS] The thermoremanent magnetization acquired by cooling in an ambient field over only a restricted temperature interval, as opposed to the entire temperature range from Curie point to room temperature. Abbreviated PTRM. { 'pär·shəl ¦thər·mō'rem·ə·nənt ,mag·nə·tə'zā·shən }

partial tide [OCEANOGR] One of the harmonic components composing the tide at any point. Also known as tidal component; tidal constituent. { 'pär·shəl 'tīd }

partial tone *See* partial. { 'pär·shəl 'tōn }

partial wetting [FL MECH] The situation in which the contact angle between a solid and a liquid is greater than zero but less than 90°. { ¦pär·shəl 'wed·iŋ }

participation crude *See* buy-back crude. { pär,tis·ə'pā·shən ,krüd }

particle [MECH] *See* material particle. [PART PHYS] *See* elementary particle. [PHYS] **1.** Any very small part of matter, such as a molecule, atom, or electron. Also known as fundamental particle. **2.** Any relatively small subdivision of matter, ranging in diameter from a few angstroms (as with gas molecules) to a few millimeters (as with large raindrops). { 'pärd·ə·kəl }

particle accelerator [NUCLEO] A device which accelerates electrically charged atomic or subatomic particles, such as electrons, protons, or ions, to high energies. Also known as accelerator; atom smasher. { 'pärd·ə·kəl ik'sel·ə,rād·ər }

particle beam [PHYS] A concentrated, nearly unidirectional flow of particles. { 'pärd·ə·kəl ,bēm }

particle board [MATER] Construction board made with wood particles impregnated with low-molecular-weight resin and then cured. { 'pärd·ə·kəl ,bórd }

particle counting [ANALY CHEM] Microscopic or photomicrographic technique for the visual counting of the numbers of particles in a known quantity of a solid-liquid suspension. { 'pärd·ə·kəl ,kaùnt·iŋ }

particle derivative [FL MECH] The rate of change of a quantity with respect to time, measured at a point that moves along with a particle of a fluid. { ¦pärd·ə·kəl də'riv·ə·div }

particle detector [NUCLEO] A device used to indicate the presence of fast-moving charged atomic or nuclear particles by observation of the electrical disturbance created by a particle as it passes through the device. Also known as radiation detector. { 'pärd·ə·kəl di,tek·tər }

particle diameter [GEOL] The diameter of a sedimentary particle considered as a sphere. { 'pärd·ə·kəl dī,am·əd·ər }

particle displacement velocimetry [FL MECH] A method of flow-field measurement in which a two-dimensional sheet of a flow field seeded with fluorescent particles is illuminated by a pulsed laser and particle displacements are recorded by a camera. { ¦pär·də·kəl di¦pläs·mənt ˌvel·ə'sim·ə·trē }

particle distribution function [STAT MECH] A function whose value is the number of particles per unit volume of phase space. { 'pärd·ə·kəl ˌdi·strə'byü·shən ˌfəŋk·shən }

particle dynamics [MECH] The study of the dependence of the motion of a single material particle on the external forces acting upon it, particularly electromagnetic and gravitational forces. { 'pärd·ə·kəl dī̇ˌnam·iks }

particle electrophoresis [PHYS CHEM] Electrophoresis in which the particles undergoing analysis are of sufficient size to be viewed either with the naked eye or with the assistance of an optical microscope. { 'pärd·ə·kəl iˌlek·trō·fə'rē·səs }

particle emission [NUC PHYS] The ejection of a particle other than a photon from a nucleus, in contrast to gamma emission. { 'pärd·ə·kəl iˌmish·ən }

particle energy [MECH] For a particle in a potential, the sum of the particle's kinetic energy and potential energy. [RELAT] For a relativistic particle the sum of the particle's potential energy, kinetic energy, and rest energy; the last is equal to the product of the particle's rest mass and the square of the speed of light. { 'pärd·ə·kəl ˌen·ər·jē }

particle flow [FL MECH] The transport of particles in fluids and gases. { 'pärd·i·kəl ˌflō }

particle horizon [RELAT] The spatial boundary beyond which, in certain universe models, it is impossible for an observer at a given time to receive a signal. { 'pärd·ə·kəl həˌrīz·ən }

particle image velocimetry [ENG] A method of measuring local fluid velocities at thousands of locations in a fluid flow by optically observing large numbers of particles that are suspended in the fluid and move with it, using a photograph of the flow illuminated by two or more successive pulses of light or continuously for a known time interval. Also known as particle tracking velocimetry. { ˌpärd·i·kəl ˌim·ij ˌvel·ə'säm·ə·trē }

particle-induced x-ray emission [ANALY CHEM] A method of trace analysis in which a beam of ions is directed at a thin foil on which the sample to be analyzed has been deposited, and the energy spectrum of the resulting x-rays is measured. { 'pärd·ə·kəl in¦düst 'eksˌrā iˌmish·ən }

particle lens [PHYS] An electric or magnetic field, or a combination thereof, which acts upon an electron beam in a manner analogous to that in which an optical lens acts upon a light beam. { 'pärd·ə·kəl ˌlenz }

particle mechanics [MECH] The study of the motion of a single material particle. { 'pärd·ə·kəl miˌkan·iks }

particle multiplet See isospin multiplet. { 'pärd·ə·kəl 'məl·tə·plət }

particle-oriented paper [MATER] A chart paper that has a magnetic coating which is produced by combining microscopic magnetic flakes with oil to form droplets and then forming these particles into an emulsion that can be applied to the surface of ordinary bond paper or to a clear plastic substrate; the magnetic field of a small-diameter recording head rotates the magnetic flakes so that they absorb or scatter incident light to give a visible dark trace that can also be read magnetically. { 'pärd·ə·kəl ˌor·ē¦ent·əd 'pā·pər }

particle physics [PHYS] The branch of physics concerned with understanding the properties and behavior of elementary particles, especially through study of collisions or decays involving energies of hundreds of megaelectronvolts or more. Also known as high-energy physics. { 'pärd·ə·kəl ¦fiz·iks }

particle properties [PARTIC PHYS] The various quantities which characterize the behavior of an elementary particle, such as mass, charge, baryon number, spin, parity, hypercharge, and isospin. { 'pärd·ə·kəl ˌpräp·ərd·ēz }

particle-scattering factor [ANALY CHEM] Factor in light-scattering equations used to compensate for the loss in scattered light intensity caused by destructive interference during the analysis of macromolecular compounds. { 'pärd·ə·kəl ¦skad·ə·riŋ ˌfak·tər }

particle size [GEOL] The general dimensions of the particles or mineral grains in a rock or sediment based on the premise that the particles are spheres; commonly measured by sieving, by calculating setting velocities, or by determining areas of

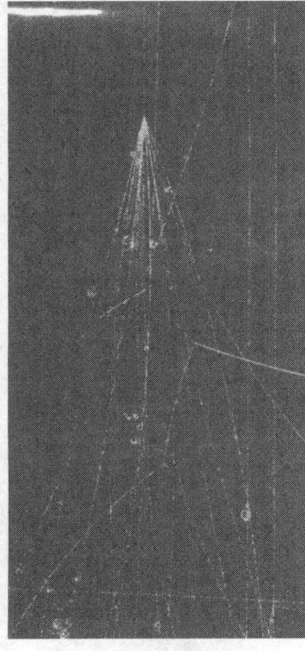

PARTICLE TRACK

Photograph of article tracks consisting of trails of bubbles in a liquid-hydrogen bubble chamber, produced by negative pions with momentum 16 GeV/c, where c is the speed of light. (CERN)

microscopic images. [MET] The average and controlling lineal dimension of an individual particle of metal powder as determined by suitable screens or other methods of analysis. { 'pärd·ə·kəl ˌsīz }

particle-size analysis [ENG] Determination of the proportion of particles of a specified size in a granular or powder sample. [GEOL] A determination of the distribution of particles in a series of size classes of a soil, sediment, or rock. Also known as size analysis; size-frequency analysis. { 'pärd·ə·kəl ¦sīz əˌnal·ə·səs }

particle-size distribution [ENG] The percentages of each fraction into which a granular or powder sample is classified, with respect to particle size, by number or weight. { 'pärd·ə·kəl ¦sīz ˌdi·strə'byü·shən }

particle spectrum See mass spectrum. { 'pärd·ə·kəl ˌspek·trəm }

particle-thickness technique [PHYS CHEM] A microscopic technique for visual measurement of the thickness of a fine particle (in the 3–100 micrometer range). { 'pärd·ə·kəl ¦thik·nəs tekˌnēk }

particle track [PHYS] Any visible phenomenon along the path of an ionizing particle, such as a trail of bubbles, water droplets, or sparks in a bubble chamber, cloud chamber, or spark chamber respectively, or of altered material in an emulsion or in glass. { 'pärd·ə·kəl ˌtrak }

particle tracking velocimetry See particle image velocimetry. { ¦pärd·ə·kəl ¦trak·iŋ ˌvel·ə'sim·ə·trē }

particle trap [PHYS] A device used to confine particles, either charged or neutral, in situations where the interaction of the particles with the wall of a container must be avoided. { 'pärd·ə·kəl ˌtrap }

particle velocity [ACOUS] The instantaneous velocity of a given infinitesimal part of a medium, with reference to the medium as a whole, due to the passage of a sound wave. [OCEANOGR] In ocean wave studies, the instantaneous velocity of a water particle undergoing orbital motion. { 'pärd·ə·kəl vəˌläs·əd·ē }

parti-colored buoy [NAV] A buoy whose visible surface is painted more than one color. { 'pär·dē¦kəl·ərd 'bȯi }

particular integral See particular solution. { pər¦tik·yə·lər 'int·ə·grəl }

particular solution [MATH] A solution to an ordinary differential equation obtained by assigning numerical values to the parameters in the general solution. Also known as particular integral. { pər¦tik·yə·lər sə'lü·shən }

particulate composite [MATER] A composite material composed of particles embedded in a matrix. { pär·tik·yə·lət kəmˌpäz·ət }

particulate mass analyzer [ENG] A unit which measures dust concentrations in emissions from furnaces, kilns, cupolas, and scrubbers. { pär¦tik·yə·lət 'mas 'an·əˌlīz·ər }

particulate matter [PHYS] matter in the form of small liquid or solid particles. { pär·tik·yə·lət ˌmad·ər }

particulates [MATER] Fine solid particles which remain individually dispersed in gases and stack emissions. { pär·tik·yə·ləts }

particulate solid See bulk solid. { pär·tik·yə·lət ˌsäl·əd }

parting [GEOL] **1.** A bed or bank of waste material dividing mineral veins or beds. **2.** A soft, thin sedimentary layer following a surface of separation between thicker strata of different lithology. **3.** A surface along which a hard rock can be readily separated or is naturally divided into layers. [MET] **1.** Recovery of gold (or occasionally another metal) from its alloys by a corrosion process. **2.** Zone of separation between cope and drag portions of mold or flask in sand casting. **3.** In sand molding, a composition to facilitate removal of the pattern. **4.** A shearing operation to produce two or more parts from a stamping. [MINERAL] Fracturing a mineral along planes weakened by deformation or twinning. { 'pärd·iŋ }

parting agent See release agent. { 'pärd·iŋ ˌā·jənt }

parting cast [GEOL] A sand-filled tension crack produced by creep along the sea floor. { 'pärd·iŋ ˌkast }

parting compound [MET] A material, such as silica or graphite, used to facilitate the separation of the cope and drag parting surfaces. { 'pärd·iŋ ˌkäm·pau̇nd }

parting line [MET] **1.** The line along which a mold is separated. **2.** A line or seam on a casting corresponding to the joint of mold parts. { 'pärd·iŋ ˌlīn }

parting lineation [GEOL] A small-scale primary sedimentary structure made up of a series of parallel ridges and grooves formed parallel to the current. Also known as current lineation. { 'pärd·iŋ ˌlin·ē'ā·shən }

parting plane lineation [GEOL] A parting lineation on a laminated surface, consisting of subparallel, linear, shallow grooves and ridges of low relief, generally less than 1 millimeter. { 'pärd·iŋ ˈplān ˌlin·ē,ā·shən }

parting sand [MET] Fine, dry sand applied to the faces of a sand mold to allow disassembly. { 'pärd·iŋ ˌsand }

parting-step lineation [GEOL] A parting lineation characterized by subparallel, steplike ridges where the parting surface cuts across several adjacent laminae. { 'pärd·iŋ ˌstep 'lin·ē,ā·shən }

parting stop [BUILD] A thin strip of wood that separates the sashes in a double-hung window. { 'pärd·iŋ ˌstäp }

parting tool [DES ENG] A narrow-bladed hand tool with a V-shaped gouge used in woodworking for cutting grooves and in wood turning for cutting a piece in two. Also known as V-tool. { 'pärd·iŋ ˌtül }

partition [BUILD] An interior wall having a height of one story or less, which divides a structure into sections. [COMPUT SCI] **1.** A reserved portion of a computer memory, sometimes used for the execution of a single computer program. **2.** One of a number of fixed portions into which a computer memory is divided in certain multiprogramming systems. [IND ENG] A slotted sheet of paperboard that can be assembled with similar sheets to form cells for holding goods during shipment. [MATH] **1.** For an integer *n*, any collection of positive integers whose sum equals *n*. **2.** For a set *A*, a collection of disjoint sets whose union is *A*. **3.** For a closed interval *I*, a finite set of closed subintervals of *I* that intersect only at their end points and whose union is *I*. { pär'tish·ən }

partition chromatography [ANALY CHEM] Chromatographic procedure in which the stationary phase is a high-boiling liquid spread as a thin film on an inert support, and the mobile phase is a vaporous mixture of the components to be separated in an inert carrier gas. { pär'tish·ən ˌkrō·mə'täg·rə·fē }

partition coefficient [ANALY CHEM] In the equilibrium distribution of a solute between two liquid phases, the constant ratio of the solute's concentration in the upper phase to its concentration in the lower phase. Symbolized K. { pär'tish·ən ˌkō·i,fish·ənt }

partitioned data set [COMPUT SCI] A single data set, divided internally into a directory and one or more sequentially organized subsections called members, residing on a direct access for each device, and commonly used for storage or program libraries. { pär'tish·ənd 'dad·ə ˌset }

partitioned display [COMPUT SCI] An electronic display that can be divided into two or more viewing areas under user or program control. Also known as split screen. { pär'tish·ənd di'splā }

partitioned file [COMPUT SCI] A file on disk storage that is divided into subdivisions, each of which constitutes a complete file. { pär'tish·ənd 'fīl }

partition function [STAT MECH] **1.** The integral, over the phase space of a system, of the exponential of $(-E/kT)$, where E is the energy of the system, k is Boltzmann's constant, and T is the temperature; from this function all the thermodynamic properties of the system can be derived. **2.** In quantum statistical mechanics, the sum over allowed states of the exponential of $(-E/kT)$. Also known as sum of states; sum over states. { pär'tish·ən ˌfəŋk·shən }

partition noise [ELECTR] Noise that arises in an electron tube when the electron beam is divided between two or more electrodes, as between screen grid and anode in a pentode. { pär'tish·ən ˌnoiz }

partition of unity [MATH] On a topological space *X*, this is a covering by open sets U_α with continuous functions f_α from *X* to [0,1], where each f_α is zero on all but a finite number of the U_α, and the sum of all these f_α at any point equals 1. { pär'tish·ən əv 'yü·nəd·ē }

partiversal [GEOL] Pertaining to formations that dip in different directions roughly as far as a semicircle. { 'pärd·ə,vər·səl }

partly cloudy [METEOROL] **1.** The character of a day's weather when the average cloudiness, as determined from frequent observations, has been from 0.1 to 0.5 for the 24-hour period. **2.** In popular usage, the state of the weather when clouds are conspicuously present, but do not completely dull the day or the sky at any moment. { 'pärt·lē 'klaůd·ē }

parton [PART PHYS] One of the very singular (or hard), small charged particles of which hadrons are proposed to be constructed, according to a theory developed to account for the scattering of very-high-energy electrons from protons at large angles and with large momentum transfers. { 'pär,tän }

part operation [COMPUT SCI] The part in an instruction that specifies the kind of arithmetical or logical operation to be performed, but not the address of the operands. { 'pärt ˌäp·ə,rā·shən }

part programming [CONT SYS] The planning and specification of the sequence of steps or events in the operation of a numerically controlled machine tool. { 'pärt ˌprō,gram·iŋ }

partridge [VERT ZOO] Any of the game birds comprising the genera *Alectoris* and *Perdix* in the family Phasianidae. { 'pär·trij }

partridgeite *See* bixbyite. { 'pär·trə,jīt }

partridgewood *See* acapau. { 'pär·trij,wůd }

parts kit [ENG] A group of parts, not all having the same basic name, used for repair or replacement of the worn broken parts of an item; it may include instruction sheets and material, such as sandpaper, tape, cement, and gaskets. { 'pärts ˌkit }

parts list [ENG] One or more printed sheets showing a manufacturer's parts or assemblies of an end item by illustration or a numerical listing of part numbers and names; it does not outline any assembly, maintenance, or operating instructions, and it may or may not have a price list cover sheet. { 'pärts ˌlist }

parturient [MED] **1.** In labor; giving birth. **2.** Of or pertaining to parturition. { pär'tůr·ē·ənt }

parturifacient [MED] **1.** Inducing labor. **2.** An agent that induces labor. { pär'tůr·ə¦fā·shənt }

parturiometer [MED] An instrument to determine the progress of labor by measuring the expulsive force of the uterus. { pär,tůr·ē'äm·əd·ər }

parturition [MED] The process of giving birth. { ˌpär·chə'rish·ən }

party line [COMMUN] A subscriber line arranged to serve more than one station, with discriminatory ringing for each station. { 'pärd·ē 'līn }

party-line bus [COMPUT SCI] Parallel input/output bus lines to which are wired all external devices, connected to a processor register by suitable logic. { 'pärd·ē ¦līn ˌbəs }

party-line carrier system [COMMUN] A single-frequency carrier telephone system in which the carrier energy is transmitted directly to all other carrier terminals of the same channel. { 'pärd·ē ¦līn 'kar·ē·ər ˌsis·təm }

party wall [BUILD] A wall providing joint service between two buildings. { 'pärd·ē ˌwol }

parulis [MED] A subperiosteal abscess arising from dental structures. { pə'rü·ləs }

parvafacies [GEOL] A body of rock constituting the part of any magnafacies that occurs between designated time-stratigraphic planes or key beds traced across the magnafacies. { ¦pär·və'fā·shēz }

Parvoviridae [VIROL] A family of deoxyribonucleic acid (DNA)–containing animal viruses characterized by an icosahedral virion containing three or four structural proteins and a single-stranded DNA genome; includes the *Dependovirus*, *Parvovirus*, and *Densovirus* genera. { ˌpär·və'vir·ə,dī }

parvovirus [VIROL] The equivalent name for picodnavirus. { ¦pär·vō'vī·rəs }

parylene [ORG CHEM] Polyparaxylylene, used in ultrathin plastic films for capacitor dielectrics, and as a pore-free coating. { 'par·ə,lēn }

parylene capacitor [ELEC] A highly stable fixed capacitor using parylene film as the dielectric; it can be operated at temperatures up to 170°C, as well as at cryogenic temperatures. { 'par·ə,lēn kə'pas·əd·ər }

PAS *See* para-aminosalicylic acid; photoacoustic spectroscopy.

pascal [MECH] A unit of pressure equal to the pressure resulting from a force of 1 newton acting uniformly over an area of 1 square meter. Symbolized Pa. { pa'skal }

Pascal [COMPUT SCI] A procedure-oriented programming language whose highly structured design facilitates the rapid location and correction of coding errors. { pa'skal }

Pascal distribution See negative binomial distribution. { pa¦skal ˌdi·strə'byü·shən }

Pascal rules [PHYS CHEM] Rules which give the diamagnetic susceptiblity of a complex molecule in terms of the sum of the susceptibilities of its constituent atoms, and a correction factor which depends on the type of bonds linking the atoms. { pa'skal ˌrülz }

Pascal's identity [MATH] The equation $C(n,r) = C(n-1, r) + C(n-1, r-1)$ where, in general, $C(n,r)$ is the number of distinct subsets of r elements in a set of n elements (the binomial coefficient). { pa¦skalz ī'den·ə·dē }

Pascal's law [FL MECH] The law that a confined fluid transmits externally applied pressure uniformly in all directions, without change in magnitude. { pa'skalz ˌló }

Pascal's limaçon See limaçon. { pa'skalz ˌlē·mə'son }

Pascal's theorem [MATH] The theorem that when one inscribes a simple hexagon in a conic, the three pairs of opposite sides meet in collinear points. { pa'skalz ˌthir·əm }

Pascal's triangle [MATH] A triangular array of the binomial coefficients, bordered by ones, where the sum of two adjacent entries from a row equals the entry in the next row directly below. Also known as binomial array. { pa'skalz 'trī·aŋ·gəl }

Paschen-Back effect [SPECT] An effect on spectral lines obtained when the light source is placed in a very strong magnetic field; the anomalous Zeeman effect obtained with weaker fields changes over to what is, in a first approximation, the normal Zeeman effect. { 'päsh·ən 'bäk iˌfekt }

Paschen bodies [PATH] Accumulations of the elementary bodies of smallpox. { 'päsh·ən 'bäd·ēz }

Paschen-Runge mounting [SPECT] A diffraction grating mounting in which the slit and grating are fixed, and photographic plates are clamped to a fixed track running along the corresponding Rowland circle. { 'päsh·ən 'rùŋ·ə ˌmaùnt·iŋ }

Paschen series [SPECT] A series of lines in the infrared spectrum of atomic hydrogen whose wave numbers are given by $R_H [(1/9) - (1/n^2)]$, where R_H is the Rydberg constant for hydrogen, and n is any integer greater than 3. { 'päsh·ən ˌsir·ēz }

Paschen's law [ELECTR] The law that the sparking potential between two parallel plate electrodes in a gas is a function of the product of the gas density and the distance between the electrodes. Also known as Paschen's rule. { 'päsh·ənz ˌló }

Paschen's rule See Paschen's law. { 'päsh·ənz ˌrül }

pascoite [MINERAL] $Ca_2V_6O_{17}\cdot11H_2O$ A dark-red-orange to yellow-orange mineral composed of hydrous vanadate of calcium. { 'pas·kəˌwīt }

Pasiphae [ASTRON] A small satellite of Jupiter with a diameter of about 35 miles (56 kilometers), orbiting with retrograde motion at a mean distance of about 1.46×10^7 miles (2.35×10^7 kilometers). Also known as Jupiter VIII. { pə'sif·əˌē }

pasmo [PL PATH] A destructive fungus disease of flax caused by *Mycosphaerella linorum*. { 'paz·mō }

pass [AERO ENG] **1.** A single circuit of the earth made by a satellite; it starts at the time the satellite crosses the equator from the Southern Hemisphere into the Northern Hemisphere. **2.** The period of time in which a satellite is within telemetry range of a data acquisition station. [COMPUT SCI] A complete cycle of reading, processing, and writing in a computer. [GEOGR] **1.** A natural break, depression, or other low place providing a passage through high terrain, such as a mountain range. **2.** A navigable channel leading to a harbor or river. **3.** A narrow opening through a barrier reef, atoll, or sand bar. [MECH ENG] **1.** The number of times that combustion gases are exposed to heat transfer surfaces in boilers (that is, single-pass, double-pass, and so on). **2.** In metal rolling, the passage in one direction of metal deformed between rolls. **3.** In metal cutting, transit of a metal cutting tool past the workpiece with a fixed tool setting. [MET] **1.** Passage of a metal bar between rolls. **2.** Open space between two grooved rolls through which metal is processed. **3.** Weld metal deposited in one trip along the axis of a weld. [MIN ENG] **1.** A mine opening through which coal or ore is delivered from a higher to a lower level. **2.** A passage left in old workings for workers to travel as they move from one level to another. **3.** A treatment of the whole ore sample in a sample divider. **4.** A passage of an excavation or grading machine. **5.** In surface mining, a complete excavation cycle in removing overburden. { pas }

passage [GEOGR] A navigable channel, especially one

through reefs or islands. [NAV] A transit from one place to another; one leg of a voyage. { 'pas·ij }

passage bed [GEOL] A stratum marking a transition from rocks of one geological system to those of another. { 'pas·ij ˌbed }

passage number [MICROBIO] The number of times that a culture has been subcultured. { 'pas·ij ˌnəm·bər }

passageway [SCI TECH] A way that permits passage between two places or points. { 'pas·ijˌwā }

Passalidae [INV ZOO] The peg beetles, a family of tropical coleopteran insects in the superfamily Scarabaeoidea. { pə'sal·əˌdē }

passband [ELECTR] A frequency band in which the attenuation of a filter is essentially zero. { 'pasˌband }

pass-by [ENG] The double-track part of any single-track system of rail transport. [MIN ENG] A passage around the working part of a shaft. { 'pasˌbī }

pass element [ELECTR] Controlled variable resistance device, either a vacuum tube or power transistor, in series with the source of direct-current power; the pass element is driven by the amplified error signal to increase its resistance when the output needs to be lowered or to decrease its resistance when the output must be raised. { 'pas ˌel·ə·mənt }

passenger [MOL BIO] A deoxyribonucleic acid segment that will be spliced into a plasmid or bacteriophage for subsequent cloning. { 'pas·ən·jər }

passenger car [ENG] **1.** A railroad car in which passengers are carried. **2.** An automobile for carrying as many as nine passengers. { 'pas·ən·jər ˌkär }

passenger ship [NAV ARCH] A ship used primarily to carry passengers. { 'pas·ən·jər ˌship }

Passeres [VERT ZOO] The equivalent name for Oscines. { 'pas·əˌrēz }

Passeriformes [VERT ZOO] A large order of perching birds comprising two major divisions: Suboscines and Oscines. { ˌpas·ə·rə'fór·mēz }

Passifloraceae [BOT] A family of dicotyledonous, often climbing plants in the order Violales; flowers are polypetalous and hypogynous with a corona, and seeds are arillate with an oily endosperm. { ˌpas·ə·flə'rās·ēˌē }

passiflorin See harman. { ˌpas·ə¦flór·ən }

passing point [MIN ENG] The point at which two vehicles, such as coal cars or mine elevators, pass each other while going in opposite directions. { 'pas·iŋ ˌpóint }

passing track [ENG] A sidetrack with switches at both ends. { 'pas·iŋ ˌtrak }

passivation [ELECTR] Growth of an oxide layer on the surface of a semiconductor to provide electrical stability by isolating the transistor surface from electrical and chemical conditions in the environment; this reduces reverse-current leakage, increases breakdown voltage, and raises power dissipation rating. [MET] To render passive; to reduce the reactivity of a chemically active metal surface by electrochemical polarization or by immersion in a passivating solution. { ˌpas·ə'vā·shən }

passivation potential [PHYS CHEM] The potential corresponding to the critical anodic current density of an electrode which behaves in an active-passive manner. { ˌpas·ə'vā·shən pəˌten·chəl }

passive accommodation [CONT SYS] The alteration in the positioning or motion of the end point of a robot manipulator that results from bending or deforming of the manipulator components in response to forces exerted on the robot. { 'pas·iv əˌkäm·ə'dā·shən }

passive-active cell [MET] An electrochemical corrosion cell established between passive and active areas on a metal surface. { 'pas·iv 'ak·tiv ˌsel }

passive-aggressive personality [PSYCH] A personality disorder characterized by the passive expression of hostility and aggressiveness, as by stubbornness, pouting, or inefficiency. { 'pas·iv ə'gres·iv ˌpərs·ən'al·əd·ē }

passive anaphylaxis [IMMUNOL] Anaphylaxis elicited by temporary sensitization with antibodies followed by injection of the corresponding sensitizing antigen. { 'pas·iv ˌan·ə·fə'lak·səs }

passive AND gate [ELECTR] See AND gate. [ENG] A fluidic device which achieves an output signal, by stream interaction, only when both of two control signals appear simultaneously. { 'pas·iv 'and ˌgāt }

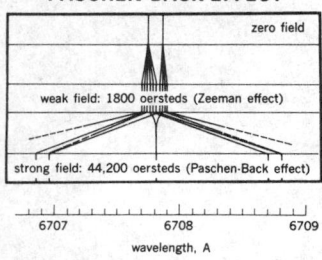

PASCHEN-BACK EFFECT

zero field

weak field: 1800 oersteds (Zeeman effect)

strong field: 44,200 oersteds (Paschen-Back effect)

6707   6708   6709
wavelength, A

Zeeman and Paschen-Back effects of red lithium doublet, whose natural separation is 0.175 angstrom.

passive antenna [ELECTROMAG] An antenna which influences the directivity of an antenna system but is not directly connected to a transmitter or receiver. { 'pas·iv an'ten·ə }

passive antiroll system [NAV ARCH] A system of antiroll tanks connected by a channel that is sized so that the flow of the liquid is out of phase with the roll of the ship. { 'pas·iv ,an·tē'rōl ,sis·təm }

passive armor [ORD] A protective device against shaped charge ammunition; designed to absorb the energy of a shaped charge. { 'pas·iv 'är·mər }

passive communications satellite [AERO ENG] A satellite that reflects communications signals between stations, without providing amplification; an example is the Echo satellite. Also known as passive satellite. { 'pas·iv kə,myü·nə'kā·shənz ,sad·əl,īt }

passive component *See* passive element. { 'pas·iv kəm'pō·nənt }

passive congestion [MED] An increased content of blood in an organ or other body part due to impaired return of venous blood. { 'pas·iv kən'jes·chən }

passive corner reflector [ELECTROMAG] A corner reflector that is energized by a distant transmitting antenna; used chiefly to improve the reflection of radar signals from objects that would not otherwise be good radar targets. { 'pas·iv 'kor·nər ri,flek·tər }

passive cutaneous anaphylaxis [IMMUNOL] The vascular reaction at the site of intradermally injected antibody when, 3 hours later, the specific antigen, usually mixed with Evans blue dye, is injected intravenously. { 'pas·iv kyü'tā·nē·əs ,an·ə·fə'lak·səs }

passive-dependent personality [PSYCH] A character disorder marked by a behavioral pattern characterized by a lack of self-confidence, indecisiveness, and a tendency to cling to and seek support from others. { 'pas·iv di'pen·dənt ,pərs·ən'al·əd·ē }

passive detection [ORD] The detection of a target or other object by means that do not reveal the position of the detecting instrument. { 'pas·iv di'tek·shən }

passive device [COMPUT SCI] A unit of a computer which cannot itself initiate a request for communication with another device, but which honors such a request from another device. { 'pas·iv di'vīs }

passive double reflector [ELECTROMAG] A combination of two passive reflectors positioned to bend a microwave beam over the top of a mountain or ridge, generally without appreciably changing the general direction of the beam. { 'pas·iv 'dəb·əl ri'flek·tər }

passive earth pressure [CIV ENG] The maximum value of lateral earth pressure exerted by soil on a structure, occurring when the soil is compressed sufficiently to cause its internal shearing resistance along a potential failure surface to be completely mobilized. { 'pas·iv 'ərth ,presh·ər }

passive electronic countermeasures [ELECTR] Electronic countermeasures that do not radiate energy, including reconnaissance or surveillance equipment that detects and analyzes electromagnetic radiation from radar and communications transmitters, and devices such as chaff which return spurious echoes to enemy radar. { 'pas·iv i,lek'trän·ik 'kaunt·ər,mezh·ərz }

passive element [ELEC] An element of an electric circuit that is not a source of energy, such as a resistor, inductor, or capacitor. Also known as passive component. [ELECTROMAG] *See* parasitic element. { 'pas·iv 'el·ə·mənt }

passive filter [ELEC] An electric filter composed of passive elements, such as resistors, inductors, or capacitors, without any active elements, such as vacuum tubes or transistors. { 'pas·iv 'fil·tər }

passive fold [GEOL] A fold in which the mechanism of folding, either flow or slip, crosses the boundaries of the strata at random. { 'pas·iv 'fōld }

passive front *See* inactive front. { 'pas·iv 'frənt }

passive glacier [HYD] A glacier with sluggish movement, generally occurring in a continental environment at a high latitude, where both accumulation and ablation are minimal. { 'pas·iv 'glā·shər }

passive guidance [NAV] Guidance of a vehicle by preset or inertial devices, without reliance on external signals or observations. { 'pas·iv 'gīd·əns }

passive homing [NAV] The homing of an aircraft or spacecraft wherein the craft directs itself toward its destination by receiving radio emission from a base station without the necessity of initiating the transmission by sending an interrogating signal from the craft. { 'pas·iv 'hōm·iŋ }

passive immunity [IMMUNOL] **1.** Immunity acquired by injection of antibodies in another individual or in an animal. **2.** Immunity acquired by the fetus by the transfer of maternal antibodies through the placenta. { 'pas·iv i'myün·əd·ē }

passive immuno-gene therapy [IMMUNOL] Immuno-gene therapy in which autologous T cells are repeatedly sensitized in culture by exposure to cytokines or by transfer of cytokine genes and then are transferred back into the patient. { ¦pas·iv ¦im·yə·nō¦jēn ,ther·ə·pē }

passive immunotherapy [IMMUNOL] Immunotherapy involving the transfer of antibodies to tumor-bearing recipients. { ¦pas·iv ,im·yə'ther·ə·pē }

passive jamming [ELECTR] Use of confusion reflectors to return spurious and confusing signals to enemy radars. Also known as mechanical jamming. { 'pas·iv 'jam·iŋ }

passive junction [ELECTROMAG] A waveguide junction that does not have a source of energy. { 'pas·iv 'jəŋk·shən }

passive margin [GEOL] A continental margin formed by rifting during continental breakup. { 'pas·iv 'mär·jən }

passive-matrix liquid-crystal display *See* supertwisted nematic liquid-crystal display. { ¦pas·iv ¦ma·triks ¦lik·wəd ¦krist·əl di'splā }

passive metal [MET] A metal on which a surface film forms by natural process or by immersion in a passivating solution, making the metal resistant to corrosion. { 'pas·iv 'med·əl }

passive method [CIV ENG] A construction method in permafrost areas in which the frozen ground near the structure is not disturbed or altered, and the foundations are provided with additional insulation to prevent thawing of the underlying ground. { 'pas·iv 'meth·əd }

passive network [ELEC] A network that has no source of energy. { 'pas·iv 'net,wərk }

passive permafrost [GEOL] Permafrost that will not refreeze under present climatic conditions after being disturbed or destroyed. Also known as fossil permafrost. { 'pas·iv 'pər·mə,fròst }

passive radar [ENG] A technique for detecting objects at a distance by picking up the microwave electromagnetic energy that is both radiated and reflected by all bodies. { 'pas·iv 'rā,där }

passive radiator [ENG ACOUS] A loudspeaker driver with no voice-coil or magnet assemblies that is mounted in a box with a woofer and exhibits a resonance that can be used to improve the low-frequency response of the system. { ¦pas·iv '¦ād·ē,ēd·ər }

passive-radiator system [ELECTR] A loudspeaker system in which the woofer is mounted in a box that also has a second speaker with no voice-coil or magnet assemblies. { ¦pas·iv 'rād·ē,ād·ər ,sis·təm }

passive reflector [ELECTROMAG] A flat reflector used to change the direction of a microwave or radar beam; often used on microwave relay towers to permit placement of the transmitter, repeater, and receiver equipment on the ground, rather than at the tops of towers. Also known as plane reflector. { 'pas·iv ri'flek·tər }

passive satellite *See* passive communications satellite. { 'pas·iv 'sad·əl,īt }

passive solar system [MECH ENG] A solar heating or cooling system that operates by using gravity, heat flows, or evaporation rather than mechanical devices to collect and transfer energy. { 'pas·iv 'sō·lər ,sis·təm }

passive sonar [ENG] Sonar that uses only underwater listening equipment, with no transmission of location-revealing pulses. { 'pas·iv 'sō,när }

passive system [ELECTR] Electronic system which emits no energy, and does not give away its position or existence. { 'pas·iv ,sis·təm }

passive termination [COMPUT SCI] The simplest means of ending a chain of peripheral devices connected to a small computer system interface (SCSI) port, suitable for chains with no more than four devices. { ,pas·iv ,tər·mə'nā·shən }

passive transducer [ELECTR] A transducer containing no internal source of power. { 'pas·iv tranz'dü·sər }

passivity [CHEM] A state of chemical inactivity, especially

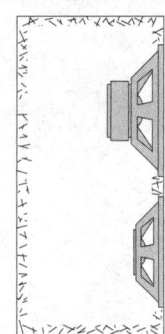

PASSIVE-RADIATOR SYSTEM

Passive-lower-driver system.

of a metal that is relatively resistant to corrosion due to loss of chemical activity. [MET] The property of a metal that has been made passive. { pə'siv·əd·ē }

pass point [GRAPHICS] A point whose horizontal or vertical position is determined from photographs by photogrammetric methods and which is intended for use as a supplemental control point in the orientation of other photographs. { 'pas ,pȯint }

passthrough [COMPUT SCI] A procedure that allows a user to communicate with a computer through the use of the operating system of a second computer. { 'pas,thrü }

password [COMPUT SCI] A unique word or string of characters that must be supplied to meet security requirements before a program, computer operator, or user can gain access to data. { 'pas,wərd }

password guessing [COMPUT SCI] A method of gaining unauthorized access to a computing system by using computers and dictionaries or large word lists to try likely passwords. { 'pas,wərd ,ges·iŋ }

past [RELAT] For an event in space-time, all events from which a signal could be emitted that could reach the event in question by traveling at speeds less than or equal to the speed of light. { 'past }

past Cauchy development [RELAT] The set of points p relative to a surface S in space-time such that every future-directed timelike or null curve through p intersects S. Symbolized D⁻(S). { ¦past 'kō·shē di,vel·əp·mənt }

paste [ELEC] In batteries, the medium in the form of a paste or jelly, containing an electrolyte; it is positioned adjacent to the negative electrode of a dry cell; in an electrolytic cell, the paste serves as one of the conducting plates. [MATER] An adhesive mixture with a characteristic plastic consistency, a high order of yield value, and a low bond strength; for example, a paste prepared by heating a starch and water mixture, then cooling the hydrolyzed product. [MET] Finely divided particles of ferromagnetic material in paste form used in the wet method of magnetic particle inspection. { pāst }

pasteboard [MATER] A type of thin cardboard made from gluing together two or more sheets of paper. { 'pāst,bȯrd }

pasted-plate storage battery See Faure storage battery. { 'pās·təd ¦plāt 'stȯr·ij ,bad·ə·rē }

paste ink [MATER] A pastelike mixture of pigment or dye with oil and other additives, such as resins, driers, tackifiers, and adhesives; used in paper and textile printing and ballpoint pens. { 'pāst ,iŋk }

pastel [MATER] A chalk or crayon made of a finely ground pigment and a minimum of nongreasy binder, such as gum tragacanth or methylcellulose; since pastels are blended on the painting itself, a larger assortment of shades and tints is required than with oil or other color mediums. { pa'stel }

pastel fixative [MATER] A thin varnish material for the simple and even application to drawings; only enough fixative should be used to keep the pastel particles from falling off the paper. { pa'stel 'fik·səd·iv }

paste mixer [ENG] Device for the blending together of solid particles and a liquid, with the final formation of a single paste phase. [FOOD ENG] See melangeur. { 'pāst ,mik·sər }

paste resin [MATER] Solventless, fluid or semisolid mixture of powdered resin and plasticizer. { 'pāst ,rez·ən }

paste solder [MET] Finely powdered solder metal combined with a flux. { 'pāst ,säd·ər }

paste-up See mechanical. { 'pāst,əp }

Pasteur-Chamberland filter [MICROBIO] A porcelain filter with small pores used in filtration sterilization. { pa'stər 'chām·bər·lənd ,fil·tər }

Pasteur effect [MICROBIO] Inhibition of fermentation by supplying an abundance of oxygen to replace anerobic conditions. { pa'stər i,fekt }

Pasteurella [MICROBIO] A genus of gram-negative, nonmotile, nonsporulating, facultatively anaerobic coccobacillary to rod-shaped bacteria which are parasitic and often pathogens in many species of mammals, birds, and reptiles; it was named to honor Louis Pasteur in 1887. { ,pas·chə'rel·ə }

pasteurellosis See hemorrhagic septicemia. { ,pa·stə·rə'lō·səs }

Pasteuriaceae [MICROBIO] Formerly a family of stalked bacteria in the order Hyphomicrobiales. { ,pa·stə·rē,ās·ē,ē }

pasteurization [SCI TECH] The application of heat to matter for a specified time to destroy harmful microorganisms or other undesirable species.

pasteurizer [ENG] An apparatus used for pasteurization of fluids. { 'pas·chə,rīz·ər }

Pasteur's salt solution [ANALY CHEM] Laboratory reagent consisting of potassium phosphate and calcium phosphate, magnesium sulfate, and ammonium tartrate in distilled water. { pa'stȯrz 'sȯlt sə,lü·shən }

past light cone [RELAT] The set of all points in space-time from which signals traveling at the speed of light reach a specified point. { 'past ¦līt ,kōn }

patagium [VERT ZOO] **1.** A membrane or fold of skin extending between the forelimbs and hindlimbs of flying squirrels, flying lizards, and other arboreal animals. **2.** A membrane or fold of skin on a bird's wing anterior to the humeral and radioulnar bones. { pə'tā·jē·əm }

patch [COMPUT SCI] **1.** To modify a program or routine by inserting a machine language correction in an object deck, or by inserting it directly into the computer through the console. **2.** The section of coding inserted in this way. [ELEC] A temporary connection between jacks or other terminations on a patch board. { pach }

patch board [ELEC] A board or panel having a number of jacks at which circuits are terminated; patch cords are plugged into the jacks to connect various circuits temporarily as required in broadcast, communication, and computer work. { 'pach ,bȯrd }

patch bolt [DES ENG] A bolt with a countersunk head having a square knob that twists off when the bolt is screwed in tightly; used to repair boilers and steel ship hulls. { 'pach ,bōlt }

patch budding [BOT] Budding in which a small rectangular patch of bark bearing the scion (a bud) is fitted into a corresponding opening in the bark of the stock. { 'pach ,bəd·iŋ }

patch cord [ELEC] A cord equipped with plugs at each end, used to connect two jacks on a patch board. { 'pach ,kȯrd }

patchouli oil [MATER] A brownish essential oil with strong camphor aroma; soluble in ether, chloroform, alcohol, and oils; derived from dried leaves of the patchouli (*Pogostemon patchouly*); main components are patchouli alcohol, eugenol, and cinnamic aldehyde; used to perfume toiletries. { pə'chü·lē ,ȯil }

patch panel See control panel; panel. { 'pach ,pan·əl }

patch reef [GEOL] **1.** A small, irregular organic reef with a flat top forming a part of a reef complex. **2.** A small, thick, isolated lens of limestone or dolomite surrounded by rocks of different facies. **3.** See reef patch. { 'pach ,rēf }

patch test [IMMUNOL] A test in which material is applied and left in contact with intact skin surfaces for 48 hours in order to demonstrate tissue sensitivity. { 'pach ,test }

patella [ANAT] A sesamoid bone in front of the knee, developed in the tendon of the quadriceps femoris muscle. Also known as kneecap. { pə'tel·ə }

Patellacea [PALEON] An extinct superfamily of gastropod mollusks in the order Aspidobranchia which developed a cap-shaped shell and were specialized for clinging to rock. { ,pad·əl'ās·ē·ə }

Patellidae [INV ZOO] The true limpets, a family of gastropod mollusks in the order Archeogastropoda. { pə'tel·ə,dē }

patent [IND ENG] A certificate of grant by a government of an exclusive right with respect to an invention for a limited period of time. Also known as letters patent. [MED] Open; exposed. { 'pat·ənt }

patent base [GRAPHICS] In letterpress technology, a metal base with slots that is used to hold unmounted electrotypes. { 'pat·ənt ,bās }

patented claim [MIN ENG] A mining claim to which a patent has been secured from the government by compliance with the laws relating to such claims. { 'pat·ənt·əd 'klām }

patent foramen ovale [MED] Persistence, usually functional, of the fetal foramen ovale after birth. { 'pat·ənt fə'rā·mən 'ōv·yül }

patenting [MET] A process used in the production of high-strength steel wire containing 0.35–0.85% carbon, in which the wire is heated to above the transformation temperature, then quenched in molten lead or molten salt, or cooled in air. { 'pat·ənt·iŋ }

patent leveling See stretcher leveling. { 'pat·ənt 'lev·ə·liŋ }

patent log [NAV] Any mechanical log; for example, an instrument that is dragged from the stem of a vessel to measure the speed or the distance traveled. { 'pat·ənt 'läg }

patent medicine [PHARM] A medicine, generally trade-marked, whose composition is incompletely disclosed. { 'pat·ənt 'med·ə·sən }

patent period [MED] The period of an infective disease during which the causative agent can be detected. { 'pat·ənt ,pir·ē·əd }

patera [ASTRON] A shallow crater with complex scalloped edges, which occurs on Mars but has not been observed on any other planet. { pə'ter·ə }

Patera process [MET] A method used in the 19th century for extracting silver from ore: the ore was roasted with sodium chloride, a solution of sodium thiosulfate was then added to leach out the silver chloride, and sodium sulfide precipitated the silver as silver sulfide. { pə'ter·ə ,prä·səs }

Paterinida [PALEON] A small extinct order of inarticulated brachiopods, characterized by a thin shell of calcium phosphate and convex valves. { ,pad·ə'rin·əd·ə }

paternity test [IMMUNOL] Identification of the blood groups of a mother, her child, and a putative father in order to establish the probability of paternity or nonpaternity; actually, only non-paternity can be established. { pə'tər·nəd·ē ,test }

paternoite See kaliborite. { ,päd·ər'nō,īt }

paternoster lake [HYD] One of a linear chain or series of small circular lakes, usually at different levels, which occupy rock basins in a glacial valley and are separated by morainal dams or riegels, but connected by streams, rapids, or waterfalls to resemble a rosary or string of beads. Also known as beaded lake; rock-basin lake; step lake. { 'päd·ər,näs·tər ,lāk }

Paterson-Kelly syndrome See Plummer-Vinson syndrome. { 'pad·ər·sən 'kel·ē ,sin,drōm }

path [COMPUT SCI] **1.** The logical sequence of instructions followed by a computer in carrying out a routine. **2.** A series of physical or logical connections between records or segments in a database management system, generally involving the use of pointers. [MATH] **1.** In a topological space, a path is a continuous curve joining two points. **2.** In graph theory, a walk whose vertices are all distinct. Also known as simple path. **3.** See walk. [NAV] A line connecting a series of points and constituting a proposed or traveled route. { path }

path attenuation [COMMUN] Power loss between transmitter and receiver, due to any cause. { 'path ə ,ten·yə'wā·shən }

path clamp [PHYSIO] An electrophysiologic recording technique for studying the function and regulation of ionic currents in cell membranes down to the level of single ion channels. { 'path ,klamp }

path coefficient [COMMUN] The ratio of the power transmitted over some designated path to that transmitted over the most direct path. { 'path ,kō·i,fish·ənt }

path computation [CONT SYS] The calculations involved in specifying the trajectory followed by a robot. { 'path ,käm·pyə,tā·shən }

path-connected set See arcwise-connected set. { 'path kə,nek·təd ,set }

path curve [MATH] A curve whose equation is given in parametric form. { 'path ,kərv }

pathergy [IMMUNOL] Either a subnormal response to an allergen or an unusually intense one in which the individual becomes sensitive not only to the specific substance but to others. { 'path·ər·jē }

path integral [QUANT MECH] An integral of a functional over function space; central to a formulation of quantum mechanics developed by R. Feynman. { 'path ,int·ə·grəl }

path length See physical path length; software path length. { 'path ,leŋkth }

pathogen [MED] A disease-producing agent; usually refers to living organisms. { 'path·ə·jən }

pathogenesis [MED] The origin and course of development of disease. { ,path·ə'jen·ə·səs }

pathogenic [MED] **1.** Producing or capable of producing bdisease. **2.** Pertaining to pathogenesis. { ,path·ə'jen·ik }

pathogenicity [MED] The ability of an organism to enter a host and cause disease. { ,path·ə·jə'nis·əd·ē }

pathogenicity island [GEN] A deoxyribonucleic acid cluster commonly containing genes associated with pathogenesis. Also known as virulence cassette. { ,path·ə·jə'nis·əd·ē ,ī·lənd }

pathognomonic [MED] Characteristic of a given disease, enabling it to be distinguished from other diseases. { ,path·əg·nō'män·ik }

pathologic anatomy [PATH] The study of structural changes caused by disease. Also known as morbid anatomy. { ,path·ə'läj·ik }

pathologist [MED] A person who specializes in the study and practice of pathology. { pə'thäl·ə·jəst }

pathology [MED] The study of the causes, nature, and effects of diseases and other abnormalities. { pə'thäl·ə·jē }

pathophobia [PSYCH] Exaggerated dread of death. { ,path·ə'fō·bē·ə }

pathopsychology [PSYCH] The branch of science dealing with mental processes, particularly as manifested by abnormal cognitive, perceptual, and intellectual functioning, during the course of mental disorders. { ,path·ō·sī'käl·ə·jē }

pathotoxin [PL PATH] A chemical of biological origin, other than an enzyme, that plays an important causal role in a plant disease. { ,path·ə'täk·sən }

pathovar [MICROBIO] A pathological variant of a nonpathological bacterial species. { 'path·ə,vär }

path plotting [ELECTROMAG] In laying out a microwave system, the plotting of the path followed by the microwave beam on a profile chart which indicates the earth's curvature. { 'path ,pläd·iŋ }

pathwise-connected set See arcwise-connected set. { 'path,wīz kə,nek·təd ,set }

Patientia [ASTRON] An asteroid with a diameter of about 153 miles (247 kilometers), mean distance from the sun of 3.06 astronomical units, and B-type (C-like) surface composition. { pä·shē'en·chə }

patina [GEOL] A thin, colored film produced on a rock surface by weathering. [MET] The greenish product, usually basic copper sulfate, formed on copper and copper-rich alloys as a result of prolonged atmospheric corrosion. { 'pat·ən·ə or pə'tē·nə }

patio process [MET] A crude chemical method of reducing silver from its ores, followed by amalgamation in low heaps with the aid of salt and copper sulfate. { 'pad·ē·ō 'prä·səs }

patroclinous inheritance [GEN] Inheritance in which the offspring more closely resembles the male parent than the female parent. { ,pa·trə,klī·nəs in'her·əd·əns }

Patroclus group See pure Trojan group. { pə'trō·kləs ,grüp }

patronite [MINERAL] A black vanadium sulfide mineral; mined as a vanadium ore in Minasragra, Peru. { 'pa·trə,nīt }

pattern [AERO ENG] The flight path flown by an aircraft, or prescribed to be flown, as in making an approach to a landing. [ENG] A form designed and used as a model for making things. [GRAPHICS] A design or form. [MATH] An equivalence class of colorings of the elements of a finite set, which are indistinguishable with respect to a group of permutations of the colors. [ORD] The distribution of a series of shots fired from one gun or a battery of guns under conditions as nearly identical as possible, the points of impact of the projectiles being dispersed about a point called the center of impact. { 'pad·ərn }

pattern analysis [COMPUT SCI] The phase of pattern recognition that consists of using whatever is known about the problem at hand to guide the gathering of data about the patterns and pattern classes, and then applying techniques of data analysis to help uncover the structure present in the data. { 'pad·ərn ə,nal·ə·səs }

pattern bombing [ORD] A method of bombing in which the bombs are made to strike the target in a certain pattern. { 'pad·ərn ,bäm·iŋ }

pattern card [TEXT] A perforated card used in jacquard weaving. { 'pad·ərn ,kärd }

pattern chain [TEXT] A device on a loom or knitting machine that controls the pattern in the material. { 'pad·ərn ,chān }

patterned ground [GEOL] Any of several well-defined, generally symmetrical forms, such as circles, polygons, and steps, that are characteristic of surficial material subject to intensive frost action. { 'pad·ərnd 'graùnd }

pattern flood [PETRO ENG] Waterflood of a petroleum reservoir in which there is an areal pattern of injection wells located so as to sweep oil from the area toward the bores of producing wells. { 'pad·ərn ,fləd }

pattern formation [EMBRYO] The embryogenic process in which the spatial differentiation of cells is specified in a structure that initially is largely homogeneous. { 'pad·ərn fər ,mā·shən }

pattern gene [GEN] A gene involved in the establishment of a particular pattern during cell differentiation. { 'pad·ərn ‚jēn }

pattern generator [ELECTR] A signal generator used to generate a test signal that can be fed into a television receiver to produce on the screen a pattern of lines having usefulness for servicing purposes. { 'pad·ərn ‚jen·ə‚rād·ər }

pattern harmonization [ORD] The adjusting of a fighter aircraft's fixed guns so that they will produce the largest uniform pattern of lethal density possible at a given range. { 'pad·ərn ‚har·mə·nə'zā·shən }

pattern recognition [COMPUT SCI] The automatic identification of figures, characters, shapes, forms, and patterns without active human participation in the decision process. { 'pad·ərn ‚rek·ig'nish·ən }

pattern-sensitive fault [COMPUT SCI] A fault that appears only in response to one pattern or sequence of data, or certain patterns or sequences. { 'pad·ərn ‚sen·səd·iv 'fȯlt }

pattern shooting [ENG] In seismic prospecting, firing of explosive charges arranged in geometric pattern. { 'pad·ərn ‚shüd·iŋ }

Patterson function [SOLID STATE] A function of three spatial coordinates, constructed in the Patterson-Harker method, which has peaks at all vectors between two atoms in a crystal, the heights of the peaks being approximately proportional to the product of the atomic numbers of the corresponding atoms. { 'pad·ər·sən ‚fəŋk·shən }

Patterson-Harker method [SOLID STATE] A method of analyzing the structure of a crystal from x-ray diffraction results; a Fourier series involving squares of the absolute values of the structure factors, which are directly observable, is used to construct a vectorial representation of interatomic distances in the crystal (Patterson map). { 'pad·ər·sən 'här·kər ‚meth·əd }

Patterson map [SOLID STATE] A contour chart of the Patterson function. { 'pad·ər·sən ‚map }

Patterson projection [SOLID STATE] A projection of the Patterson function on a section through a crystal. { 'pad·ər·sən prə‚jek·shən }

Patterson vectors [SOLID STATE] In analysis of crystal structure, the vectors of peaks relative to the origin in a Patterson function or Patterson projection. { 'pad·ər·sən ‚vek·tərz }

Pattinson process [MET] A method for separating silver from its alloys rich in lead by slow cooling of the melt so that silver-poor lead crystals separate out and are removed. { 'pat·ən·sən ‚prä·səs }

patulin [PHARM] $C_7H_6O_4$ An antibiotic derived from several fungi (Aspergillus, Penicillium species); crystalline compound soluble in water and most organic solvents; melting point is 111°C; used as an antimicrobial agent; also appears to be a potent carcinogenic mycotoxin. Also known as penicidin. { 'pach·ə·lən }

paua See abalone. { 'pau·ə }

Paucituberculata [VERT ZOO] An order of marsupial mammals in some systems of classification, including the opossum, rats, and polydolopids. { ‚pȯs·ē·tə‚bər·kyə'läd·ə }

Paul-Bunnell test See heterophile agglutination test. { 'pȯl 'bən·əl ‚test }

Pauli anomalous moment term [QUANT MECH] An additional term inserted in the Dirac equation to provide for a g-value of the particle different from 2. { 'pȯl·ē ə'näm·ə·ləs 'mō·mənt ‚tərm }

Pauli electron correlation [QUANT MECH] Correlation in space of electrons as a result of the Pauli exclusion principle. { 'pȯl·ē i'lek‚trän ‚kär·ə‚lā·shən }

Pauli exclusion principle See exclusion principle. { 'pȯl·ē ik'sklü·zhən ‚prin·sə·pəl }

Pauli-Fermi principle [QUANT MECH] The principle that each level of a quantized system can include one, two, or no electrons; if there are two electrons, they must have spins in opposite directions. { 'pȯl·ē 'fer·mē ‚prin·sə·pəl }

Pauli g-permanence rule [ATOM PHYS] For given L, S, and M_J in LS coupling, the sum, over J, of the weak-field g-factors is equal to the sum of the strong-field g-factors. { 'pȯl·ē ǀjē 'pər·mə·nəns ‚rül }

Pauli g-sum rule [ATOM PHYS] For all the states arising from a given electron configuration, the sum of the g-factors for levels with the same J value is a constant, independent of the coupling scheme. { 'pȯl·ē ǀjē ‚səm ‚rül }

---

paulin [MATER] A fabricated textile item generally used as a weather protection cover for various items or materials during storage or transit. { 'pȯl·ən }

paulingite [MINERAL] An isometric zeolite mineral consisting of an aluminosilicate of potassium, calcium, and sodium. { 'pȯl·iŋ‚īt }

Pauling rule [SOLID STATE] A rule governing the number of ions of opposite charge in the neighborhood of a given ion in an ionic crystal, in accordance with the requirement of local electrical neutrality of the structure. { 'pȯl·iŋ ‚rül }

Pauling scale [PHYS CHEM] A numerical scale of electronegativities based on bond-energy calculations for different elements joined by covalent bonds. { 'pȯl·iŋ ‚skāl }

Pauli paramagnetism See free-electron paramagnetism. { ǀpau·lē ‚par·ə'mag·nə‚tiz·əm }

Pauli spin matrices [QUANT MECH] Three anticommuting matrices, each having two rows and two columns, which represent the components of the electron spin operator:

$$\sigma_x = \begin{pmatrix} 0 & 1 \\ 1 & 0 \end{pmatrix} \quad \sigma_y = \begin{pmatrix} 0 & -i \\ j & 0 \end{pmatrix} \quad \sigma_z = \begin{pmatrix} 1 & 0 \\ 0 & -1 \end{pmatrix}$$

{ 'pȯl·ē 'spin ‚mā·trə·sēz }

Pauli spin space [QUANT MECH] A two-dimensional vector space over the complex numbers, whose vectors describe orientations of the electron spin. { 'pȯl·ē 'spin 'spās }

Pauli spin susceptibility [SOLID STATE] The susceptibility of free electrons in a metal due to the tendency of their spins to align with a magnetic field. { 'pȯl·ē ǀspin sə‚sep·tə'bil·əd·ē }

Pauli-Weisskopf equation [QUANT MECH] The equation resulting from second quantization of the Klein-Gordon equation. { 'pȯl·ē 'vīs‚kȯpf i‚kwā·zhən }

paulopost See deuteric. { 'pȯl·ə‚pōst }

Paul trap [PHYS] A device in which ions and other charged particles can be suspended by radio-frequency electric fields having a quadrupole configuration, for times limited chiefly by collisions with the background gas. { 'pȯl ‚trap }

paunch [ANAT] In colloquial usage, the cavity of the abdomen and its contents. [VERT ZOO] See rumen. { pȯnch }

paurometabolous metamorphosis [INV ZOO] A simple, gradual, direct metamorphosis in which immature forms resemble the adult except in size and are referred to as nymphs. { ‚pȯr·ē·mə'tab·ə·ləs ‚med·ə'mȯr·fə·səs }

Pauropoda [INV ZOO] A class of the Myriapoda distinguished by bifurcate antennae, 12 trunk segments with 9 pairs of functional legs, and the lack of eyes, spiracles, tracheae, and a circulatory system. { pȯ'räp·ə·də }

Paussidae [INV ZOO] The flat-horned beetles, a family of coleopteran insects in the suborder Adephaga. { 'pȯs·ə‚dē }

pavement [BUILD] A hard floor of concrete, brick, tiles, or other material. [CIV ENG] A paved surface. [GEOL] A bare rock surface that suggests a paved road surface or other pavement in smoothness, hardness, horizontality, surface extent, or close packing of units. { 'pāv·mənt }

pavement light [CIV ENG] A window built into the surface of a pavement to admit daylight to a space below ground level. { 'pāv·mənt ‚līt }

paver [MECH ENG] Any of several machines which, moving along the road, carry and lay paving material. { 'pāv·ər }

pavilion [LAP] The portion of a faceted gemstone below the girdle. Also known as base. { pə'vil·yən }

pavilion facet [LAP] A main facet on the pavilion of any fashioned gemstone. { pə'vil·yən ‚fas·ət }

paving brick [MATER] A vitrified clay brick used in the construction of pavements. { 'pāv·iŋ ‚brik }

paving-brick clay [MATER] Impure shale or fire clay with good tensile strength and plasticity; used to make paving bricks. { 'pāv·iŋ ‚brik ‚klā }

Pavlov's pouch [PHYSIO] A small portion of stomach, completely separated from the main stomach, but retaining its vagal nerve branches, which communicates with the exterior; used in the long-term investigation of gastric secretion, and particularly in the study of conditioned reflexes. { 'pav‚läfs ‚pauch }

Pavo [ASTRON] A southern constellation; right ascension 20 hours, declination 65°S. Also known as Peacock. { 'pä·vō }

pavonite [MINERAL] $AgBi_3S_5$ A mineral composed of silver bismuth sulfide. { 'pa·və‚nīt }

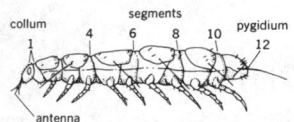

PAUROPODA

Pauropus silvaticus, showing the 9 pairs of legs and the 12 trunk segments indicated by number.

Pavy's solution [ANALY CHEM] Laboratory reagent used to determine the concentration of sugars in solution by color titration; contains copper sulfate, sodium potassium tartrate, sodium hydroxide, and ammonia in water solution. { 'pā·vēz sə‚lü·shən }

paw [VERT ZOO] The foot of an animal, especially a quadruped having claws. { pö }

pawdite [PETR] A dark-colored, fine-grained, granular hypabyssal rock composed of magnetite, titanite, biotite, hornblende, calcic plagioclase, and traces of quartz. { 'pö‚dīt }

pawl [MECH ENG] The driving link or holding link of a ratchet mechanism, permits motion in one direction only. { pöl }

PAX See private automatic exchange. { paks }

paxilla [INV ZOO] A pillarlike spine in certain starfishes that sometimes has a flattened summit covered with spinules. { pak'sil·ə }

Paxillosida [INV ZOO] An order of the Asteroidea in some systems of classification, equivalent to the Paxillosina. { ‚pak·sə'läs·əd·ə }

Paxillosina [INV ZOO] A suborder of the Phanerozonida with pointed tube feet which lack suckers, and with paxillae covering the upper body surface. { ‚pak·sə'läs·ə·nə }

payback period [IND ENG] The amount of time required for achieving an amount in profits to offset the cost of a capital expenditure, such as the cost of investment in modifications in an industrial facility for the purpose of conserving energy. { 'pā‚bak ‚pir·ē·əd }

pay dirt [MIN ENG] Profitable mineral-rich earth or ore. { 'pā ‚dərt }

paying [NAV ARCH] **1.** Filling the seams between planks with pitch, marine glue, or other material after the calking has been inserted. **2.** Slackening away on rope or chain. { 'pā·iŋ }

payload [AERO ENG] That which an aircraft, rocket, or the like carries over and above what is necessary for the operation of the vehicle in its flight. [MIN ENG] The weight of coal, ore, or mineral handled, as distinct from dirt, stone, or gangue. { 'pā‚lōd }

payload-mass ratio [AERO ENG] Of a rocket, the ratio of the effective propellant mass to the initial vehicle mass. { 'pā‚lōd ‚mas ‚rā·shō }

payoff matrix [MATH] A matrix arising from certain two-person games which gives the amount gained by a player. { 'pā‚öf ‚mā·triks }

pay ore [MIN ENG] Ore which can be mined, concentrated, or smelted at current cost of exploitation profitably at ruling market value of products. { 'pā ‚ör }

payout time [IND ENG] A measurement of profitability or liquidity of an investment, being the time required to recover the original investment in depreciable facilities from profit and depreciation; usually, but not always, calculated after income taxes. { 'pā‚aut ‚tīm }

pay sand [PETRO ENG] That portion of an oil or gas sand in which the oil or gas is found in commercial quantity. { 'pā ‚sand }

pay streak [MIN ENG] A layer of oil, ore, or other mineral that can be mined profitably. { 'pā ‚strēk }

pay television See subscription television. { 'pā 'tel·ə‚vizh·ən }

pay zone [PETRO ENG] The reservoir rock in which oil and gas are found in exploitable quantities. { 'pā ‚zōn }

Pb See lead.

P band [COMMUN] A band of radio frequencies extending from 225 to 390 megahertz, corresponding to wavelengths of 133.3 to 76.9 centimeters. { 'pē ‚band }

PBI See protein-bound iodine.

Pb-I-Pb junction See lead-I-lead junction.

PBI test See protein-bound iodine test. { ‚pē‚bē'ī ‚test }

p-block elements [CHEM] Elements of the main groups III–VII and 0 in the periodic table whose outer electronic configurations have occupied p levels. { ‚pē 'bläk 'el·ə‚mənts }

P blood group [IMMUNOL] A system of immunologically distinct, genetically determined erythrocyte antigens first defined by their reaction with anti-P, and immune rabbit antiserum, and later broadened to include related antigens. { 'pē 'bləd ‚grüp }

P-branch [SPECT] A series of lines in molecular spectra that correspond, in the case of absorption, to a unit decrease in the rotational quantum number J. { 'pē ‚branch }

PBX See private branch exchange.

PC See phosphocreatine.

PCA See polar-cap absorption.

PCB See polychlorinated biphenyl.

p-channel metal-oxide semiconductor See PMOS. { ‚pē ‚chan·əl ‚med·əl ‚äk‚sīd 'sem·i·kən‚dək·tər }

p chart [IND ENG] A chart of the fraction defective, either observed in the sample or in some production period. { 'pē ‚chärt }

P class [COMPUT SCI] The class of decision problems that can be solved in polynomial time.

PCI See peripheral component interconnect.

PCM See pulse-code modulation.

PCN See personal communications network.

PCNB See pentachloronitrobenzene.

PCP See phencyclidine; primary control program.

PCR See polymerase chain reaction.

PCS See personal communications service.

PCSB See pulse-coded scanning beam.

P Cygni star [ASTRON] An explosive variable star of spectral type B, with broad emission lines and strong absorption of violet light. { 'pē 'sig·nē ‚stär }

Pd See palladium.

PD See potential difference.

PDA See postacceleration.

PDB See PeeDee belemnite.

PDF See portable document format.

PDGF See platelet-derived growth factor.

P display See plan position indicator. { 'pē di‚splā }

PDL See page description language. { ‚pē‚dē'el }

pdl-ft See foot-poundal.

PDM See precedence diagram method; pulse-duration modulation.

PDMS See plasma desorption mass spectrometry.

PDR See precision depth recorder.

4PDT See four-pole double-throw.

PDU See power distribution unit.

pea [BOT] **1.** *Pisum sativum.* The garden pea, an annual leafy leguminous vine cultivated for its smooth or wrinkled, round edible seeds which are borne in dehiscent pods. **2.** Any of several related or similar cultivated plants. { pē }

peach [BOT] *Prunus persica.* A low, spreading, freely branching tree of the order Rosales, cultivated in less rigorous parts of the temperate zone for its edible fruit, a juicy drupe with a single large seed, a pulpy yellow or white mesocarp, and a thin firm epicarp. { pēch }

peachblossom ore See erythrite. { 'pēch‚bläs·əm ‚ör }

pea coal [GEOL] A size of anthracite that will pass through a $^{13}/_{16}$-inch (20.6-millimeter) round mesh but not through a $^9/_{16}$-inch (14.3-millimeter) round mesh. { 'pē ‚kōl }

Peacock See Pavo. { 'pē‚käk }

peacock blue [ORG CHEM] $HSO_3C_6H_4COH[C_6H_4N-(C_2H_5)CH_2C_6H_4SO_3Na]_2$ Blue pigment used in inks for multicolor printing. { 'pē‚käk 'blü }

peacock copper See peacock ore. { 'pē‚käk 'käp·ər }

peacock ore [MINERAL] A copper mineral, such as bornite, having an iridescent tarnished surface upon exposure to air. Also known as peacock copper. { 'pē‚käk ‚ör }

pea gravel [GEOL] A type of gravel whose individual particles are about the size of peas. { 'pē ‚grav·əl }

peak [GEOL] **1.** The conical or pointed top of a hill or mountain. **2.** An individual mountain or hill taken as a whole, used especially when it is isolated or has a pointed, conspicuous summit. [METEOROL] The point of intersection of the cold and warm fronts of a mature extra-tropical cyclone. [SCI TECH] The maximum instantaneous value of a quantity. { pēk }

peak amplitude [PHYS] The maximum amplitude of an alternating quantity, measured from its zero value. { 'pēk 'am·plə‚tüd }

peak analysis [SPECT] Determination of the relevant peak parameters, such as position or area, from a spectogram. { 'pēk ə‚nal·ə·səs }

peak area [ANALY CHEM] The area enclosed between the peak and the base line on a spectrogram or chromatogram. { 'pēk ‚er·ē·ə }

peak attenuation [COMMUN] The diminution of response to a modulated wave experienced on modulation crests. { 'pēk ə‚ten·yə'wā·shən }

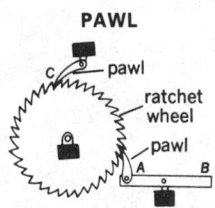

PAWL

Pawls with a ratchet wheel. Driving pawl at *A*, forced upward by lever *B*, engages teeth of ratchet wheel and rotates it counterclockwise. Holding pawl *C* prevents clockwise rotation of the wheel.

peak cathode current [ELECTR] **1.** Maximum instantaneous value of a periodically recurring cathode current. **2.** Highest instantaneous value of a randomly recurring pulse of cathode current. **3.** Highest instantaneous value of a nonrecurrent pulse of cathode current occurring under fault conditions. { 'pēk 'kath,ōd ,kə·rənt }

peak clipper *See* limiter. { 'pēk ,klip,ər }

peak clipping [ELEC] Reduction of the maximum demand for electric power from an electrical utility, often achieved by direct control of customer loads by signals directed to customer appliances. { 'pēk ,klip·iŋ }

peak detector [ELECTR] A detector whose output voltage approximates the true peak value of an applied signal; the detector tracks the signal in its sample mode and preserves the highest input signal in its hold mode. { 'pēk di,tek·tər }

peak distortion [COMMUN] Largest total distortion of telegraph signals noted during a period of observation. { 'pēk di'stȯr·shən }

peaked roof [ARCH] A roof that rises to a point or ridge. { 'pēkt 'rüf }

peak effect [SOLID STATE] In certain hard superconductors, the occurrence of a maximum in the value of the critical current as the external magnetic field is varied, near the critical magnetic field. { 'pēk i,fekt }

peak enthalpimetry [ANALY CHEM] A thermochemical analytical procedure applicable to biochemical and chemical analyses; the salient feature is rapid mixing of a reagent stream with an isothermal solvent stream into which discrete samples are intermittently injected; peak enthalpograms result which exhibit the response characteristics of genuine differential detectors. { 'pēk ¦en·thəl'pim·ə·trē }

peak envelope power [ELECTR] Of a radio transmitter, the average power supplied to the antenna transmission line by a transmitter during one radio-frequency cycle at the highest crest of the modulation envelope, taken under conditions of normal operation. { 'pēk 'en·və,lōp ,paù·ər }

peaker [ELECTR] A small fixed or adjustable inductance used to resonate with stray and distributed capacitances in a broad-band amplifier to increase the gain at the higher frequencies. { 'pēk·ər }

peak expiratory flow rate [MED] A measurement of the amount of air that leaves the lungs on forced exhalation. Abbreviated PEFR. { ¦pēk ik,spī·rə,tȯr·ē 'flō ,rāt }

peak factor *See* crest factor. { 'pēk ,fak·tər }

peak-flow monitor [IMMUNOL] A device that measures the amount of air that enters and leaves the lungs.

peak forward voltage [ELECTR] The maximum instantaneous voltage applied to an electronic device in the direction of lesser resistance to current flow. { 'pēk 'fȯr·wərd 'vōl·tij }

peak gust [METEOROL] After United States weather observing practice, the highest instantaneous wind speed recorded at a station during a specified period, usually the 24-hour observational day; therefore, a peak gust need not be a true gust of wind. { 'pēk 'gəst }

peaking circuit [ELECTR] A circuit used to improve the high-frequency response of a broad-band amplifier; in shunt peaking, a small coil is placed in series with the anode load; in series peaking, the coil is placed in series with the grid of the following stage. { 'pēk·iŋ ,sər·kət }

peaking network [ELECTR] Type of interstage coupling network in which an inductance is effectively in series (series-peaking network), or in shunt (shunt-peaking network), with the parasitic capacitance to increase the amplification at the upper end of the frequency range. { 'pēk·iŋ ,net,wərk }

peaking transformer [ELEC] A transformer in which the number of ampere-turns in the primary is high enough to produce many times the normal flux density values in the core; the flux changes rapidly from one direction of saturation to the other twice per cycle, inducing a highly peaked voltage pulse in a secondary winding. { 'pēk·iŋ tranz,fȯr·mər }

peak inverse anode voltage [ELECTR] Maximum instantaneous anode voltage in the direction opposite to that in which the tube or other device is designed to pass current. { 'pēk 'in,vərs 'an,ōd ,vōl·tij }

peak inverse voltage [ELECTR] Maximum instantaneous anode-to-cathode voltage in the reverse direction which is actually applied to the diode in an operating circuit. { 'pēk 'in,vərs ,vōl·tij }

peakless pumping [MIN ENG] Spreading the pumping load over the entire day in a mine. { 'pēk·ləs 'pəmp·iŋ }

peak limiter *See* limiter. { 'pēk ,lim·əd·ər }

peak load [ELEC] The maximum instantaneous load or the maximum average load over a designated interval of time. Also known as peak power. [ENG] The maximum quantity of a specified material to be carried by a conveyor per minute in a specified period of time. { 'pēk ,lōd }

peak overpressure [ORD] The highest overpressure resulting from a blast wave; peak overpressures near the fireball of an atomic explosion are very high but drop off rapidly as the blast wave travels along the ground outward from ground zero. { 'pēk 'ō·vər,presh·ər }

peak plain [GEOL] A high-level plain formed by a series of summits of approximately the same elevation, often described as an uplifted and fully dissected peneplain. Also known as summit plain. { 'pēk ,plān }

peak power [ELEC] *See* peak load. [ELECTROMAG] The maximum instantaneous power of a transmitted radar pulse. { 'pēk 'paù·ər }

peak pressure [ORD] The maximum pressure reached in the bore of a weapon during the burning of the propellant. { 'pēk 'presh·ər }

peak second algorithm [COMMUN] A set of mathematical procedures for attempting to predict the number of transmissions that will be carried out in a communications system during the busiest 1-second interval during some study period. { 'pēk 'sek·ənd 'al·gə,rith·əm }

peak signal level [ELECTR] Expression of the maximum instantaneous signal power or voltage as measured at any point in a facsimile transmission system; this includes auxiliary signals. { 'pēk 'sig·nəl ,lev·əl }

peak tank [NAV ARCH] A tank which is at the bow or stern end of a ship and is low in the ship, usually kept empty and dry but sometimes used to carry potable water. { 'pēk ,taŋk }

peak-to-peak amplitude [PHYS] Amplitude of an alternating quantity measured from positive peak to negative peak. { ¦pēk tə 'pēk 'am·plə,tüd }

peak-to-valley ratio [COMMUN] The ratio of the largest amplitude of a modulated wave to its smallest value. { ¦pēk tə 'val·ē ,rā·shō }

peak value [ELEC] The maximum instantaneous value of a varying current, voltage, or power during the time interval under consideration. Also known as crest value. { 'pēk 'val·yü }

peak width [ANALY CHEM] In a gas chromatogram (plot of eluent rise and fall versus time), the width of the base (time duration) of a symmetrical peak (rise and fall) of eluent. { 'pēk 'width }

peak zone [PALEON] An informal biostratigraphic zone consisting of a body of strata characterized by the exceptional abundance of some taxon (or taxa) or representing the maximum development of some taxon. { 'pēk ,zōn }

Peano continuum [MATH] A compact, connected, and locally connected metric space. { pā'än·ō kən,tin·yü·əm }

Peano curve [MATH] **1.** A continuous curve that passes through each point of the unit square. **2.** *See* Peano space. { pā'än·ō ,kərv }

Peano space [MATH] Any Hausdorff topological space that is the image of the closed unit interval under a continuous mapping. Also known as Peano curve. { pā'än·ō ,spās }

Peano's postulates [MATH] The five axioms by which the natural numbers may be formally defined; they state that (1) there is a natural number 1; (2) every natural number n has a successor n^+; (3) no natural number has 1 as its successor; (4) every set of natural numbers which contains 1 and the successor of every member of the set contains all the natural numbers; (5) if $n^+ = m^+$, then $n = m$. { pā'än·ōz 'päs·chə·ləts }

peanut [BOT] *Arachis hypogaea.* A low, branching, self-pollinated annual legume cultivated for its edible seed, which is a one-loculed legume formed beneath the soil in a pod. { 'pē·nət }

peanut oil [MATER] A yellow to greenish-yellow fatty oil obtained from peanuts; soluble in ether, chloroform, benzine, and carbon disulfide; main components are glycerides of oleic and linoleic acids; used as an edible substitute for olive oil, and in soaps and medicine. Also known as arachis oil; earthnut oil; katchung oil. { 'pē·nət ,òil }

pea ore [MINERAL] A variety of pisolitic limonite or bean ore occurring in small, rounded grains or masses about the size of a pea. { 'pē ,òr }

pear [BOT] Any of several tree species of the genus *Pyrus* in the order Rosales, cultivated for their fruit, a pome that is wider at the apical end and has stone cells throughout the flesh. { per }

pearceite [MINERAL] $Ag_{16}As_2S_{11}$ A black mineral composed of sulfide of arsenic and silver. { 'pir,sīt }

pearl [MATER] A dense, more or less round, white or light-colored concretion having various degrees of luster formed within or beneath the mantle of various mollusks by deposition of thin concentric layers of nacre about a foreign particle. [PATH] **1.** Rounded masses of concentrically arranged squamous epithelial cells, seen in some carcinomas. **2.** Mucous casts of the bronchi or bronchioles found in the sputum of asthmatic persons. { pərl }

pearl ash [MATER] An impure substance derived from potash following partial purification from wood ash. { 'pərl ,ash }

pearl essence [MATER] A brilliant, translucent, lustrous material obtained from fish scales; used in pearl lacquers and to make artificial pearls. Also known as pearl white. { 'pərl ,es·əns }

pearl hardening [INORG CHEM] Commerical name for a crystallized grade of calcium sulfate; used as a paper filler. { 'pərl ¦härd·ən·iŋ }

pearlite [GEOL] *See* perlite. [MET] A lamellar aggregate of ferrite (almost pure iron) and cementite (Fe_3C) often occurring in carbon steels and in cast iron. { 'pər,līt }

pearl moss *See* carragene. { 'pərl ,mós }

Pearl-Reed curve *See* logistic curve. { 'pərl 'rēd ,kərv }

pearl sinter *See* siliceous sinter. { 'pərl ¦sin·tər }

pearl spar [MINERAL] A crystalline carbonate having a pearly luster; an example is ankerite. { 'pərl ,spär }

pearlstone *See* perlite. { 'pərl,stōn }

pearl white *See* pearl essence.

pearly *See* nacreous. { 'pər·lē }

pearly tumor *See* cholesteatoma. { 'pər·lē 'tü·mər }

pear-shape cut [LAP] A variation of the brilliant cut, generally having 58 facets, and a pear-shaped girdle outline. { 'per ,shāp ,kət }

Pearson Type I distribution *See* beta distribution. { 'pir·sən 'tīp 'wən ,dis·trə'byü·shən }

pea-soup fog [METEOROL] Any particularly dense fog. { 'pē ,süp 'fäg }

peat [GEOL] A dark-brown or black residuum produced by the partial decomposition and disintegration of mosses, sedges, trees, and other plants that grow in marshes and other wet places. { pēt }

peat ball [ECOL] A lake ball containing an abundance of peaty fragments. { 'pēt ,bòl }

peat bed *See* peat bog. { 'pēt ,bed }

peat bog [GEOL] A bog in which peat has formed under conditions of acidity. Also known as peat bed; peat moor. { 'pēt ,bäg }

peat breccia [GEOL] Peat that has been broken up and then redeposited in water. Also known as peat slime. { 'pēt ,brech·ə }

peat coal [GEOL] A coal transitional between peat and lignite. [MATER] Artificially carbonized peat that is used as a fuel. { 'pēt ,kōl }

peat flow [ECOL] A mudflow of peat produced in a peat bog by a bog burst. { 'pēt ,flō }

peat formation [GEOCHEM] Decomposition of vegetation in stagnant water with small amounts of oxygen, under conditions intermediate between those of putrefaction and those of moldering. { 'pēt fòr'mā·shən }

peat moor *See* peat bog. { 'pēt ,mùr }

peat moss [ECOL] Moss, especially sphagnum moss, from which peat has been produced. { 'pēt ,mós }

peat-sapropel [GEOL] A product of the degradation of organic matter that is transitional between peat and sapropel. Also known as sapropel-peat. { 'pēt 'sap·rə,pel }

peat slime *See* peat breccia. { 'pēt ,slīm }

peat soil [GEOL] Soil containing a large amount of peat; it is rich in humus and gives an acid reaction. { 'pēt ,sòil }

peat tar [MATER] A peat distillate containing 2–6% tar. { 'pēt ,tär }

peat wax [MATER] A hard, waxy material extracted from

peat; it is similar to, and a substitute for, montan wax. { 'pēt ,waks }

Peaucellier linkage [MECH ENG] A mechanical linkage to convert circular motion exactly into straight-line motion. { pò'sel·yā ,liŋ·kij }

peau de soie [TEXT] Soft silk satin-textured fabric having a slight luster and faintly showing fine ribs in the filling. { 'pò· də ¦swä }

pebble [GEOL] A clast, larger than a granule and smaller than a cobble, having a diameter in the range of 0.16–2.6 inches (4–64 millimeters). Also known as pebblestone. [MINERAL] *See* rock crystal. { 'peb·əl }

pebble armor [GEOL] A desert armor made up of rounded pebbles. { 'peb·əl ,är·mər }

pebble bed [GEOL] Any pebble conglomerate, especially one in which the pebbles weather conspicuously and become loose. Also known as popple rock. { 'peb·əl ,bed }

pebble-bed reactor [NUCLEO] A nuclear reactor in which the fuel consists of small spheres or pellets stacked in the core; the reaction rate is controlled by coolant flow and by loading and unloading pellets. { 'peb·əl ,bed rē,ak·tər }

pebble coal [GEOL] Coal that is transitional between peat and brown coal. { 'peb·əl ,kōl }

pebble conglomerate [PETR] A consolidated rock consisting mainly of pebbles. { 'peb·əl kən'gläm·ə·rət }

pebble dike [GEOL] **1.** A clastic dike composed largely of pebbles. **2.** A tabular body containing sedimentary fragments in an igneous matrix. { 'peb·əl ,dīk }

pebble heater [CHEM ENG] Gas-heating device (for air, hydrogen, methane, and steam) in which heat is transferred to the gas via a countercurrent movement of preheated pebbles. { 'peb·əl ,hēd·ər }

pebble mill [MECH ENG] A solids size-reduction device with a cylindrical or conical shell rotating on a horizontal axis, and with a grinding medium such as balls of flint, steel, or porcelain. { 'peb·əl ,mil }

pebble peat [GEOL] Peat that is formed in a semiarid climate by the accumulation of moss and algae, no more than 0.25 inch (6 millimeters) in thickness, under the surface pebbles of well-drained soils. { 'peb·əl ,pēt }

pebble phosphate [GEOL] A secondary phosphorite of either residual or transported origin, consisting of pebbles or concretions of phosphatic material. { 'peb·əl ,fäs,fāt }

pebbles [MATER] Grinding media for pebble mills, usually balls of hard flint or hard burned white porcelain. { 'peb·əlz }

pebblestone *See* pebble. { 'peb·əl,stōn }

pebble-weave [TEXT] Of a material, having an irregular texture produced by weaving with shrunken and twisted yarn. { 'peb·əl ,wēv }

pebbling *See* orange peel. { 'peb·liŋ }

pebbly mudstone [GEOL] A delicately laminated till-like conglomeratic mudstone. { 'peb·lē 'məd,stōn }

pebbly sand [GEOL] An unconsolidated sedimentary deposit containing at least 75% sand and up to a maximum of 25% pebbles. { 'peb·lē 'sand }

pebbly sandstone [GEOL] A sandstone that contains 10–20% pebbles. { 'peb·lē 'san,stōn }

pébrine [INV ZOO] A contagious protozoan disease of silkworms and other caterpillars caused by *Nosema bombycis.* { pā'brēn }

pecan [BOT] *Carya illinoensis.* A large deciduous hickory tree in the order Fagales which produces an edible, oblong, thin-shelled nut. { pi'kän }

peccary [VERT ZOO] Either of two species of small piglike mammals in the genus *Tayassu,* composing the family Tayassuidae. { 'pek·ə·rē }

peck [MECH] Abbreviated pk. **1.** A unit of volume used in the United States for measurement of solid substances, equal to 8 dry quarts, or $^1/_4$ bushel, or 537.605 cubic inches, or 0.00880976754172 cubic meter. **2.** A unit of volume used in the United Kingdom for measurement of solid and liquid substances, although usually the former, equal to 2 gallons, or 0.00909218 cubic meter. { pek }

pecking order [PSYCH] A social hierarchy of prestige, dominance, or authority. [VERT ZOO] A hierarchy of social dominance within a flock of poultry where each bird is allowed to peck another lower in the scale and must submit to pecking by one of higher rank. { 'pek·iŋ ,òr·dər }

Peclet number [CHEM ENG] Dimensionless group used to

PECAN

Pecan. (*a*) Leaves and fruit. (*b*) Hulled nuts

determine the chemical reaction similitude for the scale-up from pilot-plant data to commercial-sized units; incorporates heat capacity, density, fluid velocity, and other pertinent physical parameters. { pə'klā ,nəm·bər }

Pecora [VERT ZOO] An infraorder of the Artiodactyla; includes those ruminants with a reduced ulna and usually with antlers, horns, or deciduous horns. { 'pek·ə·rə }

pecten [ZOO] Any of various comblike structures possessed by animals. { 'pek·tən }

Pectenidae [INV ZOO] A family of bivalve mollusks in the order Anisomyaria; contains the scallops. { pek'ten·ə,dē }

pectic acid [BIOCHEM] A complex acid, partially demethylated, obtained from the pectin of fruits. { 'pek·tik 'as·əd }

pectin [BIOCHEM] A purified carbohydrate obtained from the inner portion of the rind of citrus fruits, or from apple pomace; consists chiefly of partially methoxylated polygalacturonic acids. { 'pek·tən }

Pectinariidae [INV ZOO] The cone worms, a family of polychaete annelids belonging to the Sedentaria. { ,pek·tə·nə'rī·ə,dē }

pectinase [BIOCHEM] An enzyme that catalyzes the transformation of pectin into sugars and galacturonic acid. { 'pek·tə,nās }

pectinesterase [BIOCHEM] An enzyme that catalyzes the hydrolytic breakdown of pectins to pectic acids. { ¦pek·tən'es·tə,rās }

pectineus [ANAT] A muscle arising from the pubis and inserted on the femur. { pek'tin·ē·əs }

Pectinibranchia [INV ZOO] An order of gastropod mollusks which contains many families of snails; respiration is by means of ctenidia, the nervous system is not concentrated, and sexes are separate. { ,pek·tə·nə'braŋ·kē·ə }

Pectobothridia [INV ZOO] A subclass of parasitic worms in the class Trematoda, characterized by caudal hooks or hard posterior suckers or both. { ,pek·tə·bä'thrid·ē·ə }

pectolite [MINERAL] NaCa₂Si₃O₈(OH) A colorless, white, or gray inosilicate, crystallizing in the monoclinic system and having a vitreous to silky luster; hardness is 5 on Mohs scale, and specific gravity is 2.75. { 'pek·tə,līt }

pectoral fin [VERT ZOO] One of the pair of fins of fishes corresponding to forelimbs of a quadruped. { 'pek·tə·rəl 'fin }

pectoral girdle [ANAT] The system of bones supporting the upper or anterior limbs in vertebrates. Also known as shoulder girdle. { 'pek·tə·rəl 'gərd·əl }

pectoralis major [ANAT] The large muscle connecting the anterior aspect of the chest with the shoulder and upper arm. { ,pek·tə'ral·əs 'mā·jər }

pectoralis minor [ANAT] The small, deep muscle connecting the third to fifth ribs with the scapula. { ,pek·tə'ral·əs 'mīn·ər }

peculiar part [ORD] A part of ordnance for which the design is controlled by a single manufacturer, and the use is restricted to items produced by a single manufacturer. { pə'kyül·yər 'pärt }

peculiar star [ASTRON] A star that does not fit into a standard spectral classification. { pə'kyül·yər 'stär }

peculiar velocity [ASTRON] Superposed on the systematic rotation of the galaxy are individual motions of the stars; each star moves in a somewhat elliptical orbit and therefore shows a velocity of its own (peculiar velocity) to the local standard of rest, the standard moving in a circular orbit around the galactic center. { pə'kyül·yər və'läs·əd·ē }

ped [GEOL] A naturally formed unit of soil structure. { ped }

pedal [BIOL] Of or pertaining to the foot. [DES ENG] A lever operated by foot. { 'ped·əl }

pedal coordinates [MATH] The coordinates r and p describing a point P on a plane curve C, where r is the distance from a fixed point O to P, and p is the perpendicular distance from O to the tangent to C at P. { 'ped·əl kō'ȯrd·ən·əts }

pedal curve [MATH] **1.** The pedal curve of a given curve C with respect to a fixed point P is the locus of the foot of the perpendicular from P to a variable tangent to C. Also known as first pedal curve; first positive pedal curve; positive pedal curve. **2.** Any curve that can be derived from a given curve C by repeated application of the procedure specified in the first definition. { 'ped·əl ,kərv }

pedal disk [INV ZOO] The broad, flat base of many sea anemones, used for attachment to a substrate. { 'ped·əl ,disk }

pedal equation [MATH] An equation that characterizes a plane curve in terms of its pedal coordinates. { 'ped·əl i,kwā·zhən }

pedalfer [GEOL] A soil in which there is an accumulation of sesquioxides; it is characteristic of a humid region. { pə'dal·fər }

pedal ganglion [INV ZOO] One of the paired ganglia supplying nerves to the foot muscles in most mollusks. { 'ped·əl 'gaŋ·glē·ən }

pedal gland See foot gland. { 'ped·əl ,gland }

pedality [GEOL] The physical nature of a soil as expressed by the features of its constituent peds. { pe'dal·əd·ē }

pedal point [MATH] **1.** The fixed point with respect to which a pedal curve is defined. **2.** The fixed point with respect to which the pedal coordinates of a curve are defined. { 'ped·əl ,pȯint }

pedal triangle [MATH] **1.** The triangle whose vertices are located at the feet of the perpendiculars from some given point to the sides of a specified triangle. **2.** In particular, the triangle whose vertices are located at the feet of the altitudes of a given triangle. { 'ped·əl 'trī,aŋ·gəl }

pedate [BIOL] **1.** Having toelike parts. **2.** Having a foot. **3.** Having tube feet. { 'pe,dāt }

pedestal [CIV ENG] **1.** The support for a column. **2.** A metal support carrying one end of a bridge truss or girder and transmitting any load to the top of a pier or abutment. [ELECTR] See blanking level. [ENG] A supporting part or the base of an upright structure, such as a radar antenna. [GEOL] A relatively slender column of rock supporting a wider rock mass and formed by undercutting as a result of wind abrasion or differential weathering. Also known as rock pedestal. { 'ped·əst·əl }

pedestal boulder [GEOL] A rock mass supported on a rock pedestal. Also known as pedestal rock. { 'ped·əst·əl ,bȯl·dər }

pedestal design [MECH ENG] A robot design centered on the vertical axis of a central pedestal, in which the motion of any workpiece is confined to a spherical working envelope. { 'ped·əst·əl di,zīn }

pedestal flooring See raised flooring. { 'ped·ə·stəl ,flȯr·iŋ }

pedestal level See blanking level. { 'ped·əst·əl ,lev·əl }

pedestal pile [CIV ENG] A concrete pile with a bulbous enlargement at the bottom. { 'ped·əst·əl ,pīl }

pedestal rock See pedestal boulder. { 'ped·əst·əl ,räk }

pedestal sight [ORD] In an aircraft gunnery system, a sight mounted on a pedestal for remote control of the guns. { 'ped·əst·əl ,sīt }

pedial class [CRYSTAL] That class in the triclinic system which has no symmetry. { 'ped·ē·əl ,klas }

pediatrician [MED] A physician who specializes in pediatrics. { ,pēd·ē·ə'trish·ən }

pediatrics [MED] The branch of medicine that deals with the growth and development of the child through adolescence, and with the care, treatment, and prevention of diseases, injuries, and defects of children. { ,ped·ē'a·triks }

pedicel [BOT] **1.** The stem of a fruiting or sporebearing organ. **2.** The stem of a single flower. [ZOO] A short stalk in an animal body. { 'ped·ə,sel }

pedicellaria [INV ZOO] In echinoids and starfishes, any of various small grasping organs in the form of a beak carried on a stalk. { ,ped·ə·sə'ler·ē·ə }

pedicellate [BIOL] Having a pedicel. { pe'dis·ə,lāt }

Pedicellinea [INV ZOO] The single order of the class Calyssozoa, including all entoproct bryozoans. { ,ped·ə·sə'lin·ē·ə }

pedicle [ANAT] A slender process acting as a foot or stalk (as the base of a tumor), or the basal portion of an organ that is continuous with other structures. { 'ped·ə·kəl }

pediculophobia [PSYCH] Abnormal fear of pediculosis. { pə,dik·yə·lə'fō·bē·ə }

pediculosis [MED] Infestation with lice, especially of the genus *Pediculus*. { pə,dik·yə'lō·səs }

pedigree [GEN] Diagrammatic representation of the ancestry of an individual. { 'ped·ə,grē }

pedigree mud [MATER] A high-chemical-content drilling mud that includes barium sulfate, caustic soda, soda ash, sodium bicarbonate, and phosphates. { 'ped·ə,grē ,məd }

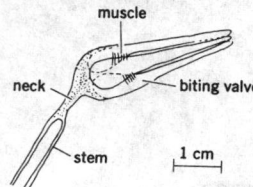

pedilidae [INV ZOO] The false ant-loving flower beetles, a family of coleopteran insects in the superfamily Tenebrionoidea. { pə'dil·ə,dē }

pediment [ARCH] A triangular face forming the gable of a two-pitched roof. [GEOL] A piedmont slope formed from a combination of processes which are mainly erosional; the surface is chiefly bare rock but may have a covering veneer of alluvium or gravel. Also known as conoplain; piedmont interstream flat. { 'ped·ə·mənt }

pedimentation [GEOL] The actions or processes by which pediments are formed. { ,ped·ə·mən'tā·shən }

pediment gap [GEOL] A broad opening formed by the enlargement of a pediment pass. { 'ped·ə·mənt ,gap }

pediment pass [GEOL] A flat, narrow tongue that extends from a pediment on one side of a mountain to join a pediment on the other side. { 'ped·ə·mənt ,pas }

Pedinidae [INV ZOO] The single family of the order Pedinoida. { pe'din·ə,dē }

Pedinoida [INV ZOO] An order of Diadematacea making up those forms of echinoderms which possess solid spines and a rigid test. { ,ped·ən'oid·ə }

pediocratic [GEOL] Pertaining to a period of time in which there is little diastrophism. { ped·ē·ə'krad·ik }

pedion [CRYSTAL] A crystal form with only one face; member of the asymmetric class of the triclinic system. { 'ped·ē·ən }

Pedionomidae [VERT ZOO] A family of quaillike birds in the order Gruiformes. { ,ped·ē·ə'näm·ə,dē }

Pedipalpida [INV ZOO] Former order of the Arachnida; these animals are now placed in the orders Uropygi and Amblypygi. { ,ped·ə'pal·pəd·ə }

pedipalpus [INV ZOO] One of the second pair of appendages of an arachnid. { 'ped·ə,pal·pəs }

pediplain [GEOL] A rock-cut erosion surface formed in a desert by the coalescence of two or more pediments. Also known as desert peneplain; desert plain; panfan. { 'ped·ə,plān }

pediplanation [GEOL] The actions or processes by which pediplanes are formed. { ,ped·ə·plə'nā·shən }

pediplane [GEOL] Any planate erosion surface formed in the piedmont area of a desert, either bare or covered with a veneer of alluvium. { 'ped·ə,plān }

pedocal [GEOL] A soil containing a concentration of carbonates, usually calcium carbonate; it is characteristic of arid or semiarid regions. { 'ped·ə,kal }

pedodontics [MED] The branch of dentistry concerned with the care of children's teeth. { ,ped·ō'dän·tiks }

pedogenesis See soil genesis. { ,ped·ō'jen·ə·səs }

pedogenics [GEOL] The study of the origin and development of soil. { ,ped·ō'jen·iks }

pedogeochemical survey [GEOCHEM] A geochemical prospecting survey in which the materials sampled are soil and till. { ped·ō,jē·ō'kem·ə·kəl 'sər,vā }

pedogeography [GEOL] The study of the geographic distribution of soils. { ped·ō·jē'äg·rə·fē }

pedography [GEOL] The systematic description of soils; an aspect of soil science. { pə'däg·rə·fē }

pedolith [GEOL] A surface formation that has undergone one or more pedogenic processes. { 'ped·ə,lith }

pedologic age [GEOL] The relative maturity of a soil profile. { ped·ō'läj·ik 'āj }

pedologic unit [GEOL] A soil considered without regard to its stratigraphic relations. { ped·ō'läj·ik 'yü·nət }

pedology [GEOL] See soil science. [MED] The science of the study of the physiological as well as the psychological aspects of childhood. { pe'däl·ə·jē }

pedometer [ENG] 1. An instrument for measuring and weighing a newborn child. 2. An instrument that registers the number of footsteps and distance covered in walking. { pə'däm·əd·ər }

pedon [GEOL] The smallest unit or volume of soil that represents or exemplifies all the horizons of a soil profile; it is usually a horizontal, hexagonal area of about 1 square meter, or possibly larger. { 'pe,dän }

pedophilia [PSYCH] Love of children by adults for sexual purposes. { ,ped·ə'fil·ē·ə }

pedophobia [PSYCH] An abnormal fear or dislike of children. { ,ped·ə'fō·bē·ə }

pedorelic [GEOL] Referring to a soil feature that is derived from a preexisting soil horizon. { ped·ō'rel·ik }

pedosphere [GEOL] That shell or layer of the earth in which soil-forming processes occur. { 'ped·ə,sfir }

pedotubule [GEOL] A soil feature consisting of skeleton grains, or skeleton grains plus plasma, and having a tubular external form (either single tubes or branching systems of tubes) characterized by relatively sharp boundaries and relatively uniform cross-sectional size and shape (circular or elliptical). { ped·ō'tüb·yül }

peduncle [ANAT] A band of white fibers joining different portions of the brain. [BOT] 1. A flower-bearing stalk. 2. A stalk supporting the fruiting body of certain thallophytes. [INV ZOO] The stalk supporting the whole or a large part of the body of certain crinoids, brachiopods, and barnacles. { 'pē,dəŋ·kəl }

pedunculate [BIOL] 1. Having or growing on a peduncle. 2. Being attached to a peduncle. { pē'dəŋ·kyə·lət }

PeeDee belemnite [GEOCHEM] Limestone from the PeeDee Formation in South Carolina (derived from the Cretaceous marine fossil *Belemnitella americana*), the carbon and oxygen isotope ratios of which are used as an international reference standard. Abbreviated PDB. { ,pē,dē bə'lem,nīt }

peek [COMPUT SCI] An instruction that causes the contents of a specific storage location in a computer to be displayed. { pēk }

peekaboo system See Batten system. { 'pēk·ə,bü ,sis·təm }

peel-back [ENG] The separation of two bonded materials, one or both of which are flexible, by stripping or pulling the flexible material from the mating surface at a 90 or 180° angle to the plane in which it is adhered. { 'pēl ,bak }

peeling [MATER] 1. Stripping or detaching a rubber coating from a metal, cloth, or other material. 2. Pulling a layer of material away from another layer, breaking one row of bonds at a time. { 'pēl·iŋ }

peel-off time [ENG] In seismic prospecting, the time correction applied to observed data to adjust them to a depressed reference datum. { 'pēl ,óf ,tīm }

peel test [ENG] A test to ascertain the adhesive strength of bonded strips of metals by peeling or pulling the metal strips back and recording the adherence values. { 'pēl ,test }

peel thrust [GEOL] A sedimentary sheet peeled off a sedimentary sequence usually along a bedding plane. { 'pēl ,thrəst }

peen [DES ENG] The end of a hammer head with a hemispherical, wedge, or other shape; used to bend, indent, or cut. { pēn }

peening [MET] Surface-hardening a piece of metal by hammering or by bombarding with hard shot. { 'pēn·iŋ }

peepdoor [MECH ENG] A small door in a furnace with a glass opening through which combustion may be observed. { 'pēp,dór }

peephole masks [COMPUT SCI] In character recognition, a set of characters (each character residing in the character reader in the form of strategically placed points) which theoretically render all input characters as being unique regardless of their style. { 'pēp,hōl ,masks }

peep sight [ORD] A rear gunsight having a small hole in which the front sight is centered in aiming; distinguished from an open sight. { 'pēp ,sīt }

peep slot [ORD] A vision slot in a combat vehicle which can readily be opened or closed from within. { 'pēp ,slät }

peer [COMMUN] A functional unit in a communications system that is in the same protocol layer as another such unit. { pir }

peer-to-peer network [COMMUN] A local-area network in which there is no central controller and all the nodes have equal access to the resources of the network. { pir tə 'pir 'net,wərk }

peesweep storm [METEOROL] An early-spring storm in Scotland and England. { 'pēz,wēp ,stórm }

PEFR See peak expiratory flow rate.

peg [ENG] 1. A small pointed or tapered piece, often cylindrical, used to pin down or fasten parts. 2. A projection used to hang or support objects. [MET] See plug. { peg }

Pegasidae [VERT ZOO] The single family of the order Pegasiformes. { pə'gas·ə,dē }

PEGASIFORMES

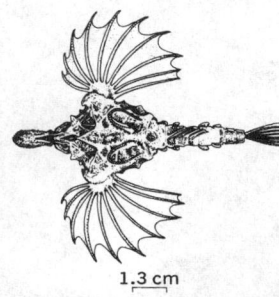

1.3 cm

Sea moth (*Pegasus draconis*).

PEGASUS

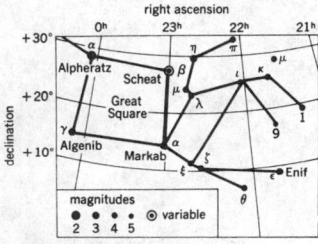

Line pattern of the constellation Pegasus. The grid lines represent the coordinates of the sky. The apparent brightness, or magnitude, of the stars is shown by the sizes of the dots, graded by appropriate numbers.

PEKING MAN

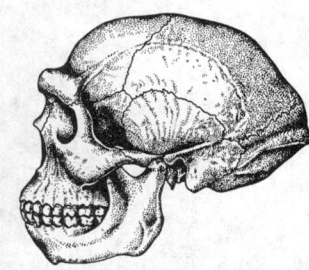

Reconstruction of female *Homo erectus pekinensis* skull showing the heavy browridges. (*After F. Weidenreich, from M. F. Ashley Montagu, An Introduction to Physical Anthropology, 2d ed., Charles C. Thomas, 1951*)

Pegasiformes [VERT ZOO] The sea moths or sea dragons, a small order of actinopterygian fishes; the anterior of the body is encased in bone, and the nasal bones are enlarged to form a rostrum that projects well forward of the mouth. { pə̇gas·ə·fȯr‚mēz }

Pegasus [ASTRON] A northern constellation; right ascension 22 hours, declination 20°N. Also known as Winged Horse. { 'peg·ə·səs }

peg count meter [ENG] A meter or register that counts the number of trunks tested, the number of circuits passed busy, the number of test failures, or the number of repeat tests completed. { 'peg ‚kau̇nt ‚mēd·ər }

pegging rammer [MET] A rod with an oblong piece of iron at its end; used to compact sand in a mold. { 'peg·iŋ ‚ram·ər }

peg graft [BOT] A graft made by driving a scion of leafless dormant wood with wedge-shaped base into an opening in the stock and sealing with wax or other material. { 'peg ‚graft }

pegmatite [PETR] Any extremely coarse-grained, igneous rock with interlocking crystals; pegmatites are relatively small, are relatively light colored, and range widely in composition, but most are of granitic composition; they are principal sources for feldspar, mica, gemstones, and rare elements. Also known as giant granite; granite pegmatite. { 'peg·mə‚tīt }

pegmatitic stage [GEOL] A stage in the normal sequence of crystallization of magma containing volatiles when the residual fluid is sufficiently enriched in volatile materials to permit the formation of coarse-grained rocks, that is pegmatites. { ¦peg·mə¦tid·ik ‚stāj }

pegmatitization [GEOL] Formation of or replacement by a pegmatite. { ‚peg·mə‚tīd·ə'zā·shən }

pegmatoid [PETR] An igneous rock that has the coarse-grained texture of a pegmatite but that lacks graphic intergrowths or typically granitic composition. { 'peg·mə‚tȯid }

pegmatolite See orthoclase. { peg'mad·əl‚īt }

peg model [GEOL] Three-dimensional model used to illustrate and study stratigraphic and structural conditions of subsurface geology; consists of a flat platform onto which vertical pegs of varying heights are mounted to represent the contours of various strata. { 'peg ‚mäd·əl }

Peierls-Nabarro force [SOLID STATE] The force required to displace a dislocation along its slip plane. { 'pā·ərlz nə'bär·ō ‚fȯrs }

Peirce stroke relationship See NOR. { 'pirs ¦strōk ri'lā·shən‚ship }

Peisidicidae [INV ZOO] A monogeneric family of polychaete annelids belonging to the Errantia. { ‚pī·sə'dis·ə‚dē }

Peking man [PALEON] *Sinanthropus pekinensis.* An extinct human type; the braincase was thick, with a massive basal and occipital torus structure and heavy browridges. { 'pē‚kiŋ 'man }

pel See pixel. { pel }

PEL See permissible exposure limit. { pel }

pelagic [GEOL] Pertaining to regions of a lake at depths of 33–66 feet (10–20 meters) or more, characterized by deposits of mud or ooze and by the absence of vegetation. Also known as eupelagic. [OCEANOGR] Pertaining to water of the open portion of an ocean, above the abyssal zone and beyond the outer limits of the littoral zone. { pə'laj·ik }

pelagic limestone [GEOL] A fine-textured limestone formed in relatively deep water by the concentration of calcareous tests of pelagic Foraminifera. { pə'laj·ik 'līm‚stōn }

pelagism See seasickness. { 'pel·ə‚jiz·əm }

pelagochthonous [GEOL] Referring to coal derived from a submerged forest or from driftwood. { ‚pel·ə'gäk·thə·nəs }

pelagosite [GEOL] A superficial calcareous crust a few millimeters thick, generally white, gray, or brownish with a pearly luster, formed in the intertidal zone by ocean spray and evaporation, and composed of calcium carbonate with higher contents of magnesium carbonate, strontium carbonate, calcium sulfate, and silica than are found in normal limy sediments. { pə'lag·ə‚sīt }

pelargonic acid [ORG CHEM] $CH_3(CH_2)_7CO_2H$ A colorless or yellowish oil, boiling at 254°C; soluble in ether and alcohol, insoluble in water; used as a chemical intermediate and flotation agent, in lacquers, pharmaceuticals, synthetic flavors and aromas, and plastics. { ¦pe‚lär¦gän·ik 'as·əd }

pelargonidin [BIOCHEM] An anthocyanidin pigment obtained by hydrolysis of pelargonin in the form of its red-brown crystalline chloride, $C_{15}H_{11}ClO_5$. { ‚pe‚lär'gän·əd·ən }

pelargonin [BIOCHEM] An anthocyanin obtained from the dried petals of red pelargoniums or blue cornflowers in the form of its red crystalline chloride, $C_{27}H_{31}ClO_{15}$. { ‚pe‚lär'gō·nən }

peldon [PETR] A very hard, smooth, compact sandstone with conchoidal fracture, occurring in coal measures. { 'pel·dən }

Pelean cloud See nuée ardente. { pə'lē·ən 'klau̇d }

Pel-Ebstein fever [MED] A relapsing fever characteristic of Hodgkin's disease. { pel 'eb‚stīn ‚fē·vər }

Pelecanidae [VERT ZOO] The pelicans, a family of aquatic birds in the order Pelecaniformes. { ‚pel·ə'kan·ə‚dē }

Pelecaniformes [VERT ZOO] An order of aquatic, fish-eating birds characterized by having all four toes joined by webs. { ‚pel·ə‚kan·ə'fȯr‚mēz }

Pelecanoididae [VERT ZOO] The diving petrels, a family of oceanic birds in the order Procellariiformes. { ‚pel·ə·kə'nȯid·ə‚dē }

Pelecinidae [INV ZOO] The pelecinid wasps, a monospecific family of hymenopteran insects in the superfamily Proctotrupoidea. { ‚pel·ə'sin·ə‚dē }

p electron [ATOM PHYS] In the approximation that each electron has a definite central-field wave function, an atomic electron that has an orbital angular momentum quantum number of unity. { 'pē i‚lek·trän }

Pelecypoda [INV ZOO] The equivalent name for Bivalvia. { ‚pel·ə'si·pə·də }

pelelith [GEOL] Vesicular or pumiceous lava in the throat of a volcano. { pə'lā‚lith }

Pele's hair [GEOL] A spun volcanic glass formed naturally by blowing out during quiet fountaining of fluid lava. Also known as capillary ejecta; filiform lapilli; lauoho o pele. { 'pā‚läz 'her }

Pele's tears [GEOL] Volcanic glass in the form of small, solidified drops which precede pendants of Pele's hair. { 'pā‚läz 'tirz }

Pelger anomaly [MED] A hereditary anomaly of granulocytes in the peripheral blood with no more than one or two nuclear lobes and unusually coarse nuclear chromatin. { 'pel·gər ə‚näm·ə·lē }

pelican [VERT ZOO] Any of several species of birds composing the family Pelecanidae, distinguished by the extremely large bill which has a distensible pouch under the lower mandible. { 'pel·ə·kən }

pelican hook [NAV ARCH] A hinged hook shaped like a pelican's beak, held closed by a ring that can be instantaneously released; used to secure chain, cable, or cargo gear on a ship. { 'pel·ə·kən ‚hu̇k }

pelite [GEOL] A sediment or sedimentary rock, such as mudstone, composed of fine, clay- or mud-size particles. Also spelled pelyte. { 'pē‚līt }

pelitic [GEOL] Pertaining to, characteristic of, or derived from pelite. { pə'lid·ik }

pelitic hornfels [PETR] A fine-grained metamorphic rock derived from pelite. { pə'lid·ik 'hȯrn‚felz }

pelitic schist [PETR] A foliated crystalline metamorphic rock derived from pelite. { pə'lid·ik 'shist }

pellagra [MED] A disease caused by nicotinic acid deficiency characterized by skin lesions, inflammation of the soft tissues of the mouth, diarrhea, and central nervous system disorders. { pə'lag·rə }

Pell equation [MATH] The diophantine equation $x^2 - Dy^2 = 1$, with D a positive integer that is not a perfect square. { 'pel i‚kwā·zhən }

pellet [AGR] A small, cylindrical, compressed mass of livestock feed. [GEOL] A fine-grained, sand-size, spherical to elliptical aggregate of clay-sized calcareous material, devoid of internal structure, and contained in the body of a well-sorted carbonate rock. [ORD] A small stone or metal ball used as a missile in firearms. [PHARM] A small pill. [SCI TECH] A small spherical or cylindrical body. [VERT ZOO] A mass of undigestible material regurgitated by a carnivorous bird. { 'pel·ət }

pellet cooler [CHEM ENG] Gas-cooled, gravity-bed device for the cooling and drying of extruded pellets and briquets. { 'pel·ət ‚kül·ər }

pellet fusion [NUCLEO] A method of controlled fusion in which the rapid implosion of a fuel pellet, produced by laser, electron, or ion beams, raises the temperature and density of the pellet core to levels at which nuclear fusion can take place before the pellet flies apart. Also known as inertial-confinement fusion. { 'pel·ət ˌfyü·zhən }

pelletierine [PHARM] C₅H₁₀N(CH₂)₂CHO An alkaloid obtained as a liquid from the root of the pomegranate; boiling point is 195°C; soluble in water, alcohol, and ether; used in medicine. { ˌpel·ə'ti͟ˌrēn }

pelleting [ENG] Method of accelerating solidification of cast explosive charges by blending precast pellets of the explosives into the molten charge. { 'pel·əd·iŋ }

pelletization [MIN ENG] Forming aggregates of about 1/2-inch (13-millimeter) diameter from finely divided ore or coal. { pel·əd·ə'zā·shən }

pelletizer [CHEM ENG] A machine for cutting bulk plastic into pellets, suitable for use as feedstock, either from solidified polymer at the end of the manufacturing process or from the molten polymer as it emerges from the die. { 'pel·ə,tīz·ər }

pellet mill [MECH ENG] Device for injecting particulate, granular or pasty feed into holes of a roller, then compacting the feed into a continuous solid rod to be cut off by a knife at the periphery of the roller. { ˌpel·ət ˌmil }

pelletron [NUCLEO] A type of electrostatic accelerator that utilizes a charging system consisting of steel cylinders joined by links of solid insulating material such as nylon to form a chain; the metal cylinders are charged as they leave a pulley at ground potential, and the charge is removed as they pass over a pulley in the high-potential terminal. { 'pelə,trän }

pellet technique See potassium bromide-disk technique. { 'pel·ət tek,nēk }

pellicle [CYTOL] A plasma membrane. [INV ZOO] A thin protective membrane, as on certain protozoans. { 'pel·ə·kəl }

pellicularia disease [PL PATH] A fungus disease of coffee and other tropical plants caused by *Pellicularia koleroga* and characterized by leaf spots. { pə,lik·yə'lar·ē·ə di,zēz }

pellicular resins [ANALY CHEM] Glass spheres coated with a thin layer of ion-exchange resin, used in liquid chromatography. { pə'lik·yə·lər 'rez·ənz }

pellicular water [HYD] Films of groundwater adhering to particles or cavities above the water table. { pə'lik·yə·lər 'wód·ər }

pell-mell structure [GEOL] A sedimentary structure characterized by absence of bedding in a coarse deposit of waterworn material; it may occur where deposition is too rapid for sorting or where slumping has destroyed the layered arrangement. { 'pel͟mel 'strək·chər }

pellodite See pelodite. { 'pel·ə,dīt }

pellotine [ORG CHEM] C₁₃H₁₉O₃N A colorless, crystalline alkaloid, derived from the dried cactus pellote, *Lophophora williamsi* (Mexico), slightly soluble in water; used as a hypnotic. { 'pel·ə,tēn }

Pelmatozoa [INV ZOO] A division of the Echinodermata made up of those forms which are anchored to the substrate during at least part of their life history. { pel,mad·ə'zō·ə }

pelmicrite [GEOL] A limestone containing less than 25% each of intraclasts and ooliths, having a volume ratio of pellets to fossils greater than 3 to 1, and with the micrite matrix more abundant than the sparry-calcite cement. { 'pel·mə,krīt }

Pelobatidae [VERT ZOO] A family of frogs in the suborder Anomocoela, including the spadefoot toads. { ,pel·ō'bad·ə,dē }

pelodite [GEOL] A lithified glacial rock flour which is composed of glacial pebbles in a silt or clay matrix and which was formed by redeposition of the fine fraction of a till. Also spelled pellodite. { 'pel·ə,dīt }

Pelodytidae [VERT ZOO] A family of frogs in the suborder Anomocoela. { ,pel·ə'did·ə,dē }

pelogloea [GEOL] Marine detrital slime from settled plankton. { ,pel·ə͟glē·ə }

Pelogonidae [INV ZOO] The equivalent name for Ochteridae. { ,pel·ə͟gän·ə,dē }

Pelomedusidae [VERT ZOO] The side-necked or hidden-necked turtles, a family of the order Chelonia. { ,pel·ō·mə'düs·ə,dē }

Pelonemataceae [MICROBIO] A family of gliding bacteria of uncertain affiliation; straight, flexuous, or spiral, unbranched filaments containing colorless, cylindrical cells. { ,pel·ō,nem·ə'tās·ē,ē }

Pelopidae [INV ZOO] A family of oribatid mites, order Sarcoptiformes. { pə'läp·ə,dē }

Peloplocaceae [MICROBIO] Formerly a family in the order Chlamydobacteriales; long, unbranched trichomes in a delicate sheath. { ,pel·ō·plə'kās·ē,ē }

Peloridiidae [INV ZOO] The single family of the homopteran series Coleorrhyncha. { ,pel·ō·rə'dī·ə,dē }

pelorus [NAV] A compass card suitably mounted and provided with vanes to permit observation of the true or relative bearings; it may not include a magnetic compass, in which case reference must be made to a local compass to give true or magnetic bearings. { pə'lór·əs }

pelorus card [NAV] The part of a pelorus on which the direction graduations are placed, usually in the form of a thin disk or annulus graduated in degrees, clockwise from 0° at the reference direction to 360°. { pə'lór·əs ,kärd }

pelotherapy [MED] Therapeutic treatment with earth or mud. { ˌpel·ō'ther·ə·pē }

pelouse See meadow. { pə'lüz }

pelphyte [GEOL] A lake-bottom deposit consisting mainly of fine, nonfibrous plant remains. { 'pel,fīt }

pelsparite [PETR] A limestone containing less than 25% each of intraclasts and ooliths, having a volume ratio of pellets to fossils greater than 3 to 1, and with the sparry-calcite cement more abundant than the micrite matrix. { 'pel,spä,rīt }

peltate [BOT] Of leaves, having the petiole attached to the lower surface instead of the base. { 'pel,tāt }

Peltier coefficient [PHYS] The ratio of the rate at which heat is evolved or absorbed at a junction of two metals in the Peltier effect to the current passing through the junction. { pel'tyā ,kō·i,fish·ənt }

Peltier effect [PHYS] The phenomenon in which heat is evolved or absorbed at the junction of two dissimilar metals carrying a small current, depending upon the direction of the current. { pel'tyā i,fekt }

Pelton turbine See Pelton wheel. { 'pel·tən ,tər·bən }

Pelton wheel [MECH ENG] An impulse hydraulic turbine in which pressure of the water supply is converted into velocity by a few stationary nozzles, and the water jets then impinge on the buckets mounted on the rim of a wheel; usually limited to high head installations, exceeding 500 feet (150 meters). Also known as Pelton turbine. { 'pel·tən ,wēl }

pelvic cavity See pelvis. { 'pel·vik ,kav·əd·ē }

pelvic fin [VERT ZOO] One of the pair of fins of fishes corresponding to the hindlimbs of a quadruped. { 'pel·vik ,fin }

pelvic girdle [ANAT] The system of bones supporting the lower limbs, or the hindlimbs, of vertebrates. { 'pel·vik ͟gərd·əl }

pelvic index [ANAT] The ratio of the anteroposterior diameter to the transverse diameter of the pelvis. { 'pel·vik ͟in,deks }

pelvis [ANAT] **1.** The main, basin-shaped cavity of the kidney into which urine is discharged by nephrons. **2.** The basin-shaped structure formed by the hipbones together with the sacrum and coccyx, or caudal vertebrae. **3.** The cavity of the bony pelvis. Also known as pelvic cavity. { 'pel·vəs }

pelviscope [MED] An endoscope for examination of the pelvic organs of the female. { 'pel·və,skōp }

Pelycosauria [PALEON] An extinct order of primitive, mammallike reptiles of the subclass Synapsida, characterized by a temporal fossa that lies low on the side of the skull. { ,pel·ə·kə'sór·ē·ə }

pelyte See pelite. { 'pe,līt }

pemphigoid [MED] An autoimmune skin disorder resembling, but histologically and clinically distinct from, pemphigus and characterized by bloody blisters, especially on the trunk and limbs. { 'pem·fə,góid }

pemphigus [MED] An acute or chronic disease of the skin characterized by the appearance of bullae, which develop in crops or in continuous succession. { 'pem·fə·gəs }

pemphigus contageosus [MED] A vesicular dermatitis endemic in tropical areas, chiefly affecting the armpits and groin. { 'pem·fə·gəs kən,tā·jē'ō·səs }

pen [ENG] **1.** A small place for confinement, storage, or protection. **2.** A device for writing with ink. [INV ZOO] The inner horny, feather-shaped, chitinous shell of a squid. Also known as gladius. { pen }

Penaeidea [INV ZOO] A primitive section of the Decapoda in the suborder Natantia; in these forms, the pleurae of the first abdominal somite overlap those of the second, the third legs

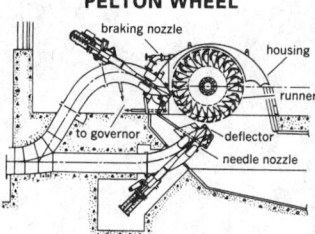

PELTON WHEEL

Cross section of a Pelton type of hydraulic-turbine installation.

are chelate, and the gills are dendrobranchiate. { ˌpen·ēˈid·ē·ə }

penalty function [MATH] A function used in treating maxima and minima problems subject to constraints. { 'pen·əl·tē ˌfəŋk·shən }

pencatite [PETR] A recrystallized limestone containing periclase or brucite and calcite in approximately equal molecular proportions. { 'peŋ·kəˌtīt }

pencil [ENG] An implement for writing or making marks with a solid substance; the three basic kinds are graphite, carbon, and colored. [MATH] **1.** In general, a family of geometric objects which share a common property. **2.** All the lines that lie in a particular plane and pass through a particular point. **3.** All the lines parallel to a particular line. **4.** All the circles that pass through two fixed points and lie in a particular plane. **5.** All the planes that include a particular line. **6.** All the spheres that include a particular circle. [OPTICS] A bundle of rays that emanate from or converge to a common point. { 'pen·səl }

pencil beam [ELECTROMAG] A beam of radiant energy concentrated in an approximately conical or cylindrical portion of space of relatively small diameter; this type of beam is used for many revolving navigational lights and radar beams. { 'pen·səl ˌbēm }

pencil beam antenna [ELECTROMAG] Unidirectional antenna designed so that cross sections of the major lobe formed by planes perpendicular to the direction of maximum radiation are approximately circular. { 'pen·səl ˌbēm anˌten·ə }

pencil cave [ENG] A driller's term for hard, closely jointed shale that caves into a well in pencil-shaped fragments. { 'pen·səl ˌkāv }

pencil cleavage [GEOL] Cleavage in which fracture produces long, slender pieces of rock. { 'pen·səl ˌklē·vij }

pencil follower [COMPUT SCI] A device for converting graphic images to digital form; the information to be analyzed appears on a reading table where a reading pencil is made to follow the trace, and a mechanism beneath the table surface transmits position signals from the pencil to an electronic console for conversion to digital form. { 'pen·səl ˌfäl·ə·wər }

pencil gneiss [GEOL] A gneiss that splits into thin, rodlike quartz-feldspar crystal aggregates. { 'pen·səl ˌnīs }

pencil ore [GEOL] Hard, fibrous masses of hematite that can be broken up into splinters. { 'pen·səl ˌór }

pencil stone See pyrophyllite. { 'pen·səl ˌstōn }

pencil tube [ELECTR] A small tube designed especially for operation in the ultra-high-frequency band; used as an oscillator or radio-frequency amplifier. { 'pen·səl ˌtüb }

pendant See roof pendant. { 'pen·dənt }

pendant atomizer See hanging-drop atomizer. { 'pen·dənt 'ad·əˌmiz·ər }

pendant cloud See tuba. { 'pen·dənt ˌklaùd }

pendant-drop melt extraction [MET] Melt extraction in which the molten metal is produced by heating the end of a rod above a disk. { 'pen·dənt ˌdräp 'melt ikˌstrakˌshən }

pendant-drop method [PHYS] Method for the measurement of liquid surface tension by the elongation of a hanging drop of the liquid. { 'pen·dənt ˌdräp ˌmeth·əd }

pendant post [BUILD] A post on a solid support and set against a wall to support a collar beam or other part of a roof. { 'pen·dənt ˌpōst }

pendeloque [LAP] A modification of the round brilliant cut, with an outline resembling that of the pear-shape cut, but with the small end longer and more pointed. { ˌpän·dəˈlōk }

pendent terrace [GEOL] A connecting ribbon of sand that joins an isolated point of rock with a neighboring coast. { 'pen·dənt ˌter·əs }

pending input/output [COMPUT SCI] An input/output operation that has been initiated but not yet carried out, so that the central processing unit either is temporarily idle or services other programs and tasks until the operation is completed. { 'pend·iŋ 'inˌpùt ˌaùtˌpùt }

pendular ring [PETRO ENG] Distribution of two nonmiscible liquids in a porous system; the pendular ring is a state of reservoir saturation in which the wetting phase is not continuous and the nonwetting phase is in contact with some of the solid surface. { 'pen·jəˌlər ˌriŋ }

pendular water [HYD] Capillary water ringing the contact points of adjacent rock or soil particles in the zone of areation. { 'pen·jəˌlər ˌwód·ər }

pendulous gyroscope [MECH] A gyroscope whose axis of rotation is constrained by a suitable weight to remain horizontal; it is the basis of one type of gyrocompass. { 'pen·jə·ləs 'jī·rəˌskōp }

pendulum [PHYS] A rigid body mounted on a fixed horizontal axis, about which it is free to rotate under the influence of gravity. Also known as compound pendulum; gravity pendulum. { 'pen·jə·ləm }

pendulum anemometer [ENG] A pressure-plate anemometer consisting of a plate which is free to swing about a horizontal axis in its own plane above its center of gravity; the angular deflection of the plate is a function of the wind speed; this instrument is not used for station measurements because of the false reading which results when the frequency of the wind gusts and the natural frequency of the swinging plate coincide. { 'pen·jə·ləm ˌan·əˈmäm·əd·ər }

pendulum clock [HOROL] A clock having a swinging pendulum as the means for producing a regularly recurring action. { 'pen·jə·ləm ˌkläk }

pendulum day [PHYS] The time required for the plane of a freely suspended (Foucault) pendulum to complete an apparent rotation about the local vertical. { 'pen·jə·ləm ˌdā }

pendulum level [ENG] A leveling instrument in which the line of sight is automatically kept horizontal by a built-in pendulum device (such as a horizontal arm and a plumb line at right angles to the arm). { 'pen·jə·ləm ˌlev·əl }

pendulum press [MECH ENG] A punch press actuated by a swinging treadle operated by the foot. { 'pen·jə·ləm ˌpres }

pendulum property [MATH] The property of a cycloid that, if a simple pendulum is hung from a cusp and made to swing between two branches, and if the length of the pendulum equals the length of the cycloid between successive cusps, then the period of the pendulum's oscillation does not depend on its amplitude, and the end of the pendulum traces out another cycloid. { 'pen·jə·ləm ˌpräp·ərd·ē }

pendulum saw [MECH ENG] A circular saw that swings in a vertical arc for crosscuts. { 'pen·jə·ləm ˌsó }

pendulum scale [ENG] Weight-measurement device in which the load is balanced by the movement of one or more pendulums from vertical (zero weight) to horizontal (maximum weight). { 'pen·jə·ləm ˌskāl }

pendulum seismograph [ENG] A seismograph that measures the relative motion between the ground and a loosely coupled inertial mass; in some instruments, optical magnification is used whereas others exploit electromagnetic transducers, photocells, galvanometers, and electronic amplifiers to achieve higher magnification. { 'pen·jə·ləm 'sīz·məˌgraf }

pendulum sextant [NAV] A sextant provided with a pendulum to indicate the horizontal. { 'pen·jə·ləm ˌsekˌstənt }

penecontemporaneous [GEOL] Of a geologic process or the structure or mineral that is formed by the process, occurring immediately following deposition but before consolidation of the enclosing rock. { ˌpēn·ē·kənˌtem·pəˈrā·nē·əs }

peneplain See base-leveled plain. { 'pēn·əˌplān }

peneplanation [GEOL] The actions or processes by which peneplains are formed. { ˌpēn·ə·pləˈnā·shən }

penesaline [ECOL] Referring to an environment intermediate between normal marine and saline, characterized by evaporitic carbonates often interbedded with gypsum or anhydrite, and by a salinity high enough to be toxic to normal marine organisms. { ˌpēn·əˈsāˌlēn }

penetrance [GEN] The proportion of individuals carrying a dominant gene in the heterozygous condition or a recessive gene in the homozygous condition in which the specific phenotypic effect is apparent. Also known as gene penetrance. { 'pen·əˌtrəns }

penetrant [INV ZOO] A large barbed nematocyst that pierces the body of the prey and injects a paralyzing agent. [MATER] A liquid with low surface tension, usually containing a dye or fluorescent chemical; when flowed over a metal surface, it is used to determine the existence and extent of cracks and other discontinuities. { 'pen·əˌtrənt }

penetrating oil [MATER] Low-viscosity oil that can penetrate between closely fitted parts, such as the leaves of springs and screw threads; used to loosen rusted parts. { 'pen·əˌtrād·iŋ ˌoil }

penetrating shower [NUC PHYS] A cosmic-ray shower, consisting mainly of muons, that can penetrate 6 to 8 inches (15 to 20 centimeters) of lead. { 'pen·əˌtrād·iŋ 'shaù·ər }

penetration [AERO ENG] That phase of the letdown from high altitude to a specified approach altitude. [MET] **1.** The distance from the original surface of the base metal to that point at which weld fusion ends. **2.** A surface defect on a casting caused by molten metal filling voids in the sand mold. [ORD] Distance to which a projectile sinks into a target. { ‚pen·ə'trā·shən }

penetration ballistics [MECH] A branch of terminal ballistics concerned with the motion and behavior of a missile during and after penetrating a target. { ‚pen·ə'trā·shən bə‚lis·tiks }

penetration depth [CRYO] The depth beneath the surface of superconductor in a magnetic field at which the magnetic field strength has fallen to $1/e$ of its value at the surface. [ELEC] In induction heating, the thickness of a layer, extending inward from a conductor's surface, whose resistance to direct current equals the resistance of the whole conductor to alternating current of a given frequency. [ENG] The greatest depth in an ultrasonic test piece at which indications can be measured. { ‚pen·ə'trā·shən ‚depth }

penetration frequency See critical frequency. { ‚pen·ə'trā·shən ‚frē·kwən·sē }

penetration funnel [GEOL] An impact crater, generally funnel-shaped, formed by a small meteorite striking the earth at a relatively low velocity and containing nearly all the impacting mass within it. { ‚pen·ə'trā·shən ‚fən·əl }

penetration gland [INV ZOO] A gland at the anterior end of certain cercariae that secretes a histolytic substance. { ‚pen·ə'trā·shən ‚gland }

penetration hardness See indentation hardness. { ‚pen·ə'trā·shən ‚härd·nəs }

penetration log [MIN ENG] A record of the speed with which a drill penetrates a bore, including such factors as hole size, bit size, mud pressure, rotation speed, and force on the bit; used to determine the thickness of coal and dirt bands in the bore. { ‚pen·ə'trā·shən ‚läg }

penetration macadam [MATER] A paving material consisting of crushed stone in two sizes bound together by asphalt or tar. { ‚pen·ə'trā·shən mə'kad·əm }

penetration number [ENG] The consistency of greases, waxes, petrolatum, and asphalt or other bituminous materials expressed as the distance that a standard needle penetrates the sample under specified American Society for Testing and Materials test conditions. { ‚pen·ə'trā·shən ‚nəm·bər }

penetration of fractures [PETRO ENG] The depth to which artificially produced fractures penetrate the fractured reservoir formation. { ‚pen·ə'trā·shən əv 'frak·chərz }

penetration phosphors [ELECTR] Phosphors of two different colors that are placed in separate layers on the screen of a cathode-ray tube to form a system for creating color displays in which a high-energy beam penetrates the first layer and excites the second, while a low-energy beam is stopped by the first layer and excites it. { ‚pen·ə'trā·shən ‚fäs·fərz }

penetration probability [QUANT MECH] The probability that a particle will pass through a potential barrier, that is, through a finite region in which the particle's potential energy is greater than its total energy. Also known as transmission coefficient. { ‚pen·ə'trā·shən ‚präb·ə‚bil·əd·ē }

penetration range See night visual range. { ‚pen·ə'trā·shən ‚ränj }

penetration rate [MECH ENG] The actual rate of penetration of drilling tools. { ‚pen·ə'trā·shən ‚rāt }

penetration speed [MECH ENG] The speed at which a drill can cut through rock or other material. { ‚pen·ə'trā·shən ‚spēd }

penetration test [ENG] A test to determine the relative values of density of noncohesive sand or silt at the bottom of boreholes. { ‚pen·ə'trā·shən ‚test }

penetration testing [COMPUT SCI] An activity that is intended to determine if there is a way to cause a computer program to fail to perform in the expected manner; it involves hypothesizing flaws that would prevent the program from enforcing security, and conducting experiments to confirm or refute the hypothesized flaws. { ‚pen·ə'trā·shən ‚test·iŋ }

penetration twin See interpenetration twin. { ‚pen·ə'trā·shən ‚twin }

penetrative [GEOL] Referring to a texture of deformation that is uniformly distributed in a rock, without notable discontinuities; for example, slaty cleavage. { 'pen·ə‚trā·div }

penetrometer [ENG] An instrument that measures the penetrating power of a beam of x-rays or other penetrating radiation. An instrument used to determine the consistency of a material by measurement of the depth to which a standard needle penetrates into it under standard conditions. { ‚pen·ə'träm·əd·ər }

Penex process [CHEM ENG] A continuous, nonregenerative petroleum-refinery process for isomerization of C_5 or C_6 fractions in the presence of hydrogen and a platinum catalyst. { 'pe‚neks ‚prä·səs }

penfieldite [MINERAL] $Pb_2(OH)Cl_3$ A white hexagonal mineral composed of basic chloride of lead, occurring in hexagonal prisms. { 'pen‚fēl‚dīt }

penguin [VERT ZOO] Any member of the avian order Sphenisciformes; structurally modified wings do not fold and they function like flippers, the tail is short, feet are short and webbed, and the legs are set far back on the body. { 'peŋ·gwən }

penicidin See patulin. { ‚pen·ə'sīd·ən }

penicillamine [PHARM] $C_5H_{11}NO_2S$ The most characteristic degradation product of penicillin-type antibiotics; used as a chelating agent, as a drug to reduce cystine excretion in cystinurea, and in the treatment of rheumatoid arthritis. { ‚pen·ə'sil·ə‚mēn }

penicillate [BIOL] Having a tuft of fine hairs. { 'pen·ə‚sil·ət }

penicillin [MICROBIO] **1.** The collective name for salts of a series of antibiotic organic acids produced by a number of *Penicillium* and *Aspergillus* species; active against most gram-positive bacteria and some gram-negative cocci. **2.** See benzyl penicillin sodium. { ‚pen·ə'sil·ən }

penicillinase [BIOCHEM] A bacterial enzyme that hydrolyzes and inactivates penicillin. { ‚pen·ə'sil·ə‚nās }

penicillin V See phenoxymethylpenicillin. { ‚pen·ə'sil·ən 'fīv or 'vē }

Peniculina [INV ZOO] A suborder of the Hymenostomatida. { pe‚nik·yə'lī·nə }

penikkavaarite [PETR] An intrusive rock composed chiefly of augite, barkevikite, and green hornblende in a feldspathic groundmass. { ‚pen·ə'ka·və‚rīt }

peninsula [GEOGR] A body of land extending into water from the mainland, sometimes almost entirely separated from the mainland except for an isthmus. { pə'nin·sə·lə }

penis [ANAT] The male organ of copulation in humans and certain other vertebrates. Also known as phallus. { 'pē·nəs }

penis envy [PSYCH] The envy of the young female child for the penis which she does not possess, or which she thinks she has lost. { 'pē·nəs ‚en·vē }

penitent See nieve penitente. { 'pen·ə·tənt }

penitent ice [HYD] A jagged spike or pillar of compacted firn caused by differential melting and evaporation; necessary for this formation are air temperature near freezing, dew point much below freezing, and strong insolation. { 'pen·ə·tənt 'īs }

penitent snow [HYD] A jagged spike or pillar of compacted snow caused by differential melting and evaporation. { 'pen·ə·tənt 'snō }

penna See contour feather. { 'pen·ə }

pennaceous [ZOO] Referring to the stiff, tightly bound portion of the feather vane on a bird. { pə'nā·shəs }

Pennales [BOT] An order of diatoms (Bacillariophyceae) in which the form is often circular, and the markings on the valves are radial. { pə'nā·lēz }

pennant [METEOROL] A means of representing wind speed in the plotting of a synoptic chart; it is a triangular flag, drawn pointing toward lower pressure from a wind-direction shaft. { 'pen·ənt }

pennantite [MINERAL] $Mn_9Al_6Si_5O_{20}(OH)_{16}$ Orange mineral composed of basic manganese aluminum silicate; member of the chlorite group; it is isomorphous with thuringite. { 'pen·ən‚tīt }

pennate [BIOL] **1.** Wing-shaped. **2.** Having wings. **3.** Having feathers. { 'pe‚nāt }

Pennatulacea [INV ZOO] The sea pens, an order of the subclass Alcyonaria; individuals lack stolons and live with their bases embedded in the soft substratum of the sea. { pə‚nach·ə'lā·shə }

Pennellidae [INV ZOO] A family of copepod crustaceans in

PENICILLIN

Basic structural formula for penicillin showing the β-lactamthiazolide ring system common to all penicillins; proper combination of substituents at R and Y results in any one of nine penicillins.

the suborder Caligoida; skin-penetrating external parasites of various marine fishes and whales. { ,pen·ə'lī·ə,dē }

penniculus [ANAT] A tuft of arterioles in the spleen. [BIOL] A brush-shaped structure. { pə'nik·yə·ləs }

Penning gage *See* Philips ionization gage. { 'pen·iŋ ,gāj }

Penning ionization [ATOM PHYS] The ionization of gas atoms or molecules in collisions with metastable atoms. { 'pen·iŋ ,ī·ə·nə'zā·shən }

Penning ion source [NUCLEO] A source of positively charged heavy ions used in accelerators, consisting of an anode chamber with cathodes at each end, into which the desired element is introduced; an arc discharge is generated in the chamber, creating a plasma, and electrons confined by the cathodes and an axial magnetic field make many ionizing collisions with the gas in the chamber. { 'pen·iŋ 'ī,än ,sȯrs }

Penning trap [ENG] A device for trapping electrons and isolating single electrons, consisting of a large, homogeneous magnetic field plus a superimposed weak parabolic electric potential created by a positive charge $+Q$ on a ring electrode and two negative charges $-Q/2$ each on two cap electrodes. { 'pen·iŋ ,trap }

Penning-trap mass spectrometer [ENG] A device for making highly accurate comparisons of the masses of charged atoms and molecules by comparing the cyclotron frequencies of single ions in a Penning trap. { pen·iŋ ,trap ,mas spek 'träm·əd·ər }

penninite [MINERAL] $(Mg,Fe,Al)_6(Si,Al)_4O_{11}(OH)_8$ An emerald-green, olive-green, pale-green, or bluish mineral of the chlorite group crystallizing in the monoclinic system, with a hardness of 2–2.5 on Mohs scale, and specific gravity of 2.6–2.85. { 'pen·ə,nīt }

Pennsylvania-base crude [MATER] A type of crude petroleum produced in Pennsylvania, New York, West Virginia, and parts of Ohio; contains a high percentage of paraffin-base lube-oil stock. { pen·səl'vā·nyə 'bās 'krüd }

Pennsylvanian [GEOL] A division of late Paleozoic geologic time, extending from 320 to 280 million years ago, varyingly considered to rank as an independent period or as an epoch of the Carboniferous period; named for outcrops of coal-bearing rock formations in Pennsylvania. { pen·səl'vā·nyən }

Pennsylvania truss [CIV ENG] A truss characterized by subdivided panels, curved top chords on through trusses, and curved bottom chords on deck spans; used on long bridge spans. { pen·səl'vā·nyə 'trəs }

pennyroyal oil [MATER] An essential oil distilled from the dried leaves and tops of the small pennyroyal plants *Hedeoma pulegioides* or *Mentha pulegium;* used as a counterirritant in liniments, in insect repellents, and for the production of methanol. { pen·ē'rȯi·əl ,ȯil }

pennyweight [MECH] A unit of mass equal to 1/20 troy ounce or to 1.55517384 grams; the term is employed in the United States and in England for the valuation of silver, gold, and jewels. Abbreviated dwt; pwt. { 'pen·ē,wāt }

Penokean *See* Animikean. { pə'nō·kē·ən }

pen recorder [ENG] A device in which the varying inputs (electrical, pneumatic, mechanical) are marked by a signal-controlled pen onto a continuous recorder chart (circular or roll chart). { 'pen ri,kȯrd·ər }

Penrose diagram [RELAT] A diagram of a space-time where the causal and infinity structure is displayed through the use of conformal transformations. Also known as conformal diagram. { 'pen,rōz ,dī·ə,gram }

Penrose-Hawking theorems [RELAT] The general relativistic theorems proving that singularities must occur in space-times, such as the universe, based on reasonable assumptions such as causality and dependent on the existence of a trapped surface. { pen,rōz 'hȯk·iŋ ,thir·əmz }

penroseite [MINERAL] $(Ni,Co,Cu)Se_2$ A lead gray, isometric mineral consisting of a selenide of nickel, copper, and cobalt; occurs in reniform masses. { 'pen,rō,zīt }

Penrose process [RELAT] A hypothetical means of extracting energy from a rotating black hole in which a particle spirals into the ergosphere of the black hole in a direction opposite to the black hole's rotation and then breaks up into two fragments, one of which escapes with an energy greater than the energy of the original particle. { 'pen,rōz ,prä·səs }

Penrose theorem [RELAT] A theorem which states that a collapsing object whose radius is smaller than its gravitational radius must collapse into a singularity. { 'pen,rōz ,thir·əm }

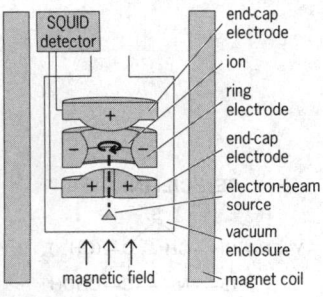

PENNING-TRAP MASS SPECTROMETER

Single-ion Penning-trap mass spectrometer. The end-cap voltages are typically 10 volts higher than the ring.

PENNSYLVANIAN

CENOZOIC	QUATERNARY	
	TERTIARY	
MESOZOIC	CRETACEOUS	
	JURASSIC	
	TRIASSIC	
PALEOZOIC	PERMIAN	
	CARBONIFEROUS	PENNSYLVANIAN
		MISSISSIPPIAN
	DEVONIAN	
	SILURIAN	
	ORDOVICIAN	
	CAMBRIAN	
	PRECAMBRIAN	

Chart showing position of the Pennsylvanian in relation to the periods and eras of geologic time.

Pensky-Martens closed tester [CHEM ENG] Device to determine the American Society for Testing and Materials flash point of fuel oils and cutback asphalts and other viscous materials and suspensions of solids. { 'pen·skē 'märt·ənz 'klōsd 'tes·tər }

penstock [CIV ENG] A valve or sluice gate for regulating water or sewage flow. [ENG] A closed water conduit controlled by valves and located between the intake and the turbine in a hydroelectric plant. { 'pen,stäk }

pentabasic [CHEM] A description of a molecule that has five hydrogen atoms that may be replaced by metals or bases. { pen·tə'bā·sik }

pentaborane [INORG CHEM] B_5H_9 Flammable liquid boiling at 48°C; ignites spontaneously in air; proposed as high-energy fuel for aircraft and missiles. { pen·tə'bȯr,ān }

pentachloride [CHEM] A molecule containing five atoms of chlorine in its structure. { pen·tə'klȯr,īd }

pentachloroethane [ORG CHEM] $CHCl_2CCl_3$ Colorless, water-insoluble liquid, boiling at 159°C; used as a solvent to degrease metals. Also known as pentalin. { pen·tə,klȯr·ō'eth,ān }

pentachloronitrobenzene [ORG CHEM] $C_6Cl_5NO_2$ Crystals of cream color with a melting point of 142–145°C; slightly soluble in alcohols; used as a fungicide and herbicide. Abbreviated PCNB. Also known as quintozene; terrachlor. { pen·tə,klȯr·ō,nī·trə'ben,zēn }

pentachlorophenol [ORG CHEM] C_6Cl_5OH A toxic white powder, decomposing at 310°C, melting at 190°C; soluble in alcohol, acetone, ether, and benzene; used as a fungicide, bactericide, algicide, herbicide, and chemical intermediate. { pen·tə,klȯr·ō·'fē,nȯl }

pentacite [MATER] Alkyd resin in which pentaerythritol is the polyhydric alcohol; used in coatings and printing inks. { 'pen·tə,sīt }

pentacosane [ORG CHEM] $C_{25}H_{52}$ A water-insoluble hydrocarbon derived from beeswax. { ,pen·tə'kō,sān }

pentacrinoid [INV ZOO] The larva of a feather star. { pen 'tak·rə,nȯid }

pentactinal [ZOO] Having five rays or branches. { pen 'tak·tə·nəl }

pentacula [INV ZOO] The five-tentacled stage in the life history of echinoderms. { pen'tak·yə·lə }

pentacyanium bis(methyl sulfate) [PHARM] $C_{29}H_{45}N_3$-O_9S_2 A water-soluble, crystalline compound obtained from ethanol solution, and melting at 173–175°C; used in medicine in the dichloride form as an antihypertensive agent. { ,pen·tə·sī'an·ē·əm bis 'meth·əl 'səl,fāt }

pentad [CLIMATOL] A period of 5 consecutive days, often preferred to the week for climatological purposes since it is an exact factor of the 365-day year. { 'pen,tad }

pentadactyl [VERT ZOO] Having five digits on the hand or foot. { pen·tə'dakt·əl }

pentadecagon [MATH] A polygon with 15 sides. { ,pen·tə'dek·ə,gän }

pentadecane [ORG CHEM] $C_{15}H_{32}$ A colorless, water-insoluble liquid, boiling at 270.5°C; soluble in alcohol; used as a chemical intermediate. { pen·tə'de,kān }

pentadecanolide [ORG CHEM] $C_{15}H_{18}O_2$ A colorless liquid with a musky odor extracted from angelica oil; soluble in 90% ethyl alcohol in equal volume; used in perfumes. { pen·tə·də'kan·əl,īd }

pentadelphous [BOT] Having the stamens in five sets with the filaments more or less united within each set. { pen·tə'del·fəs }

pentadentate ligand [INORG CHEM] A chelating agent having five groups capable of attachment to a metal ion. Also known as quinquidentate ligand. { pen·tə'den,tāt 'lī·gənd }

pentadiene [ORG CHEM] C_5H_8 Any of several straight-chain liquid diolefins: $CH_3CH_2CH=C=CH_2$, a colorless liquid boiling at 45°C, also known as ethylallene; $CH_3CH=CHCH=CH_2$, a colorless liquid boiling at 43°C; $CH_2=CHCH_2CH=CH_2$, a colorless liquid boiling at 26°C. { pen·tə'dī,ēn }

pentaerythritol [ORG CHEM] $(CH_2OH)_4C$ A white crystalline solid, melting at 261–262°C; moderately soluble in cold water, freely soluble in hot water; used to make the explosive pentaerythritol tetranitrate (PETN) and in the manufacture of alkyol resins and other coating compounds. { pen·tə·ə'rith·rə,tȯl }

pentaerythritol tetranitrate [ORG CHEM] $C(CH_2ONO_2)_4$ A white crystalline compound, melting at 139°C; explodes at 205–215°C; soluble in acetone, insoluble in water; used in medicines and explosives. Also known as penthrite; PETN. { ¦pen·tə·ə'rith·rə,tȯl ,te·trə'nī,trāt }

pentaerythritol tetrastearate [ORG CHEM] $C(CH_2OCC-C_{17}H_{35})_4$ A hard, ivory-colored wax with a softening point of 67°C; used in polishes and textile finishes. { ¦pen·tə·ə'rith·rə,tȯl ,te·trə'stir,āt }

pentafluoride [CHEM] A chemical compound onto which five fluoride atoms are bonded. { ,pen·tə'flur,īd }

pentagon [MATH] A polygon with five sides. { 'pen·tə,gän }

pentagonal dodecahedron See pyritohedron. { pen'tag·ən·əl dō·dek·ə'hē·drən }

pentagonal number [MATH] The total number, $P(n)$, of dots marking off unit segments of the sides of a set of $n − 1$ nested pentagons, given by the formula $P(n) = n(3n − 1)/2$. { pen'tag·ən·əl ¦nəm·bər }

pentagonal prism [MATH] A prism with two pentagonal sides, parallel and congruent. { pen'tag·ən·əl 'priz·əm }

pentagonal pyramid [MATH] A pyramid whose base is a pentagon. { pen'tag·ən·əl 'pir·ə,mid }

pentagrid See heptode. { 'pen·tə,grid }

pentahedron [MATH] A polyhedron with five faces. { ,pen·tə'hē·drən }

pentahydrite [MINERAL] $MgSO_4·5H_2O$ A triclinic mineral composed of hydrous magnesium sulfate; it is isostructural with chalcanthite. { ,pen·tə'hī,drīt }

pentahydroxyhexoic acid See galactonic acid. { ,pen·tə·hī¦dräk·sē·hek'sō·ik 'as·əd }

pentalin See pentachloroethane. { 'pent·əl·ən }

pentalogy [MED] Five symptoms or defects which together characterize a disease or syndrome. { pen'tal·ə·jē }

pentalogy of Fallot [MED] A syndrome consisting of the tetralogy of Fallot plus an interatrial septal defect. { pen'tal·ə·jē əv fa'lō }

Pentamerida [PALEON] An extinct order of articulate brachiopods. { ,pen·tə'mer·ə·də }

Pentameridina [PALEON] A suborder of extinct brachiopods in the order Pentamerida; dental plates associated with the brachiophores were well developed, and their bases enclosed the dorsal adductor muscle field. { ,pen·tə·mə'rid·ən·ə }

pentamerous [BOT] Having each whorl of the flower consisting of five members, or a multiple of five. { pen'tam·ə·rəs }

pentandrous [BOT] Having five stamens. { pen'tan·drəs }

n-pentane [ORG CHEM] $CH_3(CH_2)_3CH_3$ A colorless, flammable, water-insoluble hydrocarbon liquid, freezing at −130°C, boiling at 36°C; soluble in hydrocarbons and ethers; used as a chemical intermediate, solvent, and anesthetic. { ¦en 'pen,tān }

pentane candle [OPTICS] A unit of luminous intensity equal to one-tenth of the luminous intensity of a standard pentane lamp, and approximately equal to 1 candela. { 'pen,tēn 'kand·əl }

1,5-pentanediol [ORG CHEM] $HOCH_2(CH_2)_3CH_2OH$ Colorless, water-miscible liquid boiling at 242.5°C; used as a hydraulic fluid, lube-oil additive, and antifreeze, and in manufacture of polyester and polyurethane resins. { ¦wən ¦fīv ,pen,tān'dī,ȯl }

pentane insolubles [ANALY CHEM] Insoluble matter that can be separated from used lubricating oil in solution in n-pentane; may include resinous bitumens produced from the oxidation of oil and fuel; used in an American Society for Testing and Material test. { 'pen,tān in'säl·yə·bəlz }

pentane lamp [ENG] A pentane-burning lamp formerly used as a standard for photometry. { 'pen,tān ,lamp }

pentanol [ORG CHEM] $C_5H_{11}OH$ A toxic organic alcohol; 1-pentanol is n-amyl alcohol, primary; 2-pentanol is methylpropylcarbinol; 3-pentanol is diethylcarbinol; tert-pentanol is tert-amyl alcohol; pentanols are used in pharmaceuticals, as chemical intermediates, and as solvents. { 'pen·tə,nȯl }

pentanone [ORG CHEM] Either of two isomeric ketones derived from pentane: $CH_3COC_3H_7$ is a flammable, colorless, clear liquid, a mixture of methyl propyl and diethyl ketones; insoluble in water, soluble in ether and alcohol; used as a solvent. $C_2H_5COC_2H_5$ is a colorless, flammable liquid with

acetone aroma, boiling at 101°C; soluble in alcohol and ether; used in medicine and organic synthesis. { 'pen·tə,nōn }

pentaprismane [ORG CHEM] $C_{10}H_{10}$ A highly strained, saturated hydrocarbon cage structure. { ,pen·tə'priz,mān }

pentasaccharide [BIOCHEM] A carbohydrate which, on hydrolysis, yields five molecules of monosaccharides. { ,pen·tə'sak·ə,rīde }

Pentastomida [INV ZOO] A class of bloodsucking parasitic arthropods; the adult is vermiform, and there are two pairs of hooklike, retractile claws on the cephalothorax. { ,pen·tə'stäm·ə·də }

pentastyle [ARCH] Having five columns across the front. { 'pen·tə,stīl }

Pentatomidae [INV ZOO] The true stink bugs, a family of hemipteran insects in the superfamily Pentatomoidea. { ,pen·tə'täm·ə,dē }

Pentatomoidea [INV ZOO] A subfamily of the hemipteran group Pentatomorpha distinguished by marginal trichobothria and by antennae which are usually five-segmented. { ,pen·tə·tə'mȯid·ē·ə }

Pentatomorpha [INV ZOO] A large group of hemipteran insects in the subdivision Geocorisae in which the eggs are ronoperculate, a median spermatheca is present, accessory salivary glands are tubular, and the abdomen has trichobothria. { ,pen·tə·də'mȯr·fə }

pentavalent [CHEM] An atom or radical that exhibits a valency of 5. { ¦pen·tə'vā·lənt }

pentene [ORG CHEM] C_5H_{12} Colorless, flammable liquids derived from natural gasoline; isomeric forms are α-n-amylene and β-n-amylene. { 'pen,tēn }

penthouse [BUILD] 1. An enclosed space built on a flat roof to cover a stairway, elevator, or other equipment. 2. A dwelling built on top of the main roof. 3. A sloping shed or roof attached to a wall or building. { 'pent,haus }

penthrite See pentaerythritol tetranitrate. { 'pen,thrīt }

pentice [MIN ENG] 1. A rock pillar left, or a heavy timber bulkhead placed, in the bottom of a deep shaft of two or more compartments; the shaft is then further sunk through the pentice. 2. In shaft sinking, a solid rock pillar left in the bottom of the shaft for overhead protection of miners while the shaft is being extended by sinking. { 'pen·təs }

pentlandite [MINERAL] $(Fe,Ni)_9S_8$ A yellowish-bronze mineral having a metallic luster and crystallizing in the isometric system; hardness is 3.5–4 on Mohs scale, and specific gravity is 4.6–5.0; the major ore of nickel. { 'pent·lən,dīt }

pentobarbital sodium [PHARM] $C_{11}H_{17}N_2NaO_3$ A short- to intermediate-acting barbiturate; used as a hypnotic and sedative drug. Also known as sodium pentobarbitone. { ¦pen·tō'bär·bə,tȯl 'sōd·ē·əm }

pentode [ELECTR] A five-electrode electron tube containing an anode, a cathode, a control electrode, and two additional electrodes that are ordinarily grids. { 'pen,tōd }

pentode transistor [ELECTR] Point-contact transistor with four-point-contact electrodes; the body serves as a base with three emitters and one collector. { 'pen,tōd tran'zis·tər }

pentoglycan [BIOCHEM] A polysaccharide yielding a pentose sugar on hydrolysis. { ,pent·ə'glī,kan }

pentoglycerine See trimethylolethane. { ¦pen·tō'glis·ə·rən }

pentolinium tartrate [PHARM] $C_{23}H_{42}N_2O_{12}$ A white to cream-colored powder, soluble in water; used in medicine. { ,pen·tə'lin·ē·əm 'tär,trāt }

pentolite [MATER] A high explosive composed of pentaerythritol and trinitrotoluene. { 'pen·tə,līt }

pentomino [MATH] One of the 12 plane figures that can be formed by joining five unit squares along their sides. { pen'täm·ə·nō }

pentosan [BIOCHEM] A hemicellulose present in cereal, straws, brans, and other woody plant tissues; yields five-carbon-atom sugars. { 'pen·tə,san }

pentose [BIOCHEM] Any one of a class of carbohydrates containing five atoms of carbon. { 'pen,tōs }

pentose phosphate pathway [BIOCHEM] A pathway by which glucose is metabolized or transformed in plants and microorganisms; glucose-6-phosphate is oxidized to 6-phosphogluconic acid, which then undergoes oxidative decarboxylation to form ribulose-5-phosphate, which is ultimately transformed to fructose-6-phosphate. { 'pen,tōs 'fäs,fāt 'path,wā }

pentosuria [MED] The presence of pentose in the urine. { ˌpen·təsˈyu̇r·ē·ə }

pentoxide [INORG CHEM] A compound that is binary and has five atoms of oxygen; for example, phosphorus pentoxide, P_2O_5. { penˈtäkˌsīd }

pentyl See amyl. { ˈpent·əl }

pentylenetetrazol [PHARM] $C_6H_{10}N_4$ A central nervous system stimulant used as an analeptic. { ˈpent·əlˌenˈtetrəˌzól }

para-pentyloxyphenol [ORG CHEM] $C_{11}H_{16}O_2$ Compound melting at 49–50°C; used as a bactericide. { ˈpar·ə ˈpent·əlˈäk·sēˈfēˌnól }

pentyne [ORG CHEM] C_5H_8 Either of two normal isometric acetylene hydrocarbons: $HC≡C(CH_2)_2CH_3$, colorless liquid boiling at 40°C, also known as pentine, propylacetylene; $CH_3C≡CC_2H_5$, liquid boiling at 56°C. { ˈpenˌtēn }

penumbra [ASTRON] The outer, relatively light part of a sunspot. [OPTICS] That portion of a shadow illuminated by only part of a radiating source. { pəˈnəm·brə }

penumbral eclipse See lunar appulse. { pəˈnəm·brəl iˈklips }

penumbral waves [ASTRON] Waves that are often observed to propagate outward across the penumbrae of large sunspots when the penumbrae are viewed in the light of the Hα spectral line of hydrogen. { pəˈnəm·brəl ˈwāvz }

Penutian [GEOL] A North American stage of geologic time: lower Eocene (above Bulitian, below Ulatasian). { pəˈnü·shən }

PEPE See parallel element-processing ensemble. { ˈpeˌpē }

peperite [GEOL] A breccialike material in marine sedimentary rock, considered to be either a mixture of lava with sediment, or shallow intrusions of magma into wet sediment. { ˈpep·əˌrīt }

pepo [BOT] A fleshy indehiscent berry with many seeds and a hard rind; characteristic of the Cucurbitaceae. { ˈpēˌpō }

pepper [BOT] Any of several warm-season perennials of the genus *Capsicum* in the order Polemoniales, especially *C. annum* which is cultivated for its fruit, a many-seeded berry with a thickened integument. [FOOD ENG] Any of various spices and condiments obtained from the fruits of plants of the genus *Piper*. { ˈpep·ər }

peppermint [BOT] Any of various aromatic herbs of the genus *Mentha* in the family Labiatae, especially *M. piperita*. { ˈpep·ərˌmint }

peppermint camphor See menthol. { ˈpep·ərˌmint ˈkam·fər }

peppermint oil See oil of peppermint. { ˈpep·ərˌmint ˌói̇l }

pepper sludge [MATER] Fine particles of sludge produced during the acid treatment of lubricating oils and other petroleum products. { ˈpep·ər ˌsləj }

PeP reaction See proton-electron-proton reaction. { ˈpēˌēˈpē rēˌak·shən }

pepsin [BIOCHEM] A proteolytic enzyme found in the gastric juice of mammals, birds, reptiles, and fishes. { ˈpep·sən }

pepsinogen [BIOCHEM] The precursor of pepsin, found in the stomach mucosa. { pepˈsin·ə·jən }

peptic [PHYSIO] **1.** Of or pertaining to pepsin. **2.** Of or pertaining to digestion. { ˈpep·tik }

peptic ulcer [MED] An ulcer involving the mucosa, submucosa, and muscular layer on the lower esophagus, stomach, or duodenum, due in part at least to the action of acid-pepsin gastric juice. { ˈpep·tik ˈəl·sər }

peptidase [BIOCHEM] An enzyme that catalyzes the hydrolysis of peptides to amino acids. { ˈpep·təˌdās }

peptide [BIOCHEM] A compound of two or more amino acids joined by peptide bonds. { ˈpepˌtīd }

peptide bond [ORG CHEM] A bond in which the carboxyl group of one amino acid is condensed with the amino group of another to form a $-CO·NH-$ linkage. Also known as peptide linkage. { ˈpepˌtīd ˌbänd }

peptide linkage See peptide bond. { ˈpepˌtīd ˌliŋ·kij }

peptide map [CELL MOL] A fragmentation pattern generated by digestion of a particular protein with proteolytic enzymes of known specificity; used in protein identification. { ˈpepˌtīd ˌmap }

peptide synthetase [BIOCHEM] A ribosomal synthetase that catalyzes the formation of peptide bonds during protein synthesis. { ˈpepˌtīd ˈsin·thəˌtās }

peptization [CHEM] **1.** Aggregation in which a hydrophobic colloidal sol is stabilized by the addition of electrolytes (peptizing agents) which are adsorbed on the particle surfaces. **2.** Liquefaction of a substance by trace amounts of another substance. { ˌpep·təˌzā·shən }

peptized fuel [MATER] Thickened flamethrower fuel to which water or other chemical is added before mixing, to reduce mixing time and to increase storage stability. { ˈpepˌtīzd ˈfyül }

Peptococcaceae [MICROBIO] A family of gram-positive cocci; organisms can use either amino acids or carbohydrates for growth and energy. { ˌpep·təˈkäkˈsās·ēˌē }

peptone [BIOCHEM] A water-soluble mixture of proteoses and amino acids derived from albumin, meat, or milk; used as a nutrient and to prepare nutrient media for bacteriology. { ˈpepˌtōn }

per- [CHEM] Prefix meaning: **1.** Complete, as in hydrogen peroxide. **2.** Extreme, or the presence of the peroxy $(-O-O-)$ group. **3.** Exhaustive (complete) substitution, as in perchloroethylene. { pər or per }

Peracarida [INV ZOO] A superorder of the Eumalacostraca; these crustaceans have the first thoracic segment united with the head, the cephalothorax usually larger than the abdomen, and some thoracic segments free from the carapace. { ˌper·əˈkar·ə·də }

peracetic acid [ORG CHEM] CH_3COOOH A toxic, colorless liquid with strong aroma; boils at 105°C; explodes at 110°C; miscible with water, alcohol, glycerin, and ether; used as an oxidizer, bleach, catalyst, bactericide, fungicide, epoxy-resin precursor, and chemical intermediate. Also known as peroxyacetic acid. { ˈpər·əˈsēd·ik ˈas·əd }

peracid [CHEM] Acid containing the peroxy $(-O-O-)$ group, such as peracetic acid or perchloric acid. { ˈpərˈas·əd }

peralcohol [ORG CHEM] Chemical compound containing the peroxy group $(-O-O-)$, such as peracetic acid and perchromic acid. { pərˈal·kəˌhól }

peralkaline [PETR] Of igneous rock, having a molecular proportion of aluminum lower than that of sodium oxide and potassium oxide combined. { pərˈal·kəˌlīn }

peraluminous [PETR] Of igneous rock, having a molecular proportion of aluminum oxide greater than that of sodium oxide and potassium oxide combined. { ˌpər·əˈlü·mə·nəs }

Peramelidae [VERT ZOO] The bandicoots, a family of insectivorous mammals in the order Marsupialia. { ˌpər·əˈmelˌə·dē }

P-E ratio See precipitation-evaporation ratio. { ˈpēˈē ˌrā·shō }

perazine [PHARM] $C_{20}H_{25}N_3S$ A crystalline compound, melting at 51–53°C; used in medicine as a tranquilizer. { ˈper·əˌzēn }

perbenzoic acid [ORG CHEM] $C_6H_5CO_2OH$ A crystalline compound forming leaflets from benzene solution, melting at 41–45°C, freely soluble in organic solvents; used in analysis of unsaturated compounds and to change ethylinic compounds into oxides. { ˈpər·benˈzō·ik ˈas·əd }

perbituminous [GEOL] Referring to bituminous coal containing more than 5.8% hydrogen, analyzed on a dry, ash-free basis. { ˈpər·bəˈtü·mə·nəs }

percale [TEXT] A plain-weave, medium-weight fabric made of cotton or cotton and polyester. { pərˈkāl }

percaline [TEXT] Fine cotton with a glossy or moiré surface. { ˈpər·kəˌlēn }

perceived noise decibel [ACOUS] A unit of perceived noise level. Abbreviated PNdB. { pərˈsēvd ˈnói̇z ˈdes·əˌbel }

perceived noise level [ACOUS] In perceived noise decibels, the noise level numerically equal to the sound pressure level, in decibels, of a band of random noise of width one-third of one octave centered on a frequency of 1000 hertz which is judged by listeners to be equally noisy. { pərˈsēvd ˈnói̇z ˌlev·əl }

percent [MATH] A quantitative term whereby n percent of a number is n one-hundredths of the number. Symbolized %. { pərˈsent }

percentage [MATH] The result obtained by taking a given percent of a given quantity. { pərˈsen·tij }

percentage depletion [PETRO ENG] Oil- or gas-reservoir depletion allowance calculated on the basis of unit sales and initial depletable leasehold cost. { pərˈsen·tij diˈplē·shən }

percentage depth dose [NUCLEO] The percentage of the absorbed dose from ionizing radiation at a given depth within a body to the absorbed dose at a reference point, usually the position of the peak absorbed dose. { pərˈsen·tij ˈdepth ˈdōs }

percentage differential relay [ELECTR] Differential relay which functions when the difference between two quantities of the same nature exceeds a fixed percentage of the smaller quantity. Also known as biased relay; ratio-balance relay; ratio-differential relay. { pər'sen·tij ,dif·ə'ren·chəl 'rē,lā }

percentage distribution [STAT] A frequency distribution in which the individual class frequencies are expressed as a percentage of the total frequency equated to 100. Also known as relative frequency distribution; relative frequency table. { pər'sen·tij ,dis·trə'byü·shən }

percentage extraction [MIN ENG] The proportion of a coal seam or other ores which is removed from a mine. { pər'sen·tij ik'strak·shən }

percentage log [ENG] A sample log in which the percentage of each type of rock (except obvious cavings) present in each sample of cuttings is estimated and plotted. { pər'sen·tij ,läg }

percentage map [PETRO ENG] Contoured map in which the percentage of one component (such as sand) is compared with the total unit (such as sand plus shale). { pər'sen·tij ,map }

percentage modulation See percent modulation. { pər'sen·tij ,maj·ə'lā·shən }

percentage ripple [ELECTR] Ratio of the effective value of the ripple voltage to the average value of the total voltage, expressed as a percentage. { pər'sen·tij 'rip·əl }

percentage subsidence [MIN ENG] The measured amount of subsidence expressed as a percentage of the thickness of coal extracted. { pər'sen·tij səb'sīd·əns }

percentage support [MIN ENG] The percentage of the total wall area which will actually be covered by supports. { pər'sen·tij sə'pȯrt }

percent compaction [ENG] The ratio, expressed as a percentage, of dry unit weight of a soil to maximum unit weight obtained in a laboratory compaction test. { pər'sent kəm'pak·shən }

percent defective [IND ENG] The ratio of defective pieces per lot or sample, expressed as a percentage. { pər'sent di'fek·div }

percent distortion [COMMUN] The ratio of the amplitude of a harmonic component to the fundamental component multiplied by 100. { pər'sent di'stȯr·shən }

percentile [STAT] A value in the range of a set of data which separates the range into two groups so that a given percentage of the measures lies below this value. { pər'sen,tīl }

percent make [ELECTR] **1.** In pulse testing, the length of time a circuit stands closed compared to the length of the test signal. **2.** Percentage of time during a pulse period that telephone dial pulse springs are making contact. { pər'sent 'māk }

percent modulation [COMMUN] The modulation factor expressed as a percentage. Also known as percentage modulation. { pər'sent ,maj·ə'lā·shən }

percept [PSYCH] A mental image of something perceived. { 'pər,sept }

perception [PHYSIO] Recognition in response to sensory stimuli; the act or process by which the memory of certain qualities of an object is associated with other qualities impressing the senses, thereby making possible recognition of the object. { pər'sep·shən }

perceptron [COMPUT SCI] A pattern recognition machine, based on an analogy to the human nervous system, capable of learning by means of a feedback system which reinforces correct answers and discourages wrong ones. { pər'sep,trän }

perceptual constancy [PSYCH] The tendency of the perceived properties of objects (size, shape, orientation, direction) to remain intact despite variations in distance, slant, and retinal locus caused by movements of the observer. { pər'sep·chə·wəl 'kän·stən·sē }

perceptual overload [NEUROSCI] Saturation of the nervous system by an input of excess sensory information, resulting in an absence of response. { pər'sep·chə·wəl 'ō·vər,lōd }

perch [MECH] Also known as pole; rod. **1.** A unit of length, equal to 5.5 yards, or 16.5 feet, or 5.0292 meters. **2.** A unit of area, equal to 30.25 square yards, or 272.25 square feet, or 25.29285264 square meters. [NAV] A staff placed on top of a buoy, rock, or shoal as a mark for navigators; a ball or cage is sometimes placed at the top of the perch, as an identifying mark. [VERT ZOO] **1.** Any member of the family Percidae. **2.** The common name for a number of unrelated species of fish belonging to the Centrarchidae, Anabantoidei, and Percopsiformes. { pərch }

perched aquifer [HYD] An aquifer that is separated from another water-bearing stratum by an impermeable layer. { 'pərcht 'ak·wə·fər }

perched block [GEOL] A large, detached rock fragment presumed to have been transported and deposited by a glacier, and perched in a conspicuous and precarious position on the side of a hill. Also known as balanced rock; perched boulder; perched rock. { 'pərcht 'bläk }

perched boulder See perched block. { 'pərcht 'bōl·dər }

perched groundwater See perched water. { 'pərcht 'graùnd ,wȯd·ər }

perched lake [HYD] A perennial lake whose surface level lies at a considerably higher elevation than those of other bodies of water, including aquifers, directly or closely associated with the lake. { 'pərcht 'lāk }

perched rock See perched block. { 'pərcht 'räk }

perched spring [HYD] A spring that arises from a body of perched water. { 'pərcht 'spriŋ }

perched stream [HYD] A stream whose surface level is above that of the water table and that is separated from underlying groundwater by an impermeable bed in the zone of aeration. { 'pərcht 'strēm }

perched water [HYD] Groundwater that is unconfined and separated from an underlying main body of groundwater by an unsaturated zone. Also known as perched groundwater. { 'pərcht 'wȯd·ər }

perched water table [HYD] The water table or upper surface of a body of perched water. Also known as apparent water table. { 'pərcht 'wȯd·ər ,tā·bəl }

perching bed [GEOL] A body of rock, generally stratiform, that supports a body of perched water. { 'pərch·iŋ ,bed }

perchlorate [INORG CHEM] A salt of perchloric acid containing the ClO_4^- radical; for example, potassium perchlorate, $KClO_4$. { pər'klȯr,āt }

perchloric acid [INORG CHEM] $HClO_4$ Strongly oxidizing, corrosive, colorless, hygroscopic liquid, boiling at 16°C (8 mmHg, or 1067 pascals); soluble in water; unstable in pure form, but stable when diluted in water; used in medicine, electrolytic baths, electropolishing, explosives, and analytical chemistry, and as a chemical intermediate. Also known as Fraude's reagent. { pər'klȯr·ik 'as·əd }

perchloroethylene [ORG CHEM] CCl_2CCl_2 Stable, colorless liquid, boiling at 121°C; nonflammable and nonexplosive, with low toxicity; used as a dry-cleaning and industrial solvent, in pharmaceuticals and medicine, and for metal cleaning. Also known as tetrachloroethylene. { pər'klȯr·ō'eth·ə,lēn }

perchloromethyl mercaptan [ORG CHEM] $ClSCCl_3$ Poisonous, yellow oil with disagreeable aroma; decomposes at 148°C; used as a chemical intermediate, granary fumigant, and military poison gas. { pər'klȯr·ō'meth·əl mər'kap,tan }

perchloryl fluoride [INORG CHEM] $ClFO_3$ A colorless gas with a sweet odor; boiling point is −46.8°C and melting point is −146°C; used as an oxidant in rocket fuels. { pər'klȯr·əl 'flùr,īd }

Percidae [VERT ZOO] A family of fresh-water actinopterygian fishes in the suborder Percoidei; comprises the true perches. { 'pər·sə,dē }

Perciformes [VERT ZOO] The typical spiny-rayed fishes, comprising the largest order of vertebrates; characterized by fin spines, a swim bladder without a duct, usually ctenoid scales, and 17 or fewer caudal fin rays. { ,pər·sə'fȯr,mēz }

Percoidei [VERT ZOO] A large, typical suborder of the order Perciformes; includes over 50% of the species in this order. { pər'kȯid·ē,ī }

percolation [COMPUT SCI] The transfer of needed data back from secondary storage devices to main storage. [HYD] Gravity flow of groundwater through the pore spaces in rock or soil. [MIN ENG] Gentle movement of a solvent through an ore bed in order to extract a mineral. [SCI TECH] Slow movement of a liquid through a porous material. { pər·kə'lā·shən }

percolation filtration [CHEM ENG] A continuous petroleum-refinery process in which lubricating oils and waxes are percolated through a clay bed to improve color, odor, and stability. { pər·kə'lā·shən fil,trā·shən }

PERCH

Rock bass (*Ambloplites rupestris*), a perch of the family Centrarchidae.

percolation leaching [MIN ENG] The selective removal of a mineral by causing a suitable solvent to seep into and through a mass or pile of material containing the desired soluble mineral. { pər·kə'lā·shən ‚lēch·iŋ }

percolation limit [SOLID STATE] In a disordered crystalline alloy having one constituent with a magnetic moment, the concentration of the magnetic element above which the spin-glass phase is replaced by the ferromagnetic state. { pər·kə'lā·shən ‚lim·ət }

percolation network [MATH] A lattice constructed of a random mixture of conducting and nonconducting links. { ‚pər·kə'lā·shən 'net‚wərk }

percolation problem [MATH] The problem of determining the critical threshold concentration of conducting links in a pecolation network at which an infinite cluster of conducting links is formed and the lattice transforms from an insulator to a conductor. { ‚pər·kə'lā·shən ‚präb·ləm }

percolation test [CIV ENG] A test to determine the suitability of a soil for the installation of a domestic sewage-disposal system, in which a hole is dug and filled with water and the rate of water-level decline is measured. { pər·kə'lā·shən ‚test }

percolation zone [HYD] The area on a glacier or ice sheet where a limited amount of surface melting occurs, but the meltwater refreezes in the same snow layer and the snow layer is not completely soaked or brought up to the melting temperature. { pər·kə'lā·shən ‚zōn }

Percomorphi [VERT ZOO] An equivalent, ordinal name for the Perciformes. { ‚pər·kə'mór‚fī }

Percopsidae [VERT ZOO] A family of fishes in the order Percopsiformes. { pər'käp·sə‚dē }

Percopsiformes [VERT ZOO] A small order of actinopterygian fishes characterized by single, ray-supported dorsal and anal fins and a subabdominal pelvic fin with three to eight soft rays. { pər‚käp·sə'fór‚mēz }

PERCOPSIFORMES

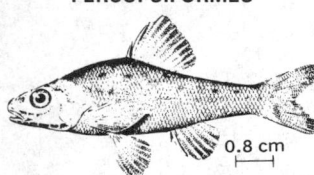

Sand roller (*Percopsis transmontana*).

percussion [MED] The act of striking or firmly tapping the surface of the body with a finger or a small hammer to elicit sounds, or vibratory sensations, of diagnostic value. [ORD] Setting off an explosive charge by a sharp blow, as of a gun hammer. { pər'kəsh·ən }

percussion bit [MECH ENG] A rock-drilling tool with chisel-like cutting edges, which when driven by impacts against a rock surface drills a hole by a chipping action. { pər'kəsh·ən ‚bit }

percussion drill [MECH ENG] A drilling machine usually using compressed air to drive a piston that delivers a series of impacts to the shank end of a drill rod or steel and attached bit. { pər'kəsh·ən ‚dril }

percussion drilling [MECH ENG] A drilling method in which hammer blows are transmitted by the drill rods to the drill bit. { pər'kəsh·ən ‚dril·iŋ }

percussion figure [CRYSTAL] Radiating lines on a crystal section produced by a sharp blow. { pər'kəsh·ən ‚fig·yər }

percussion fire [ORD] Fire with fuses set to burst on impact. { pər'kəsh·ən ‚fīr }

percussion firing mechanism [ORD] Any firing mechanism which fires the primer by percussion. { pər'kəsh·ən ‚fir·iŋ ‚mek·ə‚niz·əm }

percussion hammer firing mechanism [ORD] Firing mechanism in which a hammer, actuated by a pull of a lanyard, strikes the firing pin and fires the weapon. { pər'kəsh·ən ‚ham·ər 'fir·iŋ ‚mek·ə‚niz·əm }

percussion mark [GEOL] A small, crescent-shaped scar produced on a hard, dense pebble by a blow. { pər'kəsh·ən ‚märk }

percussion powder [MATER] Powder so composed as to ignite by a slight percussion. { pər'kəsh·ən ‚paüd·ər }

percussion side-wall sampling [PETRO ENG] A method of side-wall core sampling from wellbores in softer reservoir formations. { pər'kəsh·ən 'sīd ‚wól ‚samp·pliŋ }

percussion table *See* concussion table. { pər'kəsh·ən ‚tā·bəl }

percussion welding [MET] Resistance welding with arc heat and simultaneously applied pressure from a hammerlike blow. { pər'kəsh·ən ‚weld·iŋ }

percylite [MINERAL] $PbCuCl_2(OH)_2$ Mineral made up of a basic chloride of copper and lead and occurring as cubic blue crystals, with a hardness of 2.5. { 'pər·sē·līt }

pereletok [GEOL] A frozen layer of ground, at the base of the active layer, which may persist for one or several years. Also known as intergelisol. { ‚per·ə·lə'täk }

perennial [BOT] A plant that lives for an indefinite period, dying back seasonally and producing new growth from a perennating part. { pə'ren·ē·əl }

perennial lake [HYD] A lake that retains water in its basin throughout the year and is not usually subject to extreme water-level fluctuations. { pə'ren·ē·əl 'lāk }

perennial spring [HYD] A spring that flows continuously, as opposed to an intermittent spring or a periodic spring. { pə'ren·ē·əl 'spriŋ }

perennial stream [HYD] A stream which contains water at all times except during extreme drought. { pə'ren·ē·əl 'strēm }

perezone [GEOL] A zone in which sediments accumulate along coastal lowlands; includes lagoons and brackish-water bays. { 'per·ə‚zōn }

perfect cosmological principle [ASTRON] The assumption that the universe is homogeneous and isotropic, and does not change with time. { 'pər·fikt ‚käz·mə'läj·ə·kəl 'prin·sə·pəl }

perfect crystal [CRYSTAL] A crystal without lattice defects; it is an unattained ideal or standard. { 'pər·fikt 'krist·əl }

perfect cube [MATH] A number or polynomial which is the exact cube of another number or polynomial. { 'pər·fikt 'kyüb }

perfect dielectric *See* ideal dielectric. { 'pər·fikt ‚dī·ə'lek·trik }

perfect field [MATH] A field such that any irreducible polynomial with coefficients in this field is separable; a field whose finite extensions are all separable. { 'pər‚fikt 'fēld }

perfect flower [BOT] A flower having both stamens and pistils. { 'pər·fikt 'flaů·ər }

perfect fluid *See* inviscid fluid. { 'pər·fikt 'flü·əd }

perfect fractionation path [PHYS CHEM] On a phase diagram, a line or a path representing a crystallization sequence in which any crystal that has been formed remains inert, that is, its composition is not altered. { 'pər·fikt ‚frak·shə'nā·shən ‚path }

perfect gas *See* ideal gas. { 'pər·fikt 'gas }

perfect group [MATH] A group that is equal to its commutator subgroup. { 'pər·fikt 'grüp }

perfecting press [GRAPHICS] A printing press that is able to print both sides of the paper at once. { pər'fek·tiŋ ‚pres }

perfect lubrication [ENG] A complete, unbroken film of liquid formed over each of two metal surfaces moving relatively to one another with no contact. { 'pər·fikt ‚lü·brə'kā·shən }

perfectly diffuse radiator [OPTICS] A body that emits radiant energy in accordance with Lambert's law. { 'pər·fik·lē di'fyüs 'rād·ē‚ād·ər }

perfectly diffuse reflector [OPTICS] A body that reflects radiant energy in such a manner that the reflected energy may be treated as if it were being emitted (radiated) in accordance with Lambert's law. { 'pər·fik·lē di'fyüs ri'flek·tər }

perfectly inelastic collision [PHYS] A collision in which much translational kinetic energy is converted into internal energy of the colliding systems as is consistent with the conservation of momentum. Also known as completely inelastic collision. { 'pər·fik·lē ‚in·i'las·tik kə'lizh·ən }

perfectly mobile component [PHYS CHEM] A component whose quantity in a system is determined by its externally imposed chemical potential rather than by its initial quantity in the system. Also known as boundary value component. { 'pər·fik·lē 'mō·bəl kəm'pō·nənt }

perfectly separable space [MATH] A topological space whose topology has a countable base. Also known as completely separable space. { ‚pər·fikt·lē 'sep·rə·bəl 'spās }

perfect number [MATH] An integer which equals the sum of all its factors other than itself. { 'pər·fikt 'nəm·bər }

perfect power [MATH] A number or polynomial which equals another number or polynomial raised to an integral power greater than one. { ‚pər·fikt 'paů·ər }

perfect prognostic [METEOROL] The observed pressure pattern at the verifying time of a forecast of some element other than pressure; used in objective forecast studies in which a forecast of the element is based on a simultaneous relation between this element and the pressure pattern plus a forecast of the pressure pattern at some future time. { 'pər·fikt präg'näs·tik }

perfect set [MATH] A set in a topological space which equals its set of accumulation points. { 'pər·fikt 'set }

perfect solution [PHYS CHEM] A solution that is ideal throughout its entire compositional range. { 'pər·fikt sə'lü·shən }

perfect square [MATH] A number or polynomial which is the exact square of another number or polynomial. { 'pər·fikt 'skwer }

perfect trinomial square [MATH] A trinomial that is the exact square of a binomial. { pər·fikt trī'nō·mē·əl 'skwer }

perfect vacuum *See* absolute vacuum. { 'pər·fikt 'vak·yəm }

perfemic rock [GEOL] An igneous rock in which the ratio of salicalic to femic minerals is less than 1:7. { pər'fem·ik 'räk }

perfluorocarbon [ORG CHEM] **1.** A compound consisting of carbon and fluorine. **2.** A compound in which all the hydrogen atoms of a hydrocarbon are replaced with fluorine atoms. Abbreviated PFC. { pər'flur·ə,kär·bən }

perfluorochemical [ORG CHEM] A hydrocarbon in which all the hydrogen atoms have been replaced by fluorine. { pər'flur·ō'kem·ə·kəl }

perfoliate [BOT] Pertaining to the form of a leaf having its base united around the stem. [INV ZOO] Pertaining to the form of certain insect antennae having the terminal joints expanded and flattened to form plates which encircle the stalk. { pər'fōl·ē,āt }

perforate [SCI TECH] To pierce or puncture; particularly, to make a line or series of holes for such purposes as identification, decoration, or easy separation. { 'pər·fə,rāt }

perforated completion [PETRO ENG] Well completion in which the production casing or liner is pierced to allow passage of fluids between the wellbore and the producing formation. { 'pər·fə,rād·əd kəm'plē·shən }

perforated crust [HYD] A type of snow crust containing pits and hollows produced by ablation. { 'pər·fə,rād·əd 'krəst }

perforated film [GRAPHICS] Roll film containing perforations on one or both sides. { 'pər·fə,rād·əd 'film }

perforated metal [MATER] Sheet metal with round, square, diamond, or rectangular perforations; used for screens and for construction. { 'pər·fə,rād·əd 'med·əl }

perforated-pipe distributor [CHEM ENG] Liquid distribution device consisting of a length of piping or tubing with holes at spaced intervals along the length; used in spray columns, liquid-vapor contactors, and spray driers. Also known as a sparger. { 'pər·fə,rād·əd 'pīp di'strib·yəd·ər }

perforated plate [CHEM ENG] Flat plate with series of holes used to control fluid distribution, as in a perforated-plate (distillation) column. { 'pər·fə,rād·əd 'plāt }

perforated-plate column [CHEM ENG] Distillation column in which vapor-liquid contact is provided by perforated plates instead of bubble-cap trays. { 'pər·fə,rād·əd 'plāt 'käl·əm }

perforated-plate distributor [CHEM ENG] **1.** A perforated plate or screen used to even out liquid-flow fluctuations through flow channels. **2.** A perforated plate as used in a distillation column or liquid-liquid extraction column. { 'pər·fə,rād·əd 'plāt di'strib·yəd·ər }

perforated-plate extractor [CHEM ENG] A liquid-liquid extraction vessel in which perforated plates are used to bring about contact between the two or more liquid phases. { 'pər·fə,rād·əd 'plāt ik'strak·tər }

perforating [PETRO ENG] Special oil-well downhole procedure to make holes in tubing walls and surrounding cement; used to allow formation oil or gas to enter the wellbore tubing, or to allow water to be forced out into the formation to cause fracturing (hydraulic fracturing). { 'pər·fə,rād·iŋ }

perforation [ORD] Passage of a missile completely through an object. [SCI TECH] Any hole made by boring, punching, or piercing. { ,pər·fə'rā·shən }

perforation deposit [GEOL] An isolated kame consisting of material that accumulated in a vertical shaft which pierced a glacier and afforded no outlet for water at the bottom. { ,pər·fə'rā·shən di,päz·ət }

perforations [GRAPHICS] Small rectangular sprocket holes along the edge of film. [PETRO ENG] Downwell holes made in well tubing, usually by shot-and-explosive or shaped-charge techniques; used for oil or gas production from desired horizons, or for injection of acidizing or fracturing fluids into the formation at predetermined depths. { ,pər·fə'rā·shənz }

perforator [COMMUN] In telegraph practice, a device for punching code signals in paper tape for application to a tape transmitter. { 'pər·fə,rād·ər }

perforatorium *See* acrosome. { ,pər·fə·rə'tór·ē·əm }

perforin [IMMUNOL] An enzyme that is secreted by natural killer cells and cytotoxic T cells and destroys foreign cells by puncturing their membranes, causing leakage of the cell contents. Also known as cytolysin; pore-forming protein. { 'pər·fər·ər }

perform [COMPUT SCI] A subroutine in the COBOL programming language that allows a portion of a program to be executed on command by other portions of the same program. { pə'fórm }

performance bond [ENG] A bond that guarantees performance of a contract. { pər'fór·məns bänd }

performance characteristic [ENG] A characteristic of a piece of equipment, determined during its test or during its operation. { pər'fór·məns ,kar·ik·tə'ris·tik }

performance chart [ENG] A graph used in evaluating the performance of any device, for example, the performance of an electrical or electronic device, such as a graph of anode voltage versus anode current for a magnetron. { pər'fór·məns ,chärt }

performance curves [ENG] Graphical representations showing the abilities of rotating equipment at various operating conditions; for example, the performance curve for a compressor would include rotor speed for various intake and outlet pressures versus gas flow rate adjusted for temperature, density, viscosity, head, and other factors. { pər'fór·məns ,kərvz }

performance data [ENG] Data on the manner in which a given substance or piece of equipment performs during actual use. { pər'fór·məns ,dad·ə }

performance evaluation [IND ENG] The analysis in terms of initial objectives and estimates, and usually made on site, of accomplishments using an automatic data-processing system, to provide information on operating experience and to identify corrective actions required, if any. { pər'fór·məns i,val·yə'wā·shən }

performance failure [COMPUT SCI] Failure of a computer system in which the system operates correctly but fails to deliver the results in a timely fashion. { pər'fór·məns ,fāl·yər }

performance index [IND ENG] The ratio of standard hours to the hours of work actually used; a ratio exceeding 1.00 (or 100) indicates standard output is being exceeded. { pər'fór·məns ,in,deks }

performance measurement baseline [IND ENG] A time-phased budget plan developed for use in measuring contract performance; includes the budgets assigned to scheduled work elements and the related indirect budgets. { pər'fórm·əns ,mezh·ər·mənt 'bās,līn }

performance number [ENG] One of a series of numbers (constituting the PN, or performance-number, scale) used to convert fuel antiknock values in terms of a reference fuel into an index which is an indication of relative engine performance; used mostly to rate aviation gasolines with octane values greater than 100. { pər'fór·məns ,nəm·bər }

performance rating *See* effort rating. { pər'fór·məns ,rād·iŋ }

performance sampling [IND ENG] A technique in work measurement used to determine the leveling factor to be applied to an operator or a group of operators by short, randomly spaced observations of the performance index. { pər'fór·məns ,sam·pliŋ }

performing arts medicine [MED] A subspecialty in medicine that deals with problems specific to the activities of dancers, musicians, and vocalists. { pər'fórm·iŋ 'arts ,med·ə·sən }

perfory [COMPUT SCI] The removable edges of computer paper containing holes engaged by the pin-feed mechanism. { 'pər·fə·rē }

perfume base [MATER] Any natural or synthetic material used by perfumers as a starting point for perfume manufacture. { 'pər,fyüm ,bās }

perfume oil [MATER] Any volatile oil distilled or extracted from the leaves, flowers, gums, or woods of plant life (but occasionally of animal origin) and used in making perfume; examples are linalyl acetate (from citral), crab apple, wisteria, lavendar, and attar of rose. { 'pər,fyüm ,oil }

perfusion [PHYSIO] The pumping of a fluid through a tissue or organ by way of an artery. { pər'fyü·zhən }

PERFOLIATE

Shape of a perfoliate leaf.

pergelation [HYD] The act or process of forming permafrost. { ˌpər·jə'lā·shən }

pergelic [GEOL] Referring to a soil temperature regime in which the mean annual temperature is less than 0°C and there is permafrost. { pər'jel·ik }

pergelisol table *See* permafrost table. { pər'jel·ə,sȯl }

Pergidae [INV ZOO] A small family of hymenopteran insects in the superfamily Tenthredinoidea. { 'pər·jə,dē }

perhumid climate [CLIMATOL] As defined by C. W. Thornthwaite in his climatic classification, a type of climate which has humidity index values of +100 and above; this is his wettest type of climate (designated A), and compares closely to the "wet climate" which heads his 1931 grouping of humidity provinces. { pər'hyü·məd 'klī·mət }

perhydro- [ORG CHEM] Prefix designating a completely saturated aromatic compound, as for decalin ($C_{10}H_{18}$), also known as perhydronaphthalene. { pər'hī·drō }

perhydrous coal [GEOL] Coal that contains more than 6% hydrogen, analyzed on a dry, ash-free basis. { pər'hī·drəs 'kōl }

perianal [ANAT] Situated or occurring around the anus. { ˌper·ē'ān·əl }

perianth [BOT] The calyx and corolla considered together. { 'per·ē,anth }

periapical cyst *See* radicular cyst. { per·ē'ap·ə·kəl 'sist }

periappendicitis [MED] Inflammation of the tissue around the vermiform process, or of the serosal region of the vermiform appendix. { ˌper·ē·ə,pen·də'sīd·əs }

periapsis [ASTRON] The orbital point nearest the center of attraction of an orbiting body. { ˌper·ē'ap·səs }

periarteritis [MED] Inflammation of the outer coat of an artery and of the periarterial tissues. { ˌper·ē,ärd·ə'rīd·əs }

periarteritis nodosa *See* polyarteritis nodosa. { ˌper·ē,ärd·ə'rīd·əs nō'dō·sə }

periastron [ASTRON] The coordinates and time when the two stars of a binary star system are nearest to each other in their orbits. { ˌper·ē'as·trən }

periblast [EMBRYO] The nucleated layer of cytoplasm that surrounds the blastodisk of an egg undergoing discoidal cleavage. { 'per·ə,blast }

periblastula [EMBRYO] The blastula of a centrolecithal egg, formed by superficial segmentation. { ˌper·ə'blas·tyə·lə }

periblem [BOT] A layer of primary meristem which produces the cortical cells. { 'per·ə,blem }

periblinite [GEOL] A variety of provitrinite consisting of cortical tissue. { pə'rib·lə,nīt }

pericardial [ANAT] **1.** Of or pertaining to the pericardium. **2.** Located around the heart. { ˌper·ə'kärd·ē·əl }

pericardial cavity [ANAT] A potential space between the inner layer of the pericardium and the epicardium of the heart. { ˌper·ə'kärd·ē·əl 'kav·əd·ē }

pericardial fluid [PHYSIO] The fluid in the pericardial cavity. { ˌper·ə'kärd·ē·əl 'flü·əd }

pericardial organ [INV ZOO] One of the neurohemal organs associated with the pericardial cavity in crustaceans. { ˌper·ə'kärd·ē·əl 'ȯr·gən }

pericarditis [MED] Inflammation of the pericardium. { ˌper·ə,kär'dīd·əs }

pericardium [ANAT] The membranous sac that envelops the heart; it contains 5–20 grams of clear serous fluid. { ˌper·ə'kärd·ē·əm }

pericarp [BOT] The wall of a fruit, developed by ripening and modification of the ovarian wall. { 'per·ə,kärp }

pericenter [PHYS] That point on any orbit nearest to the center of attraction. { 'per·ə,sen·tər }

pericentric inversion [GEN] An aberration resulting from a break in each arm of a chromosome and rejoining with the middle fragment containing the centromere now inverted with respect to the two terminal fragments. { ˌper·ə'sen·trik in'vər·zhən }

pericholangitis [MED] Inflammation of the tissues around the bile ducts or the interlobular bile capillaries. { ˌper·ē,kō,-lan'jīd·əs }

perichondrium [ANAT] The fibrous connective tissue covering cartilage, except at joints. { ˌper·ə'kän·drē·əm }

periclase [MINERAL] MgO Native magnesia; a mineral occurring in granular forms or isometric crystals, with hardness of 6 on Mohs scale, and specific gravity of 3.67–3.90. Also known as periclasite. { 'per·ə,klās }

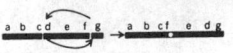

PERICENTRIC INVERSION

Schematic drawing of a pericentric inversion in a chromosome; the white dot is the chromomere.

periclasite *See* periclase. { ˌper·ə'klā,sīt }

periclinal [BOT] Pertaining to a cell layer that is parallel to the surface of a plant part. [GEOL] Referring to strata and structures that dip radially outward from, or inward toward, a center, forming a dome or a basin. { ˌper·ə'klīn·əl }

periclinal chimera [GEN] A plant composed of cells of two distinct species separated into distinctive zones. { ˌper·ə'klīn·əl kī'mir·ə }

pericline [GEOL] A fold characterized by central orientation of the dip of the beds. [MINERAL] A variety of albite elongated, and often twinned, along the *b*-axis. { 'per·ə,klīn }

pericline ripple mark [GEOL] A ripple mark arranged in an orthogonal pattern either parallel to or transverse to the current direction and having a wavelength up to 80 centimeters and amplitude up to 30 centimeters. { 'per·ə,klīn 'ripəl ,märk }

pericline twin law [CRYSTAL] A parallel twin law in triclinic feldspars, in which the *b* axis is the twinning axis and the composition surface is a rhombic section. { 'per·ə,klīn 'twin ,lȯ }

pericolitis [MED] Inflammation of the peritoneum or tissues around the colon. { ˌper·ə·kə'līd·əs }

pericondensed polycyclic [ORG CHEM] Referring to an aromatic compound in which three or more rings share common carbon atoms. { ˌper·ə·kən'denst ,päl·ē,sī·klik }

pericoronitis [MED] Inflammation of the tissue surrounding the coronal portion of the tooth, usually a partially erupted third molar. { ˌper·ə,kär·ə'nīd·əs }

pericranium [ANAT] The periosteum on the outer surface of the cranial bones. { ˌper·ə'krā·nē·əm }

pericronus [ASTRON] The nearest point of a satellite in its orbit about Saturn. Also known as perisaturnium. { 'per·ə'krō·nəs }

pericycle [BOT] The outer boundary of the stele of plants; may not be present as a distinct layer of cells. { 'per·ə,sī·kəl }

pericyclic reaction [ORG CHEM] Any one of a group of reactions that involve conjugated polyenes and proceed by single-step (concerted) reaction mechanisms. { ˌper·ə'sīk·lik rē'ak·shən }

pericynthion [ASTRON] The point in the orbit of a satellite around the moon that is nearest to the moon. { ˌper·ə'sin·thē,än }

pericystium [ANAT] The tissues surrounding a bladder. [MED] The vascular wall of a cyst. { ˌper·ə'sis·tē·əm }

pericyte [HISTOL] A mesenchymal cell found around a capillary; it may or may not be contractile. { 'per·ə,sīt }

periderm [BOT] A group of secondary tissues forming a protective layer which replaces the epidermis of many plant stems, roots, and other parts; composed of cork cambium, phelloderm, and cork. [EMBRYO] The superficial transient layer of epithelial cells of the embryonic epidermis. { 'per·ə,dərm }

peridium [BOT] The outer investment of the sporophore of many fungi. { pə'rid·ē·əm }

peridot [MINERAL] **1.** A gem variety of olivine that is transparent to translucent and pale-, clear-, or yellowish-green in color. **2.** A variety of tourmaline approaching olivine in color. { 'per·ə,dät }

peridotite [PETR] A dark-colored, ultrabasic phaneritic igneous rock composed largely of olivine, with smaller amounts of pyroxene or hornblende. { pə'rid·ə,tīt }

peridotite shell *See* upper mantle. { pə'rid·ə,tīt ,shel }

peridynamic loudspeaker [ENG ACOUS] Box-type loudspeaker baffle designed to give good bass response by minimizing acoustic standing. { ˌper·ə·də'nam·ik 'laùd,spēk·ər }

peri-endothelial cell [HISTOL] A cell that resides next to and supports the endothelial cells. { ˌpe·rē ,en·də,thē·lē·əl 'sel }

perifollicular [HISTOL] Surrounding a follicle. { ˌper·ə·fə'lik·yə·lər }

perigalacticon [ASTRON] The point in the orbit of a star that is closest to the center of the Galaxy. { ˌper·i·gə'lak·tē,kän }

perigean range [OCEANOGR] The average range of tide at the time of perigean tides, when the moon is near perigee; the perigean range is greater than the mean range. { 'per·ə,jē·ən 'rānj }

perigean tidal currents [OCEANOGR] Tidal currents of increased speed occurring at the time of perigean tides. { 'per·ə,jē·ən 'tīd·əl ,kə·rəns }

perigean tide [OCEANOGR] Tide of increased range occurring when the moon is near perigee. { ˌper·ə'jē·ən 'tīd }

perigee [ASTRON] The point in the orbit of the moon or other satellite when it is nearest the earth. { 'per·ə‚jē }

perigee-to-perigee period See anomalistic period. { 'per·ə‚jē tü 'per·ə‚jē ‚pir·ē·əd }

perigenic [GEOL] Referring to a rock constituent or mineral formed at the same time as the rock it is part of, but not formed at the specific location it now occupies in the rock. { ¦per·ə¦jen·ik }

periglacial [GEOL] Of or pertaining to the outer perimeter of a glacier, particularly to the fringe areas immediately surrounding the great continental glaciers of the geologic ice ages, with respect to environment, topography, areas, processes, and conditions influenced by the low temperature of the ice. { ¦per·ə¦glā·shəl }

periglacial climate [CLIMATOL] The climate which is characteristic of the regions immediately bordering the outer perimeter of an ice cap or continental glacier; the principal climatic feature is the high frequency of very cold and dry winds off the ice area; it is also thought that these regions offer ideal conditions for the maintenance of a belt of intense cyclonic activity. { ¦per·ə¦glā·shəl 'klī·mət }

perigon [MATH] An angle that contains 360° or 2π radians. Also known as round angle. { 'per·ə‚gän }

perigonium [BOT] The perianth of a liverwort. [INV ZOO] The sac containing the generative bodies in the gonophore of a hydroid. { ‚per·ə'gō·nē·əm }

perigynium [BOT] A fleshy cup- or tubelike structure surrounding the archegonium of various bryophytes. { ‚per·ə'jin·ē·əm }

perigynous [BOT] Bearing the floral organs on the rim of an expanded saucer- or cup-shaped receptacle or hypanthium. { pə'rij·ə·nəs }

perihelion [ASTRON] That orbital point nearest the sun when the sun is the center of attraction. { ‚per·ə¦hēl·yən }

perihepatitis [MED] Inflammation of the peritoneum and tissues surrounding the liver. { ‚per·ē‚hep·ə'tīd·əs }

peri-infarction block See intraventricular heart block. { ¦per·ē·in'färk·shən ‚bläk }

perijove [ASTRON] The nearest point of a satellite in its orbit about Jupiter. { 'per·ə‚jōv }

perikaryon [CELL MOL] A cytoplasmic mass surrounding a nucleus. [NEUROSCI] The body of a nerve cell, containing the nucleus. { ¦per·ə'kar·ē‚än }

Perilampidae [INV ZOO] A family of hymenopteran insects in the superfamily Chalcidoidea. { ‚per·ə'lam·pə‚dē }

perilla oil [MATER] A light-yellow drying oil derived from seeds of mints of the genus *Perilla*; soluble in alcohol, benzene, carbon disulfide, ether, and chloroform; used as a substitute for linseed oil, as an edible oil, and in manufacture of varnishes and artificial leather. { pə'ril·ə ‚oil }

perilymph [PHYSIO] The fluid separating the membranous from the osseous labyrinth of the internal ear. { 'per·ə‚limpf }

perimagmatic [GEOL] Referring to a hydrothermal mineral deposit located near its magmatic source. { ‚per·ə·mag'ned·ik }

perimeter [MATH] The total length of a closed curve; for example, the perimeter of a polygon is the total length of its sides. { pə'rim·əd·ər }

perimeter blasting [MIN ENG] A method of blasting in tunnels, drifts, and raises, designed to minimize overbreak and leave clean-cut solid walls; the outside holes are loaded with very light continuous explosive charges and fired simultaneously, so that they shear from one hole to the other. { pə'rim·əd·ər ‚blast·iŋ }

perimetrium [ANAT] The serous covering of the uterus. { ‚per·ə'mē·trē·əm }

perimysium [ANAT] The connective tissue sheath enveloping a muscle or a bundle of muscle fibers. { ‚per·ə'mī·sē·əm }

perineum [ANAT] **1.** The portion of the body included in the outlet of the pelvis, bounded in front by the pubic arch, behind by the coccyx and sacrotuberous ligaments, and at the sides by the tuberosities of the ischium. **2.** The region between the anus and the scrotum in the male, between the anus and the posterior commissure of the vulva in the female. { ‚per·ə'nē·əm }

perineural [ANAT] Situated around nervous tissue or a nerve. { ‚per·ə'nùr·əl }

perineural fibroblastoma See neurofibroma. { ‚per·ə'nùr·əl ‚fī·brō·bla'stō·mə }

perineural fibroma See neurofibroma. { ‚per·ə'nùr·əl fī'brō·mə }

periocular [ANAT] Surrounding the eye. { ‚per·ē'äk·yə·lər }

period [ASTRON] The average time interval for a variable star to complete a cycle of its variations. [CHEM] A family of elements with consecutive atomic numbers in the periodic table and with closely related properties; for example, chromium through copper. [GEOL] A unit of geologic time constituting a subdivision of an era; the fundamental unit of the standard geologic time scale. [MATH] **1.** A number T such that $f(x + T) = f(x)$ for all x, where $f(x)$ is a specified function of a real or complex variable. **2.** The period of an element a of a group G is the smallest positive integer n such that a^n is the identity element; if there is no such integer, a is said to be of infinite period. [NUCLEO] The time required for exponentially rising or falling neutron flux in a nuclear reactor to change by a factor of e (2.71828). [PHYS] The duration of a single repetition of a cyclic phenomenon. { 'pir·ē·əd }

periodate [INORG CHEM] A salt of periodic acid, HIO_4, for example, potassium periodate, KIO_4. { 'pər'ī‚ə‚dāt }

period doubling [PHYS] A scenario for the transition of a natural process from regular periodic motion to chaos, in which the time required for the motion of the system to repeat itself doubles again and again as a parameter describing the system is increased. { 'pir·ē·əd ‚dəb·liŋ }

periodic [SCI TECH] Repeating itself identically at regular intervals. { ‚pir·ē¦äd·ik }

periodic acid [INORG CHEM] $HIO_4 \cdot 2H_2O$ Water- and alcohol-soluble white crystals; loses water at 100°C; used as an oxidant. { ‚pir·ē¦äd·ik 'as·əd }

periodic antenna [ELECTROMAG] An antenna in which the input impedance varies as the frequency is altered. { ‚pir·ē¦äd·ik an'ten·ə }

periodic attractor [PHYS] A finite sequence of points in phase space such that certain orbits approach each of them in succession, coming closer on each approach. { ‚pir·ē¦äd·ik ə trak·tər }

periodic continued fraction See recurring continued fraction. { ‚pir·ē¦äd·ik kər‚tin·yüd 'frak·shən }

periodic current [OCEANOGR] Current produced by the tidal influence of moon and sun or by any other oscillatory forcing function. { ‚pir·ē¦äd·ik 'kə·rənt }

periodic damping [PHYS] Damping which is less than critical damping. { ‚pir·ē¦äd·ik 'damp·iŋ }

periodic decimal See repeating decimal. { ‚pir·ē¦äd·ik 'des·məl }

periodic discing [AGR] A type of soil tillage involving a series of disc-shaped plows. { ‚pir·ē¦äd·ik 'disk·iŋ }

periodic disease See familial Mediterranean fever. { ‚pir·ē¦äd·ik di‚zēz }

periodic duty [ELEC] Intermittent duty in which the load conditions are regularly recurrent. { ‚pir·ē¦äd·ik 'düd·ē }

periodic field focusing [ELECTR] Focusing of an electron beam where the electrons follow a trochoidal path and the focusing field interacts with them at selected points. { ‚pir·ē¦äd·ik 'fēld ‚fō·kə·siŋ }

periodic function [MATH] A function $f(x)$ of a real or complex variable is periodic with period T if $f(x + T) = f(x)$ for every value of x. { ‚pir·ē¦äd·ik ‚fəŋk·shən }

periodicity [MATH] The property of periodic functions. [MOL BIO] The number of base pairs in one turn of the deoxyribonucleic acid duplex. { ‚pir·ē·ə'dis·əd·ē }

periodic kiln [ENG] A kiln in which the cycle of setting ware in the kiln, heating up, "soaking" or holding at peak temperature for some time, cooling, and removing or "drawing" the ware is repeated for each batch. { ‚pir·ē¦äd·ik 'kil }

periodic lattice See lattice. { ‚pir·ē¦äd·ik 'lad·əs }

periodic law [CHEM] The law that the properties of the chemical elements and their compounds are a periodic function of their atomic weights. { ‚pir·ē¦äd·ik 'lò }

periodic line [ELEC] Line consisting of successive and identical sections, similarly oriented, the electrical properties of each section not being uniform throughout; the periodicity is in space and not in time; an example of a periodic line is the loaded line with loading coils uniformly spaced. { ‚pir·ē¦äd·ik ‚līn }

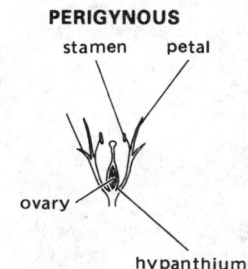

PERIGYNOUS

stamen   petal

ovary

hypanthium

Perigynous arrangement of flower parts on the receptacle.

periodic motion [MECH] Any motion that repeats itself identically at regular intervals. { ¦pir·ē¦äd·ik 'mō·shən }

periodic peritonitis See familial Mediterranean fever. { ¦pir·ē¦äd·ik ‚pər·ə·tə'nīd·əs }

periodic perturbation [ASTRON] Small deviations from the computed orbit of a planet or satellite; the deviations extend through cycles that generally do not exceed a century. [MATH] A perturbation which is periodic as a function. { ¦pir·ē¦äd·ik pər·tər'bā·shən }

periodic quantity [PHYS] Oscillating quantity, the values of which recur for equal increments of the independent variable. { ¦pir·ē¦äd·ik 'kwän·əd·ē }

periodic spring [HYD] A spring that ebbs and flows periodically, apparently due to natural siphon action. { ¦pir·ē¦äd·ik 'spriŋ }

periodic table [CHEM] A table of the elements, written in sequence in the order of atomic number or atomic weight and arranged in horizontal rows (periods) and vertical columns (groups) to illustrate the occurrence of similarities in the properties of the elements as a periodic function of the sequence. { ¦pir·ē¦äd·ik 'tā·bəl }

periodic wave [PHYS] A wave whose displacement has a periodic variation with time or distance, or both. { ¦pir·ē¦äd·ik 'wāv }

period-luminosity relation [ASTRON] Relation between the periods of Cepheid variable stars and their absolute magnitude; the absolutely brighter the star, the longer the period. { 'pir·ē·əd ‚lü·mə'näs·əd·ē ri‚lā·shən }

period of vibration [PHYS] The time for one complete cycle of a vibration. { 'pir·ē·əd əv vī'brā·shən }

periodogram [STAT] A graph used in harmonic analysis of a series that oscillates, such as a time series consisting potentially of several cycles differing in length; the square of the amplitude or intensity for each curve covering a length of time is plotted against the lengths of the various curves. { ‚pir·ē'äd·ə‚gram }

periodontal [ANAT] **1.** Surrounding a tooth. **2.** Of or pertaining to the periodontium. { ¦per·ē·ō¦dänt·əl }

periodontics [MED] The branch of dentistry devoted to diseases of the gingiva (gum tissue), alveolar bone, periodontal ligament, and cementum. { ‚per·ē·ə'dän·tiks }

periodontitis [MED] Inflammation of the periodontium. { ¦per·ē·ō‚dän'tīd·əs }

periodontium [ANAT] The tissues surrounding a tooth. { ¦per·ē·ō'dan·chəm }

periodontoclasia [MED] Any periodontal disease that results in the destruction of the periodontium. { ¦per·ē·ō‚dänt·ə'klā·zhə }

periodontosis [MED] A degenerative disturbance of the periodontium, characterized by degeneration of connective-tissue elements of the periodontal ligament and by bone resorption. { ¦per·ē·ō‚dän'tō·səs }

period parallelogram [MATH] For a doubly periodic function $f(z)$ of a complex variable, a parallelogram with vertices at z_0, $z_0 + a$, $z_0 + a + b$, and $z_0 + b$, where z_0 is any complex number, and a and b are periods of $f(z)$ but are not necessarily primitive periods. { ¦pir·ē·əd ‚par·ə'lel·ə‚gram }

perionychium [ANAT] The border of epidermis surrounding an entire nail. Also known as paronychium. { ‚per·ē·ō'nik·ē·əm }

periosteum [ANAT] The fibrous membrane enveloping bones, except at joints and the points of tendonous and ligamentous attachment. { ‚per·ē'äs·tē·əm }

periostitis [MED] Inflammation of the periosteum. { ‚per·ē‚ä'stīd·əs }

periostracum [INV ZOO] A protective layer of chitin covering the outer portion of the shell in many mollusks, especially fresh-water forms. { ‚per·ē'äs·trə·kəm }

periotic [ANAT] **1.** Situated about the ear. **2.** Of or pertaining to the parts immediately about the internal ear. { ¦per·ē¦äd·ik }

peripediment [GEOL] The segment of a pediplane extending across the younger rocks or alluvium of a basin which is always beyond but adjacent to the segment developed on the older upland rocks. { ¦per·ə'ped·ə·mənt }

peripheral [ANAT] Pertaining to or located at or near the surface of a body or an organ. [COMPUT SCI] See peripheral device. [SCI TECH] Remote from the center; marginal; on the periphery. { pə'rif·ə·rəl }

peripheral buffer [COMPUT SCI] A device acting as a temporary storage when transmission occurs between two devices operating at different transmission speeds. { pə'rif·ə·rəl 'bəf·ər }

peripheral component interconnect [COMPUT SCI] A bus standard for connecting additional input/output devices (such as graphics or modem cards) to a personal computer. Abbreviated PCI. { pə‚rif·ə·rəl kəm‚pō·nənt 'in·tər·kə·nek }

peripheral control unit [COMPUT SCI] A device which connects a unit of peripheral equipment with the central processing unit of a computer and which interprets and responds to instructions from the central processing unit. { pə'rif·ə·rəl kən'trōl ‚yü·nət }

peripheral depression See ring depression. { pə'rif·ə·rəl di'presh·ən }

peripheral device [COMPUT SCI] Any device connected internally or externally to a computer and used to enter or display data, such as the keyboard, mouse, monitor, scanner, and printer. { pə'rif·ə·rel di‚vīs }

peripheral equipment [COMPUT SCI] Equipment that works in conjunction with a computer but is not part of the computer itself. { pə'rif·ə·rəl i'kwip·mənt }

peripheral faults [GEOL] Arcuate faults bounding an elevated or depressed area such as a diapir. { pə'rif·ə·rəl 'fóls }

peripheral hemodynamics [PHYSIO] A division of hematology concerned with blood flow in regions of the body that are close to the surface, such as the extremities. { pə'rif·ə·rəl ‚hē·mō·dī'nam·iks }

peripheral initiation [ORD] Simultaneous initiation of detonation around the entire periphery of a cylindrical explosive charge; it may be accomplished from point initiation by inserting a disk of inert material, of proper dimensions, in the explosive column. { pə'rif·ə·rəl ə‚nish·ē'ā·shən }

peripheral interface channel [COMPUT SCI] A path along which information can flow between a unit of peripheral equipment and the central processing unit of a computer. { pə'rif·ə·rəl 'in·tər‚fās ‚chan·əl }

peripheral-limited [COMPUT SCI] Property of a computer system whose processing time is determined by the speed of its peripheral equipment rather than by the speed of its central processing unit. Also known as I/O-bound. { pə'rif·ə·rəl ¦lim·əd·əd }

peripheral membrane protein [MOL BIO] A protein that is associated with the plasma membrane via electrostatic interactions with integral membrane proteins or membrane lipids. Also known as extrinsic protein. { pə¦rif·ə·rəl 'mem‚brān ‚prō‚tēn }

peripheral milling [MET] Removing metal from a surface parallel to the axis of a milling cutter. { pə'rif·ə·rəl 'mil·iŋ }

peripheral nervous system [NEUROSCI] The autonomic nervous system, the cranial nerves, and the spinal nerves including their associated sensory receptors. { pə'rif·ə·rəl 'nər·vəs ‚sis·təm }

peripheral operation [COMPUT SCI] An operation in which an input or output device is used, and which is not directly controlled by a computer while the operation is being carried out. { pə'rif·ə·rəl ‚äp·ə'rā·shən }

peripheral processing [COMPUT SCI] Processing that is carried out by peripheral equipment or by an auxiliary computer. { pə'rif·ə·rəl 'prä‚ses·iŋ }

peripheral processor [COMPUT SCI] Auxiliary computer performing specific operations under control of the master computer. { pə'rif·ə·rəl 'prä‚ses·ər }

peripheral sink See rim syncline. { pə'rif·ə·rəl 'siŋk }

peripheral speed See cutting speed. { pə'rif·ə·rəl ¦spēd }

peripheral stream [HYD] A stream that flows parallel to the edge of a glacier, usually just beyond the moraine. { pə'rif·ə·rəl 'strēm }

peripheral transfer [COMPUT SCI] The transmission of data between two units of peripheral equipment or between a peripheral unit and the central processing unit of a computer. { pə'rif·ə·rəl 'tranz·fər }

peripheral units See peripheral equipment. { pə'rif·ə·rəl ‚yü·nəts }

peripheral vision [PHYSIO] The act of seeing images that fall upon parts of the retina outside the macula lutea. Also known as indirect vision. { pə'rif·ə·rəl 'vizh·ən }

periphery [MATH] The bounding curve of a surface or the surface of a solid. { pə'rif·ə·rē }

periphlebitis [MED] Inflammation of the tissues around a vein or of the adventitia of a vein. { ¦per·ə·flə'bīd·əs }

periphyton [ECOL] Sessile biotal components of a fresh-water ecosystem. { pə'rif·ə,tän }

periplasm [CYTOL] The region between the cytoplasmic and outer membranes of a cell. { 'per·ə,plaz·əm }

periplast [CYTOL] **1.** A cell membrane. **2.** A pellicle covering ectoplasm. [HISTOL] The stroma of an animal organ. [INV ZOO] The ectoplasm of a flagellate. { 'per·ə,plast }

periproct [INV ZOO] The area surrounding the anus of echinoids. { 'per·ə,präkt }

Periptychidae [PALEON] A family of extinct herbivorous mammals in the order Condylarthra distinguished by specialized, fluted teeth. { ,per·əp'tik·ə,dē }

Peripylina [INV ZOO] An equivalent name for Porulosida. { ¦per·ə'plī·nə }

perisarc [INV ZOO] The outer integument of a hydroid. { 'per·ə,särk }

perisaturnium See pericronus. { 'per·ə,sə'tər·nē·əm }

Periscelidae [INV ZOO] A family of myodarian cyclorrhaphous dipteran insects in the subsection Acalypteratae. { ,per·ə'sel·ə,dē }

Perischoechinoidea [INV ZOO] A subclass of principally extinct echinoderms belonging to the Echinoidea and lacking stability in the number of columns of plates that make up the ambulacra and interambulacra. { pə¦ris·kō,ek·ə'nóid·ē·ə }

periscope [OPTICS] **1.** An optical instrument used to provide a raised line of vision where it may not be practical or possible, as in entrenchments, tanks, or submarines; the raised line of vision is obtained by the use of mirrors or prisms within the structure of the item; it may have single or dual optical systems. **2.** A thin astigmatic lens which approximates a meniscus shape and has a base curve of ±1.25 diopters. { 'per·ə,skōp }

periscopic sextant [NAV] **1.** A sextant designed to be mounted inside an aircraft, with a tube extending vertically upward through the skin of the aircraft. **2.** A sextant designed to be used in conjunction with the periscope of a submarine. { ,per·ə'skäp·ik 'sek·stənt }

periscopic sight [ORD] Gunsight made in the form of a periscope, permitting a gunner to see over an obstacle. { ,per·ə'skäp·ik 'sīt }

perisperm [BOT] In a seed, the nutritive tissue that is derived from the nucellus and deposited on the outside of the embryo sac. { 'per·ə,spərm }

Perissodactyla [VERT ZOO] An order of exclusively herbivorous mammals distinguished by an odd number of toes and mesaxonic feet, that is, with the axis going through the third toe. { pə,ris·ō'dak·tə·lə }

peristalsis [PHYSIO] The rhythmic progressive wave of muscular contraction in tubes, such as the intestine, provided with both longitudinal and transverse muscular fibers. { ,per·ə'stäl·səs }

peristaltic charge-coupled device [ELECTR] A high-speed charge-transfer integrated circuit in which the movement of the charges is similar to the peristaltic contractions and dilations of the digestive system. { ¦per·ə¦stäl·tik 'chärj ¦kəp·əld di'vīs }

peristaltic pump [MECH ENG] A device for moving fluids by the action of multiple, equally spaced rollers, which rotate and compress a flexible tube. { ,per·ə,stal·tik 'pəmp }

peristerite [MINERAL] A gem variety of albite (An₂-An₂₄) that resembles moonstone and has a blue or bluish-white luster characterized by sharp internal reflections of blue, green, and yellow. { pə'ris·tə,rīt }

peristome [BOT] The fringe around the opening of a moss capsule. [INV ZOO] The area surrounding the mouth of various invertebrates. { 'per·ə,stōm }

peristyle [ARCH] A surrounding series of columns; the space enclosed by such columns. { 'per·ə,stīl }

peritectic [PHYS CHEM] An isothermal reversible reaction in which a liquid phase reacts with a solid phase during cooling to produce a second solid phase. { ,per·ə¦tek·tik }

peritectic point [PHYS CHEM] In a binary two-phase heteroazeotropic system at constant pressure, that point up to which the boiling point has remained constant until one of the phases has boiled away. { ,per·ə¦tek·tik ,póint }

peritectoid [PHYS CHEM] An isothermal reversible reaction in which a solid phase on cooling reacts with another solid phase to form a third solid phase. { ,per·ə'tek,tóid }

perithecial ascomycetes See Pyrenomycetes. { ,per·i,thē·shəl ,as·kə·mī'sēd,ēz }

perithecium [MYCOL] A spherical, cylindrical, or oval ascocarp which usually opens by a terminal slit or pore. { ,per·ə'thē·shəm }

peritoneal cavity [ANAT] The potential space between the visceral and parietal layers of the peritoneum. { ,per·ə·tə¦nē·əl 'kav·əd·ē }

peritoneoscope [MED] A long, slender endoscope equipped with sheath, obturator, biopsy forceps, a sphygmomanometer bulb and tubing, scissors, and a syringe; introduced through a small incision in the abdominal wall to permit visualization of the gas-inflated peritoneal cavity. { ,per·ə·tə¦nē·ə,skōp }

peritoneoscopy See laparoscopy. { ,per·ə·tə,nē'äs·kə·pē }

peritoneum [ANAT] The serous membrane enveloping the abdominal viscera and lining the abdominal cavity. { ,per·ə·tə'nē·əm }

peritonitis [MED] Inflammation of the peritoneum. { ,per·ə·tə'nīd·əs }

peritonsillar abscess [MED] An abscess forming in acute tonsillitis around one or both tonsils. { ,per·ə'täns·əl·ər 'ab,ses }

Peritrichia [INV ZOO] A specialized subclass of the class Ciliatea comprising both sessile and mobile forms. { ,per·ə'trik·ē·ə }

Peritrichida [INV ZOO] The single order of the protozoan subclass Peritrichia. { ,per·ə'trik·ə·də }

peritrichous [INV ZOO] Of certain protozoans, having spirally arranged cilia around the oral disk. [MICROBIO] Of bacteria, having a uniform distribution of flagella on the body surface. { pə'ri·trə·kəs }

perityphlitis [MED] Inflammation of the peritoneum surrounding the cecum and vermiform appendix. { ,per·ə'tif·ləd·əs }

perivitelline space [CYTOL] In mammalian ova, the space formed between the ovum and the zona pellucida at the time of maturation, into which the polar bodies are given off. { ¦per·ə'vid·əl,ēn ,spās }

periwinkle See Vinca rosea. { 'per·i,wiŋ·kəl }

Perkin reaction [ORG CHEM] The formation of unsaturated cinnamic-type acids by the condensation of aromatic aldehydes with fatty acids in the presence of acetic anhydride. { 'pər·kən rē,ak·shən }

perlèche [MED] An inflammatory condition occurring at the angles of the mouth with resultant fissuring. { pər'lesh }

Perl See Practical Extraction and Reporting Language. { pərl }

perlite [GEOL] A rhyolitic glass with abundant spherical or convolute cracks that cause it to break into small pearllike masses or pebbles, usually less than a centimeter across; it is commonly gray or green with a pearly luster and has the composition of rhyolite. Also known as pearlite; pearlstone. { 'pər,līt }

perlitic [PETR] **1.** Of the texture of a glassy igneous rock, exhibiting small spheruloids formed from cracks due to contraction during cooling. **2.** Pertaining to or characteristic of perlite. { ¦pər¦lid·ik }

perlucidus [METEOROL] A cloud variety, usually of the species stratiformis, in which distinct spaces between its elements permit the sun, moon, blue sky, or higher clouds to be seen. { pər'lü·səd·əs }

perm [PETRO ENG] A unit indicating the degree of permeability of a porous reservoir structure; the unit is expressed as bbl day⁻¹ ft⁻² psi⁻¹ ft cp or ft³ day⁻¹ ft⁻² psi⁻¹ ft cp. { pərm }

permafil [MATER] Polymerizable mixture that cures without any evaporation. { 'pər·mə,fil }

permafrost [GEOL] Perennially frozen ground, occurring wherever the temperature remains below 0°C for several years, whether the ground is actually consolidated by ice or not and regardless of the nature of the rock and soil particles of which the earth is composed. { 'pər·mə,fróst }

permafrost drilling [ENG] Boreholes drilled in subsoil and rocks in which the contained water is permanently frozen. { 'pər·mə,fróst 'dril·iŋ }

permafrost island [GEOL] A small, shallow, isolated patch of permafrost surrounded by unfrozen ground. { 'pər·mə,fróst 'ī·lənd }

permafrost line [GEOL] A line on a map representing the border of the arctic permafrost. { 'pər·mə,fróst ,līn }

PERITRICHIA

20 µm

Trichodina, a mobile peritrich.

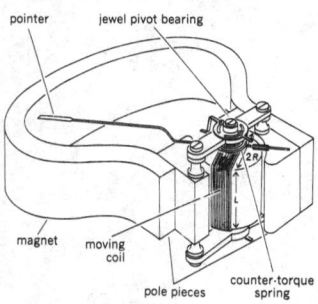

PERMANENT-MAGNET MOVING-COIL INSTRUMENT

pointer
jewel pivot bearing
magnet
moving coil
pole pieces
counter-torque spring

Mechanism of a permanent-magnet moving-coil instrument. L = length of active conductors, R = radius of action of these conductors. (*Weston Instruments, Division of Sangamo Weston, Inc.*)

permafrost table [GEOL] The upper limit of permafrost. Also known as pergelisol table. { 'pər·mə,frȯst ,tā·bəl }

permalloy [MET] A trade name for any of several highly magnetically permeable iron-base alloys containing about 45–80% nickel. { 'pər·mə,lȯi }

permanent [MATH] For a matrix with m rows and n columns, with n equal to or greater than m, the sum, over all permutations of m columns, of a product of m terms, where the ith term in the product is the term in the ith row and the permutation of the ith column. { 'pər·mə·nənt }

permanent anode [MET] A very-corrosion-resistant anode, made of a material such as a carbon, aluminum, or lead alloy, or 14.5% silicon iron; used in cathodic protection against corrosion. { 'pər·mə·nənt 'an,ōd }

permanent aurora See airglow. { 'pər·mə·nənt ȯ'rȯr·ə }

permanent axis [MECH] The axis of the greatest moment of inertia of a rigid body, about which it can rotate in equilibrium. { 'pər·mə·nənt 'ak·səs }

permanent benchmark [ENG] A readily identifiable, relatively permanent, recoverable benchmark that is intended to maintain its elevation without change over a long period of time with reference to an adopted datum, and is located where disturbing influences are believed to be negligible. { 'pər·mə·nənt 'bench,märk }

permanent-completion packer [PETRO ENG] A packer able to withstand large pressure differentials to allow for its permanent installation in a producing well. { 'pər·mə·nənt kəm'plē·shən ,pak·ər }

permanent current [OCEANOGR] A current which continues with relatively little periodic or seasonal change. { 'pər·mə·nənt 'kə·rənt }

permanent echo [ELECTROMAG] A signal reflected from an object that is fixed with respect to a radar site. { 'pər·mə·nənt 'ek·ō }

permanent emplacement See fixed emplacement. { 'pər·mə·nənt im'plās·mənt }

permanent error [COMPUT SCI] An error that occurs when a sector mark on disk pack or floppy disk is incorrectly modified by writing data over it, and that can be corrected only by clearing the entire disk and rewriting the track and sector marks. { 'pər·mə·nənt 'er·ər }

permanent extinction [GEOL] The extinction of a lake by destruction of the lake basin, because of such processes as deposition of sediments, erosion of the basin rim, filling with vegetation, or catastrophic events. { 'pər·mə·nənt ik'stiŋk·shən }

permanent fault [COMPUT SCI] A hardware malfunction that always occurs when a particular set of conditions exists, and that can be made to occur deliberately, in contrast to a sporadic fault. { 'pər·mə·nənt 'fȯlt }

permanent finish [TEXT] Any one of a number of fabric treatments used to improve glaze, hand, or performance of fabrics; generally effective for the life of the fabric in normal use. { 'pər·mə·nənt 'fin·ish }

permanent gas [THERMO] A gas at a pressure and temperature far from its liquid state. { 'pər·mə·nənt 'gas }

permanent hardness [CHEM] The hardness of water persisting after boiling. { 'pər·mə·nənt 'härd·nəs }

permanent ice foot [HYD] An ice foot that does not melt completely in summer. { 'pər·mə·nənt 'īs ,fu̇t }

permanent ink [MATER] Ink that contains up to 1% dissolved iron to prevent fading or washing away when dried. { 'pər·mə·nənt 'iŋk }

permanently convergent series [MATH] A series that is convergent for all values of the variable or variables involved in its terms. { ¦pər·mə·nənt·lē kən¦vər·jənt 'sir,ēz }

permanent magnet [ELECTROMAG] A piece of hardened steel or other magnetic material that has been strongly magnetized and retains its magnetism indefinitely. Abbreviated PM. { 'pər·mə·nənt 'mag·nət }

permanent-magnet dynamic loudspeaker See permanent-magnet loudspeaker. { 'pər·mə·nənt ¦mag·nət dī¦nam·ik 'lau̇d,spēk·ər }

permanent-magnet focusing [ELECTR] Focusing of the electron beam in a television picture tube by means of the magnetic field produced by one or more permanent magnets mounted around the neck of the tube. { 'pər·mə·nənt ¦mag·nət 'fō·kəs·iŋ }

permanent-magnet loudspeaker [ENG ACOUS] A moving-conductor loudspeaker in which the steady magnetic field is produced by a permanent magnet. Also known as permanent-magnet dynamic loudspeaker. { 'pər·mə·nənt ¦mag·nət 'lau̇d,spēk·ər }

permanent-magnet moving-coil instrument [ENG] An ammeter or other electrical instrument in which a small coil of wire, supported on jeweled bearings between the poles of a permanent magnet, rotates when current is carried to it through spiral springs which also exert a restoring torque on the coil; the position of the coil is indicated by an attached pointer. { 'pər·mə·nənt ¦mag·nət ¦mu̇v·iŋ ¦kȯil 'in·strə·mənt }

permanent-magnet moving-iron instrument [ENG] A meter that depends for its operation on a movable iron vane that aligns itself in the resultant magnetic field of a permanent magnet and adjacent current-carrying coil. { 'pər·mə·nənt ¦mag·nət ¦mu̇v·iŋ 'ī·ərn 'in·strə·mənt }

permanent-magnet stepper motor [ELEC] A stepper motor in which the rotor is a powerful permanent magnet and each stator coil is energized independently in sequence; the rotor aligns itself with the stator coil that is energized. { 'pər·mə·nənt ¦mag·nət 'step·ər ,mōd·ər }

permanent mold [MET] A reusable metal mold for the production of many castings of the same kind. { 'pər·mə·nənt 'mōld }

permanent monument [MIN ENG] A monument of a lasting character for marking a mining claim; it may be a mountain, hill, or ridge. { 'pər·mə·nənt 'män·yə·mənt }

permanent press See durable press. { 'pər·mə·nənt 'pres }

permanent-press resin [ORG CHEM] A thermosetting resin, based on chemicals such as formaldehyde and maleic anhydride, which is used to impart crease resistance to textiles and fibers. Also known as durable-press resin. { 'pər·mə·nənt ¦pres 'rez·ən }

permanent pump [MIN ENG] A pump on which the mine depends for the final disposal of its drainage. { 'pər·mə·nənt 'pəmp }

permanent set [MECH] Permanent plastic deformation of a structure or a test piece after removal of the applied load. Also known as set. { 'pər·mə·nənt 'set }

permanent-split capacitor motor [ELEC] A capacitor motor in which the starting capacitor and the auxiliary winding remain in the circuit for both starting and running. Abbreviated PSC motor. Also known as capacitor start-run motor. { 'pər·mə·nənt ¦split kə'pas·əd·ər ,mōd·ər }

permanent starch [MATER] An emulsion of polyvinyl acetate used for starching clothing and textiles; it is not removed by washing. { 'pər·mə·nənt 'stärch }

permanent stop [IND ENG] In a flexible manufacturing system a type of controlled stop where an automated guided vehicle will always halt, regardless of programming. { 'pər·mə·nənt 'stäp }

permanent storage [COMPUT SCI] A means of storing data for rapid retrieval by a computer; does not permit changing the stored data. { 'pər·mə·nənt 'stȯr·ij }

permanent teeth [ANAT] The second set of teeth of a mammal, following the milk teeth; in humans, the set of 32 teeth consists of 8 incisors, 4 canines, 8 premolars, and 12 molars. { 'pər·mə·nənt ¦tēth }

permanent thermocline See main thermocline. { 'pər·mə·nənt 'thər·mə,klīn }

permanent water [HYD] A source of water that remains constant throughout the year. { 'pər·mə·nənt 'wȯd·ər }

permanent wave [FL MECH] A wave (in a fluid) which moves with no change in streamline pattern, and which, therefore, is a stationary wave relative to a coordinate system moving with the wave. { 'pər·mə·nənt 'wāv }

permanganate [INORG CHEM] A purple salt of permanganic acid containing the MnO_4^- radical; used as an oxidizing agent and a disinfectant. { ¦pər'maŋ·gə,nāt }

permanganic acid [INORG CHEM] $HMnO_4$ An unstable acid that exists only in dilute solutions; decomposes to manganese dioxide and oxygen. { ¦pər·man'gan·ik 'as·əd }

Permasyn motor [ELEC] A synchronous motor which has permanent magnets embedded in the squirrel-cage rotor to provide an equivalent direct-current field. { 'pər·mə·sən 'mōd·ər }

permatron [ELECTR] Thermionic gas-discharge diode in

which the start of conduction is controlled by an external magnetic field. { 'pər·mə,trän }

permeability [ELECTROMAG] A factor, characteristic of a material, that is proportional to the magnetic induction produced in a material divided by the magnetic field strength; it is a tensor when these quantities are not parallel. Also known as magnetic permeability. [FL MECH] **1.** The ability of a membrane or other material to permit a substance to pass through it. **2.** Quantitatively, the amount of substance which passes through the material under given conditions. [GEOL] The capacity of a porous rock, soil, or sediment for transmitting a fluid without damage to the structure of the medium. Also known as conductivity; perviousness. [NAV ARCH] The percentage of a given space in a ship that can be occupied by water. { ,pər·mē·ə'bil·əd·ē }

permeability alloy [MET] An iron-nickel alloy having greater magnetic susceptibility than iron. { ,pər·mē·ə'bil·əd·ē 'al,ȯi }

permeability-block method [PETRO ENG] Calculation method for oil recovery from water-drive oil fields in which there are variable-permeability distributions. { ,pər·mē·ə'bil·əd·ē 'bläk ,meth·əd }

permeability coefficient [FL MECH] The rate of water flow in gallons per day through a cross section of 1 square foot under a unit hydraulic gradient, at the prevailing temperature or at 60°F (16°C). Also known as coefficient of permeability; hydraulic conductivity; Meinzer unit. { ,pər·mē·ə'bil·əd·ē ,kō·i,fish·ənt }

permeability number [ENG] A numbered value assigned to molding materials indicating the relative ease of passage of gases through them. { ,pər·mē·ə'bil·əd·ē ,nəm·bər }

permeability profile [PETRO ENG] A graphical plot of porous reservoir permeability versus distance down the wellbore. { ,pər·mē·ə'bil·əd·ē ,prō,fīl }

permeability trap [GEOL] An oil trap formed by lateral variation within a reservoir bed which seals the contained hydrocarbons through a change of permeability. { ,pər·mē·ə'bil·əd·ē ,trap }

permeability tuning [ELEC] Process of tuning a resonant circuit by varying the permeability of an inductor; it is usually accomplished by varying the amount of magnetic core material of the inductor by slug movement. { ,pər·mē·ə'bil·əd·ē ,tün·iŋ }

permeable bed [GEOL] A porous reservoir formation through which hydrocarbon fluids (oil or gas) or water (waterflood or interstitial) can flow. { 'pər·mē·ə·bəl 'bed }

permeable membrane [CHEM] A thin sheet or membrane of material through which selected liquid or gas molecules or ions will pass, either through capillary pores in the membrane or by ion exchange; used in dialysis, electrodialysis, and reverse osmosis. { 'pər·mē·ə·bəl 'mem,brān }

permeameter [ENG] **1.** A laboratory device for measurement of permeability of materials, for example, soil or rocks: consists of a powder bed of known dimension and degree of packing through which the particles are forced; pressure drop and rate of flow are related to particle size, and pressure drop is related to surface area. **2.** A device for measuring the coefficient of permeability by measuring the flow of fluid through a sample across which there is a pressure drop produced by gravity. **3.** An instrument for measuring the magnetic flux or flux density produced in a test specimen of ferromagnetic material by a given magnetic intensity, to permit computation of the magnetic permeability of the material. { ,pər·mē'am·əd·ər }

permeametry [ANALY CHEM] Determination of the average size of fine particles in a fluid (gas or liquid) by passing the mixture through a powder bed of known dimensions and recording the pressure drop and flow rate through the bed. { ,pər·mē'am·ə·trē }

permeance [ELECTROMAG] A characteristic of a portion of a magnetic circuit, equal to magnetic flux divided by magnetomotive force; the reciprocal of reluctance. Symbolized P. { 'pər·mē·əns }

permeant [CHEM] A material that permeates another material. { 'pər·mē·ənt }

permease [BIOCHEM] Any of a group of enzymes which mediate the phenomenon of active transport. { 'pər·mē,ās }

permeate [CHEM ENG] The clear fluid that passes through the membrane in a membrane filtration process. { 'pər·mē,āt }

permeation [CHEM] The movement of atoms, molecules, or ions into or through a porous or permeable substance (such as zeolite or a membrane). { ,pər·mē'ā·shən }

permeation gneiss [PETR] A gneiss formed as a result of or modified by the passage of geochemically mobile materials through or into solid rock. { ,pər·mē'ā·shən ,nīs }

permeator [CHEM ENG] A membrane assembly that performs an ion-exchange function, for example, desalting in a membrane water-desalting process. { 'pər·mē,ād·ər }

Permendur [MET] A magnetic alloy which is composed of equal parts of iron and cobalt and has an extremely high permeability when saturated. { 'pər·mən,dùr }

permenorm alloy [MET] An alloy containing 50% nickel and 50% iron; used as magnet core material and in magnetic amplifiers. { 'pər·mə,nȯrm 'al,ȯi }

Permian [GEOL] The last period of geologic time in the Paleozoic era, from 280 to 225 million years ago. { 'pər·mē·ən }

per mille [SCI TECH] Per thousand or 10^{-3}. Symbolized +. { pər 'mil }

permineralization [GEOL] A fossilization process whereby additional minerals are deposited in the pore spaces of originally hard animal parts. { pər,min·rə·lə'zā·shən }

permissible [MIN ENG] Said of equipment completely assembled and conforming in every respect with the design formally approved by the U.S. Bureau of Mines for use in gassy and dusty mines. { pər'mis·ə·bəl }

permissible dose [NUCLEO] The amount of radiation that may be safely received by an individual within a specified period. Formerly known as tolerance dose. { pər'mis·ə·bəl 'dōs }

permissible explosive [MATER] An explosive approved by the U.S. Bureau of Mines as safe for blasting in gassy and dusty mines. { pər'mis·ə·bəl ik'splō·siv }

permissible exposure limit [IND ENG] The level of air contaminants that represents an acceptable exposure level as specified in standards set by a U.S. government agency; generally expressed as 8-hour time-weighted average concentrations. Abbreviated PEL. { pər'mis·ə·bəl ik'spō·zhər ,lim·ət }

permissible lamp [MIN ENG] A lamp that meets the standards of the U.S. Bureau of Mines. { pər'mis·ə·bəl 'lamp }

permissible length [NAV ARCH] The floodable length of a ship times the factor of subdivision. { pər'mis·ə·bəl 'leŋkth }

permissible machine [MIN ENG] A machine, such as a drill, mining machine, loading machine, conveyor, or locomotive, that meets the standards of the U.S. Bureau of Mines. { pər'mis·ə·bəl mə'shēn }

permissible value [MATH] A value of a variable for which a given function is defined. { pər'mis·ə·bəl 'val·yü }

permissible velocity [CIV ENG] The highest velocity at which water is permitted to pass through a structure or conduit without excessive damage. { pər'mis·ə·bəl və'läs·əd·ē }

permissive block system [CIV ENG] A block system in which a railroad train is permitted to enter a block section already occupied by a train. { pər'mis·iv 'bläk ,sis·təm }

permissive cell [VIROL] A cell that supports replication of a virus. { pər'mis·iv 'sel }

permissive condition [GEN] An environment under which a conditional lethal mutant can survive and show the wild phenotype. { pər'mis·iv kən,dish·ən }

permissive stop [CIV ENG] A railway signal indicating the train must stop but can proceed slowly and cautiously after a specified interval, usually 1 minute. { pər'mis·iv 'stäp }

permittivity [ELEC] The dielectric constant multiplied by the permittivity of empty space, where the permittivity of empty space (ϵ_0) is a constant appearing in Coulomb's law, having the value of 1 in centimeter-gram-second electrostatic units, and of 8.854×10^{-12} farad/meter in rationalized meter-kilogram-second units. Symbolized ϵ. { ,pər·mə'tiv·əd·ē }

Permo-Carboniferous [GEOL] **1.** The Permian and Carboniferous periods considered as one unit. **2.** The Permian and Pennsylvanian periods considered as a single unit. **3.** The rock unit, or the period of geologic time, transitional between the Upper Pennsylvanian and the Lower Permian periods. { ,pər·mō,kär·bə'nif·ə·rəs }

Permo-Triassic mass extinction [EVOL] A mass extinction event marking the division between the Permian Period and

PERMIAN

CENOZOIC	QUATERNARY	
	TERTIARY	
MESOZOIC	CRETACEOUS	
	JURASSIC	
	TRIASSIC	
PALEOZOIC	PERMIAN	
	CARBONIFEROUS	PENNSYLVANIAN
		MISSISSIPPIAN
	DEVONIAN	
	SILURIAN	
	ORDOVICIAN	
	CAMBRIAN	
PRECAMBRIAN		

Chart showing position of Permian period in relation to other periods and to the eras of geologic time.

the Triassic Period as well as the border between the Paleozoic and Mesozoic eras. It is estimated to have triggered the extinction of 85% or more of all ocean species, approximately 70% of land vertebrates, and significant extinctions of plants and insects. { ¦pər·mō trī¦as·ik ¦mas ik'stiŋk·shən }

perm-plug method [PETRO ENG] Laboratory method of measuring the permeability of reservoir core samples (or plugs) by the measurement of airflow through the sample at several flow rates. { 'pərm ‚pləg ‚meth·əd }

permselective membrane [PHYS CHEM] An ion-exchange material that allows ions of one electrical sign to enter and pass through. { ¦pərm·si¦lek·tiv 'mem‚brān }

permutation [MATH] A function which rearranges a finite number of symbols; more precisely, a one-to-one function of a finite set onto itself. { ‚pər·myə'tā·shən }

permutation character [MATH] The set of all fixed points of a specified permutation. { ‚pər‚myü'tā·shən ‚kar·ik·tər }

permutation group [MATH] The group whose elements are permutations of some set of symbols where the product of two permutations is the permutation arising from successive application of the two. Also known as substitution group. { ‚pər·myə'tā·shən ‚grüp }

permutation matrix [MATH] A square matrix whose elements in any row, or any column, are all zero, except for one element that is equal to unity. { ‚pər·myə'tā·shən ‚mā‚triks }

permutation modulation [COMMUN] Proposed method of transmitting digital information by means of band-limited signals in the presence of additive white gaussian noise; pulse-code modulation and pulse-position modulation are considered simple special cases of permutation modulation. { ‚pər·myə'tā·shən ‚mäj·ə‚lā·shən }

permutation table [COMMUN] In computers, a table designed for the systematic construction of code groups; it may also be used to correct garbles in groups of code text. { ‚pər·myə'tā·shən ‚tā·bəl }

permutation tensor *See* determinant tensor. { ‚pər·myə'tā·shən ‚ten·sər }

pernetti [ENG] **1.** Small iron pins or tripods that support ware while it is being fired in a kiln. **2.** The marks left on baked pottery by these supporting pins. { pər'ned·ē }

pernicious anemia [MED] A megaloblastic macrocytic anemia resulting from lack of vitamin B_{12}, secondary to gastric atrophy and loss of intrinsic factor necessary for vitamin B_{12} absorption, and accompanied by degeneration of the posterior and lateral columns of the spinal cord. { pər'nish·əs ə'nēm·ē·ə }

pernicious malaria *See* falciparum malaria. { pər'nish·əs mə'ler·ē·ə }

perniosis [MED] Any dermatitis resulting from chilblain. { ‚pər·nē'ō·səs }

Perognathinae [VERT ZOO] A subfamily of the rodent family Heteromyidae, including the pocket and kangaroo mice. { ‚per·əg'nath·ə‚nē }

peronine [PHARM] $C_{17}H_{17}NO(OH)OC_7H_7 \cdot HCl$ A white, crystalline powder having a bitter taste; soluble in boiling water; used in medicine. { 'per·ə‚nēn }

Peronosporales [MYCOL] An order of aquatic and terrestrial phycomycetous fungi with a hyphal thallus and zoospores with two flagella. { ‚per·ə‚näs·pə'rā·lēz }

Perothopidae [INV ZOO] A small family of coleopteran insects in the superfamily Elateroidea found only in the United States. { ‚per·ə'thäp·ə‚dē }

perovskite [MINERAL] $CaTiO_3$ A natural, yellow, brownish-yellow, reddish, brown, or black mineral and a structure type which includes no less than 150 synthetic compounds; the crystal structure is ideally cubic, occurs as rounded cubes modified by the octahedral and dodecahedral forms, luster is subadamantine to submetallic, hardness is 5.5 on Mohs scale, and specific gravity is 4.0. { pə'rävz‚kīt }

peroxidase [BIOCHEM] An enzyme that catalyzes reactions in which hydrogen peroxide is an electron acceptor. { pə'räk·sə‚dās }

peroxide [CHEM] **1.** A compound containing the peroxy $(-O-O-)$ group, as in hydrogen peroxide. **2.** *See* hydrogen peroxide. { pə'räk‚sīd }

peroxide number [ANALY CHEM] Measure of millimoles of peroxide (or milliequivalents of oxygen) taken up by 1000 grams of fat or oil; used to measure rancidity. Also known as peroxide value. { pə'räk‚sīd ‚nəm·bər }

peroxide value *See* peroxide number. { pə'räk‚sīd ‚val·yü }

peroxisome [CELL MOL] Any of a subclass of microbodies that contain at least four enzymes involved in the metabolism of hydrogen peroxide. { pə'räk·sə‚sōm }

peroxyacetic acid *See* peracetic acid. { pə¦räk·sē·ə¦sēd·ik 'as·əd }

peroxydol *See* sodium perborate. { pə'räk·sə‚dȯl }

peroxynitrite [INORG CHEM] A nitrogen oxyanion containing an O-O peroxo bond that is a structural isomer of the nitrate ion. Species are generally distinguished by writing the chemical formula for peroxynitrite as $ONOO^-$ and nitrate as NO^{3-}. Other names that have been given to peroxynitrite include pernitrite and peroxonitrite; its recommended IUPAC name is oxoperoxonitrate(1-). { pə‚räk·sē'nī‚trīt }

perpend [CIV ENG] A bondstone that extends completely through a masonry wall and is exposed on each side of the wall. { 'pər‚pend }

perpendicular [MATH] Geometric objects are perpendicular if they intersect in an angle of 90°. { ¦pər·pən¦dik·yə·lər }

perpendicular axis theorem [MECH] A theorem which states that the sum of the moments of inertia of a plane lamina about any two perpendicular axes in the plane of the lamina is equal to the moment of inertia about an axis through their intersection perpendicular to the lamina. { ¦pər·pən¦dik·yə·lər 'ak·səs ‚thir·əm }

perpendicular bisector [MATH] For a line segment in a plane or in space, the line or plane that is perpendicular to this segment and passes through its midpoint. { ‚pər·pən¦dik·yə·lər 'bī‚sek·tər }

perpendicular recording *See* vertical recording. { ‚pər·pən¦dik·yə·lər ri'kȯrd·iŋ }

perpendiculars [NAV ARCH] Vertical lines through the intersections of the fore and aft ends of a ship with the design waterline. { ¦pər·pən¦dik·yə·lərz }

perpendicular slip [GEOL] The component of a fault slip measured at right angles to the trace of the fault on any intersecting surface. { ¦pər·pən¦dik·yə·lər 'slip }

perpendicular slope [GEOL] A very steep slope or precipitous face, as on a mountain. { ¦pər·pən¦dik·yə·lər 'slōp }

perpendicular throw [GEOL] The distance between two points which were formerly adjacent in a faulted bed, vein, or other surface, measured at right angles to the surface. { ¦pər·pən¦dik·yə·lər 'thrō }

perpetual calendar [ASTRON] A table or mechanical device used to determine the day of the week corresponding to any given date over a period of many years. { pər'pech·ə·wəl 'kal·ən·dər }

perpetual frost climate [CLIMATOL] The climate of the ice cap regions of the world; thus, it requires temperatures sufficiently cold so that the annual accumulation of snow and ice is never exceeded by ablation. Also known as ice-cap climate. { pər'pech·ə·wəl 'frȯst ‚klī·mət }

perpetual motion machine of the first kind [PHYS] A mechanism which, once set in motion, continues to do useful work without an input of energy, or which produces more energy than is absorbed in its operation; it violates the principle of conservation of energy. { pər'pech·ə·wəl 'mō·shən mə‚shēn əv thə 'fərst ‚kīnd }

perpetual motion machine of the second kind [PHYS] A device that extracts heat from a source and then converts this heat completely into other forms of energy; it violates the second law of thermodynamics. { pər'pech·ə·wəl 'mō·shən mə‚shēn əv thə 'sek·ənd ‚kīnd }

perpetual motion machine of the third kind [PHYS] A device which has a component that can continue moving forever; an example is a superconductor. { pər'pech·ə·wəl 'mō·shən mə‚shēn əv thə 'thərd ‚kīnd }

perpetuation [VIROL] Maintenance of a viral genome within bacterial host cells without killing them due to weakening of viral virulence. { pər‚pech·ə'wā·shən }

Perret phase [GEOL] That stage of a volcanic eruption that is characterized by the emission of much high-energy gas that may significantly enlarge the volcanic conduit. { 'per·ət ‚fāz }

Perron-Frobenius theorem [MATH] If M is a matrix with positive entries, then its largest eigenvalue λ is positive and simple; moreover, there exist vectors v and w with positive components such that $vM = \lambda v$ and $Mw = \lambda w$, if the inner product of v with w is 1, then the limit of λ^{-n} times the i,jth entry of M^n as n goes to infinity is the product of the ith

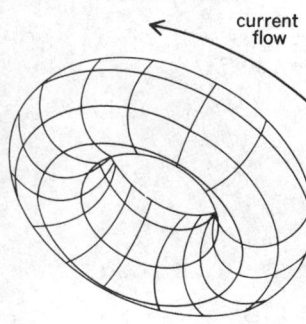

PERPETUAL MOTION MACHINE OF THE THIRD KIND

current flow

If a direct current is caused to flow in a superconducting ring, this current will continue to flow undiminished in time without application of any external force.

component of *w* and the *j*th component of *v*. { pe'rōn frō'bā·nē·ús ˌthir·əm }

Perron-Frobenius theory [MATH] The study of positive matrices and their eigenvalues; in particular, application of the Perron-Frobenius theorem. { pe'rōn frō'bā·nē·ús ˌthē·ə·rē }

perry [METEOROL] In England, a sudden, heavy fall of rain; a squall, sometimes referred to as "half a gale." { 'per·ē }

perryite [MINERAL] $(Ni,Fe)_5(Si,P)_2$ A mineral found only in meteorites. { 'per·ēˌīt }

persalic rock [GEOL] An igneous rock in which the ratio of salic to femic minerals is greater than 7:1. { pər'sal·ik 'räk }

Perseids [ASTRON] A meteor shower whose radiant lies in the constellation Perseus; it reaches a maximum about August 12. { 'pər·sē·ədz }

Perseus [ASTRON] A northern constellation; right ascension 3 hours; declination 45°N. { 'pər·sē·əs }

Perseus A [ASTRON] A strong radio source, having a redshift $z = 0.018$ and centered on the Seyfert galaxy NGC 1275, that undergoes extremely violent outbursts. { 'pər·sē·əs 'ā }

Perseus arm [ASTRON] A spiral arm of the Milky Way galaxy visible in the constellation Perseus, located (as viewed from the earth) in the direction opposite that of the galactic center. { 'pər·sē·əs ˌärm }

Perseus cluster [ASTRON] An irregular, diffuse cluster of galaxies centered on the Seyfert galaxy NGC 1275, with redshift $z = 0.018$ { 'pər·sē·əs ˌkləs·tər }

Perseus-Pisces supercluster [ASTRON] A dominant supercluster that occurs in the south galactic hemisphere and includes the Perseus cluster toward one end of a long, filamentary central condensation. { 'pər·sē·əs 'pīˌsēz 'sü·pərˌkləs·tər }

Perseus X-1 [ASTRON] The strongest known x-ray source outside the Milky Way galaxy, centered on the Seyfert galaxy NGC 1275. { 'pər·sē·əs 'eks 'wən }

Persian melon [BOT] A variety of muskmelon (*Cucumis melo*) in the order Violales; the fruit is globular and without sutures, and has dark-green skin, thin abundant netting, and firm, thick, orange flesh. { 'pər·zhən 'mel·ən }

Persian red [INORG CHEM] Red pigment made from basic lead chromate or ferric oxide. { 'pər·zhən 'red }

persic oil [MATER] A pale-yellow to red fatty oil; soluble in ether, chloroform, and carbon disulfide; taste and aroma are similar to almond oil; the oil is expressed from blanched seeds of peaches or apricots; used as a flavoring, in medicine, and as a nutrient similar to olive and almond oils. { 'pər·sik ˌoil }

persilicic *See* silicic. { 'pər·sə'lis·ik }

persistence [ELECTR] **1.** A measure of the length of time that the screen of a cathode-ray tube remains luminescent after excitation is removed; ranges from 1 for short persistence to 7 for long persistence. **2.** A faint luminosity displayed by certain gases for some time after the passage of an electric discharge. [METEOROL] With respect to the long-term nature of the wind at a given location, the ratio of the magnitude of the mean wind vector to the average speed of the wind without regard to direction. Also known as constancy; steadiness. { pər'sis·təns }

persistence forecast [METEOROL] A forecast that the future weather condition will be the same as the present condition; often used as a standard of comparison in measuring the degree of skill of forecasts prepared by other methods. { pər'sis·təns ˌfôr·kast }

persistence of vision [PHYSIO] The ability of the eye to retain the impression of an image for a short time after the image has disappeared. { pər'sis·təns əv 'vizh·ən }

persistent [BOT] Of a leaf, withering but remaining attached to the plant during the winter. { pər'sis·tənt }

persistent current [CRYO] **1.** A magnetically induced electric current that flows undiminished in a superconducting material or circuit. **2.** A superfluid current that flows undiminished around a closed path. { pər'sis·tənt 'kə·rənt }

persistent-image device [ELECTR] An optoelectronic amplifier capable of retaining an image for a definite length of time. { pər'sis·tənt ˌim·ij di,vīs }

persistent spectral holeburning [OPTICS] A process in which radiation from a narrow-band laser source sharply reduces the absorption of a solid within a frequency range that is much smaller that the linewidth of an inhomogeneously broadened transition. { pər'sis·tənt 'spek·trəl 'hōlˌbərn·iŋ }

persistent virus infection [VIROL] A covert viral infection in which a degree of equilibrium is established between the virus and the host's immune system, resulting in an infection of long duration. { pər'sis·tənt 'vī·rəs inˌfek·shən }

persistent war gas [MATER] War gas that is normally effective in the open at the point of dispersion for more than 10 minutes. { pər'sis·tənt 'wôr ˌgas }

persistron [ELECTR] A device in which electroluminescence and photoconductivity are used in a single panel capable of producing a steady or persistent display with pulsed signal input. { pər'sisˌträn }

personal communications network [COMMUN] The series of small low-power antennas that support a personal communications service, and are linked to a master telephone switch that is connected to the main telephone network. Abbreviated PCN. { ˌpərs·ən·əl kəˌmyü·nəˌkā·shənz ˌnetˌwərk }

personal communications service [COMMUN] A mobile telephone service in which pocket-sized telephones carried by the users communicate via small low-power transmitter-receiver antennas that are installed throughout a city or community. Abbreviated PCS. { ˌpərs·ən·əl kəˌmyü·nəˌkā·shənz ˌsər·vəs }

personal computer [COMPUT SCI] A computer for home or personal use. { 'pərs·ən·əl kəm'pyüd·ər }

personal digital assistant *See* handheld computer. { ˌpərs·ən·əl ˌdij·əd·əl ə'sis·tənt }

personal equation [SCI TECH] A systematic observational error due to the characteristics of the observer; the uncertainty in a reading made by an observer may be ascertained by a statistical analysis of the observer's readings. { 'pərs·ən·əl i'kwā·zhən }

personal identification code [COMPUT SCI] A special number up to six characters in length on a strip of magnetic tape embedded in a plastic card which identifies a user accessing a special-purpose computer. Abbreviated PIC. { 'pərs·ən·əl ī,den·tə·fə'kā·shən ˌkōd }

personal information manager [COMPUT SCI] Software that combines the functions of word-processing, database, and desktop accessory programs, making it possible to organize information that is relatively loosely structured. Abbreviated PIM. { 'pərs·ən·əl ˌin·fər'mā·shən ˌman·ij·ər }

personality [PSYCH] The characteristic way in which a person thinks, feels, and behaves. { ˌpərs·ən'al·əd·ē }

personality disorder [PSYCH] Any of various disorders characterized by abnormal behavior rather than by neurotic, psychotic, or mental disturbances. { ˌpərs·ən'al·əd·ē dis ˌôrd·ər }

personality theory [PSYCH] A branch of psychology concerned with developing a scientifically defensible model or view of human nature. { ˌpərs·ə'nal·əd·ē ˌthē·ə·rē }

personal locator beacon [NAV] A locator beacon capable of providing homing signals to help search-and-rescue operations. { 'pərs·ən·əl 'lō,kād·ər ˌbē·kən }

personal probability [STAT] A number between 0 and 1 assigned to an event based upon personal views concerning whether the event will occur or not; it is obtained by deciding whether one would accept a bet on the event at odds given by this number. Also known as subjective probability. { 'pərs·ən·əl ˌpräb·ə'bil·əd·ē }

personnel carrier [ORD] A motor vehicle, sometimes armored, used for transportation of troops and their equipment. { ˌpərs·ən'el ˌkar·ē·ər }

personnel monitoring [NUCLEO] Determination of the degree of radioactive contamination on individuals, using standard survey meters, and determination of the dose received by means of dosimeters. { ˌpərs·ən'el ˌmän·ə·triŋ }

Persoz's reagent [MATER] Chemical reagent used to detect the presence of silk with wool (only silk dissolves); consists of zinc chloride and zinc oxide in water. { pər'sôz·əz rē,ā·jənt }

perspective [GRAPHICS] The technique representing a figure or the space relationships of natural objects, on either a plane or curved surface, by means of projecting lines emanating from a single point, which may be infinity. { pər'spek·tiv }

perspective axis *See* axis of homology. { pər'spek·tiv ˌak·səs }

perspective center [GRAPHICS] The point of origin or termination of bundles of rays or projecting lines directed to a point object, for a photographic image or drawing, respectively. { pər'spek·tiv ˌsen·tər }

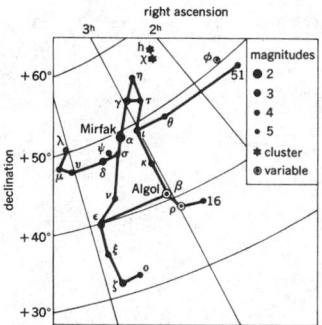

PERSEUS

Line patterns of the constellation Perseus. The grid lines in the chart represent the coordinates of the sky. The apparent brightness, or magnitude, of the stars is shown by the size of the dots, which are graded by appropriate numbers as indicated.

perspective chart [MAP] A chart on a perspective projection. { pər'spek·tiv ˌchärt }

perspective plane [GRAPHICS] Any plane containing the perspective center. { pər'spek·tiv ˌplān }

perspective projection [GRAPHICS] A projection of points by straight lines drawn through them from some given point to an intersection with the plane of projection. { pər'spek·tiv prə'jek·shən }

perspiration [PHYSIO] **1.** The secretion of sweat. **2.** See sweat. { ˌpər·spə'rā·shən }

Pers sunshine recorder [ENG] A type of sunshine recorder in which the time scale is supplied by the motion of the sun. { 'pərs 'sən,shīn ri,kȯrd·ər }

persuader [ELECTR] Element of storage tube which directs secondary emission to electron multiplier dynodes. { pər'swād·ər }

persulfate [INORG CHEM] Salt derived from persulfuric acid and containing the radical $S_2O_8^{2-}$; made by electrolysis of sulfate solutions. { ¦pər'səl,fāt }

persulfuric acid [INORG CHEM] $H_2S_2O_8$ Acid formed in lead-cell batteries by electrolyzing sulfuric acid; strong oxidizing agent. { ¦pər·səl'fyu̇r·ik 'as·əd }

PERT [SYS ENG] A management control tool for defining, integrating, and interrelating what must be done to accomplish a desired objective on time; a computer is used to compare current progress against planned objectives and give management the information needed for planning and decision making. Derived from program evaluation and review technique. { pərt }

perthite [GEOL] A parallel to subparallel intergrowth of potassium and sodium feldspar; the potassium-rich phase is usually the host from which the sodium-rich phase evolves. { 'pər,thīt }

perthitic [GEOL] Of a texture produced by perthite, exhibiting sodium feldspar as small strings, blebs, films, or irregular veinlets in a host of potassium feldspar. { pər'thid·ik }

perthosite [PETR] A light-colored syenite composed almost entirely of perthite, with less than 3% mafic minerals. { 'pər·thə,sīt }

pertinency factor [COMPUT SCI] In information retrieval, the ratio obtained in dividing the total number of relevant documents retrieved by the total number of documents retrieved. { 'pər·tə·nən·sē ,fak·tər }

perturbation [ASTRON] A deviation of an astronomical body from its computed orbit because of the attraction of another body or bodies. [MATH] A function which produces a small change in the values of some given function. [PHYS] Any effect which makes a small modification in a physical system, especially in case the equations of motion could be solved exactly in the absence of this effect. { ,pər·tər'bā·shən }

perturbation equation [PHYS] Any equation governing the behavior of a perturbation; often this will be a linear differential equation. { ,pər·tər'bā·shən i,kwā·zhən }

perturbation motion [PHYS] The motion of a disturbance (usually but not necessarily assumed infinitesimal), as opposed to the motion of the system on which the perturbation is superimposed. { ,pər·tər'bā·shən ,mō·shən }

perturbation quantity [PHYS] Any characteristic of a system which may be assumed to be a perturbation from an established value. { ,pər·tər'bā·shən ,kwän·əd·ē }

perturbation theory [MATH] The study of the solutions of differential and partial differential equations from the viewpoint of perturbation of solutions. [PHYS] The theory of obtaining approximate solutions to the equations of motion of a physical system when these equations differ by a small amount from equations which can be solved exactly. { ,pər·tər'bā·shən ,thē·ə·rē }

pertussis [MED] An infectious inflammatory bacterial disease of the air passages, caused by *Hemophilus pertussis* and characterized by explosive coughing ending in a whooping inspiration. Also known as whooping cough. { pər'təs·əs }

Peru balsam [MATER] A dark, viscous liquid with bitter taste and pleasant aroma; soluble in alcohol, ether, acetone, chloroform, benzene, and glacial acetic acid; derived from the tropical American tree *Myroxylon pereirae*; main components are esters of cinnamic and benzoic acids; used in perfumery, medicine, and chocolate manufacture. Also known as balsam of Peru; black balsam; China oil; Chinese oil; Indian balsam; Peruvian balsam. { pə'rü ¦bȯl·səm }

Peru Current [OCEANOGR] The cold ocean current flowing north along the coasts of Chile and Peru. Also known as Humboldt Current. { pə'rü 'kə·rənt }

Peru saltpeter See soda niter. { pə'rü 'sȯlt'pēd·ər }

Peruvian balsam See Peru balsam. { pə'rü·vē·ən 'bȯl·səm }

pervaporation [CHEM] A chemical separation technique in which a solution is placed in contact with a heated semipermeable membrane that selectively retains one of the components of a solution. { pər,vap·ə'rā·shən }

perveance [ELECTR] The space-charge-limited cathode current of a diode divided by the 3/2 power of the anode voltage. { 'pər·vē·əns }

perversion See lateral inversion. { pər'vər·zhən }

perviousness See permeability. { 'pər·vē·əs·nəs }

pessary [MED] An appliance of varied form placed in the vagina for uterine support or contraception. [PHARM] Any suppository or other form of medication placed in the vagina for therapeutic purposes. { 'pes·ə·rē }

pessulus [VERT ZOO] A bar composed of cartilage or bone that crosses the windpipe of a bird at its division into bronchi. { 'pes·yə·ləs }

pesticide [MATER] A chemical agent that destroys pests. Also known as biocide. { 'pes·tə,sīd }

pestilence [MED] **1.** Any epidemic contagious disease. **2.** Infection with the plague organism *Pasteurella pestis*. { 'pes·tə·ləns }

PET See polyethylene terephthalate; positron emission tomography. { ¦pē¦ē'tē or pet }

peta- [SCI TECH] A prefix that represents 10^{15}. Abbreviated P. { 'ped·ə }

petal [BOT] One of the sterile, leaf-shaped flower parts that make up the corolla. { 'ped·əl }

Petalichthyida [PALEON] A small order of extinct dorsoventrally flattened fishes belonging to the class Placodermi; the external armor is in two shields of large plates. { ,ped·ə·lik'thē·ə·də }

petaling [ORD] Condition of armor plate where the metal around the penetration is forced into a leaflike or petal form. { 'ped·əl·iŋ }

petalite [MINERAL] $LiAlSi_4O_{10}$ A white, gray, or colorless monoclinic mineral composed of silicate of lithium and aluminum, occurring in foliated masses or as crystals. { 'ped·əl,īt }

Petalodontidae [PALEON] A family of extinct cartilaginous fishes in the order Bradyodonti distinguished by teeth with deep roots and flattened diamond-shaped crowns. { ,ped·əl·ə'dänt·ə,dē }

Petaluridae [INV ZOO] A family of dragonflies in the suborder Anisoptera. { ,ped·ə'lu̇r·ə,dē }

petasma [INV ZOO] A modified endopodite of the first abdominal appendage in a male decapod crustacean. { pə'taz·mə }

petechiae [MED] Hemorrhages the size of the head of a pin. { pə'tēk·ē,ē }

peter out [ENG] To fail gradually in size, quantity, or quality; for example, a mine may be said to have petered out. { 'pēd·ər ¦au̇t }

Petersen coil See arc-suppression coil. { 'pēd·ər·sən ,kȯil }

Petersen grab [ENG] A bottom sampler consisting of two hinged semicylindrical buckets held apart by a cocking device which is released when the grab hits the ocean floor. { 'pēd·ər·sən ,grab }

Peters' formula [STAT] An approximate formula for the probable error in the value of a quantity determined from several equally careful, independent measurements of the value of the quantity. { 'pēd·ərz ,fȯr·myə·lə }

petiole [BOT] The stem which supports the blade of a leaf. { 'ped·ē,ōl }

petitgrain oil [MATER] A yellowish oil extracted from the leaves and twigs of the bitter orange tree; soluble in 70% alcohol, ether, and chloroform; used in perfumes for soaps and skin cream, and for flavoring. { 'ped·ə,grän ,ȯil }

petit mal [MED] A generalized epileptic seizure of the absence type, that is, characterized by different degrees of impaired consciousness. { pə¦tē¦mäl }

PETN See pentaerythritol tetranitrate.

petrel [VERT ZOO] A sea bird of the families Procellariidae and Hydrobatidae, generally small to medium-sized with long wings and dark plumage with white areas near the rump. { 'pe·trəl }

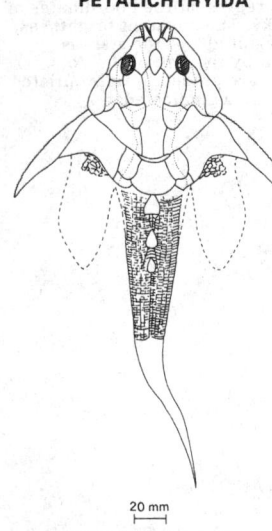

PETALICHTHYIDA

Lunaspis broilii from the Lower Devonian of Germany; restoration in dorsal aspect.

20 mm

petri dish [MICROBIO] A shallow glass or plastic dish with a loosely fitting overlapping cover used for bacterial plate cultures and plant and animal tissue cultures. { 'pē·trē ˌdish }

petrifaction [GEOL] A fossilization process whereby inorganic matter dissolved in water replaces the original organic materials, converting them to a stony substance. { ˌpe·trə'fak·shən }

petrified wood See silicified wood. { 'pe·trəˌfīd 'wůd }

Petriidae [INV ZOO] A small family of coleopteran insects in the superfamily Tenebrionoidea. { pə'trī·əˌdē }

Petri net [COMMUN] An abstract, formal model of information flow, which is used as a graphical language for modeling systems with interacting concurrent components; in mathematical terms, a structure with four parts or components: a finite set of places, a finite set of transitions, an input function, and an output function. { 'pē·trē ˌnet }

petrochemicals [ORG CHEM] Chemicals made from feedstocks derived from petroleum or natural gas; examples are ethylene, butadiene, most large-scale plastics and resins, and petrochemical sulfur. Also known as petroleum chemicals. { ¦pe·trō¦kem·ə·kəlz }

petrochemistry [GEOCHEM] An aspect of geochemistry that deals with the study of the chemical composition of rocks. [ORG CHEM] The chemistry and reactions of materials derived from petroleum, natural gas, or asphalt deposits. { ¦pe·trō 'kem·ə·strē }

petrofabric See fabric. { ¦pe·trō'fab·rik }

petrofabric analysis See structural petrology. { ¦pe·trō'fab·rik ə'nal·ə·səs }

petrofabric diagram See fabric diagram. { ¦pe·trō'fab·rik 'dī·əˌgram }

petrofabrics See structural petrology. { ¦pe·trō'fab·riks }

petrofacies See petrographic facies. { ¦pe·trō'fā·shēz }

petrogenesis [PETR] That branch of petrology dealing with the origin of rocks, particularly igneous rocks. Also known as petrogeny. { ¦pe·trō'jen·ə·səs }

petrogenic grid [PETR] A diagram whose coordinates are parameters of the rock-forming environment on which equilibrium curves are plotted indicating the limits of the stability fields of specific minerals and mineral assemblages. { ¦pe·trō¦jen·ik 'grid }

petrogeny See petrogenesis. { pə'träj·ə·nē }

petrogeometry See structural petrology. { ¦pe·trō·jē'äm·ə·trē }

petrographer [GEOL] An individual who does petrography. { pə'träg·rə·fər }

petrographic facies [GEOL] Facies distinguished principally by composition and appearance. Also known as petrofacies. { ¦pe·trə¦graf·ik 'fā·shēz }

petrographic microscope [OPTICS] A polarizing microscope used for analysis of petrographic thin sections. { ¦pe·trə¦graf·ik 'mī·krəˌskōp }

petrographic period [GEOL] The extension in time of a rock association. { ¦pe·trə¦graf·ik 'pir·ē·əd }

petrographic province [GEOL] A broad area in which similar igneous rocks are formed during the same period of igneous activity. Also known as comagmatic region; igneous province; magma province. { ¦pe·trə¦graf·ik 'präv·əns }

petrography [GEOL] The branch of geology that deals with the description and systematic classification of rocks, especially by means of microscopic examination. { pə'träg·rə·fē }

petrol See gasoline. { 'pe·trəl }

petrolatum [MATER] A smooth, semisolid blend of mineral oil with waxes crystallized from the residual type of petroleum lubricating oil; the wax molecules contain 30–70 carbon atoms and are straight chains with a few branches or naphthene rings; used as a lubricant, as a carrier in polishes and cosmetics, and as a rust preventive. { ˌpe·trə'lād·əm }

petroleum [GEOL] A naturally occurring complex liquid hydrocarbon which after distillation yields combustible fuels, petrochemicals, and lubricants; can be gaseous (natural gas), liquid (crude oil, crude petroleum), solid (asphalt, tar, bitumen), or a combination of states. { pə'trō·lē·əm }

petroleum asphalt [MATER] Asphalt recovered or made from petroleum. { pə'trō·lē·əm 'as,fȯlt }

petroleum benzin [MATER] Colorless, volatile, flammable, toxic petroleum distillate with boiling range of 35–80°C; a special grade is ligroin (petroleum ether; boiling range 20–135°C), used as an extractive solvent, especially for drugs.

Also known as benzin; benzine (archaic usage). { pə'trō·lē·əm 'benˌzēn }

petroleum chemicals See petrochemicals. { pə'trō·lē·əm ˌkem·ə·kəlz }

petroleum coke [MATER] A carbonaceous solid material made by the destructive heating of high-molecular-weight petroleum-refining residues. { pə'trō·lē·əm ˌkōk }

petroleum engineer [PETRO ENG] An engineer whose primary objective is to find and produce oil or gas from petroleum reserves. { pə'trō·lē·əm ˌen·jə'nir }

petroleum engineering [ENG] The application of almost all types of engineering to the drilling for and production of oil, gas, and liquefiable hydrocarbons. { pə'trō·lē·əm ˌen·jə'nir·iŋ }

petroleum ether [MATER] A volatile fraction of petroleum consisting chiefly of pentanes and hexanes. Also known as ligroin; white sprit. { pə'trō·lē·əm ˌē·thər }

petroleum geology [GEOL] The branch of economic geology dealing with the origin, occurrence, movement, accumulation, and exploration of hydrocarbon fuels. { pə'trō·lē·əm jē'äl·ə·jē }

petroleum isomerization process [CHEM ENG] A fixed-bed, vapor-phase petroleum-refinery process using a precious-metal catalyst and external hydrogen; feedstocks include natural gas, pentane, and hexane cuts; the product is high-octane blending stock. { pə'trō·lē·əm ī,säm·ə·rə'zā·shən ,prä·səs }

petroleum microbiology [MICROBIO] Those aspects of microbiological science and engineering of interest to the petroleum industry, including the role of microbes in petroleum formation, and the exploration, production, manufacturing, storage, and food synthesis from petroleum. { pə'trō·lē·əm ,mī·krō·bī'äl·ə·jē }

petroleum processing [CHEM ENG] The recovery and processing of various usable fractions from the complex crude oils; usable fractions include gasoline, kerosine, diesel oil, fuel oil, and asphalt. Also known as petroleum refining. { pə'trō·lē·əm 'prä,ses·iŋ }

petroleum products [MATER] Materials derived from petroleum, natural gas, or asphalt deposits; includes gasolines, diesel and heating fuels, liquefied petroleum gases (LPG and bugas), lubricants, waxes, greases, petroleum coke, petrochemicals, and (from sour crudes and natural gases) sulfur. { pə'trō·lē·əm ,präd·əks }

petroleum refining See petroleum processing. { pə'trō·lē·əm ri,fīn·iŋ }

petroleum resin [ORG CHEM] Any one of a family of polymers produced from mixed unsaturated monomers recovered from petroleum processing streams. { pə'trō·lē·əm ,rez·ən }

petroleum secondary engineering [PETRO ENG] The process of removing oil from its native reservoirs by the use of supplemental energies after the natural energies causing oil production have been depleted. { pə'trō·lē·əm 'sek·ən,der·ē ,en·jə'nir·iŋ }

petroleum seep See oil seep. { pə'trō·lē·əm ,sēp }

petroleum sulfonates [MATER] Sulfonated petroleum products made as by-products of SO_3 treatment of white oil or lubestock; used as lube-oil additives, textile-processing emulsifiers, and rust preventives. { pə'trō·lē·əm 'səl·fə,nāts }

petroleum tailings See wax tailings. { pə'trō·lē·əm 'tāl·iŋz }

petroleum tar [MATER] A viscous, black or dark-brown product of petroleum refining; yields substantial quantity of solid residue when partly evaporated or fractionally distilled. { pə'trō·lē·əm ,tär }

petroleum trap [GEOL] Stable underground formation (geological or physical) of such nature as to trap and hold liquid or gaseous hydrocarbons; usually consists of sand or porous rock surrounded by impervious rock or clay formations. { pə'trō·lē·əm ,trap }

petroleum wax [MATER] A wax occurring naturally in various fractions of crude petroleum; there are two groups: paraffin wax and microcrystalline wax. { pə'trō·lē·əm ,waks }

petroliferous [GEOL] Containing petroleum. { ,pe·trə'lif·ə·rəs }

petrologen See kerogen. { pə'träl·ə·jən }

petrologist [GEOL] An individual who studies petrology. { pə'träl·ə·jəst }

petrology [GEOL] The branch of geology concerned with the origin, occurrence, structure, and history of rocks, principally igneous and metamorphic rock. { pə'träl·ə·jē }

petromict [GEOL] Of a sediment, composed of metastable rock fragments. { 'pe·trə,mikt }

petromorph [GEOL] A speleothem or cave formation that is exposed to the surface by erosion of the limestone in which the cave was formed. { 'pe·trə,mȯrf }

petromorphology *See* structural petrology. { ¦pe·trō·mȯr'fäl·ə·jē }

Petromyzonida [VERT ZOO] The lampreys, an order of eel-like, jawless vertebrates (Agnatha) distinguished by a single, dorsal nasal opening, and the mouth surrounded by an oral disk and provided with a rasping tongue. { ,pe·trō·mī'zän·ə·də }

Petromyzontiformes [VERT ZOO] The equivalent name for Petromyzonida. { ,pe·trō·mī,zänt·ə'fȯr,mēz }

petrophysics [GEOL] Study of the physical properties of reservoir rocks. { ¦pe·trō¦fiz·iks }

petrosal nerve [NEUROSCI] Any of several small nerves passing through the petrous part of the temporal bone and usually attached to the geniculate ganglion. { pə'trō·səl ,nərv }

petrosal process [ANAT] A sharp process of the sphenoid bone located below the notch for the passage of the abducens nerve, which articulates with the apex of the petrous portion of the temporal bone and forms the medial boundary of the foramen lacerum. { pə'trō·səl ,prä·sas }

Petrosaviaceae [BOT] A small family of monocotyledonous plants in the order Triuridales characterized by perfect flowers, three carpels, and numerous seeds. { ,pe·trō,sav·ē'ās·ē,ē }

petrotectonics [GEOL] Extension of the field of structural petrology to include analysis of the movements that produced the rock's fabric. Also known as tectonic analysis. { ¦pe·trō·tek'tän·iks }

petrous [MATER] Referring to a material whose hardness resembles that of stone. { 'pe·trəs }

Petrov classification [RELAT] An algebraic classification of space-times based on eigenvalues of the curvature tensor. { 'pe,träv ,klas·ə·fə,kā·shən }

Petterson-Nansen water bottle *See* Nansen bottle. { 'ped·ər·sən 'nan·sən 'wȯd·ər ,bäd·əl }

petticoat insulator [ELEC] Insulator having an outward-flaring lower part that is hollow inside to increase the length of the surface leakage path and keep part of the path dry at all times. { 'ped·i,kōt 'in·sə,lād·ər }

Pettit truss [CIV ENG] A bridge truss in which the panel is subdivided by a short diagonal and a short vertical member, both intersecting the main diagonal at its midpoint. { 'ped·ət ,trəs }

petzite [MINERAL] Ag₃AuTe₂ A steel-gray to iron-black mineral consisting of a silver gold telluride; hardness is 2.5–3 on Moh's scale, and specific gravity is 8.7–9.0. { 'pet,sīt }

Petzval condition [OPTICS] The condition whereby an optical system will eliminate the aberration of curvature of field only if the Petzval curvature vanishes. { 'pets·väl kən,dish·ən }

Petzval curvature [OPTICS] The axial curvature of the image of a plane object produced by an optical system, equal to the sum over all the optical surfaces in the system of $R(1/n' - 1/n)$, where R is the curvature of the surface, and n and n' are the refractive indices before and after the surface. Also known as Petzval sum. { 'pets·väl 'kər·və·chər }

Petzval lens [OPTICS] A photographic objective which consists of four lenses ordered in two pairs widely separated from each other, with the first pair cemented together and the second usually having a small air space. { 'pets·väl ,lenz }

Petzval sum *See* Petzval curvature. { 'pets·väl ,səm }

Petzval surface [OPTICS] A paraboloidal surface on which point images of point objects are formed by a doublet lens whose separation is such that astigmatism is eliminated. { 'pets·väl ,sər·fəs }

peuroseite [MINERAL] (Ni,Cu,Pb)Se₂ A gray mineral composed of nickel copper lead selenide, occurring in columnar masses. { pyu'rō,zīt }

pewter [MET] An alloy that typically contained tin as the principal component and some antimony and copper; older produced pewter typically contains lead along with the other components. { 'pyüd·ər }

Peyer's patches [HISTOL] Aggregates of lymph nodules beneath the epithelium of the ileum. { 'pī·ərz ,pach·əz }

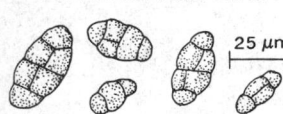

PHAEODICTYOSPORAE

Characteristic dark muriform spores of *Stemphylium*.

Peyronie's disease [MED] Scarring of the shaft of the penis; may interfere with normal erections. { ¦pā·rə¦nēz di,zēz }

pezograph *See* regmaglypt. { 'pez·ə,graf }

pf *See* power factor.

pF *See* picofarad.

Pfaffian differential equation [MATH] The first-order linear total differential equation $P(x,y,z)dx + Q(x,y,z)dy + R(x,y,z)dz = 0$, where the functions P, Q, and R are continuously differentiable. { 'faf·ē·ən ,dif·ə¦ren·chəl i'kwā·zhən }

PFC *See* perfluorocarbon.

PFE *See* photoferroelectric effect.

Pfeiffer's disease *See* infectious mononucleosis. { 'fī·fərz di,zēz }

p53 [GEN] A multifunctional tumor suppressor protein consisting of several domains that exhibit a number of biochemical activities and interact with a large variety of cellular and viral proteins; its encoding gene is the most commonly mutated gene in tumor cells. { ,pē,fif·tē'thrē }

PF key *See* programmed function key. { ,pē'ef ,kē }

PFM *See* pulse-frequency modulation.

p-form [MATH] A totally antisymmetric covariant tensor of rank p. { 'pē ,fȯrm }

Pfund series [SPECT] A series of lines in the infrared spectrum of atomic hydrogen whose wave numbers are given by $R_H[(1/25) - (1/n^2)]$, where R_H is the Rydberg constant for hydrogen, and n is any integer greater than 5. { 'funt ,sir·ēz }

PGA *See* folic acid.

PGR *See* precision depth recorder.

pH [CHEM] A term used to describe the hydrogen-ion activity of a system; it is equal to $-\log a_{H^+}$; here a_{H^+} is the activity of the hydrogen ion; in dilute solution, activity is essentially equal to concentration and pH is defined as $-\log_{10}[H^+]$, where H^+ is hydrogen-ion concentration in moles per liter; a solution of pH 0 to 7 is acid, pH of 7 is neutral, pH over 7 to 14 is alkaline. { pē'āch }

phacella *See* gastric filament. { fə'sel·ə }

phacellite *See* kaliophilite. { 'fas·əl,īt }

phacolith [GEOL] A minor, concordant, lens-shaped, and usually granitic intrusion into folded sedimentary strata. { 'fak·ə,lith }

Phaenocephalidae [INV ZOO] A monospecific family of coleopteran insects in the superfamily Cucujoidea, found only in Japan. { ,fē·nō·sə'fal·ə,dē }

Phaenothontidae [VERT ZOO] The tropic birds, a family of fish-eating aquatic forms in the order Pelecaniformes. { ,fē·nō'thänt·ə,dē }

Phaeocoleosporae [MYCOL] A spore group of the Fungi Imperfecti with dark filiform spores. { ,fē,kō·lē'äs·pə,rē }

Phaeodictyosporae [MYCOL] A spore group of the Fungi Imperfecti with dark muriform spores. { ,fē·ō,dik·tē'äs·pə,rē }

Phaeodidymae [MYCOL] A spore group of the Fungi Imperfecti with dark two-celled spores. { ,fē·ō'did·ə,mē }

Phaeodorina [INV ZOO] The equivalent name for Tripylina. { ,fē·ə·də'rī·nə }

Phaeohelicosporae [MYCOL] A spore group of the Fungi Imperfecti with dark, spirally coiled, septate spores. { ,fē·ō,hel·ə'käs·pə,rē }

Phaeophragmiae [MYCOL] A spore group of the Fungi Imperfecti with dark three- to many-celled spores. { ,fē·ō'frag·mē,ē }

Phaeophyta [BOT] The brown algae, constituting a division of plants; the plant body is multicellular, varying from a simple filamentous form to a complex, sometimes branched body having a basal attachment. { fē'äf·əd·ə }

Phaeosporae [MYCOL] A spore group of Fungi Imperfecti characterized by dark one-celled, nonfiliform spores. { fē'äs·pə,rē }

Phaeostaurosporae [MYCOL] A spore group of the Fungi Imperfecti with dark star-shaped or forked spores. { ,fē·ō·stȯ'räs·pə,rē }

phage *See* bacteriophage. { fāj }

phage cross [VIROL] Multiple infection of a single bacterium by phages that differ at one or more genetic sites, leading to the production of recombinant progeny phage. { 'fāj ,krȯs }

phage induction [VIROL] Prophage stimulation by a variety of means that induce the vegetative state. { 'fāj in,dək·shən }

phage restriction [VIROL] The inability of a phage to replicate due to an enzyme mechanism for degrading foreign deoxyribonucleic acid that enter the bacterial host cell. { 'fāj ri,strik·shən }

phagocyte [CELL MOL] An ameboid cell that engulfs foreign material. { 'fag·ə,sīt }

phagocytic vacuole *See* food vacuole. { ¦fag·ə¦sid·ik 'vak·yə,wōl }

phagocytin [BIOCHEM] A type of bactericidal agent present within phagocytic cells. { ¦fag·ə¦sīt·ən }

phagocytosis [CELL MOL] A specialized form of macropinocytosis in which cells engulf large solid objects such as bacteria and deliver the internalized objects to special digesting vacuoles; exists in certain cell types, such as macrophages and neutrophils. { ,fag·ə,sī'tō·səs }

phagolysosome [CELL MOL] An intracellular vesicle formed by fusion of a lysosome with a phagosome. { ,fag·ə'lī·sə,sōm }

phagosome [CELL MOL] A closed intracellular vesicle containing material captured by phagocytosis. { 'fag·ə,sōm }

phagotroph [INV ZOO] An organism that ingests nutrients by phagocytosis. { 'fag·ə,träf }

Phalacridae [INV ZOO] The shining flower beetles, a family of coleopteran insects in the superfamily Cucujoidea. { fə'lak·rə,dē }

Phalacrocoracidae [VERT ZOO] The cormorants, a family of aquatic birds in the order Pelecaniformes. { ,fal·ə,krō·kə'ras·ə,dē }

Phalaenidae [INV ZOO] The equivalent name for Noctuidae. { fə'len·ə,dē }

Phalangeridae [VERT ZOO] A family of marsupial mammals in which the marsupium is well developed and opens anteriorly, the hindfeet are syndactylous, and the hallux is opposable and lacks a claw. { fal·ən'jer·ə,dē }

Phalangida [INV ZOO] An order of the class Arachnida characterized by an unsegmented cephalothorax broadly joined to a segmented abdomen, paired chelate chelicerae, and paired palpi. { fə'lan·jə·də }

phalanx [ANAT] One of the bones of the fingers or toes. { 'fā,laŋks }

Phalaropodidae [VERT ZOO] The phalaropes, a family of migratory shore birds characterized by lobate toes and by reversal of the sex roles with respect to dimorphism and care of the young. { fə,ler·ə'päd·ə,de }

phallic phase [PSYCH] In psychoanalytic theory, the period from about 2½ to 6 years of age during which sexual interest, curiosity, and pleasurable experience center about the penis in boys and the clitoris in girls. { 'fal·ik ,fāz }

Phallostethidae [VERT ZOO] A family of actinopterygian fishes in the order Atheriniformes. { ,fal·ō'steth·ə,dē }

Phallostethiformes [VERT ZOO] An equivalent name for Atheriniformes. { ,fal·ō'steth·ə'fór,mēz }

phallotoxin [BIOCHEM] One of a group of toxic peptides produced by the mushroom *Amanita phalloides*. { ,fal·ō¦täk·sən }

phallus [ANAT] *See* penis. [EMBRYO] An undifferentiated embryonic structure derived from the genital tubercle that differentiates into the penis in males and the clitoris in females. { 'fal·əs }

phanerite [PETR] An igneous rock having phaneritic texture. { 'fan·ə,rīt }

phaneritic [PETR] Of the texture of an igneous rock, being visibly crystalline. Also known as coarse-grained; phanerocrystalline; phenocrystalline. { ,fan·ə'rid·ik }

phanerocryst *See* phenocryst. { 'fan·ə,rō,krist }

phanerocrystalline *See* phaneritic. { ¦fan·ə·rō¦krist·əl·ən }

phanerogam [BOT] A plant that produces seeds, for example, an angiosperm or gymnosperm. { 'fan·ə·rə,gam }

phanerophyte [ECOL] A perennial tree or shrub with dormant buds borne on aerial shoots. { 'fan·ə·rō,fīt }

Phanerorhynchidae [PALEON] A family of extinct chondrostean fishes in the order Palaeonisciformes having vertical jaw suspension. { ¦fan·ə·rō'riŋ·kə,dē }

Phanerozoic [GEOL] The part of geologic time for which there is abundant evidence of life, especially higher forms, in the corresponding rock, essentially post-Precambrian. { ¦fan·ə·rō¦zō·ik }

Phanerozonida [INV ZOO] An order of the Asteroidea in which the body margins are defined by two conspicuous series

of plates and in which pentamerous symmetry is generally constant. { ,fan·ə·rō'zän·ə·də }

Phanodermatidae [INV ZOO] A family of free-living nematodes in the superfamily Enoploidea. { ,fa·nō·dər'mad·ə,dē }

phanotron [ELECTR] A hot-filament diode rectifier tube utilizing an arc discharge in mercury vapor or an inert gas, usually xenon. { 'fan·ə,trän }

phantastran [ELECTR] A solid-state phantastron. { fan 'tas,trän }

phantastron [ELECTR] A monostable pentode circuit used to generate sharp pulses at an adjustable and accurately timed interval after receipt of a triggering signal. { 'fan'tas,trän }

phantom [GEOL] A bed or member that is absent from a specific stratigraphic section but is usually present in a characteristic position in a sequence of similar geologic age. [NUCLEO] A volume of material approximating as closely as possible the density and effective atomic number of living tissue, used in biological experiments involving radiation. [PETR] *See* ghost. { 'fan·təm }

phantom bottom [OCEANOGR] A false bottom indicated by an echo sounder, some distance above the actual bottom; such an indication, quite common in the deeper parts of the ocean, is due to large quantities of small organisms. { 'fan·təm 'bäd·əm }

phantom circuit [COMMUN] A communication circuit derived from two other communication circuits or from one other circuit and ground, with no additional wire lines. { 'fan·təm 'sər·kət }

phantom-circuit loading coil [ELEC] Loading coil for introducing a desired amount of inductance into a phantom circuit, and a minimum amount of inductance into its constituent circuits. { 'fan·təm ¦sər·kət ¦lōd·iŋ ,kȯil }

phantom-circuit repeating coil [ELEC] Repeating coil used at a terminal of a phantom circuit, in the terminal circuit extending from the midpoints of the associated side-circuit repeating coils. { 'fan·təm ¦sər·kət ri'pēd·iŋ ,kȯil }

phantom crystal [CRYSTAL] A crystal containing an earlier stage of crystallization outlined by dust, minute inclusions, or bubbles. Also known as ghost crystal. { 'fan·təm 'krist·əl }

phantom group [ELEC] **1.** Group of four open-wire conductors suitable for the derivation of a phantom circuit. **2.** Three circuits which are derived from simplexing two physical circuits to form a phantom circuit. { 'fan·təm 'grüp }

phantom horizon [GEOL] In seismic reflection prospecting, a line constructed so that it is parallel to the nearest actual dip segment at all points along a profile. { 'fan·təm hə'rīz·ən }

phantom repeating coil [ELEC] A side-circuit repeating coil or a phantom-circuit repeating coil when discrimination between these two types is not necessary. { 'fan·təm ri'pēd·iŋ ,kȯil }

phantom signals [ELECTR] Signals appearing on the screen of a cathode-ray-tube indicator, the cause of which cannot readily be determined and which may be caused by circuit fault, interference, propagation anomalies, jamming, and so on. { 'fan·təm 'sig·nəlz }

phantom target *See* echo box. { 'fan·təm 'tär·gət }

Pharetronida [INV ZOO] An order of calcareous sponges in the subclass Calcinea characterized by a leuconoid structure. { ¦far·ə'trän·ə·də }

pharmaceutical [PHARM] A chemical produced industrially (medicinal drug), which is useful in preventive or therapeutic treatment of a physical, mental, or behavioral condition. { ,fär·mə'süd·i·kəl }

pharmaceutical biotechnology [PHARM] A field that uses micro- and macroorganisms and hybridomas to create pharmaceuticals that are safer and more cost-effective than conventionally produced pharmaceuticals. { ,fär·mə,süd·i·kəl ,bī·ō·tek'näl·ə·jē }

pharmaceutical chemistry [CHEM ENG] The chemistry of drugs and of medicinal and pharmaceutical products. { ,fär·mə'süd·ə·kəl 'kem·ə·strē }

pharmaceutics *See* pharmacy. { ,fär·mə'süd·iks }

pharmacodynamics [PHARM] The science that deals with the actions of drugs. { ¦fär·mə·kō·dī'nam·iks }

pharmacogenetics [GEN] The science of genetically determined variations in drug responses. { ¦fär·mə·kō·jə'ned·iks }

pharmacognosy [PHARM] A subfield of pharmacology

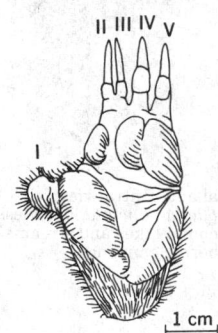

PHALANGERIDAE

Hindfoot of *Trichosurus*, representing the five-toed, syndactylous condition (digits II and III are bound in one web of skin).

which studies the biological and chemical components of medically useful substances that occur naturally (primarily those synthesized by plants). { ¦fär·mə'käg·nə·sē }

pharmacokinetics [PHARM] The study of the way that drugs move through the body after they are swallowed or injected. { ¦fär·mə·kō·ki'ned·iks }

pharmacolite [MINERAL] CaH(AsO₄)·2H₂O A white to grayish monoclinic mineral composed of hydrous acid arsenate of calcium, occurring in fibrous form. { fär'mak·ə‚līt }

pharmacologic pyrogen [PHARM] A naturally occurring pharmacologic agent, such as serotonin or a catecholamine, that controls body temperature; it can cause fever when injected under experimental conditions. { ¦fär·mə·kə'läj·ik 'pī·rə·jən }

pharmacology [CHEM] The science dealing with the nature and properties of drugs, particularly their actions. { ‚fär·mə'käl·ə·jē }

pharmacophobia [PSYCH] Abnormal fear of medicine. { ‚fär·mə·kə'fō·bē·ə }

pharmacopoeia [PHARM] A book containing a selected list of medicinal substances and their dosage forms, providing also a description and the standards for purity and strength for each. { ‚fär·mə·kə'pē·ə }

pharmacosiderite [MINERAL] Fe₃(AsO₄)₂(OH)₃·5H₂O Green or yellowish-green mineral composed of a hydrous basic iron arsenate and commonly found in cubic crystals. Also known as cube ore. { ¦fär·mə·kō'sīd·ə‚rīt }

pharmacotherapy [MED] The treatment of disease by means of drugs. { ¦fär·mə·kō'ther·ə·pē }

pharmacy Also known as pharmaceutics. [MED] **1.** The art and science of the preparation and dispensation of drugs. **2.** A place where drugs are dispensed. { 'fär·mə·sē }

pharyngeal aponeurosis [ANAT] The fibrous submucous layer of the pharynx. { fə'rin·jē·əl ‚ap·ō·nú'rō·səs }

pharyngeal bursa [EMBRYO] A small pit caudal to the pharyngeal tonsil, resulting from the ingrowth of epithelium along the course of the degenerating tip of the notochord of the vertebrate embryo. { fə'rin·jē·əl 'bər·sə }

pharyngeal cleft [EMBRYO] One of the paired open clefts on the sides of the embryonic pharynx between successive visceral arches in vertebrates. { fə'rin·jē·əl 'kleft }

pharyngeal plexus [ANAT] A plexus of veins situated at the side of the pharynx. [NEUROSCI] A nerve plexus innervating the pharynx. { fə'rin·jē·əl 'plek·səs }

pharyngeal pouch [EMBRYO] One of the five paired sacculations in the lateral aspect of the pharynx in vertebrate embryos. Also known as visceral pouch. { fə'rin·jē·əl 'paùch }

pharyngeal tonsil See adenoid. { fə'rin·jē·əl 'tän·səl }

pharyngeal tooth [VERT ZOO] A tooth developed on the pharyngeal bone in many fishes. { fə'rin·jē·əl 'tüth }

pharyngitis [MED] Inflammation of the pharynx. { ‚far·ən'jīd·əs }

Pharyngobdellae [INV ZOO] A family of leeches in the order Arhynchobdellae that is distinguished by the lack of jaws. { fə‚rin‚gäb'del·ə‚dē }

pharyngology [MED] The science of the pharyngeal mechanism, functions, and diseases. { ‚far·in'gäl·ə·jē }

pharyngoscope [MED] An instrument for examining the pharynx. { fə'rin‚gə‚skōp }

pharyngo-tonsillar diphtheria [MED] A type of diphtheria that is characterized by a sore throat, difficulty in swallowing, and low-grade fever. { fə‚rin‚gō 'täns·əl·ər dif'thir·ē·ə }

pharynx [ANAT] A chamber at the oral end of the vertebrate alimentary canal, leading to the esophagus. { 'far·inks }

phase [ASTRON] One of the cyclically repeating appearances of the moon or other orbiting body as seen from earth. [CHEM] Portion of a physical system (liquid, gas, solid) that is homogeneous throughout, has definable boundaries, and can be separated physically from other phases. [MATH] An additive constant in the argument of a trigonometric function. [MET] A constituent of an alloy that is physically distinct and is homogeneous in chemical composition. [PHYS] **1.** The fractional part of a period through which the time variable of a periodic quantity (alternating electric current, vibration) has moved, as measured at any point in time from an arbitrary time origin; usually expressed in terms of angular measure, with one period being equal to 360° or 2π radians. **2.** For a sinusoidally varying quantity, the phase (first definition) with the time origin located at the last point at which the quantity passed through

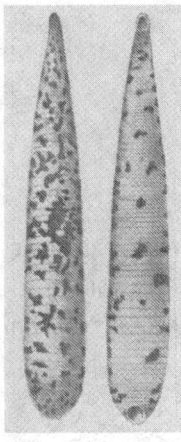

PHARYNGOBDELLAE

Dorsal and ventral view of *Erpobdella punctata*, a jawless leech common in lakes and streams in the Northern Hemisphere.

PHASE ANGLE

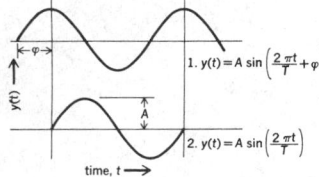

1. $y(t) = A \sin\left(\frac{2\pi t}{T} + \varphi\right)$

2. $y(t) = A \sin\left(\frac{2\pi t}{T}\right)$

An illustration of the meaning of phase for a sinusoidal wave, $y(t)$. The difference in phase between waves 1 and 2 is φ and is called the phase angle. For each wave, A is the amplitude and T is the period.

a zero position from a negative to a positive direction. **3.** The argument of the trigonometric function describing the space and time variation of a sinusoidal disturbance, $y = A \cos [(2\pi/\lambda)(x - vt)]$, where x and t are the space and time coordinates, v is the velocity of propagation, and λ is the wavelength. [THERMO] The type of state of a system, such as solid, liquid, or gas. { fāz }

phase advancer [ELEC] Phase modifier which supplies leading reactive volt-amperes to the system to which it is connected; may be either synchronous or asynchronous. { 'fāz id‚van·sər }

phase age See age of phase inequality. { 'fāz ‚āj }

phase-alternation line system [COMMUN] A color television system used in Europe, in which the phase of the color subcarrier is changed from scanning line to scanning line, requiring transmission of a line switching signal as well as a color burst. Abbreviated PAL system. { 'fāz ‚ól·tər¦nā·shən ‚līn ‚sis·təm }

phase angle [PHYS] The difference between the phase of a sinusoidally varying quantity and the phase of a second quantity which varies sinusoidally at the same frequency. Also known as phase difference. { 'fāz ‚an·gəl }

phase-angle meter See phase meter. { 'fāz ¦an·gəl ‚mēd·ər }

phase-balance relay [ELEC] Relay which functions by reason of a difference between two quantities associated with different phases of a polyphase circuit. { 'fāz ¦bal·əns 'rē‚lā }

phase behavior [PETRO ENG] The equilibrium relationships between water, liquid hydrocarbons, and dissolved or free gas, either in reservoirs or as liquids and gases are separated above ground in gas-oil separator systems. { 'fāz bi‚hāv·yər }

phase boundary [PHYS] The interface between two or more separate phases, such as liquid-gas, liquid-solid, gas-solid, or, for immiscible materials, liquid-liquid or solid-solid. { 'fāz ‚baùn·drē }

phase change [PHYS] **1.** The metamorphosis of a material or mixture from one phase to another, such as gas to liquid, solid to gas. **2.** See phase shift. { 'fāz ‚chānj }

phase-change coefficient See phase constant. { ¦fāz ‚chānj ‚kō·i‚fish·ənt }

phase-change material [ENG] A material which is used to store the latent heat absorbed in the material during a phase transition. { 'fāz ‚chānj mə‚tir·ē·əl }

phase-change recording [COMPUT SCI] An optical recording technique that uses a laser to alter the crystalline structure of a metallic surface to create bits that reflect or absorb light when they are illuminated during the read operation. { ¦fāz ‚chānj ri'kòrd·iŋ }

phase coherence [PHYS] The existence of a statistical or time coherence between the phases of two or more waves. { 'fāz kō‚hir·əns }

phase comparator [COMPUT SCI] A comparator that accepts two radio-frequency input signals of the same frequency and provides two video outputs which are proportional, respectively, to the sine and cosine of the phase difference between the two inputs. { 'fāz kəm‚par·əd·ər }

phase-comparison relaying [ELEC] A method of detecting faults in an electric power system in which signals are transmitted from each of two terminals every half cycle so that a continuous signal is received at an intermediate point if there is no fault between the terminals, while a periodic signal is received if there is a fault. { 'fāz kəm‚par·ə·sən 'rē‚lā·iŋ }

phase conductor [ELEC] In a polyphase circuit, any conductor other than the neutral conductor. { 'fāz kən‚dək·tər }

phase conjugate system [OPTICS] An adaptive optics system in which the wavefront to be corrected is measured directly, using either a geometric or interferometric test. { 'fāz ¦kän·jə·gət ‚sis·təm }

phase constant [ELECTROMAG] A rating for a line or medium through which a plane wave of a given frequency is being transmitted; it is the imaginary part of the propagation constant, and is the space rate of decrease of phase of a field component (or of the voltage or current) in the direction of propagation, in radians per unit length. Also known as phase-change coefficient; wavelength constant. { 'fāz ‚kän·stənt }

phase-contrast microscope [OPTICS] A compound microscope that has an annular diaphragm in the front focal plane of the substage condenser and a phase plate at the rear focal plane of the objective, to make visible differences in phase or optical path in transparent or reflecting media. 　{ 'fāz ¦kän‚trast 'mī·krə‚skōp }

phase control [ELECTR] **1.** A control that changes the phase angle at which the alternating-current line voltage fires a thyratron, ignitron, or other controllable gas tube. Also known as phase-shift control. **2.** *See* hue control. 　{ 'fāz kən‚trōl }

phase converter [ELEC] A converter that changes the number of phases in an alternating-current power source without changing the frequency. 　{ 'fāz kən‚vərd·ər }

phase-correcting network *See* phase equalizer. 　{ 'fāz kə‚rek·tiŋ 'net‚wərk }

phase correction [COMMUN] Process of keeping synchronous telegraph mechanisms in substantially correct phase relationship. 　{ 'fāz kə‚rek·shən }

phase crossover [CONT SYS] A point on the plot of the loop ratio at which it has a phase angle of 180°. 　{ 'fāz 'krós‚ō·vər }

phased array [ELECTROMAG] An array of dipoles on a radar antenna in which the signal feeding each dipole is varied so that antenna beams can be formed in space and scanned very rapidly in azimuth and elevation. 　{ 'fāzd ə'rā }

phase delay [COMMUN] Ratio of the total phase shift (radians) of a sinusoidal signal in transmission through a system or transducer, to the frequency (radians/second) of the signal. 　{ 'fāz di‚lā }

phase detector [ELECTR] A circuit that provides a direct-current output voltage which is related to the phase difference between an oscillator signal and a reference signal, for use in controlling the oscillator to keep it in synchronism with the reference signal. Also known as phase discriminator. 　{ 'fāz di‚tek·tər }

phase deviation [COMMUN] The peak difference between the instantaneous angle of a modulated wave and the angle of the sine-wave carrier. 　{ 'fāzd ‚dē·vē'ā·shən }

phase diagram [MET] *See* constitution diagram. [PHYS CHEM] A graphical representation of the equilibrium relationships between phases (such as vapor-liquid, liquid-solid) of a chemical compound, mixture of compounds, or solution. [THERMO] **1.** A graph showing the pressures at which phase transitions between different states of a pure compound occur, as a function of temperature. **2.** A graph showing the temperatures at which transitions between different phases of a binary system occur, as a function of the relative concentrations of its components. 　{ 'fāz ‚dī·ə‚gram }

phase difference *See* phase angle. 　{ 'fāz 'dif·rəns }

phase discriminator *See* phase detector. 　{ 'fāz di‚skrim·ə‚nād·ər }

phase distortion [COMMUN] **1.** The distortion which occurs in an instrument when the relative phases of the input signal differ from those of the output signal. **2.** *See* phase-frequency distortion. 　{ 'fāz di‚stór·shən }

phase encoding [COMPUT SCI] A method of recording data on magnetic tape in which a logical 1 is defined as the transition from one magnetic polarity to another positioned at the center of the bit cell, and 0 is defined as the transition in the opposite direction, also at the center of the cell. Also known as Manchester coding. 　{ 'fāz in'kōd·iŋ }

phase equalizer [ELECTR] A network designed to compensate for phase-frequency distortion within a specified frequency band. Also known as phase-correcting network. 　{ 'fāz 'ē‚kwə‚liz·ər }

phase equilibria [PHYS CHEM] The equilibrium relationships between phases (such as vapor, liquid, solid) of a chemical compound or mixture under various conditions of temperature, pressure, and composition. 　{ 'fāz ‚ē·kwə'lib·rē·ə }

phase excursion [COMMUN] In angle modulation, the difference between the instantaneous angle of the modulated wave and the angle of the carrier. 　{ 'fāz ik‚skər·zhən }

phase factor [ELEC] *See* power factor. [SOLID STATE] The argument (phase) of a structure factor; it cannot be directly observed. 　{ 'fāz 'fak·tər }

phase-frequency distortion [COMMUN] Distortion occurring because phase shift is not proportional to frequency over the frequency range required for transmission. Also known as phase distortion. 　{ 'fāz 'fre·kwən‚sē di‚stór·shən }

phase front [PHYS] A surface of constant phase (or phase angle) of a propagating wave disturbance. 　{ 'fāz 'frənt }

phase function [OPTICS] The angular distribution of light reflected from an object when it is illuminated by light from a specified direction. 　{ 'fāz ‚fəŋk·shən }

phase generator [ELECTR] An instrument that accepts single-phase input signals over a given frequency range, or generates its own signal, and provides continuous shifting of the phase of this signal by one or more calibrated dials. 　{ 'fāz ‚jen·ə‚rād·ər }

phase inequality [OCEANOGR] Variations in the tide or tidal currents associated with changes in the phase of the moon. 　{ 'fāz ‚in·i'kwäl·əd·ē }

phase integra *See* action. 　{ 'fāz ¦int·ə‚grəl }

phase integral method *See* Wentzel-Kramers-Brillouin method. 　{ 'fāz ¦int·ə‚grəl ‚meth·əd }

phase inversion [ELECTR] Production of a phase difference of 180° between two similar wave shapes of the same frequency. 　{ 'fāz in‚vər·zhən }

phase inverter [ELECTR] A circuit or device that changes the phase of a signal by 180°, as required for feeding a push-pull amplifier stage without using a coupling transformer, or for changing the polarity of a pulse; a triode is commonly used as a phase inverter. Also known as inverter. 　{ 'fāz in‚vərd·ər }

phase jitter [ELECTR] Jitter that undesirably shortens or lengthens pulses intermittently during data processing or transmission. 　{ 'fāz jid·ər }

phase lag [OCEANOGR] Angular retardation of the maximum of a constituent of the observed tide behind the corresponding maximum of the same constituent of the hypothetical equilibrium tide. Also known as tidal epoch. [PHYS] *See* lag angle. 　{ 'fāz ‚lag }

phase lead *See* lead angle. 　{ 'fāz ‚lēd }

phase localizer [NAV] A localizer used by aircraft for landing guidance in which the on-course line is defined by the phase reversal of the emission from the side-band antenna system; phase is determined by comparison with a carrier signal which is transmitted for this purpose. 　{ 'fāz ‚lō·kə‚līz·ər }

phase lock [ELECTR] Technique of making the phase of an oscillator signal follow exactly the phase of a reference signal by comparing the phases between the two signals and using the resultant difference signal to adjust the frequency of the reference oscillator. 　{ 'fāz ‚läk }

phase-locked communication [COMMUN] Systems in which oscillators at the receiver and transmitter are locked in phase. 　{ 'fāz ¦läkt kə‚myü·nə'kā·shən }

phase-locked loop [ELECTR] A circuit that consists essentially of a phase detector which compares the frequency of a voltage-controlled oscillator with that of an incoming carrier signal or reference-frequency generator; the output of the phase detector, after passing a loop filter, is fed back to the voltage-controlled oscillator to keep it exactly in phase with the incoming or reference frequency. Abbreviated PLL. 　{ 'fāz ¦läkt ‚lüp }

phase-locked oscillator *See* parametron. 　{ 'fāz ¦läkt ‚äs·ə‚läd·ər }

phase-locked subharmonic oscillator *See* parametron. 　{ 'fāz ¦läkt ‚səb·här'män·ik 'äs·ə‚läd·ər }

phase-locked system [ENG] A radar system, having a stable local oscillator, in which information regarding the target is gained by measuring the phase shift of the echo. 　{ 'fāz ¦läkt ‚sis·təm }

phase magnet [COMMUN] Magnetically operated latch used to phase a facsimile transmitter or recorder. Also known as trip magnet. 　{ 'fāz ‚mag·nət }

phase margin [CONT SYS] The difference between 180° and the phase of the loop ratio of a stable system at the gain-crossover frequency. 　{ 'fāz ‚mär·jən }

phase matching [OPTICS] A condition in which the polarization wave produced by two or more beams of incident radiation in a nonlinear medium has the same phase velocity as a freely propagating wave of the same frequency; the amplitude of the polarization wave is then greatly enhanced. 　{ 'fāz ‚mach·iŋ }

phase meter [ENG] An instrument for the measurement of electrical phase angles. Also known as phase-angle meter. 　{ 'fāz ‚mēd·ər }

phase modifier [ELEC] Machine whose chief purpose is to supply leading or lagging reactive volt-amperes to the system to which it is connected; may be either synchronous or asynchronous. 　{ 'fāz ‚mäd·ə‚fī·ər }

phase modulation [COMMUN] Modulation in which the linearly increasing angle of a sine wave has added to it a phase

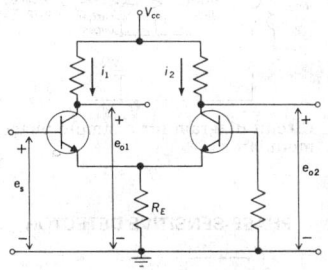

PHASE INVERTER

Circuit diagram for an emitter-coupled phase inverter.
V_{cc} = collector supply voltage, e_s = input signal voltage. When emitter resistance R_E is much larger than impedance seen looking into the emitter of each transistor, current i_1 equals i_2, and output voltages e_{O1} and e_{O2} are equal and opposite, and therefore 180° out of phase.

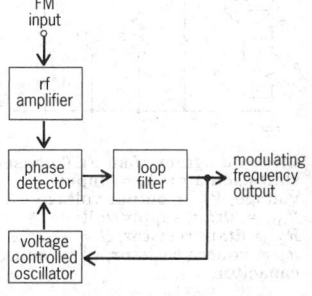

PHASE-LOCKED LOOP

Basic configuration of phase-locked loop.

PHASE MODULATOR

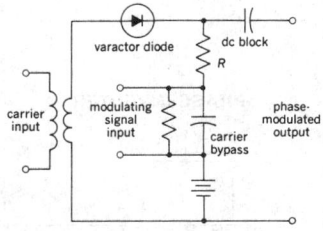

Circuit diagram for a simple phase modulator.

PHASE-SENSITIVE DETECTOR

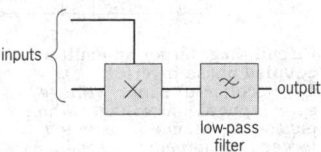

Diagram of a detector.

PHASE-SHIFT OSCILLATOR

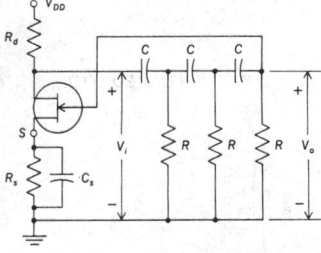

Circuit diagram of an FET phase-shift oscillator; V_i = input voltage, V_O = output voltage, V_{DD} = drain supply voltage, R_d = drain resistor, S = source, R_s = source resistor, C_s = source capacitor.

angle that is proportional to the instantaneous value of the modulating signal (message to be communicated). Abbreviated PM. { 'fāz ˌmäj·ə‚lā·shən }

phase-modulation detector [ELECTR] A device which recovers or detects the modulating signal from a phase-modulated carrier. { 'fāz ˌmäj·ə‚lā·shən di‚tek·tər }

phase-modulation transmitter [ELECTR] A radio transmitter used to broadcast a phase-modulated signal. { 'fāz ˌmäj·ə‚lā·shən tranz‚mid·ər }

phase modulator [ELECTR] An electronic circuit that causes the phase angle of a modulated wave to vary (with respect to an unmodulated carrier) in accordance with a modulating signal. { 'fāz ˌmäj·ə‚lād·ər }

phase plane analysis [CONT SYS] A method of analyzing systems in which one plots the time derivative of the system's position (or some other quantity characterizing the system) as a function of position for various values of initial conditions. { 'fāz ‚plān ə'nal·ə‚səs }

phase plate [OPTICS] In a polarizing microscope, a plate of doubly refracting material that changes the relative phase of the polarized light's components. { 'fāz ‚plāt }

phase portrait [CONT SYS] A graph showing the time derivative of a system's position (or some other quantity characterizing the system) as a function of position for various values of initial conditions. { 'fāz ‚pȯr·trət }

phase problem [CRYSTAL] The problem that arises in determining the electron density function of a crystal from x-ray diffraction data, namely that a complete determination requires knowledge of both the magnitudes and phases of the structure factors, but experimental measurements yield only the magnitudes. { 'fāz ‚präb·ləm }

phase quadrature See quadrature. { 'fāz ‚kwäd·rə‚chər }

phaser [COMMUN] Facsimile device for adjusting equipment so the recorded elemental area bears the same relation to the record sheet as the corresponding transmitted elemental area bears to the subject copy in the direction of the scanning line. [ELECTROMAG] Microwave ferrite phase shifter employing a longitudinal magnetic field along one or more rods of ferrite in a waveguide. { 'fāz·ər }

phase ratio [ANALY CHEM] In chromatography, the ratio of the volume of the mobile phase to that of the stationary phase in a chromatographic column. { 'fāz ‚rā·shō }

phase resonance [PHYS] The frequency at which the angular phase difference between the fundamental components of an oscillation and of the applied agency is 90° ($\pi/2$ radians). Also known as velocity resonance. { 'fāz ‚rez·ən·əns }

phase response [ELECTR] A graph of the phase shift of a network as a function of frequency. { 'fāz ri‚späns }

phase reversal [PHYS] A change of 180°, or one half-cycle, in phase. { 'fāz ri‚vər·səl }

phase reversal modulation [COMMUN] Form of pulse modulation in which reversal of signal phase serves to distinguish between the two binary states used in data transmission. { 'fāz ri‚vər·səl ˌmäj·ə'lā·shən }

phase-rotation relay See phase-sequence relay. { 'fāz rō‚tā·shən 'rē‚lā }

phase rule See Gibbs phase rule. { 'fāz rül }

phase-sensitive detector [ELECTR] An electronic circuit that consists essentially of a multiplier and a low-pass circuit and that produces a direct-current output signal that is proportional to the product of the amplitudes of two alternating-current input signals of the same frequency and to the cosine of the phase between them. { 'fāz ‚sen·səd·iv di‚tek·tər }

phase-sequence relay [ELEC] Relay which functions according to the order in which the phase voltages successively reach their maximum positive values. Also known as phase-rotation relay. { 'fāz ‚sē·kwəns 'rē‚lā }

phase shift [ELECTR] The phase angle between the input and output signals of a network or system. [PHYS] **1.** A change in the phase of a periodic quantity. Also known as phase change. **2.** A change in the phase angle between two periodic quantities. [QUANT MECH] For a partial wave of a particle scattered by a spherically symmetric potential, the phase shift is the difference between the phase of the wave function far from the scatterer and the corresponding phase of a free particle. { 'fāz ‚shift }

phase-shift circuit [ELECTR] A network that provides a voltage component which is shifted in phase with respect to a reference voltage. { 'fāz ‚shift ‚sər·kət }

phase-shift control See phase control. { 'fāz ‚shift kən‚trōl }

phase-shift discriminator [ELECTR] A discriminator that uses two similarly connected diodes, fed by a transformer that is tuned to the center frequency; when the frequency-modulated or phase-modulated input signal swings away from this center frequency, one diode receives a stronger signal than the other; the net output of the diodes is then proportional to the frequency displacement. Also known as Foster-Seeley discriminator. { 'fāz ‚shift di‚skrim·ə‚nād·ər }

phase shifter [ELEC] A device used to change the phase relation between two alternating-current values. { 'fāz ‚shif·tər }

phase-shifting transformer [ELEC] A transformer which produces a difference in phase angle between two circuits. { 'fāz ‚shif·tiŋ tranz‚fȯr·mər }

phase-shift keying [COMMUN] A form of phase modulation in which the modulating function shifts the instantaneous phase of the modulated wave between predetermined discrete values. Abbreviated PSK. { 'fāz ‚shift ‚kē·iŋ }

phase-shift mutation See frameshift mutation. { 'fāz ‚shift myü‚tā·shən }

phase-shift oscillator [ELECTR] An oscillator in which a network having a phase shift of 180° per stage is connected between the output and the input of an amplifier. { 'fāz ‚shift ‚äs·ə‚lād·ər }

phase solubility [PHYS CHEM] The different solubilities of a sample's solid constituents (phases) in a selected solvent. { 'fāz ‚säl·yə‚bil·əd·ē }

phase-solubility analysis [ANALY CHEM] Solvent technique used to determine the amount and number of components in a solid substance; the weight of sample added to the solvent is plotted against the weight of sample dissolved, with breakpoints in the curve occurring with each progressive saturation of the solvent with respect to each of the components; can be combined with extraction and recrystallization procedures. { 'fāz ‚säl·yə‚bil·əd·ē ə‚nal·ə·səs }

phase space [MATH] In a dynamical system or transformation group, the topological space whose points are being moved about by the given transformations. [STAT MECH] For a system with n degrees of freedom, a euclidean space with $2n$ dimensions, one dimension for each of the generalized coordinates and one for each of the corresponding momenta. { 'fāz ‚spās }

phase speed See phase velocity. { 'fāz ‚spēd }

phase splitter [ELEC] A circuit that takes a single input alternating voltage and produces two or more output alternating voltages that differ in phase from one another. { 'fāz ‚splid·ər }

phase stability [NUCLEO] A principle governing the stability of motion of particles in a synchrotron; the charged particle must be accelerated in each cycle at a time slightly earlier than the peak value of the accelerating potential. { 'fāz stə‚bil·əd·ē }

phase titration [ANALY CHEM] Analysis of a binary mixture of miscible liquids by titrating with a third liquid that is miscible with only one of the components, using the ternary phase diagram to determine the end point. { 'fāz tī‚trā·shən }

phase transfer catalysis [ORG CHEM] Enhancement of the reaction rate of a two-phase organic-water system by addition of a catalyst which alters the rate of transfer of water-soluble reactant across the interface to the organic phase. { 'fāz ‚trans‚fər kə'tal·ə·səs }

phase transformation [ELEC] A change of polyphase power from three-phase to six-phase, from three-phase to twelve-phase, and so forth, by use of transformers. [PHYS] See phase transition. { 'fāz ‚tranz·fər‚mā·shən }

phase transformer [ELEC] A transformer for changing a two-phase current to a three-phase current, or vice versa. { 'fāz tranz‚fȯr·mər }

phase transition [PHYS] A change of a substance from one phase to another. Also known as phase transformation. { 'fāz tran‚zish·ən }

phase undervoltage relay [ELEC] Relay which functions by reason of the reduction of one phase voltage in a polyphase circuit. { 'fāz ‚ən·dər‚vōl·tij 'rē‚lā }

phase velocity [PHYS] The velocity of a point that moves with a wave at constant phase. Also known as celerity; phase speed; wave celerity; wave speed; wave velocity. { 'fāz və‚läs·əd·ē }

phase winding [ELEC] One of the individual windings on the armature of a polyphase motor or generator. { 'fāz ˌwīnd·iŋ }

Phasianidae [VERT ZOO] A family of game birds in the order Galliformes; typically, members are ground feeders, have bare tarsi and copious plumage, and lack feathers around the nostrils. { ˌfāz·ē'an·ə,dē }

phasing See framing. { 'fāz·iŋ }

phasing line [ELECTR] That portion of the length of scanning line set aside for the phasing signal in a television or facsimile system. { 'fāz·iŋ ˌlīn }

phasing signal [ELECTR] A signal used to adjust the picture position along the scanning line in a facsimile system. { 'fāz·iŋ ˌsig·nəl }

phasitron [ELECTR] An electron tube used to frequency-modulate a radio-frequency carrier; internal electrodes are designed to produce a rotating disk-shaped corrugated sheet of electrons; audio input is applied to a coil surrounding the glass envelope of the tube, to produce a varying axial magnetic field that gives the desired phase or frequency modulation of the rf carrier input to the tube. { 'fāz·ə,trän }

phasmajector See monoscope. { 'faz·mə,jek·tər }

phasmid [INV ZOO] One of a pair of lateral caudal pores which function as chemoreceptors in certain nematodes. { 'faz·məd }

Phasmidae [INV ZOO] A family of the insect order Orthoptera including the walking sticks and leaf insects. { 'faz·mə,dē }

Phasmidea [INV ZOO] An equivalent name for Secernentea. { faz'mid·ē·ə }

Phasmidia [INV ZOO] An equivalent name for Secernentea. { faz'mid·ē·ə }

phasor [PHYS] **1.** A rotating line used to represent a sinusoidally varying quantity; the length of the line represents the magnitude of the quantity, and its angle with the x-axis at any instant represents the phase. **2.** Any quantity (such as impedance or admittance) which is a complex number. [SOLID STATE] A low-energy collective excitation of the conduction electrons in a metal, corresponding to a slowly varying phase modulation of a charge-density wave. { 'fāz·ər }

phasotron See cyclotron. { 'fāz·ə,trän }

pheasant [VERT ZOO] Any of various large sedentary game birds with long tails in the family Phasianidae; sexual dimorphism is typical of the group. { 'fez·ənt }

pH electrode [ANALY CHEM] Membrane-type glass electrode used as the hydrogen-ion sensor of most pH meters; the pH-response electrode surface is a thin membrane made of a special glass. { ˌpē'āch i'lek,trōd }

α-phellandrene [ORG CHEM] $C_{10}H_{16}$ A colorless oil soluble in ether; boiling point of d-optical isomer is 66–68°C, of l-optical isomer is 58–59°C; used in flavoring and perfumes. { ¦al·fə fə'lan,drēn }

phellem [BOT] Cork; the outer tissue layer of the periderm. { 'fel·əm }

phelloderm [BOT] Layers of parenchymatous cells formed as inward derivatives of the phellogen. { 'fel·ə,dərm }

phellogen [BOT] The meristematic portion of the periderm, consisting of one layer of cells that initiate formation of the cork and secondary cortex tissue. { 'fel·ə·jən }

phenacaine hydrochloride [PHARM] White crystals with a slight bitter taste and a melting point of 190°C; soluble in boiling water, chloroform, and alcohol; used in medicine. { 'fen·ə,kān ,hī·drə'klór,īd }

phenacemide [PHARM] $C_6H_5CH_2CONHCONH_2$ A white to cream-colored, crystalline solid with a melting point of 212–216°C; soluble in chloroform, benzene, ether, and alcohol; used in medicine. { fə'nas·ə·məd }

phenacite See phenakite. { 'fen·ə,sīt }

Phenacodontidae [PALEON] An extinct family of large herbivorous mammals in the order Condylarthra. { fə,nak·ə'dänt·ə,dē }

phenakite [MINERAL] Be_2SiO_4 A colorless, white, wine-yellow, pink, blue, or brown glassy mineral that crystallizes in the rhombohedral system; used as a minor gemstone. Also spelled phenacite. { 'fen·ə,kīt }

phenanthrene [ORG CHEM] $C_{14}H_{10}$ A colorless, crystalline hydrocarbon; melts at about 100°C; the nucleus is produced by the degradation of certain alkaloids; used in the synthesis of dyes and drugs. { fə'nan,thrēn }

phenanthroline [ORG CHEM] $C_{12}H_8N_2$ Any of three nitrogen bases related to phenanthrene; the ortho form is an oxidation-reduction indicator, turning faint blue when oxidized. { fə'nan·thrə,lēn }

phenanthroline indicator [ANALY CHEM] A sensitive, red-colored specific reagent for iron. { fə'nan·thrə,lēn 'in·də,kād·ər }

phenarsazine chloride [ORG CHEM] $C_{12}H_9AsClN$ A yellow, crystalline compound obtained as a precipitate from carbon tetrachloride solutions; it sublimes readily, and is slightly soluble in xylene, benzene, and carbon tetrachloride; used as a war gas. Also known as adamsite. { fə'när·sə,zēn 'klór,īd }

phenazine [ORG CHEM] $C_6H_4N_2C_6H_4$ Yellow crystals, melting at 170°C; slightly soluble in water, soluble in alcohol and ether; used as chemical intermediate and to make dyes. { 'fen·ə,zēn }

phencyclidine [PHARM] $C_{17}H_{25}N$ An addicting drug originally used as an intravenous anesthetic that was subsequently removed from medical use in humans because it produces hallucinations and delusions. It has been illegally manufactured and sold on the street. sometimes causing serious adverse reactions. Also known as angel dust; PCP. { ,fen·ə'sī·klə,dēn }

phenethyl acetate [ORG CHEM] $C_6H_5CH_2CH_2OOCCH_3$ A colorless liquid with a peachlike odor and a boiling point of 226°C; soluble in alcohol, ether, and fixed oils; used in perfumes. Also known as phenylethyl acetate. { fen'eth·əl 'as·ə,tāt }

phenethyl alcohol [ORG CHEM] $C_8H_{10}O$ A liquid with a floral odor found in many natural essential oils; soluble in 50% alcohol; used in perfumes and flavors, and in medicine as an antibacterial agent in diseases of the eye. { fen'eth·əl 'al·kə,hól }

phenethyl isobutyrate [ORG CHEM] $(CH_3)_2CHCOO-C_2H_4C_6H_5$ A colorless liquid, soluble in alcohol and ether; used in perfumes and flavoring. { fen'eth·əl ,ī·sō'byüd·ə,rāt }

phenetidine [ORG CHEM] $NH_2C_6H_4OC_2H_5$ Either of two toxic, oily liquids that darken when exposed to light and air; soluble in alcohol, insoluble in water; the ortho form boils at 228–230°C, is used to make dyes, and is also known at 2-aminophenetole; the para form boils at 253–255°C, is used to make dyes and in pharmaceuticals. { fə'ned·ə,dēn }

Phengodidae [INV ZOO] The fire beetles, a New World family of coleopteran insects in the superfamily Cantharoidea. { fen'gäd·ə,dē }

phengophobia [PSYCH] An abnormal fear of daylight. { ,feŋ·gə'fō·bē·ə }

phenicochroite See phoenicochroite. { ,fen·ə'käk·rə,wīt }

phenindione [PHARM] $C_{15}H_{10}O_2$ Pale yellow crystals, soluble in alcohol, ether, acetone, and benzene; used as an anticoagulant. { ,fen·ən'dī,ōn }

pheniramine maleate [PHARM] $C_{16}H_{20}N_2·C_4H_4O_4$ A white, crystalline powder with a melting point of 104–108°C; soluble in alcohol and water; used as an antihistamine. { fə'nir·ə,mēn 'mal·ē,āt }

phenobarbital [PHARM] $C_{12}H_{12}N_2O_3$ A crystalline compound, 5-ethyl-5-phenylbarbituric acid, with a slightly bitter taste, melting at 174–178°C; soluble in water, alcohol, chloroform, and ether; used in medicine as a long-acting sedative, anticonvulsant, and hypnotic. { ¦fē·nō'bär·bə,tól }

phenoclastic rock [PETR] A nonuniformly sized clastic rock containing phenoclasts. { ¦fen·ə¦klas·tik 'räk }

phenoclasts [PETR] The larger, conspicuous fragments in a sediment or sedimentary rock, such as cobbles in a conglomerate. { 'fen·ə,klasts }

phenocoll hydrochloride [PHARM] $C_2H_5OC_6H_4NHCO-CH_2NH_2·HCl$ A white, crystalline powder with a melting point of 95°C; soluble in warm alcohol and in water; used in medicine. { 'fen·ə,käl ,hī·drə'klór,īd }

phenocopy [GEN] The nonhereditary alteration of a phenotype to a form imitating a mutant trait; caused by external conditions during development. { 'fen·ə¦käp·ē }

phenocritical period [EMBRYO] During development, that period during which a gene's effect can most readily be influenced by external factors. { ¦fē·nə,krid·ə·kəl 'pir·ē·əd }

phenocryst [PETR] A large, conspicuous crystal in a porphyritic rock. Also known as phanerocryst. { 'fen·ə,krist }

phenocrystalline See phaneritic. { ¦fen·ə'krist·əl·ən }

phenogenetics [GEN] The study of the phenotypic effects

PHASMIDAE

Walking stick (*Diapheromera femorata*).

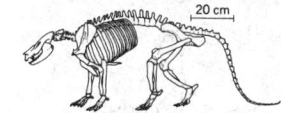

PHENACODONTIDAE

Skeleton of *Ectoconus*, an early Paleocene phenacodont condylarth.

PHENOL

Structural formula for phenol.

of the genetic material. Also known as physiological genetics. { ¦fēn·ə·jə'ned·iks }

phenol [ORG CHEM] **1.** C_6H_5OH White, poisonous, corrosive crystals with sharp, burning taste; melts at 43°C, boils at 182°C; soluble in alcohol, water, ether, carbon disulfide, and other solvents; used to make resins and weed killers, and as a solvent and chemical intermediate. Also known as carbolic acid; phenylic acid. **2.** A chemical compound based on the substitution product of phenol, for example, ethylphenol ($C_2H_4C_4H_5OH$), the ethyl substitute of phenol. { 'fē,nól }

phenolate process [CHEM ENG] A process which employs sodium phenolate to remove hydrogen sulfide from gas. { 'fēn·əl,āt ,prä·səs }

phenol coefficient [ANALY CHEM] Number scale for comparison of antiseptics, using the efficacy of phenol as unity. { 'fē,nól ,kō·i,fish·ənt }

phenol-coefficient method [CHEM] A method for evaluating water-miscible disinfectants in which a test organism is added to a series of dilutions of the disinfectant; the phenol coefficient is the number obtained by dividing the greatest dilution of the disinfectant killing the test organism by the greatest dilution of phenol showing the same result. { 'fē,nól ,kō·i,fish·ənt ,meth·əd }

phenol extraction [CHEM ENG] Petroleum-refinery solvent-extraction process using phenol as the solvent to remove aromatic, unsaturated and naphthenic constituents from lubricating-oil stocks. { 'fē,nól ik,strak·shən }

phenol-formaldehyde resin [ORG CHEM] Thermosetting resin made by the reaction of phenol and formaldehyde; has good strength and chemical resistance and low cost; used as a molding material for mechanical and electrical parts. Originally known as Bakelite. { 'fē,nól fər'mal·də,hīd ,rez·ən }

phenol-furfural resin [ORG CHEM] A phenolic resin characterized by the ability to be fabricated by injection molding since it hardens after curing conditions are reached. { 'fē,nól 'fər·fə,ral ,rez·ən }

phenolic laminate [MATER] Canvas, linen, kraft paper, glass fiber, or other substrate impregnated with 30% or more of thermosetting phenolic resin and cured; used for structural, mechanical, and electrical purposes. { fə'näl·ik 'lam·ə·nət }

phenolic plastic [MATER] A thermosetting plastic material available in many combinations of phenol and formaldehyde, often with added fillers to provide a broad range of physical, electrical, chemical, and molding properties. { fə'näl·ik 'plas·tik }

phenological shift [ECOL] A change in the timing of growth and breeding events in the life of an individual organism. { ,fēn·ə,läj·i·kəl 'shift }

phenology [CLIMATOL] The science which treats of periodic biological phenomena with relation to climate, especially seasonal changes; from a climatologic viewpoint, these phenomena serve as bases for the interpretation of local seasons and the climatic zones, and are considered to integrate the effects of a number of bioclimatic factors. { fə'näl·ə·jē }

phenolphthalein [ORG CHEM] $(C_6H_4OH)_2COC_6H_4CO$ Pale-yellow crystals; soluble in alcohol, ether, and alkalies, insoluble in water; used as an acid-base indicator (carmine-colored to alkalies, colorless to acids) for titrations, as a laxative and dye, and in medicine. { ¦fē,nól'thal·ē·ən }

phenol process [CHEM ENG] A single-solvent petroleum-refining process in which phenol is the selective solvent. { 'fē,nól ,prä·səs }

phenol red See phenolsulfonphthalein. { 'fē,nól 'red }

phenolsulfonic acid [ORG CHEM] $C_6H_5SO_3H$ Water- and alcohol-soluble mixture of ortho- and para-phenolsulfonic acids; yellowish liquid that turns brown when exposed to air; used as a chemical intermediate and in water analysis. Also known as sulfocarbolic acid. { ¦fē,nól·səl'fän·ik 'as·əd }

phenolsulfonphthalein [ORG CHEM] $C_{19}H_{14}O_5S$ A bright-red, crystalline compound, soluble in water, alcohol, and acetone; used as a pH indicator, and to test for kidney function in dogs. Also known as phenol red. { ¦fē,nól¦səl,fón·'thal·ē·ən }

phenomenal gem [LAP] A gemstone exhibiting an optical phenomenon, such as asterism, chatoyancy, or play of color. { fə'näm·ən·əl 'jem }

phenoplast [PETR] A large rock fragment in a rudaceous rock that was plastic at the time of its incorporation into the matrix. { 'fē·nə,plast }

PHENYLALANINE

Structural formula of phenylalanine.

phenothiazine [ORG CHEM] $C_{12}H_9N$ A yellow, crystalline compound, forming rhomboid leaflets or diamond-shaped plates, obtained from toluene or butanol solution; soluble in hot acetic acid, benzene, and ether; used as an insecticide and in pharmaceutical manufacture. { 'fē·nə'thī·ə,zēn }

phenotole [ORG CHEM] $C_6H_5OC_2H_5$ Combustible, colorless liquid, boiling at 172°C; soluble in alcohol and ether, insoluble in water. { 'fē·nə,tōl }

phenotype [GEN] The observable characters of an organism, dependent upon genotype and environment. { 'fē·nə,tīp }

phenotypic lag [GEN] Delay in the expression of a newly acquired character. { ¦fē·nə¦tip·ik 'lag }

phenotypic masking [MICROBIO] Masking of the phenotype in strains of bacteria that are drug-dependent. { ¦fē·nə,tip·ik 'mask·iŋ }

phenotypic mixing [VIROL] The production of virus particles having different structural components in the protein coats, each synthesized under the direction of a different genome. { ¦fē·nə,tip·ik 'mik·siŋ }

phenotypic plasticity [GEN] The range of genotype expression in different environments. { ¦fē·nə,tip·ik plas'tis·əd·ē }

phenotypic sex determination [BIOL] Control of the development of gonads by environmental stimuli, such as temperature. { ¦fē·nə,tip·ik 'seks di,tər·mə,nā·shən }

phenotypic suppression [BIOL] Prevention of mutant phenotype expression. { ¦fē·nə,tip·ik sə'presh·ən }

phenoxyacetic acid [ORG CHEM] $C_6H_5OCH_2COOH$ A light tan powder with a melting point of 98°C; soluble in ether, water, carbon disulfide, methanol, and glacial acetic acid; used in the manufacture of pharmaceuticals, pesticides, fungicides, and dyes. { fə¦näk·sē·ə¦sēd·ik 'as·əd }

phenoxybenzamine hydrochloride [ORG CHEM] $C_{18}H_{22}$-ONCl·HCl White crystals, slightly soluble in water, melting at 139°C; used in medicine. { fə¦näk·sē'ben·zə,mēn ,hī-drə'klór,īd }

2-phenoxyethanol [ORG CHEM] $C_6H_5OCH_2CH_2OH$ An oily liquid with a faint aromatic odor; melting point is 14°C; soluble in water; used in perfumes as a fixative, in organic synthesis, as an insect repellent, and as a topical anesthetic. { ¦tü fə¦näk·sē'eth·ə,nól }

phenoxymethylpenicillin [PHARM] $C_{16}H_{18}N_2O_5S$ A white, crystalline powder, soluble in alcohol and acetone; used as an oral antibiotic. Also known as penicillin V. { fə¦näk·sē,meth·əl,pen·ə'sil·ən }

phenoxypropanediol [ORG CHEM] $C_9H_{12}O_3$ A white, crystalline solid with a melting point of 53°C; soluble in water, alcohol, glycerin, and carbon tetrachloride; used in medicine and as a plasticizer. { fə¦näk·sē,prō·pān'dī,ól }

phenoxy resin [ORG CHEM] A high-molecular-weight thermoplastic polyether resin based on bisphenol-A and epichlorohydrin with bisphenol-A terminal groups; used for injection molding, extrusion, coatings, and adhesives. { fə¦näk·sē 'rez·ən }

phentolamine hydrochloride [ORG CHEM] $C_{17}H_{19}ON_2$·HCl White, water-soluble crystals, melting at 240°C; a sympatholytic; used in medicine. { fen'täl·ə,mēn ,hī·drə'klór,īd }

phenyl [ORG CHEM] C_6H_5- A functional group consisting of a benzene ring from which a hydrogen has been removed. { 'fen·əl }

phenylacetaldehyde [ORG CHEM] C_8H_8O A colorless liquid with a boiling point of 193–194°C; soluble in ether and fixed oils; used in perfumes and flavoring. Also known as α-toluic aldehyde. { 'fen·əl,as·ə'täl·də,hīd }

phenylacetic acid [ORG CHEM] $C_8H_8O_2$ White crystals with a boiling point of 262°C; soluble in alcohol and ether; used in perfumes, medicine, and flavoring and in the manufacture of penicillin. Also known as α-toluic acid. { ¦fen·əl·ə'sēd·ik 'as·əd }

phenylalanine [BIOCHEM] $C_9H_{11}O_2N$ An essential amino acid, obtained in the levo form by hydrolysis of proteins (as lactalbumin); converted to tyrosine in the normal body. Also known as α-aminohydrocinnamic acid; α-amino-β-phenyl-propionic acid; β-phenylalanine. { 'fen·əl'al·ə,nēn }

β-phenylalanine See phenylalanine. { 'bäd·ə ¦fen·əl'al·ə,nēn }

phenylaniline See diphenylamine. { ¦fen·əl'an·ə·lən }

N-phenylanthranilic acid [ORG CHEM] $(C_6H_5NH)C_6H_4$-COOH A crystalline compound, soluble in hot alcohol;

decomposes at 183–184°C; used to detect vanadium in steel. { ¦en ¦fen·əl‚an·thrə¦nil·ik 'as·əd }

phenylbenzene See biphenyl. { ¦fen·əl'ben‚zēn }

phenylbutazone [ORG CHEM] $C_{19}H_{20}O_2N_2$ White or light-yellow powder with aromatic aroma and bitter taste; melts at 107°C; slightly soluble in water, soluble in acetone; used in medicine as an analgesic and antipyretic. Also known as butazolidine. { ¦fen·əl'byüd·ə‚zōn }

phenyl cyanide See benzonitrile. { ¦fen·əl 'sī·ə‚nīd }

phenylcyclohexane [ORG CHEM] $C_{12}H_{16}$ A colorless, oily liquid with a boiling point of 237.5°C; soluble in alcohol, benzene, castor oil, carbon tetrachloride, xylene, and hexane; used as a high-boiling solvent and a penetrating agent. { ¦fen·əl‚sī·klō'hek‚sān }

phenyldichloroarsine [ORG CHEM] $C_6H_5AsCl_2$ A liquid which becomes a microcrystalline mass at −20°C (melting point) and decomposes in water; soluble in alcohol, ether, and benzene; used as a poison gas. { ¦fen·əl·dī‚klór·ō'är‚sēn }

phenyl diglycol carbonate [ORG CHEM] $C_{18}H_{18}O_7$ A colorless solid with a melting point of 40°C; soluble in organic solvents; used as a plasticizer. { ¦fen·əl dī¦glī‚kól 'kär·bə‚nāt }

phenylene blue See indamine. { ¦fən·əl‚ēn 'blü }

phenylenediamine [ORG CHEM] $C_6H_4(NH_2)_2$ Also known as diaminobenzene. Any of three toxic isomeric crystalline compounds that are diamino derivatives of benzene; the ortho form, toxic colorless crystals melting at 102–104°C and soluble in alcohol, ether, water, and chloroform, is used to manufacture dyes, in photographic developers, and as a chemical intermediate; the meta form, colorless crystals unstable in air, melting at 63°C, and soluble in alcohol, ether, and water, is used to manufacture dyes, in textile dyeing, and as a nitrous acid detector; the para form, white to purple crystals melting at 147°C, soluble in alcohol and ether, and irritating to the skin, is used to manufacture dyes, in chemical analysis, and in photographic developers. { ¦fen·əl‚ēn'dī·ə‚mēn }

phenylephrine [PHARM] $C_9H_{13}NO_2$ A sympathomimetic amine, used in its hydrochloride salt form as a vasoconstrictor. { ¦fen·əl'ef·rən }

phenyl ether See diphenyl oxide. { ¦fen·əl 'eth·ər }

phenylethyl acetate See phenethyl acetate. { ¦fen·əl¦eth·əl 'as·ə‚tāt }

phenylethylene See styrene. { ¦fen·əl¦eth·ə‚lēn }

phenyl fluoride See fluorobenzene. { ¦fen·əl 'flür‚īd }

N-phenylglycine [ORG CHEM] $C_6H_5NHCH_2COOH$ A crystalline compound, moderately soluble in water, melting at 127–128°C; used in dye manufacture (indigo). { ¦en ¦fen·əl'glī‚sēn }

phenylglyoxylonitriloxime O,O-diethyl phosphorothioate [ORG CHEM] $(H_5C_2O)_2PSONCCNC_6H_5$ A yellow liquid with a boiling point of 102°C at 0.01 mmHg (1.333 pascals); solubility in water is 7 parts per million at 20°C; used as an insecticide for stored products. Also known as phoxim. { ¦fen·əl·glī¦äk·sē¦län·ə·trəl'äk‚sēm ¦ō¦ō dī'eth·əl ‚fäs·fə·rō'thī·ə‚wāt }

phenylhydrazine [ORG CHEM] $C_6H_5NHNH_2$ Poisonous, oily liquid, boiling at 244°C; soluble in alcohol, ether, chloroform, and benzene, slightly soluble in water; used in analytical chemistry to detect sugars and aldehydes, and as a chemical intermediate. Also known as hydrazinobenzene. { ¦fen·əl'hī·drə‚zēn }

phenyl ketone See benzophenone. { ¦fen·əl 'kē‚tōn }

phenylketonuria [MED] A hereditary disorder of metabolism, transmitted as an autosomal recessive, in which there is a lack of the enzyme phenylalanine hydroxylase, resulting in excess amounts of phenylalanine in the blood and of excess phenylpyruvic and other acids in the urine. Abbreviated PKU. Also known as phenylpyruvic oligophrenia. { ¦fen·əl‚kēd·ə'nyur·ē·ə }

phenyl mercaptan See thiophenol. { ¦fen·əl mər'kap‚tan }

phenylmercuric acetate [ORG CHEM] $C_8H_8O_2Hg$ White to cream-colored prisms with a melting point of 148–150°C; soluble in alcohol, benzene, and glacial acetic acid; used as an antiseptic, fungicide, herbicide, and mildewcide. { ¦fen·əl·mər¦kyúr·ik 'as·ə‚tāt }

phenylmercuric chloride [ORG CHEM] C_6H_5HgCl White crystals with a melting point of 251°C; soluble in benzene and ether; used as an antiseptic and fungicide. { ¦fen·əl·mər¦kyúr·ik 'klór‚īd }

phenylmercuric hydroxide [ORG CHEM] C_6H_5HgOH

White to cream-colored crystals with a melting point of 197–205°C; soluble in acetic acid and alcohol; used as a fungicide, germicide, and alcohol denaturant. { ¦fen·əl·mər¦kyúr·ik hi'dräk‚sīd }

phenylmercuric oleate [ORG CHEM] $C_{41}H_{21}O_2Hg$ A white, crystalline powder with a melting point of 45°C; soluble in organic solvents; used in paints as a mildew-proofing agent, and as a fungicide. { ¦fen·əl·mər¦kyúr·ik 'ō·lē‚āt }

phenylmercuric propionate [ORG CHEM] $C_9H_{10}O_2Hg$ A white, waxlike powder with a melting point of 65–70°C; used in paints as a fungicide and bactericide. { ¦fen·əl·mər¦kyúr·ik 'prō·pē·ə‚nāt }

phenylmercuriethanolammonium acetate [ORG CHEM] $C_{10}H_{15}O_3NHg$ A white, water-soluble, crystalline solid; used as an insecticide and fungicide. { ¦fen·əl·mər¦kyúr·ik¦eth·ə‚nól·ə mō·nē·əm 'as·ə‚tāt }

phenylmethane See toluene. { ¦fen·əl'meth‚ān }

phenylmethanol See benzyl alcohol. { ¦fen·əl'meth·ə‚nól }

phenylmethyl acetate See benzyl acetate. { ¦fen·əl¦meth·əl 'as·ə‚tāt }

N-phenylmorpholine [ORG CHEM] $C_{10}H_{13}NO$ A white, water-soluble solid with a melting point of 57°C; used in the manufacture of dyestuffs, corrosion inhibitors, and photographic developers, and as an insecticide. { ¦en ¦fen·əl'mór·fə‚lēn }

phenyl mustard oil [ORG CHEM] C_6H_5NCS A pale yellow or colorless liquid with a boiling point of 221°C; soluble in alcohol and ether; used in medicine. { ¦fen·əl 'məs·tərd ‚óil }

phenylphenol [ORG CHEM] $C_6H_5C_6H_4OH$ Almost white crystals, soluble in alcohol, insoluble in water; the ortho form, melting at 56–58°C, is used to manufacture dyes, as germicide and fungicide, and in the rubber industry, and is also known as 2-hydroxybiphenyl, ortho-xenol; the para form, melting at 164–165°C, is used to manufacture dyes, resins, and rubber chemicals, and as a fungicide. { ¦fen·əl'fē‚nól }

N-phenylpiperazine [ORG CHEM] $C_{10}H_{14}N_2$ A pale yellow oil with a boiling point of 286.5°C; soluble in alcohol and ether; used for pharmaceuticals and in the manufacture of synthetic fibers. { ¦en ¦fen·əl·pī¦per·ə‚zēn }

phenylpropane See propyl benzene. { ¦fen·əl'prō‚pān }

1-phenyl-1-propanol [ORG CHEM] $C_6H_5CH(OH)CH_2CH_3$ An oily liquid that has a weak esterlike odor; miscible with methanol, ethanol, ether, benzene, and toluene; used in industry as a heat transfer medium, in the manufacture of perfumes, and as a choleretic in medicine. { ¦wən ¦fen·əl ¦wən 'prō·pə‚nól }

phenylpropanolamine hydrochloride [PHARM] C_9H_{13}ON·HCl A white, crystalline powder with a melting point of 198–199°C; soluble in alcohol and water; used in medicine. { ¦fen·əl‚prō·pə'näl·ə‚mēn ‚hī·drə'klór‚īd }

phenylpropyl alcohol [ORG CHEM] $C_9H_{12}O$ A colorless liquid with a floral odor and a boiling point of 219°C; soluble in 70% alcohol; used in perfumes and flavoring. Also known as hydrocinnamic alcohol. { ¦fen·əl¦prō·pəl 'al·kə‚hól }

phenylpropyl aldehyde [ORG CHEM] $C_9H_{10}O$ A colorless liquid with a floral odor; soluble in 50% alcohol; used in perfumes and flavoring. Also known as hydrocinnamic aldehyde. { ¦fen·əl¦prō·pəl 'al·də‚hīd }

1-phenyl-3-pyrazolidinone [ORG CHEM] $C_9H_{10}N_2O$ A crystalline compound soluble in dilute aqueous solutions of acids and alkalies; melting point is 121°C; used as a high-contrast photographic developer. { ¦wən ¦fen·əl ¦thrē ‚pī·rə·zə'lid·ən‚ōn }

phenylpyruvic acid [BIOCHEM] $C_6H_5CH_2·CO·COOH$ A keto acid, occurring as a metabolic product of phenylalanine. { ¦fen·əl·pī'rü·vik 'as·əd }

phenylpyruvic oligophrenia See phenylketonuria. { ¦fen·əl·pī'rü·vik ‚äl·ə·gə'frē·nē·ə }

phenyl salicylate See salol. { ¦fen·əl sə'lis·ə‚lāt }

phenylthiourea [ORG CHEM] $C_6H_5NHCSNH_2$ A crystalline compound that has either a bitter taste or is tasteless, depending on the heredity of the taster; used in human genetics studies. { ¦fen·əl¦thī·ō·yủ'rē·ə }

ortho-phenyl tolyl ketone [ORG CHEM] $CH_3C_6H_4COC_6H_5$ An oily liquid with a boiling point of 309–311°C; soluble in alcohol, oils, and organic solvents; used as a fixative in perfumery. { ¦ór·thō ¦fen·əl 'tä‚lil 'kē‚tōn }

phenyltrichlorosilane [ORG CHEM] $C_6H_5SiCl_3$ A liquid with a sharp odor [boiling point 201°C (410°F), melting point

−31°C (−23.8°F)] used in the manufacture of various silicone oligomers and polymers.　{ ˌfen·əlˌtrīˌklȯr·ōˈsīˌlan }

phenytoin [PHARM] $C_{15}H_{12}N_2O_2$ A powder with a melting point of 295–298°C; used as an anticonvulsant and antiepileptic in humans.　{ fəˈnid·ə·wən }

pheochromoblast [HISTOL] A precursor of a pheochromocyte.　{ ˌfē·ōˈkrō·məˌblast }

pheochromocytoma [MED] A tumor of the sympathetic nervous system composed principally of chromaffin cells; found most often in the adrenal medulla.　{ ˌfē·ōˌkrō·mōˈsī ˈtō·mə }

pheoplast [CYTOL] A plastid containing brown pigment and found in diatoms, dinoflagellates, and brown algae.　{ ˈfē· əˌplast }

pheromone [PHYSIO] Any substance secreted by an animal which influences the behavior of other individuals of the same species.　{ ˈfer·əˌmōn }

phialospore [MYCOL] One of a chain of spores produced successively on phialides.　{ ˈfī·ə·ləˌspȯr }

phi function See Euler's phi function.　{ fī or ˈfē ˌfəŋkˈshən }

phi grade scale [GEOL] A logarithmic transformation of the Wentworth grade scale in which the diameter value of the particle is replaced by the negative logarithm to the base 2 of the particle diameter (in millimeters).　{ ˈfī ˈgrād ˌskāl }

Philadelphia chromosome [PATH] An abnormally small G-group chromosome found in the hematopoietic cells of most patients with chronic granulocytic leukemia.　{ ˌfil·əˈdel·fyə ˈkrō·məˌskōp }

Philippine fowl disease See Newcastle disease.　{ ˈfil·əˌpēn ˈfau̇l diˌzēz }

Philips hot-air engine [MECH ENG] A compact hot-air engine that is a Philips Research Lab (Holland) design; it uses only one cylinder and piston, and operates at 3000 revolutions per minute, with hot-chamber temperature of 1200°F (650°C), maximum pressure of 50 atmospheres (5.07 megapascals), and mean effective pressure of 14 atmospheres (1.42 megapascals).　{ ˈfil·əps ˈhät ˌer ˌen·jən }

Philips ionization gage [ELECTR] An ionization gage in which a high voltage is applied between two electrodes, and a strong magnetic field deflects the resulting electron stream, increasing the length of the electron path and thus increasing the chance for ionizing collisions of electrons with gas molecules. Abbreviated pig. Also known as cold-cathode ionization gage; Penning gage.　{ ˈfil·əps ˌī·ə·nəˈzā·shən ˌgāj }

philipstadite [MINERAL] $Ca_2(Fe,Mg)_5(Si,Al)_8O_{22}(OH)_2$ Monoclinic mineral composed of basic silicate of calcium, iron, magnesium, and aluminum; member of the amphibole group.　{ ˈfil·əpˌstäˌdīt }

phillipsite [MINERAL] $(K_2,Na_2CA)Al_2Si_4O_{12}·H_2O$ A white or reddish zeolite mineral crystallizing in the orthorhombic system; occurs in complex fibrous crystals, which make up a large part of the red-clay sediments in the Pacific Ocean.　{ ˈfil·əpˌsīt }

Phillips relation [ASTRON] A correlation between the relatively small variation in the peak brightness of type Ia supernovae and their rates of decline, whose use enables the distances of type Ia supernovae to be estimated.　{ ˈfil·ips riˌlā·shən }

Phillips screw [DES ENG] A screw having in its head a recess in the shape of a cross; it is inserted or removed with a Phillips screwdriver that automatically centers itself in the screw.　{ ˈfil·əps ˌskrü }

Philomycidae [INV ZOO] A family of pulmonate gastropods composed of slugs.　{ ˌfil·əˈmīs·əˌdē }

philopatry [ECOL] A dispersal method in which reproductive particles remain near their point of origin. [PSYCHOL] The drive to stay on or near the site of birth.　{ ˌfī·əˈpa·trē }

Philopteridae [INV ZOO] A family of biting lice (Mallophaga) that are parasitic on most land birds and water birds.　{ ˌfil·əpˈter·əˌdē }

philosopher's wool See zinc oxide.　{ fəˈläs·ə·fərz ˈwu̇l }

-philous [SCI TECH] Suffix meaning having an affinity for.　{ fil·əs }

philtrum-otobasion inferior [ANTHRO] A measure of the distance from the midpoint of the philtrum to the otobasion inferior.　{ ˈfil·trəm ˌōd·ōˈbā·sē·ən inˈfir·ē·ər }

phi meson [PARTIC PHYS] A neutral vector meson resonance, having a mass of about 1019.4 MeV, a width of about 4.6 MeV, and negative charge parity and G parity.　{ ˈfī ˈmāˌsän }

phimosis [MED] Elongation of the prepuce and constriction of the orifice, so that the foreskin cannot be retracted to uncover the glans penis.　{ fəˈmō·səs }

phlebite [PETR] Roughly banded or veined metamorphite or migmatite.　{ ˈfleˌbīt }

phlebitis [MED] Inflammation of a vein.　{ fləˈbīd·əs }

phlebography [MED] 1. X-ray photography of a vein or veins following intravenous injection of a radiopaque substance. 2. Recording of venous pulsations.　{ fləˈbäg·rə·fē }

phlebolith [MED] A calculus in a vein.　{ ˈfleb·əˌlith }

phlebosclerosis [MED] 1. Sclerosis of a vein. 2. Chronic phlebitis.　{ ˌfle·bōˈsklə·rō·səs }

phlebotaxis [BIOL] Movement of a simple motile organism in response to the presence of blood.　{ ˌfle·bəˌtak·səs }

phlebothrombosis [MED] A venous thrombus not associated with inflammation of the vein.　{ ˌfle·bō·thrämˈbō·səs }

phlebotomus fever [MED] An acute viral infection, transmitted by the fly *Phlebotomus papatosii* and characterized by fever, pains in the head and eyes, inflammation of the conjunctiva, leukopenia, and general malaise. Also known as Chitral fever; pappataci fever; sandfly fever; three-day fever.　{ fləˈbäd·ə·məs ˌfēv·ər }

phlebotomy [MED] The withdrawal of blood from a vein.　{ fləˈbäd·ə·mē }

Phlebovirus [VIROL] A genus of the family Bunyaviridae that causes sandfly fever.　{ ˈflē·bəˌvī·rəs }

phleger corer [ENG] A device for obtaining ocean bottom cores up to about 4 feet (1.2 meters) in length; consists of an upper tube, main body weight, and tailfin assembly with a check valve that prevents the flow of water into the upper section and a consequent washing out of the core sample while hoisting the corer.　{ ˈflej·ər ˌkȯr·ər }

phlegm [PHYSIO] A viscid, stringy mucus, secreted by the mucosa of the air passages.　{ flem }

phlegmasia alba dolens [MED] A painful swelling of the leg usually seen postpartum, due to femoral vein thrombophlebitis or lymphatic obstruction. Also known as milk leg.　{ flegˈmä·zhə ˈal·bə ˈdō·lənz }

phleomycin [MICROBIO] An antibacterial antibiotic produced by *Streptomyces verticillatus;* antitumor activity has also been demonstrated.　{ ˈflē·əˌmīs·ən }

Phloeidae [INV ZOO] The bark bugs, a small neotropical family of hemipteran insects in the superfamily Pentatomoidea.　{ ˈflē·əˌdē }

phloem [BOT] A complex, food-conducting vascular tissue in higher plants; principal conducting cells are sieve elements. Also known as bast; sieve tissue.　{ ˈflō·əm }

phloem necrosis [PL PATH] A pathological state in a plant in which the phloem undergoes brown discoloration and disintegration.　{ ˈflō·əm nəˈkrō·səs }

phlogopite [MINERAL] $K_2[Mg,Fe(II)]_6(Si_6,Al_2)O_{20}(OH)_4$ A yellow-brown to copper mineral of the mica group occurring in disseminated flakes, foliated masses, or large crystals; hardness is 2.5–3.0 on Mohs scale, and specific gravity is 2.8–3.0. Also known as bronze mica; brown mica.　{ ˈfläg·əˌpīt }

phloridzin [ORG CHEM] $C_{21}H_{24}O_{19}·2H_2O$ A glycoside extracted from the root bark of apple, plum, and pear trees; white needles with a melting point of 109°C; soluble in alcohol and hot water; used in medicine.　{ fləˈrid·zən }

phloroglucinol [ORG CHEM] $C_6H_3(OH)_3·2H_2O$ White to yellow crystals with a melting point of 212–217°C when heated rapidly and 200–209°C when heated slowly; soluble in alcohol and ether; used as a bone decalcifying agent, as a floral preservative, and in the manufacture of pharmaceuticals.　{ ˈflȯr· əˈglüs·ənˌȯl }

pH measurement [ANALY CHEM] Determination of the hydrogen-ion concentration in an ionized solution by means of an indicator solution (such as phenolphthalein) or a pH meter.　{ ˌpēˈach ˌmezh·ər·mənt }

pH meter [ENG] An electronic voltmeter using a pH-responsive electrode that gives a direct conversion of voltage differences to differences of pH at the temperature of the measurement.　{ ˌpēˈach ˌmēd·ər }

phobia [PSYCH] A disproportionate, obsessive, persistent, and unrealistic fear of an external situation or object, symbolically taking the place of an internal unconscious conflict.　{ ˈfō·bē·ə }

phobic neurosis [PSYCH] A neurotic disorder characterized by a persistent phobia which frequently interferes with the individual's activities and creates tension and anxiety, sometimes accompanied by physical manifestation. { 'fō·bik nú'rō·səs }

phobophobia [PSYCH] An abnormal fear of developing a phobia. { ˌfō·bə'fō·bē·ə }

Phobos [ASTRON] A satellite of Mars; it is the larger of the two satellites, with a diameter of about 15 miles (24 kilometers). { 'fō,bós }

Phocaenidae [VERT ZOO] The porpoises, a family of marine mammals in the order Cetacea. { fō'sē·nə,dē }

Phocidae [VERT ZOO] The seals, a pinniped family of carnivoran mammals in the superfamily Canoidea. { 'fō·sə,dē }

phocomelia [MED] A congenital or inherited condition in which the proximal part of a limb is missing. { ˌfō·kə'mēl·yə }

Phodilidae [VERT ZOO] A family of birds in the order Strigiformes; the bay owl (*Pholidus bodius*) is the single species. { fō'dil·ə,dē }

Phoebe [ASTRON] A satellite of the planet Saturn; its diameter is judged to be about 190 miles (320 kilometers); it has an eccentric orbit and retrograde revolution. { 'fē·bē }

phoenicite *See* phoenicochroite. { 'fē·nə,sīt }

phoenicochroite [MINERAL] Pb_2CrO_5 A red mineral composed of basic chromate of lead, occurring in crystals and masses. Also known as beresovite; phenicochroite; phoenicite. { ˌfēn·ə'kä·krə,wīt }

Phoenicopteridae [VERT ZOO] The flamingos, a family of long-legged, long-necked birds in the order Ciconiiformes. { ˌfēn·ə·käp'ter·ə,dē }

Phoenicopteriformes [VERT ZOO] An order comprising the flamingos in some systems of classification. { ˌfē·nə,käp·tə·rə'fōr,mēz }

Phoeniculidae [VERT ZOO] The African wood hoopoes, a family of birds in the order Coraciiformes. { ˌfēn·ə'kyü·lə,dē }

Phoenix [ASTRON] A southern constellation; right ascension 1 hour, declination 50°S. { 'fē·niks }

Pholadidae [INV ZOO] A family of bivalve mollusks in the subclass Eulamellibranchia; individuals may have one or more dorsal accessory plates, and the visceral mass is attached to the valves in the dorsal portion of the body. { fō'lad·ə,dē }

Pholidophoridae [PALEON] A generalized family of extinct fishes belonging to the Pholidophoriformes. { fə,lid·ə'fōr·ə,dē }

Pholidophoriformes [PALEON] An extinct actinopterygian group composed of mostly small fusiform marine and freshwater fishes of an advanced holostean level. { fə,lid·ə,fōr·ə'fōr,mēz }

Pholidota [VERT ZOO] An order of mammals comprising the living pangolins and their fossil predecessors; characterized by an elongate tubular skull with no teeth, a long protrusive tongue, strong legs, and five-toed feet with large claws. { ˌfäl·ə'dōd·ə }

Phomaceae [MYCOL] The equivalent name for Sphaerioidaceae. { fō'mās·ē,ē }

Phomales [MYCOL] The equivalent name for Sphaeropsidales. { fō'mā·lēz }

phon [ACOUS] A unit of loudness level; the loudness level, in phons, of a sound is numerically equal to the sound pressure level, in decibels, of a 1000-hertz reference tone which is judged by listeners to be equally loud to the sound under evaluation. { fän }

phonagnosia [PSYCH] A disturbance in the recognition of familiar voices in which the affected individual has good comprehension of what is spoken, but the speaker cannot be identified. { ˌfän·ag'nō·zhə }

phonation [LING] Production of speech sounds. { fō'nā·shən }

phone [ENG ACOUS] *See* headphone. [LING] Any sound made during speech. { fōn }

phoneme [LING] A speech sound (phone) that is contrastive, that is, is perceived as being different from all other speech sounds. { 'fō,nēm }

phoneme model [ACOUS] A succinct representation of the acoustic signal that corresponds to a phoneme, usually embedded in an utterance. { 'fō,nēm ,mäd·əl }

phonemic synthesizer [ENG ACOUS] A voice response system in which each word is abstractly represented as a sequence of expected vowels and consonants, and speech is composed

by juxtaposing the expected phonemic sequence for each word with the sequences for the preceding and following words. { fə'nē·mik 'sin·thə,sīz·ər }

phone patch [ELECTR] A device connecting an amateur or citizens'-band transceiver temporarily to a telephone system. { 'fōn ,pach }

phone plug [ELEC] A standard plug having a $^3/_4$-inch-diameter (19-millimeter) shank, used with headphones, microphones, and other audio equipment; usually designed for use with either two or three conductors. Also known as telephone plug. { 'fōn ,pləg }

phonetic alphabet [COMMUN] A list of standard words used for positive identification of letters in a voice message transmitted by radio or telephone. { fə'ned·ik 'al·fə,bet }

phonetics [LING] The study of the production or articulation and perception of speech as well as the acoustic characteristics of the sounds produced. { fə'ned·iks }

phonetic search [COMPUT SCI] A method of locating information in a file in which an algorithm is used to locate combinations of characters that sound similar to a specified combination. { fə'ned·ik 'sərch }

phonic motor [ELEC] A small synchronous motor which is driven by the current of an accurate oscillator, such as a crystal oscillator, and whose frequency is thus constant to a high degree of accuracy; used in astronomical instruments where a driving speed of great accuracy is required. { 'fän·ik 'mōd·ər }

phonoatomic effect [ACOUS] An effect that can be observed when a sound source of sufficiently high frequency is placed in liquid helium maintained at a very low temperature, less than 0.1 kelvin above absolute zero, wherein sound quanta, when they arrive at the liquid surface, have sufficient energy to knock helium atoms out of the liquid. { ˌfō·nō·ə,täm·ik i'fekt }

phonocardiograph [MED] An instrument that provides a graphic record of heart murmurs and other sounds. { ˌfō·nō'kärd·ē·ə,graf }

phonoelectrocardioscope [MED] An electronic medical instrument that uses a double-beam cathode-ray oscilloscope to show simultaneously the waveforms of two different quantities related to the heart. { ˌfō·nō·i,lek·trə'kärd·ē·ə,skōp }

phonograph [ENG ACOUS] An instrument for recording or reproducing acoustical signals, such as voice or music, by transmission of vibrations from or to a stylus that is in contact with a groove in a rotating disk. { 'fō·nə,graf }

phonograph cartridge *See* phonograph pickup. { 'fō·nə,graf ,kär·trij }

phonograph cutter *See* cutter. { 'fō·nə,graf ,kəd·ər }

phonograph needle *See* stylus. { 'fō·nə,graf ,nēd·əl }

phonograph pickup [ENG ACOUS] A pickup that converts variations in the grooves of a phonograph record into corresponding electric signals. Also known as cartridge; phonograph cartridge. { 'fō·nə,graf ,pik,əp }

phonograph record [ENG ACOUS] A shellac-composition or vinyl-plastic disk, usually 7 or 12 inches (18 or 30 centimeters) in diameter, on which sounds have been recorded as modulations in grooves. Also known as disk; disk recording. { 'fō·nə,graf ,rek·ərd }

phono jack [ELECTR] A jack designed to accept a phono plug and provide a ground connection for the shield of the conductor connected to the plug. { 'fō·nō ,jak }

phonolite [PETR] A light-colored, aphanitic rock of volcanic origin, composed largely of alkali feldspar, feldspathoids, and smaller amounts of mafic minerals. { 'fō·nə,līt }

phonology [LING] The study of the sound components of a spoken language. { fə'näl·ə·jē }

phonon [SOLID STATE] A quantum of an acoustic mode of thermal vibration in a crystal lattice. { 'fō,nän }

phonon-drag effect *See* Gurevich effect. { 'fō,nän ,drag i,fekt }

phonon-electron interaction [SOLID STATE] An interaction between an electron and a vibration of a lattice, resulting in a change in both the momentum of the particle and the wave vector of the vibration. { 'fō,nän i'lek,trän ,in·tər,ak·shən }

phonon emission [SOLID STATE] The production of a phonon in a crystal lattice, which may result from the interaction of other phonons via anharmonic lattice forces, from scattering of electrons in the lattice, or from scattering of x-rays or particles which bombard the crystal. { 'fō,nän i,mish·ən }

phonon friction [MECH] Friction that arises when atoms

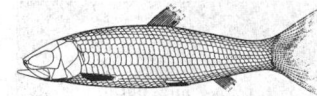

PHOLIDOPHORIDAE

Pholidophorus bechei. Lower Jurassic of England; length to 8 inches (20 centimeters).

PHONOATOMIC EFFECT

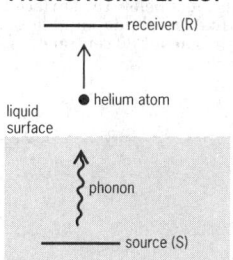

Experimental demonstration of the effect in liquid helium. Phonons from the source have sufficient energy to knock helium atoms out of the liquid, and these atoms are detected by the receiver.

close to a surface are set into motion by the sliding action of atoms in an opposing surface, and the mechanical energy needed to slide one surface over the other is thereby converted to the energy of atomic lattice vibrations (phonons) and is eventually transformed into heat. { 'fō,nän ,frik·shən }

phonon wind [SOLID STATE] A stream of nonthermal phonons that is effective in propelling electron-hole droplets through a crystal. { 'fō,nän ,wind }

phonophobia [PSYCH] An abnormal fear of sound. { ‚fō·nə'fō·bē·ə }

phono plug [ELECTR] A plug designed for attaching to the end of a shielded conductor, for feeding audio-frequency signals from a phonograph or other audio-frequency source to a mating phono jack on a preamplifier or amplifier. { 'fō·nō ,pləg }

phonoreception [PHYSIO] The perception of sound through specialized sense organs. { ¦fō·nō·ri'sep·shən }

phonotelemeter [ENG] A device consisting essentially of a stopwatch, for estimating the distance of guns in action by measuring the interval between the flash and the arrival of the sound waves from the discharge. { ¦fō·nō·tə'lem·əd·ər }

phony disease [PL PATH] A virus disease of the peach characterized by dwarfing, abnormal darkening of leaves, and a light crop of small, highly colored fruit; the tree stops bearing fruit after a few years. { 'fō·nē di‚zēz }

phony mine [ORD] Harmless object used to simulate a mine or to give false signals in detectors; used in phony mine fields. { 'fō·nē 'mīn }

phorate [ORG CHEM] $C_7H_{17}O_2PS_2$ A clear liquid with slight solubility in water; used as an insecticide for a wide range of insects on a wide range of crops. { 'fòr,āt }

phoresy [ECOL] A relationship between two different species of organisms in which the larger, or host, organism transports a smaller organism, the guest. { 'fòr·ə·sē }

Phoridae [INV ZOO] The hump-backed flies, a family of cyclorrhaphous dipteran insects in the series Aschiza. { 'fòr·ə,dē }

phorogenesis [GEOL] The shifting or slipping of the earth's crust relative to the mantle. { ‚fòr·ə'jen·ə·səs }

Phoronida [INV ZOO] A small, homogeneous group, or phylum, of animals having an elongate body, a crown of tentacles surrounding the mouth, and the anus occurring at the level of the mouth. { fə'rän·ə·də }

phosgene [ORG CHEM] $COCl_2$ A highly toxic, colorless gas that condenses at 0°C to a fuming liquid; used as a war gas and in manufacture of organic compounds. { 'fäz‚jēn }

phosgenite [MINERAL] $Pb_2Cl_2(CO_3)$ A white, yellow, or grayish mineral that crystallizes in the tetragonal system, has adamantine luster, hardness of 3 on Mohs scale, and specific gravity of 6–6.3. Also known as cromfordite; horn lead. { 'fäz·jə,nīt }

phosphatase [BIOCHEM] An enzyme that catalyzes the hydrolysis and synthesis of phosphoric acid esters and the transfer of phosphate groups from phosphoric acid to other compounds. { 'fäs·fə‚tās }

phosphate [CHEM] **1.** Generic term for any compound containing a phosphate group ($PO_4{}^{3-}$), such as potassium phosphate, K_3PO_4. **2.** Generic term for a phosphate-containing fertilizer material. [MINERAL] A mineral compound characterized by a tetrahedral ionic group of phosphate and oxygen, $PO_4{}^{3-}$. { 'fä,sfāt }

phosphate anion [INORG CHEM] $PO_4{}^{3-}$ The negative ion of phosphoric acid. { 'fä,sfāt 'an,ī·ən }

phosphate buffer [ANALY CHEM] Laboratory pH reference solution made of KH_2PO_4 and Na_2HPO_4; when 0.025 molal (equimolal of the potassium and sodium salts), the pH is 6.865 at 25°C. { 'fä,sfāt 'bəf·ər }

phosphate coating [MET] A conversion coating on metal, usually steel, produced by dipping in an aqueous solution of zinc or manganese acid phosphate; used to furnish a black finish to small arms, artillery, or automotive components to provide resistance to corrosion. { 'fä,sfāt ,kōd·iŋ }

phosphate desulfurization [CHEM ENG] A continuous, regenerative petroleum-refinery process using a tripotassium phosphate solution to remove hydrogen sulfide from natural gas, refinery gas, or liquid hydrocarbons. { 'fä,sfāt dē‚səl·fə·rə'zā·shən }

phosphate fertilizer [MATER] Fertilizer compound or mixture containing available (soluble) phosphate; examples are

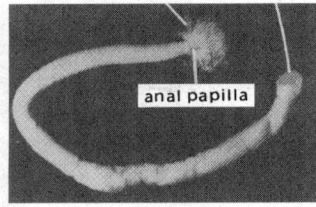

PHORONIDA

Phoronopsis harmeri removed from its tube, just below the surface in intertidal or subtidal mud flats. Length is about 20 centimeters.

phosphate rock (phosphorite), superphosphates or triple superphosphates, nitrophosphate, potassium phosphates, or N-P-K mixtures. { 'fä,sfāt 'fərd·əl‚īz·ər }

phosphate glass [MATER] Glass in which phosphorus pentoxide is a major component; resistant to hydrofluoric acid. { 'fä,sfāt ,glas }

phosphate recovery process [MIN ENG] A process developed by the U.S. Bureau of Mines for recovering phosphate from low-grade phosphorus-bearing shales. { 'fä,sfāt ri'kəv·ə·rē ,prä·səs }

phosphate rock [MATER] **1.** A rock that is naturally high enough in phosphorus to be used directly in fertilizer manufacturing. **2.** The beneficiated concentrate of a phosphate deposit. { 'fä,sfāt 'räk }

phosphatic nodule [GEOL] A dark, usually black, earthy mass or pebble of variable size and shape, having a hard shiny surface and occurring in marine strata. { fä'sfad·ik 'naj·yül }

phosphatide *See* phospholipid. { 'fäs·fə‚tīd }

phosphating [MET] Forming a phosphate coating on a metal. Also known as phosphatizing. { 'fäs‚fād·iŋ }

phosphatization [GEOCHEM] Conversion to a phosphate or phosphates; for example, the diagenetic replacement of limestone, mudstone, or shale by phosphate-bearing solutions, producing phosphates of calcium, aluminum, or iron. { ‚fäs·fəd·ə'zā·shən }

phosphatizing *See* phosphating. { 'fäs·fə‚tīz·iŋ }

phosphaturia *See* hyperphosphaturia. { ‚fäs·fə'tùr·ē·ə }

phosphene [PSYCHOL] Sensation of spots of light in the visual field due to a stimulus other than light, such as pressure on the eyeball or an electrical stimulus to the retina or visual pathway. { 'fäs‚fēn }

phosphide [INORG CHEM] Binary compound of trivalent phosphorus, as in Na_3P. { 'fä,sfīd }

phosphine [INORG CHEM] PH_3 Poisonous, colorless, spontaneously flammable gas with garlic aroma; soluble in alcohol, slightly soluble in cold water; boils at −85°C; used in organic reactions. Also known as hydrogen phosphide; phosphoretted hydrogen. { 'fä,sfēn }

phosphinic acid [ORG CHEM] Organic derivative of hypophosphorous acid; contains the radical −H_2PO_2 or =HPO_2; examples are methylphosphinic acid, CH_3HPOOH, and dimethyl phosphinic acid, $(CH_3)_2POOH$. { fä'sfin·ik 'as·əd }

phosphite [INORG CHEM] Salt of phosphorous acid; contains the radical $PO_3{}^{3-}$; an example is normal sodium phosphite, Na_3PO_3. { 'fä,sfīt }

phosphocreatine [BIOCHEM] $C_4H_{10}N_3O_5P$ Creatine phosphate, a phosphoric acid derivative of creatine which contains an energy-rich phosphate bond; it is present in muscle and other tissues, and during the anaerobic phase of muscular contraction it hydrolyzes to creatine and phosphate and makes energy available. Abbreviated PC. { ‚fäs·fō'krē·ə‚tēn }

phosphoenolpyruvic acid [BIOCHEM] $CH_2=O(OPO_3H_2)$-COOH A high-energy phosphate formed by dehydration of 2-phosphoglyceric acid; it reacts with adenosine diphosphate to form adenosine triphosphate and enolpyruvic acid. { ¦fäs-fō¦ē‚nòl·pī'rü·vik 'as·əd }

phosphoferrite [MINERAL] $(Fe,Mn)_3(PO_4)_2·3H_2O$ A white or greenish orthorhombic mineral composed of hydrous phosphate of ferrous iron manganese phosphate; exhibits micalike cleavage. { ¦fäs·fō'fe‚rīt }

phosphofructokinase [BIOCHEM] A glycolytic enzyme that functions in carbohydrate metabolism by catalyzing the phosphorylation of fructose phosphate. { ‚fäs·fə‚frük‚tō 'kī‚nās }

phosphoglucoisomerase [BIOCHEM] An enzyme that catalyzes the conversion of galactose-1-phosphate to glucose-1-phosphate. { ¦fäs·fō¦glü·kō·ī'säm·ə‚rās }

phosphoglucomutase [BIOCHEM] An enzyme that catalyzes the conversion of glucose-1-phosphate to glucose-6-phosphate. { ¦fäs·fō¦glü·kō'myü‚tās }

phosphoglycolate [BOT] A two-carbon by-product of photorespiration. { ‚fäs·fə'glī·kə‚lāt }

phosphohexose isomerase [BIOCHEM] An enzyme found in muscle and yeast that catalyzes the interconversion of glucose-6-phosphate and fructose-6-phosphate. { ‚fäs·fə‚hek‚sōs ī'säm·ə‚rās }

phospholan [ORG CHEM] $C_6H_{14}O_3PNS_2$ A colorless to yellow solid with a melting point of 37–45°C; used as an insecticide and miticide for cotton. { 'fä·sfə‚lan }

phospholipase [BIOCHEM] An enzyme that catalyzes a hydrolysis of a phospholipid, especially a lecithinase that acts in this manner on a lecithin. { ¦fäs·fō'lī¸pās }

phospholipase A₁ [BIOCHEM] A principal phospholipase that releases fatty acids from the first position (F1) of phospholipids. Abbreviated PLA1. { ¸fäs·fə'lī¸pās ¦ā'wən }

phospholipase A₂ [BIOCHEM] A principal phospholipase that cleaves fatty acids (principally arachidonic acid) from the second position (F2) of phospholipids. Abbreviated PLA2. { ¸fäs·fə'lī¸pās ¦ā'tü }

phospholipase C [BIOCHEM] A widely distributed enzyme of considerable physiological significance that hydrolyzes phospholipids at the phosphodiester bond to produce diacylglycerol and a phosphorylated head group, as found in phosphocholine or phosphoinositol. Abbreviated PLC. { ¸fäs·fə'lī¸pās ¦sē }

phospholipase D [BIOCHEM] An enzyme found in almost all mammalian cells that hydrolyzes phosphatidylcholine to produce the signaling molecule phosphatidic acid, which acts on many regulatory enzymes and other proteins in the cell. Abbreviated PLD. { ¸fäs·fə'lī¸pās 'dē }

phospholipid [BIOCHEM] Any of a class of esters of phosphoric acid containing one or two molecules of fatty acid, an alcohol, and a nitrogenous base. Also known as phosphatide. { ¦fäs·fō'lip·əd }

phosphomolybdic acid [INORG CHEM] $H_3PO_4·12MoO_3·xH_2O$ Yellow crystals; soluble in alcohol, ether, and water; used as an alkaloid reagent and a pigment. Abbreviated PMA. { ¦fä·sfō·mə'lib·dik 'as·əd }

phosphomonoesterase [BIOCHEM] An enzyme catalyzing hydrolysis of phosphoric acid esters containing one ester linkage. { ¦fäs·fō¸män·ō'es·tə¸rās }

phosphonic acid [ORG CHEM] $ROP(OH)_2$, where R is an organic radical such as $C_6H_5^-$, as in phenylphosphonic acid. { fä'sfän·ik 'as·əd }

phosphophyllite [MINERAL] $Zn_2(FeMn)(PO_4)_2·4H_2O$ Colorless to pale-blue mineral composed of hydrous zinc ferrous iron manganese phosphate; exhibits micalike cleavage. { ¸fäs·fō'fi¸līt }

phosphor See luminophor. { 'fäs·fər }

phosphor bronze [MET] A hard copper-base alloy containing several percent tin, and sometimes smaller percentages of lead, deoxidized with phosphorus. { 'fäs·fər 'bränz }

phosphor dot [ELECTR] One of the tiny dots of phosphor material that are used in groups of three, one group for each primary color, on the screen of a color television picture tube. { 'fäs·fər ¸dät }

phosphorescence [ATOM PHYS] **1.** Luminescence that persists after removal of the exciting source. Also known as afterglow. **2.** Luminescence whose decay, upon removal of the exciting source, is temperature-dependent. { ¸fäs·fə'res·əns }

phosphorescent paint [MATER] A luminous paint containing phosphors or phosphorogens which requires activation from an outside source of light, depending upon the ability of the chemical to absorb light energy, and to emit it in the form of photons of light. { ¸fäs·fə'res·ənt pānt }

phosphoretted hydrogen See phosphine. { 'fäs·fə¸red·əd 'hī·drə·jən }

phosphoric acid [INORG CHEM] H_3PO_4 Water-soluble, transparent crystals, melting at 42°C; used as a fertilizer, in soft drinks and flavor syrups, pharmaceuticals, water treatment, and animal feeds and to pickle and rust-proof metals. Also known as orthophosphoric acid. { fä'sför·ik 'as·əd }

phosphoric acid polymerization [CHEM ENG] A petroleum-refinery process using phosphoric acid catalyst to convert propylene, butylene, or both, into high-octane gasoline or petrochemical polymers. { fä'sför·ik 'as·əd pə¸lim·ə·rə'zā·shən }

phosphoric anhydride [INORG CHEM] P_2O_5 A flammable, dangerous, soft-white deliquescent powder; used as a dehydrating agent, in medicine and sugar refining, and as a chemical intermediate and analytical reagent. Also known as anhydrous phosphoric acid; phosphoric oxide; phosphorus pentoxide. { fä'sför·ik an'hī¸drīd }

phosphoric oxide See phosphoric anhydride. { fä'sför·ik 'äk¸sīd }

phosphorimetry [ANALY CHEM] Low-temperature, analytical procedure related to fluorometry; based on the nature and intensity of the phosphorescent light emitted by an appropriately excited molecule. { ¸fäs·fə'rim·ə·trē }

phosphorite [PETR] A sedimentary rock composed chiefly of phosphate minerals. { 'fäs·fə¸rīt }

phosphorization [GEOCHEM] Impregnation or combination with phosphorus or a compound of phosphorus; for example, the diagenetic process of phosphatization. { ¸fäs·fə·rə'zā·shən }

phosphorized copper [MET] A phosphorus deoxidized copper. { 'fäs·fə¸rīzd 'käp·ər }

phosphorogen [PHYS] A substance that promotes phosphorescence in another substance, as manganese does in zinc sulfide. { fä'sför·ə·jən }

phosphorolysis [BIOCHEM] A reaction by which elements of phosphoric acid are incorporated into the molecule of a compound. { ¸fäs·fə'räl·ə·səs }

phosphorous acid [INORG CHEM] H_3PO_3 Alcohol- and water-soluble deliquescent white or yellowish crystals; decomposes at 200°C; used as an analytical reagent and reducing agent. { 'fäs·fə·rəs 'as·əd }

phosphorroesslerite [MINERAL] $MgH(PO_4)·7H_2O$ A yellowish, monoclinic mineral consisting of a hydrated acid magnesium phosphate. { ¦fäs·fə'res·lə¸rīt }

phosphor tin [MET] A master alloy of tin and phosphorus, usually containing up to 5% phosphorus; used to make phosphor bronze. { 'fäs·fər 'tin }

phosphorus [CHEM] A nonmetallic element, symbol P, atomic number 15, atomic weight 30.97376; used to manufacture phosphoric acid, in phosphor bronzes, incendiaries, pyrotechnics, matches, and rat poisons; the white (or yellow) allotrope is a soft waxy solid melting at 44.5°C, is soluble in carbon disulfide, insoluble in water and alcohol, and is poisonous and self-igniting in air; the red allotrope is an amorphous powder subliming at 416°C, igniting at 260°C, is insoluble in all solvents, and is nonpoisonous; the black allotrope comprises lustrous crystals similar to graphite, and is insoluble in most solvents. { 'fäs·fə·rəs }

phosphorus nitride [INORG CHEM] P_3N_5 Amorphous white solid that decomposes in hot water; insoluble in cold water, soluble in organic solvents; used to dope semiconductors. { 'fäs·fə·rəs 'nī¸trīd }

phosphorus-nitrogen ratio [OCEANOGR] The proportion, by weight, of phosphorus to nitrogen in seawater or in plankton; the ratio is approximately 7:1 { 'fäs·fə·rəs 'nī·trə·jən 'rā¸shō }

phosphorus oxide [INORG CHEM] An oxygen compound of phosphorus; examples are phosphorus monoxide (P_2O), phosphorus trioxide (P_2O_3), phosphorus suboxide (P_4O). { 'fäs·fə·rəs 'äk¸sīd }

phosphorus oxychloride [INORG CHEM] $POCl_3$ Toxic, colorless, fuming liquid with pungent aroma; boils at 107°C; decomposes in water or alcohol; causes skin burns; used as a catalyst, chlorinating agent, and in manufacture of various anhydrides. Also known as phosphoryl chloride. { 'fäs·fə·rəs 'äk·sē'klör¸īd }

phosphorus pentabromide [INORG CHEM] PBr_5 Yellow crystals, decomposing at 106°C and in water; used in organic synthesis. { 'fäs·fə·rəs ¸pen·tə'brō¸mīd }

phosphorus pentachloride [INORG CHEM] PCl_5 Toxic, yellowish crystals with irritating aroma; an eye irritant; sublimes on heating, but will melt at 148°C under pressure; soluble in carbon disulfide; decomposes in water; used as a catalyst and chlorinating agent. { 'fäs·fə·rəs ¸pen·tə'klör¸īd }

phosphorus pentasulfide [INORG CHEM] P_2S_5 Flammable, hygroscopic, yellow crystals, melting at 281°C; decomposes in moist air; soluble in alkali hydroxides; used to make lube-oil additives, rubber additives, and flotation agents. { 'fäs·fə·rəs ¸pen·tə'səl¸fīd }

phosphorus pentoxide See phosphoric anhydride. { 'fäs·fə·rəs pen'täk¸sīd }

phosphorus sesquisulfide [INORG CHEM] P_4S_3 Flammable, yellow crystals, melting at 172°C; decomposed by hot water, insoluble in water, soluble in carbon disulfide; used as chemical intermediate and to make matches. Also known as tetraphosphorus trisulfide. { 'fäs·fə·rəs ¸ses·kwē'səl¸fīd }

phosphorus thiochloride [INORG CHEM] $PSCl_3$ Yellow liquid, boiling at 125°C; used to make insecticides and oil additives. { 'fäs·fə·rəs ¸thī·ə'klör¸īd }

phosphorus tribromide [INORG CHEM] PBr_3 A corrosive, fuming, colorless liquid with penetrating aroma; soluble in acetone, alcohol, carbon disulfide, and hydrogen sulfide;

decomposes in water; used as an analytical reagent to test for sugar and oxygen. { 'fäs·fə·rəs trī'brō,mīd }

phosphorus trichloride [INORG CHEM] PCl_3 A colorless, fuming liquid that decomposes rapidly in moist air and water; soluble in ether, benzene, carbon disulfide, and carbon tetrachloride; boils at 76°C; used as a chlorinating agent, phosphorus solvent, and in saccharin manufacture. { 'fäs·fə·rəs trī'klór,īd }

phosphorus triiodide [INORG CHEM] PI_3 Hygroscopic, red crystals, melting at 61°C; soluble in alcohol and carbon disulfide; decomposes in water; used in organic syntheses. { 'fäs·fə·rəs trī'ī·ə,dīd }

phosphorus trisulfide [INORG CHEM] P_2S_3 or P_4S_6 Grayish-yellow, tasteless, odorless solid that burns in air; soluble in alcohol, carbon disulfide, and ether; melts at 290°C; used as an analytical reagent. { 'fäs·fə·rəs trī'səl,fīd }

phosphorylase [BIOCHEM] An enzyme that catalyzes the formation of glucose-1-phosphate (Cori ester) from glycogen and inorganic phosphate; it is widely distributed in animals, plants, and microorganisms. { fäs'fòr·ə,lās }

phosphorylation [ORG CHEM] The esterification of compounds with phosphoric acid. { ,fäs,fór·ə'lā·shən }

phosphoryl chloride *See* phosphorus oxychloride. { 'fäs·fə·rəl 'klór,īd }

phosphosiderite [MINERAL] $FePO_4·2H_2O$ A pinkish-red mineral crystallizing in the monoclinic system, dimorphous with strengite and isomorphous with metavariscite. { ¦fä·sfō'sīd·ə,rīt }

phosphotransacetylase [BIOCHEM] An enzyme that catalyzes the reversible transfer of an acetyl group from acetyl coenzyme A to a phosphate, with formation of acetyl phosphate. { ¦fä·sfō,tranz·ə'sed·əl,ās }

phosphotungstic acid [INORG CHEM] $H_3PO_4·12WO_3·xH_2O$ Heavy-greenish, water- and alcohol-soluble crystals; used as an analytical reagent and in the manufacture of organic pigments. Also known as heavy acid; phosphowolframic acid; PTA. { ¦fä·sfō'təŋ·stik 'as·əd }

phosphotungstic pigment [ORG CHEM] A green or blue pigment prepared by precipitating solutions of phosphotungstic or phosphomolybdic acid with malachite green, Victoria blue, and other basic dyestuffs; used in printing inks, paints, and enamels. Also known as tungsten lake. { ¦fä·sfō'təŋ·stik 'pig·mənt }

phosphowolframic acid *See* phosphotungstic acid. { ¦fä·sfō·wúl'fram·ik 'as·əd }

phosphuranylite [MINERAL] $(UO_2)(PO_4)_2·6H_2O$ A yellow secondary mineral composed of hydrous uranyl phosphate, occurring in powder form; it is phosphorescent when exposed to radium emanations. { ,fäs·fyə'ran·əl,īt }

phot [OPTICS] A unit of illumination equal to the illumination of a surface, 1 square centimeter in area, on which there is a luminous flux of 1 lumen, or the illumination on a surface all points of which are at a distance of 1 centimeter from a uniform point source of 1 candela. Also known as centimeter-candle (deprecated usage). { fōt }

photic zone [ECOL] The uppermost layer of a body of water (approximately the upper 330 feet or 100 meters) that receives enough sunlight to permit the occurrence of photosynthesis. { 'fōd·ik }

Photidae [INV ZOO] A family of amphipod crustaceans in the suborder Gammaridea. { 'fäd·ə,dē }

photino [PARTIC PHYS] A hypothetical counterpart of the photon, postulated to exist in supersymmetry theories. { fō'tē·nō }

photoabsorption [PHYS] A process in which a photon transfers all its energy to an atom, molecule, or nucleus. { ,fōd·ō·əb'sórp·shən }

photoacoustic spectroscopy [SPECT] A spectroscopic technique for investigating solid and semisolid materials, in which the sample is placed in a closed chamber filled with a gas such as air and illuminated with monochromatic radiation of any desired wavelength, with intensity modulated at some suitable acoustic frequency; absorption of radiation results in a periodic heat flow from the sample, which generates sound that is detected by a sensitive microphone attached to the chamber. Abbreviated PAS. Also known as optoacoustic spectroscopy. { ¦fōd·ō·ə¦kü·stik spek'träs·kə·pē }

photoaddition [PHYS CHEM] A bimolecular photochemical

process in which a single product is formed by electronically excited unsaturated molecules. { ¦fōd·ō·ə'dish·ən }

photoalidade [ENG] A photogrammetric instrument which has a telescopic alidade, a plateholder, and a hinged ruling arm and is mounted on a tripod frame; used for plotting lines of direction and measuring vertical angles to selected features appearing on oblique and terrestrial photographs. { ¦fōd·ō'al·ə,dād }

photoautotroph [ECOL] An autotroph that uses energy from light to produce organic molecules. { ,fōd·ō'ód·ō,träf }

photoautotrophic [BIOL] Pertaining to organisms which derive energy from light and manufacture their own food. { ¦fōd·ō,ód·ō'träf·ik }

photobiont [ECOL] A photosynthetic partner of a symbiotic pair, such as the algal component of the fungal-algal association in lichens. { ¦fōd·ō'bī,änt }

photobleach [PHYS CHEM] Upon exposure to light, to decrease in absorbance intensity or, for fluorescent compounds, to decrease in emission intensity. { ¦fōd·ō,blēch }

photocapacitive effect [ELEC] A change in the capacitance of a bulk semiconductor or semiconductor surface film upon exposure to light. { ,fōd·ō·kə'pas·ə,tā·tiv i,fekt }

photocatalysis [PHYS CHEM] The phenomenon by which a relatively small amount of light-absorbing material, called a photocatalyst, changes the rate of chemical reaction without itself being consumed. { ,fōd·ō·kə'tal·ə·səs }

photocatalyst [PHYS CHEM] A light-absorbing substance which, when added to a reaction, facilitates the reaction, while remaining unchanged at the end of the reaction. { ,fōd·ō·'kad·əl·ist }

photocathode [ELECTR] A photosensitive surface that emits electrons when exposed to light or other suitable radiation; used in phototubes, television camera tubes, and other light-sensitive devices. { ¦fōd·ō'kath,ōd }

photocell [ELECTR] A solid-state photosensitive electron device whose current-voltage characteristic is a function of incident radiation. Also known as electric eye; photoelectric cell. { 'fōd·ə,sel }

photocell relay [ELECTR] A relay actuated by a signal received when light falls on, or is prevented from falling on, a photocell. { 'fōd·ə,sel 're,lā }

photochemical oxidant [CHEM] Any of the chemicals which enter into oxidation reactions in the presence of light or other radiant energy. { ¦fōd·ō'kem·ə·kəl 'äk·sə·dənt }

photochemical reaction [PHYS CHEM] A chemical reaction influenced or initiated by light, particularly ultraviolet light, as in the chlorination of benzene to produce benzene hexachloride. { ¦fōd·ō'kem·ə·kəl rē'ak·shən }

photochemical reduction *See* photoreduction. { ,fōd·ō¦kem·ə·kəl ri'dək·shən }

photochemical smog [METEOROL] Chemical pollutants in the atmosphere resulting from chemical reactions involving hydrocarbons and nitrogen oxides in the presence of sunlight. { ¦fōd·ō'kem·ə·kəl 'smäg }

photochemistry [PHYS CHEM] The study of the effects of light on chemical reactions. { ¦fōd·ō'kem·ə·strē }

photochromic compound [CHEM] A chemical compound that changes in color when exposed to visible or near-visible radiant energy; the effect is reversible; used to produce very-high-density microimages. { ¦fōd·ō¦krō·mik 'käm,paúnd }

photochromic glass [MATER] A glass that darkens when exposed to light but regains its original transparency a few minutes after light is removed; the rate of clearing increases with temperature. { ¦fōd·ō¦krō·mik 'glas }

photochromic reaction [CHEM] A chemical reaction that produces a color change. { ¦fōd·ō¦krō·mik rē'ak·shən }

photochromism [CHEM] The ability of a chemically treated plastic or other transparent material to darken reversibly in strong light. { ¦fōd·ō'krō,miz·əm }

photochromism method [FL MECH] A flow visualization method in which a laser or other light source is used to convert a photochromic compound in the liquid under study from a transparent to an opaque state. { ¦fōd·ō'krō,miz·əm 'meth·əd }

photoclinometer [ENG] A directional surveying instrument which records photographically the direction and magnitude of well deviations from the vertical. { ¦fōd·ō·klə'näm·əd·ər }

photoclinometry [GEOL] A technique for ascertaining slope information from an image brightness distribution, used especially for studying the amount of slope to a lunar crater wall or ridge by measuring the density of its shadow. { ¦fōd·ō·klə'näm·ə·trē }

photocoagulator [MED] An instrument that uses a xenon flash lamp and an associated train of optics to focus an intense beam of light on a detached retina for the purpose of inducing coagulation and a lesion that welds the retina back into position. { ¦fōd·ō·kō'ag·yə¸lād·ər }

photocomposition [COMPUT SCI] Composition of type using electrophotographic techniques such as phototypesetters and laser printers. [GRAPHICS] Reproduction of type images by photographic means. { ¦fōd·ō¸käm·pə'zish·ən }

photoconduction [SOLID STATE] An increase in conduction of electricity resulting from absorption of electromagnetic radiation. { ¦fōd·ō·kən'dək·shən }

photoconductive cell [ELECTR] A device for detecting or measuring electromagnetic radiation by variation of the conductivity of a substance (called a photoconductor) upon absorption of the radiation by this substance. Also known as photoresistive cell; photoresistor. { ¦fōd·ō·kən'dək·tiv 'sel }

photoconductive device [ELECTR] A photoelectric device which utilizes the photoinduced change in electrical conductivity to provide an electrical signal. { fōd·ō·kən'dək·tiv di'vīs }

photoconductive film [ELECTR] A film of material whose current-carrying ability is enhanced when illuminated. { fōd·ō·kən'dək·tiv 'film }

photoconductive gain factor [ELECTR] The ratio of the number of electrons per second flowing through a circuit containing a cube of semiconducting material, whose sides are of unit length, to the number of photons per second absorbed in this volume. { fōd·ō·kən'dək·tiv 'gān ¸fak·tər }

photoconductive meter [ELECTR] An exposure meter in which a battery supplies power through a photoconductive cell to a milliammeter. { fōd·ō·kən'dək·tiv 'med·ər }

photoconductivity [SOLID STATE] The increase in electrical conductivity displayed by many nonmetallic solids when they absorb electromagnetic radiation. { ¦fōd·ō¸kän¸dək'tiv·əd·ē }

photoconductivity gain [ELECTR] The number of charge carriers that circulate through a circuit involving a photoconductor for each charge carrier generated by light. { fōd·ō ¸kän¸dək'tiv·əd·ē ¸gān }

photoconductor [SOLID STATE] A nonmetallic solid whose conductivity increases when it is exposed to electromagnetic radiation. { ¦fōd·ō·kən'dək·tər }

photoconductor diode See photodiode. { fōd·ō·kən'dək·tər 'dī¸ōd }

photocopying process [GRAPHICS] Any of the means by which a copy is created on a sensitized surface (generally paper, film, or metal plate) by the action of radiant energy. { ¦fōd·ō'käp·ē·iŋ ¸prä·səs }

photocoupler See optoisolator. { ¦fōd·ō'kəp·lər }

photocurrent [PHYS CHEM] An electric current induced at an electrode by radiant energy. { ¦fōd·ō'kə·rənt }

photodarlington [ELECTR] A Darlington amplifier in which the input transistor is a phototransistor. { ¦fōd·ō'där·liŋ·tən }

photodegradation [ORG CHEM] Chemical changes resulting from the absorption of light that reduce the useful properties of materials, particularly polymers. The chemical changes can include bond scission (especially of the molecular backbone), color formation, crosslinking, and chemical rearrangements. { ¦fōd·ō¸deg·rə'dā·shən }

photodetachment [PHYS CHEM] The removal of an electron from a negative ion by absorption of a photon, resulting in a neutral atom or molecule. { ¦fōd·ō·di'tach·mənt }

photodetector [ELECTR] A detector that responds to radiant energy; examples include photoconductive cells, photodiodes, photoresistors, photoswitches, phototransistors, phototubes, and photovoltaic cells. Also known as light-sensitive cell; light-sensitive detector; light sensor photodevice; photodevice; photoelectric detector; photosensor. { ¦fōd·ō·di'tek·tər }

photodevice See photodetector. { ¦fōd·ō·di¸vīs }

photodichroic material [OPTICS] A material which exhibits photoinduced dichroism and birefringence. { ¦fōd·ō·dī'krō·ik mə'tir·ē·əl }

photodiffusion effect See Dember effect. { ¦fōd·ō·di'fyü·zhən i¸fekt }

photodimerization [PHYS CHEM] A bimolecular photochemical process involving an electronically excited unsaturated molecule that undergoes addition with an unexcited molecule of the same species. { ¦fōd·ō¸dī·mə·rə'zā·shən }

photodiode [ELECTR] A semiconductor diode in which the reverse current varies with illumination; examples include the alloy-junction photocell and the grown-junction photocell. Also known as photoconductor diode. { ¦fōd·ō'dī¸ōd }

photodisintegration [NUC PHYS] The breakup of an atomic nucleus into two or more fragments as a result of bombardment by gamma radiation. Also known as Chadwick-Goldhaber effect. { ¦fōd·ō·di¸sin·tə'grā·shən }

photodissociation [PHYS CHEM] The removal of one or more atoms from a molecule by the absorption of a quantum of electromagnetic energy. { ¦fōd·ō·di¸sō·shē'ā·shən }

photodosimetry [NUCLEO] Determination of the cumulative dose of ionizing radiation by use of photographic film. { ¦fōd·ō·dō'sim·ə·trē }

photodraft [DES ENG] A photographic reproduction of a master layout or design on a specially prepared emulsion-coated piece of sheet metal; used as a master in a tool-construction department. { 'fōd·ō¸draft }

photo echo [OPTICS] A coherent pulse of light generated in a nonlinear medium at a characteristic time after two other pulses, separated by a certain time interval, have entered the medium. { 'fōd·ō ¸ek·ō }

photoecology [ENG] The application of air photography to ecology, integrated land resource studies, and forestry. { ¦fōd·ō·i'käl·ə·jē }

photoelastic effect [OPTICS] Changes in optical properties of a transparent dielectric when it is subjected to mechanical stress, such as mechanical birefringence. Also known as photoelasticity. { ¦fōd·ō·i¸las·tik i'fekt }

photoelasticity [OPTICS] **1.** An experimental technique for the measurement of stresses and strains in material objects by means of the phenomenon of mechanical birefringence. **2.** See photoelastic effect. { ¦fōd·ō¸i¸las'tis·əd·ē }

photoelectret [SOLID STATE] An electret produced by the removal of light from an illuminated photoconductor in an electric field. { ¦fōd·ō·i'lek·trət }

photoelectric [ELECTR] Pertaining to the electrical effects of light, such as the emission of electrons, generation of voltage, or a change in resistance when exposed to light. { ¦fōd·ō·i'lek·trik }

photoelectric absorption [ELECTR] Absorption of photons in one of the several photoelectric effects. { ¦fōd·ō·i'lek·trik əb'sórp·shən }

photoelectric absorption analysis [ANALY CHEM] Type of activation analysis in which the γ-photon gives all of its energy to an electron in the crystal under analysis, generating a maximum-sized pulse for that particular γ-energy. { ¦fōd·ō·i'lek·trik əb'sórp·shən ə¸nal·ə·səs }

photoelectric cell See photocell. { ¦fōd·ō·i'lek·trik 'sel }

photoelectric color comparator See color comparator. { ¦fōd·ō·i'lek·trik 'kəl·ər kəm¸par·əd·ər }

photoelectric colorimeter [ENG] A colorimeter that uses a phototube or photocell, a set of color filters, an amplifier, and an indicating meter for quantitative determination of color. { ¦fōd·ō·i'lek·trik ¸kəl·ə'rim·əd·ər }

photoelectric colorimetry [ANALY CHEM] Measurement of the colorant concentration in a solution by means of the tristimulus values of three primary light filter-photocell combinations. { ¦fōd·ō·i'lek·trik ¸kəl·ə'rim·ə·trē }

photoelectric constant [ELECTR] The ratio of the frequency of radiation causing emission of photoelectrons to the voltage corresponding to the energy absorbed by a photoelectron; equal to Planck's constant divided by the electron charge. { ¦fōd·ō·i'lek·trik 'kän·stənt }

photoelectric control [ELECTR] Control of a circuit or piece of equipment by changes in incident light. { ¦fōd·ō·i'lek·trik kən'trōl }

photoelectric counter [ELECTR] A photoelectrically actuated device used to record the number of times a given light path is intercepted by an object. { ¦fōd·ō·i'lek·trik 'kaúnt·ər }

photoelectric cutoff register control [ELECTR] Use of a photoelectric control system as a longitudinal position regulator to maintain the position of the point of cutoff with respect to a repetitive pattern of moving material. { ¦fōd·ō·i'lek·trik ¦kət óf ¦rej·ə·stər kən¸trōl }

photoelectric densitometer [ENG] An electronic instrument used to measure the density or opacity of a film or other material; a beam of light is directed through the material, and the amount of light transmitted is measured with a photocell and meter. { ¦fōd·ō·i'lek·trik ¦den·sə'täm·əd·ər }

photoelectric detector See photodetector. { ¦fōd·ō·i'lek·trik di'tek·tər }

photoelectric device [ELECTR] A device which gives an electrical signal in response to visible, infrared, or ultraviolet radiation. { ¦fōd·ō·i'lek·trik di'vīs }

photoelectric door opener [CONT SYS] A control system that employs a photocell or other photo device, used to open and close a power-operated door. { ¦fōd·ō·i'lek·trik 'dȯr ¸ȯp·ə·nər }

photoelectric effect See photoelectricity. { ¦fōd·ō·i'lek·trik i¸fekt }

photoelectric electron-multiplier tube See multiplier phototube. { ¦fōd·ō·i'lek·trik i¸lek¸trän 'məl·tə¸plī·ər ¸tüb }

photoelectric flame-failure detector [CONT SYS] A photoelectric control that cuts off fuel flow when the fuel-consuming flame is extinguished. { ¦fōd·ō·i'lek·trik 'flām ¸fāl·yər di¸tek·tər }

photoelectric fluorometer [ENG] Device using a photoelectric cell to measure fluorescence in a chemical sample that has been excited (one or more electrons have been raised to higher energy level) by ultraviolet or visible light; used for analysis of chemical mixtures. { ¦fōd·ō·i'lek·trik flü'räm·əd·ər }

photoelectric imaging [GRAPHICS] The process of storing an image in a ferroelectric material by utilizing either the intrinsic or extrinsic photosensitivity in conjuction with the ferroelectric properties of the material. { ¦fōd·ō·i¦lek·trik 'im·ə·jiŋ }

photoelectric infrared radiation See near-infrared radiation. { ¦fōd·ō·i'lek·trik ¸in·frə¸red ¸rā·dē'ā·shən }

photoelectric intrusion detector [ELECTR] A burglar-alarm system in which interruption of a light beam by an intruder reduces the illumination on a phototube and thereby closes an alarm circuit. { ¦fōd·ō·i'lek·trik in'trü·zhən di¸tek·tər }

photoelectricity [ELECTR] The liberation of an electric charge by electromagnetic radiation incident on a substance; includes photoemission, photoionization, photoconduction, the photovoltaic effect, and the Auger effect (an internal photoelectric process). Also known as photoelectric effect; photoelectric process. { ¦fōd·ō¸i¸lek'tris·əd·ē }

photoelectric lighting control [ELECTR] Use of a photoelectric relay actuated by a change in illumination in a given area or at a given point. { ¦fōd·ō·i'lek·trik 'līd·iŋ kən¸trōl }

photoelectric liquid-level indicator [ENG] A level indicator in which rising liquid interrupts the light beam of a photoelectric control system; used in a tank or process vessel. { ¦fōd·ō·i'lek·trik ¦lik·wəd ¦lev·əl 'in·də¸kād·ər }

photoelectric loop control [CONT SYS] A photoelectric control system used as a position regulator for a loop of material passing from one strip-processing line to another that may travel at a different speed. Also known as loop control. { ¦fōd·ō·i'lek·trik 'lüp kən¸trōl }

photoelectric magnitude [ASTRON] The magnitude of a celestial object, as measured by a photoelectric photometer attached to a telescope. { ¦fōd·ō·i'lek·trik 'mag·nə¸tüd }

photoelectric photometer [ENG] A photometer that uses a photocell, phototransistor, or phototube to measure the intensity of light. Also known as electronic photometer. { ¦fōd·ō·i'lek·trik fə'täm·əd·ər }

photoelectric photometry [OPTICS] In contrast to the methods of visual photometry, an objective approach to the problems of photometry, wherein any of several types of photoelectric devices are used to replace the human eye as the sensing element. { ¦fōd·ō·i'lek·trik fə'täm·ə·trē }

photoelectric plethysmograph [MED] A medical instrument for measuring and recording ear opacity by means of a tiny phototube and lamp clipped to the ear, as a measure of the state of fullness of blood vessels; also worn by aircraft pilots during high-altitude flights, as an alarm indicating the need for more oxygen. { ¦fōd·ō·i'lek·trik plə'thiz·mə¸graf }

photoelectric process See photoelectricity. { ¦fōd·ō·i'lek·trik 'prä·səs }

photoelectric pyrometer [ENG] An instrument that measures high temperatures by using a photoelectric arrangement

to measure the radiant energy given off by the heated object. { ¦fōd·ō·i'lek·trik pī'räm·əd·ər }

photoelectric reader [COMPUT SCI] A device for reading information stored on paper tape or cards; data are read by sensing the presence or absence of holes. { ¦fōd·ō·i'lek·trik 'rēd·ər }

photoelectric reflectometer [ENG] A reflectometer that uses a photocell or phototube to measure the diffuse reflection of surfaces, powders, pastes, and opaque liquids. { ¦fōd·ō·i'lek·trik ¸rē¸flek'täm·əd·ər }

photoelectric register control [CONT SYS] A register control using a light source, one or more phototubes, a suitable optical system, an amplifier, and a relay to actuate control equipment when a change occurs in the amount of light reflected from a moving surface due to register marks, dark areas of a design, or surface defects. Also known as photoelectric scanner. { ¦fōd·ō·i'lek·trik 'rej·ə·stər kən¸trōl }

photoelectric relay [ELECTR] A relay combined with a phototube and amplifier, arranged so changes in incident light on the phototube make the relay contacts open or close. Also known as light relay. { ¦fōd·ō·i'lek·trik 'rē¸lā }

photoelectric scanner [COMPUT SCI] A device that scans punched cards by photoelectric means, as opposed to the standard brushes or "feelers," or mechanical plungers. [CONT SYS] See photoelectric register control. { ¦fōd·ō·i'lek·trik 'skan·ər }

photoelectric smoke-density control [CONT SYS] A photoelectric control system used to measure, indicate, and control the density of smoke in a flue or stack. { ¦fōd·ō·i'lek·trik 'smōk ¸den·səd·ē kən¸trōl }

photoelectric sorter [CONT SYS] A photoelectric control system used to sort objects according to color, size, shape, or other light-changing characteristics. { ¦fōd·ō·i'lek·trik 'sȯrd·ər }

photoelectric transmissometer [ENG] A device to measure the runway visibility at an airport by measuring the degree to which a light beam falling on a photocell is obscured by clouds or fog. { ¦fōd·ō·i'lek·trik ¸tranz·mə'säm·əd·ər }

photoelectric tube See phototube. { ¦fōd·ō·i'lek·trik 'tüb }

photoelectric turbidimeter [ENG] Device for measurement of solution turbidity by use of photocells to detect the loss of intensity of light beamed through the solution. { ¦fōd·ō·i'lek·trik ¸tər·bə'dim·əd·ər }

photoelectrolysis [PHYS CHEM] The process of using optical energy to assist or effect electrolytic processes that ordinarily require the use of electrical energy. { ¦fōd·ō¸i¸lek'träl·ə·səs }

photoelectromagnetic effect [ELECTR] The effect whereby, when light falls on a flat surface of an intermetallic semiconductor located in a magnetic field that is parallel to the surface, excess hole-electron pairs are created, and these carriers diffuse in the direction of the light but are deflected by the magnetic field to give a current flow through the semiconductor that is at right angles to both the light rays and the magnetic field. { ¦fōd·ō·i¦lek·trō·mag'nedik i'fekt }

photoelectromotive force [ELECTR] Electromotive force caused by photovoltaic action. { ¦fōd·ō·i¦lek·trō'mōd·iv 'fȯrs }

photoelectron [ELECTR] An electron emitted by the photoelectric effect. { ¦fōd·ō·i'lek¸trän }

photoelectron holography [ATOM PHYS] A technique for three-dimensional imaging of surface atoms in which electron waves produce holograms that are subjected to numerical image processing to yield computer displays of individual atoms. { ¦fōd·ō·i'lek¸trän hō'läg·rə·fē }

photoelectron spectroscopy [SPECT] The branch of electron spectroscopy concerned with the energy analysis of photoelectrons ejected from a substance as the direct result of bombardment by ultraviolet radiation or x-radiation. { ¦fōd·ō·i'lek¸trän spek'träs·kə·pē }

photoemission [ELECTR] The ejection of electrons from a solid (or less commonly, a liquid) by incident electromagnetic radiation. Also known as external photoelectric effect. { ¦fōd·ō·i'mish·ən }

photoemission threshold [ELECTR] The energy of a photon which is just sufficient to eject an electron from a solid or liquid in photoemission. { ¦fōd·ō·i'mish·ən 'thresh¸hōld }

photoemissive cell [ELECTR] A device which detects or

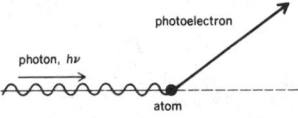

PHOTOEMISSION

Albert Einstein's approach to photoemission. Light beam behaves like a stream of photons, each of energy $h\nu$, where h is Planck's constant, and ν the frequency of the photon. When a photon interacts with an electron, the electron absorbs entire photon energy and is ejected from emitter if this energy exceeds a well-defined minimum value.

measures radiant energy by measurement of the resulting emission of electrons from the surface of a photocathode. { ¦fōd·ō·i'mis·iv 'sel }

photoemissive tube photometer [ENG] A photometer which uses a tube made of a photoemissive material; it is highly accurate, but requires electronic amplification, and is used mainly in laboratories. { ¦fōd·ō·i'mis·iv ¦tüb fə'täm·əd·ər }

photoemissivity [ELECTR] The property of a substance that emits electrons when struck by light. { ¦fōd·ō,ē·mə'siv·əd·ē }

photoemitter [SOLID STATE] A material that emits electrons when sufficiently illuminated. { ¦fōd·ō·i'mid·ər }

photoengraving [GRAPHICS] The technique of producing relief plates such as halftones or zinc etchings by photography; a metal plate is coated with a photosensitive emulsion and exposed to light under a reversed positive; the picture is developed by dissolving away the portion of the emulsion not acted upon by the light, and the plate is etched. { ¦fōd·ō·in'grāv·iŋ }

photoengraving zinc [MET] Pure zinc mixed with a small amount of iron to reduce grain size, and alloyed with a maximum of 0.2% each of cadmium, manganese, and magnesium; used for printing plates. { ¦fōd·ō·in'grāv·iŋ ,ziŋk }

photoenlarger See enlarger. { ¦fōd·ō·in'lär·jər }

photofabrication [ELECTR] In manufacturing circuit boards and integrated circuits, a process in which the etching pattern is placed over the circuit board or semiconductor material, the board or chip is placed in a special solution, and the assembly is exposed to light. { ¦fōd·ō,fab·rə'kā·shən }

photoferroelectric effect [SOLID STATE] An effect observed in ferroelectric ceramics such as PLZT materials, in which light at or near the band-gap energy of the material has an effect on the electric field in the material created by an applied voltage, and, at a certain value of the voltage, also influences the degree of ferroelectric remanent polarization. Abbreviated PFE. { ¦fōd·ō¦fer·ō·i'lek·trik i,fekt }

photo finishing [GRAPHICS] The commercial processing of photographs. { 'fōd·ō 'fin·ish·iŋ }

photofission [NUC PHYS] Fission of an atomic nucleus that results from absorption by the nucleus of a high-energy photon. { ¦fōd·ō'fish·ən }

photoflash bomb [ENG] A missile dropped from aircraft; it contains a photoflash mixture and a means for ignition at a distance above the ground, to produce a brilliant light of short duration for photographic purposes. { 'fōd·ə,flash ,bäm }

photoflash composition [MATER] A pyrotechnic material which, when loaded in a suitable casing and ignited, produces a flash of sufficient intensity and duration for photographic purposes; used as the filler in photoflash bombs and cartridges. { 'fōd·ə,flash ,käm·pə'zish·ən }

photoflash lamp [ELEC] A lamp consisting of a glass bulb filled with finely shredded aluminum foil in an atmosphere of oxygen; when the foil is ignited by a low-voltage dry cell, it burns with a burst of high-intensity light of short time duration and with definitely regulated time characteristics. { 'fōd·ə,flash ,lamp }

photoflash unit [ELECTR] A portable electronic light source for photographic use, consisting of a capacitor-discharge power source, a flash tube, a battery for charging the capacitor, and sometimes also a high-voltage pulse generator to trigger the flash. { 'fōd·ə,flash ,yü·nət }

photoflood lamp [ELEC] An incandescent lamp used in photography which has a high-temperature filament, so that it gives high illumination and high color temperature for a short lifetime. { 'fōd·ə,fləd ,lamp }

photofluorography See fluorography. { ¦fōd·ō·flù'räg·rə·fē }

photogelatin printing [GRAPHICS] A method of printing that produces extremely fine tones on uncoated papers by means of a photogelatin printing plate. { ¦fōd·ō'jel·ət·ən ,print·iŋ }

photogelatin printing plate [GRAPHICS] An aluminum or heavy-glass base coated with a layer of light-sensitive gelatin, exposed to light through an unscreened photographic negative of the copy to be reproduced; after treatment, the printing image varies in thickness, moisture content, and ability to hold or repel printing ink. Also known as albertype; artotype; collotype; heliotype; phototype. { ¦fōd·ō'jel·ət·ən 'print·iŋ ,plāt }

photogelatin process [GRAPHICS] Photomechanical process for producing prints directly from a hardened colloid film; the sensitized film is exposed under a reversed negative, desensitized, and then soaked in glycerin and salt water so that the parts that have not been exposed to light undergo swelling; the

swelled parts are ink-repellent and the unswelled parts ink-receptive, thereby forming a printing surface of the lithographic sort. Also known as collotype. { ¦fōd·ō'jel·ət·ən ,prä·səs }

photogeologic anomaly [GEOL] Any systematic deviation of a photogeologic factor from the expected norm in a given area. { ¦fōd·ō,jē·ə'läj·ik ə'näm·ə·lē }

photogeologic map [GEOL] A compilation of interpretations of a series of aerial photographs, including annotations of geologic features. { ¦fōd·ō,jē·ə'läj·ik 'map }

photogeology [GEOL] The geologic interpretation of landforms by means of aerial photographs. { ¦fōd·ō,jē'äl·ə·jē }

photogeomorphology [GEOL] The study of landforms by means of aerial photographs. { ¦fōd·ō,jē·ō·mòr'fäl·ə·jē }

photoglow tube [ELECTR] Gas-filled phototube used as a relay by making the operating voltage sufficiently high so that ionization and a flow discharge occur, with considerable current flow, when a certain illumination is reached. { 'fōd·ō,glō ,tüb }

photoglycine See glycin. { ¦fōd·ō'glī,sēn }

photogoniometer [ENG] A goniometer that uses a phototube or photocell as a sensing device for studying x-ray spectra and x-ray diffraction effects in crystals. { ¦fōd·ō,gō·nē'äm·əd·ər }

photogram [GRAPHICS] A design or pattern produced on regular photographic paper without the use of a negative or a lens system; opaque or transparent objects are assembled on the paper, and the paper is exposed to light and processed in the usual way. { 'fōd·ə,gram }

photogrammetry [ENG] **1.** The science of making accurate measurements and maps from aerial photographs. **2.** The practice of obtaining surveys by means of photography. { ,fōd·ə'gram·ə·trē }

photograph [GRAPHICS] A positive or negative image obtained by photography. { 'fōd·ə,graf }

photographic barograph [ENG] A mercury barometer arranged so that the position of the upper or lower meniscus may be measured photographically. { ¦fōd·ə¦graf·ik 'bar·ə,graf }

photographic emulsion [GRAPHICS] Microscopic grains of light-sensitive silver halide suspended in a gelatin surface on paper, plastic, metal, or glass; used to coat photographic film. { ¦fōd·ə¦graf·ik i'məl·shən }

photographic field [OPTICS] The area covered or "seen" by the lens of a camera. { ¦fōd·ə¦graf·ik 'fēld }

photographic film [GRAPHICS] Sensitized material (emulsion) coated on a flexible support, usually a transparent plastic material. { ¦fōd·ə¦graf·ik 'film }

photographic fixing [GRAPHICS] The process by which the unexposed and unreduced silver halide in a negative is removed in an exposed film; sodium thiosulfate (hypo) is the chemical usually used. { ¦fōd·ə¦graf·ik 'fiks·iŋ }

photographic interpretation See photointerpretation. { ¦fōd·ə¦graf·ik in,tər·prə'tā·shən }

photographic magnitude [ASTRON] The magnitude of a star, as obtained by measuring the apparent size of a star's image on a photographic emulsion sensitive to blue light at wavelengths between 400 and 500 nanometers. { ¦fōd·ə¦graf·ik 'mag·nə,tüd }

photographic meteor [ASTRON] A meteor which has been photographed for the purpose of determining its origin, velocity, and other characteristics. { ¦fōd·ə¦graf·ik 'mēd·ē·ər }

photographic objective [OPTICS] A camera lens designed to form sharp real images of objects on a photographic film. { ¦fōd·ə¦graf·ik əb'jek·tiv }

photographic photometry [SPECT] The use of a comparator-densitometer to analyze a photographed spectrograph spectrum by emulsion density measurements. { ¦fōd·ə¦graf·ik fə'täm·ə·trē }

photographic plane [GRAPHICS] **1.** The flat area on which a document is photographed. **2.** The surface at which an image is formed by a lens for recording on a photosensitive material. { ¦fōd·ə¦graf·ik 'plān }

photographic recording [COMMUN] Facsimile recording in which a photosensitive surface is exposed to a signal-controlled light beam or spot. { ¦fōd·ə¦graf·ik ri'kòrd·iŋ }

photographic sound recorder [ELECTR] A sound recorder having means for producing a modulated light beam and means for moving a light-sensitive medium relative to the beam to give a photographic recording of sound signals. Also known as optical sound recorder. { ¦fōd·ə¦graf·ik 'saùnd ri,kòrd·ər }

photographic sound reproducer [ELECTR] A sound reproducer in which an optical sound record on film is moved through a light beam directed at a light-sensitive device, to convert the recorded optical variations back into audio signals. Also known as optical sound reproducer. { ¦fōd·ə¦graf·ik 'saund ¦rē·prə‚düs·ər }

photographic surveying [ENG] Photographing of plumb bobs, clinometers, or magnetic needles in borehole surveying to provide an accurate permanent record. { ¦fōd·ə¦graf·ik sər'vā·iŋ }

photographic zenith tube [OPTICS] A type of zenith telescope in which light is reflected from a pool of mercury, and the photographic plate is held in a carriage just below the objective that is alternately rotated through 180° and moved slowly across the field of view to follow a star image; used for the accurate determination of time. { ¦fōd·ə¦graf·ik 'zēn·əth ‚tüb }

photograph nadir [OPTICS] The point at which a vertical line through the perspective center of a camera lens pierces the plane of the photograph. Also known as nadir point. { ¦fōd·ə¦graf·ik 'nā‚dir }

photography [GRAPHICS] The process of forming visible images directly or indirectly by the action of light or other forms of radiation on sensitive surfaces. { fə'täg·rə·fē }

photogravure [GRAPHICS] A method of making intaglio engravings in copper from photographs in which a gelatin image is used as an acid resist; it is considered high quality for halftones, but not very good for type. { ¦fōd·ə·grə¦vyùr }

photo-Hall effect [PHYS] An effect in which the illumination of a semiconductor in a magnetic field produces a change in its Hall resistance. { ¦fōd·ō 'hòl i‚fekt }

photoheliograph [OPTICS] A refracting telescope specially designed to photograph the sun's disk. { ¦fōd·ō·hē·lē·ə‚graf }

photohomolysis [PHYS CHEM] A homolysis reaction in which bond breaking is caused by radiant energy. { ¦fōd·ō·hə'mäl·ə·səs }

photoinhibition [BOT] Damage to the light-harvesting reactions of the photosynthetic apparatus caused by excess light energy trapped by the chloroplast. { ¦fōd·ō‚in·ə'bish·ən }

photoinitiated polymerization [PHYS CHEM] A chain reaction of monomer to polymer initiated by a photogenerated radical or ion. { ‚fōd·ō·ə¦nish·ē‚ād·əd pə‚lim·ə·rə'zā·shən }

photoinitiator [PHYS CHEM] A substance (other than reactant) which, on absorption of light, generates a reactive species (ion or radical), initiates a chemical reaction or transformation, and is consumed. { ‚fōd·ō·ə'nish·ē‚ād·ər }

photointerpretation [ENG] The science of identifying and describing objects in a photograph, such as deducing the topographic significance or the geologic structure of landforms on an aerial photograph. Also known as photographic interpretation. { ¦fōd·ō‚in‚ter·prə'tā·shən }

photoionization [PHYS CHEM] The removal of one or more electrons from an atom or molecule by absorption of a photon of visible or ultraviolet light. Also known as atomic photoelectric effect. { ¦fōd·ō‚ī·ə·nə'zā·shən }

photoisland grid [ELECTR] Photosensitive surface in the storage-type, Farnsworth dissector tube for television cameras. { 'fōd·ō‚ī·lənd ‚grid }

photoisolator See optoisolator. { ¦fōd·ō'ī·sə‚lād·ər }

photoisomer [PHYS CHEM] An isomer produced by photolysis. { ¦fōd·ō'ī·sə·mər }

photojunction battery [NUCLEO] A nuclear-type battery in which a radioactive material such as promethium-147 irradiates a phosphor which converts nuclear energy into light; the light is then converted to electrical energy by a small silicon junction. { ¦fōd·ō'jəŋk·shən ‚bad·ə·rē }

photolithography [GRAPHICS] Lithography in which photographically produced plates or masks are used. { ¦fōd·ō·li'thäg·rə·fē }

photology [OPTICS] The scientific study of light. { fō'täl·ə·jē }

photoluminescence [ATOM PHYS] Luminescence stimulated by visible, infrared, or ultraviolet radiation. { ¦fōd·ō‚lü·mə'nes·əns }

photolysis [PHYS CHEM] The use of radiant energy to produce chemical changes. { fō'täl·ə·səs }

photomacrography [GRAPHICS] Making large pictures of small subjects by using a short-focal-length lens on a long-bellows camera. { ¦fōd·ō·ma'kräg·rə·fē }

photomagnetic effect [PHYS] **1.** The direct effect of light on the magnetic susceptibility of certain substances. **2.** Paramagnetism displayed by certain substances when they are in a phosphorescent state. [NUC PHYS] Photodisintegration that results from the action of the magnetic field component of electromagnetic radiation. { ¦fōd·ō·mag¦ned·ik i'fekt }

photomagnetoelectric effect [ELECTROMAG] The generation of a voltage when a semiconductor material is positioned in a magnetic field and one face is illuminated. { ¦fōd·ō·mag¦ned·ō·i'lek·trik i‚fekt }

photomap [MAP] An aerial photograph or a controlled mosaic of rectified photographs to which have been added a reference grid, scale, place names, marginal information, and other pertinent data or map symbols. { 'fōd·ō‚map }

photomask [ELECTR] A film or glass negative that has many high-resolution images, used in the production of semiconductor devices and integrated circuits. { 'fōd·ō‚mask }

photomechanical [GRAPHICS] Pertaining to any platemaking process in which photographic negatives and positives are exposed onto plates or cylinders that have been coated with photosensitive substances. { ¦fōd·ō·mi'kan·ə·kəl }

photomechanochemistry [PHYS CHEM] A branch of polymer sciences that deals with photochemical conversion of chemical energy into mechanical energy. { ¦fōd·ō·mə‚kan·ō'kem·ə·strē }

photometer [ENG] An instrument used for making measurements of light or electromagnetic radiation, in the visible range. { fō'täm·əd·ər }

photometric binary See eclipsing variable star. { ¦fōd·ō¦me·trik 'bī‚ner·ē }

photometric parallax [ASTRON] The annual parallax of a star too far away for its parallax to be measured directly, as calculated from its apparent magnitude and its absolute magnitude inferred from its spectral type. { ¦fōd·ə¦me·trik 'par·ə‚laks }

photometric titration [ANALY CHEM] A titration in which the titrant and solution cause the formation of a metal complex accompanied by an observable change in light absorbance by the titrated solution. { ¦fōd·ə¦me·trik tī'trā·shən }

photometry [OPTICS] The calculation and measurement of quantities describing light, such as luminous intensity, luminous flux, luminous flux density, light distribution, color, absorption factor, spectral distribution, and the reflectance and transmittance of light; sometimes taken to include measurement of near-infrared and near-ultraviolet radiation as well as visible light. { fō'täm·ə·trē }

photomicrograph [GRAPHICS] A micrograph produced by photography. { ¦fōd·ō'mī·krə‚graf }

photomicrography [GRAPHICS] The photography of the image formed by the microscope. { ¦fōd·ō·mī'kräg·rə·fē }

photomorphogenesis [BOT] The control exerted by light over growth, development, and differentiation of plants that is independent of photosynthesis. { ¦fōd·ō‚mòr·fō'jen·ə·səs }

photomosaic [GRAPHICS] Composite photograph (usually aerial) made up of individual, small-area photographs placed side by side. { ¦fōd·ō·mō'zā·ik }

photomultiplier See multiplier phototube. { ¦fōd·ō'məl·tə‚plī·ər }

photomultiplier cell [ELECTR] A transistor whose *pn*-junction is exposed so that it conducts more readily when illuminated. { ¦fōd·ō'məl·tə‚plī·ər ‚sel }

photomultiplier counter [ELECTR] A scintillation counter that has a built-in multiplier phototube. { ¦fōd·ō'məl·tə‚plī·ər ¦kaunt·ər }

photomultiplier tube See multiplier phototube. { ¦fōd·ō'məl·tə‚plī·ər ‚tüb }

photon [OPTICS] See troland. [QUANT MECH] A massless particle, the quantum of the electromagnetic field, carrying energy, momentum, and angular momentum. Also known as light quantum. { 'fō‚tän }

photon antibunching [OPTICS] A quantum phenomenon that occurs in certain types of light emission such as resonance fluorescence, in which the emission of one photon reduces the probability that another photon will be emitted immediately afterward. { 'fō‚tän 'an·ti‚bənch·iŋ }

photon bunching [OPTICS] The tendency of photoelectric pulses from an illuminated photodetector to occur in bunches rather than at random. { 'fō‚tän ‚bənch·iŋ }

photon coupled isolator [ELECTR] Circuit coupling device, consisting of an infrared emitter diode coupled to a photon detector over a short shielded light path, which provides extremely high circuit isolation. { 'fō₁tän ¦kəp·əld 'ī·sə₁lād·ər }

photon coupling [ELECTR] Coupling of two circuits by means of photons passing through a light pipe. { 'fō₁tän ₁kəp·liŋ }

photon curve [PETRO ENG] A graphical plot of depth versus gamma radiation (photon) scatter during the radioactive logging of a well bore; used to detect differences in density at various reservoir depths. { 'fō₁tän ₁kərv }

photonegative [ELECTR] Having negative photoconductivity, hence decreasing in conductivity (increasing in resistance) under the action of light; selenium sometimes exhibits photonegativity. { ¦fōd·ō'neg·ə·tiv }

photon emission spectrum [PHYS] The relative numbers of optical photons emitted by a scintillator material per unit wavelength as a function of wavelength; the emission spectrum may also be given in alternative units such as wave number, photon energy, or frequency. { 'fō₁tän i¦mish·ən ₁spek·trəm }

photonephelometer [ENG] A nephelometer that uses a phototocell or phototube to measure the amount of light transmitted by a suspension of particles. { ¦fōd·ō₁nef·ə'läm·əd·ər }

photoneutrino [PARTIC PHYS] A member of a neutrino-antineutrino pair that is produced in the collision of a high-energy photon with an electron. { ¦fōd·ō'nü'trē·nō }

photoneutron [NUC PHYS] A neutron released from a nucleus in a photonuclear reaction. { ¦fōd·ō'nü₁trän }

photon flux [OPTICS] The number of photons in a light beam reaching a surface, such as the surface of the photocathode of a photomultiplier tube, in a unit of time. { 'fō₁tän ₁fləks }

photon gas [STAT MECH] An electromagnetic field treated as a collection of photons; it behaves as any other collection of bosons, except that the particles are emitted or absorbed without restriction on their number. { 'fō₁tän ₁gas }

photon-gated material [MATER] A material in which persistent spectral holeburning with a narrow-band laser occurs only in the presence of a second enabling light beam. { 'fō₁tän ¦gād·əd mə'tir·ē·əl }

photonic band-gap material See photonic crystal. { fə₁tän·ik 'band₁gap mə₁tir·ē·əl }

photonic crystal [OPTICS] A macroscopic, periodic dielectric structure that possesses spectral gaps (stop bands) for electromagnetic waves, in analogy with the energy bands and gaps in regular semiconductors. Also known as photonic band-gap material. { fə₁tän·ik 'krist·əl }

photonics [ELECTR] The electronic technology involved with the practical generation, manipulation, analysis, transmission, and reception of electromagnetic energy in the visible, infrared, and ultraviolet portions of the light spectrum. It contributes to many fields, including astronomy, biomedicine, data communications and storage, fiber optics, imaging, optical computing, optoelectronics, sensing, and telecommunications. Also known as optoelectronics. { fō'tän·iks }

photon microscope See optical microscope. { 'fō₁tän ₁mī·krə₁skōp }

photon sail See solar sail. { 'fō₁tän ₁sāl }

photon theory [QUANT MECH] A theory of photoemission developed by Einstein, according to which a light beam behaves like a stream of particles (called photons) when it delivers energy to a substance displaying photoemission, the particles each having an energy equal to Planck's constant times the frequency of the light. { 'fō₁tän ₁thē·ə·rē }

photonuclear reaction [NUC PHYS] A nuclear reaction resulting from the collision of a photon with a nucleus. { ¦fōd·ō'nü·klē·ər rē'ak·shən }

photooxidation [PHYS CHEM] **1.** The loss of one or more electrons from a photoexcited chemical species. **2.** The reaction of a substance with oxygen and light. When oxygen remains in the product, the reaction is also known as photooxygenation. { ¦fōd·ō₁äk·sə'dā·shən }

photooxygenation See photooxidation. { ₁fōd·ō₁äk·sə·jə'nā·shən }

photoperiodism [PHYSIO] The physiological responses of an organism to the length of night or day or both. { ¦fōd·ō'pir·ē·ə₁diz·əm }

photophilic [BIOL] Thriving in full light. { ¦fōd·ō¦fil·ik }

photophobia [PSYCH] An abnormal fear of light. { ₁fōd·ə'fō·bē·ə }

photophobic [BIOL] **1.** Avoiding light. **2.** Exhibiting negative phototropism. { ₁fōd·ə'fō·bik }

photophobic response [PHYSIO] A transient alteration in swimming direction or velocity when a motile organism is exposed to a sudden change in light intensity. { ₁fōd·ō₁fō·bik ri'späns }

photophore gland [VERT ZOO] A highly modified integumentary gland which develops into a luminous organ composed of a lens and a light-emitting gland; occurs in deep-sea teleosts and elasmobranchs. { 'fōd·ə₁fōr ₁gland }

photophoresis [PHYS] Production of unidirectional motion in a collection of very fine particles, suspended in a gas or falling in a vacuum by a powerful beam of light. { ₁fōd·ə·fə'rē·səs }

photophosphorylase [BIOCHEM] An enzyme that is associated with the surface of a thylakoid membrane and is involved in the final stages of adenosine triphosphate production by photosynthetic phosphorylation. { fōd·ō₁fä'sfōr·ə₁lās }

photophosphorylation [BIOCHEM] Phosphorylation that is induced by light energy in photosynthesis. { ¦fōd·ō₁fä·sfə·rə'lā·shən }

photophygous [BIOL] Thriving in shade. { fə'täf·ə·gəs }

photopic vision See foveal vision. { fō'täp·ik 'vizh·ən }

photopigment [BIOCHEM] A pigment that is unstable in the presence of light of appropriate wavelengths, such as the chromophore pigment which combines with opsins to form rhodopsin in the rods and cones of the vertebrate eye. { ¦fōd·ō¦pig·mənt }

photopolymer [PHYS CHEM] Any polymer which, on exposure to light, undergoes a spontaneous and permanent change in physical properties, such as crosslinking or depolymerization. { ₁fōd·ō¦päl·ə·mər }

photopositive [ELECTR] Having positive photoconductivity, hence increasing in conductivity (decreasing in resistance) under the action of light; selenium ordinarily has photopositivity. { ¦fōd·ō'päz·əd iv }

photoproton [NUC PHYS] A proton released from a nucleus in a photonuclear reaction. { ¦fōd·ō'prō₁tän }

photoreactive chlorophyll [BIOCHEM] Chlorophyll molecules which receive light quanta from antenna chlorophyll and constitute a photoreaction center where light energy conversion occurs. { ¦fōd·ō·rē'ak·tiv 'klōr·ə₁fil }

photoreactivation See photoreversal. { ₁fōd·ō₁rē·ak·tə'vā·shən }

photoreception [PHYSIO] The process of absorption of light energy by plants and animals and its utilization for biological functions, such as photosynthesis and vision. { ¦fōd·ō·ri'sep·shən }

photoreceptor [PHYSIO] A highly specialized, light-sensitive cell or group of cells containing photopigments. { ¦fōd·ō·ri'sep·tər }

photoreduction [CHEM] A chemical reduction that is produced by electromagnetic radiation. Also known as photochemical reduction. [PHYS CHEM] **1.** Addition of one or more electrons to a photoexcited chemical species. **2.** Photochemical hydrogenation of a substance. { ₁fōd·ō·ri'dək·shən }

photorefractive effect [OPTICS] An effect displayed by many electrooptic materials in which a change in the index of refraction is induced by the presence of light, and this change is retained for a time after the light exposure ceases. { ₁fōd·ō·r'frak·tiv i'fekt }

photorefractive material [MATER] A material whose index of refraction changes when it is exposed to light. { ₁fōd·ō·ri'frak·tiv mə'tir·ē·əl }

photorelief map [MAP] **1.** A map consisting of a photograph of a relief model of the area under study and showing salient physical features. **2.** A diagrammatic map that simulates or gives the impression of a photograph of a relief model of the area under study. { 'fōc·ō·ri₁lēf ₁map }

photoresist [GRAPHICS] A light-sensitive coating that is applied to a substrate or board, exposed, and developed prior to chemical etching; the exposed areas serve as a mask for selective etching. { 'fōd·ō·ri₁zist }

photoresistive cell See photoconductive cell. { ¦fōd·ō·ri'zis·tiv 'sel }

photoresistor See photoconductive cell. { ¦fōd·ō·ri'zis·tər }

photorespiration [BIOCHEM] Respiratory activity taking place in plants during the light period; CO_2 is released and O_2

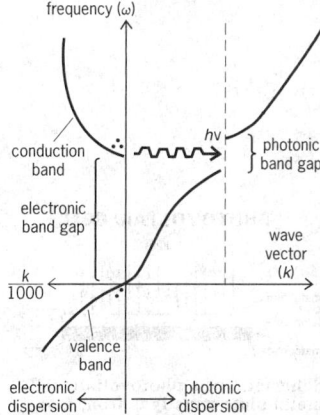

PHOTONIC CRYSTAL

Electromagnetic dispersion of a photonic crystal (right side), together with the electron wave dispersion typical of a direct-gap semiconductor (left side). The dots represent electrons and holes. *(After E. Yablonovich, Photonic band-gap structures, J. Opt. Soc. Amer., B, 10: 283–295, 1993)*

is taken up, but no useful form of energy, such as adenosinetri-phosphate, is derived. { ¦fōd·ō,res·pə'rā·shən }

photoreversal [BIOPHYS] An enzymatic repair system that uses short-wavelength visible (violet and blue) or long-wave-length ultraviolet light to reconstitute the deoxyribonucleic acid of cells that have been irradiated by ultraviolet light.   Also known as photoreactivation. { ,fōd·ō·ri'vər·səl }

photoscanner [ENG] A scanner used to make a film record of gamma rays passing through tissue from an injected radioac-tive material. { 'fōd·ō,skan·ər }

photo-SCR See light-activated silicon controlled rectifier. { ¦fōd·ō 'es¦sē'är }

photosensitive See light-sensitive. { fōd·ō·'sen·səd·iv }

photosensitive glass [GRAPHICS] Glass containing submi-croscopic metallic particles; when ultraviolet light passes through a negative on the glass, it precipitates the particles, with shadowed areas of the negative permitting deeper penetra-tion into the glass than highlight areas, giving the picture three dimensions and color; photograph is developed by heating the glass to 1000°F (538°C). { ¦fōd·ō'sen·səd·iv 'glas }

photosensitizer [PHYS CHEM] A light-absorbing substance that initiates a photochemical or photophysical reaction in another substance (molecule), and is not consumed in the reac-tion. { ¦fōd·ō'sen·sə,tīz·ər }

photosensor See photodetector. { ¦fōd·ō'sen·sər }

photosphere [ASTRON] The intensely bright portion of the sun visible to the unaided eye; it is a shell a few hundred miles in thickness marking the boundary between the dense interior gases of the sun and the more diffuse cooler gases in the outer portions of the sun. { 'fōd·ə,sfir }

photospheric granulation See granulation. { ¦fōd·ə'sfir·ik ,gran·yə'lā·shən }

photostabilize [ORG CHEM] To incorporate stabilizers in polymers, such as ultraviolet absorbers, to prevent photodegra-dation. { ,fōd·ō'stā·bə,līz }

photostriction [PHYS] The changes in the dimensions of piezoelectric materials that also exhibit one of the photoelectric effects when they are illuminated by light. { ,fōd·ə'strik·shən }

photosynthesis [BIOCHEM] Synthesis of chemical com-pounds in light, especially the manufacture of organic com-pounds (primarily carbohydrates) from carbon dioxide and a hydrogen source (such as water), with simultaneous liberation of oxygen, by chlorophyll-containing plant cells. { ¦fōd·ō'sin·thə·səs }

photosystem I [BIOCHEM] One of two reaction sequences of the light phase of photosynthesis in green plants that involves a pigment system which is excited by wavelengths shorter than 700 nanometers and which transfers this energy to energy carriers such as NADPH that are subsequently utilized in carbon dioxide fixation. { 'fōd·ō,sis·təm 'wən }

photosystem II [BIOCHEM] One of two reaction sequences of the light phase of photosynthesis in green plants which involves a pigment system excited by wavelengths shorter than 685 nanometers and which is directly involved in the splitting or photolysis of water. { 'fōd·ō,sis·təm 'tü }

phototaxis [BIOL] Movement of a motile organism or free plant part in response to light stimulation. { ¦fōd·ə¦tak·səs }

phototelegraphy See facsimile. { ¦fōd·ō·tə'leg·rə·fē }

phototheodolite [ENG] A ground-surveying instrument used in terrestrial photogrammetry which combines the func-tions of a theodolite and a camera mounted on the same tripod. { ¦fōd·ō·thē'äd·əl,īt }

photothermoelasticity [OPTICS] Changes in optical proper-ties of a transparent dielectric when it is subjected to mechanical stress, which is, in turn, induced by temperature gradients. { ¦fōd·ō,thər·mō,i,las'tis·əd·ē }

photothyristor See light-activated silicon controlled rectifier. { ¦fōd·ō·thī'ris·tər }

phototopography [ENG] The science of mapping and sur-veying in which details are plotted entirely from photographs taken at suitable ground stations. { ¦fōd·ō·tə'päg·rə·fē }

phototransistor [ELECTR] A junction transistor that may have only collector and emitter leads or also a base lead, with the base exposed to light through a tiny lens in the housing; collector current increases with light intensity, as a result of amplification of base current by the transistor structure. { ¦fōd·ō·tran'zis·tər }

phototriangulation [ENG] The extension of horizontal or

vertical control points, or both, by photogrammetric methods, whereby the measurements of angles and distances on overlap-ping photographs are related into a spatial solution using the perspective principles of the photographs. { ¦fōd·ō·trī,aŋ·gyə'lā·shən }

phototronic photocell See photovoltaic cell. { ¦fōd·ō¦trän·ə 'fōd·ə,sel }

phototroph [BIOL] An organism that utilizes light as a source of metabolic energy. { 'fōd·ə,träf }

phototrophic bacteria [MICROBIO] Primarily aquatic bacte-ria comprising two principal groups: purple bacteria and green sulfur bacteria; all contain bacteriochlorophylls. { ¦fōd·ə¦träf·ik bak'tir·ē·ə }

phototropism [BOT] A growth-mediated response of a plant to stimulation by visible light. [SOLID STATE] A reversible change in the structure of a solid exposed to light or other radiant energy, accompanied by a change in color. Also known as phototropy. { fō'tä·trə,piz·əm }

phototropy See phototropism. { fō'tä·trə·pē }

phototube [ELECTR] An electron tube containing a photo-cathode from which electrons are emitted when it is exposed to light or other electromagnetic radiation. Also known as electric eye; light-sensitive tube; photoelectric tube. { 'fōd·ō,tüb }

phototube cathode [ELECTR] The photoemissive surface which is the most negative element of a phototube. { 'fōd·ō,tüb 'kath,ōd }

phototube current meter [ENG] A device for measuring the speed of water currents in which a perforated disk, which rotates with the current by means of a propeller, is placed in the path of a beam of light that is then reflected from a mirror onto a phototube. { 'fōd·ō,tüb 'kə·rənt ,mēd·ər }

phototube relay [ELECTR] A photoelectric relay in which a phototube serves as the light-sensitive device. { 'fōd·ō,tüb 'rē,lā }

phototype See photogelatin printing plate. { 'fōd·ə,tīp }

phototypesetter [GRAPHICS] Machine for placing individ-ual characters on photographic film. { ¦fōd·ō'tīp,sed·ər }

phototypesetting [GRAPHICS] A method of composing text matter by successively projecting the images of characters on light-sensitive film or on photographic paper. { ¦fōd·ō'tīp,sed·iŋ }

photovaristor [ELECTR] Varistor in which the current-volt-age relation may be modified by illumination, for example, one in which the semiconductor is cadmium sulfide or lead telluride. { ¦fōd·ō·və'ris·tər }

photoviscoelasticity [OPTICS] Changes in optical proper-ties of a transparent, viscoelastic substance when it is subjected to stress. { ¦fōd·ō,vis·gō,i,las'tis·əd·ē }

photovisual magnitude [ASTRON] The magnitude of a star, obtained by measuring the size of the star's image on an isochro-matic photographic emulsion, using a filter transmitting only the longer wavelengths between 500 and 600 nanometers; nearly identical with visual magnitude. { ¦fōd·ō'vizh·ə·wəl 'mag·nə,tüd }

photovoltaic [ELECTR] Capable of generating a voltage as a result of exposure to visible or other radiation. { ¦fōd·ō·vōl'tā·ik }

photovoltaic cell [ELECTR] A device that detects or meas-ures electromagnetic radiation by generating a potential at a junction (barrier layer) between two types of material, upon absorption of radiant energy. Also known as barrier-layer cell; barrier-layer photocell; boundary-layer photocell; photronic photocell. { ¦fōd·ō·vōl'tā·ik ,sel }

photovoltaic effect [ELECTR] The production of a voltage in a nonhomogeneous semiconductor, such as silicon, or at a junction between two types of material, by the absorption of light or other electromagnetic radiation. { ¦fōd·ō·vōl'tā·ik i,fekt }

photovoltaic meter [ELECTR] An exposure cell in which a photovoltaic cell produces a current proportional to the light falling on the cell, and this current is measured by a sensitive microammeter. { ¦fōd·ō·vōl'tā·ik ,mēd·ər }

photox cell [ELECTR] Type of photovoltaic cell in which a voltage is generated between a copper base and a film of cuprous oxide during exposure to visible or other radiation. { 'fō,täks ,sel }

photronic cell [ELECTR] Type of photovoltaic cell in which

PHOTOVOLTAIC CELL

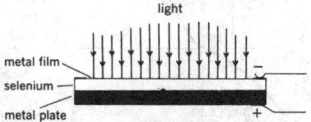

Schematic of a photovoltaic cell; metal plate usually is iron, and thin metal film is either gold or platinum.

PHOTOVOLTAIC METER

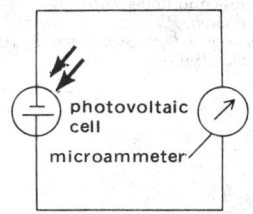

Circuit diagram of photovoltaic type of exposure meter. Heavy arrows represent light falling on cell.

a voltage is generated in a layer of selenium during exposure to visible or other radiation. { fō'trän·ik ,sel }

photronic photocell *See* photovoltaic cell. { fō'trän·ik 'fōd·ə,sel }

Phoxichilidiidae [INV ZOO] A family of marine arthropods in the subphylum Pycnogonida; typically, chelifores are present, palpi are lacking, and ovigers have five to nine joints in males only. { ,fäk·sə,kil·ə'dī·ə,dē }

phoxim *See* phenylglyoxylonitriloxime *O,O*-diethyl phosphorothioate. { 'fäk,sim }

Phoxocephalidae [INV ZOO] A family of amphipod crustaceans in the suborder Gammaridea. { ,fäk·sō·sə'fal·ə,dē }

Phractolaemidae [VERT ZOO] A family of tropical African fresh-water fishes in the order Gonorynchiformes. { ,frak·tə'lē·mə,dē }

phragmacone *See* phragmocone. { 'frag·mə,kōn }

Phragmobasidiomycetes [MYCOL] An equivalent name for Heterobasidiomycetidae. { ¦frag·mō·bə¦sid·ē·ō·mī'sēd·ēz }

phragmocone [INV ZOO] The siphuncular tube of the chambered part of the shell of certain mollusks. Also spelled phragmacone. { 'frag·mə,kōn }

phragmoid [BOT] Having septae perpendicular to the long axis, as the conidia of certain fungi. { 'frag,mȯid }

phragmoplast [CYTOL] A thin barrier which is formed across the spindle equator in late cytokinesis in plant cells and within which the cell wall is laid down. { 'frag·mə,plast }

phragmosome [CYTOL] A differentiated cytoplasmic partition in which the phragmoplast and cell plate develop during cell division in plant cells. { 'frag·mə,sōm }

Phragmosporae [MYCOL] A spore group of the Fungi Imperfecti with three- to many-celled spores. { frag'mäs·pə,rē }

phrase name *See* metavariable. { 'frāz ,nām }

phreatic [GEOL] Of a volcanic explosion of material such as steam or mud, not being incandescent. { frē'ad·ik }

phreatic cycle [HYD] The period of time during which the water table rises and then falls. { frē'ad·ik 'sī·kəl }

phreatic gas [GEOL] A gas formed by the contact of atmospheric or surface water with ascending magma. { frē'ad·ik 'gas }

phreatic surface *See* water table. { frē'ad·ik 'sər·fəs }

phreatic water [HYD] Groundwater in the zone of saturation. { frē'ad·ik 'wȯd·ər }

phreatic-water discharge *See* groundwater discharge. { frē'ad·ik ¦wȯd·ər 'dis,chärj }

phreatic zone *See* zone of saturation. { frē'ad·ik ,zōn }

Phreatoicidae [INV ZOO] A family of isopod crustaceans in the suborder Phreatoicoidea in which only the left mandible retains a lacinia mobilis. { frē,ad·ō'īs·ə,dē }

Phreatoicoidea [INV ZOO] A suborder of the Isopoda having a subcylindrical body that appears laterally compressed, antennules shorter than the antennae, and the first thoracic segment fused with the head. { frē,ad·ō·i'kȯid·ē·ə }

phreatomagmatic [GEOL] Pertaining to a volcanic explosion that extrudes both magmatic gases and steam; it is caused by the contact of the magma with groundwater or ocean water. { frē¦ad·ō·mag'mad·ik }

phreatophyte [ECOL] A plant with a deep root system which obtains water from the groundwater or the capillary fringe above the water table. { frē'ad·ə,fīt }

phrenectomy [MED] Resection of a section of a phrenic nerve or removal of an entire phrenic nerve. { frə'nek·tə·mē }

phrenic nerve [NEUROSCI] A nerve, arising from the third, fourth, and fifth cervical (cervical plexus) segments of the spinal cord; innervates the diaphragm. { 'fren·ik 'nərv }

phrynoderma [MED] Dryness of the skin with follicular hyperkeratosis, caused by vitamin A deficiency. { ,frī·nə'dər·mə }

Phrynophiurida [INV ZOO] An order of the Ophiuroidea in which the vertebrae usually articulate by means of hourglass-shaped surfaces, and the arms are able to coil upward or downward in the vertical plane. { ,frī·nə'fyür·ə·də }

pH(S) *See* pH standard. { }

pH standard [ANALY CHEM] Five standard laboratory solutions available from the U. S. National Bureau of Standards, each solution having a known pH value; the standards cover pH ranges from 3.557 to 8.833. Abbreviated pH(S). { ,pē'āch ¦stan·dərd }

phthalate [ORG CHEM] A salt of phthalic acid; contains the radical $C_6H_4(CCO)_2^{2-}$; an example is dibutylphthalate, $C_{16}H_{22}O_4$; used as a plasticizer in plastics, and as a buffer in standard laboratory solutions. { 'tha,lāt }

phthalate buffer [ANALY CHEM] Laboratory pH reference solution made of potassium hydrogen phthalate, $KHC_8H_4O_4$; at 0.05 molal, the pH is 4.008 at 25°C. { 'tha,lāt 'bəf·ər }

phthalate ester [ORG CHEM] Any of a group of plastics plasticizers made by the direct action of alcohol on phthalic anhydride; generally characterized by moderate cost, good stability, and good general properties. { 'tha,lāt 'es·tər }

phthalazine [ORG CHEM] $C_6H_4CHN_2CH$ Colorless crystals, melting at 91°C; soluble in alcohol. { 'thal·ə,zēn }

phthalic acid [ORG CHEM] $C_5H_4(CO_2H)_2$ Any of three isomeric benzene dicarboxylic acids; the ortho form is usually called phthalic acid, comprises alcohol-soluble, colorless crystals decomposing at 191°C, slightly soluble in water and ether, is used to make dyes, medicine, and synthetic perfumes, and as a chemical intermediate, and is also known as benzene orthodicarboxylic acid; the para form, known as terephthalic acid, is used to make polyester resins (Dacron) and as poultry feed additives; the meta form is isophthalic acid. { 'thal·ik 'as·əd }

meta-**phthalic acid** *See* isophthalic acid. { ¦med·ə 'thal·ik 'as·əd }

ortho-**phthalic acid** *See* phthalic acid. { ¦ȯr·thō 'thal·ik 'as·əd }

para-**phthalic acid** *See* terephthalic acid. { ¦par·ə 'thal·ik 'as·əd }

phthalic anhydride [ORG CHEM] $C_6H_4(CO)_2O$ White crystals, melting at 131°C; sublimes when heated; slightly soluble in ether and hot water, soluble in alcohol; used to make dyes, resins, plasticizers, and insect repellents. { 'thal·ik an'hī,drīd }

phthalimide [ORG CHEM] $C_8H_5NO_2$ The product made by heating phthalic anhydride with ammonia; used in Gabriel's synthesis of primary amines, amino acids, and anthranilic acid (*o*-aminobenzoic acid). { 'thal·ə,mīd }

phthalocyanine pigments [ORG CHEM] A group of lightfast organic pigments with four isoindole groups, $(C_6H_4)C_2N$, linked by four nitrogen atoms to form a conjugated chain; included are phthalocyanine (blue-green), copper phthalocyanine (blue), chlorinated copper phthalocyanine (green), and sulfonated copper phthalocyanine (green); used in enamels, plastics, linoleum, inks, wallpaper, and rubber goods. { ¦thal·ō'sī·ə·nən 'pig·məns }

phthalocyanine Q switching [OPTICS] Laser Q switching in which a solution of metal-organic compounds known as phthalocyanines is placed in a cell between an uncoated ruby laser crystal and a high-reflectivity mirror; when the incident ruby light reaches a certain level, the solution suddenly becomes almost perfectly transparent to this light, permitting the release of all the energy stored in the ruby as a giant pulse. { ¦thal·ō'sī·ə·nən 'kyü ,swich·iŋ }

phthalonitrile [ORG CHEM] $C_6H_4(CN)_2$ Buff-colored crystals with a melting point of 138°C; soluble in acetone and benzene; used in organic synthesis and as an insecticide. { ¦thal·ō'nī·trəl }

phthanite *See* chert. { 'tha,nīt }

phugoid [AERO ENG] Pertaining to variations in the longitudinal motion or course of the center of mass of an aircraft. { 'fü,gȯid }

Phycitinae [INV ZOO] A large subfamily of moths in the family Pyralididae in which the frenulum of the female is a simple spine rather than a bundle of bristles. { fī'sīt·ən,ē }

phycobilin [BIOCHEM] Any of various protein-bound pigments which are open-chain tetrapyrroles and occur in some groups of algae. { ,fī·kō'bī·lən }

phycobiliprotein [BIOCHEM] A water-soluble photosynthetic membrane protein that covalently binds with phycobilins (photosynthetic pigments) in some groups of algae. { ,fī·kō,bil·ē'prō,tēn }

phycobilisome [BIOCHEM] A light-harvesting structure containing aggregates of photosynthetic accessory pigments that is located on the surface of thylakoid membranes in all cyanobacteria and red algae. { ,fī·kō'bil·ē,sōm }

phycobiliviolin [BIOCHEM] A yellow-light (575-nanometer) absorbing pigment found in all cryptophytes but in only a few

PHREATOICOIDEA

Onchotelson brevicaudatus.

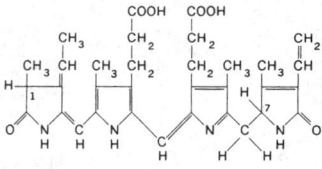

PHYCOCYANOBILIN

Structural formula of
phycocyanobilin.

PHYLLOBRANCHIATE GILL

The phyllobranchiate gill of the
mud shrimp (*Thalassinia*).

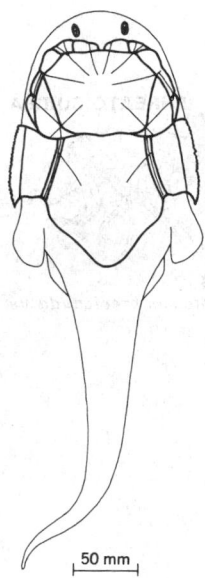

PHYLLOLEPIDA

Phyllolepis orvini from the Upper
Devonian of Greenland. The
restoration of the dermal armor is
shown in dorsal aspect. The
outline of the front of the head
and the body is hypothetical, but
shows how the fish might have
appeared in life.

cyanobacteria. Also known as cryptoviolin. { ‚fī·kō‚bil·ē'vī·ə·lən }

phycobiont [BOT] The algal component of a lichen, commonly the green unicell of the genus *Trebouxia*. { ‚fī·kō'bī‚änt }

phycocyanin [BIOCHEM] A blue phycobilin. { ¦fī·kō'sī·ə·nən }

phycocyanobilin [BIOCHEM] $C_{31}H_{38}O_2N_4$ Phycobilin with an ethylidene side chain ($=CH-CH_3$) and only one asymmetric carbon atom (C_1). { ¦fī·kō‚sī·ə·nō'bī·lən }

phycoerythrin [BIOCHEM] A red phycobilin. { ¦fī·kō·ə'rith·rən }

phycoerythrobilin [BIOCHEM] $C_{31}H_{38}O_2N_4$ Phycobilin with seven conjugated double bonds, an ethylidine side chain ($=CH-CH_3$), and two asymmetric carbon atoms (C_1 and C_7). { ¦fī·kō·ə‚rith·rə'bī·lən }

phycology See algology. { fī'käl·ə·jē }

Phycomycetes [MYCOL] A primitive class of true fungi belonging to the Eumycetes; they lack regularly spaced septa in the actively growing portions of the plant body, and have the sporangiospore, produced in the sporangium by cleavage, as the fundamental, asexual reproductive unit. { ‚fī·kō‚mī'sēd·ēz }

Phycosecidae [INV ZOO] A small family of coleopteran insects of the superfamily Cucujoidea, including five species found in New Zealand, Australia, and Egypt. { ¦fī·kō'sē·sə‚dē }

phycourobilin [BIOCHEM] A blue-green light (495-nanometer) absorbing pigment found in some cyanobacteria and red algae. { ‚fī·kō·yù'rō·bi·lin }

Phylactolaemata [INV ZOO] A class of fresh-water ectoproct bryozoans; individuals have lophophores which are U-shaped in basal outline, and relatively short, wide zooecia. { ‚fə‚lak·tō'lē·məd·ə }

phyletic evolution [EVOL] The gradual evolution of population without separation into isolated parts. { fī'led·ik ‚ev·ə'lü·shən }

phyletic gradualism See gradualism. { fī'led·ik 'gra·jə·wə‚liz·əm }

phyllary [BOT] A bract of the involucre of a composite plant. { 'fil·ə·rē }

phyllite [PETR] A metamorphic rock intermediate in grade between slate and schist, and derived from argillaceous sediments; has a silky sheen on the cleavage surface. { 'fi‚līt }

Phyllobothrioidea [INV ZOO] The equivalent name for Tetraphyllidea. { ‚fil·ō‚bäth·rē'òid·ē·ə }

phyllobranchiate gill [INV ZOO] A type of decapod crustacean gill with flattened branches, or lamellae usually arranged in two opposite series. { ‚fil·ō'braŋ·kē·ət 'gil }

phylloclade [BOT] A flattened stem that fulfills the same functions as a leaf. { 'fil·ə‚klād }

phyllode [BOT] A broad, flat petiole that replaces the blade of a foliage leaf. [INV ZOO] A petal-shaped group of ambulacra near the mouth of certain echinoderms. { 'fi‚lōd }

Phyllodocidae [INV ZOO] A leaf-bearing family of errantian annelids in which the species are often brilliantly iridescent and are highly motile. { ‚fil·ə'däs·ə‚dē }

phyllofacies [GEOL] A facies differentiated on the basis of stratification characteristics, especially the stratification index. { ‚fil·ō'fā·shēz }

Phyllogoniaceae [BOT] A family of mosses in the order Isobryales in which the leaves are equitant. { ‚fil·ə‚gō·nē'ās·ē‚ē }

Phyllolepida [PALEON] A monogeneric order of placoderms from the late Upper Devonian in which the armor is broad and low with a characteristic ornament of concentric and transverse ridges on the component plates. { ‚fil·ə'lep·ə·də }

phyllomorphic stage [GEOL] The most advanced geochemical stage of diagenesis, characterized by authigenic development of micas, feldspars, and chlorites at the expense of clays. { ¦fil·ə¦mòr·fik ‚stāj }

phyllonite [PETR] A metamorphic rock occupying an intermediate position between phyllite and mylonite. { 'fil·ə‚nīt }

Phyllophoridae [INV ZOO] A family of dendrochirotacean holothurians in the order Dendrochirotida having a rather naked skin and a complex calcareous ring. { ‚fil·ə'fòr·ə‚dē }

phyllosilicate [MINERAL] A structural type of silicate mineral in which flat sheets are formed by the sharing of three of the four oxygen atoms in each tetrahedron with neighboring

tetrahedrons. Also known as layer silicate; sheet mineral; sheet silicate. { ‚fil·ō'sil·ə·kət }

phyllosoma [INV ZOO] A flat, transparent, long-legged larval stage of various spiny lobsters. { ‚fil·ə'sō·mə }

phyllospondylous [VERT ZOO] Of vertebrae, having a hypocentrum but no pleurocentra; the neural arch extends ventrad to enclose the notochord and form transverse processes which articulate with the ribs. { ‚fil·ə'spän·də·ləs }

Phyllostictales [MYCOL] An equivalent name for Sphaeropsidales. { ‚fil·ə‚stik'tā·lēz }

Phyllostomatidae [VERT ZOO] The New World leaf-nosed bats (Chiroptera), a large tropical and subtropical family of insect- and fruit-eating forms with narrow, pointed ears. { ‚fil·ə‚stō'mad·ə‚dē }

phyllotaxy [BOT] The arrangement of leaves on a stem. { ¦fil·ə¦tak·sē }

Phylloxerinae [INV ZOO] A subfamily of homopteran insects in the family Chermidae in which the sexual forms lack mouthparts, and the parthenogenetic females have a beak but the digestive system is closed, and no honeydew is produced. { ‚fi‚läk'ser‚ə‚nē }

phylogenetic tree See evolutionary tree. { ‚fī·lō·jə¦ned·ik 'trē }

phylogeny [EVOL] The evolutionary or ancestral history of organisms. { fə'läj·ə·nē }

phylum [SYST] A major taxonomic category in classifying animals (and plants in some systems), composed of groups of related classes. { 'fī·ləm }

Phymatidae [INV ZOO] A family of carnivorous hemipteran insects characterized by strong, thick forelegs. { fī'mad·ə‚dē }

Phymosomatidae [INV ZOO] A family of echinacean echinoderms in the order Phymosomatoida with imperforate crenulate tubercles; one surviving genus is known. { ‚fī·mə·sō'mad·ə‚dē }

Phymosomatoida [INV ZOO] An order of Echinacea with a stirodont lantern and diademoid ambulacral plates. { ‚fī·mə·sə·mə'tòid·ē·ə }

physa [INV ZOO] The rounded basal portion of the body of certain sea anemones. { 'fī·sə }

Physalopteridae [INV ZOO] A family of parasitic nematodes in the superfamily Spiruroidea. { ‚fī·sə·läp'ter·ə‚dē }

Physalopteroidea [INV ZOO] A superfamily of parasitic nematodes in the order Spirurida, characterized by two large lateral lips generally provided with teeth on their inner surfaces, reduced stoma, and a reduced or absent inner whorl of circumoral sensilla and an external circle with four fused sensilla. { 'fī·sə‚läp·tə'ròid·ē·ə }

Physaraceae [MYCOL] A family of slime molds in the order Physarales. { ‚fī·sə'rās·ē‚ē }

Physarales [MYCOL] An order of Myxomycetes in the subclass Myxogastromycetidae. { ‚fī·sə'rā·lēz }

physiatrics [MED] A branch of medicine dealing with treatment, prevention, and diagnosis of disorders and disabilities through the use of physical therapy, physical agents such as electricity, light, heat, and water, and mechanical apparatus. { ‚fiz·ē'a·triks }

physiatrist [MED] A physician specializing in physiatrics. { ‚fiz·ē'a·trəst }

physical adsorption [PHYS CHEM] Reversible adsorption in which the adsorbate is held by weak physical forces. { 'fiz·ə·kəl ad'sòrp·shən }

physical anthropology [ANTHRO] The science that deals with the biological aspects of humankind and their relation to historical or cultural aspects. Also known as biological anthropology. { 'fiz·ə·kəl ‚an·thrə'päl·ə·jē }

physical chemistry [CHEM] The branch of chemistry that deals with the interpretation of chemical phenomena and properties in terms of the underlying physical processes, and with the development of techniques for their investigation. { 'fiz·ə·kəl 'kem·ə·strē }

physical climate [CLIMATOL] The actual climate of a place, as distinguished from a hypothetical climate, such as the solar climate or mathematical climate. { 'fiz·ə·kəl 'klī·mət }

physical climatology [CLIMATOL] The major branch of climatology, which deals with the explanation of climate, rather than with presentation of it (climatography). { 'fiz·ə·kəl ‚klī·mə'täl·ə·jē }

physical compatibility [ENG] The ability of two or more

materials, substances, or chemicals to be used together without ill effect. { 'fiz·ə·kəl kəm,pad·ə'bil·əd·ē }

physical constant [PHYS] A physical quantity which has a fixed and unchanging numerical value. { 'fiz·ə·kəl 'kän·stənt }

physical data independence [COMPUT SCI] A file structure such that the physical structure of the data can be modified without changing the logical structure of the file. { 'fiz·ə·kəl ¦dad·ə ,in·di'pen·dəns }

physical data structure [COMPUT SCI] The manner in which data are physically arranged on a storage medium, including various indices and pointers. { 'fiz·ə·kəl 'dad·ə ,strək·chər }

physical device table [COMPUT SCI] A table associated with a physical input/output unit containing such information as the device type, an indication of data paths that may be used to transfer information to and from the device, status information on whether the device is busy, the input/output operation currently pending on the device, and the availability of any storage contained in the device. { 'fiz·ə·kəl di¦vīs ,tā·bəl }

physical drive [COMPUT SCI] An operational hard disk, which may be formatted to include more than one logical drive. { 'fiz·i·kəl ¦drīv }

physical electronics [ELECTR] The study of physical phenomena basic to electronics, such as discharges, thermionic and field emission, and conduction in semiconductors and metals. { 'fiz·ə·kəl i,lek'trän·iks }

physical exfoliation [GEOL] A type of exfoliation caused by physical forces; for example, by the freezing of water that has penetrated fine cracks in rock or by the removal of overburden concealing deeply buried rocks. { 'fiz·ə·kəl eks,fō·lē'ā·shən }

physical forecasting *See* numerical forecasting. { 'fiz·ə·kəl 'fȯr,kast·iŋ }

physical geography [GEOGR] The study of the earth's surface features and associated processes. { 'fiz·ə·kəl jē'äg·rə·fē }

physical geology [GEOL] That branch of geology concerned with understanding the composition of the earth and the physical changes occurring in it, based on the study of rocks, minerals, and sediments, their structures and formations, and their processes of origin and alteration. { 'fiz·ə·kəl jē'äl·ə·jē }

physical input/output control system *See* PIOCS. { 'fiz·ə·kəl ¦in,pu̇t ¦au̇t,pu̇t kən'trōl ,sis·təm }

physical law [PHYS] A property of a physical phenomenon, or a relationship between the various quantities or qualities which may be used to describe the phenomenon, that applies to all members of a broad class of such phenomena, without exception. { 'fiz·ə·kəl 'lȯ }

physical libration of the moon *See* lunar libration. { 'fiz·ə·kəl lī'brā·shən əv <u>th</u>ə 'mün }

physical measurement [PHYS] Quantitative information on a physical condition, property, or relation, generally in the form of the ratio of the measured quantity to a standard quantity, or to some fixed multiple or fraction thereof. { 'fiz·ə·kəl 'mezh·ər·mənt }

physical medicine [MED] A consultative, diagnostic, and therapeutic medical specialty, coordinating and integrating the use of physical and occupational therapy and physical reconditioning in the professional management of the diseased and injured. { 'fiz·ə·kəl 'med·ə·sən }

physical metallurgy [MET] The branch of metallurgy concerned with physical and mechanical properties of metals as affected by composition, mechanical working, and heat treatment. { 'fiz·ə·kəl 'med·əl·ər·jē }

physical meteorology [METEOROL] That branch of meteorology which deals with optical, electrical, acoustical, and thermodynamic phenomena of the atmosphere, its chemical composition, the laws of radiation, and the explanation of clouds and precipitation. { 'fiz·ə·kəl ,mēd·ē·ə'räl·ə·jē }

physical modeling synthesis [ENG ACOUS] A method of synthesizing the sounds of a musical instrument that uses computational algorithms that are based directly on the mathematical physics of the instrument. { ,fiz·i·kəl 'mäd·əl·iŋ ,sin·thə·səs }

physical network [COMPUT SCI] A system of computers that communicate via cabling, modems, or other hardware, and may include more than one logical network or form part of a logical network. { 'fiz·i·kəl ¦net,wərk }

physical oceanography [OCEANOGR] The study of the physical aspects of the ocean, the movements of the sea, and the variability of these factors in relationship to the atmosphere and the ocean bottom. { 'fiz·ə·kəl ,ō·shə'näg·rə·fē }

physical optics [OPTICS] The study of the interaction of electromagnetic waves in the optical frequency range with material systems. { 'fiz·ə·kəl 'äp·tiks }

physical organic chemistry [ORG CHEM] The study of the scope and limitations of the various rules, effects, and generalizations in use in organic chemistry by application of physical and mathematical means. { ¦fiz·ə·kəl ȯr¦gan·ik 'kem·ə·strē }

physical path length [COMPUT SCI] The physical distance that an electronic signal must travel between two points. Also known as path length. { 'fiz·ə·kəl ¦path ,leŋkth }

physical property [CHEM] Property of a compound that can change without involving a change in chemical composition; examples are the melting point and boiling point. { 'fiz·ə·kəl 'präp·ərd·ē }

physical realizability [CONT SYS] For a transfer function, the possibility of constructing a network with this transfer function. { 'fiz·ə·kəl ,rē·ə,līz·ə'bil·əd·ē }

physical record [COMPUT SCI] A set of adjacent data characters recorded on some storage medium, physically separated from other physical records that may be on the same medium by means of some indication that can be recognized by a simple hardware test. Also known as record block. { 'fiz·ə·kəl 'rek·ərd }

physical residue [GEOL] A residue which results from physical, as opposed to chemical, weathering processes. { 'fiz·ə·kəl 'rez·ə,dü }

physical stratigraphy [GEOL] Stratigraphy based on the physical aspects of rocks, especially the sedimentologic aspects. { 'fiz·ə·kəl strə'tig·rə·fē }

physical system *See* causal system. { 'fiz·ə·kəl 'sis·təm }

physical testing [ENG] Determination of physical properties of materials based on observation and measurement. { 'fiz·ə·kəl 'test·iŋ }

physical theory [PHYS] An attempt to explain a certain class of physical phenomena by deducing them as necessary consequences of some primitive assumptions. { 'fiz·ə·kəl 'thē·ə·rē }

physical therapy [MED] The treatment of disease and injury by physical means. { 'fiz·ə·kəl 'ther·ə·pē }

physical time [GEOL] Geologic time as measured by some physical process, such as the radioactive decay of elements. { 'fiz·ə·kəl 'tīm }

physical vapor deposition [MATER] A thin-film deposition process in which a material (metal, alloy, compound, cermet, or composite) is either evaporated or sputtered onto a substrate in a vacuum. Abbreviated PVD. { ,fiz·i·kəl 'vā·pər ,dep·ə,zish·ən }

physical weathering *See* mechanical weathering. { 'fiz·ə·kəl 'we<u>th</u>·ə·riŋ }

physician [MED] An individual authorized to practice medicine. { fi'zish·ən }

physicist [PHYS] A person who does research in physics. { 'fiz·ə,sist }

physics [SCI TECH] The study of those aspects of nature which can be understood in a fundamental way in terms of elementary principles and laws. { 'fiz·iks }

physiognomy [PSYCH] The prediction of personality functioning from facial appearances and expression. { ,fiz·ē'äg·nə·mē }

physiographic diagram [GEOL] A small-scale map showing landforms by the systematic application of a standardized set of simplified pictorial symbols that represent the appearance such forms would have if viewed obliquely from the air at an angle of about 45°. Also known as landform map; morphographic map. { ¦fiz·ē·ə¦graf·ik 'dī·ə,gram }

physiographic feature [GEOL] A prominent or conspicuous physiographic form or noticeable part thereof. { ¦fiz·ē·ə¦graf·ik 'fē·chər }

physiographic form [GEOL] A landform considered with regard to its origin, cause, or history. { ¦fiz·ē·ə¦graf·ik 'fȯrm }

physiographic province [GEOL] A region having a pattern of relief features or landforms that differs significantly from that of adjacent regions. { ¦fiz·ē·ə¦graf·ik 'präv·əns }

physiological acoustics [ACOUS] The study of the responses to acoustic stimuli that take place in the ear or in the associated central neural auditory pathways of humans and animals.   { ˌfiz·ē·ə¦läj·ə·kəl ə'kü·stiks }

physiological biophysics   [BIOPHYS]   An area of biophysics concerned with the use of physical mechanisms to explain the behavior and the functioning of living organisms or parts thereof, and with the response of living organisms to physical forces.   { ˌfiz·ē·ə'läj·ə·kəl ˌbī·ō'fiz·iks }

physiological dead space   *See* dead space.   { ˌfiz·ē·ə'läj·ə·kəl 'ded ˌspās }

physiological ecology   [ECOL]   The study of biophysical, biochemical, and physiological processes used by animals to cope with factors of their physical environment, or employed during ecological interactions with other organisms.   { ˌfiz·ē·ə'läj·ə·kəl ē'käl·ə·jē }

physiological genetics   *See* phenogenetics.   { ˌfiz·ē·ə'läj·ə·kəl jə'ned·iks }

physiological homeostasis   [PHYSIO]   Maintenance of the body's natural resistance to disease.   { ˌfiz·ē·ə¦läj·i·kəl ˌhō·mē·ō'stās·əs }

physiological psychology   [PSYCH]   The study of the physiological mechanisms or correlates of behavior.   { ˌfiz·ē·ə'läj·ə·kəl sī'käl·ə·jē }

physiological saline   *See* normal saline.   { ˌfiz·ē·ə'läj·ə·kəl 'sā,lēn }

physiological salt solution   *See* normal saline.   { ˌfiz·ē·ə'läj·ə·kəl 'sólt sə,lü·shən }

physiological sodium chloride solution   *See* normal saline.   { ˌfiz·ē·ə'läj·ə·kəl 'sōd·ē·əm 'klór,īd sə,lü·shən }

physiologic diplopia   [PHYSIO]   A normal phenomenon in which there is formation of images in noncorresponding retinal points, giving a perception of depth. Also known as introspective diplopia.   { ˌfiz·ē·ə'läj·ik di'plō·pē·ə }

physiologic tremor   [PHYSIO]   A tremor in normal individuals, caused by fatigue, apprehension, or overexposure to cold.   { ˌfiz·ē·ə'läj·ik 'trem·ər }

physiology   [BIOL]   The study of the basic activities that occur in cells and tissues of living organisms by using physical and chemical methods.   { ˌfiz·ē·ē'äl·ə·jē }

physisorption   [PHYS CHEM]   A physical adsorption process in which there are van der Waals forces of interaction between gas or liquid molecules and a solid surface.   { ¦fiz·ə'sórp·shən }

Physopoda   [INV ZOO]   The equivalent name for Thysanoptera.   { fī'säp·ə·də }

Physosomata   [INV ZOO]   A superfamily of amphipod crustaceans in the suborder Hyperiidea; the eyes are small or rarely absent, and the inner plates of the maxillipeds are free at the apex.   { ˌfī·sə'säm·əd·ə }

physostigmine   [ORG CHEM] $C_{15}H_{21}O_2N_3$ An alkaloid; poisonous, colorless-to-pinkish crystals; soluble in alcohol and dilute acids; melts at 86°C; used as a source of salicylate and sulfate forms. Also known as calabarine; eserine.   { ˌfī·sə'stig·mēn }

physostigmine salicylate   [ORG CHEM] $C_{15}H_{21}O_2N_3 \cdot C_7H_6O_3$ Poisonous, colorless-to-yellow crystals; soluble in water, alcohol, and chloroform; melts at 182°C; used for medicines.   { ˌfī·sə'stig,mēn sə'lis·ə,lāt }

physostigmine sulfate   [ORG CHEM] $(C_{15}H_{21}O_2N_3)_2 \cdot H_2SO_4$ Poisonous, white crystals; soluble in water, alcohol, and chloroform; melts at 150°C; used for medicines.   { ˌfī·sə'stig,mēn 'səl,fāt }

Phytalmiidae   [INV ZOO]   A family of myodarian cyclorrhaphous dipteran insects in the subsection Acalypteratae.   { ˌfīd·əl'mī·ə,dē }

phytal zone   [ECOL]   The part of a lake bottom covered by water shallow enough to permit the growth of rooted plants.   { 'fīd·əl ˌzōn }

Phytamastigophorea   [INV ZOO]   A class of the subphylum Sarcomastigophora, including green and colorless phytoflagellates.   { ¦fīd·ə,ma·stə¦gäf·ə'rē·ə }

phytane   [ORG CHEM] $C_{20}H_{42}$ A hydrocarbon derivative of chlorophyll that is found in rock specimens 2.5–3 × 10⁹ years old; frequently associated with Precambrian fossil plant matter.   { 'fī,tān }

phytase   [BIOCHEM]   An enzyme occurring in plants, especially cereals, which catalyzes hydrolysis of phytic acid to inositol and phosphoric acid.   { 'fī,tās }

phyteral   [GEOL]   Morphologically recognizable forms of vegetal matter in coal.   { 'fīd·ə·rəl }

phytic acid   [ORG CHEM] $C_6H_6[OPO(OH)_2]_6$ An acid found in seeds of plants as the insoluble calcium magnesium salt (phytin); derived from corn steep liquor; inhibits calcium absorption in intestine; used to treat hard water, to remove iron and copper from wines, and to inactivate trace-metal contaminants in animal and vegetable oils.   { 'fīd·ik 'as·əd }

phytoalexin   [BIOCHEM]   A natural substance that is toxic to fungi and is synthesized by a plant as a response to fungal infection.   { ¦fīd·ō·ə'lek·sən }

phytochemistry   [BOT]   The study of the chemistry of plants, plant products, and processes taking place within plants.   { 'fīd·ō,kem·ə·strē }

phytochorology   *See* plant geography.   { ¦fīd·ō·kó'räl·ə·jē }

phytochrome   [BIOCHEM]   A protein plant pigment which serves to direct the course of plant growth and development in response variously to the presence or absence of light, to photoperiod, and to light quality.   { 'fīd·ə,krōm }

phytoclimatology   [CLIMATOL]   The study of the microclimate in the air space occupied by plant communities, on the surfaces of the plants themselves and, in some cases, in the air spaces within the plants.   { ¦fīd·ō,klī·mə'täl·ə·jē }

phytocoenosis   [ECOL]   The entire plant population of a particular habitat.   { ¦fīd·ō·sē'nō·səs }

phytocollite   [GEOL]   A black, gelatinous, nitrogenous humic body occurring beneath or within peat deposits.   { fī'täk·ə,līt }

phytogenic dam   [ECOL]   A natural dam consisting of plants and plant remains.   { ¦fīd·ə¦jen·ik 'dam }

phytogenic dune   [ECOL]   Any dune in which the growth of vegetation influences the form of the dune, for example, by arresting the drifting of sand.   { ¦fīd·ə¦jen·ik 'dün }

phytogeography   *See* geobotany; plant geography.   { ¦fīd·ō·jē'äg·rə·fē }

phytohemagglutinin   *See* phytolectin.   { ¦fīd·ō,hē·mə'glüt·ən·ən }

phytohormone   *See* plant hormone.   { ¦fīd·ō'hór,mōn }

phytol   [ORG CHEM] $C_{20}H_{40}O$ A liquid with a boiling point of 202–204°C; soluble in organic solvents; used in the synthesis of vitamins E and K.   { 'fī,tól }

Phytolacca americana   *See* pokeweed.   { ˌfīd·ō,lak·ə ə,mer·ə'kän·ə }

phytolectin   [BIOCHEM]   A lectin found in plants. Also known as phytohemagglutinin.   { ˌfīd·ə'lek·tən }

phytolith   [PALEON]   A fossilized part of a living plant that secreted mineral matter.   { 'fīd·ə,lith }

Phytomastigina   [INV ZOO]   The equivalent name for Phytamastigophorea.   { ˌfīd·ō·mas·tə'jī·nə }

phytometer   [ENG]   A device for measuring transpiration, consisting of a vessel containing soil in which one or more plants are rooted and sealed so that water can escape only by transpiration from the plant.   { fī'täm·əd·ər }

Phytomonadida   [INV ZOO]   The equivalent name for Volvocida.   { ˌfīd·ō·mō'näd·ə·də }

phytonadione   [ORG CHEM] $C_{31}H_{46}O_2$ A yellow, viscous liquid soluble in benzene, chloroform, and vegetable oils; used in medicine and as a food supplement. Also known as vitamin K_1.   { fī,tän·ə'dī,ōn }

phytopathogen   [ECOL]   An organism that causes a disease in a plant.   { ¦fīd·ō'path·ə·jən }

phytophagous   [ZOO]   Feeding on plants.   { fī'täf·ə·gəs }

Phytophthra citrophthora   [MYCOL]   A water mold that causes citrus gummosis.   { fī,däf·thrə ,si·trəf'thór·ə }

phytoplankton   [ECOL]   Planktonic plant life.   { ¦fīd·ə'plaŋk·tən }

phytoremediation   [AGR]   The use of green plants to manage or reduce high levels of soil and groundwater contaminants.   { ˌfīd·ō·ri,mēd·ē'ā·shən }

Phytosauria   [PALEON]   A suborder of Late Triassic long-snouted aquatic thecodonts resembling crocodiles but with posteriorly located external nostrils, absence of a secondary palate, and a different structure of the pelvic and pectoral girdles.   { ˌfīd·ə'sór·ē·ə }

Phytoseiidae   [INV ZOO]   A family of the suborder Mesostigmata.   { ˌfīd·ō·sē'ī·ə,dē }

phytosociology   [ECOL]   A broad study of plants that includes the study of all phenomena affecting their lives as social units.   { ¦fīd·ō,sō·sē'äl·ə·jē }

phytosterol   [BIOCHEM]   Any of various sterols obtained from plants, including ergosterol and stigmasterol.   { fī'täs·tə,ról }

PHYTOL

CH₃   CH₃   CH₃   CH₂OH

CH₃

CH₃

Structural formula of phytol.

PHYTOSAURIA

Nicrosaurus, a quadrupedal thecodont. (*After E. H. Colbert, Evolution of Vertebrates, 2d ed., copyright © 1969 by John Wiley & Sons, Inc.; reprinted by permission*)

phytotoxin [BIOCHEM] **1.** A substance toxic to plants. **2.** A toxin produced by plants. { ¦fīd·ō'täk·sən }

phytotron [BOT] A research tool used to study whole plants; contains a large number of individually controlled environments that provide the means of studying the effect of each environmental factor, such as temperature or light, at many levels simultaneously. { 'fīd·ə,trän }

pi [MATH] The irrational number which is the ratio of the circumference of any circle to its diameter; an approximation is 3.14159. Symbolized π. { pī }

pia arachnoid [VERT ZOO] The outer meninx of certain submammalian forms having two membranes covering the brain and spinal cord. { 'pī·ə ə'rak,nȯid }

Piacention See Plaisancian. { ¸pē·ə'sen·chən }

pia mater [ANAT] The vascular membrane covering the surface of the brain and spinal cord. { 'pē·ə ¸mād·ər }

piano wire [MET] High-tensile-strength, 0.75 to 0.85% carbon steel wire cold-drawn to uniform thickness. { pē'an·ō ¸wīr }

piastrenemia See thrombocytosis. { pē¸as·trə'nē·mē·ə }

pi attenuator [ELEC] An attenuator consisting of a pi network whose impedances are all resistances. { 'pī ə'ten·yə,wād·ər }

pibal See pilot-balloon observation. { 'pī,bal }

pi bonding [PHYS CHEM] Covalent bonding in which the greatest overlap between atomic orbitals is along a plane perpendicular to the line joining the nuclei of the two atoms. { 'pī ¸bänd·iŋ }

PIC See personal identification code. { ¦pē¦ī¦sē or pik }

pica [GRAPHICS] A printer's unit of measurement, 0.166044 inch (approximately $1/6$ inch) and equal to 12 points. [MED] Craving for substances not normally used as food; an abnormal appetite; sometimes seen in hysterical patients or during pregnancy. { 'pī·kə }

Picard method [MATH] A method of successive substitution for solving differential equations. { pi'kär ¸meth·əd }

Picard's big theorem [MATH] The image of every neighborhood of an essential singularity of a complex function is dense in the complex plane. Also known as Picard's second theorem. { pi'kärz 'big ¦thir·əm }

Picard's first theorem See Picard's little theorem. { pi'kärz 'fərst ¸thir·əm }

Picard's little theorem [MATH] A nonconstant entire function of the complex plane assumes every value save at most one. Also known as Picard's first theorem. { pi'kärz 'lid·əl ¦thir·əm }

Picard's second theorem See Picard's big theorem. { pi'kärz 'sek·ənd 'thir·əm }

Picatinny test [ENG] An impact test used in the United States for evaluating the sensitivity of high explosives; a small sample of the explosive is placed in a depression in a steel die cup and capped by a thin brass cover, a cylindrical steel plug is placed in the center of the cover, and a 2-kilogram weight is dropped from varying heights on the plug; the reported sensitivity figure is the minimum height, in inches, at which at least 1 firing results from 10 trials. { pik·ə tin·ē ¸test }

Piche evaporimeter [ENG] A porous-paper-wick atmometer. { 'pēsh i,vap·ə'rim·əd·ər }

Picidae [VERT ZOO] The woodpeckers, a large family of birds in the order Piciformes; adaptive modifications include a long tongue and hyoid mechanism, and stiffened tail feathers. { 'pis·ə,dē }

Piciformes [VERT ZOO] An order of birds characterized by the peculiar arrangement of the tendons of the toes. { ¸pis·ə'fȯr,mēz }

Picinae [VERT ZOO] The true woodpeckers, a subfamily of the Picidae. { 'pis·ə,nē }

pick [COMPUT SCI] To select the next card from an input stack for feeding into a card machine. [DES ENG] **1.** The steel cutting points used on a coal-cutter chain. **2.** A miner's steel or iron digging tool with sharp points at each end. [ENG] **1.** To dress the sides of a shaft or other excavation. **2.** To remove shale, dirt, and such from coal. [TEXT] See filling. { pik }

pick-a-back conveyor [MIN ENG] A short conveyor that advances with a loader or continuous miner at the face of a mine and loads coal on the main haulage system. { 'pik·ə,bak kən,vā·ər }

pick-and-pick [TEXT] A method of weaving fabric in which one type of crosswise thread is alternated with another, for example, cotton or acetate alternated with elastic yarn. { 'pik ən 'pik }

pick-and-place robot [CONT SYS] A simple robot, often with only two or three degrees of freedom and little or no trajectory control, whose sole function is to transfer items from one place to another. { 'pik ən 'plās 'rō,bät }

Pickard core barrel [MIN ENG] A type of double-tube core barrel; the distinguishing feature of the barrel is that when blocked the inner barrel slides upward into the head, closing the water ports and stopping the flow of the circulating liquid, without irreparably damaging the bit until the barrel is pulled and the blocked inner tube cleared. { 'pik·ərd 'kȯr ¸bar·əl }

pickax [DES ENG] A pointed steel or iron tool mounted on a wooden handle and used for breaking earth and stone. { 'pik,aks }

pick device See pointing device. { 'pik di,vīs }

picker [MIN ENG] **1.** An employee who picks or discards slate and other foreign matter from the coal in an anthracite breaker or at a picking table. **2.** A mechanical arrangement for removing slate from coal. [TEXT] A machine used to pull apart and separate cotton fibers. { 'pik·ər }

pickeringite [MINERAL] $MgAl_2(SO_4)_4 \cdot 22H_2O$ A white or faintly colored mineral composed of hydrous sulfate of magnesium and aluminum, occurring in fibrous masses. { 'pik·riŋ,īt }

Pickering series [SPECT] A series of spectral lines of singly ionized helium, observed in very hot O-type stars, associated with transitions between the level with principal quantum number $n = 4$ and higher energy levels. { 'pik·riŋ ,sir·ēz }

picker knives [COMPUT SCI] The narrow edges of a moving slide which will pick the bottom card from a stack and feed it to a card reader. { 'pik·ər ,nīvz }

picket ship [NAV ARCH] A radar-equipped ship, generally anchored, that is used to extend radar early-warning coverage seaward. { 'pik·ət ,ship }

pick glass [TEXT] Calibrated magnifying glass used to count warp and filling threads in a square inch of fabric. { 'pik ,glas }

pick hammer [DES ENG] A hammer with a point at one end of the head and a blunt surface at the other end. { 'pik ,ham·ər }

picking [COMPUT SCI] Identification of information displayed on a screen for subsequent computer processing, by pointing to it with a lightpen. [GRAPHICS] The lifting of the paper surface during printing. [MIN ENG] **1.** Removal of waste material from an ore. **2.** Extraction of the lightest-grade ore from a mine. **3.** Emission of particles from the roof of a mine on the verge of collapse. [TEXT] **1.** In yarn manufacture, the removal of extraneous matter from the face of woolen fabric. **2.** In the weaving process, the movement of the shuttle along the warp of the fabric. { 'pik·iŋ }

picking conveyor [MIN ENG] A continuous belt or apron conveyor used to carry a relatively thin bed of material past pickers who hand-sort or pick the material being conveyed. { 'pik·iŋ kən,vā·ər }

picking table [MIN ENG] A flat or slightly inclined platform on which the coal or ore is run to be picked free from slate or gangue. { 'pik·iŋ ,tā·bəl }

pick lacing [DES ENG] The pattern to which the picks are set in a cutter chain. { 'pik ,lās·iŋ }

pickle liquor [MET] A spent pickling solution. { 'pik·əl ,lik·ər }

pickle patch [MET] A coating of oxide or scale that remains adherent after pickling. { 'pik·əl ,pach }

pickle stain [MET] Discoloration of a metal surface due to chemical cleaning without adequate washing and drying. { 'pik·əl ,stān }

pickling [CHEM ENG] A method of preparing hides for tanning by immersion in a salt solution with a pH of 2.5 or less. [FOOD ENG] A method of preserving food by using salt, sugar, spices, and acetic acid. [MET] Preferential removal of oxide or mill scale from the surface of a metal by immersion usually in an acidic or alkaline solution. { 'pik·liŋ }

pickling acid [CHEM] Any of the acids used in pickling solutions, such as hydrochloric, sulfuric, nitric, phosphoric, or hydrofluoric acid. { 'pik·liŋ ,as·əd }

pick miner [MIN ENG] **1.** In anthracite and bituminous coal mining, one who uses hand tools to extract coal in underground working places. **2.** One who cuts out a channel under the bottom of the working face of coal with a pick. { 'pik ˌmīn·ər }

pickoff [ELECTR] A device used to convert mechanical motion into a proportional electric signal. [MECH ENG] A mechanical device for automatic removal of the finished part from a press die. { 'pikˌóf }

Pick's disease [MED] **1.** A form of presenile dementia characterized by severe atrophy of the frontal and temporal lobes of the cerebrum. **2.** A recurrent or progressive form of ascites with little or no edema. **3.** *See* constrictive pericarditis. **4.** *See* polyserositis. { 'piks diˌzēz }

pickup [AERO ENG] A potentiometer used in an automatic pilot to detect the motion of the airplane around the gyro and initiate corrective adjustments. [ELEC] **1.** A device that converts a sound, scene, measurable quantity, or other form of intelligence into corresponding electric signals, as in a microphone, phonograph pickup, or television camera. **2.** The minimum current, voltage, power, or other value at which a relay will complete its intended function. **3.** Interference from a nearby circuit or system. [MET] Transfer of metal from the work to the tool, or from the tool to the work, during a forming operation. [NUC PHYS] A type of nuclear reaction in which the incident particle takes a nucleon from the target nucleus and proceeds with this nucleon bound to itself. { 'pikˌəp }

pickup tube *See* camera tube. { 'pikˌəp ˌtüb }

pickup voltage [ELEC] Of a magnetically operated device, the voltage at which the device starts to operate. { 'pikˌəp ˌvōl·tij }

pico- [MATH] A prefix meaning 10^{-12}; used with metric units. Abbreviated p. { 'pē·kō }

picoammeter [ENG] An ammeter whose scale is calibrated to indicate current values in picoamperes. { ˌpē·kō'amˌēd·ər }

picoampere [ELEC] A unit of current equal to 10^{-12} ampere, or one-millionth of a microampere. Abbreviated pA. { ˌpē·kō'amˌpir }

picodnavirus [VIROL] A group of deoxyribonucleic acid-containing animal viruses including the adeno-satellite viruses. { pē'kädˌnəˌvī·rəs }

picofarad [ELEC] A unit of capacitance equal to 10^{-12} farad, or one-millionth of a microfarad. Abbreviated pF. { ˌpē·kō'farˌəd }

picoline [ORG CHEM] $C_5H_4N(CH_3)$ Family of colorless liquid isomers, soluble in water and alcohol; the alpha form, boiling at 129°C, is used as a solvent and chemical intermediate, and is also known as 2-methyl pyridine; the beta form, boiling at 143.5°C, is used as a solvent for chemical synthesis reactions, to make nicotinic acid, and in fabric waterproofing, and is also known as 3-methyl pyridine; the gamma form, boiling at 143.1°C, is used as a solvent for chemical synthesis reactions and in fabric waterproofing. { 'pik·əˌlēn }

picolinic acid [ORG CHEM] $C_{10}H_8N_4O_5$ An alcohol-soluble crystalline compound, forming yellow leaflets that melt at 116–117°C; used as a reagent in phenylalanine, tryptophan, and alkaloids production, and for the quantitative detection of calcium. { ˌpik·əˌlin·ik 'as·əd }

Picornaviridae [VIROL] A viral family made up of the small (18–30 nanometers) ether-sensitive viruses that lack an envelope and have a Togaviridae genome; contains the genera *Enterovirus* (human polio), *Cardiovirus* (mengo), *Rhinovirus* (common cold), and *Aphtovirus* (foot-and-mouth disease). { pēˌkór·nə'virˌəˌdī }

picornavirus [VIROL] A viral group made up of small (18–30 nanometers), ether-sensitive viruses that lack an envelope and have a ribonucleic acid genome; among subgroups included are enteroviruses and rhinoviruses, both of human origin. { pē'kór·nəˌvī·rəs }

picosecond [MECH] A unit of time equal to 10^{-12} second, or one-millionth of a microsecond. Abbreviated ps; psec. { ˌpē·kō'sek·ənd }

picotite [MINERAL] A dark-brown variety of hercynite that contains chromium and is commonly found in dunites. Also known as chrome spinel. { 'pik·əˌtīt }

picowatt [MECH] A unit of power equal to 10^{-12} watt, or one-millionth of a microwatt. Abbreviated pW. { 'pē·kəˌwät }

picramic acid [ORG CHEM] $C_6H_5N_3O_5$ A crystalline acid, forming dark red needles from alcohol solutions, melting at

169–170°C; used in dye manufacture and as a reagent in tests for albumin. { pi'kram·ik 'as·əd }

picratol [MATER] A binary explosive composed of 52% ammonium picrate and 48% TNT (trinitrotoluene); it can be melt-loaded; less sensitive than TNT, it was developed for use in armor-piercing bombs. { 'pik·rəˌtól }

picric acid [ORG CHEM] $C_6H_2(NO_2)_3OH$ Poisonous, explosive, highly oxidative yellow crystals with bitter taste; soluble in water, alcohol, chloroform, benzene, and ether; melts at 122°C; used in explosives, in external medicines; to make dyes, matches, and batteries, and to etch copper. { 'pik·rik 'as·əd }

picrite [PETR] A medium- to fine-grained igneous rock composed chiefly of olivine, with smaller amounts of pyroxene, hornblende, and plagioclase felspar. { 'piˌkrīt }

Picrodendraceae [BOT] A small family of dicotyledonous plants in the order Juglandales characterized by unisexual flowers borne in catkins, four apical ovules in a superior ovary, and trifoliate leaves. { ˌpik·rō·den'drāsˌēˌē }

picrolite *See* antigorite. { 'pik·rəˌlīt }

picromerite [MINERAL] $K_2Mg(SO_4)_2 \cdot 6H_2O$ A white mineral composed of hydrous sulfate of magnesium and potassium, occurring as crystalline encrustations. { pi'kräm·əˌrīt }

picropharmacolite [MINERAL] $(Ca,Mg)_3(AsO_4)_2 \cdot 6H_2O$ Mineral composed of hydrous calcium magnesium arsenate. { ˌpik·rō·fär'mak·əˌlīt }

picrotoxin [BIOCHEM] $C_{30}H_{34}O_{13}$ A poisonous, crystalline plant alkaloid found primarily in *Cocculus indicus*; used as a stimulant and convulsant drug. Also known as cocculin. { ˌpik·rə'täk·sən }

Pictet's liquid [MATER] Liquid mixture of carbon dioxide and sulfur dioxide; used to produce low temperatures. { pik 'tāz ˌlik·wəd }

Pictor [ASTRON] A southern constellation; right ascension 6 hours, declination 55°S. Also known as Easel. { 'pikˌtór }

picture [COMPUT SCI] In COBOL, a symbolic description of each data element or item according to specified rules concerning numerals, alphanumerics, location of decimal points, and length. [COMMUN] The image on the screen of a television receiver. [GRAPHICS] The image produced by a photographic process. { 'pik·chər }

picture black *See* black signal. { 'pik·chər ˌblak }

picture carrier [COMMUN] A carrier frequency located 1.25 megahertz above the lower frequency limit of a standard National Television Systems Committee television signal; in color television, it is used for transmitting color information. Also known as luminance carrier. { 'pik·chər ˌkar·ē·ər }

picture compression [COMPUT SCI] The elimination of redundant information from a digital picture through the use of efficient encoding techniques in which frequently occurring gray levels or blocks of gray levels are represented by short codes and infrequently occurring ones by longer codes. { 'pik·chər kəmˌpresh·ən }

picture element [ELECTR] **1.** That portion, in facsimile, of the subject copy which is seen by the scanner at any instant; it can be considered a square area having dimensions equal to the width of the scanning line. **2.** In television, any segment of a scanning line, the dimension of which along the line is exactly equal to the nominal line width; the area which is being explored at any instant in the scanning process. Also known as critical area; elemental area; pixel; recording spot; scanning spot. { 'pik·chər ˌel·ə·mənt }

picture frequency [COMMUN] A frequency that results solely from scanning of subject copy in a facsimile system. [ELECTR] *See* frame frequency. { 'pik·chər ˌfrē·kwən·sē }

picture grammar [COMPUT SCI] A formalism for carrying out computations on pictures and describing picture structure. { 'pik·chər ˌgram·ər }

picture processing *See* image processing. { 'pik·chər ˌprä·sesˌiŋ }

picture segmentation [COMPUT SCI] The division of a complex picture into parts corresponding to regions or objects, so that the picture can then be described in terms of the parts, their properties, and their spatial relationships. Also known as scene analysis; segmentation. { 'pik·chər ˌseg·mən'tā·shən }

picture signal [COMMUN] The signal resulting from the scanning process in television or facsimile. { 'pik·chər ˌsigˌnəl }

picture synchronizing pulse *See* vertical synchronizing pulse. { 'pik·chər 'siŋ·krə‚nīz·iŋ ‚pəls }

picture transmission [COMMUN] Electric transmission of a picture having a gradation of shade values. { 'pik·chər tranz 'mish·ən }

picture transmitter *See* visual transmitter. { 'pik·chər tranz ‚mid·ər }

picture tube [ELECTR] A cathode-ray tube used in television receivers to produce an image by varying the electron-beam intensity as the beam is deflected from side to side and up and down to scan a raster on the fluorescent screen at the large end of the tube. Also known as kinescope; television picture tube. { 'pik·chər ‚tüb }

picture-tube brightener [ELECTR] A small step-up transformer that can be inserted between the socket and base of a picture tube to increase the heater voltage and thereby increase picture brightness to compensate for normal aging of the tubes. { 'pik·chər ‚tüb ‚brīt·ən·ər }

picture white *See* white signal. { 'pik·chər ‚wīt }

picture window [BUILD] A large window framing an exterior view. { 'pik·chər 'win·dō }

Picumninae [VERT ZOO] The piculets, a subfamily of the avian family Picidae. { pi'kyüm·nə‚nē }

Pidgeon process [MET] A method for producing magnesium from calcined dolomite by reduction with ferrosilicon. Also known as ferrosilicon process; silicothermic process. { 'pij·ən ‚präs·əs }

piebaldism [MED] A pigmentary disorder characterized by patterns of white spots on the skin and caused by an inherited absence of melanocytes. { 'pī‚bȯl‚diz·əm }

piece [ORD] An artillery weapon, a machine gun, a rifle, or any firearm. { pēs }

piece mark [ENG] Identification number for an individual part, subassembly, or assembly; shown on the drawing, but not necessarily on the part. { 'pēs‚märk }

piecemeal stoping [GEOL] Magmatic stoping in which only isolated blocks of roof rock are assimilated. { 'pēs‚mēl ‚stōp·iŋ }

piece rate [IND ENG] Wages paid per unit of production. { 'pēs ‚rāt }

piece-root grafting [BOT] Grafting in which each piece of a cut seedling root is used as a stock. { 'pēs ‚rüt ‚graft·iŋ }

piecewise-continuous function [MATH] A function defined on a given region, which can be divided into a finite number of pieces such that the function is continuous on the interior of each piece and its value approaches a finite limit as the argument of the function in the interior approaches a boundary point of the piece. { 'pēs‚wīz kən‚tin·yə·wəs 'fəŋk·shən }

piecewise-linear [MATH] A continuous curve or function obtained by joining a finite number of linear pieces. { 'pēs‚wīz ‚lin·ē·ər }

piecewise linear system [CONT SYS] A system for which one can divide the range of values of input quantities into a finite number of intervals such that the output quantity is a linear function of the input quantity within each of these intervals. { 'pēs‚wīz ‚lin·ē·ər ‚sis·təm }

piecewise linear topology *See* combinatorial topology. { 'pēs‚wīz ‚lin·ē·ər tə'päl·ə·jē }

piecewise-smooth curve [MATH] The range of a function from a closed interval to a Euclidean space such that each of the Cartesian coordinates of the image point is a continuously differentiable function on the closed interval, except at a finite set of points where the function is differentiable on the left and on the right. { 'pēs‚wīz ‚smüth ‚kərv }

piecework [IND ENG] Work paid for in accordance with the amount done rather than the hours taken. { 'pēs‚wərk }

pie chart [MATH] A circle divided by several radii into sectors whose relative areas represent the relative magnitudes of quantities or the relative frequencies of items in a frequency distribution. Also known as circle graph; sectorgram. { 'pī ‚chärt }

piedmont [GEOL] Lying or formed at the base of a mountain or mountain range, as a piedmont terrace or a piedmont pediment. { 'pēd‚mänt }

piedmont alluvial plain *See* bajada. { 'pēd‚mänt ə'lüv·ē·əl ‚plān }

piedmont angle [GEOL] The sharp break of slope between a hill and a plain, such as the angle at the junction of a mountain front and the pediment at its base. { 'pēd‚mänt ‚aŋ·gəl }

piedmont bench *See* piedmont step. { 'pēd‚mänt ‚bench }

piedmont benchland [GEOL] One of several successions or systems of piedmont steps. Also known as piedmont stairway; piedmont treppe. { 'pēd‚mänt 'bench‚land }

piedmont bulb [HYD] The lobe or fan of ice formed when a glacier spreads out on a plain at the lower end of a valley. { 'pēd‚mänt ‚bəlb }

piedmont flat *See* piedmont step. { 'pēd‚mänt ‚flat }

piedmont glacier [HYD] A thick, continuous ice sheet formed at the base of a mountain range by the spreading out and coalescing of valley glaciers from higher mountain elevations. { 'pēd‚mänt ‚glā·sī·ər }

piedmont gravel [GEOL] Coarse gravel derived from high ground by mountain torrents and spread out on relatively flat ground where the velocity of the water is decreased. { 'pēd‚mänt ‚grav·əl }

piedmont ice [HYD] An ice sheet formed by the joining of two or more glaciers on a comparatively level plain at the base of the mountains down which the glaciers descended; it may be partly afloat. { 'pēd‚mänt ‚īs }

piedmont interstream flat *See* pediment. { 'pēd‚mänt 'in·tər‚strēm ‚flat }

piedmontite *See* piemontite. { 'pēd‚män‚tīt }

piedmont lake [HYD] An oblong lake occupying a partly overdeepened basin excavated from rock by a piedmont glacier, or dammed by a glacial moraine. { 'pēd‚mänt ‚lāk }

piedmont plain *See* bajada. { 'pēd‚mänt ‚plān }

piedmont plateau [GEOL] A plateau lying between the mountains and the plains or the ocean. { 'pēd‚mänt pla'tō }

piedmont scarp [GEOL] A small, low cliff formed in alluvium on a piedmont slope at the foot of a steep mountain range; due to dislocation of the surface, especially by faulting. Also known as scarplet. { 'pēd‚mänt ‚skärp }

piedmont slope *See* bajada. { 'pēd‚mänt ‚slōp }

piedmont stairway *See* piedmont benchland. { 'pēd‚mänt 'ster‚wā }

piedmont step [GEOL] A terracelike or benchlike piedmont feature that slopes outward or downvalley. Also known as piedmont bench; piedmont flat. { 'pēd‚mänt ‚step }

piedmont treppe *See* piedmont benchland. { 'pēd‚mänt 'trep·ə }

pi electron [PHYS CHEM] An electron which participates in pi bonding. { 'pī i' ‚ek‚trän }

piemontite [MINERAL] $Ca_2(Al,Mn^{3+},Fe)_3Si_3O_{12}(OH)$ Reddish-brown epidote mineral that contains manganese. Also known as manganese epidote; piedmontite. { 'pē ‚män‚tīt }

pier [BUILD] A concrete block that supports the floor of a building. [CIV ENG] **1.** A vertical, rectangular or circular support for concentrated loads from an arch or bridge superstructure. **2.** A structure with a platform projecting from the shore into navigable waters for mooring vessels. { pir }

piercement *See* diapir. { 'pirs·mənt }

piercement dome *See* diapir. { 'pirs·mənt ‚dōm }

Pierce oscillator [ELECTR] Oscillator in which a piezoelectric crystal unit is connected between the grid and the plate of an electron tube, in what is basically a Colpitts oscillator, with voltage division provided by the grid-cathode and plate-cathode capacitances of the circuit. { 'pirs 'äs·ə‚lād·ər }

Pierce's disease [PL PATH] A virus disease of grapes in which there is mottling between the veins of leaves, early defoliation, and early ripening and withering of the fruit. { 'pirs·əz di‚zēz }

piercing *See* fusion piercing. { 'pirs·iŋ }

piercing fold *See* diapir. { 'pirs·iŋ ‚fōld }

piercing gripper [CONT SYS] A robot component that first punctures a material such as cloth, rubber, or porous sheets, or soft plastic in order to lift and handle it. { 'pirs·iŋ ‚grip·ər }

piercing point *See* trace. { 'pirs·iŋ ‚pȯint }

pier foundation *See* caisson foundation. { 'pir ‚fau̇n‚dā·shən }

pierhead line [CIV ENG] The line in navigable waters beyond which construction is prohibited; open-pier construction may extend outward from the bulkhead line to the pierhead line. { 'pir‚hed ‚līn }

Pieridae [INV ZOO] A family of lepidopteran insects in the superfamily Papilionoidea including white, sulfur, and orange-tip butterflies; characterized by the lack of a prespiracular bar at the base of the abdomen. { pī'er·ə‚dē }

Piesmatidae [INV ZOO] The ash-gray leaf bugs, a family of

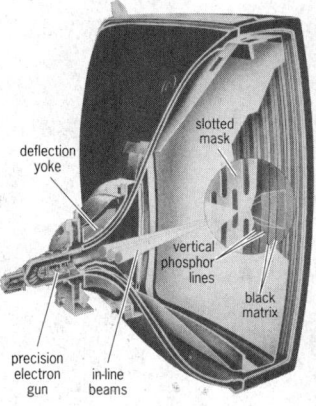

PICTURE TUBE

deflection yoke

slotted mask

vertical phosphor lines

black matrix

precision electron gun

in-line beams

Cutaway view of a color picture tube.

hemipteran insects belonging to the Pentatomorpha. { pēz 'mad·ə,dē }

Piesmidae [INV ZOO] A small family of hemipteran insects in the superfamily Lygaeoidea. { 'pez·mə,dē }

pièze [MECH] A unit of pressure equal to 1 sthène per square meter, or to 1000 pascals. Abbreviated pz. { pē'ez }

piezocaloric effect [SOLID STATE] The production of entropy in a crystal that is subjected to mechanical stress. { pē,ā·zō·kə'lór·ik i,fekt }

piezoceramic [MATER] A ceramic, such as lead zirconate titanate, that converts an electrical field to a mechanical strain or a mechanical strain to an electrical charge. In smart structures, piezoceramics are used as sensors, actuators, or both, for vibration suppression applications. Also known as piezoelectric ceramic. { pē¦ā·zō·sə'ram·ik }

piezochemistry [CHEM] The field of chemical reactions under high pressures. { pē¦ā·zō'kem·ə·strē }

piezocomposite [MATER] A composite material that has piezoelectric properties and is fabricated by interleaving a cut or preshaped piezoceramic with a passive polymer or epoxy host matrix compound. { pē¦ā·zō·kəm'päz·ət }

piezocrystallization [GEOL] Crystallization of a magma under pressure, such as the pressure associated with orogeny. { pē¦ā·zō,krist·əl·ə'zā·shən }

piezoelectric [SOLID STATE] Having the ability to generate a voltage when mechanical force is applied, or to produce a mechanical force when a voltage is applied, as in a piezoelectric crystal. { pē¦ā·zō·ə'lek·trik }

piezoelectric ceramic See piezoceramic. { pē¦ā·zō·ə'lek·trik sə'ram·ik }

piezoelectric crystal [SOLID STATE] A crystal which exhibits the piezoelectric effect; used in crystal loudspeakers, crystal microphones, and crystal cartridges for phono pickups. { pē¦ā·zō·ə'lek·trik 'krist·əl }

piezoelectric detector [ENG] A seismic detector constructed from a stack of piezoelectric crystals with an inertial mass mounted on top and intervening metal foil to collect the charges produced on the crystal faces when the crystals are strained. { pē¦ā·zō·ə'lek·trik di'tek·tər }

piezoelectric effect [SOLID STATE] **1.** The generation of electric polarization in certain dielectric crystals as a result of the application of mechanical stress. **2.** The reverse effect, in which application of a voltage between certain faces of the crystal produces a mechanical distortion of the material. { pē¦ā·zō·ə'lek·trik i'fekt }

piezoelectric element [ELECTR] A piezoelectric crystal used in an electric circuit, for example, as a transducer to convert mechanical or acoustical signals to electric signals, or to control the frequency of a crystal oscillator. { pē¦ā·zō·ə'lek·trik 'el·ə·mənt }

piezoelectric gage [ENG] A pressure-measuring gage that uses a piezoelectric material to develop a voltage when subjected to pressure; used for measuring blast pressures resulting from explosions and pressures developed in guns. { pē¦ā·zō·ə'lek·trik 'gāj }

piezoelectric hysteresis [SOLID STATE] Behavior of a piezoelectric crystal whose electric polarization depends not only on the mechanical stress to which the crystal is subjected, but also on the previous history of this stress. { pē¦ā·zō·ə'lek·trik ,his·tə'rē·səs }

piezoelectricity [SOLID STATE] Electricity or electric polarization resulting from the piezoelectric effect. { pē¦ā·zō·ə,lek'tris·əd·ē }

piezoelectric loudspeaker See crystal loudspeaker. { pē¦ā·zō·ə'lek·trik 'laůd,spēk·ər }

piezoelectric microphone See crystal microphone. { pē¦ā·zō·ə'lek·trik 'mī·krə,fōn }

piezoelectric oscillator See crystal oscillator. { pē¦ā·zō·ə'lek·trik 'äs·ə,lād·ər }

piezoelectric pickup See crystal pickup. { pē¦ā·zō·ə'lek·trik 'pik,əp }

piezoelectric polymer See piezopolymer. { pē¦ā·zō·ə'lek·trik 'päl·ə·mər }

piezoelectric resonator See crystal resonator. { pē¦ā·zō·ə'lek·trik 'rez·ən,ād·ər }

piezoelectric semiconductor [SOLID STATE] A semiconductor exhibiting the piezoelectric effect, such as quartz, Rochelle salt, and barium titanate. { pē¦ā·zō·ə'lek·trik 'sem·i·kən,dək·tər }

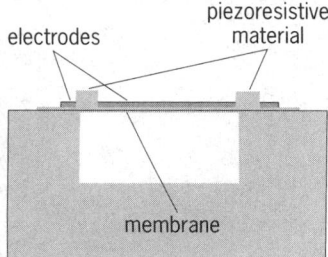

PIEZORESISTIVE MICROPHONE

Cross section of a piezoresistive microphone.

piezoelectric transducer [ELECTR] A piezoelectric crystal used as a transducer, either to convert mechanical or acoustical signals to electric signals, as in a microphone, or vice versa, as in ultrasonic metal inspection. { pē¦ā·zō·ə'lek·trik tranz'dü·sər }

piezoelectric vibrator [SOLID STATE] An element cut from piezoelectric material, usually in the form of a plate, bar, or ring, with electrodes attached to or supported near the element to excite one of its resonant frequencies. { pē¦ā·zō·ə'lek·trik 'vī,brād·ər }

piezogene [GEOL] Pertaining to the formation of minerals primarily under the influence of pressure. { pē'ā·zō,jēn }

piezoglypt See regmaglypt. { pē'ā·zō,glipt }

piezojunction effect [ELECTR] A change in the current-voltage characteristic of a pn junction that is produced by a mechanical stress. { pē,ā·zō'jəŋk·shən i,fekt }

piezomagnetism [SOLID STATE] Stress dependence of magnetic properties. { pē¦ā·zō'mag·nə,tiz·əm }

piezometer [ENG] **1.** An instrument for measuring fluid pressure, such as a gage attached to a pipe containing a gas or liquid. **2.** An instrument for measuring the compressibility of materials, such as a vessel that determines the change in volume of a substance in response to hydrostatic pressure. { ,pē·ə'zäm·əd·ər }

piezometer opening See pressure tap. { ,pē·ə'zäm·əd·ər ,ō·pən·iŋ }

piezometric surface See potentiometric surface. { pē¦ā·zō¦me·trik 'sər·fəs }

piezooptical effect [OPTICS] The change produced in the index of refraction of a light-transmitting material by externally applied stress. { pē¦ā·zō'äp·tə·kəl i,fekt }

piezopolymer [ORG CHEM] A polymeric film that has the ability to reversibly convert heat and pressure to electricity. Also known as piezoelectric polymer. { pē¦ā·zō'päl·ə·mər }

piezoresistance effect [SOLID STATE] The change in the electrical resistance of a metal or semiconductor that is produced by mechanical stress. { pē,ā·zō·ri'zis·təns i,fekt }

piezoresistive material [MATER] A metal or semiconductor in which a change in electrical resistance occurs in response to changes in the applied stress. { pē¦ā·zō·ri¦zis·tiv mə'tir·ē·əl }

piezoresistive microphone [ENG ACOUS] A microphone in which a piezoresistive material is deposited on the edges of a membrane, and variations in the resistance of this material resulting from motion of the membrane are sensed, typically in a Wheatstone bridge. { pē¦ā·zō·ri¦zis·tiv 'mī·krə,fōn }

piezoresistive sensor [ENG] A transducer which converts variations in mechanical stress into an electrical output; it consists of an element of piezoresistive material that is connected to a Wheatstone bridge circuit and is placed on a highly stressed part of a suitable mechanical structure, usually attached to a cantilever or other beam configuration. { pē¦ā·zō·ri¦zis·tiv 'sen·sər }

piezotransistor accelerometer [ENG] An accelerometer in which a seismic mass supported by a stylus transmits a concentrated force to the upper diode surface of a transistor and acceleration is determined from the resulting change in current across the pn junction of the transistor. { pē¦ā·zō·tran'zis·tər ak,sel·ə'räm·əd·ər }

piezotropic [FL MECH] Characterized by piezotropy. { pē¦ā·zō¦träp·ik }

piezotropy [FL MECH] The property of a fluid in which processes are characterized by a functional dependence of the thermodynamic functions of state: $dp/dt = b(dp/dt)$, where ρ is the density, p the pressure, and b a function of the thermodynamic variables, called the coefficient of piezotropy. { ,pē·ə'zä·trə·pē }

pi filter [ELECTR] A filter that has a series element and two parallel elements connected in the shape of the Greek letter pi (π). { 'pī ,fil·tər }

pig [ELECTR] **1.** An ion source based on the same principle as the Philips ionization gage. **2.** See Philips ionization gage. [ENG] In-line scraper (brush, blade cutter, or swab) forced through pipelines by fluid pressure; used to remove scale, sand, water, and other foreign matter from the interior surfaces of the pipe. [MET] A crude metal casting prepared for storage, transportation, or remelting. [NUCLEO] A heavily shielded container, usually lead, used to ship or store radioisotopes and other radioactive materials. [VERT ZOO] Any wild or domestic mammal of the superfamily Suoidea in the order Artiodactyla; toes terminate in nails which are modified into hooves, the tail is short, and the body is covered sparsely with hair which is frequently bristlelike. { pig }

pigeon [VERT ZOO] Any of various stout-bodied birds in the family Columbidae having short legs, a bill with a horny tip, and a soft cere. { 'pij·ən }

pigeonhole principle [MATH] The principle, that if a very large set of elements is partitioned into a small number of blocks, then at least one block contains a rather large number of elements. Also known as Dirichlet drawer principle. { 'pij·ən‚hōl ‚prin·sə·pəl }

pigeonite [MINERAL] $(Mg,Fe^{2+},Ca)(MgFe^{2+})Si_{20}$ Clinopyroxene mineral species intermediate in composition between clinoenstatite and diopside, found in basic igneous rocks. { 'pij·ə‚nīt }

pigeon milk [PHYSIO] A milky glandular secretion of the crop of pigeons that is regurgitated to feed newly hatched young. { 'pij·ən ‚milk }

piggyback board [ELECTR] A small printed circuit board that is mounted on a larger board to provide additional circuitry. { 'pig·ē‚bak ‚bȯrd }

piggyback twistor [ELECTR] Electrically alterable nondestructive-readout storage device that uses a thin narrow tape of magnetic material wound spirally around a fine copper conductor to store information; another similar tape is wrapped on top of the first, piggyback fashion, to sense the stored information; a binary digit or bit is stored at the intersection of a copper strap and a pair of these twistor wires. { 'pig·ē‚bak ‚twis·tər }

pig iron [MET] **1.** Crude, high-carbon iron produced by reduction of iron ore in a blast furnace. **2.** Cast iron in the form of pigs. { 'pig ‚ī·ərn }

pigment [CELL MOL] Any coloring matter in plant or animal cells. [MATER] A solid that reflects light of certain wavelengths while absorbing light of other wavelengths, without producing appreciable luminescence; used to impart color to other materials. { 'pig·mənt }

pigmentation [PHYSIO] The normal color of the body and its organs, resulting from a summation of the natural color of the tissue, the pigments deposited therein, and the pigments carried through the blood bathing the tissue. { ‚pig·mən'tā·shən }

pigment cell [CYTOL] Any cell containing deposits of pigment. { 'pig·mənt ‚sel }

pigment epithelium [HISTOL] A heavily pigmented layer of epithelial cells interposed between the photoreceptors of the vertebrate retina and their blood supply, the choroid. It forms a barrier regulating the exchange of fluids and substances between the blood and outer layer of the retina; it also plays a number of roles in support of photoreceptor structure and function in vertebrates. { ‚pig·mənt ‚ep·ə'thē·lē·əm }

pigment print [TEXT] A fabric that has been printed with a paste made of an insoluble pigment mixed with a binder and a thickener. { 'pig·mənt ‚print }

pigtail [ELEC] A short, flexible wire, usually stranded or braided, used between a stationary terminal and a terminal having a limited range of motion, as in relay armatures. { 'pig‚tāl }

pigtail splice [ELEC] A splice made by twisting together the bared ends of parallel conductors. { 'pig‚tāl ‚splīs }

pika [VERT ZOO] Any member of the family Ochotonidae, which includes 14 species of lagomorphs resembling rabbits but having a vestigial tail and short, rounded ears. { 'pī·kə }

pike [GEOL] A mountain or hill which has a peaked summit. [VERT ZOO] Any of about five species of predatory fish which compose the family Esocidae in the order Clupeiformes; the body is cylindrical and compressed, with cycloid scales that have deeply scalloped edges. { pīk }

pike pole [ENG] **1.** A pole with a sharp metal point in one end that is used to hold utility poles upright while they are being installed. **2.** See fire hook. { 'pīk ‚pōl }

Pilacraceae [MYCOL] A family of Basidiomycetes. { ‚pil·ə'krās·ē‚ē }

Pilargidae [INV ZOO] A family of small, short, depressed errantian polychaete annelids. { pə'lär·jə‚dē }

pilaster [CIV ENG] A vertical rectangular architectural member that is structurally a pier and architecturally a column. { pə'las·tər }

pilchard oil [MATER] Pale-yellow oil expressed from pickled pilchards (of the herring family); used to make paints and potash soft soap. { 'pil·chərd ‚ȯil }

pile [ENG] A long, heavy timber, steel, or reinforced concrete post that has been driven, jacked, jetted, or cast vertically into the ground to support a load. [NUCLEO] See nuclear reactor. [TEXT] Loops on a fabric surface. { pīl }

pile beacon [NAV] In marine operation, a beacon formed of one or more piles. { 'pīl ‚bē·kən }

pile bent [CIV ENG] A row of timber or concrete bearing piles with a pile cap forming that part of a trestle which carries the adjacent ends of timber stringers or concrete slabs. { 'pīl ‚bent }

pile cap [CIV ENG] A mass of reinforced concrete cast around the head of a group of piles to ensure that they act as a unit to support the imposed load. { 'pīl ‚kap }

pile dike [CIV ENG] A dike consisting of a group of piles braced and lashed together along a riverbank. { 'pīl ‚dīk }

pile dolphin [NAV] A minor light structure consisting of a number of piles driven into the offshore bottom in a circular pattern and drawn together with a light mounted at the top. { 'pīl ‚däl·fən }

pile driver [MECH ENG] A hoist and movable steel frame equipped to handle piles and drive them into the ground. { 'pīl ‚drīv·ər }

pile extractor [MECH ENG] **1.** A pile hammer which strikes the pile upward so as to loosen its grip and remove it from the ground. **2.** A vibratory hammer which loosens the pile by high-frequency jarring. { 'pīl ik‚strak·tər }

pile formula [MECH] An equation for the forces acting on a pile at equilibrium: $P = pA + tS + Sn \sin \phi$, where P is the load, A is the area of the pile point, p is the force per unit area on the point, S is the embedded surface of the pile, t is the force per unit area parallel to S, n is the force per unit area normal to S, and ϕ is the taper angle of the pile. { 'pīl ‚fȯr·myə·lə }

pile foundation [CIV ENG] A substructure supported on piles. { 'pīl faun‚dā·shən }

pile hammer [MECH ENG] The heavy weight of a pile driver that depends on gravity for its striking power and is used to drive piles into the ground. Also known as drop hammer. { 'pīl ‚ham·ər }

pile lighthouse [NAV] A lighthouse built on piles. { 'pīl 'līt‚haus }

pile shoe [CIV ENG] A cast-iron point on the foot of a timber or concrete driven pile to facilitate penetration of the ground. { 'pīl ‚shü }

pileum [VERT ZOO] The top of a bird's head, from the nape to the bill. { 'pil·ē·əm }

pileup [ELECTR] A set of moving and fixed contacts, insulated from each other, formed as a unit for incorporation in a relay or switch. Also known as stack. { 'pīl‚əp }

pileus [BIOL] The umbrella-shaped upper cap of mushrooms and other basidiomycetous fungi. [METEOROL] An accessory cloud of small horizontal extent, often cirriform, in the form of a cap, hood, or scarf, which occurs above or attached to the top of a cumuliform cloud that often pierces it; several pileus clouds fairly often are observed above each other. Also known as scarf cloud. { 'pil·ē·əs }

pile weave [TEXT] Type of weave made by using two warp yarns and one filling yarn, or one warp yarn and two filling yarns. { 'pīl ‚wēv }

Pilger tube-reducing process [MET] A tube-reducing process in which pierced billets are forced over a mandrel between two rolls with inclined axes and given a rotary forging treatment; the tube is advanced during the gap in each revolution. { 'pil·gər 'tüb ri‚düs·iŋ ‚prä·səs }

Pilidae [INV ZOO] A family of fresh-water snails in the order Pectinibranchia. { 'pil·ə‚dē }

Pilifera [VERT ZOO] Collective designation for animals with hair, that is, mammals. { pī'lif·ə·rə }

piling [GRAPHICS] Caking of ink on the rollers, plate, or blanket of a printing press or an accumulation of paper coating on the blanket of an offset press. { 'pīl·iŋ }

pill [ELECTROMAG] A microwave stripline termination. [PHARM] A small, solid dosage form of a globular, ovoid, or lenticular shape, containing one or more medicinal substances. { pil }

pillar [CIV ENG] A column for supporting part of a structure.

PIKA

A pika, a rock-dwelling diurnal mammal which grows to the size of a guinea pig.

PIKE

Esox lucius.

[GEOL] **1.** A natural formation shaped like a pillar. **2.** A joint block produced by columnar jointing. **3.** *See* stalacto-stalagmite. [MIN ENG] An area of coal or ore left to support the overlying strata or hanging wall in a mine. { 'pil·ər }

pillar-and-breast system [MIN ENG] A system of coal mining in which the working places are rectangular rooms usually five or ten times as long as they are broad, opened on the upper side of the gangway. { 'pil·ər ən 'brest ,sis·təm }

pillar-and-room system [MIN ENG] A system of mining whereby solid blocks of coal are left on either side of working places to support the roof until first-mining has been completed, when the pillar coal is then recovered. { 'pil·ər ən 'rüm ,sis·təm }

pillar-and-stall system [MIN ENG] A system of working coal and other minerals where the first stage of excavation is accomplished with the roof sustained by coal or ore. { 'pil·ər ən 'stȯl ,sis·təm }

pillar bolt [DES ENG] A bolt projecting from a part so as to support it. { 'pil·ər ,bōlt }

pillar buoy [NAV] A buoy composed of a tall central structure mounted on a broad flat base; not used in United States waters. { 'pil·ər ,bȯi }

pillar burst [MIN ENG] A failure of a pillar, by crushing. { 'pil·ər ,bərst }

pillar crane [MECH ENG] A crane whose mechanism can be rotated about a fixed pillar. { 'pil·ər ,krān }

pillar drive [MIN ENG] A wide irregular drift or entry, in firm dry ground, in which the roof is supported by pillars of the natural earth, or by artificial pillars of stone, no timber being used. { 'pil·ər ,drīv }

pillar extraction [MIN ENG] Removal of the pillars of coal left over from mining by the pillar-and-stall method. Also known as pillar mining. { 'pil·ər ik'strak·shən }

pillaring [MIN ENG] The process of extracting pillars. Also called pillar robbing; pulling pillars; robbing pillars. [ORD] The rapid vertical movement of smoke which sometimes results, for instance, from the explosion of a white phosphorus bomb or projectile; the effect is undesirable because it does not produce obscuration over a desirably large area. { 'pil·ə·riŋ }

pillar ladder [NAV ARCH] A ladder formed by fitting rungs extending out from a pillar or stanchion; commonly used as a means of providing access to cargo holds. { 'pil·ər ,lad·ər }

pillar line [MIN ENG] Air currents which have definitely coursed through an inaccessible abandoned panel or area or which have ventilated a pillar line or a pillar area, regardless of the methane content, or absence of methane, in such air. { 'pil·ər ,līn }

pillar man *See* pack builder. { 'pil·ər ,man }

pillar mining *See* pillar extraction. { 'pil·ər ,mīn·iŋ }

pillar press [MECH ENG] A punch press framed by two upright columns; the driving shaft passes through the columns, and the slide operates between them. { 'pil·ər ,pres }

pillar robbing *See* pillaring. { 'pil·ər ,räb·iŋ }

pillar split [MIN ENG] An opening or crosscut driven through a pillar in the course of extraction of ore. { 'pil·ər ,split }

pillbox [ORD] Small, low fortification that houses machine guns or antitank weapons; usually made of concrete, steel, or sandbags. { 'pil,bäks }

pillbox antenna [ELECTROMAG] Cylindrical parabolic reflector enclosed by two plates perpendicular to the cylinder, spaced to permit the propagation of only one mode in the desired direction of polarization. { 'pil,bäks an'ten·ə }

pilling [TEXT] The formation of little fuzzy balls on a fabric surface caused by the rubbing off of loose ends of fiber too long or strong to break away entirely. { 'pil·iŋ }

pillotina [MICROBIO] A large spirochete that contains microtubules and lives symbiotically in the hind gut of termites. { ,pil·ə'tē·nə }

pillow breccia [PETR] A deposit of pillow structures and fragments of lava in a matrix of tuff. { 'pil·ō ,brech·ə }

pillow lace *See* bobbin lace. { 'pil·ō ,lās }

pillow lava [GEOL] Any lava characterized by pillow structure and presumed to have formed in a subaqueous environment. Also known as ellipsoidal lava. { 'pil·ō ,läv·ə }

pillow structure [GEOL] A primary sedimentary structure that resembles a pillow in size and shape. Also known as mammillary structure. [PETR] A pillow-shaped structure

visible in some extrusive lavas attributed to the congealment of lava under water. { 'pil·ō ,strək·chər }

pilmer [METEOROL] In England, a heavy shower of rain. { 'pil·mər }

pilocarpine [ORG CHEM] $C_{11}H_{16}N_2O_2$ An alkaloid, in either oil or crystal form, melting at 34°C; soluble in chloroform, water, and alcohol; used in medicine. { pī·lə'kär,pēn }

pilomotor nerve [NEUROSCI] A nerve causing contraction of one of the arrectoris pilorum muscles. { ¦pī·lō'mōd·ər ,nərv }

pilomotor reflex [PHYSIO] Erection of the hairs of the skin (gooseflesh) in response to chilling or irritation of the skin or to an emotional stimulus. { ¦pī·lō'mōd·ər 'rē,fleks }

pilosebaceous [ANAT] Pertaining to the hair follicles and sebaceous glands, as the pilosebaceous apparatus, comprising the hair follicle and its attached gland. { ¦pī·lō·sə'bā·shəs }

pilosis [MED] The abnormal or excessive development of hair. { pī'lō·səs }

pilot [AERO ENG] **1.** A person who handles the controls of an aircraft or spacecraft from within the craft, and guides or controls the craft in flight. **2.** A mechanical system designed to exercise control functions in an aircraft or spacecraft. [COMMUN] **1.** In a transmission system, a signal wave, usually single frequency, transmitted over the system to indicate or control its characteristics. **2.** Instructions, in tape relay, appearing in routing line, relative to the transmission or handling of that message. [COMPUT SCI] A model of a computer system designed to test its design, logic, and data flow under operating conditions. [DES ENG] A bullet-nosed cylindrical component used in a die that enters prepunched holes of a metal strip advancing through a series of operations to assure precise registration at each station. [MECH ENG] A cylindrical steel bar extending through, and about 8 inches (20 centimeters) beyond the face of, a reaming bit; it acts as a guide that follows the original unreamed part of the borehole and hence forces the reaming bit to follow, and be concentric with, the smaller-diameter, unreamed portion of the original borehole. [NAV] **1.** A person who directs the movements of a vessel through pilot waters, usually a person who has demonstrated extensive knowledge of channels, aids to navigation, dangers to navigation, and so on, in a particular area and is licensed for that area. **2.** A book of sailing directions; for waters of the United States and its possessions, the books are prepared by the U.S. Coast and Geodetic Survey, and are called coast pilots. **3.** The person who flies aircraft. { 'pī·lət }

PILOT [COMPUT SCI] A programming language designed for applications to computer-aided instruction and the question-and-answer type of interaction that occurs in that environment. { 'pī·lət }

pilotage [NAV] The procedure of using landmarks, such as cities, towns, rivers, railroads, and prominent highways, to guide an aircraft to its destination { 'pī·lət·ij }

pilotage waters *See* pilot waters. { 'pī·lət·ij ,wȯd·ərz }

pilotaxitic [GEOL] Pertaining to the texture of the groundmass of a holocrystalline igneous rock in which lath-shaped microlites (usually of plagioclase) are arranged in a glass-free felty mesh, often aligned along the flow lines. { ¦pī·lō·tak'sid·ik }

pilot balloon [ENG] A small balloon whose ascent is followed by a theodolite in order to obtain data for the computation of the speed and direction of winds in the upper air. { 'pī·lət bə,lün }

pilot-balloon observation [METEOROL] A method of winds-aloft observation, that is, the determination of wind speeds and directions in the atmosphere above a station; involves reading the elevation and azimuth angles of a theodolite while visually tracking a pilot balloon. Also known as pibal. { 'pī·lət bə,lün ,äb·zər'vā·shən }

pilot bit [DES ENG] A noncoring bit with a cylindrical diamond-set plug of somewhat smaller diameter than the bit proper, set in the center and projecting beyond the main face of the bit. { 'pī·lət ,bit }

pilot boat [NAV] A small vessel used by a pilot to go to or from a vessel employing the pilot's services. Also known as pilot vessel. { 'pī·lət ,bōt }

pilot briefing [METEOROL] Oral comment on the observed and forecast weather conditions along a route, given by a forecaster to the pilot, navigator, or other air crew member prior to takeoff. Also known as briefing; flight briefing; flight-weather briefing. { 'pī·lət ,brēf·iŋ }

pilot cell [ELEC] Selected cell of a storage battery whose temperature, voltage, and specific gravity are assumed to indicate the condition of the entire battery. { 'pī·lət ,sel }

pilot channel [CIV ENG] One of a series of cutoffs for converting a meandering stream into a straight channel of greater slope. { 'pī·lət ,chan·əl }

pilot chart [NAV] A chart of a major ocean area published, for the benefit of mariners, by the U.S. Naval Oceanographic Office in cooperation with the U.S. Weather Bureau; these charts contain information required in planning safe routes, including ocean currents, ice at sea, wind roses, storm tracks, isotherms, magnetic variation, and recommended routes or steamer tracks. { 'pī·lət ,chärt }

pilot chute [AERO ENG] A small parachute canopy attached to a larger canopy to actuate and accelerate the opening of the load-bearing canopy. { 'pī·lət ,shüt }

pilot drill [MECH ENG] A small drill to start a hole to ensure that a larger drill will run true to center. { 'pī·lət ,dril }

piloted ignition [ENG] The accidental initiation of combustion by means of contact of gaseous material with an external high-energy source, such as a flame, spark, electrical arc, or glowing wire. { ¦pīl·əd·əd ig'nish·ən }

pilot flood [PETRO ENG] A test waterflood operation (water-injection well) designed to evaluate procedures and to give advance information prior to instituting an extensive, multipoint waterflood. { 'pī·lət ,fləd }

pilot hole [DES ENG] In metal-forming operations, a prepunched hole in a metal strip into which the pilot component of the die enters in order to assure precise registration of the strip at each work station. [ENG] A small hole drilled ahead of a larger borehole. [MECH ENG] A hole drilled in a piece of wood to serve as a guide for a nail or a screw or for drilling a larger hole. { 'pī·lət ,hōl }

pilothouse [NAV ARCH] A compartment on or near the bridge of a ship that contains the steering wheel and other controls, compass, charts, navigating equipment, and means of communicating with the engine room and other parts of the ship. Also known as wheelhouse. { 'pī·lət,haus }

piloting [NAV] The navigation of a vehicle, particularly a marine craft, by determining position relative to external reference points, usually fixed points on the earth. { 'pī·ləd·iŋ }

pilot lamp [ELEC] A small lamp used to indicate that a circuit is energized. Also known as pilot light. { 'pī·lət ,lamp }

pilotless aircraft [AERO ENG] An aircraft adapted to control by or through a preset self-reacting unit or a radio-controlled unit, without the benefit of a human pilot. { 'pī·lət·ləs 'er,kraft }

pilot light [ELEC] See pilot lamp. [ENG] A small, constantly burning flame used to ignite a gas burner. { 'pī·lət ,līt }

pilot lightship [NAV] A lightship which also serves as a pilot station. { 'pī·lət 'līt,ship }

pilot line operation [IND ENG] Minimum production of an item in order to preserve or develop the art of its production. { 'pī·lət 'līn ,äp·ə,rā·shən }

pilot materials [IND ENG] A minimum quantity of special materials, partially finished components, forgings, and castings, identified with specific production equipment and processes and required for the purpose of proofing, tooling, and testing manufacturing processes to facilitate later reactivation. { 'pī·lət mə,tir·ē·əlz }

pilot model [IND ENG] An early production model of a product used to debug the manufacturing process. { 'pī·lət ,mäd·əl }

pilot motor [ELEC] A small motor used in the automatic control of an electric current. { 'pī·lət ,mōd·ər }

pilot plant [IND ENG] A small version of a planned industrial plant, built to gain experience in operating the final plant. { 'pī·lət 'plant }

pilot production [PETRO ENG] Limited or test production of oil or gas from a field to determine reservoir and product characteristics before commencing full-scale recovery operations. { 'pī·lət prə,dək·shən }

pilot reaming bit See reaming bit. { 'pī·lət 'rēm·iŋ ,bit }

pilot relaying [ELEC] A system for protecting transmission consisting of protective relays at line terminals and a communication channel between relays which is used by the relays to determine if a fault is within the protected line section, in which case all terminals are tripped simultaneously at high speed, or outside it, in which case tripping is blocked. { 'pī·lət rē,lā·iŋ }

pilot report [METEOROL] A report of in-flight weather by an aircraft pilot or crew member; a complete pilot report includes the following information in this order: location or extent of reported weather phenomena, time of observation, description of phenomena, altitude of phenomena, type of aircraft (only with reports of turbulence or icing). Also known as aircraft report; pirep. { 'pī·lət ri,pórt }

pilot rules [NAV] Regulations supplementing the Inland Rules of the Road. { 'pī·lət ,rülz }

pilot-scale chemical reaction [CHEM ENG] Small-scale chemical reaction used to test operating conditions and product yields; used as a pilot for design of large-scale reaction systems. { 'pī·lət ¦skāl ¦kem·ə·kəl rē'ak·shən }

pilot station [NAV] At seaports, the office or headquarters of marine pilots; the place where the services of a pilot may be obtained. [NAV ARCH] Position on the bridge of a ship where the pilot stands to steer or to give directions for steering a ship into and out of a harbor. { 'pī·lət ,stā·shən }

pilot streamer [GEOPHYS] A relatively slow-moving, nonluminous lightning streamer, the existence of which has been postulated to help account for the observed mode of advance of a stepped leader as it initiates a lightning discharge. { 'pī·lət ,strēm·ər }

pilot system [COMPUT SCI] A system for evaluating new procedures for handling data in which a sample that is representative of the data to be handled is processed. { 'pī·lət ,sis·təm }

pilot test [COMPUT SCI] A test of a computer system under operating conditions and in the environment for which the system was designed. { 'pī·lət ,test }

pilot tone [COMMUN] Single frequency transmitted over a channel to operate an alarm or automatic control. { 'pī·lət ,tōn }

pilot tunnel [ENG] A small tunnel or shaft excavated in advance of the main drivage in mining and tunnel building to gain information about the ground, create a free face, and thus simplify the blasting operations. { 'pī·lət ,tən·əl }

pilot vessel See pilot boat. { 'pī·lət ,ves·əl }

pilot waters [NAV] 1. Areas in which the services of a marine pilot are essential. 2. Waters in which navigation is by piloting. Also known as pilotage waters. { 'pī·lət ,wód·ərz }

pilot wire regulator [CONT SYS] Automatic device for controlling adjustable gains or losses associated with transmission circuits to compensate for transmission changes caused by temperature variations, the control usually depending upon the resistance of a conductor or pilot wire having substantially the same temperature conditions as the conductors of the circuits being regulated. { 'pī·lət ¦wīr 'reg·yə,lād·ər }

Piltdown man [PALEON] An alleged fossil man based on fragments of a skull and mandible that were eventually discovered to constitute a skillful hoax. { 'pilt,daun ,man }

pilus [ANAT] A hair. [BIOL] A fine, slender, hairlike body. [MICROBIO] Any filamentous appendage other than flagella on certain gram-negative bacteria. Also known as fimbria. { 'pī·ləs }

PIM See personal information manager. { ¦pē¦ī'em or pim }

pima cotton [TEXT] Fine-quality, long-staple cotton fibers or fine-combed cottons. { 'pē·mə ,kät·ən }

pimaricin [ORG CHEM] $C_{33}H_{47}NO_{13}$ A compound crystallizing from a methanol-water solution, decomposing at about 200°C; soluble in water and organic solvents; used in medicine as an antifungal agent for *Candida albicans* vaginitis. { pə'mar·ə·sən }

pimelic acid [ORG CHEM] $HOOC(CH_2)_5COOH$ Crystals melting at 105°C; slightly soluble in water, soluble in alcohol and ether; used in biochemical research. { pə'mel·ik 'as·əd }

pimenta oil [MATER] A yellow to brownish essential oil with spicy aroma and pungent taste; derived from allspice (*Pimenta officinalis*); main components are eugenol, cineol, and phellandrene; used in medicine and flavors. Also known as allspice oil; pimento oil. { pə'ment·ə ,óil }

pimento [BOT] *Capsicum annuum*. A type of pepper in the order Polemoniales grown for its thick, sweet-fleshed red fruit. { pə'ment·ō }

pimento oil See pimenta oil. { pə'ment·ō ,óil }

pi meson [PARTIC PHYS] 1. Collective name for three semistable mesons which have charges of +1, 0, and −1 times the proton charge, and form a charge multiplet, with an approximate mass of 138 megaelectronvolts, spin 0, negative parity,

PIMARICIN

Structural formula of pimaricin.

negative *G* parity, and positive charge parity (for the neutral meson). Also known as pion. Symbolized π. **2.** Any meson belonging to an isospin triplet with hypercharge 0, negative *G* parity, and positive charge parity (for the neutral meson). { 'pī 'mā,sän }

pi mode [ELECTR] Of a magnetron, the mode of operation for which the phases of the fields of successive anode openings facing the interaction space differ by pi radians. { 'pī ,mōd }

pimple [MATER] A small, conical elevation on the surface of a plastic. [MED] A small pustule or papule. { 'pim·pəl }

pimple mound [GEOL] A low, flattened, roughly circular or elliptical dome consisting of sandy loam that is entirely distinct from the surrounding soil; peculiar to the Gulf coast of eastern Texas and southwestern Louisiana. { 'pim·pəl ,maůnd }

pimple plain [GEOL] A plain distinguished by the presence of numerous, conspicuous pimple mounds. { 'pim·pəl ,plān }

pi-mu atom *See* pionium. { 'pī 'myü 'ad·əm }

pin [DES ENG] **1.** A cylindrical fastener made of wood, metal, or other material used to join two members or parts with freedom of angular movement at the joint. **2.** A short, pointed wire with a head used for fastening fabrics, paper, or similar materials. [ELECTR] A terminal on an electron tube, semiconductor, integrated circuit, plug, or connector. Also known as base pin; prong. { 'pin }

piñ [BOT] A fiber obtained from the large leaves of the pineapple plant. Also known as pineapple fiber. { 'pēn·yə }

pinacocyte [INV ZOO] A flattened polygonal cell occurring in the dermal epithelium of sponges, and lining the exhalant canals. { 'pin·ə·kō,sīt }

pinacoid [CRYSTAL] An open crystal form that comprises two parallel faces. { 'pin·ə,ko̊id }

pinacoidal class [CRYSTAL] That crystal class in the triclinic system having only a center of symmetry. { ¦pin·ə¦ko̊id·əl ¦klas }

pinacoidal cleavage [CRYSTAL] A type of crystal cleavage that is parallel to one of the crystal's pinacoidal surfaces. { ¦pin·ə¦ko̊id·əl ¦klē·vij }

pinakiolite [MINERAL] $Mg_3Mn_3B_2O_{10}$ A black mineral composed of borate of magnesium and manganese; it is polymorphous with orthopinakiolite. { pə'näk·ē·ə,līt }

Pinales [BOT] An order of gymnospermous woody trees and shrubs in the class Pinopsida, including pine, spruce, fir, cypress, yew, and redwood; the largest plants are the conifers. { pī'nā·lēz }

Pinatae *See* Pinopsida. { pī'näd·ē }

pin bar [MET] A small-diameter, case-hardened steel rod used for making dowel pins. { 'pin ,bär }

pinboard [COMPUT SCI] A board or panel containing an array of uniform holes into which pins may be inserted to control the operation of equipment. { 'pin,bo̊rd }

pincer [INV ZOO] A grasping apparatus, as on the anterior legs of a lobster, consisting of two hinged jaws. { 'pin·sər }

pinch [ENG] The closing-in of borehole walls before casing is emplaced, resulting from rock failure when drilling in formations having a low compressional strength. [GEOL] Thinning of a rock layer, as where a vein narrows. [MIN ENG] *See* horseback; squeeze. { pinch }

pinch-and-swell structure [GEOL] A structural condition common in pegmatites and veins of quartz in metamorphosed rocks; the vein is pinched at frequent intervals, leaving expanded parts between. { ¦pinch ən 'swel ,strək·chər }

pinch bar [DES ENG] A pointed lever, used somewhat like a crowbar, to roll heavy wheels. { 'pinch ,bär }

pinch effect [ELEC] Manifestation of the magnetic self-attraction of parallel electric currents, such as constriction of ionized gas in a discharge tube, or constriction of molten metal through which a large current is flowing. Also known as cylindrical pinch; magnetic pinch; rheostriction. { 'pinch i,fekt }

pinch graft [MED] A small, full-thickness graft lifted from the donor area by a needle, and cut free with a razor. { 'pinch ,graft }

pinch grasp [IND ENG] A grasp by the human hand that involves the thumb and the facing side of the index finger at the knuckle; used to apply a large force to a small object. Also known as key grasp. { 'pinch ,grasp }

pinch-off *See* cutoff. { 'pinch,o̊f }

pinch-off blades [ENG] In blow molding, the part that compresses the parison to seal it prior to blowing, and to allow easy cooling and removal of flash. { 'pinch,o̊f ,blādz }

pinch-off voltage [ELECTR] Of a field-effect transistor, the voltage at which the current flow between source and drain is blocked because the channel between these electrodes is completely depleted. { 'pinch,o̊f ,vōl·tij }

pinch pass [MET] A cold rolling of sheet metal to effect a very small reduction in thickness and to produce a piece of accurate dimensions. { 'pinch ,pas }

pinch point [IND ENG] A point in a plant layout or on an automated guided vehicle such that the distance between the automated guided vehicle and the surrounding equipment and structures is so small that it represents a safety hazard to personnel. { 'pinch ,po̊int }

pinch resistor [ELECTR] A silicon integrated-circuit resistor produced by diffusing an *n*-type layer over a *p*-type resistor; this narrows or pinches the resistive channel, thereby increasing the resistance value. { 'pinch ri'zis·tər }

pinch roller [ELECTR] A small, freely turning wheel that presses the magnetic tape against the capstan in order to move the tape. { 'pinch ,rō·lər }

pinch trimming [MET] Trimming a tubular or hollow part by pinching the flange or lip over the cutting edge of a punch. { 'pinch ,trim·iŋ }

pinch-tube process [ENG] A plastics blow-molding process in which the extruder drops a tube between mold halves, and the tube is pinched off when the mold closes. { 'pinch ,tüb ,prä·səs }

pincushion distortion [ELECTR] Distortion in which all four sides of a received television picture are concave (curving inward). [OPTICS] Aberration in which the magnification produced by an optical system increases with the distance of the object point from the optical axis, so that the image of a square has concave sides. { 'pin,kůsh·ən di,stor·shən }

pin diode [ELECTR] A diode consisting of a silicon wafer containing nearly equal *p*-type and *n*-type impurities, with additional *p*-type impurities diffused from one side and additional *n*-type impurities from the other side; this leaves a lightly doped intrinsic layer in the middle, to act as a dielectric barrier between the *n*-type and *p*-type regions. Also known as power diode. { 'pin 'dī,ōd }

pine [BOT] Any of the cone-bearing trees composing the genus *Pinus*; characterized by evergreen leaves (needles), usually in tight clusters of two to five. { 'pīn }

pineal body [ANAT] An unpaired, elongated, club-shaped, knoblike or threadlike organ attached by a stalk to the roof of the vertebrate forebrain. Also known as conarium; epiphysis. { 'pin·ē·əl ,bäd·ē }

pineapple [BOT] *Ananas sativus*. A perennial plant of the order Bromeliales with long, swordlike, usually rough-edged leaves and a dense head of small abortive flowers; the fruit is a sorosis that develops from the fleshy inflorescence and ripens into a solid mass, covered by the persistent bracts and crowned by a tuft of leaves. { 'pī,nap·əl }

pineapple fiber *See* piña. { 'pī,nap·əl ,fī·bər }

pinene [ORG CHEM] $C_{10}H_{16}$ Either of two colorless isomeric unsaturated bicyclic terpene hydrocarbon liquids derived from sulfate wood turpentine; 95% of the alpha form boils in the range 156–160°C, and of the beta form boils in the range 164–169°C; used as solvents for coatings and wax formulations, as chemical intermediates for resins, and as lube-oil additives. Also known as nopinene. { 'pī,nēn }

pine needle oil [MATER] An essential oil derived from various pines; colorless to yellowish oil with balsamic aroma; soluble in alcohol; used in perfumes and medicines. Also known as Douglas fir oil; fir wood oil. { 'pīn ,nēd·əl ,oil }

pinene hydrochloride *See* terpene hydrochloride. { 'pī,nēn ,hi·drə'klo̊r,īd }

pine nut [BOT] The edible seed borne in the cone of various species of pine (*Pinus*), such as stone pine (*P. pinea*) and piñon pine (*P. cembroides* var. *edulis*). { 'pīn ,nət }

pine oil [MATER] Any of a group of volatile essential oils with pinaceous aromas distilled from cones, needles, or stumps of various pine or other conifer species; used as solvents, emulsifying agents, wetting agents, deodorants, germicides, and sources of chemicals. { 'pīn ,oil }

pine oleoresin [MATER] Fused solid blend of turpentine and rosin. { 'pīn ¦ō·lē·ō'rez·ən }

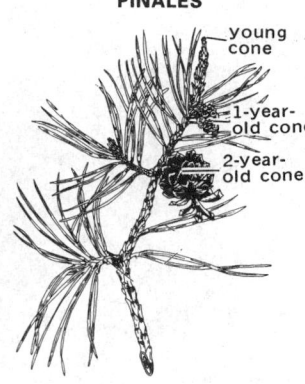

PINALES

young cone

1-year-old cone

2-year-old cone

Branch of pine with leaves and ovulate cones. *(From J. B. Hill et al., Botany, 3d ed., McGraw-Hill, 1960)*

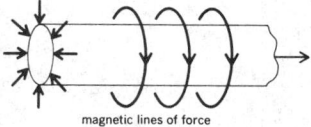

PINCH EFFECT

magnetic lines of force

Example of pinch effect. Current *I*, uniformly distributed in cylindrical wire, gives rise to compression force indicated by arrows at left.

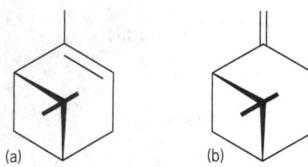

PINENE

(a)   (b)

Two principal terpenes of the oil of southern pines. (*a*) α-Pinene. (*b*) β-Pinene.

piner [METEOROL] In England, a rather strong breeze from the north or northeast. { 'pīn·ər }

pine tar [MATER] A viscous black mass obtained as a by-product in the distillation of pine wood; used for roofing. { 'pīn ,tär }

pine tar pitch [MATER] Viscous residue resulting from distillation of volatile oils from pine tar. { 'pīn ,tär ,pich }

pine terpene [MATER] Any terpene in the essential oils obtained from various *Pinus* species. { 'pīn 'tər,pēn }

pine-tree array [ELECTROMAG] Array of dipole antennas aligned in a vertical plane known as the radiating curtain, behind which is a parallel array of dipole antennas forming a reflecting curtain. { 'pīn ,trē ə ,rā }

pi network [ELEC] An electrical network which has three impedance branches connected in series to form a closed circuit, with the three junction points forming an output terminal, an input terminal, and a common output and input terminal. { 'pī ,net,wərk }

pin expansion test [MET] A test for determining tube expandability or for revealing longitudinal weaknesses by forcing a tapered pin into the open end of the tube. { 'pin ik'span·shən ,test }

pinfeather [VERT ZOO] A young, underdeveloped feather, especially one still enclosed in a cylindrical horny sheath which is afterward cast off. { 'pin,feth·ər }

pin-feed printer [COMPUT SCI] A computer printer in which the paper is aligned and advanced by protrusions on two wheels which engage evenly spaced holes along the edges of the paper. Also known as tractor-feed printer. { 'pin ,fēd 'print·ər }

ping [ELECTR] A sonic or ultrasonic pulse sent out by an echo-ranging sonar. { piŋ }

pinger [ENG ACOUS] A battery-powered, low-energy source for an echo sounder. { 'piŋ·ər }

pingo [HYD] A frost mound resembling a volcano, being a relatively large and conical mound of soil-covered ice, elevated by hydrostatic pressure of water within or below the permafrost of arctic regions. { 'piŋ·gō }

pingo ice [HYD] Clear or relatively clear ice that occurs in permafrost; originates from groundwater under pressure. { 'piŋ·gō ,īs }

pingo remnant [GEOL] A rimmed depression formed by the rupturing of a pingo summit which results in the exposure of the ice core to melting followed by partial or total collapse. Also known as pseudokettle. { 'piŋ·gō 'rem·nənt }

ping-pong [COMMUN] To switch a transmission so that it travels in the opposite direction. [COMPUT SCI] The programming technique of using two magnetic tape units for multiple reel files and switching automatically between the two units until the complete file is processed. { 'piŋ,päŋ }

pinguecula [ANAT] A small patch of yellowish-white connective tissue located on the conjunctiva, between the cornea and the canthus of the eye. { piŋ'gwek·yə·lə }

pinguite *See* nontronite. { 'piŋ,gwīt }

pinhole [MET] A material fault resulting from small blisters that have burst in a casting or that have formed during electroplating. { 'pin,hōl }

pinhole camera [OPTICS] A camera which has no lenses, but consists essentially of a darkened box with a small hole in one side, so that an inverted image of outside objects is projected on the opposite side where it is recorded on photographic film. { 'pin,hōl 'kam·rə }

pinhole chert [PETR] Chert containing weathered pebbles which are pierced by minute holes or pores. { 'pin,hōl 'chərt }

pinhole detector [ENG] A photoelectric device that detects extremely small holes and other defects in moving sheets of material. { 'pin,hōl di,tek·tər }

pinic acid [ORG CHEM] $C_9H_{14}O_4$ A crystalline dicarboxylic acid derived from α-pinene; used to make diesters for plasticizers and lubricants. { 'pī·nik 'as·əd }

Pinicae [BOT] A large subdivision of the Pinophyta, comprising woody plants with a simple trunk and excurrent branches, simple, usually alternative, needlelike or scalelike leaves, and wood that lacks vessels and usually has resin canals. { 'pī·nə,sē }

pinion [MECH ENG] The smaller of a pair of gear wheels or the smallest wheel of a gear train. [VERT ZOO] The distal portion of a bird's wing. { 'pin·yən }

pinite [MINERAL] A compact gray, green, or brown mica, chiefly muscovite derived from other minerals such as cordierite. { 'pē,nīt }

pin jack [ELEC] Single conductor jack having an opening for the insertion of a plug of very small diameter. { 'pin ,jak }

pin joint [DES ENG] A joint made with a pin hinge which has a removable pin. { 'pin ,jóint }

pin junction [ELECTR] A semiconductor device having three regions: *p*-type impurity, intrinsic (electrically pure), and *n*-type impurity. { 'pin ,jəŋk·shən }

pink disease [PL PATH] A fungus disease of the bark of rubber, cacao, citrus, coffee, and other trees caused by *Corticium salmonicolor* and characterized by a pink covering of hyphae on the stems and branches. { 'pink di,sēz }

pinkeye [MED] **1.** A contagious, mucopurulent conjunctivitis. **2.** *See* catarrhal conjunctivitis. { 'pink,ī }

pink noise [ACOUS] Noise whose intensity is inversely proportional to frequency over a specified range, to give constant energy per octave. { 'piŋk ,nóiz }

pink knot [MATER] A small knot with a diameter of 0.5 inch (1.3 centimeters) or less occurring in a wood board. { 'pink ,nät }

pink root [PL PATH] A fungus disease of onion and garlic caused by various organisms, especially species of *Phoma* and *Fusarium*; marked by red discoloration of the roots. { 'pink ,rüt }

pink rot [PL PATH] **1.** A fungus disease of potato tubers caused by *Phytophtora erythroseptica* and characterized by wet rot and pink color of the cut surfaces of the tuber upon exposure to air. **2.** A rot disease of apples caused by the fungus *Tricothecium roseum*. **3.** A watery soft rot of celery caused by the fungus *Sclerotinia sclerotiorum*. { 'pink ,rät }

pink salmon [VERT ZOO] *Oncorhynchus*. Weighing less than 2 kilograms (5 pounds), it is the smallest but typically most abundant of the salmon. Also known as humpback salmon. { ,pink 'sam·ən }

pin metal [MET] Brass with a composition of 63% copper and 37% zinc, used as cold-drawn wire for making ordinary dressmaking pins. { 'pin ,med·əl }

pinna [ANAT] The cartilaginous, projecting flap of the external ear of vertebrates Also known as auricle. { 'pin·ə }

pinnacle [ARCH] A projection on the highest point of the roof of a building. [GEOL] **1.** A sharp-pointed rock rising from the bottom, which may extend above the surface of the water, and may be a hazard to surface navigation; due to the sheer rise from the sea floor, no warning is given by sounding. **2.** Any high tower or spire-shaped pillar of rock, alone or cresting a summit. { 'pin·ə·kəl }

pinnacled iceberg [OCEANOGR] An iceberg weathered in such manner as to produce spires or pinnacles. Also known as irregular iceberg; pyramidal iceberg. { 'pin·ə·kəld 'īs,bərg }

pinnate [BOT] Having parts arranged like a feather, branching from a central axis. { 'pi,nāt }

pinnate drainage [HYD] A dendritic drainage pattern in which the main stream receives many closely spaced, subparallel tributaries that join it at acute angles; resembles a feather in plan view. { 'pi,nāt 'drā·nij }

pinnate joint *See* feather joint. { 'pi,nāt ,jóint }

pinnately compound leaf [BOT] A leaf with leaflets that are borne on the continuation of the petiole. { pi,nāt·lē ,käm,paund 'lēf }

pinnate muscle [ANAT] A muscle having a central tendon onto which many short, diagonal muscle fibers attach at rather acute angles. { 'pi,nāt ,məs·əl }

pinning [SOLID STATE] The hindering of motion of dislocations in a solid, and the consequent hardening of the solid, by impurities which collect near the dislocations, resulting in a large energy barrier being imposed against the motion of the dislocations. { 'pin·iŋ }

Pinnipedia [VERT ZOO] A suborder of aquatic mammals in the order Carnivora, including walruses and seals. { ,pin·ə'pē·dē·ə }

pinnoite [MINERAL] $Mg(BO_2)_2·3H_2O$ A yellow mineral composed of hydrous borate of magnesium, occurring in nodular masses. { 'pin·ə,wīt }

Pinnotheridae [INV ZOO] The pea crabs, a family of decapod crustaceans belonging to the Brachygnatha. { ,pin·ə'ther·ə,dē }

pinnulate [BIOL] Having pinnules. { 'pin·yə·lāt }

PINNATE MUSCLE

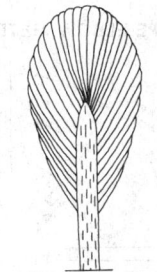

Diagram of pinnate muscle showing central tendon.

pinnule [BIOL] The secondary branch of a plumelike or pinnate organ. { 'pin,yül }

pinocytosis [CYTOL] Deprecated term formerly used to describe the process of uptake or internalization of particles, macromolecules, and fluid droplets by living cells; the process is now termed endocytosis. { ¦pin·ō·sī'tō·səs }

pinolite [PETR] A metamorphic rock containing magnesite (breunnerite) as crystals and as granular aggregates in a schistose matrix (phyllite or talc schist). { 'pin·əl,īt }

Pinophyta [BOT] The gymnosperms, a division of seed plants characterized as vascular plants with roots, stems, and leaves, and with seeds that are not enclosed in an ovary but are borne on cone scales or exposed at the end of a stalk. { pə'näf·əd·ə }

Pinopsida [BOT] A class of gymnospermous plants in the subdivision Pinicae characterized by entire-margined or slightly toothed, narrow leaves. { pə'näp·səd·ə }

pinosome [CYTOL] A closed intracellular vesicle containing material captured by pinocytosis. { 'pin·ə,sōm }

pinout [ELECTR] A graphic or text description of the function of electronic signals transmitted through each pin and receptacle in a connector. { 'pin,aüt }

pinpoint [NAV] **1.** A precisely identified point, especially on the ground, that locates a very small target, a reference point for rendezvous or for other purposes; the coordinates that define this point. **2.** The ground position of aircraft determined by direct observation of the ground. **3.** To establish (position) with great accuracy. { 'pin,póint }

pinpoint gate [ENG] In plastics molding, an orifice of 0.030 inch (0.76 millimeter) or less in diameter through which molten resin enters a mold cavity. { 'pin,póint ,gāt }

pin register [GRAPHICS] The use of accurately positioned holes and special pins to ensure exact superimposition of copy, film, plates, and presses during makeup and printing. { 'pin ,rej·ə·stər }

pin rod [DES ENG] A rod designed to connect two parts so they act as one. { 'pin ,räd }

pin sensing [COMPUT SCI] Device using a punched card, sensing the opening and closing of switches to generate digital data. { 'pin ,sens·iŋ }

pint [MECH] Abbreviated pt. **1.** A unit of volume, used in the United States for measurement of liquid substances, equal to 1/8 U.S. gallon, or 231/8 cubic inches, or 4.73176473 × 10^{-4} cubic meter. Also known as liquid pint (liq pt). **2.** A unit of volume used in the United States for measurement of solid substances, equal to 1/64 U.S. bushel, or 107,521/3200 cubic inches, or approximately 5.50610 × 10^{-4} cubic meter. Also known as dry pint (dry pt). **3.** A unit of volume, used in the United Kingdom for measurement of liquid and solid substances, although usually the former, equal to 1/8 imperial gallon, or 5.6826125 × 10^{-4} cubic meter. Also known as imperial pint. { pīnt }

pinta [MED] A disease of the skin seen most frequently in tropical America, characterized by dyschromic changes and hyperkeratosis in patches of the skin; caused by the spirochete *Treponema carateum*. Also known as carate; mal de pinto; piquite; purupuru; quitiqua. { 'pēnt·ə }

pintadoite [MINERAL] $Ca_2V_2O_7·9H_2O$ A green mineral consisting of a hydrated calcium vanadate; occurs as an efflorescence. { ¦pin·tə'dō,īt }

pin timbering [MIN ENG] A method of mine roof support in which bolts are driven up into strong material, thus supporting lower weak layers. { 'pin ,tim·bə·riŋ }

pintle [DES ENG] A vertical pivot pin, as on a rudder or a gun carriage. { 'pint·əl }

pintle center [ORD] An assumed center of a weapon on which all firing data computations are based. { 'pint·əl ,sen·tər }

pintle chain [DES ENG] A chain with links held together by pivot pins; used with sprocket wheels. { 'pint·əl ,chān }

pintle hitch [ORD] A frame, secured to the rear of a tank or combat vehicle, that carries a quick-release pintle assembly with a cable running over a pulley into the tank; used for towing trailers. { 'pint·əl ,hich }

pin-type mill [MECH ENG] Solids pulverizer in which protruding pins on high-speed rotating disk provide the breaking energy. { 'pin ,tīp ,mil }

pinulus [INV ZOO] A sponge spicule, usually with five rays, one of which develops numerous small spines. { 'pin·yə·ləs }

pinworm [INV ZOO] *Enterobius vermicularis*. A phasmid nematode of the superfamily Oxyuroidea; causes enterobiasis. Also known as human threadworm; seatworm. { 'pin,wərm }

Piobert lines See Lüders' lines. { pyō'ber ,līnz }

PIOCS [COMPUT SCI] An extension of the hardware, constituting an interface between programs and data channels; opposed to LIOCS, logical input/output control system. Derived from physical input/output control system. { 'pī,äks }

pion See pi meson. { 'pī,än }

pion bremsstrahlung [NUC PHYS] The emission of pions in the collision of two heavy nuclei in a manner analogous to the radiation that occurs when a charged particle is accelerated or decelerated. { 'pī,än brem'shträ·ləŋ }

pion condensate [NUC PHYS] A state of nuclear matter compressed to abnormally high densities, in which great numbers of pairs of particles, each consisting of a positive pion and a negative pion, are generated, and interact strongly with the nucleons, causing them to form a coherent spin-isospin structure. { 'pī,än 'känd·ən,sāt }

pion double-charge exchange [NUC PHYS] A nuclear reaction in which a positive pion interacts with a nucleus by a two-step process and a negative pion emerges, accompanied by the conversion of two neutrons in the nucleus to two protons, or the inverse reaction. { 'pī,än ¦dəb·əl ¦chärj iks'chānj }

pioneer [ECOL] An organism that is able to establish itself in a barren area and begin an ecological cycle. { ,pī·ə'nir }

pioneer tunnel [MIN ENG] A small tunnel parallel to but ahead of a main tunnel and used to make crosscuts to the path that the main tunnel will follow. { ,pī·ə'nir ,tən·əl }

pionium [PARTIC PHYS] **1.** An exotic atom consisting of a muon orbiting about an oppositely charged pion. Also known as pi-mu atom. **2.** An exotic atom consisting of an electron in orbit about an oppositely charged pion. { pī'ō·nē·əm }

Piophilidae [INV ZOO] The skipper flies, a family of myodarian cyclorrhaphous dipteran insects in the subsection Acalypteratae. { ,pī·ə'fil·ə,dē }

piotine See saponite. { 'pī·ə,tēn }

pip See blip. { pip }

pipe [COMPUT SCI] Any software-controlled technique for transfering data fron one program or task to another during processing. [DES ENG] A tube made of metal, clay, plastic, wood, or concrete and used to conduct a fluid, gas, or finely divided solid. [GEOL] **1.** A vertical, cylindrical ore body. Also known as chimney; neck; ore chimney; ore pipe; stock. **2.** A tubular cavity of varying depth in calcareous rocks, often filled with sand and gravel. **3.** A vertical conduit through the crust of the earth below a volcano, through which magmatic materials have passed. Also known as breccia pipe. [MET] **1.** The central cavity in an ingot or casting formed by contraction of the metal during solidification. **2.** An extrusion defect caused by the oxidized surface of the billet flowing toward the center of the rod at the back end. { pīp }

pipe amygdule [GEOL] An elongate amygdule occurring toward the base of a lava flow, probably formed by the generation of gases or vapor from the underlying material. { 'pīp ə'mig,dyül }

pipe bit [DES ENG] A bit designed for attachment to standard coupled pipe for use in socketing the pipe in bedrock. { 'pīp ,bit }

pipebox [ENG] In a pipework installation, a casing packed with loose insulation to enclose a set of pipes. { 'pīp,bäks }

pipe clamp [DES ENG] A device similar to a casing clamp, but used on a pipe to grasp it and facilitate hoisting or suspension. { 'pīp ,klamp }

pipe clay [GEOL] A mass of fine clay, usually lens-shaped, which forms the surface of bedrock and upon which often rests the gravel of old river beds. { 'pīp ,klā }

pipe culvert [CIV ENG] A buried pipe for carrying a watercourse below ground level. { 'pīp ,kəl·vərt }

pipe cutter [DES ENG] A hand tool consisting of a clamplike device with three cutting wheels which are forced inward by screw pressure to cut into a pipe as the tool is rotated around the pipe circumference. { 'pīp ,kəd·ər }

pipe elbow meter [ENG] A variable-head meter for measuring flow around the bend in a pipe. { 'pīp 'el·bō ,mēd·ər }

pipe fitter [ENG] A technician who fits, threads, installs, and repairs pipes in a pipework system. { 'pīp ,fid·ər }

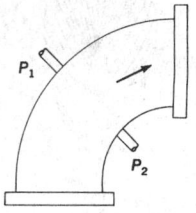

PIPE ELBOW METER

The meter measures flow rate by using the difference in pressure ($P_1 - P_2$) between pressure connections on the inside and outside of the bend in a pipe elbow.

pipe fitting [ENG] A piece, such as couplings, unions, nipples, tees, and elbows for connecting lengths of pipes. { 'pīp ,fid·iŋ }

pipe flow [ENG] Conveyance of fluids in closed conduits. { 'pīp ,flō }

pipe laying [ENG] The placing of pipe into position in a trench, as with buried pipelines for oil, water, or chemicals. { 'pīp ,lā·iŋ }

pipeline [ENG] A line of pipe connected to valves and other control devices, for conducting fluids, gases, or finely divided solids. { 'pīp,līn }

pipelining [COMPUT SCI] A procedure for processing instructions in a computer program more rapidly, in which each instruction is divided into numerous small stages, and a population of instructions are in various stages at any given time. { 'pīp,līn·iŋ }

pipe pile [CIV ENG] A steel pipe 6–30 inches (15–76 centimeters) in diameter, usually filled with concrete and used for underpinning. { 'pīp ,pīl }

Piperaceae [BOT] A family of dicotyledonous plants in the order Piperales characterized by alternate leaves, a solitary ovule, copious perisperm, and scanty endosperm. { ,pip·ə'rās·ē,ē }

Piperales [BOT] An order of dicotyledonous herbaceous plants marked by ethereal oil cells, uniaperturate pollen, and reduced crowded flowers with orthotropous ovules. { 'pip·ə'rā·lēz }

piperazine [ORG CHEM] $C_4H_{10}N_2$ A cyclic compound; colorless, deliquescent crystals, melting at 104–107°C; soluble in water, alcohol, glycerol, and glycols; absorbs carbon dioxide from air; used in medicine. { pī'par·ə,zēn }

piperazine dihydrochloride [ORG CHEM] $C_4H_{10}N_2·2HCl$ White, water-soluble needles; used for insecticides and pharmaceuticals. { pī'par·ə,zēn dī,hī·drə'klór,īd }

piperazine hexahydrate [ORG CHEM] $C_4H_{10}N_2·6H_2O$ White crystals with a melting point of 44°C; soluble in alcohol and water; used for pharmaceuticals and insecticides. { pī'par·ə,zēn ,hek·sə'hī,drāt }

piperidine [ORG CHEM] $C_5H_{11}N$ A cyclic compound and strong base; colorless liquid with pepper aroma; boils at 106°C; soluble in water, alcohol, and ether; used as a chemical intermediate and rubber accelerator, and in medicine. { pī'per·ə,dēn }

piperine [ORG CHEM] $C_{17}H_{19}NO_3$ A crystalline compound that is found in black pepper; melting point is 130°C; soluble in benzene and acetic acid; used to give a pungent taste to brandy and as an insecticide. { 'pip·ə,rēn }

pipernoid texture [GEOL] The eutaxitic texture of certain extrusive igneous rocks in which dark patches and stringers occur in a light-colored groundmass. { 'pī·pər,nóid ,teks·chər }

piperocaine hydrochloride [ORG CHEM] $C_{16}H_{23}NO_2·HCl$ A white, crystalline powder with a bitter taste and a melting point of 172–175°C; soluble in water, chloroform, and alcohol; used in medicine. { pī'per·ə,kān ,hī·drə'klór,īd }

pipe rock [PETR] A marine sandstone containing abundant scolites. { 'pīp ,räk }

piperonal [ORG CHEM] $C_8H_6O_3$ White crystals with a floral odor and a melting point of 35.5–37°C; soluble in alcohol and ether; used in medicine, perfumes, suntan preparations, and mosquito repellents. Also known as heliotropin. { pə'per·ə,nal }

piperoxan [PHARM] $C_{14}H_{19}NO_2$ An adrenergic blocking agent that has been used, as the hydrochloride salt, for diagnosis of pheochromocytoma. { 'pī·pər,äk·sən }

pipe run [ENG] The path followed by a piping system. { 'pīp ,rən }

pipe scale [ENG] Rust and corrosion products adhering to the inner surfaces of pipes; serve to decrease ability to transfer heat and to increase the pressure drop for flowing fluids. { 'pīp ,skāl }

pipe sleeve [ENG] A hollow, cylindrical insert placed in a form for a concrete wall at the position where a pipe is to penetrate in order to prevent flow of concrete into the opening. { 'pīp ,slēv }

pipe still [CHEM ENG] A petroleum-refinery still in which heat is applied to the oil while it is being pumped through a coil or pipe arranged in a firebox, the oil then running to a fractionator with continuous removal of overhead vapor and liquid bottoms. { 'pīp ,stil }

pipestone [PETR] A pink or mottled argillaceous stone; carved by the Indians into tobacco pipes. { 'pīp,stōn }

pipet [CHEM] Graduated or calibrated tube which may have a center reservoir (bulb); used to transfer known volumes of liquids from one vessel to another; types are volumetric or transfer, graduated, and micro. { pī'pet }

pipe tap [ENG] A small threaded hole or entry made into the wall of a pipe; used for sampling of pipe contents, or connection of control devices or pressure-drop-measurement devices. { 'pīp ,tap }

pipe tee [DES ENG] A T-shaped pipe fitting with two outlets, one at 90° to the connection to the main line. { 'pīp ,tē }

pipe thread [DES ENG] Most commonly, a 60° thread used on pipes and tubes, characterized by flat crests and roots and cut with 3/4-inch taper per foot (about 1.9 centimeters per 30 centimeters). Also known as taper pipe thread. { 'pīp ,thred }

pipe-thread protector See thread protector. { 'pīp ,thred prə,tek·tər }

pipe tongs [ENG] Heavy tongs that are hung on a cable and used for screwing pipe and tool joints. { 'pīp ,täŋz }

pipe-to-soil potential [ELEC] The voltage potential (emf) generated between a buried pipe and its surrounding soil, the result of electrolytic action and a cause of electrolytic corrosion of the pipe. { 'pīp tə 'sóil pə,ten·chəl }

pipe train [ENG] In the extrusion of plastic pipe, the entire equipment assembly used to fabricate the pipe (such as the extruder, die, cooling bath, haul-off, and cutter). { 'pīp ,trān }

pipe vesicle [GEOL] A slender vertical cavity, a few centimeters or tens of centimeters in length, extending upward from the base of a lava flow. { 'pīp ,ves·ə·kəl }

pipework See piping. { 'pīp,wərk }

pipe wrench [DES ENG] A tool designed to grip and turn a pipe or rod about its axis in one direction only. { 'pīp ,rench }

Pipidea [VERT ZOO] A family of frogs sometimes included in the suborder Opisthocoela, but more commonly placed in its own suborder, Aglossa; a definitive tongue is lacking, and free ribs are present in the tadpole but they fuse to the vertebrae in the adult. { pə'pid·ē·ə }

piping [ENG] A system of pipes provided to carry a fluid. Also known as pipework. [HYD] Erosive action of water passing through or under a dam, which may result in leakage or failure. { 'pīp·iŋ }

pipkrake [HYD] A small, thin needlelike crystal of ice formed just below ground level and growing perpendicular to the soil surface. Also known as needle ice. { 'pip,krāk }

pi point [ELEC] Frequency at which the insertion phase shift of an electric structure is 180° or an integral multiple of 180°. { 'pī ,póint }

pipper [OPTICS] A small hole in the reticle of an optical sight or computing sight. { 'pip·ər }

pipper image [OPTICS] A spot of light projected through the pipper in an optical or computing sight, used in aiming. { 'pip·ər ,im·ij }

Pipridae [VERT ZOO] The manakins, a family of colorful, neotropical suboscine birds in the order Passeriformes. { 'pip·rə,dē }

piptoblast [INV ZOO] A statoblast that is free but has no float. { 'pip·tə,blast }

piqué [TEXT] A cotton, polyester and cotton, rayon, or silk fabric with ribs running lengthwise, sometimes forming a waffle weave. { pē'kā }

piquite See pinta. { pē'kēt }

piracy See capture. { 'pī·rə·sē }

Pirani gage [PHYS] A thermal conductivity gage (where the thermal conductivity of a gas heated by a hot wire varies with pressure) connected to a Wheatstone bridge to measure the resistance of the hot wire, thus the gas pressure; used to measure pressure from 1 to 10^{-5} mmHg (133.32 to 0.13332 pascals). { pə'rän·ē ,gāj }

pirep See pilot report. { 'pī,rep }

piriform [MATH] A plane curve whose equation in cartesian coordinates x and y is $y^2 = ax^3 - bx^4$, where a and b are constants. { 'pir·ə,fórm }

piriformis [ANAT] A muscle arising from the front of the sacrum and inserted into the greater trochanter of the femur. { ,pir·ə'fór·məs }

pirimiphosethyl [ORG CHEM] $C_{13}H_{24}N_3O_3PS$ A straw-colored liquid which decomposes at 130°C; used as an insecticide

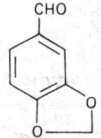

PIPERINE

Structural formula of piperine.

PIPERONAL

Structural formula of piperonal.

PIPTOBLAST

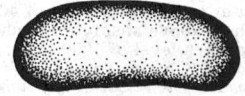

Piptoblast of *Fredericella sultana*.

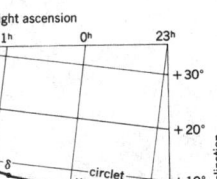

PISCES

Line pattern in the constellation Pisces. The grid lines represent the coordinates of the sky. The apparent brightness, or magnitude, of the various stars is shown by the sizes of the dots, which are graded by appropriate numbers as indicated.

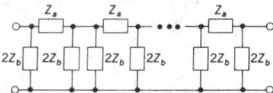

PI SECTION FILTER

Schematic diagram of pi section filter. Z_a and Z_b represent impedances. These are chosen so that $|Z_a| \gg |Z_b|$ in the frequency range which is to be a stop band and $|Z_a| \ll |Z_b|$ in the desired pass band.

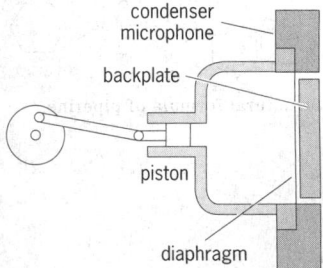

PISTONPHONE

condenser microphone
backplate
piston
diaphragm

Sectional view.

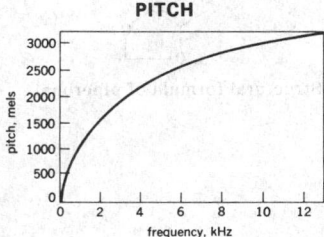

PITCH

Pitch level of sounds of different frequency, in mels. A tone of 1000 hertz at an intensity level of 40 decibels above absolute threshold has a pitch level of 1000 mels. *(After S. S. Stevens and J. Volkman, The relation of pitch to frequency, Amer. J. Psychol., 53(3):329–353, 1940)*

for the control of soil insects in vegetables and other crops. { ¦pir·əm·fäs'eth·əl }

Piroplasmea [INV ZOO] A class of parasitic protozoans in the superclass Sarcodina; includes the single genus *Babesia*. { ¦pir·ō'plaz·mē·ə }

pirssonite [MINERAL] $Na_2Ca(CO_3)_2 \cdot 2H_2O$ A colorless or white orthorhombic mineral composed of hydrous carbonate of sodium and calcium. { 'pirs·ən¸īt }

pisanite [MINERAL] $(Fe,Cu)SO_4 \cdot 7H_2O$ A blue mineral composed of hydrous sulfate of copper and iron; it is isomorphous with kirovite and melanterite. { pə'zä¸nīt }

Pisces [ASTRON] A northern constellation; right ascension 1 hour, declination 15°N. Also know as Fishes. [VERT ZOO] The fish and fishlike vertebrates, including the classes Agnatha, Placodermi, Chondrichthyes, and Osteichthyes. { 'pī·sēz }

piscicide [MATER] A substance capable of killing fish. { 'pis·ə¸sīd }

Piscis Australis [ASTRON] A southern constellation; right ascension 22 hours, declination 30°S. Also known as Southern Fish. { 'pis·kəs 'ós·trə·ləs }

α Piscis Australis [ASTRON] The brightest star in the southern constellation Piscis Australis. Also known as Fomalhaut. { ¦al·fə 'pis·kəs 'ós·trə·ləs }

Piscis Volans *See* Volans. { 'pis·kəs 'vō·lənz }

piscivorous [ZOO] Feeding on fishes. { pə'siv·ə·rəs }

pi section filter [ELEC] An electric filter made of several pi networks connected in series. { 'pī ¦sek·shən ¸fil·tər }

Pisionidae [INV ZOO] A small family of errantian polychaete annelids; allies of the scale bearers. { ¸pī·sē'än·ə¸dē }

pisolite [PETR] A sedimentary rock composed principally of pisoliths. { 'pī·zə¸līt }

pisolith [GEOL] Small, more or less spherical particles found in limestones and dolomites, having a diameter of 2–10 millimeters and often formed of calcium carbonate. { 'pī·zə¸lith }

pisolitic [PETR] Pertaining to pisolite or to the characteristic texture of such a rock. { ¦pī·zə¦lid·ik }

pisolitic tuff [GEOL] Of a tuff, composed of accretionary lapilli or pisolites. { ¦pī·zə¦lid·ik 'təf }

pisoparite [PETR] A limestone which contains at least 25% pisoliths and no more than 25% intraclasts and in which the sparry-calcite cement is more abundant than the carbonate-mud matrix (micrite). { pī'zäp·ə¸rīt }

pistachio [BOT] *Pistacia vera.* A small, spreading dioecious evergreen tree with leaves that have three to five broad leaflets, and with large drupaceous fruit; the edible seed consists of a single green kernel covered by a brown coat and enclosed in a tough shell. { pə'stash·ē¸ō }

pistil [BOT] The ovule-bearing organ of angiosperms; consists of an ovary, a style, and a stigma. { 'pist·əl }

pistillate [BOT] **1.** Having a pistil. **2.** Having pistils but no stamens. { 'pist·əl¸āt }

pistol [ORD] A short automatic or semiautomatic firearm aimed and fired from one hand, using the force of recoil to eject the empty shell and to insert a new round into the firing chamber. { 'pist·əl }

pistol lanyard [ORD] An assembly of a cord, slides, and a fastening device, generally used by military police; it is worn looped over the shoulder with the end attached to the pistol. { 'pist·əl ¸lan·yərd }

piston [ELECTROMAG] A sliding metal cylinder used in waveguides and cavities for tuning purposes or for reflecting essentially all of the incident energy. Also known as plunger; waveguide plunger. [ENG] *See* force plug. [MECH ENG] A sliding metal cylinder that reciprocates in a tubular housing, either moving against or moved by fluid pressure. { 'pis·tən }

piston attenuator [ELECTROMAG] A microwave attenuator inserted in a waveguide to introduce an amount of attenuation that can be varied by moving an output coupling device along its longitudinal axis. { 'pis·tən ə'ten·yə¸wād·ər }

piston blower [MECH ENG] A piston-operated, positive-displacement air compressor used for stationary, automobile, and marine duty. { 'pis·tən ¸blō·ər }

piston corer [MECH ENG] A steel tube which is driven into the sediment by a free fall and by lead attached to the upper end, and which is capable of recovering undistorted vertical sections of sediment. { 'pis·tən ¦kór·ər }

piston displacement [MECH ENG] The volume which a piston in a cylinder displaces in a single stroke, equal to the distance the piston travels times the internal cross section of the cylinder. { 'pis·tən di¸splās·mənt }

piston drill [MECH ENG] A heavy percussion-type rock drill mounted either on a horizontal bar or on a short horizontal arm fastened to a vertical column; drills holes to 6 inches (15 centimeters) in diameter. Also known as reciprocating drill. { 'pis·tən ¸dril }

piston engine [MECH ENG] A type of engine characterized by reciprocating motion of pistons in a cylinder. Also known as displacement engine; reciprocating engine. { 'pis·tən ¸en·jən }

piston flow [FL MECH] Two-phase (vapor-liquid) flow in which the gas flows as large plugs; occurs for gas superficial velocities from about 2 to 30 feet per second (60 to 900 centimeters per second). Also known as plug flow; slug flow. { 'pis·tən ¸flō }

piston gage *See* free-piston gage. { 'pis·tən ¸gāj }

piston head [MECH ENG] That part of a piston above the top ring. { 'pis·tən ¸hed }

piston meter [ENG] A variable-area, constant-head fluid-flow meter in which the position of the piston, moved by the buoyant force of the liquid, indicates the flow rate. Also known as piston-type area meter. { 'pis·tən ¦mēd·ər }

pistonphone [ENG ACOUS] A small chamber equipped with a reciprocating piston having a measurable displacement and used to establish a known sound pressure in the chamber, as for testing microphones. { 'pis·tən¸fōn }

piston pin [MECH ENG] A cylindrical pin that connects the connecting rod to the piston. Also known as wrist pin. { 'pis·tən ¸pin }

piston pump [MECH ENG] A pump in which motion and pressure are applied to the fluid by a reciprocating piston in a cylinder. Also known as reciprocating pump. { 'pis·tən ¦pəmp }

piston ring [DES ENG] A sealing ring fitted around a piston and extending to the cylinder wall to prevent leakage. Also known as packing ring. { 'pis·tən ¸riŋ }

piston rod [MECH ENG] The rod which is connected to the piston, and moves or is moved by the piston. { 'pis·tən ¸räd }

piston skirt [MECH ENG] That part of a piston below the piston pin bore. { 'pis·tən ¸skərt }

piston speed [MECH ENG] The total distance a piston travels in a given time; usually expressed in feet per minute. { 'pis·tən ¸spēd }

piston-type area meter *See* piston meter. { 'pis·tən ¸tīp 'er·ē·ə ¸mēd·ər }

piston valve [MECH ENG] A cylindrical type of steam engine slide valve for admission and exhaust of steam. { 'pis·tən ¦valv }

piston viscometer [ENG] A device for the measurement of viscosity by the timed fall of a piston through the liquid being tested. { 'pis·tən vi'skäm·əd·ər }

pi symbol *See* dingbat. { 'pī ¸sim·bəl }

pit [BOT] **1.** A cavity in the secondary wall of a plant cell, formed where secondary deposition has failed to occur, and the primary wall remains uncovered; two main types are simple pits and bordered pits. **2.** The stone of a drupaceous fruit. [MET] A small hole in the surface of a metal; usually caused by corrosion or formed during electroplating operations. [MIN ENG] **1.** A coal mine; the term is not commonly used by the coal industry, except in reference to surface mining where the workings may be known as a strip pit. **2.** Any quarry, mine, or excavation area worked by the open-cut method to obtain material of value. { pit }

pitch [ACOUS] That psychological property of sound characterized by highness or lowness, depending primarily upon frequency of the sound stimulus, but also upon its sound pressure and waveform. [ARCH] The ratio of the rise of a roof to its span. [CELL MOL] The distance between two adjacent turns of double-stranded deoxyribonucleic acid. [COMPUT SCI] The distance between the centerlines of adjacent rows of hole positions in punched paper tape. [DES ENG] The distance between similar elements arranged in a pattern or between two points of a mechanical part, as the distance between the peaks of two successive grooves on a disk recording or on a screw. [GEOL] *See* plunge. [GRAPHICS] The number of characters printed per horizontal inch on a typewriter or computer printer. [MATER] A dark heavy liquid or solid substance obtained as a residue after distillation of tar, oil, and such materials; occurs naturally as asphalt. [MECH] **1.** Of an aerospace vehicle, an angular displacement about an axis parallel to the lateral axis of the vehicle. **2.** The rising and falling motion of the bow of a ship or the tail of an airplane as the craft oscillates about a transverse axis. [SCI TECH] The inclination or degree of slope of an object or structure. { pich }

pitch acceleration [MECH] The angular acceleration of an aircraft or missile about its lateral, or Y, axis. { 'pich ik,sel·ə,rā·shən }

pitch attitude [MECH] The attitude of an aircraft, rocket, or other flying vehicle, referred to the relationship between the longitudinal body axis and a chosen reference line or plane as seen from the side. { 'pich ,ad·ə,tüd }

pitch axis [MECH] A lateral axis through an aircraft, missile, or similar body, about which the body pitches. Also known as pitching axis. { 'pich ,ak·səs }

pitchblende [MINERAL] A massive, brown to black, and fine-grained, amorphous, or microcrystalline variety of uraninite which has a pitchy to dull luster and contains small quantities of uranium. Also known as nasturan; pitch ore. { 'pich,blend }

pitch circle [DES ENG] In toothed gears, an imaginary circle concentric with the gear axis which is defined at the thickest point on the teeth and along which the tooth pitch is measured. { 'pich ,sər·kəl }

pitch coal See bituminous lignite. { 'pich ,kōl }

pitch coke [MATER] Coke made from coal tar pitch, characterized by high carbon and low ash content, and used mainly for production of electrode carbon. { 'pich ,kōk }

pitch cone [DES ENG] A cone representing the pitch surface of a bevel gear. { 'pich ,kōn }

pitch cylinder [DES ENG] A cylinder representing the pitch surface of a spur gear. { 'pich ,sil·ən·dər }

pitch diameter [DES ENG] The diameter of the pitch circle of a gear. { 'pich dī,am·əd·ər }

pitched roof [BUILD] **1.** A roof that has one or more surfaces with a slope greater than 10°. **2.** A roof that has two slopes meeting at a central ridge. { ¦picht 'rüf }

pitcher plant [BOT] Any of various insectivorous plants of the families Sarraceniaceae and Nepenthaceae; the leaves form deep pitchers in which water collects and insects are drowned and digested. { 'pich·ər ,plant }

pitch indicator [AERO ENG] An instrument for indicating the existence and approximate magnitude of the angular velocity about the lateral axis of an airframe. { 'pich ,in·də,kād·ər }

pitching axis See pitch axis. { 'pich·iŋ ,ak·səs }

pitching fold See plunging fold. { 'pich·iŋ ,fōld }

pitching moment [MECH] A moment about a lateral axis of an aircraft, rocket, or airfoil. { 'pich·iŋ ,mō·mənt }

pitch line See cam profile. { 'pich ,līn }

pitch mining [MIN ENG] Mining coal beds with steep slopes. { 'pich ,mīn·iŋ }

pitch opal [MINERAL] A yellowish to brownish inferior quality of common opal displaying a luster resembling that of pitch. { 'pich ,ō·pəl }

pitch ore See pitchblende. { 'pich ,ór }

pitchover [AERO ENG] The programmed turn from the vertical that a rocket under power takes as it describes an arc and points in a direction other than vertical. { 'pich,ō·vər }

pitch-row [COMPUT SCI] The distance between two adjacent holes in a paper tape. { 'pich ,rō }

pitchstone [GEOL] A type of volcanic glass distinguished by a waxy, dull, resinous, pitchy luster. Also known as fluolite. { 'pich,stōn }

pit furnace [MET] A low-temperature furnace in which steel is tempered. { 'pit ,fər·nəs }

pith [BOT] A central zone of parenchymatous tissue that occurs in most vascular plants and is surrounded by vascular tissue. { pith }

pi theorem See Buckingham's π theorem. { 'pī ,thir·əm }

pith knot [MATER] A small knot with a diameter of 0.25 inch (0.64 centimeter) or less in a wood board that has a pith hole. { 'pith ,nät }

pith ray See medullary ray. { 'pith ,rā }

pit limits [MIN ENG] The vertical and lateral extent to which the mining of a mineral deposit by open pitting may be carried economically. { 'pit ,lim·əts }

pitman [ENG] **1.** A worker in or near a pit, as in a quarry, mine, garage, or foundry. **2.** On a pumping unit, an arm connecting the crank with the walking beam for converting rotary motion to reciprocating motion. [MECH ENG] In an automotive steering system, the arm that is connected to the shaft of the steering gear sector and the tie rod, and swings back and forth as the steering wheel is turned. Also known as pitman arm. { 'pit·mən }

pitman arm See pitman. { 'pit·mən ,ärm }

pitometer [ENG] Reversed pitot-tube-type flow-measurement device with one pressure opening facing upstream and the other facing downstream. { pə'täm·əd·ər }

pitometer log [ENG] A log consisting essentially of a pitot tube projecting into the water, and suitable registering devices. { pə'täm·əd·ər ,läg }

pitot pressure [FL MECH] Pressure at the open end of a pitot tube. { pē'tō ,presh·ər }

pitot tube [ENG] An instrument that measures the stagnation pressure of a flowing fluid, consisting of an open tube pointing into the fluid and connected to a pressure-indicating device. Also known as impact tube. { pē'tō ,tüb }

pitot-tube anemometer [ENG] A pressure-tube anemometer consisting of a pitot tube mounted on the windward end of a wind vane and a suitable manometer to measure the developed pressure, and calibrated in units of wind speed. { pē'tō ,tüb ,an·ə'mäm·əd·ər }

pitot-venturi flow element [ENG] Liquid-flow measurement device in which a pair of concentric venturi elements replaces the pitot-tube probe. { pē'tō ven'tùr·ē ¦flō ,el·ə·mənt }

pit-run gravel [GEOL] A natural deposit of a mixture of gravel, sand, and foreign materials. { 'pit ,rən ,grav·əl }

pit sampling [MIN ENG] Using small untimbered pits to gain access to shallow alluvial deposits or ore dumps for purpose of testing or valuation. { 'pit ,sam·pliŋ }

pit slope [MIN ENG] The angle at which the wall of an open pit or cut stands as measured along an imaginary plane extended along the crests of the berms or from the slope crest to its toe. { 'pit ,slōp }

pitted outwash plain [GEOL] An outwash plain characterized by numerous depressions such as kettles, shallow pits, and potholes. { 'pid·əd 'aut,wäsh ,plān }

pitted pebble [GEOL] A pebble having marked concavities not related to the texture of the rock in which it appears or to differential weathering. { 'pid·əd 'peb·əl }

pitticite [MINERAL] A mineral of varying color composed of a hydrous sulfate-arsenate of iron. { 'pid·ə,sīt }

Pittidae [VERT ZOO] The pittas, a homogeneous family of brightly colored suboscine birds with an erectile crown of feathers, in the suborder Tyranni. { 'pid·ə,dē }

pitting [MED] **1.** The formation of pits; in the fingernails, a consequence and sign of psoriasis. **2.** The preservation for a short time of indentations on the skin made by pressing with the finger; seen in pitting edema. [MET] Selective localized formation of rounded cavities in a metal surface due to corrosion or to nonuniform electroplating. [MIN ENG] The act of digging or sinking a pit. { 'pid·iŋ }

pitting edema [MED] Edema of such degree that the skin can be temporarily indented by pressure with the fingers. { 'pid·iŋ ə'dē·mə }

pitting potential [MET] The electrochemical potential in a given environment above which, but not below, a corrosion pit initiates in a metal surface. { 'pid·iŋ pə,ten·chəl }

pi-T transformation See Y-delta transformation. { ¦pī ¦tē ,tranz·fər,mā·shən }

pituicyte [HISTOL] The characteristic cell of the neurohypophysis; these cells are pigmented and fusiform and are probably derived from neuroglial cells. { pə'tü·ə,sīt }

pituitary [ANAT] Of or pertaining to the hypophysis. [PHYSIO] Secreting phlegm or mucus (archaic usage). { pə'tü·ə,ter·ē }

pituitary dwarfism [MED] Stunted growth due to a deficiency of the primary growth hormone; characterized clinically by growth failure in early life, and in older persons by deficient subcutaneous fat with loose, wrinkled skin and precocious senility. { pə'tü·ə,ter·ē 'dwór fiz·əm }

pituitary gland See hypophysis. { pə'tü·ə,ter·ē ,gland }

Pityaceae [PALEOBOT] A family of fossil plants in the order Cordaitales known only as petrifactions of branches and wood. { ,pid·ē'ās·ē,ē }

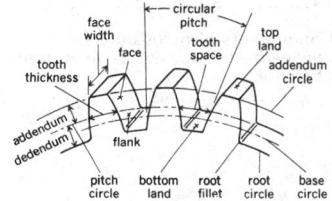

PITCH CIRCLE

Drawing of principal features of gear teeth showing the pitch circle.

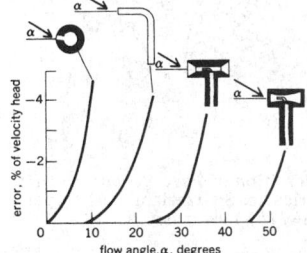

PITOT TUBE

Effect of the flow alignment of a pitot tube with streamline on accuracy of measurement. Here α = angle between direction of flow and direction of tube opening.

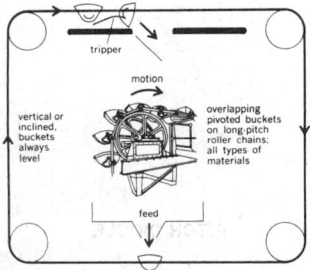

PIVOT-BUCKET CONVEYOR-ELEVATOR

Components of pivot-bucket conveyor-elevator and typical path of travel.

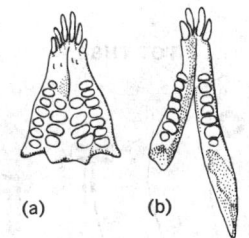

PLACODONTIA

Dentition of *Paraplacodus broilii*, Triassic, Switzerland. *(a)* Upper jaw. *(b)* Lower jaw.

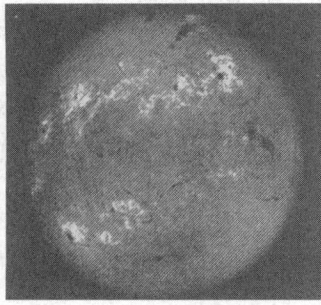

PLAGE

The disk of the sun photographed in H-α light, showing bright plages in centers of activity and dark filaments (prominences). *(Sacramento Peak Observatory, operated by the Association of Universities for Research in Astronomy, Inc.)*

pityriasis [MED] A fine, branny desquamation of the skin. { ˌpid·ə'rī·ə·səs }

Pitzer equation [PHYS CHEM] Equation for the approximation of data for heats of vaporization for organic and simple inorganic compounds; derived from temperature and reduced temperature relationships. { 'pit·sər iˌkwā·zhən }

pivot [MECH] A short, pointed shaft forming the center and fulcrum on which something turns, balances, or oscillates. { 'piv·ət }

pivotal condensation [MATH] A method of evaluating a determinant that is convenient for determinants of large order, especially when digital computers are used, involving a repeated process in which a determinant of order n is reduced to the product of one of its elements raised to a power and a determinant of order $n - 1$. { ˌpiv·əd·əl ˌkän·dən'sā·shən }

pivotal fault See rotary fault. { 'piv·əd·əl 'fȯlt }

pivotal method [STAT] A technique for passing from one set of double inequalities to another in order to find a confidence interval for a parameter. { ˌpiv·əd·əl 'meth·əd }

pivot anchor [DES ENG] An anchor that permits a pipe to swivel around a fixed point. { 'piv·ət ˌaŋ·kər }

pivot bridge [CIV ENG] A bridge in which a span can open by pivoting about a vertical axis. { 'piv·ət 'brij }

pivot-bucket conveyor-elevator [MECH ENG] A bucket conveyor having overlapping pivoted buckets on long-pitch roller chains; buckets are always level except when tripped to discharge materials. { 'piv·ət ˌbək·ət kən'vā·ər el·ə,vād·ər }

pivoted window [BUILD] A window having a section which is pivoted near the center so that the top of the section swings in and the bottom swings out. { 'piv·əd·əd 'win·dō }

pivoting [MATH] In the solution of a system of linear equations by elimination, a method of choosing a suitable equation to eliminate at each step so that certain difficulties are avoided. { 'piv·əd·iŋ }

pivoting point [NAV ARCH] The point about which a ship pivots when turning, usually somewhat forward of amidships. { 'piv·əd·iŋ ˌpȯint }

pivot joint [ANAT] A diarthrosis that permits a rotation of one bone around another; an example is the articulation of the atlas with the axis. Also known as trochoid. { 'piv·ət ˌjȯint }

PIXE See proton-induced x-ray emission. { 'pik·sē }

pixel [COMPUT SCI] The smallest part of an electronically coded picture image. [ELECTR] The smallest addressable element in an electronic display; a short form for picture element. Also known as pel. { pik'sel }

pk See peck.

pK [CHEM] The logarithm (to base 10) of the reciprocal of the equilibrium constant for a specified reaction under specified conditions.

PK See psychokinesis.

PKA See cyclic AMP-dependent protein kinase; primary knocked-on atom.

PKC See protein kinase C.

PKU See phenylketonuria.

PI See poiseuille.

PLA See programmed logic array.

placanticline [GEOL] A gentle, anticlinallike uplift of the continental platform, usually asymmetric and without a typical outline. { plak'ant·iˌklīn }

place [MATH] A position corresponding to a given power of the base in positional notation. Also known as column. { plās }

placebo [MED] A preparation, devoid of pharmacologic effect, given to patients for psychologic effect, or as a control in evaluating a medicinal believed to have a pharmacologic action. { plä'chā·bō or plə'sē·bō }

placeholder [COMPUT SCI] A section of computer storage reserved for information that will be provided later. { 'plās ˌhōl·dər }

placenta [BOT] A plant surface bearing a sporangium. [EMBRYO] A vascular organ that unites the fetus to the wall of the uterus in all mammals except marsupials and monotremes. { plə'sent·ə }

placenta accreta [MED] A placenta that has partially grown into the myometrium of the uterus. { plə'sent·ə ə'krēd·ə }

placental barrier [EMBRYO] The tissues intervening between the maternal and the fetal blood of the placenta, which prevent or hinder certain substances or organisms from passing from mother to fetus. { plə'sent·əl 'bar·ē·ər }

placenta previa [MED] A pregnancy disorder in which the placenta is abnormally located near or over the cervix. { plə,sen·tə 'prē·vē·ə }

placentation [BOT] The attachment of ovules along the inner ovarian wall by means of the placenta. [EMBRYO] The formation and fusion of the placenta to the uterine wall. { plas·ən'tā·shən }

placer [GEOL] A mineral deposit at or near the surface of the earth, formed by mechanical concentration of mineral particles from weathered debris. Also known as ore of sedimentation. { 'plās·ər }

placer claim [MIN ENG] A mining claim located upon gravel or ground whose mineral contents are extracted by the use of water, as by sluicing, or hydraulicking. { 'plās·ər ˌklām }

placer dredge [MIN ENG] A dredge for mining metals from placer deposits; it consists of a chain of closely connected buckets passing over an idler tumbler and an upper or driving tumbler, mounted on a structural-steel ladder which carries a series of rollers. { 'plās·ər ˌdrej }

placer location [MIN ENG] Location of a tract of land for the sake of loose mineral-bearing or other valuable deposits on or near its surface, rather than within lodes or veins in rock in place. { 'plās·ər lōˌkā·shən }

placer mining [MIN ENG] **1.** The extraction and concentration of heavy metals from placers. **2.** Mining of gold by washing the sand, gravel, or talus. { 'plās·ər ˌmīn·iŋ }

place value [MATH] The value given to a digit by virtue of its location in a numeral. { 'plās ˌval·yü }

placic horizon [GEOL] A black to dark red soil horizon that is usually cemented with iron and is not very permeable. { 'plā·sik hə'rīz·ən }

placode [EMBRYO] A platelike epithelial thickening, frequently marking, in the embryo, the anlage of an organ or part. { 'plaˌkōd }

Placodermi [PALEON] A large and varied class of Paleozoic fishes characterized by a complex bony armor covering the head and the front portion of the trunk. { ˌpla·kə'dər·mē }

Placodontia [PALEON] A small order of Triassic marine reptiles of the subclass Euryapsida characterized by flat-crowned teeth in both the upper and lower jaws and on the palate. { ˌplā·kə'dän·chə }

Placothuriidae [INV ZOO] A family of holothurian echinoderms in the order Dendrochirotida; individuals are invested in plates and have a complex calcareous ring mechanism. { ˌpla·kə·tho'rī·əˌdē }

plage [ASTRON] One of the luminous areas that appear in the vicinity of sunspots or disturbed areas on the sun; they may be seen distinctively in spectroheliograms taken in the calcium K line. { pläzh }

Plaggept [GEOL] A suborder of the soil order Inceptisol, with very thick surface horizons of mixed mineral and organic materials resulting from manure or human wastes added over long periods of time. { 'plä·gept }

plagiaplite [PETR] An aplite composed chiefly of plagioclase (oligoclase to andesine), possibly green hornblende, and accessory quartz, biotite, and muscovite. { 'plä·jē·əˌplīt }

Plagiaulacida [PALEON] A primitive, monofamilial suborder of multituberculate mammals distinguished by their dentition (dental formula I 3/0 C 0/0 Pm 5/4 M 2/2), having cutting premolars and two rows of cusps on the upper molars. { 'plä·jē·ə·yü'läs·əˌdə }

Plagiaulacidae [PALEON] The single family of the extinct mammalian suborder Plagiaulacida. { ˌplä·jē·ə·yü'läs·əˌdē }

plagiocephaly [MED] A type of strongly asymmetric cranial deformation, in which the anterior portion of one side and the posterior portion of the opposite side of the skull are developed more than their counterparts so that the maximum length of the skull is not in the midline but on a diagonal. { ˌplä·jē·ō'sef·ə·lē }

plagioclase [MINERAL] **1.** A type of triclinic feldspars having the general formula $(Na,Ca)Al(Si,Al)Si_2O_8$; they are common rock-forming minerals. **2.** A series in the plagioclase group which can be divided into a number of varieties based on the relative proportion of the solid solution end members, albite and anorthite (An): albite (An 0–10) oligoclase (An 10–30), andesine (An 30–50), labradorite (An 50–70), bytownite (An 70–90), and anorthite (An 90–100). Also known as sodium-calcium feldspar. { 'plä·jē·əˌklās }

plagioclimax [ECOL] A plant community which is in equilibrium under present conditions, but which has not reached its natural climax, or has regressed from it, due to biotic factors such as human intervention. { ¦plā·jē·ō'klī̇,maks }

plagiodont [VERT ZOO] Of a snake, having obliquely set, or two converging series of, palatal teeth. { 'plā·jē·ə,dänt }

plagiogravitropism [BOT] A response of root and shoot branches to gravity where growth is at different angles from the vertical. { ¦plā·jē·ō,grav·ə'trō·piz·əm }

plagiohedral [CRYSTAL] Pertaining to obliquely arranged spiral faces; in particular, to a member of a group in the isometric system with 13 axes but no center or planes. { ¦plā·jē·ō¦hē·drəl }

plagionite [MINERAL] $Pb_5Sb_8S_7$ A lead-gray mineral with metallic appearance, composed of sulfide of lead and antimony. { 'plā·jē·ə,nīt }

Plagiosauria [PALEON] An aberrant Triassic group of labyrinthodont amphibians. { ¦plā·jē·ə'sór·ē·ə }

plagiosere [ECOL] A plant succession deflected from its normal course by biotic factors. { 'plā·jē·ə,sir }

plague [MED] **1.** An infectious bacterial disease of rodents and humans caused by *Pasteurella pestis,* transmitted to humans by the bite of an infected flea (*Xenopsylla cheopis*) or by inhalation. Also known as black death; bubonic plague. **2.** Any contagious, malignant, epidemic disease. { plāg }

plain [GEOGR] An extensive, broad tract of level or rolling, almost treeless land with a shrubby vegetation, usually at a low elevation. [GEOL] A flat, gently sloping region of the sea floor. Also known as submarine plain. { plān }

plain arch [FOREN SCI] A fingerprint pattern in which the ridges enter on one side of the impression and flow or tend to flow out the other side with a rise or wave in the center. { 'plān ¦ärch }

plain concrete [CIV ENG] Concrete without reinforcement but often with light steel to reduce shrinkage and temperature cracking. { 'plān kän'krēt }

plain-laid [DES ENG] Pertaining to a rope whose strands are twisted together in a direction opposite to that of the twist in the strands. { 'plān,lād }

plain milling cutter [DES ENG] A cylindrical milling cutter with teeth on the periphery only; used for milling plain or flat surfaces. Also known as slab cutter. { 'plān ¦mil·iŋ ,kəd·ər }

plain of denudation [GEOL] A surface that has been reduced to sea level or to just above sea level by the agents of erosion (usually considered to be of subaerial origin). { 'plān əv ,dē·nü¦dā·shən }

plain of lateral planation [GEOL] An extensive, smooth, apronlike surface developed at the base of a mountain or escarpment by the widening of valleys and the coalescence of floodplains as a result of lateral planation. { 'plān əv ¦lad·ə·rəl plā'nā·shən }

plain of marine denudation [GEOL] A plane or nearly plane surface worn down by the gradual encroachment of ocean waves upon the land; or a plane or nearly plane imaginary surface representing such a plain after uplift and partial subaerial erosion. Also known as plain of submarine denudation. { 'plān əv mə¦rēn ,dē·nü¦dā·shən }

plain of marine erosion [GEOL] A theoretical platform representing a plane surface of unlimited width produced below sea level by the complete cutting away of the land by marine processes acting over a very long period of stillstand. { 'plān əv mə¦rēn i'rō·zhən }

plain of submarine denudation See plain of marine denudation. { 'plān əv ¦səb·mə,rēn ,dē·nü¦dā·shən }

plain-sawing See backsawing. { 'plān ¦só·iŋ }

plains-type fold [GEOL] An anticlinal or domelike structure of the continental platform which has no typical outline and for which there is no corresponding synclinal structure. { 'plānz ¦tīp ,fōld }

plaintext [COMMUN] The form of a message in which it can be generally understood, before it has been transformed by a code or cipher into a form in which it can be read only by those privy to the secrets of the cipher. [COMPUT SCI] Data that are to be encrypted. { 'plān,tekst }

plain tract [GEOL] The lower part of a stream, characterized by a low gradient and a wide floodplain. { 'plān ¦trakt }

plain turning [MECH ENG] Lathe operations involved when machining a workpiece between centers. { 'plān ¦tərn·iŋ }

plain vanilla See vanilla. { 'plān və'nil·ə }

plain weave [TEXT] A weave made by passing the filling yarns over one warp yarn and under the next, continuing alternately across each row. { 'plān ¦wēv }

plain whorl [FOREN SCI] A whorl fingerprint pattern that has two deltas and at least one ridge making a complete circuit, which may be spiral, oval, circular, or any variant of a circle. An imaginary line drawn between the two deltas must touch or cross at least one of the recurving ridges within the inner pattern area. { 'plān ,worl }

Plaisancian [GEOL] A European stage of geologic time: lower Pliocene (above Pontian of Miocene, below Astian). Also known as Piacentian; Plaisanzian. { plā'zän·chən }

Plaisanzian See Plaisancian. { plā'zän·zhən }

plaiting [GEOL] A texture in some schists that results from the intersection of relict bedding planes with well-developed cleavage planes. Also known as gaufrage. { 'plād·iŋ }

plait point [CHEM] Composition conditions in which the three coexisting phases of partially soluble components of a three-phase liquid system approach each other in composition. { 'plāt ,póint }

plan [GRAPHICS] **1.** An orthographic drawing on a horizontal plane, as of an instrument, a horizontal section, or a layout. **2.** A large-scale map or chart of a small area. { plan }

planar [MATH] Lying in or pertaining to a euclidean plane. { 'plā·nər }

planar area [COMPUT SCI] In computer graphics, an object with boundaries, such as a circle or polygon. { 'plān·ər ,er·ē·ə }

planar array [ELECTR] An array of ultrasonic transducers that can be mounted in a single plane or sheet, to permit closer conformation with the hull design of a sonar-carrying ship. { 'plā·nər ə¦rā }

planar-array antenna [ELECTROMAG] An array antenna in which the centers of the radiating elements are all in the same plane. { 'plā·nər ə¦rā an'ten·ə }

planar ceramic tube [ELECTR] Electron tube having parallel planar electrodes and a ceramic envelope. { 'plā·nər sə¦ram·ik 'tüb }

planar cross-bedding [GEOL] Cross-bedding characterized by planar surfaces of erosion in the lower bounding surface. { 'plā·nər 'krós ,bed·iŋ }

planar device [ELECTR] A semiconductor device having planar electrodes in parallel planes, made by alternate diffusion of *p*- and *n*-type impurities into a substrate. { 'plā·nər di,vīs }

planar diode [ELECTR] A diode having planar electrodes in parallel planes. { 'plā·nər 'dī,ōd }

planar flow structure See platy flow structure. { 'plā·nər 'flō ,strək·chər }

planar graph [MATH] A graph that can be drawn in a plane without any lines crossing. { 'plā·nər ,graf }

planaria [INV ZOO] Any flatworm of the turbellarian order Tricladida; the body is broad and dorsoventrally flattened, with anterior lateral projections, the auricles, and a pair of eyespots on the anterior dorsal surface. { plə'ner·ē·ə }

planar laser-induced fluorescence [FL MECH] A flow visualization technique in which a thin sheet of monochromatic laser radiation is used to excite a particular molecular species in a flow, and the resulting fluorescence emission of this species gives some indication of its number density. Abbreviated PLIF. { ¦plā·nər ¦lā·zər in,düst flə'res·əns }

planar linkage [MECH ENG] A linkage that involves motion in only two dimensions. { 'plā·nər 'liŋ·kij }

planar map [MATH] A plane or sphere divided into connected regions by a topological graph. { 'plān·ər ,map }

planar photodiode [ELECTR] A vacuum photodiode consisting simply of a photocathode and an anode; light enters through a window sealed into the base, behind the photocathode. { 'plā·nər ¦fōd·ō'dī,ōd }

planar point [MATH] A point on a surface at which the curvatures of all the normal sections vanish. { 'plā·nər or 'plā·när ,póint }

planar process [ENG] A silicon-transistor manufacturing process in which a fractional-micrometer-thick oxide layer is grown on a silicon substrate; a series of etching and diffusion steps is then used to produce the transistor inside the silicon substrate. { 'plā·nər ,prä·səs }

planar transistor [ELECTR] A transistor constructed by an etching and diffusion technique in which the junction is never exposed during processing, and the junctions reach the surface

PLAIN ARCH

Plain arch fingerprint pattern. (*Federal Bureau of Investigation*)

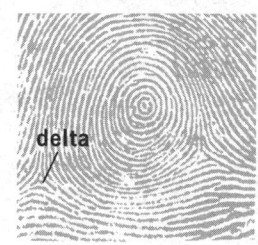

PLAIN WHORL

delta

Plain whorl fingerprint pattern, O tracing. (*Federal Bureau of Investigation*)

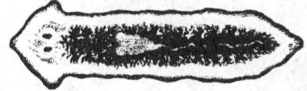

PLANARIA

Planaria, a bilaterally symmetrical, dorsoventrally flattened, triploblastic organism.

in one plane; characterized by very low leakage current and relatively high gain. { 'plā·nər tran'zis·tər }

planate [GEOL] Referring to a surface that has been flattened or leveled by planation. { 'plā͵nāt }

planation [GEOL] Erosion resulting in flat surfaces, caused by meandering streams, waves, ocean currents, wind, or glaciers. { plā'nā·shən }

planation stream piracy [HYD] Capture effected by the lateral planation of a stream invading and diverting the upper part of a smaller stream. { plā'nā·shən ͵strēm ͵pī·rə·sē }

planchet [ENG] A small metal container or sample holder; usually used to hold radioactive materials that are being checked for the degree of radioactivity in a proportional counter or scintillation detector. [MATER] A milled metal disk ready for coining. { 'plan·chət }

planck [PHYS] A unit of action equal to the product of an energy of 1 joule and a time of 1 second. { pläŋk }

Planck distribution law See Planck radiation formula. { 'pläŋk ͵dis·trə'byü·shən ͵lò }

Planck era [ASTRON] The epoch in the early universe when the gravitational interaction between particles was as strong as the other interactions. { 'pläŋk ͵ir·ə }

Planck function [THERMO] The negative of the Gibbs free energy divided by the absolute temperature. { 'pläŋk ͵fəŋk·shən }

Planckian locus [OPTICS] The locus of points on a chromaticity diagram that represents blackbody radiators at various temperatures. { 'pläŋ·kē·ən 'lō·kəs }

Planck length [PHYS] The length $\sqrt{Gh/2\pi c^3}$ (where G is the gravitational constant, h is Planck's constant, and c is the speed of light) at which quantum fluctuations are believed to dominate the geometry of space-time; it is equal to 1.6162×10^{-35} m. { 'pläŋk ͵leŋkth }

Planck mass [PHYS] The mass $\sqrt{hc/2\pi G}$, where h is Planck's constant, c is the speed of light, and G is the gravitational constant; equivalently, the mass of a particle whose reduced Compton wavelength equals the Planck length; it is equal to 21.764 micrograms or 1.2209×10^{19} GeV/c^2. { 'pläŋk ͵mas }

Planck oscillator [QUANT MECH] An oscillator which can absorb or emit energy only in amounts which are integral multiples of Planck's constant times the frequency of the oscillator. Also known as radiation oscillator. { 'pläŋk ͵äs·ə͵lād·ər }

Planck radiation formula [STAT MECH] A formula for the intensity of radiation emitted by a blackbody within a narrow band of frequencies (or wavelengths), as a function of frequency, and of the body's temperature. Also known as Planck distribution law; Planck's law. { 'pläŋk ͵rād·ē'ā·shən ͵fòr·myə·lə }

Planck's constant [QUANT MECH] A fundamental physical constant, the elementary quantum of action; the ratio of the energy of a photon to its frequency, it is equal to 6.62606876 ± 0.00000052 × 10^{-34} joule-second. Symbolized h. Also known as quantum of action. { 'pläŋks ͵kän·stənt }

Planck's law [QUANT MECH] A fundamental law of quantum theory stating that energy associated with electromagnetic radiation is emitted or absorbed in discrete amounts which are proportional to the frequency of radiation. [STAT MECH] See Planck radiation formula. { 'pläŋks ͵lò }

Planck time [PHYS] The constant $(Gh/2\pi c^5)^{1/2}$ with dimensions of "time" formed from Planck's constant h, the gravitational constant G, and the speed of light c; approximately 10^{-43} second. { 'pläŋk ͵tīm }

Planctomyces [MICROBIO] A genus of appendaged bacteria; spherical, oblong, or pear-shaped cells with long, slender stalks; reproduce by budding. { ͵plaŋk·tə'mī·sēz }

plane [ELECTR] Screen of magnetic cores; planes are combined to form stacks. [DES ENG] A tool consisting of a smooth-soled stock from the face of which extends a wide-edged cutting blade for smoothing and shaping wood. [MATH] **1.** A surface such that a straight line that joins any two of its points lies entirely in that surface. **2.** In projective geometry, a triple of sets (P,L,I) where P denotes the set of points, L the set of lines, and I the incidence relation on points and lines, such that (1) P and L are disjoint sets, (2) the union of P and L is nonnull, and (3) I is a subset of $P \times L$, the cartesian product of P and L. { plān }

plane angle [MATH] An angle between lines in the euclidean plane. { 'plān ͵aŋ·gəl }

plane atmospheric wave [METEOROL] An atmospheric wave represented in two-dimensional rectangular cartesian coordinates, in contrast to a wave considered on the spherical earth. { 'plān ͵at·mə͵sfir·ik 'wāv }

plane bed [GEOL] A sedimentary bed without elevations or depressions larger than the maximum size of the bed material. { 'plān ͵bed }

plane correction [ENG] A correction applied to observed surveying data to reduce them to a common reference plane. { 'plān kə͵rek·shən }

plane curve [MATH] Any curve lying entirely within a plane. { 'plān ͵kərv }

plane cyclic curve See cyclic curve. { 'plān 'sī·klik ͵kərv }

plane defect [CRYSTAL] A type of crystal defect that occurs along the boundary plane of two regions of a crystal, or between two grains. { 'plān di͵fekt }

plane dendrite See plane-dendritic crystal. { 'plān ͵den͵drīt }

plane-dendritic crystal [CRYSTAL] An ice crystal exhibiting an elaborately branched (dendritic) structure of hexagonal symmetry, with its much larger dimension lying perpendicular to the principal (c-axis) of the crystal. Also known as plane dendrite; stellar crystal. { 'plān den͵drid·ik 'krist·əl }

plane earth [ELECTROMAG] Earth that is considered to be a plane surface as used in ground-wave calculations. { 'plān ͵ərth }

plane-earth attenuation [ELECTROMAG] Attenuation of an electromagnetic wave over an imperfectly conducting plane earth in excess of that over a perfectly conducting plane. { 'plān ͵ərth ə͵ten·yə'wā·shən }

plane field See field of planes on a manifold. { 'plān ͵fēld }

plane geometry [MATH] The geometric study of the figures in the euclidean plane such as lines, triangles, and polygons. { 'plān jē'äm·ə·trē }

plane group [CRYSTAL] The group of operations (rotations, reflections, translations, and combinations of these) which leave a regular, periodic structure in a plane unchanged. [MATH] One of 17 two-dimensional patterns which can be produced by one asymmetric motif that is repeated by symmetry operations to produce a pattern unit which then is repeated by translation to build up an ordered pattern that fills any two-dimensional area. Also known as plane symmetry group. { 'plān ͵grüp }

plane jet [HYD] A stream flow pattern characteristic of hyperpycnal inflow, in which the inflowing water spreads as a parabola whose width is about three times the square root of the distance downstream from the mouth. { 'plān ͵jet }

plane lamina [MECH] A body whose mass is concentrated in a single plane. { 'plān 'lam·ə·nə }

plane lattice [CRYSTAL] A regular, periodic array of points in a plane. { 'plān 'lad·əs }

plane mirror [OPTICS] A mirror whose surface lies in a plane; it forms an image of an object such that the mirror surface is perpendicular to and bisects the line joining all corresponding object-image points. { 'plān 'mir·ər }

plane of departure [MECH] Vertical plane containing the path of a projectile as it leaves the muzzle of the gun. { 'plān əv di'pär·chər }

plane of fire [MECH] Vertical plane containing the gun and the target, or containing a line of site. { 'plān əv 'fīr }

plane of flotation [FL MECH] The plane in which the surface of a liquid intersects a stationary floating body. { 'plān əv flō'tā·shən }

plane of incidence [PHYS] A plane containing the direction of propagation of a wave striking a surface and a line perpendicular to the surface. Also known as incidence plane. { 'plān əv 'in·səd·əns }

plane of maximum shear stress [MECH] Either of two planes that lie on opposite sides of and at angels of 45° to the maximum principal stress axis and that are parallel to the intermediate principal stress axis. { 'plān əv ͵mak·si·məm 'shir ͵stres }

plane of mirror symmetry Also known as mirror plane of symmetry; plane of symmetry; reflection plane; symmetry plane. [CRYSTAL] In certain crystals, a symmetry element whereby reflection of the crystal through a certain plane leaves the crystal unchanged. [MATH] An imaginary plane which divides an object into two halves, each of which is the mirror image of the other in this plane. { 'plān əv 'mir·ər ͵sim·ə·trē }

plane of polarization [ELECTROMAG] Plane containing the electric vector and the direction of propagation of electromagnetic wave. { 'plān əv ¸pō·lə·rə'zā·shən }

plane of reflection [CRYSTAL] See plane of mirror symmetry. [MATH] See plane of mirror symmetry. [OPTICS] A plane containing the direction of propagation of radiation reflected from a surface, and the normal to the surface. Also known as reflection plane. { 'plān əv ri'flek·shən }

plane of saturation See water table. { 'plān əv ¸sach·ə'rā·shən }

plane of support [MATH] Relative to a convex body in a three-dimensional space, a plane that contains at least one point of the body but is such that the half-space on one side of the plane contains no points of the body. { 'plān əv sə'pȯrt }

plane of symmetry See plane of mirror symmetry. { 'plān əv 'sim·ə·trē }

plane of work [IND ENG] The plane in which most of a worker's motions occur in the performance of a task. { 'plān əv 'wərk }

plane of yaw [MECH] The plane determined by the tangent to the trajectory of a projectile in flight and the axis of the projectile. { 'plān əv 'yȯ }

plane-parallel resonator [OPTICS] A beam resonator that consists of a pair of plane mirrors which are perpendicular to the axis of the beam. { 'plān 'par·ə¸lel 'rez·ən¸ād·ər }

plane parallel texture [PETR] The parallel texture of a rock in which the constituents are parallel to a plane, but not to a line as in linear parallel texture. { 'plān 'par·ə¸lel 'teks·chər }

plane Poiseuille flow [FL MECH] Rheological (viscosity) measurement in which the fluid of interest is propelled through a narrow slot, and the volumetric flow rate and the pressure gradient are measured simultaneously to determine viscosity. { 'plān pwä'zə·ē ¸flō }

plane polarization See linear polarization. { 'plān ¸pō·lə·rə'zā·shən }

plane-polarized wave [ELECTROMAG] An electromagnetic wave whose electric field vector at all times lies in a fixed plane that contains the direction of propagation through a homogeneous isotropic medium. { 'plān 'pō·lə¸rīzd ¸wāv }

plane polygon [MATH] A polygon lying in the Euclidean plane. { 'plān 'päl·ē¸gän }

plane quadrilateral [MATH] A four-sided polygon lying in the Euclidean plane. { 'plān ¸kwä·drə'lad·ə·rəl }

plan equation [MECH ENG] The mathematical statement that horsepower = $plan/33,000$, where p = mean effective pressure (pounds per square inch), l = length of piston stroke (feet), a = net area of piston (square inches), and n = number of cycles completed per minute. { 'plan i¸kwā·zhən }

planer [MECH ENG] A machine for the shaping of long, flat, or flat contoured surfaces by reciprocating the workpiece under a stationary single-point tool or tools. [MIN ENG] A fixed-blade device for continuous longwall mining of narrow seams of friable coal; the machine is operated along the coal face, planing a narrow cut from the solid coal as it travels. { 'plān·ər }

plane reflector See passive reflector. { 'plān ri'flek·tər }

plane sailing [NAV] A method of solving the various problems involving course, distance, difference of latitude, and departure, in which the earth or a smaller area is considered a plane. { 'plān 'sāl·iŋ }

plane section [MATH] The intersection of a plane with a surface or a solid. Also known as section. { 'plān 'sek·shən }

plane strain [MECH] A deformation of a body in which the displacements of all points in the body are parallel to a given plane, and the values of these displacements do not depend on the distance perpendicular to the plane. { 'plān 'strān }

plane stress [MECH] A state of stress in which two of the principal stresses are always parallel to a given plane and are constant in the normal direction. { 'plān ¸stres }

plane surveying [ENG] Measurement of areas on the assumption that the earth is flat. { 'plān sər'vā·iŋ }

plane symmetry group See plane group. { 'plān 'sim·ə·trē ¸grüp }

planet [ASTRON] A relatively small celestial body moving in orbit around the sun or another star. { 'plan·ət }

plane table [ENG] A surveying instrument consisting of a drawing board mounted on a tripod and fitted with a compass and a straight-edge ruler; used to graphically plot survey lines directly from field observations. { 'plān ¸tā·bəl }

plane-table method [MIN ENG] A method of measuring areas of mine roadways; a drawing board is set up on a tripod in the plane of the mine section to be measured; the distance from a central point on the board to the perimeter of the roadway is measured with a tape along various offsets; the distance measured is scaled on the drawing board along the proper offset line. { 'plān ¸tā·bəl ¸meth·əd }

planetarium [ASTRON] **1.** A projection device which accurately portrays the position of the stars and planets at any time in the past, present, or future from any point on the earth or the near region of space; the modern planetarium instrument is a mechanical-electrical analog of space. **2.** The name given to the building and gear associated with this device. { ¸plan·ə'ter·ē·əm }

planetary aberration [OPTICS] The apparent displacement of an object in the solar system that results from the fact that light takes a certain time to travel from the object to earth, during which time the object travels a certain distance in its orbit. { 'plan·ə¸ter·ē ¸ab·ə'rā·shən }

planetary atmosphere [ASTRON] The outer shell of gas around some planets. { 'plan·ə¸ter·ē 'at·mə¸sfir }

planetary boundary layer [METEOROL] That layer of the atmosphere from the earth's surface to the geostrophic wind level, including, therefore, the surface boundary layer and the Ekman layer; above this layer lies the free atmosphere. { 'plan·ə¸ter·ē 'baȯn·drē ¸lā·ər }

planetary circulation See general circulation. { 'plan·ə¸ter·ē ¸sər·kyə'lā·shən }

planetary electron See orbital electron. { 'plan·ə¸ter·ē i'lek¸trän }

planetary gear train [MECH ENG] An assembly of meshed gears consisting of a central gear, a coaxial internal or ring gear, and one or more intermediate pinions supported on a revolving carrier. { 'plan·ə¸ter·ē 'gir ¸trān }

planetary geology [GEOL] A science that applies geologic principles and techniques to the study of planets and their natural satellites. Also known as planetary geoscience. { 'plan·ə¸ter·ē jē äl·ə·jē }

planetary geoscience See planetary geology. { 'plan·ə¸ter·ē ¸jē·ō'sī·əns }

planetary nebula [ASTRON] An oval or round nebula of expanding concentric rings of gas associated with a hot central star. { 'plan·ə¸ter·ē 'neb·yə·lə }

planetary nebula symbiotic See subdwarf symbiotic. { 'plan·ə¸ter·ē 'neb·yə·lə ¸sim·bē'äd·ik }

planetary orbit [ASTRON] The path that a planet has as it revolves about the sur. { 'plan·ə¸ter·ē 'ȯr·bət }

planetary perturbation [ASTRON] A deviation of a planet from its computed orbit because of the attraction of another celestial body or bodies. { 'plan·ə¸ter·ē ¸pər·tə'bā·shən }

planetary physics [ASTROPHYS] The study of the structure, composition, and physical and chemical properties of the planets of the solar system, including their atmospheres and immediate cosmic environment. { 'plan·ə¸ter·ē 'fiz·iks }

planetary precession [ASTRON] A comparatively small eastward motion of the equinoxes caused by the action of other planets in altering the plane of the earth's orbit. { 'plan·ə¸ter·ē pri'sesh·ən }

planetary vorticity effect [GEOPHYS] The effect of the variation of the earth's vorticity with latitude in altering the relative vorticity of a flow with a meridional component; a fluid with a free surface in a rotating cylinder exhibits a corresponding effect, owing to the shrinking or stretching of radially displaced columns. { 'plan·ə¸ter·ē vȯr'tis·əd·ē i¸fekt }

planetary wave See long wave; Rossby wave. { 'plan·ə¸ter·ē 'wāv }

planetary wind [METEOROL] Any wind system of the earth's atmosphere which owes its existence and direction to solar radiation and to the rotation of the earth. { 'plan·ə¸ter·ē 'wind }

planet carrier [MECH ENG] A fixed member in a planetary gear train that contains the shaft upon which the planet pinion rotates. { 'plan·ət ¸kar·ē·ər }

planetesimal [ASTRON] One of the rocky bodies, of the order of 1 mile (1.6 kilometer) in diameter, that are believed to have formed in the protosolar nebula, and whose accretion

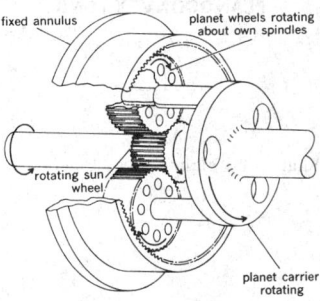

formed the rocky cores of the larger planets. { ˌplan·ə'tes·ə·məl }

planet gear [MECH ENG] A pinion in a planetary gear train. { 'plan·ət ˌgir }

planetocentric coordinates [ASTRON] Coordinates that indicate the position of a point on the surface of a planet, determined by the direction of a line joining the center of the planet to the point. { plə¦ned·ō¦sen·trik kō'ȯrd·ən·əts }

planetographic coordinates [ASTRON] Coordinates that indicate the position of a point on the surface of a planet, determined by the direction of a perpendicular to the mean surface at the point. { plə¦ned·ō¦graf·ik kō'ȯrd·ən·əts }

planetography [ASTRON] The descriptive science of the physical features of planets. { ˌplan·ə'täg·rə·fē }

planetoid See asteroid. { 'plan·əˌtȯid }

planetology [ASTRON] Scientific study of the planets, in particular their surface markings. { ˌplan·ə'täl·ə·jē }

planet pinion [MECH ENG] One of the gears in a planetary gear train that meshes with and revolves around the sun gear. { 'plan·ət ˌpin·yən }

plane triangulation [MATH] The process of adding arcs between pairs of verticles of a planar graph to producer another planar graph, each of whose regions is bounded by three sides. { 'plān trī·aŋ·gyə'lā·shən }

plane trigonometry [MATH] The study of triangles in the euclidean plane with the use of functions defined by the ratios of sides of right triangles. { 'plān trig·ə'näm·ə·trē }

plane wave [PHYS] Wave in which the wavefront is a plane surface; a wave whose equiphase surfaces form a family of parallel planes. { 'plān ˌwāv }

plane-wave initiation [ORD] Simultaneous initiation at all points of the rear surface of the main explosive charge by a flat detonation wave, usually accomplished by a composite explosive charge of proper dimensions. { 'plān ˌwāv iˌnish·ē'ā·shən }

planform [AERO ENG] The shape or form of an object, such as an airfoil, as seen from above, as in a plan view. [GEOGR] A body of water's outline or morphology as defined by the still water line. { 'planˌfȯrm }

planidium [INV ZOO] A first-stage legless larva of various insects in the orders Diptera and Hymenoptera. { plə'nid·ē·əm }

planigraphy See sectional radiography. { plə'nig·rə·fē }

planimeter [ENG] A device used for measuring the area of any plane surface by tracing the boundary of the area. { plə'nim·əd·ər }

planimetric map [MAP] A map indicating only the horizontal positions of features, without regard to elevation, in contrast with a topographic map, which indicates both horizontal and vertical positions. Also known as line map. { ¦plan·ə¦me·trik 'map }

planimetric method [MET] A method of measuring grain size by counting the number of grains in a given area. { ¦plan·ə¦me·trik 'meth·əd }

planimetry [MAP] **1.** The measurement of plane surfaces; for example, the determination of horizontal distances, angles, and areas on a map. **2.** The plan details of a map; the natural and cultural features of a region (excluding relief) as shown on a map. { plə'nim·ə·trē }

planing [ENG] Smoothing or shaping the surface of wood, metal, or plastic workpieces. { 'plān·iŋ }

planing boat See hydroplane. { 'plān·iŋ ˌbōt }

planing hull [NAV ARCH] A hull form with straight buttock lines, designed to develop positive hydrodynamic pressures on its bottom so that its draft decreases with increasing speed, enabling it to rise higher on the wave that it is generating. { 'plān·iŋ ˌhəl }

Planipennia [INV ZOO] A suborder of insects in the order Neuroptera in which the larval mandibles are modified for piercing and for sucking. { ˌplan·ə'pen·ē·ə }

planisaic [MAP] A photomap in which the planimetric detail is shown by color overprints. { ˌplan·ə'zā·ik }

planishing [MECH ENG] Smoothing the surface of a metal by a rapid series of overlapping, light hammerlike blows or by rolling in a planishing mill. { 'plan·ish·iŋ }

planisphere [MAP] A representation, on a plane, of the celestial sphere, especially one on a polar projection, with means provided for making certain measurements such as altitude and azimuth. { 'plan·əˌsfir }

plank [MATER] A heavy board with thickness of 2–4 inches (5–10 centimeters) and a width of at least 8 inches (20 centimeters). { plaŋk }

planking [NAV ARCH] The wood decks and outside planks in wood or composite ships. { 'plaŋk·iŋ }

plankton [ECOL] Passively floating or weakly motile aquatic plants and animals. { 'plaŋk·tən }

planktonic [ECOL] Free-floating. { plaŋk'tän·ik }

plankton net [ENG] A net for collecting plankton. { 'plaŋk·tən ˌnet }

planning by abstraction [COMPUT SCI] A computer problem-solving method in which the task to be accomplished is simplified; the simplified task is solved; and the solution is used as a guide. { 'plan·iŋ bī ab'strak·shən }

planning chart [NAV] A chart designed for use in planning voyages or flight operations, or investigating areas of marine or aviation activities. { 'plan·iŋ ˌchärt }

planning horizon [IND ENG] In a materials-requirements planning system, the time from the present to some future date for which plans are being generated for acquisition of materials. { 'plan·iŋ həˌrīz·ən }

planoblast [INV ZOO] The medusa form of a hydrozoan. { 'plan·əˌblast }

planocaine base See procaine base. { 'plan·əˌkän ˌbās }

planoclastic rock [PETR] An even-grained or uniformly sized clastic rock. { ¦plan·ə¦kla·stik 'räk }

planoconcave lens [OPTICS] A lens for which one surface is plane and the other is concave. { ¦plā·nō'kän,kāv 'lenz }

planoconformity [GEOL] The relation between conformable strata that are approximately uniform in thickness and sensibly parallel throughout. { ¦plā·nō·kən'fȯr·məd·ē }

planoconvex lens [OPTICS] A lens for which one surface is plane and the other is convex. { ¦plā·nō'kän,veks 'lenz }

planoconvex spotlight [ELEC] A light that can be used as a sharply defined spotlight or for soft-edged lighting; ranges in power from 100 to 2000 watts. { ¦plā·nō'kän,veks 'spätˌlīt }

planocylindrical lens [OPTICS] A lens, one of whose surfaces is a portion of a plane, while the other is a portion of a cylinder. { ¦plā·nō·sə'lin·drə·kəl 'lenz }

planographic process [GRAPHICS] Nonrelief printing process (such as collotype and lithography), in which the areas of the plate to receive ink are on the same level or plane as those that remain uninked. { ¦plā·nə¦graf·ik 'prä·səs }

planomycin See fervenulin. { ˌplan·ə'mīs·ən }

Planosol [GEOL] An intrazonal, hydromorphic soil having a clay pan or hardpan covered with a leached surface layer; developed in a humid to subhumid climate. { 'plan·əˌsȯl }

planospiral [INV ZOO] Having the shell coiled in one plane, used particularly of foraminiferans and mollusks. { ¦plā·nō'spī·rəl }

plan position indicator [ELECTR] A radarscope display in which echoes from various targets appear as bright spots at the same locations as they would on a circular map of the area being scanned, the radar antenna being at the center of the map. Abbreviated PPI. Also known as P display. { 'plan pə'zish·ən 'in·dəˌkād·ər }

plan position indicator repeater [ELECTR] Unit which repeats a plan position indicator (PPI) at a location remote from the radar console. Also known as remote plan position indicator. { 'plan pə'zish·ən 'in·dəˌkād·ər riˌpēd·ər }

plansifter [FOOD ENG] A stack of sieves of decreasing mesh size that separate particles by size; used in milling operations. { 'planˌsift·ər }

plant [BOT] Any organism belonging to the kingdom Plantae, generally distinguished by the presence of chlorophyll, a rigid cell wall, and abundant, persistent, active embryonic tissue, and by the absence of the power of locomotion. [COMPUT SCI] To place a number or instruction that has been generated in the course of a computer program in a storage location where it will be used or obeyed at a later stage of the program. [IND ENG] The land, buildings, and equipment used in an industry. { plant }

Plantae [BOT] The plant kingdom. { 'planˌtē }

Plantaginaceae [BOT] The single family of the plant order Plantaginales. { ˌplan·tə·jə'nās·ēˌē }

Plantaginales [BOT] An order of dicotyledonous herbaceous plants in the subclass Asteridae, marked by small hypogynous flowers with a persistent regular corolla and four petals. { ˌplan·tə·jə'nā·lēz }

PLANOCONCAVE LENS

Shape of a planoconcave lens.

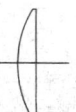

PLANOCONVEX LENS

Shape of a planoconvex lens.

azimuth

PLAN POSITION INDICATOR

Plan position indicator type of radar display; range is measured radially from center.

plantar [ANAT] Of or relating to the sole of the foot. { 'plan·tər }

plantaris [ANAT] A small muscle of the calf of the leg; origin is the lateral condyle of the femur, and insertion is the calcaneus; flexes the knee joint. { plan'tar·əs }

plantar reflex [PHYSIO] Flexion of the toes in response to stroking of the outer surface of the sole, from heel to little toe. { 'plan·tər 'rē,fleks }

plant decomposition [CONT SYS] The partitioning of a large-scale control system into subsystems along lines of weak interaction. { 'plant dē,käm·pə'zish·ən }

Plante cell [ELEC] A type of lead-acid cell in which the active material is formed on the plates by electrochemical means during repeated charging and discharging, instead of being applied as a prepared paste. { plän'tā ,sel }

plant extract *See* crude drug. { 'plant 'ek,strakt }

plant factor [ELEC] The ratio of the average power load of an electric power plant to its rated capacity. Also known as capacity factor. { 'plant ,fak·tər }

plant fermentation [BIOCHEM] A form of plant metabolism in which carbohydrates are partially degraded without the consumption of molecular oxygen. { 'plant ,fər·mən'tā·shən }

plant geography [BOT] A major division of botany, concerned with all aspects of the spatial distribution of plants. Also known as geographical botany; phytochorology; phytogeography. { 'plant jē,äg·rə·fē }

plant hormone [BIOCHEM] An organic compound that is synthesized in minute quantities by one part of a plant and translocated to another part, where it influences physiological processes. Also known as phytohormone. { 'plant ,hȯr ,mōn }

plantigrade [VERT ZOO] Pertaining to walking with the whole sole of the foot touching the ground. { 'plan·tə,grād }

plant key [BOT] An analytical guide to the identification of plants, based on the use of contrasting characters to subdivide a group under study into branches. { 'plant kē }

plant kingdom [BOT] The worldwide array of plant life constituting a major division of living organisms. { 'plant ,kiŋ·dəm }

plant layout [IND ENG] The location of equipment and facilities in a manufacturing plant. { 'plant ,lā,aút }

plant pathology [BOT] The branch of botany concerned with diseases of plants. { 'plant pə'thäl·ə·jē }

plant physiology [BOT] The branch of botany concerned with the processes which occur in plants. { 'plant ,fiz·ē'äl·ə·jē }

plant propagation [BOT] The deliberate, directed reproduction of plants using seeds or spores (sexual propagation), or using vegetative cells, tissues, or organs (asexual reproduction). { 'plant ,präp·ə,gā·shən }

plant protection [IND ENG] That portion of industrial security which concerns the safeguarding of industrial installations, resources, utilities, and materials by physical measures such as guards, fences, and lighting designation of restricted areas. { 'plant prə,tek·shən }

plant respiration [BOT] A biochemical process in plants whereby specific substrates are oxidized with a subsequent release of carbon dioxide, CO_2. { 'plant ,res·pə,rā·shən }

plant societies [ECOL] Assemblages of plants which constitute structural parts of plant communities. { 'plant sə,sī·əd·ēz }

plantula [INV ZOO] A small, cushionlike structure on the ventral surface of the segments of insect tarsi. { 'plan·chə·lə }

plant virus [VIROL] A virus that replicates only within plant cells. { 'plant ,vī·rəs }

planula [INV ZOO] The ciliated, free-swimming larva of coelenterates. { 'plan·yə·lə }

Planuloidea [INV ZOO] The equivalent name for Moruloidea. { ,plan·yə'lȯid·ē·ə }

plan view [GRAPHICS] A drawing that shows the appearance of an object when seen from above. { 'plan ,vyü }

plaque [MED] **1.** A patch, or an abnormal flat area on any internal or external body surface. **2.** A localized area of atherosclerosis. [VIROL] A clear area representing a colony of viruses on a plate culture formed by lysis of the host cell. { plak }

plash [HYD] A shallow, standing, usually short-lived pool or small pond resulting from a flood, heavy rain, or melting snow. { plash }

plasma [GEOL] The part of a soil material that can be, or has been, moved, reorganized, or concentrated by soil-forming processes. [HISTOL] The fluid portion of blood or lymph. [MINERAL] A faintly translucent or semitranslucent and bright green, leek green, or nearly emerald green variety of chalcedony, sometimes having white or yellowish spots. [PL PHYS] **1.** A highly ionized gas which contains equal numbers of ions and electrons in sufficient density so that the Debye shielding length is much smaller than the dimensions of the gas. **2.** A completely ionized gas, composed entirely of a nearly equal number of positive and negative free charges (positive ions and electrons). { 'plaz·mə }

plasma accelerator [PL PHYS] An accelerator that forms a high-velocity jet of plasma by using a magnetic field, an electric arc, a traveling wave, or other similar means. { 'plaz·mə ak'sel·ə,rād·ər }

plasma-arc cutting [ENG] Metal cutting by melting a localized area with an arc followed by removal of metal by high-velocity, high-temperature ionized gas. { 'plaz·mə ¦ärk 'kəd·iŋ }

plasma-arc welding [MET] Welding metal in a gas stream heated by a tungsten arc to temperatures approaching 60,000°F (33,315°C). { 'plaz·mə ¦ärk 'weld·iŋ }

plasma cathode [ELECTR] A cathode in which the source of electrons is a gas plasma rather than a solid. { 'plaz·mə 'kath,ōd }

plasma cell *See* plasmacyte. { 'plaz·mə ,sel }

plasma centrifuge [NUCLEO] A device for separating isotopes in which a cylinder of ionized matter is contained by a magnetic field and set into rotation by application of an electromagnetic body force, so that centrifugal force causes heavier ions (or isotopes) to move nearer the periphery of the rotating plasma. { 'plaz·mə 'sen·trə,fyüj }

plasma cloud [ASTROPHYS] An aggregate of electrically charged particles that is embedded in the solar wind. { 'plaz·mə ,klaúd }

plasmacyte [HISTOL] A fairly large, generally ovoid cell with a small, eccentrically placed nucleus; the chromatin material is adherent to the nuclear membrane and the cytoplasm is agranular and deeply basophilic everywhere except for a clear area adjacent to the nucleus in the area of the cytocentrum. Also known as plasma cell. { 'plaz·mə,sīt }

plasma desorption mass spectrometry [SPECT] A technique for analysis of nonvolatile molecules, particularly heavy molecules with atomic weight over 2000, in which heavy ions with energies on the order of 100 MeV penetrate and deposit energy in thin films, giving rise to chemical reactions that result in the formation of molecular ions and shock waves that result in the ejection of these ions from the surface; the ions are then analyzed in a mass spectrometer. Abbreviated PDMS. { 'plaz·mə dē¦sȯrp·shən 'mas spek'träm·ə·trē }

plasma diode [ELECTR] A diode used for converting heat directly into electricity; it consists of two closely spaced electrodes serving as cathode and anode, mounted in an envelope in which a low-pressure cesium vapor fills the interelectrode space; heat is applied to the cathode, causing emission of electrons. { 'plaz·mə 'dī,ōd }

plasma display [ELECTR] A display in which sets of parallel conductors at right angles to each other are deposited on glass plates, with the very small space between the plates filled with a gas; each intersection of two conductors defines a single cell that can be energized to produce a gas discharge forming one element of a dot-matrix display. { 'plaz·mə di'splā }

plasma engine [AERO ENG] An engine for space travel in which neutral plasma is accelerated and directed by external magnetic fields that interact with the magnetic field produced by current flow through the plasma. Also known as plasma jet. { 'plaz·mə ,en·jən }

plasma etching [ELECTR] A method of forming integrated-circuit patterns on a surface, in which charged species in a plasma formed above a masked surface are directed to impact the nonmasked regions of the surface and knock out substrate atoms. Also known as dry plasma etching. { 'plaz·mə 'ech·iŋ }

plasma frequency *See* Langmuir plasma frequency. { 'plaz·mə ,frē·kwən·sē }

plasmagel [CELL MOL] The outer, gelated zone of protoplasm in a pseudopodium. { 'plaz·mə,jel }

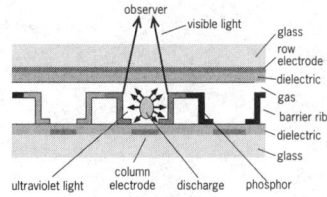

PLASMA DISPLAY

Section through three pixels of an alternating-current color plasma display panel structure. Each pixel is located where a column electrode and a row electrode cross. (*Thomson Plasma*)

plasmagene [CELL MOL] A cytoplasmic particle or substance, which may be present in bodies such as plastids or mitochondria, and which can reproduce and pass on inherited qualities to daughter cells. { 'plaz·mə‚jēn }

plasma generator [ELECTR] Any device that produces a high-velocity plasma jet, such as a plasma accelerator, engine, oscillator, or torch. { 'plaz·mə 'jen·ə‚rād·ər }

plasma gun [ELECTR] A machine, such as an electric-arc chamber, that will generate very high heat fluxes to convert neutral gases into plasma. [ELECTROMAG] An electromagnetic device which creates and accelerates bursts of plasma. { 'plaz·mə ‚gən }

plasma instability [PL PHYS] A sudden change in the quasistatic distribution of positions or velocities of particles constituting a plasma, and a sudden change in the accompanying electromagnetic field. { 'plaz·mə ‚in·stə'bil·əd·ē }

plasma jet See plasma engine. { 'plaz·mə ‚jet }

plasma-jet excitation [SPECT] The use of a high-temperature plasma jet to excite an element to provide measurable spectra with many ion lines similar to those from spark-excited spectra. { 'plaz·mə ‚jet ‚ek·sə'tā·shən }

plasmalemma See cell membrane. { ‚plaz·mə'lem·ə }

plasmalogen [BIOCHEM] Any of a group of glycerol-based phospholipids in which a fatty acid group is replaced by a fatty aldehyde. { plaz'mal·ə·jən }

plasmals [BIOCHEM] Aldehydic components of lipids which give positive color tests with reagents used for detecting aldehydes in tissues. { 'plaz·məlz }

plasma mantle [GEOPHYS] A thick layer of plasma just inside the magnetopause characterized by a tailward bulk flow with a speed of 60 to 120 miles (100 to 200 kilometers) per second and by a gradual decrease of density, temperature, and speed as the depth inside the magnetosphere increases. { 'plaz·mə 'mant·əl }

plasma membrane See cell membrane. { 'plaz·mə 'mem‚brān }

plasma oscillations [PL PHYS] Various vibrations and wave motions of the electrons and ions in a plasma. { 'plaz·mə ‚äs·ə'lā·shənz }

plasmapause [GEOPHYS] The sharp outer boundary of the plasmasphere, at which the plasma density decreases by a factor of 100 or more. { 'plaz·mə‚pòz }

plasmapheresis [MED] The withdrawal of blood from a donor to obtain plasma, its components, or the nonerythrocytic formed elements of blood, followed by the return of the erythrocytes to the donor. { ‚plaz·mə·fə'rē·səs }

plasma physics [PHYS] The study of highly ionized gases. { 'plaz·mə 'fiz·iks }

plasma pinch [PL PHYS] Application of the pinch effect to plasma in attempts to produce controlled nuclear fusion. { 'plaz·mə ‚pinch }

plasma processing [ENG] Methods and technologies that utilize a plasma to treat and manufacture materials, generally through etching, deposition, or chemical alteration at a surface inside or at the boundary of the plasma. { 'plaz·mə prä‚ses·iŋ }

plasma propulsion [AERO ENG] Propulsion of spacecraft and other vehicles by using electric or magnetic fields to accelerate both positively and negatively charged particles (plasma) to a very high velocity. { 'plaz·mə prə'pəl·shən }

plasma radiation [PL PHYS] Electromagnetic radiation emitted from a plasma, primarily by free electrons undergoing transitions to other free states or to bound states of atoms and ions, but also by bound electrons as they undergo transitions to other bound states. { 'plaz·mə ‚rād·ē'ā·shən }

plasma rocket [AERO ENG] A rocket that is accelerated by means of a plasma engine. { 'plaz·mə ‚räk·ət }

plasma sheath [ELECTR] An envelope of ionized gas that surrounds a spacecraft or other body moving through an atmosphere at hypersonic velocities; affects transmission, reception, and diffraction of radio waves. { 'plaz·mə ‚shēth }

plasma sheet [GEOPHYS] A region of relatively hot plasma outside the plasmasphere, which reaches, during quiet times, from an altitude of about 30,000 miles (50,000 kilometers) to at least past the moon's orbit in a long tail extending away from the sun; composed of particles with typical thermal energies of 2 to 4 kiloelectronvolts. { 'plaz·mə ‚shēt }

plasmasol [CELL MOL] The inner, solated zone of protoplasm in a pseudopodium. { 'plaz·mə‚sòl }

plasma-source ion implantation [ENG] A method of ion implantation in which the workpiece is placed in a plasma containing the appropriate ion species and is repetitively pulse-biased to a high negative potential so that positive plasma ions are accelerated to the surface and implant in the bulk material. Abbreviated PSII. { ‚plaz·mə ‚sòrs 'ī·ən ‚im·plan‚tā·shən }

plasmasphere [GEOPHYS] A region of relatively dense, cold plasma surrounding the earth and extending out to altitudes of approximately 2 to 6 earth radii, composed predominantly of electrons and protons, with thermal energies not exceeding several electronvolts. { 'plaz·mə‚sfir }

plasma spraying [MET] In thermal spraying, melting and transference of a metal coating to a workpiece by use of a nontransferred arc. { 'plaz·mə 'sprā·iŋ }

plasma tail [ASTRON] A comet tail that is composed primarily of electrons and molecular ions, the dominant visible ion being positively ionized carbon monoxide, and that is generally straight, with a length in the range from 0.62×10^7 to 0.62×10^8 miles (1×10^7 to 1×10^8 kilometers). { 'plaz·mə 'tāl }

plasma thromboplastin antecedent See factor XI. { 'plaz·mə ‚thräm·bō‚plas·tən 'ant·i‚sēd·ənt }

plasma thromboplastin component See Christmas factor. { 'plaz·mə ‚thräm·bō‚plas·tən kəm‚pō·nənt }

plasma torch [ENG] A torch in which temperatures as high as 50,000°C are achieved by injecting a plasma gas tangentially into an electric arc formed between electrodes in a chamber; the resulting vortex of hot gases emerges at very high speed through a hole in the negative electrode, to form a jet for welding, spraying of molten metal, and cutting of hard rock or hard metals. { 'plaz·mə ‚tòrch }

plasmatron [ELECTR] A gas-discharge tube in which independently generated plasma serves as a conductor between a hot cathode and an anode; the anode current is modulated by varying either the conductivity or the effective cross section of the plasma. { 'plaz·mə‚trän }

Plasmaviridae [VIROL] A family of enveloped deoxyribonucleic acid (DNA) phages that are characterized by rounded pleomorphic particles with a small, densely stained center and a supercoiled, double-stranded, circular DNA genome. { ‚plaz·mə'vir·ə‚dī }

Plasmavirus [VIROL] The sole genus of the deoxyribonucleic acid-containing phage family Plasmaviridae. { 'plaz·mə‚vī·rəs }

plasma wave [PL PHYS] A disturbance of a plasma involving oscillation of its constituent particles and of an electromagnetic field, which propagates from one point in the plasma to another without net motion of the plasma. { 'plaz·mə ‚wāv }

plasmid [GEN] An extrachromosomal genetic element found among various strains of Escherichia coli and other bacteria. { 'plaz·məd }

plasmid cloning vector [GEN] A plasmid that accepts foreign deoxyribonucleic acid (DNA) and is therefore used in recombinant DNA experiments. { 'plaz·mid ‚klōn·iŋ ‚vek·tər }

plasmid donation [GEN] The transfer of a nonconjugative plasmid from a donor cell to a recipient cell by way of a contact function provided by a conjugative plasmid. { 'plaz·mid dō‚nā·shən }

plasmin [BIOCHEM] A proteolytic enzyme in plasma which can digest many proteins through the process of hydrolysis. Also known as fibrinolysin. { 'plaz·mən }

plasminogen [BIOCHEM] The inert precursor, or zymogen, of plasmin. Also known as profibrinolysin. { plaz'min·ə·jən }

plasmodesma [CYTOL] An intercellular bridge, thought to be strands of cytoplasm connecting two cells. { ‚plaz·mə'dez·mə }

plasmodesmata [BOT] Strands of cytoplasm connecting the protoplasts of two contiguous plant cells. { ‚plaz·mə·dez'mäd·ə }

Plasmodiidae [INV ZOO] A family of parasitic protozoans in the suborder Haemosporina inhabiting the erythrocytes of the vertebrate host. { ‚plaz·mə'dī·ə‚dē }

Plasmodiophorida [INV ZOO] An order of the protozoan subclass Mycetozoia occurring as endoparasites of plants. { ‚plaz·mō‚dī·ə'fòr·ə·də }

Plasmodiophoromycetes [MYCOL] A class of the Fungi. { ‚plaz·mō·dī‚äf·ə·rō·mī'sē·dēz }

plasmoditrophoblast *See* syncytiotrophoblast. { ˌplaz·mō·dī'träf·ə,blast }

plasmodium [MICROBIO] The noncellular, multinucleate, jellylike, ameboid, assimilative stage of the Myxomycetes. { plaz'mō·dē·əm }

Plasmodroma [INV ZOO] A subphylum of the Protozoa, including Mastigophora, Sarcodina, and Sporozoa, in some taxonomic systems. { plaz'mä·drə·mə }

plasmogamy [INV ZOO] Fusion of protoplasts, without nuclear fusion, to form a multinucleate mass; occurs in certain protozoans. { plaz'mäg·ə·mē }

plasmoid [PHYS] An isolated collection of electrons, ions, and neutral particles which holds together for a duration many times as long as the collision times between particles. { 'plaz,mȯid }

plasmolysis [PHYSIO] Shrinking of the cytoplasm away from the cell wall due to exosmosis by immersion of a plant cell in a solution of higher osmotic activity. { plaz'mäl·ə·səs }

plasmon [GEN] The cytoplasmic genetic system in eukaryotes consisting primarily of mitochondrial deoxyribonucleic acid (DNA) and chloroplast DNA. [SOLID STATE] A quantum of a collective longitudinal wave in the electron gas of a solid. { 'plaz,män }

plasmosome *See* nucleolus. { 'plaz·mə,sōm }

plasmotomy [INV ZOO] Subdivision of a plasmodium into two or more parts. { plas'mäd·ə·mē }

plaster [MATER] A plastic mixture of various materials, such as lime or gypsum, and water which sets to a hard, coherent solid. { 'plas·tər }

plaster bat [GRAPHICS] Basic working surface on which clay is turned or modeled. { 'plas·tər ¦bat }

plasterboard [MATER] A large, thin sheet of pulpboard, paper, or felt bonded to a hardened gypsum plaster core and used as a wall backing or as a substitute for plaster. { 'plas·tər,bȯrd }

plaster coat [BUILD] A thin layer of plaster lining walls in buildings. { 'plas·tər ¦kōt }

plaster conglomerate [GEOL] A conglomerate composed entirely of boulders derived from a partially exhumed monadnock forming a wedgelike mass of its flank. { 'plas·tər kən'gläm·ə·rət }

plaster ground [BUILD] A piece of wood used as a gage to control the thickness of a plaster coat placed on a wall; usually put around windows and doors and at the floor. { 'plas·tər ¦graund }

plaster of paris [INORG CHEM] White powder consisting essentially of the hemihydrate of calcium sulfate ($CaSO_4$·$^1/_2H_2O$ or $2CaSO_4$·H_2O), produced by calcining gypsum until it is partially dehydrated; forms with water a paste that quickly sets; used for casts and molds, building materials, and surgical bandages. Also known as calcined gypsum. { 'plas·tər əv 'par·əs }

plaster shooting [ENG] A surface blasting method used when no rock drill is necessary or one is not available; consists of placing a charge of gelignite, primed with safety fuse and detonator, in close contact with the rock or boulder and covering it completely with stiff damp clay. { 'plas·tər 'shüd·iŋ }

plastic [MATER] A polymeric material (usually organic) of large molecular weight which can be shaped by flow; usually refers to the final product with fillers, plasticizers, pigments, and stabilizers included (versus the resin, the homogeneous polymeric starting material); examples are polyvinyl chloride, polyethylene, and urea-formaldehyde. [MECH] Displaying, or associated with, plasticity. { 'plas·tik }

plasticate [ENG] To soften a material by heating or kneading. Also known as plastify. { 'plas·tə,kāt }

Plastic Ball [NUCLEO] A large gamma-ray detector used at colliding-beam accelerators, and consisting of about 1400 lead-glass elements for recording photons that completely surround the point of collision. { 'plas·tik 'bȯl }

plastic bonding [ENG] The joining of plastics by heat, solvents, adhesives, pressure, or radio frequency. { 'plas·tik 'bänd·iŋ }

plastic bronze [MET] A copper alloy containing lead, usually on the order of 30%, of sufficient plasticity to make a good bearing. { 'plas·tik 'bränz }

plastic cement [MATER] A plastic material used to seal narrow openings in buildings. { 'plas·tik si'ment }

plastic clay [MATER] Fireclay which forms a moldable mass when mixed with water. { 'plas·tik 'klā }

plastic collision [MECH] A collision in which one or both of the colliding bodies suffers plastic deformation and mechanical energy is dissipated. { 'plas·tik kə'lizh·ən }

plastic deformation [MECH] Permanent change in shape or size of a solid body without fracture resulting from the application of sustained stress beyond the elastic limit. { 'plas·tik ,dē,fȯr'mā·shən }

plastic design *See* ultimate-load design. { 'plas·tik di'zīn }

plastic dielectric [MATER] A plastic used in an application in which its high resistance, dielectric strength, or other electrical properties are important, such as for electrical insulation or in a capacitor. { 'plas·tik ,dī·ə'lek·trik }

plastic equilibrium [GEOL] State of stress within a soil mass or a portion thereof that has been deformed to such an extent that its ultimate shearing resistance is mobilized. { 'plas·tik ,ē·kwə'lib·rē·əm }

plastic explosive *See* high-explosive plastic. { 'plas·tik ik'splō·siv }

plastic film [MATER] Film with thickness from 0.0015 to 0.006 inch (0.0038 to 0.015 centimeter); made from polyvinyl chloride, polyethylene, polypropylene, polystyrene, Mylar, and other resins; used for wrapping, sealing, garment waterproofing, and coating wood, paper, or fabric. { 'plas·tik 'film }

plastic film capacitor [ELEC] A capacitor constructed by stacking, or forming into a roll, alternate layers of foil and a dielectric which consists of a plastic, such as polystyrene or Mylar, either alone or as a laminate with paper. { 'plas·tik 'film kə'pas·əd·ər }

plastic flow [PHYS] Rheological phenomenon in which flowing behavior of the material occurs after the applied stress reaches a critical (yield) value, such as with putty. { 'plas·tik 'flō }

plastic foam *See* expanded plastic. { 'plas·tik 'fōm }

plasticity [MECH] The property of a solid body whereby it undergoes a permanent change in shape or size when subjected to a stress exceeding a particular value, called the yield value. { plas'tis·əd·ē }

plasticity index [GEOL] The percent difference between moisture content of soil at the liquid and plastic limits. { plas'tis·əd·ē ,in,deks }

plasticize [ENG] To soften a material to make it plastic or moldable by adding a plasticizer or by using heat. { 'plas·tə,sīz }

plasticizer *See* flexibilizer. { 'plas·tə,sīz·ər }

plasticizing oil [MATER] Coal tar distillate or solvent naphthas distilling in a wide range above 300°C; used with plastics as a plasticizer. { 'plas·tə,sīz·iŋ ,ȯil }

plasticlast [GEOL] An intraclast consisting of calcareous mud that has been torn up while still soft. { 'plas·tə,klast }

plastic limit [GEOL] The water content of a sediment, such as a soil, at the point of transition between the plastic and semisolid states. { 'plas·tik 'lim·ət }

plasticorder [ENG] Laboratory device used to predict the performance of a plastic material by measurement of temperature, viscosity, and shear-rate relationships. Also known as plastigraph. { 'plas·tə,kȯrd·ər }

plasticoviscosity [MECH] Plasticity in which the rate of deformation of a body subjected to stresses greater than the yield stress is a linear function of the stress. { ¦plas·tə·kō·vi skäs·əd·ē }

plastic paint [MATER] Paint composed of a plastic (such as vinyl or nitrocellulose) in a solvent. { 'plas·tik ,pānt }

plastic plate [ELECTR] A plate of plastic dielectric material used as a base for a semiconductor device. [GRAPHICS] A direct printing plate formed on a plastic base. { 'plas·tik 'plāt }

plastic relief map [MAP] A topographic map printed on plastic and then molded by heat and pressure into a three-dimensional form to emphasize the relief. { 'plas·tik ri'lēf ,map }

plastic semiconductor [MATER] An organic plastic resin with a conjugated double-bond structure, such as polyacetylene; the material is a semiconductor due to resistance of electrons to transfer from one molecule to another. { 'plas·tik 'sem·i·kən,dək·tər }

plastic shading *See* hill shading. { 'plas·tik 'shād·iŋ }

plastic surgery [MED] Surgical repair, replacement, or alteration of lost, injured, or deformed parts of the body by transfer of tissue. { 'plas·tik 'sər·jə·rē }

plastic viscosity [FL MECH] A measure of the internal resistance to fluid flow of a Bingham plastic, expressed as the tangential shear stress in excess of the yield stress divided by the resulting rate of shear. { 'plas·tik vi'skäs·əd·ē }

plastic wood [MATER] Wood flour or wood cellulose compounded with a synthetic resin of high molecular weight; it is adhesive but does not penetrate wood, and is used to fill cavities or seams in wood products. { 'plas·tik 'wu̇d }

plastic zone [GEOL] A region located adjacent to the rupture zone of an explosion crater and at an increased distance from the shot site, differing from the rupture zone by having less fracturing and only small permanent deformations. { 'plas·tik ˌzōn }

plastid [CELL MOL] One of the specialized cell organelles containing pigments or protein materials, often serving as centers of special metabolic activities; examples are chloroplasts and leukoplasts. { 'plas·təd }

plastify See plasticate. { 'plas·tə,fī }

plastigel [MATER] A plastisol with gellike flow properties achieved by adding a thixotropic agent (such as bentonite) to the plastisol. { 'plas·tə,jel }

plastigraph See plasticorder. { 'plas·tə,graf }

plastisol [MATER] A vinyl resin dissolved in a plasticizer to make a pourable liquid. { 'plas·tə,sȯl }

plastizymes [PHYS CHEM] Artificial enzymes (artificial polymeric materials with molecule-shaped pores) that possess catalytic properties. { 'plas·ti,zīmz }

plastogene [CELL MOL] A cytoplasmic factor, controlled by or interacting with the nucleus, which determines differentiation of a plastid. { 'plas·tə,jēn }

plastometer [ENG] Instrument used to determine the flow properties of a thermoplastic resin by forcing molten resin through a specified die opening or orifice at a given pressure and temperature. { pla'stäm·əd·ər }

plastoquinone [BIOCHEM] Any of a group of quinones that are involved in electron transport in chloroplasts during photosynthesis. { ˌplas·tə·kwə'nōn }

plastron [INV ZOO] The ventral plate of the cephalothorax of spiders. [VERT ZOO] The ventral portion of the shell of tortoises and turtles. { 'plas,trän }

plat [MAP] A plan that shows land ownership, boundaries, and subdivisions together with data for description and identification of various parts. { plat }

Platanaceae [BOT] A small family of monoecious dicotyledonous plants in which flowers have several carpels which are separate, three or four stamens, and more or less orthotropous ovules, and leaves are stipulate. { ˌplat·ən'ās·ē,ē }

Plataspidae [INV ZOO] A family of shining, oval hemipteran insects in the superfamily Pentatomoidea. { plə'tas·pə,dē }

plate [BUILD] **1.** A shoe or base member, such as of a partition or other kind of frame. **2.** The top horizontal member of a row of studs used in a frame wall. [DES ENG] A rolled, flat piece of metal of some arbitrary minimum thickness and width depending on the type of metal. [ELEC] **1.** One of the conducting surfaces in a capacitor. **2.** One of the electrodes in a storage battery. [ELECTR] See anode. [GEOL] **1.** A smooth, thin, flat fragment of rock, such as a flagstone. **2.** A large rigid, but mobile, block involved in plate tectonics; thickness ranges from 30 to 150 miles (50 to 250 kilometers) and includes both crust and a portion of the upper mantle. [GRAPHICS] **1.** In etching, the piece of copper, zinc, or other metal that constitutes the base from which prints are made. **2.** In photography, a sheet of glass coated with a sensitized emulsion. **3.** In printing, the reproduction of type or cuts in metal or other material; a plate may bear a relief, intaglio, or planographic printing surface. [MET] A thick flat particle of metal powder. { plāt }

plate amalgamation [MET] Use of copper-alloy plates or copper coated with mercury in order to trap gold from crushed ore pulp as it flows over the plates. { 'plāt ə,mal·gə'mā·shən }

plate anemometer See pressure-plate anemometer. { 'plāt ,an·ə'mäm·əd·ər }

plateau [ELECTR] The portion of the plateau characteristic of a counter tube in which the counting rate is substantially independent of the applied voltage. [GEOGR] An extensive, flat-surfaced upland region, usually more than 45–90 meters (150–300 feet) in elevation and considerably elevated above the adjacent country and limited by an abrupt descent on at least one side. [GEOL] A broad, comparatively flat and poorly defined elevation of the sea floor, commonly over 60 meters (200 feet) in elevation. { pla'tō }

plateau basalt [GEOL] One or a succession of high-temperature basaltic lava flows from fissure eruptions which accumulate to form a plateau. Also known as flood basalt. { pla'tō bə'sȯlt }

plateau characteristic [ELECTR] The relation between counting rate and voltage for a counter tube when radiation is constant, showing a plateau after the rise from the starting voltage to the Geiger threshold. Also known as counting rate-voltage characteristic. { pla'tō ˌkar·ik·tə'ris·tik }

plateau glacier [HYD] A highland glacier that overlies a generally flat mountain tract; usually overflows its edges in hanging glaciers. { pla'tō ˈglā·shər }

plateau gravel [GEOL] A sheet, spread, or patch of surficial gravel, often compacted, occupying a flat area on a hilltop, plateau, or other high region at a height above that normally occupied by a stream terrace gravel. { pla'tō 'grav·əl }

plateau level [PETRO ENG] The peak production level reached by an oil field. { pla'tō 'lev·əl }

plateau mountain [GEOL] A pseudomountain produced by the dissection of a plateau. { pla'tō ˈmau̇nt·ən }

plateau plain [GEOL] An extensive plain surmounted by a sublevel summit area and bordered by escarpments. { pla'tō ˌplān }

plateau problem [MATH] The problem of finding a minimal surface having as boundary a given curve. { pla'tō ˈpräb·ləm }

plateau ring structure [ASTRON] A lunar crater whose floor is significantly higher than the surrounding surface. { pla'tō 'riŋ ,strək·chər }

Plateau's sphere [FL MECH] A small drop of liquid which follows a larger drop that breaks away and falls. { plaˈtōz 'sfir }

plate bearing test [ENG] Former method to estimate the bearing capacity of a soil; a rigid steel plate about 1 foot (30 centimeters) square was placed on the foundation level and then loaded until the foundation failed, as evidenced by rapid sinking of the plate. { 'plāt 'ber·iŋ ,test }

plate-belt feeder See apron feeder. { 'plāt ,belt ,fēd·ər }

plate budding [BOT] Plant budding by inserting a rectangular scion with a bud under a flap of bark on the stock in such a manner that the exposed wood on the stock is covered. { 'plāt ,bəd·iŋ }

plate burning [GRAPHICS] After a negative has been prepared for an engraving, it is placed on the plate in a vacuum frame; the rays from the light source "burn" the plate, passing through the transparent areas of the negative and hardening the plate surface coating; this coating is later developed and the unexposed areas washed away; plates are also burned for offset printing. { 'plāt ,bərn·iŋ }

plate cam [MECH ENG] A flat, open cam that imparts a sliding motion. { 'plāt ,kam }

plate center [ASTRON] The point used as the origin of coordinates for measuring positions of stars on a photographic plate in photographic astrometry, ideally located on the optical axis of the telescope. { 'plāt ,sen·tər }

plate circuit See anode circuit. { 'plāt ˈsər·kət }

plate-circuit detector See anode-circuit detector. { 'plāt ˈsər·kət di,tek·tər }

plate coil [MECH ENG] Heat-transfer device made from two metal sheets held together, one or both plates embossed to form passages between them for a heating or cooling medium to flow through. Also known as panel coil. { 'plāt ˈkȯil }

plate constants [ASTRON] Coefficients that appear in linear equations used to derive the standard coordinates of the position of a star on a photographic plate from the measured coordinates of the star's image on the plate. { 'plāt ,kän·stəns }

plate conveyor [MECH ENG] A conveyor with a series of steel plates as the carrying medium; each plate is a short trough, all slightly overlapped to form an articulated band, and attached to one center chain or to two side chains; the chains join rollers running on an angle-iron framework and transmit the drive from the driveheads, installed at intermediate points and sometimes also at the head or tail ends. { 'plāt kən,vā·ər }

plate count [MICROBIO] The number of bacterial colonies

that develop on a medium in a petri dish seeded with a known amount of inoculum. { 'plāt ,kau̇nt }

plate crystal [HYD] An ice crystal exhibiting typical hexagonal (rarely triangular) symmetry and having comparatively little thickness parallel to its principal axis (c axis); as such crystals fall through the clouds in which they form, they may encounter conditions causing them to develop dendritic extensions, that is, to become plane-dendritic crystals. { 'plāt ,krist·əl }

plate current See anode current. { 'plāt ,kə·rənt }

plate cut [BUILD] The cut made in a rafter to rest on the plate. { 'plāt ,kət }

plated circuit [ELECTR] A printed circuit produced by electrodeposition of a conductive pattern on an insulating base. Also known as plated printed circuit. { 'plād·əd 'sər·kət }

plate detector See anode detector. { 'plāt di,tek·tər }

plate dissipation See anode dissipation. { 'plāt ,dis·ə,pā·shən }

plated printed circuit See plated circuit. { 'plād·əd 'print·əd 'sər·kət }

plated wire memory [COMPUT SCI] A nonvolatile magnetic memory utilizing small zones of thin films plated on wires; such memories are characterized by very fast access and nondestructive readout. { 'plād·əd 'wīr 'mem·rē }

plate efficiency [CHEM ENG] The equilibrium produced by an actual plate of a distillation column or countercurrent tower extractor compared with that of a perfect plate, expressed as a ratio. [ELECTR] See anode efficiency. { 'plāt i,fish·ən·sē }

plate feeder See apron feeder. { 'plāt ,fēd·ər }

plate-fin exchanger [MECH ENG] Heat-transfer device made up of a stack or layers, with each layer consisting of a corrugated fin between flat metal sheets sealed off on two sides by channels or bars to form passages for the flow of fluids. { 'plāt ,fin iks,chān·jər }

plate finish [GRAPHICS] A smooth, hard finish given to paper by a calendering machine. { 'plāt ,fin·ish }

plate girder [CIV ENG] A riveted or welded steel girder having a deep vertical web plate with a pair of angles along each edge to act as compression and tension flanges. { 'plāt ,gərd·ər }

plate girder bridge [CIV ENG] A fixed bridge consisting, in its simplest form, of two flange plates welded to a web plate in the overall shape of an I. { 'plāt ,gərd·ər ,brij }

plate glass [MATER] Flat, high-quality glass with plane, parallel surfaces. { 'plāt ,glas }

plate ice See pancake ice. { 'plāt ,īs }

plate impedance See anode impedance. { 'plāt im,pēd·əns }

plate input power See anode input power. { 'plāt 'in,pu̇t ,pau̇·ər }

platelet [HISTOL] See thrombocyte. [HYD] A small ice crystal which, when united with other such crystals, forms a layer of floating ice, especially sea ice, and serves as seed crystals for further thickening of the ice cover. { 'plāt·lət }

platelet-activating factor [IMMUNOL] A phospholipid released by leukocytes that causes aggregation of platelets and other effects, such as an increase in vascular permeability, and bronchoconstriction. Abbreviated PAF. { 'plāt·lət 'ak·tə,vād·iŋ ,fak·tər }

platelet-derived growth factor [CELL MOL] A glycolytic protein released by platelets and other cells that stimulates growth of cells of mesenchymal origin, for example, bone cartilage, vascular tissue, and connective tissue. { 'plāt·lət də,rīvd 'grōth ,fak·tər }

plate-load impedance See anode impedance. { 'plāt ,lōd im,pēd·əns }

plate-making [GRAPHICS] The forming of printing plates; the plate may be an offset plate, gravure plate, or a photogelatin plate. { 'plāt,māk·iŋ }

plate modulation See anode modulation. { 'plāt ,mäj·ə,lā·shən }

plate modulus [MECH] The ratio of the stress component T_{xx} in an isotropic, elastic body obeying a generalized Hooke's law to the corresponding strain component S_{xx}, when the strain components S_{yy} and S_{zz} are 0; the sum of the Poisson ratio and twice the rigidity modulus. { 'plāt ,mäj·ə·ləs }

platen [ENG] **1.** A flat plate against which something rests or is pressed. **2.** The rubber-covered roller of a typewriter against which paper is pressed when struck by the typebars. [MECH ENG] A flat surface for exchanging heat in a boiler or heat exchanger which may have extended heat transfer surfaces. { 'plat·ən }

plate neutralization See anode neutralization. { 'plāt ,nü·trə·lə,zā·shən }

platen press [GRAPHICS] A type of printing press with a flat surface bearing the inked type; another flat surface, bearing the paper, is pressed against the type; small hand presses are ordinarily of this sort. { 'plat·ən ,pres }

plate proof [GRAPHICS] A proof obtained from a printing plate. { 'plāt ,prüf }

plate pulse modulation See anode pulse modulation. { 'plāt 'pəls ,mäj·ə,lā·shən }

plate resistance See anode resistance. { 'plāt ri,zis·təns }

plate saturation See anode saturation. { 'plāt ,sach·ə,rā·shən }

plate scale [ASTRON] The ratio of the angular distance between two stars to the linear distance between their images on a photographic plate. { 'plāt ,skāl }

plate-shear test [ENG] A method used to get true shear data on a honeycomb core by bonding the core between two thick steel plates and subjecting the core to shear by displacing the plates relative to each other by loading in either tension or compression. { 'plāt 'shir ,test }

plate tectonics [GEOL] Global tectonics based on a model of the earth characterized by a small number (10–25) of semirigid plates which float on some viscous underlayer in the mantle; each plate moves more or less independently and grinds against the others, concentrating most deformation, volcanism, and seismic activity along the periphery. Also known as raft tectonics. { 'plāt tek'tän·iks }

plate theory [ANALY CHEM] In gas chromatography, the theory that the column operates similarly to a distillation column; for example, chromatographic columns are considered as consisting of a number of theoretical plates, each performing a partial separation of components. { 'plāt ,thē·ə·rē }

plate tower [CHEM ENG] A distillation tower along the internal height of which is a series of transverse plates (bubble-cap or sieve) to force intimate contact between downward flowing liquid and upward flowing vapor. { 'plāt ,tau̇·ər }

plate-type exchanger [MECH ENG] Heat-exchange device similar to a plate-and-frame filter press; fluids flow between the frame-held plates, transferring heat between them. { 'plāt ,tīp iks,chān·jər }

plate vibrator [ENG] A mechanically operated tamper fitted with a flat base. { 'plāt vī'brād·ər }

plate wave [ACOUS] A type of ultrasonic vibration generated in a thin solid, such as a sheet of metal having a thickness of less than one wavelength, and usually consisting of a variety of simultaneous modes having different velocities; it is used in metal inspection. Also known as Lamb wave. { 'plāt ,wāv }

platform [COMPUT SCI] The hardware system and the system software used by a computer program. [GEOL] **1.** Any level or almost level surface; a small plateau. **2.** A continental area covered by relatively flat or gently tilted, mainly sedimentary strata which overlay a basement of rocks consolidated during earlier deformations; platforms and shields together constitute cratons. [MIN ENG] A wooden floor on the side of a gangway at the bottom of an inclined seam, to which the coal runs by gravity, and from which it is shoveled into mine cars. [ORD] **1.** Temporary or permanent solid bed on which artillery pieces are supported to give greater stability. **2.** Metal stand at the base of some types of guns upon which the gun crew stands while serving the gun. { 'plat,form }

platform balance [ENG] A weighing device with a flat plate mounted above a balanced beam. { 'plat,form ,bal·əns }

platform beach [GEOL] A looped bar or ridge of sand and gravel formed on a wave-cut platform. { 'plat,form ,bēch }

platform blowing [ENG] Special technique for blow-molding large parts made of plastic without sagging of the part being formed. { 'plat,form ,blō·iŋ }

platform conveyor [MECH ENG] A single- or double-strand conveyor with plates of steel or hardwood forming a continuous platform on which the loads are placed. { 'plat,form kən,vā·ər }

platform deck [NAV ARCH] A partial deck fitted in the hold of a ship. { 'plat,form ,dek }

platform erection [NAV] In the alignment of inertial navigation equipment, the alignment of the stable platform vertical axis with the local vertical. { 'plat,form i,rek·shən }

platform facies See shelf facies. { 'plat,fȯrm ,fā·shēz }

platform framing [BUILD] A construction method in which each floor is framed independently by nailing the horizontal framing member to the top of the wall studs. { 'plat,fȯrm ,frām·iŋ }

platform reef [GEOL] An organic reef, generally small but more extensive than a patch reef, with a flat upper surface. { 'plat,fȯrm ,rēf }

platina [MET] A white brittle brass containing 75% zinc and 25% copper; used for jewelry. { plə'tēn·ə }

plating [MET] Forming a thin, adherent layer of metal on an object. Also known as metal plating. { 'plād·iŋ }

plating rack [MET] A fixture that holds, and conducts current to, a piece of work during electrodeposition. { 'plād·iŋ ,rak }

platinic chloride See chloroplatinic acid. { plə'tin·ik 'klȯr,īd }

platinic sodium chloride See sodium chloroplatinate. { plə'tin·ik 'sōd·ē·əm 'klȯr,īd }

platinic sulfate See platinum sulfate. { plə'tin·ik 'səl,fāt }

platiniridium [MINERAL] A silver-white cubic mineral composed of platinum, iridium, and related metals, occurring in grains. { ¦plat·ən·ə'rid·ē·əm }

platinite See platynite. { 'plat·ən,īt }

platinochloride See chloroplatinate. { ¦plat·ən·ō'klȯr,īd }

platinocyanide [INORG CHEM] A double salt of platinous cyanide and another cyanide, such as $K_2Pt(CN)_4$; used in photography and fluorescent x-ray screens. Also known as cyanoplatinate. { ¦plat·ən·ō'sī·ə,nīd }

platinoid [MET] **1.** Resembling or related to platinum. **2.** A copper-nickel-zinc alloy used for electrical resistance wire. { 'plat·ən,ȯid }

platinotron [ELECTR] A microwave tube that may be used as a high-power saturated amplifier or oscillator in pulsed radar applications; requires permanent magnet just as does a magnetron. { plə'tin·ə,trän }

platinous chloride See platinum dichloride. { 'plat·ən·əs 'klȯr,īd }

platinous iodide See platinum iodide. { 'plat·ən·əs 'ī·ə,dīd }

platinum [CHEM] A chemical element, symbol Pt, atomic number 78, atomic weight 195.08. [MET] A soft, ductile, malleable, grayish white noble metal with relatively high electric resistance; used in alloys, in electrical and electronic devices, and in jewelry. { 'plat·ən·əm }

platinum bichloride See platinum dichloride. { 'plat·ən·əm bī'klȯr,īd }

platinum black [MET] Black-colored, finely divided metallic platinum; soluble in aqua regia; used as a catalyst, as an absorbent for gases (hydrogen, oxygen), and for gas ignition. Also known as platinum Mohr. { 'plat·ən·əm 'blak }

platinum chloride [INORG CHEM] $PtCl_4$ or $PtCl_4·5H_2O$ A brown solid or red crystals; soluble in alcohol and water; decomposes when heated (loses $4H_2O$ at 100°C); used as an analytical reagent. { 'plat·ən·əm 'klȯr,īd }

platinum dichloride [INORG CHEM] $PtCl_2$ Water-insoluble, green-gray powder; decomposes to platinum at red heat; used to make platinum salts; platinum bichloride. { 'plat·ən·əm dī'klȯr,īd }

platinum diiodide See platinum iodide. { 'plat·ən·əm dī'ī·ə,dīd }

platinum electrode [PHYS CHEM] A solid platinum wire electrode used during voltammetric analyses of electrolytes. { 'plat·ən·əm i'lek,trōd }

platinum iodide [INORG CHEM] PtI_2 Water- and alkali-insoluble black powder; slightly soluble in hydrochloric acid; decomposes at 300–350°C. Also known as platinous iodide; platinum diiodide. { 'plat·ən·əm 'ī·ə,dīd }

platinum-iridium alloy [MET] An alloy with 1–30% iridium; as concentration of iridium increases, so do hardness, chemical resistance, and melting point; used in jewelry, electrical contacts, and hypodermic needles. { 'plat·ən·əm ə'rid·ē·əm 'al,ȯi }

platinum metal [CHEM] A group of transition metals that includes ruthenium, osmium, rhodium, iridium, palladium, and platinum. { 'plat·ən·əm 'med·əl }

platinum Mohr See platinum black. { 'plat·ən·əm 'mȯr }

platinum oxide [INORG CHEM] An oxide of platinum; examples are platinum monoxide (or platinous oxide), PtO, and

platinum dioxide (or platinic oxide), PtO_2. { 'plat·ən·əm 'äk,sīd }

platinum potassium chloride See potassium chloroplatinate. { 'plat·ən·əm pə'tas·ē·əm 'klȯr,īd }

platinum resistance thermometer [ENG] The basis of the International Practical Temperature Scale of 1968 from 259.35° to 630.74°C; used in industrial thermometers in the range 0 to 650°C; capable of high accuracy because platinum is noncorrosive, ductile, and nonvolatile, and can be obtained in a very pure state. Also known as Callendar's thermometer. { 'plat·ən·əm ri¦zis·təns thər'mäm·əd·ər }

platinum-rhodium alloy [MET] An alloy with up to 40% rhodium; as concentration of rhodium increases, so do chemical resistance and hardness (although less hard than for platinum-iridium alloys); used as a catalyst to make nitric acid, in thermocouples, and in rayon spinnerets. { 'plat·ən·əm 'rō·dē·əm 'al,ȯi }

platinum sodium chloride See sodium chloroplatinate. { 'plat·ən·əm 'sōd·ē·əm 'klȯr,īd }

platinum sponge [MET] Porous, grayish-black mass of finely divided platinum; soluble in aqua regia; used as a catalyst, and for ignition of combustible gases. { 'plat·ən·əm 'spənj }

platinum sulfate [INORG CHEM] $Pt(SO_4)_2$ A hygroscopic, dark mass; soluble in alcohol, ether, water, and dilute acids; used in microanalysis for halogens. Also known as platinic sulfate. { 'plat·ən·əm 'səl,fāt }

platonic solid See regular polyhedron. { plə,tän·ik 'säl·əd }

platonic year See great year. { plə'tän·ik 'yir }

platte [GEOL] A resistant knob of rock in a glacial valley or rising in the midst of an existing glacier, often causing a glacier to split near its snout. { 'plad·ə }

platter [COMPUT SCI] One of the disks in a hard-disk drive or disk pack. { 'plad·ər }

plattnerite [MINERAL] PbO_2 An iron-black mineral consisting of lead dioxide, occurring in masses with submetallic luster. { 'plat·nə,rīt }

Plattner's process [MET] A process for extracting gold in which a charge of gold-bearing pulp is placed in a revolving iron drum lined with lead, and a stream of chlorine gas is conducted through the pulp, producing chloride of gold, which is soluble in water. { 'plat·nərz ,prä·səs }

platy [GEOL] **1.** Referring to a sedimentary particle whose length is more than three times its thickness. **2.** Referring to a sandstone or limestone that splits into laminae having thicknesses in the range of 2 to 10 millimeters. { 'plad·ē }

Platyasterida [INV ZOO] An order of Asteroidea in which traces of metapinnules persist, the ossicles of the arm skeleton being arranged in two growth gradient systems. { ¦plad·ē·a'ster·ə·də }

Platybelondoninae [PALEON] A subfamily of extinct elephantoid mammals in the family Gomphotheriidae consisting of species with digging specializations of the lower tusks. { ,plad·ē,bel·ən'dän·ə,nē }

platycelous [VERT ZOO] Of a vertebra, having a flat or concave ventral surface and a convex dorsal surface. { ¦plad·ē¦sē·ləs }

Platycephalidae [VERT ZOO] The flatheads, a family of perciform fishes in the suborder Cottoidei. { ,plad·ē·sə'fal·ə,dē }

Platyceratacea [PALEON] A specialized superfamily of extinct gastropod mollusks which adapted to a coprophagous life on crinoid calices. { ,plad·ē,ser·ə'tās·ē·ə }

platycnemic index [ANTHRO] The ratio, multiplied by 100, of the anteroposterior diameter to the lateral diameter of the shinbone. { ¦plad·ik¦nē·mik 'in,deks }

Platycopa [INV ZOO] A suborder of ostracod crustaceans in the order Podocopida including marine forms with two pairs of thoracic legs. { plə'tik·ə·pə }

Platycopina [INV ZOO] The equivalent name for Platycopa. { ,plad·ə'käp·ə·nə }

Platyctenea [INV ZOO] An order of the ctenophores whose members are sedentary or parasitic; adults often lack ribs and are flattened due to shortening of the main axis. { ,plad·ik'tē·nē·ə }

Platyctenida [ZOO] An order of the phylum Ctenophora comprising four families (Ctenoplanidae, Coeloplanidae, Tjalfiellidae, Savangiidae) and six genera. { ,plad·ē'ten·əd·ə }

platy flow structure [PETR] Structure of an igneous rock characterized by tabular sheets which suggest stratification, and formation by contraction during cooling. Also known

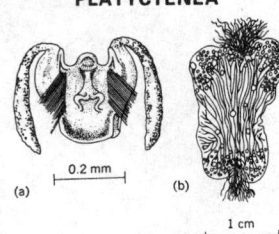

PLATYCTENEA

0.2 mm

(a)   (b)

1 cm

Coeloplana bocki. (a) Larva. (b) Adult.

as linear flow structure; planar flow structure. { 'plad·ē 'flō ,strək·chər }

Platygasteridae [INV ZOO] A family of hymenopteran insects in the superfamily Proctotrupoidea. { ,plad·ē·ga'ster·ə,dē }

Platyhelminthes [INV ZOO] A phylum of invertebrates composed of bilaterally symmetrical, nonsegmented, dorsoventrally flattened worms characterized by lack of coelom, anus, circulatory and respiratory systems, and skeleton. { ¦plad·ē·hel'min,thēz }

platykurtic distribution [STAT] A distribution of a data set which is relatively flat. { ,plad·ə¦kərd·ik ,dis·trə'byü·shən }

platymeric [ANTHRO] Of a thighbone, being flattened laterally, with a platymeric index of 75 to 85. { ¦plad·ē'mer·ik }

platymeric index [ANTHRO] The index, multiplied by 100, of the anteroposterior diameter to the lateral diameter of the femur. { ¦plad·ē'mer·ik 'in,deks }

platymyarian [INV ZOO] In nematodes, pertaining to flat muscle cells with the fibrillar region limited to a basal zone. { ¦plad·ē,mī'a·rē·ən }

platynite [MINERAL] $PbBi_2(Se,S)_3$ An iron-black mineral composed of selenide and sulfide of lead and bismuth; occurs in thin metallic plates resembling graphite. Also spelled platinite. { 'plad·ə,nīt }

platypellic [ANTHRO] Referring to or descriptive of a broad pelvis having a pelvic index of less than 90. { ,plad·ē¦pel·ik }

Platypodidae [INV ZOO] The ambrosia beetles, a family of coleopteran insects in the superfamily Curculionoidea. { ,plad·ə'päd·ə,dē }

Platypsyllidae [INV ZOO] The equivalent name for Leptinidae. { ,plad·ē'sil·ə,dē }

platypus [VERT ZOO] *Ornithorhynchus anatinus.* A monotreme, making up the family Ornithorhynchidae, which lays and incubates eggs in a manner similar to birds, and retains some reptilian characters; the female lacks a marsupium. Also known as duckbill platypus. { 'plad·ə,pùs }

platysma [ANAT] A subcutaneous muscle of the neck, extending from the face to the clavicle; muscle of facial expression. { plə'tiz·mə }

Platysomidae [PALEON] A family of extinct palaeonisciform fishes in the suborder Platysomoidei; typically, the body is laterally compressed and rhombic-shaped, with long dorsal and anal fins. { ,plad·ē'säm·ə,dē }

Platysomoidei [PALEON] A suborder of extinct deep-bodied marine and fresh-water fishes in the order Palaeonisciformes. { ,plad·ē·sə'mòid·ē,ī }

platyspondylia [MED] A rare congenital skeletal defect marked by abnormally shaped vertebrae. { ,plad·ē·spän'dil·yə }

Platysternidae [VERT ZOO] The big-headed turtles, a family of Asiatic fresh-water Chelonia with a single species (*Platysternon megacephalum*), characterized by a large head, hooked mandibles, and a long tail. { ,plad·ē'stər·nə,dē }

Plaut-Vincent's infection See Vincent's infection. { 'plaùt 'vin·səns in,fek·shən }

play [MECH ENG] Free or unimpeded motion of an object, such as the motion between poorly fitted or worn parts of a mechanism. { plā }

playa [GEOL] **1.** A low, essentially flat part of a basin or other undrained area in an arid region. **2.** A small, generally sandy land area at the mouth of a stream or along the shore of a bay. **3.** A flat, alluvial coastland, as distinguished from a beach. { 'plī·ə }

playa lake [HYD] A shallow temporary sheet of water covering a playa in the wet season. { 'plī·ə ,lāk }

playback [ENG ACOUS] Reproduction of a sound recording. { 'plā,bak }

playback head [ELECTR] A head that converts a changing magnetic field on a moving magnetic tape into corresponding electric signals. Also known as reproduce head. { 'plā,bak ,hed }

playback robot [CONT SYS] A robot that repeats the same sequence of motions in all its operations, and is first instructed by an operator who puts it through this sequence. { 'plā,bak 'rō,bät }

Playfair's law [GEOL] The law that each stream cuts its own valley, the valley being proportional in size to its stream, and the stream junctions in the valley are accordant in level. { 'plā,ferz ,lò }

play for position [IND ENG] The prepositioning of an object by a worker for a subsequent operation in the performance of a task. { 'plā fòr pə'zish·ən }

play of color [OPTICS] An optical phenomenon consisting of a rapid succession of flashes of a variety of prismatic colors as certain minerals or cabochon-cut gems are moved about; caused by diffraction of light from spherical particles of amorphous silica stacked in an orderly three-dimensional pattern. Also known as schiller. { 'plā əv 'kəl·ər }

play therapy [PSYCH] A form of treatment, used particularly with children, in which a child's play, as with dolls in the presence of a therapist, is used as a medium for expression and communication. { 'plā ,ther·ə·pē }

pleasure principle [PSYCH] The instinctive attempt to avoid pain, discomfort, or unpleasant situations; the desire to obtain maximum gratification with minimum effort. { 'plezh·ər ,prin·sə·pəl }

pleated cartridge [DES ENG] A filter cartridge made into a convoluted form that resembles the folds of an accordion. { 'plēd·əd 'kär·trij }

Plecoptera [INV ZOO] The stoneflies, an order of primitive insects in which adults differ only slightly from immature stages, except for wings and tracheal gills. { plə'käp·tə·rə }

plectane [INV ZOO] A cuticular plate supporting papillae in some nematodes. { 'plek,tān }

Plectascales [MYCOL] An equivalent name for Eurotiales. { ,plek·tə'skā·lēz }

Plectognathi [VERT ZOO] The equivalent name for Tetraodontiformes. { plek'täg·nə,thī }

Plectoidea [INV ZOO] A superfamily of small, free-living nematodes characterized by simple spiral amphids or variants thereof, elongate cylindroconoid stoma, and reflexed ovaries. { plek'tòid·ē·ə }

Plectomycetes [MYCOL] A class of the subdivision Ascomycotina; members produce a well-developed mycelium on which both sexual (asci) and asexual (conidia) states occur. { ,plek·to·mī'sē,dēz }

plectostele [BOT] A protostele that has the xylem divided into plates. { 'plek·tə,stēl }

Pleiades [ASTRON] An open cluster of a few hundred stars in the constellation Taurus; six of the stars are easily visible to the naked eye. { 'plē·ə,dēz }

Pleidae [INV ZOO] A family of hemipteran insects in the superfamily Pleoidea. { 'plē·ə,dē }

pleiomorphism [GEN] The occurrence of variable phenotypes in a group of organisms with the same genotype. { ,plē·ə'mór,fiz·əm }

pleiotropic [GEN] Referring to a gene or mutation that has multiple effects. { ,plī·ə'träp·ik }

pleiotropy [GEN] The quality of a gene having more than one phenotypic effect. { plī'ä·trə·pē }

Pleistocene [GEOL] The older of the two epochs of the Quaternary Period, spanning about 1.8 million to 10,000 years ago. It represents the interval of geological time (and rocks accumulated during that time) extending from the end of the Pliocene Epoch (and the end of Tertiary Period) to the start of the Holocene Epoch. It is commonly characterized as an epoch when the earth entered its most recent phase of widespread glaciation. Also known as Ice Age; Oiluvium. { 'plī·stə,sēn }

Plemelj formulas [MATH] Formulas for the limits of the Cauchy integrals of an arc with respect to a point z as z approaches the arc from either side. { 'plā·mə·lē ,fòr·myə·ləz }

plenum [ENG] A condition in which air pressure within an enclosed space is greater than that in the outside atmosphere. { 'plen·əm }

plenum blower assembly [MECH ENG] In an automotive air-conditioning system, the assembly through which air passes on its way to the evaporator or heater core. { 'plē·nəm 'blō·ər ə,sem·blē }

plenum chamber [ENC] An enclosed space in which a plenum condition exists; air is forced into it for slow distribution through ducts. { 'plen·əm ,chām·bər }

plenum system [MECH ENG] A heating or air conditioning system in which air is forced through a plenum chamber for distribution to ducts. { 'plen·əm ,sis·təm }

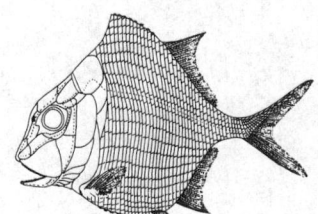

PLATYSOMIDAE

Platysomus parvulus, an Upper Carboniferous platysomid from Britain which attained a length of 5 inches (13 centimeters).

PLEISTOCENE

PRECAMBRIAN	PALEOZOIC									MESOZOIC		CENOZOIC	
					CARBON-IFEROUS								
	CAMBRIAN	ORDOVICIAN	SILURIAN	DEVONIAN	Mississippian	Pennsylvanian	PERMIAN	TRIASSIC	JURASSIC	CRETACEOUS	TERTIARY	QUATERNARY	

TERTIARY					QUATERNARY	
Paleocene	Eocene	Oligocene	Miocene	Pliocene	Pleistocene	Recent

Chart showing the relationship of the Pleistocene epoch to the periods and eras of geologic time.

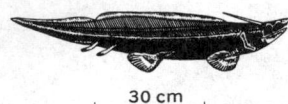

PLEURACANTHODII

Xenacanthus (Pleuracanthus), Carboniferous and Permian sharklike form; note the spine projecting from the posterior braincase.

PLEURONECTIFORMES

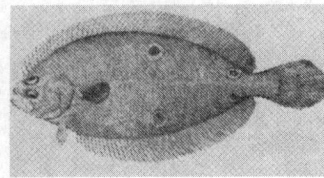

Fourspot flounder (*Paralichthys oblongus*), of the Pleuronectiformes.

pleochroic halos [OPTICS] Halos of color or color differences that are sometimes observed around inclusions in minerals, resulting from irradiation by alpha particles. { ¦plē·ə¦krō·ik 'hā·lōz }

pleochroism [OPTICS] Phenomenon exhibited by certain transparent crystals in which light viewed through the crystal has different colors when it passes through the crystal in different directions. Also known as polychroism. { plē'äk·rə‚wiz·əm }

pleocytosis [MED] Increase of cells in the cerebrospinal fluid. { ‚plē·ō‚sī'tō·səs }

pleodont [VERT ZOO] Having solid teeth. { 'plē·ə‚dänt }

Pleoidea [INV ZOO] A superfamily of suboval hemipteran insects belonging to the subdivision Hydrocoriseae. { plē'oid·ē·ə }

pleomorphism [BIOL] The occurrence of more than one distinct form of an organism in a single life cycle. [CRYSTAL] *See* polymorphism. { ‚plē·ō'mór‚fiz·əm }

pleonaste *See* ceylonite. { 'plē·ə‚nast }

pleopod [INV ZOO] An abdominal appendage in certain crustaceans that is modified for swimming. { 'plē·ə‚päd }

Pleosporales [BOT] The equivalent name for the lichenized Pseudosphaeriales. { ‚plē·ə·spə'rā·lēz }

plerocercoid [INV ZOO] The infective metacestode of certain cyclophyllidean tapeworms; distinguished by a solid body. { ‚plir·ə'sər‚kóid }

plerome [BOT] Central core of primary meristem which gives rise to all cells of the stele from the pericycle inward. { 'pli‚rōm }

plerotic water [HYD] That part of subsurface water that forms the zone of saturation, including underground streams. { plə'räd·ik 'wód·ər }

plesioaster [INV ZOO] A type of poriferan microscleric monaxonic spicule. { ‚plē·sē·ō'as·tər }

Plesiocidaroida [PALEON] An extinct order of echinoderms assigned to the Euechinoidea. { ‚plē·sē·ō‚sik·ə'róid·ə }

plesiomorph [EVOL] The original character of a branching phyletic lineage, found in the ancestral forms. { 'plē·sē·ə‚mórf }

Plesiosauria [PALEON] A group of extinct reptiles in the order Sauropterygia constituting a highly specialized offshoot of the nothosaurs. { ‚plē·sē·ə'sór·ē·ə }

plesiotype [SYST] A specimen or specimens on which subsequent descriptions are based. { 'plē·sē·ə‚tīp }

Plessy's green [INORG CHEM] $CrPO_4 \cdot xH_2O$ Deep-green pigment made of chromium phosphate mixed with chromium oxide and calcium phosphate. { ple'sēz 'grēn }

Plethodontidae [VERT ZOO] A large family of salamanders in the suborder Salamandroidea characterized by the absence of lungs and the presence of a fine groove from nostril to upper lip. { ‚pleth·ə'dänt·ə‚dē }

plethora [MED] An excess of blood in an organ or the circulatory system. { 'pleth·ə·rə }

plethysmograph [MED] An instrument for measuring changes in the size of a part of the body by measuring changes in the amount of blood in that part. { plə'thiz·mə‚graf }

pleura [ANAT] The serous membrane covering the lung and lining the thoracic cavity. { 'plùr·ə }

Pleuracanthodii [PALEON] An order of Paleozoic sharklike fishes distinguished by two-pronged teeth, a long spine projecting from the posterior braincase, and direct backward extension of the tail. { plù‚rak·ən'thō·dē‚ī }

pleural cavity [ANAT] The potential space included between the parietal and visceral layers of the pleura. { 'plùr·əl ¦kav·əd·ē }

pleural effusion [MED] Abnormal accumulation of fluid in the area between the membranes lining the lungs and the chest cavity (the pleural space). { ‚plùr·əl i'fyü·zhən }

pleural rib *See* ventral rib. { 'plùr·əl ¦rib }

pleurapophysis [ANAT] One of the lateral processes of a vertebra, corresponding morphologically to a rib. { ‚plùr·ə'päf·ə·səs }

pleurisy [MED] Inflammation of the pleura. Also known as pleuritis. { 'plùr·ə·sē }

pleuritis *See* pleurisy. { plù'rīd·əs }

pleurobranchia [INV ZOO] A gill that arises from the lateral wall of the thorax in certain arthropods. { ‚plùr·ə'braŋ·kē·ə }

pleurocarpous [BOT] Having the sporophyte in leaf axils

along the side of the stem or on lateral branches; refers specifically to mosses. { ¦plùr·ə¦kär·pəs }

Pleuroceridae [INV ZOO] A family of fresh-water snails in the order Pectinibranchia. { ‚plùr·ə'ser·ə‚dē }

Pleurocoelea [INV ZOO] An extinct superfamily of gastropod mollusks of the order Opisthobranchia in which the shell, mantle cavity, and gills were present. { ‚plùr·ə'sē·lē·ə }

Pleurodira [VERT ZOO] A suborder of turtles (Chelonia) distinguished by spines on the posterior cervical vertebrae so that the head is retractile laterally. { ‚plùr·ə'dī·rə }

pleurodontia [VERT ZOO] Attachment of the teeth to the inner surface of the jawbone. { ‚plùr·ə'dän·chə }

pleurodynia [MED] Severe paroxysmal pain and tenderness of the intercostal muscles. { ‚plùr·ə'dī·nē·ə }

pleurolophocercous cercaria [INV ZOO] A larval digenetic trematode distinguished by a long, powerful tail with a pair of fin folds, a protrusible oral sucker, and pigmented dorsal eyespots. { ‚plùr·ə¦läf·ə¦sər·kəs sər'kar·ē·ə }

Pleuromeiaceae [PALEOBOT] A family of plants in the order Pleuromiales, but often included in the Isoetales due to a phylogenetic link. { ‚plùr·ō·mē'ās·ē‚ē }

Pleuromeiales [PALEOBOT] An order of Early Triassic lycopods consisting of the genus *Pleuromeia*; the upright branched stem had grasslike leaves and a single terminal strobilus. { ‚plùr·ō·mē'ā·lēz }

pleuron [INV ZOO] The lateral portion of a single thoracic segment in arthropods. { 'plùr‚än }

Pleuronectiformes [VERT ZOO] The flatfishes, an order of actinopterygian fishes distinguished by the loss of bilateral symmetry. { ‚plùr·ō‚nek·tə'fór·mēz }

Pleuronematina [INV ZOO] A suborder of the Hymenostomatida. { ‚plùr·ō‚nem·ə'tī·nə }

pleuroperitoneal cavity [VERT ZOO] The body cavity containing both the lungs and the abdominal viscera in all pulmonate vertebrates except mammals. { ‚plùr·ō¦per·ə·tə¦nē·əl 'kav·əd·ē }

pleuropneumonia [MED] Combined pleurisy and pneumonia. [VET MED] An infectious disease of cattle producing pleural and lung inflammation, caused by *Mycoplasma* species. { ‚plùr·ō·nù'mō·nyə }

pleuropneumonialike organism [MICROBIO] A former classification for a poorly defined group of microorganisms classified in the order Mycoplasmatales, including the smallest organisms capable of independent life, and comparable in size to the large filterable viruses. Abbreviated PPLO. { ‚plùr·ō·nù¦mō·nyə‚lik 'ór·gə‚niz·əm }

Pleurostigmophora [INV ZOO] A subclass of the centipedes, in some taxonomic systems, distinguished by lateral spiracles. { ‚plùr·ə·stig'mäf·ə·rə }

Pleurotomariacea [PALEON] An extinct superfamily of gastropod mollusks in the order Aspidobranchia. { ‚plùr·əd·ə‚mar·ē'ās·ē·ə }

plexus [ANAT] A network of interlacing nerves or anastomosing vessels. [GEOL] An area on a subglacial deposit that encloses a giant's kettle. { 'plek·səs }

pli [MECH] A unit of line density (mass per unit length) equal to 1 pound per inch, or approximately 17.8580 kilograms per meter. { plē }

plica [BIOL] A fold, as of skin or a leaf. { 'plī·kə }

plication [GEOL] Intense, small-scale folding. { plī'kā·shən }

plied yarn [TEXT] A yarn composed of two or more strands. { 'plīd ¦yärn }

Pliensbachian [GEOL] A European stage of geologic time: Lower Jurassic (above Sinemurian, below Toarcian). { plēnz'bäk·ē·ən }

pliers [DES ENG] A small instrument with two handles and two grasping jaws, usually long and roughened, working on a pivot; used for holding small objects and cutting, bending, and shaping wire. { 'plī·ərz }

PLIF *See* planar laser-induced fluorescence. { ¦pē¦el¦ī'ef or plif }

Plimsoll mark [NAV ARCH] **1.** A circular disk with a horizontal line placed on the side of a ship to indicate the summer load line. **2.** A series of marks on the side of a ship indicating the load line for various seasons and geographical areas. { 'plim·səl ‚märk }

Plinian eruption *See* Vulcanian eruption. { 'plin·ē·ən i'rəp·shən }

plinth [ARCH] **1.** The block forming the lowest member or base of a column or pedestal. **2.** A slight widening at the base of a column or wall. [GEOL] The lower and outer part of a seif dune, beyond the slip-face boundaries, that has never been subjected to sand avalanches. { plinth }

plinth block *See* skirting block. { 'plinth ,bläk }

plinthite [GEOL] In a soil, a material consisting of a mixture of clay and quartz with other diluents, that is rich in sesquioxides, poor in humus, and highly weathered. { 'plin,thīt }

Pliocene [GEOL] The youngest of the five geological epochs of the Tertiary Period. The Pliocene represents the interval of geological time (and rocks deposited during that time) extending from the end of the Miocene Epoch to the beginning of the Pleistocene Epoch of the Quaternary Period. Modern time scales assign the duration of 5.0 million to 1.8 million years ago to the Pliocene. { 'plī·ə,sēn }

Pliohyracinae [PALEON] An extinct subfamily of ungulate mammals in the family Procaviidae. { ,plī·ō·hī'ras·ə,nē }

pliothermic [GEOL] Pertaining to a period in geologic history characterized by more than average climatic warmth. { ¦plī·ō¦thər·mik }

pliotron [ELECTR] Any hot-cathode vacuum tube having one or more grids. { 'plī·ə,trän }

plissé [TEXT] Thin cotton fabric, soft or crisp, with puckered stripes or patterns forming an allover blister effect; produced either by weaving with yarns having different degrees of shrinkage in finishing or by chemical treatment. { plē'sā }

PLL *See* phase-locked loop.

ploidy [GEN] Number of complete chromosome sets in a nucleus: haploid (N), diploid (2N), triploid (3N), tetraploid (4N), and so on. { 'plòid·ē }

Plokiophilidae [INV ZOO] A small family of predacious hemipteran insects in the superfamily Cimicoidea; individuals live in the webs of spiders and embiids. { ,pläk·ē·ō'fil·ə,dē }

PL/1 [COMPUT SCI] A multipurpose programming language, developed by IBM for the Model 360 systems, which can be used for both commercial and scientific applications. { ¦pē¦el'wən }

plosive [LING] A primary type of speech sound of the major languages that is characterized by the complete interception of airflow at one or more places along the vocal tract. For example, the English words par, bar, tar, and car begin with plosives. { 'plō·siv }

plot [CIV ENG] A measured piece of land. { plät }

Plotosidae [VERT ZOO] A family of Indo-Pacific salt-water catfishes (Siluriformes). { plə'täs·ə,dē }

plotter [ENG] A visual display or board on which a dependent variable is graphed by an automatically controlled pen or pencil as a function of one or more variables. { 'pläd·ər }

plotting board [ENG] The surface portion of a plotter, on which graphs are recorded. Also known as plotting table. { 'pläd·iŋ ,bórd }

plotting chart [NAV] A chart designed primarily for plotting lines of position obtained from dead reckoning, celestial observations, or radio aids, and other sources of navigational data. { 'pläd·iŋ ,chärt }

plotting head *See* reflection plotter. { 'pläd·iŋ ,hed }

plotting interval [NAV] The elapsed time between the first and last readings of radar range and bearing of an object being tracked. { 'pläd·iŋ ,in·tər·vəl }

plotting paper [GRAPHICS] Drawing paper laminated to both sides of an aluminum sheet or foil to ensure better dimensional stability; it is often used in drawing the color separations used in mapmaking. { 'pläd·iŋ ,pā·pər }

plotting sheet [NAV] A blank chart, usually on the Mercator projection, showing only the graticule and a compass rose, so that the plotting sheet can be used for any longitude. [ORD] An item having lines forming rectangular grids printed on one side; the grid lines are approximately parallel with and extending to the edges of the sheet, and are approximately centered thereon; used in graphical determination of artillery ranges and deflection angles to targets from plotted gun position. { 'pläd·iŋ ,shēt }

plotting table *See* plotting board. { 'pläd·iŋ ,tā·bəl }

plough [ENG] A groove cut lengthwise with the grain in a piece of wood. [MIN ENG] **1.** A continuous mining machine in which cutting blades, moved over the face being worked, bite into the coal as they are pulled along and discharge it on an accompanying conveyor. **2.** A V-shaped scraper that presses against the return belt of a conveyor, removing coal and debris from it. { plaù }

plough deflector [MIN ENG] A steel plate at the end of a cutter-loader to deflect cut coal onto the face conveyor. { 'plaù di,flek·tər }

ploughed-and-tongued joint *See* feather joint. { ¦plaùd ən 'təŋd ,jòint }

plough wind *See* plow wind. { 'plaù ,wind }

plow [AGR] An implement consisting of a share, moldboard, and landside attached to a frame; used to cut, lift, turn, and pulverize soil in preparation for a seedbed. { plaù }

plowshare [DES ENG] The pointed part of a moldboard plow, which penetrates and cuts the soil first. [HYD] A wedge-shaped feature developed on a snow surface by further ablation of foam crust. { 'plaù,sher }

plow sole [GEOL] A pressure pan representing a layer of soil compacted by repeated plowing to the same depth. { 'plaù ,sōl }

plow steel [MET] High-quality, high-strength steel with 0.5 to 0.95% carbon content, used for wire rope. { 'plaù ,stēl }

plow wind [METEOROL] A term used in the midwestern United States to describe strong, straight-line winds associated with squall lines and thunderstorms; resulting damage is usually confined to narrow zones like that caused by tornadoes; however, the winds are all in one direction. Also spelled plough wind. Also known as derecho. { 'plaù ,wind }

plucking [GEOL] A process of glacial erosion which involves the penetration of ice or rock wedges into subglacial niches, crevices, and joints in the bedrock; as the glacier moves, it plucks off pieces of jointed rock and incorporates them. Also known as glacial plucking; quarrying. { 'plək·iŋ }

plug [ELEC] The half of a connector that is normally movable and is generally attached to a cable or removable subassembly; inserted in a jack, outlet, receptacle, or socket. [GEOL] **1.** A vertical pipelike magmatic body representing the conduit to a former volcanic vent. **2.** A crater filling of lava, the surrounding material of which has been removed by erosion. **3.** A mass of clay, sand, or other sediment filling the part of a stream channel abandoned by the formation of a cutoff. [MET] **1.** A rod or mandrel over which a pierced tube is forced, or that fills a tube as it is drawn through a die. **2.** A punch or mandrel over which a cup is drawn. **3.** A protruding portion of a die impression for forming a corresponding recess in the forging. **4.** A false bottom in a die. Also known as peg. [MIN ENG] A watertight seal in a shaft formed by removing the lining and inserting a concrete dam, or by placing a plug of clay over ordinary debris used to fill the shaft up to the location of the plug. [SCI TECH] **1.** A piece of material used to fill a hole. **2.** A small segment of material removed from a larger object. { pləg }

plug adapter lamp holder [ELEC] A device that can be inserted in a lamp holder to act as a lamp holder and one or more receptacles. Also known as current tap. { 'pləg ə,dap·tər 'lamp ,hōld·ər }

plug-and-feather hole [ENG] A hole drilled in quarries for the purpose of splitting a block of stone by the plug-and-feather method. { ¦pləg ən 'feth·ər ,hōl }

plug-and-feather method [MIN ENG] A method of breaking large quarry stones into smaller blocks; a row of holes is drilled in the stone along a line where the break is desired; a pair of feathers (semicircular cross-section rods) is inserted in each hole; a plug (steel wedge) is inserted between each feather pair; the plugs are hammered in succession until the stone fractures. { ¦pləg ən 'feth·ər ,meth·əd }

plug back [PETRO ENG] To place cement or a mechanical plug in a well bottom for the purpose of excluding bottom water, sidetracking, or producing from a formation already drilled through. { 'pləg ,bak }

plug bit *See* noncoring bit. { 'pləg ,bit }

plugboard *See* control panel. { 'pləg,bórd }

plugboard chart *See* plugging chart. { 'pləg,bórd ,chärt }

plug cock *See* plug valve. { 'pləg ,käk }

plug-compatible hardware [COMPUT SCI] A piece of equipment which can be immediately connected to a computer manufactured by another company. { 'pləg kəm,pad·ə·bəl 'här·dwer }

plug cutter [DES ENG] A device for boring out short dowels or plugs from wood that exactly match standard drill sizes. { 'pləg ,kəd·ər }

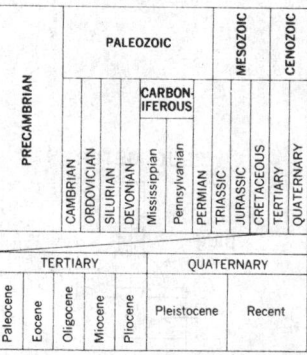

PLIOCENE

PRECAMBRIAN	PALEOZOIC									MESOZOIC		CENOZOIC
	CAMBRIAN	ORDOVICIAN	SILURIAN	DEVONIAN	Mississippian	Pennsylvanian	PERMIAN	TRIASSIC	JURASSIC	CRETACEOUS	TERTIARY	QUATERNARY
					CARBON-IFEROUS							

TERTIARY					QUATERNARY	
Paleocene	Eocene	Oligocene	Miocene	Pliocene	Pleistocene	Recent

Chart showing the relationship of the Pliocene to the eras and periods of geologic time.

plug die See floating plug. { 'pləg ‚dī }

plug dome [GEOL] A volcanic dome characterized by an upheaved, consolidated conduit filling. { 'pləg ‚dōm }

plug drawing [MET] Drawing tubing over a plug or mandrel and through a die simultaneously to reduce diameter and thickness and to produce a smooth symmetrical bore surface. { 'pləg ‚dro·iŋ }

plug flow See piston flow. { 'pləg ‚flō }

plug forming [ENG] Thermoforming process for plastics molding in which a plug or male mold is used to partially preform the part before forming is completed, using vacuum or pressure. { 'pləg ‚fȯrm·iŋ }

plug fuse [ELEC] A fuse designed for use in a standard screw-base lamp socket. { 'pləg ‚fyüz }

plug gage [DES ENG] A steel gage that is used to test the dimension of a hole; may be straight or tapered, plain or threaded, and of any cross-sectional shape. { 'pləg ‚gāj }

plugging [ELEC] Braking an electric motor by reversing its connections, so it tends to turn in the opposite direction; the circuit is opened automatically when the motor stops, so the motor does not actually reverse. [ENG] The formation of a barrier (plug) of solid material in a process flow system, such as a pipe or reactor. [MIN ENG] See blinding. [PETRO ENG] The act or process of stopping the flow of water, oil, or gas in strata penetrated by a borehole or well so that fluid from one stratum will not escape into another or to the surface; especially the sealing up of a well that is dry and is to be abandoned. { 'pləg·iŋ }

plugging agent [PETRO ENG] A chemical used to plug or block off selected permeable zones of a reservoir formation; used during formation acidizing to direct the acid to the tighter (less permeable) zones; examples are viscous gels, suspensions of graded solids, and finely ground vegetable material. { 'pləg·iŋ ‚ā·jənt }

plugging chart [COMPUT SCI] A printed chart of the sockets in a plugboard on which may be shown the jacks or wires connecting these sockets. Also known as plugboard chart. { 'pləg·iŋ ‚chärt }

plughole [MIN ENG] **1.** A passageway left open while an old portion of a mine is sealed off, to help maintain normal ventilation; it is sealed when the work is finished. **2.** A hole for an explosive charge or for a bolt. { 'pləg‚hōl }

plug-in [COMPUT SCI] A small software application that extends the capabilities (such as multimedia, audio, or video) of a browser. { 'pləg‚in }

plug-in unit [ELEC] A component or subassembly having plug-in terminals so all connections can be made simultaneously by pushing the unit into a suitable socket. { 'pləg·in ‚yü·nət }

plug meter [ENG] A variable-area flowmeter in which a tapered plug, located in an orifice and raised until the resulting opening is sufficient to handle the fluid flow, is used to measure the flow rate. { 'pləg ‚mēd·ər }

plug nozzle [AERO ENG] A nozzle that is obtained by truncating a full-length spike nozzle, eliminating a significant portion of the spike nozzle without undue loss in propulsive thrust. { 'pləg ‚näz·əl }

plug program patching [COMPUT SCI] A relatively small auxiliary plugboard patched with a specific variation of a portion of a program and designed to be plugged into a relatively larger plugboard patched with the main program. { 'pləg 'prō ‚gram ‚pach·iŋ }

plug reef [GEOL] A small, triangular reef that grows with its apex pointing seaward through openings between linear shelf-edge reefs. { 'pləg ‚rēf }

plug-to-plug compatibility [COMPUT SCI] Property of a peripheral device that can be made to operate with a computer merely by attachment of a plug or a relatively small number of cables. { 'pləg tə 'pləg kəm‚pad·ə'bil·əd·ē }

plug valve [MECH ENG] A valve fitted with a plug that has a hole through which fluid flows and that is rotatable through 90° for operation in the open or closed position. Also known as plug cock. { 'pləg ‚valv }

plug weld [MET] A circular fusion weld made in the hole of a slotted lap or tee joint. { 'pləg ‚weld }

plum [BOT] Any of various shrubs or small trees of the genus *Prunus* that bear smooth-skinned, globular to oval, drupaceous stone fruit. [GEOL] A clast embedded in a matrix of a different kind, especially a pebble in a conglomerate. { pləm }

plumaceous [VERT ZOO] Referring to the portion of a feather vane near the base that lacks hooklets and is loosely bound. { plü'mā·shəs }

plumage [VERT ZOO] The entire covering of feathers of a bird. { 'plü·mij }

Plumatellina [INV ZOO] The single order of the ectoproct bryozoan class Phylactolaemata. { ‚plü·mə·tə'lī·nə }

plumb [ENG] Pertaining to an object or structure in true vertical position as determined by a plumb bob. { pləm }

Plumbaginaceae [BOT] The leadworts, the single family of the order Plumbaginales. { ‚pləm·bə·jə'nās·ē‚ē }

Plumbaginales [BOT] An order of dicotyledonous plants in the subclass Caryophyllidae; flowers are pentamerous with fused petals, trinucleate pollen, and a compound ovary containing a single basal ovule. { ‚pləm·bə·jə'nā·lēz }

plumbago See graphite. { ‚pləm'bā·gō }

plumb bob [ENG] A weight suspended on a string to indicate the direction of the vertical. { 'pləm ‚bäb }

plumb bond [CIV ENG] A masonry bond in which corresponding joints (for example, on alternate courses) are aligned. { 'pləm ‚bänd }

plumbing [CIV ENG] The system of pipes and fixtures concerned with the introduction, distribution, and disposal of water in a building. [ELECTROMAG] Slang term for the pipelike waveguide circuit elements used in microwave radio and radar equipment. { 'pləm·iŋ }

plumbism [MED] Lead poisoning. { 'pləm‚biz·əm }

plumb line [ENG] The string on which a plumb bob hangs. [GEOPHYS] A continuous curve to which the direction of gravity is everywhere tangential. { 'pləm ‚līn }

plum blotch [PL PATH] A fungus disease of plums caused by *Phyllosticta congesta* and characterized by minute brown or gray angular leaf spots and brown or gray blotches on the fruit. { 'pləm ‚bläch }

plumboferrite [MINERAL] $PbFe_4O_7$ A dark hexagonal mineral composed of lead iron oxide. { ‚pləm·bō'fe‚rīt }

plumbogummite [MINERAL] **1.** $PbAl_3(PO_4)_2(OH)_5 \cdot H_2O$ A mineral composed of hydrous basic lead aluminum phosphate. **2.** A group of isostructural minerals, that includes gorceixite, goyazite, crandallite, deltaite, florencite, and dussertite, as well as plumbogummite. { ‚pləm·bō'gə‚mīt }

plumbojarosite [MINERAL] $PbFe_6(SO_4)_4(OH)_{12}$ A mineral composed of basic lead iron sulfate; it is isostructural with jarosite. { ‚pləm·bō·jə'rō‚sīt }

plumbous oxide See lead monoxide. { 'pləm·bəs 'äk‚sīd }

plumbous sulfide See lead sulfide. { 'pləm·bəs 'səl‚fīd }

plumbum [CHEM] Latin name for lead; source of the element symbol Pb. { 'pləm·bəm }

plume See column. { plüm }

plume structure [GEOL] On the surface of a master joint, a ridgelike tracing in a plumelike pattern, usually oriented parallel to the upper and lower surfaces of the constituent rock unit. Also known as plumose structure. { 'plüm ‚strək·chər }

plumicome [BIOL] A spicule with plumelike tufts. { 'plü·mə‚kōm }

Plummer-Vinson syndrome [MED] Dysphagia koilonychia, gastric achlorhydria, glossitis, and hypochromic microcytic anemia caused by iron deficiency. Also known as Paterson-Kelly syndrome; sideropenic dysphagia. { 'pləm·ər 'vin·sən ‚sin‚drōm }

plummet [ENG] A loose-fitting metal plug in a tapered rotameter tube which moves upward (or downward) with an increase (or decrease) in fluid flow rate upward through the tube. Also known as float. { 'pləm·ət }

plumose [VERT ZOO] Having feathers or plumes. { 'plü ‚mōs }

plumose structure See plume structure. { 'plü‚mōs ‚strək·chər }

plum pocket [PL PATH] A mild fungus disease of plums, caused either by *Taphrina pruni* or *T. communia*, in which the stone of the fruit is aborted. { 'pləm ‚päk·ət }

plumule [BOT] The primary bud of a plant embryo. [VERT ZOO] A down feather. { 'plü·myül }

plunge [ENG] **1.** To set the horizontal cross hair of a theodolite in the direction of a grade when establishing a grade between two points of known level. **2.** See transit. [GEOL] The inclination of a geologic structure, especially a fold axis, measured by its departure from the horizontal. Also known as pitch; rake. { plənj }

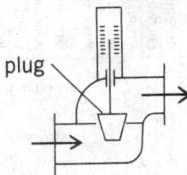

PLUG METER

Plug-type variable-area flowmeter.

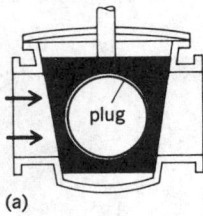

PLUG VALVE

(a)

(b)

Plug valve. (a) Closed. (b) Open. Arrows show direction of fluid flow.

PLUM

Fruit on branch of Damson plum.

plunge basin [GEOL] A deep, large hollow or cavity scoured in the bed of a stream at the foot of a waterfall or cataract by the force and eddying effect of the falling water. { 'plənj ˌbās·ən }

plunge grinding [MECH ENG] Grinding in which the wheel moves radially toward the work. { 'plənj ˌgrīnd·iŋ }

plunge point [OCEANOGR] The point at which a plunging wave curls over and falls as it moves toward the shore. { 'plənj ˌpȯint }

plunge pool [HYD] **1.** The water in a plunge basin. **2.** A deep, circular lake occupying a plunge basin after the waterfall has ceased to exist or the stream has been diverted. Also known as waterfall lake. **3.** A small, deep plunge basin. { 'plənj ˌpül }

plunger [DES ENG] A wooden shaft with a large rubber suction cup at the end, used to clear plumbing traps and waste outlets. [ELECTROMAG] *See* piston. [ENG] *See* force plug. [MECH ENG] The long rod or piston of a reciprocating pump. { 'plən·jər }

plunger jig washer [MIN ENG] A machine for washing ore, coal, or stones in which water is forced alternately up or down by a plunger. { 'plən·jər ˌjig ˌwȧsh·ər }

plunger lift [PETRO ENG] A method of lifting oil by using compressed gas to drive a free piston from the lower end of the tubing string to the surface. { 'plən·jər ˌlift }

plunger overtravel [PETRO ENG] Excessive upward or downward movement in a reciprocating-plunger-type sucker-rod oil-well pump. { 'plən·jər 'ō·vər,trav·əl }

plunger pump [MECH ENG] A reciprocating pump where the packing is on the stationary casing instead of the moving piston. { 'plən·jər ˌpəmp }

plunger-type instrument [ENG] Moving-iron instrument in which the pointer is attached to a long and specially shaped piece of iron that is drawn into or moved out of a coil carrying the current to be measured. { 'plən·jər ˌtīp 'in·strə·mənt }

plunging breaker [OCEANOGR] A breaking wave whose crest curls over and collapses suddenly. Also known as spilling breaker; surging breaker. { 'plənj·iŋ 'brāk·ər }

plunging cliff [GEOL] A sea cliff bordering directly on deep water, having a base that lies well below water level. { 'plənj·iŋ 'klif }

plunging fire [ORD] Gunfire that strikes the earth's surface at a high angle of fall, that is, greater than 45°. { 'plənj·iŋ 'fīr }

plunging fold [GEOL] A fold having a relatively steep plunge. Also known as pitching fold. { 'plənj·iŋ 'fōld }

plural scattering [PHYS] A change in direction of a particle or photon because of a small number of collisions. { ˈplu̇r·əl 'skad·ə·riŋ }

plurilocular sporangium [BOT] A multicelled, compartmentalized sporangium, such as is found in some brown algae. { ˌplu̇r·ə'läk·yə·lər spə'ran·jē·əm }

pluripotent cell [EMBRYO] A cell capable of differentiating into most cell types found in an organism but not capable of forming a functional organism. { ˈplu̇r·əˌpōt·ent ˌsel }

plus [MATH] A mathematical symbol; *A* plus *B*, where *A* and *B* are mathematical quantities, denotes the quantity obtained by taking their sum in an appropriate context. { pləs }

plush [TEXT] Warp pile fabric with silk or wool pile longer than that of velvet. { pləsh }

plush copper ore *See* chalcotrichite. { 'pləsh 'käp·ər ˌȯr }

plus-90 orientation [COMPUT SCI] In optical character recognition, that determinate position which indicates that the line elements of an inputted source document appear perpendicular with the leading edge of the optical reader. { ˈpləs 'nīn·tē ˌȯr·ē·ənˌtā·shən }

plus sieve [MET] That portion of a powder sample that is retained by a standard sieve of a specified number. { 'pləs ˌsiv }

plus sign *See* addition sign. { 'pləs ˌsīn }

plus strand [MOL BIO] A parental ribonucleic acid (RNA) strand in RNA bacteriophages that is used as a template for a complementary RNA strand (minus strand) produced with the formation of a double-stranded (double-helix) RNA. { 'pləs ˌstrand }

plus zone [COMPUT SCI] The bit positions in a computer code which represent the algebraic plus sign. { 'pləs ˌzōn }

pluteus [INV ZOO] The free-swimming, bilaterally symmetrical, easel-shaped larva of ophiuroids and echinoids. { 'plüd·ē·əs }

plutino [ASTRON] A member of the Kuiper Belt that, like Pluto, is protected from close encounters with Neptune because its period is 3/2 Neptune's period. { plü'tē·nō }

Pluto [ASTRON] The most distant planet in the solar system; mean distance to the sun is about 3.7×10^9 miles (5.9×10^9 kilometers); it has no known satellite, and its sidereal revolution period is 248 years. { 'plüd·ō }

plutology [GEOL] The study of the interior of the earth. { plü'täl·ə·jē }

pluton [GEOL] **1.** An igneous intrusion. **2.** A body of rock formed by metasomatic replacement. { 'plü,tän }

plutonian *See* plutonic. { plü'tō·nē·ən }

plutonic [GEOL] Pertaining to rocks formed at a great depth. Also known as abyssal; deep-seated; plutonian. { plü'tän·ik }

plutonic breccia [GEOL] Breccia consisting of older annular rock fragments enclosed in younger plutonic rock. { plü'tän·ik 'brech·ə }

plutonic metamorphism [GEOL] Deep-seated regional metamorphism at high temperatures and pressures, often accompanied by strong deformation. { plü'tän·ik ˌmed·ə'mȯr,fiz·əm }

plutonic rock [GEOL] A rock formed at considerable depth by crystallization of magma or by chemical alteration. { plü'tän·ik 'räk }

plutonic theory *See* volcanic theory. { plü'tän·ik 'thē·ə·rē }

plutonic water [HYD] Juvenile water in magma, or derived from magma, at a considerable depth, probably several kilometers. { plü'tän·ik 'wȯd·ər }

plutonism [GEOL] **1.** Pertaining to the processes associated with pluton formation. **2.** The theory that the earth formed by solidification of a molten mass. [MED] A disease caused by exposure to plutonium, manifested in experimental animals by graying of the hair, liver degeneration, and tumor formation. { 'plüt·ən,iz·əm }

plutonium [CHEM] A reactive metallic element, symbol Pu, atomic number 94, in the transuranium series of elements; the first isotope to be identified was plutonium-239; used as a nuclear fuel, to produce radioactive isotopes for research, and as the fissile agent in nuclear weapons. { plü'tō·nē·əm }

plutonium-238 [NUC PHYS] The first synthetic isotope made of plutonium; similar chemically to uranium and neptunium; atomic number 94; formed by bombardment of uranium with deuterons. { plü'tō·nē·əm ˌtü,thər·dē'āt }

plutonium-239 [NUC PHYS] A synthetic isotope chemically similar to uranium and neptunium; atomic number 94; made by bombardment of uranium-238 with slow neutrons in a nuclear reactor; used as nuclear reactor fuel and an ingredient for nuclear weapons. { plü'tō·nē·əm ˌtü,thər·dē'nīn }

plutonium bomb [ORD] An atomic bomb using plutonium. { plü'tō·nē·əm 'bäm }

plutonium oxide [INORG CHEM] PuO_2 A radioactively poisonous pyrophoric oxide of plutonium; particles may be easily airborne. { plü'tō·nē·əm 'äk,sīd }

plutonium reactor [NUCLEO] A nuclear reactor in which plutonium is the principal fissionable material. { plü'tō·nē·əm rē'ak·tər }

pluvial [GEOL] Of a geologic process or feature, effected by rain action. [METEOROL] Pertaining to rain, or more broadly, to precipitation, particularly to an abundant amount thereof. { 'plü·vē·əl }

pluvial lake [GEOL] A lake formed during a period of exceptionally heavy rainfall; specifically, a Pleistocene lake formed during a period of glacial advance and now either extinct or only a remnant. { 'plü·vē·əl 'lāk }

pluviilignosa [ECOL] A tropical rain forest. { ˌplü·vē·ilə g'nō·sə }

pluviofluvial [GEOL] Pertaining to the combined action of rainwater and streams. { ˌplü·vē·ōˌflü·vē·əl }

pluviograph *See* recording rain gage. { 'plü·vē·əˌgraf }

pluviometer *See* rain gage. { ˌplü·vē'äm·əd·ər }

pluviometric coefficient [METEOROL] For any month at a given station, the ratio of the monthly normal precipitation to one-twelfth of the annual normal precipitation. Also known as hyetal coefficient. { ˌplü·vē·əˌme·trik ˌkō·i'fish·ənt }

ply [MATER] A thin sheet of wood or other material bonded to one or more additional thin sheets, as in plywood. [TEXT]

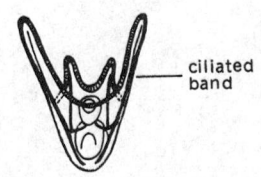

PLUTEUS

ciliated band

Drawing of a pluteus larva.

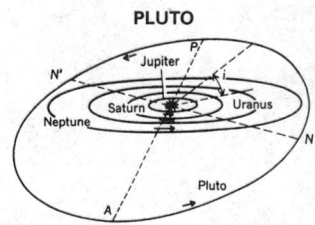

PLUTO

Orbit of Pluto, in a perspective view to show the inclination *i* and eccentricity of the orbit. A, aphelion; *P*, perihelion; *NN'*, line of nodes. (*From L. Rudaux and G. de Vaucouleurs, Astronomie, Larousse, 1948*)

A strand of yarn made by twisting together two or more strands. { plī }

plymetal [MATER] **1.** A material consisting of layers of dissimilar metals bonded together. **2.** Plywood faced with aluminum on both sides. { 'plī,med·əl }

plywood [MATER] A material composed of thin sheets of wood glued together, with the grains of adjacent sheets oriented at right angles to each other. { 'plī,wůd }

PLZT See lead lanthanum zirconate titanate.

Pm See promethium.

PM See permanent magnet; phase modulation.

PMA See phosphomolybdic acid; pyromellitic acid.

PMDA See pyromellitic dianhydride.

p.m.f. See probability mass function.

PMLCD See supertwisted nematic liquid-crystal display.

PMOS [ELECTR] Metal-oxide semiconductors that are made on *n*-type substrates, and whose active carriers are holes that migrate between *p*-type source and drain contacts. Derived from *p*-channel metal-oxide semiconductor. { 'pē,môs }

PMR See projection microradiography.

PMS notation [COMPUT SCI] A notation that provides a clear, concise description of the physical structure of computer systems, and that contains only a few primitive components, namely symbols for memory, link, switch, data operation, control unit, and transducer. Acronym for processor-memory-switch notation. { ¦pē¦em¦es nō'tā·shən }

pNa [CHEM] Logarithm of the sodium-ion concentration in a solution; that is, pNa = $-\log a_{Na^+}$, where a_{Na^+} is the sodium-ion concentration.

PN code See pseudorandom noise code. { ¸pē'en 'kōd }

PNdB See perceived noise decibel.

pneumatic [ENG] Pertaining to or operated by air or other gas. { nů'mad·ik }

pneumatic atomizer [MECH ENG] An atomizer that uses compressed air to produce drops in the diameter range of 5–100 micrometers. { nů'mad·ik 'ad·ə,mīz·ər }

pneumatic caisson [CIV ENG] A caisson having a chamber filled with compressed air at a pressure equal to the pressure of the water outside. { nů'mad·ik 'kā,sän }

pneumatic controller [MECH ENG] A device for the mechanical movement of another device (such as a valve stem) whose action is controlled by variations in pneumatic pressure connected to the controller. { nů'mad·ik kən'trōl·ər }

pneumatic control valve [MECH ENG] A valve in which the force of compressed air against a diaphragm is opposed by the force of a spring to control the area of the opening for a fluid stream. { nů'mad·ik kən'trōl ,valv }

pneumatic conveyor [MECH ENG] A conveyor which transports dry, free-flowing, granular material in suspension, or a cylindrical carrier, within a pipe or duct by means of a high-velocity airstream or by pressure of vacuum generated by an air compressor. Also known as air conveyor. { nů'mad·ik kən'vā·ər }

pneumatic deception device [ORD] A dummy tank, vehicle, or weapon made of inflatable material; used for deceiving the enemy intelligence as to location of friendly installations. { nů'mad·ik di'sep·shən di,vīs }

pneumatic drill [MECH ENG] Compressed-air drill worked by reciprocating piston, hammer action, or turbo drive. { nů'mad·ik 'dril }

pneumatic drilling [MECH ENG] Drilling a hole when using air or gas in lieu of conventional drilling fluid as the circulating medium; an adaptation of rotary drilling. { nů'mad·ik 'dril·iŋ }

pneumatic filling [MIN ENG] A filling method using compressed air to blow filling material into the mined-out stope. { nů'mad·ik 'fil·iŋ }

pneumatic gun charger [ORD] A gun charger that operates by compressed air or gas. { nů'mad·ik 'gən ,chär·jər }

pneumatic hammer [MECH ENG] A hammer in which compressed air is utilized for producing the impacting blow. Also known as air hammer; jack hammer. { nů'mad·ik 'ham·ər }

pneumatic hoist See air hoist. { nů'mad·ik 'hóist }

pneumatic injection [MIN ENG] A method for fighting underground coal fires, developed by the U.S. Bureau of Mines; this air-blowing technique involves the injection of incombustible mineral, like rock wool or dry sand, through 6-inch (15-centimeter) boreholes drilled from the surface to intersect

underground passageways in the mines. { nů'mad·ik in'jek·shən }

pneumatic lighting [MIN ENG] Lighting of underground chambers by a compressed-air turbomotor driving a small dynamo. { nů'mad·ik 'līd·iŋ }

pneumatic loudspeaker [ENG ACOUS] A loudspeaker in which the acoustic ouput results from controlled variation of an airstream. { nů'mad·ik 'laůd,spēk·ər }

pneumatic method [MIN ENG] A method of flotation in which gas is introduced near the bottom of the flotation vessel. { nů'mad·ik 'meth·əd }

pneumatic riveter [MECH ENG] A riveting machine having a rapidly reciprocating piston driven by compressed air. { nů'mad·ik 'riv·əd·ər }

pneumatics [FL MECH] Fluid statics and behavior in closed systems when the fluid is a gas. { nů'mad·iks }

pneumatic servo See valve positioner.

pneumatic servomechanism [CONT SYS] A servomechanism in which power is supplied and transmission of signals is carried out through the medium of compressed air. { nů'mad·ik ¦sər·vō 'mek·ə,niz·əm }

pneumatic steelmaking [MET] Any steelmaking process which employs air or oxygen and for which all heat is derived from the initial heat content of the charge materials and from the thermal energy of the refining reactions. { nů'mad·ik 'stēl ,māk·iŋ }

pneumatic stowing [MIN ENG] A method of filling used mine cavities with crushed rock; which is forced by compressed air into the cavity. { nů'mad·ik 'stō·iŋ }

pneumatic tank switcher [PETRO ENG] Pneumatic actuated valving system for oil-field tanks to shut off crude-oil flow to a filled tank and then to direct incoming crude flow to the next available empty tank. { nů'mad·ik 'taŋk ,swich·ər }

pneumatic telemetering [ENG] The transmission of a pressure impulse by means of pneumatic pressure through a length of small-bore tubing; used for remote transmission of signals from primary process-unit sensing elements for pressure, temperature, flow rate, and so on. { nů'mad·ik 'tel·ə,mēd·ə·riŋ }

pneumatic test [ENG] Pressure testing of a process vessel by the use of air pressure. { nů'mad·ik 'test }

pneumatic transmission lag [ELEC] The time delay in a pneumatic transmission line between the generation of an impulse at one end and the resultant reaction at the other end. { nů'mad·ik 'tranz,mish·ən ,lag }

pneumatic weighing system [ENG] A system for weight measurement in which the load is detected by a nozzle and balanced by modulating the air pressure in an opposing capsule. { nů'mad·ik 'wā·iŋ ,sis·təm }

pneumatocele [MED] **1.** Herniation of the lung. **2.** A sac or tumor containing gas; especially the scrotum filled with gas. { 'nü·məd·ō,sēl }

pneumatocodon [INV ZOO] Exumbrellar surface of the float or pneumatophore of siphonophorans. { ¸nü·məd·ō'kō,dän }

pneumatogenic [GEOL] Referring to a rock or mineral deposit formed by a gaseous agent. { ¦nü·məd·ō¦jen·ik }

pneumatolysis [GEOL] Rock alteration or mineral crystallization effected by gaseous emanations from solidifying magma. { ¸nü·mə'täl·ə·səs }

pneumatolytic [GEOL] Formed by gaseous agents. { ¦nü·məd·ō¦lid·ik }

pneumatolytic metamorphism [PETR] Contact metamorphism by the chemical action of magmatic gases. { ¦nü·məd·ō¦lid·ik ,med·ə'mór,fiz·əm }

pneumatolytic stage [GEOL] The stage in the cooling of a magma in which the solid and gaseous phases are in equilibrium. { ¦nü·məd·ō¦lid·ik 'stāj }

pneumatophore [BOT] **1.** An air bladder in marsh plants. **2.** A submerged or exposed erect root that functions in the respiration of certain marsh plants. [INV ZOO] The air sac of a siphonophore. { 'nü·məd·ə,fôr }

pneumatosaccus [INV ZOO] Subumbrellar surface of the float or pneumatophore of siphonophorans. { ¸nü·mə·dō¦sak·əs }

pneumatosis [MED] The presence of air or gas in abnormal situations in the body. { ¸nü·mə'tō·səs }

pneumaturia [MED] The voiding of urine containing free gas. { ¸nü·mə'tůr·ē·ə }

pneumoangiography [MED] The outline of the vessels of

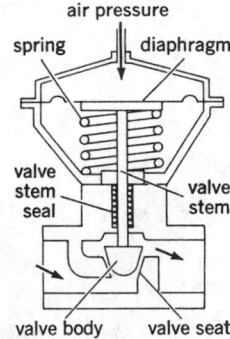

PNEUMATIC CONTROL VALVE

air pressure

spring　diaphragm

valve stem seal

valve stem

valve body　valve seat

Simplified cross section of typical pneumatic control valve. Arrows indicate direction of fluid stream.

the lung by means of radiopaque material, for roentgenographic visualization. { ˌnü·mōˌan·jē'äg·rə·fē }

pneumobacillus *See* Klebsiella pneumoniae. { ˌnü·mō·bə'sil·əs }

pneumoconiosis [MED] Any lung disease caused by dust inhalation. { ˌnü·mōˌkō·nē'ō·səs }

Pneumocystis carinii pneumonia [MED] A lung infection in humans caused by the protozoan *Pneumocystis carinii*. Also known as interstitial plasma-cell pneumonia. { ˌnü·məˌsis·təs kə'rin·ē·ˌī nü'mō·nyə }

pneumoencephalography [MED] A method of visualizing the ventricular system and subarachnoid pathways of the brain by roentgenography after removal of spinal fluid followed by the injection of air or gas into the subarachnoid space. { ˌnü·mō·inˌsef·ə'läg·rə·fē }

pneumoenteritis [MED] Inflammation of the lungs and of the intestine. { ˌnü·mōˌent·ə'rīd·əs }

pneumography [MED] **1.** Roentgenography of the lung. **2.** The recording of the respiratory excursions. { nü'mäg·rə·fē }

pneumohemothorax [MED] The presence of air or gas and blood in the thoracic cavity. { ˌnü·mōˌhē·mə'thȯr·aks }

pneumolithiasis [MED] The occurrence of calculi or concretions in a lung. { ˌnü·mō·li'thī·ə·səs }

pneumomycosis [MED] Any disease of the lungs caused by a fungus. { ˌnü·mō·mī'kō·səs }

pneumonectomy [MED] Surgical removal of an entire lung. { ˌnü·mə'nek·tə·mē }

pneumonia [MED] An acute or chronic inflammation of the lungs caused by numerous microbial, immunological, physical, or chemical agents, and associated with exudate in the alveolar lumens. { nü'mō·nyə }

pneumonic plague [MED] A virulent type of plague in humans, with lung involvement. { nü'män·ik ˌplāg }

pneumonitis [MED] Inflammation of the lung. { ˌnü·mə'nīd·əs }

pneumonolysis [MED] The loosening of any portion of lung adherent to the chest wall; a form of collapse therapy used in the treatment of pulmonary tuberculosis. { ˌnü·mə'näl·ə·səs }

pneumopericardium [MED] The presence of air in the pericardial cavity. { ˌnü·mōˌper·ə'kärd·ē·əm }

pneumoperitoneum [MED] **1.** The presence of air or gas in the peritoneal cavity. **2.** Injection of a gas into the peritoneal cavity as a diagnostic or therapeutic measure. { ˌnü·mōˌper·ə·tə'nē·əm }

pneumostome [INV ZOO] The respiratory aperture of gastropod mollusks. { 'nü·məˌstōm }

pneumotectic [GEOL] Referring to processes and products of magmatic consolidation affected to some degree by the gaseous constituents of the magma. { ˌnü·mō'tek·tik }

pneumothorax [MED] The presence of air or gas in the pleural cavity. { ˌnü·mō'thȯr·aks }

pn hook transistor *See* hook collector transistor. { 'pēˌen 'hu̇k tranˌzis·tər }

pnicogen [CHEM] Any member of the nitrogen family of elements, group V in the periodic table. { 'nī·kə·jən }

pnictide [CHEM] A simple compound of a pnicogen and an electropositive element. { 'nikˌtīd }

pnigophobia [PSYCH] An abnormal fear of choking. { ˌnī·gə'fō·bē·ə }

pnip transistor [ELECTR] An intrinsic junction transistor in which the intrinsic region is sandwiched between the *n*-type base and the *p*-type collector. { 'pēˌenˌi'pē tranˌzis·tər }

pn junction [ELECTR] The interface between two regions in a semiconductor crystal which have been treated so that one is a *p*-type semiconductor and the other is an *n*-type semiconductor; it contains a permanent dipole charge layer. { 'pēˌen ˌjəŋk·shən }

pnpn diode [ELECTR] A semiconductor device consisting of four alternate layers of *p*-type and *n*-type semiconductor material, with terminal connections to the two outer layers. Also known as *npnp* diode. { 'pēˌen'pēˌen ˌdī·ōd }

pnpn transistor *See* npnp transistor. { 'pēˌen'pēˌen tranˌzis·tər }

pnp transistor [ELECTR] A junction transistor having an *n*-type base between a *p*-type emitter and a *p*-type collector. { 'pēˌen'pē tranˌzis·tər }

Po *See* polonium.

Poaceae [BOT] The equivalent name for Gramineae. { pō'ās·ē·ˌē }

Poales [BOT] The equivalent name for Cyperales. { pō'ā·lēz }

pochoir *See* stencil printing. { pō'shwär }

pock [MED] A pustule of an eruptive fever, especially of smallpox. { päk }

Pockels cell [OPTICS] A crystal that exhibits the Pockels effect, such as potassium dihydrogen phosphate, which is placed between crossed polarizers and has ring electrodes bonded to two faces to allow application of an electric field; used to modulate light beams, especially laser beams. { 'päk·əlz ˌsel }

Pockels effect [OPTICS] Changes in the refractive properties of certain crystals in an applied electric field, which are proportional to the first power of the electric field strength. { 'päk·əlz iˌfekt }

Pockels equation [MATH] A partial differential equation which states that the Laplacian of an unknown function, plus the product of the value of the function with a constant, is equal to 0; it arises in finding solutions of the wave equation that are products of time-independent and space-independent functions. { 'päk·əlz iˌkwā·zhən }

Pockels readout optical modulator [ELECTR] A device for storing data in the form of images; it consists of bismuth silicon oxide crystal coated with an insulating layer of parylene and transparent electrodes evaporated on the surfaces; a blue laser is used for writing and a red laser is used for nondestructive readout or processing. Abbreviated PROM. { 'päk·əlz ˌrēdˌau̇t ˌäp·tə·kəl 'mäj·əˌlād·ər }

pocket [BUILD] A recess in a wall designed to receive a folding or sliding door in the open position. [CIV ENG] A recess made in masonry to receive the end of a beam. [COMPUT SCI] One of the several receptacles into which punched cards are fed by a card sorter. [GEOL] **1.** A cavity that contains a deposit such as a gas or an ore. **2.** An enclosed or sheltered place along a coast, such as a reentrant between rocky, cliffed headlands or a bight on a lee shore. [MIN ENG] A receptacle from which coal, ore, or waste is loaded into wagons or cars. { 'päk·ət }

pocket beach [GEOL] A small, narrow beach formed in a pocket, commonly crescentic in plan, with the concave edge toward the sea, and displaying well-sorted sands. { 'päk·ət ˌbēch }

pocket gopher *See* gopher. { 'päk·ət ˌgō·fər }

pocket valley [GEOL] A valley whose head is enclosed by steep walls at the base of which underground water emerges as a spring. { 'päk·ət ˌval·ē }

pod [AERO ENG] An enclosure, housing, or detachable container of some kind on an airplane or space vehicle, as an engine pod. [BOT] A dry dehiscent fruit; a legume. [DES ENG] **1.** The socket for a bit in a brace. **2.** A straight groove in the barrel of a pod auger. [GEOL] An orebody of elongate, lenticular shape. Also known as podiform orebody. { päd }

POD analysis [ANALY CHEM] A precision laboratory distillation procedure used to separate low-boiling hydrocarbon fractions quantitatively for analytical purposes. Also known as Podbielniak analysis. { 'pēˌō'dē əˌnal·ə·səs }

Podargidae [VERT ZOO] The heavy-billed frogmouths, a family of Asian and Australian birds in the order Caprimulgiformes. { pə'där·jəˌdē }

Podbielniak analysis *See* POD analysis. { päd'bēl·nē·ˌak əˌnal·ə·səs }

Podbielniak extractor [CHEM ENG] A solvent-extraction device in which centrifugal action enhances liquid-liquid contact and increases resultant separation efficiency. { päd'bēl·nē·ˌak ik·strak·tər }

pod blight [PL PATH] A fungus disease of legumes caused by *Diaporthe* species. { 'päd ˌblīt }

podded propulsion [MAR ENG] Ship propulsion by electric motors mounted in pods that are underneath the hull of a ship and attached to the ship by a strut. { 'päd·əd prə'pəl·shən }

podiatrist *See* chiropodist. { pə'dī·ə·trəst }

Podicipedidae [VERT ZOO] The single family of the avian order Podicipediformes. { ˌpäd·ə·sə'ped·əˌdē }

Podicipediformes [VERT ZOO] The grebes, an order of swimming and diving birds distinguished by dense, silky plumage, a rudimentary tail, and toes that are individually broadened and lobed. { ˌpäd·ə·sə·ped·ə'fȯrˌmēz }

PODDED PROPULSION

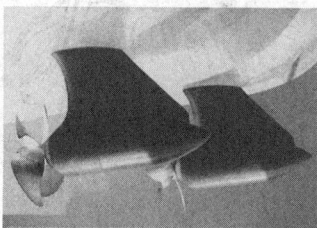

Podded propulsors mounted on the ship hull. (*Kamewa Group*)

Podicipitiformes [VERT ZOO] The equivalent name for Podicipediformes. { ˌpäd·ə·sə͵pid·ə′fȯr͵mēz }

podiform orebody *See* pod. { ′päd·ə͵fȯrm ′ȯr͵bäd·ē }

podite [INV ZOO] A segment of a limb of an arthropod. { ′pä͵dīt }

podium [INV ZOO] The terminal portion of a body wall appendage in certain echinoderms. { ′pō·dē·əm }

podobranch [INV ZOO] A gill of a crustacean attached to the basal segment of a thoracic limb. { ′päd·ə͵braŋk }

Podocopa [INV ZOO] A suborder of fresh-water ostracod crustaceans in the order Podocopida in which the inner lamella has a calcified rim joining the outer lamella along a chitinous zone of concrescence, and the two valves fit together firmly. { pə′däk·ə·pə }

Podocopida [INV ZOO] An order of the Ostracoda; contains all fresh-water ostracods and is divided into the suborders Podocopa, Metacopina, and Platycopina. { ˌpäd·ə′käp·ə·də }

Podocopina [INV ZOO] The equivalent name for Podocopa. { ˌpäd·ə·kə′pī·nə }

podocyst [INV ZOO] A sinus in the foot of certain gastropod mollusks. { ′päd·ə͵sist }

Podogona [INV ZOO] The equivalent name for Ricinuleida. { pə′däg·ə·nə }

podophyllin [PHARM] A light yellow to greenish-yellow or light brown powder with a bitter taste; soluble in alcohol, ether, and ammonium hydroxide; used in medicine. Also known as podophyllum resin. { ˌpäd·ə′fil·ən }

podophyllinic acid lactone *See* podophyllotoxin. { ˌpäd·ə·fə′lin·ik ′as·əd ′lak͵tōn }

podophyllotoxin [PHARM] $C_{22}H_{22}O_8$ A crystalline compound with a melting point of 114–118°C; soluble in alcohol, chloroform, acetone, and warm benzene; used as an antineoplastic agent in medicine. Also known as podophyllinic acid lactone. { ˌpäd·ə͵fil·ə′täk·sən }

podophyllum resin *See* podophyllin. { ˌpäd·ə′fil·əm ′rez·ən }

Podosphaera leucotricha [MYCOL] A fungal plant pathogen that causes apple powdery mildew. { ˌpäd·ə͵sfī·rə ͵lü·kə′trik·ə }

Podostemaceae [BOT] The single family of the order Podostemales. { pə͵däs·tə′mās·ē͵ē }

Podostemales [BOT] An order of dicotyledonous plants in the subclass Rosidae; plants are submerged aquatics with modified, branching shoots and small, perfect flowers having a superior ovary and united carpels. { ˌpäd·ə·stə′mā·lēz }

Podoviridae [VIROL] A family of linear double-stranded deoxyribonucleic acid–containing bacterial viruses (bacteriophages) characterized by a short contractile tail and an icosahedral head; contains the T7 coliphage. { ˌpäd·ə′vir·ə͵dī }

pod rot [PL PATH] A fungus disease of cacao caused by *Monilia roreri* and characterized by lesions on the pods. { ′päd ͵rät }

Podzol [GEOL] A soil group characterized by mats of organic matter in the surface layer and thin horizons of organic minerals overlying gray, leached horizons and dark-brown illuvial horizons; found in coal forests to temperate coniferous or mixed forests. { ′päd͵zȯl }

podzolic soil *See* red-yellow podzolic soil. { päd′zäl·ik ′sȯil }

podzolization [GEOL] The process by which a soil becomes more acid because of the depletion of bases, and develops surface layers that have been leached of clay. { ˌpäd·zə·lə′zā·shən }

Poeciliidae [VERT ZOO] A family of fishes in the order Atheriniformes including the live-bearers, such as guppies, swordtails, and mollies. { ˌpē·sə′lī·ə͵dē }

Poecilosclerida [INV ZOO] An order of sponges of the class Demospongiae in which the skeleton includes two or more types of megascleres. { ˌpē·sə·lō′skler·ə·də }

Poeobiidae [INV ZOO] A monotypic family of spioniform worms (*Poeobius meseres*) belonging to the Sedentaria and found in the North Pacific Ocean. { ˌpē·ə′bī·ə͵dē }

Poetsch process [MIN ENG] Shaft sinking in which brine at subzero temperature is circulated through boreholes to freeze running water through which a shaft or tunnel is to be driven, during development of a waterlogged mine. { ′pech ͵prä·səs }

Poggendorff's first method *See* constant-current dc potentiometer. { ′päg·ən͵dȯrfs ′first ͵meth·əd }

Poggendorff's second method *See* constant-resistance dc potentiometer. { ′päg·ən͵dȯrfs ′sek·ənd ͵meth·əd }

pogonip *See* ice fog. { ′päg·ə͵nip }

pogonochore [BOT] A type of plant that produces plumed disseminules. { pə′gän·ə͵chȯr }

Pogonophora [INV ZOO] The single class of the phylum Brachiata; the elongate body consists of three segments, each with a separate coelom; there is no mouth, anus, or digestive canal, and sexes are separate. { ˌpō·gə′näf·ə·rə }

Pogson scale [ASTRON] An index of brightness used in star catologs; it is the ratio of 2.512 to 1 between the brightness of successive magnitudes. { ′päg·sən ͵skāl }

Pohlé air lift pump [MECH ENG] A pistonless pump in which compressed air fills the annular space surrounding the uptake pipe and is free to enter the rising column at all points of its periphery. { pō′lā ′er ͵lift ͵pəmp }

poidometer [ENG] An automatic weighing device for use on belt conveyors. { pȯi′däm·əd·ər }

poikilitic [PETR] Of the texture of an igneous rock, having small crystals of one mineral randomly scattered without common orientation in larger crystals of another mineral. { ˌpȯi·kə′lid·ik }

poikiloblast [GEOL] A large crystal (xenoblast) formed by recrystallization during metamorphism and containing numerous inclusions of small idioblasts. { pȯi′kil·ə͵blast }

poikiloblastic [PETR] Of a metamorphic texture, simulating the poikilitic texture of igneous rocks in having small idioblasts of one constituent lying within larger xenoblasts. Also known as sieve texture. { pȯi͵kil·ə͵blas·tik }

poikilocrystallic *See* poikilotopic. { pȯi͵kil·ə·kri′stal·ik }

poikilocytosis [MED] A condition in which erythrocytes are distorted in shape. { pȯi͵kil·ə·sī′tō·səs }

poikilophitic [GEOL] Referring to ophitic texture characterized by lath-shaped feldspar crystals completely included in large, anhedral pyroxene crystals. { pȯi͵kil·ə′fid·ik }

poikilotherm [ZOO] An animal, such as reptiles, fishes, and invertebrates, whose body temperature varies with and is usually higher than the temperature of the environment; a cold-blooded animal. { pȯi′kil·ə͵thərm }

poikilotope [GEOL] A large crystal enclosing smaller crystals of another mineral in a sedimentary rock showing poikilotopic fabric. { pȯi′kil·ə͵tōp }

poikilotopic [GEOL] Referring to the fabric of a crystalline sedimentary rock in which the constituent crystals are multisized and larger crystals enclose smaller crystals of another mineral. Also known as poikilocrystallic. { pȯi͵kil·ə′täp·ik }

Poincaré-Birkhoff fixed-point theorem [MATH] The theorem that a bijective, continuous, area-preserving mapping of the ring between two concentric circles onto itself that moves one circle in the positive sense and the other in the negative sense has at least two fixed points. { ˌpwän·kə͵rä ′bərk͵hȯf ′fikst ͵pȯint ′thir·əm }

Poincaré conjecture [MATH] The question as to whether a compact, simply connected three-dimensional manifold without boundary must be homeomorphic to the three-dimensional sphere. { ˌpwän͵kä′rä kən͵jek·chər }

Poincaré electron [ELECTROMAG] A classical model of the electron in which nonelectromagnetic forces hold the electron together so that it has zero self-stress; it is unstable and has infinite self-energy in the case of a point electron. { ˌpwän ͵kä′rä ′lek͵trän }

Poincaré recurrence theorem [MATH] **1.** A volume-preserving homeomorphism T of a finite dimensional Euclidean space will have, for almost all points x, infinitely many points of the form $T^i(x)$, $i = 1, 2, \ldots$ within any open set containing x. **2.** A measure preserving transformation on a space with finite measure is recurrent. { ˌpwän͵kä′rä ri′kə·rəns ͵thir·əm }

Poincaré surface of section [MECH] A method of displaying the character of a particular trajectory without examining its complete time development, in which the trajectory is sampled periodically, and the rate of change of a quantity under study is plotted against the value of that quantity at the beginning of each period. Also known as surface of section. { ˌpwän͵kä′rä ′sər·fəs əv ′sek·shən }

Poinsot ellipsoid *See* inertia ellipsoid. { pwän′sō ə′lip͵sȯid }

Poinsot motion [MECH] The motion of a rigid body with a point fixed in space and with zero torque or moment acting on the body about the fixed point. { pwän′sō ′mō·shən }

Poinsot's central axis [MECH] A line through a rigid body

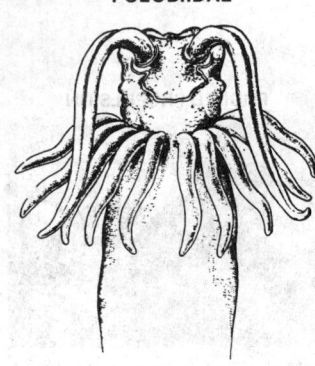

POEOBIIDAE

Anterior end of *Poeobius* (Poeobiidae) in dorsal view.

which is parallel to the vector sum **F** of a system of forces acting on the body, and which is located so that the system of forces is equivalent to the force **F** applied anywhere along the line, plus a couple whose torque is equal to the component of the total torque **T** exerted by the system in the direction **F**. { ¦pwän·sōz ¦sen·trəl ¦ak·səs }

Poinsot's method [MECH] A method of describing Poinsot motion, by means of a geometrical construction in which the inertia ellipsoid rolls on the invariable plane without slipping. { pwän'sōz ¦meth·əd }

Poinsot's spiral [MATH] Either of two plane curves whose equations in polar coordinates (r,θ) are $r \cosh n\theta = a$ and $r \sinh n\theta = a$, where a is a constant and n is an integer. { pwän 'sōz ¦spī·rəl }

point [GEOGR] A tapering piece of land projecting into a body of water; it is generally less prominent than a cape. [GRAPHICS] A printer's unit of measurement, equivalent to 0.013837 inch (approximately $^1/_{72}$ inch) or $^1/_{12}$ pica; six lines of 12-point type measure 1 inch. Also known as printer's point; typography point. [LAP] A unit of mass, used in measuring precious stones, equal to 0.01 metric carat, or to 2 milligrams. [MATH] **1.** An element in a topological space. **2.** One of the basic undefined elements of geometry possessing position but no nonzero dimension. **3.** In positional notation, the character or the location of an implied symbol that separates the integral part of a numerical expression from its fractional part; for example, it is called the binary point in binary notation and the decimal point in decimal notation. [NAV] In marine operation, one thirty-second of a circle, or $11^1/_4$ degrees. { pȯint }

point angle [DES ENG] The angle at the point or edge of a cutting tool. { ¦pȯint ¦aŋ·gəl }

point at infinity [MATH] **1.** A single point that is adjoined to the complex plane so that it corresponds to the pole of a stereographic projection of the Riemann sphere onto the complex plane, giving the complex plane a compact topology. **2.** See ideal point. { pȯint at in'fin·əd·ē }

point bar [GEOL] One of a series of low, arcuate sand and gravel ridges formed on the inside of a growing meander by the gradual addition of accretions. Also known as meander bar. { pȯint ¦bär }

point-bearing pile See end-bearing pile. { pȯint¦ber·iŋ ¦pīl }

point biserial correlation coefficient [STAT] A modification of the biserial correlation coefficient in which one variable is dichotomous and the other is continuous; a product moment correlation coefficient. { ¦pȯint ¦bī¦sir·ē·əl ¦kär·ə'lā·shən ¦kō· ə¦fish·ənt }

point-blank range [MECH] Distance to a target that is so short that the trajectory of a bullet or projectile is practically a straight, rather than a curved, line. { ¦pȯint¦blaŋk 'rānj }

point characteristic function [OPTICS] The integral between two points of $n\,ds$ along some path, where ds is the arc length of an infinitesimal piece of the path and n is the refractive index; according to Fermat's principle, it is a maximum or minimum with respect to nearby paths for the actual path of a light ray. { ¦pȯint ¦kar·ik·tə'ris·tik ¦fəŋk·shən }

point contact [ELECTR] A contact between a specially prepared semiconductor surface and a metal point, usually maintained by mechanical pressure but sometimes welded or bonded. { ¦pȯint 'kän¦takt }

point-contact diode [ELECTR] A semiconductor rectifier that uses the barrier formed between a specially prepared semiconductor surface and a metal point to produce the rectifying action. { ¦pȯint 'kän¦takt ¦dī·ōd }

point-contact silicon cell [ELECTR] A type of solar cell whose efficiency is enhanced by a combination of tiny doped-silicon dots scattered across the lower surface of the silicon crystal and fine aluminum threads that penetrate the silicon layer to collect current from each point. { ¦pȯint ¦kän¦takt 'sil· ə·kən ¦sel }

point-contact transistor [ELECTR] A transistor having a base electrode and two or more point contacts located near each other on the surface of an n-type semiconductor. { ¦pȯint ¦kän¦takt tran¦zis·tər }

point defect [CRYSTAL] A departure from crystal symmetry which affects only one, or, in some cases, two lattice sites. { ¦pȯint di¦fekt }

point diagram [PETR] A fabric diagram in which a point represents the preferred orientation of each individual fabric

element. Also known as scatter diagram. { ¦pȯint ¦dī· ə¦gram }

point distal flow [MATH] A transformation group on a compact metric space for which there exists a distal point with a dense orbit. { ¦pȯint ¦dis·təl 'flō }

pointed bracket See angle bracket. { ¦pȯint·əd ¦brak·ət }

pointer [COMPUT SCI] The part of an instruction which contains the address of the next record to be accessed. [ENG] The needle-shaped rod that moves over the scale of a meter. { ¦pȯint·ər }

Pointers [ASTRON] The stars α and β Ursae Majoris, which appear to point toward the north celestial pole and Polaris. { ¦pȯint·ərz }

point estimates [STAT] Estimates which produce a single value of the population. { ¦pȯint ¦es·tə·məts }

point-finite family of subsets [MATH] A family of subsets of a particular set, S, such that any member of S is a member of at most a finite number of these subsets. { ¦pȯint ¦fī¦nīt ¦fam·lē əv ¦səb¦sets }

point fire [ORD] Concentrated fire from a number of guns, directed at a single point or small area. { ¦pȯint ¦fīr }

point function [MATH] A function whose values are points. [PHYS] A quantity whose value depends on the location of a point in space, such as an electric field, pressure, temperature, or density. { ¦pȯint ¦fəŋk·shən }

point group [CRYSTAL] A group consisting of the symmetry elements of an object having a single fixed point; 32 such groups are possible. { ¦pȯint ¦grüp }

point harmonization [ORD] The adjusting of a fighter aircraft's fixed guns so that they converge upon a point at a given range. { ¦pȯint ¦här·mə·nə'zā·shən }

pointing [CIV ENG] **1.** Finishing a mortar joint. **2.** Pressing mortar into a raked joint. [MATER] The material (mortar) used in pointing. [MET] Reducing the diameter and tapering a short length at the end of a rod, wire, or tube. Also known as metal pointing. [ORD] The operation of giving the piece a designated elevation and direction. { ¦pȯint·iŋ }

pointing device [COMPUT SCI] A handheld device, such as a mouse, puck, or stylus, that controls a position indicator on a display screen. Also known as pick device. { ¦pȯint· iŋ di¦vīs }

pointing stick [COMPUT SCI] A small rubberized device located in the center of a computer keyboard, which is moved with a finger tip to position a pointer. { ¦pȯint·iŋ ¦stik }

pointing trowel [ENG] A tool used to apply pointing to the joints between bricks. { ¦pȯint·iŋ ¦traül }

point initiation [ENG] Application of the initial impulse from the detonator to a single point on the main charge surface; for a cylindrical charge this point is usually the center of one face. { ¦pȯint i¦nish·ē'ā·shən }

point jammer [ELECTR] Any electronic jammer directed against a specific enemy installation operating on a specific frequency. { ¦pȯint ¦jam·ər }

point-junction transistor [ELECTR] Transistor having a base electrode and both point-contact and junction electrodes. { ¦pȯint ¦jəŋk·shən tran¦zis·tər }

point localization [PHYSIO] The ability to locate the point on the skin that has been touched. { ¦pȯint ¦lō·kə·lə'zā·shən }

point-mode display [COMPUT SCI] A method of representing information in the form of dots on the face of a cathode-ray tube. { ¦pȯint ¦mōd di splā }

point mutation [GEN] Mutation of a single gene due to addition, loss, replacement, or change of sequence in one or more base pairs of the deoxyribonucleic acid of that gene. { ¦pȯint myü'tā·shən }

point of arrival [NAV] The position which a craft will reach after following specified courses for specified distances from a point of departure. { ¦pȯint əv ə'rīv·əl }

point of contraflexure [MECH] A point at which the direction of bending changes. Also known as point of inflection. { ¦pȯint əv ¦kän·trə'flek shər }

point of control [IND ENG] Fraction defective in those lots that have a probability of .50 of acceptance according to a specific sampling acceptance plan. { ¦pȯint əv kən'trōl }

point of departure See departure. { ¦pȯint əv di'pär·chər }

point of destination [NAV] The point at which the final course from the point of departure ends, exclusive of the courses needed to reach a berth or runway. { ¦pȯint əv ¦des·tə'nā· shən }

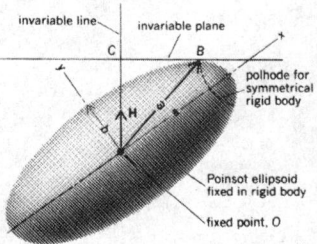

POINSOT'S METHOD

invariable line — invariable plane
polhode for symmetrical rigid body
Poinsot ellipsoid fixed in rigid body
fixed point, O

Inertia ellipsoid has axes with lengths a, b, and c (not shown) along principal axes of body, x, y, and z (not shown). OA = invariable line drawn in direction of fixed vector angular momentum **H**. Inertia ellipsoid rolls on invariable plane which is perpendicular to OA and intersects it at C. Ellipsoid is tangent to plane at B, where OB = vector angular momentum ω.

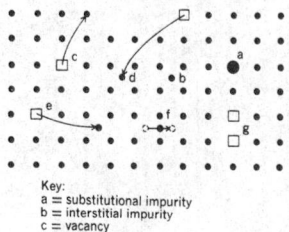

POINT DEFECT

Key:
a = substitutional impurity
b = interstitial impurity
c = vacancy
d = interstitial
e = Frenkel pair
f = interstitial impurity moving from one site through a saddle point to a neighboring site
g = divacancy

Point defects in a lattice. The squares indicate positions that were formerly occupied by atoms displaced as shown by the arrows.

point of division [MATH] The point that divides the line segment joining two given points in a given ratio. { 'pȯint əv di'vizh·ən }

point of fall [MECH] The point in the curved path of a falling projectile that is level with the muzzle of the gun. Also known as level point. { 'pȯint əv 'fȯl }

point of frog [CIV ENG] The place of intersection of the gage lines of the main track and a turnout. { 'pȯint əv 'frȯg }

point of impact [ORD] The point at which a bullet, bomb, or projectile strikes. { 'pȯint əv 'im,pakt }

point of inflection [MATH] A point where a plane curve changes from the concave to the convex relative to some fixed line; equivalently, if the function determining the curve has a second derivative, this derivative changes sign at this point. Also known as inflection point. [MECH] *See* point of contraflexure. { 'pȯint əv in'flek·shən }

point of intersection [CIV ENG] The point at which two straight sections or tangents to a road curve or rail curve meet when extended. { 'pȯint əv ,in·tər'sek·shən }

point of no return [NAV] A point along an aircraft track beyond which the aircraft's fuel supply will not permit it to return to its own or some other associated base but must continue in attempt to reach its destination. { 'pȯint əv 'nō ri'tərn }

point of operation [MET] That portion of a metal-forming press in which the material is positioned and the work is performed. { 'pȯint əv ,äp·ə'rā·shən }

point-of-origin system [COMPUT SCI] A computer system in which data collection occurs at the point where the data are actually created, as in a point-of-sale terminal. { 'pȯint əv 'är·ə·jən ,sis·təm }

point of osculation *See* double cusp. { 'pȯint əv ,äs·kyə'lā·shən }

point-of-sale terminal [COMPUT SCI] A computer-connected terminal used in place of a cash register in a store, for customer checkout and such added functions as recording inventory data, transferring funds from the customer's bank account to the merchant's bank account, and checking credit on charged or charge-card purchases; the terminals can be modified for many nonmerchandising applications, such as checkout of books in libraries. Abbreviated POS terminal. { 'pȯint əv 'sāl 'term·ən·əl }

point of switch [CIV ENG] That place in a track where a car passes from the main track to a turnout. { 'pȯint əv 'swich }

point of tangency [CIV ENG] The point at which a road curve or railway curve becomes straight or changes its curvature. Also known as tangent point. { 'pȯint əv 'tan·jən·sē }

point-placement [AGR] Positioning of fertilizer or some other agricultural chemical within the length of the seed row or in the specific location where the seed is planted. { 'pȯint ,plās·mənt }

point projection electron microscope [ELECTR] An electron microscope in which a real or virtual point source of electrons produces a highly magnified shadow. { 'pȯint prə'jek·shən i'lek,trän 'mī·krə,skōp }

point rainfall [METEOROL] The rainfall during a given time interval (or often one storm) measured in a rain gage, or an estimate of the amount which might have been measured at a given point. { 'pȯint 'rān,fȯl }

point set [MATH] A collection of points in a geometrical or topological space. { 'pȯint ,set }

point-set topology [MATH] *See* general topology. { 'pȯint ,set tə'päl·ə·jē }

point-slope form [MATH] The equation of a straight line in the form $y - y_1 = m(x - x_1)$, where m is the slope of the line and (x_1, y_1) are the coordinates of a given point on the line in a cartesian coordinate system. { 'pȯint ,slōp ,fȯrm }

points of apparent equality [PSYCH] Values of stimuli that elicit no response because a change in one aspect compensates for a change in another. { 'pȯins əv ə,par·ənt ē'kwäl·əd·ē }

points of the compass *See* compass points. { 'pȯins əv thə 'käm·pəs }

point source [CIV ENG] A municipal or industrial wastewater discharge through a discrete pipe or channel. [PHYS] A source of radiation having definite position but no extension in space; this is an ideal which is a good approximation for distances from the source sufficiently large compared to the dimensions of the source. { 'pȯint ,sȯrs }

point-source light [ELEC] A special lamp in which the radiating element is concentrated in a small physical area. { 'pȯint ,sȯrs ,līt }

point spectrum [MATH] Those eigenvalues in the spectrum of a linear operator between Banach spaces whose corresponding eigenvectors are nonzero and of finite norm. { 'pȯint ,spek·trəm }

point system [IND ENG] **1.** A system of job evaluation wherein job requirements are rated according to a scale of point values. **2.** A wage incentive plan based on points instead of man-minutes. { 'pȯint ,sis·təm }

point target [ELECTROMAG] In radar, an object which returns a target signal by reflection from a relatively simple discrete surface; such targets are ships, aircraft, projectiles, missiles, and buildings. [ORD] A precise target of small dimensions. { 'pȯint ,tär·gət }

point-to-point communication [COMMUN] Radio communication between two fixed stations. { 'pȯint tə 'pȯint kə,myü·nə'kā·shən }

point-to-point programming [CONT SYS] A method of programming a robot in which each major change in the robot's path of motion is recorded and stored for later use. { 'pȯint tə 'pȯint 'prō,gram·iŋ }

Point-to-Point Protocol [COMMUN] A standard governing dial-up connections of computers to the Internet via a telephone modem. Abbreviated PPP. { ,pȯin·tü ,pȯint 'prōd·ə,kȯl }

point transposition [ELEC] Transposition, usually in an open-wire line, which is executed within a distance comparable to the wire separation, without material distortion of the normal wire configuration outside this distance. { 'pȯint ,tranz·pə,zish·ən }

pointwise convergence [MATH] A sequence of functions $f_1, f_2, \ldots$ defined on a set S converges pointwise to a function f if the sequence $f_1(x), f_2(x), \ldots$ converges to $f(x)$ for each x in S. { 'pȯint,wīz kən'vər·jəns }

pointwise equicontinuous family of functions [MATH] A family of functions defined on a common domain D with the property that for any point x in D and for any $\epsilon > 0$ there is a $\delta > 0$ such that, whenever y is in D and $|x - y| < \delta$, $|f(x) - f(y)| < \epsilon$ for every function f in the family. { 'pȯint,wīz ,ek·wi·kən'tin·yə·wəs 'fam·lē əv 'fəŋk·shənz }

poise [FL MECH] A unit of dynamic viscosity equal to the dynamic viscosity of a fluid in which there is a tangential force 1 dyne per square centimeter resisting the flow of two parallel fluid layers past each other when their differential velocity is 1 centimeter per second per centimeter of separation. Abbreviated P. { pȯiz }

poised stream [HYD] A stream that is neither eroding nor depositing sediment. { 'pȯizd 'strēm }

poiseuille [FL MECH] A unit of dynamic viscosity of a fluid in which there is a tangential force of 1 newton per square meter resisting the flow of two parallel layers past each other when their differential velocity is 1 meter per second per meter of separation; equal to 10 poise; used chiefly in France. Abbreviated Pl. { pwä'zə·ē }

Poiseuille flow [FL MECH] The steady flow of an incompressible fluid parallel to the axis of a circular pipe of infinite length, produced by a pressure gradient along the pipe. { pwä'zə·ē ,flō }

Poiseuille's law [FL MECH] The law that the volume flow of an incompressible fluid through a circular tube is equal to $\pi/8$ times the pressure differences between the ends of the tube, times the fourth power of the tube's radius divided by the product of the tube's length and the dynamic viscosity of the fluid. { pwä'zə·ēz ,lȯ }

poison [ATOM PHYS] A substance which reduces the phosphorescence of a luminescent material. [CHEM] A substance that exerts inhibitive effects on catalysts, even when present only in small amounts; for example, traces of sulfur or lead will poison platinum-based catalysts. [ELECTR] A material which reduces the emission of electrons from the surface of a cathode. [MATER] A substance that in relatively small doses has an action that either destroys life or impairs seriously the functions of organs or tissues. [NUCLEO] A substance that absorbs neutrons without any fission resulting, and thereby lowers the reactivity of a nuclear reactor. { 'pȯiz·ən }

poison gas [MATER] A substance employed in chemical warfare to disable enemy troops; may be a gas, or a liquid or solid that gives off a gas. An example is mustard gas.　{ 'pȯiz·ən 'gas }

poison gland [VERT ZOO] Any of various specialized glands in certain fishes and amphibians which secrete poisonous mucuslike substances.　{ 'pȯiz·ən ‚gland }

poison hemlock [BOT] *Conium maculatum.* A branching biennial poisonous herb that contains a volatile alkaloid, coniine, in its fruits and leaves.　{ 'pȯiz·ən 'hem‚läk }

poison ivy [BOT] Any of several climbing, shrubby, or arborescent plants of the genus *Rhus* in the sumac family (Anacardiaceae); characterized by ternate leaves, greenish flowers, and white berries that produce an irritating oil.　{ 'pȯiz·ən 'ī·vē }

poison oak [BOT] Any of several bushy poison ivy plants or shrubby poison sumacs.　{ 'pȯiz·ən 'ōk }

poisonous plant [BOT] Any of about 400 species of vascular plants containing principles which initiate pathological conditions in man and animals.　{ 'pȯiz·ən·əs 'plant }

poison sumac [BOT] *Rhus vernix.* A tall bush of the sumac family (Anacardiaceae) bearing pinnately compound leaves with 7–13 entire leaflets, and drooping, axillary clusters of white fruits that produce an irritating oil.　{ 'pȯiz·ən 'sü‚mak }

Poisson binomial trials model See generalized binomial trials model.　{ pwä'sōn bī'nō·mē·əl ‚trīlz ‚mäd·əl }

Poisson bracket [MECH] For any two dynamical variables, *X* and *Y*, the sum, over all degrees of freedom of the system, of $(\partial X/\partial q)(\partial Y/\partial p) - (\partial X/\partial p)(\partial Y/\partial q)$, where *q* is a generalized coordinate and *p* is the corresponding generalized momentum.　{ pwä'sōn ‚brak·ət }

Poisson constant [PHYS] The ratio *k* of the gas constant *R* to the specific heat at constant pressure C_p.　{ pwä'sōn ‚kän·stənt }

Poisson density functions [STAT] Density functions corresponding to Poisson distributions.　{ pwä'sōn 'den·səd·ē ‚faŋk·shənz }

Poisson distribution [STAT] A probability distribution whose mean and variance have a common value *k*, and whose frequency is $f(x) = k^x e^{-k}/x!$, for $x = 0,1,2,\ldots$.　{ pwä'sōn ‚dis·trə'byü·shən }

Poisson effect [FL MECH] The deflection of a spinning projectile with right-handed spin to the right, and vice versa. Also known as cushion effect.　{ pwä'sōn i‚fekt }

Poisson formula [MATH] If the infinite series of functions $f(2\pi k + t)$, *k* ranging from $-\infty$ to ∞, converges uniformly to a function of bounded variation, then the infinite series with term $f(2\pi k)$, *k* ranging from $-\infty$ to ∞, is identical to the series with term the integral of $f(x)e^{-iks}dx$, *k* ranging from $-\infty$ to ∞.　{ pwä'sōn ‚fȯr·myə·lə }

Poisson index of dispersion [STAT] An index used for events which follow a Poisson distribution and should have a chi-square distribution.　{ ‚pwä'sōn ‚in‚deks əv di'spər·zhən }

Poisson integral formula [MATH] This formula gives a solution function for the Dirichlet problem in terms of integrals; an integral representation for the Bessel functions.　{ pwä'sōn 'int·ə·grəl ‚fȯr·myə·lə }

Poisson number [MECH] The reciprocal of the Poisson ratio.　{ pwä'sōn ‚nəm·bər }

Poisson process [STAT] A process given by a discrete random variable which has a Poisson distribution.　{ pwä'sōn ‚prä·səs }

Poisson ratio [MECH] The ratio of the transverse contracting strain to the elongation strain when a rod is stretched by forces which are applied at its ends and which are parallel to the rod's axis.　{ pwä'sōn ‚rā·shō }

Poisson relation [GEOPHYS] A model of elastic behavior used in experimental structural geology that takes the Poisson ratio as equal to 0.25.　{ pwä'sōn ri‚lā·shən }

Poisson's equation [MATH] The partial differential equation which states that the Laplacian of an unknown function is equal to a given function.　{ pwä'sōnz i‚kwā·zhən }

Poisson transform [MATH] An integral transform which transforms the function $f(t)$ to the function

$$F(x) = (2/\pi)\int_0^\infty [t/(x^2 + t^2)]f(t)dt$$

Also known as potential transform.　{ pwä'sōn 'tranz‚fȯrm }

poke [COMPUT SCI] An instruction that causes a value in a storage location in a microcomputer's main storage to be replaced.　{ pōk }

pokeweed [BOT] *Phytolacca americana.* A poisonous garden weed that is found throughout the United States but is native to the eastern and central areas. Symptoms of poisoning include an immediate burning bitter taste, salivation, vomiting, diarrhea, and possible shock.　{ 'pōk‚wēd }

poke welding See push welding.　{ 'pōk 'weld·iŋ }

polacke [METEOROL] A cold, dry, northeasterly katabatic wind in Bohemia descending from the Sudeten Mountains (from the direction of Poland).　{ pō'läk·ə }

polar [ASTRON] A member of a class of cataclysmic variable stars whose light displays strong circular polarization. Also known as AM Herculis star. [MATH] **1.** For a conic section, the polar of a point is the line that passes through the points of contact of the two tangents drawn to the conic from the point. **2.** For a quadric surface, the polar of a point is the plane that passes through the curve which is the locus of the points of contact of the tangents drawn to the surface from the point. **3.** For a quadric surface, the polar of a line is the line of intersection of the planes which are tangent to the surface at its points of intersection with the original line.　{ 'pō·lər }

polar air [METEOROL] A type of air whose characteristics are developed over high latitudes; there are two types: continental polar air and maritime polar air.　{ 'pō·lər 'er }

polar angle [MATH] The angular coordinate θ in a polar coordinate system whose value at a given point *p* is the angle that a line from the origin to *p* makes with the polar axis.　{ 'pō·lər 'aŋ·gəl }

polar anticyclone See arctic high; subpolar high.　{ 'pō·lər ‚ant·i'sī‚klōn }

polar automatic weather station [METEOROL] An automatic weather station which measures meteorological elements and transmits them by radio; the station is designed to function primarily in frigid or polar climates in order to fill the need for weather reports from inaccessible regions where manned stations are not practicable; since the equipment is designed to operate on ice or slush, the main structure is in the form of a sled with external pontoons for added stability.　{ 'pō·lər ‚ȯd·ə‚mad·ik 'weth·ər ‚stā·shən }

polar axis [CRYSTAL] An axis of crystal symmetry which does not have a plane of symmetry perpendicular to it. [MATH] The directed straight line relative to which the angle is measured for a representation of a point in the plane by polar coordinates.　{ 'pō·lər 'ak·səs }

polar bear [VERT ZOO] *Thalarctos maritimus.* A large aquatic carnivore found in the polar regions of the Northern Hemisphere.　{ 'pō·lər ‚ber }

polar body [CYTOL] One of the small bodies cast off by the oocyte during maturation.　{ 'pō·lər ‚bäd·ē }

polar cap [ASTRON] Any of the bright areas covering the poles of Mars, believed to be composed of frozen carbon dioxide and water-ice. [HYD] An ice sheet centered at one of the poles of the earth.　{ 'pō·lər ‚kap }

polar-cap absorption [GEOPHYS] Very strong attenuation of radio waves over the polar regions during strong solar flares, due to extremely heavy ionization of the upper atmosphere. Abbreviated PCA.　{ 'pō·lər ‚kap əb‚sȯrp·shən }

polar-cap ice See polar ice.　{ 'pō·lər ‚kap 'īs }

polar chart [MAP] **1.** A chart of polar areas. **2.** A chart on a polar projection; the projections most used for polar charts are the gnomonic, stereographic, azimuthal equidistant, transverse Mercator, and modified Lambert conformal.　{ 'pō·lər 'chärt }

polar circle [GEOD] A parallel of latitude whose distance from the pole is equal to the obliquity of the ecliptic (approximately 23°27′).　{ 'pō·lər 'sər·kəl }

polar-class icebreaker [NAV ARCH] A ship that has a reinforced hull, a specially designed bow, and a system that allows rapid shifting of ballast to increase the effectiveness of icebreaking. It serves in the Arctic or the Antarctic as a platform for polar and ocean science and research.　{ ‚pō·lər ‚klas 'īs‚brāk·ər }

polar climate [CLIMATOL] The climate of a geographical polar region, most commonly taken to be a climate which is too cold to support the growth of trees. Also known as arctic climate; snow climate.　{ 'pō·lər 'klī·mət }

polar compound [CHEM] Molecules which contain polar covalent bonds; they can ionize when dissolved or fused; polar compounds include inorganic acids, bases, and salts.　{ 'pō·lər 'käm‚paȯnd }

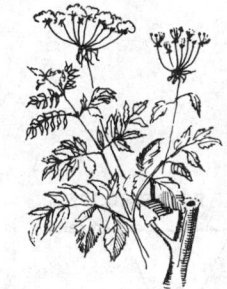

POISON HEMLOCK

Flowering branch of the poison hemlock.

POISON IVY

Three-foliolate leaf of poison ivy.

polar continental air [METEOROL] Air of an air mass that originates over land or frozen ocean areas in the polar regions; characterized by low temperature, stability, low specific humidity, and shallow vertical extent. { 'pō·lər ‚kant·ən·ent·əl 'er }

polar convergence [OCEANOGR] The line of convergence of polar and subpolar water masses in the ocean. { 'pō·lər kən'vər·jəns }

polar-coordinate navigation system [NAV] A system in which one or more signals are emitted from a facility (or co-located facilities) to produce simultaneous indication of bearing and distance. { 'pō·lər kō¦ȯrd·ən·ət ‚nav·ə'gā·shən ‚sis·təm }

polar coordinates [MATH] A point in the plane may be represented by coordinates (r,θ), where θ is the angle between the positive x-axis and the ray from the origin to the point, and r the length of that ray. { 'pō·lər kō'ȯrd·ən·əts }

polar covalent bond [PHYS CHEM] A bond in which a pair of electrons is shared in common between two atoms, but the pair is held more closely by one of the atoms. { 'pō·lər kō'vā·lənt 'bänd }

polar crystal See ferroelectric crystal. { 'pō·lər 'krist·əl }

polar cyclone See polar vortex. { 'pō·lər 'sī‚klōn }

polar desert [GEOGR] A high-latitude desert where the existing moisture is frozen in ice sheets and is thus unavailable for plant growth. Also known as arctic desert. { 'pō·lər 'dez·ərt }

polar developable [MATH] The envelope of the normal planes of a space curve. { ¦pō·lər di¦vel·əp·ə·bəl }

polar diagram [PHYS] A diagram employing polar coordinates to show the magnitude of a quantity in some or all directions from a point; examples include directivity patterns and radiation patterns. { 'pō·lər 'dī·ə‚gram }

polar distance [ASTRON] Angular distance from a celestial pole; the arc of an hour circle between a celestial pole, usually the elevated pole, and a point on the celestial sphere, measured from the celestial pole through 180°. { 'pō·lər 'dis·təns }

polar easterlies [METEOROL] The rather shallow and diffuse body of easterly winds located poleward of the subpolar low-pressure belt; in the mean in the Northern Hemisphere, these easterlies exist to an appreciable extent only north of the Aleutian low and Icelandic low. { 'pō·lər 'ēs·tər‚lēz }

polar-easterlies index [METEOROL] A measure of the strength of the easterly wind between the latitudes of 55° and 70°N; the index is computed from the average sea-level pressure difference between these latitudes and is expressed as the east to west component of geostrophic wind in meters and tenths of meters per second. { 'pō·lər 'ēs·tər‚lēz 'in‚deks }

polar electrojet [GEOPHYS] An intense current that flows in a relatively narrow band of the auroral zone ionosphere during disturbances of the magnetosphere. { 'pō·lər i'lek·trə‚jet }

polar equation [MATH] An equation expressed in polar coordinates. { ¦pō·lər i¦kwä·zhən }

polar firn [HYD] Firn formed at low temperatures with no melting or liquid water present. Also known as dry firn. { 'pō·lər 'fərn }

polar form [MATH] A complex number $x + iy$ has as polar form $re^{i\theta}$, where (r,θ) are the polar coordinates corresponding to the point of the plane with rectangular coordinates (x,y), that is, $r = \sqrt{x^2 + y^2}$ and $\theta = \arctan y/x$. { 'pō·lər 'fȯrm }

polar front [METEOROL] The semipermanent, semicontinuous front separating air masses of tropical and polar origin; this is the major front in terms of air mass contrast and susceptibility to cyclonic disturbance. { 'pō·lər 'frənt }

polar-front theory [METEOROL] A theory whereby a polar front, separating air masses of polar and tropical origin, gives rise to cyclonic disturbances which intensify and travel along the front, passing through various phases of a characteristic life history. { 'pō·lər ¦frənt ‚thē·ə·rē }

polar glacier [HYD] A glacier whose temperature is below freezing throughout its mass, and on which there is no melting during any season. { 'pō·lər 'glā·shər }

polar high See arctic high; subpolar high. { 'pō·lər 'hī }

polar ice [OCEANOGR] Sea ice that is more than 1 year old; the thickest form of sea ice. Also known as polar-cap ice. { 'pō·lər 'īs }

polarimeter [OPTICS] An instrument used to determine the rotation of the plane of polarization of plane polarized light when it passes through a substance; the light is linearly polarized by a polarizer (such as a Nicol prism), passes through the

material being analyzed, and then passes through an analyzer (such as another Nicol prism). { ‚pō·lə'rim·əd·ər }

polarimetric analysis [ANALY CHEM] A method of chemical analysis based on the optical activity of the substance being determined; the measurement of the extent of the optical rotation of the substance is used to identify the substance or determine its quantity. { pō¦lar·ə¦me·trik ə'nal·ə·səs }

polarimetry [OPTICS] The science of determining the polarization state of electromagnetic radiation (x-rays, light, or radio waves). { ‚pō·lə'rim·ə·trē }

Polaris [ASTRON] A creamy supergiant star of stellar magnitude 2.0, spectral classification F8, in the constellation Ursa Minor; marks the north celestial pole, being about 1° from this point; the star Ursae Minoris. Also known as North Star; Pole Star. { pə'lar·əs }

polariscope [OPTICS] Any of several instruments used to determine the effects of substances on polarized light, in which linearly or elliptically polarized light passes through the substance being studied, and then through an analyzer. { pə'lar·ə‚skōp }

Polaris correction [NAV] A correction to be applied to the corrected sextant altitude of Polaris to obtain latitude; this correction for the offset of Polaris from the north celestial pole varies with the local hour angle of Aries, latitude, and date. { pə'lar·əs kə‚rek·shən }

Polaris missile [ORD] A U.S. Navy surface-to-surface intermediate-range ballistic missile designed to be launched from submarines and surface ships for accurate bombardment of small target areas with conventional or nuclear warheads at ranges up to 2500 nautical miles (4600 kilometers). { pə'lar·əs 'mis·əl }

polariton [SOLID STATE] A coupled mode of motion in an ionic crystal due to the coupling between the electromagnetic field and transverse optical phonons of long wavelength. { pə'lar·ə‚tän }

polarity [COMMUN] **1.** The direction in which a direct current flows, in a teletypewriter system. **2.** The sense of the potential of a portion of a television picture signal representing a dark area of a scene relative to the potential of a portion of the signal representing a light area. [MATH] Property of a line segment whose two ends are distinguishable. [CELL MOL] The orientation of a strand of polynucleotide with respect to its partner, expressed in terms of nucleotide linkages. [PHYS] Property of a physical system which has two points with different (usually opposite) characteristics, such as one which has opposite charges or electric potentials, or opposite magnetic poles. { pə'lar·əd·ē }

polarity effect [ELECTR] An effect for which the breakdown voltage across a vacuum separating two electrodes, one of which is pointed, is much higher when the pointed electrode is the anode. { pə'lar·əd·ē i‚fekt }

polarity epoch [GEOPHYS] A period of time during which the earth's magnetic field was predominantly of a single polarity. { pə'lar·əd·ē ‚ep·ək }

polarity event [GEOPHYS] A period of no more than about 100,000 years when the earth's magnetic polarity was opposite to the predominant polarity of that polarity epoch. { pə'lar·əd·ē i‚vent }

polarity zone [GEOL] In stratigraphy, a material unit that is defined in terms of magnetic polarity, that is, reversals of the earth's magnetic field. { pə'lar·əd·ē ‚zōn }

polarizability [ELEC] The electric dipole moment induced in a system, such as an atom or molecule, by an electric field of unit strength. { ‚pō·lə‚rīz·ə'bil·əd·ē }

polarizability catastrophe [ELEC] According to a theory using the Lorentz field concept, the phenomenon where, at a certain temperature, the dielectric constant of a material becomes infinite. { ‚pō·lə‚rīz·ə'bil·əd·ē kə'tas·trə·fē }

polarizability ellipsoid See index ellipsoid. { ‚pō·lə‚rīz·ə'bil·əd·ē i'lip‚sȯid }

polarization [ELEC] **1.** The process of producing a relative displacement of positive and negative bound charges in a body by applying an electric field. **2.** A vector quantity equal to the electric dipole moment per unit volume of a material. Also known as dielectric polarization; electric polarization. **3.** A chemical change occurring in dry cells during use, increasing the internal resistance of the cell and shortening its useful life. [PHYS] **1.** Phenomenon exhibited by certain electromagnetic waves and other transverse waves in which the direction of the

POLARISCOPE

screen

lens

analyzer

quarter-wave plate

model

quarter-wave plate

polarizer

lens

light

A polariscope used in measuring photoelastic stress.

electric field or the displacement direction of the vibrations is constant or varies in some definite way. Also known as wave polarization. **2.** The direction of the electric field or the displacement vector of a wave exhibiting polarization (first definition). **3.** The process of bringing about polarization (first definition) in a transverse wave. **4.** Property of a collection of particles with spin, in which the majority have spin components pointing in one direction, rather than at random. { ˌpō·lə·rə′zā·shən }

polarization charge See bound charge. { ˌpō·lə·rə′zā·shən ˌchärj }

polarization diversity [COMMUN] A method of transmission and reception used to minimize the effects of selective fading of the horizontal and vertical components of a radio signal; it is usually accomplished through the use of separate vertically and horizontally polarized receiving antennas. { ˌpō·lə·rə′zā·shən də′vər·səd·ē }

polarization division multiple access [COMMUN] A technique for allowing multiple users at geographically dispersed locations to gain access to a shared communications channel by assigning them electric fields of different polarization. { ˌpō·lə·rə′zā·shən də‚vizh·ən ′məl·tə·pəl ′ak‚ses }

polarization division multiplexing [COMMUN] The sharing of a communications channel among multiple users by assigning them electric fields of different polarization. { ˌpō·lə·rə′zā·shən di‚vizh·ən ′məl·tə‚pleks·iŋ }

polarization ellipse [PHYS] The ellipse traced out by the tip of the electric field vector or the displacement vector of a polarized wave at a fixed point in space in the course of time. { ˌpō·lə·rə′zā·shən i‚lips }

polarization error [NAV] An error in a radio direction-finder bearing or the course indicated by a radio beacon because of a change in the polarization of the radio waves between the transmitter and receiver on being reflected from the ionosphere. Formerly referred to as night effect because it occurs principally during the night. { ˌpō·lə·rə′zā·shən ‚er·ər }

polarization fading [COMMUN] Fading as the result of changes in the direction of polarization in one or more of the propagation paths of waves arriving at a receiving point. { ˌpō·lə·rə′zā·shən ‚fād·iŋ }

polarization isocline [METEOROL] A locus of all points at which the inclination to the vertical of the plane of polarization of the diffuse sky radiation has the same value. { ˌpō·lə·rə′zā·shən ′īs·ə‚klīn }

polarization optical coherence tomography [OPTICS] A variation of optical coherence tomography which uses polarization optics in the arms of the fiber-optic Michelson interferometer to determine the sample birefringence from the magnitude of the back-reflected light. { ˌpō·lə·rə′zā·shən ‚äp·tə·kəl kō‚hir·əns tə′mäg·rə·fē }

polarization potential [ELECTROMAG] One of two vectors from which can be derived, by differentiation, an electric scalar potential and magnetic vector potential satisfying the Lorentz condition. Also known as Hertz vector. [PHYS CHEM] The reverse potential of an electrolytic cell which opposes the direct electrolytic potential of the cell. { ˌpō·lə·rə′zā·shən pə‚ten·chəl }

polarization spectroscopy [SPECT] A type of saturation spectroscopy in which a circularly polarized saturating laser beam depletes molecules with a certain orientation preferentially, leaving the remaining ones polarized; the latter are detected through their induction of elliptical polarization in a probe beam, allowing the beam to pass through crossed linear polarizers. { ˌpō·lə·rə′zā·shən spek′träs·kə·pē }

polarized ceramics [MATER] A substance, such as lead zirconate and barium titanate, having high electromechanical conversion efficiency and used as a transducer element in an ultrasonic system. { ′pō·lə‚rīzd sə′ram·iks }

polarized electrolytic capacitor [ELEC] An electrolytic capacitor in which the dielectric film is formed adjacent to only one metal electrode; the impedance to the flow of current is then greater in one direction than in the other. { ′pō·lə‚rīzd i‚lek·trə′lid·ik kə′pas·əd·ər }

polarized electromagnetic radiation [ELECTROMAG] Electromagnetic radiation in which the direction of the electric field vector is not random. { ′pō·lə‚rīzd i‚lek·trō·mag‚ned·ik ‚rād·ē′ā·shən }

polarized ion source [ELECTR] A device that generates ion beams in such a manner that the spins of the ions are aligned in some direction. { ′pō·lə‚rīzd ′ī‚än ‚sórs }

polarized light [OPTICS] Polarized electromagnetic radiation whose frequency is in the optical region. { ′pō·lə‚rīzd ′līt }

polarized meter [ENG] A meter having a zero-center scale, with the direction of deflection of the pointer depending on the polarity of the voltage or the direction of the current being measured. { ′pō·lə‚rīzd ′mēd·ər }

polarized neutrons [PHYS] A collection of neutrons in which the majority have spin pointing in one direction rather than at random. { ′pō·lə‚rīzd ′nü ‚tränz }

polarized plug [ELEC] A plug that can be inserted in its receptacle only when in a predetermined position. { ′pō·lə‚rīzd ′pləg }

polarized receptacle [ELEC] A receptacle designed for use with a polarized plug, to ensure that the grounded side of an alternating-current line or the positive side of a direct-current line is always connected to the same terminal on a piece of equipment. { ′pō·lə‚rīzd ri′sep·tə·kəl }

polarized relay [ELEC] Relay in which the movement of the armature depends upon the direction of the current in the circuit controlling the armature. Also known as polar relay. { ′pō·lə‚rīzd ′rē‚lā }

polarized scattering [PHYS CHEM] In a quasi-elastic light scattering experiment performed with polarizers, the type of scattering produced when the polarizers select both the incident and final polarizations perpendicular to the scattering plane. { ′pō·lə‚rīzd ′skad·ə·riŋ }

polarized-vane ammeter [ENG] An ammeter of only moderate accuracy in which the current to be measured passes through a small coil, distorting the field of a circular permanent magnet, and an iron vane aligns itself with the axis of the distorted field, the deflection being roughly proportional to the current. { ′pō·lə‚rīzd ‚vān ′am‚ēd·ər }

polarizer [OPTICS] A device which produces polarized light, such as a Nicol prism or Polaroid sheet. { ′pō·lə‚rīz·ər }

polarizing angle See Brewster's angle. { ′pō·lə‚rīz·iŋ ‚aŋ·gəl }

polarizing filter [OPTICS] A device which selectively absorbs components of electromagnetic radiation passing through it, so that light emerging from it is plane-polarized. { ′pō·lə‚rīz·iŋ ‚fil·tər }

polarizing microscope [OPTICS] A microscope in which an object is viewed in polarized light. { ′pō·lə‚rīz·iŋ ′mī·krə‚skōp }

polarizing pyrometer [ENG] A type of pyrometer, such as the Wanner optical pyrometer, in which monochromatic light from the source under investigation and light from a lamp with filament maintained at a constant but unknown temperature are both polarized and their intensities compared. { ′pō·lə‚rīz·iŋ pī′räm·əd·ər }

polar keying [COMMUN] Telegraph signal in which circuit current flows in one direction for spacing. { ′pō·lər ′kē·iŋ }

polar lake [HYD] A lake whose surface temperature never exceeds 4°C. { ′pō·lər ′lāk }

polar line [MATH] For a point on a space curve, the line that is normal to the osculating plane of the curve and passes through the center of curvature at that point. { ′pō·lər ‚līn }

polar low See polar vortex. { ′pō·lər ′lō }

polar maritime air [METEOROL] Air of an air mass that originates in the polar regions and is then modified by passing over a relatively warm ocean surface; characterized by moderately low temperature, moderately high surface specific humidity, and a considerable degree of vertical instability. { ′pō·lər ′mar·ə‚tīm ′er }

polar meteorology [METEOROL] The application of meteorological principles to a study of atmospheric conditions in the earth's high latitudes or polar-cap regions, northern and southern { ′pō·lər ‚mē·dē·ə′räl·ə·jē }

polar migration See polar wandering. { ′pō·lər mī′grā·shən }

polar modulation [COMMUN] Amplitude modulation in which the positive excursions of the carrier are modulated by one signal and the negative excursions by another. { ′pō·lər ‚mäj·ə′lā·shən }

polar molecule [PHYS CHEM] A molecule having a permanent electric dipole moment. { ′pō·lər ′mäl·ə‚kyül }

polar mutations [GEN] A class of mutations in the genes of an operon that affect the expression not only of the gene in which the mutation resides, but also of the genes located to one side of the mutated gene. { 'pō·lər myü'tā·shənz }

polar navigation [NAV] Navigation in polar regions, where unique considerations and techniques are applied; no definite limit for these regions is recognized, but polar navigation techniques are usually used from about latitude 70° to the nearest pole, north or south. { 'pō·lər ,nav·ə'gā·shən }

polar night [ASTRON] The period of winter darkness in the polar regions, both northern and southern. { 'pō·lər 'nīt }

polar normal [MATH] For a given point on a plane curve, the segment of the normal between the given point and the intersection of the normal with the radial line of a polar coordinate system that is perpendicular to the radial line to the given point. { 'pō·lər 'nor·məl }

polar nucleus [BOT] One of the two nuclei in the center of the embryo sac of a seed plant which fuse to form the endosperm nucleus. { 'pō·lər 'nü·klē·əs }

polarogram [ANALY CHEM] Plotted output (current versus electrode voltage) for polarographic analysis of an electrolyte. { pə'lar·ə,gram }

polarographic analysis [ANALY CHEM] An electroanalytical technique in which the current through an electrolysis cell is measured as a function of the applied potential; the apparatus consists of a potentiometer for adjusting the potential, a galvanometer for measuring current, and a cell which contains two electrodes, a reference electrode whose potential is constant and an indicator electrode which is commonly the dropping mercury electrode. Also known as polarography. { pō¦lar·ə¦graf·ik ə'nal·ə·səs }

polarographic cell [ANALY CHEM] Device for polarographic (voltammetric) analysis of an electrolyte solution; a known voltage is applied to the solution, and the ensuing current that passes through the cell (to an electrode) is measured. { pō¦lar·ə¦graf·ik 'sel }

polarographic maximum [ANALY CHEM] A deceptively high voltage buildup on an electrode during polarographic analysis of an electrolyte; caused by a reduction or oxidation process at the electrode. { pō¦lar·ə¦graf·ik 'mak·sə·məm }

polarography See polarographic analysis. { ,pō·lə'räg·rə·fē }

polaron [SOLID STATE] An electron in a crystal lattice together with a cloud of phonons that result from the deformation of the lattice produced by the interaction of the electron with ions or atoms in the lattice. { 'pō·lə,rän }

polar orbit [AERO ENG] A satellite orbit running north and south, so the satellite vehicle orbits over both the North Pole and the South Pole. { 'pō·lər 'or·bət }

polar outbreak [METEOROL] The movement of a cold air mass from its source region; almost invariably applied to a vigorous equatorward thrust of cold polar air, a rapid equatorward movement of the polar front. Also known as cold-air outbreak. { 'pō·lər 'aut,brāk }

polar plumes [ASTRON] Columnlike plumes of hot coronal gas that are concentrated at the sun's magnetic poles. { 'pō·lər 'plümz }

polar projection [MAP] A map projection centered on a pole. { 'pō·lər prə'jek·shən }

polar radiation pattern [ELECTROMAG] Diagram showing the relative strength of the radiation from an antenna in all directions in a given plane. [ENG ACOUS] Diagram showing the strength of sound waves radiated from a loudspeaker in various directions in a given plane, or a similar response pattern for a microphone. { 'pō·lər ,rād·ē'ā·shən ,pad·ərn }

polar reciprocal convex bodies [MATH] Any two convex bodies, each containing the origin in its interior, such that the Minkowski distance function of each is the support function of the other. { 'pō·lər ri¦sip·rə·kəl ¦kän,veks 'bäd,ēz }

polar-reciprocal curves [MATH] Two curves configured so that the polar of every point of one of them, with respect to a particular conic, is tangent to the other curve. { 'pō·lər ri,sip·rə·kəl 'kərvz }

polar-reciprocal triangles [MATH] Two triangles configured so that the vertices of each triangle are the poles of the sides of the other with respect to some conic. { 'pō·lər ri,sip·rə·kəl 'trī,aŋ·gəlz }

polar regions [GEOGR] The regions near the geographic poles; no definite limit for these regions is recognized. { 'pō·lər ,rē·jənz }

polar relay See polarized relay. { 'pō·lər 'rē,lā }

polar resolution [COMPUT SCI] Given the x and y components of a vector, the process of finding the magnitude of the vector and the angle it makes with the x axis. { 'pō·lər ,rez·ə'lü·shən }

polar sequence [ASTRON] A compilation of 96 brightness-standard stars within 2° of the North Pole. { 'pō·lər 'sē·kwəns }

polar solvent [MATER] A solvent in whose molecules there is either a permanent separation of positive and negative charges, or the centers of positive and negative charges do not coincide; these solvents have high dielectric constants, are chemically active, and form coordinate covalent bonds; examples are alcohols and ketones. { 'pō·lər 'säl·vənt }

polar subnormal [MATH] For a given point on a plane curve, the projection of the polar normal on the radial line of the polar coordinate system that is perpendicular to the radial line to the given point. { 'pō·lər səb'nor·məl }

polar subtangent [MATH] For a given point on a plane curve, the projection of the polar tangent on the radial line of the polar coordinate system that is perpendicular to the radial line to the given point. { 'pō·lər səb'tan·jənt }

polar symmetry [CRYSTAL] A type of crystal symmetry in which the two ends of the central crystallographic axis are not symmetrical. { 'pō·lər 'sim·ə·trē }

polar tangent [MATH] For a given point on a plane curve, the segment of the tangent between the given point and the intersection of the tangent with the radial line of a polar coordinate system that is perpendicular to the radial line to the given point. { 'pō·lər 'tan·jənt }

polar telescope [OPTICS] A telescope which uses rotating mirrors to enable celestial objects to be observed through a fixed eyepiece. { 'pō·lər 'tel·ə,skōp }

polar timing diagram [MECH ENG] A diagram of the events of an engine cycle relative to crankshaft position. { 'pō·lər 'tīm·iŋ ,dī·ə,gram }

polar transmission [COMMUN] **1.** A method of signaling in teletypewriter transmission in which direct currents flowing in opposite directions represent a mark and a space respectively, and absence of current indicates a no-signal condition. **2.** By extension, any system of signaling that uses three conditions, representing a mark, a space, or a no-signal condition. { 'pō·lər tranz'mish·ən }

polar triangle [MATH] A triangle associated to a given spherical triangle obtained from three directed lines perpendicular to the planes associated with the sides of the original triangle. { 'pō·lər 'trī,aŋ·gəl }

polar trough [METEOROL] In tropical meteorology, a wave trough in the circumpolar westerlies having sufficient amplitude to reach the tropics in the upper air; at the surface it is reflected as a trough in the tropical easterlies, but at moderate elevations it is characterized by westerly winds. { 'pō·lər 'trof }

polar variation [GEOPHYS] A small movement of the earth's axis of rotation relative to the geoid, the resultant of the Chandler wobble and other smaller movements. { 'pō·lər ,ver·ē'ā·shən }

polar vector See vector. { 'pō·lər 'vek·tər }

polar vortex [METEOROL] The large-scale cyclonic circulation in the middle and upper troposphere centered generally in the polar regions; specifically, the vortex has two centers in the mean, one near Baffin Island and another over northeastern Siberia; the associated cyclonic wind system comprises the westerlies of middle latitudes. Also known as Antarctic vortex; circumpolar whirl; polar cyclone; polar low. { 'pō·lər 'vor,teks }

polar wandering [GEOL] Migration during geologic time of the earth's poles of rotation and magnetic poles. Also known as Chandler motion; polar migration. { 'pō·lər 'wan·də·riŋ }

polar westerlies See westerlies. { 'pō·lər 'wes·tər,lēz }

polder [CIV ENG] Land reclaimed from the sea or other body of water by the construction of an embankment to restrain the water. { 'pōl·dər }

pole [CRYSTAL] **1.** A direction perpendicular to one of the faces of a crystal. **2.** One of the points at which normals to crystal faces or planes intersect a reference sphere at whose center the crystal is located. [ELEC] **1.** One of the electrodes in an electric cell. **2.** An output terminal on a switch; a double-pole switch has two output terminals. [MATH] **1.** An isolated singular point z_0 of a complex function whose Laurent series expansion about z_0 will include finitely many terms of form

$a_n(z - z_0)^{-n}$. **2.** For a great circle on a sphere, the pole of the circle is a point of intersection of the sphere and a line that passes through the center of the sphere and is perpendicular to the plane of the circle. **3.** For a conic section, the pole of a line is the intersection of the tangents to the conic at the points of intersection of the conic with the line. **4.** For a quadric surface, the pole of a plane is the vertex of the cone which is tangent to the surface along the curve where the plane intersects the surface. **5.** The origin of a system of polar coordinates on a plane. **6.** The origin of a system of geodesic polar coordinates on a surface. [MECH] **1.** A point at which an axis of rotation or of symmetry passes through the surface of a body. **2.** *See* perch. [OPTICS] **1.** The geometric center of a convex or concave mirror. **2.** *See* optical center. { 'pōl }

pole blight [PL PATH] A destructive disease of white pines characterized by shortening of the needle-bearing stems, yellowing and shortening of needles, and copious flow of resin. { 'pōl ,blīt }

pole-changing control [ELECTROMAG] A method of obtaining two or more running speeds of a three-phase motor by making changes in the number of magnetic poles, usually by making changes in the coil connections at the winding terminals. { 'pōl ¦chānj·iŋ kən,trōl }

pole-dipole array [ENG] An electrode array used in a lateral search conducted during a resistivity or induced polarization survey, or in drill hole logging, in which one current electrode is placed at infinity while another current electrode and two potential electrodes in proximity are moved across the structure to be investigated. { 'pōl 'dī,pōl ə,rā }

pole dominance [PARTIC PHYS] Property of a scattering amplitude, analytically continued to complex values of energy and scattering angle, whose behavior is dominated by the term or terms of negative power in the Laurent series of a nearby pole. { 'pōl ,däm·ə·nəns }

pole face [ELECTROMAG] The end of a magnetic core that faces the air gap in which the magnetic field performs useful work. { 'pōl ,fās }

pole-face winding [ELECTROMAG] Winding in the pole face of a motor or generator used to neutralize the cross-magnetizing armature reaction under the pole faces, which would otherwise cause a nonuniform distribution of voltage between commutator segments. Also known as compensated winding. { 'pōl ,fās ,wīnd·iŋ }

pole horn [ELECTROMAG] The part of a pole piece or pole shoe in an electrical machine that projects circumferentially beyond the pole core. { 'pōl ,hȯrn }

pole lathe [MECH ENG] A simple lathe in which the work is rotated by a cord attached to a treadle. { 'pōl ,lāth }

Polemoniaceae [BOT] A family of autotrophic dicotyledonous plants in the order Polemoniales distinguished by lack of internal phloem, corolla lobes that are convolute in the bud, three carpels, and axile placentation. { ,päl·ə,mō·nē'ās·ē,ē }

Polemoniales [BOT] An order of dicotyledonous plants in the subclass Asteridae, characterized by sympetalous flowers, a regular, usually five-lobed corolla, and stamens equal in number and alternate with the petals. { ,päl·ə,mō·nē'ā·lēz }

pole of inaccessibility *See* ice pole. { 'pōl əv ,in·ak,ses·ə'bil·əd·ē }

pole piece [ELECTROMAG] A piece of magnetic material forming one end of an electromagnet or permanent magnet, shaped to control the distribution of magnetic flux in the adjacent air gap. { 'pōl ,pēs }

pole-pole array [ENG] An electrode array, used in lateral search or in logging, in which one current electrode and the other potential electrode are kept in proximity and traversed across the structure. { 'pōl ,pōl ,ā·rā }

pole-positioning [CONT SYS] A design technique used in linear control theory in which many or all of a system's closed-loop poles are positioned as required, by proper choice of a linear state feedback law; if the system is controllable, all of the closed-loop poles can be arbitrarily positioned by this technique. { 'pōl pə,zish·ən·iŋ }

pole shoe [ELECTROMAG] Portion of a field pole facing the armature of the machine; it may be separable from the body of the pole. { 'pōl ,shü }

Pole Star *See* Polaris. { 'pōl ,stär }

polestar recorder [ENG] An instrument used to determine approximately the amount of cloudiness during the dark hours; consists of a fixed long-focus camera positioned so that Polaris is permanently within its field of view; the apparent motion of the star appears as a circular arc on the photograph and is interrupted as clouds come between the star and the camera. { 'pōl,stär ri,kȯrd·ər }

pole strength *See* magnetic pole strength. { 'pōl ,streŋkth }

pole tide [OCEANOGR] An ocean tide, theoretically 6 millimeters in amplitude, caused by the Chandler wobble of the earth; has a period of 428 days. { 'pōl ,tīd }

pole-zero configuration [CONT SYS] A plot of the poles and zeros of a transfer function in the complex plane; used to study the stability of a system, its natural motion, its frequency response, and its transient response. { 'pōl ¦zir·ō kən,fig·yə'rā shən }

polhode [MECH] For a rotating rigid body not subject to external torque, the closed curve traced out on the inertia ellipsoid by the intersection with this ellipsoid of an axis parallel to the angular velocity vector and through the center. { 'pä,lōd }

polhode cone *See* body cone. { 'pä,lōd ,kōn }

polian vesicle [INV ZOO] Interradial reservoirs connecting with the ring vessel in most asteroids and holothuroids. { 'pō·lē·ən 'ves·ə·kəl }

poling [ELEC] Adjustment of polarity; specifically, in wire-line practice, the use of transpositions between transposition sections of open wire or between lengths of cable, to cause the residual cross-talk couplings in individual sections or lengths to oppose one another. [MET] A technique used in the refining of copper that consists of the thrusting of green-wood poles into the molten metal in order to generate the reducing gases that react with the oxides in the metal. [MIN ENG] The act or process of temporarily protecting the face of a level, drift, or cut by driving poles or planks along the sides of the yet unbroken ground. { 'pōl·iŋ }

poling back [MIN ENG] Carrying out excavation behind timbering already in place. { 'pōl·iŋ 'bak }

poling board [CIV ENG] A timber plank driven into soft soil to support the sides of an excavation. { 'pōl·iŋ ,bȯrd }

polioencephalomyelitis [MED] A disease in which both the gray matter of the brain and the spinal cord are inflamed. { ¦pō·lē·ō·in¦sef·ə·lō,mī·ə līd·əs }

poliomyelitis [MED] An acute infectious viral disease which in its most serious form involves the central nervous system and, by destruction of motor neurons in the spinal cord, produces flaccid paralysis. Also known as Heine-Medin disease; infantile paralysis. { ¦pō·lē·ō,mī·ə'līd·əs }

poliovirus vaccine [IMMUNOL] A vaccine prepared from one or all three types of polioviruses in a live or attenuated state. { ¦pō·lē·ō vī·rəs vak'sēn }

polish [MATER] A powder, liquid, or semiliquid used to give smoothness, surface protection, or decoration to finishes; for example, finely ground red oxide (rouge) is used to polish plate glass, mirror backs, and optical glass; solvent-wax liquids and pastes are used to protect and enhance leather and wood surfaces; nitrocellulose lacquers are used to paint finger- and toenails. { 'päl·ish }

polished cotton [TEXT] Cotton fabric, usually of plain weave, with a glazed finish produced with resin. { 'päl·isht 'kät·ən }

polished-joint hanger [PETRO ENG] A type of tubing hanger that is slipped over or assembled around the top tubing joint in an oil well tubing string. { 'päl·isht ¦jȯint ,haŋ·ər }

polishing [CHEM ENG] In petroleum refining, removal of final traces of impurities, as for a lubricant, by clay adsorption or mild hydrogen treating. [MECH ENG] Smoothing and brightening a surface such as a metal or a rock through the use of abrasive materials. { 'päl·ish·iŋ }

polishing roll [MECH ENG] A roll or series of rolls on a plastics mold; has highly polished chrome-plated surfaces; used to produce a smooth surface on a plastic sheet as it is extruded. { 'päl·ish·iŋ ,rōl }

polishing wheel [DES ENG] An abrasive wheel used for polishing. { 'päl·ish·iŋ ,wēl }

Polish notation [COMPUT SCI] **1.** A notation system for digital-computer or calculator logic in which there are no parenthetical expressions and each operator is a binary or unary operator in the sense that it operates on not more than two operands. Also known as Lukasiewicz notation; parenthesis-free notation. **2.** The version of this notation in which operators precede the operands with which they are associated. Also known as prefix notation. { 'pō·lish nō'tā·shən }

Polish space [MATH] A separable metric space which is homeomorphic to a complete metric space. { 'pōl·ish 'spās }

pollen [BOT] The small male reproductive bodies produced in pollen sacs of the seed plants. { 'päl·ən }

pollen count [BOT] The number of grains of pollen that collect on a specified area (often taken as 1 square centimeter) in a specified time. { 'päl·ən ‚kaůnt }

pollen sac [BOT] In the anther of angiosperms and gymnosperms, a cavity that contains microspores. { 'päl·ən ‚sak }

pollen tube [BOT] The tube produced by the wall of a pollen grain which enters the embryo sac and provides a passage through which the male nuclei reach the female nuclei. { 'päl·ən ‚tüb }

pollination [BOT] The transfer of pollen from a stamen to a pistil; fertilization in flowering plants. { ‚päl·ə'nā·shən }

polling [COMMUN] A process that involves interrogating in succession every terminal on a shared communications line to determine which of the terminals require service. { 'pōl·iŋ }

polling list [COMMUN] A roster of transmitting devices sequentially scanned in a time-sharing system. { 'pōl·iŋ ‚list }

pollinosis See hay fever. { ‚päl·ə'nō·səs }

pollucite [MINERAL] $(Cs,Na)_2Al_2Si_4O_{12}·H_2O$ A colorless, transparent zoolite mineral composed of hydrous silicate of cesium, sodium, and aluminum, occurring massive or in cubes; used as a gemstone. Also known as pollux. { pə'lü‚sīt }

pollution [ECOL] Destruction or impairment of the purity of the environment. [PHYSIO] Emission of semen at times other than during coitus. { pə'lü·shən }

pollux See pollucite. { 'päl·əks }

Pollux [ASTRON] A giant orange-yellow star with visual brightness of 1.16, a little less than 35 light-years from the sun, spectral classification K0-III, in the constellation Gemini; the star β Geminorum. { 'päl·əks }

polonium [CHEM] A chemical element, symbol Po, atomic number 84; all polonium isotopes are radioactive; polonium-210 is the naturally occurring isotope found in pitchblende. { pə'lō·nē·əm }

polonium-210 [NUC PHYS] Radioactive isotope of polonium; mass 210, half-life 140 days, α-radiation; used to calibrate radiation counters, and in oil well logging and atomic batteries. Also known as radium F. { pə'lō·nē·əm 'tü‚ten }

poly- [ORG CHEM] A chemical prefix meaning many; for example, a polymer is made of a number of single molecules known as monomers, as polyethylene is made from ethylene. { 'päl·ē, 'päl·ə, 'päl·i }

polyabolo [MATH] A plane figure formed by joining isosceles right triangles along their edges. { 'päl·ē'ab·ə‚lō }

polyacetals See acetal resins. { 'päl·ē'as·ə‚talz }

Polya counting formula [MATH] A formula which counts the number of functions from a finite set D to another finite set, with two functions f and g assumed to be the same if some element of a fixed group of complete permutations of D takes f into g. { 'pōl·yə 'kaůn·tiŋ ‚fȯr·myə·lə }

polyacrylamide [ORG CHEM] $(CH_2CHCONH_2)_x$ A white, water-soluble high polymer based on acrylamide; used as a thickening or suspending agent in water-base formulations. { ¦päl·ē·ə¦kril·ə·məd }

polyacrylate [ORG CHEM] A polymer of an ester or salt of acrylic acid. { ¦päl·ē'ak·rə‚lāt }

polyacrylic acid [ORG CHEM] $(CH_2CHCOOH)_x$ An acrylic or acrylate resin formed by the polymerization of acrylic acid; water-soluble; used as a suspending and textile-sizing agent, and in adhesives, paints, and hydraulic fluids. { ¦päl·ē·ə¦kril·ik 'as·əd }

polyacrylic fiber [ORG CHEM] Continuous-strand fiber extruded from an acrylate resin. { ¦päl·ē·ə¦kril·ik 'fī·bər }

polyacrylonitrile [ORG CHEM] Polymer of acrylonitrile; semiconductive; used like an inorganic oxide catalyst to dehydrogenate tert-butyl alcohol to produce isobutylene and water. { ¦päl·ē¦ak·rə·lō'nī·trəl }

polyadelphous [BOT] Pertaining to stamens that are united by their filaments into several sets or bundles. { ¦päl·ē·ə¦del·fəs }

polyadenylation [MOL BIO] The addition of adenine nucleotides to the 3′ end of messenger ribonucleic acid molecules during posttranscriptional modification. { ¦päl·ē·ə‚den·ə'lā·shən }

Polya-Eggenberger distribution [STAT] A discrete frequency distribution that was originally considered in connection with contagious distributions. { ¦pōl·yə 'eg·ən‚bərg·ər ‚dis·trə‚byü·shən }

polyalcohol See polyhydric alcohol. { ¦päl·ē'al·kə‚hȯl }

polyalgorithm [MATH] A set of algorithms together with a strategy for choosing and switching among them. { ¦päl·ē'al·gə‚rith·əm }

polyallomer [ORG CHEM] A copolymer of propylene with other olefins. { ¦päl·ē'al·ə·mər }

polyalphabetic substitution cipher [COMMUN] A cipher that uses several substitution alphabets in turn. { ¦päl·ē‚al·fə'bed·ik ‚səb·stə'tü·shən ‚sī·fər }

polyamide [ORG CHEM] Any member of a class of polymers in which individual structural units are joined by amide bonds. { ¦päl·ē'am·əd }

polyamide resin [ORG CHEM] Product of polymerization of amino acid or the condensation of a polyamine with a polycarboxylic acid; an example is the nylons. { ¦päl·ē'am·əd 'rez·ən }

Polyangiaceae [MICROBIO] A family of bacteria in the order Myxobacterales; vegetative cells and myxospores are cylindrical with blunt, rounded ends; the slime capsule is lacking; sporangia are sessile or stalked. { ‚päl·ē‚an·jē'ās·ē‚ē }

polyargyrite [MINERAL] $Ag_{24}Sb_2S_{15}$ A gray to black mineral composed of antimony silver sulfide. { ‚päl·ē'är·jə‚rīt }

polyarteritis [MED] Inflammation of several arteries simultaneously. Also known as panarteritis. { ‚päl·ē‚ärd·ə'rīd·əs }

polyarteritis nodosa [MED] A systemic disease characterized by widespread inflammation of small and medium-sized arteries in which some of the foci are nodular. Also known as disseminated necrotizing periarteritis; periarteritis nodosa. { ‚päl·ē‚ärd·ə'rīd·əs nō'dō·sə }

polyarthritis [MED] Inflammation of several joints simultaneously. { ‚päl·ē·är'thrīd·əs }

polyatomic ion [CHEM] An electrically charged species formed by covalent bonding of atoms of two or more different elements, usually nonmetals, for example, the ammonium ion (NH_4^+). { ‚päl·ē·ə¦täm·ik 'ī·ən }

polyatomic molecule [CHEM] A chemical molecule with three or more atoms. { ‚päl·ē·ə¦täm·ik 'mäl·ə‚kyül }

polyaxon [INV ZOO] A spicule that is laid down along several axes. { ¦päl·ē'ak‚sän }

polybasic [CHEM] A chemical compound in solution that yields two or more H^- ions per molecule, such as sulfuric acid, H_2SO_4. { ¦päl·i‚bā·sik }

polybasite [MINERAL] $(Ag,Cu)_{16}Sb_2S_{11}$ An iron-black to steel-gray metallic-looking mineral; an ore of silver. { ‚päl·i'bā‚sīt }

polyblend [MATER] A mechanical mixture of two or more polymers, such as polystyrene and rubber. { 'päl·i‚blend }

Polybrachiidae [INV ZOO] A family of sedentary marine animals in the order Thecanephria. { ‚päl·i·brə'kī·ə‚dē }

polybutadiene [ORG CHEM] Oil-extendable synthetic elastomer polymer made from butadiene; resilience is similar to natural rubber; it is blended with natural rubber for use in tire and other rubber products. Also known as butadiene rubber. { ‚päl·i‚byüd·ə'dī‚ēn }

polybutene [ORG CHEM] A polymer of isobutene, $(CH_3)_2CCH_2$; made in varying chain lengths to give a wide range of properties from oily to solid; used as a lube-oil additive, in adhesives, and in rubber products. { ‚päl·i'byü‚tēn }

polybutylene [ORG CHEM] A polymer of one or more butylenes whose consistency ranges from a viscous liquid to a rubbery solid. { ‚päl·i'byüd·ə‚lēn }

polycapillary optics See Kumakhov optics. { ‚päl·ē‚kap·ə‚ler·ē 'äp·tiks }

polycarbonate [ORG CHEM] $[OC_6H_4C(CH_3)_2C_6H_4OCO]_x$ A linear polymer of carbonic acid which is a thermoplastic synthetic resin made from bisphenol and phosgene; used in emulsion coatings with glass fiber reinforcement. { ¦päl·i'kär·bə·nət }

polycarboxylic [ORG CHEM] Prefix for a compound containing two or more carboxyl (−COOH) groups. { ¦päl·i¦kär‚bäk¦sil·ik }

Polychaeta [INV ZOO] The largest class of the phylum Annelida, distinguished by paired, lateral, fleshy appendages (parapodia) provided with setae, on most segments. { ‚päl·i'kēd·ə }

polychlorinated biphenyl [ORG CHEM] Any member of the

group of chlorinated isomers of biphenyl. Abbreviated PCB. { ¦päl·i¦klȯr·ə‚nād·əd bī'fen·əl }

polychroism *See* pleochroism.

polychromatic radiation [ELECTROMAG] Electromagnetic radiation that is spread over a range of frequencies. { ¦päl·i‚krō'mad·ik ‚rād·ē'ā·shən }

Polycirrinae [INV ZOO] A subfamily of polychaete annelids in the family Terebellidae. { ‚päl·i'sir·ə‚nē }

polycistronic messenger [BIOCHEM] In ribonucleic acid viruses, messenger ribonucleic acid that contains the amino acid sequence for several proteins. { ¦päl·i·sis'trän·ik 'mes·ən‚jər }

Polycladida [INV ZOO] A class of marine Turbellaria whose leaflike bodies have a central intestine with radiating branches, many eyes, and tentacles in most species. { ‚päl·i'klad·əd·ə }

polyclimax [ECOL] A climax community under the controlling influence of many environmental factors, including soils, topography, fire, and animal interactions. { ¦päl·i'klī‚maks }

polyclinal fold [GEOL] One of a group of adjacent folds, the axial surfaces of which are oriented randomly, but which have similar surface axes. { ¦päl·i¦klīn·əl 'fōld }

polyclonal [IMMUNOL] Pertaining to cells or molecules that arise from more than one clone. { ‚päl·i'klō·nəl }

polyclonal mixed cryoglobulin [BIOCHEM] A cryoglobulin made of heterogeneous immunoglobin molecules belonging to two or more different classes, and sometimes additional serum proteins. { ¦päl·i¦klōn·əl 'mikst ¦krī·ə'gläb·yə·lən }

polycondensation [ORG CHEM] A chemical condensation leading to the formation of a polymer by the linking together of molecules of a monomer and the releasing of water or a similar simple substance. { ¦päl·i‚kän·dən'sā·shən }

polyconic chart [MAP] A chart on the polyconic projection. { ¦päl·i¦kän·ik 'chärt }

polyconic projection [MAP] A conic map projection in which the surface of a sphere or spheroid, such as the earth, is conceived as developed on a series of tangent cones, which are then spread out to form a plane; a separate cone is used for each small zone. { ¦päl·i¦kän·ik prə'jek·shən }

Polycopidae [INV ZOO] The single family of the suborder Cladocopa. { ‚päl·i'käp·ə‚dē }

polycrase [MINERAL] (Y,Ca,Ce,U,Th)(Ti,Cb,Ta)$_2$O$_6$ Black mineral composed of titanate, columbate, and tantalate of yttrium-group metals; it is isomorphous with euxenite and occurs in granite pegmatites. { 'päl·i‚krās }

polycrystal [MATER] A polycrystalline solid. { 'päl·i'krist·əl }

polycrystalline [MATER] **1.** Pertaining to a material composed of aggregates of individual crystals. **2.** Characterized by variously oriented crystals. { ‚päl·i'krist·əl·ən }

Polyctenidae [INV ZOO] A family of hemipteran insects in the superfamily Cimicoidea; the individuals are bat ectoparasites which resemble bedbugs but lack eyes and have ctenidia and strong claws. { ‚päl·ək'ten·ə‚dē }

polycyclic [ORG CHEM] A molecule that contains two or more closed atomic rings; can be aromatic (such as DDT), aliphatic (bianthryl), or mixed (dicarbazyl). { ‚päl·i'sī·klik }

polycyclic aromatic hydrocarbon [ORG CHEM] A compound containing two or more fused benzene rings such as naphthalene or anthracene. { ‚päl·ē‚sī·klik ‚ar·ə‚mad·ik 'hī·drə‚kär·bən }

polycyclic hydrocarbon *See* polynuclear hydrocarbon. { ‚päl·i'sī·klik 'hī·drə‚kär·bən }

polycyesis [MED] Multiple pregnancy. { ¦päl·i·sī'ē·səs }

polycystic kidney [MED] A usually hereditary, congenital, and bilateral disease in which a large number of cysts are present on the kidney. { ‚päl·i'sis·tik 'kid·nē }

polycystic ovarian syndrome [MED] A disorder in which the ovaries are bilaterally enlarged with multiple follicular cysts due to abnormal regulation of the hypothalamic-pituitary-ovarian axis. Symptoms include amenorrhea, menstrual abnormalities, infertility, and hirsutism. { ‚päl·ē‚sis·tik ō‚ver·ē·ən 'sin‚drōm }

polycythemia [MED] A condition characterized by an increased number of erythrocytes in the circulation. { ‚päl·i‚sī'thē·mē·ə }

polycythemia vera [MED] An absolute increase in all blood cells derived from bone marrow, especially erythrocytes. { ‚päl·i‚sī'thē·mē·ə 'vir·ə }

polydactyly [MED] The condition of having supernumerary fingers or toes. { ‚päl·i'dak·təl·ē }

polydent [ORG CHEM] Pertaining to a chemical species whose molecules possess more than two reactive sites. Also known as multident. { 'päl·ə‚dent }

polydipsia [MED] Excessive thirst. { ‚päl·i'dip·sē·ə }

polydisperse colloidal system [CHEM] A colloidal system in which the suspended particles have various sizes and shapes. { ¦päl·i·di'spərs kə'lȯid·əl 'sis·təm }

polydispersity [CHEM] Molecular-weight nonhomogeneity in a polymer system; that is, there is some molecular-weight distribution throughout the body of the polymer. { ¦päl·i·di'spər·səd·ē }

Polydolopidae [PALEON] A Cenozoic family of rodentlike marsupial mammals. { ‚päl·i·də'läp·ə‚dē }

polydymite [MINERAL] Ni$_3$S$_4$ A mineral of the linnaeite group consisting of nickel sulfide. { pə'lid·ə‚mīt }

polyelectrolyte [ORG CHEM] A natural or synthetic electrolyte with high molecular weight, such as proteins, polysaccharides, and alkyl addition products of polyvinyl pyridine; can be a weak or strong electrolyte; when dissociated in solution, it does not give un form distribution of positive and negative ions (the ions of one sign are bound to the polymer chain while the ions of the other sign diffuse through the solution). { ¦päl·ē·ə'lek·trə‚līt }

polyembryony [ZOO] A form of sexual reproduction in which two or more offspring are derived from a single egg. { ¦päl·ē·im'brī·ə‚nē }

polyene [ORG CHEM] Compound containing many double bonds, such as the carotenoids. { 'päl·ē‚ēn }

polyester fiber [TEXT] A fiber filament made from a material that is 85% or more thermoplastic polyester resin. { 'päl·ē‚es·tər 'fī·bər }

polyester film [MATER] Thin film made of polyester resin; used for packaging food and other products. { 'päl·ē‚es·tər 'film }

polyester laminate [MATER] Glass fabric or fiber mat impregnated with a polyester resin slurry, and cured; used to make sheets, bars, and structural shapes. { 'päl·ē‚es·tər 'lam·ə·nət }

polyester-reinforced urethane [MATER] A poromeric material which may have a urethane impregnation or a silicone coating; used for shoe uppers and as a substitute for industrial leathers. { 'päl·ē‚es·tər ‚rē·in‚fȯrst 'yùr·ə‚thān }

polyester resin [ORG CHEM] A thermosetting or thermoplastic synthetic resin made by esterification of polybasic organic acids with polyhydric acids; examples are Dacron and Mylar; the resin has high strength and excellent resistance to moisture and chemicals when cured. { 'päl·ē‚es·tər 'rez·ən }

polyester rubber *See* polyurethane rubber. { 'päl·ē‚es·tər 'rəb·ər }

polyestrous [PHYSIO] Having several periods of estrus in a year. { ¦päl·ē'es·trəs }

polyether [ORG CHEM] Any compound whose molecular structure contains linked ethers, R−O−R', where R and R' represent functional groups. { 'päl·ē‚ē·thər }

polyether resin [ORG CHEM] Any member of a large group of thermoplastic or thermosetting polymers that contain the typical polyether linkages in the polymer chain. { 'päl·ē‚ē·thər 'rez·ən }

polyethylene *See* ethylene resin. { ‚päl·ē'eth·ə‚lēn }

polyethylene glycol [ORG CHEM] Any of a family of colorless, water-soluble liquids with molecular weights from 200 to 6000; soluble also in aromatic hydrocarbons (not aliphatics) and many organic solvents; used to make emulsifying agents and detergents, and as plasticizers, humectants, and water-soluble textile lubricants. { ‚päl·ē eth·ə‚lēn 'glī‚kȯl }

polyethylene glycol distearate *See* polyglycol distearate. { ‚päl·ē'eth·ə‚lēn 'glī‚kȯl dī'stir‚āt }

polyethylene resin *See* ethylene resin. { ‚päl·ē'eth·ə‚lēn 'rez·ən }

polyethylene terephthalate [ORG CHEM] A thermoplastic polyester resin made from ethylene glycol and terephthalic acid; melts at 265°C; used to make films or fibers. Abbreviated PET. { ¦päl·ē'eth·ə‚lēn ‚ter·ə'tha‚lāt }

polyformaldehyde *See* polyoxymethylene. { ‚päl·ē·fȯr'mal·də‚hīd }

polyforming [CHEM ENG] A noncatalytic, petroleum-refinery process charging C$_3$ and C$_4$ gases with naphtha or gas oil

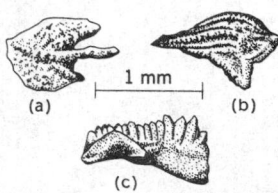

(a)

(b)

1 mm

(c)

Toothlike shapes of polygnaths show their platformlike appearance. *(a) Ancyrodella. (b) Palmatolepis. (c) Polygnathus.*

Polygonum hydropiper, eastern American smartweed. *(Photograph by A.W. Ambler, from National Audubon Society)*

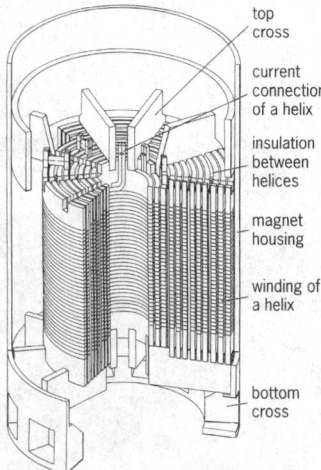

top cross

current connection of a helix

insulation between helices

magnet housing

winding of a helix

bottom cross

Diagram showing the geometry of the helical coils. Cooling water flows at high pressure between the helices.

at high temperature to produce high-quality gasoline and fuel oil; mostly replaced by catalytic reforming; the product is known as polyformdistillate. { 'päl·ē‚form·iŋ }

Polygalaceae [BOT] A family of dicotyledonous plants in the order Polygalales distinguished by having a bicarpellate pistil and monadelphous stamens. { ‚päl·i·gə'lās·ē‚ē }

polygalacturonase [BIOCHEM] An enzyme that catalyzes the hydrolysis of glycosidic linkage of polymerized galacturonic acids. { ‚päl·i‚ga‚lak'tûr·ə‚nās }

Polygalales [BOT] An order of dicotyledonous plants in the subclass Rosidae characterized by its simple leaves and usually irregular, hypogynous flowers. { ‚päl·i·gə'lā·lēz }

polygamous [BOT] Having both perfect and imperfect flowers on the same plant. [VERT ZOO] Having more than one mate at one time. { pə'lig·ə·məs }

polygen See polyvalent. { 'päl·i·jən }

polygene [GEN] One of a group of nonallelic genes that collectively control a quantitative character. [GEOL] An igneous rock composed of two or more minerals. Also known as polymere. { 'päl·i‚jēn }

polygenetic [GEOL] **1.** Resulting from more than one process of formation or derived from more than one source, or originating or developing at various places and times. **2.** Consisting of more than one type of material, or having a heterogeneous composition. Also known as polygenic. { ¦päl·i·jə'ned·ik }

polygenic See polygenetic. { ¦päl·i¦jen·ik }

polygenic inheritance [GEN] The phenotypic expression of a trait involving the interaction of many genes. { ‚päl·i‚jen·ik in'her·əd·əns }

polygeosyncline [GEOL] A geosynclinal-geoanticlinal belt that lies along the continental margin and receives sediments from a borderland on its oceanic side. { ‚päl·i‚jē·ō'sin‚klīn }

polyglycol [ORG CHEM] A dihydroxy ether derived from the dehydration (removal of a water molecule) of two or more glycol molecules; an example is diethylene glycol, CH_2OH-$CH_2OCH_2CH_2OH$. { 'päl·i‚glī‚kȯl }

polyglycol distearate [ORG CHEM] $(C_{17}H_{35})_2CO_2CO(CH_2$-$CH_2O)_x$ An off-white, soft solid with a melting point of 43°C; soluble in chlorinated solvents, acetone, and light esters; used as a resin plasticizer. Also known as polyethylene glycol distearate. { 'päl·i‚glī‚kȯl dī'stir‚āt }

polyglycolic acid polymer [MED] A synthetic biodegradable polymer material used as dissolvable sutures and tissue engineering scaffolds. { ‚päl·ē·glī¦käl·ik ¦as·əd 'päl·i·mər }

Polygnathidae [PALEON] A family of Middle Silurian to Cretaceous conodonts in the suborder Conodontiformes, having platforms with small pitlike attachment scars. { pal·ig'nath·ə‚dē }

polygon [MATH] A figure in the plane given by points p_1, $p_2, \ldots, p_n$ and line segments $p_1p_2, p_2p_3, \ldots, p_{n-1}p_n, p_np_1$. { 'päl·i‚gän }

Polygonaceae [BOT] The single family of the order Polygonales. { pə‚lig·ə'nās·ē‚ē }

Polygonales [BOT] An order of dicotyledonous plants in the subclass Caryophyllidae characterized by well-developed endosperm, a unilocular ovary, and often trimerous flowers. { pə‚lig·ə'nā·lēz }

polygonal ground [GEOL] A ground surface consisting of polygonal arrangements of rock, soil, and vegetation formed on a level or gently sloping surface by frost action. Also known as cellular soil. { pə'lig·ən·əl 'graúnd }

polygonal karst [GEOL] A karst pattern that is characteristic of tropical types such as cone karsts, with the surface completely divided into a polygonal network. { pə'lig·ən·əl 'kärst }

polygonal method [MIN ENG] A method of estimating ore reserves in which it is assumed that each drill hole has an area of influence extending halfway to the neighboring drill holes. { pə'lig·ən·əl 'meth·əd }

polygonal region [MATH] The union of the interior of a polygon with some, all, or none of the polygon itself. { pə'lig·ən·əl ‚rē·jən }

polygonal ring structure [ASTRON] A lunar crater whose wall approximates a polygon in shape. { pə'lig·ən·əl 'riŋ ‚strək·chər }

polygonization [SOLID STATE] A phenomenon observed during the annealing of plastically bent crystals in which the edge dislocations created by cold working organize themselves

vertically above each other so that polygonal domains are formed. { pə‚lig·ə·nə'zā·shən }

polygon of vectors [MATH] A polygon all but one of whose sides represent vectors to be added, directed in the same sense along the perimeter, and whose remaining side represents the sum of these vectors, directed in the opposite sense. { ¦päl·i‚gän əv 'vek·tərz }

polygon wall See tilt boundary. { 'päl·i‚gän ‚wȯl }

polygraph See lie detector. { 'päl·i‚graf }

polyhalite [MINERAL] $K_2MgCa_2(SO_4)_4 \cdot 2H_2O$ A sulfate mineral usually found in fibrous brick-red masses due to iron. { ‚päl·i·ha‚līt }

polyhaloalkane [ORG CHEM] An alkane derivative in which two or more hydrogen atoms have been replaced by halogen atoms. { ‚päl·ē‚ha·lō'al‚kān }

polyhalogeno compound [ORG CHEM] An organic compound containing more than one halogen atom. { ‚päl·ē‚hə'läj·ə·nō ‚käm‚paúnd }

polyhedral angle [MATH] The shape formed by the lateral faces of a polyhedron which have a common vertex. { ¦päl·i¦hē·drəl 'aŋ·gəl }

polyhedral disease See polyhedrosis. { ¦päl·i¦hē·drəl di'zēz }

polyhedric projection [MAP] A projection for large-scale topographic maps in which a small quadrangle on the sphere or spheroid is projected onto a plane trapezoid, the rectilinear parallels and meridians corresponding closely to arc distances on the sphere or spheroid. { ¦päl·i¦hē·drik prə'jek·shən }

polyhedron [MATH] **1.** A solid bounded by planar polygons. **2.** The set of points that belong to the simplexes of a simplicial complex. **3.** See triangulable space. { ‚päl·i¦hē·drən }

polyhedrosis [INV ZOO] Any of several virus diseases of insect larvae characterized by the breakdown of tissues and presence of polyhedral granules. Also known as polyhedral disease. { ‚päl·i‚hē'drō·səs }

polyhelix magnet [ELECTROMAG] A normal-conductor air-cored magnet constructed from a series of concentric monolayer coils. { ‚päl·ē‚hē·liks 'mag·nət }

polyhex [MATH] A plane figure formed by joining a finite number of regular hexagons along their sides. { 'päl·i‚heks }

polyhidrosis See hyperhidrosis. { ¦päl·i·hī'drō·səs }

polyhydramnios [MED] An excessive volume of amniotic fluid. Also known as hydramnios. { ‚päl·i·hī'dram‚nē‚ōs }

polyhydric alcohol [ORG CHEM] An alcohol with many hydroxyl ($-OH$) radicals, such as glycerol, $C_3H_5(OH)_3$. Also known as polyalcohol; polyol. { ‚päl·i¦hī·drək 'al·kə‚hȯl }

polyhydric phenol [ORG CHEM] A phenolic compound containing two or more hydroxyl groups, such as diphenol, $C_6H_4(OH)_2$. { ‚päl·i¦hī·drək 'fē‚nȯl }

polyiamond [MATH] A plane figure formed by joining a finite number of equilateral triangles along their sides. { 'päl·ē·ə‚mänd }

polyimide [CHEM ENG] A group of polymers that contain a repeating imide group ($-CONHCO-$). Aromatic polyimides are noted for their resistance to high temperatures, wear, and corrosion. { ‚päl·ē'ī‚mīd }

polyimide resin [ORG CHEM] An aromatic polyimide made by reacting pyromellitic dianhydride with an aromatic diamine; has high resistance to thermal stresses; used to make components of internal combustion engines. { ¦päl·ē'ī‚mīd ‚rez·ən }

polyisoprene [ORG CHEM] $(C_5H_8)_x$ The basis of natural rubber, balata, gutta-percha, and other rubberlike materials; can also be made synthetically; the stereospecific forms are cis-1,4- and trans-1,4-polyisoprene; the polymer is thermoplastic. { ‚päl·ē'īs·ə‚prēn }

polykaryocyte [MED] See syncytium. { ‚päl·i'kar·ē·ə‚sīt }

polyking See polyplet. { 'päl·i‚kiŋ }

polylactic resin [ORG CHEM] A soft, elastic resin made by the heat reaction of lactic acid with castor oil or other fatty oils; used to produce tough, water-resistant coatings. { ¦päl·i¦lak·tik ‚rez·ən }

polyLED See polymer light-emitting diode. { ‚päl·ē¦el¦ē'dē }

polyligated atom [PHYS CHEM] An atom that is bonded to more than one other atom. { ‚päl·ē¦lī‚gäd·əd 'ad·əm }

polyline [COMPUT SCI] In computer graphics, a series of connected line segments and arcs that are treated as a single entity. { 'päl·ē‚līn }

polyliner [ENG] A perforated sleeve with longitudinal ribs that is used inside the cylinder of an injection-molding machine. { 'päl·i‚līn·ər }

polymenorrhea *See* metrorrhagia. ｛ ¦päl·i‚men·ə'rē·ə ｝

polymer [ORG CHEM] Substance made of giant molecules formed by the union of simple molecules (monomers); for example polymerization of ethylene forms a polyethylene chain, or condensation of phenol and formaldehyde (with production of water) forms phenol-formaldehyde resins. ｛ 'päl·ə·mər ｝

Polymera [INV ZOO] Formerly a subphylum of the Vermes; equivalent to the phylum Annelida. ｛ pə'lim·ə·rə ｝

polymerase [BIOCHEM] An enzyme that links nucleotides together to form polynucleotide chains. ｛ pə'lim·ə‚rās ｝

polymerase chain reaction [MOL BIO] A technique for copying and amplifying the complementary strands of a target deoxyribonucleic acid molecule. Abbreviated PCR. ｛ pə¦lim·ə‚rās 'chān rē‚ak·shən ｝

polymer battery [MATER SCI] An electrochemical device in which the cathodes, anodes, and electrolytes are made of conductive polymers. ｛ 'päl·ə·mər ‚bad·ə·rē ｝

polymer blend [ORG CHEM] A homogeneous mixture of two or more different polymers. ｛ 'päl·ə·mər ‚blend ｝

polymer-dispersed liquid-crystal display [ELECTR] An electronic display in which the display elements have micrometer-sized diameter, have nearly spherical liquid-crystal droplets surrounded by a solid polymer, and the display is switched from a white opaque appearance to a clear transparent appearance by applying an electric field. ｛ ¦päl·ə·mər di‚spərst ‚lik·wəd ‚krist·əl di'splā ｝

polymere *See* polygene. ｛ 'päl·ə‚mir ｝

polymer gasoline [MATER] A product of polymerization of normally gaseous hydrocarbons to form high-octane liquid hydrocarbons boiling in the gasoline range. ｛ 'päl·ə·mər 'gas·ə‚lēn ｝

polymeric [CHEM] Made of repeating subunits. ｛ ‚päl·ə'mer·ik ｝

polymerization [CHEM] **1.** The bonding of two or more monomers to produce a polymer. **2.** Any chemical reaction that produces such a bonding. ｛ pə‚lim·ə·rə'zā·shən ｝

polymer light-emitting diode [ORG CHEM] An organic polymeric material that emits light in response to the application of an electric field. It may be an organic semiconductor sandwiched between metals of high and low work functions or a heterostructure made of two polymers, which increases the likelihood of radiative electron-hole recombination because of the energy-band structure. Also known as light-emitting polymer; polyLED. ｛ ¦päl·ə·mər ‚līt·i‚mid·iŋ 'dī‚ōd ｝

polymer paint [MATER] A paint made of acrylic resin or vinyl resin, or a combination of both resins, in a liquid form with water as the base; it spreads out in a layer, and the water evaporates to leave a continuous, flexible, and waterproof film of plastic. ｛ 'päl·ə·mər 'pānt ｝

polymer plastic [MATER] The product of a high polymer with or without additives, such as plasticizers, autooxidants, colorants, or fillers; can be sprayed, shaped, molded, extruded, cast or foamed, depending on whether it is thermoplastic or thermosetting. ｛ 'päl·ə·mər 'plas·tik ｝

polymetamorphic diaphthoresis [GEOL] Retrograde changes during a second phase of metamorphism that is clearly separated from a previous, higher-grade metamorphic period. ｛ ¦päl·i‚med·ə'mór·fik dī‚af·thə'rē·səs ｝

polymetamorphism [GEOL] Polyphase or multiple metamorphism whereby two or more successive metamorphic events have left their imprint upon the same rocks. ｛ ¦päl·i‚med·ə'mór‚fiz·əm ｝

polymethyl methacrylate [ORG CHEM] A thermoplastic polymer that is derived from methyl methacrylate, $CH_2 = C(CH_3)COOCH_3$; transparent solid with excellent optical qualities and water resistance; used for aircraft domes, lighting fixtures, optical instruments, and surgical appliances. ｛ 'päl·i‚meth·əl mə'thak·rə‚lāt ｝

polymictic [HYD] Pertaining to or characteristic of a lake having no stabile thermal stratification. [PETR] Of a clastic sedimentary rock, being made up of many rock types or of more than one mineral species. ｛ 'päl·i¦mik·tik ｝

polymignite *See* polymignyte. ｛ ‚päl·i'mig‚nīt ｝

polymignyte [MINERAL] $(Ca,Fe,Y,Zr,Th)(Nb,Ti,Ta)O_4$ A black mineral composed of niobate, titanate, and tantalate of cerium-group metals, with calcium and iron. Also spelled polymignite. ｛ ‚päl·i'mig‚nīt ｝

polymodal distribution [STAT] A frequency distribution characterized by two or more localized modes, each having a higher frequency of occurrence than other immediately adjacent individuals or classes. ｛ ¦päl·i'mōd·əl ‚dis·trə'byü·shən ｝

polymolecular assembly [CHEM] The spontaneous association of a large number of components into a specific phase (films, layers, membranes, vesicles, micelles, mesophases, surfaces, solids, and so on). ｛ ¦päl·ē·mə‚lek·yə·lər ə'sem·blē ｝

polymorph [BIOL] An organism that exhibits polymorphism. [CRYSTAL] One of the crystal forms of a substance displaying polymorphism. Also known as polymorphic modification. [HISTOL] *See* granulocyte. ｛ 'päl·i‚mórf ｝

polymorphic modification *See* polymorph. ｛ ¦päl·i¦mór·fik ‚mäd·ə·fə'kā·shən ｝

polymorphic system [COMPUT SCI] A computer system that is organized around a central pool of shared software modules which are selected as they are needed for processing. ｛ ¦päl·i¦mór·fik 'sis·təm ｝

polymorphism [BIOL] **1.** Occurrence of different forms of individual in a single species. **2.** Occurrence of different structural forms in a single individual at different periods in the life cycle. [COMPUT SCI] A property of object-oriented programming that allows many different types of objects to be treated in a uniform manner by invoking the same operation on each object. [CRYSTAL] The property of a chemical substance crystallizing into two or more forms having different structures, such as diamond and graphite. Also known as pleomorphism. [GEN] The coexistence of genetically determined distinct forms in the same population, even the rarest of them being too common to be maintained solely by mutation; human blood groups are an example. ｛ ‚päl·i'mór‚fiz·əm ｝

polymorphonuclear leukocyte *See* granulocyte. ｛ ¦päl·i‚mór·fō'nü·klē·ər 'lü·kə‚sīt ｝

polymyarian [INV ZOO] Referring to the cross-sectional appearance of muscle cells in a nematode, having many cells in each quadrant. ｛ ¦päl·i‚mī'ar·ē·ən ｝

polymyositis [MED] Inflammation of many muscles simultaneously. ｛ ¦päl·i‚mī·ə'sīd·əs ｝

polymyxin [MICROBIO] Any of the basic polypeptide antibiotics produced by certain strains of *Bacillus polymyxa*. ｛ ¦päl·i'mik·sən ｝

Polynemidae [VERT ZOO] A family of perciform shore fishes in the suborder Mugiloidei. ｛ ‚päl·i'nem·ə‚dē ｝

polyneuritis [MED] Degenerative or inflammatory lesions of several nerves simultaneously, usually symmetrical. Also known as multiple neuritis. ｛ ¦päl·i·nü'rīd·əs ｝

Polynoidae [INV ZOO] The largest family of polychaetes, included in the Errantia and having a body of varying size and shape that is covered with elytra. ｛ ‚päl·ə'nói‚dē ｝

polynomial [MATH] A polynomial in the quantities $x_1, x_2, \ldots, x_n$ is an expression involving a finite sum of terms of form $bx_1^{p_1} x_2^{p_2} \ldots x_n^{p_n}$, where b is some number, and $p_1, \ldots, p_n$ are integers. ｛ ‚päl·ə¦nō·mē·əl ｝

polynomial equation [MATH] An equation in which a polynomial in one or more variables is set equal to zero. ｛ ‚päl·ə¦nō·mē·əl i'kwē·zhən ｝

polynomial function [MATH] A function whose values can be found by substituting the value (or values) of the independent variable (or variables) in a polynomial. ｛ ‚päl·ə¦nō·mē·əl 'fəŋk·shən ｝

polynomial time [COMPUT SCI] The property of the time required to solve a problem on a computer for which there exist constants c and k such that, if the input to the problem can be specified in N bits, the problem can be solved in $c \times N^k$ elementary operations. ｛ ‚päl·ə¦nō·mē·əl 'tīm ｝

polynomial trend [STAT] A trend line which is best approximated by a polynomial function; used in time series analysis. ｛ ‚päl·i¦nō·mē·əl ‚trend ｝

polynuclear hydrocarbon [ORG CHEM] Hydrocarbon molecule with two or more closed rings; examples are naphthalene, $C_{10}H_8$, with two benzene rings side by side, or diphenyl, $(C_6H_5)_2$, with two bond-connected benzene rings. Also known as polycyclic hydrocarbon. ｛ ¦päl·ə nü·klē·ər 'hī·drə‚kär·bən ｝

polynucleotide [BIOCHEM] A linear sequence of nucleotides. ｛ ¦päl·ə'nü·klē·ə‚tīd ｝

polyn'ya [OCEANOGR] A Russian term for a water area, other than a lead, lane, or crack, which is surrounded by sea ice; the term "window" is sometimes used for a similar open area in river ice. Also known as ice clearing. ｛ ‚päl·ən‚yä ｝

POLYNOIDAE

Harmothoe of the Polynoidae, dorsal view.

Polyodontidae [INV ZOO] A family of tubicolous, often large-bodied errantian polychaetes with characteristic cephalic and parapodial structures. { ˌpäl·ē·ōˈdänt·ə‚dē }

polyol *See* polyhydric alcohol. { ˈpäl·ē‚ól }

polyolefin [ORG CHEM] A resinous material made by the polymerization of olefins, such as polyethylene from ethylene, polypropylene from propylene, or polybutene from butylene. { ¦päl·ēˈōl·ə·fən }

polyolefin fiber [MATER] Continuous-strand fiber made from a polyolefin. { ¦päl·ēˈōl·ə·fən ˈfī·bər }

polyoma virus [VIROL] A small deoxyribonucleic acid virus normally causing inapparent infection in mice, but experimentally capable of producing parotid tumors and a wide variety of other tumors. { ˌpäl·ēˈō·mə ˈvī·rəs }

polyomino [MATH] A plane figure formed by joining a finite number of unit squares along their sides. { ˌpäl·ēˈäm·ə·nō }

polyopia [MED] A condition in which more than one image of an object is formed upon the retina. { ˌpäl·ēˈō·pē·ə }

Polyopisthocotylea [INV ZOO] An order of the trematode subclass Monogenea having a solid posterior holdfast bearing suckers or clamps. { ¦päl·ē·ō‚pisˈthə‚kad·əlˈē·ə }

polyorrhymenitis *See* polyserositis. { ¦päl·ē·óˌrim·əˈnīd·əs }

polyoxyalkylene resin [ORG CHEM] Condensation polymer produced from an oxyalkene, such as polyethylene glycol from oxyethylene or ethylene glycol. { ¦päl·ē¦akˈsēˈal·kə‚len ˈrez·ən }

polyoxyethylene (8) stearate *See* polyoxyl (8) stearate. { ¦päl·ē‚akˈsēˈeth·ə‚len ˈāt ˈstir‚āt }

polyoxyl (8) stearate [ORG CHEM] A cream-colored, soft, waxy solid at 25°C; soluble in toluene, acetone, ether, and ethanol; used in bakery products as an emulsifier. Also known as polyoxyethylene (8) stearate. { ˌpäl·ē‚akˈsəl ¦āt ˈstir‚āt }

polyoxymethylene [ORG CHEM] (OCH₂)ₙ A polymer of formaldehyde that has excellent mechanical and high-temperature properties. Also know as polyacetal; polyformaldehyde. { ˌpäl·ē‚akˈsēˈmeth·ə‚len }

polyp [INV ZOO] A sessile cnidarian individual having a hollow, somewhat cylindrical body, attached at one end, with a mouth surrounded by tentacles at the free end; may be solitary (hydra) or colonial (coral). [MED] A smooth, rounded or oval mass projecting from a membrane-covered surface. { ˈpäl·əp }

polypectomy [MED] Surgical excision of a polyp. { ˌpäl·iˈpek·tə·mē }

polypeptide [BIOCHEM] A chain of amino acids linked together by peptide bonds but with a lower molecular weight than a protein; obtained by synthesis, or by partial hydrolysis of protein. { ¦päl·iˈpep‚tīd }

polypetalous [BOT] Having distinct petals, in reference to a flower or a corolla. Also known as choripetalous. { ¦päl·iˈped·əl·əs }

Polyphaga [INV ZOO] A suborder of the order Coleoptera; members are distinguished by not having the hind coxae fused to the metasternum and by lacking notopleural sutures. { pəˈlif·ə·gə }

polyphagous [ZOO] Feeding on many different kinds of plants or animals. { pəˈlif·ə·gəs }

polyphase [ELEC] Having or utilizing two or more phases of an alternating-current power line. { ˈpäl·iˌfāz }

polyphase circuit [ELEC] Group of alternating-current circuits (usually interconnected) which enter (or leave) a delimited region at more than two points of entry; they are intended to be so energized that, in the steady state, the alternating currents through the points of entry, and the alternating potential differences between them, all have exactly equal periods, but have differences in phase, and may have differences in waveform. { ˈpäl·iˌfāz ˈsər·kət }

polyphase meter [ENG] An instrument which measures some electrical quantity, such as power factor or power, in a polyphase circuit. { ˈpäl·iˌfāz ˈmēd·ər }

polyphase rectifier [ELECTR] A rectifier which utilizes two or more diodes (usually three), each of which operates during an equal fraction of an alternating-current cycle to achieve an output current which varies less than that in an ordinary half-wave or full-wave rectifier. { ˈpäl·iˌfāz ˈrek·tə‚fī·ər }

polyphase synchronous generator [ELEC] Generator whose alternating-current circuits are so arranged that two or more symmetrical alternating electromotive forces with definite phase relationships are produced at its terminals. { ˈpäl·iˌfāz ˈsiŋ·krə·nəs ˈjen·ə‚rād·ər }

polyphase transformer [ELEC] A transformer with multiple sets of primary and secondary windings on a single core; used in a polyphase circuit. { ˈpäl·iˌfāz tranzˈfór·mər }

polyphase wattmeter [ENG] An instrument that measures electric power in a polyphase circuit. { ˈpäl·iˌfāz ˈwät‚med·ər }

polyphenol oxidase [BIOCHEM] A copper-containing enzyme that catalyzes the oxidation of phenol derivatives to quinones. { ¦päl·iˈfē‚nól ˈäk·sə‚dās }

polyphenyl [ORG CHEM] Any of a group of direct colors used to dye cotton and wool. { ¦päl·iˈfen·əl }

polyphenylene oxide [ORG CHEM] A polyether resin of 2,6-dimethylphenol, (CH₃)₂C₆H₃OH; useful temperature range is −275 to 375°F (−168 to 191°C), with intermittent use possible up to 400°F (204°C). { ˌpäl·iˈfen·əl‚ēn ˈäk‚sīd }

polyphobia [PSYCH] An abnormal fear of many different things. { ˌpäl·iˈfō·bē·ə }

polyphosphazene [ORG CHEM] A high-molecular-weight, essentially linear polymer with alternating phosphorus and nitrogen atoms in the skeleton and two side groups attached to each phosphorus. { ˌpäl·iˈfä·sfə‚zēn }

polyphosphoric acid [INORG CHEM] H₆P₄O₁₃ Viscous, water-soluble, hygroscopic, water-white liquid; used wherever concentrated phosphoric acid is needed. { ¦päl·i·fäˈsfór·ik ˈas·əd }

polyphyodont [VERT ZOO] Having teeth which may be constantly replaced. { ¦päl·iˈfī·ə‚dänt }

polypide [INV ZOO] The internal contents of an ectoproct bryozoan zooid. { ˈpäl·i‚pīd }

Polyplacophora [INV ZOO] The chitons, an order of mollusks in the class Amphineura distinguished by an elliptical body with a dorsal shell that comprises eight calcareous plates overlapping posteriorly. { ˌpäl·i·plaˈkäf·ə·rə }

polyplet [MATH] A plane figure formed by joining squares either along their sides or at their corners. Also known as polyking. { ˈpäl·ə‚plət }

polyploidy [GEN] The occurrence of related species possessing three, four, or larger multiples of the haploid set of chromosomes. { ˈpäl·iˌplóid·ē }

Polypodiales [BOT] The true ferns; the largest order of modern ferns, distinguished by being leptosporangiate and by having small sporangia with a definite number of spores. { ˌpäl·i‚pädˈē·ā‚lēz }

Polypodiatae *See* Polypodiopsida. { ˌpäl·i·pəˈdī·ə‚tē }

Polypodiophyta [BOT] The ferns, a division of the plant kingdom having well-developed roots, stems, and leaves that contain xylem and phloem and show well-developed alternation of generations. { ˌpäl·i‚pädˈē·äf·əd·ē }

Polypodiopsida [BOT] A class of the division Polypodiophyta; stems of these ferns bear several large, spirally arranged, compound leaves with sporangia grouped in sori on their undermargins. { ˌpäl·i‚pädˈē·äp·səd·ə }

polypore [MYCOL] Any member of the Basidiomycetes having basidia that line the numerous tubes or pores of the basidiocarp. { ˈpäl·i‚pór }

polypropylene [ORG CHEM] (C₃H₆)ₓ A crystalline, thermoplastic resin made by the polymerization of propylene, C₃H₆; the product is hard and tough, resists moisture, oils, and solvents, and withstands temperatures up to 170°C; used to make molded articles, fibers, film, rope, printing plates, and toys. { ˌpäl·əˈprō·pə‚lēn }

polypropylene glycol [ORG CHEM] CH₃CHOH(CH₂OCH-CH₃)ₓCH₂OH Polymeric material similar to polyethylene glycol, but with greater oil solubility and less water solubility; used as a solvent for vegetable oils, waxes, and resins, in hydraulic fluids and as a chemical intermediate. { ˌpäl·əˈprō·pə‚lēn ˈglī‚kól }

Polypteridae [VERT ZOO] The single family of the order Polypteriformes. { ˌpäl·əpˈter·ə‚dē }

Polypteriformes [VERT ZOO] An ancient order of actinopterygian fishes distinguished by thick, rhombic, ganoid scales with an enamellike covering, a slitlike spiracle behind the eye, a symmetrical caudal fin, and a dorsal series of free, spinelike finlets. { ˌpäl·əp‚ter·əˈfór‚mēz }

polyribosome *See* polysome. { ˌpäl·iˈrī·bə‚sōm }

polyrod antenna [ELECTROMAG] End-fire directional

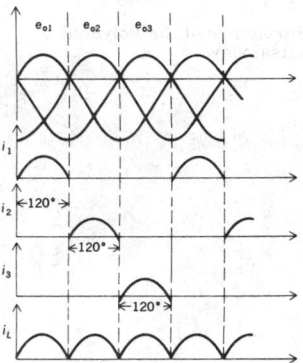

POLYPHASE RECTIFIER

Graphs of transformer voltages e_{o1}, e_{o2}, e_{o3}, diode currents i_1, i_2, i_3, and load current i_L versus time, in a three-phase half-wave rectifier.

dielectric antenna consisting of a polystyrene rod energized by a section of waveguide. { 'päl·i,räd an'ten·ə }

polysaccharide [BIOCHEM] A carbohydrate composed of many monosaccharides. Also known as glycan. { 'päl·i'sak·ə,rīd }

polysaccharide vaccine [IMMUNOL] A noninfectious vaccine that contains the polysaccharide coats, or capsules, of encapsulated bacteria. { ,päl·ē,sak·ə,rīd vak'sēn }

polysaprobic [ECOL] Referring to a body of water in which organic matter is decomposing rapidly and free oxygen either is exhausted or is present in very low concentrations. { ,päl·ə·sə'prō·bik }

polysepalous [BOT] Having separate sepals. Also known as chorisepalous. { päl·ə'sep·ə·ləs }

polyserositis [MED] Widespread, chronic fibrosing inflammation of serous membranes, especially in the upper abdomen. Also known as chronic hyperplastic perihepatitis; Concato's disease; multiple serositis; Pick's disease; polyorrhymenitis. { ,päl·i,ser·ə'sīd·əs }

polysiloxane [ORG CHEM] (R₂SiO)ₙ A polymer in which the chain contains alternate silicon and oxygen atoms; in the formula, R can be H or an alkyl or aryl group; commercially, the R is usually CH_3 (the methylsiloxanes); properties vary with molecular weight, from oils to greases to rubbers to plastics. { ,päl·i·si'läk,sān }

polysome [CELL MOL] A complex of ribosomes bound together by a single messenger ribonucleic acid molecule. Also known as polyribosome. { 'päl·i,sōm }

polysomy [GEN] The occurrence in a nucleus of an extra copy of one or more individual chromosomes. { 'päl·ē,sō·mē }

polysorbate [ORG CHEM] Any compound that is an ester of sorbitol. { ,päl·ē'sór,bāt }

polyspermy [PHYSIO] Penetration of the egg by more than one sperm. { 'päl·ē,spər·mē }

polyspore [BOT] In certain red algae, an asexual spore, of which there are 12 to 16. { 'päl·ē,spór }

polystele [BOT] A stele consisting of vascular units in the parenchyma. { 'päl·i,stēl }

Polystomatoidea [INV ZOO] A superfamily of monogeneid trematodes characterized by strong suckers and hooks on the posterior end. { ,päl·i,stō·mə'tóid·ē·ə }

Polystylifera [INV ZOO] A suborder of the Hoplonemertini distinguished by many stylets. { ,päl·i·stə'lif·ə·rə }

polystyrene [ORG CHEM] (C₆H₅CHCH₂)ₓ A water-white, tough synthetic resin made by polymerization of styrene; soluble in aromatic and chlorinated hydrocarbon solvents; used for injection molding, extrusion or casting for electrical insulation, fabric lamination, and molding of plastic objects. { 'päl·i'stī,rēn }

polystyrene capacitor [ELEC] A capacitor that uses film polystyrene as a dielectric between rolled strips of metal foil. { ,päl·i'stī,rēn kə'pas·əd·ər }

polystyrene dielectric [ELEC] Polystyrene used in applications where its very high resistivity, good dielectric strength, and other electrical properties are important, such as for electrical insulation or in dielectrics. { ,päl·i'stī,rēn ,dī·ə'lek·trik }

polysulfide rubber [ORG CHEM] A synthetic polymer made by the reaction of sodium polysulfide with an organic dichloride; resistant to light, oxygen, oils, and solvents; impermeable to gases; poor tensile strength and abrasion resistance. { ,päl·i'səl,fīd 'rəb·ər }

polysulfide treating [CHEM ENG] A petroleum-refinery process used to remove elemental sulfur from refinery liquids by contacting them with a nonregenerable solution of sodium polysulfide. { ,päl·i'səl,fīd 'trēd·iŋ }

polysulfone resin [MATER] A thermoplastic polymer containing the sulfone linkage (O=S=O); these resins have exceptional high-temperature, low-creep, and arc-resistance properties, and are self-extinguishing. { ,päl·i'səl,fōn ,rez·ən }

polysynthetic twinning [CRYSTAL] Repeated twinning that involves three or more individual crystals according to the same twin law and on parallel twin planes. { ,päl·i·sin'thed·ik 'twin·iŋ }

polytene chromosome [GEN] A giant, multistranded, chromosome produced by multiple rounds of endoreduplication and thus composed of many copies of the parental chromosome pair having their chromomeres in register. Also known as Balbiani chromosome. { 'päl·i,tēn 'krō·mə,sōm }

polyterpene resin [ORG CHEM] A thermoplastic resin or viscous liquid from polymerization of turpentine; used in paints, polishes, and rubber plasticizers, and to cure concrete and impregnate paper. { ,päl·i'tər,pēn ,rez·ən }

polytetrafluoroethylene [ORG CHEM] (CF₂CF₂)ₙ A highly crystalline perfluorinated polymer that is characteristically resistant to heat and chemicals. { ,päl·ē,te·trə,flür·ō'eth·ə,lēn }

polythene [ORG CHEM] Common name for polyethlylene in Great Britain. { 'päl·i,thēn }

polytope [MATH] A finite region in n-dimensional space (n = 2, 3, 4, . . .), enclosed by a finite number of hyperplanes; it is the n-dimensional analog of a polygon (n = 2) and a polyhedron (n = 3). { 'päl·i,tōp }

Polytrichales [BOT] An order of ascocarpous perennial mosses; rigid, simple stems are highly developed and arise from a prostrate subterranean rhizome. { pə,li·trə'kā·lēz }

polytrifluorochloroethylene resin See chlorotrifluoroethylene polymer. { ,päl·i·t,flür·ō,klór·ō'eth mbə,lēn ,rez·ən }

polytropic atmosphere [METEOROL] A model atmosphere in hydrostatic equilibrium with a constant nonzero lapse rate. { ,päl i,träp·ik at·mə,sfir }

polytropic compression curve [PHYS] Graphical relationship between pressure p and volume V for various values of specific-heat ratios n in the compression formula $pV^n = K$. { ,päl·i,träp·ik kəm'presh·ən ,kərv }

polytropic process [THERMO] An expansion or compression of a gas in which the quantity pV^n is held constant, where p and V are the pressure and volume of the gas, and n is some constant. { ,päl·i,träp·ik 'prä·səs }

polytype [CRYSTAL] A type of polymorph whose different forms are due to more than one possible mode of atomic packing. { 'päl·i,tīp }

polytypic [SYST] A taxon that contains two or more taxa in the immediately subordinate category. { 'päl·i,tip·ik }

polytypism [CRYSTAL] The ability of a mineral to crystallize into more than one form, because of more than one possible mode of atomic packing. { 'päl·i'ti,piz·əm }

polyunsaturated acid [ORG CHEM] A fatty acid with two or more double bonds per molecule, such as linoleic or linolenic acid. { ,päl·ē,ən'sach·ə,rād·əd 'as·əd }

polyunsaturated fat [MATER] A fat or oil based on fatty acids such as linoleic or linolenic acids which have two or more double bonds in each molecule; corn oil and safflower oil are examples. { ,päl·ē ən sach·ə,rād·əd 'fat }

polyurethane foam [MATER] A solid or spongy cellular material produced by the reaction of a polyester (such as glycerin) with a diisocyanate (such as toluene diisocyanate) while carbon dioxide is liberated by the reaction of a carboxyl with the isocyanate; used for thermal insulation, soundproofing, and padding. { ,päl·ē'yùr·ə,thän 'fōm }

polyurethane resin [ORG CHEM] Any resin resulting from the reaction of diisocyanates (such as toluene diisocyanate) with a phenol, amine, or hydroxylic or carboxylic compound to produce a polymer with free isocyanate groups; used as protective coatings, potting or casting resins, adhesives, rubbers, and foams, and in paints, varnishes, and adhesives. { ,päl·ē'yùr·ə,thän 'rez·ən }

polyurethane rubber [ORG CHEM] A synthetic polyurethane-resin elastomer made by the reaction of a diisocyanate to a polyester (such as the glycol-adipic acid ester); has high resistance to abrasion, oil, ozone, and high temperatures. Also known as polyester rubber. { ,päl·ē'yùr·ə,thän 'rəb·ər }

polyuria [MED] The passage of copious amounts of urine. { ,päl·ē'yùr·ē·ə }

polyvalent [CHEM] Pertaining to an ion with more than one valency, such as the sulfate ion, $SO_4{}^{2-}$. Also known as multivalent; polygen. [IMMUNOL] 1. Of antigens, having many combining sites or determinants. 2. Pertaining to vaccines composed of mixtures of different organisms, and to the resulting mixed antiserum. { ,päl·i'vā·lənt }

polyvalent number [COMPUT SCI] A number, consisting of several figures, used for description, wherein each figure represents one of the characteristics being described. { ,päl·i'vā·lənt 'nəm·bər }

polyvinyl acetal resin See vinyl acetal resin. { ,päl·i'vīn·əl 'as·ə,tal 'rez·ən }

polyvinyl acetate [ORG CHEM] (H₂CCHOOCCH₃)ₓ A thermoplastic polymer; insoluble in water, gasoline, oils, and

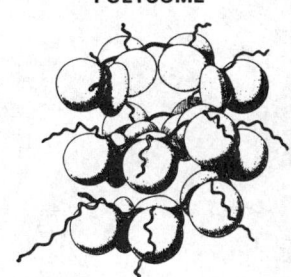
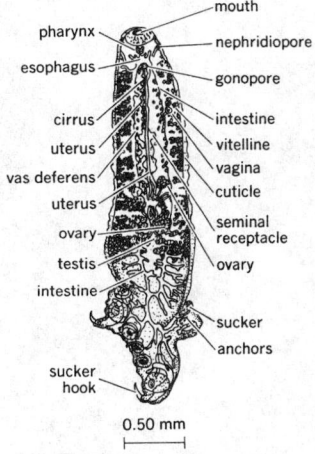

fats, soluble in ketones, alcohols, benzene, esters, and chlorinated hydrocarbons; used in adhesives, films, lacquers, inks, latex paints, and paper sizes. Abbreviated PVA; PVAc. { ¦päl·i'vīn·əl 'as·ə,tāt }

polyvinyl alcohol [ORG CHEM] Water-soluble polymer made by hydrolysis of a polyvinyl ester (such as polyvinyl acetate); used in adhesives, as textile and paper sizes, and for emulsifying, suspending, and thickening of solutions. Abbreviated PVA. { ¦päl·i'vīn·əl 'al·kə,hól }

polyvinyl carbazole [ORG CHEM] Thermoplastic resin made by reaction of acetylene with carbazole; softens at 150°C; has good electrical properties and heat and chemical stabilities; used as a paper-capacitor impregnant and as a substitute for electrical mica. { ¦päl·i'vīn·əl 'kär·bə,zōl }

polyvinyl chloride [ORG CHEM] $(H_2CCHCl)_x$ Polymer of vinyl chloride; tasteless, odorless; insoluble in most organic solvents; a member of the family of vinyl resins; used in soft flexible films for food packaging and in molded rigid products such as pipes, fibers, upholstery, and bristles. Abbreviated PVC. { ¦päl·i'vīn·əl 'klór,īd }

polyvinyl chloride acetate [ORG CHEM] Thermoplastic copolymer of vinyl chloride, CH_2CHCl, and vinyl acetate, $CH_3COOCH=CH_2$; colorless solid with good resistance to water, concentrated acids, and alkalies; compounded with plasticizers, it yields a flexible material superior to rubber in aging properties; used for cable and wire coverings and protective garments. { ¦päl·i'vīn·əl 'klór,īd 'as·ə,tāt }

polyvinyl dichloride [ORG CHEM] A high-strength polymer of chlorinated polyvinyl chloride; it is self-extinguishing and has superior chemical resistance; used for pipes carrying hot, corrosive materials. Abbreviated PVDC. { ¦päl·i'vīn·əl dī'klór,īd }

polyvinyl ether *See* polyvinyl ethyl ether. { ¦päl·i'vīn·əl 'ē·thər }

polyvinyl ethyl ether [ORG CHEM] $[-CH(OC_2H_5)CH_2-]_x$ A viscous gum to rubbery solid, soluble in organic solvents; used for pressure-sensitive tape. Also known as polyvinyl ether. { ¦päl·i'vīn·əl 'eth·əl 'ē·thər }

polyvinyl fluoride [ORG CHEM] $(-H_2CCHF-)_x$ Vinyl fluoride polymer; has superior resistance to weather, chemicals, oils, and stains, and has high strength; used for packaging (but not of food) and electrical equipment. { ¦päl·i'vīn·əl 'flúr,īd }

polyvinyl formate resin [ORG CHEM] $(CH_2=CHOOCH)_x$ Clear-colored resin that is hard and solvent-resistant; used to make clear, hard plastics. { ¦päl·i'vīn·əl 'fór,māt ,rez·ən }

polyvinylidene chloride [ORG CHEM] Thermoplastic polymer of vinylidene chloride, $H_2C=CCl_2$; white powder softening at 185–200°C; used to make soft-flexible to rigid products. { ¦päl·i·ə·vī'nil·ə,dēn 'klór,īd }

polyvinylidene fluoride [ORG CHEM] Fluorocarbon polymer made from vinylidene fluoride, $H_2C=CF_2$; has good tensile and compressive strength and high impact strength; used in chemical equipment for gaskets, impellers, and other pump parts, and for drum linings and protective coatings. { ¦päl·i·vī'nil·ə,dēn 'flúr,īd }

polyvinylidene resin *See* vinylidene resin. { ¦päl·i·vī'nil·ə,dēn 'rez·ən }

polyvinyl isobutyl ether [ORG CHEM] $[-CH_2CHOCH_2-CH(CH_3)_2-]_x$ An odorless synthetic resin; elastomer to viscous liquid depending on molecular weight; soluble in hydrocarbons, esters, ethers, and ketones, insoluble in water; used in adhesives, waxes, plasticizers, lubricating oils, and surface coatings. Abbreviated PVI. { ¦päl·i'vīn·əl ¦ī·sə ¦byüd·əl 'e·thər }

polyvinyl methyl ether [ORG CHEM] $(-CH_2CHOCH_3-)_x$ A colorless, tacky liquid, soluble in organic solvents, except aliphatic hydrocarbons, and in water below 32°C; used for pressure-sensitive adhesives, as a heat sensitizer for rubber latex, and as a pigment binder in inks and textile finishing. Abbreviated PVM. { ¦päl·i'vīn·əl 'meth·əl 'ē·thər }

polyvinyl pyrrolidone [ORG CHEM] $(C_6H_9NO)_x$ A water-soluble, white, resinous solid; used in pharmaceuticals, cosmetics, detergents, and foods, and as a synthetic blood plasma. Abbreviated PVP. { ¦päl·i'vīn·əl pə'räl·ə,dōn }

polyvinyl resin [ORG CHEM] Any resin or polymer derived from vinyl monomers. Also known as vinyl plastic. { ¦päl·i'vīn·əl 'rez·ən }

Polyzoa [INV ZOO] The equivalent name for Bryozoa. { ¦päl·ə·ə'zō·ə }

polzenite [PETR] **1.** A group of lamprophyres characterized by the presence of olivine and melilite. **2.** Any rock in this group. { 'päl·zə,nīt }

pomace [FOOD ENG] **1.** In the preparation of cider, the material derived from apples or similar fruits by crushing or grinding. **2.** The residue of grape skins, seeds, and stems after wine grapes have been pressed and the juice has been separated. { 'päm·əs }

Pomacentridae [VERT ZOO] The damselfishes, a family of perciform fishes in the suborder Percoidei. { ,pō·mə'sen·trə,dē }

Pomadasyidae [VERT ZOO] The grunts and sweetlips, a family of perciform fishes in the suborder Percoidei. { ,pō·mə·də'sī·ə,dē }

Pomatiasidae [INV ZOO] A family of land snails in the order Pectinibranchia. { ,pō·mə·tī'as·ə,dē }

Pomatomidae [VERT ZOO] A monotypic family of the Perciformes containing the bluefish (*Pomatomus saltatrix*). { ,pō·mə'täm·ə,dē }

pomegranate [BOT] *Punica granatum.* A small, deciduous ornamental tree of the order Myrtales cultivated for its fruit, which is a reddish, pomelike berry containing numerous seeds embedded in crimson pulp. { 'päm·ə,gran·ət }

Pomeranchuk cooling [CRYO] A method of attaining temperatures as low as 1 millikelvin in which helium-3 is cooled by adiabatic compression at temperatures below 0.3 K. { 'päm·ə¦ran·chək 'kúl·iŋ }

Pomeranchuk pole *See* Pomeron. { 'päm·ə¦rän·chək ,pōl }

Pomeranchuk theorem [PARTIC PHYS] The theorem that if the total cross section both for scattering of a particle by a given target particle and for scattering of its antiparticle by the same target particle, approach a limit at high energies, and do so sufficiently rapidly, then these limits must be the same. { 'päm·ə¦rän·chək ,thir·əm }

Pomeron [PARTIC PHYS] A Regge pole which is located at +1 in the angular momentum plane when the momentum transfer in the crossed channel equals zero, corresponding to the fact that total cross sections of reactions are observed to approach constants at high energies. Also known as Pomeranchuk pole. { 'päm·ə,rän }

Pompe disease [MED] A hereditary glycogen storage disease in humans arising from deficiency of a lysosomal enzyme and characterized by weakness, enlargement of the heart and cardiac failure, enlargement of the tongue, and moderate enlargement of the liver. { 'pämp di,zēz }

pompeii worm [INV ZOO] *Alvinella pompejana.* A polychaetous annelid that lives in sea-floor hydrothermal vent chimneys and may experience extreme thermal gradients between its anterior (80°C; 176°F) and posterior (22°C; 72°F) ends. { päm'pā ,wərm }

Pompilidae [INV ZOO] The spider wasps, the single family of the superfamily Pompiloidea. { pam'pil·ə,dē }

Pompiloidea [INV ZOO] A monofamilial superfamily of hymenopteran insects in the suborder Apocrita with oval abdomen and strong spinose legs. { ,päm·pə'lóid·ē·ə }

pom-pom [ORD] **1.** A rack of antiaircraft cannon, usually mounted in fours, as on the deck of a ship. **2.** An automatic cannon. { 'päm,päm }

PONA analysis [ENG] American Society for Testing and Materials analysis of paraffins (P), olefins (O), naphthenes (N), and aromatics (A) in gasolines. { 'pō·nə *or* ¦pē¦ō¦en'ā a,nal·ə·səs }

poncelet [PHYS] A unit of power equal to the power delivered by a force of 100 kilograms-force when the point at which the force is applied is moved at a rate of 1 meter per second in the direction of the force; equal to 980.665 watts. { 'päns·lət }

Ponchon-Savarit method [CHEM ENG] Graphical solution on an enthalpy-concentration diagram of liquid-vapor equilibrium values between trays of a distillation column. { ,pón·shón ,sav·ə'rē ,meth·əd }

pond [GEOGR] A small natural body of standing fresh water filling a surface depression, usually smaller than a lake. [MECH] *See* gram-force. { pänd }

pondage [HYD] Water held in a reservoir for short periods to regulate natural flow, usually for hydroelectric power. { 'pän·dij }

pondage land [GEOL] Land on which water is stored as dead water during flooding, and which does not contribute to the downstream passage of flow. Also known as flood fringe. { 'pän·dij ,land }

ponded stream [HYD] A stream in which a pond forms due to an interruption of the normal streamflow. { 'pän·dəd ,strēm }

ponderomotive force [ATOM PHYS] The part of the interaction of light with atoms that exerts a force on the atoms, rather than coupling with their internal structure. { ,pän·dər,mōd·iv 'fȯrs }

ponderosa pine [BOT] *Pinus ponderosa*. A hard pine tree of western North America; attains a height of 150–225 feet (46–69 meters) and has long, dark-green leaves in bundles of two to five and tawny, yellowish bark. { ,pän·də,rō·sə 'pīn }

ponding [BUILD] An accumulation of water on a flat roof because of clogged or inadequate drains. [CIV ENG] **1.** The impoundment of stream water to form a pond. **2.** Covering the surface of newly poured concrete with a thin layer of water to promote curing. [HYD] The natural formation of a pond in a stream by an interruption of the normal streamflow. { 'pänd·iŋ }

ponente [METEOROL] A west wind on the French Mediterranean coast, the northern Roussillon region, and Corsica. { pȯ'nȯnt }

Ponerinae [INV ZOO] A subfamily of tropical carnivorous ants (Formicidae) in which pupae characteristically form in cocoons. { pō'ner·ə,nē }

pongee [TEXT] A rough-textured, plain-woven fabric of raw silk with irregular filling yarns and generally an ecru color. { pän'jē }

Pongidae [VERT ZOO] A family of anthropoid primates in the superfamily Hominoidea; includes the chimpanzee, gorilla, and orangutan. { 'pän·jə,dē }

poniente [METEOROL] The west wind in the Straits of Gibraltar. { ,pō·nē'en·tē }

pons [ANAT] **1.** A process or bridge of tissue connecting two parts of an organ. **2.** A convex white eminence located at the base of the brain; consists of fibers receiving impulses from the cerebral cortex and sending fibers to the contralateral side of the cerebellum. { pänz }

Pontian [GEOL] A European stage of geologic time in the uppermost Miocene, above the Sarmatian and below the Plaisancian of the Pliocene; it has also been regarded as the lowermost Pliocene. { 'pän·chən }

pontianak gum [MATER] A grayish-white copal from various jelutongs of the genus *Dyera*, indigenous to Malacca and Borneo; used in rubber manufacture, chewing gum, adhesives, paints, and varnishes. Also known as jelutong. { ,pän·tē'ä·nək ,gəm }

pontic [GEOL] Pertaining to sediments or facies deposited in comparatively deep and motionless water, such as an association of black shales and dark limestones deposited in a stagnant basin. { 'pän·tik }

pontine flexure [EMBRYO] A flexure in the embryonic brain concave dorsally, occurring in the region of the myelencephalon. { 'pän,tēn 'flek·shər }

Pontodoridae [INV ZOO] A monotypic family of pelagic polychaetes assigned to the Errantia. { ,pän·tə'dȯr·ə,dē }

pontoon [AERO ENG] A float on an airplane. [NAV ARCH] **1.** A low, flat-bottomed ship, similar to a barge, carrying cranes, capstans, and other machinery used to lift weights, to lean ships on their sides for repairs, and to perform other such operation. **2.** A wooden flat-bottomed boat or other float, used particularly to make temporary bridges. { pän'tün }

pontoon bridge [CIV ENG] A fixed floating bridge supported by pontoons. { pänj'tün 'brij }

pontoon-tank roof [ENG] A type of floating tank roof, supported by buoyant floats on the liquid surface of a tank; the roof rises and falls with the liquid level in the tank; used to minimize vapor space above the liquid, thus reducing vapor losses during tank filling and emptying. { pänh'tün ,taŋk ,rüf }

Pontryagin's maximum principle [MATH] A theorem giving a necessary condition for the solution of optimal control problems: let $\theta(\tau)$, $\tau_0 \leq \tau \leq T$ be a piecewise continuous vector function satisfying certain constraints; in order that the scalar function $S = \Sigma c_i x_i(T)$ be minimum for a process described by the equation $\partial x_i/\partial \tau = (\partial H/\partial z_i)[z(\tau), x(\tau), \theta(\tau)]$ with given initial conditions $x(\tau_0) = x^0$ it is necessary that there exist a nonzero continuous vector function $z(\tau)$ satisfying $dz_i/d\tau = -(\partial H/\partial x_i)$,

$[z(\tau), x(\tau), \theta(\tau)]$, $z_i(T) = -c_i$, and that the vector $\theta(\tau)$ be so chosen that $H[z(\tau), x(\tau), \theta(\tau)]$ is maximum for all τ, $\tau_0 \leq \tau \leq T$. { ,pän·trē'ä·gənz 'mak·sə·məm ,prin·sə·pəl }

pony set [MIN ENG] A small timber set or frame incorporated in the main sets of a haulage level to accommodate an ore chute or other equipment from above or below. { 'pō·nē ,set }

pony truss [CIV ENG] A truss too low to permit overhead braces. { 'pō·nē trəs }

poodle cloth [TEXT] Loopy bouclé or knotted yarn cloth resembling the coat of a french poodle; originally wool, but now of various fibers and mixtures in knitted as well as woven versions. { 'püd·əl ,klȯth }

pool [CIV ENG] A body of water contained in a reservoir, by a dam, or by the gates of a lock. [GEOL] Underground accumulation of petroleum. [HYD] A small deep body of water, often fed by a spring. [MIN ENG] **1.** To wedge for splitting in quarrying or mining. **2.** To undermine or undercut. { pül }

pool boiling [PHYS CHEM] Boiling of a liquid whose flow results from natural convection. { 'pül ,bȯil·iŋ }

pool cathode [ELECTR] A cathode at which the principal source of electron emission is a cathode spot on a liquid-metal electrode, usually mercury. { 'pül ,kath,ōd }

pool-cathode mercury-arc rectifier [ELECTR] A pool tube connected in an electric circuit; its rectifying properties result from the fact that only the mercury-pool cathode, and not the anode, can emit electrons. Also known as mercury-pool rectifier. { 'pül ,kath,ōd 'mər·kyə·rē ,ärk 'rek·tə,fī·ər }

pool-cathode tube *See* pool tube. { 'pül ,kath,ōd ,tüb }

Poole-Frenkel effect [ELEC] An increase in the electrical conductivity of insulators and semiconductors in strong electric fields. { 'pül 'freŋ·kəl i,fekt }

pooling of error [STAT] A method used in the analysis of variance to secure more degrees of freedom for estimating the error variance; the sums of squares of several sets of data considered to be generated under the same model are added together and divided by the sum of the degrees of freedom in the several sets of data. { 'pül·iŋ əv 'er·ər }

pool reactor [NUCLEO] A research nuclear reactor in which the core is suspended in a large pool of water that serves as moderator, reflector, coolant, and radiation shield. Also known as swimming-pool reactor. { 'pül rē,ak·tər }

pool stage [HYD] As used along the Ohio and upper Mississippi Rivers of the United States, a low-water condition with the navigation dams up so that the river is a series of shallow pools; when this condition exists, the river is said to be "in pool"; river depth is regulated by the dams so as to be adequate for navigation. { 'pül ,stāj }

pool tube [ELECTR] A gas-discharge tube having a mercury-pool cathode. Also known as mercury tube; pool-cathode tube. { 'pül ,tüb }

poop bulkhead [NAV ARCH] A partition wall that extends between the upper and lower poopdecks at the poop's forward end. { 'püp 'bəlk,hed }

poop deck [NAV ARCH] A partial deck above the main deck in a ship's afterpart. { 'püp ,dek }

poor cluster [ASTRON] A galaxy cluster that has relatively few member galaxies. { 'pür ,kləs·tər }

pop [COMPUT SCI] To obtain information from the top of a stack and then reset a pointer to the next item in the stack. [MIN ENG] A drill hole blasted to reduce larger pieces of rock or to trim a working face. Also known as pop hole; pop shot. { päp }

POP *See* Post Office Protocol. { päp *or* 'pē,ō'pē }

pop action [MECH ENG] The action of a safety valve as it opens under steam pressure when the valve disk is lifted off its seat. { 'päp ,ak·shən }

popcorn noise [ELECTR] Noise that is produced by erratic jumps of bias current between two levels at random intervals in operational amplifiers and other semiconductor devices. { 'päp,kȯrn }

pop hole *See* pop. { 'päp ,hōl }

poplar [BOT] Any tree of the genus *Populus*, family Salicaceae, marked by simple, alternate leaves, scaly buds, bitter bark, and flowers and fruits in catkins. { 'päp·lər }

poplin [TEXT] A corded fabric made of cotton, polyester, wool, nylon, or blend of fibers. { 'päp·lən }

popliteal artery [ANAT] A continuation of the femoral artery

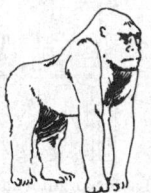

PONGIDAE

Typical anthropoid ape of the family Pongidae.

in the posterior portion of the thigh above the popliteal space and below the buttock. { päp'lid·ē·əl 'ärd·ə·rē }

popliteal nerve [NEUROSCI] Either of two branches of the sciatic nerve in the lower part of the thigh; the larger branch continues as the tibial nerve, and the smaller branch continues as the peroneal nerve. { päp'lid·ē·əl 'nərv }

popliteal space [ANAT] A diamond-shaped area behind the knee joint. { päp'lid·ē·əl 'spās }

popliteal vein [ANAT] A vein passing through the popliteal space, formed by merging of the tibial veins and continuing to become the femoral vein. { päp'lid·ē·əl 'vān }

popliteus [ANAT] **1.** The ham or hinder part of the knee joint. **2.** A muscle on the back of the knee joint. { päp'lid·ē·əs }

Popov's stability criterion [CONT SYS] A frequency domain stability test for systems consisting of a linear component described by a transfer function preceded by a nonlinear component characterized by an input-output function), with a unity gain feedback loop surrounding the series connection. { pä'pófs stə'bil·əd·ē krī,tir·ē·ən }

poppet [CIV ENG] One of the timber and steel structures supporting the fore and aft ends of a ship for launching from sliding ways. [DES ENG] A spring-loaded ball engaging a notch; a ball latch. [MIN ENG] A pulley frame or the headgear over a mine shaft. { 'päp·ət }

POPPET VALVE

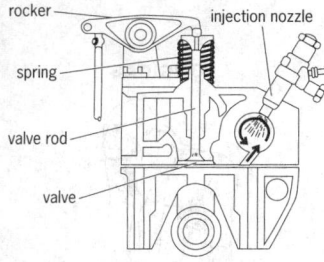

rocker
injection nozzle
spring
valve rod
valve

Poppet valve for an internal combustion engine. *(After E. A. Avallone and T. Baumeister III, eds., Marks' Standard Handbook for Mechanical Engineers, 10th ed., McGraw-Hill, 1996)*

poppet valve [MECH ENG] A cam-operated or spring-loaded reciprocating-engine mushroom-type valve used for control of admission and exhaust of working fluid; the direction of movement is at right angles to the plane of its seat. { 'päp·ət ,valv }

popping [COMPUT SCI] The deletion of the top element of a stack. [MIN ENG] Exploding a stick of dynamite on a boulder so as to break it for easy removal from a quarry or opencast mine. { 'päp·iŋ }

popping pressure [MECH ENG] In compressible fluid service, the inlet pressure at which a safety valve disk opens. { 'päp·iŋ ,presh·ər }

popple rock *See* pebble bed. { 'päp·əl ,räk }

poppy [BOT] Any of various ornamental herbs of the genus *Papaver*, family Papaveraceae, with large, showy flowers; opium is obtained from the fruits of the opium poppy (*P. somniferum*). { 'päp·ē }

poppy oil [MATER] Golden-yellow drying oil with pleasant taste and aroma; soluble in carbon disulfide, ether, and chloroform; expressed from poppy seeds, especially of the opium poppy; used as a food oil; in artist's colors, soaps, varnishes, and lubricants. Also known as poppy-seed oil. { 'päp·ē ,oil }

poppy-seed oil *See* poppy oil. { 'päp·ē ,sēd ,oil }

pop shot *See* pop. { 'päp ,shät }

populate [COMPUT SCI] To add electronic components, such as memory chips, to a circuit board. { 'päp·yə,lāt }

population [BIOL] A group of organisms occupying a specific geographic area or biome. [COMPUT SCI] A collection of records in a data base that share one or more characteristics in common. [ELECTR] The set of electronic components on a printed circuit board. [STAT] A specified set of objects or outcomes to be measured or observed. { ,päp·yə'lā·shən }

population I [ASTRON] A class of stars which are relatively young, have relatively low peculiar velocities, and are found chiefly in the spiral arms of galaxies. Also known as arm population. { ,päp·yə'lā·shən 'wən }

population II [ASTRON] A class of stars which are relatively old and evolved, have low metallic content and high peculiar velocities from 60 to 300 miles (100 to 500 kilometers) per second, and are found chiefly in the spheroidal halo of a galaxy. Also known as halo population. { ,päp·yə'lā·shən 'tü }

population III [ASTRON] A class of stars that condensed from the gas formed in the nucleosynthesis of the big bang, and consist entirely of hydrogen and helium. { ,päp·yə'lā·shən 'thrē }

population bottleneck [EVOL] Genetic drift that occurs as a result of a drastic reduction in population by an event having little to do with the usual forces of natural selection. { ,päp·yə'lā·shən 'bäd·əl,nek }

population correlation coefficient [STAT] The ratio of the covariance of two random variables to their standard deviations. { ,päp·yə'lā·shən ,kär·ə'lā·shən ,kō·ə,fish·ənt }

population covariance [STAT] The number $(1/N)[(v_1 - \bar{v}) \cdot (w_1 - \bar{w}) + \cdots + (v_N - \bar{v})(w_N - \bar{w})]$, where v_i and w_i, $i = 1, 2, \ldots, N$, are the values obtained from two populations, and

$\bar{v}$ and $\bar{w}$ are the respective means. { ,päp·yə'lā·shən kō'ver·ē·əns }

population density [ECOL] The size of the population within a particular unit of space. { ,päp·yə'lā·shən 'den·səd·ē }

population dispersal [BIOL] The process by which groups of living organisms expand the space or range within which they live. { ,päp·yə'lā·shən di'spər·səl }

population dispersion [BIOL] The spatial distribution at any particular moment of the individuals of a species of plant or animal. { ,päp·yə'lā·shən di'spər·zhən }

population dynamics [BIOL] The aggregate of processes that determine the size and composition of any population. { ,päp·yə'lā·shən dī'nam·iks }

population ecology [ECOL] The study of the vital statistics of populations, and the interactions within and between populations that influence survival and reproduction. { ,päp·yə,lā·shən ē'käl·ə·jē }

population genetics [GEN] The study of both experimental and theoretical consequences of Mendelian heredity on the population level; includes studies of gene frequencies, genotypes, phenotypes, mating systems, selection, and migration. { ,päp·yə'lā·shən jə'ned·iks }

population inversion [ATOM PHYS] The condition in which a higher energy state in an atomic system is more heavily populated with electrons than a lower energy state of the same system. { ,päp·yə'lā·shən in,vər·zhən }

population mean [STAT] The average of the numbers obtained for all members in a population by measuring some quantity associated with each member. { ,päp·yə'lā·shən ,mēn }

population multiple linear regression equation [STAT] An equation relating the conditional mean of the dependent variable to each one of the independent variables under the assumption that this relationship is linear; for the multivariate, normal distribution linearity always exists. { ,päp·yə'lā·shən ¦məl·tə·pəl ¦lin·ē·ər ri'gresh·ən i,kwā·zhən }

population of levels [STAT MECH] The number of members of an ensemble which are in each of the allowed energy states of a system. { ,päp·yə'lā·shən əv 'lev·əlz }

population variance [STAT] The arithmetic average of the numbers $(v_1 - \bar{v})^2, \ldots, (v_N - \bar{v})^2$, where v_i are numbers obtained from a population with N members, one for each member, and $\bar{v}$ is the population mean. { ,päp·yə'lā·shən 'ver·ē·əns }

p orbital [ATOM PHYS] The orbital of an atomic electron with an orbital angular momentum quantum number of unity. { 'pē 'ór·bəd·əl }

porcelain [MATER] A high-grade ceramic ware characterized by high strength, a white color, very low absorption, good translucency, and a hard glaze. Also known as European porcelain; hard paste porcelain; true porcelain. { 'pórs·lən }

porcelain capacitor [ELEC] A fixed capacitor in which the dielectric is a high grade of porcelain, molecularly fused to alternate layers of fine silver electrodes to form a monolithic unit that requires no case or hermetic seal. { 'pórs·lən kə'pas·əd·ər }

porcelain cement [MATER] A cement for bonding porcelain to porcelain, such as a mixture of gutta-percha and shellac. { 'pórs·lən si,ment }

porcelain clay [MATER] A clay suitable for use in the manufacture of porcelain, specifically kaolin. Also known as porcelain earth. { 'pórs·lən ,klā }

porcelain earth *See* porcelain clay. { 'pórs·lən ,ərth }

porcelain enamel *See* vitreous enamel. { 'pórs·lən i'nam·əl }

porcelain insulator [MATER] An electrical insulator made from porcelain; the porcelain is often made in a one-fire process, the glaze being applied to the green or unfired ware, in contrast to the two-fire process used in making ordinary porcelain. { 'pórs·lən 'in·sə,lād·ər }

porcelainite *See* mullite. { 'pór·slə,nīt }

porcelain jasper [GEOL] A hard, naturally baked, impure clay (or porcellanite) which because of its red color had long been considered a variety of jasper. { 'pór·slən 'jas·pər }

porcelaneous [GEOL] Resembling unglazed porcelain. { 'pór·sə¦lā·nē·əs }

porcelaneous chert [PETR] A hard, opaque to subtranslucent smooth chert, having a smooth fracture surface and a typically china-white appearance resembling chinaware and glazed porcelain. { ¦pór·sə¦lā·nē·əs 'chərt }

Porcellanasteridae [INV ZOO] A family of essentially deep-water forms in the suborder Paxillosina. { pȯr‚sel·ə·nə'ster·ə‚dē }

Porcellanidae [INV ZOO] The rock sliders, a family of decapod crustaceans of the group Anomura which resemble true crabs but are distinguished by the reduced, chelate fifth pereiopods and the well-developed tail fan. { ‚pȯr·sə'lan·ə‚dē }

porcellanite [PETR] A hard, dense siliceous rock, such as impure chert or indurated clay or shale. { pȯr'sel·ə‚nīt }

porcupine [VERT ZOO] Any of about 26 species of rodents in two families (Hystricidae and Erethizontidae) which have spines or quills in addition to regular hair. { 'pȯr·kyə‚pīn }

porcupine boiler [MECH ENG] A boiler having dead end tubes projecting from a vertical shell. { 'pȯr·kyə‚pīn ‚bȯil·ər }

pore [ASTRON] A very small, dark area on the sun formed by the separation of adjacent flocculi. [BIOL] Any minute opening by which matter passes through a wall or membrane. [GEOL] An opening or channelway in rock or soil. [MET] A minute cavity in a powder compact, metal casting, or electroplated coating. { pȯr }

pore compressibility [GEOL] The fractional change in reservoir-rock pore volume with a unit change in pressure upon that rock. { 'pȯr kəm‚pres·ə'bil·əd·ē }

pore diameter [DES ENG] The average or effective diameter of the openings in a membrane, screen, or other porous material. { 'pȯr dī‚am·əd·ər }

pore diffusion [FL MECH] The movement of fluids (gas or liquid) into the interstices of porous solids or membranes; occurs in membrane separation, zeolite adsorption, dialysis, and reverse osmosis. { 'pȯr di‚fyü·zhən }

pore-forming protein See perforin. { 'pȯr‚fȯrm·iŋ ‚prō‚tēn }

pore fungus [MYCOL] The common name for members of the families Boletaceae and Polyporaceae in the group Hymenomycetes; sporebearing surfaces are characteristically within tubes or pores. { 'pȯr ‚fəŋ·gəs }

pore ice [HYD] Ice which fills or partially fills pore spaces in permafrost; forms by freezing soil water in place, with no addition of water. { 'pȯr ‚īs }

porencephaly [MED] A condition in which the cavity of a lateral ventricle extends to the surface of the cerebral hemisphere; may result from brain tissue destruction or maldevelopment. { ‚pȯr·ən'sef·ə·lē }

pore pressure See neutral stress. { 'pȯr ‚presh·ər }

pore-size distribution [GEOL] Variations in pore sizes in reservoir formations; each type of rock has its own typical pore size and related permeability. { 'pȯr‚sīz ‚dis·trə'byü·shən }

pore space [GEOL] The pores in a rock or soil considered collectively. Also known as pore volume. { 'pȯr ‚spās }

pore volume See pore space. { 'pȯr ‚väl·yəm }

pore-water pressure See neutral stress. { 'pȯr ‚wȯd·ər ‚presh·ər }

poriaz [METEOROL] Violent northeast winds on the Black Sea near the Bosporus. { 'pȯr·ē‚äz }

Porifera [INV ZOO] The sponges, a phylum of the animal kingdom characterized by the presence of canal systems and chambers through which water is drawn in and released; tissues and organs are absent. { pə'rif·ə·rə }

Porlezzina [METEOROL] An east wind on Lake Lugano (Italy and Switzerland), blowing from the Gulf of Porlezza. { ‚pȯr·let'sē·nə }

porocyte [INV ZOO] One of the perforated, tubular cells which constitute the wall of the incurrent canals in certain Porifera. { 'pȯr·ə‚sīt }

porogamy [BOT] Passage of the pollen tube through the micropyle of an ovule in a seed plant. { pō'räg·ə·mē }

poromeric material [TEXT] A fabric made of polyurethane strengthened by polyester. { ‚pȯr·ō‚me·trik mə'tir·ē·əl }

porosimeter [ENG] Laboratory compressed-gas device used for measurement of the porosity of reservoir rocks. { ‚pȯr·ə'sim·əd·ər }

porosis [MED] Condition characterized by increased porosity, as of bone. { pə'rō·səs }

porosity [PHYS] 1. Property of a solid which contains many minute channels or open spaces. 2. The fraction as a percent of the total volume occupied by these channels or spaces; for example, in petroleum engineering the ratio (expressed in percent) of the void space in a rock to the bulk volume of that rock. { pə'räs·əd·ē }

porosity feet [PETRO ENG] Reservoir porosity fraction multiplied by net pay in feet, where porosity fraction is the portion of the reservoir that is porous, and net pay is the depth and areal extent of the hydrocarbons-containing reservoir. { pə'räs·əd·ē 'fēt }

porosity trap See stratigraphic trap. { pə'räs·əd·ē ‚trap }

porous [MATER] 1. Filled with pores. 2. Capable of absorbing liquids. { 'pȯr·əs }

porous alum See aluminum sodium sulfate. { 'pȯr·əs 'al·əm }

porous bearing [DES ENG] A bearing made from sintered metal powder impregnated with oil by a vacuum treatment. { 'pȯr·əs 'ber·iŋ }

porous carbon [MATER] Plates, tubes, or disks of uniform carbon particles pressed together without a binder; used for the filtration of corrosive liquids and gases. { 'pȯr·əs 'kär·bən }

porous graphite [MATER] Plates, tubes, or disks of uniform graphite particles pressed together without a binder; more resistant to oxidation but lower in strength than porous carbon. { 'pȯr·əs 'gra‚fīt }

porous metals [MET] Metals, made by powder metallurgy, having uniformly distributed controlled pore sizes, in the form of sheets, tubes, and shapes; used for filtering liquids and gases at elevated temperatures. { 'pȯr·əs 'med·əlz }

porous mold [ENG] A plastic-forming mold made from bonded or fused aggregates (such as powdered metal or coarse pellets) so that the resulting mass contains numerous open interstices through which air or liquids can pass. { 'pȯr·əs 'mōld }

porous reservoir model [PETRO ENG] Scaled laboratory model of porous reservoir used for the study of reservoir areal waterflood efficiencies. { 'pȯr·əs 'rez·əv‚wär ‚mäd·əl }

porous wheel [DES ENG] A grinding wheel having a porous structure and a vitrified or resinoid bond. { 'pȯr·əs 'wēl }

Poroxylaceae [PALEOBOT] A monogeneric family of extinct plants included in the Cordaitales. { pə‚räk·sə'lās·ē‚ē }

porpezite [MINERAL] A mineral consisting of a native alloy of palladium (5–10%) and gold. Also known as palladium gold. { 'pȯr·pə‚zīt }

porphin [BIOCHEM] A heterocyclic ring consisting of four pyrrole rings linked by methine (−CH=) bridges; the basic structure of chlorophyll, hemoglobin, the cytochromes, and certain other related substances. { 'pȯr·fən }

porphobilinogen [BIOCHEM] $C_{10}H_{14}O_4N_2$ Dicarboxylic acid derived from pyrrole; a product of hemoglobin breakdown that gives the urine a Burgundy-red color. { 'pȯr·fō·bə'lin·ə·jən }

porphrite See porphyry. { 'pȯr‚frīt }

porphyria [MED] A usually hereditary, pathologic disorder of porphyrin metabolism characterized by porphyrinuria and photosensitivity. { pȯr'fir·ē·ə }

porphyrin [BIOCHEM] A class of red-pigmented compounds with a cyclic tetrapyrrolic structure in which the four pyrrole rings are joined through their α-carbon atoms by four methene bridges (=C−); the porphyrins form the active nucleus of chlorophylls and hemoglobin. { 'pȯr·fə·rən }

porphyrinuria [MED] The excretion of large quantities of porphyrin in the urine. { ‚pȯr·fə·rə'nyùr·ē·ə }

porphyritic [PETR] Pertaining to or resembling porphyry. { ‚pȯr·fə'rid·ik }

porphyroblast [PETR] A relatively large crystal formed in a metamorphic rock. { 'pȯr·fə·rə‚blast }

porphyroblastic [PETR] Pertaining to the texture of recrystallized metamorphic rock having large idioblasts of minerals possessing high form energy in a finer-grained crystalloblastic matrix. { ‚pȯr·fir·ə'blas·tik }

porphyrocrystallic See porphyrotopic. { ‚pȯr·fir·ō·kri'stal·ik }

porphyroclastic structure See mortar structure. { ‚pȯr·fir·ō'klas·tik 'strək·chər }

porphyrogranulitic [PETR] Referring to ophitic texture characterized by large phenocrysts of feldspar and augite or olivine in a groundmass of smaller lath-shaped feldspar crystals and irregular augite grains; a combination of porphyritic and intergranular textures. { ‚pȯr·fir·ō‚gran·yə'lid·ik }

porphyroid [PETR] 1. A blastoporphyritic, or sometimes porphyroblastic, metamorphic rock of igneous origin. 2. A feldspathic metasedimentary rock having the appearance of a porphyry. { 'pȯr·fə‚rȯid }

porphyroskelic [GEOL] Pertaining to an arrangement in a soil fabric whereby the plasma occurs as a dense matrix in

PORCUPINE

The Canadian porcupine (*Erethizon dorsatum*), about 3½ feet (1 meter) long.

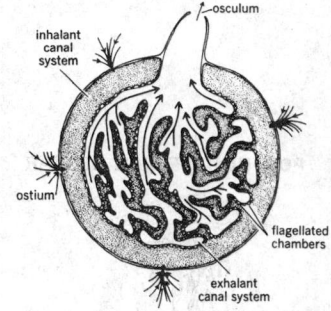

PORIFERA

Diagram of the canal system of a young fresh-water sponge.

which skeleton grains are set like phenocrysts in a porphyritic rock. { 'pȯr¦fir·ə¦skel·ik }

porphyrotope [GEOL] A large crystal enclosed in a finer-grained matrix in a sedimentary rock showing porphyrotopic fabric. { pȯr'fir·ə,tōp }

porphyrotopic [GEOL] Referring to the fabric of a crystalline sedimentary rock in which the constituent crystals are of more than one size and in which larger crystals are enclosed in a finer-grained matrix. Also known as porphyrocrystallic. { pȯr¦fir·ə¦täp·ik }

porphyry [PETR] An igneous rock in which large phenocrysts are enclosed in a very-fine-grained to aphanitic matrix. Formerly known as porphrite. { 'pȯr·fə·rē }

porpoise [VERT ZOO] Any of several species of marine mammals of the family Phocaenidae which have small flippers, a highly developed sonar system, and smooth, thick, hairless skin. { 'pȯr·pəs }

porpoise oil [MATER] A pale-yellow fatty oil obtained from blubber of the brown porpoise; soluble in ether, benzene, carbon disulfide, and chloroform; used as a lubricant, leather dressing, and illuminating oil, and in soap stock. { 'pȯr·pəs ,ȯil }

Porro prism [OPTICS] One of two identical prisms used in the Porro prism erecting system; it is a right-angle prism with the corners rounded to minimize breakage and simplify assembly. { 'pȯ·rō ,priz·əm }

Porro-prism erecting system [OPTICS] A compound erecting system, designed by M. Porro, in which there are four reflections to completely erect the image; two Porro prisms are employed; the line of sight is bent through 360°, is displaced, but is not deviated; used in prism binoculars and some telescope systems. { 'pȯ·rō ,priz·əm i'rek·tiŋ ,sis·təm }

port [COMPUT SCI] An interface between a communications channel and a unit of computer hardware. [ELEC] An entrance or exit for a network. [ELECTROMAG] An opening in a waveguide component, through which energy may be fed or withdrawn, or measurements made. [ENG] The side of a ship or airplane on the left of a person facing forward. [ENG ACOUS] An opening in a bass-reflex enclosure for a loudspeaker, designed and positioned to improve bass response. [GEOGR] *See* harbor. [NAV ARCH] An opening in a vessel to provide access for passengers, cargo handling, discharging water, and so forth. [NUCLEO] An opening in a research reactor through which objects are inserted for irradiation or from which beams of radiation emerge for experimental use. { pȯrt }

portability [COMPUT SCI] Property of a computer program that is sufficiently flexible to be easily transferred to run on a computer of a type different from the one for which it was designed. { ,pȯrd·ə'bil·əd·ē }

portable [ENG] Capable of being easily and conveniently transported. { 'pȯrd·ə·bəl }

portable audio terminal [COMPUT SCI] A lightweight, self-contained computer terminal with a typewriter keyboard, which can be attached to a telephone line by placing the telephone handset in a receptacle in the terminal. { 'pȯrd·ə·bəl 'ȯd·ē·ō ,tərm·ən·əl }

portable data terminal [COMPUT SCI] A computer terminal that can be carried about by hand to collect data from remote locations and to transfer this data to a computer system. { 'pȯrd·ə·bəl 'dad·ə ,tər·mən·əl }

portable document format [COMPUT SCI] A computer file format for publishing and distributing electronic documents (text, image, or multimedia) with the same layout, formatting, and font attributes as in the original. The files can be opened and viewed on any computer or operating system; however, special software is required. Abbreviated PDF. { 'pȯrd·ə·bəl ,däk·yə·mənt 'fȯr,mat }

porta hepatis [ANAT] The transverse fissure of the liver through which the portal vein and hepatic artery enter the liver and the hepatic ducts leave. { 'pȯrd·ə he'pad·əs }

portal [ANAT] **1.** Of or pertaining to the porta hepatis. **2.** Pertaining to the portal vein or system. [ENG] A redundant frame consisting of two uprights connected by a third member at the top. [MIN ENG] **1.** An entrance to a mine. **2.** The rock face at which a tunnel is started. { 'pȯrd·əl }

portal circulation [PHYSIO] The passage of venous blood through a portal system. { 'pȯrd·əl ,sər·kyə'lā·shən }

portal cirrhosis [MED] Replacement of normal liver structure by abnormal lobules of liver cells, often hyperplastic,

delimited by bands of fibrous tissue, giving the gross appearance of a finely nodular surface. Also known as Laennec's cirrhosis. { 'pȯrd·əl sə'rō·səs }

portal crane [MECH ENG] A jib crane carried on a four-legged portal built to run on rails. { 'pȯrd·əl 'krān }

portal hypertension [MED] Portal venous pressure in excess of 20 mmHg (2666 pascals), resulting from intrahepatic or extrahepatic portal venous compression or occlusion. { 'pȯrd·əl ,hī·pər'ten·shən }

portal system [ANAT] A system of veins that break into a capillary network before returning the blood to the heart. { 'pȯrd·əl 'sis·təm }

portal vein [ANAT] Any vein that terminates in a network of capillaries. { 'pȯrd·əl ,vān }

ported system *See* vented-box system. { 'pȯrt·əd ,sis·təm }

Porterfield [GEOL] A North American geologic stage of the Middle Ordovician, forming the lower division of the Mohawkian, and lying above Ashby and below Wilderness. { 'pȯrd·ər,fēld }

Portevin-Le Chatelier effect [SOLID STATE] The effect of foreign atoms on the deformation curve of a material, in which steps appear in what was initially a smooth curve. { ¦pȯrt,van lə,shat·lē'ā i,fekt }

port expander [COMPUT SCI] Equipment that connects links to several other devices to one port in a computer. { 'pȯrt ik,span·dər }

porthole [DES ENG] The opening or passageway connecting the inside of a bit or core barrel to the outside and through which the circulating medium is discharged. [ENG] A circular opening in the side of a ship or airplane, usually serving as a window and containing one or more panes of glass. { 'pȯrt,hōl }

porthole die [MET] An extrusion die having two or more sections in which metal is extruded separately in each section and welded before leaving the die to form intricate hollow shapes. { 'pȯrt,hōl ,dī }

Portia [ASTRON] A satellite of Uranus orbiting at a mean distance of 41,070 miles (66,100 kilometers) with a period of 12 hours 21 minutes, and with a diameter of about 68 miles (110 kilometers). { 'pȯr·shə }

portico [ARCH] A colonnade or other sheltered place to walk in. { 'pȯrd·ə,kō }

porting [COMPUT SCI] The process of converting software to run on a computer other than the one for which it was originally written. { 'pȯrd·iŋ }

portland cement [MATER] A hydraulic cement made of pulverized, calcined argillaceous and calcareous materials; the proper name for ordinary cement. { 'pȯrt·lənd si'ment }

Portlandian [GEOL] A European geologic stage of the Upper Jurassic, above Kimmeridgian, below Berriasian of Cretaceous. { pȯrt'land·ē·ən }

portlandite [MINERAL] Ca(OH)₂ A colorless, hexagonal mineral consisting of calcium hydroxide; occurs as minute plates. { 'pȯrt·lən,dīt }

portland-pozzolana cement [MATER] Portland cement to which pozzolana has been added, in the amount of about 20%, to reduce the liability of leaching. { 'pȯrt·lənd ,pät·sə'län·ə si'ment }

portland stone [MATER] A type of oolitic limestone that consists of fossils cemented together with lime; found only on the Isle of Portland, Dorset, England. { 'pȯrt·lənd 'stōn }

port of entry [CIV ENG] A location for clearance of foreign goods and citizens through a customhouse. { 'pȯrt əv 'en·trē }

port operations service [COMMUN] Maritime mobile communications service in or near a port, between coast stations and ship stations, or between ship stations, in which messages are restricted to those relating to the movement and safety of ships and, in an emergency, to the safety of persons. { 'pȯrt ,äp·ə'rā·shənz ,sər·vəs }

portrait [GRAPHICS] A printing orientation in which the printed lines are parallel to the narrow side of the paper. { 'pȯr·trət }

Portuguese man-of-war [INV ZOO] Any of several brilliantly colored tropical siphonophores in the genus *Physalia* which possess a large float and extremely long tentacles. { ,pȯr·chə'gēz ¦man əv 'wȯr }

Portulacaceae [BOT] A family of dicotyledonous plants in

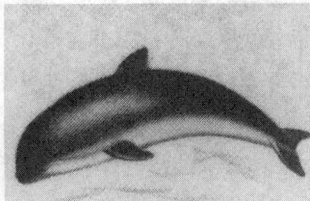
PORPOISE

The common porpoise.

PORTUGUESE MAN-OF-WAR

Float and tentacles of a Portuguese man-of-war. *(From T. I. Storer and R. L. Usinger, General Zoology, 3d ed., McGraw-Hill, 1957)*

the order Caryophyllales distinguished by a syncarpous gynoecium, few, cyclic tepals and stamens, two sepals, and two to many ovules. { ‚pȯr·chə·lə'kās·ē‚ē }

Portunidae [INV ZOO] The swimming crabs, a family of the Brachyura having the last pereiopods modified as swimming paddles. { pȯr'tü·nə‚dē }

port-wine stain [MED] A congenital hemangioma characterized by one or more red to purplish patches, usually on the face. { 'pȯrt ‚wīn ‚stān }

Porulosida [INV ZOO] An order of the protozoan subclass Radiolaria in which the central capsule shows many pores. { ‚pȯr·yə'lä·səd·ə }

Poseidon [ORD] A submarine-launched multiple-warhead nuclear missile that replaces the Polaris missile in nuclear submarines. { pə'sīd·ən }

poset *See* partially ordered set. { 'pō‚set }

posistor [ELECTR] A thermistor having a large positive resistance-temperature characteristic. { pä'zis·tər }

position [NAV] A point defined by stated or implied coordinates, usually on the surface of the earth. { pə'zish·ən }

positional astronomy [ASTRON] The branch of astronomy that deals with the determination of the positions of celestial objects. { pə'zish·ən·əl ə'strän·ə·mē }

positional cloning [GEN] The identification of a gene and its isolation as a cloned deoxyribonucleic acid fragment, starting from knowledge of its position on a genetic map or chromosome. { pə'zish·ən·əl 'klōn·iŋ }

positional-error constant [CONT SYS] For a stable unity feedback system, the limit of the transfer function as its argument approaches zero. { pə'zish·ən·əl ¦er·ər ‚kän·stənt }

positional isomer [CHEM] One of a set of structural isomers which differ only in the point at which a side-chain group is attached. [ORG CHEM] Constitutional isomer having the same functional group located in different positions along a chain or in a ring. { pə'zish·ən·əl 'ī·sə·mər }

positional notation [MATH] Any of several numeration systems in which a number is represented by a sequence of digits in such a way that the significance of each digit depends on its position in the sequence as well as its numeric value. Also known as notation. { pə'zish·ən·əl nō'tā·shən }

positional parameter [COMPUT SCI] One of a number of parameters in a group, whose significance is determined by its position within the group. { pə'zish·ən·əl pə'ram·əd·ər }

positional servomechanism [CONT SYS] A feedback control system in which the mechanical position (as opposed to velocity) of some object is automatically maintained. { pə'zish·ən·əl ¦sər·vō'mek·ə‚niz·əm }

position-analog unit [ENG] A device employed in machining operations to transmit analog information about the positions of machine parts to a servoamplifier which then compares it with input data. { pə'zish·ən ¦an·ə‚läg ‚yü·nət }

position angle [ASTRON] **1.** The angle formed by the great circle running through two celestial objects and the hour circle running through one of the objects. **2.** In measuring double stars, the angle formed between the great circle running through both components and the hour circle going through the primary measured from the north through the east from 0 to 360°. [NAV] That angle of the navigational triangle at the celestial body having the hour circle and the vertical circle at its sides. Also known as parallactic angle. { pə'zish·ən 'aŋ·gəl }

position approximate [NAV] Of inexact position; used principally on charts to indicate that the position of a wreck, shoal, and so on, has not been accurately determined or does not remain fixed. { pə'zish·ən ə'präk·sə·mət }

position blocks [MIN ENG] Unproved mining claims that are in a position to contain a lode if the lode continues in the direction in which it has been proved in other claims. { pə'zish·ən ‚bläks }

position buoy [NAV] A buoy attached to the stern of a boat which allows the vessel behind to maintain the appropriate or prescribed distance, especially under conditions of low visibility. Also known as fog buoy; towing spar. { pə'zish·ən ‚bȯi }

position circle [NAV] **1.** A circle covering all possible points within which a search objective can be located. **2.** The line of position produced by a distance measurement such as those produced by a DME (distance-measuring equipment). { pə'zish·ən ‚sər·kəl }

position-contouring system [CONT SYS] A numerical control system that exerts contouring control in two dimensions and position control in a third. { pə'zish·ən 'kän‚tür·iŋ ‚sis·təm }

position control [CONT SYS] A type of automatic control in which the input commands are the desired position of a body. { pə'zish·ən kən‚trōl }

position correction [ORD] Correction applied to firing data to compensate for difference in location of individual pieces in a battery. { pə'zish·ən kə‚rek·shən }

position correction grid [ORD] Transparent template superimposed on a large-scale plot of a battery position used to determine individual range and deflection corrections within the battery in order to obtain the type of sheaf (planned planes of fire) desired. { pə'zish·ən kə‚rek·shən ‚grid }

position doubtful [NAV] Of uncertain position; used principally on charts to indicate that a wreck, shoal, and so on, has been reported in various positions and not definitely determined in any. { pə'zish·ən 'daut·fül }

positioned weld [MET] A weld made in a joint in which members have been positioned to facilitate welding. { pə'zish·ənd 'weld }

position effect [GEN] **1.** Change in expressivity of a gene associated with changes in location on a chromosome, as from euchromatin to heterochromatin. **2.** Inherent gene expression as influenced by neighboring genes. { pə'zish·ən i‚fekt }

position error [NAV] **1.** The error of a position. **2.** That error of an instrument reading due to location or orientation. { pə'zish·ən ‚er·ər }

position finder [ORD] Optical or electrical instrument used in finding the range and position of a target. { pə'zish·ən ‚fīnd·ər }

position finding [ORD] The process of determining the position of a target with relation to the battery, and the determination of a future position upon which to direct the fire. { pə'zish·ən ‚fīnd·iŋ }

position firing [ORD] A method of defensive gunnery used by bombers, especially during World War II, in which definite amounts of deflection are prescribed for firing at attacking fighter planes. { pə'zish·ən ‚fīr·iŋ }

position fixing [NAV] The process of determining a craft's present position without reference to a previous position, by either radio (electronic) or astronomical (celestial) means. { pə'zish·ən ‚fiks·iŋ }

position indicator [ENG] An electromechanical dead-reckoning computer, either an air-position indicator or a ground-position indicator. { pə'zish·ən in·də‚kād·ər }

positioning [MECH ENG] A tooling function concerned with manipulating the workpiece in relationship to the working tools. { pə'zish·ən·iŋ }

positioning action [CONT SYS] Automatic control action in which there is a predetermined relation between the value of a controlled variable and the position of a final control element. { pə'zish·ən·iŋ ‚ak·shən }

positioning band [ORD] A metal band on some recoilless ammunition, placed to ensure the proper positioning of the round inside the chamber and tube. { pə'zish·ən·iŋ ‚band }

positioning time [COMPUT SCI] The time required for a storage medium such as a disk to be positioned and for read/write heads to be properly located so that the desired data can be read or written. [MECH ENG] The time required to move a machining tool from one coordinate position to the next. { pə'zish·ən·iŋ ‚tīm }

position line *See* line of position. { pə'zish·ən ‚līn }

position operator [QUANT MECH] The quantum-mechanical operator corresponding to the classical position variable of a particle. { pə'zish·ən ‚äp·ə‚rād·ər }

position pulse *See* commutator pulse. { pə'zish·ən ‚pəls }

position report [NAV] A radio message containing specified information regarding the position and progress of a craft. { pə'zish·ən ri‚pȯrt }

position representation [QUANT MECH] A representation in which the state functions are eigenfunctions of the position operator. Also known as Schrödinger representation. { pə'zish·ən ‚rep·ri·zən'tā·shən }

position sensor [ENG] A device for measuring a position and converting this measurement into a form convenient for transmission. Also known as position transducer. { pə'zish·ən ‚sen·sər }

position telemetering [ENG] A variation of voltage telemetering in which the system transmits the measurand by positioning a variable resistor or other component in a bridge circuit so as to produce relative magnitudes of electrical quantities or phase relationships. { pə'zish·ən ¦tel·ə'mēd·ə·riŋ }

position transducer See position sensor. { pə'zish·ən tranz ¦dü·sər }

position vector [MATH] The position vector of a point in euclidean space is a vector whose length is the distance from the origin to the point and whose direction is the direction from the origin to the point. Also known as radius vector. { pə'zish·ən ¦vek·tər }

positive [ELEC] Having fewer electrons than normal, and hence having ability to attract electrons. [GRAPHICS] Having the same rendition of light and shade as in the original scene. [MATH] Having value greater than zero. { 'päz·əd·iv }

positive acceleration [MECH] **1.** Accelerating force in an upward sense or direction, such as from bottom to top, or from seat to head; **2.** The acceleration in the direction that this force is applied. { 'päz·əd·iv ak,sel·ə'rā·shən }

positive afterimage [PHYSIO] An afterimage persisting after the eyes are closed or turned toward a dark background, and of the same color as the stimulating light. { 'päz·əd·iv 'af·tər,im·ij }

positive allosteric control See allosteric activation. { ¦päz·əd·iv ,al·ə,stir·ik kən'trōl }

positive angle [MATH] The angle swept out by a ray moving in a counterclockwise direction. { 'päz·əd·iv 'aŋ·gəl }

positive area See positive element. { 'päz·əd·iv 'er·ē·ə }

positive axis [MATH] The segment of an axis arising from a cartesian coordinate system which is realized by positive values of the coordinate variables. [METEOROL] In tropical synoptic analysis, a locus of maximum streamline curvature in an easterly wave; used primarily in the analysis of waves that span the equatorial trough (equatorial waves); a positive axis corresponds to a trough line in the Northern Hemisphere and a ridge line in the Southern Hemisphere. { 'päz·əd·iv 'ak·səs }

positive bias [ELECTR] A bias such that the control grid of an electron tube is positive with respect to the cathode. { 'päz·əd·iv 'bī·əs }

positive birefringence [OPTICS] Birefringence in which the velocity of the ordinary ray is greater than that of the extraordinary ray. { 'päz·əd·iv ,bi·ri'frin·jəns }

positive charge [ELEC] The type of charge which is possessed by protons in ordinary matter, and which may be produced in a glass object by rubbing with silk. { 'päz·əd·iv 'chärj }

positive click adjustment [IND ENG] A means of adjusting dials or push buttons to incorporate audible clicks or their tactile counterparts at predetermined positions in order to provide appropriate motor-sensory feedback to the operator. { ¦päz·əd·iv ¦klik ə'jəz·mənt }

positive clutch [MECH ENG] A clutch designed to transmit torque without slip. { 'päz·əd·iv 'kləch }

positive column [ELECTR] The luminous glow, often striated, that occurs between the Faraday dark space and the anode in a glow-discharge tube. Also known as positive glow. { 'päz·əd·iv 'käl·əm }

positive correlation [STAT] A relation between two quantities such that when one increases the other does also. { 'päz·əd·iv ,kä·rə'lā·shən }

positive crystal [OPTICS] **1.** Uniaxial anisotropic crystal having the ordinary index of refraction greater than the extraordinary index. **2.** Biaxial anisotropic crystal having the intermediate index of refraction beta closer in value to alpha, and with Z the acute bisectrix. { 'päz·əd·iv 'krist·əl }

positive definite [MATH] **1.** A square matrix A of order n is positive definite if

$$\sum_{i,j=1}^{n} A_{ij}x_i\bar{x}_j > 0$$

for every choice of complex numbers $x_1, x_2, \ldots, x_n$, not all equal to 0, where $\bar{x}_j$ is the complex conjugate of x_j. **2.** A linear operator T on an inner product space is positive definite if $\langle Tu,u\rangle$ is greater than 0 for all nonzero vectors u in the space. { 'päz·əd·iv 'def·ə·nət }

positive derail [MIN ENG] A device installed in or on a mine track to derail runaway cars or trips. { 'päz·əd·iv di'rāl }

positive-displacement compressor [MECH ENG] A compressor that confines successive volumes of fluid within a closed space in which the pressure of the fluid is increased as the volume of the closed space is decreased. { 'päz·əd·iv dis¦plās·mənt kəm,pres·ər }

positive-displacement meter [ENG] A fluid quantity meter that separates and captures definite volumes of the flowing stream one after another and passes them downstream, while counting the number of operations. { 'päz·əd·iv dis¦plās·mənt ,mēd·ər }

positive-displacement pump [MECH ENG] A pump in which a measured quantity of liquid is entrapped in a space, its pressure is raised, and then it is delivered; for example, a reciprocating piston-cylinder or rotary-vane, gear, or lobe mechanism. { 'päz·əd·iv dis¦plās·mənt ,pəmp }

positive draft [MECH ENG] Pressure in the furnace or gas passages of a steam-generating unit which is greater than atmospheric pressure. { 'päz·əd·iv 'draft }

positive drive belt See timing belt. { 'päz·əd·iv 'drīv ,belt }

positive electrode See anode. { 'päz·əd·iv i'lek,trōd }

positive electron See positron. { 'päz·əd·iv i'lek,trän }

positive element [GEOGR] A large structural feature of the earth's crust characterized by long-term upward movement (uplift, emergence) or subsidence less rapid than that of adjacent negative elements. Also known as archibole; positive area. { 'päz·əd·iv 'el·ə·mənt }

positive estuary [HYD] An estuary in which there is a measurable dilution of seawater by land drainage. { 'päz·əd·iv 'es·chə,wer·ē }

positive feedback [CONT SYS] Feedback in which a portion of the output of a circuit or device is fed back in phase with the input so as to increase the total amplification. Also known as reaction (British usage); regeneration; regenerative feedback; retroaction (British usage). { 'päz·əd·iv 'fēd,bak }

positive gene control [MOL BIO] Enhancement of gene expression through binding of specific expressor molecules to promoter sites. { ¦päz·əd·iv 'jēn kən,trōl }

positive glow See positive column. { 'päz·əd·iv 'glō }

positive-grid oscillator See retarding-field oscillator. { 'päz·əd·iv ¦grid 'äs·ə,lād·ər }

positive image [GRAPHICS] A picture as normally seen on a television picture tube or in a photograph, having the same rendition of light and shade as in the original scene. { 'päz·əd·iv 'im·ij }

positive integer [MATH] An integer greater than zero; one of the numbers 1, 2, 3, ... { ¦päz·əd·iv 'int·i·jər }

positive interference [GEN] The reduction of the likelihood of another crossover in the vicinity of a nearby crossover. { 'päz·əd·iv ,in·tər'fir·əns }

positive ion [CHEM] An atom or group of atoms which by loss of one or more electrons has acquired a positive electric charge; occurs on ionization of chemical compounds as H⁺ from ionization of hydrochloric acid, HCl. { 'päz·əd·iv 'ī,än }

positive-ion sheath [ELECTR] Collection of positive ions on the control grid of a gas-filled triode tube. { 'päz·əd·iv ¦ī,än ,shēth }

positive landform [GEOL] An upstanding topographic form, such as a mountain, hill, plateau, or cinder cone. { 'päz·əd·iv 'land,fòrm }

positive lens See converging lens. { 'päz·əd·iv ,lenz }

positive linear functional [MATH] A linear functional on some vector space of real-valued functions which takes every nonnegative function into a nonnegative number. { ¦päz·əd·iv ¦lin·ē·ər 'fəŋk·shən·əl }

positive logic [ELECTR] Logic circuitry in which the more positive voltage (or current level) represents the 1 state; the less positive level represents the 0 state. { 'päz·əd·iv 'läj·ik }

positive meniscus lens [OPTICS] A lens having one convex (bulging) and one concave (depressed) surface, with the radius of curvature of the convex surface smaller than that of the concave surface. { 'päz·əd·iv mə'nis·kəs ,lenz }

positive mirror See converging mirror. { 'päz·əd·iv 'mir·ər }

positive modulation [ELECTR] In an amplitude-modulated television system, that form of television modulation in which an increase in brightness corresponds to an increase in transmitted power. { 'päz·əd·iv ,mäj·ə'lā·shən }

positive mold [ENG] A plastics mold designed to trap all of the molding resin when the mold closes. { 'päz·əd·iv 'mōld }

positive motion [MECH ENG] Motion transferred from one machine part to another without slippage. { 'päz·əd·iv 'mō·shən }

positive movement [GEOL] **1.** Uplift or emergence of the earth's crust relative to an adjacent area of the crust. **2.** A relative rise in sea level with respect to land level. { 'päz·əd·iv 'müv·mənt }

positive number [MATH] A real number that is greater than 0. { ¦päz·əd·iv 'nəm·bər }

positive ore [MIN ENG] Ore exposed on four sides in blocks of a size variously prescribed. { 'päz·əd·iv 'ȯr }

positive part [MATH] For a real-valued function f, this is the function, denoted f^+, for which $f^+(x) = f(x)$ if $f(x) \geq 0$ and $f^+(x) = 0$ if $f(x) < 0$. { 'päz·əd·iv 'pärt }

positive pedal curve See pedal curve. { 'päz·əd·iv 'ped·əl ,kərv }

positive phase sequence [ELEC] The phase sequence that corresponds to the normal order of phases in a polyphase system. { 'päz·əd·iv ¦fāz ,sē·kwəns }

positive-phase-sequence relay [ELEC] Relay which functions in conformance with the positive-phase-sequence component of the current, voltage, or power of the circuit. { 'päz·əd·iv ¦fāz ,sē·kwəns ,rē·lā }

positive photokinesis [PHYSIO] The faster movement of an organism upon entering an illuminated area. { päz·əd·iv ,fōd·ōkə'nē·səs }

positive phototaxis [PHYSIO] The orientation and movement of an organism toward the source of a light stimulus. { ¦päz·əd·iv ,fōd·ō'tak·səs }

positive pole See north pole. { 'päz·əd·iv 'pōl }

positive ray [ELECTR] A stream of positively charged atoms or molecules, produced by a suitable combination of ionizing agents, accelerating fields, and limiting apertures. { 'päz·əd·iv 'rā }

positive real function [MATH] An analytic function whose value is real when the independent variable is real, and whose real part is positive or zero when the real part of the independent variable is positive or zero. { 'päz·əd·iv 'rēl 'fəŋk·shən }

positive semidefinite [MATH] Also known as nonnegative semidefinite. **1.** A square matrix A is positive semidefinite if

$$\sum_{i,j=1}^{n} A_{ij} x_i \bar{x}_j \geq 0$$

for every choice of complex numbers $x_1, x_2, \ldots, x_n$, where $\bar{x}_j$ is the complex conjugate of x_j. **2.** A linear operator T on an inner product space is positive semidefinite if $\langle Tu,u \rangle$ is equal to or greater than 0 for all vectors u in the space. { 'päz·əd·iv ¦sem·i¦def·ə·nət }

positive series [MATH] A series whose terms are all positive real numbers. { 'päz·əd·iv 'sir,ēz }

positive shoreline See shoreline of submergence. { 'päz·əd·iv 'shȯr,līn }

positive sign [MATH] The symbol +, used to indicate a positive number. { 'päz·əd·iv ,sīn }

positive skewness [STAT] Property of a unimodal distribution with a longer tail in the direction of higher values of the random variable. { 'päz·əd·iv 'skü·nəs }

positive temperature coefficient [THERMO] The condition wherein the resistance, length, or some other characteristic of a substance increases when temperature increases. { 'päz·əd·iv 'tem·prə·chər ,kō·i,fish·ənt }

positive terminal [ELEC] The terminal of a battery or other voltage source toward which electrons flow through the external circuit. { 'päz·əd·iv 'tərm·ən·əl }

positive transfer [PSYCH] Rapid learning in a new situation because the stimuli or responses are similar to those learned in an earlier situation. { 'päz·əd·iv 'tranz·fər }

positive transmission [COMMUN] Transmission of television signals in such a way that an increase in initial light intensity causes an increase in the transmitted power. { 'päz·əd·iv tranz'mish·ən }

positive with respect to a signed measure [MATH] A set A is positive with respect to a signed measure m if, for every measurable set B, the intersection of A and B, $A \cap B$, is measurable and $m(A \cap B) \geq 0$. { 'päz·əd·iv with ri,spekt tü ə ¦sīnd 'mezh·ər }

positive zero [COMPUT SCI] The zero value reached by counting down from a positive number in the binary system. { 'päz·əd·iv 'zir·ō }

positron [PARTIC PHYS] An elementary particle having mass equal to that of the electron, and having the same spin and statistics as the electron, but a positive charge equal in magnitude to the electron's negative charge. Also known as positive electron. { 'päz·ə,trän }

positron camera [ENG] An instrument that uses photomultiplier tubes in combination with scintillation counters to detect oppositely directed gamma-ray pairs resulting from the annihilation of electrons of positrons emitted by short-lived radioisotopes used as tracers in the human body. { 'päz·ə,trän ,kam·rə }

positron depth profiling [SOLID STATE] A technique in which the spread in stopping depths from a low-energy monoenergetic positron beam is measured and used to obtain information on the presence and depth of various crystal defects below the surface. { 'päz·ə,trän 'depth ,prō,fīl·iŋ }

positron emission [NUC PHYS] A β-decay process in which a nucleus ejects a positron and a neutrino. { 'päz·ə,trän i,mish·ən }

positron emission spectroscopy [SPECT] A technique in which a solid surface is bombarded with a low-energy monoenergetic positron beam and the energies of positrons emitted from the surface are measured to determine the amounts of energy lost to molecules adsorbed on the surface. { 'päz·ə,trän i,mish·ən spek'träs·kə·pē }

positron emission tomography [MED] A technique that uses measurements of the back-to-back emission of gamma rays from the annihilation of positrons emitted by radioactive tracers to map the distribution of these tracers in the human body. Abbreviated PET. { 'päz·ə,trän i,mish·ən tō'mäg·rə·fē }

positronium [PARTIC PHYS] The bound state of an electron and a positron. { ,päz·ə'trō·nē·əm }

positronium velocity spectroscopy [SPECT] A technique in which a solid surface is bombarded with a low-energy monoenergetic positron beam and the velocities of the emitted positronium atoms are measured to determine the energy and momentum spectrum of the density of electron states near the surface. { ,päz·ə'trō·nē·əm və'läs·əd·ē spek'träs·kə·pē }

possible ore [MIN ENG] A class of ore whose existence is a reasonable possibility, based upon geologic-mineralogic relationships and the extent of ore bodies already developed. Also known as extension ore. { 'päs·ə·bəl ¦ȯr }

possible reserves [PETRO ENG] Primary petroleum reserves that may exist, but available data do not confirm their presence. { 'päs·ə·bəl ri'zərvz }

post [CIV ENG] **1.** A vertical support such as a pillar, upright, or fence stake. **2.** A pole used as a boundary marker. [COMPUT SCI] To add or update records in a file. [MIN ENG] **1.** A mine timber, or any upright timber, more commonly the uprights which support the roof crosspieces. **2.** The support fastened between the roof and floor of a coal seam, used with certain types of mining machines or augers. **3.** A pillar of coal or ore. [NAV] A small beacon used for marking channels; it is usually more substantial than a perch. { pōst }

postaccelerating electrode See intensifier electrode. { ,pōst·ak'sel·ə,rād·iŋ i'lek,trōd }

postacceleration [ELECTR] Acceleration of beam electrons after deflection in an electron-beam tube. Also known as postdeflection acceleration (PDA). { ,pōst·ak,sel·ə'rā·shən }

postaccelerator [NUCLEO] A particle accelerator that gives particles additional energy after they have passed through another particle accelerator. { ,pōst·ik'sel·ə,rād·ər }

post-and-beam construction [BUILD] A type of wall construction using posts instead of studs. { ,pōst ən 'bēm kən,strək·shən }

post-and-lintel [ARCH] Pertaining to construction employing vertical supports and horizontal beams instead of arches or vaults. { 'pōst ən 'lint·əl }

postauricular hearing aid [ENG ACOUS] A hearing aid that fits behind the ear and has a sound tip attached to plastic tubing that conducts sound through an ear mold to the ear canal. { ,pōst·ō,rik·yə·lər 'hēr·iŋ ,ād }

post brake [MECH ENG] A brake occasionally fitted on a

steam winder or haulage, and consisting of two upright posts mounted on either side of the drum that operate on brake paths bolted to the drum cheeks. { 'pōst ‚brāk }

postcard paper [MATER] Lightweight Bristol board, made from soda and sulfite pulps, that has a smooth, firm surface for writing with pen or pencil, or for ordinary printing. { 'pōs‚kärd ‚pā·pər }

postcentral gyrus [ANAT] The cerebral convolution that lies immediately posterior to the central sulcus and extends from the longitudinal fissure above the posterior ramus of the lateral sulcus. { pōst'sen·trəl 'jī·rəs }

postcentral sulcus [ANAT] The first sulcus of the parietal lobe of the cerebrum, lying behind and roughly parallel to the central sulcus. { pōst'sen·trəl 'səl·kəs }

postcollision interaction [ATOM PHYS] A phenomenon that arises near a threshold of photoionization in which a slowly receding photoelectron transfers energy to an Auger electron. { ‚pōst·kə‚lizh·ən ‚in·tər'ak·shən }

postcure bonding [ENG] A method of postcuring at elevated temperatures of parts previously subjected to autoclave or press in order to obtain higher heat-resistant properties of the adhesive bond. { 'pōst‚kyùr 'bänd·iŋ }

postcure finish [TEXT] A durable-press finish applied to a fabric by the mill but heat-set after the garment is completed. { 'pōst‚kyùr 'fin·ish }

postdecrementing See autodecrement addressing. { ‚pōst'dek·rə‚ment·iŋ }

postdeflection accelerating electrode See intensifier electrode. { ‚pōst·di'flek·shən ak'sel·ə‚rād·iŋ i'lek‚trōd }

postdeflection acceleration See postacceleration. { ‚pōst·di'flek·shən ak'sel·ə‚rā·shən }

post drill [ENG] An auger or drill supported by a post. { 'pōst‚dril }

postedit [COMPUT SCI] To edit the output data of a computer. { 'pōst‚ed·ət }

postemphasis See deemphasis. { ‚pōst'em·fə·səs }

postencephalitic parkinsonism [MED] The parkinsonian syndrome occurring as a sequel to lethargic encephalitis within a variable period, from days to many years, after the acute process. { ‚pōst·in‚sef·ə'lid·ik 'pärk·ən·sə‚niz·əm }

postequalization See deemphasis. { ‚pōst‚ē·kwə·lə'zā·shən }

poster board [MATER] A stiff cardboard used for show cards, posters, display advertising, and signs; it may be white or colored on one side. { 'pōs·tər ‚bòrd }

posterior [ZOO] **1.** The hind end of an organism. **2.** Toward the back, or hinder end, of the body. { pä'stir·ē·ər }

posterior chamber [ANAT] The space in the eye between the posterior surface of the iris and the ciliary body, and the lens. { pä'stir·ē·ər 'chäm·bər }

posterior distribution [STAT] A probability distribution on the values of an unknown parameter that combines prior information about the parameter contained in the observed data to give a composite picture of the final judgments about the values of the parameter. { pä‚stir·ē·ər ‚dis·trə'byü·shən }

posterior horn [NEUROSCI] The dorsal column of gray matter in the spinal cord containing the axons of sensory (afferent) neurons. { pä‚stir·ē·ər 'hòrn }

posterior probabilities [STAT] Probabilities of the outcomes of an experiment after it has been performed and a certain event has occurred. { pä'stir·ē·ər präb·ə'bil·əd·ēz }

POS terminal See point-of-sale terminal. { ‚pē‚ō'es ‚term·ən·əl }

poster paint [MATER] A water paint with a gum binder, which is brilliant, opaque, and fast-drying, and usually sold in jars. Also known as show-card color. { 'pōs·tər ‚pānt }

poster paper [MATER] A strong, waterproofed paper used for billboard poster work; it is white or colored with nonfading pigments, and does not curl when paste is applied to it. { 'pōs·tər ‚pā·pər }

postfix notation See reverse Polish notation. { 'pōst‚fiks nō'tā·shən }

postforming [ENG] Forming, bonding, or shaping of heated, flexible thermoset laminates before the final thermoset reaction has occurred; upon cooling, the formed shape is held. { 'pōst‚fòrm·iŋ }

post-Galilean transformation [RELAT] A modification of the Lorentz transformation that includes first-order corrections associated with the effects of general relativity. { ‚pōst ‚gal·ə'lē·ən ‚tranz·fər'mā·shən }

Postglacial See Holocene. { pōst'glā·shəl }

postheating [MET] Application of heat after thermal spraying, brazing, or welding to control cooling rate. { ‚pōst'hēd·iŋ }

posthitis [MED] Inflammation of the prepuce. { päs'thīd·əs }

posthole [CIV ENG] A hole bored in the ground to hold a fence post. { 'pōst‚hōl }

posthumous structure [GEOL] Folds, faults, and other structural features in covering strata which revive or mimic the structure of older underlying rocks that are generally more deformed. { 'päs·chə·məs 'strək·chər }

postignition [CHEM] Surface ignition after the passage of the normal spark. { ‚pōst·ig'nish·ən }

postincrementing See autoincrement addressing. { ‚pōst'in·krə‚ment·iŋ }

postindexing [COMPUT SCI] Operation in which the contents of a register indicated by the index bits of an indirect address are added to the indirect address to form the effective address. { pōst'in‚dek·siŋ }

posting interpreter See transfer interpreter. { 'pōst·iŋ in'tər·prəd·ər }

postmagmatic [GEOL] Pertaining to geologic reactions or events occurring after the bulk of the magma has crystallized. { ‚pōst·mag'mad·ik }

postmeiotic fusion [CELL MOL] The union of two identical haploid nuclei produced by mitotic division of the egg nucleus, resulting in restoration of the diploid state. { ‚pōst·mē‚äd·ik 'fyü·zhən }

postmeridian [SCI TECH] After noon, or the period of time between noon (1200) and midnight (2400). { ‚pōst·mə'rid·ē·ən }

postmortem [COMPUT SCI] Any action taken after an operation is completed to help analyze that operation. [MED] Occurring after death. { pōst'mòrd·əm }

postmortem dump [COMPUT SCI] **1.** The printout showing the state of all registers and the contents of main memory, taken after a computer run terminates normally or terminates owing to fault. **2.** The program which generates this printout. { pōst'mòrd·əm 'dəmp }

postmortem program See postmortem routine. { pōst'mòrd·əm 'prō·grəm }

postmortem routine [COMPUT SCI] A computer routine designed to provide information about the operation of a program after the program is completed. Also known as postmortem program. { pōst'mòrd·əm rü‚tēn }

postmultiplication [MATH] In multiplying a matrix or operator **B** by another matrix or operator **A**, the operation that results in the matrix or operator **BA**. Also known as multiplication on the right. { ‚pōst·məl·tə·plə'kā·shən }

postnatal [MED] Subsequent to birth. { pōs'nād·əl }

postnecrotic cirrhosis [MED] Cirrhosis, usually due to toxic agents or viral hepatitis, characterized by necrosis of liver cells, regenerating nodules of hepatic tissue, and the presence of large bands of connective tissue. { ‚pōs·nə'kräd·ik sə'rō·səs }

post-Newtonian effects [RELAT] The first-order corrections of general relativity to classical Newtonian mechanics. { ‚pōst nü'tō·nē·ən i'feks }

postnova [ASTRON] A nova that has faded to the brightness it had before its outburst. { pōs'nō·ə }

postobsequent stream [HYD] A strike stream developed after the obsequent stream into which it flows. { ‚pōst·äb'sē·kwənt 'strēm }

post office [COMPUT SCI] The software and files in an electronic mail system that receive messages and deliver them to recipients. { 'pōst ‚òf·əs }

Post Office Protocol [COMPUT SCI] An Internet standard for delivering e-mail from a server to an e-mail client on a personal computer. Abbreviated POP. { ‚pōst ‚òf·əs 'prōd·ə‚kòl }

postoperative hernia See incisional hernia. { pōst'äp·rəd·iv 'hər‚nē·ə }

postorogenic [GEOL] Of a geologic process or event, occurring after a period of orogeny. { ‚pōst‚òr·ə'jen·ik }

postpartum [MED] Following childbirth. { pōs'pärd·əm }

postpartum depression [PSYCH] Any acute depression occurring in approximately 3 months following childbirth. { ‚pōs‚pärd·əm di'presh·ən }

postpartum pituitary necrosis See Sheehan's syndrome. { ‚pòst‚pärd·əm pə‚tü·ə‚ter·ē nə'krō·səs }

postpartum psychosis [PSYCH] A psychotic reaction in a woman following childbirth. { ‚pōs‚pärd·əm sī'kō·səs }

postprandial [MED] After a meal. { pōs'pran·dē·əl }

postprecipitation [CHEM] Precipitation of an impurity from a supersaturated solution onto the surface of an already present precipitate; used for analytical laboratory separations. { ‚pōs·pri‚sip·ə'tā·shən }

postprocessor [COMPUT SCI] A program that converts graphical output data to a form that can be used by computing equipment. { 'pōst'prä‚ses·ər }

postsynchronizing studio See ADR studio. { ‚pōst‚siŋ·krə‚nīz·iŋ 'stüd·ē·ō }

posttandem machine [NUCLEO] A particle accelerator that gives particles additional energy after they have passed through a tandem accelerator. { ‚pōst'tan·dəm mə'shēn }

posttensioning [ENG] Compressing of cast concrete beams or other structural members to impart the characteristics of prestressed concrete. { pōs'ten·shən·iŋ }

posttranslational modification [MOL BIO] Any polypeptide alteration that occurs after synthesis of the chain. { ‚pōs‚tranz'lā·shən·əl ‚mäd·ə·fə'kā·shən }

posttraumatic hernia See incisional hernia. { ‚pōs·trō'mad·ik 'hər·nē·ə }

posttraumatic stress disorder [PSYCH] An anxiety disorder in some individuals who have experienced an extremely stressful and traumatic event. It is marked by periodic and persistent reexperiencing of the event, persistent avoidance of events related to the trauma, psychological numbing that was not present prior to the trauma, and enduring symptoms of anxiety and arousal. { ‚pos·trō‚mad·ik 'stres dis‚ord·ər }

posttuning drift [ELECTR] In a frequency-agile source such as the fast-tuning oscillators used in set-on jammers for electronic warfare equipment, the increase in frequency brought about by the drop in temperature of the varactor after warmup time, settling time, and the time when the oscillator has reached a new frequency. Abbreviated PTD. { 'pōs‚tün·iŋ 'drift }

postulate See axiom. { 'päs·chə·lət }

postural edema [MED] A condition resulting from an increased hydrostatic pressure in the capillaries caused by fluid collection in the subcutaneous tissues of the feet and ankles that occurs when an individual has been standing motionless for a long period of time. { ‚päs·chə·rəl ə'dē·mə }

postweld interval [MET] The elapsed time between the end of a resistance welding current and the start of hold time. { 'pōst‚weld 'in·tər·vəl }

pot See potentiometer; pothole. { pät }

potable [SCI TECH] Suitable for drinking. { 'pōd·ə·bəl }

potamic [SCI TECH] Pertaining to rivers or river navigation. { pə'tam·ik }

Potamogalinae [VERT ZOO] An aberrant subfamily of West African tenrecs (Tenrecidae). { ‚päd·ə·mō'gal·ə‚nē }

potamogenic rock [PETR] A sedimentary rock formed by precipitation from river water. { 'päd·ə·mō'jen·ik räk }

Potamogetonaceae [BOT] A large family of monocotyledonous plants in the order Najadales characterized by a solitary, apical or lateral ovule, usually two or more carpels, flowers in spikes or racemes, and four each of tepals and stamens. { ‚päd·ə·mō‚jēd·ə'nās·ē‚ē }

Potamogetonales [BOT] The equivalent name for Najadales. { ‚päd·ə·mō‚jēd·ə'nā·lēz }

potamology [HYD] The scientific study of rivers. { ‚päd·ə'mäl·ə·jē }

Potamonidae [INV ZOO] A family of fresh-water crabs included in the Brachyura. { ‚päd·ə'män·ə‚dē }

potamoplankton [BIOL] Plankton found in rivers. { 'päd·ə·mō'plaŋk·tən }

potarite [MINERAL] PdHg A silver-white isometric mineral composed of palladium and mercury alloy. Also known as palladium amalgam. { pə'tä‚rīt }

potash See potassium carbonate. { 'päd‚ash }

potash alum See kalinite. { 'päd‚ash 'al·əm }

potash bentonite See potassium bentonite. { 'päd‚ash 'bent·ən‚īt }

potash blue [INORG CHEM] A pigment made by oxidizing ferrous ferrocyanide; used in making carbon paper. { 'päd‚ash 'blü }

potash feldspar See potassium feldspar. { 'päd‚ash 'fel‚spär }

potash kettle See giant's kettle. { 'päd‚ash 'ked·əl }

potash lake [HYD] An alkali lake whose waters contain a high content of dissolved potassium salts. { 'päd‚ash 'lāk }

potash mica See muscovite. { 'päd‚ash 'mī·kə }

potash regulations [MIN ENG] Rules governing the prospecting and exploitation of land containing potash. { 'päd‚ash ‚reg·yə'lā·shənz }

potassic [PETR] Referring to a rock which contains a significant amount of potassium. { pə'tas·ik }

potassium [CHEM] A chemical element, symbol K, atomic number 19, atomic weight 39.0983; an alkali metal. Also known as kalium. { pə'tas·ē·əm }

potassium-40 [NUC PHYS] A radioactive isotope of potassium having a mass number of 40, a half-life of approximately 1.31×10^9 years, and an atomic abundance of 0.000122 gram per gram of potassium. { pə'tas·ē·əm 'fȯr·dē }

potassium-42 [NUC PHYS] Radioactive isotope with mass number of 42; half-life is 12.4 hours, with β- and γ-radiation; radiotoxic; used as radiotracer in medicine. { pə'tas·ē·əm 'fȯr·dē‚tü }

potassium acetate [ORG CHEM] $KC_2H_3O_2$ White, deliquescent solid; soluble in water and alcohol, insoluble in ether; melts at 292°C; used as analytical reagent, dehydrating agent, in medicine, and in crystal glass manufacture. { pə'tas·ē·əm 'as·ə‚tāt }

potassium acid carbonate See potassium bicarbonate. { pə'tas·ē·əm 'as·əd 'kär·bə‚nāt }

potassium acid fluoride See potassium bifluoride. { pə'tas·ē·əm 'as·əd 'flür‚īd }

potassium acid oxalate See potassium binoxalate. { pə'tas·ē·əm 'as·əd 'äk·sə‚lāt }

potassium acid phosphate See potassium phosphate. { pə'tas·ē·əm 'as·əd 'fäs‚fāt }

potassium acid phthalate See potassium biphthalate. { pə'tas·ē·əm 'as·əd 'tha‚lāt }

potassium acid saccharate [ORG CHEM] HOOC-(CHOH)₄COOK An off-white powder, soluble in hot water, acid, or alkaline solutions; used in rubber formulations, soaps, and detergents, and for metal plating. { pə'tas·ē·əm 'as·əd 'sak·ə‚rāt }

potassium acid sulfate See potassium bisulfate. { pə'tas·ē·əm 'as·əd 'səl‚fāt }

potassium acid sulfite See potassium bisulfite. { pə'tas·ē·əm 'as·əd 'səl‚fīt }

potassium acid tartrate See potassium bitartrate. { pə'tas·ē·əm 'as·əd 'tär‚trāt }

potassium alginate [ORG CHEM] $(C_6H_7O_6K)_n$ A hydrophilic colloid occurring as filaments, grains, granules, and powder; used in food processing as a thickener and stabilizer. Also known as potassium polymannuronate. { pə'tas·ē·əm 'al·jə‚nāt }

potassium alum See potassium aluminum sulfate. { pə'tas·ē·əm 'al·əm }

potassium aluminate [INORG CHEM] $K_2Al_2O_4 \cdot 3H_2O$ Water-soluble, alcohol-insoluble, lustrous crystals; used as a dyeing and printing mordant, and as a paper sizing. { pə'tas·ē·əm ə'lüm·ə‚nāt }

potassium aluminum fluoride [INORG CHEM] K_3AlF_6 A toxic, white powder used as an insecticide. { pə'tas·ē·əm ə'lüm·ə·nəm 'flür‚īd }

potassium aluminum sulfate [INORG CHEM] $KAl(SO_4)_2 \cdot 12H_2O$ White, odorless crystals that are soluble in water; used in medicines and baking powder, in dyeing, papermaking, and tanning. Also known as alum; aluminum potassium sulfate; potassium alum. { pə'tas·ē·əm ə'lüm·ə·nəm 'səl‚fāt }

potassium antimonate [INORG CHEM] $KSbO_3$ White, water-soluble crystals. Also known as potassium stibnate. { pə'tas·ē·əm 'ant·ə·mə‚nāt }

potassium antimonyl tartrate See tartar emetic. { pə'tas·ē·əm 'ant·ə·mə‚nil 'tär‚trāt }

potassium argentocyanide See silver potassium cyanide. { pə'tas·ē·əm 'är·jən·tō'sī·ə‚nīd }

potassium-argon dating [GEOL] Dating of archeological, geological, or organic specimens by measuring the amount of argon accumulated in the matrix rock through decay of radioactive potassium. { pə'tas·ē·əm 'är‚gän 'dād·iŋ }

potassium arsenate [INORG CHEM] K_3AsO_4 Poisonous, colorless crystals; soluble in water, insoluble in alcohol; used as an insecticide, analytical reagent, and in hide preservation and textile printing. Also known as Macquer's salt. { pə'tas·ē·əm 'ärs·ən,āt }

potassium arsenite [INORG CHEM] $KH(AsO_2)_2$ Poisonous, hygroscopic, white powder; soluble in alcohol; decomposes slowly in air; used in medicine, on mirrors, and as an analytical reagent. Also known as potassium metarsenite. { pə'tas·ē·əm 'ärs·ən,īt }

potassium aurichloride See potassium gold chloride. { pə'tas·ē·əm ˌȯr·ə'klȯr,īd }

potassium bentonite [GEOL] A clay of the illite group that contains potassium and is formed by alteration of volcanic ash. Also known as K bentonite; potash bentonite. { pə'tas·ē·əm 'bent·ən,īt }

potassium bicarbonate [INORG CHEM] $KHCO_3$ A white powder or granules, or transparent colorless crystals; used in baking powder and in medicine as an antacid. Also known as potassium acid carbonate. { pə'tas·ē·əm bī'kär·bə,nāt }

potassium bichromate See potassium dichromate. { pə'tas·ē·əm bī'krō,māt }

potassium bifluoride [INORG CHEM] KHF_2 Colorless, corrosive, poisonous crystals; soluble in water and dilute alcohol; used to etch glass and as a metallurgy flux. Also known as Fremy's salt; potassium acid fluoride. { pə'tas·ē·əm bī'flur,īd }

potassium binoxalate [ORG CHEM] $KHC_2O_4·H_2O$ A poisonous, white, odorless, crystalline compound; used to clean wood and remove ink stains, as a mordant in dyeing, and in photography. Also known as potassium acid oxalate; sal acetosella; salt of sorrel. { pə'tas·ē·əm bə'näk·sə,lāt }

potassium biphthalate [ORG CHEM] $HOOCC_6H_4COOK$ A crystalline compound, soluble in 12 parts of water; used as a buffer in pH determinations and as a primary standard for preparation of volumetric alkali solutions. Also known as acid potassium phthalate; potassium acid phthalate; potassium hydrogen phthalate. { pə'tas·ē·əm bī'tha,lāt }

potassium bismuth tartrate [ORG CHEM] A white, granular powder with a sweet taste; soluble in water; used in medicine. Also known as bismuth potassium tartrate. { pə'tas·ē·əm 'biz·məth 'tär,trāt }

potassium bisulfate [INORG CHEM] $KHSO_4$ Water-soluble, colorless crystals, melting at 214°C; used in winemaking, fertilizer manufacture, and as a flux and food preservative. Also known as acid potassium sulfate; potassium acid sulfate. { pə'tas·ē·əm bī'səl,fāt }

potassium bisulfite [INORG CHEM] $KHSO_3$ White, water-soluble powder with sulfur dioxide aroma; insoluble in alcohol; decomposes when heated; used as an antiseptic and reducing chemical, and in analytical chemistry, tanning, and bleaching. Also known as potassium acid sulfite. { pə'tas·ē·əm bī'səl,fīt }

potassium bitartrate [ORG CHEM] $KHC_4H_4O_6$ White, water-soluble crystals or powder; used in baking powder, for medicine, and as an acid and buffer in foods. Also known as cream of tartar; potassium acid tartrate. { pə'tas·ē·əm bī'tär,trāt }

potassium borohydride [INORG CHEM] KBH_4 A white, crystalline powder, soluble in water, alcohol, and ammonia; used as a hydrogen source and a reducing agent for aldehydes and ketones. { pə'tas·ē·əm ˌbȯr·ō'hī,drīd }

potassium bromate [INORG CHEM] $KBrO_3$ Water-soluble, white crystals, melting at 434°C; insoluble in alcohol; strong oxidizer and a fire hazard; used in analytical chemistry and as an additive for permanent-wave compounds. { pə'tas·ē·əm 'brō,māt }

potassium bromide [INORG CHEM] KBr White, hygroscopic crystals with bitter taste; soluble in water and glycerin, slightly soluble in alcohol and ether; melts at 730°C; used in medicine, soaps, photography, and lithography. { pə'tas·ē·əm 'brō,mīd }

potassium bromide–disk technique [ANALY CHEM] Method of preparing an infrared spectrometry sample by grinding it and mixing it with a dry powdered alkali halide (such as KBr), then compressing the mixture into a tablet or pellet. Also known as pellet technique; pressed-disk technique. { pə'tas·ē·əm 'brō,mīd ‖disk tek,nēk }

potassium cadmium iodide See potassium tetraiodocadmate. { pə'tas·ē·əm 'kad·mē·əm 'ī·ə,dīd }

potassium carbonate [INORG CHEM] K_2CO_3 White, water-soluble, deliquescent powder, melting at 891°C; insoluble

in alcohol; used in brewing, ceramics, explosives, fertilizers, and as a chemical intermediate. Also known as potash; salt of tartar. { pə'tas·ē·əm 'kär·bə,nāt }

potassium chlorate [INORG CHEM] $KClO_3$ Transparent, colorless crystals or a white powder with a melting point of 356°C; soluble in water, alcohol, and alkalies; used as an oxidizing agent, for explosives and matches, and in textile printing and paper manufacture. { pə'tas·ē·əm 'klȯr,āt }

potassium chloride [INORG CHEM] KCl Colorless crystals with saline taste; soluble in water, insoluble in alcohol; melts at 776°C; used as a fertilizer and in photography and pharmaceutical preparations. Also known as potassium muriate. { pə'tas·ē·əm 'klȯr,īd }

potassium chloroaurate See potassium gold chloride. { pə'tas·ē·əm ˌklȯr·ō'ȯr,āt }

potassium chloroplatinate [INORG CHEM] K_2PtCl_6 Orange-yellow crystals or powder which decomposes when heated (250°C); used in photography. Also known as platinum potassium chloride; potassium platinichloride. { pə'tas·ē·əm ˌklȯr·ō'plat·ən,āt }

potassium chromate [INORG CHEM] K_2CrO_4 Yellow crystals, melting at 971°C; soluble in water, insoluble in alcohol; used as an analytical reagent and textile mordant, in enamels, inks, and medicines, and as a chemical intermediate. { pə'tas·ē·əm 'krō,māt }

potassium chromium sulfate See chrome alum. { pə'tas·ē·əm 'krō·mē·əm 'səl,fāt }

potassium citrate [ORG CHEM] $K_3C_6H_5O_7·H_2O$ Odorless crystals with saline taste; soluble in water and glycerol, deliquescent and insoluble in alcohol; decomposes about 230°C; used in medicine. { pə'tas·ē·əm 'sī,trāt }

potassium cobaltinitrite See cobalt potassium nitrite. { pə'tas·ē·əm kō¦bȯl·tə'nī,trāt }

potassium cyanate [INORG CHEM] $KOCN$ Colorless, water-soluble crystals; used as an herbicide and for the manufacture of drugs and organic chemicals. { pə'tas·ē·əm 'sī·ə,nāt }

potassium cyanide [INORG CHEM] KCN Poisonous, white, deliquescent crystals with bitter almond taste; soluble in water, alcohol, and glycerol; used for metal extraction, for electroplating, for heat-treating steel, and as an analytical reagent and insecticide. { pə'tas·ē·əm 'sī·ə,nīd }

potassium cyanoargentate See silver potassium cyanide. { pə'tas·ē·əm ˌsī·ə·nō'är·jən,tāt }

potassium cyanoaurite See potassium gold cyanide. { pə'tas·ē·əm ˌsī·ə·nō'ȯr,īt }

potassium dichloroisocyanurate [INORG CHEM] White, crystalline powder or granules; strong oxidant used in dry household bleaches, detergents, and scouring powders. { pə'tas·ē·əm dī¦klȯr·ō,ī'sō,sī'an·yur,āt }

potassium dichromate [INORG CHEM] $K_2Cr_2O_7$ Poisonous, yellowish-red crystals with metallic taste; soluble in water, insoluble in alcohol; melts at 396°C, decomposes at 500°C; used as an oxidizing agent and analytical reagent, and in explosives, matches, and electroplating. Also known as potassium bichromate; red potassium chromate. { pə'tas·ē·əm dī'krō,māt }

potassium dihydrogen phosphate See potassium phosphate. { pə'tas·ē·əm dī'hī·drə·jən 'fäs,fāt }

potassium diphosphate See potassium phosphate. { pə'tas·ē·əm dī'fäs,fāt }

potassium feldspar [MINERAL] Any alkali feldspar (orthoclase, microcline, sonidine, adularia) containing the molecule $KAlSi_3O_8$. Incorrectly known as K feldspar; potash feldspar. { pə'tas·ē·əm 'fel,spär }

potassium ferric oxalate [INORG CHEM] $K_3Fe(C_2O_4)_3·3H_2O$ Green crystals decomposing at 230°C, soluble in water and acetic acid; used in photography and blueprinting. { pə'tas·ē·əm 'fer·ik 'äk·sə,lāt }

potassium ferricyanide [INORG CHEM] $K_3Fe(CN)_6$ Poisonous, water-soluble, bright-red crystals; decomposes when heated; used in calico printing and wool dyeing. Also known as red potassium prussiate; red prussiate of potash. { pə'tas·ē·əm ˌfer·ə'sī·ə,nīd }

potassium ferrocyanide [INORG CHEM] $K_4Fe(CN)_6·3H_2O$ Yellow crystals with saline taste; soluble in water, insoluble in alcohol; loses water at 60°C; used in medicine, dry colors, explosives, and as an analytical reagent. Also known as yellow prussiate of potash. { pə'tas·ē·əm ˌfer·ō'sī·ə,nīd }

potassium fluoborate [INORG CHEM] KBF_4 White powder or gelatinous crystals that decompose at high temperatures; slightly soluble in water and hot alcohol; used as a sand agent to cast magnesium and aluminum, and in electrochemical processes. { pə'tas·ē·əm ¦flü·ə'bôr,āt }

potassium fluoride [INORG CHEM] KF or $KF·2H_2O$ Poisonous, white, deliquescent crystals with saline taste; soluble in water and hydrofluoric acid, insoluble in alcohol; melts at 846°C; used to etch glass and as a preservative and insecticide. { pə'tas·ē·əm 'flùr,īd }

potassium fluosilicate [INORG CHEM] K_2SiF_6 An odorless, white crystalline compound; slightly scluble in water; used in vitreous frits, synthetic mica, metallurgy, and ceramics. Also known as potassium silicofluoride. { pə'tas·ē·əm ¦flü·ə'sil·ə·kət }

potassium gluconate [ORG CHEM] $KC_6H_{11}O_7$ An odorless, white crystalline compound with salty taste; soluble in water, insoluble in alcohol and benzene; used in medicine. { pə'tas·ē·əm 'glü·kə,nāt }

potassium glutamate [ORG CHEM] $KOOC(CH_2)_2CH(NH_2)COOH·H_2O$ White, hygroscopic, water-soluble powder; used as a flavor enhancer and salt substitute. Also known as monopotassium L-glutamate. { pə'tas·ē·əm 'glüd·ə,māt }

potassium glycerinophosphate See potassium glycerophosphate. { pə'tas·ē·əm 'glis·ə·rə·nō'fäs,fāt }

potassium glycerophosphate [ORG CHEM] $K_2C_3H_5O_2·H_2PO_4·3H_2O$ Pale yellow, syrupy liquid, soluble in alcohol; used in medicine and as a dietary supplement. Also known as potassium glycerinophosphate. { pə'tas·ē·əm ¦glis·ə·rō'fäs,fāt }

potassium gold chloride [INORG CHEM] $KAuCl_4·2H_2O$ Yellow crystals, soluble in water, ether, and alcohol; used in photography and medicine. Also known as gold potassium chloride; potassium aurichloride; potassium chloroaurate. { pə'tas·ē·əm 'gōld 'klòr,īd }

potassium gold cyanide [INORG CHEM] $KAu(CN)_2$ A white, water-soluble, crystalline powder; used in medicine and for gold plating. Also known as gold potassium cyanide; potassium cyanoaurite. { pə'tas·ē·əm 'gōld 'sī·ə,nīd }

potassium hydrate See potassium hydroxide. { pə'tas·ē·əm 'hī,drāt }

potassium hydrogen phosphate See potassium phosphate. { pə'tas·ē·əm 'hī·drə·jən 'fäs,fāt }

potassium hydrogen phthalate See potassium biphthalate. { pə'tas·ē·əm 'hī·drə·jən 'tha,lāt }

potassium hydroxide [INORG CHEM] KOH Toxic, corrosive, water-soluble, white solid, melting at 360°C; used to make soap and matches, and as an analytical reagent and chemical intermediate. Also known as caustic potash; potassium hydrate. { pə'tas·ē·əm hī'dräk,sīd }

potassium hyperchlorate See potassium perchlorate. { pə'tas·ē·əm ,hī·pər'klòr,āt }

potassium hypophosphite [INORG CHEM] KH_2PO_2 White, opaque crystals or powder, soluble in water and alcohol; used in medicine. { pə'tas·ē·əm ,hī·pō'fäs,fīt }

potassium iodate [INORG CHEM] KIO_3 Odorless, white crystals; soluble in water, insoluble in alcohol; melts at 560°C; used as an analytical reagent and in medicine. { pə'tas·ē·əm 'ī·ə,dāt }

potassium iodide [INORG CHEM] KI Water- and alcohol-soluble, white crystals with saline taste; melts at 686°C; used in medicine and photography, and as an analytical reagent. { pə'tas·ē·əm 'ī·ə,dīd }

potassium linoleate [ORG CHEM] $C_{17}H_{31}COOK$ Light-tan, water-soluble paste; used as an emulsifying agent. { pə'tas·ē·əm li'nō·lē,āt }

potassium manganate [INORG CHEM] K_2MnO_4 Water-soluble dark-green crystals, decomposing at 190°C; used as an analytical reagent, bleach, oxidizing agent, disinfectant, mordant for dyeing wool and in photography, printing, and water purification. { pə'tas·ē·əm 'maŋ·gə,nāt }

potassium metabisulfite [INORG CHEM] $K_2S_2O_5$ White granules or powder, decomposing at 150–190°C; used as an antiseptic, for winemaking, food preservation, and process engraving, and as a source for sulfurous acid. Also known as potassium pyrosulfite. { pə'tas·ē·əm ,med·ə·bī'səl,fīt }

potassium metarsenite See potassium arsenite. { pə'tas·ē·əm ¦med·ə'ärs·ən,īt }

potassium monophosphate See potassium phosphate. { pə'tas·ē·əm ,män·ō'fäs,fāt }

potassium muriate See potassium chloride. { pə'tas·ē·əm 'myùr·ē,āt }

potassium nitrate [INORG CHEM] KNO_3 Flammable, water-soluble, white crystals with saline taste; melts at 337°C; used in pyrotechnics, explosives, and matches, as a fertilizer, and as an analytical reagent. Also known as niter. { pə'tas·ē·əm 'nī,trāt }

potassium nitrite [INORG CHEM] KNO_2 White, deliquescent prisms, melting at 297–450°C; soluble in water, insoluble in alcohol; strong oxidizer, exploding at over 550°C; used as an analytical reagent, in medicine, organic synthesis, pyrotechnics, and explosives. { pə'tas·ē·əm 'nī,trīt }

potassium oxalate [ORG CHEM] $K_2C_2O_4·H_2O$ Odorless, efflorescent, water-soluble, colorless crystals; decomposes when heated; used in analytical chemistry and photography and as a bleach and oxalic acid source. { pə'tas·ē·əm 'äk·sə,lāt }

potassium oxide [INORG CHEM] K_2O Gray, water-soluble crystals; melts at red heat; forms potassium hydroxide in water. { pə'tas·ē·əm 'äk,sīd }

potassium percarbonate [INORG CHEM] $K_2C_2O_6·H_2O$ White, granular, water-soluble mass with a melting point of 200–300°C; used in microscopy, photography, and textile printing. { pə'tas·ē·əm pər'kär·bə,nāt }

potassium perchlorate [INORG CHEM] $KClO_4$ Explosive, oxidative, colorless crystals; soluble in water, insoluble in alcohol; decomposes at 400°C; used in explosives, medicine, pyrotechnics, analysis, and as a reagent and oxidizing agent. Also known as potassium hyperchlorate. { pə'tas·ē·əm pər'klòr,āt }

potassium permanganate [INORG CHEM] $KMnO_4$ Highly oxidative, water-soluble, purple crystals with sweet taste; decomposes at 240°C; and explodes in contact with oxidizable materials; used as a disinfectant and analytical reagent, in dyes, bleaches, and medicines, and as a chemical intermediate. Also known as purple salt. { pə'tas·ē·əm pər'maŋ·gə,nāt }

potassium peroxide [INORG CHEM] K_2O_2 Yellow mass with a melting point of 490°C; decomposes with oxygen evolution in water; used as an oxidizing and bleaching agent. { pə'tas·ē·əm pə'räk,sīd }

potassium peroxydisulfate See potassium persulfate. { pə'tas·ē·əm pə¦räk·sē·dī'səl,fāt }

potassium persulfate [INORG CHEM] $K_2S_2O_8$ White, water-soluble crystals, decomposing below 100°C; used for bleaching and textile desizing, as an oxidizing agent and antiseptic, and in the manufacture of soap and pharmaceuticals. Also known as potassium peroxydisulfate. { pə'tas·ē·əm pər'səl,fāt }

potassium phosphate [INORG CHEM] Any one of three orthophosphates of potassium. The monobasic form, KH_2PO_4, consists of colorless, water-soluble crystals melting at 253°C; used in sonar transducers, optical modulation, medicine, baking powders, and nutrient solutions; also known as potassium acid phosphate, potassium dihydrogen phosphate (KDP), potassium diphosphate, potassium orthophosphate. The dibasic form, K_2HOP_4, consists of white, water-soluble crystals; used in medicine, fermentation, and nutrient solutions; also known as potassium hydrogen phosphate, potassium monophosphate. The tribasic form, K_3PO_4, is a water-soluble, hygroscopic white powder, melting at 1340°C; used to purify gasoline, to soften water, and to make liquid soaps and fertilizers; also known as neutral potassium phosphate, tripotassium orthophosphate. { pə'tas·ē·əm 'fäs,fāt }

potassium platinichloride See potassium chloroplatinate. { pə'tas·ē·əm ¦plat·ən·ə'klòr,īd }

potassium polymannuronate See potassium alginate. { pə'tas·ē·əm ¦päl·ē·man'yùr·ə,nāt }

potassium polymetaphosphate [INORG CHEM] $(KPO_3)_n$ White powder with a molecular weight up to 500,000; used in foods as a fat emulsifier and moisture-retaining agent. { pə'tas·ē·əm ¦päl·e,med·ə'fäs,fāt }

potassium pyrophosphate [INORG CHEM] $K_4P_2O_7·3H_2O$ Water-soluble, colorless crystals; dehydrates below 300°C, melts at 1090°C; used in tin plating, china-clay purification, dyeing, oil-drilling muds, and synthetic rubber production. Also known as normal potassium pyrophosphate; tetrapotassium pyrophosphate. { pə'tas·ē·əm ,pī·rō'fäs,fāt }

potassium pyrosulfite See potassium metabisulfite.   { pə'tas·ē·əm ˌpī·rō'səlˌfīt }

potassium silicate [INORG CHEM] $SiO_2 = K_2O$ A compound existing in two forms, solution and solid (glass); as a solution, it is colorless to turgid in water, and is used in paints and coatings, as an arc-electrode binder and catalyst and in detergents; as a solid, it is colorless and water-soluble solid, and is used in glass manufacture and for dyeing and bleaching.   { pə'tas·ē·əm 'sil·əˌkāt }

potassium silicofluoride See potassium fluosilicate.   { pə'tas·ē·əm ˌsil·ə·kō'flúrˌīd }

potassium sodium ferricyanide [INORG CHEM] K_2Na-$Fe(CN)_6$ Red, water-soluble crystals; used for blueprint paper and in photography.   { pə'tas·ē·əm 'sōd·ē·əm ˌfer·ə'sī·əˌnīd }

potassium sodium tartrate [INORG CHEM] $KNaC_4H_4O_6$·$4H_2O$ Colorless, water-soluble, efflorescent crystals or white powder with a melting point of 70–80°C; used in medicine and as a buffer and sequestrant in foods. Also known as Rochelle salt; Seignette salt.   { pə'tas·ē·əm 'sōd·ē·əm 'tärˌtrāt }

potassium sorbate [ORG CHEM] $C_6H_7KO_2$ A crystalline compound, more soluble in water than in alcohol; decomposes above 270°C; used to inhibit mold and yeast growth in food.   { pə'tas·ē·əm 'sórˌbāt }

potassium stannate [INORG CHEM] K_2SnO_3·$3H_2O$ White crystals; soluble in water, insoluble in alcohol; used in textile printing and dyeing, and in tin-plating baths.   { pə'tas·ē·əm 'stanˌāt }

potassium stibnate See potassium antimonate.   { pə'tas·ē·əm 'stibˌnīt }

potassium sulfate [INORG CHEM] K_2SO_4 Colorless crystals with bitter taste; soluble in water, insoluble in alcohol; melts at 1072°C; used as an analytical reagent, medicine, and fertilizer, and in aluminum and glass manufacture. Also known as salt of Lemery.   { pə'tas·ē·əm 'səlˌfāt }

potassium sulfide [INORG CHEM] K_2S Moderately flammable, water-soluble, deliquescent red crystals; melts at 840°C; used in analytical chemistry, medicine, and depilatories. Also known as fused potassium sulfide; hepar sulfuris; potassium sulfuret.   { pə'tas·ē·əm 'səlˌfīd }

potassium sulfite [INORG CHEM] K_2SO_3·$2H_2O$ Water-soluble, white crystals; used in medicine and photography.   { pə'tas·ē·əm 'səlˌfīt }

potassium sulfuret See potassium sulfide.   { pə'tas·ē·əm 'səl·fəˌret }

potassium tetraiodocadmate [INORG CHEM] $K_2(CdI_4)$·$2H_2O$ A crystalline compound; used in analytical chemistry for alkaloids, amines, and other compounds. Also known as cadmium potassium iodide; potassium cadmium iodide.   { pə'tas·ē·əm ˌte·trəˌī·ə·dō'kadˌmāt }

potassium thiocyanate [INORG CHEM] $KCNS$ Water- and alcohol-soluble, colorless, odorless hygroscopic crystals with saline taste; decomposes at 500°C; used as an analytical reagent and in freezing mixtures, chemicals manufacture, textile printing and dyeing, and photographic chemicals.   { pə'tas·ē·əm ˌthī·ō'sī·əˌnāt }

potassium undecylenate [ORG CHEM] CH_2:$CH(CH_2)_8$-$COOK$ A white, water-soluble powder, decomposing at about 250°C; used in pharmaceuticals and cosmetics as a fungistat and bacteriostat.   { pə'tas·ē·əm ˌən·des·ə'leˌnāt }

potassium xanthate [ORG CHEM] KC_2H_5OCSS Water- and alcohol-soluble, yellow crystals; used as an analytical reagent and soil-treatment fungicide.   { pə'tas·ē·əm 'zanˌthāt }

potato [BOT] Solanum tuberosum. An erect herbaceous annual that has a round or angular aerial stem, underground lateral stems, pinnately compound leaves, and white, pink, yellow, or purple flowers occurring in cymose inflorescences; produces an edible tuber which is a shortened, thickened underground stem having nodes (eyes) and internodes. Also known as Irish potato; white potato.   { pə'tā·dō }

potato stone [GEOL] A potato-shaped geode, especially one consisting of hard, silicified limestone with an internal lining of quartz crystals.   { pə'tā·dō ˌstōn }

potato virus X group See Potexvirus.   { pəˌtā·dō ˌvī·rəs 'eks ˌgrüp }

potato virus Y [VIROL] A species in the genus Potyvirus that is a common pathogen of potato plants. Symptoms of infection can include shortening of the stem internodes, spotting

and severe malformation of the upper leaves, defoliation, and early plant death. Abbreviated PVY.   { pəˌtā·dō ˌvī·rəs 'wī }

pot clay [MATER] Refractory clay used to make the pots in which glass is produced.   { 'pät ˌklā }

pot core [ELECTROMAG] A ferrite magnetic core that has the shape of a pot, with a magnetic post in the center and a magnetic plate as a cover; the coils for a choke or transformer are wound on the center post.   { 'pät ˌkór }

pot die forming [MECH ENG] Forming sheet or plate metal through a hollow die by the application of pressure which causes the workpiece to assume the contour of the die.   { 'pät 'dī ˌfórm·iŋ }

pot earth See potter's clay.   { 'pät ˌərth }

potential [ELEC] See electric potential. [PHYS] A function or set of functions of position in space, from whose first derivatives a vector can be formed, such as that of a static field intensity.   { pə'ten·chəl }

potential barrier [PHYS] The potential in a region in a field of force where the force exerted on a particle is such as to oppose the passage of the particle through the region. Also known as barrier; potential hill.   { pə'ten·chəl 'bar·ē·ər }

potential density [PHYS] The density that would be reached by a compressible fluid if it were adiabatically compressed or expanded to a standard pressure of the bar.   { pə'ten·chəl 'denˌsəd·ē }

potential difference [ELEC] Between any two points, the work which must be done against electric forces to move a unit charge from one point to the other. Abbreviated PD.   { pə'ten·chəl 'dif·rəns }

potential divider See voltage divider.   { pə'ten·chəl di'vīd·ər }

potential drop [ELEC] The potential difference between two points in an electric circuit. [FL MECH] The difference in pressure head between one equipotential line and another.   { pə'ten·chəl ˌdräp }

potential electrolyte [PHYS CHEM] A solid material composed of uncharged molecules that can react chemically with a solvent to yield some ions in solution.   { pə'ten·chəl i'lek·trəˌlīt }

potential energy [MECH] The capacity to do work that a body or system has by virtue of its position or configuration.   { pə'ten·chəl 'en·ər·jē }

potential evaporation See evaporative power.   { pə'ten·chəl iˌvap·ə'rā·shən }

potential evapotranspiration [HYD] Generally, the amount of moisture which, if available, would be removed from a given land area by evapotranspiration; expressed in units of water depth.   { pə'ten·chəl iˌvap·ōˌtranz·pə'rā·shən }

potential flow [FL MECH] Flow in which the velocity of flow is the gradient of a scalar function, known as the velocity potential.   { pə'ten·chəl 'flō }

potential flow analyzer See electrolytic tank.   { pə'ten·chəl ˌflō 'an·əˌlīz·ər }

potential gradient [ELEC] Difference in the values of the voltage per unit length along a conductor or through a dielectric.   { pə'ten·chəl 'grād·ē·ənt }

potential head See elevation head.   { pə'ten·chəl 'hed }

potential hill See potential barrier.   { pə'ten·chəl 'hil }

potential index of refraction [METEOROL] An atmospheric index of refraction so formulated that it would have no height variation in an adiabatic atmosphere. Also known as potential refractive index.   { pə'ten·chəl 'inˌdeks əv ri'frak·shən }

potential instability See convective instability.   { pə'ten·chəl ˌin·stə'bil·əd·ē }

potentially hazardous asteroid [ASTRON] An asteroid whose orbit approaches within 0.05 astronomical unit of the earth's orbit, and which is brighter than an absolute visual magnitude of 22.0, corresponding to a diameter of at least 110–240 meters (360–800 feet).   { pəˌten·chə·lē ˌhaz·ərd·əs 'as·təˌróid }

potential ore [MIN ENG] **1.** As yet undiscovered mineral deposits. **2.** A known mineral deposit for which recovery is not yet economically feasible.   { pə'ten·chəl ˌór }

potential refractive index See potential index of refraction.   { pə'ten·chəl ri'frak·tiv 'inˌdeks }

potential scattering [QUANT MECH] Scattering of a particle which can be treated as the effect of a potential, representing the particle's potential energy, on the particle's Schrödinger wave function.   { pə'ten·chəl 'skad·əˌriŋ }

POTATO

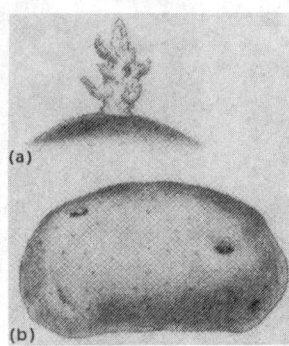

(a)

(b)

Irish potato. (a) Flowering stem. (b) Tuber.

potential sputtering [ELECTR] The ejection of mainly neutral atoms from the surface of a solid insulator due to the impact of slow, multiply charged ions whose kinetic energy alone is incapable of initiating sputtering. { pə'ten·chəl ˌspəd·ə·riŋ }

potential temperature [THERMO] The temperature that would be reached by a compressible fluid if it were adiabatically compressed or expanded to a standard pressure, usually 1 bar. { pə'ten·chəl 'tem·prə·chər }

potential theory [MATH] The study of the functions arising from Laplace's equation, especially harmonic functions. { pə'ten·chəl ˌthē·ə·rē }

potential transform See Poisson transform. { pə'ten·chəl 'tranzˌfȯrm }

potential transformer See voltage transformer. { pə'ten·chəl tranz'fȯr·mər }

potential transformer phase angle [ELEC] Angle between the primary voltage vector and the secondary voltage vector reversed; this angle is conveniently considered as positive when the reversed, secondary voltage vector leads the primary voltage vector. { pə'ten·chəl tranz'fȯr·mər 'fāz ˌaŋ·gəl }

potential vorticity [FL MECH] The product of the absolute vorticity and the static stability, conservative in adiabatic flow, given by the expression $(\eta/\theta)(\partial\theta/\partial p)$, where η is the absolute vorticity of a fluid parcel, θ the potential temperature, and p the pressure. Also known as absolute potential vorticity. { pə'ten·chəl vȯr'tis·əd·ē }

potential well [PHYS] For an object in a conservative field of force, a region in which the object has a lower potential energy than in all the surrounding regions. { pə'ten·chəl ˌwel }

potentiator [FOOD ENG] See flavor enhancer. [PHARM] A drug that augments the response of a second drug so that the total response is greater than the predictible additive effect. { pə'ten·chē·ād·ər }

potentiometer [ELEC] A resistor having a continuously adjusted sliding contact that is generally mounted on a rotating shaft; used chiefly as a voltage divider. Also known as pot (slang). [ENG] A device for the measurement of an electromotive force by comparison with a known potential difference. { pəˌten·chē'äm·əd·ər }

potentiometric cell [ANALY CHEM] Container for the two electrodes and the electrolytic solution being titrated potentiometrically. { pəˌten·chē·ə¦me·trik 'sel }

potentiometric controller [CONT SYS] A controller that operates on the null balance principle, in which an error signal is produced by balancing the sensor signal against a set-point voltage in the input circuit; the error signal is amplified for use in keeping the load at a desired temperature or other parameter. { pəˌten·chē·ə¦me·trik kən'trōl·ər }

potentiometric electrode [ELEC] An electrode that produces a voltage logarithmically dependent on the concentration of a selected ionic substance. { pəˌten·chē·ə¦me·trik i'lek ˌtrōd }

potentiometric map [HYD] A map showing the elevation of a potentiometric surface of an aquifer by means of contour lines or other symbols. Also known as pressure-surface map. { pəˌten·chē·ə¦me·trik 'map }

potentiometric model study [PETRO ENG] Analogic electrical-resistance model (electrolyte in a contoured container) of an underground reservoir based on Darcy's law, that is, the steady-state flow of liquids through porous media is analogous to the flow of current through an electrical conductor; used to predict conditions in gas-condensate oil reservoirs. { pəˌten·chē·ə¦me·trik 'mäd·əl ˌstəd·ē }

potentiometric surface [HYD] An imaginary surface that represents the static head of groundwater and is defined by the level to which water will rise. Also known as isopotential level; piezometric surface; pressure surface. { pəˌten·chē·ə¦me·trik 'sər·fəs }

potentiometric titration [ANALY CHEM] Solution titration in which the end point is read from the electrode-potential variations with the concentrations of potential-determining ions, following the Nernst concept. Also known as constant-current titration. { pəˌten·chē·ə¦me·trik tī'trā·shən }

potentiometry [ELEC] Use of a potentiometer to measure electromotive forces, and the applications of such measurements. { pəˌten·chē'äm·ə·trē }

potentiostat [ENG] An automatic laboratory instrument that controls the potential of a working electrode to within certain limits during coulometric (electrochemical reaction) titrations. { pə'ten·chē·əˌstat }

Potexvirus [VIROL] A genus of plant viruses characterized by flexuous helical rods containing one molecule of linear, positive-sense, single-stranded ribonucleic acid. Also known as potato virus X group. { pōt'eksˌvī·rəs }

pot furnace [ENG] **1.** A furnace containing several pots in which glass is melted. **2.** A furnace in which the charge is contained in a pot or crucible. { 'pät ˌfər·nəs }

pothole [CIV ENG] A pot-shaped hole in a pavement surface. [GEOL] **1.** A shaftlike cave opening upward to the surface. **2.** Any bowl-shaped, cylindrical, or circular hole formed by the grinding action of a stone in the rocky bed of a river or stream. Also known as churn hole; colk; eddy mill; evorsion hollow; kettle; pot. **3.** A vertical, or nearly vertical shaft in limestone. Also known as aven; cenote. **4.** A small depression with steep sides in a coastal marsh; contains water at or below low-tide level. Also known as rotten spot. [HYD] See moulin. { 'pätˌhōl }

Potier diagram [ELEC] Vector diagram showing the voltage and current relations in an alternating-current generator. { pō'tyā ˌdī·əˌgram }

pot life [CHEM ENG] See work life. [ENG] The period of time during which paint remains useful after its original package has been opened or after a catalyst or other additive has been incorporated. Also known as spreadable life; usable life. { 'pät ˌlīf }

potometer [ENG] A device for measuring transpiration, consisting of a small vessel containing water and sealed so that the only escape of moisture is by transpiration from a leaf, twig, or small plant with its cut end inserted in the water. { pō'täm·əd·ər }

potomology [CIV ENG] The systematic study of the factors affecting river channels to provide the basis for predictions of the effects of proposed engineering works on channel characteristics. { ˌpäd·ə'mäl·ə·jē }

pot plunger [ENG] A plunger used to force softened plastic molding material into the closed cavity of a transfer mold. { 'pät ˌplən·jər }

potrero [GEOL] An elongate, islandlike beach ridge, surrounded by mud flats and separated from the coast by a lagoon and barrier island, made up of a series of accretionary dune ridges. { pə'trer·ō }

pot still [FOOD ENG] A type of still in which the heat from a flame is applied directly to the pot containing the wash; used in the distillation of Irish grain whiskey and Scotch malt whiskey. { 'pät ˌstil }

potted circuit [ELEC] A pulse-forming network immersed in oil and enclosed in a metal container. { 'päd·əd 'sər·kət }

potted line [ELEC] Pulse-forming network immersed in oil and enclosed in a metal container. { 'päd·əd 'līn }

Potter-Bucky grid [MED] An assembly of lead strips resembling an open venetian blind, placed between a patient being x-rayed and the screen or film, to reduce the effects of scattered radiation. Also known as Bucky diaphragm; grid. { 'päd·ər 'bək·ē ˌgrid }

potter's clay [MATER] A plastic clay, free from iron and devoid of fissility, suitable for modeling or pottery making or adapted for use on a potter's wheel. Also known as argil; pot earth; potter's earth. { 'päd·ərz 'klā }

potter's earth See potter's clay. { 'päd·ərz 'ərth }

potter's wheel [ENG] A revolving horizontal disk that turns when a treadle is operated; used to shape clay by hand. { 'päd·ərz 'wēl }

pottery [MATER] Objects made of clay which may be nonvitreous, porous, opaque, and glazed or unglazed; also included is earthenware such as stoneware. { 'päd·ə·rē }

Pottiales [BOT] An order of mosses distinguished by erect stems, lanceolate to broadly ovate or obovate leaves, a strong, mostly percurrent or excurrent costa, and a cucullate calyptra. { ˌpäd·ē'ā·lēz }

potting [ELECTR] Process of filling a complete electronic assembly with a thermosetting compound for resistance to shock and vibration, and for exclusion of moisture and corrosive agents. { 'päd·iŋ }

Pott's disease [MED] Abnormal backward curvature of the spine caused by tuberculous osteitis. { 'päts diˌzēz }

Potyvirus [VIROL] A genus of plant viruses characterized by virions that are flexuous helical rods containing one molecule of

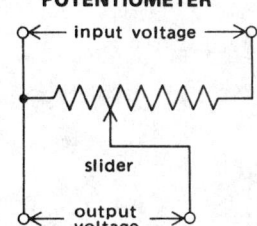

POTENTIOMETER

Schematic of potentiometer.

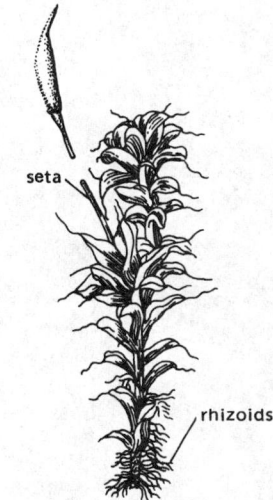

POTTIALES

The entire plant of the moss *Tortula ruralis.*

linear, positive-sense, single-stranded ribonucleic acid. Potato virus Y is the most common member. { ¦pōt'wī¸vī·rəs }

Pougachev maneuver *See* cobra maneuver. { 'pü·gə¸chef mə¸nü·vər }

Poulter seismic method [GEOPHYS] A type of air shooting in which the explosive is set on poles above the ground. { 'pōl·tər 'sīz·mik ¸meth·əd }

poultice [MED] A soft mass of hot, moist material applied as an external counterirritant, analgesic, or antiseptic. { 'pōl·təs }

poultry [AGR] Domesticated fowl grown for their meat and eggs. { 'pōl·tre }

poumar [TEXT] A unit of line density (mass per unit length) used chiefly in the textile industry to measure yarn fiber, equal to 1 pound per 1,000,000 yards, or approximately 4.96055×10^{-7} kilogram per meter. { pü'mär }

pounce [MATER] Pumice in the form of a very fine powder, used for preparing parchment and tracing cloth. { paůns }

pounce wheel [GRAPHICS] An instrument having a toothed wheel at the tip, used to follow the lines in a master drawing in pouncing. { 'paůns ¸wēl }

pouncing [GRAPHICS] A method of making copies of patterns and signs; a master drawing is made on paper of the design to be copied, and the lines are traced with a pounce wheel; the teeth perforate the paper with 12 to 24 holes per inch; the master drawing is then placed on the permanent support, and powder is forced through the holes, effectively transferring the drawing. { 'paůns·iŋ }

pouncing paper [MATER] Paper coated with pumice for polishing felt hats. { 'paůns·iŋ ¸pā·pər }

pound [MECH] **1.** A unit of mass in the English absolute system of units, equal to 0.45359237 kilogram. Abbreviated lb. Also known as avoirdupois pound; pound mass. **2.** A unit of force in the English gravitational system of units, equal to the gravitational force experienced by a pound mass when the acceleration of gravity has its standard value of 9.80665 meters per second per second (approximately 32.1740 ft/s^2) equal to 4.4482216152605 newtons. Abbreviated lb. Also spelled Pound (Lb). Also known as pound force (lbf). **3.** A unit of mass in the troy and apothecaries' systems, equal to 12 troy or apothecaries' ounces, or 5760 grains, or 5760/7000 avoirdupois pound, or 0.3732417216 kilogram. Also known as apothecaries' pound (abbreviated lb ap in the United States or lb apoth in the United Kingdom); troy pound (abbreviated lb t in the United States, or lb tr or lb in the United Kingdom). { paůnd }

poundal [MECH] A unit of force in the British absolute system of units equal to the force which will impart an acceleration of 1 ft/s^2 to a pound mass, or to 0.138254954376 newton. { 'paůnd·əl }

poundal-foot *See* foot-poundal. { 'paůnd·əl 'fůt }

pound-foot *See* foot-pound. { 'paůnd 'fůt }

pound force *See* pound. { 'paůnd 'fòrs }

pound mass *See* pound. { 'paůnd 'mas }

pound per square foot [MECH] A unit of pressure equal to the pressure resulting from a force of 1 pound applied uniformly over an area of 1 square foot. Abbreviated psf. { 'paůnd pər ¦skwer 'fůt }

pound per square inch [MECH] A unit of pressure equal to the pressure resulting from a force of 1 pound applied uniformly over an area of 1 square inch. Abbreviated psi. { 'paůnd pər ¦skwer 'inch }

Pound-Rebka experiment [RELAT] A terrestrial experiment demonstrating the gravitational redshift of light. { ¦paůnd 'reb·kə ik¸sper·ə·mənt }

pounds per square inch absolute [MECH] The absolute, thermodynamic pressure, measured by the number of pounds-force exerted on an area of 1 square inch. Abbreviated lbf in.$^{-2}$ abs; psia. { 'paůns pər ¦skwer 'inch 'ab·sə¸lüt }

pounds per square inch differential [ENG] The difference in pressure between two points in a fluid-flow system, measured in pounds per square inch. Abbreviated psid. { 'paůns pər ¦skwer 'inch dif·ə'ren·chəl }

pounds per square inch gage [MECH] The gage pressure, measured by the number of pounds-force exerted on an area of 1 square inch. Abbreviated psig. { 'paůns pər ¦skwer 'inch ¸gāj }

Poupart's ligament *See* inguinal ligament. { pü'pärz ¸lig·ə·mənt }

Pourbaix diagram [MET] A plot of potential versus pH used to predict the thermodynamic tendency of a metal to corrode. { ¦púr¸bā 'dī·ə¸gram }

pour depressant *See* pour-point depressant. { 'pór di¸pres·ənt }

pouring [MET] Transferring molten metal to a mold or ladle. { 'pór·iŋ }

pouring rope *See* asbestos joint runner. { 'pór·iŋ ¸rōp }

pour-plate culture [MICROBIO] A technique for pure-culture isolation of bacteria; liquid, cooled agar in a test tube is inoculated with one loopful of bacterial suspension and mixed by rolling the tube between the hands; subsequent transfers are made from this to a second test tube, and from the second to a third; contents of each tube are poured into separate petri dishes; pure cultures can be isolated from isolated colonies appearing on the plates after incubation. { 'pór ¸plāt ¸kəl·chər }

pour point [FL MECH] Lowest test temperature at which a liquid will flow. [MET] Temperature at which a molten alloy is cast. { 'pór ¸póint }

pour-point depressant [MATER] An additive that lowers the pour point of a wax-containing petroleum-base lubricating oil by reducing the tendency of the wax to solidify. Also known as pour depressant; pour-point inhibitor. { 'pór ¸póint di¸pres·ənt }

pour-point inhibitor *See* pour-point depressant. { 'pór ¸póint in'hib·əd·ər }

pour reversion [MATER] The difference between the original American Society for Testing and Materials (ASTM) pour point of a lubricating oil and the relatively high solidification temperature observed in the field. { 'pór ri¸vər·zhən }

pour stability [MATER] Ability of a pour-depressant-treated petroleum lubricating oil to maintain its original American Society for Testing and Materials (ASTM) pour point at low temperatures approximating winter conditions. { 'pór stə¸bil·əd·ē }

Pourtalesiidae [INV ZOO] A family of exocyclic Euechinoidea in the order Holasteroida, including those forms with a bottle-shaped test. { ¸pórd·əl·ə'sī·ə¸dē }

pour test [ENG] The chilling of a liquid under specified test conditions to determine the American Society for Testing and Materials (ASTM) pour point. { 'pór ¸test }

powder [MATER] **1.** A general term for explosives. **2.** A loose grouping or aggregation of solid particles, usually smaller than 1000 micrometers. **3.** *See* bulk solid. { 'paůd·ər }

powder avalanche [GEOL] Loose powder snow rapidly descending a mountainside. { ¦paůd·ər ¦av·ə¸lanch }

powder box [MIN ENG] A wooden box used by miners to store explosive powder and blasting caps. { 'paůd·ər ¸bäks }

powder clutch [MECH ENG] A type of electromagnetic disk clutch in which the space between the clutch members is filled with dry, finely divided magnetic particles; application of a magnetic field coalesces the particles, creating friction forces between clutch members. { 'paůd·ər ¸kləch }

powder coating [MATER] A dry coating method in which fine clear or pigmented powder particles, containing resin, modifiers, and possibly a curing agent, are electrostatically sprayed onto a substrate and heated (melted) in a oven to form a continuous film. { 'paůd·ər ¸kōt·iŋ }

powder diffraction camera [CRYSTAL] A metal cylinder having a window through which an x-ray beam of known wavelength is sent by an x-ray tube to strike a finely ground powder sample mounted in the center of the cylinder; crystal planes in this powder sample diffract the x-ray beam at different angles to expose a photographic film that lines the inside of the cylinder; used to study crystal structure. Also known as x-ray powder diffractometer. { 'paůd·ər di'frak·shən ¸kam·rə }

powdered-iron core *See* ferrite core. { ¦paůd·ərd 'ī·ərn ¸kór }

powder flowmeter [ENG] A device used to measure the flow rate of a metal powder. { 'paůd·ər 'flō¸mēd·ər }

powder house [CIV ENG] A magazine for the temporary storage of explosives. { 'paůd·ər ¸haůs }

powder insulation [MATER] Thermal insulation material made up of a finely divided solid held between two surfaces (one hot, one cold); the powder reduces both convection and radiation heat flow between the surfaces. { 'paůd·ər ¸in·sə¸lā·shən }

powder keg [ENG] A small metal keg for black blasting powder. { 'paůd·ər ¸keg }

powder lubricant [MET] An agent mixed with a powder metal to facilitate formation and ejection of a compact. { 'paùd·ər 'lü·brə·kənt }

powderman [MIN ENG] An individual in charge of explosives in an operation of any nature requiring their use. { 'paùd·ər·mən }

powder metal [MET] Finely divided particles of a metal. { 'paùd·ər ,med·əl }

powder metallurgy [MET] A metalworking process used to fabricate parts of simple or complex shape from a wide variety of metals and alloys in the form of powders. The process involves shaping of the powder and subsequent bonding of its individual particles by heating or mechanical working. { 'paùd·ər 'med·əl,ər·jē }

powder method [SOLID STATE] A method of x-ray diffraction analysis in which a collimated, monochromatic beam of x-rays is directed at a sample consisting of an enormous number of tiny crystals having random orientation, producing a diffraction pattern that is recorded on film or with a counter tube. Also known as x-ray powder method. { 'paùd·ər ,meth·əd }

powder mine [MIN ENG] An excavation filled with powder for the purpose of blasting rocks. { 'paùd·ər ,mīn }

powder-moisture test [ENG] Determination of moisture in a propellant by drying under prescribed conditions; expressed as percentage by weight. { 'paùd·ər ,mòis·chər ,test }

powder molding [ENG] Generic term for plastics-molding techniques to produce objects of varying sizes and shapes by melting polyethylene powder, usually against the heated inside of a mold. { 'paùd·ər ,mōld·iŋ }

powder pattern [CRYSTAL] In the powder method of x-ray diffraction analysis, the display of lines made on film by the Debye-Scherrer method or on paper by a recording diffractometer. [ELECTROMAG] The pattern created by very fine powders or colloidal particles, spread over the surface of a magnetic material; reveals the magnetic domains in a single crystal of such material. { 'paùd·ər ,pad·ərn }

powder silk [TEXT] Special silk fabric formerly used in making propellant bags; it left no burning residue when the propellant burned; silk has now been largely replaced by other materials, and the fabric is called cartridge cloth. Also known as cartridge silk. { 'paùd·ər ,silk }

powder snow [HYD] A cover of dry snow that has not been compacted in any way. { 'paùd·ər ,snō }

powder train [ENG] **1.** Train, usually of compressed black powder, used to obtain time action in older fuse types. **2.** Train of explosives laid out for destruction by burning. { 'paùd·ər ,trān }

powdery mildew [MYCOL] A fungus characterized by production of abundant powdery conidia on the host; a member of the family Erysiphaceae or the genus *Oidium*. [PL PATH] A plant disease caused by a powdery mildew fungus. { 'paùd·ə·rē 'mil,dü }

powdery scab [PL PATH] A fungus disease of potato tubers caused by *Spongospora subterranea* and characterized by nodular discolored lesions, which burst and expose masses of powdery fungus spores. { 'paùd·ə·rē ,skab }

powellite [MINERAL] Ca(WMo)O₄ A commercially important tungsten mineral, crystallizing in the tetragonal system; isomorphous with scheelite (CaWO₄). { 'paù·ə,līt }

power [MATH] **1.** The value that is assigned to a mathematical expression and its exponent. **2.** The power of a set is its cardinality. **3.** For a point, with reference to a circle, the quantity $(x - a)^2 + (y - b)^2 - r^2$, where x and y are the coordinates of the point, a and b are the coordinates of the center of the circle, and r is the radius of the circle. **4.** For a point, with reference to a sphere, the quantity $(x - a)^2 + (y - b)^2 + (z - c)^2 - r^2$, where x, y, and z are the coordinates of the point; a, b, and c are the coordinates of the center of the sphere; and r is the radius of the sphere. [OPTICS] *See* focal power. [PHYS] The time rate of doing work. [STAT] One minus the probability that a given test causes the acceptance of the null hypothesis when it is false due to the validity of an alternative hypothesis; this is the same as the probability of rejecting the null hypothesis by the test when the alternative is true. { 'paù·ər }

power-actuated pressure relief valve [MECH ENG] A pressure relief valve connected to and controlled by a device which utilizes a separate energy source. { 'paù·ər ,ak·chə,wād·əd 'presh·ər ri,lēf ,valv }

power amplification *See* power gain. { 'paù·ər ,am·plə·fə'kā·shən }

power amplifier [ELECTR] The final stage in multistage amplifiers, such as audio amplifiers and radio transmitters, designed to deliver maximum power to the load, rather than maximum voltage gain, for a given percent of distortion. { 'paù·ər 'am·plə,fī·ər }

power amplifier tube *See* power tube. { 'paù·ər 'am·plə,fī·ər ,tüb }

power attenuation *See* power loss. { 'paù·ər ə,ten·yə'wā·shən }

power bandwidth [COMMUN] The frequency range for which half the rated power of an audio amplifier is available at rated distortion. { 'paù·ər ,band,width }

power barker *See* barker. { 'paù·ər ,bärk·ər }

power brake [MECH ENG] An automotive brake with engine-intake-manifold vacuum used to amplify the atmospheric pressure on a piston operated by movement of the brake pedal. { 'paù·ər ,brāk }

power breeder [NUCLEO] A nuclear reactor designed to produce both useful power and fuel. { 'paù·ər ,brēd·ər }

power car [AERO ENG] A suspended structure on an airship that houses an engine. [MECH ENG] **1.** A railroad car with equipment for furnishing heat and electric power to a train. **2.** A railroad car with controls, which can be operated by itself or as part of a train. { 'paù·ər ,kär }

power check [COMPUT SCI] An automatic suspension of computer operations resulting from a significant fluctuation in internal electric power. { 'paù·ər ,chek }

power circuit [ELEC] The wires that carry current to electric motors and other devices that use electric power. { 'paù·ər ,sər·kət }

power component *See* active component. { 'paù·ər kəm,pō·nənt }

power control rod [NUCLEO] A control rod that produces only a small change in reactivity of a nuclear reactor, as required for controlling power level. { 'paù·ər kən'trōl ,räd }

power control valve [MECH ENG] A safety relief device operated by a power-driven mechanism rather than by pressure. { 'paù·ər kən'trōl ,valv }

power cord *See* line cord. { 'paù·ər ,kòrd }

power curve [STAT] The graph of the power of a test for various alternatives. { 'paù·ər ,kərv }

power cylinder [CONT SYS] A linear actuator consisting of a piston in a cylinder, driven by pneumatic or hydraulic fluid under high pressure. { 'paù·ər ,sil·ən·dər }

power dam [CIV ENG] A dam designed to raise the level of a stream to create or concentrate hydrostatic head for power purposes. { 'paù·ər ,dam }

power density [ELECTROMAG] The amount of power per unit area in a radiated microwave or other electromagnetic field, usually expressed in units of watts per square centimeter. [NUCLEO] The power generation per unit volume of a nuclear-reactor core. { 'paù·ər ,den·səd·ē }

power-density spectrum *See* frequency spectrum. { 'paù·ər ,den·səd·ē ,spek·trəm }

power detection [ELECTR] Form of detection in which the power output of the detecting device is used to supply a substantial amount of power directly to a device such as a loudspeaker or recorder. { 'paù·ər di,tek·shən }

power detector [ELECTR] Detector capable of handling strong input signals without appreciable distortion. { 'paù·ər di,tek·tər }

power diode *See* pin diode. { 'paù·ər ,dī,ōd }

power distribution unit [COMPUT SCI] Equipment located in or near a computer room which breaks down electric power from a high-voltage source to appropriate levels for distribution to the central processing unit and peripheral devices. Abbreviated PDU. { 'paù·ər ,di·strə'byü·shən ,yü·nət }

power divider [ELECTROMAG] A device used to produce a desired distribution of power at a branch point in a waveguide system. { 'paù·ər di,vīd·ər }

power down [COMPUT SCI] To exit from any running programs and remove floppy- and hard-disk cartridges before switching the computer off. { 'paù·ər ,daun }

power drill [MECH ENG] A motor-driven drilling machine. { 'paù·ər ,dril }

power-driven [MECH ENG] Of a component or piece of

equipment, moved, rotated, or operated by electrical or mechanical energy, as in a power-driven fan or power-driven turret. { 'pau̇·ər ,driv·ən }

power efficiency [STAT] The probability of rejecting a statistical hypothesis when it is false. { 'pau̇·ər i,fish·ən·sē }

power equation [MIN ENG] The relationship indicating that the natural ventilating power plus the power required to force air through a mine is equal to the power used in lifting water out of the mine plus the power lost in the kinetic energy of air leaving the mine plus the power converted to heat in overcoming friction. { 'pau̇·ər i,kwā·zhən }

power excursion [NUCLEO] A sudden increase in the power level of a nuclear reactor, caused by a sudden increase in reactivity. { 'pau̇·ər ik,skər·zhən }

power factor [ELEC] The ratio of the average (or active) power to the apparent power (root-mean-square voltage times rms current) of an alternating-current circuit. Abbreviated pf. Also known as phase factor. { 'pau̇·ər ,fak·tər }

power-factor controller [ELECTR] A solid-state electronic device that reduces excessive energy waste in alternating-current induction motors by holding constant the phase angle between current and voltage. { 'pau̇·ər ,fak·tər kən,trōl·ər }

power-factor meter [ENG] A direct-reading instrument for measuring power factor. { 'pau̇·ər ,fak·tər ,mēd·ər }

power-factor regulator [ELEC] Regulator which functions to maintain the power factor of a line or an apparatus at a predetermined value, or to vary it according to a predetermined plan. { 'pau̇·ər ,fak·tər ,reg·yə,lād·ər }

power flow [ELECTROMAG] The rate at which energy is transported across a surface by an electromagnetic field. { 'pau̇·ər ,flō }

power frequency [ELEC] The frequency at which electric power is generated and distributed; in most of the United States it is 60 hertz. { 'pau̇·ər ,frē·kwən·sē }

power function [MATH] A function whose value is the product of a constant and a power of the independent variable. [STAT] The function that indicates the probability of rejecting the null hypothesis for all possible values of the population parameter for a given critical region. { 'pau̇·ər ,fəŋk·shən }

power gain [ELECTR] The ratio of the power delivered by a transducer to the power absorbed by the input circuit of the transducer. Also known as power amplification. [ELECTROMAG] An antenna ratio equal to 4π (12.57) times the ratio of the radiation intensity in a given direction to the total power delivered to the antenna. { 'pau̇·ər ,gān }

power generator [ELEC] A device for producing electric energy, such as an ordinary electric generator or a magnetohydrodynamic, thermionic, or thermoelectric power generator. { 'pau̇·ər ,jen·ə,rād·ər }

power grasp See power grip. { 'pau̇·ər ,grasp }

power grip [IND ENG] A basic grasp whereby the fingers are wrapped around an object and the thumb placed against it; used, for example, in certain hammering operations. Also known as power grasp. { 'pau̇·ər ,grip }

power grizzly [MIN ENG] Power-operated machine for removing dirt and fine particles from ore before it is crushed. { 'pau̇·ər ,griz·lē }

power-law fluid [FL MECH] A fluid in which the shear stress at any point is proportional to the rate of shear at that point raised to some power. { 'pau̇·ər ,lȯ ,flü·əd }

power-law profile [METEOROL] A formula for the variation of wind with height in the surface boundary layer. { 'pau̇·ər ,lȯ ,prō,fīl }

power level [ELEC] The ratio of the amount of power being transmitted past any point in an electric system to a reference power value; usually expressed in decibels. [NUCLEO] The power production of a nuclear reactor in watts. { 'pau̇·ər ,lev·əl }

power line [ELEC] Two or more wires conducting electric power from one location to another. Also known as electric power line. { 'pau̇·ər ,līn }

power-line carrier [ELEC] The use of transmission lines to transmit speech, metering indications, control impulses, and other signals from one station to another, without interfering with the lines' normal function of transmitting power. { 'pau̇·ər ,līn ,kar·ē·ər }

power-line filter See line filter. { 'pau̇·ər ,līn ,fil·tər }

power-line interference [COMMUN] Interference caused by radiation from high-voltage power lines. { 'pau̇·ər ,līn ,in·tər,fir·əns }

power-line monitor [ELECTR] A device that continuously observes and records levels of electric power on a power line. { 'pau̇·ər ,līn 'män·əd·ər }

power loader [MIN ENG] Power-operated machine for loading ore, coal, or other material into a car, conveyor, or other collector. { 'pau̇·ər ,lōd·ər }

power loss [ELECTR] The ratio of the power absorbed by the input circuit of a transducer to the power delivered to a specified load; usually expressed in decibels. Also known as power attenuation. { 'pau̇·ər ,lȯs }

power meter See electric power meter. { 'pau̇·ər ,mēd·ər }

power of the continuum [MATH] The cardinality of the set of real numbers. { 'pau̇·ər əv thə kən,tin·yə·wəm }

power oil [MATER] Fluid used to actuate (power) hydraulic pumps, motors, and other power-type equipment; can be based on petroleum oils or synthetic materials. { 'pau̇·ər ,ȯil }

power output [ELECTR] The alternating-current power in watts delivered by an amplifier to a load. { 'pau̇·ər ¦au̇t,pu̇t }

power output tube See power tube. { 'pau̇·ər ¦au̇t,pu̇t ,tüb }

power pack [ELECTR] Unit for converting power from an alternating- or direct-current supply into an alternating- or direct-current power at voltages suitable for supplying an electronic device. { 'pau̇·ər ,pak }

power package [MECH ENG] A complete engine and its accessories, designed as a single unit for quick installation or removal. { 'pau̇·ər ,pak·ij }

power plant [MECH ENG] Any unit that converts some form of energy into electrical energy, such as a hydroelectric or steam-generating station, a diesel-electric engine in a locomotive, or a nuclear power plant. Also known as electric power plant. { 'pau̇·ər ,plant }

power rating [ELEC] The power available at the output terminals of a component or piece of equipment that is operated according to the manufacturer's specifications. { 'pau̇·ər ,rād·iŋ }

power ratio [ELECTROMAG] The ratio of the maximum power to the minimum power in a waveguide that is improperly terminated. { 'pau̇·ər ,rā·shō }

power reactor [NUCLEO] A nuclear reactor designed to provide useful power, as for submarines, aircraft, ships, vehicles, and power plants. { 'pau̇·ər rē,ak·tər }

power rectifier [ELEC] A device which converts alternating current to direct current and operates at high power loads. { 'pau̇·ər 'rek·tə,fī·ər }

power relay [ELEC] Relay that functions at a predetermined value of power; may be an overpower relay, an underpower relay, or a combination of both. { 'pau̇·ər 'rē,lā }

power resistor [ELEC] A resistor used in electric power systems, ranging in size from 5 watts to many kilowatts, and cooled by air convection, air blast, or water. { 'pau̇·ər ri,zis·tər }

power saw [MECH ENG] A power-operated woodworking saw, such as a bench or circular saw. { 'pau̇·ər ,sȯ }

power semiconductor [ELECTR] A semiconductor device capable of dissipating appreciable power (generally over 1 watt) in normal operation; may handle currents of thousands of amperes or voltages up into thousands of volts, at frequencies up to 10 kilohertz. { 'pau̇·ər 'sem·i·kən,dək·tər }

power series [MATH] An infinite series composed of functions having nth term of the form $a_n(x - x_0)^n$, where x_0 is some point and a_n some constant. { 'pau̇·ər ,sir·ēz }

power set [MATH] The set consisting of all subsets of a given set. { 'pau̇·ər ,set }

power shovel [MECH ENG] A power-operated shovel that carries a short boom on which rides a movable dipper stick carrying an open-topped bucket; used to excavate and remove debris. { 'pau̇·ər ,shəv·əl }

power-shovel mining [MIN ENG] A technique utilizing power shovels to mine ores by mining or stripping and taking away overburden. { 'pau̇·ər ,shəv·əl ,mīn·iŋ }

power slips See automatic slips. { 'pau̇·ər ,slips }

power spectrum See frequency spectrum. { 'pau̇·ər ,spek·trəm }

power station See generating station. { 'pau̇·ər ,stā·shən }

power steering [MECH ENG] A steering control system for a propelled vehicle in which an auxiliary power source assists the driver by providing the major force required to direct the road wheels. { 'paù·ər ˌstir·iŋ }

power stroke [MECH ENG] The stroke in an engine during which pressure is applied to the piston by expanding steam or gases. { 'paù·ər ˌstrōk }

power supply [ELECTR] A source of electrical energy, such as a battery or power line, employed to furnish the tubes and semiconductor devices of an electronic circuit with the proper electric voltages and currents for their operation. Also known as electronic power supply. { 'paù·ər sə,plī }

power supply circuit [ELEC] An electrical network used to convert alternating current to direct current. { 'paù·ər sə,plī ,sər·kət }

power-supply rejection ratio [ELECTR] The ratio between the gain of an amplifier for difference signals between the input terminals, and the gain for variations of the power-supply voltages. Abbreviated PSRR. { ¦paù·ər sə,plī ri'jek·shən ˌrā·shō }

power switch [ELEC] An electric switch which energizes or deenergizes an electric load; ranges from ordinary wall switches to load-break switches and disconnecting switches in power systems operating at voltages of hundreds of thousands of volts. { 'paù·ər ˌswich }

power switchboard [ELEC] Part of a switch gear which consists of one or more panels upon which are mounted the switching control, measuring, protective, and regulatory equipment; the panel or panel supports may also carry the main switching and interrupting devices together with their connection. { 'paù·ər 'swich,bòrd }

power switching [ELEC] Switching between supplies of electrical energy at high levels of current and voltage. { 'paù·ər ˌswich·iŋ }

power train [MECH ENG] The part of a vehicle connecting the engine to propeller or driven axle; may include drive shaft, clutch, transmission, and differential gear. Also known as drive train. { 'paù·ər ˌtrān }

power transfer equation [ELEC] An equation for the power flow across a transmission line in terms of the relative magnitudes and phases of the terminal voltages, and the inductive reactance component and resistive component of the line. { ¦paù·ər 'tranz·fər iˌkwā·zhən }

power transfer theorem [ELEC] The theorem that, in an electrical network which carries direct or sinusoidal alternating current, the greatest possible power is transferred from one section to another when the impedance of the section that acts as a load is the complex conjugate of the impedance of the section that acts as a source, where both impedances are measured across the pair of terminals at which the power is transferred, with the other part of the network disconnected. { ¦paù·ər 'tranz·fər ,thir·əm }

power transformer [ELEC] An iron-core transformer having a primary winding that is connected to an alternating-current power line and one or more secondary windings that provide different alternating voltage values. { 'paù·ər tranz,fòr·mər }

power transistor [ELECTR] A junction transistor designed to handle high current and power; used chiefly in audio and switching circuits. { 'paù·ər tran,zis·tər }

power transmission line [ELEC] The facility in an electric power system used to transfer large amounts of power from one location to a distant location; distinguished from a subtransmission or distribution line by higher voltage, greater power capability, and greater length. Also known as electric main; main (both British usages). { 'paù·ər tranz'mish·ən ,līn }

power transmission tower [ELEC] A rigid steel tower supporting a high-voltage electric power transmission line, having a large enough spacing between conductors, and between conductors and ground, to prevent corona discharge. { 'paù·ər tranz'mish·ən ,taù·ər }

power traverse [ORD] Turning of a gun to change the direction of fire by means of a power-driven mechanism, as in a tank, aircraft, or ship turret. { 'paù·ər trə,vərs }

power tube [ELECTR] An electron tube capable of handling more current and power than an ordinary voltage-amplifier tube; used in the last stage of an audio-frequency amplifier or in high-power stages of a radio-frequency amplifier. Also known as power amplifier tube; power output tube. { 'paù·ər ,tüb }

power typing [COMPUT SCI] A word-processing technique that allows the automatic typing of repetitious text, such as appears in a form letter. { 'paù·ər ,tüp·iŋ }

power up [COMPUT SCI] To check that the computer memory, peripherals, and input/output channels are working properly before the operating system is loaded. { 'paù·ər ,əp }

power winding [ELEC] In a saturable reactor, a winding to which is supplied the power to be controlled; commonly the functions of the output and power windings are accomplished by the same winding, which is then termed the output winding. { 'paù·ər ,wind·iŋ }

pox [MED] A vesicular or pustular exanthematic disease that may leave pit scars. { päks }

Poxviridae [VIROL] A family of deoxyribonucleic acid–containing animal viruses that is characterized by its ability to replicate in the cell cytoplasm; includes the subfamilies Chordopoxviridae and Entomopoxviridae. { päks'vir·ə,dī }

poxvirus [VIROL] A deoxyribonucleic acid-containing animal virus group including the viruses of smallpox, molluscum contagiosum, and various animal pox and fibromas. { 'päks,vī·rəs }

Poynting effect [MECH] The effect of torsion of a very long cylindrical rod on its length. { 'pòin·tiŋ iˌfekt }

Poynting-Robertson effect [ASTRON] The gradual decrease in orbital velocity of a small particle such as a micrometeorite in orbit about the sun due to the absorption and reemission of radiant energy by the particle. { 'pòint·iŋ 'räb·ərt·sən iˌfekt }

Poynting's law [THERMO] A special case of the Clapeyron equation, in which the fluid is removed as fast as it forms, so that its volume may be ignored. { 'pòint·iŋz ,lò }

Poynting theorem [ELECTROMAG] A theorem, derived from Maxwell's equations, according to which the rate of loss of energy stored in electric and magnetic fields within a region of space is equal to the sum of the rate of dissipation of electrical energy as heat and the rate of flow of electromagnetic energy outward through the surface of the region. { 'pòint·iŋ ,thir·əm }

Poynting vector [ELECTROMAG] A vector, equal to the cross product of the electric-field strength and the magnetic-field strength (mks units) whose outward normal component, when integrated over a closed surface, gives the outward flow of electromagnetic energy through that surface. { 'pòint·iŋ ,vek·tər }

pozzolan [GEOL] A finely ground burnt clay or shale resembling volcanic dust, found near Pozzuoli, Italy; used in cement because it hardens underwater. [MATER] Cement made by mixing and grinding together slaked lime and pozzolan without burning; sometimes used for concrete not exposed to the air. { 'pät·sə·lən }

PPI See plan position indicator.

pp junction [ELECTR] A region of transition between two regions having different properties in p-type semiconducting material. { ¦pē¦pē ,jəŋk·shən }

PPLO See pleuropneumonialike organism.

PPM See pulse-position modulation.

PPP See Point-to-Point Protocol.

p-process [NUC PHYS] The synthesis of certain nuclides in stars through capture of protons or ejection of neutrons by gamma rays. { 'pē ,prä·səs }

PP reaction See proton-proton reaction. { ¦pē¦pē rē,ak·shən }

P pulse See commutator pulse. { 'pē ,pəls }

Pr See praseodymium.

practical astronomy [ASTRON] That part of astronomy concerned with the use of information acquired by an observer in the solution of problems determining latitude and longitude on sea or land and directions on the earth's surface by the help of celestial objects. { 'prak·ti·kəl ə'strän·ə·mē }

practical entropy See virtual entropy. { 'prak·ti·kəl 'en·trə·pē }

Practical Extraction and Reporting Language [COMPUT SCI] A scripting language often used for creating CGI programs. Abbreviated Perl. { ¦prak·ti·kəl ik,strak·shən and ri'pòrt·iŋ ,laŋ·gwij }

practical system See meter-kilogram-second-ampere system. { 'prak·ti·kəl 'sis·təm }

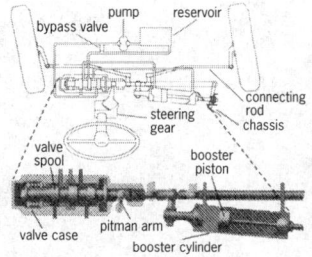
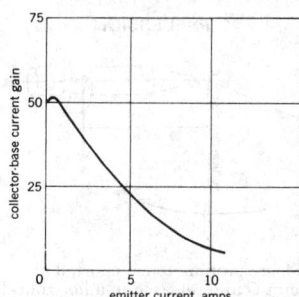

practical units [ELECTROMAG] The units of the meter-kilogram-second-ampere system. { 'prak·ti·kəl ‚yü·nəts }

practice ammunition [ORD] Ammunition used for target practice or similar types of training; for gun and rocket types of weapons, practice ammunition contains a propelling charge, and either an inert filler or a spotting charge in the projectile; other types of practice ammunition, such as bombs or mines, usually contain a spotting charge or some form of charge to indicate functioning. { 'prak·təs ‚am·yə‚nish·ən }

practice mine [ORD] Imitation land mine used in training; it may contain a smoke-producing agent for maneuvers or for practice in observing the effects of mines against vehicles, or it may be simply a block of wood, metal, or concrete for practice in laying mine fields. { 'prak·təs ‚mīn }

Prader-Willi syndrome [MED] A genetic disorder that is caused by defects on the paternally derived chromosome 15, causing mild mental retardation, neonatal hypotonia, hypogonadism, compulsive overeating, childhood onset obesity, and mild facial dysmorphism. { 'präd·ər ‚wil·ē 'sin‚drōm }

praecipitatio [METEOROL] Precipitation falling from a cloud and apparently reaching the earth's surface; this supplementary cloud feature is mostly encountered in altostratus, nimbostratus, stratocumulus, stratus, cumulus, and cumulonimbus. { prē‚sip·ə‚tä·shō }

Praesepe [ASTRON] A cluster of faint stars in the center of the constellation Cancer. Also known as Beehive; Manger. { prē'sə‚pē }

pragma [COMPUT SCI] A directive inserted into a computer program to prevent the automatic execution of certain error checking and reporting routines which are no longer necessary when the program has been perfected. { 'prag·mə }

pragmatics [COMMUN] The branch of semiotics that treats the relation of symbols to behavior and the meaning received by the listener or reader of a statement. [COMPUT SCI] The fourth and final phase of natural language processing, following contextual analysis, that takes into account the speaker's goal in uttering a particular thought in a particular way in determining what constitutes an appropriate response. { prag'mad·iks }

prairie [GEOGR] An extensive level-to-rolling treeless tract of land in the temperate latitudes of central North America, characterized by deep, fertile soil and a cover of coarse grass and herbaceous plants. { 'prer·ē }

prairie climate See subhumid climate. { 'prer·ē ‚klī·mət }

prairie dog [VERT ZOO] The common name for three species of stout, fossorial rodents belonging to the genus *Cynomys* in the family Sciuridae; all have a short, flat tail, small ears, and short limbs terminating in long claws. { 'prer·ē ‚dòg }

prairie soil [GEOL] A group of zonal soils having a surface horizon that is dark or grayish brown, which grades through brown soil into lighter-colored parent material; it is 2–5 feet (0.6–1.5 meters) thick and develops under tall grass in a temperate and humid climate. { 'prer·ē ‚sòil }

prairie wolf See coyote. { 'prer·ē ‚wùlf }

Prandtl-Glauert rule [FL MECH] The rule that the pressure coefficient at any point in the subsonic flow of a fluid about a slender body is equal to the pressure coefficient at that point in the corresponding incompressible fluid flow, divided by $\sqrt{1 - M^2}$, where M is the Mach number far from the body. { 'pränt·əl 'glaù·ərt ‚rül }

Prandtl-Meyer flow [FL MECH] A two-dimensional, supersonic fluid flow in which an initially uniform flow passes a sharp, convex corner in a boundary, resulting in expansion of the fluid. { 'pränt·əl 'mī·ər ‚flō }

Prandtl number [FL MECH] A dimensionless number used in the study of diffusion in flowing systems, equal to the kinematic viscosity divided by the molecular diffusivity. Symbolized Pr_m. Also known as Schmidt number 1 (N_{Sc}). [THERMO] A dimensionless number used in the study of forced and free convection, equal to the dynamic viscosity times the specific heat at constant pressure divided by the thermal conductivity. Symbolized N_{Pr}. { 'pränt·əl ‚nəm·bər }

prase [MINERAL] **1.** A translucent and dull leek green or light-grayish yellow-green variety of chalcedony. **2.** Crystalline quartz containing a multitude of green hairlike crystals of actinolite. Also known as mother-of-emerald. { prāz }

praseodymium [CHEM] A chemical element, symbol Pr, atomic number 59, atomic weight 140.9077; a metallic element of the rare-earth group. { ‚prā·zē·ō'dim·ē·əm }

prase opal See prasopal. { 'prāz 'ō·pəl }

prasinite [PETR] A greenschist in which the proportions of the hornblende-chlorite-epidote assemblage are more or less equal. { 'prāz·ən‚īt }

Prasinovolvocales [BOT] An order of green algae in which there are lateral appendages in the flagellum. { ‚prāz·ən‚ō‚väl·və'kā·lēz }

prasopal [MINERAL] A green variety of common opal containing chromium. Also spelled prase opal. { 'prāz‚ō·pəl }

pratincolous [ECOL] **1.** Living in meadows. **2.** Living in low grass. { prə'tiŋ·kə·ləs }

Pratt truss [CIV ENG] A truss having both vertical and diagonal members between the upper and lower chords, with the diagonals sloped toward the center. { 'prat ‚trəs }

preadaptation [EVOL] Possession by an organism or group of organisms, specialized to one mode of life, of characters which favor easy adaptation to a new environment. { ‚prē‚ad·əp'tā·shən }

prealpine facies [GEOL] A geosynclinal facies characteristic of neritic areas, displaying thick limestone deposits and coarse terrigenous material and resembling epicontinental platform sediments. { prē'al‚pīn 'fā·shēz }

preamble [COMMUN] The portion of a commercial radio telegraph message that is sent first, containing the message number, office of origin, date, and other numerical data not part of the following message text. { 'prē‚am·bəl }

preamplifier [ELECTR] An amplifier whose primary function is boosting the output of a low-level audio-frequency, radio-frequency, or microwave source to an intermediate level so that the signal may be further processed without appreciable degradation of the signal-to-noise ratio of the system. Also known as preliminary amplifier. { prē'am·plə‚fī·ər }

preassembled [ENG] Assembled beforehand. { ‚prē·ə'sem·bəld }

preatmospheric speed [ASTRON] The speed of a meteoroid just before it enters the earth's atmosphere. { ‚prē‚at·mə‚sfir·ik 'spēd }

prebiotic molecule [ORG CHEM] A molecule that is believed to be involved in the processes leading to the origin of life. { ‚pre·bī‚äd·ik 'mäl·ə‚kyül }

prebreaker [MECH ENG] Device used to break down large masses of solids prior to feeding them to a crushing or grinding device. { 'prē‚brāk·ər }

Precambrian [GEOL] All geologic time prior to the beginning of the Paleozoic era (before 600,000,000 years ago); equivalent to about 90% of all geologic time. { prē'kam·brē·ən }

precancerous [PATH] Pertaining to any pathological condition of a tissue which is likely to develop into cancer. { prē'kan·sə·rəs }

precast concrete [MATER] Concrete components which are cast and partly matured in a factory or on the site before being lifted into their final position on a structure. { 'prē‚kast 'kän‚krēt }

precedence [COMPUT SCI] The order in which operators are processed in a programming language. { 'pres·əd·əns }

precedence diagram method [IND ENG] A technique for constructing a network in which the activities are represented by symbols that are connected by lines to indicate the logical relationships between them. Abbreviated PDM. { ‚pres·əd·əns 'dī·ə‚gram ‚meth·əd }

precedence effect [ACOUS] The ability of the auditory system to process sound that reaches the ears directly from a source even when significant reflected sounds reach the ears shortly afterward. { prə'sēd·əns i‚fekt or 'pre·səd·əns }

precedence relation [COMPUT SCI] A rule stating that, in a given programming language, one of two operators is to be applied before the other in any mathematical expression. { 'pres·əd·əns ri‚lā·shən }

preceding limb [ASTRON] The half of the limb of a celestial body with an observable disk that appears to precede the body in its apparent motion across the field of view of a fixed telescope. { prē'sēd·iŋ 'lim }

precentral gyrus [ANAT] The cerebral convolution that lies between the precentral sulcus and the central sulcus and extends from the superomedial border of the hemisphere to the posterior ramus of the lateral sulcus. { 'prē‚sen·trəl 'jī·rəs }

precession [MECH] The angular velocity of the axis of spin of a spinning rigid body, which arises as a result of external torques acting on the body. { prē'sesh·ən }

PRECEDENCE EFFECT

Paths traveled by sound from a ringing telephone. The fact that humans process sounds from sources in reflective environments suggests that the first-arriving sound from the source takes precedence over those arriving later from reflections.

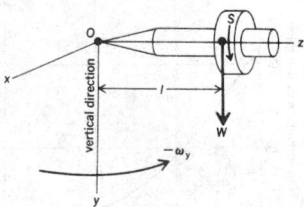

PRECESSION

A fast-spinning top supported at point O and released in a horizontal plane precesses about the vertical y axis of symmetry; x = axis perpendicular to y and z; S = spin velocity of top about z axis; W = weight of top; l = distance from O to top's center of mass; ω_y = angular velocity of precession of top about y axis, given by $\omega_y = Wl/I_zS$, where I_z = top's moment of inertia about z axis.

precessional torque [MECH] A torque which causes a rotating body to precess. { prē¦sesh·ən·əl 'tȯrk }

precession camera [CRYSTAL] An x-ray diffraction camera used in the Buerger precession method for recording the diffractions of an individual crystal. { prē'sesh·ən ˌkam·rə }

precession in declination [ASTRON] The component of general precession of the earth along a celestial meridian, amounting to about 20″ per year. { prē'sesh·ən in ˌdek·lə'nā·shən }

precession in right ascension [ASTRON] The component of general precession of the earth along the celestial equator, amounting to about 46.1″ per year. { prē'sesh·ən in 'rīt ə'sen·shən }

precession of nodes [ASTRON] The gradual change in direction of the orbital plane of a binary system. { prē'sesh·ən əv 'nōdz }

precession of the equinoxes [ASTRON] A slow conical motion of the earth's axis about the vertical to the plane of the ecliptic, having a period of 26,000 years, caused by the attractive force of the sun, moon, and other planets on the equatorial protuberance of the earth; it results in a gradual westward motion of the equinoxes. { prē'sesh·ən əv thə 'ē·kwə,näk·səz }

precharge [MET] The pressure introduced into the cavity of a mold before forming a part. { prē'chärj }

prechlorination [CIV ENG] Chlorination of water before filtration. { ¦prē,klȯr·ə'nā·shən }

precious metal [MET] A relatively scarce, valuable metal, such as gold, silver, and members of the platinum group. { 'presh·əs 'med·əl }

precious stone [MINERAL] **1.** Any genuine gemstone. **2.** A gemstone of high commercial value because of its beauty, rarity, durability, and hardness; examples are diamond, ruby, sapphire, and emerald. { 'presh·əs 'stōn }

precipice [GEOL] A very steeply inclined, vertical, or overhanging wall or surface of rock. { 'pres·ə·pəs }

precipitable water [METEOROL] The total atmospheric water vapor contained in a vertical column of unit cross-sectional area extending between any two specified levels, commonly expressed in terms of the height to which that water substance would stand if completely condensed and collected in a vessel of the same unit cross section. Also known as precipitable water vapor. { pri'sip·əd·ə·bəl 'wȯd·ər }

precipitable water vapor See precipitable water. { pri'sip·əd·ə·bəl 'wȯd·ər ˌvā·pər }

precipitant [CHEM] A chemical or chemicals that cause a precipitate to form when added to a solution. { prə'sip·ət·əns }

precipitate [CHEM] **1.** A substance separating, in solid particles, from a liquid as the result of a chemical or physical change; **2.** To form a precipitate. { prə'sip·ə,tāt }

precipitation [CHEM] The process of producing a separable solid phase within a liquid medium; represents the formation of a new condensed phase, such as a vapor or gas condensing to liquid droplets; a new solid phase gradually precipitates within a solid alloy as a result of slow, inner chemical reaction; in analytical chemistry, precipitation is used to separate a solid phase in an aqueous solution. [IMMUNOL] Aggregation of soluble antigen by an antibody. [METEOROL] **1.** Any or all of the forms of water particles, whether liquid or solid, that fall from the atmosphere and reach the ground. **2.** The amount, usually expressed in inches of liquid water depth, of the water substance that has fallen at a given point over a specified period of time. { prə,sip·ə'tā·shən }

precipitation area [METEOROL] **1.** On a synoptic surface chart, an area over which precipitation is falling. **2.** In radar meteorology, the region from which a precipitation echo is received. { prə,sip·ə'tā·shən ,er·ē·ə }

precipitation attenuation [ELECTROMAG] Loss of radio energy due to the passage through a volume of the atmosphere containing precipitation; part of the energy is lost by scattering, and part by absorption. { prə,sip·ə'tā·shən ə,ten·yə'wā·shən }

precipitation ceiling [METEOROL] After United States weather observing practice, a ceiling classification applied when the ceiling value is the vertical visibility upward into precipitation; this is necessary when precipitation obscures the cloud base and prevents a determination of its height. { prə,sip·ə'tā·shən ,sēl·iŋ }

precipitation cell [METEOROL] In radar meteorology, an element of a precipitation area over which the precipitation is more or less continuous. { prə,sip·ə'tā·shən ,sel }

precipitation clutter suppression [ELECTR] Technique of reducing, by one of the various devices integral to the radar system, clutter caused by rain in the radar range. { prə,sip·ə'tā·shən ¦kləd·ər sə,presh·ən }

precipitation current [METEOROL] The downward transport of charge, from cloud region to earth, that occurs in a fall of electrically charged rain or other hydrometeors. { prə,sip·ə'tā·shən ,kə·rənt }

precipitation echo [METEOROL] A type of radar echo returned by precipitation. { prə,sip·ə'tā·shən ,ek·ō }

precipitation effectiveness See precipitation-evaporation ratio. { prə,sip·ə'tā·shən i¦fek·tiv·nəs }

precipitation electricity [GEOPHYS] **1.** That branch of the study of atmospheric electricity concerned with the electric charges carried by precipitation particles and with the manner in which these charges are acquired. **2.** The electric charge borne by precipitation particles. { prə,sip·ə'tā·shən ə,lek'tris·əd·ē }

precipitation-evaporation ratio [CLIMATOL] For a given locality and month, an empirical expression devised for the purpose of classifying climates numerically on the basis of precipitation and evaporation. Abbreviated P-E ratio. Also known as precipitation effectiveness. { prə,sip·ə'tā·shən i,vap·ə'rā·shən ,rā·shō }

precipitation excess [HYD] The volume of water from precipitation that is available for direct runoff. { prə,sip·ə'tā·shən 'ek,ses }

precipitation facies [GEOL] Facies characteristics that provide evidence of depositional conditions; revealed mainly by sedimentary structure (such as cross-bedding and ripple marks) and by primary constituents (especially fossils). { prə,sip·ə'tā·shən ,fā·shēz }

precipitation gage [ENG] Any device that measures the amount of precipitation; principally, a rain gage or snow gage. { prə,sip·ə'tā·shən ,gāj }

precipitation-generating element [METEOROL] In radar meteorology, a relatively small volume of supercooled cloud droplets in which ice crystals form and grow much more rapidly than in a lower, larger cloud mass. { prə,sip·ə'tā·shən 'jen·ə,rād·iŋ ,el·ə·mənt }

precipitation hardening See age hardening. { prə,sip·ə'tā·shən ,härd·ən·iŋ }

precipitation indicator [ANALY CHEM] In a titration, a substance that precipitates from solution in a clearly visible form at the end point. { prə,sip·ə'tā·shən ,in·də,kād·ər }

precipitation intensity [METEOROL] The rate of precipitation, expressed in inches or millimeters per hour. Also known as rainfall intensity. { prə,sip·ə'tā·shən in,ten·səd·ē }

precipitation inversion [METEOROL] As found in some mountain areas, a decrease of precipitation with increasing elevation of ground above sea level. Also known as rainfall inversion. { prə,sip·ə'tā·shən in,vər·zhən }

precipitation noise [ELECTR] Noise generated in an antenna circuit, generally in the form of a relaxation oscillation, caused by the periodic discharge of the antenna or conductors in the vicinity of the antenna into the atmosphere. { prə,sip·ə'tā·shən ,nȯiz }

precipitation number [ANALY CHEM] The number of milliliters of asphaltic precipitate formed when 10 milliliters of petroleum-lubricating oil is mixed with 90 milliliters of a special-quality petroleum naphtha, then centrifuged according to American Society for Testing and Materials test conditions; used to determine the quantity of asphalt in petroleum-lubricating oil. { prə,sip·ə'tā·shən ,nəm·bər }

precipitation physics [METEOROL] The study of the formation and precipitation of liquid and solid hydrometeors from clouds; a branch of cloud physics and of physical meteorology. { prə,sip·ə'tā·shən ,fiz·iks }

precipitation static [COMMUN] Static interference due to the discharge of large charges built up on an aircraft or other object by rain, sleet, snow, or electrically charged clouds. { prə,sip·ə'tā·shən ,stad·ik }

precipitation station [METEOROL] A station at which only precipitation observations are made. { prə,sip·ə'tā·shən ,stā·shən }

precipitation titration [ANALY CHEM] Amperometric titration in which the potential of a suitable indicator electrode is measured during the titration. { prə‚sip·ə'tā·shən tī‚trā·shən }

precipitation trails *See* virga. { prə‚sip·ə'tā·shən ‚trālz }

precipitation trajectory [METEOROL] In radar meteorology, a characteristic echo observed on range-height indicator scopes and time-height sections which represents the height-range pattern of snow falling from isolated precipitation-generating elements of a few miles in diameter. Also known as mare's tail. { prə‚sip·ə'tā·shən trə‚jek·trē }

precipitator *See* electrostatic precipitator. { prə'sip·ə‚tād·ər }

precipitin [IMMUNOL] An antibody that chemically interacts with an antigen to form a precipitate. { prə'sip·ə·tən }

precipitin test [IMMUNOL] An immunologic test in which a specific reaction between antigen and antibody results in a visible precipitate. { prə'sip·ə·tən ‚test }

precipitous terrain [NAV] In air operations, terrain characterized by steep or abrupt slopes. { prə'sip·əd·əs tə'rān }

precision [MATH] The number of digits in a decimal fraction to the right of the decimal point. [NAV] In air operations, pertaining to a navigational facility which provides a combined azimuth and glide slope guidance to a runway. [SCI TECH] The measure of the range of values of a set of measurements; indicates reproducibility of the observations. { prə'sizh·ən }

precision adjustment [ORD] A deliberate adjustment of the fire of one weapon for the purpose of placing the center of impact accurately on the target. { prə'sizh·ən ə‚jəs·mənt }

precision agriculture [AGR] The application of technologies and agronomic principles to manage spatial and temporal variability associated with all aspects of agricultural production for the purpose of improving crop performance and environmental quality. { prə‚sizh·ən 'ag·rə‚kəl·chər }

precision approach radar [NAV] A radar system located on an airfield for observation of the position of an aircraft with respect to an approach path, and specifically intended to provide guidance to the aircraft during its approach to the field; the system consists of a ground radar equipment which is alternately connected to two antenna systems; one antenna system sweeps a narrow beam over a 20° sector in the horizontal plane; the second sweeps a narrow beam over a 7° sector in the vertical plane; course correction is transmitted to the aircraft from the ground. Abbreviated PAR. { prə'sizh·ən ə‚prōch 'rā‚där }

precision attribute [COMPUT SCI] A set of one or more integers that denotes the number of symbols used to represent a given number and positional information for determining the base point of the number. { prə'sizh·ən 'a·trə‚byüt }

precision-balanced hybrid circuit [ELEC] Circuit used to interconnect a four-wire telephone circuit to a particular two-wire circuit, in which the impedance of the balancing network is adjusted to give a relatively high degree of balance. { prə'sizh·ən ‚bal·ənst 'hī·brəd 'sər·kət }

precision block *See* gage block. { prə'sizh·ən ‚bläk }

precision bombing [ORD] Horizontal bombing done with the appropriate precision instruments and equipment so as to strike a target of comparatively small bulk or area. { prə'sizh·ən ‚bäm·iŋ }

precision casting [MET] A metal casting of accurately reproducible dimensions. { prə'sizh·ən ‚kast·iŋ }

precision depth recorder [ENG] A machine that plots sonar depth soundings on electrosensitive paper; can plot variations in depth over a range of 400 fathoms (730 meters) on a paper 18.85 inches (47.9 centimeters) wide. Abbreviated PDR. Also known as precision graphic recorder (PGR). { prə'sizh·ən 'depth ri‚kórd·ər }

precision fire [ORD] Fire in which the center of impact is accurately placed on a limited target; fire which is based on precision adjustment. { prə'sizh·ən ‚fīr }

precision graphic recorder *See* precision depth recorder. { prə'sizh·ən 'graf·ik ri'kórd·ər }

precision grinding [MECH ENG] Machine grinding to specified dimensions and low tolerances. { prə'sizh·ən ‚grīnd·iŋ }

precision net [ELEC] In a four-wire terminating set or similar device employing a hybrid coil, an artificial line designed and adjusted to provide an accurate balance for the loop and subscribers set or line impedance. { prə'sizh·ən ‚net }

precision sweep [ELECTR] Delayed expanded radar sweep for high resolution and range accuracy. { prə'sizh·ən ‚swēp }

precoat [MET] In casting, thin coating of refractory slurry applied to an expendable wax or plastic pattern as a base for the application of the main slurry. { ‚prē'kōt }

precoat filter [ENG] A device designed to filter solid particles from a liquid-solid slurry after a precoat of builtup solid material (filter aid or filtered solid) has been applied to the inner surface of the filter medium. { ‚prē'kōt 'fil·tər }

precoating [ENG] The depositing of an inert material, such as filter aid, onto the filter medium prior to the filtration of suspended solids from a solid-liquid slurry. { ‚prē'kōd·iŋ }

precognition [PSYCH] A form of extrasensory perception involving foreknowledge of a future event. { ‚prē·käg'nish·ən }

pre-cold-frontal squall line *See* prefrontal squall line. { ‚prē'kōld ‚frənt·əl 'skwól ‚līn }

precollagenous fiber *See* reticular fiber. { ‚prē·kə'laj·ə·nəs 'fī·bər }

precombustion chamber [MECH ENG] A small chamber before the main combustion space of a turbine or reciprocating engine in which combustion is initiated. { ‚prē·kəm'bəs·chən ‚chām·bər }

precompact set [MATH] A set in a metric space which can always be covered by open balls of any diameter about some finite number of its points. Also known as totally bounded set. { prē'käm‚pakt ‚set }

precompiled module [COMPUT SCI] A standardized subroutine that is separately developed and compiled for use in many different computer programs. { ‚prē·kəm'pīld 'mäj·yül }

precompiler [COMPUT SCI] A computer program that indentifies syntax errors and other problems in a program before it is converted to machine language by a compiler. { ‚prē·kəm'pīl·ər }

precomputation [NAV] The process of making navigational solutions in advance, applied particularly to the determination of computed altitude and azimuth before making a celestial observation for a line of position; when this is done, the observation must be made at the time used for the computation, or a correction applied. { ‚prē‚käm·pyə'tā·shən }

precomputed altitude [ASTRON] The altitude of a celestial body computed before observation with the sextant. Altitude corrections are included in the calculations but are applied with reversed sign. { 'prē·kəm‚pyüd·əd 'al·tə‚tüd }

precomputed curve [NAV] A graphical representation of the azimuth or altitude of a celestial body which is plotted against time for a given assumed position or positions, and which is computed for subsequent use with celestial observations. { 'prē·kəm‚pyüd·əd 'kərv }

preconduction current [ELECTR] Low value of plate current flowing in a thyratron or other grid-controlled gas tube prior to the start of conduction. { 'prē·kən'dək·shən ‚kə·rənt }

preconsolidation pressure [GEOL] The greatest effective stress exerted on a soil; result of this pressure from overlying materials is compaction. Also known as prestress. { 'prē·kən‚säl·ə'dā·shən ‚presh·ər }

precooler [MECH ENG] A device for reducing the temperature of a working fluid before it is used by a machine. { 'prē'kül·ər }

precure finish [TEXT] A durable-press finish applied to a fabric and heat-set by the mill. { prē'kyùr 'fin·əsh }

predation [BIOL] The killing and eating of an individual of one species by an individual of another species. { prə'dā·shən }

predator [ECOL] An animal that preys on other animals as a source of food. { 'pred·əd·ər }

predazzite [PETR] A recrystallized limestone that resembles pencatite, but contains less brucite than calcite. { 'pred·ə‚zīt }

predecessor [MATH] For a vertex *a* in a directed graph, any vertex *b* for which there is an arc between *a* and *b* directed from *b* to *a*. { 'pred·ə‚ses·ər }

predecessor job [COMPUT SCI] A job whose output is used as input to another job, and which must therefore be completed before the second job is started. { 'pred·ə‚ses·ər ‚jäb }

predefined function [COMPUT SCI] A sequence of instructions that is identified by name in a computer program but is built into the high-level programming language from which the program is complied or is retrieved from somewhere outside the program, such as a subroutine library. { 'prē·di'fīnd 'fəŋk·shən }

predetection combining [ELECTR] Method used to produce

an optimum signal from multiple receivers involved in diversity reception of signals. { ¦prē·di'tek·shən kəm'bīn·iŋ }

predicate [COMPUT SCI] A statement in a computer program that evaluates an expression in order to arrive at a true or false answer. [MATH] **1.** To affirm or deny, in mathematical logic, one or more subjects. Also known as logical function; propositional function. **2.** *See* propositional function. { 'pred·ə‚kāt }

predicate calculus [MATH] The mathematical study of logical statements relating to arbitrary sets of objects and involving predicates and quantifiers as well as propositional connectives. { 'pred·ə·kət ‚kal·kyə·ləs }

predict *See* forecast. { pri'dikt }

predictability [NAV] The measure of the accuracy with which the system can define the location of a point in terms of geographic coordinates rather than the lattice peculiar to that system. { prə‚dik·tə'bil·əd·ē }

predicted barrage *See* antiaircraft barrage. { prə'dik·təd bə'räzh }

predicted firing [ORD] Firing at the point at which a moving target is expected to be when the projectile reaches it, according to predictions based on observation. { prə'dik·təd 'fīr·iŋ }

predicted ground speed [NAV] The speed computed by applying estimated wind direction and speed to course and estimated airspeed. { prə'dik·təd 'graünd ‚spēd }

predicted-wave signaling [COMMUN] Communications system in which detection is optimized in the presence of severe noise by using mechanical resonator filters and other circuits in the detector to take advantage of known information on the arrival and completion times of each pulse, as well as on pulse shape, pulse frequency and spectrum, and possible data content. { prə'dik·təd ‚wāv 'sig·nəl·iŋ }

predicting dead time [ORD] The time allowed for calculation and applying firing data, from the time of observation to the instant of firing. { prə'dik·tiŋ 'ded ‚tīm }

predicting interval [ORD] The interval between successive predictions of future positions of the target. { prə'dik·tiŋ 'in·tər·vəl }

prediction [METEOROL] **1.** The act of making a weather forecast. **2.** The forecast itself. { prə'dik·shən }

predictive coder [COMMUN] Any technique for compressing audio or video signals in which a synthesizer at the receiver is controlled by signal parameters extracted at the transmitter to remake the signal. Also known as predictive encoder. { prə‚dik·tiv 'kō·dər }

predictive coding [COMMUN] In data compression, a method of coding information in which a sample value is presented as the error term formed by the difference between the sample and its prediction. { prə¦dik·tiv 'kōd·iŋ }

predictor-corrector methods [MATH] Methods of calculating numerical solutions of differential equations that employ two formulas, the first of which predicts the value of the solution function at a point x in terms of the values and derivatives of the function at previous points where these have already been calculated, enabling approximations to the derivatives at x to be obtained, while the second corrects the value of the function at x by using the newly calculated values. { prə¦dik·tər kə'rek·tər ‚meth·ədz }

predictive encoder *See* predictive coder. { prə‚dik·tiv in'kō·dər }

predissociation [PHYS CHEM] The dissociation of a molecule that has absorbed energy before it can lose energy by radiation. { ¦prē·di‚sō·sē'ā·shən }

prednisolone [BIOCHEM] $C_{21}H_{28}O_5$ A glucocorticoid that is a dehydrogenated analog of hydrocortisone. { pred'nis·ə‚lōn }

prednisone [PHARM] $C_{21}H_{26}O_5$ An adrenocortical steroid drug, obtained in crystalline form, that is an analog of cortisone. { 'pred·nə‚sōn }

predozzite [PETR] Limestone rich in periclase and brucite. { 'pred·ə‚zīt }

preece [ELEC] A unit of electrical resistivity equal to 10^{13} times the product of 1 ohm and 1 meter. { prēs }

preeclampsia [MED] A toxemia occurring in the latter half of pregnancy, characterized by an acute elevation of blood pressure and usually by edema and proteinuria, but without convulsions or coma. Also known as toxemia of pregnancy. { ¦prē·i'klamp·sē·ə }

preedit [COMPUT SCI] To edit data before feeding it to a computer. { prē'ed·ət }

preemphasis [ELECTR] A process which increases the magnitude of some frequency components with respect to the magnitude of others to reduce the effects of noise introduced in subsequent parts of the system. { prē'em·fə·səs }

preemphasis network [ELECTR] An RC (resistance-capacitance) filter inserted in a system to emphasize one range of frequencies with respect to another. Also known as emphasizer. { prē'em·fə·səs ‚net‚wərk }

preemptive multitasking [COMPUT SCI] A method of running more than one program on a computer at a time, in which control of the processor is decided by the operating system, which allocates each program a recurring time segment. { prē¦emp·tiv 'məl·tē‚task·iŋ }

preen gland *See* uropygial gland. { 'prēn ‚gland }

preferential shop [IND ENG] An establishment in which preference is given to union members in hiring, layoffs, and dismissals, with the understanding that nonunion workers may be employed without being required to join the union when the union cannot supply workers. { ‚pref·ə'ren·chəl 'shäp }

preferred numbers [ELECTR] A series of numbers adopted by the Electronic Industries Association and the military services for use as nominal values of resistors and capacitors, to reduce the number of different sizes that must be kept in stock for replacements. Also known as preferred values. { pri'fərd 'nəm·bərz }

preferred orientation [PETR] The nonrandom orientation of planar or linear fabric elements in structural petrology. { pri'fərd ‚ór·ē·ən'tā·shən }

preferred values *See* preferred numbers. { pri'fərd 'val·yüz }

prefilter [ENG] Filter used to remove gross solid contaminants before the liquid stream enters a separator-filter. { prē 'fil·tər }

prefix notation *See* Polish notation. { 'prē‚fiks nō‚tā·shən }

preflashing [GRAPHICS] The uniform exposure of a photographic plate to a small amount of light before its exposure to the image to be photographed. { prē'flash·iŋ }

prefocus lamp [ELEC] A light bulb whose filaments are precisely positioned with respect to the lamp socket. { prē'fō·kəs ‚lamp }

prefoliation *See* vernation. { ‚prē‚fō·lē'ā·shən }

preform [ENG] **1.** A preshaped fibrous reinforcement. **2.** A compact mass of premixed plastic material that has been prepared for convenient handling and control of uniformity during the mold loading process. [ENG ACOUS] The small slab of record stock material that is loaded into a press to be formed into a disk recording. Also known as biscuit (deprecated usage). { prē'fórm }

preforming [MET] **1.** Initial pressing of a powder metal to form a compact. **2.** Preliminary shaping of a refractory metal compact after presintering. **3.** In wire rope manufacturing, a process that sets a crimp in the strands, providing a permanent set and controlling the flexibility of the rope. { prē'fórm·iŋ }

prefrontal [ANAT] Situated in the anterior part of the frontal lobe of the brain. [VERT ZOO] **1.** Of or pertaining to a bone of some vertebrate skulls, located anterior and lateral to the frontal bone. **2.** Of, pertaining to, or being a scale or plate in front of the frontal scale on the head of some reptiles and fishes. { prē'frənt·əl }

prefrontal lobotomy *See* lobotomy. { prē'frənt·əl lə'bäd·ə·mē }

prefrontal squall line [METEOROL] A squall line or instability line located in the warm sector of a wave cyclone, about 50 to 300 miles (80 to 480 kilometers) in advance of the cold front, usually oriented roughly parallel to the cold front, and moving in about the same manner as the cold front. Also called nonfrontal squall line; pre-cold-frontal squall line. { prē'frənt·əl 'skwól ‚līn }

pregenital phase [PSYCH] In psychoanalytic theory, the period of early childhood that incorporates both the oral and anal phases, that is, before the genitals have begun to exert a predominant influence in the organization or patterning of sexual behavior. { prē¦jen·əd·əl 'fāz }

preglacial [GEOL] **1.** Pertaining to the geologic time immediately preceding the Pleistocene epoch. **2.** Of material, underlying glacial deposits. { prē'glā·shəl }

Pregl procedure [ANALY CHEM] Microanalysis technique in

which the sample is decomposed thermally, with subsequent oxidation of decomposition products. { 'prä·gəl prə,sē·jər }

pregnancy [MED] The state of being pregnant, from conception to childbirth. { 'preg·nən·sē }

pregnancy test [PATH] Any biologic or chemical procedure used to diagnose pregnancy. { 'preg·nən·sē ,test }

pregnanediol [BIOCHEM] $C_{21}H_{36}O_2$ A metabolite of progesterone, present in urine during the progestational phase of the menstrual cycle and also during pregnancy. { preg,nan·ə'dī,ól }

pregnenolone [BIOCHEM] $C_{21}H_{32}O_2$ A steroid ketone that is formed as an oxidation product of cholesterol, stigmasterol, and certain other steroids. { preg'nen·ə,lōn }

preheat current [MET] Current impulses which occur prior to and apart from the electric current in a resistance welding process. { 'prē,hēt ,kə·rənt }

preheater [MECH ENG] A device for preliminary heating of a material, substance, or fluid that will undergo further use or treatment by heating. { prē'hēd·ər }

preheat fluorescent lamp [ELECTR] A fluorescent lamp in which a manual switch or thermal starter is used to preheat the cathode for a few seconds before high voltage is applied to strike the mercury arc. { 'prē,hēt flú̇'res·ənt 'lamp }

preheating [ASTRON] A scenario suggested by detailed calculations of the inflationary universe cosmology in which the order parameter, as it relaxes about a new potential minimum following inflation, exhibits vibrations that can release energy in certain nonthermal modes. { prē'hēd·iŋ }

preheat roll [ENG] In plastic-extrusion coating, the heated roll between the pressure roll and the unwind roll; used to heat the substrate before it is coated. { 'prē,hēt ,rōl }

prehensile [VERT ZOO] Adapted for seizing, grasping, or plucking, especially by wrapping around some object. { prē'hen·səl }

prehension [PHYSIO] A movement that involves holding, seizing, or grasping. { prē'hen·shən }

pre-Hilbert space See inner-product space. { prē'hil·bərt ,spās }

prehnite [MINERAL] $Ca_2Al_2Si_3O_{10}(OH)_2$ A light-green to white mineral sorosilicate crystallizing in the orthorhombic system and generally found in reniform and stalactitic aggregates with crystalline surface; it has a vitreous luster, hardness is 6–6.5 on Mohs scale, and specific gravity is 2.8–2.9. { 'prä,nīt }

preignition [MECH ENG] Ignition of the charge in the cylinder of an internal combustion engine before ignition by the spark. { prē'ig'nish·ən }

pre-image [MATH] **1.** For a point y in the range of a function f, the set of points x in the domain of f for which $f(x) = y$. **2.** For a subset A of the range of a function f, the set of points x in the domain of f for which $f(x)$ is a member of A. Also known as inverse image. { prē'im·ij }

pre-Imbrian [ASTRON] Pertaining to the oldest lunar topographic features and lithologic map units constituting a system of rocks that appear in the mountainous terrae and are well displayed in the southern part of the visible lunar surface and over much of the reverse side. { prē'im·brē·ən }

preimpregnation [ENG] The mixing of a plastic resin with reinforcing material or substrate before molding takes place. { prē·im,preg'nā·shən }

preindexing [COMPUT SCI] Operation in which the address bits of a word are added to the contents of a specified register to determine the pointer address. { prē'in,deks·iŋ }

preionization See autoionization. { prē,ī·ə·nə'zā·shən }

preliminary amplifier See preamplifier. { pri'lim·ə,ner·ē 'am·plə,fī·ər }

preliminary waves [GEOPHYS] The body of waves of an earthquake, including both P waves and S waves. { pri'lim·ə,ner·ē 'wāvz }

preloading [ENG] For back-pressure-control gas valves, a weight or spring device to control the gas pressure at which the valve will open or close. { prē'lōd·iŋ }

prelogging [FOR] Cutting down and removing small trees before large trees are logged. { 'prē,läg·iŋ }

premaxilla [ANAT] Either of two bones of the upper jaw of vertebrates located in front of and between the maxillae. { prē'mak'sil·ə }

premessenger ribonucleic acid [MOL BIO] A giant molecule of ribonucleic acid that is transcribed from a cistron. { prē'mes·ən·jər rī·bō,nü,klē·ik 'as·əd }

premitosis [CYTOL] In certain protozoans, a kind of mitosis in which there is an intranuclear centriole and the mitotic apparatus is located within the nuclear membrane. { prē·mī'tō·səs }

premium motor oil [MATER] A lubricating oil with improved oxidation stability and having corrosion-preventive and detergent properties; used in internal combustion engines operating under severe conditions. { 'prē·mē·əm 'mōd·ər ,óil }

premix [ENG] In plastics molding, materials in which the resin, reinforcement, extenders, fillers, and so on have been premixed before molding. { 'prē,miks }

premix gas burner [ENG] Fuel (gas or oil) burner in which fuel and air are premixed prior to ignition in the combustion chamber. { 'prē,miks 'gas ,bər·nər }

premolar [ANAT] In each quadrant of the permanent human dentition, one of the two teeth between the canine and the first molar; a bicuspid. { prē'mō·lər }

premultiplication [MATH] In multiplying a matrix or operator **B** by another matrix or operator **A**, the operation that results in the matrix or operator **AB**. Also known as multiplication on the left. { prē·məl·tə·plə'kā·shən }

premutation [GEN] A heritable change, such as a trinucleotide repeat expansion, that has no phenotypic effect but greatly increases the likelihood of a further change at the altered site, resulting in a characteristic phenotype. { prē·myü'tā·shən }

prenatal [MED] Existing or occurring before birth. { prē'nād·əl }

prenova [ASTRON] A star that is destined to become a nova, but whose outburst has not yet taken place. { prē'nō·və }

preorogenic [GEOL] The initial phase of an orogenic cycle during which geosynclines form. { prē,ór·ə'jen·ik }

preparation fire [ORD] Fire delivered on a target or predetermined point preparatory to an assault on the target; may be naval, ground, or air. { prep·ə'rā·shən ,fīr }

preparatory fire [ORD] Antiaircraft fire to determine or check corrections of firing data prior to conducting fire for effect. { 'prep·rə,tór·ē ,fīr }

preparatory work [MIN ENG] Various excavations within a deposit so that actual mining can begin; includes inclines, drives between levels, crosscuts, and chutes. { 'prep·rə,tór·ē ,wərk }

preparing salt See sodium stannate. { prə'per·iŋ ,sólt }

prepattern [EMBRYO] The organization in a developing organism before a definite organizational pattern is established. { 'prē,pad·ərn }

preplastication [ENG] Premelting of injection-molding powders in a chamber separate from the injection cylinder. { prē,plas·tə'kā·shən }

preplumbed system [ELECTROMAG] Fixed nontunable waveguides or coaxial transmission lines. { prē'pləmd ,sis·təm }

prepolymer [ORG CHEM] A reactive low-molecular-weight macromolecule or an oligomer, capable of further polymerization. { prē'päl·i·mər }

prepolymer molding [ENG] A urethane-foam-producing system in which a portion of the polyol is prereacted with the isocyanate to form a liquid prepolymer with a pumpable viscosity; when combined with a second blend containing more polyol, catalyst, or blowing agent, the two components react and a foamed plastic results. { prē'päl·i·mər 'mōld·iŋ }

prepositioned [ORD] Actual placement of a weapon at the desired point of detonation and preparation for firing by some form of remote control or a timer mechanism. { prē·pə'zish·ənd }

prepreg [ENG] A reinforced-plastics term for the reinforcing material that contains or is combined with the full complement of resin before the molding operation. { 'prē,preg }

prepress proofs [GRAPHICS] Proofs which are made by photographic techniques in order to avoid the expense of press proofs. { 'prē,pres ,prüfs }

preprocessor [COMPUT SCI] A program that converts data into a format suitable for computer processing. { 'prē'prä,ses·ər }

preprogrammed robot [CONT SYS] A robot that cannot adapt itself to the task it is carrying out, and must follow a built-in program. Also known as sequence robot. { ¦prē'prō ‚gramd 'rō‚bät }

preprogramming [COMPUT SCI] The prerecording of instructions or commands for a machine, such as an automated tool in a factory. { prē'prō‚gram·iŋ }

prepsychotic [PSYCH] Of or pertaining to the mental state that precedes or is potentially capable of precipitating a psychotic disorder. { ‚prē‚sī'käd·ik }

prepuce [ANAT] **1.** The foreskin of the penis, a fold of skin covering the glans penis. **2.** A similar fold over the glans clitoridis. { 'prēp·əs }

preputial gland See Tyson's gland. { prē'pyü·shəl ‚gland }

preread head [COMPUT SCI] A read head that is placed near another read head in such a way that it can read data stored on a moving medium such as a tape or disk before these data reach the second head. { 'prē‚rēd ‚hed }

prerinse [GRAPHICS] A water bath used prior to development of photographic film. { prē'rins }

presbycusis [MED] A condition of diminished auditory acuity associated with old age. { ‚prez·bə'kyü·səs }

presbyophrenia [PSYCH] A variety of senile dementia in which apparent mental alertness is combined with disorientation of place and loss of memory. { ‚prez·bē·ə'frē·nē·ə }

presbyopia [MED] Diminished ability to focus the eye on near objects due to gradual loss of elasticity of the crystalline lens with age. { ‚prez·bē'ō·pē·ə }

prescaler [ELECTR] A scaler that extends the upper frequency limit of a counter by dividing the input frequency by a precise amount, generally 10 or 100. { 'prē‚skāl·ər }

prescutum [INV ZOO] The anterior part of the tergum of a segment of the thorax in insects. { prē'skyüd·əm }

preselection [COMPUT SCI] A technique for saving computation time in buffered computers in which a block of data is read into computer storage from the next input tape to be called upon before the data are required in the computer; the selection of the next input tape is determined by instructions to the computer. { ¦prē·si'lek·shən }

preselector [ELEC] Device in automatic switching which performs its selecting operation before seizing an idle trunk. [ELECTR] A tuned radio-frequency amplifier stage used ahead of the frequency converter in a superheterodyne receiver to increase the selectivity and sensitivity of the receiver. { ¦prē·si'lek·tər }

presence [ACOUS] The impression, as created by a recording or radio receiver, that the original program source is in the room. { 'prez·əns }

presensitized plate [GRAPHICS] In a photomechanical process, a metal or paper plate that has been precoated with a light-sensitive substance. { prē'sen·sə‚tīzd 'plāt }

present angular height [ORD] The element of data pertaining to the present position of the target; that is, the position at the instant the gun is fired. { 'prez·ənt 'aŋ·gyə·lər 'hīt }

presentation See radar display. { ‚prez·ən'tā·shən }

presentation graphics program [COMPUT SCI] An application program for creating and enhancing the visual appeal and understandability of charts and graphs, with the aid of a library or predrawn images that can be combined with other artwork. { ‚prez·ən¦tā·shən 'graf·iks ‚prō·grəm }

present elevation [ORD] Elevation corresponding to the present position of the target. { 'prez·ənt ‚el·ə'vā·shən }

present position [ORD] Position of a moving target at the instant of firing. { 'prez·ənt pə'zish·ən }

present range [ORD] The range between a gun and the present position of its target. { 'prez·ənt 'rānj }

present value [MIN ENG] The sum of money which, if expended on a mine for purchase, development, and equipment, would produce over the life of the mine a return of the original investment plus a commensurate profit. { 'prez·ənt 'val·yü }

present-worth factor See discount factor. { 'prez·ənt ¦wərth ‚fak·tər }

preservative [MATER] A chemical added to foodstuffs to prevent oxidation, fermentation, or other deterioration, usually by inhibiting the growth of bacteria. { pri'zər·vəd·iv }

preserve [ECOL] An area that is maintained for game or fish, especially for sport, and may have limited access requiring a permit for entry. { prə'zərv }

preset [COMPUT SCI] **1.** Of a variable, having a value established before the first time it is used. **2.** To initialize a value

of a variable before the value of the variable is used or tested. { 'prē‚set }

preset guidance [ENG] Guidance in which a predetermined path is set into the guidance mechanism of a craft, drone, or missile and is not altered after launching. { 'prē‚set 'gīd·əns }

preset parameter [COMPUT SCI] In computers, a parameter which is fixed for each problem at a value set by the programmer. { 'prē‚set pə'ram·əd·ər }

preset tool [MECH ENG] A machine tool that is used to set an initial value of a parameter controlling another device. { 'prē‚set 'tül }

preshrunk [TEXT] Pertaining to a fabric that has been shrunk during manufacture in order to minimize later shrinkage. { prē'shrəŋk }

presintering [MET] Heating a compact to a temperature lower than the final sintering temperature to facilitate handling or to remove a binder or lubricant. { prē'sint·ə·riŋ }

presoma [INV ZOO] The anterior portion of an invertebrate that lacks a definitive head structure. { prē'sō·mə }

presort [COMPUT SCI] **1.** The first part of a sort program in which data items are arranged into strings that are equal to or greater than some prescribed length. **2.** The sorting of data on off-line equipment before it is processed by a computer. { prē'sȯrt }

presque isle [GEOGR] A promontory or peninsula extending into a lake, nearly or almost forming an island; its head or end section is connected with the shore by a sag or low gap only slightly above water level or by a strip of lake bottom exposed as a land surface by a drop in lake level. { ‚pres'kīl }

press [MECH ENG] Any of various machines by which pressure is applied to a workpiece, by which a material is cut or shaped under pressure, by which a substance is compressed, or by which liquid is expressed. { pres }

press bonding [ENG] A method of bonding structures or materials through the application of pressure by a platen press or other tool. { 'pres ‚bänd·iŋ }

press camera [OPTICS] A folding camera, usually of 4- by 5-inch (10.2- by 12.7-centimeter) format, once widely used in newspaper photography. { 'pres ‚kam·rə }

press drip [MATER] Oil that drips from the wax press after pressed petroleum distillate has been removed. { 'pres ‚drip }

pressed brick [MATER] Brick subjected to pressure before burning to eliminate imperfections of shape and texture. { 'prest 'brik }

pressed density [MET] The density of a metal powder compact before sintering. { 'prest 'den·səd·ē }

pressed-disk technique See potassium bromide-disk technique. { 'prest ¦disk tek‚nēk }

pressed distillate [MATER] Oil recovered when refinery paraffin distillate is pressed to separate the liquid from the solid wax. { 'prest 'dis·tə·lət }

pressed glass [MATER] Glass shaped by being poured into a mold under pressure or pressed into a mold in a plastic state. { 'prest 'glas }

pressed loading [ENG] A loading operation in which bulk material, such as an explosive in granular form, is reduced in volume by the application of pressure. { 'prest 'lōd·iŋ }

press fit [ENG] An interference or force fit assembled through the use of a press. Also known as force fit. { 'pres ‚fit }

press forging [MET] Forging hot metal between dies in a press. { 'pres ‚fȯrj·iŋ }

press forming [MET] A metal-forming operation performed with a mechanical or hydraulic press. { 'pres ‚fȯrm·iŋ }

pressing [ENG ACOUS] A phonograph record produced in a record-molding press from a master or stamper. [MET] **1.** Shallow-drawing metal sheet or plate. **2.** Using compressive force to form a metal powder compact. { 'pres·iŋ }

pressolution See pressure solution. { 'pres·ə‚lü·shən }

pressolved [GEOL] Referring to a sedimentary bed or rock in which the grains have undergone pressure solution. { pri'zälvd }

press polish [ENG] High-sheen finish on plastic sheet stock produced by contact with a smooth metal under heat and pressure. { 'pres ‚päl·ish }

press proof [GRAPHICS] Proof removed from the printing press to inspect line and color values and overall quality; this is the last proof before the complete printing is done. { 'pres ‚prüf }

press slide [MECH ENG] The reciprocating member of a power press on which the punch and upper die are fastened. { 'pres ‚slīd }

press teletype network [COMMUN] A large teletypewriter network employed by a press association or other news distributing organization, usually employing modern carrier telegraph circuits operating over both wire and radio facilities, and transmitting to as many as 2000 stations simultaneously. { 'pres 'tel·ə‚tīp ‚net‚wərk }

press-to-talk switch [ELECTR] A switch mounted directly on a microphone to provide a convenient means for switching two-way radiotelephone equipment or electronic dictating equipment to the talk position. { 'pres tə 'tȯk ‚swich }

pressure [MECH] A type of stress which is exerted uniformly in all directions; its measure is the force exerted per unit area. { 'presh·ər }

pressure altimeter [ENG] A highly refined aneroid barometer that precisely measures the pressure of the air at the altitude an aircraft is flying, and converts the pressure measurement to an indication of height above sea level according to a standard pressure-altitude relationship. Also known as barometric altimeter. { 'presh·ər al'tim·əd·ər }

pressure altitude [METEOROL] The height above sea level at which the existing atmospheric pressure would be duplicated in the standard atmosphere; atmospheric pressure expressed as height according to a standard scale. { 'presh·ər ‚al·tə‚tüd }

pressure-altitude variation [METEOROL] The pressure difference, in feet or meters, between mean sea level and the standard datum plane. { 'presh·ər ‚al·tə‚tüd ‚ver·ē'ā·shən }

pressure angle [MECH ENG] The angle that the line of force makes with a line at right angles to the center line of two gears at the pitch points. { 'presh·ər ‚aŋ·gəl }

pressure bag [ENG] A bag made of rubber, plastic, or other impermeable material that provides a flexible barrier between the pressure medium and the part being bonded. { 'presh·ər ‚bag }

pressure bar [MECH ENG] A bar that holds the edge of a metal sheet during press operations, such as punching, stamping, or forming, and prevents the sheet from buckling or becoming crimped. { 'presh·ər ‚bär }

pressure-base factor [CHEM ENG] Factor used in orifice pressure-drop calculations to allow for conditions where the pressure base used for calculating the orifice factor is not 14.73 pounds per square inch absolute (101.56 megapascals); calculated as $F_{pb} = 14.73$/pressure base (absolute). { 'presh·ər ‚bās ‚fak·tər }

pressure block [MIN ENG] The pressure on pillars, walls, and other supports in a mine caused by removal of surrounding formations from masses of rocks or by natural geological formations. { 'presh·ər ‚bläk }

pressure bomb [PETRO ENG] Pipe-and-valve device used to capture downhole pressurized gas samples from oil wells; used to measure downhole pressure. { 'presh·ər ‚bäm }

pressure breccia See tectonic breccia. { 'presh·ər ‚brech·ə }

pressure broadening [SPECT] A spreading of spectral lines when pressure is increased, due to an increase in collision broadening. { 'presh·ər ‚brȯd·ən·iŋ }

pressure bulb [CIV ENG] The zone in a loaded soil mass bounded by an arbitrarily selected isobar of stress. { 'presh·ər ‚bəlb }

pressure bump [MIN ENG] Sudden failure of a coal pillar overloaded by the weight of the rock above it. { 'presh·ər ‚bəmp }

pressure burst [MIN ENG] A rockburst produced under stresses exceeding the elastic strength of the rock. { 'presh·ər ‚bərst }

pressure cable [ELEC] A cable in which a fluid such as oil or gas, at greater than atmospheric pressure, surrounds the conductors and insulation and keeps their temperature down. { 'presh·ər ‚kā·bəl }

pressure carburetor See injection carburetor. { 'presh·ər ‚kär·bə'rād·ər }

pressure casting [MET] Making castings of molten or plastic metal in metal molds under applied pressures. { 'presh·ər ‚kast·iŋ }

pressure center [METEOROL] **1.** On a synoptic chart (or on a mean chart of atmospheric pressure), a point of local minimum or maximum pressure; the center of a low or high. **2.**

A center of cyclonic or anticyclonic circulation. { 'presh·ər ‚sen·tər }

pressure chamber [ENG] A chamber in which an artificial environment is established at low or high pressures to test equipment under simulated conditions of operation. [MIN ENG] An enclosed space that seals off a part of a mine and in which the air pressure can be raised or lowered. { 'presh·ər ‚chām·bər }

pressure-change chart [METEOROL] A chart indicating the change in atmospheric pressure of a constant-height surface over some specified interval of time. Also known as pressure-tendency chart. { 'presh·ər ‚chānj ‚chärt }

pressure coefficient [THERMO] The ratio of the fractional change in pressure to the change in temperature under specified conditions, usually constant volume. { 'presh·ər ‚kō·i‚fish·ənt }

pressure-containing member [MECH ENG] The part of a pressure-relieving device which is in direct contact with the pressurized medium in the vessel being protected. { 'presh·ər kən‚tān·iŋ ‚mem·bər }

pressure contour [METEOROL] A line connecting points of equal height of a given barometric pressure; the intersection of a constant pressure surface by a plane parallel to mean sea level. { 'presh·ər ‚kän‚tùr }

pressure control [ENG] Any device or system able to maintain, raise, or lower the pressure in a vessel or processing system as desired. { 'presh·ər kən‚trōl }

pressure cooker [ENG] An autoclave designed for high-temperature cooking. { 'presh·ər ‚kùk·ər }

pressure decline [PETRO ENG] The loss or decline in reservoir pressure resulting from pressure drawdown during the production of gas or oil. Also known as pressure depletion. { 'presh·ər di‚klīn }

pressure deflection [ENG] In a Bourdon or bellows-type pressure gage, the deflection or movement of the primary sensing element when pressured by the fluid being measured. { 'presh·ər di‚flek·shən }

pressure depletion See pressure decline. { 'presh·ər di‚plē·shən }

pressure depth [OCEANOGR] The depth at which an ocean sample was taken, as inferred from the difference in readings on protected and unprotected thermometers on the sampler; the higher reading is on the unprotected thermometer due to the effect of pressure on the mercury column at the sampling depth. { 'presh·ər ‚depth }

pressure distillate [MATER] Light, gasoline-bearing distillate product from petroleum-refinery pressure stills; the product is cracked, as contrasted to virgin or straight-run stock; includes pressure naphtha. { 'presh·ər ‚dis·tə·lət }

pressure distribution [PETRO ENG] The relative pressures (pressure gradients) between various portions of a producing reservoir zone; lowest pressures are nearest the producing wellbores. { 'presh·ər ‚dis·trə‚byü·shən }

pressure drag See pressure resistance. { 'presh·ər ‚drag }

pressure drawdown [PETRO ENG] The drop in reservoir pressure related to the withdrawal of gas from a producing well; for low-permeability formations, pressures near the wellbores can be much lower than in the main part of the reservoir; leads to pressure decline in the reservoir and ultimate pressure depletion. { 'presh·ər ‚drȯ‚daùn }

pressure drop [FL MECH] The difference in pressure between two points in a flow system, usually caused by frictional resistance to a fluid flowing through a conduit, filter media, or other flow-conducting system. { 'presh·ər ‚dräp }

pressure-drop manometer [ENG] Manometer device (liquid-filled U tube) open at both ends, each end connected by tubing to a different location in a flow system (such as fluid- or gas-carrying pipe) to measure the drop in system pressure between the two points. { 'presh·ər ‚dräp ma'näm·əd·ər }

pressure dye test [ENG] A leak detection method in which a pressure vessel is filled with liquid dye and is pressurized under water to make possible leakage paths visible. { 'presh·ər 'dī ‚test }

pressure effect [SPECT] The effect of changes in pressure on spectral lines in the radiation emitted or absorbed by a substance; namely, pressure broadening and pressure shift. { 'presh·ər i‚fekt }

pressure elements [ENG] Those portions of a pressure-measurement gage which are moved or temporarily deformed by the gas or liquid of the system to which the gage is connected; the amount of movement or deformation is proportional to the pressure and is indicated by the position of a pointer or movable needle. { 'presh·ər ˌel·ə·məns }

pressure-enthalpy chart [PHYS] A graph of the pressure versus the enthalpy of a substance at various values of temperature, specific volume, and entropy; especially useful in refrigeration calculations. Also known as enthalpy-pressure chart. { 'presh·ər 'en͵thal·pē ˌchärt }

pressure-fall center [METEOROL] A point of maximum decrease in atmospheric pressure over a specified interval of time; on synoptic charts, a point of greatest negative pressure tendency. Also known as center of falls; isallobaric low; isallobaric minimum; katallobaric center. { 'presh·ər ǀfȯl ͵sen·tər }

pressure fan [MIN ENG] A fan that forces fresh air into a mine as distinguished from one that exhausts air from the mine. { 'presh·ər ͵fan }

pressure field [OCEANOGR] A representation of a pressure gradient as isobar contours, parallel to which ocean currents flow. { 'presh·ər ͵fēld }

pressure force [FL MECH] The force due to differences of pressure within a fluid mass; the (vector) force per unit volume is equal to the pressure gradient $-\nabla p$, and the force per unit mass (specific force) is equal to the product of the volume force and the specific volume $-\alpha\nabla p$. [METEOROL] *See* pressure-gradient force. { 'presh·ər ͵fȯrs }

pressure forming [ENG] A plastics thermoforming process using pressure to push the plastic sheet to be formed against the mold surface, as opposed to using vacuum to suck the sheet flat against the mold. { 'presh·ər ǀfȯrm·iŋ }

pressure fringe *See* pressure shadow. { 'presh·ər ͵frinj }

pressure front *See* shock front. { 'presh·ər ͵frənt }

pressure gage [ENG] An instrument having metallic sensing element (as in a Bourdon pressure gage or aneroid barometer) or a piezoelectric crystal (as in a quartz pressure gage) to measure pressure. { 'presh·ər gāj }

pressure gradient [FL MECH] The rate of decrease (that is, the gradient) of pressure in space at a fixed time; sometimes loosely used to denote simply the magnitude of the gradient of the pressure field. Also known as barometric gradient. [METEOROL] The change in atmospheric pressure per unit horizontal distance, usually measured along a line perpendicular to the isobars. { 'presh·ər ͵grād·ē·ənt }

pressure-gradient force [METEOROL] The force due to differences of pressure within the atmosphere; it usually refers only to the horizontal component of the force. Also known as pressure force. { 'presh·ər ǀgrād·ē·ənt ͵fȯrs }

pressure head [FL MECH] Also known as head. **1.** The height of a column of fluid necessary to develop a specific pressure. **2.** The pressure of water at a given point in a pipe arising from the pressure in it. { 'presh·ər ͵hed }

pressure hydrophone [ENG ACOUS] A pressure microphone that responds to waterborne sound waves. { 'presh·ər 'hī·drə͵fōn }

pressure ice [OCEANOGR] Ice, especially sea ice, which has been deformed or altered by the lateral stresses of any combination of wind, water currents, tides, waves, and surf; may include ice pressed against the shore, or one piece of ice upon another. { 'presh·ər ͵īs }

pressure ice foot [HYD] An ice foot formed along a shore by the freezing together of stranded pressure ice. { 'presh·ər ǀīs ͵fut }

pressure interface [PETRO ENG] The interrelation of several individual reservoir pressures whose productions are supported by water influx from a common aquifer; pressure depletion from withdrawal of oil from one reservoir will affect the position of the common aquifer and thus affect the pressures and gas-oil or water-oil contacts in the other reservoirs. { 'presh·ər 'in·tər͵fās }

pressure ionization [ASTROPHYS] A condition found in white dwarfs and other degenerate matter in which electron orbits overlap to the point that electrons in higher quantum levels are no longer associated with any particular nucleus and must be regarded as free. { 'presh·ər ͵ī·ə·nə′zā·shən }

pressure jump [METEOROL] A steady-state propagation of a sudden finite change of inversion height, in analogy to the shock wave in a compressible fluid or to a hydraulic jump; the prefrontal squall line has been interpreted as a pressure jump, with the cold front providing the initial pistonlike impetus. { 'presh·ər ͵jəmp }

pressure-jump line [METEOROL] A fast-moving line of sudden rise in atmospheric pressure, followed by a higher pressure level than that which preceded the jump; under suitable moisture conditions, sudden instability of the atmosphere conducive to the formation of thunderstorms can result. { 'presh·ər ǀjəmp ͵līn }

pressure line of position [NAV] A line of position parallel to the effective air path and at a distance perpendicular to it equal to the lateral displacement, determined by pressure pattern formulas. { 'presh·ər 'līn əv pə′zish·ən }

pressure maintenance [PETRO ENG] The maintenance of gas pressure in a reservoir by an active water drive, water injection, gas injection, or a combination of the foregoing. { 'presh·ər ′mānt·ən·əns }

pressure mark [GRAPHICS] A defect found in processed film which may appear as reduced or increased density. { 'presh·ər ͵märk }

pressure measurement [ENG] Measurement of the internal forces of a process vessel, tank, or piping caused by pressurized gas or liquid; can be for a static or dynamic pressure, in English or metric units, either absolute (total) or gage (absolute minus atmospheric) pressure. { 'presh·ər ͵mezh·ər·mənt }

pressure melting [PHYS] The melting of ice due to applied pressure. { 'presh·ər ǀmelt·iŋ }

pressure melting temperature [PHYS] The temperature at which ice can melt at a given pressure. { 'presh·ər ǀmelt·iŋ ͵tem·prə·chər }

pressure microphone [ENG ACOUS] A microphone whose output varies with the instantaneous pressure produced by a sound wave acting on a diaphragm; examples are capacitor, carbon, crystal, and dynamic microphones. { 'presh·ər 'mī·krə͵fōn }

pressure naphtha [MATER] Petroleum naphtha made by cracking, as contrasted to virgin or straight-run naphtha; a special grade of pressure distillate. { 'presh·ər 'naf·thə }

pressure pad [ENG] A steel reinforcement in the face of a plastics mold to help the land absorb the closing pressure. [ENG ACOUS] A felt pad mounted on a spring arm, used to hold magnetic tape in close contact with the head on some tape recorders. { 'presh·ər ͵pad }

pressure pan [GEOL] An induced soil pan which has a higher bulk density and a lower total porosity than the soil directly above or below it and is produced as a result of pressure applied by normal tillage operations or by other artificial means. { 'presh·ər ͵pan }

pressure pattern [METEOROL] The general geometric characteristics of atmospheric pressure distribution as revealed by isobars on a constant-height chart; usually applied to cyclonic-scale features of a surface chart. { 'presh·ər ͵pad·ərn }

pressure pattern displacement *See* lateral displacement. { 'presh·ər ͵pad·ərn dis͵plās·mənt }

pressure pattern flying *See* aerologation. { 'presh·ər ǀpad·ərn ͵flī·iŋ }

pressure penitente [GEOL] A nieve penitente composed of brilliantly white ice which is shaped into a slender ridge by lateral pressure of converging morainal streams and by melting of the adjacent debris-covered ice. { 'presh·ər ͵pen·ə′ten·tā }

pressure pickup [ELECTR] A device that converts changes in the pressure of a gas or liquid into corresponding changes in some more readily measurable quantity such as inductance or resistance. { 'presh·ər ǀpik͵əp }

pressure pillow [ENG] A mechanical-hydraulic snow gage consisting of a circular rubber or metal pillow filled with a solution of antifreeze and water, and containing either a pressure transducer or a riser pipe to record increase in pressure of the snow. { 'presh·ər ͵pil·ō }

pressure plate [MECH ENG] The part of an automobile disk clutch that presses against the flywheel. { 'presh·ər ͵plāt }

pressure-plate anemometer [ENG] An anemometer which measures wind speed in terms of the drag which the wind exerts on a solid body; may be classified according to the means by which the wind drag is measured. Also known as plate anemometer. { 'presh·ər ǀplāt ͵an·ə′mäm·əd·ər }

pressure plateau [GEOL] An uplifted area of a thick lava flow, measuring up to 10 or 13 feet (3 or 4 meters), the uplift of which is due to the intrusion of new lava from below that does not reach the surface. { 'presh·ər pla͵tō }

pressure point [PHYSIO] A point of marked sensibility to pressure or weight, arranged like the temperature spots, and showing a specific end apparatus arranged in a punctate manner and connected with the pressure sense. { 'presh·ər ‚pȯint }

pressure process [CHEM ENG] Treatment of timber to prevent decay by forcing a preservative such as creosote and zinc chloride into the cells of the wood. { 'presh·ər ‚prä·səs }

pressure radius [PETRO ENG] The effective radius of increased reservoir pressure surrounding a water-injection well. { 'presh·ər ‚rād·ē·əs }

pressure rating [ENG] The operating (allowable) internal pressure of a vessel, tank, or piping used to hold or transport liquids or gases. { 'presh·ər ‚rād·iŋ }

pressure-regulating valve [ENG] A valve that releases or holds process-system pressure (that is, opens or closes) either by preset spring tension or by actuation by a valve controller to assume any desired position between full open and full closed. { 'presh·ər ¦reg·yə‚läd·iŋ ‚valv }

pressure regulator [ENG] Open-close device used on the vent of a closed, gas-pressured system to maintain the system pressure within a specified range. { 'presh·ər ‚reg·yə‚läd·ər }

pressure release [GEOPHYS] The outward-expanding force of pressure which is released within rock masses by unloading, as by erosion of superincumbent rocks or by removal of glacial ice. { 'presh·ər ri‚lēs }

pressure-release jointing [GEOL] Exfoliation that occurs in once deeply buried rock that erosion has brought nearer the surface, thus releasing its confining pressure. { 'presh·ər ri¦lēs ‚jȯint·iŋ }

pressure relief [ENG] A valve or other mechanical device (such as a rupture disk) that eliminates system overpressure by allowing the controlled or emergency escape of liquid or gas from a pressured system. { 'presh·ər ri‚lēf }

pressure-relief device [MECH ENG] **1.** In pressure vessels, a device designed to open in a controlled manner to prevent the internal pressure of a component or system from increasing beyond a specified value, that is, a safety valve. **2.** A spring-loaded machine part which will yield, or deflect, when a predetermined force is exceeded. { 'presh·ər ri‚lēf di‚vīs }

pressure-relief valve [MECH ENG] A valve which relieves pressure beyond a specified limit and recloses upon return to normal operating conditions. { 'presh·ər ri‚lēf ‚valv }

pressure resistance [FL MECH] In fluid dynamics, a normal stress caused by acceleration of the fluid, which results in a decrease in pressure from the upstream to the downstream side of an object acting perpendicular to the boundary. Also known as pressure drag. { 'presh·ər ri‚zis·təns }

pressure-retaining member [MECH ENG] That part of a pressure-relieving device loaded by the restrained pressurized fluid. { 'presh·ər ri¦tān·iŋ ‚mem·bər }

pressure ridge [GEOL] **1.** A seismic feature resulting from transverse pressure and shortening of the land surface. **2.** An elongate upward movement of the congealing crust of a lava flow. **3.** A ridge of glacier ice. [OCEANOGR] A ridge or wall of hummocks where one ice floe has been pressed against another. { 'presh·ər ‚rij }

pressure ring [MIN ENG] A ring about a large excavated area, evidenced by distortion of the openings near the main excavation. { 'presh·ər ‚riŋ }

pressure-rise center [METEOROL] A point of maximum increase in atmospheric pressure over a specified interval of time; on synoptic charts, a point of maximum positive pressure tendency. Also known as anallobaric center; center of rises; isallobaric high; isallobaric maximum. { 'presh·ər ¦rīz ‚sen·tər }

pressure roll [ENG] In plastics-extrusion coating, the roll that with the chill roll applies pressure to the substrate and the molten extruded web. { 'presh·ər ‚rōl }

pressure seal [ENG] A seal used to make pressure-proof the interface (contacting surfaces) between two parts that have frequent or continual relative rotational or translational motion. { 'presh·ər ‚sēl }

pressure-sensitive adhesive [MATER] An adhesive that develops maximum bonding power when applied by a light pressure only. { 'presh·ər ¦sen·səd·iv ad'hē·siv }

pressure-sensitive paint [FL MECH] A flow visualization technique in which ultraviolet light is used to excite specific molecules in a special paint affixed to a test surface positioned in a wind tunnel flow. The resulting phosphorescence of these molecules indicates the amount of oxygen in contact with the paint and, thereby, the spatial distribution of surface pressure. { ‚presh·ər ‚sen·səd·iv 'pānt }

pressure shadow [PETR] In structural petrology, an area adjoining a porphyroblast, characterized by a growth fabric rather than a deformation fabric, as seen in a section perpendicular to the *b* axis of the fabric. Also known as pressure fringe; strain shadow. { 'presh·ər ‚shad·ō }

pressure shift [SPECT] An increase in the wavelength at which a spectral line has maximum intensity, which takes place when pressure is increased. { 'presh·ər ‚shift }

pressure solution [PETR] In a sedimentary rock, solution occurring preferentially at the grain boundary surfaces. Also known as pressolution. { 'presh·ər sə‚lü·shən }

pressure-stabilized [AERO ENG] Referring to membrane-type structures that require internal pressure for maintenance of a stable structure. { 'presh·ər 'stā·bə‚līzd }

pressure still [CHEM ENG] A continuous-flow, petroleum-refinery still in which heated oil (liquid and vapor) is kept under pressure so that it will crack (decompose into smaller molecules) to produce lower-boiling products (pressure distillate or pressure naphtha). { 'presh·ər ‚stil }

pressure storage [ENG] The storage of a volatile liquid or liquefied gas under pressure to prevent evaporation. { 'presh·ər ‚stȯr·ij }

pressure suit [AERO ENG] A garment designed to provide pressure upon the body so that respiratory and circulatory functions may continue normally, or nearly so, under low-pressure conditions such as occur at high altitudes or in space without benefit of a pressurized cabin. { 'presh·ər ‚süt }

pressure suppression *See* vapor suppression. { 'presh·ər sə‚presh·ən }

pressure surface *See* potentiometric surface. { 'presh·ər ‚sər·fəs }

pressure-surface map *See* potentiometric map. { 'presh·ər ¦sər·fəs ‚map }

pressure survey [MIN ENG] A study to determine the pressure distribution or pressure losses along consecutive lengths or sections of a ventilation circuit. [PETRO ENG] The measurement of static bottomhole pressures in an oil field with producing wells shut in for a time interval sufficient for reservoir pressure buildup to stabilize. { 'presh·ər ‚sər‚vā }

pressure switch [ELEC] A switch that is actuated by a change in pressure of a gas or liquid. { 'presh·ər ‚swich }

pressure system [ENG] Any system of pipes, vessels, tanks, reactors, and other equipment, or interconnections thereof, operating with an internal pressure greater than atmospheric. [METEOROL] An individual cyclonic-scale feature of atmospheric circulation, commonly used to denote either a high or a low, less frequently a ridge or a trough. { 'presh·ər ‚sis·təm }

pressure tank [CHEM ENG] A pressurized tank into which timber is inserted for impregnation with preservative. [CIV ENG] An airtight water tank in which air is compressed to exert pressure on the water and which is used in connection with a water distribution system. { 'presh·ər ‚taŋk }

pressure tap [ENG] A small perpendicular hole in the wall of a pressurized, fluid-containing pipe or vessel; used for connection of pressure-sensitive elements for the measurement of static pressures. Also known as piezometer opening; static pressure tap. { 'presh·ər ‚tap }

pressure tendency [METEOROL] The character and amount of atmospheric pressure change for a 3-hour or other specified period ending at the time of observation. Also known as barometric tendency. { 'presh·ər ‚ten·dən·sē }

pressure-tendency chart *See* pressure-change charty. { 'presh·ər ¦ten·dən·sē ‚chärt }

pressure tensor [PL PHYS] A tensor which plays a role in magnetohydrodynamics analogous to that of the pressure in ordinary fluid mechanics. { 'presh·ər ‚ten·sər }

pressure thrust [AERO ENG] In rocketry, the product of the cross-sectional area of the exhaust jet leaving the nozzle exit and the difference between the exhaust pressure and the ambient pressure. { 'presh·ər ‚thrəst }

pressure topography *See* height pattern. { 'presh·ər tə‚päg·rə·fē }

pressure transducer [ENG] An instrument component that detects a fluid pressure and produces an electrical signal related to the pressure. Also known as electrical pressure transducer. { 'presh·ər tranz‚dü·sər }

pressure-travel curve [MECH] Curve showing pressure plotted against the travel of the projectile within the bore of the weapon. { 'presh·ər ¦trav·əl ‚kərv }

pressure traverse [PETRO ENG] Measurement of reservoir pressures at progressive depths. { 'presh·ər trə‚vərs }

pressure treater [CHEM ENG] Any chemical treating device operated at higher-than-atmospheric pressure, as in the chemical and petroleum industries. { 'presh·ər ‚trēd·ər }

pressure tube [HYD] A deep, slender, cylindrical hole formed in a glacier by the sinking of an isolated stone that has absorbed more solar radiation than the surrounding ice. { 'presh·ər ‚tüb }

pressure-tube anemometer [ENG] An anemometer which derives wind speed from measurements of the dynamic wind pressures; wind blowing into a tube develops a pressure greater than the static pressure, while wind blowing across a tube develops a pressure less than the static; this pressure difference, which is proportional to the square of the wind speed, is measured by a suitable manometer. { 'presh·ər ¦tüb ‚an·ə'mäm·əd·ər }

pressure-tube reactor [NUCLEO] A nuclear reactor in which the fuel elements are located inside numerous tubes containing coolant circulating at high pressure; the tube assembly is surrounded by a tank containing the moderator at low pressure. { 'presh·ər ¦tüb rē'ak·tər }

pressure tunnel [CIV ENG] A waterway tunnel under pressure because the hydraulic gradient lies above the tunnel crown. { 'presh·ər ‚tən·əl }

pressure ulcer See decubitus ulcer. { 'presh·ər ‚əl·sər }

pressure vector [IND ENG] A stress on the human body produced at the interface between the operator and the equipment during the use of hand tools or other equipment, and described in terms of direction and magnitude. { 'presh·ər ‚vek·tər }

pressure vessel [ENG] A metal container, generally cylindrical or spheroid, capable of withstanding bursting pressures. { 'presh·ər ‚ves·əl }

pressure viscosity [FL MECH] Property of petroleum lubricating oils to increase in viscosity when subjected to pressure. { 'presh·ər vi‚skäs·əd·ē }

pressure wave [METEOROL] A wave or periodicity which exists in the variation of atmospheric pressure on any time scale, usually excluding normal diurnal or seasonal trends. [PHYS] See compressional wave. { 'presh·ər ‚wāv }

pressure welding [MET] Welding of metal surfaces by the application of pressure; examples are percussion welding, resistance welding, seam welding, and spot welding. { 'presh·ər ‚weld·iŋ }

pressurization [ENG] **1.** Use of an inert gas or dry air, at several pounds above atmospheric pressure, inside the components of a radar system or in a sealed coaxial line, to prevent corrosion by keeping out moisture, and to minimize high-voltage breakdown at high altitudes. **2.** The act of maintaining normal atmospheric pressure in a chamber subjected to high or low external pressure. { ‚presh·ə·rə'zā·shən }

pressurize [ENG] To maintain normal atmospheric pressure in a chamber subjected to high or low external pressures. { 'presh·ə‚rīz }

pressurized blast furnace [ENG] A blast furnace operated under pressure above the ambient; pressure is obtained by throttling the off-gas line, which permits a greater volume of air to be passed through the furnace at a lower velocity, and results in increase in smelting rate. { 'presh·ə‚rīzd 'blast ‚fər·nəs }

pressurized cabin [AERO ENG] The occupied portion of an aircraft in which the air pressure has been raised above that of the ambient atmosphere by the compression of the atmosphere into this space. { 'presh·ə‚rīzd 'kab·ən }

pressurized stoppings [MIN ENG] Stoppings which are erected in the intake and return roadways of a district to isolate an open fire or spontaneous heating and in which the pressures on both sides of each stopping are made equal by the use of auxiliary fans. { 'presh·ə‚rīzd 'stäp·iŋz }

pressurized water reactor [NUCLEO] A nuclear reactor in which water is circulated under enough pressure to prevent it from boiling, while serving as moderator and coolant for the uranium fuel; the heated water is then used to produce steam for a power plant. Abbreviated PWR. { 'presh·ə‚rīzd 'wod·ər rē‚ak·tər }

presswork [ENG] The entire range of bending and drawing operations in the cold forming of sheet metal products. [GRAPHICS] In printing, the actual operation of putting ink on paper; this activity is preceded by composition and perhaps platemaking, and is followed by binding. { 'pres‚wərk }

prestage [AERO ENG] A phase in the process of igniting a large liquid-fuel rocket in which the initial partial flow of propellants into the thrust chamber is ignited, and the combustion is satisfactorily established before the main stage is ignited. { 'prē‚stāj }

prester [METEOROL] A whirlwind or waterspout accompanied by lightning in the Mediterranean Sea and Greece. { 'pres·tər }

prestore [COMPUT SCI] To store a quantity in an available computer location before it is required in a routine. { ¦prē'stor }

prestress [ENG] To apply a force to a structure to condition it to withstand its working load more effectively or with less deflection. [GEOL] See preconsolidation pressure. { ¦prē'stres }

prestressed concrete [MATER] Concrete compressed with heavily loaded wires or bars to reduce or eliminate cracking and tensile forces. { ¦prē'strest 'kän‚krēt }

presumptive address See address constant. { pri'zəm·tiv ə'dres }

presumptive instruction See basic instruction. { pri'zəm·tiv in'strək·shən }

presuppression [GEOPHYS] In seismic prospecting, the suppression of the early events on a seismic record for control of noise and reflections on that portion of the record. { ¦prē·sə'presh·ən }

pretensioning [ENG] Process of precasting concrete beams with tensioned wires embedded in them. Also known as Hoyer method of prestressing. { prē'ten·shən·iŋ }

pretersonics See acoustoelectronics. { ¦prēd·ər'sän·iks }

pretransfer ribonucleic acid [MOL BIO] The primary transcription product of transfer ribonucleic acid-encoding genes. { ¦prē'tranz·fər ‚rī·bō·nü‚klē·ik 'as·əd }

pre-transmit-receive tube See pre-TR tube. { ¦prē 'tranz‚mit ri'sēv ‚tüb }

pretravel [CONT SYS] The distance or angle through which the actuator of a switch moves from the free position to the operating position. { 'prē‚trav·əl }

pretrigger [ELECTR] Trigger used to initiate sweep ahead of transmitted pulse. { ¦prē'trig·ər }

pre-TR tube [ELECTR] Gas-filled radio-frequency switching tube used in some radar systems to protect the transmit-receive tube from excessively high power and the receiver from frequencies other than the fundamental. Derived from pre-transmit-receive tube. { ¦prē'tē·¦är ‚tüb }

prevailing current [OCEANOGR] The ocean current most frequently observed during a given period, such as a month, a season, or a year. { pri'vāl·iŋ ‚kə·rənt }

prevailing visibility [METEOROL] In United States weather observing practice, the greatest horizontal visibility equaled or surpassed throughout half of the horizon circle; in the case of rapidly varying conditions, it is the average of the prevailing visibility while the observation is being taken. { pri'vāl·iŋ ‚viz·ə'bil·əd·ē }

prevailing westerlies [METEOROL] The prevailing westerly winds on the poleward sides of the subtropical high-pressure belts. { pri'vāl·iŋ 'wes·tər‚lēz }

prevailing wind See prevailing wind direction. { pri'vāl·iŋ 'wind }

prevailing wind direction [METEOROL] The wind direction most frequently observed during a given period; the periods most often used are the observational day, month, season, and year. Also known as prevailing wind. { pri'vāl·iŋ 'wind di‚rek·shən }

prevalence [GEN] The frequency with which a medical condition is found in a specific population at a specific time. { 'prev·ə·ləns }

preventive maintenance [ENG] A procedure of inspecting, testing, and reconditioning a system at regular intervals according to specific instructions, intended to prevent failures in service or to retard deterioration. { pri'ven·tiv 'mānt·ən·əns }

previewing [COMPUT SCI] In character recognition, a process of attempting to gain prior information about the characters that appear on an incoming source document; this information,

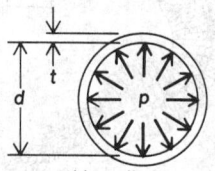

PRESSURE VESSEL

thin-walled vessel
$t < d/10$

thick-walled vessel
$t > d/10$

Pressure vessels for moderate and for high pressures; t = wall thickness; d = vessel diameter; p = pressure; d_i, d_o = inside and outside diameters.

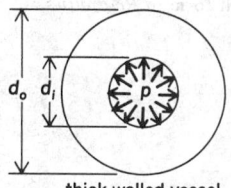

PRESSURIZED BLAST FURNACE

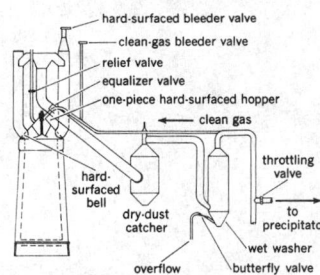

hard-surfaced bleeder valve
clean-gas bleeder valve
relief valve
equalizer valve
one-piece hard-surfaced hopper
clean gas
hard-surfaced bell
throttling valve
dry-dust catcher
to precipitator
wet washer
overflow
butterfly valve

Flow diagram of a pressurized blast furnace.

which may include the range of ink density, relative positions, and so forth, is used as an aid in the normalization phase of character recognition. { 'prē,vyü·iŋ }

previous element coding [COMMUN] System of signal coding, used for digital television transmission, whereby each transmitted picture element is dependent upon the similarity of the preceding picture element. { 'prē·vē·əs 'el·ə·mənt }

previtrain [GEOL] The woody lenses in lignite that are equivalent to vitrain in coal of higher rank. { prē'vi,trān }

Prevost's theory [THERMO] A theory according to which a body is constantly exchanging heat with its surroundings, radiating an amount of energy which is independent of its surroundings, and increasing or decreasing its temperature depending on whether it absorbs more radiation than it emits, or vice versa. { prā'vōz ,thē·ə·rē }

preweld interval [MET] In resistance spot welding, elapsed time between the end of squeeze time and the beginning of welding current. { prē'weld 'in·tər·vəl }

prewhitening filter See whitening filter. { prē'wīt·ən·iŋ ,fil·tər }

PRF See pulse repetition rate.

pri See primary winding. { prī }

Priabonian [GEOL] A European stage of geologic time in the upper Eocene, believed to consist of Auversian and Bartonian. { ,prē·ə'bō·nē·ən }

priapism [MED] Persistent erection of the penis, usually unaccompanied by sexual desire, as seen in certain pathologic conditions. { 'prī·ə,piz·əm }

Priapulida [INV ZOO] A minor phylum of wormlike marine animals; the body is made up of three distinct portions (proboscis, trunk, and caudal appendage) and is often covered with spines and tubercles, and the mouth is surrounded by concentric rows of teeth. { ,prī·ə'pyül·əd·ə }

Priapuloidea [INV ZOO] An equivalent name for Priapulida. { prī,ap·ə'lóid·ē·ə }

Pribnow box [MOL BIO] In prokaryotes, a highly conserved sequence element located upstream from the transcriptional start site to which binds the sigma subunit of the ribonucleic acid polymerase. { 'prib,nō ,bäks }

price index [STAT] A statistic used primarily in economics to indicate an average level of prices in a time series; combines several series of price data into one index. { 'prīs ,in,deks }

priceite [MINERAL] $Ca_4B_{10}O_{19} \cdot 7H_2O$ A snow-white earthy mineral composed of hydrous calcium borate, occurring as a massive. Also known as pandermite. { 'prī,sīt }

Price meter [ENG] The ocean current meter in use in the United States: six conical cups, mounted around a vertical axis, rotate and cause a signal in a set of headphones with each rotation; tail vanes and a heavy weight stabilize the instrument. { 'prīs ,mēd·ər }

price relative [STAT] The ratio of the price of certain goods in a specified period to the price of the same goods in the base period. { 'prīs 'rel·ə·tiv }

pricker See needle. { 'prik·ər }

prickly heat See miliaria. { 'prik·lē 'hēt }

prick punch [DES ENG] A tool that has a sharp conical point ground to an angle of 30–60°C; used to make a slight indentation on a workpiece to locate the intersection of centerlines. { 'prik ,pənch }

priest [OPTICS] The Z tristimulus value. { prēst }

prill [CHEM ENG] To form pellet-sized crystals or agglomerates of material by the action of upward-blowing air on falling hot solution; used in the manufacture of ammonium nitrate and urea fertilizers. [MATER] Spherical particles about the size of buckshot. [MIN ENG] **1.** The best ore after cobbing. **2.** A circular particle about the size of buckshot. **3.** Compressed and sized explosives such as ammonium nitrate. { pril }

primaquine [PHARM] $C_{15}H_{21}N_3O$ An ether-soluble viscous liquid, used as the diphosphate salt in medicine to cure malaria. { 'prī·mə,kwēn }

primary [ASTRON] **1.** A planet with reference to its satellites, or the sun with reference to its planets. **2.** The brighter star of a double star system. [CHEM] A term used to distinguish basic compounds from similar or isomeric forms; in organic compounds, for example, RCH_2OH is a primary alcohol, R_1R_2CHOH is a secondary alcohol, and $R_1R_2R_3COH$ is a tertiary alcohol; in inorganic compounds, for example, NaH_2PO_4 is primary sodium phosphate, Na_2HPO_4 is the secondary form,

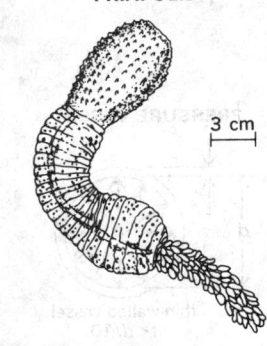

PRIAPULIDA

3 cm

Adult form of *Priapulus*.

and Na_3PO_4 is the tertiary form. [ELEC] **1.** One of the high-voltage conductors of a power distribution system. **2.** See primary winding. [GEOL] **1.** A young shoreline whose features are produced chiefly by nonmarine agencies. **2.** Of a mineral deposit, unaffected by supergene enrichment. [MET] Of a metal, obtained directly from ore. [VERT ZOO] Of or pertaining to quills on the distal joint of a bird wing. { 'prī,mer·ē }

primary air [MECH ENG] That portion of the combustion air introduced with the fuel in a burner. { 'prī,mer·ē 'er }

primary alcohol [ORG CHEM] An alcohol whose molecular structure may be written as RCH_2OH, rather than as R_1R_2CHOH (secondary) or $R_1R_2R_3COH$ (tertiary). { 'prī,mer·ē 'al·kə,hòl }

primary amine [ORG CHEM] An amine whose molecular structure may be written as RNH_2, instead of R_1R_2NH (secondary) or $R_1R_2R_3N$ (tertiary). { 'prī,mer·ē 'am,ēn }

primary arc [GEOL] **1.** A curved segment of elongated mountain zones that are the areas of the earth's major and most recent tectonic activity. **2.** See internides. { 'prī,mer·ē 'ärk }

primary area [NAV] In terminal operations, an area within a segment in which full obstacle clearance is applied. { 'prī,mer·ē 'er·ē·ə }

primary atypical pneumonia See Eaton agent pneumonia. { 'prī,mer·ē ā'tip·ə·kəl nù'mō·nyə }

primary basalt [PETR] Theoretically, the original magma from which all other rock types are supposedly obtained by various processes. { 'prī,mer·ē bə'sòlt }

primary battery [ELEC] A battery consisting of one or more primary cells. { 'prī,mer·ē 'bad·ə·rē }

primary biliary cirrhosis [MED] A slowly progressive disease primarily of middle-aged women, caused by an autoimmune destruction of bile ducts that begins as inflammation in and around larger intrahepatic bile ducts and eventually results in liver cell damage. { 'prīm·ə·rē ,bil·ē·er·ē sə'rō·səs }

primary body [ASTRON] The celestial body or central force field about which a satellite or other body orbits, or from which it is escaping, or toward which it is falling. { 'prī,mer·ē 'bäd·ē }

primary breaker [MECH ENG] A machine which takes over the work of size reduction from blasting operations, crushing rock to maximum size of about 2-inch (5-centimeter) diameter; may be a gyratory crusher or jaw breaker. Also known as primary crusher. { 'prī,mer·ē 'brāk·ər }

primary cache [COMPUT SCI] A cache memory located within a microprocessor chip itself. Also known as internal cache; level 1 cache. { 'prī,mer·ē 'kash }

primary carbon atom [ORG CHEM] A carbon atom in a molecule that is singly bonded to only one other carbon atom. { 'prī,mer·ē 'kär·bən ,ad·əm }

primary cell [ELEC] A cell that delivers electric current as a result of an electrochemical reaction that is not efficiently reversible, so that the cell cannot be recharged efficiently. { 'prī,mer·ē 'sel }

primary center [COMMUN] A telephone office having lower rank than a sectional center and higher rank than a toll center; connects toll centers and may also serve as a toll center for nearby end offices. { 'prī,mer·ē 'sen·tər }

primary circle See primary great circle. { 'prī,mer·ē 'sər·kəl }

primary circuit [ELEC] One of a collection of coupled coils or circuits that receives electric power from a source and transfers it to the secondary circuit by electromagnetic induction. { 'prī,mer·ē 'sər·kət }

primary circulation [METEOROL] The prevailing fundamental atmospheric circulation on a planetary scale which must exist in response to radiation differences with latitude, to the rotation of the earth, and to the particular distribution of land and oceans, and which is required from the viewpoint of conservation of energy. { 'prī,mer·ē ,sər·kyə'lā·shən }

primary clay See residual clay. { 'prī,mer·ē 'klā }

primary coil [ELEC] The input coil in an induction coil or transformer. { 'prī,mer·ē 'kòil }

primary colors [OPTICS] **1.** Three colors, red, yellow, and blue, which can be combined in various proportions to produce any other color. **2.** Any three colors that can be mixed in proper proportions to specify other colors; they need not be physically realizable. { 'prī,mer·ē 'kəl·ərz }

primary constriction See centromere. { 'prī,mer·ē kən'strik·shən }

primary consumer [ECOL] In an ecosystem, an animal that feeds on plants (producers) directly. Also known as a herbivore. { 'prī,mer·ē kən'sü·mər }

primary control program [COMPUT SCI] The program which provides the sequential scheduling of jobs and basic operating systems functions. Abbreviated PCP. { 'prī,mer·ē kən'trōl ,prō·gram }

primary cosmic rays See cosmic rays. { 'prī,mer·ē ¦käz·mik 'rāz }

primary crater [GEOL] **1.** An impact crater produced directly by the high-velocity impact of a meteorite or other projectile. **2.** See true crater. { 'prī,mer·ē 'krād·ər }

primary creep [MECH] The initial high strain-rate region in a material subjected to sustained stress. { 'prī,mer·ē 'krēp }

primary crusher See primary breaker. { 'prī,mer·ē 'krəsh·ər }

primary culture [BIOL] A tissue culture started from cells, tissues, or organs taken directly from the organism. { 'prī,mer·ē 'kəl·chər }

primary cyclone [METEOROL] Any cyclone (or low), especially a frontal cyclone, within whose circulation one or more secondary cyclones have developed. Also known as primary low. { 'prī,mer·ē 'sī,klōn }

primary decomposition [MATH] A primary decomposition of a submodule N of a module M is an expression of N as a finite intersection of primary submodules of M. { 'prī,mer·ē dē,käm·pə'zish·ən }

primary detector See sensor. { 'prī,mer·ē di'tek·tər }

primary dip [GEOL] The slight dip assumed by a bedded deposit at its moment of deposition. Also known as depositional dip; initial dip; original dip. { 'prī,mer·ē 'dip }

primary drilling [ENG] The process of drilling holes in a solid rock ledge in preparation for a blast by means of which the rock is thrown down. { 'prī,mer·ē 'dril·iŋ }

primary electron [ELECTR] An electron which bombards a solid surface, causing secondary emission. { 'prī,mer·ē i'lek,trän }

primary emission [ELECTR] Emission of electrons due to primary causes, such as heating of a cathode, and not to secondary effects, such as electron bombardment. { 'prī,mer·ē i'mish·ən }

primary energy [ENG] Energy that exists in a naturally occurring form, such as coal, before being converted into an end-use form. { 'prī,mer·ē 'en·ər·jē }

primary excavation [ENG] Digging performed in undisturbed soil. { 'prī,mer·ē ,eks·kə'vā·shən }

primary explosive [MATER] Explosive or explosive mixture sensitive to shock and friction; used in primers and detonators to initiate explosion. Also known as initiating explosive. { 'prī,mer·ē ik'splō·siv }

primary extinction [SOLID STATE] A weakening of the stronger beams produced in x-ray diffraction by a very perfect crystal, as compared with the weaker. { 'prī,mer·ē ik'stiŋk·shən }

primary fabric See apposition fabric. { 'prī,mer·ē 'fab·rik }

primary fault [ELEC] In an electric circuit, the initial breakdown of the insulation of a conductor, usually followed by a flow of power current. { 'prī,mer·ē 'fȯlt }

primary fission products See fission fragments. { 'prī,mer·ē 'fish·ən ,präd·əks }

primary flat joint [GEOL] An approximately horizontal joint plane in igneous rocks. Also known as L joint. { 'prī,mer·ē 'flat ,jȯint }

primary flow [ELECTR] The current flow that is responsible for the major properties of a semiconductor device. { 'prī,mer·ē 'flō }

primary focus [OPTICS] In an astigmatic system, a line at which some of the bundle of rays from an off-axis point meet; this line is perpendicular to a plane which contains the point and the optical axis, and has a smaller image distance than the secondary focus. Also known as meridional focus; tangential focus. { 'prī,mer·ē 'fō·kəs }

primary frequency [COMMUN] Frequency assigned for normal use on a particular circuit or communications channel. { 'prī,mer·ē 'frē·kwən·sē }

primary-frequency standard [COMMUN] One of the standards of frequency maintained by various governments; the operating frequency of a radio station is determined by comparison with multiples of this standard frequency. { 'prī,mer·ē 'frē·kwən·sē ,stan·dərd }

primary front [METEOROL] The principal, and usually original, front in any frontal system in which secondary fronts are found. { 'prī,mer·ē 'frənt }

primary fuel cell [ELEC] A fuel cell in which the fuel and oxidant are continuously consumed. { 'prī,mer·ē 'fyül ,sel }

primary geosyncline See orthogeosyncline. { 'prī,mer·ē 'jē·ō'sin,klīn }

primary gneiss [PETR] A rock that exhibits planar or linear structures characteristic of metamorphic rocks but lacks observable granulation or recrystallization and is therefore considered to be of igneous origin. { 'prī,mer·ē 'nīs }

primary gneissic banding [PETR] A kind of banding developed in certain igneous (plutonic) rocks of heterogeneous composition, produced by the admixture of two magmas only partly miscible or by magma intimately admixed with country rock into which it has been injected along planes of bedding or foliation. { 'prī,mer·ē ¦nī,sik 'band·iŋ }

primary great circle [GEOD] A great circle used as the origin of measurement of a coordinate; particularly, such a circle 90° from the poles of a system of spherical coordinates, as the equator. Also known as fundamental circle; primary circle. { 'prī,mer·ē 'grāt 'sər·kəl }

primary growth [BOT] Plant growth that originates in apical meristematic tissue of shoots and roots, giving rise to primary tissue. { 'prī,mer·ē 'grōth }

primary gun [ORD] Principal or main gun, especially of a tank or other armored vehicle. { 'prī,mer·ē 'gən }

primary haulage [MIN ENG] A short haul in which there is no secondary-or main-line haulage. Also known as face haulage. { 'prī·mer·ē 'hȯl·ij }

primary high explosive [MATER] An explosive which is extremely sensitive to heat and shock and is normally used to initiate a secondary high explosive; examples are mercury fulminate, lead azide, lead styphnate, and tetracene. { 'prī,mer·ē 'hī ik'splō·siv }

primary hydrogen atom [ORG CHEM] A hydrogen atom that is bonded to a primary carbon atom. { 'prī,mer·ē 'hī·drə·jən ,ad·əm }

primary hypertension See essential hypertension. { 'prī,mer·ē ,hī·pər'ten·chən }

primary hypothermia [MED] A decrease in internal body temperature caused by environmental stress that overwhelms the body's thermoregulation capability. { ¦prī·mər·ē ,hī·pōthər·mē·ə }

primary immune response [IMMUNOL] The activation and response of lymphocytes specific for a newly encountered antigen; generally slower and weaker than the secondary immune response. { ¦prī·mə·rē i'myün ri,späns }

primary index [COMPUT SCI] An index that holds the values of primary keys, in sequence. { 'prī,mer·ē 'in,deks }

primary instrument [ENG] A measuring instrument that can be calibrated without reference to another instrument. { 'prī,mer·ē 'in·strə·mənt }

primary interstice See original interstice. { 'prī,mer·ē in'tər·stəs }

primary key [COMPUT SCI] A key that identifies a record or portion of a record and determines the sequence of records in a file or other data structure. { 'prī,mer·ē 'kē }

primary knocked-on atom [PHYS] An atom in a solid that recoils from a collision with an energetic particle coming from outside the solid, rather than with another knocked-on atom. Abbreviated PKA. { 'prī,mer·ē 'näkt ,ȯn 'ad·əm }

primary lateral sclerosis [MED] A sclerotic disease of the crossed pyramidal tracts of the spinal cord, characterized by paralysis of the limbs, with rigidity, increased tendon reflexes, and absence of sensory and nutritive disorders. Also known as lateral sclerosis. { 'prī,mer·ē 'lad·ə·rəl sklə'rō·səs }

primary lead [MET] Lead recovered from ore, as contrasted with recycled scrap (secondary) lead. { 'prī,mer·ē 'led }

primary lesion [MED] **1.** In syphilis, tuberculosis, and cowpox, a chancre. **2.** In dermatology, the earliest clinically recognizable manifestation of cutaneous disease, such as a macule, papule, vesicle, pustule, or wheal. { 'prī,mer·ē 'lē·zhən }

primary lights [OPTICS] Any three lights used in a system of tristimulus colorimetric analysis of solutions. { 'prī,mer·ē 'līts }

primary low See primary cyclone. { 'prī,mer·ē 'lō }

primary magma [GEOL] A magma that originates below the earth's crust. { 'prī,mer·ē 'mag·mə }

primary measuring element [ENG] The portion of a measuring or sensing device that is in direct contact with the variables being measured (such as temperature, pressure, pH, or velocity). { 'prī,mer·ē 'mezh·ə·riŋ ,el·ə·mənt }

primary meristem [BOT] Meristem which is derived directly from embryonic tissue and which gives rise to epidermis, vascular tissue, and the cortex. { 'prī,mer·ē 'mer·ə,stem }

primary mineral [MINERAL] A mineral that is formed at the same time as the rock in which it is contained, and that retains its original form and composition. { 'prī,mer·ē 'min·rəl }

primary optic axis [OPTICS] One of two optic axes in a crystal that are perpendicular to the circular sections of the indicatrix and along which all light rays travel with equal velocity. { 'prī,mer·ē 'äp·tik ,ak·səs }

primary orogeny [GEOL] Orogeny that is characteristic of the internides and that involves deformation, regional metamorphism, and granitization. { 'prī,mer·ē ȯ'räj·ə·nē }

primary phase [THERMO] The only crystalline phase capable of existing in equilibrium with a given liquid. { 'prī,mer·ē 'fāz }

primary phase region [THERMO] On a phase diagram, the locus of all compositions having a common primary phase. { 'prī,mer·ē 'fāz ,rē·jən }

primary phloem [BOT] Phloem derived from apical meristem. { 'prī,mer·ē 'flō·əm }

primary photocurrent [ELECTR] A photocurrent resulting from nonohmic contacts unable to replenish charge carriers which pass out of the opposite contact, and whose maximum gain is unity. { 'prī,mer·ē 'fōd·ō,kə·rənt }

primary plasticizer [MATER] A plasticizer material for plastics formulations that has sufficient affinity to a polymer or resin so that it is considered compatible and therefore may be used as the sole plasticizer. { 'prī,mer·ē 'plas·tə,sīz·ər }

primary pollutant [METEOROL] A pollutant that enters the air directly from a source. { 'prī,mer·ē pə'lüt·ənt }

primary porosity [GEOL] Natural porosity in petroleum reservoir sands or rocks. { 'prī,mer·ē pə'räs·əd·ē }

primary power cable [ELEC] Power service cables connecting the outside power source to the main-office switch and metering equipment. { 'prī,mer·ē 'pau̇·ər ,kā·bəl }

primary producer [ECOL] In an ecosystem, an organism (primarily green photosynthetic plants) that utilizes the energy of the sun and inorganic molecules from the environment to synthesize organic molecules. { 'prī,mer·ē prə'dü·sər }

primary production [ECOL] The total amount of new organic matter produced by photosynthesis. { 'prī,mer·ē prə'dək·shən }

primary radar [ENG] Radar in which the incident beam is reflected from the target to form the return signal. Also known as primary surveillance radar (PSR). { 'prī,mer·ē 'rā,där }

primary radiation [PHYS] Radiation arriving directly from its source without interaction with matter. { 'prī,mer·ē ,rād·ē'ā·shən }

primary rainbow [OPTICS] The most common of the principal rainbow phenomena, which appears as an arc of angular radius of about 42° about the observer's antisolar point; it is the inner of two rainbows, whose light undergoes only one internal reflection, and which is narrower and brighter than the outer, or secondary, rainbow. { 'prī,mer·ē 'rān,bō }

primary reference fuel [MATER] **1.** Gasoline; isooctane, *n*-heptane, or mixtures thereof used in the American Society for Testing and Materials-Cooperative Fuel Research gasoline test engine to determine the octane rating of commercial gasoline. **2.** Diesel fuel; cetane, α-methylnaphthalene, or mixtures thereof used in ASTM-CFR diesel test engines to rate the cetane number of commercial diesel fuels. { 'prī,mer·ē 'ref·rəns ,fyül }

primary register [COMPUT SCI] A general-purpose register in a central processing unit that is available for direct utilization by computer programs. { 'prī,mer·ē 'rej·ə,stər }

primary relay [ELEC] Relay that produces the initial action in a sequence of operations. { 'prī,mer·ē 'rē,lā }

primary reserve [PETRO ENG] Petroleum reserve recoverable commercially at current prices and costs by conventional methods and equipment as a result of the natural energy inherent in the reservoir. { 'prī,mer·ē ri'zərv }

primary rocks [PETR] Rocks whose constituents are newly formed particles that have never been constituents of previously formed rocks and that are not the products of alteration or replacement, such as limestones formed by precipitation from solution. { 'prī,mer·ē 'räks }

primary root [BOT] The first plant root to develop; derived from the radicle. { 'prī,mer·ē 'rüt }

primary scattering [PHYS] Any scattering process in which radiation is received at a detector, such as the eye, after having been scattered just once; distinguished from multiple scattering. { 'prī,mer·ē 'skad·ə·riŋ }

primary sedimentary structure [GEOL] A sedimentary structure produced during deposition, such as ripple marks and graded bedding. { 'prī,mer·ē ,sed·ə'men·trē ,strək·chər }

primary service area [COMMUN] The area in which the ground wave of a broadcast station is not subject to objectionable interference or fading. { 'prī,mer·ē 'sər·vəs ,er·ē·ə }

primary sewage sludge [CIV ENG] A semiliquid waste resulting from sedimentation with no additional treatment. { 'prī,mer·ē 'sü·ij ,sləj }

primary skip zone [ELECTROMAG] Area around a transmitter beyond the ground wave but within the skip distance. { 'prī,mer·ē 'skip ,zōn }

primary standard [SCI TECH] A unit directly defined and established by some authority, against which all secondary standards are calibrated; for example, in analytical chemistry, reference substances or solutions of known chemical purity and concentration are used to standardize laboratory solutions prior to volumetric analysis or titration. { 'prī,mer·ē 'stan·dərd }

primary storage [COMPUT SCI] Main internal storage of a computer. { 'prī,mer·ē 'stȯr·ij }

primary stratification [GEOL] Stratification which develops when sediments are first deposited. Also known as direct stratification. { 'prī,mer·ē ,strad·ə·fə'kā·shən }

primary stratigraphic trap [GEOL] A stratigraphic trap formed by the deposition of clastic materials (such as shoestring sands, lenses, sand patches, bars, or cocinas) or through chemical deposition (such as organic reefs or biostromes). { 'prī,mer·ē ,strad·ə¦graf·ik 'trap }

primary stress [MECH] A normal or shear stress component in a solid material which results from an imposed loading and which is under a condition of equilibrium and is not self-limiting. { 'prī,mer·ē 'stres }

primary stress field *See* ambient stress field. { 'prī,mer·ē 'stres ,fēld }

primary structure [AERO ENG] The main framework, of an aircraft including fittings and attachments; any structural member whose failure would seriously impair the safety of the missile is a part of the primary structure. [BIOCHEM] The sequence of amino acids in the molecule of a protein or a peptide. [GEOL] A structure, in an igneous rock, that formed at the same time as the rock, but before its final consolidation. [ORG CHEM] The chemical structure of a polymer chain. { 'prī,mer·ē 'strək·chər }

primary submodule [MATH] A submodule N of a module M over a commutative ring R such that $M \neq N$ and, for any a in R, the principal homomorphism of the factor module M/N associated with a, $a_{M/N}$, is either injective or nilpotent. { 'prī,mer·ē səb'mä·jəl }

primary succession *See* prisere. { 'prī,mer·ē sək'sesh·ən }

primary surveillance radar *See* primary radar. { 'prī,mer·ē sər'vā·ləns ,rā,där }

primary syphilis [MED] The first stage of the venereal disease, characterized clinically by a painless ulcer, or chancre, at the point of infection and painless, discrete regional adenopathy. { 'prī,mer·ē 'sif·ə·ləs }

primary tectonite [PETR] A tectonite with depositional fabric. { 'prī,mer·ē 'tek·tə,nīt }

primary tissue [BOT] Plant tissue formed during primary growth. [HISTOL] Any of the four fundamental tissues composing the vertebrate body. { 'prī,mer·ē 'tish·ü }

primary transcript [MOL BIO] The initial transcription of a ribonucleic acid molecule from deoxyribonucleic acid. { 'prī,mer·ē 'tran,skript }

primary treatment [CIV ENG] Removal of floating solids and suspended solids, both fine and coarse, from raw sewage. { 'prī,mer·ē 'trēt·mənt }

primary voltage [ELEC] The voltage applied to the terminals of the primary winding of a transformer. { 'prī,mer·ē 'vōl·tij }

primary wave [COMMUN] A radio wave traveling by a direct path, as contrasted with skips. [GEOPHYS] The first seismic wave that reaches a station from an earthquake. { 'prī¦mer·ē 'wāv }

primary weapon [ORD] The principal arm of a combat unit; the rifle is the primary or basic weapon for an infantry rifle company, as compared with grenades or chemical projectiles, which are secondary or auxiliary weapons in such an organization. { 'prī¦mer·ē 'wep·ən }

primary winding [ELEC] The transformer winding that receives signal energy or alternating-current power from a source. Also known as primary. Abbreviated pri. Symbolized P. { 'prī¦mer·ē 'wīnd·iŋ }

primary xylem [BOT] Xylem derived from apical meristem. { 'prī¦mer·ē 'zī·ləm }

primase See DNA primase. { 'prī¦mās }

Primates [VERT ZOO] The order of mammals to which man belongs; characterized in terms of evolutionary trends by retention of a generalized limb structure and dentition, increasing digital mobility, replacement of claws by flat nails, development of stereoscopic vision, and progressive development of the cerebral cortex. { prī'mād·ēz }

prime [ENG] **1.** Main or primary, as in prime contractor. **2.** In blasting, to place a detonator in a cartridge or charge of explosive. **3.** To treat wood with a primer or penetrant primer. **4.** To add water to a pump to enable it to begin pumping. [MATH] See prime element. { prīm }

prime coat See primer. { 'prīm 'kōt }

prime contractor [ENG] A contractor having a direct contract for an entire project; the contractor may in turn assign portions of the work to subcontractors. { 'prīm 'kän,trak·tər }

primed lymphocyte [IMMUNOL] A lymphocyte that is from an immunized individual or that has been exposed to antigen in cell culture and is therefore sensitized. { 'prīmd 'lim·fə,sīt }

prime element [MATH] A member of an integral domain that is not a unit and is not the product of two members that are not units. Also known as prime. { 'prīm 'el·ə·mənt }

prime factor [MATH] A prime number or prime polynomial that exactly divides a given number or polynomial. { 'prīm 'fak·tər }

prime fictitious meridian [NAV] The reference meridian (real or fictitious) used as the origin for measurement of fictitious longitude. { 'prīm fik'tish·əs mə,rid·ē·ən }

prime field [MATH] For a field K with multiplicative unit element e, the field consisting of elements of the form $(ne)(me)^{-1}$, where $m \neq 0$ and n are integers. { 'prīm 'fēld }

prime focus [OPTICS] The position in a reflecting telescope at which light from celestial objects is focused by the main mirror, located on the axis of the mirror near the open end of the tube. { 'prīm 'fō·kəs }

prime grid meridian [NAV] The reference meridian of a grid; in polar regions it is usually the 180-0° geographic meridian, used as the origin for measuring grid longitude. { 'prīm 'grid mə,rid·ē·ən }

prime ideal [MATH] A principal ideal of a ring given by a single element that has properties analogous to those of the prime numbers. { 'prīm ī'dēl }

prime inverse meridian See prime transverse meridian. { 'prīm in'vərs mə,rid·ē·ən }

prime meridian [GEOD] The meridian of longitude 0°, used as the origin for measurement of longitude; the meridian of Greenwich, England, is almost universally used for this purpose. { 'prīm mə,rid·ē·ən }

prime mover [ANAT] A muscle that produces a specific motion or maintains a specific posture. [MECH ENG] **1.** The component of a power plant that transforms energy from the thermal or the pressure form to the mechanical form. **2.** A tractor or truck, usually with four-wheel drive, used for hauling tasks. { 'prīm 'müv·ər }

prime navaids facilities [NAV] Navigational aid facilities that, under certain operational environments (for example, weather or mission) are considered to be essential to the safe completion of departure and approach-to-land maneuvering by aircraft; these facilities may include RAPCON, GCA, ILS, TACAN, VOR, and associated critical air-traffic control communications equipments. { 'prīm 'na,vādz fə,sil·əd·ēz }

prime number [MATH] A positive integer having no divisors except itself and the integer 1. { 'prīm 'nəm·bər }

prime number theorem [MATH] The theorem that the limit of the quantity $[\pi(x)]$ (ln x)/x as x approaches infinity is 1, where $\pi(x)$ is the number of prime numbers not greater than

x and ln x is the natural logarithm of x. { 'prīm ¦nəm·bər ,thir·əm }

prime oblique meridian [NAV] The reference fictitious meridian of an oblique graticule. { 'prīm ō'blēk mə,rid·ē·ən }

prime polynomial [MATH] A polynomial whose only factors are itself and constants. { 'prīm ¦päl·i'nō·mē·əl }

primer [CELL MOL] **1.** A short ribonucleic acid (RNA) sequence that is complementary to a sequence of deoxyribonucleic acid (DNA) and has a 3'-OH terminus at which a DNA polymerase begins synthesis of a DNA chain. **2.** A short sequence of DNA that is complementary to a messenger RNA (mRNA) sequence and enables reverse transcriptase to begin copying the neighboring sequences of mRNA. **3.** A transfer RNA whose elongation starts RNA-directed DNA synthesis in retroviruses. [ENG] In general, a small, sensitive initial explosive train component which on being actuated initiates functioning of the explosive train, and will not reliably initiate high explosive charge; classified according to the method of initiation, for example, percussion primer, electric primer, or friction primer. [MATER] A prefinishing coat applied to surfaces that are to be painted or otherwise finished. Also known as prime coat. { 'prīm·ər }

primer cup [ENG] A small metal cup, into which the primer mixture is loaded. { 'prīm·ər ,kəp }

primer-detonator [ENG] A unit, in a metal housing, in which are assembled a primer, a detonator and when indicated, an intervening delay charge. { 'prīm·ər 'det·ən,ād·ər }

prime register [COMPUT SCI] One of the registers that is inactive at any given time in a central processing unit with duplicate general-purpose registers. { 'prīm 'rej·ə·stər }

prime ring [MATH] For a field K with multiplicative unit element e, the ring consisting of elements of the form ne, where n is an integer. { 'prīm 'riŋ }

primer leak [ENG] Defect in a cartridge which allows partial escape of the hot propelling gases in a primer, caused by faulty construction or an excessive charge. { 'prīm·ər ,lēk }

primer mixture [MATER] An explosive mixture containing a sensitive explosive and other ingredients, used in a primer. { 'prīm·ər ,miks·chər }

primer pouch [ORD] Container that holds the primer used in firing with separate-loading ammunition. { 'prīm·ər ,pauch }

primer seat [ORD] The chamber, in the breech mechanism of a gun that uses separate-loading ammunition, into which the primer is set. { 'prīm·ər ,sēt }

primer setback [ORD] The backward movement of a primer cup in a cartridge case which occurs when the base of the cup is not properly supported by the bolt face or breechblock. { 'prīm·ər 'set,bak }

primes [MET] High-quality metal products, particularly sheet and plate, that are free from visible defects. { prīmz }

prime transverse meridian [NAV] The reference meridian of a transverse graticule. Also known as prime inverse meridian. { 'prīm tranz'vərs mə,rid·ē·ən }

prime vertical circle [GEOD] The vertical circle through the east and west points of the horizon. { 'prīm 'vərd·ə·kəl 'sər·kəl }

prime-white kerosine [MATER] A kerosine with an off-white color, between water-white and standard-white kerosine. { 'prīm ¦wīt 'ker·ə,sēn }

primibrach [INV ZOO] In crinoids, the brachials of the unbranched arm. { 'prī·mə,brak }

priming [MECH ENG] In a boiler, the excessive carryover of fine water particles along with the steam because of insufficient steam space, faulty boiler design, or faulty operating conditions. { 'prīm·iŋ }

priming composition [MATER] A physical mixture of materials that is very sensitive to impact or percussion and, when so exploded, undergoes very rapid autocombustion, producing hot gases and incandescent solid particles; priming compositions are used for the ignition of primary high explosives, black powder igniter charges, and propellants in small arms ammunition. { 'prīm·iŋ ,kəm·pə,zish·ən }

priming of the tides [OCEANOGR] The acceleration in the times of occurrence of high and low tides when the sun's tidal effect comes before that of the moon. { 'prīm·iŋ əv thə 'tīdz }

priming pump [MECH ENG] A device on motor vehicles and tanks, providing a means of injecting a spray of fuel into the engine to facilitate starting. { 'prīm·iŋ ,pəmp }

Primitiopsacea [PALEON] A small dimorphic superfamily

of extinct ostracodes in the suborder Beyrichicopina; the velum of the male was narrow and uniform, but that of the female was greatly expanded posteriorly.   { prī¦mid·ē·äp'sā·shə }

primitive [COMPUT SCI]   A sketchy specification, omitting details, of some action in a computer program.   [CONT SYS]   A basic operation of a robot, initialized by a single command statement in the program that controls the robot.   { 'prim·əd·iv }

primitive abstract data type [COMPUT SCI]   A simple abstract data type that is typically implemented directly in a high-level programming language; examples include integers and real numbers (with appropriate arithmetic operators), booleans (with appropriate logical operators), text strings, and pointers.   { 'prim·əd·iv 'ab,strakt 'dad·ə ,tīp }

primitive abundant number [MATH]   An integer that is an abundant number and has no proper divisors that are also abundant numbers.   { ¦prim·əd·iv ə¦bən·dənt 'nəm·bər }

primitive cell [CRYSTAL]   A parallelepiped whose edges are defined by the primitive translations of a crystal lattice; it is a unit cell of minimum volume.   { 'prim·əd·iv 'sel }

primitive circle [MATH]   The stereographic projection of the great circle whose plane is perpendicular to the diameter of the projected sphere that passes through the point of projection.   { 'prim·əd·iv 'sər·kəl }

primitive element [MATH]   A member of a finite number field from which all the other members can be generated by repeated multiplication.   { 'prim·əd·iv 'el·ə·mənt }

primitive equations [FL MECH]   The Eulerian equations of motion of a fluid in which the primary dependent variables are the fluid's velocity components; these equations govern a wide variety of fluid motions and form the basis of most hydrodynamical analysis; in meteorology, these equations are frequently specialized to apply directly to the cyclonic-scale motions by the introduction of filtering approximations.   { 'prim·əd·iv i'kwā·zhənz }

primitive gut [EMBRYO]   The tubular structure in embryos which differentiates into the alimentary canal.   { 'prim·əd·iv 'gət }

primitive lattice [CRYSTAL]   A crystal lattice in which there are lattice points only at its corners.   Also known as simple lattice.   { 'prim·əd·iv 'lad·əs }

primitive period [MATH]   **1.** A period a of a simply periodic function $f(x)$ such that any period of $f(x)$ is an integral multiple of a.   **2.** Either of two periods a and b of a doubly periodic function $f(x)$ such that any period of $f(x)$ is of the form $ma + nb$, where m and n are integers.   { 'prim·əd·iv 'pir·ē·əd }

primitive period parallelogram [MATH]   For a doubly periodic function $f(z)$ of a complex variable, a parallelogram with vertices at $z_0, z_0 + a, z_0 + a + b$, and $z_0 + b$, where z_0 is any complex number and a and b are primitive periods of $f(z)$.   { ¦prim·əd·iv ¦pir·ē·əd ,par·ə'lel·ə,gram }

primitive plane [MATH]   A partial plane in which every line passes through at least two points.

primitive polynomial [MATH]   A polynomial with integer coefficients which have 1 as their greatest common divisor.   { 'prim·əd·iv ,päl·i'nō·mē·əl }

primitive pseudoperfect number [MATH]   An integer that is a pseudoperfect number and has no proper divisors that are also pseudoperfect numbers.   { ¦prim·əd·iv ¦süd·ō¦pər,fikt 'nəm·bər }

primitive root [MATH]   An nth root of unity that is not an mth root of unity for any m less than n.   { 'prim·əd·iv 'rüt }

primitive streak [EMBRYO]   A dense, opaque band of ectoderm in the bilaminar blastoderm associated with the morphogenetic movements and proliferation of the mesoderm and notochord; indicates the first trace of the vertebrate embryo.   { 'prim·əd·iv 'strēk }

primitive translation [CRYSTAL]   For a space lattice, one of three translations which can be repeatedly applied to generate any translation which leaves the lattice unchanged.   { 'prim·əd·iv tranz'lā·shən }

primitive water [HYD]   Water that has been imprisoned in the earth's interior, in either molecular or dissociated form, since the formation of the earth.   { 'prim·əd·iv 'wȯd·ər }

primordial black holes [ASTRON]   Hypothetical black holes which may have formed in the early, highly compressed stages of the universe immediately following the big bang.   { prī'mȯrd·ē·əl 'blak 'hōlz }

primordial germ cell [EMBRYO]   Embryonic cell giving rise

to a germ cell from which a gamete (egg or sperm) develops.   { prī¦mȯrd·ē·əl 'jərm ,sel }

primordial gut See archenteron.   { prī'mȯrd·ē·əl 'gət }

primordium See anlage.   { prī'mȯrd·ē·əm }

Primulaceae [BOT]   A family of dicotyledonous plants in the order Primulales characterized by a herbaceous habit and capsular fruit with two to many seeds.   { ,prim·yə'lās·ē,ē }

Primulales [BOT]   An order of dicotyledonous plants in the subclass Dilleniidae distinguished by sympetalous flowers, stamens located opposite the corolla lobes, and a compound ovary with a single style.   { ,prim·yə'lā·lēz }

principal axis [CRYSTAL]   The longest axis in a crystal.   [ENG ACOUS]   A reference direction for angular coordinates used in describing the directional characteristics of a transducer; it is usually an axis of structural symmetry or the direction of maximum response.   [MATH]   **1.** One of a set of perpendicular axes such that a quadratic function can be written as a sum of squares of coordinates referred to these axes.   **2.** For a conic, a straight line that passes through the midpoints of all the chords perpendicular to it.   **3.** For a quadric surface, the intersection of two principal planes.   [MECH]   One of three perpendicular axes in a rigid body such that the products of inertia about any two of them vanish.   [OPTICS]   See optic axis.   { 'prin·sə·pəl 'ak·səs }

principal axis of strain [MECH]   One of the three axes of a body that were mutually perpendicular before deformation.   Also known as strain axis.   { 'prin·sə·pəl 'ak·səs əv 'strān }

principal axis of stress [MECH]   One of the three mutually perpendicular axes of a body that are perpendicular to the principal planes of stress.   Also known as stress axis.   { 'prin·sə·pəl 'ak·səs əv 'stres }

principal branch [MATH]   For complex valued functions such as the logarithm which are multiple-valued, a selection of values so as to obtain a genuine single-valued function.   { 'prin·sə·pəl 'branch }

principal curvatures [MATH]   For a point on a surface, the absolute maximum and absolute minimum values attained by the normal curvature.   { 'prin·sə·pəl 'kər·və·chərz }

principal diagonal [MATH]   For a square matrix, the diagonal extending from the upper left-hand corner to the lower right-hand corner of the matrix, that is, the diagonal containing the elements a_{ij} for which $i = j$.   Also known as main diagonal.   { 'prin·sə·pəl dī'ag·ən·əl }

principal directions [MATH]   For a point on a surface, the directions in which the normal curvature attains its absolute maximum and absolute minimum values.   [MECH]   The directions of the principal axes of a strain.   { 'prin·sə·pəl di'rek·shənz }

principal distance [GRAPHICS]   The perpendicular distance from the internal perspective center to the plane of a particular finished negative or print, or of a drawing.   { 'prin·sə·pəl 'dis·təns }

principal E plane [ELECTROMAG]   Plane containing the direction of radiation of electromagnetic waves and arranged so that the electric vector everywhere lies in the plane.   { 'prin·sə·pəl 'ē ,plān }

principal focus See focal point.   { 'prin·sə·pəl 'fō·kəs }

principal function [MECH]   The integral of the Lagrangian of a system over time; it is involved in the statement of Hamilton's principle.   { 'prin·sə·pəl 'fəŋk·shən }

principal homomorphism [MATH]   Let a be an element of a ring R and M be a module over R; the principal homomorphism of M associated with a, denoted a_M, is the mapping which takes each element x in M into ax.   { 'prin·sə·pəl ¦hō·mō'mȯr,fiz·əm }

principal H plane [ELECTROMAG]   Plane that contains the direction of radiation and the magnetic vector, and is everywhere perpendicular to the E plane.   { 'prin·sə·pəl 'āch ,plān }

principal ideal [MATH]   The smallest ideal of a ring which contains a given element of the ring.   { 'prin·sə·pəl ī'dēl }

principal ideal ring [MATH]   A commutative ring with a unit element in which every ideal is a principal ideal.   { ¦prin·sə·pəl ī¦del ,riŋ }

principal item [ENG]   Item which, because of its major importance, requires detailed analysis and examination of all factors affecting its supply and demand, as well as an unusual degree of supervision; its selection is based upon such criteria as strategic importance, high monetary value, unusual complexity of issue, and procurement difficulties.　{ 'prin·sə·pəl 'īd·əm }

principal line [GRAPHICS] The trace of the principal plane upon a photograph. [SPECT] That spectral line which is most easily excited or observed.　{ 'prin·sə·pəl 'līn }

principal lobe [PHYS] The lobe of a radiation pattern or directivity pattern that lies on the axis of symmetry of an acoustic or electromagnetic transmitter or receptor.　{ 'prin·sə·pəl 'lōb }

principal meridian [CIV ENG] One of the meridians established by the United States government as a reference for subdividing public land.　{ 'prin·sə·pəl mə'rid·ē·ən }

principal mode *See* fundamental mode.　{ 'prin·sə·pəl 'mōd }

principal moments [PHYS CHEM] The three moments of inertia of a rigid molecule calculated with respect to the principal axes.　{ 'prin·sə·pəl 'mō·məns }

principal normal [MATH] The line perpendicular to a space curve at some point which also lies in the osculating plane at that point.　{ 'prin·sə·pəl 'nór·məl }

principal normal indicatrix [MATH] For a space curve, all the end points of those radii of a unit sphere that are parallel to the positive directions of the principal normals to the curve. Also known as spherical indicatrix of the principal normal. { ¦prin·sə·pəl ¦nór·məl in'dik·ə‚triks }

principal normal section [MATH] For a point on a surface, a normal section in a direction in which the curvature of this section has a maximum or minimum value.　{ ¦prin·sə·pəl ¦nór·məl 'sek·shən }

principal part [MATH] 1. The principal part of an analytic function $f(z)$ defined in an annulus about a point z_0 is the sum of terms in its Laurent expansion about z_0 with negative powers of $(z - z_0)$. 2. A principal part of a triangle is one of its sides or one of its interior angles.　{ 'prin·sə·pəl 'pärt }

principal plane [MATH] For a quadric surface, a plane that passes through the midpoints of all the chords perpendicular to it. [OPTICS] 1. Two planes perpendicular to the optical axis such that objects in one plane form images in the other with a lateral magnification of unity. 2. The vertical plane passing through the internal perspective center and containing the perpendicular from that center to the plane of a tilted photograph. 3. *See* principal section.　{ 'prin·sə·pəl 'plān }

principal plane of stress [MECH] For a point in an elastic body, a plane at that point across which the shearing stress vanishes.　{ 'prin·sə·pəl 'plān əv 'stres }

principal point [OPTICS] The intersection of a principal plane with the optical axis.　{ 'prin·sə·pəl 'póint }

principal quantum number [ATOM PHYS] A quantum number for orbital electrons, which, together with the orbital angular momentum and spin quantum numbers, labels the electron wave function; the energy level and the average distance of an electron from the nucleus depend mainly upon this quantum number.　{ 'prin·sə·pəl 'kwän·təm ‚nəm·bər }

principal radii [MATH] The radii of curvature of the normal sections with maximum and minimum curvature at a given point on a surface; the reciprocals of the principal curvatures.　{ 'prin·sə·pəl 'rād·ē‚ī }

principal radii of normal curvature [MATH] The reciprocals of the principal curvatures of a surface at a point.　{ 'prin·sə·pəl ¦rād·ē‚ī əv ¦nór·məl 'kər·və·chər }

principal ray [OPTICS] 1. The one ray within a bundle of incident rays that, upon entering an optical instrument from any given point of the object, passes through the optical center of the lens. 2. *See* principal visual ray.　{ 'prin·sə·pəl 'rā }

principal root [MATH] The positive real root of a positive number, or the negative real root in the case of odd roots of negative numbers.　{ 'prin·sə·pəl 'rüt }

principal section [MATH] A normal section at a given point on a surface whose curvature has a maximum or minimum value. [OPTICS] A plane in a crystal that contains the crystal's optic axis and the ray of light under consideration. Also known as principal plane.　{ 'prin·sə·pəl 'sek·shən }

principal series [SPECT] A series occurring in the line spectra of many atoms and ions with one, two, or three electrons in the outer shell, in which the total orbital angular momentum quantum number changes from 1 to 0.　{ 'prin·sə·pəl 'sir·ēz }

principal strain [MECH] The elongation or compression of one of the principal axes of strain relative to its original length.　{ 'prin·sə·pəl 'strān }

principal stress [MECH] A stress occurring at right angles to a principal plane of stress.　{ 'prin·sə·pəl 'stres }

principal submatrix [MATH] An $m \times m$ matrix, P, is an $m \times m$ principal submatrix of an $n \times n$ matrix, A, if P is obtained from A by removing any $n - m$ rows and the same $n - m$ columns.　{ ¦prin·sə·pəl səb'mā·triks }

principal value [MATH] 1. The numerically smallest value of the arc sine, arc cosine, or arc tangent of a number, the positive value being chosen when there are values that are numerically equal but opposite in sign. 2. *See* Cauchy principal value.　{ 'prin·sə·pəl 'val·yü }

principal vertical circle [GEOD] The vertical circle through the north and south points of the horizon, coinciding with the celestial meridian.　{ 'prin·sə·pəl 'vərd·ə·kəl ‚sər·kəl }

principal visual ray [OPTICS] A perpendicular extending from a station point to a perspective plane and theoretically passing exactly along the visual axis of a viewing eye. Also known as principal ray.　{ 'prin·sə·pəl 'vish·ə·wəl 'rā }

principle [SCI TECH] A scientific law which is highly general or fundamental, and from which other laws are derived.　{ 'prin·sə·pəl }

principle of coincidence [ENG] The principle of operation of a vernier, according to which the fraction of the smallest division of the main scale is determined by the division of the vernier which is exactly in line with a division of the main scale.　{ 'prin·sə·pəl əv kō'in·sə·dəns }

principle of covariance [RELAT] 1. In classical physics and in special relativity, the principle that the laws of physics take the same mathematical form in all inertial reference frames. 2. In general relativity, the principle that the laws of physics take the same mathematical form in all conceivable curvilinear coordinate systems.　{ 'prin·sə·pəl əv kō'ver·ē·əns }

principle of dichotomy *See* law of the excluded middle.　{ ¦prin·sə·pəl əv dī'käd·ə·mē }

principle of duality *See* duality principle.　{ 'prin·sə·pəl əv dü'al·əd·ē }

principle of dynamical similarity [MECH] The principle that two physical systems which are geometrically and kinematically similar at a given instant, and physically similar in constitution, will retain this similarity at later corresponding instants if and only if the Froude number 1 for each independent type of force has identical values in the two systems. Also known as similarity principle.　{ 'prin·sə·pəl əv di¦nam·ə·kəl ‚sim·ə'lar·əd·ē }

principle of equivalence *See* equivalence principle.　{ 'prin·sə·pəl əv i'kwiv·ə·ləns }

principle of inaccessibility *See* Carathéodory's principle.　{ 'prin·sə·pəl əv ‚in·ak‚ses·ə'bil·əd·ē }

principle of insufficient reason [STAT] The principle that cases are equally likely to occur unless reasons to the contrary are known.　{ 'prin·sə·pəl əv ‚in·sə‚fish·ənt 'rēz·ən }

principle of least action [MECH] The principle that, for a system whose total mechanical energy is conserved, the trajectory of the system in configuration space is that path which makes the value of the action stationary relative to nearby paths between the same configurations and for which the energy has the same constant value. Also known as least-action principle.　{ 'prin·sə·pəl əv ‚lēst 'ak·shən }

principle of optimality [CONT SYS] A principle which states that for optimal systems, any portion of the optimal state trajectory is optimal between the states it joins.　{ 'prin·sə·pəl əv ‚äp·tə'mal·əd·ē }

principle of reciprocity *See* reciprocity theorem.　{ 'prin·sə·pəl əv ‚res·ə'präs·əd·ē }

principle of superposition [ELEC] 1. The principle that the total electric field at a point due to the combined influence of a distribution of point charges is the vector sum of the electric field intensities which the individual point charges would produce at that point if each acted alone. 2. The principle that, in a linear electrical network, the voltage or current in any element resulting from several sources acting together is the sum of the voltages or currents resulting from each source acting alone. Also known as superposition theorem. [MECH] The principle that when two or more forces act on a particle at the same time, the resultant force is the vector sum of the two. [PHYS] Also known as superposition principle. 1. A general principle applying to many physical systems which states that if a number of independent influences act on the system, the resultant influence is the sum of the individual influences acting separately. **2.** In all theories characterized by linear homogeneous differential equations, such as optics, acoustics, and quantum theory, the principle that the sum of any number of solutions to the equations is another solution. { 'prin·sə·pəl əv ˌsü·pər·pə'zish·ən }

principle of the maximum [MATH] The principle that for a nonconstant complex analytic function defined in a domain, the absolute value of the function cannot attain its maximum at any interior point of the domain. { 'prin·sə·pəl əv thə 'mak·sə·məm }

principle of the minimum [MATH] The principle that for a nonvanishing nonconstant complex analytic function defined in a domain, the absolute value of the function cannot attain its minimum at any interior point of the domain. { 'prin·sə·pəl əv thə 'min·ə·məm }

principle of uniformity See uniformitarianism. { 'prin·sə·pəl əv ˌyü·nə'fôr·məd·ē }

principle of virtual work [MECH] The principle that the total work done by all forces acting on a system in static equilibrium is zero for any infinitesimal displacement from equilibrium which is consistent with the constraints of the system. Also known as virtual work principle. { 'prin·sə·pəl əv 'vər·chə·wəl ˌwərk }

print [GRAPHICS] A photographic copy made by placing a negative or transparency in contact with a sensitized surface or by projecting the image on a screen or sensitized photographic medium, and then developing the result. { print }

printability gage [GRAPHICS] A standard pattern of lines with graduated spaces between them, used by printers to determine the size of symbol that can be printed with fidelity and to measure the amount of spread in the width of printed lines inherent in the particular printing process. { ˌprint·ə'bil·əd·ē ˌgāj }

print driver [COMPUT SCI] The portion of a computer program that directs output to a printer and usually also controls printer functions such as pagination and the setting of the margins and page headers. { 'print ˌdrī·vər }

printed circuit [ELECTR] A conductive pattern that may or may not include printed components, formed in a predetermined design on the surface of an insulating base in an accurately repeatable manner. { 'print·əd 'sər·kət }

printed circuit board [ELECTR] A flat board whose front contains slots for integrated circuit chips and connections for a variety of electronic components, and whose back is printed with electrically conductive pathways between the components. Also known as circuit board. { 'print·əd 'sər·kət ˌbórd }

printed-wiring armature [ELEC] An armature in which the conductors consist of printed-wiring strips on both sides of a thin insulating disk, to give a low-inertia armature for servomotors and other variable high-speed applications. { 'print·əd ¦wīr·iŋ 'ärm·ə·chùr }

printed wiring board [ELECTR] A copper-clad dielectric material with conductors etched on the external or internal layers. { ¦print·əd 'wīr·iŋ ˌbórd }

printer [COMPUT SCI] A computer output mechanism that prints characters one at a time or one line at a time. { 'print·ər }

printer file [COMPUT SCI] **1.** A file that contains the information that the printer driver needs in order to generate the codes required by the printer. **2.** A document in print image format. { 'prin·tər ˌfīl }

printer's point See point. { 'print·ərz ˌpóint }

print film [GRAPHICS] A fine-grain, high-resolving-power, yellow-dyed positive photographic film used for continuous contact printing. { 'print ˌfilm }

print hammer [GRAPHICS] A device on certain kinds of printers which, upon receiving the proper signal, strikes the paper, bringing it in contact with the character to be printed. { 'print ˌham·ər }

print head [COMPUT SCI] The mechanism that generates the characters to be reproduced by a character printer. { 'print ˌhed }

print image format [COMPUT SCI] The format of a document that has been prepared for output on the printer. { 'print ˌim·ij ˌfôr·mat }

printing calculator [COMPUT SCI] A desk-model electronic calculator that provides a printed record on paper tape with or without a digital display. { 'print·iŋ 'kal·kyə,lād·ər }

printing element [COMPUT SCI] The part of the print head

mechanism that comes into contact with the paper to print characters or other images. { 'print·iŋ ˌel·ə·mənt }

printing ink [MATER] Ink generally made from carbon black, lampblack, or other pigment suspended in an oil vehicle, with a resin, solvent, adhesive, and drier; available in many variations. { 'print·iŋ ˌiŋk }

printing press [GRAPHICS] A machine for the production of printed impressions on paper and other materials. { 'print·iŋ ˌpres }

printing-telegraph code [COMMUN] A five- or seven-unit code used for operation of a teleprinter, teletypewriter, and similar telegraph printing devices. { 'print·iŋ ¦tel·ə,graf ˌkōd }

printing telegraphy [COMMUN] Method of telegraph operation in which the received signals are automatically recorded in printed characters. { 'print·iŋ tə'leg·rə·fē }

print member [COMPUT SCI] The part of a computer printer that determines the form of a printed character, such as a print wheel or type bar. { 'print ˌmem·bər }

printout [COMPUT SCI] A printed output of a data-processing machine or system. { 'print,aút }

printout paper [GRAPHICS] A paper used to make prints by light alone, without any processing involved; it has a reddish color, is often used as a proof, and can be made permanent by toning. { 'print,aút ˌpā·pər }

print position [COMPUT SCI] One of the positions on a printer at which a character can be printed. { 'print pə,zish·ən }

print processes [GRAPHICS] Three general print processes: relief process, in which the printing areas stand above the surface of the plate; intaglio process, in which the printing areas fall below the surface of the plate; and planographic process, in which there is no appreciable difference between the level of the printing areas and the general surface. { 'print ,prä,ses·əz }

print queue [COMPUT SCI] A prioritized list, maintained by the operating system, of the output from a computer system waiting on a spool file to be printed. { 'print ,kyü }

print server [COMPUT SCI] A computer controlling a series of printers. { 'print ,sər·vər }

printthrough [ELECTR] Transfer of signals from one recorded layer of magnetic tape to the next on a reel. { 'print,thrü }

print train [COMPUT SCI] **1.** The chain in a chain printer or the drum in a drum printer that holds the type slugs used to make impressions on paper. **2.** The electronic character set that serves a similar function in a laser printer. { 'print ,trān }

print wheel [COMPUT SCI] A disk which has around its rim the letters, numerals, and other characters that are used in printing in a wheel printer. { 'print ,wēl }

prion [BIOCHEM] Any of a group of infectious proteins that cause fatal neurodegenerative diseases in humans and animals, including scrapie and bovine spongiform encephalopathy in animals and Creutzfeldt-Jakob disease and Gerstmann-Straussler-Scheinker disease in humans. { 'prī,än }

prion diseases [MED] A group of invariably fatal disorders affecting humans and animals that are clinically characterized by neurological and behavioral degeneration caused by the cerebral accumulation of an abnormal prion protein which is resistant to proteolytic enzymes and, in contrast to other infectious agents, does not require nucleic acid for replication. The diseases are transmissible either genetically (for example, Creutzfeldt-Jakob disease) or via infection (new variant Creutzfeldt-Jakob disease and mad cow disease) or can occur spontaneously (classical or sporadic Creutzfeldt-Jakob disease). Also known as spongiform encephalopathies. { 'prī,än di,zēz·əz }

Prioniodidae [PALEON] A family of conodonts in the suborder Conodontiformes having denticulated bars with a large denticle at one end. { ˌprī·ə,nī'äd·ə,dē }

Prioniodinidae [PALEON] A family of conodonts in the suborder Conodontiformes characterized by denticulated bars or blades with a large denticle in the middle third of the specimen. { ˌprī·ə,nī·ə'din·ə,dē }

prionodont [VERT ZOO] Having many simple, similar teeth set in a row like sawteeth. { prī'än·ə,dänt }

prior-art search [ENG] **1.** A search for prior art which may possibly anticipate an invention which is being considered for patentability. **2.** A similar search but for the purpose of

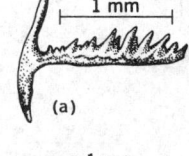

PRIONIODIDAE

Three examples of Prioniodidae; *(a) Ligonodina. (b) Hibbardella. (c) Hindeodella.*

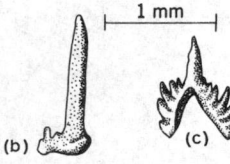

PRIONIODINIDAE

Two examples of the Prioniodinidae: *(a) Bryantodus; (b) Lewistownella.*

determining what the status of existing technology is before going ahead with new research; it is done to avoid unwittingly retracing new steps taken by other workers in the field. { 'prī·ər ,ärt ,sərch }

prior distribution [STAT] A probability distribution on the set of all possible values of an unknown parameter of a statistical model that describes information available from sources other than a statistical investigation, in particular, expert judgment, past experience, or prior belief. { 'prī·ər ,dis·trə'byü·shən }

priorite [MINERAL] $(Y,Ce,Th)(Ti,Nb)_2O_6$ A mineral composed of titanoniobate of rare-earth metals; it is isomorphous with eschynite. Also known as blomstrandine. { 'prī·ə,rīt }

priority-arbitration circuit [COMPUT SCI] A logic circuit which combines all interrupts but allows only the highest-priority request to enable its active flipflop. { prī'är·əd·ē ,är·bə'trā·shən ,sər·kət }

priority indicator [COMMUN] Data attached to a message to indicate its relative priority and hence the order in which it will be transmitted. [COMPUT SCI] Data attached to a computer program or job which are used to determine the order in which it will be processed by the computer. { prī'är·əd·ē 'in·də,kād·ər }

priority interrupt [COMPUT SCI] An interrupt procedure in which control is passed to the monitor, the required operation is initiated, and then control returns to the running program, which never knows that it has been interrupted. { prī'är·əd·ē 'int·ə,rəpt }

priority phase [COMPUT SCI] Phase consisting of execution of operations in response to instruments or process interrupts other than clock interrupts. { prī'är·əd·ē 'fāz }

priority polling [COMMUN] In a data communications network, a system in which nodes with high activity are interrogated more frequently than those with only occasional traffic. { prī'är·əd·ē 'pōl·iŋ }

priority processing [COMPUT SCI] A method of computer time-sharing in which the order in which programs are processed is determined by a system of priorities, involving such factors as the length, nature, and source of the programs. { prī'är·əd·ē 'prä,ses·iŋ }

prior probabilities [STAT] Probabilities of the outcomes of an experiment before the experiment has been performed. { 'prī·ər ,präb·ə'bil·əd·ēz }

priority queueing [COMPUT SCI] The arrangement of jobs to be carried out in a list according to their relative importance, with the most important first. { prī'är·əd·ē 'kyü·iŋ }

priscol See tolazoline hydrochloride. { 'pris,kól }

prisere [ECOL] The ecological succession of vegetation that occurs in passing from barren earth or water to a climax community. Also known as primary succession. { 'prī,sir }

prism [CRYSTAL] A crystal which has three, four, six, eight, or twelve faces, with the face intersection edges parallel, and which is open only at the two ends of the axis parallel to the intersection edges. [GEOL] A long, narrow, wedge-shaped sedimentary body with a width-thickness ratio greater than 5 to 1 but less than 50 to 1. [MATH] A polyhedron with two parallel, congruent faces and all other faces parallelograms. [OPTICS] An optical system consisting of two or more usually plane surfaces of a transparent solid or embedded liquid at an angle with each other. Also known as optical prism. { 'priz·əm }

prismatic astrolabe [ENG] A surveying instrument that makes use of a pan of mercury forming an artificial horizon, and a prism mounted in front of a horizontal telescope to determine the exact times at which stars reach a fixed altitude, and thereby to establish an astronomical position. { priz'mad·ik 'as·trə,lāb }

prismatic binoculars See prism binoculars. { priz'mad·ik bə'näk·yə·lərz }

prismatic cleavage [CRYSTAL] A type of crystal cleavage that occurs parallel to the faces of a prism. { priz'mad·ik 'klē·vij }

prismatic coefficient See longitudinal coefficient. { priz 'mad·ik ,kō·i'fish·ənt }

prismatic compass [ENG] A hand compass used by surveyors which is equipped with a prism that allows the compass to be read while the site is being taken. { priz'mad·ik 'kəm·pəs }

prismatic error [OPTICS] That error due to lack of parallelism of the two faces of an optical element, such as a mirror or a shade glass. { priz'mad·ik 'er·ər }

prismatic jointing See columnar jointing. { priz'mad·ik 'jóint·iŋ }

prismatic plane [MET] Any plane parallel to the principal c axis in noncubic metals. { priz'mad·ik 'plān }

prismatic structure See columnar jointing. { priz'mad·ik 'strək·chər }

prismatic surface [MATH] A surface generated by moving a straight line which always meets a broken line lying in a given plane and which is always parallel to some given line not in that plane. { priz'mad·ik 'sər·fəs }

prismatoid [MATH] A polyhedron whose vertices all are in one or the other of two parallel planes. { 'priz·mə,tóid }

prism binoculars [OPTICS] A type of binoculars, each half of which is a Kepler telescope that employs a Porro prism erecting system both to erect the image and to reduce the length of the instrument. Also known as prismatic binoculars. { 'priz·əm bə'näk·yə·lərz }

prism crack [GEOL] A mud crack that develops in regular or irregular polygonal patterns on the surface of drying mud puddles and that breaks the sediment into prisms standing normal to the bedding. { 'priz·əm ,krak }

prism diopter [OPTICS] A unit used in measuring the deviating power of a prism; this power in prism diopters is 100 times the tangent of the angle of deviation of a ray of light. { 'priz·əm dī äp·tər }

prism joint [MECH ENG] A robotic articulation that has only one degree of freedom, in sliding motion only. { 'priz·əm ,jóint }

prism level [ENG] A surveyor's level with prisms that allow the levelman to view the level bubble without moving his eye from the telescope. { 'priz·əm 'lev·əl }

prismoid [MATH] A prismatoid whose two parallel faces are polygons having the same number of sides while the other faces are trapezoids or parallelograms. { 'priz,móid }

prismoidal formula [MATH] A formula that gives the volume of a prismatoid as $(1/6)h(A_1 + 4A_m + A_2)$, where h is the altitude, A_1 and A_2 are the areas of the bases, and A_m is the area of a plane section halfway between the bases. { priz'móid·əl 'fór·myə·lə }

prism spectrograph [SPECT] Analysis device in which a prism is used to give two different but simultaneous light wavelengths derived from a common light source; used for the analysis of materials by flame photometry. { 'priz·əm 'spek·trə,graf }

prism transit See broken-back transit. { 'priz·əm 'trans·ət }

pristane [ORG CHEM] $C_{19}H_{40}$ A liquid soluble in such organic solvents as ether, petroleum ether, benzene, chloroform, and carbon tetrachloride; used as a lubricant, as an oil in transformers, and as an anticorrosion agent. Also known as norphytane. { 'pri,stān }

Pristidae [VERT ZOO] The sawfishes, a family of modern sharks belonging to the batoid group. { 'pris·tə,dē }

Pristiophoridae [VERT ZOO] The saw sharks, a family of modern sharks often grouped with the squaloids which have a greatly extended rostrum with enlarged denticles along the margins. { ,pris·tē·ə 'fór·ə,dē }

privacy system [COMMUN] A device or method for scrambling overseas telephone conversations handled by radio links in order to make them unintelligible to outside listeners. Also known as privacy transformation; secrecy system. { 'prī·və·sē ,sis·təm }

privacy transformation See privacy system. { 'prī·və·sē ,tranz·fər'mā·shən }

private aid to navigation [NAV] In marine operation, an aid to navigation authorized by the U.S. Coast Guard but established, operated, and maintained by private interests in the navigable waters of the United States. { 'prī·vət ,ād tə ,nav·ə gā·shən }

private automatic branch exchange [COMMUN] A private branch exchange in which connections are made by remote-controlled switches. Abbreviated PABX. { 'prī·vət ,öd·ə,mad·ik 'branch iks,chānj }

private automatic exchange [COMMUN] A private telephone exchange in which connections are made by remote-controlled switches. Abbreviated PAX. { 'prī·vət ,öd·ə,mad·ik iks,chānj }

private branch exchange [COMMUN] A telephone exchange serving a single organization, having a switchboard and associated equipment, usually located on the customer's premises; provides for switching calls between any two extensions served by the exchange or between any extension and the national telephone system via a trunk to a central office. Abbreviated PBX. { 'prī·vət 'branch iks,chānj }

private branch exchange access line [ELEC] Circuit that connects a main private branch exchange (PBX) to a switching center. { 'prī·vət 'branch iks,chānj 'ak,ses ,līn }

private data [COMPUT SCI] Data that are open to a single user only. { 'prī·vət 'dad·ə }

private exchange [COMMUN] Telephone exchange serving a single organization and having no means for connecting to a public telephone system. { 'prī·vət iks'chānj }

private library [COMPUT SCI] An organized collection of programs and other software that is the property of a single user of a computer system and is not generally available to other users. { 'prī·vət 'lī,brer·ē }

private line [COMMUN] A line, channel, or service reserved solely for one user. { 'prī·vət 'līn }

private line arrangement [COMPUT SCI] The structure of a computer system in which each input/output device has a set of lines leading to the central processing unit for the device's own private use. Also known as radial selector. { 'prī·vət |līn ə,rānj·mənt }

private line service [COMMUN] Service provided by United States common carriers engaged in domestic or international wire, radio, and cable communications for the intercity communications purposes of a customer; this service is provided over integrated communications pathways, including facilities or local channels, which are integrated components of intercity private line services, and station equipment between specified locations for a continuous period or for regularly recurring periods at stated hours. { 'prī·vət |līn ,sər·vəs }

private pack [COMPUT SCI] A disk pack assigned exclusively to one application or one user so that the operating system does not try to allocate space on the device to others. { 'prī·vət 'pak }

private stream [HYD] Any stream which diverts part or all of the drainage of another stream. { 'prī·vət 'strēm }

privileged direction [OPTICS] One of two mutually perpendicular directions for the plane of polarization of a beam of plane-polarized light falling on a plate of anisotropic material such that the light which emerges from the plate is also plane-polarized. { 'priv·ə·lijd də'rek·shən }

privileged instruction [COMPUT SCI] A class of instructions, usually including storage protection setting, interrupt handling, timer control, input/output, and special processor status-setting instructions, that can be executed only when the computer is in a special privileged mode that is generally available to an operating or executive system, but not to user programs. { 'priv·ə·lijd in'strək·shən }

privileged mode See master mode. { 'priv·ə·lijd ,mōd }

PRML technique See partial-response maximum-likelihood technique. { |pē|är|em'el tek,nēk }

proaccelerin [BIOCHEM] A labile procoagulant in normal plasma but deficient in the blood of patients with parahemophilia; essential for rapid conversion of prothrombin to thrombin. Also known as factor V; labile factor. { |prō·ak'sel·ə·rən }

proactive interference [PSYCH] The situation in which old memories inhibit the learning of new memories. { prō,ak·tiv ,in·tər'fir·əns }

proamnion [EMBRYO] The part of the embryonic area at the sides and in front of the head of the developing amniote embryo, which remains without mesoderm for a considerable period. { prō'am·nē,än }

Proanura [PALEON] Triassic forerunners of the Anura. { prō'an·yə·rə }

probabilistic automaton [COMPUT SCI] A device, with a finite number of internal states, which is capable of scanning input words over a finite alphabet and responding by successively changing its internal state in a probabilistic way. Also known as stochastic automaton. { ,präb·ə·bə'lis·tik ō'täm·ə,tän }

probabilistic sampling [STAT] A process in which the laws of probability determine which elements are to be included in a sample. { ,präb·ə'lis·tik 'sam·pliŋ }

probabilistic sequential machine [COMPUT SCI] A probabilistic automaton that has the capability of printing output words probabilistically, over a finite output alphabet. Also known as stochastic sequential machine. { ,präb·ə·bə'lis·tik si'kwen·chəl mə'shēn }

probability [STAT] The probability of an event is the ratio of the number of times it occurs to the large number of trials that take place; the mathematical model of probability is a positive measure which gives the measure of the space the value 1. { ,präb·ə'bil·əd·ē }

probability amplitude See Schrödinger wave function. { ,präb·ə'bil·əd·ē 'am·plə,tüd }

probability current density [QUANT MECH] A vector whose component normal to a surface gives the probability that a particle will cross a unit area of the surface during a unit time. { ,präb·ə'bil·əd·ē |kə·rənt ,den·səd·ē }

probability density [QUANT MECH] The square of the absolute value of the Schrödinger wave function for a particle at a given point; gives the probability per unit volume of finding the particle at that point. { ,präb·ə'bil·əd·ē ,den·səd·ē }

probability density function [STAT] A real-valued function whose integral over any set gives the probability that a random variable has values in this set. Also known as density function; frequency function. { ,präb·ə'bil·əd·ē |den·səd·ē ,fəŋk·shən }

probability deviation See probable error. { ,präb·ə'bil·əd·ē ,dē·vē,ā·shən }

probability distribution See distribution. { ,präb·ə'bil·əd·ē ,dis·trə,byü·shən }

probability forecast [METEOROL] A forecast of the probability of occurrence of one or more of a mutually exclusive set of weather contingencies, as distinguished from a series of categorical statements. { ,präb·ə'bil·əd·ē ,fór,kast }

probability mass function [STAT] A function which gives the relative frequency of each possible value of the random variable in an experiment involving a discrete set of outcomes. Abbreviated p.m.f. { ,präb·ə'bil·əd·ē 'mas ,fəŋk·shən }

probability measure [MATH] The measure on a probability space. { ,präb·ə'bil·əd·ē ,mezh·ər }

probability paper [STAT] Graph paper with one axis specially ruled to transform the distribution function of a specified function to a straight line when it is plotted against the variate as the abscissa. { ,präb·ə'bil·əd·ē ,pā·pər }

probability ratio test [STAT] Testing a simple hypothesis against a simple alternative by using the ratio of the probability of each simple event under the alternative to the probability of the event under the hypothesis. { ,präb·ə'bil·əd·ē 'rā·shō ,test }

probability sampling [STAT] A method of sampling from a finite population where the probability of each set of units being selected is known. { ,präb·ə'bil·əd·ē ,sam·pliŋ }

probability space [MATH] A measure space such that the measure of the entire space equals 1. { ,präb·ə'bil·əd·ē ,spās }

probability theory [MATH] The study of the mathematical structures and constructions used to analyze the probability of a given set of events from a family of outcomes. { ,präb·ə'bil·əd·ē ,thē·ə·rē }

probable [ORD] **1.** An instance in which a hostile aircraft is probably destroyed. **2.** The hostile aircraft so designated. { 'präb·ə·bəl }

probable error [STAT] The error that is exceeded by a variable with a probability of 1/2. Also known as probability deviation. { 'präb·ə·bəl 'er·ər }

probable maximum precipitation [METEOROL] The theoretically greatest depth of precipitation for a given duration that is physically possible over a particular drainage area at a certain time of year; in practice, this is derived over flat terrain by storm transposition and moisture adjustment to observed storm patterns. { 'präb·ə·bəl 'mak·sə·məm pri,sip·ə'tā·shən }

probable ore [MIN ENG] **1.** A mineral deposit adjacent to a developed ore but not yet proved by development. **2.** See indicated ore. { 'präb·ə·bəl 'ór }

probable reserves [PETRO ENG] Primary petroleum reserves based on limited evidence, but not proved by a commercial oil-production rate. { 'präb·ə·bəl ri'zərvz }

proband [GEN] The clinically affected individual through whom a family is found that can be used to study the genetics of a particular disorder. Also known as propositius. { 'prō,band }

probe [AERO ENG] An instrumented vehicle moving through the upper atmosphere or space or landing upon another celestial body in order to obtain information about the specific environment. [BIOL] A biochemical substance labeled with a radioactive isotope or tagged in some other way and used to identify or isolate a gene, a gene product, or a protein. [COMMUN] To determine a radio interference by obtaining the relative interference level in the immediate area of a source by the use of a small, insensitive antenna in conjunction with a receiving device. [ELECTROMAG] A metal rod that projects into but is insulated from a waveguide or resonant cavity; used to provide coupling to an external circuit for injection or extraction of energy or to measure the standing-wave ratio. Also known as waveguide probe. [ENG] A small tube containing the sensing element of electronic equipment, which can be lowered into a borehole to obtain measurements and data. [PHYS] A small device which can be brought into contact with or inserted into a system in order to make measurements on the system; ordinarily it is designed so that it does not significantly disturb the system. { prōb }

probe coil [ELECTROMAG] In eddy-current nondestructive tests, a type of test coil which is placed on the surface of an object. { 'prōb ,kȯil }

probe gas [ENG] Tracer gas emitted from a small orifice for impingement on a restricted area being tested for leaks. { 'prōb ,gas }

probertite [MINERAL] $NaCaB_5O_9 \cdot 5H_2O$ A colorless mineral crystallizing in the monoclinic system, consisting of hydrous sodium calcium borate. { 'präb·ər,tīt }

probe-type liquid-level meter [ENG] Device to sense or measure the level of liquids in storage or process vessels by means of an immersed electrode or probe. { 'prōb ,tīp 'lik·wəd ¦lev·əl ,mēd·ər }

probe-type microelectrode [NEUROSCI] A microelectrode consisting of one or more long thin shanks projecting from a larger carrier area that is designed to facilitate the investigation of processing in three-dimensional neural networks on the level of individual neurons. { 'prōb,tīp ,mī·krō·i'lek,trōd }

probit [STAT] A procedure used in dosage-response studies to avoid obtaining negative response values to certain dosages; five is added to the values of the standardized variate which is assumed to be normal; the term is a contraction of probability unit. { 'prō·bət }

problem check [COMPUT SCI] One or more tests used to assist in obtaining the correct machine solution to a problem. { 'präb·ləm ,chek }

problem-defining language [COMPUT SCI] A programming language that literally defines a problem and may specifically define the input and output, but does not define the method of transforming one to the other. Also known as problem specification language. { 'präb·ləm di¦fīn·iŋ ,laŋ·gwij }

problem definition [COMPUT SCI] The art of compiling logic in the form of general flow charts and logic diagrams which clearly explain and present the problem to the programmer in such a way that all requirements involved in the run are presented. { 'präb·ləm ,def·ə,nish·ən }

problem-describing language [COMPUT SCI] A programming language that describes, in the most general way, the problem to be solved, but gives no indication of the problem's detailed characteristics or its solution. { 'präb·ləm di¦skrīb·iŋ ,laŋ·gwij }

problème des ménages [MATH] The problem of seating a specified number (greater than two) of married couples at a circular table so that the sexes alternate and no husband and wife sit side by side. Also known as ménage problem. { prȯ·blem de mā'nazh }

problème des recontres [MATH] The problem of determining the number of derangements of a specified number of distinct objects. { prȯ·blem de rə'kän·trə }

problem file See run book. { 'präb·ləm ,fīl }

problem folder See run book. { 'präb·ləm ,fōld·ər }

problem mode [COMPUT SCI] A condition of computer operation in which, in contrast to supervisor mode, the privileged instructions cannot be executed, preventing the program from upsetting the supervisor program or any other program. { 'präb·ləm ,mōd }

problem of nontaking rooks [MATH] The problem of

determining the number of ways that a specified number of rooks can be placed on a chessboard of specified size so that no rook can capture another rook (that is, no two rooks are in same row or in the same column). Also known as rook problem. { ¦präb·ləm əv ¦nän,tāk·iŋ 'rüks }

problem of type [MATH] The problem of determining whether a given simply connected Riemann surface is of hyperbolic, parabolic, or elliptic type. { ¦präb·ləm əv 'tīp }

problem-oriented language [COMPUT SCI] A language designed to facilitate the accurate expression of problems belonging to specific sets of problem types. { 'präb·ləm ,ȯr·ē¦ent·əd ,laŋ·gwij }

problem-solving language [COMPUT SCI] A programming language that can be used to specify a complete solution to a problem. { 'präb·ləm sälv·iŋ ,laŋ·gwij }

problem space [PSYCH] A mental representation of a problem that contains knowledge of the initial state and the goal state of the problem as well as possible intermediate states that must be searched in order to link up the beginning and the end of the task. { 'präb·ləm ,spās }

problem-specification language See problem-defining language. { 'präb·ləm ,spes·ə·fə¦kā·shən ,laŋ·gwij }

Proboscidea [VERT ZOO] An order of herbivorous placental mammals characterized by having a proboscis, incisors enlarged to become tusks, and pillarlike legs with five toes bound together on a broad pad. { ,prō·bə'sid·ē·ə }

proboscis [INV ZOO] A tubular organ of varying form and function on a large number of invertebrates, such as insects, annelids, and tapeworms. [VERT ZOO] The flexible, elongated snout of certain mammals. { prə'bäs·kəs }

Proca equations [QUANT MECH] A set of equations, analogous to Maxwell's equations, relating a four-vector potential and a second-rank tensor field describing a particle of spin 1 and nonzero mass. { 'prō·kə i,kwā·zhənz }

procaine See procaine base. { 'prō,kān }

procaine base [ORG CHEM] $C_6H_4NH_2COOCH_2CH_2$- $N(C_2H_5)_2$ Water-insoluble, light-sensitive, odorless, white powder, melting at 60°C; soluble in alcohol, ether, chloroform, and benzene; used in medicine as a local anesthetic. Also known as planocaine base; procaine. { 'prō,kān ,bās }

procaine penicillin G [ORG CHEM] $C_{29}H_{38}N_4O_6S \cdot H_2O$ White crystals or powder, fairly soluble in chloroform; used as an antibiotic in animal feed. { 'prō,kān ,pen·ə'sil·ən 'jē }

procambium [BOT] The part of the apical meristematic tissue from which primary vascular tissues are derived. { prō 'käm·bē·əm }

Procampodeidae [INV ZOO] A family of the insect order Diplura. { prō,kam·pə'dē·ə,dē }

procarp [BOT] The reproductive structure of the female gametophyte that is found in certain red algae. { 'prō,kärp }

Procaviidae [VERT ZOO] A family of mammals in the order Hyracoidea including the hyraxes. { ,prō·kə'vī·ə,dē }

Procaviinae [VERT ZOO] A subfamily of ungulate mammals in the family Procaviidae. { ,prō·kə'vī·ə,nē }

procedural memory [PSYCH] The memory of motor, perceptual, and cognitive skills. { prə,sēj·ə·rəl 'mem·rē }

procedural programming [COMPUT SCI] A list of instructions telling a computer, step-by-step, what to do, usually having a linear order of execution from the first statement to the second and so forth with occasional loops and branches. Procedural programming languages include C, C++, Fortran, Pascal, and Basic. { prə sē·jə·rəl 'prō,gram·iŋ }

procedural representation [COMPUT SCI] The representation of certain concepts in a computer by procedures or programs in some appropriate language, rather than by static data items such as numbers or lists. { prə'sē·jə·rəl ,rep·rə·zen'tā·shən }

procedure [COMPUT SCI] **1.** A sequence of actions (or computer instructions) which collectively accomplish some desired task. **2.** In particular, a subroutine that causes an effect external to itself. { prə'sē·jər }

procedure declaration [COMPUT SCI] A statement that causes a procedure to be given a name and written as a segment of a computer program. { prə'sē·jər ,dek·lə,rā·shən }

procedure division [COMPUT SCI] The section of a program (written in the COBOL language) in which a programmer specifies the operations to be performed with the data names appearing in the program. { prə'sē·jər di,vizh·ən }

procedure library [COMPUT SCI] A collection of job control language routines that are stored on a disk file and can be executed by entering a command naming the routine. Abbreviated PROCLIB. { prə'sē·jər ˌlī·brer·ē }

procedure-oriented language [COMPUT SCI] A language designed to facilitate the accurate description of procedures, algorithms, or routines belonging to a certain set of procedures. { prə'sē·jər ˌȯr·ē|ent·əd ˌlaŋ·gwij }

procedure track [NAV] The path specified for making an instrument approach to an airport and for pull-up in case of a missed approach; this information is indicated on instrument approach charts by lines both in plan and profile. { prə'sē·jər ˌtrak }

procedure turn [NAV] A constant-rate turn of an aircraft in flight; used for computing the radius of turn and time required for its execution when very accurate navigation is required in controlling time or maintaining accurate, predetermined tracks. { prə'sē·jər ˌtərn }

procedure work log system [NAV] In air operations, a system utilized for logging the flight procedures in airspace data and establishing the source of documentation for automatic data processing. Abbreviated PROWL. { prə'sē·jər 'wərk ˌläg ˌsis·təm }

proceed-to-select signal [COMMUN] Signal returned from distant automatic equipment over the backward signaling path, in response to a calling signal, to indicate that selecting information can be transmitted; in certain signaling systems, both signals can be the same. { prə'sēd tə si'lekt ˌsig·nəl }

proceed-to-transmit signal [COMMUN] Signal returned from a distant manual switchboard over the backward signaling path, in response to a calling signal, to indicate that the teleprinter of the distant operator is connected to the circuit. { prə'sēd tə tranz'mit ˌsig·nəl }

Procellarian [GEOL] Pertaining to lunar lithologic map units and topographic forms constituting, or closely associated with, the maria. { ˌprō·sə'lar·ē·ən }

Procellariidae [VERT ZOO] A family of birds in the order Procellariiformes comprising the petrels, fulmars, and shearwaters. { ˌprō·sə·lə'rī·ə,dē }

Procellariiformes [VERT ZOO] An order of oceanic birds characterized by tubelike nostril openings, webbed feet, dense plumage, compound horny sheath of the bill, and, often, a peculiar musky odor. { ˌprō·sə·lə,rī·ə'fȯr,mēz }

procephalon [INV ZOO] The part of an insect's head that lies anteriorly to the segment in which the mandibles are located. { prō'sef·ə,län }

procercoid [INV ZOO] The solid parasitic larva of certain eucestodes, such as pseudophyllideans, that develops in the body of the intermediate host. { prō'sər,kȯid }

process [ANAT] A projection from the central mass of an organism. [COMPUT SCI] **1.** To assemble, compile, generate, interpret, compute, and otherwise act on information in a computer. **2.** A program that is running on a computer. [ENG] A system or series of continuous or regularly occurring actions taking place in a predetermined or planned manner to produce a desired result. { 'prä,ses }

process analytical chemistry [ANALY CHEM] A branch of analytical chemistry concerned with quantitative and qualitative information about a chemical process. { 'prä,səs ˌan·əl'it·i·kəl 'kem·ə·strē }

process analyzer [CHEM ENG] An instrument for determining the chemical composition of the substances involved in a chemical process directly, or for measuring the physical parameters indicative of composition. { 'prä,səs ˌan·ə,līz·ər }

process annealing [MET] Softening a ferrous alloy by heating to a temperature close to but below the lower limit of the transformation range and then cooling. { 'prä,səs ə,nēl·iŋ }

process-bound program *See* CPU-bound program. { 'prä,ses ˈbaŭnd 'prō·grəm }

process camera [OPTICS] Large camera used to produce materials for reproduction in printing; permits a large range of enlargement and reduction. { 'prä,səs ˌkam·rə }

process chart [IND ENG] A graphic representation of events occurring during a series of actions or operations. { 'prä,səs ˌchärt }

process color [GRAPHICS] Method of reproducing full-color originals such as paintings and color photographs; four-color process plates print in yellow, magenta, cyan, and black. { 'prä,səs ˌkəl·ər }

process control [ENG] Manipulation of the conditions of a process to bring about a desired change in the output characteristics of the process. { 'prä,səs kən,trōl }

process control chart [IND ENG] A tabulated graphical arrangement of test results and other pertinent data for each production assembly unit, arranged in chronological sequence for the entire assembly. { 'prä,səs kən,trōl ˌchärt }

process control engineering [ENG] A field of engineering dealing with ways and means by which conditions of continuous processes are automatically kept as close as possible to desired values or within a required range. { 'prä,səs kən,trōl ˌen·jə,nir·iŋ }

process control system [CONT SYS] The automatic control of a continuous operation. { 'prä,səs kən,trōl ˌsis·təm }

process dynamics [ENG] The dynamic response interrelationships between components (units) of a complex system, such as in a chemical process plant. { 'prä,səs dī,nam·iks }

process engineering [ENG] A service function of production engineering that involves selection of the processes to be used, determination of the sequence of all operations, and requisition of special tools to make a product. { 'prä,səs ˌen·jə,nir·iŋ }

process furnace [CHEM ENG] Furnace used to heat process-stream materials (liquids, gases, or solids) in a chemical-plant operation; types are direct-fired, indirect-fired, and pebble heaters. { 'prä,səs ˌfər·nəs }

process heater [CHEM ENG] Equipment for the heating of chemical process streams (gases, liquids, or solids); usually refers to furnaces, in contrast to heat exchangers. { 'prä,səs ˌhēd·ər }

process heat reactor [NUCLEO] A nuclear reactor that produces heat for use in manufacturing processes. { 'prä,səs ˈhēt rē,ak·tər }

processing [COMMUN] Further handling, manipulation, consolidation, compositing, and so on, of information to convert it from one format to another or to reduce it to manageable or intelligible information. [ENG] The act of converting material from one form into another desired form. { 'prä,ses·iŋ }

processing interrupt [COMPUT SCI] The interruption of the batch processing mode in a real-time system when live data are entered in the system. { 'prä,ses·iŋ 'int·ə,rəpt }

processing program [COMPUT SCI] Any computer program that is not a control program, such as an application program, or a noncontrolling part of the operating system, such as a sort-merge program or language translator. { 'prä,ses·iŋ ˌprō,gram }

processing section [COMPUT SCI] The computer unit that does the actual changing of input into output; includes the arithmetic unit and intermediate storage. { 'prä,ses·iŋ ˌsek·shən }

process lapse rate [METEOROL] The rate of decrease of the temperature of an air parcel as it is lifted, expressed as $-dT/dz$, where z is the altitude, or occasionally dT/dp, where p is pressure; the concept may be applied to other atmospheric variables, such as the process lapse rate of density. { 'prä,səs 'laps ˌrāt }

process layout [IND ENG] In a processing plant, the layout of machines, equipment, and locations which groups the same or similar operations. { 'prä,səs ˌlā,aŭt }

process lens [OPTICS] A highly corrected, apochromatic lens used for precise color-separation work. { 'prä,səs ˌlenz }

process-limited *See* processor-limited. { 'prä,səs ˈlim·əd·əd }

process metallurgy [MET] The branch of metallurgy concerned with the extraction of metals from ore, and with the refining of metals; usually synonymous with extractive metallurgy. { 'prä,səs ˌmed·əl,ər·jē }

process monitoring [CHEM ENG] The observation of chemical process variables by means of pressure, temperature, flow, and other types of indicators; usually occurs in a central control room. { 'prä,səs ˌmän·ə·triŋ }

processor [COMPUT SCI] **1.** A device that performs one or many functions, usually a central processing unit. Also known as engine. **2.** A program that transforms some input into some output, such as an assembler, compiler, or linkage editor. { 'prä,ses·ər }

processor complex [COMPUT SCI] The central portion of a very large computer consisting of several central processing units working in concert. { 'prä,ses·ər ˌkäm,pleks }

processor error interrupt [COMPUT SCI] The interruption

of a computer program because a parity check indicates an error in a word that has been transferred to or within the central processing unit. { 'prä,ses·ər ¦er·ər ¦int·ə,rəpt }

processor-limited [COMPUT SCI] Property of a computer system whose processing time is determined by the speed of its central processing unit rather than by the speed of its peripheral equipment. Also known as process-limited. { 'prä,ses·ər ,lim·əd·əd }

processor-memory-switch notation See PMS notation. { 'prä,ses·ər 'mem·rē ,swich nō,tā·shən }

processor stack pointer [COMPUT SCI] A programmable register used to access all temporary-storage words related to an interrupt-service routine which was halted when a new service routine was called in. { 'prä,ses·ər 'stak ,póint·ər }

processor status word [COMPUT SCI] A word comprising a set of flag bits and the interrupt-mask status. { 'prä,ses·ər 'stad·əs ,wərd }

process piping [ENG] In an industrial facility, pipework whose function is to convey the materials used for the manufacturing processes. { 'prä,ses ,pīp·iŋ }

process planning [IND ENG] Determining the conditions necessary to convert material from one state to another. { 'prä,ses ,plan·iŋ }

process printing [GRAPHICS] The printing from a series of two or more halftone plates to produce intermediate colors and shades. { 'prä,ses ,print·iŋ }

process reengineering [SYS ENG] The study, capture, and modification of the internal mechanisms or functionality of an existing process or systems-engineering life cycle in order to reconstitute it in a new form and with new functional and nonfunctional features, often to take advantage of newly emerged or desired organizational or technological capabilities without changing the inherent purpose of the process that is being reengineered. { ,prä'səs ,rē,en·jə'nir·iŋ }

process research [SCI TECH] Applied research with a new or improved process in view. { 'prä,ses ri,sərch }

process schizophrenia [PSYCH] Schizophrenia having a slow, insidious onset; the person exhibits poor adjustment before hospitalization. { 'prä,ses ,skit·sə'frē·nē·ə }

process sequencing [IND ENG] Specification of the appropriate order for the processes required to manufacture a part. { 'prä,ses ,sē·kwəns·iŋ }

process simulation [COMPUT SCI] The use of computer programming, computer vision, and feedback to simulate manufacturing techniques. { 'prä,ses ,sim·yə,lā·shən }

process time [IND ENG] **1.** Time needed for completion of the machine-controlled portion of a work cycle. **2.** Time required for completion of an entire process. { 'prä,ses ,tīm }

process variable [CHEM ENG] Any of those varying operational and physical conditions associated with a chemical processing operation, such as temperature, pressure, flowrate, density, pH, viscosity, or chemical composition. { 'prä,ses ,ver·ē·ə·bəl }

prochirality [ORG CHEM] The property displayed by a molecule or atom which contains (or is bonded to) two constitutionally identical ligands; Also known as prostereoisomerism. { ,prō·kī'ral·əd·ē }

PROCLIB See procedure library. { 'präk,lib }

procoagulant [BIOCHEM] Any of blood clotting factors V to VIII; accelerates the conversion of prothrombin to thrombin in the presence of thromboplastin and calcium. { ¦prō·kō'ag·yə·lənt }

Procoela [VERT ZOO] A suborder of the Anura characterized by a procoelous vertebral column and a free coccyx articulating with a double condyle. { prō'sēl·ə }

procoelous [VERT ZOO] The form of a vertebra that is concave anteriorly and convex posteriorly. { prō'sēl·əs }

procollagen [BIOCHEM] A high-molecular-weight form of collagen that is found in intracellular spaces and is believed to be the precursor of collagen. { prō'käl·ə·jən }

Procolophonia [PALEON] A subclass of extinct cotylosaurian reptiles. { ,präk·ə·lə'fō·nē·ə }

proctiger [INV ZOO] The cone-shaped, reduced terminal segment of the abdomen of an insect which contains the anus. { 'präk·tə·jər }

proctitis [MED] Inflammation of the anus or rectum. { präk'tīd·əs }

proctodone [INV ZOO] An insect hormone that causes diapause to end. { 'präk·tə,dōn }

proctology [MED] A branch of medicine concerned with the structure and disease of the anus, rectum, and sigmoid colon. { präk'täl·ə·jē }

proctoscope [MED] An instrument for inspecting the anal canal and rectum. { 'präk·tə,skōp }

proctosigmoidectomy [MED] The abdominoperineal excision of the anus and rectosigmoid, usually with the formation of an abdominal colostomy. { ¦präk·tō,sig,mói'dek·tə·mē }

Proctotrupidae [INV ZOO] A family of hymenopteran insects in the superfamily Proctotrupoidea. { ,präk·tə'trü·pə,dē }

Proctotrupoidea [INV ZOO] A superfamily of parasitic Hymenoptera in the suborder Apocrita. { ,präk·tə·trə'póid·ē·ə }

procumbent [BOT] Having stems that lie flat on the ground but do not root at the nodes. [SCI TECH] **1.** Lying stretched out. **2.** Slanting forward. **3.** Lying face down. { prō'kəm·bənt }

procurement [ORD] **1.** The complete action or process of acquiring or obtaining personnel, materiel, services, or property from outside a military service by means authorized in pertinent directives. **2.** More specifically, the action or process of acquiring or obtaining materiel, property, or services at the operational level, for example, purchasing, contracting, and negotiating directly with the source of supply. { prə'kyur·mənt }

procurement lead time [ORD] The time elapsing between the initiation of procurement action and the receipt into the supply system of the materiel procured. { prə'kyur·mənt ¦lēd ,tīm }

Procyon [ASTRON] A star of magnitude 0.3, of spectral type F5, and 11 light-years (1.04×10^{17} meters) from earth; one of a binary. Also known as α Canis Minoris. { 'prō·sē,än }

Procyonidae [VERT ZOO] A family of carnivoran mammals in the superfamily Canoidea, including raccoons and their allies. { ,prō·sē'än·ə,dē }

prod See test prod. { präd }

prod cast [GEOL] The cast of a prod mark. Also known as impact cast. { 'präd ,kast }

prodelta [GEOL] The part of a delta lying beyond the delta front, and sloping gently down to the basin floor of the delta; it is entirely below the water level. { 'prō,del·tə }

prodelta clay [GEOL] Fine sand, silt, and clay transported by the river and deposited on the floor of a sea or lake beyond the main body of a delta. { 'prō,del·tə ,klā }

Prodinoceratinae [PALEON] A subfamily of extinct herbivorous mammals in the family Uintatheriidae; animals possessed a carnivorelike body of moderate size. { ¦präd·ən·ō·sə'rat·ən,ē }

prod mark [GEOL] A short tool mark oriented parallel to the current and gradually deepening downcurrent. Also known as impact mark. { 'präd ,märk }

prodrome [MED] **1.** An early or premonitory manifestation of impending disease before the specific symptoms begin. **2.** An aura. { 'prō,drōm }

prodrug [PHYSIOL] An inactive precursor of a drug that is activated via the body's metabolism. { 'prō,drəg }

producer [ECOL] An autotrophic organism of the ecosystem; any of the green plants. { prə'dü·sər }

producer gas [MATER] Fuel gas high in carbon monoxide and hydrogen, produced by burning a solid fuel with a deficiency of air or by passing a mixture of air and steam through a bed of incandescent fuel; used as a cheap, low-Btu industrial fuel. { prə'dü·sər ,gas }

producer's risk [IND ENG] The probability that in an acceptance sampling plan, material of an acceptable quality level will be rejected. { prə'dü·sərz ,risk }

producing gas-oil ratio [PETRO ENG] The ratio of gas to oil (GOR, or gas-oil ratio) from a producing well; an increase in GOR is a danger signal in the efficient control of reservoir performance. { prə'düs·iŋ ¦gas ¦oil 'rā·shō }

producing horizon [PETRO ENG] A reservoir bed within the stratigraphic series of an oil province from which gas or liquid hydrocarbons can be obtained by drilling a well. { prə'düs·iŋ hə,rīz·ən }

producing reserves [PETRO ENG] Developed (proved) petroleum reserves to be produced by existing wells in that portion of a reservoir subjected to full-scale secondary-recovery operations. { prə'düs·iŋ ri,zərvz }

PROCUMBENT

The procumbent stem of purslane.

product [CHEM] A substance formed as a result of a chemical reaction. [CHEM ENG] *See* discharge liquor. [IND ENG] **1.** An item or goods made by an industrial firm. **2.** The total of such items or goods. [MATH] **1.** For two integers, m and n, the number of objects in the set formed by combining m sets, each of which has n objects. **2.** For two rational numbers, a/b and c/d, where a, b, c, and d are integers, the number $(ac)/(bd)$. **3.** For any two real numbers, which are the limits of sequences of rational numbers p_n and q_n respectively, the limit of the sequence $p_n q_n$. **4.** The product of two algebraic quantities is the result of their multiplication relative to an operation analogous to multiplication of real numbers. **5.** The product of a collection of sets A_1, A_2, ..., A_n is the set of all elements of the form $(a_1, a_2, ..., a_n)$ where each a_i is an element of A_i for each $i = 1, 2, ..., n$. **6.** For two transformations, the transformation that results from their successive application. **7.** For two fuzzy sets A and B, with membership functions m_A and m_B, the fuzzy set whose membership function $m_{A \cdot B}$ satisfies the equation $m_{A \cdot B}(x) = m_A(x) \cdot m_B(x)$ for every element x. **8.** The product AB of two matrices A and B, where the number n of columns in A equals the number of rows in B, is the matrix whose element c_{ij} in row i and column j is the sum over $k = 1, 2, ..., n$ of the product of the elements a_{ik} in A and b_{kj} in B. { 'präd·əkt }

product bundle [MATH] A bundle whose total space is the cartesian product of the base space B and a topological space F and whose projection map sends (b,a) to b. { 'präd·əkt ˌbən·dəl }

product demodulator [ELECTR] A receiver demodulator whose output is the product of the input signal voltage and a local oscillator signal voltage at the input frequency. Also known as product detector. { 'präd·əkt di ˌmäj·ə ˌlād·ər }

product design [DES ENG] The determination and specification of the parts of a product and their interrelationship so that they become a unified whole. { 'präd·əkt di ˌzīn }

product detector *See* product demodulator. { 'präd·əkt di ˌtek·tər }

Productinida [PALEON] A suborder of extinct articulate brachiopods in the order Strophomenida characterized by the development of spines. { ˌprä'dək'tin·ə·də }

production [COMPUT SCI] **1.** The processing of useful work by a computer system, excluding the development and testing of new programs. **2.** A rule in a grammar of a formal language that describes how parts of a string (or word, phrase, or construct) can be replaced by other strings. Also known as rule of inference. [ENG] Output, such as units made in a factory, oil from a well, or chemicals from a processing plant. { prə'dək·shən }

production control [IND ENG] The procedure for planning, routing, scheduling, dispatching, and expediting the flow of materials, parts, subassemblies, and assemblies within a plant, from the raw state to the finished product, in an orderly and efficient manner. { prə'dək·shən kən ˌtrōl }

production-decline curve [PETRO ENG] A graphical means to estimate the ultimate recovery (oil or gas) from a reservoir; cumulative production is plotted against time, the curve being extrapolated to an end point (that is, ultimate recovery). { prə'dək·shən di ˌklīn ˌkərv }

production ecology *See* ecological energetics. { prə'duk·shən ē'käl·ə·jē }

production engineering [IND ENG] The planning and control of the mechanical means of changing the shape, condition, and relationship of materials within industry toward greater effectiveness and value. { prə'dək·shən ˌen·jə'nir·iŋ }

production model [IND ENG] A model in its final mechanical and electrical form of final production design made by production tools, jigs, fixtures, and methods. { prə'dək·shən ˌmäd·əl }

production packer [PETRO ENG] A downhole tool used to assist in the efficient production of oil and gas from a well having one or more productive horizons; the function is to provide a seal between the outside of the tubing and the inside of the casing to prevent movement of fluids past that point. Also known as packer. { prə'dək·shən ˌpak·ər }

production program [COMPUT SCI] A proprietary program used primarily for internal processing in a business and not generally made available to third parties for profit. { prə'dək·shən ˌprō ˌgram }

production reactor [NUCLEO] A nuclear reactor designed primarily for large-scale production of transmutation products, such as plutonium. { prə'dək·shən rē ˌak·tər }

production requirements [IND ENG] The sum of authorized stock levels and pipeline needs less stocks expected to become available, stock on hand, stocks due in, returned stocks, and stocks from salvage, reclamation, rebuild, and other sources. { prə'dək·shən ri ˌkwīr·məns }

production standard *See* standard time. { prə'dək·shən ˌstan·dərd }

production tank *See* lease tank. { prə'dək·shən ˌtaŋk }

production test [COMPUT SCI] A test of a computer system with actual data in the environment where it will be used. { prə'dək·shən ˌtest }

production time [COMPUT SCI] Good computing time, including occasional duplication of one case for a check or rerunning of the test run; also including duplication requested by the sponsor, any reruns caused by misinformation or bad data supplied by sponsor, and error studies using different intervals, covergence criteria, and so on. { prə'dək·shən ˌtīm }

production track [ENG ACOUS] A sound track which is either prerecorded or recorded directly on the set, and which exists in the film at that time when the music breakdown for scoring is about to begin. { prə'dək·shən ˌtrak }

production tubing [PETRO ENG] The final string of pipe that is placed in an oil well after the oil has stopped flowing naturally. { prə'dək·shən ˌtüb·iŋ }

productive cough [MED] A cough accompanied by expectoration. { prə'dək·tiv 'kof }

productive time [IND ENG] Time during which useful work is performed in an operation or process. { prə'dək·tiv 'tīm }

productivity [AGR] The yield of a given crop per unit of land. [IND ENG] The ratio of output production to input effort; it is an indicator of the efficiency with which an enterprise converts its resources (inputs) into finished goods or services (outputs). [PETRO ENG] Measure of an oil well's ability to produce liquid or gaseous hydrocarbons; categories include relative, specific, ultimate, and fractured-well productivity. { ˌprä ˌdək'tiv·əd·ē }

productivity index [PETRO ENG] The number of barrels of oil produced per day per decline in well bottom-hole pressure in pounds per square inch. { ˌprä ˌdək'tiv·əd·ē ˌin ˌdeks }

productivity ratio [PETRO ENG] **1.** The amount of damage or improvement to reservoir formation permeability adjacent to the borehole (due to invasion or reduction of drilling mud present, drilling-fluid filtrate water, swollen clay particles, or salt or wax deposition). **2.** The ratio of permeability calculated from the productivity index to the permeability calculated from reservoir buildup pressure. { ˌprä ˌdək'tiv·əd·ē ˌrā·shō }

productivity test [PETRO ENG] Graphical relation of bottomhole static pressure (calculated or measured) versus producing pressure for various gas flow rates; used to predict future oil well behavior. { ˌprä ˌdək'tiv·əd·ē ˌtest }

product life-cycle [IND ENG] All the phases, from conception and scale-up, through production, growing use, maturity, and obsolescence of a product. { 'präd·əkt 'līf ˌsī·kəl }

product line [IND ENG] **1.** The range of products offered by a firm. **2.** A group of basically similar products, differentiated only by such characteristics as color, style, or size. { 'präd·əkt ˌlīn }

product measure [MATH] A measure on a product of measure spaces constructed from the measures on each of the individual spaces by taking the measure of the product of a finite number of measurable sets, one from each of the measure spaces in the product, to be the product of the measures of these sets. { 'präd·əkt ˌmezh·ər }

product model [STAT] A model for independent repetition of an experiment, or independent performance of several experiments, obtained by taking the cartesian product of the probability spaces representing the experiments. { 'präd·əkt ˌmäd·əl }

product modulator [ELECTR] Modulator whose modulated output is substantially equal to the carrier and the modulating wave; the term implies a device in which intermodulation between components of the modulating wave does not occur. { 'präd·əkt ˌmäj·ə ˌlād·ər }

product-moment coefficient *See* sample correlation coefficient. { 'präd·əkt 'mō·mənt ˌkō·i'fish·ənt }

product of inertia [MECH] Relative to two rectangular axes, the sum of the products formed by multiplying the mass (or, sometimes, the area) of each element of a figure by the product of the coordinates corresponding to those axes. { 'prä·dəkt əv i'nȯr·shə }

product reengineering [SYS ENG] The study, capture, and modification of the internal mechanisms or functionality of an existing system or product in order to reconstitute it in a new form with new features, often to take advantage of newly emerged technologies without major change to the inherent functionality and purpose of the system. { ¦präd·əkt ¦rē¸en 'jə'nir·iŋ }

product research [SCI TECH] Applied research with a new or improved product in view. { 'prä·dəkt ri¸sərch }

product topology [MATH] A topology on a product of topological spaces whose open sets are constructed from cartesian products of open sets from the individual spaces. { 'prä¸dəkt tə'päl·ə·jē }

product water [CHEM ENG] Fresh water that is produced by a desalination process; Also known as converted water. { 'prä¸dəkt ¸wȯd·ər }

proenzyme See zymogen. { prō'en¸zīm }

proerythroblast of Ferrata See pronormoblast. { ¦prō·ə¦rith·rə¸blast əv fe'räd·ə }

proestrus [PHYSIO] The beginning of the follicular phase of estrus. { prō'es·trəs }

profibrinolysin See plasminogen. { ¦prō¸fī·brə'näl·ə·sən }

profile [GEOL] **1.** The outline formed by the intersection of the plane of a vertical section and the ground surface. Also known as topographic profile. **2.** Data recorded by a single line of receivers from one shot point in seismic prospecting. [GEOPHYS] A graphic representation of the variation of one property, such as gravity, usually as ordinate, with respect to another property, usually linear, such as distance. [HYD] A vertical section of a potentiometric surface, such as a water table. [PETR] In structural petrology, a cross section of a homoaxial structure. { 'prō¸fīl }

profile chart [ELECTROMAG] A vertical cross-section drawing of a microwave path between two stations, indicating terrain, obstructions, and antenna height requirements. { 'prō¸fīl ¸chärt }

profile die [ENG] A plastics extrusion die used to produce continuous shapes, but not tubes or sheets. { 'prō¸fīl ¸dī }

profiled keyway [DES ENG] A keyway for a straight key formed by an end-milling cutter. Also known as end-milled keyway. { 'prō¸fīld 'kē¸wā }

profile drag [FL MECH] That part of the airfoil drag that results from the skin friction and the shape of the airfoil as indicated by the airfoil profile. { 'prō¸fīl ¸drag }

profile line [GEOL] The top line of a profile section, representing the intersection of a vertical plane with the surface of the ground. { 'prō¸fīl ¸līn }

profile of equilibrium [GEOL] **1.** The slope of the floor of a sea, ocean, or lake, taken in a vertical plane, when deposition of sediment is balanced by erosion. **2.** The longitudinal profile of a graded stream. Also known as equilibrium profile; graded profile. { 'prō¸fīl əv ¸ē·kwə'lib·rē·əm }

profile section [GEOL] A diagram or drawing that shows along a given line the configuration or slope of the surface of the ground as it would appear if it were intersected by a vertical plane. { 'prō¸fīl ¸sek·shən }

profile thickness [AERO ENG] The maximum distance between the upper and lower contours of an airfoil, measured perpendicular to the mean line of the profile. { 'prō¸fīl ¸thik·nəs }

profiling [ENG] Electrical exploration wherein the transmitter and receiver are moved in unison across a structure to obtain a profile of mutual impedance between transmitter and receiver. Also known as lateral search. { 'prō¸fīl·iŋ }

profiling machine [MECH ENG] A machine used for milling irregular profiles; the cutting tool is guided by the contour of a model. { 'prō¸fīl·iŋ mə¸shēn }

profiling snow gage [HYD] A type of radioactive gage for measuring the water equivalent and density/depth distribution of a snowpack, consisting of a radioactive source and a radioactivity detector which move up and down in two adjacent vertical pipes surrounded by snow. Also known as nuclear twin-probe gage. { 'prō¸fīl·iŋ 'snō ¸gāj }

profilograph [ENG] An instrument for measuring and recording roughness of the surface over which it travels. { prō'fīl·ə¸graf }

profilometer [ENG] An instrument for measuring the roughness of a surface by means of a diamond-pointed tracer arm attached to a coil in an electric field; movement of the arm across the surface induces a current proportional to surface roughness. { ¸prō·fə'läm·əd·ər }

profit in sight [MIN ENG] Probable gross profit from a mine's ore reserves, as distinct from the ground that is still to be blocked out. { 'präf·ət in ¸sīt }

profit sharing [IND ENG] Sharing of company profits with the employees. { 'präf·ət ¸sher·iŋ }

proflavine sulfate [ORG CHEM] $C_{13}H_{11}N_3 \cdot H_2SO_4$ A reddish-brown, crystalline powder, soluble in alcohol and water; used in medicine. { prō'flā¸vēn 'səl¸fāt }

profunda [ANAT] Deep-seated; applied to certain arteries. { prō'fən·də }

profundal zone [ECOL] The region occurring below the limnetic zone and extending to the bottom in lakes deep enough to develop temperature stratification. { prō'fənd·əl ¸zōn }

Proganosauria [PALEON] The equivalent name for Mesosauria. { ¸prō¸gan·ə'sȯr·ē·ə }

progenitor cell [CYTOL] A precursor cell that completes a series of cell divisions to produce a distinct cell lineage. { prə'jen·əd·ər ¸sel }

progeny [BIOL] Offspring, descendants. { 'präj·ə·nē }

progeny test [GEN] The assessment of parental genotype by study of its progeny under controlled conditions. { 'präj·ə·nē ¸test }

progeria [MED] An abnormal childhood state of premature senescence, characterized by wrinkled skin, gray hair, lack of pubic or facial hair, development of atherosclerosis, and a short life span. Also known as Hutchinson-Gilford syndrome. { ¸prō'jir·ē·ə }

progestational hormone [BIOCHEM] **1.** The natural hormone progesterone, which induces progestational changes of the uterine mucosa. **2.** Any derivative or modification or progesterone having similar actions. { ¦prō¸je'stā·shən·əl 'hȯr¸mōn }

progesterone [BIOCHEM] $C_{21}H_{30}O_2$ A steroid hormone produced in the corpus luteum, placenta, testes, and adrenals; plays an important physiological role in the luteal phase of the menstrual cycle and in the maintenance of pregnancy; it is an intermediate in the biosynthesis of androgens, estrogens, and the corticoids. { prō'jes·tə¸rōn }

proglacial [GEOL] Of streams, deposits, and other features, being immediately in front of or just beyond the outer limits of a glacier or ice sheet, and formed by or derived from glacier ice. { prō'glā·shəl }

proglottid [INV ZOO] One of the segments of a tapeworm. { prō'gläd·əd }

prognathic [ANTHRO] A condition of the upper jaw in which it projects anteriorly with respect to the profile of the facial skeleton, when the skull is oriented on the Frankfort horizontal plane; having a gnathic index of 103.0 or more. { präg 'nath·ik }

prognosis [MED] A prediction as to the course and outcome of a disease, injury, or developmental abnormality. { präg'nō·səs }

prognostic chart [METEOROL] A chart showing, principally, the expected pressure pattern (or height pattern) of a given synoptic chart at a specified future time; usually, positions of fronts are also included, and the forecast values of other meteorological elements may be superimposed. { präg'näs·tik ¦chär· }

prognostic equation [METEOROL] Any equation governing a system which contains a time derivative of a quantity and therefore can be used to determine the value of that quantity at a later time when the other terms in the equation are known (for example, the vorticity equation). { präg'näs·tik i'kwā·zhər }

progradation [GEOL] Seaward buildup of a beach, delta, or fan by nearshore deposition of sediments transported by a river, by accumulation of material thrown up by waves, or by material moved by longshore drifting. { ¦prō·grə'dā·shən }

prograde metamorphism [GEOL] Metamorphic changes in response to a higher pressure or temperature than that to which the rock was last adjusted. { ¦prō'grād ¸med·ə'mȯr¸fiz·əm }

prograde motion [ASTRON] **1.** The apparent motion of a planet around the sun in the direction of the sun's rotation. **2.** See prograde orbit. { prō'grād 'mō·shən }

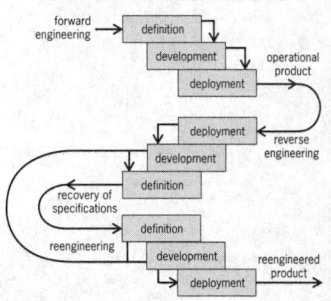

PRODUCT REENGINEERING

Product reengineering as a sequence of forward, reverse, and forward engineering.

PROGESTERONE

Structural formula of progesterone.

prograde orbit [ASTRON] Orbital motion in the usual direction of celestial bodies within a given system; specifically, of a satellite, motion in the direction of rotation of the primary. Also known as prograde motion. { ¦prō'grād 'ȯr·bət }

prograding shoreline [GEOL] A shoreline that is being built seaward by accumulation or deposition. { ¦prō'grād·iŋ 'shȯr,līn }

program [AERO ENG] In missile guidance, the planned flight path events to be followed by a missile in flight, including all the critical functions, preset in a program device, which control the behavior of the missile. [COMMUN] A sequence of audio signals alone, or audio and video signals, transmitted for entertainment or information. [COMPUT SCI] A detailed and explicit set of directions for accomplishing some purpose, the set being expressed in some language suitable for input to a computer, or in machine language. [IND ENG] An undertaking of significant scope that is enduring rather than occurring within a limited time span. { 'prō·grəm or 'prō,gram }

program analysis [COMPUT SCI] The process of determining the functions to be carried out by a computer program. { 'prō·grəm ə,nal·ə·səs }

program block [COMPUT SCI] A division or section of a computer program that functions to a large extent as if it were a separate program. { 'prō·grəm ,bläk }

program check [COMPUT SCI] A built-in check system in a program to determine that the program is running correctly. { 'prō·grəm ,chek }

program compatibility [COMPUT SCI] The type of compatibility shared by two computers that can process the identical program or programs written in the same source language or machine language. { 'prō·grəm kəm,pad·ə'bil·əd·ē }

program control [CONT SYS] A control system whose set point is automatically varied during definite time intervals in order to make the process variable vary in some prescribed manner. { 'prō·grəm kən,trōl }

program conversion [COMPUT SCI] The changing of the source language of a computer program from one dialect to another, or the modification of the program to operate with a different operating system or data-base management system. { 'prō·grəm kən,vər·zhən }

program counter See instruction counter. { 'prō·grəm ,kaunt·ər }

program design [COMPUT SCI] The phase of computer program development in which the hardware and software resources needed by the program are identified and the logic to be used by the program is determined. { 'prō·grəm di,zīn }

program development time [COMPUT SCI] The total time taken on a computer to produce operating programs, including the time taken to compile, test, and debug programs, plus the time taken to develop and test new procedures and techniques. { 'prō·grəm di'vel·əp·mənt ,tīm }

program device [CONT SYS] In missile guidance, the automatic device used to control time and sequence of events of a program. { 'prō·grəm di,vīs }

program editor [COMPUT SCI] A computer routine used in time-sharing systems for on-line modification of computer programs. { 'prō·grəm ,ed·ə·tər }

program element [COMPUT SCI] Part of a central computer system that carries out the instruction sequence scheduled by the programmer. { 'prō·grəm ,el·ə·mənt }

program evaluation and review technique See PERT. { 'prō·grəm i,val·yə'wā·shən ən ri'vyü tek,nēk }

program failure alarm [COMMUN] Signal-operated radio or television relay that gives a visual and aural alarm when the program fails on the line being monitored; a time delay is provided to prevent the relay from operating and giving a false alarm during station identification periods or other short periods of silence in program continuity. { 'prō·grəm 'fāl·yər ə,lärm }

program generator [COMPUT SCI] A program that permits a computer to write other programs automatically. { 'prō·grəm ,jen·ə,rād·ər }

program level [ENG ACOUS] The level of the program signal in an audio system, expressed in volume units. { 'prō·grəm ,lev·əl }

program library [COMPUT SCI] An organized set of computer routines and programs. { 'prō·grəm ,lī·brer·ē }

program listing [COMPUT SCI] A list of the statements in a computer program, usually produced as a by-product of the compilation of the program. { 'prō·grəm ,list·iŋ }

program logic [COMPUT SCI] A particular sequence of instructions in a computer program. { 'prō·grəm ,läj·ik }

programmable calculator [COMPUT SCI] An electronic calculator that has some provision for changing its internal program, usually by inserting a new magnetic card on which the desired calculating program has been stored. { prō'gram·ə·bəl 'kal·kyə,lād·ər }

programmable controller [CONT SYS] A control device, normally used in industrial control applications, that employs the hardware architecture of a computer and a relay ladder diagram language. Also known as programmable logic controller. { prō'gram·ə·bəl kən'trōl·ər }

programmable counter [ELECTR] A counter that divides an input frequency by a number which can be programmed into decades of synchronous down counters; these decades, with additional decoding and control logic, give the equivalent of a divide-by-N counter system, where N can be made equal to any number. { prō'gram·ə·bəl 'kaunt·ər }

programmable decade resistor [ELECTR] A decade box designed so that the value of its resistance can be remotely controlled by programming logic as required for the control of load, time constant, gain, and other parameters of circuits used in automatic test equipment and automatic controls. { prō'gram·ə·bəl 'de,kād ri,zis·tər }

programmable device [COMPUT SCI] Any device whose operation is controlled by a stored program that can be changed or replaced. { prō'gram·ə·bəl di'vīs }

programmable electronic system [SYS ENG] A system based on a computer and connected to sensors or actuators for the purpose of control, protection, or monitoring. { prō'gram·ə·bəl i'lek,trän·ik ,sis·təm }

programmable logic array See field-programmable logic array. { prō'gram·ə·bəl ,läj·ik ə,rā }

programmable logic controller See programmable controller. { prō'gram·ə·bəl ,läj·ik kən,trōl·ər }

programmable power supply [ELEC] A power supply whose output voltage can be changed by digital control signals. { prō'gram·ə·bəl 'pau·ər sə,plī }

programmable read-only memory [COMPUT SCI] An integrated-circuit memory chip which can be programmed only once by the user after which the information stored in the chip cannot be altered. Abbreviated PROM. { prō'gram·ə·bəl ¦rēd ¦ōn·lē 'mem·rē }

program maintenance [COMPUT SCI] The updating of computer programs both by error correction and by alteration of programs to meet changing needs. { 'prō·grəm 'mānt·ən·əns }

programmatic interface See application program interface. { ,prō·grə¦mad·ik 'in·tər,fās }

programmed check [COMPUT SCI] **1.** An error-detecting operation programmed by instructions rather than built into the hardware. **2.** A computer check in which a sample problem with known answer, selected for having a program similar to that of the next problem to be run, is put through the computer. { 'prō,gramd 'chek }

programmed dump [COMPUT SCI] A storage dump which results from an instruction in a computer program at a particular point in the program. { 'prō,gramd 'dəmp }

programmed function key [COMPUT SCI] A key on the keyboard of a computer terminal that lacks a predefined function but can be assigned a function by a computer program. Abbreviated PF key. { 'prō,gramd 'fəŋk·shən ,kē }

programmed halt [COMPUT SCI] A halt that occurs deliberately as the result of an instruction in the program. Also known as programmed stop. { 'prō,gramd 'hȯlt }

programmed logic array [ELECTR] An array of AND/OR logic gates that provides logic functions for a given set of inputs programmed during manufacture and serves as a read-only memory. Abbreviated PLA. { 'prō,gramd ¦läj·ik ə,rā }

programmed marginal check [COMPUT SCI] Computer program that varies its own voltage to check some piece of electronic computer equipment during a preventive maintenance check. { 'prō,gramd 'mär·jən·əl 'chek }

programmed operators [COMPUT SCI] Computer instructions which enable subroutines to be accessed with a single programmed instruction. { 'prō,gramd 'äp·ə,rād·ərz }

programmed stop See programmed halt. { 'prō,gramd 'stäp }

programmed turn [AERO ENG] The automatically controlled turn of a ballistic missile into the curved path that will lead to the correct velocity and vector for the final portion of the trajectory. { 'prō,gramd 'tərn }

programmer [COMPUT SCI] A person who prepares sequences of instructions for a computer, without necessarily converting them into the detailed codes. [CONT SYS] A device used to control the motion of a missile in accordance with a predetermined plan. { 'prō,gram·ər }

programmer analyst [COMPUT SCI] A person who both writes computer programs and analyzes and designs information systems. { 'prō,gram·ər 'an·əl,ist }

programmer-defined macroinstruction [COMPUT SCI] A macroinstruction which is equivalent to a set of ordinary instructions as specified by the programmer for use in a particular computer program. { 'prō,gram·ər di¦fīnd ¦ma·krō·in'strak·shən }

programmer's tool kit [COMPUT SCI] A collection of programs designed to help programmers in developing software, usually oriented toward a particular programming language. { 'prō,gram·ərz 'tül ,kit }

programming [COMPUT SCI] Preparing a detailed sequence of operating instructions for a particular problem to be run on a digital computer. Also known as computer programming. [ENG] In a plastics process, extruding a parison whose thickness differs longitudinally in order to equalize wall thickness of the blown container. { 'prō,gram·iŋ }

programming language [COMPUT SCI] The language used by a programmer to write a program for a computer. { 'prō,gram·iŋ ,laŋ·gwij }

programming panel [CONT SYS] A device used to edit a program or insert and monitor it in a programmable controller. { 'prō,gram·iŋ ,pan·əl }

programming unit See manual control unit. { 'prō,gram·iŋ ,yü·nət }

program module [COMPUT SCI] A logically self-contained and discrete part of a larger computer program, for example, a subroutine or a coroutine. { 'prō·gram ,mäj·yül }

program monitor [COMMUN] A monitor used to observe the quality of a radio or television broadcast. { 'prō·gram 'män·əd·ər }

program parameter [COMPUT SCI] In computers, an adjustable parameter in a subroutine which can be given a different value each time the subroutine is used. { 'prō·gram pə'ram·əd·ər }

program register [COMPUT SCI] The register in the control unit of a digital computer that stores the current instruction of the program and controls the operation of the computer during the execution of that instruction. Also known as computer control register. { 'prō·gram ,rej·ə·stər }

program scan [CONT SYS] The span of time during which a programmable controller processor executes all the instructions of a given program. { 'prō·gram ,skan }

program-sensitive fault [COMPUT SCI] A hardware malfunction that appears only in response to a particular sequence (or kind of sequence) of program instructions. { 'prō·gram ¦sen·səd·iv 'fȯlt }

program specification [COMPUT SCI] A statement of the precise functions which are to be carried out by a computer program, including descriptions of the input to be processed by the program, the processing needed, and the output from the program. { 'prō·gram ,spes·ə·fə'kā·shən }

program star [ASTRON] A star whose properties are observed or measured during a specified series of observations. { 'prō·gram ,stär }

program state [COMPUT SCI] The mode of operation of a computer during the execution of instructions in an application program. { 'prō·gram ,stāt }

program status word [COMPUT SCI] An internal register to the central processing unit denoting the state of the computer at a moment in time. { 'prō·gram 'stad·əs ,wərd }

program step [COMPUT SCI] In computers, some part of a program, usually one instruction. { 'prō·gram ,step }

program stop [COMPUT SCI] An instruction built into a computer program that will automatically stop the machine under certain conditions, or upon reaching the end of processing or completing the solution of a program. Also known as halt instruction; stop instruction. { 'prō·gram ,stäp }

program storage [COMPUT SCI] Portion of the internal storage reserved for the storage of programs, routines, and subroutines; in many systems, protection devices are used to prevent inadvertent alteration of the contents of the program storage; contrasted with temporary storage. { 'prō·gram ,stȯr·ij }

program tape [COMPUT SCI] Tape containing the sequence of computer instructions for a given problem. { 'prō·gram tāp }

program test [COMPUT SCI] A system of checking before running any problem in which a sample problem of the same type with a known answer is run. { 'prō·gram ,test }

program testing time [COMPUT SCI] The machine time expended for program testing, debugging, and volume and compatibility testing. { 'prō·gram 'test·iŋ ,tīm }

program time [COMPUT SCI] The phase of computer operation when an instruction is being interpreted so that it can be carried out. { 'prō·gram ,tīm }

progress chart [IND ENG] A graphical representation of the degree of completion of work in progress. { 'präg·rəs ,chärt }

progression [MATH] A sequence or series of mathematical objects or quantities, each entry determined from its predecessors by some algorithm. [MET] The fixed dimension between adjacent stations in a progressive die and thus the precise distance the strip must advance between successive cycles of the press. { prə'gresh·ən }

progressive aging [MET] Aging of metals achieved by increasing the temperature in stages or by continuous elevation of the temperature. { prə'gres·iv 'āj·iŋ }

progressive block sequence [MET] A welding sequence in which the joint is completed in sections from one end to the other or from the center alternately to both ends. { prə'gres·iv 'bläk 'sē·kwəns }

progressive bonding [ENG] A method of curing a resin adhesive wherein heat and pressure are applied in successive steps. Also known as progressive gluing. { prə'gres·iv 'bänd·iŋ }

progressive die [MET] A die in which two or more operations are performed sequentially at different positions. { prə'gres·iv 'dī }

progressive forming [MET] Sequential forming at consecutive stations either with a single die or with separate dies. { prə'gres·iv 'form·iŋ }

progressive gluing See progressive bonding. { prə'gres·iv 'glü·iŋ }

progressive granulation [MATER] Propellant granulation in which the surface area of a grain increases during burning. { prə'gres·iv ,gran·yə'lā·shən }

progressive lateral sclerosis [MED] Amyotrophic lateral sclerosis with primary involvement of the pyramidal tracts. { prə'gres·iv 'lad·ə·rəl sklə'rō·səs }

progressive lens See omnifocal lens. { prə'gres·iv 'lenz }

progressive metamorphism [GEOL] Systematic change in metamorphic grade from lower to higher in any metamorphic terrain. { prə'gres·iv ,med·ə'mȯr,fiz·əm }

progressive muscular dystrophy [MED] Chronic progressive dystrophy of the skeletal muscles. { prə'gres·iv 'məs·kyə·lər 'dis·trə·fē }

progressive overflow [COMPUT SCI] Retrieval of a randomly stored overflow record by a forward serial search from the home address. { prə'gres·iv 'ō·vər,flō }

progressive powder [MATER] A slow-burning explosive. { prə'gres·iv 'paud·ər }

progressive proofs [GRAPHICS] The proofs from color plates showing each color alone, and also in combination with each succeeding color in printing rotation; the order is yellow, magenta, cyan, and black. { prə'gres·iv 'prüfs }

progressive sand wave [GEOL] A sand wave characterized by downcurrent migration. { prə'gres·iv 'sand ,wāv }

progressive scanning [COMMUN] Scanning all lines in sequence, without interlace, so all picture elements are included during one vertical sweep of the scanning beam. Also known as sequential scanning. { prə'gres·iv 'skan·iŋ }

progressive sorting [GEOL] Sorting of sedimentary particles in the downcurrent direction, resulting in a systematic downcurrent decrease in the mean grain size of the sediment. { prə'gres·iv 'sȯrd·iŋ }

progressive wave [METEOROL] A wave or wavelike disturbance which moves relative to the earth's surface. [PHYS] A wave which transfers energy from one part of a medium to another, in contrast to a standing wave. Also known as free-traveling wave. { prə'gres·iv 'wāv }

progressive-wave antenna *See* traveling-wave antenna. { prə'gres·iv ¦wāv an'ten·ə }

Progymnospermopsida [PALEON] A class of plants intermediate between ferns and gymnosperms; comprises the Devonian genus *Archaeopteris*. { prō¦jim·nō,spər'mäp·səd·ə }

prohaptor [INV ZOO] The anterior attachment organ of a typical monogenetic trematode. { prō'hap·tər }

prohibited area [NAV] **1.** An airspace of defined dimensions identified by an area on the surface of the earth within which flight is prohibited, except as authorized by governmental authority. **2.** An area within which no vessels may be navigated, except as authorized by governmental authority. { prō 'hib·əd·əd 'er·ē·ə }

prohormone [BIOCHEM] The precursor of a hormone. { 'prō,hȯr,mōn }

Projapygidae [INV ZOO] A family of wingless insects in the order Diplura. { ¦prä·jə'pij·əd·ē }

project [ENG] A specifically defined task within a research and development field, which is established to meet a single requirement, either stated or anticipated, for research data, an end item of material, a major component, or a technique. { 'prä,jekt }

project development methodology [COMPUT SCI] A structured set of procedures designed to control the development of computer programs in a large organization. { 'prä,jekt di¦vel·əp·mənt ,meth·ə'däl·ə·jē }

projected planform [AERO ENG] The contour of the planform as viewed from above. { prə'jek·təd 'plan,fȯrm }

projected-scale instrument [ENG] An indicating instrument in which a light beam projects an image of the scale on a screen. { prə'jek·təd ¦skāl ,in·strə·mənt }

projected window [BUILD] A window having one or more rotatable sashes which swing either inward or outward. { prə'jek·təd 'win,dō }

project engineering [ENG] **1.** The engineering design and supervision (coordination) aspects of building a manufacturing facility. **2.** The engineering aspects of a specific project, such as development of a product or solution to a problem. { 'prä,jekt en·jə'nir·iŋ }

projectile [AERO ENG] **1.** Any object, especially a missile, that is fired, thrown, launched, or otherwise projected. **2.** Originally, an object, such as a bullet or artillery shell, projected by an applied external force. { prə'jek·təl }

projectile ogive [ORD] A hollow, conical, metallic item, designed to enclose the forward portion of a projectile to reduce air resistance during flight. { prə'jek·təl 'ō,jīv }

projecting cylinder [MATH] A cylinder whose elements pass through a given curve and are perpendicular to one of the three coordinate planes. { prə¦jekt·iŋ 'sil·ən·dər }

projecting plane [MATH] A plane that contains a given straight line in space and is perpendicular to one of the three coordinate planes. { prə¦jekt·iŋ 'plān }

projection [MAP] A system for presenting on a plane surface the spherical surface of the earth or the celestial sphere; some of these systems are conic, cylindrical, gnomonic, Mercator, orthographic, and stereographic. Also known as map projection. [MATH] **1.** The continuous map for a fiber bundle. **2.** Geometrically, the image of a geometric object or vector superimposed on some other. **3.** A linear map *P* from a linear space to itself such that *P* ∘ *P* is equal to *P*. [PSYCH] Ascribing one's motives to someone else to disguise a source of conflict in oneself. { prə'jek·shən }

projection area [ANAT] An area of the cortex connected with lower centers of the brain by projection fibers. { prə'jek·shən ,er·ē·ə }

projection cathode-ray tube [ELECTR] A television cathode-ray tube designed to produce an intensely bright but relatively small image that can be projected onto a large viewing screen by an optical system. { prə'jek·shən ¦kath,ōd 'rā ,tüb }

projection chamber *See* projection spark chamber. { prə'jek·shən ,chām·bər }

projection display [ELECTR] An electronic system in which an image is generated on a high-brightness cathode-ray tube or similar electronic image generator and then optically projected onto a larger screen. { prə'jek·shən di'splā }

projection fibers [ANAT] Fibers joining the cerebral cortex to lower centers of the brain, and vice versa. { prə'jek·shən ,fī·bərz }

projection microradiography [PHYS] Microradiography in which an electron beam, focused into an extremely fine pencil, generates a point source of x-rays, and enlargement is achieved by placing the sample very near this source, and several centimeters from the recording material. Abbreviated PMR. Also known as shadow microscopy; x-ray projection microscopy. { prə'jek·shən ¦mī·krō,rad·ē'äg·rə·fē }

projection microscope [PHYS] An x-ray microscope which magnifies by image projection, either in contact microradiography or in projection microradiography. { prə'jek·shən 'mī·krə,skōp }

projection net *See* net. { prə'jek·shən ,net }

projection optics *See* Schmidt system. { prə'jek·shən ,äp·tiks }

projection plan position indicator [ELECTR] Unit in which the image of a 4-inch (10-centimeter) dark-trace cathode-ray tube is projected on a 24-inch (61-centimeter) horizontal plotting surface; the echoes appear as magenta-colored arcs on white background. { prə'jek·shən 'plan pə'zish·ən 'in·də,kād·ər }

projection printer [OPTICS] An optical, image-enlarging device, used in enlarging photographs. { prə'jek·shən ,print·ər }

projection printing [GRAPHICS] The production of photographic prints with the use of an enlarger. { prə'jek·shən ,print·iŋ }

projection slide [GRAPHICS] A positive transparent image on glass or film intended for projection onto a reflecting screen or the rear of a diffuse transmitting screen. Also known as diapositive; lantern slide. { prə'jek·shən ,slīd }

projection spark chamber [NUCLEO] A spark chamber in which the track of the particle is perpendicular, or nearly so, to the electric field, so that each electron of the track produces a streamer across the gap; the resulting curtain contains information only as to the projection of the track perpendicular to the electric field. Also known as projection chamber. { prə'jek·shən 'spärk ,chām·bər }

projection thermography [ENG] A method of measuring surface temperature in which thermal radiation from a surface is imaged by an optical system on a thin screen of luminescent material, and the pattern formed corresponds to the heat radiation of the surface. { prə'jek·shən thər'mäg·rə·fē }

projection welding [MET] Resistance welding in which the welds are localized at projections, intersections, and overlaps on the parts. { prə'jek·shən ,weld·iŋ }

projective geometry [MATH] The study of those properties of geometric objects which are invariant under projection. { prə'jek·tiv jē'äm·ə·trē }

projective group [MATH] A group of transformations arising in the general theory of projective geometry. { prə'jek·tiv 'grüp }

projective line [MATH] The line obtained from the stereographic projection of the circle. { prə'jek·tiv 'līn }

projective plain curve [MATH] The set of all points in the projective plane for which a particular homogeneous polynomial in the coordinates equals zero. { prə'jek·tiv ,plän 'kərv }

projective plane [MATH] **1.** The topological space obtained from the two-dimensional sphere by identifying antipodal points; the space of all lines through the origin in Euclidean space. **2.** More generally, a plane (in the sense of projective geometry) such that (1) every two points lie on exactly one line, (2) every two lines pass through exactly one point, and (3) there exists a four-point. { prə'jek·tiv 'plän }

projective point [MATH] The point from which a projection by rays is performed, as in stereographic projection. { prə'jek·tiv 'pȯint }

projective space [MATH] The topological space obtained from the *n*-dimensional sphere under identification of antipodal points. { prə'jek·tiv 'spās }

projective technique [PSYCH] A procedure used to identify and evaluate an individual's characteristic modes of thought and behavior, personality traits, attitudes, and motivation, by means of an objective test. { prə'jek·tiv tek'nēk }

projective test [PSYCH] Observation of a subject's responses to various test materials presented in a relatively unstructured, yet standard situation. { prə'jek·tiv 'test }

projective topology [MATH] The finest topology on the tensor product of two locally convex topological vector spaces such that the function that maps each element of the Cartesian product of the two spaces to the corresponding element of their

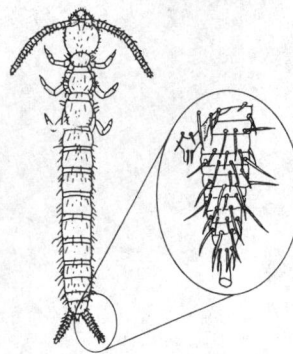

PROJAPYGIDAE

Anajapyx vesiculosus showing the cerci which contain a silk gland.

PROJECTION MICROSCOPE

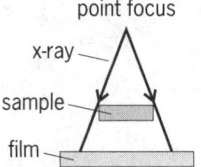

Principle of the projection x-ray microscope.

tensor product is a continuous function. { prə‚jek·tiv tə'päl·ə·jē }

project life *See* economic life. { 'prä·jikt ‚līf }

projector [ENG ACOUS] **1.** A horn designed to project sound chiefly in one direction from a loudspeaker. **2.** An underwater acoustic transmitter. [MATH] One of the lines or rays in a central projection. [OPTICS] *See* optical projection system. [ORD] **1.** Any apparatus for launching a projectile, such as a gun or rocket launcher. **2.** Smooth-bore-type barrel or other unrifled weapon from which pyrotechnic signals, grenades, and certain mortar projectiles are fired. **3.** A rack for launching target rockets. **4.** Special type of gun for projecting antisubmarine projectiles. { prə'jek·tər }

projector compass [NAV] A magnetic compass in which the lubber's line and compass card, or a portion thereof, are viewed as an image projected through a system of lenses upon a screen adjacent to the helmsman's position. { prə'jek·tər ‚kəm·pəs }

Prokaryotae [BIOL] A superkingdom of predominantly unicellular microorganisms lacking a membrane-bound nucleus containing chromosomes and having asexual reproduction by binary fission; it includes the kingdom Monera, and viruses, which are acellular, are included by some. { prō‚kar·ē'ō‚dē }

prokaryote [CYTOL] **1.** A primitive nucleus, where the deoxyribonucleic acid-containing region lacks a limiting membrane. **2.** Any cell containing such a nucleus, such as the bacteria and the blue-green algae. { prō'kar‚ē‚ōt }

Prolacertiformes [PALEON] A suborder of extinct terrestrial reptiles in the order Eosuchia distinguished by reduction of the lower temporal arcade. { prō‚las·ər·də'fòr‚mēz }

prolactin [BIOCHEM] A protein hormone produced by the adenohypophysis; stimulates lactation and promotes functional activity of the corpus luteum. Also known as lactogenic hormone; luteotropic hormone; mammary-stimulating hormone; mammogen; mammogenic hormone; mammotropin. { prō'lak·tən }

prolamellar body [BOT] An accumulation of vesicles formed by the invagination of the proplastid membrane during etiolation. { ‚prō·lə‚mel·ər 'bäd·ē }

prolamin [BIOCHEM] Any of the simple proteins, such as zein, found in plants; soluble in strong alcohol, insoluble in absolute alcohol and water. { prō'lam·ən }

prolapse [MED] The falling or sinking down of a part or organ. { 'prō‚laps }

prolapsed bedding [GEOL] Bedding characterized by a series of flat folds with near-horizontal axial planes contained entirely within a bed which has undisturbed boundaries. { 'prō‚lapst 'bed·iŋ }

prolate [SCI TECH] Pertaining to a body that is extended or elongated in the direction of an axis of symmetry. { 'prō‚lāt }

prolate cycloid [MATH] A trochoid in which the distance from the center of the rolling circle to the point describing the curve is greater than the radius of the circle. { 'prō‚lāt 'sī‚klòid }

prolate ellipsoid *See* prolate spheroid. { 'prō‚lāt i'lip‚sòid }

prolate spheroid [MATH] The ellipsoid or surface obtained by revolving an ellipse about one of its axes so that the equatorial circle has a diameter less than the length of the axis of revolution. Also known as prolate ellipsoid. { 'prō‚lāt 'sfir‚òid }

prolate spheroidal coordinate system [MATH] A three-dimensional coordinate system whose coordinate surfaces are the surfaces generated by rotating a plane containing a system of confocal ellipses and hyperbolas about the major axis of the ellipses, together with the planes passing through the axis of rotation. { ‚prō‚lāt sfir‚òid·əl kō'òrd·ən·ət ‚sis·təm }

proliferative arthritis *See* rheumatoid arthritis. { prō'lif·ə‚rād·iv ärth'rīd·əs }

proline [BIOCHEM] $C_5H_9O_2$ A heterocyclic amino acid occurring in essentially all proteins, and as a major constituent in collagen proteins. { 'prō‚lēn }

prolinemia [MED] A rare hereditary disease caused by absence of the degradative enzymes that convert proline to glutamic acid, and characterized by a high content of proline in blood and urine with consequent mental retardation and renal malfunction. { ‚prō·lə'nē·mē·ə }

PROLOG [COMPUT SCI] A programming language that is for artificial intelligence applications, and uses problem descriptions to reach solutions, based on precise rules. { 'prō‚läg }

prolonge [ORD] Rope, with a hook or loop at one end, with which soldiers can move a vehicle or gun carriage into position. { prō'länj }

proluvium [GEOL] A complex, friable, deltaic sediment accumulated at the foot of a slope as a result of an occasional torrential washing of fragmental material. { prō'lü·vē·əm }

PROM *See* Pockels readout optical modulator; programmable read-only memory. { präm }

promazine hydrochloride [ORG CHEM] $C_{17}H_{20}N_2S\cdot HCl$ A white to slightly yellow, crystalline powder, melting at 172–182°C; used in medicine and as a food additive. { 'präm·ə‚zēn ‚hī·drə'klòr‚īd }

PROM burner [COMPUT SCI] A special device used to write on a programmable read-only memory (PROM). { 'präm ‚bər·nər }

promenade deck [NAV ARCH] An upper deck, or part of a deck, of a passenger ship where passengers walk about. Also known as hurricane deck. { ‚präm·ə'nād ‚dek }

prometaphase [CYTOL] A stage between prophase and metaphase in mitosis in which the nuclear membrane disappears and the spindle forms. { prō'med·ə‚fāz }

Prometheus [ASTRON] A satellite of Saturn which orbits at a mean distance of 139,000 kilometers (86,000 miles), just inside the F ring; together with Pandora, it holds this ring in place. { prə'mē·thē·əs }

promethium [CHEM] A chemical element, symbol Pm, atomic number 61, produced artificially in nuclear reactors; atomic weight of the most abundant separated isotope is 147; a member of the rare-earth group of metals. { prə'mē·thē·əm }

promethium-147 [NUC PHYS] Artificially produced isotope with atomic number 61 and mass 147; produced during fission of ^{235}U. Also known as florentium; illinium. { prə'mē·thē·əm ‚wən'fòrd·ē'sev·ən }

promethium cell [NUCLEO] A nuclear energy cell in which beta particles from promethium-147 cause a phosphor to glow; the light output is converted to electric energy by photocells. { prə'mē·thē·əm ‚sel }

prominence [ASTROPHYS] A volume of luminous, predominantly hydrogen gas that appears on the sun above the chromosphere; occurs only in the region of horizontal magnetic fields because these fields support the prominences against solar gravity. { 'präm·ə‚nəns }

promontory [GEOL] **1.** A high, prominent projection or point of land, or a rock cliff, jutting out boldly into a body of water. **2.** A cape, either low-lying or of considerable height, with a bold termination. **3.** A bluff or prominent hill overlooking or projecting into a lowland. { 'präm·ən‚tòr·ē }

promoter [CHEM] A chemical which itself is a feeble catalyst, but greatly increases the activity of a given catalyst. [GEN] The site on deoxyribonucleic acid to which ribonucleic acid polymerase binds preparatory to initiating transcription of a gene or an operon. { prə'mōd·ər }

PROM programmer [ELECTR] A device that holds several programmable read-only memory (PROM) chips and writes instructions and data into them by melting connections in their circuitry. { 'präm 'prō‚gram·ər }

prompt [COMPUT SCI] A message or format displayed on the screen of a computer terminal that requires the user to respond in some way before processing can continue. { prämpt }

prompt critical [NUCLEO] Capable of sustaining a chain reaction without the aid of delayed neutrons. { 'prämpt ‚krid·ə·kəl }

prompt neutron [NUC PHYS] A neutron released coincident with the fission process, as opposed to neutrons subsequently released. { 'prämpt ‚nü‚trän }

prompt radiation [NUC PHYS] Radiation emitted within a time too short for measurement, including γ-rays, characteristic x-rays, conversion and Auger electrons, prompt neutrons, and annihilation radiation. { 'prämpt ‚rād·ē‚ā·shən }

promyelocyte [HISTOL] The earliest myelocyte stage derived from the myeloblast. { prō'mī·ə·lə‚sīt }

pronate [ANAT] **1.** To turn the forearm so that the palm of the hand is down or toward the back. **2.** To turn the sole of the foot outward with the lateral margin of the foot elevated; to evert. [CONT SYS] To orient a robot toward a position in which the back or protected side of a manipulator faces up and is exposed. { 'prō‚nāt }

pronator [PHYSIO] A muscle which pronates, as the muscles of the forearm attached to the ulna and radius. { 'prō‚nād·ər }

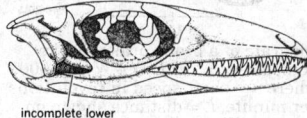

PROLACERTIFORMES

incomplete lower
temporal arcade

Lateral view of *Prolacerta* skull, Lower Triassic. (*From A. S. Romer, Vertebrate Paleontology, 3d ed., University of Chicago Press, 1966*)

pronephros [EMBRYO] One of the anterior pair of renal organs in higher vertebrate embryos; the pair initiates formation of the archinephric duct. { prō′ne·frəs }

prong *See* pin. { präŋ }

pronghorn [VERT ZOO] *Antilocapra americana.* An antelopelike artiodactyl composing the family Antilocapridae; the only hollow-horned ungulate with branched horns present in both sexes. { 'präŋ,hȯrn }

prong reef [GEOL] A wall reef that has developed irregular buttresses normal to its axis in both leeward and (to a smaller degree) seaward directions. { präŋ ,rēf }

pronormoblast [HISTOL] A nucleated erythrocyte precursor with scanty basophilic cytoplasm without hemoglobin. Also known as lymphoid hemoblast of Pappenheim; macroblast of Naegeli; megaloblast of Sabin; proerythroblast of Ferrata; prorubricyte; rubriblast; rubricyte. { prō′nȯr·mə,blast }

pronucleus [CYTOL] One of the two nuclear bodies of a newly fertilized ovum, the male pronucleus and the female pronucleus, the fusion of which results in the formation of the germinal (cleavage) nucleus. { prō′nü·klē·əs }

prony brake [MECH ENG] An absorption dynamometer that applies a friction load to the output shaft by means of wood blocks, a flexible band, or other friction surface. { 'prō·nē ,brāk }

proof [ENG] Reproduction of a die impression by means of a cast. [FOOD ENG] **1.** The strength of the ethyl alcohol in distilled spirits; in the United States, each degree of proof is equal to 0.5% of alcohol by volume. **2.** To activate yeast by mixing it with water or milk and letting the mixture stand for a given period of time. [GRAPHICS] The inked impression of composed type or a plate; used for inspection purposes or for pasting up with other artwork. [MATH] A deductive demonstration of a mathematical statement. { prüf }

proof by contradiction *See* reductio ad absurdum. { 'prüf bī ,kän·trə′dik·shən }

proof by descent *See* mathematical induction. { 'prüf bī di′sent }

proof firing [ORD] The firing of certain rounds for the purpose of testing the serviceability of a weapon or its mount. { 'prüf ,fīr·iŋ }

proof load [ENG] A predetermined test load, greater than the service load, to which a specimen is subjected before acceptance for use. { 'prüf ,lōd }

proof mark [ORD] Distinguishing mark on a weapon to indicate inspection and proof firing. { 'prüf ,märk }

proof plane [ELEC] A small metal plane supported by an insulating handle and used to transfer a small fraction of the electric charge on a body to an electrometer to investigate the charge distribution on the body. { 'prüf ,plān }

proofreading [MOL BIO] Any mechanism for correcting errors in replication, transcription, or translation that involves monitoring of individual units after they have been added to the chain. Also known as editing. { 'prüf,rēd·iŋ }

proof resilience [MECH] The tensile strength necessary to stretch an elastomer from zero elongation to the breaking point, expressed in foot-pounds per cubic inch of original dimension. { 'prüf ri,zil·yəns }

proof stress [MECH] **1.** The stress that causes a specified amount of permanent deformation in a material. **2.** A specified stress to be applied to a member or structure in order to assess its ability to support service loads. { 'prüf ,stres }

proof total [COMPUT SCI] One of a group of totals which are compared with each other to check their consistency. { 'prüf ,tōd·əl }

prop [MIN ENG] Underground supporting post set across the lode, seam, bed, or other opening. { präp }

propadiene *See* allene. { ,präp·ə′dī,ēn }

propaedeutic stratigraphy *See* prostratigraphy. { ,prō·pi′düd·ik strə′tig·rə·fē }

propagated blast [ENG] A blast of a number of unprimed charges of explosives plus one hole primed, generally for the purpose of ditching, where each charge is detonated by the explosion of the adjacent one, the shock being transmitted through the wet soil. { 'präp·ə,gād·əd 'blast }

propagated error [COMPUT SCI] An error which takes place in one operation and spreads through succeeding operations. { 'präp·ə,gād·əd 'er·ər }

propagation *See* wave motion. { ,präp·ə′gā·shən }

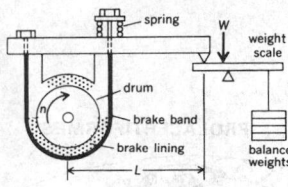

PRONY BRAKE

spring
W
weight scale
drum
brake band
brake lining
balance weights
L

Diagram of a prony brake. Brake horsepower = $2\pi nL(W - W_0)/33{,}000$, where n = shaft speed in revolutions per minute, L = distance shown on drawing, W = scale weight with brake operating, and W_0 = scale weight with brake free.

propagation anomaly [PHYS] Change in propagation characteristics due to a resonance in the medium of propagation. { ,präp·ə′gā·shən ə,näm·ə·le }

propagation constant [ELECTROMAG] A rating for a line or medium along or through which a wave of a given frequency is being transmitted; it is a complex quantity; the real part is the attenuation constant in nepers per unit length, and the imaginary part is the phase constant in radians per unit length. { ,präp·ə′gā·shən ,kän·stənt }

propagation delay [ELECTR] The time required for a signal to pass through a given complete operating circuit; it is generally of the order of nanoseconds, and is of extreme importance in computer circuits. { ,präp·ə′gā·shən di,lā }

propagation forecasting [METEOROL] Forecasting in which the known or predicted vertical distribution of the index of refraction over an area is used to forecast the propagation performance of radars or any microwave radio equipment operating in that area. { ,präp·ə′gā·shən 'fȯr,kast·iŋ }

propagation loss [COMMUN] The attenuation of signals passing between two points of a transmission path. { ,präp·ə′gā·shən ,lȯs }

propagation mode [ELECTROMAG] A form of propagation of electromagnetic radiation in a periodic beamguide in which the field distributions over cross sections of the beam are identical at positions separated by one period of the guide. { ,präp·ə′gā·shən ,mōd }

propagation notice [COMMUN] A forecast of propagation conditions for long-distance radio communications, broadcast at regular intervals over radio stations operated by the National Institute of Standards and Technology. { ,präp·ə′gā·shən ,nōd·əs }

propagation path [COMMUN] A path between receiver and transmitter including direct tropospheric scatter, ionospheric scatter, E-layer skip, and F_1-layer and F_2-layer skip and echo. { ,präp·ə′gā·shən ,path }

propagation rate [CHEM] The speed at which a flame front progresses through the body of a flammable fuel-oxidizer mixture, such as gas and air. { ,präp·ə′gā·shən ,rāt }

propagation step [CHEM] In a chain reaction, one of the fundamental steps that take place repeatedly until the reaction is complete. { ,präp·ə′gā·shən ,step }

propagation time delay [COMMUN] The time required for a wave to travel between two points of a transmission path. { ,präp·ə′gā·shən 'tīm di,lā }

propagation velocity [ELECTROMAG] Velocity of electromagnetic wave propagation in the medium under consideration. { ,präp·ə′gā·shən və,läs·əd·ē }

propagator [QUANT MECH] The probability amplitude for a particle to move or propagate to some new point of space and time when its amplitude at some point of origination is known. { 'präp·ə,gād·ər }

propagule [BOT] **1.** A reproductive structure of brown algae. **2.** A propagable shoot. { 'präp·ə,gyül }

Propalticidae [INV ZOO] A family of coleopteran insects of the superfamily Cucujoidea found in Old World tropics and Pacific islands. { ,prō·pəl′sid· əd·ē }

propane [ORG CHEM] $CH_3CH_2CH_3$ A heavy, colorless, gaseous petroleum hydrocarbon gas of the paraffin series; boils at −44.5°C; used as a solvent, refrigerant, and chemical intermediate. { 'prō,pān }

propane bubble chamber [NUCLEO] A bubble chamber whose liquid is propane; used when it is desirable to have a dense, heavy target in experiments such as detection of neutrinos or measurements of the decay products of neutral pions, not usually visible in a hydrogen bubble chamber. { 'prō,pān 'bəb·əl ,chām·bər }

propane deasphalting [CHEM ENG] Petroleum-refinery solvent process using propane to remove and precipitate asphalt from petroleum stocks, such as for lubricating oils. { 'prō,pān dē′as,fȯld·iŋ }

propane decarbonizing [CHEM ENG] Petroleum-refinery solvent process using propane to recover catalytic-cracking feedstock from heavy-fuel residues; when butane or butane-propane solvent is used, the process is called solvent decarbonizing. { 'prō,pān dē′kär·bə,nīz·iŋ }

propane dewaxing [CHEM ENG] Petroleum-refinery solvent process using propane to remove waxes from lubricating oils to lower the lube-oil pour point. { 'prō,pān dē′waks·iŋ }

propane fractionation [CHEM ENG] Continuous, petroleum-refinery solvent process using liquid propane to segregate long-vacuum residue into two or more grades of lube-oil stock (such as heavy neutral stock or bright stock) and asphalt. { 'prō,pān ,frak·shə'nā·shən }

1-propanethiol *See* n-propyl mercaptan. { |wən|prō,pān 'thī,ól }

propanoic acid *See* propionic acid. { |prō·pə|nō·ik 'as·əd }

propanol *See* propyl alcohol. { 'prō·ə,nól }

2-propanone *See* acetone. { |tü 'prō·pə,nōn }

propargyl alcohol [ORG CHEM] HCCCH$_2$OH Colorless, water- and alcohol-soluble liquid, boiling at 114°C; used as a chemical intermediate, stabilizer, and corrosion inhibitor. Also known as 2-propyn-1-ol. { prō'pär·jəl 'al·kə,hól }

propargyl bromide [ORG CHEM] C$_3$H$_3$Cl A flammable liquid with a boiling point range of 56.0–57.1°C; used as a soil fumigant. { prō'pär·jəl 'brō,mīd }

propargyl chloride [ORG CHEM] C$_3$H$_3$Cl A liquid miscible with benzene, carbon tetrachloride, ethanol, and ethylene glycol; used as an intermediate in organic synthesis. { prō'pär·jəl 'klór,īd }

prop-crib timbering [MIN ENG] Shaft timbering with cribs kept apart at the proper distance by means of props. { 'präp ,krib 'tim·bə·riŋ }

propellant [MATER] A combustible substance that produces heat and supplies ejection particles as in a rocket engine. { prə'pel·ənt }

propellant 23 *See* fluoroform. { prə'pel·ənt |twen·tē'thrē }

propellant-actuated device [ENG] A device that employs the energy supplied by the gases produced by burning propellants to accomplish or initiate a mechanical action other than propelling a projectile. { prə'pel·ənt |ak·chə wäd·əd di,vīs }

propellant-acutated exactor [ORD] A propellant-actuated device intended to release the safety mechanism of another propellant-actuated device. { prə'pel·ənt |ak·chə,wäd·əd ig'zak·tər }

propellant additive [MATER] Any material added to the basic formulation of a solid propellant to accomplish some special purpose such as to increase or decrease the rate of burning. { prə'pel·ənt 'ad·əd·iv }

propellant binder [MATER] An elastomeric fuel used in a composite propellant so that the propellant may be cast directly into the combustion chamber, where the binder cures to a rubber and the propellant grain is then supported by adhesion to the walls; used in missiles. { prə'pel·ənt 'bīnd·ər }

propellant fouling [ORD] Bits of unburned or partially burned propellant left in the bore after firing. { prə'pel·ənt 'faúl·iŋ }

propellant grain [MATER] An elongated molding or extrusion, often of intricate shape, of solid propellant for a rocket, regardless of size. { prə'pel·ənt ,grān }

propellant injector [AERO ENG] A device for injecting propellants, which include fuel and oxidizer, into the combustion chamber of a rocket engine. { prə'pel·ənt in,ek·tər }

propellant mass fraction *See* propellant mass ratio. { prə'pel·ənt |mas ,frak·shən }

propellant mass ratio [AERO ENG] Of a rocket, the ratio of the effective propellant mass to the initial vehicle mass. Also known as propellant mass fraction. { prə'pel·ənt |mas ,rā·shō }

propellant powder [MATER] A low explosive of fine granulation which, through burning, produces gases at a controlled rate to provide the energy for propelling a projectile. { prə'pel·ənt ,paúd·ər }

propellant weight fraction [AERO ENG] The weight of the solid propellant charge divided by weight of the complete solid propellant propulsion unit. { prə'pel·ənt |wāt frak·shən }

propeller [MECH ENG] A bladed device that rotates on a shaft to produce a useful thrust in the direction of the shaft axis. { prə'pel·ər }

propeller anemometer [ENG] A rotation anemometer which is encased in a strong glass outer shell that protects it against hydrostatic pressure. { prə'pel·ər ,an·ə'mäm·əd·ər }

propeller blade [DES ENG] One of two or more plates radiating out from the hub of a propeller and normally twisted to form part of a helical surface. { prə'pel·ər ,blād }

propeller boss [DES ENG] The central portion of the screw propeller which carries the blades, and forms the medium of attachment to the propeller shaft. Also known as propeller hub. { prə'pel·ər ,bós }

propeller cavitation [FL MECH] Formation of vapor-filled and air-filled bubbles or cavities in water at or on the surface of a rotating propeller, occurring when the pressure falls below the vapor pressure of water. { prə'pel·ər ,kav·ə,tā·shən }

propeller efficiency [MECH ENG] The ratio of the thrust horsepower delivered by the propeller to the shaft horsepower as delivered by the engine to the propeller. { prə'pel·ər i,fish·ən·sē }

propeller fan [MECH ENG] An axial-flow blower, with or without a casing, using a propeller-type rotor to accelerate the fluid. { prə'pel·ə·'fan }

propeller horsepower [NAV ARCH] The horsepower delivered to the propeller of a ship. { prə'pel·ər 'hórs,paú·ər }

propeller hub *See* propeller boss. { prə'pel·ər ,həb }

propeller meter [ENG] A quantity meter in which the flowing stream rotates a propellerlike device and revolutions are counted. { prə'pel·ər 'mēd·ər }

propeller post [NAV ARCH] The forward post of a stern frame on vessels having a center-line propeller. { prə'pel·ər ,pōst }

propeller pump *See* axial-flow pump. { prə'pel·ər 'pəmp }

propeller racing [NAV ARCH] The sudden increase in the number of revolutions made by the engine when the propeller blades are lifted clear of the water, or nearly so, due to the roll or pitch of the ship. { prə'pel·ər rās·iŋ }

propeller shaft [MECH ENG] A shaft, carrying a screw propeller at its end, that transmits power from an engine to the propeller. { prə'pel·ər ,shaft }

propeller slip angle [MECH ENG] The angle between the plane of the blade face and its direction of motion. { prə'pel·ər 'slip ,aŋ·gəl }

propeller strut *See* strut. { prə'pel·ər ,strət }

propeller thrust [NAV ARCH] The effort delivered by a propeller in pushing a vessel ahead. { prə'pel·ər ,thrəst }

propeller tip clearance [NAV ARCH] Generally, the shortest distance between the skin of a vessel and the circle swept by the propeller tips. { prə'pel·ər 'tip ,klir·əns }

propeller tip speed [MECH ENG] The speed in feet per minute swept by the propeller tips. { prə'pel·ər 'tip ,spēd }

propeller turbine [MECH ENG] A form of reactive-type hydraulic turbine using an axial-flow propeller rotor. { prə'pel·ər 'tər·bən }

propeller windmill [MECH ENG] A windmill that extracts wind power from horizontal air movements to rotate the blades of a propeller. { prə'pel·ər 'win,mil }

propenyl guaethol [ORG CHEM] C$_{11}$H$_{14}$O$_2$ A white powder with a vanilla flavor and a melting point of 85–86°C; soluble in fats, essential oils, and edible solvents; used for artificial vanilla flavoring. { 'prō·pə,nil 'gwē,thól }

properdin [IMMUNOL] A macroglobin of normal plasma capable of killing various bacteria and viruses in the presence of complement and magnesium ions. { 'prō·pər·dən }

proper divisor [MATH] A proper divisor of a positive integer *n* is any divisor other than 1 and *n*. { 'präp·ər di'vīz·ər }

proper face [MATH] **1.** For a simplex, a face whose dimension is strictly less than that of the simplex. **2.** For a convex polytope, the intersection of the convex polytope with one of the hyperplanes enclosing it. { 'präp·ər 'fās }

proper fraction [MATH] **1.** A fraction a/b where the absolute value of *a* is less than the absolute value of *b*. **2.** The quotient of two polynomials in which the degree of the numerator is less than the degree of the denominator. { 'präp·ər 'frak·shən }

proper function *See* eigenfunction. { 'präp·ər 'fəŋk·shən }

proper Lorentz transformation [RELAT] A Lorentz transformation which can be represented by a matrix whose determinant is +1. { 'präp·ər 'lór·əns ,tranz·fər,mā·shən }

properly divergent series [MATH] A series whose partial sums become either arbitrarily large or arbitrarily small (algebraically). { 'präp·ər·lē də'vər·jənt 'sir,ēz }

proper motion [ASTRON] That component of the space motion of a celestial body perpendicular to the line of sight, resulting in the change of a star's apparent position relative to that of other stars; expressed in angular units. { 'präp·ər 'mō·shən }

proper orthogonal transformation [MATH] An orthogonal transformation such that the determinant of its matrix is +1. { 'präp·ər ór'thäg·ən·əl ,tranz·fər'mā·shən }

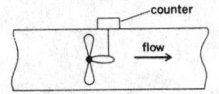

PROPELLER METER

Schematic of a propeller meter, a type of quantity meter.

proper rational function [MATH] The quotient of a polynomial P by a polynomial Q whose order is greater than P. { 'präp·ər ¦rash·ən·əl ¦faŋk·shən }

proper subset [MATH] A set X is a proper subset of a set Y if there is an element of Y which is not in X while X is a subset of Y. { 'präp·ər 'səb,set }

proper time [RELAT] The time measured by an ideal clock that is carried along with a specified particle, and is based on the invariant timelike space-time intervals between points along the particle's trajectory. { 'präp·ər 'tīm }

property detector [COMPUT SCI] In character recognition, that electronic component of a character reader which processes the normalized signal for the purpose of extracting from it a set of characteristic properties on the basis of which the character can be subsequently identified. { 'präp·ərd·ē di,tek·tər }

property list [COMPUT SCI] A list for describing some object or concept, in which odd-numbered items name a property or attribute of a relevant class of objects, and the item following the property name is the property's value for the described objects. { 'präp·ərd·ē ,list }

proper value See eigenvalue. { 'präp·ər 'val·yü }

propfan [AERO ENG] An advanced turboprop with very thin, highly swept blades to reduce both compressibility losses and propeller noise during high-speed cruise. { 'präp,fan }

prop-free [MIN ENG] A face with no posts between the coal and the conveyor used to remove it in longwall mining of a coal seam. { 'präp ,frē }

prop-free front [MIN ENG] In coal mining, longwall working in which support to the roof is given by roof beams cantilevered from behind the working face. { 'präp ,frē ,frənt }

prophage [VIROL] Integrated unit formed by union of the provirus into the bacterial genome. { 'prō,fāj }

propham [ORG CHEM] $C_{10}H_{13}NO_2$ A light brown solid with a melting point of 87–88°C; slightly soluble in water; used as a pre- and postemergence herbicide for vegetable crops. Abbreviated IPC (isopropyl-N-phenylcarbamate). { 'prō ,fam }

prophase [CYTOL] The initial stage of mitotic or meiotic cell division in which chromosomes are condensed from the nuclear material and split logitudinally to form pairs. { 'prō,fāz }

prophylactic vaccination [IMMUNOL] Vaccination occurring before exposure to pathogens. { ,prō·fə¦lak·tic ,vak·sə'nā·shən }

prophylaxis [MED] The prevention of disease. { ,prō·fə'lak·səs }

propiodal [PHARM] $C_5H_{13}ONI_2$ A white, crystalline solid with a melting point of 275°C; soluble in water; used in medicine for iodine therapy. { 'prō·pē·ə,dal }

β-propiolactone [ORG CHEM] $C_3H_4O_2$ Water-soluble liquid that decomposes rapidly at boiling point (155°C); miscible with ethanol, acetone, chloroform, and ether; reacts with alcohol; used as a chemical intermediate. { ¦bād·ə ,prō·pē·ə'lak,tōn }

propionaldehyde [ORG CHEM] C_2H_5CHO Flammable, water-soluble, water-white liquid, with suffocating aroma; boils at 48.8°C; used to manufacture acetals, plastics, and rubber chemicals, and as a disinfectant and preservative. { ¦prō·pē·ən'al·də,hīd }

propionate [ORG CHEM] A salt of propionic acid, CH_3CH_2COOH; an example is sodium propionate, CH_3CH_2COONa. { 'prō·pē·ə,nāt }

Propionibacteriaceae [MICROBIO] A family of bacteria related to the actinomycetes; gram-positive, anaerobic to aerotolerant rods or filaments; ferment carbohydrates, with propionic acid as the principal product. { ¦prō·pē,än·ə,bak·tir·ē'ās·ē,ē }

propionic acid [ORG CHEM] CH_3CH_2COOH Water- and alcohol-soluble, clear, colorless liquid with pungent aroma; boils at 140.7°C; used to manufacture various propionates, in nickel-electroplating solutions, for perfume esters and artificial flavors, for pharmaceuticals, and as a cellulosics solvent. Also known as methylacetic acid; propanoic acid. { ¦prō·pē¦än·ik 'as·əd }

propionic anhydride [ORG CHEM] $(CH_3CH_2CO)_2O$ A colorless liquid with a boiling point of 167–169°C; soluble in ether, alcohol, and chloroform; used as an esterifying agent and for dyestuffs and pharmaceuticals. { ¦prō·pē¦än·ik an'hī,drīd }

propionic ether See ethyl propionate. { ¦prō·pē¦än·ik 'ē·thər }

propionitrile See ethyl cyanide. { ,prō·pē'än·ə,tril }

proplastid [BOT] Precursor body of a cell plastid. { prō 'plas·təd }

proplatinum [MET] A nickel-silver-bismuth alloy used as a substitute for platinum. { prō'plat·ən·əm }

propleuron [INV ZOO] A pleuron of the prothorax in insects. { prō'plur,än }

proplyd [ASTRON] A disk of dense gas and dust surrounding a young star. Derived from protoplanetary disk. { 'präp,lid }

propodite [INV ZOO] The sixth leg joint of certain crustaceans. Also known as propodus. { 'präp·ə,dīt }

propodus See propodite. { 'präp·əd·əs }

proportion [MATH] **1.** The proportion of two quantities is their ratio. **2.** The statement that two ratios are equal. { prə'pór·shən }

proportional band [ACOUS] One of a series of frequency bands whose members have equal band ratios. [CONT SYS] The range of values of the controlled variable that will cause a controller to operate over its full range. { prə'pór·shən·əl 'band }

proportional control [CONT SYS] Control in which the amount of corrective action is proportional to the amount of error; used, for example, in chemical engineering to control pressure, flow rate, or temperature in a process system. { prə'pór·shən·əl kən'trōl }

proportional controller [CONT SYS] A controller whose output is proportional to the error signal. { prə'pór·shən·əl kən'trōl·ər }

proportional counter [NUCLEO] A radiation counter consisting of a proportional counter tube and its associated circuits; resembles a Geiger-Müller counter, but with a different counting gas (argon methane) and a lower voltage on the tube; used to measure α, β, and x-rays; has low sensitivity for γ-radiation. { prə'pór·shən·əl 'kaunt·ər }

proportional counter tube [NUCLEO] A radiation-counter tube operated at voltages high enough to produce ionization by collision and adjusted so the total ionization per count is proportional to the ionization produced by the initial ionizing event. { prə'pór·shən·əl 'kaunt·ər ,tüb }

proportional dividers [DES ENG] Dividers with two legs, pointed at both ends, and an adjustable pivot; distances measured by the points at one end can be marked off in proportion by the points at the other end. { prə'pór·shən·əl di'vīd·ərz }

proportional elastic limit [MECH] The greatest stress intensity for which stress is still proportional to strain. { prə'pór·shən·əl i'las·tik ,lim·ət }

proportional ionization chamber [ELECTR] An ionization chamber in which the initial ionization current is amplified by electron multiplication in a region of high electric-field strength, as in a proportional counter; used for measuring ionization currents or charges over a period of time, rather than for counting. { prə'pór·shən·əl ,ī·ə·nə'zā·shən ,chām·bər }

proportional limit [MECH] The greatest stress a material can sustain without departure from linear proportionality of stress and strain. { prə'pór·shən·əl 'lim·ət }

proportional navigation [NAV] Homing guidance of a guided missile toward a moving target in which the missile turning rate is proportional to the rate of change of line of sight between missile and target. { prə'pór·shən·əl ,nav·ə'gā·shən }

proportional parts [MATH] Numbers in the same proportion as a set of given numbers; such numbers are used in an auxiliary interpolation table based on the assumption that the tabulated quantity and entering arguments differ in the same proportion. { prə'pór·shən·əl 'pärts }

proportional-plus-derivative control [CONT SYS] Control in which the control signal is a linear combination of the error signal and its derivative. { prə'pór·shən·əl ,pləs də'riv·əd·iv kən,trōl }

proportional-plus-integral control [CONT SYS] Control in which the control signal is a linear combination of the error signal and its integral. { prə'pór·shən·əl ,pləs 'int·ə·grəl kən,trōl }

proportional-plus-integral-plus-derivative control [CONT SYS] Control in which the control signal is a linear combination of the error signal, its integral, and its derivative. { prə'pór·shən·əl ,pləs 'int·ə·grəl ,pləs də'riv·əd·iv kən,trōl }

proportional reducer [GRAPHICS] A solution that is used to lower density in a negative or positive image in proportion to the silver present, with a certain amount of loss in shadow detail. { prə'pȯr·shən·əl ri'dü·sər }

proportional region [NUCLEO] The range of applied voltages in a radiation counter tube in which the gas amplification is greater than 1 and does not depend on the charge liberated in the initial ionizing event. { prə'pȯr·shən·əl 'rē·jən }

proportional spacing [GRAPHICS] Spacing in which each character occupies space according to its width. { prə'pȯr·shən·əl 'spās·iŋ }

proportional-speed control See floating control. { prə'pȯr·shən·əl 'spēd kən,trōl }

proportioning probe [ENG] A leak-testing probe capable of changing the air-tracer gas ratio without changing the amount of flow it transmits to the testing device. { prə'pȯr·shən·iŋ ,prōb }

proportioning pump See metering pump. { prə'pȯr·shən·iŋ ,pəmp }

proportioning reactor [ELECTROMAG] A saturable-core reactor used for regulation and control; increasing the input control current from zero to rated value makes output current increase in proportion from cutoff up to full load value. { prə'pȯr·shən·iŋ rē,ak·tər }

proposition [MATH] **1.** Any problem or theorem. **2.** A statement that makes an assertion that is either false or true or has been designated as false or true. { ,präp·ə'zish·ən }

propositional algebra [MATH] The study of finite configurations of symbols and the interrelationships between them. { ,prä·pə'zish·ən·əl 'al·jə·brə }

propositional calculus [MATH] The mathematical study of logical connectives between propositions and deductive inference. Also known as sentential calculus. { ,präp·ə'zish·ən·əl 'kal·kyə·ləs }

propositional connectives [MATH] The symbols ~, ∧, ∨, → or ⊃, and ↔ or ≡, denoting logical relations that may be expressed by the phrases "it is not the case that," "and," "or," "if . . . , then," and "if and only if." Also known as sentential connectives. { ,prä·pə'zish·ənəl kə'nek·tivz }

propositional function [MATH] An expression that becomes a proposition when the values of certain symbols in the expression are specified, and that is either true or false depending on these values. Also known as logical function; open sentence; open statement; predicate; sentential function; statement function. { ,präp·ə'zish·ən·əl 'fəŋk·shən }

propositus See proband. { prə'päz·əd·əs }

proppant See propping agent. { 'präp·ənt }

propped cantilever [CIV ENG] A beam having one built-in support and one simple support. { 'präpt 'kant·əl,ē·vər }

propping agent [PETRO ENG] Sand, gravel, or particles of other material (such as sintered bauxite or ceramic beads) suspended in drilling fluid during formation fracturing to keep (prop) open the cracks in the rock when the fluid is withdrawn. Also known as proppant. { 'präp·iŋ ,ā·jənt }

proprietary program [COMPUT SCI] **1.** A computer program that is owned by someone, and whose use may thus be restricted in some manner or entail payment of a fee. Also known as owned program. **2.** More narrowly, a program that is exploited commercially as a separate product. { prə'prī·ə,ter·ē 'prō·grəm }

proprioception [PHYSIO] The reception of internal stimuli. [PSYCH] Sensory awareness of one's location with regard to the external environment. { ,prō·prē·ə'sep·shən }

proprioceptive defect [PSYCH] A distorted self-perception of body parts and of contact with the environment. { ,prō·prē·ə,sep·tiv di'fekt }

proprioceptor [CONT SYS] A device that senses the position of an arm or other computer-controlled articulated mechanism of a robot and provides feedback signals. [PHYSIO] A sense receptor that signals spatial position and movements of the body and its members, as well as muscular tension. { ,prō·prē·ə'sep·tər }

prop root [BOT] A root that serves to support or brace the plant. Also known as brace root. { 'präp ,rüt }

propterygium [VERT ZOO] The anterior of the three principal basal cartilages forming a support of one of the paired fins of sharks, rays, and certain other fishes. { ,präp·tə'rij·ē·əm }

proptosis [MED] A falling downward or forward, especially of an eyeball. { präp'tō·səs }

propulsion [MECH] The process of causing a body to move by exerting a force against it. { prə'pəl·shən }

propulsion system [MECH ENG] For a vehicle moving in a fluid medium, such as an airplane or ship, a system that produces a required change in momentum in the vehicle by changing the velocity of the air or water passing through the propulsive device or engine; in the case of a rocket-propelled vehicle operating without a fluid medium, the required momentum change is produced by using up some of the propulsive device's own mass, called the propellant. { prə'pəl·shən ,sis·təm }

propulsive coefficient [NAV ARCH] The ratio between the effective horsepower (ehp) and the shaft horsepower (shp) at any given speed. { prə'pəl·siv ,kō·ə'fish·ənt }

propyl- [ORG CHEM] The $CH_3CH_2CH_2-$ radical, derived from propane; found, for example, in 1-propanol. { 'prō·pəl }

n-**propyl acetate** [ORG CHEM] $C_3H_7OOCCH_3$ Colorless liquid with pleasant aroma, miscible with alcohols, ketones, esters, and hydrocarbons; boils at 96–102°C; used for flavors and perfumes, in organic synthesis, and as a solvent. { ¦en 'prō·pəl 'as·ə,tāt }

propylacetone See methyl butyl ketone. { ¦prō·pəl'as·ə,tōn }

propyl alcohol [ORG CHEM] $CH_3CH_2CH_2OH$ A colorless liquid made by oxidation of aliphatic hydrocarbons; boils at 97°C; used as a solvent and chemical intermediate. Also known as ethyl carbinol; propanol. { 'prō·pəl 'al·kə,hȯl }

n-**propylamine** [ORG CHEM] $C_3H_7NH_2$ Colorless, flammable liquid, boiling at 46–51°C; used as a sedative. { ¦en ,prō 'pil·ə,mēn }

propyl benzene [ORG CHEM] $C_6H_5C_3H_7$ Water-insoluble, colorless liquid, boiling at 158°C. Also known as phenylpropane. { 'prō·pəl 'ben,zēn }

propylene [ORG CHEM] $CH_3CH=CH_2$ Colorless unsaturated hydrocarbon gas, with boiling point of −47°C; used to manufacture plastics and as a chemical intermediate. Also known as methyl ethylene; propene. { 'prō·pə,lēn }

propylene aldehyde See crotonaldehyde. { 'prō·pə,lēn 'al·də,hīd }

propylene carbonate [ORG CHEM] $C_3H_6CO_3$ Odorless, colorless liquid, boiling at 242°C; miscible with acetone, benzene, and ether; used as a solvent, extractant, plasticizer, and chemical intermediate. { 'prō·pə,lēn 'kär·bə,nāt }

propylene dichloride [ORG CHEM] $CH_3CHClCH_2Cl$ Water-insoluble, colorless, moderately flammable liquid, with chloroform aroma; boils at 96.3°C; miscible with most common solvents; used as a solvent, dry-cleaning fluid, metal degreaser, and fumigant. { 'prō·pə,lēn dī'klȯr,īd }

propylene glycol [ORG CHEM] $CH_3CHOHCH_2OH$ A viscous, colorless liquid, miscible with water, alcohol, and many solvents; boils at 188°C; used as a chemical intermediate, antifreeze, solvent, lubricant, plasticizer, and bactericide. { 'prō·pə,lēn 'glī,kȯl }

propylene glycol alginate [ORG CHEM] $C_9H_{14}O_7$ A white, water-soluble powder; used as a stabilizer, thickener, and emulsifier. { 'prō·pə,lēn 'glī kȯl al·jə,nāt }

propylene glycol monomethyl ether [ORG CHEM] $C_4H_{10}O_2$ A colorless liquid with a boiling point of 120.1°C; soluble in water, methanol, and ether; used as a solvent for cellulose, dyes, and inks. { 'prō·pə,lēn 'glī,kȯl ¦män·ō'meth·əl 'ē·thər }

propylene glycol monoricinoleate [ORG CHEM] $C_{21}H_{30}O_4$ A pale yellow, moderately viscous oily liquid, soluble in organic solvents; used as a plasticizer and lubricant and in dye solvents and cosmetics. { 'prō·pə,lēn 'glī,kȯl ¦män·ō,ris·ən'ō·lē,āt }

propyleneimine [ORG CHEM] C_3H_7N A clear, colorless liquid with a boiling point of 66–67°C; soluble in water and organic solvents; used as an intermediate in organic synthesis. { ,prō·pə'lēn·ə,mēn }

propylene oxide [ORG CHEM] C_3H_6O Colorless, flammable liquid, with etherlike aroma; soluble in water, alcohol, and ether; boils at 33.9°C; used as a solvent and fumigant, in lacquers, coatings, and plastics, and as a petrochemical intermediate. { 'prō·pə,lēn 'äk,sīd }

propylene tetramer See dodecane. { 'prō·pə,lēn 'te·trə·mər }

propyl formate [ORG CHEM] $C_4H_8O_2$ A flammable liquid with a boiling point of 81.3°C; used for flavoring. { 'prō·pəl 'fȯr,māt }

n-**propyl furoate** [ORG CHEM] $C_8H_{10}O_3$ A colorless, fragrant liquid with a boiling point of 210.9°C; soluble in alcohol and ether; used for flavoring. { ¦en 'prō·pəl 'fyùr·ə,wāt }

propyl gallate [ORG CHEM] $C_3H_7OOCC_6H_2(OH)_3$ Colorless crystals with a melting point of 150°C; used to prevent or retard rancidity in edible fats and oils. { 'prō·pəl 'ga,lāt }

propylhexedrine [PHARM] $C_{10}H_{21}N$ A clear, colorless liquid with a boiling point of 202–206°C; soluble in dilute acids; used in medicine. { ,prō·pəl,hek'sed·rən }

propyliodone [ORG CHEM] $C_{10}H_{11}O_3NI_2$ A white, crystalline powder with a melting point of 187–190°C; soluble in alcohol, acetone, and ether; used in medicine as a radiopaque medium. { ,prō·pəl'ī·ə,dōn }

propylite [PETR] A modified andesite, altered by hydrothermal processes, resembling a greenstone and consisting of calcite, epidote, serpentine, quartz, pyrite, and iron ore. { 'prō·pə,līt }

propylization [PETR] A hydrothermal process by which propylite is formed from andesite by the introduction of or replacement by an assemblage of minerals. { ,prō·pəl·ə'zā·shən }

n-propyl mercaptan [ORG CHEM] C_3H_7SH A liquid with an offensive odor and a boiling range of 67–73°C; used as a herbicide. Also known as 1-propanethiol. { ¦en 'prō·pəl mər'kap,tan }

N-propyl nitrate [ORG CHEM] $C_3H_7NO_3$ A white to straw-colored liquid with a boiling range of 104–127°C; used as a monopropellant rocket fuel. { ¦en 'prō·pəl 'nī,trāt }

propylparaben [ORG CHEM] $C_{10}H_{12}O_3$ Colorless crystals or white powder with a melting point of 95–98°C; soluble in acetone, ether, and alcohol; used in medicine and as a food preservative and fungicide. { ,prō·pəl'par·ə·bən }

1-propylphosphonic acid [ORG CHEM] $C_3H_9O_3P$ A white solid with a melting point of 68–69°C; soluble in water; used as a growth regulator for herbaceous and woody species. { ¦wən ¦prō·pəl·fä'sfän·ik 'as·əd }

propylpiperidine See coniine. { 'prō·pəl·pi'per·ə,dēn }

propylthiopyrophosphate [ORG CHEM] $C_{12}H_{28}P_2S_2O$ A straw-colored to dark amber liquid with a boiling point of 148°C; used as an insecticide for chinch bugs in lawns and turf. { 'prō·pəl¦thī·ō,pī·rə'fä,sfāt }

propylthiouracil [PHARM] $C_7H_{10}N_2OS$ White, crystalline powder with a melting point of 218–221°C; soluble in ammonia and alkali hydroxides; used in medicine. { ,prō·pəl¦thī·ō'yūr·ə,sil }

2-propyn-1-ol See propargyl alcohol. { ¦tü 'prō·pən ,wən ,ol }

Prorastominae [PALEON] A subfamily of extinct dugongs (Dugongidae) which occur in the Eocene of Jamaica. { ,prór·ə'stäm·ə,nē }

prorennin See renninogen. { prō'ren·ən }

Prorhynchidae [INV ZOO] A family of turbellarians in the order Alloeocoela. { prō'riŋ·kə,dē }

prorubricyte See pronormoblast. { prō'rü·brə,sīt }

Prosauropoda [PALEON] A division of the extinct reptilian suborder Sauropodomorpha; they possessed blunt teeth, long forelimbs, and extremely large claws on the first finger of the forefoot. { ,prä·só'räp·əd·ə }

Prosobranchia [INV ZOO] The largest subclass of the Gastropoda; generally, respiration is by means of ctenidia, an operculum is present, there is one pair of tentacles, and the sexes are separate. { ,prä·sə'braŋ·kē·ə }

prosodus [INV ZOO] A canal leading from an incurrent canal to a flagellated chamber in Porifera. { 'präs·əd·əs }

prosoma [INV ZOO] The anterior part of the body of mollusks and other invertebrates; primitive segmentation is not apparent. { prō'sō·mə }

prosopagnosia [PSYCH] The inability to recognize familiar faces. { ,präs·ō·pag'nōzh·yə }

prosopite [MINERAL] $CaAl_2(F,OH)_8$ A colorless mineral composed of basic calcium aluminum fluoride. { 'präs·ə,pīt }

Prosopora [INV ZOO] An order of the class Oligochaeta comprising mesonephridiostomal forms in which there are male pores in the segment of the posterior testes. { prə'säp·ə·rə }

prosopyle [INV ZOO] The opening into a flagellated chamber from an inhalant canal in sponges. { 'präs·ə,pīl }

prospect [MIN ENG] **1.** To search for minerals or oil by looking for surface indications, by drilling boreholes, or both. **2.** A plot of ground believed to be mineralized enough to be of economic importance. { 'prä,spekt }

prospecting seismology [PETRO ENG] The application of seismology to the exploration for natural resources, especially gas and oil. { 'prä,spek·tiŋ sīz'mäl·ə·jē }

prospector [MIN ENG] A person engaged in exploring for valuable minerals, or in testing supposed discoveries of the same. { 'prä,spek·tər }

Prospector [AERO ENG] A specific uncrewed spacecraft designed to make a soft landing on the moon to take measurements, photographs, and soil samples, and then return to earth. { 'prä,spek·tər }

prospect pit [MIN ENG] A pit excavated for the purpose of prospecting mineral-bearing ground. { 'prä,spekt ,pit }

prospect shaft [MIN ENG] A shaft constructed for the purpose of excavating mineral-bearing ground. { 'prä,spekt ,shaft }

Prospero [ASTRON] A small satellite of Uranus in a retrograde orbit with a mean distance of 10,250,000 miles (16,500,000 kilometers), eccentricity of 0.324, and sidereal period of 5.50 years. { 'präs·pə·rō }

prostaglandin [BIOCHEM] Any of various physiologically active compounds containing 20 carbon atoms and formed from essential fatty acids; found in highest concentrations in normal human semen; activities affect the nervous system, circulation, female reproductive organs, and metabolism. { ,präs·tə'glan·dən }

prostate [ANAT] A glandular organ that surrounds the urethra at the neck of the urinary bladder in the male. { 'prä,stāt }

prostatectomy [MED] Surgical removal of all or part of the prostate. { ,präs·tə'tek·tə·mē }

prostate specific antigen [IMMUNOL] A glycoprotein with 240 amino acids that is expressed exclusively by human prostate epithelial cells and that exhibits protease activity; elevated levels in the blood are detected in individuals with prostate cancer. Abbreviated PSA. { ¦prä,stāt spə,sif·ik 'ant·i·jən }

prostatitis [MED] Inflammation of the prostate. { ,präs·tə'tīd·əs }

prostereoisomerism See prochirality. { prō¦ster·ē·ō·ī'säm·ə,riz·əm }

prosthecae [MICROBIO] Appendages that are part of the wall in bacteria in the genus Caulobacter. { präs'thē·sē }

prosthecate bacteria [MICROBIO] Single-celled microorganisms that differ from typical unicellular bacteria in having one or more appendages which extend from the cell surface; the best-known genus is Caulobacter. { 'präs·thə,kāt bak'tir·ē·ə }

prosthesis [MED] An artificial substitute for a missing part of the body, such as a substitute hand, leg, eye, or denture. { präs'thē·səs }

prosthetic group [BIOCHEM] A characteristic nonamino acid substance that is strongly bound to a protein and necessary for the protein portion of an enzyme to function; often used to describe the function, as in hemeprotein for hemoglobin. { präs'thed·ik 'grüp }

prosthodontics [MED] The science and practice of replacement of missing dental and oral structures. { ¦präs·thə¦dän·tiks }

Prostigmata [INV ZOO] The equivalent name for Trombidiformes. { ,prō,stig'mäd·ə }

prostomium [INV ZOO] The portion of the head anterior to the mouth in annelids and mollusks. { prō'stō·mē·əm }

prostratigraphy [GEOL] Preliminary stratigraphy, including lithologic and paleontologic studies, without consideration of the time factor. Also known as propaedeutic stratigraphy; protostratigraphy. { ,prä·strə'tig·rə·fē }

protactinium [CHEM] A chemical element, symbol Pa, atomic number 91; the third member of the actinide group of elements; all the isotopes are radioactive; the longest-lived isotope is protactinium-231. { ¦prōd,ak'tin·ē·əm }

protactinium-ionium age method [GEOL] A method of calculating the ages of deep-sea sediments formed during the last 150,000 years from measurements of the ratio of protactinium-231 to ionium (thorium-230), based on the gradual change of this ratio over time because of the difference in half-lives. { ¦prōd,ak'tin·ē·əm ī'ō·nē·əm 'āj ,meth·əd }

protalus rampart [GEOL] An arcuate ridge consisting of boulders and other coarse debris marking the downslope edge of an existing or melted snowbank. { prō'tal·əs 'ram,pärt }

protamine [BIOCHEM] Any of the simple proteins that are combined with nucleic acid in the sperm of certain fish, and that upon hydrolysis yield basic amino acids; used in medicine to control hemorrhage, and in the preparation of an insulin form to control diabetes. { 'prōd·ə,mēn }

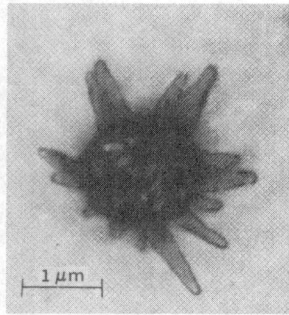

PROSTHECATE BACTERIA

Photograph of a bacterial cell taken with the electron microscope. Note the 14 prosthecae extending from the cell and the transparent gas vesicles inside the cell.

1 µm

protandry [PHYSIO] That condition in which an animal is first a male and then becomes a female. { prō'tan·drē }

protanomaly See protanopia. { ¸prŏd·ə'näm·ə·lē }

protanopia [MED] Partial color blindness in which there is defective red vision; green sightedness. Also known as protanomaly. { ¸prŏd·ə'nō·pē·ə }

Proteaceae [BOT] A large family of dicotyledonous plants in the order Proteales, notable for often having a large cluster of small or reduced flowers. { prō'tās·ē¸ē }

Proteales [BOT] An order of dicotyledonous plants in the subclass Rosidae marked by its strongly perigynous flowers, a four-lobed, often corolla-like calyx, and reduced or absent true petals. { ¸prŏd·ē'ā·lēz }

protease [BIOCHEM] An enzyme that digests proteins. { 'prŏd·ē¸ās }

proteasome [CELL MOL] A large proteolytic particle found in the cytoplasm and nucleus of all eukaryotic cells that is the site for degradation of most intracellular proteins. { 'prŏd·ē·ə¸sōm }

protected format [COMPUT SCI] Parts of a computer display that cannot be altered by typing from the keyboard. { prə'tek·təd 'fòr¸mat }

protected location [COMPUT SCI] A storage cell arranged so that access to its contents is denied under certain circumstances, in order to prevent programming accidents from destroying essential programs and data. { prə'tek·təd lō'kā·shən }

protected-logic module [COMPUT SCI] A module that stores selected computer programs that must remain unaltered. { prə'tek·təd ¦läj·ik 'mäj·yül }

protected subnetwork See domain. { prə'tek·təd səb'net¸wərk }

protected thermometer [ENG] A reversing thermometer which is encased in a strong glass outer shell that protects it against hydrostatic pressure. { prə'tek·təd thər'mäm·əd·ər }

protection [NUCLEO] Any provision to reduce exposure of persons to radiation; for example, protective barriers to reduce external radiation or measures to prevent inhalation of radioactive materials. { prə'tek·shən }

protection code [COMPUT SCI] A component of a task descriptor that specifies the protection domain of the task, that is, the authorizations it has to perform certain actions. { prə'tek·shən ¸kōd }

protection key [COMPUT SCI] An indicator, usually 1 to 6 bits in length, associated with a program and intended to grant the program access to those sections of memory which the program can use but to deny the program access to all other parts of memory. { prə'tek·shən ¸kē }

protection profile [COMPUT SCI] A structure for defining the security and functionality requirements of a computing system. { prə'tek·shən ¸prō·fīl }

protectite [PETR] A rock formed by the crystallization of a primary magma. { prə'tek¸tīt }

protective action guide [NUCLEO] The absorbed dose of ionizing radiation in individuals in the general population which would warrant protective action following a contaminating event, such as a nuclear explosion. { prə'tek·tiv 'ak·shən ¸gīd }

protective atmosphere [MET] A substance such as inert or combusted fuel gas which surrounds a workpiece to be heat-treated, welded, brazed, or thermally sprayed under controlled conditions. { prə'tek·tiv 'at·mə¸sfir }

protective clothing [NUCLEO] Special clothing worn by a radiation worker to prevent contamination of the body or personal clothing. { prə'tek·tiv 'klōth·iŋ }

protective colloid [PHYS CHEM] A colloidal substance that protects other colloids from the coagulative effect of electrolytes and other agents. { prə'tek·tiv 'kä¸lóid }

protective coloration [ZOO] A color pattern that blends with the environment and increases the animal's probability of survival. { prə'tek·tiv ¸kəl·ə'rā·shən }

protective device See electric protective device. { prə'tek·tiv di'vīs }

protective finish [ENG] A coating applied to equipment to protect it from corrosion and wear; many substances, including metals, glass, and ceramics, are used. { prə'tek·tiv 'fin·ish }

protective fire [ORD] Fire delivered by supporting guns and directed against the enemy to hinder his fire or movement against friendly forces. { prə'tek·tiv 'fīr }

protective grounding [ELEC] Grounding of the neutral conductor of a secondary power-distribution system, and of all metal enclosures for conductors, to protect persons from dangerous currents. { prə'tek·tiv 'graùnd·iŋ }

protective relay [ELEC] A relay whose principal function is to protect service from interruption or to prevent or limit damage to apparatus. { prə'tek·tiv 'rē¸lā }

protective resistance [ELECTR] Resistance used in series with a gas tube or other device to limit current flow to a safe value. { prə'tek·tiv ri'zis·təns }

protective survey [NUCLEO] An evaluation of the radiation hazards incidental to the production, use, or existence of radioactive materials or other sources of radiation under a specific set of conditions. { prə'tek·tiv 'sər¸vā }

protector [ELEC] Device to protect equipment or personnel from high voltages or currents. { prə'tek·tər }

protector block [ELEC] Rectangular piece of carbon with an insulated metal insert, or porcelain with a carbon insert, constituting an element of a protector; it forms a gap which will break down and provide a path to ground for voltages over 350 volts. { prə'tek·tər ¸bläk }

protector gap [ELEC] A device designed to limit or equalize voltage in order to protect telephone and telegraph equipment; consists of two carbon blocks with an air gap between them, which are brought into contact when there is a steady-state discharge across the gap. Also known as gap. { prə'tek·tər ¸gap }

protector tube [ELECTR] A glow-discharge cold-cathode tube that becomes conductive at a predetermined voltage, to protect a circuit against overvoltage. { prə'tek·tər ¸tüb }

protectoscope [OPTICS] Device in a tank or armored car, similar to the periscope of a submarine; it enables a soldier to see around a shield without exposing himself to enemy gunfire directed at the ports of the vehicle. { prə'tek·tə¸skōp }

Proteeae [MICROB IO] Formerly a tribe of the Enterobacteriaceae comprising the genus *Proteus*; included organisms which were characteristically motile, fermented dextrose with gas production, and produced urease. { ¸prō'tē·ē¸ē }

Proteida [VERT ZOO] A suborder coextensive with Proteidae in some classification systems. { prō'tē·əd·ə }

Proteidae [VERT ZOO] A family of the amphibian suborder Salamandroidea; includes the neotenic, aquatic *Necturus* and *Proteus* species. { ɔrō'tē·ə¸dē }

protein [BIOCHEM] Any of a class of high-molecular-weight polymer compounds composed of a variety of α-amino acids joined by peptide linkages. { 'prō¸tēn }

proteinaceous [MATER] **1.** Pertaining to any material having a protein base. **2.** Pertaining to adhesive materials having a protein base such as animal glue, casein, and soya. { ¦prōt·ən'ā·shəs }

proteinase [BIOCHEM] A type of protease which acts directly on native proteins in the first step of their conversion to simpler substances. { 'prōt·ən¸ās }

protein-bound iodine [BIOCHEM] Iodine bound to blood protein. Abbreviated PBI. { 'prō¸tēn ¸baùnd 'ī·ə¸dīn }

protein-bound iodine test [PATH] A test of thyroid function that reflects the level of circulating thyroid hormone by determination of the level of protein-bound iodine in the blood. Abbreviated PBI test. { 'prō¸tēn ¸baùnd 'ī·ə¸dīn ¸test }

protein coat See capsid. { 'prō¸tēn ¸kōt }

protein engineering [MOL BIO] The design and construction of new proteins or enzymes with novel or desired functions by modifying amino acid sequences by using recombinant deoxyribonucleic acid technology. { 'prō¸tēn ¸en·jə'nir·iŋ }

protein kinase [BIOCHEM] An enzyme that exerts regulatory effects on growth and malignant transformation by phosphorylating proteins. { ¦prō¸tēn 'kī¸nās }

protein kinase A See cyclic AMP-dependent protein kinase. { ¦prō¸tēn ¦kī¸nās 'ā }

protein kinase C [CELL MOL] A Ca²⁺-dependent enzyme mediating the phosphorylation of certain cellular proteins, thereby regulating important physiological functions, such as cell growth, ion channel activity, secretion, and synaptic transmission. Abbreviated PKC. { ¦prō¸tēn ¦kī¸nās 'sē }

proteinometer See hand sugar refractometer. { ¸prōt·ən'äm·əd·ər }

protein translocator [CELL MOL] A protein that mediates

the transport of other proteins across the membrane of an organelle such as the endoplasmic reticulum, nucleus, mitochondria, and chloroplast. { ¦prō̇tēn tranz'lō̇kād·ər }

proteinuria [MED] The presence of protein in the urine. { ˌprōt·ən'ür·ē·ə }

Proteocephalidae [INV ZOO] A family of tapeworms in the order Proteocephaloidea in which the reproductive organs are within the central mesenchyme of the segment. { ˌprōd·ē·ō·sə'fal·ə,dē }

Proteocephaloidea [INV ZOO] An order of tapeworms of the subclass Cestoda in which the holdfast organ bears four suckers and, frequently, a suckerlike apical organ. { ˌprōd·ē·ō,sef·ə'lȯid·ē·ə }

proteoglycan [BIOCHEM] A high-molecular-weight polyanionic substance covalently linked by numerous heteropolysaccharide side chains to a polypeptide chain backbone. { ˌprōd·ē·ō'glī·kən }

proteolysin [BIOCHEM] A lysin that produces proteolysis. { ˌprōd·ē'äl·ə·sən }

proteolysis [BIOCHEM] Fragmentation of a protein molecule by addition of water to the peptide bonds. { ˌprōd·ē'äl·ə·səs }

proteolytic enzyme [BIOCHEM] Any enzyme that catalyzes the breakdown of protein. { ¦prōd·ē·ə¦lid·ik 'en,zīm }

proteome [MOL BIO] The complete set of proteins present in the various cells of an organism. { 'prōd·ē,ōm }

proteomics [MOL BIO] The separation, identification, and characterization of the complete set of proteins present in the various cells of an organism. { ˌprōd·ē'äm·iks }

Proteomyxida [INV ZOO] The single order of the Proteomyxidia. { ¦prōd·ē·ə¦mik·səd·ə }

Proteomyxidia [INV ZOO] A subclass of Actinopodea including protozoan organisms which lack protective coverings or skeletal elements and have reticulopodia, or filopodia. { ¦prōd·ē·ə·mik'sid·ē·ə }

proteoplast [CYTOL] A type of cell plastid containing crystalline, fibrillar, or amorphous masses of protein. { 'prōd·ē·ə,plast }

proteose [BIOCHEM] One of a group of derived proteins intermediate between native proteins and peptones; soluble in water, not coagulable by heat, but precipitated by saturation with ammonium or zinc sulfate. { 'prōd·ē,ōs }

Proterostomia [ZOO] That part of the animal kingdom in which cleavage of the egg is of the determinate type; includes all bilateral phyla except Echinodermata, Chaetognatha, Pogonophora, Hemichordata, and Chordata. { ˌpräd·ə·rō'stō·mē·ə }

Proterosuchia [PALEON] A suborder of moderate-sized thecodont reptiles with lightly built triangular skulls, downturned snouts, and palatal teeth. { ˌpräd·ə·rō'sü·kē·ə }

Proterotheriidae [PALEON] A group of extinct herbivorous mammals in the order Litopterna which displayed an evolutionary convergence with the horses in their dentition and in reduction of the lateral digits of their feet. { ˌpräd·ə·rō·thə'rī·ə,dē }

Proterozoic [GEOL] Geologic time between the Archean and Paleozoic eras, that is, from 2500 million to 550 million years ago. Also known as Algonkian. { ˌpräd·ə·rə¦zō·ik }

Proteus [ASTRON] A satellite of Neptune orbiting at a mean distance of 73,100 miles (117,600 kilometers) with a period of 26.9 hours, and with a diameter of about 250 miles (400 kilometers). [COMPUT SCI] *See* advanced signal-processing system. { 'prōd·ē·əs }

Proteutheria [VERT ZOO] A group of primatelike insectivores that contains the living tree shrews. { ˌprōd·ē·yü'thir·ē·ə }

prothallium [BOT] The gametophyte of a pteridophyte in the form of a flat green thallus with thizoids. { prō'thal·ē·əm }

prothoracic gland [INV ZOO] One of the paired glands in the prothorax of insects which produce ecdysone. { ¦prō·thə'ras·ik 'gland }

prothorax [INV ZOO] The first thoracic segment of an insect; bears the first pair of legs. { prō'thȯr,aks }

prothrombin [BIOCHEM] An inactive plasma protein precursor of thrombin. Also known as factor II; thrombinogen. { prō'thräm·bən }

prothrombin factor *See* vitamin K. { prō'thräm·bən ,fak·tər }

prothrombin time [PATH] A one-stage clotting test based on the time required for clotting to occur after the addition of tissue thromboplastin and calcium to decalcified plasma. { prō'thräm·bən ,tīm }

prothymocyte [IMMUNOL] A T-cell precursor. { prō'thī·mə,sīt }

proticity [BIOCHEM] In oxidative phosphorylation, the flowing of protons in the proton circuit from high to low protic potential. { prō'tis·əd·ē }

Protista [BIOL] A proposed kingdom to include all unicellular organisms lacking a definite cellular arrangement, such as bacteria, algae, diatoms, and fungi. { prə'tis·tə }

protium [NUC PHYS] The lightest hydrogen isotope, having a mass number of 1 and consisting of a single proton and electron. Also known as light hydrogen. { 'prōd·ē·əm }

Protoariciinae [INV ZOO] A subfamily of polychaete annelids in the family Orbiniidae. { ¦prōd·ō,ar·ə'sī·ə,nē }

protobitumen [MATER] Any of the fats, oils, waxes, or resins which are present as unaltered or nearly unaltered plant and animal products from which fossil bitumens are formed. { ¦prōd·ō·bə'tü·mən }

Protobranchia [INV ZOO] A small and primitive order in the class Bivalvia; the hinge is taxodont in all but one family, there is a central ligament pit, and the anterior and posterior adductor muscles are nearly equal in size. { ˌprōd·ō'braŋ·kē·ə }

Protoceratidae [PALEON] An extinct family of pecoran ruminants in the superfamily Traguloidea. { ¦prōd·ō·sə'rad·ə,dē }

Protochordata [INV ZOO] The equivalent name for Hemichordata. { ¦prōd·ō·kȯr'dad·ə }

protoclastic [PETR] Of igneous rocks, characterized by granulation and deformation of the earlier-formed minerals due to differential flow of the magma before solidification. { ¦prōd·ō¦klas·tik }

Protococcaceae [BOT] A monogeneric family of green algae in the suborder Ulotrichineae in which reproduction is entirely vegetative. { ˌprōd·ō·käk'sās·ē,ē }

Protococcida [INV ZOO] A small order of the protozoan subclass Coccidia; all are invertebrate parasites, and only sexual reproduction is known. { ˌprōd·ō'käk·səd·ə }

protocol [COMPUT SCI] **1.** A set of hardware and software interfaces in a terminal or computer which allows it to transmit over a communications network, and which collectively forms a communications language. **2.** *See* communication protocol. [SCI TECH] A procedure that must be used when performing specified measurements or related operations in order for results to be acceptable to the specifying agency. { 'prōd·ə,kȯl }

protocol-level timer [COMMUN] A time-measuring unit within a communicating device that issues high-priority interrupts which synchronize and set deadlines for protocol-related activities. { 'prōd·ə,kȯl ¦lev·əl 'tīm·ər }

Protocucujidae [INV ZOO] A small family of coleopteran insects in the superfamily Cucujoidea found in Chile and Australia. { ˌprōd·ō·kə'kü·yə,dē }

protoderm *See* dermatogen. { 'prōd·ə,dərm }

protodolomite [MINERAL] A crystalline calcium-magnesium carbonate with a disordered lattice in which the metallic ions occur in the same crystallographic layers instead of in alternate layers as in the dolomite mineral. { ¦prōd·ō'dō·lə,mīt }

Protodonata [PALEON] An extinct order of huge dragonflylike insects found in Permian rocks. { ¦prōd·ō·də'näd·ə }

Protodrilidae [INV ZOO] A family of annelids belonging to the Archiannelida. { ˌprōd·ō'dril·ə,dē }

protoenstatite [MINERAL] An artificial, unstable, altered form of MgSiO₃ produced by thermal decomposition of talc; convertible to enstatite by grinding or heating to a high temperature. { ¦prōd·ō'en·stə,tīt }

Protoeumalacostraca [PALEON] The stem group of the crustacean series Eumalacostraca. { ¦prōd·ō,yü·mə·lə'käs·trə·kə }

protogalaxy [ASTRON] A clump of matter in an early stage of galaxy formation in which the matter has started to collapse under its own gravity to form a galaxy. { ¦prōd·ō'gal·ik·sē }

protogenic [CHEM] Strongly acidic. { ¦prōd·ə¦jen·ik }

protogyny [PHYSIO] A condition in hermaphroditic or dioecious organisms in which the female reproductive structures mature before the male structures. { prō'täj·ə·nē }

protointraclast [GEOL] A limestone component that resulted from a premature attempt at resedimentation while it was still in an unconsolidated and viscous or plastic state, and that never existed as a free clastic entity. { ¦prōd·ō'in·trə,klast }

protolith [PETR] The original, unmetamorphosed rock from which a given metamorphic rock is formed. { 'prōd·ə,lith }

Protomastigida [INV ZOO] The equivalent name for Kinetoplastida. { ,prōd·ō·ma'stij·ə·də }

protomer [BIOCHEM] One of the polypeptide chains composing an oligomeric protein. Also known as subunit. { 'prōd·ə·mər }

Protomonadina [INV ZOO] An order of flagellates, subclass Mastigophora, with one or two flagella, including many species showing protoplasmic collars ringing the base of the flagellum. { ,prōd·ō,män·ə'dī·nə }

Protomonida [INV ZOO] The equivalent name for Protomonadina. { ,prōd·ō'män·ə·də }

protomylonite [PETR] A mylonitic rock that develops from contact-metamorphosed rock; granulation and flowage are caused by overthrusts following the contact surfaces between the intrusion and the country rock. { ,prōd·ō'mī·lə,nīt }

Protomyzostomidae [INV ZOO] A family of parasitic polychaetes belonging to the Myzostomaria and known for three species from Japan and the Murman Sea. { ,prōd·ō,mī·zə'stäm·ə,dē }

proton [PHYS] An elementary particle that is the positively charged constituent of ordinary matter and, together with the neutron, is a building block of all atomic nuclei; its mass is approximately 938 megaelectronvolts and spin 1/2. { 'prō,tän }

proton accelerator [NUCLEO] A particle accelerator which accelerates protons to high energies, as opposed to one which accelerates heavier ions or electrons. { 'prō,tän ak'sel·ə,rād·ər }

proton acid See Brönsted acid. { 'prō,tän 'as·əd }

protonate [CHEM] To add protons to a base by a proton source. { 'prōt·ən,āt }

proton capture [NUC PHYS] A nuclear reaction in which a proton combines with a nucleus. { 'prō,tän 'kap·chər }

proton drip-line [NUC PHYS] On a chart of the nuclides, which plots proton number versus neutron number, the boundary beyond which proton-rich nuclei are unstable against proton emission. { 'prō,tän 'drip ,līn }

proton-electron-proton reaction [NUC PHYS] A nuclear reaction in which two protons and an electron react to form a deuteron and a neutrino; it is an important source of detectable neutrinos from the sun. Abbreviated PeP reaction. { 'prō,tän i'lek,trän 'prō,tän rē,ak·shən }

protonema [BOT] A green, filamentous structure that originates from an asexual spore of mosses and some liverworts and that gives rise by budding to a mature plant. { ,prōt·ən'ē·mə }

protonephridium [INV ZOO] **1.** A primitive excretory tube in many invertebrates. **2.** The duct of a flame cell. { ,prōd·ō·nə'frid·ē·əm }

protonic acid See Brönsted acid. { prō'tän·ik 'as·əd }

proton-induced x-ray emission [ANALY CHEM] A method of elemental analysis in which the energy of the characteristic x-rays emitted when a sample is bombarded with a beam of energetic protons is used to identify the elements present in the sample. Abbreviated PIXE. { 'prō,tän in,düst 'eks,rā i,mish·ən }

protonium [ATOM PHYS] A bound state of a proton and an antiproton. { prō'tō·nē·əm }

proton magnetometer [ELECTROMAG] A highly sensitive magnetometer which measures the frequency of the proton resonance in ordinary water. { 'prō,tän ,mag·nə'täm·əd·ər }

proton microscope [ELECTR] A microscope that is similar to the electron microscope but uses beams of protons instead of electrons as the charged particles. { 'prō,tän 'mī·krə,skōp }

proton moment [NUC PHYS] The magnetic dipole moment of the proton, a physical constant equal to $(1.410606633 \pm 0.000000058) \times 10^{-23}$ joule per tesla. { 'prō,tän ,mō·mənt }

proton number See atomic number. { 'prō,tän ,nəm·bər }

proton-proton chain [NUC PHYS] An energy-releasing nuclear reaction chain which is believed to be of major importance in energy production in hydrogen-rich stars. Also known as deuterium cycle. { 'prō,tän 'prō,tän ,chān }

proton-proton reaction [NUC PHYS] The initiating reaction in the proton-proton chain, in which two protons react to form a deuteron, a positron, and a neutrino. Abbreviated PP reaction. { 'prō,tän 'prō,tän rē,ak·shən }

proton-proton scattering [NUC PHYS] A collision of a proton with another proton, usually the nucleus of a hydrogen atom. { 'prō,tän 'prō,tän 'skad·ər·iŋ }

proton radioactivity [NUC PHYS] A process by which an unstable nucleus spontaneously decays by the emission of a proton. { 'prō,tän ,rād·ē·ō·ak'tiv·əd·ē }

proton-recoil counter [NUCLEO] A counter for measuring fast neutrons. { 'prō,tän 'rē,kȯil ,kaȯnt·ər }

proton resonance [SPECT] A phenomenon in which protons absorb energy from an alternating magnetic field at certain characteristic frequencies when they are also subjected to a static magnetic field; this phenomenon is used in nuclear magnetic resonance quantitative analysis technique. { 'prō,tän 'rez·ən·əns }

proton-rich nucleus [NUC PHYS] An atomic nucleus in which the ratio of proton number to neutron number is much larger than that of nuclei found in nature. { 'prō,tän ,rich 'nü·klē·əs }

proton scattering microscope [SOLID STATE] A microscope in which protons produced in a cold-cathode discharge are accelerated and focused on a crystal in a vacuum chamber; protons reflected from the crystal strike a fluorescent screen to give a visual and photographable display that is related to the structure of the target crystal. { 'prō,tän 'skad·ə·riŋ 'mī·krə,skōp }

proton stability constant [PHYS CHEM] The reciprocal of the dissociation constant of a weak base in solution. { 'prō,tän stə'bil·əd·ē ,kän·stənt }

proton storage ring [NUCLEO] A machine consisting of magnets and vacuum chambers in which beams of high-energy protons can be stored. { 'prō,tän 'stȯr·ij ,riŋ }

proton synchrotron [NUCLEO] A device for accelerating protons in circular orbits in a time-varying magnetic field, in which the orbit radius is kept constant. { 'prō,tän 'siŋ·krə,trän }

proton vector magnetometer [ELECTROMAG] A type of proton magnetometer with a system of auxiliary coils that permits measurement of horizontal intensity or vertical intensity as well as total intensity. { 'prō,tän ,vek·tər ,mag·nə'täm·əd·ər }

proto-oncogene [GEN] The normal-functioning precursor of an oncogene; mutation of this gene to produce an oncogene causes excessive activity of the gene product, leading to cancer. { ,prōd·ō 'äŋ·kō,jēn }

protophilic [CHEM] Strongly basic. { ,prōd·ō'fil·ik }

protophloem [BOT] The primary phloem developed from the procambium. { ,prōd·ə'flō·əm }

Protophyta [BOT] A division of the plant kingdom, according to one system of classification, set up to include the bacteria, the blue-green algae, and the viruses. { prə'täf·əd·ə }

protoplanet [ASTRON] A precursor of one of the giant planets, which is believed to have formed, along with its satellites, from a minisolar nebula in a manner similar to that of the formation of the sun and planets. { ,prōd·ō,plan·ət }

protoplanetary disk See proplyd. { ,prōd·ō,plan·ä,ter·ē 'disk }

protoplasm [CYTOL] The colloidal complex of protein that composes the living material of a cell. { 'prōd·ə,plaz·əm }

protoplast [CELL MOL] **1.** The living portion of a cell considered as a unit; includes the cytoplasm, the nucleus, and the plasma membrane. **2.** Plant, fungal, or bacterial cell that has had its cell wall removed. { 'prōd·ə,plast }

protoplast fusion [GEN] A technique by which two protoplasts are fused to form hybrid cells that can grow into mature hybrid organisms; usually performed on plants. { 'prōd·ə,plast ,fyü·zhən }

protopodite [INV ZOO] The basal segment of a crustacean limb bearing an endopodite or exopodite, or both, at its distal extremity. { prə'täp·ə,dīt }

Protopteridales [PALEOBOT] An extinct order of ferns, class Polypodiatae. { ,prōd·ō,ter·ə'dā·lēz }

protoquartzite [PETR] A well-sorted sandstone that is intermediate in composition between subgraywacke and orthoquartzite, consisting of 75–95% quartz and chert, with less than 15% detrital clay matrix and 5–25% unstable materials in which there is a greater abundance of rock fragments than feldspar grains. Also known as quartzose subgraywacke. { ,prōd·ō'kwȯrt,sīt }

PROTON MAGNETOMETER

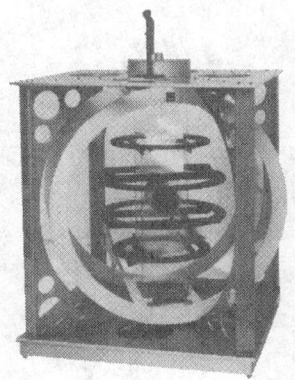

The water container and biasing coil assembly of the proton vector magnetometer. Instrument measures frequency of voltage induced in coil by the protons in water. (*U. S. Coast and Geodetic Survey*)

PROTON SYNCHROTRON

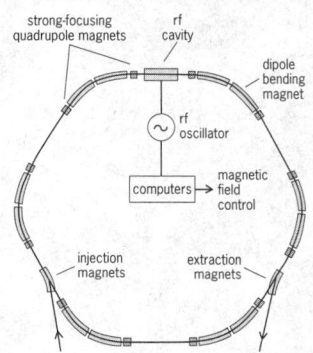

Idealized proton synchrotron.

protore [MIN ENG] **1.** A primary mineral deposit which, through enrichment, can be modified to form an economic ore. **2.** A deposit which could become economically workable if technological change occurred or prices were increased. { 'prōd‚ōr }

Protosireninae [PALEON] An extinct superfamily of sirenian mammals in the family Dugongidae found in the middle Eocene of Egypt. { ‚prōd·ō·sə'ren·ə‚nē }

Protospondyli [VERT ZOO] An equivalent name for Semionotiformes. { ‚prōd·ō'spän·də‚lī }

protostar [ASTRON] A dense condensation of material that is still in the process of accreting matter to form a star. { 'prōd·ə‚stär }

protostele [ASTRON] A stele consisting of a solid rod of xylem surrounded by phloem { 'prōd·ə‚stēl }

Protostomia [INV ZOO] A major division of bilateral animals; includes most worms, arthropods, and mollusks. { ‚prōd·ə'stō·mē·ə }

protostratigraphy *See* prostratigraphy. { ¦prōd·ō·strə'tig·rə·fē }

Protosuchia [PALEON] A suborder of extinct crocodilians from the Late Triassic and Early Jurassic. { ¦prōd·ō'sü·kē·ə }

protosun [ASTRON] The condensation of material that lay at the center of the solar nebula and accreted material from it to form the sun. { 'prōd·ō‚sən }

Prototheria [VERT ZOO] A small subclass of Mammalia represented by a single order, the Monotremata. { ¦prōd·ō'thir·ē·ə }

prototroch [INV ZOO] The band of cilia characteristic of a trochophore larva. { 'prōd·ə‚träk }

prototroph [MICROBIO] A microorganism that has the ability to synthesize all of its amino acids, nucleic acids, vitamins, and other cellular constituents from inorganic nutrients. { 'prōd·ə‚träf }

prototrophic [MICROBIO] Pertaining to bacteria with the nutritional properties of the wild type, or the strains found in nature. { ¦prōd·ō¦träf·ik }

prototropy [ORG CHEM] A reversible interconversion of structural isomers that involves the transfer of a proton. { prō'tä·trə·pē }

Prototrupoidea [INV ZOO] A superfamily of the Hymenoptera. { ‚prōd·ō·trə'pòid·ē·ə }

prototype [ENG] A model suitable for use in complete evaluation of form, design, and performance. { 'prōd·ə‚tīp }

protoxin [BIOCHEM] A chemical compound that is a precursor to a toxin. { 'prō¦täk·sən }

protoxylem [BOT] The part of the primary xylem that differentiates from the procambium and is formed during elongation of an embryonic plant organ. { ¦prōd·ō'zī·ləm }

Protozoa [INV ZOO] A diverse phylum of eukaryotic microorganisms; the structure varies from a simple uninucleate protoplast to colonial forms, the body is either naked or covered by a test, locomotion is by means of pseudopodia or cilia or flagella, there is a tendency toward universal symmetry in floating species and radial symmetry in sessile types, and nutrition may be phagotrophic or autotrophic or saprozoic. { ¦prōd·ə¦zō·ə }

protozoology [INV ZOO] That branch of biology which deals with the Protozoa. { ¦prōd·ō·zō'äl·ə·jē }

Protrachaeta [INV ZOO] The equivalent name for Onychophora. { prō·trə'kēd·ə }

protractor [ENG] A semicircular instrument used to construct and measure angles formed by intersecting lines of a plane; the midpoint of the diameter of the semicircle is marked and serves as the vertex of angles constructed or measured. { 'prō‚trak·tər }

protropic [CHEM] Pertaining to chemical reactions that are influenced by protons. { prō'träp·ik }

protrypsin *See* trypsinogen. { prō'trip·sən }

Protura [INV ZOO] An order of primitive wingless insects belonging to the subclass Apterygota; individuals are elongate and eyeless, lack antennae, and are from pale amber to white in color; anamorphosis is characteristic of the group. { prə'tyùr·ə }

proustite [MINERAL] Ag₃AsS₃ A cochineal-red mineral that crystallizes in the rhombohedral system, consists of silver arsenic sulfide, is isomorphous with pyrargyrite, and occurs massively and in crystals. Also known as light-red silver ore; light-ruby silver. { 'prü‚sīt }

Prout's hypothesis [PHYS CHEM] The hypothesis that all atoms are built up from hydrogen atoms. { 'praùts hī‚päth·ə·səs }

proved ore [MIN ENG] Ore in which there is practically no risk of failure of continuity. { 'prüvd ‚or }

proved reserves [PETRO ENG] The estimated quantities of crude-oil liquids which with reasonable certainty can be recovered in future years from delineated reservoirs under existing economic and operating conditions. { 'prüvd ri'zərvz }

provenance [GEOL] The location, topography, and composition of the source area for any sedimentary rock. Also known as source area; sourceland. { 'präv·ə·nəns }

proventriculus [INV ZOO] **1.** A sac anterior to the gizzard in earthworms. **2.** A dilation of the foregut anterior to the midgut of Mandibulata. [VERT ZOO] The true stomach of a bird, usually separated from the gizzard by a constriction. { ‚prō·vən'trik·yə·ləs }

province [OCEANOGR] An area composed of a grouping of like bathymetric elements whose features are in obvious contrast with surrounding regions. { 'präv·əns }

provincial series [GEOL] A time-stratigraphic series recognized only in a particular region and involving a major division of time within a period. { prə'vin·chəl 'sir·ēz }

proving [COMPUT SCI] Testing whether a computer is free of faults and capable of functioning normally, usually by having it carry out a check routine or diagnostic routine. { 'prüv·iŋ }

proving ring [DES ENG] A ring used for calibrating test machines; the diameter of the ring changes when a force is applied along a diameter. { 'prüv·iŋ ‚riŋ }

provirus [VIROL] The phage genome. { prō'vī·rəs }

provitamin [BIOCHEM] A vitamin precursor; assumes vitamin activity upon activation or chemical change. { prō'vīd·ə·mən }

provitrain [GEOL] Vitrain in which some plant structure can be discerned by microscope. Also known as telain. { prō'vi‚trān }

provitrinite [GEOL] A variety of vitrinite characteristic of provitrain and including the varieties periblinite, suberinite, and xylinite. { prō'vi·trə‚nīt }

prow [NAV ARCH] The bow of a ship, especially the part above the waterline. { praù }

PROWL *See* procedure work log system. { praùl }

Proxima Centauri [ASTRON] The star that is the sun's nearest neighbor; stellar magnitude is ll, and it is 2° from the bright star α Centauri. { 'präk·sə·mə sen'tór·ē }

proximal [ANAT] Near the body or the median line of the body. [CONT SYS] Located close to the base or pedestal and away from the end effector of a robot. [GEOL] Of a sedimentary deposit, composed of coarse clastics and formed near the source. { 'präk·sə·məl }

proximal convoluted tubule [ANAT] The convoluted portion of the vertebrate nephron lying between Bowman's capsule and the loop of Henle; functions in the resorption of sugar, sodium and chloride ions, and water. { 'präk·sə·məl ‚kän·və'lüd·əd 'tüb‚yül }

proximate analysis [CHEM ENG] A technique that separates and identifies categories of compounds in a mixture; reported are moisture and ash content, the extracts of the mixture made with alcohol, petroleum ether, water, hydrochloric acid and resins, starches, reducing sugars, proteins, fats, esters, free acids, and so on; this type of analysis of solid fuels allows a prediction to be made as to how the fuel will behave in a furnace. { 'präk·sə·mət ə'nal·ə·səs }

proximity detector [ENG] A sensing device that produces an electrical signal when approached by an object or when approaching an object. { präk'sim·əd·ē di‚tek·tər }

proximity effect [ELEC] Redistribution of current in a conductor brought about by the presence of another conductor. { präk'sim·əd·ē i‚fekt }

proximity-effect microbridge [CRYO] A Josephson junction formed by overcoating a few micrometers of thin superconducting film with normal metal, thereby weakening the superconductivity in the film beneath the metal. Also known as Notarys-Mercereau microbridge. { präk'sim·əd·ē i¦fekt 'mī·krō‚brij }

proximity-focused tube [ELECTR] A type of image tube in which electrons are rapidly accelerated across a narrow gap, 1.5 to 3.5 millimeters wide, between the photocathode and

PROTURA

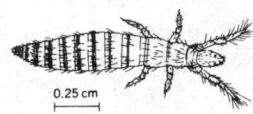

0.25 cm

Acerentulus barberi. (From H. S. Ewing, Ann. Entomol. Soc. Amer., 33(3):497, 1940)

the phosphor screen, both deposited on plane-parallel optical windows. { präk'sim·əd·ē 'fō·kəst 'tüb }

proximity fuse [ORD] A fuse that detonates a warhead when the target is within some specified region near the fuse; radio, radar, photoelectric, or other devices may be used as activating elements. Also known as influence fuse; variable-time fuse; vt fuse. { präk'sim·əd·ē ,fyüz }

proximity sensor [CONT SYS] Any device that measures short distances within a robotic system. Also known as non-contact sensor. { präk'sim·əd·ē 'sen·sər }

proximity warning indicator [NAV] An airborne instrument which produces a warning signal indicating the approach of an aircraft on a possible collision course. Abbreviated PWI. { präk'sim·əd·ē ¦wȯrn·iŋ 'in·də,kād·ər }

proximoceptor [PHYSIO] An exteroceptor involved in taste or cutaneous sensations. { ¦präk·sə·mō¦sep·tər }

proxy server [COMPUT SCI] Software for caching and filtering Web content to reduce network traffic on intranets, and for increasing security by filtering content and restricting access. { 'präk·sē ,sər·vər }

PRR *See* pulse repetition rate.

prudent limit of endurance [AERO ENG] The time during which an aircraft can remain airborne and still retain a given safety margin of fuel. { 'prüd·ənt 'lim·ət əv in'dür·əns }

Prüfer domain [MATH] An integral domain in which every nonzero finitely generated ideal is invertible. { 'prüf·ər də,mān }

pruritus [MED] Localized or generalized itch due to irritation of sensory nerve endings. { prü'rīd·əs }

Prussian blue [INORG CHEM] $Fe_4[Fe(CN)_6]_3$ Ferric ferrocyanide, used as a blue pigment and in the removal of hydrogen sulfide from gases. { 'prəsh·ən 'blü }

prussic acid *See* hydrocyanic acid. { 'prəs·ik 'as·əd }

pyrrolidine [ORG CHEM] C_4H_9N A colorless to pale yellow liquid with a boiling point of 87°C; soluble in water and alcohol; used in the manufacture of pharmaceuticals, insecticides, and fungicides. { pə'ral·ə,dēn }

ps *See* picosecond.

PSA *See* prostate specific antigen.

psalterium *See* omasum. { sȯl'tir·ē·əm }

Psamment [GEOL] A suborder of the soil order Entisol, characterized by a texture of loamy fine sand or coarser sand, and by a coarse fragment content of less than 35. { 'sa,ment }

Psammettidae [INV ZOO] A family of Psamminida, with a strongly built test, haphazardly arranged xenophyae, no specialized surface layer, and no large openings in the test. { sə'med·ə,dē }

Psamminida [INV ZOO] An order of Xenophyophorea distinguished by the absence of linellae in the test and, in general, rigidity of the body. { ,sam·ə'nī·də }

Psamminidae [INV ZOO] A family of Psamminida, with a solid, sometimes fragile test and external xenophyae arranged in a distinct surface layer. { sə'min·ə,dē }

psammite *See* arenite. { 'sa,mīt }

psammitic *See* arenaceous. { sə'mid·ik }

Psammodontidae [PALEON] A family of extinct cartilaginous fishes in the order Bradyodonti in which the upper and lower dentitions consisted of a few large quadrilateral plates arranged in two rows meeting in the midline. { ,sam·ə'dänt·ə,dē }

Psammodrilidae [INV ZOO] A small family of spioniform worms belonging to the Sedentaria. { ,sam·ə'dril·ə,dē }

psammoma [MED] A tumor, usually a meningioma, which contains psammoma bodies. { sa'mäm·ə }

psammoma bodies [PATH] Laminated calcific spherical structures found in certain benign and malignant tumors. { sa'mäm·ə ,bäd·ēz }

psammomatus papilloma *See* serous cystadenoma. { sa'mäm·əd·əs ,pap·ə'lō·mə }

psammon [ECOL] **1.** In a body of fresh water, that part of the environment composed of a sandy beach and bottom lakeward from the water line. **2.** Organisms which inhabit the interstitial water in the sands on a lake shore. { 'sa,män }

psammophilic [ECOL] Pertaining to an organism found in sand. { ¦sam·ə¦fil·ik }

psammophyte [ECOL] Thriving (as a plant) on sandy soil. { 'sam·ə,fīt }

psammosere [ECOL] Stages in plant succession which begin in sandy soil. { 'sam·ə,sir }

PSC motor *See* permanent-split capacitor motor. { ¦pē¦es¦sē 'mōd·ər }

psec *See* picosecond.

Pselaphidae [INV ZOO] The ant-loving beetles, a large family of coleopteran insects in the superfamily Staphylinoidea. { sə laf·ə,dē }

Psephenidae [INV ZOO] The water penny beetles, a small family of coleopteran insects in the superfamily Dryopoidea. { sə'fen·ə,dē }

psephicity [GEOL] A coefficient of roundability of a pebble- or sand-size mineral fragment, expressed as the ratio of specific gravity to hardness (as measured in the air) or the quotient of specific gravity minus one divided by hardness (as measured in water). { sə'fis·əd·ē }

psephite [GEOL] A sediment or sedimentary rock composed of fragments that are coarser than sand and which are set in a qualitatively and quantitatively varying matrix; equivalent to a rudite or, generally, a conglomerate. { sē,fīt }

psephyte [GEOL] A lake-bottom deposit consisting mainly of coarse, fibrous plant remains. { sē,fīt }

p series [MATH] The series $1 + (1/2)^p + (1/3)^p + \cdots$, where p is a real number. { 'pē ,sir·ēz }

Pseudaliidae [INV ZOO] A family of roundworms belonging to the Strongyloidea which occur as parasites of whales and porpoises. { süd·ə·də'lī·ə,dē }

pseudaposematic [ECOL] Pertaining to an imitation in coloration or form by an organism of another organism that possesses dangerous or disagreeable characteristics. { ¦süd·ə,pōz·ə'mad·ik }

pseudoadiabat [METEOROL] On a thermodynamic diagram, a line representing a pseudoadiabatic expansion of an air parcel; in practice, approximate computations are employed, and the resulting lines represent, ambiguously, pseudoadiabats and saturation adiabats. { ¦sü·dō ad·ē·ə,bat }

pseudoadiabatic chart *See* Stuve chart. { ¦sü·dō,ad·ē·ə'bad·ik 'chärt }

pseudoadiabatic expansion [GEOPHYS] A saturation-adiabatic process in which the condensed water substance is removed from the system, and which therefore is best treated by the thermodynamics of open systems; meteorologically, this process corresponds to rising air from which the moisture is precipitating. { ¦sü·dō,ad·ē·ə'bad·ik ik'span·shən }

pseudoalleles [GEN] Closely linked genes of similar function that can be separated from each other by crossing over. { ¦sü·cō·ə'lēlz }

pseudoallochem [GEOL] An object resembling an allochem but produced in place within a calcareous sediment by a secondary process such as recrystallization. { ¦sü·dō'al·ə,kem }

pseudoanalog display [ELECTR] An electronic display consisting of a dedicated arrangement of discrete pixels used to present analog or quantitative information. { ¦süd·ō¦an·ə,läg di'splā }

pseudoautosomal [GEN] Pertaining to segments of the X and Y chromosomes that undergo obligatory crossing-over so that they show an autosomal pattern of inheritance instead of the typical X- or Y-linked pattern. { ¦süd·ō,ȯd·ō'sō·məl }

Pseudoborniales [PALEOBOT] An order of fossil plants found in Middle and Upper Devonian rocks. { ¦sü·dō,bȯr·nē'ā·lēz }

pseudobreccia [PETR] Limestone that is partially and irregularly dolomitized and is characterized by a mottled, breccialike appearance. Also known as recrystallization breccia. { ¦sü·dō'brech·ə }

pseudobrookite [MINERAL] Fe_2TiO_5 A brown or black mineral consisting of iron titanium oxide and occurring in orthorhombic crystals; specific gravity is 4.4–4.98. { ¦sü·dō'brù,kīt }

pseudocannel coal [GEOL] Cannel coal that contains much humic matter. Also known as humic-cannel coal. { ¦sü·dō¦kan·əl 'kōl }

pseudocarburizing *See* blank carburizing. { ¦sü·dō'kär·bə,rīz·iŋ }

pseudocentrum [VERT ZOO] A centrum formed by fusion of the dorsal or dorsal and ventral arcualia, as in tailed amphibians. { ¦sü·dō'sen·trəm }

pseudocercus *See* urogomphus. { ¦sü·dō'sər·kəs }

pseudochrysolite *See* moldavite. { ¦sü·dō'kris·ə,līt }

PSEUDOBORNIALES

Pseudobornia node with a whorl of three leaves.

1.5 cm

pseudocode [COMPUT SCI] In software engineering, an outline of a program written in English or the user's natural language; it is used to plan the program, and also serves as a source for test engineers doing software maintenance; it cannot be compiled. { 'süd·ō‚kōd }

pseudocoele [INV ZOO] A space between the body wall and internal organs that is not formed by gastrulation and lacks a cellular lining. { 'süd·ə‚sēl }

pseudocoelocyte [INV ZOO] In nematodes, a mesenchymal cell in the pseudocoelom. { ¦sü·dō'sel·ə‚sīt }

Pseudocoelomata [INV ZOO] A group comprising the animal phyla Entoprocta, Aschelminthes, and Acanthocephala; characterized by a pseudocoelom. { ¦sü·dō‚sē·lə'mäd·ə }

pseudocol [GEOL] A landform represented by a constriction of a stream valley diverted by a glacial ponding, formed by the cutting through of a cover of drift and subsequent exposure of a former col. { 'süd·ə‚kól }

pseudo cold front See pseudo front. { 'sü·dō 'kōld ‚frənt }

pseudocoloring [COMPUT SCI] A method of assigning arbitrary colors to the gray levels of a black-and-white image. It is popular in thermography (the imaging of heat), where hotter objects (with high pixel values) are assigned one color (for example, red), and cool objects (with low pixel values) are assigned another color (for example, blue), with other colors assigned to intermediate values. { ‚süd·ō'kəl·ər·iŋ }

pseudocolumella [INV ZOO] In anthozoans, a type of axial structure. { ¦sü·dō‚kal·yə'mel·ə }

pseudoconcretion [GEOL] A subspherical, secondary sedimentary structure resembling a true concretion but not formed by orderly precipitation of mineral matter in the pores of a sediment. { ¦sü·dō·kän'krē·shən }

pseudoconformity See paraconformity. { ¦sü·dō·kən'fȯr·məd·ē }

pseudoconglomerate [GEOL] A rock that resembles, or may easily be mistaken for, a true or normal (sedimentary) conglomerate. { ¦sü·dō·kən'gläm·ə·rət }

pseudocotunnite [MINERAL] K_2PbCl_4 A yellow or yellowish-green, orthorhombic mineral consisting of a potassium lead chloride. { ¦sü·dō·kə'tə‚nīt }

pseudocritical properties [CHEM] Effective (empirical) values for the critical properties (such as temperature, pressure, and volume) of a multicomponent chemical system. { ¦sü·dō'krid·ə·kəl 'präp·ərd·ēz }

pseudo cross-bedding [GEOL] **1.** An inclined bedding produced by deposition in response to ripple-mark migration and characterized by foreset beds that appear to dip into the current. **2.** A structure resembling cross-bedding, caused by distortion-free slumping and sliding of a semiconsolidated mass of sediments (such as sandy shales). { 'sü·dō 'krós ‚bed·iŋ }

pseudocrystal [CRYSTAL] A substance that appears to be crystalline but does not have a true crystalline diffraction pattern. { ¦sü·dō'krist·əl }

pseudocumene [ORG CHEM] C_9H_{12} Water-insoluble, hydrocarbon liquid, boiling at 168°C; soluble in alcohol, benzene, and ether; used to manufacture perfumes and dyes, and as a catgut sterilant. Also known as pseudocumol; uns-trimethylbenzene. { ¦sü·dō'kyü·mēn }

pseudocumol See pseudocumene. { ¦sü·dō'kyü‚mól }

Pseudocycnidae [INV ZOO] A family of the Caligoida which comprises external parasites on the gills of various fishes. { ¦sü·dō'sik·nə‚dē }

pseudocyesis [MED] A condition characterized by amenorrhea, enlargement of the abdomen, and other symptoms simulating gestation, due to an emotional disorder. { ¦sü·dō·sī'ē·səs }

Pseudodiadematidae [INV ZOO] A family of Jurassic and Cretaceous echinoderms in the order Phymosomatoida which had perforate crenulate tubercles. { ¦sü·dō‚dī·ə·də'mad·ə‚dē }

pseudodiffusion [GEOL] Mixing of thin superpositioned layers of slowly accumulated marine sediments by the action of water motion or subsurface organisms. { ¦sü·dō·di'fyü·zhən }

pseudoequivalent temperature See equivalent temperature. { ¦sü·dō·i'kwiv·ə·lənt 'tem·prə·chər }

pseudofault [GEOL] A faultlike feature resulting from weathering along joint, shrinkage, or bedding planes. { 'süd·ə‚fȯlt }

pseudofibrous peat [GEOL] Peat that is fibrous in texture but is plastic and incoherent. { ¦sü·dō'fī·brəs 'pēt }

pseudo front [METEOROL] A small-scale front, formed in association with organized severe convective activity, between a mass of rain-cooled air from the thunderstorm clouds and the warm surrounding air. Also known as pseudo cold front. { 'sü·dō 'frənt }

pseudogalena See sphalerite. { ¦sü·do·gə'lē·nə }

pseudogamy See pseudomixis. { sü'däg·ə·mē }

pseudogene [GEN] A sequence of deoxyribonucleic acid resembling but not functioning like a gene; usually produced by gene duplication followed by mutations that alter or abolish function. { ¦sü·dō‚jēn }

pseudoglanders [VET MED] An infectious bacterial lymphangitis of horses and other equines caused by *Corynebacterium pseudotuberculosis* and characterized by ulcerating nodules of the lymph nodes in the legs. { ¦sü·dō'glan·dərz }

pseudogley [GEOL] A densely packed, silty soil that is alternately waterlogged and rapidly dried out. { ¦sü·dō‚glā }

pseudo-Goldstone bosons [PARTIC PHYS] Goldstone bosons which accompany the breakdown of approximate, accidental symmetries in certain unified gauge theories of weak and electromagnetic interactions. { ¦sü·dō 'gōl‚stōn 'bō‚sänz }

pseudogradational bedding [GEOL] A structure in metamorphosed sedimentary rock in which the original textural graduation (coarse at the base, finer at the top) appears to be reversed because of the formation of porphyroblasts in the finer-grained part of the rock. { ¦sü·dō·grā'dā·shən·əl 'bed·iŋ }

pseudograph [MATH] A graph with at least one loop. { 'süd·ə‚graf }

pseudohalogen [CHEM] Any one of a group of molecules that exhibit significant similarity to the halogens, for example, cyanogen (NCCN). { ‚süd·ō'hal·ə·jən }

pseudohermaphroditism [PHYSIO] A condition in humans which simulates hermaphroditism, with gynandry in females and androgyny in males. { ¦sü·dō·hər'maf·rə·də‚tiz·əm }

pseudoinstruction [COMPUT SCI] **1.** A symbolic representation in a compiler or interpreter. **2.** See quasi-instruction. { 'sü·dō·in‚strək·shən }

pseudoionone [ORG CHEM] $C_{13}H_{20}O$ A pale yellow liquid with a boiling point of 143–145°C; soluble in alcohol and ether; used for perfumes and cosmetics. { ¦sü·dō'ī·ə‚nōn }

pseudokarst [GEOL] A topography that resembles karst but that is not formed by the dissolution of limestone; usually a rough-surfaced lava field in which ceilings of lava tubes have collapsed. { ¦sü·də‚kärst }

pseudokettle See pingo remnant. { ¦sü·dō'ked·əl }

pseudoleucite [MINERAL] A pseudomorph after leucite consisting of a mixture of nepheline, orthoclase, and analcime. { ¦sü·dō'lü‚sīt }

pseudoliquid density [PETRO ENG] For reservoir studies, the calculated value of a pseudodensity for reservoir liquid at atmospheric conditions, followed by application of suitable correction factors to obtain an approximate value for actual liquid density in the reservoir. { ¦sü·dō'lik·wəd 'den·səd·ē }

pseudolite [NAV] A ground-based reference station in a differential global positioning system, situated at a known location, which broadcasts a signal that has the same structure as a satellite signal and also contains differential corrections for the signals of satellites in view. { 'süd·ə‚līt }

pseudomalachite [MINERAL] $Cu_5(PO_4)_2(OH)_4·H_2O$ An emerald green to dark green and blackish-green, monoclinic mineral consisting of a hydrated basic copper phosphate. Also known as tagilite. { ¦sü·dō'mal·ə‚kīt }

pseudometric See semimetric. { ¦süd·ə¦me·trik }

pseudomicroseism [GEOPHYS] A microseism due to instrumental effects. { ‚sü·dō'mī·krə‚sīz·əm }

pseudomixis [BIOL] Formation of an embryo from the fusion of vegetative cells instead of gametes. Also known as pseudogamy; somatogamy. { ¦süd·ō'mik·səs }

Pseudomonadaceae [MICROBIO] A family of gram-negative, aerobic, rod-shaped bacteria; cells are straight or curved and motile by polar flagella. { sü‚däm·ə·nə'dās·ē‚ē }

Pseudomonadales [MICROBIO] Formerly an order of ovoid, rod-shaped, comma-shaped, or spiral bacteria in the class Schizomycetes; cells characterized as rigid and motile by means of polar flagella. { sü‚däm·ə·nə'dā·lēz }

Pseudomonadineae [MICROBIO] Formerly a suborder of bacteria of the order Pseudomonadales including those families whose cells lacked photosynthetic pigments. { sü‚däm·ə·nə'dī·nē‚ē }

Pseudomonas [MICROBIO] A genus of gram-negative, motile, non-spore-forming, rod-shaped bacteria that cause a variety of infectious diseases in animals and humans (such as glanders and melioidosis) and in plants. { ‚süd·ə'mōn·əs }

Pseudomonas aeruginosa [MICROBIO] An opportunistic pathogen that is the most significant cause of hospital-acquired infections, particularly in predisposed patients with metabolic, hematologic, and malignant diseases. It produces toxic factors such as lipase, esterase, lecithinase, elastase, and endotoxin, some of which may contribute to its pathogenesis. { ‚süd·ə‚mōn·əs ‚ar·ə·jə'nō·sə }

Pseudomonas mallei [MICROBIO] A mammalian parasite that is the causative agent of glanders, an infectious disease of horses that is occasionally transmitted to humans by direct contact. { ‚süd·ə‚mōn·əs 'mal·ē‚ī }

Pseudomonas pseudomallei [MICROBIO] A bacteria that is the causative agent of melioidosis, an endemic glanders-like disease of humans and animals that occurs most frequently in southeastern Asia and northern Australia. { ‚süd·ə‚mōn·əs ‚süd·ə'mal·ē‚ī }

pseudomorph [MINERAL] An altered mineral whose crystal form has the outward appearance of another mineral species. Also known as false form. { 'süd·ə‚mórf }

pseudomountain [GEOL] A mountain formed by differential erosion, in contrast to one produced by uplift. { 'sü·dō‚maúnt·ən }

pseudomucinous cystadenoma [MED] A benign ovarian tumor composed of columnar mucin-producing cells lining multilocular cysts filled with mucinous material. { ‚süd·ə'myüs·ən·əs ‚sist·ad·ən'ō·mə }

pseudonodule [GEOL] A primary sedimentary structure consisting of a ball-like mass of sandstone enclosed in shale or mudstone; characterized by a rounded base with upturned or inrolled edges and resulting from the settling of sand into underlying clay or mud which has welled up between isolated sand masses. Also known as sand roll. { 'sü·dō'näj‚ül }

pseudonoise code See pseudorandom noise code. { 'süd·ō'nóiz ‚kōd }

pseudo-oolith [GEOL] A spherical or roundish pellet or particle (generally less than 1 millimeter in diameter) in a sedimentary rock, externally resembling an oolith in size or shape but of secondary origin and amorphous or crypto- or microcrystalline, and lacking the radial or concentric internal structure of an oolith. Also known as false oolith. { ‚sü·dō'ō‚ō‚lith }

pseudo-operation [COMPUT SCI] An operation which is not part of the computer's operation repertoire as realized by hardware; hence, an entension of the set of machine operations. { 'sü·dō‚äp·ə'rā·shən }

pseudoparalysis [MED] An apparent motor paralysis that is caused by voluntary inhibition of motor impulses because of pain or other organic or psychic causes. { 'sü·dō·pə'ral·ə·səs }

pseudoparasite [MED] Something in the blood that is mistaken for a parasite. { 'sü·dō'par·ə‚sīt }

pseudoparkinsonism [MED] A reversible syndrome resembling parkinsonism that may result from the dopamine-blocking action of antipsychotic drugs. Also known as drug-induced parkinsonism. { ‚süd·ō'pär·kən·sə‚niz·əm }

pseudoperfect number [MATH] An integer that is equal to the sum of some of its proper divisors. { ‚süd·ō‚pər·fikt 'nəm·bər }

Pseudophoracea [INV ZOO] An extinct superfamily of gastropod mollusks in the order Aspidobranchia. { ‚süd·ə·fə'rā·shə }

Pseudophyllidea [INV ZOO] An order of tapeworms of the subclass Cestoda, parasitic principally in the intestine of cold-blooded vertebrates. { ‚süd·ə·fə'lid·ē·ə }

pseudoplasmodium [INV ZOO] An aggregate of amebas resembling a plasmodium. { 'sü·dō·plaz'mō·dē·əm }

pseudoplastic [BIOL] Referring to an organism which lacks the capacity for major modification or for evolutionary differentiation. { 'sü·dō'plas·tik }

pseudoplastic fluid [FL MECH] A fluid whose apparent viscosity or consistency decreases instantaneously with an increase in shear rate. { 'sü·dō'plas·tik 'flü·əd }

pseudopodium [BOT] A slender, leafless branch of the gametophyte in certain Bryatae. [CYTOL] Temporary projection of the protoplast of ameboid cells in which cytoplasm streams actively during extension and withdrawal. [INV ZOO] Foot of a rotifer. { ‚süd·ə'pōd·ē·əm }

pseudoporphyritic [PETR] Pertaining to a rock that is not a true porphyry, but resembles one because of rapid growth of some of the crystals. { ‚sü·dō‚pór·fə'rid·ik }

pseudopotential [SOLID STATE] The common effective potential for electrons in a crystal lattice that is calculated in the orthogonalized plane-wave method and in the pseudopotential method, and that is relatively weak (except for diffracted electrons) because the electrons are moving rapidly past the atoms in the lattice. { ‚sü·dō·pə'ten·chəl }

pseudopotential method [SOLID STATE] A method of approximating the energy states of electrons in a crystal lattice in which the electrons are assumed to move in a common effective potential that is calculated from the experimentally determined energy levels and the effective masses of the electrons. { ‚sü·dō·pə‚ten·chəl 'meth·əd }

pseudorandom noise code [COMMUN] A method of transmitting messages in the presence of interference or noise, in which each binary digit in the original message is encoded by a long series of binary digits with desirable autocorrelation properties. Also known as pseudonoise code. Abbreviated PN code. { ‚süd·ō'ran·dəm 'nóiz ‚kōd }

pseudorandom numbers [COMPUT SCI] Numbers produced by a definite arithmetic process, but satisfying one or more of the standard tests for randomness. { ‚süd·ō'ran·dəm 'nəm·bərz }

pseudoreduced compressibility [CHEM] The compressibility factor for a multicomponent gaseous system, calculated at reduced conditions using the pseudoreduced properties of the mixture. { ‚sü·dō·ri'düst kəm‚pres·ə'bil·əd·ē }

pseudoreduced properties [CHEM] Reduced-state relationships (such as reduced pressure, reduced temperature, and reduced volume) calculated for multicomponent chemical systems by using pseudocritical properties. { ‚sü·dō·ri'düst 'präp·ərd·ēz }

pseudo ripple mark [GEOL] A bedding-plane feature that resembles a ripple mark but is formed by lateral pressure caused by slumping or by local, small-scale tectonic deformation. { sü·dō 'rip·əl ‚märk }

pseudorotation [ORG CHEM] Twist conformation in aliphatic rings or ring structures containing five or more atoms. { süd·ō·rō'tā·shən }

pseudorotaxane [CHEM] A supramolecular species consisting of a linear molecular component (without bulky end groups) encircled by a macrocyclic component. { ‚süd·ō·rō·tak‚sān }

pseudoscalar [PHYS] A quantity which has magnitude only, and which acts, under Lorentz transformation, like a scalar but with a sign change under space reflection or time reflection, or both. { 'sü·dō'skāl·ər }

pseudoscalar coupling [PARTIC PHYS] A type of interaction postulated between a nucleon and a pion in which the interaction energy is a product of the pion's pseudoscalar field and a bilinear pseudoscalar function of the nucleon fields. { 'sü·dō'skāl·ər 'kəp·iŋ }

pseudoscalar meson [PARTIC PHYS] A meson, such as the pion, which has spin 0 and negative parity, and may be described by a field which is a pseudoscalar quantity. Also known as pseudoscalar particle. { ‚sü·dō'skāl·ər 'mā‚sän }

pseudoscalar particle See pseudoscalar meson. { 'sü·dō'skāl·ər 'pärd·ə·kəl }

pseudoscope [OPTICS] A device that produces reversed stereoscopic effects, for example, by transposing the pictures of a stereoscope. { 'süd·ə‚skōp }

Pseudoscorpionida [INV ZOO] An order of terrestrial Arachnida having the general appearance of miniature scorpions without the postabdomen and sting. { ‚sü·dō‚skór·pē'än·ə·də }

Pseudosphaeriales [BOT] An order of the class Ascolichenes, shared by the class Ascomycetes; the ascocarp is flask-shaped and lined with a layer of interwoven, branched pseudoparaphyses. { ‚süd·ō‚sfir·ē'ā·lēz }

pseudospharolith [MINERAL] A spherulite consisting of two minerals, one with parallel and one with inclined extinction, growing from the same center. { ‚süd·ō'sfar·ə‚lith }

pseudosphere [MATH] The pseudospherical surface generated by revolving a tractrix about its asymptote. { 'süd·ə‚sfir }

pseudospherical surface [MATH] A surface whose total

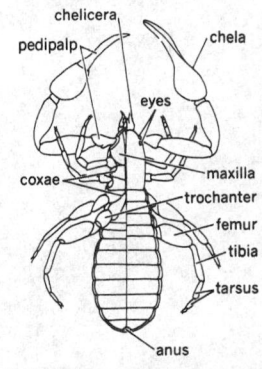

PSEUDOSCORPIONIDA

Adult Pseudoscorpionida: ventral view, left side; dorsal view, right side. *(From H. S. Pratt, A Manual of the Common Invertebrate Animals, rev. ed., McGraw-Hill, 1951)*

curvature has a constant negative value. { ¦süd·ō¦sfer·ə·kəl 'sər·fəs }

Pseudosporidae [INV ZOO] A family of the protozoan subclass Proteomyxidia; flagellated stages invade Volvocidae and filamentous algae and become amebas. { ¦sü·dō'spór·ə͵dē }

pseudostatic SP *See* pseudostatic spontaneous potential. { ¦sü·dō¦stad·ik ¦es'pē }

pseudostatic spontaneous potential [PETRO ENG] Theoretical maximum spontaneous potential current that can be measured in a downhole, mud-column log in shaly sand. Abbreviated pseudostatic SP; PSP. { ¦sü·dō¦stad·ik spän'tā·nē·əs pə'ten·chəl }

pseudo steady-state pressure distribution [PETRO ENG] The condition when the declining pressure distribution within a reservoir system closed at one boundary is declining at a uniform rate everywhere (or, there is constant pressure gradient within the reservoir). { ¦sü·dō 'sted·ē ͵stāt 'presh·ər ͵dis·trə͵byü·shən }

pseudostem [BOT] A false stem composed of concentric rolled or folded blades and sheaths that surround the growing point. { 'süd·ō͵stem }

pseudostratification *See* sheeting structure. { ¦sü·dō͵strad·ə·fə'kā·shən }

PSEUDOSTRATIFIED EPITHELIUM

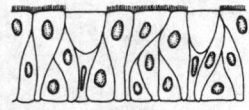

Arrangement of cells in pseudostratified epithelium.

pseudostratified epithelium [HISTOL] A type of epithelium in which all cells reach to the basement membrane but some extend toward the surface only part way, while others reach the surface. { ¦sü·dō'strad·ə͵fīd ͵ep·ə'thē·lē·əm }

Pseudosuchia [PALEON] A suborder of extinct reptiles of the order Thecodontia comprising bipedal, unarmored or feebly armored forms which resemble dinosaurs in many skull features but retain a primitive pelvis. { ¦sü·dō'sü·kē·ə }

pseudosymmetry [CRYSTAL] Apparent symmetry of a crystal, resembling that of another system; generally due to twinning. { ¦sü·dō'sim·ə·trē }

pseudotachylite [PETR] A black rock that resembles tachylite; carries fragmental enclosures and shows evidence of having been at high temperature. { ¦sü·dō'tak·ə͵līt }

pseudotensor [PHYS] **1.** A quantity which transforms as a tensor under space rotations, but which transforms as a tensor, together with a change in sign, under space inversion. **2.** A quantity which transforms as a tensor under Lorentz transformations, but with an additional sign change under space reflection or time reflection or both. { ¦sü·dō'ten·sər }

Pseudothelphusidae [INV ZOO] A family of fresh-water crabs belonging to the Brachyura. { ¦sü·dō·thel'fyüz·ə͵dē }

pseudotillite [GEOL] A nonglacial tillite-like rock, such as a pebbly mudstone, formed on land by the flow of nonglacial mud or deposited by a subaqueous turbidity flow. { ¦sü·dō'däd·əl͵īt }

Pseudotriakidae [VERT ZOO] The false catsharks, a family of galeoids in the carcharinid line. { ¦sü·dō·trī'ak·ə͵dē }

pseudotuberculosis [MED] A bacterial infection in humans and many animals caused by *Pasteurella pseudotuberculosis*; may be severe in humans with septicemia and symptoms resembling typhoid fever. { ¦sü·dō·tə͵bər·kyə'lō·səs }

pseudounconformity [GEOL] A stratigraphic relationship that appears unconformable but is characterized by a superabundance or an excess accumulation of sediment, due to factors like submarine slumping which occurs penecontemporaneously with sedimentation off the sides of a rising anticline or dome. { ¦sü·dō·kən'fór·məd·ē }

pseudovector [PHYS] **1.** A quantity which transforms as a vector under space rotations but which transforms as a vector, together with a change in sign, under a space inversion. Also known as axial vector. **2.** A quantity which transforms as a four-vector under Lorentz transformations, but with an additional sign change under space reflection or time reflection or both. { ¦sü·dō'vek·tər }

pseudovector coupling [PARTIC PHYS] A type of interaction postulated between a nucleon and another particle in which the expression for the interaction energy contains a bilinear pseudovector function of nucleon fields. { ¦sü·dō'vek·tər 'kəp·liŋ }

pseudovector meson [PARTIC PHYS] A meson which has spin quantum number 1 and positive parity, and may be described by a field which is a pseudovector quantity. { ¦sü·dō'vek·tər 'mä͵sän }

pseudovitrinite [GEOL] A maceral of coal that is superficially similar to vitrinite but that is higher in reflectance from

polished surfaces in oil immersion and has slitted structure, remnant cellular structures, uncommon fracture patterns, higher relief, and paucity or absence of pyrite inclusions. { ¦sü·dō'vi·trə͵nīt }

pseudovitrinoid [GEOL] Pseudovitrinite occurring in bituminous coal. { ¦sü·dō'vi·trə͵nóid }

pseudovolcano [GEOL] A large crater or circular hollow believed not to be associated with recent volcanic activity, such as a crater which is the result of cauldron subsidence or of a phreatic explosion in the distant past. { ¦sü·dō·väl'kā·nō }

pseudo-wild type [GEN] A wild phenotype due to a second (suppressor) mutation of a mutant. { ¦sü·dō' wīld ͵tīp }

psf *See* pound per square foot.

P shell [ATOM PHYS] The sixth layer of electrons about the nucleus of an atom, having electrons whose principal quantum number is 6. { 'pē ͵shel }

psi *See* pound per square inch.

psia *See* pounds per square inch absolute.

psid *See* pounds per square inch differential.

psi function [MATH] The special function of a complex variable which is obtained from differentiating the logarithm of the gamma function. [QUANT MECH] *See* Schrödinger wave function. { 'sī ͵fəŋk·shən }

psig *See* pounds per square inch gage.

PSII *See* plasma-source ion implantation.

psilate [BOT] Lacking ornamentation; generally applied to pollen. { 'sī͵lāt }

Psilidae [INV ZOO] The rust flies, a family of myodarian cyclorrhaphous dipteran insects in the subsection Acalyptratae. { 'sil·ə͵dē }

psilomelane [MINERAL] $BaMn_9O_{16}(OH)_4$ A massive, hard, black, botryoidal manganese oxide mineral mixture with a specific gravity ranging from 3.7 to 4.7. { ͵sī·lō'me͵län }

Psilophytales [PALEOBOT] A group formerly recognized as an order of fossil plants. { ͵sī·lō͵fī'tā·lēz }

Psilophytineae [PALEON] The equivalent name for Rhyniopsida. { ͵sī·lō͵fī'tin·ē͵ē }

Psilopsida [BOT] A subdivision of the Tracheophyta. { sī'läp·səd·ə }

Psilorhynchidae [VERT ZOO] A small family of actinopterygian fishes belonging to the Cyprinoidei. { ͵sī·lō'riŋ·kə͵dē }

psilosis [MED] Falling out of the hair. { sī'lō·səs }

Psilotales [BOT] The equivalent name for Psilotophyta. { ͵sī·lō'tā·lēz }

Psilotatae [BOT] A class of the Psilotophyta. { sī'läd·ə͵dē }

Psilotophyta [BOT] A division of the plant kingdom represented by three living species; the life cycle is typical of the vascular cryptogams. { ͵sī·lō'täf·əd·ə }

psi particle *See* J particle. { 'sī ͵pärd·ə·kəl }

psi-prime particle [PARTIC PHYS] A neutral meson which has a mass of 3684 megaelectronvolts, spin quantum number 1, and negative parity and charge parity; it has an anomalously long lifetime. Symbolized ψ′. { 'sī ͵prīm ͵pärd·ə·kəl }

Psittacidae [VERT ZOO] The single family of the Psittaciformes. { sə'tas·ə͵dē }

Psittaciformes [VERT ZOO] The parrots, a monofamilial order of birds that exhibit zygodactylism and have a strong hooked bill. { sə͵tas·ə'fór͵mēz }

psittacinite *See* mottramite. { sə'tas·ə͵nīt }

psittacosis [MED] Pneumonia and generalized infection of man and of birds caused by agents of the PLT-Bedsonia group; transmitted to humans by psittacine birds. { ͵sid·ə'kō·səs }

PSK *See* phase-shift keying.

psoas [ANAT] Either of two muscles: psoas major which arises from the bodies and transverse processes of the lumbar vertebrae and is inserted into the lesser trochanter of the femur, and psoas minor which arises from the bodies and transverse processes of the lumbar vertebrae and is inserted on the pubis. { 'sō·əs }

Psocoptera [INV ZOO] An order of small insects in which wings may be present or absent, tarsi are two- or three-segmented, cerci are absent, and metamorphosis is gradual. { sō'käp·tə·rə }

Psolidae [INV ZOO] A family of echinoderms in the order Dendrochirotida characterized by a ventral adhesive sucker and a U-shaped gut, with the mouth and anus opening upward on the adoral surface. { 'säl·ə‚dē }

Psophiidae [VERT ZOO] The trumpeters, a family of birds in the order Gruiformes. { sō'fil·ə‚dē }

psophometer [ENG] An instrument for measuring noise in electric circuits; when connected across a 600-ohm resistance in the circuit under study, the instrument gives a reading that by definition is equal to half of the psophometric electromotive force actually existing in the circuit. { sō'fäm·əd·ər }

psophometric electromotive force [ELECTR] The true noise voltage that exists in a circuit. { ‚säf·ə‚me·trik i‚lek·trə‚mōd·iv 'fórs }

psophometric voltage [ELECTR] The noise voltage as actually measured in a circuit under specified conditions. { ‚säf·ə‚me·trik 'vōl·tij }

psoriasis [MED] A usually chronic, often acute inflammatory skin disease of unknown cause; characterized by dull red, well-defined lesions covered by silvery scales which when removed disclose tiny capillary bleeding points. { sə'rī·ə‚səs }

psorosis [PL PATH] A virus disease of tangerine, grapefruit, and sweet orange trees characterized by scaly bark, a gummy exudate, retarded growth, small yellow leaves, and dieback of twigs. Also known as scaly bark. { sə'rō·səs }

PSP *See* pseudostatic spontaneous potential.

P spot [ASTRON] One of a pair of sunspots that appears to precede or lead the other across the face of the sun, or whose magnetic polarity is that which is normally found in such a sunspot during that sunspot cycle and in the hemisphere of the sun. { 'pē ‚spät }

PSR *See* primary radar.

PSRR *See* power-supply rejection ratio.

PSTN *See* public switched telephone network.

psyche [PSYCH] The mind or self as a functional entity. { 'sī·kē }

Psyche [ASTRON] An asteroid with a diameter of about 155 miles (249 kilometers), mean distance from the sun of 2.92 astronomical units, and unusual (M-type) surface composition; it may be made of solid metal. { 'sī·kē }

psychedelic [PSYCH] Of, pertaining to, or producing a psychic state. { ‚sī·kə'del·ik }

psychiatric social worker [PSYCH] A specialist who utilizes the techniques of both social work and psychiatry to serve the community. { ‚sī·kē'a·trik 'sō·shəl ‚wər·ər }

psychiatrist [MED] A person who specializes in psychiatry; a licensed physician trained in psychiatry. { sə'kī·ə·trəst }

psychiatry [MED] The medical science that deals with the origins, diagnosis, and treatment of mental and emotional disorders. { sə'kī·ə·trē }

Psychidae [INV ZOO] The bagworms, a family of lepidopteran insects in the superfamily Tineoidea; males are large, hairy moths, but females are degenerate, wingless, and legless and live in bag-shaped cases. { 'sī·kə‚dē }

psychoacoustics [PSYCH] The study of the psychological interactions that take place between humans or animals and the world of sound, including studies of the perception of sound and studies of the production of speech. { ‚sī·kō·ə'küs·tiks }

psychoanalysis [PSYCH] A technique used in the treatment of neuroses and other emotional disorders which relies upon free associations of the patient to bring ideas and experiences from the unconscious to the conscious divisions of the psyche. { ‚sī·kō·ə'nal·ə·səs }

psychobiology [PSYCH] The school of psychiatry and psychology in which the individual is considered as the sum of his environment as well as being considered a physical organism. { ‚sī·kō·bī'äl·ə·jē }

Psychodidae [INV ZOO] The moth flies, a family of orthorrhaphous dipteran insects in the series Nematocera. { sī'käd·ə‚dē }

psychodynamics [PSYCH] The study of human behavior from the point of view of motivation and drives, depending largely on the functional significance of emotion, and based on the assumption that an individual's total personality and reactions at any given time are the product of the interaction between his genetic constitution and his environment. { ‚sī·kō·dī'nam·iks }

psychogalvanic reflex [PHYSIO] A variation in the electric conductivity of the skin in response to emotional stimuli, which cause changes in blood circulation, secretion of sweat, and skin temperature. { ‚sī·kō·gal'van·ik 'rē‚fleks }

psychogalvanometer [ENG] An instrument for testing mental reaction by determining how skin resistance changes when a voltage is applied to electrodes in contact with the skin. { ‚sī·kō‚gal·və'näm·əd·ər }

psychogenic [MED] Of psychic origin. { ‚sī·kō'jen·ik }

psychointegroammeter *See* lie detector. { ‚sī·kō‚in·tə·grō'am‚ēd·ər }

psychokinesis [PSYCH] The alleged ability of an individual to exert a mental influence on physical events in advance of their occurrence. Abbreviated PK. { ‚sī·kō·kə'nē·səs }

psychologist [PSYCH] **1.** An individual who has made a professional study of, and usually thereafter professionally engages in, psychology. **2.** Specifically, an individual with the minimum professional qualifications set forth by an intraprofessionally recognized psychological association. { sī'käl·ə·jəst }

psychology [BIOL] **1.** The science that deals with the functions of the mind and the behavior of an organism in relation to its environment. **2.** The mental activity characteristic of a person or a situation. { sī'käl·ə·jē }

psychometrics [PSYCH] The mathematical and statistical treatment of psychological data. { ‚sī·kō‚me·triks }

psychomotor [PSYCH] Pertaining to both mental and motor activity. { ‚sī·kə'mōd·ər }

psychomotor excitement [PSYCH] Generalized physical and emotional overactivity that is marked by nonproductivity and repetitiveness and associated with feelings of inner tension. { ‚sī·kə‚mōd·ər ik'sīt·mənt }

psychomotor performance [IND ENG] The degree of skill demonstrated by an operator in the completion of a task. { ‚sī·kə‚mōd·ər pər'fór·məns }

psychomotor task [IND ENG] An aspect of a job that requires the operator to use controlled movements of the body. { ‚sī·kə‚mōd·ər ‚task }

psychopathic personality [PSYCH] An emotionally immature individual characterized by pronounced defects in judgment and prone to impulsive, generally amoral or antisocial behavior. { ‚sī·kə‚path·ik ‚pər·sə'nal·əd·ē }

psychopathology [PSYCH] The systematic study of mental diseases. { ‚sī·kō·pə'thäl·ə·jē }

psychopharmacology [PSYCH] The science that deals with the action of drugs on mental function. { ‚sī·kō‚fär·mə'käl·ə·jē }

psychophysics [PSYCH] **1.** The study of mental processes by physical methods. **2.** The study of the relations of stimuli to the sensations they produce. { ‚sī·kō'fiz·iks }

psychophysiologic nervous system reaction *See* neurasthenic neurosis. { ‚sī·kō‚fiz·ē·ō'läj·ik 'nər·vəs ‚sis·təm rē‚ak·shən }

psychosexual development [PSYCH] In psychoanalytic theory, a series of four developmental stages (oral, anal, phallic, and Oedipal), relatively fixed in time, that are determined by the interaction between a person's biological drives and the environment. { ‚sī·kō‚sek·chə·wəl di'vel·əp·mənt }

psychosis [PSYCH] An impairment of mental functioning to the extent that it interferes grossly with an individual's ability to meet the ordinary demands of life, characterized generally by severe affective disturbance, profound introspection, and withdrawal from reality, formation of delusions or hallucinations, and regression presenting the appearance of personality disintegration. { sī'kō·səs }

psychosomatic [MED] Of or pertaining to the interrelationship between mental processes and somatic functions. { ‚sī·kō·sə'mad·ik }

psychosomatograph [ENG] An instrument for recording muscular action currents or physical movements during tests of mental-physical coordination. { sī·kō·sə'mad·ə‚graf }

psychosurgery [MED] The branch of medicine that deals with the treatment of various psychoses, severe neuroses, and chronic painful conditions by means of operative procedures on the brain. { ‚sī·kō'sər·jə·rē }

psychotaxis [PSYCH] Involuntary adjustment of a person's thoughts and behavior in order to retain the agreeable and avoid the disagreeable; an ego defense mechanism. { ‚sī·kō'tak·səs }

psychotherapy [PSYCH] The use of psychological means in

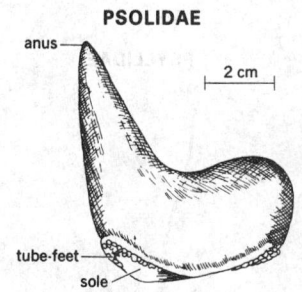

PSOLIDAE

anus

2 cm

tube-feet

sole

Psolus phantapus.

PSYCHIDAE

(a)

(b)

Thyridopteryx ephemeraeformis Haworth. (a) Male. (b) Female case.

PSYLLIDAE

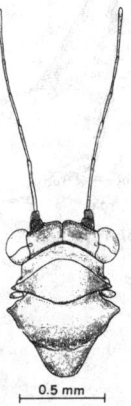

0.5 mm

Anterior dorsum of a psyllid.

PTERIDOSPERMAE

1 m

Reconstruction of a seed fern. Note appearance of adventitious roots on the lower portion of the stem.

PTEROBRANCHIA

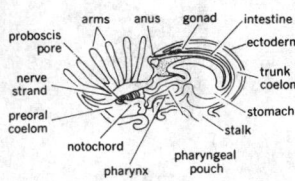

Cephalodiscus.

the treatment of emotional and mental disorders. { ¦sī·kō'ther·ə·pē }

psychotomimetic [PSYCH] **1.** Mimicking a psychotic disorder. **2.** Pertaining to any drug or compound, such as lysergic acid diethylamide or mescaline, which can induce a psychotic-like state. { sī¦käd·ō·mi'med·ik }

psychotropic [PSYCH] Pertaining to any drug or agent having a particular affinity for or effect on the psyche. { ¦sī·kə'träp·ik }

psychromatic ratio [THERMO] Ratio of the heat-transfer coefficient to the product of the mass-transfer coefficient and humid heat for a gas-vapor system; used in calculation of humidity or saturation relationships. { sī·krə'mad·ik 'rā·shō }

psychrometer [ENG] A device comprising two thermometers, one a dry bulb, the other a wet or wick-covered bulb, used in determining the moisture content or relative humidity of air or other gases. Also known as wet and dry bulb thermometer. { sī'kräm·əd·ər }

psychrometric calculator [ENG] A device for quickly computing certain psychrometric data, usually the dew point and the relative humidity, from known values of the dry- and wet-bulb temperatures and the atmospheric pressure. { ¦sī·krə¦me·trik 'kal·kyə,lād·ər }

psychrometric chart [THERMO] A graph each point of which represents a specific condition of a gas-vapor system (such as air and water vapor) with regard to temperature (horizontal scale) and absolute humidity (vertical scale); other characteristics of the system, such as relative humidity, wet-bulb temperature, and latent heat of vaporization, are indicated by lines on the chart. { ¦sī·krə¦me·trik 'chärt }

psychrometric formula [THERMO] The semiempirical relation giving the vapor pressure in terms of the barometer and psychrometer readings. { ¦sī·krə¦me·trik 'fȯr·myə·lə }

psychrometric tables [THERMO] Tables prepared from the psychrometric formula and used to obtain vapor pressure, relative humidity, and dew point from values of wet-bulb and dry-bulb temperature. { ¦sī·krə¦me·trik 'tā·bəlz }

psychrometry [ENG] The science and techniques associated with measurements of the water vapor content of the air or other gases. { sī'kräm·ə·trē }

psychrophile [BIOL] An organism that thrives at low temperatures. { 'sī·krə,fīl }

psychrophilic [PHYSIO] Relating to the ability to live and grow at low temperatures. { ¦sī·krə'fil·ik }

psychrophobia [PHYSIO] Abnormal sensitivity to cold. [PSYCH] An abnormal fear of the cold. { ¦sī·krə'fō·bē·ə }

psychrophyte [ECOL] A plant adapted to the climatic conditions of the arctic or alpine regions. { 'sī·krə,fīt }

psychrosphere [OCEANOGR] The cold deep layer of the ocean, 100–700 meters (330–2300 feet) below the surface, where the water temperature is typically less than 10°C (50°F). { 'sī·krə,sfir }

Psyllidae [INV ZOO] The jumping plant lice, a family of the Homoptera in the series Sternorrhyncha in which adults have a transverse head with protuberant eyes and three ocelli, 6- to 10-segmented antennae, and wings with reduced but conspicuous venation. { 'sil·ə,dē }

pt *See* pint.

Pt *See* platinum.

PTA *See* factor XI; phosphotungstic acid.

ptarmigan [VERT ZOO] Any of various birds of the genus *Lagopus* in the family Tetraonidae; during the winter, plumage is white and hairlike feathers cover the feet. { 'tar·mə·gən }

PT boat *See* motor torpedo boat. { ¦pē'tē ,bōt }

PTC *See* Christmas factor.

PTD *See* posttuning drift.

Pteraspidomorphi [VERT ZOO] The equivalent name for Diplorhina. { tə,ras·pə·də'mȯr,fī }

Pterasteridae [INV ZOO] A family of deep-water echinoderms in the order Spinulosida distinguished by having webbed spine fins. { ,ter·ə'ster·ə,dē }

Pteridophyta [BOT] The equivalent name for Polypodiophyta. { ,ter·ə'däf·əd·ə }

Pteridophytic *See* Paleophytic. { tə¦rid·ə¦fid·ik }

Pteridospermae [PALEOBOT] Seed ferns, a class of the Cycadicae comprising extinct plants characterized by naked seeds borne on large fernlike fronds. { ,ter·ə·dō'spər,mē }

Pteridospermophyta [PALEOBOT] The equivalent name for Pteridospermae. { ,ter·ə·dō·spər'mäf·əd·ē }

Pteriidae [INV ZOO] Pearl oysters, a family of bivalve mollusks which have nacreous shells. { tə'rī·ə,dē }

pterinophore [CYTOL] A yellow to orange chromatophore that contains pterine pigment. { tə'rin·ə,fȯr }

pterion [ANTHRO] The region surrounding the sphenoparietal suture where the frontal bone, parietal bone, squama temporalis, and greater wing of the sphenoid bone come together most closely. { 'ter·ē,än }

Pterobranchia [INV ZOO] A group of small or microscopic marine animals regarded as a class of the Hemichordata; all are sessile, tubicolous organisms with a U-shaped gut and three body segments. { ,ter·ə'braŋ·kē·ə }

pterochore [BOT] A type of plant that produces winged disseminules. { 'ter·ə,kȯr }

Pteroclidae [VERT ZOO] The sandgrouse, a family of granivorous birds in the order Columbiformes; mainly an Afro-Asian group resembling pigeons and characterized by cryptic coloration, usually corresponding with the soil color of the habitat. { tə'räk·lə,dē }

pterodactyl [PALEON] The common name for members of the extinct reptilian order Pterosauria. { ¦ter·ə¦dak·təl }

Pterodactyloidea [PALEON] A suborder of Late Jurassic and Cretaceous reptiles in the order Pterosauria distinguished by lacking tails and having increased functional wing length due to elongation of the metacarpels. { ¦ter·ə¦dak·tə'lȯid·ē·ə }

pteroic acid [BIOCHEM] $C_{14}H_{12}N_6O_3$ A crystalline amino acid formed by hydrolysis of folic acid or other pteroylglutamic acids. { tə'rō·ik 'as·əd }

Pteromalidae [INV ZOO] A family of hymenopteran insects in the superfamily Chalcidoidea. { ,ter·ə'mal·ə,dē }

Pteromedusae [INV ZOO] A suborder of hydrozoan cnidarians in the order Trachylina characterized by a modified, bipyramidal medusae. { ,ter·ə·mə'dü,sē }

Pterophoridae [INV ZOO] The plume moths, a family of the lepidopteran superfamily Pyralidoidea in which the wings are divided into featherlike plumes, maxillary palpi are lacking, and the legs are long. { ,ter·ə'fȯr·ə,dē }

Pteropidae [INV ZOO] The fruit bats, a large family of the Chiroptera found in Asia, Australia, and Africa. { tə'räp·ə,dē }

Pteropoda [INV ZOO] The sea butterflies, an order of pelagic gastropod mollusks in the subclass Opisthobranchia in which the foot is modified into a pair of large fins and the shell, when present, is thin and glasslike. { tə'räp·ə·də }

Pteropodidae [VERT ZOO] A family of fruit-eating bats in the suborder Megachiroptera, characterized by primitive ears and by shoulder joints. { ,ter·ə'päd·ə,dē }

pteropod ooze [GEOL] A pelagic sediment containing at least 45% calcium carbonate in the form of tests of marine animals, particularly pteropods. { 'ter·ə,päd 'üz }

Pteropsida [BOT] A large group of vascular plants characterized by having parenchymatous leaf gaps in the stele and by having leaves which are thought to have originated in the distant past as branched stem systems. { tə'räp·səd·ə }

Pterosauria [PALEON] An extinct order of flying reptiles of the Mesozoic era belonging to the subclass Archosauria; the wing resembled that of a bat, and a large heeled sternum supported strong wing muscles. { ,ter·ə'sȯr·ē·ə }

pterostigma [INV ZOO] Opaque thickened spot occurring on the costal margin of an insect wing. { ,ter·ə'stig·mə }

pteroylglutamic acid *See* folic acid. { ¦ter·ə·wəl·glü'tam·ik 'as·əd }

pterygium [MED] **1.** A triangular mass of mucous membrane growing on the conjunctiva, usually near the inner canthus. **2.** Overgrowth of the cuticle forward on the nail. [VERT ZOO] A generalized vertebrate limb. { tə'rij·ē·əm }

pterygoid bone [VERT ZOO] A rodlike bone or group of bones forming a portion of the palatoquadrate arch in lower vertebrates. { 'ter·ə,gȯid }

pterygopalatine fossa [ANAT] The gap between the pterygoid process of the sphenoid bone and the maxilla and palatine bone. { ¦ter·ə·gō¦pal·ə,tēn 'fäs·ə }

pterygoquadrate [EMBRYO] Of, pertaining to, or being the first branchial arch in lower vertebrate embryos; gives rise to most of the upper jaw. { ¦ter·ə·gō¦kwä,drāt }

Ptiliidae [INV ZOO] The feather-winged beetles, a family of coleopteran insects in the superfamily Staphylinoidea. { 'til·ə,dē }

Ptilodactylidae [INV ZOO] The toed-winged beetles, a family of the Coleoptera in the superfamily Dryopoidea. { ¦til·ō·dak'til·ə,dē }

Ptilodontoidea [PALEON] A suborder of extinct mammals in the order Multituberculata. { ¦til·ō·dän'tóid·ē·ə }

ptilolite See mordenite. { 'til·ə,līt }

Ptinidae [INV ZOO] The spider beetles, a family of coleopteran insects in the superfamily Bostrichoidea. { 'tin·ə,dē }

PTM See pulse-time modulation.

Ptolemaic system [ASTRON] The movements of the solar system according to Clandius Ptolemy; supposedly, the earth was a fixed center, with the sun and moon revolving about it in circular orbits; planets revolved in small circles (epicycles) whose centers revolved about the earth in larger circles (deferents). { ¦täl·ə¦mā·ik 'sis·təm }

Ptolemy's theorem [MATH] The theorem that a necessary and sufficient condition for a convex quadrilateral to be inscribed in a circle is that the sum of the products of the two pairs of opposite sides equal the product of the diagonals. { 'täl·ə·mēz ,thir·əm }

ptosis [MED] Prolapse, abnormal depression, or falling down of an organ or part; applied especially to drooping of the upper eyelid, from paralysis of the third cranial nerve. { 'tō·səs }

PTRM See partial thermoremanent magnetization.

p27 [CELL MOL] A cyclin-dependent kinase inhibitor in normal and neoplastic cells. { ,pē,twen·tē 'sev·ən }

ptyalase See ptyalin. { 'tī·ə,lās }

ptyalin [BIOCHEM] A diastatic enzyme found in saliva which catalyzes the hydrolysis of starch to dextrin, maltose, and glucose, and the hydrolysis of sucrose to glucose and fructose. Also known as ptyalase; salivary amylase; salivary diastase. { 'tī·ə·lən }

Ptychodactiaria [INV ZOO] An order of the zoantharian anthozoans of the phylum Cnidaria known only from two genera, *Ptychodactis* and *Dactylanthus*. { ¦tī·kō,dak·tē'ar·ē·ə }

Ptychomniaceae [BOT] A family of mosses in the order Isobryales distinguished by an eight-ribbed capsule. { tī,käm·nē'ās·ē,ē }

ptychotis oil See ajowan oil. { tī'kōd·əs ,oil }

Ptyctodontida [PALEON] An order of Middle and Upper Devonian fishes of the class Placodermi in which both the head and trunk shields are present, and the joint between them is a well-differentiated and variable structure. { ,tik·tə'dänt·əd·ə }

ptygma [GEOL] Pegmatitic material with migmatite or gneiss, resembling disharmonic folds. Also known as ptygmatic fold. { 'tig·mə }

ptygmatic fold See ptygma. { tig'mad·ik 'fōld }

p-type conductivity [ELECTR] The conductivity associated with holes in a semiconductor, which are equivalent to positive charges. { 'pē ¦tīp ,kän,dək'tiv·əd·ē }

p-type crystal rectifier [ELECTR] Crystal rectifier in which forward current flows when the semiconductor is positive with respect to the metal. { 'pē ¦tīp 'krist·əl 'rek·tə,fī·ər }

p-type semiconductor [ELECTR] An extrinsic semiconductor in which the hole density exeeds the conduction electron density. { 'pē ¦tīp 'sem·i·kən,dək·tər }

p⁺-type semiconductor [ELECTR] A *p*-type semiconductor in which the excess mobile hole concentration is very large. { 'pē¦pləs ,tīp 'sem·i·kən,dək·tər }

p-type silicon [ELECTR] Silicon to which more impurity atoms of acceptor type (with valence of 3, such as boron) than of donor type (with valence of 5, such as phosphorus) have been added, with the result that the hole density exceeds the conduction electron density. { 'pē ¦tīp 'sil·ə,kän }

Pu See plutonium.

puberty [PHYSIO] The period at which the generative organs become capable of exercising the function of reproduction; signalized in the boy by a change of voice and discharge of semen, in the girl by the appearance of the menses. { 'pyü·bərd·ē }

puberulent [BOT] Having a surface covered with very fine downlike hairs. { pyü'ber·yə·lənt }

puberulic acid [BIOCHEM] $(HO)_3(C_7H_2O)COOH$ A keto acid formed as a metabolic product of certain species of *Penicillium;* has some germicidal activity against gram-positive bacteria. { pyü'bər·yə·lik 'as·əd }

pubic arch [ANAT] The arch formed by the conjoined rami of the pubis and ischium. { 'pyü·bik 'ärch }

pubic crest [ANAT] The crest extending from the pubic tubercle to the medial extremity of the pubis. { 'pyü·bik krest }

pubic height [ANTHRO] A measure of the vertical distance from the pubis to the floor taken when the subject is in a standing position. { 'pyü·bik 'hīt }

pubic symphysis [ANAT] The fibrocartilaginous union of the pubic bones. Also known as symphysis pubis. { 'pyü·bik 'sim·fə·səs }

pubis [ANAT] The pubic bone, the portion of the hipbone forming the front of the pelvis. { 'pyü·bəs }

public address system See sound-reinforcement system. { 'pəb·lik ə'dres ,s.s·təm }

public area [BUILD] The total nonrentable area of a building, such as public conveniences and rest rooms. { 'pəb·lik 'er·ē·ə }

public communications service [COMMUN] Telephone or telegraph service provided for the transmission of unofficial communications for the public. { 'pəb·lik kə,myü·nə'kā·shənz ,sər·vəs }

public correspondence [COMMUN] Any telecommunications which offices and stations at the disposal of the public must accept for transmission. { 'pəb·lik ,kär·ə'spän·dəns }

public data [COMPUT SCI] Data that are open to all users, with no security measures necessary as far as reading is concerned. { 'pəb·lik 'dad·ə }

public health [MED] **1.** The state of health of a community or of a population. **2.** The art and science dealing with the protection and improvement of community health. { 'pəb·lik 'helth }

public health dent stry [MED] The science of promoting dental health through community effort; it comprises research, education, prevention, diagnosis, prescription, and the evaluation of community dental care. { ,pəb·lik ,helth 'dent·ə·strē }

public-key algorithm [COMMUN] A cryptographic algorithm in which one key (usually the enciphering key) is made public and a different key (usually the deciphering key) is kept secret; it must not be possible to deduce the private key from the public key. { 'pəb·lik 'kē 'al·gə,rith·əm }

public network [COMMUN] A communications network that can be used by anyone, usually on a fee basis. { 'pəb·lik 'ret,wərk }

public pack [COMPUT SCI] A disk pack that can be used by any program and any application in a computer system. { 'pəb·lik 'pak }

public radio communications services [COMMUN] Land, mobile, and fixed services, the stations of which are open to public correspondence. { 'pəb·lik 'rād·ē·ō kə,myü·nə'kā·shənz ,sər·və·səz }

public-safety frequency bands [COMMUN] Radio-frequency bands allocated in the United States for communication on land between base stations and mobile stations or between mobile stations by police, fire, highway, forestry, and emergency services. { 'pəb·lik ¦sāf·tē 'frē·kwən·sē ,banz }

public-safety radio service [COMMUN] Any service of radio communication essential to either the discharge of nonFederal governmental functions relating to public safety responsibilities or the alleviation of an emergency endangering life or property, the radio transmitting facilities of which are defined as fixed, land, or mobile stations. { 'pəb·lik ¦sāf·tē 'rād·ē·ō ,sər·vəs }

public switched telephone network [COMMUN] The worldwide voice telephone network. Abbreviated PSTN. { 'pəb·lik ,swicht 'tel·ə,fōn ,ret,wərk }

public utility [IND ENG] A business organization considered by law to be vested with public interest and subject to public regulation. { 'pəb·lik yü'til·əd·ē }

public works [IND ENG] Government-owned and financed works and improvements for public enjoyment or use. { 'pəb·lik 'wərks }

Puccinia asparagi [MYCOL] An autoecious fungus of the order Uredinales; the causative agent of asparagus rust. { pü,sin·ē·ə ə'spär·ə,gē }

Puccinia graminis [MYCOL] A macrocylic heteroecious

PTYCTODONTIDA

Rhamphodopsis threiplandi from the Middle Devonian of Scotland. Restoration of skeleton with outline of body in lateral aspect.

fungus of the order Uredinales; the causative agent of black stem rust of cereal grains. { pü,sin·ē·ə 'gram·ə,nəs }

Puccinia malvacearum [MYCOL] A microcyclic fungus of the order Uredinales; the causative agent of hollyhock rust. { pü,sin·ē·ə ,mal·və'sē·ə·rəm }

pucherite [MINERAL] BiVO₄ A reddish-brown orthorhombic mineral composed of bismuth vanadate, occurring as small crystals. { 'pü·kə,rīt }

Puck [ASTRON] A satellite of Uranus orbiting at a mean distance of 53,440 miles (86,010 kilometers) with a period of 18 hours 20 minutes, and with a diameter of about 96 miles (154 kilometers). { pək }

puckering [MET] Corrugations in metal parts resulting from pressing or drawing. { 'pək·ə·riŋ }

pudding ball See armored mud ball. { 'pùd·iŋ ,ból }

puddingstone [GEOL] In Great Britain, a conglomerate consisting of rounded pebbles whose colors are in marked contrast with the matrix, giving a section of the rock the appearance of a raisin pudding. { 'pùd·iŋ,stōn }

puddle [ENG] To apply water in order to settle loose dirt. [MET] A batch of molten iron within the puddling furnace. { pəd·əl }

puddling [MET] A process for the production of wrought iron by agitation of a bath of molten pig iron with iron oxide in order to reduce the carbon, silicon, phosphorus, and manganese content. { 'pəd·liŋ }

puddling furnace [MET] A coal-fired reverberatory furnace for puddling pig iron. { 'pəd·liŋ ,fər·nəs }

puelche [METEOROL] An east wind which has crossed the Andes; the Andean foehn of the South American west coast. { 'pwel·chē }

puerperal sepsis [MED] A toxic condition caused by infection in the birth canal, occurring as a complication or sequel of pregnancy. { pyü'ər·prəl 'sep·səs }

puerperium [MED] **1.** The state of a woman during labor or immediately after delivery. **2.** The period from delivery to the time when the uterus returns to normal size, about 6 weeks. { ,pyü·ər'pir·ē·əm }

puff [ELEC] See picofarad. [MECH ENG] A small explosion within a furnace due to combustion conditions. { pəf }

puff ball [BOT] A spherical basidiocarp that retains spores until fully mature and, when disturbed, releases them as puffs of fine dust. { 'pəf ,ból }

puff cone See mud cone. { 'pəf ,kōn }

puffer [MIN ENG] A small stationary engine used in coal mines for hoisting material. { 'pəf·ər }

puffing gun [FOOD ENG] A device for puffing food particles in which they are rapidly heated and pressurized up to 200 pounds per square inch (1.38 megapascals) in a sealed vessel and then suddenly depressurized to atmospheric pressure, causing rapid volatilization of internal moisture. { 'pəf·iŋ ,gən }

puff of wind [METEOROL] A slight local breeze which causes a patch of ripples on the surface of the sea. { 'pəf əv 'wind }

pug mill [MECH ENG] A machine for mixing and tempering a plastic material by the action of blades revolving in a drum or trough. { 'pəg ,mil }

puking [CHEM ENG] In a distillation column, the foaming and rising of liquid so that part of it is driven out of the vessel through the vapor line. { 'pyük·iŋ }

pulaskite [PETR] A light-colored, feldspathoid-bearing, granular or trachytoid alkali syenite composed chiefly of orthoclase, soda pyroxene, arfvedsonite, and nepheline. { pü'las,kīt }

pulegium oil See hedeoma oil. { pyü'lē·jē·əm ,oil }

Pulfrich refractometer [OPTICS] A critical angle refractometer in which the material to be tested rests on a prism of material of known higher index of refraction and angle, and the angle of refraction of light which is directed at the interface between the two materials at grazing incidence is observed. { 'pùl·frik ,rē,frak'täm·əd·ər }

pull-apart [GEOL] A precompaction sedimentary structure having the appearance of boudinage and consisting of beds that have been stretched and pulled apart into relatively short slabs. { 'pùl ə,pärt }

pull crack [MET] A crack in a casting caused by contraction strains during cooling and resulting from the shape of the casting. { 'pùl ,krak }

pull-down menu [COMPUT SCI] A list of options for action that appears near the top of a display screen, usually overlaying

PULLEY

Some typical examples of pulleys. (*a*) Multigroove V-belt sheave; (*b*) solid-hub crown face pulley; (*c*) brass pulley for round belt. (*Boston Gear INCOM International Inc.*)

the current contents of the screen without disrupting them, and usually in response to an indicator being pointed at an icon. { 'pùl ¦daùn 'men·yü }

puller [MECH ENG] A lever-operated chain or wire-rope hoist for lifting or pulling at any angle, which has a reversible ratchet mechanism in the lever permitting short-stroke operation for both tensioning and relaxing, and which holds the loads with a Weston-type friction brake or a releasable ratchet. Also known as come-along. { 'pùl·ər }

pulley [DES ENG] A wheel with a flat, round, or grooved rim that rotates on a shaft and carries a flat belt, V-belt, rope, or chain to transmit motion and energy. { 'pùl·ē }

pulley lathe [MECH ENG] A lathe for turning pulleys. { 'pùl·ē ,lāth }

pulley stile [BUILD] The upright part of a window frame which holds the pulley and guides the sash. { 'pùl·ē ,stīl }

pulley top [MECH ENG] A top with a long shank used to tap setscrew holes in pulley hubs. { 'pùl·ē ,täp }

pulling [ELECTR] An effect that forces the frequency of an oscillator to change from a desired value; causes include undesired coupling to another frequency source or the influence of changes in the oscillator load impedance. [PETRO ENG] Withdrawing sucker rods and production tubing from a pumping well prior to cleaning out or replacing parts of the pump. { 'pùl·iŋ }

pulling figure [ELECTR] The total frequency change of an oscillator when the phase angle of the reflection coefficient of the load impedance varies through 360°, the absolute value of this reflection coefficient being constant at 0.20. { 'pùl·iŋ ,fig·yər }

pulling pillars See pillaring. { 'pùl·iŋ ,pil·ərz }

pull-in torque [MECH ENG] The largest steady torque with which a motor will attain normal speed after accelerating from a standstill. { 'pùl,in ,tórk }

pullorum disease [VET MED] A highly fatal disease of chickens and other birds caused by *Salmonella pullorum*, characterized by weakness, lassitude, lack of appetite, and whitish or yellowish diarrhea. Also known as bacillary white diarrhea; white diarrhea. { pə'lòr·əm di,zēz }

pull-out torque [MECH ENG] Th largest torque under which a motor can operate without sharply losing speed. { 'pūl,aùt ,tórk }

pull rope [MIN ENG] **1.** The rope that pulls a journey of loaded cars on a haulage plane. **2.** The rope that pulls the loaded scoop or bucket in a scraper loader layout. Also known as main rope. { 'pùl ,rōp }

pullshovel See backhoe. { 'pùl,shəv·əl }

pull strength [MECH] A unit in tensile testing; the bond strength in pounds per square inch. { 'pùl ,streŋkth }

pull tube [PETRO ENG] Tube used in rod-type, traveling-barrel oil well pumps to connect the pump plunger with the seating anchor. { 'pùl ,tüb }

pulmonary anthrax [MED] A form of anthrax in humans caused by the inhalation of dust containing *Bacillus anthracis* spores. { 'pùl·mə,ner·ē 'an,thraks }

pulmonary artery [ANAT] A large artery that conducts venous blood from the heart to the lungs of tetrapods. { 'pùl·mə,ner·ē 'ärd·ə·rē }

pulmonary circulation [PHYSIO] The circulation of blood through the lungs for the purpose of oxygenation and the release of carbon dioxide. Also known as lesser circulation. { 'pùl·mə,ner·ē ,sər·kyə'lā·shən }

pulmonary embolism [MED] Obstruction of the pulmonary artery or a branch of it by a free-floating blood clot (embolus) usually originating from a vein in the leg or pelvic area. { ,pùl·mə,ner·ē 'em·bə,liz·əm }

pulmonary edema [MED] An effusion of fluid into the alveoli and interstitial spaces of the lungs. { 'pùl·mə,ner·ē i'dē·mə }

pulmonary plexus [NEUROSCI] A nerve plexus composed chiefly of vagal fibers situated on the anterior and posterior aspects of the bronchi and accompanying them into the substance of the lung. { 'pùl·mə,ner·ē 'plek·səs }

pulmonary stenosis [MED] Narrowing of the orifice of the pulmonary artery. { 'pùl·mə,ner·ē stə'nō·səs }

pulmonary valve [ANAT] A valve consisting of three semilunar cusps situated between the right ventricle and the pulmonary trunk. { 'pùl·mə,ner·ē 'valv }

pulmonary vein [ANAT] A large vein that conducts oxygenated blood from the lungs to the heart in tetrapods. { 'pùl·mə,ner·ē 'vān }

pulmonary ventilation [PHYSIO] The volume of gas entering and exiting the lungs per unit time of respiration. { 'pùl·mə,ner·ē ,vent·əl'ā·shən }

Pulmonata [INV ZOO] A subclass of the gastropod mollusks which contains the "lung"-bearing snails; the gills have been lost and in their place the mantle cavity has become a pulmonary sac. { ,pùl·mə'näd·ə }

pulmonate [VERT ZOO] Possessing lungs or lunglike organs. { 'pùl·mə,nāt }

pulp [ANAT] A mass of soft spongy tissue in the interior of an organ. [BOT] The soft succulent portion of a fruit. [ENG] See slime. [MATER] The cellulosic material produced by reducing wood mechanically or chemically and used in making paper and cellulose products. Also known as wood pulp. { pəlp }

pulpboard [MATER] Chipboard to which is added a percentage of mechanical wood pulp. { 'pəlp,bórd }

pulp cavity [ANAT] The space within the central part of a tooth containing the dermal pulp and made up of the pulp chamber and a root canal. { 'pəlp ,kav·əd·ē }

pulp chamber [ANAT] The coronal portion of the central cavity of a tooth. { 'pəlp ,chām·bər }

pulper [MECH ENG] A machine that converts materials to pulp, for example, one that reduces paper waste to pulp. { 'pəlp·ər }

pulping [ENG] Reducing wood to pulp. { 'pəlp·iŋ }

pulp molding [ENG] A plastics-industry process in which a resin-impregnated pulp material is preformed by application of a vacuum, after which it is oven-cured and molded. { 'pəlp ,mōld·iŋ }

pulpotomy [MED] Surgical removal of the pulp of a tooth. { pəl'päd·ə·mē }

pulpstone [MATER] A block of sandstone cut into wheels for grinding, especially wood pulp in paper manufacture. { 'pəlp,stōn }

pulpwood [MATER] Any wood that can be reduced to pulp. { 'pəlp,wùd }

pulque [FOOD ENG] The sap of the agave plant after natural fermentation; it is distilled to make tequila. { 'pül,kā }

pulsar [ASTROPHYS] A celestial radio source, emitting intense short bursts of radio emission; the periods of known pulsars range between 33 milliseconds and 3.75 seconds, and pulse durations range from 2 to about 150 milliseconds with longer-period pulsars generally having a longer pulse duration. { 'pəl,sär }

pulsatance [PHYS] Angular velocity in radians, equal to 2π times frequency in hertz. { 'pəl·səd·əns }

pulsating combustion [CHEM] Combustion that is accompanied by spontaneous pressure oscillations, which occur if the Rayleigh criterion is satisfied. { 'pəl,sād·iŋ kəm'bus·chən }

pulsating current [ELEC] Periodic direct current. { 'pəl,sād·iŋ 'kə·rənt }

pulsating electromotive force [ELEC] Sum of a direct electromotive force and an alternating electromotive force. Also known as pulsating voltage. { 'pəl,sād·iŋ i,lek·trə'mōd·iv 'fòrs }

pulsating flow [ENG] Irregular fluid flow in a piping system often resulting from the pressure variations of reciprocating compressors or pumps within the system. { 'pəl,sād·iŋ 'flō }

pulsating star [ASTRON] Variable star whose luminosity fluctuates as the star expands and contracts; the variation in brightness is thought to come from the periodic change of radiant energy to gravitational energy and back. { 'pəl,sād·iŋ 'stär }

pulsating voltage See pulsating electromotive force. { 'pəl,sād·iŋ 'vōl·tij }

pulsation [PHYSIO] A beating or throbbing, usually rhythmic, as of the heart or an artery. { pəl'sā·shən }

pulsation dampening [ENG] Device installed in a fluid piping system (gas or liquid) to eliminate or even out the fluid-flow pulsations caused by reciprocating compressors, pumps, and such. { pəl'sā·shən ,dam·pən·iŋ }

pulsation welding See multiple-impulse welding. { pəl'sā·shən ,weld·iŋ }

pulse [PHYS] A variation in a quantity which is normally constant; has a finite duration and is usually brief compared to the time scale of interest. [PHYSIO] **1.** The regular, recurrent, palpable wave of arterial distention due to the pressure of the blood ejected with each contraction of the heart. **2.** A single wave. { pəls }

pulse altimeter [ENG] A device which is used to measure the distance of an aircraft above the ground by sending out radar signals in short pulses and measuring the time delay between the leading edge of the transmitted pulse and that of the pulse returned from the ground. { pəls al'tim·əd·ər }

pulse amplifier [ELEC] An amplifier designed specifically to amplify electric pulses without appreciably changing their waveforms. { 'pəls ,am·plə,fī·ər }

pulse amplitude [PHYS] The peak, average, effective, instantaneous, or other magnitude of a pulse, usually with respect to the normal constant value; the exact meaning should be specified when giving a numerical value. { 'pəls ,am·plə,tüd }

pulse-amplitude discriminator [ENG] Electronic instrument used to investigate the amplitude distribution of the pulses produced in a nuclear detector. { 'pəls ,am·plə,tüd di'skrim·ə,nād·ər }

pulse-amplitude modulation [COMMUN] Amplitude modulation of a pulse carrier. Abbreviated PAM. { 'pəls ,am·plə,tüd ,mäj·ə,lā·shən }

pulse-amplitude modulation-frequency modulation [COMMUN] System in which pulse-amplitude-modulated subcarriers are used to frequency-modulate a second carrier; binary digits are formed by the absence or presence of a pulse in an assigned position. { 'pəls ,am·plə,tüd ,mäj·ə,lā·shən 'frē·kwən·sē ,mäj·ə,lā·shən }

pulse analyzer [ELECTR] An instrument used to measure pulse widths and repetition rates, and to display on a cathode-ray screen the waveform of a pulse. { 'pəls ,an·ə,līz·ər }

pulse bandwidth [COMMUN] The bandwidth outside of which the amplitude of a pulse-frequency spectrum is below a prescribed fraction of the peak amplitude. { 'pəls 'band,width }

pulse cable [COMMUN] A communications cable, capable of transmitting pulses without unacceptable distortion. { 'pəls ,kā·bəl }

pulse carrier [COMMUN] A pulse train used as a carrier. { 'pəls ,kar·ē·ər }

pulse circuit [ELECTR] An active electrical network designed to respond to discrete pulses of current or voltage. { 'pəls ,sər·kət }

pulse code [COMMUN] A code consisting of various combinations of pulses, such as the Morse code, Baudot code, and the binary code used in computers. { 'pəls ,kōd }

pulse-coded scanning beam [NAV] **1.** A radio or radar beam which is swept over a sector of space and is accompanied by a repeated pattern of pulses that is varied to indicate the position of the beam in space. **2.** A system of ground equipment that generates such beams at microwave frequencies to furnish guidance to aircraft making microwave landings. Abbreviated PCSB. { 'pəls ,kōd·əd 'skan·iŋ ,bēm }

pulse-code modulation [COMMUN] Modulation in which the peak-to-peak amplitude range of the signal to be transmitted is divided into a number of standard values, each having its own code; each sample of the signal is then transmitted as the code for the nearest standard amplitude. Abbreviated PCM. { 'pəls ,kōd ,mäj·ə'lā·shən }

pulse-code modulation television [COMMUN] A television system in which digital signals using a pulse code are transmitted, rather than analog signals. { 'pəls ,kōd ,mäj·ə'lā·shən 'tel·ə,vizh·ən }

pulse coder See coder. { 'pəls ,kōd·ər }

pulse coding and correlation [COMMUN] A general technique concerning a variety of methods used to change the transmitted waveform and then decode upon its reception; pulse compression is a special form of pulse coding and correlation. { 'pəls ,kōd·iŋ ən ,kär·ə'lā·shən }

pulse column [CHEM ENG] Continuous-phase process column (such as liquid only or gas only) in which the flow-through is pulsating; used to increase mass-transfer rates, as in a liquid-liquid extraction operation. { 'pəls ,käl·əm }

pulse communication [COMMUN] Radio communication using pulse modulation. { 'pəls kə,myü·nə,kā·shən }

pulse compression [ELECTR] A matched filter technique

used to discriminate against signals which do not correspond to the transmitted signal. { 'pəls kəm‚presh·ən }

pulse-compression radar [ENG] A radar system in which the transmitted signal is linearly frequency-modulated or otherwise spread out in time to reduce the peak power that must be handled by the transmitter; signal amplitude is kept constant; the receiver uses a linear filter to compress the signal and thereby reconstitute a short pulse for the radar display. { 'pəls kəm‚presh·ən 'rā‚där }

pulse counter [ELECTR] A device that indicates or records the total number of pulses received during a time interval. { 'pəls ‚kaůnt·ər }

pulsed-bed sorption [CHEM ENG] Solid-liquid countercurrent adsorption process (such as an ion-exchange process) in which the granulated solids bed and the solution flow alternately, in opposite directions. { 'pəlst ‚bed 'sȯrp·shən }

pulse decay time [COMMUN] The interval of time required for the trailing edge of a pulse to decay from 90% to 10% of the peak pulse amplitude. { 'pəls di'kā ‚tīm }

pulse-delay network [ELECTR] A network consisting of two or more components such as resistors, coils, and capacitors, used to delay the passage of a pulse. { 'pəls di'lā 'net‚wərk }

pulse demodulator [ELECTR] A device that recovers the modulating signal from a pulse-modulated wave. { 'pəls dē¦mäj·ə‚lād·ər }

pulse-density modulation See pulse-frequency modulation. { 'pəls ‚den·sət·ē ‚mäj·ə'lā·shən }

pulsed fast neutron analysis [ENG] A technique for detecting contraband materials, in which a pulsed beam of high-energy neutrons is scanned up and down in a raster pattern while the object under inspection is conveyed through the beam; characteristic gamma rays emitted by materials in the object are detected in order to analyze and image these materials with the help of time-of-flight measurements. { 'pəlst ‚fast 'nü‚trän ə‚nal·əsəs }

pulsed-gel electrophoresis [CELL MOL] A technique in which rare cutting restriction enzymes are used to generate very large fragments of deoxyribonucleic acid (up to 1 million base pairs long), which are separated in gels by applying alternating cycles of electric fields in different directions. { 'pəlst ‚jel ¦‚lek·trə·fə'rē·səs }

pulse discriminator [ELECTR] A discriminator circuit that responds only to a pulse having a particular duration or amplitude. { 'pəls di‚skrim·ə‚nād·ər }

pulsed laser [OPTICS] A laser in which a pulse of coherent light is produced at fixed time intervals, as required for ranging and tracking applications or to permit higher output power than can be obtained with continuous operation. { 'pəlst 'lā·zər }

pulsed light [OPTICS] A beam of light whose intensity is modulated in some prescribed manner; analogous to a radar pulse. { 'pəlst 'līt }

pulsed-light ceilometer See pulsed-light cloud-height indicator. { 'pəlst 'līt sē'läm·əd·ər }

pulsed-light cloud-height indicator [ENG] An instrument used for the determination of cloud heights; it operates on the principle of pulse radar, employing visible light rather than radio waves. Also known as pulsed-light ceilometer. { 'pəlst 'līt 'klaůd ‚hīt 'in·də‚kād·ər }

pulse-Doppler radar [ENG] Pulse radar that uses the Doppler effect to obtain information about the velocity of a target. { 'pəls ¦däp·lər 'rā‚där }

pulsed oscillator [ELECTR] An oscillator that generates a carrier-frequency pulse or a train of carrier-frequency pulses as the result of self-generated or externally applied pulses. { 'pəlst 'äs·ə‚lād·ər }

pulse dot soldering iron [ENG] A soldering iron that provides heat to the tip for a precisely controlled time interval, as required for making a good soldered joint without overheating adjacent parts. { 'pəls ‚dät 'säd·ə·riŋ ‚ī·ərn }

pulsed reactor [NUCLEO] A research nuclear reactor in which continual short, intense surges of power and radiation can be produced; the neutron flux during the surge is much higher than could be tolerated during steady-state operation. { 'pəlst rē'ak·tər }

pulse droop [ELECTR] A distortion of an otherwise essentially flat-topped rectangular pulse, characterized by a decline of the pulse top. { 'pəls ‚drüp }

pulsed ruby laser [OPTICS] A laser in which ruby is used as the active material; the extremely high pumping power

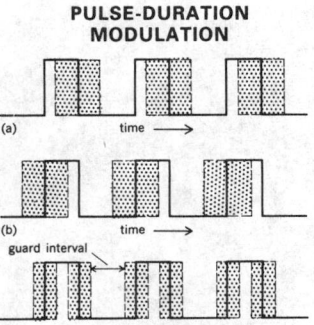

PULSE-DURATION MODULATION

Types of pulse-duration modulation. *(a)* Trailing edge modulated, leading edge fixed. *(b)* Leading edge modulated, trailing edge fixed. *(c)* Both edges modulated. Solid lines indicate duration of unmodulated pulses, shaded areas limits of maximum modulation. *(From H. S. Black, Modulation Theory, Van Nostrand, 1953)*

required is obtained by discharging a bank of capacitors through a special high-intensity flash tube, giving a coherent beam that lasts for about 0.5 millisecond. { 'pəlst 'rü·bē 'lā·zər }

pulsed transfer function [CONT SYS] The ratio of the z-transform of the output of a system to the z-transform of the input, when both input and output are trains of pulses. Also known as discrete transfer function; z-transfer function. { 'pəlst 'tranz·fər ‚faŋk·shən }

pulse duration [COMMUN] The time interval between the first and last instants at which the instantaneous amplitude reaches a stated fraction of the peak pulse amplitude. Also known as pulse length; pulse width (both deprecated usages). { 'pəls dů'rā·shən }

pulse-duration coder See coder. { 'pəls dů¦rā·shən 'kōd·ər }

pulse-duration discriminator [ELECTR] A circuit in which the sense and magnitude of the output are a function of the deviation of the pulse duration from a reference. { 'pəls dů¦rā·shən di'skrim·ə‚nād·ər }

pulse-duration modulation [COMMUN] Modulation of a pulse carrier wherein the value of each instantaneous sample of a modulating wave produces a pulse of proportional duration by varying the leading, trailing, or both edges of a pulse. Abbreviated PDM. Also known as pulse-length modulation; pulse-width modulation. { 'pəls dů¦rā·shən ‚mäj·ə‚lā·shən }

pulse - duration modulation - frequency modulation [COMMUN] System in which pulse-duration-modulated subcarriers are used to frequency-modulate a second carrier. Also known as pulse-width modulation-frequency modulation. { 'pəls dů¦rā·shən ‚mäj·ə‚lā·shən 'frē·kwən·sē ‚mäj·ə‚lā·shən }

pulsed video thermography [ENG] A method of nondestructive testing in which a source of heat is applied to an area of a specimen for a very short time duration, and an infrared detection system reveals anomalously hot or cold regions that then appear close to defects. { 'pəlst 'vid·ē·ō thər'mäg·rə·fē }

pulse-echo method [MET] A nondestructive test in which pulses of energy are directed into a part, and the time for the echo to return from one or more surfaces is measured. { 'pəls ¦ek·ō ‚meth·əd }

pulse form [PHYS] The amplitude of a pulse plotted as a function of time. { 'pəls ‚fȯrm }

pulse-forming network [ELECTR] A network used to shape the leading or trailing edge of a pulse. { 'pəls ¦fȯrm·iŋ 'net‚wərk }

pulse-frequency modulation [COMMUN] A form of pulse-time modulation in which the pulse repetition rate is the characteristic varied. Abbreviated PFM. { 'pəls ¦frē·kwən·sē ‚mäj·ə‚lā·shən }

pulse-frequency spectrum See pulse spectrum. { 'pəls ¦frē·kwən·sē ‚spek·trəm }

pulse generator [ELEC] See impulse generator. [ELECTR] A generator that produces repetitive pulses or signal-initiated pulses. { 'pəls ‚jen·ə‚rād·ər }

pulse group See pulse train. { 'pəls ‚grüp }

pulse hardening [MET] A surface-hardening process performed by heating to the required temperature in a span of several milliseconds by using an energy and time-controlled pulse of very high power at a very high frequency, about 27 megahertz. { 'pəls ‚härd·ən·iŋ }

pulse height [ELECTR] The strength or amplitude of a pulse, measured in volts. { 'pəls ‚hīt }

pulse-height analyzer [NUCLEO] An instrument capable of indicating the number of occurrences of pulses falling within each of one or more specified amplitude ranges; used to obtain the energy spectrum of nuclear radiations. Also known as kick-sorter (British usage); multichannel analyzer. { 'pəls ‚hīt 'an·ə‚līz·ər }

pulse-height discriminator [ELECTR] A circuit that produces a specified output pulse when and only when it receives an input pulse whose amplitude exceeds an assigned value. Also known as amplitude discriminator. { 'pəls ‚hīt di'skrim·ə‚nād·ər }

pulse-height selector [ELECTR] A circuit that produces a specified output pulse only when it receives an input pulse whose amplitude lies between two assigned values. Also known as amplitude selector; diffractional pulse-height discriminator. { 'pəls ‚hīt si'lek·tər }

pulse-height spectrum [PHYS] Distribution of various pulse wavelengths and strengths (heights) developed during activation analysis. { 'pəls ‚hīt 'spek·trəm }

pulse improvement threshold [COMMUN] In a constant-amplitude pulse-modulation system, the condition in which the peak pulse voltage is greater than twice the peak noise voltage, after selection and before nonlinear processes such as amplitude clipping and limiting. { 'pəls im¦prüv·mənt 'thresh,hōld }

pulse integrator [ELECTR] An RC (resistance-capacitance) circuit which stretches in time duration a pulse applied to it. { 'pəls ¦int·ə‚grād·ər }

pulse interference eliminator [ELECTR] Device which removes pulsed signals which are not precisely on the radar operating frequency. { 'pəls ‚in·tər¦fir·əns i‚lim·ə‚nād·ər }

pulse interference separator and blanker [ELECTR] Automatic interference blanker that will blank all video signals not synchronous with the radar pulse-repetition frequency. { 'pəls ‚in·tər¦fir·əns 'sep·ə‚rād·ər ən 'blaŋk·ər }

pulse interleaving [COMMUN] A process in which pulses from two or more sources are combined in time-division multiplex for transmission over a common path. { 'pəls ‚in·tər'lēv·iŋ }

pulse interval *See* pulse spacing. { 'pəls ‚in·tər·vəl }

pulse-interval modulation *See* pulse-spacing modulation. { 'pəls ¦in·tər·vəl ‚mäj·ə‚lā·shən }

pulse ionization chamber [NUCLEO] An ionization chamber that detects individual ionizing events. Also known as counting ionization chamber. { 'pəls ‚ī·ə·nə'zā·shən ‚chām·bər }

pulsejet engine [AERO ENG] A type of compressorless jet engine in which combustion occurs intermittently so that the engine is characterized by periodic surges of thrust; the inlet end of the engine is provided with a grid to which are attached flap valves; these can be sucked inward by a negative differential pressure to allow a regulated amount of air to flow inward to mix with the fuel. Also known as aeropulse engine. { 'pəls‚jet ‚en·jən }

pulse jitter [COMMUN] A relatively small variation of the pulse spacing in a pulse train; the jitter may be random or systematic, depending on its origin, and is generally not coherent with any pulse modulation imposed. { 'pəls ‚jid·ər }

pulse length *See* pulse duration. { 'pəls ‚leŋkth }

pulse-length modulation *See* pulse-duration modulation. { 'pəls ¦leŋkth ‚mäj·ə‚lā·shən }

pulse-link repeater [ELECTR] Arrangement of apparatus used in telephone signaling systems for receiving pulses from one E and M signaling circuit, and retransmitting corresponding pulses into another E and M signaling circuit. { 'pəls ¦liŋk ri'pēd·ər }

pulse-mode multiplexing [COMMUN] A type of time-division multiplexing employing pulse-amplitude modulation in which a sequence of pulses is repeatedly transmitted, and the amplitude of each pulse in the sequence is modulated by a different communication channel. { 'pəls ¦mōd 'məl·tə‚pleks·iŋ }

pulse-modulated jamming [COMMUN] Use of jamming pulses of various widths and repetition rates. { 'pəls ‚mäj·ə¦läd·əd 'jam·iŋ }

pulse-modulated radar [ENG] Form of radar in which the radiation consists of a series of discrete pulses. { 'pəls ‚mäj·ə¦läd·əd 'rā‚där }

pulse modulation [COMMUN] A system of modulation in which the amplitude, duration, position, or mere presence of discrete pulses may be so controlled as to represent the message to be communicated. { 'pəls ‚mäj·ə‚lā·shən }

pulse modulator [ELECTR] A device for carrying out the pulse modulation of a radio-frequency carrier signal. { 'pəls ‚mäj·ə‚läd·ər }

pulse-numbers modulation [COMMUN] Modulation in which a pulse carrier's pulse density per unit time varies in accordance with a modulating wave, by making systematic omissions without changing the phase or amplitude of the transmitted pulses; as an example, the omission of every other pulse could correspond to zero modulation; the reinsertion of some or all pulses then corresponds to positive modulation, and the omission of more than every other pulse corresponds to negative modulation. { 'pəls ‚nəm·bərz ‚mäj·ə‚lā·shən }

pulse operation [ELECTR] For microwave tubes, a method of operation in which the energy is delivered in pulses. { 'pəls ‚äp·ə‚rā·shən }

pulse period [COMMUN] In telephony, time required for one opening and closing of the loop of a calling telephone; for example, the time required to open and close the dial pulse springs once. Also known as impulse period. { 'pəls ‚pir·ē·əd }

pulse-phase modulation *See* pulse-position modulation. { 'pəls ‚fāz ‚mäj·ə‚lā·shən }

pulse-position modulation [COMMUN] Modulation of a pulse carrier wherein the value of each instantaneous sample of a modulating wave varies the position in time of a pulse relative to its unmodulated time of occurrence. Abbreviated PPM. Also known as pulse-phase modulation. { 'pəls pə¦zish·ən ‚mäj·ə‚lā·shən }

pulse pressure [PHYSIO] The difference between the systolic and diastolic blood pressure. { 'pəls ‚presh·ər }

pulser [CHEM ENG] Device used to create a pulsating fluid flow through a process vessel, such as a liquid-liquid or vapor-liquid extraction tower; used to increase contact and mass transfer rates. [ELECTR] A generator used to produce high-voltage, short-duration pulses, as required by a pulsed microwave oscillator or a radar transmitter. { 'pəl·sər }

pulse radar [ENG] Radar in which the transmitter sends out high-power pulses that are spaced far apart in comparison with the duration of each pulse; the receiver is active for reception of echoes in the interval following each pulse. { 'pəls 'rā‚där }

pulse radiolysis [PHYS CHEM] A method of studying fast chemical reactions in which a sample is subjected to a pulse of ionizing radiation, and the products formed by the resulting reactions are studied spectroscopically. { 'pəls ‚rād·ē'äl·ə·səs }

pulse rate [PHYSIO] The number of pulsations of an artery per minute. { 'pəls ‚rāt }

pulse-rate telemetering [ELECTR] Telemetering in which the number of pulses per unit time is proportional to the magnitude of the measured quantity. { 'pəls ‚rāt ‚tel·ə‚mēd·ə·riŋ }

pulse recurrence rate *See* pulse repetition rate. { 'pəls ri'kə·rəns ‚rāt }

pulse recurrence time [COMMUN] Time elapsing between the start of one transmitted pulse and the next pulse; the reciprocal of the pulse repetition rate. { 'pəls ri'kə·rəns ‚tīm }

pulse regeneration [ELECTR] The process of restoring pulses to their original relative timings, forms, and magnitudes. { 'pəls ri‚jen·ə‚rā·shən }

pulse repeater [ELECTR] Device used for receiving pulses from one circuit and transmitting corresponding pulses into another circuit; it may also change the frequencies and waveforms of the pulses and perform other functions. { 'pəls ri‚pēd·ər }

pulse repetition frequency *See* pulse repetition rate. { 'pəls ‚rep·ə¦tish·ən ‚frē·kwən·sē }

pulse repetition rate [ELECTR] The number of times per second that a pulse is transmitted. Abbreviated PRR. Also known as pulse recurrence rate; pulse repetition frequency (PRF). { 'pəls ‚rep·ə¦tish·ən ‚rāt }

pulse rise time [COMMUN] The interval of time required for the leading edge of a pulse to rise from 10% to 90% of the peak pulse amplitude. { 'pəls ‚rīz ‚tīm }

pulse scaler [ELECTR] A scaler that produces an output signal when a prescribed number of input pulses has been received. { 'pəls ‚skāl·ər }

pulse selector [ELECTR] A circuit or device for selecting the proper pulse from a sequence of telemetering pulses. { 'pəls si‚lek·tər }

pulse shaper [ELECTR] A transducer used for changing one or more characteristics of a pulse, such as a pulse regenerator or pulse stretcher. { 'pəls ‚shāp·ər }

pulse spacing [PHYS] Time between corresponding points of successive pulses. Also known as pulse interval. { 'pəls ‚spās·iŋ }

pulse-spacing modulation [COMMUN] A form of pulse-time modulation in which the pulse spacing is varied. Also known as pulse-interval modulation. { 'pəls ¦spās·iŋ ‚mäj·ə‚lā·shən }

pulse spectrum [PHYS] The frequency distribution of the sinusoidal components of a pulse in relative amplitude and in relative phase. Also known as pulse-frequency spectrum. { 'pəls ‚spek·trəm }

pulse stretcher [ELECTR] A pulse shaper that produces an output pulse whose duration is greater than that of the input pulse and whose amplitude is proportional to the peak amplitude of the input pulse. { 'pəls ‚strech·ər }

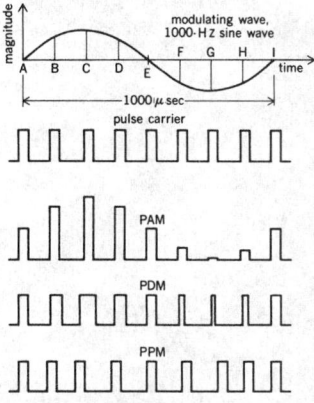

PULSE MODULATION

Examples of pulse modulation. PAM = pulse-amplitude modulation, PDM = pulse-duration modulation, PPM = pulse-position modulation. (*From H. B. Black, Modulation Theory, Van Nostrand, 1953*)

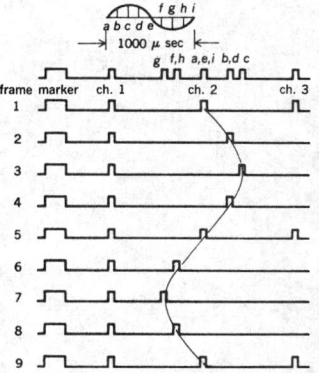

PULSE-POSITION MODULATION

Modulation of channel 2 pulse by a sine wave. Successive diagrams show relative position in time of channel 2 pulse for nine successive samples, at phases of sine wave indicated by letters *a–i*. (*From H. S. Black, Modulation Theory, Van Nostrand, 1953*)

pulse subcarrier [COMMUN] One of a number of frequency-modulation carriers modulating a radio-frequency carrier, each of which is in turn pulse-modulated. { 'pəls 'səb,kar·ē·ər }

pulse synthesizer [ELECTR] A circuit used to supply pulses that are missing from a sequence due to interference or other causes. { 'pəls ,sin·thə,sīz·ər }

pulse-time-modulated radiosonde [ENG] A radiosonde which transmits the indications of the meteorological sensing elements in the form of pulses spaced in time; the meteorological data are evaluated from the intervals between the pulses. Also known as time-interval radiosonde. { 'pəls ¦tīm ¦mäj·ə,lād·əd 'rād·ē·ō,sänd }

pulse-time modulation [COMMUN] Modulation in which the time of occurrence of some characteristic of a pulse carrier is varied from the unmodulated value; examples include pulse-duration, pulse-interval, and pulse-position modulation. Abbreviated PTM. { 'pəls ¦tīm ,mäjə,lā·shən }

pulse tracking system [ENG] Tracking system which uses a high-energy, short-duration pulse radiated toward the target from which the velocity, direction, and range are determined by the characteristics of the reflected pulse. { 'pəls 'trak·iŋ ,sis·təm }

pulse train [PHYS] A series of regularly recurrent pulses having similar characteristics. Also known as pulse group. { 'pəls ,trān }

pulse-train analysis [COMMUN] A Fourier analysis of a pulse train. { 'pəls ¦trān ə,nal·ə·səs }

pulse transformer [ELECTR] A transformer capable of operating over a wide range of frequencies, used to transfer nonsinusoidal pulses without materially changing their waveforms. { 'pəls tranz,för·mər }

pulse transmitter [ELECTR] A pulse-modulated transmitter whose peak-power-output capabilities are usually large with respect to the average-power-output rating. { 'pəls tranz ,mid·ər }

pulse-type altimeter See radar altimeter. { 'pəls ¦tīp al'tim·əd·ər }

pulse-type telemetering [COMMUN] Signal transmission system with pulses as a function of time, but independent of electrical magnitude; in a pulse-counting system the number of pulses per unit time corresponds to the measured variable; in pulse-width or pulse-duration types, the length of the pulse is controlled by the measured variable. { 'pəls ¦tīp ,tel·ə,mēd·ə·riŋ }

pulse voltage See impulse voltage. { 'pəls ,vōl·tij }

pulse wave [PHYSIO] A wave of increased pressure over the arterial system, started by contraction of septum and valves in the heart. { 'pəls ,wāv }

pulse width See pulse duration. { 'pəls ,width }

pulse-width discriminator [ELECTR] Device that measures the pulse length of video signals and passes only those whose time duration falls into some predetermined design tolerance. { 'pəls ¦width di'skrim·ə,nād·ər }

pulse-width modulated static inverter [ELEC] A variation of the quasi-square-wave static inverter, operating at high frequency, in which the pulse width, and not the amplitude, of the square wave is adjusted to approximate the sine wave. { 'pəls ¦width ¦mäj·ə,lād·əd 'stad·ik in,vərd·ər }

pulse-width modulation See pulse-duration modulation. { 'pəls ¦width ,mäj·ə,lā·shən }

pulse-width modulation-frequency modulation See pulse-duration modulation-frequency modulation. { 'pəls ¦width ,mäj·ə,lā·shən 'frē·kwən·sē ,mäj·ə,lā·shən }

pulse window See mean profile. { 'pəls ,win·dō }

pulsing key [COMMUN] **1.** Method of passing voice frequency pulses over the line under control of a key at the original office; used with E and M supervision on intertoll dialing. **2.** System of signaling where numbered keys are depressed instead of using a dial. { 'pəls·iŋ ,kē }

pulsing transformer [ELEC] Transformer designed to supply pulses of voltage or current. { 'pəls·iŋ tranz,för·mər }

pulsometer [MECH ENG] A simple, lightweight pump in which steam forces water out of one of two chambers alternately. { pəl'säm·əd·ər }

pultrusion [ENG] A process for producing continuous fibers for advanced composites which involves pulling reinforcements through tanks of thermoset resins, a preformer, and then a die, where the product is formed into its final shape. { pùl'trü·zhən }

pulverite [PETR] A sedimentary rock composed of silt- or clay-sized aggregates of nonclastic origin with a texture simulating a lutite of clastic origin. { 'pəl·və,rīt }

pulverization See comminution. { ,pəl·və·rə'zā·shən }

pulverizer [MECH ENG] Device for breaking down of solid lumps into a fine material by cleavage along crystal faces. { 'pəl·və,rīz·ər }

pulvillus [INV ZOO] A small cushion or cushionlike pad, often covered with short hairs, on an insect's foot between the claws of the last segment. { ,pəl'vil·əs }

pulvinus [BOT] A cushionlike enlargement of the base of a petiole which functions in turgor movements of leaves. { ,pəl'vī·nəs }

puma [VERT ZOO] *Felis concolor.* A large, tawny brown wild cat (family Felidae) once widespread over most of the Americas. Also known as American lion; catamount; cougar; mountain lion. { 'pü·mə }

pumice [GEOL] A rock froth, formed by the extreme puffing up of liquid lava by expanding gases liberated from solution in the lava prior to and during solidification. Also known as foam; pumice stone; pumicite; volcanic foam. { 'pəm·əs }

pumice fall [GEOL] Pumice falling from a volcano eruption cloud. { 'pəm·əs ,fȯl }

pumiceous [GEOL] Pertaining to the texture of a pyroclastic rock, such as pumice, characterized by numerous small cavities presenting a spongy, frothy appearance. { pyü'mish·əs }

pumice stone See pumice. { 'pəm·əs ,stōn }

pumicite See pumice. { 'pəm·ə,sīt }

pumilith [GEOL] A lithified deposit of volcanic ash. { 'pəm·ə,lith }

pump [ELECTR] Of a parametric device, the source of alternating-current power which causes the nonlinear reactor to behave as a time-varying reactance. [MECH ENG] A machine that draws a fluid into itself through an entrance port and forces the fluid out through an exhaust port. { pəmp }

pumpability [MATER] **1.** The property of a lubricating grease that causes it to flow under pressure through lines, nozzles, and fittings. **2.** The ability of any liquid, slurry, or suspension to be moved through a flow conduit by pressure from a pump. { ,pəm·pə'bil·əd·ē }

pumpability test [ENG] Standard test to ascertain the lowest temperature at which a petroleum fuel oil may be pumped. { ,pəm·pə'bil·əd·ē ,test }

pumparound [CHEM ENG] A system or process vessel that moves liquid out of and back into the vessel at a new location; for example, in a bubble tower, the withdrawing of liquid from a plate or tray, followed by cooling, and returning to another plate to induce condensation of vapors. { 'pəm·pə,raùnd }

pump bob [MECH ENG] A device such as a crank that converts rotary motion into reciprocating motion. { 'pəmp ,bäb }

pump-down time [ENG] The length of time required to evacuate a leak-tested vessel. { 'pəmp 'daùn ,tīm }

pumped hydroelectric storage [ELEC] A method of energy storage in which excess electrical energy produced at times of low demand is used to pump water into a reservoir, and this water is released at times of high demand to operate hydroelectric generators. { 'pəmpt ¦hī·drō·i'lek·trik 'stȯr·ij }

pumped tube [ELECTR] An electron tube that is continuously connected to evacuating equipment during operation; large pool-cathode tubes are often operated in this manner. { 'pəmpt ,tüb }

pumpellyite [MINERAL] $Ca_2Al_3Si_3O_{12}(OH)$ A greenish epidote-like mineral that is probably related to clinozoisite. Also known as lotrite; zonochlorite. { ,pəm'pel·ē,īt }

pumpellyite-prehnite-quartz facies [PETR] A variety of low-temperature, moderate-pressure metamorphism. { ¦pəm 'pel·ē,īt ¦prā,nīt ¦kwȯrts 'fā·shēz }

pumphouse [CIV ENG] A building in which are housed pumps that supply an irrigation system, a power plant, a factory, a reservoir, a farm, a home, and so on. { 'pəmp,haùs }

pumping [FL MECH] Unsteadiness of the mercury in the barometer, caused by fluctuations of the air pressure produced by a gusty wind or due to the motion of a vessel. [PHYS] **1.** The application of optical, infrared, or microwave radiation of appropriate frequency to a laser or maser medium so that absorption of the radiation increases the population of atoms or molecules in higher energy states. Also known as electronic pumping. **2.** The removal of gases and vapors from a vacuum system. { 'pəmp·iŋ }

pumping frequency [ELECTR] Frequency at which pumping is provided in a maser, quadrupole amplifier, or other amplifier requiring high-frequency excitation. { 'pəmp·iŋ ˌfrē·kwən·sē }

pumping level See dynamic level. { 'pəmp·iŋ ˌlev·əl }

pumping loss [MECH ENG] Power consumed in purging a cylinder of exhaust gas and sucking in fresh air instead. { 'pəmp·iŋ ˌlós }

pumping pressure [PETRO ENG] Pressure required to inject (pump) water, gas, or acid into a pressurized petroleum reservoir. { 'pəmp·iŋ ˌpresh·ər }

pumping radiation [PHYS] Electromagnetic radiation applied to a laser or maser in the process of pumping. { 'pəmp·iŋ ˌrād·ē,ā·shən }

pumping station [CIV ENG] A building in which two or more pumps operate to supply fluid flowing at adequate pressure to a distribution system. { 'pəmp·iŋ ˌstā·shən }

pumping well [PETRO ENG] A producing oil well in which liquid products are recovered from the reservoir by means of a pump, rather than by gas lift. { 'pəmp·iŋ ˌwel }

pumpkin [BOT] Any of several prickly vines with large lobed leaves and yellow flowers in the genus Cucurbita of the order Violales; the fruit is orange-colored and large, with a firm rind. { 'pəm·kən }

pump off [PETRO ENG] In an oil well, to pump so rapidly that the oil level drops below the pump's standing valve. { 'pəmp ˌóf }

pump oscillator [ELECTR] Alternating-current generator that supplies pumping energy for maser and parametric amplifiers; operates at twice or some higher multiple of the signal frequency. { 'pəmp ˌäs·ə,lād·ər }

puna [ECOL] An alpine biological community in the central portion of the Andes Mountains of South America characterized by low-growing, widely spaced plants that lack much green color most of the year. { 'pü·nə }

punch [DES ENG] See nail set. [MECH ENG] A tool that forces metal into a die for extrusion or similar operations. { pənch }

punch card [COMPUT SCI] A medium by means of which data are fed into a computer in the form of rectangular holes punched in the card. Also known as punched card. { 'pənch ˌkärd }

punched card See punch card. { 'pəncht ˌkärd }

punched-plate screen [ENG] Flat, perforated plate with round, square, hexagonal, or elongated openings; used for screening (size classification) of crushed or pulverized solids. { 'pəncht ˌplāt ˌskrēn }

punching [ENG] **1.** A piece removed from a sheet of metal or other material by a punch press. **2.** A method of extrusion, cold heading, hot forging, or stamping in a machine for which mating die sections determine the shape or contour of the work. { 'pənch·iŋ }

punch press [MECH ENG] **1.** A press consisting of a frame in which slides or rams move up and down, of a bed to which the die shoe or bolster plate is attached, and of a source of power to move the slide. Also known as drop press. **2.** Any mechanical press. { 'pənch ˌpres }

punch radius [DES ENG] The radius on the bottom end of the punch over which the metal sheet is bent in drawing. { 'pənch ˌrād·ē·əs }

punch-through [ELECTR] An emitter-to-collector breakdown which can occur in a junction transistor with very narrow base region at sufficiently high collector voltage when the space-charge layer extends completely across the base region. { 'pənch,thrü }

punctate [BIOL] Dotted; full of minute points. { 'pəŋk ˌtāt }

punctuated equilibria [EVOL] In the fossil record, long periods of little change in lineages interspersed with brief periods of relatively rapid change.

punctuated evolution See punctuated equilibrium. { ˌpəŋk·chə,wād·əd ˌev·ə'lü·shən }

punctuation bit [COMPUT SCI] A binary digit used to indicate the beginning or end of a variable-length record. { ˌpəŋk·chə'wā·shən ˌbit }

punctum remotum See far point. { 'pəŋk·təm ri'mōd·əm }

puncture [ELEC] Disruptive discharge through insulation involving a sudden and large increase in current through the insulation due to complete failure under electrostatic stress. [SCI TECH] To pierce or indent. { 'pəŋk·chər }

puncture-sealing tire [ENG] A tire whose interior surface is coated with a plastic material that is forced into a puncture by high-pressure air inside the tire and subsequently hardens to seal the puncture. { 'pəŋk·chər ˌsēl·iŋ 'tīr }

puncture voltage [ELEC] The voltage at which a test specimen is electrically punctured. { 'pəŋk·chər ˌvōl·tij }

punt [NAV ARCH] **1.** A heavily built boat of rectangular shape used by workmen employed in painting, cleaning, or repairing a ship's topsides when in sheltered waters. **2.** A square-ended boat used on shallow rivers and lakes, often propelled by poles. { pənt }

pupa [INV ZOO] The quiescent, intermediate form assumed by an insect that undergoes complete metamorphosis; it follows the larva and precedes the adult stages and is enclosed in a hardened cuticle or a cocoon. { 'pyü·pə }

pupate [INV ZOO] **1.** To develop into a pupa. **2.** To pass through a pupal stage. { 'pyü,pāt }

pupil [ANAT] The contractile opening in the iris of the vertebrate eye. { 'pyü·pəl }

pupillary reflex [PHYSIO] **1.** Contraction of the pupil in response to stimulation of the retina by light. Also known as Whytt's reflex. **2.** Contraction of the pupil on accommodation for close vision, and dilation of the pupil on accommodation for distant vision. **3.** Contraction of the pupil on attempted closure of the eye. Also known as Westphal-Pilcz reflex; Westphal's pupillary reflex. { 'pyü·pə,ler·ē ˌrē,fleks }

Pupin coil See loading coil. { 'pyü'pēn ˌkóil }

Pupipara [INV ZOO] A section of cyclorrhaphous dipteran insects in the Schizophora series in which the young are born as mature maggots ready to become pupae. { pyü'pip·ə·rə }

pup jack See tip jack. { 'pəp ˌjak }

Puppis [ASTRON] A southern constellation; right ascension 8 hours, declination 40° south. Also known as Stern. { 'pəp·əs }

Puppis A [ASTRON] An extended, nonthermal radio source, the remnant of a supernova that exploded about 10,000 years ago, about 1.8 kiloparsecs (3.5×10^{16} miles or 5.6×10^{16} kilometers) from the earth. { 'pəp·əs 'ā }

Purbeckian [GEOL] A stage of geologic time in Great Britain: uppermost Jurassic (above Bononian, below Cretaceous). { pər'bek·ē·ən }

purebred breed [AGR] A group of plants or animals that possess certain inherited characteristics, such as color or markings, which are deliberately chosen for using selective breeding techniques. { 'pyúr,bred ˌbred }

pure coal See vitrain. { 'pyúr ˌkōl }

pure culture [MICROBIO] A culture that contains cells of one kind, all progeny of a single cell. { 'pyúr ˌkəl·chər }

pure forest [FOR] A forest in which one species makes up 80% or more of the total number of trees. { 'pyúr ˌfär·əst }

pure geometry [MATH] Geometry studied from the standpoint of its axioms and postulates rather than its objects. { 'pyúr jē'äm·ə·trē }

pure imaginary number [MATH] A complex number $z = x + iy$, where $x = 0$ { 'pyúr i,maj·ə,ner·ē ˌnəm·bər }

pure inverse scattering theory [PHYS] The branch of inverse scattering theory that treats the case in which the data consist of pure, noisefree information about the scattering amplitude. { 'pyúr in,vərs ˌskad·ə·riŋ ˌthē·ə·rē }

purely inseparable [MATH] An element a is said to be purely inseparable over a field F with characteristic p greater than 0 if it is algebraic over F and if there exists a nonnegative integer n such that a^{p^n} lies in F. { 'pyúr·lē in'sep·rə·bəl }

purely inseparable extension [MATH] A purely inseparable extension E of a field F is an algebraic extension of F whose separable degree over F equals 1 or, equivalently, an algebraic extension of F in which every element is purely inseparable over F. { 'pyúr·lē in,sep·rə·bəl ik sten·chən }

pure mathematics [MATH] The intrinsic study of mathematical structures, with no consideration given as to the utility of the results for practical purposes. { 'pyúr ˌmath·ə'mad·iks }

pure procedure [COMPUT SCI] A procedure that never modifies any part of itself during execution. { 'pyúr prə,sē·jər }

pure projective geometry [MATH] The axiomatic study of geometric systems which exhibit invariance relative to a notion of projection. { 'pyúr prə,jek·tiv jē'äm·ə·trē }

pure research See basic research. { 'pyúr ri'sərch }

pure shear [MECH] A particular example of irrotational strain or flattening in which a body is elongated in one direction and shortened at right angles to it as a consequence of differential displacements on two sets of intersecting planes. { 'pyur 'shir }

pure strategy [MATH] In game theory, a predetermined plan covering all possible situations in a game and not involving the use of random devices. { 'pyur 'strad·ə·jē }

pure substance [CHEM] A sample of matter, either an element or a compound, that consists of only one component with definite physical and chemical properties and a definite composition. { 'pyur 'səb·stəns }

pure surd [MATH] A surd, all of whose terms are irrational numbers. { 'pyur 'sərd }

pure tone See simple tone. { 'pyur 'tōn }

pure Trojan group [ASTRON] The group of Trojan planets which lies near the Lagrangian point 60° behind Jupiter. Also known as Patroclus group. { 'pyur 'trō·jən ˌgrüp }

pure vanilla See vanilla. { 'pyur və'nil·ə }

purga [METEOROL] A severe storm similar to the blizzard and buran, which rages in the tundra regions of northern Siberia in winter. { 'pur·gə }

purge [COMPUT SCI] To remove data from computer storage so that space occupied by the data can be reused. { pərj }

purge date [COMPUT SCI] The date after which data are released and the storage area can be used for storing other data. { 'pərj ˌdāt }

purge meter interlock [MECH ENG] A meter to maintain airflow through a boiler furnace at a specific level for a definite time interval; ensures that the proper air-fuel ratio is achieved prior to ignition. { 'pərj ˌmēd·ər 'in·tərˌläk }

purging [ENG] Replacing the atmosphere in a container by an inert substance to prevent formation of explosive mixtures. [MED] The condition in which there is rapid and continuous evacuation of the bowels. [SCI TECH] The act or process of cleaning and purifying. { 'pərj·iŋ }

purify [COMPUT SCI] To remove errors from data. [ENG] To remove unwanted constituents from a substance. { 'pyur·əˌfī }

purine [BIOCHEM] A heterocyclic compound containing fused pyrimidine and imidazole rings; adenine and guanine are the purine components of nucleic acids and coenzymes. { 'pyuˌrēn }

purity [CHEM] The degree to which the content of impurity can be detected by an analytical procedure in a sample of matter that is classified as a pure substance; the grade of purity is in inverse proportion to the amount of impurity present. Also known as chemical purity. [OPTICS] The degree to which a primary color is pure and not mixed with the other two primary colors. { 'pyur·əd·ē }

purity coil [ELECTR] A coil mounted on the neck of a color picture tube, used to produce the magnetic field needed for adjusting color purity; the direct current through the coil is adjusted to a value that makes the magnetic field orient the three individual electron beams so each strikes only its assigned color of phosphor dots. { 'pyur·əd·ē ˌkȯil }

purity control [ELECTR] A potentiometer or rheostat used to adjust the direct current through the purity coil. { 'pyur·əd·ē kənˌtrōl }

purity magnet [ELECTR] An adjustable arrangement of one or more permanent magnets used in place of a purity coil in a color television receiver. { 'pyur·əd·ē ˌmag·nət }

purity of state [STAT MECH] Property of a system which is definitely in a certain quantum state, rather than having a certain probability of being in any of several quantum states. { 'pyur·əd·ē əv 'stāt }

Purkinje cell [HISTOL] Any of the cells of the cerebral cortex with large, flask-shaped bodies forming a single cell layer between the molecular and granular layers. { pər'kin·jē ˌsel }

Purkinje effect [PHYSIO] When illumination is reduced to a low level, slowly enough to allow adaptation by the eye, the sensation produced by the longer-wave stimuli (red, orange) decreases more rapidly than that produced by shorter-wave stimuli (violet, blue). Also spelled Parkinje effect. { pər'kin·jē iˌfekt }

Purkinje fibers [HISTOL] Modified cardiac muscle fibers composing the terminal portion of the conducting system of the heart. { pər'kin·jē ˌfī·bərz }

PURINE

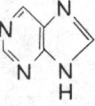

Structural formula of purine.

purl [HYD] A swirling or eddying stream or rill, moving swiftly around obstructions. { pərl }

purlin [BUILD] A horizontal roof beam, perpendicular to the trusses or rafters; supports the roofing material or the common rafters. { 'pər·lən }

puromycin [MICROBIO] $C_{22}H_{29}O_5N_7$ A colorless, crystalline broad-spectrum antibiotic produced by a strain of *Streptomyces*. { ˌpur·ə'mīs·ən }

purple bacteria [MICROBIO] Any of various photosynthetic bacteria that contain bacteriochlorophyll, distinguished by purplish or reddish-brown pigments. { 'pər·pəl bak'tir·ē·ə }

purple blende See kermesite. { 'pər·pəl 'blend }

purple blotch [PL PATH] A fungus disease of onions, garlic, and shallots caused by *Alternaria porri* and characterized by small white spots which become large purplish blotches. { 'pər·pəl 'bläch }

purple boundary [OPTICS] A straight line connecting the ends of the spectrum locus on the chromaticity diagram. { 'pər·pəl 'baun·drē }

purple lakes [MATER] A class of lake (pigment) used in printing inks; derived from a combination of such compounds as β-hydroxynaphthoic acid and 2-diazonaphthalene-1-sulfonic acid. { 'pər·pəl 'lāks }

purple light [GEOPHYS] The faint purple glow observed on clear days over a large region of the western sky after sunset and over the eastern sky before sunrise. { 'pər·pəl 'līt }

purple nonsulfur bacteria [MICROBIO] Any of various purple photosynthetic bacteria, especially members of the family Athiorhodaceae, that utilize organic hydrogen donor compounds. { 'pər·pəl ˌnän'səl·fər bak'tir·ē·ə }

purple of Cassius See gold tin purple. { 'pər·pəl əv 'kash·əs }

purple plague [ELECTR] A compound formed by intimate contact of gold and aluminum, which appears on silicon planar devices and integrated circuits using gold leads bonded to aluminum thin-film contacts and interconnections, and which seriously degrades the reliability of semiconductor devices. { 'pər·pəl 'plāg }

purple salt See potassium permanganate. { 'pər·pəl 'sȯlt }

purple sulfur bacteria [MICROBIO] Any of various anaerobic photosynthetic purple bacteria, especially in the family Thiorhodaceae, that utilize H_2S and other inorganic sulfur compounds as a source of hydrogen, while the carbon source can be carbon monoxide. { 'pər·pəl 'səl·fər bak,tir·ē·ə }

purple-top [PL PATH] A virus disease of potato plants characterized by purplish or chlorotic discoloration of the top shoots, swelling of axillary branches, and severe wilting. { 'pər·pəlˌtäp }

purpura [MED] Spontaneous hemorrhages into tissues such as joints, skin, and mucosal surfaces. { 'pər·pyə·rə }

purpurin [ORG CHEM] $C_{14}H_8O_5$ A compound crystallizing as long orange needles from dilute alcohol solutions; used in the manufacture of dyes, and as a reagent for the detection of boron. Also known as natural red. { 'pər·pyə·rən }

purpurite [MINERAL] $(Mn,Fe)PO_4$ A dark-red or purple mineral composed of ferric-manganic phosphate; it is isomorphous with heterosite. { 'pər·pyəˌrīt }

purpurogallin [ORG CHEM] $C_{11}H_8O_5$ A red, crystalline compound, the aglycon of several glycosides from nutgalls; decomposes at 274-275°C; soluble in boiling alcohol, methanol, and acetone; used as an antioxidant or to retard metal contamination in hydrocarbon fuels or lubricants. { ˌpər·pyə·rō'gal·ən }

purse seine [ENG] A net that can be dropped by two boats to encircle a school of fish, then pulled together at the bottom and raised, thereby catching the fish. { 'pərs ˌsān }

purse seiners [NAV ARCH] Fishing boats equipped to fish with a purse seine. { 'pərs ˌsān·ərz }

purulent [MED] Consisting of, containing, or forming pus. { 'pyur·ə·lənt }

purupuru See pinta. { ˌpur·ü'pur·ü }

pus [MED] A viscous, creamy, pale-yellow or yellow-green fluid produced by liquefaction necrosis in a neutrophil-rich exudate. { pəs }

push [COMPUT SCI] To add an item to a stack. { push }

push-bar conveyor [MECH ENG] A type of chain conveyor in which two endless chains are cross-connected at intervals by push bars which propel the load along a stationary bed or trough of the conveyor. { 'push ˌbar kən‚vā·ər }

push bench [MECH ENG] A machine used for drawing tubes of moderately heavy gage by cupping metal sheet and applying pressure to the inside bottom of the cup to force it through a die. { 'push ˌbench }

push button [COMPUT SCI] A small rectangle on a graphical user interface whose selection initiates immediate action. { 'push ˌbət·ən }

push-button dialing [ELECTR] Dialing a number by pushing buttons on the telephone rather than turning a circular wheel; each depressed button causes a transistor oscillator to oscillate simultaneously at two different frequencies, generating a pair of audio tones which are recognized by central-office (or PBX) switching equipment as digits of a telephone number. Also known as tone dialing; touch call. { 'push ˌbət·ən 'dī·liŋ }

push-button switch [ELEC] A master switch that is operated by finger pressure on the end of an operating button. { 'push ˌbət·ən 'swich }

push-button tuner [ELECTR] A device that automatically tunes a radio receiver or other piece of equipment to a desired frequency when the button assigned to that frequency is pressed. { 'push ˌbət·ən 'tün·ər }

push development [GRAPHICS] The use of extended development times to improve the speed and detective quantum efficiency of a photographic plate. { 'push di‚vel·əp·mənt }

push-down automaton [COMPUT SCI] A nondeterministic, finite automaton with an auxiliary tape having the form of a push-down storage. { 'push‚daun ȯ'täm·ə‚tän }

push-down list [COMPUT SCI] An ordered set of data items so constructed that the next item to be retrieved is the item most recently stored; in other words, last-in, first-out (LIFO). { 'push‚daun ‚list }

push-down storage [COMPUT SCI] A computer storage in which each new item is placed in the first location in the storage and all the other items are moved back one location; it thus follows the principle of a push-down list. Also known as cellar; nesting storage; running accumulator. { 'push‚daun ‚stȯr·ij }

pusher furnace [MET] A type of continuous furnace in which the stock to be heated is charged at one end, carried through a sequence of one or more heating zones, and discharged at the opposite end. { 'push·ər ‚fər·nəs }

pusher grade See helper grade. { 'push·ər ‚grād }

pusher tractor [MIN ENG] A bulldozer exerting pressure on the rear of a scraper-loader while the loader is digging and loading unconsolidated ground during opencast mining. { 'push·ər ‚trak·tər }

push fit [DES ENG] A hand-tight sliding fit between a shaft and a hole. { 'push ‚fit }

pushing [COMPUT SCI] The placing of a data element at the top of a stack. [ELEC] A change in the resonant frequency of a circuit due to changes in the applied voltages. { 'push·iŋ }

push moraine [GEOL] A broad, smooth, arc-shaped ridge consisting of material mechanically pushed or shoved along by an advancing glacier. Also known as push-ridge moraine; shoved moraine; thrust moraine; upsetted moraine. { 'push mə‚rān }

push nipple [MECH ENG] A short length of pipe used to connect sections of cast iron boilers. { 'push ‚nip·əl }

push-pull amplifier [ELECTR] A balanced amplifier employing two similar electron tubes or equivalent amplifying devices working in phase opposition. { 'push ‚pul 'am·plə‚fī·ər }

push-pull currents See balanced currents. { 'push ‚pul 'kə·rəns }

push-pull electret transducer [ELECTR] A type of transducer in which a foil electret is sandwiched between two electrodes and is specially treated or arranged so that the electrodes exert forces in opposite directions on the diaphragm, and the net force is a linear function of the applied voltage. { 'push ‚pul i'lek·trət tranz'dü·sər }

push-pull magnetic amplifier [ELECTR] A realization of a push-pull amplifier using magnetic amplifiers. { 'push ‚pul mag'ned·ik 'am·plə‚fī·ər }

push-pull oscillator [ELECTR] A balanced oscillator employing two similar electron tubes or equivalent amplifying devices in phase opposition. { 'push ‚pul 'äs·ə‚lād·ər }

push-pull sound track [ENG ACOUS] A sound track having two recordings so arranged that the modulation in one is 180° out of phase with that in the other. { 'push ‚pul 'saun ‚trak }

push-pull transformer [ELECTR] An audio-frequency transformer having a center-tapped winding and designed for use in a push-pull amplifier. { 'push ‚pul tranz'fȯr·mər }

push-pull transistor [ELECTR] **1.** A realization of a push-pull amplifier using transistors. **2.** A Darlington circuit in which the two transistors required for a push-pull amplifier exist in a single substrate. { 'push ‚pul tran'zis·tər }

push-pull voltages See balanced voltages. { 'push ‚pul 'vȯl·tij·əz }

push-push amplifier [ELECTR] An amplifier employing two similar electron tubes with grids connected in phase opposition and with anodes connected in parallel to a common load; usually used as a frequency multiplier to emphasize even-order harmonics; transistors may be used in place of tubes. { 'push ‚push 'am·plə‚fī·ər }

push-ridge moraine See push moraine. { 'push ‚rij mə‚rān }

push rod [MECH ENG] A rod, as in an internal combustion engine, which is actuated by the cam to open and close the valves. { 'push ‚räd }

push-to-talk circuit [ELEC] Simplex circuit in which changeover from the receive to transmit state is accomplished by depressing a single spring-return switch, and releasing the switch returns the circuit to the receive state; the push-to-talk switch is located on microphones and telephone handsets; it is most often applied to radio circuits. { 'push tə 'tȯlk ‚sər·kət }

push-up [ENG] Concave bottom contour of a plastic container; allows an even bearing surface on the outer edge and prevents the container from rocking. { 'push‚əp }

push-up list [COMPUT SCI] An ordered set of data items so constructed that the next item to be retrieved will be the item that was inserted earliest in the list, resulting in a first-in, first-out (FIFO) structure. { 'push‚əp ‚list }

push welding [MET] Spot or projection welding in which the force is applied manually by one electrode; the work takes the place of the other electrode. Also known as poke welding. { 'push ‚weld·iŋ }

pustule [MED] A small, circumscribed, pus-filled elevation on the skin. [PL PATH] A blisterlike mark on a leaf due to rupture of surface tissues overlying spore masses of a parasitic fungus. { 'pəs·chül }

Pustulosa [PALEON] An extinct suborder of echinoderms in the order Phanerozonida found in the Paleozoic. { ‚pəs·chə'lō·sə }

pusule [INV ZOO] A noncontractile fluid-filled vacuole emptied by means of a duct; found in dinoflagellates. { 'pəs·yül }

put [COMPUT SCI] A programming instruction that causes data to be written from computer storage into a file. { 'püt }

putlog [CIV ENG] A crosspiece in a scaffold or formwork; supports the soffits and is supported by the ledgers. { 'püt‚läg }

putrefaction [BIOCHEM] Decomposition of organic matter, particularly the anaerobic breakdown of proteins by bacteria, with the production of foul-smelling compounds. { ‚pyü·trə'fak·shən }

putty [MATER] A cement of dough consistency made of whiting and boiled linseed oil and used in fastening glass in sashes and sealing crevices in woodwork. { 'pəd·ē }

putty knife [DES ENG] A knife with a broad flexible blade, used to apply and smooth putty. { 'pəd·ē ‚nīf }

putty oil [MATER] Petroleum oil that is added to putty; serves to lubricate the putty and keep it soft after the linseed oil dries. { 'pəd·ē ‚oil }

puy [GEOL] A small, remnant volcanic cone. { pwē }

PVA See polyvinyl acetate; polyvinyl alcohol.

PVAc See polyvinyl acetate.

PVC See polyvinyl chloride.

PVD See physical vapor deposition.

PVDC See polyvinyl dichloride.

PVI See polyvinyl isobutyl ether.

PVM See polyvinyl methyl ether.

PVP See polyvinyl pyrrolidone.

PVY See potato virus Y.

pW See picowatt.

P wave [GEOPHYS] A body wave that can pass through all

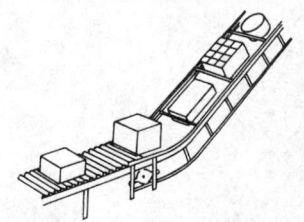

Push-bar conveyor, a basic type of chain conveyor.

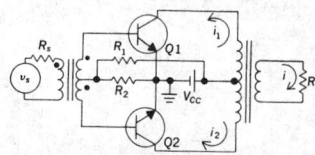

Circuit diagram of two transistors, Q_1 and Q_2, in a push-pull arrangement. Here, v_s = source voltage, R_s = source resistance, R_L = load resistance, V_{CC} = collector supply voltage.

layers of the earth. It is fastest of all seismic waves, traveling at a velocity of 3–4 miles (5–7 kilometers) per second in the crust and 5–6 miles (8–9 kilometers) per second in the upper mantle. Also known as compressional wave; longitudinal wave; primary wave. { 'pē ‚wāv }

PWI *See* proximity warning indicator.

PWR *See* pressurized water reactor.

pwt *See* pennyweight.

pyarthrosis [MED] Suppuration involving a joint. { ¦pī· är'thrō·səs }

pycnidiospore [MYCOL] A conidium produced by a pycnidium. { pik'nid·ē·ō‚spór }

pycnidium [MYCOL] A cavity that bears pycnidiospores in certain fungi. { pik'nid·ē·əm }

pycniospore [MYCOL] A haploid spore of a rust fungus that fuses with a haploid hypha of opposite sex to produce dikaryotic aeciospores. { 'pik·nē·ə‚spór }

pycnite [MINERAL] A variety of topaz occurring in massive columnar aggregations. { 'pik‚nīt }

pycnium [MYCOL] A flask-shaped fruit body of a rust fungus formed in clusters just beneath the surface of a host tissue. { 'pik·nē·əm }

pycnocline [GEOPHYS] A change in density of ocean or lake water or rock with displacement in some direction, especially a rapid change in density with vertical displacement. [OCEANOGR] A region in the ocean where water density increases relatively rapidly with depth. { 'pik·nə‚klīn }

Pycnodontiformes [PALEON] An extinct order of specialized fishes characterized by a laterally compressed, disk-shaped body, long dorsal and anal fins, and an externally symmetrical tail. { ‚pik·nə‚dänt·ə'fór‚mēz }

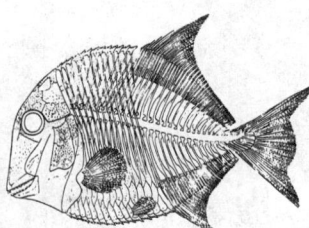

Coelodus costae, a pycnodont from Lower Cretaceous of Italy; length to 4 inches (10 centimeters).

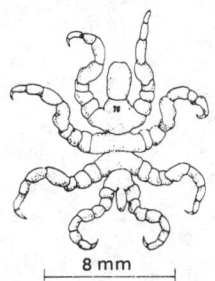

8 mm

Pycnogonum stearnsi.

Pycnogonida [INV ZOO] The sea spiders, a subphylum of marine arthropods in which the body is reduced to a series of cylindrical trunk somites supporting the appendage. { ‚pik· nə'gän·əd·ə }

Pycnogonidae [INV ZOO] A family of the Pycnogonida lacking both chelifores and palpi and having six to nine jointed ovigers in the male only. { ‚pik·nə'gän·ə‚dē }

pycnometer [ENG] A container whose volume is precisely known, used to determine the density of a liquid by filling the container with the liquid and then weighing it. Also spelled pyknometer. { pik'näm·əd·ər }

pycnometry [PHYS] The determination of liquid density by weighing the liquid in a container (pycnometer) of known volume. { pik'näm·ə·trē }

pyelitis [MED] Inflammation of the renal pelvis. { ‚pī· ə'līd·əs }

pyelolithotomy [MED] Excision of a renal calculus through an incision in the renal pelvis. { ¦pī·ə·lō·li'thäd·ə·mē }

pyelonephritis [MED] The disease process resulting from the effects of infections of the parenchyma and the pelvis of the kidney. Also known as interstitial nephritis. { ¦pī·ə·lō· ne'frīd·əs }

pyemia [MED] A disease state due to the presence of pyogenic microorganisms in the blood and the formation, wherever these organisms lodge, of embolic or metastatic abscesses. { pī'ē·mē·ə }

Pygasteridae [PALEON] The single family of the extinct order Pygasteroida. { ‚pī·gə'ster·ə‚dē }

Pygasteroida [PALEON] An order of extinct echinoderms in the superorder Diadematacea having four genital pores, noncrenulate tubercles, and simple ambulacral plates. { ¦pī·gə· stə'róid·ə }

pygidium [INV ZOO] **1.** A caudal shield on the abdomen of some Arthropoda. **2.** The terminal body segment of many invertebrates. { pī'jid·ē·əm }

Pygopodidae [VERT ZOO] The flap-footed lizards, a family of the suborder Sauria. { ‚pī·gə'päd·ə‚dē }

pygostyle [VERT ZOO] A specialized bone in birds which is formed by a number of fused tail vertebrae. { 'pīg·ə‚stīl }

pyknometer *See* pycnometer. { pik'näm·əd·ər }

pyknosis [PATH] The polymerization and contraction of the nuclear chromosomal components. { pik'nō·səs }

pylephlebitis [MED] Inflammation of the portal vein. { ¦pī·lə·flə'bīd·əs }

pylome [INV ZOO] An aperture for emission of pseudopodia and intake of food in some Sarcodina. { 'pī‚lōm }

pylon [AERO ENG] A suspension device externally installed under the wing or fuselage of an aircraft; it is aerodynamically

designed to fit the configuration of specific aircraft, thereby creating an insignificant amount of drag; it includes means of attaching to accommodate fuel tanks, bombs, rockets, torpedoes, rocket motors, or the like. [CIV ENG] **1.** A massive structure, such as a truncated pyramid, on either side of an entrance. **2.** A tower supporting a wire over a long span. **3.** A tower or other structure marking a route for an airplane. { 'pī‚län }

pyloric caecum [INV ZOO] **1.** One of the tubular pouches that open into the ventriculus of an insect. **2.** One of the paired tubes having lateral glandular diverticula in each ray of a starfish. [VERT ZOO] One of the tubular pouches that open from the pyloric end of the stomach into the alimentary canal of most fishes. { pə'lór·ik 'sē·kəm }

pyloric sphincter [ANAT] The thickened ring of circular smooth muscle at the lower end of the pyloric canal of the stomach. { pə'lór·ik 'sfiŋk·tər }

pyloric stenosis [MED] Obstruction of the pyloric opening of the stomach due to hypertrophy of the pyloric sphincter. { pə'lór·ik stə'nō·səs }

pylorospasm [MED] Spasm of the pylorus. { pī'lór· ə‚spaz·əm }

pylorus [ANAT] The orifice of the stomach communicating with the small intestine. { pə'lór·əs }

pyobacillosis [VET MED] A bacterial infection of sheep, swine, or rarely cattle caused by *Corynebacterium pyogenes;* usually marked by abscess formation, but in sheep takes the form of chronic purulent pneumonia. { ‚pī·ō‚bas·ə'lō·səs }

pyocyanin [MICROBIO] $C_{13}H_{10}N_2O$ An antibiotic substance forming blue crystals, produced by *Pseudomonas aeruginosa;* active against many bacteria and fungi. { ‚pī·ō'sī·ə·nən }

pyoderma [MED] Any pus-producing skin lesion or lesions, used in reference to groups of furuncles, pustules, or even carbuncles. { ‚pī·ō'dər·mə }

pyogenesis [MED] The formation of pus. { ¦pī·ō'jen·ə· səs }

pyonephritis [MED] Suppurative inflammation of a kidney. { ¦pī·ō·ne'frīd·əs }

pyonephrosis [MED] Replacement of renal tissue by abscesses. { ¦pī·ō·nē'frō·səs }

pyorrhea [MED] A purulent discharge. { ‚pī·ə'rē·ə }

pyosalpinx [MED] A reproductive system disorder in which the Fallopian tubes are distended with pus. { ‚pī·ə'sal‚piŋks }

pyracetic acid *See* pyroligneous acid. { ‚pī·rə‚sēd·ik 'as·əd }

Pyralidae [INV ZOO] The equivalent name for Pyralididae. { pə'ral·ə‚dē }

Pyralididae [INV ZOO] A large family of moths in the lepidopteran superfamily Pyralidoidea; the labial palpi are well developed, and the legs are usually long and slender. { ‚pir· ə'lid·ə‚dē }

Pyralidinae [INV ZOO] A subfamily of the Pyralididae. { ‚pir·ə'lin·ə‚nē }

Pyralidoidea [INV ZOO] A superfamily of the Lepidoptera belonging to the Heteroneura and including long-legged, slender-bodied moths with well-developed maxillary palpi. { ‚pir· ə·lə'dóid·ē·ə }

pyramid [CRYSTAL] An open crystal having three, four, six, eight, or twelve nonparallel faces that meet at a point. [MATH] A polyhedron with one face a polygon and all other faces triangles with a common vertex. { 'pir·ə‚mid }

pyramidal area *See* motor area. { 'pir·ə¦mid·əl *or* pə'ram·ə· dəl 'er·ē·ə }

pyramidal cleavage [CRYSTAL] A type of crystal cleavage that occurs parallel to the faces of a pyramid. { ¦pir·ə¦mid·əl 'klē·vij }

pyramidal horn [ENG] Horn whose sides form a pyramid. { ¦pir·ə¦mid·əl 'hórn }

pyramidal iceberg *See* pinnacled iceberg. { ¦pir·ə¦mid·əl 'īs‚bərg }

pyramidal molecule [CHEM] A molecular structure in the shape of a pyramid in which the central atom at the peak possesses either three or four valence bonds that are directed to the other atoms, which form the base of the pyramid. { ¦pir· ə‚mid·əl 'mäl·ə‚kyül }

pyramidal numbers [MATH] The numbers 1, 4, 10, 20, 35, . . . , which are the number of dots in successive pyramidal arrays and are given by $(1/6)n(n + 1) (n + 2)$, where $n = 1$, 2, 3, ... { ¦pir·ə‚mid·əl 'nəm·bərz }

pyramidal surface [MATH] A surface generated by a line

passing through a fixed point and moving along a broken line in a plane not containing that point. { ¦pir·ə¦mid·əl 'sər·fəs }

pyramidal system [ANAT] The corticospinal and cortico-bulbar tracts. { ¦pir·ə¦mid·əl 'sis·təm }

Pyramidellidae [INV ZOO] A family of gastropod mollusks in the order Tectibranchia; the operculum is present in this group. { pə¸ram·ə'del·ə¸dē }

pyramid of numbers [ECOL] The concept that an organism making up the base of a food chain is numerically abundant while each succeeding member of the chain is represented by successively fewer individuals; uses feeding relationship as a basis for the quantitative analysis of an ecological system. { 'pir·ə¸mid əv 'nəm·bərz }

pyranometer [ENG] An instrument used to measure the combined intensity of incoming direct solar radiation and diffuse sky radiation; compares heating produced by the radiation on blackened metal strips with that produced by an electric current. Also known as solarimeter. { ¦pir·ə'näm·əd·ər }

pyranose [BIOCHEM] A sugar whose cyclic or ring structure resembles that of pyran. { 'pī·rə¸nōs }

pyranoside [BIOCHEM] A glycoside whose cyclic sugar component resembles that of pyran. { pī'ran·ə¸sīd }

pyrargyrite [MINERAL] Ag_3SbS_3 A deep ruby-red to black mineral, crystallizing in the hexagonal system, occurring in massive form and in disseminated grains, and having an adamantine luster; hardness is 2.5 on Mohs scale, and specific gravity is 5.85; an important silver ore. Also known as dark-red silver ore; dark ruby silver. { pī'rär·jə¸rīt }

Pyraustinae [INV ZOO] A large subfamily of the Pyralididae containing relatively large, economically important moths. { pə'ròs·tə¸nē }

pyrazinamide [PHARM] $C_5H_5N_3O$ A crystalline compound with a melting point of 189–191°C; used as a drug in the treatment of tuberculosis. { ¸pir·ə'zin·ə¸mīd }

pyrazolone dye [ORG CHEM] An acid dye containing both −N=N− and =C=C= chromophore groups, such as tartrazine; used for silk and wool. { pə'raz·ə¸lōn ¸dī }

Pyrenean orogeny [GEOL] A short-lived orogeny that occurred during the late Eocene, between the Bartonian and Ludian stages. { ¸pir·ə'nē·ən ò'räj·ə¸nē }

pyrenoid [BOT] A colorless body found within the chromatophore of certain algae; a center for starch formation and storage. { 'pir·ə¸nòid }

Pyrenolichenes [BOT] The equivalent name for Pyrenulales. { ¸pī·rə·nō·lī'kē·nēz }

Pyrenomycetes [MYCOL] The largest class in the subdivision Ascomycotina, distinguished by a single-walled ascus and the coiled branches that form on the hyphae to initiate ascocarp formation. Also known as perithecial ascomycetes. { pī¸rē·nō·mī'sēd¸ēz }

Pyrenulaceae [BOT] A family of the Pyrenulales; all species are crustose and most common on tree bark in the tropics. { pī¸ren·yə'lās·ē¸ē }

Pyrenulales [BOT] An order of the class Ascolichenes including only those lichens with perithecia that contain true paraphyses and unitunicate asci. { pī¸ren·yə'lā·lēz }

pyrethrum [MATER] A toxicant obtained in the form of dried powdered flowers of the plant of the same name; mixed with petroleum distillates, it is used as an insecticide. { pī'rē·thrəm }

pyrexia [MED] Elevation of temperature above the normal; fever. { pī'rek·sē·ə }

pyrgeometer [ENG] An instrument for measuring radiation from the surface of the earth into space. { ¦pīr·jē'äm·əd·ər }

Pyrgotidae [INV ZOO] A family of myodarian cyclorrhaphous dipteran insects in the subsection Acalyptratae. { pər'gäd·ə¸dē }

pyrheliometer [ENG] An instrument for measuring the total intensity of direct solar radiation received at the earth. { ¦pir¸hē·lē'äm·əd·ər }

pyridine [ORG CHEM] C_5H_5N Organic base; flammable, toxic yellowish liquid, with penetrating aroma and burning taste; soluble in water, alcohol, ether, benzene, and fatty oils; boils at 116°C; used as an alcohol denaturant, solvent, in paints, medicine, and textile dyeing. { 'pir·ə¸dēn }

2,3-pyridinedicarboxylic acid [BIOCHEM] $C_7H_5NO_4$ An odorless, crystalline compound with a melting point of 190°C; soluble in water; inhibits glucose synthesis. Also known as quinolinic acid. { tü ¦thrē ¦pir·ə¸dēn¸kär·bäk'sil·ik 'as·əd }

pyridostigmine bromide [PHARM] $C_9H_{13}O_2NBr$ A white, crystalline powder with a melting point of 154–157°C; soluble in water, alcohol, and chloroform; used in medicine. { ¦pir·ə·dō¦stig¸mēn 'brō¸mīd }

pyridoxal hydrochloride See pyridoxine hydrochloride. { ¸pir·ə¦däk·səl ¸hī·drə'klòr¸īd }

pyridoxal phosphate See codecarboxylase. { ¸pir·ə¦däk·səl 'fä¸sfāt }

pyridoxine hydrochloride [BIOCHEM] $C_8H_{11}NO_3\cdot HCl$ A crystalline compound, decomposing at about 208°C; used in medicine in vitamin therapy. Also known as pyridoxal hydrochloride; vitamin B_6 hydrochloride. { ¸pir·ə¦däk¸sēn ¸hī·d·ə'klòr¸īd }

pyrimidine [BIOCHEM] $C_4H_4N_2$ A heterocyclic organic compound containing nitrogen atoms at positions 1 and 3; naturally occurring derivatives are components of nucleic acids and coenzymes. { pə'rim·ə¸dēn }

pyrite [MINERAL] FeS_2 A hard, brittle, brass-yellow mineral with metallic luster, crystallizing in the isometric system; hardness is 6–6.5 on Mohs scale, and specific gravity is 5.02. Also known as common pyrite; fool's gold; iron pyrites; mundic. { 'pī¸rīt }

pyrite roasting [MIN ENG] Thermal processing of iron pyrite (FeS_2, iron disulfide) in the presence of air to produce iron oxide sinter (used in steel mills) and elemental sulfur. { 'pī¸rīt 'rōst·iŋ }

pyritization [GEOL] A common process of hydrothermal alteration involving introduction of or replacement by pyrite. { ¸pī¸rīd·ə'zā·shən }

pyritobitumen [GEOL] Any of various dark-colored, relatively hard, nonvolatile hydrocarbon substances often associated with mineral matter, which decompose upon heating to yield bitumens. Also known as pyrobitumen. { pə'rīd·ō·bə'tü·mən }

pyritohedron [CRYSTAL] A dodecahedral crystal with 12 irregular pentagonal faces; it is characteristic of pyrite. Also known as pentagonal dodecahedron; pyritoid; regular dodecahedron { pə¸rīd·ō'hē·drən }

pyritoid See pyritohedron. { 'pī¸rīd¸òid }

pyro- [CHEM] A chemical prefix for compounds formed by heat, such as pyrophosphoric acid, an inorganic acid formed by the loss of one water molecule from two molecules of an ortho acid. { 'pī¸rō, 'pī·rə }

pyroaurite [MINERAL] $Mg_6Fe_2(OH)_{16}\cdot CO_3\cdot 4H_2O$ A gold-like or brownish rhombohedral mineral composed of hydrous basic magnesium iron carbonate. { ¦pī·rō'ò¸rīt }

pyrobelonite [MINERAL] $PbMn(VO_4)(OH)$ A fire-red to deep brilliant-red mineral composed of basic vanadate of manganese and lead, occurring as crystal needles. { ¦pī·rō'bel·ə¸nīt }

pyrobiolite [PETR] An organic rock containing organic remains that have been altered by volcanic action. { ¦pī·rō'bī·ə¸līt }

pyrobitumen See pyritobitumen. { ¦pī·rō·bə'tü·mən }

pyroborate See borax. { ¦pī·rō'bò¸rāt }

pyrocatechuic acid See catechol. { ¦pī·rō¦kad·ə¦chü·ik 'as·əd }

pyrocellulose [ORG CHEM] Highly nitrated cellulose; used to make explosives; originally called guncotton in the United States, cordite in England. { ¦pī·rō'sel·yə¸lōs }

pyrochlore [MINERAL] $(Na,Ca)_2(Nb,Ta)_2O_6(OH,F)$ Pale-yellow, reddish, brown, or black mineral, crystallizing in the isometric system, and occurring in pegmatites derived from alkalic igneous rocks. Also known as pyrrhite. { 'pī·rə¸klòr }

pyrochroite [MINERAL] $Mn(OH)_2$ A hexagonal mineral composed of naturally occurring manganese hydroxide; it is white when fresh, but darkens upon exposure. { ¸pī·rə'krō¸īt }

pyroclast [GEOL] An individual pyroclastic fragment or clast. { 'pī·rə¸klast }

pyroclastic flow [GEOL] Ash flow not involving high-temperature conditions. { ¦pī·rə¦klas·tik 'flō }

pyroclastic-flow deposit See ignimbrite. { ¸pī·rə¸klas·tik 'flō dī päz·ət }

pyroclastic ground surge [GEOL] The relatively thin mantle of rock found around a volcanic vent; the thickness is not uniform, the internal stratification is not parallel to the top and bottom of the layer, and the extent is a few kilometers from the source. { ¦pī·rə¦klas·tik 'graùnd ¸sərj }

PYRIMIDINE

Structural formula of pyrimidine.

pyroclastic rock [PETR] A rock that is composed of fragmented volcanic products ejected from volcanoes in explosive events. { ¦pī·rə¦klas·tik ′räk }

pyroconductivity [SOLID STATE] Electrical conductivity that develops in a material only at high temperature, chiefly at fusion, in solids that are practically nonconductive at atmospheric temperatures. { ¦pī·rō,kän·dək¦tiv·əd·ē }

pyroelectric crystal [SOLID STATE] A crystal exhibiting pyroelectricity, such as tourmaline, lithium sulfate monohydrate, cane sugar, and ferroelectric barium titanate. { ¦pī·rō·i¦lek·trik ′krist·əl }

pyroelectricity [SOLID STATE] **1.** The property of certain crystals to produce a state of electrical polarity by a change of temperature. **2.** An electric charge released as the result of a temperature change. { ¦pī·rō·i,lek′tris·əd·ē }

pyrogallic acid [ORG CHEM] $C_6H_3(OH)_3$ Lustrous, light-sensitive white crystals, melting at 133°C; soluble in alcohol, ether, and water; used for photography, dyes, drugs, medicines, and process engravings, and as an analytical reagent and protective colloid. Also known as pyrogallol. { ¦pī·rō′gal·ik ′as·əd }

pyrogallol See pyrogallic acid. { ,pī·rō′ga,lól }

pyrogallolphthalein See gallein. { ¦pī·rō¦gal·ō′thal·ē·ən }

pyrogen [BIOCHEM] A group of substances thought to be polysaccharides of microbial origin that produce an increase in body temperature when injected into humans and some animals. { ′pī·rə,jən }

pyrogenesis [GEOL] The intrusion and extrusion of magma and its derivatives. { ,pī·rō′jen·ə·səs }

pyrogenetic mineral [MINERAL] An anhydrous mineral of an igneous rock, usually crystallized at high temperature in a magma containing relatively few volatile components. { ¦pī·rō·jə′ned·ik ′min·rəl }

pyrogenic distillation [CHEM ENG] A cracking process that runs at high temperatures, high pressures, or both, resulting in greater yields of the light hydrocarbon components of gasoline. { ¦pī·rō¦jen·ik ,dist·əl′ā·shən }

pyroheliometer [METEOROL] An instrument that measures the sun's radiation output. { ,pī·rō,hē·lē′äm·əd·ər }

pyroligneous [CHEM ENG] Referring to a substance obtained by the destructive distillation of wood. { ¦pī·rō′lig·nē·əs }

pyroligneous acid [ORG CHEM] An impure acetic acid derived from destructive distillation of wood or pine tar. Also known as pyracetic acid; wood vinegar. { ¦pī·rō′lig·nē·əs ′as·əd }

pyrolithic acid See cyanuric acid. { ¦pī·rō¦lith·ik ′as·əd }

pyrolusite [MINERAL] MnO_2 An iron-black mineral that crystallizes in the tetragonal system and is the most important ore of manganese; hardness is 1–2 on Mohs scale, and specific gravity is 4.75. { ,pī·rə′lü,sīt }

pyrolysate [CHEM] Any product of pyrolysis. { pī′räl·ə,zāt }

pyrolysis [CHEM] The breaking apart of complex molecules into simpler units by the use of heat, as in the pyrolysis of heavy oil to make gasoline. Also known as thermolysis. { pə′räl·ə·səs }

pyromagma [GEOL] A highly mobile lava, oversaturated with gases, that exists at shallower depths than hypomagma. { ,pī·rō′mag·mə }

pyromania [PSYCH] A monomania for setting or watching fires. { ,pī·rō′mā·nē·ə }

pyromelane See brookite. { ¦pī·rō′me,lān }

pyromellitic acid [ORG CHEM] $C_6H_2(COOH)_4$ A white powder with a melting point of 257–265°C; used as an intermediate for polyesters and polyamides. Abbreviated PMA. { ,pī·rō·mə′lid·ik ′as·əd }

pyromellitic dianhydride [ORG CHEM] $C_6H_2(C_2O_3)_2$ A white powder with a melting point of 286°C; soluble in some organic solvents; used for curing epoxy resins. Abbreviated PMDA. { ,pī·rō·mə′lid·ik ′dī·an′hī,drīd }

pyrometallurgy [MET] Processes that use chemical reactions at elevated temperatures for the extraction of metals from raw materials such as ores and concentrates, or for the treatment of recycled scrap. { ¦pī·rō′med·əl,ər·jē }

pyrometamorphism [PETR] Contact metamorphism at temperatures near the melting points of the component minerals. { ¦pī·rō,med·ə′mór,fiz·əm }

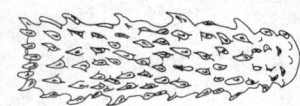

PYROSOMIDA

Colony of *Pyrosoma atlanticum*.

pyrometasomatism [PETR] Forming of contact-metamorphic mineral deposits at high temperatures by emanations from the intrusive rock, involving replacement of the enclosing rock with the addition of materials. { ¦pī·rō,med·ə′sō·mə,tiz·əm }

pyrometer [ENG] Any of a broad class of temperature-measuring devices; they were originally designed to measure high temperatures, but some are now used in any temperature range; includes radiation pyrometers, thermocouples, resistance pyrometers, and thermistors. { pī′räm·əd·ər }

pyrometric cone See Seger cone. { ¦pī·rə¦me·trik ′kōn }

pyrometry [THERMO] The science and technology of measuring high temperatures. { pī′räm·ə·trē }

pyromorphite [MINERAL] $Pb_5(PO_4)_3Cl$ A green, yellow, brown, gray, or white mineral of the apatite group, crystallizing in the hexagonal system; a minor ore of lead. Also known as green lead ore. { ¦pī·rō′mór,fīt }

pyromucic acid See furoic acid. { ¦pī·rō′myü·sik ′as·əd }

pyron [PHYS] A unit of area-density of power, equal to the area-density of power resulting from a power of one international table calorie per minute acting uniformly over an area of 1 square centimeter; equal to 697.8 watts per square meter. { ′pī,rän }

pyrone detector [ELECTR] Crystal detector in which rectification occurs between iron pyrites and copper or other metallic points. { ′pī,rōn di,tek·tər }

pyrope [MINERAL] $Mg_3Al_2(SiO_4)_3$ A mineral species of the garnet group characterized by a deep fiery-red color and occurring in basic and ultrabasic igneous rocks. { ′pī,rōp }

pyrophane See fire opal. { ′pī·rə,fān }

pyrophanite [MINERAL] $MnTiO_3$ A blood-red rhombohedral mineral consisting of manganese titanate; it is isomorphous with ilmenite. { pī′räf·ə,nīt }

pyrophobia [PSYCH] An abnormal fear of fires. { ,pī·rə′fō·bē·ə }

pyrophoric alloy [MET] **1.** An alloy such as ferrocerium that produces a spark when struck with metal (steel) at an angle; used for automatic cigarette lighters. **2.** An alloy in powder form that spontaneously oxidizes in air, reaching high temperatures. { ¦pī·rə¦fór·ik ′al,ói }

pyrophoric material [CHEM] A material that spontaneously ignites in air below 113°C (45°C), such as fine metal powder, alkali metal, partially or fully alkylated metal or nonmetal hydride, and metal carbonyl. { ,pī·rə,fór·ik mə′tir·ē·əl }

pyrophoric propellant [MATER] A propellant combination of a liquid fuel and a fluid oxidizer (usually air) that will quickly react when brought into intimate contact and achieve ignition temperature. { ¦pī·rə¦fór·ik prə′pel·ənt }

pyrophosphatase [BIOCHEM] An enzyme catalyzing hydrolysis of esters containing two or more molecules of phosphoric acid to form a simpler phosphate ester. { ,pī·rō′fä·sfə,tās }

pyrophosphoric acid [INORG CHEM] $H_4P_2O_7$ Water-soluble, syrupy liquid melting at 61°C; used as a catalyst and to make organic phosphate esters. { ¦pī·rō·fä′sfór·ik ′as·əd }

pyrophyllite [MINERAL] $AlSi_2O_5(OH)$ A white, greenish, gray, or brown phyllosilicate mineral that resembles talc and occurs in a foliated form or in compact masses in quartz veins, granites, and metamorphic rocks. Also known as pencil stone. { ,pī·rō′fi,līt }

pyroretinite [MINERAL] A type of retinite found in the brown coals of Aussig (Usti and Labem), in Bohemian Czechoslovakia. { ,pī·rō′ret·ən,īt }

pyroschist [PETR] A schist or shale that has a sufficiently high carbon content to burn with a bright flame or to yield volatile hydrocarbons when heated. { ′pī·rə,shist }

pyrosin See tetraiodofluorescein. { ,pī·rə·sən }

pyrosmalite [MINERAL] $(Mn,Fe)_4Si_3O_7(OH,Cl)_6$ A colorless, pale-brown, gray, or gray-green mineral composed mainly of basic iron manganese silicate with chlorine. { pī′räz·mə,līt }

Pyrosomida [INV ZOO] An order of pelagic tunicates in the class Thaliacea in which species form tubular swimming colonies and are often highly luminescent. { ,pī·rə′säm·əd·ə }

pyrosphere [GEOL] The zone of the earth below the lithosphere, consisting of magma. Also known as magmosphere. { ′pī·rə,sfir }

pyrostat [ENG] **1.** A sensing device that automatically actuates a warning or extinguishing mechanism in case of fire. **2.** A high-temperature thermostat. { ′pī·rə,stat }

pyrostibite See kermesite. { ˌpī·rə'sti,bīt }

pyrostilpnite [MINERAL] Ag_3SbS_3 A hyacinth-red mineral composed of silver antimony sulfide, occurring in monoclinic crystal tufts; it is polymorphous with pyrargerite. { ˌpī·rə'stilp,nīt }

pyrotechnic code [COMMUN] Significant arrangement of the various colors and patterns of fireworks, signal lights, or signal smokes used for communication between units or between ground and air. { ¦pī·rə¦tek·nik 'kōd }

pyrotechnic pistol [ENG] A single-shot device designed specifically for projecting pyrotechnic signals. { ¦pī·rə¦tek·nik 'pis·təl }

pyrotechnics [ENG] Art and science of preparing and using fireworks. [MATER] Items which are used for both military and nonmilitary purposes to produce a bright light for illumination, or colored lights or smoke for signaling, and which are consumed in the process. { ¦pī·rə¦tek·niks }

pyrotechnic signal [COMMUN] Signal designed for military use to produce a colored light or smoke, for the purpose of transmitting information. { ¦pī·rə¦tek·nik 'sig·nəl }

Pyrotheria [PALEON] An extinct monofamilial order of primitive, mastodonlike, herbivorous, hoofed mammals restricted to the Eocene and Oligocene deposits of South America. { ˌpī·rō'thir·ē·ə }

Pyrotheriidae [PALEON] The single family of the Pyrotheria. { ˌpī·rō·thə'rī·ə,dē }

pyroxene [MINERAL] A family of diverse and important rock-forming minerals having infinite (Si_2O_6) single inosilicate chains as their principal motif; colors range from white through yellow and green to brown and greenish black; hardness is 5.5–6 on Mohs scale, and specific gravity is 3.2–4.0. { pə'räk,sēn }

pyroxene alkali syenite [GEOL] A quartz-poor (less than 20%) member of the charnockite series, characterized by the presence of microperthite. { pə'räk,sēn 'al·kə,lī 'sī·ə,nīt }

pyroxene monzonite [GEOL] A quartz-poor (less than 20%) member of the charnockite series, containing approximately equal amounts of microperthite and plagioclase. { pə'räk,sēn 'män·zə,nīt }

pyroxene syenite [GEOL] A quartz-poor (less than 20%) member of the charnockite series, containing more microperthite than plagioclase. { pə'räk,sēn 'sī·ə,nīt }

pyroxenite [PETR] A heavy, dark-colored, phaneritic igneous rock composed largely of pyroxene with smaller amounts of olivine and hornblende, and formed by crystallization of gabbroic magma. { pə'räk·sə,nīt }

pyroxenoids [MINERAL] A mineral group (including wollastonite and rhodonite) compositionally similar to pyroxene, but SiO_4 tetrahedrons are connected in rings rather than chains. { pə'räk·sə,nóidz }

pyroxylin [ORG CHEM] $[C_{12}H_{16}O_6(NO_3)_4]_x$ Any member of the group of commercially available nitrocelluloses that are used for properties other than their combustibility; the term is commonly used to identify products that are principally made from nitrocellulose, such as pyroxylin plastic or pyroxylin lacquer. Also known as collodion cotton; soluble guncotton; soluble nitrocellulose. { pə'räk·sə·lən }

pyroxylin cement [MATER] A solution of nitrocellulose in a chemical solvent, compounded with a resin, or plasticized with a gum or synthetic; dries by evaporation of the solvent. { pə'räk·sə·lən si'ment }

pyrrhite See pyrochlore. { 'pi,rīt }

Pyrrhocoridae [INV ZOO] A family of hemipteran insects belonging to the superfamily Pyrrhocoroidea. { ˌpir·ə'kór·ə,dē }

Pyrrhocoroidea [INV ZOO] A superfamily of the Pentatomorpha. { ˌpir·ə·kə'róid·ē·ə }

Pyrrhophyta [BOT] A small division of motile, generally unicellular flagellate algae characterized by the presence of yellowish-green to golden-brown plastids and by the general absence of cell walls. { pə'räf·əd·ə }

pyrrhotite [MINERAL] $Fe_{1-x}S$ A common reddish-brown to brownish-bronze mineral that occurs as rounded grains to large masses, more rarely as tabular pseudohexagonal crystals and rosettes; hardness is 4 on Mohs scale, and specific gravity is 4.6 (for the composition Fe_7S_8). { 'pir·ə,tīt }

pyrrobutamine phosphate [PHARM] $C_{20}H_{22}NCl·2H_3PO_4$ A cream-colored, water-soluble powder with a melting range of 127–131°C; used in medicine. { ˌpi·rō'byüd·ə,mēn 'fä,sfāt }

pyrrole [ORG CHEM] C_4H_5N Water-insoluble, yellowish oil, with pungent taste; soluble in alcohol, ether, and dilute acids; boils at 130°C; polymerizes in light; used to make drugs. { 'pi,rōl }

pyrrole ring [ORG CHEM] A five-member heterocycle containing one nitrogen atom and four carbon atoms in the ring; frequently found in structures of natural products occurring as joined rings or attached to straight chains. { 'pī,rōl ,rin }

2-pyrrolidone [ORG CHEM] C_4H_7ON Combustible, light-yellow liquid, boiling at 245°C; soluble in ethyl alcohol, water, chloroform, and carbon disulfide; used as a plasticizer and polymer solvent, in insecticides and specialty inks, and as a nylon-4 precursor. { ˌtü pə'räl·ə,dēn }

pyrrone [ORG CHEM] A polyimidazopyrrolone synthesized from dianhydrides and tetramines; soluble in sulfuric acid; resists temperatures to 600°C. { 'pi,rōn }

pyruvate [BIOCHEM] Salt of pyruvic acid, such as sodium pyruvate, $NaOOCCOCH_3$. { pī'rü,vāt }

pyruvic acid [BIOCHEM] Important intermediate in protein and carbohydrate metabolism; liquid with acetic-acid aroma; melts at 11.8°C; miscible with alcohol, ether, and water; used in biochemical research. { pī'rü·vik 'as·əd }

pyruvic carboxylase [BIOCHEM] An enzyme found in yeast, bacteria, and plants that catalyzes the decarboxylation of pyruvate to acetaldehyde and carbon dioxide. { pī,rü·vik kär'bäk·sə,lās }

pyruvic decarboxylase [BIOCHEM] An enzyme found in yeast that, along with thiamine pyrophosphate and magnesium ions, catalyzes the removal of a carboxyl group from pyruvic acid to produce acetaldehyde and carbon dioxide. { pī,rü·vik ,dē·kär'bäk·sə,lās }

Pythagorean numbers [MATH] Positive integers x, y, and z which satisfy the equation $x^2 + y^2 = z^2$. Also known as Pythagorean triple. { pə,thag·ə'rē·ən 'nəm·bərz }

Pythagorean scale [ACOUS] A musical scale such that the frequency intervals are represented by the ratios of integral powers of the numbers 2 and 3. { pə,thag·ə'rē·ən 'skāl }

Pythagorean theorem [MATH] In a right triangle the square of the length of the hypotenuse equals the sum of the squares of the lengths of the other two sides. { pə,thag·ə'rē·ən 'thir·əm }

Pythagorean triple See Pythagorean numbers. { pə,thag·ə'rē·ən 'trip·əl }

Pythidae [INV ZOO] An equivalent name for Salpingidae. { 'pith·ə,dē }

python [VERT ZOO] The common name for members of the reptilian subfamily Pythoninae. { 'pī,thän }

Pythoninae [VERT ZOO] A subfamily of the reptilian family Boidae distinguished anatomically by the skull structure and the presence of a pair of vestigial hindlegs in the form of stout, movable spurs. { pī'thän·ə,nē }

pyuria [MED] The presence of pus in the urine. { pī'yùr·ē·ə }

pyxidium [BOT] A capsular fruiting body dehiscing around its circumference, thus causing the upper part to fall off. { pik'sid·ē·əm }

pyxis [BOT] A capsule that dehisces by a transverse fissure around the circumference. { 'pik·səs }

Pyxis [ASTRON] A southern constellation; right ascension 9 hours, declination 30° south. Also known as Malus. { 'pik·səs }

pz See pièze.

PZT See lead zirconate titanate.

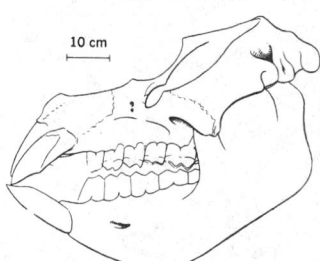

PYROTHERIA

10 cm

Skull and jaw of *Pyrotherium sorondi*, an early Oligocene pyrothere from South America.

PYRROLE

Structural formula for pyrrole.

Q [NUC PHYS] *See* disintegration energy. [PHYS] A measure of the ability of a system with periodic behavior to store energy equal to 2π times the average energy stored in the system divided by the energy dissipated per cycle. Also known as *Q* factor; quality factor; storage factor. [THERMO] A unit of heat energy, equal to 10^{18} British thermal units, or approximately 1.055×10^{21} joules.

QAM *See* quadrature amplitude modulation.

Q band [ELECTROMAG] A radio-frequency band of 36 to 46 gigahertz. { 'kyü ,band }

QBE *See* query by example.

Q branch [SPECT] A series of lines in molecular spectra that correspond to changes in the vibrational quantum number with no change in the rotational quantum number. { 'kyü ,branch }

QCD *See* quantum chromodynamics.

Q factor [ORD] A correction factor applied to a bombsight setting to help account for the differing winds between flight altitude and the ground; this factor corrects only for that component of the differential ballistic wind that is parallel to the actual wind at flight level. [PHYS] *See* Q. { 'kyü ,fak·tər }

Q fever [MED] An acute, febrile infectious disease of humans, characterized by sudden onset and patchy pneumonitis, and caused by a bacterialike organism, *Coxiella burneti*. { 'kyü ,fē·vər }

Q machine [PL PHYS] A device in which a highly ionized, magnetically confined plasma is created by contact ionization of atoms and thermionic emission of electrons. { 'kyü mə,shēn }

Q-machine plasma [PL PHYS] A plasma column in a magnetic field created by surface ionization of a cesium beam on a hot tungsten plate. { 'kyü mə,shēn 'plaz·mə }

Q magnitude [ASTRON] The magnitude of a celestial object based on observations in the infrared at a wavelength of 19.5 micrometers. { 'kyü ,mag·nə,tüd }

Q meter [ENG] A direct-reading instrument which measures the *Q* of an electric circuit at radio frequencies by determining the ratio of inductance to resistance, and which has also been developed to measure many other quantities. Also known as quality-factor meter. { 'kyü ,mēd·ər }

Q multiplier [ELECTR] A filter that gives a sharp response peak or a deep rejection notch at a particular frequency, equivalent to boosting the *Q* of a tuned circuit at that frequency. { 'kyü 'məl·tə,plī·ər }

Q point *See* quiescent operating point. { 'kyü ,point }

QPRK *See* quadrature partial-response keying.

QPSK *See* quadrature phase-shift keying.

qr *See* quarter.

QRS complex [MED] The electrocardiographic deflection representing ventricular depolarization; the initial downward deflection is termed a *Q* wave; the initial upward deflection, an *R* wave; and the downward deflection following the *R* wave, an *S* wave. Also known as ventricular depolarization complex. { ¦kyü¦är'es ,käm,pleks }

qr tr *See* quarter.

Q signal [COMMUN] A three-letter abbreviation starting with Q, used in the International List of Abbreviations for radio-telegraphy to represent complete sentences. [ELECTR] The quadrature component of the chrominance signal in color television, having a bandwidth of 0 to 0.5 megahertz; it consists of $+0.48(R - Y)$ and $+0.41(B - Y)$, where Y is the luminance signal, R is the red camera signal, and B is the blue camera signal. { 'kyü ,sig·nəl }

QSO *See* quasar.

QSS *See* quasi-stellar radio source.

Q-switched laser [OPTICS] A laser whose *Q* factor is kept at a low value while an ion population inversion is built up, and then is suddenly switched to a high value just before instability occurs, resulting in a very high rate of stimulated emission. Also known as giant pulse laser. { 'kyü ,swich 'lā·zər }

qt *See* quart.

quad [ELEC] A series of four separately insulated conductors, generally twisted together in pairs. [ELECTR] A series-parallel combination of transistors, used to obtain increased reliability through double redundancy, because the failure of one transistor will not disable the entire circuit. [GRAPHICS] One of the small pieces of metal used in typesetting to space or to fill out a line of characters; used mostly to fill space when indenting the first line and to fill out the last line of type. [THERMO] A unit of heat energy, equal to 10^{15} British thermal units, or approximately 1.055×10^{18} joules. { kwäd }

quadded cable [ELEC] Cable in which at least some of the conductors are arranged in the form of quads. { 'kwäd·əd 'kā·bəl }

quadded redundancy [COMPUT SCI] A form of redundancy in which each logic gate is quadruplicated, and the outputs of one stage are interconnected to the inputs of the succeeding stage by a connection pattern so that errors made in earlier stages are overridden in later stages, where the original correct signals are restored. { 'kwäd·əd ri'dən·dən·sē }

quad density [COMPUT SCI] A format for floppy-disk storage that holds four times as much data as would normally be contained. { 'kwäd 'den·səd·ē }

quad in-line [ELECTR] An integrated-circuit package that has two rows of staggered pins on each side, spaced closely enough together to permit 48 or more pins per package. Abbreviated QUIL. { 'kwäd ,in'līn }

quadrangle [CIV ENG] 1. A four-cornered, four-sided courtyard, usually surrounded by buildings. 2. The buildings surrounding such a courtyard. 3. A four-cornered, four-sided building. [MAP] A four-sided tract of land, defined by parallels of latitude and meridians of longitude, used as an area unit in systematic mapping. [MATH] A geometric figure bounded by four straight-line segments called sides, each of which intersects each of two adjacent sides in points called vertices, but fails to intersect the opposite sides. Also known as quadrilateral. { 'kwä,draŋ·gəl }

quadrangular prism [MATH] A prism whose bases are quadrangles { kwə¦draŋ·gyə·lər 'priz·əm }

quadrangular pyramid [MATH] A pyramid whose base is a quadrangle. { kwə¦draŋ·gyə·lər 'pir·ə,mid }

quadrant [ANAT] One of the four regions into which the abdomen may be divided for purposes of physical diagnosis. [ELECTROMAG] *See* international henry. [ENG] 1. An instrument for measuring altitudes, used, for example, in astronomy, surveying, and gunnery; employs a sight that can be moved through a graduated 90° arc. 2. A lever that can move through a 90° arc. [MATH] 1. A quarter of a circle; either an arc of 90° or the area bounded by such an arc and the two radii. 2. Any of the four regions into which the plane is divided by a pair of coordinate axes. [MECH ENG] A device for converting horizontal reciprocating motion to vertical reciprocating motion. [NAV] One of the four areas between consecutive equisignal zones of a four course radio range station. [NAV ARCH] A casting, forging, or built-up frame in the shape

Q
R

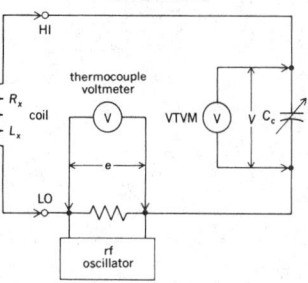

Q METER

Simplified measurement circuit of a *Q* meter. Coil being measured, with inductance L_x and resistance R_x, is connected into the circuit by external terminals HI and LO. Calibrated capacitor C_c is turned to bring the coil into resonance. An input voltage *e* is supplied by a radio-frequency oscillator and measured by a thermocouple voltmeter. A vacuum-tube voltmeter (VTVM) measure voltage *V* across calibrated capacitor. *Q* of coil is determined from the equation $V/e = (1 + Q^2)^{1/2}$.

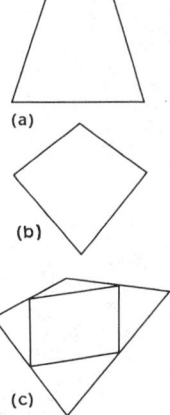

QUADRANGLE

(a)

(b)

(c)

Quadrangles. (*a*) Isosceles trapezoid. (*b*) Kite. (*c*) Parallelogram in quadrangle.

of a sector of a circle attached to the rudder stock and through which the steering gear leads turn the rudder. [OPTICS] A double-reflecting instrument for measuring angles, used primarily for measuring altitudes of celestial bodies; the instrument was replaced by the sextant. [PHYSIO] A sector of one-fourth of the field of vision of one or both eyes. { 'kwä·drənt }

quadrantal angle [MATH] An angle equal to 90° or π/2 radians multiplied by a positive or negative integer or zero. { kwä¦drant·əl 'aŋ·gəl }

quadrantal correctors [NAV] Masses of soft iron placed near a magnetic compass to correct for quadrantal deviation. { kwä'drant·əl kə'rek·tərz }

quadrantal deviation [NAV] Deviation which changes its sign (E or W) approximately each 90° change of heading; caused by induced magnetism in horizontal soft iron present near the compass. { kwä'drant·əl ¦dē·vē'ā·shən }

quadrantal error [NAV] The angular error in a measured bearing that is due to the presence of metal in the vicinity of the direction-finding antenna, such as the metal structure and engines of an airplane or the hull of a ship. { kwä'drant·əl ¦er·ər }

quadrantal point *See* intercardinal point. { kwä'drant·əl ¦point }

quadrantal spheres [NAV] Two hollow spheres of soft iron placed near a magnetic compass to correct for quadrantal deviation. { kwä'drant·əl ¦sfirz }

quadrantal spherical triangle [MATH] A spherical triangle that has one and only one right angle. { kwä¦drant·əl ¦sfir·ə·kəl 'trī¸aŋ·gəl }

quadrant angle of fall [MECH] The vertical acute angle at the level point, between the horizontal and the line of fall of a projectile. { 'kwä·drənt 'aŋ·gəl əv 'fȯl }

quadrant electrometer [ENG] An instrument for measuring electric charge by the movement of a vane suspended on a wire between metal quadrants; the charge is introduced on the vane and quadrants in such a way that there is a proportional twist to the wire. { 'kwä·drənt i¸lek'träm·əd·ər }

quadrant elevation [ORD] Vertical angle between the base and the axis of the bore of a gun which exists just prior to firing. { 'kwä·drənt ¸el·ə'vā·shən }

Quadrantids [ASTRON] A meteor shower whose radiant-right ascension of 15 hours and declination of +48° is in the constellation Boötes; velocity is 27 miles (43 kilometers) per second, and the strength is medium. { kwä'dran·tidz }

quadrant mount [ORD] A device on a gun that holds the gunner's quadrant while the gun is being laid in elevation. { 'kwä·drənt ¸maunt }

quadrant sight [ORD] Sighting instrument on a gun that is used in laying the gun in elevation; used in conjunction with a telescope, which lays the gun for direction. { 'kwä·drənt ¸sīt }

quadraphonic sound system [ENG ACOUS] A system for reproducing sound by means of four loudspeakers properly situated in the listening room, usually at the four corners of a square, with each loudspeaker being fed its own identifiable segment of the program signal. Also known as four-channel sound system. { ¦kwä·drə¦fän·ik 'saund }

quadrate bone [VERT ZOO] A small element forming part of the upper jaw joint on each side of the head in vertebrates below mammals. { 'kwä¸drāt ¸bōn }

quadratic [MATH] Any second-degree expression. { kwä'drad·ik }

quadratic congruence [MATH] A statement that two polynomials of second degree have the same remainder on division by a given integer. { kwä¦drad·ik kən'grü·əns }

quadratic equation [MATH] Any second-degree polynomial equation. { kwä'drad·ik i'kwā·zhən }

quadratic form [MATH] Any second-degree, homogeneous polynomial. { kwä'drad·ik 'fȯrm }

quadratic formula [MATH] A formula giving the roots of a quadratic equation in terms of the coefficients; for the equation $ax^2 + bx + c = 0$, the roots are $x = (-b \pm \sqrt{b^2 - 4ac})/2a$. { kwä'drad·ik 'fȯr·myə·lə }

quadratic function [MATH] A function whose value is given by a quadratic polynomial in the independent variable. { kwə¸drad·ik 'fəŋk·shən }

quadratic inequality [MATH] An inequality in which one side is a quadratic polynomial and the other side is zero. { kwə¸drad·ik ¸in·i'kwal·əd·ē }

quadratic performance index [CONT SYS] A measure of

system performance which is, in general, the sum of a quadratic function of the system state at fixed times, and the integral of a quadratic function of the system state and control inputs. { kwä'drad·ik pər'fȯr·məns ¸in¸deks }

quadratic polynomial [MATH] A polynomial where the highest degree of any of its terms is 2. { kwä'drad·ik ¸päl·ə'nō·mē·əl }

quadratic programming [MATH] A body of techniques developed to find extremal points for systems of quadratic inequalities. { kwä'drad·ik 'prō¸gram·iŋ }

quadratic reciprocity law [MATH] The law that, if *p* and *q* are distinct odd primes, then

$$(p \mid q)(q \mid p) = (-1)^{(1/4)(p-1)(q-1)}$$

(the vertical line inside parentheses is Legendre's symbol). { kwuä¦drad·ik ¸res·ə'präs·əd·ē ¸lȯ }

quadratic residue [MATH] A residue of order 2. { kwä¦drad·ik 'rez·ə·dü }

quadratic Stark effect [ATOM PHYS] A splitting of spectral lines of atoms in an electric field in which the energy levels shift by an amount proportional to the square of the electric field, and all levels shift to lower energies; observed in lines resulting from the lower energy states of many-electron atoms. { kwä'drad·ik 'stärk i¸fekt }

quadratic surd [MATH] A square root of a rational number that is itself an irrational number. { kwä'drad·ik ¸sərd }

quadratic Zeeman effect [ATOM PHYS] A splitting of spectral lines of atoms in a magnetic field in which the energy levels shift by an amount proportional to the square of the magnetic field. { kwä'drad·ik 'zē·mən i¸fekt }

quadratojugal [VERT ZOO] A small bone connecting the quadrate and jugal bones on each side of the skull in many lower vertebrates. { kwä¦drā·dō'jü·gəl }

quadratrix of Hippias [MATH] A plane curve whose equation in cartesian coordinates *x* and *y* is $y = x \cot [\pi x/(2a)]$, where *a* is a constant. { 'kwäd·rə¸triks əv 'hip·ē·əs }

quadrature [ASTRON] The right-angle physical alignment of the sun, moon, and earth. [MATH] **1.** The construction of a square whose area is equal to that of a given surface. **2.** The process of calculating a definite integral. [PHYS] State of being separated in phase by 90°, or one quarter-cycle. Also known as phase quadrature. { 'kwä·drə·chər }

quadrature amplifier [ELECTR] An amplifier that shifts the phase of a signal 90°; used in a color television receiver to amplify the 3.58-megahertz chrominance subcarrier and shift its phase 90° for use in the *Q* demodulator. { 'kwä·drə·chər ¸am·plə¸fī·ər }

quadrature amplitude modulation [COMMUN] Abbreviated QAM. **1.** Quadrature modulation in which the two carrier components are amplitude-modulated. **2.** A digital modulation technique in which digital information is encoded in bit sequences of specified length and these bit sequences are represented by discrete amplitude levels of an analog carrier, by a phase shift of the analog carrier from the phase that represented the previous bit sequence by a multiple of 90°, or by both. { ¦kwäd·rə·chər ¸am·plə¸tüd ¸mäj·ə'lā·shən }

quadrature component [ELEC] **1.** A vector representing an alternating quantity which is in quadrature (at 90°) with some reference vector. **2.** *See* reactive component. { 'kwä·drə·chər kəm¸pō·nənt }

quadrature current *See* reactive current. { 'kwä·drə·chər ¸kə·rənt }

quadrature modulation [COMMUN] Modulation of two carrier components 90° apart in phase by separate modulating functions. { 'kwä·drə·chər ¸mäj·ə'lā·shən }

quadrature partial-response keying [COMMUN] A modulation technique in which two orthogonally phased carriers are combined; each carrier is modulated by one of the digital bit streams to one of three levels; used to convey chrominance information in color television, and considered for use in terrestrial radio. Abbreviated QPRK. { 'kwä·drə·chər ¦pär·shəl ri'späns ¸kē·iŋ }

quadrature phase-shift keying [COMMUN] Phase-shift keying in which four different phase angles are used, usually spaced 90° apart. Abbreviated QPSK. Also known as quadriphase; quaternary phase-shift keying. { ¦kwäd·rə·chər 'fāz ¸shift ¸kē·iŋ }

quadric cone [MATH] A conical surface whose directrices are conic curves. { 'kwä·drik ¦kōn }

quadric curve [MATH] An algebraic curve whose equation is of the second degree. { 'kwäd·rik 'kərv }

quadric quantic [MATH] A quantic of the second degree. { 'kwä·drik ¦kwän·tik }

quadriceps [ANAT] Four-headed, as a muscle. { 'kwä·drə,seps }

quadriceps femoris [ANAT] The large extensor muscle of the thigh, combining the rectus femoris and vastus muscles. { 'kwä·drə,seps 'fem·ə·rəs }

quadrics [MATH] Homogeneous, second-degree expressions. { 'kwä·driks }

quadric surface [MATH] A surface whose equation is a second-degree algebraic equation. { 'kwä·drik 'sər·fəs }

quadricycle [MECH ENG] A four-wheeled human-powered land vehicle, usually propelled by the action of the rider's feet on the pedals. { 'kwäd·rə,sī·kəl }

quadridentate ligand [CHEM] A group which forms a chelate and has four points of attachment. { ¦kwä·drə'den,tāt 'līg·ənd }

quadrigeminal body *See* corpora quadrigemina. { ¦kwä·drə¦jem·ə·nəl 'bäd·ē }

Quadrijugatoridae [PALEON] A monomorphic family of extinct ostracods in the superfamily Hollinacea. { ¦kwä·drə,jü·gə'tór·ə,dē }

quadrilateral *See* quadrangle. { ¦kwä·drə'lad·ə·rəl }

quadrille paper [MATER] A good-quality white ledger paper with light-blue lines ruled on it. { 'kwä'dril ,pā·pər }

quadrillion [MATH] **1.** The number 10^{15}. **2.** In British and German usage, the number 10^{24}. { kwə'dril·yən }

quadrilocular [BIOLOGY] Having four cells or cavities. { ¦kwä·drə'läk·yə·lər }

quadrinomial distribution [STAT] A multinomial distribution with four possible outcomes. { ¦kwä·drə'nō·mē·əl ,di·strə'byü·shən }

quadriphase *See* quadrature phase-shift keying. { 'kwäd·rə,fāz }

quadriplegia [MED] Paralysis affecting the four extremities of the body; may be spastic or flaccid. { ¦kwä·drə'plē·jə }

quadripuntal [COMPUT SCI] Having four punches; specifically, having four random punches on an IBM or Hollerith-type punched card. { ¦kwä·drə'pənt·əl }

quadruped [VERT ZOO] An animal that has four legs. { 'kwä·drə,ped }

quadruple point [PHYS CHEM] Temperature at which four phases are in equilibrium, such as a saturated solution containing an excess of solute. { kwə'drüp·əl 'point }

quadruple thread [DES ENG] A multiple thread having four separate helices equally spaced around the circumference of the threaded member; the lead is equal to four times the pitch of the thread. { kwə'drüp·əl 'thred }

quadruple vector product [MATH] **1.** For any four vectors, the dot product of two derived vectors, one of which is the cross product of two of the original vectors, and the other of which is the cross product of the other two. **2.** For any four vectors, the cross product of two derived vectors, one of which is the cross product of two of the original vectors, and the other of which is the cross product of the other two. { kwə'drüp·əl 'vek·tər ,präd·əkt }

quadruplex circuit [ELEC] Telegraph circuit designed to carry two messages in each direction at the same time. { 'kwä·drə,pleks ,sər·kət }

quadrupole [ELECTROMAG] A distribution of charge or magnetization which produces an electric or magnetic field equivalent to that produced by two electric or magnetic dipoles whose dipole moments have the same magnitude but point in opposite directions, and which are separated from each other by a small distance. [MATH] A mass distribution that has unequal components of the moment-of-inertia tensor along the three principal directions. { 'kwä·drə,pōl }

quadrupole amplifier [ELECTR] A low-noise parametric amplifier consisting of an electron-beam tube in which quadrupole fields act on the fast cyclotron wave of the electron beam to produce high amplification at frequencies in the range of 400–800 megahertz. { 'kwä·drə,pōl 'am·plə,fī·ər }

quadrupole field [ELECTROMAG] **1.** An electric or magnetic field equivalent to that produced by two electric or magnetic dipoles whose dipole moments have the same magnitude but point in opposite directions, and which are separated from each other by a small distance. **2.** The field produced by a quadrupole lens. { 'kwä·drə,pōl ,fēld }

quadrupole lens [ELECTROMAG] A device for focusing beams of charged particles which has four electrodes or magnetic poles of alternating sign arranged in a circle about the beam; used in instruments such as electron microscopes and particle accelerators. { 'kwä·drə,pōl ,lenz }

quadrupole moment [ELECTROMAG] A quantity characterizing a distribution of charge or magnetization; it is given by integrating the product of the charge density or divergence of magnetization density, the second power of the distance from the origin, and a spherical harmonic Y^*_{2m} over the charge or magnetization distribution. { 'kwä·drə,pōl ,mō·mənt }

quadrupole spectrometer [ANALY CHEM] A type of mass spectroscope in which ions pass along a line of symmetry between four parallel cylindrical rods; an alternating potential superimposed on a steady potential between pairs of rods filters out all ions except those of a predetermined mass. Also known as Massenfilter. { 'kwä·drə,pōl spek'träm·əd·ər }

quad word [COMPUT SCI] A word 16 bytes long. { 'kwäd ,wərd }

quagmire *See* bog. { 'kwäg,mīr }

quail [VERT ZOO] Any of several migratory game birds in the family Phasianidae. { kwāl }

quake sheet [GEOL] A well-defined bed resembling a slump sheet but produced by an earthquake and resulting in the formation of a load cast without horizontal slip. { kwāk ,shēt }

quaking bog [GEOL] A peat bog floating or growing over water-saturated land which shakes or trembles when walked on. { 'kwāk·iŋ ,bäg }

qualification test [ENG] A formally defined series of tests by which the functional, environmental, and reliability performance of a component or system may be evaluated in order to satisfy the engineer, contractor, or owner as to its satisfactory design and construction prior to final approval and acceptance. { ,kwäl·ə·fə'kā·shən test }

qualified name [COMPUT SCI] A name that is further identified by associating it with additional names, usually the names of things that contain the thing being named. { 'kwäl·ə,fīd ¦nām }

qualifier [COMPUT SCI] A name that is associated with another name to give additional information about the latter and distinguish it from other things having the same name. { 'kwäl·ə,fī·ər }

qualitative analysis [ANALY CHEM] The analysis of a gas, liquid, or solid sample or mixture to identify the elements, radicals, or compounds composing the sample. { 'kwäl·ə,tād·iv ə,nal·ə·səs }

quality analysis [IND ENG] Examination of the quality goals of a product or service. { 'kwäl·əd·ē ə,nal·ə·səs }

quality assurance [IND ENG] A series of planned or systematic actions required to provide adequate confidence that a product or service will satisfy given needs. { 'kwäl·əd·ē ə'shúr·əns }

quality control [IND ENG] The operational techniques and the activities that sustain the quality of a product or service in order to satisfy given requirements. It consists of quality planning, data collection, data analysis, and implementation, and is applicable to all phases of the product life cycle: design, development, manufacturing, delivery and installation, and operation and maintenance. { 'kwäl·əd·ē kən,trōl }

quality-control chart [IND ENG] A control chart used to indicate and control the quality of a product. { 'kwäl·əd·ē kən¦trōl ,chärt }

quality factor [NUCLEO] The factor by which absorbed dose is to be multiplied to obtain a quantity that expresses on a common scale, for all ionizing radiations, the irradiation incurred by exposed persons. [PHYS] *See* Q. { 'kwäl·əd·ē ,fak·tər }

quality-factor meter *See* Q meter. { 'kwäl·əd·ē ,fak·tər ,mēd·ər }

quality of snow [METEOROL] The amount of ice in a snow sample expressed as a percent of the weight of the sample. Also known as thermal quality of snow. { 'kwäl·əd·ē əv 'snō }

quality of sound *See* timbre. { 'kwäl·əd·ē əv 'saúnd }

quality program [COMPUT SCI] A computer program that is correct, reliable, efficient, maintainable, flexible, testable, portable, and reusable. { ¦kwäl·əd·ē 'prō·grəm }

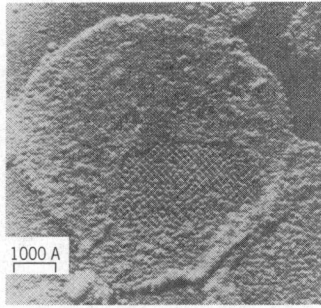

QUANTASOME

Membranes containing chlorophyll taken from a spinach chloroplast. This chromium-shadowed preparation shows that the membrane is composed of a highly ordered array of units, or quantasomes. Scale bar is 1000 angstroms. (*After R. B. Park, courtesy of Science, 144 (3621), 1964*)

QUANTIZED ELECTRONIC STRUCTURE

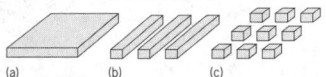

Geometric configurations of quantized electronic structures: (*a*) quantum well; (*b*) quantum wires; (*c*) quantum dots.

quantal response [STAT] Response to treatment which has only two outcomes, all or none. { 'kwänt·əl ri,späns }

quantasome [CYTOL] One of the highly ordered array of units that has a "cobblestone" appearance in electron micrographs of the lamella of chloroplasts, and thought to be the most probable site of the light reaction in photosynthesis. { 'kwän·tə,sōm }

quantic [MATH] A homogeneous algebraic polynomial with more than one variable. { 'kwän·tik }

quantification [SCI TECH] The act of quantifying, that is, of giving a numerical value to a measurement of something, as in computer applications, psychology, or market research. { ,kwän·tə·fə'kā·shən }

quantifier [MATH] Either of the phrases "for all" and "there exists"; these are symbolized respectively by an inverted A and a backward E. { 'kwän·tə,fī·ər }

quantile [STAT] A value which divides a set of data into equal proportions; examples are quartile and decile. { 'kwän,tīl }

quantitative analysis [ANALY CHEM] The analysis of a gas, liquid, or solid sample or mixture to determine the precise percentage composition of the sample in terms of elements, radicals, or compounds. { 'kwän·ə·tād·iv ə'nal·ə·səs }

quantitative genetics [GEN] The study of continuously varying traits, such as height or milk yield. { 'kwän·ə·tād·iv jə'ned·iks }

quantitative geomorphology [GEOL] The assignment of dimensions of mass, length, and time to all descriptive parameters of landform geometry and geomorphic processes, followed by the derivation of empirical mathematical relationships and formulation of rational mathematical models relating these parameters. { 'kwän·ə·tād·iv jē·ō·mór'fäl·ə·jē }

quantitative inheritance [GEN] The acquisition of characteristics which show a quantitative and continuous type of variation. { 'kwän·ə·tād·iv in'her·əd·əns }

quantitative structure-activity relationships [BIOCHEM] The establishment of statistical correlations between the potencies of a series of structurally related compounds and one or more quantitative structural parameters, such as lipophilicity, polarity, and molecular size, by using multilinear regression analysis. { 'kwän·ə,tād·iv ¦strək·chər ak¦tiv·əd·ē ri'lā·shən,ships }

quantitative trait [GEN] A trait that is under the control of many factors, both genetic and environmental, each of which contributes only a small amount to the total variability of the trait. { ,kwänt·ə,tād·iv 'trāt }

quantitative trait locus [GEN] The location of a gene that affects a quantitative trait. { ,kwänt·ə,tād·iv 'trāt 'lō·kəs }

quantity [COMPUT SCI] In computers, a positive or negative real number in the mathematical sense; the term quantity is preferred to the term number in referring to numerical data; the term number is used in the sense of natural number and reserved for "the number of digits," the "number of operations," and so forth. [MATH] Any expression which is concerned with value rather than relations. { 'kwän·əd·ē }

quantity-distance tables [ORD] The regulations pertaining to the amounts and kinds of explosives that can be stored and the proximity of such storage to buildings, highways, railways, magazines, or other installations. { 'kwän·əd·ē 'dis·təns ,tā·bəlz }

quantity meter [ENG] A type of fluid meter used to measure volume of flow. { 'kwän·əd·ē ,mēd·ər }

quantity of electricity *See* charge. { 'kwän·əd·ē əv ,i,lek'tris·əd·ē }

quantization [COMMUN] Division of the range of values of a wave into a finite number of subranges, each of which is represented by an assigned or quantized value within the subrange. [QUANT MECH] **1.** The restriction of an observable quantity, such as energy or angular momentum, associated with a physical system, such as an atom, molecule, or elementary particle, to a discrete set of values. **2.** The transition from a description of a system of particles or fields in the classical approximation where canonically conjugate variables commute, to a description where these variables are treated as noncommuting operators; quantization (first definition) is a result of this procedure. [SCI TECH] The restriction of a variable to a discrete number of possible values; thus the age of a person is usually quantized as a whole number of years. { ,kwän·tə'zā·shən }

quantization distortion [COMMUN] Inherent distortion introduced in the process of quantization of a waveform. Also known as quantization noise; quantumization distortion; quantumization noise. { ,kwän·tə'zā·shən di,stór·shən }

quantization level [COMMUN] Discrete value of the output designating a particular subrange of the input. { ,kwän·tə'zā·shən ,lev·əl }

quantization noise *See* quantization distortion. { ,kwän·tə'zā·shən ,nóiz }

quantized electronic structure [ELECTR] A material that confines electrons in such a small space that their wave-like behavior becomes important and their properties are strongly modified by quantum-mechanical effects. { ¦kwän,tīzd i ¦lek·¦trän·ik 'strək·chər }

quantized frequency modulation [COMMUN] Frequency modulation that involves quantization; it uses time and frequency redundancy within a voice frequency channel during each transmitted symbol; used to combat distortion due to multipath, selection fading, and noise spikes. { 'kwän,tīzd 'frē·kwən·sē ,mäj·ə,lā·shən }

quantized Hall conductance [PHYS] The reciprocal of the von Klitzing constant, equal to e^2/h, where e is the charge of the electron and h is Planck's constant. { ¦kwän,tīzd 'hól kən,dək·təns }

quantized Hall resistance *See* von Klitzing constant. { ¦kwän,tīzd 'hól ri,zis·təns }

quantized pulse modulation [COMMUN] Pulse modulation that involves quantization, such as pulse-numbers modulation and pulse-code modulation. { 'kwän,tīzd 'pəls ,mäj·ə,lā·shən }

quantized Rabi oscillations [ATOM PHYS] Rabi oscillations that occur when only a small number of photons are present at discrete frequencies determined by the number of photons. { ¦kwän,tīzd 'rä·bē ,äs·ə,lā·shənz }

quantized spin wave *See* magnon. { 'kwän,tīzd 'spin ,wāv }

quantized vortex [CRYO] A circular flow pattern observed in superfluid helium and type II superconductors, in which a superfluid flows about a normal (nonsuperfluid) cylindrical region or core which has the form of a thin line, and either the circulation or the magnetic flux is quantized. { 'kwän,tīzd 'vór,teks }

quantizer [ELECTR] A device that measures the magnitude of a time-varying quantity in multiples of some fixed unit, at a specified instant or specified repetition rate, and delivers a proportional response that is usually in pulse code or digital form. { kwän'tīz·ər }

quantum [COMMUN] One of the subranges of possible values of a wave which is specified by quantization and represented by a particular value within the subrange. [QUANT MECH] **1.** For certain physical quantities, a unit such that the values of the quantity are restricted to integral multiples of this unit; for example, the quantum of angular momentum is Planck's constant divided by 2π. **2.** An entity resulting from quantization of a field or wave, having particlelike properties such as energy, mass, momentum and angular momentum; for example, the photon is the quantum of an electromagnetic field, and the phonon is the quantum of a lattice vibration. { 'kwän·təm }

quantum acoustics [ACOUS] The study of the properties of propagating sound waves that are directly attributable to the underlying quantum-mechanical nature of the medium. { 'kwän·təm ə'kü·stiks }

quantum anomaly [QUANT MECH] A phenomenon whereby a quantity that vanishes according to the dynamical rules of classical physics acquires a finite value when quantum rules are used. { ,kwän·təm ə'näm·ə·lē }

quantum cascade laser [OPTICS] A semiconductor laser whose light is generated by electronic transitions between bound states created by quantum confinement in alternating ultrathin layers of semiconductor material. { ¦kwänt·əm ,kas,kād 'lā·zər }

quantum chaos [QUANT MECH] The dynamics of quantum systems whose classical counterparts exhibit chaotic behavior. { ,kwänt·əm 'kā,äs }

quantum chemistry [PHYS CHEM] A branch of physical chemistry concerned with the explanation of chemical phenomena by means of the laws of quantum mechanics. { 'kwän·təm 'kem·ə·strē }

quantum chromodynamics [PART PHYS] A gauge theory of

the strong interactions among quarks; the mathematical structure of the theory resembles that of quantum electrodynamics, with color as the conserved charge. Abbreviated QCD. { 'kwän·təm ¦krō·mō·dī'nam·iks }

quantum computer [COMPUT SCI] A computer in which the time evolution of the state of the individual switching elements of the computer is governed by the laws of quantum mechanics. { 'kwän·təm kəm¦pyüd·ər }

quantum defect [ATOM PHYS] The difference between the principal quantum number of an atomic energy level and the effective quantum number obtained by fitting with a Rydberg formula the energy required to ionize the atom from that level. { 'kwän·təm 'dē‚fekt }

quantum detector [PHYS] A detector of electromagnetic radiation which converts a quantum of the radiation into a proportionate signal by some process which is insensitive to quanta of less than a certain energy; examples include photographic emulsions, photoelectric cells, and Geiger counters. { 'kwän·təm di‚tek·tər }

quantum discontinuity [QUANT MECH] The emission or absorption of a definite amount of energy that accompanies a quantum jump. { 'kwän·təm ‚dis‚känt·ən'ü·əd·ē }

quantum dot [ELECTR] A quantized electronic structure in which electrons are confined with respect to motion in all three dimensions. { ‚kwänt·əm 'dät }

quantum dot laser [OPTICS] A laser that has a dense array of equal-sized quantum dots in the active region, each with only a few thousand atoms of semiconductor material, and emits light from electronic transitions between the discrete energy levels of these quantum dots. { 'kwän·təm 'dät ‚lā·zər }

quantum efficiency [ELECTR] The average number of electrons photoelectrically emitted from a photocathode per incident photon of a given wavelength in a phototube. { 'kwän·təm i‚fish·ən·sē }

quantum electrodynamics [QUANT MECH] The quantum theory of electromagnetic radiation, synthesizing the wave and corpuscular pictures, and of the interaction of radiation with electrically charged matter, in particular with atoms and their constituent electrons. Also known as quantum theory of light; quantum theory of radiation. { 'kwän·təm i¦lek·trō·dī'nam·iks }

quantum electronics [ELECTR] The branch of electronics associated with the various energy states of matter, motions within atoms or groups of atoms, and various phenomena in crystals; examples of practical applications include the atomic hydrogen maser and the cesium atomic-beam resonator. { 'kwän·təm i‚lek'trän·iks }

quantum entanglement [QUANT MECH] The property of two particles with a common origin whereby a measurement on one of the particles determines not only its quantum state but the quantum state of the other particle as well. { ‚kwän·təm in'taŋ·gəl·mənt }

quantum evolution [EVOL] A special but extreme case of phyletic evolution; the rapid evolution that takes place when relatively sudden and drastic change occurs in the environment or when organisms spread into new habitats where conditions differ from those to which they are adapted; the organisms must then adapt quickly to the new conditions if they are to survive. { 'kwän·təm ‚ev·ə'lü·shən }

quantum field theory [QUANT MECH] Quantum theory of physical systems possessing an infinite number of degrees of freedom, such as the electromagnetic field, gravitation field, or wave fields in a medium. { 'kwän·təm 'fēld ‚thē·ə·rē }

quantum gravitation [QUANT MECH] Also known as quantum gravity. **1.** The quantum theory of the gravitational field. **2.** The study of quantum fields in a curved space-time. { 'kwän·təm ‚grav·ə'tā·shən }

quantum gravity See quantum gravitation. { 'kwän·təm 'grav·əd·ē }

quantum Hall effect [ELECTR] A phenomenon exhibited by certain semiconductor devices at low temperatures and high magnetic fields, whereby the Hall resistance becomes precisely equal to $(h/e^2)/n$, where h is Planck's constant, e is the electronic charge, and n is either an integer or a rational fraction. Also known as von Klitzing effect. { 'kwän·təm 'hȯl i‚fekt }

quantum Hall liquid See quantum Hall state.

quantum Hall state [CRYO] A kind of incompressible liquid state obtained by placing a two-dimensional electron gas, confined on the interface of two different semiconductors, in a strong magnetic field at low temperature. Also known as quantum Hall liquid. { ‚kwän·təm 'hȯl ‚stāt }

quantum hydrodynamics [CRYO] The mechanics of a superfluid, such as helium II, investigating phenomena such as the fountain effect and second sound. { 'kwän·təm ‚hī·drō·dī'nam·iks }

quantum hypothesis [QUANT MECH] A hypothesis that some physical quantity can assume only a certain discrete set of values; examples are Planck's law, and the condition in the Bohr-Sommerfeld theory that the action integral of a system must be an integral multiple of Planck's constant. { 'kwän·təm hī'päth·ə·səs }

quantumization distortion See quantization distortion. { ‚kwän·tə·mə'zā·shən di‚stȯr·shən }

quantumization noise See quantization distortion. { ‚kwän·tə·mə'zā·shən ‚nȯiz }

quantum jump [QUANT MECH] The transition of a quantum system from one stationary state to another, accompanied by emission or absorption of energy. { 'kwän·təm 'jəmp }

quantum limit [SPECT] The shortest wavelength present in a continuous x-ray spectrum. Also known as boundary wavelength; end radiation. { 'kwän·təm 'lim·ət }

quantum measurement paradox [QUANT MECH] A paradox that arises because, at the atomic level where the quantum formalism has been directly tested, the most natural interpretation implies that when two or more different outcomes are possible it is not necessarily true that one or the other is actually realized, whereas at the everyday level such a state of affairs seems to conflict with direct experience. { 'kwänt·əm 'mezh·ə·mənt ‚par·ə‚däks }

quantum-mechanical operator [QUANT MECH] A linear, Hermitian operator associated with some physical quantity; for a physical system in any state, the expectation value of the physical quantity equals the integral over configuration space of $\psi^*(A\psi)$, where $A\psi$ is the result of the operator acting on the wave function of the system, and ψ^* is the complex conjugate of the wave function. { 'kwän·təm mi'kan·ə·kəl ‚äp·ə‚rād·ər }

quantum mechanics [PHYS] The modern theory of matter, of electromagnetic radiation, and of the interaction between matter and radiation; it differs from classical physics, which it generalizes and supersedes, mainly in the realm of atomic and subatomic phenomena. Also known as quantum theory. { 'kwän·təm mi'kan·iks }

quantum mineralogy [MINERAL] A branch of mineralogy concerned with the application of quantum mechanics to mineralogical systems. { 'kwän·təm ‚min·ə'räl·ə·jē }

quantum nondemolition measurement [QUANT MECH] A measurement of a physical observable of some system without altering its value. { 'kwän·təm ‚nän‚dem·ə'lish·ən 'mezh·ər·mənt }

quantum number [QUANT MECH] One of the quantities, usually discrete with integer or half-integer values, needed to characterize a quantum state of a physical system; they are usually eigenvalues of quantum-mechanical operators or integers sequentially assigned to these eigenvalues. { 'kwän·təm ‚nən·bər }

quantum of action See Planck's constant. { 'kwän·təm əv 'ak·shən }

quantum size effects [SOLID STATE] Unusual properties of extremely small crystals that arise from confinement of electrons to small regions of space in one, two, or three dimensions. { 'kwänt·əm 'sīz i‚feks }

quantum solid [SOLID STATE] A solid whose atoms or molecules undergo large zero-point motion even in the quantum ground state (at absolute zero temperature) as a result of their small mass and the weak attractive part of their interaction potential. { 'kwän·təm ‚säl·əd }

quantum state [QUANT MECH] **1.** The condition of a physical system as described by a wave function; the function may be simultaneously an eigenfunction of one or more quantum-mechanical operators; the eigenvalues are then the quantum numbers that label the state. **2.** See energy state. { 'kwän·təm ‚stāt }

quantum statistics [STAT MECH] The statistical description of particles or systems of particles whose behavior must be described by quantum mechanics rather than classical mechanics. { 'kwän·təm stə'tis·tiks }

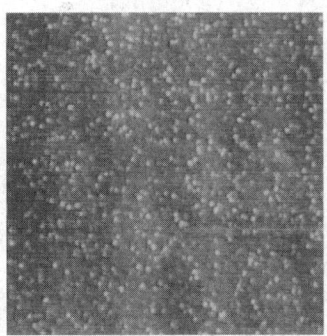

QUANTUM DOT

Atomic force microscope surface image (1×1 micrometer) of quantum dots formed after the deposition of 2.5 atomic layers of indium phosphide and gallium indium phosphide.

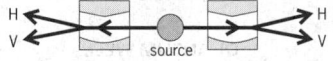

QUANTUM ENTANGLEMENT

Quantum entanglement of two photons emitted by a source. Each photon travels to its own two-channel polarizer, which can be rotated around the respective beam direction. Each detector (H or V) has the same probability of registering a photon. If the two polarizers are oriented parallel, the two photons will always be registered in different detectors.

quantum teleportation [QUANT MECH] The replication of a quantum state at a distant location by utilizing the concept of quantum entanglement. { ¦kwän·təm ¦tel·ə·pór′tā·shən }

quantum theory See quantum mechanics. { ′kwän·təm ¦thē-ə·rē }

quantum theory of heat capacity [STAT MECH] Application of quantum statistics to calculate heat capacities of various substances; an important result of the theory is the decrease of specific heats at low temperatures to values smaller than their classical values as a result of energy quantization. { ¦kwän·təm ¦thē·ə·rē əv ′hēt kə,pas·əd·ē }

quantum theory of light See quantum electrodynamics. { ¦kwän·təm ¦thē·ə·rē əv ′līt }

quantum theory of matter [QUANT MECH] The microscopic explanation of the properties of condensed matter, that is, solids and liquids, based on the fundamental laws of quantum mechanics. { ¦kwänt·əm ¦thē·ə·rē əv ′mad·ər }

quantum theory of measurement [QUANT MECH] The attempt to reconcile the counterintuitive features of quantum mechanics with the hypothesis that it is in principle a complete description of the physical world, even at the level of everyday objects. { ¦kwänt·əm ¦thē·ə·rē əv ′mezh·ər·mənt }

quantum theory of radiation [QUANT MECH] **1.** The theory of heat radiation based on Planck's law; its principal result is the Planck radiation formula. **2.** See quantum electrodynamics. { ¦kwän·təm ¦thē·ə·rē əv ′rād·ē′ā·shən }

quantum theory of spectra [QUANT MECH] The contemporary theory of spectra, based on the idea that an atom, molecule, or nucleus can exist only in certain allowed energy states, that it emits or absorbs energy as it changes from one state to another, and that the frequency of the associated electromagnetic radiation equals the difference in energies of two states divided by Planck's constant. { ¦kwän·təm ¦thē·ə·rē əv ′spek·trə }

quantum theory of valence [PHYS CHEM] The theory of valence based on quantum mechanics; it accounts for many experimental facts, explains the stability of a chemical bond, and allows the correlation and prediction of many different properties of molecules not possible in earlier theories. { ¦kwän·təm ¦thē·ə·rē əv ′vā·ləns }

quantum turbulence [CRYO] A phenomenon observed in a channel filled with superfluid and subjected to a heat flux which exceeds a certain critical value, in which the superfluid becomes filled with a tangled mass of quantized vortex lines. { ¦kwän·təm ′tər·byə·ləns }

quantum-wave equation [QUANT MECH] A partial differential equation which relates the spatial and time dependences of the wave function of a system of one or more atomic or subatomic particles; examples are the Schrödinger equation in nonrelativistic quantum mechanics, and the Klein-Gordon, Dirac, Rarita-Schwinger and Proca equations in relativistic quantum mechanics. { ¦kwän·təm ¦wāv i,kwā·zhən }

quantum well [ELECTR] A thin layer of material (typically between 1 and 10 nanometers thick) within which the potential energy of an electron is less than outside the layer, so that the motion of the electron perpendicular to the layer is quantized. { ¦kwän·təm ′wel }

quantum well infrared photodetector [ELECTR] A detector of infrared radiation composed of numerous alternating layers of controlled thickness of gallium arsenide and aluminum gallium arsenide; the spectral response of the device can be tailored within broad limits by adjusting the aluminum-to-gallium ratio and the thicknesses of the layers during growth. Abbreviated QWIP. { ¦kwänt·əm ¦wel ,in·frə′red ¦fōd·ō·di′tek·tər }

quantum well injection transit-time diode [ELECTR] An active microwave diode that employs resonant tunneling through a gallium arsenide quantum well located between two aluminum gallium arsenide barriers to inject electrons into an undoped gallium arsenide drift region. Abbreviated QWITT diode. { ¦kwän·təm ¦wel in,jek·shən ¦tranz·it ¦tīm ′dī,ōd }

quantum wire [ELECTR] A strip of conducting material about 10 nanometers or less in width and thickness that displays quantum-mechanical effects such as the Aharanov-Bohm effect and universal conductance fluctuations. { ′kwän·təm ′wīr }

quantum yield [PHYS CHEM] For a photochemical reaction, the number of moles of a stated reactant disappearing, or the number of moles of a stated product produced, per einstein of light of the stated wavelength absorbed. { ′kwän·təm ′yēld }

quaquaversal [GEOL] Of strata and geologic structures, dipping outward in all directions away from a central point. { ¦kwä·kwə¦vər·səl }

quarantine [MED] Limitation of freedom of movement of susceptible individuals who have been exposed to communicable disease, for a period of time equal to the incubation period of the disease. { ′kwär·ən,tēn }

quarantine anchorage [CIV ENG] An area where a vessel anchors when satisfying quarantine regulations. { ′kwär·ən,tēn ,aŋ·kər·ij }

quarantine buoy [NAV] A buoy marking the location of a quarantine anchorage. { ′kwär·ən,tēn ,bói }

quark [PARTIC PHYS] One of the hypothetical basic particles, having charges whose magnitudes are one-third or two-thirds of the electron charge, from which many of the elementary particles may, in theory, be built up; for example, nucleons may be formed from three quarks and mesons from quark-antiquark combinations; no experimental evidence for the actual existence of free quarks has been found. { kwärk }

quark confinement [PART PHYS] The phenomenon wherein quarks can never be removed from the hadrons they compose, even though the interactions between them are relatively weak. { ′kwärk kən,fīn·mənt }

quark-gluon plasma [NUC PHYS] A state of nuclear matter postulated by quantum chromodynamics to exist at extremely high temperatures and densities in which the neutrons and protons lose their identities and the quarks and gluons form an unstructured collection of particles. { ′kwärk ′glü,än ,plaz·mə }

quarkonium [PARTIC PHYS] A meson that is made up of a heavy quark and its antiparticle, an antiquark. { kwär′kō·nē·əm }

quark star [ASTRON] A hypothetical star so dense that the nucleons have lost their identity and stability is derived from degenerate quarks. { ′kwärk ,stär }

quarry [ENG] An open or surface working or excavation for the extraction of building stone, ore, coal, gravel, or minerals. { ′kwär·ē }

quarry bar [ENG] A horizontal bar with legs at each end, used to carry machine drills. { ′kwär·ē ,bär }

quarry face [MIN ENG] The freshly split face of ashlar, squared off for the joints only and used for massive work. { ′kwär·ē ,fās }

quarrying [ENG] The surface exploitation and removal of stone or mineral deposits from the earth's crust. [GEOL] See plucking. { ′kwär·ē·iŋ }

quarrying machine [MECH ENG] Any machine used to drill holes or cut tunnels in native rock, such as a gang drill or tunneling machine; most commonly, a small locomotive bearing rock-drilling equipment operating on a track. { ′kwär·ē·iŋ mə,shēn }

quarry powder [MATER] Ammonium nitrate dynamites used in quarrying where blasts of several tons of explosives are needed. { ′kwär·ē ,paůd·ər }

quarry sap See quarry water. { ′kwär·ē ,sap }

quarry water [ENG] Subsurface water retained in freshly quarried rock. Also known as quarry sap. { ′kwär·ē ,wód·ər }

quart [MECH] Abbreviated qt. **1.** A unit of volume used for measurement of liquid substances in the United States, equal to 2 pints, or $\frac{1}{4}$ gallon, or $57\frac{3}{4}$ cubic inches, or $9.46352946 \times 10^{-4}$ cubic meter. **2.** A unit of volume used for measurement of solid substances in the United States, equal to 2 dry pints, or $\frac{1}{32}$ bushel, or 107,521/1600 cubic inches, or approximately 1.10122×10^{-3} cubic meter. **3.** A unit of volume used for measurement of both liquid and solid substances, although mainly the former, in the United Kingdom and Canada, equal to 2 U.K. pints, or $\frac{1}{4}$ U.K. gallon, or 1.1365225×10^{-3} cubic meter. { kwórt }

quartation See inquartation. { kwór′tā·shən }

quarter [MECH] **1.** A unit of mass in use in the United States, equal to $\frac{1}{4}$ short ton, or 500 pounds, or 226.796185 kilograms. **2.** A unit of mass used in troy measure, equal to $\frac{1}{4}$ troy hundredweight, or 25 troy pounds, or 9.33104304 kilograms. Abbreviated qr tr. **3.** A unit of mass used in the United Kingdom, equal to $\frac{1}{4}$ hundredweight, or 28 pounds, or 12.70058636 kilograms. Abbreviated qr. **4.** A unit of volume used in the United Kingdom for measurement of liquid and solid substances, equal to 8 bushels, or 64 gallons, or 0.29094976 cubic

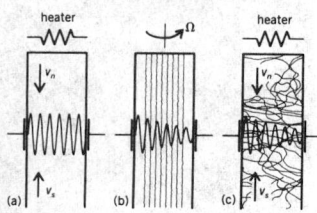

QUANTUM TURBULENCE

Wide-channel counterflow experiments. (*a*) Heat flux produced by the heater is subcritical, producing smooth counterflow of the superfluid (velocity v_s) and normal fluid (velocity v_n). (*b*) Heater is turned off and the channel is rotated at an angular velocity Ω to produce a uniform array of quantized vortices. (*c*) Heat flux is supercritical, producing a tangle of quantized vortices (quantum turbulence).

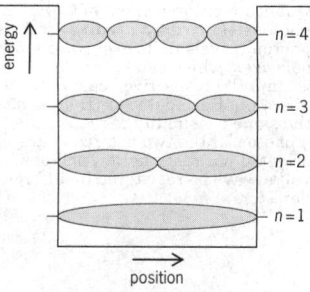

QUANTUM WELL

Electron energy levels and electron standing waves for $n = 1, 2, 3,$ and 4 quantum states in a quantum well.

meter. [NAV ARCH] Portions of a vessel's sides about midway between the stem and the middle and between the middle and the stern. { 'kwȯrd·ər }

quarter deck [NAV ARCH] The after portion of a weather deck. { 'kwȯrd·ər ,dek }

quartering machine [MECH ENG] A machine that bores parallel holes simultaneously in such a way that the center lines of adjacent holes are 90° apart. { 'kwȯrd·ə·riŋ mə,shēn }

quartering sea [NAV] Waves moving in a direction approximately 45° from a vessel's heading, striking the vessel on the quarter. { 'kwȯrd·ə·riŋ 'sē }

quarternary phase-shift keying [ELECTR] Modulation of a microwave carrier with two parallel streams of nonreturn-to-zero data in such a way that the data is transmitted as 90° phase shifts of the carrier; this gives twice the message channel capacity of binary phase-shift keying in the same bandwidth. Abbreviated QPSK. { 'kwät·ə,ner·ē 'fāz ,shift ,kē·iŋ }

quarter-phase *See* two-phase. { 'kwȯrd·ər ,fāz }

quarterpolymer [CHEM] A polymer in which the repeating groups comprise four species of monomer. { 'kwȯrd·ər¦päl·i·mər }

quarter-sawed [MATER] The grain pattern that is produced when hardwood is cut so that the annular rings are at an angle of 45° or less with the board's surface. { 'kwȯrd·ər ,sȯd }

quarter-square multiplier [COMPUT SCI] A device used to carry out function multiplication in an analog computer by implementing the algebraic identity $xy = \frac{1}{4}[(x + y)^2 - (x - y)^2]$. { 'kwȯrd·ər ,skwer 'məl·tə,plī·ər }

quarter-turn drive [MECH ENG] A belt drive connecting pulleys whose axes are at right angles. { 'kwȯrd·ər ,tərn 'drīv }

quarter-wave [ELECTROMAG] Having an electrical length of one quarter-wavelength. { 'kwȯrd·ər ,wāv }

quarter-wave antenna [ELECTROMAG] An antenna whose electrical length is equal to one quarter-wavelength of the signal to be transmitted or received. { 'kwȯrd·ər ,wāv an'ten·ə }

quarter-wave attenuator [ELECTROMAG] Arrangement of two wire gratings, spaced an odd number of quarter-wavelengths apart in a waveguide, used to attenuate waves traveling through in one direction. { 'kwȯrd·ər ,wāv ə'ten·yə,wād·ər }

quarter-wave line *See* quarter-wave stub. { 'kwȯrd·ər ,wāv ,līn }

quarter-wave matching section *See* quarter-wave transformer. { 'kwȯrd·ər ,wāv 'mach·iŋ ,sek·shən }

quarter-wave plate [OPTICS] A thin sheet of mica or other doubly refracting crystal material of such thickness as to introduce a phase difference of one quarter-cycle between the ordinary and the extraordinary components of light passing through; such a plate converts circularly polarized light into plane-polarized light. { 'kwȯrd·ər ,wāv ,plāt }

quarter-wave stub [ELECTROMAG] A section of transmission line that is one quarter-wavelength long at the fundamental frequency being transmitted; when shorted at the far end, it has a high impedance at the fundamental frequency and all odd harmonics, and a low impedance for all even harmonics. Also known as quarter-wave line; quarter-wave transmission line. { 'kwȯrd·ər ,wāv ¦stəb }

quarter-wave termination [ELECTROMAG] Metal plate and a wire grating spaced about one-fourth of a wavelength apart in a waveguide, with the plate serving as the termination of the guide; waves reflected from the metal plate are canceled by waves reflected from the grating so that all energy is absorbed (none is reflected) by the quarter-wave termination. { 'kwȯrd·ər ,wāv tər·mə'nā·shən }

quarter-wave transformer [ELECTROMAG] A section of transmission line approximately one quarter-wavelength long, used for matching a transmission line to an antenna or load. Also known as quarter-wave matching section. { 'kwȯrd·ər ,wāv tranz'fȯr·mər }

quarter-wave transmission line *See* quarter-wave stub. { 'kwȯrd·ər ,wāv tranz'mish·ən ,līn }

quartic *See* biquadratic. { 'kwȯrd·ik }

quartic equation [MATH] Any fourth-degree polynomial equation. Also known as biquadratic equation. { 'kwȯrd·ik i'kwā·zhən }

quartic quantic [MATH] A quantic of the fourth degree. { ¦kwȯr·tik 'kwän·tik }

quartic surd [MATH] A fourth root of a rational number that is itself an irrational number. { 'kwärd·ik ,sərd }

quartile [STAT] The value of any of the three random variables which separate the frequency of a distribution into four equal parts. { 'kwȯr,tīl }

quartile deviation [STAT] One-half of the difference between the upper and lower, that is, the third and first, quartiles. Also known as semi-interquartile range. { ¦kwȯr,tīl ,dē·vē'ā·shən }

quartz [MINERAL] SiO_2 A colorless, transparent rock-forming mineral with vitreous luster, crystallizing in the trigonal trapezohedral class of the rhombohedral subsystem; hardness is 7 on Mohs scale, and specific gravity is 2.65; the most abundant and widespread of all minerals. { kwȯrts }

quartzarenite [PETR] A quartz-rich sandstone with framework grains separated predominantly by cement rather than matrix; essentially an orthoquartzite. { kwȯrt'sar·ə,nīt }

quartz basalt [PETR] An igneous rock with more than 5% quartz. { 'kwȯrts bə'sȯlt }

quartz-bearing diorite *See* quartz diorite. { 'kwȯrts ,ber·iŋ 'dī·ə,rīt }

quartz claim [MIN ENG] In the United States, a mining claim containing ore in veins or lodes, as contrasted with placer claims carrying mineral, usually gold, in alluvium. { 'kwȯrts ¦klām }

quartz clock [HOROL] A clock using the piezoelectric property of a quartz crystal, in which the crystal is introduced into an oscillating electric circuit having a frequency nearly equal to the natural frequency of vibration of the crystal. { 'kwȯrts ¦kläk }

quartz crystal [ELECTR] A natural or artificially grown piezoelectric crystal composed of silicon dioxide, from which thin slabs or plates are carefully cut and ground to serve as a crystal plate. [MINERAL] *See* rock crystal. { 'kwȯrts ¦krist·əl }

quartz-crystal filter [ELECTR] A filter which utilizes a quartz crystal; it has a small bandwidth, a high rate of cutoff, and a higher unloaded Q than can be obtained in an ordinary resonator. { 'kwȯrts ,krist·əl 'fil·tər }

quartz-crystal resonator [ELECTR] A quartz plate whose natural frequency of vibration is used to control the frequency of an oscillator. Also known as quartz resonator. { 'kwȯrts ,krist·əl 'rez·ən,ād·ər }

quartz delay line [ELECTR] An acoustic delay line in which quartz is used as the medium of sound transmission. { 'kwȯrts di'lā ,līn }

quartz diorite [PETR] A group of plutonic rocks having the composition of diorite but with large amounts of quartz (greater than 20%). Also known as quartz-bearing diorite; tonalite. { 'kwȯrts 'dī·ə,rīt }

quartz fiber [ENG] An extremely fine and uniform quartz filament that may be used as a torsion thread or as an indicator in an electroscope or cosimeter. { 'kwȯrts 'fī·bər }

quartz-fiber dosimeter [ENG] A dosimeter in which radiation dose is determined from the deflection of a quartz fiber that is initially charged, repelling it from its metal support, and has its charge reduced by ionizing radiation, causing a proportional reduction in its deflection. { ¦kwȯrts ¦fī·bər dō'sim·əd·ər }

quartz-fiber electroscope [ELECTR] Electroscope in which a gold-plated quartz fiber serves the same function as the gold leaf of a conventional electroscope. { ¦kwȯrts ¦fī·bər i'lek·trə,skōp }

quartz-fiber manometer *See* decrement gage. { ¦kwȯrts ¦fī·bər mə'näm·əd·ər }

quartz-flooded limestone [PETR] A limestone characterized by an abundance of quartz particles that had been imported suddenly from a nearby source by wind or water currents, but that gradually become sparser in an upward direction and completely disappear within a few centimeters. { 'kwȯrts ,fləd·əd 'līm,stōn }

quartz graywacke [PETR] A graywacke containing abundant grains of quartz and chert and less than 10% each of feldspars and rock fragments. { 'kwȯrts 'grā,wak·ə }

quartz horizontal magnetometer [ENG] A type of relative magnetometer used as a geomagnetic field instrument and as an observatory instrument for routine calibration of recording equipment. { 'kwȯrts ,här·ə'zänt·əl ,mag·nə'täm·əd·ər }

quartz-iodine lamp [ELECTR] An electric lamp having a tungsten filament and a quartz envelope filled with iodine vapor. { 'kwȯrts 'ī·ə,dīn ,lamp }

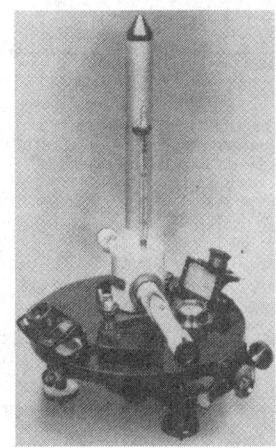

QUARTZ HORIZONTAL MAGNETOMETER

Photograph of quartz horizontal magnetometer. (*U. S. Coastal and Geodetic Survey*)

quartzite [PETR] A granoblastic metamorphic rock consisting largely or entirely of quartz; most quartzites are formed by metamorphism of sandstone. { 'kwȯrt‚sīt }

quartzitic sandstone [PETR] Sandstone consisting of 100% quartz grains cemented with silica. { 'kwȯrt'sid·ik 'san‚stōn }

quartz lamp [ELECTR] A mercury-vapor lamp having a transparent envelope made from quartz instead of glass; quartz resists heat, permitting higher currents, and passes ultraviolet rays that are absorbed by ordinary glass. { 'kwȯrts ‚lamp }

quartz lattice See rhyodacite. { 'kwȯrts 'lad·əs }

quartz mine [MIN ENG] A mine in which a valuable constituent, such as gold, is found in veins rather than in placers; so named because quartz is the chief accessory mineral in such deposits. { 'kwȯrts ‚mīn }

quartz monzonite [PETR] Granitic rock in which 10–50% of the felsic constituents are quartz, and in which the ratio of alkali feldspar to total feldspar is between 35% and 65%. Also known as adamellite. { 'kwȯrts 'män·zə‚nīt }

quartz oscillator [ELECTR] An oscillator in which the frequency of the output is determined by the natural frequency of vibration of a quartz crystal. { 'kwȯrts 'äs·ə‚lād·ər }

quartzose [GEOL] Referring to a substance which contains quartz as a principal constituent. { 'kwȯrt‚sōs }

quartzose arkose [PETR] A sandstone containing 50–85% quartz, chert, and metamorphic quartzite, 15–25% feldspars and feldspathic crystalline rock fragments, and 0–25% micas and micaceous metamorphic rock fragments. { 'kwȯrt‚sōs 'är‚kōs }

quartzose chert [PETR] A vitreous, sparkly, shiny chert, which under high magnification shows a heterogeneous mixture of pyramids, prisms, and faces of quartz, but also includes chert in which the secondary quartz is largely anhedral. { 'kwȯrt‚sōs 'chərt }

quartzose graywacke [PETR] **1.** A sandstone containing 50–85% quartz, chert, and metamorphic quartzite, 15–25% micas and micaceous metamorphic rock fragments, and 0–25% feldspars and feldspathic crystalline rock fragments. **2.** A graywacke that has lost its micaceous constituents through abrasion and thus tends to approach an orthoquartzite. { 'kwȯrt‚sōs 'grā‚wak·ə }

quartzose sandstone [PETR] Sandstone consisting of more than 95% clear quartz grains and less than 5% matrix. Also known as quartz sandstone. { 'kwȯrt‚sōs 'san‚stōn }

quartzose shale [PETR] A green or gray shale composed predominantly of rounded quartz grains of silt size, commonly associated with highly mature sandstones (orthoquartzites), representing the reworking of residual clays as transgressive seas encroached on old land areas. { 'kwȯrt‚sōs 'shāl }

quartzose subgraywacke See protoquartzite. { 'kwȯrt‚sōs 'səb‚grā‚wak·ə }

quartz-pebble conglomerate See orthoquartzitic conglomerate. { 'kwȯrts ‚peb·əl kən'gläm·ə·rət }

quartz plate See crystal plate. { 'kwȯrts ‚plāt }

quartz porphyry [PETR] A porphyritic extrusive or hypabyssal rock containing quartz and alkali feldspar phenocrysts embedded in a microcrystalline or cryptocrystalline matrix. Also known as granite porphyry. { 'kwȯrts 'pȯr·fə·rē }

quartz pressure gage [ENG] A pressure gage that uses a highly stable quartz crystal resonator whose frequency changes directly with applied pressure. { 'kwȯrts 'presh·ər ‚gāj }

quartz resonator See quartz-crystal resonator. { 'kwȯrts 'rez·ən‚ād·ər }

quartz resonator force transducer [ENG] A type of accelerometer which measures the change in the resonant frequency of a small quartz plate with a longitudinal slot, forming a double-ended tuning fork, when a longitudinal force associated with acceleration is applied to the plate. { 'kwȯrts 'rez·ən‚ād·ər 'fȯrs tranz‚dü·sər }

quartz sandstone See quartzose sandstone. { 'kwȯrts 'san‚stōn }

quartz schist [PETR] A schist whose foliation is due mainly to streaks and lenticles of nongranular quartz. { 'kwȯrts 'shist }

quartz strain gage [ELECTR] A device used to measure small deformations of a substance by determining the resulting voltage that develops in a quartz attached to it. { 'kwȯrts 'strän ‚gāj }

quartz syenite [PETR] A group of plutonic rocks having the characteristics of syenite but with a greater amount of quartz (5–20%). { 'kwȯrts 'sī·ə‚nīt }

quartz thermometer [ENG] A thermometer based on the sensitivity of the resonant frequency of a quartz crystal to changes in temperature. { 'kwȯrts thər‚mäm·əd·ər }

quartz topaz See citrine. { 'kwȯrts 'tō‚paz }

quartz wedge [OPTICS] A very thin wedge of quartz cut parallel to an optic axis; used to determine the sign of double refraction of biaxial crystals, and in other applications involving polarized light and its interaction with matter. { 'kwȯrts 'wej }

quasar [ASTRON] Quasi-stellar astronomical object, often a radio source; all quasars have large red shifts; they have small optical diameter, but may have large radio diameter. Also known as quasi-stellar object (QSO). { 'kwā‚zär }

quasi-acoustical holography [ACOUS] An optical technique by which images produced by B-mode ultrasonic medical holography are assembled in a single three-dimensional image volume which, when reconstructed with visible light, renders a realistic three-dimensional image of the target under investigation. Also known as holographic multiplexing. { ‚kwä·zē ə'kü·stə·kəl hō'läg·rə·fē }

quasi-atom [ATOM PHYS] A system formed by two colliding atoms whose nuclei approach each other so closely that, for a very short time, the atomic electrons arrange themselves as if they belonged to a single atom whose atomic number equals the sum of the atomic numbers of the colliding atoms. { ‚kwä·zē 'ad·əm }

quasi-cratonic [GEOL] Pertaining to a part of oceanic crust marginal to the continent which is considered to be former continental material that stretched and foundered during expansion. Also known as semicratonic. { ‚kwä·zē krə'tän·ik }

quasi-crystal [CRYSTAL] A phase of solid matter that, like a crystal, exhibits long-range orientational order and translational order but whose atoms and clusters repeat in a sequence defined by a sum of periodic functions whose periods are in an irrational ratio. { ‚kwä·zē 'krist·əl }

quasi-equilibrium [GEOL] The state of balance or grade in a stream cross section, whereby conditions of approximate equilibrium tend to be established in a reach of the stream as soon as a rather smooth longitudinal profile has been established in that reach, even though downcutting may go on. { ‚kwä·zē ‚ē·kwə'lib·rē·əm }

quasi-fission See deep inelastic collision. { ‚kwä·zē 'fish·ən }

quasi-F martingale [MATH] A stochastic process which is the sum of an F martingale and an F process having bounded variation on every finite time interval. { ‚kwä·zē ‚ef 'märt·ən‚gäl }

quasi-free-electron theory [SOLID STATE] A modification of the free-electron theory of metals to take into account the periodic variation of the potential acting on a conduction electron, in which these electrons are assigned an effective scalar mass which differs from their real mass. { ‚kwä·zē ‚frē i'lek‚tran ‚thē·ə·rē }

quasi-geostationary satellites [AERO ENG] A constellation of satellites that simulates an object hovering over a particular location on the earth by having one member of the constellation moving slowly over a nearby location at all times. { ‚kwäz·ē ‚jē·ō‚stā·shən·er·ē 'sad·əl‚īts }

quasi-hydrostatic approximation [METEOROL] The use of the hydrostatic equation as the vertical equation of motion, thus implying that the vertical accelerations are small without constraining them to be zero. Also known as quasi-hydrostatic assumption. { ‚kwä·zē ‚hī·drə‚stad·ik ə‚präk·sə'mā·shən }

quasi-hydrostatic assumption See quasi-hydrostatic approximation. { ‚kwä·zē ‚hī·drə‚stad·ik ə'səm·shən }

quasi-instruction [COMPUT SCI] An expression in a source program which resembles an instruction in form, but which does not have a corresponding machine instruction in the object program, and is directed to the assembler or compiler. Also known as pseudoinstruction. { ‚kwä·zē in'strək·shən }

quasi-linear feedback control system [CONT SYS] Feedback control system in which the relationships between the pertinent measures of the system input and output signals are substantially linear despite the existence of nonlinear elements. { ‚kwä·zē 'lin·ē·ər 'fēd‚bak kən'trol ‚sis·təm }

quasi-linear system [CONT SYS] A control system in which the relationships between the input and output signals are substantially linear despite the existence of nonlinear elements. { ‚kwä·zē 'lin·ē·ər 'sis·təm }

QUASI-CRYSTAL

Example of a quasi-crystal lattice, formed from a square and a rhombus, with a disallowed (eightfold) symmetry.

quasi-molecule [ATOM PHYS] The structure formed by two colliding atoms when their nuclei are close enough for the atoms to interact, but not so close as to form a quasi-atom. { ¦kwä·zē 'mäl·ə‚kyül }

quasi-parallel execution [COMPUT SCI] The execution of a collection of coroutines by a single processor that can work on only one coroutine at a time; the order of execution is arbitrary and each coroutine is executed independently of the rest. { ¦kwä·zē 'par·ə‚lel ‚ek·sə'kyü·shən }

quasi-particle [PHYS] An entity used in the description of a system of many interacting particles which has particlelike properties such as mass, energy, and momentum, but which does not exist as a free particle; examples are phonons and other elementary excitations in solids, and "dressed" helium-3 atoms in Landau's theory of liquid helium-3. { ¦kwä·zē ¦pard·ə·kəl }

quasi-particle detector [ENG] A detector of electromagnetic radiation at wavelengths close to 1 millimeter, based on the tunneling of single electrons (more precisely, quasi-particles) through a tunnel junction consisting of an oxide barrier between two superconductors, with a responsivity of one tunneling electron for each microwave photon absorbed. { ¦kwä·zē ¦pard·ə·kəl di‚tek·tər }

quasi-perfect number [MATH] An integer that is 1 less than the sum of all its factors other than itself. { ¦kwäz·ē ¦pər·fikt 'nəm·bər }

quasi-periodic motion [PHYS] Motion at two frequencies simultaneously, where the ratio of the frequencies is not a rational number. { ¦kwä·zē ‚pir·ē'äd·ik 'mō·shən }

quasi-random code generator [COMMUN] High-speed coded information source used in the design and evaluation of wide-band communications links by providing a means of closed-loop testing. { ¦kwä·zē 'ran·dəm 'kōd ‚jen·ə‚rād·ər }

quasi-reflection [OPTICS] A term applied to the very strong return of light produced by dust particles and other suspensoids whose diameters are large compared to the wavelength of the incident radiation. { ¦kwä·zē ri'flek·shən }

quasi-square-wave static inverter [ELEC] A static inverter that generates two square waves superimposed on one another to approximate an ac sine wave, using a silicon-controlled rectifier bridge and control circuit to control the pulse width and amplitude of the resulting wave, thereby achieving regulation. { ¦kwä·zē 'skwer ‚wāv 'stad·ik in'vərd·ər }

quasi-stable elementary particle [PARTIC PHYS] A term formerly (before the discovery of charmed particles) used for elementary particles that cannot decay into other particles through strong interactions and that have lifetimes longer than 10^{-20} second. Also known as semistable elementary particle. { ¦kwä·zē 'stā·bəl 'el·ə‚men·trē 'pärd·ə·kəl }

quasi-static process See reversible process. { ¦kwä·zē 'stad·ik 'prä·səs }

quasi-stationary front [METEOROL] A front which is stationary or nearly so; conventionally, a front which is moving at a speed less than about 5 knots (0.26 meter per second) is generally considered to be quasi-stationary. Commonly known as stationary front. { ¦kwä·zē 'stā·shə‚ner·ē 'frənt }

quasi-stellar object See quasar. { ¦kwä·zē 'stel·ər 'äb‚jekt }

quasi-stellar radio source [ASTRON] A quasar that emits a significant fraction of its energy at radio frequencies ranging from 30 megahertz to 100 gigahertz. Abbreviated QSS. { ¦kwä·zē 'stel·ər 'rād·ē·ō ‚sórs }

Quaternary [GEOL] The second period of the Cenozoic geologic era, following the Tertiary, and including the last 2–3 million years. { 'kwät·ən‚er·ē }

quaternary alloy [MET] An alloy containing four principal elements apart from accidental impurities. { 'kwät·ən‚er·ē 'al‚ói }

quaternary ammonium base [ORG CHEM] Ammonium hydroxide (NH_4OH) with the ammonium hydrogens replaced by organic radicals, such as (CH_3)$_4$NOH. { 'kwät·ən‚er·ē ə¦mō·nē·əm 'bās }

quaternary ammonium salt [ORG CHEM] A nitrogen compound in which a central nitrogen atom is joined to four organic radicals and one acid radical, for example, hexamethonium chloride; used as an emulsifying agent, corrosion inhibitor, and antiseptic. { 'kwät·ən‚er·ē ə¦mō·nē·əm 'sólt }

quaternary carbon atom [ORG CHEM] A carbon atom bonded to four other carbon atoms with single bonds. { 'kwät·ən‚er·ē 'kär·bən 'ad·əm }

quaternary phase equilibria [PHYS CHEM] The solubility relationships in any liquid system with four nonreactive components with varying degrees of mutual solubility. { 'kwät·ən‚er·ē ¦fāz ‚ē·kwə'lib·rē·ə }

quaternary phase-shift keying See quadrature phase-shift keying. { 'kwät·ər‚ner·ē 'fāz ‚shift ‚kē·iŋ }

quaternary quantic [MATH] A quantic that contains four variables. { 'kwät·ən‚er·ē 'kwän·tik }

quaternary signaling [COMMUN] An electrical communications mode in which information is passed by the presence and absence, or plus and minus variations of four discrete levels of one parameter of the signaling medium. { 'kwät·ən‚er·ē 'sig·nə·liŋ }

quaternary system [PHYS CHEM] An equilibrium relationship between a mixture of four (four phases, four components, and so on). { 'kwät·ən‚er·ē 'sis·təm }

quaternion [MATH] The division algebra over the real numbers generated by elements i, j, k subject to the relations $i^2 = j^2 = k^2 = -1$ and $ij = -ji = k$, $jk = -kj = i$, and $ki = -ik = j$. Also known as hypercomplex number. { kwə'ter·nē·ən }

quatrefoil [MATH] A multifoil consisting of four congruent arcs of a circle arranged around a square. { 'kwä·trə‚fóil }

quay [CIV ENG] A solid embankment or structure parallel to a waterway; used for loading and unloading ships. { kē }

qubit [COMPUT SCI] In quantum computation, a superposition of the ground state and the excited state of an elementary two-level quantum system (such as a two-level atom or a nuclear spin), corresponding to a classical bit that is either 0 (corresponding to the ground state) or 1 (corresponding to the excited state). { 'kyü·bit }

quebracho [BOT] Any of a number of South American trees in different genera in the order Sapindales, but all being a valuable source of wood, bark, and tannin. [MATER] A drilling-fluid additive used for thinning or dispersing in order to control viscosity and thixotropy; made from an extract of the quebracho tree and consisting essentially of tannic acid. { kā'brä·chō }

quebracho bark [BOT] Bark of the white quebracho tree *Aspidosperma quebracho* of Chile and Argentina; main components are aspidospermine, tannin, and quebrachine; used in medicine and tanning. { kā'brä·chō ‚bärk }

queen [INV ZOO] A mature, fertile female in a colony of ants, bees, or termites, whose function is to lay eggs. { kwēn }

queen closer [CIV ENG] In masonry work, a brick that has been cut in half along its length and is used at the end of a course. { 'kwēn ‚klōs·ər }

queen post [CIV ENG] Either of two vertical members, one on each side of the apex of a triangular truss. { 'kwēn ‚pōst }

queenslandite See Darwin glass. { 'kwēnz·lən‚dīt }

Queensland tick typhus [MED] A benign infectious disease of humans found in rural northeastern Australia, caused by the bacterialike microorganism *Rickettsia australis* and presumed to be carried by the tick *Ixodes holocyclus*. { 'kwēnz·lənd ¦tik 'tī·fəs }

queen's metal [MET] An alloy consisting principally of tin to which antimony, zinc, and lead or copper are added. { 'kwēnz ‚med·əl }

Queenston shale [GEOL] A red bed series from the Ordovician found in Niagara Gorge; it is composed of deltaic red shale. { 'kwēnz·tən shāl }

queenstownite See Darwin glass. { 'kwēn·stə‚nīt }

quellung reaction [MICROBIO] Swelling of the capsule of a bacterial cell, caused by contact with serum containing antibodies capable of reacting with polysaccharide material in the capsule; applicable to *Pneumococcus*, *Klebsiella*, and *Hemophilus*. { 'kwel·əŋ rē‚ak·shən }

quench aging [MET] Aging of metal induced by rapid cooling from solution heat-treatment temperatures. { 'kwench ‚āj·iŋ }

quench annealing [MET] Annealing an austenitic ferrous alloy by heating followed by quenching from solution temperatures. { 'kwench ə‚nēl·iŋ }

quench bath [ENG] A liquid medium, such as oil, fused salt, or water, into which a material is plunged for heat-treatment purposes. { 'kwench ‚bath }

quenched spark gap [ELEC] A spark gap having provisions for rapid deionization; one form consists of many small gaps between electrodes that have relatively large mass and are good

	QUATERNARY	
CENOZOIC	TERTIARY	
MESOZOIC	CRETACEOUS	
	JURASSIC	
	TRIASSIC	
PALEOZOIC	PERMIAN	
	CARBONIFEROUS	PENNSYLVANIAN
		MISSISSIPPIAN
	DEVONIAN	
	SILURIAN	
	ORDOVICIAN	
	CAMBRIAN	
PRECAMBRIAN		

Chart showing the position of the Quaternary in relation to other periods and to the eras of geologic time.

QUEBRACHO

Quebracho, branch with fruit.

radiators of heat; the electrodes serve to cool the gaps rapidly and thereby stop conduction. { 'kwencht 'spärk ,gap }

quenched water [OCEANOGR] Ocean water which produces an abnormally large propagation loss in the sound passing through it; it is usually found in shallow water or near shores where there are strong currents accompanied by considerable turbulence. { 'kwencht 'wȯd·ər }

quench frequency [ELECTR] Number of times per second that a circuit is caused to go in and out of oscillation. { 'kwench ,frē·kwən·sē }

quench hardening [MET] The hardening of a ferrous alloy by quenching from a temperature above the transformation range. { 'kwench ,hard·ən·iŋ }

quenching [ATOM PHYS] Phenomenon in which a very strong electric field, such as a crystal field, causes the orbit of an electron in an atom to precess rapidly so that the average magnetic moment associated with its orbital angular momentum is reduced to zero. [ELECTR] **1.** The process of terminating a discharge in a gas-filled radiation-counter tube by inhibiting reignition. **2.** Reduction of the intensity of resonance radiation resulting from deexcitation of atoms, which would otherwise have emitted this radiation, in collisions with electrons or other atoms in a gas. [ENG] Shock cooling by immersing liquid or molten material into a cooling medium (liquid or gas); used in metallurgy, plastics forming, and petroleum refining. [IMMUNOL] An adaptation of immunofluorescence that uses two fluorochromes, one of which absorbs light emitted by the other; one fluorochrome labels that antigen, another the antibody, and the antigen-antibody complexes retain both; the initially emitted light is absorbed and so quenched by the second compound. [MECH ENG] Rapid removal of excess heat from the combustion chamber of an automotive engine. [SOLID STATE] Reduction in the intensity of sensitized luminescence radiation when energy migrating through a crystal by resonant transfer is dissipated in crystal defects or impurities rather than being reemitted as radiation. { 'kwench·iŋ }

quenching frequency [ELECTR] The frequency of an alternating voltage that is applied to a superregenerative detector stage to prevent sustained oscillation. { 'kwench·iŋ ,frē·kwən·sē }

quenching oil [MET] Animal, vegetable, or mineral oil, such as fish oil, cottonseed oil, or lard, used in quenching baths for carbon and alloy steels; removes heat from the steel more slowly and uniformly than water. { 'kwench·iŋ ,ȯil }

quenching stress [MET] Internal stresses set up in a metal as a result of quenching. { 'kwench·iŋ ,stres }

quench oscillator [ELECTR] Circuit in a superregenerative receiver which produces the frequency signal. { 'kwench ,äs·ə,lād·ər }

quench-tank extrusion [ENG] Plastic-film or metal extrusion that is cooled in a quenching medium. { 'kwench ,taŋk ik'strü·zhən }

quench temperature [ENG] The temperature of the medium used for quenching. { 'kwench ,tem·prə·chər }

quenite [PETR] A fine-grained, dark-colored hypabyssal rock composed of anorthite, chrome diopside, with less olivine and a small amount of bronzite. { 'kwe,nīt }

quenselite [MINERAL] $PbMnO_2(OH)$ A pitch black mineral consisting of an oxide of lead and manganese; occurs in tabular form. { 'kwens·əl,īt }

quenstedtite [MINERAL] $Fe_2(SO_4)_3 \cdot 10H_2O$ A pale violet to reddish-violet, triclinic mineral consisting of hydrated ferric sulfate; occurs in aggregates of crystals. { 'kwen,ste,tīt }

quercetin [BIOCHEM] $C_{15}H_5O_2(OH)_5$ A yellow, crystalline flavonol obtained from oak bark and Douglas-fir bark; used as an antioxidant and absorber of ultraviolet rays, and in rubber, plastics, and vegetable oils. { 'kwer·sə·tən }

quercimelin See quercitrin. { ,kwer'sim·ə·lən }

quercite See quercitol. { 'kwer,sīt }

quercitol [PHARM] $C_6H_7(OH)_5$ Colorless, water-soluble, sweet-tasting crystals with a melting point of 234°C; used in medicine. Also known as acorn sugar; quercite. { 'kwer·sə,tȯl }

quercitrin [ORG CHEM] $C_{21}H_{20}O_{11}$ The 3-rhamnoside of quercitin, forming yellow crystals from dilute ethanol or methanol solution, melting at 176–179°C, soluble in alcohol; used as a textile dye. Also known as quercimelin; quercitroside. { 'kwer·sə·trən }

quercitroside See quercitrin. { kwer'si·trə,sīd }

QUERCETIN

Structural formula.

query [COMPUT SCI] A computer instruction to interrogate a data base. { 'kwir·ē }

query by example [COMPUT SCI] A software product used to search a data base for information having formats or ranges of values specified by English-like statements that indicate the desired results. Abbreviated QBE. { 'kwir·ē bī ig'zam·pəl }

query language [COMPUT SCI] A generalized computer language that is used to interrogate a data base. { 'kwir·ē ,laŋ·gwij }

query layer [COMPUT SCI] A program that mediates between data sources on the World Wide Web and a user's query by breaking the query into subqueries against each information source and then gathering together the results for presentation to the user. { 'kwir·ē ,lā·ər }

query program [COMPUT SCI] A computer program that allows a user to retrieve information from a data base and have it displayed on a terminal or printed out. { 'kwir·ē ,prō·grəm }

QUEST See quantized electronic structure. { kwest }

question-answering system [COMPUT SCI] An information retrieval system in which a direct answer is expected in response to a submitted query, rather than a set of references that may contain the answers. { 'kwes·chən 'an·sə·riŋ ,sis·təm }

quetsch [TEXT] **1.** A vat containing rollers, in which chemical solutions are applied to yarns or fabrics. **2.** One of the rollers in such a vat. { kvech }

queue [COMPUT SCI] **1.** A list of items waiting for attention in a computer system, generally ordered according to some criteria. **2.** A linear list whose elements are inserted and deleted in a first-in-first-out order. [IND ENG] See waiting line. { kyü }

queued access method [COMPUT SCI] A set of precedures controlled by queues for efficient transfer of data between a computer and input-output devices. { 'kyüd 'ak,ses ,meth·əd }

queue-driven system [COMPUT SCI] A software system that uses many queues for tasks in various phases of processing. { 'kyü ,driv·ən ,sis·təm }

queueing [ENG] The movement of discrete units through channels, such as programs or data arriving at a computer, or movement on a highway of heavy traffic. { 'kyü·iŋ }

queueing network model [COMPUT SCI] A model that represents a computer system by a network of devices through which customers (such as transactions, processes, or server requests) flow, and queues may form at each device due to its finite service rate. { 'kyü·iŋ ,net,wərk ,mäd·əl }

queueing theory [MATH] The area of stochastic processes emphasizing those processes modeled on the situation of individuals lining up for service. { 'kyü·iŋ ,thē·ə·rē }

Quevenne scale [CHEM] Arbitrary scale used with hydrometers or lactometers in the determination of the specific gravity of milk; degrees Quevenne = 1000 (specific gravity − 1). { kə'ven ,skāl }

quibinary [COMPUT SCI] A numeration system, used in data processing, in which each decimal digit is represented by seven binary digits, a group of five which are coefficients of 8, 6, 4, 2, and 0, and a group of two which are coefficients of 1 and 0. { 'kwib·ə,ner·ē }

quick [GEOL] **1.** Referring to a sediment that, when mixed with or absorbing water, becomes extremely soft, incoherent, or loose, and is capable of flowing easily under load or by force of gravity. **2.** Referring to a soil in which a decrease in effective stress allows water to flow upward with sufficient velocity to reduce significantly the soil's bearing capacity. **3.** Referring to a highly porous soil that readily absorbs heat. [MIN ENG] Referring to an economically valuable or productive mineral deposit. { kwik }

quick-break fuse [ELEC] A fuse designed to draw out the arc and break the circuit rapidly when the fuse wire melts, generally by separating the broken ends with a spring. { 'kwik ,brāk 'fyüz }

quick-break switch [ELEC] A switch that breaks a circuit rapidly, independently of the rate at which the switch handle is moved, to minimize arcing. { 'kwik ,brāk 'swich }

quick-change gearbox [MECH ENG] A cluster of gears on a machine tool, the arrangement of which allows for the rapid change of gear ratios. { 'kwik ,chānj 'gir,bäks }

quick clay [GEOL] Clay that loses its shear strength after being disturbed. { 'kwik ,klā }

quick flashing light [NAV] In marine operations, a light showing short flashes at the rate of not less than 60 per minute; it has a cautionary significance. { 'kwik ¦flash·iŋ 'līt }

quicklime *See* calcium oxide. { 'kwik,līm }

quick-make switch [ELEC] Switch or circuit breaker which has a high contact-closing speed, independent of the operator. { 'kwik ¦māk 'swich }

quick malleable iron [MET] Malleable iron containing 2.2% carbon, 1.5% silicon, 0.30–0.60% manganese, and 0.75–1% copper. { 'kwik 'mal·yə·bəl 'ī·ərn }

quickmatch [ENG] Fast-burning fuse made from a cord impregnated with black powder. { 'kwik,mach }

quickness [ORD] General term, expressing the mass rate of gas evolution of a propellant in a quantitative sense; basically a function of the propellant geometry. { 'kwik·nəs }

quick return [MECH ENG] A device used in a reciprocating machine to make the return stroke faster than the power stroke. { 'kwik ri'tərn }

quicksand [GEOL] A highly mobile mass of fine sand consisting of smooth, rounded grains with little tendency to mutual adherence, usually thoroughly saturated with upward-flowing water; tends to yield under pressure and to readily swallow heavy objects on the surface. Also known as running sand. [MATER] A loose sand mixture with a high proportion of water, thus having a low bearing pressure. { 'kwik,sand }

quicksilver *See* mercury. { 'kwik,sil·vər }

quicksilver vermilion *See* mercuric sulfide. { 'kwik,sil·vər vər'mil·yən }

quickstone [PETR] A consolidated rock that flowed under the influence of gravity before lithification. { 'kwik,stōn }

quickwater [HYD] The part of a stream characterized by a strong current. { 'kwik,wȯd·ər }

quiesce [COMPUT SCI] To prevent a computer system from starting new jobs so that the system gradually winds down as current jobs are completed, usually in preparation for a planned outage. { kwē'es }

quiescence [CELL MOL] A period in which a cell is not increasing its mass or going through the cell cycle. { kwē'es·əns }

quiescent [ELECTR] Pertaining to a circuit element which has no input signal, so that it does not perform its active function. [ENG] Pertaining to a body at rest, or inactive, such as an undisturbed liquid in a storage or process vessel. [MED] Inactive, latent, or dormant, referring to a disease or pathological process. { kwē'es·ənt }

quiescent-carrier telephony [COMMUN] A radiotelephony system in which the carrier is suppressed whenever there are no voice signals to be transmitted. { kwē'es·ənt 'kar·ē·ər tə'lef·ə·nē }

quiescent center [BOT] A central region (usually hemispherical) of many root stems that consists of cells which rarely divide or synthesize deoxyribonucleic acid and have less ribonucleic acid and protein than adjacent cells. { kwē'es·ənt ¦sen·tər }

quiescent operating point [ELECTR] The currents and voltages in an electronic circuit when the input signal is replaced by its average value, so that all currents and voltages can be approximated by series expansions around this point. Also known as Q point. { kwē,es·ənt 'äp·ə,rād·iŋ ,pȯint }

quiescent period [COMMUN] Resting period, or the period between pulse transmissions. { kwē'es·ənt 'pir·ē·əd }

quiescent point [ELECTR] The point on the characteristic curve of an amplifier representing the conditions that exist when the input signal equals zero. { kwē'es·ənt ¦pȯint }

quiescent prominence [ASTROPHYS] A vertical sheet of cool gas that is suspended in the solar corona for a period of days to months. { kwē'es·ənt ¦präm·ə·nəns }

quiescent push-pull [ELECTR] Push-pull output stage so arranged in a radio receiver that practically no current flows when no signal is being received. { kwē'es·ənt ¦pu̇sh ¦pu̇l }

quiescent value [ELECTR] The voltage or current value for an electron-tube electrode when no signals are present. { kwē'es·ənt 'val·yü }

quiet automatic volume control *See* delayed automatic gain control. { 'kwī·ət ¦ȯd·ə¦mad·ik 'väl·yəm kən,trōl }

quiet battery [ELECTR] Source of energy of special design or with added filters which is sufficiently quiet and free from interference that it may be used for speech transmission. Also known as talking battery. { 'kwī·ət 'bad·ə·rē }

quieting sensitivity [ELECTR] Minimum signal input to a

frequency-modulated receiver which is required to give a specified output signal-to-noise ratio under specified conditions. { 'kwī·əd·iŋ ,sen·sə,tiv·əd·ē }

quiet sun [ASTROPHYS] The sun when it is free from unusual radio wave or thermal radiation such as that associated with sunspots. { 'kwī·ət 'sən }

quiet-sun noise [ASTROPHYS] Electromagnetic noise originating in the sun at a time when there is little or no sunspot activity. { 'kwī·ət ¦sən ,nȯiz }

quiet tuning [ELECTR] Circuit arrangement for silencing the output of a radio receiver, except when it is accurately tuned to an incoming carrier wave. { 'kwī·ət 'tün·iŋ }

QUIL *See* quad in-line. { kwil }

quill [DES ENG] A hollow shaft into which another shaft is inserted in mechanical devices. [TEXT] A shaft or spool on which filling yarn is wound before insertion into a shuttle. [VERT ZOO] The hollow, horny shaft of a large stiff wing or tail feather. { kwil }

quill drive [MECH ENG] A drive in which the motor is mounted on a nonrotating hollow shaft surrounding the driving-wheel axle; pins on the armature mesh with spokes on the driving wheels, thereby transmitting motion to the wheels; used on electric locomotives. { 'kwil ,drīv }

quill gear [MECH ENG] A gear mounted on a hollow shaft. { 'kwil gir }

quillwort [BOT] The common name for plants of the genus *Isoetes*. { 'kwil,wȯrt }

quilted surface [GEOL] A land surface characterized by broad, rounded, uniformly convex hills separating valleys that are comparatively narrow. { 'kwil·təd 'sər·fəs }

quinacrine [PHARM] $C_{23}H_{30}ClN_3O$ Formerly an important antimalarial drug but now used in the treatment of giardiasis, tapeworm infections, amebiasis, and a variety of other conditions. { 'kwin·ə·krən }

quinalbarbitone *See* secobarbital. { ,kwin·əl'bär·bə,tōn }

quinaldine [ORG CHEM] $C_9H_6NCH_3$ A colorless, oily liquid with a boiling point of 246–247°C; soluble in alcohol, chloroform, and ether; used in medicine as an antimalarial. Also known as chinaldine. { 'kwin·əl,dēn }

quinalizarin [ORG CHEM] $C_{14}H_8O_6$ A red, crystalline compound, soluble in water solutions of alkalies, and in acetic and sulfuric acid; used to dye cottons. { ,kwin·ə'liz·ə·rən }

quinaphthol [PHARM] $C_{43}H_{40}N_2O_{10}S_2$ A yellow, crystalline powder with a melting point of 185–186°C; used in medicine. { kwi'naf,thȯl }

quinary code [COMPUT SCI] A code based on five possible combinations for representing digits. { 'kwī·nə·rē ,kōd }

quince [BOT] *Cydonia oblonga*. A deciduous tree of the order Rosales characterized by crooked branching, leaves that are densely hairy on the underside and solitary white or pale-pink flowers; fruit is an edible pear- or apple-shaped tomentose pome. { kwins }

Quincke tube *See* Herschel-Quincke tube. { 'kviŋ·kə ,tüb }

quinhydrone [ORG CHEM] $C_6H_4O_2 \cdot C_6H_4(OH)_2$ Green, water-soluble powder, subliming at 171°C; a compound of quinone and hydroquinone dissociating in solution. { kwin'hī,drōn }

quinhydrone electrode [ANALY CHEM] A platinum wire in a saturated solution of quinhydrone; used as a reversible electrode standard in pH determinations. { kwin'hī,drōn i'lek,trōd }

quinic acid [ORG CHEM] $C_6H_7(OH)_4COOH \cdot H_2O$ Ether-insoluble, white crystals with acid taste; melts at 162°C; soluble in alcohol, water, and glacial acetic acid; used in medicine. Also known as chinic acid; kinic acid. { 'kwin·ik 'as·əd }

quinidine [ORG CHEM] $C_{20}H_{24}N_2O_2$ A crystalline alkaloid that melts at 174–175°C (345–347°F) and that may be derived from the bark of cinchona used as the salt in medicine. Also known as β-quinine. { 'kwin·ə,dēn }

quinine [ORG CHEM] $C_{20}H_{24}N_2O_2 \cdot 3H_2O$ White powder or crystals, soluble in alcohol, ether, carbon disulfide, chloroform, and glycerol; an alkaloid derived from cinchona bark; used as an antimalarial drug and in beverages. { 'kwī,nīn }

β-quinine *See* quinidine. { ¦bād·ə 'kwī,nīn }

quinoa [BOT] *Chenopodium quinoa*. An annual herb of the family Chenopodiaceae grown at high altitudes in South America for the highly nutritious seeds. { kwi'nō·ə }

quinoidine [ORG CHEM] A brownish-black mass consisting of a mixture of alkaloids which remain in solution after

QUILLWORT

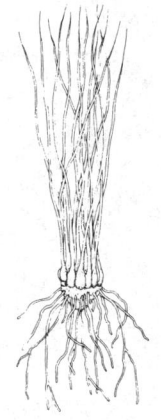

A quillwort (*Isoetes nuttallii*).
(From A. W. Haupt, Plant Morphology, McGraw-Hill, 1953)

extracting crystallized alkaloids from cinchona bark; soluble in dilute acids, alcohol, and chloroform; used in medicine. Also known as chinoidine. { kwi′nō·ə‚dēn }

quinol *See* hydroquinone. { ′kwi‚nȯl }

quinoline [ORG CHEM] C_9H_7N Water-soluble, aromatic nitrogen compound; colorless, hygroscopic liquid; also soluble in alcohol, ether, and carbon disulfide; boils at 238°C; used in medicine and as a chemical intermediate. Also known as chinoline; leucoline; leukol. { ′kwin·ə‚lēn }

quinoline blue *See* cyanine dye. { ′kwin·ə‚lēn ′blü }

8-quinolinol *See* 8-hydroxyquinoline; oxine. { ¦āt kwi′näl· ə‚nȯl }

quinone [ORG CHEM] $CO(CHCH)_2CO$ Yellow crystalline compound with irritating aroma; melts at 116°C; soluble in alcohol, alkalies, and ether; used to make dyes and hydroquinone. Also known as benzoquinone; chinone. { ′kwi‚nōn }

quinoprotein [BIOCHEM] A member of a class of proteins that uses pyrroloquinoline quinone as a cofactor. { ‚kwin· ə′prō‚tēn }

quinoxaline [ORG CHEM] $C_8H_6N_2$ Bicyclic organic base; colorless powder, soluble in water and organic solvents; melts at 30°C; used in organic synthesis. { kwi′näk·sə‚lēn }

N′-2-quinoxalysulfanilimide [ORG CHEM] $C_{14}H_{12}N_4SO_2$ Crystals with a melting point of 247°C; almost insoluble in water; used as a rodenticide. Also known as sulfaquinoxaline. { ¦en‚prīm ¦tü kwi¦näk·sə·lē‚sȯl·fə′nil·ə‚mīd }

quinquefoliate [BOT] Of a leaf, having five leaflets. { ‚kwin·kə′fō·lē‚āt }

quintal *See* hundredweight; metric centner. { ′kwint·əl }

quintic [MATH] A fifth-degree expression. { ′kwin·tik }

quintic equation [MATH] A fifth-degree polynomial equation. { ′kwin·tik i′kwā·zhən }

quintic quantic [MATH] A quantic of the fifth degree. { ′kwin·tik ¦kwän·tik }

quintic surd [MATH] A fifth root of a rational number that is itself an irrational number. { ′kwin·tik ‚sərd }

quintillion [MATH] **1.** The number 10^{18}. **2.** In British and German usage, the number 10^{30}. { kwin′til·yən }

quintozene *See* pentachloronitrobenzene. { ′kwin·tə‚zēn }

quintuplet [BIOL] One of five children who have been born at one birth. { kwin′təp·lət }

QWIP *See* quantum well infrared photodetector.

quire [MATER] Twenty-five sheets of paper, or one-twentieth of a 500-sheet ream. { kwīr }

quirk [BUILD] **1.** An indentation separating one element from another, as between moldings. **2.** A V groove in the finish-coat plaster where it abuts the return on a door or window. { kwərk }

quirk bead [BUILD] **1.** A bead with a quirk on one side only, as on the edge of a board. Also known as bead and quirk. **2.** A bead that is flush with the adjoining surface and separated from it by a quirk on each side. Also known as bead and quirk; double-quirked bead; flush bead; recessed bead. **3.** A bead located at a corner with quirks at either side at right angles

to each other. Also known as bead and quirk; return bead. **4.** A bead with a quirk on its face. Also known as bead and quirk. { kwərk ‚bēd }

quitclaim [MIN ENG] Legal release of a claim, right, title, or interest by one person or estate to another. { ′kwit‚klām }

quiteron [ELECTR] A three-terminal superconducting switch that consists of three superconducting layers separated by two ultrathin insulating layers, and operates by heavy injection of quasiparticles into a common, thin middle electrode to create a nonequilibrium state and cause the properties of the electrode to become those of a normal metal. { ′kwid·ə‚rän }

quitiqua *See* pinta. { ′kēt·ə‚kä }

QWITT diode *See* quantum well injection transit-time diode. { ¦kyü¦dəb·əl‚yü¦i¦tē¦tē ′dī‚ōd }

Q unit [THERMO] A unit of energy, used in measuring the heat energy of fuel reserves, equal to 10^{18} British thermal units, or approximately 1.055×10^{21} joules. { ′kyü ‚yü·nət }

quoin [BUILD] One of the members forming an outside corner or exterior angle of a building, and differentiated from the wall by color, texture, size, or projection. [GRAPHICS] An expandable device used to secure type and printing plates in a chase. { kȯin }

quoin post [CIV ENG] The vertical member at the jointed end of a gate in a navigation lock. { kȯin ‚pōst }

quotient [MATH] The result of dividing one quantity by another. { ′kwō·shənt }

quotient field [MATH] The smallest field containing a given integral domain; obtained by formally introducing all quotients of elements of the integral domain. { ′kwō·shənt ‚fēld }

quotient group [MATH] A group G/H whose elements are the cosets gH of a given normal subgroup H of a given group G, and the group operation is defined as $g_1H \cdot g_2H \equiv (g_1 \cdot g_2)H$. Also known as factor group. { ′kwō·shənt ‚grüp }

quotient ring [MATH] A ring R/I whose elements are the cosets rI of a given ideal I in a given ring R, where the additive and multiplicative operations have the form: $r_1I + r_2I = (r_1 + r_2)I$ and $r_1I \cdot r_2I \equiv (r_1 \cdot r_2)I$. Also known as factor ring; residue class ring. { ′kwō·shənt ‚riŋ }

quotient set [MATH] The set of all the equivalence classes relative to a given equivalence relation on a given set. { ′kwō-shənt ‚set }

quotient space [MATH] The topological space Y which is the set of equivalence classes relative to some given equivalence relation on a given topological space X; the topology of Y is canonically constructed from that of X. Also known as factor space. { ′kwō·shənt ‚spās }

quotient topology [MATH] If X is a topological space, X/R the quotient space by some equivalence relation on X, the quotient topology on X/R is the smallest topology which makes the function which assigns to each element of X its equivalence class in X/R a continuous function. { ′kwō·shənt tə′päl·ə·jē }

Q value *See* disintegration energy. { ′kyü ‚val·yü }

qwerty keyboard [ENG] A keyboard containing the standard arrangement of letters so named after the first letters on the top alphabetic row. { ′kwərd·ē kē‚bȯrd }

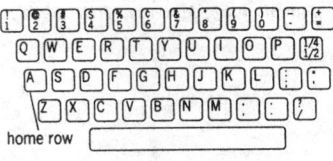

QWERTY KEYBOARD

home row

Layout of the qwerty keyboard.

R

r *See* roentgen.

R *See* roentgen.

r$_s$ *See* Spearman's rank correlation coefficient.

Ra *See* radium.

R.A. *See* right ascension.

Raabe's convergence test [MATH] An infinite series with positive terms a_n where, for each n, $a_{n+1}/a_n = 1/(1 + b_n)$ will converge if, after a certain term, nb_n always exceeds a fixed number greater than 1 and will diverge if nb_n always is less than a fixed number less than or equal to 1. { 'räb·əz kən'vər·jəns ,test }

rabal [METEOROL] A method of winds-aloft observation, that is, the determination of wind speeds and directions in the atmosphere above a station; it is accomplished by recording the elevation and azimuth angles of the balloon at specified time intervals while visually tracking a radiosonde balloon with the theodolite. { 'rā,bäl }

rabbet [ENG] **1.** A groove cut into a part. **2.** A strip applied to a part as, for example, a stop or seal. **3.** A joint formed by fitting one member into a groove, channel, or recess in the face or edge of a second member. { 'rab·ət }

rabbet plane [DES ENG] A plane with the blade extending to the outer edge of one side that is open. { 'rab·ət ,plān }

rabbit [NUCLEO] A small container that is propelled, usually pneumatically or hydraulically, through a tube into a nuclear reactor; used to expose samples to the radiation, especially neutron flux, then remove them rapidly for measurements of radioactive atoms having short half-lives. Also known as shuttle. [PETRO ENG] A small plug driven by pressure through a flow line to clean the line or to check that it is unobstructed. [VERT ZOO] Any of a large number of burrowing mammals in the family Leporidae. { 'rab·ət }

rabbit ear [MET] A recess in the corner of a die allowing wrinkling or folding of the blank. { 'rab·ət ,ir }

rabbit sequence [MATH] A sequence of binary numbers that is recursively generated by the rules 0 → 1 (young rabbits grow old) and 1 → 10 (old rabbits stay old and beget young ones); beginning with a single 1, successive generations are 1, 10, 101, 10110, 10110101, and so forth. { 'rab·ət ,sē·kwəns }

rabble [MET] An iron bar for skimming the bath in a smelting or refining furnace or for stirring the ore in a roasting furnace either manually or mechanically. { 'rab·əl }

rabbling [ENG] Stirring a molten charge, as of metal or ore. { 'rab·liŋ }

rabies [VET MED] An acute, encephalitic viral infection transmitted to humans by the bite of a rabid animal. Also known as hydrophobia. { 'rā·bēz }

Rabi oscillation [ATOM PHYS] The periodic exchange of energy between atoms in an electromagnetic field and a single mode of the field. { 'rä·bē ,äs·ə'lā·shən }

Racah coefficient [QUANT MECH] A coefficient that appears in the transformation between the modes of coupling eigenfunctions of three angular momenta; they differ only by, at most, a sign from the six-j coefficients. Also known as W coefficient. { 'rä·kä ,kō·i'fish·ənt }

raccoon [VERT ZOO] Any of 16 species of carnivorous nocturnal mammals belonging to the family Procyonidae; all are arboreal or semiarboreal and have a bushy, long ringed tail. { ra'kün }

race [ANTHRO] **1.** A distinctive human type possessing characteristic traits that are transmissible by descent. **2.** Descendants of a common ancestor. [BIOL] **1.** An infraspecific taxonomic group of organisms, such as subspecies or microspecies. **2.** A fixed variety or breed. [DES ENG] Either of the concentric pair of steel rings of a ball bearing or roller bearing. [ENG] A channel transporting water to or away from hydraulic machinery, as in a power house. [OCEANOGR] A rapid current, or a constricted channel in which such a current flows; the term is usually used only in connection with a tidal current, which may be called a tide race. { rās }

race condition [ELEC] An ambiguous condition occurring in control counters when one flip-flop changes to its next state before a second one has had sufficient time to latch. { 'rās kən,dish·ən }

racemase [BIOCHEM] Any of group of enzymes that catalyze racemization reactions. { 'ras·ə,mās }

racemate [ORG CHEM] An equimolar mixture of the two enantiomers (+ and −, or R and S) of a substance; it is optically inactive. { 'ras·ə,māt }

raceme [BOT] An inflorescence on which flowers are borne on stalks of equal length on an unbranched main stalk that continues to grow during flowering. { rā'sēm }

racemic acid [ORG CHEM] $C_2H_4O_2(COOH)_2 \cdot H_2O$ Colorless crystals, melting at 205°C; soluble in water, slightly soluble in alcohol; used as a chemical intermediate. Also known as inactive tartaric acid. { rə'sēm·ik 'as·əd }

racemic compound [ORG CHEM] Crystals containing an equimolar, random and ordered mixture of enantiomers (heterochiral crystals). { rə,sēm·ik 'käm,paund }

racemic conglomerate [ORG CHEM] Spontaneous resolution of a racemate, through crystallization, into a mixture of pure enantiomers (homochiral crystals). { rə,sēm·ik kən'gläm·ə·rət }

racemic mixture [ORG CHEM] According to the IUPAC, this usage is strongly discouraged; racemate is preferred. { rə,sēm·ik 'miks·chər }

racemic modification *See* racemic mixture. { rā,sēm·ik ,mäd·ə·fə'kā·shən }

racemization [ORG CHEM] A process by which an optically active form of a substance is converted into a racemic mixture. { ,rā·sə·mə'zā·shən }

racemose [ANAT] Of a gland, compound and shaped like a bunch of grapes, with freely branching ducts that terminate in acini. [BOT] Bearing, or occurring in the form of, a raceme. { 'ras·ə,mōs }

race track [NUCLEO] An assembly of several Calutron isotope separators in the shape of a race track, having a common magnetic field. Also known as track. { 'rās ,trak }

raceway [ELEC] A channel used to hold and protect wires, cables, or busbars. Also known as electric raceway. { 'rās,wā }

rachiglossate radula [INV ZOO] A radula of certain gastropod mollusks which has one or three longitudinal series of teeth, each of which may bear many cusps. { ,rā·kə'glä,sāt 'raj·ə·lə }

rachilla [BOT] The axis of a grass spikelet. { rə'kil·ə }

rachis [ANAT] The vertebral column. [BIOL] An axial structure such as the axis of an inflorescence, the central petiole of a compound leaf, or the central cord of an ovary in Nematoda. { 'rā·kəs }

rack [AERO ENG] A suspension device permanently fixed to an aircraft; it is designed for attaching, arming, and releasing one or more bombs; it may also be utilized to accommodate other items such as mines, rockets, torpedoes, fuel tanks, rescue

---

RABBIT

The snowshoe hare (*Lepus americanus*).

RACCOON

A semiarboreal raccoon.

RACHIGLOSSATE RADULA

Rachiglossate radula, from *Murex* (marine gastropod). *(From R. R. Shrock and W. H. Twenhofel, Principles of Invertebrate Paleontology, 2d ed., McGraw-Hill, 1953)*

equipment, sonobuoys, and flares. [CIV ENG] A fixed screen composed of parallel bars placed in a waterway to catch debris. [DES ENG] *See* relay rack. [ENG] A frame for holding or displaying articles. [MECH ENG] A bar containing teeth on one face for meshing with a gear. [MIN ENG] An inclined trough or table for washing or separating ore. { rak }

rack and pinion [MECH ENG] A gear arrangement consisting of a toothed bar that meshes with a pinion. { 'rak ən 'pin·yən }

rack-and-pinion steering [MECH ENG] A steering system in which the rotation of pinion gear at the end of the steering column moves a toothed bar (the rack) left or right to transmit steering movements. { 'rak ən ¦pin·yən 'stir·iŋ }

rack and snail [HOROL] A mechanism in a striking timepiece that permits the hands to be advanced without waiting for the striking to be completed. { 'rak ən 'snāl }

racking [CIV ENG] Setting back the end of each course of brick or stone from the end of the preceding course. [MET] Suspending work from a frame that holds and conducts current to one or more electrodes for electroplating and other electrochemical operations. [PETRO ENG] During oil well drilling, placing stands of pipe in an orderly fashion in the derrick while hoisting pipe out of the well bore. [TEXT] Lateral movements of the needles or needle bed of a knitting machine in order to compact a fabric and enable it to hold its shape. { 'rak·iŋ }

rack panel [ELECTR] A panel designed for mounting on a relay rack; its width is 19 inches (48.26 centimeters), height is a multiple of $1^3/_4$ inches (4.445 centimeters), and the mounting notches are standardized as to size and position. { 'rak ¦pan·əl }

rack railway [CIV ENG] A railway with a rack between the rails which engages a gear on the locomotive; used on steep grades. { 'rak 'rāl¦wā }

racon *See* radar beacon. { 'rā¦kän }

rad [NUCLEO] A special unit of absorbed dose, equal to energy absorption of 100 ergs per gram (0.01 joule per kilogram); equal to 0.01 gray. { rad }

radappertization [FOOD ENG] The use of radiation for sterilizing foods. { rā·dap·ərd·ə'zā·shən }

radar [ENG] **1.** A system using beamed and reflected radio-frequency energy for detecting and locating objects, measuring distance or altitude, navigating, homing, bombing, and other purposes; in detecting and ranging, the time interval between transmission of the energy and reception of the reflected energy establishes the range of an object in the beam's path. Derived from radio detection and ranging. **2.** *See* radar set. { 'rā¦där }

radar-absorbing material [MATER] A material that is designed to reduce the reflection of electromagnetic radiation by a conducting surface in the frequency range from approximately 100 megahertz to 100 gigahertz. { 'rā¦där əb¦sòrb·iŋ mə¦tir·ē·əl }

radar advisory [COMMUN] The term used to indicate that the provision of advice and information is based on radar observation. { 'rā¦där id'vīz·ə·rē }

radar altimeter [NAV] A radio altimeter, useful at altitudes much greater than the 5000-foot (1500-meter) limit of frequency-modulated radio altimeters, in which simple pulse-type radar equipment is used to send a pulse straight down from an aircraft and to measure its total time of travel to the surface and back to the aircraft. Also known as high-altitude radio altimeter; pulse-type altimeter. { 'rā¦där al'tim·əd·ər }

radar altimetry area [NAV] A large and comparatively level terrain area with a defined elevation which can be used in determining the altitude of airborne equipment by the use of radar. { 'rā¦där al'tim·ə·trē ¦er·ē·ə }

radar altitude [NAV] The altitude of an aircraft or spacecraft as determined by a radio altimeter; thus, the actual vertical distance from the terrain. Also known as radio altitude. { 'rā¦där 'al·tə¦tüd }

radar and television aid to navigation [NAV] Device which converts a circular scan radar presentation to a horizontally scanned television presentation; it provides a continuous bright display with target trails for course and speed indications of moving targets. { 'rā¦där ən 'tel·ə¦vizh·ən 'ād tə ¦nav·ə'gā·shən }

radar antenna [ELECTROMAG] A device which radiates radio-frequency energy in a radar system, concentrating the transmitted power in the direction of the target, and which

provides a large area to collect the echo power of the returning wave. { 'rā¦där an'ten·ə }

radar antijamming [ELECTR] Measures taken to counteract radar jamming. { 'rā¦där ¦ant·i'jam·iŋ }

radar approach control [NAV] A facility providing radar approach control service by use of airport surveillance radar and precision approach radar equipment. { 'rā¦där ə'prōch kən¦trōl }

radar astronomy [ASTRON] The study of astronomical bodies and the earth's atmosphere by means of radar pulse techniques, including tracking of meteors and the reflection of radar pulses from the moon and the planets. { 'rā¦där ə'strän·ə·mē }

radar attenuation [ELECTROMAG] Ratio of the power delivered by the transmitter to the transmission line connecting it with the transmitting antenna, to the power reflected from the target which is delivered to the receiver by the transmission line connecting it with the receiving antenna. { 'rā¦där ə¦ten·yə'wā·shən }

radar beacon [NAV] A radar receiver-transmitter that transmits a strong coded radar signal whenever its radar receiver is triggered by an interrogating radar on an aircraft or ship; the coded beacon reply can be used by the navigator to determine his own position in terms of bearing and range from the beacon. Also known as racon; radar transponder. { 'rā¦där ¦bē·kən }

radar beam [ELECTROMAG] The movable beam of radio-frequency energy produced by a radar transmitting antenna; its shape is commonly defined as the loci of all points at which the power has decreased to one-half of that at the center of the beam. { 'rā¦där ¦bēm }

radar bearing [NAV] A bearing obtained by radar. { 'rā¦där ¦ber·iŋ }

radar bombing [ORD] Bombing in which radar is used to locate the target or aiming point, to aid in positioning the bombing aircraft at the proper release point for bombing, or to release bombs automatically, especially under conditions of poor visibility. { 'rā¦där ¦bäm·iŋ }

radar bombsight [ENG] An airborne radar set used to sight the target, solve the bombing problem, and drop bombs. { 'rā¦där ¦bäm¦sīt }

radar buoy [NAV] A buoy having corner reflectors designed into the superstructure, but maintaining the characteristic shape of the buoy in order to differentiate a radar buoy from a buoy on which a corner reflector is mounted. { 'rā¦där ¦bói }

radar camera [OPTICS] A special manual or automatic camera used to photograph images on a radarscope. Also known as radarscope camera. { 'rā¦där ¦kam·rə }

radar camouflage [ORD] The use of special coverings or surfaces that reduce the reflection of radio-frequency energy back to a radar set, to minimize chances for detection of the object by an enemy radar set. { 'rā¦där ¦kam·ə'flazh }

radar cell [ELECTROMAG] Volume whose dimensions are one radar pulse length by one radar beam width. { 'rā¦där ¦sel }

radar chart [NAV] A special map used in radar navigation that emphasizes objects which give prominent radar echoes. { 'rā¦där ¦chärt }

radar climatology [CLIMATOL] The statistics in time and space of radar weather echoes. { 'rā¦där ¦klī·mə'täl·ə·jē }

radar clutter *See* clutter. { 'rā¦där ¦kləd·ər }

radar command guidance [ENG] A missile guidance system in which radar equipment at the launching site determines the positions of both target and missile continuously, computes the missile course corrections required, and transmits these by radio to the missile as commands. { 'rā¦där kə'mand ¦gīd·əns }

radar confusion reflector *See* confusion reflector. { 'rā¦där kən'fyü·zhən ri¦flek·tər }

radar conspicuous object [ELECTROMAG] An object which returns a strong radar echo. { 'rā¦där kən'spik·yə·wəs 'äb·jekt }

radar constant [ELECTR] One of those terms of the radar equation or radar storm-detection equation which are functions of the particular radar to which the equations are applied; these include peak power, antenna gain or aperture, beam width, pulse length, pulse repetition frequency, wavelength, polarization, and noise level of the receiver. { 'rā¦där ¦kän·stənt }

radar contact [ENG] Recognition and identification of an echo on a radar screen; an aircraft is said to be on radar contact when its radar echo can be seen and identified on a PPI (plan-position indicator) display. { 'rā‚där ‚kän‚takt }

radar control [ELECTR] Guidance, direction, or employment exercised over an aircraft, guided missile, gun battery, or the like, by means of, or with the aid of, radar. { 'rā‚där kən‚trōl }

radar controller [NAV] An air-traffic controller or other responsible person proficient in the use and interpretation of radar and capable of performing one or more of the following functions: surveillance controller, traffic director, or final controller. { 'rā‚där kən‚trōl‚ər }

radar countermeasure [ELECTR] An electronic countermeasure used against enemy radar, such as jamming and confusion reflectors. Abbreviated RCM. { 'rā‚där 'kaunt‚ər‚mezh‚ər }

radar coverage [ENG] The limits within which objects can be detected by one or more radar stations. { 'rā‚där ‚kəv‚rij }

radar coverage indicator [ENG] Device that shows how far a given aircraft should be tracked by a radar station, and also provides a reference (detection) range for quality control; takes into account aircraft size, altitude, screening angle, site elevation, type radar, antenna radiation pattern, and antenna tilt. { 'rā‚där ‚kəv‚rij ‚in‚də‚kād‚ər }

radar cross section *See* echo area. { 'rā‚där ‚krös ‚sek‚shən }

radar data [NAV] Data originating from a radar facility as distinct from air-traffic control data facility or other source. [ORD] Either long-range or gap-filler radar information modified to a form which a computer can use; the two types of data are as follows: correlated, associated with a track; and uncorrelated, not associated with a track. { 'rā‚där ‚dad‚ə }

radar data filtering [ELECTR] Quality analysis process that causes the computer to reject certain radar data and to alert personnel of mapping and surveillance consoles to the rejection. { 'rā‚där 'dad‚ə ‚fil‚triŋ }

radar deception [ORD] Radiation, reradiation, alteration, absorption, or reflection of electromagnetic energy in a manner intended to mislead an enemy in interpretation or use of information received by his radar systems. { 'rā‚där di‚sep‚shən }

radar decoy [ORD] A reflecting object used in radar deception, having the same reflective characteristics as the target. { 'rā‚där 'dē‚koi }

radar display [ELECTR] The pattern representing the output data of a radar set, generally produced on the screen of a cathode-ray tube. Also known as presentation; radar presentation. { 'rā‚där di‚splā }

radar distribution switchboard [ELECTR] Switching panel for connecting video, trigger, and bearing from any one of five systems, to any or all of 20 repeaters; also contains order lights, bearing cutouts, alarms, test equipment, and so forth. { 'rā‚där ‚dis‚trə'byü‚shən ‚swich‚bórd }

radar dome [ENG] Weatherproof cover for a primary radiating element of a radar or radio device which is transparent to radio-frequency energy, and which permits active operation of the radiating element, including mechanical rotation or other movement as applicable. { 'rā‚där ‚dōm }

radar echo *See* echo. { 'rā‚där ‚ek‚ō }

radar equation [ELECTROMAG] An equation that relates the transmitted and received powers and antenna gains of a primary radar system to the echo area and distance of the radar target. { 'rā‚där i‚kwā‚zhən }

radar fire [ORD] Gunfire aimed at a target which is tracked by radar. { 'rā‚där ‚fīr }

radar fire control [ORD] Fire control by means of radar. { 'rā‚där 'fīr kən‚trōl }

radar fix [NAV] A determination of position by means of radar. { 'rā‚där ‚fiks }

radar frequency band [ELECTROMAG] A frequency band of microwave radiation in which radar operates. { 'rā‚där 'frē‚kwən‚sē ‚band }

radar gun-layer [ENG] A radar device which tracks a target and aims a gun or guns automatically. { 'rā‚där 'gən ‚lā‚ər }

radar homing [ENG] Homing in which a missile-borne radar locks onto a target and guides the missile to that target. [NAV] Homing on the source of a radar beam. { 'rā‚där ‚hōm‚iŋ }

radar horizon [NAV] The distance to which a radar's operation is limited by the quasi-optical characteristics of the radio waves employed. { 'rā‚där hə‚rīz‚ən }

radar image [ELECTR] The image of an object which is produced on a radar screen. { 'rā‚där ‚im‚ij }

radar indicator [ELECTR] A cathode-ray tube and associated equipment used to provide a visual indication of the echo signals picked up by a radar set. { 'rā‚där ‚in‚də‚kād‚ər }

radar intelligence [COMMUN] **1.** Intelligence concerning radar or intelligence derived from the use of radar equipment. **2.** Organization or activity that deals with such intelligence. { 'rā‚där in‚tel‚ə‚jəns }

radar intelligence item [ELECTR] A feature which is radar significant but which cannot be identified exactly at the moment of its appearance as homogeneous. { 'rā‚där in‚tel‚ə‚jəns ‚īd‚əm }

radar interferometry [GEOPHYS] A microwave remote sensing method for combining imagery collected over time by radar systems on board airplane or satellite platforms to map the elevations, movements, and changes of the earth's surface. Such detectable changes include earthquakes, volcanoes, glaciers, landslides, and underground explosions, as well as fires, floods, forestry operations, moisture changes, and vegetation growth. { 'rā‚där ‚in‚tər‚fə'räm‚ə‚trē }

radar jamming [ELECTR] Radiation, reradiation, or reflection of electromagnetic waves so as to impair the usefulness of radar used by the enemy. { 'rā‚där ‚jam‚iŋ }

radar line of position [NAV] A line of position determined by radar. { 'rā‚där 'līn əv pə'zish‚ən }

radar marker [ENG] A fixed facility which continuously emits a radar signal so that a bearing indication appears on a radar display. { 'rā‚där ‚mär‚kər }

radar meteorological observation [METEOROL] Evaluation of the echoes appearing on the indicator of a weather radar, in terms of orientation, coverage, intensity, tendency of intensity, height, movement, and unique characteristics of echoes, that may be indicative of certain types of severe storms (such as hurricanes, tornadoes, or thunderstorms) and of anomalous propagation. Also known as radar weather observation. { 'rā‚där ‚mēd‚ē‚ə‚rə'läj‚ə‚kəl ‚äb‚sər'vā‚shən }

radar meteorology [METEOROL] The study of the scattering of radar waves by all types of atmospheric phenomena and the use of radar for making weather observations and forecasts. { 'rā‚där ‚mēd‚ē‚ə'räl‚ə‚jē }

radar mile [ELECTROMAG] The time for a radar pulse to travel from the radar to a target 1 mile (1.61 kilometers) distant and return, equal to 10.75 microseconds. { 'rā‚där 'mīl }

radar nautical mile [ELECTROMAG] The time interval of approximately 12.355 microseconds that is required for the radio-frequency energy to travel 1 radar pulse to travel 1 nautical mile (1852 meters) and return. { 'rā‚där 'nód‚ə‚kəl ‚mīl }

radar netting [ENG] The linking of several radars to a single center to provide integrated target information. { 'rā‚där ‚ned‚iŋ }

radar netting station [ENG] A center which can receive data from radar tracking stations and exchange these data among other radar tracking stations, thus forming a radar netting system. { 'rā‚där 'ned‚iŋ ‚stā‚shən }

radar netting unit [ELECTR] Optional electronic equipment that converts the operations central of certain air defense fire distribution systems to a radar netting station. { 'rā‚där 'ned‚iŋ ‚yü‚nət }

radar paint [MATER] A coating that absorbs radar waves. { 'rā‚där ‚pānt }

radar photography [GRAPHICS] Making a photograph of a radar display. { 'rā‚där fə‚täg‚rə‚fē }

radar picket [ENG] A ship or aircraft equipped with early-warning radar and operating at a distance from the area being protected, to extend the range of radar detection. { 'rā‚där ‚pik‚ət }

radar picket escort ship [NAV ARCH] Any of the escort ships modified to give an increased combat information center, electronic countermeasures, and electronic search facilities. { 'rā‚där ‚pik‚ət 'es‚kórt ‚ship }

radar prediction [ENG] A graphic portrayal of the estimated radar intensity, persistence, and shape of the cultural and natural features of a specific area. { 'rā‚där pri'dik‚shən }

radar presentation *See* radar display. { 'rā‚där ‚prē‚zen'tā‚shən }

radar pulse [ELECTROMAG] Radio-frequency radiation emitted with high power by a pulse radar installation for a period of time which is brief compared to the interval between such pulses. { 'rā‚där ‚pəls }

radar range [ELECTROMAG] The maximum distance at

which a radar set is ordinarily effective in detecting objects. { 'rā‚där ‚rānj }

radar range equation [ELECTROMAG] An equation which expresses radar range in terms of transmitted power, minimum detectable signal, antenna gain, and the target's radar cross section. { 'rā‚där ‚rānj i‚kwā·zhən }

radar range marker See distance marker. { 'rā‚där 'rānj ‚mär·kər }

radar receiver [ELECTR] A high-sensitivity radio receiver that is designed to amplify and demodulate radar echo signals and feed them to a radarscope or other indicator. { 'rā‚där ri‚sēv·ər }

radar receiver-transmitter [ELECTR] A single component having the dual functions of generating electromagnetic energy for transmission, and of receiving, demodulating, and sometimes presenting intelligence from the reflected electromagnetic energy. { 'rā‚där ri‚sēv·ər tranz'mid·ər }

radar reconnaissance [ORD] Reconnaissance by means of radar to determine the location, disposition, and strength of enemy forces, and to determine the nature of the terrain. { 'rā‚där ri'kän·ə·səns }

radar reflection [ELECTROMAG] The return of electromagnetic waves, generated by a radar installation, from an object on which the waves are incident. { 'rā‚där ri‚flek·shən }

radar reflection interval [ELECTROMAG] The time required for a radar pulse to travel from the source to the target and return to the source, taking the velocity of radio propagation to be equal to the velocity of light. { 'rā‚där ri‚flek·shən ‚in·tər·vəl }

radar reflectivity [ELECTROMAG] The fraction of electromagnetic energy generated by a radar installation which is reflected by an object. { 'rā‚där ‚rē‚flek'tiv·əd·ē }

radar reflector [ELECTROMAG] A device that reflects or deflects radar waves. { 'rā‚där rē‚flek·tər }

radar relay [ENG] **1.** Equipment for relaying the radar video and appropriate synchronizing signal to a remote location. **2.** Process or system by which radar echoes and synchronization data are transmitted from a search radar installation to a receiver at a remote point. { 'rā‚där 'rē‚lā }

radar repeater [ELECTR] A cathode-ray indicator used to reproduce the visible intelligence of a radar display at a remote position; when used with a selector switch, the visible intelligence of any one of several radar systems can be reproduced. { 'rā‚där ri‚pēd·ər }

radar report [METEOROL] The encoded and transmitted report of a radar meteorological observation; these reports usually give the azimuth, distance, altitude, intensity, shape and movement, and other characteristics of precipitation echoes observed by the radar. Also known as rain area report. Abbreviated RAREP. { 'rā‚där ri‚pȯrt }

radar return [NAV] The signal indication of an object which has reflected energy that was transmitted by a primary radar. Also known as radio echo. { 'rā‚där ri‚tərn }

radar scanning [ENG] The process or action of directing a radar beam through a space search pattern for the purpose of locating a target. { 'rā‚där ‚skan·iŋ }

radarscope [ELECTR] Cathode-ray tube, serving as an oscilloscope, the face of which is the radar viewing screen. Also known as scope. { 'rā‚där‚skōp }

radarscope camera See radar camera. { 'rā‚där‚skōp ‚kam·rə }

radarscope overlay [ENG] A transparent overlay placed on a radarscope for comparison and identification of radar returns. { 'rā‚där‚skōp 'ō·vər‚lā }

radar selector switch [ELECTR] Manual or motor-driven switch which transfers a plan-position indicator repeater from one system to another, switching video, trigger, and bearing data. { 'rā‚där si'lek·tər ‚swich }

radar set [ENG] A complete assembly of radar equipment for detecting and ranging, consisting essentially of a transmitter, antenna, receiver, and indicator. Also known as radar. { 'rā‚där ‚set }

radar shadow [ELECTROMAG] A region shielded from radar illumination by an intervening reflecting or absorbing medium such as a hill. { 'rā‚där ‚shad·ō }

radar signal film [GRAPHICS] The film on which is recorded all the reflected signals acquired by a coherent radar, and which must be viewed or processed through an optical correlator to permit interpretation. { 'rā‚där ‚sig·nəl ‚film }

radar signal spectrograph [ELECTR] An electronic device in the form of a scanning filter which provides a frequency analysis of the amplitude-modulated back-scattered signal. { 'rā‚där ¦sig·nəl 'spek·trə‚graf }

radar silence [ORD] A period of time during which radar transmission is stopped, generally for security reasons. { 'rā‚där ‚sī·ləns }

radarsonde [ENG] **1.** An electronic system for automatically measuring and transmitting high-altitude meteorological data from a balloon, kite, or rocket by pulse-modulated radio waves when triggered by a radar signal. **2.** A system in which radar techniques are used to determine the range, elevation, and azimuth of a radar target carried aloft by a radiosonde. { 'rā‚där‚sänd }

radar station [ENG] The place, position, or location from which, or at which, a radar set transmits or receives signals. { 'rā‚där ‚stā·shən }

radar station pointer [NAV] A transparent plotting chart inscribed with radial lines from 0° to 360°; used to plot radar echoes to the scale of the chart in use; it is also used to assist in identifying radar responses with charted features. { 'rā‚där ‚stā·shən ‚pȯint·ər }

radar storm detection [METEOROL] The detection of certain storms or stormy conditions by means of radar; liquid or frozen water drops within the storm reflect radar echoes. { 'rā‚där 'stȯrm di‚tek·shən }

radar storm-detection equation [METEOROL] The equation which relates the variables involved in the radar detection of precipitation. { 'rā‚där 'stȯrm di‚tek·shən i‚kwā·zhən }

radar surveying [ENG] Surveying in which airborne radar is used to measure accurately the distance between two ground radio beacons positioned along a baseline; this eliminates the need for measuring distance along the baseline in inaccessible or extremely rough terrain. { 'rā‚där sər'vā·iŋ }

radar target [ELECTROMAG] An object belonging to a desired class which reflects back a signal sufficient to produce a fluorescent mark on the radar screen. { 'rā‚där ‚tär·gət }

radar telescope [ENG] A large radar antenna and associated equipment used for radar astronomy. { 'rā‚där 'tel·ə‚skōp }

radar theodolite [ENG] A theodolite that uses radar to obtain azimuth, elevation, and slant range to a reflecting target, for surveying or other purposes. { 'rā‚där thē'äd·əl‚īt }

radar threshold limit [ENG] For a given radar and specified target, the point in space relative to the focal point of the antenna at which initial detection criteria can be satisfied. { 'rā‚där 'thresh‚hōld ‚lim·ət }

radar tracking [ENG] Tracking a moving object by means of radar. { 'rā‚där ‚trak·iŋ }

radar tracking information [NAV] Information issued to alert an aircraft to any radar targets observed on the controller's radar display which may be in a dangerous situation. { 'rā‚där ¦trak·iŋ ‚in·fər‚mā·shən }

radar tracking station [ENG] A radar facility which has the capability of tracking moving targets. { 'rā‚där 'trak·iŋ ‚stā·shən }

radar transmitter [ELECTR] The transmitter portion of a radar set. { 'rā‚där tranz‚mid·ər }

radar transponder See radar beacon. { 'rā‚där tranz'pän·dər }

radar triangulation [ENG] A radar system of locating targets, usually aircraft, in which two or more separate radars are employed to measure range only; the target is located by automatic trigonometric solution of the triangle composed of a pair of radars and the target in which all three sides are known. { 'rā‚där trī‚aŋ·gyə'lā·shən }

radar upper band See upper bright band. { 'rā‚där 'əp·ər ‚band }

radar volume [ELECTROMAG] The volume in space that is irradiated by a given radar; for a continuous-wave radar it is equivalent to the antenna radiation pattern; for a pulse radar it is a function of the cross-section area of the beam of the antenna and the pulse length of the transmitted pulse. { 'rā‚där ‚väl·yəm }

radar weather observation See radar meteorological observation. { 'rā‚där 'weth·ər ‚äb·zər‚vā·shən }

radar wind [METEOROL] Wind of which the movement, speed, and direction is observed or determined by a radar tracking of a balloon carrying a radiosonde, a radio transmitter, or a radar reflector. { 'rā‚där ‚wind }

radar wind system [ENG] Apparatus in which radar techniques are used to determine the range, elevation, and azimuth of a balloon-borne target, and hence to compute upper-air wind data. { 'rā‚där ¦wind ‚sis·təm }

radechon [ELECTR] A storage tube having a single electron gun and a dielectric storage medium consisting of a sheet of mica sandwiched between a continuous metal backing plate and a fine-mesh screen; used in simple delay schemes, signal-to-noise improvement, signal comparison, and conversion of signal-time bases. Also known as barrier-grid storage tube. { 'rad·ə‚kän }

Rademacher functions [MATH] The functions f_n, with $n = 1, 2, 3, \ldots$, defined on the closed interval $[0, 1]$ by the equation $f_n(x) = \text{sgn}\ [\sin\ (2^n\pi x)]$, where sgn represents the signum function and sin represents the sine function. { 'räd·ə‚mäk·ər ‚fəŋk·shənz }

radiac [NUCLEO] Detection, identification, and measurement of the intensity of nuclear radiation in an area. Derived from radioactivity detection, identification, and computation. { 'rā·dē‚ak }

radiac instrument See radiac set. { 'rā·dē‚ak ‚in·strə·mənt }

radiacmeter See radiac set. { 'rā·dē‚ak‚mēd·ər }

radiac set [NUCLEO] A complete system for detecting, identifying, and measuring radioactivity. Also known as radiac instrument; radiacmeter. { 'rā·dē‚ak ‚set }

radial [GRAPHICS] A line or direction from the center (principal point, isocenter, nadir point, or substitute center) to any point on a photograph. [MATH] For a plane curve C, the locus of end points of lines, drawn from a fixed point, that are equal and parallel to the radius of curvature of C. [NAV] One of a number of radial lines of position defined by an azimuthal radio navigation facility and identified in terms of the bearing (usually magnetic) of all points on that line from the facility. [SCI TECH] Directed or diverging from a center. { 'rād·ē·əl }

radial acceleration See centripetal acceleration. { 'rād·ē·əl ak‚sel·ə'rā·shən }

radial artery [ANAT] A branch of the brachial artery in the forearm; principal branches are the radial recurrent and the main artery of the thumb. { 'rād·ē·əl 'ärd·ə·rē }

radial assumption [MAP] The assumption, forming the basis of planimetric mapping by photogrammetric methods, that on a nearly vertical photograph all displacements resulting from tilt and relief are radial from the photograph's principal point. { 'rād·ē·əl ə‚səm·shən }

radial astigmatism [OPTICS] Astigmatism which affects the imaging of points that lie off the axis of an optical system, due to oblique incidence of rays from these points. Also known as oblique astigmatism. { 'rād·ē·əl ə'stig·mə‚tiz·əm }

radial band pressure [MECH] The pressure which is exerted on the rotating band by the walls of the gun tube, and hence against the projectile wall at the band seat, as a result of the engraving of the band by the gun rifling. { 'rād·ē·əl ¦band ‚presh·ər }

radial-beam tube [ELECTR] A vacuum tube in which a radial beam of electrons is rotated past circumferentially arranged anodes by an external rotating magnetic field; used chiefly as a high-speed switching tube or commutator. { 'rād·ē·əl ¦bēm ‚tüb }

radial bearing [MECH ENG] A bearing with rolling contact in which the direction of action of the load transmitted is radial to the axis of the shaft. { 'rād·ē·əl 'ber·iŋ }

radial canal [INV ZOO] **1.** One of the numerous canals that radiate from the spongocoel in certain Porifera. **2.** Any of the canals extending from the coelenteron to the circular canal in the margin of the umbrella in jellyfishes. **3.** A canal radiating from the circumoral canal along each ambulacral area in many echinoderms. { 'rād·ē·əl kə'nal }

radial chromatography [ANALY CHEM] A circular disk of absorbent paper which has a strip (wick) cut from edge to center to dip into a solvent; the solvent climbs the wick, touches the sample, and resolves it into concentric rings (the chromatogram). Also known as circular chromatography; radial paper chromatography. { 'rād·ē·əl ‚krō·mə'täg·rə·fē }

radial cleavage [EMBRYO] A cleavage pattern characterized by formation of a mass of cells that show radial symmetry. { 'rād·ē·əl 'klē·vij }

radial deviation [BIOPHYS] A position of the human hand in which the wrist is bent toward the thumb. { ¦rād·ē·əl ‚dē·vē'ā·shən }

radial displacement [GRAPHICS] On vertical photographs, the apparent "leaning out," or the apparent displacement of the top of any object having height in relation to its base. { 'rād·ē·əl di'splās·mənt }

radial distribution function [MATH] A function $F(r)$ equal to the average of a given function of the three coordinates over a sphere of radius r centered at the origin of the coordinate system. [PHYS CHEM] A function $\rho(r)$ equal to the average over all directions of the number density of molecules at distance r from a given molecule in a liquid. { 'rād·ē·əl ‚dis·trə'byü·shən ‚fəŋk·shən }

radial Doppler effect [ELECTROMAG] The part of the optical Doppler effect which depends on the direction of the relative velocity of source and observer, and is analogous to the acoustical Doppler effect, in contrast to the transverse Doppler effect. { 'rād·ē·əl 'däp·lər ə‚fekt }

radial drainage pattern [GEOL] A drainage pattern characterized by radiating streams diverging from a high central area. Also known as centrifugal drainage pattern. { 'rād·ē·əl 'drān·ij ‚pad·ərn }

radial draw forming [MECH ENG] A metal-forming method in which tangential stretch and radial compression are applied gradually and simultaneously. { 'rād·ē·əl 'drò ‚fòrm·iŋ }

radial drill [MECH ENG] A drilling machine in which the drill spindle can be moved along a horizontal arm which itself can be rotated about a vertical pillar. { 'rād·ē·əl 'dril }

radial drilling [ENG] The drilling of several holes in one plane, all radiating from a common point. { 'rād·ē·əl 'dril·iŋ }

radial engine [MECH ENG] An engine characterized by radially arranged cylinders at equiangular intervals around the crankshaft. { 'rād·ē·əl 'en·jən }

radial error [ORD] An error associated with delivery of munitions on a target; it is the distance between the desired point of impact and actual point of impact, both points projected and measured on an imaginary plane drawn perpendicular to the flight path of the munition. { 'rād·ē·əl 'er·ər }

radial faults [GEOL] Faults arranged like the spokes of a wheel, radiating from a central point. { 'rād·ē·əl 'fòls }

radial-flow [ENG] Having the fluid working substance flowing along the radii of a rotating tank. [PETRO ENG] Pertaining to spokelike flow of reservoir fluids radially inward toward a wellbore focal area. { 'rād·ē·əl ¦flō }

radial-flow turbine [MECH ENG] A turbine in which the gases flow primarily in a radial direction. { 'rād·ē·əl ¦flō ¦ər·bən }

radial force [MECH ENG] In machining, the force acting on the cutting tool in a direction opposite to depth of cut. { 'rād·ē·əl 'fòrs }

radial gate See Tainter gate. { 'rād·ē·əl 'gāt }

radial grating [ELECTROMAG] Conformal wire grating consisting of wires arranged radially in a circular frame, like the spokes of a wagon wheel, and placed inside a circular waveguide to obstruct E waves of zero order while passing the corresponding H waves. { 'rād·ē·əl 'grād·iŋ }

radial heat flow [THERMO] Flow of heat between two coaxial cylinders maintained at different temperatures; used to measure thermal conductivities of gases. { 'rād·ē·əl 'hēt ‚flō }

radial lead [ELEC] A wire lead coming from the side of a component rather than axially from the end. { 'rād·ē·əl 'lēd }

radial load [MECH ENG] The load perpendicular to the bearing axis. { 'rād·ē·əl 'lōd }

radial locating [MECH ENG] One of the three locating problems in tooling to maintain the desired relationship between the workpiece, the cutter, and the body of the machine tool; the other two locating problems are concentric and plane locating. { 'rād·ē·əl 'lō‚kād·iŋ }

radial loop [FOREN SCI] A loop fingerprint pattern which flows in the direction of the radius bone. { ‚rād·ē·əl 'lüp }

radially related figures See homothetic figures. { 'rād·ē·ə·lē ri‚lād·əd 'fig·yərz }

radial motion [MECH] Motion in which a body moves along a line connecting it with an observer or reference point; for example, the motion of stars which move toward or away from the earth without a change in apparent position. { 'rād·ē·əl 'mō·shən }

radial nerve [NEUROSCI] A large nerve that arises in the brachial plexus and branches to enervate the extensor muscles and skin of the posterior aspect of the arm, forearm, and hand. { 'rād·ē·əl 'nərv }

RADIAL CLEAVAGE

Radial cleavage pattern found in invertebrates.

RADIAL LOOP

delta—

Radial loop fingerprint pattern, right hand. (*Federal Bureau of Investigation*)

radial paper chromatography *See* radial chromatography. { 'rād·ē·əl 'pā·pər ,krō·mə'täg·rə·fē }

radial-parallel search [NAV] Search by ship or aircraft on a series of radial course lines to a given line, beyond which parallel course lines are followed. { 'rād·ē·əl 'par·ə,lel ,sərch }

radial percussive coal cutter [MIN ENG] A heavy coal cutter having a percussive drill, with extension rods; used in headings and rooms in pillar methods of working and for drilling shot-firing holes. { 'rād·ē·əl pər'kəs·iv 'kōl ,kəd·ər }

radial-ply [DES ENG] Pertaining to the construction of a tire in which the cords run straight across the tire, and an additional layered belt of fabric is placed around the circumference between the plies and the tread. { ¦rād·ē·əl ¦plī }

radial-ply tire *See* radial tire. { ¦rād·ē·əl ¦plī 'tīr }

radial rake [MECH ENG] The angle between the cutter tooth face and a radial line passing through the cutting edge in a plane perpendicular to the cutter axis. { 'rād·ē·əl 'rāk }

radial road [CIV ENG] One of a group of roads leading outward from the center of a city in a pattern similar to spokes on a wheel. { 'rād·ē·əl ,rōd }

radial saw [MECH ENG] A power saw that has a circular blade suspended from a transverse head mounted on a rotatable overarm. { 'rād·ē·əl 'sȯ }

radial search [NAV] Search by ship or aircraft on a series of radial course lines. { 'rād·ē·əl 'sərch }

radial selector [ENG] *See* omnibearing selector; private line arrangement. { 'rād·ē·əl si'lek·tər }

radial shear interferometer [OPTICS] An interferometer in which a wavefront is interfered with by an expanded version of itself, resulting in fringes along which the radial slope of the wavefront is constant. { 'rād·ē·əl ¦shir ,in·tər·fə'räm·əd·ər }

radial stress [MECH] Tangential stress at the periphery of an opening. { 'rād·ē·əl 'stres }

radial symmetry [SCI TECH] An arrangement of usually similar parts in a regular pattern around a central axis. { 'rād·ē·əl 'sim·ə·trē }

radial tire [ENG] A pneumatic tire constructed with a layer of fabric between the tread and the plies (cords), which run straight across the tire. Also known as radial-ply tire. { ¦rād·ē·əl 'tīr }

radial triangulation [MAP] A triangulation procedure in which radials of overlapping vertical or oblique photographs to control points imaged on the photographs are measured and used for horizontal-control extension by successive intersection and resection of such lines. { 'rād·ē·əl trī,aŋ·gyə'lā·shən }

radial velocity [MECH] The component of the velocity of a body that is parallel to a line from an observer or reference point to the body; the radial velocities of stars are valuable in determining the structure and dynamics of the Galaxy. Also known as line-of-sight velocity. { 'rād·ē·əl və'läs·əd·ē }

radial wave equation [MECH] Solutions to wave equations with spherical symmetry can be found by separation of variables; the ordinary differential equation for the radial part of the wave function is called the radial wave equation. { 'rād·ē·əl ¦wāv i,kwā·zhən }

radian [MATH] The central angle of a circle determined by two radii and an arc joining them, all of the same length. { 'rād·ē·ən }

radiance [OPTICS] The radiant flux per unit solid angle per unit of projected area of the source; the usual unit is the watt per steradian per square meter. Also known as steradiancy. { 'rād·ē·əns }

radiancy *See* radiant emittance. { 'rād·ē·ən·sē }

radian frequency *See* angular frequency. { 'rād·ē·ən 'frē·kwən·sē }

radian length [PHYS] Distance, in a sinusoidal wave, between phases differing by an angle of 1 radian; it is equal to the wavelength divided by 2π. { 'rād·ē·ən 'leŋkth }

radiant [ASTRON] **1.** A point on the celestial sphere through which pass the backward extensions of the trail of a meteor as observed at various locations, or the backward extensions of trails of a number of meteors traveling parallel to each other. **2.** A point on the celestial sphere toward which the stars in a moving cluster appear to travel. [PHYS] **1.** Pertaining to motion of particles or radiation along radii from a common point or a small region. **2.** A point, region, substance, or entity from which particles or radiations are emitted. { 'rād·ē·ənt }

radiant density [PHYS] The instantaneous amount of radiant energy contained in a unit volume of propagation medium. { 'rād·ē·ənt 'den·səd·ē }

radiant efficiency [OPTICS] The ratio of the radiant flux emitted by a radiation source to the power consumed by the source. { 'rād·ē·ənt i'fish·ən·sē }

radiant emittance [ELECTROMAG] The radiant flux per unit area that emerges from a surface. Also known as radiancy; radiant exitance. { 'rād·ē·ənt i'mit·əns }

radiant energy *See* radiation. { 'rād·ē·ənt 'en·ər·jē }

radiant-energy thermometer *See* radiation pyrometer. { 'rād·ē·ənt ¦en·ər·jē thər'mäm·əd·ər }

radiant exitance *See* radiant emittance. { 'rād·ē·ənt 'ek·sit·əns }

radiant exposure [OPTICS] A measure of the total radiant energy incident on a surface per unit area; equal to the integral over time of the radiant flux density. Also known as exposure. { 'rād·ē·ənt ik'spō·zhər }

radiant flux [OPTICS] The time rate of flow of radiant energy. { 'rād·ē·ənt 'fləks }

radiant flux density [ELECTROMAG] The amount of radiant power per unit area that flows across or onto a surface. Also known as irradiance. { 'rād·ē·ənt ¦fləks 'den·səd·ē }

radiant heating [ENG] Any system of space heating in which the heat-producing means is a surface that emits heat to the surroundings by radiation rather than by conduction or convection. { 'rād·ē·ənt 'hēd·iŋ }

radiant intensity [ELECTROMAG] The energy emitted per unit time per unit solid angle about the direction considered; usually expressed in watts per steradian. { 'rād·ē·ənt in'ten·səd·ē }

radiant power [ELECTROMAG] The energy carried across or onto a surface by electromagnetic radiation per unit time, or the total radiant energy emitted by a source of electromagnetic radiation per unit time. { 'rād·ē·ənt 'pau̇·ər }

radiant quantities [OPTICS] Physical quantities used in photometry, such as radiant flux and radiance, which are based on the energy carried by light, and are thus independent of the response of the human eye. { 'rād·ē·ənt 'kwän·əd·ēz }

radiant reflectance [ELECTROMAG] Ratio of reflected radiant power to incident radiant power. { 'rād·ē·ənt ri'flek·təns }

radiant superheater [MECH ENG] A superheater designed to transfer heat from the products of combustion to the steam primarily by radiation. { 'rād·ē·ənt 'sü·pər,hēd·ər }

radiant transmittance [ELECTROMAG] Ratio of transmitted radiant power to incident radiant power. { 'rād·ē·ənt tranz'mit·əns }

radiant-type boiler [MECH ENG] A water-tube boiler in which boiler tubes form the boundary of the furnace. { 'rād·ē·ənt ¦tīp 'bȯil·ər }

Radiata [INV ZOO] Members of the Eumetazoa which have a primary radial symmetry; includes the Cnidaria and Ctenophora. { ,rād·ē'äd·ə }

radiated interference [COMMUN] Interference which is transmitted through the atmosphere according to the laws of electromagnetic wave propagation; the term is generally considered to include the transfer of interfering energy in inductive or capacitive coupling. { 'rād·ē,ād·əd ,in·tər'fir·əns }

radiated power [ELECTROMAG] The total power emitted by a transmitting antenna. { 'rād·ē,ād·əd 'pau̇·ər }

radiating curtain [ELECTROMAG] Array of dipoles in a vertical plane, positioned to reinforce each other; it is usually placed one-fourth wavelength ahead of a reflecting curtain of corresponding half-wave reflecting antennas. { 'rād·ē,ād·iŋ 'kərt·ən }

radiating element [ELECTROMAG] Basic subdivision of an antenna which in itself is capable of radiating or receiving radio-frequency energy. { 'rād·ē,ād·iŋ 'el·ə·mənt }

radiating guide [ELECTROMAG] Waveguide designed to radiate energy into free space; the waves may emerge through slots or gaps in the guide, or through horns inserted in the wall of the guide. { 'rād·ē,ād·iŋ 'gīd }

radiating power *See* emittance. { 'rād·ē,ād·iŋ 'pau̇·ər }

radiating scattering [PHYS] The diversion of radiation (thermal, electromagnetic, or nuclear) from its orginal path as a result of interactions or collisions with atoms, molecules, or larger particles in the atmosphere or other media between the source of radiation (for example, a nuclear explosion) and a point at some distance away. { 'rād·ē,ād·iŋ 'skad·ə·riŋ }

radiation [ENG] A method of surveying in which points are

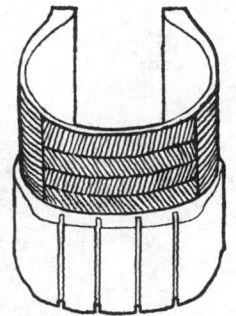
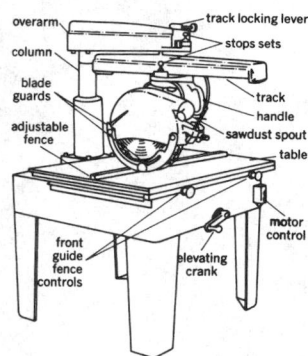

located by knowledge of their distances and directions from a central point. [PHYS] **1.** The emission and propagation of waves transmitting energy through space or through some medium; for example, the emission and propagation of electromagnetic, sound, or elastic waves. **2.** The energy transmitted by waves through space or some medium; when unqualified, usually refers to electromagnetic radiation. Also known as radiant energy. **3.** A stream of particles, such as electrons, neutrons, protons, α-particles, or high-energy photons, or a mixture of these. { ¸rād·ē'ā·shən }

radiation accident [NUCLEO] Any accident resulting in the spread of radioactive materials or in the exposure of individuals to radiation. { ¸rād·ē'ā·shən ¸ak·sə·dənt }

radiational cooling [METEOROL] The cooling of the earth's surface and adjacent air, accomplished (mainly at night) whenever the earth's surface suffers a net loss of heat due to terrestrial radiation. { ¸rād·ē'ā·shən·əl 'kül·iŋ }

radiational inversion [METEOROL] An inversion at the land surface resulting from rapid radiational cooling of lower air; usually occurs on cold winter nights. { rad·ē'ā·shən·əl in'vər·zhən }

radiation angle [ELECTROMAG] The vertical angle between the line of radiation emitted by a directional antenna and the horizon. { ¸rād·ē'ā·shən ¸aŋ·gəl }

radiation area [NUCLEO] Any accessible area in which the level of radiation is such that a major portion of an individual's body could receive in any 1 hour a dose in excess of 5 millirem or in any 5 consecutive days a dose in excess of 150 millirem. { ¸rād·ē'ā·shən ¸er·ē·ə }

radiation biochemistry [BIOCHEM] The study of the response of the constituents of living matter to radiation. { ¸rād·ē'ā·shən ¸bī·ō'kem·ə·strē }

radiation biology See radiobiology. { ¸rād·ē'ā·shən bī'äl·ə·jē }

radiation biophysics [BIOPHYS] The study of the response of organisms to ionizing radiations and to ultraviolet light. { ¸rād·ē'ā·shən ¸bī·ō'fiz·iks }

radiation-bounded nebula [ASTRON] An emission nebula whose central star is not hot enough to ionize the entire cloud. { ¸rād·ē'ā·shən ¸baün·dəd 'neb·yə·lə }

radiation budget [GEOPHYS] A quantitative statement of the amounts of radiation entering and leaving a given region of the earth. { ¸rād·ē'ā·shən ¸bəj·ət }

radiation burn [MED] A burn caused by overexposure to radiant energy. { ¸rād·ē'ā·shən ¸bərn }

radiation catalysis [CHEM] The use of radiation (such as gamma, neutron, proton, electron, or x-ray) to activate or speed up a chemical or physical change; for example, radiation alone can initiate polymerization without heat, pressure, or chemical catalysts. { ¸rād·ē'ā·shən kə'tal·ə·səs }

radiation cataract See irradiation cataract. { ¸rād·ē'ā·shən 'kad·ə¸rakt }

radiation characteristic [COMMUN] One of the identifying features of a radiating signal, such as frequency and pulse width. { ¸rād·ē'ā·shən ¸kar·ik·tə'ris·tik }

radiation chart [GEOPHYS] Any chart or diagram which permits graphical solution of the (generally unintegrable) flux integrals arising in problems of atmospheric infrared radiation transfer. { ¸rād·ē'ā·shən ¸chärt }

radiation chemistry [NUCLEO] The branch of chemistry that is concerned with the chemical effects, including decomposition, of energetic radiation or particles on matter. { ¸rād·ē'ā·shən 'kem·ə·strē }

radiation cooling [ELECTR] Cooling of an electrode resulting from its emission of heat radiation. [PHYS] The cooling of gases to very low temperatures by means of the resonant radiation pressure of intense laser light. { ¸rād·ē'ā·shən ¸kül·iŋ }

radiation correction See cooling correction. { ¸rād·ē'ā·shən kə¸rek·shən }

radiation corrosion [MET] Accelerated corrosion of a metal caused by radiation. { ¸rād·ē'ā·shən kə¸rō·zhən }

radiation counter [NUCLEO] An instrument used for detecting or measuring nuclear radiation by counting the resultant ionizing events; examples include Geiger counters and scintillation counters. Also known as counter. { ¸rād·ē'ā·shən ¸kaünt·ər }

radiation counter tube See counter tube. { ¸rād·ē'ā·shən ¸kaünt·ər ¸tüb }

radiation cytology [CYTOL] An aspect of biology that deals with the effects of radiations on living cells. { ¸rād·ē'ā·shən sī'täl·ə·jē }

radiation damage [NUCLEO] Harmful changes in the properties of liquics, gases, and solids caused by any type of radiation. { ¸rād·ē'ā·shən ¸dam·ij }

radiation damping [ELECTROMAG] Damping of a system which loses energy by electromagnetic radiation. [QUANT MECH] Damping which arises in quantum electrodynamics from the virtual interaction of a particle with its zero point field. { ¸rād·ē'ā·shən ¸damp·iŋ }

radiation decontamination [NUCLEO] The removal of unwanted radioactive material. { ¸rād·ē'ā·shən ¸dē·kən¸tam·ə'nā·shən }

radiation dermatitis See radiodermatitis. { ¸rād·ē'ā·shən ¸dər·mə'tīd·əs }

radiation detection instrument [NUCLEO] Any device that detects and records the characteristics of ionizing radiation. { ¸rād ē'ā·shən di'tek·shən ¸in·strə·mənt }

radiation detector See particle detector. { ¸rād·ē'ā·shən di¸tek·tər }

radiation dose [NUCLEO] The total amount of ionizing radiation absorbed by material or tissues, in the sense of absorbed dose (expressed in rads), exposure dose (expressed in roentgens), or dose equivalent (expressed in rems). { ¸rād·ē'ā·shən ¸dōs }

radiation dose rate [NUCLEO] The radiation dosage absorbed per unit of time; a radiation dose rate can be set at some particular unit of time (for example, H-hour plus 1 hour) and would be called H-hour plus 1 radiation dose rate. { ¸rād·ē'ā·shən 'dōs ¸rāt }

radiation dosimetry See dosimetry. { ¸rād·ē'ā·shən dō'sim·ə·trē }

radiation effects [BIOL] The harmful effects of ionizing radiation on humans and other animals, such as production of cancers, cataracts, and radiation ulcers, loss of hair, reddening of skin, sterilization, nausea, vomiting, mucous or bloody diarrhea, purpura, epilation, and agranulocytic infections. { ¸rād·ē'ā·shən i¸feks }

radiation efficiency [ELECTROMAG] Of an antenna, the ratio of the power radiated to the total power supplied to the antenna at a given frequency. { ¸rād·ē'ā·shən i¸fish·ən·sē }

radiation-enhanced diffusion [ELECTR] A mechanism for ion-beam mixing of a film and a substrate in which lattice defects that are formed by the atomic displacements produced by ion bombardment result in an increase in interdiffusion coefficients. { ¸rād·ē'ā·shən in¸hanst də'fyü·zhən }

radiation era [ASTRON] The period in the early universe, lasting from roughly 20 seconds to 10^5 years after the big bang, when photons dominated the universe. { ¸rād·ē'ā·shən ¸ir·ə }

radiation field [ELECTROMAG] The electromagnetic field that breaks away from a transmitting antenna and radiates outward into space as electromagnetic waves; the other type of electromagnetic field associated with an energized antenna is the induction field. { ¸rād·ē'ā·shən ¸fēld }

radiation filter [ELECTROMAG] Selectively transparent body, which transmits only certain wavelength ranges. { ¸rād·ē'ā·shən ¸fil·tər }

radiation fog [METEOROL] A major type of fog, produced over a land area when radiational cooling reduces the air temperature to or below its dew point; thus, strictly, a nighttime occurrence, although the fog may begin to form by evening twilight and often does not dissipate until after sunrise. { ¸rād·ē'ā·shən ¸fäg }

radiation gage [NUCLEO] An instrument for measuring radiation quantity and intensity. { ¸rād·ē'ā·shən ¸gāj }

radiation hardening [ENG] Improving the ability of a device or piece of equipment to withstand nuclear or other radiation; applies chiefly to dielectric and semiconductor materials. { ¸rād·ē'ā·shən 'härd·ən·iŋ }

radiation hazard [MED] Health hazard arising from exposure to ionizing radiation. { ¸rād·ē'ā·shən ¸haz·ərd }

radiation hybrid panel [GEN] A group of interspecific hybrid cell lines each containing a different set of very tiny fragments of the deoxyribonucleic acid of one parental species; those used in mapping human genes are produced by massive radiation of human cells before hybridizing them with rodent cells. { ¸rā·dē'ā·shən ¸hī·brəd ¸pan·əl }

radiation impedance See radiation resistance. { ¦rād·ē'ā·shən im¸pēd·əns }

radiation intensity [ELECTROMAG] The power radiated from an antenna per unit solid angle in a given direction. [NUCLEO] The quantity of radiant energy passing perpendicularly through a specified location of unit area in unit time; reported as a number of particles or photons per square centimeter per second, or in energy units such as ergs per square centimeter per second. { ¦rād·ē'ā·shən in¸ten·səd·ē }

radiation ionization [PHYS] Ionization of the atoms or molecules of a gas or vapor by the action of electromagnetic radiation. { ¦rād·ē'ā·shən ¸ī·ə·nə'zā·shən }

radiation laws [PHYS] **1.** The four physical laws which, together, fundamentally describe the behavior of blackbody radiation, Kirchhoff's law, Planck's law, Stefan-Boltzmann law, and Wien's law. **2.** All of the more inclusive assemblage of empirical and theoretical laws describing all manifestations of radiative phenomena. { ¦rād·ē'ā·shən ¸lóz }

radiation length [NUCLEO] The mean path length required to reduce the energy of relativistic charged particles by the factor $1/e$, or 0.368, as they pass through matter. Also known as cascade unit; radiation unit. { ¦rād·ē'ā·shən ¸leŋkth }

radiationless transition [PHYS] A transition of a system between two energy states in which energy is given to or taken up from another system or particle, rather than being emitted or absorbed in electromagnetic radiation; examples include internal conversion, the Auger effect, and excitation or deexcitation of atoms or molecules in collisions with other atoms or molecules. { ¦rād·ē'ā·shən·ləs tran'zish·ən }

radiation lobe See lobe. { ¦rād·ē'ā·shən ¸lōb }

radiation loss [MECH ENG] Boiler heat loss to the atmosphere by conduction, radiation, and convection. { ¦rād·ē'ā·shən ¸lós }

radiation microbiology [MICROBIO] A field of basic and applied radiobiology concerned chiefly with the damaging effects of radiation on microorganisms. { ¦rād·ē'ā·shən ¦mī·krō·bī'äl·ə·jē }

radiation monitoring [NUCLEO] Continuous or periodic determination of the amount of radiation present in a given area. { ¦rād·ē'ā·shən ¸män·ə·triŋ }

radiation noise See electromagnetic noise. { ¦rād·ē'ā·shən ¸nóiz }

radiation oscillator See Planck oscillator. { ¦rād·ē'ā·shən ¸äs·ə¸läd·ər }

radiation oven [ENG] Heating chamber relying on tungsten-filament infrared lamps with reflectors to create temperatures up to 600°F (315°C); used to dry sheet and granular material and to bake surface coatings. { ¦rād·ē'ā·shən ¸əv·ən }

radiation pattern [ELECTROMAG] Directional dependence of the radiation of an antenna. Also known as antenna pattern; directional pattern; field pattern. { ¦rād·ē'ā·shən ¸pad·ərn }

radiation physics [PHYS] The study of ionizing radiation and its effects on matter. { ¦rād·ē'ā·shən 'fiz·iks }

radiation preservation [FOOD ENG] Exposure of food products to ionizing radiation, such as electrons, x-rays, and γ-rays, in order to destroy microorganisms and thereby aid preservation. { ¦rād·ē'ā·shən ¸prez·ər'vā·shən }

radiation pressure [ACOUS] The average pressure exerted on a surface or interface between two media by a sound wave. [ELECTROMAG] The pressure exerted by electromagnetic radiation on objects on which it impinges. { ¦rād·ē'ā·shən ¸presh·ər }

radiation protection [NUCLEO] **1.** Legislation and regulations to protect the public and laboratory or industrial workers against radiation. **2.** Measures to reduce exposure to radiation. { ¦rād·ē'ā·shən prə¸tek·shən }

radiation protection guide [NUCLEO] The officially determined radiation doses which should not be exceeded without careful consideration of the reasons for doing so; these standards, established by the Federal Radiation Council, are equivalent to what was formerly called the maximum permissible dose or maximum permissible exposure. { ¦rād·ē'ā·shən prə¸tek·shən ¸gīd }

radiation pyrometer [ENG] An instrument which measures the temperature of a hot object by focusing the thermal radiation emitted by the object and making some observation on it; examples include the total-radiation, optical, and ratio pyrometers. Also known as noncontact thermometer; radiant-energy

thermometer; radiation thermometer. { ¦rād·ē'ā·shən pī'räm·əd·ər }

radiation quality [PHYS] The spectrum of radiant energy produced by a given radiation source with respect to its penetration or its suitability for a specific application. { ¦rād·ē'ā·shən ¸kwäl·əd·ē }

radiation resistance [ACOUS] For a medium, the acoustic impedance of a plane wave in that medium. Also known as radiation impedance. [ELECTROMAG] The total radiated power of an antenna divided by the square of the effective antenna current measured at the point where power is supplied to the antenna. { ¦rād·ē'ā·shən ri¸zis·təns }

radiation safety [NUCLEO] Protection of personnel against harmful effects of ionizing radiation by taking steps to ensure that people will not receive excessive doses of radiation and by monitoring all sources of radiation to which they may be exposed. { ¦rād·ē'ā·shən ¸sāf·tē }

radiation scattering [PHYS] The diversion of radiation (thermal, electromagnetic, or nuclear) from its original path as a result of interactions or collisions with atoms, molecules, or larger particles in the atmosphere or other media. { ¦rād·ē'ā·shən 'skad·ə·riŋ }

radiation shelter See fallout shelter. { ¦rād·ē'ā·shən ¸shel·tər }

radiation shield [ENG] A shield or wall of material interposed between a source of radiation and a radiation-sensitive body, such as a person, radiation-detection instrument, or photographic film, to protect the latter. { ¦rād·ē'ā·shən ¸shēld }

radiation sickness [MED] **1.** Illness, usually manifested by nausea and vomiting, resulting from the effects of therapeutic doses of radiation. **2.** Radiation injury following exposure to excessive doses of radiation, such as the explosion of an atomic bomb. { ¦rād·ē'ā·shən ¸sik·nəs }

radiation source [NUCLEO] Usually a sealed capsule containing an artificial radioisotope, used in teletherapy, radiography, as a power source for batteries, or in various types of industrial gages; machines such as accelerators, and radioisotopic generators and natural radionuclides may also be considered as sources. { ¦rād·ē'ā·shən ¸sórs }

radiation standards [NUCLEO] Exposure standards, permissible concentrations, rules for safe handling, regulations for transportation, regulations for industrial control of radiation, and control of radiation exposure by legislative means. { ¦rād·ē'ā·shən ¸stan·dərdz }

radiation sterilization [NUCLEO] Exposure of a material, object, or body to ionizing radiation in order to destroy microorganisms. { ¦rād·ē'ā·shən ¸ster·ə·lə'zā·shən }

radiation survey meter [NUCLEO] Portable device to measure the intensity of nuclear radiation in a given region, in such applications as health physics (atomic radiation safety) or supervision of radioactively hot areas. { ¦rād·ē'ā·shən 'sər¸vā ¸mēd·ər }

radiation therapy [MED] The use of ionizing radiation or radioactive substances to treat disease. Also known as actinotherapy; radiotherapy. { ¦rād·ē'ā·shən 'ther·ə·pē }

radiation thermocouple [ELEC] An infrared detector consisting of several thermocouples connected in series, arranged so that the radiation falls on half of the junctions, causing their temperature to increase so that a voltage is generated. { ¦rād·ē'ā·shən 'thər·mə¸kəp·əl }

radiation thermometer See radiation pyrometer. { ¦rād·ē'ā·shən thər'mäm·əd·ər }

radiation unit See radiation length. { ¦rād·ē'ā·shən ¸yü·nət }

radiation vacuum gage [ENG] Vacuum (reduced-pressure) measurement device in which gas ionization from an alpha source of radiation varies measurably with changes in the density (molecular concentration) of the gas being measured. { ¦rād·ē'ā·shən 'vak·yəm ¸gāj }

radiation warning symbol [NUCLEO] A standard symbol used on posters displayed in locations where radiation hazards exist; consists of a magenta trefoil printed on a yellow background. { ¦rād·ē'ā·shən 'wórn·iŋ ¸sim·bəl }

radiation well logging See radioactive well logging. { ¦rād·ē'ā·shən 'wel ¸läg·iŋ }

radiation zone See Fraunhofer region. { ¦rād·ē'ā·shən ¸zōn }

radiative braking [ASTRON] Deceleration of a star's rotation due to emission of electromagnetic radiation. { 'rād·ē¸ād·iv 'brāk·iŋ }

radiative capture [NUC PHYS] A nuclear capture process

whose prompt result is the emission of electromagnetic radiation only. { 'rād·ē,ād·iv 'kap·chər }

radiative collision [PHYS] A collision between two charged particles in which part of the kinetic energy of the particles is converted into electromagnetic radiation. { 'rād·ē,ād·iv kə'lizh·ən }

radiative correction [QUANT MECH] The change produced in the value of some physical quantity, such as the mass or charge of a particle, as the result of the particle's interactions with various fields. { 'rād·ē,ād·iv kə'rek·shən }

radiative diffusivity [METEOROL] A characteristic property of a given layer of the atmosphere which governs the rate at which that layer will warm or cool as a result of the transfer, within it, of infrared radiation; the radiative diffusivity is dependent upon the temperature and water-vapor content of the layer of air and upon the pressure within the layer. { 'rād·ē,ād·iv ,di,fyü'siv·əd·ē }

radiative equilibrium [ASTROPHYS] The energy transfer through a star by radiation, absorption, and reradiation at a rate such that each section of the star is maintained at the appropriate temperature. { 'rād·ē,ād·iv ,ē·kwə'lib·rē·əm }

radiative forcing [METEOROL] The relative effectiveness of greenhouse gases to restrict long-wave radiation from escaping back into space. For a particular greenhouse gas, radiative forcing is measured as the change in average net radiation (in watts per square meter) at the top of the troposphere, and depends on the wavelength at which the gas absorbs the radiation, the strength of absorption per molecule, and the concentration of the gas. { ,rād·ē,ād·iv 'fòrs·iŋ }

radiative recombination [PHYS] Recombination of parts of an atom or an electron and a hole in a semiconductor during which electromagnetic radiation is emitted. { 'rād·ē,ād·iv ri,käm·bə'nā·shən }

radiative transfer [PHYS] The propagation of energy by radiative processes, involving emission, absorption, and scattering of electromagnetic radiation. Also known as radiative transport. { 'rād·ē,ād·iv 'tranz·fər }

radiative transition [QUANT MECH] A change of a quantum-mechanical system from one energy state to another in which electromagnetic radiation is emitted. { 'rād·ē,ād·iv tran'zish·ən }

radiative transport See radiative transfer. { 'rād·ē,ād·iv 'tranz,pórt }

radiator [ACOUS] A vibrating element of a transducer which radiates sound waves. [ELECTROMAG] **1.** The part of an antenna or transmission line that radiates electromagnetic waves either directly into space or against a reflector for focusing or directing. **2.** A body that emits radiant energy. [ENG] Any of numerous devices, units, or surfaces that emit heat, mainly by radiation, to objects in the space in which they are installed. [PHYS] **1.** In general, a body which emits particles or radiation in any form. **2.** A body placed in a beam of ionizing radiation which, as a result, emits radiation of another kind. { 'rād·ē,ād·ər }

radiator temperature drop [MECH ENG] In internal combustion engines, the difference in temperature of the coolant liquid entering and leaving the radiator. { 'rād·ē,ād·ər 'tem·prə·chər ,dräp }

radiatus [METEOROL] A cloud variety whose elements are arranged in straight parallel bands; owing to the effect of perspective, these bands seem to converge toward a point on the horizon, or, when the bands cross the entire sky, toward two opposite points. { ,rād·ē'läd·əs }

radical [BOT] **1.** Of, pertaining to, or proceeding from the root. **2.** Arising from the base of a stem or from an underground stem. [MATH] **1.** In a ring, the intersection of all maximal ideals. Also known as Jacobson radical. **2.** An indicated root of a quantity. Symbolized √. **3.** See nilradical. [ORG CHEM] See free radical. { 'rad·ə·kəl }

radical axis [MATH] The line passing through the two points of intersection of a pair of circles. { 'rad·ə·kəl 'ak·səs }

radical center [MATH] **1.** For three circles, the point at which the three radical axes of pairs of the circles intersect. **2.** For four spheres, the point at which the six radical planes of pairs of the spheres intersect. { 'rad·ə·kəl 'sen·tər }

radical equation See irrational equation. { 'rad·ə·kəl i'kwā·zhən }

radical ion [CHEM] A charged compound that has an unpaired electron; it may be either a radical cation (positively charged) or radical anion (negatively charged). { ,rad·ə·kəl 'ī·ən }

radical mastectomy [MED] Surgical removal of the breast, subcutaneous fat, muscle, lymph glands, and a wide area of skin for cancer. { 'rad·ə·kəl ,ma'stek·tə·mē }

radical plane [MATH] The plane containing the circle of intersection of a pair of spheres. { 'rad·ə·kəl 'plān }

radical scavenger [CHEM] One of a group of molecules that combines with free radicals in a chemical or biochemical system to render them less active chemically. { 'rad·ə·kəl ,skav·ən·jər }

radical sign [MATH] The symbol √, indicating that a root of a quantity is to be taken. { 'rad·ə·kəl ,sīn }

radicand [MATH] The quantity that appears under a radical sign. { 'rad·ə,kand }

radicidation [FOOD ENG] Destruction by radiation of microorganisms in food that are significant to public health; an example is Salmonella species. { ,rād·ə·sə'dā·shən }

radicle [BOT] The embryonic root of a flowering plant. { 'rad·ə·kəl }

radicofunctional name [ORG CHEM] A name for an organic compound that uses two key words; the first word corresponds to the group or groups involved and the second word indicates the functional group—for example, alkyl halide. { ,rad·ə·kō'fəŋk·shən·əl 'nām }

radicotomy See rhizotomy. { ,rad·ə'käd·ə·mē }

radicular cyst [MED] A cyst arising from a granuloma of the root of a tooth. Also known as periapical cyst. { rə'dik·yə·lər 'sist }

radiculitis [MED] Inflammation of a nerve root. { rə,dik·yə'līd·əs }

radiculoneuropathy [MED] Disease of the peripheral spinal nerves and their roots. { ra,dik·yə·lō·nù'räp·ə·thē }

radio [COMMUN] The transmission of signals through space by means of electromagnetic waves. [ELECTR] See radio receiver. { 'rād·ē·ō }

radio- [ELECTROMAG] A prefix denoting the use of radiant energy, particularly radio waves. [NUCLEO] Chemical prefix designating radiation or radioactivity; used to designate radioactive elements (such as radiocarbon) and substances containing them (such as radiochemicals, radiocolloids, or radio compounds). { 'rād·ē·ō }

radioacoustic position finding See radioacoustic ranging. { ¦rād·ē·ō·ə¦küs·tik pə'zish·ən ,find·iŋ }

radioacoustic ranging [ENG] A method for finding the position of a vessel at sea; a bomb is exploded in the water, and the sound of the explosion transmitted through water is picked up by the vessel and by shore stations, other vessels, or buoys whose positions are known; the received sounds are transmitted instantaneously by radio to the surveying vessel, and the elapsed times are proportional to the distances to the known positions. Abbreviated RAR. Also known as radioacoustic position finding; radioacoustic sound ranging. { ¦rād·ē·ō·ə¦küs·tik 'ranj·iŋ }

radioacoustics [COMMUN] Study of the production, transmission, and reproduction of sounds carried from one place to another by radiotelephony. { ¦rād·ē·ō·ə¦küs·tiks }

radioacoustic sound ranging See radioacoustic ranging. { ¦rād·ē·ō·ə¦küs·tik 'saùnd ,ranj·iŋ }

radioactinium [NUC PHYS] Conventional name for the isotope of thorium which has mass number 227 and is in the actinium series. Symbolized RdAc. { ¦rād·ē·ō·ak'tin·ē·əm }

radioactive [NUC PHYS] Exhibiting radioactivity or pertaining to radioactivity. { ¦rād·ē·ō'ak·tiv }

radioactive age determination See radiometric dating. { ¦rād·ē·ō'ak·tiv 'āj di,tər·mə,nā·shən }

radioactive beam [NUCLEO] A collimated stream of accelerated radioactive nuclei, of sufficient intensity to be utilized for a variety of scientific enterprises such as studies of nuclear structure, nuclear astrophysics, and materials science. { ,rād·ē·ō,ak·tiv 'bēm }

radioactive carbon dating See carbon-14 dating. { ¦rād·ē·ō'ak·tiv ¦kär·bən ,dād·iŋ }

radioactive chain See radioactive series. { ¦rād·ē·ō'ak·tiv 'chān }

radioactive clock [NUC PHYS] A radioactive isotope such as potassium-40 which spontaneously decays to a stable end product at a constant rate, allowing absolute geologic age to be determined. { ¦rād·ē·ō'ak·tiv 'kläk }

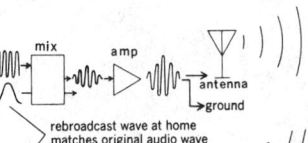

RADIO

Transmission of audio information by radio.

radioactive cloud [NUCLEO] A mass of air and vapor in the atmosphere carrying radioactive debris from a nuclear explosion. { ¦rād·ē·ō¦ak·tiv ˈklau̇d }

radioactive cobalt [NUC PHYS] Radioactive form of cobalt, such as cobalt-60 with a half-life of 5.3 years. { ¦rād·ē·ō¦ak·tiv ˈkō‚bȯlt }

radioactive collision [NUC PHYS] A nuclear reaction in which a neutron is absorbed by a nucleus and a gamma ray is emitted. { ‚rād·ē·ō¦ak·tiv kəˈlizh·ən }

radioactive contaminant [NUCLEO] A radioactive material which has spread to places where it may harm persons, spoil experiments, or make products or equipment unsuitable or unsafe for some specific purpose. { ¦rād·ē·ō¦ak·tiv kənˈtam·ə·nənt }

radioactive dating See radiometric dating. { ¦rād·ē·ō¦ak·tiv ˈdād·iŋ }

radioactive debris [NUCLEO] Radioactive material which is carried through the air from the site of a nuclear explosion. { ¦rād·ē·ō¦ak·tiv dəˈbrē }

radioactive decay [NUC PHYS] The spontaneous transformation of a nuclide into one or more different nuclides, accompanied by either the emission of particles from the nucleus, nuclear capture or ejection of orbital electrons, or fission. Also known as decay; nuclear spontaneous reaction; radioactive disintegration; radioactive transformation; radioactivity. { ¦rād·ē·ō¦ak·tiv diˈkā }

radioactive decay constant See decay constant. { ¦rād·ē·ō¦ak·tiv di¦kā ‚kän·stənt }

radioactive decay product See daughter. { ¦rād·ē·ō¦ak·tiv di¦kā ‚präd·əkt }

radioactive decay series See radioactive series. { ¦rād·ē·ō¦ak·tiv di¦kā ‚sir·ēz }

radioactive disintegration See radioactive decay. { ¦rād·ē·ō¦ak·tiv di‚sin·tə¦grā·shən }

radioactive displacement law [NUC PHYS] The statement of the changes in mass number *A* and atomic number *Z* that take place during various nuclear transformations. Also known as displacement law. { ¦rād·ē·ō¦ak·tiv diˈsplās·mənt ‚lȯ }

radioactive element [NUC PHYS] An element all of whose isotopes spontaneously transform into one or more different nuclides, giving off various types of radiation; examples include promethium, radium, thorium, and uranium. { ¦rād·ē·ō¦ak·tiv ˈel·ə·mənt }

radioactive emanation [NUC PHYS] A radioactive gas given off by certain radioactive elements; all of these gases are isotopes of the element radon. Also known as emanation. { ¦rād·ē·ō¦ak·tiv ‚em·əˈnā·shən }

radioactive equilibrium [NUC PHYS] In radioactivity, the condition of equilibrium in which the rate of decay of the parent isotope is exactly matched by the rate of decay of every intermediate daughter isotope. { ¦rād·ē·ō¦ak·tiv ‚ē·kwəˈlib·rē·əm }

radioactive fallout See fallout. { ¦rād·ē·ō¦ak·tiv ˈfȯl‚au̇t }

radioactive half-life See half-life. { ¦rād·ē·ō¦ak·tiv ˈhaf‚līf }

radioactive heat [THERMO] Heat produced within a medium as a result of absorption of radiation from decay of radioisotopes in the medium, such as thorium-232, potassium-40, uranium-238, and uranium-235. { ¦rād·ē·ō¦ak·tiv ˈhēt }

radioactive isotope See radioisotope. { ¦rād·ē·ō¦ak·tiv ˈī·sə‚tōp }

radioactive material [NUCLEO] A material having one or more constituents that exhibit significant radioactivity. { ¦rād·ē·ō¦ak·tiv məˈtir·ē·əl }

radioactive metal [NUC PHYS] A luminous metallic element, such as actinium, radium, or uranium, that spontaneously and continuously emits radiation capable in some degree of penetrating matter impervious to ordinary light. { ¦rād·ē·ō¦ak·tiv ˈmed·əl }

radioactive mineral [MINERAL] Any mineral species that contains uranium or thorium as an essential part of the chemical composition; examples are uraninite, pitchblende, carnotite, coffinite, and autunite. { ¦rād·ē·ō¦ak·tiv ˈmin·rəl }

radioactive paint [MATER] A luminous paint that gives off light without being activated. { ¦rād·ē·ō¦ak·tiv ˈpānt }

radioactive salt [NUCLEO] A salt whose molecules have radioactive atoms, such as radium bromide or mesothorium bromide; used in trace amounts to energize self-luminous paints. { ¦rād·ē·ō¦ak·tiv ˈsȯlt }

radioactive series [NUC PHYS] A succession of nuclides, each of which transforms by radioactive disintegration into the next until a stable nuclide results. Also known as decay chain; decay family; decay series; disintegration chain; disintegration family; disintegration series; radioactive chain; radioactive decay series; series decay; transformation series. { ¦rād·ē·ō¦ak·tiv ˈsir·ēz }

radioactive snow gage [ENG] A device which automatically and continuously records the water equivalent of snow on a given surface as a function of time; a small sample of a radioactive salt is placed in the ground in a lead-shielded collimator which directs a beam of radioactive particles vertically upward; a Geiger-Müller counting system (located above the snow level) measures the amount of depletion of radiation caused by the presence of the snow. { ¦rād·ē·ō¦ak·tiv ˈsnō ‚gāj }

radioactive source [NUCLEO] Any quantity of radioactive material intended for use as a source of ionizing radiation. { ¦rād·ē·ō¦ak·tiv ˈsȯrs }

radioactive standard [NUCLEO] A sample of radioactive material which contains a known number and type of radioactive atoms at some definite time; used to calibrate radiation measuring instruments. { ¦rād·ē·ō¦ak·tiv ˈstan·dərd }

radioactive thickness gage [NUCLEO] An instrument for measuring the thickness of the metal wall of a pipe or tank from one side, by directing a beam of γ-rays through the wall at an angle and measuring the amount of backscattering with a radiation detector. { ¦rād·ē·ō¦ak·tiv ˈthik·nəs ‚gāj }

radioactive tracer [NUCLEO] A radioactive isotope which, when attached to a chemically similar substance or injected into a biological or physical system, can be traced by radiation detection devices, permitting determination of the distribution or location of the substance to which it is attached. Also known as radiotracer. { ¦rād·ē·ō¦ak·tiv ˈtrā·sər }

radioactive transformation See radioactive decay. { ¦rād·ē·ō¦ak·tiv ‚tranz·fərˈmā·shən }

radioactive waste [NUCLEO] Liquid, solid, or gaseous waste resulting from mining of radioactive ore, production of reactor fuel materials, reactor operation, processing of irradiated reactor fuels, and related operations, and from use of radioactive materials in research, industry, and medicine. { ¦rād·ē·ō¦ak·tiv ˈwāst }

radioactive-waste disposal [NUCLEO] The disposal of waste radioactive materials and of equipment contaminated by radiation; the two basic disposal methods are concentration for burial underground or in the sea, and dilution for controlled dispersion; reprocessing of reactor fuel is a major source of radioactive waste. { ¦rād·ē·ō¦ak·tiv ˈwāst diˈspō·zəl }

radioactive well logging [ENG] The recording of the differences in radioactive content (natural or neutron-induced) of the various rock layers found down an oil well borehole; types include γ-ray, neutron, and photon logging. Also known as radiation well logging; radioactivity prospecting. { ¦rād·ē·ō¦ak·tiv ˈwel ‚läg·iŋ }

radioactivity [NUC PHYS] **1.** A particular type of radiation emitted by a radioactive substance, such as alpha radioactivity. **2.** See radioactive decay. **3.** See activity. { ‚rād·ē·ō·akˈtiv·əd·ē }

radioactivity analysis See activation analysis. { ‚rād·ē·ō·akˈtiv·əd·ē əˈnal·ə·səs }

radioactivity concentration guide [NUCLEO] The concentration of radioactive material in an environment which would result in doses equal, over a period of time, to those in the radiation protection guide; this Federal Radiation Council term replaces the former maximum permissible concentration. { ‚rād·ē·ō·akˈtiv·əd·ē ‚käns·ən¦trā·shən ‚gīd }

radioactivity equilibrium [NUC PHYS] A condition which may arise in the decay of a radioactive parent with short-lived descendants, in which the ratio of the activity of a parent to that of a descendant remains constant. { ‚rād·ē·ō·akˈtiv·əd·ē ‚ē·kwəˈlib·rē·əm }

radioactivity log [ENG] Record of radioactive well logging. { ‚rād·ē·ō·akˈtiv·əd·ē ‚läg }

radioactivity prospecting See radioactive well logging. { ‚rād·ē·ō·akˈtiv·əd·ē ˈprä‚spekt·iŋ }

radio aid to navigation [ELECTR] An aid to navigation which utilizes the propagation characteristics of radio waves to furnish navigation information. { ˈrād·ē·ō ˈād tə ‚nav·əˈgā·shən }

radio altimeter [ENG] An absolute altimeter that depends on the reflection of radio waves from the earth for the determination of altitude, as in a frequency-modulated radio altimeter and a radar altimeter. Also known as electronic altimeter; reflection altimeter. { 'rād·ē·ō al'tim·əd·ər }

radio altitude See radar altitude. { 'rād·ē·ō 'al·tə,tüd }

radio and wire integration [COMMUN] The combining of wire circuits with radio facilities. { 'rād·ē·ō ən 'wīr ,int·ə'grā·shən }

radio antenna See antenna. { 'rād·ē·ō an'ten·ə }

radio approach aids [NAV] Equipment that uses radio or radar to furnish guidance to an aircraft with required accuracy from the time it is in the vicinity of an airfield until it reaches a position from which a landing can be made. { 'rād·ē·ō ə'prōch ,ādz }

radioassay [ANALY CHEM] An assay procedure involving the measurement of the radiation intensity of a radioactive sample. { ¦rād·ē·ō'a,sā }

radio astronomy [ASTRON] The study of celestial objects by measurement and analysis of their emitted electromagnetic radiation in the wavelength range from roughly 1 millimeter to 30 millimeters. { 'rād·ē·ō ə'strän·ə·mē }

radio atmometer [ENG] An instrument designed to measure the effect of sunlight upon evaporation from plant foliage; consists of a porous-clay atmometer whose surface has been blackened so that it absorbs radiant energy. { 'rād·ē·ō at'mäm·əd·ər }

radio attenuation [ELECTROMAG] For one-way propagation, the ratio of the power delivered by the transmitter to the transmission line connecting it with the transmitting antenna to the power delivered to the receiver by the transmission line connecting it with the receiving antenna. { 'rād·ē·ō ə,ten·yə'wā·shən }

radio aurora See artificial radio aurora. { 'rād·ē·ō ə'rȯr·ə }

radioautography See autoradiography. { ¦rād·ē·ō,ȯ'täg·rə·fē }

radio autopilot coupler [ENG] Equipment providing means by which an electrical navigational signal operates an automatic pilot. { 'rād·ē·ō 'ȯd·ō,pī·lət 'kəp·lər }

radio B battery [ELEC] A B-type battery used in a radio set, usually consisting of 15 to 30 permanently connected cells. { 'rād·ē·ō 'bē ,bad·ə·rē }

radio beacon [NAV] A nondirectional radio transmitting station in a fixed geographic location, emitting a characteristic signal from which bearing information can be obtained by a radio direction finder on a ship or aircraft. Also known as aerophare; radiophare. { 'rād·ē·ō 'bē·kən }

radio-beacon buoy [NAV] A buoy equipped with a marker radio beacon; such a buoy is usually used to mark an important entrance to a channel; the beacon is of low power and provides a signal for a short range. { 'rād·ē·ō ¦bē·kən ,bȯi }

radio-beacon monitor station [COMMUN] A station which monitors the signal from one or more remotely located marine radio beacons. { 'rād·ē·ō ¦bē·kən 'man·ə·tər ,stā·shən }

radio beam [ELECTROMAG] A concentrated stream of radio-frequency energy as used in radio ranges, microwave relays, and radar. { 'rād·ē·ō ,bēm }

radio bearing [NAV] The bearing of a radio transmitter from a receiver as determined by a radio direction finder. { 'rād·ē·ō ,ber·iŋ }

radiobioassay [BIOL] The analysis of the kind, concentration, and location of radioactive material in the human body by direct measurement (in vivo counting) or the evaluation of materials removed (or excreted). { ,rād·ē·ō,bī·ō'as,ā }

radiobiology [BIOL] Study of the scientific principles, mechanisms, and effects of the interaction of ionizing radiation with living matter. Also known as radiation biology. { 'rād·ē·ō·bī'äl·ə·jē }

radio blackout [COMMUN] A fadeout that may last several hours or more at a particular frequency. Also known as blackout. { 'rād·ē·ō 'blak,aut }

radio broadcasting [COMMUN] Radio transmission intended for general reception. { 'rād·ē·ō 'brȯd,kast·iŋ }

radio button [COMPUT SCI] In a graphical user interface, one of a group of small circles that represent a set of choices (indicated by text next to the circles) from which only one can be selected; the selected choice is indicated by a partly filled circle. { 'rād·ē·ō ,bət·ən }

radiocarbon See carbon-14. { ¦rād·ē·ō'kär·bən }

radiocarbon dating See carbon-14 dating. { ¦rād·ē·ō'kär·bən 'dād·iŋ }

radiocardiogram [MED] An x-ray recording of the variation with time of the concentration of a radioisotope in a heart chamber; usually the radioactive material is injected intravenously. { ¦rād·ē·ō'kärd·ē·ə,gram }

radiocesium See cesium-137. { ¦rād·ē·ō'sē·zē·əm }

radiochemical laboratory [CHEM] A specially equipped and shielded chemical laboratory designed for conducting radiochemical studies without danger to the laboratory personnel. { ¦rād·ē·ō'kem·ə·kəl 'lab·rə,tȯr·ē }

radiochemistry [CHEM] That area of chemistry concerned with the study of radioactive substances. { ¦rād·ē·ō'kem·ə·strē }

radiochromatography [ANALY CHEM] An analytic process for quantitative or qualitative determination of radioactive substances in a mixture by measuring the radioactivity of various zones in the chromatogram. { ¦rād·ē·ō,krō·mə'täg·rə·fē }

radiochronology [GEOL] An absolute-age dating method based on the existing ratio between radioactive parent elements (such as uranium-238) and their radiogenic daughter isotopes (such as lead-206). { ¦rād·ē·ō·krə'näl·ə·jē }

radio climatology [CLIMATOL] The study of regional and seasonal variations in the manner of propagation of radio energy through the atmosphere. { 'rād·ē·ō 'klī·mə'täl·ə·jē }

radiocolloid [CHEM] A colloid having a component that consists of radioactive atoms. { ,rād·ē·ō'kä,lȯid }

radio command [ELECTR] A radio control signal to which a guided missile or other remote-controlled vehicle or device responds. { 'rād·ē·ō kə,mand }

radio communication [COMMUN] Communication by means of radio waves, such as by radio facsimile, radiotelegraph, radiotelephone, and radioteletypewriter. { 'rād·ē·ō kə,myü·nə'kā·shən }

radiocommunication service [COMMUN] A service involving the emission, transmission, or reception of radio waves for specific telecommunications purposes. { ,rād·ē·ō·kə,myü·nə'kā·shən ,sər·vəs }

radio communications guard See radio guard. { 'rād·ē·ō kə,myü·nə'kā·shənz ,gärd }

radio compass See automatic direction finder. { 'rād·ē·ō 'kam·pəs }

radio control [ELECTR] The control of stationary or moving objects by means of signals transmitted through space by radio. { 'rād·ē·ō kən'trōl }

radio countermeasures [ELECTR] Electrical or other techniques depriving the enemy of the benefits which would ordinarily accrue to him through the use of any technique employing the radiation of radio waves; it includes benefits derived from radar and intercept services. { 'rād·ē·ō 'kaunt·ər,mezh·ərz }

radio deception [COMMUN] The use of radio to deceive the enemy, as by sending false dispatches or using enemy call signs. { 'rād·ē·ō di'sep·shən }

radiodermatitis [MED] Degenerative changes in the skin following excessive exposure to ionizing radiation. Also known as radiation dermatitis. { ¦rād·ē·ō,dər·mə'tīd·əs }

radio detection [ENG] The detection of the presence of an object by radiolocation without precise determination of its position. { 'rād·ē·ō di'tek·shən }

radio detection and location [ENG] Use of an electronic system to detect, locate, and predict future positions of earth satellites. { 'rād·ē·ō di'tek·shən ən lō'kā·shən }

radio detection and ranging See radar. { 'rād·ē·ō di'tek·shən ən 'rānj·iŋ }

radiodetermination satellite service [COMMUN] A system that employs at least two geosynchronous satellites, a central ground station, and hand-held or vehicle-mounted transceivers to enable users to determine and transmit their precise position. Abbreviated RDSS. { 'rād·ē·ōdi,tər·nə'nā·shən 'sad·əl,īt ,sər·vəs }

radio direction finder [NAV] A radio aid to navigation that uses a rotatable loop or other highly directional antenna arrangement to determine the direction of arrival of a radio signal. Abbreviated RDF. Also known as direction finder. { 'rād·ē·ō di'rek·shən ,fīn·dər }

radio direction-finder station [COMMUN] A land-based radio station equipped with special apparatus for determining the direction of radio signals transmitted by ships and other stations. { 'rād·ē·ō di'rek·shən ,fīn·dər ,stā·shən }

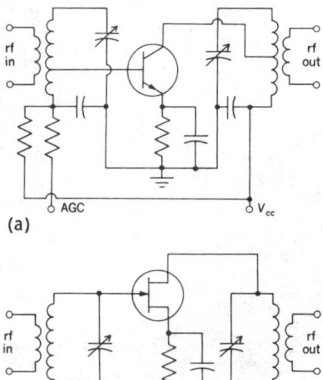

RADIO-FREQUENCY AMPLIFIER

Typical radio-frequency amplifier circuits with (a) bipolar transistor and (b) field-effect transistor. AGC = automatic gain control; V_{CC} = collector supply voltage; V_{DD} = drain supply voltage.

radio direction finding [NAV] A procedure for determining the bearing at a receiving point of the source of a radio signal by observing the direction of arrival of the wave front. { 'rād·ē·ō di'rek·shən ˌfīnd·iŋ }

radio Doppler [ENG] Direct determination of the radial component of the relative velocity of an object by an observed frequency change due to such velocity. { 'rād·ē·ō 'däp·lər }

radio duct [GEOPHYS] An atmospheric layer, typically shallow and almost horizontal, in which radio waves propagate in an anomalous fashion; ducts occur when, due to sharp inversions of temperature or humidity, the vertical gradient of the radio index of refraction exceeds a critical value. { 'rād·ē·ō ˌdəkt }

radio echo See radar return. { 'rād·ē·ō ˌek·ō }

radio echo observation [ENG] A method of determining the distance of objects in the atmosphere or outer space, in which a radar pulse is directed at the object and the time that elapses from transmission of the pulse to reception of a reflected pulse is measured. { 'rād·ē·ō ˌekō ˌäb·zər'vā·shən }

radioecology [ECOL] The interdisciplinary study of organisms, radionuclides, ionizing radiation, and the environment. { ˌrād·ē·ō·ē'käl·ə·jē }

radioelectric meteorology See radio meteorology. { ˌrād·ē·ō·i'lek·trik ˌmēd·ē·ə'räl·ə·jē }

radio element [NUC PHYS] A radioactive isotope of an element, or a sample consisting of one or more radioactive isotopes of an element. { 'rād·ē·ō 'el·ə·mənt }

radio emission [ELECTROMAG] The emission of radio-frequency electromagnetic radiation by oscillating charges or currents. { 'rād·ē·ō i,mish·ən }

radio energy [ELECTROMAG] The energy carried by radio-frequency electromagnetic radiation. { 'rād·ē·ō ,en·ər·jē }

radio engineering [ENG] The field of engineering that deals with the generation, transmission, and reception of radio waves and with the design, manufacture, and testing of associated equipment. { 'rād·ē·ō ,en·jə'nir·iŋ }

radio facility chart See enroute chart. { 'rād·ē·ō fə'sil·əd·ē ,chärt }

radio facsimile system [COMMUN] A facsimile system in which signals are transmitted by radio rather than by wire. { 'rād·ē·ō fak'sim·ə·lē ,sis·təm }

radio fadeout [COMMUN] Increased absorption of radio waves passing through the lower layers of the ionosphere due to a sudden and abnormal increase in ionization in these regions; signals at receivers then fade out or disappear. Also known as fadeout. { 'rād·ē·ō 'fād,aut }

radio fan-marker beacon See fan-marker beacon. { 'rād·ē·ō 'fan ,mär·kər ,bē·kən }

radio field intensity [ELECTROMAG] Electric or magnetic field intensity at a given location associated with the passage of radio waves. { 'rād·ē·ō 'fēld in,ten·səd·ē }

radio field-to-noise ratio [ELECTROMAG] Ratio, at a given location, of the radio field intensity of the desired wave to the noise field intensity. { 'rād·ē·ō 'fēld tə 'nȯiz ,rā·shō }

radio fix [COMMUN] Determination of the position of the source of radio signals by obtaining cross bearings on the transmitter with two or more radio direction finders in different locations, then computing the position by triangulation. [NAV] 1. Determination of the position of a vessel or aircraft equipped with direction-finding equipment by ascertaining the direction of radio signals received from two or more transmitting stations of known location and then computing the position by triangulation. 2. Determination of position of an aircraft in flight by identification of a radio beacon or by locating the intersection of two radio beams. { 'rād·ē·ō ,fiks }

radio fixing aid [NAV] Equipment making use of radio to assist in the determination of a geographical position. { 'rād·ē·ō 'fiks·iŋ ,ād }

radio frequency [ELECTROMAG] A frequency at which coherent electromagnetic radiation of energy is useful for communication purposes; roughly the range from 10 kilohertz to 100 gigahertz. Abbreviated rf. { 'rād·ē·ō ,frē·kwən·sē }

radio-frequency alternator [ELEC] A rotating-type alternator designed to produce high power at frequencies above power-line values but generally lower than 100,000 hertz; used chiefly for high-frequency heating. { 'rād·ē·ō ˌfrē·kwən·sē 'ȯl·tə,nād·ər }

radio-frequency amplifier [ELECTR] An amplifier that amplifies the high-frequency signals commonly used in radio communications. { 'rād·ē·ō ˌfrē·kwən·sē 'am·plə,fī·ər }

radio-frequency bandwidth [COMMUN] Band of frequencies comprising 99% of the total radiated power of the signal transmission extended to include any discrete frequency on which the power is at least 0.25% of the total radiated power. { 'rād·ē·ō ˌfrē·kwən·sē 'band,width }

radio-frequency cable [ELECTROMAG] A cable having electric conductors separated from each other by a continuous homogeneous dielectric or by touching or interlocking spacer beads; designed primarily to conduct radio-frequency energy with low losses. Also known as RG line. { 'rād·ē·ō ˌfrē·kwən·sē ,kā·bəl }

radio-frequency cavity preselector [ELECTROMAG] A tunable cavity resonator in an ultra-high-frequency circuit, which is similar in function to a tuned resonant circuit. { 'rād·ē·ō ˌfrē·kwən·sē 'kav·əd·ē ,prē·si'lek·tər }

radio-frequency choke [ELEC] A coil designed and used specifically to block the flow of radio-frequency current while passing lower frequencies or direct current. { 'rād·ē·ō ˌfrē·kwən·sē ,chōk }

radio-frequency component [COMMUN] Portion of a signal or wave which consists only of the radio-frequency alternations, and not including its audio rate of change in amplitude frequency. { 'rād·ē·ō ˌfrē·kwən·sē kəm,pō·nənt }

radio-frequency current [ELEC] Alternating current having a frequency higher than 10,000 hertz. { 'rād·ē·ō ˌfrē·kwən·sē ,kə·rənt }

radio-frequency filter [ELECTR] An electric filter which enhances signals at certain radio frequencies or attenuates signals at undesired radio frequencies. { 'rād·ē·ō ˌfrē·kwən·sē ,fil·tər }

radio-frequency generator [ELECTR] A generator capable of supplying sufficient radio-frequency energy at the required frequency for induction or dielectric heating. { 'rād·ē·ō ˌfrē·kwən·sē 'jen·ə,rād·ər }

radio-frequency head [ENG] Unit consisting of a radar transmitter and part of a radar receiver, the two contained in a package for ready removal and installation. { 'rād·ē·ō ˌfrē·kwən·sē 'hed }

radio-frequency heating See electronic heating. { 'rād·ē·ō ˌfrē·kwən·sē 'hēd·iŋ }

radio-frequency interference [COMMUN] Interference from sources of energy outside a system or systems, as contrasted to electromagnetic interference generated inside systems. Abbreviated RFI. { 'rād·ē·ō ˌfrē·kwən·sē ,in·tər'fir·əns }

radio-frequency line See radio-frequency transmission line. { 'rād·ē·ō ˌfrē·kwən·sē ,līn }

radio-frequency measurement [ELECTR] The precise measurement of frequencies above the audible range by any of various techniques, such as a calibrated oscillator with some means of comparison with the unknown frequency, a digital counting or scaling device which measures the total number of events occurring during a given time interval, or an electronic circuit for producing a direct current proportional to the frequency of its input signal. { 'rād·ē·ō ˌfrē·kwən·sē 'mezh·ər·mənt }

radio-frequency oscillator [ELECTR] An oscillator that generates alternating current at radio frequencies. { 'rād·ē·ō ˌfrē·kwən·sē 'äs·ə,lād·ər }

radio-frequency power supply [ELECTR] A high-voltage power supply in which the output of a radio-frequency oscillator is stepped up by an air-core transformer to the high voltage required for the second anode of a cathode-ray tube, then rectified to provide the required high direct-current voltage; used in some television receivers. { 'rād·ē·ō ˌfrē·kwən·sē 'pau·ər sə,plī }

radio-frequency preheating [ENG] Preheating of plastics-molding materials by radio frequencies of 10-100 megahertz per second to facilitate the molding operation or to reduce the molding-cycle time. Abbreviated rf preheating. { 'rād·ē·ō ˌfrē·kwən·sē ,prē'hēd·iŋ }

radio-frequency pulse [COMMUN] A radio-frequency carrier that is amplitude-modulated by a pulse; the amplitude of the modulated carrier is zero before and after the pulse. Also known as radio pulse. { 'rād·ē·ō ˌfrē·kwən·sē ,pəls }

radio-frequency quadrupole [NUCLEO] A special vane-type accelerating structure that is used as an injector in some proton linear accelerators and also provides quadrupole focusing by electric fields near the axis. { ¦rād·ē·ō ¦frē·kwən·sē 'kwä·drə‚pōl }

radio-frequency reactor [ELECTR] A reactor used in electronic circuits to pass direct current and offer high impedance at high frequencies. { ¦rād·ē·ō ¦frē·kwən·sē rē'ak·tər }

radio-frequency resistance See high-frequency resistance. { ¦rād·ē·ō ¦frē·kwən·sē ri'zis·təns }

radio-frequency sensor [ENG] A device that uses radio signals to determine the position of objects to be manipulated by a robotic system. { ¦rād·ē·ō ¦frē·kwən·sē ‚sen·sər }

radio-frequency shielding [ELECTROMAG] Enclosure of a physical space or an object with a shield that prevents radio-frequency electromagnetic radiation from leaving or entering. { ¦rād·ē·ō ¦frē·kwən·sē ‚shēld·iŋ }

radio-frequency shift See frequency shift. { ¦rād·ē·ō ¦frē·kwən·sē ‚shift }

radio-frequency signal generator [ELECTR] A test instrument that generates the various radio frequencies required for alignment and servicing of radio, television, and electronic equipment. Also known as service oscillator. { ¦rād·ē·ō ¦frē·kwən·sē 'sig·nəl ‚jen·ə‚rād·ər }

radio-frequency spectrometer [SPECT] An instrument which measures the intensity of radiation emitted or absorbed by atoms or molecules as a function of frequency at frequencies from 10^5 to 10^9 hertz; examples include the atomic-beam apparatus, and instruments for detecting magnetic resonance. { ¦rād·ē·ō ¦frē·kwən·sē spek'träm·əd·ər }

radio-frequency spectroscopy [SPECT] The branch of spectroscopy concerned with the measurement of the intervals between atomic or molecular energy levels that are separated by frequencies from about 10^5 to 10^9 hertz, as compared to the frequencies that separate optical energy levels of about 6×10^{14} hertz. { ¦rād·ē·ō ¦frē·kwən·sē spek'träs·kə·pē }

radio-frequency spectrum See radio spectrum. { ¦rād·ē·ō ¦frē·kwən·sē 'spek·trəm }

radio-frequency SQUID [ELECTR] A type of SQUID which has only one Josephson junction in a superconducting loop; its state is determined from radio-frequency measurements of the impedance of the ring. { ¦rād·ē·ō ¦frē·kwən·sē 'skwid }

radio-frequency transformer [ELECTROMAG] A transformer having a tapped winding or two or more windings designed to furnish inductive reactance or to transfer radio-frequency energy from one circuit to another by means of a magnetic field; may have an air core or some form of ferrite core. Also known as radio transformer. { ¦rād·ē·ō ¦frē·kwən·sē tranz'fòr·mər }

radio-frequency transmission line [ELECTROMAG] A transmission line designed primarily to conduct radio-frequency energy, consisting of two or more conductors supported in a fixed spatial relationship along their own length. Also known as radio-frequency line. { ¦rād·ē·ō ¦frē·kwən·sē tranz'mish·ən ‚līn }

radio-frequency welding See high-frequency welding. { ¦rād·ē·ō ¦frē·kwən·sē 'weld·iŋ }

radio galaxy [ASTROPHYS] A galaxy that is emitting much energy in radio frequencies often from regions devoid of visible matter. { ¦rād·ē·ō 'gal·ik·sē }

radiogenic [NUC PHYS] Pertaining to a material produced by radioactive decay, as the production of lead from uranium decay. { ¦rād·ē·ō¦jen·ik }

radiogenic age determination See radiometric dating. { ¦rād·ē·ō¦jen·ik 'āj di‚tərm·ə‚nā·shən }

radiogenic argon [NUC PHYS] Argon occurring in rocks and minerals that is the result of in-place decay of potassium-40 since the formation of the earth. { ¦rād·ē·ō¦jen·ik 'är‚gän }

radiogenic dating See radiometric dating. { ¦rād·ē·ō¦jen·ik 'dād·iŋ }

radiogenic isotope [NUC PHYS] An isotope which was produced by the decay of a radioisotope, but which itself may or may not be radioactive. { ¦rād·ē·ō¦jen·ik 'ī·sə‚tōp }

radiogenic lead [NUC PHYS] Stable, end-product lead (Pb-206, Pb-207, and Pb-208) occurring in rocks and minerals that is the result of in-place decay of uranium and thorium since the formation of the earth. { ¦rād·ē·ō¦jen·ik 'led }

radiogenic strontium [NUC PHYS] Strontium-87 occurring in rocks and minerals that is the direct result of in-place decay of rubidium-87 since the formation of the earth. { ¦rād·ē·ō¦jen·ik 'strän·chəm }

radiogeology [GEOCHEM] The study of the distribution patterns of radioactive elements in the earth's crust and the role of radioactive processes in geologic phenomena. { ¦rād·ē·ō·jē'äl·ə·jē }

radioglaciology [GEOPHYS] The study of glacier ice by means of radar, especially the sounding of ice depth. { ¦rād·ē·ō‚glā·sē'äl·ə·jē }

radiogoniometer [ELECTR] A goniometer used as part of a radio direction finder. { ¦rād·ē·ō gō·nē'äm·əd·ər }

radiogoniometry [ENG] Science of locating a radio transmitter by means of taking bearings on the radio waves emitted by such a transmitter. { ¦rād·ē·ō‚gō·nē'äm·ə·trē }

radiogram [COMMUN] A message transmitted by radio. { 'rād·ē·ə‚gram }

radiograph [GRAPHICS] The photographic image produced in radiography. Also known as shadowgraph. { 'rād·ē·ə‚graf }

radiographic equivalence factor [NUCLEO] The reciprocal of the thickness of a specified material having the same radiographic absorption as a unit thickness of a standard material. { ¦rād·ē·ə¦graf·ik i'kwiv·ə·ləns ‚fak·tər }

radiographic film [GRAPHICS] The photographic film used in radiography, which must be properly selected for contrast, latitude, and sensitivity. { ¦rād·ē·ə¦graf·ik 'film }

radiographic sensitivity [NUCLEO] A measure of radiographic quality whereby the minimum discontinuity that may be detected on the film is expressed as a percentage of the base thickness. { ¦rād·ē·ə¦graf·ik ‚sen·sə'tiv·əd·ē }

radiography [GRAPHICS] The technique of producing a photographic image of an opaque specimen by transmitting a beam of x-rays or γ-rays through it onto an adjacent photographic film; the image results from variations in thickness, density, and chemical composition of the specimen; used in medicine and industry. { ‚rād·ē'äg·rə·fē }

radio guard [COMMUN] Ship, aircraft, or radio station designed to listen for and record transmission, and to handle traffic on a designated frequency for a certain unit or units. Also known as radio communications guard. { 'rād·ē·ō ‚gärd }

radio guidance [ELECTR] Guidance of a flight-borne missile or other vehicle from a ground station by means of radio signals. { 'rād·ē·ō 'gīd·əns }

radio guided bomb [ORD] A bomb, such as the azon, guided by radio control from outside the missile. { 'rād·ē·ō 'gīd·əd 'bäm }

radio hole [GEOPHYS] Strong fading of the radio signal at some position in space along an air-to-air or air-to-ground path; the effect is caused by the abnormal refraction of radio waves. { 'rād·ē·ō ‚hōl }

radio homing aid [NAV] Radio equipment which permits an aircraft to reach a way point or its destination by maintaining constant some parameter which is derived through the use of the equipment. { 'rād·ē·ō 'hōm·iŋ ‚ād }

radio homing beacon See homing beacon. { 'rād·ē·ō 'hōm·iŋ ‚bē·kən }

radio horizon [COMMUN] The locus of points at which direct rays from a transmitter become tangential to the surface of the earth; the distance to the radio horizon is affected by atmospheric refraction. { 'rād·ē·ō hə'rīz·ən }

radiohumeral index [ANTHRO] The ratio, multiplied by 100, of the length of the radius to the length of the humerus of the human arm. { ¦rād·ē·ō'hyüm·ə·rəl 'in‚deks }

radiohumeral joint [ANAT] The joint in the elbow between the radius and the humerus bones. { ¦rād·ē·ō'hyüm·ə·rəl 'jòint }

radiohydrology [NUCLEO] The study of the hydrologic relationships of extraction processing, and use (including use in hydrologic investigations) of radioactive materials and the disposal of associated waste products. { ¦rād·ē·ō·hī'dräl·ə·jē }

radioimmunoassay [IMMUNOL] A sensitive method for determining the concentration of an antigenic substance in a sample by comparing its inhibitory effect on the binding of a radioactivity-labeled antigen to a limited amount of a specific antibody with the inhibitory effect of known standards. { ¦rād·ē·ō‚im·yə·nō'a‚sā }

radio-inertial guidance system [ENG] A command type of missile guidance system consisting essentially of a radar

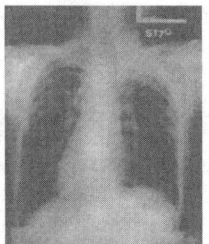

tracking unit; a computer that accepts missile position and velocity information from the tracking system and furnishes to the command link appropriate signals to steer the missile; the command link, which consists of a transmitter on the ground and an antenna and receiver on the missile; and an inertial system for partial guidance in case of radio guidance failure. { ¦rād·ē·ō i¦nər·shəl 'gīd·əns ˌsis·təm }

radio intelligence [COMMUN] Information regarding the enemy obtained by interception and interpretation of enemy radio transmissions. { 'rād·ē·ō in'tel·ə·jəns }

radio interception [COMMUN] Tuning in on a radio message not intended for the listener. { 'rād·ē·ō ˌin·tər'sep·shən }

radio interference *See* interference. { 'rād·ē·ō ˌin·tər'fir·əns }

radio interferometer [ENG] Radiotelescope or radiometer employing a separated receiving antenna to measure angular distances as small as 1 second of arc; records the result of interference between separate radio waves from celestial radio sources. { 'rād·ē·ō ˌin·tər·fə'räm·əd·ər }

radioiodine [NUC PHYS] Any radioactive isotope of iodine, especially iodine-131; used as a tracer to determine the activity and size of the thyroid gland, and experimentally, to destroy the thyroid glands of animals. { ¦rād·ē·ō'ī·ə¸dīn }

radioisotope [NUC PHYS] An isotope which exhibits radioactivity. Also known as radioactive isotope; unstable isotope. { ¦rād·ē·ō'ī·sə¸tōp }

radioisotope assay [ANALY CHEM] An analytical technique including procedures for separating and reproducibly measuring a radioactive tracer. { ¦rād·ē·ō'ī·sə¸tōp 'aˌsā }

radioisotope battery *See* nuclear battery. { ¦rād·ē·ō'ī·sə¸tōp 'bad·ə·rē }

radioisotope heating unit [NUCLEO] A unit that contains a small amount of plutonium and may be carried on board a spacecraft to heat nearby components. Abbreviated RHU. { ¦rād·ē·ōˌī·sə¸tōp 'hēd·iŋ ¸yü·nət }

radioisotope thermoelectric generator [NUCLEO] A device for converting nuclear energy to electrical energy in which the heat produced by radioactivity of a radioisotope is used to produce a voltage in a thermocouple circuit; chief use has been in space vehicles and in instruments left on the lunar surface. Abbreviated RTG. { ¦rād·ē·ō'ī·sə¸tōp ¦thər·mō·i'lek·trik 'jen·ə¸rād·ər }

radioisotopic generator *See* nuclear battery. { ¦rād·ē·ōˌī·sə'täp·ik 'jen·ə¸rād·ər }

radio knife [MED] A surgical knife that uses a high-frequency electric arc at its tip to cut tissue and simultaneously sterilize the edges of the wound. { 'rād·ē·ō ¸nīf }

radio landing aid [NAV] Equipment using radio waves to assist aircraft in carrying out their actual landings. { 'rād·ē·ō 'land·iŋ ¸ād }

radio landing beam [NAV] A radio beam used for vertical guidance of aircraft during descent to a landing surface. { 'rād·ē·ō 'land·iŋ ¸bēm }

Radiolaria [INV ZOO] A subclass of the protozoan class Actinopodea whose members are noted for their siliceous skeletons and characterized by a membranous capsule which separates the outer from the inner cytoplasm. { ¸rād·ē·ō'lar·ē·ə }

radiolarian chert [GEOL] A homogeneous cryptocrystalline radiolarite with a well-developed matrix. { ¦rād·ē·ō¦lar·ē·ən 'chərt }

radiolarian earth [GEOL] A porous, unconsolidated siliceous sediment formed from the opaline silica skeletal remains of Radiolaria; formed from radiolarian ooze. { ¦rād·ē·ō¦lar·ē·ən 'ərth }

radiolarian ooze [GEOL] A siliceous ooze containing the skeletal remains of the Radiolaria. { ¦rād·ē·ō¦lar·ē·ən 'üz }

radiolarite [GEOL] **1.** A whitish, hard, consolidated equivalent of radiolarian earth. **2.** Radiolarian ooze that has been indurated. { ¸rād·ē·ō'la¸rīt }

radiole [INV ZOO] A spine on a sea urchin. { 'rād·ē·ōl }

radio line of position [NAV] A line of position determined by any of a series of radio navigation aids. { 'rād·ē·ō 'līn əv pə'zish·ən }

radio link [COMMUN] A radio system used to provide a communication or control channel between two specific points. { 'rād·ē·ō ¸liŋk }

radiolitic [PETR] **1.** Pertaining to the texture of an igneous rock, characterized by radial, fanlike groupings of acicular crystals, resembling sectors of spherulites. **2.** Referring to

limestones in which the components radiate from central points, with the cement making up less than 50% of the total rock. { ¦rād·ē·ō¦lid·ik }

radiolocation [ENG] Determination of relative position of an object by means of equipment operating on the principle that propagation of radio waves is at a constant velocity and rectilinear. { ¦rād·ē·ō·lō'kā·shən }

radio log [COMMUN] A log of radio messages sent and received, together with other pertinent information, maintained by radio operators. { 'rād·ē·ō ¸läg }

radiological [NUCLEO] Pertaining to nuclear radiation, radioactivity, and atomic weapons. { ¦rād·ē·ə¦läj·ə·kəl }

radiological agent [NUCLEO] Any of a family of substances that produce casualties by emitting radiation. { ¦rād·ē·ə¦läj·ə·kəl 'ā·jənt }

radiological defense [ORD] The methods, plans, and procedures involved in establishing and exercising defensive measures against the radiation effects of an attack by atomic weapons or radiological warfare agents. { ¦rād·ē·ə¦läj·ə·kəl di'fens }

radiological dose [NUCLEO] The total amount of ionizing radiation absorbed by an individual exposed to any radiating source. { ¦rād·ē·ə¦läj·ə·kəl 'dōs }

radiological survey [NUCLEO] Determination of the distribution and dose rates of radiation in an area. { ¦rād·ē·ə¦läj·ə·kəl 'sər¸vā }

radiological warfare [ORD] The employment of agents or weapons to produce residual radioactive contamination, as distinguished from the initial effects of a nuclear explosion (blast, thermal, and initial nuclear radiation). { ¦rād·ē·ə¦läj·ə·kəl 'wȯr¸fer }

radiologist [MED] A physician who specializes in the use of radiant energy in the diagnosis and treatment of disease. { ¸rād·ē·äl·ə·jəst }

radiology [MED] The medical science concerned with radioactive substances, x-rays, and other ionizing radiations, and the application of the principles of this science to diagnosis and treatment of disease. { ¸rād·ē·äl·ə·jē }

radiolucent [ELECTROMAG] Transparent to x-rays and radio waves. { ¦rād·ē·ō'lüs·ənt }

radioluminescence [PHYS] Luminescence produced by x-rays or γ-rays, or by particles emitted in radioactive decay. { ¦rād·ē·ō¸lü·mə'nes·əns }

radiolysis [PHYS CHEM] The dissociation of molecules by radiation; for example, a small amount of water in a reactor core dissociates into hydrogen and oxygen during operation. { ¸rād·ē'äl·ə·səs }

radio magnetic indicator [NAV] An aircraft instrument which presents a combined display of omnibearing vehicle heading and relative bearings of a radio station. { 'rād·ē·ō mag'ned·ik 'in·də¸kād·ər }

radio marker beacon station [NAV] Station marking a definite location on the ground as an aid to aircraft at altitude; the International Civil Aviation Organization beacons operate at a single frequency of 75 megahertz. { 'rād·ē·ō 'mär·kər ¸bē·kən ¸stā·shən }

radio mast [ENG] A tower, pole, or other structure for elevating an antenna. { 'rād·ē·ō 'mast }

radio metal locator *See* metal detector. { 'rād·ē·ō 'med·əl 'lō¸kād·ər }

radio meteor [ASTRON] A meteor which has been detected by the reflection of a radio signal from the meteor trail of relatively high ion density (ion column); such an ion column is left behind a meteoroid when it reaches the region of the upper atmosphere between about 50 and 75 miles (80 and 120 kilometers), although occasionally radio meteors are detected at higher altitudes. { 'rād·ē·ō 'mēd·ē·ər }

radio-meteorograph [COMMUN] A device for the automatic radio transmission of the indications of a set of meteorological instruments. { 'rād·ē·ō ¸mēd·ē'ȯr·ə¸graf }

radio meteorology [METEOROL] That branch of the science of meteorology which embraces the propagation of radio energy through the atmosphere, and the use of radio and radar equipment in meteorology; this is the most general term and includes radar meteorology. Also known as radioelectric meteorology. { 'rād·ē·ō ¸mēd·ē·ə'räl·ə·jē }

radiometer [ELECTR] A receiver for detecting microwave thermal radiation and similar weak wide-band signals that resemble noise and are obscured by receiver noise; examples include the Dicke radiometer, subtraction-type radiometer, and

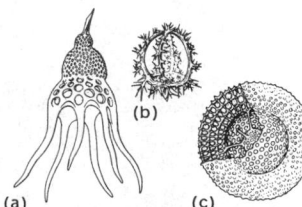

RADIOLARIA

Radiolaria. Skeletons representing (*a, b*) certain Monopylina (or Nasellina) and (*c*) certain Periphylina (or Spumellina).

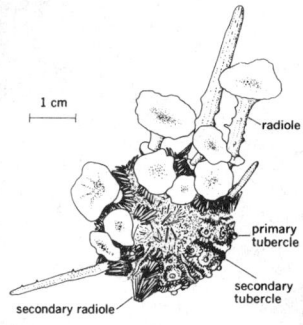

RADIOLE

Morphological features of *Goniocidaris parasol* showing radioles.

two-receiver radiometer. Also known as microwave radiometer; radiometer-type receiver. [ENG] An instrument for measuring radiant energy; examples include the bolometer, microradiometer, and thermopile. { ‚rād·ē'äm·əd·ər }

radiometer effect [PHYS] The effect of the temperature of a surface on the pressure exerted on it in a low vacuum, due to the effect on the momentum transferred to gas molecules colliding with the surface. { ‚rād·ē'äm·əd·ər i‚fekt }

radiometer-type receiver *See* radiometer. { ‚rād·ē'äm·əd·ər ¦tīp ri'sē·vər }

radiometric age [GEOL] Geologic age expressed in years determined by quantitatively measuring radioactive elements and their decay products. { ¦rād·ē·ō¦me·trik 'āj }

radiometric analysis [ANALY CHEM] Quantitative chemical analysis that is based on measurement of the absolute disintegration rate of a radioactive component having a known specific activity. { ¦rād·ē·ō¦me·trik ə'nal·ə·səs }

radiometric dating [ARCHEO] A dating method that utilizes the radioactive decay of certain long-lived, naturally occurring parent isotopes to stable daughter isotopes. [NUCLEO] A technique for measuring the age of an object or sample of material by determining the ratio of the concentration of a radioisotope to that of a stable isotope in it; for example, the ratio of carbon-14 to carbon-12 reveals the approximate age of bones, pieces of wood, and other archeological specimens. Also known as isotopic age determination; nuclear age determination; radioactive age determination; radioactive dating; radiogenic age determination; radiogenic dating. { ‚rād·ē·ə‚me·trik 'dād·iŋ }

radiometric magnitude [ASTRON] A celestial body's magnitude as calculated from the total amount of radiant energy of all the wavelengths that reach the earth's surface. { ¦rād·ē·ō¦me·trik 'mag·nə‚tüd }

radiometric titration [ANALY CHEM] Use of radioactive indicator to track the transfer of material between two liquid phases in equilibrium, such as titration of $^{110}AgNO_3$ (silver nitrate, with the silver atom having mass number 110) against potassium chloride. { ¦rād·ē·ō¦me·trik tī'trā·shən }

radiometry [PHYS] The detection and measurement of radiant electromagnetic energy, especially that associated with infrared radiation. { ‚rād·ē'äm·ə·trē }

radiomicrometer *See* microradiometer. { ¦rād·ē·ō·mī'kräm·əd·ər }

radiomimetic activity [BIOL] The radiationlike effects of certain chemicals, such as nitrogen mustard, urethane, and fluorinated pyrimidines. { ¦rād·ē·ō·mi¦med·ik ak'tiv·əd·ē }

radiomimetic substances [CHEM] Chemical substances which cause biological effects similar to those caused by ionizing radiation. { ¦rād·ē·ō·mi¦med·ik 'səb·stəns·əz }

radio mirage [ELECTROMAG] The detection of radar targets at phenomenally long range due to radio ducting. { ¦rād·ē·ō mə'räzh }

radiomutation [GEN] A chromosomal aberration which is the result of exposure of living tissue to ionizing radiation. { ¦rād·ē·ō·myü'tā·shən }

radio navigation [NAV] The use of apparatus operating at radio frequencies to determine parameters useful for navigation; it includes radio direction finding, radio ranges, radio compasses, radio homing beacons, and loran. { rād·ē·ō ‚nav·ə'gā·shən }

radio nebula [ASTROPHYS] A nebula that emits nonthermal radio-frequency radiation; derives its luminosity from collisions with the surrounding interstellar medium, or from processes associated with the magnetic fields presumably involved within the nebula; examples are the network nebulae in Cygnus and NGC 443. { 'rād·ē·ō 'neb·yə·lə }

radionecrosis [PATH] Destruction of living tissue by radiation. { ¦rād·ē·ō·ne'krō·səs }

radio net [COMMUN] System of radio stations operating with each other; a military net usually consists of a radio station of a superior unit and stations of all subordinate or supporting units. { 'rād·ē·ō ‚net }

radio noise [ELECTROMAG] Electromagnetic noise having radio frequencies. { 'rād·ē·ō ‚nȯiz }

radio-noise field strength [ELECTROMAG] A quantity which is proportional, or related in a known manner, to the field strength of electromagnetic waves of an interfering character at a point, such as a radio receiver. Also known as noise field strength. { 'rād·ē·ō ¦nȯiz 'fēld ‚streŋkth }

radionuclide [NUC PHYS] A nuclide that exhibits radioactivity. { ¦rād·ē·ō'nü‚klīd }

radionuclide content [NUC PHYS] The sum of the number of radioactive nuclei before and after a nuclear reaction as a fraction of the total number of nuclei involved (reactant nuclei plus reaction-product nuclei). { ¦rād·ē·ō'nü‚klīd 'kän‚tent }

radio-paging system [COMMUN] A system consisting of personal paging receivers, radio transmitters, and an encoding device, designed to alert an individual, or group of individuals, and deliver a short message. { 'rād·ē·ō ¦pāj·iŋ ‚sis·təm }

radiopaque [ELECTROMAG] Not appreciably penetrable by x-rays or other forms of radiation. { ¦rād·ē·ō'pāk }

radiopaque agent [MED] A substance, such as barium sulfate, which is opaque to x-rays, administered orally to a patient to provide contrast in x-ray photographs of his gastrointestinal system. { ¦rād·ē·ō'pāk ‚ā·jənt }

radiopasteurization [ENG] Pasteurization by surface treatment with low-energy irradiation. { ¦rād·ē·ō‚pas·chúr·ə'zā·shən }

radiophare *See* radio beacon. { 'rād·ē·ō‚fer }

radiopharmaceutical [PHARM] A radioactive drug used for medicinal purposes, either diagnostic or therapeutic. { ‚rād·ē·ō‚fär·mə·süd·ə·kəl }

radiophone *See* radiotelephone. { 'rād·ē·ə‚fōn }

radiophoto *See* facsimile. { ¦rād·ē·ō'fōd·ō }

radiophotoluminescence [PHYS] Luminescence exhibited by minerals such as fluorite and kunzite as a result of irradiation with β- and γ-rays followed by exposure to light. { ¦rād·ē·ō¦fōd·ō‚lü·mə'nes·əns }

radio pill [ELECTR] A device used in biotelemetry for monitoring the physiologic activity of an animal, such as pH values of stomach acid; an example is the Heidelberg capsule. { 'rād·ē·ō ‚pil }

radio position finding [ENG] Process of locating a radio transmitter by plotting the intersection of its azimuth as determined by two or more radio direction finders. { 'rād·ē·ō pə'zish·ən ‚fīnd·iŋ }

radio-positioning land station [COMMUN] Station in the radiolocation service, other than a radio-navigation station, not intended for operation while in motion. { 'rād·ē·ō pə¦zish·ən·iŋ 'land ‚stā·shən }

radio-positioning mobile station [COMMUN] Station in the radiolocation service, other than a radio-navigation station, intended to be used while in motion or during halts at unspecified points. { 'rād·ē·ō pə¦zish·ən·iŋ 'mō·bəl ‚stā·shən }

radio prospecting [ENG] Use of radio and electric equipment to locate mineral or oil deposits. { 'rād·ē·ō 'prä‚spek·tiŋ }

radio proximity fuse [ORD] A proximity fuse that contains a miniature radio transmitter and uses radio echoes from a target to trigger the fuse within predetermined limits of distance from the target. { 'rād·ē·ō präk'sim·əd·ē ‚fyüz }

radio pulse [COMMUN] *See* radio-frequency pulse. [ELECTROMAG] An intense burst of radio-frequency energy lasting for a fraction of a second. { 'rād·ē·ō ‚pəls }

radio range finding [NAV] Radiolocation in which the distance of an object is determined by measuring the time of arrival of its radio emissions, whether independent, reflected, or retransmitted on the same or other wavelength. { 'rād·ē·ō 'rānj ‚fīnd·iŋ }

radio range station [NAV] Radio-navigation land station in the aeronautical, radio-navigation service providing radio equisignal zones. { 'rād·ē·ō 'rānj ‚stā·shən }

radio receiver [ELECTR] A device that converts radio waves into intelligible sounds or other perceptible signals. Also known as radio; radio set; receiving set. { 'rād·ē·ō ri‚sēv·ər }

radio recognition [COMMUN] Determination by radio means of the friendly or enemy character, or the individuality, of another radio station. { 'rād·ē·ō ‚rek·ig'nish·ən }

radio recombination line [SPECT] A radio-frequency spectral line that results from an electron transition between energy levels in an atom or ion having a large principal quantum number n, greater than 50. { 'rād·ē·ō rē‚käm·bə'nā·shən ‚līn }

radio relay satellite *See* communications satellite. { 'rād·ē·ō 'rē‚lā ‚sad·əl‚īt }

radio relay system [COMMUN] A radio transmission system in which intermediate radio stations or radio repeaters receive and retransmit radio signals. Also known as relay system. { 'rād·ē·ō 'rē‚lā ‚sis·təm }

radio repeater [COMMUN] A repeater that acts as an intermediate station in transmitting radio communications signals or radio programs from one fixed station to another; serves to extend the reliable range of the originating station; a microwave repeater is an example. { 'rād·ē·ō ri‚pēd·ər }

radioresistance [BIOL] The resistance of organisms or tissues to the harmful effects of various radiations. { ¦rād·ē·ō·ri'zis·təns }

radio scanner *See* scanning radio. { 'rād·ē·ō 'skan·ər }

radio scattering *See* scattering. { 'rād·ē·ō 'skad·ə·riŋ }

radiosensitive [MATER] Sensitive to damage by radiant energy. { ¦rād·ē·ō 'sen·səd·iv }

radio set *See* radio transmitter. { 'rād·ē·ō ‚set }

radio sextant [ELECTROMAG] An antenna with a high-resolution beam pattern that measures the angle between local direction references and an astronomical radio signal source such as an artificial satellite, the sun, the moon, or a radio star. { 'rād·ē·ō 'sek·stənt }

radio shielding [ELEC] Metallic covering over all electric wiring and ignition apparatus, which is grounded at frequent intervals for the purpose of eliminating electric interference with radio communications. { 'rād·ē·ō ‚shēld·iŋ }

radio signal [COMMUN] A signal transmitted by radio. { 'rād·ē·ō 'sig·nəl }

radio-signal reporting code [COMMUN] A code for reporting the quality of radiotelephone or radiotelegraph transmission, consisting of a code word followed by a group of numbers rating various characteristics. Also known as signal reporting code. { 'rād·ē·ō ‚sig·nəl ri'pórd·iŋ ‚kōd }

radio silence [COMMUN] Period during which all or certain radio equipment capable of radiation is kept inoperative. { 'rād·ē·ō 'sī·ləns }

radiosonde [ENG] A balloon-borne instrument for the simultaneous measurement and transmission of meteorological data; the instrument consists of transducers for the measurement of pressure, temperature, and humidity, a modulator for the conversion of the output of the transducers to a quantity which controls a property of the radio-frequency signal, a selector switch which determines the sequence in which the parameters are to be transmitted, and a transmitter which generates the radio-frequency carrier. { 'rād·ē·ō‚sänd }

radiosonde balloon [AERO ENG] A balloon used to carry a radiosonde aloft; it is considerably larger than a pilot balloon or a ceiling balloon. { 'rād·ē·ō‚sänd bə‚lün }

radiosonde commutator [ELECTR] A component of a radiosonde consisting of a series of alternate electrically conducting and insulating strips; as these are scanned by a contact, the radiosonde transmits temperature and humidity signals alternately. { 'rād·ē·ō‚sänd 'käm·yə‚täd·ər }

radiosonde observation [METEOROL] An evaluation in terms of temperature, relative humidity, and pressure aloft, of radio signals received from a balloon-borne radiosonde; the height of each mandatory and significant pressure level of the observation is computed from these data. Also known as raob. { 'rād·ē·ō‚sänd ‚äb·zər‚vā·shən }

radiosonde-radio-wind system [ENG] An apparatus consisting of a standard radiosonde and radiosonde ground equipment to obtain upper-air data on pressure, temperature, and humidity, and a self-tracking radio direction finder to provide the elevation and azimuth angles of the radiosonde so that the wind vectors may be obtained. { 'rād·ē·ō‚sänd ¦rād·ē·ō ‚wind ‚sis·təm }

radiosonde set [ENG] A complete set for automatically measuring and transmitting high-altitude meteorological data by radio from such carriers as a balloon or rocket. { 'rād·ē·ō‚sänd ‚set }

radio sonobuoy *See* sonobuoy. { 'rād·ē·ō 'sän·ə‚bói }

radio source [ASTROPHYS] A source of extragalactic or interstellar electromagnetic emission in radio wavelengths. { 'rād·ē·ō ‚sórs }

radio spectrum [COMMUN] The entire range of frequencies in which useful radio waves can be produced, extending from the audio range to about 300,000 megahertz. Also known as radio-frequency spectrum. { 'rād·ē·ō 'spek·trəm }

radio spectrum allocation [COMMUN] The specification of the frequencies of the radio spectrum which are available for use by the various radio services. { 'rād·ē·ō 'spek·trəm ‚al·ə'kā·shən }

RADIO TELESCOPE

Installation of electronics at the focus of an 85-foot-diameter (26-meter) radio telescope.

radio star [ASTROPHYS] A discrete celestial radio source. { 'rād·ē·ō ‚stär }

radio station [COMMUN] A station equipped to engage in radio communication or radio broadcasting. { 'rād·ē·ō ‚stā·shən }

radio storm [ASTROPHYS] A prolonged period of disturbed emission or reception that lasts for periods of hours up to days. { 'rād·ē·ō ‚stórm }

radio sun [ASTROPHYS] The sun as defined by its electromagnetic radiation in the radio portion of the spectrum. { 'rād·ē·ō ¦sən }

radio tail object [ASTROPHYS] An extragalactic object that displays a strong tail or jet at radio frequencies. { 'rād·ē·ō ¦tāl ‚äb·jəkt }

radiotelegraphy [COMMUN] Telegraphy involving the use of radio waves in place of wire lines. { ¦rād·ē·ō·tə'leg·rə·fē }

radiotelemetry [COMMUN] The reception of data at a location remote from the source of the data, using radio-frequency electromagnetic radiation as the means of transmission. { ¦rād·ē·ō·tə'lem·ə·trē }

radiotelephone [COMMUN] **1.** Pertaining to telephony over radio channels. **2.** A radio transmitter and radio receiver used together for two-way telephone communication by radio. Also known as radiophone. { ¦rād·ē·ō'tel·ə‚fōn }

radiotelephony [COMMUN] Two-way transmission of sounds by means of modulated radio waves, without interconnecting wires. { ¦rād·ē·ō·tə'lef·ə·nē }

radio telescope [ENG] An astronomical instrument used to measure the amount of radio energy coming from various directions in the sky, consisting of a highly directional antenna and associated electronic equipment. { 'rād·ē·ō 'tel·ə‚skōp }

radioteletype [COMMUN] A teletypewriter and the associated equipment needed for operation over a radio channel rather than over wires. Abbreviated RTTY. { ¦rād·ē·ō'tel·ə‚tīp }

radioteletypewriter [COMMUN] A teletypewriter and the associated equipment needed for operation over a radio channel rather than over wires. { ¦rād·ē·ō‚tel·ə'tīp‚wrīd·ər }

radiotherapy *See* radiation therapy. { ¦rād·ē·ō'ther·ə·pē }

radiothorium [NUC PHYS] Conventional name of the isotope of thorium which has mass number 228. Symbolized RdTh. { ¦rād·ē·ō'thór·ē·əm }

radio time signal [COMMUN] A time signal sent by radio broadcast. { 'rād·ē·ō 'tīm ‚sig·nəl }

radio tower [COMMUN] A tower, usually several hundred meters tall, either guyed or freestanding, on which a transmitting antenna is mounted to increase the range of radio transmission; in some cases, the tower itself may be the antenna. { 'rād·ē·ō ‚taú·ər }

radiotoxicity [MED] A radioactive compound that is toxic to living cells or tissues, causing radiation sickness. { ‚rād·ē·ō·täk'sis·əd·ē }

radiotracer *See* radioactive tracer. { ¦rād·ē·ō'trā·sər }

radio tracking [ENG] The process of keeping a radio or radar beam set on a target and determining the range of the target continuously. { 'rād·ē·ō 'trak·iŋ }

radio transmission [COMMUN] The transmission of signals through space at radio frequencies by means of radiated electromagnetic waves. { 'rād·ē·ō tranz'mish·ən }

radio transmitter [ELECTR] The equipment used for generating and amplifying a radio-frequency carrier signal, modulating the carrier signal with intelligence, and feeding the modulated carrier to an antenna for radiation into space as electromagnetic waves. Also known as radio set; transmitter. { 'rād·ē·ō 'tranz‚mid·ər }

radio transponder [ELECTR] A transponder which receives and transmits radio waves, in contrast to a sonar transponder, which receives and transmits acoustic waves. { 'rād·ē·ō tran'spän·dər }

radio tube *See* electron tube. { 'rād·ē·ō ‚tüb }

radio watch *See* watch. { 'rād·ē·ō ‚wäch }

radio wave [ELECTROMAG] An electromagnetic wave produced by reversal of current in a conductor at a frequency in the range from about 10 kilohertz to about 300,000 megahertz. { 'rād·ē·ō ‚wāv }

radio wavefront distortion [ELECTROMAG] Change in the direction of advance of a radio wave. { 'rād·ē·ō 'wāv‚frənt di‚stór·shən }

radio-wave propagation [ELECTROMAG] The transfer of energy through space by electromagnetic radiation at radio frequencies. { 'rād·ē·ō ¦wāv ‚präp·ə‚gā·shən }

radio window [GEOPHYS] A band of frequencies extending from about 6 to 30,000 megahertz, in which radiation from the outer universe can enter and travel through the atmosphere of the earth. { 'rād·ē·ō ‚win·dō }

radish [BOT] *Raphanus sativus.* **1.** An annual or biennial crucifer belonging to the order Capparales. **2.** The edible, thickened hypocotyl of the plant. { 'rad·ish }

radist [NAV] Radio-navigation system in which the comparison of arrival times of transmitted pulses, at three or more ground stations, indicates the position of the vehicle. { 'rā‚dist }

radium [CHEM] **1.** A radioactive member of group II, symbol Ra, atomic number 88; the most abundant naturally occurring isotope has mass number 226 and a half-life of 1620 years. **2.** A highly toxic solid that forms water-soluble compounds; decays by emission of α, β, and γ-radiation; melts at 700°C, boils at 1140°C; turns black in air; used in medicine, in industrial radiography, and as a source of neutrons and radon. { 'rād·ē·əm }

radium age [NUCLEO] The age of a mineral as calculated from the numbers of radium atoms present originally, now, and when equilibrium is established with ionium. { 'rād·ē·əm ‚āj }

radium bromide [INORG CHEM] $RaBr_2$ Water-soluble, poisonous, radioactive white powder, corrosive to skin or flesh; melts at 728°C; used in medicine, physical research, and luminous paint. { 'rād·ē·əm 'brō‚mīd }

radium carbonate [INORG CHEM] $RaCO_3$ Water-insoluble, poisonous, radioactive, white powder; used in medicine. { 'rād·ē·əm 'kär·bə‚nāt }

radium cell [NUCLEO] A sealed thin-wall tube containing radium. { 'rād·ē·əm ‚sel }

radium chloride [INORG CHEM] $RaCl_2$ Water- and alcohol-soluble, poisonous, radioactive, yellow-white crystals; corrosive effect on skin and flesh; melts at 1000°C; used in medicine, physical research, and luminous paint. { 'rād·ē·əm 'klór‚īd }

radium F *See* polonium-210. { 'rād·ē·əm 'ef }

radium needle [NUCLEO] A radium cell in the form of a needle, usually of platinum-iridium or gold alloy, designed primarily for insertion in tissue. { 'rād·ē·əm ‚nēd·əl }

radium plaque [NUCLEO] A radium container in which the radium is distributed over a surface; the shielding is usually small in one direction so as to permit transmission of β-rays as well as γ-rays. { 'rād·ē·əm ‚plak }

radium sulfate [INORG CHEM] $RaSO_4$ Water-insoluble, radioactive, poisonous, white crystals; used in medicine. { 'rād·ē·əm 'səl‚fāt }

radium therapy [MED] Radiotherapy using the radiations from radium. { 'rād·ē·əm ‚ther·ə·pē }

radius [ANAT] The outer of the two bones of the human forearm or of the corresponding part in vertebrates other than fish. [MATH] **1.** A line segment joining the center and a point of a circle or sphere. **2.** The length of such a line segment. { 'rād·ē·əs }

radius cutter [MECH ENG] A formed milling cutter with teeth ground to produce a radius on the workpiece. { 'rād·ē·əs ‚kəd·ər }

radius of action [ENG] The maximum distance a ship, aircraft, or other vehicle can travel away from its base along a given course with normal load and return without refueling, but including the fuel required to perform those maneuvers made necessary by all safety and operating factors. { 'rād·ē·əs əv 'ak·shən }

radius of convergence [MATH] The positive real number corresponding to a power series expansion about some number a with the property that if $x - a$ has absolute value less than this number the power series converges at x, and if $x - a$ has absolute value greater than this number the power series diverges at x. { 'rād·ē·əs əv kən'vər·jəns }

radius of curvature [MATH] The radius of the circle of curvature at a point of a curve. { 'rād·ē·əs əv 'kər·və·chər }

radius of damage [ORD] The distance from ground zero of a nuclear blast at which there is a 0.50 probability of achieving the desired damage. { 'rād·ē·əs əv 'dam·ij }

radius of geodesic curvature [MATH] For a point on a curve lying on a surface, the reciprocal of the geodesic curvature at the point. { ¦rād·ē·əs əv ¦jē·ə¦des·ik 'kər·və·chər }

radius of geodesic torsion [MATH] The reciprocal of the geodesic torsion of a surface at a point in a given direction. { ¦rād·ē·əs əv ¦jē·ə¦des·ik 'tór·shən }

radius of gyration [MATH] The square root of the ratio of the moment of inertia of a plane figure about a given axis to its area. [MECH] The square root of the ratio of the moment of inertia of a body about a given axis to its mass. { 'rād·ē·əs əv ji'rā·shən }

radius of normal curvature [MATH] The reciprocal of the normal curvature of a surface at a point and in a given direction. { ¦rād·ē·əs əv ¦nór·məl 'kər·və·chər }

radius of protection [ENG] The radius of the circle within which a lightning discharge will not strike, due to the presence of an elevated lightning rod at the center. { 'rād·ē·əs əv prə'tek·shən }

radius of rupture [ORD] Greatest distance from the center of an underground explosive charge at which the explosion will be destructive. { 'rād·ē·əs əv 'rəp·chər }

radius of safety [ORD] The horizontal distance from ground area beyond which the weapon effects on friendly troops are acceptable. { 'rād·ē·əs əv 'sāf·tē }

radius of torsion [MATH] The reciprocal of the torsion of a space curve at a point. { ¦rād·ē·əs əv 'tór·shən }

radius of total curvature [MATH] The quantity $\sqrt{-1/C}$, where C is the total curvature of a surface at a point. { ¦rād·ē·əs əv ¦tōd·əl 'kər·və·chər }

radius of visibility [NAV] The radius of a circle limiting the area in which an object can be seen under specified conditions. { 'rād·ē·əs əv ‚viz·ə'bil·əd·ē }

radius ratio [PHYS CHEM] The ratio of the radius of a cation to the radius of an ion; relative ionic radii are pertinent to crystal lattice structure, particularly the determination of coordination number. { 'rād·ē·əs ‚rē·shō }

radius rod [ENG] A rod which restricts movement of a part to a given arc. { 'rād·ē·əs ‚räd }

radius vector [ASTRON] A line joining the center of an orbiting body with the focus of its orbit located near its primary. [MATH] The coordinate r in a polar coordinate system, which gives the distance of a point from the origin. { 'rād·ē·əs ‚vek·tər }

radix *See* root. { 'rād·iks }

radix approximation [MATH] The approximation of a number by a number that can be expressed by a specified finite number of digits in radix notation. { 'rād·iks ə‚präk·sə'mā·shən }

radix complement [MATH] A numeral in positional notation that can be derived from another by subtracting the original numeral from the numeral of highest value with the same number of digits, and adding 1 to the difference. Also known as complement; true complement. { 'rād·iks 'käm·plə·mənt }

radix fraction [MATH] A generalization of a decimal fraction given by an expression of the form $(a/r) + (b/r^2) + (c/r^3) + \ldots$, where r is an integer and $a, b, c, \ldots$ are integers that are less than r. { 'rād·iks ‚frak·shən }

radix-minus-one complement [MATH] A numeral in positional notation of base (or radix) B derived from a given numeral by subtracting the latter from the highest numeral with the same number of digits, that is, from $B-1$; it is 1 less than the radix complement. { 'rād·iks ¦mī·nəs ¦wən 'käm·plə·mənt }

radix notation [MATH] A positional notation in which the successive digits are interpreted as coefficients of successive integral powers of a number called the radix or base; the represented number is equal to the sum of this power series. Also known as base notation. { 'rād·iks nō‚tā·shən }

radix point [MATH] A dot written either on or slightly above the line, used to mark the point at which place values change from positive to negative powers of the radix in a number system; a decimal point is a radix point for radix 10. { 'rād·iks ‚póint }

radix transformation [COMPUT SCI] A method of transformation that involves changing the radix or base of the original key and either discarding excess high-order digits (that is, digits in excess of the number desired in the key) or extracting some part of the transformed number. { 'rād·iks ‚tranz·fər'mā·shən }

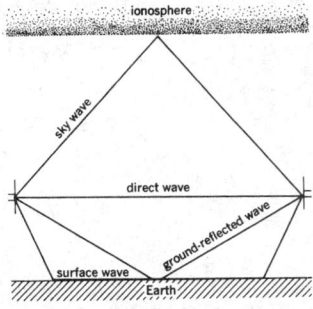

Possible transmission paths of electromagnetic radiation at radio frequencies.

Radlac accelerator [NUCLEO] A linear induction accelerator in which air-core cavities excited by external fields are used to generate the accelerating field. { 'rad,lak ik'sel·ə,rād·ər }

radome [ELECTROMAG] A strong, thin shell, made from a dielectric material that is transparent to radio-frequency radiation, and used to house a radar antenna, or a space communications antenna of similar structure. { 'rā,dōm }

radon [CHEM] A chemical element, symbol Rn, atomic number 86; all isotopes are radioactive, the longest half-life being 3.82 days for mass number 222; it is the heaviest element of the noble-gas group, produced as a gaseous emanation from the radioactive decay of radium. [NUC PHYS] The conventional name for radon-222. Symbolized Rn. { 'rā,dän }

radon-220 [NUC PHYS] The isotope of radon having mass number 220, symbol ^{220}Rn, which is a radioactive member of the thorium series with a half-life of 56 seconds. { 'rā,dän ,tü'twen·tē }

radon-222 [NUC PHYS] The isotope of radon having mass number 222, symbol ^{222}Rn, which is a radioactive member of the uranium series with a half-life of 3.82 days. { 'rā,dän ,tü'twen·tē,tü }

radon breath analysis [MED] Examination of exhaled air for the presence of radon to determine the presence and quantity of radium in the human body. { 'rā,dän 'breth ə'nal·ə·səs }

Radon measure See regular Borel measure. { 'rā,dän ,mezh·ər }

Radon theorem [MATH] The theorem that a set in an n-dimensional Euclidean space with at least $n + 2$ points is the union of two sets whose convex spans are disjoint. { 'rā,dän ,thir·əm }

Radon transform [MATH] A mathematical operation that is roughly equivalent to finding the projection of a function along a given line; useful in computerized tomography. { 'rā,dän 'tranz,fȯrm }

Radstockian [GEOL] A European stage of geologic time forming the upper Upper Carboniferous, above Staffordian and below Stephanian, equivalent to uppermost Westphalian. { rad'stäk·ē·ən }

radula [INV ZOO] A filelike ribbon studded with horny or chitinous toothlike structures, found in the mouth of all classes of mollusks except Bivalvia. { 'raj·ə·lə }

radzimir [TEXT] A firm, lustrous fabric with embedded cross-ribs, woven of silk. Also spelled rhadzimer. { 'rad·zə,mir }

rafaelite [PETR] A nepheline-free orthoclase-bearing hypabyssal rock that also contains analcime and calcic plagioclase. { 'raf·ē·ə,līt }

raffiche [METEOROL] In the Mediterranean region, gusts from the mountains; violent gusts of the bora. { rä'fēsh }

raffinase [BIOCHEM] An enzyme that hydrolyzes raffinose, yielding fructose in the reaction. { 'raf·ə,nās }

raffinate [CHEM ENG] In solvent refining, that portion of the treated liquid mixture that remains undissolved and is not removed by the selective solvent. Also known as good oil to petroleum-refinery operators. { 'raf·ə,nāt }

raffinose [BIOCHEM] $C_{18}H_{32}O_{16}·5H_2O$ A white, crystalline trisaccharide found in sugarbeets, cottonseed meal, and molasses; yields glucose, fructose, and galactose on complete hydrolysis. Also known as gossypose; melitose; melitriose. { 'raf·ə,nōs }

Rafflesiales [BOT] A small order of dicotyledonous plants; members are highly specialized, nongreen, rootless parasites which grow from the roots of the host. { re,flē·zhē'ā·lēz }

rafos [NAV] A long-range marine navigation system in which the ship measures the arrival times of sound waves from explosive charges released at fixed shore stations at known times. { 'rā,fōs }

raft [ENG] A quantity of timber or lumber secured together by means of ropes, chains, or rods and used for transportation by floating. [GEOL] **1.** A rock fragment caught up in a magma and drifting freely, more or less vertically. **2.** See float coal. [HYD] An accumulation or jam of floating logs, driftwood, dislodged trees, or other debris, formed naturally in a stream by caving of the banks. { raft }

rafted ice [OCEANOGR] A form of pressure ice composed of overlying pieces of ice floe. { 'raf·təd ,īs }

rafter [BUILD] A roof-supporting member immediately beneath the roofing material. { 'raf·tər }

rafter dam [CIV ENG] A dam made of horizontal timbers that meet in the center of the stream like rafters in a roof. { 'raf·tər ,dam }

raft foundation [CIV ENG] A continuous footing that supports an entire structure, such as a floor. Also known as foundation mat. { 'raft fau̇n,dā·shən }

rafting [GEOL] Transporting of rock by floating ice or floating organic materials (such as logs) to places not reached by water currents. [OCEANOGR] The process of forming rafted ice. { 'raft·iŋ }

raft lake [HYD] A relatively short-lived body of water impounded along a stream by a raft. { 'raft ,lāk }

raft tectonics See plate tectonics. { 'raft tek,tän·iks }

rag [PETR] Any of various hard, coarse, rubbly, or shell rocks that weather with a rough, irregular surface, such as a flaggy sandstone or limestone used as a building stone. Also known as ragstone. { rag }

rag bolt See barb bolt. { 'rag ,bōlt }

ragged ceiling See indefinite ceiling. { 'rag·əd'sēl·iŋ }

raggiatura [METEOROL] Land squalls descending with great force from ravines and valleys in high land in Italy; they extend only a short distance off the west coast. { ,rä·jə'tu̇r·ə }

raggle [BUILD] **1.** A manufactured masonry unit, frequently made of terra cotta, having a slot or groove to receive a metal flashing. Also known as flashing block; raggle block. **2.** A groove cut into masonry to receive adjoining material. { 'rag·əl }

raggle block See raggle. { 'rag·əl ,bläk }

raglanite [PETR] A nepheline syenite composed of oligoclase, nepheline, and corundum with minor amounts of mica, calcite, magnetite, and apatite. { 'rag·lə,nīt }

rag papers [MATER] The most expensive papers, made wholly or partly from cotton or linen rags; rag content is expressed as 25, 50, 75, or 100%; pure rag papers are the strongest and most resistant to the discoloration and deterioration due to age. { 'rag ,pāpərs }

ragstone See rag. { 'rag,stōn }

RAID [COMPUT SCI] A group of hard disks that operate together to improve performance or provide fault tolerance and error recovery through data striping, mirroring, and other techniques. Derived from redundant array of inexpensive disks. { rād }

rail [ENG] **1.** A bar extending between posts or other supports as a barrier or guard. **2.** A steel bar resting on the crossties to provide track for railroad cars and other vehicles with flanged wheels. [MECH ENG] A high-pressure manifold in some fuel injection systems. { rāl }

rail anchor [CIV ENG] A device that prevents tracks from moving longitudinally and maintains the proper gap between sections of rail. { 'rāl ,aŋ·kər }

rail bender [ENG] A portable appliance for bending rails for track or for straightening bent or curved rails. { 'rāl ,ben·dər }

rail capacity [CIV ENG] The maximum number of trains which can be planned to move in both directions over a specified section of track in a 24-hour period. { 'rāl kə,pas·əd·ē }

rail clip [CIV ENG] **1.** A plate that holds a rail at its base. **2.** A device used to fasten a derrick or crane to the rails of a track to prevent tipping. **3.** A support on a track rail, used for holding a detector bar. { 'rāl ,klip }

rail crane See locomotive crane. { 'rāl ,krān }

rail-fence jammer See continuous-wave jammer. { 'rāl ¦fens ,jam·ər }

railhead [CIV ENG] **1.** The topmost part of a rail, supporting the wheels of railway vehicles. **2.** A point at which railroad traffic originates and terminates. **3.** The temporary ends of a railroad line under construction. { 'rāl,hed }

railing [CIV ENG] A barrier consisting of a rail and supports. [ELECTR] Radar pulse jamming at high recurrence rates (50 to 150 kilohertz); it results in an image on a radar indicator resembling fence railing. { 'rāl·iŋ }

rail joint [CIV ENG] A rigid connection of the ends of two sections of railway track. { 'rāl ,jȯint }

Raillietiellidae [INV ZOO] A small family of parasitic arthropods in the order Cephalobaenida. { ,rāl·yə'tyel·ə,dē }

railroad [CIV ENG] A permanent line of rails forming a route for freight cars and passenger cars drawn by locomotives. { 'rāl,rōd }

railroad engineering [CIV ENG] That part of transportation engineering involved in the planning, design, development, operation, construction, maintenance, use, or economics of facilities for transportation of goods and people in wheeled units of rolling stock running on, and guided by, rails normally supported on crossties and held to fixed alignment. Also known as railway engineering. { 'rāl‚rōd ‚en·jə'nir·iŋ }

railroad ferry [NAV ARCH] A ship having a deck with railroad tracks, for carrying railroad cars between two ports. { 'rāl‚rōd ‚fer·ē }

railroad jack [MECH ENG] **1.** A hoist used for lifting locomotives. **2.** A portable jack for lifting heavy objects. **3.** A hydraulic jack, either powered or lever-operated. { 'rāl‚rōd ‚jak }

railroad thermit [MET] Red thermit to which is added up to 16% nickel, manganese, and steel. { 'rāl‚rōd 'thər·mət }

rail steel [MET] Steel used to make rail track. { 'rāl ‚stēl }

railway artillery [ORD] Heavy artillery, given mobility by mounting on railway mounts or carriages; no longer in use by U.S. forces. { 'rāl‚wā är'til·ə·rē }

railway carriage See railway mount. { 'rāl‚wā ‚kar·ij }

railway dry dock [CIV ENG] A railway dock consisting of tracks built on an incline on a strong foundation, and extending from a sufficient distance in shore to allow a vessel to be hauled out of the water. { 'rāl‚wā 'drī ‚däk }

railway end-loading ramp [CIV ENG] A sloping platform situated at the end of a track and rising to the level of the floor of the railcars (wagons). { 'rāl‚wā 'end ‚lōd·iŋ ‚ramp }

railway engineering See railroad engineering. { 'rāl‚wā ‚en·jə'nir·iŋ }

railway mount [ORD] Mount for railway artillery; no longer in use by U.S. forces, but still of interest for possible future application. Also known as railway carriage. { 'rāl‚wā ‚maůnt }

RAIM See receiver autonomous integrity monitoring. { räm or 'är‚ā‚ī'em }

rain [METEOROL] Precipitation in the form of liquid water drops with diameters greater than 0.5 millimeter, or if widely scattered the drops may be smaller; the only other form of liquid precipitation is drizzle. { rān }

rain and snow mixed [METEOROL] Precipitation consisting of a mixture of rain and wet snow; usually occurs when the temperature of the air layer near the ground is slightly above freezing. { 'rān ən 'snō 'mikst }

rain area report See radar report. { 'rān ‚er·ē·ə ri‚pȯrt }

rain attenuation [COMMUN] Attenuation of radio waves when passing through moisture-bearing cloud formations or areas in which rain is falling; increases with the density of the moisture in the transmission path. { 'rān ə‚ten·yə‚wā·shən }

rainbow [ELECTR] Technique which applies pulse-to-pulse frequency changing to identifying and discriminating against decoys and chaff. [OPTICS] Colored arc seen in the sky when the sun or moon is illuminating large numbers of falling raindrops. [PETRO ENG] Chromatic iridescence observed in drilling fluid that has been circulated in a well, indicating contamination or contact with fresh hydrocarbons. { 'rān‚bō }

rainbow granite [PETR] A type of granite having either a black or dark-green background with pink, yellowish, or reddish mottling, or a pink background with dark mottling. { 'rān‚bō 'gran·ət }

rainbow roof [ARCH] A pitched roof with slightly arched slopes. Also known as whaleback roof. { 'rān‚bō 'rüf }

rainbow scattering [PHYS] The scattering of particles by a potential that contains both attractive and repulsive parts and whose width is much greater than the de Broglie wavelength of the particles; analogous to scattering of light by liquid droplets, which produces a rainbow. { 'rān‚bō ‚skad·ər·iŋ }

rain cloud [METEOROL] Any cloud from which rain falls; a popular term having no technical denotation. { 'rān ‚klaůd }

rain crust [HYD] A type of snow crust, formed by refreezing after surface snow crystals have been melted and wetted by liquid precipitation; composed of individual ice particles such as firn. { 'rān ‚krəst }

rain desert [ECOL] A desert in which rainfall is sufficient to maintain a sparse general vegetation. { 'rān ‚dez·ərt }

raindrop [METEOROL] A drop of water of diameter greater than 0.5 millimeter falling through the atmosphere. { 'rān‚dräp }

raindrop impressions See rain prints. { 'rān‚dräp im'presh·ənz }

raindrop imprints See rain prints. { 'rān‚dräp 'im‚prins }

Rainey's corpuscle [INV ZOO] The sickle-shaped spore of an encysted sarcosporidian. { 'rā‚nēz ‚kȯr·pə·səl }

rain factor [HYD] A coefficient designed to measure the combined effect of temperature and moisture on the formation of soil humus; it is obtained by dividing the annual rainfall (in millimeters) by the mean annual temperature (in degrees Celsius). { 'rān ‚fak·tər }

rainfall [METEOROL] The amount of precipitation of any type; usually taken as that amount which is measured by means of a rain gage (thus a small, varying amount of direct condensation is included). { 'rān‚fȯl }

rainfall frequency [CLIMATOL] The number of times, during a specified period of years, that precipitation of a certain magnitude or greater occurs or will occur at a station; numerically, the reciprocal of the frequency is usually given. { 'rān‚fȯl ‚frē·kwən·sē }

rainfall intensity See precipitation intensity. { 'rān‚fȯl in'ten·səd·ē }

rainfall inversion See precipitation inversion. { 'rān‚fȯl in‚vər·zhən }

rainfall penetration [HYD] The depth below the soil surface to which water from a given rainfall has been able to infiltrate. { 'rān‚fȯl ‚pen·ə‚trā·shən }

rainfall regime [CLIMATOL] The character of the seasonal distribution of rainfall at any place; the chief rainfall regimes, as defined by W. G. Kendrew, are equatorial, tropical, monsoonal, oceanic and continental westerlies, and Mediterranean. { 'rān‚fȯl rə‚zhēm }

rainforest [ECOL] A forest of broad-leaved, mainly evergreen, trees found in continually moist climates in the tropics, subtropics, and some parts of the temperate zones. { 'rān‚fär·əst }

rainforest climate See wet climate. { 'rān‚fär·əst ‚klī·mət }

rain gage [ENG] An instrument designed to collect and measure the amount of rain that has fallen. Also known as ombrometer; pluviometer; udometer. { 'rān ‚gāj }

rain-gage shield [ENG] A device which surrounds a rain gage and acts to maintain horizontal flow in the vicinity of the funnel so that the catch will not be influenced by eddies generated near the gage. Also known as wind shield. { 'rān ‚gāj ‚shēld }

rain gush See cloudburst. { 'rān ‚gəsh }

rain gust See cloudburst. { rān ‚gəst }

raininess [METEOROL] Generally, the quantitative character of rainfall for a given place. { 'rān·ē·nəs }

rain-intensity gage [ENG] An instrument which measures the instantaneous rate at which rain is falling on a given surface. Also known as rate-of-rainfall gage. { 'rān in'ten·səd·ē ‚gāj }

rainmaking [METEOROL] Popular term applied to all activities designed to increase, through any artificial means, the amount of precipitation released from a cloud. { 'rān‚māk·iŋ }

rain pillar [GEOL] A minor landform consisting of a column of soil or soft rock capped and protected by pebbles or concretions, produced by the differential erosion from the impact of falling rain. { 'rān ‚pil·ər }

rain prints [GEOL] Small, shallow depressions formed in soft sediment or mud by the impact of falling raindrops. Also known as raindrop impressions; raindrop imprints. { 'rān ‚prins }

rain rot [VET MED] Weeping dermatitis accompanied by swelling of the skin and loss of hair in sheep exposed to prolonged periods of rain. { 'rān ‚rät }

rain shadow [METEOROL] An area of diminished precipitation on the lee side of mountains or other topographic obstacles. { 'rān ‚shad·ō }

rainsquall [METEOROL] A squall associated with heavy convective clouds, frequently the cumulonimbus type; usually sets in shortly before the thunderstorm rain, blowing outward from the storm and generally lasting only a short time. Also known as thundersquall. { 'rān‚skwȯl }

rain stage [METEOROL] The thermodynamic process of condensation of water from moist air in an idealized saturation-adiabatic or pseudoadiabatic lifting, at temperatures above the freezing point; begins at the condensation level. { 'rān ‚stāj }

rainwash [GEOL] **1.** The washing away of loose surface material by rainwater after it has reached the ground but before it has been concentrated into definite streams. **2.** Material

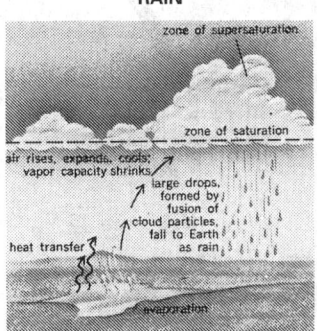

RAIN

Diagram of the steps in the formation of rain.

transported and accumulated, or washed away, by rainwater. { 'rān‚wäsh }

rainwater [HYD] Water that has fallen as rain and is quite soft, as it has not yet collected soluble matter from the soil. { 'rān‚wȯd·ər }

rainy climate [CLIMATOL] In W. Koppen's climatic classification, any climate type other than the dry climates; however, it is generally understood that this refers principally to the tree climates and not the polar climates. { 'rān·ē 'klī·mət }

rainy season [CLIMATOL] In certain types of climate, an annually recurring period of one or more months during which precipitation is a maximum for that region. Also known as wet season. { 'rān·ē ‚sēz·ən }

raise [MIN ENG] A shaftlike mine opening, driven upward from a level to connect with a level above, or to the surface. { rāz }

raise boring machine [MIN ENG] A machine that is used to drill pilot holes between levels in a mine, and to ream the pilot hole to the finished dimension of the raise. { rāz ¦bȯr·iŋ mə‚shēn }

raised beach [GEOL] An ancient beach raised to a level above the present shoreline by uplift or by lowering of the sea level; often bounded by inland cliffs. { 'rāzd 'bēch }

raised bog [ECOL] An area of acid, peaty soil, especially that developed from moss, in which the center is relatively higher than the margins. { 'rāzd 'bäg }

raised flooring [CIV ENG] A flooring system having removable panels supported on adjustable pedestals or stringers to allow convenient access to the space below. Also known as access flooring; elevated flooring; pedestal flooring. { 'rāzd 'flȯr·iŋ }

raise drill [MIN ENG] A circular raise driving machine which bores a pilot hole and reams it to finished raise diameter. { 'rāz ‚dril }

raise driller [MIN ENG] A person who works in a raise. Also known as raiseman. { 'rāz ‚dril·ər }

raiseman See raise driller. { 'rāz·mən }

raising plate See wall plate. { 'rāz·iŋ ‚plāt }

Rajakaruna engine [MECH ENG] A rotary engine that uses a combustion chamber whose sides are pin-jointed together at their ends. { 'rä·jä·kə'rün·ə ‚en·jən }

Rajidae [VERT ZOO] The skates, a family of elasmobranchs included in the batoid group. { 'raj·ə‚dē }

Rajiformes [VERT ZOO] The equivalent name for Batoidea. { ‚raj·ə'fȯr‚mēz }

rake [BUILD] The exterior finish and trim applied parallel to the sloping end walls of a gabled roof. [DES ENG] A hand tool consisting of a long handle with a row of projecting prongs at one end; for example, the tool used for gathering leaves or grass on the ground. [ENG] The angle between an inclined plane and the vertical. [GEOL] See plunge. [MECH ENG] The angle between the tooth face or a tangent to the tooth face of a cutting tool at a given point and a reference plane or line. [NAV ARCH] The angle between the vertical direction and a part of a ship, such as a mast, funnel, bow, stern, rudder, or sternpost. [ORD] To sweep a target, especially a ship or a column of troops, with gun or cannon fire. { rāk }

rake blade [ENG] A blade on a bulldozer in the form of spaced tines that point down. { 'rāk ‚blād }

raked joint [CIV ENG] A mortar, or masonry, joint from which the mortar has been scraped out to about 3/4 inch (20 millimeters). { 'rākt ‚jȯint }

râle [MED] An abnormal sound accompanying the normal sounds of respiration within the air passages and heard on auscultation of the chest. { ral }

Rallidae [VERT ZOO] A large family of birds in the order Gruiformes comprising rails, gallinules, and coots. { 'ral·ə‚dē }

ralstonite [MINERAL] $NaMgAl_5F_{12}(OH)_6 \cdot 3H_2O$ A colorless, white, or yellowish mineral composed of hydrous basic sodium magnesium aluminum fluoride, occurring in octahedral crystals. { 'rȯl·stə‚nīt }

ram [AERO ENG] The forward motion of an air scoop or air inlet through the air. [HYD] An underwater ledge or projection from an ice wall, ice front, iceberg, or floe, usually caused by the more intensive melting and erosion of the unsubmerged part. Also known as apron; spur. [MECH ENG] A plunger,

weight, or other guided structure for exerting pressure or drawing something by impact. [MIN ENG] See barney. [VERT ZOO] A male sheep or goat. { ram }

Ram See Aries. { ram }

RAM See random-access memory. { ram }

ram accelerator [AERO ENG] A launching device in which a projectile is fired into a stationary tube filled with a combustible gas mixture that accelerates the projectile to hypersonic speed in a manner analogous to a ramjet engine. { ¦ram ik'sel·ə‚rād·ər }

Raman effect [OPTICS] A phenomenon observed in the scattering of light as it passes through a transparent medium; the light undergoes a change in frequency and a random alteration in phase due to a change in rotational or vibrational energy of the scattering molecules. Also known as Raman scattering. { 'räm·ən i‚fekt }

Raman-induced Kerr effect [OPTICS] The birefringence of an observation beam whose frequency differs from that of a pumping beam by a characteristic frequency of the medium (molecular vibration or rotation). { ¦räm·ən in¦düst 'kər i‚fekt }

Raman lidar [OPTICS] A type of lidar that measures the scattered signal at the Raman-shifted wavelength in order to determine atmospheric density, temperature, and water vapor concentration. { 'räm·ən 'lī‚där }

Raman-Rayleigh ratio [OPTICS] The ratio of Raman scattering (of a light beam passing through a transparent medium) to Rayleigh scattering (of a light beam horizontal to the medium). { 'räm·ən 'rā‚lē ‚rā·shō }

Raman scattering See Raman effect. { 'räm·ən ‚skad·ə·riŋ }

Raman spectroscopy [SPECT] Analysis of the intensity of Raman scattering of monochromatic light as a function of frequency of the scattered light; the information obtained is useful for determining molecular structure. { 'räm·ən spek'träs·kə‚pē }

Raman spectrum [SPECT] A display, record, or graph of the intensity of Raman scattering of monochromatic light as a function of frequency of the scattered light. { 'räm·ən ‚spek·trəm }

Ramapithecinae [PALEON] A subfamily of Hominidae including the protohominids of the Miocene and Pliocene. { ‚räm·ə·pə'thes·ən‚ē }

Ramapithecus [PALEON] The genus name given to a fossilized upper jaw fragment found in the Siwalik hills, India; closely related to the human family. { ‚räm·ə'pith·ə·kəs }

ramark [NAV] A radio marker beacon which continuously transmits pulsed signals appearing as radial lines on a ship's PPI (plan position indicator); the line indicates the direction to the beacon. { 'rā‚märk }

ramate [BIOL] Having branches. { 'ra‚māt }

rambla [GEOL] A dry ravine, or the dry bed of an ephemeral stream. { 'ram·blə }

Rambus dynamic random-access memory [COMPUT SCI] High-performance memory that can transfer data at rates of 800 megahertz and higher. Abbreviated RDRAM. { ¦ram‚bəs dī¦nam·ik ‚ran·dəm 'ak‚ses ‚mem·rē }

RAM disk See RAM drive. { 'ram ‚disk }

ramdohrite [MINERAL] $Pb_3Ag_2Sb_6S_{13}$ A dark-gray mineral composed of a lead silver antimony sulfur compound. { 'räm‚dō‚rīt }

RAM drive [COMPUT SCI] A portion of a computer's random-access memory (RAM) that is made to simulate a disk drive. Also known as RAM disk. { 'ram ‚drīv }

ram effect [MECH ENG] The increased air pressure in a jet engine or in the manifold of a piston engine, due to ram. { 'ram i‚fekt }

ramentum [BOT] A thin brownish scale consisting of a single layer of cells and occurring on the leaves and young shoots of many ferns. { rə'men·təm }

ramicolous [BOT] Living on twigs. { rə'mik·ə·ləs }

ramie [BOT] Boehmeria nivea. A shrub or half-shrub of the nettle family (Urticaceae) cultivated as a source of a tough, strong, durable, lustrous natural woody fiber resembling flax, obtained from the phloem of the plant; used for high-quality papers and fabrics. Also known as China grass; rhea. { 'rā·mē }

ramiform [BOT] Branched or comblike. { 'ram·ə‚fȯrm }

ramjet engine [AERO ENG] A type of jet engine with no mechanical compressor, consisting of a specially shaped tube

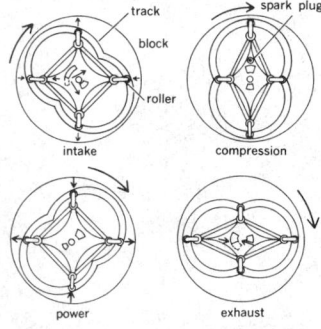

RAJAKARUNA ENGINE

Cross sections of the Rajakaruna engine, a type of revolving block rotary engine, showing the four stages of one combustion cycle.

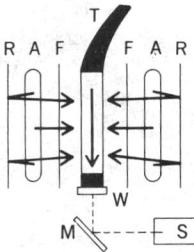

RAMAN EFFECT

Arrangement for excitation of the Raman effect with a noncoherent source. Mercury vapor arcs A are surrounded by appropriate reflectors R to increase intensity of available radiation. Light from arcs passes through filters F which transmit one of several monochromatic frequencies of mercury radiation. Monochromatic radiation enters circular cylindrical tube T in which liquid or vapor is to be studied. Light scattered in direction of vertical arrow passes through window W and to the spectrograph S through optical system M.

or duct open at both ends, the air necessary for combustion being shoved into the duct and compressed by the forward motion of the engine; the air passes through a diffuser and is mixed with fuel and burned, the exhaust gases issuing in a jet from the rear opening. { 'ram,jet ,en·jən }

ramjet exhaust nozzle [AERO ENG] The discharge nozzle in a ramjet engine; hot gas is ejected rearward through this nozzle. { 'ram,jet ig'zȯst ,näz·əl }

rammelsbergite [MINERAL] $NiAs_2$ A gray mineral composed of nickel diarsenide; it is dimorphous with pararammelsbergite. Also known as white nickel. { 'ram·əlz,bər,gīt }

rammer [ENG] An instrument for driving something, such as wood or stones, into another material with force. Also known as beetle; maul. [ORD] **1.** Device for driving a projectile into position in a gun; may be hand- or power-operated. **2.** Tool used to remove a live projectile from the bore of a gun. { 'ram·ər }

ramming [ENG] Packing a powder metal or sand into a compact mass. { 'ram·iŋ }

ramoff [MET] A defect in a casting due to improper ramming of the sand. { 'ram,ȯf }

Ramon flocculation test [IMMUNOL] A method of standardizing antitoxins; a toxin-antitoxin flocculation that is a precipitin reaction in which the end point is the zone of optimal proportion; that is, the zone in which there is no uncombined antigen or antibody. { rə'mōn ,fläk·yə'lā·shən ,test }

ramose [BIOL] Having lateral divisions or branches. { 'rā,mōs }

ramp [ENG] **1.** A uniformly sloping platform, walkway, or driveway. **2.** A stairway which gives access to the main door of an airplane. [HYD] An accumulation of snow forming an inclined plane between land or land ice and sea ice or shelf ice. Also known as drift ice foot. [MIN ENG] A slope between levels in open-pit mining. { ramp }

rampage through core [COMPUT SCI] Action of a computer program that writes data in incorrect locations or otherwise alters storage locations improperly, because of a program error. { 'ram,pāj thrü 'kȯr }

rampart [GEOL] **1.** A narrow, wall-like ridge, 3–7 feet (1–2 meters) high, built up by waves along the seaward edge of a reef flat, and consisting of boulders, shingle, gravel, or reef rubble, commonly capped by dune sand. **2.** A wall-like ridge of unconsolidated material formed along a beach by the action of strong waves and current. **3.** A crescentic or ringlike deposit of pyroclastics around the top of a volcano. { 'ram,pärt }

rampart wall [GEOL] A rimming wall formed along the outer or seaward margin of a terrace, as on various high limestone Pacific islands. { 'ram,pärt ,wȯl }

ram penetrometer See ramsonde. { 'ram ,pen·ə'träm·əd·ər }

ramp generator [ELECTR] A circuit that generates a sweep voltage which increases linearly in value during one cycle of sweep, then returns to zero suddenly to start the next cycle. { 'ramp ,jen·ə,rād·ər }

Ramphastidae [VERT ZOO] The toucans, a family of birds with large, often colorful bills in the order Piciformes. { ,ram 'fas·tə,dē }

ramphoid cusp [MATH] A cusp of a curve which has both branches of the curve on the same side of the common tangent. Also known as single cusp of the second kind. { 'ram,fȯid ,kəsp }

ramping [ENG] In the production of parts fabricated from composite materials, a gradual and programmed sequence of changes in temperature or pressure that control curing and cooling. { 'ramp·iŋ }

ramp mining [MIN ENG] The development of moderately inclined accessways from the surface to mining levels for haulage of ore, materials, waste, workers, and equipment. { 'ramp ,mīn·iŋ }

ram pressure [FL MECH] The pressure that is exerted by a fluid as a result of its motion. { 'ram ,presh·ər }

ram pressure stripping [ASTRON] A process in which the gas in a galaxy interacts with the hot x-ray-emitting gas filling the cluster to which the galaxy belongs and is stripped away. { 'ram ,presh·ər ,strip·iŋ }

RAMPS See resource allocation in multiproject scheduling. { ramps }

ramp sight [ORD] Type of metallic gun sight in which the

aperture is raised or lowered by moving it forward or backward on an inclined ramp. { 'ramp ,sīt }

ramp valley [GEOL] A trough between faults, forced downward by lateral pressure. { 'ramp ,val·ē }

ramp weight [AERO ENG] The static weight of a mission aircraft determined by adding operating weight, payload, flight plan fuel load, and fuel required for ground turbine power unit, taxi, runup, and takeoff. { 'ramp ,wāt }

ram recovery See recovery. { 'ram ri'kəv·ə·rē }

RAM resident [COMPUT SCI] A program that remains stored in a computer's random-access memory (RAM) at all times. Also known as terminate and stay resident (TSR). { ¦ram 'rez·ə·dənt }

ram rocket [AERO ENG] **1.** A rocket motor mounted coaxially in the open front end of a ramjet, used to provide thrust at low speeds and to ignite the ramjet fuel. **2.** The entire unit or power plant consisting of the ramjet and such a rocket. { 'ram ,räk·ət }

Ramsauer effect [ATOM PHYS] The vanishing of the scattering cross section of electrons from atoms of a noble gas at some value of the electron energy, always below 25 electronvolts. { 'räm,zau·ər i,fekt }

Ramsay-Shields-Eötvös equation [THERMO] An elaboration of the Eötvös rule which states that at temperatures not too near the critical temperature, the molar surface energy of a liquid is proportional to $t_c - t - 6$ K, where t is the temperature and t_c is the critical temperature. { 'ram·zē 'shēlz 'öt·vȯsh i,kwā·zhən }

Ramsay-Young method [THERMO] A method of measuring the vapor pressure of a liquid, in which a thermometer bulb is surrounded by cotton wool soaked in the liquid, and the pressure, measured by a manometer, is reduced until the thermometer reading is steady. { ¦ram·zē 'yəŋ ,meth·əd }

Ramsay-Young rule [THERMO] An empirical relationship which states that the ratio of the absolute temperatures at which two chemically similar liquids have the same vapor pressure is independent of this vapor pressure. { 'ram·zē 'yəŋ ,rül }

ramsdellite [MINERAL] MnO_2 An orthorhombic mineral composed of manganese dioxide; it is dimorphous with pyrolusite. { 'ramz·de,līt }

Ramsden circle [OPTICS] A sharp, bright circle of light which appears on a sheet of white paper held near the eyepiece of a telescope focused for infinity and pointed at a bright sky. Also known as Ramsden disk. { 'ramz·dən ,sər·kəl }

Ramsden disk See Ramsden circle. { 'ramz·dən ,disk }

Ramsden eyepiece [OPTICS] An eyepiece consisting of two planoconvex lenses with their plane sides facing outward, having the same power and focal length, and separated by a distance equal to their common focal length. { 'ramz·dən 'ī,pēs }

Ramsey fringes [PHYS] Oscillations in the number of transitions in a molecular beam that passes through two separated radio-frequency fields, as a function of L/v, where L is the separation of the field and v is the speed of the beam. { 'ram·zē ,frin·jəz }

Ramsey number [MATH] For any two positive integers, p and q, the smallest integer, $R(p,q)$, that has the (p,q)-Ramsey property. { 'ram·zē ,nəm·bər }

Ramsey property [MATH] For any two positive integers, p and q, a positive integer r is said to have the (p,q)-Ramsey property if in any set of r people there is either a subset of p people who are all mutual acquaintances or a set of q people who are all mutual strangers. { 'ram·zē ,präp·ərd·ē }

Ramsey theorem [MATH] The theorem that for any two positive integers p and q there is a positive integer r that has the (p,q)-Ramsey property. { ¦ram·zē ¦thir·əm }

Ramsey theory [MATH] The theory of order that must exist in subsets of sufficiently large sets, as illustrated by Ramsey's theorem and van der Waerden's theorem. { 'ram·zē ,thē·ə·rē }

ramsonde [ENG] A cone-tipped metal rod or tube that is driven downward into snow to measure its hardness. Also known as ram penetrometer. { 'ram,sänd }

ram travel [ENG] In injection or transfer molding, the distance moved by the injection ram when filling the mold. { 'ram ,trav·əl }

ram-type turret lathe [MECH ENG] A horizontal turret lathe in which the turret is mounted on a ram or slide which rides on a saddle. { 'ram ¦tər·ət ,lāth }

ramus [ANAT] A slender bone process branching from a large bone. [VERT ZOO] The barb of a feather. [ZOO] The

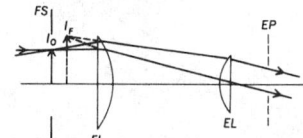

RAMSDEN EYEPIECE

Ramsden eyepiece. FL = field lens; EL = eye lens; FS = field stop; EP = exit pupil or eye point; I_O = image formed by the preceding system; I_F = image formed by the preceding system and the field lens.

branch of a structure such as a blood vessel, nerve, arthropod appendage, and so on. { 'rä·məs }

Rancholabrean [GEOL] A stage of geologic time in southern California, in the upper Pleistocene, above the Irvingtonian. { ˌran·chō·lə'brä·ən }

randannite [MINERAL] An earthy form of opal. { ran 'da,nīt }

randkluft [HYD] A crevasse at the head of a mountain glacier, separating the moving ice and snow from the surrounding rock wall of the valley, where no ice apron is present. { 'ränt,klůft }

random access [COMPUT SCI] **1.** The ability to read or write information anywhere within a storage device in an amount of time that is constant regardless of the location of the information accessed and of the location of the information previously accessed. Also known as direct access. **2.** A process in which data are accessed in nonsequential order and possibly at irregular intervals of time. Also known as single reference. { 'ran·dəm 'ak,ses }

random-access discrete address [COMMUN] Communications technique in which radio users share one wide band instead of each user getting an individual narrow band. { 'ran·dəm ¦ak,ses di'skrēt ə'dres }

random-access disk file [COMPUT SCI] A file which is contained on a disk having one head per track and in which consecutive records are not necessarily in consecutive locations. { 'ran·dəm ¦ak,ses 'disk ,fīl }

random-access input/output [COMPUT SCI] A technique which minimizes seek time and overlaps with processing. { 'ran·dəm ¦ak,ses 'in,půt 'aůt,půt }

random-access memory [COMPUT SCI] A data storage device having the property that the time required to access a randomly selected datum does not depend on the time of the last access or the location of the most recently accessed datum. Abbreviated RAM. Also known as direct-access memory; direct-access storage; random-access storage; random storage; uniformly accessible storage. { 'ran·dəm ¦ak,ses 'mem·rē }

random-access programming [COMPUT SCI] Programming without regard for the time required for access to the storage positions called for in the program, in contrast to minimum-access programming. { 'ran·dəm ¦ak,ses 'prō,gram·iŋ }

random-access storage See random-access memory. { 'ran·dəm ¦ak,ses 'stȯr·ij }

random coil [PHYS CHEM] Any of various irregularly coiled polymers that can occur in solution. Also known as cyclic coil. { 'ran·dəm 'kȯil }

random copolymer [ORG CHEM] Resin copolymer in which the molecules of each monomer are randomly arranged in the polymer backbone. { 'ran·dəm kō'päl·i·mər }

random diffusion chamber See reverberation chamber. { 'ran·dəm di¦fyü·zhən ,chām·bər }

random digit [STAT] Digit taken from a table of random numbers according to some specified probability rule. { 'ran·dəm 'dij·ət }

random error [STAT] An error that can be predicted only on a statistical basis. { 'ran·dəm 'er·ər }

random experiments [STAT] Experiments which do not always yield the same result when repeated under the same conditions. { 'ran·dəm ik'sper·ə·məns }

random forecast [METEOROL] A forecast in which one of a set of meteorological contingencies is selected on the basis of chance; it is often used as a standard of comparison in determining the degree of skill of another forecast method. { 'ran·dəm 'fȯr,kast }

random function [MATH] A function whose domain is an interval of the extended real numbers and has range in the set of random variables on some probability space; more precisely, a mapping of the cartesian product of an interval in the extended reals with a probability space to the extended reals so that each section is a random variable. { 'ran·dəm 'fəŋk·shən }

random interstratification [SOLID STATE] A crystalline structure in which two or more types of layers alternate in a random fashion. { 'ran·dəm ,in·tər,strad·i·fə'kā·shən }

randomization [STAT] Assigning subjects to treatment groups by use of tables of random numbers. { ˌran·də·mə'zā·shən }

randomized blocks [STAT] An experimental design in which the various treatments are reproduced in each of the blocks and are randomly assigned to the units within the blocks,

permitting unbiased estimates of error to be made. { 'ran·də,mīzd 'bläks }

randomized jitter [ELECTR] Jitter by means of noise modulation. { 'ran·də,mīzd 'jid·ər }

randomized test [STAT] Acceptance or rejection of the null hypothesis by use of a random variable to decide whether an observation causes rejection or acceptance. { 'ran·də,mīzd 'test }

randomizing scheme [COMPUT SCI] A technique of distributing records among storage modules to ensure even distribution and seek time. { 'ran·də,mīz·iŋ ,skēm }

random length [ENG] One of a group of various lengths of pipe as delivered by the manufacturer, usually 13–23 feet (4–7 meters) long. Also known as mill length. { 'ran·dəm 'leŋkth }

random line [ENG] A trial surveying line that is directed as closely as circumstances permit toward a fixed terminal point that cannot be seen from the initial point. Also known as random traverse. { 'ran·dəm 'līn }

random mating [GEN] A mating system in which there is an equal opportunity for all male and female gametes to join in fertilization. { 'ran·dəm ,mād·iŋ }

random matrices [MATH] Collections of large matrices, chosen at random from some ensemble. { ˌran·dəm 'mā·tri,sēz }

random noise [MATH] A form of random stochastic process arising in control theory. [PHYS] Noise characterized by a large number of overlapping transient disturbances occurring at random, such as thermal noise and shot noise. Also known as fluctuation noise. { 'ran·dəm 'nȯiz }

random number generator [COMPUT SCI] **1.** A mathematical program which generates a set of numbers which pass a randomness test. **2.** An analog device that generates a randomly fluctuating variable, and usually operates from an electrical noise source. { 'ran·dəm 'nəm·bər ,jen·ə,rād·ər }

random numbers [MATH] A listing of numbers which is nonrepetitive and satisfies no algorithm. { 'ran·dəm 'nəm·bərz }

random ordered sample [STAT] An ordered sample of size s drawn from a population of size N such that the probability of any particular ordered sample is the reciprocal of the number of permutations of N things taken s at a time. { 'ran·dəm ¦ȯr·dərd 'sam·pəl }

random process See stochastic process. { 'ran·dəm 'prä·səs }

random pulsing [COMMUN] Continuous, varying, pulse-repetition rate, accomplished by noise modulation or continuous frequency change. { 'ran·dəm 'pəls·iŋ }

random sampling [STAT] A sampling from some population where each entry has an equal chance of being drawn. { 'ran·dəm 'sam·pliŋ }

random-sampling voltmeter [ENG] A sampling voltmeter which takes samples of an input signal at random times instead of at a constant rate; the synchronizing portions of the instrument can then be simplified or eliminated. { 'ran·dəm ¦sam·pliŋ 'vōlt,mēd·ər }

random sequence [MET] A longitudinal sequence of weld beads deposited in random increments. { 'ran·dəm 'sē·kwəns }

random start [STAT] In a systematic sample, the random selection of a starting point in the first sample block followed by taking that value in the same position in every succeeding block. { 'ran·dəm 'stärt }

random storage See random-access memory. { 'ran·dəm 'stȯr·ij }

random structure [CRYSTAL] A crystal structure in which different types of atoms are associated with the various points in a crystal lattice in a random fashion. { 'ran·dəm 'strək·chər }

random superimposed coding [COMPUT SCI] A system of coding in which a set of random numbers is assigned to each concept to be encoded; with punched cards, each number corresponds to some one hole to be punched in a given field. { 'ran·dəm ¦sü·pər·im'pōzd 'kōd·iŋ }

random traverse See random line. { 'ran·dəm trə'vərs }

random variable [MATH] A measurable function on a probability space; usually real valued, but possibly with values in a general measurable space. Also known as chance variable; stochastic variable; variate. { 'ran·dəm 'ver·ē·ə·bəl }

random vector See diverse vector. { 'ran·dəm 'vek·tər }

random vibration [MECH] A varying force acting on a mechanical system which may be considered to be the sum of a large number of irregularly timed small shocks; induced typically by aerodynamic turbulence, airborne noise from rocket jets, and transportation over road surfaces. { 'ran·dəm vī'brā·shən }

random walk [MATH] A succession of movements along line segments where the direction and possibly the length of each move is randomly determined. { 'ran·dəm 'wȯk }

random winding [ELEC] A coil winding in which the turns are positioned haphazardly rather than in layers. { 'ran·dəm 'wīnd·iŋ }

Raney nickel [MET] A nickel powder prepared from an alloy of nickel and aluminum in equal parts by preferentially dissolving the aluminum in a warm solution of sodium hydroxide. { 'rā·nē ‚nik·əl }

rang [PETR] A unit of subdivision in the C.I.P.W. (Cross-Iddings-Pirsson-Washington) classification of igneous rocks. { räŋ }

range [CIV ENG] Any series of contiguous townships of the U.S. Public Land Survey system. [COMMUN] **1.** In printing telegraphy, that fraction of a perfect signal element through which the time of selection may be varied to occur earlier or later than the normal time of selection without causing errors while signals are being received; the range of a printing telegraph receiving device is commonly measured in percent of a perfect signal element by adjusting the indicator. **2.** Upper and lower limits through which the index arm of the range-finder mechanism of a teletypewriter may be moved and still receive correct copy. [CONT SYS] **1.** The maximum distance a robot's arm or wrist can travel. Also known as reach. **2.** The volume comprising the locations to which a robot's arm or wrist can travel. [ECOL] The area or region over which a species is distributed. [ENG] **1.** The distance capability of an aircraft, missile, gun, radar, or radio transmitter. **2.** A line defined by two fixed landmarks, used for missile or vehicle testing and other test purposes. [MATH] The range of a function f from a set X to a set Y consists of those elements y in Y for which there is an x in X with $f(x) = y$. [MECH] The horizontal component of a projectile displacement at the instant it strikes the ground. [NAV] **1.** A line of bearing defined by a radio range. **2.** *See* radio range. [NUCLEO] The distance that a given ionizing particle can penetrate a given medium before its energy drops to the point that the particle no longer produces ionization. [PHYS] The greatest distance between two particles at which a given force between them is appreciable. [STAT] The difference between the maximums and minimums of a variable quantity. { rānj }

range adjustment [ORD] Correction of firing data so that the impact or burst will be on the target with respect to range. { 'rānj ə‚jəs·mənt }

range-amplitude display [ELECTR] Radar display in which a time base provides the range scale from which echoes appear as deflections normal to the base. { 'rānj 'am·plə‚tüd di‚splā }

range angle [ORD] Angle between the aircraft-target line and the vertical line from the aircraft to the ground at the instant a bomb is released. Also known as dropping angle. { 'rānj ‚aŋ·gəl }

range arithmetic *See* interval arithmetic. { 'rānj ə‚rith·mə·tik }

range attenuation [ELECTROMAG] In radar terminology, the decrease in power density (flux density) caused by the divergence of the flux lines with distance, this decrease being in accordance with the inverse-square law. { 'rānj ə‚ten·yə‚wā·shən }

range-bearing display *See* B display. { 'rānj 'ber·iŋ di‚splā }

range calibration [ENG] Adjustment of a radar set so that when on target the set will indicate the correct range. { 'rānj ‚kal·ə‚brā·shən }

range calibrator [ELECTR] **1.** A device with which the operator of a transmitter calculates the distance over which the signal will extend intelligibly. **2.** A device for adjusting radar range indications by use of known range targets or delayed signals, so when on target the set will indicate the correct range. { 'rānj ‚kal·ə‚brād·ər }

range check [COMPUT SCI] A method of checking the validity of input data by determining whether the values fall within an expected range. { 'rānj ‚chek }

range coding [ENG] Method of coding a radar transponder beacon response so that it appears as a series of illuminated bars on a radarscope; the coding provides identification. { 'rānj ‚kōd·iŋ }

range comprehension [ELECTR] In a frequency-modulation sonar system, valves between the maximum and the minimum ranges. { 'rānj ‚käm·pri‚hen·shən }

range control [AERO ENG] The operation of an aircraft to obtain the optimum flying time. { 'rānj kən‚trōl }

range control chart [AERO ENG] A graph kept in flight on which actual fuel consumption is plotted against distance flown for comparison with planned fuel consumption. { 'rānj kən‚trōl ‚chärt }

range correction [ORD] Changes of firing data necessary to allow for deviations of range due to weather, material, or ammunition. { 'rānj kə‚rek·shən }

range correction board [ORD] Device with which the correction to be applied to a gun is computed mechanically; the correction obtained allows for all nonstandard conditions, such as variations in weather and ammunition. { 'rānj kə‚rek·shən ‚bȯrd }

range corrector setting [ENG] Degree to which the range scale of a position-finding apparatus must be adjusted before use. { 'rānj kə‚rek·tər ‚sed·iŋ }

range delay [ELECTROMAG] A control used in radars which permits the operator to present on the radarscope only those echoes from targets which lie beyond a certain distance from the radar; by using range delay, undesired echoes from nearby targets may be eliminated while the indicator range is increased. { 'rānj di‚lā }

range determination [ORD] Process of finding the distance between a gun and a target, usually by firing the gun, by estimating with the eye, by using a range-finding instrument, or by plotting. { 'rānj di‚tər·mə‚nā·shən }

range deviation [MECH] Distance by which a projectile strikes beyond, or short of, the target; the distance as measured along the gun-target line or along a line parallel to the gun-target line. { 'rānj ‚dē·vē‚ā·shən }

range discrimination *See* distance resolution. { 'rānj di‚skrim·ə‚nā·shən }

range disk [ORD] Graduated disk, used for range setting, connected mechanically with the elevating mechanism of a gun. { 'rānj ‚disk }

range dispersion diagram [ORD] Chart indicating the expected percentage of shots fired with the same data which will fall into each of eight areas within the dispersion pattern for ranges. { 'rānj di‚spər·zhən ‚dī·ə‚gram }

range drum [ORD] Graduated indicator of cylinder type, connected mechanically with the elevating mechanism of a gun, used for range setting. { 'rānj ‚drəm }

range error [ORD] Difference between the range to the point of impact of a particular projectile and the range to the center of impact of a group of shots fired with the same data. { 'rānj ‚er·ər }

range estimation [ORD] A rough estimate of the distance to a target or other object. { 'rānj ‚es·tə‚mā·shən }

rangefinder [COMMUN] A movable, calibrated unit of the receiving mechanism of a teletypewriter by means of which the selecting interval may be moved with respect to the start signal. [ELECTR] A device which determines the distance to an object by measuring the time it takes for a radio wave to travel to the object and return. [ENG] *See* optical rangefinder. { 'rānj ‚fīnd·ər }

range flag [ORD] Red flag displayed on or near a target range during firing practice as a warning that firing is being conducted. { 'rānj ‚flag }

range gate capture [ELECTR] Electronic countermeasure technique using a spoofer radar transmitter to produce a false target echo that can make a fire-control tracking radar move off the real target and follow the false one. { 'rānj ‚gāt ‚kap·chər }

range gating [ELECTR] The process of selecting those radar echoes that lie within a small range interval. { 'rānj ‚gād·iŋ }

range-height indicator [ENG] A scope which simultaneously indicates range and height of a radar target; this presentation is commonly used by height finders. { 'rānj 'hīt ‚in·də‚kad·ər }

range-height indicator display [ELECTR] A radar display that presents visually the scalar distance between a reference point and a target, along with the vertical distance between a reference plane and the target. Abbreviated RHI display. { 'rānj 'hīt 'in·də,kād·ər di,splā }

range-imaging sensor [ENG] A robotic device that makes precise measurements, by using the principles of algebra, trigonometry, and geometry, of the distance from a robot's end effector to various parts of an object, in order to form an image of the object. { 'rānj ¦im·ij·iŋ ,sen·sər }

range ladder [ORD] A naval term used to describe a method of adjusting gunfire by firing successive volleys; starting with a range which is assuredly beyond or short and applying small uniform range corrections to the successive volleys until the target is crossed. { 'rānj ,lad·ər }

range lights [NAV] Two or more lights in the same horizontal direction, and installed as aids to nautical navigation, particularly those lights placed to mark any line of importance to vessels, as a channel. { 'rānj ,līts }

range marker *See* distance marker. { 'rānj ,mär·kər }

range mark offset [ELECTR] Displacement of range mark on a type B indicator. { 'rānj ¦märk 'óf,set }

range of a loop [COMPUT SCI] The set of instructions contained between the opening and closing statements of a do loop. { 'rānj əv ə 'lüp }

range of motion [BIOPHYS] The degree of movement that can occur in a joint. { 'rānj əv 'mō·shən }

range of tide [OCEANOGR] The difference in height between consecutive high and low tides at a place. { 'rānj əv 'tīd }

range of visibility [NAV] The extreme distance at which an object or light can be seen. { 'rānj əv ,viz·ə'bil·əd·ē }

range oil [MATER] Kerosine similar in properties to a No. 1 distillate heating oil; used for space heating. { 'rānj ,óil }

range pole *See* range rod. { 'rānj ,pōl }

range probable error [ORD] Error in range that a gun or other weapon may be expected to exceed as often as not; given in the firing tables for the gun, and may be taken as an index of accuracy of the piece. { 'rānj ,präb·ə·bəl ,er·ər }

range rake [ORD] A T-shaped device with pegs set in the cross; the distance between pegs subtends a definite angle at the base of the T; by sighting with a range rake, a flank observer can get a quick angular measurement of range deviation. { 'rānj ,rāk }

range rate [ELECTR] The rate at which the distance from the measuring equipment to the target or signal source that is being tracked is changing with respect to time. { 'rānj ,rāt }

range recorder [ENG] An item which makes a permanent representation of distance, expressed as range, versus time. [ENG ACOUS] A display used in sonar in which a stylus sweeps across a paper moving at a constant rate and chemically treated so that it is darkened by an electrical signal from the stylus; the stylus starts each sweep as a sound pulse is emitted so that the distance along the trace at which the echo signal appears is a measure of the range to the target. { 'rānj ri,kórd·ər }

range resolution *See* distance resolution. { 'rānj ,rez·ə,lü·shən }

range ring [ELECTR] Accurate, adjustable, ranging mark on a plan position indicator corresponding to a range step on a type M indicator. { 'rānj ,riŋ }

range rod [ENG] A long (6–8 feet or 1.8–2.4 meters) rod fitted with a sharp-pointed metal shoe and usually painted in 1-foot (30-centimeter) bands of alternate red and white; used for sighting points and lines in surveying or for showing the position of a ground point. Also known as line rod; lining pole; range pole; ranging rod; sight rod. { 'rānj ,räd }

range safety [ORD] On a rocket test range, the planning and execution of each test so as to ensure the maximum safety of all personnel and property within the range boundaries. { 'rānj ,sāf·tē }

range scale [ORD] **1.** Scale on the arm of a plotting board where the observed range of a moving target is recorded in finding firing data. **2.** Graduated scale on the sight or mount of a gun, used to show the elevation of the gun. **3.** Table of firing data giving elevation settings corresponding to various ranges for the standard charges. { 'rānj ,skāl }

range selection [ELECTR] Control on a radar indicator for selection of range scale. { 'rānj si,lek·shən }

range sensing [ENG] The precise measurement of the distance of a device from a robot's end effector. [ORD] Observing the location of the striking or bursting point of a projectile with respect to range, and reporting it as a hit, over, short, lost,

or doubtful; range sensing does not include accurate determination of distances. { 'rānj ,sens·iŋ }

range spotting [ORD] Watching the burst or impact of shots to note their deviation beyond, or short of, the target. { 'rānj ,späd·iŋ }

range step [ELECTR] Vertical displacement on M-indicator sweep to measure range. { 'rānj ,step }

range strobe [ELECTROMAG] An index mark which may be displayed on various types of radar indicators to assist in the determination of the exact range of a target. { 'rānj ,strōb }

range surveillance [ENG] Surveillance of a missile range by means of electronic and other equipment. { 'rānj sər,vā·ləns }

range sweep [ELECTR] A sweep intended primarily for measurement of range. { 'rānj ,swēp }

range table [ORD] Prepared table that gives elevations corresponding to ranges for a gun or other weapon, under various conditions; part of a firing table. { 'rānj ,tā·bəl }

range-tracking element [ELECTR] An element in a radar set which measures range and its time derivative; by means of the latter, a range gate is actuated slightly before the predicted instant of signal reception. { 'rānj ¦trak·iŋ ,el·ə·mənt }

range unit [ELECTR] Radar system component used for control and indication (usually counters) of range measurements. { 'rānj ,yü·nət }

range wind [ORD] Horizontal component of true wind in the vertical plane through the line of fire of an artillery piece. { 'rānj ,wind }

range zero [ELECTR] Alignment of start sweep trace with zero range. { 'rānj ,zir·ō }

range zone [GEOL] Formal biostratigraphic zone made up of a body of strata comprising the total horizontal (geographic) and vertical (stratigraphic) range of occurrence of a specified taxon of a group of taxa. { 'rānj ,zōn }

ranging [ORD] **1.** Wide-scale scouting, especially by aircraft, designed to search an area systematically. **2.** Locating an enemy gun by watching its flash, listening to its report, or other similar means. { 'rānj·iŋ }

ranging oscillator [ELECTR] Oscillator circuit containing an LC (inductor-capacitor) resonant combination in the cathode circuit, usually used in radar equipment to provide range marks. { 'rānj·iŋ 'äs·ə,lād·ər }

ranging rod *See* range rod. { 'rānj·iŋ ,räd }

Ranidae [VERT ZOO] A family of frogs in the suborder Diplasiocoela including the large, widespread genus *Rana*. { 'ran·ə,dē }

rank [GEOL] **1.** A coal classification based on degree of metamorphism. **2.** *See* stack. [MATH] **1.** The rank of a matrix is its maximum number of linearly independent rows. **2.** The rank of a system of homogeneous linear equations equals the rank of the matrix of its coefficients. **3.** A tensor in an n-dimensional space is of rank r if it has n^r components. **4.** The rank of a group G is the number of elements in the basis of the quotient group of G over the subgroup consisting of all elements of G having finite period. **5.** The rank of a place or valuation is equal to the number of proper prime ideals in its valuation ring. **6.** The rank of a prime ideal P is the largest number n for which there exists a sequence $P_0 = P, P_1, P_2, \ldots, P_n$ of prime ideals such that P_i is a subset of P_{i-1}. [MECH ENG] The number of rotational joints belonging to a robot. [STAT] The number assigned to an observation if a collection of observations is ordered from smallest to largest and each observation is given the number corresponding to its place in the order. { raŋk }

rank correlation [STAT] A nonparametric test of statistical dependence for a random sample of paired observations. { 'raŋk ,kä·rə'lā·shən }

ranked p₀ set [MATH] A partially ordered set for which there is a function, r, defined on the elements of the set, such that $r(x) = 0$ if x is a minimal element, and $r(x) = r(y) + 1$ if x covers y. { 'raŋkt ,pē'zir·ō *or* 'pē,səb¦zir·ō ,set }

rank estimator [NAV ARCH] A parameter that is designed to estimate a vessel's seakeeping performance, and is calculated from the ship's dimensions and its waterplane and prismatic coefficients. { 'raŋk ,es·tə,mād·ər }

Rankine body [FL MECH] A fluid flow pattern formed by combining a uniform stream with a source and a sink of equal strengths, with the line joining the source and sink along the stream direction. { 'raŋ·kən ,bäd·ē }

Rankine cycle [THERMO] An ideal thermodynamic cycle

consisting of heat addition at constant pressure, isentropic expansion, heat rejection at constant pressure, and isentropic compression; used as an ideal standard for the performance of heat-engine and heat-pump installations operating with a condensable vapor as the working fluid, such as a steam power plant. Also known as steam cycle. { 'raŋ·kən ˌsī·kəl }

Rankine efficiency [MECH ENG] The efficiency of an ideal engine operating on the Rankine cycle under specified conditions of steam temperature and pressure. { 'raŋ·kən iˌfish·ən·sē }

Rankine-Hugoniot equations [THERMO] Equations, derived from the laws of conservation of mass, momentum, and energy, which relate the velocity of a shock wave and the pressure, density, and enthalpy of the transmitting fluid before and after the shock wave passes. { 'raŋ·kən yü'gō·nē·ō iˌkwā·zhənz }

Rankine temperature scale [THERMO] A scale of absolute temperature; the temperature in degrees Rankine (°R) is equal to 9/5 of the temperature in kelvins and to the temperature in degrees Fahrenheit plus 459.67. { 'raŋ·kən 'tem·prə·chər ˌskāl }

Rankine vortex [FL MECH] A vortex with a vertical axis and circular motion, in which the motion is that of a rotating solid cylinder inside some fixed radius, and the circulation is constant outside this radius. { 'raŋ·kən vȯr'teks }

ranking method [IND ENG] A system of job evaluation wherein each job as a whole is given a rank with respect to all the other jobs, and no attempt is made to establish a measure of value. { 'raŋk·iŋ ˌmeth·əd }

rankinite [MINERAL] $Ca_3Si_2O_7$ A monoclinic mineral composed of calcium silicate. { 'raŋ·kəˌnīt }

RANK ligand [BIOCHEM] A local paracrine factor that originates from osteoblasts and mediates the effects of most, if not all, agents that are known to impact osteoclast development in bone. Also known as osteoprotegerin ligand. { ¦raŋk 'līg·ənd }

rank of an observation [STAT] The number assigned to an observation if a collection of observations is ordered from smallest to largest and each observation is given the number corresponding to its place in the order. { 'raŋk əv ən 'äb·zər'vā·shən }

rank-order statistics [STAT] Statistics computed from rankings of the observations rather than from the observations themselves. { ¦raŋk 'ȯr·dər stə'tis·tiks }

rank tests [STAT] Tests which use the ranks of observations with respect to one another rather than the observations themselves. { 'raŋk ˌtests }

Ranney oil-mining system [PETRO ENG] A method used to get oil from oil sands that involves driving mine galleries from shafts communicating to the surface in impermeable strata above and below the oil strata; holes drilled at short intervals along the galleries into the oil sands drain the oil or gas through pipes sealed in the drill holes into tanks from which the gas or oil is pumped to the surface. { 'ran·ē 'ȯil ˌmīn·iŋ ˌsis·təm }

Ranney well [CIV ENG] A well that has a center caisson with horizontal perforated pipes extending radially into an aquifer; particularly applicable to the development of thin aquifers at shallow depths. { 'ran·ē ˌwel }

Ranque effect [FL MECH] An effect whereby turbulent flow in a tube supplied with air through a tangential nozzle at high pressure produces warming near the walls of the tube and cooling at the axis. { 'räŋk iˌfekt }

RANS analysis See Reynolds-averaged Navier-Stokes analysis. { 'ranz əˌnal·ə·səs }

ransomite [MINERAL] $Cu(Fe,Al)_2(SO_4)_4 \cdot 7H_2O$ A sky-blue mineral composed of hydrous copper iron aluminum sulfate. { 'ran·səˌmīt }

ranula [MED] A retention cyst of a salivary gland. { 'ran·yə·lə }

Ranunculaceae [BOT] A family of dicotyledonous herbs in the order Ranunculales distinguished by alternate leaves with net venation, two or more distinct carpels, and numerous stamens. { rəˌnəŋ·kyə'lās·ē,ē }

Ranunculales [BOT] An order of dicotyledons in the subclass Magnoliidae characterized by its mostly separate carpels, triaperturate pollen, herbaceous or only secondarily woody habit, and frequently numerous stamens. { rəˌnəŋ·kyə'lā·lēz }

raob See radiosonde observation. { 'rāˌäb }

Raoult's law [PHYS CHEM] The law that the vapor pressure of a solution equals the product of the vapor pressure of the pure solvent and the mole fraction of solvent. { rä'ülz ˌlȯ }

rapakivi [PETR] Granite or quartz monzonite characterized by orthoclase phenocrysts mantled with plagioclase. Also known as wiborgite. { ¦rä·pə'kē·vē }

rapakivi texture [PETR] An igneous and metamorphic rock texture in which spherical potassium feldspar crystals are surrounded by a rim of sodium feldspar, both within a finer-grained matrix. { ¦rä·pə'kē·vē 'teks·chər }

rape [BOT] *Brassica napus.* A plant of the cabbage family in the order Capparales; the plant does not form a compact head, the leaves are bluish-green, deeply lobed, and curled, and the small flowers produce black seeds; grown for forage. { rāp }

rape oil [MATER] A fatty, nondrying or semidrying, viscous, dark-brown to yellow oil with unpleasant taste and aroma, obtained from the seed of rape and turnip; soluble in ether, carbon disulfide, and chloroform; solidifies at −2 to −10°C; used to make lubricants and rubber substitutes, as an illuminant, and in steel heat treatment. Also known as colza oil; rapeseed oil. { 'rāp ˌȯil }

rape-seed oil See rape oil. { 'rāp ˌsēd ˌȯil }

raphania [MED] A disease thought to be due to chronic ingestion of the poison in seeds of the wild radish. { rə'fan·yə }

raphe [ANAT] A broad seamlike junction between two lateral halves of an organ or other body part. [BOT] **1.** The part of the funiculus attached along its full length to the integument of an anatropous ovule, between the chalaza and the attachment to the placenta. **2.** The longitudinal median line or slit on a diatom valve. { 'rā·fē }

Raphidae [VERT ZOO] A family of birds in the order Columbiformes that included the dodo (*Raphus cucullatus*); completely extirpated during the 17th and early 18th centuries. { 'raf·əˌdē }

raphide [BOT] One of the long, needle-shaped crystals, usually consisting of calcium oxalate, occurring as a metabolic by-product in certain plant cells. { 'rāfˌīd }

rapid [HYD] A portion of a stream in swift, disturbed motion, but without cascade or waterfall; usually used in the plural. { 'rap·əd }

rapid access loop [COMPUT SCI] A small section of storage, particularly in drum, tape, or disk storage units, which has much faster access than the remainder of the storage. { 'rap·əd 'ak,ses ˌlüp }

rapid-curing asphalt [MATER] A liquid asphalt composed of asphalt cement and a gasoline- or naphtha-type diluent. Abbreviated RC asphalt. { 'rap·əd ¦kyür·iŋ 'as,fȯlt }

rapid-eye-movement sleep [PSYCH] That part of the sleep cycle during which the eyes move rapidly, accompanied by a loss of muscle tone and a low-amplitude encephalogram recording; most dreaming occurs during this stage of sleep. Abbreviated REM sleep. { 'rap·əd ¦ī müv·mənt ˌslēp }

rapid fire [ORD] Rate of firing small arms or automatic weapons, faster than slow fire, but slower than quick fire. { 'rap·əd 'fīr }

rapid flow [HYD] Water flow whose velocity exceeds the velocity of propagation of a long surface wave in still water. Also known as supercritical flow. { 'rap·əd 'flō }

rapid memory See rapid storage. { 'rap·əd 'mem·rē }

rapid proton capture process [ASTROPHYS] A mode of explosive nucleosynthesis in which each of the nuclei lighter than iron captures many protons, populating nuclides near the proton drip line, and these subsequently undergo a series of beta decays back to one of the stable nuclides. Also known as rp-process. { ¦rap·id ¦prō·tän 'kap·chər ¦prä·ses }

rapid prototyping [IND ENG] A modeling process used in product design in which a CAD drawing of a part is processed to create a file of the part in slices, and then a part is built by depositing layer (slice) upon layer of material; includes stereolithography, selective laser sintering, or fused deposition modeling. { ¦rap·əd ¦prōd·əˌtīp·iŋ }

rapid quenching [MET] Superfast cooling ($1–5 \times 10^6$ K per second) of a molten metal to produce new and amorphous alloys and new crystalling material with improved properties. { 'rap·əd 'kwench·iŋ }

rapid sand filter [CIV ENG] A system for purifying water,

RANUNCULACEAE

Colorado columbine (*Aquilegia coerulea*). (*U. S. Forest Service photograph by C. A. Kulzleb*)

which is forced through layers of sand and gravel under pressure. { 'rap·əd 'sand ,fil·tər }

rapid selector [COMPUT SCI] A device which scans codes recorded on microfilm; microimages of the documents associated with the codes may also be recorded on the film. { 'rap·əd si'lek·tər }

rapid sequence camera [OPTICS] A conventional camera in most respects except that it is designed to permit a number of photographs to be obtained in rapid succession with one winding of the shutter. { 'rap·əd |sē·kwəns ,kam·rə }

rapid storage [COMPUT SCI] In computers, storage with a very short access time; rapid access is generally gained by limiting storage capacity. Also known as high-speed storage; rapid memory. { 'rap·əd 'stȯr·ij }

rapid traverse [MECH ENG] A machine tool mechanism which rapidly repositions the workpiece while no cutting takes place. { 'rap·əd trə'vərs }

rapid wasting syndrome [INV ZOO] A coral reef disease that is characterized by a rapid loss of tissue and destruction of the underlying skeleton. { |rap·əd 'wāst·iŋ ,sin,drōm }

raptorial [ZOO] **1.** Living on prey. **2.** Adapted for snatching or seizing prey, as birds of prey. { rap'tȯr·ē·əl }

rapture of the deep *See* nitrogen narcosis. { 'rap·chər əv thə 'dēp }

rare-earth alloy [MET] An alloy containing rare-earth materials. { 'rer ,ərth 'al,ȯi }

rare-earth chelate laser *See* chelate laser. { 'rer ,ərth 'kē,lāt 'lā·zər }

rare-earth-doped fiber amplifier [COMMUN] An optical fiber amplifier whose fiber core is lightly doped with trivalent rare-earth ions, which absorb light at certain pump wavelengths and emit it at some signal wavelength through stimulated emission. { |rär ,ərth ,dōpt ,fī·bər 'am·plə,fī·ər }

rare-earth element [CHEM] The name given to any of the group of chemical elements with atomic numbers 58 to 71; the name is a misnomer since they are neither rare nor earths; examples are cerium, erbium, and gadolinium. { 'rer ,ərth 'el·ə·mənt }

rare-earth garnet [MATER] A synthetic garnet having the general structure of grossularite, but with calcium replaced by a rare-earth metal, and aluminum and silicon replaced by iron; used for electronic applications. { 'rer ,ərth 'gär·nət }

rare-earth magnet [ELECTROMAG] Any of several types of magnets made with rare-earth elements, such as rare-earth-cobalt magnets, which have coercive forces up to ten times that of ordinary magnets; used for computers and signaling devices. { 'rer ,ərth 'mag·nət }

rare-earth mineral [MINERAL] A mineral containing lanthanides and yttrium as essential constituents. The total atomic ratio of lanthanides and yttrium is greater than any other element within at least one crystallographic site. Examples are monazite, xenotime, and bastnaesite. { 'rer ,ərth 'min·rəl }

rare-earth salts [INORG CHEM] Salts derived from monazite, and with rare earths in similar proportions as in monazite; contains La, Ce, Pr, Nd, Sm, Gd, and Y as acetates, carbonates, chlorides, fluorides, nitrates, sulfates, and so on. { 'rer ,ərth 'sȯls }

rarefaction [ACOUS] The instantaneous, local reduction in density of a gas resulting from passage of a sound wave, or the region in which the density is reduced at some instant. Also known as rarefraction. { |rer·ə|fak·shən }

rarefaction wave [FL MECH] A pressure wave or rush of air or water induced by rarefaction; it travels in the opposite direction to that of a shock wave directly following an explosion. Also known as suction wave. { |rer·ə|fak·shən ,wāv }

rarefied gas [FL MECH] A gas whose pressure is much less than atmospheric pressure. { 'rer·ə,fīd 'gas }

rarefraction *See* rarefaction. { ,rer·ə'frak·shən }

rare gas *See* noble gas. { 'rer 'gas }

rare metal [MET] Any metal that is difficult to extract from ore and is rare and expensive commercially; includes masurium, alabamine, and virginium. { 'rer 'med·əl }

RAREP *See* radar report. { 'rer,ep }

rare set *See* nowhere dense set. { |rär 'set }

Rarita-Schwinger equation [QUANT MECH] A partial differential equation, similar in form to the Dirac equation, relating the spatial and time dependence of a 16-component wave function describing a free relativistic particle with intrinsic spin 3/2, and its antiparticle. { 'rä·rē·tä 'shviŋ·ər i,kwā·shən }

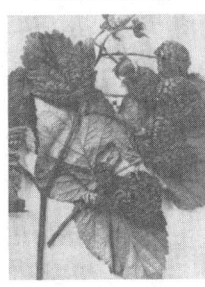

RASPBERRY

A red raspberry branch, Loudon variety. (*USDA*)

raschel knitting [TEXT] A type of warp knitting that produces fabric similar to tricot but usually of coarser texture and with openwork patterns. { 'rash·əl ,nid·iŋ }

Raschig process [CHEM ENG] A method for production of phenol that begins with a first-stage chlorination of benzene, using an air-hydrochloric acid mixture. { 'rä·shik 'prä·səs }

Raschig ring [CHEM ENG] A type of packing in the shape of a short pipe; used in columns for absorption operations, and to a limited extent for distillation operations. { 'rä·shik ,riŋ }

rash [MED] A lay term for nearly any skin eruption, but more commonly for acute inflammatory dermatoses. [MIN ENG] Very impure coal, so mixed with waste material as to be unsalable. { rash }

Ra-Shalom [ASTRON] An Aten asteroid that has a period of 0.759 year and an eccentricity of 0.436. { 'rä shə'lȯm }

rashing [MIN ENG] A soft, friable, and flaky or scaly rock (shale or clay) immediately beneath a coal seam, often containing much carbonaceous material and readily mixed with the coal in mining. { 'rash·iŋ }

RA size [ENG] One of a series of sizes to which untrimmed paper is manufactured; for reels of paper, the standard sizes in millimeters are 430, 610, 860, and 1220; for sheets of paper, the sizes are RA0, 860 × 1220; RA1, 610 × 860; RA2, 430 × 610; RA sizes correspond to A sizes when trimmed. { |är'ā ,sīz }

rasorial [ZOO] Adapted for scratching for food; applied to birds. { rə'sȯr·ē·əl }

rasorite *See* kernite. { 'rä·zə,rīt }

rasp [DES ENG] A metallic tool with a rough surface of small points used for shaping and finishing metal, plaster, stone, and wood; designed in a number of useful curved shapes. { rasp }

raspberry [BOT] Any of several species of upright shrubs of the genus *Rubus*, with perennial roots and prickly biennial stems, in the order Rosales; the edible black or red juicy berries are aggregate fruits, and when ripe they are easily separated from the fleshy receptacle. { 'raz,ber·ē }

raspite [MINERAL] PbWO₄ A yellow or brownish-yellow mineral composed of lead tungstate, occurring as monoclinic crystals. { 'ra,spīt }

Ras protein [CELL MOL] A GTPase that is part of a signaling cascade that begins with external chemical signals and controls cell growth; it turns itself off by hydrolyzing guanosine triphosphate (GTP) to guanosine diphosphate (GDP); however, if left in its active (GTP-bound) form, it can lead to uncontrollable cell growth, or cancer. { |är|ā'es ,prō,tēn }

R association [ASTRON] A grouping of stars in a reflection nebula. { 'är ə,sō·sē,ā·shən }

raster [ELECTR] A predetermined pattern of scanning lines that provides substantially uniform coverage of an area; in television the raster is seen as closely spaced parallel lines, most evident when there is no picture. { 'ras·tər }

raster graphics [COMPUT SCI] A computer graphics coding technique which codes each picture element of the picture area in digital form. Also known as bit-mapped graphics. { 'ras·tər |graf·iks }

raster image processor [GRAPHICS] A computer that accepts digital files and creates a print-ready file. { ,ras·tər ,im·ij 'prä,ses·ər }

rasterization [COMPUT SCI] The conversion of graphics objects composed of vectors or line segments into dots for transmission to raster graphics displays and to dot matrix and laser printers. { ,ras·tə·rə'zā·shən }

raster scanning [ELECTR] Radar scan very similar to electron-beam scanning in an ordinary television set; horizontal sector scan that changes in elevation. { 'ras·tər |skan·iŋ }

Rast method [ANALY CHEM] The melting-point depression method often used for the determination of the molecular weight of organic compounds. { 'rast ,meth·əd }

rat [VERT ZOO] The name applied to over 650 species of mammals in several families of the order Rodentia; they differ from mice in being larger and in having teeth modified for gnawing. { rat }

RAT *See* rocket-assisted torpedo. { rat }

rataria larva [INV ZOO] The second, hourglass-shaped, free-swimming larva of the siphonophore *Velella*. { rə'tar·ē·ə ,lär·və }

rat-bite fever [MED] Either of two diseases transmitted by the bite of a rat: spirillary rat-bite fever and streptobacillary fever. { 'rat ,bīt ,fē·vər }

ratchet [DES ENG] A wheel, usually toothed, operating with a catch or a pawl so as to rotate in only a single direction. { 'rach·ət }

ratchet coupling [MECH ENG] A coupling between two shafts that uses a ratchet to allow the driven shaft to be turned in one direction only, and also to permit the driven shaft to overrun the driving shaft. { 'rach·ət ,kəp·liŋ }

ratchet jack [DES ENG] A jack operated by a ratchet mechanism. { 'rach·ət ,jak }

ratchet tool [DES ENG] A tool in which torque or force is applied in one direction only by means of a ratchet. { 'rach·ət ,tül }

rat distillate [CHEM ENG] A refinery designation for gasoline and other fuels as they come from the condenser, before undesirable substances are removed by further processing. { 'rat ,dist·əl·ət }

rate [SCI TECH] The amount of change of some quantity during a time interval divided by the length of the time interval. { rāt }

rate action See derivative action. { 'rāt ,ak·shən }

Rateau formula [FL MECH] A formula, $m = A_2 p_1 (16.367 - 0.96 \log p_1)/1000$, for determining the discharge m of saturated steam in pounds per second through a well-rounded convergent orifice; A_2 is the area of the orifice in square inches, and p_1 the reservoir pressure in pounds per square inch. { rä'tō ,for·myə·lə }

rate climb [AERO ENG] The climb of an aircraft to higher altitudes at a constant rate. { 'rāt ,klīm }

rate constant [PHYS CHEM] Numerical constant in a rate-of-reaction equation; for example, $r_A = kC_A^a C_B^b C_C^c$, where C_A, C_B, and C_C are reactant concentrations, k is the rate constant (specific reaction rate constant), and a, b, and c are empirical constants. { 'rāt ,kän·stənt }

rate control [CONT SYS] A form of control in which the position of a controller determines the rate or velocity of motion of a controlled object. Also known as velocity control. { 'rāt kən,trōl }

rated capacity [MECH ENG] The maximum capacity for which a boiler is designed, measured in pounds of steam per hour delivered at specified conditions of pressure and temperature. { 'rād·əd kə'pas·əd·ē }

rated engine speed [MECH ENG] The rotative speed of an engine specified as the allowable maximum for continuous reliable performance. { 'rād·əd 'en·jən ,spēd }

rate descent [AERO ENG] An aircraft descent from higher altitudes at a constant rate. { 'rāt di,sent }

rate-determining step [CHEM] In a multistep chemical reaction, the step with the lowest velocity, which determines the rate of the overall reaction. { 'rāt di,tər·mən·iŋ ,step }

rated flow [ENG] **1.** Normal operating flow rate at which a fluid product is passed through a vessel or piping system. **2.** Flow rate for which a vessel or process system is designed. { 'rād·əd 'flō }

rated horsepower [MECH ENG] The normal maximum, allowable, continuous power output of an engine, turbine motor, or other prime mover. { 'rād·əd 'hòrs,pau̇·ər }

rated load [MECH ENG] The maximum load a machine is designed to carry. { 'rād·əd 'lōd }

rated relieving capacity [DES ENG] The measured relieving capacity for which the pressure relief device is rated in accordance with the applicable code or standard. { 'rād·əd ri'lēv·iŋ kə,pas·əd·ē }

rated speed [COMPUT SCI] The maximum operating speed that can be sustained by a data-processing device or communications line, not allowing for periodic pauses for various reasons such as carriage return on a printer. { 'rād·əd 'spēd }

rate effect [ELECTR] The phenomenon of a *pnpn* device switching to a high-conduction mode when anode voltage is applied suddenly or when high-frequency transients exist. { 'rāt i,fekt }

rate feedback [ELECTR] The return of a signal, proportional to the rate of change of the output of a device, from the output to the input. { 'rāt 'fēd,bak }

rate-grown transistor [ELECTR] A junction transistor in which both impurities (such as gallium and antimony) are placed in the melt at the same time and the temperature is suddenly raised and lowered to produce the alternate *p*-type and *n*-type layers of rate-grown junctions. Also known as graded-junction transistor. { 'rāt ,grōn tran'zis·tər }

rate gyroscope [MECH ENG] A gyroscope that is suspended in just one gimbal whose bearings form its output axis and which is restrained by a spring; rotation of the gyroscope frame about an axis perpendicular to both spin and output axes produces precession of the gimbal within the bearings proportional to the rate of rotation. { 'rāt 'jī·rə,skōp }

rate integrating gyroscope [MECH ENG] A single-degree-of-freedom gyro having primarily viscous restraint of its spin axis about the output axis; an output signal is produced by gimbal angular displacement, relative to the base, which is proportional to the integral of the angular rate of the base about the input axis. { 'rāt ,int·ə,grād·iŋ 'jī·rə,skōp }

rate meter See counting rate meter. { 'rāt ,mēd·ər }

rate multiplier [COMPUT SCI] An integrator in which the quantity to be integrated is held in a register and is added to the number standing in an accumulator in response to pulses which arrive at a constant rate. { 'rāt ,məl·tə,plī·ər }

rate of approach [AERO ENG] The relative speed of two aircraft when the distance between them is decreasing. { 'rāt əv ə'prōch }

rate of change See derivative. { 'rāt əv 'chānj }

rate-of-change map [GEOL] A derived stratigraphic map that shows the rate of change of structure, thickness, or composition of a given stratigraphic unit. { 'rāt əv 'chānj ,map }

rate of change of acceleration [MECH] Time rate of change of acceleration; this rate is a factor in the design of some items of ammunition that undergo large accelerations. { 'rāt əv 'chānj əv ik,sel·ə'rā·shən }

rate of climb [AERO ENG] Ascent of aircraft per unit time, usually expressed as feet per minute. { 'rāt əv 'klīm }

rate-of-climb indicator [AERO ENG] A device used to indicate changes in the vertical position of an aircraft by comparing the actual outside air pressure to a reference volume that lags the outside pressure because a calibrated restrictor imposes a lag-time constant to the reference pressure volume. Also known as rate-of-descent indicator; vertical speed indicator. { 'rāt əv 'klīm 'in·də,kād·ər }

rate of departure [AERO ENG] The relative speed of two aircraft when the distance between them is increasing. { 'rāt əv di'pär·chər }

rate-of-descent indicator See rate-of-climb indicator. { 'rāt əv di'sent 'in·də,kād·ər }

rate of detonation [ORD] Rate at which detonation of an explosive progresses; usually expressed in meters or yards per second. { 'rāt əv ,det·ən'ā·shən }

rate of flow See flow rate. { 'rāt əv 'flō }

rate-of-flow control valve See flow control valve. { 'rāt əv 'flō kən,trōl ,valv }

rate-of-rainfall gage See rain-intensity gage. { 'rāt əv 'rān,fòl ,gāj }

rate of reaction [CHEM] A measurement based on the mass of reactant consumed in a chemical reaction during a given period of time. { 'rāt əv rē'ak·shən }

rate of return [AERO ENG] Aircraft relative to its base, either fixed or moving. { 'rāt əv ri'tərn }

rate of rise [ENG] The time rate of pressure increase during an isolation test for leaks. { 'rāt əv 'rīz }

rate of sedimentation [GEOL] The amount of sediment accumulated in an aquatic environment over a given period of time, usually expressed as thickness of accumulation per unit time. Also known as sedimentation rate. { 'rāt əv ,sed·ə·mən'tā·shən }

rate of strain hardening [MET] Rate of change of true stress with respect to true strain in the plastic range. Also known as modulus of strain hardening. { 'rāt əv 'strān 'härd·ən·iŋ }

rate process [PHYS] Any process in which the derivatives with respect to time of one or more variables, evaluated at any given time t_0, depend on the values of the variables at time t_0 and possibly at times earlier than t_0. { 'rāt ,prä·ses }

rate receiver [ELECTROMAG] A guidance antenna that receives a signal from a launched missile as to its rate of speed. { 'rāt ri,sē·vər }

rate response [ENG] Quantitative expression of the output rate of a control system as a function of its input signal. { 'rāt ri,späns }

rate servomechanism See velocity servomechanism. { 'rāt ,sər·vō mek·ə,niz·əm }

rate test [COMPUT SCI] A test that verifies that the time

RATCHET

(a)

(b)   driver   follower

Ratchet mechanisms. (a) Toothed ratchet is driven in direction of arrow by catch when arm moves to left; pawl holds ratchet during return stroke of catch. (b) In roller ratchet, rollers become wedged between driver and follower when driver turns faster than follower in direction of arrow.

constants of the integrators are correct; used in analog computers. { 'rāt ,test }

rate transmitter [ELECTR] A transmitter in a missile being launched, used with a ground receiver to indicate the rate of speed increase. { 'rāt tranz,mid·ər }

ratfish [VERT ZOO] The common name for members of the chondrichthyan order Chimaeriformes. { 'rat,fish }

rat guard [NAV ARCH] A circular or conical metal shield placed on a mooring line to keep rats from boarding or leaving a vessel. { 'rat ,gärd }

rathite [MINERAL] $Pb_{13}As_{18}S_{40}$ A dark-gray mineral with metallic luster composed of sulfide of lead and arsenic; occurs as orthorhombic crystals. { 'rä,tīt }

Rathke's pouch *See* craniobuccal pouch. { 'rät,kēz ,pauch }

rathole [MIN ENG] A shallow, small-diameter, auxiliary hole alongside the main borehole, drilled at an angle to the main hole; after core drilling is completed, the rathole is reamed out and the larger-size hole is advanced, usually by some noncoring method. { 'rat,hōl }

rating [ENG] A designation of an operating limit for a machine, apparatus, or device used under specified conditions. { 'rād·iŋ }

rating curve [HYD] For a given point on a stream, a graph of discharge versus stage. { 'rād·iŋ ,kərv }

rating nut [HOROL] A nut used to vary the effective length of a pendulum in adjusting the rate of a clock. { 'rād·iŋ ,nət }

ratio [MATH] A ratio of two quantities or mathematical objects *A* and *B* is their quotient or fraction *A/B*. { 'rā·shō }

ratio arm circuit [ELEC] Two adjacent arms of a Wheatstone bridge, designed so they can be set to provide a variety of indicated resistance ratios. { 'rā·shō ,ärm ,sər·kət }

ratio-balance relay *See* percentage differential relay. { 'rā·shō ,bal·əns ,rē,lā }

ratio control system [CONT SYS] Control system in which two process variables are kept at a fixed ratio, regardless of the variation of either of the variables, as when flow rates in two separate fluid conduits are held at a fixed ratio. { 'rā·shō kən'trōl ,sis·təm }

ratio delay study *See* work sampling. { 'rā·shō di'lā ,stəd·ē }

ratio detector [ELECTR] A frequency-modulation detector circuit that uses two diodes and requires no limiter at its input; the audio output is determined by the ratio of two developed intermediate-frequency voltages whose relative amplitudes are a function of frequency. { 'rā·shō di,tek·tər }

ratio deviation *See* modulation index. { 'rā·shō ,dē·vē'ā·shən }

ratio-differential relay *See* percentage differential relay. { 'rā·shō ,dif·ə¦ren·chəl 'rē,lā }

ratio estimator [STAT] A ratio of two random variables that is used as an estimator. { 'rā·shō ,es·tə,mād·ər }

ratio map [GEOL] A facies map that depicts the ratio of thicknesses between rock types in a given stratigraphic unit. { 'rā·shō ,map }

ratio meter [ENG] A meter that measures the quotient of two electrical quantities; the deflection of the meter pointer is proportional to the ratio of the currents flowing through two coils. { 'rā·shō ,mēd·ər }

rational algebraic expression [MATH] An algebraic expression that equals a quotient of polynomials. { ¦rash·ən·əl ¦al·jə,brā·ik ik'spresh·ən }

rational formula [HYD] The expression of peak discharge as equal to the product of rainfall, drainage area, and a runoff coefficient depending on drainage-basin characteristics. { 'rash·ən·əl 'fōr·myə·lə }

rational fraction [MATH] **1.** A fraction whose numerator and denominator are both rational numbers. **2.** A fraction whose numerator and denominator are both polynomials. { 'rash·ən·əl 'frak·shən }

rational function [MATH] A function which is a quotient of polynomials. { 'rash·ən·əl 'fəŋk·shən }

rational horizon *See* celestial horizon. { 'rash·ən·əl hə'rīz·ən }

rationalization [PSYCH] A defense mechanism against difficult and unpleasant situations in which the individual attempts to use plausible means to justify or defend the unacceptable situations. { ,rash·ən·əl·ə'zā·shən }

rationalize [MATH] **1.** To carry out operations on an algebraic equation that remove radicals containing the varible. **2.** To multiply the numerator and denominator of a fraction by a

quantity that removes the radicals in the denominator. **3.** To make a substitution in an integral that removes the radicals in the integrand. { 'rash·ən·əl,īz }

rationalized units [ELEC] A system of electrical units, such as occurs in the International System, in which the factor of 4π is removed from the field equations and appears instead in the explicit expressions for the fields of a point charge and current element. { 'rash·ən·əl,īzd 'yü·nəts }

rational number [MATH] A number which is the quotient of two integers. { 'rash·ən·əl 'nəm·bər }

rational root theorem [MATH] The theorem that, if a rational number *p/q*, where *p* and *q* have no common factors, is a root of a polynomial equation with integral coefficients, then the coefficient of the term of highest order is divisible by *q* and the coefficient of the term of lowest order is divisible by *p*. { 'rash·ən·əl ,rüt ,thir·əm }

rational synthesis [CHEM] The production of a compound using a sequence of chemical reaction steps strategically chosen. { 'rash·ən·əl 'sin·thə·səs }

ratio of expansion [MECH ENG] The ratio of the volume of steam in the cylinder of an engine when the piston is at the end of a stroke to that when the piston is in the cutoff position. { 'rā·shō əv ik'span·shən }

ratio of reduction [ENG] The ratio of the maximum size of the stone which will enter a crusher, to the size of its product. { 'rā·shō əv ri'dək·shən }

ratio of rise [OCEANOGR] The ratio of the height of tide at two places. { 'rā·shō əv 'rīz }

ratio of similitude [MATH] The ratio of the lengths of corresponding line segments of similar figures. Also known as homothetic ratio. { ¦rā·shō əv sə'mil·ə,tüd }

ratio of specific heats [PHYS CHEM] The ratio of specific heat at constant pressure to specific heat at constant volume, $\gamma = C_p/C_v$. { 'rā·shō əv spə'sif·ik 'hēts }

ratio of transformation [ELEC] Ratio of the secondary voltage of a transformer to the primary voltage under no-load conditions, or the corresponding ratio of currents in a current transformer. { 'rā·shō əv ,tranz·fər'mā·shən }

ratio of transformer [ELEC] Ratio of the number of turns in one winding of a transformer to the number of turns in the other, unless otherwise specified. { 'rā·shō əv tranz'fōr·mər }

ratio paper *See* semilogarithmic coordinate paper. { 'rā·shō ,pā·pər }

ratio print [GRAPHICS] A print the scale of which has been changed from that of the negative by photographic enlargement or reduction. { 'rā·shō ,print }

ratio resistor [ELEC] One of the resistors in a Wheatstone or Kelvin bridge whose resistances appear in a pair of ratios which are equal in a balanced bridge. { 'rā·shō ri,zis·tər }

ratio scale [STAT] A rule or system for assigning numbers to objects which has all the properties of an interval scale and, in addition, has a natural origin, so that ratios of numbers assigned to different objects have meaning. { 'rā·shō ,skāl }

ratio test *See* Cauchy ratio test. { 'rā·shō ,test }

ratites [VERT ZOO] A group of flightless, mostly large, running birds comprising several orders and including the emus, cassowaries, kiwis, and ostriches. { 'ra,tīts }

rato [AERO ENG] A rocket system providing additional thrust for takeoff of an aircraft. Derived from rocket-assisted takeoff. { 'rād·ō }

rat race [ELECTR] A particular type of radar waveguide configuration which allows the handling of greater power. { 'rat ,rās }

rattail [MET] A small irregular line marking a minor buckle on the surface of a casting. { 'rat,tāl }

rattail file [DES ENG] A round tapering file used for smoothing or enlarging holes. { 'rat,tāl 'fīl }

rattan [BOT] Any of several long-stemmed, climbing palms, especially of the genera *Calanius* and *Daemonothops*; stem material is used to make walking sticks, wickerwork, and cordage. { ra'tan }

rattlesnake [VERT ZOO] Any of a number of species of the genera *Sistrurus* or *Crotalus* distinguished by the characteristic rattle on the end of the tail. { 'rad·əl,snāk }

rattlesnake ore [GEOL] A gray, black, and yellow mottled ore of carnotite and vanoxite; its spotted appearance resembles that of a rattlesnake. { 'rad·əl,snāk ,ōr }

rattle stone [GEOL] A concretion composed of concentric laminae of different compositions, in which the more soluble

RATTLESNAKE

Rattlesnake (*Crotalus horridus*), one of the crotalid vipers.

layers have been removed by solution, leaving the central part detached from the outer part, such as a concretion of iron oxide filled with loose sand that rattles on shaking. Also known as klapperstein. { 'rad·əl ˌstōn }

rat typhus See murine typhus. { 'rat 'tī·fəs }

rauhaugite [PETR] A carbonatite that contains ankerite. { raù'haù·gīt }

Raunkiaer system [BOT] A classification system for plant life-forms based on the position of perennating buds in relation to the soil surface. { 'raùn·kē·ir ˌsis·təm }

Rauracian [GEOL] A substage of Upper Jurassic geologic time in Great Britain forming the middle Lusitanian, above the Argovian and below the Sequanian. { raù'rā·shən }

Rauschelback rotor [ENG] A free-turning S-shaped propeller used to measure ocean currents; the number of rotations per unit time is proportional to the flow. { 'raùsh·əl,bak ˌrōd·ər }

Rauwolfia [BOT] A genus of mostly poisonous, tropical trees and shrubs of the dogbane family (Apocynaceae); certain species yield substances used as emetics and cathartics, while *R. serpentina* is a source of alkaloids used as tranquilizers. { raù'wùl·fē·ə }

ravelly ground [GEOL] Rock that breaks into small pieces when drilled and tends to cave or slough into the hole when the drill string is pulled, or binds the drill string by becoming wedged or locked between the drill rod and the borehole wall. { 'rav·lē 'graùnd }

ravine [GEOGR] A small and narrow valley with steeply sloping sides. { rə'vēn }

ravinement [GEOL] **1.** The formation of a ravine or ravines. **2.** An irregular junction which marks a break in sedimentation, such as an erosion line occurring where shallow-water marine deposits have cut down into slightly eroded underlying beds. { rə'vēn·mənt }

raw [METEOROL] Colloquially descriptive of uncomfortably cold weather, usually meaning cold and damp, but sometimes cold and windy. { rȯ }

raw data [SCI TECH] Data that have not been processed; may be in machine-readable form. { 'rȯ 'dad·ə }

rawhide [MATER] Untanned animal hide that has been dried or treated with a preservative. { 'rȯ,hīd }

raw humus See ectohumus. { 'rȯ 'hyü·məs }

rawin [METEOROL] A method of winds-aloft observation, that is, the determination of wind speeds and directions in the atmosphere above a station; accomplished by tracking a balloon-borne radar target, responder, or radiosonde transmitter with either radar or a radio direction finder. { 'rā,win }

rawinsonde [METEOROL] A method of upper-air observation consisting of an evaluation of the wind speed and direction, temperature, pressure, and relative humidity aloft by means of a balloon-borne radiosonde tracked by a radar or radio direction finder. { 'rā·wən,sänd }

rawin target [ELECTROMAG] A special type of radar target tied beneath a free balloon, and designed to be an efficient reflector of radio energy; such targets usually consist of a corner reflector and are made of some reflecting material stretched over light wooden or metal struts. { 'rā·wən ,tär·gət }

raw map [GEOPHYS] A seismic map in which the z coordinate is time. { 'rȯ 'map }

raw material [IND ENG] A crude, unprocessed or partially processed material used as feedstock for a processing operation; for example, crude petroleum, raw cotton, or steel scrap. Also known as crude material. { 'rȯ mə'tir·ē·əl }

raw score [STAT] Any number as it originally appears in an experiment; for example, in evaluating test results the raw scores express the number of correct answers, uncorrected for position in the reference population. { 'rȯ 'skȯr }

raw sewage [CIV ENG] Untreated waste materials. { 'rȯ 'sü·ij }

raw silk [TEXT] A stage in the production of silk from the cocoons of cultivated silkworms when the silk filament consists of 80% fibroin and 20% sericin. { 'rȯ 'silk }

raw sludge [CIV ENG] Sewage sludge preliminary to primary and secondary treatment processes. { 'rȯ 'sləj }

raw water [CIV ENG] Water that has not been purified. { 'rȯ 'wȯd·ər }

ray [ASTRON] One of the broad streaks that radiate from some craters on the moon, especially Copernicus and Tycho; they consist of material of high reflectivity and are seen from earth best at full moon. [MATH] A straight-line segment emanating from a point. Also known as half line. [OPTICS] A curve whose tangent at any point lies in the direction of propagation of a light wave. [PHYS] A moving particle or photon of ionizing radiation. [VERT ZOO] Any of about 350 species of the elasmobranch order Batoidea having flattened bodies with large pectoral fins attached to the side of the head, ventral gill slits, and long, spikelike tails. { rā }

ray acoustics [ACOUS] The study of the behavior of sound under the assumption that sound traversing a homogeneous medium travels along straight lines or rays. Also known as geometrical acoustics. { 'rā ə,küs·tiks }

ray center See homothetic center. { 'rā ,sen·tər }

Raychuraduri equation [RELAT] An equation of general relativity theory, useful in proving singularity theorems, that relates the expansion, convergence, and shear of a congruence of time-like or null curves to the amount of matter present. { ,rā·chùr,ə,dùr·ē i,kwā·zhən }

ray crater [ASTRON] A large, relatively young lunar crater with visible rays. { 'rā ,krād·ər }

ray diagram [OPTICS] A diagram showing the paths of selected rays through an optical system. { 'rā ,dī·ə,gram }

Raydist [NAV] A radio navigation system used in hydrographic and geophysical surveying and ships' trials which employs phase comparison techniques. { 'rā,dist }

ray ellipsoid See Fresnel ellipsoid. { 'rā i'lip,sȯid }

rayfin fish [VERT ZOO] The common name for members of the Actinopterygii. { 'rā,fin ,fish }

ray flower [BOT] One of the small flowers with a strap-shaped corolla radiating from the margin of the head of a capitulum. { 'rā ,flaù·ər }

ray initial [BOT] One of the cells of the cambium which divide to produce new phloem and xylem ray cells. { 'rā i,nish·əl }

Raykin fender [CIV ENG] Sandwich-type fender buffer to protect docks from the impact of mooring ships; made of a connected series of steel plates cemented to layers of rubber. { 'rā·kən ,fen·dər }

rayl [ACOUS] A unit of specific acoustical impedance, equal to a sound pressure of 1 dyne per square centimeter divided by a sound particle velocity of 1 centimeter per second. Also known as specific acoustical ohm (Ω_s); unit-area acoustical ohm. { rāl }

rayleigh [OPTICS] A unit of brightness, used to measure the brightness of the night sky and aurorae, equal to $10^{10}/4\pi$ quanta per square meter per second per steradian. { 'rā·lē }

Rayleigh atmosphere [METEOROL] An idealized atmosphere consisting of only those particles, such as molecules, that are smaller than about one-tenth of the wavelength of all radiation incident upon that atmosphere; in such an atmosphere, simple Rayleigh scattering would prevail. { 'rā·lē ,at·mə,sfir }

Rayleigh balance [ELECTROMAG] An apparatus for assigning the value of the ampere in which the force exerted on a movable circular coil by larger circular coils above and below, but coaxial with, the movable coil is compared with the gravitational force on a known mass. { 'rā·lē ,bal·əns }

Rayleigh-Bènard convection [PHYS] Convection of a fluid heated from below, characterized by a regular array of usually hexagonal cells. { 'rā·lē bā'när kən,vek·shən }

Rayleigh criterion [CHEM] The criterion for spontaneous pressure oscillations to accompany combustion, namely, that combustion progresses more rapidly or efficiently during the compression phase of the pressure oscillation than during the rarefaction phase. [OPTICS] A criterion for the resolving power of an optical instrument which states that the images of two point objects are resolved when the principal maximum of the diffraction pattern of one falls exactly on the first minimum of the diffraction pattern of the other. { 'rā·lē krī'tir·ē·ən }

Rayleigh cycle [ELECTROMAG] A cycle of magnetization that does not extend beyond the initial portion of the magnetization curve, between zero and the upward bend. { 'rā·lē ,sī·kəl }

Rayleigh disk [ACOUS] An acoustic radiometer, used to measure particle velocity, consisting of a thin disk set at an angle of 45° to a sound beam; the particle velocity is calculated from the resulting torque on the disk. { 'rā·lē ,disk }

Rayleigh distance [PHYS] For electromagnetic, acoustic, or elastic waves emitted from a uniformly excited planar array

RAY

Devil ray, which is named for the anterior cephalic fins, extensions of the pectoral fins.

transmitting a sinusoidal signal, the distance from the array at which there is a transition from a near-field region, in which the radiated energy is confined to a cylindrical region, to a far-field region, in which the wave field exhibits spherical spreading and the field amplitude varies inversely with range. { 'rā·lē ˌdis·təns }

Rayleigh distribution [STAT] A normal distribution of two uncorrelated variates with the same variance. { 'rā·lē ˌdis·trəˌbyü·shən }

Rayleigh flow [FL MECH] An idealized type of gas flow in which heat transfer may occur, satisfying the assumptions that the flow takes place in constant-area cross section and is frictionless and steady, that the gas is perfect and has constant specific heat, that the composition of the gas does not change, and that there are no devices in the system which deliver or receive mechanical work. { 'rā·lē ˌflō }

Rayleigh interferometer [OPTICS] An optical interferometer in which two rays of light, emanating from a single slit, are collimated by a lens, pass through separate slits and cells, and are brought to focus by a second lens so that interference fringes become visible. Also known as Rayleigh refractometer. { 'rā·lē ˌin·tər·fə'räm·əd·ər }

Rayleigh-Jansen method [FL MECH] A method for solving equations for compressible fluid flow past a body, in which the velocity potential for the difference between the fluid velocity and the velocity distant from the body (V) is expressed as a power series in the square of the Mach number corresponding to V. { ¦rā·lē 'jan·sən ˌmeth·əd }

Rayleigh-Jeans law [STAT MECH] A law giving the intensity of radiation emitted by a blackbody within a narrow band of wavelengths; it states that this intensity is proportional to the temperature divided by the fourth power of the wavelength; it is a good approximation to the experimentally verified Planck radiation formula only at long wavelengths. { 'rā·lē 'jēnz ˌlò }

Rayleigh law [ELECTROMAG] **1.** For small values of the magnetic field strength H, the normal permeability of a material is approximately by $a + bH$, where a is the initial permeability and b is a constant. **2.** In a magnetic material subject to cyclic magnetization, with maximum magnetic field strength small compared with the coercive force, the hysteresis loss per cycle is proportional to the cube of the maximum value of the magnetic induction. [OPTICS] In Rayleigh scattering, the intensity of light scattered in a direction making an angle θ with the incident direction is proportional to $1 + \cos^2 θ$ and inversely proportional to the fourth power of the wavelength of the incident radiation. { 'rā·lē ˌlò }

Rayleigh lidar [OPTICS] A type of lidar that is designed to measure the Rayleigh scattering of laser light from molecules in the atmosphere, thereby determining atmospheric density. { 'rā·lē 'lī·där }

Rayleigh line [MECH] A straight line connecting points corresponding to the initial and final states on a graph of pressure versus specific volume for a substance subjected to a shock wave. [SPECT] Spectrum line in scattered radiation which has the same frequency as the corresponding incident radiation. { 'rā·lē ˌlīn }

Rayleigh loop [ELECTROMAG] A parabolic approximation to a magnetic hysteresis loop. { 'rā·lē ˌlüp }

Rayleigh number 1 [FL MECH] A dimensionless number used in studying the breakup of liquid jets, equal to Weber number 2. Symbolized N_{Ra1}. { 'rā·lē ¦nəm·bər 'wən }

Rayleigh number 2 [THERMO] A dimensionless number used in studying free convection, equal to the product of the Grashof number and the Prandtl number. Symbolized R'_2. { 'rā·lē ¦nəm·bər 'tü }

Rayleigh number 3 [THERMO] A dimensionless number used in the study of combined free and forced convection in vertical tubes, equal to Rayleigh number 2 times the Nusselt number times the tube diameter divided by its entry length. Symbolized Ra_3. { 'rā·lē ¦nəm·bər 'thrē }

Rayleigh prism [OPTICS] A system of prisms used to produce greater dispersion of light than would be produced by a single prism. { 'rā·lē ˌpriz·əm }

Rayleigh ratio [OPTICS] Light-scattering relationship defined by the ratio of intensities of incident and scattered light at a specified distance; used in photometric and refractometric analyses. { 'rā·lē ˌrā·shō }

Rayleigh reciprocity theorem [ELECTROMAG] Reciprocal relationship for an antenna when it is transmitting or receiving;

the effective heights, radiation resistance, and the radiation pattern are alike, whether the antenna is transmitting or receiving. { 'rā·lē ˌres·ə'präs·əd·ē ˌthir·əm }

Rayleigh refractometer See Rayleigh interferometer. { 'rā·lē ˌrē‚frak'täm·əd·ər }

Rayleigh-Ritz method [MATH] An approximation method for finding solutions of functional equations in terms of finite systems of equations. { 'rā·lē 'rits ˌmeth·əd }

Rayleigh scattering [ELECTROMAG] Scattering of electromagnetic radiation by independent particles which are much smaller than the wavelength of the radiation. { 'rā·lē ˌskad·ə·riŋ }

Rayleigh's dissipation function [MECH] A function which enters into the equations of motion of a system undergoing small oscillations and represents frictional forces which are proportional to velocities; given by a positive definite quadratic form in the time derivatives of the coordinates. Also known as dissipation function. { 'rā·lē ˌdis·ə'pā·shən ˌfəŋk·shən }

Rayleigh-Taylor instability [FL MECH] The instability of the interface separating two fluids having different densities when the lighter fluid is accelerated toward the heavier fluid. { ¦rā·lē 'tā·lər ˌin·stə'bil·əd·ē }

Rayleigh wave [GEOPHYS] In seismology, a surface wave with a retrograde, elliptical motion at the free surface. Also known as R wave. [MECH] A wave which propagates on the surface of a solid; particle trajectories are ellipses in planes normal to the surface and parallel to the direction of propagation. Also known as surface wave. { 'rā·lē ˌwāv }

Raymond concrete pile [CIV ENG] A pile made by driving a thin steel shell into the ground with a tapered mandrel and filling it with concrete. { 'rā·mənd ¦kän¦krēt ˌpīl }

Raynaud's disease [MED] A usually bilateral disease of blood vessels, especially of the extremities; excited by cold or emotion, characterized by intermittent pallor, cyanosis, and redness, and generally accompanied by pain. { rā'nōz di‚zēz }

rayon [TEXT] A fiber made from regenerated cellulose by the viscose or cuprammonium process. { 'rā‚än }

rayon coning oil [MATER] Lubricant oil used to reduce static in yarns being wound on cones; composed of low-viscosity mineral oils emulsifiable in water. { 'rā‚än 'kōn·iŋ ˌoil }

rayon staple [TEXT] Rayon fibers of spinnable lengths; continuous filaments from a number of spinnerets are collected into a loose, ropelike form and cut into suitable lengths. { 'rā‚än ‚stā·pəl }

rayon tow [TEXT] Large number of continuous rayon filaments that give the appearance of untwisted rope; when cut into suitable lengths, this material becomes rayon staple. { 'rā‚än ˌtō }

ray parameter [GEOPHYS] A function p that is constant along a given seismic ray, given by $p = rv^{-1} \sin i$, where r is the distance from the center O of the earth, v is the velocity, and i is the angle that the ray at a point P makes with the radius OP. { 'rā pə‚ram·əd·ər }

ray path [COMMUN] Geometric path between the transmitting and receiving locations. [PHYS] An imaginary path along which travels the energy associated with a point on a wavefront. { 'rā ˌpath }

ray shake [BOT] A radial crack in wood caused by wounds in a tree along the barrier zone. { 'rā ˌshāk }

ray surface [OPTICS] The locus of points reached in a unit time in an anisotropic medium by an electromagnetic disturbance that starts from the origin. { 'rā ˌsər·fəs }

ray system [ASTRON] The bright streaks radiating outward from a lunar crater. { 'rā ˌsis·təm }

ray tracing [COMPUT SCI] The creation of reflections, refractions, and shadows in a graphics image by following a series of rays from a light source and determining the effect of light on each pixel in the image. [OPTICS] Calculation of the paths followed by rays of light through an optical system, using Snell's law and trigonometrical formulas. { 'rā ˌtrās·iŋ }

Razin effect [PL PHYS] An effect whereby electrons in a cool, collisionless plasma strongly reduce the intensity of synchrotron radiation. { 'räz·ən i‚fekt }

razon [ORD] A kind of glide bomb having movable control surfaces in the tail adjusted by radio signals to control the bomb in range and in azimuth. { 'rā‚zän }

razorback [GEOL] A sharp, narrow ridge. { 'rā·zər‚bak }

razor stone See novaculite. { 'rā·zər ˌstōn }

Rb See rubidium.

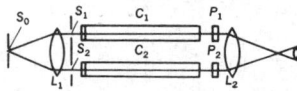

RAYLEIGH INTERFEROMETER

Schematic drawing of the Rayleigh interferometer. S_1, S_2 are slits illuminated with parallel light from lens L_1 whose focus is a single narrow slit S_0. Second lens L_2 brings the two beams to a focus, where interference fringes become visible. Cells (C_1, C_2) are placed in front of each slit; material of interest is placed in one cell and reference material in other. One of the plates of glass (P_1, P_2) is tilted until effect of cells on the interference fringes is cancelled.

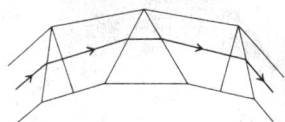

RAYLEIGH PRISM

Rayleigh prism system; the arrow shows the path of light through the prism.

RBE *See* relative biological effectiveness.

R-branch [SPECT] A series of lines in molecular spectra that correspond, in the case of absorption, to a unit increase in the rotational quantum number *J*. { 'är ,branch }

RBS *See* Rutherford backscattering spectrometry.

R-C amplifier *See* resistance-capacitance coupled amplifier. { ¦är¦sē am·plə,fī·ər }

RC asphalt *See* rapid-curing asphalt. { ¦är¦sē 'as,fȯlt }

R-C circuit *See* resistance-capacitance circuit. { ¦är¦sē 'sər·kət }

R-C constant *See* resistance-capacitance constant. { ¦är¦sē 'kän·stənt }

R-C coupled amplifier *See* resistance-capacitance coupled amplifier. { ¦är¦sē ¦kəp·əld am·plə,fī·ər }

R-C coupling *See* resistance coupling. { ¦är¦sē 'kəp·liŋ }

R center [SOLID STATE] A color center whose absorption band lies between the F band and M band, and which is produced by prolonged irradiation with light in the F band or prolonged x-ray exposure at room temperature. Also known as D center; E center. { 'är ,sen·tər }

RCM *See* radar countermeasure.

R-C network *See* resistance-capacitance network. { ¦är¦sē 'net,wərk }

r-combination [MATH] An *r*-combination of a set is an selection of *r* elements of the set. { 'är ,käm·bə,nā·shən }

R Coronae Borealis star [ASTROPHYS] A rare type of irregular variable star which has long periods of maximum brightness followed by a sudden, unpredictable reduction in brightness of several magnitudes, and a slower, sometimes erratic return to the original brightness. { 'är kə¦rō,nē ,bȯr·ē'al·əs ,stär }

R-C oscillator *See* resistance-capacitance oscillator. { ¦är¦sē 'äs·ə,lād·ər }

rd *See* rutherford.

RdAc *See* radioactinium.

R-DAT system *See* rotary digital audio tape system. { 'är ,dat ,sis·təm *or* ¦är ¦dē¦ā'tē ,sis·təm }

RDC extractor *See* rotary-disk contactor. { ¦är¦dē'sē ik'strak·tər }

RDF *See* radio direction finder.

RDGE *See* resorcinol diglycidyl ether.

R display [ELECTR] Radar display, essentially an expanded A display, in which an echo can be expanded for more detailed examination. { 'är di,splā }

RDRAM *See* Rambus dynamic random-access memory. { ¦är¦dē'ram }

RDS *See* respiratory distress syndrome of newborn.

RDSS *See* radiodetermination satellite service.

RdTh *See* radiothorium.

RDX *See* cyclonite.

Re *See* rhenium.

reach [CIV ENG] A portion of a waterway between two locks or gages. [CONT SYS] *See* range. [ENG] The length of a channel, uniform with respect to discharge, depth, area, and slope. [GEOGR] **1.** A continuous, unbroken surface of land or water. Also known as stretch. **2.** A bay, estuary, or other arm of the sea extending up into the land. [HYD] A straight, continuous, or extended part of a river, stream, or restricted waterway. { rēch }

reach rod [MECH ENG] A rod motion in a link used to transmit motion from the reversing rod to the lifting shaft. { 'rēch ,räd }

reactance [ELEC] The imaginary part of the impedance of an alternating-current circuit. { rē'ak·təns }

reactance amplifier *See* parametric amplifier. { rē'ak·təns 'am·plə,fī·ər }

reactance drop [ELEC] The component of the phasor representing the voltage drop across a component or conductor of an alternating-current circuit which is perpendicular to the current. { rē'ak·təns ,dräp }

reactance frequency multiplier [ELECTR] Frequency multiplier whose essential element is a nonlinear reactor. { rē'ak·təns 'frē·kwən·sē 'məl·tə,plī·ər }

reactance grounded [ELEC] Grounded through a reactance. { rē'ak·təns ,grauṅ·dəd }

reactance relay [ELEC] Form of impedance relay, the operation of which is a function of the reactance of a circuit. { rē'ak·təns ,rē,lā }

reactance tube [ELECTR] Vacuum tube operated in a way that it presents almost a pure reactance to the circuit. { rē'ak·təns ,tüb }

reactance-tube modulator [ELECTR] An electron-tube circuit, used to produce phase or frequency modulation, in which the reactance is varied in accordance with the instantaneous amplitude of the modulating voltage. { rē'ak·təns ,tüb 'mäj·ə,lād·ər }

reactant [CHEM] A substance that reacts with another one to produce a new set of substances (products). { rē'ak·tənt }

reactant ratio [AERO ENG] The ratio of the weight flow of oxidizer to fuel in a rocket engine. { rē'ak·tənt ,rā·shō }

reaction [CONT SYS] *See* positive feedback. [MECH] The equal and opposite force which results when a force is exerted on a body, according to Newton's third law of motion. [NUC PHYS] *See* nuclear reaction. { rē'ak·shən }

reaction border *See* reaction rim. { rē'ak·shən ,bȯrd·ər }

reaction boundary *See* reaction line. { rē'ak·shən ,baůn·drē }

reaction curve *See* reaction line. { rē'ak·shən ,kərv }

reaction energy *See* disintegration energy. { rē'ak·shən ,en·ər·jē }

reaction engine [AERO ENG] An engine that develops thrust by its reaction to a substance ejected from it; specifically, such an engine that ejects a jet or stream of gases created by the addition of energy to the gases in the engine. Also known as reaction motor. { rē'ak·shən ,en·jən }

reaction enthalpy number [PHYS CHEM] A dimensionless number used in the study of interphase transfer in chemical reactions, equal to the enthalpy of reaction per unit mass of a specified compound produced in a reaction, times the mass fraction of that compound, divided by the product of the specific heat at constant pressure and the temperature change during the reaction. { rē'ak·shən 'en,thal·pē ,nəm·bər }

reaction fin [NAV ARCH] A multiple fin or stationary propeller installed ahead of a ship's propeller, with fins warped so that water flowing into the propeller rotates in the direction opposite, resulting in elimination of race rotation aft of the propeller, and thereby in increased efficiency. { rē'ak·shən ,fin }

reaction flux [MATER] Soldering flux which reacts chemically with the base metal and has a rapid fluxing action when heated. { rē'ak·shən ,fləks }

reaction formation [PSYCH] A defense mechanism, characterized by the development of conscious, socially acceptable activity which is the antithesis of repressed or rejected unconscious desires. { rē'ak·shən fȯr,mā·shən }

reaction injection molding [ENG] A plastics fabrication process in which two streams of highly reactive, low-molecular-weight, low-viscosity resin systems are combined to form a solid material. { rē'ak·shən in'jek·shən 'mōl·diŋ }

reaction kinetics *See* chemical kinetics. { rē'ak·shən ki'ned·iks }

reaction line [PHYS CHEM] In a ternary system, a special case of the boundary line along which one of the two crystalline phases present reacts with the liquid, as the temperature is decreased, to form the other crystalline phase. Also known as reaction boundary; reaction curve. { rē'ak·shən ,līn }

reaction mechanism [CHEM] The sequence of steps during which a chemical reaction occurs, including the transition state during which the reactants are converted into products. { rē'ak·shən ,mek·ə,niz·əm }

reaction motor [AERO ENG] *See* reaction engine. [ELEC] A synchronous motor whose rotor contains salient poles but which has no windings and no permanent magnets. { rē'ak·shən ,mōd·ər }

reaction pair [MINERAL] Any two minerals, one of which is formed at the expense of the other by reaction with liquid. { rē'ak·shən ,per }

reaction path *See* mechanism. { rē'ak·shən ,path }

reaction principle [MINERAL] The concept of a reaction series for the principal rock-forming minerals. { rē'ak·shən ,prin·sə·pəl }

reaction propulsion [AERO ENG] Propulsion by means of reaction to a jet of gas or fluid projected rearward, as by a jet engine, rocket engine, or rocket motor. { rē'ak·shən prə,pəl·shən }

reaction rim [PETR] A surficial rim around one mineral produced by the reaction of the core mineral with the surrounding magma. Also known as reaction border. { rē'ak·shən ,rim }

REACTION SERIES

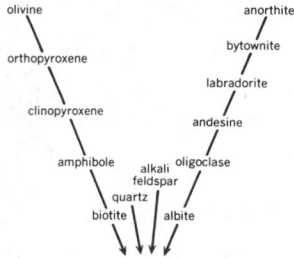

Continuous reaction series showing the course of crystallization of the common magmas; early crystals are at top of diagram.

REACTIVE POWER

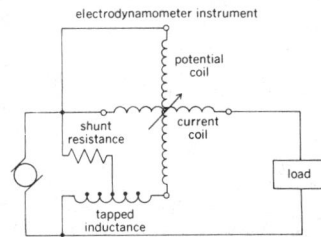

Diagram of a circuit for single-phase reactive power measurement which uses a tapped inductance. This inductance and shunt resistance are adjusted until currents in potential coil and current coil are in quadrature.

reaction series [MINERAL] Any series of minerals in which early formed varieties react with the melt to yield new minerals; two different types of reaction series exist, continuous and discontinuous. { rē'ak·shən ¦sir·ēz }

reactions inventory [IND ENG] A summary of the various possible responses of an individual to a stimulus or group of stimuli. { rē'ak·shənz 'in·ven¦tȯr·ē }

reaction time [PHYSIO] The interval between application of a stimulus and the beginning of the response. { rē'ak·shən ¦tīm }

reaction turbine [MECH ENG] A power-generation prime mover utilizing the steady-flow principle of fluid acceleration, where nozzles are mounted on the moving element. { rē'ak·shən ¦tər·bən }

reaction wheel [MECH ENG] A device capable of storing angular momentum which may be used in a space ship to provide torque to effect or maintain a given orientation. { rē'ak·shən ¦wēl }

reaction wood [BOT] An abnormal development of a tree and therefore its wood as the result of unusual forces acting on it, such as an atypical gravitational pull. { rē'ak·shən ¦wu̇d }

reaction zone [CHEM ENG] In a catalytic reactor vessel, the location or zone within the vessel where the bulk of the chemical reaction takes place. { rē'ak·shən ¦zōn }

reactive [ELEC] Pertaining to either inductive or capacitance reactance; a reactive circuit has a high value of reactance in comparison with resistance. { rē'ak·tiv }

reactive bond [CHEM] A bond between atoms that is easily invaded (reacted to) by another atom or radical; for example, the double bond in $CH_2=CH_2$ (ethylene) is highly reactive to other ethylene molecules in the reaction known as polymerization to form polyethylene. { rē'ak·tiv 'bänd }

reactive component [ELEC] In the phasor representation of quantities in an alternating-current circuit, the component of current, voltage, or apparent power which does not contribute power, and which results from inductive or capacitive reactance in the circuit, namely, the reactive current, reactive voltage, or reactive power. Also known as idle component; quadrature component; wattless component. { rē'ak·tiv kəm'pō·nənt }

reactive current [ELEC] In the phasor representation of alternating current, the component of the current perpendicular to the voltage, which contributes no power but increases the power losses of the system. Also known as idle current; quadrature current; wattless current. { rē'ak·tiv 'kə·rənt }

reactive dye [MATER] Dye that reacts with the textile fiber to produce both a hydroxyl and an oxygen linkage, the chlorine combining with the hydroxyl to form a strong ether linkage; gives fast, brilliant colors. { rē'ak·tiv 'dī }

reactive factor [ELEC] The ratio of reactive power to apparent power. { rē'ak·tiv 'fak·tər }

reactive fluid [PETRO ENG] Any fluid that alters the internal geometry of a reservoir's porosity; for example, water is a reactive fluid when it causes swelling of clays and consequent changes in porosity. { rē'ak·tiv 'flü·əd }

reactive intermediate [CHEM] An unstable compound formed as an intermediate during a chemical reaction. { rē'ak·tiv ˌin·tər'mē·dē·ət }

reactive ion etching [ELECTR] A directed chemical etching process used in integrated circuit fabrication in which chemically active ions are accelerated along electric field lines to meet a substrate perpendicular to its surface. { rē'ak·tiv 'ī·än ˌech·iŋ }

reactive load [ELEC] A load having inductive or capacitive reactance. { rē'ak·tiv 'lōd }

reactive muffler [ENG] A muffler that attenuates by reflecting sound back to the source. Also known as nondissipative muffler. { rē'ak·tiv 'məf·lər }

reactive power [ELEC] The power value obtained by multiplying together the effective value of current in amperes, the effective value of voltage in volts, and the sine of the angular phase difference between current and voltage. Also known as wattless power. { rē'ak·tiv 'pau̇·ər }

reactive voltage [ELEC] In the phasor representation of alternating current, the voltage component that is perpendicular to the current. { rē'ak·tiv 'vōl·tij }

reactive volt-ampere See volt-ampere reactive. { rē'ak·tiv 'vōlt 'am¦pir }

reactive volt-ampere hour See var hour. { rē'ak·tiv 'vōlt 'am¦pir 'au̇·ər }

reactive volt-ampere meter See varmeter. { rē'ak·tiv 'vōlt 'am¦pir ˌmēd·ər }

reactivity [CHEM] The relative capacity of an atom, molecule, or radical to combine chemically with another atom, molecule, or radical. [NUCLEO] A measure of the deviation of a nuclear reactor from the critical state at any instant of time such that positive and negative values correspond to reactors above and below critical, respectively; measured in percent k, millikays, dollars, or in-hours. { ˌrē·ak'tiv·əd·ē }

reactor [CHEM ENG] Device or process vessel in which chemical reactions (catalyzed or noncatalyzed) take place during a chemical conversion type of process. [ELEC] A device that introduces either inductive or capacitive reactance into a circuit, such as a coil or capacitor. Also known as electric reactor. [NUC PHYS] See nuclear reactor. { rē'ak·tər }

reactor cavity cooling system [NUCLEO] A passive heat-removal system incorporated in the design of the modular high-temperature gas-cooled reactor, in which ambient air is heated by reactor decay heat and returned to the environment. { rē'ak·tər ¦kav·əd·ē 'kül·iŋ ˌsis·təm }

reactor fuel See nuclear fuel. { rē'ak·tər ˌfyül }

reactor fuel cycle [NUCLEO] The processes of preparing fuel elements and assemblies for use in a reactor, using these elements in reactor operation, recovering radioactive by-products from spent fuel, and reprocessing remaining fissionable material into new fuel elements. Also known as fuel cycle; nuclear fuel cycle. { rē'ak·tər 'fyül ˌsī·kəl }

reactor fuel element See nuclear fuel element. { rē'ak·tər 'fyül ˌel·ə·mənt }

reactor fuel pellet See nuclear fuel pellet. { rē'ak·tər 'fyül ˌpel·ət }

reactor period [NUCLEO] The time required for the power of a nuclear reactor to increase by a factor of $e = 2.72$ for a given multiplication constant. { rē'ak·tər ˌpir·ē·əd }

reactor physics [NUCLEO] The science of the interaction of the elementary particles and radiations characteristic of nuclear reactors with matter in bulk. { rē'ak·tər ˌfiz·iks }

reactor vessel [NUCLEO] A large tanklike structure built to prevent radioactive materials from escaping from the reactor and associated equipment. { rē'ak·tər ˌves·əl }

read [COMPUT SCI] **1.** To acquire information, usually from some form of storage in a computer. **2.** To convert magnetic spots, characters, or punched holes into electrical impulses. [ELECTR] To generate an output corresponding to the pattern stored in a charge storage tube. { rēd }

read-around number See read-around ratio. { 'rēd ə¦rau̇nd ˌnəm·bər }

read-around ratio [COMPUT SCI] The number of times that a particular bit in electrostatic storage may be read without seriously affecting nearby bits. Also known as read-around number. { 'rēd ə¦rau̇nd ˌrā·shō }

read-back check See echo check. { 'rēd¦bak ˌchek }

Read diode [ELECTR] A high-frequency semiconductor diode consisting of an avalanching pn junction, biased to fields of several hundred thousand volts per centimeter, at one end of a high-resistance carrier serving as a drift space for the charge carriers. { 'rēd ˌdī·ōd }

reader [COMPUT SCI] A device that converts information from one form to another, as from punched paper tape to magnetic tape. [GRAPHICS] A projection device for viewing an enlarged microimage with the unaided eye. { 'rēd·ər }

reader-interpreter [COMPUT SCI] A service routine that reads an input string, stores programs and data on random-access storage for later processing, identifies the control information contained in the input string, and stores this control information separately in the appropriate control lists. { 'rēd·ər in'tər·prəd·ər }

reader-punch equipment [COMPUT SCI] An input/output unit which can punch computer results on cards and read card data into the computer. { 'rēd·ər 'pənch i¦kwip·mənt }

read error [COMPUT SCI] A condition in which the content of a storage device cannot be electronically identified. { 'rēd ˌer·ər }

read head [COMPUT SCI] A device that converts digital information stored on a magnetic tape, drum, or disk into electrical signals usable by the computer arithmetic unit. { 'rēd ˌhed }

read-in [COMPUT SCI] To sense information contained in

some source and transmit this information to an internal storage. { 'rēd¦in }

readiness review [COMPUT SCI] An on-site examination of the adequacy of preparations for effective utilization upon installation of a computer, and to identify any necessary corrective actions. { 'red·i·nəs ri,vyü }

readiness time [ENG] The length of time required to obtain a stabilized system ready to perform its intended function (readiness time includes warm-up time); the time is measured from the point when the system is unassembled or uninstalled to such time as it can be expected to perform as accurately as at any later time; maintenance time is excluded from readiness time. { 'red·i·nəs ,tīm }

reading [ENG] **1.** The indication shown by an instrument. **2.** Observation of the readings of one or more instruments. [MOL BIO] A linear process by which amino acid sequences are recognized by the protein-synthesizing system of a cell from messenger ribonucleic codes. { 'rēd·iŋ }

reading frame [MOL BIO] A nucleotide sequence that starts with an initiation codon, partitions the subsequent nucleotides into a series of amino acid-encoding triplets, and ends with a termination codon. { 'rēd·iŋ ,frām }

reading microscopes [OPTICS] A set of microscopes used to read the division circle of a transit circle in order to precisely determine the inclination of the telescope. { 'rēd·iŋ ,mī·krə,skōps }

reading mistake [MOL BIO] The incorrect placement of one or more amino acid residues in a polypeptide chain during genetic translation. { 'rēd·iŋ mə'stāk }

reading point See breakpoint. { 'rēd·iŋ ,pȯint }

reading rate [COMPUT SCI] Number of characters, words, fields, or blocks sensed by an input sensing device per unit of time. { 'rēd·iŋ ,rāt }

read-in program [COMPUT SCI] Computer program that can be put into a computer in a simple binary form and allows other programs to be read into the computer in more complex forms. { 'rēd ,in ,prō·grəm }

read-only memory [COMPUT SCI] A device for storing data in permanent, or nonerasable, form; usually a static electronic or magnetic device allowing extremely rapid access to data. Abbreviated ROM. Also known as read-only storage. { 'rēd ¦ōn·lē 'mem·rē }

read-only storage See read-only memory. { 'rēd ¦ōn·lē 'stȯr·ij }

read-only terminal [COMPUT SCI] A peripheral device, such as a printer, that can only receive signals. { 'rēd ¦ōn·lē 'tər·mən·əl }

readout [COMPUT SCI] **1.** The presentation of output information by means of a display, printout, or other methods. **2.** To sense information contained in some computer internal storage and transmit this information to a storage external to the computer. { 'rēd,aút }

readout station [COMMUN] A recording or receiving radio station at which data are received. { 'rēd,aút ,stā·shən }

read screen [COMPUT SCI] In optical character recognition (OCR), the transparent component part of most character readers through which appears the input document to be recognized. { 'rēd ,skrēn }

readthrough [GEN] Transcription beyond a termination sequence due to failure of ribonucleic acid polymerase to recognize the termination codon. { 'rēd,thrü }

read time [COMPUT SCI] The time interval between the instant at which information is called for from storage and the instant at which delivery is completed in a computer. { 'rēd ,tīm }

read-while-writing [COMPUT SCI] The reading of a record or group of records into storage from tape at the same time another record or group of records is written from storage to tape. { 'rēd ,wīl 'rīd·iŋ }

read/write channel [COMPUT SCI] A path along which information is transmitted between the central processing unit of a computer and an input, output, or storage unit under the control of the computer. { 'rēd 'rīt ,chan·əl }

read/write check indicator [COMPUT SCI] A device incorporated in certain computers to indicate upon interrogation whether or not an error was made in reading or writing; the machine can be made to stop, retry the operation, or follow a special subroutine, depending upon the result of the interrogation. { 'rēd 'rīt 'chek ,in·də,kād·ər }

read/write comb [COMPUT SCI] The set of arms mounted with magnetic heads that reach between the disks of a disk storage device to read and record information. { 'rēd 'rīt ,kōm }

read/write head [COMPUT SCI] A magnetic head that both senses and records data. Also known as combined head. { 'rēd 'rīt ,hed }

read/write memory [COMPUT SCI] A computer storage in which data may be stored or retrieved at comparable intervals. { 'rēd 'rīt ,mem·rē }

read/write random-access memory [COMPUT SCI] A random access memory in which data can be written into memory as well as read out of memory. { 'rēd 'rīt 'ran·dəm 'ak,ses ,mem·rē }

ready [ORD] Of a weapon, aimed, loaded, and prepared to fire. { 'red·ē }

ready-mixed concrete [MATER] Concrete mixed away from the construction site and delivered in readiness for placing. { 'red·ē ¦mikst kän'krēt }

ready-to-receive signal [COMMUN] Signal sent back to a facsimile transmitter to indicate that a facsimile receiver is ready to accept the transmission. { 'red·ē tə ri'sēv ,sig·nəl }

reagent [ANALY CHEM] A substance, chemical, or solution used in the laboratory to detect, measure, or otherwise examine other substances, chemicals, or solutions; grades include ACS (American Chemical Society standards), reagent (for analytical reagents), CP (chemically pure), USP (U.S. Pharmacopeia standards), NF (National Formulary standards), and purified, technical (for industrial use). [CHEM] The compound that supplies the molecule, ion, or free radical which is arbitrarily considered as the attacking species in a chemical reaction. { rē'ā·jənt }

reagent chemicals [ANALY CHEM] High-purity chemicals used for analytical reactions, for testing of new reactions where the effect of impurities are unknown, and, in general, for chemical work where impurities must either be absent or at a known concentration. { rē'ā·jənt ,kem·ə·kəlz }

reagin [IMMUNOL] The class of immunoglobulin that mediates allergic reactions such as asthma and urticaria. Also known as IgE; reaginic antibody. { rē'ā·jən }

reaginic antibody See reagin. { ,rē·ə,jin·ik 'an·tə,bäd·ē }

real axis [MATH] The horizontal axis of the Cartesian coordinate system for the euclidean or complex plane. { 'rēl 'ak·səs }

real closed field [MATH] A real field which has no algebraic extensions other than itself. { ¦rēl ¦klōzd 'fēld }

real closure [MATH] A real closure of a real field F is a real closed field which is an algebraic extension of F. { ¦rēl 'klō·zhər }

real continuum See real number system. { ¦rēl kən'tin·yə·wəm }

real crystal [CRYSTAL] A crystal for which the finite extent of the crystal and its various imperfections and defects are taken into account. { 'rēl 'krist·əl }

real data type [COMPUT SCI] A scalar data type which contains a normalized fraction (mantissa) and an exponent (characteristic) and is used to represent floating-point data, usually decimal. { 'rēl 'dad·ə ,tīp }

real fluid flow [FL MECH] The flow in which effects of tangential or shearing forces are taken into account; these forces give rise to fluid friction, because they oppose the sliding of one particle past another. { 'rēl 'flü·əd ,flō }

realgar [MINERAL] AsS A red to orange mineral crystallizing in the monoclinic system, having a resinous luster and found in short, vertical striated crystals; specific gravity is 3.48, and hardness is 1.5–2 on Mohs scale. Also known as red arsenic; red orpiment; sandarac. { rē'al,gär }

real gas [THERMO] A gas, as considered from the viewpoint in which deviations from the ideal gas law, resulting from interactions of gas molecules, are taken into account. Also known as imperfect gas. { 'rēl 'gas }

real image [OPTICS] An optical image such that all the light from a point on an object that passes through an optical system actually passes close to or through a point on the image. { 'rēl 'im·ij }

reality principle [PSYCH] The concept that the pleasure principle is normally modified by the demands of the external environment and that the individual adjusts to these inescapable

REAL IMAGE

Formation of real image. Rays leaving object point Q and passing through the refracting surface separating media n and n' are brought to a focus at the image point Q'. Modified from F. A. Jenkins and H. E. White, *Fundamentals of Optics*, 3d ed., McGraw-Hill, 1957)

requirements so that he ultimately secures satisfaction of his instinctual wishes. { rē'al·əd·ē ,prin·sə·pəl }

realizability [CONT SYS] Property of a transfer function that can be realized by a network that has only resistances, capacitances, inductances, and ideal transformers. { ,rē·ə,līz·ə'bil·əd·ē }

realization [STAT] For a stochastic process, a probability space whose points are sample paths of the stochastic process and whose probability is obtained from the joint probability distributions of the random variables in the process. { ,rē·ə·lə'zā·shən }

real line [MATH] A straight line of infinite extent upon which the real numbers are plotted according to their distance in a positive or negative direction from a point arbitrarily chosen as zero. Also known as number line. { rēl 'līn }

real linear group [MATH] The group of all nonsingular linear transformations of a real vector space whose group operation is composition. { ¦rēl ¦lin·ē·ər 'grüp }

real number [MATH] Any member of the real number system. { 'rēl 'nəm·bər }

real number system [MATH] The unique (to within isomorphism) complete ordered field; the field of real numbers. Also known as real continuum. { ¦rēl 'nəm·bər ,sis·təm }

real object [OPTICS] A collection of points which actually serves as a source of light rays in an optical system. { 'rēl 'äb·jikt }

real orthogonal group [MATH] The group composed of orthogonal matrices having real number entries. { 'rēl ȯr'thäg·ən·əl ,grüp }

real part [MATH] The real part of a complex number $z = x + iy$ is the real number x. { 'rēl ¦pärt }

real plane [MATH] A plane whose points are assigned ordered pairs of real numbers for coordinates. { 'rēl ,plān }

real power [ELEC] The component of apparent power that represents true work; expressed in watts, it is equal to volt-amperes multiplied by the power factor. { 'rēl 'paů·ər }

real precession [NAV] In marine gyroscopes, that component of the total precession caused by bearing friction, torque unbalance, and other manufacturing or design defects. Also known as induced precession. { 'rēl prē'sesh·ən }

real source [ACOUS] A source of sound consisting of a macroscopic body that is composed of materials different from those of the medium in which the sound propagates and has sharply delineated physical extent, and which generates sound by executing complex motions while immersed in the medium. { 'rēl 'sȯrs }

real-space-transfer transistor [ELECTR] A transistor that utilizes the effect of the increase in electron energy and temperature in high electric fields. { ,rēl ,spās 'tranz·fər tran,zis·tər }

real storage [COMPUT SCI] Actual physical storage of data and instructions. { 'rēl 'stȯr·ij }

real-time [COMPUT SCI] Pertaining to a data-processing system that controls an ongoing process and delivers its outputs (or controls its inputs) not later than the time when these are needed for effective control; for instance, airline reservations booking and chemical processes control. { 'rēl ,tīm }

real-time clock [COMPUT SCI] A pulse generator which operates at precise time intervals to determine time intervals between events and initiate specific elements of processing. { 'rēl ,tīm 'kläk }

real-time control system [COMPUT SCI] A computer system which controls an operation in real time, such as a rocket flight. { 'rēl ,tīm kən'trōl ,sis·təm }

real-time holographic interferometry [OPTICS] The study of the interference fringes generated when a hologram is made of an object and is later placed back into its original position relative to the object, now slightly deformed, so that there is interference between the object and its hologram. { 'rēl ,tīm ¦hō·lə¦graf·ik ,in·tər·fə'räm·ə·trē }

real-time operation [COMPUT SCI] **1.** Of a computer or system, an operation or other response in which programmed responses to an event are essentially simultaneous with the event itself. **2.** An operation in which information obtained from a physical process is processed to influence or control the physical process. { 'rēl ,tīm ,äp·ə'rā·shən }

real-time processing [COMPUT SCI] The handling of input data at a rate sufficient to ensure that the instructions generated by the computer will influence the operation under control at the required time. { 'rēl ,tīm 'prä,ses·iŋ }

real-time programming [COMPUT SCI] Programming for a situation in which results of computations will be used immediately to influence the course of ongoing physical events. { 'rēl ,tīm 'prō,gram·iŋ }

real-time system [COMPUT SCI] A system in which the computer is required to perform its tasks within the time restraints of some process or simultaneously with the system it is assisting. { 'rēl ,tīm 'sis·təm }

real unimodular group [MATH] The group of all square $n \times n$ matrices with real number entries and of determinant 1. { 'rēl ,yün·i'mäj·ə·lər ,grüp }

real-valued function [MATH] A function whose values are real numbers. { 'rēl ¦val·yüd 'fəŋk·shən }

real variable [MATH] A variable that assumes real numbers for its values. { 'rēl 'ver·ē·ə·bəl }

ream [ENG] To enlarge or clean out a hole. [MATER] **1.** A layer of nonhomogeneous material in flat glass. **2.** Five hundred sheets of paper; a printer's ream consists of 516 sheets. { rēm }

reamed extrusion ingot [MET] A hollow extrusion ingot whose original inside surface has been removed by reaming. { 'rēmd ik'strü·zhən ,iŋ·gət }

reamer [DES ENG] A tool used to enlarge, shape, smooth, or otherwise finish a hole. { 'rēm·ər }

reaming bit [DES ENG] A bit used to enlarge a borehole. Also known as broaching bit; pilot reaming bit. { 'rēm·iŋ ,bit }

rear area [ORD] The area in the rear of the combat and forward areas. { 'rir 'er·ē·ə }

rearming [ORD] **1.** An operation that replenishes the prescribed stores of ammunition, bombs, and other armament items for an aircraft, naval ship, tank, or armored vehicle, including replacement of defective ordnance equipment, in order to make it ready for combat service. **2.** Resetting the fuse on a bomb, or on an artillery, mortar, or rocket projectile, so that it will detonate at the desired time. { rē'ärm·iŋ }

rear-projection [ELECTR] Pertaining to television system in which the picture is projected on a ground-glass screen for viewing from the opposite side of the screen. { 'rir prə'jek·shən }

rearrangement reaction [NUC PHYS] A nuclear reaction in which nucleons are exchanged between nuclei. [ORG CHEM] A chemical reaction involving a change in the bonding sequence within a molecule. Also known as molecular rearrangement. { ,rē·ə'rānj·mənt rē,ak·shən }

rear response [ENG ACOUS] The maximum pressure within 60° of the rear of a transducer in decibels relative to the pressure on the acoustic axis. { 'rir ri,späns }

rear sight [ORD] An item attached to the breech end and integral to a carbine, machine gun, pistol, rifle, or the like; it may be a fixed or adjustable cross blade with a U- or V-shaped notch or aperture, or it may have elevation and windage adjustment knobs, slides, and graduated scales and be provided with aperture disks. { 'rir ,sīt }

reasonableness [COMPUT SCI] A measure of the extent to which data processed by a computer falls within an acceptable allowance for errors, as determined by quantitative tests. { 'rēz·nə·bəl·nəs }

reassortant virus [VIROL] A virion containing deoxyribonucleic acid from one virus species and a protein coat from another. { ,rē·ə¦sȯrt·ənt 'vī·rəs }

Réaumur temperature scale [THERMO] Temperature scale where water freezes at 0°R and boils at 80°R. { 'rā·ō¦myůr 'tem·prə·chər ,skāl }

rebar [CIV ENG] A steel bar or rod used to reinforce concrete. { 'rē,bär }

rebat [METEOROL] The lake breeze of Lake Geneva, Switzerland; it blows from about 10 a.m. to 4 p.m. { re'bä }

rebecca [NAV] An electronic navigation system that has at least one radio transmitter and one radio receiver; the transmitter emits pulses which travel over a single path to a transponder and return to the interrogation receiver. { ri'bek·ə }

rebecca-eureka system [NAV] An aircraft radar homing system in which an airborne interrogator-responsor (rebecca) homes on a ground radar beacon (eureka) that has been dropped or set up in advance; the system can also give the distance from the rebecca radar to the eureka beacon. { ri'bek·ə yů'rē·kə ,sis·təm }

reboiler [CHEM ENG] An auxiliary heating unit for a fractionating tower designed to supply additional heat to the lower portion of the tower; liquid withdrawn from the side or bottom of the tower is reheated by heat exchange, then reintroduced into the tower. { rē'bȯil·ər }

reboot [COMPUT SCI] To reload systems software into a computer so that it makes a new start. { rē'büt }

rebound [GEOL] The isostatic readjustment upward of a landmass depressed by glacial loading. { 'rē,baȯnd }

rebound clip [DES ENG] A clip surrounding the back and one or two other leaves of a leaf spring, to distribute the load during rebounds. { 'rē,baȯnd ,klip }

rebound leaf [DES ENG] In a leaf spring, a leaf placed over the master leaf to limit the rebound and help carry the load imposed by it. { 'rē,baȯnd ,lēf }

reboyo [METEOROL] A persistent (day-long) storm from the southwest during the rainy season on the Brazilian coast. { rə'bȯi·ō }

rebreather [ENG] A closed-loop oxygen supply system consisting of gas supply and face mask. { rē'brēth·ər }

rebroadcast [COMMUN] Repetition of a radio or television program at a later time. { rē'brȯd,kast }

rebuild [ENG] To restore to a condition comparable to new by disassembling the item to determine the condition of each of its component parts, and reassembling it, using serviceable, rebuilt, or new assemblies, subassemblies, and parts. { rē'bild }

recalescence [MET] Brightening (reglowing) of iron on cooling through the gamma- to alpha-phase transformation temperature caused by liberation of the latent heat of transformation. { ,rē·kə'les·əns }

recall [COMMUN] A flashing signal to the attendant's switchboard; the operator may be recalled by the subscriber operating the switch hook of the subscriber's set. { 'rē,kȯl }

recall factor [COMPUT SCI] A measure of the efficiency of an information retrieval system, equal to the number of retrieved relevant documents divided by the total number of relevant documents in the file. { 'rē,kȯl ,fak·tər }

recapitulation theory [BIOL] The biological theory that an organism passes through developmental stages resembling various stages in the phylogeny of its group; ontogeny recapitulates phylogeny. Also known as biogenetic law; Haeckel's law. { ,rē·kə,pich·ə'lā·shən ,thē·ə·rē }

recarburize [MET] **1.** To increase the carbon content of molten steel or cast iron. **2.** To carburize a metal part, making up for any loss of carbon during processing. { rē'kär·bə,rīz }

receding leg [ORD] That portion of the target's course line in which the slant range increases for successive target positions. { ri'sēd·iŋ 'leg }

received power [ELECTROMAG] **1.** The total power received at an antenna from a signal, such as a radar target signal. **2.** In a mobile communications system, the root-mean-square value of power delivered to a load which properly terminates an isotropic reference antenna. { ri'sēvd 'paȯ·ər }

receive-only [COMMUN] A teleprinter which has no keyboard, and thus can receive but not transmit. Abbreviated RO. { ri'sēv 'ōn·lē }

receiver [CHEM ENG] Vessel, container, or tank used to receive and collect liquid material from a process unit, such as the distillate receiver from the overhead condenser of a distillation column. [ELECTR] The complete equipment required for receiving modulated radio waves and converting them into the original intelligence, such as into sounds or pictures, or converting to desired useful information as in a radar receiver. [MECH ENG] An apparatus placed near the compressor to equalize the pulsations of the air as it comes from the compressor to cause a more uniform flow of air through the pipeline and to collect moisture and oil carried in the air. { ri'sē·vər }

receiver autonomous integrity monitoring [NAV] A method of providing consistency for the Global Positioning System (GPS) within the user avionics by checking the consistency of range measurements from redundant observations on GPS satellites. Abbreviated RAIM. { ri,sē·vər ȯ'tän·ə·məs in'teg·rəd·ē ,män·ə·triŋ }

receiver bandwidth [ELECTR] Spread, in frequency, between the halfpower points on the receiver response curve. { ri'sē·vər 'band,width }

receiver gating [ELECTR] Application of operating voltages to one or more stages of a receiver only during that part of a cycle of operation when reception is desired. { ri'sē·vər ,gād·iŋ }

receiver incremental tuning [ELECTR] Control feature to permit receiver tuning (of a transceiver) up to 3 kilohertz to either side of the transmitter frequency. { ri'sē·vər ,in·krə,ment·əl 'tün·iŋ }

receiver lockout system *See* lockout. { ri'sē·vər 'läk,aȯt ,sis·təm }

receiver noise threshold [ELECTR] External noise appearing at the front end of a receiver, plus the noise added by the receiver itself, determines a noise threshold that has to be exceeded by the minimum discernible signal. { ri'sē·vər 'nȯiz ,thresh,hōld }

receiver primary *See* display primary. { ri'sē·vər 'prī,mer·ē }

receiver radiation [ELECTROMAG] Radiation of interfering electromagnetic fields by the oscillator of a receiver. { ri'sē·vər ,rād·ē'ā·shən }

receiver synchro *See* synchro receiver. { ri'sē·vər 'siŋ·krō }

receiving antenna [ELECTROMAG] An antenna used to convert electromagnetic waves to modulated radio-frequency currents. { ri'sēv·iŋ an,ten·ə }

receiving area [ELECTROMAG] The factor by which the power density must be multiplied to obtain the received power of an antenna, equal to the gain of the antenna times the square of the wavelength divided by 4π. { ri'sēv·iŋ ,er·ē·ə }

receiving gage [ENG] A fixed gage designed to inspect a number of dimensions and also their reaction to each other. { ri'sēv·iŋ ,gāj }

receiving house [CHEM ENG] A building where liquid streams from petroleum-refining-process condensers are observed through a look box, and samples are taken for testing, and also where products are diverted to storage tanks or to other processing units. { ri'sēv·iŋ ,haȯs }

receiving loop loss [COMMUN] In telephones, that part of the repetition equivalent assignable to the station set, subscriber line, and battery supply circuit that are on the receiving end. { ri'sēv·iŋ ,lüp ,lȯs }

receiving set *See* radio receiver. { ri'sēv·iŋ ,set }

receiving station [MECH ENG] The location or device on conveyor systems where bulk material is loaded or otherwise received onto the conveyor. { ri'sēv·iŋ ,stā·shən }

receiving tank *See* rundown tank. { ri'sēv·iŋ ,taŋk }

receiving tube [ELECTR] A low-voltage and low-power vacuum tube used in radio receivers, computers, and sensitive control and measuring equipment. { ri'sēv·iŋ ,tüb }

recent [EVOL] Referring to taxa which still exist; the antonym of fossil. { 'rē·sənt }

Recent *See* Holocene. { 'rē·sənt }

receptacle [BOT] The pointed end of a pedicel or peduncle from which the flower parts grow. [ELEC] *See* outlet. { ri'sep·tə·kəl }

reception [COMMUN] The conversion of modulated electromagnetic waves or electric signals, transmitted through the air or over wires or cables, into the original intelligence, or into desired useful information (as in radar), by means of antennas and electronic equipment. { ri'sep·shən }

receptive aphasia *See* sensory aphasia. { ri'sep·tiv ə'fā·zhə }

receptor [BIOCHEM] A site or structure in a cell which combines with a drug or other biological to produce a specific alteration of cell function. [PHYSIO] A sense organ. { ri'sep·tər }

recess [ENG] A surface groove or depression. [GEOL] **1.** An indentation occurring in a surface, bounded by a straight line. **2.** An area having the axial traces of folds concave toward the outer edge of the folded belt. { 'rē,ses }

recessed bead *See* quirk bead. { 'rē,sest ,bēd }

recessed tube wall [MECH ENG] A boiler furnace wall which has openings to partially expose waterwall tubes to the radiant combustion gases. { 'rē,sest 'tüb ,wȯl }

recession [GEOL] **1.** The backward movement, or retreat, of an eroded escarpment. **2.** A continuing landward movement of a shoreline or beach undergoing erosion. Also known as retrogression. **3.** The withdrawal of a body of water (as a sea or lake), thereby exposing formerly submerged areas. [HYD] The gradual upstream retreat of a waterfall. { ri'sesh·ən }

recessional moraine [GEOL] **1.** An end moraine formed during a temporary halt in the final retreat of a glacier. **2.** A moraine formed during a minor readvance of the ice front during a period of glacial recession. Also known as stadial moraine. { ri'sesh·ən·əl mə'rān }

recession curve [HYD] A hydrograph showing the decrease of the runoff rate after rainfall or the melting of snow. { ri 'sesh·ən ‚kərv }

recession of galaxies [ASTROPHYS] The increase in the velocity of recession (red shift) of galaxies with distance from an observer on earth. { ri'sesh·ən əv 'gal·ik·sēz }

recessive [GEN] **1.** An allele that is not expressed phenotypically when present in the heterozygous condition. **2.** An organism homozygous for a recessive gene. { ri'ses·iv }

rechamber [ORD] To rebore or otherwise alter the chamber of a small arm, normally for the purpose of adapting it to cartridges for which not originally designed. { rē'chām·bər }

recharge [ELEC] To restore a cell or battery to a charged condition by sending a current through it in a direction opposite to that of the discharging current. [HYD] **1.** The processes involved in the replenishment of water to the zone of saturation. **2.** The amount of water added or absorbed. Also known as groundwater increment; groundwater recharge; groundwater replenishment; increment; intake. { rē'chärj }

rechargeable battery See storage battery. { rē'chär·jə·bəl'bad·ə·rē }

recharge area [HYD] An area in which water is absorbed that eventually reaches the zone of saturation in one or more aquifers. Also known as intake area. { 'rē‚chärj ‚er·ē·ə }

recharge basin [CIV ENG] A basin constructed in sandy material to collect water, as from storm drains, for the purpose of replenishing groundwater supply. { 'rē‚chärj ‚bās·ən }

recharge well [HYD] A well used as a source of water in the process of artificial recharge. Also known as injection well. { 'rē‚chärj ‚wel }

reciprocal [MATH] The reciprocal of a number A is the number $1/A$. { ri'sip·rə·kəl }

reciprocal bearing See back bearing. { ri'sip·rə·kəl 'ber·iŋ }

reciprocal curve [MATH] For a particular curve, C, a curve that includes a point with cartesian coordinates (x, y) if C has a point with cartesian coordinates $(x, 1/y)$. { ri‚sip·rə·kəl 'kərv }

reciprocal differences [MATH] An interpolation technique using successive quotients of a function with its values so as to obtain a continued fraction expansion approximating the given function by a rational function. { ri'sip·rə·kəl 'dif·rən·səz }

reciprocal ellipsoid See index ellipsoid. { ri'sip·rə·kəl i'lip‚sóid }

reciprocal equation [MATH] An algebraic equation in one variable whose roots are unchanged when the unknown is replaced by its reciprocal. { rə'sip·rə·kəl i'kwā·zhən }

reciprocal ferrite switch [ELECTROMAG] A ferrite switch that can be inserted in a waveguide to switch an input signal to either of two output waveguides; switching is done by a Faraday rotator when acted on by an external magnetic field. { ri'sip·rə·kəl 'fe‚rīt ‚swich }

reciprocal impedance [ELEC] Two impedances Z_1 and Z_2 are said to be reciprocal impedances with respect to an impedance Z (invariably a resistance) if they are so related as to satisfy the equation $Z_1 Z_2 = Z^2$. { ri'sip·rə·kəl im'pēd·əns }

reciprocal inhibition [PHYSIO] In muscular movement, the simultaneous relaxation of one muscle and the contraction of its antagonist. [PSYCH] The modification of a behavioral pattern by the conditioning of responses that are incompatible with the response to be eliminated. { ri'sip·rə·kəl ‚in·ə'bish·ən }

reciprocal junction [ELECTROMAG] A waveguide junction in which the transmission coefficient from the ith port to the jth port is the same as that from the jth port to the ith port; that is, the S matrix is symmetrical. { ri'sip·rə·kəl 'jəŋk·shən }

reciprocal lattice [CRYSTAL] A lattice array of points formed by drawing perpendiculars to each plane (hkl) in a crystal lattice through a common point as origin; the distance from each point to the origin is inversely proportional to spacing of the specific lattice planes; the axes of the reciprocal lattice are perpendicular to those of the crystal lattice. { ri'sip·rə·kəl 'lad·əs }

reciprocal laying [ORD] Method of making the planes of fire of two guns parallel by pointing the guns in parallel direction; in reciprocal laying, the two guns sight on each other,

then swing out through supplementary angles to produce equal deflections from the base line connecting the two pieces. { ri'sip·rə·kəl 'lā·iŋ }

reciprocal leveling [CIV ENG] A variant of straight differential leveling applied to long distances in which levels are taken on two points, and the average of the two elevation differences is the true difference. { ri'sip·rə·kəl 'lev·ə·liŋ }

reciprocal matrix See inverse matrix. { ri¦sip·rə·kəl 'mā·triks }

reciprocal ohm See siemens. { ri'sip·rə·kəl 'ōm }

reciprocal ohm centimeter See roc. { ri'sip·rə·kəl 'ōm 'sent‚i‚mēd·ər }

reciprocal ohm meter See rom. { ri'sip·rə·kəl 'ōm ‚mēd·ər }

reciprocal polar figures [MATH] Two plane figures consisting of lines and their points of intersection such that the points of each of them are the poles of the lines of the other with respect to a given conic. { ri¦sip·rə·kəl ¦pō·lər 'fig·yərz }

reciprocal ratio See inverse ratio. { ri¦sip·rə·kəl 'rā·shō }

reciprocal recombination [GEN] In dihybrid gametes, the generation, by meiotic crossing over, of a new combination of alleles unlike those of the maternal and paternal homologues. { ri'sip·rə·kəl rē‚käm·bə'nā·shən }

reciprocal series [MATH] A series whose terms are reciprocals of the corresponding terms of a given series. { rə'sip·rə·kəl 'sir‚ēz }

reciprocal space See wave-vector space. { ri'sip·rə·kəl ‚spās }

reciprocal spiral See hyperbolic spiral. { ri'sip·rə·kəl ‚spī·rəl }

reciprocal strain ellipsoid [MECH] In elastic theory, an ellipsoid of certain shape and orientation which under homogeneous strain is transformed into a set of orthogonal diameters of the sphere. { ri'sip·rə·kəl ¦strān i'lip‚sóid }

reciprocal substitution [MATH] The substitution of a new variable for the reciprocal of the original variable. { ri¦sip·rə·kəl ‚səb·stə'tü·shən }

reciprocal theorem [MATH] **1.** In plane geometry, a theorem (which may be true or false) that is obtained from a given theorem by exchanging points and lines, angles and sides, and so forth. **2.** See dual theorem. { ri'sip·rə·kəl 'thir·əm }

reciprocal transducer [ELECTR] Transducer which satisfies the principle of reciprocity. { ri'sip·rə·kəl tranz'dü·sər }

reciprocal translocation [CELL MOL] The special case of translocation in which two segments exchange positions. { ri'sip·rə·kəl ‚tranz·lō'kā·shən }

reciprocal vectors [CRYSTAL] For a set of three vectors forming the primitive translations of a lattice, the vectors that form the primitive translations of the reciprocal lattice. [MATH] For a set of three linearly independent vectors, a second set of three vectors, each of which is perpendicular to two of the original vectors and has a scalar product of unity with the third. { ri'sip·rə·kəl 'vek·tərz }

reciprocal velocity region [NUC PHYS] The energy region in which the capture cross section for neutrons by a given element is inversely proportional to neutron velocity. { ri'sip·rə·kəl və'läs·əd·ē ‚rē·jən }

reciprocal wavelength See wave number. { ri'sip·rə·kəl 'wāv‚leŋkth }

reciprocating compressor [MECH ENG] A positive-displacement compressor having one or more cylinders, each fitted with a piston driven by a crankshaft through a connecting rod. { ri'sip·rə‚kād·iŋ kəm'pres·ər }

reciprocating drill See piston drill. { ri'sip·rə‚kād·iŋ 'dril }

reciprocating engine See piston engine. { ri'sip·rə‚kād·iŋ 'en·jən }

reciprocating flight conveyor [MECH ENG] A reciprocating beam or beams with hinged flights that advance materials along a conveyor trough. { ri'sip·rə‚kād·iŋ 'flīt kən‚vā·ər }

reciprocating-plate column See reciprocating-plate extractor. { ri'sip·rə‚kād·iŋ ¦plāt 'käl·əm }

reciprocating-plate extractor [CHEM ENG] A liquid-liquid contactor in which equally spaced perforated plates (as in a distillation column) move up and down rapidly over a short distance to cause liquid agitation and mixing. Also known as reciprocating-plate column. { ri'sip·rə‚kād·iŋ ¦plāt ik'strak·tər }

reciprocating-plate feeder [MECH ENG] A back-and-forth shaking tray used to feed abrasive materials, such as pulverized coal, into process units. { ri'sip·rə‚kād·iŋ ¦plāt 'fēd·ər }

reciprocating pump *See* piston pump. { ri'sip·rə,kād·iŋ 'pəmp }

reciprocating screen [MECH ENG] Horizontal solids-separation screen (sieve) oscillated back and forth by an eccentric gear; used for solids classification. { ri'sip·rə,kād·iŋ 'skrēn }

reciprocation [ELECTR] In electronics, a process of deriving a reciprocal impedance from a given impedance, or finding a reciprocal network for a given network. { ri,sip·rə'kā·shən }

reciprocity calibration [ENG ACOUS] A measurement of the projector loss and hydrophone loss of a reversible transducer by means of the reciprocity theorem and comparisons with the known transmission loss of an electric network, without knowing the actual value of either the electric power or the acoustic power. { ,res·ə'präs·əd·ē ,kal·ə,brā·shən }

reciprocity failure [GRAPHICS] A phenomenon wherein the usual combination of light intensity and exposure time (following the reciprocity law) required to produce a specified density in a photographic emulsion changes due to especially long exposure time or low light intensity. { ,res·ə'präs·əd·ē ,fāl·yər }

reciprocity law [GRAPHICS] The law that, with standard development, the optical density of an exposed emulsion is proportional to the exposure time and to the illumination. { ,res·ə'präs·əd·ē ,lō }

reciprocity theorem Also known as principle of reciprocity. [ACOUS] The theorem that, in an acoustic system consisting of a fluid medium with boundary surfaces and subject to no impressed body forces, if p_1 and p_2 are the pressure fields produced respectively by the components of the fluid velocities V_1 and V_2 normal to the boundary surfaces, then the integral over the boundary surfaces of $p_1V_2 - p_2V_1$ vanishes. [ELEC] **1.** The electric potentials V_1 and V_2 produced at some arbitrary point, due to charge distributions having total charges of q_1 and q_2 respectively, are such that $q_1V_2 = q_2V_1$. **2.** In an electric network consisting of linear passive impedances, the ratio of the electromotive force introduced in any branch to the current in any other branch is equal in magnitude and phase to the ratio that results if the positions of electromotive force and current are exchanged. [ELECTROMAG] Given two loop antennas, a and b, then $I_{ab}/V_a = I_{ba}/V_b$, where I_{ab} denotes the current received in b when a is used as transmitter, and V_a denotes the voltage applied in a; I_{ba} and V_b are the corresponding quantities when b is the transmitter, a the receiver; it is assumed that the frequency and impedances remain unchanged. [ENG ACOUS] The sensitivity of a reversible electroacoustic transducer when used as a microphone divided by the sensitivity when used as a source of sound is independent of the type and construction of the transducer. [PHYS] In general, any theorem that expresses various reciprocal relations for the behavior of some physical systems, in which input and output can be interchanged without altering the response of the system to a given excitation. { ,res·ə'präs·əd·ē ,thir·əm }

recirculating-ball steering [MECH ENG] A steering system that transmits steering movements by means of steel balls placed between a worm gear and a nut. { rē'sər·kyə,lād·iŋ ,bol 'stir·iŋ }

recirculator [ENG] A self-contained underwater breathing apparatus that recirculates an oxygen supply (mix-gas or pure) to the diver until the oxygen is depleted. { rē'sər·kyə,lād·ər }

reclaimed oil [MATER] Used lubricating oil that is collected, reprocessed, and sold for reuse. Also known as recovered oil. { 'rē,klāmd 'oil }

reclaimed rubber [MATER] Scrap rubber (natural or synthetic) prepared for reuse; fragmented scrap is digested in hot caustic solution to which reclaiming agents have been added; reclaimed rubber is used to blend with virgin rubber, or for low-grade rubber products. { 'rē,klāmd 'rəb·ər }

reclaimer [COMPUT SCI] A device that performs dynamic storage allocation, periodically searching memory to locate cells whose contents are no longer useful for computation, and making them available for other uses. { rē'klām·ər }

reclaim rinse [MET] A nonflowing rinse used to recover dragout in electroplating operations. { 'rē,klām ,rins }

reclamation [CIV ENG] **1.** The recovery of land or other natural resource that has been abandoned because of fire, water, or other cause. **2.** Reclaiming dry land by irrigation. { ,rek·lə'mā·shən }

reclassify [ORD] To change the security classification of a document, piece of equipment, or the like. { rē'klas·ə,fī }

reclinate [BOT] Vernation in which the upper part of the leaf is bent down on the lower part. { 'rek·lə,nāt }

reclined fold *See* recumbent fold. { ,ri'klīnd 'fōld }

reclosing relay [ELEC] Form of voltage, current, power, or other type of relay which functions to reclose a circuit. { 'rē,klōz·iŋ 'rē,lā }

recognition [COMPUT SCI] The act or process of identifying (or associating) an input with one of a set of possible known alternatives, as in character recognition and pattern recognition. { ,rek·ig'nish·ən }

recognition differential [ACOUS] For a specified listener, the amount by which the signal level exceeds the noise level reaching the ear when there is a 50% probability of detection of the signal. { ,rek·ig'nish·ən ,dif·ə'ren·chəl }

recognition gate [COMPUT SCI] A logic circuit used to select devices identified by a binary address code. Also known as decoding gate. { rek·ig'nish·ən ,gāt }

recognition sequence [MOL BIO] A specific sequence of nucleotides at which a restriction endonuclease cleaves a deoxyribonucleic acid molecule. { ,rek·ig'nish·ən ,sē·kwəns }

recognition site [MOL BIO] The nucleotide sequence in duplex deoxyribonucleic acid (DNA) to which a restriction endonuclease binds initially and within which the endonuclease cuts the DNA. { ,rek·ig'nish·ən ,sīt }

recoil *See* gun reaction. { 'rē,koil }

recoil adapter [ORD] A device fastened between a gun, especially an aircraft machine gun, and its mount to adapt the gun for mounting and to absorb the recoil. { 'rē,koil ə,dap·tər }

recoil booster [ORD] Component of a machine gun which traps some of the gas from the barrel and acts to ensure positive recoil action when the gun is fired at angles other than the usual horizontal. { 'rē,koil ,büs·tər }

recoil brake [ORD] That part of the recoil mechanism that actually absorbs the energy of recoil and stops the rearward movement of the recoiling parts. { 'rē,koil ,brāk }

recoil click [HOROL] A device in a timepiece that prevents a mainspring from being wound too tightly; uses a pawl that recoils after winding. { 'rē,koil ,klik }

recoil electron [PHYS] An electron that has been set into motion by a collision. { 'rē,koil i,lek,trän }

recoil escapement *See* anchor escapement. { 'rē,koil i,skāp·mənt }

recoil implantation [ELECTR] A mechanism for ion-beam mixing of a film and a substrate in which atoms are driven from the film into the substrate as a result of direct collisions with incident ions. { 'rē,koil ,im·plan'tā·shən }

recoiling mass [ORD] The mass of the recoiling parts of a weapon. { ri'koil·iŋ 'mas }

recoiling parts [ORD] Those parts of a weapon which move in recoil, usually including the tube, breech housing, breechblock assembly, and parts of the recoil mechanism. { ri'koil·iŋ 'pärts }

recoil ion spectroscopy [ATOM PHYS] A method of studying highly ionized and highly excited atomic states, in which relatively light atoms in a gaseous target are bombarded by highly ionized, fast, heavy projectiles, resulting in single collisions in which the target atoms are raised to very high states of ionization and excitation while incurring relatively small recoil velocities. { 'rē,koil 'ī,än ,spek'träs·kə·pē }

recoilless [ORD] Built so as to eliminate or cancel out recoil; most recoilless guns are designed to let part of the propellant gases escape to the rear. { ri'koil·ləs }

recoilless ammunition [ORD] Ammunition intended for use in recoilless rifles; provision is made in the ammunition for release of propellant gases in the manner and quantity necessary to produce the recoilless action. { ri'koil·ləs ,am·yə'nish·ən }

recoilless gun [ORD] A smooth-bore, open-breech, launcher-type artillery weapon constructed of lightweight metals and employing a muzzle-inserted propellant; it is designed with a firing mechanism activated electrically or mechanically by remote control. { ri'koil·ləs 'gən }

recoilless rifle [ORD] A weapon consisting of a light artillery tube of the recoilless type and a very light mount. { ri'koil·ləs 'rif·əl }

recoil mechanism [ORD] A hydraulic-, pneumatic-, or spring-type shock absorber that decreases the energy of the recoil gradually and so avoids violent movement of the gun. { 'rē,koil ,mek·ə,niz·əm }

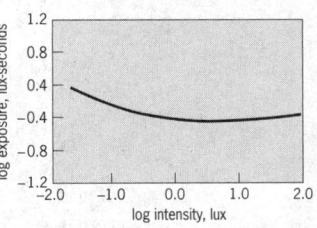

RECIPROCITY FAILURE

Result of reciprocity failure for a hypothetical monochrome emulsion. The graph shows the amount of exposure required to produce a constant density at different levels of illumination for a single photographic emulsion type. Loss of sensitivity at very high or very low levels of illumination is reflected in the greater exposures required at these extreme levels.

recoil milking [NUCLEO] A technique for detecting transmutation recoil atoms knocked out of a target by heavy-ion bombardment, in which the atoms come to rest in a stream of helium or other gas which carries them through an orifice to a rough vacuum where they are adsorbed on a surface and their radioactivity detected. { 'rē̩kȯil ˌmilk·iŋ }

recoil oil [MATER] A neutral, constant-viscosity oil used in hydropneumatic and hydrospring recoil systems. { 'rē̩kȯil ˌȯil }

recoil particle [PHYS] A particle that has been set into motion by a collision or by a process involving the ejection of another particle. { 'rē̩kȯil ˌpärd·ə·kəl }

recoil pit [ORD] Pit dug near the breech of a gun to provide space for the breech when it moves backward during recoil. { 'rē̩kȯil ˌpit }

recoil velocity [ORD] Velocity in recoil of the recoiling parts of a gun. { 'rē̩kȯil və̩läs·əd·ē }

recombinant [GEN] Any new cell, individual, or molecule that is produced in the laboratory by recombinant deoxyribonucleic acid technology or that arises naturally as a result of recombination. { rē'käm·bə·nənt }

recombinant DNA [CELL MOL] Deoxyribonucleic acid (DNA) that has been altered (caused to recombine) by rearrangement of its sequence, addition or deletion of DNA segments, or introduction of foreign DNA segments (for example, a gene from another organism introduced into an organism's genome). { ˌrē'käm·bə·nənt ˌdē'en·ā }

recombinant technology [GEN] **1.** In genetic engineering, laboratory techniques used to join deoxyribonucleic acid (DNA) from different sources to produce novel DNA. Also known as gene splicing. **2.** In genetic engineering, laboratory techniques used to join ribonucleic acid (RNA) from different sources to produce novel RNA. { ri'käm·bə·nənt tek'näl·ə·jē }

recombination [GEN] **1.** The occurrence of gene combinations in the progeny that differ from those of the parents as a result of independent assortment and crossing-over. **2.** The production of genetic information in which some elements of one line of descent are replaced or added to by those of another line. [PHYS] The combination and resultant neutralization of particles or objects having unlike charges, such as a hole and an electron or a positive ion and a negative ion. { ˌrē̩käm·bə'nā·shən }

recombination coefficient [ELECTR] The rate of recombination of positive ions with electrons or negative ions in a gas, per unit volume, divided by the product of the number of positive ions per unit volume and the number of electrons or negative ions per unit volume. { ˌrē̩käm·bə'nā·shən ˌkō·i̩fish·ənt }

recombination electroluminescence *See* injection electroluminescence. { ˌrē̩käm·bə'nā·shən i̩lek·trō̩lü·mə'nes·əns }

recombination energy [PHYS] The energy released when two oppositely charged portions of an atom or molecule rejoin to form a neutral atom or molecule. { ˌrē̩käm·bə'nā·shən ˌen·ər·jē }

recombination frequency [GEN] The number of recombinants divided by the total number of progeny. { rē'käm·bə'nā·shən ˌfrē·kwən·sē }

recombination mosaic [GEN] A mosaic produced as the result of somatic crossing-over. { ˌrē̩käm·bə'nā·shən mō̩zā·ik }

recombination radiation [SOLID STATE] The radiation emitted in semiconductors when electrons in the conduction band recombine with holes in the valence band. { ˌrē̩käm·bə'nā·shən ˌrād·ē̩ā·shən }

recombination repair [MOL BIO] A repair mechanism involving exchange of correct for incorrect segments between two damaged deoxyribonucleic acid molecules. { rē̩käm·bə'nā·shən ri̩per }

recombination velocity [ELECTR] On a semiconductor surface, the ratio of the normal component of the electron (or hole) current density at the surface to the excess electron (or hole) charge density at the surface. { ˌrē̩käm·bə'nā·shən və̩läs·əd·ē }

recombinator [MOL BIO] Any nucleotide sequence that stimulates genetic recombination at neighboring sites. { rē'käm·bə̩nād·ər }

recompletion [PETRO ENG] Redrilling an oil well to a new producing zone (new depth) when the current zone is depleted. { ˌrē·kəm'plē·shən }

recomposed granite [PETR] An arkose composed of consolidated feldspathic residue that has been reworked and decomposed so slightly that upon cementation the rock resembles granite except that its grain is less even and it contains a greater percentage of quartz. Also known as reconstructed granite. { ˌrē·kəm'pōzd 'gran·ət }

recomposed rock [PETR] A rock produced in place by the cementation of the fragmental products of surface weathering; for example, a recomposed granite. { ˌrē·kəm'pōzd 'räk }

recomputed point of turn [NAV] An altered dead-reckoning position of an aircraft at a turning point, determined after wind has been established by drift observations made before and after the turn. { 'rē·kəm̩pyüd·əd ̩pȯint əv 'tərn }

reconditioned carrier reception [ELECTR] Method of reception in which the carrier is separated from the sidebands to eliminate amplitude variations and noise, and is then added at an increased level to the sideband, to obtain a relatively undistorted output. { ˌrē·kən'dish·ənd 'kar·ē·ər ri̩sep·shən }

reconditioning [ENG] Restoration of an object to a good condition. { ˌrē·kən'dish·ən·iŋ }

reconnaissance [ENG] A mission to secure data concerning the meteorological, hydrographic, or geographic characteristics of a particular area. [ORD] A mission undertaken to obtain, by visual observation or other detection methods, information about the activities and resources of an enemy or potential enemy. { ri'kän·ə·səns }

reconnaissance drone [AERO ENG] An uncrewed aircraft guided by remote control, with photographic or electronic equipment for providing information about an enemy or potential enemy. { ri'kän·ə·səns ̩drōn }

reconnaissance map [MAP] A map based on the information obtained in a reconnaissance survey. { ri'kän·ə·səns ̩map }

reconnaissance spacecraft [AERO ENG] A satellite put into orbit about the earth and containing electronic equipment designed to pick up and transmit back to earth information pertaining to activities such as military. { ri'kän·ə·səns 'spās̩kraft }

reconnaissance survey [ENG] A preliminary survey, usually executed rapidly and at relatively low cost, prior to mapping in detail and with greater precision. { ri'kän·ə·səns ̩sər̩vā }

reconnection [ASTRON] The rejoining of solar magnetic field lines that have been severed at a neutral region. { ˌrē·kə'nek·shən }

reconstituted mica [MATER] Mica sheets or shaped objects made by breaking up scrap natural mica, combining with a binder, and pressing into forms suitable for use as electrical insulating material. { rē'kän·stə̩tüd·əd 'mī·kə }

reconstitution [COMPUT SCI] The conversion of tokens back to the keywords they represent in a programming language, before generation of the output of an interpreted program. [GEOL] The formation of new chemicals, minerals, or structures under the influence of metamorphism. { rē̩kän·stə'tü·shən }

reconstructed coal [MATER] Coal formed from crushed or powdered, briquetted lignite or coal, waterproofed with a coating of pitch. { ˌrē·kən'strək·təd 'kōl }

reconstructed granite *See* recomposed granite. { ˌrē·kən'strək·təd 'gran·ət }

reconstructed stone [LAP] A gem material made by the fusing or sintering of small particles of the genuine stone. { ˌrē·kən'strək·təd 'stōn }

reconstruction [SOLID STATE] A process in which atoms at the surface of a solid displace and form bands different from those existing in the bulk solid. { ˌrē·kən'strək·shən }

reconstructive processing [INORG CHEM] The spinning of an inorganic compound of an organic support or binder subsequently removed by oxidation or volatilization to form an inorganic polymer. { ˌrē·kən'strək·tiv 'prä̩ses·iŋ }

reconstructive transformation [CRYSTAL] A type of crystal transformation that involves the breaking of either first- or second-order coordination bonds. { ˌrē·kən'strək·tiv ̩tranz·fər'mā·shən }

recontrol time *See* deionization time. { ˌrē·kən'trōl ̩tīm }

record [COMPUT SCI] A group of adjacent data items in a computer system, manipulated as a unit. Also known as entity.

[SCI TECH] **1.** To preserve for later reproduction or reference. **2.** *See* recording.   { 'rek·ərd }

record block *See* physical record.   { 'rek·ərd ‚bläk }

record changer [ENG ACOUS] A record player that plays a number of records automatically in succession.   { 'rek·ərd ‚chānj·ər }

record density *See* bit density; character density.   { 'rek·ərd ‚den·səd·ē }

recorder *See* recording instrument.   { ri'kórd·ər }

record gap [COMPUT SCI] An area in a storage medium, such as magnetic tape or disk, which is devoid of information; it delimits records, and, on tape, allows the tape to stop and start between records without loss of data. Also known as interrecord gap (IRG).   { 'rek·ərd ‚gap }

record head *See* recording head.   { ri'kórd ‚hed }

recording [SCI TECH] **1.** Any process for preserving signals, sounds, data, or other information for future reference or reproduction, such as disk recording, facsimile recording, ink-vapor recording, magnetic tape or wire recording, and photographic recording. **2.** The end product of a recording process, such as the recorded magnetic tape, disk, or record sheet. Also known as record.   { ri'kórd·iŋ }

recording balance [ANALY CHEM] An analytical balance equipped to record weight results by electromagnetic or servomotor-driven accessories.   { ri'kórd·iŋ ‚bal·əns }

recording-completing trunk [ELEC] Trunk for extending a connection from a local line to a toll operator, used for recording the call and for completing the toll connection.   { ri'kórd·iŋ kəm'plēd·iŋ ‚trəŋk }

recording density [COMPUT SCI] The amount of data that can be stored in a unit length of magnetic tape, usually expressed in bits per inch or characters per inch.   { ri'kórd·iŋ ‚den·səd·ē }

recording head [ELECTR] A magnetic head used only for recording. Also known as record head. [ENG ACOUS] *See* cutter.   { ri'kórd·iŋ ‚hed }

recording instrument [ENG] An instrument that makes a graphic or acoustic record of one or more variable quantities. Also known as recorder.   { ri'kórd·iŋ ‚in·strə·mənt }

recording lamp [ELECTR] A lamp whose intensity can be varied at an audio-frequency rate, for exposing variable-density sound tracks on motion picture film and for exposing paper or film in photographic facsimile recording.   { ri'kórd·iŋ ‚lamp }

recording level [ELECTR] Amplifier output level required to secure a satisfactory recording.   { ri'kórd·iŋ ‚lev·əl }

recording noise [ELECTR] Noise that is introduced during a recording process.   { ri'kórd·iŋ ‚nóiz }

recording optical tracking instrument [ENG] Optical system used for recording data in connection with missile flights.   { ri'kórd·iŋ ‚äp·tə·kəl 'trak·iŋ ‚in·strə·mənt }

recording rain gage [ENG] A rain gage which automatically records the amount of precipitation collected, as a function of time. Also known as pluviograph.   { ri'kórd·iŋ 'rān ‚gāj }

recording spot *See* picture element.   { ri'kórd·iŋ ‚spät }

recording storage tube [ELECTR] Type of cathode-ray tube in which the electric equivalent of an image can be stored as an electrostatic charge pattern on a storage surface; there is no visual display, but the stored information can be read out at any later time as an electric output signal.   { ri'kórd·iŋ 'stór·ij ‚tüb }

recording thermometer *See* thermograph.   { ri'kórd·iŋ thər‚mäm·əd·ər }

recording trunk [ELEC] Trunk extending from a local central office or private branch exchange to a toll office, which is used only for communications with toll operators and not for completing toll connections.   { ri'kórd·iŋ ‚trəŋk }

record layout [COMPUT SCI] A form showing how fields are positioned within a record, usually with information about each field.   { 'rek·ərd ‚lā‚aut }

record length [COMPUT SCI] The number of characters required for all the information in a record.   { 'rek·ərd ‚leŋkth }

record locking [COMPUT SCI] Action of a computer system that makes a record that is being processed by one user unavailable to other users, to prevent more than one user from attempting to update the same information simultaneously.   { 'rek·ərd ‚läk·iŋ }

record mark [COMPUT SCI] A symbol that signals a record's beginning or end.   { 'rek·ərd ‚märk }

record observation [METEOROL] A type of aviation weather observation; the most complete of all such observations and usually taken at regularly specified and equal intervals (hourly, usually on the hour). Also known as hourly observation.   { 'rek·ərd ‚äb·zər‚vā·shən }

record player [ENG ACOUS] A motor-driven turntable used with a phonograph pickup to obtain audio-frequency signals from a phonograph record.   { 'rek·ərd ‚plā·ər }

record storage mark [COMPUT SCI] A special character which appears only in the record storage unit of the card reader to limit the length of the record read into storage.   { 'rek·ərd 'stór·ij ‚märk }

record variable [COMPUT SCI] A group of related but dissimilar data items that can be worked on as a single unit. Also known as structured variable.   { 'rek·ərd ‚ver·ē·ə·bəl }

recoupling [QUANT MECH] A transformation between eigenfunctions of total angular momentum resulting from coupling eigenfunctions of three or more angular momenta in some order, and eigenfunctions of total angular momentum resulting from coupling of the same eigenfunctions in a different order.   { rē'kəp·liŋ }

recoverable shear [FL MECH] Measure of the elastic content of a fluid, related to elastic recovery (mechanicallike property of elastic recoil); found in unvulcanized, unfilled natural rubber, and certain polymer solutions, soap gels, and biological fluids.   { ri'kəv·rə·bəl 'shir }

recovered oil *See* reclaimed oil.   { ri'kəv·ərd 'óil }

recovery [AERO ENG] **1.** The procedure or action that obtains when the whole of a satellite, or a section, instrumentation package, or other part of a rocket vehicle, is retrieved after a launch. **2.** The conversion of kinetic energy to potential energy, such as in the deceleration of air in the duct of a ramjet engine. Also known as ram recovery. **3.** In flying, the action of a lifting vehicle returning to an equilibrium attitude after a nonequilibrium maneuver. [HYD] The rise in static water level in a well, occurring upon the cessation of discharge from that well or a nearby well. [MECH] The return of a body to its original dimensions after it has been stressed, possibly over a considerable period of time. [MET] **1.** The percentage of valuable material obtained from a processed ore. **2.** Reduction or elimination of work-hardening effects, usually by heat treatment. [MIN ENG] The proportion or percentage of coal or ore mined from the original seam or deposit. [PETRO ENG] The removal (recovery) of oil or gas from reservoir formations.   { ri'kəv·ə·rē }

recovery area [AERO ENG] An area in which a satellite, satellite package, or spacecraft is recovered after reentry.   { ri'kəv·ə·rē ‚er·ē·ə }

recovery capsule [AERO ENG] A space capsule designed to be recovered after reentry.   { ri'kəv·ə·rē ‚kap·səl }

recovery factor [PETRO ENG] The ratio of recoverable oil reserves to the oil in place in a reservoir.   { ri'kəv·ə·rē 'fak·tər }

recovery interrupt [COMPUT SCI] A type of interruption of program execution which provides the computer with access to subroutines to handle an error and, if successful, to continue with the program execution.   { ri'kəv·ə·rē 'int·ə‚rəpt }

recovery package [AERO ENG] A package attached to a reentry or other body designed for recovery, containing devices intended to locate the body after impact.   { ri'kəv·ə·rē ‚pak·ij }

recovery party [ORD] A form of contact party whose purpose is the recovery of disabled ordnance materiel from predesignated collecting points, and the transportation of this materiel to the ordnance shops for repairs.   { ri'kəv·ə·rē ‚pärd·ē }

recovery room [MED] A hospital room in which surgical patients are kept during the period immediately following an operation for care and recovery from anesthesia.   { ri'kəv·ə·rē ‚rüm }

recovery routine [COMPUT SCI] A computer routine that attempts to resolve automatically conditions created by errors, without causing the computer system to shut down or otherwise do serious damage.   { ri'kəv·ə·rē rü‚tēn }

recovery ship [NAV ARCH] A naval vessel designed to participate in the retrieval of a satellite, instrument package, or spaceship after it has reentered the atmosphere and landed in the ocean.   { ri'kəv·ə·rē ‚ship }

recovery system [COMPUT SCI] A system for recognizing a malfunction in a data-base management system, reporting it, reconstructing the damaged part of the data base, and resuming processing. { ri'kəv·ə·rē ,sis·təm }

recovery temperature *See* adiabatic recovery temperature. { ri'kəv·ə·rē ,tem·prə·chər }

recovery time [ELECTR]   **1.** The time required for the control electrode of a gas tube to regain control after anode-current interruption.   **2.** The time required for a fired TR (transmit-receive) or pre-TR tube to deionize to such a level that the attenuation of a low-level radio-frequency signal transmitted through the tube is decreased to a specified value.   **3.** The time required for a fired ATR (anti-transmit-receive) tube to deionize to such a level that the normalized conductance and susceptance of the tube in its mount are within specified ranges.   **4.** The interval required, after a sudden decrease in input signal amplitude to a system or component, to attain a specified percentage (usually 63%) of the ultimate change in amplification or attenuation due to this decrease.   **5.** The time required for a radar receiver to recover to half sensitivity after the end of the transmitted pulse, so it can receive a return echo. [NUCLEO]   The minimum time from the start of a counted pulse to the instant a succeeding pulse can attain a specific percentage of the maximum value of the counted pulse in a Geiger counter. { ri'kəv·ə·rē ,tīm }

recovery vehicle [MECH ENG]   A special-purpose vehicle equipped with winch, hoist, or boom for recovery of vehicles. { ri'kəv·ə·rē ,vē·ə·kəl }

recruitment [PHYSIO]   A serial discharge from neurons innervating groups of muscle fibers. { ri'krüt·mənt }

recrystallization [CHEM]   Repeated crystallization of a material from fresh solvent to obtain an increasingly pure product. [CRYSTAL]   A change in the structure of a crystal without a chemical alteration. [MET]   A process which takes place in metals and alloys following distortion and fragmentation of constituent crystals by severe mechanical deformation, in which some fragments grow at the expense of others, so that larger, strain-free grains are formed; it progresses slowly at room temperature, but is greatly speeded by annealing. [PETR]   The formation of new mineral grains in crystalline form in a rock under the influence of metamorphic processes. { rē,krist·əl·ə'zā·shən }

recrystallization annealing [MET]   Producing a new grain structure without phase change by annealing cold-worked metal. { rē,krist·əl·ə'zā·shən ə,nēl·iŋ }

recrystallization breccia *See* pseudobreccia. { rē,krist·əl·ə'zā·shən ,brech·ə }

recrystallization flow [GEOL]   Flow in which there is molecular rearrangement by solution and redeposition, solid diffusion, or local melting. { rē,krist·əl·ə'zā·shən ,flō }

recrystallization temperature [MET]   The minimum temperature at which complete recrystallization occurs in an annealed cold-worked metal within a specified time. { rē,krist·əl·ə'zā·shən ,tem·prə·chər }

rectangle [MATH]   A plane quadrilateral having four interior right angles and opposite sides of equal length. { 'rek,taŋ·gəl }

rectangular Cartesian coordinate system *See* Cartesian coordinate system. { rek'taŋ·gyə·lər kär¦tē·zhən kō'órd·ən·ət ,sis·təm }

rectangular cavity [ELECTROMAG]   A resonant cavity having the shape of a rectangular parallelepiped. { rek'taŋ·gyə·lər 'kav·əd·ē }

rectangular chart [MAP]   **1.** A chart in a rectangular shape. **2.** A chart on the rectangular map projection. { rek'taŋ·gyə·lər 'chärt }

rectangular coordinates *See* Cartesian coordinates. { rek'taŋ·gyə·lər kō'órd·ən·əts }

rectangular cross ripple mark [GEOL]   An oscillation cross ripple mark consisting of two sets of ripples which intersect at right angles, enclosing a rectangular pit. { rek'taŋ·gyə·lər ¦krós 'rip·əl ,märk }

rectangular distribution *See* uniform distribution. { rek'taŋ·gyə·lər ,dis·trə'byü·shən }

rectangular drainage pattern [GEOL]   A drainage pattern characterized by many right-angle bends in both the main streams and their tributaries. Also known as lattice drainage pattern. { rek'taŋ·gyə·lər 'drān·ij ,pad·ərn }

rectangular game *See* matrix game. { rek'taŋ·gyə·lər 'gām }

rectangular graph *See* bar graph. { rek¦taŋ·gyə·lər 'graf }

rectangular hyperbola [MATH]   A hyperbola whose major and minor axes are equal. { rek'taŋ·gyə·lər hī'pər·bə·lə }

rectangular mesh [TEXT]   Cloth with a different mesh count in the fill than in the warp. Also known as oblong mesh. { rek'taŋ·gyə·lər 'mesh }

rectangular parallelepiped [MATH]   A parallelepiped with bases as rectangles all perpendicular to its lateral faces. Also known as cuboid; rectangular solid. { rek'taŋ·gyə·lər ,par·ə,lel·ə'pī,ped }

rectangular projection [MAP]   A cylindrical map projection with uniform spacing of the parallels; used for the star chart in the Air Almanac. { rek'taŋ·gyə·lər prə'jek·shən }

rectangular pulse [ELECTR]   A pulse in which the wave amplitude suddenly changes from zero to another value at which it remains constant for a short period of time, and then suddenly changes back to zero. { rek'taŋ·gyə·lər 'pəls }

rectangular scanning [ELECTR]   Two-dimensional sector scanning in which a slow sector scanning in one direction is superimposed on a rapid sector scanning in a perpendicular direction. { rek'taŋ·gyə·lər 'skan·iŋ }

rectangular search [NAV]   A search of three legs from a moving point, the first and third legs being perpendicular to the base course of the moving point, and the second leg being parallel to it. { rek'taŋ·gyə·lər 'sərch }

rectangular solid *See* rectangular parallelepiped. { rek'taŋ·gyə·lər 'säl·əd }

rectangular wave [ELECTR]   A periodic wave that alternately and suddenly changes from one to the other of two fixed values. Also known as rectangular wave train. { rek'taŋ·gyə·lər 'wāv }

rectangular waveguide [ELECTROMAG]   A waveguide having a rectangular cross section. { rek'taŋ·gyə·lər 'wāv,gīd }

rectangular wave train *See* rectangular wave. { rek'taŋ·gyə·lər 'wāv ,trān }

rectangular weir [CIV ENG]   A weir with a rectangular notch at top for measurement of water flow in open channels; it is simple, easy to make, accurate, and popular. { rek'taŋ·gyə·lər 'wer }

Rectenna [ELECTR]   A device that converts microwave energy in direct-current power; consists of a number of small dipoles, each having its own diode rectifier network, which are connected to direct-current buses. { rek'ten·ə }

Recticornia [INV ZOO]   A family of amphipod crustaceans in the superfamily Genuina containing forms in which the first antennae are straight, arise from the anterior margin of the head, and have few flagellar segments. { ,rek·tə'kór·nē·ə }

rectifiable curve [MATH]   A curve whose length can be computed and is finite. { 'rek·tə,fī·ə·bəl 'kərv }

rectification [CIV ENG]   A new alignment to correct a deviation of a stream channel or bank. [ELEC]   The process of converting an alternating current to a unidirectional current. [GEOL]   The simplification and straightening of the outline of an initially irregular and crenulate shoreline through the cutting back of headlands and offshore islands by marine erosion, and through deposition of waste from erosion or of sediment brought down by neighboring rivers. [GRAPHICS]   The transformation of a photograph onto a horizontal plane so as to remove or correct displacements (distortions in perspective) by tilt. { ,rek·tə·fə'kā·shən }

rectification distillation [CHEM ENG]   A distillation technique in which a rectifying column is used. { ,rek·tə·fə'kā·shən ,dis·tə'lā·shən }

rectification factor [ELECTR]   Quotient of the change in average current of an electrode by the change in amplitude of the alternating sinusoidal voltage applied to the same electrode, the direct voltages of this and other electrodes being maintained constant. { ,rek·tə·fə'kā·shən ,fak·tər }

rectified value [ELEC]   For an alternating quantity, the average of all the positive (or negative) values of the quantity during an integral number of periods. { 'rek·tə,fīd 'val·yü }

rectifier [ELEC]   A nonlinear circuit component that allows more current to flow in one direction than the other; ideally, it allows current to flow in one direction unimpeded but allows no current to flow in the other direction. { 'rek·tə,fī·ər }

rectifier filter [ELECTR]   An electric filter used in smoothing out the voltage fluctuation of an electron tube rectifier, and generally placed between the rectifier's output and the load resistance. { 'rek·tə,fī·ər ,fil·tər }

rectifier instrument [ENG]   Combination of an instrument sensitive to direct current and a rectifying device whereby alternating current (or voltages) may be rectified for measurement. { 'rek·tə,fī·ər ,in·strə·mənt }

rectifier rating [ELECTR] A performance rating for a semiconductor rectifier, usually on the basis of the root-mean-square value of sinusoidal voltage that it can withstand in the reverse direction and the average current density that it will pass in the forward direction. { 'rek·tə,fī·ər ,rād·iŋ }

rectifier stack [ELECTR] A dry-disk rectifier made up of layers or stacks of disks of individual rectifiers, as in a selenium rectifier or copper-oxide rectifier. { 'rek·tə,fī·ər ,stak }

rectifier transformer [ELECTR] Transformer whose secondary supplies energy to the main anodes of a rectifier. { 'rek·tə,fī·ər tranz'fȯr·mər }

rectifying column [CHEM ENG] Portion of a distillation column above the feed tray in which rising vapor is enriched by interaction with a countercurrent falling stream of condensed vapor; contrasted to the stripping column section below the column feed tray. { 'rek·tə,fī·iŋ ,käl·əm }

rectifying developable [MATH] The envelope of the rectifying planes of a space curve. { 'rek·tə,fī·iŋ də'vel·əp·ə·bəl }

rectifying plane [MATH] The plane that contains the tangent and binormal to a curve at a given point on the curve. { 'rek·tə,fī·iŋ ,plān }

rectilinear [MATH] Consisting of or bounded by lines. { ¦rek·tə'lin·ē·ər }

rectilinear generators [MATH] Straight lines which generate ruled surfaces. { ¦rek·tə'lin·ē·ər 'jen·ə,rād·ərz }

rectilinear motion [MECH] A continuous change of position of a body so that every particle of the body follows a straight-line path. Also known as linear motion. { ¦rek·tə'lin·ē·ər 'mō·shən }

rectilinear scanning [ELECTR] Process of scanning an area in a predetermined sequence of narrow parallel strips. { ¦rek·tə'lin·ē·ər 'skan·iŋ }

rectilinear shoreline [GEOL] A long, relatively straight shoreline. { ¦rek·tə'lin·ē·ər 'shȯr,līn }

rectilinear system See orthoscopic system. { ¦rek·tə'lin·ē·ər 'sis·təm }

rectiliner lens [OPTICS] A lens that is free from distortion, imaging straight lines onto straight lines regardless of their orientation. { ¦rek·tə'lin·ē·ər 'lenz }

recto [GRAPHICS] A right-hand book page, which always bears an odd number. { 'rek·tō }

rectocele [MED] Bulging or herniation of the rectum into the vagina. Also known as vaginal protocele. { 'rek·tə,sēl }

rectorite [GEOL] A white clay-mineral mixture with a regular interstratification of two mica layers (pyrophyllite and vermiculite) and one or more water layers. Also known as allevardite. { 'rek·tə,rīt }

rectrix [VERT ZOO] One of the stiff tail feathers used by birds to control direction of flight. { 'rek·triks }

rectum [ANAT] The portion of the large intestine between the sigmoid flexure and the anus. { 'rek·təm }

rectus [ANAT] Having a straight course, as certain muscles. { 'rek·təs }

rectus abdominis [ANAT] The long flat muscle of the anterior abdominal wall which, as vertical fibers, arises from the pubic crest and symphysis, and is inserted into the cartilages of the fifth, sixth, and seventh ribs. { 'rek·təs ab'däm·ə·nəs }

rectus femoris [ANAT] A division of the quadriceps femoris inserting in the patella and ultimately into the tubercle of the tibia. { 'rek·təs 'fem·ə·rəs }

rectus oculi [ANAT] Any of four muscles (superior, inferior, lateral, and medial) of the eyeball, running forward from the optic foramen and inserted into the sclerotic coat. { 'rek·təs 'äk·yə·lī }

recumbent [BOT] Of or pertaining to a plant or plant part that tends to rest on the surface of the soil. { ri'kəm·bənt }

recumbent fold [GEOL] An overturned fold with a nearly horizontal axial surface. Also known as reclined fold. { ri'kəm·bənt 'fōld }

recuperability [COMMUN] Ability to continue to operate after a partial or complete loss of the primary communications facility resulting from sabotage, enemy attack, or other disaster. { rē,küp·rə'bil·əd·ē }

recuperative air heater [ENG] An air heater in which the

heat-transferring metal parts are stationary and form a separating boundary between the heating and cooling fluids. { rē'küp·rəd·iv 'er ,hēd·ər }

recuperator [ENG] An apparatus in which heat is conducted from the combustion products to incoming cooler air through a system of thin-walled ducts. { rē'kü·pə,rād·ər }

recurrence formula methods [MATH] Methods of calculating numerical solutions of differential equations in which the equation is written in the form of a recurrence relation between values of the solution function at successive points by replacing the derivatives with corresponding finite difference expressions. { ri'kər·əns 'fȯr·myə·lə ,meth·ədz }

recurrence interval [HYD] The average time interval between occurrences of a hydrologic event, such as a flood, of a given or greater magnitude. { ri'kər·əns 'int·ər·vəl }

recurrence rate See repetition rate. { ri'kər·əns ,rāt }

recurrence relation [MATH] An equation relating a term in a sequence to one or more of its predecessors in the sequence. { ri'kər·əns ri,lā·shən }

recurrent backcrossing [GEN] Repetitive sexual crossing of hybrids to one parent, used to eliminate all but the desired alleles and traits of the donor parent. { ri'kər·ənt 'bak,krȯs·iŋ }

recurrent folding [GEOL] A type of folding due to periodic deformation or subsidence and characterized by thinning or possible disappearance of formations at the crest. Also known as revived folding. { ri'kər·ənt 'fōld·iŋ }

recurrent nova [ASTRON] A binary star that undergoes outbursts every few decades in which the brightness increases roughly 100–1000 times, as the result of nuclear explosions in matter that has accreted on a white dwarf component star from a neighboring red giant component. { ri'kər·ənt 'nō·və }

recurrent parent [GEN] In recurrent backcrossing, the parent that is crossed with the first and the subsequent generations. Also known as backcross parent. { ri'kər·ənt 'per·ənt }

recurrent transformation [MATH] 1. A measurable function from a measure space T to itself such that for every measurable set A in the space and every point x in A there is a positive integer n such that $T^n(x)$ is also in A. 2. A continuous function from a topological space T to itself such that for every open set A in the space and every point x in A there is a positive integer n such that $T^n(x)$ is also in A. { ri'kər·ənt ,tranz·fər'mā·shən }

recurring continued fraction [MATH] A continued fraction in which a finite sequence of terms is repeated indefinitely. Also known as periodic continued fraction. { ri¦kər·iŋ kən¦tin·yüd 'frak·shən }

recurring decimal See repeating decimal. { ri'kər·iŋ 'des·məl }

recurring demand [IND ENG] A request made periodically or anticipated to be repetitive by an authorized requisitioner for material for consumption or use, or for stock replenishment. { ri'kər·iŋ di'mand }

recurring issue [ORD] An issue made on a cyclic basis to replenish materiel consumed or worn out through fair wear and tear in operations, with each issue being made to a consignee eligible to request further replenishment when required, in the foreseeable future. { ri'kər·iŋ 'ish·ü }

recursion [COMPUT SCI] A technique in which an apparently circular process is used to perform an iterative process. { ri'kər·zhən }

recursion formula [MATH] An algorithm allowing computation of a succession of quantities. Also known as recursion relation. { ri'kər·zhən ,fȯr·myə·lə }

recursion relation See recursion formula. { ri'kər·zhən ri,lā·shən }

recursive [MATH] Pertaining to a process that is inherently repetitive, with the results of each repetition usually depending upon those of the previous repetition. { ri'kər·siv }

recursive filter [ELECTR] A digital filter that has feedback; that is, its output depends not only on present and past input values but on past output values as well. { ri,kər·siv 'fil·tər }

recursive functions [MATH] Functions that can be obtained by a finite number of operations, computations, or algorithms. { ri'kər·siv 'fəŋk·shənz }

recursive macro call [COMPUT SCI] A call to a macroinstruction already called when used in conjunction with conditional assembly. { ri'kər·siv ¦mak·rō ,kȯl }

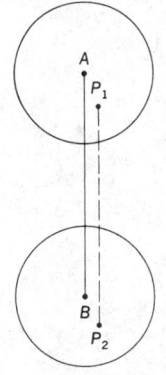

RECTILINEAR MOTION

Diagram of rectilinear motion. Motion of center of mass of a body from A to B is along the straight line connecting these points. Path of any particle in the body is a straight line parallel to or coinciding with AB, such as the straight line connecting P_1 and P_2.

recursive procedure [COMPUT SCI] A method of calculating a function by deriving values of it which become more accurate at each step; recursive procedures are explicitly outlawed in most systems with the exception of a few which use languages such as ALGOL and LISP. { ri'kər·siv prə'sē·jər }

recursive subroutine [COMPUT SCI] A reentrant subroutine whose partial results are stacked, with a processor stack pointer advancing and retracting as the subroutine is called and completed. { ri'kər·siv 'səb·rü,tēn }

recurvature [METEOROL] With respect to the motion of severe tropical cyclones (hurricanes and typhoons), the change in direction from westward and poleward to eastward and poleward; such recurvature of the path frequently occurs as the storm moves into middle latitudes. { rē'kər·və·chər }

recurved [SCI TECH] Curving inward to backward. { rē'kərvd }

recurved spit See hook. { rē'kərvd 'spit }

recycle [AERO ENG] **1.** To stop the count in a countdown and to return to an earlier point. **2.** To give a rocket or other object a completely new checkout. { rē'sī·kəl }

recycle base [ORD] A base used by returning mission aircraft for servicing and maintenance, before the next mission launch. { rē'sī·kəl ,bās }

recycle mixing [CHEM ENG] The mixing of a portion of a product stream (fluid or solid) from a processing unit with incoming raw feed. { rē'sī·kəl ,miks·iŋ }

recycle ratio [CHEM ENG] In a continuous chemical process, the ratio of recycle stock to fresh feed. { rē'sī·kəl ,rā·shō }

recycle stock [CHEM ENG] That portion of a feedstock that has passed through a processing unit and is recirculated (recycled) back through the process. { rē'sī·kəl ,stäk }

recycling [ELECTR] Returning to an original condition, as to 0 or 1 in a counting circuit. [ENG] The extraction and recovery of valuable materials from scrap or other discarded materials. [NUCLEO] Reuse of fissionable nuclear reactor fuel by chemical processing, reenriching, and refabricating into new fuel elements. { rē'sīk·liŋ }

recycling endosome [CELL MOL] An intracellular vesicle through which internalized receptors pass during their transport from early endosomes back to the plasma membrane for reuse. { rē¦sīk·liŋ 'en·də,sōmz }

red [OPTICS] The hue evoked in an average observer by monochromatic radiation having a wavelength in the approximate range from 622 to 770 nanometers; however, the same sensation can be produced in a variety of other ways. { red }

red acetate See mordant rouge. { 'red 'as·ə,tāt }

red algae [BOT] The common name for members of the phylum Rhodophyta. { 'red 'al·jē }

red antimony See kermesite. { 'red 'an·tə,mō·nē }

red arsenic See realgar. { 'red 'ärs·ən·ik }

red azimuth tables [NAV] Popular title for Hydrographic Office Publication No. 260, "Azimuths of the Sun," used in navigation. { 'red 'az·ə·məth ,tā·bəlz }

redbed [GEOL] Continentally deposited sediment composed principally of sandstone, siltsone, and shale; red in color due to the presence of ferric oxide (hematite). Also known as red rock. { 'red,bed }

red blood cell See erythrocyte. { 'red 'bləd ,sel }

red brass [MET] Brass containing 85% copper, 5% zinc, 5% tin, and 5% lead. { 'red 'bras }

red charcoal [MATER] An impure charcoal made by heating wood to 300°C; much of the oxygen and hydrogen is retained. { 'red 'chär,kōl }

red clay [GEOL] A fine-grained, reddish-brown pelagic deposit consisting of relatively large proportions of windblown particles, meteoric and volcanic dust, pumice, shark teeth, manganese nodules, and debris transported by ice. Also known as brown clay. { 'red 'klā }

red cobalt See erythrite. { 'red 'kō,bȯlt }

red copper ore See cuprite. { 'red 'käp·ər ,ȯr }

Reddish-Brown Lateritic soil [GEOL] One of a zonal, lateritic group of soils developed from a mottled red parent material and characterized by a reddish-brown surface horizon and underlying red clay. { 'red·ish ¦braün ,lad·ə'rid·ik 'sȯil }

Reddish-Brown soil [GEOL] A group of zonal soils having a reddish, light brown surface horizon overlying a heavier, more reddish horizon and a light-colored lime horizon. { 'red·ish ¦braün 'sȯil }

red dwarf star [ASTRON] A red star of low luminosity, so designated by E. Hertzsprung; dwarf stars are commonly those main-sequence stars fainter than an absolute magnitude of +1, and red dwarfs are the faintest and coolest of the dwarfs. { 'red ¦dwȯrf 'stär }

red earth [GEOL] Leached, red, deep, clayey soil that is characteristic of a tropical climate. Also known as red loam. { 'red ¦ərth }

redefine [COMPUT SCI] A procedure used in certain programming languages to specify different utilizations of the same storage area at different times. { ¦rē·di'fīn }

redeposition [GEOL] Formation into a new accumulation, such as the deposition of sedimentary material that is picked up and moved (reworked) from the place of its original deposition, or the solution and reprecipitation of mineral matter. { rē,dep·ə'zish·ən }

red giant star [ASTRON] A star whose evolution has progressed to the point where hydrogen core burning has been completed, the helium core has become denser and hotter than originally, and the envelope has expanded to perhaps 100 times its initial size. { 'red 'jī·ənt 'stär }

red giant tip [ASTRON] The upper tip of the red giant branch in the Hertzsprung-Russell diagram that represents stars undergoing a flash process. { 'red ¦jī·ənt 'tip }

red glass [MATER] Soda-zinc glass to which small amounts of cadmium and selenium are added. { 'red 'glas }

red gum See eucalyptus gum. { 'red ¦gəm }

red-hardness [MET] In reference to high-speed steel and other cutting tool materials, the property of being hard enough to cut metals even when heated to a dull-red color. { 'red ¦härd·nəs }

red hematite See hematite. { 'red 'hē·mə,tīt }

redia [INV ZOO] A larva produced within the miracidial sporocyst of certain digenetic trematodes which may give rise to daughter rediae or to cercariae. { 'rē·dē·ə }

redifferentiation [PHYSIO] The return to a position of greater specialization in actual and potential functions, or the developing of new characteristics. { rē,dif·ə,ren·chē'ā·shən }

redingtonite [MINERAL] (Fe,Mg,Ni)(Cr,Al)$_2$(SO$_4$)$_4$·22H$_2$O A pale-purple mineral composed of a hydrous sulfate of iron, magnesium, nickel, chromium, and aluminum. { 'red·iŋ·tə,nīt }

red iron ore See hematite. { 'red 'ī·ərn ,ȯr }

redistribution [ELECTR] The alteration of charges on an area of a storage surface by secondary electrons from any other area of the surface in a charge storage tube or television camera tube. { rē,dis·trə'byü·shən }

red lake C pigment [MATER] An organic azo pigment; made by coupling the diazonium salt (barium or sodium) of ortho-chloro-meta-toluidine-para-sulfonic acid with β-naphthol; used to color inks, plastics, and rubbers. { 'red ¦lāk 'sē ,pig·mənt }

red lead See lead tetroxide. { 'red 'led }

red lead ore See crocoite. { 'red 'led ,ȯr }

red leaf [PL PATH] Any of various nonparasitic plant diseases marked by red discoloration of the foliage. { 'red ,lēf }

Redler conveyor [MECH ENG] A conveyor in which material is dragged through a duct by skeletonized or U-shaped impellers which move the material in which they are submerged because the resistance to slip through the element is greater than the drag against the walls of the duct. { 'red·lər kən'vā·ər }

red liquor See mordant rouge. { 'red 'lik·ər }

red loam See red earth. { 'red 'lōm }

red magnetism [GEOPHYS] The magnetism of the north-seeking end of a freely suspended magnet; this is the magnetism of the earth's south magnetic pole. { 'red 'mag·nə,tiz·əm }

red mercury sulfide See mercuric sulfide. { 'red 'mər·kyə·rē 'səl,fīd }

red metal [MET] A copper matte having a copper content of about 48%. { 'red 'med·əl }

red mud [GEOL] A reddish terrigenous mud composed of up to 25% calcium carbonate and deriving its color from the presence of ferric oxide; found on the sea floor near deserts and near the mouths of large rivers. [MET] An iron oxide-rich residue obtained in purifying bauxite in the Bayer process. { 'red 'məd }

red nitric acid [MATER] A type of liquid bipropellant with boiling point 104°F (40°C), freezing point −80°F (−62°C), and density 1.58 grams per milliliter; used to supply power for jet propulsion. { 'red 'nī·trik 'as·əd }

red nucleus [HISTOL] A mass of reticular fibers in the gray

REDBED

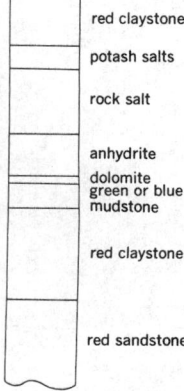

red claystone

potash salts

rock salt

anhydrite

dolomite
green or blue
mudstone

red claystone

red sandstone

Sequence of redbeds with evaporites, characteristic of Permian formations in southwestern United States.

matter of the tegmentum of the mesencephalon of higher vertebrates; it receives fibers from the cerebellum of the opposite side and gives rise to rubrospinal tract fibers of the opposite side. { 'red 'nü·klē·əs }

red ocher *See* ferric oxide; hematite. { 'red 'ō·kər }

red oil [MATER] **1.** Any intermediate-grade petroleum lubricating oil that is red in color by transmitted light; includes so-called red engine oils, bearing oils, and machinery oils. **2.** *See* oleic acid. { 'red 'öil }

red orpiment *See* realgar. { 'red 'ör·pə·mənt }

redox blemish [GRAPHICS] A microscopic formation on a silver-gelatin film caused by air pollution or improper handling. Also known as aging blemish; microspot. { 'rē,däks ,blem·ish }

redox cell [ELEC] Cell designed to convert the energy of reactants to electrical energy; an intermediate reductant, in the form of liquid electrolyte, reacts at the anode in a conventional manner; it is then regenerated by reaction with a primary fuel. { 'rē,däks ,sel }

red oxide of zinc *See* zincite. { 'red 'äk,sīd əv 'ziŋk }

redoxomorphic stage [GEOCHEM] The earliest geochemical stage of diagenesis characterized by mineral changes primarily due to oxidation and reduction reactions. { ri'däk·sə¦mór·fik ,stāj }

redox polymer [ORG CHEM] A polymer whose structure contains functional groups that can be reversibly reduced or oxidized. Also known as electron exchanger. { 'rē,däks ¦päl·ə·mər }

redox potential [PHYS CHEM] Voltage difference at an inert electrode immersed in a reversible oxidation-reduction system; measurement of the state of oxidation of the system. Also known as oxidation-reduction potential. { 'rē,däks pə,ten·chəl }

redox potentiometry [ANALY CHEM] Use of neutral electrode probes to measure the solution potential developed as the result of an oxidation or reduction reaction. { 'rē,däks pə,ten·chē'äm·ə·trē }

redox system [CHEM] A chemical system in which reduction and oxidation (redox) reactions occur. { 'rē,däks ,sistəm }

redox titration [ANALY CHEM] A titration characterized by the transfer of electrons from one substance to another (from the reductant to the oxidant) with the end point determined colorimetrically or potentiometrically. { 'rē,däks tī'trā·shən }

red phosphorus [CHEM] An allotropic form of the element phosphorus; violet-red, amorphous powder subliming at 416°C, igniting at 260°; insoluble in all solvents,; nonpoisonous. { 'red 'fä·sfə·rəs }

red potassium chromate *See* potassium dichromate. { 'red pə'tas·ē·əm 'krō,māt }

red potassium prussiate *See* potassium ferricyanide. { 'red pə'tas·ē·əm 'prəs·ē,āt }

red precipitate *See* mercuric oxide. { 'red prə'sip·ə,tāt }

red prussiate of potash *See* potassium ferricyanide. { 'red 'prəs·ē,āt əv 'päd,ash }

red prussiate of soda *See* sodium ferricyanide. { 'red 'prəs·ē,āt əv 'sōd·ə }

red ring [PL PATH] A nematode disease of the coconut palm caused by *Aphelenchoides cocophilus* and marked by the appearance of red rings in the stem cross section. { 'red ,riŋ }

red rock *See* redbed. { 'red ,räk }

red rot [PL PATH] Any of several fungus diseases of plants characterized by red patches on stems or leaves; common in sugarcane, sisal, and various evergreen and deciduous trees. { 'red ,rät }

red rust [PL PATH] An algal disease of certain subtropical plants, such as tea and citrus, caused by the green alga *Cephaleuros virescens* and characterized by a rusty appearance of the leaves or twigs. { 'red 'rəst }

redruthite *See* chalcocite. { 'red¦rü,thīt }

Red Sea [GEOGR] A body of water that lies between Arabia and northeastern Africa, about 1200 miles (2000 kilometers) long, 180 miles (300 kilometers) wide, and a maximum depth of about 7600 feet (2300 meters). { 'red 'sē }

red sector [NAV] A sector of the circle of visibility of a navigational light in which a red light is exhibited; such sectors are designated by their limiting bearing, as observed at some point other than the light; red sectors are often located so that they warn of danger to vessels. { 'red ,sek·tər }

redshift [ASTROPHYS] A systematic displacement toward longer wavelengths of lines in the spectra of distant galaxies and also of the continuous portion of the spectrum; increases with distance from the observer. Also known as Hubble effect. { 'red,shift }

red snow [HYD] A snow surface of reddish color caused by the presence within it of certain microscopic algae or particles of red dust. { 'red 'snō }

Red Spot [ASTRON] A semipermanent marking of the planet Jupiter; some fluctuations in visibility exist; it does not rotate uniformly with the planet, indicating that it is a disturbance of Jupiter's atmosphere. { 'red ,spät }

red stele [PL PATH] A fatal fungus disease of strawberries caused by *Phytophthora fragariae* invading the roots, producing redness and rotting with consequent dwarfing and wilting of the plant. { 'red ¦stēl }

redstone [PETR] **1.** Any reddish sedimentary rock, such as red-colored sandstone. **2.** A deep-red, clayey sandstone or silt-stone representing a floodplain micaceous arkose. { 'red,stōn }

red stripe [PL PATH] **1.** A fungus decay of timber caused by *Polyporus vaporarius* and characterized by reddish streaks. **2.** A bacterial disease of sugarcane caused by *Xanthomonas rubrilineans* and characterized by red streaks in the young leaves, followed by invasion of the vascular system. { 'red ¦strīp }

red-tape operation *See* bookkeeping operation. { 'red ¦tāp ,äp·ə,rā·shən }

red tetrazolium *See* triphenyltetrazolium chloride. { 'red ,tetrə'zäl·ē·əm }

red thermit [MET] Thermit made with red iron oxide. { 'red 'thər·mət }

red thread [PL PATH] A fungus disease of turf grasses caused by *Cortecium fuciforme* and characterized by the appearance of red stromata on the pinkish hyphal threads. { 'red ¦thred }

red tide [BIOL] A reddish discoloration of coastal surface waters due to concentrations of certain toxin-producing dinoflagellates. Also known as red water. { 'red 'tīd }

redtop grass [BOT] One of the bent grasses. *Agrostis alba* and its relatives, which grow on a wide variety of soils; it is a perennial, spreads slowly by rootstocks, and has top growth 2–3 feet (60–90 centimeters) tall. { 'red,täp ,gras }

reduce [ORD] To clear a stoppage in a weapon. { ri'düs }

reduced characteristic equation [MATH] The polynomial equation of lowest degree that is satisfied by a given matrix. Also known as minimal equation. { ri'düst ,kär·ik·tə'ris·tik i,kwā·zhən }

reduced Compton wavelength [QUANT MECH] The Compton wavelength of a particle divided by 2π. { ri'düst 'käm·tən 'wāv,leŋkth }

reduced crude [MATER] In petroleum refining, a residual product remaining after removal by distillation or other means of an appreciable quantity of the more volatile components of crude oil. { ri'düst 'krüd }

reduced cubic equation [MATH] A cubic equation in a variable x, where the coefficient of x^2 is zero. { ri'düst ¦kyü·bik i'kwā·zhən }

reduced distance [OPTICS] A distance in a medium divided by the medium's index of refraction. { ri'düst 'dis·təns }

reduced echelon matrix [MATH] A matrix in which the first nonzero term in a row is the only nonzero term in its column. { ri'düst ,esh·ə,län 'mā,triks }

reduced equation *See* auxiliary equation. { ri'düst i'kwā·zhən }

reduced equation of state [PHYS] An equation relating the reduced pressure, reduced volume, and reduced temperature of a substance. { ri'düst i'kwā·zhən əv 'stāt }

reduced form [MATH] A lambda expression that has no subexpressions of the form (λxMA), where M and A are lambda expressions, is said to be in reduced form. { ri'düst 'förm }

reduced frequency *See* Strouhal number. { ri'düst 'frē·kwən·sē }

reduced inspection [IND ENG] The decrease in the number of items inspected from that specified in the original sampling plan because the quality of the item has consistently improved. { ri'düst in'spek·shən }

reduced instruction set computer [COMPUT SCI] A computer in which the compiler and hardware are interlocked, and the compiler takes over some of the hardware functions of

Appearance of Red Spot on Jupiter.

conventional computers and translates high-level-language programs directly into low-level machine code. Abbreviated RISC. { ri¦düst in'strək·shən ˌset kəm'pyüd·ər }

reduced latitude [GEOD] The angle at the center of a sphere tangent to a reference ellipsoid along the equator, between the plane of the equator and a radius to the point intersected on the sphere by a straight line perpendicular to the plane of the equator. Also known as geometric latitude; parametric latitude. { ri'düst 'lad·ə,tüd }

reduced mass [MECH] For a system of two particles with masses m_1 and m_2 exerting equal and opposite forces on each other and subject to no external forces, the reduced mass is the mass m such that the motion of either particle, with respect to the other as origin, is the same as the motion with respect to a fixed origin of a single particle with mass m acted on by the same force; it is given by $m = m_1 m_2 / (m_1 + m_2)$. { ri'düst 'mas }

reduced nickel [MET] Nickel obtained by the precipitation of nickel hydroxide or nickel carbonate onto kieselguhr, then reducing the precipitate by heating it with hydrogen. { ri'düst 'nik·əl }

reduced oil [MATER] 1. Oil rerun in a vacuum or steam still from an oil that is already distilled. 2. An oil made from the residue in the still after another product has been distilled from the crude oil. { ri'düst 'öil }

reduced-order controller [CONT SYS] A control algorithm in which certain modes of the structure to be controlled are ignored, to enable control commands to be computed with sufficient rapidity. { ri'düst ¦ör·dər kən'trōl·ər }

reduced pressure [METEOROL] The calculated value of atmospheric pressure at mean sea level or some other specified level, as derived (reduced) from station pressure or actual pressure; thus, sea level pressure is nearly always a reduced pressure. [THERMO] The ratio of the pressure of a substance to its critical pressure. { ri'düst 'presh·ər }

reduced-pressure distillation See vacuum distillation. { ri'düst ¦presh·ər ,dis·tə'lā·shən }

reduced proper motion [ASTRON] The proper motion of a star expressed as a linear velocity. { ri'düst 'präp·ər 'mō·shən }

reduced property See reduced value. { ri'düst 'präp·ərd·ē }

reduced residue system modulo n [MATH] A set of integers that includes those members of a complete residue system modulo n that are relatively prime to n. { ri¦düst 'res·ə,dü ¦sis·təm ¦mäj·ə,lō 'en }

reduced telemetry [COMMUN] Raw telemetry data transformed into a usable form. { ri'düst tə'lem·ə·trē }

reduced temperature [THERMO] The ratio of the temperature of a substance to its critical temperature. { ri'düst 'tem·prə·chər }

reduced value [THERMO] The actual value of a quantity divided by the value of that quantity at the critical point. Also known as reduced property. { ri'düst 'val·yü }

reduced viscosity [ENG] In plastics processing, the ratio of the specific viscosity to concentration. { ri'düst vi'skäs·əd·ē }

reduced volume [THERMO] The ratio of the specific volume of a substance to its critical volume. { ri'düst 'väl·yəm }

reducer [BIOL] See decomposer. [CHEM] See reducing agent. [DES ENG] A fitting having a larger size at one end than at the other and threaded inside, unless specifically flanged or for some special joint. [GRAPHICS] A solution capable of dissolving silver; used to cut down the contrast or density of a negative or positive image. { ri'dü·sər }

reducible configuration [MATH] A graph such that the four-colorability of any planar graph containing the configuration can be deduced from the four-colorability of planar graphs with fewer vertices. { ri'dü·sə·bəl kən,fig·yə'rā·shən }

reducible curve [MATH] A curve that can be shrunk to a point by a continuous deformation without passing outside a given region. { ri'düs·ə·bəl 'kərv }

reducible polynomial [MATH] A polynomial relative to some field which can be written as the product of two polynomials of degree at least 1. { ri'dü·sə·bəl ,päl·i'nō·mē·əl }

reducible representation of a group [MATH] A representation of a group as a family of linear operators on a vector space V such that there is a proper closed subspace of V that is invariant under these operators. { ri¦düs·ə·bəl ,rep·rə·zen'tā·shən əv ə ,grüp }

reducible transformation [MATH] A linear transformation T on a vector space V that can be completely specified by describing its effect on two subspaces, M and N, that are each transformed into themselves by T and are such that any vector of V can be uniquely represented as the sum of a vector of M and a vector of N. { ri¦düs·ə·bəl ,tranz·fər'mā·shən }

reducing agent [CHEM] Also known as reducer. 1. A material that adds hydrogen to an element or compound. 2. A material that adds an electron to an element or compound, that is, decreases the positiveness of its valence. { ri'düs·iŋ ,ā·jənt }

reducing atmosphere [CHEM] An atmosphere of hydrogen (or other substance that readily provides electrons) surrounding a chemical reaction or physical device; the effect is the opposite to that of an oxidizing atmosphere. { ri'düs·iŋ 'at·mə,sfir }

reducing coupling [ENG] A coupling used to connect a smaller pipe to a larger one. { ri'düs·iŋ ,kəp·liŋ }

reducing flame [CHEM] A flame having excess fuel and being capable of chemical reduction, such as extracting oxygen from a metallic oxide. { ri'düs·iŋ ,flām }

reducing glass [OPTICS] A double-concave lens that reduces the apparent size of objects viewed through it; used by illustrators and painters to create an artificial sense of distance from their work. { ri'düs·iŋ ,glas }

reducing sugar [ORG CHEM] Any of the sugars that because of their free or potentially free aldehyde or ketone groups, possess the property of readily reducing alkaline solutions of many metallic salts such as copper, silver, or bismuth; examples are the monosaccharides and most of the disaccharides, including maltose and lactose. { ri'düs·iŋ ,shug·ər }

reductant [MET] Coal or other reducing materials introduced in a smelting process to remove oxygen from ores or concentrates. { ri'dək·tənt }

reductio ad absurdum [MATH] A method of proof in which it is first supposed that the fact to be proved is false, and then it is shown that this supposition leads to the contradiction of accepted facts. Also known as indirect proof; proof by contradiction. { ri¦dək·tē·ō äd ab'sərd·əm }

reduction [ANALY CHEM] Preparation of one or more subsamples from a sample of material that is to be analyzed chemically. [CHEM] 1. Reaction of hydrogen with another substance. 2. Chemical reaction in which an element gains an electron (has a decrease in positive valence). [COMPUT SCI] Any process by which data are condensed, such as changing the encoding to eliminate redundancy, extracting significant details from the data and eliminating the rest, or choosing every second or third out of the totality of available points. [GEOL] The lowering of a land surface by erosion. [NAV] The process of substituting for an observed value one derived therefrom, as the calculation of the corresponding meridian altitude from an observation of a celestial body near the meridian, or the derivation from a celestial observation of the information needed for establishing a line of position. { ri'dək·shən }

reduction cell [CHEM] A vessel in which aqueous solutions of salts or fused salts are reduced electrolytically. { ri'dək·shən ,sel }

reduction formula [MATH] 1. An equation that expresses an integral as the sum of certain functions and a simpler integral. 2. An identity that expresses the values of a trigonometric function of an angle greater than 90° in terms of a function of an angle less than 90°. { ri'dək·shən ,för·myə·lə }

reduction gear [MECH ENG] A gear train which lowers the output speed. { ri'dək·shən ,gir }

reduction index [GEOL] The rate of wear of a sedimentary particle subject to abrasion, expressed as the difference between the mean weight of the particle before and after transport divided by the product of mean weight before transport and the distance traveled. { ri'dək·shən ,in,deks }

reduction of area [MET] In tensile testing, the percentage of decrease in cross-sectional area of a specimen at the point of rupture. { ri'dək·shən əv 'er·ē·ə }

reduction of observations [SCI TECH] The mathematical analysis of data from observations to obtain the desired information. { ri'dək·shən əv ,äb·zər'vā·shənz }

reduction of star places [ASTRON] The computation of mean places of stars from observations of their apparent places. { ri'dək·shən əv 'stär ,plās·əz }

reduction of tidal current [OCEANOGR] The processing of observed tidal current data to obtain mean values of tidal current constants. { ri'dək·shən əv 'tīd·əl ,kə·rənt }

reduction of tides [OCEANOGR] The processing of observed tidal data to obtain mean values of tidal constants. { ri'dək·shən əv 'tīdz }

reduction potential [PHYS CHEM] The potential drop involved in the reduction of a positively charged ion to a neutral form or to a less highly charged ion, or of a neutral atom to a negatively charged ion. { ri'dək·shən pə,ten·chəl }

reduction ratio [ENG] Ratio of feed size to product size for a mill (crushing or grinding) operation; measured by lump and sieve sizes. { ri'dək·shən ,rā·shō }

reduction roll [MET] A roller used to reduce the thickness of a piece of metal. { ri'dək·shən ,rōl }

reduction rule [COMPUT SCI] The principal computation rule in the lambda calculus; it states that an operator-operand combination of the form $(\lambda x MA)$ may be transformed into the expression $S^x_A M$, obtained by substituting the lambda expression A for all instances of x in M, provided there are no conflicts of variable names. Also known as beta rule. { ri'dək·shən ,rül }

reduction sequence [MATH] A sequence of applications of the reduction rule to a lambda expression. { ri'dək·shən ,sē·kwəns }

reduction sphere [GEOL] A white, leached, spheroidal mass produced in a reddish or brownish sandstone by a localized reducing environment, commonly surrounding an organic nucleus or a pebble and ranging in size from a poorly defined speck to a large, perfect sphere more than 10 inches (25 centimeters) in diameter. { ri'dək·shən ,sfir }

reduction to sea level [ENG] The application of a correction to a measured horizontal length on the earth's surface, at any altitude, to reduce it to its projected or corresponding length at sea level. { ri'dək·shən tə 'sē ,lev·əl }

reduction to the meridian [NAV] The process of applying a correction to an altitude observed when a body is near the celestial meridian of the observer, for the purpose of finding the altitude at meridian transit. { ri'dək·shən tü t̲h̲ə mə'rid·ē·ən }

reduction to the sun [ASTRON] In the spectroscopic determination of a star's radial motion referred to the sun, the correction that is needed to be applied to the observed radial velocity of the star to compensate for the motion of the earth with respect to the sun. { ri'dək·shən tü t̲h̲ə 'sən }

reductive grammar [COMPUT SCI] A set of syntactic rules for the analysis of strings to determine whether the strings exist in a language. { ri'dək·tiv 'gram·ər }

redundancy [COMPUT SCI] Any deliberate duplication or partial duplication of circuitry or information to decrease the probability of a system or communication failure. [COMMUN] In the transmission of information, the fraction of the gross information content of a message which can be eliminated without loss of essential information. [GEN] **1.** Repetition of a deoxyribonucleic acid sequence in a nucleus. **2.** Multiplicity of codons for individual amino acids. [MATH] A repetitive statement. [MECH] A statically indeterminate structure. { ri'dən·dən·sē }

redundancy bit [COMPUT SCI] A bit which carries no information but which is added to the information-carrying bits of a character or stream of characters to determine their accuracy. { ri'dən·dən·sē ,bit }

redundancy check [COMPUT SCI] A forbidden-combination check that uses redundant digits called check digits to detect errors made by a computer. { ri'dən·dən·sē ,chek }

redundant array of inexpensive disks See RAID. { ri,dən·dənt ə¦rā əv ,in·ik,spen·siv 'disks }

redundant bombing [ORD] The bombing of targets which have already been made inoperative or useless by attacks on other targets upon which the former are dependent for their operation. { ri'dən·dənt 'bäm·iŋ }

redundant character [COMPUT SCI] A character specifically added to a group of characters to ensure conformity with certain rules which can be used to detect computer malfunction. { ri'dən·dənt 'kar·ik·tər }

redundant code [COMMUN] A code which uses more signal elements than are needed to represent the information it transmits. { ri'dən·dənt 'kōd }

redundant digit [COMPUT SCI] Digit that is not necessary for an actual computation but serves to reveal a malfunction in a digital computer. { ri'dən·dənt 'dij·it }

redundant equation [MATH] An equation with roots that have been introduced in the process of solving another equation but that are not solutions of the equation to be solved. { ri'dən·dənt i'kwā·zhən }

redundant number See abundant number. { ri'dən·dənt ¦nəm·bər }

redundant system See duplexed system. { ri'dən·dənt ,sis·təm }

Reduviidae [INV ZOO] The single family of the hemipteran group Reduvioidea; nearly all have a stridulatory furrow on the prosternum, ocelli are generally present, and the beak is three-segmented. { ,rej·ə'vī·ə,dē }

Reduvioidea [INV ZOO] The assassin bugs or conenose bugs, a monofamilial group of hemipteran insects in the subdivision Geocorisae. { ,rej·ə,vē'oid·ē·ə }

reduzate [GEOL] A sediment accumulated under reducing conditions and consequently rich in organic carbon and in iron sulfide minerals; examples are coal and black shale. { 'rej·yə,zāt }

red water [BIOL] See red tide. [VET MED] **1.** A babesiasis of cattle characterized by hematuria following release of hemoglobin by destruction of erythrocytes. **2.** A chronic disease of cattle attributed to oxalic acid in the forage; hematuria results from escape of blood from lesions in the bladder. **3.** An acute febrile septicemia of cattle, and sometimes horses, sheep, and swine, caused by the bacterium *Clostridium hemolyticum* and characterized by hemoglobinuria and sometimes intestinal hemorrhages. { 'red 'wöd·ər }

red white-dwarf star [ASTRON] A star type that is considered an anomaly; these are objects 10,000 times fainter than the sun, with surface temperature below 4000 K so that surface radiation has cooled the star at an unexpectedly rapid rate. { 'red 'wīt ¦dwórf ,stär }

redwood [BOT] *Sequoia sempervirens.* An evergreen tree of the pine family; it is the tallest tree in the Americas, attaining 350 feet (107 meters); its soft heartwood is a valuable building material. { 'red,wud }

REDWOOD

Leaves and cone of redwood (*Sequoia sempervirens*).

Redwood viscometer [ENG] A standard British-type viscometer in which the viscosity is determined by the time, in seconds, required for a certain quantity of liquid to pass out through the orifice under given conditions; used for determining viscosities of petroleum oils. { 'red 'wud vi'skäm·əd·ər }

Red-Yellow Podzolic soil [GEOL] Any of a group of acidic, zonal soils having a leached, light-colored surface layer and a subsoil containing clay and oxides of aluminum and iron, varying in color from red to yellowish red to a bright yellowish brown. { 'red 'yel·ō päd'zäl·ik 'soil }

red zinc ore See zincite. { 'red 'ziŋk ,ór }

reed [BOT] Any tall grass characterized by a slender jointed stem. [ENG] A thin bar of metal, wood, or cane that is clamped at one end and set into transverse elastic vibration, usually by wind pressure; used to generate sound in musical instruments, and as a frequency standard, as in a vibrating-reed frequency meter. [TEXT] A comblike loom attachment that keeps the warp yarns apart and pushes the filling thread against the woven fabric. { rēd }

reed frequency meter See vibrating-reed frequency meter. { 'rēd 'frē·kwən·sē ,mēd·ər }

reed horn [ENG ACOUS] A horn that produces sound by means of a steel reed vibrated by air under pressure. { 'rēd ,hórn }

reeding [ENG] Corrugating or serrating, as in coining or embossing. { 'rēd·iŋ }

reedmergnerite [MINERAL] $NaBSi_3O_8$ A colorless, triclinic borate mineral that represents the boron analog of albite. { rēd'mər·nyə,rīt }

reed pen [GRAPHICS] A short section of bamboo sharpened to a point at both ends and used for ink drawing. { 'rēd ,pen }

reed relay [ELECTROMAG] A relay having contacts mounted on magnetic reeds sealed into a length of small glass tubing; an actuating coil is wound around the tubing or wound on an auxiliary ferrite-core structure, to provide the magnetic field required for relay operation. { 'rēd ,rē,lā }

Reed-Solomon code [COMMUN] A linear, block-based error-correcting code with wide-ranging applications, which is based on the mathematics of finite fields. { ¦rēd 'säl·ə·mən ,kōd }

Reed-Sternberg cell [PATH] An anaplastic reticuloendothelial cell constituting the characteristic microscopic feature of Hodgkin's disease. { 'rēd 'stərn,bərg sel }

reed switch [ELECTROMAG] A switch that has contacts mounted on ferromagnetic reeds sealed in a glass tube, designed for actuation by an external magnetic field. Also known as magnetic reed switch. { 'rēd ,swich }

reef [GEOL] **1.** A ridge- or moundlike layered sedimentary rock structure built almost exclusively by organisms. **2.** An offshore chain or range of rock or sand at or near the surface of the water. [MIN ENG] A major ore trend or ore body. { rēf }

reef breccia [PETR] A rock formed by the consolidation of limestone fragments broken off from a reef by the action of waves and tides. { 'rēf 'brech·ə }

reef cap [GEOL] A deposit of fossil-reef material overlying or covering an island or mountain. { 'rēf ,kap }

reef cluster [GEOL] A group of reefs of wholly or partly contemporaneous growth, found within a circumscribed area or geologic province. { 'rēf ,kləs·tər }

reef complex [GEOL] The solid reef core and the heterogeneous and contiguous fragmentary material derived from it by abrasion. { 'rēf ,käm,pleks }

reef conglomerate See reef talus. { 'rēf kən,gläm·ə·rət }

reef core [GEOL] The rock mass constructed in place, and within the rigid growth lattice formed by reef-building organisms. { 'rēf ,kor }

reef debris See reef detritus. { 'rēf də,brē }

reef detritus [GEOL] Fragmental material derived from the erosion of an organic reef. Also known as reef debris. { 'rēf di,trīd·əs }

reef edge [GEOL] The seaward margin of the reef flat, commonly marked by surge channels. { 'rēf ,ej }

reef flank [GEOL] The part of the reef that surrounds, interfingers with, and locally overlies the reef core, often indicated by massive or medium beds of reef talus dipping steeply away from the reef core. { 'rēf ,flaŋk }

reef flat [GEOL] A flat expanse of dead reef rock which is partly or entirely dry at low tide; shallow pools, potholes, gullies, and patches of coral debris and sand are features of the reef flat. { 'rēf ,flat }

reef front [GEOL] The upper part of the outer or seaward slope of a reef, extending to the reef edge from above the dwindle point of abundant living coral and coralline algae. { 'rēf ,frənt }

reef-front terrace [GEOL] A shelflike or benchlike eroded surface, sometimes veneered with organic growth, sloping seaward to a depth of 8–15 fathoms (15–27 meters). { 'rēf ¦frənt ,ter·əs }

reef knoll [GEOL] **1.** A bioherm or fossil coral reef represented by a small, prominent, rounded hill, up to 330 feet (100 meters) high, consisting of resistant reef material, being either a local exhumation of an original reef feature or a feature produced by later erosion. **2.** A present-day reef in the form of a knoll; a small reef patch developed locally and built upward rather than outward. { 'rēf ,nōl }

reef limestone [PETR] Limestone composed of the remains of sedentary organisms such as sponges, and of sediment-binding organic constituents such as calcareous algae. Also known as coral rock. { 'rēf 'līm,stōn }

reef milk [GEOL] A very-fine-grained matrix material of the back-reef facies, consisting of white, opaque microcrystalline calcite derived from abrasion of the reef core and reef flank. { 'rēf ,milk }

reef patch [GEOL] A single large colony of coral formed independently on a shelf at depths less than 220 feet (70 meters) in the lagoon of a barrier reef or of an atoll. Also known as patch reef. { 'rēf ,pach }

reef pinnacle [GEOL] A small, isolated spire of rock or coral, especially a small reef patch. { 'rēf ,pin·ə·kəl }

reef rock [PETR] A hard, unstratified rock composed of sand, shale, and the calcareous remains of sedentary organisms, cemented by calcium carbonate. { 'rēf ,räk }

reef segment [GEOL] A part of an organic reef lying between passes, gaps, or channels. { 'rēf ,seg·mənt }

reef slope [GEOL] The face of a reef rising from the sea floor. { 'rēf ,slōp }

reef talus [GEOL] Massive inclined strata composed of reef detritus deposited along the seaward margin of an organic reef. Also known as reef conglomerate. { 'rēf ,tā·ləs }

reef tufa [GEOL] Drusy, prismatic, fibrous calcite deposited directly from supersaturated water upon the void-filling internal sediment of the calcite mudstone of a reef knoll. { 'rēf ,tüf·ə }

reef wall [GEOL] A wall-like upgrowth of living coral and the skeletal remains of dead coral and other reef-building organisms, which reaches an intertidal level and acts as a partial barrier between adjacent environments. { 'rēf ,wol }

reel [DES ENG] A revolving spool-shaped device used for storage of hose, rope, cable, wire, magnetic tape, and so on. { rēl }

reel and bead See bead and reel. { 'rēl ən 'bēd }

reeled silk [TEXT] The raw silk strand made by combining several filaments from separate cocoons. { 'rēld 'silk }

reel locomotive [MIN ENG] A trolley locomotive with a wire-rope reel for drawing mining cars out of rooms. { 'rēl ,lō·kə,mōd·iv }

reel number [COMPUT SCI] A number identifying a reel of magnetic tape in a file containing more than one reel and indicating the order in which the reel is to be used. Also known as reel sequence number. { 'rēl ,nəm·bər }

reel sequence number See reel number. { 'rēl 'sē·kwəns ,nəm·bər }

reengineering [SYS ENG] The application of technology and management science to the modification of existing systems, organizations, processes, and products in order to make them more effective, efficient, and responsive. { ,rē·en·jə'nir·iŋ }

reenterable [COMPUT SCI] The attribute that describes a program or routine which can be shared by several tasks concurrently. { rē'en·trə·bəl }

reentrant [ENG] Having one or more sections directed inward, as in certain types of cavity resonators. [GEOL] A prominent, generally angular indentation into a coastline. { rē'en·trənt }

reentrant angle [CRYSTAL] The angle between two plane surfaces on a crystalline solid, in which the external angle is less than 180°. [MATH] An interior angle of a polygon that is greater than 180°. { rē'en·trənt ,aŋ·gəl }

reentrant code See reentrant program. { rē'en·trənt ,kōd }

reentrant program [COMPUT SCI] A subprogram in a time-sharing or multiprogramming system that can be shared by a number of users, and can therefore be applied to a given user program, interrupted and applied to some other user program, and then reentered at the point of interruption of the original user program. Also known as reentrant code. { rē'en·trənt ,prō,gram }

reentrant winding [ELEC] Armature winding that returns to its starting point, thus forming a closed circuit. { rē'en·trənt ,wīnd·iŋ }

reentry [AERO ENG] The event when a spacecraft or other object comes back into the sensible atmosphere after being in space. { rē'en·trē }

reentry angle [AERO ENG] That angle of the reentry body trajectory and the sensible atmosphere at which the body reenters the atmosphere. { rē'en·trē ,aŋ·gəl }

reentry body [AERO ENG] That part of a space vehicle that reenters the atmosphere after flight above the sensible atmosphere. { rē'en·trē ,bäd·ē }

reentry nose cone [AERO ENG] A nose cone designed especially for reentry, consists of one or more chambers protected by a heat sink. { rē'en·trē 'nōz ,kōn }

reentry point [COMPUT SCI] The instruction in a computer program at which execution is resumed after the program has jumped to another place. { rē'en·trē ,point }

reentry system [COMPUT SCI] In character recognition, a system in which the input data to be read are printed by the computer with which the reader is associated. [ORD] That portion of a ballistic missile or spacecraft designed to place one or more reentry vehicles on terminal trajectories so as to arrive at selected targets; penetration aids, spacers, deployment modules, and associated programming, and control and sensing devices are included in the reentry system. { rē'en·trē ,sis·təm }

reentry trajectory [AERO ENG] That part of a rocket's trajectory that begins at reentry and ends at the target or at the surface. { rē'en·trē trə,jek·trē }

reentry vehicle [AERO ENG] Any payload-carrying vehicle designed to leave the sensible atmosphere and then return through it to earth. { rē'en·trē ,vē·ə·kəl }

reentry window [AERO ENG] The area, at the limits of the earth's atmosphere, through which a spacecraft in a given trajectory can pass to accomplish a successful reentry for a landing in a desired region. { rē'en·trē ,win·dō }

reepithelialization [MED] **1.** Regrowth of epithelial tissue over a denuded surface. **2.** Surgical placement of a graft of epithelium over a denuded surface. { rē,ep·ə,thē·lē·ə·lə'zā·shən }

reeve [NAV ARCH] **1.** To pass a rope through a hole or opening in a block, ringbolt, or similar device. **2.** To fasten something by passing a rope or line through or around it. { rēv }

reevesite [MINERAL] $Na_6Fe_2(OH)_{16}(CO_3)\cdot4H_2O$ Hydrous oxide mineral known only in meteorites. { 'rēv,zīt }

refer [ORD] To bring the gunsights on a chosen aiming point without moving an artillery piece which has been laid for direction. { ri'fər }

reference acoustic pressure [ACOUS] Magnitude of any complex sound that will produce a sound-level meter reading equal to that produced by a sound pressure of 20 micropascals at 1000 hertz. Also known as reference sound level. { 'ref·rəns ə'küs·tik 'presh·ər }

reference address See address constant. { 'ref·rəns 'ad,res }

reference angle [ELECTROMAG] Angle formed between the center line of a radar beam as it strikes a reflecting surface and the perpendicular drawn to that reflecting surface. { 'ref·rəns ,aŋ·gəl }

reference block [COMPUT SCI] A block within a computer program governing a numerically controlled machine which has enough data to allow resumption of the program following an interruption. { 'ref·rəns ,bläk }

reference burst See color burst. { 'ref·rəns ,bərst }

reference craft [NAV] A craft to which relative movement of other craft is referred; on a maneuvering board this first craft is placed at the center; it may be a reference ship, reference aircraft, or reference vehicle, depending upon the type of circumstances. { 'ref·rəns ,kraft }

reference dimension [DES ENG] In dimensioning, a dimension without tolerance used for informational purposes only, and does not govern machining operations in any way; it is indicated on a drawing by writing the abbreviation REF directly following or under the dimension. { 'ref·rəns di,men·shən }

reference dipole [ELECTROMAG] Straight half-wave dipole tuned and matched for a given frequency, and used as a unit of comparison in antenna measurement work. { 'ref·rəns 'dī,pōl }

reference direction [NAV] A direction used as a basis for comparison of other directions. { 'ref·rəns di,rek·shən }

reference electrode [PHYS CHEM] A nonpolarizable electrode that generates highly reproducible potentials; used for pH measurements and polarographic analyses; examples are the calomel electrode, silver-silver chloride electrode, and mercury pool. { 'ref·rəns i'lek,trōd }

reference ellipsoid [MAP] A reference surface used to represent the size and shape of the earth for cartography. { 'ref·rəns i'lip,sòid }

reference frame See frame of reference. { 'ref·rəns ,frām }

reference frequency [COMMUN] Frequency having a fixed and specified position with respect to the assigned frequency. { 'ref·rəns ,frē·kwən·sē }

reference fuel [MATER] One of the standardized laboratory engine fuels, blends of which are used to determine the octane numbers of motor gasoline and the cetane numbers of diesel fuels. { 'ref·rəns ,fyül }

reference level [ENG] See datum plane. [ENG ACOUS] The level used as a basis of comparison when designating the level of an audio-frequency signal in decibels or volume units. Also known as reference signal level. [OCEANOGR] **1.** Level of no motion. **2.** A level for which current is known; allows determination of absolute current from relative current. { 'ref·rəns ,lev·əl }

reference line [NAV] A line from which angular or linear measurements are reckoned. Also known as datum line. { 'ref·rəns ,līn }

reference listing [COMPUT SCI] A list printed by a compiler showing the instructions in the machine language program which it generates. { 'ref·rəns ,list·iŋ }

reference locality [GEOL] A locality containing a reference section, established to supplement the type locality. { 'ref·rəns lō'kal·əd·ē }

reference lot [IND ENG] A lot of select components, used as a standard. { 'ref·rəns ,lät }

reference mark [ELECTR] One of the marks used in a design of a printed circuit, giving scale dimensions and indicating the edges of the circuit board. { 'ref·rəns ,märk }

reference material [ANALY CHEM] A material or substance whose properties are sufficiently well established to be used in calibrating an apparatus, assessing a measurement method, or assigning values to other materials. { 'ref·rəns mə,tir·ē·əl }

reference monitor [COMPUT SCI] A means of checking that a particular user is allowed access to a specified object in a computing system. Also known as access control mechanism; reference validation mechanism. { 'ref·rəns ,män·əd·ər }

reference noise [ELECTR] The power level used as a basis of comparison when designating noise power expressed in decibels above reference noise (dBrn); the reference usually used is 10^{-12} watt (-90 decibels above 1 milliwatt; dBm) at 1000 hertz. { 'ref·rəns ,nòiz }

reference plane [ENG] See datum plane. [MECH ENG] The plane containing the axis and the cutting point of a cutter. { 'ref·rəns ,plān }

reference point [NAV] A point to which other points, lines, and so forth are referred, usually in terms of distance or direction, or both. { 'ref·rəns ,pòint }

reference range [ENG] Range obtained from the radar coverage indicator for a given penetrating aircraft. { 'ref·rəns ,rānj }

reference record [COMPUT SCI] Output of a compiler that lists the operations and their positions in the final specific routine and contains information describing the segmentation and storage allocation of the routine. { 'ref·rəns ,rek·ərd }

reference rounds [ORD] Ammunition rounds of known performance which are fired during ballistic tests of ammunition for comparative purposes. { 'ref·rəns ,raùnz }

reference section [GEOL] A rock section, or group of sections, designated to supplement the type section, or sometimes to supplant it (as where the type section is no longer exposed), and to afford a standard for correlation for a certain part of the geologic column. { 'ref·rəns ,sek·shən }

reference seismometer [ENG] In seismic prospecting, a detector placed to record successive shots under similar conditions, to permit overall time comparisons. { 'ref·rəns sīz'mäm·əd·ər }

reference signal level See reference level. { 'ref·rəns 'sig·nəl ,lev·əl }

reference sound level See reference acoustic pressure. { 'ref·rəns 'saùnd ,lev·əl }

reference spheroid [GEOD] An ellipsoid of revolution, chosen to approximate the geoid, on which geodetic triangulation measurements are computed. { 'ref·rəns 'sfir,òid }

reference station [OCEANOGR] **1.** A place for which independent daily predictions are given in the tide or current tables, from which corresponding predictions are obtained for other stations by means of differences or factors. **2.** A place for which tidal or tidal current constants have been determined and which is used as a standard for the comparison of simultaneous observations at a second station. Also known as standard station. { 'ref·rəns stā·shən }

reference supply [ELECTR] A source of stable and constant voltage, such as a Zener diode, used in analog computers, regulated power supplies, and a variety of other circuits for comparison with a varying voltage. { 'ref·rəns sə,plī }

reference time [COMPUT SCI] The instant near the beginning of switching that is chosen as a reference for time measurements in a digital computer. { 'ref·rəns ,tīm }

reference tone [ENG] Stable tone of known frequency continuously recorded on one track of multitrack signal recordings and intermittently recorded on signal track recordings by the collection equipment operators for subsequent use by the data analysts as a frequency reference. { 'ref·rəns ,tōn }

reference validation mechanism See reference monitor. { 'ref·rəns ,val·ə'dā·shən ,mek·ə,niz·əm }

reference vehicle [NAV] A vehicle serving as a reference craft. { 'ref·rəns ,vē·ə·kəl }

reference voltage [ELEC] An alternating-current voltage used for comparison, usually to identify an in-phase or out-of-phase condition in an ac circuit. { 'ref·rəns ,vōl·tij }

reference volume [ACOUS] The audio volume level that gives a reading of 0 VU (volume units) on a standard volume indicator; the sensitivity of the volume indicator is adjusted so reference volume or 0 is read when the instrument is connected across a 600-ohm resistance to which is delivered a power of 1 milliwatt at 1000 hertz. { 'ref·rəns ˌväl·yəm }

reference white [COMMUN] **1.** In a scene viewed by television cameras, the color of light from a nonselective diffuse reflector that is lighted by the normal illumination of the scene. **2.** The color by which this color is simulated on a television screen or other display device. { 'ref·rəns ˌwīt }

reference white level [ELECTR] In television, the level at the point of observation corresponding to the specified maximum excursion of the picture signal in the white direction. { 'ref·rəns 'wīt ˌlev·əl }

referencing [ENG] The process of measuring the horizontal (or slope) distances and directions from a survey station to nearby landmarks, reference marks, and other permanent objects which can be used in the recovery or relocation of the station. { 'ref·rən·siŋ }

referred pain [MED] Pain felt in one area but originating in another area. { ri'fird 'pān }

refine [ENG] To free from impurities, as the separation of petroleum, ores, or chemical mixtures into their component parts. { ri'fīn }

refined kerosine See deodorized kerosine. { ri'fīnd 'ker·ə,sēn }

refined lard [FOOD ENG] A purified lard, produced by removing with alkali the free fatty acids, coloring matter, and mucilaginous gums from rendered lard. { ri'fīnd 'lärd }

refined lecithin See lecithin. { ri'fīnd 'les·ə·thən }

refined oil [MATER] A class of petroleum oil used for home lighting and cooking purposes. Also known as burning oil. { ri'fīnd 'oil }

refined paraffin wax [MATER] A grade of paraffin wax; a hard, crystalline hydrocarbon wax derived from mixed-base or paraffin-base crude oils. { ri'fīnd 'par·ə·fən ˌwaks }

refined tar [MATER] A tar from which water has been extracted by evaporation or distillation. { ri'fīnd 'tär }

refinement [MATH] A tower that can be obtained by inserting a finite number of subsets in a given tower. { ri'fīn·mənt }

refinery [CHEM ENG] System of process units used to convert crude petroleum into fuels, lubricants, and other petroleum-derived products. [MET] System of process units used to convert nonferrous-metal ores into pure metals, such as copper or zinc. { ri'fīn·rē }

refinery gas [MATER] Gas produced in petroleum refineries by cracking, reforming, and other processes; principally methane, ethane, ethylene, butanes, and butylenes. { ri'fīn·rē ˌgas }

refining temperature [MET] The temperature just above the transformation range employed in the heat treatment of steel in order to refine grain size. { ri'fīn·iŋ ˌtem·prə·chər }

reflectance [COMPUT SCI] In optical character recognition, the relative brightness of the inked area that forms the printed or handwritten character; distinguished from background reflectance and brightness. [ELEC] See reflection factor. [PHYS] See reflectivity. { ri'flek·təns }

reflectance spectrophotometry [SPECT] Measurement of the ratio of spectral radiant flux reflected from a light-diffusing specimen to that reflected from a light-diffusing standard substituted for the specimen. { ri'flek·təns ˌspek·trə·fə'täm·ə·trē }

reflected binary [COMPUT SCI] A particular form of gray code which is constructed according to the following rule: Let the first 2^N code patterns be given, for any N greater than 1; the next 2^N code patterns are derived by changing the $(N + 1)$-th bit from the right from 0 to 1 and repeating the original 2^N patterns in reverse order in the N rightmost positions. Also known as reflected code. { ri'flek·təd 'bī,ner·ē }

reflected buried structure [GEOL] The distortion of surface beds that reflect a similar structural distortion of underlying formations. { ri'flek·təd 'ber·ēd 'strək·chər }

reflected code See reflected binary. { ri'flek·təd 'kōd }

reflected impedance [ELEC] **1.** Impedance value that appears to exist across the primary of a transformer due to current flowing in the secondary. **2.** Impedance which appears at the input terminals as a result of the characteristics of the impedance at the output terminals. { ri'flek·təd im'pēd·əns }

reflected pressure [PHYS] The pressure from an explosion (especially an airburst bomb), which is reflected from a solid object or surface, rather than dissipated in the air. { ri'flek·təd 'presh·ər }

reflected ray [PHYS] A ray extending outward from a point of reflection. { ri'flek·təd 'rā }

reflected resistance [ELEC] Resistance value that appears to exist across the primary of a transformer when a resistive load is across the secondary. { ri'flek·təd ri'zis·təns }

reflected signal indicator [ENG] Pen recorder which presents the radar signals within frequency gates; these recordings enable the operator to determine that an airborne object has penetrated the Doppler link and its direction of penetration. { ri'flek·təd ˌsig·nəl 'in·də,kād·ər }

reflected ultraviolet method [GRAPHICS] A method of ultraviolet photography in which an ultraviolet source is used and the camera is provided with a filter which permits only reflected ultraviolet light to reach the film. { ri'flek·təd ˌəl·trə'vī·lət ˌmeth·əd }

reflected wave [PHYS] A wave reflected from a surface, discontinuity, or junction of two different media, such as the sky wave in radio, the echo wave from a target in radar, or the wave that travels back to the source end of a mismatched transmission line. { ri'flek·təd 'wāv }

reflecting antenna [ELECTROMAG] An antenna used to achieve greater directivity or desired radiation patterns, in which a dipole, slot, or horn radiates toward a larger reflector which shapes the radiated wave to produce the desired pattern; the reflector may consist of one or two plane sheets, a parabolic or paraboloidal sheet, or a paraboloidal horn. { ri'flek·tiŋ an'ten·ə }

reflecting curtain [ELECTROMAG] A vertical array of half-wave reflecting antennas, generally used one quarter-wavelength behind a radiating curtain of dipoles to form a high-gain antenna. { ri'flek·tiŋ 'kərt·ən }

reflecting electrode [ELECTR] Tabular outer electrode or the repeller plate in a microwave oscillator tube, corresponding in construction but not in function to the plate of an ordinary triode; used for generating extremely high frequencies. { ri'flek·tiŋ i'lek,trōd }

reflecting galvanometer See mirror galvanometer. { ri'flek·tiŋ ˌgal·və'näm·əd·ər }

reflecting grating [ELECTROMAG] Arrangement of wires placed in a waveguide to reflect one desired wave while allowing one or more other waves to pass freely. { ri'flek·tiŋ 'grād·iŋ }

reflecting microscope [OPTICS] A microscope whose objective is composed of two mirrors, one convex and the other concave; its imaging properties are independent of the wavelength of light, allowing it to be used even for infrared and ultraviolet radiation. { ri'flek·tiŋ 'mī·krə,skōp }

reflecting nephoscope See mirror nephoscope. { ri'flek·tiŋ 'nef·ə,skōp }

reflecting prism [OPTICS] A prism used in place of a mirror for deviating light, usually designed so that there is no dispersion of light; the light undergoes at least one internal reflection. { ri'flek·tiŋ 'priz·əm }

reflecting sign [CIV ENG] A road sign painted with reflective paint so as to be easily visible in the light of a headlamp. { ri'flek·tiŋ 'sīn }

reflecting spectrograph [OPTICS] A solar spectrograph in which the collimator and camera element are long-focus concave mirrors. { ri'flek·tiŋ 'spek·trə,graf }

reflecting telescope [OPTICS] A telescope in which a concave parabolic mirror gathers light and forms a real image of an object. Also known as reflector telescope. { ri'flek·tiŋ 'tel·ə,skōp }

reflection [MATH] **1.** The reflection of a configuration in a line, in a plane, or in the origin of a coordinate system is the replacement of each point in the configuration by a point that is symmetric to the given point with respect to the line, plane, or origin. **2.** Two permutations, a and b, of the same objects are reflections of each other if the first object in a is the last object in b, the second object in a is the next-to-last object in b, and so forth, with the last object in a being the first object in b. [PHYS] The return of waves or particles from surfaces on which they are incident. { ri'flek·shən }

reflection altimeter See radio altimeter. { ri'flek·shən al'tim·əd·ər }

reflection angle See angle of reflection. { ri'flek·shən ˌaŋ·gəl }

reflection coefficient [PHYS] The ratio of the amplitude of a wave reflected from a surface to the amplitude of the incident

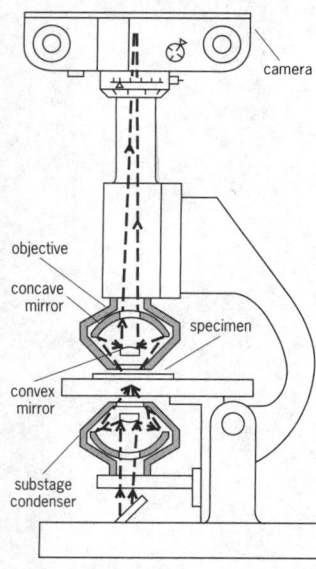

camera

objective

concave mirror

specimen

convex mirror

substage condenser

Reflecting microscope arranged for photomicrography.

REFLECTING PRISM

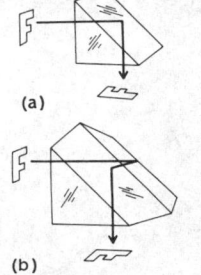

(a)

(b)

Examples of reflecting prism. (a) Right-angle. (b) Amici roof prism.

wave. Also known as coefficient of reflection. { ri'flek·shən ,kō·i,fish·ənt }

reflection copy [GRAPHICS] Material for illustrations, such as photographs, which must be viewed or photographed by surface-reflected light. { ri'flek·shən ,käp·ē }

reflection density [OPTICS] The common logarithm of the ratio of the luminance of a nonabsorbing perfect diffuser to that of the surface under consideration, when both are illuminated at an angle of 45° to the normal and the direction of measurement is perpendicular to the surface. { ri'flek·shən ,den·səd·ē }

reflection diffraction [PHYS] Type of electron diffraction analysis in which the electron beam grazes the sample surface. { ri'flek·shən di,frak·shən }

reflection factor [ELEC] Ratio of the load current that is delivered to a particular load when the impedances are mismatched to that delivered under conditions of matched impedances. Also known as mismatch factor; reflectance; transition factor. { ri'flek·shən ,fak·tər }

reflection goniometer [ENG] A goniometer that measures the angles between crystal faces by reflection of a parallel beam of light from successive crystal faces. { ri'flek·shən ,gō·nē'äm·əd·ər }

reflection HEED [PHYS] A form of HEED (high-energy electron diffraction) in which electrons are incident at small angles (0.5–4°) to the surface, and are reflected from it. Abbreviated RHEED. { ri'flek·shən ,hēd }

reflection law [PHYS] When a wave, such as electromagnetic radiation or sound, is reflected from a surface in a sharply defined direction, the reflected and incident waves travel in directions that make the same angle with a perpendicular to the surface and lie in a common plane with it. Also known as law of reflection. { ri'flek·shən ,lo }

reflection lobes [ELECTROMAG] Three-dimensional sections of the radiation pattern of a directional antenna, such as a radar antenna, which results from reflection of radiation from the earth's surface. { ri'flek·shən ,lōbz }

reflection loss [ELEC] **1.** Reciprocal of the ratio, expressed in decibels, of the scalar values of the volt-amperes delivered to the load to the volt-amperes that would be delivered to a load of the same impedance as the source. **2.** Apparent transmission loss of a line which results from a portion of the energy being reflected toward the source due to a discontinuity in the transmission line. { ri'flek·shən ,los }

reflection nebula [ASTRON] A type of bright diffuse nebula composed mainly of cosmic dust; it is visible because of starlight from nearby stars or nebula stars that is scattered by the dust particles. { ri'flek·shən ,neb·yə·lə }

reflection plane See plane of mirror symmetry; plane of reflection. { ri'flek·shən ,plān }

reflection plotter [NAV] An attachment fitted to a radar plan position indicator which provides a plotting surface permitting plotting without parallax errors; any mark made on the plotting surface will be reflected on the radarscope directly below. Also known as plotting head. { ri'flek·shən ,pläd·ər }

reflection profile [ENG] A seismic profile obtained by designing the spread geometry in such a manner as to enhance reflected energy. { ri'flek·shən ,prō,fīl }

reflection rainbow [OPTICS] A rainbow formed by light rays which have been reflected from an extended water surface; not to be confused with a reflected rainbow whose image may be seen in a still body of water. { ri'flek·shən ,rān,bō }

reflection seismology See reflection shooting. { ri'flek·shən sīz'mäl·ə·jē }

reflection shooting [ENG] A procedure in seismic prospecting based on the measurement of the travel times of waves which, originating from an artificially produced disturbance, have been reflected to detectors from subsurface boundaries separating media of different elastic-wave velocities; used primarily for oil and gas exploration. Also known as reflection seismology. { ri'flek·shən ,shüd·iŋ }

reflection spectrum [PHYS] The spectrum seen when incident waves are selectively altered by a reflecting substance. { ri'flek·shən ,spek·trəm }

reflection survey [ENG] Study of the presence, depth, and configuration of underground formations; a ground-level explosive charge (shot) generates vibratory energy (seismic rays) that strike formation interfaces and are reflected back to ground-level sensors. Also known as seismic survey. { ri'flek·shən ,sər,vā }

reflection twin [CRYSTAL] A crystal twin whose symmetry is formed by an apparent mirror image across a plane. { ri'flek·shən ,twin }

reflection x-ray microscopy [ENG] A technique for producing enlarged images in which a beam of x-rays is successively reflected at grazing incidence, from two crossed cylindrical surfaces; resolution is about 0.5–1 micrometer. { ri'flek·shən ¦eks,rā mi'kräs·kə·pē }

reflective binary code See reflected binary. { ri'flek·tiv 'bī,ner·ē 'kōd }

reflective coating See mirror coating. { ri'flek·tiv 'kōd·iŋ }

reflective code See Gray code. { ri'flek·tiv 'kōd }

reflective insulation [MATER] An insulating material used to retard the flow of heat by reflecting heat radiation; usually made of aluminum foil or sheets, although coated steel sheets, aluminized paper, gold and silver surfaces, and refractory metals at higher temperatures are also used. { ri'flek·tiv ,in·sə'lā·shən }

reflective spot [COMPUT SCI] A piece of metallic foil that is embedded in a magnetic tape to indicate the end of a reel. { ri'flek·tiv ,spät }

reflectivity [PHYS] The ratio of the energy carried by a wave which is reflected from a surface to the energy carried by the wave which is incident on the surface. Also known as reflectance. { ,rē,flek'tiv·əd·ē }

reflectometer [ELECTROMAG] See microwave reflectometer. [ENG] A photoelectric instrument for measuring the optical reflectance of a reflecting surface. { ,rē,flek'täm·əd·ər }

reflector [ELECTR] See repeller. [ELECTROMAG] **1.** A single rod, system of rods, metal screen, or metal sheet used behind an antenna to increase its directivity. **2.** A metal sheet or screen used as a mirror to change the direction of a microwave radio beam. [GEOPHYS] A layer or horizon that reflects seismic waves. [NUCLEO] A layer of water, graphite, beryllium, or other scattering material placed around the core of a nuclear reactor to reduce the loss of neutrons. Also known as tamper. { ri'flek·tər }

reflector characteristic [ELECTR] A chart of power output and frequency deviation of a reflex klystron as a function of reflector voltage. { ri'flek·tər ,kar·ik·tə'ris·tik }

reflector compass [NAV] A magnetic compass in which the image of the compass card is viewed by direct reflection in a mirror adjacent to the helmsman's position. { ri'flek·tər ,kam·pəs }

reflector microphone [ENG ACOUS] A highly directional microphone which has a surface that reflects the rays of impinging sound from a given direction to a common point at which a microphone is located, and the sound waves in the speech-frequency range are in phase at the microphone. { ri'flek·tər ,mī·krə,fōn }

reflector plate [OPTICS] A transparent mirror in a computing gunsight, or in some types of optical gunsights and bombsights, that reflects the reticle image or images to the eye. { ri'flek·tər ,plāt }

reflector satellite [AERO ENG] Satellite so designed that radio or other waves bounce off its surface. { ri'flek·tər ,sad·əl,īt }

reflector telescope See reflecting telescope. { ri'flek·tər ,tel·ə,skōp }

reflector voltage [ELECTR] Voltage between the reflector electrode and the cathode in a reflex klystron. { ri'flek·tər ,vōl·tij }

reflex [PHYSIO] An automatic response mediated by the nervous system. { rē,fleks }

reflex angle [MATH] An angle greater than 180° and less than 360°. { 'rē,fleks ,aŋ·gəl }

reflex arc [NEUROSCI] A chain of neurons composing the anatomical substrate or pathway of the unconditioned reflex. { 'rē,fleks ,ärk }

reflex baffle [ENG ACOUS] A loudspeaker baffle in which a portion of the radiation from the rear of the diaphragm is propagated forward after controlled shift of phase or other modification, to increase the overall radiation in some portion of the audio-frequency spectrum. Also known as vented baffle. { 'rē,fleks ,baf·əl }

reflex bladder [ANAT] A urinary bladder controlled only by the simple reflex arc through the sacral part of the spinal cord. { 'rē,fleks ,blad·ər }

reflex bunching [ELECTR] The bunching that occurs in an

REFLECTION NEBULA

Reflection nebula associated with the Pleiades star cluster. (*Lick Observatory Photograph*)

REFLECTION X-RAY MICROSCOPY

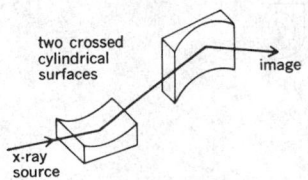

Principle for reflection x-ray microscopy.

REFLECTOR MICROPHONE

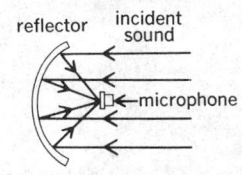

Schematic view of a parabolic reflector microphone.

electron stream which has been made to reverse its direction in the drift space. { 'rē,fleks ,bənch·iŋ }

reflex camera [OPTICS] A camera in which a mirror is used to reflect a full-size image of a scene on a ground glass so that the composition and focus may be judged. { 'rē,fleks ,kam·rə }

reflex circuit [ELECTR] A circuit in which the signal is amplified twice by the same amplifier tube or tubes, once as an intermediate-frequency signal before detection and once as an audio-frequency signal after detection. { 'rē,fleks ,sər·kət }

reflexed [BOT] Turned abruptly backward. { 'rē,flekst }

reflex epilepsy [MED] Seizure brought on by specific sensory stimuli. { 'rē,fleks 'ep·ə,lep·sē }

reflexive Banach space [MATH] A Banach space B such that, for every continuous linear functional F on the conjugate space B^*, there corresponds a point x_0 of B such that $F(f) = f(x_0)$ for each element f of B^*. Also known as regular Banach space. { ri¦flek·siv 'bä,näk ,spās }

reflexive processing [COMPUT SCI] Information processing in which two or more computers connected by communications channels run identical programs and take the same actions at the same time, so that users in different locations can work on the same programs at the same time. { ri'flek·siv 'prä,ses·iŋ }

reflexive relation [MATH] A relation among the elements of a set such that every element stands in that relation to itself. { ri'flek·siv ri,lā·shən }

reflex klystron [ELECTR] A single-cavity klystron in which the electron beam is reflected back through the cavity resonator by a repelling electrode having a negative voltage; used as a microwave oscillator. Also known as reflex oscillator. { 'rē,fleks 'klī,strän }

reflex oscillator See reflex klystron. { 'rē,fleks 'äs·ə,lād·ər }

reflex printing [GRAPHICS] A method of copying in which a sensitized film or paper is placed face down on the material to be copied; light passes through the base of the sensitized material and is reflected from the light and dark portions of the original back to the emulsion. { 'rē,fleks ,print·iŋ }

reflex sight [OPTICS] An optical or computing sight that reflects a reticle image or images onto a reflector plate for superimposition on the target by the eye. { 'rē,fleks ,sīt }

reflex time [NEUROSCI] The time required for the nerve impulse to travel in a reflex action. { 'rē,fleks ,tīm }

reflowing [ENG] Melting and resolidifying an electrodeposited or other type coating. { rē'flō·iŋ }

reflux [CHEM ENG] In a chemical process, that part of the product stream that may be returned to the process to assist in giving increased conversion or recovery, as in distillation or liquid-liquid extraction. { 'rē,fləks }

reflux condenser [CHEM ENG] An auxiliary vessel for a distillation column that constantly condenses vapors and returns liquid to the column. { 'rē,fləks kən,den·sər }

reflux esophagitis [MED] An inflammation of the lower esophagus that results from excessive acid reflux or regurgitation from the stomach; symptoms include heartburn and indigestion. { ,rē,fləks i,säf·ə'jīd·əs }

reflux ratio [CHEM ENG] The quantity of liquid reflux per unit quantity of product removed from the process unit, such as a distillation tower or extraction column. { 'rē,fləks ,rā·shō }

refolding [GEOL] A process by which folds of one generation are subjected to and stressed by a force of different orientation. { rē'fōld·iŋ }

refoliation [GEOL] A foliation that is subsequent to and oriented differently from an earlier foliation. { ri,fō·lē'ā·shən }

reforestation [FOR] Establishment of a new forest by seeding or planting seedlings on forest land that fails to restock naturally. { rē,fär·ə'stā·shən }

reformat [COMPUT SCI] To change the arrangement of data in a storage device. { rē'fòr·mat }

reformate [MATER] Product from a petroleum-refinery reforming process; types are thermal reformate (from thermal reforming), and cat or catalytic reformate (from catalytic reforming). { rē'fòr,māt }

Reformatsky reaction [ORG CHEM] A condensation-type reaction between ketones and α-bromoaliphatic acids in the presence of zinc or magnesium, such as $R_2CO + BrCH_2COOR + Zn \rightarrow (ZnO \cdot HBr) + R_2C(OH)CH_2COOR$. { rif·ər'mat·skē rē,ak·shən }

reformed gas [MATER] A lower-thermal-value fuel gas made by pyrolysis and steam decomposition of high-thermal-value natural and refinery gases. { 'rē,fòrmd 'gas }

reformed gasoline [MATER] Gasoline made by a catalytic or thermal reforming process. { 'rē,fòrmd 'gas·ə,lēn }

reforming [CHEM ENG] The thermal or catalytic conversion of petroleum naphtha into more volatile products of higher octane number; represents the total effect of numerous simultaneous reactions, such as cracking, polymerization, dehydrogenation and isomerization. { ¦rē'fòrm·iŋ }

refracted ray [PHYS] A ray extending onward from the point of refraction. { ri'frak·təd 'rā }

refracted wave [PHYS] That portion of an incident wave which travels from one medium into a second medium. Also known as transmitted wave. { ri'frak·təd 'wāv }

refracting angle See apical angle. { ri'frak·tiŋ ,aŋ·gəl }

refracting edge [OPTICS] The intersection of the two refracting faces of a prism. { ri'frak·tiŋ ,ej }

refracting sphere [OPTICS] A sphere made of a transparent material whose index of refraction differs from the medium surrounding it, so that it refracts light passing through it. { ri'frak·tiŋ ,sfir }

refracting telescope [OPTICS] A telescope in which a lens gathers light and forms a real image of an object. Also known as refractor telescope. { ri'frak·tiŋ 'tel·ə,skōp }

refraction [ELECTROMAG] The change in direction of lines of force of an electric or magnetic field at a boundary between media with different permittivities or permeabilities. [PHYS] The change of direction of propagation of any wave, such as an electromagnetic or sound wave, when it passes from one medium to another in which the wave velocity is different, or when there is a spatial variation in a medium's wave velocity. { ri'frak·shən }

refraction coefficient [OCEANOGR] The square root of the ratio of the spacing between orthogonals in deep water and in shallow water; it is a measure of the effect of refraction in diminishing wave height by increasing the length of the wave crest. { ri'frak·shən ,kō·i,fish·ənt }

refraction diagram [OCEANOGR] A chart showing the position of the wave crests at a particular time, or the successive positions of a particular wave crest as it moves shoreward. { ri'frak·shən ,dī·ə,gram }

refraction error [NAV] An error due to refraction, particularly such an error in a sextant altitude, due to atmospheric refraction. { ri'frak·shən ,er·ər }

refraction loss [ELECTROMAG] Portion of the transmission loss that is due to refraction resulting from nonuniformity of the medium. { ri'frak·shən ,lòs }

refraction process [ENG] Seismic (reflection) survey in which the distance between the explosive shot and the receivers (sensors) is large with respect to the depths to be mapped. { ri'frak·shən ,prä·səs }

refraction profile [ENG] A seismic profile obtained by designing the spread geometry in such a manner as to enhance refracted energy. { ri'frak·shən ,prō,fīl }

refraction shooting [ENG] A type of seismic shooting based on the measurement of seismic energy as a function of time after the shot and of distance from the shot, by determining the arrival times of seismic waves which have traveled nearly parallel to the bedding in high-velocity layers, in order to map the depth of such layers. { ri'frak·shən ,shüd·iŋ }

refractive constant See index of refraction. { ri'frak·tiv 'kän·stənt }

refractive index See index of refraction. { ri'frak·tiv ,in,deks }

refractive modulus See modified index of refraction. { ri'frak·tiv 'mäj·ə·ləs }

refractivity [ELECTROMAG] **1.** Some quantitative measure of refraction, usually a measure of the index of refraction. **2.** The index of refraction minus 1. { ,rē,frak'tiv·əd·ē }

refractometer [ENG] An instrument used to measure the index of refraction of a substance in any one of several ways, such as measurement of the refraction produced by a prism, measurement of the critical angle, observation of an interference pattern produced by passing light through the substance, and measurement of the substance's dielectric constant. { ,rē,frak'täm·əd·ər }

refractometry [OPTICS] The measurement of the index of refraction of a substance; it is an important tool in analytical chemistry. { ,rē,frak'täm·ə·trē }

refractor telescope *See* refracting telescope. { ri'frak·tər 'tel·ə,skōp }

refractory [MATER] **1.** A material of high melting point. **2.** The property of resisting heat. [MED] Not readily yielding to treatment. { ri'frak·trē }

refractory cement [MATER] Any of a variety of mixtures, such as fireclay-silica-ganister mixture, or fireclay mixed with crushed brick, or fireclay and silica sand, with a refractory range of 2600–2800°F (1412–1523°C); used for furnace and oven linings and for fillers. { ri'frak·trē si'ment }

refractory clay [GEOL] Clay with a melting point above 1600°C; used to make firebrick and linings for furnaces and reactors. { ri'frak·trē ,klā }

refractory coating [MATER] A coating composed of a refractory material. { ri'frak·trē ,kōd·iŋ }

refractory concrete [MATER] Heat-resistant concrete made with high-alumina or calcium-aluminate cement and a refractory aggregate. { ri'frak·trē 'kän,krēt }

refractory enamel [MATER] An enamel for coating and protecting metals against attack by hot gases. { ri'frak·trē i'nam·əl }

refractory hard metals [CHEM] True chemical compounds composed of two or more metals in the crystalline form, and having a very high melting point and high hardness. { ri'frak·trē 'härd 'med·əlz }

refractory-lined firebox boiler [MECH ENG] A horizontal fire-tube boiler with the front portion of the shell located over a refractory furnace; the rear of the shell contains the first-pass tubes, and the second-pass tubes are located in the upper part of the shell. { ri'frak·trē ,līnd 'fīr,bäks ,bȯil·ər }

refractory metal [MET] A metal or alloy that is heat-resistant, having a high melting point. { ri'frak·trē ,med·əl }

refractory period [NEUROSCI] A brief period of time following the stimulation of a nerve during which the nerve will not respond to a second stimulus. { ri'frak·trē ,pir·ē·əd }

refractory sand [MATER] Sand used for refractory which is capable of resisting high temperatures. { ri'frak·trē ,sand }

refrangible [PHYS] Capable of being refracted. { ri'fran·jə·bəl }

refresh [COMPUT SCI] A process of periodically replacing data to prevent the data from decaying, as on a cathode-ray-tube display or in a dynamic random-access memory. { ri'fresh }

refrigerant [MATER] A substance that by undergoing a change in phase (liquid to gas, gas to liquid) releases or absorbs a large latent heat in relation to its volume, and thus effects a considerable cooling effect; examples are ammonia, sulfur dioxide, ethyl or methyl chloride (these are no longer widely used), and the fluorocarbons, such as Freon, Ucon, and Genetron. { ri'frij·ə·rənt }

refrigerant 23 *See* fluoroform. { ri'frij·ə·rənt ,twen·tē'thrē }

refrigerated truck [MECH ENG] An insulated truck equipped and used as a refrigerator to transport fresh perishable or frozen products. { ri'frij·ə,rād·əd 'trək }

refrigeration [MECH ENG] The cooling of a space or substance below the environmental temperature. { ri,frij·ə'rā·shən }

refrigeration condenser [MECH ENG] A vapor condenser in a refrigeration system, where the refrigerant is liquefied and discharges its heat to the environment. { ri,frij·ə'rā·shən kən,den·sər }

refrigeration cycle [THERMO] A sequence of thermodynamic processes whereby heat is withdrawn from a cold body and expelled to a hot body. { ri,frij·ə'rā·shən ,sī·kəl }

refrigeration oil [MATER] A mineral oil with all moisture and wax removed; used for lubricating refrigerating machinery. { ri,frij·ə'rā·shən ,ȯil }

refrigeration system [MECH ENG] A closed-flow system in which a refrigerant is compressed, condensed, and expanded to produce cooling at a lower temperature level and rejection of heat at a higher temperature level for the purpose of extracting heat from a controlled space. { ri,frij·ə'rā·shən ,sis·təm }

refrigerator [MECH ENG] An insulated, cooled compartment. { ri'frij·ə,rād·ər }

refrigerator car [MECH ENG] An insulated freight car constructed and used as a refrigerator. { ri'frij·ə,rād·ər ,kär }

Refsum's disease [MED] A familial disorder characterized by visual disturbances, ataxia, neuritic changes, and cardiac damage, associated with high blood level of phytinic acid. { 'ref·səmz di,zēz }

Refugian [GEOL] A North American stage of geologic time in the Eocene and Oligocene, above the Fresnian and below the Zemorrian. { rə'fyü·jē·ən }

refugium [ECOL] An area which has escaped the great changes which occurred in the region as a whole, often providing conditions in which relic colonies can survive; for example, a driftless area which has escaped the effects of glaciation because it projected above the ice. { rə'fyü·jē·əm }

reg [GEOL] An extensive, nearly level, low desert plain from which fine sand has been removed by wind, leaving a sheet of coarse, smoothly angular, wind-polished gravel and small stones lying on an alluvial soil, strongly cemented by mineralized solutions to form a broad desert pavement. Also known as gravel desert. { reg }

RE galaxy [ASTRON] A type of ring galaxy that consists of a single, relatively empty, ringlike structure, without any prominent condensation or nucleus. { ,rē'ē ,gal·ik·sē }

regelation [HYD] Phenomenon in which ice melts at the bottom of droplets of highly concentrated saline solution that are trapped in ice which has frozen over polar waters, and freezes at the top of these droplets, so that the droplets move downward through the ice, leaving it hard and clear. [THERMO] Phenomenon in which ice (or any substance which expands upon freezing) melts under intense pressure and freezes again when this pressure is removed; accounts for phenomena such as the slippery nature of ice and the motion of glaciers. { ,rē·jə'lā·shən }

regenerant [CHEM] A solution whose purpose is to restore the activity of an ion-exchange bed. { rē'jen·ə·rənt }

regenerate [CHEM ENG] To clean of impurities and make reusable as in regeneration of a catalytic cracking catalyst by burning off carbon residue, regeneration of clay adsorbent by washing free of adherents, or regeneration of a filtration system by cleaning off the filter media. [ELECTR] **1.** To restore pulses to their original shape. **2.** To restore stored information to its original form in a storage tube in order to counteract fading and disturbances. { rē'jen·ə,rāt }

regenerated cellulose [MATER] **1.** Rayon in which the raw cellulose is changed physically but not chemically, such as viscose, cuprammonium, and nitrocellulose rayons. **2.** A transparent cellulose plastic material made by mixing cellulose xanthate with a dilute sodium hydroxide solution to form a viscose. { rē'jen·ə,rād·əd 'sel·yə,lōs }

regenerated fiber [TEXT] Textile fiber produced by dissolving a natural material (such as cellulose), then regenerating it by extrusion and precipitation, as with viscose. { rē'jen·ə,rād·əd 'fī·bər }

regenerated flow control [HYD] Control of glacial drainage by modified morainal features, resulting from the readvance of a previously stagnant glacier. { rē'jen·ə,rād·əd 'flō kən,trōl }

regenerated glacier [HYD] A glacier that becomes active after a period of stagnation. { rē'jen·ə,rād·əd glā·shər }

regeneration [BIOL] The replacement by an organism of tissues or organs which have been lost or severely injured. [CHEM] Restoration of the activity of a deactivated catalyst. [CONT SYS] *See* positive feedback. [ELECTR] Replacement or restoration of charges in a charge storage tube to overcome decay effects, including loss of charge by reading. [NUCLEO] Restoration of contaminated nuclear fuel to a usable condition. { rē jen·ə'rā·shən }

regeneration electrode [NEUROSCI] A silicon-based microelectrode with holes through which severed peripheral nerve fibers can regenerate and then reinnervate on the target tissue or organ. { rē jen·ə'rā·shən i,lek,trōd }

regeneration system [MECH ENG] A system within a gas turbine that recovers waste heat from the turbine exhaust and uses it for the compression cycle. { rē jen·ə'rā·shən ,sis·təm }

regenerative air heater [MECH ENG] An air heater in which the heat-transferring members are alternately exposed to heat-surrendering gases and to air. { rē'jen·rəd·iv 'er ,hēd·ər }

regenerative amplifier [ELECTR] An amplifier that uses positive feedback to give increased gain and selectivity. { rē'jen·rəd·iv 'am·plə,fī·ər }

regenerative braking [ELEC] A system of dynamic braking in which the electric drive motors are used as generators and return the kinetic energy of the motor armature and load to the electric supply system. { rē'jen·rəd·iv 'brāk·iŋ }

regenerative chaff [ORD] An electronic countermeasure technique in which a missile or powered decoy sequentially ejects clouds of chaff. { rē'jen·rəd·iv 'chaf }

regenerative clipper [ELECTR] A type of monostable multivibrator which is a modification of a Schmitt trigger; used for pulse generation. { rē'jen·rəd·iv 'klip·ər }

regenerative cooling [ENG] A method of cooling gases in which compressed gas is cooled by allowing it to expand through a nozzle, and the cooled expanded gas then passes through a heat exchanger where it further cools the incoming compressed gas. { rē'jen·rəd·iv 'kül·iŋ }

regenerative cycle [MECH ENG] *See* bleeding cycle. [THERMO] An engine cycle in which low-grade heat that would ordinarily be lost is used to improve the cyclic efficiency. { rē'jen·rəd·iv ˌsī·kəl }

regenerative detector [ELECTR] A vacuum-tube detector circuit in which radio-frequency energy is fed back from the anode circuit to the grid circuit to give positive feedback at the carrier frequency, thereby increasing the amplification and sensitivity of the circuit. { rē'jen·rəd·iv di'tek·tər }

regenerative divider [ELECTR] Frequency divider which employs modulation, amplification, and selective feedback to produce the output wave. { rē'jen·rəd·iv di'vīd·ər }

regenerative engine [AERO ENG] **1.** A jet or rocket engine that utilizes the heat of combustion to preheat air or fuel entering the combustion chamber. **2.** Specifically, to a type of rocket engine in which one of the propellants is used to cool the engine by passing through a jacket prior to combustion. { rē'jen·rəd·iv 'en·jən }

regenerative feedback *See* positive feedback. { rē'jen·rəd·iv 'fēd‚bak }

regenerative fuel cell [ELEC] A fuel cell in which the reaction product is processed to regenerate the reactants. { rē'jen·rəd·iv 'fyül ‚sel }

regenerative pump [MECH ENG] Rotating-vane device that uses a combination of mechanical impulse and centrifugal force to produce high liquid heads at low volumes. Also known as turbine pump. { rē'jen·rəd·iv 'pəmp }

regenerative reactor [NUCLEO] A nuclear reactor that produces fissionable material in addition to energy; when loaded with fissionable uranium-235, and nonfissionable uranium-238 or thorium, it converts the uranium-238 or thorium into fissionable materials which are then used as fuel in the core of the reactor. { rē'jen·rəd·iv rē'ak·tər }

regenerative read [COMPUT SCI] A read operation in which the data are automatically written back into the locations from which they are taken. { rē'jen·rəd·iv 'rēd }

regenerative receiver [ELECTR] A radio receiver that uses a regenerative detector. { rē'jen·rəd·iv ri'sē·vər }

regenerative repeater [COMMUN] A repeater that performs pulse regeneration to restore the original shape of a pulse signal used in teletypewriter and other code circuits. { rē'jen·rəd·iv ri'pēd·ər }

regenerative storage [COMPUT SCI] A storage unit, such as a delay line or storage tube, in which the stored data must be constantly read and restored to prevent decay or loss. { rē'jen·rəd·iv 'stȯr·ij }

regenerative track [COMPUT SCI] Track on a magnetic drum with interconnected reading and writing heads; information stored on these tracks is continuously read from the drum, transmitted round a closed circuit, and recorded back on the drum; consequently, access times to these data are short; operation is analogous to that of the acoustic delay line. { rē'jen·rəd·iv 'trak }

regenerator [CHEM ENG] Device or system used to return a system or a component of it to full strength in a chemical process; examples are a furnace to burn carbon from a catalyst, a tower to wash impurities from clay, and a flush system to clean off the surface of filter media. [ELECTR] **1.** A circuit that repeatedly supplies current to a display or memory device to prevent data from decaying. **2.** *See* repeater. [MECH ENG] A device used with hot-air engines and gas-burning furnaces which transfers heat from effluent gases to incoming air or gas. { rē'jen·ə‚rād·ər }

Reggeism [PARTIC PHYS] An attempt to account for and correlate hadron resonances and the asymptotic behavior of scattering amplitudes of hadrons at high energies in terms of Regge poles. { 'reg·ə‚iz·əm }

Regge pole [PARTIC PHYS] A pole singularity of a scattering amplitude in the complex angular momentum plane; the scattering amplitude is formed by continuing partial wave amplitudes from positive integer values of the angular momentum to the complex plane. { 'reg·ə ‚pōl }

Regge recurrence [PARTIC PHYS] One of a sequence of hadrons, with successive hadrons increasing by one in spin and also increasing in mass, but with the same values of other quantum numbers, except for parity, charge parity and G parity, which alternate in sign; it is believed that they are rotationally excited states of a particle, and that they alternate between two Regge trajectories. { 'reg·ə ri‚kər·əns }

Regge trajectory [PARTIC PHYS] **1.** The path followed by a Regge pole in the complex angular momentum plane as the center-of-mass energy is varied. **2.** The relationship between the spin and mass of a sequence of hadrons, with successive hadrons increasing by 2 in spin and also increasing in mass, but with the same values of other quantum numbers; the hadrons are thought to correspond to energies at which a Regge pole passes near positive integers (or half integers). { 'reg·ə trə‚jek·trē }

regime [GEOL] The existence in a stream channel of a balance between erosion and deposition over a period of years. { rə'zhēm }

regimen [HYD] **1.** The behavior characteristic of the total amount of water involved in a drainage basin. **2.** Analysis of the total volume of water involved with a lake, including water losses and gains, over a period of a year. **3.** The flow characteristics of a stream with respect to velocity, volume, form of and alterations in the channel, capacity to transport sediment, and the amount of material supplied for transportation. { 'rej·ə·mən }

region [COMPUT SCI] A group of machine addresses which refer to a base address. [MATH] **1.** The union of an open connected set with a subset of its boundary points (which may be the entire boundary or the empty set). **2.** *See* domain. { 'rē·jən }

regional address [COMPUT SCI] An address of a machine instruction within a series of consecutive addresses; for example, R18 and R19 are specific addresses in an R region of N consecutive addresses, where all addresses must be named. { 'rēj·ən·əl ə'dres }

regional anatomy [ANAT] The detailed study of the anatomy of a part or region of the body of an animal. { 'rēj·ən·əl ə'nad·ə·mē }

regional anesthesia *See* regional block anesthesia. { 'rēj·ən·əl ‚an·əs'thē·zhə }

regional block anesthesia [MED] Anesthesia limited to a part or region of the body by injecting an anesthetic around the nerves that supply the area to block conduction from the area. Also known as regional anesthesia. { 'rēj·ən·əl ‚bläk ‚an·əs'thē·zhə }

regional center [COMMUN] A long-distance telephone office which has the highest rank in routing of telephone calls; there are 12 such offices in the United States and Canada. { 'rēj·ən·əl 'sen·tər }

regional dip [GEOL] The nearly uniform and generally low-angle inclination of strata over a wide area. Also known as normal dip. { 'rēj·ən·əl 'dip }

regional enteritis [MED] Inflammation of isolated segments of the small intestine, with involved parts becoming thick-walled and rigid. Also known as regional enterocolitis; regional ileitis. { 'rēj·ən·əl ‚ent·ə'rīd·əs }

regional enterocolitis *See* regional enteritis. { 'rēj·ən·əl ‚ent·ə·rō·kə'līd·əs }

regional forecast *See* area forecast. { 'rēj·ən·əl 'fȯr‚kast }

regional geology [GEOL] The geology of a large region, treated from the viewpoint of the spatial distribution and position of stratigraphic units, structural features, and surface forms. { 'rēj·ən·əl jē'äl·ə·jē }

regional ileitis *See* regional enteritis. { 'rēj·ən·əl ‚il·ē'īd·əs }

regional metamorphism [GEOL] Geological metamorphism affecting an extensive area. { 'rēj·ən·əl ‚med·ə'mȯr‚fiz·əm }

regional metasomatism [GEOL] Metasomatic processes affecting extensive areas whereby the introduced material may be derived from partial fusion of the rocks involved from deep-seated magmatic sources. { 'rēj·ən·əl ‚med·ə·sō·mə‚tiz·əm }

regional migration [PETRO ENG] Horizontal movement of gas or oil through a reservoir formation as a result of artificial pressure differences created by withdrawal of gas or oil at well sites. { 'rēj·ən·əl mī'grā·shən }

regional slope [GEOL] The generally uniform dip of rock strata or land surface over a wide area. { 'rēj·ən·əl 'slōp }

regional slope deposit [GEOL] A sedimentary deposit widely distributed as a thin sheet over a regional slope. { 'rēj·ən·əl 'slōp di,päz·ət }

regional snowline [HYD] The level above which, averaged over a large area, snow accumulation exceeds ablation year after year. { 'rēj·ən·əl 'snō,līn }

regional unconformity [GEOL] A continuous unconformity extending throughout a wide region that may be nearly continentwide, and usually represents a long period of time. { 'rēj·ən·əl ,ən·kən'fòr·məd·ē }

region of escape See exosphere. { 'rē·jən əv i'skāp }

regioselective [ORG CHEM] Pertaining to a chemical reaction which favors a single positional or structural isomer, leading to its yield being greater than that of the other products in the reaction. Also known as regiospecific. { ¦rē·jē·ō·si'lek·tiv }

regioselectivity [BIOCHEM] Control over the location of reaction in a complex molecule. { ,rē·jē·ō,si·lek'tiv·əd·ē }

regiospecific See regioselective. { ¦rē·jē·ō·spə'sif·ik }

register [COMPUT SCI] The computer hardware for storing one machine word. [COMMUN] Part of an automatic switching telephone system which receives and stores the dialing pulses which control the further operations necessary in establishing a telephone connection. [ENG] Also known as registration. **1.** The accurate matching or superimposition of two or more images, such as the three color images on the screen of a color television receiver, or the patterns on opposite sides of a printed circuit board, or the colors of a design on a printed sheet. **2.** The alignment of positions relative to a specified reference or coordinate, such as hole alignments in punched cards, or positioning of images in an optical character recognition device. [GRAPHICS] **1.** Exact agreement in the position of printed material on both sides of a sheet or on all pages of a book or pamphlet. **2.** Exact overprinting of colorplates, or other subsequent plates, so that all printed detail is correctly combined; proper color overprinting is checked by the exact superimposition of the register marks that are printed with each color run. **3.** In flat preparation, the exact agreement between color or complementary flats. [MECH ENG] The portion of a burner which directs the flow of air used in the combustion process. [ORD] **1.** To adjust fire on a visible point, called a check point, and compute accurate adjusted data so that firing data for later targets may be computed with reference to that check point. **2.** To adjust fire on several selected points in order that they may serve later as auxiliary targets. { 'rej·ə·stər }

register breadth [NAV ARCH] The breadth of beam dimension of a vessel; specifically, the breadth of the broadest part on the outside of the vessel. { 'rej·ə·stər ¦bredth }

register capacity [COMPUT SCI] The upper and lower limits of the numbers which may be processed in a register. { 'rej·ə·stər kə'pas·əd·ē }

register circuit [ELECTR] A switching circuit with memory elements that can store from a few to millions of bits of coded information; when needed, the information can be taken from the circuit in the same code as the input, or in a different code. { 'rej·ə·stər ,sər·kət }

register control [CONT SYS] Automatic control of the position of a printed design with respect to reference marks or some other part of the design, as in photoelectric register control. { 'rej·ə·stər kən,trōl }

register glass [GRAPHICS] In photography, a glass plate at the focal plane, against which the film is pressed during exposure. { 'rej·ə·stər ,glas }

register length [COMPUT SCI] The number of digits, characters, or bits, which a register can store. { 'rej·ə·stər ¦leŋkth }

register-level compatibility [COMPUT SCI] Property of hardware components that are totally compatible, having registers with the same type, size, and names. { ¦rej·ə·stər ,lev·əl kəm,pad·ə'bil·əd·ē }

register mark [ENG] A mark or line printed or otherwise impressed on a web of material for use as a reference to maintain register. [GRAPHICS] One of a set of marks, usually cross-shaped, placed on the margin of colored originals to be photographed for colorplates; the marks are used to register the images when the plates are printed. { 'rej·ə·stər ,märk }

register-sender [COMMUN] A unit that generates and recognizes the supervisory signals to make connection to a circuit switching unit. { 'rej·ə·stər 'sen·dər }

register ton See ton. { 'rej·ə·stər 'tən }

register tonnage [NAV ARCH] The available gross tonnage of a ship for transport of freight or passengers. { 'rej·ə·stər 'tən ij }

register variable [COMPUT SCI] A variable in a computer program that is assigned to a register in the central processing unit instead of to a location in main storage. { 'rej·ə·stər ,ver·ē·ə·bəl }

registration See register. { ,rej·ə'strā·shən }

registration fire [ORD] That artillery fire delivered to obtain corrections for increasing the accuracy of subsequent artillery. { ,rej·ə'strā·shən ,fīr }

registration mark [COMPUT SCI] In character recognition, a preprinted indication of the relative position and direction of various elements of the source document to be recognized. { ,rej·ə'strā·shən ,märk }

regmagenesis [GEOL] Diastrophic production of regional strike-slip displacements. { ¦reg·mə'jen·ə·səs }

regmaglypt [GEOL] Any of various small, well-defined, characteristic indentations or pits on the surface of meteorites, frequently resembling the imprints of fingertips in soft clay. Also known as pezograph; piezoglypt. { 'reg·mə,glipt }

regolith [GEOL] The layer rock or blanket of unconsolidated rocky debris of any thickness that overlies bedrock and forms the surface of the land. Also known as mantle rock. { 'reg·ə,lith }

Regosol [GEOL] In early United States soil classification systems, one of an azonal group of soils that form from deep, unconsolidated deposits and have no definite genetic horizons. { 'reg·ə,säl }

regradation [GEOL] The formation by a stream of a new profile of equilibrium, as when the former profile, after gradation, became deformed by crustal movements. { ¦rē·grā'dā·shən }

regression [GEOL] The theory that some rivers have sources on the rainier sides of mountain ranges and gradually erode backward until the ranges are cut through. [OCEANOGR] Retreat of the sea from land areas, and the consequent evidence of such withdrawal. [PSYCH] A mental state and a mode of adjustment to difficult and unpleasant situations, characterized by behavior of a type that had been satisfying and appropriate at an earlier stage of development but which no longer befits the age and social status of the individual. [STAT] Given two stochastically dependent random variables, regression functions measure the mean expectation of one relative to the other. { ri'gresh·ən }

regression analysis [STAT] The description of the nature of the relationship between two or more variables; it is concerned with the problem of describing or estimating the value of the dependent variable on the basis of one or more independent variables. { ri'gresh·ən ə,nal·ə·səs }

regression coefficient [STAT] The coefficient of the independent variables in a regression equation. { ri'gresh·ən ,kō·ə,fish·ənt }

regression conglomerate [GEOL] A coarse sedimentary deposit formed during a retreat (recession) of the sea. { ri'gresh·ən kən,gläm·ə·rət }

regression curve [STAT] A plot of a regression equation; for two variables, the independent variable is plotted as the abscissa and the dependent variable as the ordinate; for three variables, a solid model can be constructed or the representation can be reduced by an isometric chart or stereogram. { ri'gresh·ən ,kərv }

regression estimate [STAT] An estimate of one variable obtained by substituting the known value of another variable in a regression equation calculated on sample values of the two variables. { ri'gresh·ən ,es·tə·mət }

regression line [STAT] A linear regression equation with two or more variables. { ri'gresh·ən ,līn }

regression of nodes [ASTRON] The westward movement of the nodes of the moon's orbit; one cycle is completed in about 18.6 years. { ri'gresh·ən əv 'nōdz }

regressive overlap See offlap. { ri'gres·iv 'ō·vər,lap }

regressive reef [GEOL] One of a series of nearshore reefs or bioherms superimposed on basinal deposits during the rising

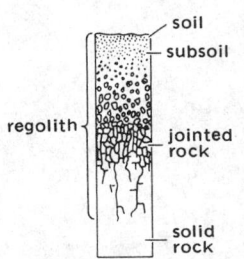

REGOLITH

Cross section through the regolith showing the components.

(labels: soil, subsoil, regolith, jointed rock, solid rock)

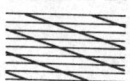

REGULAR LAY

Drawing of wire rope wound in regular lay showing position of the wires and strands.

of a landmass or the lowering of the sea level, and developed more or less parallel to the shore. { ri'gres·iv 'ref }

regressive ripple [GEOL] An asymmetric ripple mark formed by a current but oriented in a direction opposite to the general movement of current flow (steep side facing upcurrent). { ri'gres·iv 'rip·əl }

regressive sediment [GEOL] A sediment deposited during the retreat or withdrawal of water from a land area or during the emergence of the land, and characterized by an offlap arrangement. { ri'gres·iv 'sed·ə·mənt }

regret criterion See Savage principle. { ri'gret krī,tir·ē·ən }

regula falsi [MATH] A method of calculating an unknown quantity by first making an estimate and then using this and the properties of the unknown to obtain it. Also known as rule of false position. { 'reg·yə·lə 'fäl·sē }

regular [BOT] Having radial symmetry, referring to a flower. [ELECTROMAG] In a definite direction; not diffused or scattered, when applied to reflection, refraction, or transmission. { 'reg·yə·lər }

regular analytic curve [MATH] An analytic curve for which the derivatives of the analytic functions that represent the coordinates of points on the curve do not all vanish at any point. { ¦reg·yə·lər ¦an·ə,lid·ik 'kərv }

regular Baire measure [MATH] A Baire measure such that the measure of any Baire set E is equal to both the greatest lower bound of measures of open Baire sets containing E, and to the least upper bound of closed, compact sets contained in E. { 'reg·yə·lər 'bär ,mezh·ər }

regular Banach space See reflexive Banach space. { 'reg·yə·lər 'bä,näk ,späs }

regular Borel measure [MATH] A Borel measure such that the measure of any Borel set E is equal to both the greatest lower bound of measures of open Borel sets containing E, and to the least upper bound of measures of compact sets contained in E. Also known as Radon measure. { 'reg·yə·lər bə'rel ,mezh·ər }

regular connective tissue [HISTOL] Connective tissue in which the fibers are arranged in definite patterns. { 'reg·yə·lər kə'nek·tiv 'tish·ü }

regular cluster [ASTRON] A galaxy cluster that shows a smooth, centrally concentrated distribution of galaxies and an overall symmetric shape. { ¦reg·yə·lər 'kləs·tər }

regular curve [MATH] A curve that has no singular points. { 'reg·yə·lər 'kərv }

regular definition [MATH] A definition of the sum of a divergent series which yields the ordinary sum when applied to a convergent series. { ¦reg·yə·lər ,def·ə'nish·ən }

regular dodecahedron [CRYSTAL] See pyritohedron. [MATH] A regular polyhedron of 12 faces. { 'reg·yə·lər dō,dek·ə'hē·drən }

regular element [IND ENG] An element that occurs with a fixed frequency in each work cycle. Also known as repetitive element. { 'reg·yə·lər 'el·ə·mənt }

regular expression [COMPUT SCI] A formal description of a language acceptable by a finite automaton or for the behavior of a sequential switching circuit. { 'reg·yə·lər ik'spresh·ən }

regular extension [MATH] An extension field K of a field F such that F is algebraically closed in K and K is separable over F; equivalently, an extension field K of a field F such that K and $\overline{F}$ are linearly disjoint over F, where $\overline{F}$ is the algebraic closure of F. { 'reg·yə·lər ik'sten·chən }

regular function [MATH] An analytic function of one or more complex variables. { 'reg·yə·lər 'fəŋk·shən }

regular graph [MATH] A graph whose vertices all have the same degree. { 'reg·yə·lər ¦graf }

Regularia [INV ZOO] An assemblage of echinoids in which the anus and periproct lie within the apical system; not considered a valid taxon. { ,reg·yə'lar·ē·ə }

regular icosahedron [MATH] A 20-sided regular polyhedron, having five equilateral triangles meeting at each face. { 'reg·yə·lər ī,käs·ə'hē·drən }

regularization [QUANT MECH] A formal procedure used to eliminate ambiguities which arise in evaluating certain integrals in a quantized field theory; corresponds to adding extra fields whose masses are allowed to approach infinity. { ,reg·yə·lər·ə'zā·shən }

regular lay [DES ENG] The lay of a wire rope in which the wires in the strand are twisted in directions opposite to the direction of the strands. { 'reg·yə·lər 'lā }

regular-lay left twist See left-laid. { 'reg·yə·lər ¦lā 'left 'twist }

regular map See normal map. { 'reg·yə·lər 'map }

regular motor oil [MATER] A petroleum lubricating oil suitable for use in internal combustion engines under normal operating conditions. { 'reg·yə·lər 'mōd·ər ,oil }

regular octahedron [MATH] A regular polyhedron of eight faces. { 'reg·yə·lər ,äk·tə'hē·drən }

regular parameter [MATH] The independent variable that parametrizes a regular analytic curve. { ¦reg·yə·lər pə'ram·əd·ər }

regular permutation group [MATH] A permutation group of order n on n objects, where n is a positive integer. { 'reg·yə·lər ,pər·myə'tā·shən ,grüp }

regular point [MATH] **1.** Any point or a surface that is not a singular point. **2.** See ordinary point. { ¦reg·yə·lər 'póint }

regular polygon [MATH] A polygon with congruent sides and congruent interior angles. { 'reg·yə·lər 'päl·i,gän }

regular polyhedron [MATH] A polyhedron all of whose faces are regular polygons, and whose polyhedral angles are congruent. Also known as platonic solid. { 'reg·yə·lər ,päl·i'hē·drən }

regular polymer [CHEM] A polymer whose molecules possess only one kind of constitutional unit in a single sequential structure. { 'reg·yə·lər 'päl·ə·mər }

regular polytope [MATH] A geometric object in multidimensional Euclidean space that is analogous to the regular polygons (in two-dimensional space) and the regular polyhedra (in three-dimensional space). { ¦reg·yə·lər 'päl·ə,tōp }

regular prism [MATH] A right prism whose bases are regular polygons. { ¦reg·yə·lər 'priz·əm }

regular pyramid [MATH] A pyramid whose base is a regular polygon and whose lateral faces are inclined at equal angles to the base. { ¦reg·yə·lər 'pir·ə,mid }

regular reflection See specular reflection. { 'reg·yə·lər ri'flek·shən }

regular reflector See specular reflector. { 'reg·yə·lər ri'flek·tər }

regular representation [MATH] A regular representation of a finite group is an isomorphism of it with a group of permutations. { 'reg·yə·lər ,rep·rə·zən'tā·shən }

regular sampling [MIN ENG] The continuous or intermittent sampling of the same coal or coke received regularly at a given point. { 'reg·yə·lər 'samp·liŋ }

regular sequence See Cauchy sequence. { ¦reg·yə·lər 'sē·kwəns }

regular singular point [MATH] A regular singular point of a differential equation is a singular point of the equation at which none of the solutions has an essential singularity. { 'reg·yə·lər ¦siŋ·gyə·lər 'póint }

regular space [MATH] A topological space such that any neighborhood of a point in the space contains the closure of another neighborhood of the same point. { ¦reg·yə·lər 'späs }

regular tetrahedron [MATH] A regular polyhedron of four faces. { 'reg·yə·lər ,te·trə'hē·drən }

regular topological space [MATH] A topological space where any point and a closed set not containing it can be enclosed in disjoint open sets. { 'reg·yə·lər ¦täp·ə¦läj·ə·kəl 'späs }

regular variable star [ASTRON] A variable star whose variation in brightness is repeated with a uniform period and light curve from one cycle to the next. { 'reg·yə·lər 'ver·ē·ə·bal 'stär }

regular ventilating circuit [MIN ENG] All places in the mine through which there is a positive natural flow of air. { 'reg·yə·lər 'vent·əl,ād·iŋ ,sər·kət }

regulated item [ORD] An item requiring special control in handling distribution and use, because of its highly technical, costly, or hazardous nature, or because of some other specified concern. { 'reg·yə,lād·əd 'īd·əm }

regulated power supply [ELEC] A power supply containing means for maintaining essentially constant output voltage or output current under changing load conditions. { 'reg·yə,lād·əd 'pau·ər sə,plī }

regulated split [MIN ENG] In mine ventilation, a split where it is necessary to control the volumes in certain low-resistance splits to cause air to flow into the splits of high resistance. { 'reg·yə,lād·əd 'split }

regulating reservoir [CIV ENG] A reservoir that regulates the flow in a water-distributing system. { 'reg·yə,lād·iŋ 'rez·əv,wär }

regulating rod [NUCLEO] A control rod intended to accomplish rapid, fine, and sometimes continuous adjustment of the reactivity of a nuclear reactor; it usually can move much more rapidly than a shim rod but makes a smaller change in reactivity. { 'reg·yə,lād·iŋ ,räd }

regulating station [ORD] A command agency established to control all movements of personnel and supplies into or out of a given area. { 'reg·yə,lād·iŋ ,stā·shən }

regulating system See automatic control system. { 'reg·yə,lād·iŋ ,sis·təm }

regulating transformer [ELEC] Transformer having one or more windings excited from the system circuit or a separate source and one or more windings connected in series with the system circuit for adjusting the voltage or the phase relation or both in steps, usually without interrupting the load. { 'reg·yə,lād·iŋ tranz,fȯr·mər }

regulating winding [ELEC] Of a transformer, a supplementary winding connected in series with one of the main windings to change the ratio of transformation or the phase relation, or both, between circuits. { 'reg·yə,lād·iŋ ,wīnd·iŋ }

regulation [CONT SYS] The process of holding constant a quantity such as speed, temperature, voltage, or position by means of an electronic or other system that automatically corrects errors by feeding back into the system the condition being regulated; regulation thus is based on feedback, whereas control is not. [ELEC] The change in output voltage that occurs between no load and full load in a transformer, generator, or other source. [ELECTR] The difference between the maximum and minimum tube voltage drops within a specified range of anode current in a gas tube. { ,reg·yə'lā·shən }

regulation of constant-current transformer [ELEC] Maximum departure of the secondary current from its rated value expressed in percent of the rated secondary current, with rated primary voltage and frequency applied. { ,reg·yə'lā·shən əv ¦kän·stənt ¦kə·rənt tranz'fȯr·mər }

regulative egg [EMBRYO] An egg in which unfertilized fragments develop as complete, normal individuals. { 'reg·yə,lād·iv 'eg }

regulator [CONT SYS] A device that maintains a desired quantity at a predetermined value or varies it according to a predetermined plan. [MIN ENG] An opening in a wall or door in the return airway of a district to increase its resistance and reduce the volume of air flowing. { 'reg·yə,lād·ər }

regulator gene [GEN] A gene that controls the rate of transcription of one or more other genes. { 'reg·yə,lād·ər ,jēn }

regulator pin [HOROL] A pin in a watch to adjust the effective length of the hairspring and thereby adjust the speed of the watch. { 'reg·yə,lād·ər ,pin }

regulator problem See linear regulator problem. { 'reg·yə,lād·ər ,präb·ləm }

regulatory control function [CONT SYS] That level in the functional decomposition of a large-scale control system which interfaces with the plant to implement the decisions of the optimizing controller inputted in the form of set points, desired trajectories, or targets. Also known as direct control function. { 'reg·yə·lə,tȯr·ē kən'trōl ,fəŋk·shən }

regulatory sequence [CELL MOL] A sequence of deoxyribonucleic acid to which gene regulatory proteins bind to control the rate of transcription. { 'reg·yə·lə,tȯr·ē ,sē·kwəns }

regulatory site [BIOCHEM] A site on an enzyme, other than the active site, at which certain molecules bind, thereby affecting the enzyme's activity. { 'reg·yə·lə,tȯr·ē ,sit }

regulon [GEN] In bacteria, a system of genes, formed by one or more operons, which regulate enzyme induction and whose activity is controlled by a single repressor substance. { 'reg·yə,län }

regulus [MET] Impure metal formed beneath the slag during smelting or reduction of ores. { 'reg·yə·ləs }

Regulus [ASTRON] A star of stellar magnitude 1.34, about 67 light-years from the sun, spectral classification B8, in the constellation Leo; the star α Leonis. Also known as Little Ruler. [ORD] A United States surface-to-surface, jet-powered guided missile; it is equipped with a nuclear warhead,

and launched from a surfaced submarine or a cruiser. { 'reg·yə·ləs }

regur [GEOL] One of a group of calcareous, intrazonal soils characterized by dark color and a high clay content. Also known as black cotton soil. { 'reg·ər }

regurgitation [MED] Reverse circulation of blood in the heart due to defective functioning of the valves. [PHYSIO] Bringing back into the mouth undigested food from the stomach. { ri gər·jə'tā·shən }

rehabilitation [MED] The restoration to a disabled individual of maximum independence commensurate with his limitations by developing his residual capacity. { ,rē·ə,bil·ə'tā·shən }

rehabilitation engineering [ENG] The use of technology to make disabled persons as independent as possible by providing assistive devices to compensate for disability. { ,rē·ə,bil·ə'tā·shən ,en·jə,nir·iŋ }

Rehbinder effect [PHYS] The reduction in the hardness and ductility of a material by a surface-active molecular film. { 'rā,bin·dər i fekt }

Rehbock weir formula [FL MECH] Probably the most accurate formula for the rate of flow of water over a rectangular suppressed weir; it includes a correction for the velocity of approach for normal, or fairly uniform, velocity distribution in the upstream channel; the formula is

$$Q = [3.234 + 5.347/(320h - 3) + 0.428h/d_0]lh^{3/2}$$

where Q is the flow rate in cubic feet per second, l is the width of the weir in feet, h is the head of water above the crest of the weir in feet, and d_0 is the height of weir or depth of water at zero head in feet. { 'rā,bäk 'wer ,fȯr·myə·lə }

reheating [ASTRON] A scenario suggested by the original calculations of the inflationary universe cosmology in which, following inflation, the universe quickly thermalizes to a temperature that is comparable to the energy density stored in the original symmetric pre-inflationary phase of matter. [THERMO] A process in which the gas or steam is reheated after a partial isentropic expansion to reduce moisture content. Also known as resuperheating. { rē'hēd·iŋ }

Rehfuss tube [MED] A stomach tube designed for the removal of specimens of gastric contents for analysis after administration of a test meal. { 'rā·fūs ,tüb }

Reichenbach's lamellae [GEOL] Thin, platy inclusions of foreign minerals (usually troilite, schreibersite, or chromite) occuring in iron meteorites. { 'rī·kən,bäks lə'mel·ē }

Reichert-Meissl number [ANALY CHEM] An indicator of the measure of volatile soluble fatty acids. { 'rī·kərt 'mīs·əl ,rəm·bər }

Reichert's cartilage [EMBRYO] The cartilage of the hyoid arch in a human embryo. { 'rī·kərts ,kärt·lij }

Reich process [CHEM ENG] Process to purify carbon dioxide produced during fermentation; organic impurities in the gas are oxidized and absorbed, then the gas is dehydrated. { 'rīk ,prä·səs }

Reid equation [PETRO ENG] Relation of gas-well flow rate to pitot-tube readings for various impact pressures. { 'rēd i,kwā·zhən }

Reid vapor pressure [ENG] A measure in a test bomb of the vapor pressure in pounds pressure of a sample of gasoline at 100°F (37.8°C). { 'rēd 'vā·pər ,presh·ər }

Reighardiidae [INV ZOO] A monotypic family of arthropods in the order Cephalobaenida; the posterior end of the organism is rounded, without lobes, and the cuticula is covered with minute spines. { ,rī·gär'dī·ə,dē }

reimbursed time [COMPUT SCI] The machine time which is loaned or rented to another office, agency, or organization, either on a reimbursable or reciprocal basis. { 'rē·əm,bərst 'tīm }

Reimer-Tiemann reaction [ORG CHEM] Formation of phenolic aldehydes by reaction of phenol with chloroform in the presence of an alkali. { 'rīm·ər ¦tē·mən rē,ak·shən }

Reinartz crystal oscillator [ELECTR] Crystal-controlled vacuum-tube oscillator in which the crystal current is kept low by placing a resonant circuit in the cathode lead tuned to half the crystal frequency; the resulting regeneration at the crystal frequency improves efficiency without the danger of uncontrollable oscillation at other frequencies. { 'rīn,ärts 'krist·əl 'äs·ə,lād·ər }

REINDEER

Reindeer do not display sexual dimorphism: males are larger than females.

reindeer [VERT ZOO] *Rangifer tarandus.* A migratory ruminant of the deer family (Cervidae) which inhabits the Arctic region and has a circumpolar distribution; characteristically, both sexes have antlers and are brown with yellow-white areas on the neck and chest. { 'rān,dir }

Reinecke's salt [ANALY CHEM] [(NH$_3$)$_2$Cr(SCN)$_4$]NH$_4$· H$_2$O A reagent to detect mercury (gives a red color or a precipitate), and to isolate organic bases (such as proline or histidine). { 'rīn·ə·kēz ,sȯlt }

reinfection [MED] A second infection after recovery from an earlier infection with the same kind of organism. { ,rē·ən'fek·shən }

reinforced beam [CIV ENG] A concrete beam provided with steel bars for longitudinal tension reinforcement and sometimes compression reinforcement and reinforcement against diagonal tension. { ¦rē·ən'fȯrst 'bēm }

reinforced brickwork [CIV ENG] Brickwork strengthened by expanded metal, steel-wire mesh, hoop iron, or thin rods embedded in the bed joints. { ¦rē·ən'fȯrst 'brik,wərk }

reinforced column [CIV ENG] **1.** A long concrete column reinforced with longitudinal bars with ties or circular spirals. **2.** A composite column. **3.** A combination column. { ¦rē·ən'fȯrst 'käl·əm }

reinforced concrete [CIV ENG] Concrete containing reinforcing steel rods or wire mesh. { ¦rē·ən'fȯrst 'kän,krēt }

reinforced molding compound [MATER] A compound containing polymer or resin and a reinforcing filler, supplied in the form of ready-to-use material as distinguished from premix. { ¦rē·ən'fȯrst 'mōld·iŋ ,käm,paůnd }

reinforced plastic [MATER] High-strength filled plastic product used for mechanical, construction, and electrical products, automotive components, and ablative coatings; filling can be whiskers of glass, metal, boron, or other materials. { ¦rē·ən'fȯrst 'plas·tik }

reinforcement [CIV ENG] Strengthening concrete, plaster, or mortar by embedding steel rods or wire mesh in it. [MATER] A strong inert material bonded to a plastic to enhance its strength, stiffness, and resistance to impact. { ,rē·ən'fȯrs·mənt }

reinforcement conditioning See operant conditioning. { ,rē·ən'fȯrs·mənt kən'dish·ən·iŋ }

reinforcement of weld [MET] Weld metal that extends beyond the surface or plane of the weld joint. { ,rē·ən'fȯrs·mənt əv 'weld }

reinforcing bars [CIV ENG] Steel rods that are embedded in building materials such as concrete for reinforcement. { ¦rē·ən'fȯrs·iŋ ,bärz }

reinitialize [COMPUT SCI] To return a computer program to the condition it was in at the start of processing, so that nothing remains from previous executions of the program. { ¦rē·i'nish·əl,īz }

Reinsch test [ANALY CHEM] A test for detecting small amounts of arsenic, silver, bismuth, and mercury. { 'rīnsh ,test }

reinserter See direct-current restorer. { ¦rē·ən'sərd·ər }

reinsertion of carrier [ELECTR] Combining a locally generated carrier signal in a receiver with an incoming signal of the suppressed carrier type. { ¦rē·ən'sər·shən əv 'kar·ē·ər }

reishi mushroom See Ganoderma lucidum. { 'rā·ē·shē ,məsh,rüm }

Reissner-Nordstrom solution [RELAT] The unique solution of general relativity theory describing a nonrotating, charged black hole. { 'rīs·nər 'nȯrd·strəm sə,lü·shən }

Reissner's membrane [ANAT] The anterior wall of the cochlear duct, which separates the cochlear duct from the scala vestibuli. Also known as vestibular membrane of Reissner. { 'rīs·nərz ,mem,brān }

Reiter's syndrome [MED] The triad of idiopathic nongonococcal urethritis, conjunctivitis, and subacute or chronic polyarthritis. Also known as arthritis urethritica; idiopathic arthritis; infectious uroarthritis. { 'rīd·ərz ,sin,drōm }

rejected recharge [HYD] Water that infiltrates to the water table but then discharges because the aquifer is full and cannot accept it. { ri'jek·təd 'rē,chärj }

rejection [IMMUNOL] Destruction of a graft by the immune system of the recipient. { ri'jek·shən }

rejection band Also known as stop band. [ELECTROMAG] The band of frequencies below the cutoff frequency in a uniconductor waveguide. [PHYS] A frequency band within which electrical or electromagnetic signals are reduced or eliminated. { ri'jek·shən ,band }

rejection number [IND ENG] A predetermined number of defective items in a batch which, if not exceeded, requires acceptance of the batch. { ri'jek·shən ,nəm·bər }

rejector See trap. { ri'jek·tər }

rejector circuit See band-stop filter. { ri'jek·tər ,sər·kət }

rejector impedance See dynamic impedance. { ri'jek·tər im,pēd·əns }

rejuvenate [GEOL] The act of stimulating a stream to renewed erosive activity either by tectonic uplift or a drop in sea level. { ri'jü·və,nāt }

rejuvenated fault scarp [GEOL] A fault scarp revived by renewed movement along an old fault line after partial dissection or erosion of the initial scarp. Also known as revived fault scarp. { ri'jü·və,nād·əd 'fȯlt ,skärp }

rejuvenated stream [HYD] A mature stream that has reverted to the behavior and forms of a more youthful stage due to rejuvenation, usually as a result of uplift. Also known as revived stream. { ri'jü·və,nād·əd 'strēm }

rejuvenated water [HYD] Water returned to the terrestrial water supply as a result of compaction and metamorphism. { ri'jü·və,nād·əd 'wȯd·ər }

rejuvenation [GEOL] The restoration of youthful features to fluvial landscapes; the renewal of youthful vigor to low-gradient streams is usually caused by regional upwarping of broad areas formerly at or near base level. [HYD] **1.** The stimulation of a stream to renew erosive activity. **2.** The renewal of youthful vigor in a mature stream. { ri,jü·və'nā·shən }

rejuvenation head See knickpoint. { ri,jü·və'nā·shən ,hed }

rel [ELECTROMAG] Unit of reluctance equal to 1 ampere-turn per magnetic line of force. { rel }

relapsing fever [MED] An acute infectious disease caused by various species of the spirochete *Borrelia,* characterized by episodes of fever which subside spontaneously and recur over a period of weeks. { ri'laps·iŋ ,fē·vər }

related angle [MATH] The acute angle at which trigonometric functions have the same absolute values as at a given angle outside the first quadrant. { ri'lād·əd ,aŋ·gəl }

relation [COMPUT SCI] A two-dimensional table in which data are arranged in a relational data structure. [MATH] A set of ordered pairs. Also known as correspondence. { ri'lā·shən }

relational algebraic language [COMPUT SCI] A low-level procedural language for carrying out fundamental algebraic operations on a database of relations. { ri'lā·shən·əl 'al·jə,brā·ik ,laŋ·gwij }

relational calculus language [COMPUT SCI] A higher-level nonprocedural language for operating on a data base of relations, containing statements that can be mapped to the fundamental algebraic operations on the database. { ri'lā·shən·əl 'kal·kyə·ləs ,laŋ·gwij }

relational capability [COMPUT SCI] Property of two or more data files that can be joined together for viewing, editing, or creation of reports. { ri'lā·shən·əl ,kāp·ə'bil·əd·ē }

relational database See relational system. { ri'lā·shən·əl 'dad·ə,bās }

relational data structure [COMPUT SCI] A type of data structure in which data are represented as tables in which no entry contains more than one value. { ri'lā·shən·əl 'dad·ə ,strək·chər }

relationally complete [COMPUT SCI] Property of a programming language that provides for the construction of all relations derivable from some set of base relations by the application of the primitive algebraic operations. { ri'lā·shən·əl·ē kəm'plēt }

relational operator [COMPUT SCI] An operator that indicates whether one quantity is equal to, greater than, or less than another. { ri'lā·shən·əl 'äp·ə,rād·ər }

relational spreadsheet [COMPUT SCI] A spreadsheet whose data are stored in a central database and are copied from the database into the spreadsheet when the spreadsheet is called up. { ri'lā·shən·əl 'spred,shēt }

relational system [COMPUT SCI] A database management system in which a relational data structure is used. Also known as relational database. { ri'lā·shən·əl 'sis·təm; }

relative [NAV] **1.** Related to a moving point; apparent, as relative wind, relative movement. **2.** Related to or measured from the heading, as relative bearing. { 'rel·əd·iv }

relative address [COMPUT SCI] The numerical difference between a desired address and a known reference address. { 'rel·əd·iv ə'dres }

relative age [GEOL] The geologic age of a fossil organism, rock, or geologic feature or event defined relative to other organisms, rocks, or features or events rather than in terms of years. { 'rel·əd·iv 'āj }

relative atomic mass See atomic weight. { 'rel·əd·iv ə'täm·ik 'mas }

relative attenuation [ELECTR] The ratio of the peak output voltage of an electric filter to the voltage at the frequency being considered. { 'rel·əd·iv ə,ten·yə'wā·shən }

relative azimuth [NAV] Azimuth relative to heading of a craft. { 'rel·əd·iv 'az·ə·məth }

relative bandwidth [ELECTR] For an electric filter, the ratio of the bandwidth being considered to a specified reference bandwidth, such as the bandwidth between frequencies at which there is an attenuation of 3 decibels. { 'rel·əd·iv 'band,width }

relative bearing [NAV] Bearing relative to heading or to the craft; the horizontal direction of a terrestrial point from a craft expressed as the angular difference between the heading and the direction. { 'rel·əd·iv 'ber·iŋ }

relative biological effectiveness [NUCLEO] A factor used to compare the biological effectiveness of different types of ionizing radiation; it is the inverse ratio of the amount of absorbed radiation required to produce a given effect to a standard (or reference) radiation required to produce the same effect. Abbreviated RBE. { 'rel·əd·iv ¦bī·ə¦läj·ə·kəl i'fek·tiv·nəs }

relative byte address [COMPUT SCI] A relative address expressed as the number of bytes from a point of reference to the desired address. { 'rel·ə·tiv 'bīt ,ad,res }

relative chronology [GEOL] Geochronology in which the time order is based on superposition or fossil content rather than on an age expressed in years. { 'rel·əd·iv krə'näl·ə·jē }

relative coding [COMPUT SCI] A form of computer programming in which the address part of an instruction indicates not the desired address but the difference between the location of the instruction and the desired address. { 'rel·əd·iv 'kōd·iŋ }

relative compaction [ENG] The percentage ratio of the field density of soil to the maximum density as determined by standard compaction. { 'rel·əd·iv kəm'pak·shən }

relative contour See thickness line. { 'rel·əd·iv 'kän,tùr }

relative coordinates [MATH] Coordinates given as offsets from some point whose location can be adjusted. { 'rel·əd·iv kō'órd·ən·əts }

relative coordinate system [PHYS] Any coordinate system which is moving with respect to an inertial coordinate system. { 'rel·əd·iv kō'órd·ən·ət ,sis·təm }

relative current [OCEANOGR] The current which is a function of the dynamic slope of an isobaric surface and which is determined from an assumed layer of no motion. { 'rel·əd·iv 'kə·rənt }

relative damping ratio See damping ratio. { 'rel·əd·iv 'damp·iŋ ,rā·shō }

relative dating [GEOL] The proper chronological placement of a feature, object, or happening in the geologic time scale without reference to its absolute age. { 'rel·əd·iv 'dād·iŋ }

relative deflection See astrogeodetic deflection.

relative density See specific gravity. { 'rel·əd·iv 'den·səd·ē }

relative-density bottle See specific-gravity bottle. { 'rel·əd·iv ¦den·səd·ē ,bäd·əl }

relative dielectric constant See dielectric constant. { 'rel·əd·iv ¦dī·i'lek·trik 'kän·stənt }

relative direction [NAV] Horizontal direction expressed as angular distance from heading. { 'rel·əd·iv dī rek·shən }

relative distance [NAV] Distance relative to a specified reference point, usually one in motion. { 'rel·əd·iv 'dis·təns }

relative divergence See development index. { 'rel·əd·iv di'vər·jəns }

relative efficiency [STAT] **1.** Of an estimator, the comparative efficiency of the two estimators of the same parameter. **2.** For experimental design, the number of replications each design requires to reach the same precision. { 'rel·ə·tiv i'fish·ən·sē }

relative error [MATH] The absolute error in estimating a quantity divided by its true value. { 'rel·əd·iv 'er·ər }

relative erythrocytosis [MED] A form of erythrocytosis that occurs when the concentration of erythrocytes in the circulating blood increases through loss of plasma. { 'rel·əd·iv ə¦rith·rō,sī'tō·səs }

relative force [ENG] Ratio of the force of a test propellant to the force of a standard propellant, measured at the same initial temperature and loading density in the same closed chamber. { 'rel·əd·iv 'fòrs }

relative frequency [STAT] The ratio of the number of occurrences of a given type of event or the number of members of a population in a given class to the total number of events or the total number of members of the population. { 'rel·əd·iv frē·kwən·sē }

relative frequency distribution See percentage distribution. { ¦rel·ə·tiv 'frē·kwən·sē ,dis·trə,byü·shən }

relative frequency table See percentage distribution. { 'rel·ə·tiv 'frē·kwən·sē tā·bəl }

relative fugacity [PHYS CHEM] The ratio of the fugacity in a given state to the fugacity in a defined standard state. { 'rel·əd·iv fyü'gas·əd·ē }

relative gain [ELECTROMAG] The gain of an antenna in a given direction when the reference antenna is a half-wave, loss-free dipole isolated in space whose equatorial plane contains the given direction. { 'rel·əd·iv ¦gān }

relative gain array [CONT SYS] An analytical device used in process control multivariable applications, based on the comparison of single-loop control to multivariable control; expressed as an array (for all possible input-output pairs) of the ratios of a measure of the single-loop behavior between an input-output variable pair, to a related measure of the behavior of the same input-output pair under some idealization of multivariable control. { 'rel·əd·iv ¦gān ə,rā }

relative geologic time [GEOL] Nonabsolute geological time in which events may be placed relatively to one another. { 'rel·əd·iv ¦jē·ə,läj·ik 'tīm }

relative gravity instrument [ENG] Any device for measuring the differences in the gravity force or acceleration at two or more points. { 'rel·əd·iv 'grav·əd·ē ,in·strə·mənt }

relative humidity [METEOROL] The (dimensionless) ratio of the actual vapor pressure of the air to the saturation vapor pressure. Abbreviated RH. { 'rel·əd·iv hyü'mid·əd·ē }

relative hypsography See thickness pattern. { 'rel·əd·iv hip 'sig·rə·fē }

relative index of refraction [OPTICS] The ratio of the velocity of light in one medium to that in another medium. { 'rel·əd·iv 'in,deks əv ri'frak·shən }

relative interference effect [ENG ACOUS] Of a single-frequency electric wave in an electroacoustic system, the ratio, usually expressed in decibels, of the amplitude of a wave of specified reference frequency to that of the wave in question when the two waves are equal in interference effects. { 'rel·əd·iv ,in·tər'fir·əns i,fekt }

relative ionospheric opacity meter See riometer. { 'rel·əd·iv ī¦än·ə,sfir·ik ō'pas·əd·ē ,mēd·ər }

relative isohypse See thickness line. { 'rel·əd·iv 'ī·sə,hips }

relative luminosity factor See luminosity function. { 'rel·əd·iv ,lü·mə'näs·əd·ē ,fak·tər }

relatively closed set [MATH] A subset of a topological space is relatively closed if it is a closed set in some relative topology of a subset. { 'rel·ə,tiv·lē ¦klōzd ,set }

relatively compact set See conditionally compact set. { 'rel·ə,tiv·lē ¦käm,pakt ,set }

relatively open set [MATH] A subset of a topological space is relatively open if it is an open set in some relative topology of a subset. { 'rel·ə,tiv·lē ¦ō·pən ,set }

relatively prime [MATH] Integers m and n are relatively prime if there are integers p and q so that $pm + qn = 1$; equivalently, if they have no common factors other than 1. { 'rel·ə,tiv·lē 'prīm }

relative Mach number See Mach number. { 'rel·əd·iv 'mäk ,nəm·bər }

relative magnetometer [ENG] Any magnetometer which must be calibrated by measuring the intensity of a field whose strength is accurately determined by other means; opposed to absolute magnetometer. { 'rel·əd·iv ,mag·nə'täm·əd·ər }

relative maximum [MATH] A value of a function at a point x_0 which is equal to or greater than the values of the function at all points in some neighborhood of x_0. { 'rel·ə·tiv 'mak·sə·məm }

relative minimum [MATH] A value of a function at a point x_0 which is equal to or less than the values of the function at all points in some neighborhood of x_0. { 'rel·ə·tiv 'min· ə·məm }

relative molecular mass *See* molecular weight. { 'rel·əd·iv mə'lek·yə·lər 'mas }

relative momentum [MECH] The momentum of a body in a reference frame in which another specified body is fixed. { 'rel·əd·iv mə'men·təm }

relative motion [MECH] The continuous change of position of a body with respect to a second body or to a reference point that is fixed. Also known as apparent motion. { 'rel·əd·iv 'mō·shən }

relative movement line [NAV] A line connecting successive positions of a maneuvering craft relative to a reference craft. { 'rel·əd·iv ¦müv·mənt ¦līn }

relative orbit [ASTRON] The closed path described by the apparent position of the fainter member of a binary system relative to the brighter member. { 'rel·əd·iv 'ȯr·bət }

relative permeability [ELECTROMAG] The ratio of the permeability of a substance to the permeability of a vacuum at the same magnetic field strength. [GEOL] Specific permeability of a porous rock formation to a particular phase (oil, water, gas) at a particular saturation and a particular saturation distribution; for example, ratio of effective permeability to a specified phase to the rock's absolute permeability. { 'rel·əd· iv ¸pər·mē·ə'bil·əd·ē }

relative permittivity *See* dielectric constant. { 'rel·əd·iv ¸pər· mə'tiv·əd·ē }

relative plot [NAV] A plot of the successive positions of a craft relative to a reference point, which is usually in motion. { 'rel·əd·iv 'plät }

relative position [NAV] A point defined with reference to another position, either fixed or moving; the coordinates of such a point are usually bearing, true or relative, and distance from an identified reference point. { 'rel·əd·iv pə'zish·ən }

relative power gain [ELECTROMAG] Of one transmitting or receiving antenna over another, the measured ratio of the signal power one produces at the receiver input terminals to that produced by the other, the transmitting power level remaining fixed. { 'rel·əd·iv 'paů·ər ¸gān }

relative pressure response [ENG ACOUS] The amount, in decibels, by which the acoustic pressure induced by a projector under some specified condition exceeds the pressure induced under a reference condition. { 'rel·əd·iv ¦presh·ər ri¸späns }

relative primes [MATH] Two positive integers with no common positive divisor other than 1. { 'rel·ə·tiv 'prīmz }

relative quickness [ORD] Ratio of the quickness of a test propellant to the quickness of a standard propellant, measured at the same initial temperature and loading density in the same closed chamber. { 'rel·əd·iv 'kwik·nəs }

relative refractory period [NEUROSCI] A period of a few milliseconds following the absolute refractory period during which the excitation threshold of neural tissue is raised and a stronger-than-normal stimulus is required to initiate an action potential. { 'rel·əd·iv ri'frak·trē ¸pir·ē·əd }

relative relief *See* local relief. { 'rel·əd·iv ri'lēf }

relative resistance [ELEC] The ratio of the resistance of a piece of a material to the resistance of a piece of specified material, such as annealed copper, having the same dimensions and temperature. { 'rel·əd·iv ri'zis·təns }

relative response [ELECTR] In a transducer, the amount (in decibels) by which the response under some particular condition exceeds the response under a reference condition. { 'rel·əd· iv ri'späns }

relative roughness factor [FL MECH] Roughness of pipe-wall interior (distance from peaks to valleys) divided by pipe internal diameter; used to modify Reynolds number calculations for fluid flow through pipes. { 'rel·əd·iv ¦rəf·nəs ¸fak·tər }

relative scatter intensity [OPTICS] For scattering of radiation under any given set of physical conditions, the ratio of the radiant intensity scattered in any given direction to the radiant intensity scattered in the direction of the incident beam. { 'rel· əd·iv 'skad·ər in¸ten·səd·ē }

relative search [NAV] A search in which the area to be searched is defined relative to a point which is moving over the surface of the earth. { 'rel·əd·iv 'sərch }

relative sector search [NAV] A search of three legs, the first and third being on courses such as to produce relative movement along radial lines of equal length from a moving

point, and the middle leg connecting the other two. { 'rel·əd· iv 'sek·tər ¸sərch }

relative square search [NAV] Search by a series of course lines such as to produce relative movement lines of increasing length, each change of course being such as to produce a change of 90° in the direction of relative movement, all changes being in the same direction (right or left), so that the pattern of the search is an expanding square relative to a moving point. { 'rel·əd·iv 'skwer ¸sərch }

relative stability test [ANALY CHEM] A color test using methylene blue that indicates when the oxygen present in a sewage plant's effluent or polluted water is exhausted. { 'rel· əd·iv stə'bil·əd·ē ¸test }

relative-state interpretation *See* Everett-Wheeler interpretation. { ¦rel·əd·iv ¸stāt in¸tər·prə'tā·shən }

relative sunspot number [ASTRON] A measure of sunspot activity, computed from the formula $R = k(10g + f)$, where R is the relative sunspot number, f the number of individual spots, g the number of groups of spots, and k a factor that varies with the observer (his or her personal equation), the seeing, and the observatory (location and instrumentation). Also known as sunspot number; sunspot relative number; Wolf number; Wolf-Wolfer number; Zurich number. { 'rel·əd·iv 'sən¸spät ¸nəm·bər }

relative time [GEOL] Geologic time determined by the placing of events in a chronologic order of occurrence, especially time as determined by organic evolution or superposition. { 'rel·əd·iv 'tīm }

relative topography *See* thickness pattern. { 'rel·əd·iv tə'päg·rə·fē }

relative topology [MATH] In a topological space X any subset A has a topology on it relative to the given one by intersecting the open sets of X with A to obtain open sets in A. { 'rel·əd· iv tə'päl·ə·jē }

relative transmitting response [ENG ACOUS] In a sonar projector, the ratio of the transmitting response for a given bearing and frequency to the transmitting response for a specified bearing and frequency. { 'rel·əd·iv tranz'mid·iŋ ri¸späns }

relative triple precision [COMPUT SCI] The retention of three times as many digits of a quantity as the computer normally handles; for example, a computer whose basic word consists of 10 decimal digits is called upon to handle 30 decimal digit quantities. { 'rel·əd·iv 'trip·əl prə'sizh·ən }

relative vector [COMPUT SCI] In computer graphics, a vector whose end points are given in relative coordinates. { 'rel·əd· iv 'vek·tər }

relative velocity [MECH] The velocity of a body with respect to a second body; that is, its velocity in a reference frame where the second body is fixed. { 'rel·əd·iv və'läs·əd·ē }

relative volatility [CHEM] The volatility of a standard material whose relative volatility is by definition equal to unity. { 'rel·əd·iv ¸väl·ə'til·əd·ē }

relative wind [NAV] The speed and relative direction from which the wind appears to blow with reference to a moving point. Also known as apparent wind. { 'rel·əd·iv 'wind }

relativistic beam [RELAT] A beam of particles traveling at a speed comparable with the speed of light. { ¸rel·ə·tə'vis· tik 'bēm }

relativistic bremsstrahlung *See* gravitational bremsstrahlung. { ¸rel·ə·tə'vis·tik 'brem¸shträ·ləŋ }

relativistic electrodynamics [ELECTROMAG] The study of the interaction between charged particles and electric and magnetic fields when the velocities of the particles are comparable with that of light. { ¸rel·ə·tə'vis·tik i¦lek·trō·dī'nam·iks }

relativistic jet model [ASTRON] The accepted explanation of superluminal motion, wherein a feature moves from the nucleus of a quasar at relativistic speed and at a small angle to the line of sight to the earth. { ¸rel·ə·tə¸vis·tik 'jet ¸mäd·əl }

relativistic kinematics [RELAT] A description of the motion of particles compatible with the special theory of relativity, without reference to the causes of motion. { ¸rel·ə·tə'vis·tik ¸kin·ə'mad·iks }

relativistic mass [RELAT] The mass of a particle moving at a velocity exceeding about one-tenth the velocity of light; it is significantly larger than the rest mass. { ¸rel·ə·tə'vis·tik 'mas }

relativistic mechanics [RELAT] **1.** Any form of mechanics compatible with either the special or the general theory of relativity. **2.** The nonquantum mechanics of a system of particles or of a fluid interacting with an electromagnetic field, in the case when some of the velocities are comparable with the speed of light. { ,rel·ə·tə'vis·tik mi'kan·iks }

relativistic particle [RELAT] A particle moving at a speed comparable with the speed of light. { ,rel·ə·tə'vis·tik 'pärd·ə·kəl }

relativistic quantum theory [QUANT MECH] The quantum theory of particles which is consistent with the special theory of relativity, and thus can describe particles moving arbitrarily close to the speed of light. { ,rel·ə·tə'vis·tik 'kwänt·əm ,thē·ə·rē }

relativistic theory [PHYS] Any theory which is consistent with the special or general theory of relativity. { ,rel·ə·tə'vis·tik 'thē·ə·rē }

relativistic velocity [RELAT] A velocity comparable to the speed of light. { ,rel·ə·tə'vis·tik və'läs·əd·ē }

relativity [PHYS] Theory of physics which recognizes the universal character of the propagation speed of light and the consequent dependence of space, time, and other mechanical measurements on the motion of the observer performing the measurements; it has two main divisions, the special theory and the general theory. { ,rel·ə'tiv·əd·ē }

relaxation [GEOL] In experimental structural geology, the diminution of applied stress with time, as the result of any of various creep processes. [MATH] *See* relaxation method. [MECH] **1.** Relief of stress in a strained material due to creep. **2.** The lessening of elastic resistance in an elastic medium under an applied stress resulting in permanent deformation. [PHYS] A process in which a physical system approaches a steady state after conditions affecting it have been suddenly changed, and in which the presence of dissipative agents prevents the system from overshooting and then oscillating about this state. { ,rē,lak'sā·shən }

relaxation circuit [ELECTR] Circuit arrangement, usually of vacuum tubes, reactances, and resistances, which has two states or conditions, one, both, or neither of which may be stable; the transient voltage produced by passing from one to the other, or the voltage in a state of rest, can be used in other circuits. { ,rē,lak'sā·shən ,sər·kət }

relaxation inverter [ELECTR] An inverter that uses a relaxation oscillator circuit to convert direct-current power to alternating-current. { ,rē,lak'sā·shən in,vərd·ər }

relaxation kinetics [PHYS CHEM] A branch of kinetics that studies chemical systems by disturbing their states of equilibrium and making observations as they return to equilibrium. { ,rē,lak'sā·shən ki,ned·iks }

relaxation method [MATH] A successive approximation method for solving systems of equations where the errors from an initial approximation are viewed as constraints to be minimized or relaxed within a toleration limit. Also known as relaxation. { ,rē,lak'sā·shən ,meth·əd }

relaxation oscillations [PHYS] Oscillations having a sawtooth waveform in which the displacement increases to a certain value and then drops back to zero, after which the cycle is repeated. { ,rē,lak'sā·shən ,äs·ə,lā·shənz }

relaxation oscillator [ELECTR] An oscillator whose fundamental frequency is determined by the time of charging or discharging a capacitor or coil through a resistor, producing waveforms that may be rectangular or sawtooth. { ,rē,lak'sā·shən ,äs·ə,lād·ər }

relaxation test [ENG] A creep test in which the decrease of stress with time is measured while the total strain (elastic and plastic) is maintained constant. { ,rē,lak'sā·shən ,test }

relaxation time [PHYS] For many physical systems undergoing relaxation, a time τ such that the displacement of a quantity from its equilibrium value at any instant of time t is the exponential of $-t/\tau$. [SOLID STATE] The travel time of an electron in a metal before it is scattered and loses its momentum. { ,rē,lak'sā·shən ,tīm }

relaxed circular deoxyribonucleic acid [CELL MOL] A form of circular deoxyribonucleic acid in which the circle of one strand is broken. Also known as open-circle deoxyribonucleic acid. { ri'lakst 'sər·kyə·lər dē¸äk·sē¸rī·bō·nü'klē·ik 'as·əd }

relaxed peak process *See* deep inelastic collision. { ri'lakst ¦pēk ,prä·səs }

relaxin [BIOCHEM] A hormone found in the serum of humans and certain other animals during pregnancy; probably acting

with progesterone and estrogen, it causes relaxation of pelvic ligaments in the guinea pig. { ri'lak·sən }

relay [COMMUN] A microwave or other radio system used for passing a signal from one radio communication link to another. [ELEC] A device that is operated by a variation in the conditions in one electric circuit and serves to make or break one or more connections in the same or another electric circuit. Also known as electric relay. { 'rē,lā }

relay center [COMMUN] A switching center in which messages are automatically routed according to data contained in the messages or message headers. { 'rē,lā ,sen·tər }

relay contact [ELEC] One of the pair of contacts that are closed or opened by the movement of the armature of a relay. { 'rē,lā ,kän,takt }

relay control system [CONT SYS] A control system in which the error signal must reach a certain value before the controller reacts to it, so that the control action is discontinuous in amplitude. { 'rē,lā kən'trōl ,sis·təm }

relay haulage [MIN ENG] Single-track, high-speed mine haulage from one relay station to another. Also known as intermediate haulage. { 'rē,lā ,hȯl·ij }

relay intercropping [AGR] A form of intercropping in which two or more crops grow simultaneously during part of the life cycle of each; that is, a second crop is planted before the first crop matures. { 'rē,lā 'in·tər,kräp·iŋ }

relay rack [DES ENG] A standardized steel rack designed to hold 19-inch (48.25-centimeter) panels of various heights, on which are mounted radio receivers, amplifiers, and other units of electronic equipment. Also known as rack. { 'rē,lā ,rak }

relay satellite *See* communications satellite. { 'rē,lā ,sad·əl,īt }

relay selector [ELEC] Relay circuit associated with a selector, consisting of a magnetic impulse counter, for registering digits and holding a circuit. { 'rē,lā si,lek·tər }

relay station *See* repeater station. { 'rē,lā ,stā·shən }

relay system [COMMUN] *See* radio relay system. [ELEC] Dial-switching equipment that does not use mechanical switches, but is made up principally of relays. { 'rē,lā ,sis·təm }

release [MECH ENG] A mechanical arrangement of parts for holding or freeing a device or mechanism as required. { ri'lēs }

release adiabat [MECH] A curve or locus of points which defines the succession of states through which a mass that has been shocked to a high-pressure state passes while monotonically returning to zero pressure. { ri'lēs 'ad·ē·ə,bat }

release agent [MATER] A lubricant, such as wax or silicone oil, used to coat a mold cavity to prevent the molded piece from sticking when removed. Also known as mold release; parting agent. { ri'lēs ,ā·jənt }

release altitude [AERO ENG] Altitude of an aircraft above the ground at the time of release of bombs, rockets, missiles, tow targets, and so forth. { ri'lēs ,al·tə,tüd }

released mineral [MINERAL] A mineral formed during the crystallization of a magma due to failure of an earlier phase to react with the liquid portion of the magma. { ri'lēst 'min·rəl }

release factor [MOL BIO] Any protein that responds to termination codons in messenger ribonucleic acid and causes the release of the finished polypeptide. { ri'lēs ,fak·tər }

release fracture [GEOL] A fracture formed as a result of a decrease in the maximum principal stress. { ri'lēs ,frak·chər }

release joint *See* sheeting structure. { ri'lēs ,jȯint }

releaser stimulus [ZOO] A stimulus which affects an animal by initiating an instinctual behavior pattern. { ri'lēs·ər ,stim·yə·ləs }

reliability [ENG] The probability that a component part, equipment, or system will satisfactorily perform its intended function under given circumstances, such as environmental conditions, limitations as to operating time, and frequency and thoroughness of maintenance for a specified period of time. [STAT] **1.** The amount of credence placed in a result. **2.** The precision of a measurement, as measured by the variance of repeated measurements of the same object. { ri,lī·ə'bil·əd·ē }

relic [GEOL] **1.** A landform that remains intact after decay or disintegration or that remains after the disappearance of the major portion of its substance. **2.** A vestige of a particle in a sedimentary rock, such as a trace of a fossil fragment. { 'rel·ik }

RELAY

A mulitcontrol telephone-type auxiliary relay. When appled current or voltage in electric coil exceeds a threshold value, the coil activates an armature which opens or closes contacts. (*C. P. Clare and Co.*)

relict [BIOL] A persistent, isolated remnant of a once-abundant species. [GEOL] **1.** Referring to a topographic feature that remains after other parts of the feature have been removed or have disappeared. **2.** Pertaining to a mineral, structure, or feature of a rock which represents features of an earlier rock and which persists in spite of processes tending to destroy it, such as metamorphism. { 'rel·ikt }

relict dike [GEOL] In a granitized mass, a tabular, crystalloblastic body that represents a dike which was emplaced prior to, and which was relatively resistant to, the granitization process. { 'rel·ikt 'dīk }

relict glacier [HYD] A remnant of an older and larger glacier. { 'rel·ikt 'glā·shər }

reliction [HYD] The slow and gradual withdrawal or recession of the water in a sea, a lake, or a stream, leaving the former bottom as permanently exposed and uncovered dry land. { rə'lik·shən }

relict lake [HYD] A lake that survives in an area formerly covered by the sea or a larger lake, or a lake that represents a remnant resulting from a partial extinction of the original body of water. { 'rel·ikt 'lāk }

relict mineral [MINERAL] A mineral of a rock that persists from an earlier rock. { 'rel·ikt 'min·rəl }

relict permafrost [GEOL] Permafrost formed in the past which persists in areas where it would not form today. { 'rel·ikt 'pər·mə,frȯst }

relict sediment [GEOL] A sediment which was in equilibrium with its environment when first deposited but which is unrelated to its present environment even though it is not buried by later sediments, such as a shallow-marine sediment on the deep ocean floor. { 'rel·ikt 'sed·ə·mənt }

relict soil [GEOL] A soil formed on a preexisting landscape but not subsequently buried under younger sediments. { 'rel·ikt 'sȯil }

relict texture [GEOL] In mineral deposits, an original texture that persists after partial replacement. { 'rel·ikt 'teks·chər }

relief [CRYSTAL] The apparent topography exhibited by minerals in thin section as a consequence of refractive index. [GEOD] The configuration of a part of the earth's surface, with reference to altitude and slope variations and to irregularities of the land surface. [MECH ENG] **1.** A passage made by cutting away one side of a tailstock center so that the facing or parting tool may be advanced to or almost to the center of the work. **2.** Clearance provided around the cutting edge by removal of tool material. { ri'lēf }

relief angle [MECH ENG] The angle between a relieved surface and a tangential plane at a cutting edge. { ri'lēf ,aŋ·gəl }

relief frame [MECH ENG] A frame placed between the slide valve of a steam engine and the steam chest cover; reduces pressure on the valve and thereby reduces friction. { ri'lēf ,frām }

relief hole [ENG] Any of the holes fired after the cut holes and before the lifter holes in breaking ground for tunneling or shaft sinking. { ri'lēf ,hōl }

relief limonite [MINERAL] Indigenous limonite that is porous and cavernous in texture. { ri'lēf 'līm·ə,nīt }

relief map [MAP] A map of an area showing the topographic relief. { ri'lēf ,map }

relief model [MAP] A three-dimensional relief map. { ri'lēf ,mäd·əl }

relief press See letterpress. { ri'lēf ,pres }

relief printing [GRAPHICS] A method of printing in which the type or other images stand above the printing surface; even though the lower areas may have ink in them, they do not print because the paper does not contact them. { ri'lēf ,print·iŋ }

relief shading See hill shading. { ri'lēf ,shād·iŋ }

relief valve See pressure-relief valve. { ri'lēf ,valv }

relief well [CIV ENG] A well that drains a pervious stratum, to relieve waterlogging at the surface. [PETRO ENG] A directional well which is drilled to intersect a well that is blowing out, and down which heavy drilling fluid is pumped to kill the blow-out well. { ri'lēf ,wel }

relieving [MECH ENG] Treating an embossed metal surface with an abrasive to reveal the base-metal color on the elevations or highlights of the surface. { ri'lēv·iŋ }

relieving anode [ELECTR] Of a pool-cathode tube, an auxiliary anode which provides an alternative conducting path for reducing the current to another electrode. { ri'lēv·iŋ ,an,ōd }

relieving arch See discharging arch. { ri'lēv·iŋ ,ärch }

relieving platform [CIV ENG] A deck on the land side of a retaining wall to transfer loads vertically down to the wall. { ri'lēv·iŋ ,plat,fȯrm }

relighter flame safety lamp [MIN ENG] A locked spirit-burning lamp fitted with an internal relighting device. { rē'līd·ər ,flām 'sāf·tē ,lamp }

reline [ORD] To replace a worn liner of a gun to give it the ballistics of a new weapon. { rē'līn }

relish [ENG] The shoulder of a tenon, used in a mortise and tenon system. { 'rel·ish }

Relizean stage [GEOL] A subdivision of the Miocene in the California-Oregon-Washington area. { rə'lē·zē·ən ,stāj }

reloadable control storage [COMPUT SCI] The control storage made up of unit control words necessary for channel multiplexing. { rē'lōd·ə·bəl kən'trōl ,stȯr·ij }

relocatable code [COMPUT SCI] A code generated by an assembler or compiler, and in which all memory references needing relocation are either specially marked or relative to the current program-counter reading. { ¦rē·lō¦kād·ə·bəl 'kōd }

relocatable emulator [COMPUT SCI] An emulator which does not require a stand-alone machine but executes in a multi-programming environment. { ¦rē·lō¦kād·ə·bəl 'em·yə,lād·ər }

relocatable program [COMPUT SCI] A program coded in such a way that it may be located and executed in any part of memory. { ¦rē·lō¦kād·ə·bəl 'prō,gram }

relocate [COMPUT SCI] To establish or change the location of a program routine while adjusting or modifying the address references within the instructions to correctly indicate the new locations. { rē'lō,kāt }

relocating loader [COMPUT SCI] A loader in which some of the addresses in the program to be loaded are expressed relative to the start of the program rather than in absolute form. { ¦rē·lō¦kād·iŋ 'lōd·ər }

relocation hardware [COMPUT SCI] Equipment in a multi-programming system which allows a computer program to be run in any available space in memory. { ,rē·lō'kā·shən ,härd,wer }

relocation register [COMPUT SCI] A hardware element that holds a constant to be added to the address of each memory location in a computer program running in a multiprogramming system, as determined by the location of the area in memory assigned to the program. { ,rē·lō'kā·shən ,rej·ə·stər }

relogging [FOR] An operation in which small trees are salvaged, often for pulpwood, after the large trees are logged. { rē'läg·iŋ }

reluctance [ELECTROMAG] A measure of the opposition presented to magnetic flux in a magnetic circuit, analogous to resistance in an electric circuit; it is equal to magnetomotive force divided by magnetic flux. Also known as magnetic reluctance. { ri'lək·təns }

reluctance microphone See magnetic microphone. { ri'lək·təns ,mī·krə,fōn }

reluctance motor [ELEC] A synchronous motor, similar in construction to an induction motor, in which the member carrying the secondary circuit has salient poles but no direct-current excitation; it starts as an induction motor but operates normally at synchronous speed. { ri'lək·təns ,mōd·ər }

reluctance pickup See variable-reluctance pickup. { ri'lək·təns ,pik,əp }

reluctance pressure transducer [ENG] Pressure-measurement transducer in which pressure changes activate equivalent magnetic-property changes. { ri'lək·təns 'presh·ər tranz,dü·sər }

reluctivity [PHYS] The reciprocal of magnetic permeability; the reluctivity of empty space is unity. Also known as magnetic reluctivity; specific reluctance. { ,rē,lək'tiv·əd·ē }

rem [NUCLEO] A unit of ionizing radiation, equal to the amount that produces the same damage to humans as 1 roentgen of high-voltage x-rays. Derived from roentgen equivalent man. { rem }

remainder [MATH] **1.** The remaining integer when a division of an integer by another is performed; if $l = m \cdot p + r$, where l, m, p, and r are integers and r is less than p, then r is the remainder when l is divided by p. **2.** The remaining polynomial when division of a polynomial is performed; if $l = m \cdot p + r$, where l, m, p, and r are polynomials, and the degree of r is less than that of p, then r is the remainder when l is divided by p. **3.** The remaining part of a convergent infinite series after a computation, for some *n,* of the first *n* terms. { ri'mān·dər }

remainder formula [MATH] A formula by which the remainder resulting from an approximation of a function by a partial sum of a power series can be computed or analyzed. { ri'mān·dər ‚för·myə·lə }

remainder theorem [MATH] Dividing a polynomial $p(x)$ by $(x - a)$ gives a remainder equaling the number $p(a)$. { ri'mān·dər ‚thir·əm }

remaining velocity [MECH] Speed of a projectile at any point along its path of fire. { ri'mān·iŋ və'läs·əd·ē }

Remak's ganglion [NEUROSCI] A ganglion near the junction of the coronary sinus and the right atrium. { 'rā‚mäks ‚gaŋ·glē‚än }

remanence [ELECTROMAG] The magnetic flux density that remains in a magnetic circuit after the removal of an applied magnetomotive force; if the magnetic circuit has an air gap, the remanence will be less than the residual flux density. { 'rem·ə·nəns }

remanent magnetization [GEOPHYS] That component of a rock's magnetization whose direction is fixed relative to the rock and which is independent of moderate, applied magnetic fields. { 'rem·ə·nənt ‚mag·nə·tə'zā·shən }

remedial maintenance *See* corrective maintenance. { ri'mēd·ē·əl 'mānt·ən·əns }

remedial operation [CHEM ENG] In a chemical process operation, the revision of operating conditions so as to correct the overall operation and bring the product into desired rate or specification limits. Also known as corrective operation. { ri'mēd·ē·əl ‚äp·ə'rā·shən }

remember condition [ELECTR] Condition of a flip-flop circuit in which no change takes place between a given internal state and the next state. { ri'mem·bər kən‚dish·ən }

remex *See* flight feather. { 'rē‚meks }

remiges [VERT ZOO] Wing feathers of a bird. { 'rem·ə‚jēz }

remineralization [PHYSIO] The continual reforming of tooth mineral that occurs at the surface of teeth, chiefly from constituents of saliva. { rē‚min·rə·lə'zā·shən }

remodulator [ELECTR] A circuit that converts amplitude modulation to audio frequency-shift modulation for transmission of facsimile signals over a voice-frequency radio channel. Also known as converter. { rē'mäj·ə‚lād·ər }

remoistening adhesive [MATER] Any adhesive material, such as dextrin, animal glue, or gum arabic, which is reactivated with the application of water upon the adhesive surface. { rē'mòis·ən·iŋ ad‚hē·ziv }

remolded soil [GEOL] Soil that has had its natural internal structure modified or disturbed by manipulation so that it lacks shear strength and gains compressibility. { rē'mōl·dəd 'sòil }

remolding index [GEOL] The ratio of the modulus of deformation of a soil in the undisturbed state to that of a soil in the remolded state. { rē'mōld·iŋ ‚in‚deks }

remote access [COMPUT SCI] Ability to gain entry to a computer system from a location some distance away. { ri'mōt 'ak‚ses }

remote-access admittance [CONT SYS] A special piece of hardware, with built-in sensors and actuators, that is used by a robot to carry out the last stages of assembling several parts into a piece of equipment. { ri'mōt ‚ak‚ses ad'mit·əns }

remote batch computing [COMPUT SCI] The running of programs, usually during nonprime hours, or whenever the demands of real-time or time-sharing computing slacken sufficiently to allow less pressing programs to be run. { ri'mōt 'bach kəm‚pyüd·iŋ }

remote batch processing [COMPUT SCI] Batch processing in which an input device is located at a distance from the main installation and has access to a computer through a communication link. { ri'mōt 'bach ‚prä‚ses·iŋ }

remote calculator [COMPUT SCI] A keyboard device that can be connected to the central processing unit of a distant computer over an ordinary telephone channel, enabling the user to present programs to the computer. { ri'mōt 'kal·kyə‚lād·ər }

remote communications software [COMPUT SCI] Software that allows a microcomputer to control or duplicate the operation of another microcomputer at a distant location, using the standard telephone system. { ri'mōt kə‚myü·nə'kā·shənz }

remote-center compliance [MECH ENG] A compliant device that allows a part that is gripped by a robot or other automatic machinery to rotate about the tip of the robot end effector or to translate without rotation when it is pushed, thereby easing the mechanical assembly of parts. { ri'mōt ‚sen·tər kəm'plī·əns }

remote computing system [COMPUT SCI] A data-processing system that has terminals distant from the central processing unit, from which users can communicate with the central processing unit and compile, debug, test, and execute programs. { ri'mōt kəm'pyüd·iŋ ‚sis·təm }

remote computing system exchange [COMPUT SCI] A device that handles communications between the central processing unit and remote consoles of a remote computing system, and enables several remote consoles to operate at the same time without interfering with each other. { ri'mōt kəm'pyüd·iŋ ‚sis·təm iks‚chānj }

remote computing system language [COMPUT SCI] A computer language used for communications between the central processing unit and remote consoles of a remote computer system, generally incorporating a procedure-oriented language such as FORTRAN, but also containing operating statements, such as instructions to debug or execute programs. { ri'mōt kəm'pyüd·iŋ ‚sis·təm ‚laŋ·gwij }

remote computing system log [COMPUT SCI] A record of the volumes of data transmitted and of the frequency of various types of events during the operation of remote consoles in a remote computing system. { ri'mōt kəm'pyüd·iŋ ‚sis·təm ‚läg }

remote console [COMPUT SCI] A terminal in a remote computing system that has facilities for communicating with, and exerting control over, the central processing unit, and which may have devices for reading and punching cards or paper tape, any of various types of display units, printers, and a keyboard device for direct communication with the central processing unit. { ri'mōt 'kän‚sōl }

remote control [CONT SYS] Control of a quantity which is separated by an appreciable distance from the controlling quantity; examples include master-slave manipulators, telemetering, telephone, and television. { ri'mōt kən'trōl }

remote-cutoff tube *See* variable-mu tube. { ri'mōt ‚kəd‚òf ‚tüb }

remote debugging [COMPUT SCI] **1.** The testing and correction of computer programs at a remote console of a remote computing system. **2.** *See* remote testing. { ri'mōt dē'bəg·iŋ }

remote gun control [ORD] Pointing a gun in azimuth and elevation by means of a remote control system, which automatically keeps the gun pointed according to the firing data. { ri'mōt 'gən kən‚trōl }

remote-indicating compass [NAV] A compass equipped with one or more indicators to repeat at a distance the readings of the master compass. { ri'mōt ‚in·də‚kād·iŋ ‚käm·pəs }

remote indicator [ELECTR] **1.** An indicator located at a distance from the data-gathering sensing element, with data being transmitted to the indicator mechanically, electrically over wires, or by means of light, radio, or sound waves. **2.** *See* repeater. { ri'mōt 'in·də kād·ər }

remote inquiry [COMPUT SCI] Interrogation of the content of an automatic data-processing equipment storage unit from a device remotely displaced from the storage unit site. { ri'mōt 'in‚kwə·rē }

remotely operated vehicle [OCEANOGR] A crewless submersible vehicle that is tethered to a vessel on the surface by a cable; it has a video camera, lights, thrusters that generally provide three-dimensional maneuverability, depth sensors, and a wide array of manipulative and acoustic devices, as well as special instrumentation to perform a variety of work tasks. Abbreviated ROV. { rə‚mōt·lē ‚äp·ə‚rād·əd 've·ə·kəl }

remotely piloted vehicle [AERO ENG] A robot aircraft, controlled over a two-wave radio link from a ground station or mother aircraft that can be hundreds of miles away; electronic guidance is generally supplemented by remote control television cameras feeding monitor receivers at the control station. Abbreviated RPV. { ri'mōt·lē ‚pī·ləd·əd 've·ə·kəl }

remote manipulation [ENG] Use of mechanical equipment controlled from a distance to handle materials, such as radioactive materials. Also known as teleoperation. { ri'mōt mə‚nip·yə'lā·shən }

remote manipulator [ENG] A mechanical, electromechanical, or hydromechanical device that enables a person, directly controlling the device through handles or switches, to perform manual operations while separated from the site of the work. Also known as manipulator; teleoperator. { ri'mōt mə'nip·yə,lād·ər }

remote metering *See* telemetering. { ri'mōt 'mēd·ə·riŋ }

remote pickup [COMMUN] Picking up a radio or television program at a remote location and transmitting it to the studio or transmitter over wire lines or a shortwave or microwave radio link. { ri'mōt 'pik,əp }

remote plan position indicator *See* plan position indicator repeater. { ri'mōt ¦plan pə¦zish·ən 'in·də,kād·ər }

remote sensing [ELEC] Sensing, by a power supply, of voltage directly at the load, so that variations in the load lead drop do not affect load regulation. [ENG] The gathering and recording of information without actual contact with the object or area being investigated. { ri'mōt 'sens·iŋ }

remote subscriber [COMMUN] Subscriber to a network that does not have direct access to the switching center, but has access to the circuit through a facility such as a base message center. { ri'mōt səb'skrīb·ər }

remote terminal [COMPUT SCI] A computer terminal which is located away from the central processing unit of a data-processing system, at a location convenient to a user of the system. { ri'mōt 'tər·mən·əl }

remote testing [COMPUT SCI] A method of testing and correcting computer programs; programmers do not go to the computer center but provide detailed instructions to be carried out by computer operators along with the programs and associated test data. Also known as remote debugging. { ri'mōt 'test·iŋ }

removable discontinuity [MATH] A point where a function is discontinuous, but it is possible to redefine the function at this point so that it will be continuous there. { ri'müv·ə·bəl ¦dis,känt·ən'ü·əd·ē }

removable medium [COMPUT SCI] A data storage medium, such as magnetic tape or floppy disk, that can be physically removed from the drive that reads and writes on it. { ri'müv·ə·bəl 'mē·dē·əm }

removable plugboard *See* detachable plugboard. { ri'müv·ə·bəl 'pləg,bȯrd }

removable prosthodontics [MED] A subdivision of prosthodontics that focuses on the replacement of missing teeth by dentures. { ri,müv·ə·bəl ,präs·thə'dän·tiks }

REM sleep *See* rapid-eye-movement sleep. { 'rem ,slēp }

REM statement [COMPUT SCI] A statement in a computer program that consists of remarks or comments that document the program, and contains no executable code. { 'rem ,stāt·mənt }

renaissance lace [TEXT] A type of lace in which woven tape motifs are joined by a variety of flat stitches. { ,ren·ə'säns 'lās }

renal artery [ANAT] A branch of the abdominal or ventral aorta supplying the kidneys in vertebrates. { 'rēn·əl 'ärd·ə·rē }

renal calculus [MED] A concretion in the kidney. { 'rēn·əl 'kal·kyə·ləs }

renal-cell carcinoma [MED] A malignant renal tumor composed principally of large, often hyalin, polygonal cells. Also known as clear-cell carcinoma; Grawitz's tumor; hypernephroma. { 'rēn·əl 'sel ,kärs·ən'ō·mə }

renal clearance [PHYSIO] The volume of blood plasma completely cleared of a particular substrate by the kidneys per unit time; a measure of kidney function. { 'rēn·əl klir·əns }

renal corpuscle [ANAT] The glomerulus together with its Bowman's capsule in the renal cortex. Also known as Malpighian corpuscle. { 'rēn·əl 'kȯr·pə·səl }

renal dwarfism [MED] Dwarfism due to any of various chronic diseases of the kidney in children. { 'rēn·əl 'dwȯr,fiz·əm }

renal failure [MED] Severe malfunction of the kidneys, producing uremia and the resulting constitutional symptoms. { 'rēn·əl 'fāl·yər }

renal papilla [ANAT] A fingerlike projection into the renal pelvis through which the collecting tubules discharge. { 'rēn·əl pə'pil·ə }

renal pyramid [ANAT] Any of the conical masses composing

the medullary substance of the kidney. Also known as Malpighian pyramid. { 'rēn·əl 'pir·ə·məd }

renal rickets [MED] The metabolic bone disease due to increased bone resorption resulting from the acidosis and secondary hyperparathyroidism of renal insufficiency. { 'rēn·əl 'rik·əts }

renal threshold [PHYSIO] A concentration of a substance within the blood which, when reached, causes the substance to appear in the urine. { 'rēn·əl 'thresh,hōld }

renal tubular acidosis [MED] Defective hydrogen-ion excretion in the renal tubules, resulting in hyperchloremic acidosis and inadequate acidification of the urine. { 'rēn·əl 'tüb·yə·lər ,as·ə'dō·səs }

renal tubule [ANAT] One of the glandular tubules which elaborate urine in the kidneys. { 'rēn·əl 'tü·byül }

renal vein [ANAT] A vein which returns blood from the kidney to the vena cava. { 'rēn·əl 'vān }

renaming rule [MATH] A transformation rule in the lambda calculus that allows conflicts of variables to be eliminated; it states that a bound variable x in a lambda expression M may be uniformly replaced by some other bound variable y, provided y does not occur in M. Also known as alpha rule. { rē'nām·iŋ ,rül }

renardite [MINERAL] $Pb(UO_2)_4(PO_4)_2(OH)_4·7H_2O$ A yellow mineral composed of hydrous basic lead uranyl phosphate. { rə'när,dīt }

renaturation [BIOCHEM] The process of restoring denatured proteins to their original condition. { rē,nach·ə'rā·shən }

rendering [GRAPHICS] Methods or techniques that are used to display realistic-looking three-dimensional images on a two-dimensional medium such as a computer display. { 'ren·dər·iŋ }

rendezvous [AERO ENG] **1.** The event of two or more objects meeting with zero relative velocity at a preconceived time and place. **2.** The point in space at which such an event takes place, or is to take place. { 'rän·də,vü }

Rendoll [GEOL] A suborder of the soil order Mollisol, formed in highly calcareous parent materials, mostly restricted to humid, temperate regions; the soil profile consists of a dark upper horizon grading to a pale lower horizon. { 'ren,däl }

Rendzina [GEOL] One of an intrazonal, calcimorphic group of soils characterized by a brown to black, friable surface horizon and a light-gray or yellow, soft underlying horizon; found under grasses or forests in humid to semiarid climates. { rent'sin·ə }

renette [INV ZOO] An excretory cell found in certain nematodes. { re'net }

renewable energy source [ENG] A form of energy that is constantly and rapidly renewed by natural processes such as solar, ocean wave, and wind energy. { ri,nü·ə·bəl 'en·ər·jē ,sȯrs }

renewable resources [CHEM ENG] Agricultural materials used as feedstocks for industrial processes. { ri'nü·ə·bəl ri'sȯr·səs }

reniform [SCI TECH] Bean- or kidney-shaped, as describing the structure of a crystal in which rounded masses occur at the ends of radiating crystals, or certain structures in animals and plants. { 'rēn·ə,fȯrm }

renin [BIOCHEM] A proteolytic enzyme produced in the afferent glomerular arteriole which reacts with the plasma component hypertensinogen to produce angiotensin II. { 'rēn·ən }

Renner-Teller effect [PHYS CHEM] The splitting, into two, of the potential function along the bending coordinate in degenerate electronic states of linear triatomic or polyatomic molecules. { 'ren·ər 'tel·ər i,fekt }

rennet [VERT ZOO] The lining of the stomach of certain animals, especially the fourth stomach in ruminants. { 'ren·ət }

rennin [BIOCHEM] An enzyme found in the gastric juice of the fourth stomach of calfs; used for coagulating milk casein in cheesemaking. Also known as chymosin. { 'ren·ən }

Renninger effect [PHYS] A phenomenon observed in the analysis of thick crystals with x-rays or neutrons, in which a strong diffracted beam acts as a primary beam and can undergo further diffraction. { 'ren·iŋ·ər i,fekt }

renninogen [BIOCHEM] The zymogen of rennin. Also known as prorennin. { rə'nin·ə·jən }

Renn-Walz process [MET] A method of reclaiming iron and other metals from the waste materials produced in the smelting of zinc and lead ores; this material is brought up to 1000°C in

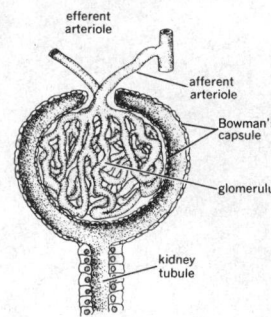

RENAL CORPUSCLE

efferent arteriole
afferent arteriole
Bowman's capsule
glomerulus
kidney tubule

Drawing of renal corpuscle.
(From C. K. Weichert, Anatomy of the Chordates, 4th ed., McGraw-Hill, 1970)

the preheating zone of the kiln by the countercurrent gases, and the oxidized metal vapors are carried off in the flue gases, from which they are subsequently filtered. { 'ren 'välts ‚prä·səs }

renormalizability [QUANT MECH] The property of some quantum field theories whereby all infinite quantities can be absorbed into a renormalization of physical parameters such as mass and charge. { rē‚nòr·mə‚līz·ə'bil·əd·ē }

renormalization [QUANT MECH] In certain quantum field theories, a procedure in which nonphysical bare values of certain quantities such as mass and charge are eliminated and the corresponding physically observable quantities are introduced. { rē‚nòr·mə·lə'zā·shən }

renormalization group methods [STAT MECH] Methods for treating the behavior of substances near critical points, in which the canonical ensemble is generalized by dividing a substance into cells of arbitrary size and forming an ensemble consisting of all microscopic configurations consistent with specified values of the thermodynamic variables in each of these cells. { rē‚nòr·mə·lə'zā·shən ‚grüp ‚meth·ədz }

renormalization transformation [MATH] A transformation of a mathematical function involving a change of scale. { rē‚nòr·mə·lə'zā·shən ‚tranz·fər‚mā·shən }

renovation [ORD] **1.** Process of restoring materiel to, or nearly to, its original condition by cleaning, painting, or similar methods. **2.** Restoration of ammunition to serviceable condition by operations more extensive or hazardous than routine ammunition maintenance; normally involves replacement of components. { ‚ren·ə'vā·shən }

rensselaerite [PETR] A soft, compact, fibrous talc pseudomorphous after pyroxene and found in Canada and northern New York. { 'ren·sə·lə‚rīt }

reorder cycle [IND ENG] The interval between successive reorder (procurement) actions. { re'òr·dər ‚sī·kəl }

reorder point [IND ENG] An arbitrary level of stock on hand plus stock due in, at or below which routine requisitions for replenishment purposes are submitted in accordance with established requisitioning schedules. { re'òr·dər ‚pòint }

Reoviridae [VIROL] A family of ribonucleic acid (RNA)–containing viruses characterized by an unenveloped icosahedral virion containing a genome of 10–12 segments (depending on the genus) of double-stranded RNA; hosts include vertebrates, invertebrates, and plants. { ‚rē·ə'vir·ə‚dī }

reovirus [VIROL] A group of ribonucleic acid-containing animal viruses, including agents of encephalitis and phlebotomus fever. { 'rē·ō‚vī·rəs }

rep [NUCLEO] A unit of ionizing radiation, equal to the amount that causes absorption of 93 ergs per gram of soft tissue. Derived from roentgen equivalent physical. Also known as parker; tissue roentgen. [TEXT] A type of fabric characterized by distinct, round, padded ribs running from selvage to selvage. { rep }

repair [ENG] To restore that which is unserviceable to a serviceable condition by replacement of parts, components, or assemblies. { ri'per }

repair cycle [ENG] The period that elapses from the time the item is removed in a reparable condition to the time it is returned to stock in a serviceable condition. { ri'per ‚sī·kəl }

repair dock [CIV ENG] A graving dock or floating dry dock built primarily for ship repair. { ri'per ‚däk }

repair forecast [ENG] The quantity of items estimated to be repaired or rebuilt for issue during a stated future period. { ri'per ‚fòr‚kast }

repair kit [ENG] A group of parts and tools, not all having the same basic name, used for repair or replacement of the worn or broken parts of an item; it may include instruction sheets and material, such as sandpaper, tape, cement, gaskets, and the like. { ri'per ‚kit }

repair parts list [ENG] List approved by designated authorities, indicating the total quantities of repair parts, tools, and equipment necessary for the maintenance of a specified number of end items for a definite period of time. { ri'per ‚pärts ‚list }

repair synthesis [MOL BIO] Enzymatic excision and replacement of regions of damaged deoxyribonucleic acid, as in repair of thymine dimers by ultraviolet irradiation. { ri'per ‚sin·thə·səs }

repand [BOT] Having a margin that undulates slightly, referring to a leaf. { rə'pand }

repeatability [CONT SYS] The ability of a robot to reposition itself at a location to which it is directed or at which it is commanded to stop. [NAV] In a navigation system, the measure of the accuracy with which the system permits the operator to return to a specific point as defined only in terms of the lattice peculiar to that system. { ri‚pēd·ə'bil·əd·ē }

repeat accuracy [CONT SYS] The variations in the actual position of a robot manipulator from one cycle to the next when the manipulator is commanded to repeatedly return to the same point or position. { ri'pēt 'ak·yə·rə·sē }

repeated load [MECH] A force applied repeatedly, causing variation in the magnitude and sometimes in the sense, of the internal forces. { ri'pēd·əd 'lōd }

repeated measurements model [STAT] A product model in which each factor is the same. { ri'pēd·əd ‚mezh·ər·məns ‚mäd·əl }

repeated reflection *See* multiple reflection. { ri'pēd·əd ri'flek·shən }

repeated root *See* multiple root. { ri'pēd·əd 'rüt }

repeated twinning [CRYSTAL] Crystal twinning that involves more than two simple crystals. { ri'pēd·əd 'twin·iŋ }

repeater [ELEC] *See* repeating coil. [ELECTR] **1.** An amplifier or other device that receives weak signals and delivers corresponding stronger signals with or without reshaping of waveforms; may be either a one-way or two-way repeater. Also known as regenerator. **2.** An indicator that shows the same information as is shown on a master indicator. Also known as remote indicator. { ri'pēd·ər }

repeater compass *See* compass repeater. { ri'pēd·ər ‚käm·pəs }

repeater jammer [ELECTR] A jammer that intercepts an enemy radar signal and reradiates the signal after modifying it to incorporate erroneous data on azimuth, range, or number of targets. { ri'pēd·ər ‚jam·ər }

repeater station [COMMUN] A station containing one or more repeaters. Also known as relay station. { ri'pēd·ər ‚stā·shən }

repeat glass [OPTICS] Used by textile and wallpaper designers, a device consisting of four lenses formed in one piece of glass; when a single drawing or pattern is viewed, the subject matter is repeated four times. { ri'pēt ‚glas }

repeating coil [ELEC] A transformer used to provide inductive coupling between two sections of a telephone line when a direct connection is undesirable. Also known as repeater. { ri'pēd·iŋ ‚kòil }

repeating-coil bridge cord [ELEC] In telephony, a method of connecting the common office battery to the cord circuits by connecting the battery to the midpoints of a repeating coil, bridged across the cord circuit. { ri‚pēd·iŋ ‚kòil 'brij ‚kòrd }

repeating decimal [MATH] A decimal that is either finite or infinite with a finite block of digits repeating indefinitely. Also known as periodic decimal; recurring decimal. { ri'pēd·iŋ 'des·məl }

repeating unit [ORG CHEM] The group of atoms that is derived from a monomer and repeats throughout a polymer. Also known as monomeric unit. { ri'pēd·iŋ ‚yü·nət }

repeat key [COMPUT SCI] A key on a typewriter or computer keyboard that, when depressed at the same time as a character key, causes repeated printing or generation of the character until one of the keys is released. { ri'pēt ‚kē }

repeat operator [COMPUT SCI] A pseudo instruction using two arguments, a count p and an increment n: the word immediately following the instruction is repeated p times, with the values $0, n, 2n, \ldots, (p-1)n$ added to the successive words. { ri'pēt ‚äp·ə‚rād·ər }

repellency [CHEM] Ability to repel water, or being hydrophobic; opposite to water wettability. { ri'pel·ən·sē }

repeller [ELECTR] An electrode whose primary function is to reverse the direction of an electron stream in an electron tube. Also known as reflector. { ri'pel·ər }

repent [BOT] Of a stem, creeping along the ground and rooting at the nodes. { 'rē·pent }

reperforation [PETRO ENG] Creation of new perforations (holes) in oil well tubing opposite to oil-bearing reservoir zones; creates more opportunity for fluid to drain from the formation into the wellbore. { rē‚pər·fə'rā·shən }

repetition [GEOL] The duplication of certain stratigraphic beds at the surface or in any specified section owing to disruption and displacement of the beds by faulting or intense folding. { ‚rep·ə'tish·ən }

repetition equivalent [COMMUN] In a complete telephone connection, a measure of the grade of transmission experienced by the subscribers using the connection; it includes the combined effects of volume, distortion, noise, and all other subscriber reactions and usages. { ,rep·ə'tish·ən i'kwiv·ə·lənt }

repetition frequency [COMMUN] *See* repetition rate. [CELL MOL] The number of copies of a given nucleotide sequence present in the haploid genome. { ,rep·ə'tish·ən ,frē·kwən·sē }

repetition instruction [COMPUT SCI] An instruction that causes one or more other instructions to be repeated a specified number of times, usually with systematic address modification occurring between repetitions. { ,rep·ə'tish·ən in,strək·shən }

repetition priming [PSYCH] The faster processing or easier identification of studied stimuli as compared with unstudied stimuli. { ,rep·ə,tish·ən 'prīm·iŋ }

repetition rate [COMMUN] The rate at which recurrent signals are produced or transmitted. Also known as recurrence rate; repetition frequency. { ,rep·ə'tish·ən ,rāt }

repetitious deoxyribonucleic acid [CELL MOL] Nucleotide sequences occurring repeatedly in chromosomal deoxyribonucleic acid. { rep·ə'tish·əs dē,äk·sē¦rī·bō·nü,klē·ik 'as·əd }

repetitive addressing [COMPUT SCI] A system used on some computers in which, under certain conditions, an instruction is written without giving the address of the operand, and the operand address is automatically that of the location addressed by the last previous instruction. { rə'ped·əd·iv ə'dres·iŋ }

repetitive analog computer [COMPUT SCI] An analog computer which repeatedly carries out the solution of a problem at a rapid rate (10 to 60 times a second) while an operator may vary parameters in the problem. { rə'ped·əd·iv 'an·ə,läg kəm'pyüd·ər }

repetitive element *See* regular element. { rə'ped·əd·iv 'el·ə·mənt }

repetitive statement [COMPUT SCI] A statement in a computer program that is repeatedly executed for a specified number of times or for as long as a specified condition holds true. { ri'ped·əd·iv 'stāt·mənt }

repetitive time method [IND ENG] A technique where the stopwatch is read and simultaneously returned to zero at each break point. Also known as snapback method. { ri'ped·əd·iv 'tīm ,meth·əd }

repetitive unit [COMPUT SCI] A type of circuit which appears more than once in a computer. { rə'ped·əd·iv yü·nət }

Repettian [GEOL] A North American stage of lower Pliocene geologic time, above the Delmontian and below the Venturian. { rə'pesh·ən }

repi [HYD] A lake, pond, or other standing water body associated with a sink or subsidence of land surface. { 'rep·ē }

replacement [GEOL] Growth of a new or chemically different mineral in the body of an old mineral by simultaneous capillary solution and deposition. [PALEON] Substitution of inorganic matter for the original organic constituents of an organism during fossilization. { ri'plās·mənt }

replacement bit *See* reset bit. { ri'plās·mənt ,bit }

replacement demand [ENG] A demand representing replacement of items consumed or worn out. { ri'plās·mənt di,mand }

replacement deposit [MINERAL] A mineral deposit formed by the in-position replacement of one mineral for another. { ri'plās·mənt di,päz·ət }

replacement dike [GEOL] A dike which is made by gradual transformation of wall rock by solutions along fractures or permeable zones. { ri'plās·mənt ,dīk }

replacement factor [ENG] The estimated percentage of equipment or repair parts in use that will require replacement during a given period. { ri'plās·mənt ,fak·tər }

replacement study [IND ENG] An economic analysis involving the comparison of an existing facility and a proposed replacement facility. { ri'plās·mənt ,stəd·ē }

replacement texture [GEOL] The texture exhibited where one mineral has replaced another. { ri'plās·mənt ,teks·chər }

replacement transfusion *See* exchange transfusion. { ri'plās·mənt tranz,fyü·zhən }

replacement vein [GEOL] A mineral vein formed by the gradual transformation of an original vein by secondary fluids. { ri'plās·mənt vān }

replenisher [ORD] A cylinder containing recoil oil and a spring-actuated piston, which allows for expansion of oil in the recoil system which has become heated during firing of the weapon; it also returns oil to the recoil system after firing, when the temperature (and volume) of the oil is down. { ri'plen·ish·ər }

replenishment [GEOL] The stage in development of a cavern in which the presence of air in the passages allows the deposition of speleothems. { ri'plen·ish·mənt }

replica [ENG] A thin plastic or inorganic film which is formed on a surface and then removed from it for study in an electron microscope. { 'rep·lə·kə }

replica grating [OPTICS] A diffraction grating made by flowing a plastic solution over an original grating, evaporating the solvent, and removing the resulting film, which has the lines of the original grating impressed on it. { 'rep·lə·kə ,grād·iŋ }

replica master [MECH ENG] A robotlike machine whose motions are duplicated by another robot when the machine is moved by a human operator. { 'rep·lə·kə ,mas·tər }

replica plating [MICROBIO] A method for the isolation of nutritional mutants in microorganisms; colonies are grown from a microorganism suspension previously exposed to a mutagenic agent, on a complete medium in a petri dish; a velour surface is used to transfer the impression of all these colonies to a petri dish containing a minimal medium; colonies that do not grow on the minimal medium are the mutants. { 'rep·lə·kə ,plād·iŋ }

replicating fork [CELL MOL] The Y-shaped region of a chromosome that is a growing point in the replication of deoxyribonucleic acid. { 'rep·lə,kād·iŋ ¦fork }

replication [ANALY CHEM] The formation of a faithful mold or replica of a solid that is thin enough for penetration by an electron microscope beam; can use plastic (such as collodion) or vacuum deposition (such as of carbon or metals) to make the mold. [CELL MOL] Duplication, as of a nucleic acid, by copying from a molecular template. [STAT] In experimental design, the repetition of an experiment or parts of an experiment to secure more data as an aid to determining the experimental error and to arrive at better estimates of the effects of various treatments with smaller standard errors. [VIROL] Multiplication of phage in a bacterial cell. { ,rep·lə'kā·shən }

replication bubble *See* replication eye. { ,rep·lə'kā·shən ,bəb·əl }

replication eye [CELL MOL] A replicated region of deoxyribonucleic acid contained within a longer, unreplicated region and presented in the shape of an eye. Also known as replication bubble. { ,rep·lə'kā·shən ī }

replicon [GEN] Each deoxyribonucleic segment that acts as a unit of replication. { 'rep·lə,kän }

replum [BOT] A thin wall separating the two valves or chambers of certain fruits. { 'rep·ləm }

reply [COMMUN] A radio-frequency signal or combination Of signals transmitted by a transponder in response to an interrogation. Also known as response. { ri'plī }

report [ACOUS] Sharp explosive sound, as of a shot, bursting bomb, or projectile. [COMPUT SCI] An output document prepared by a data-processing system. { ri'port }

reporter gene [GEN] A transfected gene that produces a signal, such as green fluorescence, when it is expressed; it is typically included in a larger cloned gene that is introduced into an organism to study its temporal and spatial pattern of expression. { ri'pord·ər ¦jēn }

report generator [COMPUT SCI] A routine which produces a complete data-processing report, given only a description of the desired content and format, plus certain information concerning the input file. Also known as report writer. { ri'port jen·ə,rād·ər }

reporting point [NAV] **1.** A specified point in relation to which a craft reports its position. **2.** In air operations, a geographical point established for use by air-traffic control in the movement and separation of aircraft. { ri'pord·iŋ ,point }

reporting time interval [COMMUN] The time for transmission of data or a report from the originating terminal to the end receiver. { ri'pord·iŋ 'tīm ,in·tər·vəl }

report of survey [ORD] An official report prepared on a standardized form, which records the circumstance concerning the loss, damage, or destruction of property, and which serves as, or supports, a voucher for dropping the property from property records. { ri'port əv 'sər,vā }

report program [COMPUT SCI] A program that prints out an analysis of a file of records, usually arranged by keys, each analysis or total being produced when a key change takes place. { ri'pȯrt ‚prō‚gram }

report program generator [COMPUT SCI] A nonprocedural programming language that provides a convenient method of producing a wide variety of reports. Abbreviated RPG. { ri'pȯrt ‚prō‚gram ‚jen·ə‚rād·ər }

report writer See report generator. { ri'pȯrt ‚rīd·ər }

repoussage [GRAPHICS] In etching, when a plate is repaired, scraped, and rubbed, there is a hollow left that will print a gray smudge; repoussage involves hammering out the hollow by putting the plate face down on a flat steel surface and tapping it gently from the back with a repoussage hammer. { rə‚pü ‚säzh }

repoussé [GRAPHICS] A method of producing a relief design on a thin sheet of metal or leather by hammering or otherwise working the reverse side. { rə‚pü'sā }

Reppe process [CHEM ENG] A family of high-pressure, catalytic acetylene-reaction processes yielding (depending upon what the acetylene reacts with) butadiene, allyl alcohol, acrylonitrile, vinyl ethers and derivatives, acrylic acid esters, cyclooctatraene, and resins. { 'rep·ə ‚prä·səs }

representation [MATH] A representation of a group is given by a homomorphism of it onto some group either of matrices or unitary operators of a Hilbert space. { ‚rep·ri‚zen'tā·shən }

representation condition [COMPUT SCI] The condition that, if one software entity is less than another entity in terms of a selected attribute, then any software metric for that attribute must associate a smaller number to the first entity than it does to the second entity. { ‚rep·rə·zen'tā·shən kən‚dish·ən }

representation theory [MATH] **1.** The study of groups by the use of their representations. **2.** The determination of representations of specific groups. [QUANT MECH] Quantum-mechanical device in which one selects the common eigenfunctions of a complete set of quantum-mechanical operators as a basis of vectors in a Hilbert space, and expresses wave functions and operators in terms of column matrices and square matrices, respectively, which correspond to this basis. { ‚rep·ri‚zen'tā·shən ‚thē·ə·rē }

representative calculating time [COMPUT SCI] The time required to perform a specified operation or series of operations. { ‚rep·ri‚zen·təd·iv 'kal·kyə‚lād·iŋ ‚tīm }

representative fraction See natural scale. { ‚rep·ri‚zen·təd·iv 'frak·shən }

representative sample [MIN ENG] In testing or valuation of a mineral deposit, a sample so large and average in composition as to be considered representative of a specified volume of the surrounding ore body. [STAT] A sample whose characteristics reflect those of the population from which it is drawn. { ‚rep·ri‚zen·təd·iv 'sam·pəl }

repressing [BIOCHEM] The termination of enzyme synthesis when the products of the reaction catalyzed by the enzyme reach a critical concentration. [CELL MOL] Inhibition of transcription or translation due to binding of a repressor to an operator on a deoxyribonucleic acid molecule or to a specific messenger ribonucleic acid site. [MET] Applying pressure to a pressed and sintered compact to improve some physical property. { ri'pres·iŋ }

repression [BIOCHEM] The termination of enzyme synthesis when the products of the reaction catalyzed by the enzyme reach a critical concentration. [CELL MOL] Inhibition of transcription or translation due to binding of a repressor to an operator on a deoxyribonucleic acid molecule or to a specific messenger ribonucleic acid site. [PSYCH] A defense mechanism whereby ideas, feelings, or desires, in conflict with the individual's conscious self-image or motives, are unconsciously dismissed from consciousness. { ri'presh·ən }

repressor [BIOCHEM] An end product of metabolism which represses the synthesis of enzymes in the metabolic pathway. [GEN] The product of a regulator gene that acts to repress the transcription of another gene. { ri'pres·ər }

repressuring [PETRO ENG] Forcing gas or water under pressure into an oil reservoir with the intention of increasing the recovery of crude oil. { rē'presh·ə‚riŋ }

repro See reproduction proof. { 'rē‚prō }

reprocessed wool [TEXT] Yarn manufactured from previously processed fibers derived from an unused wool product, such as cloth cuttings. { rē'prä‚sest 'wu̇l }

reproduce head See playback head. { rē·prə'düs ‚hed }

reproducer [COMPUT SCI] A punched-card machine that reads a punched card and duplicates part or all of its contents by punching another card. Also known as punched-card reproducer. { rē·prə'düs·ər }

reproducing stylus See stylus. { ‚rē·prə‚düs·iŋ ‚stī·ləs }

reproducing system See sound-reproducing system. { ‚rē·prə‚düs·iŋ ‚sis·təm }

reproducing unit [COMPUT SCI] An electromechanical device which will duplicate a deck of cards. { ‚rē·prə‚düs·iŋ ‚yü·nət }

reproduction [BIOL] The mechanisms by which organisms give rise to other organisms of the same kind. { ‚rē·prə·dək·shən }

reproduction proof [GRAPHICS] A page proof made with great care on smooth paper and used for photographic reproduction; it smudges easily and must be handled with care. Also known as repro; slick paper proof. { ‚rē·prə‚dək·shən ‚prüf }

reproduction speed [COMMUN] Area of copy recorded per unit time in facsimile transmission. { ‚rē·prə‚dək·shən ‚spēd }

reproductive behavior [ZOO] The behavior patterns in different types of animals by means of which the sperm is brought to the egg and the parental care of the resulting young insured. { ‚rē·prə‚dək·tiv bi'hā·vyər }

reproductive distribution [ECOL] The range of areas where conditions are favorable to maturation, spawning, and early development of marine animals. { ‚rē·prə‚dək·tiv ‚dis·rə'byü·shən }

reproductive system [ANAT] The structures concerned with the production of sex cells and perpetuation of the species. { ‚rē·prə‚dək·tiv ‚sis·təm }

reproductive technology [MED] Any procedure undertaken to aid in conception, intrauterine development, and birth when natural processes do not function normally. { ‚rē·prə‚dək·tiv tek'näl·ə·jē }

reproductive toxicity [MED] Adverse effects on the male and/or female reproductive systems caused by exposure to a toxic chemical. It may be expressed as alterations in sexual behavior, decreases in fertility, or fetal loss during pregnancy. Developmental toxicity may also be included. { ‚rē·prə‚dək·tiv täk'sis·əd·ē }

reprographics [GRAPHICS] The generation of graphic images by using an interactive network linking several mechanical, electronic, and electromechanical devices, such as typewriters, optical character readers, magnetic disks, and computers capable of graphical outputs. { ‚rē·prə'graf·iks }

Reptantia [INV ZOO] A suborder of the crustacean order Decapoda including all decapods other than shrimp. { rep'tan·chē·ə }

reptile [VERT ZOO] Any member of the class Reptilia. { 'rep‚tīl }

Reptilia [VERT ZOO] A class of terrestrial vertebrates composed of turtles, tuatara, lizards, snakes, and crocodileans; characteristically they lack hair, feathers, and mammary glands, the skin is covered with scales, they have a three-chambered heart, and the pleural and peritoneal cavities are continuous. { rep'til·yə }

repulsion [MECH] A force which tends to increase the distance between two bodies having like electric charges, or the force between atoms or molecules at very short distances which keeps them apart. Also known as repulsive force. { ri'pəl·shən }

repulsion-induction motor [ELEC] A repulsion motor that has a squirrel-cage winding in the rotor in addition to the repulsion-motor winding. { ri'pəl·shən in'dək·shən ‚mōd·ər }

repulsion motor [ELEC] An alternating-current motor having stator windings connected directly to the source of ac power and rotor windings connected to a commutator; brushes on the commutator are short-circuited and are positioned to produce the rotating magnetic field required for starting and running. { ri'pəl·shən ‚mōd·ər }

repulsion-start induction motor [ELEC] An alternating-current motor that starts as a repulsion motor; at a predetermined speed the commutator bars are short-circuited to give the equivalent of a squirrel-cage winding for operation as an induction motor with constant-speed characteristics. { ri'pəl·shən ‚stärt in'dək·shən ‚mōd·ər }

repulsive force See repulsion. { ri'pəl·siv 'fȯrs }

request/grant logic [COMPUT SCI] Logic circuitry which, in

REPULSION MOTOR

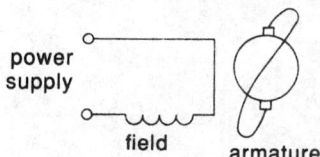

Schematic of a repulsion motor showing stator winding, which is connected to the ac power, and the rotor or armature winding, which is short-circuited through a commutator and brushes.

effect, selects the interrupt line with highest priority. { ri 'kwest 'grant ˌläj·ik }

request repeat system [COMMUN] System using an error-detecting code, and so arranged that a signal detected as being in error automatically initiates a request for retransmission. { ri'kwest riˌpēt ˌsis·təm }

required navigation performance [NAV] A method of regulating the performance of air navigation equipment, in which airspace requirements can be satisfied independently of the methods (that is, the sensors) by which they are achieved. Abbreviated RNP. { rēˌkwī·ord ˌnav·ə'gā·shən pərˌform·əns }

required thickness [DES ENG] The thickness calculated by recognized formulas for boiler or pressure vessel construction before corrosion allowance is added. { ri'kwīrd 'thik·nəs }

requirements engineering [SYS ENG] The process of identifying and articulating needs for a new technology and applications. { riˌkwīr·məns ˌen·jə'nir·iŋ }

reradiation [COMMUN] Undesirable radiation of signals generated locally in a radio receiver, causing interference or revealing the location of the receiver. { rē·rā·dē'ā·shən }

rerailer [ENG] A small, lightweight Y-shaped device, used to retrack railroad cars and locomotives; as the car is pulled across the device, the derailed wheels are channeled back onto the tracks. Also known as retracker. { rē'rāl·ər }

rerun [CHEM ENG] To distill a liquid material that has already been distilled; usually implies taking a large proportion of the charge stock overhead. [COMPUT SCI] To run a program or a portion of it again on a computer. Also known as rollback. { 'rēˌrən }

rerun point [COMPUT SCI] A location in a program from which the program may be started anew after an interruption of the computer run. { 'rēˌrən ˌpóint }

rerun routine [COMPUT SCI] A routine designed to be used in the wake of a computer malfunction or a coding or operating mistake to reconstitute a routine from the last previous rerun point. { 'rēˌrən ˌrü·tēn }

RES See reticuloendothelial system.

resaw [ENG] To cut lumber to boards of final thickness. { rē'só }

resbenzophenone See benzoresorcinol. { rezˌben'zäf·ə ˌnōn }

rescap [ELEC] A capacitor and resistor assembly manufactured as a packaged encapsulated circuit. Also known as capacitor-resistor unit; capristor; packaged circuit; resistor-capacitor unit. { 'resˌkap }

rescinnamine [PHARM] $C_{35}H_{42}N_2O_2$ A white to buff-colored, crystalline powder with a melting point of 238°C (in vacuum); soluble in acetic acid and chloroform; an alkaloid extracted from a few species of *Rauwolfia* and used in medicine. { rə'sin·əˌmēn }

rescue dump [COMPUT SCI] The copying of the entire contents of a computer memory into auxiliary storage devices, carried out periodically during the course of a computer program so that in case of a machine failure the program can be reconstituted at the last point at which this operation was executed. { 'resˌkyü ˌdəmp }

resealing pressure [MECH ENG] The inlet pressure at which leakage stops after a pressure relief valve is closed. { rē'sēl·iŋ ˌpresh·ər }

research [SCI TECH] Scientific investigation aimed at discovering and applying new facts, techniques, and natural laws. { ri'sərch }

research method [ENG] A standard test to determine the research octane number (or rating) of fuels for use in spark-ignition engines. { ri'sərch ˌmeth·əd }

research octane number [ENG] An expression for the antiknock rating of a motor gasoline as a guide to how vehicles will operate under mild conditions associated with low engine speeds. { ri'sərch 'äkˌtān ˌnəm·bər }

research reactor [NUCLEO] A reactor primarily designed to supply neutrons or other ionizing radiation for experimental purposes; it may also be used for training, materials testing, and production of radioisotopes. Also known as teaching reactor. { ri'sərch rēˌak·tər }

research rocket [AERO ENG] A rocket-propelled vehicle used to collect scientific data. { ri'sərch ˌräk·ət }

research ship [NAV ARCH] A ship used to carry out oceanographic research; conventional research ships carry on multidiscipline research, either on a single cruise or on successive cruises, while other research ships carry out special tasks, such as fisheries research ships, ships to tend and carry deep-submergence vehicles, and research drilling ships. Also known as oceanographic ship. { ri'sərch ˌship }

réseau [ASTRON] A grid that is photographed by a separate exposure on the same plate as images of celestial objects. [METEOROL] The term adopted by the World Meteorological Organization for the worldwide network of meteorological stations which have been chosen to represent the meteorology of the globe (*réseau mondial*). { rā'zō }

resection [ENG] **1.** A method in surveying by which the horizontal position of an occupied point is determined by drawing lines from the point to two or more points of known position. **2.** A method of determining a plane-table position by orienting along a previously drawn foresight line and drawing one or more rays through the foresight from previously located stations. [MED] The surgical removal of a section or segment of an organ or other structure. { ri'sek·shən }

resectoscope [MED] A tubelike instrument containing an optical device and a sliding knife; permits excision of tissue in body cavities without an opening other than that made by the instrument itself. { ri'sek·təˌskōp }

Resedaceae [BOT] A family of dicotyledonous herbs in the order Capparales having irregular, hypogynous flowers. { ˌres·ə'dās·ēˌē }

resedimentation [GEOL] **1.** Sedimentation of material derived from a preexisting sedimentary rock, that is, redeposition of sedimentary material. **2.** Mechanical deposition of material in cavities of postdepositional age, such as the deposition of carbonate muds and silts by internal mechanical erosion or solution of a limestone. **3.** The general process of subaqueous, downslope movement of sediment under the influence of gravity, such as the formation of a turbidity-current deposit. { rēˌsed·ə·mən'tā·shən }

resequent [GEOL] Referring to a geologic or topographic feature that resembles or agrees with a consequent feature but that developed from the feature at a later date. { rē'sē·kwənt }

resequent fault-line scarp [GEOL] A fault-line scarp which faces in the same direction as the original fault scarp or in which the downthrown block is topographically lower than the upthrown block. { rē'sē·kwənt 'fólt ˌlīn ˌskärp }

resequent stream [HYD] A stream whose direction follows an original consequent stream but is generally lower; resequent streams are generally tributary to a subsequent stream. { rē'sē·kwənt 'strēm }

reserpine [PHARM] $C_{33}H_{40}N_2O_9$ An alkaloid extracted from certain species of *Rauwolfia* and used as a sedative and antihypertensive. { rə'sər·pēn }

reserve [COMPUT SCI] To assign portions of a computer memory and of input/output and storage devices to a specific computer program in a multiprogramming system. { ri'zərv }

reserve aircraft [AERO ENG] Those aircraft which have been accumulated in excess of immediate needs for active aircraft and are retained in the inventory against possible future needs. { ri'zərv 'erˌkraft }

reserve battery [ELEC] A battery which is inert until an operation is performed which brings all the cell components into the proper state and location to become active. { ri'zərv 'bad·ə·rē }

reserve buoyancy [NAV ARCH] That part of the volume of a ship which is above the water surface and is watertight, so that it will increase buoyancy if the ships sinks deeper into the water. { ri'zərv 'bói·ən·sē }

reserve cell [HISTOL] **1.** One of the small, undifferentiated epithelial cells at the base of the stratified columnar lining of the bronchial tree. **2.** A chromophobe cell. { ri'zərv 'sel }

reserved minerals [MIN ENG] Economic minerals that belong to the state, which confers the right to prospect for and to mine them on any applicant. { ri'zərvd 'min·rəlz }

reserved word [COMPUT SCI] A word which cannot be used in a programming language to represent an item of data because it has some particular significance to the compiler, or which can be used only in a particular context. { ri'zərvd 'wərd }

reserves [MIN ENG] The quantity of workable mineral or of gas or oil which is calculated to lie within given boundaries. { ri'zərvz }

reserves-decline relationship [PETRO ENG] Relationship between production-rate decline over a period of time to the total remaining hydrocarbon reserves in a reservoir. { ri'zərvz di¦klīn ri'lā·shən,ship }

reservoir [CIV ENG] A pond or lake built for storage of water, usually by the construction of a dam across a river. [GEOL] **1.** A subsurface accumulation of crude oil or natural gas under adequate trap conditions. **2.** An area covered by névé where snow collects to form a glacier. **3.** A space within the earth that is occupied by magma. [SCI TECH] A container for storage of a liquid material, for example, a tank. { 'rez·əv,wär }

reservoir cycling [PETRO ENG] Repressuring of an oil reservoir by reinjection of dry gas (gas with liquids stripped out) into the formation. { 'rez·əv,wär ,sīk·liŋ }

reservoir drive mechanism [PETRO ENG] The physical action by which hydrocarbons (gas or liquid) are moved through the porous reservoir structure; for example, gas drive or water drive. Also known as oil-well drive. { 'rez·əv,wär 'drīv ,mek·ə,niz·əm }

reservoir dynamics [PETRO ENG] Fluid-flow performance within an oil or gas reservoir. { 'rez·əv,wär dī'nam·iks }

reservoir fluid [GEOL] The subterranean fluid trapped by a reservoir formation; can include natural gas, liquid and vapor petroleum hydrocarbons, and interstitial water. { 'rez·əv,wär ,flü·əd }

reservoir pressure [GEOL] **1.** The pressure on fluids (water, oil, gas) in a subsurface formation. Also known as formation pressure. **2.** The pressure under which fluids are confined in rocks. { 'rez·əv,wär ,presh·ər }

reservoir rock [GEOL] Friable, porous sandstone containing deposits of oil or gas. { 'rez·əv,wär ,räk }

reset See clear. { 'rē,set }

reset action [CONT SYS] Floating action in which the final control element is moved at a speed proportional to the extent of proportional-position action. { 'rē,set ,ak·shən }

reset bit [DES ENG] A diamond bit made by reusing diamonds salvaged from a used bit and setting them in the crown attached to a new bit blank. Also known as replacement bit. { 'rē,set bit }

reset condition [ELECTR] Condition of a flip-flop circuit in which the internal state of the flip-flop is reset to zero. { 'rē,set kən,dish·ən }

reset cycle [COMPUT SCI] The return of a cycle index counter to its initial value. { 'rē,set ,sī·kəl }

reset input [COMPUT SCI] The act of resetting the original conditions of a problem after a program is run on an analog computer. { 'rē,set 'in,pút }

reset mode [COMPUT SCI] The phase of operation of an analog computer during which the required initial conditions are entered into the system and the computing units are inoperative. Also known as initial condition mode. { 'rē,set ,mōd }

reset pulse [ELECTR] **1.** A drive pulse that tends to reset a magnetic cell in the storage section of a digital computer. **2.** A pulse used to reset an electronic counter to zero or to some predetermined position. { 'rē,set ,pəls }

reset rate [ENG] The number of times per minute that the effect of the proportional-position action upon the final control element is repeated by the proportional-speed floating action. { 'rē,set ,rāt }

resettability [ELECTR] The ability of the tuning element of an oscillator to retune the oscillator to the same operating frequency for the same set of input conditions. { ri,sed·ə'bil·əd·ē }

reshabar [METEOROL] A strong, very turbulent, dry northeast wind of bora type which blows down mountain ranges in southern Kurdistan in Persia; it is dry and hot in summer and cold in winter. { ¦rä·shə¦bär }

resid See residual oil. { rə'zid }

residence time [CHEM ENG] The average length of time a particle of reactant spends within a process vessel or in contact with a catalyst. [NUCLEO] The time during which radioactive material remains in the atmosphere following the detonation of a nuclear explosive; it is usually expressed as a half-time, since the time for all material to leave the atmosphere is not well known. { 'rez·ə·dəns ,tīm }

resident executive [COMPUT SCI] The portion of the executive program (sometimes called monitor system) which is permanently stored in core. Also known as resident monitor. { 'rez·ə·dənt ig'zek·yəd·iv }

resident module See resident routine. { 'rez·ə·dənt 'mä·jəl }

resident monitor See resident executive. { 'rez·ə·dənt 'män·əd·ər }

resident routine [COMPUT SCI] Any computer routine which is stored permanently in the memory, such as the resident executive. Also known as resident module. { 'rez·ə·dənt rü'tēn }

residual [GEOL] **1.** Of a mineral deposit, formed by either mechanical or chemical concentration. **2.** Pertaining to a residue left in place after weathering of rock. **3.** Of a topographic feature, representing the remains of a formerly great mass or area and rising above the surrounding surface. { rə'zij·ə·wəl }

residual air See residual volume. { rə'zij·ə·wəl 'er }

residual anticline [GEOL] In salt tectonics, a relative structural high resulting from the depression of two adjacent rim synclines. Also known as residual dome. { rə'zij·ə·wəl 'ant i,klīn }

residual charge [ELEC] The charge remaining on the plates of a capacitor after initial discharge. { rə'zij·ə·wəl 'chärj }

residual clay [GEOL] Very finely divided clay material formed in place by weathering of rock. Also known as primary clay. { rə'zij·ə·wəl 'klā }

residual compaction [GEOL] The difference between the amount of compaction that will ultimately occur for a given increase in applied stress, and that which has occurred at a specified time. { rə'zij·ə·wəl kəm'pak·shən }

residual current [ELECTR] Current flowing through a thermionic diode when there is no anode voltage, due to the velocity of the electrons emitted by the heated cathode. { rə'zij·ə·wəl ¦kə·rənt }

residual deviation [NAV] Deviation of a magnetic compass after adjustment or compensation. { rə'zij·ə·wəl ,dē·vē'ā·shən }

residual dome See residual anticline. { rə'zij·ə·wəl ¦dōm }

residual elements [MET] Elements present in small amounts in a metal or alloy, not added intentionally. { rə'zij·ə·wəl 'el·ə·məns }

residual error rate See undetected error rate. { rə'zij·ə·wəl 'er·ər ,rāt }

residual error ratio [PHYS] The difference between an optimum result derived from experience or experiment and a supposedly exact result derived from theory. { rə'zij·ə·wəl 'er·ər ,rā·shō }

residual field [ELECTROMAG] The magnetic field left in an iron core after excitation has been removed. { rə'zij·ə·wəl ¦fēld }

residual flux density [ELECTROMAG] The magnetic flux density at which the magnetizing force is zero when the material is in a symmetrically and cyclically magnetized condition. Also known as residual induction; residual magnetic induction; residual magnetism. { rə'zij·ə·wəl 'fläks ,den·səd·ē }

residual free gas [PETRO ENG] Free gas-cap gas in equilibrium with residual liquid hydrocarbons in a depleted reservoir, such as a reservoir at the end of its primary or economic producing life. { rə'zij·ə·wəl 'frē ,gas }

residual fuel oil [MATER] Topped crude petroleum or viscous residuums from refinery operations; commercial grades of burner oils Nos. 5 and 6 are residual oils, and include the bunker fuels. { rə'zij·ə·wəl 'fyül ,òi }

residual induction See residual flux density. { rə'zij·ə·wəl in'dək·shən }

residual intensity [SPECT] The intensity of radiation at some wavelength in a spectral line divided by the intensity in the adjacent continuum. { rə'zij·yə·wəl in'ten·səd·ē }

residual ionization [PHYS] Ionization of air or other gas in a closed chamber, not accounted for by recognizable neighboring agencies; now attributed to cosmic rays. { rə'zij·ə·wəl ,ī·ə·nə'zā·shən }

residual kame [GEOL] A ridge or mound of sand or gravel formed by the denudation of glaciofluvial material that had been deposited in glacial lakes or on the flanks of hills of till. { rə'zij·ə·wəl 'kām }

residual liquid [GEOL] The volatile components of a magma that remain in the magma chamber after much crystallization has taken place. { rə'zij·ə·wə·'lik·wəd }

residual liquor See rest magma. { rə'zij·ə·wəl 'lik·ər }

residual magnetic induction See residual flux density. { rə'zij·ə·wəl mag'ned·ik in'dək·shən }

residual magnetism See residual flux density. { rə'zij·ə·wəl 'mag·nə,tiz·əm }

residual map [GEOL] A stratigraphic map that displays the small-scale variations (such as local features in the sedimentary environment) of a given stratigraphic unit. { rə'zij·ə·wəl ¦map }

residual material [GEOL] Unconsolidated or partly weathered parent material of a soil, presumed to have developed in place (by weathering) from the consolidated rock on which it lies. { rə'zij·ə·wəl mə'tir·ē·əl }

residual method [MET] Magnetic particle inspection in which particles are supplied to a specimen after the magnetizing force has been removed. { rə'zij·ə·wəl ¦meth·əd }

residual mineral [GEOL] A mineral that has been concentrated in place by weathering and leaching of rock. { rə'zij·ə·wəl ¦min·rəl }

residual mode [CONT SYS] A characteristic motion of a structure which is deliberately ignored in the control algorithm of an active control system for the structure in the process of model reduction. { rə'zij·ə·wəl ¦mōd }

residual modulation See carrier noise. { rə'zij·ə·wəl ,maj·ə'lā·shən }

residual ochre [GEOL] An earthy, red, yellow, or brownish iron oxide powder of iron oxide (usually the mineral limonite) produced during chemical weathering. { rə'zij·ə·wəl 'ō·kər }

residual oil [MATER] Petroleum-refinery term for combustible, viscous or semiliquid bottoms product from crude oil distillation; used in adhesives, roofing compounds, asphalt manufacture, low-grade fuel oils, and sealants. Also known as liquid asphalt; resid; residuum; tailings. { rə'zij·ə·wəl 'oil }

residual penetration [ORD] As pertains to shaped charge ammunition, penetration of the jet into a backing target of some standard material after passage through a thickness of target material under test; this penetration is a measure of the effectiveness of the tested material against the jet. { rə'zij·ə·wəl ,pen·ə'trā·shən }

residual radiation [NUCLEO] Nuclear radiation emitted by radioactive material deposited after an atomic burst, including fission products, unfissioned nuclear material, and material in which radioactivity may have been induced by neutron bombardment. Also known as reststrahlen. [OPTICS] The nearly monochromatic radiation resulting from several reflections of light or other radiation from polished surfaces of certain substances such as quartz and rock salt, due to high reflectivity of these substances in certain bands of wavelengths. { rə'zij·ə·wəl ,rād·ē'ā·shən }

residual resistance [SOLID STATE] The value to which the electrical resistance of a metal drops as the temperature is lowered to near absolute zero, caused by imperfections and impurities in the metal rather than by lattice vibrations. { rə'zij·ə·wəl ri'zis·təns }

residual sediment See resistate. { rə'zij·ə·wəl 'sed·ə·mənt }

residual set [MATH] In a topological space, the complement of a set which is a countable union of nowhere dense sets. { rə'zij·ə·wəl 'set }

residual spectrum [MATH] Those members λ of the spectrum of a linear operator A on a Banach space X for which $(A − λI)^{-1}$, I being the identity operator, is unbounded with domain not dense in X. { rə'zij·ə·wəl 'spek·trəm }

residual stress See internal stress. { rə'zij·ə·wəl 'stres }

residual stress field See ambient stress field. { rə'zij·ə·wəl 'stres ,fēld }

residual sum of squares See error sum of squares. { rə'zij·ə·wəl ¦səm əv 'skwerz }

residual swelling [GEOL] The difference between the original prefreezing level of the ground and the level reached by the settling after the ground is completely thawed. { rə'zij·ə·wəl 'swel·iŋ }

residual tack See aftertack. { rə'zij·ə·wəl 'tak }

residual valley [GEOL] An intervening trough between uplifted mountains. { rə'zij·ə·wəl 'val·ē }

residual variance [STAT] In analysis of variance and regression analysis, that part of the variance which cannot be attributed to specific causes. { rə'zij·ə·wəl 'ver·ē·əns }

residual vibration See zero-point vibration. { rə'zij·ə·wəl vī'brā·shən }

residual voltage [ELEC] Vector sum of the voltages to ground of the several phase wires of an electric supply circuit. { rə'zij·ə·wəl 'vōl·tij }

residual volume [PHYSIO] Air remaining in the lungs after the most complete expiration possible; it is elevated in diffuse obstructive emphysema and during an attack of asthma. Also known as residual air. { rə'zij·ə·wəl 'väl·yəm }

residuary resistance [NAV ARCH] The sum of wavemaking resistance and eddy resistance opposing the motion of a ship through the water; the resistance which remains when frictional resistance is subtracted from total fluid resistance or drag. { rə'zij·ə,wer·ē ri'zis·təns }

residue [CHEM ENG] **1.** The substance left after distilling off all but the heaviest components from crude oil in petroleum refinery operations. Also known as bottoms; residuum. **2.** Solids deposited onto the filter medium during filtration. Also known as cake; discharged solids. [GEOL] The in-place accumulation of rock debris which remains after weathering has removed all but the least soluble constituent. [MATH] **1.** The residue of a complex function $f(z)$ at an isolated singularity z_0 is given by $(1/2\pi i) \int f(z)dz$ along a simple closed curve interior to an annulus about z_0; equivalently, the coefficient of the term $(z − z_0)^{-1}$ in the Laurent series expansion of $f(z)$ about z_0. **2.** In general, a coset of an ideal in a ring. **3.** A residue of m of order n, where m and n are integers, is a remainder that results from raising some integer to the nth power and dividing by m. { 'rez·ə,dü }

residue check See modulo N check. { 'rez·ə,dü ,chek }

residue class [MATH] A set of numbers satisfying a congruency relation. { 'rez·ə,dü ,klas }

residue class ring See quotient ring. { 'rez·ə,dü ¦klas ,riŋ }

residue system [COMPUT SCI] A number system in which each digit position corresponds to a different radix, all pairs of radices are relatively prime, and the value of a digit with radix r for an integer A is equal to the remainder when A is divided by r. { 'rez·ə,dü ,sis·təm }

residue theorem [MATH] The value of the integral of a complex function, taken along a simple closed curve enclosing at most a finite number of isolated singularities, is given by $2\pi i$ times the sum of the residues of the function at each of the singularities. { 'rez·ə,dü ,thir·əm }

residuum See residual oil; residue. { rə'zij·ə·wəm }

resilience [COMPUT SCI] The ability of computer software to be used for long periods of time. [MECH] **1.** Ability of a strained body, by virtue of high yield strength and low elastic modulus, to recover its size and form following deformation. **2.** The work done in deforming a body to some predetermined limit, such as its elastic limit or breaking point, divided by the body's volume. { rə'zil·yəns }

resin [ORG CHEM] Any of a class of solid or semisolid organic products of natural or synthetic origin with no definite melting point, generally of high molecular weight; most resins are polymers. { 'rez·ən }

resin duct [BOT] A canal (intercellular space) lined with secretory cells that release resins into the canal; common in gymnosperms. { 'rez·ən ,dəkt }

resin emulsion [MATER] Stable emulsion of a resin in a solvent carrier, such as the latex emulsions used in water-based latex paints. { 'rez·ən i,məl·shən }

resin finish [TEXT] A synthetic-resin finish produced by impregnating the fiber with resin and then baking it. { 'rez·ən ,fin·ish }

resin-in-pulp ion exchange [CHEM ENG] Combination of coarse anion-exchange resin with a slurry of finely ground uranium ore in an acid-leach liquor. { 'rez·ən in 'pəlp 'ī,än iks,chānj }

resinite [GEOL] A variety of exinite composed of resinous compounds, often in elliptical or spindle-shaped bodies. { 'rez·ən,īt }

resin matrix [PHYS CHEM] The molecular network of an ion exchange material that carries the ionogenic groups. { 'rez·ən ,mā,triks }

resin of copper See cuprous chloride. { 'rez·ən əv 'käp·ər }

resinography [CHEM] Science of resins, polymers, plastics, and their products; includes study of morphology, structure, and other characteristics relatable to composition or treatment. { ,rez·ən'äg·rə·fē }

resinoid [ORG CHEM] A thermosetting synthetic resin either in its initial (temporarily fusible) or in its final (infusible) state. { 'rez·ən,oid }

resinoid wheel [DES ENG] A grinding wheel bonded with a synthetic resin. { 'rez·ən,oid 'wēl }

resinol [MATER] Heat- and oxidation-sensitive, benzene-soluble coal tar fraction containing phenols; insoluble in light petroleum. { 'rez·ən,ȯl }

resin opal [MINERAL] A wax-, honey-, or ocher-yellow variety of common opal with a resinous luster or appearance. { 'rez·ən 'ō·pəl }

resinous cement [MATER] An acid-proof cement with a base of synthetic resin. { 'rez·ən·əs si'ment }

resinous coal [GEOL] Coal in which large proportions of resinous material are contained in the attritus. { 'rez·ən·əs 'kōl }

resinous luster [GEOL] The luster on the fractured surfaces of certain minerals (such as opal, sulfur, amber, and sphalerite) and rocks (such as pitchstone) that resemble the appearance of resin. { 'rez·ən·əs 'ləs·tər }

resin roof bolting [MIN ENG] The fixation of metal roof bolts in rock holes with a bonding resin. { 'rez·ən 'rüf ,bōlt·iŋ }

resin tin See rosin tin. { 'rez·ən ,tin }

resist [GRAPHICS] A protective layer applied to the image, or other parts of a plate, to protect that portion of the metal from the action of an etching bath or a sandblasting operation. [MATER] An acid-resistant nonconducting coating used to protect desired portions of a wiring pattern from the action of the etchant during manufacture of printed wiring boards. [MET] An insulating material, for example lacquer, applied to the surface of work to prevent electroplating or electrolytic action at the coated area. Also known as stopoff. { ri'zist }

resistance [ACOUS] See acoustic resistance. [ELEC] **1.** The opposition that a device or material offers to the flow of direct current, equal to the voltage drop across the element divided by the current through the element. Also known as electrical resistance. **2.** In an alternating-current circuit, the real part of the complex impedance. [FL MECH] See fluid resistance. [MECH] In damped harmonic motion, the ratio of the frictional resistive force to the speed. Also known as damping coefficient; damping constant; mechanical resistance. { ri'zis·təns }

resistance box [ELEC] A box containing a number of precision resistors connected to panel terminals or contacts so that a desired resistance value can be obtained by withdrawing plugs (as in a post-office bridge) or by setting multicontact switches. { ri'zis·təns ,bäks }

resistance brazing [MET] Brazing employing the heat developed by an electric current, the joint being part of the electric circuit. { ri'zis·təns ,brāz·iŋ }

resistance bridge See Wheatstone bridge. { ri'zis·təns ,brij }

resistance-capacitance circuit [ELEC] A circuit which has a resistance and a capacitance in series, and in which inductance is negligible. Abbreviated R-C circuit. { ri'zis·təns kə'pas·əd·əns ,sər·kət }

resistance-capacitance constant [ELEC] Time constant of a resistive-capacitive circuit, equal in seconds to the resistance value in ohms multiplied by the capacitance value in farads. Abbreviated R-C constant. { ri'zis·təns kə'pas·əd·əns ,kän·stənt }

resistance-capacitance coupled amplifier [ELECTR] An amplifier in which a capacitor provides a path for signal currents from one stage to the next, with resistors connected from each side of the capacitor to the power supply or to ground; it can amplify alternating-current signals but cannot handle small changes in direct currents. Also known as R-C amplifier; R-C coupled amplifier; resistance-coupled amplifier. { ri'zis·təns kə'pas·əd·əns 'kəp·əld 'am·plə,fī·ər }

resistance-capacitance network [ELEC] Circuit containing resistances and capacitances arranged in a particular manner to perform a specific function. Abbreviated R-C network. { ri'zis·təns kə'pas·əd·əns 'net,wərk }

resistance-capacitance oscillator [ELECTR] Oscillator in which the frequency is determined by resistance and capacitance elements. Abbreviated R-C oscillator. { ri'zis·təns kə'pas·əd·əns 'äs·ə,lād·ər }

resistance coefficient 1 [FL MECH] A dimensionless number used in the study of flow resistance, equal to the resistance force in flow divided by one-half the product of fluid density, the square of fluid velocity, and the square of a characteristic length. Symbolized cf_{ff}. { ri'zis·təns ,kō·i,fish·ənt 'wən }

resistance coefficient 2 See Darcy number 1. { ri'zis·təns,kō·i,fish·ənt 'tü }

resistance commutation [ELEC] Commutation of an electric rotating machine in which brushes with relatively high resistance span at least one commutator segment, in order to achieve a linear variation of current with time, and thereby minimize self-inductive voltage in the coils. { ri'zis·təns ,kam·yə'tā·shən }

resistance-coupled amplifier See resistance-capacitance coupled amplifier. { ri'zis·təns ,kəp·əld 'am·plə,fī·ər }

resistance coupling [ELECTR] Coupling in which resistors are used as the input and output impedances of the circuits being coupled; a coupling capacitor is generally used between the resistors to transfer the signal from one stage to the next. Also known as R-C coupling; resistance-capacitance coupling; resistive coupling. { ri'zis·təns ,kəp·liŋ }

resistance drop [ELEC] The voltage drop occurring between two points on a conductor due to the flow of current through the resistance of the conductor; multiplying the resistance in ohms by the current in amperes gives the voltage drop in volts. Also known as IR drop. { ri'zis·təns ,dräp }

resistance element [ELEC] An element of resistive material in the form of a grid, ribbon, or wire, used singly or built into groups to form a resistor for heating purposes, as in an electric soldering iron. { ri'zis·təns ,el·ə·mənt }

resistance factor See R factor. { ri'zis·təns ,fak·tər }

resistance furnace [ENG] An electric furnace in which the heat is developed by the passage of current through a suitable internal resistance that may be the charge itself, a resistor embedded in the charge, or a resistor surrounding the charge. Also known as electric resistance furnace. { ri'zis·təns ,fər·nəs }

resistance gage [ENG] An instrument for determining high pressures from the change in the electrical resistance of manganin or mercury produced by these pressures. { ri'zis·təns ,gāj }

resistance grounding [ELEC] Electrical grounding in which lines are connected to ground by a resistive (totally dissipative) impedance. { ri'zis·təns ,graùnd·iŋ }

resistance heating [ELEC] The generation of heat by electric conductors carrying current; degree of heating is proportional to the electrical resistance of the conductor; used in electrical home appliances, home or space heating, and heating ovens and furnaces. { ri'zis·təns ,hēd·iŋ }

resistance lamp [ELEC] Electric lamp used to prevent the current in a circuit from exceeding a desired limit. { ri'zis·təns ,amp }

resistance loss [ELEC] Power loss due to current flowing through resistance; its value in watts is equal to the resistance in ohms multiplied by the square of the current in amperes. { ri'zis·təns ,lȯs }

resistance magnetometer [ENG] A magnetometer that depends for its operation on variations in the electrical resistance of a material immersed in the magnetic field to be measured. { ri'zis·təns ,mag·nə'täm·əd·ər }

resistance material [ELEC] Material having sufficiently high resistance per unit length or volume to permit its use in the construction of resistors. { ri'zis·təns mə'tir·ē·əl }

resistance measurement [ELEC] The quantitative determination of that property of an electrically conductive material, component, or circuit called electrical resistance. { ri'zis·təns ,mezh·ər·mənt }

resistance meter [ENG] Any instrument which measures electrical resistance. Also known as electrical resistance meter. { ri'zis·təns ,mēd·ər }

resistance methanometer [ENG] A catalytic methanometer, with platinum used as the filament, which both heats the detecting element and acts as a resistance-type thermometer. { ri'zis·təns ,meth·ə'näm·əd·ər }

resistance noise See thermal noise. { ri'zis·təns ,nȯiz }

resistance pyrometer See resistance thermometer. { ri'zis·təns pī'räm·əd·ər }

resistance-rate flowmeter See resistive flowmeter. { ri'zis·təns ,rāt 'flō,mēd·ər }

resistance seam welding [MET] Resistance welding process which produces a series of individual spot welds, overlapping spot welds, or a continuous nugget weld made by circular or wheel-type electrodes. { ri'zis·təns 'sēm ,weld·iŋ }

resistance spot welding [MET] Resistance welding process in which the parts are lapped and held in place under pressure;

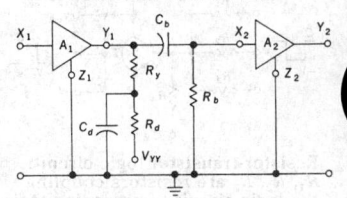

RESISTANCE-CAPACITANCE COUPLED AMPLIFIER

Circuit diagram of resistance-capacitance coupled amplifier. Amplifier stages A_1 and A_2 have inputs X_1 and X_2 and outputs Y_1 and Y_2. Y_1 is coupled to X_2 by blocking capacitor C_b. For vacuum-tube amplifier, Z_1 and Z_2 = cathode of A_1 and A_2; R_b = grid-leak resistor; R_y = plate resistor; V_{YY} = plate supply voltage. Decoupling filter R_d and C_d is used for compensation.

RESISTANCE THERMOMETER

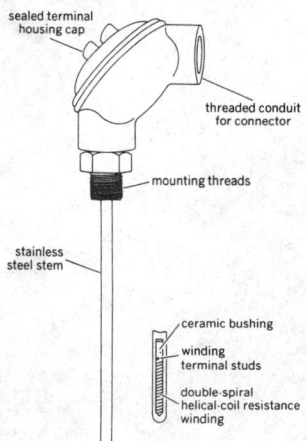

sealed terminal
housing cap

threaded conduit
for connector

mounting threads

stainless
steel stem

ceramic bushing

winding
terminal studs

double-spiral
helical-coil resistance
winding

Industrial-type resistance thermometer. *(From D. M. Considine, ed., Process Instruments and Controls Handbook, McGraw-Hill, 1957)*

RESISTIVE SUPERCONDUCTING FAULT-CURRENT LIMITER

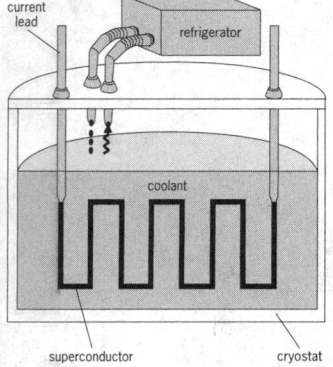

current
lead

refrigerator

coolant

superconductor   cryostat

Schematic diagram of the limiter.

RESISTOR-TRANSISTOR LOGIC

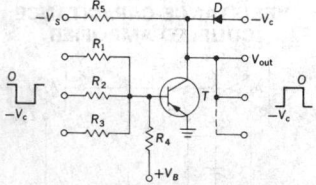

Resistor-transistor logic circuit: R_1, R_2, R_3 are resistors coupling the logic circuit to preceding logic circuits. With resistor R_4 and positive supply voltage $+V_B$ they form the gate circuit. D is the diode and T the transistor. V_s = input signal voltage; V_c = collector voltage; V_{out} = output voltage.

the size and shape of the electrodes (usually circular) control the size and shape of the welds. { ri'zis·təns 'spät ˌweld·iŋ }

resistance-start motor [ELEC] A split-phase motor having a resistance connected in series with the auxiliary winding; the auxiliary circuit is opened when the motor attains a predetermined speed. { ri'zis·təns ˌstärt ˌmōd·ər }

resistance strain gage [ELECTR] A strain gage consisting of a strip of material that is cemented to the part under test and that changes in resistance with elongation or compression. { ri'zis·təns 'strān ˌgāj }

resistance thermometer [ENG] A thermometer in which the sensing element is a resistor whose resistance is an accurately known function of temperature. Also known as electrical resistance thermometer; resistance pyrometer. { ri'zis·təns thər'mäm·əd·ər }

resistance transfer factor [GEN] A carrier of genetic information in bacteria which is considered to control the ability of self-replication and conjugal transfer of R factors. Abbreviated RTF. { ri'zis·təns 'tranz·fər ˌfak·tər }

resistance welding [MET] Joining metals together under pressure by making use of heat developed by an electric current, the work being part of the electrical circuit. { ri'zis·təns ˌweld·iŋ }

resistance wire [MET] Wire made from a metal or alloy having high resistance per unit length, such as Nichrome; used in wire-wound resistors and heating elements. { ri'zis·təns ˌwīr }

resistate [GEOL] A sediment consisting of minerals that are chemically resistant and are enriched in the residues of weathering processes. Also known as residual sediment. { ri'zis·ˌtāt }

resist-dyeing [TEXT] A cross-dyeing method in which a chemical is applied to certain yarns before weaving so that, when the material is dyed, only the untreated yarns take the dye, producing a colorful pattern; the chemical is later removed from the fabric. { ri'zist ˌdī·iŋ }

resisting moment [MECH] A moment produced by internal tensile and compressive forces that balances the external bending moment on a beam. { ri'zist·iŋ ˌmō·mənt }

resistive coupling *See* resistance coupling. { ri'zis·tiv 'kəp·liŋ }

resistive flowmeter [ENG] Liquid flow-rate measurement device in which flow rates are read electrically as the result of the rise or fall of a conductive differential-pressure manometer fluid in contact with a resistance-rod assembly. Also known as resistance-rate flowmeter. { ri'zis·tiv 'flō,mēd·ər }

resistive load [ELEC] A load whose total reactance is zero, so that the alternating current is in phase with the terminal voltage. Also known as nonreactive load. { ri'zis·tiv 'lōd }

resistive superconducting fault-current limiter [ELEC] A fault-current limiter in which a superconductor is directly connected in series to the line to be protected and is immersed in a coolant which is chilled by a refrigerant, and the connection from the line at room temperature to the superconductor is provided by special current leads, which are designed to minimize the heat transfer to the coolant. { ri¦zis·tiv ¦sü·pər·kən,dək·tiŋ 'fölt, kər·ənt ˌlim·əd·ər }

resistive unbalance [ELEC] Unequal resistance in the two wires of a transmission line. { ri'zis·tiv ən'bal·əns }

resistivity *See* electrical resistivity. { ˌrē,zis'tiv·əd·ē }

resistivity factor *See* formation factor. { ˌrē,zis'tiv·əd·ē ˌfak·tər }

resistivity index [PETRO ENG] Ratio of the true electrical resistivity of a rock system at a specified water saturation, to the resistivity of the rock itself; used for calculation of electrical well-logging data. { ˌrē,zis'tiv·əd·ē ˌin,deks }

resistivity method [ENG] Any electrical exploration method in which current is introduced in the ground by two contact electrodes and potential differences are measured between two or more other electrodes. { ˌrē,zis'tiv·əd·ē ˌmeth·əd }

resistivity well logging [PETRO ENG] The measurement of subsurface electrical resistivities (normal and lateral to the borehole) during electrical logging of oil wells. { ˌrē,zis'tiv·əd·ē 'wel ˌläg·iŋ }

resistojet [AERO ENG] A type of propulsion system in which propellant is heated by passing it over resistively heated surfaces and then through a converging-diverging nozzle. { ri'zis·tə,jet }

resistor [ELEC] A device designed to have a definite amount

of resistance; used in circuits to limit current flow or to provide a voltage drop. Also known as electrical resistor. { ri'zis·tər }

resistor bulb [ENG] A temperature-measurement device inside of which is a resistance winding; changes in temperature cause corresponding changes in resistance, varying the current in the winding. { ri'zis·tər ˌbəlb }

resistor-capacitor-transistor logic [ELECTR] A resistor-transistor logic with the addition of capacitors that are used to enhance switching speed. { ri'zis·tər kə'pas·əd·ər tran'zis·tər ˌläj·ik }

resistor-capacitor unit *See* rescap. { ri'zis·tər kə'pas·əd·ər ˌyü·nət }

resistor color code [ELEC] Code adopted by the Electronic Industries Association to mark the values of resistance on resistors in a readily recognizable manner; the first color represents the first significant figure of the resistor value, the second color the second significant figure, and the third color represents the number of zeros following the first two figures; a fourth color is sometimes added to indicate the tolerance of the resistor. { ri'zis·tər 'kəl·ər ˌkōd }

resistor core [ELEC] Insulating support on which a resistor element is wound or otherwise placed. { ri'zis·tər ˌkor }

resistor element [ELEC] That portion of a resistor which possesses the property of electric resistance. { ri'zis·tər ˌel·ə·mənt }

resistor furnace [ENG] An electric furnace in which heat is developed by the passage of current through distributed resistors (heating units) mounted apart from the charge. { ri'zis·tər ˌfər·nəs }

resistor network [ELEC] An electrical network consisting entirely of resistances. { ri'zis·tər 'net,wərk }

resistor oven [ENG] Heating chamber relying on an electrical-resistance element to create temperatures of up to 800°F (430°C); used for drying and baking. { ri'zis·tər 'əv·ən }

resistor termination [ELECTR] A thick-film conductor pad overlapping and contacting a thick-film resistor area. { ri'zis·tər ˌtər·mə'nā·shən }

resistor-transistor logic [ELECTR] One of the simplest logic circuits, having several resistors, a transistor, and a diode. Abbreviated RTL. { ri'zis·tər tran'zis·tər ˌläj·ik }

resite *See* C stage. { 're,zīt }

resnatron [ELECTR] A microwave-beam tetrode containing cavity resonators, used chiefly for generating large amounts of continuous power at high frequencies. { 'rez·nə,trän }

resolution [CONT SYS] The smallest increment in distance that can be distinguished and acted upon by an automatic control system. [ELECTR] In television, the maximum number of lines that can be discerned on the screen at a distance equal to tube height; this ranges from 350 to 400 for most receivers. [ELECTROMAG] In radar, the minimum separation between two targets, in angle or range, at which they can be distinguished on a radar screen. Also known as resolving power. [MATH] For a vector, the determination of vectors parallel to specified (usually perpendicular) axes such that their sum equals the given vector. [OPTICS] *See* resolving power. [ORG CHEM] The process of separating a racemic mixture into the two component optical isomers. [PHYS] **1.** For a measurement of energy or momentum of a collection of particles, the difference between the highest and lowest energies at which the response of an instrument to a beam of monoenergetic particles is at least half its maximum value, divided by the energy of the particles. **2.** The procedure of breaking up a vectorial quantity into its components. [SPECT] *See* resolving power. { ˌrez·ə'lü·shən }

resolution chart [COMMUN] *See* test pattern. [OPTICS] A device to test resolving power; usually alternate black and white lines of equal width arranged in groups of decreasing line width, identified as the number of line pairs per millimeter. { ˌrez·ə'lü·shən ˌchärt }

resolution error [COMPUT SCI] An error of an analog computing unit that results from its inability to respond to changes of less than a given magnitude. { ˌrez·ə'lü·shən ˌer·ər }

resolution factor [COMPUT SCI] In information retrieval, the ratio obtained in dividing the total number of documents retrieved (whether relevant or not to the user's needs) by the total number of documents available in the file. { ˌrez·ə'lü·shən ˌfak·tər }

resolution in azimuth [ENG] The angle by which two targets must be separated in azimuth in order to be distinguished by a radar set when the targets are at the same range. { ‚rez·ə'lü·shən in 'az·ə·məth }

resolution in range [ENG] Distance by which two targets must be separated in range in order to be distinguished by a radar set when the targets are on the same azimuth line. { ‚rez·ə'lü·shən in 'ränj }

resolution of the identity [MATH] A family of linear projection operators on a Banach space used in studying the spectra of linear operators. { ‚rez·ə'lü·shən əv the ə'den·əd·ē }

resolution reading [OPTICS] A number indicating how many lines per millimeter are contained in the finest group which can be distinguished on a resolution chart. { ‚rez·ə'lü·shən ‚rēd·iŋ }

resolution wedge [COMMUN] On a television test pattern, a group of gradually converging lines used to measure resolution. { ‚rez·ə'lü·shən ‚wej }

resolvable balanced incomplete block design [MATH] A balanced incomplete block design such that the blocks themselves are partitioned into r families of v/k blocks, such that every element occurs in exactly one block of each of these families. { ri¦zäl·və·bəl ¦bal·ənst ‚iŋ·kəm‚plēt 'bläk di‚zīn }

resolve motion-rate control [CONT SYS] A form of robotic control in which the controlled variables are the velocity vectors of the end points of a manipulator, and the angular velocities of the joints are determined to obtain the desired results. { ri'zolv 'mō·shən ¦rāt kən‚trōl }

resolvent [MATH] For a linear operator T on a Banach space, the function, defined on the complement of the spectrum of T given by $(T - \lambda I)^{-1}$ for each λ in this complement, where I is the identity operator; this enables a study of T relative to its eigenvalues. { ri'zäl·vənt }

resolvent kernel [MATH] A function appearing as an integrand in an integral representation for a solution of a linear integral equation which often completely determines the solutions. { ri'zäl·vənt 'kər·nəl }

resolvent set [MATH] Those scalars λ for which the operator $T - \lambda I$ has a bounded inverse, where T is some linear operator on a Banach space, and I is the identity operator. { ri'zäl·vənt 'set }

resolver [ELEC] A synchro or other device whose rotor is mechanically driven to translate rotor angle into electrical information corresponding to the sine and cosine of rotor angle; used for interchanging rectangular and polar coordinates. Also known as sine-cosine generator; synchro resolver. [ELECTR] **1.** A synchro or other device whose input is the angular position of an object, such as the rotor of an electric machine, and whose output is electric signals, usually proportional to the sine and cosine of an angle, and often in digital form; used to interchange rectangular and polar coordinates, and in servomechanisms to report the orientation of controlled objects. Also known as angular resolver. **2.** A device that accepts a single vector-valued analog input and produces for output either analog or digital signals proportional to two or three orthogonal components of the vector. Also known as vector resolver. { ri'zäl·vər }

resolving cell [ELECTROMAG] In radar, volume in space whose diameter is the product of slant range and beam width, and whose length is the pulse length. { ri'zälv·iŋ ‚sel }

resolving power [ELECTROMAG] **1.** The reciprocal of the beam width of a unidirectional antenna, measured in degrees. **2.** See resolution. [OPTICS] A quantitative measure of the ability of an optical instrument to produce separable images of different points on an object; usually, the smallest angular or linear separation of two object points for which they may be resolved according to the Rayleigh criterion. Also known as resolution. [PHYS] A measure of the ability of a mass spectroscope to separate particles of different masses, equal to the ratio of the average mass of two particles whose mass spectrum lines can just be completely separated, to the difference in their masses. [SPECT] A measure of the ability of a spectroscope or interferometer to separate spectral lines of nearly equal wavelength, equal to the average wavelength of two equally strong spectral lines whose images can barely be separated, divided by the difference in wavelengths; for spectroscopes, the lines must be resolved according to the Rayleigh criterion; for interferometers, the wavelengths at which the lines have half of maximum intensity must be equal. Also known as resolution. { ri'zälv·iŋ ‚paů·ər }

resolving time [COMPUT SCI] In computers, the shortest permissible period between trigger pulses for reliable operation of a binary cell. [ENG] Minimum time interval, between events, that can be detected; resolving time may refer to an electronic circuit, to a mechanical recording device, or to a counter tube. { ri'zälv·iŋ ‚tīm }

resonance [ELEC] A phenomenon exhibited by an alternating-current circuit in which there are relatively large currents near certain frequencies, and a relatively unimpeded oscillation of energy from a potential to a kinetic form; a special case of the physics definition. [PHYS] **1.** A phenomenon exhibited by a physical system acted upon by an external periodic driving force, in which the resulting amplitude of oscillation of the system becomes large when the frequency of the driving force approaches a natural free oscillation frequency of the system. **2.** In general, any phenomenon which is greatly enhanced at frequencies or energies that are at or very close to a given characteristic value. [PHYS CHEM] A feature of the valence-bond method that accounts for the anomalies in certain molecules by representing their structures with approximate resonance hybrid formulas no single electronic formula conforms both to the observed properties and to the octet rule. Also known as mesomerism. [QUANT MECH] **1.** An enhanced coupling between quantum states with the same energy. **2.** See resonance absorption. **3.** See resonance level. { 'rez·ən·əns }

resonance absorption [NUCLEO] The absorption of neutrons having a narrow range of energies corresponding to a nuclear resonance level of the absorber in a nuclear reactor. [QUANT MECH] The absorption of electromagnetic radiation by a quantum-mechanical system at a characteristic frequency satisfying the Bohr frequency condition. Also known as resonance. { 'rez·ən·əns əb‚sórp·shən }

resonance bridge [ELEC] A four-arm alternating-current bridge used to measure inductance, capacitance, or frequency; the inductor and the capacitor, which may be either in series or in parallel, are tuned to resonance at the frequency of the source before the bridge is balanced. { 'rez·ən·əns ‚brij }

resonance capture [NUC PHYS] The combination of an incident particle and a nucleus in a resonance level of the resulting compound nucleus, characterized by having a large cross section at and very near the corresponding resonance energy. { 'rez·ən·əns ‚kap·chər }

resonance curve [ELEC] Graphical representation illustrating the manner in which a tuned circuit responds to the various frequencies in and near the resonant frequency. { 'rez·ən·əns ‚kərv }

resonance energy [PHYS] The characteristic energy at which, or very close to which, the amplitude of a resonance phenomenon is greatly enhanced. { 'rez·ən·əns ‚en·ər·jē }

resonance fluorescence [ATOM PHYS] See resonance radiation. [NUC PHYS] Resonant scattering from an atomic nucleus. { 'rez·ən·əns flu‚res·əns }

resonance fluorescence lidar [OPTICS] A type of lidar in which the laser wavelength is tuned to the resonance absorption wavelength of a specific molecular species whose resonant backscatter cross section is measured in order to determine its density in the upper atmosphere. { ¦rez·ən·əns flù'res·əns ‚lī‚cär }

resonance frequency [PHYS] A frequency at which some measure of the response of a physical system to an external periodic driving force is a maximum; three types are defined, namely, phase resonance, amplitude resonance, and natural resonance, but they are nearly equal when dissipative effects are small. Also known as resonant frequency. [QUANT MECH] A characteristic frequency, satisfying the Bohr frequency condition, at which a quantum-mechanical system absorbs radiation. { 'rez·ən·əns ‚frē·kwən·sē }

resonance hybrid [CHEM] A molecule that may be considered an intermediate between two or more valence bond structures. { 'rez·ən·əns ‚hī·brəd }

resonance ionization spectroscopy [SPECT] A technique capable of detecting single atoms or molecules of a given element or compound in a gas, in which an atom or molecule in its ground state is excited to a bound state when a photon is absorbed from a laser beam at a very well-controlled wavelength that is resonant with the excitation energy; a second photon removes the excited electron from the atom or molecule, and this electron is then accelerated by an electric field and collides with the gas molecules, creating additional ionization which is detected by a proportional counter.   Abbreviated RIS. { 'rez·ən·əns ‚ī·ə·nə'zā·shən spek'träs·kə·pē }

resonance Kuiper Belt object [ASTRON]   A member of the Kuiper Belt that lies in a stable region where it is protected from close encounters with the planet Neptune because its period of revolution is a simple fraction of Neptune's period. { ‚rez·ən·əns 'kī·pər ‚belt ‚äb‚jekt }

resonance lamp [ATOM PHYS]   An evacuated quartz bulb containing mercury, which acts as a source of radiation at the wavelength of the pure resonance line of mercury when irradiated by a mercury-arc lamp. { 'rez·ən·əns ‚lamp }

resonance level [QUANT MECH]   An unstable state of a compound system capable of being formed in a collision between two particles, and associated with a peak in a graph of cross section versus energy for the scattering of the particles.   Also known as resonance. { 'rez·ən·əns ‚lev·əl }

resonance line [SPECT]   The line of longest wavelength associated with a transition between the ground state and an excited state. { 'rez·ən·əns ‚līn }

resonance luminescence See resonance radiation. { 'rez·ən·əns ‚lü·mə‚nes·əns }

resonance method [ELEC]   A method of determining the impedance of a circuit element, in which resonance frequency of a resonant circuit containing the element is measured. [ENG]   In ultrasonic testing, a method of measuring the thickness of a metal by varying the frequency of the beam transmitted to excite a maximum amplitude of vibration. { 'rez·ən·əns ‚meth·əd }

resonance radiation [ATOM PHYS]   The emission of radiation by a gas or vapor as a result of excitation of atoms to higher energy levels by incident photons at the resonance frequency of the gas or vapor; the radiation is characteristic of the particular gas or vapor atom but is not necessarily the same frequency as the absorbed radiation.   Also known as resonance fluorescence; resonance luminescence. { 'rez·ən·əns ‚rād·ē‚ā·shən }

resonance reaction [NUC PHYS]   A nuclear reaction that takes place only when the energy of the incident particles is at or very close to a characteristic value. { 'rez·ən·əns rē‚ak·shən }

resonance scattering [NUC PHYS]   A peak in the cross section of a nucleus for elastic scattering of neutrons at energies near a resonance level, accompanied by an anomalous phase shift in the scattered neutrons. { 'rez·ən·əns ‚skad·ə·riŋ }

resonance spectrum [SPECT]   An emission spectrum resulting from illumination of a substance (usually a molecular gas) by radiation of a definite frequency or definite frequencies. { 'rez·ən·əns ‚spek·trəm }

resonance structure [ORG CHEM]   Any of two or more possible structures of the same compound that have identical geometry but different arrangements of their paired electrons; none of the structures has physical reality or adequately accounts for the properties of the compound, which exists as an intermediate form. { 'rez·ən·əns ‚strək·chər }

resonance transformer [ELEC]   A high-voltage transformer in which the secondary circuit is tuned to the frequency of the power supply. [ELECTR]   An electrostatic particle accelerator, used principally for acceleration of electrons, in which the high-voltage terminal oscillates between voltages which are equal in magnitude and opposite in sign. { 'rez·ən·əns tranz‚fòr·mər }

resonance trough [METEOROL]   A large-scale pressure trough which forms at an appropriate wavelength away from a dominant trough; for example, the mean trough over the Mediterranean in winter is often considered a resonance trough between the two more dynamically active troughs along the east coasts of North America and Asia. { 'rez·ən·əns ‚tròf }

resonance vibration [MECH]   Forced vibration in which the frequency of the disturbing force is very close to the natural frequency of the system, so that the amplitude of vibration is very large. { 'rez·ən·əns vī‚brā·shən }

resonant antenna [ELECTROMAG]   An antenna for which there is a sharp peak in the power radiated or intercepted by the antenna at a certain frequency, at which electric currents in the antenna form a standing-wave pattern. { 'res·ən·ənt an'ten·ə }

resonant capacitor [ELEC]   A tubular capacitor that is wound to have inductance in series with its capacitance. { 'res·ən·ənt kə'pas·əd·ər }

resonant cavity See cavity resonator. { 'res·ən·ənt 'kav·əd·ē }

resonant-cavity maser [PHYS]   A maser in which the paramagnetic active material is placed in a cavity resonator. { 'res·ən·ənt ‚kav·əd·ē 'mā·zər }

resonant chamber See cavity resonator. { 'res·ən·ənt 'chām·bər }

resonant-chamber switch [ELECTROMAG]   Waveguide switch in which a tuned cavity in each waveguide branch serves the functions of switch contacts; detuning of a cavity blocks the flow of energy in the associated waveguide. { 'res·ən·ənt ‚chām·bər ‚swich }

resonant circuit [ELEC]   A circuit that contains inductance, capacitance, and resistance of such values as to give resonance at an operating frequency. { 'res·ən·ənt 'sər·kət }

resonant coupling [ELEC]   Coupling between two circuits that reaches a sharp peak at a certain frequency. { 'res·ən·ənt 'kəp·liŋ }

resonant detector [PHYS]   A detector of electromagnetic radiation which is responsive to radiation only at certain frequencies at which resonance is created in the detector. { 'res·ən·ənt di'tek·tər }

resonant diaphragm [ELECTROMAG]   Diaphragm, in waveguide technique, so proportioned as to introduce no reactive impedance at the design frequency. { 'res·ən·ənt 'dī·ə‚fram }

resonant element See cavity resonator. { 'res·ən·ənt 'el·ə·mənt }

resonant frequency See resonance frequency. { 'res·ən·ənt 'frē·kwən·sē }

resonant gate transistor [ELECTR]   Surface field-effect transistor incorporating a cantilevered beam which resonates at a specific frequency to provide high-Q-frequency discrimination. { 'res·ən·ənt 'gāt tran‚zis·tər }

resonant helix [ELECTROMAG]   An inner helical conductor in certain types of transmission lines and resonant cavities, which carries currents with the same frequency as the rest of the line or cavity. { 'res·ən·ənt 'hē·liks }

resonant ionization mass spectrometry [SPECT]   An instrumental technique for quantitative identification of trace impurities (at or below the part-per-billion level); it begins with laser-induced or ion-induced desorption, followed by resonant laser ionization (usually from two or three lasers), and then analysis by time-of-flight mass spectrometry.   Abbreviated RIMS. { ‚rez·ən·ənt ‚ī·ə·nə‚zā·shən ‚mas spek'träm·ə·trē }

resonant iris [ELECTROMAG]   A resonant window in a circular waveguide; it resembles an optical iris. { 'res·ən·ənt 'ī·rəs }

resonant jet [AERO ENG]   A pulsejet engine, exhibiting intensification of power under the rhythm of explosions and compression waves within the engine. { 'res·ən·ənt 'jet }

resonant line [ELECTROMAG]   A transmission line having values of distributed inductance and distributed capacitance so as to make the line resonant at the frequency it is handling. { 'res·ən·ənt 'līn }

resonant-line oscillator [ELECTR]   Oscillator in which one or more sections of transmission lines are employed as resonant elements. { 'res·ən·ənt ‚līn 'äs·ə‚lād·ər }

resonant-line tuner [ELECTR]   A television tuner in which resonant lines are used to tune the antenna, radio-frequency amplifier, and radio-frequency oscillator circuits; tuning is achieved by moving shorting contacts that change the electrical lengths of the lines. { 'res·ən·ənt ‚līn 'tün·ər }

resonant-mass antenna [ENG]   A detector of gravitational radiation, consisting of a mass of several tons of aluminum or other metal, in the shape of a cylinder or a truncated icosahedron, and attached electromechanical transducers that convert deformations of the mass to electronic signals. { ‚rez·ən·ənt ‚mas an'ten·ə }

resonant-mode power supply [ELECTR]   An electronic power supply in which the current and voltage waveforms are shaped to sinusoids by a small inductor and capacitor inserted in the current path. { ‚rez·ən·ənt ‚mōd 'pau·ər sə‚plī }

resonant Raman effect [ATOM PHYS]   A process in which a photon whose energy is exactly matched to the transition energy between two atomic energy levels (within the natural linewidth) promotes an atomic electron to an excited state, which decays in the same step. { 'rez·ən·ənt 'rä‚män i‚fekt }

resonant reaction [NUC PHYS]   A nuclear reaction whose

probability is enhanced at an energy corresponding to an energy level of one of the nuclei. { 'rez·ən·ənt rē‚ak·shən }

resonant-reed relay [ELEC] A reed relay in which the reed switch closes only when the required frequency is applied to the operating coil, to make one of the reeds vibrate until its amplitude is sufficient to make contact with the other reed; used in selective paging systems. { 'res·ən·ənt ‚rēd 'rē‚lā }

resonant resistance [ELEC] Resistance value to which a resonant circuit is equivalent. { 'res·ən·ənt ri'zis·təns }

resonant scattering [QUANT MECH] Scattering of a photon by a quantum-mechanical system (usually an atom or nucleus) in which the system first absorbs the photon by undergoing a transition from one of its energy states to one of higher energy, and subsequently reemits the photon by the exact inverse transition. { 'res·ən·ənt 'skad·ə·riŋ }

resonant ultrasound spectroscopy [ACOUS] An experimental technique for obtaining a complete set of elastic constants of a material in which a sample of the material of well-defined shape, usually a sphere or a rectangular parallelepiped, is held between two transducers, one of which excites the sample while the other measures its response, and the spectrum of free mechanical resonances of the sample is measured. { ‚rez·ən·ənt ‚əl·trə‚saund spek'träs·kə·pē }

resonant voltage step-up [ELEC] Ability of an inductor and a capacitor in a series resonant circuit to deliver a voltage several times greater than the input voltage of the circuit. { 'res·ən·ənt 'vōl·tij 'step‚əp }

resonant wavelength [ELECTROMAG] The wavelength in free space of electromagnetic radiation having a frequency equal to a natural resonance frequency of a cavity resonator. { 'res·ən·ənt 'wāv‚leŋkth }

resonant window [ELECTROMAG] A parallel combination of inductive and capacitive diaphragms, used in a waveguide structure to provide transmission at the resonant frequency and reflection at other frequencies. { 'res·ən·ənt 'win‚dō }

resonate [ELEC] To bring to resonance, as by tuning. { 'rez·ən‚āt }

resonating cavity [ELECTROMAG] Short piece of waveguide of adjustable length, terminated at either or both ends by a metal piston, an iris diaphragm, or some other wave-reflecting device; it is used as a filter, as a means of coupling between guides of different diameters, and as impedance networks corresponding to those used in radio circuits. { 'rez·ən‚ād·iŋ 'kav·əd·ē }

resonator [PHYS] A device that exhibits resonance at a particular frequency, such as an acoustic resonator or cavity resonator. { 'rez·ən‚ād·ər }

resonator grid [ELECTR] Grid that is attached to a cavity resonator in velocity-modulated tubes to provide coupling between the resonator and the electron beam. { 'rez·ən‚ād·ər ‚grid }

resonator wavemeter [ELECTROMAG] Any resonant circuit used to determine wavelength, such as a cavity-resonator frequency meter. { 'rez·ən‚ād·ər 'wāv‚mēd·ər }

resorbed reef [GEOL] A reef characterized by embayed margins and by the numerous isolated patches of reef that are closely distributed about the main mass. { rē'sórbd 'rēf }

resorcin See resorcinol. { rə'zórs·ən }

resorcinol [ORG CHEM] $C_6H_4(OH)_2$ Sweet-tasting, white, toxic crystals; soluble in water, alcohol, ether, benzene, and glycerol; melts at 111°C; used for resins, dyes, pharmaceuticals, and adhesives, and as a chemical intermediate. Also known as resorcin. { rə'zórs·ən‚ól }

resorcinol acetate [ORG CHEM] $HOC_6H_4OCOCH_3$ A viscous, combustible, yellow to amber liquid with burning taste; soluble in alcohol; boils at 283°C; used in cosmetics and medicine. Also known as resorcinol monoacetate. { rə'zórs·ən‚ól 'as·ə‚tāt }

resorcinol diglycidyl ether [ORG CHEM] $C_{12}H_{14}O_2$ A straw yellow liquid with a boiling point of 172°C (at 0.8 mmHg or 100 pascals); used for epoxy resins. Abbreviated RDGE. { rə'zórs·ən‚ól di'glis·ə‚dil ‚ē·thər }

resorcinol-formaldehyde resin [ORG CHEM] A phenol-formaldehyde resin, soluble in water, ketones, and alchol; used to make fast-curing adhesives for wood gluing. { rə'zórs·ən‚ól fór'mal·də‚hīd 'rez·ən }

resorcinol monoacetate See resorcinol acetate. { rə'zórs·ən‚ól ‚män·ō'as·ə‚tāt }

β-resorcylic acid [ORG CHEM] $(OH)_2C_6H_3COOH$ Combustible, white needles; soluble in alcohol and ether, very slightly soluble in water; decomposes at 220°C; used as a dyestuff and a pharmaceutical intermediate, and in the manufacture of fine chemicals. { ‚bād·ə ‚rē·zór‚sil·ik 'as·əd }

resorption [PETR] The process by which a magma redissolves previously crystallized minerals. [PHYS] Absorption or, less commonly, adsorption of material by a body or system from which the material was previously released. { rē'sórp·shən }

resource [SCI TECH] A reserve source of supply, such as a material or mineral. { 'rē‚sórs }

resource allocation in multiproject scheduling [IND ENG] A system that employs network analysis as an aid in making the best assignment of resources which must be stretched over a number of projects. Abbreviated RAMPS. { 'rē‚sórs ‚al·ə'kā·shən in ‚məl·ti‚prä·jekt 'sked·jə·liŋ }

respiration [PHYSIO] **1.** The processes by which tissues and organisms exchange gases with their environment. **2.** The act of breathing with the lungs, consisting of inspiration and expiration. { ‚res·pə'rā·shən }

respirator [ENG] A device for maintaining artificial respiration to protect the respiratory tract against irritating and poisonous gases, fumes, smoke, and dusts, with or without equipment supplying oxygen or air; some types have a fitting which covers the nose and mouth. { 'res·pə‚rād·ər }

respiratory arrest [MED] Sudden cessation of spontaneous respiration due to failure of the respiratory center. { 'res·prə‚tór·ē ə 'rest }

respiratory center [PHYSIO] A large area of the brain involved in regulation of respiration. { 'res·prə‚tór·ē ‚sen·tər }

respiratory dead space [PHYSIO] That part of the respiratory system which has no alveoli and in which little or no exchange of gas between air and blood takes place. { 'res·prə‚tór·ē 'ded ‚spās }

respiratory distress syndrome of newborn [MED] A disease occurring during the first days of life, characterized by respiratory distress and cyanosis; a hyaline membrane lines the alveoli when the disease persists for more than several hours. Abbreviated RDS. { 'res·prə‚tór·ē di'stres ‚sin‚drōm əv 'nü‚bórn }

respiratory epithelium [HISTOL] The ciliated pseudostratified epithelium lining the respiratory tract. { 'res·prə‚tór·ē ‚ep·ə'hē·lē·əm }

respiratory minute volume [PHYSIO] The total amount of air which moves in and out of the lungs in a minute. { 'res·prə‚tór·ē 'min·ət 'väl·yəm }

respiratory pigment [BIOCHEM] Any of various conjugated proteins that function in living organisms to transfer oxygen in cellular respiration. { 'res·prə‚tór·ē ‚pig·mənt }

respiratory quotient [PHYSIO] The ratio of volumes of carbon dioxide evolved and oxygen consumed during a given period of respiration. Abbreviated RQ. { 'res·prə‚tor·ē ‚kwō·shənt }

respiratory syncytial virus [VIROL] An enveloped, single-stranded RNA animal virus belonging to the Paramyxoviridae genus *Pneumovirus*; associated with a large proportion of respiratory illnesses in very young children, particularly bronchiolitis and pneumonia. { 'res·prə‚tór·ē sin'sish·əl 'vī·rəs }

respiratory system [ANAT] The structures and passages involved with the intake, expulsion, and exchange of oxygen and carbon dioxide in the vertebrate body. { 'res·prə‚tór·ē ‚sis·təm }

respiratory tree [ANAT] The trachea, bronchi, and bronchioles. [INV ZOO] Either of a pair of branched tubular appendages of the cloaca in certain holothurians that is thought to have a respiratory function. { 'res·prə‚tor·ē ‚trē }

respirometer [ENG] **1.** An instrument for studying respiration. **2.** A diver's helmet containing a compressed air supply for replenishing oxygen used by the diver. { ‚res·pə'räm·əd·ər }

responder [ELECTR] The transmitter section of a radar beacon { ri'spän·dər }

responder beacon [ELECTR] The radar beacon that serves to emit the signals of the responder in a transponder. { ri'spän·dər‚bē·kən }

response [COMMUN] *See* reply. [CONT SYS] A quantitative expression of the output of a device or system as a function

RESONANT ULTRASOUND SPECTROSCOPY

Resonant ultrasound spectroscopy transducer with a mounted sample. (*Dynamic Resonance Systems Inc.*)

of the input. Also known as system response. [STAT] The value of some measurable quantity after a treatment has been applied. { ri'späns }

response characteristic [CONT SYS] The response as a function of an independent variable, such as direction or frequency, often presented in graphical form. { ri'späns ,kar·ik·tə,ris·tik }

response time [COMPUT SCI] The delay experienced in time sharing between request and answer, a delay which increases when the number of users on the system increases. [CONT SYS] The time required for the output of a control system or element to reach a specified fraction of its new value after application of a step input or disturbance. [ELEC] The time it takes for the pointer of an electrical or electronic instrument to come to rest at a new value, after the quantity it measures has been abruptly changed. { ri'späns ,tīm }

responsor [ELECTR] The receiving section of an interrogator-responsor. { ri'spän·sər }

restart [AERO ENG] The act of firing a stage of a rocket after a previous powered flight. [COMPUT SCI] To go back to a specific planned point in a routine, usually in the case of machine malfunction, for the purpose of rerunning the portion of the routine in which the error occurred; the length of time between restart points in a given routine should be a function of the mean free error time of the machine itself. { 'rē,stärt }

rest density [RELAT] The density of a small portion of a fluid in a Lorentz frame in which that portion of the fluid is at rest. { 'rest ,den·səd·ē }

rest energy [RELAT] The energy equivalent to the rest mass m_0 of a particle or body; that is, the quantity of m_0c^2, where c is the speed of light; often expressed in electronvolts. { 'rest ,en·ər·jē }

rest frame [RELAT] The Lorentz frame in which the total momentum of a system equals zero; for an accelerated system, the rest frame varies from instant to instant. { 'rest ,frām }

rest hardening [GEOL] The increase of strength, with time, of a clay subsequent to its deposition, remolding, or modification by the application of shear stress. { 'rest ,härd·ən·iŋ }

restiform body See inferior cerebellar peduncle. { 'res·tə,form ,bäd·ē }

resting cell [CELL MOL] An interphase cell. { 'rest·iŋ ,sel }

resting frequency See carrier frequency. { 'rest·iŋ ,frē·kwən·sē }

resting metabolism [PHYSIO] The metabolism of a person at rest while seated or standing in a normal position. { 'rest·iŋ me,tab·ə,liz·əm }

resting potential [PHYSIO] The potential difference between the interior cytoplasm and the external aqueous medium of the living cell. { 'rest·iŋ pə,ten·chəl }

resting spore [BIOL] A spore that remains dormant for long periods before germination, withstanding adverse conditions; usually invested in a thickened cell wall. { 'rest·iŋ ,spor }

Restionaceae [BOT] A large family of monocotyledonous plants in the order Restionales characterized by unisexual flowers, wholly cauline leaves, unilocular anthers, and a more or less open inflorescence. { ,res·tē·ə'nās·ē,ē }

Restionales [BOT] An order of monocotyledonous plants in the subclass Commelinidae having reduced flowers and a single, pendulous, orthotropous ovule in each of the one to three locules of the ovary. { ,res·tē·ə'nā·lēz }

restitution coefficient See coefficient of restitution. { ,res·tə'tü·shən ,kō·i,fish·ənt }

restless legs syndrome [MED] A condition that is characterized by intense disagreeable feelings in the legs at rest and repose with compulsion to move the legs to get relief from these symptoms; peak onset usually occurs during middle age, and the disorder tends to become more severe with age. { ,rest·ləs 'legz ,sin,drōm }

rest magma [GEOL] The part of magma that remains after many minerals have crystallized from it during a long series of differentiations. Also known as residual liquor. { 'rest ,mag·mə }

rest mass [RELAT] The mass of a particle in a Lorentz reference frame in which it is at rest. { 'rest ,mas }

restoration [ECOL] A conservation measure involving the correction of past abuses that have impaired the productivity of the resources base. { ,res·tə'rā·shən }

restoration ecology [ECOL] The application of ecological principles and field methodologies to the successful restoration of damaged ecosystems. { ,res·tə,rā·shən ē'käl·ə·jē }

restorative dentistry [MED] A branch of dentistry that focuses on the preservation and restoration of decayed, defective, missing, and traumatized teeth. { rə,stir·əd·iv 'dent·ə·strē }

restorative immunotherapy [IMMUNOL] Immunotherapy that involves the direct and indirect restoration of deficient immunological function through any means other than the direct transfer of cells. { rə,stor·əd·iv ,im·yə·nō'ther·ə·pē }

restore [COMPUT SCI] In computers, to regenerate, to return a cycle index or variable address to its initial value, or to store again. [ELECTR] Periodic charge regeneration of volatile computer storage systems. { ri'stor }

restorer See direct-current restorer. { ri'stor·ər }

restorer pulses [ELECTR] In computers, pairs of complement pulses, applied to restore the coupling-capacitor charge in an alternating-current flip-flop. { ri'stor·ər ,pəls·əz }

restoring logic [ELECTR] Circuitry designed so that even with an imperfect input pulse a standard output occurs at the exit of each successive logic gate. { ri'stor·iŋ ,läj·ik }

rest point [ENG] On a balance, the position of the pointer with respect to the pointer scale when the beam has ceased moving. { 'rest ,point }

rest potential [ELEC] Residual potential difference remaining between an electrode and an electrolyte after the electrode has become polarized. { 'rest pə,ten·chəl }

restrainer [GRAPHICS] The ingredient of a developer which prevents too rapid development and chemical fog. { ri'strān·ər }

restraint of loads [ENG] The process of binding, lashing, and wedging items into one unit onto or into its transporter in a manner that will ensure immobility during transit. { ri'strānt əv 'lōdz }

restricted [GEOL] Referring to tectonic transport or movement in which elongation of particles is transverse to the direction of movement. { ri'strik·təd }

restricted adhesive [MATER] An adhesive which for any reason cannot satisfactorily pass its evaluation test; as a result, the maximum time required for curing, that is, its usable life, cannot be assigned; it cannot be used for structural bonding. { ri'strik·təd ad'hē·ziv }

restricted air cargo [IND ENG] Cargo which is not highly dangerous under normal conditions, but which possesses certain qualities which require extra precautions in packing and handling. { ri'strik·təd 'er ,kär·gō }

restricted area [NAV] **1.** An airspace of defined dimensions identified by an area on the surface of the earth within which the flight of aircraft is subject to restrictions. **2.** An area within which vessels may be navigated subject to prohibitions with respect to anchoring, trawling, fishing, and so on. { ri'strik·təd 'er·ē·ə }

restricted basin [GEOL] A depression in the floor of the ocean in which the water circulation is topographically restricted and therefore generally is oxygen-depleted. Also known as barred basin; silled basin. { ri'strik·təd 'bās·ən }

restricted function [COMPUT SCI] A function of the operating system that cannot be used by application programs. { ri'strik·təd 'fəŋk·shən }

restricted gate [ENG] Small opening between runner and cavity in an injection or transfer mold which breaks cleanly when the piece is ejected. { ri'strik·təd 'gāt }

restricted internal rotation [PHYS CHEM] Restrictions on the rotational motion of molecules or parts of molecules in some substances, such as solid methane, at certain temperatures. { ri'strik·təd in¦tərn·əl rō'tā·shən }

restricted job [IND ENG] A task whose performance time is governed by a machine, a process, another task, or the nature of the job itself, rather than being under the control of the worker. { ri'strik·təd 'jäb }

restricted limit See limit inferior. { ri¦strik·təd 'lim·ət }

restricted propellant [AERO ENG] A solid propellant having only a portion of its surface exposed for burning while the other surfaces are covered by an inhibitor. { ri'strik·təd prə'pel·ənt }

restricted proper motion [ASTRON] The rate of change of a star's apparent position relative to surrounding stars, corrected for precession, nutation, and aberration. { ri'strik·təd 'präp·ər 'mō·shən }

restricted waters [NAV] Areas which for navigational reasons, such as the presence of shoals or other dangers, confine the movements of shipping within narrow limits. { ri'strik·təd 'wȯd·ərz }

restricted work [IND ENG] Manual or machine work where the work pace is only partially under the control of the worker. { ri'strik·təd 'wərk }

restriction [CELL MOL] The degradation of foreign deoxyribonucleic acid by restriction endonucleases capable of recognizing particular patterns of specificity. { ri¦strik·shən }

restriction endonuclease [CELL MOL] Any of the specific endonucleases that recognizes a short specific sequence within a deoxyribonucleic acid molecule and then catalyzes double-stranded cleavage of that molecule. Also known as endodeoxyribonuclease. { ri'strik·shən ¦en·dō'nü·klē‚ās }

restriction endonuclease analysis [CELL MOL] A technique in which deoxyribonucleic acid (DNA) fragments obtained from digestion with restriction enzymes are compared to construct a restriction map showing the position of specific sites along a sequence of DNA. { ri‚strik·shən ‚en·dō'nü·klē‚ās ə‚nal·ə·səs }

restriction fragment [CELL MOL] Any of the individual polynucleotide sequences produced by digestion of deoxyribonucleic acid with a restriction endonuclease. { ri'strik·shən ‚frag·mənt }

restriction fragment length polymorphism [CELL MOL] Variations in the length of restriction fragments resulting from action by a specific endonuclease. { ri¦strik·shən ¦frag·mənt ‚leŋkth ‚päl·i'mȯr‚fiz·əm }

restriction map [CELL MOL] A diagram of a deoxyribonucleic acid molecule showing sites at which restriction endonucleases produce cleavage. { ri'strik·shən ‚map }

restriction of ego [PSYCH] A defense mechanism for escaping anxiety by avoiding situations consciously perceived as dangerous or uncomfortable. { ri'strik·shən əv 'ē·gō }

restriction point [CELL MOL] In the mammalian cell cycle, a time late in the G1 phase at which the cell commits to the replication of its deoxyribonucleic acid (DNA). { ri'strik·shən ‚pȯint }

restriction site [CELL MOL] A sequence in a deoxyribonucleic acid molecule that can be cleaved with a specific restriction endonuclease. { ri'strik·shən ‚sīt }

restrictive condition [GEN] An environmental condition under which a conditional lethal mutant either cannot grow or shows the mutant phenotype. { ri¦strik·tiv kən¦dish·ən }

rest stick See mahlstick. { 'rest ‚stik }

reststrahlen See residual radiation. { 'rest‚sträl·ən }

resue [MIN ENG] To mine a very narrow vein by first stoping the rock wall on one side and then removing the ore. { rə'sü }

resultant [MATH] **1.** For a set of polynomial equations, a function of the coefficients which equals zero if the equations have at least one solution. Also known as eliminant. **2.** See vector sum. { ri'zəl·tənt }

resultant of forces [MECH] A system of at most a single force and a single couple whose external effects on a rigid body are identical with the effects of the several actual forces that act on that body. { ri'zəl·tənt əv 'fȯrs·əz }

resultant rake [MECH ENG] The angle between the face of a cutting tooth and an axial plane through the tooth point measured in a plane at right angles to the cutting edge. { ri'zəlt·ənt 'rāk }

resultant wind [CLIMATOL] The vectorial average of all wind directions and speeds for a given level at a given place for a certain period, such as a month. { ri'zəlt·ənt 'wind }

resuperheating See reheating. { ‚rē'sü·pər'hēd·iŋ }

resupinate [BOT] Inverted, usually through 180°, so as to appear upside down or reversed. { rē'sü·pə‚nāt }

resupply [IND ENG] The act of replenishing stocks in order to maintain required levels of supply. { ¦rē·sə'plī }

resurgence [HYD] The point where an underground stream reappears at the surface to become a surface stream. Also known as emergence; exsurgence; rise. { ri'sər·jəns }

resurgent [GEOL] Referring to magmatic water or gases that were derived from sources on the earth's surface, from its atmosphere, or from country rock of the magma. { ri'sər·jənt }

resurgent cauldron [GEOL] A cauldron in which the cauldron block has been uplifted following subsidence, usually in the form of a structural dome. { ri'sər·jənt 'kȯl·drən }

resurrected [GEOL] Pertaining to a surface, landscape, or feature (such as a mountain, peneplain, or fault scarp) that has been restored by exhumation to its previous status in the existing relief. Also known as exhumed. [HYD] Pertaining to a stream that follows an earlier drainage system after a period of brief submergence has slightly masked the old course by a thin film of sediments. Also known a palingenetic. { ‚rez·ə¦rek·təd }

resuscitation [MED] Restoration of consciousness or life functions after apparent death. { ri‚səs·ə'tā·shən }

resuscitator [ENG] A device for supplying oxygen to and inducing breathing in asphyxiation victims. { ri'səs·ə‚tād·ər }

ret [CHEM] The reduction or digestion of fibers (usually linen) by enzymes. { ret }

retained water [HYD] The water remaining in rock or soil after gravity groundwater has been drained out. { ri'tānd 'wȯd·ər }

retainer [ENG] A device that holds a mechanical component in place. { ri'tān·ər }

retainer plate [ENG] The plate on which removable mold parts (such as a cavity or ejector pin) are mounted during molding. { ri'tān·ər ‚plāt }

retainer wall [ENG] A wall, usually earthen, around a storage tank or an area of storage tanks (tank farm); used to hold (retain) liquid in place if one or more tanks begin to leak. { ri'tān·ər ‚wȯl }

retaining ring [DES ENG] **1.** A shoulder inside a reaming shell that prevents the core lifter from entering the core barrel. **2.** A steel ring between the races of a ball bearing to maintain the correct distribution of the balls in the races. { ri'tān·iŋ ‚riŋ }

retaining wall [CIV ENG] A wall designed to maintain differences in ground elevations by holding back a bank of material. { ri'tān·iŋ ‚wȯl }

retard [CIV ENG] A permeable bank-protection structure, situated at and parallel to the toe of a slope and projecting into a stream channel, designed to check stream velocity and induce silting or accretion. { ri'tärd }

retardant See retarder. { ri'tärd·ənt }

retardation [MED] Slow mental or physical functioning. [NAV] The amount of time or phase angle introduced by the resistivity of the surface over which the radio wave in radio navigation is passing. [OCEANOGR] The amount of time by which corresponding tidal phases grow later day by day, averaging approximately 50 minutes. [OPTICS] In interference microscopy, the difference in optical path between the light passing through the specimen and the light bypassing the specimen. Also known as optical-path difference. { ‚rē‚tär'dā·shən }

retardation coil [ELECTROMAG] A high-inductance coil used in telephone circuits to permit passage of direct current or low-frequency ringing current while blocking the flow of audio-frequency currents. { ‚rē‚tär‚dā·shən ‚kȯil }

retardation plate See wave plate. { ‚rē‚tär'dā·shən ‚plāt }

retardation sheet See wave plate. { ‚rē‚tär'dā·shən ‚shēt }

retardation theory [OPTICS] General methods of calculating the effect of one or more wave plates on light which is normally incident on the plates and which is initially polarized in some fashion. { ‚rē‚tär'dā·shən ‚thē·ə·rē }

retarded acid [PETRO ENG] Oil well acidizing solution whose reactivity is slowed by addition of artificial gums and thickening agents, so that the acid penetrates deeper into the formation before being spent. { ri'tärd·əd 'as·əd }

retarded field [ELECTROMAG] An electric or magnetic field strength as found from the retarded potentials. { ri'tärd·əd fēld }

retarded potentials [ELECTROMAG] The electromagnetic potentials at an instant in time t and a point in space r as a function of the charges and currents that existed at earlier times at points on the past light cone of the event r,t. { ri'tärd·əd pə'ten·chəlz }

retarder [MATER] A material that inhibits the action of another substance, such as flameproofing agents or substances added to cement to retard setting time. Also known as retardant. [MECH ENG] **1.** A braking device used to control the speed of railroad cars moving along the classification tracks in a hump yard. **2.** A strip inserted in a tube of a fire-tube boiler to increase agitation of the hot gases flowing therein. { ri'tärd·ər }

retarding basin [CIV ENG] A basin designed and operated

to provide temporary storage and thus reduce the peak flood flows of a stream. { ri'tärd·iŋ ,bäs·ən }

retarding conveyor [MECH ENG] Any type of conveyor used to restrain the movement of bulk materials, packages, or objects where the incline is such that the conveyed material tends to propel the conveying medium. { ri'tärd·iŋ kən,vā·ər }

retarding-field oscillator [ELECTR] An oscillator employing an electron tube in which the electrons oscillate back and forth through a grid that is maintained positive with respect to both the cathode and anode; the field in the region of the grid exerts a retarding effect through the grid in either direction. Also known as positive-grid oscillator. { ri'tärd·iŋ ¦fēld 'äs·ə ,lād·ər }

retarding potential [PHYS] A potential which causes the speed of a moving particle to be reduced. { ri'tärd·iŋ pə,ten·chəl }

retard transmitter [ELECTR] Transmitter in which a delay period is introduced between the time of actuation and the time of transmission. { ,ri'tärd tranz,mid·ər }

rete cord [EMBRYO] One of the deep, anastomosing strands of cells of the medullary cords of the vertebrate embryo that form the rete testis or the rete ovarii. { 'rēd·ē ,kȯrd }

rete mirabile [VERT ZOO] A network of small blood vessels that are formed by the branching of a large vessel and that usually reunite into a single trunk; believed to have an oxygen-storing function in certain aquatic fauna. { 'rēd·ē mi'räb·ə·lē }

retene [ORG CHEM] $C_{18}H_{18}$ A cyclic hydrocarbon, melting at 100.5–101°C, soluble in benzene and hot ethanol; used in organic syntheses. { 'rē,tēn }

retention cyst [MED] A cyst caused by obstructed outflow of secretion from a gland. { ri'ten·chən ,sist }

retention index [ANALY CHEM] In gas chromatography, the relationship of retention volume with arbitrarily assigned numbers to the compound being analyzed; used to indicate the volume retention behavior during analysis. { ri'ten·chən ,in,deks }

retention period [COMPUT SCI] The length of time that data must be kept on a reel of magnetic tape before it can be destroyed. { ri'ten·chən ,pir·ē·əd }

retention time [ANALY CHEM] In gas chromatography, the time at which the center, or maximum, of a symmetrical peak occurs on a gas chromatogram. [ELECTR] The maximum time between writing into a storage tube and obtaining an acceptable output by reading. Also known as storage time. { ri'ten·chən ,tīm }

retention volume [ANALY CHEM] In gas chromatography, the product of retention time and flow rate. { ri'ten·chən ,väl·yəm }

retentivity [ELECTROMAG] The residual flux density corresponding to the saturation induction of a magnetic material. { ,rē,ten'tiv·əd·ē }

rete ovarii [ANAT] Vestigial tubules or cords of cells near the hilus of the ovary, corresponding with the rete testis, but not connected with the mesonephric duct. { 'rēd·ē ō'var·ē,ī }

rete testis [ANAT] The network of anastomosing tubules in the mediastinum testis. { 'rēd·ē 'tes·təs }

retgersite [MINERAL] $NiSO_4·6H_2O$ A deep emerald green, tetragonal mineral consisting of a hydrated nickel sulfate. { 'ret·gər,sīt }

Retgers' law [SOLID STATE] The law that the properties of crystalline mixtures of isomorphous substances are continuous functions of the percentage composition. { 'ret·gərz,lȯ }

rethrolone [ORG CHEM] A generic name for the five-member ring portion of a pyrethrin. { 'reth·rə,lōn }

reticle [OPTICS] A series of intersecting fine lines, wires, or the like which are placed in the focus of the objective of an optical instrument to aid in measurement of angles or distances. { 'red·ə·kəl }

reticle image [OPTICS] A light image of the reticle in a computing gunsight or in certain types of optical gunsights and bombsights, cast on a reflector plate and superimposed on the target. { 'red·ə·kəl ,im·ij }

reticular See reticulate. { re'tik·yə·lər }

reticular cell See reticulocyte. { re'tik·yə·lər 'sel }

reticular degeneration [PATH] Rupture of epidermal cells with formation of multilocular bullae due to intracellular edema. { re'tik·yə·lər di,jen·ə'rā·shən }

reticular density [MATH] The number of points per unit area

in a two-dimensional lattice, such as the plane of a crystal lattice. { re'tik·yə·lər 'den·səd·ē }

reticular fiber [HISTOL] Any of the delicate, branching argentophile fibers conspicuous in the connective tissue of lymphatic tissue, myeloid tissue, the red pulp of the spleen, and most basement membranes. Also known as argentaffin fiber; argyrophil lattice fiber; precollagenous fiber. { re'tik·yə·lər 'fī·bər }

reticular formation [NEUROSCI] The portion of the central nervous system which consists of small islands of gray matter separated by fine bundles of nerve fibers running in every direction. { re'tik·yə·lər fȯr¦mā·shən }

Reticulariaceae [MYCOL] A family of plasmodial slime molds in the order Liceales. { rə¦tik·yə,lar·ē'ās·ē,ē }

reticular system See reticuloendothelial system. { re'tik·yə·lər ,sis·təm }

reticular tissue [HISTOL] Connective tissue having reticular fibers as the principal element. { re'tik·yə·lər ,tish·ü }

reticulate [BIOL] Having or resembling a network of fibers, veins, or lines. [GEOL] **1.** Referring to a vein or lode with netlike texture. **2.** Referring to rock texture in which crystals are partly altered to a secondary material, forming a network that encloses the remnants of the original mineral. Also known as mesh texture; reticular; reticulated. { rə'tik·yə·lət }

reticulated See reticulate. { rə'tik·yə,lād·əd }

reticulated bar [GEOL] One of a group of slightly submerged sandbars in two sets, both of which are diagonal to the shoreline, forming a crisscross pattern. { rə'tik·yə,lād·əd 'bär }

reticulated glass [MATER] Ornamental glassware containing interlacing sets of lines. { rə'tik·yə,lād·əd 'glas }

reticulate venation [BOT] A branching vascular system with successively thinner veins diverging as branches from the thicker veins. { rə,tik·yə·lət ve'nā·shən }

reticulation [GRAPHICS] The contraction or wrinkling of a film emulsion because of sudden changes of temperature during processing. { rə,tik·yə'lā·shən }

reticulin [BIOCHEM] A protein isolated from reticular fibers. { rə'tik·yə·lən }

reticulocyte [HISTOL] Also known as reticular cell. **1.** A large, immature red blood cell, having a reticular appearance when stained due to retention of portions of the nucleus. **2.** A cell of reticular tissue. { rə'tik·yə·lə,sīt }

reticulocytopenia [MED] A decrease in the normal number of circulating reticulocytes. Also known as reticulopenia. { re,tik·yə·lō,sīd·ə'pē·nyə }

reticulocytosis [MED] An increase in the normal number of circulating reticulocytes. { rə,tik·yə·lō·sə'tō·səs }

reticuloendothelial granulomatosis [MED] A group of rare diseases characterized by generalized reticuloendothelial hyperplasia with or without intracellular lipid deposition. { rə¦tik·yə·lō,en·dō'thē·lē·əl ¦gran·yə·lō·mə'tō·səs }

reticuloendothelial system [ANAT] The macrophage system, including all phagocytic cells such as histiocytes, macrophages, reticular cells, monocytes, and microglia, except the granular white blood cells. Abbreviated RES. Also known as hematopoietic system; reticular system. { rə¦tik·yə·lō,en·dō'thē·lē·əl ,sis·təm }

reticuloendothelium [HISTOL] The cells making up the reticuloendothelial system. { rə,tik·yə·lō,en·də'thē·lē·əm }

reticulohistiocytoma [MED] A solitary skin nodule composed of large multinucleated histiocytes that contain glycolipids. { rə,tik·yə·lō,his·tē·ō·sə'tō·mə }

reticulopenia See reticulocytopenia. { re,tik·yə·lō'pē·nyə }

reticulopodia [INV ZOO] Pseudopodia in the form of a branching network. { rə¦tik·yə·lō'päd·ē·ə }

Reticulosa [PALEON] An order of Paleozoic hexactinellid sponges with a branching form in the subclass Hexasterophora. { rə,tik·yə'lō·sə }

reticulosis [MED] An increase in the number of histiocytes, monocytes, or other reticuloendothelial elements. { rə,tik·yə'lō·səs }

reticulospinal tract [NEUROSCI] Nerve fibers descending from large cells of the reticular formation of the pons and medulla into the spinal cord. { rə¦tik·yə·lō'spīn·əl 'trakt }

reticulum [BIOL] A fine network. [VERT ZOO] The second stomach in ruminants. { rə'tik·yə·ləm }

Reticulum [ASTRON] A southern constellation, right ascension 4 hours, declination 60° south. Also known as Net. { rə'tik·yə·ləm }

Reticulum system [ASTRON] A globular cluster or dwarf galaxy near the Large Magellanic Cloud { ri'tik·yə·ləm ,sis·təm }

retina [COMPUT SCI] In optical character recognition, a scanning device. [NEUROSCI] The photoreceptive layer and terminal expansion of the optic nerve in the dorsal aspect of the vertebrate eye. { 'ret·ən·ə }

retina character reader [COMPUT SCI] A character reader that operates in the manner of the human retina in recognizing identical letters in different type fonts. { 'ret·ən·ə 'kar·ik·tər ,rēd·ər }

retinaculum [INV ZOO] **1.** A clasp on the forewing of certain moths for retaining the frenulum of the hindwing. **2.** An appendage on the third abdominal somite of springtails that articulates with the furcula. { ,ret·ən'ak·yə·ləm }

retinal [BIOCHEM] A carotenoid, produced as an intermediate in the bleaching of rhodopsin and decomposition to vitamin A. Also known as vitamin A aldehyde. { 'ret·ən·əl }

retinal astigmatism [MED] Astigmatism due to changes in the localization of the fixation point. { 'ret·ən·əl ə'stig·mə,tiz·əm }

retinal detachment [MED] An eye disorder characterized by the separation of the sensory layers of the retina from their supporting foundations. { ,ret·ən·əl di'tach·mənt }

retinal illuminance [OPTICS] A psychophysiological quantity which is a measure of the brightness of a visual sensation; it is measured in trolands. { 'ret·ən·əl i'lü·mə·nəns }

retinalite [MINERAL] A massive, honey-yellow or greenish serpentine mineral with a waxy or resinous luster; a variety of chrysolite. { 'ret·ən·əl,īt }

retinal pigment See rhodopsin. { 'ret·ən·əl 'pig·mənt }

retinal retinitis See vascular retinopathy. { 'ret·ən·əl ,ret·ən'īd·əs }

retinasphalt [MINERAL] A light-brown variety of retinite usually found with lignite. { ,ret·ən'as,sfolt }

retinene [BIOCHEM] A pigment extracted from the retina, which turns yellow by the action of light; the chief carotenoid of the retina. { 'ret·ən,ēn }

retinite [MINERAL] A fossil resin, such as glessite, krantzite, muckite, and ambrite, composed of 6–15% oxygen, lacking succinic acid, and found in brown coals and peat. { 'ret·ən,īt }

retinitis [MED] Inflammation of the retina. { ,ret·ən'īd·əs }

retinitis pigmentosa [MED] A hereditary affection inherited as a sex-linked recessive and characterized by slowly progressing atrophy of the retinal nerve layers, and clumping of retinal pigment, followed by attenuation of the retinal arterioles and waxy atrophy of the optic disks. { ,ret·ən'īd·əs ,pig·mən'tō·sə }

retinoblastoma [MED] A malignant tumor of the sensory layer of the retina. { ,ret·ən·ō·bla'stō·mə }

retinochoroiditis [MED] Inflammation of the retina and choroid. { ,ret·ən·ō,kór,ói'dīd·əs }

retinoid [BIOCHEM] The set of molecules composing vitamin A and its synthetic analogs, such as retinal or retinyl acetate. { 'ret·ən,óid }

retinoid receptor [CELL MOL] Any member of a family of nuclear receptors that mediate the actions of natural and synthetic analogs of vitamin A (retinoids) by regulating the transcription of retinoid-responsive genes in the cell nucleus. { 'ret·ən,óid ri,sep·tər }

retinol See vitamin A. { 'ret·ən,ól }

retinopathy [MED] Any pathologic condition involving the retina. { ,ret·ən'äp·ə·thē }

retinoschisis [MED] **1.** Separation with hole formation of the layers composing the retina. **2.** A congenital anomaly characterized by cleavage of the retina. { ,ret·ən·ō'ski·səs }

retinula [INV ZOO] The receptor element at the inner end of the ommatidium in a compound eye. { rə'tin·yə·lə }

retire [NAV] To move a line of position back, parallel to itself, along a course line to obtain a line of position at an earlier time. { ri'tīr }

retired line of position [NAV] A line of position which has been moved backward along the course line to correspond with a time previous to that at which the line was established. { ri'tīrd 'līn əv pə'zish·ən }

retort [CHEM ENG] **1.** A closed refractory chamber in which coal is carbonized for manufacture of coal gas. **2.** A vessel for the distillation or decomposition of a substance. { ri'tórt }

Retortamonadida [INV ZOO] An order of parasitic flagellate protozoans belonging to the class Zoomastigophorea, having two or four flagella and a complex blepharoplast-centrosome-axostyle apparatus. { ri'tòr·də·mə'näd·ə·də }

retouch colors [GRAPHICS] Colors used to correct defects in black-and-white and color photographs; they adhere without crawling and can be used with brushes or airbrush on matte and glossy prints. { 'rē,toch ,kəl·ərz }

retrace See flyback. { 'rē,trās }

retrace blanking [ELECTR] Blanking a television picture tube during vertical retrace intervals to prevent retrace lines from showing on the screen. { 'rē,trās ,blaŋk·iŋ }

retrace line [ELECTR] The line traced by the electron beam in a cathode-ray tube in going from the end of one line or field to the start of the next line or field. Also known as return line. { 'rē,trās ,līn }

retracker See rerailer. { rē'trak·ər }

retract [MATH] A subset R of a topological space X is a retract of X if there is a continuous map f from X to R, with $f(r) = r$ for all points r of R. { 'rē,trakt }

retractor [ANAT] A muscle that draws a limb or other body part toward the body. [MED] A clawlike instrument for holding tissues away from the surgical field. { ri'trak·tər }

retransmission unit [ELECTR] Control unit used at an intermediate station for feeding one radio receiver-transmitter unit for two-way communication. { 'rē·tranz'mish·ən ,yü·nət }

retreat [MIN ENG] Workings in the opposite direction of advance work which, when completed, will permit the area to be abandoned as finished. { ri'trēt }

retreater [ENG] A defective maximum thermometer of the liquid-in-glass type in which the mercury flows too freely through the constriction; such a thermometer will indicate a maximum temperature that is too low. { ri'trēd·ər }

retreat gun See evening gun. { ri'trēt ,gən }

retrievable inner barrel [ENG] The inner barrel assembly of a wire-line core barrel, designed for removing core from a borehole without pulling the rods. { ri'trēv·ə·bəl 'in·ər 'bar·əl }

retrieve [COMPUT SCI] To find and select specific information. { ri'trēv }

retroaction See positive feedback. { 're·trō'ak·shən }

retroactive interference [PSYCH] Type of interference in which information that is learned later interferes with information that was learned earlier. { ,re·trō,ak·tiv ,in·tər'fir·əns }

retroactive refit See retrofit. { ,re·trō'ak·tiv 'rē,fit }

retrocerebral gland [INV ZOO] Any of various endocrine glands located behind the brain in insects which function in postembryonic development and metamorphosis. { 're·trō·sə'rē·brəl 'gland }

retrodirective mirror [OPTICS] **1.** An optical system consisting of two mutually perpendicular plane mirrors; it reflects any beam of light which lies in a plane perpendicular to the mirrors into a direction antiparallel to its original direction. **2.** An optical system consisting of three mutually perpendicular plane mirrors; it reflects any beam of light into a direction antiparallel to its original direction. { 're·tro·di'rek·tiv 'mir·ər }

retrofire time [AERO ENG] The computed starting time and duration of firing of retrorockets to decrease the speed of a recovery capsule and make it reenter the earth's atmosphere at the correct point for a planned landing. { 're·trō,fīr ,tīm }

retrofit [ENG] A modification of equipment to incorporate changes made in later production of similar equipment; it may be done in the factory or field. Derived from retroactive refit. { 're·trō,fit }

retroflexion [ANAT] The state of being bent backward. [MED] A condition in which the uterus is bent backward on itself, producing a sharp angle in its longitudinal axis at the junction of the cervix and the fundus. { ,re·trə'flek·shən }

retrogradation [CHEM] **1.** Generally, a process of deterioration; a reversal or retrogression to a simpler physical form. **2.** A chemical reaction involving vegetable adhesives, which revert to a simpler molecular structure. { ,re·trō·grā'dā·shən }

retrograde amnesia [MED] Loss of memory for events occurring prior to, but not after, the onset of a current disease or trauma. { 're·trə,grād am'nē·zha }

retrograde condensation [ORG CHEM] Phenomenon associated with the behavior of a hydrocarbon mixture in the critical region wherein, at constant temperature, the vapor phase in

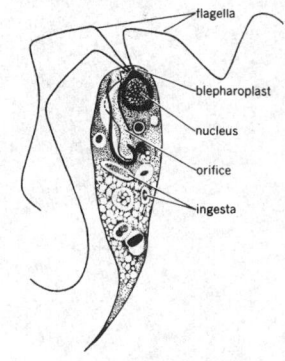

RETORTAMONADIDA

flagella
blepharoplast
nucleus
orifice
ingesta

Chilomastix aulastomi.

contact with the liquid may be condensed by a decrease in pressure; or at constant pressure, the vapor is condensed by an increase in temperature. { 're·trə‚grād ‚kän·dən'sā·shən }

retrograde evaporation [ORG CHEM] Phenomenon associated with the behavior of a hydrocarbon mixture in the critical region wherein, at constant temperature, the liquid phase in contact with the vapor may be vaporized by an increase in pressure; or at constant pressure, the liquid is evaporated by a decrease in temperature. { 're·trə‚grād i‚vap·ə'rā·shən }

retrograde gas-condensate reservoir *See* dew-point reservoir. { 're·trə‚grād ¦gas 'kän·dən‚sät 'rez·əv‚wär }

retrograde metamorphism [PETR] Formation of metamorphic minerals of a lower grade of metamorphism at the expense of minerals which are characteristic of a higher grade. Also known as diaphthoresis; retrogressive metamorphism. { 're·trə‚grād ‚med·ə'mȯr‚fiz·əm }

retrograde motion [ASTRON] **1.** An apparent backward motion of a planet among the stars resulting from the observation of the planet from the planet earth which is also revolving about the sun at a different velocity. Also known as retrogression. **2.** *See* retrograde orbit. { 're·trə‚grād 'mō·shən }

retrograde orbit [ASTRON] Motion in an orbit opposite to the usual orbital direction of celestial bodies within a given system; specifically, of a satellite, motion in a direction opposite to the direction of rotation of the primary. Also known as retrograde motion. { 're·trə‚grād 'ȯr·bət }

retrograde reservoir [GEOL] Hydrocarbon reservoir in which hydrocarbons are initially in the vapor phase; as pressure is reduced, the bubble-point line is passed and liquids are formed; upon further pressure reduction, a vapor phase is again formed. { 're·trə‚grād 'rez·əv‚wär }

retrograde wave [METEOROL] An atmospheric wave which moves in a direction opposite to that of the flow in which the wave is embedded; retrogression of a particular wave on daily charts is rarely seen, but is frequently observed on 4-day or monthly mean charts. { 're·trə‚grād 'wāv }

retrograding shoreline [GEOL] A shoreline that is being moved landward by wave erosion. { 're·trə‚grād·iŋ 'shȯr‚līn }

retrogression [ASTRON] *See* retrograde motion. [GEOL] *See* recession. [MED] Going backward, as in degeneration or atrophy of tissues. [METEOROL] The motion of an atmospheric wave or pressure system in a direction opposite to that of the basic flow in which it is embedded. [PSYCH] Return to infantile behavior. { ‚re·trə'gresh·ən }

retrogressive metamorphism *See* retrograde metamorphism. { ¦re·trə¦gres·iv ‚med·ə'mȯr‚fiz·əm }

retrolental fibroplasia [MED] An oxygen-induced disease of the retina in premature infants characterized by formation of an opaque membrane behind the lens of the eye. { ¦re·trə¦lent·əl ‚fī·brə'plā·zhə }

retroposon [GEN] A mobile genetic element that transposes to or from a chromosomal site by reverse transcription from a ribonucleic acid intermediate. { ‚re·trə'pō‚zän }

retroreflection [PHYS] Reflection wherein the reflected rays of radiation return along paths parallel to those of their corresponding incident rays. { ¦re·trō·ri'flek·shən }

retroreflector [PHYS] Any instrument used to cause reflected radiation to return along paths parallel to those of their corresponding incident rays; one type, the corner reflector, is an efficient radar target. { ¦re·trō·ri'flek·tər }

retrorocket [AERO ENG] A rocket fitted on or in a spacecraft, satellite, or the like to produce thrust opposed to forward motion. Also known as braking rocket. { ¦re·trō'räk·ət }

retrorse [BIOL] Bent downward or backward. { ri'trȯrs }

retrostalsis [PHYSIO] Reverse peristalsis. { ¦re·trō'stäl·səs }

retrosynthetic analysis [ORG CHEM] A method for planning an organic chemical synthesis in which the desired product molecule is considered first, and then steps are considered one at a time leading back to the appropriate starting materials. { ‚re·trō·sin¦thed·ik ə'nal·ə·səs }

retrotransposon [CELL MOL] A small, mobile deoxyribonucleic acid (DNA) sequence that can retrotranspose, that is, move from one genomic location to another by producing ribonucleic acid (RNA) that is transcribed by reverse transcriptase back into DNA which is then inserted at a new site. { ¦re·trō·tranz'pō‚zän }

retroussage [GRAPHICS] The method of bringing ink up from incised lines in an intaglio plate; dragging a soft cloth

across the ink-filled lines prior to printing makes the lines wider and renders certain passages darker and richer. { ‚re·trə'säzh }

retroversion [ANAT] A turning back. [MED] A condition in which the uterus is tilted backward without any change in the angle of its longitudinal axis. { ‚re·trə'vər·zhən }

Retroviridae [VIROL] A family of ribonucleic acid (RNA)–containing animal viruses characterized by spherical enveloped virions containing two single-stranded RNA molecules and reverse transcriptase; includes the subfamilies Oncovirinae, Spumavirinae, and Lentivirinae. { ‚re·trə'vir·ə‚dī }

retrovirus [VIROL] A family of ribonucleic acid viruses distinguished by virions which possess reverse transcriptase and which have two proteinaceous structures, a dense core, and an envelope that surrounds the core. { 're·trō‚vī·rəs }

retry [COMPUT SCI] When a central processing unit error is detected during execution of an instruction, the computer will execute this instruction unless a register was altered by the operation. { 'rē‚trī }

retting [CHEM ENG] Soaking vegetable stalks to decompose the gummy material and release the fibers. { 'red·iŋ }

Rett syndrome [PSYCH] An inherited developmental disorder observed only in females that is characterized by a short period of normal development, followed by loss of developmental skills (particularly purposeful hand movements) and marked psychomotor retardation. A brief autistic-like phase may be observed during the preschool period, but the subsequent course and clinical features are markedly different from autism. { 'ret ‚sin‚drōm }

return [BUILD] The continuation of a molding, projection, member, cornice, or the like, in a different direction, usually at a right angle. [COMPUT SCI] **1.** To return control from a subroutine to the calling program. **2.** To go back to a planned point in a computer program and rerun a portion of the program, usually when an error is detected; rerun points are usually not more than 5 minutes apart. [ELECTR] *See* echo. [GEOPHYS] Any of those surface waves on the record of a large earthquake which have traveled around the earth's surface by the long (greater than 180°) arc between epicenter and station, or which have passed the station and returned after traveling the entire circumference of the earth. { ri'tərn }

return address [COMPUT SCI] The address in storage to which a computer program is directed upon completion of a subroutine. { ri'tərn 'ad‚res }

return bead *See* quirk bead. { ri'tərn ‚bēd }

return bend [DES ENG] A pipe fitting, equal to two ells, used to connect parallel pipes so that fluid flowing into one will return in the opposite direction through the other. { ri'tərn ‚bend }

return busy tone [COMMUN] A signal returned to the register-sender that, in turn, returns a busy indication to the calling station. { ri'tərn 'biz·ē ‚tōn }

return code [COMPUT SCI] An indicator that is issued by a computer upon completion of a subroutine or function, or of the entire program, that indicates the result of the processing and, in particular, whether the processing was successful or ended abnormally because of an error. { ri'tərn ‚kōd }

return connecting rod [MECH ENG] A connecting rod whose crankpin end is located on the same side of the crosshead as the cylinder. { ri'tərn kə'nek·tiŋ ‚räd }

return difference [CONT SYS] The difference between 1 and the loop transmittance. { ri'tərn 'dif·rəns }

return flow [HYD] Irrigation water not consumed by evapotranspiration but returned to its source or to another body of ground or surface water. Also known as return water. { ri'tərn ‚flō }

return-flow burner [MECH ENG] A mechanical oil atomizer in a boiler furnace which regulates the amount of oil to be burned by the portion of oil recirculated to the point of storage. { ri'tərn ¦flō ‚bər·nər }

return idler [MECH ENG] The idler or roller beneath the cover plates on which the conveyor belt rides after the load which it was carrying has been dumped. { ri'tərn ‚īd·lər }

return interval [ELECTR] Interval corresponding to the direction of sweep not used for delineation. { ri'tərn ‚in·tər·vəl }

return jump [COMPUT SCI] A jump instruction in a subroutine which passes control to the first statement in the program which follows the instruction called the subroutine. { ri'tərn ‚jəmp }

return key [COMPUT SCI] A key on a typewriter or a computer keyboard that, when depressed, causes a print mechanism or cursor to move to the beginning of the next line. { ri'tərn ‚kē }

return line *See* retrace line. { ri'tərn ‚līn }

return loss [COMMUN] **1.** The difference between the power incident upon a discontinuity in a transmission system and the power reflected from the discontinuity. **2.** The ratio in decibels of the power incident upon a discontinuity to the power reflected from the discontinuity. { ri'tərn ‚lòs }

return streamer [GEOPHYS] The intensely luminous streamer which propagates upward from earth to cloud base in the last phase of each lightning stroke of a cloud-to-ground discharge. Also known as main stroke; return stroke. { ri'tərn ‚strēm·ər }

return stroke *See* return streamer. { ri'tərn ‚strōk }

return to zero mode [COMPUT SCI] Computer readout mode in which the signal returns to zero between each bit indication. { ri'tərn tə 'zir·ō ‚mōd }

return trace *See* flyback. { ri'tərn ‚trās }

return wall [BUILD] An interior wall of about the same height as the outside wall of a building; distinct from a partition or a low wall. { ri'tərn ‚wòl }

return water [HYD] *See* return flow. [PETRO ENG] In a water-injection operation (waterflood) for an oil reservoir, the reinjection of salt water that is produced along with the oil. { ri'tərn ‚wòd·ər }

return wire [ELEC] The ground wire, common wire, or negative wire of a direct-current power circuit. { ri'tərn ‚wīr }

retuse [BOT] Having a rounded apex with a slight, central notch. { ri'tüs }

retzian [MINERAL] $Mn_2Y(AsO_4)(OH)_4$ A chocolate brown to chestnut brown, orthorhombic mineral consisting of a basic arsenate of calcium, rare earths, and manganese. { 'ret·sē·ən }

Reuleaux triangle [MATH] A closed plane curve, not actually a triangle, that consists of three arcs, each of which joins two vertices of an equilateral triangle and is part of a circle centered at the remaining vertex. { ‚re‚lō 'trī‚aŋ·gəl }

reusable [COMPUT SCI] Of a program, capable of being used by several tasks without having to be reloaded; it is a generic term, including reenterable and serially reusable. { rē'yü·zə·bəl }

reveal [BUILD] **1.** The side of an opening for a door or window, doorway, or the like, between the doorframe or window frame and the outer surface of the wall. **2.** The distance from the face of a door to the face of the frame on the pivot side. { ri'vēl }

reveille gun [ORD] The firing of a gun at the first note of reveille or at sunrise. Also known as morning gun. { 'rev·ə·lē ‚gən }

revenue cutter [NAV ARCH] A ship used by the government mainly for enforcement of revenue laws and prevention of smuggling. { 'rev·ə‚nü ‚kəd·ər }

reverberant sound [ACOUS] The portion of the room impulse response consisting of sound that arrives at a listener's location more than 150 milliseconds after the first direct sound, and that has been reflected against walls and ceilings many times. { ri‚vər·bə·rənt 'saùnd }

reverberation [ACOUS] The prolongation of sound at a given point after direct reception from the source has ceased, due to such causes as reflections from bounding surfaces, scattering from inhomogeneities in a medium, and vibrations excited by the original sound. { ri‚vər·bə'rā·shən }

reverberation chamber [ACOUS] An enclosure with heavy surfaces which randomly reflect as great an amount of sound as possible; used in acoustic measurements. Also known as random diffusion chamber. { ri‚vər·bə'rā·shən ‚chām·bər }

reverberation time [ACOUS] The time in seconds required for the average sound-energy density at a given frequency to reduce to one-millionth of its initial steady-state value after the sound source has been stopped; this corresponds to a decrease of 60 decibels. { ri‚vər·bə'rā·shən ‚tīm }

reverberatory furnace [ENG] A furnace in which heat is supplied by burning of fuel in a space between the charge and the low roof. { ri'vər·brə‚tòr·ē 'fər·nəs }

reversal film [GRAPHICS] A type of film designed to yield a positive image directly when reversal processed. { ri'vər·səl ‚film }

reversal of dip [GEOL] Change in the dip direction of bedding near a fault such that the beds curve toward the fault surface in a direction exactly opposite that of the drag folds. Also known as dip reversal. { ri'vər·səl əv 'dip }

reversal process [GRAPHICS] A method in which a positive is produced from the plate or film exposed in a camera; the negative and positive stages are often produced from the same emulsion layer, with no separate printing material required; only one copy can be made at a time. { ri'vər·səl ‚präs·əs }

reversal spectrum [SPECT] A spectrum which may be observed in intense white light which has traversed luminous gas, in which there are dark lines where there were bright lines in the emission spectrum of the gas { ri'vər·səl ‚spek·trəm }

reversal speed [AERO ENG] The speed of an aircraft above which the aeroelastic loads will exceed the control surface loading of a given flight control system; the resultant load will act in the reverse direction from the control surface loading, causing the control system to act in a direction opposite to that desired. { ri'vər·səl ‚spēd }

reversal temperature [SPECT] The temperature of a blackbody source such that, when light from this source is passed through a luminous gas and analyzed in a spectroscope, a given spectral line of the gas disappears, whereas it appears as a bright line at lower blackbody temperatures, and a dark line at higher temperatures. { ri'vər·səl ‚em·prə·chər }

reverse bias [ELECTR] A bias voltage applied to a diode or a semiconductor junction with polarity such that little or no current flows; the opposite of forward bias. { ri'vərs 'bī·əs }

reverse-blocking tetrode thyristor *See* silicon controlled switch. { ri'vərs ‚bläk·iŋ 'te‚trōd thī'ris·tər }

reverse-blocking triode thyristor *See* silicon controlled rectifier. { ri'vərs ‚bläk·iŋ 'trī‚ōd thī'ris·tər }

reverse bonded-phase chromatography [ANALY CHEM] A technique of bonded-phase chromatography in which the stationary phase is nonpolar and the mobile phase is polar. { ri'vərs 'bän·dəd ‚fāz ‚krō·mə'täg·rə·fē }

reverse Brayton cycle [THERMO] A refrigeration cycle using air as the refrigerant but with all system pressures above the ambient. Also known as dense-air refrigeration cycle. { ri'vərs 'brāt·ən ‚sī·kəl }

reverse Carnot cycle [THERMO] An ideal thermodynamic cycle consisting of the processes of the Carnot cycle reversed and in reverse order, namely, isentropic expansion, isothermal expansion, isentropic compression, and isothermal compression. { ri'vərs kär'nō ‚sī·kəl }

reverse cell [METEOROL] A circulating fluid system in which the circulation in a vertical plane is thermally indirect; that is, cooler air rises relative to warmer air. { ri'vərs 'sel }

reverse circulation drilling [MIN ENG] **1.** A variation of the rotary drilling method in which the cuttings are pumped up and out of the drill pipe, an advantage in certain large diameter holes. **2.** Diamond core drilling in which the water is injected through a stuffing box into the annular space around the drill rods and thus forced up special large drill rods. { ri'vərs ‚sər·kyə'lā·shən ‚dril·iŋ }

reverse code dictionary [COMPUT SCI] Alphabetic or alphanumeric arrangement of codes associated with their corresponding English words or terms. { ri'vərs ‚kōd 'dik·shə‚ner·ē }

reverse current [ELECTR] Small value of direct current that flows when a semiconductor diode has reverse bias. { ri'vərs 'kə·rənt }

reverse-current cleaning *See* anodic cleaning. { ri'vərs 'kə·rənt ‚klēn·iŋ }

reverse-current protection [ELEC] A device which senses when there is a reversal in the normal direction of current in an electric power system, indicating an abnormal condition of the system, and which initiates appropriate action to prevent damage to the system. { ri'vərs 'kə·rənt prə‚tek·shən }

reverse-current relay [ELEC] Relay that operates whenever current flows in the reverse direction. { ri'vərs 'kə·rənt 'rē‚lā }

reverse curve [MATH] An S-shaped curve, that is, one having two arcs with their centers on opposite sides of the curve. Also known as S curve. { ri'vərs 'kərv }

reversed *See* overturned. { ri'vərst }

reversed air-blast process [CHEM ENG] A gasmaking process in which, after a short period of the ordinary blow, the air blast is reversed so as to enter the top of the superheater, and

REVERSE BRAYTON CYCLE

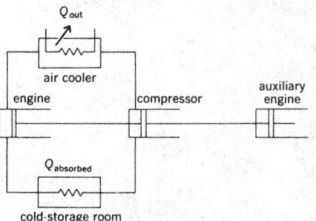

Schematic arrangement of the reverse Brayton cycle; air undergoes isentropic compression followed by reversible constant-pressure cooling; high-pressure air next expands reversibly in engine and exhausts at low temperature; cooled air passes through the cold storage chamber, picks up heat at constant pressure, and then returns to suction side of compressor. Q = heat.

REVERSE-CURRENT PROTECTION

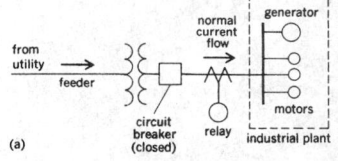

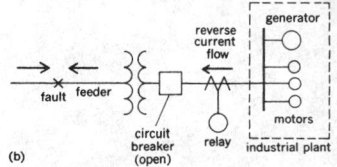

Example of reverse-current protection, in which utility supplies power to industrial plant having some generation of its own. (*a*) Normal load conditions. (*b*) Internal fault condition; relay trips circuit breaker under reverse-current condition, isolating plant from utility and preventing excessive burden on plant generator.

passes back to the top of the generator and down. { ri'vərst 'er‚blast ‚prä·səs }

reversed arc [GEOL] A curved belt of islands which is concave toward the open ocean, the opposite of most island arcs. { ri'vərst 'ärk }

reversed consequent stream [HYD] A consequent stream whose direction of flow is contrary to that normally consistent with the geologic structure. { ri'vərst 'kän·sə·kwənt 'strēm }

reverse deionization [CHEM] A process in which an ion-exchange unit and a cation-exchange unit are used in sequence to remove all ions from a solution. { ri'vərs dē‚ī·ə·nə'zā·shən }

reversed image [GRAPHICS] **1.** A mirror image in which the right and left sides of the picture are interchanged. **2.** See negative. [OPTICS] See inverted image. { ri'vərst 'im·ij }

reverse direction See inverse direction. { ri'vərs di'rek·shən }

reverse-direction flow [COMPUT SCI] A logical path that runs upward or to the left on a flowchart. { ri'vərs di‚rek·shən ‚flō }

reversed-phase partition chromatography [ANALY CHEM] Paper chromatography in which the low-polarity phase (such as paraffin, paraffin jelly, or grease) is put onto the support (paper) and the high-polarity phase (such as water, acids, or organic solvents) is allowed to flow over it. { ri'vərst ‚fāz pär'tish·ən ‚krō·mə'täg·rə·fē }

reversed polarity [GEOPHYS] Natural remanent magnetism opposite that of the present geomagnetic field. { ri'vərst pə'lar·əd·ē }

reverse drawing [MET] Drawing for a second time, in a direction opposite to the original drawing. { ri'vərs 'drò·iŋ }

reversed stream [HYD] A stream whose direction of flow has been reversed, as by glacial action, landsliding, gradual tilting of a region, or capture. { ri'vərst 'strēm }

reversed tide [OCEANOGR] An oceanic tide that is out of phase with the apparent motions of the tide-producing body, so that low tide is directly under the tide-producing body and is accompanied by a low tide on the opposite side of the earth. Also known as inverted tide. { ri'vərst 'tīd }

reverse engineering [ENG] The analysis of a completed system in order to isolate and identify its individual components or building blocks. { ri'vərs ‚en·jə'nir·iŋ }

reverse fault See thrust fault. { ri'vərs 'fòlt }

reverse feedback See negative feedback. { ri'vərs 'fēd‚bak }

reverse flange [ENG] A flange made by shrinking. { ri'vərs 'flanj }

reverse-flowage fold [GEOL] A fold in which flow from deformation has thickened the anticlinal crests and thinned the synclinal troughs, contrary to the normal flow pattern of a flow fold. { ri'vərs ‚flō·ij 'fōld }

reverse genetics [CELL MOL] An experimental method in which information from cloned deoxyribonucleic acid (DNA) or protein sequences is used to find or to produce mutations that help identify the function of a gene or protein (in contrast to classical genetics in which a known function or trait is traced back to a particular gene. { ri‚vərs jə'ned·iks }

reverse graft [BOT] A plant graft made by inserting the scion in an inverted position. { ri'vərs 'graft }

reverse key [ELEC] Key used in a circuit to reverse the polarity of that circuit. { ri'vərs 'kē }

reverse lay [DES ENG] The lay of a wire rope with strands alternating in a right and left lay. { ri'vərs 'lā }

reverse mutation [GEN] A mutation in a mutant allele which makes it capable of producing the nonmutant phenotype; may actually restore the original deoxyribonucleic acid sequence of the gene or produce a new one which has a similar effect. Also known as back mutation. { ri'vərs myü'tā·shən }

reverse osmosis [CHEM ENG] A technique used in desalination and waste-water treatment; pressure is applied to the surface of a saline (or waste) solution, forcing pure water to pass from the solution through a membrane (hollow fibers of cellulose acetate or nylon) that will not pass sodium or chloride ions. { ri'vərs äs'mō·səs }

reverse passive anaphylaxis [IMMUNOL] Hypersensitivity produced when the antigen is injected first, then followed in several hours by the specific antibody, causing shock. { ri'vərs 'pas·iv ‚an·ə·fə'lak·səs }

reverse pinocytosis See emiocytosis. { ri'vərs ‚pī·nō·sī'tō·səs }

reverse pitch [MECH ENG] A pitch on a propeller blade producing thrust in the direction opposite to the normal one. { ri'vərs 'pich }

reverse polarity [MET] An arc-welding circuit in which the electrode is connected to the positive terminal. { ri'vərs pə'lar·əd·ē }

reverse Polish notation [COMPUT SCI] The version of Polish notation, used in some calculators, in which operators follow the operators with which they are associated. Abbreviated RPN. Also known as postfix notation; suffix notation. { ri'vərs 'pō·lish nō'tā·shən }

reverse power [ELEC] Transmission of electric energy through a circuit in a direction opposite to the usual direction. { ri'vərs 'paù·ər }

reverse-printout typewriter [ENG] An automatic typewriter that eliminates conventional carriage return by typing one line from left to right and the next line from right to left. { ri'vərs ‚print‚aùt 'tīp‚rīd·ər }

reverse process [GRAPHICS] The conversion of a negative to a positive image by chemical means. { ri'vərs ‚prä·səs }

reverse-roll coating [ENG] Substrate coating that is premetered between rolls and then wiped off on the web; amount of coating is controlled by the metering gap and the rotational speed of the roll. { ri'vərs ‚rōl 'kōd·iŋ }

reverse saddle [GEOL] A mineral deposit associated with the trough of a synclinal fold and following the bedding plane. Also known as trough reef. { ri'vərs ‚sad·əl }

reverse similar fold [GEOL] A fold whose strata are thickened on the limbs and thinned on the axes, contrary to the pattern of a similar fold. { ri'vərs 'sim·ə·lər 'fōld }

reverse slip fault See thrust fault. { ri'vərs 'slip ‚fòlt }

reverse slope [GEOL] A hill descending away from a ridge. { ri'vərs ‚slōp }

reverse transcript [MOL BIO] A deoxyribonucleic acid sequence obtained from a ribonucleic acid sequence by means of reverse transcription. { ri‚vərs 'tran‚skript }

reverse transcriptase [GEN] A polymerase that mediates deoxyribonucleic acid synthesis by using a ribonucleic acid template. { ri'vərs tran'skrip‚tās }

reverse video [COMPUT SCI] An electronic display mode in which the normal properties of the display are reversed; for example, normally white characters on a black background will appear as black characters on a white background. Also known as inverse video. { ri'vərs 'vid·ē·ō }

reverse voltage [ELEC] In the case of two opposing voltages, voltage of that polarity which produces the smaller current. { ri'vərs 'vōl·tij }

reversibility principle [OPTICS] The principle that if a beam of light is reflected back on itself, it will traverse the same path or paths as it did before reversal. [STAT MECH] See microscopic reversibility. { ri‚vər·sə'bil·əd·ē ‚prin·sə·pəl }

reversible booster [ELEC] Booster capable of adding to and subtracting from the voltage of a circuit. { ri'vər·sə·bəl 'büs·tər }

reversible capacitance [ELECTR] Limit, as the amplitude of an applied sinusoidal capacitor voltage approaches zero, of the ratio of the amplitude of the resulting in-phase fundamental-frequency component of transferred charge to the amplitude of the applied voltage, for a given constant bias voltage superimposed on the sinusoidal voltage. { ri'vər·sə·bəl kə'pas·əd·əns }

reversible chemical reaction [CHEM] A chemical reaction that can be made to proceed in either direction by suitable variations in the temperature, volume, pressure, or quantities of reactants or products. { ri'vər·sə·bəl 'kem·ə·kəl rē'ak·shən }

reversible counter [COMPUT SCI] A counter which stores a number whose value can be decreased or increased in response to the appropriate control signal. { ri'vər·sə·bəl 'kaùnt·ər }

reversible electrode [PHYS CHEM] An electrode that owes its potential to unit charges of a reversible nature, in contrast to electrodes used in electroplating and destroyed during their use. { ri'vər·sə·bəl i'lek‚trōd }

reversible engine [THERMO] An ideal engine which carries out a cycle of reversible processes. { ri'vər·sə·bəl 'en·jən }

reversible motor [ELEC] A motor in which the direction of rotation can be reversed by means of a switch that changes motor connections when the motor is stopped. { ri'vər·sə·bəl 'mōd·ər }

reversible path [THERMO] A path followed by a thermodynamic system such that its direction of motion can be reversed at any point by an infinitesimal change in external conditions; thus the system can be considered to be at equilibrium at all points along the path. { ri'vər·sə·bəl 'path }

reversible-pitch propeller [MECH ENG] A type of controllable-pitch propeller; of either controllable or constant speed, it has provisions for reducing the pitch to and beyond the zero value, to the negative pitch range. { ri vər·sə·bəl ¦pich prə'pel·ər }

reversible process [THERMO] An ideal thermodynamic process which can be exactly reversed by making an indefinitely small change in the external conditions. Also known as quasistatic process. { ri'vər·sə·bəl 'präs·əs }

reversible steering gear [MECH ENG] A steering gear for a vehicle which permits road shock and wheel deflections to come through the system and be felt in the steering control. { ri'vər·sə·bəl 'stir·iŋ ˌgir }

reversible tramway See jig back. { ri'vər·sə·bəl 'tram,wā }

reversible transducer [ELECTR] Transducer whose loss is independent of transmission direction. { ri'vər·sə·bəl tranz 'düs·ər }

reversible transit circle [ENG] A transit circle that can be lifted out of its bearings and rotated through 180°, enabling systematic errors in both orientations to be determined. { ri'vər·sə·bəl 'tran·zət ˌsər·kəl }

reversing current [OCEANOGR] Any current that changes direction, with a period of slack water at each reversal of direction. { ri'vərs·iŋ ˌkə·rənt }

reversing dune [GEOL] A dune that tends to develop unusual height but migrates only a limited distance because seasonal shifts in dominant wind direction cause it to move alternately in nearly opposite directions. { ri'vərs·iŋ ˌdün }

reversing layer [ASTROPHYS] A layer of relatively cool gas forming the lower part of the sun's chromosphere, just above the photosphere, that gives rise to absorption lines in the sun's spectrum. { ri'vərs·iŋ ˌlā·ər }

reversing mill [MET] A rolling mill in which the workpiece is passed forward and backward through a given pair of rolls. { ri'vərs·iŋ ˌmil }

reversing motor [ELEC] A motor for which the direction of rotation can be reversed by changing electric connections or by other means while the motor is running at full speed; the motor will then come to a stop, reverse, and attain full speed in the opposite direction. { ri'vərs·iŋ ˌmōd·ər }

reversing switch [ELEC] A switch intended to reverse the connections of one part of a circuit. { ri'vərs·iŋ ˌswich }

reversing thermometer [ENG] A mercury-in-glass thermometer which records temperature upon being inverted and thereafter retains its reading until returned to the first position. { ri'vərs·iŋ thər'mäm·əd·ər }

reversing water bottle See Nansen bottle. { ri vərs·iŋ 'wod·ər ˌbäd·əl }

reversion [CHEM ENG] In rubber manufacture, a decrease in rubber modulus or viscosity caused by overworking. [MATH] For a series, the process of constructing a new series in which the dependent and independent variables of the original series are interchanged. { ri'vər·zhən }

revet-crag [GEOL] One of a series of narrow, pointed outliers or ridges of eroded strata inclined like a revetment against a mountain spur. { rə'vet ˌkrag }

revetment [CIV ENG] A facing made on a soil or rock embankment to prevent scour by weather or water. [ORD] A retaining wall with a facing such as concrete or stone, commonly used for fortifications or to protect against explosions. { rə'vet·mənt }

revived fault scarp See rejuvenated fault scarp. { ri'vīvd 'fȯlt ˌskärp }

revived folding See recurrent folding. { ri'vīvd 'fōld·iŋ }

revived stream See rejuvenated stream. { ri'vīvd 'strēm }

revolute [BOT] Rolled backward and downward. { 'rev·ə,lüt }

revolute-coordinate robot See jointed-arm robot. { 'rev·ə,lüt kō¦ȯrd·ən·ət 'rō,bät }

revolute joint [MECH ENG] A robotic articulation consisting of a pin with one degree of freedom. { 'rev·ə,lüt ˌjȯint }

revolution [GEOL] A little-used term to describe a time of profound crustal movements, on a continentwide or worldwide

scale, which led to abrupt geographic, climatic, and environmental changes that were related to changes in forms of life. [MECH] The motion of a body around a closed orbit. { ˌrev·ə'lü·shən }

revolution counter [ENG] An instrument for registering the number of revolutions of a rotating machine. Also known as revolution indicator. { ˌrev·ə'lü·shən ˌkaunt·ər }

revolution indicator See revolution counter. { ˌrev·ə'lü·shən ˌin·də,kād·ər }

revolution per minute [MECH] A unit of angular velocity equal to the uniform angular velocity of a body which rotates through an angle of 360° (2π radians), so that every point in the body returns to its original position, in 1 minute. Abbreviated rpm. { ˌrev·ə'lü·shən pər 'min·ət }

revolution per second [MECH] A unit of angular velocity equal to the uniform angular velocity of a body which rotates through an angle of 360° (2π radians), so that every point in the body returns to its original position, in 1 second. Abbreviated rps. { ˌrev·ə'lü·shən pər 'sek·ənd }

revolution table [NAV ARCH] A table giving the number of shaft revolutions corresponding to various speeds of a vessel. { ˌrev·ə'lü·shən ˌtā·bəl }

revolver [NAV] The pair of horizontal angles between three points, as observed at any place on the circle defined by the three points; this is the one situation in which such angles do not establish a fix. Also known as swinger. [ORD] A firearm with a cylinder of several chambers so arranged as to revolve on an axis and be discharged in succession by the same lock. { ri'väl·vər }

revolving-block engine [MECH ENG] Any of various engines which combine reciprocating piston motion with rotational motion of the entire engine block. { ri'välv·iŋ ¦bläk 'en·jən }

revolving door [BUILD] A door consisting of four leaves that revolve together on a central vertical axis within a circular vestibule. { ri'välv iŋ 'dȯr }

revolving shovel [MECH ENG] A digging machine, mounted on crawlers or on rubber tires, that has the machinery deck and attachment on a vertical pivot so that it can swing freely. { ri'välv·iŋ 'shəv·əl }

revolving storm [METEOROL] A cyclonic storm, or one in which the wind revolves about a central low-pressure area. { ri'välv·iŋ 'stȯrm }

rewind [ELECTR] 1. The components on a magnetic tape recorder that serve to return the tape to the supply reel at high speed. 2. To return a magnetic tape to its starting position. { 'rē,wīnd }

rework [GEOL] Any geologic material that has been removed or displaced by natural agents from its origin and incorporated in a younger formation. { 'rē,wərk }

rewrite [COMPUT SCI] The process of restoring a storage device to its state prior to reading; used when the information-storing state may be destroyed by reading. { 'rē,rīt }

Reychler's acid See d-camphorsulfonic acid. { 'rī·klərz ˌas·əd }

Reye's syndrome [MED] An uncommon liver disorder primarily occurring in infants and young children; characterized by convulsions, hypoglycemia, and a liver showing diffuse microvacuolar fatty metamorphosis. { 'rīz ˌsin drōm }

reyn [FL MECH] A unit of dynamic viscosity equal to the dynamic viscosity of a fluid in which there is a tangential force of 1 poundal per square foot resisting the flow of two parallel fluid layers past each other when their differential velocity is 1 foot per second per foot of separation; equal to approximately 14.8816 poise. { ren }

Reynier's isolator [ENG] A mechanical barrier made of steel that surrounds the area in which germ-free vertebrates and accessory equipment are housed; has electricity for light and power, an exit-entry opening with a steam barrier, a means for sterile air exchange, glass viewing port, and neoprene gloves which allow handling of the animals. { rān'yās 'īs·ə,lād·ər }

Reynolds analogy [CHEM ENG] Relationship showing the similarity between the transfer of mass, heat, and momentum. { 'ren·əlz ə,nal·ə·jē }

Reynolds-averaged Navier-Stokes analysis [FL MECH] The process of determining numerical solutions of the Navier-Stokes equations for a fluid flow, using time averaging of flow variables and modeling of turbulent stresses to simplify the

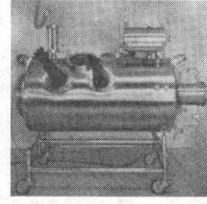

REYNIER'S ISOLATOR

Basic germ-free Reynier's isolator showing gloves used for working in the isolator.

calculations. Abbreviated RANS analysis. { ¦ren·əlz ¦av·rijd ¦näv·ē¸ā 'stōks ə¸nal·ə·səs }

Reynolds criterion [FL MECH] The principle that the type of fluid motion, that is, laminar flow or turbulent flow, in geometrically similar flow systems depends only on the Reynolds number; for example, in a pipe, laminar flow exists at Reynolds numbers less than 2000, turbulent flow at numbers above about 3000. { 'ren·əlz ¸krī¸tir·ē·ən }

Reynolds effect [METEOROL] A process of drop growth in clouds which involves net evaporation from cloud drops warmer than others and net condensation on the cooler drops. { 'ren·əlz i¸fekt }

Reynolds equation [FL MECH] A form of the Navier-Stokes equation which is

$$\rho \partial u/\partial t = (\partial/\partial x)(p_{xx} - \rho u^2) + (\partial/\partial y) \cdot (p_{xy} - \rho uv) + (\partial/\partial z)(p_{xz} - \rho uw)$$

where ρ is the fluid density, u, v, and w are the components of the fluid velocity, and p_{xx}, p_{xy}, and p_{xz} are normal and shearing stresses. { 'ren·əlz i¸kwā·zhən }

Reynolds model [OCEANOGR] A laboratory model of ocean currents in which inertial forces and frictional forces predominate, and in which the Reynolds number is used extensively in calculations. { 'ren·əlz ¸mäd·əl }

Reynolds number [FL MECH] A dimensionless number which is significant in the design of a model of any system in which the effect of viscosity is important in controlling the velocities or the flow pattern of a fluid; equal to the density of a fluid, times its velocity, times a characteristic length, divided by the fluid viscosity. Symbolized N_{Re}. Also known as Damköhler number V (DaV). { 'ren·əlz ¸nəm·bər }

Reynolds stress [FL MECH] The net transfer of momentum across a surface in a turbulent fluid because of fluctuations in fluid velocity. Also known as eddy stress. { 'ren·əlz ¸stres }

Reynolds stress tensor [FL MECH] A tensor whose components are the components of the Reynolds stress across three mutually perpendicular surfaces. { 'ren·əlz 'stres ¸ten·sər }

rezbanyite [MINERAL] $Pb_3Cu_2Bi_{10}S_{19}$ A metallic-gray mineral composed of sulfide of lead, copper, and bismuth. { rez'ban¸yīt }

rf See radio frequency.

Rf See rutherfordium.

R factor [GEN] A self-replicating, infectious agent that carries genetic information and transmits drug resistance from bacterium to bacterium by conjugation of cell. Also known as resistance factor. { 'är ¸fak·tər }

RFI See radio-frequency interference.

rf preheating See radio-frequency preheating. { ¦är¦ef prē 'hēd·iŋ }

RFQ See radio-frequency quadrupole.

R galaxy [ASTRON] A galaxy that displays rotational symmetry but lacks a clearly defined rotational or elliptical structure. { 'är ¸gal·ik·sē }

RGB monitor [COMPUT SCI] A video display screen that requires separate red, green, and blue signals from a computer or other source. { ¦är¦jē¦bē 'män·əd·ər }

RG line See radio-frequency cable. { ¦är'jē ¸līn }

RGU system [ASTRON] A system for obtaining a complete assessment of a star's magnitude, based on measurements of the star's brightness when viewed through red, green, and ultraviolet filters. { ¦är¦jē'yü ¸sis·təm }

Rh See rhodium.

RH See relative humidity.

Rhabdiasoidea [INV ZOO] An order or superfamily of parasitic nematodes. { ¸rab·dē·ə'sȯid·ē·ə }

rhabdion [INV ZOO] One of the sclerotized segments lining the buccal cavity of nematodes. { 'rab·dē¸än }

rhabdite [INV ZOO] A small rodlike or fusiform body secreted by epidermal or parenchymal cells of certain turbellarians and trematodes. [MINERAL] See schreibersite. { 'rab¸dīt }

Rhabditia [INV ZOO] A subclass of nematodes in the class Secernentea. { rab'dish·ə }

Rhabditidia [INV ZOO] An order of nematodes in the subclass Rhabditia including parasites of humans and domestic animals. { ¸rab·də'tid·ē·ə }

Rhabditoidea [INV ZOO] A superfamily of small to moderate-sized nematodes in the order Rhabditidia with small, porelike, anteriorly located amphids, and esophagus with corpus, isthmus, and valvulated basal bulb. { ¸rab·də'tȯid·ē·ə }

Rhabdocoela [INV ZOO] Formerly an order of the Turbellaria, and now divided into three orders, Catenulida, Macrostomida, and Neorhabdocoela. { ¸rab·də'sē·lə }

rhabdoglyph [PALEON] A trace fossil consisting of a presumable worm trail appearing on the undersurface of flysch beds (sandstones) as a nearly straight bulge with little or no branching. { 'rab·də¸glif }

rhabdolith [BOT] A minute coccolith having a shield surmounted by a long stem and found at all depths in the ocean, from the surface to the bottom. { 'rab·də¸lith }

rhabdome [INV ZOO] The central translucent cylinder in the retinula of a compound eye. { 'rab¸dōm }

rhabdomyoblastoma See rhabdomyosarcoma. { ¦rab·dō¸mī·ō·bla'stō·mə }

rhabdomyoma [MED] A benign tumor of skeletal muscle. { ¦rab·dō·mī'ō·mə }

rhabdomyosarcoma [MED] A malignant tumor of skeletal muscle in the extremities composed of anaplastic muscle cells. Also known as malignant rhabdomyoma; rhabdomyoblastoma. { ¦rab·dō¸mī·ō·sär'kō·mə }

rhabdophane [MINERAL] $(Ce,Y,La,Di)(PO_4)\cdot H_2O$ A brown, pinkish, or yellowish-white mineral consisting of a hydrated phosphate of cerium, yttrium, and rare earths. { 'rab·də¸fān }

Rhabdophorina [INV ZOO] A suborder of ciliates in the order Gymnostomatida. { ¸rab·dō·fə'rī·nə }

rhabdosome [INV ZOO] A colonial graptolite that develops from a single individual. { 'rab·də¸sōm }

Rhabdoviridae [VIROL] A large family of negative-strand ribonucleic acid (RNA) viruses characterized by an enveloped bullet-shaped virion containing nonfragmented single-stranded RNA. They infect mammals, birds, fish, insects, and plants. The family includes the genera Vesiculovirus (vesicular stomatitis) and Lyssavirus (rabies). { ¸rab·dō'vir·ə¸dī }

rhabdovirus [VIROL] A group of ribonucleic acid-containing animal viruses, including rabies virus and certain infective agents of fish and insects. { ¦rab·dō'vī·rəs }

rhabdus [INV ZOO] A uniaxial sponge spicule. { 'rab·dəs }

Rhachitomi [PALEON] A group of extinct amphibians in the order Temnospondyli in which pleurocentra were retained. { rə'kid·ə¸mī }

rhachitomous [VERT ZOO] Being, having, or pertaining to vertebrae with centra whose parts do not fuse. { rə'kid·ə·məs }

Rhacophoridae [VERT ZOO] A family of arboreal frogs in the suborder Diplasiocoela. { ¸rak·ō'fȯr·ə¸dē }

Rhacopilaceae [BOT] A family of mosses in the order Isobryales generally having dimorphous leaves with smaller dorsal leaves and a capsule that is plicate when dry. { ¸rak·ō·pə'lās·ē¸ē }

rhadzimer See radzimir. { 'rad·zə¸mir }

Rhaetian [GEOL] A European stage of geologic time; the uppermost Triassic (above Norian, below Hettangian of Jurassic). Also known as Rhaetic. { 'rē·shən }

Rhaetic See Rhaetian. { 'rēd·ik }

Rhagionidae [INV ZOO] The snipe flies, a family of predatory orthorrhaphous dipteran insects in the series Brachycera that are brownish or gray with spotted wings. { ¸rag·ē'än·ə¸dē }

rhagon [INV ZOO] A pyramid-shaped, colonial sponge having an osculum at the apex and flagellated chambers in the upper wall only. { 'rā¸gän }

Rhamnaceae [BOT] A family of dicotyledonous plants in the order Rhamnales characterized by a solitary ovule in each locule, free stamens, simple leaves, and flowers that are hypogynous to perigynous or epigynous. { ram'nās·ē¸ē }

Rhamnales [BOT] An order of dicotyledonous plants in the subclass Rosidae having a single set of stamens, opposite the petals, usually a well-developed intrastamenal disk, and two or more locules in the ovary. { ram'nā·lēz }

rhamnose [BIOCHEM] $C_6H_{12}O_5$ A deoxysugar occurring free in poison sumac, and in glycoside combination in many plants. Also known as isodulcitol. { 'ram¸nōs }

rhamphoid [BIOL] Beak-shaped. { 'ram¸fȯid }

Rhamphorhynchoidea [PALEON] A Jurassic suborder of the Pterosauria characterized by long, slender tails with an expanded tip. { ¦ram·fə·riŋ¦kȯid·ē·ə }

rhampotheca [VERT ZOO] The horny sheath covering a bird's beak. { ¦ram·fə'thē·kə }

Rh antigen *See* Rh factor. { ¦är¦ach 'ant·i·jən }

Rh blocking serum [IMMUNOL] A serum that reacts with Rh-positive blood without causing agglutination, but which blocks the action of anti-Rh serums that are subsequently introduced. { ¦är¦ach ¦bläk·iŋ ¸sir·əm }

Rh blocking test [IMMUNOL] A test for the detection of Rh antibody in plasma wherein erythrocytes having the Rh antigen are incubated in the patient's serum so that the antibodies may be adsorbed on these cells, which are then employed in the antiglobulin test. Also known as indirect Coombs test; indirect developing test. { ¦är¦ach ¦bläk·iŋ ¸test }

Rh blood group [IMMUNOL] The extensive, genetically determined system of red blood cell antigens defined by the immune serum of rabbits injected with rhesus monkey erythrocytes, or by human antisera. Also known as rhesus blood group. { ¦är¦ach 'bləd ¸grüp }

rhe [FL MECH] **1.** A unit of dynamic fluidity, equal to the dynamic fluidity of a fluid whose dynamic viscosity is 1 centipoise. **2.** A unit of kinematic fluidity, equal to the kinematic fluidity of a fluid whose kinematic viscosity is 1 centistoke. { rē }

rhea [BOT] *See* ramie. [VERT ZOO] The common name for members of the avian order Rheiformes. { 'rē·ə }

Rhea [ASTRON] A satellite of Saturn, with estimated diameter of 950 miles (1530 kilometers). { 'rē·ə }

RHEED *See* reflection HEED. { 'är¸hēd }

rhegmagenesis [GEOL] Orogeny characterized by the development of large-scale strike-slip faults. { ¦reg·mə'jen· ə·səs }

rhegmatogenous retinal detachment [MED] Retinal detachment that is due to a retinal hole or tear; it occurs spontaneously or following trauma. { ¸reg·mə¦täj·ə·nəs ¸ret·ən·əl di'tach·mənt }

rheid [GEOL] A substance (below its melting point) which deforms by viscous flow during applied stress at an order of magnitude at least three times that of elastic deformation under similar circumstances. { 'rē·əd }

Rheidae [VERT ZOO] The single family of the avian order Rheiformes. { 'rē·ə¸dē }

rheid fold [GEOL] A fold whose strata deform by viscous flow as if they were fluid. { 'rē·əd ¸fōld }

rheidity [GEOL] Relaxation time of a substance, divided by 1000. { rē'id·əd·ē }

Rheiformes [VERT ZOO] The rheas, an order of South American running birds; called American ostriches, they differ from the true ostrich in their smaller size, feathered head and neck, three-toed feet, and other features. { ¸rē·ə'fȯr¸mēz }

Rheinberg illumination [OPTICS] An illumination technique used in optical microscopes that is a modification of the dark-field method; the central disk is transparent and colored; an annulus of a complementary color fills the remaining condenser aperture; the specimen is seen in the color of the annulus against the background of the central disk. Also known as optical staining. { 'rīn¸bərg i¸lü·mə¸nā·shən }

Rhenanida [PALEON] An order of extinct marine fishes in the class Placodermi distinguished by mosaics of small bones between the large plates in the head shield. { re'nan·ə·də }

rhenium [CHEM] A metallic element, symbol Re, atomic number 75, atomic weight 186.207; a transition element. { 'rē·nē·əm }

rhenium halide [INORG CHEM] Halogen compound of rhenium; examples are $ReCl_3$, $ReCl_4$, ReF_4, and ReF_6. { 'rē·nē· əm 'ha¸līd }

rheobase [PHYSIO] The intensity of the steady current just sufficient to excite a tissue when suddenly applied. { 'rē· ō¸bās }

rheocasting [MET] A process in which a liquid metal is vigorously agitated during initial stages of solidification to produce a globular semisolid structure which remains highly fluid when more than 60% solidification has occurred. { 'rē· ō¸kast·iŋ }

rheoelectroencephalograph [MED] Electroencephalograph for measuring differential blood flow in both sides of the brain and in any other part of the body. { ¦rē·ō·i¸lek·trō· in'sef·ə·lə¸graf }

rheogoniometry [MECH] Rheological tests to determine the various stress and shear actions on Newtonian and non-Newtonian fluids. { ¦rē·ə·gō·nē'äm·ə·trē }

rheoignimbrite [GEOL] An ignimbrite, on the slope of a volcanic crater, that has developed secondary flowage due to high temperatures. { ¦rē·ō'ig·nim¸brīt }

rheology [MECH] The study of the deformation and flow of matter, especially non-Newtonian flow of liquids and plastic flow of solids. { rē'äl·ə·jē }

rheometer [ENG] An instrument for determining flow properties of solids by measuring relationships between stress, strain, and time. { rē'äm·əd·ər }

rheomorphic intrusion [PETR] The injection of country rock that has become mobilized into the igneous intrusion that caused the rheomorphism. { ¦rē·ə¦mȯr·fik in'trü·zhən }

rheomorphism [PETR] Mobilization of a rock by at least partial fusion accompanied by, and sometimes promoted by, addition of new material by diffusion. { ¦rē·ə¦mȯr¸fiz·əm }

rheopectic fluid [FL MECH] A fluid for which the structure builds up on shearing; this phenomenon is regarded as the reverse of thixotropy. { ¦rē·ə¦pek·tik 'flü·əd }

rheopexy [PHYS CHEM] A property of certain sols, having particles shaped like rods or plates, which set to gel form more quickly when mechanical means are used to hasten the orientation of the particles. { 'rē·ə¸pek·sē }

rheophile [ECOL] Living or thriving in running water. { 'rē·ə¸fīl }

rheophilous bog [ECOL] A bog which draws its source of water from drainage. { rē'äf·ə·ləs ¸bäg }

rheoplankton [ECOL] Plankton found in flowing water. { 'rē·ō¸plaŋk·tən }

rheostat [ELEC] A resistor constructed so that its resistance value may be changed without interrupting the circuit to which it is connected. Also known as variable resistor. { 'rē· ə¸stat }

rheostatic braking [ENG] A system of dynamic braking in which direct-current drive motors are used as generators and convert the kinetic energy of the motor rotor and connected load to electrical energy, which in turn is dissipated as heat in a braking rheostat connected to the armature. { ¦rē·ə¦stad·ik 'brāk·iŋ }

rheostatic control [ELEC] A method of controlling the speed of electric motors that involves varying the resistance or reactance in the armature or field circuit; used in motors that drive elevators. { ¦rē·ə¦stad·ik kən'trōl }

rheostriction *See* pinch effect. { 'rē·ə¸strik·shən }

rheotaxial growth [ENG] A chemical vapor deposition technique for producing silicon diodes and transistors on a fluid layer having high surface mobility. { ¦rē·ə¦tak·sē·əl 'grōth }

rheotaxis [BIOL] Movement of a motile cell or organism in response to the direction of water currents. { ¦rē·ə¦tak·səs }

rheotron *See* betatron. { 'rē·ə¸trän }

rheotropic brittleness [MET] A low-temperature or high-strain-rate brittleness that may be eliminated by prestraining under milder conditions. { ¦rē·ə¦träp·ik 'brid·əl·nəs }

rheotropism [BIOL] Orientation response of an organism to the stimulus of a flowing fluid, as water. { rē'ä·trə¸piz·əm }

rhesus blood group *See* Rh blood group. { 'rē·səs 'bləd ¸grüp }

rhesus factor *See* Rh factor. { 'rē·səs ¸fak·tər }

rhesus macaque *See* rhesus monkey. { 'rē·səs mə'kak }

rhesus monkey [VERT ZOO] *Macaque mulatta.* An agile, gregarious primate found in southern Asia and having a short tail, short limbs of almost equal length, and a stocky build. Also known as rhesus macaque. { 'rē·səs ¸mäŋ·kē }

rheumatic arteritis [MED] A type of allergic arteritis associated with acute rheumatic fever. { rü'mad·ik ¸ärd·ə'rīd·əs }

rheumatic carditis [MED] Inflammation of the heart resulting from rheumatic fever. { rü'mad·ik kär'dīd·əs }

rheumatic encephalopathy [MED] An inflammatory reaction of the brain and the smaller arteries of the cerebral cortex associated with rheumatic fever. { rü'mad·ik in¸sef·ə'läp·ə· thē }

rheumatic endocarditis [MED] Inflammation of the endocardium in acute rheumatic fever, usually involving heart valves. { rü'mad·ik ¸en·dō·kär'dīd·əs }

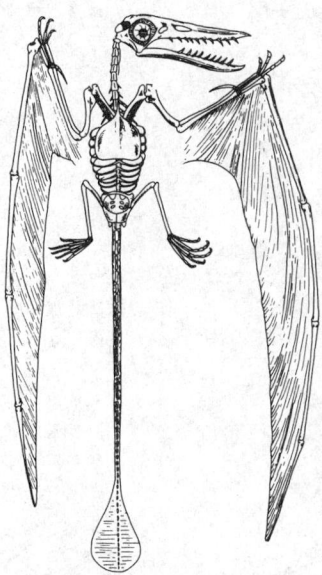

Restoration of Jurassic pterosaur skeleton of *Rhamphorhynchus*; length about 30 centimeters.

rheumatic fever [MED] A febrile disease occurring in childhood as a delayed sequel of infection by *Streptococcus hemolyticus*, group A; characterized by arthritis, carditis, nosebleeds, and chorea. { rü'mad·ik 'fē·vər }

rheumatic pneumonia [MED] Pneumonia associated with acute rheumatic fever. { rü'mad·ik nů'mō·nyə }

rheumatism [MED] Any combination of muscle or joint pain, stiffness, or discomfort arising from nonspecific disorders. { 'rü·mə,tiz·əm }

rheumatoid arthritis [MED] A chronic systemic inflammatory disease of connective tissue in which symptoms and changes predominate in articular and related structures. Also known as atrophic arthritis; chronic infectious arthritis; proliferative arthritis. { 'rü·mə,tóid är'thrīd·əs }

rheumatoid factor [IMMUNOL] The immunoglobulin in the class IgM that is detected in the synovial fluid of individuals with rheumatoid arthritis. { 'rü·mə,tóid ,fak·tər }

rheumatoid nodules [MED] Subcutaneous lateral foci of fibrinoid degeneration or necrosis surrounded by mononuclear cells in a regular palisade arrangement, occurring usually in association with rheumatoid arthritis or rheumatic fever. { 'rü·mə,tóid 'näj·ülz }

rheumatoid spondylitis [MED] A chronic progressive arthritis of young men, affecting mainly the spine and sacroiliac joints, leading to fusion and deformity. Also known as Marie-Strümpell disease. { 'rü·mə,tóid ,spän·də'līd·əs }

rhexistasy [GEOL] The mechanical breaking up and transport of old soils or other surface residual materials. { rek'sis·tə·sē }

Rh factor [IMMUNOL] Any of several red blood cell antigens originally identified in the blood of rhesus monkeys. Also known as Rh antigen; rhesus factor. { ¦är¦ach ,fak·tər }

RHI display *See* range-height indicator display. { ¦är¦ach'ī di,splā }

Rhigonematidae [INV ZOO] A family of nematodes in the superfamily Oxyuroidea. { ,rig·ō·nə'mad·ə,dē }

Rhincodontidae [VERT ZOO] The whale sharks, a family of essentially tropical galeoid elasmobranchs in the isurid line. { ,riŋ·kə'dänt·ə,dē }

Rh incompatibility [MED] A condition in which red blood cells of the fetus become coated with immunoglobulin G antibody (Rh antibody) of maternal origin which is directed against Rh-D antigen of paternal origin that is present on fetal cells. { ¦är¦ach ,in·kəm,pad·ə'bil·əd·ē }

rhinencephalon [ANAT] The anterior olfactory portion of the vertebrate brain. { ¦rīn·in'sef·ə,län }

rhinestone [MATER] A clear, colorless imitation of diamond, made of glass, paste, or gem quartz, backed with metallic foil. { 'rīn,stōn }

rhinitis [MED] Inflammation of the mucous membranes in the nose. { rī'nīd·əs }

Rhinobatidae [VERT ZOO] The guitarfishes, a family of elasmobranchs in the batoid group. { ,rī·nō'bad·ə,dē }

Rhinoceratidae [VERT ZOO] A family of perissodactyl mammals in the superfamily Rhinoceratoidea, comprising the living rhinoceroses. { rī,näs·ə'räd·ə,dē }

Rhinoceratoidea [VERT ZOO] A superfamily of perissodactyl mammals in the suborder Ceratomorpha including living and extinct rhinoceroses. { rī,näs·ə·rə'tóid·ē·ə }

rhinocerebral mucormycosis [MED] A mold infection of the sinus that spreads rapidly to the eye and brain. { ,rīn·ō·sə,rēb·rəl ,myü·kō·mī'kō·səs }

rhinoceros [VERT ZOO] The common name for the odd-toed ungulates composing the family Rhinoceratidae, characterized by massive, thick-skinned limbs and bodies, and one or two horns which are composed of a solid mass of hairs attached to the bony prominence of the skull. { rī'näs·ə·rəs }

Rhinochimaeridae [VERT ZOO] A family of ratfishes, order Chimaeriformes, distinguished by an extremely elongate rostrum. { ,rīn·ō·ki'mer·ə,dē }

Rhinocryptidae [VERT ZOO] The tapaculos, a family of ground-inhabiting suboscine birds in the suborder Tyranni characterized by a large, movable flap which covers the nostrils. { ,rīn·ə'krip·tə,dē }

rhinogenous [MED] Originating in the nose. { rī'näj·ə·nəs }

rhinolaryngology [MED] The science of the anatomy, physiology, and pathology of the nose and larynx. { ¦rī·nō,lar·iŋ'gäl·ə·jē }

RHINOCEROS

Black rhinoceros (*Diceros bicornis*).

rhinolaryngoscope [MED] A scope containing a mirror and a light, used to examine the nose and larynx. { ¦rī·nō·lə'rin·jə,skōp }

rhinology [MED] The science of the anatomy, functions, and diseases of the nose. { rī'näl·ə·jē }

Rhinolophidae [VERT ZOO] The horseshoe bats, a family of insect-eating chiropterans widely distributed in the Eastern Hemisphere and distinguished by extremely complex, horseshoe-shaped nose leaves. { ,rīn·ə·ə'läf·ə,dē }

rhinopharyngitis [MED] Inflammation of the nose and pharynx or of the nasopharynx. { ¦rī·nō,far·ən'jīd·əs }

rhinophore [INV ZOO] An olfactoreceptor of certain land mollusks, usually borne on a tentacle. { 'rīn·ə,fór }

rhinoplasty [MED] A plastic operation on the nose. { 'rīn·ə,plas·tē }

Rhinopomatidae [VERT ZOO] The mouse-tailed bats, a small family of insectivorous chiropterans found chiefly in arid regions of northern Africa and southern Asia and characterized by long, wirelike tails and rudimentary nose leaves. { ,rīn·ə·pə'mad·ə,dē }

Rhinopteridae [VERT ZOO] The cow-nosed rays, a family of batoid sharks having a fleshy pad at the front end of the head and a well-developed poison spine. { ,rī·näp'ter·ə,dē }

rhinorrhea [MED] **1.** A mucous discharge from the nose. **2.** Escape of cerebrospinal fluid through the nose. { ,rin·ə'rē·ə }

rhinoscleroma [MED] A chronic infectious bacterial disease caused by *Klebsiella rhinoscleromatis* and characterized by hard nodules and plaques of inflamed tissue in the nose and adjacent areas. { ,rīn·ə·sklə'rō·mə }

rhinoscope [MED] An instrument for examining the nasal cavities. { 'rīn·ə,skōp }

Rhinotermitidae [INV ZOO] A family of lower termites of the order Isoptera. { ¦rī·nō·tər'mad·ə,dē }

rhinotheca [VERT ZOO] The horny sheath on the upper part of a bird's bill. { ¦rīn·ə'thē·kə }

rhinovirus [VIROL] A subgroup of the picornavirus group including small, ribonucleic acid-containing forms which are not inactivated by ether. { ¦rīn·ə'vī·rəs }

Rhipiceridae [INV ZOO] The cedar beetles, a family of coleopteran insects in the superfamily Elateroidea. { ,rip·ə'ser·ə,dē }

Rhipidistia [VERT ZOO] The equivalent name for Osteolepiformes. { ,rip·ə'dis·tē·ə }

rhipidium [BOT] A fan-shaped inflorescence with cymose branching in which branches lie in the same plane and are suppressed alternately on each side. { ri'pid·ē·əm }

Rhipiphoridae [INV ZOO] The wedge-shaped beetles, a family of coleopteran insects in the superfamily Meloidea. { ,rip·ə'fór·ə,dē }

rhizanthous [BOT] Producing flowers directly from the root. { rī'zan·thəs }

rhizautoicous [BOT] Of mosses, having the antheridial branch and the archegonial branch connected by rhizoids. { ¦rīz,ó¦tói·kəs }

rhizic water *See* soil water. { 'rīz·ik ,wód·ər }

rhizine [BOT] The rhizoid of a lichen. { 'rī,zēn }

Rhizobiaceae [MICROBIO] A family of gram-negative, motile, aerobic rods; utilize carbohydrates and produce slime on carbohydrate media. { rī,zō·bē'ās·ē,ē }

Rhizobium [MICROBIO] A genus of rod-shaped, gram-negative, aerobic, and nitrogen-fixing bacteria which form symbiotic nodules on the roots of leguminous plants, such as clover and beans. { rī'zō·bē·əm }

rhizocarpous [BOT] Pertaining to perennial herbs having perennating underground parts from which stems and foliage arise annually. { ¦rī·zō¦kär·pəs }

Rhizocephala [INV ZOO] An order of crustaceans which parasitize other crustaceans; adults have a thin-walled sac enclosing the visceral mass and show no trace of segmentation, appendages, or sense organs. { ,rī·zō'sef·ə·lə }

Rhizochloridina [INV ZOO] A suborder of flagellate protozoans in the order Heterochlorida. { ,rī·zō,klór·ə'dī·nə }

rhizoconcretion *See* root cast. { ¦rī·zō·kän'krē·shən }

rhizoctol *See* methylarsinic sulfide. { rī'zäk,tól }

Rhizodontidae [PALEON] An extinct family of lobefin fishes in the order Osteolepiformes. { ,rī·zō'dänt·ə,dē }

rhizoid [BOT] A rootlike structure which helps to hold the plant to a substrate; found on fungi, liverworts, lichens, mosses, and ferns. { 'rī‚zòid }

Rhizomastigida [INV ZOO] An order of the protozoan class Zoomastigophorea; all species are microscopic and ameboid, and have one or two flagella. { ‚rī·zō·mas'tij·əd·ə }

Rhizomastigina [INV ZOO] The equivalent name for Rhizomastigida. { 'rī·zō·mas·tə'jī·nə }

rhizome [BOT] An underground horizontal stem, often thickened and tuber-shaped, and possessing buds, nodes, and scalelike leaves. { 'rī‚zōm }

rhizomorph [BOT] A rootlike structure, characteristic of many basidiomycetes, consisting of a mass of densely packed and intertwined hyphae. { 'rī·zə‚mórf }

Rhizophagidae [INV ZOO] The root-eating beetles, a family of minute coleopteran insects in the superfamily Cucujoidea. { ‚rī·zō'fā·jə‚dē }

Rhizophoraceae [BOT] A family of dicolyledonous plants in the order Cornales distinguished by opposite, stipulate leaves, two ovules per locule, folded or convolute bud petals, and a berry fruit. { rī‚zàf·ə'rās·ē‚ē }

Rhizophorales [BOT] An order of dicotyledonous flowering plants, class Magnoliopsida; mostly tanniferous trees and shrubs with leaves opposite, simple, and entire, and flowers regular, mostly perfect, and variously perigynous or epigynous. { ‚rīz·ə·fó'rā·lēz }

rhizophore [BOT] A leafless, downward-growing dichotomous *Selaginella* shoot that has tufts of adventitious roots at the apex. { 'rī·zə‚fór }

rhizoplast [CYTOL] A delicate fiber or thread running between the nucleus and the blepharoplast in cells bearing flagella. { 'rī·zə‚plast }

rhizopod [INV ZOO] An anastomosing rootlike pseudopodium. { 'rī·zə‚päd }

Rhizopodea [INV ZOO] A class of the protozoan superclass Sarcodina in which pseudopodia may be filopodia, lobopodia, or reticulopodia, or may be absent. { ‚rī·zə'pō·dē·ə }

rhizosphere [GEOL] The soil region subject to the influence of plant roots and characterized by a zone of increased microbiological activity. { 'rī·zə‚sfir }

Rhizostomeae [INV ZOO] An order of the class Scyphozoa having the umbrella generally higher than it is wide with the margin divided into many lappets but not provided with tentacles. { ‚rī·zə'stō·mē‚ē }

rhizotomy [MED] Surgical division of any root, as of a nerve. Also known as radicotomy. { rī'zäd·ə·mē }

rhizotron [BOT] An underground laboratory system designed for examining plant root growth; contains enclosed columns of soil with transparent plastic windows which permit viewing, measuring, and photographing. { 'rīz·ə‚trän }

RHM *See* roentgen per hour at one meter.

Rh negative [MED] Absence of the Rh-D antigen on the surface of red blood cells. { ‚är'āch 'neg·əd·iv }

Rh null [MED] Total absence of all Rh antigens on the surface of red blood cells. { ‚är'āch 'nəl }

rhodamine B [ORG CHEM] $C_{28}H_{31}ClN_2O_3$ Red, green, or reddish-violet powder, soluble in alcohol and water; forms bluish red, fluorescent solution in water; used as red dye for paper, wool, and silk, and as an analytical reagent and biological stain. { 'rōd·ə‚mēn 'bē }

rhodamine toner [MATER] Rhodamine dye and phosphotungstic or phosphomolybdic acid; red to maroon; used in printing inks. { 'rōd·ə‚mēn 'tōn·ər }

Rhodanian orogeny [GEOL] A short-lived orogeny that occurred at the end of the Miocene Period. { rō'dān·ē·ən ó'räj·ə·nē }

rhodanic acid *See* thiocyanic acid. { rō'dan·ik 'as·əd }

rhodanine [ORG CHEM] $C_3H_3NOS_2$ A pale-yellow crystalline compound that may decompose violently when heated, giving off toxic by-products; used in organic synthesis. { 'rōd·ə‚nīn }

L-rhodeose *See* L-fucose. { el 'rōd·ē‚ōs }

Rhodesian man [PALEON] A type of fossil man inhabiting southern and central Africa during the late Pleistocene; the skull was large and low, marked by massive browridges, with a cranial capacity of 1300 cubic centimeters or less. { rō'dē·zhən 'man }

Rhodesian trypanosomiasis [MED] A fulminating form of African sleeping sickness caused by *Trypanosoma rhodesiense*, transmitted by the tsetse fly, and characterized by parasitemia,

edema, lymphadenitis, and myocarditis. Also known as East African sleeping sickness. { rō'dē·zhən trə‚pan·ə·sō'mī·ə·səs }

Rhodininae [INV ZOO] A subfamily of limivorous worms in the family Maldanidae. { rō'din·ə‚nē }

rhodinol [MATER] Colorless, combustible liquid mixture of terpene alcohols with rose scent; soluble in mineral oil and alcohol; derived from geranium oil used in perfumes and flavors. { 'rōd·ən‚ól }

rhodinyl acetate [MATER] Terpene-alcohol-acetates mixture; colorless-to-yellow, combustible liquid with rose scent; soluble in mineral oil, alcohol, and glycerin; used in perfumes and flavors. { 'rōd·ən·əl 'as·ə‚tāt }

rhodite [MINERAL] A mineral consisting of a native alloy of rhodium (about 40) and gold. { 'rō‚dīt }

rhodium [CHEM] A chemical element, symbol Rh, atomic number 45, atomic weight 102.9055. [MET] A silver-white metal in the platinum family; sometimes alloyed with platinum for thermocouples or used as a tarnish-resistant electrode posit. { 'rōd·ē·əm }

rhodium chloride [INORG CHEM] $RhCl_3$ Water-insoluble, brown-red powder, soluble in cyanides and alkalies; decomposes at 450–500°C. Also known as rhodium trichloride. { 'rōd·ē·əm 'klór‚īd }

rhodium trichloride *See* rhodium chloride. { 'rōd·ē·əm trī'klór‚īd }

rhodizite [MINERAL] $CsAl_4Be_4B_{11}O_{25}(OH)_4$ A white mineral composed of a basic borate of cesium, aluminum, and beryllium, occurring as isometric crystals. { 'rōd·ə‚zīt }

Rhodobacterlineae [MICROBIO] Formerly a suborder of the order Pseudomonadales comprising all of the photosynthetic, or phototrophic, bacteria except those of the genus *Rhodomicrobium*. { ‚rō·dō·bak‚tir·ə'l·nē‚ē }

rhodochrosite [MINERAL] $MnCO_3$ A rose-red to pink or gray mineral form of manganese carbonate with hexagonal symmetry but occurring in massive or columnar form; isomorphous with calcite and siderite, has a hardness of 3.5–4 on Mohs scale, and a specific gravity of 3.7; a minor ore of manganese. { ‚rōd·ə'krō‚sīt }

rhodolite [MINERAL] A violet-red garnet species composed of a mixture of almandite and pyrope in about a 3:1 ratio. { 'rōd·əl‚īt }

rhodonea *See* rose. { ‚rōd·ən'ē·ə }

rhodonite [MINERAL] $MnSiO_3$ A pink or brown mineral inosilicate crystallizing in the triclinic system and commonly found in cleavable to compact masses or in embedded grains; luster is vitreous, hardness is 5.5–6 on Mohs scale, and specific gravity is 3.4–3.7. { 'rōd·ən‚īt }

Rhodophyceae [BOT] A class of algae belonging to the division or subphylum Rhodophyta. { ‚rōd·ə'fīs·ē‚ē }

Rhodophyta [BOT] The red algae, a large diverse phylum or division of plants distinguished by having an abundance of the pigment phycoerythrin. { rō'däf·əd·ə }

rhodoplast [BOT] A reddish chromatophore occurring in red algae. { 'rōd·ə‚plast }

rhodopsin [BIOCHEM] A deep-red photosensitive pigment contained in the rods of the retina of marine fishes and most higher vertebrates. Also known as retinal pigment; visual purple. { rō'däp·sən }

Rhodospirillaceae [MICROBIO] A family of bacteria in the suborder Rhodospirillineae; cells are motile by flagella, multiplication is by budding or binary fission, and photosynthetic membranes are continuous with the cytoplasmic membrane. { ‚rō·dō‚spī·rə'lās·ē‚ē }

Rhodospirillales [MICROBIO] The single order of the phototrophic bacteria; cells are spherical, rod-shaped, spiral, or vibrio-shaped, and all contain bacteriochlorophylls and carotenoid pigments. { ‚rō·dō‚spī·rə'lā·lēz }

Rhodospirillineae [MICROBIO] The purple bacteria, a suborder of the order Rhodospirillales; contain bacteriochlorophyll a or b, located on internal membranes. { ‚rō·dō‚spī·rə'lin·ē‚ē }

rhodoxanthin [BIOCHEM] $C_{40}H_{50}O_2$ A xanthophyll carotenoid pigment. { ‚rō·dō'zan·thən }

rhohelos [ECOL] A stream-crossed, nonalluvial marsh typical of filled lake areas. { rō'hē‚lōs }

Rhoipteleaceae [BOT] A monotypic family of dicotyledonous plants in the order Juglandales having pinnately compound leaves, and flowers in triplets with four sepals and six stamens, and the lateral flowers female but sterile. { ‚róip‚tē·lē'ās·ē‚ē }

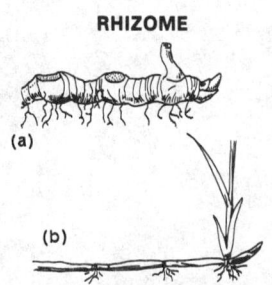

RHIZOME

The underground stems of (*a*) Solomon's seal and (*b*) grass.

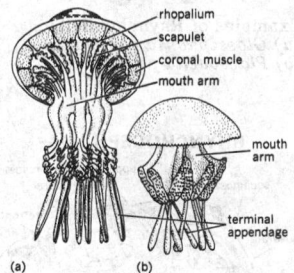

RHIZOSTOMEAE

Rhizostomeae. (*a*) *Rhizostoma*; the bell may reach 2 feet (61 centimeters) in diameter. (*b*) *Mastigias*. (From L. Hyman, *The Invertebrates, vol. 1*, McGraw-Hill, 1940)

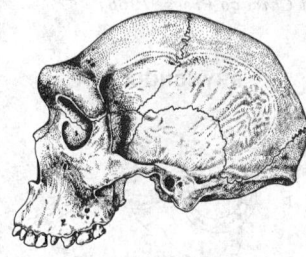

RHODESIAN MAN

Skull of Rhodesian man (*Homo erectus rhodesiensis*), late Pleistocene. (*From M. F. Ashley Montagu, An Introduction to Physical Anthropology, 2d ed., Charles C. Thomas, 1951*)

rhomb *See* rhombohedron; rhombus. { räm }

rhombencephalon [EMBRYO] The most caudal of the primary brain vesicles in the vertebrate embryo. Also known as hindbrain. { ˈräm‚benˈsef·ə‚län }

rhombic antenna [ELECTROMAG] A horizontal antenna having four conductors forming a diamond or rhombus; usually fed at one apex and terminated with a resistance or impedance at the opposite apex. Also known as diamond antenna. { ˈräm‚bik anˈten·ə }

rhombic dodecahedron [CRYSTAL] A crystal form in the cubic system that is a dodecahedron whose faces are equal rhombuses. { ˈräm‚bik ‚dōˈdek·ə‚hē·drən }

rhombic lattice *See* orthorhombic lattice. { ˈräm‚bik ˈlad·əs }

rhombic sulfur [CHEM] Crystalline sulfur with three unequal axes, all at right angles. { ˈräm‚bik ˈsəl·fər }

rhombic system *See* orthorhombic system. { ˈräm‚bik ˈsis‚təm }

Rhombifera [PALEON] An extinct order of Cystoidea in which the thecal canals crossed the sutures at the edges of the plates, so that one-half of any canal lay in one plate and the other half on an adjoining plate. { rämˈbif·ə‚rə }

rhombochasm [GEOL] A parallel-sided gap in the sialic crust occupied by simatic crust, probably caused by spreading and separation. { ˈräm·bə‚kaz·əm }

rhomboclase [MINERAL] HFe³⁺(SO₄)₂·4H₂O A colorless mineral composed of hydrous acid ferric sulfate, occurring in rhombic plates. { ˈräm·bə‚klās }

rhombogen [INV ZOO] A form of reproductive individual of the mesozoan order Dicyemida found in the sexually mature host which arises from nematogens and gives rise to free-swimming infusorigens. { ˈräm·bə‚jən }

rhombohedral [CRYSTAL] **1.** Of or pertaining to the rhombohedral system. **2.** Of or pertaining to crystal cleavage in or a centered lattice of the hexagonal system. { ˈräm·bōˈhē·drəl }

rhombohedral close packing *See* rhombohedral packing. { ˈräm·bōˈhē·drəl ˈklōs ˈpak·iŋ }

rhombohedral iron ore *See* hematite; siderite. { ˈräm·bōˈhē·drəl ˈī·ərn ‚ȯr }

rhombohedral lattice [CRYSTAL] A crystal lattice in which the three axes of a unit cell are of equal length, and the three angles between axes are the same, and are not right angles. Also known as trigonal lattice. { ˈräm·bōˈhē·drəl ˈlad·əs }

rhombohedral packing [CRYSTAL] The tightest manner of systematic arrangement of uniform solid spheres in a clastic sediment or crystal lattice, characterized by a unit cell of six planes passed through eight sphere centers situated at the corners of a regular rhombohedron. Also known as rhombohedral close packing. { ˈräm·bōˈhē·drəl ˈpak·iŋ }

rhombohedral system [CRYSTAL] A division of the trigonal crystal system in which the rhombohedron is the basic unit cell. { ˈräm·bōˈhē·drəl ˈsis‚təm }

rhombohedron [CRYSTAL] A trigonal crystal form that is a parallelepiped, the six identical faces being rhombs. Also known as rhomb. [MATH] A prism with six parallelogram faces. { ˈräm·bōˈhē·drən }

rhomboid [MATH] A parallelogram whose adjacent sides are not equal. { ˈräm‚bȯid }

rhomboidal prism [OPTICS] A prism with four parallel sides and two slanting, or oblique, parallel ends; it will divert the path of light entering its ends without changing the form of the light. { rämˈbȯid·əl ˈpriz·əm }

rhomboid ripple mark [GEOL] An aqueous current ripple mark characterized by a reticular arrangement of diamond-shaped tongues of sand, with each tongue having two acute angles, one pointing upcurrent and the other pointing downcurrent. { ˈräm‚bȯid ˈrip·əl ‚märk }

rhomboporoid cryptostome [PALEON] Any of a group of extinct bryozoans in the order Cryptostomata that built twiglike colonies with zooecia opening out in all directions from the central axis of each branch. { ˈräm·bōˈpȯr‚ȯid ˈkrip·tə‚sōm }

rhomb-porphyry [PETR] A porphyritic alkaline syenite composed of an alkali feldspar groundmass with augites having rhombohedral cross sections as the principal phenocryst minerals. { ˈräm‚pȯr·fə‚rē }

rhombus [MATH] A parallelogram with all sides equal. Also known as rhomb. { ˈräm‚bəs }

rho meson [PARTIC PHYS] Collective name for vector meson resonances belonging to a charge multiplet with total isospin 1, hypercharge 0, negative charge conjugation parity, positive

g-parity, mass of about 770 MeV, and width of about 146 MeV. Designated ρ(770). { ˈrō ˈmā‚sän }

Rhopalidae [INV ZOO] A family of pentatomorphan hemipteran insects in the superfamily Coreoidea. { rōˈpāl·ə‚dē }

rhopalium [INV ZOO] A sense organ found on the margin of a discomedusan. { rōˈpāl·ē·əm }

Rhopalocera [INV ZOO] Formerly a suborder of Lepidoptera comprising those forms with clubbed antennae. { ‚rō·pəˈläs·ə·rə }

rhopalocercous cercaria [INV ZOO] A free-swimming digenetic trematode larva distinguished by a very wide tail. { ‚rō·pə·lōˈsər·kəs sərˈkar·ē·ə }

Rhopalodinidae [INV ZOO] A family of holothurian echinoderms in the order Dactylochirotida in which the body is flask-shaped, the mouth and anus lying together. { ‚rō·pə·lōˈdin·ə‚dē }

Rhopalosomatidae [INV ZOO] A family of hymenopteran insects in the superfamily Scolioidea. { ‚rō·pə·lōˈsō‚mad·ə‚dē }

rho-theta navigation *See* omnibearing-distance navigation. { ˈrōˈthād·ə ‚nav·ə'gā·shən }

rhourd [GEOL] A pyramid-shaped sand dune, formed by the intersection of other dunes. { rȯrd }

Rh positive [MED] Presence of the Rh-D antigen on the surface of red blood cells. { ˈärˈäch ˈpäz·əd·iv }

RHU *See* radioisotope heating unit.

rhubarb [BOT] *Rheum rhaponticum.* A herbaceous perennial of the order Polygoniales grown for its thick, edible petioles. { ˈrü‚bärb }

rhumbatron *See* cavity resonator. { ˈrəm·bə‚trän }

rhumb bearing [NAV] The direction of a rhumb line through two terrestrial points, expressed as angular distance from a reference direction; usually measured from 000° at the reference direction clockwise through 360°. Also known as Mercator bearing. { ˈrəm ‚ber·iŋ }

rhumb direction *See* Mercator direction. { ˈrəm diˈrek·shən }

rhumb line [MAP] A line on the surface of the earth making the same oblique angle with all meridians. Also known as loxodrome. { ˈrəm ‚līn }

rhumb-line course [NAV] The direction of the rhumb line from the point of departure to the destination, expressed as the angular distance from a reference direction, usually north. Also known as Mercator course. { ˈrəm ‚līn ‚kȯrs }

rhumb-line distance [NAV] Distance along a rhumb line, usually expressed in nautical miles. { ˈrəm ‚līn ‚dis·təns }

rhumb-line error [NAV] An acceleration error, such as in a bubble sextant reading, due to an aircraft being in curved flight in space when proceeding along a rhumb line on the earth's surface. { ˈrəm ‚līn ‚er·ər }

rhumb-line sailing [NAV] Any method of solving the various problems involving course, distance, difference of latitude, difference of longitude, and departure, as they are related to a rhumb line. { ˈrəm ‚līn ‚sāl·iŋ }

rhyacolite *See* sanidine. { rīˈak·ə‚līt }

Rhynchobdellae [INV ZOO] An order of the class Hirudinea comprising leeches that possess an eversible proboscis and lack hemoglobin in the blood. { ‚riŋˈkäbˈde·lē }

Rhynchocephalia [VERT ZOO] An order of lepidosaurian reptiles represented by a single living species, *Sphenodon punctatus,* and characterized by a diapsid skull, teeth fused to the edges of the jaws, and an overhanging beak formed by the upper jaw. { ‚riŋ·kō·səˈfāl·yə }

rhynchocoel [INV ZOO] A cavity that holds the inverted proboscis in nemertinean worms. { ˈriŋ·kō‚sēl }

Rhynchocoela [INV ZOO] A phylum of bilaterally symmetrical, unsegmented, ribbonlike worms having an eversible proboscis and a complete digestive tract with an anus. { ‚riŋ·kōˈsē·lə }

rhynchodaeum [INV ZOO] The part of the proboscis lying anterior to the brain in nemertinean worms. { riŋˈkō·dē·əm }

Rhynchodina [INV ZOO] A suborder of ciliate protozoans in the order Thigmotrichida. { ‚riŋ·kəˈdī·nə }

Rhynchonellida [INV ZOO] An order of articulate brachiopods; typical forms are dorsibiconvex, the posterior margin is curved, the dorsal interarea is absent, and the ventral one greatly reduced. { ‚riŋ·kəˈnel·əd·ə }

rhynchophorous [ZOO] Having a beak. { riŋˈkäf·ə·rəs }

Rhynchosauridae [PALEON] An extinct family of generally

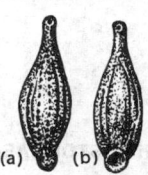
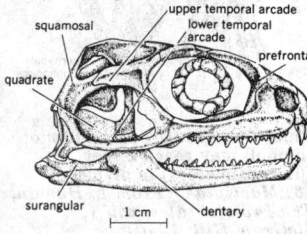
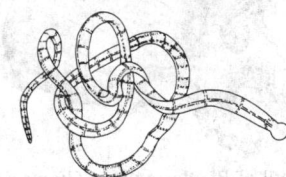

large, stout, herbivorous lepidosaurian reptiles in the order Rhynchocephalea. { ,riŋ·kə'sȯr·ə,dē }

Rhynchotheriinae [PALEON] A subfamily of extinct elephantoid mammals in the family Gomphotheriidae comprising the beak-jawed mastodonts. { ,riŋ·kə·thə'rī·ə,nē }

Rhyniatae See Rhyniopsida. { rī'nī·ə,dē }

Rhyniophyta [PALEOBOT] A subkingdom of the Embryobionta including the relatively simple, uppermost Silurian-Devonian vascular plants. { ,rī·nē'äf·əd·ə }

Rhyniopsida [PALEOBOT] A class of extinct plants in the subkingdom Rhyniophyta characterized by leafless, usually dichotomously branched stems that bore terminal sporangia. { ,rī·nē'äp·səd·ə }

Rhynochetidae [VERT ZOO] A monotypic family of gruiform birds containing only the kagu of New Caledonia. { ,rī·nə'kēd·ə,dē }

rhyodacite [PETR] A group of extrusive porphyritic igneous rocks containing quartz, plagioclase, and biotite phenocrysts in a fine-grained to glassy groundmass composed of alkali feldspar and silica minerals. Also known as dellenite; quartz lattice. { rī'äd·ə,sīt }

rhyolite [PETR] A light-colored, aphanitic volcanic rock composed largely of alkali feldspar and free silica with minor amounts of mafic minerals; the extrusive equivalent of granite. { 'rī·ə,līt }

rhyolitic glass [GEOL] Volcanic glass that is chemically equivalent to rhyolite. { ¦rī·ə¦lid·ik 'glas }

rhyolitic lava [GEOL] A highly viscous, silica-rich lava. { ¦rī·ə¦lid·ik 'lä·və }

rhyolitic magma [PETR] A type of magma formed by differentiation from basaltic magma in combination with assimilation of siliceous material, or by melting of portions of the earth's sialic layer. { ¦rī·ə¦lid·ik 'mag·mə }

rhyolitic tuff [GEOL] A tuff composed of fragments of rhyolitic lava. { ¦rī·ə¦lid·ik 'təf }

Rhysodidae [INV ZOO] The wrinkled bark beetles, a family of coleopteran insects in the suborder Adephaga. { rī'säd·ə,dē }

rhythmic accumulations [GEOL] Regular patterns of ripples and cusps in sediment on the beach or the sea floor, formed by currents and waves. { 'rith·mik ə,kyü·mə lā·shənz }

rhythmic crystallization [PETR] In igneous rocks, a phenomenon in which different minerals crystallize in concentric layers, giving rise to orbicular texture. { 'rith·mik ,krist· əl'ā·shən }

rhythmic driving [MIN ENG] Driving carried out between two shifts; that is, the drilling, loading, and blasting are carried out in one shift and the mucking and transportation in the following one. { 'rith·mik 'drīv·iŋ }

rhythmic layering [GEOL] A type of layering in an igneous intrusion which is easily observable and in which there is repetition of zones of varying composition. { 'rith·mik 'lā· ər·iŋ }

rhythmic sedimentation [GEOL] A repetitious, regular sequence of rock units formed by sedimentary succession and indicating a frequent, predictable recurrence of the same sequence of conditions. { 'rith·mik ,sed·ə·men'tā·shən }

rhythmic stratification [GEOL] The occurrence of sediment layers in repetitive patterns, such as a regular alternation of layers of lime and clay. { 'rith·mik ,strad·ə·fə'kā·shən }

rhythmic succession [GEOL] A succession of rock units showing continual and repeated changes of lithology. { 'rith· mik sək'sesh·ən }

rhythmite [GEOL] An independent unit of a rhythmic succession or of beds that were developed by rhythmic sedimentation. { 'rith·mīt }

ria [GEOGR] **1.** Any broad, estuarine river mouth. **2.** A long, narrow coastal inlet, except a fjord, whose depth and width gradually and uniformly diminish inland. { 'rē·ə }

RIAA curve [ENG ACOUS] **1.** Recording Industry Association of America curve representing standard recording characteristics for long-play records. **2.** The corresponding equalization curve for playback of long-play records. { ¦är¦ī̯a¦ā 'kərv }

ria coast [GEOGR] A coast with several parallel rias extending far inland and alternating with ridgelike promontories. { 'rē·ə 'kōst }

ria shoreline [GEOGR] A type of coastline developed along a drowning landmass in which numerous long and narrow

arms of the sea extend inland parallel with one another and perpendicular to the coastline. { 'rē·ə 'shȯr,līn }

rib [AERO ENG] A transverse structural member that gives cross-sectional shape and strength to a portion of an airfoil. [ANAT] One of the long curved bones forming the wall of the thorax in vertebrates. [BOT] A primary vein in a leaf. [GEOL] A layer or dike of rock forming a small ridge on a steep mountainside. [MIN ENG] **1.** A solid pillar of coal or ore left for support. **2.** A thin stratum in a seam of coal. [TEXT] A straight, raised cord in a fabric, formed by a heavy thread in any direction. { rib }

rib-and-furrow [GEOL] The bedding-plane expression for micro-cross-bedding, consisting of sets of small, transverse arcuate markings confined to long, narrow, parallel grooves oriented parallel to the current flow and separated by narrow ridges. { 'rib ən 'fər·ō }

riband jasper See ribbon jasper. { 'rib·ənd 'jas·pər }

rib arch [CIV ENG] An arch consisting of ribs placed side by side and extending from the springings on one end to those on the other end. { 'rib ,ärch }

ribbed-clamp coupling [DES ENG] A rigid coupling which is split longitudinally and bored to shaft diameter, with a shim separating the two halves. { 'ribd ¦klamp 'kəp·liŋ }

ribbed moraine [GEOL] One of a group of irregularly subparallel, locally branching, generally smoothly rounded and arcuate ridges that are convex in the downstream direction of a glacier but that curve upstream adjacent to eskers. { 'ribd mə'rān }

ribble See ripple till. { 'rib·əl }

ribbon [BUILD] A horizontal piece of wood nailed to the face of studs; usually used to support the floor joists. [GRAPHICS] A narrow band of inked fabric in a typewriter or other printing machine with which type is printed by striking keys against it. [MATH] The plane figure generated by a straight line which moves so that it is always perpendicular to the path traced by its middle point. [PETR] One of a set of parallel bands in a rock or mineral. { 'rib·ən }

ribbon banding [PETR] A banding produced in the bedding of a sedimentary rock by thin strata of contrasting colors, giving the rock an appearance which suggests bands of ribbons. { 'rib·ən ,band·iŋ }

ribbon bomb [GEOL] An elongate and flattened volcanic bomb derived from ropes of lava. { 'rib·ən ,bäm }

ribbon cable [ELEC] A cable made of normal, round, insulated wires arranged side by side and fastened together by a cohesion process to form a flexible ribbon. { 'rib·ən ,kā·bəl }

ribbon cartridge [GRAPHICS] An inked ribbon for a typewriter or computer printer enclosed in a housing that can be snapped into place. { 'rib·ən ,kär·trij }

ribbon conductor [ELEC] A thin, flat piece of metal suitable for carrying electric current. { 'rib·ən kən,dək·tər }

ribbon conveyor [MECH ENG] A type of screw conveyor which has an open space between the shaft and a ribbon-shaped flight, used for wet or sticky materials which would otherwise build up on the spindle. { 'rib·ən kən'vā·ər }

ribbon diagram [GEOL] A continuous geologic cross section that is drawn in perspective along a curved or sinuous line. { 'rib·ən 'dī·ə,gram }

ribbon jasper [GEOL] Banded jasper with parallel, ribbonlike stripes of alternating colors or shades of color. Also known as riband jasper. { 'rib·ən ,jas·pər }

ribbon lightning [GEOPHYS] Ordinary streak lightning that appears to be spread horizontally into a ribbon of parallel luminous streaks when a very strong wind is blowing at right angles to the observer's line of sight; successive strokes of the lightning flash are then displaced by small angular amounts and may appear to the eye or camera as distinct paths. Also known as band lightning; fillet lightning. { 'rib·ən ,līt·niŋ }

ribbon microphone [ENG ACOUS] A microphone whose electric output results from the motion of a thin metal ribbon mounted between the poles of a permanent magnet and driven directly by sound waves; it is velocity-actuated if open to sound waves on both sides, and pressure-actuated if open to sound waves on only one side. { 'rib·ən 'mī·krə,fōn }

ribbon mixer [MECH ENG] Device for the mixing of particles, slurries, or pastes of solids by the revolution of an elongated helicoid (spiral) ribbon of metal. { 'rib·ən 'mik·sər }

ribbon parachute [AERO ENG] A type of parachute having a canopy consisting of an arrangement of closely spaced tapes;

this parachute has high porosity with attendant stability and slight opening shock. { 'rib·ən 'par·ə‚shüt }

ribbon reef [GEOL] A linear reef within the Great Barrier Reef off the northeast coast of Australia, having inwardly curved extremities, and forming a festoon along the precipitous edge of the continental shelf. { 'rib·ən ‚rēf }

ribbon rock [PETR] A rock showing a succession of thin layers of differing composition or appearance. { 'rib·ən ‚räk }

ribbon slate [PETR] Slate produced by incomplete metamorphism of clearly visible residual bedding planes that cut across the cleavage surface. { 'rib·ən ‚slāt }

ribbon structure [GEOL] A succession of thin layers of different mineralogy and texture often contorted and deformed. { 'rib·ən ‚strək·chər }

ribbon vein *See* banded vein. { 'rib·ən ‚vān }

ribbon windows [ARCH] Windows in a continuous horizontal band. { 'rib·ən ‚win·dōz }

rib hole [MIN ENG] One of the final holes fired in blasting ground at the sides of a shaft or tunnel. Also known as trimmer. { 'rib ‚hōl }

ribitol *See* adonitol. { 'rī·bə‚tòl }

riblet [DES ENG] Any of the small, longitudinal striations, with spacing on the order of 0.002 inch or 50 micrometers, that are made on the surfaces of ships or aircraft to reduce the drag of turbulent flow. { 'rib·lət }

Riblet coupler *See* three-decibel coupler. { 'rib·lət ‚kəp·lər }

riboflavin [BIOCHEM] $C_{17}H_{20}N_4O_6$ A water-soluble, yellow orange fluorescent pigment that is essential to human nutrition as a component of the coenzymes flavin mononucleotide and flavin adenine dinucleotide. Also known as lactoflavin; vitamin B₂; vitamin G. { 'rī·bə‚flā·vən }

riboflavin 5'-phosphate [BIOCHEM] $C_{17}H_{21}N_4O_9P$ The phosphoric acid ester of riboflavin. Also known as flavin phosphate; flavin mononucleotide; FMN; isoalloxazine mononucleotide; vitamin B₂ phosphate. { 'rī·bə‚flā·vən ¦fīv‚prīm 'fä‚sfāt }

D-riboketose *See* ribulose. { ¦dē ¦rī·bō¦kē‚tōs }

ribonuclease [BIOCHEM] $C_{587}H_{909}N_{171}O_{197}S_{12}$ An enzyme that catalyzes the depolymerization of ribonucleic acid. { ¦rī·bō'nü·klē‚ās }

ribonucleic acid [BIOCHEM] A long-chain, usually single-stranded nucleic acid consisting of repeating nucleotide units containing four kinds of heterocyclic, organic bases: adenine, cytosine, quanine, and uracil; they are conjugated to the pentose sugar ribose and held in sequence by phosphodiester bonds; involved intracellularly in protein synthesis. Abbreviated RNA. { ¦rī·bō¦nü¦klē·ik 'as·əd }

ribonucleic acid polymerase [BIOCHEM] An enzyme that transcribes a ribonucleic acid (RNA) molecule from one strand of a deoxyribonucleic acid (DNA) molecule. Also known as transcriptase. { ¦rī·bō¦nü¦klē·ik 'as·əd pə'lim·ə‚rās }

ribonucleoprotein [BIOCHEM] Any of a large group of conjugated proteins in which molecules of ribonucleic acid are closely associated with molecules of protein. { ¦rī·bō¦nü·klē·ō'prō‚tēn }

ribonucleotide [BIOCHEM] A ribose-containing nucleotide, the structural unit of ribonucleic acid. { ¦rī·bō'nü·klē·ə‚tīd }

ribose [BIOCHEM] $C_5H_{10}O_5$ A pentose sugar occurring as a component of various nucleotides, including ribonucleic acid. { 'rī‚bōs }

riboside [BIOCHEM] Any glycoside containing ribose as the sugar component. { 'rī·bə‚sīd }

ribosomal ribonucleic acid [BIOCHEM] Any of three large types of ribonucleic acid found in ribosomes: 5S RNA, with molecular weight 40,000; 14–16S RNA, with molecular weight 600,000; and 18–22S RNA with molecular weight 1,200,000. Abbreviated r-RNA. { ¦rī·bō¦sō·məl ¦rī·bō¦nü¦klē·ik 'as·əd }

ribosome [CYTOL] One of the small, complex particles composed of various proteins and three molecules of ribonucleic acid which synthesize proteins within the living cell. { 'rī·bə‚sōm }

ribozyme [BIOCHEM] A ribonucleic acid molecule that can catalyze, or lower the activation energy for, specific biochemical reactions. { 'rīb·ə‚zīm }

rib pillar [MIN ENG] A pillar whose length is large compared with its width. { 'rib ‚pil·ər }

rib rifling [ORD] Rifling of the bore of a gun in which the lands and grooves are of equal width. { 'rib ‚rīf·liŋ }

ribulose [BIOCHEM] $C_5H_{10}O_5$ A pentose sugar that exists only as a syrup; synthesized from arabinose by isomerization with pyridine; important in carbohydrate metabolism. Also known as D-erythropentose; D-riboketose. { 'rī·byə‚lōs }

ribulose 1,5-bisphosphate carboxylase/oxygenase [BIOCHEM] A key enzyme in carbon fixation during photosynthesis by plant chloroplasts and their cyanobacterial relatives, as well as by photosynthetic proteobacteria. Also known as rubisco; RuBP carboxylase/oxygenase. { ¦rī·byə‚lōs ¦wən ¦fiv ‚bis¦fäs‚fāt kär¦bäk·sə‚lās 'äk·sə·jə‚nās }

ribulose diphosphate [BIOCHEM] $C_5H_{12}O_{11}P_2$ The phosphate ester of ribulose. { 'rī·byə‚lōs dī'fä‚sfāt }

ribut [METEOROL] Sharp, short squalls during comparatively calm winds from May to November in Malaya. { ri'bət }

Riccati-Bessel functions [MATH] Solutions of a second-order differential equation in a complex variable which have the form $zf(z)$, where $f(z)$ is a function in terms of polynomials and cos (z), sin (z). { ri'käd·ē 'bes·əl ‚fəŋk·shənz }

Riccati equation [MATH] **1.** A first-order differential equation having the form $y' = A_0(x) + A_1(x)y + A_2(x)y^2$; every second-order linear differential equation can be transformed into an equation of this form. **2.** A matrix equation of the form $dP(t)/dt + P(t)F(t) + F^T(t)P(t) - P(t)G(t)R^{-1}(t)G^T(t)P(t) + Q(t) = 0$, which frequently arises in control and estimation theory. { ri'käd·ē i‚kwā·zhən }

Ricci equations [MATH] Equations relating the components of the Ricci tensor, the curvature tensor, and an arbitrary tensor of a Riemann space. Also known as Ricci identities. { 'rē‚chē i‚kwā·zhənz }

Ricci identities *See* Ricci equations. { 'rē‚chē ī‚den·ə·dēz }

Ricci tensor *See* contracted curvature tensor. { 'rē‚chē ‚ten·sər }

Ricci theorem [MATH] The covariant derivative vanishes for either of the fundamental tensors of a Riemann space. { 'rē‚chē ‚thir·əm }

rice [BOT] *Oryza sativa.* An annual cereal grass plant of the order Cyperales, cultivated as a source of human food for its carbohydrate-rich grain. { rīs }

rice bran oil [MATER] Clear, combustible liquid, derived by solvent-extraction of oil from fresh rice bran; used to make soaps and animal feeds, salad and cooking oils, and hydrogenated shortening. { 'rīs 'bran ‚oil }

rice coal [GEOL] Anthracite that will pass through circular holes in a screen, the holes measuring 5/16 inch (7.9 millimeters), but not 3/16 inch (4.8 millimeters), in diameter. { 'rīs ‚kōl }

rice glue [MATER] A paste made from ground rice boiled in soft water; used in molded objects such as statuary. { 'rīs ‚glü }

rice grains [ASTRON] Bright patches that stand out against the darker background of the surface of the sun; they are short-lived, and the pattern changes in a matter of minutes. { 'rīs ‚grānz }

rice neutralization [ELECTR] Development of voltage in the grid circuit of a vacuum tube in order to nullify or cancel feedback through the tube. { 'rīs ‚nü·trə·lə'zā·shən }

rice neutralizing circuit [ELECTR] Radio-frequency amplifier circuit that neutralizes the grid-to-plate capacitance of an amplifier tube. { 'rīs 'nü·trə‚līz·iŋ ‚sər·kət }

rice paper [MATER] **1.** A product, not a true paper and not made from rice, but manufactured from the pith of a tree grown in Taiwan; tissue-thin sheets of the pith are peeled away as a cylindrical section of the wood rotates against a knife. **2.** Any of various oriental papers used in block printing. { 'rīs ‚pā·pər }

Rice's bromine solution [ANALY CHEM] Analytical reagent for the quantitative analysis of urea; has 12.5% bromine and sodium bromide in aqueous solution. { 'rīs·əz 'brō‚mēn sə‚lü·shən }

Richardson automatic scale [ENG] An automatic weighing and recording machine for flowable materials carried on a conveyor; weighs batches from 200 to 1000 pounds (90 to 450 kilograms). { 'rich·ərd·sən ¦òd·ə¦mad·ik 'skāl }

Richardson-Dushman equation [ELECTR] An equation for the current density of electrons that leave a heated conductor in thermionic emission. Also known as Dushman equation. { 'rich·ərd·sən 'dəsh·mən i‚kwā·zhən }

Richardson effect *See* thermionic emission. { 'rich·ərd·sən i‚fekt }

Richardson number [FL MECH] A dimensionless number used in studying the stratified flow of multilayer systems; equal

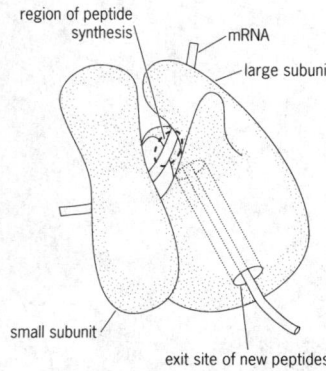

RIBOSOME

region of peptide synthesis
mRNA
large subunit
small subunit
exit site of new peptides

Three-dimensional models of *Escherichia coli* ribosome.

RIBULOSE

$$CH_2OH$$
$$|$$
$$C=O$$
$$|$$
$$HCOH$$
$$|$$
$$HCOH$$
$$|$$
$$CH_2OH$$

Structural formula of ribulose.

to the acceleration of gravity times the density gradient of a fluid, divided by the product of the fluid's density and the square of its velocity gradient at a wall. Symbolized N_{Ri}. { 'rich·ərd·sən ,nəm·bər }

Richardson plot [ELECTR] A graph of log (J/T^2) against $1/T$, where J is the current density of electrons leaving a heated conductor in thermionic emission, and T is the temperature of the conductor; according to the Richardson-Dushman equation, this is a straight line. { 'rich·ərd·sən ,plät }

Richard's solder [MET] A yellow brass containing 3% aluminum and 3% phosphor tin. { 'rich·ərdz ,säd·ər }

rich cluster [ASTRON] A galaxy cluster that has relatively many member galaxies. { ¦rich ¦kläs·tər }

rich concrete [MATER] Concrete with a high cement content. { ¦rich ¦kän,krēt }

richellite [MINERAL] $Ca_3Fe_{10}(PO_4)_8(OH,F)_{12}·nH_2O$ A yellow mineral composed of hydrous basic iron calcium fluophosphate; occurs in masses. { rə'she,līt }

rich mixture [CHEM] An air-fuel mixture that is high in its concentration of combustible component. { ¦rich ¦miks·chər }

Richmondian [GEOL] A North American stage of geologic time: Upper Ordovician (above Maysvillian, below Lower Silurian). { rich'mən·dē·ən }

rich oil [MATER] Natural-gasoline-plant absorption oil containing dissolved natural-gasoline fractions. { ¦rich ¦ȯil }

rich ore [MIN ENG] Relatively high grade ore. { ¦rich ¦ȯr }

richterite [MINERAL] $(Na,K)_2(Mg,Mn,Ca)_6Si_8O_{22}(OH)_2$ A brown, yellow, or rose-red monoclinic mineral composed of basic silicate of sodium, potassium, magnesium, manganese, and calcium; a member of the amphibole group. { 'rik·tə,rīt }

Richter scale [GEOPHYS] A scale of numerical values of earthquake magnitude ranging from 1 to 9. { 'rik·tər ,skāl }

ricin [MATER] White, poisonous powder derived from pressed castor oil bran. { 'rīs·ən }

Ricinidae [INV ZOO] A family of bird lice, order Mallophaga, which occur on numerous land and water birds. { rə'sin·ə,dē }

ricinoleic acid [ORG CHEM] $C_{18}H_{34}O_3$ Unsaturated fatty acid; a combustible, water-insoluble, viscous liquid; soluble in most organic solvents; boils at 226°C (10 mmHg); used as a chemical intermediate, in soaps and Turkey red oils, and for textile finishing. Also known as castor oil acid. { ¦ris·ən·ō¦lē·ik 'as·əd }

ricinoleyl alcohol [ORG CHEM] $C_{18}H_{36}O_2$ Fatty alcohol of ricinoleic acid; a combustible, colorless, nondrying liquid, boiling at 170–328°C; used as a chemical intermediate, in protective coatings, surface-active agents, pharmaceuticals, and plasticizers. { ¦ris·ən·ō¦lē·əl 'al·kə,hȯl }

Ricinuleida [INV ZOO] An order of rare, ticklike arachnids in which the two anterior pairs of appendages are chelate, and the terminal segments of the third legs of the male are modified as copulatory structures. { ,ris·ən·yü'lē·ə·də }

rickardite [MINERAL] Cu_4Te_3 A deep-purple mineral composed of copper telluride, occurring in masses. { 'rik·ər,dīt }

rickets [MED] A disorder of calcium and phosphorus metabolism affecting bony structures, due to vitamin D deficiency. { 'rik·əts }

Rickettsiaceae [MICROBIO] A family of the order Rickettsiales; small, rod-shaped, coccoid, or diplococcoid cells often found in arthropods; includes human and animal parasites and pathogens. { ri,ket·sē'ās·ē,ē }

Rickettsiales [MICROBIO] An order of prokaryotic microorganisms; gram-negative, obligate, intracellular animal parasites (may be grown in tissue cultures); many cause disease in humans and animals. { ri,ket·sē'ā·lēz }

rickettsialpox [MED] An acute febrile disease caused by the organism *Rickettsia akari* and transmitted from the mouse to humans by the mite *Allodermanyssus sanguineus;* characterized by rash, a primary ulcer, and often swelling of glands. { ri¦ket·sē·əl¦päks }

Rickettsieae [MICROBIO] A tribe of the family Rickettsiaceae; cells are occasionally filamentous; infect arthropods and some vertebrates and are pathogenic for humans, most frequently an incidental host. { ri'ket·sē,ē }

rickettsiosis [MED] Any disease caused by rickettsiae. { ri,ket·sē'ō·səs }

ricolettaite [PETR] A dark-colored syenite-gabbro containing anorthite as the plagioclase, along with olivine and augite. { ,rik·ə'led·ə,īt }

rictus [VERT ZOO] The mouth aperture in birds. { 'rik·təs }

riddle [DES ENG] A sieve used for sizing or for removing foreign material from foundry sand or other granular materials. { 'rid·əl }

rideau [GEOL] A small ridge or mound of earth, or a slightly elevated piece of ground. { ri'dō }

rider [GRAPHICS] A piece of type that is raised higher than surrounding type, and tends to make a heavier impression on the paper. [MIN ENG] A steel or iron crossbeam which slides between the guides in a sinking shaft; it is carried by the hoppit and serves to guide and steady the hoppit during its movement up and down the shaft. { 'rīd·ər }

rider coal *See* rider coal seam. { 'rīd·ər ,kōl }

rider coal seam [MIN ENG] In a coal mine, a small coal seam above the main seam. Also known as rider coal. { 'rīd·ər ¦kōl ,sēm }

ridge [ARCH] The line on which the sides of a sloping roof meet. [GEOL] An elongate, narrow, steep-sided elevation of the earth's surface or the ocean floor. [METEOROL] An elongated area of relatively high atmospheric pressure, almost always associated with, and most clearly identified as, an area of maximum anticyclonic curvature of wind flow. Also known as wedge. { rij }

ridge aloft *See* upper-level ridge. { 'rij ə'lȯft }

ridge board [BUILD] A horizontal board placed on edge at the apex of the roof. { 'rij ,bȯrd }

ridge cap [BUILD] Wood or metal cap which is placed over the angle of the ridge. { 'rij ,kap }

ridged ice [OCEANOGR] Sea ice having readily observed surface features in the form of one or more pressure ridges. { 'rijd 'īs }

ridge fault [GEOL] A fault structure that is a set of two faults bounding a horst. { 'rij ,fȯlt }

ridge pole [BUILD] The horizontal supporting member placed along the ridge of a roof. { 'rij ,pōl }

ridge regression analysis [STAT] A form of regression analysis in which damping factors are added to the diagonal of the correlation matrix prior to inversion, a procedure which tends to orthogonalize interrelated variables; study of the robustness of the regression coefficients with changes in the damping factors is then used to determine sets of variables that should be removed. Also known as damped regression analysis. { 'rij ri'gresh·ən ə,nal·ə·səs }

ridge-top trench [GEOL] A trench, occasionally found at or near the crest of high, steep-sided mountain ridges, formed by the creep displacement of a large slab of rock along shear surfaces more or less parallel with the side slope of the ridge. { 'rij ,täp ,trench }

ridge waveguide [ELECTROMAG] A circular or rectangular waveguide having one or more longitudinal internal ridges that serve primarily to increase transmission bandwidth by lowering the cutoff frequency. { 'rij 'wāv,gīd }

ridging [OCEANOGR] A form of deformation of floating ice, caused by lateral pressure, whereby ice is forced or piled haphazardly to form ridged ice. { 'rij·iŋ }

riding boom *See* boat boom. { 'rīd·iŋ ,büm }

riebeckite [MINERAL] $Na_2(Fe,Mg)_5Si_8O_{22}(OH)_2$ A blue or black monoclinic amphibole occurring as a primary constituent in some acid- or sodium-rich igneous rocks. { 'rē,be,kīt }

riebungsbreccia [GEOL] A breccia developed during folding. { 'rē·bəŋz,brech·ə }

Riecke's principle [MINERAL] The principle that solution of a mineral occurs most readily at points of greatest external pressure, and crystallization occurs most readily at points of least external pressure; applied to recrystallization in metamorphic rock. { 'rē·kəz ,prin·sə·pəl }

Riedel's disease [MED] A form of chronic thyroiditis with irregular localized areas of stony, hard fibrosis. { 'rēd·əlz di,zēz }

riedenite [PETR] An igneous rock composed of large tabular biotite crystals in a granular groundmass of nosean, biotite, pyroxene, and small amounts of sphene and apatite. { 'rēd·ən,īt }

Riefler clock [HOROL] A clock employing a pin escapement and a mercurial pendulum; used as a standard. { 'rēf·lər ,kläk }

riegel [GEOL] A low, traverse ridge of bedrock on the floor of a glacial valley. Also known as rock bar; threshold; verrou. { 'rē·gəl }

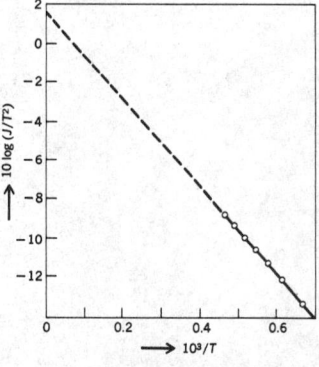

RICHARDSON PLOT

Richardson plot for tungsten, an important thermionic emitter. *(After G. Herrmann and S. Wagener, The Oxide-Coated Cathode, vol. 2, Chapman and Hall, 1951)*

Riegler's test [ANALY CHEM] Analytical technique for nitrous acid; uses sodium naphthionate and β-naphthol. { 'rēg·lərz ‚test }

Rieke diagram [ELECTR] A chart showing contours of constant power output and constant frequency for a microwave oscillator, drawn on a Smith chart or other polar diagram whose coordinates represent the components of the complex reflection coefficient at the oscillator load. { 'rē·kə ‚dī·ə‚gram }

Riemann-Christoffel tensor [MATH] The basic tensor used for the study of curvature of a Riemann space; it is a fourth-rank tensor, formed from Christoffel symbols and their derivatives, and its vanishing is a necessary condition for the space to be flat. Also known as curvature tensor. { 'rē‚män 'kris·tə·fəl ‚ten·sər }

Riemann function [MATH] A type of Green's function used in solving the Cauchy problem for a real hyperbolic partial differential equation. { 'rē‚män ‚fəŋk·shən }

Riemann hypothesis [MATH] The conjecture that the only zeros of the Riemann zeta function with positive real part must have their real part equal to $^1/_2$. { 'rē‚män hī‚päth·ə·səs }

Riemannian curvature [MATH] A general notion of space curvature at a point of a Riemann space which is directly obtained from orthonormal tangent vectors there. { rē'män·ē·ən 'kər·və·chər }

Riemannian geometry See elliptic geometry. { rē'män·ē·ən jē'äm·ə·trē }

Riemannian manifold [MATH] A differentiable manifold where the tangent vectors about each point have an inner product so defined as to allow a generalized study of distance and orthogonality. { rē'män·ē·ən 'man·ə‚fōld }

Riemannian space-time [RELAT] The space-time of general relativity, having the mathematical structure of a four-dimensional Riemann space. { rē'män·ē·ən 'spās‚tīm }

Riemann integral [MATH] The Riemann integral of a real function $f(x)$ on an interval (a,b) is the unique limit (when it exists) of the sum of $f(a_i)(x_i - x_{i-1})$, $i = 1, \ldots, n$, taken over all partitions of (a,b), $a = x_0 < a_1 < x_1 < \cdots < a_n < x_n = b$, as the maximum distance between x_i and x_{i-1} tends to zero. { 'rē‚män ‚int·ə·grəl }

Riemann-Lebesgue lemma [MATH] If the absolute value of a function is integrable over the interval where it has a Fourier expansion, then its Fourier coefficients a_n tend to zero as n goes to infinity. { 'rē‚män lə'beg ‚lem·ə }

Riemann mapping theorem [MATH] Any simply connected domain in the plane with boundary containing more than one point can be conformally mapped onto the interior of the unit disk. { 'rē‚män 'map·iŋ ‚thir·əm }

Riemann method [MATH] A method of solving the Cauchy problem for hyperbolic partial differential equations. { 'rē‚män ‚meth·əd }

Riemann P function [MATH] A scheme for exhibiting the singular points of a second-order ordinary differential equation, and the orders at these points of solutions of the equation. { 'rē‚män 'pē ‚fəŋk·shən }

Riemann space [MATH] A Riemannian manifold or subset of a euclidean space where tensors can be defined to allow a general study of distance, angle, and curvature. { 'rē‚män ‚spās }

Riemann sphere [MATH] The two-sphere whose points are identified with all complex numbers by a stereographic projection. Also known as complex sphere. { 'rē‚män ‚sfir }

Riemann-Stieltjes integral See Stieltjes integral. { ‚rē‚män 'stēl·tyəs ‚int·i·grəl }

Riemann surfaces [MATH] Sheets or surfaces obtained by analyzing multiple-valued complex functions and the various choices of principal branches. { 'rē‚män ‚sər·fə·səz }

Riemann tensors [MATH] Various types of tensors used in the study of curvature for a Riemann space. { 'rē‚män ‚ten·sərz }

Riemann zeta function [MATH] The complex function $\zeta(z)$ defined by an infinite series with nth term $e^{-z \log n}$. Also known as zeta function. { 'rē‚män 'zād·ə ‚fəŋk·shən }

Riesz-Fischer theorem [MATH] The vector space of all real- or complex-valued functions whose absolute value squared has a finite integral constitutes a complete inner product space. { 'rēsh 'fish·ər ‚thir·əm }

rifampicin [MICROBIO] An antibacterial and antiviral antibiotic; action depends upon its preferential inhibition of bacterial

ribonucleic acid polymerase over animal-cell RNA polymerase. { rə'fam·pə·sən }

riffle [HYD] **1.** A shallows across a stream bed over which water flows swiftly and is broken into waves by submerged obstructions. **2.** Shallow water flowing over a riffle. { 'rif·əl }

riffler [DES ENG] A small, curved rasp or file for filing interior surfaces or enlarging holes. { 'rif·lər }

rifle [DES ENG] A drill core that has spiral grooves on its outside surface. [ENG] A borehole that is following a spiral course. [ORD] A firearm having spiral grooves upon the surface of its bore to impart rotary motion to a projectile, thereby stabilizing the projectile and ensuring greater accuracy of impact and longer range; it may fire projectiles automatically or semiautomatically, or successive rounds may be manually loaded. { 'rī·fəl }

rifle bracket [ORD] Metal clamp for holding the rifle in easy accessibility, usually on a motor vehicle. { 'rī·fəl ‚brak·ət }

rifle grenade [ORD] A grenade especially designed or adapted to be fired or launched from the muzzle of a rifle or carbine. { 'rī·fəl grə‚nād }

rifle grenade ogive [ORD] A hollow metallic item designed for attachment to the forward end of a practice rifle grenade; it cushions the impact and permits reuse of the grenade. { 'rī·fəl grə‚nād 'ō‚jīv }

rifle range [ORD] Place for practice in shooting with a rifle. { 'rī·fəl ‚rānj }

rifling [MECH ENG] The technique of cutting helical grooves inside a rifle barrel to impart a spinning motion to a projectile around its long axis. { 'rīf·liŋ }

rift [GEOL] **1.** A narrow opening in a rock caused by cracking or splitting. **2.** A high, narrow passage in a cave. { rift }

rift-block mountain [GEOL] A mountain range which is a horst block bounded by normal faults. { 'rift ‚bläk 'maunt·ən }

rift-block valley [GEOL] A valley which occupies a graben. { 'rift ‚bläk 'val·ē }

rift lake See sag pond. { 'rift ‚läk }

rift saw [DES ENG] **1.** A saw for cutting wood radially from the log. **2.** A circular saw divided into toothed arms for sawing flooring strips from cants. { 'rift ‚sò }

rift valley [GEOL] A deep, central cleft with a mountainous floor in the crest of a midoceanic ridge. Also known as central valley; midocean rift. { 'rift ‚val·ē }

Rift Valley fever [MED] A toxic generalized febrile virus disease of humans and animals in South and East Africa, transmitted by a mosquito, and characterized by headache, photophobia, myalgia, and anorexia. { 'rift ‚val·ē 'fē·vər }

rift-valley lake See sag pond. { 'rift ‚val·ē ‚läk }

rig [MECH ENG] A tripod, derrick, or drill machine complete with auxiliary and accessory equipment needed to drill. [NAV ARCH] The method according to which spars and sails are designed and fitted. { rig }

Rigel [ASTRON] A multiple star of stellar magnitude 0.08, 650 light-years from the sun, spectral classification B8-Ia, in the constellation Orion; the star β Orionis. { 'rī·jəl }

rigging [AERO ENG] The shroud lines attached to a parachute. [NAV ARCH] Collectively, all the ropes and chains employed to support and work the masts, yards, booms, and sails of a vessel. { 'rig·iŋ }

Righi experiment [OPTICS] An experiment in which a rotating Nicol prism, a Fresnel mirror, a quarter-wave plate, and a fixed Nicol prism are used to produce effects in light beams similar to beats between sounds with slightly different frequencies. { 'rē·gē ik‚sper·ə·mənt }

Righi-Leduc effect [PHYS] The phenomenon wherein, if a magnetic field is applied at right angles to the direction of a temperature gradient in a conductor, a new temperature gradient is produced perpendicular to both the direction of the original temperature gradient and to the magnetic field. Also known as Leduc effect. { 'rē·gē lə'dük i‚fekt }

right-and-left-hand chart [IND ENG] A graphic symbolic representation of the motions made by one hand in relation to those made by the other hand. { ‚rīt ən ‚left ‚hand 'chärt }

right angle [MATH] An angle of 90°. { 'rīt 'aŋ·gəl }

right-angle prism [OPTICS] A type of prism used to turn a beam of light through a right angle (90°); it will invert (turn upside-down) or will revert (turn right for left), according to the position of the prism, any light reflected by it. { 'rīt 'aŋ·gəl 'priz·əm }

right ascension [ASTRON] A celestial coordinate; the angular distance taken along the celestial equator from the vernal equinox eastward to the hour circle of a given celestial body. Abbreviated R.A. { 'rīt ə'sen·chən }

right astern *See* dead astern. { 'rīt ə'stərn }

right bank [NAV] That bank of a stream or river on the right of the observer when he is facing in the direction of flow, or downstream. { 'rīt ¦baŋk }

right circular cone [MATH] A circular cone whose axis is perpendicular to its base. { 'rīt ¦sər·kyə·lər ¦kōn }

right circular cylinder [MATH] A solid bounded by two parallel planes and by a cylindrical surface consisting of the straight lines perpendicular to the planes and passing through a circle in one of them. { 'rīt ¦sər·kyə·lər 'sil·ən·dər }

right coset [MATH] A right coset of a subgroup H of a group G is a subset of G consisting of all elements of the form ha, where a is a fixed element of G and h is any element of H. { 'rīt 'kō¦set }

right-cut tool [DES ENG] A single-point lathe tool which has the cutting edge on the right side when viewed face up from the point end. { 'rīt ¦kət ¦tül }

right-hand cutting tool [DES ENG] A cutter whose flutes twist in a clockwise direction. { 'rīt ¦hand ¦kəd·iŋ ¦tül }

right-handed [CRYSTAL] **1.** Having a crystal structure with a mirror-image relationship to a left-handed structure. **2.** [DES ENG] **1.** Pertaining to screw threads that allow coupling only by turning in a clockwise direction. **2.** *See* right-laid. { 'rīt ¦han·dəd }

right-handed coordinate system [MATH] **1.** A three-dimensional rectangular coordinate system such that when the thumb of the right hand extends in the positive direction of the first (or x) axis the fingers fold in the direction in which the second (or y) axis could be rotated about the first axis to coincide with the third (or z) axis. **2.** A Riemann space which has negative scalar density function. { 'rīt ¦han·dəd kō'órd·ən,ət ¦sis·təm }

right-handed curve [MATH] A space curve whose torsion is negative at a given point. Also known as dextrorse curve; dextrorsum. { 'rīt ¦han·dəd ¦kərv }

right-hand helicity [QUANT MECH] Property of a particle whose spin is parallel to its momentum. { 'rīt ¦hand he'lis·əd·ē }

right-hand limit *See* limit on the right. { 'rīt ¦hand 'lim·ət }

right-hand polarization [ELECTROMAG] In elementary particle discussions, circular or elliptical polarization of an electromagnetic wave in which the electric field vector at a fixed point in space rotates in the right-hand sense about the direction of propagation; in optics, the opposite convention is used; in facing the source of the beam, the electric vector is observed to rotate clockwise. { 'rīt ¦hand ¦pō·lə·rə'zā·shən }

right-hand rule [ELECTROMAG] **1.** For a current-carrying wire, the rule that if the fingers of the right hand are placed around the wire so that the thumb points in the direction of current flow, the fingers will be pointing in the direction of the magnetic field produced by the wire. Also known as hand rule. **2.** For a moving wire in a magnetic field, such as the wire on the armature of a generator, if the thumb, first, and second fingers of the right hand are extended at right angles to one another, with the first finger representing the direction of magnetic lines of force and the second finger representing the direction of current flow induced by the wire's motion, the thumb will be pointing in the direction of motion of the wire. Also known as Fleming's rule. { 'rīt ¦hand 'rül }

right-hand screw [DES ENG] A screw that advances when turned clockwise. { 'rīt ¦hand 'skrü }

right-hand taper [ELEC] Taper in which there is greater resistance in the clockwise half of the operating range of a rheostat or potentiometer (looking from the shaft end) than in the counterclockwise half. { 'rīt ¦hand 'tā·pər }

right helicoid [MATH] The surface that is swept out by a ray that originates at an axis and remains perpendicular to this axis while the ray is rotated about the axis and is translated in the direction of the axis, both at a constant rate. { 'rīt 'hel·ə,kóid }

right identity [MATH] In a set on which a binary operation ∘ is defined, an element e with the property that $a \circ e = a$ for every element a in the set. { 'rīt ī'den·əd·ē }

righting arm [NAV ARCH] The horizontal distance between the center of gravity and a vertical line through the center of buoyancy of a ship that is displaced from the upright position;

knowledge of this quantity is necessary to determine the righting moment. { 'rīd·iŋ ,ärm }

righting lever [FL MECH] The horizontal distance from the center of mass of a floating body, slightly displaced from the equilibrium position, to a vertical line passing through the center of buoyancy. { 'rīd·iŋ ,lev·ər }

righting moment [NAV ARCH] The torque which tends to restore a vessel heeled over to its upright position; it is the product of the righting arm and the weight of the vessel. { 'rīd·iŋ ,mō·mənt }

right inverse [MATH] For a set S with a binary operation $x \circ y$ that has an identity element e, the right inverse of a member, x, of S is another member, $\bar{x}$, of S for which $x \circ \bar{x} = e$. { 'rīt 'in,vers }

right-invertible element [MATH] An element x of a groupoid with a unit element e for which there is an element x such that $x \circ \bar{x} = e$. { 'rīt in,vərd·ə·bəl 'el·ə·mənt }

right-justify [COMPUT SCI] To shift the contents of a register so that the right or least significant digit is at some specified position. [GRAPHICS] To adjust the printing positions of characters on a page so that the right margin is lined up. Also known as flush right. { 'rīt 'jəs·tə,fī }

right-laid [DES ENG] Rope or cable construction in which strands are twisted counterclockwise. Also known as right-handed. { 'rīt ¦lād }

right lang lay [DES ENG] Rope or cable in which the individual wires or fibers and the strands are twisted to the right. { 'rīt ¦laŋ ¦lā }

right-lateral fault *See* dextral fault. { 'rīt ¦lad·ə·rəl 'fólt }

right-lateral slip fault *See* dextral fault. { 'rīt ¦lad·ə·rəl 'slip ,fólt }

right module [MATH] A module over a ring in which the product of a member x of the module and a member a of the ring is written xa. { 'rīt ,mäj·əl }

right-of-way [CIV ENG] **1.** Areas of land used for a road and along the side of the roadway. **2.** A thoroughfare or path established for public use. **3.** Land occupied and used by a railroad or a public utility. { 'rīt əv ¦wā }

right parallelepiped [MATH] A parallelepiped whose lateral faces are perpendicular to its bases. { 'rīt ,par·ə,lel·ə'pī·pəd }

right prism [MATH] A prism whose lateral edges are perpendicular to the bases. { 'rīt ,priz·əm }

right rudder [NAV] The operation of moving the rudder to starboard, and consequently turning the bow of the ship to the right. { 'rīt 'rəd·ər }

right section [MATH] A plane section by a plane perpendicular to the elements of a given cylinder, or to the lateral faces of a given prism. { 'rīt 'sek·shən }

right side up *See* right way up. { 'rīt 'sīd 'əp }

right-slip fault *See* dextral fault. { 'rīt ,slip ,fólt }

right sphere [ASTRON] The appearance of the celestial sphere as seen by an observer at the earth's equator. { 'rīt sfir }

right spherical triangle [MATH] A spherical triangle that has at least one right angle. { 'rīt ¦sfir·ə·kəl 'trī,aŋ·gəl }

right strophoid [MATH] A plane curve derived from a straight line L and a point called the pole, consisting of the locus of points on a rotating line L' passing through the pole whose distance from the intersection of L and L' is equal to the distance of this intersection from the foot of the perpendicular from the pole to L. { 'rīt 'strä,fóid }

right triangle [MATH] A triangle one of whose angles is a right angle. { 'rīt 'trī,aŋ·gəl }

right truncated prism [MATH] A truncated prism, in which one of the cutting planes is perpendicular to the lateral edges. { 'rīt ¦trəŋ·kād·əd 'priz·əm }

right value [COMPUT SCI] The actual data content of a symbolic variable in a computer program; it is one of two components of the symbolic variable, the other being the memory address. Abbreviated rvalue. { 'rīt 'val·yü }

right way up [GEOL] The state of strata where the present upward succession of layers is the original (normal) order of deposition. Also known as right side up. { 'rīt 'wā ¦əp }

rigid body [MECH] An idealized extended solid whose size and shape are definitely fixed and remain unaltered when forces are applied. { 'rij·id 'bäd·ē }

rigid-body dynamics [MECH] The study of the motions of a rigid body under the influence of forces and torques. { 'rij·id ¦bäd·ē dī'nam·iks }

rigid copper coaxial line [ELECTROMAG] A coaxial cable in which the central conductor and outer conductor are formed by joining rigid pieces of copper. { 'rij·id 'käp·ər kó'ak·sē·əl ,līn }

rigid coupling [MECH ENG] A mechanical fastening of shafts connected with the axes directly in line. { 'rij·id 'kəp·liŋ }

rigid frame [BUILD] A steel skeleton frame in which the end connections of all members are rigid so that the angles they make with each other do not change. { 'rij·id 'frām }

rigid insulation [ELEC] Electrical insulation that is part of a rigid structure, and must provide mechanical strength and stability of form as well as a dielectric barrier; mica, glass, porcelain, and thermosetting resins are the principal materials used. { 'rij·id ,in·sə'lā·shən }

rigidity [ASTROPHYS] The ratio of the momentum of a cosmic-ray particle to its electric charge, in units of the electron charge. [MECH] The quality or state of resisting change in form. { ri'jid·əd·ē }

rigidity modulus See modulus of elasticity in shear. { ri'jid·əd·ē ,mäj·ə·ləs }

rigidizer [ENG] A supporting structure providing ridigity to an instrument that might otherwise be subject to undesirable vibrations. { ,ri·jə'dīz·ər }

rigid pavement [CIV ENG] A thick portland cement pavement on a gravel base and subbase, with steel reinforcement and often with transverse joints. { 'rij·əd 'pāv·mənt }

rigid resin [MATER] A resin with a modulus of 10,000 pounds per square inch (6.895 × 10⁷ pascals) or greater. { 'rij·əd 'rez·ən }

Rigil Kent See Alpha Centauri. { 'rī·jəl 'kent }

Rigil Kentaurus See Alpha Centauri. { 'rī·jəl ken'tór·əs }

rigor [MED] **1.** Stiffness. **2.** A chill associated with muscular contraction and tremor. { 'rig·ər }

rigor mortis [PATH] Stiffening and rigidity of the musculature occurring after death, beginning within 5–10 hours, and disappearing after 3–4 days. { 'rig·ər 'mórd·əs }

RIHANS See River and Harbor Aid to Navigation System. { 'rī,hanz }

Riley-Day syndrome See dysautonomia. { ¦rī·lē ¦dā 'sin,drōm }

rill [ASTRON] A crooked, narrow crack on the moon's surface; may be a kilometer or more in width and a few to several hundred kilometers in length. Also spelled rille. [GEOL] A small, transient runnel. [HYD] A small brook or stream. { ril }

rill crater [ASTRON] A lunar crater that forms part of a rill. { 'ril ,krād·ər }

rille See rill. { ril }

rillenstein [GEOL] A pattern of tiny solution grooves of about 1 millimeter or less in width, formed on the limestone surface of a karstic region. { 'ril·ən,stīn }

rill erosion [GEOL] The formation of numerous, closely spaced rills due to the uneven removal of surface soil by streamlets of running water. Also known as rilling; rill wash; rillwork. { 'ril i'rō·zhən }

rill flow [HYD] Surface runoff flowing in small irregular channels too small to be considered rivulets. { 'ril ,flō }

rilling See rill erosion. { 'ril·iŋ }

rill mark [GEOL] A small, dendritic channel formed on beach mud or sand by a rill, especially if on the lee side of a partially buried obstruction. { 'ril ,märk }

rillstone See ventifact. { 'ril,stōn }

rill wash See rill erosion. { 'ril ,wäsh }

rillwork See rill erosion. { 'ril,wərk }

rim [DES ENG] **1.** The outer part of a wheel, usually connected to the hub by spokes. **2.** An outer edge or border, sometimes raised or projecting. { rim }

rima [GEOL] A long, narrow aperture, cleft, or fissure. { 'rī·mə }

rim-bearing swing bridge [CIV ENG] A swing bridge that is supported by a cylindrical girder on rollers. { ¦rim ,ber·iŋ 'swiŋ ,brij }

rim blight [PL PATH] A fungus disease of tea caused by members of the genus *Cladosporium* and characterized by yellow discoloration of the leaf margins followed by browning. { 'rim ,blīt }

rim cement [GEOL] A thin layer of calcium carbonate, hematite, or silica developed on the surface of detrital grains during diagenesis. { 'rim si,mənt }

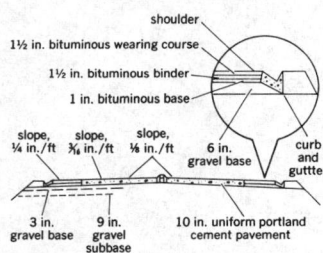

RIGID PAVEMENT

shoulder
1½ in. bituminous wearing course
1½ in. bituminous binder
1 in. bituminous base
slope, ¼ in./ft slope, ¾₆ in./ft slope, ⅛ in./ft 6 in. gravel base curb and guttter
3 in. gravel base 9 in. gravel subbase 10 in. uniform portland cement pavement

Rigid main roadway with flexible shoulders.

rim clutch [MECH ENG] A frictional contact clutch having surface elements that apply pressure to the rim either externally or internally. { 'rim ,kləch }

rim drive [ENG ACOUS] A phonograph or sound recorder drive in which a rubber-covered drive wheel is in contact with the inside of the rim of the turntable. { 'rim ,drīv }

rime [HYD] A white or milky and opaque granular deposit of ice formed by the rapid freezing of supercooled water drops as they impinge upon an exposed object; composed essentially of discrete ice granules, and has densities as low as 0.2–0.3 gram per cubic centimeter. { rīm }

rime fog See ice fog. { 'rīm ,fäg }

rimfire [ORD] Of a cartridge, having the primer mixture in the rim of the cartridge case base. { 'rim,fīr }

rim gypsum [GEOCHEM] Gypsum in thin films between anhydrite crystals, believed to have been introduced in solution rather than produced by replacement. { 'rim ,jip·səm }

rimless [ORD] Pertaining to a cartridge case in which the extracting groove is machined into the body of the case; that is, no part of the case extends beyond the body. { 'rim·ləs }

rimmed [ORD] Pertaining to a cartridge case in which an extractor rim projects beyond the body of the case. { rimd }

rimmed kettle [GEOL] A morainal depression with raised edges. { 'rimd 'ked·əl }

rimmed solution pool [GEOL] A pool in rock with a hardened rim resulting from deposition of lime during evaporation at low tide. { 'rimd sə'lü·shən ,pül }

rimmed steel [MET] Low-carbon steel, partially deoxidized, which on cooling continuously, evolves sufficient carbon monoxide to form a case or rim of metal virtually free of voids. { 'rimd 'stēl }

rimming wall [GEOL] A steep, ridgelike erosional remnant of continuous layers of porous, permeable, poorly cemented, detrital limestones, believed to form under tropical or subtropical conditions by surface-controlled secondary cementation of an original steep slope and followed by differential erosion that brings the cemented zone into relief. { 'rim·iŋ ,wól }

rimpylite [MINERAL] A group name for several green and brown hornblendes with high contents of (Al,Fe)₂O₃. { 'rim·pə,līt }

rim ridge [GEOL] A minor ridge of till defining the edge of a moraine plateau. { 'rim ,rij }

rimrock [GEOL] A top layer of resistant rock on a plateau outcropping with vertical or near vertical walls. { 'rim,räk }

RIMS See resonant ionization mass spectrometry. { rimz or ¦är¦ī¦em'es }

rimstone [GEOL] A calcium-containing deposit ringing an overflowing basin such as a hot spring. { 'rim,stōn }

rim syncline [GEOL] In salt tectonics, a local depression that develops as a border around a salt dome, as the salt in the underlying strata is displaced toward the dome. Also known as peripheral sink. { 'rim 'sin,klīn }

rincon [GEOL] **1.** A small, secluded valley. **2.** A bend in a stream. { riŋ'kōn }

rind [BOT] **1.** The bark of a tree. **2.** The thick outer covering of certain fruits. { rīnd }

rinderpest [VET MED] An acute, contagious, and often fatal virus disease of cattle, sheep, and goats which is characterized by fever and the appearance of ulcers on the mucous membranes of the intestinal tract. { 'rin·dər,pest }

R indicator See R scope. { 'är ,in·də,kād·ər }

ring [COMPUT SCI] A cyclic arrangement of data elements, usually including a specified entry pointer. [DES ENG] A tie member or chain link; tension or compression applied through the center of the ring produces bending moment, shear, and normal force on radial sections. [MATH] **1.** An algebraic system with two operations called multiplication and addition; the system is a commutative group relative to addition, and multiplication is associative, and is distributive with respect to addition. **2.** A ring of sets is a collection of sets where the union and difference of any two members is also a member. [ORG CHEM] A closed loop of bonded atoms in a chemical structure, for example, benzene or cyclohexane. { riŋ }

ring A [ASTRON] The bright outer ring of Saturn, having an outside diameter of 169,000 miles (272,000 kilometers) and an inner diameter of 150,000 miles (242,000 kilometers). { 'riŋ 'ā }

ring-and-ball test [CHEM ENG] A test for determining the melting point of asphalt, waxes, and paraffins in which a small ring is fitted with a test sample upon which a small ball is then placed; the melting point is that temperature at which the sample softens sufficiently to allow the ball to fall through the ring. Also known as ball and ring method. { 'riŋ ən ¦böl ˌtest }

ring-and-bead sight [ORD] Type of gunsight in which the front sight is a bead or post and the rear sight a ring. { 'riŋ ən ¦bēd ˌsīt }

ring-and-circle shear [DES ENG] A rotary shear designed for cutting circles and rings where the edge of the metal sheet cannot be used as a start. { 'riŋ ən ¦sər·kəl ˌshir }

ring-around [COMMUN] **1.** Improper routing of a call back through a switching center already trying to complete the same call, thus tying up the trunks by repeating the cycle. **2.** Oscillation of a repeater caused by leakage of the transmitter signal into the receiver. { 'riŋ ə,raund }

ring B [ASTRON] The brightest of Saturn's rings, with an outer diameter of 146,000 miles (235,000 kilometers) and an inner diameter of 114,000 miles (183,000 kilometers). { 'riŋ 'bē }

ringbolt [DES ENG] An eyebolt with a ring passing through the eye. { 'riŋ,bōlt }

ring bus [ELEC] A substation switching arrangement that may consist of four, six, or more breakers connected in a closed loop, with the same number of connection points. { 'riŋ ,bəs }

ring C [ASTRON] A faint ring of Saturn inside ring B having an outer diameter of 114,000 miles (183,000 kilometers) and an inner diameter of 91,000 miles (146,000 kilometers). Also known as crepe ring. { 'riŋ 'sē }

ring canal [INV ZOO] In echinoderms, the circular tube of the water-vascular system that surrounds the esophagus. { 'riŋ kə,nal }

ring circuit [ELECTROMAG] In waveguide practice, a hybrid T junction having the physical configuration of a ring with radial branches. { 'riŋ ,sər·kət }

ring closure [ORG CHEM] A chemical reaction in which one part of an open chain of a molecule reacts with another part to form a ring. { 'riŋ ,klō·zhər }

ring complex [GEOL] An association of two ring-shaped igneous intrusive forms, ring dikes and cone sheets. { 'riŋ ,käm,pleks }

ring counter [ELECTR] A loop of binary scalers or other bistable units so connected that only one scaler is in a specified state at any given time; as input signals are counted, the position of the one specified state moves in an ordered sequence around the loop. { 'riŋ ,kaunt·ər }

ring crusher [MECH ENG] Solids-reduction device with a rotor having loose crushing rings held outwardly by centrifugal force, which crush the feed by impact with the surrounding shell. { 'riŋ ,krəsh·ər }

ring current [GEOPHYS] A westward electric current which is believed to circle the earth at an altitude of several earth radii during the main phase of geomagnetic storms, resulting in a large worldwide decrease in the geomagnetic field horizontal component at low latitudes. { 'riŋ ,kə·rənt }

ring D [ASTRON] A very faint ring of Saturn that is fainter than ring C and lies between ring C and the planet's surface. { 'riŋ 'dē }

ring data structure [COMPUT SCI] Stored data that is organized by a chain of pointers so that the last pointer is directed back to the beginning of the chain. { 'riŋ 'dad·ə ,strək·chər }

ring deoxyribonucleic acid See circular deoxyribonucleic acid. { 'riŋ dē¦äk·sē¦rī·bō¦nü¦klē·ik 'as·əd }

ring depression [GEOL] The annular, structurally depressed area surrounding the central uplift of a cryptoexplosion structure; faulting and folding may be involved in its formation. Also known as peripheral depression; ring syncline. { 'riŋ di,presh·ən }

ring dike [GEOL] A roughly circular dike that is vertical or inclined away from the center of the arc. Also known as ring-fracture intrusion. { 'riŋ ,dīk }

ring discharge [ELECTR] A ring-shaped discharge generated by a high-frequency oscillating electromagnetic field produced by an external coil. Also known as toroidal discharge. { 'riŋ 'dis,chärj }

ringdown [COMMUN] In telephone switching, that method of signaling an operator in which telephone ringing current is sent over the line to operate a drop or a self-locking relay and lamp. { 'riŋ,daun }

ring-dyed fiber [TEXT] A synthetic fiber that is dipped in

plastic to coat the surface fiber and to provide a surface on which the dye will adhere. { 'riŋ ¦dīd 'fī·bər }

ring E [ASTRON] A diffuse, faint outer ring of Saturn that extends from inside the orbit of Enceladus at about 112,000 miles (181,000 kilometers) from Saturn to outside the orbit of Dione at about 300,000 miles (480,000 kilometers) from the planet. { 'riŋ 'ē }

Ringelmann chart [ENG] A chart used in making subjective estimates of the amount of solid matter emitted by smoke stacks; the observer compares the grayness of the smoke with a series of shade diagrams formed by horizontal and vertical black lines on a white background. { 'riŋ·gəl,män ,chärt }

ringent [BOT] Having widely separated, gaping lips. [ZOO] Gaping irregularly. { 'rin·jənt }

Ringer's solution [CHEM] A solution of 0.86 gram sodium chloride, 0.03 gram potassium chloride, and 0.033 gram calcium chloride in boiled, purified water, used topically as a physiological salt solution. { 'riŋ·ərz sə,lü·shən }

ring F [ASTRON] A narrow ring of Saturn just outside ring A that consists of more than five separate strands and is held in place by two small satellites. { 'riŋ 'ef }

ring fault [GEOL] **1.** A fault that bounds a rift valley. **2.** A steep-sided fault pattern that is cylindrical in outline and associated with cauldron subsidence. Also known as ring fracture. { 'riŋ ,fölt }

ring fissure [GEOL] A roughly circular desiccation crack formed on a playa around a point source (generally a phreatophyte). { 'riŋ ,fish·ər }

ring fracture See ring fault. { 'riŋ ,frak·chər }

ring-fracture intrusion See ring dike. { 'riŋ ¦frak·chər in,trü·zhən }

ring-fracture stoping [GEOL] Large-scale magmatic stoping that is associated with cauldron subsidence. { 'riŋ ¦frak·chər ,stōp·iŋ }

ring G [ASTRON] A tenuous ring of Saturn with an average diameter of 211,000 miles (340,000 kilometers), and lying 21,000 miles (34,000 kilometers) outside ring A. { 'riŋ 'jē }

ring gage [DES ENG] A cylindrical ring of steel whose inside diameter is finished to gage tolerance and is used for checking the external diameter of a cylindrical object. { 'riŋ ,gāj }

ring galaxy [ASTRON] A class of galaxy whose ringlike structure has clumps of ionized hydrogen clouds on its periphery, may have a nucleus of stars, and is usually accompanied by a small galaxy; probably formed when a small galaxy crashes through the disk of a spiral galaxy. { 'riŋ ,gal·ik·sē }

ring gate [CIV ENG] A type of gate used to regulate and control the discharge of a morning-glory spillway; like a drum gate, it offers a minimum of interference to the passage of ice or drift over the gate and requires no external power for operation. [ENG] An annular opening through which plastics enter the cavity of an injection or transfer mold. { 'riŋ ,gāt }

ring gear [MECH ENG] The ring-shaped gear in an automobile differential that is driven by the propeller shaft pinion and transmits power through the differential to the line axle. { 'riŋ ,gir }

ring head [ELECTR] A recording and playback head in a magnetic recording system which has the form of a ring with a gap at one point, and on which the coils are wound. { 'riŋ ,hed }

ring holes [MIN ENG] The group of boreholes radially drilled from a common-center setup { 'riŋ ,hōlz }

ringing [COMMUN] The production of an audible or visible signal at a station or switchboard by means of an alternating or pulsating current. [CONT SYS] An oscillatory transient occurring in the output of a system as a result of a sudden change in input. { 'riŋ·iŋ }

ringing circuit [ELECTR] A circuit which has a capacitance in parallel with a resistance and inductance, with the whole in parallel with a second resistance; it is highly underdamped and is supplied with a step or pulse input. { 'riŋ·iŋ ,sər·kət }

ringing time [ENG] In an ultrasonic testing unit, the length of time that the vibrations in a piezoelectric crystal remain after the generation of ultrasonic waves ceases. { 'riŋ·iŋ ,tīm }

ring isomerism [ORG CHEM] A type of geometrical isomerism in which bond lengths and bond angles prevent the existence of the trans structure if substituents are attached to alkenic carbons which are part of a cyclic system, the ring of which contains fewer than eight members; for example, 1,2-dichlorocyclohexene. { 'riŋ ī'säm·ə,riz·əm }

RINGING CIRCUIT

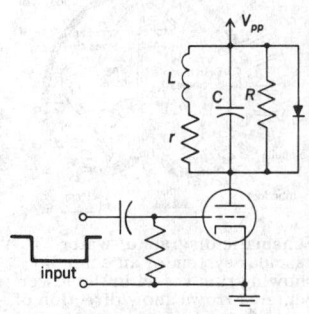

Circuit diagram of a ringing circuit used as a trigger source. V_{pp} = plate supply voltage, L = inductance, C = capacitance, R, r = resistances.

ring isomorphism [MATH] An isomorphism between rings. { 'riŋ ‚ī·sō'mȯr‚fiz·əm }

ringite [GEOL] An igneous rock formed by the mixing of silicate and carbonatite magmas. { 'riŋ‚īt }

ring jewel [DES ENG] A type of jewel used as a pivot bearing in a time-keeping device, gyro, or instrument. { 'riŋ ‚jül }

ring job [MECH ENG] Installation of new piston rings on a piston. { 'riŋ ‚jäb }

ring laser *See* laser gyro. { 'riŋ ‚lā·zər }

ring lifter *See* split-ring core lifter. { 'riŋ ‚lif·tər }

ring-lifter case *See* lifter case. { 'riŋ ‚lif·tər ‚kās }

ringlock nail [DES ENG] A nail ringed with grooves to provide greater holding power. { 'riŋ‚läk ‚nāl }

ring micrometer [OPTICS] A flat, thin ring in the focal plane of a telescope; used to measure differences in right ascension and declination. { 'riŋ mi'kräm·əd·ər }

ring modulator [ELECTR] A modulator in which four diode elements are connected in series to form a ring around which current flows readily in one direction; input and output connections are made to the four nodal points of the ring; used as a balanced modulator, demodulator, or phase detector. { 'riŋ 'mäj·ə‚lād·ər }

Ring Nebula [ASTRON] A nebula in the summer constellation Lyra; it is an example of the planetary type of gaseous nebulae. { 'riŋ 'neb·yə·lə }

ring network [COMMUN] A communications network in which the nodes can be considered to be on a circle, around which messages must be routed. Also known as loop network. { 'riŋ 'net‚wərk }

ring of operators *See* von Neumann algebra. { ¦riŋ əv 'äp·ə‚rād·ərz }

ring-oil [MECH ENG] To oil (a bearing) by conveying the oil to the point to be lubricated by means of a ring, which rests upon and turns with the journal, and dips into a reservoir containing the lubricant. { 'riŋ ‚ȯil }

ring permutation [MATH] An arrangement of objects about a ring whose orientation is unspecified. { 'riŋ ‚pər·myə‚tā·shən }

ring plain [ASTRON] A lunar crater of exceptionally large diameter and with a relatively smooth interior. { 'riŋ ‚plān }

ring power transmission line [ELEC] A power transmission line that is closed upon itself to form a ring; provides two paths between the power station and any customer, and enables a faulty section of the line to be disconnected without interrupting service to customers. { 'riŋ 'paů·ər tranz'mish·ən ‚līn }

ring road *See* beltway. { 'riŋ ‚rōd }

ring-roller mill [MECH ENG] A grinding mill in which material is fed past spring-loaded rollers that apply force against the sides of a revolving bowl. Also known as roller mill. { 'riŋ ¦rōl·ər ‚mil }

ring-rolling [MET] Producing a thin, large-diameter ring from a thicker, smaller-diameter ring by placing the ring between two rotating rolls. { 'riŋ ‚rōl·iŋ }

ring rot [PL PATH] **1.** A fungus disease of the sweet potato root caused by *Rhizopus stolonifer* and marked by rings of dry rot. **2.** A bacterial disease of potatoes caused by *Corynebacterium sepedonicum* and characterized by brown discoloration of the annular vascular tissue. { 'riŋ ‚rät }

rings and brushes [OPTICS] An interference pattern produced by ordinary and extraordinary rays when a uniaxial crystal is placed between two polarizers. { ¦riŋz ən 'brəsh·əz }

ring shift *See* cyclic shift. { 'riŋ ¦shift }

ring sight [ORD] A sight, especially a gunsight, in the shape of a ring or concentric rings, through which aim is taken and range is estimated. { 'riŋ ‚sīt }

ring silicate *See* cyclosilicate. { 'riŋ 'sil·ə‚kət }

rings of Saturn [ASTRON] Circular rings that encircle the planet Saturn at its equator; there are four main regions to the ring system; theory and observations indicate that the rings are composed of separate particles which move independently in the four series of circular coplanar orbits. { 'riŋz əv 'säd·ərn }

ring spinner [TEXT] A spinning machine in which a loop slides on a bobbin ring to control the twist of the yarn. { 'riŋ ‚spin·ər }

ring spot [PL PATH] Any of various virus and fungus diseases of plants characterized by the appearance of a discolored, annular lesion. { 'riŋ ‚spät }

ring stress [MIN ENG] The zone of stress in rock surrounding all development excavations. { 'riŋ ‚stres }

ring structure [COMPUT SCI] A chained file organization such that the end of the chain points to its beginning. [GEOL] A formation on the surface of the earth, moon, or a planet, having a ring-shaped trace in plan. [ORG CHEM] A cyclic chemical structure consisting of a chain whose ends are connected by bonds. { 'riŋ ‚strək·chər }

ring syncline *See* ring depression. { 'riŋ 'sin‚klīn }

ring system [ORG CHEM] Arbitrary designation of certain compounds as closed, circular structures, as in the six-carbon benzene ring; common rings have four, five, and six members, either carbon or some combination of carbon, nitrogen, oxygen, sulfur, or other elements. { 'riŋ ‚sis·təm }

ringtail *See* cacomistle. { 'riŋ‚tāl }

ring test [IMMUNOL] The simplest of the precipitin tests for antigen-antibody reaction; the solution containing antigen is layered on a solution containing antibody; a white disk or precipitate forms at the point where the two solutions diffuse until optimum concentration for precipitation is reached. { 'riŋ ‚test }

ring theory [MATH] The study of the structure of rings in algebra. { 'riŋ ‚thē·ə·rē }

ring time [ELECTR] The length of time in microseconds required for a pulse of energy transmitted into an echo box to die out; a measurement of the performance of the radar. { 'riŋ ‚tīm }

ring vessel [INV ZOO] A part of the water-vascular system in echinoderms; it is the circular canal around the mouth into which the stone canal empties, and from which a radial water vessel traverses to each of five radii. { 'riŋ ‚ves·əl }

ring vortex *See* vortex ring. { 'riŋ ‚vȯr‚teks }

ringwall structure [ASTRON] A lunar crater whose center lies on the wall, or on the line of the wall, of a larger crater. { 'riŋ‚wȯl ‚strək·chər }

ring whizzer [INORG CHEM] A fluxional molecule frequently encountered in organometallic chemistry in which rapid rearrangements occur by migrations about unsaturated organic rings. { 'riŋ ‚wiz·ər }

ringworm [MED] A fungus infection of skin, hair, or nails producing annular lesions with elevated margins. Also known as tinea. { 'riŋ‚wərm }

rinkite *See* mosandrite. { 'riŋ‚kīt }

rinkolite *See* mosandrite. { 'riŋ·kə‚līt }

rinneite [MINERAL] NaK₃FeCl₆ A colorless, pink, violet, or yellow mineral composed of sodium potassium iron chloride, occurring in granular masses. { 'rin·ē‚īt }

rinse [GRAPHICS] To wash off; a liquid bath to remove foreign matter. { rins }

Riodininae [INV ZOO] A subfamily of the lepidopteran family Lycaenidae in which prothoracic legs are nonfunctional in the male. { ‚rī·ə'din·ə‚dē }

riometer [ENG] An instrument that measures changes in ionospheric absorption of electromagnetic waves by determining and recording the level of extraterrestrial cosmic radio noise. Derived from relative ionospheric opacity meter. { rī'äm·əd·ər }

riot-control agent [MATER] A chemical that produces temporary irritating or disabling effects when in contact with the eyes or when inhaled. { 'rī·ət kən¦trōl ā·jənt }

riot grenade [ORD] Grenade of plastic or other nonfragmenting material, containing a charge of tear gas and a detonating fuse with short delay; the gas is released by a bursting action. { 'rī·ət grə‚nād }

riot gun [ORD] Any shotgun with a short barrel, especially a short-barreled shotgun used in guard duty or to scatter rioters; usually has a 20-inch (50.8-centimeter) cylinder barrel. { 'rī·ət ‚gən }

Rio Tinto process [MIN ENG] Heap leaching of copper-containing sulfides that have been oxidized to sulfates by prolonged atmospheric weathering. { 'rē·ō 'tin·tō ‚prä·səs }

rip [ENG] To saw wood with the grain. [MIN ENG] To break down the roof in mine roadways to increase the headroom for haulage, traffic, and ventilation. [OCEANOGR] A turbulent agitation of water generally caused by the interaction of currents and wind. { rip }

RIP *See* raster image processor. { ¦är¦ī¦pē *or* rip }

riparian [BIOL] Living or located on a riverbank. { rə'per·ē·ən }

riparian water loss [HYD] Discharge of water through

RING NEBULA

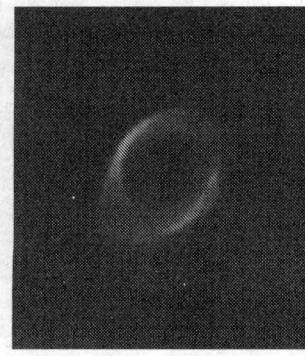

Ring Nebula NGC 6720.

RINGS OF SATURN

Telescopic appearance of Saturn at maximum inclination of ring system.

RING VESSEL

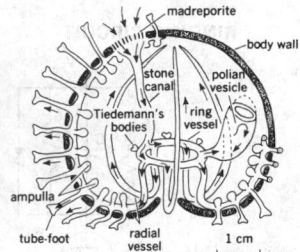

Schematic diagram of water-vascular system of an echinoid showing ring vessel in the lower center. Arrows show direction of water flow.

evapotranspiration along a watercourse, especially water transpired by vegetation growing along the watercourse. { rə'per·ē·ən 'wȯd·ər ˌlȯs }

riparian zone [ECOL] The part of the watershed immediately adjacent to the stream channel. { rī'per·ē·ən ˌzōn }

ripbit See detachable bit. See jackbit. { 'rip‚bit }

rip channel [GEOL] A channel, often more than 2 meters (6.6 feet) deep, carved on the shore by a rip current. { 'rip ˌchan·əl }

rip current [OCEANOGR] The return flow of water piled up on shore by incoming waves and wind. { 'rip ˌkə·rənt }

ripe [BOT] Of fruit, fully developed, having mature seed and so usable as food. [FOR] Of timber or a forest, ready to be cut. [GEOL] Referring to peat, in an advanced state of decay. [HYD] Descriptive of snow that is in a condition to discharge meltwater; ripe snow usually has a coarse crystalline structure, a snow density near 0.5, and a temperature near 32°F (0°C). { rīp }

ripidolite [MINERAL] $(Mg,Fe^{2+})_9Al_5Si_5O_{20}(OH)_{16}$ A mineral of the chlorite group; consists of basic magnesium iron aluminum silicate. Also known as aphrosiderite. { rə'pid·əlˌīt }

rip panel [AERO ENG] A part of a crewed free balloon; it is the panel to which the ripcord is attached and extends about one-fourth to one-fifth of the circumference of the balloon along one of its meridians; it is torn open when the ripcord is pulled so that all the gas in the balloon escapes. { 'rip ˌpan·əl }

ripping bar [DES ENG] A steel bar with a chisel at one end and a curved claw for pulling nails at the other. Also known as claw bar; wrecking bar. { 'rip·iŋ ˌbär }

ripping face support [MIN ENG] A timber or steel support at the ripping lip. { 'rip·iŋ ˌfās sə‚pȯrt }

ripping lip [MIN ENG] The end of the enlarged roadway section where work is proceeding. { 'rip·iŋ ˌlip }

ripping punch [DES ENG] A tool with a rectangular cutting edge, used in a punch press to crosscut metal plates. { 'rip·iŋ ˌpənch }

ripple [ELEC] The alternating-current component in the output of a direct-current power supply, arising within the power supply from incomplete filtering or from commutator action in a dc generator. [FL MECH] See capillary wave. [GEOL] A very small ridge of sand resembling or suggesting a ripple of water and formed on the bedding surface of a sediment. [OCEANOGR] A small curling or undulating wave controlled to a significant degree by both surface tension and gravity. { 'rip·əl }

ripple bedding [GEOL] A bedding surface characterized by ripple marks. { 'rip·əl ˌbed·iŋ }

ripple biscuit [GEOL] A bedding structure produced by lenticular lamination of sand in a bay or lagoon. { 'rip·əl ˌbis·kət }

ripple-carry adder [COMPUT SCI] A device for addition of two n-bit binary numbers, formed by connecting n full adders in cascade, with the carry output of each full adder feeding the carry input of the following full adder. { 'rip·əl ˌkar·ē ˌad·ər }

ripple drift [GEOL] A pattern of cross-lamination formed by sedimentary deposits on both sides of a migrating ripple. { 'rip·əl ˌdrift }

ripple filter [ELECTR] A low-pass filter designed to reduce ripple while freely passing the direct current obtained from a rectifier or direct-current generator. Also known as smoothing circuit; smoothing filter. { 'rip·əl ˌfil·tər }

ripple index [GEOL] On a rippled surface, the ratio of the crest-to-crest distance to the crest-to-trough distance. { 'rip·əl ˌin‚deks }

ripple lamina [GEOL] An internal sedimentary structure formed in sand or silt by currents or waves, as opposed to a ripple mark formed externally on a surface. { 'rip·əl ˌlam·ə·nə }

ripple load cast [GEOL] A load cast of a ripple mark showing evidence of penecontemporaneous deformation in the accumulation of its trough and crest and in the oversteepening of the component laminae. { 'rip·əl 'lōd ˌkast }

ripple mark [GEOL] **1.** A surface pattern on incoherent sedimentary material, especially loose sand, consisting of alternating ridges and hollows formed by wind or water action. **2.** One of the ridges on a ripple-marked surface. { 'rip·əl ˌmärk }

ripple quantity [PHYS] Alternating component of a pulsating quantity when this component is small relative to the constant component. { 'rip·əl ˌkwän·əd·ē }

ripple scour [GEOL] A shallow, linear trough with transverse ripple marks. { 'rip·əl ˌskaur }

ripple symmetry index [GEOL] A measure of the degree of symmetry of a ripple mark, equal to the ratio of the length of the gentle (upcurrent) side to the steep (downcurrent) side. { 'rip·əl 'sim·ə·trē ˌin‚deks }

ripple tank [PHYS] A shallow tray containing a liquid and equipped with means for generating surface waves; used to illustrate several types of wave phenomena, such as interference and diffraction. { 'rip·əl ˌtaŋk }

ripple till [GEOL] A till sheet containing low, winding smooth-topped ridges lying at right angles to the direction of ice movement, and grouped into narrow belts up to 48 miles (80 kilometers) long that are generally parallel to the direction of ice movement. Also known as ribble. { 'rip·əl ˌtil }

ripple voltage [ELEC] The alternating component of the unidirectional voltage from a rectifier or generator used as a source of direct-current power. { 'rip·əl ˌvōl·tij }

riprap [CIV ENG] A foundation or revetment in water or on soft ground made of irregularly placed stones or pieces of boulders; used chiefly for river and harbor work, for roadway filling, and on embankments. { 'rip‚rap }

rips [OCEANOGR] A turbulent agitation of water, generally caused by the interaction of currents and wind; in nearshore regions they may be currents flowing swiftly over an irregular bottom; sometimes referred to erroneously as tide rips. { rips }

ripsaw [MECH ENG] A heavy-tooth power saw used for cutting wood with the grain. { 'rip‚so }

ripstop fabric [TEXT] A type of tear-resistant fabric that is woven with a double thread at given intervals so that small tears do not spread. { 'rip‚stäp ˌfab·rik }

RIS See resonance ionization spectroscopy.

RISC See reduced instruction set computer. { risk }

rise [ASTRON] Of a celestial body, to cross the visible horizon while ascending. [GEOL] A long, broad elevation which rises gently from its surroundings, such as the sea floor. [HYD] See resurgence. [SCI TECH] The increase in the height or the value of something, such as a rise of tide or a rise of temperature. { rīz }

rise and run [CIV ENG] The pitch of an inclined surface or member, usually expressed as the ratio of the vertical rise to the horizontal span. { ˌrīz ən 'rən }

rise of bottom See dead rise. { ˌrīz əv 'bäd·əm }

rise of floor See dead rise. { ˌrīz əv 'flȯr }

rise of tide [OCEANOGR] Vertical distance from the chart datum to a higher water datum. { ˌrīz əv 'tīd }

rise pit [GEOL] A pit through which an underground stream rises to the surface with a calm and steady flow. { ˌrīz ˌpit }

riser [CHEM ENG] That portion of a bubble-cap assembly in a distillation tower that channels the rising vapor and causes it to flow downward to pass through the liquid held on the bubble plate. [CIV ENG] **1.** A board placed vertically beneath the tread of a step in a staircase. **2.** A vertical steam, water, or gas pipe. [GEOL] A steplike topographic feature, such as a steep slope between terraces. [MET] See feedhead. [PETRO ENG] In an offshore drilling facility, a system of piping extending from the hole and terminating at the rig. [TEXT] A raised spot in a weaving pattern where the warp traverses the weft or the filling. { 'rīz·ər }

riser-angle indicator [PETRO ENG] In an offshore drilling operation, an acoustic or electronic device that monitors the angle of the flex joint on a floating drilling rig. Also known as azimuth-angle indicator. { 'rī·zər ˌaŋ·gəl ˌin·dəˌkād·ər }

riser plate [CIV ENG] A plate used to support a tapering switch rail above the base of the rail; used with a railroad gage or tie plate to maintain minimum gage. { 'rīz·ər ˌplāt }

rise time [CONT SYS] The time it takes for the output of a system to change from a specified small percentage (usually 5 or 10) of its steady-state increment to a specified large percentage (usually 90 or 95). [ELEC] The time for the pointer of an electrical instrument to make 90% of the change to its final value when electric power suddenly is applied from a source whose impedance is high enough that it does not affect damping. { 'rīz ˌtīm }

rising factorial polynomials [MATH] The polynomials

$[x]^n = x(x + 1)(x + 2) \cdots (x + n - 1)$. { ‚rīz·iŋ fak¦tȯr·ē·əl ‚päl·ə'nō·mē·əlz }

rising hinge [BUILD] A hinge that raises a door slightly as it is opened. { 'rīz·iŋ 'hinj }

rising limb [HYD] The rising portion of the hydrograph resulting from runoff of rainfall or snowmelt. { 'rīz·iŋ 'lim }

rising-sun magnetron [ELECTR] A multicavity magnetron in which resonators having two different resonant frequencies are arranged alternately for the purpose of mode separation; the cavities appear as alternating long and short radial slots around the perimeter of the anode structure, resembling the rays of the sun. { 'rīz·iŋ ¦sən 'mag·nə‚trän }

rising tide [OCEANOGR] The portion of the tide cycle between low water and the following high water. { 'rīz·iŋ 'tīd }

risk [ENG] The potential realization of undesirable consequences from hazards arising from a possible event. { risk }

risk analysis [ENG] The scientific study of risk. { 'risk ə‚nal·ə·səs }

risk management [ENG] The overall systematic approach to analyzing risk and implementing risk controls. { 'risk ‚man·ij·mənt }

Risley prism system [OPTICS] A type of dispersing prism used to test ocular convergence in ophthalmology; consists of two thin prisms mounted so that they can be rotated simultaneously in opposite directions. { 'riz·lē 'priz·əm ‚sis·təm }

Riss [GEOL] **1.** A European stage of geologic time: Pleistocene (above Mindel, below Würm). **2.** The third stage of glaciation of the Pleistocene in the Alps. { ris }

Rissoacea [PALEON] An extinct superfamily of gastropod mollusks. { ‚ris·ə'wäs·ē·ə }

Riss-Würm [GEOL] The third interglacial stage of the Pleistocene in the Alps, following the Riss glaciation and preceding the Würm glaciation. { 'ris'virm }

Ritchey-Chrétien optics [OPTICS] A modification of the Cassegrain optical system used in large optical telescopes; it has a hyperbolic image-forming primary mirror, no spherical aberration, and no coma; it has a larger usable field than either Newtonian or Cassegrain optical systems. { 'rich·ē 'krā·chən ‚äp·tiks }

Ritchie's experiment [THERMO] An experiment that uses a Leslie cube and a differential air thermometer to demonstrate that the emissivity of a surface is proportional to its absorptivity. { 'rich·ēz ik‚sper·ə·mənt }

Ritchie wedge [OPTICS] A photometer in which a test source and a standard source of light illuminate two perpendicular white, diffusing surfaces which intersect in a movable wedge, and these surfaces are viewed from a direction perpendicular to a line connecting the sources. { 'rich·ē ‚wej }

Ritter reaction [ORG CHEM] A procedure for the preparation of amides by reacting alkenes or tertiary alcohols with nitriles in an acidic medium. { 'rid·ər rē‚ak·shən }

Rittinger's law [MECH ENG] The law that energy needed to reduce the size of a solid particle is directly proportional to the resultant increase in surface area. { 'rit·ən·jərz ‚lȯ }

Ritz formula [ATOM PHYS] A particular expansion of an equation used in studying the spectra of atoms. { 'ritz ‚fȯr·myə·lə }

Ritz method [MATH] A method of solving boundary value problems based upon reformulating the given problem as a minimization problem. { 'ritz ‚meth·əd }

Ritz's combination principle [SPECT] The empirical rule that sums and differences of the frequencies of spectral lines often equal other observed frequencies. Also known as combination principle. { 'rit·səz ‚käm·bə'nā·shən ‚prin·sə·pəl }

riveling See wrinkling. { 'riv·əl·iŋ }

river [HYD] A large, natural freshwater surface stream having a permanent or seasonal flow and moving toward a sea, lake, or another river in a definite channel. [LAP] A pure-white diamond of very high grade. { 'riv·ər }

River and Harbor Aid to Navigation System [NAV] An all-weather system for precision navigation in harbors and harbor entrances, which provides sufficient accuracy for vessels to proceed as if visual aids did not exist, with a 50-foot (15-meter) position radius accuracy requirement at a 95% confidence level. Abbreviated RIHANS. { ¦riv·ər ən ¦här·bər 'ād tə ‚nav·ə'gā·shən ‚sis·təm }

river bar [GEOL] A ridgelike accumulation of alluvium in the channel, along the banks, or at the mouth of a river. { 'riv·ər ‚bär }

river basin [GEOL] The area drained by a river and all of its tributaries. { 'riv·ər ‚bās·ən }

riverbed [GEOL] The channel which contains, or formerly contained, a river. { 'riv·ər‚bed }

riverboat [NAV ARCH] A vessel used on a river, such as a shallow-draft passenger boat, barge, towboat, or rowboat. { 'riv·ər‚bōt }

river bottom [GEOL] The low-lying alluvial land along a river. Also known as river flat. { 'riv·ər ‚bäd·əm }

river breathing [HYD] Fluctuation of the water level of a river. { 'riv·ər ‚brēth·iŋ }

river buoy [NAV] A lightweight nun or can buoy especially designed to withstand strong currents. { 'riv·ər ‚bȯi }

river capture See capture. { 'riv·ər ‚kap·chər }

river-delta marsh [EVOL] A brackish or fresh-water marsh bordering the mouth of a distributary stream. { 'riv·ər ¦del·tə ‚märsh }

river-deposition coast [GEOL] A deltaic coast characterized by lobate seaward bulges crossed by river distributaries and bordered by lowlands. { 'riv·ər ‚dep·ə'zish·ən ‚kōst }

river drift [GEOL] Rock material deposited by a river in one place after having been moved from another. { 'riv·ər ‚drift }

river end [HYD] The lowest point of a river with no outlet to the sea, situated where its water disappears by percolation or evaporation. { 'riv·ər ‚end }

river engineering [CIV ENG] A branch of transportation engineering consisting of the physical measures which are taken to improve a river and its banks. { 'riv·ər ‚en·jə‚nir·iŋ }

river flat See river bottom. { 'riv·ər ‚flat }

river forecast [HYD] A forecast of the expected stage or discharge at a specified time, or of the total volume of flow within a specified time interval, at one or more points along a stream. { 'riv·ər ‚fȯr‚kast }

river gage [ENG] A device for measuring the river stage; types in common use include the staff gage, the water-stage recorder, and wire-weight gage. Also known as stream gage. { 'riv·ər ‚gāj }

river ice [HYD] Any ice formed in or carried by a river. { 'riv·ər ‚īs }

river mining [MIN ENG] Mining or excavating beds of existing rivers after deflecting their course, or by dredging without changing the flow of water. { 'riv·ər ‚mīn·iŋ }

river morphology [GEOL] The study of the channel pattern and the channel geometry at several points along a river channel, including the network of tributaries within the drainage basin. Also known as channel morphology; fluviomorphology; stream morphology. { 'riv·ər mȯr'fäl·ə·jē }

river piracy See capture. { 'riv·ər ‚pī·rə·sē }

river plain See alluvial plain. { 'riv·ər ‚plān }

River Po See Eridanus. { 'riv·ər ‚pō }

river radar [NAV] A marine set especially designed for river pilotage, generally characterized by a high degree of resolution and a wide selection of range scales. { 'riv·ər 'rā‚där }

river run gravel [GEOL] Natural gravel as found in deposits that have been subjected to the action of running water. { 'riv·ər ¦rən ‚grav·əl }

river system [HYD] The aggregate of stream channels draining a river basin. { 'riv·ər ‚sis·təm }

river terrace See stream terrace. { 'riv·ər ‚ter·əs }

river tide [HYD] A tide that occurs in rivers emptying directly into the sea, showing three characteristic modifications of ocean tides: the speed at which the tide travels upstream depends on the depth of the channel, the further upstream the longer the duration of the falling tide and shorter the duration of the rising tide, and the range of the tide decreases with distance upstream. { 'riv·ər ‚tīd }

riverwash [GEOL] **1.** Soil material that has been transported and deposited by rivers. **2.** An alluvial deposit in a river bed or flood channel, subject to erosion and deposition during recurring flood periods. { 'riv·ər‚wäsh }

river water [HYD] Water having carbonate, sulfate, and calcium as its main dissolved constituents; distinguished from seawater by its chloride and sodium content. { 'riv·ər ‚wȯd·ər }

Rivest-Shamir-Adleman algorithm [COMMUN] A public-key algorithm whose strength is based on the fact that factoring large composite prime numbers into their prime factors involves

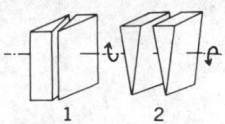

RISLEY PRISM SYSTEM

Diagram of Risley prism system. In orientation 1, combined deviation of prisms is zero; when both have been rotated 90° as in orientation 2, combined deviation is at a maximum.

an overwhelming amount of computation. Abbreviated RSA algorithm. { ri'vest shə'mir 'ad·əl·mən ˌal·gə,rith·əm }

rivet [DES ENG] A short rod with a head formed on one end; it is inserted through aligned holes in parts to be joined, and the protruding end is pressed or hammered to form a second head. { 'riv·ət }

riveting [ENG] The permanent joining of two or more machine parts or structural members, usually plates, by means of rivets. { 'riv·əd·iŋ }

riveting hammer [MECH ENG] A hammer used for driving rivets. { 'riv·əd·iŋ ˌham·ər }

riveting stake [HOROL] A block of steel with holes, used in watchmaking; arbors are inserted in the holes, and pinions or collets are placed on the edges of the holes for riveting. { 'riv·əd·iŋ ˌstāk }

rivet pitch [ENG] The center-to-center distance of adjacent rivets. { 'riv·ət ˌpich }

rivet weld [MET] A weld shaped like a countersunk rivet. { 'riv·ət ˌweld }

riving [GEOL] The splitting off, cracking, or fracturing of rock, especially by frost action. { 'rīv·iŋ }

rivulet [HYD] A small stream; a brook. { 'riv·yə·lət }

rivulose [BOT] Marked by irregular, narrow lines. { 'riv·yə,lōs }

RK galaxy [ASTRON] A type of ring galaxy which consists of a ringlike structure with a large, bright condensation or knot within the ring itself. { 'är'kā ˌgal·ik·sē }

RLL code *See* run-length-limited code. { 'är'el'el ˌkōd }

R loop hybridization [CELL MOL] A process by which ribonucleic acid is annealed with double-stranded deoxyribonucleic acid (DNA), thus displacing a single DNA strand. { 'är ˌlüp ˌhī·brid·ə'zā·shən }

R magnitude [ASTRON] The wavelength of a celestial object based on observations at a wavelength of 680 nanometers. { 'är ˌmag·nə,tüd }

R meter [NUCLEO] An ionization instrument calibrated to indicate the intensity of γ-rays, x-rays, and other ionizing radiation in roentgens. { 'är ˌmēd·ər }

R Monocerotis [ASTRON] An irregular variable star at the tip of the small fan-shaped emission nebula NGC 2261. { 'är ˌmän·ə'ser·əd·əs }

rms value *See* root-mean-square value. { 'är'em'es ˌval·ü }

Rn *See* radon.

RNA *See* ribonucleic acid.

RNA editing [CELL MOL] Alteration in the nucleotide sequence of messenger RNA after it has been transcribed from DNA, changing the nature of its protein product from what was encoded; may entail insertion, deletion, or enzyme-catalyzed chemical alteration of nucleotides. { 'är'en'ā 'ed·əd·iŋ }

RNAi *See* RNA interference. { 'är'en'ā'ī }

RNA interference [CELL MOL] The process by which foreign, double-stranded RNA is recognized and degraded by specialized protein complexes within many eukaryotic cells; believed to be an evolutionarily conserved defense mechanism against RNA viruses and transposable elements. Abbreviated RNAi. { 'är'en'ā ˌin·tər'fir·əns }

RNA polymerase I [MOL BIO] An enzyme found in the nucleolus that copies ribosomal genes to produce ribosomal ribonucleic acid. { 'är'en'ā pə'lim·ə,rās 'wən }

RNA polymerase II [MOL BIO] An enzyme that transcribes messenger ribonucleic acid sequences in the chromatin. { 'är'en'ā pə'lim·ə,rās 'tü }

RNA polymerase III [MOL BIO] An enzyme that directs the synthesis of small ribonucleic acid (RNA) molecules such as the amino acyl transfer RNAs. { 'är'en'ā pə'lim·ə,rās 'thrē }

RNA primer [CELL MOL] A short strand of RNA that is synthesized along single-stranded DNA during replication, initiating DNA polymerase-catalyzed synthesis of the complementary strand. { 'är'en'ā 'prī·mər }

RNAV *See* area navigation. { 'är,nav }

RN galaxy [ASTRON] A type of ring galaxy which consists of a ringlike structure with a nucleus somewhere within it, the nucleus being somewhat like those seen in ordinary spiral galaxies but typically lying off the center of the ring. { 'är'en ˌgal·ik·sē }

RNP *See* required navigation performance.

RO *See* receive-only.

roach [INV ZOO] An insect of the family Blattidae; the body is wide and flat, the anterior part of the thorax projects over the head, and antennae are long and filiform, with many segments. Also known as cockroach. { 'rōch }

road [CIV ENG] An open way for travel and transportation. [GEOL] One of a series of erosional terraces in a glacial valley, formed as the water level dropped in an ice-dammed lake. [MIN ENG] Any mine passage or tunnel. { 'rōd }

roadbed [CIV ENG] The earth foundation of a highway or a railroad. { 'rōd,bed }

road capacity [CIV ENG] The maximum traffic flow obtainable on a given roadway, using all available lanes, usually expressed in vehicles per hour or vehicles per day. { 'rōd kə,pas·əd·ē }

road grade [CIV ENG] The level and gradient of a road, measured along its center way. { 'rōd ,grād }

road map [MAP] A map showing the roadways of a specified area. { 'rōd ,map }

road net [CIV ENG] The system of roads available within a particular locality or area. { 'rōd ,net }

road octane number [ENG] A numerical value for automotive antiknock properties of a gasoline; determined by operating a car over a stretch of level road or on a chassis dynamometer under conditions simulating those encountered on the highway. { 'rōd 'äk,tān ,nəm·bər }

road oil [MATER] A heavy residual petroleum oil, usually one of the slow-curing grades of liquid asphalt. { 'rōd ,öil }

roadstead [GEOGR] An area near the shore, where vessels can anchor in safety; usually a shallow indentation in the coast. { 'rōd,sted }

road test [ENG] A motor-vehicle test conducted on the highway or on a chassis dynamometer to determine the performance of fuels or lubricants or the performance of the vehicle. { 'rōd ,test }

roadway [CIV ENG] The portion of the thoroughfare over which vehicular traffic passes. { 'rōd,wā }

roaring forties [METEOROL] A popular nautical term for the stormy ocean regions between 40° and 50° latitude; it usually refers to the Southern Hemisphere, where there is an almost completely uninterrupted belt of ocean with strong prevailing westerly winds. { 'rör·iŋ 'för·dēz }

roaring sand [GEOL] A sounding sand, found on a desert dune, that sets up a low roaring sound that sometimes can be heard for a distance of 1200 feet (400 meters). { 'rör·iŋ 'sand }

roast [MET] To heat ore to effect some chemical change that will facilitate smelting. { 'rōst }

roaster [ENG] Equipment for the heating of materials, such as in pyrite roasting; a furnace. { 'rōs·tər }

roasting regeneration [CHEM ENG] Regeneration of a processing (treating) clay by heating or burning it in contact with air to remove combustible impurities adsorbed onto the surface. { 'rōst·iŋ rē,jen·ə'rā·shən }

roast sintering *See* blast roasting. { 'rōst ,sin·tə·riŋ }

rob [MIN ENG] To take out ore or coal from a mine with a view to immediate product, and not to subsequent working. { 'räb }

robber [MET] An extra cathode that reduces current density at local areas of the work being electroplated for the purpose of producing a more uniform thickness coating. { 'räb·ər }

robbery *See* capture. { 'räb·ə·rē }

robbing pillars *See* pillaring. { 'räb·iŋ ,pil·ərs }

Robertinacea [INV ZOO] A superfamily of marine, benthic foraminiferans in the suborder Rotaliina characterized by a trochospiral or modified test with a wall of radial aragonite, and having bilamellar septa. { rä,bərd·ə'nās·ē·ə }

Roberts evaporator *See* short-tube vertical evaporator. { 'räb·ərts i'vap·ə,rād·ər }

Roberts' linkage [MECH ENG] A type of approximate straight-line mechanism which provided, early in the 19th century, a practical means of making straight metal guides for the slides in a metal planner. { 'räb·ərts ,liŋ·kij }

Robertson-Walker solutions [RELAT] A class of relativistic models for a homogeneous, isotropic universe that are conventionally accepted as describing the real universe. { 'räb·ərt·sən 'wök·ər sə,lü·shənz }

Robin Hood's wind [METEOROL] In Great Britain, saturated air with temperatures near freezing; it is raw and penetrating. { 'räb·ən 'hüdz 'wind }

Robin law [PHYS] The law that an increase in pressure on a system in chemical or physical equilibrium favors the system

flat   countersunk   countersunk   button   high
      flat   round   button

Some types of standard rivet heads.

ROBERTS' LINKAGE

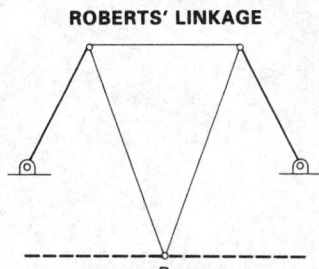

Roberts' linkage, an example of straight-line mechanisms. Point *P* moves along a straight line.

formed with a decrease in volume, and conversely, a change in pressure does not affect a system formed with no change in volume. { 'räb·ən ¦lo̅ }

Robins-Messiter system [MECH ENG] A stacking conveyor system in which material arrives on a conveyor belt and is fed to one or two wing conveyors. { 'räb·ənz 'mes·ə·tər ‚sis·təm }

robinsonite [MINERAL] $Pb_7Sb_{12}S_{25}$ A mineral composed of lead antimony sulfide. { 'räb·ən·sə‚nīt }

Robitzsch actinograph [ENG] A pyranometer whose design utilizes three bimetallic strips which are exposed horizontally at the center of a hemispherical glass bowl; the outer strips are white reflectors, and the center strip is a blackened absorber; the bimetals are joined in such a manner that the pen of the instrument deflects in proportion to the difference in temperature between the black and white strips, and is thus proportional to the intensity of the received radiation; this instrument must be calibrated periodically. { 'ro̅‚bitsh ak'tin·ə‚graf }

robot [CONT SYS] A mechanical device that can be programmed to perform a variety of tasks of manipulation and locomotion under automatic control. { 'ro̅‚bät }

robotics [IND ENG] The study of problems associated with the design, application, and control and sensory systems of self-controlled devices. { ro̅'bäd·iks }

robustness [STAT] The property of statistical procedures that are insensitive to small departures from the assumptions on which they depend, such as the assumption that certain distributions are normal. { ro̅'bəst·nəs }

robust program [COMPUT SCI] **1.** A computer program using an iterative process that converges rapidly to the solution being sought. **2.** A computer program that performs well even under unusual conditions. { ¦ro̅·bəst 'pro̅·grəm }

roc [ELEC] A unit of electrical conductivity equal to the conductivity of a material in which an electric field of 1 volt per centimeter gives rise to a current density of 1 ampere per square centimeter. Derived from reciprocal ohm centimeter. { räk }

Roccilaceae [BOT] A family of fruticose species of Hysteriales that grow profusely on trees and rocks along the coastlines of Portugal, California, and western South America. { ‚räs·ə'lās·ē‚ē }

rocdrumlin See rock drumlin. { 'räk¦drəm·lən }

Rochelle-electric See ferroelectric. { ro̅'shel·i‚lek·trik }

Rochelle salt See potassium sodium tartrate. { ro̅'shel‚so̅lt }

Roche lobes [MECH] **1.** Regions of space surrounding two massive bodies revolving around each other under their mutual gravitational attraction, such that the gravitational attraction of each body dominates the lobe surrounding it. **2.** In particular, the effective potential energy (referred to a system of coordinates rotating with the bodies) is equal to a constant V_0 over the surface of the lobes, and if a particle is inside one of the lobes and if the sum of its effective potential energy and its kinetic energy is less than V_0, it will remain inside the lobe. { 'ro̅ch ‚lo̅bz }

roche moutonnée [GEOL] A small, elongate hillock of bedrock sculptured by a large glacier so that its long axis is oriented in the direction of ice movement; the upstream side is gently inclined, smoothly rounded, but striated, and the downstream side is steep, rough, and hackly. { 'ro̅ch ¦müt·ən¦ā }

Roche's limit [ASTROPHYS] The limiting distance below which a satellite orbiting a celestial body would be disrupted by the tidal forces generated by the gravitational attraction of the primary; the distance depends on the relative densities of the bodies and on the orbit of the satellite; it is computed by $R = 2.45(Lr)$, where L is a factor that depends on the relative densities of the satellite and the body, R is the radius of the satellite's orbit measured from the center of the primary body, and r is the radius of the primary body; if the satellite and the body have the same density, the relationship is $R = 2.45r$. { 'ro̅·shəz ‚lim·ət }

Rochon polarizing prism [OPTICS] A device for producing linearly polarized beams of light, consisting of two adjacent quartz wedges, the first of which has its optic axis parallel to the beam, while the second has its optic axis perpendicular to the beam; one of the beams (the ordinary ray) is undeviated, and is therefore not spread into a spectrum. { ro̅'sho̅n 'po̅·lə‚rīz·iŋ ‚priz·əm }

rock [PETR] **1.** A consolidated or unconsolidated aggregate of mineral grains consisting of one or more mineral species and having some degree of chemical and mineralogic constancy.

2. In the popular sense, a hard, compact material with some coherence, derived from the earth. { räk }

rockair [AERO ENG] A high-altitude sounding system consisting of a small solid-propellant research rocket carried aloft by an aircraft; the rocket is fired while the aircraft is in vertical ascent. { 'räk‚er }

rock asphalt See asphalt rock. { räk 'as‚fo̅lt }

rock association [PETR] A group of igneous rocks within a petrographic province that are related chemically and petrographically, generally in a systematic manner such that chemical data for the rocks plot as smooth curves on variation diagrams. Also known as rock kindred. { räk ə‚so̅·shē¦ā·shən }

rock awash [OCEANOGR] In U.S. Coast and Geodetic Survey terminology, a rock exposed at any stage of the tide between the datum of mean high water and the sounding datum, or one just bare at these data. { räk ə'wäsh }

rock-a-well [PETRO ENG] The procedure of bleeding pressure alternately from the casing of a well and from the tubing until the well starts flowing. { 'räk·ə‚wel }

rock bar See riegel. { räk ‚bär }

rock-basin lake See paternoster lake. { räk ¦bas·ən ‚lāk }

rock bench See structural bench. { räk ‚bench }

rock bit [ENG] Any one of many different types of roller bits used on rotary-type drills for drilling large-size holes in soft to medium-hard rocks. { räk ‚bit }

rockbolt [ENG] A bar, usually constructed of steel, which is inserted into predrilled holes in rock and secured for the purpose of ground control. { 'räk‚bo̅lt }

rock bolting [ENG] A method of securing or strengthening closely jointed or highly fissured rocks in mine workings, tunnels, or rock abutments by inserting and firmly anchoring rock bolts oriented perpendicular to the rock face or mine opening. { räk ‚bo̅lt·iŋ }

rockbridgeite [MINERAL] $Fe^{2+}Fe_6^{3+}(PO_4)_4(OH)_8$ A basic phosphate mineral containing iron; isomorphous with frondelite. { 'räk‚bri‚jīt }

rock-bulk compressibility [GEOL] One of three types of rock compressibility (matrix, bulk, and pore); the fractional change in volume of the bulk volume of the rock with a unit change in pressure. { räk ¦bəlk kəm‚pres·ə'bil·əd·ē }

rock bump [MIN ENG] The sudden release of the weight of the rocks over a coal seam or of enormous lateral stresses. { räk ‚bəmp }

rockburst [MIN ENG] A sudden and violent rock failure around a mining excavation on a sufficiently large scale to be considered a hazard. { räk ‚bərst }

rock candy [FOOD ENG] Large, transparent, hydrated crystals of cane sugar. { räk ‚kan·dē }

rock cave See shelter cave. { räk ‚kāv }

rock channeler [MECH ENG] A machine used in quarrying for cutting an artificial seam in a mass of stone. { räk ‚chan·əl·ər }

rock cleavage [PETR] The capacity of a rock to split along certain parallel surfaces more easily than along others. { räk ‚klē·vij }

rock control [GEOL] The influences of differences in earth materials on development of landforms. { räk kən‚tro̅l }

rock cork See mountain cork. { räk ‚ko̅rk }

rock creep [GEOL] A form of slow flowage in rock materials evident in the downhill bending of layers of bedded or foliated rock and in the slow downslope migration of large blocks of rock away from their parent outcrop. { räk ‚krēp }

rock crystal [MINERAL] A transparent, colorless form of quartz with low brilliance; used for lenses, wedges, and prisms in optical instruments. Also known as berg crystal; crystal; mountain crystal; pebble; quartz crystal. { räk ‚krist·əl }

rock cycle [GEOL] The interrelated sequence of events by which rocks are initially formed, altered, destroyed, and reformed as a result of magmatism, erosion, sedimentation, and metamorphism. { räk ‚sī·kəl }

rock-defended terrace [GEOL] **1.** A river terrace having a ledge or outcrop of resistant rock at its base which serves as protection against undermining. **2.** A marine terrace having a mass of resistant rock at the base of the cliff which protects against wave erosion. { räk di¦fen·dəd 'ter·əs }

rock desert [GEOL] An upland desert in which bedrock is either exposed or is covered with a thin veneer of coarse rock fragments. { räk ‚dez·ərt }

rock drill [MECH ENG] A machine for boring relatively short holes in rock for blasting purposes; motive power may be compressed air, steam, or electricity. { 'räk ,dril }

rock drum See rock drumlin. { 'räk ¦drəm }

rock drumlin [GEOL] A smooth, streamlined hill modeled by glacial erosion, which has a core of bedrock usually veneered with a layer of glacial till and which resembles a true drumlin in outline and form but is generally less symmetrical and less regularly shaped. Also known as drumlinoid; false drumlin; rocdrumlin; rock drum. { 'räk ¦drəm·lən }

rock dust distributor See rock duster. { 'räk ¦dəst di,strib· yəd·ər }

rock duster [MIN ENG] A machine that distributes rock dust over the interior surfaces of a coal mine by means of air to prevent coal dust explosions. Also known as rock dust distributor. { 'räk ¦dəs·tər }

rock element [PETR] The coherent, intact piece of rock that is the basic constituent of the rock system and which has physical, mechanical, and petrographic properties that can be described or measured by laboratory tests. { 'räk ¦el·ə·mənt }

rocker [CIV ENG] A support at the end of a truss or girder which permits rotation and horizontal movement to allow for expansion and contraction. [GRAPHICS] A type of chisel, with a sharp beveled edge, used in mezzotint engraving; it is set on the surface of a copper plate and rocked back and forth to produce a rough grain, which, when printed, produces a velvety black. [MIN ENG] A small digging bucket mounted on two rocker arms in which auriferous alluvial sands are agitated by oscillation, in water, to collect gold. [ORD] Movable, built-in support in a field gun carriage, between the trail and the cradle, that allows changes in elevation to be made without disturbing the angle of position setting. { 'räk·ər }

rocker arm [MECH ENG] In an internal combustion engine, a lever that is pivoted near its center and operated by a pushrod at one end to raise and depress the valve stem at the other end. { 'räk·ər ,ärm }

rocker bearing [CIV ENG] A bridge support that is free to rotate but cannot move horizontally. { 'räk·ər ,ber·iŋ }

rocker bent [CIV ENG] A bent used on a bridge span; hinged at one or both ends to provide for the span's expansion and contraction. { 'räk·ər ,bent }

rocker cam [MECH ENG] A cam that moves with a rocking motion. { 'räk·ər ,kam }

rocker dump car [MIN ENG] A small-capacity mining car; the most popular and most widely used are the gravity dump types, designed so that the weight of the load tips the body when a locking latch is released by hand. { 'räk·ər 'dəmp ,kär }

rocker panel [ENG] The part of the paneling on a passenger vehicle located below the passenger compartment doorsill. { 'räk·ər ,pan·əl }

rocker shovel [MIN ENG] A digging and loading machine consisting of a bucket attached to a pair of semicircular runners; lifts and dumps the bucket load into a car or another materials-transport unit behind the machine. { 'räk·ər ,shəv·əl }

rocket [AERO ENG] 1. Any kind of jet propulsion capable of operating independently of the atmosphere. 2. A complete vehicle driven by such a propulsive system. { 'räk·ət }

rocket airplane [AERO ENG] An airplane using a rocket or rockets for its chief or only propulsion. { 'räk·ət 'er,plān }

rocket ammunition [ORD] Any type of ammunition incorporating rockets. { 'räk·ət ,am·yə,nish·ən }

rocket antenna [ELECTROMAG] An antenna carried on a rocket, to receive signals controlling the rocket or to transmit measurements made by instruments aboard the rocket. { 'räk· ət an,ten·ə }

rocket assist [AERO ENG] An assist in thrust given an airplane or missile by use of a rocket motor or rocket engine during flight or during takeoff. { 'räk·ət ə,sist }

rocket-assisted takeoff See rato. { 'räk·ət ə¦sis·təd 'tāk,óf }

rocket-assisted torpedo [ORD] A torpedo designed to be fired into the air by rocket and to drop into the water by parachute; the torpedo then seeks its underwater target by a special homing device. Abbreviated RAT. { 'räk·ət ə¦sis·təd tór'pēd·ō }

rocket astronomy [ASTRON] The discipline comprising measurements of the electromagnetic radiation from the sun, planets, stars, and other bodies, of wavelengths that are almost completely absorbed below the 150-mile (250-kilometer) level, by using a rocket to carry instruments above 150 miles to measure these phenomena. { 'räk·ət ə,strän·ə·mē }

rocket chamber [AERO ENG] A chamber for the combustion of fuel in a rocket; in particular, that section of the rocket engine in which combustion of propellants takes place. { 'räk· ət ,chäm·bər }

rocket electrophoresis [IMMUNOL] A variant of crossed electrophoresis in which the medium contains only one antibody; test substances are driven directly into the medium that contains the antibody, forming rocket-shaped (inverted V) trails of precipitation. { 'räk·ət i,lek·trō·fə'rē·səs }

rocket engine [AERO ENG] A reaction engine that contains within itself, or carries along with itself, all the substances necessary for its operation or for the consumption or combustion of its fuel, not requiring intake of any outside substance and hence capable of operation in outer space. Also known as rocket motor. { 'räk·ət ,en·jər }

rocket fuel [MATER] Any of the substances or mixtures of substances that can burn rapidly with controlled combustion to produce large volumes of gas at high pressures and temperatures; includes monopropellants (hydrogen peroxide and hydrazine), liquid bipropellant fuels (organic fuel and oxidizer), and solid propellants (mixed oxidizer-fuel in a propellant grain). { 'räk·ət ,fyül }

rocket igniter [AERO ENG] An igniter for the propellant in a rocket. { 'räk·ət ig nīd·ər }

rocket launcher [AERO ENG] A device for launching a rocket, wheel-mounted, motorized, or fixed for use on the ground, rocket launchers are mounted on aircraft, as under the wings, or are installed below or on the decks of ships. { 'räk· ət ,lón·chər }

rocket lightning [GEOPHYS] A rare form of lightning whose luminous channel seems to advance through the air with only the speed of a skyrocket. { 'räk·ət ,līt·niŋ }

rocket missile [ORD] A missile using rocket propulsion. { 'räk·ət ,mis·əl }

rocket motor See rocket engine. { 'räk·ət ,mōd·ər }

rocket nose section [AERO ENG] The extreme forward portion of a rocket, designed to contain instrumentation, spotting charges, fusing or arming devices, and the like, but does not contain the payload. { 'räk·ət ,nōz ,sek·shən }

rocket ogive [ORD] A hollow, conical, metallic shell covering for a rocket warhead, designed to reduce air resistance during flight. { 'räk·ət ,ō,jīv }

rocket propellant [MATER] 1. Any agent which is used for consumption or combustion in a rocket, and from which the rocket derives its thrust, such as a fuel, oxidizer, and additive. 2. The ejected fluid in a nuclear rocket. { 'räk·ət prə,pel·ənt }

rocket propulsion [AERO ENG] Reaction propulsion by a rocket engine. { 'räk·ət prō,pəl·shən }

rocket ramjet [AERO ENG] A ramjet engine having a rocket mounted within the ramjet duct, the rocket being used to bring the ramjet up to the necessary operating speed. Also known as ducted rocket. { 'räk·ət 'ram,jet }

rocketry [AERO ENG] 1. The science or study of rockets, embracing theory, research, development, and experimentation. 2. The art and science of using rockets, especially rocket ammunition { 'räk·ə·trē }

rocket sled [AERO ENG] A sled that runs on a rail or rails and is accelerated to high velocities by a rocket engine; the sled is used in determining G tolerances and for developing crash survival techniques. { 'räk·ət ,sled }

rocket-sled testing [AERO ENG] A method of subjecting structures and devices to high accelerations or decelerations and aerodynamic flow phenomena under controlled conditions; the test object is mounted on a sled chassis running on precision steel rails and accelerated by rockets or decelerated by water scoops. { 'räk·ət ¦sled test·iŋ }

rocketsonde See meteorological rocket. { 'räk·ət,sänd }

rocket staging [AERO ENG] The use of successive rocket sections or stages, each having its own engine or engines; each stage is a complete rocket vehicle in itself. { 'räk·ət ,stāj·iŋ }

rocket station [ENG] A life-saving station equipped with line-carrying rocket apparatus. { 'räk·ət ,stā·shən }

rocket thrust [AERO ENG] The thrust of a rocket engine. { 'räk·ət ,thrəst }

rocket tube [AERO ENG] 1. A launching tube for rockets. 2. A tube or nozzle through which rocket gases are ejected. { 'räk·ət ,tüb }

rockeye [ORD] A free-fall cluster bomb designed for use

against tanks and armored vehicles with antipersonnel secondary effects. { 'räk₁ī }

rock fabric See fabric. { 'räk ,fab·rik }

rock failure [GEOL] Fracture of a rock that has been stressed beyond its ultimate strength. { 'räk ,fāl·yər }

rockfall [GEOL] **1.** The fastest-moving landslide; free fall of newly detached bedrock segments from a cliff or other steep slope; usually occurs during spring thaw. **2.** The rock material moving in or moved by a rockfall. { 'räk,fòl }

rock fan [GEOL] A fan-shaped bedrock surface whose apex is where a mountain stream debouches upon a piedmont slope, and which occupies an area where a pediment meets the mountain slope. { 'räk ,fan }

rock-fill [CIV ENG] Composed of large, loosely placed rocks. { 'räk ,fil }

rock-fill dam [CIV ENG] A dam constructed of loosely placed rock or stone. { 'räk ,fil ,dam }

rock-floor robbing [GEOL] A form of sheetflood erosion in which sheetfloods remove crumbling debris from rock surfaces in desert mountains. { 'räk ,flòr ,räb·iŋ }

rock flour [GEOL] A fine, chemically unweathered powder of rock-forming minerals produced by pulverization of rock fragments during natural transport or crushing. Also known as glacial flour. { 'räk ,flaù·ər }

rock flowage See flow. { 'räk ,flō·ij }

rockforming [GEOL] Referring to any minerals which commonly occur in important proportions in common rocks. { 'räk,fòrm·iŋ }

rock fragment [PETR] A component of a sedimentary rock consisting of polymineralic or polygranular sand grains that are abraded particles of igneous, sedimentary, or metamorphic rocks. { 'räk ,frag·mənt }

rock glacier [GEOL] Boulders and fine material cemented by ice about a meter below the surface. Also known as talus glacier. { 'räk ,glā·shər }

rock-glacier creep [GEOL] A rapid talus creep of tongues of debris in a cold region, caused by the expansive force of the alternate freeze and thaw of ice in the interstices of the debris. { 'räk ,glä·shər ,krēp }

rock gypsum [MINERAL] Massive, coarsely crystalline to earthy, finely granular type of gypsum found in gyp rock. { 'räk ,jip·səm }

rocking bar [HOROL] **1.** The plate that holds the winding wheels in a watch; rocks to alternately engage the winding mechanism and the hand-setting mechanism. **2.** The part in an alarm timepiece that engages the time and alarm mainsprings. { 'räk·iŋ ,bär }

rocking furnace [MECH ENG] A horizonal, cylindrical melting furnace that is rolled back and forth on a geared cradle. { 'räk·iŋ ,fər·nəs }

rocking pier [CIV ENG] A pier that is hinged to allow for longitudinal expansion or contraction of the bridge. { 'räk·iŋ ,pir }

rocking stone [GEOL] A stone or boulder, often of great size, so finely poised upon its foundation (as on the side of a hill or cliff) that it can be moved slightly backward and forward with little force (as with the hand) and still retain its original position. Also known as roggan. { 'räk·iŋ ,stōn }

rocking the sextant See swinging the arc. { 'räk·iŋ thə 'sek·stənt }

rocking valve [MECH ENG] An engine valve in which a disk or cylinder turns in its seat to permit fluid flow. { 'räk·iŋ ,valv }

rock island See meander core. { 'räk ,ī·lənd }

rock kindred See rock association. { 'räk ,kin·drəd }

rocklath [MATER] A sheet of gypsum used as a base for plaster. { 'räk,lath }

rock loader [MIN ENG] Any device or machine used for loading slate or rock inside a mine. { 'räk ,lōd·ər }

rock magnetism [GEOPHYS] The natural remanent magnetization of igneous, metamorphic, and sedimentary rocks resulting from the presence of iron oxide minerals. { 'räk 'mag·nə,tiz·əm }

rock matrix compressibility [GEOL] One of three types of rock compressibility (matrix, bulk, and pore); the fractional change in volume of the solid rock material (grains) with a unit change in pressure. { 'räk ¦mā·triks kəm,pres·ə'bil·əd·ē }

rock meal See rock milk. { 'räk ,mēl }

rock mechanics [GEOPHYS] Application of the principles of mechanics and geology to quantify the response of rock when it is acted upon by environmental forces, particularly when human-induced factors alter the original ambient forces. { 'räk mi,kan·iks }

rock milk [MINERAL] A soft, white, earthy or powdery variety of calcite. Also known as agaric mineral; bergmehl; forril farina; rock meal. { 'räk ,milk }

rockoon [AERO ENG] A high-altitude sounding system consisting of a small solid-propellant research rocket carried aloft by a large plastic balloon. { rä'kün }

rock pedestal See pedestal. { 'räk ,ped·ə·stəl }

rock pediment [GEOL] A pediment formed on the surface of bedrock. { 'räk ,ped·ə·mənt }

rock permeability [GEOL] The ability of a rock to receive, hold, or pass fluid materials (oil, water, and gas) by nature of the interconnections of its internal porosity. { 'räk ,pər·mē·ə'bil·əd·ē }

rock phosphate See phosphorite. { 'räk ,fä,sfāt }

rock pillar [GEOL] **1.** A column of rock produced by differential weathering or erosion, as along a joint plane. **2.** In a cave, a pillar-type structure that is residual bedrock rather than a stalactostalagmite. { 'räk ,pil·ər }

rock pool [GEOL] A tidal pool formed along a rocky shoreline. { 'räk ,pül }

rock pressure [GEOPHYS] **1.** Stress in underground geologic material due to weight of overlying material, residual stresses, and pressures resulting from swelling clays. **2.** See ground pressure. { 'räk ,presh·ər }

rock river [GEOL] A very long and narrow rock stream. { 'räk ,riv·ər }

rock salt See halite. { 'räk ,sòlt }

rock saw [MIN ENG] A type of mechanical miner that is used to remove large blocks of material; cuts narrow slots or channels by the action of a moving steel band or blade and a slurry of abrasive particles (sometimes diamonds) rather than teeth; small flame jets are also used. { 'räk ,sò }

rockshaft [MIN ENG] A shaft through which rock can be brought into a mine for filling, stopes, or other excavations. { 'räk,shaft }

rock shell [INV ZOO] The common name for a large number of gastropod mollusks composing the family Muricidae and characterized by having conical shells with various sculpturing. { 'räk ,shel }

rock shelter [GEOL] A cave that is formed by a ledge of overhanging rock. { 'räk ,shel·tər }

rock silk [MINERAL] A silky variety of asbestos. { 'räk ,silk }

Rocksite program [MIN ENG] A U.S. Navy program concerned with undersea mining or consolidated mineral deposits; studied direct sea-floor access at remote sites through shafts drilled in the sea floor. { 'räk,sīt ,prō,gram }

rockslide [GEOL] The sudden, rapid downward movement of newly detached bedrock segments over a surface of weakness, such as of bedding, jointing, or faulting. Also known as rock slip. { 'räk,slīd }

rock slip See rockslide. { 'räk ,slip }

rock stack [GEOL] A rocky crag that has been uplifted from an old sea floor. { 'räk ,stak }

rock step See knickpoint. { 'räk ,step }

rock-stratigraphic unit [GEOL] A lithologically homogeneous body of strata characterized by certain observable physical features, or by the dominance of a certain rock type or combination of rock types; rock-stratigraphic units include groups, formations, members, and beds. Also known as geolith; lithologic unit; lithostratic unit; lithostratigraphic unit; rock unit. { 'räk ,strad·ə¦graf·ik 'yü·nət }

rock stream [GEOL] Rocks moving (or already moved) in a mass down a slope under the influence of their own weight. { 'räk ,strēm }

rock system [GEOPHYS] In rock mechanics, all natural environmental factors that can influence the behavior of that portion of the earth's crust that will become part of an engineering structure. { 'räk ,sis·təm }

rock terrace [GEOL] A stream terrace on the side of a valley composed of resistant bedrock which remains during erosion of weaker overlying and underlying beds. { 'räk ,ter·əs }

rock type [PETR] **1.** One of the three major rock groups: igneous, sedimentary, metamorphic. **2.** A rock having a

---

ROCK SHELL

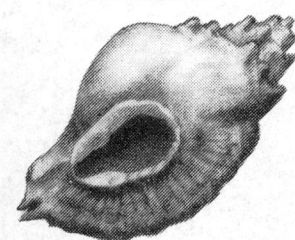

The festive rock shell (*Murex festivus*).

ROCKSITE PROGRAM

Proposed method of sea-floor access in the Rocksite program. 300 feet = 90 meters, 10 miles = 16 kilometers.

unique, identifiable set of characters, such as basalt. { 'räk ‚tīp }

rock unit *See* rock-stratigraphic unit. { 'räk ‚yü·nət }

rock varnish [GEOL] A dark coating on rock surfaces exposed to the atmosphere. It is composed of about 30% manganese and iron oxides, up to 70% clay minerals, and over a dozen trace and rare-earth minerals. Although found in all terrestrial environments, it is mostly developed and best preserved in arid regions. Also know as desert varnish. { 'räk ‚vär·nəsh }

Rockwell hardness [ENG] A measure of hardness of a material as determined by the Rockwell hardness test. { 'räk‚wel 'härd·nəs }

Rockwell hardness test [ENG] One of the arbitrarily defined measures of resistance of a material to indentation under static or dynamic load; depth of indentation of either a steel ball or a 120° conical diamond with rounded point, $^1/_{16}$, $^1/_8$, $^1/_4$, or $^1/_2$ inch (1.5875, 3.175, 6.35, 12.7 millimeters) in diameter, called a brale, under prescribed load is the basis for Rockwell hardness; 60, 100, 150 kilogram load is applied with a special machine, and depth of impression under initial minor load is indicated on a dial whose graduations represent hardness number. { 'räk‚wel 'härd·nəs ‚test }

rock wood *See* mountain wood. { 'räk ‚wůd }

rock wool *See* mineral wool. { 'räk ‚wůl }

Rocky Mountain spotted fever [MED] An acute, infectious, typhuslike disease of man caused by the rickettsial organism *Rickettsia rickettsi* and transmitted by species of hard-shelled ticks; characterized by sudden onset of chills, headache, fever, and an exanthem on the extremities. Also known as American spotted fever; tick fever; tick typhus. { 'räk·ē 'maůnt·ən 'späd·əd 'fē·vər }

rocky point effect [ELECTR] Transient but violent discharges between electrodes in high-voltage transmitting tubes. { 'räk·ē ‚pöint i‚fekt }

rod [DES ENG] **1.** A bar whose end is slotted, tapered, or screwed for the attachment of a drill bit. **2.** A thin, round bar of metal or wood. [GEOL] A rodlike sedimentary particle characterized by a width-length ratio less than 2/3 and a thickness-width ratio more than 2/3. Also known as roller. [HISTOL] One of the rod-shaped sensory bodies in the retina which are sensitive to dim light. [MECH] *See* perch. [NUCLEO] A relatively long and slender body of material used in, or in conjunction with, a nuclear reactor; may contain fuel, absorber, or fertile material or other material in which activation or transmutation is desired. { räd }

rod bit [DES ENG] A bit designed to fit a reaming shell that is threaded to couple directly to a drill rod. { 'räd ‚bit }

rod coupling [DES ENG] A double-pin-thread coupling used to connect two drill rods together. { 'räd ‚kəp·liŋ }

rodding [ENG] An operation in which a rod is passed through a length of tubing such as a rifle or pipework to determine if the bore is clear. [PETR] In metamorphic rocks, a linear structure in which the stronger parts, such as vein quartz or quartz pebbles, have been shaped into parallel rods. { 'räd·iŋ }

rod dope [MATER] Grease or other material used to protect or lubricate drill rods. Also called gunk; rod grease. { 'räd ‚dōp }

rodent [VERT ZOO] The common name for members of the order Rodentia. { 'röd·ənt }

Rodentia [VERT ZOO] An order of mammals characterized by a single pair of ever-growing upper and lower incisors, a maximum of five upper and four lower cheek teeth on each side, and free movement of the lower jaw in an anteroposterior direction. { rō'den·chə }

rodenticide [MATER] A chemical agent used to kill rodents. { rō'den·tə‚sīd }

rod gap [ELEC] **1.** A device that is usually formed of two $^1/_2$-square-inch (3-square-centimeter) rods, one grounded and the other connected to the line conductor, but may also have the shape of rings or horns, used to limit the magnitude of transient overvoltages on an electrical system as a result of lightning strikes. **2.** Spark gap in which the electrodes are two coaxial rods, with ends between which the discharge takes place, cut perpendicularly to the axis. { 'räd ‚gap }

rod grease *See* rod dope. { 'räd ‚grēs }

rodingite [PETR] A medium- to coarse-grained, commonly

calcium-enriched gabbroic rock containing grossular and diallage as essential minerals. { 'rōc·iŋ‚gīt }

rodite *See* diogenite. { 'rō‚dīt }

rod level [ENG] A spirit level attached to a level rod or stadia rod to ensure the vertical position of the rod prior to instrument reading. { 'räd ‚lev·əl }

rod mill [MECH ENG] A pulverizer operated by the impact of heavy metal rods. [MET] A mill for making metal rods. { 'räd ‚mil }

rod pump [PETRO ENG] Type of oil well sucker-rod pump that can be inserted into or removed from oil well tubing without moving or disturbing the tubing itself. Also known as insert pump. { 'räd ‚pəmp }

Rodrigues formula [MATH] **1.** The equation giving the *n*th function in a class of special functions in terms of the *n*th derivatives of some polynomial. **2.** The formula $d\mathbf{n} + k \, d\mathbf{r} = 0$, expressing the difference $d\mathbf{n}$ in the unit normals to a surface at two neighboring points on a line of curvature, in terms of the difference $d\mathbf{r}$ in the position vectors of the two points and the principal curvature k. **3.** The formula for a matrix that is used to transform the cartesian coordinates of a vector in three-space under a rotation through a specified angle about an axis with specified direction cosines. { rə'drē·gəs ‚fór·myə·lə }

rod slide *See* slide. { 'räd ‚slīd }

rod string [MECH ENG] Drill rods coupled to form the connecting link between the core barrel and bit in the borehole and the drill machine at the collar of the borehole. { 'räd ‚striŋ }

rod stuffing box [ENG] An annular packing gland fitting between the drill rod and the casing at the borehole collar; allows the rod to rotate freely but prevents the escape of gas or liquid under pressure. { 'räd ‚stəf·iŋ ‚bäks }

rod thermistor [ELECTR] A type of thermistor that has high resistance, long time constant, and moderate power dissipation; it is extruded as a long vertical rod 0.250–2.0 inches (0.63–5.1 centimeters) long and 0.050–0.110 inch (0.13–0.28 centimeter) in diameter, of oxide-binder mix and sintered; ends are coated with conducting paste and leads are wrapped on the coated area. { 'räd thər'mis·tər }

rod weeder [AGR] A type of equipment used to prepare the soil during harrowing; it is a power-driven rod, usually square in cross section, which also operates below the surface of loose soil, killing weeds and maintaining the soil in loose mulched condition; adapted to large operations and used in dry areas in the northwestern United States. { 'räd ‚wēd·ər }

roedderite [MINERAL] $(Na,K)_2(Mg,Fe)_5Si_{12}O_{30}$ A silicate meteorite mineral. { 'räd·ə‚rīt }

roemerite [MINERAL] $FeFe_2(SO_4)_4 \cdot 14H_2O$ A rust-brown to yellow mineral composed of hydrous ferric and ferrous iron sulfate. { 'räm·ə‚rīt }

roentgen [NUCLEO] An exposure dose of x- or γ-radiation such that the electrons and positrons liberated by this radiation produce, in air, when stopped completely, ions carrying positive and negative charges of 2.58×10^{-4} coulomb per kilogram of air. Abbreviated R (formerly r). Also spelled röntgen. { 'rent·gən }

roentgen current [ELEC] An electric current arising from the motion of polarization charges, as in the rotation of a dielectric in a charged capacitor. { 'rent·gən ‚kər·ənt }

roentgen diffractometry *See* x-ray crystallography. { 'rent·gən də‚frak'täm·ə·trē }

roentgen equivalent man *See* rem. { 'rent·gən i'kwiv·ə·lənt man }

roentgen equivalent physical *See* rep. { 'rent·gən i'kwiv·ə·lənt fiz·i·kəl }

roentgen meter [NUCLEO] A meter for measuring the cumulative quantity of x-rays or γ-rays, without reference to time. { 'rent·gən ‚mēd·ər }

roentgenography [PHYS] Radiography by means of x-rays. { ‚rent·gə'näg·rə·fē }

roentgenoluminescence [PHYS] Luminescence which can be produced by x-rays. { ‚rent·gə·nō‚lü·mə'nes·əns }

roentgen optics *See* x-ray optics. { 'rent·gən 'äp·tiks }

roentgenotherapy *See* x-ray therapy. { ‚rent·gə·nō'ther·ə·pē }

roentgen per hour at one meter [NUCLEO] A unit of γ-ray source strength, corresponding to a dose rate of 1 roentgen per hour at a distance of 1 meter in air. Symbolized RHM. { 'rent·gən pər ‚aůr at ‚wən 'mēd·ər }

roentgen rays *See* x-rays. { 'rent·gən ‚rāz }

roentgen spectrometry *See* x-ray spectrometry. { 'rent·gən spek'träm·ə·trē }

Roese-Gottlieb method [ANALY CHEM] A solvent extraction method used to obtain an accurate determination of the fat content of milk. { 'rez·ə 'gät,lēb ,meth·əd }

roesslerite [MINERAL] $MgH(AsO_4)·7H_2O$ A monoclinic mineral composed of hydrous acid magnesium arsenate; it is isomorphous with phosphorroesslerite. { 'res·lə,rīt }

roestone *See* oolite. { 're,stōn }

rofla [GEOL] An extremely narrow, tortuous gorge, frequently formed by meltwater streams flowing from a glacier. { 'rō·flə }

ROFOR [METEOROL] An international code word used to indicate a route forecast (along an air route). { 'rō,fôr }

ROFOT [METEOROL] An international code word used to indicate a route forecast, with units in the English system. { 'rō,fät }

Rogallo wing [AERO ENG] A glider folded inside a spacecraft; to be deployed during the spacecraft's reentry like a parachute, gliding the spacecraft to a landing. { rō'gäl·ō ,wiŋ }

rogenstein [GEOL] An oolite in which the ooliths are united by argillaceous cement. { 'rō·gən,stīn }

Roget's spiral [ELEC] A spiral wire, suspended vertically with the lower end in mercury, that is made to go through a cycle in which an electric current passing through the wire produces mutual attraction between the coils, causing the wire to lift out of the mercury and breaking the current; the spiral then expands under its own weight, so that the lower end drops back into the mercury and the current is reestablished. { rō'zhäz 'spī·rəl }

roggan *See* rocking stone. { 'räg·ən }

Rogowski coil [ENG] A device for measuring alternating current without making contact with the current-carrying conductor, which consists of an air-core coil placed around the conductor in a toroidal fashion so that the alternating magnetic field produced by the current induces a voltage in the coil. { rə'gäv·skē ,kóil }

Rohrback solution [MATER] Toxic, clear, yellow liquid used for specific-gravity separation of minerals and microchemical detection of alkaloids. { 'rôr,bak sə,lü·shən }

roil [HYD] A small section of a stream, characterized by swiftly flowing, turbulent water. { 'róil }

roily oil [MATER] Crude petroleum oil that is more or less emulsified with water. { 'róil·ē 'óil }

roily water [HYD] **1.** Muddy or sediment-filled water. **2.** Turbulent, agitated, or swirling water. { 'róil·ē 'wód·ər }

rolamite mechanism [MECH ENG] An elemental mechanism consisting of two rollers contained by two parallel planes and bounded by a fixed S-shaped band under tension. { 'rō·lə,mīt ,mek·ə,niz·əm }

role indicator [COMPUT SCI] In information retrieval, a code assigned to a key word to indicate its part of speech, nature, or function. { 'rōl ,in·də,kād·ər }

roll [GEOL] A primary sedimentary structure produced by deformation involving subaqueous slump or vertical foundering. [MECH] Rotational or oscillatory movement of an aircraft or similar body about a longitudinal axis through the body; it is called roll for any degree of such rotation. [MECH ENG] A cylinder mounted in bearings; used for such functions as shaping, crushing, moving, or printing work passing by it. [MIN ENG] *See* horseback. [TEXT] A continuous strand made by rolling, rubbing, or twisting fibers. { rōl }

roll acceleration [MECH] The angular acceleration of an aircraft or missile about its longitudinal or X axis. { 'rōl ik,sel·ə,rā·shən }

roll axis [MECH] A longitudinal axis through an aircraft, rocket, or similar body, about which the body rolls. { 'rōl ,ak·səs }

rollback *See* rerun. { 'rōl,bak }

roll bar [DES ENG] A metal bar installed overhead on a roofless automotive vehicle in order to protect the occupants if the car rolls over. { 'rōl ,bär }

roll cage [DES ENG] A frame of metal bars that is installed in a racing car around the driver's seat to protect the driver in the event of an accident. { 'rōl ,kāj }

roll cloud *See* rotor cloud. { 'rōl ,klaůd }

roll compacting [MET] Compacting a metal powder by using a rolling mill. { 'rōl kəm,pak·tiŋ }

roll control [ENG] The exercise of control over a missile so as to make it roll to a programmed degree, usually just before pitchover. { 'rōl kən,trōl }

roll crusher [MECH ENG] A crusher having one or two toothed rollers to reduce the material. { 'rōl ,krəsh·ər }

rolled glass [MATER] Thick flat glass made by passing a roller over the molten glass. { 'rōld 'glas }

rolled gold [MET] A metal or alloy of low value covered by a layer of gold alloy where the proportion of gold alloy to total weight of the article may be less than 1:20 and fineness of the gold alloy may not be less than 10 karats. { 'rōld 'gōld }

rolled joint [ENG] A joint made by expanding a tube in a tube sheet hole by use of an expander. { 'rōld 'jóint }

roller [DES ENG] A cylindrical device for transmitting motion and force by rotation. [GEOL] *See* rod. [OCEANOGR] A long, massive wave which usually retains its form without breaking until it reaches the beach or a shoal. { 'rō·lər }

roller analyzer [ENG] Device for quantitative separation of fine particles (down to 5 micrometers) by use of the graduated lift of a variable-rate pneumatic stream. { 'rō·lər ,an·ə,līz·ər }

roller bearing [MECH ENG] A shaft bearing characterized by parallel or tapered steel rollers confined between outer and inner rings. { 'rō·lər ,ber·iŋ }

roller bit *See* cone rock bit. { 'rō·lər ,bit }

roller cam follower [MECH ENG] A follower consisting of a rotatable wheel at the end of the shaft. { 'rō·lər 'kam ,fäl·ə·wər }

roller chain [MECH ENG] A chain drive assembled from roller links and pin links. { 'rō·lər ,chān }

roller coating [ENG] The application of paints, lacquers, or other coatings onto raised designs or letters by means of a roller. { 'rō·lər ,kōd·iŋ }

roller cone bit [ENG] A drilling bit containing two to four cutters (cones) mounted on very rugged bearings. Also known as bit cone; rock bit. { 'rōl·ər 'kōn ,bit }

roller conveyor [MECH ENG] A gravity conveyor having a track of parallel tubular rollers set at a definite grade, usually on antifriction bearings, at fixed locations, over which package goods which are sufficiently rigid to prevent sagging between rollers are moved by gravity or propulsion. { 'rō·lər kən,vā·ər }

roller drying [CHEM ENG] A method used to dry milk for purposes other than human consumption; concentrated milk is fed between two heated and narrowly spaced stainless steel rollers, the adhering thin film of milk dries as the rollers turn and is scraped off the roller by a doctor blade. { 'rō·lər ,drī·iŋ }

roller gate [CIV ENG] A cylindrical, usually hollow crest gate that is raised and lowered by large toothed wheels running on sloping racks. { 'rō·lər ,gāt }

roller-hearth kiln [ENG] A type of tunnel kiln through which the ware is conveyed on ceramic rollers. { 'rō·lər 'härth ,kil }

roller leveling [MECH ENG] Leveling flat stock by passing it through a machine having a series of rolls whose axes are staggered about a mean parallel path by a decreasing amount. { 'rō·lər 'lev·ə·liŋ }

roller mill *See* ring-roller mill. { 'rō·lər ,mil }

roller printing [TEXT] Printing patterns on fabrics by running them through copper rollers to which inks are applied. { 'rō·lər ,print·iŋ }

roller pulverizer [MECH ENG] A pulverizer operated by the crushing action of rotating rollers. { 'rō·lər 'pəl·və,rīz·ər }

rollers [OCEANOGR] Swells coming from a great distance and forming large breakers on exposed coasts. { 'rō·lərz }

roller stamping die [MECH ENG] An engraved roller used for stamping designs and other markings on sheet metal. { 'rō·lər 'stamp·iŋ ,dī }

roller stripping [GRAPHICS] In lithography, referring to the problem of ink failing to adhere to the metal ink rollers of a press. { 'rō·lər 'strip·iŋ }

Rolle's theorem [MATH] If a function $f(x)$ is continuous on the closed interval $[a,b]$ and differentiable on the open interval (a,b) and if $f(a) = f(b)$, then there exists x_0, $a < x_0 < b$, such that $f'(x_0) = 0$. { 'rōlz ,thir·əm }

roll film [GRAPHICS] A long strip of plastic material with a sensitive emulsion on one side and a layer of gelatin on the other; the gelatin layer reduces the tendency of the base material to curl; the film is attached to opaque backing paper that protects the film from light before and after it is exposed in the camera. { 'rōl ,film }

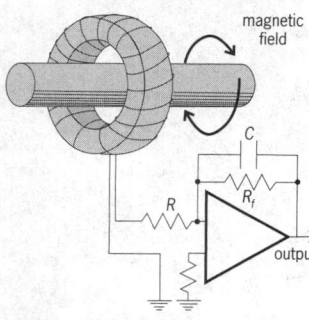

ROGOWSKI COIL

magnetic field

C

R

R_f

output

Arrangement of a Rogowski coil and electronic integrator to give a complete transducer. R = resistor; R_f = resistor of forward current; C = capacitor.

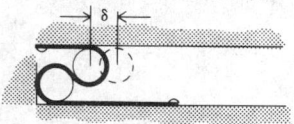

ROLAMITE MECHANISM

δ

δ represents deflection of upper roller during motion of mechanism.

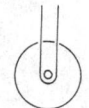

ROLLER CAM FOLLOWER

Drawing of roller cam follower.

ROLLER CHAIN

Section of a single-strand roller chain.

roll flattening *See* flattening.   { 'rōl ‚flat·ən·iŋ }

roll forging   [MET]   Forging metal by using grooved rotating dies.   { 'rōl ‚fȯrj·iŋ }

roll forming   [MET]   Metal forming by using contoured rolls.   { 'rōl ‚fȯrm·iŋ }

roll in   [COMPUT SCI]   To restore to main memory a section of program or data that had previously been rolled out.   { 'rōl ‚in }

Rollin film *See* helium film.   { 'räl·ən ‚film }

rolling   [MECH]   Motion of a body across a surface combined with rotational motion of the body so that the point on the body in contact with the surface is instantaneously at rest.   [MET]   Reducing or changing the cross-sectional area of a workpiece by the compressive forces exerted by rotating rolls. Also known as metal rolling.   [NAV ARCH]   The oscillating motion of a vessel from side to side due to ground swell, heavy sea, or other causes.   { 'rōl·iŋ }

rolling barrage   [ORD]   An artillery barrage that precedes infantry troops at a predetermined rate in their advance during the attack to protect them and facilitate their advance.   { 'rōl·iŋ bə'räzh }

rolling beach   [GEOL]   At the base of a sea cliff, the upper part of an accumulation of boulder sand pebbles which is being ground to sand and finer particles.   { 'rōl·iŋ 'bēch }

rolling chock *See* bilge keel.   { 'rōl·iŋ 'chäk }

rolling contact   [MECH]   Contact between bodies such that the relative velocity of the two contacting surfaces at the point of contact is zero.   { 'rōl·iŋ 'kän‚takt }

rolling-contact bearing   [MECH ENG]   A bearing composed of rolling elements interposed between an outer and inner ring.   { 'rōl·iŋ ‚kän‚takt 'ber·iŋ }

rolling door   [ENG]   A door that moves up and down or from side to side by means of wheels moving along a track.   { 'rōl·iŋ 'dȯr }

rolling friction   [MECH]   A force which opposes the motion of any body which is rolling over the surface of another.   { 'rōl·iŋ 'frik·shən }

rolling lift bridge   [CIV ENG]   A bridge having on the shore end of the lifting portion a segmental bearing that rolls on a flat surface.   { 'rōl·iŋ 'lift ‚brij }

rolling radius   [DES ENG]   For an automotive vehicle, the distance from the center of an axle to the ground.   { 'rōl·iŋ ‚rād·ē·əs }

rolling recoil   [ORD]   System formerly used for absorbing the recoil energy of some railway guns; with the brakes set, the entire car is allowed to roll back when the gun is fired.   { 'rōl·iŋ 'rē‚kȯil }

rolling transposition   [ELEC]   Transposition in which the conductors of an open wire circuit are physically rotated in a substantially helical manner; with two wires, a complete transposition is usually executed in two consecutive spans.   { 'rōl·iŋ ‚tranz·pə'zish·ən }

roll mill   [MECH ENG]   A series of rolls operating at different speeds for grinding and crushing.   { 'rōl ‚mil }

roll-off   [ELECTR]   Gradually increasing loss or attenuation with increase or decrease of frequency beyond the substantially flat portion of the amplitude-frequency response characteristic of a system or transducer.   { 'rōl ‚ȯf }

roll-out   [COMPUT SCI]   **1.** To make available additional main memory for one task by copying another task onto auxiliary storage.   **2.** To read a computer register or counter by adding a one to each digit column simultaneously until all have returned to zero, with a signal being generated at the instant a column returns to zero.   { 'rōl ‚aut }

rollover   [COMPUT SCI]   A keyboard feature that allows more than one key to be depressed simultaneously, enabling the keys to be depressed more rapidly in sequence.   { 'rōl‚ō·vər }

roll resistance spot welding   [MET]   Resistance spot welding using rotating circular electrodes.   { 'rōl ri‚zis·təns 'spät ‚weld·iŋ }

roll roofing   [MATER]   Composition sheet roofing supplied in rolls from which it is laid in overlapping strips.   { 'rōl ‚rüf·iŋ }

roll set   [ENG]   A series of paired convex and concave contoured rolls in a roll forming machine that progressively form a workpiece of uniform cross section.   { 'rōl ‚set }

roll straightening   [ENG]   Unbending of metal stock by passing it through staggered rolls in different planes.   { 'rōl 'strāt·ən·iŋ }

roll threading   [MECH ENG]   Threading a metal workpiece by rolling it either between grooved circular rolls or between grooved straight lines.   { 'rōl ‚thred·iŋ }

roll-tube technique   [MICROBIO]   A pure-culture technique, employed chiefly in tissue culture, in which, during incubation, the test tubes are held in a wheellike instrument at an angle of about 15° from the horizontal and the wheel is rotated vertically about once every 2 minutes.   { 'rōl ‚tüb tek‚nēk }

roll welding   [MET]   Forge welding by heating in a furnace and applying pressure with rolls.   { 'rōl ‚weld·iŋ }

roll your own *See* user program.   { 'rōl yər 'ōn }

rom   [ELEC]   A unit of electrical conductivity, equal to the conductivity of a material in which an electric field of 1 volt per meter gives rise to a current density of 1 ampere per square meter. Derived from reciprocal ohm meter.   { räm }

ROM *See* read-only memory.   { räm }

ROMable code   [COMPUT SCI]   A computer program developed to be stored permanently in a read-only memory (ROM).   { 'räm·ə‚bəl 'kōd }

Romanche trench   [GEOL]   A 24,320-foot-deep (7370-meter) trench in the Mid-Atlantic Ridge near the equator.   { rō'mänsh 'trench }

Romberg's sign   [MED]   **1.** A sign for obturator hernia in which there is pain radiating to the knee.   **2.** A sign for loss of position sense in which the patient cannot maintain equilibrium when standing with feet together and eyes closed.   { 'räm‚bərgz ‚sīn }

romeite   [MINERAL] $(Ca,Fe,Mn,Na)_2(Sb,Ti)_2O_6(O,OH,F)$ A honey-yellow to yellowish-brown mineral composed of oxide of calcium, iron, manganese, sodium, antimony, and titanium, occurring in minute octahedrons.   { 'rō·mē‚īt }

Römer method   [OPTICS]   A method of measuring the speed of light, in which apparent changes in the periods of satellites of another planet, such as Jupiter, whose distance from the earth is known, are observed throughout the year.   { 'rem·ər ‚meth·əd }

ROMET   [METEOROL]   An international code word denoting route forecast, with units in the metric system.   { 'rō‚met }

Ronchi test   [OPTICS]   An improvement on the Foucault knife-edge test for testing curved mirrors, in which the knife edge is replaced with a transmission grating with 15–80 lines per centimeter, and the pinhole source is replaced with a slit or a section of the same grating.   { 'raŋ·kē ‚test }

rondada   [METEOROL]   In Spain, a wind that shifts diurnally from northwest through north, east, south, and west.   { rȯn'däd·ə }

rongstockite   [GEOL]   A medium- to fine-grained plutonic rock composed of zoned plagioclase, orthoclase, some cancrinite, augite, mica, hornblende, magnetite, sphene, and apatite.   { raŋ'stä‚kīt }

röntgen *See* roentgen.   { 'rent·gən }

rood   [MECH]   A unit of area, equal to $^1/_4$ acre, or 10,890 square feet, or 1011.7144056 square meters.   { rüd }

Rood-Sastry classification   [ASTRON]   A classification scheme for galaxy clusters that differentiates between a number of basic cluster morphologies.   { 'rüd ‚sas·trē ‚klas·ə·fə'kā·shən }

roof   [BUILD]   The cover of a building or similar structure.   [GEOL]   **1.** The rock above an orebody.   **2.** The country rock bordering the upper surface of an igneous intrusion.   [MIN ENG]   The rock immediately above a coal seam; corresponds to a hanging wall in metal mining.   { rüf }

roof beam   [BUILD]   A load-bearing member in the roof structure.   { 'rüf ‚bēm }

roof bolt   [MIN ENG]   One of the long steel bolts driven into walls or roofs of underground excavations to strengthen the pinning of rock strata.   { 'rüf ‚bōlt }

roof control   [MIN ENG]   The study of rock behavior when undermined by mining operations, and the most effective measures to control movements.   { 'rüf kən‚trōl }

roof cut   [MIN ENG]   A machine cut made with a turret coal cutter in the roof immediately above the seam.   { 'rüf ‚kət }

roof drain   [BUILD]   A drain for receiving water that has collected on the surface of a roof and discharging it into a downspout.   { 'rüf ‚drān }

roofed dike   [GEOL]   A dike that has an upward termination.   { 'rüft ‚dīk }

roof filter   [ELECTR]   Low-pass filter used in carrier telephone systems to limit the frequency response of the equipment to frequencies needed for normal transmission, thereby blocking

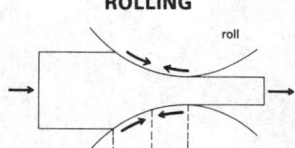

ROLLING

Direction of friction forces in the roll gap during the rolling process. At the neutral point (arrows meet), roll and metal strip have the same velocity; to the left of this point (entry side) the strip moves slower than the roll; and to the right of this point (exit) it moves faster.

unwanted higher frequencies induced in the circuit by external sources; improves runaround cross-talk suppression and minimizes high-frequency singing. { 'rüf ,fil·tər }

roof foundering [GEOL] Collapse of overlying rock into a magma chamber following excavation of a large quantity of magma. { 'rüf ,faùn·driŋ }

roofing [MATER] Material used in roof construction, such as tar, tar paper, shingles, slate, and tin. { 'rüf·iŋ }

roofing copper [MET] Copper that has been hot-rolled to sheets in 14- to 32-ounce (400- to 900-gram) weights. { 'rüf·iŋ ,käp·ər }

roofing felt [MATER] Thick asphalt-impregnated paper used for roofing. { 'rüf·iŋ ,felt }

roofing granules [MATER] Graded particles of crushed rock, slate, slag, porcelain, or tile, used as surfacing on asphalt roofing and shingles. { 'rüf·iŋ ,gran·yülz }

roofing nail [DES ENG] A nail used for attaching paper or shingle to roof boards; usually short with a barbed shank and a large flat head. { 'rüf·iŋ ,nāl }

roofing putty [MATER] Heavy consistency asphalt solution with asbestos fibers; used for caulking metal roofs. { 'rüf·iŋ ,pəd·ē }

roofing slate [MATER] Hard varieties of slate varying in size from 12 × 6 inches (30 × 15 centimeters) to 24 × 14 inches (60 × 35 centimeters), and from $^1/_8$ to $^3/_4$ inch (3 to 19 millimeters) in thickness. { 'rüf·iŋ ,slāt }

roof jack [MIN ENG] A screw- or pump-type extension post used as a temporary roof support. { 'rüf ,jak }

roof pendant [GEOL] Downward projection or sag into an igneous intrusion of the country rock of the roof. Also known as pendant. { 'rüf ,pen·dənt }

roof stringer [MIN ENG] A lagging bar running parallel with the working place above the header in a weak or scaly top in narrow rooms or entries which have short life. { 'rüf ,striŋ·ər }

roof truss [BUILD] A truss used in roof construction; it carries the weight of roof deck and framing and of wind loads on the upper chord; an example is a Fink truss. { 'rüf ,trəs }

rookery [ZOO] A location used by birds for breeding and nesting. { 'rùk·ə·rē }

rook polynomial [MATH] A polynomial in which the coefficient of xk is the number of ways the k rooks can be placed on a chessboard of specified size so that no rook can capture another rook (that is, so that no two rooks are in the same row or the same column). { 'rùk ,päl·ə'nō·mē·əl }

rook problem See problem of nontaking rooks. { 'rùk ,präb·ləm }

room [BUILD] A partitioned-off area inside a building or dwelling. [GEOL] An open area in a cave. [MIN ENG] **1.** Space driven off an entry in which coal is produced. **2.** Working place in a flat mine. { rüm }

room acoustics [ACOUS] The study of the behavior of sound waves in an enclosed room. { 'rüm ə,küs·tiks }

room-and-pillar [MIN ENG] A system of mining in which the coal or ore is mined in rooms separated by narrow ribs or pillars; pillars are subsequently worked. { 'rüm ən 'pil·ər }

room conveyor [MIN ENG] Any conveyor which carries coal from the face of a room toward the mouth. { 'rüm kən,vā·ər }

room crosscut See breakthrough. { 'rüm 'kròs,kət }

room impulse response [ACOUS] The sound pressure in a room that results from a very short sound pulse, usually measured as a function of the time after the arrival of the first direct sound from the source to a listener's location. { ¦rüm 'im,pəls ri,späns }

room noise [COMMUN] Ambient noise in a telephone station. { 'rüm ,nòiz }

root planing [MED] A periodontal procedure that removes bacterial deposits (calculus and plaque) from the crowns and roots of affected teeth. Also known as scaling. { 'rüt ,plān·iŋ }

room power [ELECTR] The electric power that is fed to the machinery in a computer room after passing through a power distribution unit, motor-generator set, or other conditioning and isolating device. { 'rüm ,paú·ər }

rooseveltite [MINERAL] $BiAsO_4$ A gray mineral consisting of bismuth arsenate; occurs as thin botryoidal crusts. { 'rōz·vəl,tīt }

rooster [AGR] An adult male domestic fowl. [VERT ZOO] An adult male of certain birds and fowl, such as pheasants and ptarmigans. { 'rüs·tər }

rooster tail [HYD] A plumelike form of water and sometimes spray that occurs at the intersection of two crossing waves. { 'rüs·tər ,tāl }

root [COMPUT SCI] The origin or most fundamental point of a tree diagram. Also known as base. [BOT] The absorbing and anchoring organ of a vascular plant; it bears neither leaves nor flowers and is usually subterranean. [CIV ENG] The portion of a dam which penetrates into the ground where the dam joins the hillside. [DES ENG] The bottom of a screw thread. [GEOL] **1.** The lower limit of an ore body. Also known as bottom. **2.** The part of a fold nappe that was originally linked to its root zone. [MATH] **1.** A root of a given real or complex number is a number which when raised to some exponent equals that number. Also known as radix. **2.** A root of a polynomial $p(x)$ is a number a such that $p(a) = 0$. **3.** A root of an equation is a number or quantity that satisfies that equation. [MET] See root of weld. { rüt }

root canal [ANAT] The cavity within the root of a tooth, occupied by pulp, nerves, and vessels. { 'rüt kə,nal }

root cap [BOT] A thick, protective mass of parenchymal cells covering the meristematic tip of the root. { 'rüt ,kap }

root cast [GEOL] A slender, tubular, near-vertical, and commonly downward-branching sedimentary structure formed by the filling of a tubular opening left by a root. Also known as rhizoconcretion. { 'rüt ,kast }

root circle [DES ENG] A hypothetical circle defined at the bottom of the tooth spaces of a gear. { 'rüt ,sər·kəl }

root clay See underclay. { 'rüt ,klā }

root component See root symbol. { 'rüt kəm,pō·nənt }

root crack [MET] A crack in the weld or in the heat-affected zone at the root of the weld. { 'rüt ,krak }

root directory [COMPUT SCI] The starting point in a hierarchical file system, where the system operates when it is first started. { 'rüt di,rek·trē }

rooted ordered tree [MATH] A rooted tree in which the order of the subtrees formed by deleting the root vertex is significant. { 'rüd·əd 'ór·dərd 'trē }

rooted tree [MATH] A directed tree graph in which one vertex has no predecessor, and each of the remaining vertices has a unique predecessor. { 'rüd·əd 'trē }

rooter [ENG] A heavy plowing device equipped with teeth and used for breaking up the ground surface; a towed scarifier. { 'rüd·ər }

root face [MET] The part of a fusion face that is not beveled in a welding operation. { 'rüt ,fās }

root field See Galois field. { 'rüt ,fēld }

root fillet [DES ENG] The rounded corner at the angle of a gear tooth flank and the bottom land. { 'rüt ,fil·ət }

root hair [BOT] One of the hairlike outgrowths of the root epidermis that function in absorption. { 'rüt ,her }

root knot [PL PATH] Any of various plant diseases caused by root-knot nematodes which produce gall-like enlargements on the roots. { 'rüt ,nät }

root-knot nematode [INV ZOO] A plant-parasitic nematode species that induces galls or knots to form on roots. { ,rüt ,nät 'nēm·ə,tōd }

rootless vent [GEOL] A source of lava that is not directly connected to a volcanic vent or magma source. { 'rüt·ləs 'vent }

root locus plot [CONT SYS] A plot in the complex plane of values at which the loop transfer function of a feedback control system is a negative number. { 'rüt ¦lō·kəs ,plät }

root-mean-square current See effective current. { 'rüt ,mēn 'skwer 'kə·rənt }

root-mean-square deviation [STAT] The square root of the sum of squared deviations from the mean divided by the number of observations in the sample. { ¦rüt ¦mēn ,skwer ,der·ə'vā·shən }

root-mean-square error [STAT] The square root of the second moment corresponding to the frequency function of a random variable. { ¦rüt ,mēn 'skwer 'er·ər }

root-mean-square sound pressure See effective sound pressure. { 'rüt ,mēn 'skwer 'saùnd ,presh·ər }

root-mean-square value Abbreviated rms value. [PHYS] The square root of the time average of the square of a quantity; for a periodic quantity the average is taken over one complete cycle. Also known as effective value. [STAT] The square

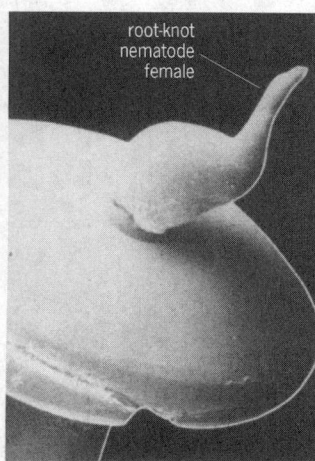

ROOT-KNOT NEMATODE

root-knot nematode female

Adult, pear-shaped, female root-knot nematode on a pinhead. (*Courtesy of J. D. Eisenback*)

root of the average of the squares of a series of related values. { 'rüt ¦mēn ¦skwer 'val·yü }

root nodule [PL PATH] An abnormal nodular growth on a plant root system caused by the establishment of symbiotic nitrogen-fixing bacteria in the host tissue. { 'rüt 'näj¸ül }

root of joint [MET] The area of closest proximity between members of a joint to be welded. { 'rüt əv 'jȯint }

root of unity [MATH] A root of unity in a field F is an element a in F such that $a^n = 1$ for some positive integer n. { ¦rüt əv 'yü·nəd·ē }

root of weld [MET] The points at which the bottom of the weld and the base metal surfaces intersect. Also known as root. { 'rüt əv 'weld }

root opening [MET] In welding, the distance between members at the root of the joint. { 'rüt 'ō·pən·iŋ }

root pass [MET] The first weld bead deposited in a multiple pass weld. Also known as root sealer bead. { 'rüt ¸pas }

root penetration [MET] The depth of penetration of the weld metal into the root of a joint. { 'rüt ¸pen·ə¸trā·shən }

root planning [MED] A periodontal procedure that removes bacterial deposits (calculus and plaque) from the crowns and roots of affected teeth. Also know as scaling. { 'rüt ¸plān· iŋ }

root rot [PL PATH] Any of various plant diseases characterized by decay of the roots. { 'rüt ¸rät }

Roots blower [MECH ENG] A compressor in which a pair of hourglass-shaped members rotate within a casing to deliver large volumes of gas at relatively low pressure increments. { 'rüts ¸blō·ər }

root sealer bead See root pass. { 'rüt ¸sēl·ər ¸bēd }

root segment [COMPUT SCI] The master or controlling segment of an overlay structure which always resides in the main memory of a computer. { 'rüt ¸seg·mənt }

root sheath [GEOL] A hollow root cast. { 'rüt ¸shēth }

root-squaring methods [MATH] Methods of solving algebraic equations which involve calculating the coefficients in a sequence of equations, each of which has roots which are the squares of the roots in the previous equation. { 'rüt ¸skwer· iŋ ¸meth·ədz }

rootstock [BOT] A root or part of a root used as the stock for grafting. { 'rüt¸stäk }

root-sum-square value [PHYS] The square root of the sum of the squares of a series of related values; commonly used to express total harmonic distortion. { 'rüt ¸səm 'skwər 'val·ü }

root symbol [COMPUT SCI] An element of a formal language, generally unique, that is not derivable from other language elements. Also known as root component. { 'rüt ¸sim·bəl }

root task [COMPUT SCI] The initial program on a parallel machine from which one or more child processes branch out in the fork-join model. { 'rüt ¸task }

root test [MATH] An infinite series of nonnegative terms a_n converges if, after some term, the ith root of a_i is less than a fixed number smaller than 1. { 'rüt ¸test }

root vertex [MATH] The vertex of a rooted tree that has no predecessor. { 'rüt 'vər¸teks }

rootworm [INV ZOO] **1.** An insect larva that feeds on plant roots. **2.** A nematode that infests the roots of plants. { 'rüt¸wərm }

root zone [GEOL] **1.** The area where a low-angle thrust fault steepens and descends into the crust. **2.** The source of the root of a fold nappe. { 'rüt ¸zōn }

ropak [OCEANOGR] An ice cake standing on edge as a result of excessive pressure. Also known as turret ice. { 'rō¸pak }

rope [MATER] A long, flexible object which consists of many strands of wire, plastic, or vegetable fiber such as manila. [ORD] A form of confusion reflector consisting of long strips of metal foil, sometimes used with tiny parachutes to reduce the rate of fall. { rōp }

rope-and-button conveyor [MECH ENG] A conveyor consisting of an endless wire rope or cable with disks or buttons attached at intervals. { 'rōp ən ¦bət·ən kən¸vā·ər }

rope boring [ENG] A method similar to rod drilling except that rigid rods are replaced by a steel rope to which the boring tools are attached and allowed to fall by their own weight. { 'rōp ¸bȯr·iŋ }

rope chaff [ORD] Chaff that contains one or more rope elements. { 'rōp ¸chaf }

rope cutter See hook tender. { 'rōp ¸kəd·ər }

rope drive [MECH ENG] A system of ropes running in grooved pulleys or sheaves to transmit power over distances too great for belt drives. { 'rōp ¸drīv }

rope-lay conductor [ELEC] Cable composed of a central core surrounded by one or more layers of helically laid groups of wires. { 'rōp ¦lā kən¸dək·tər }

rope rider See trip rider. { 'rōp ¸rīd·ər }

rope sheave [DES ENG] A grooved wheel, usually made of cast steel or heat-treated alloy steel, used for rope drives. { 'rōp ¸shēv }

rope socket [DES ENG] A drop-forged steel device, with a tapered hole, which can be fastened to the end of a wire cable or rope and to which a load may be attached. { 'rōp ¸säk·ət }

ropewalk [ENG] A long walkway down which a worker carries and lays rope in a manufacturing plant. { 'rōp¸wȯk }

ropeway [ENG] One or a pair of steel cables between several supporting towers which serve as tracks for transporting materials in mountainous areas or at sea. { 'rōp¸wā }

Roproniidae [INV ZOO] A small family of hymenopteran insects in the superfamily Proctotrupoidea. { ¸räp·rə'nī·ə¸dē }

ropy lava See pahoehoe. { 'rō·pē 'lä·və }

Rorschach test [PSYCH] A projective psychologic test in which the subject describes what he imagines seeing in a series of 10 standard inkblots of varying designs and colors. { 'rȯr ¸shäk ¸test }

Rosa and Dorsey method [ELECTROMAG] A method of measuring the speed of light by comparing the capacitance of a capacitor in electromagnetic units, as measured experimentally, with values of currents determined from a current balance, to the capacitance of the same capacitor in electrostatic units, as determined from its geometrical dimensions. { 'rō·sə ən 'dȯr· sē ¸meth·əd }

rosacea [MED] A chronic skin disorder of middle age characterized by redness, papules, and oiliness. { rō'zā·shē·ə }

Rosaceae [BOT] A family of dicotyledonous plants in the order Rosales typically having stipulate leaves and hypogynous, slightly perigynous, or epigynous flowers, numerous stamens, and several or many separate carpels. { rō'zās·ē¸ē }

Rosales [BOT] A morphologically diffuse order of dicotyledonous plants in the subclass Rosidae. { rō'zā·lēz }

Rosalind [ASTRON] A satellite of Uranus orbiting at a mean distance of 43,450 miles (69,930 kilometers) with a period of 13 hours 26 minutes, and with a diameter of about 36 miles (58 kilometers). { 'raz·lind }

rosaniline See fuchsin. { rōz'an·ə·lən }

rosasite [MINERAL] $(Cu,Zn)_2(OH)_2(CO_3)$ A green to bluish-green and sky blue mineral consisting of a carbonate-hydroxide of copper and zinc. { 'rō·zə¸sīt }

roscherite [MINERAL] $(Ca,Mn,Fe)_3Al(PO_4)(OH)·2H_2O$ A dark-brown mineral composed of hydrous basic phosphate of aluminum, calcium, manganese, and iron, occurring as monoclinic crystals. { 'räsh·ə¸rīt }

roscoelite [MINERAL] $K(V,Al,Mg)_3Si_3O_{10}(OH)_2$ Tan, grayish-brown, or greenish-brown vanadium-bearing mica mineral occurring in minute scales or flakes. { 'rä¸skō¸īt }

rose [BOT] A member of the genus *Rosa* in the rose family (Rosaceae); plants are erect, climbing, or trailing shrubs, generally prickly stemmed, and bear alternate, odd-pinnate single leaves. [MATH] A graph consisting of loops shaped like rose petals arising from the equations in polar coordinates $r = a \sin n\theta$ or $r = a \cos n\theta$. Also known as rhodonea. { rōz }

rose absolute [MATER] First filtrate from the cooled alcohol solvent solution of rose concrete (after removal of waxes) during perfume manufacture; pure oil of rose. { 'rōz 'ab·sə¸lüt }

rose bit [DES ENG] A hardened steel or alloy noncore bit with a serrated face to cut or mill out bits, casing, or other metal objects lost in the hole. { 'rōz ¸bit }

rosebloom See false blossom. { 'rōz¸blüm }

rose chucking reamer [DES ENG] A machine reamer with a straight or tapered shank and a straight or spiral flute; cutting is done at the ends of the teeth only; produces a rough hole since there are few teeth. { 'rōz ¸chək·iŋ ¸rē·mər }

rose concrete [MATER] Semisolid mixture of essential oils and waxes solvent-extracted from rose flower petals, leaves, and bark. { 'rōz 'kän¸krēt }

rose diagram [GEOL] A circular graph indicating values in several classes of vector properties of rocks such as cross-bedding direction. { 'rōz 'dī·ə¸gram }

rose flower oil See rose oil. { 'rōz ¦flaü·ər ¸ȯil }

ROSACEAE

A common eastern American species of wild rose (*Rosa carolina*). (*Photograph by A. W. Ambler, from National Audubon Society*)

rose hip [BOT] The ripened false fruit of a rose plant. { 'rōz ˌhip }

roselite [MINERAL] (Ca,Co)₂(Co,Mg)(AsO₄)₂·2H₂O A pink or rose-colored, monoclinic mineral consisting of a hydrated arsenate of calcium, cobalt, and magnesium. { 'rōz·ə‚līt }

rosemary [BOT] *Rosmarinus officinalis.* A fragrant evergreen of the mint family from France, Spain, and Portugal; leaves have a pungent bitter taste and are used as an herb and in perfumes. { 'rōz‚mer·ē }

rosemary oil [MATER] Pungent, combustible, colorless to yellow essential oil with camphorlike aroma; soluble in ether, glacial acetic acid, and alcohol; derived from flowers of rosemary (*Rosmarinus officinalis*); used in flavors, perfumes, and medicines. { 'rōz‚mer·ē ‚oil }

Rosenberg crossed-field generator [ELEC] A type of dynamoelectric amplifier which is self-regulating and can operate while the rotor varies in speed, the current never rising above a certain value. { 'rōz·ən‚bərg ¦krȯst ¦fēld 'jen·ə‚rād·ər }

Rosenmueller's organ *See* epoophoron. { 'rōz·ən‚mül·ərz ‚ȯr·gən }

Rosenmund reaction [ORG CHEM] Catalytic hydrogenation of an acid chloride to form an aldehyde; the reaction is in the presence of sulfur to prevent the subsequent hydrogenation of the aldehyde. { 'rōz·ən‚münd rē‚ak·shən }

rose oil [MATER] Transparent, combustible, yellow-to-green or red essential oil with fragrant scent and sweet taste; solidifies at 18–37°C; steam-distilled from rose flowers; used in flavors, perfumes, and medicines. Also known as attar of roses; otto of rose oil; rose flower oil. { 'rōz ‚oil }

roseola infantum *See* exanthem subitum. { ‚rō·zē'ō·lə in'fan·təm }

roseola typhosa [MED] The rose-colored eruption characteristic of typhus or typhoid fever. { ‚rō·zē'ō·ə tī'fō·sə }

rose opal [MINERAL] An opaque variety of common opal having a fine red color. { 'rōz 'ō·pəl }

rose point lace [TEXT] A needlepoint lace worked in relief. Also known as gros point; venetian rose point. { 'rōz ¦pȯint 'lās }

rose quartz [MINERAL] A pink variety of crystalline quartz; commonly massive and used as a gemstone. Also known as Bohemian ruby. { 'rōz 'kwȯrts }

rose reamer [DES ENG] A reamer designed to cut on the beveled leading ends of the teeth rather than on the sides. { 'rōz 'rē·mər }

Rose's metal [MET] An alloy of bismuth tin and lead; melts at 94°C. { 'rōz·əz ‚med·əl }

rosette [BIOL] Any structure or marking resembling a rose. [MET] **1.** Rounded constituents in a microstructure arranged in whorls. **2.** Strain gages arranged to indicate at a single position the strains in three different directions. [MINERAL] Rose-shaped, crystalline aggregates of barite, marcasite, or pyrite formed in sedimentary rock. [PL PATH] Any of various plant diseases in which the leaves become clustered in the form of a rosette. { rō'zet }

Rosette Nebula [ASTRON] A nebula classified as NGC 2237; this nebula contains numerous small dense clouds that have been photographed with large telescopes. { rō'zet 'neb·yə·lə }

rose water [MATER] Aqueous solution with rose scent (from steam distillation of fresh flowers of rose plants); used in lotions, flavors, and perfumes. { 'rōz ‚wȯd·ər }

Rosidae [BOT] A large subclass of the class Magnoliatae; most have a well-developed corolla with petals separate from each other, binucleate pollen, and ovules usually with two integuments. { 'rōz·ə‚dē }

rosieresite [MINERAL] A yellow to brown mineral composed of hydrous aluminum phosphate containing lead and copper, occurring in stalactitic masses. { ‚rō‚zē'er·ə‚sīt }

rosin [MATER] A translucent yellow, umber, or reddish resinous residue from the distillation of crude turpentine from the sap of pine trees (gum rosin) or from an extract of the stumps and other parts of the tree (wood rosin); used in varnishes, lacquers, printing inks, adhesives, and soldering fluxes, in medical ointments, and as a preservative. { 'räz·ən }

rosin-core solder [MATER] Solder made up in tubular or other hollow form, with the inner space filled with noncorrosive rosin flux. { 'räz·ən ¦kȯr 'säd·ər }

rosin essence [MATER] That part of rosin that can be distilled off at a temperature below 360°C. { 'räz·ən 'es·əns }

rosin ester *See* ester gum. { 'räz·ən 'es·tər }

rosin-extended rubber [MATER] Cold rubber with up to 50% rosin. { 'räz·ən ik‚sten·dəd 'rəb·ər }

rosin joint [ELEC] A soldered joint in which one of the wires is surrounded by an almost invisible film of insulating rosin, making the joint intermittently or continuously open even though it looks good. { 'räz·ən ‚jȯint }

rosin oil [MATER] Viscous, water-insoluble, white-to-brown liquid; soluble in ether, chloroform, carbon disulfide, and fatty oils; distilled from rosin; used as a lubricant and in adhesives, inks, and linoleum. { 'räz·ən ‚oil }

rosin size [MATER] An alkali-treated rosin used as a dry powder or emulsion to surface-size paper products. { 'räz·ən ‚sīz }

rosin tin [MINERAL] A red or yellow variety of cassiterite. Also known as resin tin. { 'räz·ən ‚tin }

Rosiwal analysis [PETR] A quantitive method of estimating the volume percentages of the minerals in a rock, in which thin sections of a rock are examined under a microscope which has a micrometer to measure the linear intercepts of each mineral along a particular set of lines. { 'räz·ə‚wȯl ə‚nal·ə·səs }

Ross Barrier [OCEANOGR] A wall of shelf ice bordering on the Ross Sea. { 'rȯs 'bar·ē·ər }

Rossby diagram [THERMO] A thermodynamic diagram, named after its designer, with mixing ratio as abscissa and potential temperature as ordinate; lines of constant equivalent potential temperature are added. { 'rȯs·bē ‚dī·ə‚gram }

Rossby number [FL MECH] The nondimensional ratio of the inertial force to the Coriolis force for a given flow of a rotating fluid, given as $R_0 = U/fL$, where U is a characteristic velocity, f the Coriolis parameter (or, if the system is cylindrical rather than spherical, twice the system's rotation rate), and L a characteristic length. { 'rȯs·bē ‚nəm·bər }

Rossby parameter [FL MECH] The northward variation of the Coriolis parameter, arising from the sphericity of the earth. Also known as Rossby term. { 'rȯs·bē pə‚ram·əd·ər }

Rossby regime [FL MECH] A type of flow pattern in a rotating fluid with differential radial heating in which the major radial transport of shear and momentum is effected by horizontal eddies of low wave-number; this regime occurs for low values of the Rossby number (of the order of 0.1). { 'rȯs·bē ri‚zhēm }

Rossby term *See* Rossby parameter. { 'rȯs·bē ‚tərm }

Rossby wave [METEOROL] A large, slow-moving, planetary-scale wave generated in the troposphere by ocean-land temperature contrasts and topographic forcing (winds flowing over mountains), and affected by the Coriolis effect due to the earth's rotation. Rossby waves have also been observed in the ocean. Also known as planetary wave. { 'rȯs·bē ‚wāv }

Ross effect [GRAPHICS] A shrinking of the blackened areas of a photographic layer upon drying. Also known as gelatin effect. { 'rȯs i‚fekt }

Rosseland mean absorption coefficient [OPTICS] A coefficient of opacity that is the inverse of a weighted mean of the transmission coefficient over all frequencies. { 'rȯs·lənd 'mēn əb'sȯrp·shən ‚kō·i‚fish·ənt }

Rossel Current [OCEANOGR] A seasonal Pacific Ocean current flowing westward and north-westward along both the southern and northeastern coasts of New Guinea, the southern part flowing through Torres Strait and losing its identity in the Arafura Sea, and the northern part curving northeastward to join the equatorial countercurrent of the Pacific Ocean. { 'rȯs·əl ‚kə·rənt }

Ross feeder [MECH ENG] A chute for conveying bulk materials by means of a screen of heavy endless chains hung on a sprocket shaft; rotation of the shaft causes materials to slide. { 'rȯs 'fēd·ər }

rossite [MINERAL] CaV₂O₆·4H₂O A yellow, triclinic mineral consisting of a hydrated calcium vanadate. { 'rȯ‚sīt }

Rossman drive [ENG] A method used to provide speed control of alternating-current motors; an induction motor stator is mounted on trunnion bearings and driven with an auxiliary motor, to provide the desired change in slip between the stator and rotor. { 'rȯs·mən ‚drīv }

Ross objective [OPTICS] A type of wide-field lens objective in cameras used for astrometric work. { 'rȯs əb‚jek·tiv }

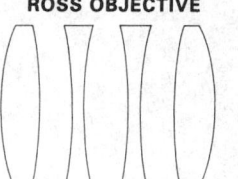

ROSS OBJECTIVE

The four component lenses of the Ross objective.

Ross Sea [GEOGR] Arm of the South Pacific Ocean off Antarctica. { 'rös 'sē }

rostellum [BIOL] The anterior, flattened region of the scolex of armed tapeworms. { rä'stel·əm }

rosterite See vorobyevite. { 'rä·stə,rīt }

rosthornite [MINERAL] A brown to garnet-red variety of retinite with a low (4.5) oxygen content, found in lenticular masses in coal. { 'räs·thər,nīt }

Rostratulidae [VERT ZOO] A small family of birds in the order Charadriiformes containing the painted snipe; females are more brightly colored than males. { ,rä·strə'tyü·lə,dē }

rostrum [BIOL] A beak or beaklike process. { 'rä·strəm }

rot [MATER] See curl. [PL PATH] Any plant disease characterized by breakdown and decay of plant tissue. { rät }

Rotaliacea [INV ZOO] A superfamily of foraminiferans in the suborder Rotaliina characterized by a planispiral or trochospiral test having apertural pores and composed of radial calcite, with secondarily bilamellar septa. { rō,tal·ē'ā·shə }

rotameter [ENG] A variable-area, constant-head, rate-of-flow volume meter in which the fluid flows upward through a tapered tube, lifting a shaped weight to a position where upward fluid force just balances its weight. { rō'tam·əd·ər }

rotary [MECH ENG] **1.** A rotary machine, such as a rotary printing press or a rotary well-drilling machine. **2.** The turntable and its supporting and rotating assembly in a well-drilling machine. { 'röd·ə·rē }

rotary abutment meter [ENG] A type of positive displacement meter in which two displacement rotating vanes interleave with cavities on an abutment rotor in such a way that the three elements are geared together. { 'röd·ə·rē ə'bət·mənt ,mēd·ər }

rotary actuator [MECH ENG] A device that converts electric energy into controlled rotary force; usually consists of an electric motor, gear box, and limit switches. { 'röd·ə·rē 'ak·chə,wād·ər }

rotary air heater [MECH ENG] A regenerative air heater in which heat-transferring members are moved alternately through the gas and air streams. { 'röd·ə·rē 'er ,hēd·ər }

rotary amplifier See rotating magnetic amplifier. { 'röd·ə·rē 'am·plə,fī·ər }

rotary annular extractor [MECH ENG] Vertical, cylindrical shell with an inner, rotating cylinder; liquids to be contacted flow countercurrently through the annular space between the rotor and shell; used for liquid-liquid extraction processes. { 'röd·ə·rē 'an·yə·lər ik'strak·tər }

rotary atomizer [MECH ENG] A hydraulic atomizer having the pump and nozzle combined. { 'röd·ə·rē 'ad·ə,mīz·ər }

rotary beam [ELECTROMAG] Short-wave antenna system highly directional in azimuth and altitude, mounted in such a manner that it can be rotated to any desired position, either manually or by an electric motor drive. { 'röd·ə·rē 'bēm }

rotary belt cleaner [MECH ENG] A series of blades symmetrically spaced about the axis of rotation and caused to scrape or beat against the conveyor belt for the purpose of cleaning. { 'röd·ə·rē 'belt ,klē·nər }

rotary blasthole drilling [MIN ENG] A term applied to two types of drilling: in quarrying and open pit mining it implies rotary drilling with roller-type bits, using compressed air for cuttings removal, either conventional rotary table drive or hydraulic motor to produce rotation, with hydraulic or wireline mechanisms to add part of the weight of the drill to the weight of the tools to increase bit pressure; and in underground mining and sometimes aboveground, it implies the drilling of small-diameter blastholes with a diamond drill, using either coring or noncoring diamond bits. { 'röd·ə·rē 'blast,hōl ,dril·iŋ }

rotary blower [MECH ENG] Positive-displacement, rotating-impeller, air-movement device; can be straight-lobe, screw, sliding-vane, or liquid-piston type. { 'röd·ə·rē 'blō·ər }

rotary boring [MECH ENG] A system of boring in which rock penetration is achieved by the rotation of the hollow cutting tool. { 'röd·ə·rē 'bör·iŋ }

rotary breaker [MIN ENG] A breaking machine for coal or ore; consists of a trommel screen with a heavy steel shell fitted with lifts which raise and convey the coal and stone forward and break it; as the material is broken, the undersize passes through the apertures. { 'röd·ə·rē 'brāk·ər }

rotary bucket [MECH ENG] A 12- to 96-inch-diameter (30- to 244-centimeter) posthole augerlike device, the bottom end of which is equipped with cutting teeth used to rotary-drill large-diameter shallow holes to obtain samples of soil lying above the groundwater level. { 'röd·ə·rē 'bək·ət }

rotary calculator [COMPUT SCI] A type of mechanical desk calculator that is distinguished from key-driven calculators by virtue of a latching selection keyboard in which the multiplicand or divisor could be indexed and then repeatedly transferred to the accumulator, positively or negatively by turning a crank; the accumulator and a cycle-counting register were mounted in a carriage that could be shifted right or left, relative to the selection keyboard, to accommodate the successive digits of the multiplier or quotient. { 'röd·ə·rē 'kal·kyə,lād·ər }

rotary camera [GRAPHICS] A microfilm camera that photographs documents while they are being moved by a transporting device; the document transport mechanism is connected to a film transport mechanism, so there is no difference in the rate of movement between film and image. { 'röd·ə·rē 'kam·rə }

rotary-combustion engine See Wankel engine. { 'röd·ə·rē kəm bəs·chən ,en·jən }

rotary compressor [MECH ENG] A positive-displacement machine in which compression of the fluid is effected directly by a rotor and without the usual piston, connecting rod, and crank mechanism of the reciprocating compressor. { 'röd·ə·rē kəm'pres·ər }

rotary converter See dynamotor. { 'röd·ə·rē kən'vərd·ər }

rotary coupler See rotating joint. { 'röd·ə·rē 'kəp·lər }

rotary crane [MECH ENG] A crane consisting of a boom pivoted to a fixed or movable structure. { 'röd·ə·rē 'krān }

rotary crusher [MECH ENG] Solids-reduction device in which a high-speed rotating cone on a vertical shaft forces solids against a surrounding shell. { 'röd·ə·rē 'krəsh·ər }

rotary-cup oil burner [ENG] Oil burner that uses centrifugal force to spray fuel oil from a rotary fuel atomizing cup into the combustion chamber. { 'röd·ə·rē 'kəp 'öil ,bər·nər }

rotary current [OCEANOGR] A current with the direction of flow rotating through all points of the compass. { 'röd·ə·rē 'kə·rənt }

rotary cutter [MECH ENG] Device used to cut tough or fibrous materials by the shear action between two sets of blades, one set on a rotating holder, the other stationary on the surrounding casing. { 'röd·ə·rē 'kəd·ər }

rotary digital audio tape system [ELECTR] A digital audio tape system that uses the helical-scan technology developed for video systems, with a rotating drum containing two metal-in-gap heads. Abbreviated R-DAT system. { 'röd·ə·rē 'dij·əd·əl ,öd·ē·ō 'tāp ,sis·təm }

rotary-disk contactor [CHEM ENG] Liquid-liquid contactor, having a vertical cylindrical shell with vertical rotating shaft upon which are mounted a spaced series of flat disks; spinning of the disks forces liquid into shell-mounted baffles, causing mixing; used for liquid-liquid extraction processes. Also known as RDC extractor. { 'röd·ə·rē 'disk 'kän,tak·tər }

rotary dispersion [OPTICS] The change in the angle through which an optically active substance rotates the plane of polarization of plane polarized light as the wavelength of the light is varied. Also known as rotatory dispersion. { 'röd·ə·rē di'spər·zhən }

rotary drill [MECH ENG] Any of various drill machines that rotate a rigid, tubular string of rods to which is attached a rock cutting bit, such as an oil well drilling apparatus. { 'röd·ə·rē 'dril }

rotary drilling [MECH ENG] The act or process of drilling a borehole by means of a rotary-drill machine, such as in drilling an oil well. { 'röd·ə·rē 'dril·iŋ }

rotary dryer [MECH ENG] A cylindrical furnace slightly inclined to the horizontal and rotated on suitable bearings; moisture is removed by rising hot gases. { 'röd·ə·rē 'drī·ər }

rotary dump car [MIN ENG] A small mine car in which the car body is mounted on a rotary dumper. { 'röd·ə·rē 'dəmp ,kär }

rotary dumper [MIN ENG] A steel structure on which a mine car revolves and discharges the contents, usually sideways. { 'röd·ə·rē 'dəm·pər }

rotary engine [MECH ENG] A positive displacement engine (such as a steam or internal combustion type) in which the thermodynamic cycle is carried out in a mechanism that is entirely rotary and without the more customary structural elements of a reciprocating piston, connecting rods, and crankshaft. { 'röd·ə·rē 'en·jən }

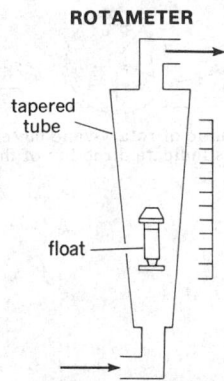

ROTAMETER

Diagram of a rotameter. Arrows indicate direction of fluid flow.

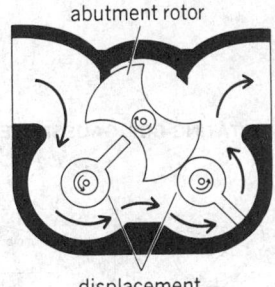

ROTARY ABUTMENT METER

Rotary abutment positive-displacement flowmeter.

rotary excavator See bucket-wheel excavator. { 'rōd·ə·rē 'ek·skə,vād·ər }

rotary fault [GEOL] A fault in which displacement is downward at one point and upward at another point. Also known as pivotal fault; rotational fault. { 'rōd·ə·rē 'fȯlt }

rotary feeder [MECH ENG] Device in which a rotating element or vane discharges powder or granules at a predetermined rate. { 'rōd·ə·rē 'fēd·ər }

rotary filter See drum filter. { 'rōd·ə·rē 'fil·tər }

rotary furnace [MECH ENG] A heat-treating furnace of circular construction which rotates the workpiece around the axis of the furnace during heat treatment; workpieces are transported through the furnace along a circular path. { 'rōd·ə·rē 'fər·nəs }

rotary gap See rotary spark gap. { 'rōd·ə·rē 'gap }

rotary joint See rotating joint. { 'rōd·ə·rē 'jȯint }

rotary kiln [ENG] A long cylindrical kiln lined with refractory, inclined at a slight angle, and rotated at a slow speed. { 'rōd·ə·rē 'kil }

rotary lock See star valve. { ¦rōd·ə·rē ¦läk }

rotary-percussive drill [MECH ENG] Drilling machine which operates as a rotary machine by the action of repeated blows to the bit. { 'rōd·ə·rē pər'kəs·iv 'dril }

rotary phase converter [ELEC] Machine which converts power from an alternating-current system of one or more phases to an alternating-current system of a different number of phases, but of the same frequency. { 'rōd·ə·rē 'fāz kən,vərd·ər }

rotary polarization See optical activity. { 'rōd·ə·rē ,pō·lə·rə'zā·shən }

rotary power source [ELEC] An uninterruptible power system in which a battery driven dc motor mechanically drives an ac generator in the event of a power outage. { 'rōd·ə·rē 'paü·ər ,sȯrs }

rotary press [GRAPHICS] A press utilizing two cylinders, one of which supports the paper while the other one prints on it; large rotary presses are web-fed and accept continuous strips of paper from large rolls; this web of paper is not cut or trimmed until after it is printed. { 'rōd·ə·rē 'pres }

rotary pump [MECH ENG] A displacement pump that delivers a steady flow by the action of two members in rotational contact. { 'rōd·ə·rē 'pəmp }

rotary reflection axis See rotoreflection axis. { 'rōd·ə·rē ri'flek·shən ,ak·səs }

rotary rig [PETRO ENG] The collective equipment used with a rotary drill; includes prime movers or engines, derrick or mast, hoisting and rotating equipment, drill pipe, drill collars and bit, and the mud system used to circulate drilling fluid. { 'rōd·ə·rē 'rig }

rotary roughening [MECH ENG] A metal preparation technique in which the workpiece surface is roughened by a cutting tool. { 'rōd·ə·rē 'rəf·ə·niŋ }

rotary shear [MECH ENG] A sheet-metal cutting machine having two rotary-disk cutters mounted on parallel shafts and driven in unison. { 'rōd·ə·rē 'shir }

rotary shot drill [MECH ENG] A rotary drill used to drill blastholes. { 'rōd·ə·rē 'shät ,dril }

rotary solenoid [ELECTROMAG] A solenoid in which the armature is rotated when actuated; the rotary stroke, ranging from 25 to 95°, is usually converted to linear motion to give a longer stroke than is possible with a conventional plunger-type solenoid. { 'rōd·ə·rē 'sō·lə,nȯid }

rotary spark gap [ELEC] A spark gap in which sparks occur between one or more fixed electrodes and a number of electrodes projecting outward from the circumference of a motor-driven metal disk. Also known as rotary gap. { 'rōd·ə·rē 'spärk ,gap }

rotary stepping relay See stepping relay. { 'rōd·ə·rē 'step·iŋ ,rē,lā }

rotary stepping switch See stepping relay. { 'rōd·ə·rē 'step·iŋ ,swich }

rotary swager [MECH ENG] A machine for reducing diameter or wall thickness of a bar or tube by delivering hammerlike blows to the surface of the work supported on a mandrel. { 'rōd·ə·rē 'swā·jər }

rotary switch [ELEC] A switch that is operated by rotating its shaft. { 'rōd·ə·rē 'swich }

rotary system [COMMUN] A telephone switching system that uses unidirectional, rotary switches that carry ten sets of

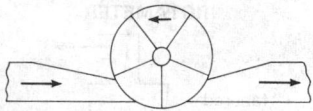

ROTARY-VANE METER

Schematic of rotary-vane meter. Arrows indicate direction of fluid flow.

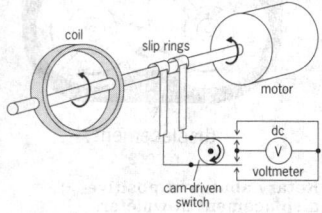

ROTATING-COIL GAUSSMETER

Components of the gaussmeter. (M. Lush)

brushes (wipers), only one of which is tripped as part of the control and selection process. { 'rōd·ə·rē 'sis·təm }

rotary table [MECH ENG] A milling machine attachment consisting of a round table with T-shaped slots and rotated by means of a handwheel actuating a worm and worm gear. [PETRO ENG] A circular unit on the floor of a derrick which rotates the drill pipe and bit. { 'rōd·ə·rē 'tā·bəl }

rotary transformer [ELEC] A rotating machine used to transform direct-current power from one voltage to another. { 'rōd·ə·rē tranz'fȯr·mər }

rotary vacuum filter See drum filter. { 'rōd·ə·rē 'vak·yəm ,fil·tər }

rotary valve [MECH ENG] A valve for the admission or release of working fluid to or from an engine cylinder where the valve member is a ported piston that turns on its axis. { 'rōd·ə·rē 'valv }

rotary-vane attenuator [ELECTROMAG] Device designed to introduce attenuation into a waveguide circuit by varying the angular position of a resistive material in the guide. { 'rōd·ə·rē ¦vān ə'ten·yə,wād·ər }

rotary-vane meter [ENG] A type of positive-displacement rate-of-flow meter having spring-loaded vanes mounted on an eccentric drum in a circular cavity; each time the drum rotates, a fixed volume of fluid passes through the meter. { 'rōd·ə·rē ¦vān 'mēd·ər }

rotary vibrating tippler [MIN ENG] A tippler designed to overcome the tendency for coal or dirt to stick to the bottom of the tubs so that when the tippler is inverted, the car rests upon a vibrating frame which frees any material tending to stick to the bottom. { 'rōd·ə·rē 'vī·brād·iŋ 'tip·lər }

rotary voltmeter [ENG] Type of electrostatic voltmeter used for measuring high voltages. { 'rōd·ə·rē 'vōlt,mēd·ər }

rotate [BOT] Of a sympetalous corolla, having a short tube and petals radiating like the spokes of a wheel. { 'rō,tāt }

rotating amplifier See rotating magnetic amplifier. { 'rō,tād·iŋ 'am·plə,fī·ər }

rotating-anode tube [ELECTR] An x-ray tube in which the anode rotates continuously to bring a fresh area of its surface into the beam of electrons, allowing greater output without melting the target. { 'rō,tād·iŋ ¦an,ōd ¦tüb }

rotating beacon [NAV] A navigational beacon with one or more beams that rotate. { 'rō,tād·iŋ 'bē·kən }

rotating-beam ceilometer [ENG] An electronic, automatic-recording meteorological device which determines cloud height by means of triangulation. { 'rō,tād·iŋ ¦bēm sē'läm·əd·ər }

rotating-coil gaussmeter [ENG] An instrument for measuring low magnetic field strengths and flux densities by measuring the voltage induced in a search coil that is rotated in the field at constant speed. { ,rō,tād·iŋ ¦kȯil 'gaüs,mēd·ər }

rotating coordinate system [MECH] A coordinate system whose axes as seen in an inertial coordinate system are rotating. { 'rō,tād·iŋ kō'ȯrd·ən·ət ,sis·təm }

rotating crystal method [SOLID STATE] Any method of studying crystalline structures by x-ray or neutron diffraction in which a monochromatic, collimated beam of x-rays or neutrons falls on a single crystal that is rotated about an axis perpendicular to the beam. { 'rō,tād·iŋ ¦krist·əl ¦meth·əd }

rotating crystal source [NUCLEO] A source of ultracold neutrons in which neutrons are slowed down by reflection from the moving surface of a rotating mica crystal; only those neutrons which satisfy the Bragg condition for reflection from crystal planes are reflected. { 'rō,tād·iŋ ¦krist·əl ,sȯrs }

rotating-cylinder method [FL MECH] A method of measuring the viscosity of a fluid in which the fluid fills the space between two concentric cylinders, and the torque on the stationary inner cylinder is measured when the outer cylinder is rotated at constant speed. { 'rō,tād·iŋ ¦sil·ən·dər ,meth·əd }

rotating-drum heat transfer [CHEM ENG] Procedure for solidifying layers of solids onto the outside surface of an inside-cooled drum that is partly immersed in a melt of the solids material. { 'rō,tād·iŋ ¦drəm ¦hēt ,tranz·fər }

rotating joint [ELECTROMAG] A joint that permits one section of a transmission line or waveguide to rotate continuously with respect to another while passing radio-frequency energy. Also known as rotary coupler; rotary joint. { 'rō,tād·iŋ 'jȯint }

rotating magnetic amplifier [ELEC] A prime-mover-driven direct-current generator whose power output can be controlled by small field input powers, to give power gain as high as

10,000. Also known as rotary amplifier; rotating amplifier. { 'rō̇,tād·iŋ mag'ned·ik 'am·plə,fī·ər }

rotating meter See velocity-type flowmeter. { 'rō̇,tād·iŋ 'mēd·ər }

rotating models [OCEANOGR] Laboratory models for studying ocean currents, the models being rotated to simulate in part the earth's rotation. { 'rō̇,tād·iŋ 'mäd·əlz }

rotating platinum electrode [ANALY CHEM] Platinum wire sealed in a soft-glass tubing and rotated by a constant-speed motor; used as the electrode in amperometric titrations. Abbreviated RPE. { 'rō̇,tād·iŋ 'plat·ən·əm i'lek,trōd }

rotating Reynolds number [FL MECH] A nondimensional number arising in problems of a rotating viscous fluid and, in particular, in problems involving the agitation of such a fluid by an impeller, equal to the product of the square of the impeller's diameter and its angular velocity divided by the kinematic viscosity of the fluid. Symbolized Re_r. { 'rō̇,tād·iŋ 'ren·əlz ,nəm·bər }

rotating spreader [ENG] Plastics-molding injection device consisting of a finned torpedo that is rotated by a shaft extending through a tubular cross-section injection ram behind it. { 'rō̇,tād·iŋ 'spred·ər }

rotating viscometer vacuum gage [ENG] Vacuum (reduced-pressure) measurement device in which the torque on a spinning armature is proportional to the viscosity (and the pressure) of the rarefied gas being measured; sensitive for absolute pressures of 1 millimeter of mercury (133.32 pascals), down to a few tens of micrometers. { 'rō̇,tād·iŋ vi'skäm·əd·ər 'vak·yəm ,gāj }

rotating wedge [OPTICS] A circular optical wedge mounted to be rotated in the path of light to divert the line of sight to a limited degree. { 'rō̇,tād·iŋ 'wej }

rotation [COMPUT SCI] An operation performed on data in a register of the central processing unit, in which all the bits in the register are shifted one position to the right or left, and the endmost bit, which is shifted out of the register, is carried around to the position at the opposite end of the register. [MATH] See curl. [MECH] Also known as rotational motion. **1.** Motion of a rigid body in which either one point is fixed, or all the points on a straight line are fixed. **2.** Angular displacement of a rigid body. **3.** The motion of a particle about a fixed point. { rō'tā·shən }

rotational bomb [GEOL] A bomb whose shape is formed by spiral motion or rotation during flight. { rō'tā·shən·əl 'bäm }

rotational casting [ENG] Method to make hollow plastic articles from plastisols and lattices using a hollow mold rotated in one or two planes; the hot mold fuses the plastisol into a gel, which is then chilled and the product stripped out. Also known as rotational molding. { rō'tā·shən·əl 'kast·iŋ }

rotational constant [PHYS CHEM] That constant inversely proportioned to the moment of inertia of a linear molecule; used in calculations of microwave spectroscopy quantums. { rō'tā·shən·əl 'kän·stənt }

rotational delay [COMPUT SCI] The time required for a disk or other random-access storage device to reach a position where the read/write head can be positioned over the desired data. { rō'tā·shən·əl di'lā }

rotational energy [MECH] The kinetic energy of a rigid body due to rotation. [PHYS CHEM] For a diatomic molecule, the difference between the energy of the actual molecule and that of an idealized molecule which is obtained by the hypothetical process of gradually stopping the relative rotation of the nuclei without placing any new constraint on their vibration, or on motions of electrons. { rō'tā·shən·əl 'en·ər·jē }

rotational fault See rotary fault. { rō'tā·shən·əl 'fȯlt }

rotational field [PHYS] A vector field whose curl does not vanish. Also known as circuital field; vortical field. { rō'tā·shən·əl 'fēld }

rotational flow [FL MECH] Flow of a fluid in which the curl of the fluid velocity is not zero, so that each minute particle of fluid rotates about its own axis. Also known as rotational motion. { rō'tā·shən·əl 'flō }

rotational impedance [MECH] A complex quantity, equal to the phasor representing the alternating torque acting on a system divided by the phasor representing the resulting angular velocity in the direction of the torque at its point of application. Also known as mechanical rotational impedance. { rō'tā·shən·əl im'pēd·əns }

rotational inertia See moment of inertia. { rō'tā·shən·əl i'nər·shə }

rotational landslide [GEOL] A landslide in which shearing takes place on a well-defined, curved shear surface, concave upward in cross section, producing a backward rotation in the displaced mass. { rō'tā·shən·əl 'an,slīd }

rotational latency [COMPUT SCI] The time required, following an order to read or write information in disk storage, for the location of the information to revolve beneath the appropriate read/write head. { rō'tā·shən·əl 'lāt·ən·sē }

rotational level [PHYS CHEM] An energy level of a diatomic or polyatomic molecule characterized by a particular value of the rotational energy and of the angular momentum associated with the motion of the nuclei. { rō'tā·shən·əl 'lev·əl }

rotational molding See rotational casting. { rō'tā·shən·əl 'mōld·iŋ }

rotational motion See rotation. See rotational flow. { rō'tā·shən·əl 'mō·shən }

rotational movement [GEOL] Apparent fault-block displacement in which the blocks have rotated relative to one another, so that alignment of formerly parallel features is disturbed. { rō'tā·shən·əl 'müv·mən }

rotational position sensing [COMPUT SCI] A fast disk search method whereby the control unit looks for a specified sector, and then receives the sector number required to access the record. { rō'tā·shən·əl pə'zish·ən ,sens·iŋ }

rotational quantum number [PHYS CHEM] A quantum number J characterizing the angular momentum associated with the motion of the nuclei of a molecule; the angular momentum is $(h/2\pi)\sqrt{J(J+1)}$ and the largest component is $(h/2\pi)J$, where h is Planck's constant. { rō'tā·shən·əl 'kwän·təm ,nəm·bər }

rotational reactance [MECH] The imaginary part of the rotational impedance. Also known as mechanical rotational reactance. { rō'tā·shən·əl rē'ak·təns }

rotational resistance [MECH] The real part of rotational impedance; it is responsible for dissipation of energy. Also known as mechanical rotational resistance. { rō'tā·shən·əl ri'zis·təns }

rotational spectrum [SPECT] The molecular spectrum resulting from transitions between rotational levels of a molecule which behaves as the quantum-mechanical analog of a rotating rigid body. { rō'tā·shən·əl 'spek·trəm }

rotational stability [MECH] Property of a body for which a small angular displacement sets up a restoring torque that tends to return the body to its original position. { rō'tā·shən·əl stə'bil·əd·ē }

rotational strain [MECH] Strain in which the orientation of the axes of strain is changed. { rō'tā·shən·əl 'strān }

rotational sum rule [SPECT] The rule that, for a molecule which behaves as a symmetric top, the sum of the line strengths corresponding to transitions to or from a given rotational level is proportional to the statistical weight of that level, that is, to $2J + 1$, where J is the total angular momentum quantum number of the level. { rō'tā·shən·əl 'səm ,rül }

rotational transform [PL PHYS] Property possessed by a magnetic field, in a system used to confine a plasma, in which magnetic lines of force do not close in on themselves after making a circuit around the system, but are rotationally displaced. { rō'tā·shən·əl 'tranz,fȯrm }

rotational transformation [CRYSTAL] A type of crystal transformation that is a change from an ordered phase to a partially disordered phase by rotation of groups of atoms. { rō'tā·shən·əl ,tranz·fər'mā·shən }

rotational transition [PHYS CHEM] A transition between two molecular energy levels which differ only in the energy associated with the molecule's rotation. { rō'tā·shən·əl tran'zish·ən }

rotational traverse [MECH ENG] The maximum angle through which a body can rotate with one point of the body remaining fixed at an axis or center. { rō'tā·shən·əl trə'vərs }

rotational viscometer See Couette viscometer. { rō'tā·shən·əl vi'skäm·əd·ər }

rotational wave See shear wave; S wave. { rō'tā·shən·əl 'wāv }

rotation anemometer [ENG] A type of anemometer in which the rotation of an element serves to measure the wind speed; rotation anemometers are divided into two classes: those in which the axis of rotation is horizontal, as exemplified by the windmill anemometer; and those in which the axis is vertical, such as the cup anemometer. { rō′tā·shən ,an·ə′mäm·əd·ər }

rotation angle [MATH] A directed angle together with a signed measure of the angle. { rō′tā·shən ,aŋ·gəl }

rotation axis [CRYSTAL] A symmetry element of certain crystals in which the crystal can be brought into a position physically indistinguishable from its original position by a rotation through an angle of 360°/n about the axis, where n is the multiplicity of the axis, equal to 2, 3, 4, or 6. Also known as symmetry axis. { rō′tā·shən ,ak·səs }

rotation camera [SOLID STATE] An instrument for studying crystalline structure by x-ray or neutron diffraction, in which a monochromatic, collimated beam of x-rays or neutrons falls on a single crystal which is rotated about an axis perpendicular to the beam and parallel to one of the crystal axes, and the various diffracted beams are registered on a cylindrical film concentric with the axis of rotation. { rō′tā·shən ,kam·rə }

rotation coefficients [MECH] Factors employed in computing the effects on range and deflection which are caused by the rotation of the earth; they are published only in firing tables involving comparatively long ranges. { rō′tā·shən ,kō·i,fish·əns }

rotation firing [ENG] Setting off explosions so that each hole throws its burden toward the space made by the preceding explosions. { rō′tā·shən ,fīr·iŋ }

rotation group [MATH] The group consisting of all orthogonal matrices or linear transformations having determinant 1. { rō′tā·shən ,grüp }

rotation-inversion axis [CRYSTAL] A symmetry element of certain crystals in which a crystal can be brought into a position physically indistinguishable from its original position by a rotation through an angle of 360°/n about the axis followed by an inversion, where n is the multiplicity of the axis, equal to 1, 2, 3, 4, or 6. Also known as inversion axis. { rō′tā·shən in′vər·zhən ,ak·səs }

rotation moment See torque. { rō′tā·shən ,mō·mənt }

rotation-reflection axis [CRYSTAL] A symmetry element of certain crystals in which a crystal can be brought into a position physically indistinguishable from its original position by a rotation through an angle of 360°/n about the axis followed by a reflection in the plane perpendicular to the axis, where n is the multiplicity of the axis, equal to 1, 2, 3, 4, or 6. { rō′tā·shən ri′flek·shən ,ak·səs }

rotation Reynolds number See rotating Reynolds number. { rō′tā·shən ′ren·əlz ,nəm·bər }

rotation spectrum [PHYS CHEM] Absorption-spectrum (absorbed electromagnetic energy) wavelengths produced if only the rotational energy of a molecule is affected during excitation. { rō′tā·shən ,spek·trəm }

rotation therapy [MED] Radiation therapy in which either the patient or the source of radiation is rotated, to permit a larger dose at the center of rotation within the patient's body than on any area of the skin. { rō′tā·shən ,ther·ə·pē }

rotation twin [CRYSTAL] A twin crystal in which the parts will coincide if one part is rotated 180° (sometimes 30, 60, or 120°). { rō′tā·shən ,twin }

rotation-vibration spectrum [PHYS CHEM] Absorption-spectrum (absorbed electromagnetic energy) wavelengths produced when both the energy of vibration and energy of rotation of a molecule are affected by excitation. { rō′tā·shən vī′brā·shən ,spek·trəm }

rotator [ANAT] A muscle that partially rotates a part of the body on the part's axis. [ELECTROMAG] A device that rotates the plane of polarization of a plane-polarized electromagnetic wave, such as a twist in a waveguide. [MECH] A rotating rigid body. [QUANT MECH] A molecule or other quantum-mechanical system which behaves as the quantum-mechanical analog of a rotating rigid body. Also known as top. { ′rō,tād·ər }

Rotatoria [INV ZOO] The equivalent name for Rotifera. { ,rōd·ə′tȯr·ē·ə }

rotatory [OPTICS] Having the capability to rotate the plane of polarization of polarized electromagnetic radiation. { ′rōd·ə,tȯr·ē }

rotatory dispersion See rotary dispersion. { ′rōd·ə,tȯr·ē di′spər·zhən }

rotatory power [OPTICS] A substance's capability to rotate the plane of polarization of polarized electromagnetic radiation.

[PHYS CHEM] The product of the specific rotation of an element or compound and its atomic or molecular weight. { ′rōd·ə,tȯr·ē ,pau·ər }

rotaxane [ORG CHEM] A compound with two or more independent portions not bonded to each other but linked by a linear portion threaded through a ring and maintained in this position by bulky end groups. { rō′tak,sān }

röteln See rubella. { ′re,teln }

rotenone [ORG CHEM] $C_{23}H_{22}O_6$ White crystals with a melting point of 163°C; soluble in ether and acetone; used as an insecticide and in flea powders and fly sprays. Also known as tubatoxin. { ′rōt·ən,ōn }

rotenturm wind [METEOROL] A warm south wind blowing through Rotenturm Pass in the Transylvanian Alps. { ′rōt·ən,türm ,wind }

Rotifera [INV ZOO] A class of the phylum Aschelminthes distinguished by the corona, a retractile trochal disk provided with several groups of cilia and located on the head. { rō′tif·ə·rə }

Rotliegende [GEOL] A European series of geologic time: Lower and Middle Permian. { ′rōt,lē·gən·də }

rotoflector [ELECTROMAG] In radar, elliptically shaped, rotating reflector used to reflect a vertically directed radar beam at right angles so that it radiates in a horizontal direction. { ′rōd·ə,flek·tər }

rotogravure [GRAPHICS] A print made by the intaglio process, using a rotary press. { ¦rōd·ə·grə¦vyůr }

rotoinversion axis [CRYSTAL] A type of crystal symmetry element that combines a rotation of 60, 90, 120, or 180° with inversion across the center. Also known as symmetry axis of rotary inversion; symmetry axis of rotoinversion. { ¦rōd·ō·in′vər·zhən ,ak·səs }

roton [PHYS] A quantum of rotational motion in a liquid, such as superfluid helium. { ′rō,tän }

rotor [AERO ENG] An assembly of blades designed as airfoils that are attached to a helicopter or similar aircraft and rapidly rotated to provide both lift and thrust. [COMMUN] **1.** Disk with a set of input contacts and a set of output contacts, connected by any prearranged scheme designed to rotate within an electrical cipher machine. **2.** Disk whose rotation produces a variation of some cryptographic element in a cipher machine usually by means of lugs (or pins) in or on its periphery. [ELEC] The rotating member of an electrical machine or device, such as the rotating armature of a motor or generator, or the rotating plates of a variable capacitor. [MECH ENG] See impeller. { ′rōd·ər }

rotor cloud [METEOROL] Turbulent, altocumulus-type cloud formation found in the lee of some large mountain barriers, particularly in the Sierra Nevadas near Bishop, California; the air in the cloud rotates around an axis parallel to the range. Also known as roll cloud. { ′rōd·ər ,klaud }

rotoreflection axis [CRYSTAL] A type of symmetry element that combines a rotation of 60, 90, 120, or 180° with reflection across the plane perpendicular to the axis. Also known as rotatory reflection axis. { ¦rōd·ō·ri¦flek·shən ,ak·səs }

rotor plate [ELEC] One of the rotating plates of a variable capacitor, usually directly connected to the metal frame. { ′rōd·ər ,plāt }

rotten ice [HYD] Any piece, body, or area of ice which is in the process of melting or disintegrating; it is characterized by honeycomb structure, weak bonding between crystals, or the presence of meltwater or sea water between grains. Also known as spring sludge. { ′rät·ən ¦īs }

rotten spot See pothole. { ′rät·ən ,spät }

rottenstone [MATER] A soft, decomposed limestone, light gray to olive in color; used in powder form as a polishing material for metal and wood. { ′rät·ən,stōn }

rotund space See strictly convex space. { rō¦tənd ′spās }

Rouché's theorem [MATH] If analytic functions $f(z)$ and $g(z)$ in a simply connected domain satisfy on the boundary $|g(z)| < |f(z)|$, then $f(z)$ and $f(z) + g(z)$ have the same number of zeros in the domain. { ′rüsh·əz ,thir·əm }

rouge [MATER] Finely divided, hydrated iron oxide, used in polishing glass, metal, or gems, and as a pigment. { ′rüzh }

rougemontite [GEOL] A coarse-grained igneous rock composed of anorthite, titanaugite, and small amounts of olivine and iron ore. { ′rüzh,män,tīt }

rough [LAP] An uncut gemstone. { rəf }

rough air [AERO ENG] An aviation term for turbulence encountered in flight. { 'rəf 'er }

rough-axed brick See axed brick. { 'rəf ¦akst 'brik }

rough bark [PL PATH] **1.** Any of various virus diseases of woody plants characterized by roughening and often splitting of the bark. **2.** A fungus disease of apples caused by *Phomopsis mali* and characterized by rough cankers on the bark. { 'rəf ¦bärk }

rough burning [AERO ENG] Pressure fluctuations frequently observed at the onset of burning and at the combustion limits of a ramjet or rocket. { 'rəf ¦bərn·iŋ }

roughcast [CIV ENG] A rough finish on a surface; in particular, a plaster made of lime and shells or pebbles, applied by throwing it against a wall with a trowel. { 'rəf,kast }

rough colony [MICROBIO] A flattened, irregular, and wrinkled colony of bacteria indicative of decreased capsule formation and virulence. { 'rəf 'käl·ə·nē }

rough cut [ENG] A heavy cut (or cuts) made before the finish cut, the primary object of which is the rapid removal of material. { 'rəf ,kət }

roughening transition [PHYS] A change in the behavior of the interface between the solid and liquid phases of a substance at a certain critical temperature, below which the surface is flat or sharp and displays distinct terraces and ledges on an atomic level, while above the critical temperature the surface is rough or rounded. { 'rəf·ən·iŋ tran'zish·ən }

rougher cell [MIN ENG] Flotation cells in which the bulk of the gangue is removed from the ore. { 'rəf·ər ,sel }

rough grinding [MECH ENG] Preliminary grinding without regard to finish. { 'rəf 'grīnd·iŋ }

rough hardware [ENG] Utility items such as nails, sash balances, and studs, without attractive finished appearance. { 'rəf 'härd,wer }

rough ice [HYD] An expanse of ice having an uneven surface caused by formation of pressure ice or by growlers frozen in place. { 'rəf 'īs }

roughing [ENG] The start of evacuation of a vacuum system under test for leaks. { 'rəf·iŋ }

roughing stand [MET] The first stand of rolls, or the last stand before the finishing rolls, through which a preheated billet is passed. { 'rəf·iŋ ,stand }

roughing tool [ENG] A single-point cutting tool having a sharp or small-radius nose, used for deep cuts and rapid material removal from the workpiece. { 'rəf·iŋ ,tül }

rough lumber [MATER] Sawed, undressed lumber. { 'rəf 'ləm·bər }

rough machining [MECH ENG] Preliminary machining without regard to finish. { 'rəf mə'shēn·iŋ }

roughness [FL MECH] Distance from peaks to valleys in pipe-wall irregularities; used to modify Reynolds number calculations for fluid flow through pipes. { 'rəf·nəs }

roughness elements [OCEANOGR] Structures attached to laboratory models to simulate the roughness of the ocean floor. { 'rəf·nəs ,el·ə·məns }

roughness factor [FL MECH] A correction factor used in fluid-flow calculations to allow for flow resistance caused by the roughness of the surface over which the fluid must flow. { 'rəf·nəs ,fak·tər }

roughness length See dynamic roughness. { 'rəf·nəs ,leŋkth }

roughness-width cutoff [MECH ENG] The maximum width of surface irregularities included in roughness height measurements. { 'rəf·nəs ¦width 'kəd,óf }

rough threading [ENG] **1.** Rapid removal of the bulk of the material in a threading operation. **2.** Roughening a surface prior to hot-metal spraying to enhance adhesions. { 'rəf 'thred·iŋ }

rough turning [MECH ENG] The removal of excess stock from a workpiece as rapidly and efficiently as possible. { 'rəf 'tərn·iŋ }

rouleau [PATH] A roll of erythrocytes resembling a stack of coins. { rü'lō }

roulette [MATH] The curve traced out by a point attached to a given curve that rolls without slipping along another given curve that remains fixed. { rü'let }

round [ENG] A series of shots fired either simultaneously or with delay periods between them. [NAV] To pass and alter direction of travel, as a vessel rounds a cape. [ORD] A single munition, missile, or device to be loaded on or in a delivery platform, vehicle, or device for purposes of expenditure; the configuration of the round may vary. { raund }

round angle See perigon. { 'raund ,aŋ·gəl }

round-face bit [DES ENG] Any bit with a rounded cutting face. { 'raund ¦fās 'bit }

round file [DES ENG] A file having a circular cross section. { 'raund 'fīl }

round-head bolt [DES ENG] A bolt having a rounded head at one end. { 'raund ¦hed ,bōlt }

round-head buttress dam [CIV ENG] A mass concrete dam built of parallel buttresses thickened at the upstream end until they meet. { 'raund ¦hed 'bə·trəs ,dam }

rounding [MATH] Dropping or neglecting decimals after some significant place. Also known as truncation. { 'raund·iŋ }

rounding error [MATH] The computational error due to always rounding numbers in a calculation. Also known as round-off error. { 'raund·iŋ ,er·ər }

round ligament [ANAT] **1.** A flattened band extending from the fovea on the head of the femur to attach on either side of the acetabular notch between which it blends with the transverse ligament. **2.** A fibrous cord running from the umbilicus to the notch in the anterior border of the liver; represents the remains of the obliterated umbilical vein. { 'raund 'lig·ə·mənt }

roundness [GEOL] The degree of abrasion of sedimentary particles; expressed as the radius of the average radius of curvature of the edges or corners to the radius of curvature of the maximum inscribed sphere. { 'raund·nəs }

roundnose chisel [DES ENG] A chisel having a rounded cutting edge. { 'raund¦nōs 'chiz·əl }

roundnose tool [DES ENG] A large-radius-nose cutting tool generally used in finishing operations. { 'raund¦nōs 'tül }

round of beam [NAV ARCH] The arch or slope from side to side of a vessel's weather deck for water drainage. Also known as camber. { 'raund əv 'bēm }

round of bearings [NAV] A group of bearings observed simultaneously, or over a short period of time with no appreciable delay between the completion of one observation and the start of the next. { 'raund əv 'ber·iŋz }

round off [MATH] To truncate the least significant digit or digits of a numeral, and adjust the remaining numeral to be as close as possible to the original number. { 'raun,dóf }

round-off error See rounding error. { 'raun,dóf ,er·ər }

round of sights [NAV] A group of celestial observations made with a sextant or other similar instrument over a short period of time, usually with no appreciable delay between the completion of one observation and the start of the next. { 'raund əv 'sīts }

round-robin scheduling [COMPUT SCI] A scheduling algorithm which repeatedly runs through a list of users, giving each user the opportunity to use the central processing unit in succession. { 'raund ¦räb·ən 'skej·ə·liŋ }

rounds complete [ORD] The term used to report that the number of rounds specified in fire for effect have been fired. { 'raunz kəm'plēt }

roundstone [GEOL] Any naturally rounded rock fragment of any size larger than a sand grain (diameter greater than 2 millimeters), such as a boulder, cobble, pebble, or granule. { 'raund,stōn }

round strand rope [DES ENG] A rope composed generally of six strands twisted together or laid to form the rope around a core of hemp, sisal, or manila, or, in a wire-cored rope, around a central strand composed of individual wires. { 'raund 'strand ¦rōp }

round-the-world echo [COMMUN] A signal occurring every $1/7$ second when a radio wave repeatedly encircles the earth at its speed of 186,000 miles (300,000 kilometers) per second. { 'raund the 'wərld 'ek·ō }

round trip [ENG] The combined operations of entering and leaving a hole during drilling operations. { 'raund ¦trip }

round-trip echoes [ELECTROMAG] Multiple reflection echoes produced when a radar pulse is reflected from a target strongly enough so that the echo is reflected back to the target where it produces a second echo. { 'raund ¦trip 'ek·ōz }

round wind [METEOROL] A wind that gradually changes direction through approximately 180° during the daylight hours. { 'raund ,wind }

round window [ANAT] A membrane-covered opening

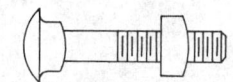

ROUND-HEAD BOLT

Drawing of round-head carriage bolt with a nut.

between the middle and inner ears in amphibians and mammals through which energy is dissipated after traveling in the membranous labyrinth. { 'raund 'win·dō }

roundworm [INV ZOO] The name applied to nematodes. { 'raund,wərm }

Rous sarcoma [VET MED] A fibrosarcoma that can be produced in chickens, pheasants, and ducklings inoculated with the filterable, ribonucleic acid Rous virus. { 'raus sär'kō·mə }

Rousseau diagram [OPTICS] A geometric construction used to determine the total luminous flux of a lamp from a number of polar diagrams which give the effective luminous intensity of the lamp in various directions. { rü,sō 'dī·ə,gram }

rout [MECH ENG] To gouge out, make a furrow, or otherwise machine a wood member. { raut }

route [NAV] The prescribed course to be traveled from a specific point of origin to a specific destination. { rüt *or* raut }

route chart [MAP] A chart showing routes between various places, usually with distances indicated. { 'rüt ,chärt }

route component [METEOROL] The average forecast wind component parallel to the flight path at flight level for an entire route; it is positive if helping (tailwind), and negative if retarding (headwind). { 'rüt kəm,pō·nənt }

route forecast [METEOROL] An aviation weather forecast for one or more specified air routes. { 'rüt ,fòr,kast }

route locking [CIV ENG] Electrically locking in position switches, movable point frogs, or derails on the route of a train, after the train has passed a proceed signal. { 'rüt ,läk·iŋ }

router [COMMUN] A device that selects an appropriate pathway for a message and routes the message accordingly. [DES ENG] **1.** A chisel with a curved point for cleaning out features such as grooves and mortises on wood members. **2.** *See* router plane. [MECH ENG] A machine tool with a rapidly rotating vertical spindle and cutter for making furrows, mortises, and similar grooves. { 'raud·ər }

router plane [DES ENG] A plane for cutting grooves and smoothing the bottom of grooves. Also known as router. { 'raud·ər ,plān }

route survey [CIV ENG] A survey for the design and construction of linear works, such as roads and pipelines. { 'rüt ,sər,vā }

Routh's procedure [MECH] A procedure for modifying the Lagrangian of a system so that the modified function satisfies a modified form of Lagrange's equations in which ignorable coordinates are eliminated. { 'rüths prə,sē·jər }

Routh's rule [MATH] The number of roots with positive real parts of an algebraic equation is equal to the number of changes of algebraic sign of a sequence whose terms are formed from coefficients of the equation in a specified manner. Also known as Routh test. { 'rauths ,rül }

Routh's rule of inertia [MECH] The moment of inertia of a body about an axis of symmetry equals $M(a^2 + b^2)/n$, where M is the body's mass, a and b are the lengths of the body's two other perpendicular semiaxes, and n equals 3, 4, or 5 depending on whether the body is a rectangular parallelepiped, elliptic cylinder, or ellipsoid, respectively. { 'rauths 'rül əv i'nər·shə }

Routh table [MATH] An array of numbers each of which is formed from coefficients of an algebraic equation in a specified manner; the first row of this array constitutes the sequence used in Routh's rule. { 'rauth ,tā·bəl }

Routh test *See* Routh's rule. { 'rauth ,test }

routine [COMPUT SCI] A set of digital computer instructions designed and constructed so as to accomplish a specified function. { rü'tēn }

routine ammunition maintenance [ORD] Maintenance operations not involving disassembly of ammunition or replacement of components, and comprising chiefly cleaning and protecting exterior surfaces of individual items, packages of ammunition, ammunition components, and explosives. { rü'tēn ,am·yə'nish·ən ,mānt·ən·əns }

routine library [COMPUT SCI] Ordered set of standard and proven computer routines by which problems or parts of problems may be solved. { rü'tēn ,lī,brer·ē }

routing [COMMUN] The assignment of a path by which a message will travel to its destination. [ENG] A manufacturing process in which wooden parts are fabricated in various configurations; in high-speed industrial applications, an overhead cutting tool drills into the workpiece and then cuts the

desired interior shape. [GRAPHICS] In letterpress printing, the removal of the nonprinting areas of a plate. { 'rüd·iŋ }

routing indicator [COMMUN] **1.** A group of letters, engineered and assigned, to identify a station within a digital communications network. **2.** A group of letters assigned to indicate the geographic location of a station; a fixed headquarters of a command, activity, or unit at a geographic location; or the general location of a tape relay or tributary station to facilitate the routing of traffic over tape relay networks. { 'rüd·iŋ ,in·də,kād·ər }

routing message [COMMUN] The function performed at a central message processor of selecting the route, or alternate route required, by which a message will proceed to the next point in reaching its destination. { 'rüd·iŋ ,mes·ij }

routivarite [GEOL] A fine-grained igneous rock containing orthoclase, plagioclase, quartz, and garnet. { ,rüd·ə¦va,rīt }

rouvillite [GEOL] A light-colored theralite composed predominantly of labradorite and nepheline, with small amounts of titanaugite, hornblende, pyrite, and apatite. { 'rüv·ə,līt }

rouvite [MINERAL] $CaU_2V_{12}O_{36}·20H_2O$ A purplish- to bluish-black mineral consisting of a hydrated vanadate of calcium and uranium; occurs as dense masses, crusts, and coatings. { 'rü,vīt }

ROV *See* remotely operated vehicle.

roving [MATER] Fibrous glass in which spun strands are woven into a tubular rope. [TEXT] Natural fiber yarns that have been drawn out and slightly twisted in preparation for spinning. { 'rōv·iŋ }

roving artillery [ORD] Artillery withdrawn from its regular position and assigned to special missions; usually moved about and fired from different positions to deceive the enemy as to position and strength. { 'rōv·iŋ är'til·ə·rē }

roving gun [ORD] Gun that is moved about and fired from different positions to mislead or harass the enemy; generally used for registration when the location of the battery position must remain secret. { 'rōv·iŋ ,gən }

row [COMPUT SCI] **1.** The characters, or corresponding bits of binary-coded characters, in a computer word. **2.** Equipment which simultaneously processes the bits of a character, the characters of a word, or corresponding bits of binary-coded characters in a word. **3.** Corresponding positions in a group of columns. { rō }

row address [COMPUT SCI] An index array entry field which contains the main storage address of a data block. { 'rō 'ad,res }

roweite [MINERAL] $(Mn,Mg,Zn)Ca(BO_2)_2(OH)_2$ A light-brown mineral composed of basic borate of calcium, manganese, magnesium, and zinc. { 'rō,īt }

rowland [SPECT] A unit of length, formerly used in spectroscopy, equal to 999.81/999.94 angstrom, or approximately 0.99987×10^{-10} meter. { 'rō,lənd }

Rowland circle [SPECT] A circle drawn tangent to the face of a concave diffraction grating at its midpoint, having a diameter equal to the radius of curvature of a grating surface; the slit and camera for the grating should lie on this circle. { 'rō,lənd ,sər·kəl }

Rowland current [ELEC] A convection current that arises when a charged capacitor plate is rotated. { 'rō·lənd ,kər·ənt }

Rowland ghost [SPECT] A false spectral line produced by a diffraction grating, arising from periodic errors in groove position. { 'rō,lənd ,gōst }

Rowland grating *See* concave grating. { 'rō,lənd ,grād·iŋ }

Rowland mounting [SPECT] A mounting for a concave grating spectrograph in which camera and grating are connected by a bar forming a diameter of the Rowland circle, and the two run on perpendicular tracks with the slit placed at their junction. { 'rō,lənd ,maunt·iŋ }

Rowland ring [ELECTROMAG] A ring-shaped sample of magnetic material, generally surrounded by a coil of wire carrying a current. { 'rō,lənd ,riŋ }

rowlock [NAV ARCH] A U-shaped fitting with shank or socket attachment to the gunwale of a boat, through which an oar is swung. { 'rō,läk }

rowlock arch [ARCH] An arch in which wedge-shaped blocks are arranged in concentric rings. { 'rō,läk 'ärch }

rowlock course [CIV ENG] A course of bricks laid on their sides so that only their ends are visible. { 'rō,läk ,kórs }

row matrix *See* row vector. { 'rō ,mā,triks }

row order [COMPUT SCI] The storage of a matrix $a(m,n)$ as

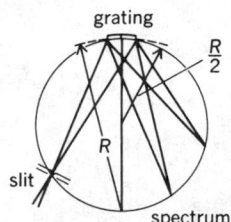

ROWLAND CIRCLE

R represents a radius of the circle of which the curved surface of the grating forms a part. *R*/2 represents a radius of the Rowland circle. The camera may be located anywhere on this circle.

$a(1,1), a(1,2), \ldots, a(1,n), a(2,1), a(2,2), \ldots$ Also known as lexicographic order. { 'rō ,or·dər }

row pitch [COMPUT SCI] The distance between the center-lines of adjacent rows of holes in punched cards and paper tapes. { 'rō ,pich }

row shooting [MIN ENG] Setting off a row of holes nearest the face first, and then other rows in succession behind it. { 'rō ,shüd·iŋ }

row vector [MATH] A matrix consisting of only one row. Also known as row matrix. { 'rō ,vek·tər }

roxarsone *See* 4-hydroxy-3-nitrobenzenearsonic acid. { 'räks,är,sōn }

royal jelly [MATER] A protein complex high in vitamin B secreted by bees to nourish the egg of the queen bee; used in face creams. { 'rȯi·əl 'jel·ē }

Roziere balloon [AERO ENG] A crewed helium balloon whose pilot uses a small burner powered by liquid petroleum gas at night to compensate for the lack of solar heating. { rōz 'yär bə,lün }

RPE *See* rotating platinum electrode.

r-permutation [MATH] An *r*-permutation of a set is an ordered selection of *r* elements of the set. { 'är ,pər·myə,tā·shən }

RPG *See* report program generator.

rpm *See* revolution per minute.

RPN *See* reverse Polish notation.

rp-process *See* rapid proton capture process. { 'är,pē 'prä,ses }

r-process [NUC PHYS] The synthesis of elements and nuclides in supernovas through rapid captures of neutrons in a matter of seconds, followed by beta decay. { 'är ,prä·səs }

rps *See* revolution per second.

RPV *See* remotely piloted vehicle.

RQ *See* respiratory quotient.

RR Lyrae stars [ASTRON] Pulsating variable stars with a period of 0.05–1.2 days in the halo population of the Milky Way Galaxy; color is white, and they are mostly stars of spectral class A. Also known as cluster cepheids; cluster variables. { 'är,är 'lī·rē ,stärz }

r-RNA *See* ribosomal ribonucleic acid.

rs *See* Spearman's rank correlation coefficient.

RSA algorithm *See* Rivest-Shamir-Adleman algorithm. { 'är,es,ā 'al·gə,rith·əm }

R scan *See* R scope. { 'är ,skan }

RS Canum Venaticorum stars [ASTRON] A group of peculiar binary stars with orbital periods between 1 and 14 days, in which the cores of the H and K lines of singly ionized calcium display strong emission, and the hotter of the two stars is of spectral type F or G. { 'är,es ,kan·ən və,nad·ə'kȯr·əm ,stärz }

R scope [ELECTR] An A scope presentation with a segment of the horizontal trace expanded near the target spot (pip) for greater accuracy in range measurement. Also known as R indicator; R scan. { 'är ,skōp }

r selection [ECOL] Selection that favors rapid population growth (r represents the intrinsic rate of increase). { 'är si,lek·shən }

R star [ASTRON] A star of spectral type R, having spectral characteristics similar to those of types G and K, except that molecular bands of molecular carbon (C_2), cyanogen radical (CN), and methyldadyne (CH) are prominent. { 'är ,stär }

RS-232 [COMMUN] A standard developed by the Electronic Industries Association that governs the interface between data processing and data communications equipment, and is widely used to connect microcomputers to peripheral devices.

R tectonite [PETR] A tectonite in which the fabric is believed to have resulted from rotation. { 'är 'tek·tə,nīt }

RTF *See* resistance transfer factor.

RTG *See* radioisotope thermoelectric generator.

r-theta navigation *See* omnibearing-distance navigation. { 'är 'thād·ə ,nav·ə'gā·shən }

RTL *See* resistor-transistor logic.

RTTY *See* radioteletype.

Ru *See* ruthenium.

rubber [ORG CHEM] A natural, synthetic, or modified high polymer with elastic properties and, after vulcanization, elastic recovery. { 'rəb·ər }

rubber accelerator [ORG CHEM] A substance that increases the speed of curing of rubber, such as thiocarbanilide. { 'rəb·ər ak'sel·ə rād·ər }

rubber adhesive [MATER] An adhesive made with a rubber base by using natural or synthetic rubber in an evaporative solvent; a tacky mixture of rubber and filler material, as used on pressure-sensitive tapes; or rubber-solvent-catalyst mixtures (usually two-part) that cure in place. { 'rəb·ər ad,hē·ziv }

rubber banding [COMPUT SCI] In computer graphics, the moving of a line or object, with one end held fixed in position. { 'rəb·ər 'band·iŋ }

rubber-base paint [MATER] A paint in which chlorinated rubber or synthetic latex is the nonvolatile vehicle. { 'rəb·ər ,bās 'pānt }

rubber belt [DES ENG] A conveyor belt that consists essentially of a rubber-covered fabric; fabric is cotton, or nylon or other synthetic fiber, with steel-wire reinforcement. { 'rəb·ər 'belt }

rubber blanket [ENG] A rubber sheet used as a functional die in rubber forming. { 'rəb·ər 'blaŋ·kət }

rubber cement [MATER] An adhesive composed of unvulcanized rubber in an organic solvent. { 'rəb·ər si,ment }

rubber-covered steel conveyor [DES ENG] A steel conveyor band with a cover of rubber bonded to the steel. { 'rəb·ər ,kəv·ərd 'stēl kər,vā·ər }

rubber fiber [MATER] A fiber composed of natural or synthetic rubber; used to make elastic yarn for clothing. { 'rəb·ər 'fī·bər }

rubber foam *See* rubber sponge. { 'rəb·ər 'fōm }

rubber hydrochloride [ORG CHEM] White, thermoplastic hydrochloric acid derivative of rubber; water-insoluble powder or clear film. soluble in aromatic hydrocarbons; softens at 110-120°C; used for protective coverings, food packaging, shower curtains, and rainwear. { 'rəb·ər ,hī·drə'klȯr,īd }

rubber ice [OCEANOGR] Newly formed sea ice which is weak and elastic. { 'rəb·ər ,īs }

rubber plating [ENG] The laying down of a rubber coating onto metals by electrodeposition or by ionic coagulation. { 'rəb·ər 'plād·iŋ }

rubber solvent [MATER] Fast-evaporating petroleum distillate used as a solvent for tackifying rubber during plying (laminating) operations, and in compounding rubber cements. { 'rəb·ər ,säl·vənt }

rubber sponge [MATER] Foamed, flexible rubber; produced by beating air into unvulcanized latex, or by incorporating a gas-producing ingredient (such as sodium bicarbonate) into a strongly masticated rubber stock; used for comfort cushioning, packaging, and shock insulation. Also known as cellular rubber; foam rubber; rubber foam; sponge rubber. { 'rəb·ər ,spənj }

rubber tree [BOT] *Hevea brasiliensis.* A tall tree of the spurge family (Euphorbiaceae) from which latex is collected and coagulated to produce rubber. { 'rəb·ər ,trē }

rubber wheel [DES ENG] A grinding wheel made with rubber as the bonding agent. { 'rəb·ər 'wēl }

rubbing [GRAPHICS] An impression made by moistening a thin, tough paper and patting it into the incised parts of a carved or modeled surface, then rubbing it with pencil, chalk, inked pad, or watercolor; an actual-size image of the original is produced on the paper. { 'rəb·iŋ }

rubbing oil [MATER] 1. A low-viscosity petroleum oil used either with or without an abrasive to polish dried surfaces, such as paint. 2. A nonviscous oil used for polishing wood furniture. { 'rəb·iŋ ,ȯil }

rubble [CIV ENG] 1. Rough, broken stones and other debris resulting from the deterioration and destruction of a building. 2. Rough stone or brick used in coarse masonry or to fill the space in a wall between the facing courses. [GEOL] 1. A loose mass of rough, angular rock fragments, coarser than sand. 2. *See* talus. [HYD] Fragments of floating or grounded sea ice in hard, roughly spherical blocks measuring 0.5–1.5 meters (1.5–4.5 feet) in diameter, and resulting from the breakup of larger ice formations. Also known as rubble ice. { 'rəb·əl }

rubble drift [GEOL] 1. A rubbly deposit (or congeliturbate) formed by solifluction under periglacial conditions. 2. A coarse mass of angular debris and large blocks set in an earthy matrix of glacial origin. { 'rəb·əl ,drift }

rubble ice *See* rubble. { 'rəb·əl ,īs }

rubble-mound structure [CIV ENG] A mound of nonselectively formed and placed stones which are protected with a covering layer of selected stones or of specially shaped concrete armored elements. { 'rəb·əl ¦maůnd ˌstrək·chər }

rubble tract [GEOL] The part of the reef flat immediately behind and on the lagoon side of the reef front, paved with cobbles, pebbles, blocks, and other coarse reef fragments. { 'rəb·əl ˌtrakt }

rubella [MED] An infectious virus disease of humans characterized by coldlike symptoms, fever, and transient, generalized pale-pink rash; its occurrence in early pregnancy is associated with congenital abnormalities. Also known as epidemic roseola; French measles; German measles; röteln. { rü'bel·ə }

rubellite [MINERAL] The red to red-violet variety of the gem mineral tourmaline; hardness is 7–7.5 on Mohs scale, and specific gravity is near 3.04. { 'rü·bəˌlīt }

rubeola See measles. { ˌrü·bē'ō·lə }

Rubiaceae [BOT] The single family of the plant order Rubiales. { ˌrü·bē'ās·ē¸ē }

Rubiales [BOT] An order of dicotyledonous plants marked by their inferior ovary, regular or nearly regular corolla, and opposite leaves with interpetiolar stipules or whorled leaves without stipules. { ˌrü·bē'ā·lēz }

rubicelle [MINERAL] A yellow or orange-red gem variety of spinel. { ¦rü·bə¦sel }

rubidium [CHEM] A chemical element, symbol Rb, atomic number 37, atomic weight 85.4678; a reactive alkali metal; salts of the metal may be used in glass and ceramic manufacture. { rü'bid·ē·əm }

rubidium bromide [INORG CHEM] RbBr Colorless, regular crystals, melting at 683°C; soluble in water; used as a nerve sedative. { rü'bid·ē·əm 'brōˌmīd }

rubidium chloride [INORG CHEM] RbCl A water-soluble, white, lustrous powder melting at 715°C; used as a source for rubidium metal, and as a laboratory reagent. { rü'bid·ē·əm 'klórˌīd }

rubidium halide [INORG CHEM] Any of the halogen compounds of rubidium; examples are RbBr, RbCl, RbF, RbIBrCl, RbBr₂Cl, and RbIBr₂. { rü'bid·ē·əm 'haˌlīd }

rubidium halometallate [INORG CHEM] Halogen-metal-containing compounds of rubidium; examples are Rb₂GeF₆ (rubidium hexafluorogermanate), Rb₂PtCl₆ (rubidium chloroplatinate), and Rb₂PdCl₅ (rubidium palladium chloride). { rü'bid·ē·əm ˌha·lō'med·əlˌāt }

rubidium magnetometer See rubidium-vapor magnetometer. { rü'bid·ē·əm ˌmag·nə'täm·əd·ər }

rubidium-strontium dating [GEOL] A method for determining the age of a mineral or rock based on the decay rate of rubidium-87 to strontium-87. { rü'bid·ē·əm 'strän·chəm 'dād·iŋ }

rubidium sulfate [INORG CHEM] Rb₂SO₄ Colorless, water-soluble rhomboid crystals, melting at 1060°C; used as a cathartic. { rü'bid·ē·əm 'səlˌfāt }

rubidium-vapor frequency standard [PHYS] An atomic frequency standard in which the frequency is established by a gas cell containing rubidium vapor and a neutral buffer gas. { rü'bid·ē·əm ¦vā·pər 'frē·kwənˌse ˌstan·dərd }

rubidium-vapor magnetometer [ENG] A highly sensitive magnetometer in which the spin precession principle is combined with optical pumping and monitoring for detecting and recording variations as small as 0.01 gamma (0.1 microoersted) in the total magnetic field intensity of the earth. Also known as rubidium magnetometer. { rü'bid·ē·əm ¦vā·pər ˌmag·nə'täm·əd·ər }

rubisco See ribulose-1,5-bisphosphate carboxylase/oxygenase. { rü'bis·kō }

RuBP carboxylase/oxygenase See ribulose 1,5-bisphosphate carboxylase/oxygenase. { ¦rü¦bē¦pē kär¦bak·səˌläs 'äk·səˌjəˌnās }

rubredoxin [BIOCHEM] A class of iron-sulfur proteins that contains one iron coordinated to the sulfur atom of four cysteine residues. { ¦rü·brə'däk·sən }

rubriblast See pronormoblast. { 'rü·brəˌblast }

rubricyte See pronormoblast. { 'rü·brəˌsīt }

ruby [MINERAL] The red variety of the mineral corundum; in its finest quality, the most valuable of gemstones. { 'rü·bē }

ruby copper ore See cuprite. { 'rü·bē 'käp·ər ¦ór }

ruby glass [MATER] Glass of a rich red color produced by adding selenium or cadmium sulfide, or copper oxide to the glass. { 'rü·bē 'glas }

ruby laser [OPTICS] An optically pumped solid-state laser

that uses a ruby crystal to produce an intense and extremely narrow beam of coherent red light. { 'rü·bē 'lā·zər }

ruby maser [PHYS] A maser that uses a ruby crystal in the cavity resonator. { 'rü·bē 'mā·zər }

ruby mica [MINERAL] The finest grade of Indian mica; used for electrical capacitors. { 'rü·bē 'mī·kə }

ruby silver [MINERAL] Either of two red silver sulfide minerals: pyrorgyrite (dark-ruby silver) and proustite (light-ruby silver). { 'rü·bē 'sil·vər }

ruby spinel [MINERAL] A clear-red gem variety of spinel, containing small amounts of chromium and having the color but none of the other attributes of true ruby. { 'rü·bē spə'nel }

ruby zinc See zincite. { 'rü·bē 'ziŋk }

rudaceous [PETR] Of or pertaining to a sedimentary rock composed of a large quantity of fragments that are larger than sand grains (diameter greater than 2 millimeters). { rü'dā·shəs }

rudder [ENG] **1.** A flat, usually foil-shaped movable control surface attached upright to the stern of a boat, ship, or aircraft, and used to steer the craft. **2.** See rudder angle. { 'rəd·ər }

rudder angle [ENG] The acute angle between a ship or plane's rudder and its fore-and-aft line. Also known as rudder. { 'rəd·ər ˌaŋ·gəl }

rudder gudgeon [NAV ARCH] One of the lugs cast or forged on the sternpost for the purpose of hanging and hinging the rudder. { 'rəd·ər ˌgə·jən }

rudder lug [NAV ARCH] A projection that is cast or fitted to the forward edge of the rudder frame for the purpose of taking the pintles. { 'rəd·ər ˌləg }

rudderpost [NAV ARCH] A second sternpost to which the rudder of a single-screw ship is joined. { 'rəd·ərˌpōst }

rudderstock [NAV ARCH] A vertical shaft of a rudder having a yoke, quadrant, or tiller fitted to its upper portion by which it may be turned. { 'rəd·ərˌstäk }

rudder stop [NAV ARCH] One of the fittings attached to the structure of the ship or to the shoulders of the sternpost, which limit the swing of the rudder to an angle of about 35°. { 'rəd·ər ˌstäp }

ruddy turnstone [VERT ZOO] *Arenaria interpes.* A member of the avian order Charadriiformes that perform transpacific flights during their migration. { 'rəd·ē 'tərnˌstōn }

ruderal [ECOL] **1.** Growing on rubbish, or waste or disturbed places. **2.** A plant that thrives in such a habitat. { 'rüd·ə·rəl }

rudistids [PALEON] Fossil sessile bivalves that formed reefs during the Cretaceous in the southern Mediterranean or the Tethyan belt. { rü'dis·tədz }

rudite [GEOL] A sedimentary rock composed of fragments coarser than sand grains. { 'rüˌdīt }

Rudzki anomaly [GEOPHYS] A gravity anomaly calculated by replacing the surface topography by its mirror image within the geoid. { 'rüdˌskē aˌnäm·ə·lē }

rue oil [MATER] Yellow, combustible essential oil, soluble in most fixed and mineral oils; derived from blooming plants of genus *Ruta;* used in perfumery and veterinary medicine, and as a chemical intermediate. { 'rü ˌóil }

Ruffini cylinder [NEUROSCI] A cutaneous nerve ending suspected as the mediator of warmth. { rü'fē·nē ˌsil·ən·dər }

ruffle [GEOL] A ripple mark produced by an eddy. { 'rəf·əl }

ruffled groove cast [GEOL] A groove cast with a feather pattern, consisting of a groove with lateral wrinkles that join the main cast in the downcurrent direction at an acute angle. { 'rəf·əld ¦grüv ˌkast }

rufous [BOT] Having a reddish-brown color. { 'rü·fəs }

ruggedization [ELECTR] Making electronic equipment and components resistant to severe shock, temperature changes, high humidity, or other detrimental environmental influences. { ˌrəg·ə·də'zā·shən }

ruggedized computer [COMPUT SCI] A computer, especially a minicomputer, built so as to reduce vibrations, resist moisture, and remain unaffected by electromagnetic interferences such as are found in factory, military, or mobile environments. { 'rəg·əˌdīzd kəm'pyüd·ər }

ruggedness number [GEOL] A dimensionless number that expresses the geometric characteristics of a drainage system; derived from the product of maximum basin relief and drainage density within the drainage basin. { 'rəg·əd·nəs ˌnəm·bər }

Rugosa [PALEON] An order of extinct corals having either simple or compound skeletons with internal skeletal structures consisting mainly of three elements, the septa, tabulae, and dissepiments. { ˌrü'gō·sə }

rugose [BIOL] Having a wrinkled surface. { 'rü,gōs }

rugose mosaic [PL PATH] A virus disease of potatoes marked by dwarfed, wrinkled, and mottled leaves and resulting in premature death. { 'rü,gōs mō'zā·ik }

ruin agate [MINERAL] A brown variety of agate displaying, on a polished surface, markings that resemble or suggest the outlines of ruins or ruined buildings. { 'rü·ən 'ag·ət }

ruin marble [PETR] A brecciated limestone that, when cut and polished, gives a mosaic effect suggesting the appearance of ruins or ruined buildings. { 'rü·ən 'mär·bəl }

rule [MATH] An antecedent condition and a consequent proposition that can support deductive processes. { rül }

Rule *See* Norma. { 'rül }

rule-based control system *See* direct expert control system. { ¦rül ¦bāst kən'trōl ˌsis·təm }

rule-based expert system [COMPUT SCI] An expert system based on a collection of rules that a human expert would follow in dealing with a problem. { ¦rül ¦bāst 'ek,spərt ˌsis·təm }

ruled surface [MATH] A surface that can be generated by the motion of a straight line. { 'rüld 'sər·fəs }

rule of 80–20 *See* Pareto's law. { ¦rül əv 'ād·ē ˌtwen·tē }

rule of approximation [MIN ENG] A rule applicable to placer mining locations and entries upon surveyed lands, to be applied on the basis of 10-acre (40,469-square-meter) legal subdivisions. { 'rül əv ə,präk·sə'mā·shən }

rule of detachment [MATH] The rule that if an implication is true and its antecedent is true, then the consequent is true. { 'rül əv di,tach·mənt }

rule of false position *See* regula falsi. { 'rül əv 'fôls pə'zish·ən }

rule of inference *See* production. { 'rül əv 'in·frəns }

rule of sixty [NAV] An air navigation rule of thumb which states that for each mile off course in a 60-mile leg, the correction to parallel the original course is 1°. { 'rül əv 'siks·tē }

rule of V's [GEOL] The outcrop of a formation that crosses a valley forms an acute angle (a V) that points in the direction in which the formation lies underneath the stream. { 'rül əv 'vēz }

ruler [ENG] A graduated strip of wood, metal, or other material, used to measure lines or as a guide in drawing lines. { 'rül·ər }

ruling [MATH] One of the positions of the straight line that generates a ruled surface. { 'rül·iŋ }

ruling engine [SPECT] A machine operated by a long micrometer screw which rules equally spaced lines on an optical diffraction grating. { 'rül·iŋ ˌen·jən }

ruling pen [GRAPHICS] A drafter's pen that has two adjustable pointed blades between which ink is contained. { 'rül·iŋ ˌpen }

rumble *See* turntable rumble. { 'rəm·bəl }

rumen [VERT ZOO] The first chamber of the ruminant stomach. Also known as paunch. { 'rü·mən }

ruminant [PHYSIO] Characterized by the act of regurgitation and rechewing of food. [VERT ZOO] A mammal belonging to the Ruminantia. { 'rü·mə·nənt }

Ruminantia [VERT ZOO] A suborder of the Artiodactyla including sheep, goats, camels, and other forms which have a complex stomach and ruminate their food. { ˌrü·mə'nan·chə }

rumination [MED] Voluntary regurgitation of food from the stomach, followed by remastication and swallowing in emotionally or mentally disturbed persons. Also known as merycism. [PHYSIO] Regurgitation and remastication of food in preparation for true digestion in ruminants. [PSYCH] An obsessional preoccupation with a single idea or system of ideas. { ˌrü·mə'nā·shən }

rumination disorder [MED] A childhood eating disorder that involves repeated regurgitation and rechewing of food. This behavior is not the result of a gastrointestinal or medical condition; the partially digested food comes back into the mouth without any observable nausea, disgust, or attempt to vomit. { ˌrü·mə'nā·shən dis,ôrd·ər }

run [BUILD] **1.** The horizontal distance from the face of a wall to the ridge of the roof. **2.** The width of a single tread in a stairway. **3.** The horizontal distance traversed by a flight of steps. **4.** The runway or track for a window. [COMPUT SCI] A single, complete execution of a computer program, or one continuous segment of computer processing, used to complete one or more tasks for a single customer or application. Also

known as machine run [CHEM ENG] **1.** The amount of feedstock processed by a petroleum refinery unit during a given time; often used colloquially in relation to the type of stock being processed, as in crude run or naphtha run. **2.** A processing-cycle or batch-treatment operation. [ENG] A portion of pipe or fitting lying in a straight line in the same direction of flow as the pipe to which it s connected. [GEOL] **1.** A ribbonlike, flat-lying, irregular orebody following the stratification of the host rock. **2.** A branching or fingerlike extension of the feeder of an igneous intrusion. [MIN ENG] *See* slant. [NAV] The distance traveled by a craft during any given time interval, or since leaving a designated place. [NAV ARCH] The underwater portion of that part of the aft end of a ship where it curves inward and upward to the stern. [ORD] **1.** Steady, level flight of an aircraft across a target to enable bombs to be dropped accurately in horizontal bombing. **2.** Passing of a moving target once across the range. [STAT] The occurrence of the same characteristic in a series of observations; can be used to test whether or not two random samples come from populations having the same frequency distribution. { rən }

run a line of soundings [ENG] To obtain a series of soundings along a course line. { 'rən ə ¦līn əv 'saůnd·iŋz }

runaround [GRAPHICS] In composition, a type area set in a measure that is adjusted to fit around a picture or another design element. [MIN ENG] A bypass driven in the shaft pillar to permit safe passage from one side of the shaft to the other. { 'rən·ə,raůnd }

runaround crosstalk [COMMUN] Crosstalk resulting from coupling between the high-level end of one repeater and the low-level end of another repeater, as at a carrier telephone repeater station. { 'rən·ə,raůnd 'krôs,tôk }

runaround scrap *See* in-house scrap. { 'rən·ə,raůnd ,skrap }

runaway effect [ELECTR] The phenomenon whereby an increase in temperature causes an increase in a collector-terminal current in a transistor, which in turn results in a higher temperature and, ultimately, failure of the transistor; the effect limits the power output of the transistor. [PL PHYS] The phenomenon whereby an electric current in a plasma produces heat through the Joule effect and the resulting temperature increase results in an increase in conductivity and, thereby, an increase in current flow. { 'rən·ə,wā i,fekt }

runaway electron [ELECTR] An electron, in an ionized gas to which an electric field is applied, that gains energy from the field faster than it loses energy by colliding with other particles in the gas. { 'rən·ə,wā i'lek,trän }

runaway gun [ORD] Automatic weapon that continues firing after the trigger is released; caused by a defect in some part of its mechanism. { 'rən·ə,wā ¦gən }

runaway star [ASTRON] A star of spectral type O or early B with an unusually high spatial velocity; believed to result from a supernova explosion in a close binary system. { 'rən·ə,wā 'stär }

runaway tape [COMPUT SCI] A tape reel that spins rapidly and out of control as the result of a hardware malfunction. { ¦rən·ə¦wā 'tāp }

runback [CHEM ENG] A pipe through which all or part of a distillation column's overhead condensate can be run back into the column, instead of being drawn off as product. [ENG] **1.** To retract the drill feed mechanism to its starting position. **2.** To drill slowly downward toward the bottom of the hole when the drill string has been lifted off-bottom for rechucking. { 'rən,bak }

run before the wind [NAV] To steer a course downwind. { 'rən bi,fôr thə 'wind }

run book [COMPUT SCI] The collection of materials necessary to document a program run on a computer. Also known as problem file; problem folder. { 'rən ,bůk }

run chart [COMPUT SCI] A flow chart for one or more computer runs which shows input, output, and the use of peripheral units, but no details of the execution of the run. Also known as run diagram. { 'rən ,chärt }

runcinate [BOT] Pinnately cut with downward-pointing lobes. { 'rən·sə,nāt }

run diagram *See* run chart. { 'rən ,dī·ə,gram }

run documentation [COMPUT SCI] Detailed instructions to the operator on how to run a particular computer program. { 'rən ,däk·yə·men'tā·shən }

run down a coast [NAV] To sail approximately parallel with the coastline. { 'rən ,daůn ə 'kōst }

rundown line [CHEM ENG] A line from a process unit that connects the look box in the receiving house with the tank in which the product is temporarily stored. { 'rən,daún ,līn }

rundown tank [CHEM ENG] A tank in which the product from a still, agitator, or other processing equipment is received, and from which the product is pumped to larger storage tanks. Also known as pan tank; receiving tank. { 'rən,daún ,taŋk }

Runge-Kutta method [MATH] A numerical approximation technique for solving differential equations. { 'rəŋ·ə 'kúd·ə ,meth·əd }

Runge vector [MECH] A vector which describes certain unchanging features of a nonrelativistic two-body interaction obeying an inverse-square law, either in classical or quantum mechanics; its constancy is a reflection of the symmetry inherent in the inverse-square interaction. { 'rəŋ·ə ,vek·tər }

run in [ENG] To lower the assembled drill rods and auxiliary equipment into a borehole. { 'rən 'in }

R unit See German R unit. See Solomon R unit. { 'är ,yü·nət }

runite See graphic granite. { 'rü,nīt }

run-length encoding [COMPUT SCI] A method of data compression that encodes strings of the same character as a single number. { 'rən ˌleŋkth in'kōd·iŋ }

run-length-limited code [COMMUN] A binary code in which a 1 is inserted after a certain number of 0's, in order to avoid long strings of 0's, which would require very accurate clocking in order to ensure that a bit was not lost. Abbreviated RLL code. { ¦rən ˌleŋkth ˌlim·əd·əd 'kōd }

run motor [ELEC] In facsimile equipment, a motor which supplies the power to drive the scanning or recording mechanisms; a synchronous motor is used to limit the speed. { 'rən ,mōd·ər }

runnel [GEOL] A troughlike hollow on a tidal sand beach which carries water drainage off the beach as the tide retreats. { 'rən·əl }

runner [BOT] A horizontally growing, sympodial stem system; adventitious roots form near the apex, and a new runner emerges from the axil of a reduced leaf. Also known as stolon. [ENG] In a plastics injection or transfer mold, the channel (usually circular) that connects the sprue with the gate to the mold cavity. [MET] **1.** The part of a casting between itself and the gate assembly of the mold. **2.** A channel through which molten metal flows from one receptacle to another. [MIN ENG] A vertical timber sheet pile used to prevent collapse of an excavation. { 'rən·ər }

runner box [MET] A box that divides the molten metal into several streams before it enters the cavity of the mold. { 'rən·ər ,bäks }

running accumulator See push-down storage. { 'rən·iŋ ə'kyü·mə,lād·ər }

running bird [VERT ZOO] Any of the large, flightless, heavy birds usually categorized as ratites. { 'rən·iŋ ,bərd }

running block See traveling block. { 'rən·iŋ ,bläk }

running bond [CIV ENG] A masonry bond involving the placing of each brick as a stretcher and overlapping the bricks in adjoining courses. { 'rən·iŋ ¦bänd }

running fit [DES ENG] The intentional difference in dimensions of mating mechanical parts that permits them to move relative to each other. { 'rən·iŋ ¦fit }

running fix [NAV] A position determined by crossing lines of position obtained at different times and advanced or retarded to a common time. { 'rən·iŋ ¦fiks }

running foot [GRAPHICS] A title that appears at the bottom of almost every page of a book or magazine, such as the title of a chapter, a section, or the publication. { 'rən·iŋ ¦fút }

running gate [MET] A gate through which molten metal enters a mold. { 'rən·iŋ ,gāt }

running gear [MECH ENG] The means employed to support a truck and its load and to provide rolling-friction contact with the running surface. { 'rən·iŋ ,gir }

running ground [MIN ENG] Insecure, incoherent ground that may be semiplastic or plastic and deforms readily under pressure. { 'rən·iŋ ¦graúnd }

running head [GRAPHICS] A title (as of a chapter, of a section, or of the book itself) which appears at the top of almost every page of a book. { 'rən·iŋ ¦hed }

running-in [ENG] The process of operating new or repaired machinery or equipment in order to detect any faults and to ensure smooth, free operation of parts before delivery. { 'rən·iŋ 'in }

running ornament [ARCH] Any molding ornament in which the design is continuous, in intertwined or flowing lines. { 'rən·iŋ 'ȯr·nə·mənt }

running sand See quicksand. { 'rən·iŋ ,sand }

running survey [MAP] A rough survey made by a vessel while coasting. { 'rən·iŋ ,sər,vā }

run-of-bank gravel See bank-run gravel. { ¦rən əv 'baŋk ,grav·əl }

runoff [HYD] **1.** Surface streams that appear after precipitation. **2.** The flow of water in a stream, usually expressed in cubic feet per second; the net effect of storms, accumulation, transpiration, meltage, seepage, evaporation, and percolation. [MIN ENG] Collapse of a coal pillar in a mine. { 'rən,ȯf }

runoff coefficient [HYD] The percentage of precipitation that appears as runoff. { 'rən,ȯf ,kō·i,fish·ənt }

runoff cycle [HYD] The part of the hydrologic cycle involving water between the time it reaches the land as precipitation and its subsequent evapotranspiration or runoff. { 'rən,ȯf ,sī·kəl }

runoff desert [ECOL] An arid region in which local rain is insufficient to support any perennial vegetation except in drainage or runoff channels. { 'rən,ȯf ,dez·ərt }

runoff intensity [HYD] The excess of rainfall intensity over infiltration capacity, usually expressed in inches of rainfall per hour. Also known as runoff rate. { 'rən,ȯf in,ten·səd·ē }

runoff pit [MIN ENG] Catchment area to which spillage from classifiers, thickeners, and slurry pumps can gravitate if it becomes necessary to dump their contents. Also known as spill pit. { 'rən,ȯf ,pit }

runoff rate See runoff intensity. { 'rən,ȯf ,rāt }

run-of-mill [MIN ENG] Ore accepted for treatment, after waste and dense media rejection. Also known as mill-head ore. { 'rən əv 'mil }

run-of-mine See mine run. { 'rən əv 'mīn }

run of the coast [GEOGR] The trend of the coast. { 'rən əv thə 'kōst }

run-on See dieseling. { 'rən,ȯn }

runout [HYD] The location where an avalanche slows down or stops, depositing the avalanche debris. [MET] **1.** Escape of molten metal from a casting mold, crucible, or furnace. **2.** Defect in a casting caused by escape of metal from a mold. { 'rən,aút }

runout table [MET] A roll table used to receive a rolled or extruded member. { 'rən,aút ,tā·bəl }

run-out time [IND ENG] Time required by machine tools after cutting time is finished before tool and material are completely free of interference and before the start of the next sequence of operation. { 'rən,aút ,tīm }

run-time data [MECH ENG] Information obtained from sensors during a machine's regular operation and used to improve its performance. { 'rən ¦tīm 'dad·ə }

run-time error [COMPUT SCI] An error in a computer program that is not detected until the program is executed, and then causes a processing error to occur. { 'rən ¦tīm 'er·ər }

run-time error handler [COMPUT SCI] A system control program that detects and diagnoses run-time errors and issues messages concerning them. { 'rən ¦tīm 'er·ər ,hand·lər }

run-time library [COMPUT SCI] A collection of general-purpose routines that form part of a language translator and allow computer programs to be run with a particular operating system. { 'rən ¦tīm 'lī,brer·ē }

run-up See swash. { 'rən,əp }

runway [CIV ENG] A straight path, often hard-surfaced, within a landing strip, normally used for landing and takeoff of aircraft. [GEOL] The channel of a stream. { 'rən,wā }

runway environment [NAV] The conditions that affect the navigability of a runway, such as the presence of a runway threshold or lighting aids which identify the runway. { 'rən,wā in,vī·ərn·mənt }

runway lights [NAV] Aeronautical ground lights arranged along a runway indicating its direction or boundaries. { 'rən,wā ,līts }

runway observation [METEOROL] An evaluation of certain meteorological elements observed at a specified point on or near an airport runway; temperature, wind speed and direction, ceiling, and visibility are among the elements frequently observed at such locations, because of the importance of these data to aircraft landing and takeoff operations. { 'rən,wā ,äb·zər'vā·shən }

RUNNING BIRD

A kiwi with typical furlike plumage, vestigial wings, and a long, slender slightly curved bill.

runway temperature [METEOROL] The temperature of the air just above the runway at an airport (usually at about 4 feet or 1.2 meters but ideally at engine or wing height), used in the determination of density altitude; therefore, runway temperature observations are made and reported at airports when critical values of density altitude prevail. { 'rən,wā ,tem·prə·chər }

runway visibility [METEOROL] The visibility along an identified runway, determined from a specified point on the runway with the observer facing in the same direction as a pilot using the runway. { 'rən,wā ,viz·ə'bil·əd·ē }

runway visual range [METEOROL] The maximum distance along the runway at which the runway lights are visible to a pilot after touchdown. { 'rən,wā 'vizh·ə·wəl 'rānj }

Rupelian [GEOL] A European stage of middle Oligocene geologic time, above the Tongrian and below the Chattian. Also known as Stampian. { rü'pel·yən }

rupicolous [ECOL] Living among or growing on rocks. { rü'pik·ə·ləs }

Rüping process [ENG] A system for preservative treatment of wood by using positive initial pressure, followed by introduction of the preservative and release of air, creating a vacuum. { 'rüp·iŋ ,prä·səs }

rupture See fracture. See hernia. { 'rəp·chər }

rupture disk See burst disk. { 'rəp·chər ,disk }

rupture disk device [MECH ENG] A nonreclosing pressure relief device which relieves the inlet static pressure in a system through the bursting of a disk. { 'rəp·chər ,disk di,vīs }

rupture zone [GEOL] The region immediately adjacent to the boundary of an explosion crater, characterized by excessive in-place crushing and fracturing where the stresses produced by the explosion have exceeded the ultimate strength of the medium. { 'rəp·chər ,zōn }

rural radio service [COMMUN] A radio service used to provide public message communication service between a central office and subscribers located in rural areas to which it is impracticable or uneconomic to run wire lines. { 'rùr·əl 'rād·ē·ō ,sər·vəs }

Rusa oil See palmarosa oil. { 'rü·sə ,óil }

Rushton-Oldshue column [CHEM ENG] A mixing unit used for continuous pipeline blending in which two-phase contacting is desired; it is a column containing separation plates, baffles, and mixing impellers. { 'rəsh·tən 'ōl,shü ,käl·əm }

Russell bodies [PATH] Hyaline eosinophilic globules 4–5 micrometers in diameter, thought to be particles of antibody globulin, occurring in the cytoplasm of plasma cells in chronic inflammatory exudates. { 'rəs·əl ,bäd·ēz }

Russell diagram See Hertzsprung-Russell diagram. { 'rəs·əl ,dī·ə,gram }

Russell effect [GRAPHICS] The formation of latent developable images on a photographic film or paper by an agent other than electromagnetic radiation, such as a resin, metal, volatile liquid, or printing ink. Also known as Vogel-Colson-Russell effect. { 'rəs·əl i,fekt }

Russell flask [PETRO ENG] Device for volumetric determination of the true volume of sand grains within a unit bulk volume of grains plus voids. { 'rəs·əl ,flask }

russellite [MINERAL] Bi_2WO_6 A pale yellow to greenish, tetragonal mineral consisting of an oxide of bismuth and tungsten; occurs as fine-grained compact masses. { 'rəs·ə,līt }

Russell mixture [ASTROPHYS] A mixture of elements with the same relative proportions as are found in the sun and other stars. { 'rəs·əl ,miks·chər }

Russell movable-wall oven [CHEM ENG] An oven for coal carbonization which cokes a 400-pound (180-kilogram) charge in a horizontal, 12-inch-wide (30-centimeter) chamber, heated from both sides, but with one side floating and balanced against scales. { 'rəs·əl ¦müv·ə·bəl ¦wól ,əv·ən }

Russell-Saunders coupling [PHYS] A process for building many-electron single-particle eigenfunctions of orbital angular momentum and spin; the orbital functions are combined to make an eigenfunction of the total orbital angular momentum, the spin functions are combined to make an eigenfunction of the total spin angular momentum, and then the results are combined into eigenfunctions of the total angular momentum of the system. Also known as LS coupling. { 'rəs·əl 'sòn·dərz ,kəp·liŋ }

Russell's paradox [MATH] The paradox concerning the concept of all sets which are not members of themselves which

forces distinctions in set theory between sets and classes. { 'rəs·əlz 'par·ə,däks }

Russell's viper See tic polonga. { 'rəs·əlz 'vī·pər }

Russell-Vogt theorem See Vogt-Russell theorem. { 'rəs·əl 'vòt 'thir·əm }

rust [MET] The iron oxides formed on corroded ferrous metals and alloys. [PL PATH] Any plant disease caused by rust fungi (Uredinales) and characterized by reddish-brown lesions on the plant parts. { rəst }

rust fungi See Urediniomycetes. { 'rəst ,fən,jī }

rusting [GEOL] The formation of red, yellow, or brown iron oxide minerals by oxidation of mineral deposits. [MET] The formation of rust on ferrous metals and alloys. { 'rəst·iŋ }

rust joint [ENG] A joint to which some oxidizing agent is applied either to cure a leak or to withstand high pressure. { 'rəst ,jóint }

rust prevention [ENG] Surface protection of ferrous structures or equipment to prevent formation of iron oxide; can be by coatings, surface treatment, plating, chemicals, cathodic arrangements, or other means. { 'rəst pri,ven·chən }

rust preventive [MATER] One of a group of products, often with petroleum thinners, used to prevent corrosion to metal surfaces. { 'rəst pri,ven·tiv }

rusty blotch [PL PATH] A fungus disease of barley caused by *Helminthosporium californicum* and characterized by brown blotches on the foliage. { 'rəs·tē 'bläch }

rusty gold [MET] Native gold that has a thin coat of iron oxide or silica that prevents it from amalgamating readily. { 'rəs·tē 'gōld }

rusty mottle [PL PATH] A virus disease of cherry characterized by retarded development of blossoms and leaves in the spring, followed by necrotic spotting and shot-holing of the foliage with considerable defoliation. { 'rəs·tē 'mäd·əl }

rut [PHYSIO] The period during which the male animal has a heightened mating drive. { rət }

rutabaga [BOT] *Brassica napobrassica*. A biennial crucifer of the order Capparales probably resulting from the natural crossing of cabbage and turnip and characterized by a large, edible, yellowish fleshy root. { ¦rüd·ə¦bā·gə }

Rutaceae [BOT] A family of dicotyledonous plants in the order Sapindales distinguished by mostly free stamens and glandular-punctate leaves. { rü'tās·ē,ē }

ruthenic chloride See ruthenium chloride. { rü'then·ik 'klòr,īd }

ruthenium [CHEM] A chemical element, symbol Ru, atomic number 44, atomic weight 101.07. [MET] A hard, brittle, grayish-white metal used as a catalyst; workable only at high temperatures. { rü'thē·nē·əm }

ruthenium chloride [INORG CHEM] $RuCl_3$ Black, deliquescent, water-insoluble solid that decomposes in hot water and above 500°C; used as a laboratory reagent. Also known as ruthenic chloride; ruthenium sesquichloride. { rü'thē·nē·əm 'klòr,īd }

ruthenium halide [INORG CHEM] Halogen compound of ruthenium; examples are $RuCl_2$, $RuCl_3$, $RuCl_4$, $RuBr_3$, and RuF_5. { rü'thē·nē·əm 'ha,līd }

ruthenium red [INORG CHEM] $Ru_2(OH)_2Cl_4 \cdot 7NH_3 \cdot 3H_2O$ A water-soluble, brownish-red powder; used as an analytical reagent and stain. { rü'thē·nē·əm 'red }

ruthenium sesquichloride See ruthenium chloride. { rü'thē·nē·əm ,ses·kwi'klòr,īd }

ruthenium tetroxide [INORG CHEM] RuO_4 A yellow, toxic solid, melting at 25°C; used as an oxidizing agent. { rü'thē·nē·əm te'träk,sīd }

rutherford [NUCLEO] Abbreviated rd. **1.** A unit used to express the decay rate of radioactive material, equal to 10^6 disintegrating atoms per second. **2.** That amount of a substance which is undergoing 10^6 disintegrations per second. { 'rəth·ər·fərd }

Rutherford backscattering spectrometry [SPECT] A method of determining the concentrations of various elements as a function of depth beneath the surface of a sample, by measuring the energy spectrum of ions which are backscattered out of a beam directed at the surface. { 'rəth·ər·fərd 'bak,skad·ə·riŋ spek'träm·ə·trē }

rutherfordine [MINERAL] $(UO_2)(CO_3)$ A yellow mineral composed of uranyl carbonate, occurring as masses of fibers. { 'rəth·ər·fər,dēn }

rutherfordium [CHEM] A chemical element, symbolized Rf,

atomic number 104, a synthetic element; the first element beyond the actinide series, and the twelfth transuranium element. { ‚rəth·ər'fȯr·dē·əm }

Rutherford nuclear atom [ATOM PHYS] A theory of atomic structure in which nearly all the mass is concentrated in a small nucleus, electrons surrounding the nucleus fill nearly all the atom's volume, the number of these electrons equals the atomic number, and the positive charge on the nucleus is equal in magnitude to the negative charge of the electrons. { 'rəth·ər·fərd ¦nü·klē·ər 'ad·əm }

Rutherford scattering [ATOM PHYS] Scattering of heavy charged particles by the Coulomb field of an atomic nucleus. { 'rəth·ər·fərd ‚skad·ə·riŋ }

rutilated quartz [MINERAL] Sagenitic quartz characterized by the presence of enclosed needlelike crystals of rutile. Also known as Venus hairstone. { 'rüd·əl‚ād·əd 'kwȯrts }

rutile [MINERAL] TiO_2 A reddish-brown tetragonal mineral common in acid igneous rocks, in metamorphic rocks, and as residual grain in beach sand. { 'rü‚tēl }

rutin [ORG CHEM] $C_{27}H_{32}O_{16}$ A hydroxyflavone glucorhamnoside derived from cowslip and other plants; yellow needles melting at 190°C; used to treat capillary disorders. { 'rüt·ən }

rutterite [PETR] A medium-grained, equigranular, dark-pink plutonic rock composed chiefly of microperthite, microcline, and albite, with small amounts of nepheline, biotite, amphibole, graphite, and magnetite. { 'rəd·ə‚rīt }

rvalue *See* right value. { 'är‚val·yü }

R-value [ENG] An index of the ability of a substance or material to retard the flow of heat; higher numerical values correspond to higher insulating ability. { 'är ‚val·yü }

R_F value [ANALY CHEM] In chromatography, a measurement based on the relative distance traveled by a sample of a substance in a specific procedure; under standard conditions it is a characteristic property of the substance. { ‚är'ef *or* ‚är səb'ef ‚val·yü }

RV Tauri stars [ASTRON] A class of stars; they are long-period pulsating variable types with periods from about 50 to 150 days; otherwise they are like the shorter-period W Virginis stars; they are found in both the Milky Way Galaxy and the globular clusters. { ‚är¦vē 'tȯr·ē ‚stärz }

RW Aurigae stars [ASTRON] A class of stars that are variable, and whose light variations are rapid and irregular. { ‚är¦dəb·əl‚yü ȯ'rī·gē ‚stärz }

R wave *See* Rayleigh wave. { 'är ‚wāv }

ry *See* rydberg.

rydberg [ATOM PHYS] A unit of energy used in atomic physics, equal to the square of the charge of the electron divided by twice the Bohr radius; equal to 13.605698 ± 0.000004 electronvolts. Symbolized ry. [SPECT] *See* kayser. { 'rid‚bərg }

Rydberg atom [ATOM PHYS] An atom whose outer electron

RYE

Spikes of Tetra Petkus, a variety of rye. (*Wisconsin Agricultural Experiment Station*)

has been excited to very high energy states, far from the nucleus. { 'rid‚bərg ‚ad·əm }

Rydberg constant [ATOM PHYS] **1.** The most accurately measured of the fundamental constants, which enters into the formulas for wave numbers of atomic spectra and serves as a universal scaling factor for any spectroscopic transition and as an important cornerstone in the determination of other constants; it is equal to $\alpha^2 mc/(2h)$ or, in International System (SI) units, to $me^4/(8h^3\varepsilon_0^2 c)$, where α is the fine-structure constant, m and e are the electron mass and charge, c is the speed of light, h is Planck's constant, and ε_0 is the electric constant; numerically, it is equal to 10,973,731.568 549 ± 0.000 083 inverse meters. Symbolized R_∞. **2.** For any atom, the Rydberg constant (first definition) divided by $1 + m/M$, where m and M are the masses of an electron and of the nucleus. { 'rid‚bərg ‚kän·stənt }

Rydberg correction [ATOM PHYS] A term inserted into a formula for the energy of a single electron in the outermost shell of an atom to take into account the failure of the inner electron shells to screen the nuclear charge completely. { 'rid‚bərg kə‚rek·shən }

Rydberg maser [PHYS] A maser that amplifies microwave radiation by means of stimulated emission of atoms whose outer electrons have been excited to very high energy states. { 'rid‚bərg ‚mā·zər }

Rydberg series formula [SPECT] An empirical formula for the wave numbers of various lines of certain spectral series such as neutral hydrogen and alkali metals; it states that the wave number of the nth member of the series is $\lambda_\infty - R/(n + a)^2$, where λ_∞ is the series limit, R is the Rydberg constant of the atom, and a is an empirical constant. { 'rid‚bərg ¦sir·ēz ‚fȯr·myə·lə }

Rydberg spectrum [SPECT] An ultraviolet absorption spectrum produced by transitions of atoms of a given element from the ground state to states in which a single electron occupies an orbital farther from the nucleus. { 'rīd‚bərg ‚spek·trəm }

rye [BOT] *Secale cereale*. A cereal plant of the order Cyperales cultivated for its grain, which contains the most desirable gluten, next to wheat. { rī }

rye buckwheat *See* tartary buckwheat. { 'rī 'bək‚wēt }

rye whiskey [FOOD ENG] One of the potable alcoholic beverages, obtained by distilling the alcohol-containing liquid resulting from the fermentation of a rye mash by yeast; the distilled product is aged for a number of years in new-charred oak barrels. { 'rī 'wis·kē }

Rynchopidae [VERT ZOO] The skimmers, a family of birds in the order Charadriiformes with a knifelike lower beak that is longer and narrower than the upper one. { ‚riŋ'käp·ə‚dē }

Rytiodinae [VERT ZOO] A subfamily of trichechiform sirenians in the family Dugongidae. { ‚rid·ē'äd·ən‚ē }

Rzeppa joint [MECH ENG] A special application of the Bendix-Weiss universal joint in which four large balls are transmitting elements, while a center ball acts as a spacer; it transmits constant angular velocity through a single universal joint. { 'zhep·ə ‚jȯint }

S

s *See* second; strange quark.

S *See* secondary winding; siemens; stoke; sulfur.

Saalic orogeny [GEOL] A short-lived orogeny that occurred early in the Permian period, between the Autunian and Saxonian stages. { 'sä·lik ȯ'räj·ə·nē }

saba [BOT] A plant (*Musa sapientum* var. *compressa*) that is common in the Philippines; the fruit is a cooking banana. [TEXT] A textile made from fibers of the saba plant. { sə'bä }

sabach *See* caliche. { ,sä,bäk }

sabadilla [MATER] Ripe seeds of the sabadilla plant (*Schoenocaulon officinale*) that have been dried; used as an insecticide on cattle. Also known as caustic barley; cevedilla. { ,sa·bə'dē·ə }

Sabathé's cycle [MECH ENG] An internal combustion engine cycle in which part of the combustion is explosive and part at constant pressure. { sa·bə'tāz ,si·kəl }

Sabattier effect [GRAPHICS] Formation of a positive image on a developed photographic plate or film exposed to diffuse light; the image is formed by exposure of silver halide in undeveloped regions of the emulsion. Also known as solarization. { sa·bə'tyā i,fekt }

Sabellariidae [INV ZOO] The sand-cementing worms, a family of polychaete annelids belonging to the Sedentaria and characterized by a compact operculum formed of setae of the first several segments. { sə,bel·ə'rī·ə,dē }

Sabellidae [INV ZOO] A family of sedentary polychaete annelids often occurring in intertidal depths but descending to great abyssal depths; one of two families that make up the feather-duster worms. { sə'bel·ə,dē }

Sabellinae [INV ZOO] A subfamily of the Sabellidae including the most numerous and largest members. { sə'bel·ə,nē }

saber saw [MECH ENG] A portable saw consisting of an electric motor, a straight saw blade with reciprocating mechanism, a handle, baseplate, and other essential parts. { 'sā·bər ,sȯ }

sabin [ACOUS] A unit of sound absorption for a surface, equivalent to 1 square foot (0.09290304 square meter) of perfectly absorbing surface. Also known as absorption unit; open window unit (OW unit); square-foot unit of absorption. { 'sā·bən }

Sabinas [GEOL] A North American (Gulf Coast) provincial series in Upper Jurassic geologic time, below the Coahuilan. { sə'bēn·əs }

Sabine formula [ACOUS] An empirical equation for the reverberation time of sound in a room; its form is identical to that of the Franklin equation. { 'sā,bēn ,fȯr·myə·lə }

Sabin vaccine [IMMUNOL] A live-poliovirus vaccine that is administered orally. { 'sā·bən vak'sēn }

sabkha *See* sebkha. { 'sab·kə }

sable [VERT ZOO] *Martes zibellina.* A carnivore of the family Mustelidae; a valuable fur-bearing animal, quite similar to the American marten. { 'sā·bəl }

sable brush [GRAPHICS] A fine artist's brush, usually round and fairly long with tapered point, made from the hair of the Siberian mink. { 'sā·bəl ,brəsh }

sablefish [VERT ZOO] *Anoplopoma fimbria.* An abundant black-skinned fish in the North Pacific. { 'sā·bəl,fish }

sabot [ORD] Lightweight carrier in which a subcaliber projectile is centered to permit firing the projectile in the larger-caliber weapon; the sabot diameter fills the bore of the weapon from which the projectile is fired. { sa'bō }

sabotage [ORD] Action by enemy agents or sympathizers with intent to stop or otherwise hinder a nation's war effort or to interfere with or obstruct the defense of a nation. { 'sab·ə,täzh }

Sabouraud's agar [MICROBIO] A peptone-maltose agar used as a culture medium for pathogenic fungi, especially the dermatophytes. { sa·bu'rōz 'ag·ər }

sabulous *See* arenaceous. { 'sab·yə·ləs }

sac [BIOL] A soft-walled cavity within a plant or animal, often containing a special fluid and usually having a narrow opening or none at all. [MAP] Indentation in the contour lines of equal depth showing submarine relief. { sak }

saccadic movement [PHYSIO] Rapid eye movement that transfers the gaze from one fixation point to another. { sə'kad·ik 'mūv·mənt }

saccate [BOT] Having a saclike or pouchlike form. { 'sa,kāt }

saccharase [BIOCHEM] An enzyme that catalyzes the hydrolysis of disaccharide to monosaccharides, specifically of sucrose to dextrose and levulose. Also known as invertase; invertin; sucrase. { 'sak·ə,rās }

saccharimeter [ENG] An instrument for measuring the amount of sugar in a solution, often by determining the change in polarization produced by the solution. { ,sak·ə'rim·əd·ər }

saccharin [ORG CHEM] $C_6H_4COSO_2NH$ A sweet-tasting, white powder, soluble in acetates, benzene, and alcohol; slightly soluble in water and ether; melts at 228°C; used as a sugar substitute for syrups, and in medicines, foods, and beverages. Also known as benzosulfimide; gluside. { 'sak·ə·rən }

saccharoidal [PETR] The texture of a rock that is crystalline or granular. Also known as sucrosic; sugary. { ¦sak·ə¦rȯid·əl }

saccharolactic acid *See* mucic acid. { ¦sak·ə·rō¦lak·tik 'as·əd }

saccharolytic [ZOO] Pertaining to an organism that metabolizes carbohydrates. { ,sak·ə·rə'lid·ik }

saccharometer [ENG] An instrument for measuring the amount of sugar in a solution, by determining either the specific gravity or the gases produced by fermentation. { ,sak·ə'räm·əd·ər }

Saccharomycetaceae [MYCOL] The single family of the order Saccharomycetales. { ¦sak·ə·rō,mī·sə'tās·ē,ē }

Saccharomycetales [MYCOL] An order of the subclass Hemiascomycetidae comprising typical yeasts, characterized by the presence of naked asci in which spores are formed by free cells. { ¦sak·ə·rō,mī·sə'tā·lēz }

Saccharomycetoideae [MYCOL] A subfamily of Saccharomycetacae in which spores may be hat-, sickle-, or kidney-shaped, or round or oval. { ¦sak·ə·rō,mī·sə'tȯid·ē,ē }

saccharopinuria [MED] An inborn error of amino acid metabolism characterized by abnormally high levels of saccharopine in the urine. { ¦sak·ə·rō·pi'nùr·ē·ə }

saccharose *See* sucrose. { 'sak·ə,rōs }

Saccoglossa [INV ZOO] An order of gastropod mollusks belonging to the Opisthobranchia. { ,sak·ə'gläs·ə }

Saccopharyngiformes [VERT ZOO] Formerly an order of actinopterygian fishes, the gulpers, now included in the Anguilliformes. { ,sak·ō·fə,rin·ə'fȯr,mēz }

Saccopharyngoidei [VERT ZOO] The gulpers, a suborder of actinopterygian fishes in the order Anguilliformes having

SABELLIDAE

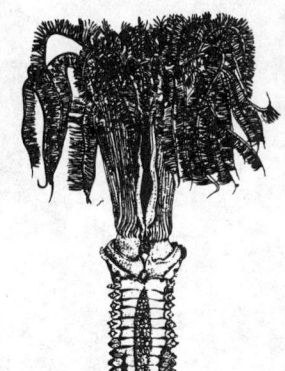

Anterior end of *Sabella* in dorsal view, showing crown used as gill and a collecting organ.

SABLE

The sable (*Marles zibellina*).

degenerative adaptations, including loss of swim bladder, opercle, branchiostegal ray, caudal fin, scales, and ribs. { ‚sak·ō‚far·əŋ'góid·ē‚ī }

saccular aneurysm [MED] A saclike arterial dilation communicating with the artery by a relatively small opening. { 'sak·yə·lər 'an·yə‚riz·əm }

sacculus [ANAT] The smaller, lower saclike chamber of the membranous labyrinth of the vertebrate ear. { 'sak·yə·ləs }

saccus See vesicle. { 'sak·əs }

sac fungus [MYCOL] The common name for members of the class Ascomycetes. { 'sak ‚fəŋ·gəs }

sack cloth [TEXT] A coarse cloth made of animal hairs, cotton, flax, or hemp. { 'sak ‚klóth }

sackungen [GEOL] Deep-seated rock creep which has produced a ridge-top trench by gradual settlement of a slablike mass into an adjacent valley. { 'sa‚kùŋ·ən }

Sackur-Tetrode equation [STAT MECH] An equation for the translational entropy of an ideal gas made up of free fermions. { 'säk·ər 'te‚tröd i‚kwā·zhən }

sacral block [MED] Anesthesia induced by injection of an anesthetic through the caudal hiatus. { 'sak·rəl 'bläk }

sacral nerve [NEUROSCI] Any of five pairs of spinal nerves in the sacral region which innervate muscles and skin of the lower back, lower extremities, and perineum, and branches to the hypogastric and pelvic plexuses. { 'sak·rəl 'nərv }

sacral vertebrae [ANAT] Three to five fused vertebrae that form the sacrum in most mammals; amphibians have one sacral vertebra, reptiles usually have two, and birds have 10–23 fused in the synsacrum. { 'sak·rəl 'vərd·ə‚brā }

sacrificial anode [PHYS CHEM] A protective coating applied to a metal surface to act as an anode and be consumed in an electrochemical reaction, thereby preventing electrolytic corrosion of the metal. { ‚sak·rə‚fish·əl 'an‚ōd }

sacrificial compliant substrate See compliant substrate. { ‚sak·rə‚fish·əl kəm‚plī·ənt 'səb‚strāt }

sacrificial metal [PHYS CHEM] A metal that can be used for a sacrificial anode. { ‚sak·rə‚fish·əl 'med·əl }

sacrococcygeus [ANAT] One of two inconstant thin muscles extending from the lower sacral vertebrae to the coccyx. { ‚sa·krō‚käk'sij·ē·əs }

sacroiliac [ANAT] Pertaining to the sacrum and the ilium. { ‚sa·krō'il·ē‚ak }

sacrospinous [ANAT] Pertaining to the sacrum and the spine of the ischium. { ‚sa·krō'spī·nəs }

sacrum [ANAT] A triangular bone, consisting in humans of five fused vertebrae, located below the last lumbar vertebra, above the coccyx, and between the hipbones. { 'sak·rəm }

saddle [DES ENG] A support shaped to fit the object being held. [GEOL] **1.** A gap that is broad and gently sloping on both sides. **2.** A relatively flat ridge that connects the peaks of two higher elevations. **3.** That part along the surface axis or axial trend of an anticline that is a low point or depression. { 'sad·əl }

saddleback [GEOL] A hill or ridge with a concave outline along its crest. [METEOROL] The cloudless air between the "towers" of two cumulus congestus or cumulonimbus clouds and above a lower cloud mass. { 'sad·əl‚bak }

saddle fold [GEOL] A flexural fold perpendicular to the parent fold and having an additional flexure at its crest. { 'sad·əl ‚fōld }

saddle leather [MATER] **1.** Tanned cattlehide used in furnishings for saddle horses. **2.** Leather resembling saddle leather used in handbags and other leather goods. { 'sad·əl ‚leth·ər }

saddle point [GEOL] See col. [MATH] **1.** A point where all the first partial derivatives of a function vanish but which is not a local maximum or minimum. **2.** For a matrix of real numbers, an element that is both the smallest element of its row and the largest element of its column, or vice versa. **3.** For a two-person, zero-sum game, an element of the payoff matrix that is the smallest element of its row and the largest element of its column, so that the corresponding strategies are optimal for each player, given the strategy chosen by the other player. { 'sad·əl ‚póint }

saddle-point azeotrope [PHYS CHEM] A rarely occurring azeotrope which is formed in ternary systems and for which the boiling point is intermediate between the highest and lowest boiling mixture in the system. { 'sad·əl ‚póint 'ā·zē·ə‚trōp }

SADDLE-TYPE TURRET LATHE

Saddle-type turret lathe, showing characteristic hexagon turret on the bedways of the lathe. (*Jones and Lamson Machine Co.*)

saddle-point method See steepest-descent method. { 'sad·əl ‚póint ‚meth·əd }

saddle-point theory [MATH] The study of differentiable functions and their derivatives from the viewpoint of saddle points, especially applicable to the calculus of variations. { 'sad·əl ‚póint ‚thē·ə·rē }

saddle reef [GEOL] A mineral deposit associated with the crest of an anticlinal fold and following the bedding plane, usually found in vertical succession. Also known as saddle vein. { 'sad·əl ‚rēf }

saddle stone See apex stone. { 'sad·əl ‚stōn }

saddle-type turret lathe [MECH ENG] A turret lathe designed without a ram and with the turret mounted directly on a support (saddle) which slides on the bedways of the lathe. { 'sad·əl ‚tīp 'tər·ət ‚lāth }

saddle vein See saddle reef. { 'sad·əl ‚vān }

saddling [MET] Forming a seamless ring by forging a pierced disk over a mandrel (or saddle). Also known as mandrel forging. { 'sad·liŋ }

sadism [PSYCH] Sexual perversion in which one derives pleasure from inflicting physical or mental cruelty upon another. { 'sä‚diz·əm }

Saefftigen's pouch [INV ZOO] An elongated pouch inside the genital sheath in many acanthocephalans. { 'zef·ti·gənz ‚pauch }

SAE number [ENG] A classification of motor, transmission, and differential lubricants to indicate viscosities, standardized by the Society of Automotive Engineers; SAE numbers do not connote quality of the lubricant. { ‚e‚sā'ē ‚nəm·bər }

safe [ORD] Pertaining to ordnance constituted and set so as not to detonate or function accidentally. { saf }

safe burst height [ORD] The height of burst at or above which the level of fallout or damage to ground installations is at a predetermined level acceptable to the military commander. { 'saf ‚bərst ‚hīt }

safelight [GRAPHICS] A darkroom light of a particular color that will not affect the sensitive material being handled; for example, blue-sensitive materials require a red, orange, or yellow safelight. { 'saf‚līt }

safe load [MECH] The stress, usually expressed in tons per square foot, which a soil or foundation can safely support. { 'saf ‚lōd }

safety [ENG] Methods and techniques of avoiding accident or disease. [ORD] A locking or cut-off device that prevents a weapon or missile from being fired accidentally. { 'saf·tē }

safety belt [ENG] A strong strap or harness used to fasten a person to an object, such as the seat of an airplane or automobile. { 'saf·tē ‚belt }

safety block [ORD] A block which, in the safe position, prevents functioning of the fuse by limiting the motion of the firing pin. { 'saf·tē ‚bläk }

safety board [PETRO ENG] A board placed in a derrick for a worker to stand on when handling drill rods at single, double, triple, or quadruple levels; the boards are placed at suitable heights to handle a stand of drill rods for that number of joints. { 'saf·tē ‚bórd }

safety bolt [CIV ENG] A bolt that can be opened from only one side of the door or gate it fastens. { 'saf·tē ‚bōlt }

safety button [NUCLEO] A device worn by workers exposed to nuclear radiation to warn of excessive exposure. { 'saf·tē ‚bət·ən }

safety cable [MIN ENG] A mining machine cable designed to cut off power when the positive conductor insulation is damaged. { 'saf·tē ‚kā·bəl }

safety cage [MIN ENG] A cage, box, or platform used for lowering and hoisting miners, tools, and equipment into and out of mines. { 'saf·tē ‚kāj }

safety can [ENG] A cylindrical metal container used for temporary storage or handling of flammable liquids, such as gasoline, naphtha, and benzine, in buildings not provided with properly constructed storage rooms; these cans are also used to transport such liquids for filling and supply purposes within local areas. { 'saf·tē ‚kan }

safety car [MIN ENG] Any mine car or hoisting cage provided with safety stops, catches, or other precautionary devices. { 'saf·tē ‚kär }

safety chain [MIN ENG] A chain connecting the first and last cars of a trip to prevent separation, if a coupling breaks. { 'saf·tē ‚chān }

safety chuck [DES ENG] Any drill chuck on which the heads of the set screws do not protrude beyond the outer periphery of the chuck. { 'sāf·tē ˌchək }

safety door [MIN ENG] An extra door ready for use in the event of damage to the existing ventilation door or in any emergency, for example, explosion or fire. { 'sāf·tē ˌdȯr }

safety engineer [IND ENG] A person who inspects all possible danger spots in a factory, mine, or other industrial building or plant. { 'sāf·tē ˌen·jə'nir }

safety engineering [IND ENG] The testing and evaluating of equipment and procedures to prevent accidents. { 'sāf·tē ˌen·jə'nir·iŋ }

safety explosive [MATER] An explosive which may be handled safely under ordinary conditions; it requires a powerful detonating force. { 'sāf·tē ik'splō·siv }

safety factor [ELEC] The amount of load, above the normal operating rating, that a device can handle without failure. [MECH] *See* factor of safety. [ORD] **1.** Increase in range or elevation that must be set on a gun so that friendly troops, over whose heads fire is to be delivered, will not be endangered. **2.** Overload factor in design to ensure safe operation. { 'sāf·tē ˌfak·tər }

safety film [GRAPHICS] Films made from cellulose acetate, polyester, and other plastics that are not readily flammable. { 'sāf·tē ˌfilm }

safety flange [DES ENG] A type of flange with tapered sides designed to keep a wheel intact in the event of accidental breakage. { 'sāf·tē ˌflanj }

safety fork [ORD] Metal clip that fits over the collar of the fuse in a land mine and prevents the mine from being set off accidentally; its function is the same as that of a safety pin. { 'sāf·tē ˌfȯrk }

safety fuse [ENG] A train of black powder which is enclosed in cotton, jute yarn, and waterproofing compounds, and which burns at the rate of 2 feet (60 centimeters) per minute; it is used mainly for small-scale blasting. { 'sāf·tē ˌfyüz }

safety gate [MIN ENG] An automatically operated gate at the top of a mine shaft or at landings both to guard the entrance and to prevent falling into the shaft. { 'sāf·tē ˌgāt }

safety glass [MATER] **1.** A glass that resists shattering (such as a glass containing a net of wire or constructed of sheets separated by plastic film). **2.** A glass that has been tempered so that when it shatters, it breaks up into grains instead of jagged fragments. { 'sāf·tē ˌglas }

safety groove [ORD] A groove incorporated in an item or component of ammunition to ensure that any possible failure will occur at a selected location and thereby will be of a less hazardous nature. { 'sāf·tē ˌgrüv }

safety hoist [MECH ENG] A hoisting gear that does not continue running when tension is released. { 'sāf·tē ˌhȯist }

safety hook [DES ENG] A hoisting hook with a spring-loaded latch that prevents the load from accidentally slipping off the hook. { 'sāf·tē ˌhük }

safety lamp [MIN ENG] In coal mining, a lamp that is relatively safe to use in atmospheres which may contain flammable gas. { 'sāf·tē ˌlamp }

safety lanes [NAV] Specified sea lanes designated for use in transit by submarines and surface ships to prevent attack by friendly forces. { 'sāf·tē ˌlānz }

safety level of supply [IND ENG] The quantity of material, in addition to the operating level of supply, required to be on hand to permit continuous operations in the event of minor interruption of normal replenishment or unpredictable fluctuations in demand. { 'sāf·tē 'lev·əl əv sə'plī }

safety lever [ORD] **1.** A metal piece forming part of a grenade fuse that is restrained either by hand or by the projection adapter after the safety pin is removed; following the release of the grenade, the lever is discarded and the fuse train is initiated by the action of the released fuse firing pin. **2.** A lever that sets the safety mechanism on certain types of automatic weapons. { 'sāf·tē ˌlev·ər }

safety lock [ORD] Locking device that prevents a weapon from being fired accidentally. { 'sāf·tē ˌläk }

safety match [ENG] A match that can be ignited only when struck against a specially made friction surface. { 'sāf·tē ˌmach }

safety pin [ORD] **1.** A device designed to fit the mechanism of a fuse in order to prevent accidental arming or functioning of the fuse and to ensure transport safety; the device is removed just before employment of the fuse. **2.** *See* safety wire. { 'sāf·tē ˌpin }

safety pinion [HOROL] A center pinion gear in a watch that protects delicate parts by unscrewing when the mainspring breaks, thereby preventing pressure from being applied to the train wheel teeth and arbor pivots. { 'sāf·tē ˌpin·yən }

safety plug [ENG] A protective device used on a heated pressure vessel (for example, a steam boiler), and containing a fusible element that melts at a predetermined safe temperature to prevent the buildup of excessive pressure. Also known as fusible plug. { 'sāf·tē ˌpləg }

safety post [MIN ENG] A timber placed near the face of workings to protect the workers. Also known as safety prop. { 'sāf·tē ˌpōst }

safety prop *See* safety post. { 'sāf·tē ˌpräp }

safety rail *See* guardrail. { 'sāf·tē ˌrāl }

safety relief valve *See* safety valve. { 'sāf·tē ri'lēf ˌvalv }

safety rod [NUCLEO] A control rod capable of shutting down a reactor quickly in case of failure of the ordinary control system using regulating rods and shim rods; a safety rod may be suspended above the core by a magnetic coupling and allowed to fall in when power reaches a predetermined level. Also known as scram rod. { 'sāf·tē ˌräd }

safety service [COMMUN] Radio communications service used permanently or temporarily for safeguarding human life and property. { 'sāf·tē ˌsər·vəs }

safety shoe [ENG] A special shoe without spark-producing nails or plates, worn by personnel working around explosives. { 'sāf·tē ˌshü }

safety stake [ORD] One of the stakes set in the ground to mark the right or left limit of safe fire of a weapon. { 'sāf·tē ˌstāk }

safety stop [MECH ENG] **1.** On a hoisting apparatus, a device by which the load may be prevented from falling. **2.** An automatic device on a hoisting engine designed to prevent overwinding. { 'sāf·tē ˌstäp }

safety time [IND ENG] The difference between the time when a certain material will be required and the time when the material will actually be in stock. { 'sāf·tē ˌtīm }

safety valve [MECH ENG] A spring-loaded, pressure-actuated valve that allows steam to escape from a boiler at a pressure slightly above the safe working level of the boiler; fitted by law to all boilers. Also known as safety relief valve. { 'sāf·tē ˌvalv }

safety wire [ORD] Wire set into the body of a fuse to lock all movable parts into safe positions so that the fuse will not be set off accidentally; it is pulled out just before firing. Also known as safety pin. { 'sāf·tē ˌwīr }

safe yield [CIV ENG] The maximum dependable draft that can be made continuously upon a source of water supply over a given period of time during which the probable driest period, and therefore period of greatest deficiency in water supply, is likely to occur. { 'sāf 'yēld }

safflorite [MINERAL] CoAs₂ A cobalt arsenide mineral that occurs in tin-white masses, and is dimorphous with smaltite; found in Canada, Morocco, and the United States. { 'saf·ləˌrīt }

safflower [BOT] *Carthamus tinctorius.* An annual thistlelike herb belonging to the composite family (Compositae); the leaves are edible, flowers yield dye, and seeds yield a cooking oil. { 'saˌflau̇·ər }

safflower oil [MATER] Nonyellowing oil derived from safflower seed and similar to linseed oil; contains high proportion of linoleic acid; used as a drying oil and in food and medicine. { 'saˌflau̇·ər ˌȯil }

saffron [BOT] *Crocus sativus.* A crocus of the iris family (Iridaceae); the source of a yellow dye used for coloring food and medicine. { 'saf·rən }

SAFI *See* semiautomatic flight inspection.

safing [ORD] As applied to weapons and ammunition, the changing from a state of readiness for initiation to a safe condition. { 'sāf·iŋ }

safranine [ORG CHEM] Any of a group of phenazine-based dyes; some are used as biological stains. { 'saf·rəˌnēn }

safrole [ORG CHEM] C₃H₅C₆H₃O₂CH₂ A toxic, water-insoluble, colorless oil that boils at 233°C; found in sassafras and camphorwood oils; used in medicine, perfumes, insecticides, and soaps, and as a chemical intermediate. { 'saˌfrōl }

Safflower (*Carthamnus tinctorius*). (*USDA*)

SAGE

Sage (*Salvia officinalis*). (*USDA*)

SAGITTARIUS

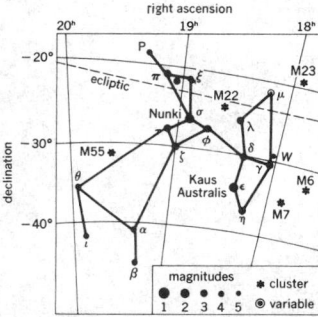

Line pattern of the constellation Sagittarius. The grid lines in the chart represent the coordinates of the sky. The apparent brightness, or magnitude, of the stars is shown by the sizes of the dots, which are graded by appropriate numbers as indicated.

SAINT JOHN'S WORT

Saint John's wort plants.

sag [ELEC] Slack introduced in an aerial cable or open-wire line to compensate for contraction during cold weather. [GEOL] **1.** A pass or gap in a ridge or mountain range shaped like a saddle. **2.** A shallow depression in a relatively flat land surface. **3.** A regional basin with gently sloping sides. [MET] Decrease in the section thickness of a casting caused by weakness of the sand mold. { sag }

sag-and-swell topography [GEOGR] An undulating surface characteristic of till sheets, for example, the landscape of the midwestern United States. { ¦sag ən ¦swel tə¦pag·rə·fē }

Sagartiidae [INV ZOO] A family of zoantharians in the order Actiniaria. { ‚sag·ər'tī·ə‚dē }

sag bolt [MIN ENG] A device to measure roof sag; a 12-foot (3.7-meter) unit installed without a bearing plate and securely anchored with the aid of a heavy nut, extending about 2 inches (5 centimeters) from the hole; three $1/2$-inch (1.3-centimeter) strips of colored tape wrapped around the extending section of the bolt, green at the roof line followed by yellow and then red, help detect roof sag at a glance. { sag ‚bōlt }

sage [BOT] *Salvia officinalis*. A half-shrub of the mint family (Labiatae); the leaves are used as a spice. { sāj }

SAGE [ORD] An air defense system in which air surveillance data are processed for transmission to computers at direction centers, where the data is further processed, evaluated, and analyzed automatically to produce weapon assignment and guidance orders. Derived from semiautomatic ground environment. { sāj }

sagebrush [BOT] Any of various hoary undershrubs of the genus *Artemisia* found on the alkaline plains of the western United States. { 'sāj‚brəsh }

sagenite [MINERAL] A variety of rutile that is acicular and occurs in reticulated twin groups of crystals crossing at 60°. { 'saj·ə‚nīt }

sagenitic [GEOL] Containing acicular minerals. { ‚saj·ə'nid·ik }

Sagenocrinida [PALEON] A large order of extinct, flexible crinoids that occurred from the Silurian to the Permian. { ¦saj·ə·nō'krī·nə·də }

sage oil *See* oil of sage. { 'sāj ‚oil }

saggar clay [MATER] A fire clay of which the case is made that is used for the firing of porcelain and pottery. Also spelled sagger clay. { 'sag·ər ‚klā }

sagger clay *See* saggar clay. { 'sag·ər ‚klā }

sagging [NAV ARCH] Deflection of the hull of a ship in which the middle of the keel is bowed downward. { 'sag·iŋ }

Saghathiinae [PALEON] An extinct subfamily of hyracoids in the family Procaviidae. { ‚sag·ə'thī·ə‚nē }

sagitta [MATH] The distance between the midpoint of an arc and the midpoint of its chord. [VERT ZOO] The larger of two otoliths in the ear of most fishes. { sə'jid·ə }

Sagitta [ASTRON] A small constellation; right ascension 20 hours, declination 10° north. Also known as Arrow. { sə'jid·ə }

sagittal [ZOO] In the median longitudinal plane of the body, or parallel to it. { 'saj·əd·əl }

sagittal coma [OPTICS] The radius of the circle formed in the focal plane by rays from an off-axis point passing near the edge of a lens that displays coma. { ¦saj·ə·təl ¦kō·mə }

sagittal focus *See* secondary focus. { 'saj·əd·əl 'fō·kəs }

sagittal plane [OPTICS] A plane that is perpendicular to the meridional plane of an optical system and contains a specified ray. Also known as equatorial plane. { ¦saj·ə·təl ¦plān }

sagittal surface [OPTICS] A surface containing the secondary foci of points in a plane perpendicular to the optical axis of an astigmatic system. { ¦saj·ə·təl ¦sər·fəs }

Sagittariidae [VERT ZOO] A family of birds in the order Falconiformes comprising a single species, the secretary bird, noted for its nuchal plumes resembling quill pens stuck behind an ear. { ‚saj·ə·tə'rī·ə‚dē }

Sagittarius [ASTRON] A constellation whose major portion lies in the Milky Way; right ascension 19 hours, declination 25° south. Also known as Archer. { ‚saj·ə'ter·ē·əs }

Sagittarius A [ASTRON] An intense radio source in the constellation Sagittarius, apparently comprising a gaseous envelope surrounding a small dense core that is believed to constitute the center of the Milky Way Galaxy. { ‚saj·ə'ter·ē·əs 'ā }

Sagittarius arm [ASTRON] A spiral arm of the Milky Way Galaxy that lies between the sun and the galactic center in the direction of the constellation Sagittarius. { ‚saj·ə'ter·ē·əs 'ärm }

Sagittarius B2 [ASTRON] The richest molecular radio source in the Galaxy, located near the galactic center and consisting of a massive, dense complex of at least seven HII regions and molecular clouds. { ‚saj·ə'ter·ē·əs ¦bē'tü }

Sagittarius star cloud [ASTRON] A large star cloud within the Milky Way; its extension is about 1500 to 6000 light-years (1.42×10^{19} to 5.68×10^{19} meters) from the sun. { ‚saj·ə'ter·ē·əs 'stär ‚klaud }

sagittate [BOT] Shaped like an arrowhead, especially referring to leaves. { 'saj·ə‚tāt }

sagittocyst [INV ZOO] A cyst in the epidermis of certain turbellarians containing a single spindle-shaped needle. { sə'jid·ə‚sist }

Sagnac effect [OPTICS] The shift in interference fringes from two coherent light beams traveling in opposite directions around a ring when the ring is rotated about an axis perpendicular to the ring. { 'sän·yäk i‚fekt }

sago [MATER] A starch obtained from the trunks of certain tropical palms, such as the sago; used as a thickening agent in food and as textile stiffening. { 'sä·gō }

sag pond [GEOL] A small body of water occupying an enclosed depression or sag formed where active or recent fault movement has impounded drainage. Also known as fault-trough lake; rift lake; rift-valley lake. { 'sag ‚pänd }

Saha ionization [STAT MECH] The ionization of a gas which exists when the gas is in thermal equilibrium at a given temperature, in the absence of external influences; it increases with increasing temperature. Also known as thermal ionization. { ‚sä‚hä ‚ī·ə·nə'zā·shən }

Saha's equation [STAT MECH] An equation for Saha ionization of a monatomic gas in terms of the temperature and pressure of the gas, the ionization potential, and statistical weights of ion, electron, and atom. { ‚sä‚häz i‚kwā·zhən }

sahel [ECOL] A region having characteristics of a savanna or a steppe and bordering on a desert. [METEOROL] A strong dust-bearing desert wind in Morocco. { sə'hel }

sahlinite [MINERAL] $Pb_{14}(AsO_4)_2O_9Cl_4$ A pale sulfur-yellow, monoclinic mineral consisting of a basic chloride-arsenate of lead; occurs in aggregates of small scales. { 'sä·lə‚nīt }

sahlite *See* salite. { 'sä‚līt }

sail [NAV ARCH] An article made of canvas and rope designed to be spread on spars in such a manner as to utilize the power of the wind in driving a vessel. { sāl }

Sail *See* Vela. { sāl }

sailboat [NAV ARCH] A boat fitted with one or more sails, and propelled by wind striking the sails. { 'sāl‚bōt }

sailcloth [TEXT] A generic term for heavy, durable fabrics with a strong canvas weave made of cotton, linen, jute, polyester, or nylon; used for sails as well as for other applications. { 'sāl‚klóth }

sailfish [VERT ZOO] Any of several large fishes of the genus *Istiophorus* characterized by a very large dorsal fin that is highest behind its middle. { 'sāl‚fish }

sailing [NAV] A method of solving the various problems involving course, distance, difference of latitude, difference of longitude, and departure; the various methods are collectively termed the sailings. { 'sāl·iŋ }

sailing chart [NAV] A small-scale nautical chart for offshore navigation. { 'sāl·iŋ ‚chärt }

sailing directions *See* coast pilot. { 'sāl·iŋ di‚rek·shənz }

sail plan [NAV ARCH] A plan drawn to show the number, arrangement, and dimensions of the sails for a sailing vessel. { 'sāl ‚plan }

Saint Elmo's fire [ELEC] A visible electric discharge, sometimes seen on the mast of a ship, on metal towers, and on projecting parts of aircraft, due to concentration of the atmospheric electric field at such projecting parts. { 'sānt 'el·mōz 'fīr }

Saint Hilaire method [NAV] The establishing of a line of position by assuming an altitude for a celestial body and then comparing this assumed altitude with that which is measured; the difference between the observed and computed altitudes and the azimuth gives the line of position. Also known as the intercept method. { sānt ē'ler ‚meth·əd }

Saint John's wort [PHARM] *Hypericum perforatum*. A herbacious perennial that has been used for millennia for its many medicinal properties, including wound healing and treatment of kidney and lung ailments, insomnia, and depression. { sānt 'jänz ˌwȯrt }

Saint Joseph retort process [MET] An electrothermic retort process for processing zinc ore and zinc from secondary sources into zinc; heat of reaction between the sintered zinc concentrate and the coke mixture is supplied by passage of heavy electric current through the resistance of the charge. { 'sānt 'jō·səf ri'tȯrt ˌprä·səs }

Saint Louis encephalitis [MED] A mosquito-borne arbovirus infection of the central nervous system, occurring in the central and western United States and in Florida. { sānt 'lü·əs inˌsef·ə'līd·əs }

Saint Peter sandstone [GEOL] An artesian aquifer of early Lower Paleozoic age which underlies part of Minnesota, Wisconsin, Iowa, Illinois, and Indiana. { sānt 'pēd·ər ˌsan·stōn }

Saint Venant's compatibility equations [MECH] Equations for the components e_{ij} of the strain tensor that follow from their integrability, namely, $(e_{ij})_{kl} + (e_{kl})_{ij} - (e_{ik})_{jl} - (e_{jl})_{ik} = 0$, where i, j, k, and l can take on any of the values x, y, and z, and subscripts outside the parentheses indicate partial differentiation. { ˌsän·və'nänz kəmˌpad·ə'bil·əd·ē iˌkwā·shənz }

Saint Venant's principle [MECH] The principle that the strains that result from application, to a small part of a body's surface, of a system of forces that are statically equivalent to zero force and zero torque become negligible at distances which are large compared with the dimensions of the part. { ˌsän·və'nänz 'prin·sə·pəl }

Saint Vitus dance [MED] Chorea associated with rheumatic fever. Also known as Sydenham's chorea. { sānt 'vīd·əs ˌdans }

Sakata-Taketani equation [QUANT MECH] A relativistic wave equation for a particle with spin 1 whose form resembles that of the nonrelativistic Schrödinger equation. { sä'käd·ə ˌtä·kə'tä·nē iˌkwā·zhən }

Sakmarian [GEOL] A European stage of geologic time; the lowermost Permian, above Stephanian of Carboniferous and below Artinskian. { säk'mär·ē·ən }

sal See sial. { sal }

salable coal [MIN ENG] Total output of a coal mine, less the tonnage rejected or consumed during preparation for market. { 'sāl·ə·bəl 'kōl }

sal acetosella See potassium binoxalate. { 'sal ˌas·ə·dō 'sel·ə }

Salado formation [GEOL] A red-bed formation from the Permian found in southeast New Mexico; contains rock salt and potash salts. { sə'lä·dō fȯrˌmā·shən }

salamander [VERT ZOO] The common name for members of the order Urodela. { 'sal·əˌman·dər }

salamander stove [ENG] A small portable stove used for temporary or emergency heat; for example, on construction sites or in greenhouses. { 'sal·əˌman·dər 'stōv }

Salamandridae [VERT ZOO] A family of urodele amphibians in the suborder Salamandroidea characterized by a long row of prevomerine teeth. { ˌsal·ə'man·drəˌdē }

Salamandroidea [VERT ZOO] The largest suborder of the Urodela characterized by teeth on the roof of the mouth posterior to the openings of the nostrils. { ˌsal·ə'man'drȯid·ē·ə }

salammoniac [MINERAL] NH₄Cl A white, isometric, crystalline mineral composed of native ammonium chloride. { ˌsal·ə'mō·nē·ak }

salammoniac cell [ELEC] Cell in which the electrolyte consists primarily of a solution of ammonium chloride. { ˌsal·ə'mō·nē·ak ˌsel }

Salam-Weinberg theory See Weinberg-Salam theory. { sə'läm 'wīnˌbərg ˌthē·ə·rē }

Salangidae [VERT ZOO] A family of soft-rayed fishes, in the suborder Galaxioidei, which live in estuaries of eastern Asia. { sə'lan·jəˌdē }

salazosulfadimidine [ORG CHEM] $C_{19}H_{17}N_5O_5S$ A brown crystalline compound that melts at 207°C; used in medicine in cases of ulcerative colitis. { ˌsal·ä·zōˌsəl·fə'dī·məˌdēn }

salband See selvage. { 'salˌband }

salcrete [GEOL] A thin, hard crust of salt-cemented sand grains, occurring on a marine beach that is occasionally or periodically saturated by saline water. { 'salˌkrēt }

Saldidae [INV ZOO] The shore bugs, a family of predacious hemipteran insects in the superfamily Saldoidea. { 'sal·dəˌdē }

Saldoidea [INV ZOO] A superfamily of the hemipteran group Leptopodoidea. { sal'dȯi·dē·ə }

saléeite [MINERAL] $Mg(UO_2)_2(PO_4)_2 \cdot 10H_2O$ A lemon-yellow mineral composed of hydrous phosphate of magnesium and uranium. { sə'lāˌīt }

Saleniidae [INV ZOO] A family of echinoderms in the order Salenioida distinguished by imperforate tubercles. { ˌsa·lə'nī·ə·dē }

Salenioida [INV ZOO] An order of the Echinacea in which the apical system includes one or several large angular plates covering the periproct. { səˌlē·nē'ȯi·də }

salesite [MINERAL] $Cu(IO_3)(OH)$ A bluish-green mineral composed of basic iodate of copper. { 'salˌzīt }

sal ethyl See ethyl salicylate. { 'sal 'eth·əl }

salfemic rock [GEOL] An igneous rock in which the ratio of salic to femic minerals is greater than 3:5 and less than 5:3. { sal'fē·mik 'räk }

salic [GEOL] A soil horizon enriched with secondary salts, at least 2 percent, and measuring at least 6 inches (15 centimeters) in thickness. [MINERAL] Pertaining to certain light-colored minerals, such as quartz and feldspars, that are rich in silica or magnesium and commonly occur in igneous rock. { 'sal·ik }

Salicaceae [BOT] The single family of the order Salicales. { ˌsal·ə'kās·ē·ē }

Salicales [BOT] A monofamilial order of dicotyledonous plants in the subclass Dilleniidae; members are dioecious, woody plants, with alternate, simple, stipulate leaves and plumose-hairy mature seeds. { ˌsal·ə'kā·lēz }

salicin [ORG CHEM] $C_{13}H_{13}O_7$ A glucoside; colorless crystals, soluble in water, alcohol, alkalies, and glacial acetic acid; melts at 199°C; used in medicine and as an analytical reagent. { 'sal·ə·sən }

salicyl alcohol [ORG CHEM] $C_7H_8O_2$ A crystalline alcohol that forms plates or powder, melting at 86–87°C; used in medicine as a local anesthetic. { 'sal·ə·səl 'al·kəˌhȯl }

salicylaldehyde [ORG CHEM] C_6H_4OHCHO Clear to dark-red oily liquid, with burning taste and almond aroma; soluble in alcohol, benzene, and ether, very slightly soluble in water; boils at 196°C; used in analytical chemistry, in perfumery, and for synthesis of chemicals. Also known as helicin. { ˌsal·ə·səl'al·dəˌhīd }

salicylamide [ORG CHEM] $C_6H_4(OH)CONH_2$ Pinkish or white crystals; soluble in alcohol, ether, chloroform, and hot water; melts at 193°C; used in medicine as an analgesic, antipyretic, and antirheumatic drug. { ˌsal·ə'sī·lə·məd }

salicylanilide [PHARM] $C_{13}H_{11}NO_2$ A crystalline compound with a melting point of 135.8–136.2°C; freely soluble in alcohol, ether, chloroform, and benzene, slightly soluble in water; used as a topical antifungal agent in humans and animals and in medicine. { ˌsal·ə·səl'an·əl·əd }

salicylate [ORG CHEM] A salt of salicylic acid with the formula $C_6H_4(OH)COOM$, where M is a monovalent metal; for example, $NaC_7H_5O_3$, sodium salicylate. { sə'lis·əˌlāt }

salicylated mercury See mercuric salicylate. { sə'lis·əˌlād·əd 'mər·kyə·rē }

salicylic acid [ORG CHEM] $C_6H_4(OH)COOH$ White crystals with sweetish taste; soluble in alcohol, acetone, ether, benzene, and turpentine, slightly soluble in water; discolored by light; melts at 158°C; used as a chemical intermediate and in medicine, dyes, perfumes, and preservatives. { ˌsal·ə'sil·ik 'as·əd }

salicylic acid ethyl ether See ethyl salicylate. { ˌsal·ə'sil·ik 'as·əd 'eth·əl 'ē·thər }

salicylic ether See ethyl salicylate. { ˌsal·ə'sil·ik 'ē·thər }

salicylism [MED] A syndrome produced by excessive doses of salicylates; characterized by dizziness, headache, and nausea. { 'sal·ə·səˌliz·əm }

salient [GEOL] **1.** A landform that projects or extends outward or upward from its surroundings. **2.** An area in which the axial traces of folds are convex toward the outer edge of the folded belt. { 'sāl·yənt }

salient angle [MATH] An interior angle of a polygon that is less than 180°. { 'sāl·yənt ˌaŋ·gəl }

Salientia [VERT ZOO] The equivalent name for Anura. { ˌsā·lē'en·chə }

salient point [MATH] A point at which two branches of a curve with different tangents meet and terminate. { 'sāl·yənt 'pȯint }

Eastern American cottonwood (*Populus deltoides*), showing female catkins with ripening capsules and cottony seeds. (*Photograph by John H. Gerard, from National Audubon Society*)

SALIENT-POLE FIELD WINDING

Salient-pole field winding on rotor of synchronous motor showing the pole core, lower right. (*Allis-Chalmers*)

SALLEN-KEY FILTER

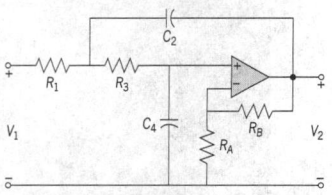

Low-pass Sallen-Key filter.

SALMON

Coho salmon (*Oncorhynchus kisutch*).

SALPIDA

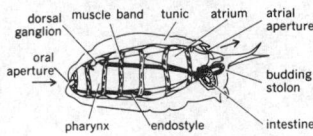

Salp, *Thalia democratica*.

salient pole [ELECTROMAG] A structure of magnetic material on which is mounted a field coil of a generator, motor, or similar device. { 'sāl·yənt 'pōl }

salient-pole field winding [ELEC] A type of field winding in electric machinery where the winding turns are concentrated around the pole core. { 'sāl·yənt ¦pōl 'fēld ¦wīnd·iŋ }

saliferous stratum [GEOL] A stratum that contains, produces, or is impregnated with salt. Also known as saliniferous stratum. { sə'lif·ə·rəs 'strad·əm }

saligenol See salicyl alcohol. { sə'lij·ə,nól }

salimeter [ENG] A hydrometer graduated to read directly the percentage of salt in a solution such as brine. { sə'lim·əd·ər }

salina [ENG] See saltworks. [GEOL] An area, such as a salt flat, in which deposits of crystalline salts are formed or found. [HYD] A body of water containing high concentrations of salt. { sə'lē·nə }

salinastone [GEOL] A sedimentary rock composed mostly of saline minerals which are usually precipitated but may be fragmental. { sə'lē·nə,stōn }

saline-alkali soil [GEOL] A salt-affected soil with a content of exchangeable sodium greater than 15, with much soluble salts, and with a pH value usually less than 9.5. { 'sā,lēn 'al·kə,lī ,sóil }

salinelle [GEOL] A mud volcano erupting saline mud. { ,sa·lə'nel }

saline soil [GEOL] A nonalkali, salt-affected soil with a high content of soluble salts, with exchangeable sodium of less than 15, and with a pH value less than 8.5. { 'sā,lēn ,sóil }

saline-water reclamation [CHEM ENG] Purification and removal of salts from brine or brackish water by ion exchange, crystallization, distillation, evaporation, and reverse osmosis. { 'sā,lēn ¦wód·ər ,rek·lə,mā·shən }

saliniferous stratum See saliferous stratum. { ,sal·ə'nif·ə·rəs 'strad·əm }

salinity [OCEANOGR] The total quantity of dissolved salts in sea water, measured by weight in parts per thousand. { sə'lin·əd·ē }

salinity current [OCEANOGR] A density current in the ocean whose flow is caused, controlled, or maintained by its relatively greater density due to excessive salinity. { sə'lin·əd·ē ,kə·rənt }

salinity logging [PETRO ENG] Technique for measurement and recording of saltwater-bearing zones in an oil or gas reservoir; uses a combination of neutron logging with a chlorine curve. { sə'lin·əd·ē ,läg·iŋ }

salinity-temperature-depth recorder [ENG] An instrument consisting of sensing elements usually lowered from a stationary ship, and a recorder on board which simultaneously records measurements of temperature, salinity, and depth. Also known as CTD recorder; STD recorder. { sə'lin·əd·ē 'tem·prə·chər 'depth ri,kórd·ər }

salinization [GEOL] In a soil of an arid, poorly drained region, the accumulation of soluble salts by the evaporation of the waters that bore them to the soil zone. { ,sal·ən·ə'zā·shən }

salinometer [ENG] An instrument that measures water salinity by means of electrical conductivity or by a hydrometer calibrated to give percentage of salt directly. { ,sal·ə'näm·əd·ər }

salinon [MATH] A plane figure bounded by a semicircle, two smaller semicircles which lie inside the larger semicircle with diameters along the diameter of the larger semicircle and are tangent to the larger semicircle, and another semicircle which is outside the larger. { 'sal·ə,nän }

Salisbury dark box [ELECTR] Isolating chamber used for test work in connection with radar equipment; the walls of the chamber are specially constructed to absorb all impinging microwave energy at a certain frequency. { 'sólz,ber·ē 'därk 'bäks }

salite [MINERAL] $(Mg,Fe)_2Si_2O_6$ A grayish-green to black mineral variety of diopside containing more magnesium than iron; member of the clinopyroxene group. Also spelled sahlite. { 'sa,līt }

salitrite [PETR] A lamprophyre composed chiefly of titanite and diopside with acmite, accessory apatite, microcline, and occasionally anorthoclase and baddeleyite. { 'sal·ə,trīt }

saliva [PHYSIO] The opalescent, tasteless secretions of the oral glands. { sə'lī·və }

salivary amylase See ptyalin. { 'sal·ə,ver·ē 'am·ə,lās }

salivary diastase See ptyalin. { 'sal·ə,ver·ē 'dī·ə,stās }

salivary gland [PHYSIO] A gland that secretes saliva, such as the sublingual or parotid. { 'sal·ə,ver·ē ,gland }

salivary gland chromosomes [CELL MOL] Polytene chromosomes found in the interphase nuclei of salivary glands in the larvae of Diptera; chromosomes in the larva undergo complete somatic pairing to form two homologous polytene chromosomes fused side by side. { 'sal·ə,ver·ē ,gland 'krō·mə,sōmz }

salivation [MED] Mild mercury poisoning suffered by workers in amalgamation plants. [PHYSIO] Excessive secretion of saliva. { ,sal·ə'vā·shən }

Salk vaccine [IMMUNOL] A killed-virus vaccine administered for active immunization against poliomyelitis. { 'sók vak,sēn }

Sallen-Key filter [ELECTR] An electric filter that uses a single amplifier of positive low gain, realized by an operational amplifier and two feedback resistors. { ¦sal·ən 'kē ,fil·tər }

sallying ship [NAV ARCH] Producing rolling motion of a vessel by running a group of people in unison from side to side. { 'sal·ē·iŋ ,ship }

sally port [ORD] Large gate or passage in a fortified place. { 'sal·ē ,pórt }

salmon [VERT ZOO] The common name for a number of fish in the family Salmonidae which live in coastal waters of the North Atlantic and North Pacific and breed in rivers tributary to the oceans. { 'sam·ən }

Salmonella [MICROBIO] A genus of gram-negative, facultatively anaerobic bacteria belonging to the family Enterobacteriaceae that cause enteric infections with or without blood invasion. Most species are motile, utilize citrate, decarboxylate ornithine, form gas from glucose, and produce hydrogen sulfide. Salmonellae do not ferment lactose, produce indole, or split urea; the Voges-Proskauer reaction is negative. { ,sal·mə'nel·ə }

Salmonelleae [MICROBIO] Formerly a tribe of the Enterobacteriaceae comprising the pathogenic genera *Salmonella* and *Shigella*. { ,sal·mə'nel·ē,ē }

salmonellosis [MED] Infection with any species of *Salmonella*. { ,sal·mə·ne'lō·səs }

Salmonidae [VERT ZOO] A family of soft-rayed fishes in the suborder Salmonoidei including the trouts, salmons, whitefishes, and graylings. { sal'män·ə,dē }

Salmoniformes [VERT ZOO] An order of soft-rayed fishes comprising salmon and their allies; the stem group from which most higher teleostean fishes evolved. { ,sal,män·ə'fòr,mēz }

Salmonoidei [VERT ZOO] A suborder of the Salmoniformes comprising forms having an adipose fin. { ,sal·mə'nóid·ē,ī }

salmon oil [MATER] Combustible, pale golden-yellow liquid with sweet taste; soluble in alcohol, ether, chloroform, and carbon disulfide; obtained from the waste in the canning of salmon; used in pet foods, soaps, and leather dressing. { 'sam·ən ,óil }

salmonsite [MINERAL] A buff-colored mineral composed of hydrous phosphate of manganese and iron occurring in cleavable masses. { 'sam·ən,zīt }

Salmopercae [VERT ZOO] An equivalent name for Percopsiformes. { ,sal·mō'pər,sē }

salol [ORG CHEM] $C_6H_4OHCOOC_6H_5$ White powder with aromatic taste and aroma; soluble in alcohol, ether, chloroform, and benzene; slightly soluble in water; melts at 42°C; used in medicinals and as a preservative. Also known as phenyl salicylate. { 'sa,lól }

salpeter process See three-alpha process. { sal'pēd·ər ,präs·əs }

Salpida [INV ZOO] An order of tunicates in the class Thaliacea including transparent forms ringed by muscular bands. { 'sal·pə·də }

Salpingidae [INV ZOO] The narrow-waisted bark beetles, a family of coleopteran insects in the superfamily Tenebrionoidea. { sal'pin·jə,dē }

salpingitis [MED] 1. Inflammation of the fallopian tube. 2. Inflammation of the eustachian tube. { ,sal·pən'jīd·əs }

salpingo-oophoritis [MED] Inflammation of the fallopian tubes and ovaries. { sal¦piŋ·gō,ō·ə·fə'rīd·əs }

sal prunella See niter balls. { 'sal prü'nel·ə }

sal soda [INORG CHEM] $Na_2CO_3 \cdot 10H_2O$ White, water-soluble crystals, insoluble in alcohol; melts and loses water at about 33°C; mild irritant to mucous membrane; used in cleansers and for washing textiles and bleaching linen and cotton. Also known as sodium carbonate decahydrate; washing soda. { 'sal ˌsō·də }

salsoline [PHARM] $C_{11}H_{15}NO_2$ A compound that crystallizes from alcohol solution, melts at 221°C, soluble in hot alcohol and chloroform; used in medicine as an antihypertensive agent. { 'sal·sə,lēn }

salt [CHEM] The reaction product when a metal displaces the hydrogen of an acid; for example, $H_2SO_4 + 2NaOH \rightarrow Na_2SO_4$ (a salt) $+ 2H_2O$. [ENG] To add an accelerator or retardant to cement. [MIN ENG] **1.** To introduce extra amounts of a valuable or waste mineral into a sample to be assayed. **2.** To artificially enrich, as a mine, usually with fraudulent intent. { sȯlt }

salt-affected soil [GEOL] A general term for a soil that is not suitable for the growth of crops because of an excess of salts, exchangeable sodium, or both. { 'sȯlt i¦fek·təd 'sȯil }

salt-and-pepper sand [GEOL] A sand composed of a mixture of light- and dark-colored grains. { 'sȯlt ən ¦pep·ər 'sand }

salt anticline [GEOL] A structure like a salt dome but with a linear salt core. Also known as salt wall. { 'sȯlt 'ant·i,klīn }

sal tartar See sodium tartrate. { 'sal 'tär·tər }

saltation [GEOL] Transport of a sediment in which the particles are moved forward in a series of short intermittent bounces from a bottom surface. { sȯl'tā·shən }

saltation load [GEOL] The part of the bed load that is bouncing along the stream bed or is moved, directly or indirectly, by the impact of bouncing particles. { sȯl'tā·shən ˌlōd }

saltatorial [ZOO] Adapted for leaping. { ¦sal·tə¦tȯr·ē·əl }

salt bath [MET] Molten salts in which steel is heated for hardening and tempering. { 'sȯlt ,bath }

salt bottom [GEOL] A flat piece of relatively low-lying ground encrusted with salt. { 'sȯlt ,bäd·əm }

saltbox [ARCH] A type of house built in colonial New England, having a frame structure with two stories in the front and one in the rear and a double-sloping roof. { 'sȯlt,bäks }

salt bridge [PHYS CHEM] A bridge of a salt solution, usually potassium chloride, placed between the two half-cells of a galvanic cell, either to reduce to a minimum the potential of the liquid junction between the solutions of the two half-cells or to isolate a solution under study from a reference half-cell and prevent chemical precipitations. { 'sȯlt ,brij }

salt burst [GEOL] Rock destruction caused by crystallization of soluble salts that enter the pores. { 'sȯlt ,bərst }

salt cake [INORG CHEM] Impure sodium sulfate; used in soaps, paper pulping, detergents, glass, ceramic glaze, and dyes. { 'sȯlt ,kāk }

salt crust [HYD] A salt deposit formed on an ice surface by crystal growth forcing salt out of young sea ice and pushing it upward. { 'sȯlt ,krəst }

salt dome [GEOL] A diapiric or piercement structure in which there is a central, equidimensional salt plug. { 'sȯlt ,dōm }

salt-dome breccia [GEOL] A breccia found in deep shale sequences and occurring as a dome-shaped mass in a broad zone surrounding a salt plug. { 'sȯlt ¦dōm 'brech·ə }

salted weapon [ORD] A nuclear weapon which has, in addition to its normal components, certain elements or isotopes which capture neutrons at the time of the explosion and produce radioactive weapon debris. { 'sȯl·təd 'wep·ən }

salt-effect distillation [CHEM ENG] A process of extractive distillation in which a salt that is soluble in the liquid phase of the system being separated is used as a separating agent. { 'sȯlt i,fekt ,dis·tə¦lā·shən }

saltern See salt garden; saltworks. { 'sȯl·tərn }

salt error [ANALY CHEM] An error introduced in an analytical determination of a saline liquid such as sea water; caused by the effect of the neutral ions in the solution on the color of the pH indicator, and hence upon the apparent pH. { 'sȯlt ,er·ər }

salt field [GEOL] An area overlying a usually workable salt deposit of economic value. { 'sȯlt ,fēld }

salt fingers [CHEM] A close-packed array of rising and sinking columns of fluid that form in a liquid when a slow-diffusing solute is separated by an interface from another, lower solute that diffuses more rapidly. { 'sȯlt ,fiŋ·gərz }

salt flat [GEOL] The level, salt-encrusted bottom of a lake or pond that is temporarily or permanently dried up. { 'sȯlt ,flat }

salt flowers See ice flowers. { 'sȯlt ,flaů·ərz }

salt-fog test [MET] An accelerated corrosion test in which a piece of metal is subjected to a fine spray of sodium chloride solution. Also known as salt-spray test. { 'sȯlt ¦fäg ,test }

salt garden [ENG] A large, shallow basin or pond where sea water is evaporated by solar heat. Also known as saltern. { 'sȯlt ,gärd·ən }

salt glacier [GEOL] A gravitational flow of salt down the slopes of a salt plug, following the preexisting structure. { 'sȯlt ,glā·shər }

salt gland [VERT ZOO] A compound tubular gland, located around the eyes and nasal passages in certain marine turtles, snakes, and birds, which copiously secretes a watery fluid containing a high percentage of salt. { 'sȯlt ,gland }

salt glaze [ENG] Glaze formed on the surface of stoneware by putting salt into the kiln during firing. { 'sȯlt ,glāz }

salt-gradient solar pond See solar pond. { 'sȯlt ¦grād·ē·ənt ¦sō·lər 'pänd }

salt grainer [CHEM ENG] Type of evaporative crystallizer in which the solution is kept hot, and supersaturation is developed by evaporation rather than by cooling. { 'sȯlt ,grān·ər }

salt haze [METEOROL] A haze created by the presence of finely divided particles of sea salt in the air, usually derived from the evaporation of sea spray. { 'sȯlt ,hāz }

salt hill [GEOL] An abrupt hill of salt, with sinkholes and pinnacles at its summit. { 'sȯlt ,hil }

Salticidea [INV ZOO] The jumping spiders, a family of predacious arachnids in the suborder Dipneumonomorphae having keen vision and rapid movements. { ,sal·tə'sid·ē·ə }

saltierra [GEOL] A deposit of salt left by evaporation of a shallow salt lake. { ,sal·tē'er·ə }

salting-out effect [CHEM ENG] The growth of crystals of a substance on heated, liquid-holding surfaces of a crystallizing evaporator as a result of the decrease in solubility of the substance with increase in temperature. { 'sȯlt·iŋ ¦aůt i,fekt }

salt lake [HYD] A confined inland body of water having a high concentration of salts, principally sodium chloride. { 'sȯlt ,lāk }

saltmarsh [ECOL] A maritime habitat found in temperate regions, but typically associated with tropical and subtropical mangrove swamps, in which excess sodium chloride is the predominant environmental feature. { 'sȯlt,märsh }

saltmarsh plain [ECOL] A salt marsh that has been raised above the level of the highest tide and has become dry land. { 'sȯlt,märsh ,plān }

salt mine [MIN ENG] A mine containing deposits of rock salt. { 'sȯlt ,mīn }

salt-mud combination log [PETRO ENG] Record of electrical logging of the mud in oil well boreholes in the presence of sodium chloride incursions from adjacent formations. { 'sȯlt 'məd ,käm·bə,nā·shən 'läg }

salt of Lemery See potassium sulfate. { 'sȯlt əv lem'rē }

salt of sorrel See potassium binoxalate. { 'sȯlt əv 'sär·əl }

salt of tartar See potassium carbonate. { 'sȯlt əv 'tär·tər }

salt pan [CHEM] A pool used for obtaining salt by the natural evaporation of sea water. [GEOL] **1.** An undrained, usually small and shallow, natural depression or hollow in which water accumulates and evaporates, leaving a salt deposit. **2.** A shallow lake of brackish water occupying such a depression. { 'sȯlt ,pan }

saltpeter See potassium nitrate. { 'sȯlt'pēd·ər }

saltpeter cave [GEOL] A cave in which there are deposits of saltpeter earth. { 'sȯlt 'pēd·ər ,kāv }

saltpeter earth [GEOL] A deposit containing calcium nitrate and found in caves. { 'sȯlt'pēd·ər ,ərth }

salt pillow [GEOL] An embryonic salt dome rising from its source bed, still at depth. { 'sȯlt ,pil·ō }

salt pit [GEOL] A pit in which sea water is received and evaporated and from which salt is obtained. { 'sȯlt ,pit }

salt plug [GEOL] The salt core of a salt dome. { 'sȯlt ,pləg }

salt polygon [GEOL] A surface of salt on a playa, having three to eight sides marked by ridges of material formed as a result of the expansive forces of crystalizing salt, and ranging in width from an inch or so to 100 feet (30 meters). { 'sȯlt 'päl·i,gän }

salt screen [GRAPHICS] An intensifying screen consisting of a chemical salt which fluoresces when bombarded by x-rays. { 'sȯlt ,skrēn }

salt-spray climax [ECOL] A climax community along exposed Atlantic and Gulf seacoasts composed of plants able to tolerate the harmful effects of salt picked up and carried by onshore winds from seawater. { 'sȯlt ¦sprā 'klī,maks }

salt-spray test See salt-fog test. { 'sȯlt ¦sprā ,test }

salt stock [GEOL] An immature salt dome comprising a pluglike salt diapir that has pierced the overlying strata. { 'sȯlt ,stäk }

salt tectonics [GEOL] The study of the structure and mechanism of emplacement of salt domes. Also known as halokinesis. { 'sȯlt tek'tän·iks }

saltus See oscillation. { 'sal·təs }

salt velocity meter [ENG] A rate-of-flow volume meter used to find the transit time of passage between two fixed points of a small quantity of salt or radioactive isotope in a flowing stream by measuring electrical conductivity or radiation level at those points. { 'sȯlt və'läs·əd·ē ,mēd·ər }

salt wall See salt anticline. { 'sȯlt ,wȯl }

salt water See seawater. { 'sȯlt ¦wȯd·ər }

salt-water front [OCEANOGR] The interface between fresh and salt water in a coastal aquifer or in an estuary. { 'sȯlt ¦wȯd·ər ,frənt }

salt-water intrusion [HYD] Displacement of fresh surface water or groundwater by salt water due to its greater density. { 'sȯlt ¦wȯd·ər in,trü·zhən }

salt-water underrun [OCEANOGR] A type of density current occurring in a tidal estuary, due to the greater salinity of the bottom water. { 'sȯlt ¦wȯd·ər 'ən·də,rən }

salt-water wedge [OCEANOGR] A wedge-shaped intrusion of salty ocean water into a fresh-water estuary or tidal river; it slopes downward in the upstream direction, and salinity increases with depth. { 'sȯlt ¦wȯd·ər ,wej }

salt-water well [PETRO ENG] A well from which salt water flows after the petroleum contents are depleted. { 'sȯlt ¦wȯd·ər ,wel }

salt weathering [GEOL] The granular disintegration or fragmentation of rock material produced by saline solutions or by salt-crystal growth. { 'sȯlt ,weth·ə·riŋ }

salt well [ENG] A bored or driven well from which brine is obtained. { 'sȯlt ,wel }

saltworks [ENG] A building or group of buildings where salt is produced commercially, as by extraction from sea water or from the brine of salt springs. Also known as salina; saltern. { 'sȯlt,wərks }

saluting gun [ORD] Cannon used for firing salutes. { sə'lüd·iŋ ,gən }

salvage procedure [ENG] The recovery, evacuation, and reclamation of damaged, discarded, condemned, or abandoned material, ships, craft, and floating equipment for reuse, repair, refabrication, or scrapping. { 'sal·vij prə,sē·jər }

salvage value [ENG] **1.** The cost that could be recovered from the sale of used equipment when removed or scrapped. **2.** The actual market value of a specific facility or equipment at a particular point in time. { 'sal·vij ,val·yü }

salvage vessel [NAV ARCH] A ship equipped to rescue or save ships or material that has been sunk, wrecked, damaged, or discarded. { 'sal·vij ,ves·əl }

salvia [MATER] The dried leaves of the sage, *Salvia officinalis;* contains volatile oil, resin, and tannin; used in food engineering as a flavoring agent and condiment, and in medicine as an antisecretory agent. { 'sal·vē·ə }

salvia oil See oil of sage. { 'sal·vē·ə ,ȯil }

Salviniales [BOT] A small order of heterosporous, leptosporangiate ferns (division Polypodiophyta) which float on the surface of the water. { ,sal,vin·ē'ā,lēz }

salvo [ORD] **1.** A simultaneous, or nearly simultaneous, discharge of shots from two or more closely placed guns or launchers against the same target. **2.** The release of several bombs or rocket missiles simultaneously, or in close train, from one or more aircraft, at a single target. **3.** The aggregate of shots, bombs, or rockets so discharged or released. { 'sal·vō }

salvor [NAV ARCH] A person who rescues maritime vessels and cargo from loss or destruction. { 'sal·vȯr }

SAM See surface-to-air missile. { sam }

samara [BOT] A dry, indehiscent, winged fruit usually containing a single seed, such as sugar maple (*Acer saccharum*). { sə'mar·ə }

samarium [CHEM] A rare-earth metal, atomic number 62,

symbol Sm; melts at 1350°C, tarnishes in air, ignites at 200–400°C. { sə'mar·ē·əm }

samarium-cobalt magnet [ELECTROMAG] A rare-earth permanent magnet that is more efficient, has lower leakage and greater resistance to demagnetization, and can be magnetized to higher levels than conventional permanent magnets. { sə'mar·ē·əm 'kō,bȯlt 'mag·nət }

samarium oxide [INORG CHEM] Sm_2O_3 A cream-colored powder with a melting point of 2300°C; soluble in acids; used for infrared-absorbing glass and as a neutron absorber. { sə'mar·ē·əm 'äk,sīd }

samarskite [MINERAL] $(Y,Ce,U,Ca,Fe,Pb,Th)(Nb,Ta,Ti,Sn)_2O_6$ A velvet-black to brown metamict orthorhombic mineral with splendent vitreous to resinous luster occurring in granite pegmatites. Also known as ampangabeite; uranotantalite. { sə'mär,skīt }

Sambonidae [INV ZOO] A family of pentastomid arthropods in the suborder Porocephaloidea of the order Porocephalida. { sam'bän·ə,dē }

sample [SCI TECH] Representative fraction of material tested or analyzed in order to determine the nature, composition, and percentage of specified constituents, and possibly their reactivity. [STAT] A selection of a certain collection from a larger collection. { 'sam·pəl }

sample-and-hold circuit [ELECTR] A circuit that measures an input signal at a series of definite times, and whose output remains constant at a value corresponding to the most recent measurement until the next measurement is made. { ¦sam·pəl ən 'hōld ,sər·kət }

sample correlation coefficient [STAT] The ratio of the sample covariance of x and y to the standard deviation of x times the standard deviation of y. Also known as product-moment coefficient. { ¦sam·pəl ,kä·rə'lā·shən ,kō·ə,fish·ənt }

sampled data [STAT] Data that are obtained at discrete rather than continuous intervals. { 'sam·pəld 'dad·ə }

sampled-data control system [CONT SYS] A form of control system in which the signal appears at one or more points in the system as a sequence of pulses or numbers usually equally spaced in time. { 'sam·pəld ¦dad·ə kən'trōl ,sis·təm }

sample design [STAT] A procedure or plan drawn up before any data are collected to obtain a sample from a given population. Also known as sampling plan; survey design. { 'sam·pəl di,zīn }

sampled grade [MIN ENG] The amount of valuable metal in the ore in place as determined by underground, surface, or drill-hole sampling. { 'sam·pəld 'grād }

sample function [STAT] A function or procedure which, when applied repeatedly to a given population, produces a collection of samples. { 'sam·pəl ,fəŋk·shən }

sampleite [MINERAL] $NaCaCu_5(PO_4)_4Cl·5H_2O$ A blue mineral composed of hydrous phosphate and chloride of sodium, calcium, and copper. { 'sam·pə,līt }

sample log [ENG] Record of core samples or drill cuttings; gives geological, visual, and hydrocarbon-content record versus depth of drilling. { 'sam·pəl ,läg }

sample path [MATH] If $\{X_t: t \text{ in } T\}$ is a stochastic process, a sample path for the process is the function on T to the range of the process which assigns to each t the value $X_t(w)$, where w is a previously given fixed point in the domain of the process. { 'sam·pəl ,path }

sampler [CONT SYS] A device, used in sampled-data control systems, whose output is a series of impulses at regular intervals in time; the height of each impulse equals the value of the continuous input signal at the instant of the impulse. [ENG] A mechanical or other device designed to obtain small samples of materials for analysis; used in biology, chemistry, and geology. { 'sam·plər }

sample size [STAT] The number of objects in the sample. { 'sam·pəl ,sīz }

sample space [STAT] A concept in probability theory which considers all possible outcomes of an experiment, game, and so on, as points in a space. { 'sam·pəl ,spās }

sample splitter [ENG] An instrument, generally constructed of acrylic resin, designed to subdivide a total sample of marine plankton while maintaining a quantitatively correct relationship between the various phyla in the sample. { 'sam·pəl ,splid·ər }

sample survey [STAT] A survey of a population made by using only a portion of the population. { 'sam·pəl ,sər·vā }

sampling [ENG] Process of obtaining a sequence of instantaneous values of a wave. [SCI TECH] The obtaining of small representative quantities of materials (gas, liquid, solid) for the purpose of analysis. [STAT] A drawing of a collection from a given population. { 'sam·pliŋ }

sampling area ratio [MIN ENG] The volume of the soil displaced in proportion to the volume of the sample; a well-designed tool has an area ratio of about 20%. { 'sam·pliŋ ‚er·ē·ə ‚rā·shō }

sampling bottle [ENG] A cylindrical container, usually closed at a chosen depth, to trap a water sample and transport it to the surface without introducing contamination. { 'sam·pliŋ ‚bäd·əl }

sampling distribution [STAT] A distribution of the estimates that can be made by each of all possible samples of a fixed size that could be taken from a universe. { 'samp·liŋ ‚dis·trə‚byü·shən }

sampling error [STAT] That portion of the difference between the value of a statistic derived from observations and the value that it is supposed to estimate; attributed to the fact that samples represent only a portion of a population. { 'samp·liŋ ‚er·ər }

sampling fraction [STAT] The ratio of the sample size to the population size. { 'sam·pliŋ ‚frak·shən }

sampling gate [ELECTR] A gate circuit that extracts information from the input waveform only when activated by a selector pulse. { 'sam·pliŋ ‚gāt }

sampling interval [CONT SYS] The time between successive sampling pulses in a sampled-data control system. { 'sam·pliŋ ‚in·tər·vəl }

sampling pipe [MIN ENG] A small pipe built into a seal to take air samples in a sealed area. { 'sam·pliŋ ‚pīp }

sampling plan [IND ENG] A plan stating sample sizes and the criteria for accepting or rejecting items or taking another sample during inspection of a group of items. [STAT] See sample design. { 'sam·pliŋ ‚plan }

sampling probe [ENG] A leak-testing probe which collects tracer gas from the test area of an object under pressure and feeds it to the leak detector at reduced pressure. { 'sam·pliŋ ‚prōb }

sampling process [ENG] The process of obtaining a sequence of instantaneous values of some quantity that varies continuously with time. { 'sam·pliŋ ‚präs·əs }

sampling rate [ENG] The rate at which measurements of physical quantities are made; for example, if it is desired to calculate the velocity of a missile and its position is measured each millisecond, then the sampling rate is 1000 measurements per second. { 'sam·pliŋ ‚rāt }

sampling risk [IND ENG] In inspection procedure, the probability, under the sampling plan used, that acceptable material will be rejected or that unsatisfactory material will be accepted. { 'sam·pliŋ ‚risk }

sampling spark chamber [NUCLEO] A spark chamber that yields as output data the coordinates of a single point on the track (or tracks) in each gap; all narrow-gap chambers are of this type. { 'sam·pliŋ 'spärk ‚chām·bər }

sampling spoon [MIN ENG] A cylinder with a spoonlike cutting edge for taking soil samples. { 'sam·pliŋ ‚spün }

sampling switch See commutator switch. { 'sam·pliŋ ‚swich }

sampling synthesis [ENG ACOUS] Any method of synthesizing musical tones that is based on playing back digitally recorded sounds. { 'sam·pliŋ ‚sin·thə·səs }

sampling techniques [STAT] The methods used in drawing samples from a population usually in such a manner that the sample will facilitate determination of some hypothesis concerning the population. { 'sam·pliŋ tek‚nēks }

sampling theorem [COMMUN] The theorem that a signal that varies continuously with time is completely determined by its values at an infinite sequence of equally spaced times if the frequency of these sampling times is greater than twice the highest frequency component of the signal. Also known as Shannon's sampling theorem. { 'sam·pliŋ ‚thir·əm }

sampling theory [STAT] The mathematical study of sampling techniques. { 'sam·pliŋ ‚thē·ə·rē }

sampling time [ENG] The time between successive measurements of a physical quantity. { 'sam·pliŋ ‚tīm }

sampling voltmeter [ENG] A special type of voltmeter that detects the instantaneous value of an input signal at prescribed times by means of an electronic switch connecting the signal to a memory capacitor; it is particularly effective in detecting high-frequency signals (up to 12 gigahertz) or signals mixed with noise. { 'sam·pliŋ 'vōlt‚mēd·ər }

samsonite [MINERAL] $Ag_4MnSb_2S_6$ A black mineral composed of sulfide of silver, manganese, and antimony occurring in monoclinic prismatic crystals. { 'sam·sə‚nīt }

samson post See king post. { 'sam·sən ‚pōst }

Samythinae [INV ZOO] A subfamily of sedentary polychaete annelids in the family Ampharetidae having a conspicuous dorsal membrane. { sə'mith·ə‚nē }

SAN See styrene-acrylonitrile resin.

sanakite [PETR] A glassy andesite composed of bronzite, augite, magnetite, and a few large plagioclase and garnet crystals. { 'san·ə‚kīt }

sanatron circuit [ELECTR] A variable time-delay circuit having two pentodes and two diodes, used to produce very short gate waveforms having time durations that vary linearly with a reference voltage. { 'san·ə ‚trän ‚sər·kət }

sanbornite [MINERAL] $BaSi_2O_5$ A white triclinic mineral composed of barium silicate. { 'san‚bȯr‚nīt }

sand [GEOL] Unconsolidated granular material consisting of mineral, rock, or biological fragments between 63 micrometers and 2 millimeters in diameter; usually produced primarily by the chemical or mechanical breakdown of older source rocks, but may also be formed by the direct chemical precipitation of mineral grains or by biological processes. { sand }

Sandalidae [INV ZOO] The equivalent name for Rhipiceridae. { san'dal·ə‚dē }

sandal oil See sandalwood oil. { 'san·dəl ‚ȯil }

sandalwood [BOT] **1.** Any species of the genus *Santalum* of the sandalwood family (Santalaceae) characterized by a fragrant wood. **2.** *S. album.* A parasitic tree with hard, close-grained, aromatic heartwood used in ornamental carving and cabinetwork. { 'san·dəl‚wu̇d }

sandalwood oil [MATER] Pale-yellow essential oil with harsh taste and faint aromatic scent; soluble in fixed oils; insoluble in glycerin; derived from the sandalwood *Santalum album;* used in medicine, perfumes, and flavors. Also known as East Indian sandalwood oil; sandal oil; santal oil; santalwood oil. { 'san·dəl‚wu̇d ‚ȯil }

sand apron [GEOL] A deposit of sand along the shore of a lagoon of a reef. { 'sand ‚ā·prən }

sandarac See realgar. { 'san·də‚rak }

sandarac gum [MATER] Yellow, brittle, water-insoluble, natural resin obtained from the African sandarac tree of Morocco; used in varnishes and lacquers. { 'san·də‚rak ‚gəm }

sand auger See dust whirl. { 'sand ‚ȯg·ər }

sand avalanche [GEOL] Movement of large masses of sand down a dune face when the angle of repose is exceeded or when the dune is disturbed. { 'sand ‚av·ə‚lanch }

sandbag [ENG] A bag filled with sand; used to build temporary protective walls. [GEOL] In the roof of a coal seam, a deposit of glacial debris formed by scour and fill subsequent to coal formation. { 'san‚bag }

sandbank [GEOL] A deposit of sand forming a mound, hillside, bar, or shoal. { 'san‚baŋk }

sandbar [GEOL] A bar or low ridge of sand bordering the shore and built up, or near, to the surface of the water by currents or wave action. Also known as sand reef. { 'san‚bär }

sandblasting [ENG] Surface treatment in which steel grit, sand, or other abrasive material is blown against an object to produce a roughened surface or to remove dirt, rust, and scale. [GEOL] Abrasion affected by the action of hard, windblown mineral grains. { 'san‚blast·iŋ }

sandblow [ECOL] A patch of coarse, sandy soil denuded of vegetation by wind action. { 'san‚blō }

sand boil See blowout. { 'san ‚bȯil }

sand-cast [MET] Made by pouring molten metal into a mold made of sand. { 'san‚kast }

sand cay See sandkey. { 'san ‚kē }

sand cone [GEOL] **1.** A cone-shaped deposit of sand, produced especially in an alluvial cone. **2.** A low debris cone whose protective veneer consists of sand. { 'san ‚kōn }

sand control [MET] A process to regulate the properties of foundry sand to produce defect-free castings. { 'san kən‚trōl }

SANDALWOOD

Sandalwood branch with foliage and fruit.

sand count [PETRO ENG] Determination of the total thickness of an oil or gas reservoir's permeable section (excluding shale streaks and other impermeable zones); can be derived from electrical logs. { 'san ,kaùnt }

sand crystal [GEOL] A large crystal loaded up to 60% with detrital sand inclusions formed in a sandstone during or as a result of cementation. { 'san ,krist·əl }

sand devil See dust whirl. { 'san ,dev·əl }

sand dike [GEOL] A sedimentary dike consisting of sand that has been squeezed or injected upward into a fissure. { 'san ,dīk }

SAND DOLLAR

The nearly circular body may reach 3 inches (7.5 centimeters) in diameter. The characteristic pattern of a five-part flower is seen on the a boral surface.

sand dollar [INV ZOO] The common name for the flat, disk-shaped echinoderms belonging to the order Clypeasteroida. { 'san ,däl·ər }

sand drain [CIV ENG] A vertical boring through a clay or silty soil filled with sand or gravel to facilitate drainage. { 'san ,drān }

sand drift [GEOL] **1.** Movement of windblown sand along the surface of a desert or shore. **2.** An accumulation of sand against the leeward side of a fixed obstruction. { 'san ,drift }

sand drip [GEOL] A rounded or crescentic surface form on a beach sand, resulting from the sudden absorption of overwash. { 'san ,drip }

sand dune [GEOL] A mound of loose windblown sand commonly found along low-lying seashores above high-tide level. { 'san ,dün }

sander [MECH ENG] **1.** An electric machine used to sand the surface of wood, metal, or other material. **2.** A device attached to a locomotive or electric rail car which sands the rails to increase friction on the driving wheels. { 'san·dər }

sandfall See slip face. { 'san,fòl }

sand filter [CIV ENG] A filter consisting of graded layers of sand and aggregate for purifying domestic water. { 'san ,fil·tər }

sand finish [ENG] A smooth finish on a plaster surface made by rubbing the sand or mortar coat. { 'san ,fin·ish }

sand flat [GEOL] A sandy tidal flat barren of vegetation. { 'san ,flat }

sand flood [GEOL] A vast body of sand moving or borne along a desert, as in the Arabian deserts. { 'san ,fləd }

sandfly [INV ZOO] Any of various small biting Diptera, especially of the genus *Phlebotomus,* which are vectors for phlebotomus (sandfly) fever. { 'san,flī }

sandfly fever See phlebotomus fever. { 'san,flī ,fē·vər }

sand gall See sand pipe. { 'san ,gòl }

sand glacier [GEOL] **1.** An accumulation of sand that is blown up the side of a hill or mountain and through a pass or saddle, and then spread out on the opposite side to form a wide, fan-shaped plain. **2.** A horizontal plateau of sand terminated by a steep talus slope. { 'san ,glā·shər }

sand-grain volume [PETRO ENG] In oil reservoir porosity calculations, the actual volume filled by sand grains, without allowance for spaces (voids) between the grains. { 'san ¦grān ,väl·yəm }

sand heap analogy See sand hill analogy. { 'sand ,hēp ə,nal·ə·jē }

sand hill [GEOL] A ridge of sand, especially a sand dune in a desert region. { 'san ,hil }

sand hill analogy [MECH] A formal identity between the differential equation and boundary conditions for a stress function for torsion of a perfectly plastic prismatic bar, and those for the height of the surface of a granular material, such as dry sand, which has a constant angle of rest. Also known as sand heap analogy. { 'sand ,hil ə,nal·ə·jē }

sandhog [ENG] A worker in compressed-air environments, as in driving tunnels by means of pneumatic caissons. { 'san,häg }

sand hole [GEOL] A small pit (7–8 millimeters in depth and a little less wide than deep) with a raised margin, formed on a beach by waves expelling air from a formerly saturated mass of sand. { 'san ,hōl }

sand hopper [INV ZOO] The common name for gammaridean crustaceans found on beaches. { 'san ,häp·ər }

sand horn [GEOL] A pointed sand deposit extending from the shore into shallow water. { 'san ,hòrn }

sanding [ENG] **1.** Covering or mixing with sand. **2.** Smoothing a surface with sandpaper or other abrasive paper or cloth. { 'sand·iŋ }

sandkey [GEOL] A small sandy island parallel with the shore. Also known as sand cay. { 'san,kē }

sand levee See whaleback dune. { 'san ,lev·ē }

sand-lime brick [MATER] **1.** Decorative brick made of sand and lime pressed in an atmosphere of steam. **2.** A firebrick made of refractory silica sand with lime as a bonding agent. { 'san 'līm 'brik }

sand line [ENG] A wire line used to raise and lower a bailer or sand pump to remove cuttings from a borehole. { 'san ,līn }

sand load [ELECTROMAG] An attenuator used as a power-dissipating terminating section for a coaxial line or waveguide; the dielectric space in the line is filled with a mixture of sand and graphite that acts as a matched-impedance load, preventing standing waves. { 'san ,lōd }

sand lobe [GEOL] A rounded sand deposit extending from the shore into shallow water. { 'san ,lōb }

Sandmeyer's reaction [CHEM] Conversion of diazo compounds (in the presence of cuprous halogen salts) into halogen compounds. { 'san,mī·ərz rē,ak·shən }

sand mill [MECH ENG] Variation of a ball-type size-reduction mill in which grains of sand serve as grinding balls. { 'san ,mil }

sandpaper [MATER] Paper with abrasive glued to the surface. { 'san,pā·pər }

sand pavement [GEOL] A sandy surface derived from coarse-grained sand ripples, developed on the lower, windward slope of a dune or rolling sand area during a period of intermittent light, variable winds. { 'san ,pāv·mənt }

sand pile [CIV ENG] A compacted filling of sand in a deep round hole formed by ramming the sand with a pile; used for foundations in soft soil. { 'san ,pīl }

sand pipe [GEOL] A pipe formed in sedimentary rocks, filled with considerable sand and some gravel. Also known as sand gall. { 'san ,pīp }

sandpiper [VERT ZOO] Any of various small birds that are related to plovers and that frequent sandy and muddy shores in temperate latitudes; bill is moderately long with a soft, sensitive tip, legs and neck are moderately long, and plumage is streaked brown, gray, or black above and is white below. { 'san,pī·pər }

sandpit [CIV ENG] An excavation dug in sand, especially as a source of sand for construction materials. { 'san,pit }

sand plain [GEOL] A small outwash plain formed by deposition of sand transported by meltwater streams flowing from a glacier. { 'san ,plān }

sand pump [MECH ENG] A pump, usually a centrifugal type, capable of handling sand- and gravel-laden liquids without clogging or wearing unduly; used to extract mud and cuttings from a borehole. Also known as sludge pump. { 'san ,pəmp }

sand reef See sandbar. { 'san ,rēf }

sand reel [MECH ENG] A drum, operated by a band wheel, for raising or lowering the sand pump or bailer during drilling operations. Also known as coring reel. { 'san ,rēl }

sand return [PETRO ENG] The return of injected sand to the wellbore following formation fracturing; it constitutes a problem. { 'san ri,tərn }

sand ridge [GEOL] **1.** Any low ridge of sand formed at some distance from the shore, and either submerged or emergent, such as a longshore bar or a barrier beach. **2.** One of a series of long, wide, extremely low, parallel ridges believed to represent the eroded stumps of former longitudinal sand dunes. **3.** A crescent-shaped landform found on a sandy beach, such as a beach cusp. **4.** See sand wave. { 'san ,rij }

sand river [GEOL] A river that deposits much of its sand load along its middle course, to be subsequently removed by the wind. { 'san ,riv·ər }

sandrock [GEOL] A field term for a sandstone that is not firmly cemented. { 'san,räk }

sand roll See pseudonodule. { 'san ,rōl }

sand run [GEOL] **1.** A fluidlike motion of dry sand. **2.** A mass of dry sand in motion. { 'san ,rən }

sands [MIN ENG] The coarser and heavier particles of crushed ore, of such size that they settle readily in water and may be leached by allowing the solution to percolate. { sanz }

sand sea [GEOL] **1.** An extensive assemblage of sand dunes of several types in an area where a great supply of sand is present; characterized by an absence of travel lines, or directional indicators, and by a wavelike appearance of dunes separated by troughs. **2.** The flat, rain-smoothed plain of volcanic ash and other pyroclastics on the floor of a caldera. { 'san 'sē }

sand shadow [GEOL] A lee-side accumulation of sand, as a small turret-shaped dune, formed in the shelter of, and immediately behind, a fixed obstruction, such as clumps of vegetation. { 'san ,shad·ō }

sandshale [GEOL] A sedimentary deposit consisting of thin alternating beds of sandstone and shale. { 'san,shāl }

sand-shale ratio [GEOL] The ratio between the thickness or percentage of sandstone and that of shale in a geologic section. { 'san 'shāl 'rā·shō }

sand shark [VERT ZOO] Any of various shallow-water predatory elasmobranchs of the family Carchariidae. Also known as tiger shark. { 'san ,shark }

sand sheet [GEOL] A thin accumulation of coarse sand or fine gravel having a flat surface. { 'san ,shēt }

sand slinger [MECH ENG] A machine which delivers sand to and fills molds at high speed by centrifugal force. { 'san ,sliŋ·ər }

sand snow [HYD] Snow that has fallen at very cold temperatures (of the order of −25°C); as a surface cover, it has the consistency of dust or light dry sand. { 'san ,snō }

sandspit [GEOL] A spit consisting principally of sand. { 'san,spit }

sand splay [GEOL] A floodplain splay consisting of coarse sand particles. { 'san ,splā }

sandstone [PETR] A detrital sedimentary rock consisting of individual grains of sand-size particles 0.06 to 2 millimeters in diameter either set in a fine-grained matrix (silt or clay) or bonded by chemical cement. { 'san,stōn }

sandstone dike [GEOL] A dike made of sandstone or lithified sand. { 'san,stōn 'dīk }

sandstone sill [GEOL] A tabular mass of sandstone that has been emplaced by sedimentary injection parallel to the structure or by bedding of preexisting rock in the manner of an igneous sill. { 'san,stōn 'sil }

sandstorm [METEOROL] A strong wind carrying sand through the air, the diameter of most particles ranging from 0.08 to 1 millimeter; in contrast to a duststorm, the sand particles are mostly confined to the lowest 7 feet (2 meters) above ground, rarely rising more than 36 feet (11 meters). { 'san,storm }

sand streak [GEOL] A low, linear ridge formed at the interface of sand and air or water, oriented parallel to the direction of flow, and having a symmetric cross section. { 'san ,strēk }

sand stream [GEOL] A small sand delta spread out at the mouth of a gully, or a deposit of sand along the bed of a small creek, formed by a torrential rain. { 'san ,strēm }

sand strip [GEOL] A long, narrow ridge of sand extending for a long distance downwind from each horn of a dune. { 'san ,strip }

sand trap [ENG] A device in a conduit for trapping sand or soil particles carried by the water. { 'san ,trap }

sandur See outwash plain. { 'san·dər }

sandwash [GEOL] A sandy or gravel stream bed, devoid of vegetation, containing water only during a sudden and heavy rainstorm. { 'san,wäsh }

sand wave [GEOL] A large, ridgelike primary structure resembling a water wave on the upper surface of a sedimentary bed that is formed by high-velocity air or water currents. Also known as sand ridge. { 'san ,wāv }

sand wedge [GEOL] A wedge-shaped accumulation of sand with the apex downward formed by the filling in of winter contraction cracks. { 'san ,wej }

sand wheel [MECH ENG] A wheel fitted with steel buckets around the circumference for lifting sand or sludge out of a sump to stack it at a higher level. { 'san ,wēl }

sandwich beam See flitch girder. { 'san,wich ,bēm }

sandwich braze [MET] A technique by which a shim is placed between materials to be brazed as a transition layer to decrease thermal stress. { 'san,wich ,brāz }

sandwich construction [DES ENG] Composite construction of alloys, plastics, wood, or other materials consisting of a foam or honeycomb layer laminated and glued between two hard outer sheets. Also known as sandwich laminate. { 'san,wich kən,strək·shən }

sandwich heating [ENG] Method for heating both sides of a

thermoplastic sheet simultaneously prior to forming or shaping. { 'san,wich ,hēd·iŋ }

sandwich laminate See sandwich construction. { 'san,wich 'lam·ə·nət }

sandwich molding See coinjection molding. { 'san,wich ,mōld·iŋ }

sandwich rolling [MET] Rolling strips of metal together to form a metallurgically bonded composite sheet. { 'san,wich ,rōl·iŋ }

sandy bentonite See arkosic bentonite. { 'san·dē 'bent·ən,īt }

sandy chert [PETR] Chert formed in sandy beds by replacement of cement, or the filling of pore spaces, with silica. { 'san·dē 'chərt }

sane [PSYCH] Of sound mind. { sān }

Sanfilippo's syndrome [MED] A hereditary metabolic disorder, transmitted as an autosomal recessive, characterized by excessive amounts of heparitin sulfate in the urine, and manifested by minor skeletal changes and slight hepatomegaly. { san·fə'lip·ōz ,sin,drōm }

Sangamon [GEOL] The third interglacial stage of the Pleistocene epoch in North America, following the Illinoian glacial and preceding the Wisconsin. { 'saŋ·gə,mən }

Sanger's reagent See 1-fluoro-2,4-dinitrobenzene. { 'saŋ·ərz rē,ā·jənt }

sanguineous [PHYSIO] Pertaining to or containing blood. { saŋ'gwin·ē·əs }

sanguivorous [ZOO] Feeding on blood. { saŋ'gwiv·ə·rəs }

sanidal [GEOL] Pertaining to the continental shelf. { 'san·əd·əl }

sanidaster [INV ZOO] A rod-shaped spicule having spines at intervals along its length. { ,san·ə,das·tər }

sanidine [MINERAL] KAlSi$_3$O$_8$ An alkali feldspar mineral occurring in clear, glassy crystals embedded in unaltered acid volcanic rocks; a high-temperature, disordered form. Also known as glassy feldspar; ice spar; rhyacolite. { 'san·ə,dēn }

sanidinite [PETR] A type of igneous rock composed chiefly of sanidine. { sə'nid·ən,īt }

sanitary engineering [CIV ENG] A field of civil engineering concerned with works and projects for the protection and promotion of public health. { 'san·ə,ter·ē ,en·jə'nir·iŋ }

sanitary landfill [CIV ENG] The disposal of garbage by spreading it in layers covered with soil or ashes to a depth sufficient to control rats, flies, and odors. { 'san·ə,ter·ē lan,fil }

sanitary nipper See latrine cleaner. { 'san·ə,ter·ē 'nip·ər }

sanitary sewer [CIV ENG] A sewer which is restricted to carrying sewage and to which storm and surface waters are not admitted. { 'san·ə,ter·ē 'sü·ər }

sanitation [CIV ENG] The act or process of making healthy environmental conditions. { ,san·ə'tā·shən }

sanitizer [MATER] Disinfectant formulated to clean food-processing equipment and dairy and eating utensils. { 'san·ə,tīz·ər }

San Joaquin Valley fever See coccidioidomycosis. { ¦san wǒ¦kēn ¦val·ē 'fē·vər }

sanmartinite [MINERAL] ZnWO$_4$ A mineral composed of zinc tungstate. { san'mart·ən,īt }

sannaite [PETR] An extrusive rock containing phenocrysts of barkevikite, pyroxene, and biotite (in order of decreasing abundance) in a fine-grained or dense groundmass of alkali feldspar, acmite, chlorite, calcite, and pseudomorphs of mica after nepheline. { 'san·ə,īt }

SA node See sinauricular node. { ¦es¦ā 'nōd }

sansar [METEOROL] A northwest wind of Persia. { 'sän·sər }

sansicl [GEOL] An unconsolidated sediment, consisting of a mixture of sand, silt, and clay, in which no component forms 50% or more of the whole aggregate. { 'san,sik·əl }

SANTA See systematic analog network testing approach. { 'san·tə }

Santa Ana [METEOROL] A hot, dry, foehnlike desert wind, generally from the northeast or east, especially in the pass and river valley of Santa Ana, California, where it is further modified as a mountain-gap wind. { 'san·tə ¦an·ə }

Santalaceae [BOT] A family of parasitic dicotyledonous plants in the order Santalales characterized by dry or fleshy indehiscent fruit, plants with chlorophyll, petals absent, and ovules without integument. { ,san·tə'lās·ē,ē }

Santalales [BOT] An order of dicotyledonous plants in the subclass Rosidae characterized by progressive adaptation to parasitism, accompanied by progressive simplification of the ovules. { ˌsan·təˈlā·lēz }

santal oil See sandalwood oil. { ˈsant·əl ˌȯil }

santalol [ORG CHEM] $C_{15}H_{24}O$ A colorless liquid with a boiling point of 300°C; derived from sandalwood oil and used for perfumes. { ˈsan·təˌlȯl }

santalwood oil See sandalwood oil. { ˈsant·əlˌwùd ˌȯil }

Santa Rosa storm [METEOROL] In Argentina, an annual storm near the end of August. { ˈsan·tə ˈrō·zə ˈstȯrm }

Santonian [GEOL] A European stage of geologic time in the Upper Cretaceous, above the Coniacian and below the Campanian. { sanˈtō·nē·ən }

santonin [ORG CHEM] $C_{15}H_{18}O_3$ A white powder with a melting point of 170–173°C; soluble in chloroform and alcohol; used in medicine. { ˈsant·ən·ən }

santorinite [PETR] **1.** A light-colored extrusive rock containing approximately 60–65% silica and calcic plagioclase (labradorite to anorthite) as the only feldspar. **2.** A hypersthene andesite containing plagioclase crystals that have labradorite cores and sodic rims and a groundmass with microlites of sodic oligoclase. { sanˈtȯr·əˌnīt }

sanukite [PETR] An andesite characterized by orthopyroxene as the mafic mineral, andesine as the plagioclase, and a glassy groundmass. { ˈsan·əˌkīt }

sap [BOT] The fluid part of a plant which circulates through the vascular system and is composed of water, gases, salts, and organic products of metabolism. { sap }

Sapele mahogany [MATER] A figured wood from *Entandrophragma cylindricum*, a big tree growing on the Ivory Coast, Ghana, and Nigeria. Also known as aboundikro; scented mahogany; West African cedar. { səˈpē·lē məˈhäg·ə·nē }

saphenous nerve [NEUROSCI] A somatic sensory nerve arising from the femoral nerve and innervating the skin of the medial aspect of the leg, foot, and knee joint. { səˈfē·nəs ˌnərv }

Sapindaceae [BOT] A family of dicotyledonous plants in the order Sapindales distinguished by mostly alternate leaves, usually one and less often two ovules per locule, and seeds lacking endosperm. { ˌsap·ənˈdās·ēˌē }

Sapindales [BOT] An order of mostly woody dicotyledonous plants in the subclass Rosidae with compound or lobed leaves and polypetalous, hypogynous to perigynous flowers with one or two sets of stamens. { ˌsap·ənˈdā·lēz }

sapling [BOT] A young tree with a trunk less than 4 inches (10 centimeters) in diameter at a point approximately 4 feet (1.2 meters) above the ground. { ˈsap·liŋ }

saponification [CHEM] The process of converting chemicals into soap; involves the alkaline hydrolysis of a fat or oil, or the neutralization of a fatty acid. { səˌpän·ə·fəˈkā·shən }

saponification equivalent [CHEM] The quantity of fat in grams that can be saponified by 1 liter of normal alkalies. { səˌpän·ə·fəˈkā·shən iˌkwiv·ə·lənt }

saponification number [ANALY CHEM] Milligrams of potassium hydroxide required to saponify the fat, oil, or wax in a 1-gram sample of a given material, using a specific American Society for Testing and Materials test method. { səˌpän·ə·fəˈkā·shən ˌnəm·bər }

saponin [ORG CHEM] Any of numerous plant glycosides characterized by foaming in water and by producing hemolysis when water solutions are injected into the bloodstream; used as beverage foam producer, textile detergent and sizing, soap substitute, and emulsifier. { ˈsap·ə·nən }

saponite [MINERAL] A soft, soapy, white or light-buff to bluish or reddish trioctahedral montmorillonitic clay mineral consisting of hydrous magnesium aluminosilicate and occurring in masses in serpentine and basaltic rocks. Also known as bowlingite; mountain soap; piotine; soapstone. { ˈsap·əˌnīt }

Sapotaceae [BOT] A family of dicotyledonous plants in the order Ebenales characterized by a well-developed latex system. { ˌsap·əˈtās·ēˌē }

sappare See kyanite. { ˈsaˌper }

sapphire [MINERAL] Any of the gem varieties of the mineral corundum, especially the blue variety, except those that have medium to dark tones of red that characterize ruby; hardness is 9 on Mohs scale, and specific gravity is near 4.00. { ˈsaˌfīr }

sapphire quartz [MINERAL] An indigo-blue opaque variety of quartz. { ˈsaˌfīr ˈkwȯrts }

sapphire whiskers See alumina fibers. { ˈsaˌfīr ˌwis·kərz }

sapphirine [MINERAL] $(MgFe)_{15}(Al,Fe)_{34}Si_7O_{80}$ A green or pale-blue mineral composed of silicate and oxide of magnesium, iron, and aluminum; usually occurs in granular form. { ˈsaf·əˌrēn }

sapping [GEOL] Erosion along the base of a cliff by the wearing away of softer layers, thus removing the support for the upper mass which breaks off into large blocks and falls from the cliff face. Also known as undermining. { ˈsap·iŋ }

Saprist [GEOL] A suborder of the soil order Histosol consisting of residues in which plant structures have been largely obliterated by decay; saturated with water most of the time. { ˈsaˌprist }

saprobe [ECOL] An organism that lives on decaying organic matter. { ˈsaˌprōb }

saprobic [BOT] Living on decaying organic matter; applied to plants and microorganisms. { səˈprō·bik }

saprogen [BIOL] An organism that lives on nonliving organic matter. { ˈsap·rə·jən }

saprogenous ooze [GEOL] Ooze formed of putrefying organic matter. { səˈpräj·ə·nəs ˈüz }

Saprolegniales [MYCOL] An order of aquatic fungi belonging to the class Phycomycetes, having a mostly hyphal thallus and zoospores with two flagella. { ˌsap·rəˌleg·nēˈā·lēz }

saprolite [GEOL] A soft, earthy red or brown, decomposed igneous or metamorphic rock that is rich in clay and formed in place by chemical weathering. Also known as saprolith; sathrolith. { ˈsap·rəˌlīt }

saprolith See saprolite. { ˈsap·rəˌlith }

sapropel [GEOL] A mud, slime, or ooze deposited in more or less open water. { ˈsap·rəˌpel }

sapropel-clay [GEOL] A sedimentary deposit in which the amount of clay is greater than that of sapropel. { ˈsap·rəˌpel ˌklā }

sapropelic coal [GEOL] Coal formed by putrefaction of organic matter under anaerobic conditions in stagnant or standing bodies of water. Also known as sapropelite. { ˈsap·rəˌpel·ik ˈkōl }

sapropelite See sapropelic coal. { ˈsap·rəˌpeˌlīt }

sapropel-peat See peat-sapropel. { ˈsap·rəˌpel ˌpēt }

saprophage [BIOL] An organism that lives on decaying organic matter. { ˈsap·rəˌfāj }

saprophyte [BOT] A plant that lives on decaying organic matter. { ˈsap·rəˌfīt }

saprovore [ZOO] A detritus-eating animal. { ˈsap·rəˌvȯr }

saprozoic [ZOO] Feeding on decaying organic matter; applied to animals. { ˈsap·rəˌzō·ik }

sapwood [BOT] The younger, softer, outer layers of a woody stem, between the cambium and heartwood. Also known as alburnum. { ˈsapˌwùd }

Sapygidae [INV ZOO] A family of hymenopteran insects in the superfamily Scolioidea. { səˈpij·əˌdē }

sarah [NAV] Radio homing device originally designed for personnel rescue and now used in spacecraft recovery operations at sea. Derived from search and rescue and homing. { ˈsa·rə }

sàrca [METEOROL] A violent north wind of Lake Garda in Italy. { ˈsär·kə }

Sarcina [MICROBIO] A genus of strictly anaerobic bacteria in the family Peptococcaceae; spherical cells occur in packets; ferment carbohydrates. { ˈsär·sə·nə }

sarcochore [BOT] A plant dispersing minute, light disseminules. { ˈsär·kəˌkȯr }

Sarcodina [INV ZOO] A superclass of Protozoa in the subphylum Sarcomastigophora in which movement involves protoplasmic flow, often with recognizable pseudopodia. { ˌsär·kəˈdī·nə }

sarcoglia [CYTOL] The protoplasm occurring at a myoneural junction. { särˈkäg·lē·ə }

sarcoidosis [MED] A disease of unknown etiology characterized by granulomatous lesions, somewhat resembling true tubercles, but showing little or no necrosis, affecting the lymph nodes, skin, liver, spleen, heart, skeletal muscles, lungs, bones in distal parts of the extremities (osteitis cystica of Jüngling), and other structures, and sometimes by hyperglobulinemia, cutaneous anergy, and hypercalcinuria. { ˌsär·kȯiˈdō·səs }

sarcolemma [HISTOL] The thin connective tissue sheath enveloping a muscle fiber. { ˌsär·kəˈlem·ə }

sarcoleukemia See leukosarcoma. { ˌsär·kə·lüˈkē·mē·ə }

sarcoma [MED] A malignant tumor arising in connective tissue and composed principally of anaplastic cells that resemble those of supportive tissues. { sär'kō·mə }

sarcoma botryoides [MED] A malignant mesenchymoma that forms grapelike structures; most common in the vagina of infants. { sär'kō·mə bä·trē'ói·dēz }

Sarcomastigophora [INV ZOO] A subphylum of Protozoa comprising forms that possess flagella or pseudopodia or both. { ¦sar·kə‚mas·tə'gäf·ə·rə }

sarcomere [HISTOL] One of the segments defined by Z disks in a skeletal muscle fibril. { 'sär·kə‚mir }

Sarcophagidae [INV ZOO] A family of the myodarian orthorrhaphous dipteran insects in the subsection Calypteratae comprising flesh flies, blowflies, and scavenger flies. { ‚sär·kə'faj·ə‚dē }

sarcoplasm [HISTOL] Hyaline, semifluid interfibrillar substance of striated muscle tissue. { 'sär·kə‚plaz·əm }

sarcoplasmic reticulum [CYTOL] Collectively, the cysternae of a single muscle fiber. { ¦sar·kə¦plaz·mik rə'tik·yə·ləm }

sarcopside [MINERAL] (Fe,Mn,Mg)₃(PO₄)₂ A mineral composed of a phosphate of manganese, magnesium, and iron. { sär'käp·səd }

Sarcopterygii [VERT ZOO] A subclass of Osteichthyes, including Crossopterygii and Dipnoi in some systems of classification. { sär‚käp·tə'rij·ē‚ī }

Sarcoptiformes [INV ZOO] A suborder of the Acarina including minute globular mites without stigmata. { sär‚käp·tə'fór‚mēz }

sarcosine [ORG CHEM] CH₃NHCH₂COOH Sweet-tasting, deliquescent crystals; soluble in water, slightly soluble in alcohol; decomposes at 210–215°C; used in toothpaste manufacture. Also known as methyl glycocol. { 'sär·kə‚sēn }

sarcosoma [INV ZOO] The fleshy portion of an anthozoan. { ‚sär·kə'sō·mə }

Sarcosporida [INV ZOO] An order of Protozoa of the class Haplosporea which comprises parasites in skeletal and cardiac muscle of vertebrates. { ‚sär·kə'spór·əd·ə }

sarcosporidiosis [VET MED] A disease of mammals other than humans caused by muscle infestation by sporozoans of the order Sarcosporida. { ‚sär·kō·spə‚rid·ē'ō·səs }

sarcostyle [INV ZOO] A fibril or column of muscular tissue. { 'sär·kə‚stīl }

sarcotubule [CYTOL] A tubular invagination of a muscle fiber. { ¦sär·kō'tü‚byül }

sard [MINERAL] A translucent brown, reddish-brown, or deep orange-red variety of chalcedony. Also known as sardine; sardius. { särd }

Sardic orogeny [GEOL] A short-lived orogeny that occurred near the end of the Cambrian period. { 'sär·dik ó'räj·ə·nē }

sardine [MINERAL] See sard. [VERT ZOO] **1.** *Sardina pilchardus.* The young of the pilchard, a herringlike fish in the family Clupeidae found in the Atlantic along the European coasts. **2.** The young of any of various similar and related forms which are processed and eaten as sardines. { sär'dēn }

sardine oil [MATER] Combustible, yellow liquid obtained from sardines; soluble in alcohol, ether, and chloroform; solidifies at about 30°C; used in soaps and pet foods, and as a lubricant. { sär'dēn ‚oil }

sardius See sard. { 'sär·dē·əs }

sardonyx [MINERAL] An onyx characterized by parallel layers of sard, a deep orange-red variety of chalcedony, and a mineral of different color. { sär'dän·iks }

Sargasso Sea [GEOGR] A region of the North Atlantic Ocean; boundaries are defined in the west and north by the Gulf Stream, in the east by longitude 40°W, and in the south by latitude 20°N. { sär'ga·sō 'sē }

Sargent curve [NUC PHYS] A graph of logarithms of decay constants of radioisotopes subject to beta-decay against logarithms of the corresponding maximum beta-particle energies; most of the points fall on two straight lines. { 'sär·jənt ‚kərv }

Sargent cycle [THERMO] An ideal thermodynamic cycle consisting of four reversible processes: adiabatic compression, heating at constant volume, adiabatic expansion, and isobaric cooling. { 'sär·jənt ‚sī·kəl }

sarking [BUILD] A layer of boards or bituminous felt placed beneath tiles or other roofing to provide thermal insulation or to prevent ingress of water. { 'särk·iŋ }

sarkinite [MINERAL] Mn₂(AsO₄)(OH) A flesh-red monoclinic mineral composed of hydrous manganese arsenate, occurring in crystals. { 'sär·kə‚nīt }

sarkomycin [MICROBIO] C₇H₈O₃ An antibiotic produced by an actinomycete which acts as a carcinolytic agent. { ‚sär·kə'mīs·ən }

Sarmatian [GEOL] A European stage of geologic time: the upper Miocene, above Tortonian, below Pontian. { sär'mā·shən }

sarmentocymarin [BIOCHEM] A cardioactive, steroid glycoside from the seeds of *Strophanthus sarmentosus*; on hydrolysis it yields sarmentogenin and sarmentose. { sär¦men·tō'sī·mə·rən }

sarmentogenin [BIOCHEM] C₂₃H₃₄O₅ The steroid aglycon of sarmentocymarin; isometric with digitoxigenin, and characterized by a hydroxyl group at carbon number 11. { sär‚men·tō'jen·ən }

sarmentose [BOT] Producing slender, prostrate stems or runners. { sär'men‚tōs }

sarmientite [MINERAL] Fe₂(AsO₄)(SO₄)(OH)·5H₂O A yellow mineral composed of basic hydrous arsenate and sulfate of iron; it is isomorphous with diadochite. { ‚sär·mē'en‚tīt }

sarnaite [GEOL] A feldspathoid-bearing syenite composed of cancrinite and acmite. { 'sär·nə‚īt }

saros [ASTRON] A cycle of time after which the centers of the sun and moon, and the nodes of the moon's orbit return to the same relative position; this period is 18 years 11⅓ days, or 18 years 10⅓ days if 5 rather than 4 leap years are included. { 'sa‚räs }

sarospatakite [GEOL] A micaceous clay mineral composed of mixed layers of illite and montmorillonite. { ‚sär·ə'späd·ə‚kīt }

Sarothriidae [INV ZOO] The equivalent name for Jacobsoniidae. { ‚sar·ə'thrī·ə‚dē }

Sarraceniaceae [BOT] A small family of dicotyledonous plants in the order Sarraceniales in which leaves are modified to form pitchers, placentation is axile, and flowers are perfect with distinct filaments. { ‚sar·ə‚sē·nē'ās·ē‚ē }

Sarraceniales [BOT] An order of dicotyledonous herbs or shrubs in the subclass Dilleniidae; plants characteristically have alternate, simple leaves that are modified for catching insects, and grow in waterlogged soils. { ‚sar·ə‚sē·nē'ā·lēz }

sarsaparilla [BOT] Any of various tropical American vines of the genus *Smilax* (family Liliaceae) found in dense, moist jungles; a flavoring material used in medicine and soft drinks is obtained from the dried roots of at least four species. { ‚sas·pə'ril·ə }

sartorite [MINERAL] PbAs₂S₄ A dark-gray monoclinic mineral, occurring in crystalline form. { 'sär·də‚rīt }

sartorius [ANAT] A large muscle originating in the anterior superior iliac spine and inserting in the tibia; flexes the hip and knee joints, and rotates the femur laterally. { sär'tór·ē·əs }

SAS See aluminum sodium sulfate; stability augmentation system.

SASAR See segmented aperture-synthetic aperture radar. { 'sā‚sär }

sash [BUILD] A frame for window glass. { sash }

sash bar [BUILD] One of the strips of wood or metal that separate the panes of glass in a window. Also known as glazing bar; muntin; window bar. { 'sash ‚bär }

sash cord [BUILD] A cord or chain used to attach a counterweight to the window sash. { 'sash ‚kórd }

Sa spiral [ASTRON] A class of spiral galaxy, including those galaxies that have the largest center sections and closely wound galactic arms. { ¦es¦ā 'spī·rəl }

sassafras [BOT] *Sassafras albidum.* A medium-sized tree of the order Magnoliales recognized by the bright-green color and aromatic odor of the leaves and twigs. { 'sas·ə‚fras }

sassafras oil See oil of sassafras. { 'sas·ə‚fras ‚oil }

sassoline See sassolite. { 'sas·ə‚lēn }

sassolite [MINERAL] H₃BO₃ A white or gray mineral consisting of native boric acid usually occurring in small pearly scales as an incrustation or as tabular triclinic crystals. Also known as sassoline. { 'sas·ə‚līt }

sastruga [HYD] A ridge of snow up to 2 inches (5 centimeters) high formed by wind erosion and aligned parallel to the wind. Also known as skavl; zastruga. { 'zas·trə·gə }

satchel charge [ORD] A number of blocks of explosive

An eastern American species of the pitcher plant (*Sarracenia purpurea*). (*Photograph of Henry M. Mayer, from National Audubon Society*)

Smilax aristolchiaefolia, which yields a flavoring material known as Mexican sarsaparilla.

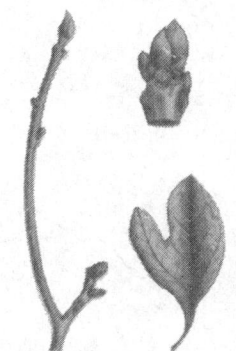

Sassafras albidum, twig, terminal bud, and leaf.

taped to a board fitted with a rope or wire loop for carrying and attaching. { 'sach·əl ,chärj }

sateen [TEXT] A satin-weave cloth, usually made of mercerized cotton and often treated with high-luster and crease-resistant finishes. { sa'tēn }

satellite [AERO ENG] See artificial satellite. [ASTRON] A small, solid body moving in an orbit around a planet; the moon is a satellite of earth. [CELL MOL] A chromosome segment distant from but attached to the rest of the chromosome by an achromatic filament. { 'sad·əl,īt }

satellite and missile surveillance [ENG] The systematic observation of aerospace for the purpose of detecting, tracking, and characterizing objects, events, and phenomena associated with satellites and inflight missiles, friendly and enemy. { ¦sad·əl,īt ən ¦mis·əl sər'vā·ləns }

satellite antenna [ELECTROMAG] Antenna on an artificial satellite to receive command signals from earth, act as a beacon for tracking, or transmit scientific or other data to earth. { 'sad·əl,īt an,ten·ə }

satellite astronomy [ASTRON] The study of astronomical objects by using detectors mounted on earth-orbiting satellites or deep-space probes; allows observations that are not obstructed by the earth's atmosphere. { 'sad·əl,īt ə'strän·ə·mē }

satellite band [CELL MOL] A fraction of the deoxyribonucleic acid (DNA) of an organism which has a different density from the rest and is therefore separable as a band in density gradient centrifugation; these bands are usually made up of highly repetitive sequences of DNA. { 'sad·əl,īt ,band }

satellite-based augmentation system [NAV] A type of wide-area DGPS, intended for aviation users in enroute, terminal, nonprecision approach, and category I (or near category I) precision approach phases of flight, in which the error corrections are broadcast from geostationary satellites and are useful throughout the geographic areas in which these satellites are visible. Abbreviated SBAS. { ¦sad·əl,īt ,bāst ,ȯg·mən'tā·shən ,sis·təm }

satellite cell [HISTOL] One of the neurilemmal cells surrounding nerve cells in the peripheral nervous system. { 'sad·əl,īt ,sel }

satellite communication [COMMUN] Communication that involves the use of an active or passive satellite to extend the range of a communications, radio, television, or other transmitter by returning signals to earth from an orbiting satellite. { 'sad·əl,īt kə,myü·nə,kā·shən }

satellite computer [COMPUT SCI] A computer which, under control of the main computer, handles the input and output routines, thereby allowing the main computer to be fully dedicated to computations. { 'sad·əl,īt kəm,pyüd·ər }

satellite deoxyribonucleic acid [CELL MOL] Any fraction, usually highly repetitious, of chromosomal deoxyribonucleic acid that differs significantly in its base composition from the majority of other fractions. { ¦sad·ə,līt dē,äk·sē'rī·bō·nü,klē·ik 'as·əd }

satellite infrared spectrometer [SPECT] A spectrometer carried aboard satellites in the Nimbus series which measures the radiation from carbon dioxide in the atmosphere at several different wavelengths in the infrared region, giving the vertical temperature structure of the atmosphere over a large part of the earth. Abbreviated SIRS. { 'sad·əl,īt ¦in·frə'red spek'träm·əd·ər }

satellite master antenna television system [COMMUN] A master antenna television system equipped with a television receive-only antenna to receive broadcasts relayed by geostationary satellites. Abbreviated SMATV system. { 'sad·əl,īt ¦mas·tər an¦ten·ə 'tel·ə,vizh·ən ,sis·təm }

satellite meteorology [METEOROL] That branch of meteorological science that employs sensing elements on meteorological satellites to define the state of the atmosphere. { 'sad·əl,īt ,mēd·ē·ə'räl·ə·jē }

satellite processor [COMPUT SCI] One of the outlying processors in a hierarchical distributed processing system, typically placed at or near point-of-transaction locations, and designed to serve the users at those locations. { 'sad·əl,īt ,prä,ses·ər }

satellite tracking [AERO ENG] Determination of the positions and velocities of satellites through radio and optical means. { 'sad·əl,īt ,trak·iŋ }

satellitic crater See secondary crater. { ¦sad·ə¦lid·ik 'krād·ər }

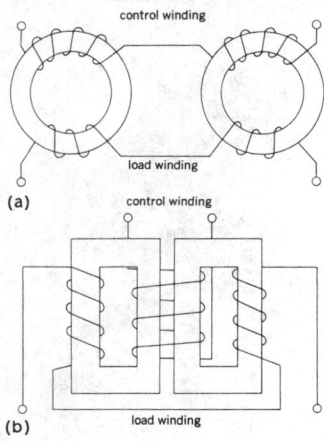

SATURABLE REACTOR

control winding
load winding
(a)

control winding
load winding
(b)

Typical construction of saturable reactors. (*a*) Two separate two-legged cores. (*b*) Three-legged core formed by placing two two-legged cores together.

satellitosis [MED] A condition, associated with inflammatory and degenerative diseases of the central nervous system, in which satellite cells increase around the nerve cells. { ,sad·əl,īd'ō·səs }

satelloid [AERO ENG] A vehicle that revolves about the earth or other celestial body, but at such altitudes as to require sustaining thrust to balance drag. { 'sad·əl,ȯid }

sathrolith See saprolite. { 'sath·rə,lith }

satin [TEXT] A closely woven fabric with a glossy face and a dull-finish back made by carrying the warp (or filling) uninterruptedly on the surface over many filling (or warp) yarns; made of silk, polyester, or other fibers. { 'sat·ən }

satin finish [MET] A finish involving soft scratch-brushing of polished metal surfaces to produce a soft sheen. Also known as Butler finish; scratch-brush finish. [TEXT] A glossy finish on any fabric. { 'sat·ən 'fin·ish }

satin ice See acicular ice. { 'sat·ən 'īs }

satin spar [MINERAL] A white, translucent, fine fibrous variety of gypsum having a silky luster. Also known as satin stone. { 'sat·ən 'spär }

satin stone See satin spar. { 'sat·ən 'stōn }

saturable absorption [OPTICS] A decrease in the absorption coefficient of certain nonlinear materials at high intensities in the incident radiation. { 'sach·rə·bəl ab'sȯrp·shən }

saturable-core magnetometer [ENG] A magnetometer that depends for its operation on the changes in permeability of a ferromagnetic core as a function of the magnetic field to be measured. { 'sach·rə·bəl ¦kȯr ,mag·nə'täm·əd·ər }

saturable-core reactor See saturable reactor. { 'sach·rə·bəl ¦kȯr rē'ak·tər }

saturable reactor [ELECTROMAG] An iron-core reactor having an additional control winding that carries direct current whose value is adjusted to change the degree of saturation of the core, thereby changing the reactance that the alternating-current winding offers to the flow of alternating current; with appropriate external circuits, a saturable reactor can serve as a magnetic amplifier. Also known as saturable-core reactor; transductor. { 'sach·rə·bəl rē'ak·tər }

saturable transformer [ELECTROMAG] A saturable reactor having additional windings to provide voltage transformation or isolation from the alternating-current supply. { 'sach·rə·bəl tranz'fȯr·mər }

saturated activity [NUCLEO] The maximum activity obtainable by activation in a definite flux in a nuclear reactor. { 'sach·ə,rād·əd ak'tiv·əd·ē }

saturated air [METEOROL] Moist air in a state of equilibrium with a plane surface of pure water or ice at the same temperature and pressure; that is, air whose vapor pressure is the saturation vapor pressure and whose relative humidity is 100. { 'sach·ə,rād·əd 'er }

saturated ammonia [CHEM] **1.** Liquid ammonia in a state in which adding heat at constant pressure causes the liquid to vaporize at constant temperature, and in which removing heat at constant pressure causes the temperature of the liquid to drop immediately. **2.** Ammonia vapor in a state in which adding heat at constant pressure causes an immediate temperature rise (superheating) and in which removing heat at constant pressure starts immediate condensation at constant temperature. { 'sach·ə,rād·əd ə'mō·nyə }

saturated calomel electrode [PHYS CHEM] A reference electrode of mercury topped by a layer of mercury (I) chloride paste with potassium chloride solution placed above; easier to assemble than the normal and the one-tenth normal (referring to the concentration of KCl) calomel electrodes. { 'sach·ə,rād·əd 'kal·ə,mel i'lek,trōd }

saturated color [OPTICS] A pure color not contaminated by white. { 'sach·ə,rād·əd 'kəl·ər }

saturated compound [ORG CHEM] An organic compound with all carbon bonds satisfied; it does not contain double or triple bonds and thus cannot add elements or compounds. { 'sach·ə,rād·əd 'käm,paůnd }

saturated diode [ELECTR] A diode that is passing the maximum possible current, so further increases in applied voltage have no effect on current. { 'sach·ə,rād·əd 'dī,ōd }

saturated hydrocarbon [ORG CHEM] A saturated carbon-hydrogen compound with all carbon bonds filled; that is, there are no double or triple bonds as in olefins and acetylenics. { 'sach·ə,rād·əd ¦hī·drə'kär·bən }

saturated interference spectroscopy [SPECT] A version

of saturation spectroscopy in which the gas sample is placed inside an interferometer that splits a probe laser beam into parallel components in such a way that they cancel on recombination; intensity changes in the recombined probe beam resulting from changes in absorption or refractive index induced by a laser saturating beam are then measured. { 'sach·ə‚rād·əd ‚in·tər‚fir·əns spek'träs·kə·pē }

saturated liquid [CHEM] A solution that contains enough of a dissolved solid, liquid, or gas so that no more will dissolve into the solution at a given temperature and pressure. { 'sach·ə‚rād·əd 'lik·wəd }

saturated mineral [MINERAL] A mineral that forms in the presence of free silica. { 'sach·ə‚rād·əd 'min·rəl }

saturated permafrost [GEOL] Permafrost that contains no more ice than the ground could hold if the water were in the liquid state. { 'sach·ə‚rād·əd 'pər·mə‚fròst }

saturated rock [PETR] An igneous rock composed principally of saturated minerals. { 'sach·ə‚rād·əd 'räk }

saturated surface *See* water table. { 'sach·ə‚rād·əd 'sər·fəs }

saturated vapor [THERMO] A vapor whose temperature equals the temperature of boiling at the pressure existing on it. { 'sach·ə‚rād·əd 'vā·pər }

saturated zone *See* zone of saturation. { 'sach·ə‚rād·əd 'zōn }

saturating signal [ELECTR] In radar, a signal of an amplitude greater than the dynamic range of the receiving system. { 'sach·ə‚rād·iŋ 'sig·nəl }

saturation [ELECTR] **1.** The condition that occurs when a transistor is driven so that it becomes biased in the forward direction (the collector becomes positive with respect to the base, for example, in a *pnp* type of transistor). **2.** *See* anode saturation. **3.** *See* temperature saturation. [ELECTROMAG] *See* magnetic saturation. [METEOROL] The maximum water vapor per unit volume that a parcel of air can contain at a given temperature. [NUCLEO] **1.** The condition in which the decay rate of a given radionuclide is equal to its rate of production in an induced nuclear reaction. **2.** The condition in which the voltage applied to an ionization chamber is high enough to collect all the ions formed by radiation but not high enough to produce ionization by collision. [OPTICS] *See* color saturation. [ORD] The striking of a target area with such numbers of missiles that no place in it remains untouched by destruction. [PHYS] **1.** The condition in which a further increase in some cause produces no further increase in the resultant effect. **2.** The property exhibited by certain forces between particles wherein each particle can interact strongly with only a limited number of other particles, as in the forces between atoms in a molecule, and between nucleons in a nucleus. [PHYS CHEM] The condition in which the partial pressure of any fluid constituent is equal to its maximum possible partial pressure under the existing environmental conditions, such that any increase in the amount of that constituent will initiate within it a change to a more condensed state. { ‚sach·ə'rā·shən }

saturation adiabat [METEOROL] On a thermodynamic diagram, a line of constant wet-bulb potential temperatures; in practice, approximate computations are usually employed, and the resulting lines represent, ambiguously, saturation adiabats and pseudoadiabats. Also known as moist adiabat; wet adiabat. { ‚sach·ə'rā·shən 'ad·ē·ə‚bat }

saturation-adiabatic lapse rate [METEOROL] A special case of process lapse rate, defined as the rate of decrease of temperature with height of an air parcel lifted in a saturation-adiabatic process through an atmosphere in hydrostatic equilibrium. Also known as moist-adiabatic lapse rate. { ‚sach·ə'rā·shən ‚ad·ē·ə‚bad·ik 'laps ‚rāt }

saturation-adiabatic process [METEOROL] An adiabatic process in which the air is maintained at saturation by the evaporation or condensation of water substance, the latent heat being supplied by or to the air respectively; the ascent of cloudy air, for example, is often assumed to be such a process. { ‚sach·ə'rā·shən ‚ad·ē·ə‚bad·ik 'prä·səs }

saturation bombing [ORD] Intense area bombing intended to leave no place in a given area free from destructive effects; it may be achieved by dropping many small bombs, or by dropping a medium number of large bombs, or by dropping a single massive bomb. { ‚sach·ə'rā·shən 'bäm·iŋ }

saturation current [ELECTR] **1.** In general, the maximum current which can be obtained under certain conditions. **2.** In a vacuum tube, the space-charge-limited current, such that further increase in filament temperature produces no specific increase

in anode current. **3.** In a vacuum tube, the temperature-limited current, such that a further increase in anode-cathode potential difference produces only a relatively small increase in current. **4.** In a gaseous-discharge device, the maximum current which can be obtained for a given mode of discharge. **5.** In a semiconductor, the maximum current which just precedes a change in conduction mode. [NUCLEO] The ionization current in a gas tube when the applied potential is large enough to collect all ions produced by ionizing radiation. { ‚sach·ə'rā·shən ‚kə·rənt }

saturation curve [GEOL] A curve showing the weight of solids per unit volume of a saturated soil mass as a function of water content. { ‚sach·ə'rā·shən ‚kərv }

saturation deficit [METEOROL] **1.** The difference between the actual vapor pressure and the saturation vapor pressure at the existing temperature. **2.** The additional amount of water vapor needed to produce saturation at the current temperature and pressure, expressed in grams per cubic meter. Also known as vapor-pressure deficit. { ‚sach·ə'rā·shən 'def·ə·sət }

saturation diving [PHYSIO] Diving in which the tissues exposed to high pressure at great ocean depths for 24 hours become saturated with gases, especially inert gases, thereby reaching a new equilibrium state. { ‚sach·ə'rā·shən 'dīv·iŋ }

saturation flux density *See* saturation induction. { ‚sach·ə'rā·shən 'fləks ‚den·səd·ē }

saturation induction [ELECTROMAG] The maximum intrinsic induction possible in a material. Also known as saturation flux density. { ‚sach·ə'rā·shən in'dək·shən }

saturation limiting [ELECTR] Limiting the minimum output voltage of a vacuum-tube circuit by operating the tube in the region of plate-current saturation (not to be confused with emission saturation). { ‚sach·ə'rā·shən 'lim·əd·iŋ }

saturation line [PETR] The line, on a variation diagram of an igneous rock series, that represents saturation with respect to silica; rocks to the right of the line are oversaturated and those to the left, undersaturated. { ‚sach·ə'rā·shən ‚līn }

saturation magnetization [ELECTROMAG] The maximum possible magnetization of a material. { ‚sach·ə'rā·shən ‚mag·nəd·ə'zā·shən }

saturation mixing ratio [METEOROL] A thermodynamic function of state; the value of the mixing ratio of saturated air at the given temperature and pressure; this value may be read directly from a thermodynamic diagram. { ‚sach·ə'rā·shən 'mik·siŋ ‚rā·shō }

saturation ratio [METEOROL] The ratio of the actual specific humidity to the specific humidity of saturated air at the same temperature. { ‚sach·ə'rā·shən ‚rā·s·ō }

saturation scale [OPTICS] A series of colors which appear to have equal differences in color saturation. { ‚sach·ə'rā·shən ‚skāl }

saturation signal [ELECTROMAG] A radio signal (or radar echo) which exceeds a certain power level fixed by the design of the receiver equipment; when a receiver or indicator is "saturated," the limit of its power output has been reached. { ‚sach·ə'rā·shən ‚sig·nəl }

saturation specific humidity [THERMO] A thermodynamic function of state; the value of the specific humidity of saturated air at the given temperature and pressure. { ‚sach·ə'rā·shən spə'sif·ik hyü'mid·əd·ē }

saturation spectroscopy [SPECT] A branch of spectroscopy in which the intense, monochromatic beam produced by a laser is used to alter the energy-level populations of a resonant medium over a narrow range of particle velocities, giving rise to extremely narrow spectral lines that are free from Doppler broadening; used to study atomic, molecular, and nuclear structure, and to establish accurate values for fundamental physical constants. { ‚sach·ə'rā·shən spek'träs·kə·pē }

saturation vapor pressure [THERMO] The vapor pressure of a thermodynamic system, at a given temperature, wherein the vapor of a substance is in equilibrium with a plane surface of that substance's pure liquid or solid phase. { ‚sach·ə'rā·shən 'vā·pər ‚presh·ər }

saturator [ENG] A device, equipment, or person that saturates one material with another; examples are a tank in which vapors become saturated with ammonia from coal (in carbonization of coal), a humidifier, and the operator of a machine for impregnating roofing felt with asphalt. { 'sach·ə‚rād·ər }

Saturn [AERO ENG] One of the very large launch vehicles

built primarily for the Apollo program; begun by Army Ordnance but turned over to the National Aeronautics and Space Administration for the manned space flight program to the moon. [ASTRON] The second largest planet in the solar system (mass is 95.2 compared to earth's 1) and the sixth in the order of distance to the sun; it is visible to the naked eye as a yellowish first-magnitude star except during short periods near its conjunction with the sun; it is surrounded by a series of rings. { 'sad·ərn }

saturnicentric coordinates [ASTRON] Coordinates that indicate the position of a point on the surface of Saturn, determined by the direction of a line joining the center of Saturn to the point. { sa'vän·sen·trik kō'ȯrd·ən,əts }

saturnigraphic coordinates [ASTRON] Coordinates that indicate the position of a point on the surface of Saturn, determined by the direction of a line perpendicular to the mean surface at the point. { sə¦tər·nə¦graf·ik kō'ȯrd·ən,əts }

Saturniidae [INV ZOO] A family of medium- to large-sized moths in the superfamily Saturnioidea including the giant silkworms, characterized by reduced, often vestigial, mouthparts and strongly bipectinate antennae. { ,sad·ər'nī·ə,dē }

Saturnioidea [INV ZOO] A superfamily of medium- to very-large-sized moths in the suborder Heteroneura having the frenulum reduced or absent, reduced mouthparts, no tympanum, and pectinate antennae. { ,sad·ər·nē'ȯid·ē·ə }

Saturn Nebula [ASTRON] A double-ring planetary nebula in the constellation Aquarius, about 700 parsecs away. { 'sa,tərn ,neb·yə·lə }

Satyrinae [INV ZOO] A large, cosmopolitan subfamily of lepidopterans in the family Nymphalidae, containing the wood nymphs, meadow browns, graylings, and arctics, characterized by bladderlike swellings at the bases of the forewing veins. { sə'tir·ə,nē }

saucer crater [ASTRON] A very shallow type of bowl crater on the moon. { 'sȯs·ər ,krād·ər }

Saucesian [GEOL] A North American stage of geologic time in the Oligocene and Miocene, above the Zemorrian and below the Relizian. { sȯ'sē·zhən }

sauconite [MINERAL] The zinc-bearing end member of the montmorillonite group; a trioctahedral clay mineral. { 'sȯ·kə,nīt }

saucyite [PETR] A glassy rhyolitic rock composed of large sanidine phenocrysts in a groundmass of orthoclase microlites and minute crystals of biotite, augite, sphene, zircon, and magnetite. { 'sȯ·sē,īt }

sault [HYD] A waterfall or rapids in a stream. { sü }

Saunders air-lift pump [MECH ENG] A device for raising water from a well by the introduction of compressed air below the water level in the well. { 'sȯn·dərz 'er ,lift ,pəmp }

Sauria [VERT ZOO] The lizards, a suborder of the Squamata, characterized generally by two or four limbs but sometimes none, movable eyelids, external ear openings, and a pectoral girdle. { 'sȯr·ē·ə }

Saurichthyidae [PALEON] A family of extinct chondrostean fishes bearing a superficial resemblance to the Aspidorhynchiformes. { ,sȯr·ək'thī·ə,dē }

Saurischia [PALEON] The lizard-hipped dinosaurs, an order of extinct reptiles in the subclass Archosauria characterized by an unspecialized, three-pronged pelvis. { sȯ'ris·kē·ə }

Sauropoda [PALEON] A group of fully quadrupedal, seemingly herbivorous dinosaurs from the Jurassic and Cretaceous periods in the suborder Sauropodomorpha; members had small heads, spoon-shaped teeth, long necks and tails, and columnar legs. { sȯ'räp·əd·ə }

Sauropodomorpha [PALEON] A suborder of extinct reptiles in the order Saurischia, including large, solid-limbed forms. { sȯ¦räp·əd·ə'mȯr·fə }

Sauropterygia [PALEON] An order of Mesozoic marine reptiles in the subclass Euryapsida. { sȯ,räp·tə'rij·ē·ə }

Saururaceae [BOT] A family of dicotyledonous plants in the order Piperales distinguished by mostly alternate leaves, two to ten ovules per carpel, and carpels distinct or united into a compound ovary. { ,sȯ·rə'rās·ē,ē }

sausage instability See kink instability. { 'sȯ·sij ,in·stə'bil·əd·ē }

saussurite [MINERAL] A white or grayish, tough, compact mineral aggregate composed chiefly of a mixture of albite or oligoclase and zoisite or epidote. { 'sȯ·sə,rīt }

saussuritization [GEOL] A metamorphic process involving

replacement of plagioclase in basalts and gabbros by a fine-grained aggregate of zoisite, epidote, albite, calcite, sericite, and zeolites. { sȯ'sùr·əd·ə'zā·shən }

sauterelle [ENG] A device used by masons for tracing and forming angles. { ,sȯd·ə'rel }

Savage principle [MATH] A technique used in decision theory; a criterion is used to construct a regret matrix in which each outcome entry represents a regret defined as the difference between best possible outcome and the given outcome; the matrix is then used as in decision making under risk with expected regret as the decision-determining quality. Also known as regret criterion. { 'sav·ij ,prin·sə·pəl }

savane armée See thornbush. { sa'vän är'mā }

savane épineuse See thornbush. { sa'vän ā·pə'nûz }

savanna [ECOL] Any of a variety of physiognomically or environmentally similar vegetation types in tropical and extratropical regions; all contain grasses and one or more species of trees of the families Leguminosae, Bombacaceae, Bignoniaceae, or Dilleniaceae. { sə'van·ə }

savanna climate See tropical savanna climate. { sə'van·ə ,klī·mət }

savanna-woodland See tropical woodland. { sə'van·ə 'wùd·lənd }

savart [ACOUS] A unit of pitch interval, such that the interval between two frequencies measured in savarts is equal to 1000 times the common logarithm of the ratio of the frequencies; one octave equals approximately 301.030 savarts. { sa'vär }

Savart plate [OPTICS] A device consisting of a pair of calcite plates having the same thickness, cut along the natural cleavage faces, and mounted with corresponding faces perpendicular to each other; used to detect polarization of light by means of interference fringes. { sa'vär ,plāt }

Savart polariscope [OPTICS] A polariscope consisting of a specially constructed double-plate polarizer and a tourmaline plate analyzer; polarized light passing through the instrument is indicated by the presence of parallel colored fringes, while unpolarized light results in a uniform field. { sa'vär pə'lar·ə,skōp }

savic orogeny [GEOL] A short-lived orogeny that occurred in late Oligocene geologic time, between the Chattian and Aquitanian stages. { 'sav·ik ȯ'räj·ə·nē }

savin oil [MATER] Pale-yellow, volatile oil, soluble in alcohol; derived from the fresh tops of the savin (*Juniperus sabina*); used in medicine. { 'sav·ən ,ȯil }

Savonius rotor [MECH ENG] A rotor composed of two offset semicylindrical elements rotating about a vertical axis. { sə'vō·nē·əs 'rōd·ər }

Savonius windmill [MECH ENG] A windmill composed of two semicylindrical offset cups rotating about a vertical axis. { sə'vō·nē·əs 'win,mil }

savory [BOT] A herb of the mint family in the genus *Satureia*; of the more than 100 species, only summer savory (*S. hortensis*) and winter savory (*S. montana*) are grown for flavoring purposes. { 'sav·ə·rē }

savory oil [MATER] A yellow to brown essential oil derived from the whole dried plant *Satureia hortensis*; used for flavoring. Also known as summer savory oil. { 'sā·və·rē ,ȯil }

saw [DES ENG] **1.** Any of various tools consisting of a thin, usually steel, blade with continuous cutting teeth on the edge. **2.** Any similar device or tool, such as a rotating disc, in which a sharp continuous edge replaces the teeth. { sȯ }

SAW See surface acoustic wave.

saw bit [DES ENG] A bit having a cutting edge formed by teeth shaped like those in a handsaw. { 'sȯ ,bit }

saw-cut [GEOL] A large canyon that cuts abruptly across a terrace, so that it is visible only from locations near its edge. { 'sȯ ,kət }

sawdust [MATER] Wood fragments made by a saw in cutting. { 'sȯ,dəst }

sawdust concrete [MATER] Concrete containing sawdust as the principal aggregate. { 'sȯ,dəst 'kän,krēt }

sawfish [VERT ZOO] Any of several elongate viviparous fishes of the family Pristidae distinguished by a dorsoventrally flattened elongated snout with stout toothlike projections along each edge. { 'sȯ,fish }

saw gumming [MECH ENG] Grinding away the punch marks in the spaces between the teeth in saw manufacture. { 'sȯ ,gəm·iŋ }

sawhorse [ENG] A wooden rack used to support wood that is being sawed. { 'so˙,hȯrs }

sawing [ENG] Cutting with a saw. { 'so˙·iŋ }

sawmill [IND ENG] A plant that houses sawing machines. [MECH ENG] A machine for cutting logs with a saw or a series of saws. { 'so˙,mil }

sawtooth barrel *See* basket. { 'so˙,tüth 'bar·əl }

sawtooth blasting [MIN ENG] The cutting of a series of slabs which, in plan, resemble sawteeth by blasting oblique, horizontal holes along a face. { 'so˙,tüth 'blast·iŋ }

sawtooth crusher [MECH ENG] Solids crusher in which feed is broken down between two sawtoothed shafts rotating at different speeds. { 'so˙,tüth 'krəsh·ər }

sawtooth floor channeling [MIN ENG] A method of channeling inclined beds of marble by removing right-angle blocks in succession from the various beds, thus giving the floor a zigzag or sawtooth appearance. { 'so˙,tüth 'flȯr ,chan·əl·iŋ }

sawtooth generator [ELECTR] A generator whose output voltage has a sawtooth waveform; used to produce sweep voltages for cathode-ray tubes. { 'so˙,tüth 'jen·ə,rād·ər }

sawtooth modulated jamming [ELECTR] Electronic countermeasure technique when a high-level jamming signal is transmitted, thus causing large automatic gain control voltages to be developed at the radar receiver that, in turn, cause target pip and receiver noise to completely disappear. { 'so˙,tüth ¦mäj·ə,lād·əd 'jam·iŋ }

sawtooth pulse [ELECTR] An electric pulse having a linear rise and a virtually instantaneous fall, or conversely, a virtually instantaneous rise and a linear fall. { 'so˙,tüth ¦pəls }

sawtooth roof [ARCH] A roof form having a succession of monitors in sawtooth shape. { 'so˙,tüth 'rüf }

sawtooth stoping [MIN ENG] In the United States, overhand stoping in which the line of advance is up the dip, and benches are advanced in a line parallel with the drift. { 'so˙,tüth 'stōp·iŋ }

sawtooth waveform [ELECTR] A waveform characterized by a slow rise time and a sharp fall, resembling a tooth of a saw. { 'so˙,tüth 'wāv,fȯrm }

sax [DES ENG] A tool for chopping away the edges of roof slates; it has a pick at one end for making nail holes. { saks }

saxicolous [ECOL] Living or growing among rocks. { sak'sik·ə·ləs }

Saxifragaceae [BOT] A family of dicotyledonous plants in the order Rosales which are scarcely or not at all succulent and have two to five carpels usually more or less united, and leaves not modified into pitchers. { ,sak·sə·frə'gās·ē,ē }

saxitoxin [BIOCHEM] A nonprotein toxin produced by the dinoflagellate *Gonyaulax catenella*. { ,sak·sə¦täk·sən }

Saxonian [GEOL] A European stage of geologic time in the Middle Permian, above the Autonian and below the Thuringian. { sak'sō·nē·ən }

saxonite [PETR] A peridotite composed chiefly of olivine and orthopyroxene. { 'sak·sə,nīt }

saxony [TEXT] A fine-quality woolen made of short-staple, botany wools of superior felting power. { 'sak·sə·nē }

saxophone [ELECTROMAG] Vertex-fed linear array antenna giving a cosecant-squared radiation pattern. { 'sak·sə,fōn }

Saybolt chromometer [OPTICS] Device used to measure the color of undyed gasolines, jet fuels, naphthas, kerosines, petroleum waxes, and pharmaceutical white oils. { 'sā,bȯlt krə'mäm·əd·ər }

Saybolt color [ENG] A color standard for petroleum products determined with a Saybolt chromometer. { 'sā,bȯlt ,kəl·ər }

Saybolt Furol viscosimeter [ENG] An instrument for measuring viscosity of very thick fluids, for example, heavy oils; similar to the Saybolt Universal viscosimeter, but with a larger-diameter tube so that the efflux time is about one-tenth that of the Universal instrument. { 'sā,bȯlt 'fyu˙,rȯl ,vis·kə'sim·əd·ər }

Saybolt Furol viscosity [FL MECH] The time in seconds for 60 milliliters of fluid to flow through a capillary tube in a Saybolt Furol viscosimeter at specified temperatures between 70 and 210°F (21 and 99°C); used for high-viscosity petroleum oils, such as transmission and gear oils, and heavy fuel oils. { 'sā,bȯlt 'fyu˙,rȯl vi'skäs·əd·ē }

Saybolt Seconds Universal [FL MECH] A unit of measurement for Saybolt Universal viscosity. Abbreviated SSU. { 'sā,bȯlt 'sek·ənz ,yü·nə'vər·səl }

Saybolt Universal viscosimeter [ENG] An instrument for measuring viscosity by the time it takes a fluid to flow through a calibrated tube; used for the lighter petroleum products and lubricating oils. { 'sā,bȯlt ,yü·nə'vər·səl ,vis·kə'sim·əd·ər }

Saybolt Universal viscosity [FL MECH] The time in seconds for 60 milliliters of fluid to flow through a capillary tube in a Saybolt Universal viscosimeter at a given temperature. { 'sā,bȯlt ,yü·nə'vər·səl vi'skäs·əd·ē }

sb *See* stilb.

Sb *See* antimony.

S band [COMMUN] A band of radio frequencies extending from 1550 to 5200 megahertz, corresponding to wavelengths of 19.37 to 5.77 centimeters. { 'es ,band }

S-band hiran *See* shiran. { 'es ,band 'hī,ran }

S-band single-access service [COMMUN] One of the services provided by the Tracking and Data Relay Satellite System, which provides return-link data rates up to 6 megabits per second for each user spacecraft and forward-link data at 300 kilobits per second. Abbreviated SSA. { ¦es ,band ,siŋ·gəl 'ak,ses ,sər·vəs }

SBAS *See* satellite-based augmentation system. { 'es,bäs *or* ¦es,bē,ā'es }

s-block element [CHEM] A chemical element whose valence shell contains s-electrons only; found in groups 1 and 2 of the periodic table. { ¦es ,bläk ,el·ə·mənt }

SBMV *See* southern bean mosaic virus.

SBR *See* styrene-butadiene rubber.

Sb spiral galaxy [ASTRON] A class of spiral galaxy characterized by smaller central bodies and more open, larger arms. { ¦es,bē 'spī·rəl 'ga·lik·sē }

Sc *See* scandium.

SC *See* sectional center.

scab [BUILD] A short, flat piece of lumber that is used to splice two pieces of wood set at right angles to each other. [MED] Crusty exudate covering a wound or ulcer during the healing process. [MET] A defect consisting of a flat, partially detached piece of metal joined to the surface of a casting or piece of rolled metal. { skab }

scabbard [ORD] A sheath with an open top; it is usually made of leather or canvas and is designed to protect edged weapons, rifles, carbines, and submachine guns from the elements and rough usage. { 'skab·ərd }

scabbing [ORD] Breaking off of fragments in the inside of a wall of hard material due to the impact or explosion of a projectile on the outside. { 'skab·iŋ }

scabies [MED] A contagious skin disorder caused by the mite *Sarcoptes scabiei* burrowing beneath the skin, causing the formation of multiform lesions with intense itching. { 'skā·bēz }

scabland [GEOL] Elevated land that is essentially flat-lying and covered with basalt and has only a thin soil cover, sparse vegetation, and usually deep, dry channels. { 'skab,land }

scabrock [GEOL] **1.** An outcropping of scabland. **2.** Weathered material of a scabland surface. { 'skab,räk }

scabrous [BIOL] Having a rough surface covered with stiff hairs or scales. { 'skab·rəs }

scacchite [MINERAL] MnCl₂ A mineral composed of native manganese chloride, found in volcanic regions. { 'ska,kīt }

SCADA *See* supervisory control and data acquisition. { 'skad·ə *or* ¦es¦sē¦ā¦dē'ā }

scaffold [CIV ENG] A temporary or movable platform supported on the ground or suspended; used for working at considerable heights above the ground. { 'ska,fōld }

scaffold protein [CELL MOL] A protein that assembles interacting signaling proteins into multimolecular signaling complexes. { 'ska,fōld 'prō,tēn }

scaglia [GEOL] A dark, very-fine-grained, somewhat calcareous shale usually developed in the Upper Cretaceous and Lower Tertiary periods of the northern Apennines. { 'skal·yə }

scalable font [GRAPHICS] A set of mathematical values that specify the outlines for each character in a font, and allow the characters to be scaled to any size through special algorithms. Also known as outline font. { ¦skāl·ə·bəl 'fänt }

scala media [ANAT] The middle channel of the cochlea, filled with endolymph and bounded above by Reissner's membrane and below by the basilar membrane. Also known as cochlear duct. { 'skā·lə 'mē·dē·ə }

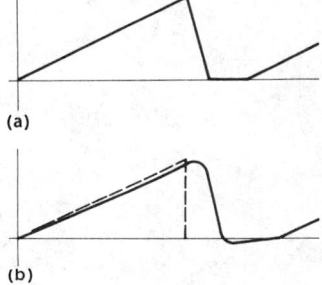

SAWTOOTH WAVEFORM

(a)

(b)

Sawtooth waveforms. (*a*) An ideal linear sawtooth. (*b*) Approximate sawtooth generated by actual circuits.

scalar [COMPUT SCI] A single value or item. [MATH] One of the algebraic quantities which form a field, usually the real or complex numbers, by which the vectors of a vector space are multiplied. [PHYS] **1.** A quantity which has magnitude only and no direction, in contrast to a vector. **2.** A quantity which has magnitude only, and has the same value in every coordinate system. Also known as scalar invariant. { 'skā‧lər }

scalar data type [COMPUT SCI] The manner in which a sequence of bits represents a single data item in a computer program. Also known as aggregate data type. { 'skā‧lər 'dad‧ə ‚tīp }

scalar field [MATH] **1.** The field consisting of the scalars of a vector space. **2.** A function on a vector space into the scalars of the vector space. [PHYS] A field which is characterized by a function of position and time whose value at each point is a scalar. { 'skā‧lər 'fēld }

scalar function [MATH] A function from a vector space to its scalar field. [PHYS] A function of position and time whose value at each point is a scalar. { 'skā‧lər 'fəŋk‧shən }

scalar gradient [MATH] The gradient of a function. { 'skā‧lər 'grā‧dē‧ənt }

scalariform [BIOL] Resembling a ladder; having transverse markings or bars. { skə'lar‧ə‚fórm }

scalar invariant See scalar. { skā‧lər in'ver‧ē‧ənt }

scalar matrix [MATH] A diagonal matrix whose diagonal elements are all equal. { ¦skā‧lər 'mā‚triks }

scalar meson [PART PHYS] A meson which has spin 0 and positive parity, and may be described by a scalar field. { 'skā‧lər 'mā‚sän }

scalar multiplication [MATH] The multiplication of a vector from a vector space by a scalar from the associated field; this usually contracts or expands the length of a vector. { 'skā‧lər ‚məl‧tə‧pli'kā‧shən }

scalar polynomial curvature singularity [RELAT] A singularity in space-time at which a scalar, formed as a polynomial in the curvature tensor, diverges. { ¦skāl‧ər ‚päl‧ə¦nō‧mē‧əl ¦kərv‧ə‧chər ‚siŋ‧gyə'lar‧əd‧ē }

scalar potential [PHYS] A scalar function whose negative gradient is equal to some vector field, at least when this field is time-independent; for example, the potential energy of a particle in a conservative force field, and the electrostatic potential. { 'skā‧lər pə'ten‧chəl }

scalar processor [COMPUT SCI] A computer that carries out computations on one number at a time. { 'skā‧lər 'prä‚ses‧ər }

scalar product [MATH] **1.** A symmetric, alternating, or Hermitian form. **2.** See inner product. { 'skā‧lər 'präd‧əkt }

scalar quantization [COMPUT SCI] A data compression technique in which a value is presented (in approximation) by the closest, in some mathematical sense, of a predefined set of allowable values. { ¦skā‧lər ‚kwän‧tə'zā‧shən }

scalar triple product [MATH] The scalar triple product of vectors v_1, v_2, and v_3 from Euclidean three-dimensional space determines the volume of the parallelepiped with these vectors as edges; it is given by the determinant of the 3×3 matrix whose rows are the components of v_1, v_2, and v_3. Also known as triple scalar product. { 'skā‧lər 'trip‧əl ‚präd‧əkt }

scala tympani [ANAT] The lowest channel in the cochlea of the ear; filled with perilymph. { 'skā‧lə tim'pan‧ē }

scala vestibuli [ANAT] The uppermost channel of the cochlea; filled with perilymph. { 'skā‧lə ve'stib‧yə‧lē }

scalded skin syndrome See toxic epidermal necrolysis. { ¦skȯl‧dəd 'skin ‚sin‚drōm }

scale [ACOUS] A series of musical notes arranged from low to high by a specified scheme of intervals suitable for musical purposes. [BOT] The bract of a catkin. [CHEM] See boiler scale. [ENG] **1.** A series of markings used for reading the value of a quantity or setting. **2.** To change the magnitude of a variable in a uniform way, as by multiplying or dividing by a constant factor, or the ratio of the real thing's magnitude to the magnitude of the model or analog of the model. **3.** A weighing device. **4.** A ruler or other measuring stick. **5.** A dense deposit bonded on the surface of a tube in a heat exchanger or on the surface of an evaporating device. [GRAPHICS] An indication of represented to actual distances on a map, chart, or drawing. [MET] A thick metallic oxide coating formed usually by heating metals in air. [PHYS] **1.** A one-to-one correspondence between numbers and the value of some physical quantity, such as the centigrade or Kelvin

temperature scales on the API (American Petroleum Institute) or Baumé scales of specific gravity. **2.** To determine a quantity at some order of magnitude by using data or relationships which are known to be valid at other (usually lower) orders of magnitude. [VERT ZOO] A flat calcified or cornified platelike structure on the skin of most fishes and of some tetrapods. { skāl }

scaleboard [MATER] Thin sheets of wood used as veneer. { 'skāl‚bórd }

scale drawing [GRAPHICS] A drawing of an object or structure showing all parts in the same proportion of their true size. { 'skāl ‚drȯ‧iŋ }

scale effect [AERO ENG] The necessary corrections applied to measurements of a model in a wind tunnel to ascertain corresponding values for a full-sized object. [FL MECH] An effect in fluid flow that results from changing the scale, but not the shape, of a body around which the flow passes; this effect is relevant to wind tunnel experiments. { 'skāl i‚fekt }

scale factor [ENG] The factor by which the reading of an instrument or the solution of a problem should be multiplied to give the true final value when a corresponding scale factor is used initially to bring the magnitude within the range of the instrument or computer. { 'skāl ‚fak‧tər }

scale height [GEOPHYS] A measure of the decrease of atmospheric pressure with height; when the atmospheric temperature is constant with height, the pressure varies exponentially with height, and the scale height is the height interval over which the pressure changes by a factor of $1/e$. Also known as e-folding height. { 'skāl ‚hīt }

scale insect [INV ZOO] Any of various small, structurally degenerate homopteran insects in the superfamily Coccoidea which resemble scales on the surface of a host plant. { 'skāl 'in‚sekt }

scale model [GRAPHICS] A three-dimensional representation of an object or structure having all parts in the same proportion of their true size. { 'skāl ¦mäd‧əl }

scalene spherical triangle [MATH] A spherical triangle no two of whose sides are equal. { ¦skā‚lēn ¦sfir‧ə‧kəl 'trī‚aŋ‧gəl }

scalene triangle [MATH] A triangle where no two angles are equal. { 'skā‚lēn 'trī‚aŋ‧gəl }

scalenohedron [CRYSTAL] A closed crystal form whose faces are scalene triangles. { skə'lē‧nō'hē‧drən }

scalenus [ANAT] One of three muscles in the neck, arising from the transverse processes of the cervical vertebrae, and inserted on the first two ribs. { skā'lē‧nəs }

scale-of-ten circuit See decade scaler. { ¦skāl əv ¦ten 'sər‧kət }

scale-of-two circuit See binary scaler. { ¦skāl əv ¦tü 'sər‧kət }

scaler [ELECTR] A circuit that produces an output pulse when a prescribed number of input pulses is received. Also known as counter; scaling circuit. { skāl‧ər }

scale scar [BOT] A mark left on a stem after bud scales have fallen off. { 'skāl ‚skär }

scales of motion [OCEANOGR] A series of increasing characteristic magnitudes of motion, ranging from tiny eddies of turbulence to oceanwide currents, each member of the series interacting with the adjacent members. { 'skālz əv 'mō‧shən }

scale symmetry [PHYS] The property of certain physical systems whereby their equations of motion in classical mechanics are unchanged when the space and time variables are rescaled; for example, for a point particle moving in a plane in a potential given by a delta function at the origin, when the vector giving the particle's position r is replaced by sr and the time t is replaced by $s^2 t$, where s is any number. { 'skāl ‚sim‧ə‧trē }

scale-up [DES ENG] Design process in which the data of an experimental-scale operation (model or pilot plant) is used for the design of a large (scaled-up) unit, usually of commercial size. [IND ENG] Transfer of a new process from a pilot plant operation to production at commercial levels. { 'skāl‚əp }

scale wax [MATER] The paraffin wax derived by sweating the greater part of the oil from slack wax; contains up to 6% oil. Also known as crude scale; paraffin scale. { 'skāl ‚waks }

Scalibregmidae [INV ZOO] A family of mud-swallowing worms belonging to the Sedentaria and found chiefly in sublittoral and great depths. { ‚skal‧ə'breg‧mə‚dē }

scaling [BIOL] The removing of scales from fishes. [ELECTR] Counting pulses with a scaler when the pulses occur too fast for direct counting by conventional means. [ENG]

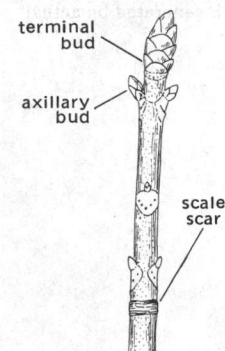

SCALE SCAR

terminal bud

axillary bud

scale scar

Position of scale scar on the twig of a buckeye tree.

Removing scale (rust or salt) from a metal or other surface. [GRAPHICS] Using a scale to measure dimensions in a scale drawing. [MECH] Expressing the terms in an equation of motion in powers of nondimensional quantities (such as a Reynolds number), so that terms of significant magnitude under conditions specified in the problem can be identified, and terms of insignificant magnitude can be dropped. [MED] *See* root planning. [MET] **1.** Forming of a thick layer of metallic oxide on metals at high temperatures. **2.** Depositing of solid inorganic solutes from water on a metal surface, such as a cooling tube or boiler. [MIN ENG] Removing loose rocks and coal from the roof, walls, or face after blasting. [NUC PHYS] A property of nuclear collisions whereby the likelihood of a nuclear reaction depends more on the ratio between energy transferred and momentum transferred than on the energy transferred between the colliding particles. { 'skāl·iŋ }

scaling circuit *See* scaler. { 'skāl·iŋ ˌsər·kət }

scaling factor [ELECTR] The number of input pulses per output pulse of a scaling circuit. Also known as scaling ratio. [ENG] Factor used in heat-exchange calculations to allow for the loss in heat conductivity of a material because of the development of surface scale, as inside pipelines and heat-exchanger tubes. [PHYS] A constant of proportionality which appears in a scaling law. { 'skāl·iŋ ˌfak·tər }

scaling law [PHYS] A law, stating that two quantities are proportional, which is known to be valid at certain orders of magnitude and is used to calculate the value of one of the quantities at another order of magnitude. { 'skāl·iŋ ˌlò }

scaling ratio [ELECTR] *See* scaling factor. [ENG] The ratio of a certain property of a laboratory model to the same property in the natural prototype. { 'skāl·iŋ ˌrā·shō }

scaling symmetry [MATH] The property of an object each part of which is identical to the whole seen at a different magnification; the property that characterizes a fractal. { 'skāl·iŋ ˌsim·ə·trē }

scallion *See* shallot. { 'skal·yən }

scallop [GEOL] *See* scalloping. [INV ZOO] Any of various bivalve mollusks in the family Pectinidae distinguished by radially ribbed valves with undulated margins. { 'skäl·əp }

scalloped upland [GEOL] The region near or at the divide of an upland into which glacial cirques have cut from opposite sides. { 'skäl·əpt 'əp·lənd }

scalloping [GEOL] A sedimentary structure superficially resembling an oscillation ripple mark, and having a concave side that is always oriented toward the top of the bed. Also known as scallop. [NAV] Cyclical variations in the apparent guidance path generated by a four-course radio range or instrument landing system, at speeds greater than can be followed by the vehicle. { 'skäl·ə·piŋ }

scalp [MET] To remove surface layers, and thereby defects, from ingots, billets, or slabs by machining. [MIN ENG] To remove undesirable fine material from broken ore, stone, or gravel. { skalp }

scalped anticline *See* breached anticline. { 'skalpt 'ant·iˌklīn }

scalped extrusion ingot [MET] A cast, solid, or hollow extrusion ingot which has been machined to remove outside surface layers. { 'skalpt ik'strü·zhən ˌiŋ·gət }

scalpel [DES ENG] A small, straight, very sharp knife (or detachable blade for a knife), used for dissecting. { 'skal·pəl }

Scalpellidae [INV ZOO] A primitive family of barnacles in the suborder Lepadomorpha having more than five plates. { skal'pel·əˌdē }

scalping chips [MET] Material removed from the surface of cast ingots to reduce surface roughness and to provide a smooth, clean surface for the rolling mill. { 'skalp·iŋ ˌchips }

scaly bark *See* psorosis. { 'skā·lē ˌbärk }

scaly leg [VET MED] A highly contagious disease of poultry caused by the sarcoptid mite *Knemidokoptes mutans*. { 'skā·lē ˌleg }

scan [COMPUT SCI] To examine information, following a systematic, predetermined sequence, for some particular purpose. [ELECTR] The motion, usually periodic, given to the major lobe of an antenna; the process of directing the radio-frequency beam successively over all points in a given region of space. [ENG] **1.** To examine an area, a region in space, or a portion of the radio spectrum point by point in an ordered sequence; for example, conversion of a scene or image to an electric signal or use of radar to monitor an airspace for detection, navigation, or traffic control purposes. **2.** One complete circular, up-and-down, or left-to-right sweep of the radar, light, or other beam or device used in making a scan. { skan }

scan converter [ELECTR] **1.** Equipment that converts radar data images at a 3 to 10 kilohertz sampling rate that can be sent over telephone line or narrow bandwidth radio circuits and converted into a slow-scan image, through a similar converter, at the receiving end. **2.** A cathode-ray tube that is capable of storing radar, television, and data displays for nondestructive readout over prolonged periods of time. { 'skan kənˌvərd·ər }

scandent [BOT] Climbing by stem-roots or tendrils. { 'skan·dənt }

scandia *See* scandium oxide. { 'skan·dē·ə }

scandium [CHEM] A transition element, symbol Sc, atomic number 21; melts at 1200°C; found associated with rare-earth elements. { 'skan·dē·əm }

scandium halide [INORG CHEM] A compound of scandium and a halogen; for example, scandium chloride, $ScCl_3$. { 'skan·dē·əm 'haˌlīd }

scandium oxide [INORG CHEM] Sc_2O_3 White powder, soluble in hot acids; used to prepare scandium. Also known as scandia. { 'skan·dē·əm 'äkˌsīd }

scandium sulfate [INORG CHEM] $Sc_2(SO_4)_3$ Water-soluble, colorless crystals. { 'skan·dē·əm 'səlˌfāt }

scandium sulfide [INORG CHEM] Sc_3S_3 Yellowish powder; decomposes in dilute acids and boiling water to give off hydrogen sulfide. { 'skan·dē·əm 'səlˌfīd }

scan head [ELECTR] A sensing device that is moved across the image being scanned. { 'skan ˌhed }

scanistor [ELECTR] Integrated semiconductor optical-scanning device that converts images into electrical signals; the output analog signal represents both amount and position of light shining on its surface. { skə'nisˌtər }

scan line [ELECTR] A horizontal row of pixels on a video screen that are examined or refreshed in succession in one sweep across the screen during the scanning process. { 'skan ˌlīn }

scanner [COMMUN] That part of a facsimile transmitter which systematically translates the densities of the elemental areas of the subject copy into corresponding electric signals. [COMPUT SCI] A device that converts an image of something outside a computer, such as text, a drawing, or a photograph, into a digital image that it sends into the computer for display or further processing. [ENG] **1.** Any device that examines an area or region point by point in a continuous systematic manner, repeatedly sweeping across until the entire area or region is covered; for example, a flying-spot scanner. **2.** A device that automatically samples, measures, or checks a number of quantities or conditions in sequence, as in process control. { 'skan·ər }

scanner selector [COMPUT SCI] An electronic device interfacing computer and multiplexers when more than one multiplexer is used. { 'skan·ər siˌlek·tər }

scanning acoustic microscope [ACOUS] A type of acoustic microscope in which a collimated beam of acoustic radiation is focused by a spherical cavity filled with coupling fluid and an object is mechanically scanned by moving it through the focus. { 'skan·iŋ əˈküs·tik 'mī·krəˌskōp }

scanning circuit *See* sweep circuit. { 'skan·iŋ ˌsər·kət }

scanning electron microscope [ELECTR] A type of electron microscope in which a beam of electrons, a few hundred angstroms in diameter, systematically sweeps over the specimen; the intensity of secondary electrons generated at the point of impact of the beam on the specimen is measured, and the resulting signal is fed into a cathode-ray-tube display which is scanned in synchronism with the scanning of the specimen. Abbreviated SEM. { 'skan·iŋ i'lekˌträn 'mī·krəˌskōp }

scanning frequency *See* stroke speed. { 'skan·iŋ ˌfrē·kwən·sē }

scanning head [ELECTR] Light source and phototube combined as a single unit for scanning a moving strip of paper, cloth, or metal in photoelectric side-register control systems. { 'skan·iŋ ˌhed }

scanning HEED [PHYS] A form of HEED in which the diffracted electrons are directly measured electronically with a sensitive detector, and the diffraction pattern is recorded either by moving the detector across it or by deflecting the diffracted

SCALLOP

Typcial ribbed valve of a scallop.

electrons across a stationary detector. Abbreviated SHEED. { 'skan·iŋ ¦hēd }

scanning line [COMMUN] **1.** In television, a single, continuous, narrow strip which is determined by the process of scanning. **2.** Path traced by the scanning or recording spot in one sweep across the subject copy or record sheet. { 'skan·iŋ ‚līn }

scanning linearity [ELECTR] In television, the uniformity of scanning speed during the trace interval. { 'skan·iŋ ‚lin·ē'ar·əd·ē }

scanning line frequency *See* stroke speed. { 'skan·iŋ ¦līn ‚frē·kwən·sē }

scanning loss [ELECTROMAG] In a radar system employing a scanning antenna, the reduction in sensitivity (usually expressed in decibels) due to scanning across the target, compared with that obtained when the beam is directed constantly at the target. { 'skan·iŋ ‚lós }

scanning near-field optical microscopy *See* near-field scanning optical microscopy. { ¦skan·iŋ ¦nir ‚fēld ‚äp·tə·kəl mī'kräs·kə·pē }

scanning proton microprobe [ENG] An instrument used for determining the spatial distribution of trace elements in samples, in which a beam of energetic protons is focused on a narrow spot which is swept over the sample, and the characteristic x-rays emitted from the target are measured. { 'skan·iŋ 'prō‚tän 'mī·krə‚skōp }

scanning radio [ELECTR] A radio receiver that automatically scans across public service, emergency service, or other radio bands and stops at the first preselected station which is on the air. Also known as radio scanner. { 'skan·iŋ 'rād·ē·ō }

scanning radiometer [ENG] An image-forming system consisting of a radiometer which, by the use of a plane mirror rotating at 45° to the optical axis, can see a circular path normal to the instrument. { 'skan·iŋ ‚rād·ē'äm·əd·ər }

scanning sequence [ENG] The order in which the points in a region are scanned; for example, in television the picture is scanned horizontally from left to right and vertically from top to bottom. { 'skan·iŋ ‚sēk·wəns }

scanning sonar [ENG] Sonar in which all targets of interest are shown simultaneously, as on a radar PPI (plan position indicator) display or sector display; the sound pulse may be transmitted in all directions simultaneously and picked up by a rotating receiving transducer, or transmitted and received in only one direction at a time by a scanning transducer. { 'skan·iŋ 'sō‚när }

scanning speed *See* spot speed. { 'skan·iŋ ‚spēd }

scanning spot *See* picture element. { 'skan·iŋ ‚spät }

scanning switch *See* commutator switch. { 'skan·iŋ ‚swich }

scanning transmission electron microscope [ELECTR] A type of electron microscope which scans with an extremely narrow beam that is transmitted through the sample; the detection apparatus produces an image whose brightness depends on atomic number of the sample. Abbreviated STEM. { 'skan·iŋ tranz'mish·ən i'lek‚trän 'mī·krə‚skōp }

scanning tunneling microscope [ELECTR] An instrument for producing surface images with atomic-scale lateral resolution, in which a fine probe tip is raster-scanned over the surface at a distance of 0.5–1 nanometer, and the resulting tunneling current, or the position of the tip required to maintain a constant tunneling current, is monitored. Also known as tunneling microscope. { 'skan·iŋ ¦tən·əl·iŋ 'mī·krə‚skōp }

scanning yoke *See* deflection yoke. { 'skan·iŋ ‚yōk }

scansorial [BOT] Adapted for climbing. { skan'sór·ē·əl }

scantlings [BUILD] Sections of timber measuring less than 8 inches (20 centimeters) wide and from 2 to 6 inches (5 to 15 centimeters) thick; used for studding. [NAV ARCH] The dimensions and material thicknesses of frames, shell plating, deck plating, and other structures of a ship, together with the suitability of the means for protecting openings and making them sufficiently watertight or weathertight. { 'skant·liŋz }

Scapanorhychidae [VERT ZOO] The goblin sharks, a family of deep-sea galeoids in the isurid line having long, sharp teeth and a long, pointed rostrum. { ‚skap·ə·nō'rik·ə‚dē }

scapha [ANAT] The furrow of the auricle between the helix and the antihelix. { 'skaf·ə }

Scaphidiidae [INV ZOO] The shining fungus beetles, a family of coleopteran insects in the superfamily Staphylinoidea. { ‚skaf·ə'dī·ə‚dē }

scaphocephaly [MED] A condition of the skull characterized by elongation and narrowing, and a projecting, keellike sagittal suture, caused by its premature closure. { ‚skaf·ə'sef·ə·lē }

scaphoid [ANAT] A boat-shaped bone of the carpus or of the tarsus. { 'skaf‚óid }

Scaphopoda [INV ZOO] A class of the phylum Mollusca in which the soft body fits the external, curved and tapering, nonchambered, aragonitic shell which is open at both ends. { skə'fäp·əd·ə }

scapolite [MINERAL] A white, gray, or pale-green complex aluminosilicate of sodium and calcium belonging to the tectosilicate group of silicate minerals; crystallizes in the tetragonal system and is vitreous; hardness is 5–6 on Mohs scale, and specific gravity is 2.65–2.74. Also known as wernerite. { 'skap·ə‚līt }

scapolitization [GEOL] Introduction of or replacement by scapolite. { skap·ə‚lid·ə'zā·shən }

scapula [ANAT] The large, flat, triangular bone forming the back of the shoulder. Also known as shoulder blade. { 'skap·yə·lə }

scapulet [INV ZOO] In some medusae, fringed outgrowths on the outer surfaces of the arms near the bell. { 'skap·yə‚let }

scapulus [INV ZOO] A modified submarginal region in some sea anemones. { 'skap·yə·ləs }

scapus [BIOL] The stem, shaft, or column of a structure. { 'skā·pəs }

scar [GEOL] **1.** A steep, rocky eminence, such as a cliff or precipice, where bare rock is well exposed. Also known as scaur; scaw. **2.** *See* shore platform. [MED] A permanent mark on the skin or other tissue, formed from connective-tissue replacement of tissue destroyed by a wound or disease process. { skär }

Scarabaeidae [INV ZOO] The lamellicorn beetles, a large cosmopolitan family of coleopteran insects in the superfamily Scarabaeoidea including the Japanese beetle and other agricultural pests. { ‚skar·ə'bē·ə‚dē }

Scarabaeoidea [INV ZOO] A superfamily of Coleoptera belonging to the suborder Polyphaga. { ‚skar·ə·bē'óid·ē·ə }

scarabiasis [MED] Invasion of the intestine by the dung beetle, characterized by anorexia, emaciation, and disturbance of the gastrointestinal tract. { ¦skar·ə'bī·ə·səs }

scarfing [MET] **1.** Cutting away of surface defects on metals by use of a gas torch. **2.** A forging process in which the ends of two pieces to be joined are tapered to avoid an enlarged joint. { 'skärf·iŋ }

scarf joint [DES ENG] A joint made by the cutting of overlapping mating parts so that the joint is not enlarged and the patterns are complementary, and securing them by glue, fasteners, welding, or other joining method. { 'skärf ‚jóint }

Scaridae [VERT ZOO] The parrotfishes, a family of perciform fishes in the suborder Percoidei which have the teeth of the jaw generally coalescent. { 'skar·ə‚dē }

scarification [MED] The operation of making numerous small, superficial incisions in skin or other tissue. { ‚skär·ə·fə'kā·shən }

scarifier [ENG] An implement or machine with downward projecting tines for breaking down a road surface 2 feet (60 centimeters) or less. { 'skär·ə‚fī·ər }

scarifying [AGR] A process that is usually preceded by hulling in which hard seeds are scratched to facilitate water absorption and speed germination. { 'skär·ə‚fī·iŋ }

scarious [BOT] Having a thin, membranous texture. { 'skar·ē·əs }

scarlet *See* scarlet red. { 'skär·lət }

scarlet fever [MED] An acute, contagious bacterial disease caused by *Streptococcus hemolyticus;* characterized by a papular, or rough, bright-red rash over the body, with fever, sore throat, headache, and vomiting occurring 2–3 days after contact with a carrier. { 'skär·lət 'fē·vər }

scarlet fever streptococcus antitoxin [IMMUNOL] A sterile aqueous solution of antitoxins obtained from the blood of animals immunized against group A beta hemolytic streptococci toxin; formerly used in the treatment of, and to produce immunity against, scarlet fever. { 'skär·lət 'fē·vər ‚strep·tə¦käk·əs ‚ant·i'täk·sən }

scarlet fever streptococcus toxin [IMMUNOL] Toxic filtrate of cultures of *Streptococcus pyogenes* responsible for the

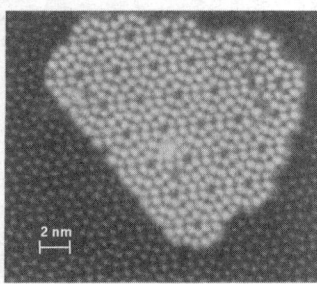

SCANNING TUNNELING MICROSCOPE

2 nm

Silicon surface imaged with the scanning tunneling microscope. *(Reprinted with permission from U. Kohler, J. E. Demuth and R. J. Hamers, Scanning tunneling microscopy study of low-temperature epitaxial growth of silicon on Si(111)-(7×7), Vacuum Sci. Technol., A7(4): 2860–2867, 1989, copyright American Vacuum Society)*

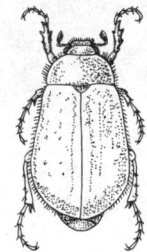

SCARABAEIDAE

A drawing of a lamellicorn beetle. *(From T. I. Storer and R. L. Usinger, General Zoology, 3d ed., McGraw-Hill, 1957)*

characteristic rash of scarlet fever; the toxin is used in the Dick test. { 'skär·lət 'fē·vər ¦strep·tə¦käk·əs 'täk·sən }

scarlet red [ORG CHEM] $CH_3C_6H_4H:NC_6H_3CH_3N:NC_{10}H_{15}OH$ A brown, water-insoluble powder, used as a dye in ointments. Also known as Biebrich red; scarlet. { 'skär·lət 'red }

scarp See escarpment. { skärp }

Scarpa's fascia [ANAT] The deep, membranous layer of the superficial fascia of the lower abdomen. { 'skär·pəz ¦fā·shə }

scarped plain [GEOL] A terrain characterized by a succession of faintly inclined or gently folded strata. { 'skärpt ¦plān }

scarp face See scarp slope. { 'skärp ¦fās }

scarp-foot spring [HYD] A spring that flows onto the land surface at or near the foot of an escarpment. { 'skärp ¦fůt ¦spriŋ }

scarpland [GEOGR] A region marked by a succession of nearly parallel cuestas separated by lowlands. { 'skärp¦lənd }

scarplet See piedmont scarp. { 'skärp·lət }

scarpline [GEOL] A relatively straight line of cliffs of considerable extent, produced by faulting or erosion along a fault. { 'skärp¦līn }

scarp slope [GEOL] The steep face of a cuesta, or asymmetric ridge, facing in an opposite direction to the dip of the strata. Also known as front slope; inface; scarp face. { 'skärp ¦slōp }

scarp stream [HYD] An obsequent stream flowing down a scarp, such as down the scarp slope of a cuesta. { 'skärp ¦strēm }

scar tissue [MED] Contracted, dense connective tissue that is formed by the healing process of a wound or diseased tissue. { 'skär ¦tish·ü }

SC asphalt See slow-curing liquid asphaltic material. { ¦es¦sē 'as¦fôlt }

Scatopsidae [INV ZOO] The minute black scavenger flies, a family of orthorrhaphous dipteran insects in the series Nematocera. { skə'täp·sə¦dē }

scatter angle See scattering angle. { 'skad·ər ¦aŋ·gəl }

scatter band [COMMUN] In pulse interrogation systems, the total bandwidth occupied by the frequency spread by numerous interrogations operating on the same nominal radio frequency. { 'skad·ər ¦band }

scatter diagram [PETR] See point diagram. [STAT] A plot of the pairs of values of two variates in rectangular coordinates. Also known as scatter gram. { 'skad·ər ¦dī·ə¦gram }

scattered [METEOROL] Descriptive of a sky cover of 0.1 to 0.5 (5 to 54%), applied only when clouds or obscuring phenomena aloft are present, not applied for surface-based obscuring phenomena. { 'skad·ərd }

scattered Kuiper Belt object [ASTRON] A member of the Kuiper belt with a very large, very elliptical orbit, whereby its greatest distance from the sun can range up to 200 astronomical units or more. { ¦skad·ərd 'kī·pər ¦äb¦jekt }

scattergram See scatter diagram. { 'skad·ər¦gram }

scattering [ELECTROMAG] Diffusion of electromagnetic waves in a random manner by air masses in the upper atmosphere, permitting long-range reception, as in scatter propagation. Also known as radio scattering. [PHYS] **1.** The change in direction of a particle or photon because of a collision with another particle or a system. **2.** Diffusion of acoustic or electromagnetic waves caused by inhomogeneity or anisotropy of the transmitting medium. **3.** In general, causing a collection of entities to assume a less orderly arrangement. { 'skad·ə·riŋ }

scattering amplitude [QUANT MECH] A quantity, depending in general on the energy and scattering angle, which specifies the wave function of particles scattered in a collision, and whose squared modulus is proportional to the number of particles scattered in a given direction. { 'skad·ə·riŋ ¦am·plə¦tüd }

scattering angle [PHYS] The angle between the initial and final directions of motion of a scattered particle. Also known as scatter angle. { 'skad·ə·riŋ ¦aŋ·gəl }

scattering coefficient [ELECTROMAG] One of the elements of the scattering matrix of a waveguide junction; that is, a transmission or reflection coefficient of the junction. [PHYS] The fractional decrease in intensity of a beam of electromagnetic radiation or particles per unit distance traversed, which results from scattering rather than absorption. Also known as dissipation coefficient. { 'skad·ə·riŋ ¦kō·i¦fish·ənt }

scattering cross section [ELECTROMAG] The power of electromagnetic radiation scattered by an antenna divided by

the incident power. [PHYS] The sum of the cross sections for elastic and inelastic scattering. { 'skad·ə·riŋ 'krós ¦sek·shən }

scattering function [ELECTROMAG] The intensity of scattered radiation in a given direction per lumen of flux incident upon the scattering material. { 'skad·ə·riŋ ¦fəŋk·shən }

scattering layer [OCEANOGR] A layer of organisms in the sea which causes sound to scatter and to return echoes. { 'skad·ə·riŋ ¦lā·ər }

scattering length [NUC PHYS] A parameter used in analyzing nuclear scattering at low energies; as the energy of the bombarding particle becomes very small, the scattering cross section approaches that of an impenetrable sphere whose radius equals this length. Also known as scattering power. { 'skad·ə·riŋ ¦leŋkth }

scattering loss [ELECTROMAG] The portion of the transmission loss that is due to scattering within the medium or roughness of the reflecting surface. { 'skad·ə·riŋ ¦lós }

scattering matrix [ELECTROMAG] A square array of complex numbers consisting of the transmission and reflection coefficients of a waveguide junction. [QUANT MECH] A matrix which expresses the initial state in a scattering experiment in terms of the possible final states. Also known as collision matrix; S matrix. { 'skad·ə·riŋ ¦mā·triks }

scattering-matrix theory See S-matrix theory. { 'skad·ə·riŋ ¦mā·triks ¦thē·ə·rē }

scattering operator [PHYS] An operator which acts in the vector space of solutions of a wave equation, transforming solutions representing incoming waves in a domain exterior to a bounded obstacle into solutions representing outgoing waves. { 'skad·ə·riŋ ¦äp·ə¦rād·ər }

scattering plane [PHYS CHEM] In a quasielastic light-scattering experiment performed with the use of polarizers, the plane containing the incident and scattered beams. { 'skad·ə·riŋ ¦plān }

scattering power See scattering length. { 'skad·ə·riŋ ¦pau·ər }

scattering theory [PHYS] The discipline that mathematically determines the amplitudes of the scattered fields in a scattering process or collision from the equations of motion of the interacting particles, including the potential energy of the interaction. Also known as direct scattering theory. { 'skad·ə·riŋ ¦thē·ə·rē }

scatter loading [COMPUT SCI] The process of loading a program into main memory such that each section or segment of the program occupies a single, connected memory area but the several sections of the program need not be adjacent to each other. { 'skad·ər ¦lōd·iŋ }

scatterometer [ENG] A microwave sensor that is essentially a radar without ranging circuits, used to measure only the reflection or scattering coefficient while scanning the surface of the earth from an aircraft or a satellite. { ¦skad·ə'räm·əd·ər }

scatter propagation [ELECTROMAG] Transmission of radio waves far beyond line-of-sight distances by using high power and a large transmitting antenna to beam the signal upward into the atmosphere and by using a similar large receiving antenna to pick up the small portion of the signal that is scattered by the atmosphere. Also known as beyond-the-horizon communication; forward-scatter propagation; over-the-horizon propagation. { 'skad·ər ¦präp·ə¦gā·shən }

scatter read [COMPUT SCI] An input operation that places various segments of an input record into noncontiguous areas in central memory. { 'skad·ər ¦rēd }

scatter reflections [ELECTROMAG] Reflections from portions of the ionosphere having different virtual heights, which mutually interfere and cause rapid fading. { 'skad·ər ri¦flek·shənz }

scaur See scar. { skär }

scavenger [CHEM] A substance added to a mixture or other system to remove or inactivate impurities. Also known as getter. [ECOL] An organism that feeds on carrion, refuse, and similar matter. [MET] A reactive metal added to a molten metal to combine with and remove dissolved gases. { 'skav·ən·jər }

scavenger system [ORD] A device for clearing smoke and gases from the chamber and bore of a firearm after firing. { 'skav·ən·jər ¦sis·təm }

scavenger well [HYD] A well located between a good well (or group of wells) and a source of potential contamination, which is pumped (or allowed to flow) as waste to prevent the contaminated water from reaching the good well. { 'skav·ən·jər ˌwel }

scavenging [MECH ENG] Removal of spent gases from an internal combustion engine cylinder and replacement by a fresh charge or air. [MET] Removal of dissolved gases from molten metal. [ORD] The sweeping out, by a blast of air, of the gaseous products resulting from the firing of a gun. { 'skav·ən·jiŋ }

scaw *See* scar. { skȯ }

Scelionidae [INV ZOO] A family of small, shining wasps in the superfamily Proctotrupoidea, characterized by elbowed, 11- or 12-segmented antennae. { ˌsel·ē'än·ə‚dē }

scenario-based design [SYS ENG] A family of techniques in which the use of a future system is concretely described at an early point in the development process, and narrative descriptions of the envisage usage episodes are then employed in a variety of ways to guide the development of the system. { sə‚ner·ē·ō ‚bāst di'zīn }

scend [ENG] **1.** The upward motion of the bow and stern of a vessel associated with pitching. **2.** The lifting of the entire vessel by waves or swell. Also known as send. { send }

scene analysis *See* picture segmentation. { 'sēn ə‚nal·ə·səs }

scene paint [MATER] A paint used in theatrical scene painting; it is a dry pigment mixed with a glue-water mixture called size water. { 'sēn ‚pānt }

scented mahogany *See* Sapele mahogany. { 'sent·əd mə'häg·ə·nē }

scent gland [VERT ZOO] A specialized skin gland of the tubuloalveolar or acinous variety which produces substances having peculiar odors; found in many mammals. { 'sent ‚gland }

scfh [FL MECH] Cubic feet per hour of gas flow at specified standard conditions of temperature and pressure.

scfm [FL MECH] Cubic feet per minute of gas flow at specified standard conditions of temperature and pressure.

Schaeffer's salt [ORG CHEM] $HOC_{10}SO_3Na$ A light-yellow to pink, water-soluble powder; the sodium salt formed from 2-naphthol-6-sulfonic acid; used as an intermediate in synthesis of organic compounds. { 'shā‚fərz ‚sȯlt }

schafarzikite [MINERAL] $Fe_5Sb_4O_{11}$ A red to brown mineral composed of iron antimony oxide. { 'shä·fər‚zi‚kīt }

schairerite [MINERAL] $Na_3(SO_4)(F,Cl)$ A colorless rhombohedral mineral composed of sodium sulfate with fluorine and chlorine, occurring in crystals. { 'shī·rə‚rīt }

schalstein [PETR] A slaty rock formed by shearing basaltic or andesitic tuff or lava. { 'shäl‚stīn }

schappe [TEXT] Yarn or fabric made of spun silk or synthetic yarn resembling silk. { 'shäp·ə }

Schardinger dextrin *See* cycloamylose. { 'shärd·ən·jər 'deks·trən }

scharnitzer [METEOROL] A cold, northerly wind of long duration in Tyrol, Austria. { 'shär·nit·sər }

Schatzki's ring [MED] A mucosal constriction at the junction of the esophagus and stomach in the presence of a hiatus hernia. Also known as lower esophageal ring. { ‚shät'skēz 'riŋ }

Schauder's fixed-point theorem [MATH] A continuous mapping from a closed, compact, convex set in a Banach space into itself has at least one fixed point. { 'shaȯd·ərz ‚fikst ‚pȯint 'thir·əm }

Scheat [ASTRON] A red giant, irregular, variable star, in the constellation Pegasus. { shē'at }

schedule [STAT] A group or list of questions used by an interviewer to obtain information directly from a subject. { 'skej·əl }

scheduled down time [COMPUT SCI] A period of time designated for closing down a computer system for preventive maintenance. { 'skej·əld 'daȯn‚tīm }

scheduled target [ORD] A planned target on which a nuclear weapon is to be delivered at a specific time during the operation of the supported force. { 'skej·əld 'tär·gət }

schedule of fire [ORD] Groups of fires or series of fires fired in a definite sequence according to a definite program. { 'skej·əl əv 'fīr }

scheduler [COMPUT SCI] A system control program that determines the sequence in which programs will be processed by a computer and automatically submits them for execution at predetermined times. { 'skej·ə·lər }

scheduling [IND ENG] A decision-making function that plays an important role in most manufacturing and service industries and often allows an organization to operate with a minimum of resources. Scheduling is applied in procurement and production, in transportation and distribution, and in information processing and communication. In manufacturing, the scheduling function coordinates the flow of parts and products through the system, and balances the workload on machines and personnel, departments, and the entire plant. { 'skej·əl·iŋ }

scheduling algorithm [COMPUT SCI] A systematic method of determining the order in which tasks will be performed by a computer system, generally incorporated into the operating system. { 'skej·ə·liŋ ‚al·gə‚rith·əm }

Scheele's green *See* copper arsenite. { 'shā·ləz 'grēn }

scheelite [MINERAL] $CaWO_4$ A yellowish-white mineral crystallizing in the tetragonal system and occurring in tabular or massive form in pneumatolytic veins associated with quartz; an ore of tungsten. { 'shā‚līt }

Scheffel engine [MECH ENG] A type of multirotor engine that uses nine approximately equal rotors turning in the same clockwise sense. { 'shef·əl ‚en·jən }

schefflerite [MINERAL] $(Ca,Mn)(Mg,Fe,Mn)\cdot Si_2O_6$ Brown to black variety of pyroxene that crystallizes in the monoclinic system and contains manganese and frequently iron. { 'shef·lə‚rīt }

Scheibel column *See* Scheibel extractor. { 'shī·bəl ‚käl·əm }

Scheibel extractor [CHEM ENG] Liquid-liquid contact vessel used in liquid-liquid extraction processes: a vertical cylinder with interspersed open spaces and wire-mesh packing along its height, with liquid agitators in the open spaces, or a vertical cylinder fully filled with wire-mesh packing. Also known as Scheibel column; Scheibel-York extractor; York-Scheibel column. { 'shī·bəl ik‚strak·tər }

Scheibel-York extractor *See* Scheibel extractor. { 'shī·bəl 'yȯrk ik‚strak·tər }

Scheie's syndrome [MED] A hereditary disease transmitted as an autosomal recessive and characterized by high levels of chondroitin sulfate B in the urine, mild distortion of the facies, hypertrichosis, clouding of the cornea, and aortic valve disease. { 'shī·əz ‚sin‚drōm }

schema [COMPUT SCI] A logical description of the data in a data base, including definitions and relationships of data. { 'skē·mə }

schematic circuit diagram *See* circuit diagram. { ski'mad·ik 'sər·kət ‚dī·ə‚gram }

schematic diagram [GRAPHICS] A presentation of the element-by-element relationship of all parts of a system. Also known as schematic drawing. { ski'mad·ik 'dī·ə‚gram }

schematic drawing *See* schematic diagram. { ski'mad·ik 'drȯ·iŋ }

scheme of maneuver [ORD] The tactical plan to be executed by a force in order to seize assigned objectives. { 'skēm əv mə'nü·vər }

schemochrome [ZOO] A feather color that originates within the feather structures, through refraction of light independent of pigments. { 'skē·mə‚krōm }

Schenck's disease *See* sporotrichosis. { 'sheŋks di‚zēz }

Schering bridge [ELEC] A four-arm alternating-current bridge used to measure capacitance and dissipation factor; bridge balance is independent of frequency. { 'sher·iŋ ‚brij }

scheteligite [MINERAL] $(Ca,Y,Sb,Mn)_2(Ti,Ta,Nb,W)_2O_6(O,OH)$ A mineral composed of oxide of calcium, rare-earth metals, antimony, manganese, titanium, columbium, and tantalum. { shə'tel·ə‚gīt }

Scheuermann's disease *See* osteochondrosis. { 'shȯi·ər‚mänz di‚zēz }

Schick test [IMMUNOL] A skin test for determining susceptibility to diphtheria performed by the intradermal injection of diluted diphtheria toxin; a positive reaction, showing edema and scaling after 5 to 7 days, indicates lack of immunity. { 'shik ‚test }

Schiff base [ORG CHEM] $RR'C=NR''$ Any of a class of derivatives of the condensation of aldehydes or ketones with primary amines; colorless crystals, weakly basic; hydrolyzed by water and strong acids to form carbonyl compounds and amines; used as chemical intermediates and perfume bases, in dyes and rubber accelerators, and in liquid crystals for electronics. { 'shif ‚bās }

SCHEIBEL EXTRACTOR

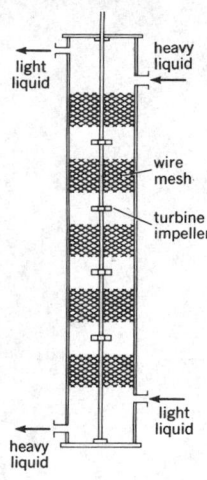

Schematic of Scheibel extractor. *(From R. E. Treybal, Mass Transfer Operations, 2d ed., McGraw-Hill, 1968)*

SCHERING BRIDGE

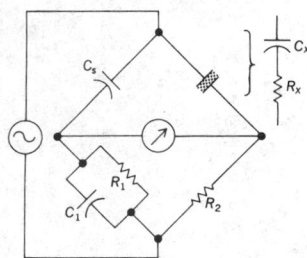

Circuit diagram of Schering bridge used to measure capacitance C_x and resistance R_x of equivalent series-circuit representation of capacitor. Standard capacitor C_s is assumed free from loss. Bridge is balanced when $C_x R_1 = C_s R_1$; $R_x C_s = R_2 C_1$.

schiffli [TEXT] A machine for putting embroidery and lace patterns on textiles. { 'shif·lē }

Schiff's reagent [ANALY CHEM] An aqueous solution of rosaniline and sulfurous acid; used in the Schiff test. { 'shifs rē,ā·jənt }

Schiff test [ANALY CHEM] A test for aldehydes by using an aqueous solution of rosaniline and sulfurous acid. { 'shif ,test }

Schilder's disease [MED] **1.** A retrogressive disease of the white matter in the central nervous system characterized by diffuse loss of myelin. **2.** Any of the progressive degenerative diseases of the white matter in the central nervous system. { 'shil·dərz di,zēz }

schiller See play of color. { 'shil·ər }

schillerization [OPTICS] Development of schiller in crystals due to the pattern of inclusions. { ,shil·ə·rə'zā·shən }

schiller layer [PHYS CHEM] One of a series of layers formed by sedimenting particles that exhibit bright colors in reflected light, because the layers are separated by approximately equal distances, with the distances being of the same order of magnitude as the wavelength of visible light. Also known as iridescent layer. { 'shil·ər ,lā·ər }

Schindleriidae [VERT ZOO] The single family of the order Schindlerioidei. { ,shind·lə'rī·ə,dē }

Schindlerioidei [VERT ZOO] A suborder of fishes in the order Perciformes composed of one monogeneric family comprising two tiny oceanic species that are transparent and neotenic. { ,shind·lə·rē'óid·ē,ī }

schindylesis [ANAT] A synarthrosis in which a plate of one bone is fixed in a fissure of another. { ,skin·də'lē·səs }

schirmerite [MINERAL] PbAg₄Bi₄S₉ A mineral composed of lead, silver, and bismuth sulfide. { 'shər·mə,rīt }

schist [GEOL] A large group of coarse-grained metamorphic rocks which readily split into thin plates or slabs as a result of the alignment of lamellar or prismatic minerals. { shist }

schist-arenite [PETR] A light-colored sandstone containing more than 20% rock fragments derived from an area of regionally metamorphosed rocks. { 'shist 'a·rə,nīt }

schistose [GEOL] Pertaining to rocks exhibiting schistosity. { 'shis,tōs }

schistosity [GEOL] A type of cleavage characteristic of metamorphic rocks, notably schists and phyllites, in which the rocks tend to split along parallel planes defined by the distribution and parallel arrangement of platy mineral crystals. { shis'täs·əd·ē }

schistosome dermatitis [MED] A dermatitis caused by penetration of the skin by certain schistosome cercariae. Also known as swamp itch; swimmer's itch. { 'shis·tə,sōm ,dər·mə'tīd·əs }

schistosomiasis [MED] A disease in which humans are parasitized by any of three species of blood flukes: *Schistosoma mansoni, S. haematobium,* and *S. japonicum;* adult worms inhabit the blood vessels. Also known as bilharzias; snail fever. { ,shis·tə·sə'mī·ə·səs }

Schistostegiales [BOT] A monospecific order of mosses; the small, slender, glaucous plants are distinguished by the luminous protonema. { ,shis·tə,stej·ē'ā·lēz }

schizaxon [NEUROSCI] An axon that divides, in its course, into equal or nearly equal branches. { skiz'ak,sän }

schizocarp [BOT] A dry fruit that separates at maturity into single-seeded indehiscent carpels. { 'skiz·ə,kärp }

Schizocoela [INV ZOO] A group of animal phyla, including Bryozoa, Brachiopoda, Phoronida, Sipunculoidea, Echiuroidea, Priapuloidea, Mollusca, Annelida, and Arthropoda, all characterized by the appearance of the coelom as a space in the embryonic mesoderm. { ¦skiz·ə¦sē·lə }

schizodont [INV ZOO] A multinucleate trophozoite that segments into merozoites. { 'skiz·ə,dänt }

schizogamy [BIOL] A form of reproduction involving division of an organism into a sexual and an asexual individual. { ski'zäg·ə·mē }

schizogenesis [BIOL] Reproduction by fission. { ,ski·zō'jen·ə·səs }

schizognathous [VERT ZOO] Descriptive of birds having a palate in which the vomer is small and pointed, the maxillopalatines are not united with each other or with the vomer, and the palatines articulate posteriorly with the rostrum. { ski'zäg·nə·thəs }

Schizogoniales [BOT] A small order of the Chlorophyta containing algae that are either submicroscopic filaments or macroscopic ribbons or sheets a few centimeters wide and attached by rhizoids to rocks. { ,skiz·ə,gō·nē'ā·lēz }

schizogony [INV ZOO] Asexual reproduction by multiple fission of a trophozoite; a characteristic of certain Sporozoa. { ski'zäg·ə·nē }

schizolite [MINERAL] A light-red variety of pectolite containing manganese. { 'skiz·ə,līt }

Schizomeridaceae [BOT] A family of green algae in the order Ulvales. { ¦skiz·ə,mer·ə'dās·ē,ē }

Schizomycetes [MICROBIO] Formerly a class of the division Protophyta which included the bacteria. { ,skiz·ə·mī'sēd·ēz }

Schizomycophyta [BOT] The designation for bacteria in those taxonomic systems that consider bacteria as plants. { ,skiz·ə·mī'käf·əd·ə }

schizont [INV ZOO] A multinucleate cell in certain members of the Sporozoa that is produced from a trophozoite in a cell of the host, and that segments into merozoites. { 'skī,zänt }

Schizopathidae [INV ZOO] A family of dimorphic zoantharians in the order Antipatharia. { ,skiz·ə'path·ə,dē }

schizopelmous [VERT ZOO] Having the two flexor tendons of the toes separate, as in certain birds. { ¦ski·zō¦pel·məs }

Schizophora [INV ZOO] A series of the dipteran suborder Cyclorrhapha in which adults possess a frontal suture through which a distensible sac, or ptilinum, is pushed to help the organism escape from its pupal case. { ski'zäf·ə·rə }

schizophrenia [PSYCH] A group of mental disorders characterized by withdrawal from reality and by alterations in thinking, feeling, and concept formations. Also known as dementia praecox. { ,skit·sə'frē·nē·ə }

Schizophyceae [MICROBIO] The blue-green algae, a class of the division Protophyta. { ,skiz·ə'fī·sē,ē }

Schizophyta [BOT] The prokaryotes, a division of the plant subkingdom Thallobionta; includes the bacteria and blue-green algae. { ski'zäf·əd·ē }

schizopod [INV ZOO] **1.** Having the limbs split so that each has an endopodite and an exopodite, as in certain crustaceans. **2.** A biramous appendage. { 'skiz·ə,päd }

Schizopteridae [INV ZOO] A family of minute ground-inhabiting hemipterans in the group Dipsocoeoidea; individuals characteristically live in leaf mold. { ,ski·zäp'ter·ə,dē }

schizorhinal [VERT ZOO] Having a deep cleft on the posterior margin of the osseous external nares, as in certain birds. { ¦skiz·ə¦rīn·əl }

schizothecal [VERT ZOO] Having the horny envelope of the tarsus divided into scalelike plates; refers to most birds. { ¦skiz·ə'thē·kəl }

schizothoracic [INV ZOO] Having a prothorax that is large and loosely articulated with the thorax. { ¦ski·zō·thə'ras·ik }

schlanite [MATER] The soluble resin extracted from anthracoxene by ether. { 'shlä,nīt }

Schleiermacher's method [THERMO] A method of determining the thermal conductivity of a gas, in which the gas is placed in a cylinder with an electrically heated wire along its axis, and the electric energy supplied to the wire and the temperatures of wire and cylinder are measured. { 'shlī·ər mäk·ərz ,meth·əd }

Schlemm's canal [ANAT] A space or series of spaces at the junction of the sclera and cornea in the eye; drains aqueous humor from the anterior chamber. { 'shlemz kə,nal }

schlempe See vinasse. { 'shlem·pə }

Schlernwind [METEOROL] East wind blowing down from the Schlern near Bozen in Tyrol, Austria. { 'shlərn,vint }

schlicht function See shlicht function. { 'shlikt ,fəŋ·shən }

Schlick vibration formula [NAV ARCH] A formula for calculation of hull vibration in ships. { 'shlik vī'brā·shən ,fōr·myə·lə }

schlieren [OPTICS] In atmospheric optics, parcels or strata of air having densities sufficiently different from that of their surroundings so that they may be discerned by means of refraction anomalies in transmitted light. [PETR] Irregular streaks with shaded borders in some igneous rocks, representing the segregation of light and dark minerals or altered inclusions, elongated by flow. { 'shlir·ən }

schlieren arch [GEOL] An intrusive igneous body with flow layers which occur along its borders but which are poorly developed or absent in its interior. { 'shlir·ən ,ärch }

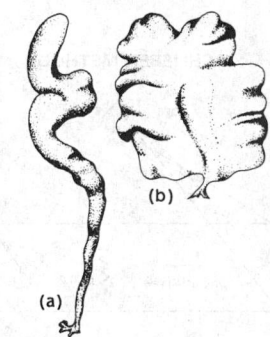

SCHIZOGONIALES

Thalli of *Prasiola* species: (a) ribbonlike; (b) sheetlike.

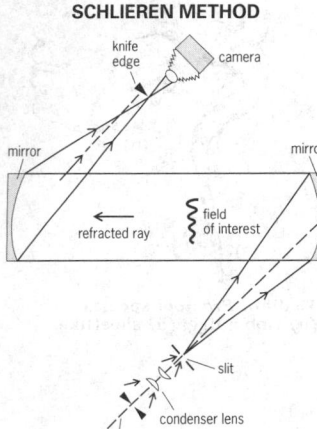

SCHLIEREN METHOD

Knife-edge method of viewing schlieren, employing the "z" configuration.

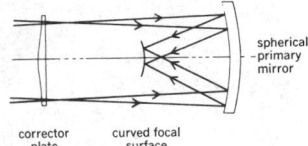

SCHMIDT SYSTEM

Optics of the Schmidt system showing axial and the extra-axial light paths.

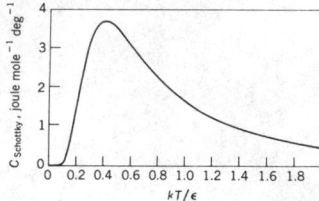

SCHOTTKY ANOMALY

Theoretical curve for the Schottky heat capacity when there are two energy levels separated by a splitting ϵ. T = temperature; k = Boltzmann constant. *(After H. M. Rosenberg, Low Temperature Solid State Physics, Clarendon Press, 1963)*

schlieren dome [GEOL] An intrusive body more or less completely outlined by flow layers which culminate in one central area. { 'shlir·ən ˌdōm }

schlieren method [OPTICS] An optical technique that detects density gradients occurring in a fluid flow; in its simplest form, light from a slit is collimated by a lens and focused onto a knife-edge by a second lens, the flow pattern is placed between these two lenses, and the diffraction pattern that results on a screen or photographic film placed behind the knife-edge is observed. { 'shlir·ən ˌmeth·əd }

Schlumberger dipmeter [ENG] An instrument that measures both the amount and direction of dip by readings taken in the borehole; it consists of a long, cylindrical body with two telescoping parts and three long, springy metal strips, arranged symmetrically round the body, which press outward and make contact with the walls of the hole. { 'shləm·bər,zhä 'dip,mēd·ər }

Schlumberger photoclinometer [ENG] An instrument that measures simultaneously the amount and direction of the deviation of a borehole; the sonde, designed to lie exactly parallel to the axis of the borehole, is fitted with a small camera on the axis of a graduated glass bowl, in which a steel ball rolls freely and a compass is mounted in gimbals; the camera is electrically operated from the surface and takes a photograph of the bowl, the steel ball marks the amount of deviation, and the position in relation to the image of the compass needle gives the direction of deviation. { 'shləm·bər,zhä ,fōd·ō·kli'näm·əd·ər }

Schmidt camera *See* Schmidt system. { 'shmit ,kam·rə }

Schmidt-Cassegrain telescope [OPTICS] A variant of the Schmidt system which uses a Schmidt corrector plate together with a pair of spheroidal or slightly aspherical mirrors arranged as in a Cassegrain telescope. { 'shmit kas·gran 'tel·ə,skōp }

Schmidt correction plate [OPTICS] In the Schmidt system, a glass plate with one face aspherical and the other aspherical and deviating from a plane in such a way that it bends light, traveling to the system's spherical mirror, so as to correct for spherical aberration and coma. Also known as Schmidt lens. { 'shmit kə'rek·shən ˌplāt }

Schmidt field balance [ENG] An instrument that operates as both a horizontal and vertical field balance and consists of a permanent magnet pivoted on a knife edge. { 'shmit 'fēld ˌbal·əns }

Schmidt lens *See* Schmidt correction plate. { 'shmit ˌlenz }

Schmidt lines [NUC PHYS] Two lines, on a graph of nuclear magnetic moment versus nuclear spin, on which points describing all nuclides should lie, according to the independent particle model; experimentally, however, points describing nuclides are scattered between the lines. { 'shmit ˌlīnz }

Schmidt-Maksutov telescope *See* meniscus-Schmidt telescope. { 'shmit mak'su·täf ˌtel·ə,skōp }

Schmidt net [GEOL] A coordinate or reference system used to plot a Schmidt projection. { 'shmit ˌnet }

Schmidt number 1 *See* Prandtl number. { 'shmit ˌnəm·bər 'wən }

Schmidt number 2 *See* Semenov number 1. { 'shmit ˌnəm·bər 'tü }

Schmidt number 3 [PHYS CHEM] A dimensionless number used in electrochemistry, equal to the product of the dielectric susceptibility and the dynamic viscosity of a fluid divided by the product of the fluid density, electrical conductivity, and the square of a characteristic length. Symbolized Sc_3. { 'shmit ˌnəm·bər 'thrē }

Schmidt objective *See* Schmidt system. { 'shmit äb,jek·tiv }

Schmidt optics *See* Schmidt system. { 'shmit ˌäp·tiks }

Schmidt projection [GEOL] A Lambert azimuthal equal-area projection of the lower hemisphere of a sphere onto the plane of a meridian; used in structural geology. { 'shmit prə,jek·shən }

Schmidt reflector [OPTICS] A telescope employing the Schmidt system. { 'shmit ri,flek·tər }

Schmidt system [OPTICS] An optical system designed to eliminate spherical aberration and coma, which, in its original form, consists of a spherical mirror, a Schmidt correction plate near the focus of the mirror, and usually a curved reflecting plate at the focus of the mirror; used in astronomical telescopes with unusually wide fields of view and in spectroscopes, and to project a television image from a cathode-ray tube onto a

screen. Also known as projection optics; Schmidt camera; Schmidt objective; Schmidt optics. { 'shmit ,sis·təm }

Schmitt circuit [ELECTR] A bistable pulse generator in which an output pulse of constant amplitude exists only as long as the input voltage exceeds a certain value. Also known as Schmitt limiter; Schmitt trigger. { 'shmit ,sər·kət }

Schmitt limiter *See* Schmitt circuit. { 'shmit 'lim·əd·ər }

Schmitt trigger *See* Schmitt circuit. { 'shmit 'trig·ər }

Schneiderian membrane [ANAT] The mucosa lining the nasal cavities and paranasal sinuses. { shnī'dir·ē·ən 'mem ,brān }

Schneider recoil system [MECH ENG] A recoil system for artillery, employing the hydropneumatic principle without a floating piston. { 'shnī·dər 'rē,kȯil ,sis·təm }

Schneider's index [MED] A test of general physical and circulatory efficiency, consisting of pulse and blood pressure observations under standard conditions of rest and exercise. { 'shnī·dərz 'in,deks }

Schoch effect [ACOUS] A shift of a few wavelengths in the position of an ultrasonic sound beam that undergoes total internal reflection from a surface. { 'shäk i,fekt }

Schoelkopf's acid [ORG CHEM] A dye of the following types: 1-naphthol-4,8-disulfonic acid, 1-naphthylamine-4,8-disulfonic acid, and 1-naphthylamine-8-sulfonic acid; may be toxic. { 'shȯl,kȯpfs ,as·əd }

Schoenbiinae [INV ZOO] A subfamily of moths in the family Pyralididae, including the genus *Acentropus,* the most completely aquatic Lepidoptera. { shən'bī·ə,nē }

Schoenherr-Hessberger process [CHEM ENG] A nitrogen-fixation process used in Norway; employs a very long (22 feet or 7 meters) alternating-current arc around which air moves in a helical path in a 746-kilowatt furnace. { 'shən·her 'hes,bərg·ər ,prä·səs }

schoepite [MINERAL] $UO_3 \cdot 2H_2O$ A yellow secondary mineral composed of hydrous uranium oxide. { 'ske,pīt }

Schönberg-Chandrasekhar limit *See* Chandrasekhar-Schönberg limit. { 'shərn,bȯrg ¦chan·drə'sä·kər ,lim·ət }

schönfelsite [PETR] A form of basalt containing embedded crystals of olivine and augite in a complex, dense fine-grained groundmass. { 'shən,fel,zīt }

Schönflies crystal symbols [CRYSTAL] Symbols denoting the 32 crystal point groups or symmetry classes; capital letters indicate the general type of class, and subscripts the multiplicity of rotation axes and the existence of additional symmetries. { 'shən,flēs 'krist·əl ,sim·bəlz }

schooner [NAV ARCH] A sailing vessel with two or more masts rigged fore and aft. { 'skün·ər }

Schoop process [ENG] A process for coating surfaces by spraying with high-velocity molten metal particles. { shōp ,prä·səs }

schorl *See* schorlite. { shȯrl }

schorlite [MINERAL] The black, iron-rich, opaque variety of tourmaline. Also known as schorl. { 'shȯr,lit }

schorlomite [MINERAL] $Ca_3(Fe,Ti)_2(Si,Ti)_3O_{12}$ A black mineral of the garnet group that has a vitreous luster and usually occurs in masses; hardness is 7–7.5 on Mohs scale, and specific gravity is 3.81–3.88. { 'shȯr·lə,mīt }

schott [GEOGR] A shallow saline lake in southern Tunisia or on the plateaus of northern Algeria, which is usually dry during the summer. { shät }

Schotten-Baumann reaction [ORG CHEM] An acylation reaction that uses an acid chloride in the presence of dilute alkali to acylate the hydroxyl and amino group of organic compounds. { 'shät·ən 'baú,män rē,ak·tən }

Schottky anomaly [SOLID STATE] A contribution to the heat capacity of a solid arising from the thermal population of discrete energy levels as the temperature is raised; the effect is particularly prominent at low temperatures. { 'shät·kē ə,näm·ə·lē }

Schottky barrier [ELECTR] A transition region formed within a semiconductor surface to serve as a rectifying barrier at a junction with a layer of metal. { 'shät·kē ,bar·ē·ər }

Schottky barrier diode [ELECTR] A semiconductor diode formed by contact between a semiconductor layer and a metal coating; it has a nonlinear rectifying characteristic; hot carriers (electrons for *n*-type material or holes for *p*-type material) are emitted from the Schottky barrier of the semiconductor and move to the metal coating that is the diode base; since majority carriers predominate, there is essentially no injection or storage of minority carriers to limit switching speeds. Also known as hot-carrier diode; Schottky diode. { 'shät·kē ¦bar·ē·ər 'dī‚ōd }

Schottky defect [SOLID STATE] **1.** A defect in an ionic crystal in which a single ion is removed from its interior lattice site and relocated in a lattice site at the surface of the crystal. **2.** A defect in an ionic crystal consisting of the smallest number of positive-ion vacancies and negative-ion vacancies which leave the crystal electrically neutral. { 'shät·kē di‚fekt }

Schottky diode *See* Schottky barrier diode. { 'shät·kē 'dī‚ōd }

Schottky-diode FET logic [ELECTR] A logic gate configuration used with gallium-arsenide field-effect transistors operating in the depletion mode, in which very small Schottky diodes at the gate input provide the logical OR function and the level shifting required to make the input and output voltage levels compatible. Abbreviated SDFL. { 'shät·kē ¦dī‚ōd ¦ef¦ē¦tē 'läj·ik }

Schottky effect [SOLID STATE] The enhancement of the thermionic emission of a conductor resulting from an electric field at the conductor surface. { 'shät·kē i‚fekt }

Schottky line [SOLID STATE] A graph of the logarithm of the saturation current from a thermionic cathode as a function of the square root of anode voltage; it is a straight line according to the Schottky theory. { 'shät·kē ‚līn }

Schottky noise *See* shot noise. { 'shät·kē ‚nóiz }

Schottky theory [SOLID STATE] A theory describing the rectification properties of the junction between a semiconductor and a metal that result from formation of a depletion layer at the surface of contact. { 'shät·kē ‚thē·ə·rē }

Schottky transistor-transistor logic [ELECTR] A transistor-transistor logic circuit in which a Schottky diode with forward diode voltage is placed across the base-collector junction of the output transistor in order to improve the speed of the circuit. { 'shät·kē tran¦zis·tər tran¦zis·tər 'läj·ik }

Schrage motor [ELEC] A type of alternating-current commutator motor whose speed is controlled by varying the position of sets of brushes on the commutator. { 'shräg·ə ‚mōd·ər }

schreibersite [MINERAL] $(Fe,Ni)_3P$ A silver-white to tin-white magnetic meteorite mineral crystallizing in the tetragonal system and occurring in tables or plates as oriented inclusions in iron meteorites. Also known as rhabdite. { 'shrī·bər‚sīt }

schreinerize [TEXT] To give a lustrous finish to a cotton fabric by passing it under rollers engraved with fine lines. { 'shrī·nə‚rīz }

schriesheimite [PETR] An amphibole peridotite that contains diopside. { 'shrē·shē·ə‚mīt }

Schrödinger equation [QUANT MECH] A partial differential equation governing the Schrödinger wave function ψ of a system of one or more nonrelativistic particles; $\hbar(\partial\psi/\partial t) = H\psi$, where H is a linear operator, the Hamiltonian, which depends on the dynamics of the system, and $\hbar$ is Planck's constant divided by 2π. { 'shräd·iŋ·ər i‚kwä·zhən }

Schrödinger-Klein-Gordon equation *See* Klein-Gordon equation. { 'shräd·iŋ·ər 'klīn 'górd·ən i‚kwä·zhən }

Schrödinger-Pauli equation [QUANT MECH] A modification of the Schrödinger equation to describe a particle with spin of $^1/_2\hbar$, where $\hbar$ is Planck's constant divided by 2π; the wave function has two components, corresponding to the particle's spin pointing in either of two opposite directions. { 'shräd·iŋ·ər 'pau·lē i‚kwä·zhən }

Schrödinger picture [QUANT MECH] A mode of description of a quantum-mechanical system in which dynamical states are represented by vectors which evolve in the course of time, and physical quantities are represented by stationary operators, in contrast to the Heisenberg picture. { 'shräd·iŋ·ər 'pik·chər }

Schrödinger representation *See* position representation. { 'shräd·iŋ·ər ‚rep·ri·zən'tā·shən }

Schrödinger's wave mechanics [QUANT MECH] The version of nonrelativistic quantum mechanics in which a system is characterized by a wave function which is a function of the coordinates of all the particles of the system and time, and obeys a differential equation, the Schrödinger equation; physical quantities are represented by differential operators which may act on the wave function, and expectation values of measurements are equal to integrals involving the corresponding operator and the wave function. Also known as wave mechanics. { 'shräd·iŋ·ərz 'wāv mi‚kan·iks }

Schrödinger wave function [QUANT MECH] A function of the coordinates of the particles of a system and of time which is a solution of the Schrödinger equation and which determines

the average result of every conceivable experiment on the system. Also known as probability amplitude; psi function; wave function. { 'shräd·iŋ·ər 'wāv ‚faŋk·shən }

schroeckingerite [MINERAL] $NaCa_3(UO_2)(CO_3)(SO_4)F\cdot10H_2O$ A yellowish secondary mineral composed of hydrous sodium calcium uranyl carbonate, sulfate, and fluoride. { 'shrek·iŋ·ə‚rīt }

Schroeder-Bernstein theorem [MATH] If a set A has at least as many elements as another set B and B has at least as many elements as A, then A and B have the same number of elements. { 'shräd·ə· 'bərn‚stīn ‚thē·ə·rəm }

Schroter effect [ASTRON] The occurrence of the dichotomy of Venus earlier than theoretically predicted when it is waning, and later than theoretically predicted when waxing. { 'shröd·ər i‚fekt }

schrötterite [MINERAL] An opaline variety of allophane that is rich in aluminum. { 'shräd·ə‚rīt }

schrund line [GEOL] The base of the bergschrund, or deep crevasse, at a late stage in the excavation of a cirque; the schrund line separates the steep slope of the cirque wall from the gentler slope below. { 'shrunt ‚līn }

Schubertellidae [PALEON] An extinct family of marine protozoans in the superfamily Fusulinacea. { ‚shü·bər'tel·ə‚dē }

Schuermann series [CHEM] A list of metals so arranged that the sulfide of one is precipitated at the expense of the sulfide of any lower metal in the series. { 'shói·ər‚män ‚sir·ēz }

Schuler pendulum [MECH] Any apparatus which swings, because of gravity, with a natural period of 84.4 minutes, that is, with the same period as a hypothetical simple pendulum whose length is the earth's radius; the pendulum arm remains vertical despite any motion of its pivot, and the apparatus is therefore useful in navigation. { 'shü·lər ‚pen·jə·ləm }

Schuler tuning [ENG] The designing of gyroscopic devices so that their periods of oscillation will be about 84.4 minutes. { 'shü·lər ‚tün·iŋ }

schultenite [MINERAL] $PbHAsO_4$ A colorless mineral composed of lead hydrogen arsenate occurring in tabular orthorhombic crystals. { 'shult·ən‚īt }

Schultz-Charlton test [IMMUNOL] A skin test for the diagnosis of scarlet fever, performed by the intradermal injection of human scarlet fever immune serum; a positive reaction consists of blanching of the rash in the area surrounding the point of injection. { 'shults 'chärl·tən ‚test }

Schultz-Dale reaction [IMMUNOL] A method for demonstrating anaphylactic hypersensitivity outside the body by suspending an excised intestinal loop or uterine strip from a sensitized animal in an oxygenated, physiological salt solution; addition of the proper allergen causes contraction of the smooth muscle. { 'shults 'dāl rē‚ak·shən }

Schultze powder [MATER] A smokeless powder propellant, consisting of wood pellets impregnated with barium nitrate and potassium nitrate. { 'shult·sə ‚paud·ər }

Schultz-Hardy rule [CHEM] The sensitivity of lyophobic colloids to coagulating electrolytes is governed by the charge of the ion opposite that of the colloid, and the sensitivity increases more rapidly than the charge of the ion. { 'shults 'här·dē ‚rül }

Schulze's reagent [ANALY CHEM] An oxidizing mixture consisting of a saturated aqueous solution of $KCIO_3$ and varying amounts of concentrated HNO_3; commonly used in palynologic macerations. { 'shult·səz rē‚ā·jənt }

Schumann plate [GRAPHICS] A type of photographic plate in which the silver halide is held to the glass, using a special method, by a layer of gelatin so thin that it is transparent to ultraviolet light with wavelengths as short as 1200 angstroms (120 nanometers), in contrast to the limit of about 2200 angstroms (220 nanometers) in a conventional photographic plate. { 'shü‚män ‚plāt }

Schumann region [OPTICS] The most extreme ultraviolet portion of the electromagnetic spectrum that will affect a photographic plate. { 'shü‚män ‚rē·jən }

Schumann resonance [GEOPHYS] A resonance created by lightning-induced electromagnetic radiation trapped in the spherical waveguide formed between the ionosphere and the earth. { 'shü·mən ‚rez·ə‚nəns }

schungite [GEOL] Amorphous carbon-rich material occurring in Precambrian schists. { 'shuŋ‚gīt }

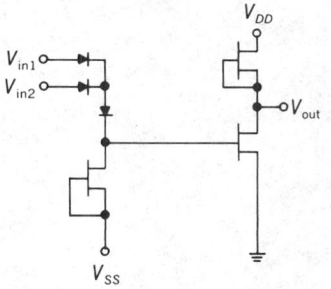
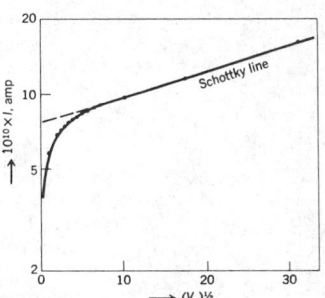

schuppen structure See imbricate structure. { 'shŭp·ən ˌstrək·chər }

Schur-Cohn test [MATH] A test to determine whether all the coefficients of a polynomial have magnitude less than one; the polynomial has this property only if each of a series of determinants formed from the coefficients of the polynomial in a specified manner is positive for determinants of even degree and negative for determinants of odd degree. { 'shür 'kŏn ˌtest }

Schur's lemma [MATH] For certain types of modules M, the ring consisting of all homomorphisms of M to itself will be a division ring. { 'shürz ˌlem·ə }

Schuster method [SPECT] A method for focusing a prism spectroscope without using a distant object or a Gauss eyepiece. { 'shŭs·tər ˌmeth·əd }

Schwagerinidae [PALEON] A family of fusulinacean protozoans that flourished during the Early and Middle Pennsylvanian and became extinct during the Late Permian. { ˌshwäg·ə'rin·ə,dē }

Schwann cell [HISTOL] One of the cells that surround peripheral axons forming sheaths of the neurilemma. { 'shwän ˌsel }

schwannoma See neurilemmoma. { shwä'nō·mə }

schwartzembergite [MINERAL] $Pb_5(IO_3)Cl_3O_3$ A mineral composed of lead iodate, chloride, and oxide. { 'shwȯrt·səm,bər,gīt }

Schwartz's theory of distributions [MATH] A theory that treats distributions as continuous linear functionals on a vector space of continuous functions which have continuous derivatives of all orders and vanish appropriately at infinity. { ˈshwȯrts ˈthē·ə·rē əv ˌdis·trə'byü·shənz }

Schwarz-Christoffel transformations [MATH] Those complex transformations which conformally map the interior of a given polygon onto the portion of the complex plane above the real axis. { 'shvärts 'kris·tə·fel ˌtranz·fər'mā·shənz }

Schwarz' inequality See Cauchy-Schwarz inequality. { 'shvärts ˌin·i,kwäl·əd·ē }

Schwarz reflection principle [MATH] To obtain the analytic continuation of a given function $f(z)$ analytic in a region R, whose boundary contains a segment of the real axis, into a region reflected from R through this segment, one takes the complex conjugate function $f(\bar{z})$. { 'shvärts ri'flek·shən ˌprin·sə·pəl }

Schwarzschild anastigmat [OPTICS] A Gregorian telescope whose surfaces are altered to reduce astigmatism. { 'shvärts,shilt ˌan·ə'stig·mat }

Schwarzschild criterion [ASTROPHYS] A criterion for determining the stability of a stellar medium against convective motion, according to which convection takes place when the temperature gradient is greater than the gradient that would exist if the medium were adiabatic. { 'shvärts,shilt krī,tir·ē·ən }

Schwarzschild radius [RELAT] Conventionally taken to be twice the black hole mass appearing in the general relativistic Schwarzschild solution times the gravitational constant divided by the square of the speed of light; the event horizon in a Schwarzschild solution is at the Schwarzschild radius. { 'shvärts,shilt ˌräd·ē·əs }

Schwarzschild singularity [RELAT] The coordinate singularity at the event horizon that exists in a certain coordinate system describing a nonrotating black hole in empty space. { ˈshvärts,shilt ˌsiŋ·gyə'lar·əd·ē }

Schwarzschild solution [RELAT] The unique solution of general relativity theory describing a nonrotating black hole in empty space. { 'shvärts,shilt sə,lü·shən }

Schwarz's lemma [MATH] If an analytic function of the unit disk to itself sends the origin to the origin then it must be distance-decreasing. { 'shvärt·səz ˌlem·ə }

Schwassman-Wachmann comet [ASTRON] A variable cometlike asteroid whose period is 16 years; its orbit is very nearly circular and lies between the orbits of Saturn and Jupiter. { 'shväs,män 'va_k,män ˌkäm·ət }

Schweitzer's reagent [CHEM] An ammoniacal solution of cupric hydroxide; used to dissolve cellulose, silk, and linen, and to test for wool. { 'shvīt·sərz rē,ā·jənt }

Schweydar mechanical detector [ENG] A seismic detector that senses and records refracted waves; a lead sphere is suspended by a flat spring, the sphere's motion is magnified by an aluminum cone that moves a bow around a spindle carrying a mirror, and this motion is then photographically recorded. { 'shwād·ər mi,kan·i·kəl di'tek·tər }

Schwinger critical field [ELEC] That electric field at which an electron is accelerated from rest to a velocity at which its kinetic energy equals its rest energy over a distance of one Compton wavelength. { ˈshviŋ·ər ˈkrid·ə·kəl 'fēld }

Schwinger's variational principle [PHYS] A method used in electromagnetic theory, or similar disciplines, to calculate an approximate value or a linear of quadratic functional, such as a scattering amplitude or reflection coefficient, when the function for which the functional is evaluated is the solution of an integral equation. { 'shviŋ·ərz ˌver·ē'ā·shən·əl ˌprin·sə·pəl }

Sciaenidae [VERT ZOO] A family of perciform fishes in the suborder Percoidei, which includes the drums. { sī'ēn·ə,dē }

sciatica [MED] Neuralgic pain in the lower extremities, hips, and back caused by inflammation or injury to the sciatic nerve. { sī'ad·ə·kə }

sciatic nerve [NEUROSCI] Either of a pair of long nerves that originate in the lower spinal cord and send fibers to the upper thigh muscles and the joints, skin, and muscles of the leg. { sī'ad·ik 'nərv }

SCIDS See severe combined immunodeficiency syndrome. { skidz or ˌesˌsēˌīˌdē'es }

science [SCI TECH] A branch of study in which facts are observed and classified, and, usually, quantitative laws are formulated and verified; involves the application of mathematical reasoning and data analysis to natural phenomena. { 'sī·əns }

scientific calculator [COMPUT SCI] An electronic calculator that has provisions for handling exponential, trigonometric, and sometimes other special functions in addition to performing arithmetic operations. { ˌsī·ən'tif·ik 'kal·kyə,lād·ər }

scientific computer [COMPUT SCI] A computer which has a very large memory and is capable of handling extremely high-speed arithmetic and a very large variety of floating-point arithmetic commands. { ˌsī·ən'tif·ik kəm'pyüd·ər }

scientific information [SCI TECH] That part of technical information concerned with the study of natural phenomena. { ˌsī·ən'tif·ik ˌin·fər'mā·shən }

scientific method [SCI TECH] The systematic collection and classification of data and, usually, the formulation and testing of hypotheses based on the data. { ˌsī·ən'tif·ik 'meth·əd }

scientific notation [COMPUT SCI] The display of numbers in which a base number, representing the significant digits, is followed by a number representing the power of 10 to which the base number is raised. { ˌsī·ən'tif·ik nō'tā·shən }

scientific system [COMPUT SCI] A system devoted principally to computations as opposed to business and data-processing systems, the main emphasis of which is on the updating of data records and files rather than the performance of calculations. { ˌsī·ən'tif·ik 'sis·təm }

scientist [SCI TECH] A person having the training, ability, and desire to seek new knowledge, new principles, and new materials in some field of science. { 'sī·ən·tist }

scillarenin [PHARM] $C_{24}H_{32}O_4$ A crystalline compound that forms prisms from a methanol solution and melts at 232–238°C; used in medicine for cardiac disease. { si'lar·ə·nən }

Scincidae [VERT ZOO] The skinks, a family of the reptilian suborder Sauria which have reduced limbs and snakelike bodies. { 'skiŋ·kə,dē }

Scinidae [INV ZOO] A family of bathypelagic, amphipod crustaceans in the suborder Hyperiidea. { 'skin·ə,dē }

scintillation [ELECTROMAG] **1.** A rapid apparent displacement of a target indication from its mean position on a radar display; one cause is shifting of the effective reflection point on the target. Also known as target glint; target scintillation; wander. **2.** Random fluctuation, in radio propagation, of the received field about its mean value, the deviations usually being relatively small. [LAP] The flashing, twinkling, or sparkling of light, or the alternating display of reflections, from the polished facets of a gemstone. [NUCLEO] A flash of light produced in a phosphor by an ionizing particle or photon. [OPTICS] **1.** Rapid changes of brightness of stars or other distant, celestial objects caused by variations in the density of the air through which the light passes. **2.** Rapid changes in the values of irradiance over the cross section of a laser beam. { ˌsint·əl'ā·shən }

scintillation camera [NUCLEO] A camera that gives a complete image of radionuclide distribution in a particular area of the human body in one exposure, in contrast to line-scanning techniques. { ‚sint·əl'ā·shən ‚kam·rə }

scintillation counter [NUCLEO] A device in which the scintillations produced in a fluorescent material by an ionizing radiation are detected and counted by a multiplier phototube and associated circuits; used in medical and nuclear research and in prospecting for radioactive ores. Also known as scintillation detector; scintillometer. { ‚sint·əl'ā·shən ‚kaúnt·ər }

scintillation-counter crystal [NUCLEO] A substance (a fluor, such as thallium-activated calcium tungstate) that emits a flash of light (scintillation) when contacted by a high-energy particle, for example, an alpha, beta, or gamma ray. { ‚sint·əl'ā·shən ‚kaúnt·ər ‚krist·əl }

scintillation detector See scintillation counter. { ‚sint·əl'ā·shən di‚tek·tər }

scintillation spectrometer [NUCLEO] A scintillation counter adapted to measuring the energy and intensity of gamma rays from radioactive elements. { ‚sint·əl'ā·shən spek'träm·əd·ər }

scintillator [NUCLEO] A material that emits optical photons in response to ionizing radiation. { 'sint·əl‚ād·ər }

scintillometer See scintillation counter. { ‚sint·əl'äm·əd·ər }

scintillon [INV ZOO] An outpocketing of the cytoplasm in dinoflagellates which contains luciferase and luciferin-binding protein and is the source of bioluminescence. { 'sint·i‚län }

Sciomyzidae [INV ZOO] A family of myodarian cyclorrhaphous dipteran insects in the subsection Acalypteratae. { ‚sī·ə'miz·ə‚dē }

scion [BOT] A section of a plant, usually a stem or bud, which is attached to the stock in grafting. { 'sī·ən }

sciophilous [ECOL] Capable of thriving in shade. { sī'äf·ə·ləs }

sciophyte [BOT] A plant that thrives at lowered light intensity. { 'sī·ə‚fīt }

scirrhous carcinoma [MED] A hard, poorly differentiated adenocarcinoma in which the anaplastic cells are surrounded by dense bundles of collagenous fibers. { 'sir·əs ‚kärs·ən'ō·mə }

scissor engine See cat-and-mouse engine. { 'siz·ər ‚en·jən }

scissoring [COMPUT SCI] In computer graphics, the deletion of those parts of an image that fall outside a window that has been placed over the original image. Also known as clipping. { 'siz·ər·iŋ }

scissor jack [MECH ENG] A lifting jack driven by a horizontal screw; the linkages of the jack are parallelograms whose horizontal diagonals are lengthened or shortened by the screw. { 'siz·ər ‚jak }

scissors bridge [CIV ENG] A light metal bridge that can be folded and carried by a military tank. { 'siz·ərz ‚brij }

scissors crossover [CIV ENG] A scissor-shaped junction between two parallel railway tracks. Also called double crossover. { 'siz·ərz 'krós‚ō·vər }

scissors fault [GEOL] A fault on which the offset or separation along the strike increases in one direction from an initial point and decreases in the other direction. Also known as differential fault. { 'siz·ərz ‚fólt }

scissors mode [NUC PHYS] A mode of nuclear collective motion in permanently deformed nuclei that is of orbital magnetic dipole character and can be explained as a scissorslike vibration of the ellipsoidally deformed bodies of protons and neutrons against each other. { 'siz·ərs ‚mōd }

scissors truss [BUILD] A roof truss in which the braces cross like scissors blades. { 'siz·ərz ‚trəs }

Scitaminales [BOT] An equivalent name for Zingiberales. { ‚sīd·ə·mə'nā·lēz }

Scitamineae [BOT] An equivalent name for Zingiberales. { ‚sīd·ə'min·ē‚ē }

Sciuridae [VERT ZOO] A family of rodents including squirrels, chipmunks, marmots, and related forms. { sī'yúr·ə‚dē }

Sciuromorpha [VERT ZOO] A suborder of Rodentia according to the classical arrangement of the order. { sī‚yùr·ə'mór·fə }

sclera [ANAT] The hard outer coat of the eye, continuous with the cornea in front and with the sheath of the optic nerve behind. { 'skler·ə }

Scleractinia [INV ZOO] An order of the subclass Zoantharia which comprises the true, or stony, corals; these are solitary or colonial anthozoans which attach to a firm substrate. { ‚sklerak'tin·ē·ə }

Scleraxonia [INV ZOO] A suborder of cnidarians in the order

Gorgonacea in which the axial skeleton has calcareous spicules. { ‚skleer·ak·sō·nē·ə }

sclereid [BOT] A thick-walled, lignified plant cell typically found in sclerenchyma. { 'sklir·ē·əd }

sclerema neonatorum [MED] A disease of the newborn, particularly the premature or the undernourished, dehydrated, and debilitated, characterized by waxy-white hardening of subcutaneous tissue. { sklə'rē·mə ‚ne·ə·nə'tór·əm }

sclerenchyma [BOT] A supporting plant tissue composed principally of sclereids whose walls are often mineralized. { sklə'ren·kə·mə }

scleriasis See scleroderma. { sklə'rī·ə·səs }

sclerite [INV ZOO] One of the sclerotized plates of the integument of an arthropod. { 'skle‚rīt }

scleroblast [INV ZOO] A spicule-secreting cell in Porifera. { 'skler·ə‚blast }

scleroblastema [EMBRYO] Embryonic tissue from which bones are formed. { ‚skler·ō·bla'stē·mə }

sclerocaulous [BOT] Having a hard, dry stem because of exceptional development of sclerenchyma. { ‚skler·ō'kól·əs }

sclerochore [BOT] A plant that disperses disseminules without apparent morphological adaptations. { 'skler·ə‚kór }

Sclerodactylidae [INV ZOO] A family of echinoderms in the order Dendrochirotida having a complex calcareous ring. { ‚skler·ō‚dak'til·ə‚dē }

scleroderma [MED] An abnormal increase in collagenous connective tissue in the skin. Also known as chorionitis; dermatosclerosis; scleriasis. { ‚skler·ō'dər·mə }

sclerodermatous [INV ZOO] Having a skeleton that is composed of scleroderm, as certain corals. [VERT ZOO] Having a hard outer covering, for example, hard plate or horny scale. { ‚skler·ō'dər·məd·əs }

Sclerogibbidae [INV ZOO] A monospecific family of the hymenopteran superfamily Bethyloidea. { ‚skler·ō'jib·ə‚dē }

sclerometer [ENG] An instrument used to determine the hardness of a material by measuring the pressure needed to scratch or indent a surface with a diamond point. { sklə'räm·əd·ər }

sclerophyllous [BOT] Characterized by thick, hard foliage due to well-developed sclerenchymatous tissue. { ‚skler·ə'fil·əs }

scleroprotein [BIOCHEM] Any one of a class of proteins, such as keratin, fibroin, and the collagens, which occur in hard parts of the animal body and serve to support or protect. Also known as albuminoid. { ‚skler·ō'prō‚tēn }

scleroscope [ENG] An instrument used to determine the hardness of a material by measuring the height to which a standard ball rebounds from its surface when dropped from a standard height. { 'skler·ə‚skōp }

scleroscope hardness test See Shore scleroscope hardness test. { 'skler·ə‚skōp 'härd·nəs ‚test }

scleroseptum [INV ZOO] A calcareous radial septum in certain corals. { ‚skler·ō'sep·təm }

sclerosing adenomatosis [MED] A form of mammary dysplasia in which ductular structures are encased in fibrous tissue, simulating invading cancerous ductular structures. Also known as fibrosing adenomatosis. { sklə'rōs·iŋ ‚ad·ən·ō·mə'tō·səs }

sclerosing hemangioma [MED] A type of benign histiocytoma marked by prominence of the capillary channels. { sklə'rōs·iŋ hi‚mān‚jē'ō·mə }

sclerosis [PATH] Hardening of a tissue, especially by proliferation of fibrous connective tissue. { sklə'rō·səs }

sclerotesta [BOT] The middle hard layer of the testa in various seeds. { ‚skler·ə'tes·tə }

sclerotic [ANAT] Pertaining to the sclera. [MED] **1.** Hard. **2.** Of or pertaining to sclerosis. { sklə'räd·ik }

sclerotinite [GEOL] A variety of inertinite composed of fungal sclerotia. { 'skler·ə·tə‚nīt }

sclerotium [MICROBIO] The hardened, resting or encysted condition of the plasmodium of Myxomycetes. [MYCOL] A hardened, resting mass of hyphae, usually black on the outside, from which fructifications may develop. { sklə'rō·shəm }

sclerotome [EMBRYO] The part of a mesodermal somite which enters into the formation of the vertebrae. [MED] A knife used in sclerotomy. [VERT ZOO] The fibrous tissue separating successive myotomes in certain lower vertebrates. { 'skler·ə‚tōm }

Solitary coral polyps, *Oulangia* species.

sclerotomy [MED] Surgical incision of the sclera. { sklə'räd·ə·mē }

scobiform [BOT] Resembling sawdust. { 'skäb·ə‚fȯrm }

Sco-Cen association [ASTRON] An association of very young stars within the Gould belt whose brightest member is Antares. { 'skō 'sen ə‚sō·sē‚ā·shən }

scolecite [MINERAL] CaAl₂Si₃O₁₀ A zeolite mineral that occurs in delicate, radiating groups of white fibrous or acicular crystals; sometimes shows wormlike motion upon heating. { 'skäl·ə‚sīt }

scolecodont [PALEON] Any of the paired, pincerlike jaws occurring as fossils of annelid worms. { skō'lē·kə‚dänt }

Scolecosporae [MYCOL] A spore group of Fungi Imperfecti characterized by filiform spores. { ‚skō·lə'käs·pə‚rē }

scolex [INV ZOO] The head of certain tapeworms, typically having a muscular pad with hooks, and two pairs of lateral suckers. { 'skō‚leks }

Scoliidae [INV ZOO] A family of the Hymenoptera in the superfamily Scolioidea. { skō'lī·ə‚dē }

Scolioidea [INV ZOO] A superfamily of hymenopteran insects in the suborder Apocrita. { ‚skō·lē'ȯid·ē·ə }

scoliosis [MED] Lateral curvature of the spine. { ‚skō·lē'ō·səs }

scolite [GEOL] Any of the small tubes in rock believed to be the fossilized burrows of worms. { 'skō‚līt }

scolop [INV ZOO] The thickened, distal tip of a vibration-sensitive organ in insects. { 'skäl·əp }

Scolopacidea [VERT ZOO] A large, cosmopolitan family of birds of the order Charadriiformes including snipes, sandpipers, curlews, and godwits. { ‚skäl·ə·pə'sīd·ē·ə }

Scolopendridae [INV ZOO] A family of centipedes in the order Scolopendromorpha which characteristically possess eyes. { ‚skäl·ə'pen·drə‚dē }

Scolopendromorpha [INV ZOO] An order of the chilopod subclass Pleurostigmophora containing the dominant tropical forms, and also the largest of the centipedes. { ¦skäl·ə‚pen·drə'mȯr·fə }

scolophore See scolopophore. { 'skäl·ə‚fȯr }

scolopophore [INV ZOO] A spindle-shaped, bipolar nerve ending in the integument of insects, believed to be auditory in function. Also known as scolophore. { skə'läp·ə‚fȯr }

Scolytidae [INV ZOO] The bark beetles, a large family of coleopteran insects in the superfamily Curculionoidea characterized by a short beak and clubbed antennae. { skə'lid·ə‚dē }

Scombridae [VERT ZOO] A family of perciform fishes in the suborder Scombroidei including the mackerels and tunas. { 'skäm·brə‚dē }

Scombroidei [VERT ZOO] A suborder of fishes in the order Perciformes; all are moderate- to large-sized shore and oceanic fishes having fixed premaxillae. { skäm'brȯi·dē‚ī }

scoop [DES ENG] **1.** Any of various ladle-, shovel-, or bucketlike utensils or containers for moving liquid or loose materials. **2.** A funnel-shaped opening for channeling a fluid into a desired path. [ELEC] See ellipsoidal floodlight. [MECH ENG] A large shovel with a scoop-shaped blade. { sküp }

scoopfish See underway sampler. { 'süp‚fish }

scopa [INV ZOO] A brushlike arrangement of short stiff hairs on the body surface of certain insects. { 'skō·pə }

scope [COMPUT SCI] For a variable in a computer program, the portion of the computer program within which the variable can be accessed (used or changed). [ELECTR] See cathode-ray oscilloscope; radarscope. [ENG] The work that will actually be done on a project as documented by the terms in a contract. { skōp }

Scopeumatidae [INV ZOO] The dung flies, a family of myodarian cyclorrhaphous dipteran insects in the subsection Calypteratae. { ‚skä·pyü'mad·ə‚dē }

Scopidae [VERT ZOO] A family of birds in the order Ciconiiformes containing a single species, the hammerhead (*Scopus umbretta*) of tropical Africa. { 'skäp·ə‚dē }

scopolamine [PHARM] C₁₇H₂₁O₄N An alkaloid derivative of several plants in the family Solanaceae, used as an anticholinergic drug; its hydrobromide salt is used as a sedative. { skə'päl·ə‚mēn }

scopoline [ORG CHEM] C₈H₁₃O₂H A white crystalline alkaloid that melts at 108–109°C, soluble in water and ethanol; used in medicine. Also known as oscine. { 'skō·pə‚lēn }

scopometer [OPTICS] An instrument used to measure the absorption or scattering of light in a solution containing solid particles by measuring the contrast between an illuminated line placed behind the solution and a field of constant brightness. { skə'päm·əd·ər }

scopophilia [PSYCH] Sexual stimulation from looking at the unclad human figure; observed chiefly in normal adolescent and adult males where it takes the aim-inhibited form of "girl watching" or looking at nude or seminude females in magazines or as part of some stage performance, or where it may be sublimated as scientific curiosity; when present to a pathologic degree, it is deviant and called voyeurism. { ‚skäp·ə'fil·ē·ə }

scopula [ZOO] A tuft of hair, as on the feet and chelicerae of certain spiders. { 'skäp·yə·lə }

scopulite [GEOL] A crystallite in the form of a rod with terminal brush or plume. { 'skäp·yə‚līt }

scorching [CHEM ENG] Premature vulcanization caused by heat during the processing of rubber. [ENG] **1.** Burning an exposed surface so as to change color, texture, or flavor without consuming. **2.** Destroying by fire. [PL PATH] Browning of plant tissues caused by heat or parasites; may also be symptomatic of disease. { 'skȯrch·iŋ }

scorch time [CHEM ENG] In rubber manufacture, the time during which a rubber compound can be worked at a given temperature before curing begins. { 'skȯrch ‚tīm }

score [GEOL] See scoring. [GRAPHICS] Indent or impress mark on paper which facilitates folding. { skȯr }

scoria [GEOL] Vesicular, cindery, dark lava formed by the escape and expansion of gases in basaltic or andesitic magma; generally denser and darker than pumice. [MATER] Refuse after melting metals or reducing ore. { 'skȯr·ē·ə }

scoria cone [GEOL] A volcanic cone composed of a vesicular, cindery crust on the surface of lava that is basaltic or andesitic in nature. { 'skȯr·ē·ə ‚kōn }

scoria mound [GEOL] A volcanic knoll composed of vesicular, cindery crust on the surface of lava that is basaltic or andesitic in nature. { 'skȯr·ē·ə ‚maùnd }

scoria tuff [GEOL] A deposit of fragmented scoria in a fine-grained tuff matrix. { 'skȯr·ē·ə ‚təf }

scorification [MET] Concentration of precious metals, such as gold and silver, in molten lead by oxidation employing appropriate fluxes. { ‚skȯr·ə·fə'kā·shən }

scoring [ENG] Scratching the surface of a material. [GEOL] **1.** The formation of parallel scratches, lines, or grooves in a bedrock surface by the abrasive action of rock fragments transported by a moving glacier. **2.** A scratch, line, or groove produced by this process. Also known as score. [MATER] See attrition. { 'skȯr·iŋ }

scoring test See L-2 test. { 'skȯr·iŋ ‚test }

scorodite [MINERAL] FeAsO₄·2H₂O A pale leek-green or liver-brown orthorhombic mineral consisting of ferric arsenate; isomorphous with mansfieldite and represents a minor ore of arsenic. { 'skȯr·ə‚dīt }

Scorpaenidae [VERT ZOO] The scorpion fishes, a family of Perciformes in the suborder Cottoidei, including many tropical shorefishes, some of which are venomous. { skȯr'pē·nə‚dē }

Scorpaeniformes [VERT ZOO] An order of fishes coextensive with the perciform suborder Cottoidei in some systems of classification. { skȯr‚pē·nə'fȯr‚mēz }

scorpioid cyme [BOT] A cyme with a curved axis and flowers arising two-ranked on alternate sides of the axis. { 'skȯr‚pē‚ȯid 'sīm }

scorpion [INV ZOO] The common name for arachnids constituting the order Scorpionida. { 'skȯr‚pē·ən }

Scorpion See Scorpius. { 'skȯr‚pē·ən }

Scorpionida [VERT ZOO] The scorpions, an order of arachnids characterized by a shieldlike carapace covering the cephalothorax and by large pedipalps armed with chelae. { ‚skȯr‚pē'än·əd·ə }

Scorpio X-1 [ASTROPHYS] The most intense celestial source of x-rays known, associated with a highly variable radio source and a variable optical source, ranging from the twelfth to fourteenth magnitude. Abbreviated Sco X-1. { 'skȯr‚pē·ō ¦eks'wən }

Scorpius [ASTRON] A southern constellation, right ascension 16 hours, declination 40° south; the bright-red star Antares is located in it. Also known as Scorpion. { 'skȯr‚pē·əs }

scorzalite [MINERAL] FeAl₂(PO₄)₂(OH)₂ A blue mineral composed of basic iron aluminum phosphate; it is isomorphous with lazulite. { 'skȯr·zə‚līt }

scotch [DES ENG] See scutch. [ENG] A wooden stopblock

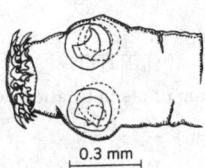
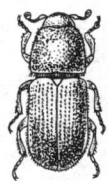
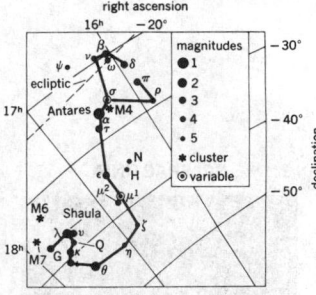

or iron catch placed under a wheel or other curved object to prevent slipping or rolling. { skäch }

scotch boiler [MECH ENG] A fire-tube boiler with one or more cylindrical internal furnaces enveloped by a boiler shell equipped with five tubes in its upper part; heat is transferred to water partly in the furnace area and partly in passage of hot gases through the tubes. Also known as dry-back boiler; scotch marine boiler (marine usage). { 'skäch 'bȯil·ər }

Scotch bond See American bond. { 'skäch 'bänd }

Scotch derrick See stiffleg derrick. { 'skäch 'der·ik }

scotch marine boiler See scotch boiler. { 'skäch mə¦rēn 'bȯil·ər }

Scotch mist [METEOROL] A combination of thick mist (or fog) and heavy drizzle occurring frequently in Scotland and in parts of England. { 'skäch 'mist }

Scotch pine [BOT] *Pinus sylvestris.* A hard pine of North America having two short, bluish needles in a cluster. { 'skäch 'pīn }

Scotch-type volcano [GEOL] A volcanic form characterized by concentric cuestas and produced by cauldron subsidence. { 'skäch ¦tīp väl'kā·nō }

Scotch whiskey [FOOD ENG] A distilled spirit made in Scotland primarily from malted barley. { 'skäch 'wis·kē }

Scotch yoke [MECH ENG] A type of four-bar linkage; it is employed to convert a steady rotation into a simple harmonic motion. { 'skäch 'yōk }

scotochromogen [MICROBIO] **1.** Any microorganism which produces pigment when grown without light as well as with light. **2.** A member of group II of the atypical mycobacteria. { ¦skäd·ə'krō·mə·jən }

scotodinia [MED] Dizziness and headache associated with the appearance of black spots before the eyes. { ¦skäd·ə'din·ē·ə }

scotoma [MED] A blind spot or area of depressed vision in the visual field. { skə'tō·mə }

scotomization [PSYCH] In psychoanalytic theory, the mind's ability to erase a traumatic or overwhelming experience. { ¦skäd·ə·mə'zā·shən }

scotophor [MATER] A solid that exhibits reversible darkening and bleaching actions of tenebrescence under suitable irradiation. { 'skäd·ə¸fȯr }

scotopic vision [PHYSIO] Vision that is due to the activity of the rods of the retina only; it is the type of vision that occurs at very low levels of illumination, and it can detect differences of brightness but not of hue. Also known as night vision. { skə'täp·ik 'vizh·ən }

scotoscope [ELECTR] A telescope which employs an image intensifier to see in the dark. { 'skäd·ə¸skōp }

Scott connection [ELECTR] A type of transformer which transmits power from two-phase to three-phase systems, or vice versa. { 'skät kə¸nek·shən }

Scott-Darcy process [CIV ENG] A chemical precipitation method used for fine solids removal in sewage plants; employs ferric chloride solution made by treating scrap iron with chlorine. { 'skät 'där·ē ¸präs·əs }

Scott top [ELEC] Transformers arranged in the Scott connection for converting electrical power from two-phase to three-phase, or vice versa. { 'skät ¸täp }

scour See tidal scour. { 'skaȯ·ər }

scour and fill [GEOL] The process of first digging out and then refilling a channel instigated by the action of a stream or tide; refers particularly to the process that occurs during a period of flood. { 'skaȯ·ər ən 'fil }

scour channel [GEOL] A large, groovelike erosional feature produced in sediments by scour. { 'skaȯ·ər ¸chan·əl }

scour depression [GEOL] A crescent-shaped hollow in the stream bed near the outside of the stream's bend, caused by water that scours below the grade of the stream. { 'skaȯ·ər di¸presh·ən }

scouring [ENG] Physical or chemical attack on process equipment surfaces, as in a furnace or fluid catalytic cracker. [GEOL] **1.** An erosion process resulting from the action of the flow of air, ice, or water. **2.** See glacial scour. [MATER] See attrition. [MECH ENG] Mechanical finishing or cleaning of a hard surface by using an abrasive and low pressure. [TEXT] **1.** Removal of grease and dirt from wool. **2.** The cleaning of fabric before the dyeing step. { 'skaȯr·iŋ }

scouring agent [TEXT] A cleaner used to remove oily substances from textile fibers; for example, natural oils and fats

from raw wool, or lubricants from rayon yarns or fabrics during an operation such as winding or weaving. { 'skaȯr·iŋ ¸ā·jənt }

scouring basin [CIV ENG] A basin containing impounded water which is released at about low water in order to maintain the desired depth in the entrance channel. Also known as sluicing pond. { 'skaȯr·iŋ ¸bas·ən }

scouring rush See horsetail. { 'skaȯr·iŋ ¸rəsh }

scouring velocity [GEOL] The velocity of water which is necessary to dislodge stranded solids from the stream bed. { 'skaȯr·iŋ və¸läs·əd·ē }

scour lineation [GEOL] A smooth, low, narrow (2–5 centimeters or 1–2 inches wide) ridge formed on a sedimentary surface and believed to result from the scouring action of a current of water. { 'skaȯ·ər ¸lin·ē ā·shən }

scour mark [GEOL] A mark produced by the cutting or scouring action of a current flowing over the bottom of a river or body of water. { 'skaȯ·ər ¸mark }

scourway [GEOL] A channel created by a powerful water current, particularly the temporary channels formed by streams on the edge of a Pleistocene ice sheet. { 'skaȯ·ər¸wā }

scout [ENG] An engineer who makes a preliminary examination of promising oil and mining claims and prospects. [NAV] **1.** To search an area by following an orderly pattern of courses. **2.** A craft engaged in search. { skaȯt }

Scout [AERO ENG] A four-stage all-solid-propellant rocket, used as a space probe and orbital test vehicle; first launched July 1, 1960, with a 150-pound (68-kilogram) payload. { skaȯt }

scout boring [MIN ENG] A bore made to test a geologic formation being prospected. { 'skaȯt ¸bȯr·iŋ }

scout car [ORD] Lightly armed and armored reconnaissance vehicle, either wheel or half-track, without turrets, adapted for high-speed operation on hard roads and for cross-country missions. { 'skaȯt ¸kär }

scout hole [MIN ENG] **1.** A borehole penetrating only the uppermost part of an ore body in order to delineate the surface configuration. **2.** A shallow borehole used to ascertain the presence of ore or to explore an area in a preliminary manner. { 'skaȯt ¸hōl }

scouting distance [NAV] The distance between adjacent scouts on a scouting line. { 'skaȯd·iŋ ¸dis·təns }

scouting front [NAV] The distance along the scouting line between the outer limits of visibility of the two outboard scouts on the scouting line. { 'skaȯd·iŋ ¸frənt }

scouting interval [NAV] The distance between consecutive scouts following the same track. { 'skaȯd·iŋ ¸in·tər·vəl }

scouting line [NAV] A line on which scouts are located while scouting in a formation suitable to cover a definite pattern. { 'skaȯd·iŋ ¸līn }

Sco X-1 See Scorpio X-1.

SCR See silicon controlled rectifier.

scram [NUCLEO] **1.** A sudden shutting down of a nuclear reactor, usually by dropping safety rods, when a predetermined neutron flux or other dangerous condition occurs. **2.** To close down a reactor by bringing about a scram. { skram }

scramble [AERO ENG] To take off as quickly as possible (usually followed by course and altitude instructions). [COMMUN] To mix, in cryptography, in random or quasi-random fashion. { 'skram·bəl }

scrambler [ELECTR] A circuit that divides speech frequencies into several ranges by means of filters, then inverts and displaces the frequencies in each range so that the resulting reproduced sounds are unintelligible; the process is reversed at the receiving apparatus to restore intelligible speech. Also known as speech inverter; speech scrambler. { 'skram·blər }

scram drive [MIN ENG] Underground drive above the tramming level, along which scrapers move ore to a discharge chute. { 'skram ¸drīv }

scramjet [AERO ENG] Essentially a ramjet engine, intended for flight at hypersonic speeds. Derived from supersonic combustion ramjet. { 'skram¸jet }

scram rod See safety rod. { 'skram ¸räd }

scram system [NUCLEO] A system which causes a scram in a nuclear reactor when dangerous conditions arise. { 'skram ¸sis·təm }

scrap [ENG] Any solid material cutting or reject of a manufacturing operation, which may be suitable for recycling as feedstock to the primary operation; for example, scrap from plastic or glass molding or metalworking. { skrap }

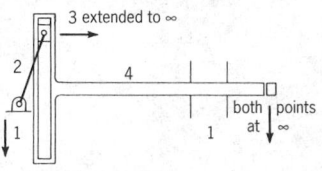

SCOTCH YOKE

Scotch yoke, a type of four-bar linkage; the numbers indicate the linkage.

scraped freezer [FOOD ENG] A refrigeration unit in which a flowing solution or slush contacts a chilled surface on which ice forms and is removed periodically by scraping. { ¦skrāpt ¦frēz·ər }

scraped-surface exchanger [CHEM ENG] A liquid-liquid heat-exchange device that has a rotating element with spring-loaded scraper blades to wipe the process-fluid exchange surfaces clean of crystals or other foulants; used in paraffin-wax processing. { 'skrāpt ¦sər·fəs iks‚chān·jər }

scraper conveyor [MECH ENG] A type of flight conveyor in which the element (chain and flight) for moving materials rests on a trough. { 'skrāp·ər kən‚vā·ər }

scraper hoist [MECH ENG] A drum hoist that operates the scraper of a scraper loader. { 'skrāp·ər ‚hȯist }

scraper loader [MECH ENG] A machine used for loading coal or rock by pulling a scoop through the material to an apron or ramp, where the load is discharged onto a car or conveyor. { 'skrāp·ər ‚lōd·ər }

scraper ring [MECH ENG] A piston ring that scrapes oil from a cylinder wall to prevent it from being burned. { 'skrāp·ər ‚riŋ }

scraper ripper [MIN ENG] A piece of strip-mine equipment with teeth on the lip to rip or break the coal and with a flight conveyor to remove the broken coal. { 'skrāp·ər ‚rip·ər }

scraper trap [ENG] A device for the insertion or recovery of pigs, or scrapers, that are used to clean the inside surfaces of pipelines. { 'skrāp·ər ‚trap }

scrapie [VET MED] A transmissible, usually fatal, virus disease of sheep, characterized by degeneration of the central nervous system. { 'skrā·pē }

scrap mica [MATER] Mica whose size, color, or quality is below specifications for sheet mica. { 'skrap ‚mī·kə }

Scraptidae [INV ZOO] An equivalent name for Melandryidae. { 'skrap·tə‚dē }

scratch [COMPUT SCI] To remove data or to set up its identifying labels so that new data can be written over it. { skrach }

scratchboard [GRAPHICS] A plain, white-coated drawing board, covered with india ink, which, when dry, is scratched with a scratch knife to produce a snow-white line; brilliant, contrasty black-and-white drawings can thus be made. { 'skrach‚bȯrd }

scratch-brush finish See satin finish. { 'skrach ‚brəsh ‚fin·ish }

scratch coat [ENG] The first layer of plaster applied to a surface; the surface is scratched to improve the bond with the next coat. { 'skrach ‚kōt }

scratch file [COMPUT SCI] A temporary file for future use, created by copying all or part of a data set to an auxiliary memory device. { 'skrach ‚fīl }

scratch filter [ENG ACOUS] A low-pass filter circuit inserted in the circuit of a phonograph pickup to suppress higher audio frequencies and thereby minimize needle-scratch noise. { 'skrach ‚fil·tər }

scratch hardness [MATER] A measure of the resistance of minerals or metals to scratching; for minerals it is defined by comparison with 10 selected minerals which are numbered in order of increasing hardness according to the Mohs scale. { 'skrach ‚härd·nəs }

scratch hardness test [MET] A hardness test in which a cutting point under given pressure is drawn across the surface of a metal, and the width of the scratch is measured. [MINERAL] A determination of the resistance of a mineral to scratching by testing it with minerals on the Mohs scale. { 'skrach ¦härd·nəs ‚test }

scratch knife [GRAPHICS] A knife point that fits in a standard penholder and makes a fine to broad line on a scratchboard; used in the graphic arts for cutting, erasing, and retouching. { 'skrach ‚nīf }

scratch-pad memory [COMPUT SCI] A very fast intermediate storage (in the form of flip-flop register or semiconductor memory) which often supplements main core memory. { 'skrach ‚pad ‚mem·rē }

scratch tape [COMPUT SCI] A reel of magnetic tape containing data that may now be destroyed. { 'skrach ‚tāp }

screaming [AERO ENG] A form of combustion instability, especially in a liquid-propellant rocket engine, of relatively high frequency, characterized by a high-pitched noise. { 'skrēm·iŋ }

scree [GEOL] **1.** A mound of loose, angular material, less than 4 inches (10 centimeters). **2.** See talus. { skrē }

screeching [AERO ENG] A form of combustion instability, especially in an afterburner, of relatively high frequency, characterized by a harsh, shrill noise. { 'skrēch·iŋ }

screed [BUILD] A long, narrow strip of plaster placed at intervals on a surface as a guide for the thickness of plaster to be applied. [CIV ENG] **1.** A straight-edged wood or metal template, fixed temporarily to a surface as a guide when plastering or concreting. **2.** An oscillating metal bar mounted on wheels and spanning a freshly placed road slab, used to strike off and smooth the surface. { skrēd }

screed wire See ground wire. { 'skrēd ‚wīr }

screen [COMPUT SCI] To make a preliminary selection from a set of entities, selection criteria being based on a given set of rules or conditions. [ELECTR] **1.** The surface on which a television, radar, x-ray, or cathode-ray oscilloscope image is made visible for viewing; it may be a fluorescent screen with a phosphor layer that converts the energy of an electron beam to visible light, or a translucent or opaque screen on which the optical image is projected. Also known as viewing screen. **2.** See screen grid. [ELECTROMAG] Metal partition or shield which isolates a device from external magnetic or electric fields. [ENG] **1.** A large sieve of suitably mounted wire cloth, grate bars, or perforated sheet iron used to sort rock, ore, or aggregate according to size. **2.** A covering to give physical protection from light, noise, heat, or flying particles. **3.** A filter medium for liquid-solid separation. { skrēn }

screen analysis [ENG] A method for finding the particle-size distribution of any loose, flowing, conglomerate material by measuring the percentage of particles that pass through a series of standard screens with holes of various sizes. { 'skrēn ə‚nal·ə·səs }

screen angle [ELECTROMAG] Vertical angle bounded by a straight line from the radar antenna to the horizon and the horizontal at the antenna assuming a $^4/_3$ earth's radius. [GRAPHICS] The angle used for a halftone screen in relation to another screen to avoid an undesirable moiré pattern in a color reproduction process. { 'skrēn ‚aŋ·gəl }

screen capture See screen shot. { 'skrēn ‚kap·chər }

screen cloth [MATER] A woven material suitable for use in a screen deck. { 'skrēn ‚klȯth }

screen deck [DES ENG] A surface provided with apertures of specified size, used for screening purposes. { 'skrēn ‚dek }

screen dissipation [ELECTR] Power dissipated in the form of heat on the screen grid as the result of bombardment by the electron stream. { 'skrēn ‚dis·ə‚pā·shən }

screen dryer See traveling-screen dryer. { 'skrēn 'drī·ər }

screen dump [COMPUT SCI] **1.** The printing of everything that appears on a computer screen. **2.** The printed copy that results from this action. { 'skrēn ‚dəmp }

screened pan [METEOROL] An evaporation pan the top of which is covered by wire-mesh screening (1/4-inch or 6-millimeter mesh); the screening reduces air circulation and insolation, and results in a pan coefficient nearer to unity than that for unscreened pans. { 'skrēnd ‚pan }

screened trailing cable [ELEC] A flexible cable provided with a protective screen of conducting material, so applied as to enclose each power core separately or to enclose together all the cores of the cable. { 'skrēnd 'trāl·iŋ ‚kā·bəl }

screen format [COMPUT SCI] The manner in which information is arranged and presented on a cathode-ray tube or other electronic display. { 'skrēn ‚fȯr‚mat }

screen formatter [COMPUT SCI] A computer program that enables the user to design and set up screen formats. Also known as screen generator; screen painter. { 'skrēn ‚fȯr‚mad·ər }

screen generator See screen formatter. { 'skrēn ‚jen·ə‚rād·ər }

screen grid [ELECTR] A grid placed between a control grid and an anode of an electron tube, and usually maintained at a fixed positive potential, to reduce the electrostatic influence of the anode in the space between the screen grid and the cathode. Also known as screen. { 'skrēn ‚grid }

screen image buffer [COMPUT SCI] A section of computer storage that contains a representation of the information that appears on an electronic display. Abbreviated SIB. { 'skrēn ¦im·ij ‚bəf·ər }

screening [ATOM PHYS] The reduction of the electric field about a nucleus by the space charge of the surrounding electrons. [ELECTROMAG] *See* electric shielding. [ENG] **1.** The separation of a mixture of grains of various sizes into two or more size-range portions by means of a porous or woven-mesh screening media. **2.** The removal of solid particles from a liquid-solid mixture by means of a screen. **3.** The material that has passed through a screen. [IND ENG] The elimination of defective pieces from a lot by inspection for specified defects. Also known as detailing. { 'skrēn·iŋ }

screening agent [ANALY CHEM] A nonchelating dye used to improve the colorimetric end point of a complexometric titration; a dye addition forms a complementary pair of colors with the metalized and unmetalized forms of the end-point indicator. { 'skrēn·iŋ ‚ā·jənt }

screening constant [ATOM PHYS] The difference between the atomic number of an element and the apparent atomic number for a given process; this difference results from screening. { 'skrēn·iŋ ‚kän·stənt }

screening factor [NUC PHYS] The actual rate of a nuclear reaction in a dense plasma divided by the rate that would prevail if there were no free electrons to screen the repulsion between the nuclei. { 'skrēn·iŋ ‚fak·tər }

screening smoke [ORD] A smoke cloud produced by chemical agents or smoke generators; used to conceal friendly troops or to deny observation to enemy troops. { 'skrēn·iŋ ‚smōk }

screen memory [COMPUT SCI] The portion of a microcomputer storage that is reserved for setting up screen formats. [PSYCH] A consciously tolerable but usually unimportant memory recalled in place of an associated important one, which would be painful and disturbing. { 'skrēn ‚mem·rē }

screen mesh [ENG] A wire network or cloth mounted in a frame for separating and classifying materials. { 'skrēn ‚mesh }

screen overlay [COMPUT SCI] **1.** An array of cells on a video display screen that allow a user to command a computer by touching buttons displayed on the screen at the locations of the cells. **2.** A window of data that is temporarily displayed on a screen, leaving the original display intact when the window is removed. **3.** *See* glare filter. { 'skrēn 'ō·vər‚lā }

screen painter *See* screen formatter. { 'skrēn ‚pān·tər }

screen pipe [ENG] Perforated pipe with a straining device in the form of closely wound wire coils wrapped around it to admit well fluids while excluding sand. { 'skrēn ‚pīp }

screen printing [GRAPHICS] A printing method in which ink is forced by a squeegee through the open areas of a resist-imaged screen (silk, nylon, or metal) onto a substrate (paper, fabric, and so on). Also known as serigraphy; silk-screen printing. { 'skrēn 'print·iŋ }

screen process [GRAPHICS] A color photography process in which analysis and synthesis of color are accomplished by use of a screen containing a mosaic of very small color filters. { 'skrēn ‚prä·səs }

screen ruling [GRAPHICS] On a contact screen or ruled glass halftone screen, the total number of lines or dots per inch. { 'skrēn ‚rül·iŋ }

screen saver [COMPUT SCI] A program that launches when a computer is not in use for a predetermined period, displaying various transient or moving images on a computer screen. Originally used to prevent computer screen damage from prolonged display of a static image, screen savers are now more of an amusement or security feature as modern monitors are less susceptible to screen burning. { 'skrēn ‚sāv·ər }

screen shot [COMPUT SCI] A digital image or file containing all or part of what is seen on a computer display. Also known as screen capture. { 'skrēn ‚shät }

screen size [MIN ENG] A standard for determining the size of diamond particles; the size of the screened particle is determined by the size of the opening through which the diamond particle will not pass. { 'skrēn ‚sīz }

screw [DES ENG] **1.** A cylindrical body with a helical groove cut into its surface. **2.** A fastener with continuous ribs on a cylindrical or conical shank and a slotted, recessed, flat, or rounded head. Also known as screw fastener. { skrü }

screw axis [CRYSTAL] A symmetry element of some crystal lattices, in which the lattice is unaltered by a rotation about the axis combined with a translation parallel to the axis and equal to a fraction of the unit lattice distance in this direction. { 'skrü ‚ak·səs }

screw blank *See* bolt blank. { 'skrü ‚blaŋk }

screw compressor [MECH ENG] A rotary-element gas compressor in which compression is accomplished between two intermeshing, counterrotating screws. { 'skrü kəm'pres·ər }

screw conveyor [MECH ENG] A conveyor consisting of a helical screw that rotates upon a single shaft within a stationary trough or casing, and which can move bulk material along a horizontal, inclined, or vertical plane. Also known as auger conveyor; spiral conveyor; worm conveyor. { 'skrü kən'vā·ər }

screw dislocation [CRYSTAL] A dislocation in which atomic planes form a spiral ramp winding around the line of the dislocation. { 'skrü ‚dis·lō‚kā·shən }

screw displacement [MECH] A rotation of a rigid body about an axis accompanied by a translation of the body along the same axis. { 'skrü di‚splās·mənt }

screw dowel [DES ENG] A metal dowel pin having a straight or tapered thread at one end. { 'skrü ‚daül }

screwdriver [DES ENG] A tool for turning and driving screws in place; a thin, wedge-shaped or fluted end enters the slot or recess in the head of the screw. { 'skrü‚drīv·ər }

screw elevator [MECH ENG] A type of screw conveyor for vertical delivery of pulverized materials. { 'skrü 'el·ə‚vād·ər }

screw fastener *See* screw. { 'skrü ‚fas·nər }

screwfeed [MECH ENG] A system or combination of gears, ratchets, and friction devices in the swivel head of a diamond drill, which controls the rate at which a bit penetrates a rock formation. { 'skrü‚fēd }

screw feeder [MECH ENG] A mechanism for handling bulk (pulverized or granulated solids) materials, in which a rotating helicoid screw moves the material forward, toward and into a process unit. { 'skrü 'fēd·ər }

screw ice [HYD] **1.** Small ice fragments in heaps or ridges, produced by the collision of ice cakes. **2.** A small formation of pressure ice. { 'skrü ‚īs }

screw jack *See* jackscrew. { 'skrü ‚jak }

screw machine [MECH ENG] A lathe for making relatively small, turned metal parts in large quantities. { 'skrü mə‚shēn }

screw pile [CIV ENG] A pile having a wide helical blade at the foot which is twisted into position, for use in soft ground or other location requiring a large supporting surface. { 'skrü ‚pīl }

screw pine *See* pandanus. { 'skrü ‚pīn }

screw plasticating injection molding [ENG] A plastic-molding technique in which plastic is converted from pellets to a viscous (plasticated) melt by an extruder screw that is an integral part of the molding machine. { 'skrü 'plas·ti‚kād·iŋ in'jek·shən ‚mōld·iŋ }

screw press [MECH ENG] A press having the slide operated by a screw mechanism. { 'skrü ‚pres }

screw propeller [MECH ENG] A marine and airplane propeller consisting of a streamlined hub attached outboard to a rotating engine shaft on which are mounted two to six blades; the blades form helicoidal surfaces in such a way as to advance along the axis about which they revolve. { 'skrü prə‚pel·ər }

screw pump [MECH ENG] A pump that raises water by means of helical impellers in the pump casing. { 'skrü ‚pəmp }

screw rivet [DES ENG] A short rod threaded along the length of the shaft that is set without access to the point. { 'skrü ‚riv·ət }

screw spike [DES ENG] A large nail with a helical thread on the upper portion of the shank; used to fasten railroad rails to the ties. { 'skrü ‚spīk }

screwstock [MECH ENG] Free-machining bar, rod, or wire. { 'skrü‚stäk }

screw thread [DES ENG] A helical ridge formed on a cylindrical core, as on fasteners and pipes. { 'skrü ‚thred }

screw-thread gage [DES ENG] Any of several devices for determining the pitch, major, and minor diameters, and the lead, straightness, and thread angles of a screw thread. { 'skrü ‚thred ‚gāj }

screw-thread micrometer [DES ENG] A micrometer used to measure pitch diameter of a screw thread. { 'skrü ‚thred mī'kräm·əd·ər }

scribe projection [COMMUN] Method of automatic information presentation; information is placed on a small metallic-coated glass slide by removing the coating with a movable fine-pointed scribe controlled by a servo system; light passing

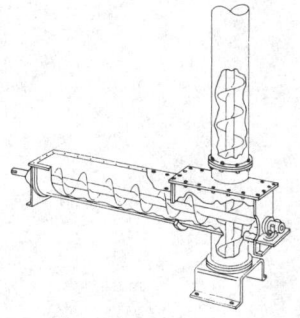

SCREW CONVEYOR

A combination of a horizontal and vertical screw conveyor.

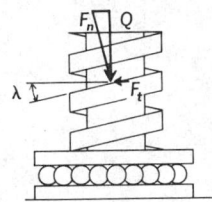

SCREW-THREAD

Diagram of a frictionless screw with square threads mounted on a ball thrust bearing, used to raise load Q. Here λ = screw lead angle, F_t = tangential force required to turn screw, F_n = normal force.

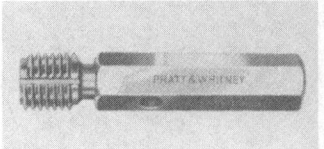

SCREW-THREAD GAGE

A plug gage, a type of screw-thread gage.

through the scribed area projects on the screen. { 'skrīb prə‚jek·shən }

scriber [DES ENG] A sharp-pointed tool used for drawing lines on metal workpieces. { 'skrī·bər }

scribing [ELECTR] Cutting a grid pattern of deep grooves with a diamond-tipped tool in a slice of semiconductor material containing a number of devices, so that the slice can be easily broken into individual chips. [GRAPHICS] A method of preparing a map or chart by cutting the lines into a prepared coating. { 'skrīb·iŋ }

Scribner log rule [FOR] A method of scaling logs to derive board-foot calculations; uses a table showing expected log output in board feet that originated from diagrams of 1-inch boards drawn to scale within cylinders of various sizes. { 'skrib·nər 'läg ‚rül }

scrim [MATER] A coarse mesh made of wire, fiberglass, or other heavy fibers and used to bridge and reinforce a joint or used as a base for painting or plastering. [TEXT] A loose plain-woven fabric, generally cotton, with fine to coarse meshes; used in various finishes in clothing, curtains, and industry. { skrim }

script [COMPUT SCI] An executable list of commands written in a programming language. { skript }

scripting language [COMPUT SCI] An interpreted language (for example, JavaScript and Perl) used to write simple programs, called scripts. { 'skrip·tiŋ ‚laŋ·gwij }

scrod [VERT ZOO] A young fish, especially cod. { skräd }

scrofula [MED] Tuberculosis of cervical lymph nodes. { 'skräf·yə·lə }

scroll [ARCH] An ornament consisting of a spirally wound band, either as a running ornament or a terminal. [COMPUT SCI] To move information in an electronic display up, down, left, or right, so that new information appears and some of the existing information is moved away. [GEOL] One of a series of crescent-shaped sediments on the inner bank of a moving channel, deposited there by the stream. { skrōl }

scroll arrow [COMPUT SCI] An arrow on a video display screen that is clicked in order to scroll the screen in the corresponding direction. { 'skrōl ‚a·rō }

scroll bar [COMPUT SCI] A horizontal or vertical bar that contains a box that is clicked and dragged up, down, left, or right in order to scroll the screen. { 'skrōl ‚bär }

scroll gear [DES ENG] A variable gear resembling a scroll with teeth on one face. { 'skrōl ‚gir }

scrolling [COMPUT SCI] The continuous movement of information either vertically or horizontally on a video screen. { 'skrōl·iŋ }

scroll meander [GEOL] A type of forced-cut meander, in which the scrolls built on the inner bank cause erosion of the outer bank. { 'skrōl mē'an·dər }

scroll saw [ENG] A saw with a narrow blade, used for cutting curves or irregular designs. { 'skrōl ‚sȯ }

Scrophulariaceae [BOT] A large family of dicotyledonous plants in the order Scrophulariales, characterized by a usually herbaceous habit, irregular flowers, axile placentation, and dry, dehiscent fruit. { ‚skräf·yə‚lar·ē'ās·ē‚ē }

Scrophulariales [BOT] An order of flowering plants in the subclass Asteridae distinguished by a usually superior ovary and, generally, either by an irregular corolla or by fewer stamens than corolla lobes, or commonly both. { ‚skräf·yə‚lar·ē'ā·lēz }

scrotum [ANAT] The pouch containing the testes. { 'skrōd·əm }

scrub [AERO ENG] To cancel a scheduled firing, either before or during countdown. [AGR] An inferior animal of nondescript breeding. [COMPUT SCI] To examine a large amount of data and eliminate duplicate or unneeded items. [ECOL] A tract of land covered with a generally thick growth of dwarf or stunted trees and shrubs and a poor soil. { skrəb }

scrubber [ENG] A device for the removal, or washing out, of entrained liquid droplets or dust, or for the removal of an undesired gas component from process gas streams. Also known as washer; wet collector. [MIN ENG] A device, such as a wash screen, wash trommel, log washer, and hydraulic jet or monitor, in which a coarse and sticky material, for example, ore or clay, is either washed free of adherents or mildly disintegrated. { 'skrəb·ər }

scrubbing oil See absorption oil. { 'skrəb·iŋ ‚ȯil }

scrub plane [DES ENG] A narrow carpenter's plane with a

blade that has a rough surface and a rounded cutting edge. { 'skrəb ‚plān }

scrub typhus See tsutsugamushi disease. { 'skrəb ‚tī·fəs }

scruff [MET] A mixture of tin oxide and iron-tin alloy formed as dross on a molten tin-coating bath. { skrəf }

scruple [PHARM] A unit of mass in apothecaries' measure, equal to 20 grains or to 1.2959782 grams. { 'skrü·pəl }

SCS See silicon controlled switch.

SCSI See small computer system interface. { 'skəz·ē }

Sc spiral galaxy [ASTRON] A class of spiral galaxy characterized by spirals with the largest and most loosely coiled arms and the smallest central portion. { ‚es‚sē 'spī·rəl 'gal·ik·sē }

SC star [ASTRON] A star of a type intermediate between carbon stars and S stars. { ‚es‚sē ‚stär }

scuba diving [ENG] Any of various diving techniques using self-contained underwater breathing apparatus. { 'skü·bə ‚dīv·iŋ }

scud [METEOROL] Ragged low clouds, usually stratus fractus; most often applied when such clouds are moving rapidly beneath a layer of nimbostratus. { skəd }

scuffing [ENG] The dull mark, sometimes the result of abrasion, on the surface of glazed ceramic or glassware. { 'skəf·iŋ }

scuffle hoe [DES ENG] A hoe having two sharp edges so that it can be pushed and pulled. { 'skəf·əl ‚hō }

sculpin [VERT ZOO] Any of several species of small fishes in the family Cottidae characterized by a large head that sometimes has spines, spiny fins, broad mouth, and smooth, scaleless skin. { 'skəl·pən }

Sculptor [ASTRON] A southern constellation, right ascension 0 hours, declination 30° south. Also known as Sculptor's Apparatus; Workshop. { 'skəlp·tər }

Sculptor Group [ASTRON] One of the nearest groups of galaxies beyond the Local Group, consisting of about five large galaxies near the south galactic pole. { 'skəlp·tər ‚grüp }

Sculptor's Apparatus See Sculptor. { 'skəlp·tərz ‚ap·ə'rad·əs }

Sculptor system [ASTRON] A dwarf, elliptical galaxy in the Local Group, about 270,000 light years away. { 'skəlp·tər ‚sis·təm }

sculpture [GRAPHICS] The art of carving or shaping stone, wood, metal, or other materials into figures, or of modeling figures in wax or clay to be cast in plaster, or bronze or other metals. { 'skəlp·chər }

scum [MATER] **1.** A film of impurities that rises to or is formed on the surface of a liquid. **2.** A slimy film formed on the surface of a solid object. { skəm }

scum chamber [CIV ENG] An enclosed compartment in an Imhoff tank, in which gas escapes from the scum which rises to the surface of sludge during sewage digestion. { 'skəm ‚chām·bər }

scumming [GRAPHICS] Ink adhering to the nonimage area of a printing plate, resulting in tinting of the nonimage area of the printed substrate. { 'skəm·iŋ }

scupper [NAV ARCH] A drain on or below the deck of a ship that guides water over or through the side. { 'skəp·ər }

scupper pipe [NAV ARCH] A pipe leading from the scuppers to the fitting in the ship's side, for draining water from the deckhouse roof. { 'skəp·ər ‚pīp }

scurf [MED] A branlike desquamation of the epidermis, especially from the scalp; dandruff. { skərf }

S curve See reverse curve. { 'es ‚kərv }

scurvy [MED] An acute or chronic nutritional disorder due to vitamin C deficiency; characterized by weakness, subcutaneous hemorrhages, and alterations of any tissue containing collagen, ground substance, dentine, intercellular cement, or osteoid. { 'skər·vē }

scutch [DES ENG] A small, picklike tool which has flat cutting edges for trimming bricks. Also known as scotch. { skəch }

scutching [TEXT] The breaking up and separating of the woody portions of flax stems from the retted fiber by means of rollers. { 'skəch·iŋ }

scute [INV ZOO] A cornified, epithelial, scalelike structure in lizards and snakes. { skyüt }

Scutechiniscidae [INV ZOO] A family of heterotardigrades in the suborder Echiniscoidea, with segmental and intersegmental thickenings of cuticle. { skü‚tek·ə'nis·ə‚dē }

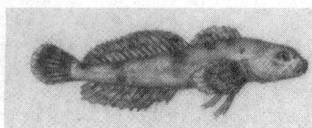

SCULPIN

The sculpin *Cottus cognatus.*

Scutelleridae [INV ZOO] The shield bugs, a family of Hemiptera in the superfamily Pentatomoidea. { ¦sküd·əl'er·ə¸dē }

scutellum [BOT] **1.** A rounded apothecium with an elevated rim found in certain lichens. **2.** The flattened cotyledon of a monocotyledonous plant embryo, such as a grass. [INV ZOO] The third of four pieces forming the upper part of the thoracic segment in certain insects. [VERT ZOO] One of the scales on the tarsi and toes of birds. { skü'tel·əm }

Scutigeromorpha [INV ZOO] The single order of notostigmophorous centipedes; members are distinguished by a dorsal respiratory opening, compound-type eyes, long flagellate multisegmental antennae, and long thin legs with multisegmental tarsi. { skü¸tij·ə·rə'mòr·fə }

scuttle [BUILD] An opening in the ceiling to provide access to the attic or roof. { 'skəd·əl }

scuttlebutt [NAV ARCH] **1.** A cask on shipboard which holds a day's supply of drinking water. **2.** A drinking fountain on shipboard or at a marine installation. { 'skəd·əl¸bət }

scutum [INV ZOO] **1.** A bony, horny, or chitinous plate. **2.** The second of four pieces forming the upper part of the thoracic segment in certain insects. **3.** One or two lower opercular valves in certain barnacles. { 'sküd·əm }

Scutum [ASTRON] A southern constellation, right ascension 19 hours, declination 10° south. Also known as Shield. { 'sküd·əm }

scuzzy See small computer system interface. { 'skəz·ē }

Scydmaenidae [INV ZOO] The antlike stone beetles, a large cosmopolitan family of the Coleoptera in the superfamily Staphylinoidea. { sid'mē·nə¸dē }

scyelite [PETR] A coarse-grained ultramafic igneous rock characterized by poikilitic texture resulting from the inclusion of olivine crystals in crystals of other minerals, especially amphiboles. { 'sī·ə¸līt }

Scyllaridae [INV ZOO] The Spanish, or shovel-nosed, lobsters, a family of the Scyllaridea. { si'lar·ə¸dē }

Scyllaridea [INV ZOO] A superfamily of decapod crustaceans in the section Macrura including the heavily armored spiny lobsters and the Spanish lobsters, distinguished by the absence of a rostrum and chelae. { ¸sil·ə'rid·ē·ə }

Scylliorhinidae [VERT ZOO] The catsharks, a family of the cacharinid group of galeoids; members exhibit the most exotic color patterns of all sharks. { ¸sil·ē·ō'rin·ə¸dē }

scyphistoma [INV ZOO] A sessile, polyploid larva of many Scyphozoa which may produce either more scyphistomae or free-swimming medusae. { sī'fis·tə·mə }

scyphomedusa [INV ZOO] A medusa of the scyphozoans. { ¸sī·fō·mə'dü·sə }

Scyphomedusae [INV ZOO] A subclass of the class Scyphozoa characterized by reduced marginal tentacles, tetramerous medusae, and medusalike polyploids. { ¸sī·fō·mə'dü¸sē }

scyphopolyp [INV ZOO] A polyp of the scyphozoans. { 'sī·fō'päl·əp }

Scyphozoa [INV ZOO] A class of the phylum Cnidaria; all members are marine and are characterized by large, well-developed medusae and by small, fairly well-organized polyps. { ¸sī·fə'zō·ə }

scythe [DES ENG] A tool with a long curved blade attached at a more or less right angle to a long handle with grips for both hands; used for cutting grass as well as grain and other crops. { sīth }

Scythian stage [GEOL] A stage in the lesser Triassic series of the alpine facies. Also known as Werfenian stage. { 'sith·ē·ən ¸stāj }

SDA See automation source data.

SDDC See sodium dimethyldithiocarbamate.

SDFL See Schottky-diode FET logic.

SDHT See high-electron-mobility transistor.

SDMA See space-division multiple access.

SDRAM See synchronous dynamic random access memory. { ¦es¦dē'ram }

SDS polyacrylamide gel electrophoresis [CELL MOL] An electrophoretic technique in which proteins are denatured by the negatively charged detergent sodium dodecyl sulfate, which masks their intrinsic electrical charge so that when the constituent polypeptide chains are run through a polyacrylamide gel, the protein molecules are separated according to size, not electrical charge. Abbreviated SDS-PAGE. { ¦es¦dē¦es ¸päl·ē·ə¦kril·ə·mīd ¸jel i¸lek·trō·fə'rē·səs }

SDS-PAGE See SDS polyacrylamide gel electrophoresis. { ¦es¦dē¦es 'pāj }

Se See selenium.

sea [GEOGR] A usually salty lake lacking an outlet to the ocean. [OCEANOGR] **1.** A major subdivision of the ocean. **2.** A heavy swell or ocean wave still under the influence of the wind that produced it. **3.** See ocean. { sē }

sea-air temperature difference correction [NAV] A correction to sextant altitude readings made necessary by abnormal terrestrial refraction occurring when there is a nonstandard density lapse rate in the atmosphere due to a difference in the temperature of the water and of the air at the surface. { 'sē 'er 'tem·prə·chər ¸dif·rəns kə¸rek·shən }

sea anchor [NAV ARCH] An object towed by a usually small vessel to keep the vessel end-on to a heavy sea or surf or to reduce drift; the usual form is a conical canvas bag whose large end is open, and, when towed with the large end in the forward position, the bag offers considerable resistance. { 'sē ¸aŋ·kər }

sea anemone [INV ZOO] Any of the 1000 marine cnidarians that constitute the order Actiniaria; the adult is a cylindrical polyp or hydroid stage with the free end bearing tentacles that surround the mouth. { 'sē ə¸nem·ə·nē }

sea arch [GEOL] An opening through a headland, formed by wave erosion or solution (as by the enlargement of a sea cave, or by the meeting of two sea caves from opposite sides), which leaves a bridge of rock over the water. Also known as marine arch; marine bridge; sea bridge. { 'sē ¸ärch }

sea ball [OCEANOGR] A spherical mass of somewhat fibrous material of living or fossil vegetation (especially algae), produced mechanically in shallow waters along a seashore by the compacting effect of wave movement. { 'sē ¸bòl }

sea bank See seawall. { 'sē ¸baŋk }

seabeach [GEOL] A beach along the margin of the sea. { 'sē¸bēch }

seabed See sea floor. { 'sē¸bed }

seaborgium [CHEM] A chemical element, symbolized Sg, atomic number 106, a synthetic element; the fourteenth transuranium element. { sē'bòrg·ē·əm }

sea bottom See sea floor. { 'sē ¸bäd·əm }

sea breeze [METEOROL] A coastal, local wind that blows from sea to land, caused by the temperature difference when the sea surface is colder than the adjacent land; it usually blows on relatively calm, sunny summer days, and alternates with the oppositely directed, usually weaker, nighttime land breeze. { 'sē ¸brēz }

sea breeze of the second kind See cold-front-like sea breeze. { 'sē ¸brēz əv thə 'sek·ənd ¸kīnd }

sea bridge See sea arch. { 'sē ¸brij }

sea buoy [NAV] The outermost buoy marking the entrance to a channel or harbor. Also known as farewell buoy, since it is the last buoy passed by a vessel proceeding out to sea. { 'sē ¸bòi }

sea-captured stream [HYD] A stream, flowing parallel to the seashore, that is cut in two as a result of marine erosion and that may enter the sea by way of a waterfall. { 'sē ¸kap·chərd ¸strēm }

sea cave [GEOL] A split or hollow opening, usually at sea level, in the base of a sea cliff, formed by waves acting on weak parts of the weathered rock. Also known as marine cave; sea chasm. { 'sē ¸kāv }

sea channel [GEOL] A long, narrow, U-shaped or V-shaped shallow depression of the sea floor, usually occurring on a gently sloping plain or fan. { 'sē ¸chan·əl }

sea chasm See sea cave. { 'sē ¸kaz·əm }

sea chest [NAV ARCH] **1.** A trunk in which a sailor stores personal property. **2.** A pipe between a ship's side and a valve in the hull for draining water. { 'sē ¸chest }

sea cliff [GEOL] An erosional landform, produced by wave action, which is either at the seaward edge of the coast or at the landward side of a wave-cut platform and which denotes the inner limit of the beach erosion. { 'sē ¸klif }

sea clutter [ELECTROMAG] A clutter on an airborne radar due to reflection of signals from the sea. Also known as sea return; wave clutter. { 'sē ¸kləd·ər }

seacoast [GEOGR] The land adjacent to the sea. { 'sē¸kōst }

seacoast artillery [ORD] Class of artillery formerly used for seacoast defense; consisted of fixed guns, howitzers, and mortars defending the harbors. { 'sē¸kōst är'til·ə·rē }

SCYPHOMEDUSA

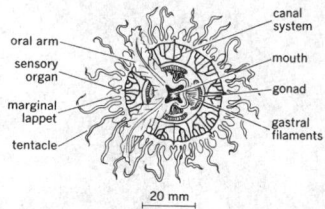

Schema of scyphomedusa, oral view.

SEA ANEMONE

Sea anemone (*Sagartia*), a translucent, sessile species that has retractile tentacles.

sea-control operations [ORD] The employment of naval forces, supported by land and air forces, as appropriate, to achieve military objectives in vital sea areas; such operations include destruction of enemy naval forces, suppression of enemy sea commerce, protection of vital sea lanes, and establishment of local military superiority in areas of naval operations. { 'sē kən,trōl ,äp·ə,rā·shənz }

sea cucumber [INV ZOO] The common name for the echinoderms that make up the class Holothuroidea. { 'sē kyü,kəm·bər }

seadrome [CIV ENG] **1.** A designated area for landing and takeoff of seaplanes. **2.** A platform at sea for landing and takeoff of land planes. { 'sē,drōm }

sea echelon [ORD] A portion of the assault shipping which withdraws from or remains out of the transport area during an amphibious landing and operates in designated areas to seaward in an on-call or unscheduled status. { 'sē ,esh·ə,län }

sea fan [GEOL] *See* submarine fan. [INV ZOO] A form of horny coral that branches like a fan. { 'sē ,fan }

sea floor [GEOL] The bottom of the ocean. Also known as seabed; sea bottom. { 'sē ,flȯr }

sea-floor spreading [GEOL] The hypothesis that the ocean floor is spreading away from the midoceanic ridges and is being conveyed landward by convective cells in the earth's mantle, carrying the continental blocks as passive passengers; the ocean floor moves away from the midoceanic ridge at the rate of 0.4 to 4 inches (1 to 10 centimeters) per year and provides the source of power in the hypothesis of plate tectonics. Also known as ocean-floor spreading; spreading concept; spreading floor hypothesis. { 'sē ,flȯr ,spred·iŋ }

sea-foam *See* sepiolite. { 'sē ,fōm }

sea fog [METEOROL] A type of advection fog formed over the ocean as a result of any of a variety of processes, as when air that has been lying over a warm water surface is transported over a colder water surface, resulting in a cooling of the lower layer of air below its dew point. { 'sē ,fäg }

sea front [GEOGR] An area partly bounded by the sea. { 'sē ,frənt }

sea frontier [ORD] The naval command of a coastal frontier, including the coastal zone in addition to the land area of the coastal frontier and the adjacent sea areas. { 'sē frən'tir }

sea gate [CIV ENG] A gate which serves to protect a harbor or tidal basin from the sea, such as one of a pair of supplementary gates at the entrance to a tidal basin exposed to the sea. [GEOGR] A way giving access to the sea such as a gate, channel, or beach. { 'sē ,gāt }

sea glow [OCEANOGR] The luminous, cobalt-blue appearance of very clear water in the open ocean, caused by upward-scattered light from which much of the red has been absorbed. { 'sē ,glō }

seagoing barge [NAV ARCH] A barge capable of sailing on the open sea. { 'sē¦gō·iŋ 'bärj }

seagoing tug [NAV ARCH] A tugboat capable of sailing on the open sea. { 'sē¦gō·iŋ ,təg }

sea grass [BOT] Marine plants which are found in shallow brackish or marine waters, are more highly organized than algae, are seed-bearing, and attain lengths of up to 8 feet (2.4 meters). { 'sē ,gras }

sea gully *See* slope gully. { 'sē ,gəl·ē }

sea horse [INV ZOO] Any of about 50 species of tropical and subtropical marine fishes constituting the genus *Hippocampus* in the family Syngnathidae; the body is compressed, the head is bent ventrally and has a tubiform snout, and the tail is tapering and prehensile. { 'sē ,hȯrs }

Sea Horse *See* Capricornus. { 'sē ,hȯrs }

sea ice [OCEANOGR] **1.** Ice formed from seawater. **2.** Any ice floating in the sea. { 'sē ,īs }

sea-ice shelf [OCEANOGR] Sea ice floating in the vicinity of its formation and separated from fast ice, of which it may have been a part, by a tide crack or a family of such cracks. { 'sē ,īs ,shelf }

seakeeping [NAV ARCH] The performance of a ship in waves, affected primarily by the ship's motion in six degrees of freedom. { 'sē,kēp·iŋ }

sea kindliness [NAV ARCH] Property of a vessel which is well adapted to being handled safely at sea in heavy weather. { 'sē ¦kīnd·lē·nəs }

sea knoll *See* knoll. { 'sē ,nōl }

seal [ENG] **1.** Any device or system that creates a nonleaking

union between two mechanical or process-system elements; for example, gaskets for pipe connection seals, mechanical seals for rotating members such as pump shafts, and liquid seals to prevent gas entry to or loss from a gas-liquid processing sequence. **2.** A tight, perfect closure or joint. [VERT ZOO] Any of various carnivorous mammals of the suborder Pinnipedia, especially the families Phoridae, containing true seals, and Otariidae, containing the eared and fur seals. { sēl }

sea-lane *See* seaway. { 'sē ,lān }

sea-launched ballistic missile [ORD] A missile launched from a submarine or surface ship. { 'sē ¦lȯncht bə'lis·tik 'mis·əl }

seal coat [MATER] A layer of bituminous material applied to bituminous macadam or concrete to seal the surface. { 'sēl ,kōt }

sealed-beam headlight [ELEC] A headlight in which the filament, reflector, and lens are contained in a single sealed unit. { ¦sēld ¦bēm 'head,līt }

sealed cabin [AERO ENG] The occupied space of an aircraft or spacecraft characterized by walls which do not allow gaseous exchange between the inner atmosphere and its surrounding atmosphere, and containing its own mechanisms for maintenance of the inside atmosphere. { ¦sēld 'kab·ən }

sealed tube [ELECTR] Electron tube which is hermetically sealed. { ¦sēld 'tüb }

sealer [MATER] A preliminary coating applied to seal the pores in a porous, uncoated surface, such as wood. { 'sēl·ər }

Seale rope [DES ENG] A wire rope with six or eight strands, each having a large wire core covered by nine small wires, which, in turn, are covered by nine large wires. { 'sēl ,rōp }

sea level [GEOL] The level of the surface of the ocean; especially, the mean level halfway between high and low tide, used as a standard in reckoning land elevation or sea depths. { 'sē ,lev·əl }

sea-level chart *See* surface chart. { 'sē ¦lev·əl ,chärt }

sea-level datum [ENG] A determination of mean sea level that has been adopted as a standard datum for heights or elevations, based on tidal observations over many years at various tide stations along the coasts. { 'sē ¦lev·əl ,dad·əm }

sea-level pressure [METEOROL] The atmospheric pressure at mean sea level, either directly measured or, most commonly, empirically determined from the observed station pressure. { 'sē ¦lev·əl ,presh·ər }

sea-level pressure chart *See* surface chart. { 'sē ¦lev·əl ¦presh·ər ,chärt }

sea lily [INV ZOO] The common name for those crinoids in which the body is flower-shaped and is carried at the tip of an anchored stem. { 'sē ,lil·ē }

sealing [MET] **1.** Impregnation of porous castings with resins to overcome porosity. **2.** Reducing porosity of an anodic oxide film on aluminum and aluminum alloys by immersion in boiling water. { 'sēl·iŋ }

sealing compound [ELEC] A compound used in dry batteries, capacitor blocks, transformers, and other components to keep out air and moisture. { 'sēl·iŋ ,käm,paȯnd }

sealing tape [MATER] Gummed tape for sealing packages. { 'sēl·iŋ ,tāp }

sealing wax [MATER] A colored, scented mixture of resins and shellac; used for sealing containers and documents. { 'sēl·iŋ ,waks }

sealing-wax structure [GEOL] A primary sedimentary flow structure produced by slumping, characterized by the lack of a sharply defined slip plane at the base or a contemporaneous erosion plane at the top, and occupying a zone of highly fluid contortion in an otherwise normal sedimentary succession. { 'sēl·iŋ ¦waks ,strək·chər }

sea lion [VERT ZOO] Any of several large, eared seals of the Pacific Ocean; related to fur seals but lack a valuable coat. { 'sē ,lī·ən }

seal off [ENG] To close off, as a tube or borehole, by using a cement or other sealant to eliminate ingress or egress. [PETRO ENG] Penetration of a drilling fluid into a formation so that the formation is prevented from producing. { 'sēl 'ȯf }

seal oil [MATER] A yellowish, liquid, fatty oil obtained from seal blubber; soluble in ether and chloroform; melts at 22–33°C; used in soapmaking, in dressing animal skins, and as a lubricant. { 'sēl ,ȯil }

seal weld [MET] A weld designed primarily for preventing leakage. { 'sēl ,weld }

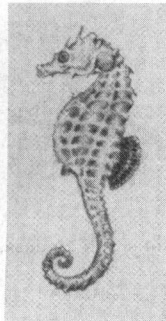

SEA HORSE

The sea horse *Hippocampus hudsonius*.

SEAL

Alaska fur seal (*Callorhinus ursinus*).

seam [ENG] **1.** A mechanical or welded joint. **2.** A mark on ceramic or glassware where matching mold parts join. **3.** A line occurring on a molded or laminated piece of plastic material that differs in appearance from the rest of the surface and is caused by a parting of the mold. Also known as mold seam. [GEOL] **1.** A stratum or bed of coal or other mineral. **2.** A thin layer or stratum of rock. **3.** A very narrow coal vein. [MET] An unwelded fold or lap which appears as a crack on the surface of a casting or wrought product. { sēm }

seamanite [MINERAL] $Mn_3(PO_4)(BO_3) \cdot 3H_2O$ A pale- to wine-yellow orthorhombic mineral that is a phosphate and borate of manganese; occurs in crystals. { 'sē·mə,nīt }

sea manners [NAV] A consideration for the other ship or craft and the exercise of good judgment under certain conditions when ships or craft meet. { 'sē ,man·ərz }

seamanship [NAV] The ability to handle, manage, and navigate a ship at sea. { 'sē·mən,ship }

seamark [NAV] An object designed as an aid to navigation and located with the express purpose of being visible from a distance to seaward; it is often erected in shoal water rather than on land. { 'sē,märk }

sea marker [ENG] A patch of color on the ocean surface produced by releasing dye; used, for example, to attract the attention of the crew of a rescue airplane. { 'sē ,märkər }

sea marsh [ECOL] A salt marsh periodically overflowed or flooded by the sea. Also known as sea meadow. { 'sē ,märsh }

seam blast [MIN ENG] A blast made by placing powdered or other explosive along and in a seam or crack between the solid wall and the stone or coal to be removed. { 'sēm ,blast }

sea meadow [ECOL] *See* sea marsh. [OCEANOGR] Any of the upper layers of the open ocean that have such an abundance of phytoplankton that they provide food for marine organisms. { 'sē ,med·ō }

sea mile [NAV] An approximate mean value of the nautical mile equal to 6080 feet (1853.184 meters) or the length of a minute of arc along the meridian at latitude 48°. { 'sē 'mīl }

seaming [MET] The joining of the edges of sheet-metal parts by interlocking folds. { 'sēm·iŋ }

sea mist *See* steam fog. { 'sē ,mist }

seamless integration [COMPUT SCI] The addition of a routine or program that works smoothly with an existing system and can be activated and used as if it had been built into the system when the system was put together. { 'sēm·ləs ,int·ə'grā·shən }

seamless ring rolling [MET] The hot-rolling of a circular blank, with a hole in the center, to form a seamless ring. { 'sēm·ləs ,riŋ ,rōl·iŋ }

seamless tubing [MET] A tubing made by extrusion or by piercing and rolling a billet. { 'sēm·ləs 'tüb·iŋ }

seamount [GEOL] A mountain rising from the ocean floor as a result of submarine volcanism. { 'sē,maunt }

seamount chain [GEOL] Several seamounts in a line with bases separated by a relatively flat sea floor. { 'sē,maunt ,chān }

seamount group [GEOL] Several closely spaced seamounts not in a line. { 'sē,maunt ,grüp }

seamount range [GEOL] Three or more seamounts having connected bases and aligned along a ridge or rise. { 'sē ,maunt ,rānj }

sea mud [GEOL] A rich, slimy deposit in a salt marsh or along a seashore, sometimes used as a manure. Also known as sea ooze. { 'sē ,məd }

seam weld [MET] **1.** A longitudinal weld joining of sheet-metal parts or in making tubing. **2.** Arc or resistance welding in which a series of overlapping spot welds is produced. { 'sēm ,weld }

sea ooze *See* sea mud. { 'sē ,üz }

sea otter [VERT ZOO] *Enhydra lutris.* A large marine otter found close to the shoreline in the North Pacific; these animals are diurnally active and live in herds. { 'sē ,äd·ər }

sea peak [GEOL] A peaked elevation of the sea floor, rising 3300 feet (1000 meters) or more from the floor. { 'sē ,pēk }

sea pen [INV ZOO] The common name for cnidarians constituting the order Pennatulacea. { 'sē ,pen }

seaplane [AERO ENG] An airplane that takes off from and alights on the water; it is supported on the water by pontoons, or floats, or by a hull which is a specially designed fuselage. Also known as airboat. { 'sē,plān }

seaport [CIV ENG] A harbor or town that has facilities for seagoing ships and is active in marine activities. { 'sē,pòrt }

seaquake [GEOPHYS] An earth tremor whose epicenter is beneath the ocean and can be felt only by ships in the vicinity of the epicenter. Also known as submarine earthquake. { 'sē,kwāk }

sear [ORD] **1.** An item that retains the firing mechanism of a gun in the cocked position. **2.** A variety of lockwork in the firing mechanism of a propellant-actuated device which prevents motion of the firing pin until released. { sir }

sea rainbow *See* marine rainbow. { 'sē ,rān,bō }

search [COMPUT SCI] To seek a desired item or condition in a set of related or similar items or conditions, especially a sequentially organized or nonorganized set, rather than a multidimensional set. [ENG] To explore a region in space with radar. [NAV] An orderly arrangement of course lines used in searching an area. { sərch }

search and rescue [ENG] The use of aircraft, surface craft, submarines, specialized rescue teams and equipment to search for and rescue personnel in distress on land or at sea. { 'sərch ən 'res,kyü }

search-and-rescue coordination center [COMMUN] A primary search and rescue facility suitably staffed by supervisory personnel and equipped for coordinating and controlling search and rescue operations. { 'sərch ən ¦res,kyü kō'órd·ən,ā·shən ,sen·tər }

search antenna [ELECTROMAG] A radar antenna or antenna system designed for search. { 'sərch an,ten·ə }

search argument [COMPUT SCI] The item or condition that is desired in a search procedure. { 'sərch ,är·gyə·mənt }

search attack unit [ORD] The designation given to one or more ships separately organized or detached from a formation as a tactical unit to search for and destroy submarines. { 'sərch ə'tak ,yü·nət }

search coil *See* exploring coil. { 'sərch ,kòil }

search engine [COMPUT SCI] **1.** Any software that locates and retrieves information in a database. **2.** A server with a stored index of Web pages that is capable of returning lists of pages that match keyword queries. { 'sərch ,en·jən }

search field [COMPUT SCI] A field in a record or segment whose value is examined in a search. { 'sərch ,fēld }

search for extraterrestrial intelligence [ASTRON] A systematic effort to discover evidence for the existence beyond the earth of other advanced civilizations. Abbreviated SETI. { ¦sərch fər ,ek·strə·tə¦res·trē·əl in¦tel·ə·jəns }

search gate [ELECTR] A gate pulse used to search back and forth over a certain range. { 'sərch ,gāt }

searching control [ENG] A mechanism that changes the azimuth and elevation settings on a searchlight automatically and constantly, so that its beam is swept back and forth within certain limits. { 'sərch·iŋ kən,trōl }

searching fire [ORD] Fire distributed in depth by successive changes in the elevation of the gun. { 'sərch·iŋ ,fīr }

searching lighting *See* horizontal scanning. { 'sərch·iŋ ,līd·iŋ }

search key [COMPUT SCI] A data item, or the value of a data item, that is used in carrying out a search. { 'sərch ,kē }

searchlight [OPTICS] A type of light projector designed to produce a beam of high intensity and minimum divergence; it usually employs a specular paraboloidal reflector to produce parallel rays from a light source located at the focus. { 'sərch,līt }

searchlight-control radar [ENG] A ground-based radar used to direct searchlights at aircraft. { 'sərch,līt kən¦trōl ,rā,där }

searchlighting [NAV] The directing of a beam of radiant energy in an oscillatory manner but continuously at a single target. { 'sərch,līd·iŋ }

searchlight-type sonar [ENG] A sonar system in which both transmission and reception are effected by the same narrow beam pattern. { 'sərch,līt ¦tīp 'sōnär }

search radar [ENG] A radar intended primarily to cover a large region of space and to display targets as soon as possible after they enter the region; used for early warning, in connection with ground-controlled approach and interception, and in air-traffic control. { 'sərch ,rā,där }

search time [COMPUT SCI] Time required to locate a particular field of data in a computer storage device; requires a comparison of each field with a predetermined standard until an identity is obtained. { 'sərch ˌtīm }

search unit [ENG] The portion of an ultrasonic testing system which incorporates sending and in some cases receiving transducers to scan the workpiece. { 'sərch ˌyü·nət }

sea return See sea clutter. { 'sē ri̇ˌtərn }

sea rim [ASTRON] The apparent horizon as actually observed at sea. { 'sē ˌrim }

searlesite [MINERAL] NaB(SiO₃)₂·H₂O A white mineral composed of hydrous sodium borosilicate occurring as spherulites. { 'sərlˌzīt }

sea room [NAV] Space in which to maneuver without danger of grounding or colliding. { 'sē ˌrüm }

sea-run [VERT ZOO] Having the habit of ascending a river from the sea, especially to spawn, as salmon and brook trout. { 'sēˌrən }

sea salt [OCEANOGR] The salt remaining after the evaporation of seawater, containing sodium and magnesium chlorides and magnesium and calcium sulfates. { 'sē ˌsȯlt }

sea-salt nucleus [OCEANOGR] A condensation nucleus of a highly hygroscopic nature produced by partial or complete desiccation of particles of sea spray or of seawater droplets derived from breaking bubbles. { 'sē ˌsȯlt ˌnü·klē·əs }

seascape [OCEANOGR] The surrounding sea as it appears to an observer. { 'sēˌskāp }

seascarp [GEOL] A submarine cliff that is relatively long, high, and straight. { 'sēˌskärp }

seashell [INV ZOO] The shell of a marine invertebrate, especially a mollusk. { 'sēˌshel }

seashore [GEOL] **1.** The strip of land that borders a sea or ocean. Also known as seaside; shore. **2.** The ground between the usual tide levels. Also known as seastrand. { 'sēˌshȯr }

seashore lake [GEOGR] A lake, containing either fresh or salt water, which lies along a seashore; it is separated from the sea by a river, a delta, or a wall of sediment. { 'sēˌshȯr ˌlāk }

seasickness [MED] Motion sickness occurring at sea. Also known as pelagism. { 'sēˌsik·nəs }

seaside See seashore. { 'sēˌsīd }

sea slope [GEOL] The slope of land toward the sea. { 'sē ˌslōp }

sea slug [INV ZOO] The common name for the naked gastropods composing the suborder Nudibranchia. { 'sē ˌsləg }

sea smoke See steam fog. { 'sē ˌsmōk }

season [CLIMATOL] A division of the year according to some regularly recurrent phenomena, usually astronomical or climatic. { 'sēz·ən }

seasonal affective disorder [PSYCH] A syndrome of annually repeating depressive symptoms (usually overeating, oversleeping, and carbohydrate craving) that are related to changes in the season and are responsive to light therapy. Also known as winter depression. { ˌsēz·ēn·əl aˌfek·tiv disˈȯrd·ər }

seasonal balancing [CHEM ENG] A seasonal adjustment of the front-end boiling range (volatility) of a motor gasoline to control engine starting characteristics by compensating for seasonal temperature changes. { 'sēz·ən·əl 'bal·əns·iŋ }

seasonal current [OCEANOGR] An ocean current which has large changes in speed or direction due to seasonal winds. { 'sēz·ən·əl 'kə·rənt }

seasonal factors [COMMUN] Factors that are used to adjust skywave absorption data for seasonal variations; these variations are due primarily to seasonal fluctuations in the heights of the ionospheric layers. { 'sēz·ən·əl 'fak·tərz }

seasonally frozen ground [GEOL] Ground that is frozen during low temperatures and remains so only during the winter season. Also known as frost zone. { 'sēz·ən·lē 'frō·zən ˌgraùnd }

seasonal recovery [HYD] Recharge of groundwater during and after a wet season, with a rise in the level of the water table. { 'sēz·ən·əl ri'kəv·ə·rē }

seasonal stream [HYD] A stream whose flow is not constant because it has water in its course only during certain seasons. { 'sēz·ən·əl 'strēm }

seasonal thermocline [OCEANOGR] A thermocline which develops in the oceans in summer at relatively shallow depths due to surface heating and downward transport of heat caused by mixing of water generated by summer winds. { 'sēz·ən·əl 'thər·məˌklīn }

seasonal variation [GEOPHYS] The variation of any parameter of the upper atmosphere with season; for example, the variation of ion densities of different parts of the ionosphere, and the resulting variation in transmission of radio signals over large distances. { 'sēz·ən·əl ˌver·ē'ā·shən }

season check [MATER] A longitudinal crack in wood, caused by uneven seasoning. { 'sēz·ən ˌchek }

season crack [MET] A stress-corrosion crack produced in a copper-base alloy subject to a residual or applied tensile stress and exposed to a specific environment such as moist air containing traces of ammonia. { 'sēz·ən ˌkrak }

season cracking [ORD] In brass cartridge cases and other brass parts, cracking which occurs because of residual internal strains from the manufacturing operations. { 'sēz·ən ˌkrak·iŋ }

seasoned lumber [MATER] Lumber which has been cured by drying to ensure a uniform moisture content. { 'sēz·ənd 'ləm·bər }

seasoning [CIV ENG] See curing. [ELECTR] Overcoming a temporary unsteadiness of a component that may appear when it is first installed. [ENG] Drying of wood either in the air or in a kiln. { 'sēz·ən·iŋ }

sea spider [INV ZOO] The common name for arthropods in the subphylum Pycnogonida. { 'sē ˌspī·dər }

sea squirt [INV ZOO] A sessile, marine tunicate of the class Ascidiacea; it squirts water from two openings in the unattached end when touched or disturbed. { 'sē ˌskwərt }

sea state [OCEANOGR] The numerical or written description of ocean-surface roughness. { 'sē ˌstāt }

seastrand See seashore. { 'sēˌstrand }

sea supremacy [ORD] That degree of sea superiority wherein the opposing force is incapable of effective interference. { 'sē səˌprem·ə·sē }

sea-surface slope [OCEANOGR] A gradual change in the level of the sea surface with distance, caused by Coriolis and wind forces. { 'sē ˌsər·fəs ˌslōp }

sea surveillance [ENG] The systematic observation of surface and subsurface sea areas by all available and practicable means primarily for the purpose of locating, identifying, and determining the movements of ships, submarines, and other vehicles, friendly and enemy, proceeding on or under the surface of seas and oceans. { 'sē sərˌvā·ləns }

seat [MECH ENG] The fixed, pressure-containing portion of a valve which comes into contact with the moving portions of that valve. [ORD] **1.** Support or holder for a mechanism, or for a part of one. **2.** To fit correctly in or on a holder, or prepared position, such as to seat a fuse in a bomb, a projectile in the bore of a gun, or a cartridge in a chamber. { sēt }

seat clay See underclay. { 'sēt ˌklā }

seat earth See underclay. { 'sēt ˌərth }

sea terrace See marine terrace. { 'sē ˌter·əs }

seating [ORD] The distance to which a projectile is rammed into the bore of a cannon, usually measured from the base of the projectile to the rear face of the breech. { 'sēd·iŋ }

seating-lock locking fastener [DES ENG] A locking fastener that locks only when firmly seated and is therefore free-running on the bolt. { 'sēd·iŋ ˌläk 'läk·iŋ 'fas·nər }

sea turn [METEOROL] A wind coming from the sea, often bringing mist; the term is limited mainly to the New England section of the United States. { 'sē ˌtərn }

sea turtle [VERT ZOO] Any of various marine turtles, principally of the families Cheloniidae and Dermochelidae, having paddle-shaped feet. { 'sē ˌtərd·əl }

seatworm See pinworm. { 'sētˌwərm }

sea urchin [INV ZOO] A marine echinoderm of the class Echinoidea; the soft internal organs are enclosed in and protected by a test or shell consisting of a number of close-fitting plates beneath the skin. { 'sē ˌər·chən }

sea valley [GEOL] A relatively shallow, wide depression with gentle slopes in the sea floor, the bottom of which grades continuously downward. { 'sē ˌval·ē }

sea van [IND ENG] Commercial or government-owned (or leased) shipping containers which are moved via ocean transportation; since wheels are not attached, they must be lifted on and off the ship. { 'sē ˌvan }

seawall [CIV ENG] A concrete, stone, or metal wall or embankment constructed along a shore to reduce wave erosion

---

SEASONAL THERMOCLINE

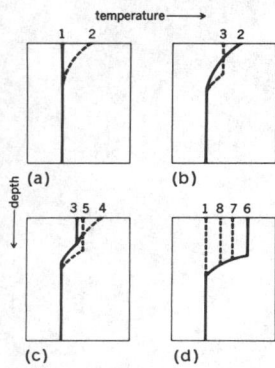

Diagrams showing temperature structure of seasonal thermocline during *(a)* first stage, *(b)* second stage, and *(c)* third stage of formation, and *(d)* breakup. Numbers 1–8 indicate sequence in this process.

SEA SQUIRT

Sea squirts; these saclike tunicates serve as scavengers and as food for higher forms.

SEA URCHIN

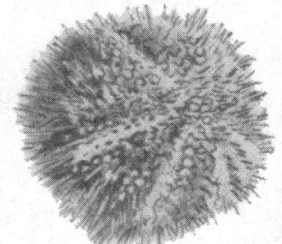

The sea urchin shell, which protects the soft internal organs.

and encroachment by the sea. Also known as sea bank. [GEOL] A steep-faced, long embankment situated by powerful storm waves along a seacoast at high-water mark. { 'sē‚wȯl }

seaward [NAV] The direction or side in which the sea is located; the direction or side away from the land. { 'sē·wərd }

seawater [OCEANOGR] Water of the seas, distinguished by high salinity. Also known as salt water. { 'sē‚wȯd·ər }

seawater thermometer [ENG] A specially designed thermometer to measure the temperature of a sample of seawater; an instrument consisting of a mercury-in-glass thermometer protected by a perforated metal case. { 'sē‚wȯd·ər thər'mäm·əd·ər }

seaway [NAV] **1.** The motion or rate of motion of a vessel. **2.** Headway of a vessel. **3.** The sea as a route of travel from one place to another; a shipping lane. Also known as sea-lane. { 'sē‚wā }

seaweed [BOT] A marine plant, especially algae. { 'sē‚wēd }

seaworthiness [NAV ARCH] The fitness of a ship to sail on the sea and meet any usual condition, or to sail on a specific voyage. { 'sē·wər‚thē·nəs }

sebaceous gland [PHYSIO] A gland, arising in association with a hair follicle, which produces and liberates sebum. { si'bā·shəs 'gland }

sebacic acid [ORG CHEM] $COOH(CH_2)_8COOH$ Combustible, white crystals; slightly soluble in water, soluble in alcohol and ether; melts at 133°C; used in perfumes, paints, and hydraulic fluids and to stabilize synthetic resins. { si'bas·ik 'as·əd }

sebacylic acid See sebacic acid. { ‚seb·ə¦sil·ik 'as·əd }

sebastianite [PETR] A plutonic rock composed of euhedral anorthite, biotite, and some augite and apatite, but without feldspathoids and quartz. { si'bas·chə‚nīt }

sebcha See sebkha. { 'seb·kə }

Sebekidae [INV ZOO] A family of pentastomid arthropods in the suborder Porocephaloidea. { si'bek·ə‚dē }

sebka See sebkha. { 'seb·kə }

sebkha [GEOL] A geologic feature, in North Africa, which is a smooth, flat, plain usually high in salt; after a rain the plain may become a marsh or a shallow lake until the water evaporates. Also known as sabkha; sebcha; sebka; sibjet. { 'seb·kə }

seborrheic dermatitis [MED] An acute inflammation of the skin, occurring usually on oily skin; characterized by scales, crusting yellowish patches, and itching. { ‚seb·ə'rē·ik ‚dər·mə'tīd·əs }

sebum [PHYSIO] The secretion of sebaceous glands, composed of fat, cellular debris, and keratin. { 'sē·bəm }

sec See secant; second; secondary winding.

SEC See secondary-electron conduction.

seca [METEOROL] A drought, or dry wind, in Brazil. { 'sä·kə }

secalose [BIOCHEM] A polysaccharide consisting of fructose units; occurs in green rye and oats, and in rye flour. { 'sek·ə‚lōs }

secant [MATH] **1.** The function given by the reciprocal of the cosine function. Abreviated sec. **2.** The secant of an angle *A* is 1/cos *A*. **3.** A line of unlimited length that intersects a given curve. { 'sē‚kant }

secant conic chart See conic chart with two standard parallels. { 'sē‚kant 'kän·ik 'chärt }

secant conic projection See conic projection with two standard parallels. { 'sē‚kant 'kän·ik prə'jek·shən }

Secchi disk [ENG] An opaque white disk used to measure the transparency or clarity of seawater by lowering the disk into the water horizontally and noting the greatest depth at which it can be visually detected. { 'sek·ē ‚disk }

Secernentea [INV ZOO] A class of the phylum Nematoda in which the primary excretory system consists of intracellular tubular canals joined anteriorly and ventrally in an excretory sinus, into which two ventral excretory gland cells may also open. { ‚se·sər'nen·chə }

sech See hyperbolic secant. { sek }

sechard [METEOROL] A dry, warm foehn wind over Lake Geneva in Switzerland. { se'chär }

seclusion [METEOROL] A special case of the process of occlusion, where the point at which the cold front first overtakes the warm front (or quasi-stationary front) is at some distance from the apex of the wave cyclone. { si'klü·zhən }

secobarbital [PHARM] $C_{12}H_{18}N_2O_3$ 5-Allyl-5-(1-methylbutyl) barbituric acid, a short-acting barbiturate; white powder with bitterish taste; very soluble in alcohol and ether, slightly soluble in water; melts at 82°C; used as a sedative and hypnotic, frequently as the sodium derivative. Also known as quinolbarbitone; Seconal (trade name). { ‚se·kō'bär·bə‚tȯl }

secodont [VERT ZOO] Having teeth adapted for cutting. { 'sek·ə‚dänt }

secohm See international henry. { 'sek‚ōm }

Seconal See secobarbital. { 'sek·ə‚nȯl }

second [MATH] A unit of plane angle, equal to 1/60 minute, or 1/3,600 degree, or π/648,000 radian. [PHYS] The fundamental unit of time equal to 9,192,631,770 periods of the radiation corresponding to the transition between the two hyperfine levels of the ground state of an atom of cesium-133. Abbreviated s: sec. { 'sek·ənd }

secondary [ELEC] Low-voltage conductors of a power distributing system. [ELECTROMAG] See secondary winding. [GEOL] A term with meanings that changed from early to late in the 19th century, when the term was confined to the entire Mesozoic era; it was finally replaced by Mesozoic era. { 'sek·ən‚der·ē }

secondary air [MECH ENG] Combustion air introduced over the burner flame to enhance completeness of combustion. { 'sek·ən‚der·ē 'er }

secondary alcohol [ORG CHEM] An organic alcohol with molecular structure R_1R_2CHOH, where R_1 and R_2 designate either identical or different groups. { 'sek·ən‚der·ē 'al·kə‚hȯl }

secondary allocation [COMPUT SCI] An area of disk storage that is assigned to a file which has become too large for the area originally assigned to it. { 'sek·ən‚der·ē ‚al·ə'kā·shən }

secondary amine [ORG CHEM] An organic compound that may be written R_1R_2NH, where R_1 and R_2 designate either identical or different groups. { 'sek·ən‚der·ē 'am‚ēn }

secondary amyloidosis [MED] Amyloidosis that usually follows chronic suppurative, inflammatory diseases, such as tuberculosis, osteomyelitis, and bronchiectasis. { 'sek·ən‚der·ē ‚am·ə‚lȯi'dō·səs }

secondary area [NAV] In air operations, that area within a segment in which ROC (required obstacle clearance) is reduced as distance from the prescribed course is increased. { 'sek·ən‚der·ē 'er·ē·ə }

secondary armament [ORD] In ships with multiple-size guns installed, that battery consisting of guns next largest to those of the main battery. { 'sek·ən‚der·ē 'är·mə·mənt }

secondary battery See storage battery. { 'sek·ən‚der·ē 'bad·ə‚rē }

secondary biliary cirrhosis [MED] A type of cirrhosis caused by obstruction of the bile duct by calculi (stones). { ‚sek·ən‚der·ē ‚bil·ē‚er·ē sə'rō·səs }

secondary blast injuries [ORD] Those injuries sustained from the indirect effects, such as falling rubble from a collapsed building, or missiles (debris or objects) which have been picked up by the generated winds and hurled against an individual; also includes injuries resulting from individuals being hurled against stationary objects. { 'sek·ən‚der·ē ‚blast ‚in·jə·rēz }

secondary bomb damage [ORD] Nonphysical bomb damage. { 'sek·ən‚der·ē ‚bäm ‚dam·ij }

secondary bow See secondary rainbow. { 'sek·ən‚der·ē 'bō }

secondary cache [COMPUT SCI] High-speed memory between the primary cache and main memory that supplies the processor with the most frequently requested data and instructions. Also known as level 2 cache. { ‚sek·ən‚der·ē 'kash }

secondary cambium [BOT] One of the tissue layers formed after the initial cambial layers in certain plant roots, and that produce a ring of tissue. { 'sek·ən‚der·ē 'kam·bē·əm }

secondary carbon atom [ORG CHEM] A carbon atom that is singly bonded to two other carbon atoms. { 'sek·ən‚der·ē 'kär·bən 'ad·əm }

secondary cell See storage cell. { 'sek·ən‚der·ē 'sel }

secondary circle See secondary great circle. { 'sek·ən‚der·ē 'sər·kəl }

secondary circuit [ELEC] The wiring connected to the secondary winding of a transformer, induction coil, or similar device. [MET] The part of a welding machine which conducts secondary current between transformer and electrodes or between electrodes and workpiece. { 'sek·ən‚der·ē 'sər·kət }

secondary clay [GEOL] A clay that has been transported

from its place of formation and redeposited elsewhere. { 'sek·ən,der·ē 'klä }

secondary coast [GEOL] A relatively stable seacoast or shoreline whose features are the result of present-day marine processes. { 'sek·ən,der·ē 'kōst }

secondary cold front [METEOROL] A front which forms behind a frontal cyclone and within a cold air mass, characterized by an appreciable horizontal temperature gradient. { 'sek·ən,der·ē 'kōld ,frənt }

secondary consequent stream [HYD] A tributary of a subsequent stream, flowing parallel to or down the same slope as the original consequent stream; it is usually developed after the formation of a subsequent stream, but in a direction consistent with that of the original consequent stream. Also known as subconsequent stream. { 'sek·ən,der·ē 'kän·sə·kwənt 'strēm }

secondary consolidation [GEOL] Consolidation of sedimentary material, at essentially constant pressure, resulting from internal processes such as recrystallization. { 'sek·ən,der·ē kən,säl·ə'dā·shən }

secondary consumer [ECOL] In an ecosystem, an animal that feeds on primary consumers. Also known as a carnivore. { ¦sek·ən,der·ē kən'sü·mər }

secondary cosmic rays [GEOPHYS] Radiation produced when primary cosmic rays enter the atmosphere and collide with atomic nuclei and electrons. { 'sek·ən,der·ē 'käz·mik 'rāz }

secondary crater [GEOL] An impact crater produced by the relatively low-velocity impact of fragments ejected from a large primary crater. Also known as satellitic crater. { 'sek·ən,der·ē 'krād·ər }

secondary creep [MECH] The change in shape of a substance under a minimum and almost constant differential stress, with the strain-time relationship a constant. Also known as steady-state creep. { 'sek·ən,der·ē 'krēp }

secondary crusher [MECH ENG] Any of a group of crushing and pulverizing machines used after the primary treatment to further reduce the particle size of shale or other rock. { 'sek·ən,der·ē 'krəsh·ər }

secondary cyclone [METEOROL] A cyclone which forms near or in association with a primary cyclone. Also known as secondary low. { 'sek·ən,der·ē 'sī,klōn }

secondary diagonal [MATH] The elements of a square matrix that lie on the straight line extending from the lower left-hand corner to the upper right-hand corner of the matrix. { ¦sek·ən,der·ē dī'ag·ən·əl }

secondary drilling [MIN ENG] The process of drilling the so-called popholes for the purpose of breaking the larger masses of rock thrown down by the primary blast. { 'sek·ən,der·ē 'dril·iŋ }

secondary electron [ELECTR] **1.** An electron emitted as a result of bombardment of a material by an incident electron. **2.** An electron whose motion is due to a transfer of momentum from primary radiation. { 'sek·ən,der·ē i'lek,trän }

secondary-electron conduction [ELECTR] Transport of charge by secondary electrons moving through the interstices of a porous material under the influence of an externally applied electric field. Abbreviated SEC. { 'sek·ən,der·ē i'lek,trän kən,dək·shən }

secondary emission [ELECTR] The emission of electrons from the surface of a solid or liquid into a vacuum as a result of bombardment by electrons or other charged particles. { 'sek·ən,der·ē i'mish·ən }

secondary enlargement [MINERAL] Overgrowth by chemical deposition on a mineral grain of additional material of identical composition in optical and crystallographic continuity with the original grain; crystal faces characteristic of the original mineral often result. Also known as secondary growth. { 'sek·ən,der·ē in'lärj·mənt }

secondary enrichment [GEOL] The addition to a vein or ore body of material that originated later in time from the oxidation of decomposed ore masses that overlie the vein. { 'sek·ən,der·ē in'rich·mənt }

secondary extinction [PHYS] Increased absorption or decreased diffraction of x-rays by a crystal lattice, due to previous reflection of the x-rays from suitably placed crystal planes. { ¦sek·ən,der·ē ik'stiŋk·shən }

secondary fermentation [FOOD ENG] A fermentation process produced by adding sugar to wine and used to create natural carbonation in, for example, champagne. { 'sek·ən,der·ē ,fər·mən'tā·shən }

secondary flow [FL MECH] A field of fluid motion which can be considered as superposed on a primary field of motion through the action of friction usually in the vicinity of solid boundaries. Also known as frictional secondary flow. { 'sek·ən,der·ē 'flō }

secondary focus [OPTICS] In an astigmatic system, a line at which some of the bundle of rays from an off-axis point meet; this line lies in a plane which contains the point and the optical axis, and has a greater image distance than the primary focus. Also known as sagittal focus. { 'sek·ən,der·ē 'fō·kəs }

secondary fragment [ORD] Secondary fragmentation resulting from either breakup of controlled or uncontrolled fragments upon impact or from the creation of fragments from the target material when it is impacted by a projectile; an example is the spalling of armor plate when it is pierced by armor-piercing shot; the fragments or spalls are broken off the back of the plate and become significant kill mechanisms within the armored enclosure. { 'sek·ən,der·ē 'frag·mənt }

secondary front [METEOROL] A front which may form within a baroclinic cold air mass which itself is separated from a warm air mass by a primary frontal system; the most common type is the secondary cold front. { 'sek·ən,der·ē 'frənt }

secondary geosyncline [GEOL] A geosyncline appearing at the culmination of or after geosynclinal orogeny. { 'sek·ən,der·ē ,jē·ō'sin,klīn }

secondary glacier [HYD] A small valley glacier that joins a larger trunk glacier as a tributary glacier. { 'sek·ən,der·ē 'glā·shər }

secondary great circle [GEOD] A great circle perpendicular to a primary great circle, as a meridian. Also known as secondary circle. { 'sek·ən,der·ē 'grāt 'sər·kəl }

secondary grid emission [ELECTR] Electron emission from a grid resulting directly from bombardment of its surface by electrons or other charged particles. { 'sek·ən,der·ē 'grid i,mish·ən }

secondary grinding [MECH ENG] A further grinding of material previously reduced to sand size. { 'sek·ən,der·ē 'grīnd·iŋ }

secondary growth See secondary enlargement. { 'sek·ən,der·ē 'grōth }

secondary hardening [MET] The hardening of certain alloy steels at moderate temperatures (250–650°C) by the precipitation of carbides; the resultant hardness is greater than that obtained by tempering the steel at some lower temperature for the same time. { 'sek·ən,der·ē 'härd·ən·iŋ }

secondary haulage [MIN ENG] That portion of the haulage system which collects coal from gathering-haulage delivery points and delivers it to the main portion of the system. { 'sek·ən,der·ē 'hól·ij }

secondary high explosive [MATER] A high explosive which is relatively insensitive to heat and shock and is usually initiated by a primary high explosive; used for boosters and bursting charges. { 'sek·ən,der·ē 'hī ik'splō·siv }

secondary hydrogen atom [ORG CHEM] A hydrogen atom that is bonded to a secondary carbon atom. { 'sek·ən,der·ē 'hī·drə·jən ,ad·əm }

secondary hypothermia [MED] A decrease in core body temperature caused by an underlying pathology that prevents the body from generating enough core heat. { ,sek·ən,der·ē ,hī·pō'thər·mē·ə }

secondary index [COMPUT SCI] An index that provides an alternate method of accessing records or portions of records in a data base or file. Also known as alternate index. { 'sek·ən,der·ē 'in,deks }

secondary interstices [GEOL] Openings in a rock that formed after the enclosing rock was formed. { 'sek·ən,der·ē in'tər·stə,sēz }

secondary ionization coefficient See Townsend second ionization coefficient. { ¦sek·ən,der·ē ,ī·ə·nə'zā·shən ,kō·ə,fish·ənt }

secondary ion mass analyzer [ENG] A type of secondary ion mass spectrometer that provides general surface analysis and depth-profiling capabilities. { 'sek·ən,der·ē 'ī,än 'mas 'an·ə,līz·ər }

secondary ion mass spectrometer [ENG] An instrument for microscopic chemical analysis, in which a beam of primary ions with an energy in the range 5–20 kiloelectronvolts bombards a small spot on the surface of a sample. and positive and negative secondary ions sputtered from the surface are analyzed in a mass spectrometer. Abbreviated SIMS. Also known as ion microprobe; ion probe. { 'sek·ən,der·ē 'ī,än 'mas spek'-tram·əd·ər }

secondary item [ORD] Any item, including end items, components, and spare parts, which has not been classified as a principal item. { 'sek·ən,der·ē 'īd·əm }

secondary key [COMPUT SCI] A key that holds the physical location of a record or a portion of a record in a file or database, and provides an alternative means of accessing data. Also known as alternate key. { 'sek·ən,der·ē 'kē }

secondary limestone [PETR] Limestone deposited from solution in cracks and cavities of other rocks. { 'sek·ən,der·ē 'līm,stōn }

secondary lobe See minor lobe. { 'sek·ən,der·ē 'lōb }

secondary low See secondary cyclone. { 'sek·ən,der·ē 'lō }

secondary meristem [BOT] Meristem developed from differentiated living tissue. { 'sek·ən,der·ē 'mer·ə,stem }

secondary metabolite [BOT] A natural chemical product of plants not normally involved in primary metabolic processes such as photosynthesis and cell respiration. Also known as secondary plant product. { 'sek·ən,der·ē mə'tab·ə,līt }

secondary metal [MET] Metal recovered from scrap by remelting and refining. { 'sek·ən,der·ē 'med·əl }

secondary mineral [MINERAL] A mineral produced in an enclosing rock after the rock was formed as a result of weathering or metamorphic or solution activity, and usually at the expense of a primary material that came into existence earlier. { 'sek·ən,der·ē 'min·rəl }

secondary oil recovery [PETRO ENG] Procedures used to increase the flow of oil from depleted or nearly depleted wells; includes fracturing, acidizing, waterflood, and gas injection. { 'sek·ən,der·ē 'óil ri,kəv·ə·rē }

secondary optic axis [OPTICS] One of two optic axes in a crystal along which all light rays travel with equal velocity. { 'sek·ən,der·ē 'äp·tik 'ak·səs }

secondary periderm [BOT] Any layer of the periderm except the first and outermost layer. { 'sek·ən,der·ē 'per·i,dərm }

secondary phloem [BOT] Phloem produced by the cambium, consisting of two interpenetrating systems, the vertical or axial and the horizontal or ray. { 'sek·ən,der·ē 'flō·əm }

secondary photocurrent [ELECTR] A photocurrent resulting from ohmic contacts that are able to replenish charge carriers which pass out of the opposite contact in order to maintain charge neutrality, and whose maximum gain is much greater than unity. { 'sek·ən,der·ē 'fōd·ō,kə·rənt }

secondary plant product See secondary metabolite. { 'sek·ən,der·ē 'plant ,präd·əkt }

secondary plasticizer [MATER] A plastics plasticizer that has insufficient affinity for a resin for it to be the sole plasticizer, and must be blended with a primary plasticizer. Also known as extender plasticizer. { 'sek·ən,der·ē 'plas·tə,sīz·ər }

secondary pollutant [METEOROL] An air pollutant produced by the reaction of a primary pollutant with some other component in the air. { 'sek·ən,der·ē pə'lüt·ənt }

secondary porosity [GEOL] The interstices that appear in a rock formation after it has formed, because of dissolution or stress distortion taking place naturally or artificially as a result of the effect of acid treatment or the injection of coarse sand. { 'sek·ən,der·ē pə'räs·əd·ē }

secondary port [CIV ENG] A port with one or more berths, normally at quays, which can accommodate oceangoing ships for discharge. { 'sek·ən,der·ē 'pórt }

secondary radar [ELECTR] Radar which receives pulses transmitted by an interrogator and makes a return transmission (usually on a different frequency) by its transponder, as opposed to a primary radar which receives pulses returned from illuminated objects. { 'sek·ən,der·ē 'rā,där }

secondary radiation [PHYS] Particles or photons produced by the action of primary radiation on matter, such as Compton recoil electrons, delta rays, secondary cosmic rays, and secondary electrons. { 'sek·ən,der·ē ,rād·ē'ā·shən }

secondary rainbow [OPTICS] A faint rainbow of angular radius about 50° which may appear outside the primary rainbow of 42° radius, and which has its colors in reverse order to those of the primary. Also known as secondary bow. { 'sek·ən,der·ē 'rān,bō }

secondary recovery [PETRO ENG] A method, such as waterflooding, for recovering additional oil from a well after it has stopped producing. { 'sek·ən,der·ē ri'kəv·ə·rē }

secondary reference fuel [MATER] A commercially produced internal combustion engine fuel which is acceptable for knock testing or cetane rating, and which has been calibrated against primary reference fuels by engine tests. { 'sek·ən,der·ē 'ref·rəns ,fyül }

secondary reflection See multiple reflection; shoot. { 'sek·ən,der·ē ri'flek·shən }

secondary rescue facilities [ENG] Local airbase-ready aircraft. crash boats, and other air, surface, subsurface, and ground elements suitable for rescue missions, including government and privately operated units and facilities. { 'sek·ən,der·ē 'res,kyü fə,sil·əd·ēz }

secondary reserves [PETRO ENG] Reserves recoverable commercially at current prices and costs as a result of artificial supplementation of the reserve's natural (gas-drive) energy. { 'sek·ən,der·ē ri'zərvz }

secondary root [BOT] A root arising from a primary root. { 'sek·ən,der·ē 'rüt }

secondary sewage sludge [CIV ENG] Sludge that includes activated sludge, mixed sludge, and chemically precipitated sludge. { 'sek·ən,der·ē 'sü·ij ,sləj }

secondary shaft [MIN ENG] The shaft which extends a mine downward from the bottom of, but not in line with, the primary shaft. { 'sek·ən,der·ē shaft }

secondary splits [MIN ENG] Splits formed by separation of the main air splits. { 'sek·ən,der·ē 'splits }

secondary standard [PHYS] **1.** A unit, as of length, capacitance, or weight, used as a standard of comparison in individual countries or localities, but checked against the one primary standard in existence somewhere. **2.** A unit defined as a specified multiple or submultiple of a primary standard, such as the centimeter. { 'sek·ən,der·ē 'stan·dərd }

secondary station [COMMUN] Any station in a radio net other than the net control station. [ORD] Observation post at the end of a base line farthest from the gun or directing point. { 'sek·ən,der·ē 'stā·shən }

secondary storage [COMPUT SCI] Any means of storing and retrieving data external to the main computer itself but accessible to the program. { 'sek·ən,der·ē 'stór·ij }

secondary stratification [GEOL] The layering that occurs when sediments that were at one time deposited are resuspended and redeposited. Also known as indirect stratification. { 'sek·ən,der·ē ,strad·ə·fə'kā·shən }

secondary stratigraphic trap See stratigraphic trap. { 'sek·ən,der·ē ,strad·ə¦graf·ik 'trap }

secondary stress [MECH] A self-limiting normal or shear stress which is caused by the constraint of a structure and which is expected to cause minor distortions that would not result in a failure of the structure. { 'sek·ən,der·ē 'stres }

secondary structure [BIOCHEM] The conformation of a protein or peptide molecule with respect to nearest-neighbor amino acids. [GEOL] A structure such as a fault, fold, or joint resulting from tectonic movement that started after the rock in which it is found was emplaced. [PALEON] A coarse structure usually between the thin sheets in the protective wall of a tintinnid. { 'sek·ən der·ē 'strək·chər }

secondary succession [ECOL] Ecological succession that occurs in habitats where the previous community has been destroyed or severely disturbed, such as following forest fire, abandonment of agricultural fields, or epidemic disease or pest attack. { ¦sek·ən,der·ē sək'sesh·ən }

secondary surveillance radar [NAV] The secondary radar that operates in conjunction with the airborne transponder of the air-traffic control radar beacon system (ATCRBS). { 'sek·ən,der·ē sər'vā·ləns ,rā,dər }

secondary target [ORD] The target against which fire is directed when the main fire mission has been accomplished, or when it has become impossible or impracticable for the gun or battery to carry out the main fire mission. { 'sek·ən,der·ē ¦tär·gət }

secondary tectonite [GEOL] A tectonite having a deformation fabric. { 'sek·ən,der·ē 'tek·tə,nīt }

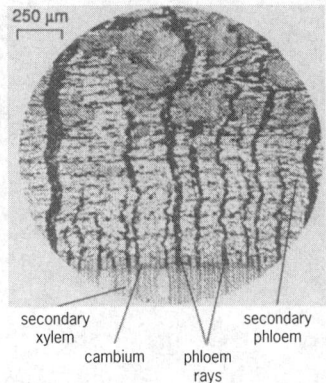

SECONDARY PHLOEM

250 μm

secondary xylem · cambium · phloem rays · secondary phloem

Photomicrograph of cross section of the secordary phloem of paper birch (*Betula papyrifera*). (*Forest Products Laboratory, USDA*)

secondary tide station [ENG] A place at which tide observations are made over a short period to obtain data for a specific purpose. { 'sek·ən,der·ē 'tīd ,stā·shən }

secondary tissue [BOT] Tissue that develops from the vascular cambium or from differentiated tissues. { 'sek·ən,der·ē 'tish·ü }

secondary twinning [CRYSTAL] Twinning of a crystal caused by an external influence, such as pressure in rock. { 'sek·ən,der·ē 'twin·iŋ }

secondary tympanic membrane [ANAT] The membrane closing the fenestra cochleae. { 'sek·ən,der·ē tim'pan·ik 'mem,brān }

secondary voltage [ELECTROMAG] The voltage across the secondary winding of a transformer. { 'sek·ən,der·ē 'vōl·tij }

secondary wall [BOT] The portion of a plant cell wall produced internal to and following deposition of the primary wall; usually consists of several anisotropic layers, and often has prominent internal rings, spirals, bars, or reticulations. { 'sek·ən,der·ē 'wol }

secondary wave [GEOPHYS] See S wave. [OPTICS] One of the waves that radiate from each point on a wavefront, according to Huygens' principle. { 'sek·ən,der·ē 'wāv }

secondary weapon [ORD] A supporting or auxiliary weapon of a unit, vehicle, position, or aircraft; generally, it is a gun of smaller caliber than the primary weapon, and its purpose is to protect or supplement the fire of the primary weapon. { 'sek·ən,der·ē 'wep·ən }

secondary winding [ELECTROMAG] A transformer winding that receives energy by electromagnetic induction from the primary winding; a transformer may have several secondary windings, and they may provide alternating-current voltages that are higher, lower, or the same as that applied to the primary winding. Abbreviated sec. Also known as secondary. { 'sek·ən,der·ē 'wind·iŋ }

secondary xylem [BOT] Xylem produced by cambium, composed of two interpenetrating systems, the horizontal (ray) and vertical (axial). { 'sek·ən,der·ē 'zī·ləm }

second boiling point [PHYS CHEM] In certain mixtures, the temperature at which a gas phase develops from a liquid phase upon cooling. { 'sek·ənd 'boil·iŋ ,point }

second bottom [GEOL] The first terrace rising over a floodplain. { 'sek·ənd 'bäd·əm }

second breakdown [ELECTR] Destructive breakdown in a transistor, wherein structural imperfections cause localized current concentrations and uncontrollable generation and multiplication of current carriers; reaction occurs so suddenly that the thermal time constant of the collector regions is exceeded, and the transistor is irreversibly damaged. { 'sek·ənd 'brāk,daùn }

second category [MATH] A set is of second category if it cannot be expressed as a countable union of nowhere dense sets. { 'sek·ənd 'kad·ə,gor·ē }

second-channel interference See alternate-channel interference. { 'sek·ənd 'chan·əl ,in·tər'fir·əns }

second-class ore [MIN ENG] An ore that needs some preliminary treatment before it is of a sufficiently high grade to be acceptable for market. Also known as milling ore. { 'sek·ənd 'klas 'or }

second countable topological space [MATH] A topological space that has a countable base. { 'sek·ənd 'kaùn·tə·bəl ,täp·ə'läj·ə·kəl 'spās }

second crop [AGR] A crop succeeding one already harvested during a growing season; either a regrowth of the harvested crop, or a newly planted crop. { 'sek·ənd 'kräp }

second curvature See torsion. { 'sek·ənd 'kər·və·chər }

second-degree burn [MED] A burn that is more severe than a first-degree burn and is characterized by blistering as well as reddening of the skin, edema, and destruction of the superficial underlying tissues. { 'sek·ənd 'di·grē 'bərn }

second derivative [MATH] The derivative of the first derivative of a function. { 'sek·ənd də'riv·əd·iv }

second-derivative map [GEOPHYS] A map of the second vertical derivative of a potential field such as the earth's gravity or magnetic field. { 'sek·ənd də'riv·əd·iv 'map }

second detector [ELECTR] The detector that separates the intelligence signal from the intermediate-frequency signal in a superheterodyne receiver. { 'sek·ənd di'tek·tər }

second-foot [HYD] A contraction of cubic foot per second (cfs), the unit of stream discharge commonly used in the United States. { 'sek·ənd 'fùt }

second-foot day [HYD] The volume of water represented by a flow of 1 cubic foot per second for 24 hours; equal to 86,400 cubic feet (approximately 2446.58 cubic meters); used extensively as a unit of runoff volume or reservoir capacity, particularly in the eastern United States. { 'sek·ənd 'fùt 'dā }

second generation See F$_2$. { 'sek·ənd jen·ə'rā·shən }

second-generation computer [COMPUT SCI] A computer characterized by the use of transistors rather than vacuum tubes, the execution of input/output operations simultaneously with calculations, and the use of operating systems. { 'sek·ənd jen·ə'rā·shən kəm'pyüd·ər }

second growth [FOR] New trees that naturally replace trees which have been removed from a forest by cutting or by fire. { 'sek·ənd 'grōth }

second hand [HOROL] The hand that indicates seconds on a timepiece. { 'sek·ənd ,hand }

second law of motion See Newton's second law. { 'sek·ənd 'lo əv 'mō·shən }

second law of thermodynamics [THERMO] A general statement of the idea that there is a preferred direction for any process; there are many equivalent statements of the law, the best known being those of Clausius and of Kelvin. { 'sek·ənd 'lo əv ,thər·mə·dī'nam·iks }

second-level controller [CONT SYS] A controller which influences the actions of first-level controllers, in a large-scale control system partitioned by plant decomposition, to compensate for subsystem interactions so that overall objectives and constraints of the system are satisfied. Also known as coordinator. { 'sek·ənd 'lev·əl kən'trōl·ər }

second mean-value theorem [MATH] The theorem that for two functions $f(x)$ and $g(x)$ that are continuous on a closed interval $[a,b]$ and differentiable on the open interval (a,b), such that $g(b) \neq g(a)$, there exists a number x_1 in (a,b) such that either $[f(b) - f(a)]/[g(b) - g(a)] = f'(x_1)/g'(x_1)$ or $f'(x_1) = g'(x_1) = 0$. Also known as Cauchy's mean-value theorem; double law of the mean; extended mean-value theorem; generalized mean-value theorem. { 'sek·ənd 'mēn 'val·yü ,thir·əm }

second messenger [CELL MOL] Any small molecule or ion that occurs in the cytoplasm of a cell, is generated in response to a hormone binding to a cell-surface receptor, and activates various kinases that regulate the activities of other enzymes. { 'sek·ənd 'mes·ən·jər }

second mining [MIN ENG] In underground coal mining, a procedure in which the pillars left for roof support are removed as mining operations move back out of the previously mined area. { 'sek·ənd 'mīn·iŋ }

second-moment closure [FL MECH] A model of the Reynolds stresses in turbulent flow which is based upon providing transport equations for the Reynolds stresses themselves, thus abandoning the concept of eddy viscosity. { 'sek·ənd ,mō·mənt 'klō·zhər }

second moment of area See geometric moment of inertia. { 'sek·ənd 'mō·mənt əv 'er·ē·ə }

second-order climatological station [CLIMATOL] A station at which observations of atmospheric pressure, temperature, humidity, winds, clouds, and weather are made at least twice daily at fixed hours, and at which the daily maximum and minimum of temperature, the daily amount of precipitation, and the duration of bright sunshine are observed. { 'sek·ənd 'or·dər ,klī·mət·ə'läj·ə·kəl 'stā·shən }

second-order difference [MATH] One of the first-order differences of the sequence formed by taking the first-order differences of a given sequence. { 'sek·ənd 'or·dər 'dif·rəns }

second-order equation [MATH] A differential equation where some term includes the second derivative of the unknown function and no derivative of higher order is present. { 'sek·ənd 'or·dər i'kwā·zhən }

second-order leveling [ENG] Spirit leveling that has less stringent requirements than those of first-order leveling, in which lines between benchmarks established by first-order leveling are run in only one direction. { 'sek·ənd 'or·dər 'lev·ə·liŋ }

second-order reaction [PHYS CHEM] A reaction whose rate of reaction is determined by the concentration of two chemical species. { 'sek·ənd 'or·dər rē'ak·shən }

second-order relief [GEOGR] Extensive relief features consisting of major mountain systems and other surface formations of subcontinental extent. { 'sek·ənd 'or·dər ri'lēf }

second-order station [METEOROL] After U.S. Weather

Bureau practice, a station operated by personnel certified to make aviation weather observations or synoptic weather observations. { 'sek·ənd ¦ȯr·dər 'stā·shən }

second-order subroutine [COMPUT SCI] A subroutine that is entered from another subroutine, in contrast to a first-order subroutine; it constitutes the second level of a two-level or higher-level routine. Also known as second-remove subroutine. { 'sek·ənd ¦ȯr·dər 'səb·rü‚tēn }

second-order transition [THERMO] A change of state through which the free energy of a substance and its first derivatives are continuous functions of temperature and pressure, or other corresponding variables. { 'sek·ənd ¦ȯr·dər tran‚zish·ən }

second pilot [AERO ENG] A pilot, not necessarily qualified on type, who is responsible for assisting the first pilot to fly the aircraft and is authorized as second pilot. { 'sek·ənd 'pī·lət }

second quadrant [MATH] **1.** The range of angles from 90 to 180°. **2.** In a plane with a system of cartesian coordinates, the region in which the x coordinate is negative and the y coordinate is positive. { 'sek·ənd 'kwä·drənt }

second quantization [QUANT MECH] A procedure in which the dependent variables of a classical field or a quantum-mechanical wave function are regarded as operators on which commutation rules are imposed; this produces a formalism in which particles may be created and destroyed. { 'sek·ənd ‚kwän·tə'zā·shən }

second radiation constant [STAT MECH] A constant appearing in the Planck radiation formula, equal to the speed of light times Planck's constant divided by Boltzmann's constant, or approximately 1.4388 degree-centimeters. Symbolized c_2; C_2. { 'sek·ənd ‚rād·ē'ā·shən ‚kän·stənt }

second-remove subroutine See second-order subroutine. { 'sek·ənd ri¦müv 'səb·rü‚tēn }

second sound [ACOUS] A transverse sound wave which propagates in smectic liquid crystals, and whose behavior resembles mathematically that of second sound in superfluid helium. [CRYO] A type of wave propagated in the superfluid phase of liquid helium (helium II), in which temperature and entropy variations propagate with no appreciable variation in density or pressure. { 'sek·ənd 'saùnd }

second species [MATH] The class of sets G_0 such that all the sets G_n are nonempty, where, in general, G_n is the derived set of G_{n-1}. { 'sek·ənd 'spē‚shēz }

seconds pendulum [HOROL] A pendulum measuring 99.353 centimeters between its center of suspension and center of oscillation, and which requires exactly 1 second to swing from one side to another at sea level and 45° latitude. { 'sek·əns ‚pen·jə·ləm }

second strike [ORD] The first counterblow of a war (generally associated with nuclear operations). { 'sek·ənd 'strīk }

second-strike capability [ORD] The ability to survive a first strike with sufficient resources to deliver an effective counterblow (generally associated with nuclear weapons). { 'sek·ənd ¦strīk ‚kā·pə'bil·əd·ē }

second-time-around echo [ELECTR] A radar echo received after an interval exceeding the pulse interval. Also known as second-trip echo. { 'sek·ənd ¦tīm ə'raùnd ‚ek·ō }

second-trip echo See second-time-around echo. { 'sek·ənd ¦trip ‚ek·ō }

second water [LAP] The quality or luster of a gemstone which is one grade below first water, such as that of a diamond that contains slight flaws or turbid patches. { 'sek·ənd 'wȯd·ər }

second-year ice [OCEANOGR] Sea ice that has survived only one summer's melt. Also known as two-year ice. { 'sek·ənd ¦yir 'īs }

secrecy system See privacy system. { 'sē·krə·sē ‚sis·təm }

secretin [BIOCHEM] A basic polypeptide hormone produced by the duodenum in response to the presence of acid; acts to excite the pancreas to activity. { si'krēt·ən }

secret-key algorithm [COMPUT SCI] A cryptographic algorithm which uses the same cryptographic key for encryption and decryption, requiring that the key first be transmitted from the sender to the recipient via a secure channel. { ‚sē·krət ‚kē 'al·gə‚rith·əm }

secretion [GEOL] A secondary structure formed of material deposited (from solution) within an empty cavity in any rock, especially a deposit formed on or parallel to the walls of the cavity, the first layer being the outer one. [PHYSIO] **1.** The act or process of producing a substance which is specialized to perform a certain function within the organism or is excreted from the body. **2.** The material produced by such a process. { si krē·shən }

secretor gene [GEN] A dominant autosomal gene in humans which controls secretion of A and B antigenic material in saliva, urine, plasma, and other body fluids; it is not linked to the ABO locus. { si'krēd·ər ‚jēn }

secretory diarrhea [MED] Diarrhea resulting from an increase in electrolyte secretion into the intestinal lumen by its epithelial cells, leading to an increased flow of water from the intestinal mucosa into the lumen; it persists during fasting. { si‚krēd·ə·rē dī·ə'rē·ə }

secretory granules [CYTOL] Accumulations of material produced within a cell for secretion outside the cell. { si'krēd·ə·rē ‚gran‚yülz }

secretory structure [BOT] Plant cells or organizations of plant cells which produce a variety of secretions. { si'krēd·ə·rē ‚strək·chər }

secret weapon [ORD] A weapon closely guarded or kept under concealment so as to be used with advantage before countermeasures can be taken against it. { 'sē·krət 'wep·ən }

sectile [MINERAL] Pertaining to a mineral whose texture is tenacious enough to be cut with a knife. { 'sek·təl }

section [CIV ENG] A piece of land usually 1 mile square (640 acres or approximately 2.58999 square kilometers) with boundaries conforming to meridians and parallels within established limits; 1 of 36 units of subdivision of a township in the U.S. Public Land survey system. [COMMUN] Each individual transmission span in a radio relay system; a system has one or more section than it has repeaters. [GEOL] **1.** An inclined or vertical surface that is uncovered either naturally (as a sea cliff or stream bank) or artificially (as a strip mine or road cut) through a part of the earth's crust. **2.** A description or scale drawing of the successive rock units or geologic structures shown by the exposed surface, or their appearance if cut through by any intersecting plane. **3.** See columnar section. **4.** See geologic section. **5.** See type section. **6.** See thin section. [MATH] **1.** For a polyhedral angle, the polygon formed by the intersection of the faces of the angle with a plane that does not pass through the vertex. **2.** See plane section. { 'sek·shən }

sectional center [COMMUN] A long-distance telephone office which connects several primary centers and which is in class number 2; only a regional center has greater importance in routing telephone calls. Abbreviated SC. { 'sek·shən·əl 'sen·tər }

sectional conveyor [MECH ENG] A belt conveyor that can be lengthened or shortened by the addition or the removal of interchangeable sections. { 'sek·shən·əl kən'vā·ər }

sectional core barrel [DES ENG] A core barrel whose length can be increased by coupling unit sections together. { 'sek·shən·əl 'kȯr ‚bar·əl }

sectional header boiler [MECH ENG] A horizontal boiler in which tubes are assembled in sections into front and rear headers; the latter, in turn, are connected to the boiler drum by vertical tubes. { 'sek·shən·əl 'hed·ər ‚bȯil·ər }

sectionalized vertical antenna [ELECTROMAG] Vertical antenna that is insulated at one or more points along its length; the insertion of suitable reactances or applications of a driving voltage across the insulated points results in a modified current distribution giving a more desired radiation pattern in the vertical plane. { 'sek·shən·əl‚īzd 'vərd·ə·kəl an‚ten·ə }

sectional radiography [ELECTR] The technique of making radiographs of plane sections of a body or an object; its purpose is to show detail in a predetermined plane of the body, while blurring the images of structures in other planes. Also known as laminography; planigraphy; tomography. { 'sek·shən·əl ‚rād·ē'äg·rə·fē }

section house [CIV ENG] A building near a railroad section for housing railroad workers, or for storing maintenance equipment for the section. { 'sek·shən ‚haùs }

section line [CIV ENG] A line representing the boundary of a section of land. [GRAPHICS] One of a series of parallel lines indicating a cut surface in a mechanical or architectural drawing. { 'sek·shən ‚līn }

section modulus [MECH] The ratio of the moment of inertia of the cross section of a beam undergoing flexure to the greatest distance of an element of the beam from the neutral axis. { 'sek·shən 'mäj·ə·ləs }

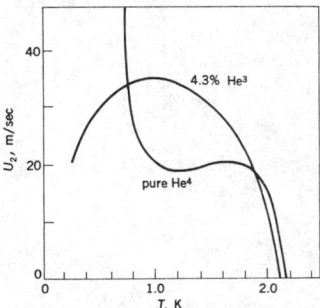

SECOND SOUND

Velocity of second sound U_2 as a function of temperature (kelvins) for pure He[4] and a 4.3 % He[3] mixture. *(From J. C. King and H. A. Fairbank, Second sound in He[3]-He[4] mixtures below 1°K, Phys. Rev., 93:21, 1954)*

sector [COMPUT SCI] **1.** A portion of a track on a magnetic disk or a band on a magnetic drum. **2.** A unit of data stored in such a portion. [CIV ENG] A clearly defined area or airspace designated for a particular purpose. [ELECTROMAG] Coverage of a radar as measured in azimuth. [MATH] A portion of a circle bounded by two radii and an arc joining their end points. [METEOROL] Something resembling the sector of a circle, as a warm sector between the warm and cold fronts of a cyclone. { 'sek·tər }

sectoral harmonic [MATH] A spherical harmonic which is 0 on a set of equally spaced meridians of a sphere with center at the origin of spherical coordinates, dividing the sphere into sectors. { ¦sek·tə·rəl här′män·ik }

sectoral horn [ELECTROMAG] Horn with two opposite sides parallel and the two remaining sides which diverge. { 'sek·tə·rəl 'hȯrn }

sector boundary [ASTROPHYS] The rapid transition from one polarity to another in the interplanetary magnetic field. { 'sek·tər ′baún·drē }

sector disk [PHYS] A device used to reduce the intensity of a beam of light or other electromagnetic radiation by an accurately known amount; in its simplest form, it consists of a circular, opaque disk with one or more sectors cut out of it, rapidly rotating in the path of the beam. { 'sek·tər ‚disk }

sector display [ELECTR] A display in which only a sector of the total service area of a radar system is shown; usually the sector is selectable. { 'sek·tər di‚splā }

sectored light [NAV] A light having sectors of different colors or of the same color in specific sectors separated by dark sectors. { 'sek·tərd 'līt }

sector gate [CIV ENG] A horizontal gate with a pie-slice cross section used to regulate the level of water at the crest of a dam; it is raised and lowered by a rack and pinion mechanism. { 'sek·tər ‚gāt }

sector gear [DES ENG] **1.** A toothed device resembling a portion of a gear wheel containing the center bearing and a part of the rim with its teeth. **2.** A gear having such a device as its chief essential feature. [MECH ENG] A gear system employing such a gear as a principal part. { 'sek·tər ‚gir }

sectorgram See pie chart. { 'sek·tər‚gram }

sector interleave [COMPUT SCI] A sequence indicating the order in which sectors are arranged on a hard disk, generally so as to minimize access times. Also known as sector map. { 'sek·tər 'in·tər‚lēv }

sector map See sector interleave. { 'sek·tər ‚map }

sector mark [COMPUT SCI] A location on each sector of each track of a disk pack or floppy disk that gives the sector's address, tells whether the sector is in use, and gives other control information. { 'sek·tər ‚märk }

sector of fire [ORD] An area which is required to be covered by fire by an individual, a weapon, or a unit. { 'sek·tər əv 'fīr }

sector scan [ELECTR] A radar scan through a limited angle, as distinguished from complete rotation. { 'sek·tər ‚skan }

sector search [NAV] A flight or sailing plan of three legs, the turning points being at equal distances along radial lines from a fixed or moving point. { 'sek·tər ‚sərch }

sector structure [ASTROPHYS] The polarity pattern of the interplanetary magnetic field observed during a solar rotation. { 'sek·tər ‚strək·chər }

sector wind [METEOROL] The average observed or computed wind (direction and speed) at flight level for a given sector of an air route; sectors for over-ocean flights usually consist of 10° of longitude. { 'sek·tər ‚wind }

secular [ENG] Of or pertaining to a long indefinite period of time. { 'sek·yə·lər }

secular acceleration [ASTRON] An apparent gradual acceleration of the moon's motion in its orbit, as measured relative to mean solar time. { 'sek·yə·lər ak‚sel·ə′rā·shən }

secular determinant [MATH] For a square matrix A, the determinant of the matrix whose off-diagonal components are equal to those of A, and whose diagonal components are equal to the difference between those of A and a parameter λ; it is equal to the characteristic polynomial in λ of the linear transformation represented by A. { 'sek·yə·lər di′tər·mən·ənt }

secular equilibrium [NUCLEO] Radioactive equilibrium in which the parent has such a small decay constant that there has been no appreciable change in the quantity of parent present by the time the decay products have reached radioactive equilibrium. { 'sek·yə·lər ‚ē·kwə′lib·rē·əm }

secular parallax [ASTRON] An apparent angular displacement of a star, resulting from the sun's motion. { 'sek·yə·lər 'par·ə‚laks }

secular perturbations [ASTROPHYS] Changes in the orbit of a planet, or of a satellite, that operates in extremely long cycles. { 'sek·yə·lər ‚pər·dər′bā·shənz }

secular trend [STAT] A concept in time series analysis that refers to a movement or trend in a series over very long periods of time. Also known as long-time trend. { 'sek·yə·lər 'trend }

secular variable [ASTRON] A star whose brightness appears to have slowly lessened or increased over a time period of centuries. { 'sek·yə·lər 'ver·ē·ə·bəl }

secular variation [ASTRON] A perturbation of the moon's motion caused by variations in the effect of the sun's gravitational attraction on the earth and moon as their relative distances from the sun vary during the synodic month. [GEOPHYS] The changes, measured in hundreds of years, in the magnetic field of the earth. Also known as geomagnetic secular variation. { 'sek·yə·lər ‚ver·ē′ā·shən }

secund [BOT] Having lateral members arranged on one side only. { 'sē‚kənd }

secundine dike [GEOL] A dike which has been intruded into hot country rock. { 'sek·ən‚dīn 'dīk }

secure [ORD] To gain possession of a position or terrain feature, with or without force, and to make such disposition as will prevent, as far as possible, its destruction or loss by enemy action. { si'kyúr }

secure visual communications [COMMUN] The transmission of an encrypted digital signal consisting of animated visual and audio information; the distance may vary from a few hundred feet to thousands of miles. { si'kyúr 'vizh·ə·wəl kə‚myü·nə′kā·shənz }

secure voice [COMMUN] Voice message that is scrambled or coded, therefore not transmitted in the clear. { si'kyúr 'vȯis }

securinine [PHARM] $C_{13}H_{15}NO_2$ A crystalline compound that forms yellow crystals from a methanol solution and melts at 142–143°C; used to make the nitrate compound for cardiac insufficiency. { si'kyúr·ə‚nēn }

securite explosive [MATER] A type of plastic explosive with a balanced oxygen content; it is built up on a nonexplosive, hydrophilic gel and contains oxygen-emitting salts, solid high explosive, and water. { si'kyúr‚īt ik‚splō·siv }

security [COMPUT SCI] The existence and enforcement of techniques which restrict access to data, and the conditions under which data may be obtained. [ELEC] The ability of an electric power system to suitably respond to disturbances arising within that system, including both local and widespread disturbances and the loss of major generation and transmission facilities. [ORD] **1.** Measures taken by a command to protect itself from espionage, observation, sabotage, annoyance, or surprise. **2.** A condition which results from the establishment and maintenance of protective measures which ensure a state of inviolability from hostile acts or influences. **3.** Protection of supplies or supply establishments against enemy attack, fire, theft, and sabotage. { si'kyúr·əd·ē }

security classification [ORD] A category or grade assigned to defense information or materiel to indicate the degree of danger to national security that would result from its unauthorized disclosure and the standard of protection required to guard against unauthorized disclosure. { si'kyúr·əd·ē ‚klas·ə·fə′kā·shən }

security clearance [ORD] A clearance given to a person to permit access to classified material, equipment, or information up to and including a given classification, provided the person can establish a need-to-know. { si'kyúr·əd·ē ‚klir·əns }

security control officer [ORD] In the United States, an officer, warrant officer, or responsible civilian official appointed in each command or agency to exercise staff supervision over the safeguarding of defense information. { si'kyúr·əd·ē kən'trōl ‚ȯf·ə·sər }

security kernel [COMPUT SCI] A portion of an operating system into which all security-related functions have been concentrated, forming a small, certifiably secure nucleus which is separate from the rest of the system. { si'kyúr·əd·ē ‚kər·nəl }

security perimeter [COMPUT SCI] A logical boundary of a distributed computer system, surrounding all the resources that are controlled and protected by the system. { sə'kyūr·əd·ē pə,rim·əd·ər }

security reporting/alerting system [COMMUN] A rapid communications procedure that integrates all U.S. Air Force bases and commands, so that a significant happening at one location, or a pattern of seemingly unrelated happenings at several locations, can serve as a basis for swift security alerting or warning throughout the system. { si'kyūr·əd·ē ri¦pórd·iŋ ə¦lərd·iŋ ,sis·təm }

security target [COMPUT SCI] A description of a product meeting the security and functionality requirements of a computing system. { sə'kyūr·əd·ē ,tär·gət }

SED See skin erythema dose.

sedation [MED] A state of lessened activity. { si'dā·shən }

sedative [PHARM] An agent or drug that has a quieting effect on the central nervous system. { 'sed·əd·iv }

Sedentaria [INV ZOO] A group of families of polychaete annelids in which the anterior, or cephalic, region is more or less completely concealed by overhanging peristomial structures, or the body is divided into an anterior thoracic and a posterior abdominal region. { ,sed·ən'tar·ē·ə }

sedentary soil [GEOL] Soil that still lies on the rock from which it was formed. { 'sed·ən,ter·ē 'sȯil }

sedifluction [GEOL] The subaquatic or subaerial movement of material in unconsolidated sediments, occurring in the primary stages of diagenesis. { ,sed·ə'flək·shən }

sediment [GEOL] **1.** A mass of organic or inorganic solid fragmented material, or the solid fragment itself, that comes from weathering of rock and is carried by, suspended in, or dropped by air, water, or ice; or a mass that is accumulated by any other natural agent and that forms in layers on the earth's surface such as sand, gravel, silt, mud, fill, or loess. **2.** A solid material that is not in solution and either is distributed through the liquid or has settled out of the liquid. { 'sed·ə·mənt }

sedimentary breccia [PETR] A rock composed of fragments that are larger than 2 millimeters in diameter and are the result of sedimentary processes; characterized by imperfect mechanical sorting of its materials and by a higher concentration of fragments from one local source or by a wide variety of materials mixed together in no particular pattern. Also known as sharpstone conglomerate. { ¦sed·ə¦men·trē 'brech·ə }

sedimentary cycle See cycle of sedimentation. { ¦sed·ə¦men·trē 'sī·kəl }

sedimentary differentiation [GEOL] The progressive separation (by erosion and transportation) of a well-defined rock mass into physically and chemically unlike products that are resorted and deposited as sediments in more or less separate areas. { ¦sed·ə¦men·trē ,dif·ə,ren·chē'ā·shən }

sedimentary dike [GEOL] A tabular mass of sedimentary material that cuts across the structure or bedding of preexisting rock in the manner of an igneous dike and that is formed by the filling of a crack or fissure by forcible injection or intrusion of sediments under abnormal pressure, or by simple infilling of sediments. { ¦sed·ə¦men·trē 'dīk }

sedimentary facies [GEOL] A stratigraphic facies differing from another part or parts of the same unit in both lithologic and paleontologic characters. { ¦sed·ə¦men·trē 'fā·shēz }

sedimentary injection See injection. { ¦sed·ə¦men·trē in'jek·shən }

sedimentary insertion [GEOL] The emplacement of sedimentary material among deposits or rocks already formed, such as by infilling, injection, or intrusion, or through localized subsidence due to solution of underlying rock. { ¦sed·ə¦men·trē in'sər·shən }

sedimentary intrusion See intrusion. { ¦sed·ə¦men·trē in'trü·zhən }

sedimentary laccolith [GEOL] An intrusion of plastic sedimentary material (such as clayey salt breccia) forced up under high pressure and penetrating parallel or nearly parallel to the bedding planes of the invaded formation; characterized by a very irregular thickness. { ¦sed·ə¦men·trē 'lak·ə,lith }

sedimentary lag [GEOL] Delay between the formation of potential sediment by weathering and its removal and deposition. { ¦sed·ə¦men·trē 'lag }

sedimentary petrography [PETR] The description and classification of sedimentary rocks. Also known as sedimentography. { ¦sed·ə¦men·trē pə'träg·rə·fē }

sedimentary petrology [PETR] The study of the composition, characteristics, and origin of sediments and sedimentary rocks. { ¦sed·ə¦men·trē pə'träl·ə·jē }

sedimentary quartzite See orthoquartzite. { ¦sed·ə¦men·trē 'kwȯrt,sīt }

sedimentary rock [PETR] A rock formed by consolidated sediment deposited in layers. Also known as derivative rock; neptunic rock, stratified rock. { ¦sed·ə¦men·trē 'räk }

sedimentary structure [GEOL] A structure in sedimentary rocks, such as cross-bedding, ripple marks, and sandstone dikes, produced either contemporaneously with deposition (primary sedimentary structures) or shortly after deposition (secondary sedimentary structures). { ¦sed·ə¦men·trē 'strək·chər }

sedimentary tectonics [GEOL] Folding and deformation in geosynclinal basins caused by subsidence and buckling of strata. { ¦sed·ə¦men·trē tek'tän·iks }

sedimentary trap [GEOL] An area in which sedimentary material accumulates instead of being transported farther, as in an area between high-energy and low-energy environments. { ¦sed·ə¦men·trē 'trap }

sedimentary tuff [GEOL] A tuff containing a small amount of nonvolcanic detrital material. { ¦sed·ə¦men·trē 'təf }

sedimentary volcanism [GEOL] The expelling, extruding, or breaking through of overlying formations by a mixture of sediment, water, and gas, driven by the gas under pressure. { ¦sed·ə¦men·trē 'väl·kə,niz·əm }

sedimentation [CHEM] The settling of suspended particles within a liquid under the action of gravity or a centrifuge. [GEOL] **1.** The act or process of accumulating sediment in layers. **2.** The process of deposition of sediment. [MET] Classification of metal powders by the rate of settling in a fluid { ,sed·ə·mən'tā·shən }

sedimentation balance [ANALY CHEM] A device to measure and record the weight of sediment (solid particles settled out of a liquid) versus time; used to determine particle sizes of fine solids. { ,sed·ə·mən'tā·shən ,bal·əns }

sedimentation basin [GEOL] A depression in the ocean floor with a wide, flat bottom in which sediment accumulates. { ,sed·ə·mən'tā·shən ,bās·ən }

sedimentation coefficient [PHYS CHEM] In the sedimentation of molecules in an accelerating field, such as that of a centrifuge, the velocity of the boundary between the solution containing the molecules and the solvent divided by the accelerating field. (In the case of a centrifuge, the accelerating field equals the distance of the boundary from the axis of rotation multiplied by the square of the angular velocity in radians per second.) { ,sed·ə·mən'tā·shən ,kō·i'fish·ənt }

sedimentation constant [PHYS CHEM] A quantity used in studying the behavior of colloidal particles subject to forces, especially centrifugal forces; it is equal to $2r^2(\rho - \rho')/9\eta$, where r is the particle's radius, ρ and ρ' are reciprocals of partial specific volumes of particle and medium respectively, and η is the medium's viscosity. { ,sed·ə·mən'tā·shən ,kän·stənt }

sedimentation curve [GEOL] A curve showing cumulatively, and in successive units of time, the amount of sediment accumulated or removed from an originally uniform suspension. { ,sed·ə·mən'tā·shən ,kərv }

sedimentation diameter [GEOL] The diameter of a sedimentary particle, determined from the measurement of a hypothetical sphere of the same gravity and settling velocity as those of a given sedimentary particle in the same fluid. { ,sed·ə·mən'tā·shən dī,am·əd·ər }

sedimentation equilibrium [ANALY CHEM] The equilibrium between the forward movement of a sample's liquid-sediment boundary and reverse diffusion during centrifugation; used in molecular-weight determinations. { ,sed·ə·mən'tā·shən ,ē·kwə,lib·rē·əm }

sedimentation radius [GEOL] One-half of the sedimentation diameter. { ,sed·ə·mən'tā·shən ,rād·ē·əs }

sedimentation rate [GEOL] See rate of sedimentation. [PATH] The rate at which red blood cells settle out of anticoagulated blood. { ,sed·ə·mən'tā·shən ,rāt }

sedimentation tank [ENG] A tank in which suspended matter is removed either by quiescent settlement or by continuous flow at high velocity and extended retention time to allow deposition. { ,sed·ə·mən'tā·shən ,taŋk }

sedimentation trend [GEOL] The direction in which sediments were laid down. { ,sed·ə·mən'tā·shən ,trend }

sedimentation trough [GEOL] A depression in the ocean floor with a narrow U- or V-shaped bottom in which sediment accumulates. { ˌsed·ə·mən'tā·shən ˌtrof }

sedimentation unit [GEOL] A sedimentary deposit formed during one distinct act of sedimentation. { ˌsed·ə·mən'tā·shən ˌyü·nət }

sedimentation velocity [ANALY CHEM] The rate of movement of the liquid-sediment boundary in the sample holder during centrifugation; used in molecular-weight determinations. { ˌsed·ə·mən'tā·shən və'läs·əd·ē }

sediment bulb [ENG] A bulb for holding sediment that settles from the liquid in a tank. { 'sed·ə·mənt ˌbəlb }

sediment charge [HYD] In a stream, the ratio of the weight or volume of sediment to the weight or volume of water passing a given cross section per unit of time. { 'sed·ə·mənt ˌchärj }

sediment concentration [HYD] The ratio of the dry weight of the sediment in a water-sediment mixture (obtained from a stream or other body of water) to the total weight of the mixture. { 'sed·ə·mənt ˌkän·sən'trā·shən }

sediment corer [ENG] A heavy coring tube which punches out a cylindrical sediment section from the ocean bottom. { 'sed·ə·mənt ˌkor·ər }

sediment-delivery ratio [GEOL] The ratio of sediment yield of a drainage basin to the total amount of sediment moved by sheet erosion and channel erosion. { 'sed·ə·mənt di¦liv·ə·rē ˌrā·shō }

sediment discharge [HYD] The amount of sediment moved by a stream in a given time, measured by dry weight or by volume. Also known as sediment-transport rate. { 'sed·ə·mənt 'dis,chärj }

sediment discharge rating [HYD] A relationship between the discharge of sediment and the total discharge of the stream. Also known as silt discharge rating. { 'sed·ə·mənt 'dis,chärj ˌrād·iŋ }

sediment load [HYD] The solid material that is transported by a natural agent, especially by a stream. { 'sed·ə·mənt ˌlōd }

sedimentography See sedimentary petrography. { ˌsed·ə·mən'täg·rə·fē }

sedimentology [GEOL] The science concerned with the description, classification, origin, and interpretation of sediments and sedimentary rock. { ˌsed·ə·mən'täl·ə·jē }

sediment-production rate [GEOL] Sediment yield per unit of drainage area, derived by dividing the annual sediment yield by the area of the drainage basin. { 'sed·ə·mənt prə'dək·shən ˌrāt }

sediment station [HYD] A vertical cross-sectional plane of a stream, usually normal to the mean direction of flow, where samples of suspended load are collected on a systematic basis for determining concentration, particle-size distribution, and other characteristics. { 'sed·ə·mənt ˌstā·shən }

sediment-transport rate See sediment discharge. { 'sed·ə·mənt 'tranz,port ˌrāt }

sediment trap [ENG] A device for measuring the accumulation rate of sediment on the floor of a body of water. { 'sed·ə·mənt ˌtrap }

sediment vein [GEOL] A sedimentary dike formed by the filling of a fissure from above with sedimentary material. { 'sed·ə·mənt ˌvān }

sediment yield [GEOL] The amount of material eroded from the land surface by runoff and delivered to a stream system. { 'sed·ə·mənt ˌyēld }

sedoheptulose [BIOCHEM] A seven-carbon ketose sugar widely distributed in plants of the Crassulaceae family; a significant intermediary compound in the cyclic regeneration of D-ribulose. { ¦sē·dō'hep·tə,lōs }

Seebeck coefficient [ELECTR] The ratio of the open-circuit voltage to the temperature difference between the hot and cold junctions of a circuit exhibiting the Seebeck effect. { 'zā,bek ˌkō·i'fish·ənt }

Seebeck effect [ELECTR] The development of a voltage due to differences in temperature between two junctions of dissimilar metals in the same circuit. [GRAPHICS] A photographic emulsion that is exposed until a faint visible image appears, and is then exposed to colored light and takes on the color of the light to which it is exposed. { 'zā,bek i,fekt }

seed [BOT] A fertilized ovule containing an embryo which forms a new plant upon germination. [CHEM] A small, single crystal of a desired substance added to a solution to induce crystallization. [COMPUT SCI] An initial number used by an algorithm such as a random number generator. [SOLID STATE] A small, single crystal of semiconductor material used to start the growth of a large, single crystal for use in cutting semiconductor wafers. { sēd }

seed charge [CHEM] A small amount of material added to a supersaturated solution to initiate precipitation. { 'sēd ,chärj }

seed coat [BOT] The envelope which encloses the seed except for a tiny pore, the micropyle. { 'sēd ,kōt }

seed core [NUCLEO] A reactor core that includes a relatively small volume of highly enriched uranium, surrounded by a much larger volume of natural uranium or thorium in a blanket; as a result of fission in the seed uranium, neutrons are supplied to the blanket, which is made to furnish a substantial fraction of the core power. Also known as spiked core. { 'sēd ,kor }

seed cotton [AGR] The material in mature bolls containing both cotton fiber and seeds. { 'sēd ,kät·ən }

seed down [AGR] To sow seeds for grass or forage legumes. { 'sēd ,daun }

seed fern [PALEOBOT] The common name for the extinct plants classified as Pteridospermae, characterized by naked seeds borne on large, fernlike fronds. { 'sēd ,fərn }

seeding [AGR] The planting of seed. [CHEM] The adding of a seed charge to a supersaturated solution, or a single crystal of a desired substance to a solution of the substance to induce crystallization. [ELECTR] The introduction of atoms with a low ionization potential into a hot gas to increase electrical conductivity. { 'sēd·iŋ }

seedling [BOT] 1. A plant grown from seed. 2. A tree younger and smaller than a sapling. 3. A tree grown from a seed. { 'sēd·liŋ }

seed nuclei [ASTROPHYS] Nuclei from which other nuclei are synthesized in stars. { 'sēd ,nü·klē,ī }

seeing [ASTRON] The clarity and steadiness of an image of a star in a telescope. { 'sē·iŋ }

seek [COMPUT SCI] 1. To position the access mechanism of a random-access storage device at a designated location or position. 2. The command that directs the positioning to take place. [ORD] To go toward a target or other object in reaction to some influence such as heat, light, sound, or other radiation emitted by the target or object. { sēk }

seek area [COMPUT SCI] An area of a direct-access storage device, such as a magnetic disk file, assigned to hold records to which rapid access is needed, and located so that the physical characteristics of the device permit such access. Also known as cylinder. { 'sēk ,er·ē·ə }

seeker [ORD] 1. Any moving object, especially a missile, that finds its target by means of a device attracted to light, heat, radio waves, sound, or other radiation emitted by the target. 2. The device used in such an object. { 'sē·kər }

seek time [COMPUT SCI] The time required for the access mechanism of a random-access storage device to be properly positioned. { 'sēk ,tīm }

Seelandian [GEOL] A European stage of geologic time in the lowermost Paleocene. { zā'län·dē·ən }

seen fire [ORD] Fire which is continuously aimed at the future position of an aircraft, the aim being derived from visual observation. { 'sēn 'fīr }

seep [GEOL] An area, generally small, where water, or another liquid such as oil, percolates slowly to the land surface. [PETRO ENG] An oil spring whose daily yield ranges from a few drops to several barrels of oil; usually located at low elevations where water has accumulated. { sēp }

seepage [FL MECH] The slow movement of water or other fluid through a porous medium. [HYD] The slow movement of water through small openings and spaces in the surface of unsaturated soil into or out of a body of surface or subsurface water. { 'sēp·ij }

seepage face [GEOL] A belt on a slope, such as the bank of a stream, along which water emerges at atmospheric pressure and flows down the slope. { 'sēp·ij ,fās }

seepage lake [HYD] 1. A closed lake that loses water mainly by seepage through the walls and floor of its basin. 2. A lake that receives its water mainly from seepage. { 'sēp·ij ,lāk }

seersucker [TEXT] A washable fabric with crinkled stripes made by altering the tension of the warp threads during weaving. { 'sir,sək·ər }

Segas process [CHEM ENG] A process for the production of low-Btu gas by the catalytic method using a fixed bed catalyst, lime-bauxite mixture bonded with bentonite. { 'sē‚gas ‚prä·səs }

Seger cone [MATER] Any of a series of conical shaped thermometric devices made of materials that deform at specified temperatures; consists of mixtures of clay, salt, and other materials in such proportions that their softening temperatures vary progressively through the series; used to indicate temperatures of furnaces, particularly in ceramic industries. Also known as pyrometric cone. { 'zā·gər ‚kōn }

segment [ANALY CHEM] A specific, demarcated portion of a lot of a substance that is to be chemically analyzed. [COMPUT SCI] **1.** A single section of an overlay program structure, which can be loaded into the main memory when and as needed. **2.** In some direct-access storage devices, a hardware-defined portion of a track having fixed data capacity. [MATH] **1.** A segment of a line or curve is any connected piece. **2.** A segment of a circle is a portion of the circle bounded by a chord and an arc subtended by the chord. **3.** A segment of a totally ordered Abelian group G is a subset D of G such that if a is in D then so are all elements b satisfying $-a \le b \le a$. [NAV] In air operations, a basic functional division of an instrument approach procedure; it bears a fixed orientation with respect to the course to be flown; it is assigned specific geometric coordinates which uniquely determine its position; the location of the segment is assigned with respect to the obstacles in the operations area. { 'seg·mənt }

segmental arch [ARCH] A round arch whose curved surface is less than a semicircle. { seg'ment·əl 'ärch }

segmental gate *See* tainter gate. { seg'ment·əl 'gāt }

segmental meter [ENG] A variable head meter whose orifice plate has an opening in the shape of a half circle. { seg'ment·əl 'mēd·ər }

segmental reflex [NEUROSCI] A reflex arc having afferent inputs by way of the spinal dorsal roots, and efferent outputs over spinal ventral roots of the same or adjacent segments. { seg'ment·əl 'rē‚fleks }

segmentation [COMMUN] The division of a long communications message into smaller messages that can be transmitted intermittently. [COMPUT SCI] **1.** The division of virtual storage into identifiable functional regions, each having enough addresses so that programs or data stored in them will not assign the same addresses more than once. **2.** The division of a large computer program into smaller units, called segments. **3.** *See* picture segmentation. [ZOO] *See* metamerism. { ‚seg·mən'tā·shən }

segmentation cavity *See* blastocoele. { ‚seg·mən'tā·shən ‚kav·əd·ē }

segment die [MET] A die made of parts that can be disassembled to facilitate removal of the workpiece. Also known as split die. { 'seg·mənt ‚dī }

segmented aperture-synthetic aperture radar [ENG] An enhancement of synthetic aperture radar that overcomes restrictions on the effective length of the receiving antenna by using a receiving antenna array composed of a set of contiguous subarrays and employing signal processing to provide the proper phase corrections for each subarray. Abbreviated SASAR. { 'seg‚ment·əd ‚ap·ə·chər sin'thed·ik ‚ap·ə·chər 'rā‚där }

segmented mirror telescope [OPTICS] A telescope consisting of many mirrors, all figured as segments of one parent paraboloidal surface. { 'seg‚ment·əd 'mir·ər 'tel·ə‚skōp }

segment mark [COMPUT SCI] A special character written on tape to separate one section of a tape file from another. { 'seg·mənt ‚märk }

segment saw [MECH ENG] A saw consisting of steel segments attached around the edge of a flange and used for cutting veneer. { 'seg·mənt ‚sȯ }

Segrè characteristic [MATH] A set of integers that are the orders of the Jordan submatrices of a classical canonical matrix, with integers that correspond to submatrices containing the same characteristic root being bracketed together. { 'se‚grā ‚kär·ik·tə‚ris·tik }

Segrè chart [NUC PHYS] A chart of the nuclides which is laid off in squares, each square displaying data about a nuclide; each column contains nuclides with a given neutron number and each row contains nuclides with a given atomic number; successive columns and rows represent successively higher numbers of neutrons and protons. { sə'grā ‚chärt }

segregated ice [HYD] Ice films, seams, lenses, rods, or layers generally 0.04 to 6 inches (1 to 150 millimeters) thick that grow in permafrost by drawing in water as the ground freezes. Also known as Taber ice. { 'seg·rə‚gād·əd 'īs }

segregated vein [GEOL] A fissure filled with mineral matter derived from country rock by the action of percolating water. Also known as exudation vein. { 'seg·rə‚gād·əd 'vān }

segregating unit [COMPUT SCI] A punched-card machine that selects from a group of punched cards those cards satisfying certain criteria; it has only two output hoppers, one for cards satisfying the criteria and one for those that do not, in contrast with a card sorter which has multiple outputs. { 'seg·rə‚gād·iŋ ‚yü·nət }

segregation [ENG] **1.** The keeping apart of process streams. **2.** In plastics molding, a close succession of parallel, relatively narrow, and sharply defined wavy lines of color on the surface of a plastic that differ in shade from surrounding areas and create the impression that the components have separated. [GEN] The separation of homologous chromosomes, and thus the alleles they carry, during meiosis in the formation of gametes. [GEOL] The formation of a secondary feature within a sediment after deposition due to chemical rearrangement of minor constituents. [MET] The nonuniform distribution of alloying elements, impurities, or microphases, resulting in localized concentrations. { ‚seg·rə'gā·shən }

segregation banding [PETR] A compositional band in gneisses that is the result of segregation of material from an originally homogeneous rock. { ‚seg·rə'gā·shən ‚band·iŋ }

segregation distorter [GEN] An abnormality of meiosis which produces a distortion of the 1:1 segregation ratio of alleles in a heterozygote. { ‚seg·rə'gā·shən dis‚tȯr·dər }

seiche [FL MECH] An oscillation of a fluid body in response to the disturbing force having the same frequency as the natural frequency of the fluid system. [OCEANOGR] A standing-wave oscillation of an enclosed or semienclosed water body, continuing pendulum-fashion after cessation of the originating force which is usually considered to be strong winds or barometric pressure changes. { sāsh }

Seidel aberrations [OPTICS] The five types of aberration of monochromatic light deduced from the Seidel theory, namely, spherical aberration, coma, astigmatism, curvature of field, and distortion. { 'zīd·əl ‚ab·ə‚rā·shənz }

Seidel method [MATH] A basic iterative procedure for solving a system of linear equations by reducing it to triangular form. Also known as Gauss-Seidel method. { 'zīd·əl ‚meth·əd }

Seidel theory [OPTICS] A theory of aberrations in which the sine of the angle which a ray makes with the optical axis is approximated by the first two terms in the sine's Taylor expansion (the first- and third-order terms), rather than the first term alone, as in a first-order theory. { 'zīd·əl ‚thē·ə·rē }

seif dune [GEOL] A large, tapering, longitudinal dune or chain of sand dunes with a sharp crest that in profile consists of a succession of peaks and cols. { 'sāf ‚dün }

Seignette-electric *See* ferroelectric. { sen'yet i'lek·trik }

Seignette salt *See* potassium sodium tartrate. { sen'yet ‚sȯlt }

seine net [ENG] A net used to catch fish by encirclement, usually by closure of the two ends and the bottom. { sān ‚net }

seismic activity *See* seismicity. { 'sīz·mik ak‚tiv·əd·ē }

seismic anisotropy [GEOPHYS] The dependence of seismic velocity on the direction of propagation. { 'sīz·mik ‚an·ə'sä·trə‚pē }

seismic area *See* earthquake zone. { 'sīz·mik ‚er·ē·ə }

seismic belt [GEOPHYS] An elongate seismic zone, such as that in the Circum-Pacific. { 'sīz·mik ‚belt }

seismic bracing [ENG] Reinforcement added to a structure to prevent collapse or deformation of building elements as a result of earthquakes. { 'sīz·mik 'brās·iŋ }

seismic constant [CIV ENG] In building codes dealing with earthquake hazards, an arbitrarily set quantity of steady acceleration, in units of acceleration of gravity, that a building must withstand. { 'sīz·mik 'kän·stənt }

seismic detector [ENG] An instrument that receives seismic impulses. { 'sīz·mik di‚tek·tər }

seismic discontinuity [GEOPHYS] **1.** A surface at which velocities of seismic waves change abruptly. **2.** A boundary between seismic layers of the earth. Also known as interface; velocity discontinuity. { 'sīz·mik ‚dis·känt·ən'ü·əd·ē }

SEGMENTAL METER

Orifice plate used in a segmental meter.

seismic efficiency [GEOPHYS] The proportion of the total available strain energy which is radiated as seismic waves. { 'sīz·mik i'fish·ən·sē }

seismic-electric effect [GEOPHYS] The variation of resistivity with elastic deformation of rocks. { 'sīz·mik i'lek·trik i‚fekt }

seismic event [GEOPHYS] An earthquake or a somewhat similar transient earth motion caused by an explosion. { 'sīz·mik i‚vent }

seismic exploration [ENG] The exploration for economic deposits by using seismic techniques, usually involving explosions, to map subsurface structures. { 'sīz·mik ‚ek·splə'rā·shən }

seismic gradient [GEOPHYS] The variation of seismic velocity with distance in a specified direction. Also known as velocity gradient. { 'sīz·mik 'grād·ē·ənt }

seismic hazard [GEOPHYS] Any physical phenomenon, such as ground shaking or ground failure, that is associated with an earthquake and that may produce adverse effects on human activities. { 'sīz·mik 'haz·ərd }

seismic intensity [GEOPHYS] The average rate of flow of seismic-wave energy through a unit section perpendicular to the direction of propagation. { 'sīz·mik in'ten·səd·ē }

seismicity [GEOPHYS] The phenomena of earth movements. Also known as seismic activity. { sīz'mis·əd·ē }

seismic load [ENG] The force on a structure caused by acceleration induced on its mass by an earthquake. { 'sīz·mik 'lōd }

seismic map [GEOPHYS] A contour map constructed from seismic data, the z coordinate of which could be either time or depth. { 'sīz·mik 'map }

seismic profiler [ENG] A continuous seismic reflection system used to study the structure beneath the sea floor to depths of 10,000 feet (3000 meters) or more, using a rotating drum to record reflections. { 'sīz·mik 'prō‚fīl·ər }

seismic prospecting [GEOPHYS] Geophysical prospecting based on the analysis of elastic waves generated in the earth by artificial means. { 'sīz·mik 'präs‚pek·tiŋ }

seismic ray [GEOL] The path along which seismic energy travels. { 'sīz·mik 'rā }

seismic reflector [GEOPHYS] A subsurface profile that is generated by seismic data and indicates a distinctive type of sediment geometry produced by sea-level changes; used to correlate stratigraphic sequences. { 'sīz·mik ri'flek·tər }

seismic risk [GEOPHYS] **1.** An assortment of earthquake effects that range from ground shaking, surface faulting, and landsliding to economic loss and casualties. **2.** The probability that social or economic consequences of earthquakes will equal or exceed specified values at a site, at several sites, or in an area, during a specified exposure time. { 'sīz·mik 'risk }

seismic shooting [ENG] A method of geophysical prospecting in which elastic waves are produced in the earth by the firing of explosives. { 'sīz·mik 'shüd·iŋ }

seismic stratigraphy [GEOL] A branch of stratigraphy in which sediments and sedimentary rocks are interpreted in a geometrical context from seismic reflectors. { 'sīz·mik strə'tig·rə·fē }

seismic survey See reflection survey. { 'sīz·mik 'sər‚vā }

seismic tomography [GEOPHYS] The estimation of seismic wave velocities throughout a region of interest from the travel times of either transmitted or reflected waves, generally through numerical models and iterative procedures. { 'sīz‚mik tō'mäg·rə·fē }

seismic velocity [GEOPHYS] The rate of propagation of an elastic wave, usually measured in kilometers per second. { 'sīz·mik və'läs·əd·ē }

seismic vertical [GEOL] **1.** The point on the earth's surface directly over the point within the earth from which an earthquake impulse originates. **2.** The vertical line between the surface point and the point of origin. { 'sīz·mik 'vərd·ə·kəl }

seismochronograph [ENG] A chronograph for determining the time at which an earthquake shock appears. { 'sīz·mə'krän·ə‚graf }

seismogram [ENG] The record made by a seismograph. { 'sīz·mə‚gram }

seismograph [ENG] An instrument that records vibrations in the earth, especially earthquakes. { 'sīz·mə‚graf }

seismology [GEOPHYS] **1.** The study of earthquakes. **2.**

The science of strain-wave propagation in the earth. { sīz'mäl·ə·jē }

seismometer [ENG] An instrument that detects movements in the earth. { sīz'mäm·əd·ər }

seismoscope [ENG] An instrument for recording only the occurrence or time of occurrence (not the magnitude) of an earthquake. { 'sīz·mə‚skōp }

Seisonacea [INV ZOO] A monofamiliar order of the class Rotifera characterized by an elongated jointed body with a small head, a long slender neck region, a thick fusiform trunk, and an elongated foot terminating in a perforated disk. { ‚sī·sə'nā·shə }

Seisonidea [INV ZOO] The equivalent name for Seisonacea. { ‚sī·sə'nīd·ē·ə }

seistan [METEOROL] A strong wind of monsoon origin which blows from between the northwest and north-northwest and sets in about the end of May or early June in the historic Seistan district of eastern Iran and Afghanistan; it continues almost without cessation until about the end of September; because of its duration it is known as the wind of 120 days (bad-i-sad-o-bistroz). { 'sā‚stän }

Seitz filter [MICROBIO] A bacterial filter made of asbestos and used to sterilize solutions without the use of heat. { 'zīts ‚fil·tər }

seizing [ENG] Abrasive damage to a metal surface caused when the surface is rubbed by another metal surface. [MET] Welding a workpiece to a die member under the combined forces of pressure and sliding friction. { 'sēz·iŋ }

seizure [MED] **1.** The sudden onset or recurrence of a disease or an attack. **2.** Specifically, an epileptic attack, fit, or convulsion. { 'sē·zhər }

Sejournet process [MET] During hot extrusion, the lubrication and insulation of a metal billet with molten glass. Also known as Ugine Sejournet process. { sə·zhùr'nā ‚prä·səs }

sejunction water [HYD] Capillary water bounded by menisci, and in static equilibrium in the soil above the capillary fringe. { sə'jəŋk·shən ‚wòd·ər }

sekaninaite [MINERAL] A violet variety of cordierite in which magnesium is largely replaced by ferrous iron. Also known as iron cordierite. { sə'kän·ən·ə‚īt }

Selachii [VERT ZOO] An order of elasmobranchs including all fossil sharks, except Cladoselachii and Pleuracanthodii. { sə'lāk·ē‚ī }

Selaginellales [BOT] The plant order of small club mosses, containing one living genus, *Selaginella;* distinguished from other lycopods in being heterosporous and in having a ligule borne on the upper base of the leaf. { sə‚laj·ə·nə'lā·lēz }

selagite [PETR] A mica trachyte characterized by abundant tabular biotite crystals in a holocrystalline groundmass of orthoclase and diopside, and possibly quartz and olivine. { 'sel·ə‚jīt }

selatan [METEOROL] Strong, dry, southerly winds of the southeast monsoon in the Netherlands East Indies and the Celebes. { sä'lä‚tän }

select [COMPUT SCI] **1.** To choose a needed subroutine from a file of subroutines. **2.** To take one alternative if the report on a condition is of one state, and another alternative if the report on the condition is of another state. **3.** To pull from a mass of data certain items that require special attention; selection of individual cards is accomplished automatically by either the sorter or collator, according to the type of selection. { si'lekt }

select bit [COMPUT SCI] The bit (or bits) in an input/output instruction word which selects the function of a specified device. Also known as subdevice bit. { si'lekt ‚bit }

selected areas See Kapetyn selected areas. { si'lek·təd 'er·ē·əz }

selected mine [ORD] A controlled underwater mine which has been connected, through the selector assembly, to the control equipment at the shore station; it is exclusive of all other mines in its group and may be fired, tested, or disarmed independently of the remainder of the group. { si'lek·təd 'mīn }

selected time [IND ENG] An observed actual time value for an element, measured by time study, which is identified as being the most representative of the situation observed. { si'lek·təd 'tīm }

selectin [CELL MOL] Any of a family of carbohydrate-binding, calcium-dependent cell adhesion molecules that play an

SELACHII

Primitive selachian *Hybodus hauffianus*, a Mesozoic shark.

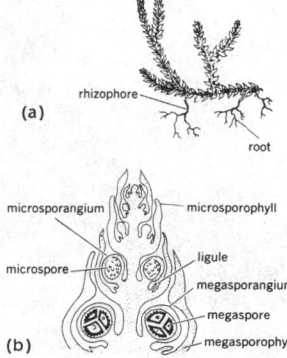

SELAGINELLALES

Selaginella, or small club moss.
(a) Plant with creeping, dichotomous stems bearing numerous, tiny leaves and leafless branches, or rhizophores. *(b)* Diagram of a longitudinal section through a cone.

important role in leukocyte-endothelial binding. { səˈlek·tən }

selecting circuit [ELEC] A simple switching circuit that receives the identity (the address) of a particular item and selects that item from among a number of similar ones. { siˈlek·tiŋ ˌsər·kət }

selection [COMMUN] The process of addressing a call to a specific station in a selective calling system. [GEN] Any natural or artificial process which favors the survival and propagation of individuals of a given phenotype in a population. { siˈlek·shən }

selection bias [STAT] A bias built into an experiment by the method used to select the subjects which are to undergo treatment. { siˈlek·shən ˌbī·əs }

selection check [COMPUT SCI] Electronic computer check, usually automatic, to verify that the correct register, or other device, is selected in the performance of an instruction. { siˈlek·shən ˌchek }

selection coefficient [GEN] A measure of the rate of transmission through successive generations of a given allele compared to the rate of transmission of another (usually the wild-type) allele. { siˈlek·shən ˌkō·iˌfish·ənt }

selection pressure [EVOL] Those factors that influence the direction of natural selection. { siˈlek·shən ˌpresh·ər }

selection rules [PHYS] Rules summarizing the changes that must take place in the quantum numbers of a quantum-mechanical system for a transition between two states to take place with appreciable probability; transitions that do not agree with the selection rules are called forbidden and have considerably lower probability. { siˈlek·shən ˌrülz }

selection sort [COMPUT SCI] A sorting routine that scans a list of items repeatedly and, on each pass, selects the item with the lowest value and places it in its final position. { siˈlek·shən ˌsórt }

selective absorption [ELECTROMAG] A greater absorption of electromagnetic radiation at some wavelengths (or frequencies) than at others. { siˈlek·tiv abˈsórp·shən }

selective acidizing [PETRO ENG] Oil-reservoir acid treatment in which the acid is injected into specific reservoir zones; contrasted with uncontrolled acidizing in which the acid solution is simply pumped down the casing and is forced into adjacent rock. { siˈlek·tiv ˈas·əˌdīz·iŋ }

selective adsorbent [CHEM ENG] Material that will selectively adsorb (or reject) one or more specific components from a multicomponent mixture of gases or liquids; common adsorbents are silica gel, carbon and activated carbon, activated alumina, and synthetic or natural zeolites (molecular sieves). { siˈlek·tiv adˈsór·bənt }

selective breeding [BIOL] Breeding of animals or plants having desirable characters. { siˈlek·tiv ˈbrēd·iŋ }

selective calling system [COMMUN] A radio communications system in which the central station transmits a coded call that activates only the receiver to which that code is assigned. { siˈlek·tiv ˌkól·iŋ ˌsis·təm }

selective circuit [ELEC] A circuit that transmits certain types of signals and fails to transmit or attenuates others. { siˈlek·tiv ˈsər·kət }

selective cracking [CHEM ENG] A refinery process in which recycled stock is distilled in equipment kept separate from that used for distillation of original stock. { siˈlek·tiv ˈkrak·iŋ }

selective dump [COMPUT SCI] An edited or nonedited listing of the contents of selected areas of memory or auxiliary storage. { siˈlek·tiv ˈdəmp }

selective fading [COMMUN] Fading that is different at different frequencies in a frequency band occupied by a modulated wave, causing distortion that varies in nature from instant to instant. { siˈlek·tiv ˈfād·iŋ }

selective filling [MIN ENG] Filling by hand so that stone or dirt is rejected and only clean coal or ore is loaded. { siˈlek·tiv ˈfil·iŋ }

selective flotation [MIN ENG] The surface or froth selecting of the valuable minerals rather than the gangue. { siˈlek·tiv flōˈtā·shən }

selective forgetting [PSYCH] Allowing memories to be shaped to better fit one's perceptions of the world and of oneself. { siˌlek·tiv fərˈged·iŋ }

selective fracturing [PETRO ENG] Procedures for obtaining multiple formation fractures in a specific reservoir zone by plugging casing perforations or by isolating (with packers) the desired zone prior to fracturing operations. { siˈlek·tiv ˈfrak·chəˌriŋ }

selective fusion [GEOL] The fusion of only a portion of a mixture, such as a rock. { siˈlek·tiv ˈfyü·zhən }

selective heating [MET] Heating only certain portions of a workpiece to impart desired properties. { siˈlek·tiv ˈhēd·iŋ }

selective identification feature [ELECTR] Airborne pulse-type transponder which provides automatic selective identification of aircraft in which it is installed to ground, shipboard, or airborne recognition installations. { siˈlek·tiv īˌden·tə·fəˈkā·shən ˌfē·chər }

selective inhibition See selective poisoning. { siˈlek·tiv ˌin·əˈbish·ən }

selective interference [COMMUN] Interference whose energy is concentrated in a narrow band of frequencies. { siˈlek·tiv ˌin·tərˈfir·əns }

selective jamming [ELECTR] Jamming in which only a single radio channel is jammed. { siˈlek·tiv ˈjam·iŋ }

selective loading [NAV ARCH] The arrangement and stowage of equipment and supplies aboard ship in a manner designed to facilitate issues to units. { siˈlek·tiv ˈlōd·iŋ }

selectively doped heterojunction transistor See high-electron-mobility transistor. { siˈlek·tiv·lē ¦dōpt ¦hed·ə·rōˈjəŋk·shən tranˈzis·tər }

selective medium [MICROBIO] A bacterial culture medium containing an individual organic compound as the sole source of carbon, nitrogen, or sulfur for growth of an organism. { siˈlek·tiv ˈmē·dē·əm }

selective mining [MIN ENG] A method of mining whereby ore of unwarranted high value is mined in such manner as to make the low-grade ore left in the mine incapable of future profitable extraction; in other words, the best ore is selected in order to make good mill returns, leaving the low-grade ore in the mine. { siˈlek·tiv ˈmīn·iŋ }

selective permeability [PHYS] The property of a membrane or other material that allows some substances to pass through it more easily than others. { siˈlek·tiv ˌpər·mē·əˈbil·əd·ē }

selective photoelectric effect [ELECTR] A resonance in the dependence of photoemission on the incident photon energy that is displayed when light is incident on a thin-metal film and the light vector has a component perpendicular to a crystal plane. Also known as spectral selective photoelectric effect; vector effect. { siˈlek·tiv ˌfōd·ō·iˈlek·trik iˌfekt }

selective plating [MET] An electrochemical process in which the base metal is masked, except the area to receive the plate, with a nonconductive material; the masked metal, with an electric current running through it, is then sprayed with a solution of plating metal which adheres only to the unmasked section. { siˈlek·tiv ˈplād·iŋ }

selective poisoning [CHEM] Retardation of the rate of one catalyzed reaction more than that of another by the use of a catalyst poison. Also known as selective inhibition. { siˈlek·tiv ˈpóiz·ən·iŋ }

selective polymerization [CHEM ENG] The polymerization of a single type of molecule in a mixture of monomers; for example, the production of diisobutylene from a mixture of butylenes. { siˈlek·tiv pəˌlim·ə·rəˈzā·shən }

selective quenching [MET] Quenching only certain portions of a piece of metal. { siˈlek·tiv ˈkwench·iŋ }

selective radiator [PHYS] An object that emits electromagnetic radiation whose spectral energy distribution differs from that of a blackbody with the same temperature. { siˈlek·tiv ˈrad·ē·ād·ər }

selective reflection [ELECTROMAG] Reflection of electromagnetic radiation more strongly at some wavelengths (or frequencies) than at others. { siˈlek·tiv riˈflek·shən }

selective replacement [GEOL] The replacement of one mineral by another, preferentially within an altered rock mass. { siˈlek·tiv riˈplās·mənt }

selective ringing [COMMUN] Telephone arrangement on party lines, in which only the bell of the called subscriber rings, with other bells on the party line remaining silent. { siˈlek·tiv ˈriŋ·iŋ }

selective scattering [ELECTROMAG] Scattering of electromagnetic radiation more strongly at some wavelengths than at others. { siˈlek·tiv ˈskad·ə·riŋ }

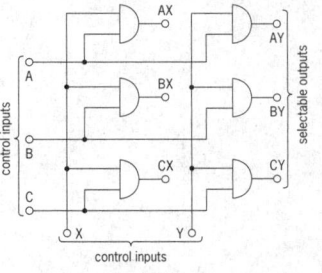

Matrix selecting circuit using AND gates.

selective solubility diffusion [CHEM ENG] The transmission of fluids through a nonporous, polymeric barrier (membrane) by an adsorption-solution-diffusion-desorption sequence. { si'lek·tiv ,säl·yə'bil·əd·ē di,fyü·zhən }

selective solvent [CHEM ENG] A solvent that, at certain temperatures and ratios with other materials, preferentially dissolves more of one component of a liquid or solids mixture than of another, thereby permitting partial separation. { si'lek·tiv 'säl·vənt }

selective trace [COMPUT SCI] A tracing routine wherein only instructions satisfying certain specified criteria are subject to tracing. { si'lek·tiv 'trās }

selective transmission [MECH ENG] A gear transmission with a single lever for changing from one gear ratio to another; used in automotive vehicles. { si'lek·tiv tranz'mish·ən }

selectivity [ANALY CHEM] The ability of a type of method or instrumentation to respond to a specified substance or constituent and not to others. [ELECTR] **1.** The ability of a radio receiver to separate a desired signal frequency from other signal frequencies, some of which may differ only slightly from the desired value. **2.** The inverse of the shape factor of a bandpass filter. { sə,lek'tiv·əd·ē }

selectivity coefficient [ANALY CHEM] Ion equilibria relationship formula for ion-exchange-resin systems. { sə,lek'tiv·əd·ē ,kō·i,fish·ənt }

selectivity diagram [CHEM ENG] A triangular plot of solubilities in a ternary liquid system; used to calculate the ability of a solvent to extract a component from a mixture (its selectivity) at various concentration combinations. { sə,lek'tiv·əd·ē 'dī·ə,gram }

selector [CIV ENG] A device that automatically connects the appropriate railroad signal to control the track selected. [COMPUT SCI] Computer device which interrogates a condition and initiates a particular operation dependent upon the report. [ELEC] An automatic or other device for making connections to any one of a number of circuits, such as a selector relay or selector switch. [ENG] **1.** A device for selecting objects or materials according to predetermined properties. **2.** A device for starting or stopping at predetermined positions. [MECH ENG] **1.** The part of the gearshift in an automotive transmission that selects the required gearshift bar. **2.** The lever with which a driver operates an automatic gearshift. [MET] A converter that separates purified copper from residue in a single operation. { si'lek·tər }

selector channel [COMPUT SCI] A unit which connects high-speed input/output devices, such as magnetic tapes, disks, and drums, to a computer memory. { si'lek·tər ,chan·əl }

selector switch [ELEC] A manually operated multiposition switch. Also called multiple-contact switch. { si'lek·tər ,swich }

s-electron [ATOM PHYS] An atomic electron that is described by a wave function with orbital angular momentum quantum number of zero in the independent particle approximation. { 'es i,lek,trän }

selenic acid [INORG CHEM] H_2SeO_4 A highly toxic, water-soluble, white solid, melting point 58°C, decomposing at 260°C. { sə'len·ik 'as·əd }

selenide [INORG CHEM] M_2Se A binary compound of divalent selenium, such as Ag_2Se, silver selenide. [ORG CHEM] An organic compound containing divalent selenium, such as $(C_2H_5)_2Se$, ethyl selenide. { 'sel·ə,nīd }

selenious acid See selenous acid. { sə'lē·nē·əs 'as·əd }

selenite [MINERAL] The clear, colorless variety of gypsum crystallizing in the monoclinic system and occurring in crystals or in crystal mass. Also known as spectacle stone. { 'sel·ə,nīt }

selenite butte [GEOL] A small tabular mound, rising 3.3–10 feet (1–3 meters) above a playa, composed of lake sediments capped with a veneer of selenite formed by deflation of the playa or by the effects of rising groundwater. { 'sel·ə,nīt 'byüt }

selenium [CHEM] A highly toxic, nonmetallic element in group 6, symbol Se, atomic number 34; steel-gray color; soluble in carbon disulfide, insoluble in water and alcohol; melts at 217°C; and boils at 690°C; used in analytical chemistry, metallurgy, and photoelectric cells, and as a lube-oil stabilizer and chemicals intermediate. { sə'lē·nē·əm }

selenium bromide [INORG CHEM] Any of three compounds of selenium and bromine: Se_2Br_2, a red liquid that melts at −46°C, also known as selenium monobromide; $SeBr_2$, a brown liquid, also known as selenium dibromide; and $SeBr_4$, orange, carbon-disulfide-soluble crystals, also known as selenium tetrabromide. { sə'lē·nē·əm 'brō,mīd }

selenium cell [ELECTR] A photoconductive cell in which a thin film of selenium is used between suitable electrodes; the resistance of the cell decreases when the illumination is increased. { sə'lē·nē·əm ,sel }

selenium dibromide See selenium bromide. { sə'lē·nē·əm dī'brō,mīd }

selenium diode [ELECTR] A small area selenium rectifier which has characteristics similar to those of selenium rectifiers used in power systems. { sə'lē·nē·əm 'dī,ōd }

selenium dioxide [INORG CHEM] SeO_2 Water- and alcohol-soluble, white to reddish, lustrous crystals; melts at 340°C; used in medicine, and as an oxidizing agent and catalyst. Also known as selenous acid anhydride; selenous anhydride; selenium oxide. { sə'lē·nē·əm dī'äk,sīd }

selenium disulfide See selenium sulfide. { sə'lē·nē·əm dī'səl,fīd }

selenium halide [INORG CHEM] A compound of selenium and a halogen, for example, Se_2Br_2, $SeBr_2$, $SeBr_4$; Se_2Cl_2, $SeCl_2$, $SeCl_4$; Se_2I_2, SeI_4. { sə'lē·nē·əm 'ha,līd }

selenium monobromide See selenium bromide. { sə'lē·nē·əm ,män·ō'brō,mīd }

selenium nitride [INORG CHEM] Se_2N_2 A water-insoluble, yellow solid that explodes at 200°C. { sə'lē·nē·əm 'nī,trīd }

selenium oxide See selenium dioxide. { sə'lē·nē·əm 'äk,sīd }

selenium rectifier [ELECTR] A metallic rectifier in which a thin layer of selenium is deposited on one side of an aluminum plate and a conductive metal coating is deposited on the selenium. { sə'lē·nē·əm 'rek·tə,fī·ər }

selenium stainless steel [MET] Stainless steel to which about 0.1 percent or more selenium is added to improve machinability. { sə'lē·nē·əm 'stān·ləs ,stēl }

selenium sulfide [PHARM] SeS_2 A bright orange powder with a melting point of 100°C; used in medicine. Also known as selenium disulfide. { sə'lē·nē·əm 'səl,fīd }

selenium tetrabromide See selenium bromide. { sə'lē·nē·əm ,te·trə'brō,mīd }

selenocentric [ASTRON] Pertaining to the moon's center. { sə,lē·nō'sen·trik }

selenodesy [ASTRON] The branch of applied mathematics that determines, by observation and measurement, the exact positions of points on the moon's surface, as well as the shape and size of the moon. { sə,lē·nə,des·ē }

selenodetic [ASTRON] Of, or pertaining to, or determined by selenodesy. { sə,lē·nə,ded·ik }

selenodont [VERT ZOO] **1.** Being or pertaining to molars having crescentic ridges on the crown. **2.** A mammal with selenodont dentition. { sə'lē·nə,dänt }

selenofault [ASTRON] A geological fault in the lunar surface. { sə'lē·nə,folt }

selenographic coordinates [ASTRON] A coordinate system for specifying positions on the moon's surface relative to the moon's center, consisting of selenographic latitude and longitude, or of a cartesian coordinate system. { sə,lē·nə,graf·ik kō'ord·ən·əts }

selenographic latitude [ASTRON] The angular distance, measured along a meridian, between a point on the moon's surface and the moon's equator. { sə,lē·nə,graf·ik 'lad·ə,tüd }

selenographic longitude [ASTRON] The angular distance, measured along the moon's equator, between the meridian passing through a point on the moon's surface and the first lunar meridian. { sə,lē·nə,graf·ik 'län·jə,tüd }

selenography [ASTRON] Studies pertaining to the physical geography of the moon; specifically, referring to positions on the moon measured in latitude from the moon's equator and in longitude from a reference meridian. { ,sel·ə'näg·rə·fē }

selenology [ASTRON] A branch of astronomy that treats of the moon, including such attributes as magnitude, motion and constitution. Also known as lunar geology. { ,sel·ə'näl·ə·jē }

selenomorphology [ASTRON] The study of landforms on the moon, including their origin, evolution, and distribution. { sə'lē·nō·mor'fäl·ə·jē }

selenone [ORG CHEM] A group of organic selenium compounds with the general formula R_2SeO_2. { 'sel·ə,nōn }

selenonic acid [ORG CHEM] Any organic acid containing

the radical $-SeO_3H$; analogous to a sulfonic acid. { ˌsel·ə'nän·ik 'as·əd }

selenosis [MED] Selenium poisoning. { ˌsel·ə'nō·səs }

selenotrope [ENG] A device used in geodetic surveying for reflecting the moon's rays to a distant point, to aid in long-distance observations. { sə'lē·nə,trōp }

selenous acid [INORG CHEM] H_2SeO_3 Colorless, transparent crystals; soluble in water and alcohol, insoluble in ammonia; decomposes when heated; used as an analytical reagent. Also spelled selenious acid. { sə'lē·nəs 'as·əd }

selenous acid anhydride See selenium dioxide. { sə'lē·nəs 'as·əd an'hī,drīd }

selenous anhydride See selenium dioxide. { sə'lē·nəs an'hī,drīd }

selenoxide [ORG CHEM] A group of organic selenium compounds with the general formula R_2SeO. { ˌsel·ə'näk,sīd }

self-absorption [NUCLEO] Absorption of ionizing radiation by the material that emits the radiation, thus reducing the radiation level against which further shielding must be provided. Also known as self-shielding. [SPECT] Reduction of the intensity of the center of an emission line caused by selective absorption by the cooler portions of the source of radiation. Also known as self-reduction; self-reversal. { ˌself əb'sórp·shən }

self-acting door [MIN ENG] A ventilation door that is constructed of two halves which move on small pulleys and which are forced apart centrally as the trams come in contact with the converging beams that operate the door. { ˌself ¦ak·tiŋ 'dór }

self-acting incline See gravity haulage. { ˌself ¦ak·tiŋ 'in,klīn }

self-action effect [OPTICS] In a medium with a third-order optical nonlinearity, the modification of the refractive index and absorption coefficient of a light field present in the medium by the strength of the light intensity, so that the light field effectively acts on itself. { ˌself 'ak·shən i,fekt }

self-adapting system [SYS ENG] A system which has the ability to modify itself in response to changes in its environment. { ˌself ə¦dap·tiŋ 'sis·təm }

self-adjoint operator [MATH] A linear operator which is identical with its adjoint operator. { ˌself ə¦jóint 'äp·ə,rād·ər }

self-adjusting communications See adaptive communications. { ˌself ə¦jəst·iŋ kə,myü·nə'kā·shənz }

self-advancing supports [MIN ENG] An assembly of hydraulically operated steel hydraulic supports, on a long-wall face, which are moved forward as a unit. Also known as walking props. { ˌself əd¦vans·iŋ sə'pórts }

self-analysis [PSYCH] The attempt to gain insight into one's own psychic state and behavior. { ˌself ə¦nal·ə·səs }

self-assembling monolayer [MATER] Monomolecular films formed by immersing an appropriate substrate into a solution of an active surfactant. { ˌself ə,sem·bliŋ 'män·ō,lā·ər }

self-bias [ELECTR] A grid bias provided automatically by the resistor in the cathode or grid circuit of an electron tube; the resulting voltage drop across the resistor serves as the grid bias. Also known as automatic C bias; automatic grid bias. { ˌself ¦bī·əs }

self-bias transistor circuit [ELECTR] A transistor with a resistance in the emitter lead that gives rise to a voltage drop which is in the direction to reverse-bias the emitter junction; the circuit can be used even if there is zero direct-current resistance in series with the collector terminal. { ˌself ¦bī·əs tran'zis·tər ,sər·kət }

self-centering chuck [MECH ENG] A drill chuck that, when closed, automatically positions the drill rod in the center of the drive rod of a diamond-drill swivel head. { ˌself ¦sen·tə·riŋ 'chək }

self-chambering [ORD] Ability of a weapon to chamber cartridges without manual aid. { ˌself ¦chām·bə·riŋ }

self-charge [QUANT MECH] A contribution to a particle's electric charge arising from the vacuum polarization in the neighborhood of the bare charge. { ˌself ¦chärj }

self-checking code [COMPUT SCI] An encoding of data so designed and constructed that an invalid code can be rapidly detected; this permits the detection, but not the correction, of almost all errors. Also known as error-checking code; error-detecting code. { ˌself ¦chek·iŋ 'kōd }

self-checking number [COMPUT SCI] A number with a suffix figure related to the figure of the number, used to check

the number after it has been transferred from one medium or device to another. { ˌself ¦chek·iŋ 'nəm·bər }

self-cleaning [ENG] Pertaining to any device that is designed to clean itself without disassembly, for example, a filter in which accumulated filter cake or sludge is removed by an internal scraper or by a blowdown or backwash action. { ˌself 'klēn·iŋ }

self-cleaning contact See wiping contact. { ˌself 'klēn·iŋ 'kän,takt }

self-complementing code [COMPUT SCI] A binary-coded-decimal code in which the combination for the complement of a digit is the complement of the combination for that digit. { ¦self ¦käm·plə,ment·iŋ 'kōd }

self-complementary graph [MATH] A simple graph that is isomorphic to its complement. { ¦self ˌkäm·plə,men·trē 'graf }

self-conjugate particle [PART PHYS] An elementary particle which is identical to its antiparticle; it must have zero charge, lepton number, baryon number, and hypercharge. { ˌself 'kän·jə·gət 'pärd·ə·kəl }

self-conjugate partition [MATH] A partition that is its own conjugate. { ¦self ˌkär·jə·gət pär'tish·ən }

self-consistent field method See Hartree method. { ¦self kən¦sis·tənt 'fēld ,meth·əd }

self-contained base-line system [ORD] System of target location whereby the target is located in azimuth and range by using a self-contained range finder. { ˌself kən¦tānd 'bās ,līn ,sis·təm }

self-contained breathing apparatus [ENG] A portable breathing unit which permits freedom of movement. { ¦self kən¦tānd 'breth·iŋ ,ap·ə rad·əs }

self-contained database management system [COMPUT SCI] A database management system that is in no way an extension of any programming language, and is usually quite independent of any language. { ¦self kən¦tānd ¦dad·ə¦bās 'man·ij·mərt ,sis·təm }

self-contained night attack [ORD] The capability of a single aircraft to accurately navigate to a target area, acquire and strike a designated target, and return to the operating base during the hours of darkness. { ¦self kən¦tānd 'nīt ə,tak }

self-contained portable electric lamp [MIN ENG] An electric lamp which is operated by an electric battery and is specifically designed to be carried about by its user. { ¦self kən¦tānd 'pord·ə·bəl i lek·trik 'lamp }

self-contained range finder [ENG] Instrument used for measuring range by direct observation, without using a base line; the two types are the coincidence range finder and the stereoscopic range finder. { ¦self kən¦tānd 'rānj ,fīn·dər }

self-defocusing [OPTICS] The action of a medium whose index of refraction decreases with increasing optical intensity on a laser beam that is more intense in the center than at the edges, whereby the profile of the refractive index corresponds to that of a negative lens, causing the beam to defocus. { ¦self dē¦fō·kə·siŋ }

self-destroying [ORD] In connection with a fuse or tracer, indicating that the projectile, rocket, or missile with which it is used will be destroyed in flight prior to ground impact in case the target is missed. { ¦self di¦strói·iŋ }

self-destroying fuse [ORD] A fuse designed to burst a projectile before the end of its flight. { ¦self di¦strói·iŋ 'fyüz }

self-destruct charge [ORD] An explosive element which operates in conjunction with that part of the guided missile which, of itself or by command, senses a catastrophic flight malfunction and destroys the missile. { ¦self di¦strəkt 'chärj }

self-destruction [ORD] The event, due to fuse or tracer action without outside stimulus when provided for in the design, in which the fuse or tracer effects projectile or missile destruction, after flight to a range greater than that of the target. { ¦self di¦strak·shən }

self-destruction equipment [ORD] Any equipment that may be destroyed by a self-contained explosive. { ¦self di strək·shən i,kwip·mənt }

self-diagnostic routine [COMPUT SCI] A test of an electronic device that is performed automatically, usually when the device is turned on. Also known as self-test. { ¦self ˌdī·əg,nas·tik rü'tēn }

self-differentiation [PHYSIO] The differentiation of a tissue, even when isolated, solely as a result of intrinsic factors after determination. { ¦self ˌdif·ə,ren·chē¦ā·shən }

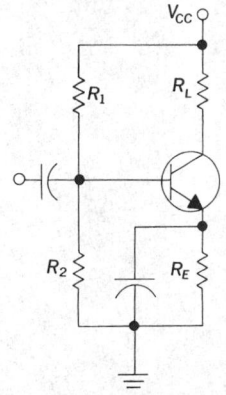

SELF-BIAS TRANSISTOR CIRCUIT

Circuit diagram of a self-bias transistor circuit. R_L = load resistance; R_E = resistance in emitter lead; V_{CC} = collector supply voltage with respect to emitter.

self-diffusion [SOLID STATE] The spontaneous movement of an atom to a new site in a crystal of its own species. { ¦self di¦fyü·zhən }

self-documenting code [COMPUT SCI] A sequence of programming statements that are simple and straightforward and can be readily implemented by another programmer. { ¦self ¦däk·yə‚ment·iŋ 'kōd }

self-dual switching function [MATH] A switching function whose value remains unchanged when the digits 0 and 1 are interchanged in each element of the domain of the function. { ¦self ‚dül 'swich·iŋ ‚faŋk·shən }

self-dumping cage [MIN ENG] A cage in which the deck is pivoted so that as the cage is lifted, toward the end of the lift, the deck tilts and the end door is lifted, discharging the coal. { ¦self ¦dəmp·iŋ 'kāj }

self-dumping car [MIN ENG] A mine car which can be side-tipped while in motion on the rail track; it is fitted with a spherically contoured wheel which engages a ramp structure and gradually tilts the car. { ¦self ¦dəmp·iŋ 'kär }

self-energizing brake [MECH ENG] A brake designed to reinforce the power applied to it, such as a hand brake. { ¦self ‚en·ər¦jīz·iŋ 'brāk }

self-energy [PHYS] **1.** Classically, the contribution to the energy of a particle that arises from the interaction between different parts of the particle. **2.** In a quantized field theory, the contribution to the energy of a particle due to virtual emission and absorption of other particles, in particular, mesons and photons. { ¦self 'en·ər·jē }

self-excited [ELEC] Operating without an external source of alternating-current power. { ¦self ik'sīd·əd }

self-excited oscillator [ELECTR] An oscillator that depends on its own resonant circuits for initiation of oscillation and frequency determination. { ¦self ik'sīd·əd 'äs·ə‚lād·ər }

self-excited vibration See self-induced vibration. { ¦self ik'sīd·əd vī'brā·shən }

self-extinguishing [MATER] The ability of a material to cease burning once the source of the flame has been removed. { ¦self ik¦stiŋ·gwə·shiŋ }

self-extracting file [COMPUT SCI] A compressed (zipped) file that unzips itself when it is executed. { ‚self ik‚strak·tiŋ 'fīl }

self-faced stone [CIV ENG] A type of stone used in masonry that splits along natural cleavage planes and does not have to be dressed. { ¦self ‚fāst 'stōn }

self-fields [ELECTROMAG] The electric and magnetic fields generated by an intense beam of charged particles, which act on the beam itself; they limit the beam intensities which can be achieved in storage rings. { ¦self ¦fēlz }

self-fluxing alloy [MET] Any alloy used in thermal spraying which does not require the addition of a flux in order to wet the substrate and coalesce when heated. { ¦self ¦flək·siŋ 'al‚ȯi }

self-focusing [OPTICS] The action of a medium whose index of refraction increases with increasing optical intensity on a laser beam that is more intense in the center than at the edges, whereby the profile of the refractive index corresponds to that of a positive lens, causing the beam to focus. { ¦self 'fō·kə·siŋ }

self-focusing fiber [OPTICS] A type of optical fiber in which the refractive index decreases continuously along the radius, but progressively more rapidly with distance from the radius, so that light rays which travel longer distances are speeded up, and nearly all light rays travel with the same net axial velocity. { ¦self 'fō·kə·siŋ 'fī·bər }

self-hardening steel See air-hardening steel. { ¦self 'härd·ən·iŋ 'stēl }

self-healing dielectric breakdown [ELECTR] A dielectric breakdown in which the breakdown process itself causes the material to become insulating again. { ¦self ¦hēl·iŋ ‚dī·ə¦lek·trik 'brāk‚daȯn }

self-impedance See mesh impedance. { ¦self im¦pēd·əns }

self-incompatibility [BOT] Pertaining to an individual flower that cannot complete fertilization with its own pollen. { ¦self ‚in·kəm‚pad·ə'bil·əd·ē }

self-induced transparency [OPTICS] A phenomenon in which a pulse of coherent light, with a certain frequency, amplitude, and duration, is transmitted by a normally opaque medium; energy absorbed from the first half of the pulse, whose frequency is at or near the average resonance peak of a band of coherent atomic two-quantum-level optical oscillators, is returned to the last half of the pulse by the medium in the form of coherently emitted light. Abbreviated SIT. { ¦self in¦düst tranz'par·ən·sē }

self-induced vibration [MECH] The vibration of a mechanical system resulting from conversion, within the system, of nonoscillatory excitation to oscillatory excitation. Also known as self-excited vibration. { ¦self in¦düst vī'brā·shən }

self-inductance [ELECTROMAG] **1.** The property of an electric circuit whereby an electromotive force is produced in the circuit by a change of current in the circuit itself. **2.** Quantitatively, the ratio of the electromotive force produced to the rate of change of current in the circuit. { ¦self in¦dəkt·əns }

self-induction [ELECTROMAG] The production of a voltage in a circuit by a varying current in that same circuit. { ¦self in¦dək·shən }

selfish deoxyribonucleic acid [CELL MOL] Any tandemly repeated or dispersed repetitive deoxyribonucleic acid sequence that has no obvious function but can spread and accumulate in the species because of its rapid replication. Also known as junk deoxyribonucleic acid. { ¦sel·fish dē‚äk·sē'rī·bō·nü‚klē·ik 'as·əd }

self-loading [MECH ENG] The capability of a powered industrial truck to pick up, transport, and deposit its load by using components that are part of its standard equipment, for example, a forklift. [ORD] Pertaining to a firearm or gun that utilizes either the explosive gases or recoil to extract the empty case and to chamber the next round; self-loading firearms or guns include both semiautomatic and full-automatic types. { ¦self ¦lōd·iŋ }

self-locking nut [DES ENG] A nut having an inherent locking action, so that it cannot readily be loosened by vibration. { 'self ¦läk·iŋ 'nət }

self-locking screw [DES ENG] A screw that locks itself in place without requiring a separate nut or lock washer. { 'self ¦läk·iŋ 'skrü }

self-optimizing communications See adaptive communications. { ¦self ¦äp·tə‚mīz·iŋ kə‚myü·nə'kā·shəns }

self-organization [CHEM] The capability of a system to spontaneously generate a well-defined supramolecular entity by self-assembling from components in a given set of conditions. { ‚self ¦ȯr·gə·nə'zā·shən }

self-organized criticality [PHYS] Property of a system that persistently operates far from equilibrium, at or near a threshold of instability, having evolved automatically to this critical state independently of external fields. { ‚self ¦ȯr·gə‚nīzd ‚krid·ə'kal·əd·ē }

self-organizing function [CONT SYS] That level in the functional decomposition of a large-scale control system which modifies the modes of control action or the structure of the control system in response to changes in system objectives, contingency events, and so forth. { ¦self ¦ȯr·gə‚nīz·iŋ 'faŋk·shən }

self-organizing system [SYS ENG] A system that is able to affect or determine its own internal structure. { ¦self ¦ȯr·gə‚nīz·iŋ 'sis·təm }

self-phase modulation [OPTICS] The temporal action of a medium whose index of refraction increases with increasing optical intensity on time-varying optical signals or pulses, whereby the rising front edge of the pulse is shifted to lower frequencies and the rear of the pulse is shifted to higher frequencies. { ¦self ‚fāz ‚mäj·ə'lā·shən }

self-poisoning [CHEM] Inhibition of a chemical reaction by a product of the reaction. Also known as autopoisoning. { ‚self ¦pȯiz·ən·iŋ }

self-pollination [BOT] Transfer of pollen from the anther to the stigma of the same flower or of another flower on the same plant. { ¦self ¦päl·ə¦nā·shən }

self-propelled [MECH ENG] Pertaining to a vehicle given motion by means of a self-contained motor. [ORD] **1.** Pertaining to a gun mounted on a vehicle that has its own motive power. **2.** Pertaining to a missile that is propelled by fuel carried by the missile itself, as in the case of a rocket. **3.** Pertaining to a military unit having self-propelled guns. { ¦self prə¦peld }

self-propelled artillery [ORD] Artillery weapons permanently installed on vehicles, which provide motive power for the piece; these weapons are fired from the vehicle. { ¦self prə¦peld är'til·ə·rē }

self-pulsing [ELECTR] Special type of grid pulsing which automatically stops and starts the oscillations at the pulsing rate by a special circuit. { 'self ¦pəls·iŋ }

self-quenched detector [ELECTR] Superregenerative detector in which the time constant of the grid leak and grid capacitor is sufficiently large to cause intermittent oscillation above audio frequencies, serving to stop normal regeneration each time just before it spills over into a squealing condition. { 'self ¦kwencht di'tek·tər }

self-quenching oscillator [ELECTR] Oscillator producing a series of short trains of radio-frequency oscillations separated by intervals of quietness. { 'self ¦kwench·iŋ 'äs·ə‚lād·ər }

self-reduction See self-absorption.

self-repair [COMPUT SCI] Any type of hardware redundancy in which faults are selectively masked and are detected, located, and subsequently corrected by the replacement of the failed unit by an unfailed replica. { 'self ri¦per }

self-rescuer [MIN ENG] A small filtering device carried by a miner underground to provide immediate protection against carbon monoxide and smoke in case of a mine fire or explosion; used for escape purposes only, because it does not sustain life in atmospheres containing deficient oxygen. { 'self 'res·kyü·wər }

self-reset [ELEC] Automatically returning to the original position when normal conditions are resumed; applied chiefly to relays and circuit breakers. { 'self ¦rē‚set }

self-resetting loop [COMPUT SCI] A loop whose termination causes the numbers stored in all locations affected by the loop to be returned to the original values which they had upon entry into the loop. { 'self ri¦sed·iŋ 'lüp }

self-reversal [GEOPHYS] Acquisition by a rock of a natural remanent magnetization opposite to the ambient magnetic field direction at the time of rock formation. [SPECT] See self-absorption. { 'self ri¦vər·səl }

self-rising ground [GEOL] The puffy, irregular, surface or near-surface zone of certain playas, formed by the effects of capillary rise of groundwater. { 'self ¦rīz·iŋ 'graůnd }

self-running droplet [FL MECH] A droplet that spontaneously coats a surface without the assistance of gravity or any external motion, due to the fact that it contains molecular species which are likely to react with the solid and cause the solvent to cease to wet the solid once it is coated with these molecules. { ¦self ‚rən·iŋ 'dräp·lət }

self-saturation [ELECTR] The connection of half-wave rectifiers in series with the output windings of the saturable reactors of a magnetic amplifier, to give higher gain and faster response. { ¦self ‚sach·ə¦rā·shən }

self-scanned image sensor [ELECTR] A solid-state device, still in the early stages of development, which converts an optical image into a television signal without the use of an electron beam; it consists of an array of photoconductor diodes, each located at the intersection of mutually perpendicular address strips respectively connected to horizontal and vertical scan generators and video coupling circuits. { 'self¦skand 'im·ij ‚sen·sər }

self-screening range [ELECTROMAG] Range at which a target can be detected by a radar in the midst of its jamming mask, with a certain specified probability. { 'self ¦skrēn·iŋ ‚rānj }

self-sealing [ENG] A fluid container, such as a fuel tank or a tire, lined with a substance that allows it to close immediately over any small puncture or rupture. { 'self ¦sēl·iŋ }

self-selection bias [STAT] Bias introduced into an experiment by having the subjects decide themselves whether or not they will receive treatment. { 'self si¦lek·shən 'bī·əs }

self-shielding [NUCLEO] **1.** Shielding of the inner portion of the fuel in a nuclear reactor by the outer portion of the fuel. **2.** See self-absorption. { 'self ¦shēld·iŋ }

self-similar flow [FL MECH] A fluid flow whose shape does not change with time, such as a spherical expansion. { ¦self 'sim·ə·lər 'flō }

self-similarity [MATH] The property whereby an object or mathematical function preserves its structure when multiplied by a certain scale factor. { 'self ‚sim·ə'lar·əd·ē }

self-starter [MECH ENG] An attachment for automatically starting an internal combustion engine. { 'self 'stär·dər }

self-starting synchronous motor [ELEC] A synchronous motor provided with the equivalent of a squirrel-cage winding, to permit starting as an induction motor. { 'self ¦stärd·iŋ 'siŋ·krə·nəs 'mōd·ər }

self-steering microwave array [ELECTROMAG] An antenna

array used with electronic circuitry that senses the phase of incoming pilot signals and positions the antenna beam in their direction of arrival. { 'self ¦stir·iŋ 'mī·krō‚wāv ə'rā }

self-supported film See film. { 'self sə‚pȯrd·əd 'film }

self-synchronous device See synchro. { 'self ¦siŋ·krə·nəs di'vīs }

self-synchronous repeater See synchro. { 'self ¦siŋ·krə·nəs ri'pēd·ər }

self-tapping screw [DES ENG] A screw with a specially hardened thread that makes it possible for the screw to form its own internal thread in sheet metal and soft materials when driven into a hole. Also known as sheet-metal screw; tapping screw. { 'self ¦tap·iŋ 'skrü }

self-test See self-diagnostic routine. { 'self 'test }

self-timer [ENG] A device that delays the tripping of a camera shutter so that the photographer can be included in the photograph. { 'self 'tīm·ər }

self-triggering program [COMPUT SCI] A computer program which automatically commences execution as soon as it is fed into the central processing unit. { 'self 'trig·ə·riŋ 'prō·grəm }

self-tuning regulator [CONT SYS] A type of adaptive control system composed of two loops, an inner loop which consists of the process and an ordinary linear feedback regulator, and an outer loop which is composed of a recursive parameter estimator and a design calculation, and which adjusts the parameters of the regulator. Abbreviated STR. { 'self 'tün·iŋ 'reg·yə‚lād·ər }

self-unloading ship [NAV ARCH] A ship equipped with endless belt or chain devices or with swinging booms, which enable it to be unloaded without the use of harbor facilities. { 'self ən¦lōd·iŋ 'ship }

Seligeriales [BOT] An order of true mosses in the class Bryopsida; members grow on rocks and may be exceedingly small to moderate size and tufted; the double structure of the peristome is distinctive. { ‚sel·ə‚jir·ē'ā·lēz }

seligmannite [MINERAL] PbCuAsS$_3$ A metallic gray orthorhombic mineral, occurring in crystals. { 'sel·əg·mə‚nīt }

Seliwanoff's test [ANALY CHEM] A color test helpful in the identification of ketoses, which develop a red color with resorcinol in hydrochloric acid. { sə'liv·ə‚nȯfs ‚test }

sellaite [MINERAL] MgF$_2$ A colorless mineral composed of magnesium fluoride occurring in tetragonal prismatic crystals. { 'sel·ə‚īt }

sella turcica [ANAT] A depression in the upper surface of the sphenoid bone in which the pituitary gland rests in vertebrates. { ‚sel·ə 'tər·kə·kə }

sellers hob [MECH ENG] A hob that turns on the centers of a lathe the work being fed to it by the lathe carriage. { 'sel·ərz 'häb }

sellite [INORG CHEM] A solution of sodium sulfite (Na$_2$SO$_3$) used in the purification of 2,4,6-trinitrotoluene to remove unsymmetrical isomers. { 'sel‚īt }

Sellmeier's equation [ELECTROMAG] An equation for the index of refraction of electromagnetic radiation as a function of wavelength in a medium whose molecules have oscillators of different frequencies. { 'zel‚mī·ərz i‚kwā·zhən }

selsyn See synchro. { 'sel·sin }

selsyn generator See synchro transmitter. { 'sel·sin ‚jen·ə‚rād·ər }

selsyn motor See synchro receiver. { 'sel·sin 'mōd·ər }

selsyn receiver See synchro receiver. { 'sel·sin ri‚sē·vər }

selsyn system See synchro system. { 'sel·sin 'sis·təm }

selsyn transmitter See synchro transmitter. { 'sel·sin tranz‚mid·ər }

selva See tropical rainforest. { 'sel·və }

selvage [PETR] The marginal zone of an igneous mass, generally characterized by a fine-grain, or sometimes glassy, texture. Also known as salband. [TEXT] Heavy reinforced outside woven edges of cloth. Also spelled selvedge. { 'sel·vij }

selvedge See selvage. { 'sel·vij }

Selwood engine [MECH ENG] A revolving-block engine in which two curved pistons opposed 180° run in toroidal tracks, forcing the entire engine block to rotate. { 'sel‚wůd ‚en·jən }

SEM See scanning electron microscope.

Semaeostomeae [INV ZOO] An order of the class Scyphozoa including most of the common medusae, characterized by

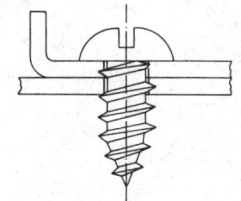
SEMAEOSTOMEAE

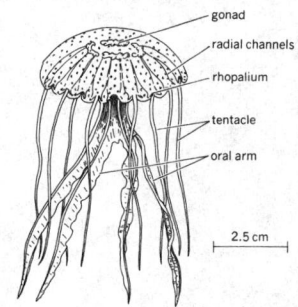

gonad
radial channels
rhopalium
tentacle
oral arm

2.5 cm

Pelagia, showing the tentacles arising between the lappets on the umbrella. *(From L. Hyman, The Invertebrates, vol. 1, McGraw-Hill, 1940)*

a flat, domelike umbrella whose margin is divided into many lappets. { sə,mē·ə'stō·mē,ē }

semanteme [COMMUN] A language element that expresses a definite idea or image, such as a word, base of a word, or data element. Also known as lexeme. { si'man,tēm }

semantic analysis [COMPUT SCI] A phase of natural language processing, following parsing, that involves extraction of context-independent aspects of a sentence's meaning, including the semantic roles of entities mentioned in the sentence, and quantification information, such as cardinality, iteration, and dependency. { si'man·tik ə'nal·ə·səs }

semantic error [COMPUT SCI] The use of an incorrect symbolic name in a computer program. { si'man·tik 'er·ər }

semantic extension [COMPUT SCI] An extension mechanism which introduces new kinds of objects into an extensible language, such as additional data types or operations. { si'man·tik ik'sten·shən }

semantic gap [COMPUT SCI] The difference between a data or language structure and the objects that it models. { si'man·tik 'gap }

semantic memory [PSYCH] Memory of generic, context-free knowledge. { sə,man·tik 'mem·rē }

semantics [COMMUN] The branch of semiotics that deals with the relations between symbols and what they stand for, and defines the meaning that is prescribed for a statement by its originator. { si'man·tiks }

semaphore [COMPUT SCI] A memory cell that is shared by two parallel processes which rely on each other for their continued operation, and that provides an elementary form of communication between them by indicating when significant events have taken place. { 'sem·ə,fôr }

sematic See aposematic. { si'mat·ik }

semeiography [MED] Description of the signs and symptoms of a disease. { ,sē·mē'äg·rə·fē }

semelparity [BIOL] Reproduction that occurs only one time during the life of an individual. { 'sem·əl,par·əd·ē }

semelparous [ZOO] Capable of breeding or reproducing only once. { ,sem·əl'par·əs }

semen [PHYSIO] The fluid that carries the male germ cells. Also known as seminal fluid. { 'sē·mən }

Semenov number 1 [PHYS CHEM] A dimensionless number used in reaction kinetics, equal to a mass transfer constant divided by a reaction rate constant. Symbolized S_m. Formerly known as Schmidt number 2. { 'se·mə,nôf 'nəm·bər 'wən }

Semenov number 2 [PHYS] The reciprocal of the Lewis number. { 'se·mə,nôf 'nəm·bər 'tü }

semiactive homing guidance [NAV] Guidance in which a craft or vehicle is directed toward a target by illuminating the target with radiation from a ground-based or shipborne radar; the missile or other vehicle is equipped with a direction-finding receiver which enables it to home on the target by use of the radar signal reflected from it. { ¦sem·ē¦ak·tiv 'hōm·iŋ ,gīd·əns }

semiactive tracking system [NAV] A tracking system used for semiactive homing guidance. { ¦sem·ē¦ak·tiv 'trak·iŋ ,sis·təm }

semialgorithm [COMPUT SCI] A procedure for solving a problem that will continue endlessly if the problem has no solution. { ,sem·ē'al·gə,rith·əm }

semianechoic room [ACOUS] A room having surfaces which reduce the reflection of sound to less than normal, although not to as great an extent as an anechoic room. { ¦sem·ē,an·e'kō·ik 'rüm }

semianthracite [GEOL] Coal which is between bituminous coal and anthracite in metamorphic rank, and which has a fixed-carbon content of 86-92%. { ,sem·ē'an·thrə,sīt }

semiarid climate See steppe climate. { ¦sem·ē'ar·əd 'klī·mət }

semiautomatic [ORD] Pertaining to a firearm or gun that utilizes a part of the force of an exploding cartridge to extract the empty case and to chamber the next round, but requires a separate pull on the trigger to fire each round; examples are the semiautomatic rifle and the semiautomatic pistol. { ¦sem·ē,öd·ə'mad·ik }

semiautomatic flight inspection [NAV] Airborne equipment that systematically records significant parameters as an aircraft flies along a previously determined route, in order to check the performance of ground-based navigation aids. Abbreviated SAFI. { ¦sem·ē,öd·ə'mad·ik 'flīt in,spek·shən }

semiautomatic ground environment See SAGE. { ¦sem·ē,öd·ə'mad·ik 'graund in,vi·rən·mənt }

semiautomatic supply [ORD] System by which certain specified items of supplies needed by units, activities, or forces are shipped by the agencies responsible for supply on the basis of periodic reports of the status of stocks on hand and en route to the using agency; all other supplies are furnished on the basis of requisitions initiated by the using agency. { ¦sem·ē,öd·ə'mad·ik sə'plī }

semiautomatic telephone system [COMMUN] Telephone system that limits automatic dialing to only those subscribers who are served by the same exchange as the calling subscriber. { ¦sem·ē,öd·ə'mad·ik 'tel·ə,fōn ,sis·təm }

semiautomatic transmission [MECH ENG] An automobile transmission that assists the driver to shift from one gear to another. { ¦sem·ē,öd·ə'mad·ik tranz'mish·ən }

semiautomatic welding [MET] An arc-welding method in which the electrode, a long length of small-diameter bare wire, usually in coil form, is positioned and advanced by the operator from a hand-held welding gun which feeds the electrode through the nozzle. { ¦sem·ē,öd·ə'mad·ik 'weld·iŋ }

semiaxis [MATH] A line segment that forms half of the axis of a geometric figure (such as an ellipse), having one end point at the center of symmetry of the figure. { ¦sem·ē'ak·səs }

semibatch chemical reactor [CHEM ENG] A reactor in which a constant liquid volume is maintained without any overflow, and with the continuous addition of one reactant, usually a gas. { ¦sem·i,bach 'kem·ə·kəl rē'ak·tər }

semibituminous coal [GEOL] Coal that is harder and more brittle than bituminous coal, has a high fuel ratio, contains 10-20% volatile matter, and burns without smoke; ranks between bituminous and semianthracite coals. { ¦sem·i·bə'tü·mə·nəs 'kōl }

semibolson [GEOL] A wide desert basin or valley whose central playa is absent or poorly developed, and which is drained by an intermittent stream that flows through canyons at each end and reaches a surface outlet. { ¦sem·i'bōls·ən }

semibright coal [GEOL] A type of banded coal defined microscopically as consisting of between 80 and 61% bright ingredients such as vitrain, clarain, and fusain, with clarodurain and durain composing the remainder. { 'sem·i,brīt 'kōl }

semicarbazide [ORG CHEM] $H_2N-NHCONH_2$ A reagent used to produce semicarbazones by reaction with aldehydes or ketones. { ¦sem·i'kär·bə,zīd }

semicarbazide hydrochloride [ORG CHEM] $CH_5ON_3 \cdot HCl$ Colorless prisms, soluble in water, decomposing at 175°C; used as an analytical reagent for aldehydes and ketones, and to recover constituents of essential oils. { ¦sem·i 'kär·bə,zīd ,hī·drə'klôr,īd }

semicarbazone [ORG CHEM] $R_2C:N_2HCONH_2$ A condensation product of an aldehyde or ketone with semicarbazide. { ,sem·i'kär·bə,zōn }

semichemical pulp [MATER] Wood which has been pulped by the process of semichemical pulping. { ¦sem·i'kem·ə·kəl 'pəlp }

semichemical pulping [CHEM ENG] A method of producing wood-fiber products in which the wood chips are merely softened by chemical treatment (neutral sodium sulfite solution), while the remainder of the pulping action is supplied by a disk attrition mill or by some similar mechanical device for separating the fibers. { ¦sem·i'kem·ə·kəl 'pəlp·iŋ }

semicircle [MATH] One of the two parts of a circle that extend from one end of a diameter to the other, and whose length is one-half that of the circle. { 'sem·i,sər·kəl }

semicircular canal [ANAT] Any of three loop-shaped tubular structures of the vertebrate labyrinth; they are arranged in three different spatial planes at right angles to each other, and function in the maintenance of body equilibrium. { ¦sem·i'sər·kyə·lər kə'nal }

semicircular deviation [NAV] A deviation of a magnetic compass which changes sign east or west approximately each 180° change of heading. { ¦sem·i'sər·kyə·lər ,dē·vē'ā·shən }

semicircumference [MATH] One-half the circumference of a circle. { ,sem·i·sər'kəm·frəns }

semiclosed-cycle gas turbine [MECH ENG] A heat engine in which a portion of the expanded gas is recirculated. { 'sem·i,klōzd,sī·kəl 'gas ,tər·bən }

semicoma [MED] A mildly or partially comatose state in

which the patient can be roused and responds to strong stimuli with some purposeful movements. { ¦sem·i'kō·mə }

semicompreg [MATER] Resin-impregnated wood compressed to a density not exceeding 1.25. { ¦sem·i'käm‚preg }

semiconducting compound [SOLID STATE] A compound which is a semiconductor, such as copper oxide, mercury indium telluride, zinc sulfide, cadmium selenide, and magnesium iodide. { ¦sem·i·kən¦dək·tiŋ ¦käm‚paúnd }

semiconducting crystal [SOLID STATE] A crystal of a semiconductor, such as silicon, germanium, or gray tin. { ¦sem·i·kən¦dək·tiŋ ¦krist·əl }

semiconductive loading tube [ENG] A loading tube for blasthole explosives which dissipates static electric charges to prevent premature blasts. { ¦sem·i·kən¦dək·tiv 'lōd·iŋ ‚tüb }

semiconductor [SOLID STATE] A solid crystalline material whose electrical conductivity is intermediate between that of a conductor and an insulator, ranging from about 10^5 mhos to 10^{-7} mho per meter, and is usually strongly temperature-dependent. { ¦sem·i·kən¦dək·tər }

semiconductor detector [NUCLEO] A particle detector which detects ionization produced by energetic charged particles in the depletion layer of a reverse-biased *pn* junction in a semiconductor, usually a very pure single crystal of silicon or germanium. { ¦sem·i·kən¦dək·tər di'tek·tər }

semiconductor device [ELECTR] Electronic device in which the characteristic distinguishing electronic conduction takes place within a semiconductor. { ¦sem·i·kən¦dək·tər di‚vīs }

semiconductor diode [ELECTR] Also known as crystal diode; crystal rectifier; diode. **1.** A two-electrode semiconductor device that utilizes the rectifying properties of a *pn* junction or a point contact. **2.** More generally, any two-terminal electronic device that utilizes the properties of the semiconductor from which it is constructed. { ¦sem·i·kən¦dək·tər 'dī‚ōd }

semiconductor-diode parametric amplifier [ELECTR] Parametric amplifier using one or more varactors. { ¦sem·i·kən¦dək·tər ¦dī‚ōd ¦par·ə¦me·trik 'am·plə‚fī·ər }

semiconductor disk [COMPUT SCI] A large semiconductor memory that imitates a disk drive in that the operating system can read and write to it as though it were an ordinary disk, but at a much faster rate. Also known as nonrotating disk. { 'sem·i·kən‚dək·tər ‚disk }

semiconductor doping *See* doping. { ¦sem·i·kən¦dək·tər 'dōp·iŋ }

semiconductor heterostructure [ELECTR] A structure of two different semiconductors in junction contact having useful electrical or electrooptical characteristics not achievable in either conductor separately; used in certain types of lasers and solar cells. { ¦sem·i·kən¦dək·tər 'hed·ə·rō‚strək·chər }

semiconductor intrinsic properties [SOLID STATE] Properties of a semiconductor that are characteristic of the ideal crystal. { ¦sem·i·kən¦dək·tər in'trin·sik 'präp·ərd·ēz }

semiconductor junction [ELECTR] Region of transition between semiconducting regions of different electrical properties, usually between *p*-type and *n*-type material. { ¦sem·i·kən¦dək·tər ‚jəŋk·shən }

semiconductor laser [OPTICS] A laser in which stimulated emission of coherent light occurs at a *pn* junction when electrons and holes are driven into the junction by carrier injection, electron-beam excitation, impact ionization, optical excitation, or other means. Also known as diode laser; laser diode. { ¦sem·i·kən¦dək·tər 'lā·zər }

semiconductor memory [COMPUT SCI] A device for storing digital information that is fabricated by using integrated circuit technology. Also known as integrated circuit memory; large-scale integrated memory; memory chip; semiconductor storage; transistor memory. { ¦sem·i·kən¦dək·tər ‚mem‚rē }

semiconductor rectifier *See* metallic rectifier. { ¦sem·i·kən¦dək·tər 'rek·tə‚fī·ər }

semiconductor storage *See* semiconductor memory. { 'sem·i·kən‚dək·tər ‚stór·ij }

semiconductor thermocouple [ELECTR] A thermocouple made of a semiconductor, which offers the prospect of operation with high-temperature gradients, because semiconductors are good electrical conductors but poor heat conductors. { ¦sem·i·kən¦dək·tər 'thər·mə‚kəp·əl }

semiconductor trap *See* trap. { ¦sem·i·kən¦dək·tər ‚trap }

semiconjugate axis [MATH] Either of the equal line segments into which the conjugate axis of a hyperbola is divided by the center of symmetry. { ¦sem·ē¦kän·jə·gət 'ak·səs }

semiconservative replication [CELL MOL] Replication of deoxyribonucleic acid by longitudinal separation of the two complementary strands of the molecule, each being conserved and acting as a template for synthesis of a new complementary strand. { ¦sem·i·kən'sər·vəd·iv rep·li'kā·shən }

semicontrolled mosaic [MAP] A mosaic which is composed of photographs of approximately the same scale laid so that major ground features match their geographical coordinates. { ¦sem·i·kən'trōld mō'zā·ik }

semiconvection [FL MECH] A partial convective mixing that causes a region to become convectively stable before complete mixing has been achieved. { ¦sem·i·kən'vek·shən }

semicratonic *See* quasi-cratonic. { ¦sem·i·krə'tän·ik }

semicubical parabola [MATH] A plane curve whose equation in Cartesian coordinates *x* and *y* is $y^2 = ax^3$, where *a* is some constant. Also known as isochrone. { ¦sem·i'kyü·bə·kəl pə'rab·ə·lə }

semicrystalline *See* hypocrystalline. { ¦sem·i'krist·əl·ən }

semidense list [COMPUT SCI] A list that can be divided into two contiguous portions, with all the cells in the larger portion filled and all the other cells empty. { ¦sem·i'dens 'list }

semidesert [ECOL] An area intermediate in character and often located between a desert and a grassland or woodland. { ¦sem·i'dez·ərt }

semidetached binary [ASTRON] A binary system whose secondary member fills its Roche lobe but whose primary member does not. { ¦sem·i·di'tacht 'bī‚ner·ē }

semidiameter [ASTRON] Measured at the observer, half the angle subtended by the visible disk of a celestial body. { ¦sem·i·dī'am·əd·ər }

semidiameter correction [NAV] The sextant altitude correction that, when applied to the observation of the upper or lower limb of a celestial body, determines the altitude of the center of that body. { ¦sem·i·dī'am·əd·ər kə'rek·shən }

semidiesel engine [MECH ENG] **1.** An internal combustion engine of a type resembling the diesel engine in using heavy oil as fuel but employing a lower compression pressure and spraying it under pressure, against a hot (uncooled) surface or spot, or igniting it by the precombustion or supercompression of a portion of the charge in a separate member or uncooled portion of the combustion chamber. **2.** A true diesel engine that uses a means other than compressed air for fuel injection. { ¦sem·i'dē·zəl 'en‚jən }

semidiurnal [ASTRON] Having a period of, occurring in, or related to approximately half a day. [METEOROL] Pertaining to a meteorological event that occurs twice a day. { ¦sem·i·dī'ərn·əl }

semidiurnal current [OCEANOGR] A tidal current in which the tidal-day current cycle consists of two flood currents and two ebb currents, separated by slack water, or of two changes in direction of 360° of a rotary current; this is the most common type of tidal current throughout the world. { ¦sem·i·dī'ərn·əl 'kə‚rənt }

semidiurnal tide [OCEANOGR] A tide having two high waters and two low waters during a tidal day. { ¦sem·i·dī'ərn·əl 'tīd }

semidormancy [BOT] Decrease in plant growth rate; may be seasonal or associated with unfavorable environmental conditions. { ¦sem·i'dór·mən·sē }

semidouble [BOT] Pertaining to a flower that has more than the usual number of petals or disk florets while it retains some pollen-bearing stamens or some perfect disk florets. { ¦sem·i'dəb·əl }

semidull coal [GEOL] A type of banded coal consisting mainly of clarodurain and durain, with from 40 to 21% bright ingredients such as vitrain, clarain, and fusain. { 'sem·i‚dəl 'kōl }

semiempirical computation [PHYS CHEM] Computation of the geometry of a molecule by using parameters that have been experimentally determined for similar molecules. { ‚sem·ē·em¦pir·ə·kəl ‚käm·pyə'tā·shən }

semifinishing [MET] The preliminary finishing operations. { ¦sem·i'fin·ish·iŋ }

semifixed ammunition [ORD] Ammunition in which the cartridge case is not permanently fixed to the projectile, so that the zone charge within the cartridge case can be adjusted to obtain the desired range; it is loaded into the weapon as a unit. { ¦sem·i'fixt ¸am·yə'nish·ən }

semifloating axle [MECH ENG] A supporting member in motor vehicles which carries torque and wheel loads at its outer end. { ¦sem·i'flōd·iŋ 'ak·səl }

semiforbidden line [SPECT] A spectral line associated with a semiforbidden transition. { ¦sem·i·fər'bid·ən 'līn }

semiforbidden transition [ATOM PHYS] An atomic transition whose probability is reduced by selection rules by a factor roughly of the order of 10^6, as compared with 10^9 for a forbidden transition. { ¦sem·i·fər'bid·ən tran'zish·ən }

semifusinite [GEOL] A coal maceral with a well-defined woody structure and optical properties intermediate between those of nitrinite and those of fusinite. { ¦sem·i'fyüz·ən¸īt }

semigel [MATER] A cohesive powder used as an explosive. { 'sem·i¸jel }

semigloss [MATER] Pertaining to a surface finish intermediate between flat and glossy; used especially of paint and varnish. { 'sem·i¸gläs }

semigroup [MATH] A set which is closed with respect to a given associative binary operation. { 'sem·i¸grüp }

semigroup theory [MATH] The formal algebraic study of the structure of semigroups. { 'sem·i¸grüp ¸thē·ə·rē }

semi-interquartile range See quartile deviation. { ¦sem·ē¸in·tər'kwȯr¸tīl ¸rānj }

semi-invariants See cumulants. { ¦sem·ē·in'ver·ē·əns }

semikilled steel [MET] Incompletely deoxidized steel containing enough dissolved oxygen to react with the carbon it contains to form carbon monoxide, the latter offsetting solidification shrinkage. { 'sem·i¸kild 'stēl }

semilate [BOT] Pertaining to a plant whose growing season is intermediate between midseason forms and late forms. { ¦sem·i'lāt }

semilethal gene [GEN] A mutant causing the death of some of the individuals of the relevant genotype, but never 100%. Also known as sublethal gene. { ¦sem·i'lēth·əl 'jēn }

semilive skid [ENG] A platform having two fixed legs at one end and two wheels at the other; used for moving bulk materials. { ¦sem·i'līv 'skid }

semilogarithmic coordinate paper [MATH] Paper ruled with two sets of mutually perpendicular, parallel lines, one set being spaced according to the logarithms of consecutive numbers, and the other set uniformly spaced. Also known as ratio paper. { ¦sem·i¦läg·ə¦rith·mik kō¦ȯrd·ən·ət 'pā·pər }

semilunar cartilage [ANAT] One of the two interarticular knee cartilages. { ¦sem·i'lü·nər 'kärt·lij }

semilunar ganglion See Gasserian ganglion. { ¦sem·i'lü·nər 'gaŋ·glē·ən }

semilunar valve [ANAT] Either of two tricuspid valves in the heart, one at the orifice of the pulmonary artery and the other at the orifice of the aorta. { ¦sem·i'lü·nər 'valv }

semimagic square See magic square. { 'sem·i¸maj·ik ¸skwer }

semimagnetic controller [ELEC] Electrical controller having only part of its basic functions performed by devices that are operated by electromagnets. { ¦sem·i'mag¸ned·ik kən'trōl·ər }

semimajor axis [MATH] Either of the equal line segments into which the major axis of an ellipse is divided by the center of symmetry. { ¸sem·ē¦mā·jər 'ak·səs }

semimat [MATER] Intermediate between glossy and mat, as photographic paper. { ¦sem·i'mat }

semimember [CIV ENG] A part in a frame or truss that ceases to bear a load when the stress in it starts to reverse. { ¦sem·i'mem·bər }

semimembranosus [ANAT] One of the hamstring muscles, arising from the ischial tuber, and inserted into the tibia. { ¦sem·i¸mem·brə'nō·səs }

semimetal See metalloid. { ¦sem·ē'med·əl }

semimetric [MATH] A real valued function $d(x,y)$ on pairs of points from a topological space which has all the same properties as a metric save that $d(x,y)$ may be zero even if x and y are distinct points. Also known as pseudometric. { ¦sem·i'me·trik }

semimicroanalysis [ANALY CHEM] A chemical analysis procedure in which the weight of the sample is between 10 and 100 milligrams. { ¦sem·i¸mī·krō·ə'nal·ə·səs }

semiminor axis [MATH] Either of the equal line segments into which the minor axis of an ellipse is divided by the center of symmetry. { ¦sem·i¦mīn·ər 'ak·səs }

semimobile artillery [ORD] Artillery weapons designed for movement but which require partial disassembly to be placed in a firing position; wheels or other suspension devices are removed from the mount so that it may rest on the ground. { ¦sem·i'mō·bəl är'til·ə·rē }

semimonocoque [AERO ENG] A fuselage structure in which longitudinal members (stringers) as well as rings or frames which run circumferentially around the fuselage reinforce the skin and help carry the stress. Also known as stiffened-shell fuselage. { ¦sem·i'män·ə¸käk }

seminal bursa [INV ZOO] A sac which retains sperm for a period of time in turbellarians. { 'sem·ən·əl 'bər·sə }

seminal fluid See semen. { 'sem·ə·nəl ¸flü·əd }

seminal fructose [PHYSIO] Fructose that is normally produced in the seminal vesicles. { 'sem·ən·əl 'frük¸tōs }

seminal groove [ZOO] A passage in many animals providing a pathway for sperm. { 'sem·ən·əl 'grüv }

seminal receptacle See spermatheca. { 'sem·ən·əl ri'sep·tə·kəl }

seminal vesicle [ANAT] A saclike, glandular diverticulum on each ductus deferens in male vertebrates; it is united with the excretory duct and serves for temporary storage of semen. { 'sem·ən·əl 'ves·i·kəl }

seminiferous tubule [ANAT] Any of the tubercles of the testes which produce spermatozoa. { ¦sem·ə¦nif·rəs 'tü¸byül }

seminivorous [ZOO] Feeding on seeds. { ¦sem·ə¦niv·ə·rəs }

seminoma [MED] A malignant tumor of the testes composed of large, uniform cells with clear cytoplasm. { ¸sem·ə'nō·mə }

seminorm [MATH] A scalar-valued function on a real or complex vector space satisfying the axioms of a norm, except that the seminorm of a nonzero vector may equal zero. { 'sem·ē¸nȯrm }

seminumerical algebraic manipulation language [COMPUT SCI] The most elementary type of algebraic manipulation language, constructed to manipulate data from rigid classes of mathematical objects possessing strictly canonical forms. { ¦sem·i·nü'mer·ə·kəl ¸al·jə'brā·ik mə¸nip·yə'lā·shən ¸laŋ·gwij }

semiochemical [PHYSIO] Any of a class of substances produced by organisms, especially insects, that participate in regulation of their behavior in such activities as aggregation of both sexes, sexual stimulation, and trail following. { ¦sem·ē·ə'kem·ə·kəl }

Semionotiformes [VERT ZOO] An order of actinopterygian fishes represented by the single living genus *Lepisosteus*, the gars; the body is characteristically encased in a heavy armor of interlocking ganoid scales. { ¸sem·ē·ə¸nō·də'fȯr¸mēz }

semiotics [COMMUN] The theory of signs and symbols, entities that represent some other thing; it includes syntactics, pragmatics, and semantics. { ¸sem·ē'äd·iks }

semioxamazide [ORG CHEM] $H_2NCOCONHNH_2$ A crystalline compound that decomposes at 220°C; soluble in hot water, acids, and alkalies; used as a reagent for aldehydes and ketones. { ¸sem·ē¸äk'sam·ə·zəd }

semipalmate [VERT ZOO] Having a web halfway down the toes. { ¦sem·i'päl¸māt }

semiparasite See hemiparasite. { ¦sem·i'par·ə¸sīt }

semiperfect number [MATH] A number which is the sum of some set of its own proper divisors. { 'sem·i¸pər·fekt ¸nəm·bər }

semiperimeter [MATH] One-half the length of a closed curve. { ¸sem·i·pə'rim·əd·ər }

semipermanent mold [MET] A reusable metal mold with expendable sand cores. { ¦sem·i'pər·mə·nənt 'mōld }

semipermeable membrane [PHYS] A membrane which allows a solvent to pass through it, but not certain dissolved or colloidal substances. { ¦sem·i'pər·mē·ə·bəl 'mem¸brān }

semiplastic explosive [MATER] An explosive in which the quantities of liquid products are insufficient to render the mixture compressible. { ¦sem·i'plas·tik ik'splō·siv }

semipositive mold [ENG] A plastics mold that allows a small amount of excess material to escape when it is closed. { ¦sem·i'päz·əd·iv 'mōld }

semiprecious [LAP] Pertaining to gemstones whose value

is lower than that of precious stones; in particular, stones whose hardness is less than 8. { ¦sem·i'presh·əs }

semiprime [MATH] A positive integer that is the product of exactly two primes. { 'sem·i‚prīm }

semiquinone [ORG CHEM] A radical ion intermediate formed in the oxidation of a hydroquinone to a quinone. { ‚sem·ē·kwə'nōn }

semirefined wax [MATER] Commercial grades of petroleum wax which are inferior to fully refined grades but which meet specified requirements as to color and oil content. { ¦sem·i·ri'fīnd 'waks }

semiregular solid See Archimedean solid. { ¦sem·i‚reg·yə·lər 'säl·əd }

semiregular variables [ASTRON] Variable red giant stars whose variation in brightness is repeated, but whose period and light curve may vary considerably from one cycle to the next; they have absolute magnitude of about 0 or −1 and quasi-periods of from about 40 to 150 days. { ¦sem·i'reg·yə·lər 'ver·ē·ə·bəlz }

semirigid plastic [MATER] A plastic that has a stiffness or apparent modulus of elasticity of between 10,000 and 100,000 pounds per square inch (6.895 × 10⁷ and 6.895 × 10⁸ pascals) under prescribed test conditions. { ¦sem·i'rij·əd 'plas·tik }

semiring of sets [MATH] A collection S of sets that includes the empty set and the intersection of any two of its members, and is such that if A and B are members of S and A is a subset of B, then $B − A$ is the union of a finite number of disjoint members of S. { ‚sem·ē'riŋ əv 'setz }

semischist [PETR] A partly metamorphosed sedimentary rock, exhibiting some foliation. { 'sem·i‚shist }

semisecant See transversal. { 'sem·i‚sē‚kant }

semiselective ringing [COMMUN] In telephone service, party line ringing wherein the bells of two stations are rung simultaneously; the differentiation is made by the number of rings. { ¦sem·i·si'lek·tiv 'riŋ·iŋ }

semisilica refractory [MATER] A silica refractory made from clay with a high silica (sand) content (over 70% total silica); characterized by dimensional stability when heated or fired. { ¦sem·i'sil·ə·kə ri'frak·trē }

semisimple module [MATH] A module which is the sum of a family of simple modules. { ¦sem·i‚sim·pəl 'mä·jəl }

semisimple representation See completely reducible representation. { ¦sem·i‚sim·pəl ‚rep·ri·zen'ta·shən }

semisimple ring [MATH] A ring in which 1 does not equal 0, and which is semisimple as a left module over itself. { ¦sem·i‚sim·pəl 'riŋ }

semispecies [SYST] 1. The species that compose a superspecies. 2. Populations that have acquired some attributes of species rank. 3. Organisms that are borderline between species and subspecies. { ¦sem·i'spē·shēz }

semispinalis [ANAT] One of the deep longitudinal muscles of the back, attached to the vertebrae. { ‚sem·i‚spī'nal·əs }

semisplint coal [GEOL] Banded coal that is intermediate between bright−banded and splint coal, and has 20–30% opaque attritus and more than 5% anthraxylon. { ¦sem·i'splint 'kōl }

semistable elementary particle See quasi-stable elementary particle. { ¦sem·i'stā·bəl 'el·ə‚men·trē 'pärd·ə·kəl }

semisteel [MET] Low-carbon steel made by replacing about one-fourth of the pig iron in the cupola with steel scrap. { ¦sem·i'stēl }

semisynchronous satellite [AERO ENG] An artificial earth satellite that makes one revolution in exactly one-half of a sidereal day (11 hours 58 minutes 2 seconds). { ‚sem·ē‚siŋ·krə·nəs 'sad·əl‚īt }

semitendinosus [ANAT] One of the hamstring muscles, arising from the ischium and inserted into the tibia. { ‚sem·ē‚ten·də'nō·səs }

semitone [ACOUS] The interval between two sounds whose frequencies have a ratio approximately equal to the twelfth root of 2. Also known as half step. { 'sem·i‚tōn }

semitrailer [ENG] A cargo-carrying piece of equipment that has one or two axles at the rear; the load is carried on these axles and on the fifth wheel of the tractor that supplies motive power to the semitrailer. { ¦sem·i'trāl·ər }

semitransparent photocathode [ELECTR] Photocathode in which radiant flux incident on one side produces photoelectric emission from the opposite side. { ¦sem·i·tranz'par·ənt ¦fōd·ō'kath‚ōd }

semitransverse axis [MATH] Either of the equal line segments into which the transverse axis of a hyperbola is divided by the center of symmetry. { ¦sem·i¦tranz‚vərs 'ak·səs }

semivariogram [STAT] A mathematical function used to quantify the dissimilarity between groups of values. { ‚sem·i'ver·ē·ə‚gram }

semivitreous [MATER] Pertaining to ceramics whose glassy content is not sufficient to reduce porosity below 0.2%. { ¦sem·i'vi·trē·əs }

semolina [FOOD ENG] A granular pale yellow flour made from milling the whole kernel of durum (hard grain) wheat. { ‚sem·ə'lē·nə }

Semper's larva [INV ZOO] A cylindrical larva in the life history of certain zoanthid corals, characterized by a hole at each end and an annular or longitudinal band of long cilia. { 'sem·pərz ‚lär·və }

Semple plunger [ORD] A centrifugal plunger which operates to maintain a fuse in a safe condition until centrifugal force unlocks and moves the firing pin into the armed position. { 'sem·pəl ‚plən·jər }

sems [DES ENG] A preassembled screw and washer combination. { semz }

semseyite [MINERAL] Pb₉Sb₈S₂₁ A gray to black mineral composed of lead antimony sulfide. { 'sem·sē‚īt }

senaite [MINERAL] (Fe,Mn,Pb)TiO₃ A black mineral consisting of a lead- and manganese-bearing ilmenite; occurs as rough crystals and rounded fragments. { 'sen·ə‚īt }

senarmontite [MINERAL] Sb₂O₃ A colorless or grayish mineral composed of native antimony trioxide occurring in masses or as octahedral crystals. { ‚sen·ər'män‚tīt }

send See scend. { send }

sender [COMMUN] Part of an automatic-switching telephone system that receives pulses from a dial or other source and, in accordance with them, controls the further operations necessary in establishing a telephone connection. { 'sen·dər }

sending-end impedance [ELEC] Ratio of an applied potential difference to the resultant current at the point where the potential difference is applied; the sending-end impedance of a line is synonymous with the driving-point impedance of the line. { 'send·iŋ ‚end im'pēd·əns }

Sendust [MATER] The trade name for an alloy consisting of approximately 85% iron, 6% aluminum, and 9% silicon; its powdered form is compacted to manufacture low-loss magnetic cores. { 'sen‚dəst }

Sendzimir mill [MET] A mill having small-diameter working rolls, each backed by a pair of supporting rolls, and each pair of these supported by a cluster of three rolls; used for cold-rolling wide sheets of metal to close tolerance. { 'sen·zə‚mir ‚mil }

Senecan [GEOL] A North American provincial series of geologic time, forming the lower part of the Upper Devonian, above the Erian and below the Chautauquan. { 'sen·i·kən }

senescence [BIOL] The study of the biological changes related to aging, with special emphasis on plant, animal, and clinical observations which may apply to humans. [GEOL] The part of the erosion cycle at which the stage of old age begins. { si'nes·əns }

senescent arthritis See degenerative joint disease. { si'nes·ənt är'thrīd·əs }

senescent lake [HYD] A lake that is approaching extinction—for example, from filling by remains of aquatic vegetation. { si'nes·ənt 'lāk }

senesland [GEOL] A land surface intermediate between a matureland and a peneplain. { 'sen·əs‚land }

Senftleben effect [PHYS] The change in the thermal conductivity of a gas in a magnetic field. { 'senft‚lā·bən i‚fekt }

sengierite [MINERAL] Cu(UO₂)₂(VO₄)₂·8-10H₂O A yellowish-green mineral composed of hydrous copper uranyl vanadate. { 'seŋ·ē·ə‚rīt }

senile [GEOL] Pertaining to the stage of senility of the cycle or erosion. [MED] Of, pertaining to, or caused by the aging process or by the infirmities of old age. { 'sē‚nīl }

senile cataract [MED] The most common type of cataract; it occurs with aging, and progressively blurred vision is the only symptom. { ‚sē‚nīl 'kad·ə‚rakt }

senile eczema [MED] A form of eczema associated with aging and caused by factors such as dryness of skin, soap sensitivity, or poor hygiene or diet. { ‚sē‚nīl 'ek·sə·mə }

senile emphysema [MED] Degenerative changes in the

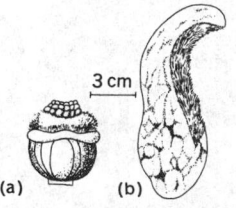

lungs and thoracic cage associated with aging. { 'sē,nīl ,em·fə'sē·mə }

senile gangrene [MED] A form of tissue death caused by deterioration of the blood supply to the extremities in the elderly. { 'sē,nīl 'gaŋ,grēn }

senile psychosis [PSYCH] A severe form of senile dementia characterized by personality deterioration, progressive memory loss, eccentricity, irritability, and sometimes delusions and hallucinations. { 'sē,nīl sī'kō·səs }

senile vaginitis [MED] Inflammation of the vagina occurring in elderly women following the chronic irritation of the thinned, atrophic mucosa. { 'sē,nīl ,vaj·ə'nīd·əs }

senility [GEOL] The stage of the cycle of erosion in which erosion of a land surface has reached a minimum, most of the hills have disappeared, and base level has been approached. [MED] Old age and its characteristics. { si'nil·əd·ē }

senna [PHARM] The dried leaflets of plants of the *Cassia* genus; used in medicine as a cathartic. { 'sen·ə }

Senn process [FOOD ENG] A butter-making process in which the cream is subjected to rapid agitation at 1500 revolutions per minute, decreasing to 20 revolutions per minute in the presence of 2–4 atmospheres (200–400 kilopascals) of carbon dioxide. { 'sen ,prä·səs }

Senonian [GEOL] A European stage of geologic time, forming the Upper Cretaceous, above the Turonian and below the Danian. { sə'nō·nē·ən }

sensation [PHYSIO] The subjective experience that results from the stimulation of a sense organ. { sen'sā·shən }

sensation level *See* level above threshold. { sen'sā·shən ,lev·əl }

sensation unit [ACOUS] A unit of loudness, no longer in use; the loudness of a sound is $20 \log_{10}(p/p_0)$ sensation units above threshold, where p is the pressure level of the sound, and p_0 is the pressure of a sound which can just be detected by the ear. { sen'sā·shən ,yü·nət }

sense [COMPUT SCI] To read punched holes in tape or cards. [ENG] To determine the arrangement or position of a device or the value of a quantity. [NAV] The general direction from which a radio signal arrives; if a radio bearing is received by a simple loop antenna, there are two possible readings approximately 180° apart; the resolving of this ambiguity is called sensing of the bearing. { sens }

sense amplifier [ELECTR] Circuit used to determine either a phase or voltage change in communications-electronics equipment and to provide automatic control function. { 'sens ,am·plə,fī·ər }

sense antenna [ELECTROMAG] An auxiliary antenna used with a directional receiving antenna to resolve a 180° ambiguity in the directional indication. Also known as sensing antenna. { 'sens an,ten·ə }

sense finder [NAV] Portion of a direction finder which permits resolution of the 180° ambiguity. { 'sens ,fīn·dər }

sense indicator *See* to-from indicator. { 'sens ,in·də,kād·ər }

sense light [COMPUT SCI] A light which can be turned on or off, its status being the determinant as to which path a program will select. { 'sens ,līt }

sense organ [PHYSIO] A structure which is a receptor for external or internal stimulation. { 'sens ,ór·gən }

sense strand [MOL BIO] The strand of a double-stranded deoxyribonucleic acid molecule that is complementary to the ribonucleic acid formed by transcription. Also known as coding strand. { 'sens ,strand }

sense switch *See* alteration switch. { 'sens ,swich }

sensibility [PHYS] The ability of a magnetic compass card to align itself with the magnetic meridian after deflection. { ,sen·sə'bil·əd·ē }

sensible atmosphere [METEOROL] That part of the atmosphere that offers resistance to a body passing through it. { 'sen·sə·bəl 'at·mə,sfir }

sensible heat [THERMO] **1.** The heat absorbed or evolved by a substance during a change of temperature that is not accompanied by a change of state. **2.** *See* enthalpy. { 'sen·sə·bəl 'hēt }

sensible-heat factor [THERMO] The ratio of space sensible heat to space total heat; used for air-conditioning calculations. Abbreviated SHF. { 'sen·sə·bəl ¦hēt ,fak·tər }

sensible-heat flow [METEOROL] In the atmosphere, the poleward transport of sensible heat (enthalpy) across a given latitude belt by fluid flow. [THERMO] The heat given up or absorbed by a body upon being cooled or heated, as the result of the body's ability to hold heat; excludes latent heats of fusion and vaporization. { 'sen·sə·bəl ¦hēt 'flō }

sensible horizon [ASTRON] That circle of the celestial sphere formed by the intersection of the celestial sphere and a plane through any point, such as the eye of an observer, and perpendicular to the zenith-nadir line. { 'sen·sə·bəl hə'rīz·ən }

sensible temperature [METEOROL] The temperature at which air with some standard humidity, motion, and radiation would provide the same sensation of human comfort as existing atmospheric conditions. { 'sen·sə·bəl 'tem·prə·chər }

sensillum [ZOO] A simple, epithelial sense organ composed of one cell or of a few cells. { sen'sil·əm }

sensing [NAV] The process of eliminating ambiguity in direction-finder bearings. [ORD] The direction of a point of burst or impact, or centers of burst or impact with respect to the target, such as over, short, air, or graze. { 'sens·iŋ }

sensing antenna *See* sense antenna. { 'sens·iŋ an,ten·ə }

sensing element *See* sensor. { 'sens·iŋ ,el·ə·mənt }

sensing signal [COMMUN] A special signal that is transmitted to alert the receiving station at the beginning of a message. { 'sens·iŋ ,sig·nəl }

sensing station *See* reading station. { 'sens·iŋ ,stā·shən }

sensing zone technique [ANALY CHEM] Particle-size measurement in a dilute solution, with fine particles passed through a small zone (opening) so that individual particles may be observed and measured by electrolytic, photic, or sonic methods. { 'sens·iŋ ¦zōn tek,nēk }

sensistor [ELECTR] Silicon resistor whose resistance varies with temperature, power, and time. { sen'zis·tər }

sensitive altimeter [ENG] An aneroid altimeter constructed to respond to pressure changes (altitude changes) with a high degree of sensitivity; it contains two or more pointers to refer to different scales, calibrated in hundreds of feet, thousands of feet, and so on. { 'sen·səd·iv al'tim·əd·ər }

sensitive clay [GEOL] A clay whose shear strength is reduced to a very small fraction of its former value on remolding at constant moisture content. { 'sen·səd·iv 'klā }

sensitive data [COMPUT SCI] Data that can be read or processed in specified transactions by a specified program, device, or user. { 'sen·səd·iv 'dad·ə }

sensitive switch *See* snap-action switch. { 'sen·səd·iv 'swich }

sensitive time [NUCLEO] The duration of supersaturation, sufficient for track formation, following expansion in a cloud chamber. { 'sen·səd·iv 'tīm }

sensitive volume [NUCLEO] The portion of a radiation-counter tube that responds to a specific radiation. { 'sen·səd·iv 'väl·yəm }

sensitivity [ELECTR] **1.** The minimum input signal required to produce a specified output signal, for a radio receiver or similar device. **2.** Of a camera tube, the signal current developed per unit incident radiation, that is, per watt per unit area. [ENG] **1.** A measure of the ease with which a substance can be caused to explode. **2.** A measure of the effect of a change in severity of engine-operating conditions on the antiknock performance of a fuel; expressed as the difference between research and motor octane numbers. Also known as spread. [GEOL] The effect of remolding on the consistency of a clay or cohesive soil, regardless of the physical nature of the causes of the change. [PHYSIO] The capacity for receiving sensory impressions from the environment. [SCI TECH] **1.** The ability of the output of a device, system, or organism to respond to an input stimulus. **2.** Mathematically, the ratio of the response or change induced in the output to a stimulus or change in the input. { ,sen·sə'tiv·əd·ē }

sensitivity function [CONT SYS] The ratio of the fractional change in the system response of a feedback-compensated feedback control system to the fractional change in an open-loop parameter, for some specified parameter variation. { ,sen·sə'tiv·əd·ē ,fəŋk·shən }

sensitivity time control [ELECTR] In a radar receiver a circuit which greatly reduces the gain at the time that the transmitter emits a pulse; following the pulse, the circuit increases the sensitivity; thus reflection from distant objects will be received and those from nearby objects will be prevented from saturating the receiver. { ,sen·sə'tiv·əd·ē 'tīm kən,trōl }

sensitization [ELECTR] *See* activation. [IMMUNOL] The

alteration of a body's responsiveness to a foreign antigen, usually an allergen, such that upon subsequent exposures to the allergen there is a heightened immune response. [PSYCH] A process in which a given behavior increases in intensity simply with repeated occurrences. { ‚sen·səd·ə'zā·shən }

sensitometer [ENG] An instrument for measuring the sensitivity of light-sensitive materials. { ‚sen·sə'täm·əd·ər }

sensor [ENG] The generic name for a device that senses either the absolute value or a change in a physical quantity such as temperature, pressure, flow rate, or pH, or the intensity of light, sound, or radio waves and converts that change into a useful input signal for an information-gathering system; a television camera is therefore a sensor, and a transducer is a special type of sensor. Also known as primary detector; sensing element. { 'sen·sər }

sensorineural deafness [MED] Deafness caused by an abnormality of the sense organ in the inner ear or the auditory nerve (cranial nerve VIII). { ‚sen·sə·rə'nür·əl 'def·nəs }

sensorium [PHYSIO] **1.** A center, especially in the brain, for receiving and integrating sensations. **2.** The entire sensory apparatus of an individual. { sen'sór·ē·əm }

sensory aphasia [MED] A form of aphasia in which the perception of sounds as language is partially preserved, but the patient is unable to comprehend the meaning of words, and in speaking, words are evoked with difficulty, are used incorrectly, and do not convey ideas correctly, resulting frequently in other forms of language impairment, particularly in agrammatism. Also known as receptive aphasia. { 'sen·sə·rē ə'fā·zhə }

sensory area [NEUROSCI] Any area of the cerebral cortex associated with the perception of sensations. { 'sen·sə·rē ‚er·ē·ə }

sensory cell [NEUROSCI] A neuron having its terminal processes connected with sensory nerve endings. [PHYSIO] A modified epithelial or connective tissue cell adapted for the reception and transmission of sensations. { 'sen·sə·rē ‚sel }

sensory control [CONT SYS] Control of a robot's actions on the basis of its sensor readings. { 'sen·sə·rē kən'trōl }

sensory controlled robot [CONT SYS] A robot whose programmed sequence of instructions can be modified by information about the environment received by the robot's sensors. { 'sen·sə·rē kən'trōld 'rō‚bät }

sensory learning [PSYCH] Learning situations in which a person or animal is trained to respond to changes in or differences between some aspects of a physical stimulus presented to one of the sense organs. { 'sen·sə·rē ‚lərn·iŋ }

sensory memory [PSYCH] The sensations that briefly continue after something has been perceived. { ‚sens·ə·rē 'mem·rē }

sensory nerve [NEUROSCI] A nerve that conducts afferent impulses from the periphery to the central nervous system. { 'sen·sə·rē ‚nərv }

sensory-neural hearing loss [PHYSIO] A type of hearing loss resulting from damage to the receptive elements in the ear or to the auditory nerve itself. { ‚sen·sə·rē ‚nür·əl 'hir·iŋ ‚lòs }

sentence [COMPUT SCI] An entire instruction in the COBOL programming language. { 'sent·əns }

sentential calculus See propositional calculus. { sen'ten·chəl 'kal·kyə·ləs }

sentential connectives See propositional connectives. { sen‚ten·chəl kə'nek·tivz }

sentential function See propositional function. { sen‚ten·chəl 'fəŋk·shən }

sentinel [COMPUT SCI] Symbol marking the beginning or end of an element of computer information such as an item or a tape. { 'sent·ən·əl }

sepal [BOT] One of the leaves composing the calyx. { 'sēp·əl }

separable degree [MATH] Let E be an algebraic extension of a field F, and let f be any embedding of F in a field L such that L is the algebraic closure of the image of F under f; the separable degree of E over F is the number of distinct embeddings of E in L which are extensions of f. { 'sep·rə·bəl di'grē }

separable element [MATH] An element a is said to be separable over a field F if it is algebraic over F and if the extension field of F generated by a is a separable extension of F. { 'sep·rə·bəl 'el·ə·mənt }

separable extension [MATH] A field extension K of a field F is separable if every element of K is a root of a separable

polynomial whose coefficients are elements of F. { 'sep·rə·bəl ik'sten·chən }

separable polynomial [MATH] A polynomial with no multiple roots. { 'sep·rə·bəl ‚päl·i'nō·mē·əl }

separable space [MATH] A topological space which has a countable subset that is dense. { 'sep·rə·bəl 'spās }

separate See soil separate. { 'sep·rət }

separated aggregate [MATER] Aggregate for concrete that has been separated into fine and coarse constituents. { 'sep·ə‚rād·əd 'eg·rə·gət }

separated ammunition [ORD] Ammunition characterized by the arrangement of the propelling charge and the projectile for loading into the gun; the propelling charge, contained in a primed cartridge case that is sealed with a closing plug, and the projectile are loaded into the gun in one operation; separated ammunition is used when the ammunition is too large to handle as fixed ammunition. { 'sep·ə‚rād·əd ‚am·yə'nish·ən }

separated-function synchrotron [NUCLEO] A proton synchrotron in which separate groups of magnets are used to focus and deflect the beam so that it follows a circular path. { 'sep·ə‚rād·əd ‚fəŋk·shən 'siŋ·krə‚trän }

separated sets [MATH] Sets A and B in a topological space are separated if both the closure of A intersected with B and the closure of B intersected with A are disjoint. { 'sep·ə‚rād·əd 'sets }

separate loading ammunition [ORD] Ammunition in which the projectile, propellant charge (bag-loaded), and primer are loaded separately into the gun; no cartridge case is utilized in this type of ammunition. { 'sep·rət ‚lōd·iŋ ‚am·yə'nish·ən }

separately excited [ELEC] Obtaining excitation from a source other than the machine or device itself. { 'sep·rət·lē ik'sīd·əd }

separate sewage system [CIV ENG] A drainage system in which sewage and groundwater are carried in separate sewers. { 'sep·rət 'sü·ij ‚sis·təm }

separating calorimeter [PHYS] A device for measuring the moisture content of steam. { 'sep·ə‚rād·iŋ ‚kal·ə'rim·əd·ər }

separating power [CHEM ENG] The measure of the ability of a system (such as a rectifying system) to separate the components of a mixture, when the components have increasingly close boiling points. { 'sep·ə‚rād·iŋ ‚paú·ər }

separating transcendence base [MATH] A transcendence base of a field E over a field F such that E is algebraic and separable over the field generated by F and the transcendence base. { 'sep·ə‚rād·iŋ tran'sen·dəns ‚bās }

separation [AERO ENG] The action of a fallaway section or companion body as it casts off from the remaining body of a vehicle, or the action of the remaining body as it leaves a fallaway section behind it. [CHEM ENG] The separation of liquids or gases in a mixture, as by distillation or extraction. [ENG] **1.** The action segregating phases, such as gas-liquid, gas-solid, liquid-solid. **2.** The segregation of solid particles by size range, as in screening. [ENG ACOUS] The degree, expressed in decibels, to which left and right stereo channels are isolated from each other. [GEOL] The apparent relative displacement on a fault, measured in any given direction. [MIN ENG] The removal of gangue from raw ores, as in frothing. { ‚sep·ə'rā·shən }

separation axioms [MATH] Properties of topological spaces such as Hausdorff, regular, and normal which reflect how points and closed sets may be enclosed in disjoint neighborhoods. { ‚sep·ə'rā·shən 'ak·sē·əmz }

separation disk [BOT] A layer of gelatinous material between two adjacent negative cells in some blue-green algae; associated with hormogonium formation. { ‚sep·ə'rā·shən ‚disk }

separation energy [NUC PHYS] The energy needed to remove a proton, neutron, or alpha particle from a nucleus. { ‚sep·ə'zā·shən ‚en·ər·jē }

separation factor [NUCLEO] The abundance ratio of two isotopes after processing, divided by their abundance ratio before processing. { 'sep·ə'rā·shən ‚fak·tər }

separation filter [ELECTR] Combination of filters used to separate one band of frequencies from another. { ‚sep·ə'rā·shən ‚fil·tər }

separation layer [BOT] A structurally distinct layer of the abscission zone of a plant containing abundant starch and dense cytoplasm. { ‚sep·ə'rā·shən ‚lā·ər }

separation negatives [GRAPHICS] The negatives made

from full-color originals and used in the preparation of colorplates; four negatives are made, for yellow, magenta, cyan, and black printing plates. { ˌsep·ə'rā·shən 'neg·əd·ivz }

separation of the first kind [MATH] A division of an ordered set into two classes in which each member of one class is greater than every member of the other, and there exists a separating element that belongs to one class or the other. { ˌsep·ə¦rā·shən əv t͟hə 'first ˌkīnd }

separation of the second kind [MATH] A division of an ordered set into two classes in which each member of one class is greater than every member of the other, and there is no least member in the class of greater elements and no greatest member in the class of lesser elements. { ˌsep·ə¦rā·shən əv t͟hə 'sek·ənd kīnd }

separation of variables [MATH] **1.** A technique where certain differential equations are rewritten in the form $f(x)dx = g(y)dy$ which is then solvable by integrating both sides of the equation. **2.** A method of solving partial differential equations in which the solution is written in the form of a product of functions, each of which depends on only one of the independent variables; the equation is then arranged so that each of the terms involves only one of the variables and its corresponding function, and each of these terms is then set equal to a constant, resulting in ordinary differential equations. Also known as product-solution method. { ˌsep·ə'rā·shən əv 'ver·ē·ə·bəlz }

separation theorem [CONT SYS] A theorem in optimal control theory which states that the solution to the linear quadratic Gaussian problem separates into the optimal deterministic controller (that is, the optimal controller for the corresponding problem without noise) in which the state used is obtained as the output of an optimal state estimator. { ˌsep·ə'rā·shən ˌthir·əm }

separative work unit [NUC PHYS] A fundamental measure of work required to separate a quantity of isotopic mixture into two component parts, one having a higher percentage of concentration of the desired isotope and one having a lower percentage. { 'sep·rəd·iv 'wərk ˌyü·nət }

separator [COMPUT SCI] A datum or character that denotes the beginning or ending of a unit of data. [ELEC] A porous insulating sheet used between the plates of a storage battery. [ELECTR] A circuit that separates one type of signal from another by clipping, differentiating, or integrating action. [ENG] **1.** A machine for separating materials of different specific gravity by means of water or air. **2.** Any machine for separating materials, as the magnetic separator. [MECH ENG] *See* cage. [PETRO ENG] *See* gas-oil separator. { 'sep·əˌrād·ər }

separator-filter [ENG] A vessel that removes solids and entrained liquid from a liquid or gas stream, using a combination of a baffle or coalescer with a screening (filtering) element. { 'sep·əˌrād·ər 'fil·tər }

separator page [COMPUT SCI] A page preceding or following a report in a computer printout giving all information needed to identify the report. { 'sep·əˌrād·ər ˌpāj }

separatory funnel [CHEM] A funnel-shaped device used for the careful and accurate separation of two immiscible liquids; a stopcock on the funnel stem controls the rate and amount of outflow of the lower liquid. { 'sep·rəˌtȯr·ē 'fən·əl }

sepatrix [CONT SYS] A curve in the phase plane of a control system representing the solution to the equations of motion of the system which would cause the system to move to an unstable point. { 'sep·əˌtriks }

sepia [MATER] A brown pigment prepared from the dried, inky exudation of a cuttlefish; used as a dye and in watercolors and ink. { 'sē·pē·ə }

sepia negative *See* vandyke. { 'sē·pē·ə 'neg·əd·iv }

Sepioidea [INV ZOO] An order of the molluscan subclass Coleoidea having a well-developed eye, an internal shell, fins separated posteriorly, and chromatophores in the dermis. { ˌsē·pē'ȯid·ē·ə }

sepiolite [MINERAL] $Mg_4(Si_2O_5)_3(OH)_2 \cdot 6H_2O$ A soft, lightweight, absorbent, white to light-gray or light-yellow clay mineral, found principally in Asia Minor; used for tobacco pipe bowls and ornamental carvings. Also known as meerschaum; sea-foam. { 'sē·pē·əˌlīt }

Sepsidae [INV ZOO] The spiny-legged flies, a family of myodarian cyclorrhaphous dipteran insects in the subsection Acalypteratae; development takes place in decaying organic matter. { 'sep·səˌdē }

sepsis [MED] **1.** Poisoning by products of putrefaction. **2.** The severe toxic, febrile state resulting from infection with pyogenic microorganisms, with or without associated septicemia. { 'sep·səs }

septal filament [INV ZOO] In anthozoans, the free edges of the septum containing gland cells and nematocysts. { 'sep·təl 'fil·ə·mənt }

septal ostium [INV ZOO] Any of the openings in septa of anthozoans. { 'sep·təl 'äs·tē·əm }

septarian [GEOL] Pertaining to the irregular polygonal pattern of internal cracks developed in septaria. { sep'tar·ē·ən }

septarian boulder *See* septarium. { sep'tar·ē·ən 'bōl·dər }

septarian nodule *See* septarium. { sep'tar·ē·ən 'näj·ül }

septarium [GEOL] A large (32–36 inches or 80–90 centimeters in diameter), spheroidal concretion, usually composed of argillaceous carbonate, characterized by internal cracking into irregular polygonal blocks that become cemented together by crystalline minerals. Also known as beetle stone; septarian boulder; septarian nodule; turtle stone. { sep'tar·ē·əm }

septate [BIOL] Having a septum. { 'sep·tāt }

septate coaxial cavity [ELECTROMAG] Coaxial cavity having a vane or septum, added between the inner and outer conductors, so that it acts as a cavity of a rectangular cross section bent transversely. { 'sep·tāt kō'ak·sē·əl 'kav·əd·ē }

septate waveguide [ELECTROMAG] Waveguide with one or more septa placed across it to control microwave power transmission. { 'sep·tāt 'wāv·gīd }

Septibranchia [INV ZOO] A small order of bivalve mollusks in which the anterior and posterior abductor muscles are about equal in size, the foot is long and slender, and the gills have been transformed into a muscular septum. { ˌsep·tə'braŋ·kē·ə }

septic [MED] Of or pertaining to sepsis. { 'sep·tik }

septic abortion [MED] An abortion complicated by acute infection of the endometrium. { 'sep·tik ə'bȯr·shən }

septic embolus [MED] An embolus formed by bacteria. { 'sep·tik 'em·bə·ləs }

septicemia [MED] A clinical syndrome in which infection is disseminated through the body in the bloodstream. Also known as blood poisoning. { ˌsep·tə'sē·mē·ə }

septicidal [BOT] A type of dehiscence exhibited by some fruit in which splitting of the pericarp occurs along the junction of component carpels. { 'sep·tə¦sīd·əl }

septic tank [CIV ENG] A settling tank in which settled sludge is in immediate contact with sewage flowing through the tank while solids are decomposed by anaerobic bacterial action. { 'sep·tik ˌtaŋk }

septillion [MATH] **1.** The number 10^{24}. **2.** In British and German usage, the number 10^{42}. { sep'til·yən }

septinary number [MATH] A number in which the quantity represented by each figure is based on a radix of 7. { 'sep·təˌner·ē 'nəm·bər }

septulum [ANAT] A small septum. { 'sep·tə·ləm }

septum [BIOL] A partition or dividing wall between two cavities. [ELECTROMAG] A metal plate placed across a waveguide and attached to the walls by highly conducting joints; the plate usually has one or more windows, or irises, designed to give inductive, capacitive, or resistive characteristics. { 'sep·təm }

septum pellucidum [ANAT] A thin translucent septum forming the internal boundary of the lateral ventricles of the brain and enclosing between its two laminas the so-called fifth ventricle. { 'sep·təm pə'lü·səd·əm }

septum primum [EMBRYO] The first incomplete interatrial septum of the embryo. { 'sep·təm 'prē·məm }

septum secundum [EMBRYO] The second incomplete interatrial septum of the embryo, containing the foramen ovale; it develops to the right of the septum primum and fuses with it to form the adult interatrial septum. { 'sep·təm si'kən·dəm }

Sequanian [GEOL] Upper Lower Jurassic (Upper Lusitanian) geologic time. Also known as Astartian. { sə'kwā·nē·ən }

sequela [MED] The abnormal aftereffects or complications of an illness, infection, or injury. { si'kwel·ə }

sequence [COMPUT SCI] To put a set of symbols into an arbitrarily defined order; that is, to select A if A is greater than or equal to B, or to select B if A is less than B. [ENG] An orderly progression of items of information or of operations in accordance with some rule. [GEOL] **1.** A sequence of geologic events, processes, or rocks, arranged in chronological order. **2.** A geographically discrete, major informal rock-stratigraphic unit of greater than group or supergroup rank. Also known as stratigraphic sequence. **3.** A body of rock deposited during a complete cycle of sea-level change. [MATH] A listing of mathematical entities $x_1, x_2 \ldots$ which is indexed by the positive integers; more precisely, a function whose domain is an infinite subset of the positive integers. Also known as infinite sequence. [METEOROL] *See* collective. { 'sē·kwəns }

sequence calling [COMPUT SCI] The instructions used for linking a closed subroutine with a main routine; that is, standard linkage and a list of the parameters. { 'sē·kwəns ˌkȯl·iŋ }

sequence check [COMPUT SCI] To verify that correct precedence relationships are obeyed, usually by checking for ascending sequence numbers. { 'sē·kwəns ˌchek }

sequence checking routine [COMPUT SCI] In computers, a checking routine which records specified data regarding the operations resulting from each instruction. { 'sē·kwəns ˌchek·iŋ rü̇ˌtēn }

sequence counter *See* instruction counter. { 'sē·kwəns ˌkau̇nt·ər }

sequence error [COMPUT SCI] An error that arises when the arrangement of items in a set, for example, a deck of punch cards, does not follow some specified order. { 'sē·kwəns ˌer·ər }

sequence monitor [COMPUT SCI] The automatic step-by-step check by a computer of the manual actions required for the starting and shutdown of a computer. { 'sē·kwəns ˌmän·əd·ər }

sequence number [COMPUT SCI] A number assigned to an item to indicate its relative position in a series of related items. { 'sē·kwəns ˌnəm·bər }

sequence of current [OCEANOGR] The order of occurrence of the tidal current strengths of a day, with special reference to whether the greater flood immediately precedes or follows the greater ebb. { 'sē·kwəns əv 'kə·rənt }

sequence of tide [OCEANOGR] The order in which the tides of a day occur, with special reference to whether the higher high water immediately precedes or follows the lower low water. { 'sē·kwəns əv 'tīd }

sequencer [COMPUT SCI] A machine which puts items of information into a particular order, for example, it will determine whether *A* is greater than, equal to, or less than *B*, and sort or order accordingly. Also known as sorter. [ENG] A mechanical or electronic device that may be set to initiate a series of events and to make the events follow in a given sequence. { 'sē·kwən·sər }

sequence register [COMPUT SCI] A counter which contains the address of the next instruction to be carried out. { 'sē·kwəns ˌrej·ə·stər }

sequence robot *See* preprogrammed robot. { 'sē·kwəns ˌrō·bät }

sequence stratigraphy [GEOL] A branch of stratigraphy that subdivides the sedimentary record along continental margins and in interior basins into a succession of depositional sequences as regional and interregional correlative units. { 'sē·kwəns strə'tig·rə·fē }

sequence-stressing loss [ENG] In posttensioning, the loss of elasticity in a stressed tendon that results from the shortening of the member as additional tendons are stressed. { 'sē·kwəns ˌstres·iŋ ˌlȯs }

sequence timer [MET] A device used in resistance welding to control the sequence and duration of all elements of the weld cycle, except weld time or heat time. { 'sē·kwəns ˌtīm·ər }

sequence weld timer [MET] A sequence timer which also controls weld time or heat time. { 'sē·kwəns 'weld ˌtīm·ər }

sequencing [IND ENG] Designating the order of performance of tasks to assure optimal utilization of available production facilities. { 'sē·kwəns·iŋ }

sequencing equipment [COMMUN] Special selecting device that permits messages received from several teletypewriter circuits to be subsequently selected and retransmitted over a reduced number of trunks or circuits. { 'sē·kwəns·iŋ iˌkwip·mənt }

sequential access [COMPUT SCI] A process that involves reading or writing data serially and, by extension, a data-recording medium that must be read serially, as a magnetic tape. { si'kwen·chəl 'akˌses }

sequential analysis [STAT] The continuous analysis of data, obtained via sampling, performed as the amount of sampling increases. { si'kwen·chəl ə'nal·ə·səs }

sequential batch operating system [COMPUT SCI] Software equipment that automatically begins running a new job on a computer system as soon as the current job is completed. { si'kwen·chəl 'bach 'äp·əˌrād·iŋ ˌsis·təm }

sequential circuit [ELEC] A switching circuit whose output depends not only upon the present state of its input, but also on what its input conditions have been in the past. { si'kwen·chəl 'sər·kət }

sequential collation of range [ENG] Spherical, long-baseline, phase-comparison trajectory-measuring system using three or more ground stations, time-sharing a single transponder, to provide nonambiguous range measurements to determine the instantaneous position of a vehicle in flight. { si'kwen·chəl kə'lā·shən əv 'rānj }

sequential color television [COMMUN] A color television system in which the primary color components of a picture are transmitted one after the other; the three basic types are the line-sequential, dot-sequential, and field-sequential color television systems. Also known as sequential system. { si'kwen·chəl ˈkäl·ər 'tel·əˌvizh·ən }

sequential control [COMPUT SCI] Manner of operating a computer by feeding orders into the computer in a given order during the solution of a problem. { si'kwen·chəl kən'trōl }

sequential cropping [AGR] A form of multiple cropping in which crops are grown in sequence on the same field, with the succeeding crop planted after the preceding crop is harvested. { si'kwen·chəl 'kräp·iŋ }

sequential landform [GEOL] One of an orderly succession of smaller landforms that are developed by the erosion, weathering, and mass wasting of larger initial landforms. { si'kwen·chəl 'landˌfȯrm }

sequential lobing [NAV] A direction-finding technique that utilizes the signal derived from partially overlapped lobes which occur in sequence. { si'kwen·chəl 'lōb·iŋ }

sequential logic element [ELECTR] A circuit element having at least one input channel, at least one output channel, and at least one internal state variable, so designed and constructed that the output signals depend on the past and present states of the inputs. { si'kwen·chəl ˈläj·ik ˌel·ə·mənt }

sequentially compact space [MATH] A topological space with the property that every sequence formed from its points has a subsequence that converges to a point in the space. { si'kwen·chə·lē ˌkämˌpakt 'spās }

sequential machine [COMPUT SCI] A mathematical model of a certain type of sequential circuit, which has inputs and outputs that can each take on any value from a finite set and are of interest only at certain instants of time, and in which the output depends on previous inputs as well as the concurrent input. { si'kwen·chəl mə'shēn }

sequential network [COMPUT SCI] An idealized model of a sequential circuit that reflects its logical but not its electronic properties. { si'kwen·chəl 'netˌwərk }

sequential operation [COMPUT SCI] The consecutive or serial execution of operations, without any simultaneity or overlap. { si'kwen·chəl ˌäp·ə'rā·shən }

sequential organization [COMPUT SCI] The write and read of records in a physical rather than a logical sequence. { si'kwen·chəl ˌȯr·gə·nə'zā·shən }

sequential processing [COMPUT SCI] Processing items in a collection of data according to some specified sequence of keys, in contrast to serial processing. { si'kwen·chəl 'präˌses·iŋ }

sequential sampling [IND ENG] A sampling plan in which an undetermined number of samples are tested one by one, accumulating the results until a decision can be made. { si'kwen·chəl 'sam·pliŋ }

sequential scanning *See* progressive scanning. { si'kwen·chəl 'skan·iŋ }

sequential scheduling system [COMPUT SCI] A first-come, first-served method of selecting jobs to be run. { si'kwen·chəl 'skej·ə·liŋ ˌsis·təm }

sequential search [COMPUT SCI] A procedure for searching a table that consists of starting at some table position (usually the beginning) and comparing the file-record key in hand with each table-record key, one at a time, until either a match is found or all sequential positions have been searched. { si'kwen·chəl 'sərch }

sequential selection [COMMUN] The selection of the elements of a message (such as letters) from a set of possible elements (such as the alphabet), one after another. { si'kwen·chəl si'lek·shən }

sequential system See sequential color television. { si'kwen·chəl 'sis·təm }

sequential trials [STAT] The outcome of each trial is known before the next trial is performed. { si'kwen·chəl 'trīlz }

sequestering agent [CHEM] A substance that removes a metal ion from a solution system by forming a complex ion that does not have the chemical reactions of the ion that is removed; can be a chelating or a complexing agent. { si'kwes·tə·riŋ ˌā·jənt }

sequestrum [MED] A piece of dead or detached bone within a cavity, abscess, or wound. { si'kwes·trəm }

Sequoia [BOT] A genus of conifers having overlapping, scalelike evergreen leaves and vertical grooves in the trunk; the giant sequoia (*Sequoia gigantea*) is the largest and oldest of all living things. { si'kwȯi·yə }

serac [HYD] A sharp ridge or pinnacle of ice among the crevasses of a glacier. { sə'rak }

serandite [MINERAL] $Na(Mn,Ca)_2Si_3O_8(OH)$ A rose-red mineral composed of a basic silicate of manganese, lime, potash, and soda occurring in monoclinic crystals. { 'ser·ən,dīt }

Serber potential [NUC PHYS] A potential between nucleons, equal to $\frac{1}{2}(1 + M)V(r)$, where $V(r)$ is a function of the distance between the nucleons, and M is an operator which exchanges the spatial coordinates of the particles but not their spins (corresponding to the Majorana force). { 'sər·bər pə,ten·chəl }

sere [ECOL] A temporary community which occurs during a successional sequence on a given site. { sir }

serein [METEOROL] The doubtful phenomenon of fine rain falling from an apparently clear sky, the clouds, if any, being too thin to be visible; frequently, fine rain is observed with a clear sky overhead, but clouds to windward clearly indicate the source of the drops. { sə'ran }

serge [TEXT] Twill weave with the diagonal prominent on both sides of the cloth. { sərj }

Sergestidae [INV ZOO] A family of decapod crustaceans including several species of prawns. { sər'jes·tə,dē }

serial [COMPUT SCI] Pertaining to the internal handling of data in sequential fashion. [IND ENG] An element or a group of elements within a series which is given a numerical or alphabetical designation for convenience in planning, scheduling, and control. { 'sir·ē·əl }

serial-access [COMPUT SCI] **1.** Pertaining to memory devices having structures such that data storage sites become accessible for read/write in time-sequential order; circulating memories and magnetic tapes are examples of serial-access memories. **2.** Pertaining to a particular process or program that accesses data items sequentially, without regard to the capability of the memory hardware. **3.** Pertaining to character-by-character transmission from an on-line real-time keyboard. { 'sir·ē·əl 'ak,ses }

serial addition [COMPUT SCI] An arithmetic operation in which two numbers are added one digit at a time. { 'sir·ē·əl ə'dish·ən }

serial bit [COMPUT SCI] Digital computer storage in which the individual bits that make up a computer word appear in time sequence. { 'sir·ē·əl ,bit }

serial communications [COMMUN] The transmission of digital data over a single channel. { 'sir·ē·əl kə,myü·nə'kā·shənz }

serial correlation [STAT] The correlation between values of events in a time series and those values ahead or behind by a fixed amount in time or space or between parts of two different time series. { 'sir·ē·əl ,kär·ə'lā·shən }

serial digital computer [COMPUT SCI] A digital computer in which the digits are handled serially, although the bits that make up a digit may be handled either serially or in parallel. { 'sir·ē·əl 'dij·əd·əl kəm'pyüd·ər }

serial dot character printer [COMPUT SCI] A computer printer in which the dot matrix technique is used to print characters, one at a time, with a movable print head that is driven back and forth across the page. { 'sir·ē·əl ¦dät 'kar·ik·tər ,print·ər }

serial file [COMPUT SCI] The simplest type of file organization, in which no subsets are defined, no directories are provided, no particular file order is specified, and a search is performed by sequential comparison of the query with identifiers of all stored items. { 'sir·ē·əl 'fīl }

serial homology [ZOO] The similarity between the members of a single series of structures, such as vertebrae, in an organism. { 'sir·ē·əl hə'mäl·ə·jē }

serial input/output [COMPUT SCI] Data that are transmitted into and out of a computer over a single conductor, one bit at a time. { 'sir·ē·əl 'in,pût 'aût,pût }

serial interface [COMPUT SCI] A link between a microcomputer and a peripheral device in which data is transmitted over a single conductor, one bit at a time. Also known as serial port. { 'sir·ē·əl 'in·tər,fās }

serialize [COMPUT SCI] To convert a signal suitable for parallel transmission into a signal suitable for serial transmission, consisting of a sequence of bits. { 'sir·ē·ə,līz }

serial learning [PSYCH] The type of association in verbal learning involved in learning the alphabet; studied in the laboratory by giving the subject serial lists to learn, where each list would consist of a number of unrelated items. { 'sir·ē·əl 'lərn·iŋ }

serially ordered set See linearly ordered set. { ¦sir·ē·ə·lē ¦ȯrd·ərd 'set }

serially reusable [COMPUT SCI] An attribute possessed by a program that can be used for several tasks in sequence without having to be reloaded into main memory for each additional use. { 'sir·ē·ə·lē rē'yü·zə·bəl }

serial memory [COMPUT SCI] A computer memory in which data are available only in the same sequence as originally stored. { 'sir·ē·əl 'mem·rē }

serial observation [OCEANOGR] The procurement of water samples and temperature readings at a number of levels between the surface and the bottom of an ocean. { 'sir·ē·əl ,äb·zər'vā·shən }

serial operation [COMPUT SCI] The flow of information through a computer in time sequence, using only one digit, word, line, or channel at a time. { 'sir·ē·əl ,äp·ə'rā·shən }

serial order See linear order. { 'sir·ē·əl ,ȯrd·ər }

serial-parallel [COMPUT SCI] **1.** A combination of serial and parallel; for example, serial by character, parallel by bits comprising the character. **2.** Descriptive of a device which converts a serial input into a parallel output. { 'sir·ē·əl 'par·ə,lel }

serial-parallel conversion [COMPUT SCI] The transformation of a serial data representation as found on a disk or drum into the parallel data representation as exists in core. { 'sir·ē·əl ¦par·ə,lel kən'vər·zhən }

serial port See serial interface. { 'sir·ē·əl ,pȯrt }

serial printer [GRAPHICS] **1.** A typewriter, or similar device, in which the paper or printing device moves back and forth, step by step to successive positions, to print one character at a time. Also known as character printer. **2.** A printer that is designed to be connected to a serial port of a computer. { 'sir·ē·əl 'print·ər }

serial processing [COMPUT SCI] Processing items in a collection of data in the order that they appear in a storage device, in contrast to sequential processing. [PSYCH] The processing of several pieces of information one at a time, in succession. { 'sir·ē·əl 'prä,ses·iŋ }

serial processor [COMPUT SCI] A computer in which data are handled sequentially by separate units of the system. { 'sir·ē·əl 'prä,ses·ər }

serial programming [COMPUT SCI] In computers, programming in which only one operation is executed at one time. { 'sir·ē·əl 'prō,gram·iŋ }

serial sampling [STAT] A method of gathering samples by a set pattern, such as a grid, to ensure randomness. { 'sir·ē·əl 'sam·pliŋ }

serial station [OCEANOGR] An oceanographic station consisting of one or more Nansen casts. { 'sir·ē·əl ,stā·shən }

serial storage [COMPUT SCI] Computer storage in which time is one of the coordinates used to locate any given bit, character, or word; access time, therefore, includes a variable waiting time, ranging from zero to many word times. { 'sir·ē·əl 'stȯr·ij }

serial transfer [COMPUT SCI] Transfer of the characters of an element of information in sequence over a single path in a digital computer. { 'sir·ē·əl 'tranz·fər }

serial transmission [COMMUN] Transmission of groups of elements of a signal in time intervals that follow each other without overlapping. { 'sir·ē·əl tranz'mish·ən }

seriate [GEOL] Having crystals that vary gradually in size. { 'sir·ē,āt }

sericeous [BOT] Of, pertaining to, or consisting of silk. { sə'rish·əs }

sericite [MINERAL] A white, fine-grained potassium mica, usually muscovite in composition, having a silky luster and found as small flakes in various metamorphic rocks. { 'ser·ə,sīt }

sericitic sandstone [PETR] A sandstone in which sericite (derived by decomposition of feldspar) intermingles with finely divided quartz and fills the voids between quartz grains. { ¦ser·ə¦sīd·ik 'san,stōn }

sericitization [GEOL] A hydrothermal or metamorphic process involving the introduction of or replacement by sericite. { ,ser·ə,sīd·ə'zā·shən }

sericulture [AGR] The raising of silkworms to produce raw silk. { 'ser·ə,kəl·chər }

series [ANALY CHEM] A group of results of repeated analyses completed by using a single analytical method on samples of a homogeneous substance. [ELEC] An arrangement of circuit components end to end to form a single path for current. [GEOL] **1.** A number of rocks, minerals, or fossils that can be arranged in a natural sequence due to certain characteristics, such as succession, composition, or occurrence. **2.** A time-stratigraphic unit, below system and above stage, composed of rocks formed during an epoch of geologic time. [MATH] An expression of the form $x_1 + x_2 + x_3 + \cdots$, where x_i are real or complex numbers. [SPECT] A collection of spectral lines of an atom or ion for a set of transitions, with the same selection rules, to a single final state; often the frequencies have the general formula $[R/(a + c_1)^2] - [R/(n + c_2)^2]$, where R is the Rydberg constant for the atom, a and c_1 and c_2 are constants, and n takes on the values of the integers greater than a for the various lines in the series. { 'sir·ēz }

series circuit [ELEC] A circuit in which all parts are connected end to end to provide a single path for current. { 'sir·ēz ,sər·kət }

series compensation [CONT SYS] *See* cascade compensation. [ELEC] The insertion of variable, controlled, high-voltage series capacitors into transmission lines in order to modify the impedance structure of a transmission network so as to adjust the power-flow distribution on individual lines and thus increase the power flow across such compensated lines. { 'sir·ēz ,käm·pən'sā·shən }

series connection [ELEC] A connection that forms a series circuit. { 'sir·ēz kə,nek·shən }

series decay *See* radioactive series. { 'sir·ēz di'kā }

series disintegration [NUC PHYS] The successive radioactive transformations in a radioactive series. Also known as chain decay; chain disintegration. { 'sir·ēz di,sin·tə'grā·shən }

series excitation [ELEC] The obtaining of field excitation in a motor or generator by allowing the armature current to flow through the field winding. { 'sir·ēz ,ek·sə'tā·shən }

series-fed vertical antenna [ELECTROMAG] Vertical antenna which is insulated from the ground and energized at the base. { 'sir·ēz ¦fed 'vərd·i·kəl an'ten·ə }

series feed [ELECTR] Application of the direct-current voltage to the plate or grid of a vacuum tube through the same impedance in which the alternating-current flows. { 'sir·ēz 'fēd }

series firing [ENG] The firing of detonators in a round of shots by passing the total supply current through each of the detonators. { 'sir·ēz 'fīr·iŋ }

series generator [ELEC] A generator whose armature winding and field winding are connected in series. Also known as series-wound generator. { 'sir·ēz 'jen·ə,rād·ər }

series loading [ELECTR] Loading in which reactances are inserted in series with the conductors of a transmission circuit. { 'sir·ēz 'lōd·iŋ }

series modulation [ELECTR] Modulation in which the plate circuits of a modulating tube and a modulated amplifier tube are in series with the same plate voltage supply. { 'sir·ēz ,mäj·ə'lā·shən }

series motor [ELEC] A commutator-type motor having armature and field windings in series; characteristics are high

starting torque, variation of speed with load, and dangerously high speed on no-load. Also known as series-wound motor. { 'sir·ēz ,mōd·ər }

series multiple [ELEC] Type of switchboard jack arrangement in which a single line circuit appears before two or more operators, all appearances being connected in series. { 'sir·ēz ¦məl·tə·pəl }

series-parallel circuit [ELEC] A circuit in which some of the components or elements are connected in parallel, and one or more of these parallel combinations are in series with other components of the circuit. { 'sir·ēz ¦par·ə,lel 'sər·kət }

series-parallel control [ELEC] A method of controlling the speed of electric motors in which the motors, or groups of motors, are connected in series at some times and in parallel at other times. { 'sir·ēz ¦par·ə,lel kən'trōl }

series-parallel firing [ENG] The firing of detonators in a round of shots by dividing the total supply current into branches, each containing a certain number of detonators wired in series. { 'sir·ēz ¦par·ə,lel ¦fīr·iŋ }

series-parallel switch [ELEC] A switch used to change the connections of lamps or other devices from series to parallel, or vice versa. { 'sir·ēz ¦par·ə,lel ¦swich }

series peaking [ELECTR] Use of a peaking coil and resistor in series as the load for a video amplifier to produce peaking at some desired frequency in the passband, such as to compensate for previous loss of gain at the high-frequency end of the passband. { 'sir·ēz 'pēk·iŋ }

series production [IND ENG] The manufacture of a product or service by a group of operations sequenced so that all materials will be routed successively through each production stage. Also known as batch production. { 'sir·ēz prə'dək·shən }

series radio tap [COMMUN] A telephone tapping procedure in which a miniature radio transmitter is inserted in series with one wire of the target pair so that the transmitter derives its power from the telephone central battery. { 'sir·ēz 'rād·ē·ō ,tap }

series reactor [ELEC] A reactor used in alternating-current power systems for protection against excessively large currents under short-circuit or transient conditions; it consists of coils of heavy insulated cable either cast in concrete columns or supported in rigid frames and mounted on insulators. Also known as current-limiting reactor. { 'sir·ēz rē,ak·tər }

series regulator [ELEC] A regulator that controls output voltage or current by automatically varying a resistance in series with the voltage source. { 'sir·ēz 'reg·yə,lād·ər }

series reliability [SYS ENG] Property of a system composed of elements in such a way that failure of any one element causes a failure of the system { 'sir·ēz ri,lī·ə'bil·əd·ē }

series repeater [ELEC] A type of negative impedance telephone repeater which is stable when terminated in an open circuit and oscillates when it is connected to a low impedance, in contrast to a shunt repeater. { 'sir·ēz ri'pēd·ər }

series resonance [ELEC] Resonance in a series resonant circuit, wherein the inductive and capacitive reactances are equal at the frequency of the applied voltage; the reactances then cancel each other, reducing the impedance of the circuit to a minimum, purely resistive value. { 'sir·ēz 'rez·ən·əns }

series resonant circuit [ELEC] A resonant circuit in which the capacitor and coil are in series with the applied alternating-current voltage. { 'sir·ēz ¦rez·ən·ənt ,sər·kət }

series shots [ENG] The connecting and firing of a number of loaded holes one after the other. { 'sir·ēz ,shäts }

series-shunt network *See* ladder network. { 'sir·ēz ¦shənt 'net,wərk }

series T junction *See* E-plane T junction. { 'sir·ēz ¦tē ¦jəŋk·shən }

series transistor regulator [ELECTR] A voltage regulator whose circuit has a transistor in series with the output voltage, a Zener diode, and a resistor chosen so that the Zener diode is approximately in the middle of its operating range. { 'sir·ēz tran'zis·tər 'reg·yə,lād·ər }

series-tuned circuit [ELEC] A simple resonant circuit consisting of an inductance and a capacitance connected in series. { 'sir·ēz ¦tünd ,sər·kət }

series ventilation [MIN ENG] A system of ventilating a number of faces consecutively by the same air current. { 'sir·ēz ,vent·əl'ā·shən }

series welding [MET] Making two or more resistance welds

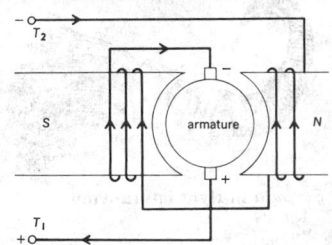

SERIES GENERATOR

Armature and field winding in a series generator. N = north; S = south; T_1, T_2 = terminals connected to external load.

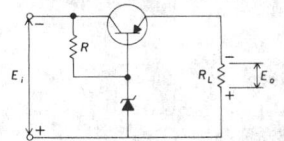

SERIES TRANSISTOR REGULATOR

Circuit diagram of a series transistor regulator. E_o = output voltage; E_i = input voltage; R = resistor; R_L = load resistance.

simultaneously by using a single welding transformer with three or more electrodes forming a series circuit. { 'sir·ēz ,weld·iŋ }

series winding [ELEC] A winding in which the armature circuit and the field circuit are connected in series with the external circuit. { 'sir·ēz ,wīnd·iŋ }

series-wound generator See series generator. { 'sir·ēz ¦waùnd ¦jen·ə,rād·ər }

series-wound motor See series motor. { 'sir·ēz ¦waùnd 'mōd·ər }

serif [GRAPHICS] Any of the short crosslines placed at the ends of the strokes of many letters appearing in some typefaces. { 'ser·əf }

serigraph [GRAPHICS] The silk-screen process when it is used as a fine-art reproduction method. { 'ser·i,graf }

serigraphy See screen printing. { sə'rig·rə·fē }

serine [BIOCHEM] $C_3H_7O_3N$ An amino acid obtained by hydrolysis of many proteins; a biosynthetic precursor of several metabolites, including cysteine, glycine, and choline. { 'se,rēn }

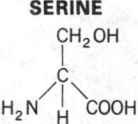

SERINE

CH₂OH

Structural formula of serine.

serine protease [CELL MOL] A family of endopeptidases whose proteolytic activity involves the hydroxy group of the serine residue; they play an important role in digestion, blood clotting, and the complement system. { 'se,rēn 'prōd·ə,ās }

serioscopy [NUCLEO] A radiographic technique enabling three-dimensional exploration by moving two of the three components of the system (tube, subject, film) in order to register the radiographic image of one plane in the object while images outside this slice have a continuous relative displacement and are blurred. { ,sē·rē'äs·kə·pē }

seritinous [ECOL] Of, pertaining to, or occurring during the latter, drier half of the summer. { ¦ser·ə'tī·nəs }

seroche See mountain sickness. { sə'rōsh }

serodiagnosis [MED] Diagnosis based upon the reaction of blood serum of a patient. { ¦si·rō,dī·əg'nō·səs }

seroepidemiology [MED] The study of the distribution of serum antibodies. { ¦si·rō,ep·ə,dē·mē'äl·ə·jē }

serofibrinous [PHYSIO] Composed of serum and fibrin. { ¦si·rō'fī·brə·nəs }

Serolidae [INV ZOO] A family of isopod crustaceans which contains greatly flattened forms that live partially buried on sandy bottoms. { sə'räl·ə,dē }

serology [BIOL] The branch of science dealing with the properties and reactions of blood sera. { sə'räl·ə·jē }

seronegative [PATH] **1.** Having a negative serologic test for some condition. **2.** Specifically, having a negative serologic test for syphilis. { ¦si·rō'neg·əd·iv }

seropositive [PATH] **1.** Having a positive serologic test for some condition. **2.** Specifically, having a positive serologic test for syphilis. { ¦si·rō'päz·əd·iv }

seropurulent [MED] Composed of serum and pus, as a seropurulent exudate. { ¦si·rō'pyùr·ə·lənt }

seroresistance [PATH] Persistent positive serologic reaction for syphilis despite prolonged intensive treatment; the patient is said to be Wassermann-fast. { ¦si·rō·ri'zis·təns }

serosa [ANAT] The serous membrane lining the pleural, peritoneal, and pericardial cavities. [EMBRYO] The chorion of reptile and bird embryos. { sə'rō·sə }

serotherapy [MED] The treatment of disease by means of human or animal serum containing antibodies. Also known as immunotherapy. { ¦si·rō'ther·ə·pē }

serotinous [BOT] Plants which flower or develop late in a season. { sə'rät·ən·əs }

serotonin [BIOCHEM] $C_{10}H_{12}ON_2$ A compound derived from tryptophan which functions as a local vasoconstrictor, plays a role in neurotransmission, and has pharmacologic properties. Also known as 5-hydroxytryptamine. { ,sir·ə'tō·nən }

serotype [MICROBIO] A serological type of intimately related microorganisms, distinguished on the basis of antigenic composition. { 'sir·ə,tīp }

serous cystadenoma [MED] A benign cystic tumor of the ovary, made up of cylindrical cells resembling those of the uterine tube; psammoma bodies often appear in the wall of the cyst. Also known as endosalpingioma; papillary adenoma of ovary; papillary cystadenoma of ovary; papillocystoma; psammomatus papilloma; serous cystoma. { 'sir·əs ,sist,ad·ən'ō·mə }

serous cystoma See serous cystadenoma. { 'sir·əs si'stō·mə }

SERPULIDAE

Serpula in right lateral view.

serous gland [PHYSIO] A structure that secretes a watery, albuminous fluid. { 'sir·əs ,gland }

serous membrane [HISTOL] A delicate membrane covered with flat, mesothelial cells lining closed cavities of the body. { 'sir·əs 'mem,brān }

serous plethora [MED] An increase in the watery part of the blood. { 'sir·əs 'pleth·ə·rə }

Serpens [ASTRON] A constellation, right ascension 17 hours, declination 0°. Also known as Serpent. { 'sər,penz }

Serpent See Serpens. { 'sər·pənt }

Serpent Bearer See Ophiucus. { 'sər·pənt 'ber·ər }

Serpentes [VERT ZOO] The snakes, a suborder of the Squamata characterized by the lack of limbs and pectoral girdle and external ear openings, immovable eyelids, and a braincase that is completely bony anteriorly. { sər'pen,tēz }

serpentine [MINERAL] $(Mg,Fe)_3Si_2O_5(OH)_4$ A group of green, greenish-yellow, or greenish-gray ferromagnesian hydrous silicate rock-forming minerals having greasy or silky luster and a slightly soapy feel; translucent varieties are used for gemstones as substitutes for jade. { 'sər·pən,tēn }

serpentine cooler See cascade cooler. { 'sər·pən,tēn 'kül·ər }

serpentine crepe [TEXT] A crepe fabric woven with a lengthwise crinkled effect. { 'sər·pən,tēn 'krāp }

serpentine curve [MATH] The curve given by the equation $x^2 y + b^2 y - a^2 x = 0$, passing through and having symmetry about the origin while being asymptotic to the x axis in both directions. { 'sər·pən,tēn 'kərv }

serpentine jade [MINERAL] A variety of the mineral serpentine resembling jade in appearance and used as an ornamental stone. { 'sər·pən,tēn 'jād }

serpentine locomotion [VERT ZOO] The wavelike or undulating movements characteristic of snakes. { 'sər·pən,tēn ,lō·kə'mō·shən }

serpentine rock See serpentinite. { 'sər·pən,tēn 'räk }

serpentine spit [GEOGR] A spit that is extended in more than one direction due to variable or periodically shifting currents. { 'sər·pən,tēn 'spit }

serpentinite [PETR] A rock composed almost entirely of serpentine minerals. Also known as serpentine rock. { 'sər·pən,tē,nīt }

serpentinization [GEOL] A hydrothermal process by which magnesium-rich silicate minerals are converted into or replaced by serpentine minerals. { ,sər·pən,tē·nə'zā·shən }

serpent kame See esker. { 'sər·pənt 'kām }

serpierite [MINERAL] $(Cu,Zn,Ca)_5(SO_4)_2(OH)_6 \cdot 3H_2O$ A bluish-green mineral composed of hydrous basic sulfate of copper, zinc, and calcium; occurs in tabular crystals and tufts. { 'sər·pē·ə,rīt }

Serpulidae [INV ZOO] A family of polychaete annelids belonging to the Sedentaria including many of the feather-duster worms which construct calcareous tubes in the earth, sometimes in such abundance as to clog drains and waterways. { sər'pyù·lə,dē }

Serranidae [VERT ZOO] A family of perciform fishes in the suborder Percoidei including the sea basses and groupers. { sə'ran·ə,dē }

serrate [BIOL] Possessing a notched or toothed edge. [GEOL] Pertaining to topographic features having a notched or toothed edge, or a saw-edge profile. { 'se,rāt }

serrated pulse [ELECTR] Vertical and horizontal synchronizing pulse divided into a number of small pulses, each of which acts for the duration of half a line in a television system. { 'se,rād·ad 'pəls }

serrate ridge See arête. { 'se,rāt 'rij }

Serratia marcescens [MICROBIO] A human pathogen that is intrinsically resistant to many antimicrobials (for example, cephalosporins, polymyxins, and nitrofurans) and occurs predominantly in hospitalized patients. { sə,rā·shē·ə mär'ses·əns }

Serratieae [MICROBIO] Formerly a tribe of the Enterobacteriaceae containing the genus *Serratia,* with soil and water forms characterized by the production of a bright-orange to deep-red pigment, prodigiosin. { sə'räsh·ē,ē }

Serret-Frenet formulas See Frenet-Serret formulas. { sə'rā fra'nā ,fór·myə·ləz }

Serridentinae [PALEON] An extinct subfamily of elephantoids in the family Gomphotheriidae. { ,ser·ə'dent·ən,ē }

Serritermitidae [INV ZOO] A family of the Isoptera which contains the single monotypic genus *Serritermes*. { ˌser·ə·tər'mid·ə‚dē }

serrodyne [ELECTR] Phase modulator using transit time modulation of a traveling-wave tube or klystron. { 'ser·ə‚dīn }

Serropalpidae [INV ZOO] An equivalent name for Melandryidae. { ˌser·ə'pal·pə‚dē }

serrulate [BIOL] Finely serrate. { 'ser·ə·lət }

Sertoli cell [HISTOL] One of the sustentacular cells of the seminiferous tubules. { sər'tō·lē ‚sel }

serum [PHYSIO] The liquid portion that remains when blood clots spontaneously and the formed and clotting elements are removed by centrifugation; it differs from plasma by the absence of fibrinogen. { 'sir·əm }

serum accident [IMMUNOL] A serious allergic reaction which immediately follows the introduction of a foreign serum into a hypersensitive individual; dyspnea and flushing occur, soon followed by shock and occasionally by fatal termination. { 'sir·əm ‚ak·sə·dənt }

serum albumin [BIOCHEM] The principal protein fraction of blood serum and serous fluids. { 'sir·əm al byü·mən }

serum globulin [BIOCHEM] The globulin fraction of blood serum. { 'sir·əm 'glä·byə·lən }

serum hepatitis [MED] A form of viral hepatitis transmitted by parenteral injection of human blood or blood products contaminated with the type B virus. { 'sir·əm ‚hep·ə'tīd·əs }

serum shock [MED] An anaphylactic reaction following the injection of foreign serum into a sensitized individual. { 'sir·əm ‚shäk }

serum sickness [MED] A syndrome manifested in 8–12 days after the administration of serum by an urticarial rash, edema, enlargement of lymph nodes, arthralgia, and fever. { 'sir·əm ‚sik·nəs }

server [COMPUT SCI] A computer or software package that sends requested information to a client or clients in a network. { 'sər·vər }

service [ENG] To perform services of maintenance, supply, repair, installation, distribution, and so on, for or upon an instrument, installation, vehicle, or territory. { 'sər·vəs }

serviceability [IND ENG] The reliability of equipment according to some objective criterion such as serviceability ratio, utilization ratio, or operating ratio. { ‚sər·və·sə'bil·əd·ē }

serviceability ratio [IND ENG] The ratio of up time to the sum of up time and down time. { ‚sər·və·sə'bil·əd·ē ‚rā·shō }

service agreement [ENG] A contract which agrees to provide mechanical maintenance of a machine for a fixed period of time at a stated charge. { 'sər·vəs ə‚grē·mənt }

service area [COMMUN] The area that is effectively served by a given radio or television transmitter, navigation aid, or other type of transmitter. Also known as coverage. { 'sər·vəs ‚er·ē·ə }

service band [COMMUN] Band of frequencies allocated to a given class of radio service. { 'sər·vəs ‚band }

service bit [COMMUN] A bit used in data transmission to monitor the transmission rather than to convey information, such as a request that part of a message be repeated. { 'sər·vəs ‚bit }

service brake [MECH ENG] The brake used for ordinary driving in an automotive vehicle; usually foot-operated. { 'sər·vəs ‚brāk }

service bureau [COMPUT SCI] An organization that offers time sharing and software services to its users who communicate with a computer in the bureau from terminals on their premises. { 'sər·vəs ‚byur·ō }

service ceiling [AERO ENG] The height at which, under standard atmospheric conditions, an aircraft is unable to climb faster than a specified rate (100 feet or 30 meters per minute in the United States, Great Britain, and Canada). { 'sər·vəs ‚sēl·iŋ }

service compartment [MIN ENG] The section of a mine shaft that houses water pipes, compressed-air pipeline, cables and telephone wires, and signaling and similar arrangements. { 'sər·vəs kəm‚pärt·mənt }

service dead load [ENG] The calculated dead load that will be supported by a member. { 'sər·vəs 'ded ‚lōd }

service engineering [ENG] The function of determining the integrity of material and services in order to measure and maintain operational reliability, approve design changes, and assure their conformance with established specifications and standards. { 'sər·vəs ‚en·jə‚nir·iŋ }

service factor [ENG] For a chemical or a petroleum processing plant or its equipment, the measure of the continuity of an operation, computed by dividing the time on-stream (actual running time) by the total elapsed time. { 'sər·vəs ‚fak·tər }

service life [ENG] The length of time during which a machine, tool, or other apparatus or device can be operated or used economically or before breakdown. { 'sər·vəs ‚līf }

service oscillator *See* radio-frequency signal generator. { 'sər·vəs 'äs·ə‚lād·ər }

service pipe [CIV ENG] A pipe linking a building to a main pipe. { 'sər·vəs ‚pīp }

service program [COMPUT SCI] A computer program that is used in a computer system to support the functioning of the system, such as a librarian or a utility program. { 'sər·vəs ‚prō‚gram }

service provider [COMPUT SCI] An organization that provides access to a wide-area network, such as the Internet. { 'sər·vəs prə‚vīd·ər }

service rating [MATER] A classification for an engine-lubricating oil that indicates the type of service for which the oil is most appropriate. { 'sər·vəs ‚rād·iŋ }

service road [CIV ENG] A small road parallel to the main road for convenient access to shops and houses. { 'sər·vəs ‚rōd }

service routine [COMPUT SCI] A section of a computer code that is used in so many different jobs that it cannot belong to any one job. { 'sər·vəs rü‚tēn }

service shaft [MIN ENG] A shaft used only for hoisting men and materials to and from underground. { 'sər·vəs ‚shaft }

service test [ORD] A test of an item, system of materiel, or technique conducted under simulated or actual operational conditions to determine whether the specified military requirements or characteristics are satisfied. { 'sər·vəs ‚test }

service time *See* machine attention time. { 'sər·vəs ‚tīm }

service valve [ENG] In a pipework system, a valve that isolates a piece of equipment from the rest of the system. { 'sər·vəs ‚valv }

service velocity [ORD] The muzzle velocity established as the velocity to be attained by a projectile of standard weight and under standard conditions of temperature, when fired from a new gun of the designated type; the range tables are computed on the basis of this velocity. { 'sər·vəs və‚läs·əd·ē }

service volume [NAV] That volume in airspace surrounding a VOR, Tacan, or Vortac facility within which a signal of usable strength exists and is free of interchannel and cochannel interference; the advertised service volume is defined as a simple cylinder of airspace, usually smaller than theoretical maximum, and is used in planning areas of air operations. { 'sər·vəs ‚väl·yəm }

service wires [ELEC] The conductors that bring the electric power into a building. { 'sər·vəs ‚wīrz }

servicing [ENG] Replacement of consumable material or items needed to keep equipment in operating condition; does not include preventive or corrective maintenance. { 'sər·vəs·iŋ }

servicing time [COMPUT SCI] Machine down-time necessary for routine testing, for machine servicing due to breakdown, or for preventive servicing measures; includes all test time (good or bad) following breakdown and subsequent repair or preventive servicing. { 'sər·vəs·iŋ ‚tīm }

serving [ELEC] A covering, such as thread or tape, that protects a winding from mechanical damage. Also known as coil serving. { 'sər·viŋ }

servo *See* servomotor. { 'sər·vō }

servo amplifier [ELECTR] An amplifier used in a servomechanism. { 'sər·vō 'am·plə‚fī·ər }

servoarm attachment [MECH ENG] A device that enhances the maximum distance over which the manipulator of a simple robot can travel. { 'sər·vō‚ärm ə‚tach·mənt }

servo brake [MECH ENG] **1.** A brake in which the motion of the vehicle is used to increase the pressure on one of the shoes. **2.** A brake in which the force applied by the operator is augmented by a power-driven mechanism. { 'sər·vō 'brāk }

servolink [CONT SYS] A power amplifier, usually mechanical, by which signals at a low power level are made to operate control surfaces requiring relatively large power inputs, for example, a relay and motor-driven actuator. { 'sər·vō‚liŋk }

servo loop *See* single-loop servomechanism. { 'sər·vō ‚lüp }

SERVOMULTIPLIER

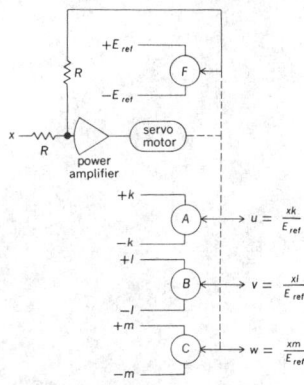

Circuit diagram of servomultiplier. Feedback control loop is used to make the shaft angle of servomotor proportional to drive signal *x*. Motor shaft is linked to potentiometers *A*, *B*, and *C* so that their output voltages *u*, *v*, and *w* are also proportional to *x*. In feedback loop servomotor moves slider shaft of an attenuator until signal from the feedback potentiometer *F* is equal (but opposite in sign) to *x* so that there is a zero error signal at the input of the power amplifier.

SESSOBLAST

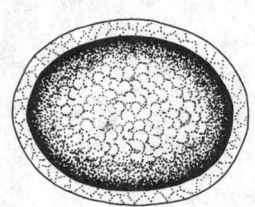

Sessoblast of *Stolella indica*.

servomechanism [CONT SYS] An automatic feedback control system for mechanical motion; it applies only to those systems in which the controlled quantity or output is mechanical position or one of its derivatives (velocity, acceleration, and so on). Also known as servo system. { ¦sər·vō'mek·ə¦niz·əm }

servomotor [CONT SYS] The electric, hydraulic, or other type of motor that serves as the final control element in a servomechanism; it receives power from the amplifier element and drives the load with a linear or rotary motion. Also known as servo. { 'sər·vō¦mōd·ər }

servomultiplier [ELECTR] An electromechanical multiplier in which one variable is used to position one or more ganged potentiometers across which the other variable voltages are applied. { ¦sər·vō'məl·tə¦plī·ər }

servonoise [ENG] Hunting action of the tracking servomechanism of a radar, which results from backlash and compliance in the gears, shafts, and structures of the mount. { 'sər·vō¦nòiz }

servo system *See* servomechanism. { 'sər·vō ¦sis·təm }

servovalve [MECH ENG] A transducer in which a low-energy signal controls a high-energy fluid flow so that the flow is proportional to the signal. { 'sər·vō¦valv }

sesame oil [MATER] A combustible, yellow, optically active, semidrying fatty oil obtained from sesame seeds; soluble in ether, benzene, and carbon disulfide, slightly soluble in alcohol; melts at 20–25°C; used in edible food products, such as shortenings, salad oils, and margarine. Also known as benne oil; gingelly oil; teel oil. { 'ses·ə·mē ¦òil }

sesamoid bone [MED] A small bone developed in a tendon subjected to much pressure. { 'ses·ə¦mòid ¦bōn }

sesquillinear form [MATH] A mapping $f(x,y)$ from $E \times F$ into R, where R is a commutative ring with an automorphism with period 2 and $E \times F$ is the cartesian product of two modules E and F over R, such that for each x in E the function which takes y into $f(x,y)$ is antilinear, and for each y in F the function which takes x into $f(x,\bar{y})$ is linear. { ¦ses·kwə¦lin·ē·ər 'fòrm }

sesquioxide [CHEM] A compound composed of a metal and oxygen in the ratio 2:3; for example, Al_2O_3. { ¦ses·kwē'äk¦sīd }

sesquisideband transmission [COMMUN] Transmission of a carrier modulated by one full sideband and half of the other sideband. { ¦ses·kwē'sīd¦band tranz'mish·ən }

sesquiterpene [ORG CHEM] Any terpene with the formula $C_{15}H_{24}$; that is, $1\frac{1}{2}$ times the terpene formula. { ¦ses·kwē'tər¦pēn }

sessile [BOT] Attached directly to a branch or stem without an intervening stalk. [ZOO] Permanently attached to the substrate. { 'ses·əl }

sessile bubble method [FL MECH] A method of measuring the surface tension of a liquid that involves measuring the dimensions of a bubble in the liquid resting under a plane or concave-downward surface. { 'ses·əl ¦bəb·əl ¦meth·əd }

sessile dislocation [MET] A dislocation in a metal lattice that is relatively immobile, offering an obstacle to the movement of other dislocations. { 'ses·əl ¦dis·lō'kā·shən }

sessile drop method [FL MECH] A method of measuring surface tension in which the depth and mass of a drop resting on a surface that it does not wet are measured; from this, the shape of the drop and, in turn, the surface tension are determined. { 'ses·əl ¦dräp ¦meth·əd }

Sessilina [INV ZOO] A suborder of ciliates in the order Peritrichida. { ¦ses·ə'lī·nə }

sessoblast [INV ZOO] A statoblast that attaches to zooecial tubes or to the substratum. { 'ses·ə¦blast }

seston [OCEANOGR] Minute living organisms and particles of nonliving matter which float in water and contribute to turbidity. { 'se¦stän }

set [ASTRON] Of a celestial body, to cross the visible western horizon while descending. [CHEM] The hardening or solidifying of a plastic or liquid substance. [COMPUT SCI] A collection of record types. [ELECTR] The placement of a storage device in a prescribed state, for example, a binary storage cell in the high or 1 state. [ENG] **1.** A combination of units, assemblies, and parts connected or otherwise used together to perform an operational function, such as a radar set. **2.** In plastics processing, the conversion of a liquid resin or adhesive into a solid state by curing or evaporation of solvent or suspending medium, or by gelling. **3.** Saw teeth bent out of the plane of the saw body, resulting in a wide cut in the

workpiece. [GEOL] A group of essentially conformable strata or cross-strata, separated from other sedimentary units by surfaces of erosion, nondeposition, or abrupt change in character. [GRAPHICS] The fixing or drying of a printing ink on a printed sheet, so that, though not completely dry, the sheet can be handled without smudging. [MATER] **1.** The hardening or firmness displayed by some materials when left undisturbed. **2.** Permanent deformation of a material, such as metal or plastic, when stressed beyond the elastic limit. [MATH] A collection of objects which has the property that, given any thing, it can be determined whether or not the thing is in the collection. [MECH] *See* permanent set. [MIN ENG] *See* frame set. [NAV] The establishment of a course. [OCEANOGR] The direction toward which an oceanic current flows. { set }

seta [BIOL] **1.** A slender, usually rigid bristle or hair. Also known as chaeta. **2.** In mosses and liverworts, the stalk of the sporophyte supporting the capsule. { 'sēd·ə }

set-associative cache [COMPUT SCI] A cache memory in which incoming data are distributed in sequence to each of two to eight areas or sets, and is generally read out in the same manner, allowing each set to prepare for the next input/output operation. { ¦set ə¦sōs·ē¦ād·iv ¦kash }

set analyzer *See* analyzer. { 'set ¦an·ə¦līz·ər }

setback [BUILD] **1.** A withdrawal of the face of a building to a line toward the rear of the building line or the rear of the wall below in order to reduce obstruction of sunlight reaching the street or the lower stories of adjacent buildings. **2.** *See* offset. [CIV ENG] The distance that a section of a building is set back from the property line as required by local zoning codes. [MECH] The relative rearward movement of component parts in a projectile, missile, or fuse undergoing forward acceleration during its launching; these movements, and the setback force which causes them, are used to promote events which participate in the arming and eventual functioning of the fuse. { 'set¦bak }

setback force [MECH] The rearward force of inertia which is created by the forward acceleration of a projectile or missile during its launching phase; the forces are directly proportional to the acceleration and mass of the parts being accelerated. { 'set¦bak ¦fòrs }

set bit [DES ENG] A bit insert with diamonds or other cutting media. { 'set ¦bit }

set casing [ENG] Introducing cement between the casing and the wall of the hole to seal off intermediate formations and prevent fluids from entering the hole. { 'set ¦kās·iŋ }

set class [COMPUT SCI] The collection of set occurrences that have been or may be created in accordance with a particular set description. { 'set ¦klas }

set composite [ELEC] Signaling circuit in which two signaling or telegraph legs may be superimposed on a two-wire, interoffice trunk by means of one of a balanced pair of high-impedance coils connected to each side of the line with an associated capacitor network. { 'set kəm¦päz·ət }

set condition [ELECTR] Condition of a flip-flop circuit in which the internal state of the flip-flop is set to 1. { 'set kən¦dish·ən }

set copper [MET] An intermediate copper product obtained at the end of the oxidizing portion of the fire-refining cycle and containing about 3–4% cuprous oxide. { 'set ¦käp·ər }

set description [COMPUT SCI] For a specified data set, a definition of the set class name, set-owner selection criteria, set-member eligibility rules, and set-member ordering rules. { 'set di¦skrip·shən }

Setebos [ASTRON] A small satellite of Uranus in a retrograde orbit with a mean distance of 11,090,000 miles (17,850,000 kilometers), eccentricity of 0.522, and sidereal period of 6.25 years. { 'sed·ə¦bōs }

set forward [MECH] Relative forward movement of component parts which occurs in a projectile, missile, or bomb in flight when impact occurs; the effect is due to inertia and is opposite in direction to setback. { 'set 'fòr·wərd }

set forward force [MECH] The forward force of inertia which is created by the deceleration of a projectile, missile, or bomb when impact occurs; the forces are directly proportional to the deceleration and mass of the parts being decelerated. Also known as impact force. { 'set 'fòr·wərd ¦fòrs }

set forward point [MECH] A point on the expected course

of the target at which it is predicted the target will arrive at the end of the time of flight. { 'set ¦fȯr·wərd ¸pȯint }

set function [MATH] A relation that assigns a value to each member of a collection of sets. { 'set ¸fəŋk·shən }

set grease soap [MATER] A soap made at room temperature by saponifying slack lime with resin oil; used to make a cold-set grease such as axle grease. { 'set ¸grēs ¸sōp }

set hammer [DES ENG] **1.** A hammer used as a shaping tool by blacksmiths. **2.** A hollow-face tool used in setting rivets. { 'set ¸ham·ər }

SETI *See* search for extraterrestrial intelligence. { 'sed·ē }

setigerous [INV ZOO] Referring to a segment with setae. { sə'tij·ə·rəs }

set mark [TEXT] A weaving defect in cloth; in particular, a transverse mark resulting from an improperly set loom or from an interruption in the operation of the loom. { 'set ¸märk }

set occurrence [COMPUT SCI] An instance of a set created in accordance with a set description. { 'set ə¸kə·rəns }

set off [GRAPHICS] In presswork, the undesirable rubbing off or marking of a printed sheet with the ink from the next sheet as it is being delivered. { 'set ¸ȯf }

set of first category *See* meager set. { ¦set əv 'fərst ¸kat·ə¸gȯr·ē }

set of second category [MATH] Any set that is not a meager set. { ¦set əv 'sek·ənd ¸kat·ə¸gȯr·ē }

setover [ENG] A device which helps move a lathe tailstock or headstock on its base so that a taper on a turned piece can be obtained. { 'set¸ō·vər }

set point [CONT SYS] The value selected to be maintained by an automatic controller. { 'set ¸pȯint }

set pressure [MECH ENG] The inlet pressure at which a relief valve begins to open as required by the code or standard applicable to the pressure vessel to be protected. { 'set ¸presh·ər }

set pulse [ELECTR] An electronic pulse designed to place a memory cell in a specified state. { 'set ¸pəls }

set screw [DES ENG] A small headless machine screw, usually having a point at one end and a recessed hexagonal socket or a slot at the other end, used for such purposes as holding a knob or gear on a shaft. { 'set ¸skrü }

set theory [MATH] The study of the structure and size of sets from the viewpoint of the axioms imposed. { 'set ¸thē·ə·rē }

setting angle [MECH ENG] The angle, usually 90°, between the straight portion of the tool shank and the machined portion of the work. { 'sed·iŋ ¸aŋ·gəl }

setting circle [ENG] A coordinate scale on an optical pointing instrument, such as a telescope or surveyor's transit. { 'sed·iŋ ¸sər·kəl }

setting gage [ENG] A standard gage for testing a limit gage or setting an adjustable limit gage. { 'sed·iŋ ¸gāj }

setting ring [ORD] Part of a mechanical fuse setter that takes hold of a fixed ring on the fuse of a projectile; it then rotates the entire projectile except a small ring, or setting element, in the fuse; this setting element is kept from turning by the adjusting ring in the fuse setter just long enough to make the desired change in the setting of the fuse. { 'sed·iŋ ¸riŋ }

setting temperature [ENG] The temperature at which a liquid resin or adhesive, or an assembly involving them, will set, that is, harden, gel, or cure. { 'sed·iŋ ¸tem·prə·chər }

setting time [ENG] The length of time that a resin or adhesive must be subjected to heat or pressure to cause them to set, that is, harden, gel, or cure. { 'sed·iŋ ¸tīm }

settleable solids test [CIV ENG] A test used in examination of sewage to help determine the sludge-producing characteristics of sewage; a measurement of the part of the suspended solids heavy enough to settle is made in an Imhoff cone. { 'sed·ə·bəl 'säl·ədz ¸test }

settled [METEOROL] Pertaining to weather, devoid of storms for a considerable period. { 'sed·əld }

settled ground [MIN ENG] Ground which has ceased to subside over the waste area of a mine, having reached a state of full subsidence. { 'sed·əld 'graund }

settled snow [HYD] An old snow that has been strongly metamorphosed and compacted. { 'sed·əld 'snō }

settlement [CIV ENG] The gradual downward movement of an engineering structure, due to compression of the soil below the foundation. [GEOL] The subsidence of surficial material (such as coastal sediments) due to compaction. [MIN ENG]

The gradual lowering of the overlying strata in a mine, due to extraction of the mined material. { 'sed·əl·mənt }

settler [ENG] A separator, such as a tub, pan, vat, or tank in which the partial separation of a mixture is made by density difference; used to separate solids from liquid or gas, immiscible liquid from liquid, or liquid from gas. { 'set·lər }

settling [ENG] The gravity separation of heavy from light materials; for example, the settling out of dense solids or heavy liquid droplets from a liquid carrier, or the settling out of heavy solid grains from a mixture of solid grains of different densities. [GEOL] The sag in outcrops of layered strata, caused by rock creep. Also known as outcrop curvature. { 'set·liŋ }

settling basin [CIV ENG] An artificial trap designed to collect suspended stream sediment before discharge of the stream into a reservoir. [IND ENG] A sedimentation area designed to remove pollutants from factory effluents. { 'set·liŋ ¸bās·ən }

settling chamber [ENG] A vessel in which solids or heavy liquid droplets settle out of a liquid carrier by gravity during processing or storage. { 'set·liŋ ¸chām·bər }

settling pond [MIN ENG] A natural or artificial pond for recovering the solids from an effluent. { 'set·liŋ ¸pänd }

settling reservoir [CIV ENG] A reservoir consisting of a series of basins connected in steps by long weirs; only the clear top layer of each basin is drawn off. { 'set·liŋ ¸rez·əv¸wär }

settling rounds [ORD] Rounds fired at varying angles of elevation to seat the spade and base plate of a gun mount firmly in the ground. { 'set·liŋ ¸raunz }

settling tank [ENG] A tank into which a two-phase mixture is fed and the entrained solids settle by gravity during storage. { 'set·liŋ ¸taŋk }

settling time *See* correction time. { 'set·liŋ ¸tīm }

settling velocity [FL MECH] The rate at which suspended solids subside and are deposited. Also known as fall velocity. [MECH] The velocity reached by a particle as it falls through a fluid, dependent on its size and shape, and the difference between its specific gravity and that of the settling medium; used to sort particles by grain size. { 'set·liŋ və¸läs·əd·ē }

setup [ELECTR] The ratio between the reference black level and the reference white level in television, both measured from the blanking level; usually expressed as a percentage. [IND ENG] The preparation of a facility or a machine for a specific work method, activity, or process. { 'sed¸əp }

setup person [CONT SYS] A person who uses a teach pendant to instruct a robot in its motions. { 'sed¸əp ¸pər·sən }

setup time [COMPUT SCI] The time before, after, and between computer machine runs, in which manual tasks are carried out, such as changing tape reels or transporting tapes, cards, or supplies to and from the computer equipment, to prepare for a new run. [CONT SYS] The total time needed to prepare a robot to carry out a task, including the time required to obtain the proper tools or end effectors and any work pieces. [IND ENG] In manufacturing operations, the time needed to perform tasks involved in starting up an operation. Also known as start-up time. [MATER] The time required for a cement or a gelatin to harden. { 'sed¸əp ¸tīm }

seven-eighths rule [NAV] A thumb rule which states that the approximate distance to an object broad on the beam equals seven-eighths of the distance traveled by a craft while the relative bearing (right or left) changes from 30 to 60° or from 120 to 150°, neglecting current and wind. { ¦sev·ən 'āths ¸rül }

seven-tenths rule [NAV] A thumb rule which states that the approximate distance to an object broad on the beam equals seven-tenths of the distance traveled by a craft while the relative bearing (right or left) changes from 22.5° to 45° or from 135° to 157.5°, neglecting current and wind. { ¦sev·ən 'tenths ¸rül }

seven-thirds rule [NAV] A thumb rule which states that the approximate distance to an object broad on the beam equals seven-thirds of the distance traveled by a craft while the relative bearing (right or left) changes from 22.5° to 26.5°, 67.5° to 90°, 90° to 112.5°, or 153.5° to 157.5°, neglecting current and wind. { ¦sev·ən 'thərdz ¸rül }

seventy-five-degree line [ORD] Imaginary line between the final bomb release line and the vulnerable area upon which antiaircraft artillery guns are located; from such positions the guns at an elevation of 75° can still engage and deliver effective fire on the final bomb release line. { ¦sev·ən·tē'fīv di¸grē ¸līn }

severe combined immunodeficiency syndrome [MED] A group of genetic disorders that result in a deficiency of B and T lymphocytes, causing great susceptibility to infection.

Abbreviated SCIDS. { sə¦vir kəm¦bīnd ¦im·yə¸nō·də'fish·ən·sē ¸sin¸drōm }

severe storm [METEOROL] In general, any destructive storm, but usually applied to a severe local storm, that is, an intense thunderstorm, hail storm, or tornado. { si'vir 'storm }

severe-storm observation [METEOROL] An observation (and report) of the occurrence, location, time, and direction of movement of severe local storms. { si'vir ¦storm ¸äb·zər'vā·shən }

severe weather [METEOROL] A more general term for severe storm. { si'vir 'weth·ər }

severity factor [CHEM ENG] A measure of the severeness or intensity of overall reaction conditions in a chemical reaction; for example, the temperature, pressure, or conversion in a catalytic cracker or reformer. { si'ver·əd·ē ¸fak·tər }

Sevier orogeny [GEOL] The deformation that occurred along the eastern edge of the Great Basin in Utah (eastern edge of the Cordilleran miogeosyncline) during times intermediate between the Nevadan orogeny to the west and the Laramide orogeny to the east, culminating early in the Late Cretaceous. { se'vyā o'räj·ə·nē }

sewage [CIV ENG] The fluid discharge from medical, domestic, and industrial sanitary appliances. Also known as sewerage. { 'sü·ij }

sewage disposal plant [CIV ENG] The land, building, and apparatus employed in the treatment of sewage by chemical precipitation or filtration, bacterial action, or some other method. { 'sü·ij di¦spōz·əl ¸plant }

sewage farm [AGR] A farm in which sewage is used for irrigation and fertilizer. { 'sü·ij ¸färm }

sewage sludge [CIV ENG] A semiliquid waste with a solid concentration in excess of 2500 parts per million, obtained from the purification of municipal sewage. Also known as sludge. { 'sü·ij ¸sləj }

sewage system [CIV ENG] A drainage system for carrying surface water and sewage for disposal. { 'sü·ij ¸sis·təm }

sewage treatment [CIV ENG] A process for the purification of mixtures of human and other domestic wastes; the process can be aerobic or anaerobic. { 'sü·ij ¸trēt·mənt }

sewer [CIV ENG] An underground pipe or open channel in a sewage system for carrying water or sewage to a disposal area. { 'sü·ər }

sewerage See sewage. { 'sü·ə·rij }

sewer gas [MATER] The gas evolved from the decomposition of municipal sewage; it has a high content of methane and hydrogen sulfide, and can be used as a fuel gas. { 'sü·ər ¸gas }

sewing machine [MECH ENG] A mechanism that stitches cloth, leather, book pages, or other material by means of a double-pointed or eye-pointed needle. { 'sō·iŋ mə¸shēn }

sewing press [GRAPHICS] A wooden fixture used to stretch cords used in hand-sewn book bindings. { 'sō·iŋ ¸pres }

sex [BIOL] **1.** The state of condition of an organism which comes to expression in the production of germ cells. **2.** To determine the sex of. { seks }

sexadecimal See hexadecimal. { ¦sek·sə'des·məl }

sexadecimal number system See hexadecimal number system. { ¦sek·sə'des·məl ¸nəm·bər ¸sis·təm }

sexagesimal [MATH] Pertaining to a multiplicity of 60 distinct alternative states or conditions or, simply, a positional numeration system to radix (or base) 60. { sek·sə¦jez·ə·məl }

sexagesimal counting table [MATH] A table for converting numbers using the 60 system into decimals, for example, minutes and seconds. { sek·sə¦jez·ə·məl 'kaunt·iŋ ¸tā·bəl }

sexagesimal measure of angles [MATH] A system of angular units in which a complete revolution is divided into 360 degrees, a degree into 60 minutes, and a minute into 60 seconds. { sek·sə¦jez·ə·məl 'mezh·ər əv 'aŋ·gəlz }

sex cell See gamete. { 'seks ¸sel }

sex chromatin See Barr body. { 'seks ¸krō·mə·tən }

sex chromosome [GEN] Either member of a pair of chromosomes responsible for sex determination of an organism. { 'seks ¸krō·mə¸sōm }

sex cords [EMBRYO] Cordlike masses of epithelial tissue that invaginate from germinal epithelium of the gonad and give rise to seminiferous tubules and rete testes in the male, and primary ovarian follicles and rete ovarii in the female. { 'seks ¸kordz }

sex determination [GEN] The mechanisms which determine

SEXTANT

A marine sextant.

whether the bipotential embryo will develop as male or female in a species. { seks di¸tər·mə¸nā·shən }

sex factor See fertility factor. { 'seks ¸fak·tər }

sex hormone [BIOCHEM] Any hormone secreted by a gonad, but also found in other tissues. { 'seks ¸hor¸mōn }

sex-influenced inheritance [GEN] That part of the inheritance pattern on which sex differences operate to promote character differences. { 'seks in¦flü·ənst in'her·ət·əns }

sexless connector See hermaphroditic connector. { 'seks·ləs kə'nek·tər }

sex-limited inheritance [GEN] Expression of a phenotype in only one sex; may be due to either a sex-linked or autosomal gene. { 'seks ¦lim·əd·əd in'her·ət·əns }

sex-linked inheritance [GEN] The transmission to successive generations of traits that are due to alleles at gene loci on a sex chromosome. { 'seks ¦liŋkt in'her·ət·əns }

sex organs [ANAT] The organs pertaining entirely to the sex of an individual, both physiologically and anatomically. { 'seks ¸or·gənz }

sex ratio [BIOL] The relative proportion of males and females in a population. { 'seks ¸rā·shō }

Sextans [ASTRON] A constellation in the southern hemisphere, right ascension 10 hours, declination 0°. Also known as Sextant. { 'sek¸stanz }

sextant [MATH] A unit of plane angle, equal to 60° or π/3 radians. [NAV] An optical instrument; a double reflecting instrument used in navigation for measuring angles, primarily altitudes of celestial bodies. { 'sek·stənt }

Sextant See Sextans. { 'sek·stənt }

sextant adjustment [NAV] In celestial navigation, the process of checking the accuracy of a sextant and removing or reducing its error. { 'sek·stənt ə¸jəs·mənt }

sextant altitude [NAV] Altitude as indicated by a sextant or similar instrument before corrections are applied. { 'sek·stənt ¸al·tə¸tüd }

sextant altitude corrections [NAV] Corrections applied to sextant altitude readings which are made necessary by external physical phenomena, such as sea-air temperature difference, wave height, and sea tilt, personal errors such as tilt and height of eye, and sextant errors. { 'sek·stənt ¦al·tə¸tüd kə¸rek·shənz }

sextant error [NAV] The error in sextant readings caused by sources other than its operator or physical phenomena; some of these errors are prismatic, collimation, perpendicularity of horizon glass, and perpendicularity of index mirror. Also known as octant error. { 'sek·stənt ¸er·ər }

sextic [MATH] Having the sixth degree or order. { 'sek·stik }

sextile aspect [ASTRON] The position of two celestial bodies when they are 60° apart. { 'sek¸stīl 'as¸pekt }

sextillion [MATH] **1.** The number 10²¹. **2.** In British and German usage, the number 10³⁶. { sek'stil·yən }

sextupole magnet [ELECTROMAG] A configuration of six magnets arranged in a circular pattern with alternating polarities, used in the control of electron beams. { 'seks·tə¸pōl ¸mag·nət }

sexual cycle [PHYSIO] A cycle of physiological and structural changes associated with sex; examples are the estrous cycle and the menstrual cycle. { 'sek·shə·wəl 'sī·kəl }

sexual dimorphism [BIOL] Diagnostic morphological differences between the sexes. { 'sek·shə·wəl dī'mor¸fiz·əm }

sexuality [BIOL] **1.** The sum of a person's sexual attributes, behavior, and tendencies. **2.** The psychological and physiological sexual impulses whose satisfaction affords pleasure. [PSYCH] The quality of being sexual, or the degree of a person's sexual attributes, attractiveness, and drives. { ¸sek·shə'wal·əd·ē }

sexually transmitted disease [MED] An infection acquired and transmitted primarily by sexual contact. Abbreviated STD. { 'sek·shə·lē trans'mid·əd di'zēz }

sexual reproduction [BIOL] Reproduction involving the paired union of special cells (gametes) from two individuals. { 'sek·shə·wəl ¸rē·prə'dək·shən }

sexual selection [EVOL] A special form of natural selection responsible for the evolution of traits that promote success in competition for mates. { ¦seksh·ə·wəl si'lek·shən }

sexual spore [BIOL] A spore resulting from conjugation of gametes or nuclei of opposite sex. { 'sek·shə·wəl ¸spor }

seybertite See clintonite. { 'sī·bər¸dīt }

Seyfert galaxy [ASTRON] A galaxy that has a small, bright nucleus from which violent explosions may occur. { 'zī·fərt ,gal·ik·sē }

Seyfert's Sextet [ASTRON] A compact collection of galaxies that surrounds the galaxy NGC 6027 and has both spiral and irregular members, most of which interact with each other. { 'zī·fərts seks'tet }

Seymouriamorpha [PALEON] An extinct group of labyrinthodont Amphibia of the Upper Carboniferous and Permian in which the intercentra were reduced. { sē,mör·ē·ə'mör·fə }

Sezary syndrome [MED] Exfoliative erythroderma with a cutaneous infiltrate of atypical mononuclear cells; similar cells are also present in the peripheral blood. { sə'zar·ē ,sin,drōm }

S factor [ORD] The deflection change in mils required to keep the burst on the observer target line when the range is changed 100 yards (91.44 meters) along that line. { 'es ,fak·tər }

SFC *See* specific fuel consumption.

sferics *See* atmospheric interference. { 'sfir·iks }

sferics fix [METEOROL] The estimated location of a source of atmospherics, presumably a lightning discharge. { 'sfir·iks ,fiks }

sferics observation [METEOROL] An evaluation, from one or more sferics receivers, of the location of weather conditions with which lightning is associated; such observations are more commonly obtained from networks of two or three widely spaced stations; simultaneous observations of the azimuth of the discharge are made at all stations, and the location of the storm is determined by triangulation. { 'sfir·iks ,äb·zər,vā·shən }

sferics receiver [ELECTR] An instrument which measures, electronically, the direction of arrival, intensity, and rate of occurrence of atmospherics; in its simplest form, the instrument consists of two orthogonally crossed antennas, whose output signals are connected to an oscillograph so that one loop measures the north-south component while the other measures the east-west component; the signals are combined vertically to give the azimuth. Also known as lightning recorder. { 'sfir·iks ri,sē·vər }

SFS *See* sodium formaldehyde sulfoxylate.

S glass [MATER] A glass containing magnesia, alumina, and a silicate. { 'es ,glas }

SGML *See* Standard Generalized Markup Language.

shackle [DES ENG] An open or closed link of various shapes with extended legs; each leg has a transverse hole to accommodate a pin, bolt, or the like, which may or may not be furnished. { 'shak·əl }

shackle bolt [DES ENG] A cylindrically shaped metal bar for connecting the ends of a shackle. { 'shak·əl ,bōlt }

shade [OPTICS] The color of a mixture of pigments or dyes which has some black pigment or dye in it. { shād }

shaded-pole motor [ELEC] A single-phase induction motor having one or more auxiliary short-circuited windings acting on only a portion of the magnetic circuit; generally, the winding is a closed copper ring embedded in the face of a pole; the shaded pole provides the required rotating field for starting purposes. { 'shād·əd ,pōl 'mōd·ər }

shaded-relief map [MAP] A map of an area whose relief is made to appear three-dimensional by the hill-shading method. { 'shād·əd ri'lēf ,map }

shade error [NAV] That error of a sextant due to refraction in the shade glasses. { shād ,er·ər }

shade glass [OPTICS] A darkened transparency that can be moved into the line of sight of an optical instrument, such as a sextant, to reduce the intensity of light reaching the eye. { shād ,glas }

shading [ELECTR] Television process of compensating for the spurious signal generated in a camera tube during trace intervals. [MAP] *See* hill shading. { 'shād·iŋ }

shading coefficient [ENG] A ratio of the solar energy transmitted through a window to the incident solar energy; used to express the effectiveness of a shading device. { 'shād·iŋ ,kō·i,fish·ənt }

shading coil *See* shading ring. { 'shād·iŋ ,kóil }

shading ring [ELECTROMAG] The copper ring used in a shaded-pole motor to produce a rotating magnetic field for starting purposes, or used around a part of the core of an alternating-current relay to prevent contact chatter. Also known as shading coil. [ENG ACOUS] A heavy copper ring

sometimes placed around the central pole of an electrodynamic loudspeaker to serve as a shorted turn that suppresses the hum voltage produced by the field coil. { 'shād·iŋ ,riŋ }

shading signal [ELECTR] Television camera signal that serves to increase the gain of the amplifier in the camera during those intervals of time when the electron beam is on an area corresponding to a dark portion of the scene being televised. { 'shād·iŋ ,sig·nəl }

shadow [OPTICS] A region of darkness caused by the presence of an opaque object interposed between such a region and a source of light. [PHYS] A region which some type of radiation, such as sound or x-rays, does not reach because of the presence of an object, which the radiation cannot penetrate, interposed between the region and the source of radiation. { 'shad·ō }

shadow attenuation [ELECTROMAG] Attenuation of radio waves over a sphere in excess of that over a plane when the distance over the surface and other factors are the same. { 'shad·ō ə,ten·yə'wā·shən }

shadow bands [ASTRON] Rippling bands of shadow that appear on every white surface of flat terrestrial objects a few minutes before the total eclipse of the sun. { 'shad·ō ,banz }

shadow batch system [COMPUT SCI] An online data collection system that initially only stores transactions in the computer system for reference, and updates the master files only at the end of the day or processing period. { 'shad·ō ,bach ,sis·təm }

shadow effect [COMMUN] Reduction in the strength of an ultra-high-frequency signal caused by some object (such as a mountain or a tall building) between the points of transmission and reception. { 'shad·ō i,fekt }

shadow factor [ELECTROMAG] The ratio of the electric-field strength that would result from propagation of waves over a sphere to that which would result from propagation over a plane under comparable conditions. [OPTICS] A multiplication factor derived from the sun's declination, the latitude of the target and the time of photography, used in determining the heights of objects from shadow length. Also known as tan alt. { 'shad·ō ,fak·tər }

shadowgram [GRAPHICS] An x-ray photograph. [PHYS] A plot or display of a shadow. { 'shad·ō,gram }

shadowgraph [GRAPHICS] **1.** A photographic image in the form of a shadow. **2.** *See* radiograph. [OPTICS] A simple method of making visible the disturbances that occur in fluid flow at high velocity, in which light passing through a flowing fluid is refracted by density gradients in the fluid, resulting in bright and dark areas on a screen placed behind the fluid. { 'shad·ō,graf }

shadow line [GRAPHICS] A thick line in a line drawing of an object illuminated by light; indicates the edge farthest from the source of light. { 'shad·ō ,līn }

shadow mask [ELECTR] A thin, perforated metal mask mounted just back of the phosphor-dot faceplate in a three-gun color picture tube; the holes in the mask are positioned to ensure that each of the three electron beams strikes only its intended color phosphor dot. Also known as aperture mask. { 'shad·ō ,mask }

shadow microscopy *See* projection microradiography. { 'shad·ō mī'kräs·kə·pē }

shadow photometer [ENG] A simple photometer in which a rod is placed in front of a screen and two light sources to be compared are adjusted in position until their shadows touch and are equal in intensity. { 'shad·ō fō'täm·əd·ər }

shadow region [ELECTROMAG] Region in which, under normal propagation conditions, the field strength from a given transmitter is reduced by some obstruction which renders effective radio reception of signals or radar detection of objects in this region improbable. { 'shad·ō ,rē·jən }

shadow scattering [QUANT MECH] Scattering that results from the interference of the incident wave and scattered waves. { 'shad·ō 'skad·ər·iŋ }

shadow-transit [ASTRON] A transit of one of Jupiter's Galilean satellites, in which the satellite casts its shadow on the planet's disk. { 'shad·ō 'tranz·ət }

shadow zone [ACOUS] A region, usually under water or in the atmosphere, which sound waves will not reach, according to ray acoustics. [GEOPHYS] The zone, between 103 and 143° from the epicenter of an earthquake, in which direct seismic waves do not arrive because of refraction and absorption by the earth's core. { 'shad·ō ,zōn }

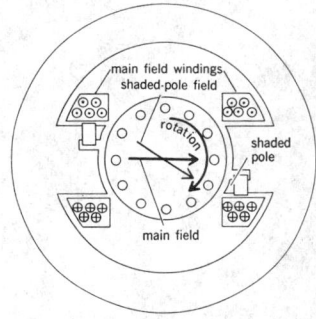

SHADED-POLE MOTOR

main field windings
shaded-pole field
rotation
shaded pole
main field

⊙ current toward reader
⊕ current away from reader

Cross-sectional view of shaded-pole motor.

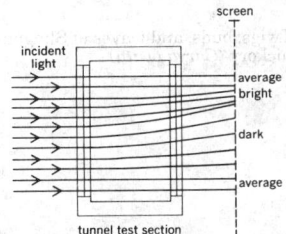

SHADOWGRAPH

screen
incident light
average bright
dark
average
tunnel test section

Shadowgraph produced by light passing through fluid in test section indicates flow pattern.

shadscale [BOT] *Atriplex confertifolia.* A small shiny shrub found in the Great Basin Desert. { 'shad,skāl }

shaft [GEOL] A passage in a cave that is vertical or nearly vertical. [MECH ENG] A cylindrical piece of metal used to carry rotating machine parts, such as pulleys and gears, to transmit power or motion. [MIN ENG] An excavation of limited area compared with its depth, made for finding or mining ore or coal, raising water, ore, rock, or coal, hoisting and lowering men and material, or ventilating underground workings; the term is often specifically applied to approximately vertical shafts as distinguished from an incline or an inclined shaft. [SCI TECH] A long, slender, usually cylindrical part. { shaft }

shaft alley [NAV ARCH] The watertight trunk in a ship through which pass the propeller shafts between the propellers and the engine room. { 'shaft ,al·ē }

shaft allowance [MIN ENG] The extra space between the excavation diameter and the finished diameter to accommodate the permanent shaft lining. { 'shaft ə,laù·əns }

shaft balancing [DES ENG] The process of redistributing the mass attached to a rotating body in order to reduce vibrations arising from centrifugal force. Also known as rotor balancing. { 'shaft ,bal·əns·iŋ }

shaft cable [MIN ENG] A specially armored cable of great mechanical strength running down a mine shaft. { 'shaft ,kā·bəl }

shaft capacity [MIN ENG] The output of ore or coal that can be expected to be raised regularly and in normal circumstances. { 'shaft kə,pas·əd·ē }

shaft column [MIN ENG] A length of pipes installed in a mine shaft for pumping, for hydraulic stowing, or for compressed air. { 'shaft ,käl·əm }

shaft coupling *See* coupling. { 'shaft ,kəp·liŋ }

shaft crusher [MIN ENG] A hard-rock crusher in a shaft, set to reduce large lumps of ore to a convenient size for delivery to the skip. { 'shaft ¦krəsh·ər }

shaft deformation bar [MIN ENG] A length of 1 1/2-inch (3.8-centimeter) pipe fitted at one end with a micrometer and at the other end with a hard-steel cone for measuring the deformation in the cross section of a shaft. { 'shaft ,dē,fòr'mā·shən ,bär }

shaft drilling [MIN ENG] The drilling of small shafts up to about 5 feet (1.5 meters) in diameter with the shot drill. { 'shaft ,dril·iŋ }

shaft furnace [ENG] A vertical, refractory-lined cylinder in which a fixed bed (or descending column) of solids is maintained, and through which an ascending stream of hot gas is forced; for example, the pig-iron blast furnace and the phosphors-from-phosphate-rock furnace. { 'shaft ¦fər·nəs }

shaft hopper [MECH ENG] A hopper that feeds shafts or tubes to grinders, threaders, screw machines, and tube benders. { 'shaft ¦häp·ər }

shaft horsepower [MECH ENG] The output power of an engine, motor, or other prime mover; or the input power to a compressor or pump. { 'shaft 'hòrs,paù·ər }

shaft house [MIN ENG] A building at the mouth of a shaft, where ore or rock is received from the mine. { 'shaft ,haùs }

shafting [MECH ENG] The cylindrical machine element used to transmit rotary motion and power from a driver to a driven element; for example, a steam turbine driving a ship's propeller. { 'shaft·iŋ }

shaft kiln [ENG] A kiln in which raw material fed into the top, moves down through hot gases flowing up from burners on either side at the bottom, and emerges as a product from the bottom; used for calcining operations. { 'shaft ¦kil }

shaft lining [MIN ENG] The timber, steel, brick, or concrete structure fixed around a shaft to support the walls. { 'shaft ,līn·iŋ }

shaft pillar [MIN ENG] A large area of a coal or ore seam which is left unworked around the shaft bottom to protect the shaft from damage by subsidence. { 'shaft ,pil·ər }

shaft plumbing [MIN ENG] The operation of orienting two plumb bobs, both at surface and at depth in order to transfer the bearing underground. { 'shaft ,pləm·iŋ }

shaft pocket [MIN ENG] Ore storage pocket, of one or more compartments, cut into the wall on one or both sides of a vertical shaft or in the hanging wall of an inclined shaft. { 'shaft ,päk·ət }

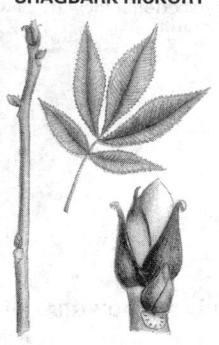

SHAGBARK HICKORY

Twigs, buds, and leaves of Shagbark hickory *(Carya ovata).*

shaft-position encoder [ELECTR] An analog-to-digital converter in which the exact angular position of a shaft is sensed and converted to digital form. { 'shaft pə¦zish·ən in'kōd·ər }

shaft siding [MIN ENG] The station or landing place arranged for buckets or tubs at the bottom of the winding shaft. { 'shaft ,sīd·iŋ }

shaft signaling [MIN ENG] The transmission of visible and audible signals between the onsetter or hitcher at the pit bottom and the banksman or hoistman at the pit top. { 'shaft ¦sig·nəl·iŋ }

shaft sinking [MIN ENG] Excavating a shaft downward, usually from the surface, to the workable coal or ore. { 'shaft ,siŋk·iŋ }

shaft-sinking drill [MIN ENG] A large-diameter drill with multiple rotary cones or cutting bits, used for shaft sinking. { 'shaft ,siŋk·iŋ ,dril }

shaft spillway [CIV ENG] A vertical shaft which has a funnel-shaped mouth and ends in an outlet tunnel, providing an overflow duct for a reservoir. Also known as morning glory spillway. { 'shaft ,spil,wā }

shaft station [MIN ENG] An enlargement of a level near a shaft from which ore, coal, or rock may be hoisted and supplies unloaded. { 'shaft ,stā·shən }

shaft strut [NAV ARCH] A term applied to a bracket supporting the after propeller shaft bearing external to the hull. { 'shaft ,strət }

shagbark hickory [FOR] *Carya ovata.* A type of hickory that grows to a height of about 120 ft (36 m) and is found in the eastern half of the United States and adjacent Canada. It is the most important species because of the commercial value of its nuts and of its wood. { ,shag,bärk 'hik·ə·rē }

shagreen [MATER] 1. A leather made by pressing grains into the hide to create indentations. 2. The skin of certain sharks and rays containing small knobs. { sha'grēn }

Shahbazian objects [ASTRON] Compact collections of the order of 10 galaxies. { shə'bä·zē·ən ,äb·jeks }

shake [MATER] 1. Separation between adjoining layers of wood, due to causes other than drying. 2. A thick hand-cut shingle. { shāk }

shake culture [MICROBIO] 1. A method for isolating anaerobic bacteria by shaking a deep liquid culture of an agar or gelatin to distribute the inoculum before solidification of the medium. 2. A liquid medium in a flask that has been inoculated with an aerobic microorganism and placed on a shaking machine; action of the machine continually aerates the culture. { 'shāk ,kəl·chər }

shakedown test [ENG] An equipment test made during the installation work. { 'shāk,daùn ,test }

shakeout [MET] Removing a casting from a sand mold. { 'shāk,aùt }

shaker [ELECTROMAG] An electromagnetic device capable of imparting known and usually controlled vibratory acceleration to a given object. Also known as electrodynamic shaker; shake table. { 'shāk·ər }

shaker conveyor [MIN ENG] A conveyor consisting of a length of metal troughs, with suitable supports, to which a reciprocating motion is imparted by drives. { 'shāk·ər kən,vā·ər }

shake table *See* shaker; vibration machine. { 'shāk ,tā·bəl }

shake-table test [ENG] A laboratory test for vibration tolerance, in which the device to be tested is placed on a shake table. { 'shāk ¦tā·bəl ,test }

shake wave *See* S wave. { 'shāk ,wāv }

shaking-out [CHEM ENG] A procedure in which a sample of crude oil is centrifuged at high speed to separate its components; used to determine sediment and water content. { 'shāk·iŋ 'aùt }

shaking screen [MECH ENG] A screen used in separating material into desired sizes; has an eccentric drive or an unbalanced rotating weight to produce shaking. { 'shāk·iŋ ,skrēn }

shaking table *See* Wilfley table. { 'shāk·iŋ ,tā·bəl }

shale [PETR] A fine-grained laminated or fissile sedimentary rock made up of silt- or clay-size particles; generally consists of about one-third quartz, one-third clay materials, and one-third miscellaneous minerals, including carbonates, iron oxides, feldspars, and organic matter. { shāl }

shale ball [GEOL] A meteorite partly or wholly converted to iron oxides by weathering. Also known as oxidite. { 'shāl ,bòl }

shale break [GEOL] A thin layer or parting of shale between harder strata or within a bed of sandstone or limestone. { 'shāl ˌbrāk }

shale clay [MATER] A clay made from ground shale. { 'shāl ˌklā }

shale crescent [GEOL] A crescent formed by the filling of a ripple-mark trough by shale. { 'shāl ˌkres·ənt }

shale ice [HYD] A mass of thin and brittle plates of river or lake ice formed when sheets of skim ice break up into small pieces. { 'shāl ˌīs }

shale naphtha [MATER] Naphtha derived from shale oil, usually containing 60–70% olefins and other hydrocarbons. { 'shāl ˌnaf·thə }

shale oil [MATER] Liquid obtained from the destructive distillation of kerogens in oil shale; further processing is needed to convert shale oil into products similar to petroleum oils. { 'shāl ˌoil }

shale reservoir [GEOL] Underground hydrocarbon reservoir in which the reservoir rock is a brittle, siliceous, fractured shale. { 'shāl 'rez·əv,wär }

shale shaker [PETRO ENG] A vibrating screen over which drilling fluid is passed to trap the drill cuttings as the fluid passes through. { 'shāl ˌshāk·ər }

shalification [GEOL] The formation of shale. { ˌshāl·ə·fə'kā·shən }

shallot [BOT] *Allium ascalonicum.* A bulbous onionlike herb. Also known as scallion. { 'shal·ət }

shallow-focus earthquake [GEOPHYS] An earthquake whose focus is located within 70 kilometers of the earth's surface. { 'shal·ō ˌfō·kəs 'ərth,kwāk }

shallow fog [METEOROL] In weather-observing terminology, low-lying fog that does not obstruct horizontal visibility at a level 6 feet (1.8 meters) or more above the surface of the earth; this is, almost invariably, a form of radiation fog. { 'shal·ō ˌfäg }

shallow fording [ORD] The ability of a vehicle or gun, equipped with built-in waterproofing with its suspension in contact with the ground, to negotiate a water obstacle without the use of a special waterproofing. { 'shal·ō 'fòrd·iŋ }

shallow inland seas [GEOL] Epeiric seas which periodically cover cratonic areas as a result of continental subsidence or eustatic rises in sea level. { 'shal·ō 'in·lənd 'sēz }

shallow marginal seas [GEOL] Epeiric seas along the cratonic margins. { 'shal·ō 'märj·ən·əl 'sēz }

shallows [HYD] A shallow place or area in a body of water, or an expanse of shallow water. { 'shal·ōz }

shallow water [HYD] Water of such a depth that bottom topography affects surface waves. { 'shal·ō 'wòd·ər }

shallow-water wave [HYD] A progressive gravity wave in water whose depth is much less than the wavelength. { 'shal·ō ˌwòd·ər ˌwāv }

shallow well [HYD] **1.** A water well, generally dug up by hand or by excavating machinery, or put down by driving or boring, that taps the shallowest aquifer in the vicinity. **2.** A well whose water level is shallow enough to permit use of a suction pump, the practical lift of which is taken as 22 feet (6.7 meters). { 'shal·ō 'wel }

shaluk [METEOROL] Any hot desert wind other than simoom. { shä'lək }

shaly [GEOL] Pertaining to, composed of, containing, or having the properties of shale, especially readily split along close-spaced bedding planes. { 'shāl·ē }

shaly bedding [GEOL] Laminated bedding varying between 2 and 10 millimeters in thickness. { 'shāl·ē 'bed·iŋ }

shamal [METEOROL] The northwest wind in the lower valley of the Tigris and Euphrates and the Persian Gulf; it may set in suddenly at any time, and generally lasts from 1 to 5 days, dying down at night and freshening again by day; but in June and early July it continues almost without cessation (the great or 40-day shamal). { shə'mäl }

shandite [MINERAL] $Ni_3Pb_2S_2$ A rhombohedral mineral composed of nickel lead sulfide, occurring in crystals. { 'shan,dīt }

shank [DES ENG] **1.** The end of a tool which fits into a drawing holder, as on a drill. **2.** *See* bit blank. { shaŋk }

shank-type cutter [DES ENG] A cutter having a shank to fit into the machine tool spindle or adapter. { 'shaŋk ˌtīp ˌkəd·ər }

shannon [COMMUN] A unit of information content, equal to the designation of one of two possible and equally likely values or states of anything used to store or convey information. { 'shan·ən }

Shannon formula [COMMUN] A theorem in information theory which states that the highest number of binary digits per second which can be transmitted with arbitrarily small frequency of error is equal to the product of the bandwidth and $\log_2(1 + R)$, where R is the signal-to-noise ratio. { 'shan·ən ˌfòr·myə·lə }

Shannon limit [COMMUN] Maximum signal-to-noise ratio improvement which can be achieved by the best modulation technique as implied by Shannon's theorem relating channel capacity to signal-to-noise ratio. { 'shan·ən ˌlim·ət }

Shannon-McMillian-Breiman theorem [MATH] Given an ergodic measure preserving transformation T on a probability space and a finite partition ζ of that space the limit as $n\rightarrow\infty$ of $1/n$ times the information function of the common refinement of ζ, $T^{-1}\zeta, \ldots, T^{-n+1}\zeta$ converges almost everywhere and in the L_1 metric to the entropy of T given ζ. { 'shan·ən mik 'mil·ən 'brī·mən ˌthir·əm }

Shannon's sampling theorem *See* sampling theorem. { 'shan·ənz 'sam·pliŋ ˌthir·əm }

Shannon's theorems [MATH] These results are foundational to the mathematical study of information; mathematically they link the concept of entropy with the amount of efficient transmittal and reception of information. { 'shan·ənz ˌthir·əmz }

shantung [GEOL] A monadnock in the process of burial by huangho deposits. [TEXT] A fabric with a plain silk weave characterized by a rough, nubbed surface. { shan'təŋ }

Shantung soil *See* Noncalcic Brown soil. { shan'təŋ ˌsoil }

shape coding [DES ENG] The use of special shapes for control knobs, to permit recognition and sometimes also position monitoring by sense of touch. { 'shāp ˌkōd·iŋ }

shaped-beam antenna [ELECTROMAG] Antenna with a directional pattern which, over a certain angular range, is of special shape for some particular use. { 'shāpt ˌbēm an 'ten·ə }

shaped-chamber manometer [ENG] A flow measurement device that measures differential pressure with a uniform flow-rate scale with a specially shaped chamber. { 'shāpt ˌchām·bər mə'näm·əd·ər }

shaped charge [ORD] An explosive charge with a shaped cavity that forces the impact of the explosion to the front so that there is an armor-piercing force. Also known as cavity charge. { 'shāpt 'chärj }

shape factor [ELEC] *See* form factor. [ELECTR] The ratio of the 60-decibel bandwidth of a bandpass filter to the 3-decibel bandwidth. [FL MECH] The quotient of the area of a sphere equivalent to the volume of a solid particle divided by the actual surface of the particle; used in calculations of gas flow through beds of granular solids. [OPTICS] For a lens, the quantity $(R_2 + R_1)/(R_2 - R_1)$, where R_1 and R_2 are the radii of the first and second surface of the lens. Also known as Coddington shape factor. { 'shāp ˌfak·tər }

shape-fill [COMPUT SCI] The filled-in areas on a graphic electronic display. { 'shāp ˌfil }

shape isomer [NUC PHYS] An excited nuclear state which has an unusually long lifetime because of its deformed shape, which differs drastically from that of the lower energy states into which it is permitted to decay. { 'shāp 'ī·sə·mər }

shape memory alloy [MET] An alloy that, after being deformed, can recover its original shape when it is heated. { 'shāp ˌmem·rē al,òi }

shaper [MECH ENG] A machine tool for cutting flat-on-flat, contoured surfaces by reciprocating a single-point tool across the workpiece. { 'shā·pər }

shape resonance [QUANT MECH] A broad resonance or peak in the cross section of a scattering process that reflects the shape of the potential between projectile and target; this shape consists typically of a barrier separating an inner deep well from a shallow, asymptotically vanishing potential at large separation distances. { 'shāp ˌrez·ən·əns }

shaping [ELECTROMAG] The adjustment of a plan-position-indicator pattern set up by a rotating magnetic field. [MECH ENG] A machining process in which a reciprocating single-point tool moves over the work in straight, parallel lines to produce a flat surface. [PSYCH] Therapeutic reinforcement of those responses that increasingly approximate sought-after behaviors. { 'shāp·iŋ }

shaping circuit See corrective network. { 'shāp·iŋ ˌsər·kət }

shaping dies [MECH ENG] A set of dies for bending, pressing, or otherwise shaping a material to a desired form. { 'shāp·iŋ ˌdīz }

shaping network See corrective network. { 'shāp·iŋ ˌnet,wərk }

shapometer [ENG] A device used to measure the shape of sedimentary particles. { shā'päm·əd·ər }

shard [GEOL] A vitric fragment in pyroclastics, having a characteristically curved surface of fracture. { shärd }

shared control unit [COMPUT SCI] A control unit which controls several devices with similar characteristics, such as tape devices. { 'sherd kən¦trōl ˌyü·nət }

shared file [COMPUT SCI] A direct-access storage device that is used by more than one computer or data-processing system. { 'sherd 'fīl }

shared load [COMPUT SCI] A workload that can be shared by more than one computer, particularly during peak periods. { 'sherd 'lōd }

shared logic [COMPUT SCI] **1.** The simultaneous use of a single computer by multiple users. **2.** An arrangement of computers or computerized equipment in which the processing capabilities of one computer, including the ability to use peripheral devices, can be distributed to the other computers. { 'sherd 'läj·ik }

shared-logic cluster word processor [COMPUT SCI] A system of terminals lacking word-processing capability and printers joined to a single computer designed to carry out word-processing functions. { 'sherd ¦läj·ik ¦kləs·tər 'wərd ˌprä,ses·ər }

shared paranoid delusion See folie à deux. { ¦sherd ¦par·əˌnȯid di'luzh·ən }

shared resource [COMPUT SCI] Peripheral equipment that is simultaneously shared by several users. { 'sherd 'rē,sȯrs }

shareware [COMPUT SCI] Copyrighted software that can be tried before buying. { 'sher,wer }

sharing device [COMPUT SCI] A small, inexpensive multiplexer that combines two independent data signals, which are then transmitted over the same communications line. { 'sher·iŋ diˌvīs }

shark [VERT ZOO] Any of about 225 species of carnivorous elasmobranchs which occur principally in tropical and subtropical oceans; the body is fusiform with a heterocercal tail and a tough, usually gray, skin roughened by tubercles, and the snout extends beyond the mouth. { shärk }

sharki See kaus. { 'shär·kē }

shark liver oil [MATER] A yellow to brown oil with strong aroma, obtained from the livers of various sharks; insoluble in water, soluble in ether, benzene, and carbon disulfide; used as a vitamin A source, in biochemical research. Also known as dogfish oil; shark oil. { shärk 'liv·ər ˌȯil }

shark oil See shark liver oil. { 'shärk ˌȯil }

sharkskin [TEXT] **1.** A smooth wool or worsted fabric with a basketweave or twill pattern in two tones. **2.** A smooth, dull-finished rayon fabric in a basketweave pattern used for dresses. { 'shärk,skin }

sharkskin pahoehoe [GEOL] A type of pahoehoe displaying numerous tiny spines or spicules on the surface. { 'shärk,skin pə'hō·ē,hō·ē }

shark-tooth projection [GEOL] Sharp pointed projections several centimeters in length, formed by the pulling apart of plastic lava. { 'shärk ¦tüth prə'jek·shən }

sharp-crested weir [CIV ENG] A weir in which the water flows over a thin, sharp edge. { 'shärp ¦kres·təd 'wer }

sharp-cutoff tube [ELECTR] An electron tube in which the control-grid openings are uniformly spaced; the anode current then decreases linearly as the grid voltage is made more negative, and cuts off sharply at a particular grid voltage. { 'shärp ¦kəd,ȯf ˌtüb }

sharp-edged gust [METEOROL] A gust that represents an instantaneous change in wind direction or speed. { 'shärp ¦ejd 'gəst }

sharpen [ENG] To give a thin keen edge or a sharp acute point to. [GRAPHICS] To decrease the size of the dots on a screened halftone. { 'shär·pən }

sharpening stone [ENG] A device such as a whetstone used for sharpening by hand. { 'shär·pə·niŋ ˌstōn }

sharp iron [ENG] A tool used to open seams for caulking. { 'shärp 'ī·ərn }

SHARK

The mackerel shark (*Lamnia nasus*).

sharpite [MINERAL] $(UO_2)(CO_3) \cdot H_2O$ A greenish-yellow mineral composed of hydrous basic uranyl carbonate. { 'shär,pīt }

sharpness function technique [OPTICS] An adaptive optics technique in which a deformable mirror or other corrective element is dithered to maximize a given function, such as the integral of the square of the irradiance in the image plane or the amount of energy within a certain region. { 'shärp·nəs ˌfəŋk·shən tek,nēk }

sharpness of resonance [ELEC] The narrowness of the frequency band around the resonance at which the response of an electric circuit exceeds an arbitrary fraction of its maximum response, often 70.7%. { 'shärp·nəs əv 'rez·ən·əns }

sharps [FOOD ENG] The bran from wheat kernels; used as cattle feed. Also known as middlings; wheat middlings. { shärps }

sharp sand [GEOL] An angular-grain sand free of clay, loam, and other foreign particles. { 'shärp 'sand }

sharp series [SPECT] A series occurring in the line spectra of many atoms and ions with one, two, or three electrons in the outer shell, in which the total orbital angular momentum quantum number changes from 0 to 1. { 'shärp 'sir·ēz }

sharpstone [GEOL] Any rock fragment having angular edges and corners and being more than 2 millimeters in diameter. { 'shärp,stōn }

sharpstone conglomerate See sedimentary breccia. { 'shärp,stōn kən'gläm·ə·rət }

sharp tuning [ELEC] Having high selectivity; responding only to a desired narrow range of frequencies. { 'shärp 'tün·iŋ }

sharp V thread [DES ENG] A screw thread having a sharp crest and root; the included angle is usually 60°. { 'shärp 'vē ˌthred }

shatter breccia [PETR] A tectonic breccia composed of angular fragments that show little rotation. { 'shad·ər 'brech·ə }

shatter cone [GEOL] A striated conical rock fragment along which fracturing has occurred. { 'shad·ər ˌkōn }

shattercrack See flake. { 'shad·ər,krak }

shattering [FL MECH] One theory to explain homogenization or globule fractionation in milk; it is the effect that occurs when whole milk under high velocity strikes a flat surface, such as an impact ring. [MECH] The breaking up into highly irregular, angular blocks of a very hard material that has been subjected to severe stresses. { 'shad·ə·riŋ }

shatterproof glass See nonshattering glass. { 'shad·ər¦prüf 'glas }

shatter zone [GEOL] An area of randomly fissured or cracked rock that may be filled by mineral deposits, forming a network pattern of veins. { 'shad·ər ˌzōn }

shattuckite [MINERAL] $Cu_5(SiO_3)_4H_2O$ A blue mineral composed of basic copper silicate, occurring in fibrous masses. { 'shad·ə,kīt }

Shaula [ASTRON] A blue-white subgiant star of stellar magnitude 1.7, spectral classification B2-IV, in the constellation Scorpius; the star λ Scorpii. { 'shaù·lə }

shave hook [DES ENG] A plumber's or metalworker's tool composed of a sharp-edged steel plate on a shank; used for scraping metal. { 'shāv ˌhùk }

shaving [ENG ACOUS] Removing material from the surface of a disk recording medium to obtain a new recording surface. [MECH ENG] **1.** Cutting off a thin layer from the surface of a workpiece. **2.** Trimming uneven edges from stampings, forgings, and tubing. { 'shāv·iŋ }

Shaw process [MET] A foundry molding process which makes use of wood or metal patterns and a refractory mold bonded with an ethyl silicate base material. { 'shȯ ˌprä·səs }

sheaf [MATH] A fiber bundle with algebraic and topological structure usually associated to a differentiable manifold M which reflects the local behavior of differentiable functions on M. [ORD] Planned planes of fire which produce a desired pattern of bursts with rounds fired by two or more weapons. { shēf }

sheaf of planes [MATH] All the planes passing through a given point. Also known as bundle of planes. { 'shēf əv 'plānz }

sheaf structure [GEOL] A bundled arrangement of crystals that is characteristic of certain fibrous minerals, such as stibnite. { 'shēf ˌstrək·chər }

shear [DES ENG] A cutting tool having two opposing blades between which a material is cut. [ENG] An apparatus for hoisting heavy loads consisting of two or more poles fastened together at their upper ends and spread apart at their lower ends, secured or steadied by a guy or guys, and provided with a tackle. Also known as shear legs. [MECH] *See* shear strain. [MIN ENG] To make vertical cuts in a coal seam that has been undercut. { 'shir }

shear angle [MECH ENG] The angle made by the shear plane with the work surface. { 'shir ,aŋ·gəl }

shear burst [MIN ENG] The explosive breaking of wall rock in a deep mining field by the occurrence of a single shear crack parallel to the face in one of the walls, causing rock behind the shear plane to expand freely into the stope and then to disrupt and fill the place with debris. { 'shir ,bərst }

shear cell [ENG] The component for holding the powder in an apparatus for making measurements of the failure properties of a sample of powder. { 'shir ,sel }

shear center *See* center of twist. { 'shir ,sen·tər }

shear cleavage *See* slip cleavage. { 'shir klē·vij }

shear diagram [MECH] A diagram in which the shear at every point along a beam is plotted as an ordinate. { 'shir ,dī·ə,gram }

shear drag *See* shear resistance. { 'shir ,drag }

shear fold [GEOL] A similar fold whose mechanism is shearing or slipping along closely spaced planes that are parallel to the fold's axial surface. Also known as glide fold; slip fold. { 'shir ,fōld }

shear fracture [MECH] A fracture resulting from shear stress. { 'shir ,frak·chər }

shear-gravity wave [GEOPHYS] A combination of gravity waves and a Helmholtz wave on a surface of discontinuity of density and velocity. { 'shir 'grav·əd·ē ,wāv }

shearing [MECH ENG] Separation of material by the cutting action of shears. [MIN ENG] The vertical side cutting which, together with holing or horizontal undercutting, constitutes the attack upon a face of coal. [TEXT] A process to remove surface irregularities in a napped fabric by passing it through a cylindrical machine with rotating spiral blades. { 'shir·iŋ }

shearing die [MECH ENG] A die with a punch for shearing the work from the stock. { 'shir·iŋ ,dī }

shearing field [PL PHYS] A special type of magnetic field, used to confine a plasma whose rotational transform angle changes with distance from the magnetic axis. { 'shir·iŋ ,fēld }

shearing forces [MECH] Two forces that are equal in magnitude, opposite in direction, and act along two distinct parallel lines. { 'sher·iŋ ,fôrs·əz }

shearing instability *See* Helmholtz instability. { 'shir·iŋ ,in·stə,bil·əd·ē }

shearing interferometer [OPTICS] An interferometer in which a wavefront is interfered with a version of itself that has been modified in some manner; examples are lateral shear interferometers and radial shear interferometers. { 'shir·iŋ ,in·tər·fə'räm·əd·ər }

shearing machine [MECH ENG] A machine for cutting cloth or bars, sheets, or plates of metal or other material. { 'shir·iŋ mə,shēn }

shearing punch [MECH ENG] A punch that cuts material by shearing it, with minimal crushing effect. { 'shir·iŋ ,pənch }

shearing strain [MECH] The distortion that results from motion of material on opposite sides of a plane in opposite directions parallel to the plane. { 'shir·iŋ ,strān }

shearing stress [MECH] A stress in which the material on one side of a surface pushes on the material on the other side of the surface with a force which is parallel to the surface. Also known as shear stress; tangential stress. { 'shir·iŋ ,stres }

shearing tool [DES ENG] A cutting tool (for a lathe, for example) with a considerable angle between its face and a line perpendicular to the surface being cut. { 'shir·iŋ ,tül }

shear joint [GEOL] A joint that is a shear fracture; it is a potential plane of shear. Also known as slip joint. { 'shir ,jóint }

shear legs *See* shear. { 'shir ,legz }

shear line [METEOROL] A line or narrow zone across which there is an abrupt change in the horizontal wind component parallel to this line; a line of maximum horizontal wind shear. { 'shir ,līn }

shear lip [MET] An area or ridge at the edge of a shear fracture surface. { 'shir ,lip }

shear mark [ENG] A crease on a piece of pressed glass; results when the piece is sheared off for pressing. { 'shir ,märk }

shear modulus *See* modulus of elasticity in shear. { 'shir ,mäj·ə·ləs }

shear moraine [GEOL] A debris-laden surface or zone found along the margin of any ice sheet or ice cap, dipping in toward the center of the ice sheet but becoming parallel to the bed at the base. { 'shir mə'rān }

shear pin [DES ENG] **1.** A pin or wire provided in a fuse design to hold parts in a fixed relationship until forces are exerted on one or more of the parts which cause shearing of the pin or wire; the shearing is usually accomplished by setback or set forward (impact) forces; the shear member may be augmented during transportation by an additional safety device. **2.** In a propellant-actuated device, a locking member which is released by shearing. **3.** In a power train, such as a winch, any pin, as through a gear and shaft, which is designed to fail at a predetermined force in order to protect a mechanism. { 'shir ,pin }

shear plane [GEOL] *See* shear surface. [HYD] A planar surface in a glacier, usually laden with rock debris, attributed to ciscontinuous shearing or overthrusting. [MECH] A confined zone along which fracture occurs in metal cutting. { 'shir ,plān }

shear rate [FL MECH] The relative velocities in laminar flow of parallel adjacent layers of a fluid body under shear force. { 'shir ,rāt }

shear resistance [FL MECH] A tangential stress caused by fluid viscosity and taking place along a boundary of a flow in the tangential direction of local motion. Also known as shear drag. { 'shir ri,zis·təns }

shear slide [GEOL] A landslide, especially a slump, produced by shear failure usually along a plane of weakness such as a bedding or cleavage plane. { 'shir ,slīd }

shear sorting [GEOL] Sorting of sediments in which the smaller grains tend to move toward the zone of greatest shear strain, and the larger grains toward the zone of least shear. { 'shir ,sórd·iŋ }

shear spinning [MECH ENG] A sheet-metal-forming process which forms parts with rotational symmetry over a mandrel with the use of a tool or roller in which deformation is carried out with a roller in such a manner that the diameter of the original blank does not change but the thickness of the part decreases by an amount dependent on the mandrel angle. { 'shir ,spin·iŋ }

shear steel [MET] A cutlery steel made from short sheared lengths of blister steel; the lengths are heated, joined by rolling or hammering, and finished by hammering. { 'shir ,stēl }

shear strain [MECH] Also known as shear. **1.** A deformation of a solid body in which a plane in the body is displaced parallel to itself relative to parallel planes in the body; quantitatively, it is the displacement of any plane relative to a second plane, divided by the perpendicular distance between planes. **2.** The force causing such deformation. { 'shir ,strān }

shear strength [MECH] **1.** The maximum shear stress which a material can withstand without rupture. **2.** The ability of a material to withstand shear stress. { 'shir ,streŋkth }

shear stress *See* shearing stress. { 'shir ,stres }

shear structure [GEOL] A local structure in which earth stresses have been relieved by many small, closely spaced fractures. { 'shir ,strək·chər }

shear surface [GEOL] A surface along which differential movement has taken place parallel to the surface. Also known as shear plane. { 'shir ,sər·fəs }

shear test [ENG] Any of various tests to determine shear strength of soil samples. { 'shir ,test }

shear thickening [FL MECH] Viscosity increase of non-Newtonian fluids (for example, complex polymers, proteins, protoplasm) that undergo viscosity increases under conditions of shear stress (that is, viscometric flow). { 'shir ,thik·ən·iŋ }

shear thinning [FL MECH] Viscosity reduction of non-Newtonian fluids (for example, polymers and their solutions, most slurries and suspensions, lube oils with viscosity-index improvers) that undergo viscosity reductions under conditions of shear stress (that is, viscometric flow). { 'shir ,thin·iŋ }

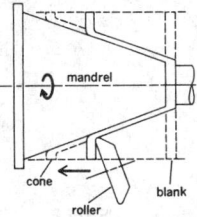

SHEAR SPINNING

Schematic of shear spinning showing the blank in three positions.

shear transformation *See* martensitic transformation. { 'shir tranz·fər'mā·shən }

shear-viscosity function [FL MECH] The expression of the viscometric flow of a purely viscous, non-Newtonian fluid in terms of velocity gradient and shear stress of the flowing fluid. { 'shir vi'skäs·əd·ē ,fəŋk·shən }

shearwater [VERT ZOO] Any of various species of oceanic birds of the genus *Puffinus* having tubular nostrils and long wings. { 'shir,wȯd·ər }

shear wave [GEOPHYS] *See* S wave. [MECH] A wave that causes an element of an elastic medium to change its shape without changing its volume. Also known as rotational wave. { 'shir ,wāv }

shear zone [GEOL] A tabular area of rock that has been crushed and brecciated by many parallel fractures resulting from shear strain; often becomes a channel for underground solutions and the seat of ore deposition. { 'shir ,zōn }

sheath [ELEC] A protective outside covering on a cable. [ELECTR] A space charge formed by ions near an electrode in a gas tube. [ELECTROMAG] The metal wall of a waveguide. [SCI TECH] A protective case or cover. { shēth }

sheathed bacteria [MICROBIO] Chains of bacterial cells, usually rod-shaped, enclosed in a hyaline envelope or sheath. { 'shēthd bak'tir·ē·ə }

sheathed explosive [ENG] A permitted explosive enveloped by a sheath containing a noncombustible powder which reduces the temperature of the resultant gases of the explosion and, therefore, reduces the risk of these hot gases causing a firedamp ignition. { 'shēthd ik'splō·siv }

sheathing board [MATER] A composition board (for example, of fiber or gypsum cement) used instead of wood sheathing. { 'shēth·iŋ ,bȯrd }

sheathing paper [MATER] A paper that is heavier and of better quality than the usual building paper. { 'shēth·iŋ ,pā·pər }

sheath-reshaping converter [ELECTROMAG] In a waveguide, a mode converter in which the change of wave pattern is achieved by gradual reshaping of the sheath of the waveguide and of conducting metal sheets mounted longitudinally in the guide. { 'shēth re̩'shāp·iŋ kən'vərd·ər }

sheave [DES ENG] A grooved wheel or pulley. { shēv }

shed [NUC PHYS] A unit of cross section, used in studying collisions of nuclei and particles, equal to 10^{-24} barn, or 10^{48} square centimeter. { shed }

SHED *See* solar heat exchanger drive. { shed }

shedding [TEXT] A process performed by the heddle frame and heddles of a loom to raise and lower certain groups of alternate warp yarns so that the filling yarns alternate in passing under one group of warp yarns and over another. { 'shed·iŋ }

shed dormer [ARCH] A dormer window which (unlike a gabled dormer) has a horizontal eave line. { 'shed 'dȯr·mər }

shed roof [ARCH] A flat roof that slopes in one direction and may lean against another wall or building. Also known as lean- to roof. { 'shed 'rüf }

SHEED *See* scanning HEED. { shed *or* 'es,hēd }

Sheehan's syndrome [MED] Pituitary necrosis caused by insufficient blood supply to the gland due to copious blood loss or intravascular clotting and hemorrhage associated with premature separation of the placenta during childbirth. Also known as postpartum pituitary necrosis. { 'shē·ənz ,sin ,drōm }

sheen [OPTICS] A subdued and often iridescent or metallic glitter which approaches, but is just short of, optical reflection and which modifies the surface luster of a mineral. { shēn }

sheep [VERT ZOO] Any of various mammals of the genus *Ovis* in the family Bovidae characterized by a stocky build and horns, when present, which tend to curl in a spiral. { shēp }

sheepsfoot roller [DES ENG] A cylindrical steel drum to which knob-headed spikes are fastened; used for compacting earth. Also known as tamping roller. { 'shēps,fút ,rōl·ər }

sheep's-head clock [HOROL] A lantern clock having one hand, a crown escapement, and large dials overlapping the clock movement. { 'shēps ,hed ,kläk }

sheepskin wheel [DES ENG] A polishing wheel made of sheepskin disks or wedges either quilted or glued together. { 'shēp,skin ,wēl }

sheer [GEOL] A steep face of a cliff. [NAV] To swerve off course to avoid collision. [NAV ARCH] The curvature of a ship's deck from bow to stern, with the lowest point of the deck usually amidships. { shir }

sheering batten [NAV ARCH] One of the long strips of wood attached to the frames of a ship to locate the strakes of the shell plating in relation to the sheer of the deck. { 'shir·iŋ ,bat·ən }

sheer line [NAV ARCH] The longitudinal curve of the rail or decks, which shows the variation in height above water or freeboard throughout the vessel's entire length. { 'shir ,līn }

sheer strake [NAV ARCH] The uppermost line of planking of a wooden ship, or the upper strake of plating at a steel ship's main deck. { 'shir ,strāk }

sheer-strake plate [NAV ARCH] A plate forming part of a sheer strake. { 'shir ,strāk ,plāt }

sheet [GEOL] 1. A thin flowstone coating of calcite in a cave. 2. A tabular igneous intrusion, especially when concordant or only slightly discordant. [HYD] *See* sheetflood. [MATER] A material in a configuration similar to a film except that its thickness is greater than 0.25 millimeter. [MATH] 1. A portion of a surface such that it is possible to travel continuously between any two points on it without leaving the surface. 2. A part of a Riemann surface such that any extension results in a multiple covering of some part of the complex plane over which the surface lies. [NAV ARCH] A rope or chain used to haul the clew of a sail out toward the yard arm or downward toward the deck and aft. { shēt }

sheet asphalt [MATER] Asphalt which provides a smooth surface and is used for continuous road surfacing. { 'shēt 'as,fȯlt }

sheet cavitation [FL MECH] A type of cavitation in which cavities form on a solid boundary and remain attached as long as the conditions that led to their formation remain unaltered. Also known as steady-state cavitation. { 'shēt ,kav·ə'tā·shən }

sheet composting [AGR] Addition of large amounts of organic residue to a soil; extra nitrogen is usually added for faster decomposition. { 'shēt kəm,pōst·iŋ }

sheet copper [MET] Copper rolled into sheets; for roofing sometimes used as it leaves the rolls, but for other purposes it is commonly employed after it has been cold-rolled to increase hardness and strength. { 'shēt ,käp·ər }

sheet crack [GEOL] A planar crack attributed to shrinkage of sediment due to dewatering. { 'shēt ,krak }

sheet deposit [GEOL] A stratiform mineral deposit that is more or less horizontal and extensive relative to its thickness. { 'shēt di,päz·ət }

sheet drift [GEOL] An evenly spread deposit of glacial drift that did not significantly alter the form of the underlying rock surface. { 'shēt ,drift }

sheeted fissure [GEOL] A closely spaced fissure. { 'shēd·əd 'fish·ər }

sheeted vein [GEOL] A vein filling a shear zone. { 'shēd·əd 'vān }

sheeted zone [GEOL] An area of mineral deposits consisting of sheeted veins. { 'shēd·əd 'zōn }

sheet erosion [GEOL] Erosion of thin layers of surface materials by continuous sheets of running water. Also known as sheetflood erosion; sheetwash; surface wash; unconcentrated wash. { 'shēt i,rō·zhən }

sheet-fed press [GRAPHICS] A printing press that accepts paper in the form of sheets. { 'shēd ,fed 'pres }

sheet feeder [COMPUT SCI] A device that feeds noncontinuous forms or sheets of paper into a printer. { 'shēt ,fēd·ər }

sheet film [GRAPHICS] Film that consists of a negative material with its emulsion coated on a base of plastic rather than glass. { 'shēd ,film }

sheetflood [HYD] A broad expanse of moving, storm-borne water that spreads as a thin, continuous, relatively uniform film over a large area for a short distance and duration. Also known as sheet; sheetwash. { 'shēt,fləd }

sheetflood erosion *See* sheet erosion. { 'shēd,fləd i,rō·zhən }

sheet flow [HYD] An overland flow or downslope movement of water taking the form of a thin, continuous film over relatively smooth soil or rock surfaces and not concentrated into channels larger than rills. { 'shēt,flō }

sheet forming [ENG] The process of producing thin, flat sections of solid materials; for example, sheet metal, sheet plastic, or sheet glass. { 'shēt ,fȯrm·iŋ }

SHEEP

(a)

(b)

Examples of some prominent breeds of sheep. (*a*) Hampshire ram. (*b*) Southdown ram (*Southdown Association*).

sheet frost [HYD] A thick coating of rime formed on windows and other surfaces. { 'shēt ‚frȯst }

sheet glass [MATER] Flat sections of glass made by drawing a continuous thin film of glass from a molten bath, then cooling and cutting it; used for common glazing. { 'shēt ‚glas }

sheet grating [ELECTROMAG] Three-dimensional grating consisting of thin, longitudinal, metal sheets extending along the inside of a waveguide for a distance of about a wavelength, and used to stop all waves except one predetermined wave that passes unimpeded. { 'shēt ‚grād·iŋ }

sheet ice [HYD] A smooth, thin layer of ice formed by rapid freezing of the surface layer of a body of water. { 'shēt ‚īs }

sheeting [GEOL] The process by which thin sheets, slabs, scales, plates, or flakes of rock are successively broken loose or stripped from the outer surface of a large rock mass in response to release of load. Also known as exfoliation. [MATER] **1.** A continuous film of a material such as plastic. **2.** Steel or wood members used to face the walls of an excavation such as a basement or a trench. { 'shēd·iŋ }

sheeting caps [MIN ENG] A row of caps put on blocks about 14 inches (36 centimeters) high which are placed on top of the drift sets when constructing the permanent floor in the stope. { 'shēd·iŋ ‚kaps }

sheeting plane [PETR] In igneous rocks, the primary cleavage plane or parting. { 'shēd·iŋ ‚plān }

sheeting structure [GEOL] A fracture or joint formed by pressure-release jointing or exfoliation. Also known as exfoliation joint; expansion joint; pseudostratification; release joint; sheet joint; sheet structure. { 'shēd·iŋ ‚strȯk·chər }

sheet joint See sheeting structure. { 'shēt ‚jȯint }

sheet lightning [GEOPHYS] A diffuse, but sometimes fairly bright, illumination of those parts of a thundercloud that surround the path of a lightning flash, particularly a cloud discharge or cloud-to-cloud discharge. Also known as luminous cloud. { 'shēt ‚līt·niŋ }

sheet line [MAP] The outermost border line of a map or chart. { 'shēt ‚līn }

sheet metal [MET] Thin sections of metal formed by rolling hot metal and usually less than 0.25 inch (6.35 millimeters) thick; when thicker than 0.25 inch, called plate. { 'shēt ‚med·əl }

sheet-metal gage [MET] A standard for expressing the thickness of metal sheets; some manufacturers, for example, Brown & Sharpe (B&S), Birmingham (BG), and Imperial, use code numbers with actual thickness in inches or millimeters. { 'shēt ¦med·əl ‚gāj }

sheet-metal screw See self-tapping screw. { 'shēt ¦med·əl ‚skrü }

sheet mica [MATER] Mica that is relatively flat and sufficiently free from structural defects to enable it to be punched or stamped into specified shapes for use by the electronic and electrical industries. { 'shēt ‚mī·kə }

sheet mineral See phyllosilicate. { 'shēt ‚min·rəl }

sheet piling [CIV ENG] Closely spaced piles of wood, steel, or concrete driven vertically into the ground to obstruct lateral movement of earth or water, and often to form an integral part of the permanent structure. { 'shēt ‚pīl·iŋ }

sheet plastic [MATER] Flat sections of extruded, molded, or cast plastic, with a thickness greater than that for film, that is, greater than 0.05 inch (1.3 millimeters). { 'shēt ‚plas·tik }

sheet polarizer [OPTICS] A mechanism for obtaining linear polarized light; there are several types, one of which is a microcrystalline polarizer in which small crystals of a dichroic material (quinine iodosulfate), oriented parallel to each other in a plastic medium, absorb one polarization and transmit the other. { 'shēt 'pō·lə‚rīz·ər }

sheet rubber [MATER] Latex that has been rolled into sheets, either smooth or ribbed. { 'shēt ¦rəb·ər }

sheet sand See blanket sand. { 'shēt ‚sand }

sheet sandstone [GEOL] A thin, blanket-shaped deposit of sandstone of regional extent. { 'shēt 'san‚stōn }

sheet separation [MET] The gap between faying surfaces surrounding the weld in spot, seam, or projection welding. { 'shēt ‚sep·ə‚rā·shən }

sheet silicate See phyllosilicate. { 'shēt 'sil·ə·kət }

sheet spar [GEOL] A sheet crack filled with spar. { 'shēt ‚spär }

sheet steel [MET] Steel rolled in the form of sheet, usually used for deep-drawing applications. { 'shēt ¦stēl }

sheet structure See sheeting structure. { 'shēt ‚strȯk·chər }

sheet train [ENG] The entire assembly needed to produce plastic sheet; includes the extruder, die, polish rolls, conveyor, draw rolls, cutter, and stacker. { 'shēt ‚trān }

sheetwash [GEOL] **1.** The detritus deposited by a sheetflood. **2.** See sheet erosion. [HYD] **1.** A wide, moving expanse of water on an arid plain; the combined result of many streams issuing from the mountains. **2.** See sheetflood. { 'shēt ‚wäsh }

sheffer stroke See NAND. { 'shef·ər ‚strōk }

Sheffield plate [MET] A cladding of silver rolled and fused on both sides of a copper sheet. { 'she‚fēld 'plāt }

Shelby tube [ENG] A thin-shelled tube used to take deep-soil samples; the tube is pushed into the undisturbed soil at the bottom of the casting of the borehole driven into the ground. { 'shel·bē ‚tüb }

shelf [GEOL] **1.** Solid rock beneath alluvial deposits. **2.** A flat, projecting ledge of rock. **3.** See continental shelf. { shelf }

shelf angle [CIV ENG] A mild steel angle section, riveted or welded to the web of an I beam to support the formwork for hollow tiles or the floor or roof units, or to form a seat for precast concrete. { 'shelf ‚aŋ·gəl }

shelf break [GEOL] An obvious steepening of the gradient between the continental shelf and the continental slope. { 'shelf ‚brāk }

shelf channel [GEOL] A valley formed in a shelf by erosion. { 'shelf ‚chan·əl }

shelf edge [GEOL] The demarcation, without dramatic change in gradient, between continental shelf and continental slope { 'shelf ‚ej }

shelf facies [GEOL] A sedimentary facies characterized by carbonate rocks and fossil shells and produced in the neritic environments of marginal shelf seas. Also known as foreland facies; platform facies. { 'shelf ‚fā·shēz }

shelf ice [HYD] The ice of an ice shelf. Also known as barrier ice. { 'shelf ‚īs }

shelf life [ENG] The time that elapses before stored food, chemicals, batteries, and other materials or devices become inoperative or unusable due to age or deterioration. { 'shelf ‚līf }

shelf sea [OCEANOGR] A shallow marginal sea located on the continental shelf, usually less than 150 fathoms (275 meters) in depth; an example is the North Sea. { 'shelf ‚sē }

shelfstone [GEOL] A speleothem formed at the water's edge as a horizontally projecting ledge. { 'shelf‚stōn }

shell [ARCH] A reinforced concrete arched or domed roof used over unpartitioned areas. [BUILD] A building without internal partitions or furnishings. [COMPUT SCI] A program that provides an interface between a user and the computer's operating system by reading commands and sending them to the operating system for execution. [DES ENG] **1.** The case of a pulley block. **2.** A thin hollow cylinder. **3.** A hollow hemispherical structure. **4.** The outer wall of a vessel or tank. [GEOL] **1.** The crust of the earth. **2.** A thin hard layer of rock. [GRAPHICS] An engraved roller made of copper and used in calico printing. [MET] **1.** The outer wall of a metal mold. **2.** The hard layer of sand and thermosetting plastic formed over a pattern and used as a mold wall in shell molding. **3.** The metal sleeve remaining when a billet is extruded with a dummy block at smaller diameter. **4.** A tubular casting used in preparing seamless drawn tubes. **5.** A pierced forging. [ORD] **1.** A hollow metal projectile designed to be projected from a gun, containing or intended to contain a high-explosive, chemical, atomic, or other charge. **2.** A shotgun cartridge or a cartridge for artillery of small arms. [PHYS] A set of energy levels with approximately the same energy in an atom or nucleus. [ZOO] **1.** A hard, usually calcareous, outer covering on an animal body, as of bivalves and turtles. **2.** The hard covering of an egg. **3.** Chitinous exoskeleton of certain arthropods. { shel }

shellac [MATER] A natural, alcohol-soluble, water-insoluble, flammable resin; made from lac resin deposited on tree twigs in India by the lac insect (Laccifer lecca) used as an ingredient of wood coatings. { shə'lak }

shell account [COMPUT SCI] A type of limited access to the Internet in which the user is connected to the Internet indirectly through a second computer on which the user has established an account. { 'shel ə‚kau̇nt }

shellac varnish [MATER] A solution of shellac in denatured alcohol; used in wood finishing where a fast-drying, light-colored, hard finish is desired. { shə'lak ˌvär·nish }

shellac wax [MATER] A hard wax with 3% shellac, from which it is extracted, and used in polishes and insulating materials. { shə'lak ˌwaks }

shellac wheel [DES ENG] A grinding wheel having the abrasive bonded with shellac. { shə'lak ˌwēl }

shell-and-tube exchanger [ENG] A device for the transfer of heat from a hot fluid to a cooler fluid; one fluid passes through a group (bundle) of tubes, the other passes around the tubes, through a surrounding shell. Also known as tubular exchanger. { ¦shel ən ¦tüb iks'chān·jər }

shell capacity [ENG] The amount of liquid that a tank car or tank truck will hold when the liquid just touches the underside of the top of the tank shell. { 'shel kəˌpas·əd·ē }

shell clearance [DES ENG] The difference between the outside diameter of a bit or core barrel and the outside set or gage diameter of a reaming shell. { 'shel ˌklir·əns }

shell core [MET] A sand core formed by shell molding. { 'shel ˌkȯr }

shell-destroying tracer [ORD] A tracer that includes an explosive element beyond the tracer element and that is designed to cause activation of the explosive by the tracer after the antiaircraft projectile has passed the target point, thus destroying the projectile to avoid impact in friendly territory. { 'shel di¦strȯi·iŋ 'trā·sər }

shell filler [ORD] Explosive or other material used to make up the filler or charge in a projectile. { 'shel ˌfil·ər }

shellfish [INV ZOO] An aquatic invertebrate, such as a mollusk or crustacean, that has a shell or exoskeleton. { 'shelˌfish }

shell-form transformer [ELECTROMAG] A transformer in which all of the widings are on the center of three legs. { 'shel ˌfȯrm tranzˌfȯr·mər }

shell gland [INV ZOO] An organ that secretes the embryonic shell in many mollusks. [VERT ZOO] A specialized structure attached to the oviduct in certain animals that secretes the eggshell material. { 'shel ˌgland }

shell ice [HYD] Ice, on a body of water, that remains as an unbroken surface when the water level drops so that a cavity is formed between the water surface and the ice. { 'shel ˌīs }

shell innage [ENG] The depth of a liquid in a tank car or tank truck shell. { 'shel ˌin·ij }

shell knocker [ENG] A device to strike the external surface of a horizontally rotating process vessel (for example, a kiln or a dryer) to loosen accumulations of solid materials from the inner walls or flights of the shell. Also known as knocker. { 'shel ˌnäk·ər }

shell marl [GEOL] A light-colored calcareous deposit formed on the bottoms of small fresh-water lakes, composed largely of uncemented mollusk shells and precipitated calcium carbonate, along with the hard parts of minute organisms. { 'shel ˌmärl }

shell membrane [CELL MOL] Either of a pair of membranes lining the inner surface of an egg shell; they allow free entry of oxygen but prevent rapid evaporation of moisture. { 'shel ˌmemˌbrān }

shell method [MATH] A method of computing the volume of a solid of revolution by integrating over the volumes of infinitesimal shell-shaped sections bounded by cylinders with the same axis of revolution as the solid. { 'shel ˌmeth·əd }

shell model [NUC PHYS] A model of the nucleus in which the shell structure is either postulated or is a consequence of other postulates; especially the model in which the nucleons act as independent particles filling a preassigned set of energy levels as permitted by the quantum numbers and Pauli principle. { 'shel ˌmäd·əl }

shell molding [MET] Forming a rigid, porous, self-supporting refractory mold by sprinkling molding sand blended with thermosetting plastic or resin over a preheated metal pattern and then curing in an oven. { 'shel ˌmōld·iŋ }

shell outage [ENG] The unfilled portion of a tank car or tank truck shell; the distance from the underside of the top of the shell to the level of the liquid in the shell. { 'shel ˌaud·ij }

shell plating [NAV ARCH] The plating forming the outer skin of a vessel; in addition to providing a watertight boundary, it contributes largely to the strength of the vessel. { 'shel ˌplād·iŋ }

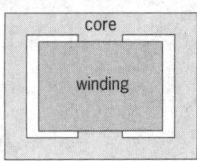

SHELL-FORM TRANSFORMER

Arrangement of winding.

shell pump [MECH ENG] A simple pump for removing wet sand or mud; consists of a hollow cylinder with a ball or clack valve at the bottom. { 'shel ˌpəmp }

shell reamer [DES ENG] A machine reamer consisting of two parts, the arbor and the replaceable reamer, with straight or spiral flutes; designed as a sizing or finishing reamer. { 'shel ˌrēm·ər }

shell roof [BUILD] A roof made of a thin, curved, platelike structure, usually of concrete but lumber and steel are also used. { 'shel ˌrüf }

shell room [ORD] A large room for the storage of projectiles. { 'shel ˌrüm }

shell sand [GEOL] A loose aggregate that is largely composed of shell fragments of sand size. { 'shel ˌsand }

shell star [ASTRON] A type of star which is believed to be surrounded by a tenuous envelope of gas, as indicated by bright emission lines in its spectrum. { 'shel ˌstär }

shell still [CHEM ENG] A distillation device formerly used in petroleum refineries; oil was charged into a closed, cylindrical shell and heat was applied to the outside of the bottom by a firebox. { 'shel ˌstil }

shell structure [NUC PHYS] Structure of the nucleus in which nucleons of each kind occupy quantum states which are in groups of approximately the same energy, called shells, the number of nucleons in each shell being limited by the Pauli exclusion principle. { 'shel ˌstrək·chər }

shell-type transformer [ELECTROMAG] Transformer in which the magnetic circuit completely surrounds the windings. { 'shel ¦tīp tranz'fȯr·mər }

shelly [GEOL] **1.** Pertaining to a sediment or sedimentary rock containing the shells of animals. **2.** Pertaining to land abounding in or covered with shells. { 'shel·ē }

shelly facies [GEOL] A nongeosynclinal sedimentary facies that is commonly characterized by abundant calcareous fossil shells, dominant carbonate rocks (limestones and dolomites), mature orthoquartzitic sandstones, and a paucity of shales. { 'shel·ē 'fā·shēz }

shelly pahoehoe [GEOL] A type of pahoehoe characterized by open tubes and blisters on the surface. { 'shel·ē pə'hō·ē,hō·ē }

shelterbelt [ECOL] A natural or planned barrier of trees or shrubs to reduce erosion and provide shelter from wind and storm activity. { 'shel·tər,belt }

shelter cave [GEOL] A cave which extends only a short way underground, and whose roof of overlying rock usually extends beyond its sides. Also known as rock cave. { 'shel·tər ˌkāv }

shelter deck [NAV ARCH] A ship's deck which is above the principal deck and is continuous and of light construction. { 'shel·tər ˌdek }

shelter porosity [GEOL] A type of primary interparticle porosity created by the sheltering effect of relatively large sedimentary particles which prevent the infilling of pore space by finer clastic particles. { 'shel·tər pə'räs·əd·ē }

shelterwood method [FOR] A method for ensuring tree reproduction; older trees are removed by successive cuttings so that the amount of light reaching the seedlings is gradually increased. { 'shel·tər,wud ,meth·əd }

Shelton loader [MIN ENG] A modified coal-cutting machine in which the picks of the cutter chain are replaced by loading flights, which push the prepared coal up a ramp on to the face conveyor. { 'shel·tən 'lōd·ər }

Shenstone effect [ELECTR] An increase in photoelectric emission of certain metals following passage of an electric current. { 'shen,stōn i,fekt }

shepherding satellite [ASTRON] A satellite that helps to hold in place a given ring of a planet. { ¦shep·ərd·iŋ 'sad·əl,īt }

Sheppard's correction [STAT] A correction for moments computed for a frequency distribution of grouped data to adjust for the error that is introduced by the assumption that all the data within a class are at the midpoint of the class; the adjustment is made by subtracting one-twelfth of the square of a grouping unit from the estimated variance. { 'shep·ərdz kəˌrek·shənz }

sherardizing [MET] Coating iron with zinc by tumbling the article in powdered zinc at about 250–375°C. { shə'rär,dīz·iŋ }

shergottite [GEOL] An achondritic stony meteorite that is composed chiefly of pigeonite and maskelynite. { 'shər·gə,tīt }

sheridanite [MINERAL] $(Mg,Al)_6(Al,Si)_4O_{18}(OH)_8$ Pale-green to colorless talclike mineral composed of basic magnesium aluminum silicate. { 'sher·ə,nīt }

Sheridan tank [ORD] The United States' light tank, M551, introduced into limited service in 1967; it can be transported by air and dropped by parachute, has a crew of four, weighs 15 tons (13,600 kilograms), and is armed with a 152-millimeter gun launcher and two machine guns. { 'sher·ə·dən ,taŋk }

sherry [FOOD ENG] A dry to sweet fortified wine with nutty flavor and ranging from pale to dark amber in color. { 'sher·ē }

sherry topaz [MINERAL] A brownish-yellow to yellow-brown variety of topaz resembling sherry wine in color. { 'sher·ē 'tō,paz }

Sherwood number See Nusselt number. { 'shər,wůd ,nəm·bər }

shetland [TEXT] A fabric made of shetland wool with a raised finish and soft hand. { 'shet·lənd }

Shetland sheep [VERT ZOO] A breed of sheep raised in the Shetland Isles of Scotland. { 'shet·lənd 'shēp }

SHF See sensible-heat factor; superhigh frequency.

shield [ENG] An iron, steel, or wood framework used to support the ground ahead of the lining in tunneling and mining. [GEOL] **1.** The very old, rigid core of relatively stable rocks within a continent around which younger sedimentary rocks have been deposited. Also known as continental shield. **2.** See palette. [NUCLEO] The material placed around a nuclear reactor, or other source of radiation, to reduce escaping radiation or particles to a permissible level. Also known as shielding. [ORD] Armor plate mounted on a gun carriage to protect the operating mechanism and gun crew from enemy fire. { shēld }

Shield See Scutum. { shēld }

shield basalt [GEOL] A basaltic lava flow from a group of small, close-spaced shield-volcano vents that coalesced to form a single unit. { shēld bə'sòlt }

shield cone [GEOL] A cone or dome-shaped volcano built up by successive outpourings of lava. { shēld ,kōn }

shielded arc welding [MET] Arc welding in which the electric arc and the weld metal are protected by gas, decomposition products of the electrode covering, or a blanket of fusible flux. { 'shēl·dəd 'ärk ,weld·iŋ }

shielded-conductor cable [ELEC] Cable in which the insulated conductor or conductors are enclosed in a conducting envelope or envelopes, constructed so that substantially every point on the surface of the insulation is at ground potential or at some predetermined potential with respect to ground. { 'shēl·dəd kən¦dək·tər 'kā·bəl }

shielded-core superconducting fault-current limiter [ELEC] A limiter which is essentially a transformer, with its primary normal conducting coil connected in series to the line to be protected, while the secondary side is a superconducting tube (that is, a one-turn coil). Also known as inductive superconducting fault-current limiter; shorted-transformer superconducting fault-current limiter. { ¦shēl·dəd ,cȯr¦sü·pər·kən,dək·tiŋ 'fȯlt ,kər·ənt ,lim·əd·ər }

shielded joint [ELEC] Cable joint having its insulation so enveloped by a conducting shield that substantially every point on the surface of the insulation is at ground potential, or at some predetermined potential with respect to ground. { 'shēl·dəd 'jȯint }

shielded line [ELECTROMAG] Transmission line, the elements of which confine the propagated waves to an essentially finite space; the external conducting surface is called the sheath. { 'shēl·dəd 'līn }

shielded metal-arc welding [MET] Arc welding in which heating with an electric arc between the electrode and the work produces fusion of the electrode covering which shields the work. { 'shēl·dəd ¦med·əl 'ärk 'weld·iŋ }

shielded pair [ELEC] A pair of wires within a cable that is individually covered by a conducting shield. { 'shēld·əd 'per }

shielded wire [ELEC] Insulated wire covered with a metal shield, usually of tinned braided copper wire. { 'shēl·dəd 'wīr }

shield factor [COMMUN] Ratio of noise (or induced current or voltage) in a telephone circuit when a source of shielding is present to the corresponding quantity when the shielding is absent. { 'shēld ,fak·tər }

shield grid [ELECTR] A grid that shields the control grid of a gas tube from electrostatic fields, thermal radiation, and deposition of thermionic emissive material; it may also be used as an additional control electrode. { 'shēld ,grid }

shield-grid thyratron [ELECTR] A thyratron having a shield grid, usually operated at cathode potential. { 'shēld ¦grid 'thī·rə,trän }

shielding [ELECTROMAG] See electric shielding. [MET] Placing a nonconducting object in an electrolytic bath during plating to alter the current distribution. [NUCLEO] **1.** Reducing the ionizing radiation reaching one region of space from another region by using a shield or other device. **2.** See shield. { 'shēld·iŋ }

shielding distance See Debye shielding length. { 'shēld·iŋ ,dis·təns }

shielding factor [GEOPHYS] The ratio of the strength of the magnetic field at a directional compass to its strength if there were no disturbing material; usually expressed as a decimal. { 'shēld·iŋ ,fak·tər }

shielding gas [MET] Gas, such as nitrogen, oxygen, and carbon dioxide, used in shielded arc welding to protect molten weld from contamination and damage by the atmosphere. { 'shēld·iŋ ,gas }

shielding layer [METEOROL] The layer of air nearest the earth, with reference to the manner in which this layer shields the earth from activity in the free atmosphere above, or vice versa. { 'shēld·iŋ ,lā·ər }

shielding ratio [ELECTROMAG] The ratio of a field in a specified region when electrical shielding is in place to the field in that region when the shielding is removed. { 'shēld·iŋ ,rā·shō }

shield volcano [GEOL] A broad, low volcano shaped like a flattened dome and built of basaltic lava. Also known as basaltic dome; lava dome. { 'shēld väl,kā·nō }

shift [COMPUT SCI] A movement of data to the right or left, in a digital-computer location, usually with the loss of characters shifted beyond a boundary. [GEOL] The relative displacement of the units affected by a fault but outside the fault zone itself. [IND ENG] The number of hours or the part of any day worked. Also known as tour. [MECH ENG] To change the ratio of the driving to the driven gears to obtain the desired rotational speed or to avoid overloading and stalling an engine or a motor. [MET] A casting defect caused by malalignment of the mold parts. [SPECT] A small change in the position of a spectral line that is due to a corresponding change in frequency which, in turn, results from one or more of several causes, such as the Doppler effect. { shift }

shifting [GEOL] The movement of the crest of a divide away from a more actively eroding stream (as on the steeper slope of an asymmetric ridge) toward a weaker stream on the gentler slope. { 'shift·iŋ }

shifting theorem [MATH] **1.** If the Fourier transform of $f(t)$ is $F(x)$, then the Fourier transform of $f(t-a)$ is $\exp(iax)F(x)$. **2.** If the Laplace transform of $f(x)$ is $F(y)$, then the Laplace transform of $f(x-a)$ is $\exp(-ay)F(y)$. { 'shift·iŋ ,thir·əm }

shift joint [BUILD] A vertical joint placed on a solid member of the course below. { 'shift ,jȯint }

shift of butts [NAV ARCH] The arrangement of butt joints in plating. { 'shift əv 'bəts }

shift register [COMPUT SCI] A computer hardware element constructed to perform shifting of its contained data. { 'shift ,rej·ə·stər }

shift-register generator [COMPUT SCI] A random-number generator which consists of a sequence of shift operations and other operations, such as no-carry addition. { 'shift ¦rej·ə·stər ¦jen·ə,rād·ər }

shift work [IND ENG] Work paid for by day wage. { 'shift ,wərk }

shiitake mushroom See Letinula edodes. { ,shē·ē,tä·kē 'məsh,rům }

shikimic acid [BIOCHEM] $C_7H_{10}O_5$ A crystalline acid that is a plant constituent, and an intermediate in the biochemical pathway from phosphoenolpyruvic acid to tyrosine. { shə'kim·ik 'as·əd }

Shillelagh [ORD] A United States weapon system including a gun launcher and a fire-control system mounted on the main battle tank and assault reconnaissance vehicle for employment against enemy armor, troops, and field fortifications; it is guided by an infrared command link. { shə'lā,lē }

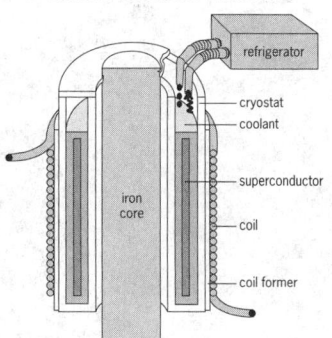

SHIELDED-CORE SUPERCONDUCTING FAULT-CURRENT LIMITER

Cross section. Only the superconducting tube is cooled by the refrigerator.

shim [ENG] **1.** In the manufacture of plywood, a long, narrow patch glued into the panel or cemented into the lumber core itself. **2.** A thin piece of material placed between two surfaces to obtain a proper fit, adjustment, or alignment. { shim }

shimmer [METEOROL] To appear tremulous or wavering, due to varying atmospheric refraction in the line of sight. { 'shim·ər }

shimmy [MECH] Excessive vibration of the front wheels of a wheeled vehicle causing a jerking motion of the steering wheel. { 'shim·ē }

shim rod [NUCLEO] A control rod used for making occasional coarse adjustments in the reactivity of a nuclear reactor. { 'shim ˌräd }

shingle [GEOL] Pebbles, cobble, and other beach material, coarser than ordinary gravel but roughly the same size and occurring typically on the higher parts of a beach. [MATER] A rectangular piece of wood, metal, or other material that is used like a tile and arranged in overlapping rows for covering roofs and walls. { 'shiŋ·gəl }

shingle barchan [GEOL] A dunelike ridge formed of shingle perpendicular to the beach in shallow water. { 'shiŋ·gəl bär'kän }

shingle beach [GEOL] A narrow beach composed of shingle and commonly having a steep slope on both its landward and seaward sides. Also known as cobble beach. { 'shiŋ·gəl ˌbēch }

shingle-block structure *See* imbricate structure. { 'shiŋ·gəl 'bläk 'strək·chər }

shingle lap [DES ENG] A lap joint in which the two surfaces are tapered, with the thinner surface lapped over the thicker one. { 'shiŋ·gəl ˌlap }

shingle nail [DES ENG] A nail about a half to a full gage thicker than a common nail of the same length. { 'shiŋ·gəl ˌnāl }

shingle rampart [GEOL] A rampart of shingle built along a reef on the seaward edge. { 'shiŋ·gəl 'ram,pärt }

shingle ridge [GEOL] A steeply sloping bank of shingle heaped upon and parallel with the shore. { 'shiŋ·gəl ˌrij }

shingles *See* herpes zoster. { 'shiŋ·gəlz }

shingle structure *See* imbricate structure. { 'shin·gəl ˌstrək·chər }

shingling *See* imbrication. { 'shiŋ·gliŋ }

shin splints [MED] An injury and an inflammation of the lower leg muscles and bones of the lower and middle third of the tibia and fibula, seen in athletes such as runners or basketball and football players. { 'shin ˌsplins }

ship [NAV ARCH] Any large vessel which travels over the seas, rivers, or lakes. { ship }

Ship *See* Argo. { ship }

ship auger [DES ENG] An auger consisting of a spiral body having a single cutting edge, with or without a screw; there is no spur at the outer end of the cutting edge. { 'ship ˌog·ər }

shipbuilding [CIV ENG] The construction of ships. { 'ship,bil·diŋ }

ship drift [OCEANOGR] A method of measuring ocean currents; the ship itself is used as a current tracer, its motions being measured by navigating equipment on board. { 'ship ˌdrift }

shipfitter [CIV ENG] A worker who builds the steel structure of a ship, including laying-off and fabricating the individual members, subassembly, and erection on the shipway. { 'ship,-fid·ər }

ship heading marker [NAV] A mark on a direction compass which indicates the position of the ship's head, such as a lubber's line, or an electronic radial sweep line on a PPI (plan position indicator). { 'ship 'hed·iŋ ˌmär·kər }

shiplap [MATER] Lumber whose edges are rabbeted in order to make a close overlapping joint. Also known as shiplap board; shiplap siding. { 'ship,lap }

shiplap board *See* shiplap. { 'ship,lap ˌbord }

shiplap siding *See* shiplap. { 'ship,lap ˌsīd·iŋ }

ship motion [ENG] Translational and rotational motions of a ship in a wave system which cause the center of gravity to deviate from simple straight-line motion; these motions are heave, surge, sway, roll, pitch, and yaw. { 'ship ˌmō·shən }

shipping [NAV ARCH] A term applied collectively to those ships which are used to transport personnel or cargo, or both; often modified to denote type, use, or force to which assigned. { 'ship·iŋ }

shipping and storage container [IND ENG] A reusable

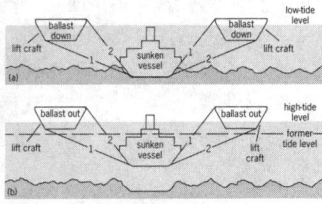

SHIP SALVAGE

Use of lift craft to refloat a vessel. *(a)* At low tide, the lift craft are ballasted down, and cables are passed beneath the vessel and pulled tight. *(b)* At high tide, the lift craft are deballasted, and the vessel is raised above the ocean bottom and towed away.

noncollapsible container of any configuration designed to provide protection for a specific item against impact, vibration, climatic conditions, and the like, during handling, shipment, and storage. { 'ship·iŋ ən 'stor·ij kən,tā·nər }

shipping control [ORD] All matters pertaining to convoy organization, routing, reporting, and diversion of shipping of all nations under charter thereto; it does not include cognizance over the general employment and allocation of shipping, harbor movements, and loading and unloading, which are functions of other agencies. { 'ship·iŋ kən,trōl }

shipping designator [COMMUN] A code word assigned to a particular overseas base, port, or area, for specific use as an address on shipments to the overseas location concerned; the code word is usually four letters and may be followed by a number to indicate a particular addressee. { 'ship·iŋ 'dez·ig,nād·ər }

shipping document [IND ENG] A document listing the items in a shipment, and showing other supply and transportation information that is required by agencies concerned in the movement of material. { 'ship·iŋ ,däk·yə·mənt }

shipping fever [VET MED] An acute, occasionally subacute, septicemic disease in cattle and sheep, probably caused by a combination of virus and *Pasteurella multocida* or *P. hemolytica*. { 'ship·iŋ ,fē·vər }

shipping lane [NAV] An established route traversed by ocean shipping. { 'ship·iŋ ,lān }

shipping time [ENG] The time elapsing between the shipment of material by the supplying activity and receipt of material by the requiring activity. { 'ship·iŋ ,tīm }

shipping ton *See* ton. { 'ship·iŋ ,tən }

ship report [METEOROL] The encoded and transmitted report of a marine weather observation. { 'ship ri,port }

ship salvage [NAV ARCH] Voluntary response, by other than a ship's own crew, to a maritime peril that threatens any type of vessel or maritime cargo. { 'ship ,sal·vij }

ship's clock [HOROL] A clock that rings to indicate the time according to the system of bells used on shipboard. { 'ships 'kläk }

ship's field error [NAV] That error in radio direction finder bearings due to the ship's directional antenna being located where the main component of the ship's field is not parallel to the ship's center line. { 'ships 'fēld ,er·ər }

ship's head [NAV] The heading of a vessel in degrees. { 'ships 'hed }

ship's lines *See* lines. { 'ships 'līnz }

ship synoptic code [METEOROL] A synoptic code for communicating marine weather observations; it is a modification of the international synoptic code. { 'ship si'näp·tik 'kōd }

ship-tended acoustic relay [NAV] An acoustic navigation system that employs ocean bottom transponders and determines the distance of the ship from these transponders by measuring the time required for a signal broadcast by an acoustic transducer below the hull of the ship to travel to the transponder, be rebroadcast there, and travel back to the ship. Abbreviated STAR. { 'ship 'ten·dəd ə'küs·tik 'rē,lā }

ship-to-shore movement [ORD] That portion of the assault phase of an amphibious operation which includes the deployment of the landing force from the assault shipping to designated landing areas. { 'ship tə 'shor 'müv·mənt }

shipway [CIV ENG] **1.** The ways on which a ship is constructed. **2.** The supports placed underneath a ship in dry dock. [NAV] A channel through which ships pass. { 'ship,wā }

shipworm [INV ZOO] Any of several bivalve mollusk species belonging to the family Teredinidae and which superficially resemble earthworms because the two valves are reduced to a pair of plates at the anterior of the animal or are used for boring into wood. { 'ship,wərm }

shipwright [CIV ENG] A worker whose responsibility is to ensure that the structure of a ship is straight and true and to the designed dimensions; the work starts with the laying down of the keel blocks and continues throughout the steelwork; applicable also to wood ship builders. { 'ship,rīt }

shipyard [CIV ENG] A facility adjacent to deep water where ships are constructed or repaired. { 'ship,yärd }

shipyard eye *See* keratoconjunctivitis. { 'ship,yärd 'ī }

shiran [ELECTR] Specially designed frequency-modulation continuous-wave distance-measuring equipment used for performing distance measurements of an accuracy comparable to first-order triangulation. Derived from S-band hiran. { 'shī,ran }

shivering [MATER] Cracks and scales on a pottery glaze caused by unequal contraction during cooling. { 'shiv·ə·riŋ }

SHM *See* harmonic motion.

shoal [GEOL] A submerged elevation that rises from the bed of a shallow body of water and consists of, or is covered by, unconsolidated material, and may be exposed at low water. { 'shōl }

shoal breccia [PETR] A breccia formed by the action of waves and tides on a shoal, and resulting from diastrophism or aggradation. { 'shōl ,brech·ə }

shoaling [OCEANOGR] The bottom effect which influences the height of waves moving from deep to shallow water. { 'shōl·iŋ }

shoal patches [OCEANOGR] Individual and scattered elevations of the bottom, with depths of 10 fathoms (18 meters) or less, but composed of any material except rock or coral. { 'shōl ,pach·əz }

shoal reef [GEOL] A reef formed in irregular masses amid submerged shoals of calcareous reef detritus. { 'shōl ,rēf }

shoal water [OCEANOGR] Shallow water; over a shoal. { 'shōl ,wȯd·ər }

shock [MECH] A pulse or transient motion or force lasting thousandths to tenths of a second which is capable of exciting mechanical resonances; for example, a blast produced by explosives. [MED] Clinical manifestations of circulatory insufficiency, including hypotension, weak pulse, tachycardia, pallor, and diminished urinary output. { 'shäk }

shock absorber [MECH ENG] A spring, a dashpot, or a combination of the two, arranged to minimize the acceleration of the mass of a mechanism or portion thereof with respect to its frame or support. { 'shäk əb,zȯr·bər }

shock action [ORD] A method of attack by mobile units in which the suddenness, violence, and massed weight of the first impact produce the main effect; tank attacks usually rely on shock action. { 'shäk ,ak·shən }

shock breccia [PETR] A fragmental rock formed by the action of shock waves, such as suevite formed by meteorite impact. { 'shäk ,brech·ə }

shock bump [MIN ENG] A rock bump resulting from the sudden collapse of a strong deposit. { 'shäk ,bəmp }

shock cells [FL MECH] Diamond-shaped regions of alternating high and low pressure in a jet flow, through which the jet exit pressure adjusts to the ambient pressure. { 'shäk ,selz }

shock diamonds [PHYS] The shock waves that appear in the exhaust stream of a rocket; they are made visible by their luminosity and describe an approximate diamond configuration in side view. { 'shäk ,dī·mənz }

shock excitation [ELEC] Excitation produced by a voltage or current variation of relatively short duration; used to initiate oscillation in the resonant circuit of an oscillator. Also known as impulse excitation. { 'shäk ,ek,sī'tā·shən }

shock front [PHYS] The outer side of a shock wave whose pressure rises from zero up to its peak value. Also known as pressure front. { 'shäk ,frənt }

shock heating [PHYS] The nonisentropic heating of a fluid which takes place when a shock wave passes through it. { 'shäk ,hēd·iŋ }

shock isolation [MECH ENG] The application of isolators to alleviate the effects of shock on a mechanical device or system. { 'shäk ,ī·sə,lā·shən }

Shockley diode [ELECTR] A *pnpn* silicon controlled switch having characteristics that permit operation as a unidirectional diode switch. { 'shäk·lē 'dī·ōd }

Shockley partial dislocation [SOLID STATE] A partial dislocation in which the Burger's vector lies in the fault plane, so that it is able to glide, in contrast to a Frank partial dislocation. Also known as glissile dislocation. { 'shäk·lē 'pär·shəl ,dis·lō'kā·shən }

shock lithification [GEOL] The conversion of originally loose fragmental materials into coherent aggregates by the action of shock waves, such as those generated by explosions or meteorite impacts. { 'shäk ,lith·ə·fə,kā·shən }

shock loading [GEOPHYS] The process of subjecting material to the action of high-pressure shock waves generated by artificial explosions or by meteorite impact. { 'shäk ,lōd·iŋ }

shock melting [GEOPHYS] Fusion of material as a result of

the high temperatures produced by the action of high-pressure shock waves. { 'shäk ,melt·iŋ }

shock metamorphism [PETR] The complete permanent changes (physical, chemical, mineralogic, morphologic) in rocks caused by transient high-pressure shock waves that act over short-time intervals, ranging from a few microseconds to a fraction of a minute. { 'shäk ,med·ə'mȯr,fiz·əm }

shock mount [MECH ENG] A mount used with sensitive equipment to reduce or prevent transmission of shock motion to the equipment. { 'shäk ,maunt }

shock organ [IMMUNOL] The organ or tissue that exhibits the most marked response to the antigen-antibody interaction in hypersensitivity, as the lungs in allergic asthma or the skin in allergic contact dermatitis. { 'shäk ,ȯr·gən }

shock resistance [ENG] The property which prevents cracking or general rupture when impacted. { 'shäk ri,zis·təns }

shock strut [AERO ENG] The primary working part of any landing gear, which supplies the force as the airplane sinks toward the ground, turning the flight path from one intersecting the ground to one parallel to the ground. { 'shäk ,strət }

shock test [ENG] The test to determine whether the armor sample will crack or spall under impact by kinetic energy or high-explosive projectiles. { 'shäk ,test }

shock therapy [PSYCH] The use of drugs, carbon dioxide, insulin, or electric current to induce coma in the treatment of psychiatric disorders. { 'shäk ,ther·ə·pē }

shock tube [FL MECH] A long tube divided into two parts by a diaphragm; the volume on one side of the diaphragm constitutes the compression chamber, the other side is the expansion chamber; a high pressure is developed by suitable means in the compression chamber, and the diaphragm ruptured; the shock wave produced in the expansion chamber can be used for the calibration of air blast gages, or the chamber can be instrumented for the study of the characteristics of the shock wave. { 'shäk ,tüb }

shock tunnel [ENG] A hypervelocity wind tunnel in which a shock wave generated in a shock tube ruptures a second diaphragm in the throat of a nozzle at the end of the tube, and gases emerge from the nozzle into a vacuum tank with Mach numbers of 6 to 25. { 'shäk ,tən·əl }

shock wave [PHYS] A fully developed compression wave of large amplitude, across which density, pressure, and particle velocity change drastically. { 'shäk ,wāv }

shock-wave lip [PHYS] The shock wave obtained from the lip of a free jet nozzle, because of failure to match the stream pressure and the ambient exhaust pressure. { 'shäk ,wāv ,lip }

shock zone [GEOL] A volume of rock in or around an impact or explosion crater in which a distinctive shock-metamorphic deformation or transformation effect is present. { 'shäk ,zōn }

shoe [ENG] In glassmaking, an open-ended crucible placed in a furnace for heating the blowing irons. [MECH ENG] **1.** A metal block used as a form or support in various bending operations. **2.** A replaceable piece used to break rock in certain crushing machines. **3.** *See* brake shoe. [MIN ENG] **1.** Pieces of steel fastened to a mine cage and formed to fit over the guides to guide it when it is in motion. **2.** The bottom wedge-shaped piece attached to tubbing when sinking through quicksand. **3.** A trough to convey ore to a crusher. **4.** A coupling of rolled, cast, or forged steel to protect the lower end of the casting or drivepipe in overburden, or the bottom end of a sampler when pressed into a formation being sampled. { shü }

shoe brake [MECH ENG] A type of brake in which friction is applied by a long shoe, extending over a large portion of the rotating drum; the shoe may be external or internal to the drum. { 'shü ,brāk }

shoemaker's knife *See* arbilos. { 'shü māk·ərz 'nīf }

shoestring [GEOL] A long, relatively straight and narrow sedimentary body having a width/thickness ratio of less than 5:1, usually 1:1. { 'shü,striŋ }

shoestring rill [GEOL] One of several long, narrow, uniform channels, closely spaced and roughly parallel with one another, that merely score the homogeneous surface of a relatively steep slope of bare soil or weak, clay-rich bedrock, and that develop wherever overland flow is intense. { 'shü,striŋ ,ril }

shoestring sand [GEOL] A shoestring composed of sand and usually buried in mud or shale, usually a sandbar or channel fill. { 'shü,striŋ ,sand }

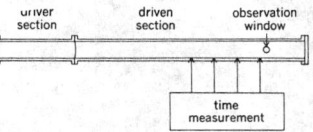

SHOCK TUBE

Schematic representation of a shock tube.

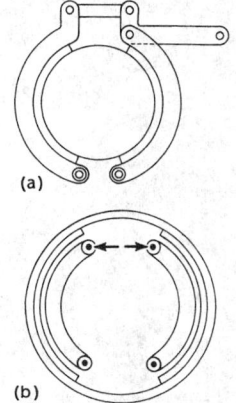

SHOE BRAKE

Two types of shoe brake. *(a)* External shoe brake; shoes are lined with frictional material. *(b)* Internal shoe brake with lining.

shonkinite [PETR] A dark-colored syenite composed principally of augite and orthoclase with some olivine, hornblende, biotite, and nepheline. { 'shäŋ·kə,nīt }

shoofly See slant. { 'shü,flī }

shoot [BOT] **1.** The aerial portion of a plant, including stem, branches, and leaves. **2.** A new, immature growth on a plant. [ENG] To detonate an explosive, used to break coal loose from a seam or in blasting operation or in a borehole. [GEOL] See ore shoot. [GEOPHYS] The energy that goes up through the strata from a seismic profiling shot and is reflected downward at the surface or at the base of the weathering; appears either as a single wave or unites with a wave train that is traveling downward. Also known as secondary reflection. [HYD] **1.** A place where a stream flows or descends swiftly. **2.** A natural or artificial channel, passage, or trough through which water is moved to a lower level. **3.** A rush of water down a steep place or a rapids. [ORD] To project a missile with force; to fire a weapon, as a gun or cannon; to strike or hit something with a missile. { shüt }

shooting board [ENG] **1.** A fixture used as a guide in planing boards; it is more accurate than a miter. **2.** A table and plane used for trimming printing plates. { 'shüd·iŋ ,bórd }

shooting star [ASTRON] A small meteor that has the brief appearance of a darting, starlike object. { 'shüd·iŋ 'stär }

shop drawing [GRAPHICS] A scale drawing to be used as a design guide in the workshop of a manufacturer. { 'shäp ,dró·iŋ }

Shope papilloma [VET MED] A transmissible, virus-induced papilloma occurring naturally on the skin of rabbits. { 'shōp ,pap·ə'lō·mə }

shop fabrication [ENG] Making parts and materials in the shop rather than at the work site. { 'shäp ,fab·rə,kā·shən }

shop lumber [MATER] Softwood lumber graded and used in the factory for general cut-up purposes; similar to factory lumber but of a lower grade. { 'shäp ,ləm·bər }

shop standards [ENG] Written criteria established to govern methods and procedures at an installation. { 'shäp ,stan·dərdz }

shop supplies [ENG] Expendable items consumed in operation and maintenance (for example, waste, oils, solvents, tape, packing, flux, or welding rod). { 'shäp sə,plīz }

shop typhus See murine typhus. { 'shäp ,tī·fəs }

shop weld [ENG] A weld made in the workshop prior to delivery to the construction site. { 'shäp ,weld }

shoran See short-range navigation. { 'shó,ran }

shore [ENG] Timber or other material used as a temporary prop for excavations or buildings; may be sloping, vertical, or horizontal. [GEOL] **1.** The narrow strip of land immediately bordering a body of water. **2.** See seashore. { shór }

shore bird [VERT ZOO] A general term applied to a large number of birds in 12 families of the suborder Charadrii which are always found near water, although the habitat and morphology is varied. Also known as wader. { 'shór ,bərd }

shore crab See Carcinus maenas. { 'shór ,krab }

shore current [HYD] A water current near a shoreline, often flowing parallel to the shore. { 'shór ,kə·rənt }

shore drift See littoral drift. { 'shór ,drift }

shore effect [ELECTROMAG] Bending of waves toward the shoreline when traveling over water near a shoreline, due to the slightly greater velocity of radio waves over water than over land; this effect causes eors in radio-direction-finder indications. { 'shór i,fekt }

shoreface [GEOL] The narrow, steeply sloping zone between the seaward limit of the shore at low water and the nearly horizontal offshore zone. { 'shór,fās }

shoreface terrace [GEOL] A wave-built terrace in the shoreface region, composed of gravel and coarse sand swept from the wave-cut bench into deeper water. { 'shór,fās ,ter·əs }

shore fire-control party [ORD] A specially trained unit for control of naval gunfire in support of troops ashore, consisting of a spotting team to adjust fire and a naval gunfire liaison team to perform liaison functions for the supported battalion commander. { 'shór 'fīr kən,trōl ,pärd·ē }

Shore hardness [ENG] A method of rating the hardness of a metal or of a plastic or rubber material. { 'shór ,härd·nəs }

shore ice [OCEANOGR] Sea ice that has been beached by wind, tides, currents, or ice pressure; it is a type of fast ice, and may sometimes be rafted ice. { 'shór ,īs }

shore lead [OCEANOGR] A lead between pack ice and fast ice or between floating ice and the shore; it may be closed by wind or currents so that only a tide crack remains. { 'shór ,lēd }

shoreline [GEOL] The intersection of a specified plane of water, especially mean high water, with the shore; a limit which changes with the tide or water level. Also known as strandline; waterline. { 'shór,līn }

shoreline cycle [GEOL] The cycle of changes through which sequential forms of coastal features pass during shoreline development, from the establishment of a water level to the time when the water can do no more work. { 'shór,līn ,sī·kəl }

shoreline-development ratio [GEOL] A ratio indicating the degree of irregularity of a lake shoreline, given as the length of the shoreline to the circumference of a circle whose area is equal to that of the lake. { 'shór,līn di'vel·əp·mənt ,rā·shō }

shoreline of depression [GEOL] A shoreline of submergence that implies an absolute subsidence of the land. { 'shór,līn əv di'presh·ən }

shoreline of elevation [GEOL] A shoreline of emergence that implies an absolute rise of the land. { 'shór,līn əv ,el·ə'vā·shən }

shoreline of emergence [GEOL] A straight or gently curving shoreline formed by the dominant relative emergence of the floor of an ocean or a lake. Also known as emerged shoreline; negative shoreline. { 'shór,līn əv i'mər·jəns }

shoreline of submergence [GEOL] A shoreline, characterized by bays, promontories, and other minor features, formed by the dominant relative submergence of a landmass. Also known as positive shoreline; submerged shoreline. { 'shór,līn əv səb'mər·jəns }

shore party [ORD] A task organization of the landing force, formed for the purpose of facilitating the landing and movement off the beaches of troops, equipment, and supplies; for the evacuation from the beaches of casualties and prisoners of war; and for facilitating the beaching, retraction, and salvaging of landing ships and craft; it comprises elements of both the naval and landing forces. { 'shór ,pärd·ē }

shore platform [GEOL] The horizontal or gently sloping surface produced along a shore by wave erosion. Also known as scar. { 'shór ,plat,fórm }

shore polyn'ya [OCEANOGR] A polyn'ya between pack ice and the coast, or between pack ice and an ice front, formed by a current or by wind. { 'shór ¦pal·ən¦yä }

shore protection [CIV ENG] Preventing erosion of the ground bordering a body of water. { 'shór prə,tek·shən }

Shore scleroscope [ENG] A device used in rebound hardness testing of rubber, metal, and plastic; consists of a small, conical hammer fitted with a diamond point and acting in a glass tube. { 'shór 'skler·ə,skōp }

Shore scleroscope hardness test [MET] A rebound hardness test in which a metal body is dropped vertically down a glass tube onto the surface of the material being tested; the height of the rebound is a measure of the hardness. Also known as scleroscope hardness test. { 'shór 'skler·ə,skōp 'härd·nəs ,test }

shoreside See onshore. { 'shór,sīd }

shore tank [PETRO ENG] A shoreside storage tank for liquid petroleum products discharged by tankers. { 'shór ,taŋk }

shore terrace [GEOL] **1.** A terrace produced along the shore by wave and current action. **2.** See marine terrace. { 'shór ,ter·əs }

shore-to-shore movement [ORD] The assault movement of personnel and materiel directly from a shore staging area to the objective, involving no further transfers between types of craft or ships incident to the assault movement. { ¦shór tə ¦shór 'müv·mənt }

shoring [ENG] Providing temporary support with shores to a building or an excavation. { 'shór·iŋ }

Shor's algorithm [COMPUT SCI] An algorithm for factoring a large number within a reasonable amount of time, using a quantum computer. { ¦shórz 'al·gə,rith·əm }

short [ELEC] See short circuit. [ENG] In plastics injection molding, the failure to fill the mold completely. Also known as short shot. [ORD] A bomb or projectile hit short of the target. { shórt }

short antenna [ELECTROMAG] An antenna shorter than about one-tenth of a wavelength, so that the current may be assumed to have constant magnitude along its length, and the antenna may be treated as an elementary dipole. { 'shȯrt an,ten·ə }

short arc [MATH] The shorter of the two arcs between the points of intersection of a chord with a circle. { 'shȯrt 'ärk }

short-baseline system [AERO ENG] A trajectory measuring system using a baseline the length of which is very small compared with the distance of the object being tracked. { 'shȯrt ¦bās,līn ,sis·təm }

short card [COMPUT SCI] A printed circuit board that is plugged into an expansion slot in a microcomputer and is only half the length of a full-size card. { 'shȯrt ¦kärd }

short circuit [ELEC] A low-resistance connection across a voltage source or between both sides of a circuit or line, usually accidental and usually resulting in excessive current flow that may cause damage. Also known as short. { 'shȯrt 'sər·kət }

short-circuit impedance [ELEC] Of a line or four-terminal network, the driving point impedance when the far-end is short-circuited. { 'shȯrt ¦sər·kət im'pēd·əns }

short-circuiting transfer [ENG] Transfer of melted material from a consumable electrode during short circuits. { 'shȯrt ¦sər·kəd·iŋ 'tranz·fər }

short-circuit transition See shunt transition. { 'shȯrt ¦sər·kət tranz'zish·ən }

short column [CIV ENG] A column in which both compression and bending is significant, generally having a slenderness ratio between 30 and 120–150. { 'shȯrt 'käl·əm }

shortcoming [DES ENG] An imperfection or malfunction occurring during the life cycle of equipment, which should be reported and which must be corrected to increase efficiency and to render the equipment completely serviceable. { 'shȯrt,kəm·iŋ }

short-contact switch [ELEC] Selector switch in which the width of the movable contact is greater than the distance between contact clips, so that the new circuit is contacted before the old one is broken; this avoids noise during switching. { 'shȯrt 'kän,tak ,swich }

short-crested wave [OCEANOGR] An ocean wave whose crest is of finite length; that is, the type actually found in nature. { 'shȯrt ¦kres·təd 'wāv }

short-day response [PHYSIO] A photoperiodic response to decreasing days and increasing nights. { ,shȯrt ,dā ri'späns }

short-delay blasting [ENG] A method of blasting by which explosive charges are detonated in a given sequence with short time intervals. { 'shȯrt di¦lā 'blas·tiŋ }

short-delay detonator See millisecond delay cap. { 'shȯrt di¦lā 'det·ən,ād·ər }

short-distance navigation [NAV] **1.** Navigation (air) served by aids located less than 200-miles (322-kilometers) distance (International Civil Aviation Organization). **2.** Navigation (marine) at distances less than 3 miles (4.8 kilometers) [International Meeting on Marine Navigation, 1958]. { 'shȯrt ¦dis·təns ,nav·ə'gā·shən }

short division [MATH] **1.** Division of numbers in which the divisor contains only one digit **2.** Division of algebraic quantities in which the divisor contains only one term. { 'shȯrt də,vizh·ən }

shorted-transformer superconducting fault-current limiter See shielded-core superconducting fault-current limiter. { ¦shȯrd·əd tranz¦fȯrm·ər ,sü·pər·kən,dək·tiŋ ¦fȯlt ,kər·ənt ,lim·əd·ər }

short flashing light [NAV] A flashing light having individual flashes of less than 2-seconds duration. { 'shȯrt ¦flash·iŋ 'līt }

short fuse [ENG] **1.** Any fuse that is cut too short. **2.** The practice of firing a blast, the fuse on the primer of which is not sufficiently long to reach from the top of the charge to the collar of the borehole; the primer, with fuse attached, is dropped into the charge while burning. { 'shȯrt 'fyüz }

short-gate gain [ELECTR] Video gain on short-range gate. { 'shȯrt ¦gāt 'gān }

short-haul [COMMUN] Pertaining to devices capable of transmitting and receiving signals over distances up to about 1 mile (1.6 kilometers). { 'shȯrt ¦hȯl }

short-haul convoy [ORD] A convoy whose voyage lies in general in coastal waters and whose terminals of departure and arrival lie in different countries. { 'shȯrt ¦hȯl ¦kän,vȯi }

short hundredweight See hundredweight. { 'shȯrt 'hən·dərd,wāt }

short ink [GRAPHICS] A type of printing ink with a butterlike viscosity that is therefore not free-flowing. { 'shȯrt 'iŋk }

short interspersed elements [GEN] Families of short deoxyribonucleic acid sequences that are individually inserted abundantly throughout the genome in mammals and other taxons; the 300-base-pair Alu short interspersed elements make up about 5% of the human genome. Abbreviated SINES. { 'shȯrt in·tər,spərst 'el·ə·məns }

shortite [MINERAL] $Na_2Ca_2(CO_3)_3$ A mineral composed of sodium and calcium carbonate. { 'shȯr,tīt }

short leg [ENG] One of the wires on an electric blasting cap, which has been shortened so that when placed in the borehole, the two splices or connections will not come opposite each other and make a short circuit. { 'shȯrt 'leg }

short-line seeking [COMPUT SCI] A method of accelerating the operation of a computer printer, in which the printer is sent directly to the beginning of the next line to be printed without going to the left margin of the paper. { 'shȯrt ¦līn 'sēk·iŋ }

short-long flashing light [NAV] A light showing a short flash of about 0.4 second, and a long flash of four times that duration, this combination recurring about six to eight times per minute. { 'shȯrt 'lȯŋ 'flash·iŋ 'līt }

shortness [MET] A form of brittleness in metal, designated as hot, cold, or red to indicate the temperature range in which brittleness occurs. { 'shȯrt·nəs }

short oil [MATER] Varnish containing a small percentage of oil. { 'shȯrt ,ȯil }

short-path principle See Hittorf principle. { 'shȯrt ¦path 'prin·sə·pəl }

short-period comet [ASTRON] A comet whose period is short enough for observations at two or more apparitions to be interrelated; usually taken to be a comet whose period is shorter than 200 years. { 'shȯrt ¦pir·ē·əd 'käm·ət }

short-precision number See single-precision number. { 'shȯrt pri¦sizh·ən 'nəm·bər }

short-pulse laser [OPTICS] A laser designed to generate a pulse of light lasting on the order of nanoseconds or less, and having very high power, such as by Q switching or mode-locking. { 'shȯrt ¦pəls 'lā·zər }

short radius See apothem. { ¦shȯrt ¦rād·ē·əs }

short-range attack missile [ORD] An air-to-surface and air-to-air solid-fuel guided missile that uses inertial guidance for complete immunity to jamming; target coordinates are fed into the missile by a computer on the aircraft just before launch. Abbreviated SRAM. { 'shȯrt ¦rānj ə'tak ,mis·əl }

short-range ballistic missile [ORD] A ballistic missile with a range capability up to about 600 nautical miles (1100 kilometers). { 'shȯrt ¦rānj bə'lis·tik ,mis·əl }

short-range force [PHYS] A force between two particles which is negligible when the distance between the particles is greater than a certain amount; in particular, nuclear forces whose range is several times 10^{-15} meter. { 'shȯrt ¦rānj 'fȯrs }

short-range forecast [METEOROL] A weather forecast made for a time period generally not greater than 48 hours in advance. { 'shȯrt ¦rānj 'fȯr,kast }

short-range navigation [NAV] Navigation employing only aids usable at short ranges. Also known as shoran. { 'shȯrt ¦rānj ,nav·ə'gā·shən }

short-range order [PHYS] A regularity in the arrangement of atoms in a disordered solid or a liquid in which the probability of a given type of atom having neighbors of a given type is greater than one would expect on a purely random basis. { 'shȯrt ¦rānj 'ȯr·dər }

short-range radar [ENG] Radar whose maximum line-of-sight range, for a reflecting target having 1 square meter of area perpendicular to the beam, is between 50 and 150 miles (80 and 240 kilometers). { 'shȯrt ¦rānj 'rā,där }

short residuum [CHEM ENG] A petroleum refinery term for residual oil from crude-oil distillation operations in which neutral oils are taken overhead with the distillate. { 'shȯrt ri'zij·ə·wəm }

short round [ORD] **1.** Defective cartridge in which the bullet has been seated too deeply. **2.** A projectile which fails to travel the expected distance or range. { 'shȯrt 'raȯnd }

short run [MET] Pertaining to a mold or casting filled only partially with molten metal. { 'shȯrt 'rən }

shorts [ENG] Oversize particles held on a screen after sieving the fines through the screen. [FOOD ENG] Wheat-husk parts that are finer than bran; used in flour milling. { 'shȯrts }

short shipment [ENG] Freight listed or manifested but not received. { 'shȯrt 'ship·mənt }

short shot *See* short. { 'shȯrt 'shät }

short-slot coupler *See* three-decibel coupler. { 'shȯrt ¦slät 'kəp·lər }

short stop [CHEM ENG] A substance added during a polymerization process to terminate the reaction. { 'shȯrt ‚stäp }

short supply [IND ENG] An item is in short supply when the total of stock on hand and anticipated receipts during a given period is less than the total estimated demand during that period. { 'shȯrt sə'plī }

short takeoff and landing [AERO ENG] The ability of an aircraft to clear a 50-foot (15-meter) obstacle within 1500 feet (450 meters) of commencing takeoff, or in landing, to stop within 1500 feet after passing over a 50-foot obstacle. Abbreviated STOL. { 'shȯrt 'tāk‚ȯf ən 'land·iŋ }

short-term exposure limit [PHYSIO] The maximum amount of harmful gas or dust to which a person may be exposed for a brief period (usually 15 minutes) without being physically harmed. Abbreviated STEL. { ¦shȯrt ¦tərm ik'spō·zhər ‚lim·ət }

short-term memory [PSYCH] Conscious, brief retention of information that is currently being processed in a person's mind. { ‚shȯrt ‚tərm 'mem·rē }

short-term predictor [COMMUN] An electric filter that removes redundancies in a signal associated with short-term correlations so that information can be transmitted more efficiently. { ‚shȯrt ‚tərm prə'dik·tər }

short-term repeatability [CONT SYS] The close agreement of positional movements of a robotic system repeated under identical conditions over a short period of time and at the same location. { 'shȯrt ‚tərm ri‚pēd·ə'bil·əd·ē }

short-time rating [ELEC] A rating defining the load that a machine, apparatus, or device can carry for a specified short time. { 'shȯrt ‚tīm 'rād·iŋ }

short ton *See* ton. { 'shȯrt 'tən }

short-tube vertical evaporator [CHEM ENG] A liquid evaporation process unit with a vertical bundle of tubes 2–3 inches (5–8 centimeters) in diameter and 4–6 feet (1.2–1.8 meters) long; the heating fluid is inside the tubes, and the liquid to be evaporated is in the shell area outside the tubes; used mainly to evaporate cane-sugar juice. Also known as calandria evaporator; Roberts evaporator; standard evaporator. { 'shȯrt ¦tüb 'vərd·ə·kəl i'vap·ə‚rād·ər }

shortwall [MIN ENG] **1.** A method of mining in which comparatively small areas are worked separately. **2.** A length of coal face between about 5 and 30 yards (4.6 and 27 meters), generally employed in pillar methods of working. { 'shȯrt‚wȯl }

shortwall coal cutter [MIN ENG] A machine for undercutting coal which has a long, rigid chain jib fixed in relation to the main body of the machine and which cuts across a heading from right to left, being drawn across by means of a steel-wire rope. { 'shȯrt‚wȯl 'kōl ‚kəd·ər }

short wave *See* deep-water wave. { 'shȯrt ‚wāv }

shortwave broadcasting [COMMUN] Radio broadcasting at frequencies in the range from about 1600 to 30,000 kilohertz, above the standard broadcast band. { 'shȯrt‚wāv 'brȯd‚kast·iŋ }

shortwave converter [ELECTR] Electronic unit designed to be connected between a receiver and its antenna system to permit reception of frequencies higher than those the receiver ordinarily handles. { 'shȯrt‚wāv kən'vərd·ər }

short waveguide isolator [ELECTR] A device that functions as an isocirculator in a miniature microwave circuit and consists of a waveguide T junction with a magnetized cylinder of ferrite at the center and an absorber on the side arm of the T. Also known as flange isolator. { 'shȯrt ¦wāv‚gīd 'ī·sə‚lād·ər }

shortwave propagation [COMMUN] Propagation of radio waves at frequencies in the range from about 1600 to 30,000 kilohertz. { 'shȯrt‚wāv ‚präp·ə'gā·shən }

shortwave radiation [ELECTROMAG] A term used loosely to distinguish radiation in the visible and near-visible portions of the electromagnetic spectrum (roughly 0.4 to 1.0 micrometer in wavelength) from long-wave radiation (infrared radiation). { 'shȯrt‚wāv ‚rād·ē'ā·shən }

short word [COMPUT SCI] The fixed word of lesser length in computers capable of handling words of two different lengths; in many computers this is referred to as a half-word because the length is exactly the half-length of the full word. { 'shȯrt 'wərd }

shoshonite [PETR] A basaltic rock composed of olivine and augite phenocrysts in a groundmass of labradorite with orthoclase rims, olivine, augite, a small amount of leucite, and some dark-colored glass. { shə'shō‚nīt }

shot [AERO ENG] An act or instance of firing a rocket, especially from the earth's surface. [ENG] **1.** A charge of some kind of explosive. **2.** Small spherical particles of steel. **3.** Small steel balls used as the cutting agent of a shot drill. **4.** The firing of a blast. **5.** In plastics molding, the yield from one complete molding cycle, including scrap. [MIN ENG] Coal broken by blasting or other methods. [ORD] **1.** A solid projectile for cannon, without a bursting charge; the term projectile is preferred for uniformity in nomenclature. **2.** A mass or load of numerous, relatively small, lead pellets used in a shotgun, as birdshot or buckshot. { shät }

shot bit [DES ENG] A short length of heavy-wall steel tubing with diagonal slots cut in the flat-faced bottom edge. { 'shät ‚bit }

shot blasting [MET] Cleaning and descaling metal by shot peening or by means of a stream of abrasive powder blown through a nozzle under air pressure in the range 30–150 pounds per square inch (200–1000 kilopascals). { 'shät ‚blast·iŋ }

shot boring [ENG] The act or process of producing a borehole with a shot drill. { 'shät ‚bȯr·iŋ }

shot break [ENG] In seismic prospecting, the electrical impulse which records the instant of explosion. { 'shät ‚brāk }

shot capacity [ENG] The maximum weight of molten resin that an accumulator can push out with one forward stroke of the ram during plastics forming operations. { 'shät kə‚pas· əd·ē }

shot copper [GEOL] Small, rounded particles of native copper, molded by the shape of vesicles in basaltic host rock, and resembling shot in size and shape. { 'shät ‚käp·ər }

shotcreting [ENG] A process of conveying mortar or concrete through a hose at high velocity onto a surface; the material bonds tenaciously to a properly prepared concrete surface and to a number of other materials. { 'shät‚krēd·iŋ }

shot depth [ENG] The distance from the surface to the charge. { 'shät ‚depth }

shot drill *See* calyx drill. { 'shät ‚dril }

shot effect *See* shot noise. { 'shät i‚fekt }

shot elevation [ENG] Elevation of the dynamite charge in the shot hole. { 'shät ‚el·ə‚vā·shən }

shot feed [MECH ENG] A device to introduce chilled-steel shot, at a uniform rate and in the proper quantities, into the circulating fluid flowing downward through the rods or pipe connected to the core barrel and bit of a shot drill. { 'shät ‚fēd }

shot-firing cable [ELEC] A two-conductor cable which leads from the exploder to the detonator wires. Also known as firing cable. { 'shät ¦fīr·iŋ ‚kā·bəl }

shot-firing circuit [ELEC] The path taken by the electric current from the exploder along the shot-firing cable, the detonator wires, and finally the detonator when a shot is detonated. { 'shät ¦fīr·iŋ ‚sər·kət }

shot-firing curtain [MIN ENG] A steel frame with chains about 6 inches (15 centimeters) apart suspended from the roof about 9 to 12 feet (2.7 to 3.7 meters) from the face of an advancing tunnel to intercept flying debris when shot-firing at the face. { 'shät ¦fīr·iŋ ‚kȯrt·ən }

shot group *See* shot pattern. { 'shät ‚grüp }

shotgun [ORD] A smooth-bore shoulder weapon; the usual classes are riot gun, skeet gun, and sporting gun. { 'shät‚gən }

shotgun cartridge [ORD] A container or capsule, usually of stiff paper with a brass base, containing primer, powder, wadding, and shot for use in a shotgun. { 'shät‚gən ‚kär·trij }

shothole [ENG] The borehole in which an explosive is placed for blasting. { 'shät‚hōl }

shothole casing [ENG] A lightweight pipe, usually about 4 inches (10 centimeters) in diameter and 10 feet (3 meters) long, with threaded connections on both ends, used to prevent the shothole from caving and bridging. { 'shät‚hōl ‚kās·iŋ }

shothole drill [MECH ENG] A rotary or churn drill for drilling shotholes. { 'shät‚hōl ‚dril }

shot mill [ENG] A high-speed, continuous mill for deagglomerating, dispersing, and milling paints, inks, dyestuffs, adhesives, food, and pharmaceuticals; consists of a chamber with rotating disks that is filled with small steel or ceramic spheres (shot), and a pump to propel material through the mill. Also known as a media mill.   { 'shät ,mil }

shot noise [ELECTR] Noise voltage developed in a thermionic tube because of the random variations in the number and the velocity of electrons emitted by the heated cathode; the effect causes sputtering or popping sounds in radio receivers and snow effects in television pictures. Also known as Schottky noise; shot effect.   { 'shät ,nȯiz }

shot pattern [ORD] Design made on a surface by all the impacts of a series of shots fired under similar conditions. Also known as shot group.   { 'shät ,pad·ərn }

shot peening [MET] Shot blasting with small steel balls driven by a blast of air.   { 'shät ,pēn·iŋ }

shot point [ENG] The point at which an explosion (such as in seismic prospecting) originates, generating vibrations in the ground.   { 'shät ,pȯint }

shot rock [ENG] Blasted rock.   { 'shät ,räk }

shotting [MET] Making shot by pouring molten metal in finely divided streams; the particles solidify during descent and are cooled in a tank of water.   { 'shäd·iŋ }

shot tongs [ORD] Device used to lift and convey heavy projectiles in a horizontal position.   { 'shät ,täŋz }

shoulder [ANAT] **1.** The area of union between the upper limb and the trunk in humans. **2.** The corresponding region in other vertebrates. [DES ENG] The portion of a shaft, a stepped object, or a flanged object that shows an increase of diameter. [ENG] A projection made on a piece of shaped wood, metal, or stone, where its width or thickness is suddenly changed. [GEOL] **1.** A short, rounded spur protruding laterally from the slope of a mountain or hill. **2.** The sloping segment below the summit of a mountain or hill. **3.** A bench on the flanks of a glaciated valley, located at the sharp change of slope where the steep sides of the inner glaciated valley meet the more gradual slope above the level of glaciation. **4.** A joint structure on a joint face produced by the intersection of plume-structure ridges with fringe joints. [GRAPHICS] That part of a plate or type that extends beyond the actual printing surface.   { 'shōl·dər }

shoulder blade *See* scapula.   { 'shōl·dər ,blād }

shoulder-elbow height [ANTHRO] A measure of the distance taken from the top of the acromion to the tip of the elbow, as the subject sits erect, with the upper arm vertical and the forearm horizontal.   { 'shōl·dər 'el·bō ,hīt }

shoulder girdle *See* pectoral girdle.   { 'shōl·dər ,gərd·əl }

shoulder guard [ORD] Any shield over the firing mechanism of a gun designed to protect the gunner from contact; particularly, such a shield on cannon mounted in tanks and in other cramped quarters.   { 'shōl·dər 'gärd }

shoulder-hand syndrome [MED] A syndrome characterized by severe constant intractable pain in the shoulder and arm, limited joint motion, diffuse swelling of the distal part of the upper extremity, fibrosis and atrophy of muscles, and decalcification of underlying bones; the cause is not well understood; it is similar to, or may be a form of, causalgia. Also known as hand-shoulder syndrome.   { 'shōl·dər 'hand 'sin ,drōm }

shoulder harness [ENG] A harness in a vehicle that fastens over the shoulders to prevent a person's being thrown forward in the seat.   { 'shōl·dər ,här·nəs }

shoulder screw [DES ENG] A screw with an unthreaded cylindrical section, or shoulder, between threads and screwhead; the shoulder is larger in diameter than the threaded section and provides an axis around which close-fitting moving parts operate.   { 'shōl·dər ,skrü }

shoulder stock [ORD] An item usually made of hardwood to which the barrel assembly and other parts of a shoulder-fired gun are attached; usually designed in one piece.   { 'shōl·dər ,stäk }

shoulder weapon [ORD] Any firearm designed to be braced on or against the shoulder when firing, as a rifle, carbine, or bazooka launcher.   { 'shōl·dər ,wep·ən }

shoved moraine *See* push moraine.   { 'shəvd mə'rān }

shovel [DES ENG] A hand tool having a flattened scoop at the end of a long handle for moving soil, aggregate, cement, or other similar material. [MECH ENG] A mechanical excavator.   { 'shəv·əl }

shovel dozer *See* tractor loader.   { 'shəv·əl ,dōz·ər }

shovel loader [MECH ENG] A loading machine mounted on wheels, with a bucket hinged to the chassis which scoops up loose material, elevates it, and discharges it behind the machine.   { 'shəv·əl ,lōd·ər }

Showalter stability index [METEOROL] A measure of the local static stability of the atmosphere, expressed as a numerical index.   { 'shō,wȯl·tər stə,bil·əd·ē ,in,deks }

show-card color *See* poster paint.   { 'shō ,kärd ,kəl·ər }

shower [METEOROL] Precipitation from a convective cloud; characterized by the suddenness with which it starts and stops, by the rapid changes of intensity, and usually by rapid changes in the appearance of the sky. [NUC PHYS] *See* cosmic-ray shower.   { 'shaù·ər }

shower meteor [ASTRON] A meteor whose direction of arrival is approximately parallel to others belonging to the same meteor shower.   { 'shaù·ər ,mēd·ē·ər }

shower unit [NUCLEO] The mean path length required to reduce the energy of relativistic charged particles to half value as they pass through matter; one shower unit is equal to 0.693 radiation length.   { 'shaù·ər ,yü·nət }

show-through [GRAPHICS] In printing, an undesirable result in which the printing is visible through the sheet under normal lighting conditions. Also known as strike-through.   { 'shō ,thrü }

Shpol'skii effect [SPECT] The occurrence of very narrow fluorescent lines in the spectra of certain compounds from molecules frozen at low temperatures.   { 'shpōl·skē i,fekt }

shrapnel [ORD] Small lead or steel balls contained in a shrapnel case which is fired from an artillery piece; the balls are projected in a forward direction upon the functioning of the fuse.   { 'shrap·nəl }

shrew [VERT ZOO] Any of more than 250 species of insectivorous mammals of the family Soricidae; individuals are small with a moderately long tail, minute eyes, a sharp-pointed snout, and small ears.   { shrü }

Shrike [ORD] An air-to-surface solid-fuel guided missile that homes in on enemy radar installations.   { shrīk }

shrimp [INV ZOO] The common name for a number of crustaceans, principally in the decapod suborder Natantia, characterized by having well-developed pleopods and by having the abdomen sharply bent in most species, producing a humped appearance.   { shrimp }

shrinkage [ENG] **1.** Contraction of a molded material, such as metal or resin, upon cooling. **2.** Contraction of a plastics casting upon polymerizing. [GEOL] The decrease in volume of soil, sediment, fill, or excavated earth due to the reduction of voids by mechanical compaction, superimposed loads, natural consolidation, or drying.   { 'shriŋ·kij }

shrinkage cavity [MET] A cavity resulting from shrinkage during casting.   { 'shriŋ·kij ,kav·əd·ē }

shrinkage crack [GEOL] A small crack produced in fine-grained sediment or rock by the loss of contained water during drying or dehydration. [MET] An irregular interdendritic crack in a casting caused by unequal contraction or inadequate feeding.   { 'shriŋ·kij ,krak }

shrinkage index [GEOL] The numerical difference between the plastic limit of a material and its shrinkage limit.   { 'shriŋ·kij ,in,deks }

shrinkage limit [GEOL] That moisture content of a soil below which a decrease in moisture content will not cause a decrease in volume, but above which an increase in moisture will cause an increase in volume.   { 'shriŋ·kij ,lim·ət }

shrinkage pore [GEOL] An irregular pore formed in muddy sediment by shrinkage.   { 'shriŋ·kij ,pȯr }

shrinkage ratio [GEOL] The ratio of a volume change to the moisture-content change above the shrinkage limit.   { 'shriŋ·kij ,rā·shō }

shrinkage rule *See* contraction rule.   { 'shriŋ·kij ,rül }

shrinkage stoping [MIN ENG] A modification of overhead stoping, involving the use of a part of the ore for the purpose of support and as a working platform. Also known as back stoping.   { 'shriŋ·kij ,stōp·iŋ }

shrink fit [DES ENG] A tight interference fit between mating parts made by shrinking-on, that is, by heating the outer member to expand the bore for easy assembly and then cooling so that the outer member contracts.   { 'shriŋk ,fit }

shrink forming [DES ENG] Forming metal wherein the piece undergoes shrinkage during cooling following the application of heat, cold upset, or pressure.   { 'shriŋk ,fȯr·miŋ }

SHREW

The Eurasian common shrew (*Sorex araneus*).

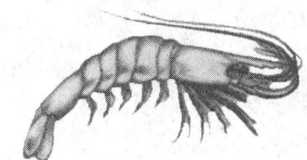

SHRIMP

Common shrimp (*Crangon vulgaris*).

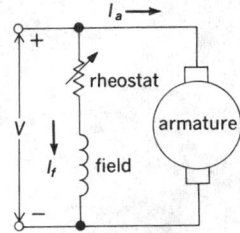

SHUNT MOTOR

Circuit diagram showing connections of shunt motor. V = total impressed voltage from line; I_a = total armature current; I_f = field current.

shrinking of the plane [MATH] A homothetic transformation in which the ratio of similitude is less than 1. Also known as shrinking transformation. { ¦shriŋk·iŋ əv thə 'plān }

shrinking space [MATH] The conjugate space of a Banach space with basis $x_1, x_2, \ldots$, which satisfies the property that, for any continuous linear functional f, the norm of f with domain restricted to the linear span of $x_{n+1}, x_{n+2}, \ldots$ approaches zero as n approaches infinity. { ¦shriŋk·iŋ 'spās }

shrinking transformation See shrinking of the plane. { 'shriŋk·iŋ ‚tranz·fər‚mā·shən }

shrink-mixed concrete [MATER] Concrete that is partially mixed before being put in a truck mixer. { 'shriŋk ¦mikst 'kän‚krēt }

shrink ring [DES ENG] A heated ring placed on an assembly of parts, which on subsequent cooling fixes them in position by contraction. { 'shriŋk ‚riŋ }

shrink rule See contraction rule. { 'shriŋk ‚rül }

shrink wrapping [ENG] A technique of packaging with plastics in which the strains in plastics film are released by raising the temperature of the film, causing it to shrink-fit over the object being packaged. { 'shriŋk ‚rap·iŋ }

shroud [ENG] A protective covering, usually of metal plate or sheet. [HOROL] The ends of lantern clock pinions that hold the pins. [NAV ARCH] A principal member of the standing rigging consisting of hemp or wire ropes which extend from or near a masthead to a vessel's side or to the rim of the top of the mast to afford lateral support for the mast. { shraúd }

shrouded propeller See ducted fan. { 'shraúd·əd prə'pel·ər }

shrub [BOT] A low woody plant with several stems. { shrəb }

shrub-coppice dune [GEOL] A small dune formed on the leeward side of bush-and-clump vegetation. { 'shrəb ¦käp·əs 'dün }

Shubnikov-de Haas effect [SOLID STATE] Oscillations of the resistance or Hall coefficient of a metal or semiconductor as a function of a strong magnetic field, due to the quantization of the electron's energy. { 'shüb·nə‚kóf də'häs i‚fekt }

Shubnikov groups [SOLID STATE] The point groups and space groups of crystals having magnetic moments. Also known as black-and-white groups; magnetic groups. { 'shüb·nə‚kóf ‚grüps }

shuga [OCEANOGR] A spongy, rather opaque, whitish chunk of ice which forms instead of pancake ice if the freezing takes place in sea water which is considerably agitated. { 'shü·gə }

shungite [GEOL] A hard, black, amorphous, coallike material composed of more than 98% carbon. { 'shəŋ‚īt }

shunt [CIV ENG] To shove or turn off to one side, as a car or train from one track to another. [ELEC] **1.** A precision low-value resistor placed across the terminals of an ammeter to increase its range by allowing a known fraction of the circuit current to go around the meter. Also known as electric shunt. **2.** To place one part in parallel with another. **3.** See parallel. [ELECTROMAG] A piece of iron that provides a parallel path for magnetic flux around an air gap in a magnetic circuit. [MED] A vascular passage by which blood is diverted from its normal circulatory path; frequently it is a surgical passage created between two blood vessels, but it may also be an anatomical feature. { shənt }

shunt-excited [ELECTROMAG] Having field windings connected across the armature terminals, as in a direct-current generator. { 'shənt ik¦sīd·əd }

shunt-excited antenna [ELECTROMAG] A tower antenna, not insulated from the ground at the base, whose feeder is connected at a point about one-fifth of the way up the antenna and usually slopes up to this point from a point some distance from the antenna's base. { 'shənt ik¦sīd·əd an'ten·ə }

shunt-fed vertical antenna [ELECTROMAG] Vertical antenna connected to the ground at the base and energized at a point suitably positioned above the grounding point. { 'shənt ¦fed ¦vərd·ə·kəl an'ten·ə }

shunt feed See parallel feed. { 'shənt ‚fed }

shunt generator [ELEC] A generator whose field winding and armature winding are connected in parallel, and in which the armature supplies both the load current and the field current. { 'shənt ¦jen·ə‚rād·ər }

shunting [ELEC] The act of connecting one device to the terminals of another so that the current is divided between the two devices in proportion to their respective admittances. { 'shənt·iŋ }

shunt loading [ELEC] Loading in which reactances are applied in shunt across the conductors. { 'shənt ¦lōd·iŋ }

shunt motor [ELEC] A direct-current motor whose field circuit and armature circuit are connected in parallel. { 'shənt ¦mōd·ər }

shunt neutralization See inductive neutralization. { 'shənt ‚nü·trə·lə'zā·shən }

shunt peaking [ELECTR] The use of a peaking coil in a parallel circuit branch connecting the output load of one stage to the input load of the following stage, to compensate for high-frequency loss due to the distributed capacitances of the two stages. { 'shənt ¦pēk·iŋ }

shunt reactor [ELEC] A reactor that has a relatively high inductance and is wound on a magnetic core containing an air gap; used to neutralize the charging current of the line to which it is connected. { 'shənt rē¦ak·tər }

shunt regulator [ELEC] A regulator that maintains a constant output voltage by controlling the current through a dropping resistance in series with the load. { 'shənt ¦reg·yə‚lād·ər }

shunt repeater [ELEC] A type of negative impedance telephone repeater which is stable when it is short-circuited, but oscillates when terminated by a high impedance, in contrast to a series repeater; it can be thought of as a negative admittance. { 'shənt ri¦pēd·ər }

shunt T junction See H-plane T junction. { 'shənt 'tē ¦jəŋk·shən }

shunt transition [ELEC] A method of changing the connection of motors from series to parallel in which one motor, or group of motors, is first short-circuited, then disconnected, and finally connected in parallel with the other motor or motors. Also known as short-circuit transition. { 'shənt tran¦zish·ən }

shunt valve [ENG] A valve that gives a fluid under pressure a more readily available escape route than the normal route. { 'shənt ‚valv }

shunt-wound [ELEC] Having armature and field windings in parallel, as in a direct-current generator or motor. { 'shənt ¦waúnd }

shut-down circuit [ENG] An electronic, electric, or pneumatic system designed to shut off and close down process systems or equipment; can be used for routine or emergency situations. { 'shət‚daún ‚sər·kət }

shut height [MECH ENG] The distance in a press between the bottom of the slide and the top of the bed, indicating the maximum die height that can be accommodated. { 'shət ‚hīt }

shut-in pressure [PETRO ENG] The equilibrated reservoir pressure measured when all the gas or oil outflow has been shut off. { 'shət¦in ‚presh·ər }

shut-in well [PETRO ENG] An oil or gas well that is closed off; the well is shut so that it does not produce a fluid product of any kind. { 'shət¦in ‚wel }

shutoff [AERO ENG] In rocket propulsion, the intentional termination of burning by command from the ground or from a self-contained guidance system. { 'shət‚óf }

shutoff head [MECH ENG] The pressure developed in a centrifugal or axial flow pump when there is zero flow through the system. { 'shət‚óf ‚hed }

shutter [NUCLEO] A movable plate of absorbing material used to cover a window or a beam hole in a reactor when radiation is not desired, or used to shut off a flow of slow neutrons. [OPTICS] A mechanical device that cuts off a beam of light by opening and closing at different rates of speed to expose film or plates; used in cameras and motion picture projectors. [ORD] A barrier in an explosive train used to stop a detonation wave; an interrupter which opens or closes as a shutter; often used to obtain fuse safety. { 'shəd·ər }

shutter dam [CIV ENG] A dam consisting of a series of pieces that can be lowered or raised by revolving them about their horizontal axis. { 'shəd·ər ‚dam }

shuttered fuse [ORD] A fuse in which inadvertent initiation of the detonator will not initiate either the booster or the burst charge. { 'shəd·ərd 'fyüz }

shuttered image converter [ELECTR] An image tube whose photoelectrons can be rapidly switched off to allow a camera to record the image on its screen. { ¦shəd·ərd 'im·ij kən‚vərd·ər }

shuttering See formwork. { 'shəd·ə·riŋ }

shutterridge [GEOL] A ridge formed by vertical, lateral, or oblique displacement of a fault traversing a ridge-and-valley topography with the displaced part of a ridge shutting in the adjacent ravine or canyon. { 'shəd·ə,rij }

shuttle [MECH ENG] A back-and-forth motion of a machine which continues to face in one direction. [NUCLEO] *See* rabbit. [TEXT] A device on a loom that moves filling yarns between the warp yarns during weaving. { 'shəd·əl }

shuttle bombing [ORD] Bombing of objectives, utilizing two bases; a bomber formation bombs its target, flies on to its second base, reloads, and returns to its home base, again bombing a target if required. { 'shəd·əl ,bäm·iŋ }

shuttle box [TEXT] **1.** A case on a loom at either end of the lay; holds the shutter after it has been moved through the shed. **2.** A compartment for quick-access storage of shuttles containing threads of various colors. { 'shəd·əl ,bäks }

shuttle car [MIN ENG] An electrically propelled vehicle on rubber tires or caterpillar treads used to transfer raw materials, such as coal and ore, from loading machines in trackless areas of a mine to the main transportation system. { 'shəd·əl ,kär }

shuttle conveyor [MECH ENG] Any conveyor in a self-contained structure movable in a defined path parallel to the flow of the material. { 'shəd·əl kən,vā·ər }

shuttle vector [MOL BIO] A deoxyribonucleic acid vector able to replicate in two different organisms, and therefore able to shuttle foreign nucleic acids between two different hosts. Also known as bifunctional vector. { 'shəd·əl ,vek·tər }

shuttling [ENG] A movement involving two or more trips or partial trips by the same motor vehicles between two points. { 'shəd·əl·iŋ }

Shwartzman phenomenon [IMMUNOL] A type of local tissue reactivity in the skin in which a preparatory injection of the endotoxin is followed by an intravenous injection of the same or another endotoxin 24 hours later, producing immediate neutropenia and thrombopenia with the development of leukocyte-platelet thrombi with subsequent hemorrhage. { 'shwörts·mənfə,näm·ə,nän }

Si *See* silicon.

SI *See* International System of Units.

Siacci method [MECH] An accurate and useful method for calculation of trajectories of high-velocity missiles with low quadrant angles of departure; basic assumptions are that the atmospheric density anywhere on the trajectory is approximately constant, and the angle of departure is less than about 15°. { sē'ä·chē ,meth·əd }

sial [PETR] A petrologic term for the silica- and alumina-rich upper rock layers of the earth's crust; gives rise to granite magma; the bulk of the continental blocks is sialic. Also known as granitic layer; sal. { 'sī,al }

sialadenitis [MED] Inflammation of a salivary gland. { sī,al·ə·də'nīd·əs }

sialagogue [PHARM] A drug producing a flow of saliva. { sī'al·ə,gäg }

sialic acid [BIOCHEM] Any of a family of amino sugars, containing nine or more carbon atoms, that are nitrogen- and oxygen-substituted acyl derivatives of neuraminic acid; as components of lipids, polysaccharides, and mucoproteins, they are widely distributed in bacteria and in animal tissues. { sī'al·ik 'as·əd }

sialogram [MED] Roentgenogram of a salivary duct system after the injection of an opaque medium. { sī'al·ə,gram }

sialography [MED] Radiographic examination of a salivary gland following injection of an opaque substance into its duct. { ,sī·ə'läg·rə·fē }

sialolith [PATH] A salivary calculus. { sī'al·ə,lith }

sialolithiasis [MED] The presence of salivary calculi. { sī,al·ə·li'thī·ə·səs }

sialomucin [BIOCHEM] An acid mucopolysaccharide containing sialic acid as the acid component. { sī,al·ə'myüs·ən }

siamese blow [ENG] In the plastics industry, the blow molding of two or more parts of a product in a single blow, then cutting them apart. { 'sī·ə,mēz 'blō }

siamese connection [ENG] A Y-shaped standpipe installed close to the ground outside a building to provide two inlet connections for fire hoses to the standpipes and to the sprinkler system. { 'sī·ə,mēz kə'nek·shən }

Siamese twins [MED] Viable conjoined twins. { sī·ə,mēz 'twinz }

SIB *See* screen image buffer.

Siberian anticyclone *See* Siberian high. { sī'bir·ē·ən 'ant·i,sī,klón }

Siberian high [METEOROL] An area of high pressure which forms over Siberia in winter, and which is particularly apparent on mean charts of sea-level pressure; centered near lake Baikal. Also known as Siberian anticyclone. { sī'bir·ē·ən 'hī }

Siberian tick typhus [MED] A relatively benign, rash- and eschar-producing spotted-fever-like disease in northern Asia, caused by *Rickettsia siberica;* transmitted by four species of *Dermacentor* and two of *Haemaphysalis.* { sī'bir·ē·ən 'tik ,tī·fəs }

siberite [MINERAL] A violet-red or purplish lithian variety of tourmaline. { sī'bi,rīt }

sibjet *See* sebkha. { 'sib·jət }

sibling rivalry [PSYCH] Competition between siblings for parental love, or for some other recognition. { 'sib·liŋ 'rī·vəl·rē }

Siboglinidae [INV ZOO] A family of pogonophores in the order Athecanephria. { ,sī·bə'glī·nə,dē }

SIC *See* dielectric constant.

sickle [AGR] The cutting mechanism of a binder, reaper, or combine. [DES ENG] A hand tool consisting of a hooked metal blade with a short handle, used for cutting grain or other agricultural products. [TEXT] A hooked arm for guiding the thread in a spinning mule. { 'sik·əl }

Sickle [ASTRON] A group of six stars in the constellation Leo that outline the head of the lion. { 'sik·əl }

sickle-cell anemia [MED] A chronic, hereditary hemolytic and thrombotic disorder in which hypoxia causes the erythrocyte to assume a sickle shape; occurs in individuals homozygous for sickle-cell hemoglobin trait. Also known as sickle-cell disease. { 'sik·əl ,sel ə'nē·mē·ə }

sickle-cell disease *See* sickle-cell anemia. { 'sik·əl ,sel di,zēz }

sickle-cell hemoglobin [PATH] The hemoglobin found in sickle-cell anemia, differing in electrophoretic mobility and other physiochemical properties from normal adult hemoglobin. Also known as hemoglobin S. { 'sik·əl ,sel 'hē·mə,glō·bən }

sicklerite [MINERAL] (Li,Mn)(PO$_4$) A dark-brown mineral composed of hydrous lithium manganese phosphate occurring in cleavable masses. { 'sik·lə,rīt }

sicula [INV ZOO] The cone-shaped chitinous skeleton of the first zooid of a graptolite colony. { 'sik·yə·lə }

SID *See* sudden ionospheric disturbance.

side [MATH] **1.** One of the line segments that bound a polygon. **2.** One of the two rays that extend from the vertex of an angle. { sīd }

side arms [ORD] Weapons that are worn at the side or in the belt when not in use; examples are the bayonet, automatic pistol, and revolver. { 'sīd ,ärmz }

sideband [ELECTROMAG] **1.** The frequency band located either above or below the carrier frequency, within which fall the frequency components of the wave produced by the process of modulation. **2.** The wave components lying within such bands. { 'sīd,band }

side bar [ENG] A bar on which molding pins are carried; operated from outside the mold. { 'sīd ,bär }

side-boom dredge [NAV ARCH] A dredge that carries the discharge in a discharge pipe hung from a boom, a distance of from 200 to 500 feet (60 to 150 meters) directly to port or starboard of the vessel, and there discharges into the atmosphere, dropping vertically from a height of about 50 feet (15 meters) onto the surface of the sea. { 'sīd ,büm 'drej }

side canyon [GEOL] A ravine or other valley smaller than a canyon, through which a tributary flows into the main stream. { 'sīd ,kan·yən }

side-centered lattice [CRYSTAL] A type of centered lattice that is centered on the side faces only. { 'sīd ,sen·tərd 'lad·əs }

side chain [ORG CHEM] A grouping of similar atoms (two or more, generally carbons, as in the ethyl radical, C$_2$H$_5$—) that branches off from a straight-chain or cyclic (for example, benzene) molecule. Also known as branch; branched chain. { 'sīd ,chān }

side-channel spillway [CIV ENG] A dam spillway in which the initial and final flow are approximately perpendicular to each other. Also known as lateral flow spillway. { 'sīd ,chan·əl 'spil ,wā }

side circuit [COMMUN] One of the circuits arranged to derive a phantom circuit. { 'sīd ,sər·kət }

side-construction tile [MATER] A type of structural clay tile designed to receive its principal stress at right angles to the axis of the cells. { 'sīd kən¦strək·shən 'til }

side-cut brick [MATER] Brick cut by taut wire along the long side, as opposed to the edge. { 'sīd ¦kət ‚brik }

side direction [MECH] In stress analysis, the direction perpendicular to the plane of symmetry of an object. { 'sīd di‚rek·shən }

side-discharge shovel [MIN ENG] A shovel loader having a 21-cubic-foot (0.59-cubic-meter) bucket, hinged to the chassis to dig, lift, and discharge the material sideways onto a scraper or a belt conveyor. { 'sīd ¦dis‚chärj 'shəv·əl }

side draw pin [ENG] Projection used to core a hole in a molded article at an angle other than the line of mold closing; must be withdrawn before the article is ejected. { 'sīd 'drȯ ‚pin }

side drift *See* adit. { 'sīd ‚drift }

side dumper [MIN ENG] An ore, rock, or coal car that can be tilted sidewise and thus emptied. { 'sīd ‚dəm·pər }

side echo [ELECTROMAG] Echo due to a side lobe of an antenna. { 'sīd ‚ek·ō }

side effect [COMPUT SCI] A consistent result of a procedure that is in addition to or peripheral to the basic result. { 'sīd i‚fekt }

side-end lines [MIN ENG] Limits of a mining claim looked on as boundary lines, especially in the case of veins originating within but extending outside the claim. { 'sīd ‚end ‚līnz }

side error [NAV] That error in the reading of a marine sextant due to nonperpendicularity of the horizon glass to the frame. { 'sīd ‚er·ər }

side-facing tool [ENG] A single-point cutting tool having a nose angle of less than 60° and used for finishing the tailstock end of work being machined between centers or the face of a workpiece mounted in a chuck. { 'sīd ‚fās·iŋ ‚tül }

sidehill bit [DES ENG] A drill bit which is set off-center so that it cuts a hole of larger diameter than that of the bit. { 'sīd‚hil ‚bit }

side hook *See* bench hook. { 'sīd ‚hu̇k }

side keelson [NAV ARCH] The fore and aft girders running along the bottom of the ship parallel, or nearly so, to the keel. { 'sīd ‚kēl·sən }

side lobe *See* minor lobe. { 'sīd ‚lōb }

side-lobe blanking [ELECTR] Radar technique which compares relative signal strengths between an omnidirectional antenna and the radar antenna. { 'sīd ¦lōb 'blaŋk·iŋ }

side-lobe cancellation [ORD] Jamming countermeasure technique which is designed to exclude or greatly reduce the strength of jamming signals introduced through the side or back lobes of a receiving system. { 'sīd ¦lōb ‚kan·sə'lā·shən }

side-looking radar [ENG] A high-resolution airborne radar having antennas aimed to the right and left of the flight path; used to provide high-resolution strip maps with photographlike detail, to map unfriendly territory while flying along its perimeter, and to detect submarine snorkels against a background of sea clutter. { 'sīd ‚lu̇k·iŋ 'rā·där }

side milling [MECH ENG] Milling with a side-milling cutter to machine one vertical surface. { 'sīd ‚mil·iŋ }

side-milling cutter [DES ENG] A milling cutter with teeth on one or both sides as well as around the periphery. { 'sīd ¦mil·iŋ ‚kəd·ər }

side oblique air photograph [GRAPHICS] An oblique photograph taken with the camera axis at right angles to the longitudinal axis of the aircraft. { 'sīd ȯ'blēk 'er 'fōd·ə‚graf }

side pinacoid [CRYSTAL] A pinacoid with Miller indices (010) in an orthorhombic, monoclinic, or triclinic crystal. { 'sīd 'pin·ə‚kȯid }

side plates [MIN ENG] In timbering, where both a cap and a sill are used, and the posts act as spreaders, the cap and the sill are termed side plates. { 'sīd ‚plāts }

sideraerolite *See* stony-iron meteorite. { ‚sid·ə·rə¦er·ə‚līt }

side rake [MECH ENG] The angle between the tool face and a reference plane for a single-point turning tool. { 'sīd ‚rāk }

side reaction [CHEM] A secondary or subsidiary reaction that takes place simultaneously with the reaction of primary interest. { 'sīd rē‚ak·shən }

sidereal [ASTRON] Referring to a quantity, such as time, to indicate that it is measured in relation to the apparent motion or position of the stars. { sī'dir·ē·əl }

sidereal clock [HOROL] A clock set at 0 hours 0 minutes 0 seconds (midnight) as the vernal equinox crosses the meridian; it is the astronomical clock. { sī'dir·ē·əl 'kläk }

sidereal day [ASTRON] The time between two successive upper transits of the vernal equinox; this period measures one sidereal day. { sī'dir·ē·əl 'dā }

sidereal hour angle [ASTRON] The angle along the celestial equator formed between the hour circle of a celestial body and the hour circle of the vernal equinox, measuring westward from the vernal equinox through 360°. { sī'dir·ē·əl 'au̇r ‚aŋ·gəl }

sidereal month [ASTRON] The time period of one revolution of the moon about the earth relative to the stars; this period varies because of perturbations, but it is a little less than 27¹/₃ days. { sī'dir·ē·əl 'mənth }

sidereal noon [ASTRON] The instant in time that the vernal equinox is on the meridian. { sī'dir·ē·əl 'nün }

sidereal period [ASTRON] The length of time required for one revolution of a celestial body about its primary, with respect to the stars. { sī'dir·ē·əl 'pir·ē·əd }

sidereal table [ORD] A device for artillery; a servo-driven table which is used to cancel out earth's rotation; the axis of the table is aligned for the particular latitude of location. { sī'dir·ē·əl 'tā·bəl }

sidereal time [ASTRON] Time based on diurnal motion of stars; it is used by astronomers but is not convenient for ordinary purposes. { sī'dir·ē·əl 'tīm }

sidereal year [ASTRON] The time period relative to the stars of one revolution of the earth around the sun; it is about 365.2564 mean solar days. { sī'dir·ē·əl 'yir }

side relief angle [DES ENG] The angle that the portion of the flanks of a cutting tool below the cutting edge makes with a plane normal to the base. { 'sīd ri'lēf ‚aŋ·gəl }

siderite [MINERAL] $FeCO_3$ A brownish, gray, or greenish rhombohedral mineral composed of ferrous carbonate; hardness is 4 on Mohs scale, and specific gravity is 3.9. Also known as chalybite; iron spar; rhombohedral iron ore; siderose; sparry iron; spathic iron; white iron ore. { 'sid·ə‚rīt }

Siderocapsaceae [MICROBIO] A family of gram-negative, chemolithotrophic bacteria; characterized by the ability to deposit iron or manganese compounds on or around the cells. { ‚sid·ə·rə‚kap'sās·ē‚ē }

siderocyte [CYTOL] An erythrocyte which contains granules staining blue with the Prussian blue reaction. { 'sid·ə·rə‚sīt }

side rod [MECH ENG] **1.** A rod linking the crankpins of two adjoining driving wheels on the same side of a locomotive; distributes power from the main rod to the driving wheels. **2.** One of the rods linking the piston-rod crossheads and the side levers of a side-lever engine. { 'sīd ‚räd }

sideroferrite [GEOL] A variety of native iron occurring as grains in petrified wood. { ‚sid·ə·rə'fe‚rīt }

siderofibrosis [MED] Fibrosis associated with deposits of iron-bearing pigment. { ‚sid·ə·rə·fī'brō·səs }

siderogel [MINERAL] A mineral consisting of truly amorphous $FeO(OH)$ and occurring in some bog iron ores. { 'sid·ə·rə‚jel }

siderograph [ENG] An instrument that keeps the time of the Greenwich longitude; consists of a clock and a navigation instrument. { 'sid·ə·rə‚graf }

siderolite *See* stony-iron meteorite. { 'sid·ə·rə‚līt }

sideromelane [MINERAL] Any iron-rich mafic mineral. { ‚sid·ə·rə'me‚lān }

sideronatrite [MINERAL] $Na_2Fe(SO_4)(OH)·3H_2O$ A yellow mineral composed of basic hydrous sodium iron sulfate occurring in fibrous masses. { ‚sid·ə·rə'nā‚trīt }

sideronitic texture [GEOL] In mineral deposits, a mesh of silicate minerals so shattered and pressed as to force out solutions and other volatiles. { ¦sid·ə·rə¦nid·ik 'teks·chər }

sideropenic dysphagia *See* Plummer-Vinson syndrome. { ¦sid·ə·rə¦pē·nik dis'fā·ja }

siderophile element [CHEM] An element with a weak affinity for oxygen and sulfur and that is readily soluble in molten iron; includes iron, nickel, cobalt, platinum, gold, tin, and tantalum. { 'sid·ə·rə‚fīl 'el·ə·mənt }

siderophore [BIOCHEM] A molecular receptor that binds and transports iron. { 'sid·ə·rə‚fȯr }

siderophyllite [MINERAL] An iron-rich variety of biotite. { ‚sid·ə·rə'fil‚īt }

siderophyre [GEOL] A stony-iron meteorite containing bronzite and tridymite crystals in a nickel-iron network. Also known as siderophyry. { 'sid·ə·rə‚fīr }

siderophyry *See* siderophyre. { ˌsid·ə'räf·ə·rē }

siderose *See* siderite. { 'sid·ə,rōs }

siderosilicosis [MED] A pneumoconiosis resulting from prolonged inhalation of silica and iron dusts. { ˌsid·ə·rə,sil·ə'kō·səs }

siderosis [MED] Pneumoconiosis due to prolonged inhalation of dust containing iron salt. Also known as arc-welder's disease. [PATH] The presence or concentration of stainable iron pigment in a tissue or organ. { ˌsid·ə'rō·səs }

siderosphere *See* inner core. { 'sid·ə·rə,sfir }

siderostat [OPTICS] A more precise model of a heliostat; the siderostat uses a modified mirror mounting so that the image of a star is kept steady while the rest of the field is in rotation about the center. { 'sid·ə·rə,stat }

siderotil [MINERAL] $(Cu,Fe)SO_4 \cdot 5H_2O$ A white to yellowish or pale greenish-white mineral consisting of ferrous sulfate pentahydrate; occurs as fibrous crusts and groups of needlelike crystals. { 'sid·ə·rə,til }

side shot [ENG] A reading or measurement from a survey station to locate a point that is off the traverse or that is not intended to be used as a base for the extension of the survey. { 'sīd ,shät }

side slope [ENG] A test course used to determine lateral stability of a vehicle as well as steering, carburetion, and other functions. { 'sīd ,slōp }

side spray [ORD] Fragments of a bursting projectile thrown sidewise from the line of flight, in contrast with base spray, thrown to the rear, and nose spray, thrown to the front. { 'sīd ,sprā }

sidestream [CHEM ENG] A liquid stream taken from an intermediate point of a liquids-processing unit, for example, a distillation or extraction tower. { 'sīd,strēm }

side stream *See* tributary. { 'sīd,strēm }

sidestream stripper [CHEM ENG] A device used to perform further distillation on a liquid stream (sidestream) from any one of the plates of a bubble tower, usually with the use of steam. { 'sīd,strēm 'strip·ər }

sideswipe [GEOPHYS] **1.** A phenomenon wherein two cross reflections come from a single seismograph, due to the almost simultaneous arrival of reflection energy from both limbs of a syncline or from two nearby, steeply dipping fault scarps. **2.** In refraction shooting, the lateral deflection of a minimum-time path to include a nearby, steeply dipping, high-velocity boundary such as a flank of a salt dome. { 'sīd,swīp }

side-tipping loader [MIN ENG] A front-end loading machine which discharges the bucket load by tipping it sideways. { 'sīd,tip·iŋ 'lōd·ər }

sidetone [COMMUN] The sound of the speaker's own voice as heard in his telephone receiver; the effect is undesirable and is usually suppressed by special circuits. { 'sīd,tōn }

sidetone level [COMMUN] The ratio of the volume of the sidetone to the volume of the speaker's voice, usually expressed in decibels. { 'sīd,tōn ,lev·əl }

sidetone ranging [COMMUN] A method of measuring time delay, and thereby range, by sending a radio signal to a satellite, in which several audio tones of different frequencies are broadcast, and the phases of the tones transmitted from the satellite are compared with the sent tone phases. { 'sīd,tōn 'rānj·iŋ }

sidetrack [CIV ENG] **1.** To move railroad cars onto a siding. **2.** *See* siding. { 'sīd,trak }

sidetracking [ENG] The deliberate act or process of deflecting and drilling a borehole away from a normal, straight course. [PETRO ENG] Drilling procedure to bypass a broken drill or casing permanently lodged in the hole being drilled for an oil well, usually by using a whipstock. { 'sīd,trak·iŋ }

sidewalk [CIV ENG] **1.** A walkway for pedestrians on the side of a street or road. **2.** A foot pavement. { 'sīd,wók }

side-wall sampling [PETRO ENG] A technique for taking rock or sand samples from the sides of oil well boreholes. { 'sīd,wól ,sam·pliŋ }

sidewall section [ENG ACOUS] A wall in a sound-recording studio with reversible panels or rotating columns that are sound-absorbent on one side and reflective on the other, used to vary the acoustic environment. { 'sīd,wól ,sek·shən }

sideways feed [COMPUT SCI] The method of placing cards in the feed hopper of a punched-card machine in which one of the long edges of the card enters the machine first, so that the columns of the card are read simultaneously. Also known as parallel feed. { 'sīd,wāz 'fēd }

Sidewinder [ORD] A Navy air-to-air missile having infrared homing guidance and a speed of over Mach 2; when launched from an airplane, the missile seeks and hits an enemy bomber or fighter target by homing on the heat emitted by the target. { 'sīd,wīnd·ər }

siding [CIV ENG] A short railroad track connected to the main track at one or more points and used to move railroad cars in order to free traffic on the main line or for temporary storage of cars. Also known as sidetrack. [MATER] Any wall cladding, except masonry or brick. { 'sīd·iŋ }

SIDS *See* sudden infant death syndrome. { sidz }

siegbahn [SPECT] A unit of length, formerly used to express wavelengths of x-rays, equal to 1/3029.45 of the spacing of the (200) planes of calcite at 18°C, or to $(1.00202 \pm 0.00003) \times 10^{-13}$ meter. Also known as x-ray unit; X-unit. Symbolized X; XU. { 'sēg,bän }

siege-howitzer [ORD] A short, heavy gun of large caliber, used for destruction of fortresses. { 'sēj ,hau·ət·sər }

siegenite [MINERAL] $(Co,Ni)_3S_4$ A mineral composed of nickel cobalt sulfide. { 'sē·gə,nīt }

siemens [ELEC] A unit of conductance, admittance, and susceptance, equal to the conductance between two points of a conductor such that a potential difference of 1 volt between these points produces a current of 1 ampere; the conductance of a conductor in siemens is the reciprocal of its resistance in ohms. Formerly known as mho (℧); reciprocal ohm. Symbolized S. { 'sē·mənz }

Siemens' electrodynamometer [ELECTROMAG] An early type of electromagnetic instrument in which current flows through all the coils in series. { 'sē·mənz i‖ek·trō,dī·nə'mäm·əd·ər }

sienna [MATER] Any of various yellowish-brown earthy substances consisting of hydrated iron oxide occurring in limonite; becomes orange-brown when burnt and is generally darker and more transparent in oils than is ocher; used as pigment for oil paints and stains. { sī'en·ə }

Sierclomorphidae [INV ZOO] A small family of hymenopteran insects in the superfamily Scolioidea. { sē‖er·ə·lō'mór·fə,dē }

sierozem [GEOL] A soil found in cool to temperate arid regions, characterized by a brownish-gray surface on a lighter layer based on a carbonate or hardpan layer. { 'sir·ə,zem }

Sierpinski gasket [MATH] A fractal which can be constructed by a recursive procedure; at each step a triangle is divided into four new triangles, only three of which are kept for further iterations. { sir'pin·skē ,gas·kət }

Sierpinski set [MATH] **1.** A set of points S on a line such that both S and its complement contain at least one point in each uncountable set on the line that is a countable intersection of open sets. **2.** A set of points in a plane that includes at least one point of each closed set of nonzero measure and does not include any subsets consisting of three collinear points. { sər'pin·skē ,set }

sierra [GEOGR] A high range of hills or mountains with irregular peaks that give a sawtooth profile. { sē'er·ə }

sieve [ENG] **1.** A meshed or perforated device or sheet through which dry loose material is refined, liquid is strained, and soft solids are comminuted. **2.** A meshed sheet with apertures of uniform size used for sizing granular materials. { siv }

sieve analysis [ENG] The size distribution of solid particles on a series of standard sieves of decreasing size, expressed as a weight percent. Also known as sieve classification; sieving. { 'siv ə,nal·ə·səs }

sieve area [BOT] An area in the wall of a sieve-tube element, sieve cell, or parenchyma cell characterized by clusters of pores through which strands of cytoplasm pass to adjoining cells. { 'siv ,er·ē·ə }

sieve cells [BOT] The food-conducting cells of phloem tissue in seedless vascular plants and gymnosperms; unlike sieve-tube members in angiosperms, they do not have sieve plates. { siv selz }

sieve classification *See* sieve analysis. { 'siv ,klas·ə·fə,kā·shən }

sieve deposition [GEOL] The formation of coarse-grained lobate masses on an alluvial fan whose material is sufficiently coarse and permeable to permit complete infiltration of water before it reaches the toe of the fan. { 'siv ,dep·ə,zish·ən }

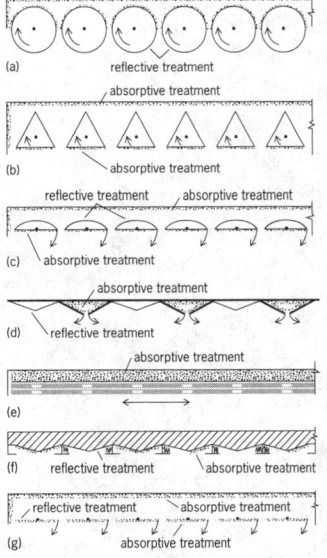

Variable sidewall treatments. *(a)* Rotatable cylinders. *(b)* Rotatable triangular prisms. *(c)* Rotatable cylindrical sections. *(d)* Folding splays. *(e)* One stationary and one sliding perforated panel. *(f)* Retractable curtains. *(g)* Rotatable flat panels.

**SIEMENS'
ELECTRODYNAMOMETER**

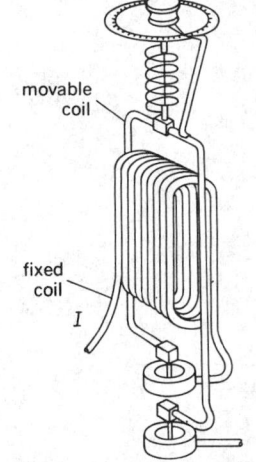

Drawing of Siemens' electrodynamometer, which can be used as ammeter. Current I flows through fixed and movable coils in series. *(Weston Instruments, Division of Sangamo Weston, Inc.)*

sieve diameter [ENG] The size of a sieve opening through which a given particle will just pass. { 'siv dī₁am·əd·ər }

sieve elements [BOT] The food-conducting cells of phloem tissue; may be either sieve cells (in seedless vascular plants and gymnosperms) or sieve-tube members (in angiosperms). { 'siv ₁el·ə·məns }

sieve fraction [ENG] That portion of solid particles which pass through a standard sieve of given number and is retained by a finer sieve of a different number. { 'siv ₁frak·shən }

sieve lobe [GEOL] A coarse-grained lobate mass produced by sieve deposition on an alluvial fan. { 'siv ₁lōb }

sieve mesh [DES ENG] The standard opening in sieve or screen, defined by four boundary wires (warp and woof); the laboratory mesh is square and is defined by the shortest distance between two parallel wires as regards aperture (quoted in micrometers or millimeters), and by the number of parallel wires per linear inch as regards mesh; 60-mesh equals 60 wires per inch. { 'siv ₁mesh }

sieve of Eratosthenes [MATH] An iterative procedure which determines all the primes less than a given number. { 'siv əv ₁er·ə'täs·thə₁nēz }

sieve plate [BOT] A perforated section of the wall of a component member of a sieve tube. [CHEM ENG] A distillation-tower tray that is perforated so that the vapor emerges vertically through the tray, passing through the liquid holdup on top of the tray; used as a replacement for bubble-cap trays in distillation. Also known as sieve tray. { 'siv ₁plāt }

sievert [NUCLEO] **1.** The International System unit of dose equivalent, equal to the dose equivalent when the absorbed dose of ionizing radiation multiplied by stipulated dimensionless factors is 1 joule per kilogram. Abbreviated Sv. **2.** A unit of radiation dose, equal to the dose delivered by a point source of 1 milligram of radium, enclosed in a platinum container with walls 0.5-millimeter thick, to a sample at a distance of 1 centimeter over a period of 1 hour; equal to approximately 8.38 roentgens. Also known as millicurie-of-intensity-hour. { 'zē·vərt }

sieve shaker [CHEM ENG] A device used to shake a stacked column of standard sieve-test trays to cause solids to sift progressively from the top (large openings) to the bottom (small openings and a final pan), according to particle size. { 'siv ₁shā·kər }

sieve texture See poikiloblastic. { 'siv ₁teks·chər }

sieve tissue See phloem. { 'siv ₁tish·ü }

sieve tray See sieve plate. { 'siv ₁trā }

sieve tube [BOT] A phloem element consisting of a series of thin-walled cells arranged end to end, in which some sieve areas are more specialized than others. { 'siv ₁tüb }

sieve-tube member [BOT] The food-conducting cells of phloem tissue in angiosperms; the individual cells have a sieve plate at each end and are arranged end to end to form a sieve tube. { 'siv ₁tüb ₁mem·bər }

sieving See sieve analysis. { 'siv·iŋ }

siffanto [METEOROL] A southwest wind of the Adriatic Sea; it is often violent. { si'fän·tō }

sift [COMPUT SCI] To extract certain desired information items from a large quantity of data. { sift }

Sigalionidae [INV ZOO] A family of scale-bearing polychaete annelids belonging to the Errantia. { ₁sig·ə·lē'än·ə₁dē }

Siganidae [VERT ZOO] A small family of herbivorous perciform fishes in the suborder Acanthuroidei having minute concealed scales embedded in the skin and strong, sharp fin spines. { si'gan·ə₁dē }

Sigatoka [PL PATH] Leaf-spot disease of banana caused by the fungus *Mycosphaerella musicola*. { ₁sig·ə'tō·kə }

sight [NAV] See celestial observation. [ORD] **1.** Mechanical or optical device for aiming a firearm or for laying a gun or launcher in position. **2.** To aim at a target or aiming point. [PHYSIO] See vision. { sīt }

sight aperture [ORD] An irregularly shaped, adjustable, mechanical item usually integral to a rear sight; it functions as a peephole through which the sight at the opposite end of a gun is brought into view in aiming at a target or object. { 'sīt ₁ap·ə·chər }

sight base [ORD] **1.** In gunnery, the distance between the eye and the rings of an optical ring sight. **2.** Mount for a gunsight. { 'sīt ₁bās }

sight bracket [ORD] Clamp used to hold a detachable sight in position when mounted on a gun. { 'sīt ₁brak·ət }

sight check [COMPUT SCI] A check that holes are punched in the same positions in two or more punched cards by superimposing the cards and looking through the holes. { 'sīt ₁chek }

sight distance [OPTICS] The distance from which an object at eye level remains visible to an observer. { 'sīt ₁dis·təns }

sight extension [ORD] Any device which raises the normal base or mount of a gunsight to provide improved or unobstructed sighting. { 'sīt ik₁sten·shən }

sight-feed [ENG] Pertaining to piping in which the flowing liquid can be observed through a transparent tube or wall. { 'sīt₁fēd }

sight glass [ENG] A glass tube or a glass-faced section on a process line or vessel; used for visual reading of liquid levels or of manometer pressures. { 'sīt ₁glas }

sighting [OPTICS] **1.** The act or procedure of aiming with the aid of a sight. **2.** The action of bringing something into view; the action of seeing something. { 'sīd·iŋ }

sighting angle [ORD] In bombing, the angle between the line of sight to the aiming point and the vertical. { 'sīd·iŋ ₁aŋ·gəl }

sighting bar [ORD] Wooden device with enlarged front and rear sight, eyepiece, and a movable target, used for training in the proper method of aiming a small-arms weapon; the eyepiece forces the student to hold the eye in proper position; because of the size of the sights, errors of aiming are very apparent. { 'sīd·iŋ ₁bär }

sighting instrument [ORD] A device or an instrument designed to aid in the pointing of a weapon. { 'sīd·iŋ ₁in·strə·mənt }

sighting shot [ORD] Trial shot, fired to find out whether the sights are properly adjusted. { 'sīd·iŋ ₁shät }

sighting station [ORD] A place from which remote-controlled weapons are sighted. { 'sīd·iŋ ₁stā·shən }

sighting system [ORD] A mechanical or optical device for aiming a firearm or for laying a gun in position. { 'sīd·iŋ ₁sis·təm }

sighting tube [ENG] A tube, usually ceramic, inserted into a hot chamber whose temperature is to be measured; an optical pyrometer is sighted into the tube to observe the interior end of the tube to give a temperature reading. { 'sīd·iŋ ₁tüb }

sight leaf [ORD] The movable hinged part of a rear sight of a gun that can be raised and set to a desired range, or snapped down when not in use. { 'sīt ₁lēf }

sight reduction [NAV] The information needed for establishing a line of position, derived from the use of a sextant or an octant. { 'sīt ri₁dək·shən }

sight-reduction tables [NAV] Tables for performing sight reduction obtained by the use of a sextant or octant, particularly those sights for determining computed altitude for comparison with the observed altitude of a celestial body to determine the altitude difference for establishing a line of position. { 'sīt ri¦dək·shən ₁tā·bəlz }

sight rod See range rod. { 'sīt ₁räd }

sight tracking line [ORD] The line of sight from a computing gunsight reticle image to the target. { 'sīt 'trak·iŋ ₁līn }

sight unit [OPTICS] A compact sighting device composed of an elbow or panoramic telescope, mount, or adapter, usually used for pointing a weapon for direct or indirect fire; it may be attached to, or used in conjunction with, a weapon rocket launcher, or the like. { 'sīt ₁yü·nət }

sigma [INV ZOO] A C-shaped spicule. { 'sig·mə }

sigma algebra [MATH] A collection of subsets of a given set which contains the empty set and is closed under countable union and complementation of sets. Also known as sigma field. { 'sig·mə 'al·jə·brə }

sigma bond [PHYS CHEM] The chemical bond resulting from the formation of a molecular orbital by the end-on overlap of atomic orbitals. { 'sig·mə ₁bänd }

sigma-delta analog-to-digital converter [ELECTR] A converter that uses an analog circuit to generate a single-valued pulse stream in which the frequency of pulses is determined by the analog source, and then uses a digital circuit to repeatedly sum the number of these pulses over a fixed time interval, converting the pulses to numeric values. { ¦sig·mə ¦del·tə ₁an·ə₁läg tü ₁dij·əd·əl kən₁vərd·ər }

sigma-delta converter [ELECTR] A class of electronic systems containing both analog and digital subsystems whose most common application is the conversion of analog signals to digital form, and vice versa, using pulse density modulation to create a high-rate stream of single-amplitude pulses in either case. Also known as delta-sigma converter. { ¦sig·mə ¦del·tə kən'vərd·ər }

sigma-delta digital-to-analog converter [ELECTR] A converter that uses a digital circuit to convert numeric values from a digital processor to a pulse stream and then uses an analog low-pass filter to produce an analog waveform. { ¦sig·mə ¦del·tə ¦dij·əd·əl tü ¦an·ə¦läg kən'vərd·ər }

sigma-delta modulator [ELECTR] The circuit used to generate a pulse stream in a sigma-delta converter. Also known as delta-sigma modulator. { ¦sig·mə ¦del·tə 'mäj·ə¦läd·ər }

sigma factor [MOL BIO] A regulatory protein in prokaryotes that combines with the ribonucleic acid (RNA) polymerase enzyme to facilitate the initiation of RNA synthesis. { 'sig·mə ¦fak·tər }

sigma field *See* sigma algebra. { 'sig·mə ¦fēld }

sigma finite [MATH] A measure is sigma finite on a space X if X is a countable disjoint union of sets each of which is measurable and has finite measure. { 'sig·mə 'fī¦nīt }

sigma function [THERMO] A property of a mixture of air and water vapor, equal to the difference between the enthalpy and the product of the specific humidity and the enthalpy of water (liquid) at the thermodynamic wet-bulb temperature; it is constant for constant barometric pressure and thermodynamic wet-bulb temperature. { 'sig·mə ¦fəŋk·shən }

sigma hyperon [PART PHYS] **1.** The collective name for three semistable baryons with charges of $+1$, 0, and -1 times the proton charge, designated Σ^+, Σ^0, and Σ^-, all having masses of approximately 1193 megaelectronvolts, spin of $^1/_2$, and positive parity; they form an isotopic spin multiplet with a total isotopic spin of 1 and a hypercharge of 0. **2.** Any baryon belonging to an isotopic spin multiplet having a total isotopic spin of 1 and a hypercharge of 0; designated $\Sigma_J{}^P(m)$, where m is the mass of the baryon in megaelectronvolts, and J and P are its spin and parity; the $\Sigma_{3/2}{}^+(1385)$ is sometimes designated Σ^*. { 'sig·mə 'hī·pə¦rän }

sigma-minus hyperonic atom [ATOM PHYS] An atom consisting of a negatively charged sigma hyperon orbiting around an ordinary nucleus. Designated Σ^- hyperonic atom. { 'sig·mə 'mī·nəs ¦hī·pə¦rän·ik 'ad·əm }

sigma phase [MET] A brittle, nonmagnetic phase of tetragonal structure occurring in many transition-metal alloys; frequently encountered in high chromium stainless steels. { 'sig·mə ¦fāz }

sigma pile [NUCLEO] An assembly of moderating material containing a neutron source, used to study the absorption cross sections and other neutron properties of the material. { 'sig·mə ¦pīl }

sigma ring [MATH] A ring of sets where any countable union of its members is also a member. { 'sig·mə ¦riŋ }

sigmaspire [INV ZOO] An S-shaped sponge spicule. { 'sig·mə¦spīr }

sigma-T [OCEANOGR] An abbreviated value of the density of a sea-water sample of temperature T and salinity S: $\sigma T = [\rho(S,T)-1] \times 10^3$, where $\rho(S,T)$ is the value of the sea-water density in centimeter-gram-second units at standard atmospheric pressure. { 'sig·mə ¦tē }

sigmatron [NUCLEO] A cyclotron and betatron operating in tandem to produce billion-volt x-rays. { 'sig·mə¦trän }

sigmatropic shift [ORG CHEM] A rearrangement reaction that consists of the migration of a sigma bond (that is, the sigma electrons) and the group of atoms that are attached to it from one position in a chain or ring into a new position. { ¦sig·mə¦träp·ik 'shift }

sigmoid [BIOL] S-shaped. { 'sig¦mȯid }

sigmoidal dune [GEOL] A dune with an S-shaped ridge crest formed by the merger of crescentic dunes. { sig'mȯid·əl 'dün }

sigmoidal fold [GEOL] A recumbent fold having an axial surface which resembles the Greek letter sigma. { sig'mȯid·əl 'fōld }

sigmoid colon [ANAT] The S-shaped portion of the colon between the descending colon and the rectum. { 'sig¦mȯid 'kō·lən }

sigmoid distortion [OPTICS] A distortion present in line-scan imagery, causing straight lines cut obliquely to appear as sigmoid curves. { 'sig¦mȯid di'stȯr·shən }

sigmoiditis [MED] Inflammation of the sigmoid flexure of the colon. { ¦sig·mȯi'dīd·əs }

sigmoidoscope [MED] An appliance for the inspection, by artificial light, of the sigmoid colon; it differs from the proctoscope in its greater length and diameter. { sig'mȯid·ə¦skōp }

sign [COMMUN] In semiotics, an entity that signifies some other thing, and may be interpreted. [MATH] **1.** A symbol which indicates whether a quantity is greater than zero or less than zero; the signs are often the marks $+$ and $-$ respectively, but other arbitrarily selected symbols are used, especially in automatic data processing. **2.** A unit of plane angle, equal to $30°$ or $\pi/6$ radians. { sīn }

signage [GRAPHICS] Environmental graphic communications whose functions include direction, identification, information or orientation, regulation, warning, or restriction. { 'sīn·ij }

signal [COMMUN] **1.** A visual, aural, or other indication used to convey information. **2.** The intelligence, message, or effect to be conveyed over a communication system. **3.** *See* signal wave. { 'sig·nəl }

signal area [NAV] That part of an airport used for the display of visual ground signals for the benefit of aircraft in flight. { 'sig·nəl ¦er·ē·ə }

signal bias [COMMUN] Form of teletypewriter signal distortion brought about by the lengthening or shortening of pulses during transmission; when marking pulses are all lengthened, a marking signal bias results; when marking pulses are all shortened, a spacing signal bias results. { 'sig·nəl ¦bī·əs }

signal carrier *See* carrier. { 'sig·nəl ¦kar·ē·ər }

signal center [COMMUN] A combination of signal communication facilities operated by the U.S. Army in the field and consisting of a communications center, telephone switching central, and appropriate means of signal communications. { 'sig·nəl ¦sen·tər }

signal channel [COMMUN] A signal path for transmitting electric signals; such paths may be separated by frequency division or time division. { 'sig·nəl ¦chan·əl }

signal conditioning [COMMUN] Processing the form or mode of a signal so as to make it intelligible to or compatible with a given device, such as a data transmission line, including such manipulation as pulse shaping, pulse clipping, digitizing, and linearizing. { 'sig·nəl kən¦dish·ən·iŋ }

signal correction [ENG] In seismic analysis, a correction to eliminate the time differences between reflection times, resulting from changes in the outgoing signal from shot to shot. { 'sig·nəl kə¦rek·shən }

signal detection theory [PSYCH] A theory which characterizes not only the acuity of an individual's discrimination but also the psychological factors that bias his judgment. { 'sig·nəl di¦tek·shən ¦thē·ə·rē }

signal distance [COMPUT SCI] The number of bits that are not the same in two binary words of equal length. Also known as hamming distance. { 'sig·nəl ¦dis·təns }

signal distortion generator [ELECTR] Instrument designed to apply known amounts of distortion on a signal for the purpose of testing and adjusting communications equipment such as teletypewriters. { 'sig·nəl di¦stȯr·shən jen·ə¦rād·ər }

signal effect [ENG] In seismology, variation in arrival times of reflections recorded with identical filter settings, as a result of changes in the outgoing signal. { 'sig·nəl i¦fekt }

signal flare [ENG] A pyrotechnic flare of distinct color and character used as a signal. { 'sig·nə ¦fler }

signal-flow graph [SYS ENG] An abbreviated block diagram in which small circles, called nodes, represent variables of the system, and the nodes are connected by lines, called branches, which represent one-way signal multipliers; an arrow on the line indicates direction of signal flow, and a letter near the arrow indicates the multiplication factor. Also known as flow graph. { 'sig·nəl ¦flō 'graf }

signal generator [ENG] An electronic test instrument that delivers a sinusoidal output at an accurately calibrated frequency that may be anywhere from the audio to the microwave range; the frequency and amplitude are adjustable over a wide range, and the output usually may be amplitude- or frequency-modulated. Also known as test oscillator. { 'sig·nəl jen·ə¦rād·ər }

signal in band [COMMUN] To send control signals at frequencies within the frequency range of the data signal. { ¦sig·nəl in ¦band }

signaling cell [PHYSIO] A cell whose products induce a specific response in target cells. { 'sig·nə·liŋ ˌsel }

signaling key *See* key. { 'sig·nə·liŋ ˌkē }

signaling rate [COMMUN] The rate at which signals are transmitted. { 'sig·nə·liŋ ˌrāt }

signal intensity [COMMUN] The electric-field strength of the electromagnetic wave transmitting a signal. { 'sig·nəl in,ten·səd·ē }

signal level [COMMUN] The difference between the level of a signal at a point in a transmission system and the level of an arbitrarily specified reference signal. { 'sig·nəl ˌlev·əl }

signal light [COMMUN] A light specifically designed for the transmission of code messages by means of visible light rays that are interrupted or deflected by electric or mechanical means. [ENG] A signal, illumination, or any pyrotechnic light used as a sign. { 'sig·nəl ˌlīt }

signal molecule [BIOCHEM] A molecule produced by a signaling cell. { 'sig·nəl ˌmäl·ə,kyül }

signal normalization *See* signal standardization. { 'sig·nəl ˌnȯr·mə·lə'zā·shən }

single nucleotide polymorphism [GEN] A single base-pair difference between two copies of a deoxyribonucleic acid sequence from two individuals. Abbreviated SNP. { ˌsiŋ·gəl ˌnü·klē·ə,tīd ˌpäl·ē'mȯr·fiz·əm }

signal out of band [COMMUN] To send control signals at frequencies outside the frequency range of the data signal. { ˌsig·nəl aút əv ˌband }

signal processing [COMMUN] The extraction of information from complex signals in the presence of noise, generally by conversion of the signals into digital form followed by analysis using various algorithms. Also known as digital signal processing (DSP). { 'sig·nəl ˌprä,ses·iŋ }

signal-recognition particle [CELL MOL] A ribonucleoprotein consisting of a ribonucleic acid (RNA) molecule and six distinct peptide chains that recognizes the signal sequence of a partially synthesized protein and guides it along with its ribosome to a signal recognition particle receptor in the endoplasmic reticulum. Abbreviated SRP. { ˌsig·nəl rek·ig'nish·ən ˌpärd·i·kəl }

signal regeneration [COMMUN] The restoration of a waveform representing a signal to approximate its original amplitude and shape. Also known as signal reshaping. { 'sig·nəl rē,jen·ə'rā·shən }

signal reporting code *See* radio-signal reporting code. { 'sig·nəl ri'pȯrd·iŋ ˌkōd }

signal reshaping *See* signal regeneration. { 'sig·nəl rē,shāp·iŋ }

signal rocket [ORD] A rocket that gives off some characteristic color or display which has a meaning according to an established code. { 'sig·nəl ˌräk·ət }

signal sequence [CELL MOL] A discrete sequence of amino acids in a protein that serves to identify it to transport mechanisms within a cell so as to guide the protein to its destination. { 'sig·nəl ˌsē·kwəns }

signal-shaping network [ELECTR] Network inserted in a telegraph circuit, usually at the receiving end, to improve the waveform of the code signals. { 'sig·nəl ˌshāp·iŋ ˌnet,wərk }

signal speed [COMMUN] The rate at which code elements are transmitted by a communications system. { 'sig·nəl ˌspēd }

signal standardization [COMMUN] The use of one signal to generate another which meets specified requirements for shape, amplitude, and timing. Also known as signal normalization. { 'sig·nəl ˌstan·dər·də'zā·shən }

signal station [COMMUN] A place on shore at which signals are made to ships at sea. { 'sig·nəl ˌstā·shən }

signal strength [ELECTROMAG] The strength of the signal produced by a radio transmitter at a particular location, usually expressed as microvolts or millivolts per meter of effective receiving antenna height. { 'sig·nəl ˌstraŋkth }

signal-strength meter [ELECTR] A meter that is connected to the automatic volume-control circuit of a communication receiver and calibrated in decibels or arbitrary S units to read the strength of a received signal. Also known as S meter; S-unit meter. { 'sig·nəl ˌstraŋkth ˌmēd·ər }

signal-to-interference ratio [ELECTR] The relative magnitude of signal waves and waves which interfere with signal-wave reception. { 'sig·nəl tü ˌin·tər'fir·əns ˌrā·shō }

signal-to-noise improvement factor *See* noise improvement factor. { 'sig·nəl tə 'nȯiz im'prüv·mənt ˌfak·tər }

signal-to-noise ratio [ELECTR] The ratio of the amplitude of a desired signal at any point to the amplitude of noise signals at that same point; often expressed in decibels; the peak value is usually used for pulse noise, while the root-mean-square (rms) value is used for random noise. Abbreviated S/N; SNR. { 'sig·nəl tə 'nȯiz ˌrā·shō }

signal tower [CIV ENG] A switch tower from which railroad signals are displayed or controlled. { 'sig·nəl ˌtaú·ər }

signal tracer [ELECTR] An instrument used for tracing the progress of a signal through a radio receiver or an audio amplifier to locate a faulty stage. { 'sig·nəl ˌtrā·sər }

signal transduction [CELL MOL] The relaying of molecular signals (for example, as contained in a hormone) or physical signals (for example, sensory stimuli) from a cell's exterior to its intracellular response mechanisms. { 'sig·nəl tranz,dək·shən }

signal voltage [ELEC] Effective (root-mean-square) voltage value of a signal. { 'sig·nəl ˌvōl·tij }

signal wave [COMMUN] A wave whose characteristics permit some intelligence, message, or effect to be conveyed. Also known as signal. { 'sig·nəl ˌwāv }

signal-wave envelope [COMMUN] Contour of a signal wave which is composed of a series of wave cycles. { 'sig·nəl ˌwāv 'en·və,lōp }

signal winding [ELEC] Control winding, of a saturable reactor, to which the independent variable (signal wave) is applied. { 'sig·nəl ˌwīnd·iŋ }

sign-and-magnitude code [COMPUT SCI] The representation of an integer X by $(-1)^{a_0}(2^{n-2} a_1 + 2^{n-3} a_2 + \cdots + a_{n-1})$, where a_0 is 0 for X positive, and a_0 is 1 for X negative, and any a_j is either 0 or 1. { 'sīn ən 'mag·nə,tüd ˌkōd }

signature [ELECTR] The characteristic pattern of a target as displayed by detection and classification equipment. [GRAPHICS] A folded, printed sheet, usually consisting of 16 or 32 pages, that forms a section of a book or a pamphlet; the sheet may have fewer pages, but is always in multiples of four. [MATH] **1.** For a quadratic or Hermitian form, the number of positive coefficients minus the number of negative coefficients when the form is reduced by a linear transformation to a sum of squares of absolute values. **2.** For a symmetric or Hermitian matrix, the number of positive entries minus the number of negative entries when the matrix is transformed to diagonal form. [NAV ARCH] The graphic record of the magnetic properties of a vessel automatically traced as the vessel passes over the sensitive element of a recording instrument; more accurately called magnetic signature. [ORD] The identifying characteristics peculiar to each type of target which enable detecting apparatus, such as certain fuses, to sense and differentiate targets. [QUANT MECH] A quantum number α that characterizes a system with the symmetry of a prolate or oblate spheroid and satisfies the equation $r = \exp(-i\pi\alpha)$, where r is the eigenvalue of the system under a rotation through 180° about an axis perpendicular to the symmetry axis. { 'sig·nə·chər }

sign bit [COMPUT SCI] A sign digit consisting of one bit. { 'sīn ˌbit }

sign check indicator [COMPUT SCI] An error checking device, indicating no sign or improper signing of a field used for arithmetic processes; the machine can, upon interrogation, be made to stop or enter into a correction routine. { 'sīn ˌchek 'in·də,kād·ər }

sign convention [OPTICS] A convention as to which quantities, such as angles, distances, and radii of curvature, are positive and which are negative in computations involving a lens or a mirror. { 'sīn kən,ven·chən }

sign digit [COMPUT SCI] A digit containing one to four binary bits, associated with a data item and used to denote an algebraic sign. { 'sīn ˌdij·ət }

signed decimal [COMPUT SCI] A form of packed decimal representation in which the low-order nibble of the last byte has a sign bit that specifies whether the number is positive or negative. { 'sīnd 'des·məl }

signed field [COMPUT SCI] A field of data that contains a number which includes a sign digit indicating the number's sign. { 'sīnd 'fēld }

signed integer [COMPUT SCI] A whole number whose value lies anywhere in a domain that extends from a negative to a positive integer, and which therefore carries a sign. { 'sīnd 'int·ə·jər }

signed measure [MATH] An extended real-valued function

m defined on a sigma algebra of subsets of a set S such that (1) the value of m on the empty set is 0, (2) the value of m on a countable union of disjoint sets is the sum of its values on each set, and (3) m assumes at most one of the values $+\infty$ and $-\infty$. { 'sīnd 'mezh·ər }

signet-ring cell [HISTOL] A cell with a large fat- or carbohydrate-filled vacuole that pushes the nucleus against the cell membrane. { 'sig·nət ,riŋ 'sel }

sign flag [COMPUT SCI] A bit in a status byte in a computer's central processing unit that indicates whether the result of an arithmetic operation is positive or negative. { 'sīn ,flag }

significance [MATH] The arbitrary rank, priority, or order of relative magnitude assigned to a given position in a number. { sig'nif·i·kəns }

significance arithmetic [COMPUT SCI] A rough technique for estimating the numbers and positions of the significant digits of the radix approximation that results when an arithmetic operation is applied to operands in radix approximation form. { sig'nif·i·kəns ə,rith·mə·tik }

significance level See level of significance. { sig'nif·i·kəns ,lev·əl }

significance probability [STAT] The probability of observing a value of a test statistic as significant as, or even more significant than, the value actually observed. { sig'nif·i·kəns ,präb·ə,bil·əd·ē }

significant digit See significant figure. { sig'nif·i·kənt ,dij·ət }

significant figure [MATH] A prescribed decimal place which determines the amount of rounding off to be done; this is usually based upon the degree of accuracy in measurement. Also known as significant digit. { sig'nif·i·kənt ,fig·yər }

significant wave [OCEANOGR] Statistically, a wave with the average height of the highest third of the waves of a given wave group. { sig'nif·i·kənt ,wāv }

signless Stirling number [MATH] The absolute value of a Stirling number of the first kind. { ¦sīn·ləs 'stər·liŋ ,nəm·bər }

sign of aggregation [MATH] One of a pair of parentheses, braces, brackets, or bars which signify that the terms they enclose are to be treated as a single term. { ¦sīn əv ,ag·rə'gā·shən }

sign of the zodiac [ASTRON] The zodiac is divided into 12 sections, called signs, in each of which the sun is situated for 1 month of the year; each sign, 30° in length, is named from a constellation with which the sign once coincided. { 'sīn əv thə 'zō·dē·ak }

sign position [COMPUT SCI] That position, always at or near the left or right end of a numeral, in which the algebraic sign of the number is represented. { 'sīn pə,zish·ən }

sign stimulus [PSYCH] A specific external stimulus that initiates certain behavioral sequences that typically occur in a fixed stereotyped fashion. { 'sīn ,stim·yə·ləs }

sign test [STAT] A test which can be used whenever an experiment is conducted to compare a treatment with a control on a number of matched pairs, provided the two treatments are assigned to the members of each pair at random. { 'sīn ,test }

signum [MATH] The function sgn(x), defined for all real values of x, where sgn(x) = 1 if $x > 0$, sgn(x) = -1 if $x < 0$, and sgn(0) = 0. { 'sig·nəm }

sigua [METEOROL] A straight-blowing monsoon gale of the Philippines. { 'sē,wä }

sikussak [OCEANOGR] Very old sea ice trapped in fjords; it resembles glacier ice because snowfall and snow drifts contribute to its formation. { sə'kü,säk }

SIL See speech interference level.

silage [AGR] Green or mature fodder that is fermented to retard spoilage and produce a succulent winter feed for livestock. { 'sī·lij }

silane [INORG CHEM] Si$_n$H$_{+2}$ A class of silicon-based compounds analogous to alkanes, that is, straight-chain, saturated paraffin hydrocarbons; they can be gaseous or liquid. Also known as silicon hydride. { 'si,lān }

silanol [CHEM] A member of the family of compounds whose structure contains a silicon atom that is bound directly to one or more hydroxyl groups. { 'sī·lə,nòl }

silcrete [GEOL] A conglomerate of sand and gravel cemented by silica. { 'sil,krēt }

silent discharge [ELECTR] An inaudible electric discharge in air that occurs at high voltage and consumes a relatively large amount of energy. { ¦sī·lənt 'dis,chärj }

silent mutation [GEN] A mutation that does not result in amino acid sequence change. { 'sī·lənt myü'tā·shən }

silent period [COMMUN] Period during each hour in which ship and shore radio stations must remain silent and listen for distress calls. { 'sī·lənt 'pir·ē·əd }

silent speed [ENG] The speed at which silent motion pictures are fed through a projector, equal to 16 frames per second (sound-film speed is 24 frames per second). { 'sī·lənt 'spēd }

silent stock support [MECH ENG] A flexible metal guide tube in which the stock tube of an automatic screw machine rotates; it is covered with a casing which deadens sound and prevents transfer of noise and vibration. { 'sī·lənt 'stäk sə,pòrt }

silex [MATER] Heat- and shock-resistant glass containing about 98% quartz. [MINERAL] A pure or finely ground quartz. { 'sī,leks }

silexite [GEOL] Chert occurring in calcareous beds. [PETR] Igneous rock composed mainly of primary quartz. { sī'lek,sīt }

silhouette target [ORD] **1.** Target whose shape is outlined against a light background, although its body features cannot be clearly seen. **2.** Practice target consisting of the dark image of a person or object outlined against a light background. { ,sil·ə'wet 'tär·gət }

silica [MINERAL] SiO$_2$ Naturally occurring silicon dioxide; occurs in five crystalline polymorphs (quartz, tridymite, cristobalite, coesite, and stishovite), in cryptocrystalline form (as chalcedony), in amorphous and hydrated forms (as opal), and combined in silicates. { 'sil·ə·kə }

silica aerogel [MATER] A colloidal silica powder whose grains have small pores; used as a low-temperature insulator. { 'sil·ə·kə 'er·ə,jel }

silica brick [MATER] A type of refractory brick formed of at least 90% silica cemented with, for example, slurried lime; used to line furnace roofs. { 'sil·ə·kə ¦brik }

silica cement [MATER] A mortar used with silica cement; it is a refractory material. { 'sil·ə·kə si'ment }

silica flour [MET] A sand additive for casting produced by pulverizing quartz sand. { 'sil·ə·kə ¦flaù·ər }

silica fume [MATER] A fine-particulate waste product of electric-arc furnaces, consisting primarily of amorphous (noncrystalline) silicon dioxide; its most important use is in the production of high-strength concrete. Also known as microsilica. { 'sil·ə·kə ,fyüm }

silica gel [INORG CHEM] A colloidal, highly absorbent silica used as a dehumidifying and dehydrating agent, as a catalyst carrier, and sometimes as a catalyst. { 'sil·ə·kə ¦jel }

silica glass [MATER] A translucent or transparent vitreous material consisting almost entirely of silica. Also known as fused silica; vitreous silica. { 'sil·ə·kə ¦glas }

silica sand [GEOL] Sand having a very high percentage of silicon dioxide; a source of silicon. { 'sil·ə·kə ¦sand }

silica stone [PETR] A sedimentary rock composed of siliceous minerals. { 'sil·ə·kə ¦stōn }

silicate [INORG CHEM] The generic term for a compound that contains silicon, oxygen, and one or more metals, and may contain hydrogen. [MINERAL] Any of a large group of minerals whose crystal lattice contains SiO$_4$ tetrahedra, either isolated or joined through one or more of the oxygen atoms. { 'sil·ə·kət }

silicate cement [MATER] The silicate of soda glue, used as an adhesive in cardboard and plywood boxes. { 'sil·ə·kət si,ment }

silicate cotton See mineral wool. { 'sil·ə·kət ¦kät·ən }

silicate grinding wheel [DES ENG] A mild-acting grinding wheel where the abrasive grain is bonded with sodium silicate and fillers. { 'sil·ə·kət ¦grīnd·iŋ ,wēl }

silicate of soda See sodium silicate. { 'sil·ə·kət əv 'sōd·ə }

silicate paint [MATER] A paint in which the vehicle is water-soluble sodium silicate; used for painting mortar. { 'sil·ə·kət ¦pānt }

silication [GEOL] The conversion to or the replacement by silicates. { ,sil·ə'kā·shən }

silicatization [MIN ENG] The sealing off of water by the injection of calcium silicate under pressure; sometimes used to reduce the leakage of water through defective lengths of tubing in a shaft. { ,sil·ə,kād·ə'zā·shən }

siliceous [PETR] Describing a rock containing abundant silica, especially free silica. { sə'lish·əs }

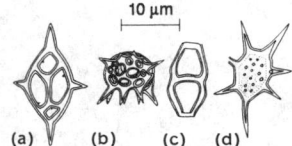

SILICOFLAGELLATA

10 μm

(a)   (b)   (c)   (d)

Examples of fossil and modern Silicoflagellata. *(a) Dictyocha,* Cretaceous to Recent; *(b) Cannopilus,* Miocene; *(c) Naviculopsis,* Eocene to Miocene; and *(d) Vallacerta,* Upper Cretaceous.

SILICON CONTROLLED RECTIFIER

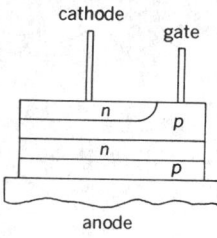

cathode

gate

n

p

n

p

anode

Diagrammatic view of typical silicon controlled rectifier showing four alternate layers of *p*-type and *n*-type material.

siliceous dust [MIN ENG] The dust arising from the dry-working of sand, sandstone, trap, granite, and other igneous rocks; the dust is not soluble in the body fluids, and often results in a form of pneumoconiosis, known as silicosis. { sə'lish·əs 'dəst }

siliceous earth [GEOL] A loose, friable, soft, porous, light-weight, fine-grained, and usually white siliceous sediment, usually derived from the remains of organisms. { sə'lish·əs 'ərth }

siliceous limestone [PETR] **1.** A dense, dark, commonly thin-bedded limestone representing an intimate admixture of calcium carbonate and chemically precipitated silica that are believed to have accumulated simultaneously. **2.** A silicified limestone, bearing evidence of replacement of calcite by silica. { sə'lish·əs 'līm,stōn }

siliceous ooze [GEOL] An ooze composed of siliceous skeletal remains of organisms, such as radiolarians. { sə'lish·əs 'üz }

siliceous sediment [GEOL] Fine-grained sediment and sedimentary rock mainly composed of the microscopic remains of the unicellular, silica-secreting plankton diatoms and radiolarians. Minor constituents include extremely small shards of sponge spicules and other microorganisms such as silicoflagellates. Siliceous sedimentary rock sequences are often highly porous and can form excellent petroleum source and reservoir rocks. { sə'lish·əs 'sed·ə·mənt }

siliceous shale [PETR] A hard, fine-grained rock with the texture of shale and with as much as 85% silica. { sə'lish·əs 'shāl }

siliceous sinter [MINERAL] A white, lightweight, porous, opaline variety of silica, deposited by a geyser or hot spring. Also known as fiorite; geyserite; pearl sinter; sinter. { sə'lish·əs 'sin·tər }

silicic [PETR] Describing magma or igneous rock rich in silica (usually at least 65); granite is a silicic rock. Also known as oversaturated; persilicic. { sə'lis·ik }

silicic acid [INORG CHEM] $SiO_2 \cdot nH_2O$ A white, amorphous precipitate; used to bleach fats, waxes, and oils. Also known as hydrated silica. { sə'lis·ik 'as·əd }

siliciclastic *See* siliclastic. { ,sil·ə·si'klas·tik }

silicide [CHEM] A binary compound in which silicon is bonded with a more electropositive element. { 'sil·ə,sīd }

silicide resistor [ELECTR] A thin-film resistor that uses a silicide of molybdenum or chromium, deposited by direct-current sputtering in an integrated circuit when radiation hardness or high resistance values are required. { 'sil·ə,sīd ri'zis·tər }

silicification [GEOL] Introduction of or replacement by silica. Also known as silification. { sə,lis·ə·fə'kā·shən }

silicified wood [GEOL] A material formed by the silicification of wood, generally in the form of opal or chalcedony, in such a manner as to preserve the original form and structure of the wood. Also known as agatized wood; opalized wood; petrified wood; woodstone. { sə'lis·ə,fīd 'wùd }

silicinate [GEOL] Pertaining to the silica cement of a sedimentary rock. { sə'lis·ən,āt }

siliclastic [PETR] Pertaining to clastic noncarbonate rocks which are almost exclusively silicon-bearing, either as forms of quartz or as silicates. Also known as siliciclastic. { ,sil·ə'klas·tik }

silicle [BOT] A many-seeded capsule formed from two united carpels, usually of equal length and width, and divided on the inside by a replum. { 'sil·ə·kəl }

silicoblast [INV ZOO] Poriferan amebocytes involved in formation of siliceous spicules. { 'sil·ə·kə,blast }

Silicoflagellata [BOT] A class of unicellular flagellates of the plant division Chrysophyta represented by a single living genus, *Dictyocha*. { ,sil·ə·kō,flaj·ə'läd·ə }

Silicoflagellida [INV ZOO] An order of marine flagellates in the class Phytamastigophorea which have an internal, siliceous, tubular skeleton, numerous yellow chromatophores, and a single flagellum. { ,sil·ə·kō·flə'jel·əd·ə }

silicomagnesiofluorite [MINERAL] $Ca_4Mg_3Si_2O_5(OH)_2F_{10}$ A mineral composed of basic calcium magnesium fluoride and silicate. { ,sil·ə·kō·mag,nē·zē·ō'flùr,īt }

silicomanganese [MET] A crude alloy made up of 65–70% manganese, 16–25% silicon, and 1–2.5% carbon; used in the manufacture of low-carbon steel. { ,sil·ə·kō'man·gə,nēs }

silicon [CHEM] A group 14 nonmetallic element, symbol Si, with atomic number 14, atomic weight 28.086; dark-brown crystals that burn in air when ignited; soluble in hydrofluoric acid and alkalies; melts at 1410°C; used to make silicon-containing alloys, as an intermediate for silicon-containing compounds, and in rectifiers and transistors. { 'sil·ə·kən }

silicon bromide *See* silicon tetrabromide. { 'sil·ə·kən 'brō,mīd }

silicon bronze [MET] An alloy of copper with 1–5% silicon; it is corrosion-resistant and has good mechanical properties. { 'sil·ə·kən 'bränz }

silicon burning [NUC PHYS] The synthesis, in stars, of elements, chiefly in the iron group, resulting from the photodisintegration of silicon-28 and other intermediate-mass nuclei; copious supplies of protons, alpha particles, and neutrons are produced, followed by the capture of these particles by other intermediate-mass nuclei. { 'sil·ə·kən 'bərn·iŋ }

silicon capacitor [ELECTR] A capacitor in which a pure silicon-crystal slab serves as the dielectric; when the crystal is grown to have a *p* zone, a depletion zone, and an *n* zone, the capacitance varies with the externally applied bias voltage, as in a varactor. { 'sil·ə·kən kə'pas·əd·ər }

silicon carbide [INORG CHEM] SiC Water-insoluble, bluish-black crystals, very hard and iridescent; soluble in fused alkalies; sublimes at 2210°C; used as an abrasive and a heat refractory, and in light-emitting diodes to produce green or yellow light. { 'sil·ə·kən 'kär,bīd }

silicon chloride *See* silicon tetrachloride. { 'sil·ə·kən 'klôr,īd }

silicon controlled rectifier [ELECTR] A semiconductor rectifier that can be controlled; it is a *pnpn* four-layer semiconductor device that normally acts as an open circuit, but switches rapidly to a conducting state when an appropriate gate signal is applied to the gate terminal. Abbreviated SCR. Also known as reverse-blocking triode thyristor. { 'sil·ə·kən kən'trōld 'rek·tə,fī·ər }

silicon controlled switch [ELECTR] A four-terminal switching device having four semiconductor layers, all of which are accessible; it can be used as a silicon controlled rectifier, gate-turnoff switch, complementary silicon controlled rectifier, or conventional silicon junction transistor. Abbreviated SCS. Also known as reverse-blocking tetrode thyristor. { 'sil·ə·kən kən'trōld 'swich }

silicon copper [MET] An alloy containing 70–80% copper and 20–30% silicon, used as an addition to molten copper or brass. { 'sil·ə·kən 'käp·ər }

silicon detector *See* silicon diode. { 'sil·ə·kən di'tek·tər }

silicon diode [ELECTR] A crystal diode that uses silicon as a semiconductor; used as a detector in ultra-high- and super-high-frequency circuits. Also known as silicon detector. { 'sil·ə·kən 'dī,ōd }

silicon dioxide [INORG CHEM] SiO_2 Colorless, transparent crystals, soluble in molten alkalies and hydrofluoric acid; melts at 1710°C; used to make glass, ceramic products, abrasives, foundry molds, and concrete. { 'sil·ə·kən dī'äk,sīd }

silicone [MATER] A fluid, resin, or elastomer; can be a grease, a rubber, or a foamable powder; the group name for heat-stable, water-repellent, semiorganic polymers of organic radicals attached to the silicones, for example, dimethyl silicone; used in adhesives, cosmetics, and elastomers. { 'sil·ə,kōn }

silicon fluoride *See* silicon tetrafluoride. { 'sil·ə·kən 'flùr,īd }

silicon halide [INORG CHEM] A compound of silicon and a halogen; for example, $SiBr_4$, Si_2Br_6, $SiCl_4$, Si_2Cl_6, Si_3Cl_8, SiF_4, Si_2F_6, SiI_4, and Si_2F_6. { 'sil·ə·kən 'ha,līd }

silicon homojunction *See* bipolar junction transistor. { ,sil·ə·kən 'hä·mə,jəŋk·shən }

silicon hydride *See* silane. { 'sil·ə·kən 'hī,drīd }

silicon image sensor [ELECTR] A solid-state television camera in which the image is focused on an array of individual light-sensitive elements formed from a charge-coupled-device semiconductor chip. Also known as silicon imaging device. { 'sil·ə·kən 'im·ij ,sen·sər }

silicon imaging device *See* silicon image sensor. { 'sil·ə·kən 'im·ij·iŋ di,vīs }

siliconized graphite [MATER] A graphite material whose surface has been chemically converted to silicon carbide. { ,sil·ə·kə,nīzd 'gra,fīt }

siliconizing [MET] Diffusing silicon into solid metal at an elevated temperature. { 'sil·ə·kə,nīz·iŋ }

silicon monoxide [INORG CHEM] SiO A hard, abrasive,

amorphous solid used as thin surface films to protect optical parts, mirrors, and aluminum coatings. { 'sil·ə·kən mə'näk,sīd }

silicon nitride [INORG CHEM] Si_3N_4 A white, water-insoluble powder, resistant to thermal shock and to chemical reagents; used as a catalyst support and for stator blades of high-temperature gas turbines. { 'sil·ə·kən 'nī,trīd }

silicon-on-insulator [ELECTR] A semiconductor manufacturing technology in which thin films of single-crystalline silicon are grown over an electrically insulating substrate. { 'sil·ə·kən ȯn 'in·sə,lād·ər }

silicon-on-sapphire [ELECTR] A semiconductor manufacturing technology in which metal oxide semiconductor devices are constructed in a thin single-crystal silicon film grown on an electrically insulating synthetic sapphire substrate. Abbreviated SOS. { 'sil·ə·kən ȯn 'sa,fīr }

silicon pn junction detector [NUCLEO] A type of junction detector made by diffusing an *n*-type dopant, usually phosphorus, about 2 micrometers into the surface of a *p*-type silicon base. { 'sil·ə·kən ¦pē¦en ¦jəŋk·shən di'tek·tər }

silicon rectifier [ELECTR] A metallic rectifier in which rectifying action is provided by an alloy junction formed in a high-purity silicon slab. { 'sil·ə·kən 'rek·tə,fī·ər }

silicon resistor [ELECTR] A resistor using silicon semiconductor material as a resistance element, to obtain a positive temperature coefficient of resistance that does not appreciably change with temperature; used as a temperature-sensing element. { 'sil·ə·kən ri'zis·tər }

silicon retina [ELECTR] An analog very large scale integrated circuit chip that performs operations which resemble some of the functions performed by the retina of the human eye. { ¦sil·ə,kän 'ret·ən·ə }

silicon solar cell [ELECTR] A solar cell consisting of *p* and *n* silicon layers placed one above the other to form a *pn* junction at which radiant energy is converted into electricity. { 'sil·ə·kən 'sō·lər 'sel }

silicon steel [MET] A steel that contains 0.5–4.5% silicon, used in electric transformer coils. { 'sil·ə·kən 'stēl }

silicon surface barrier detector [NUCLEO] A type of junction detector made from a wafer of *n*-type silicon which is subjected to etching and surface treatments to create a thin layer of *p*-type material and then receives a thin layer of gold evaporated onto the surface. { 'sil·ə·kən sər·fəs ¦bar·ē·ər di,tek·tər }

silicon-symmetrical switch [ELECTR] Thyristor modified by adding a semiconductor layer so that the device becomes a bidirectional switch; used as an alternating-current phase control, for synchronous switching and motor speed control. { 'sil·ə·kən si'me·trə·kəl 'swich }

silicon tetrabromide [INORG CHEM] $SiBr_4$ A fuming, colorless liquid that yellows in air; disagreeable aroma; boils at 153°C. Also known as silicon bromide. { 'sil·ə·kən ¦te·trə'brō,mīd }

silicon tetrachloride [INORG CHEM] $SiCl_4$ A clear, corrosive, fuming liquid with suffocating aroma; decomposes in water and alcohol; boils at 57.6°C; used in warfare smoke screens, to make ethyl silicate and silicones, and as a source of pure silicon and silica. Also known as silicon chloride; tetrachlorosilane. { 'sil·ə·kən ¦te·trə'klȯr,īd }

silicon tetrafluoride [INORG CHEM] SiF_4 A colorless, suffocating gas absorbed readily by water, in which it decomposes; boiling point, −86°C; used in chemical analysis and to make fluosilicic acid. Also known as silicon fluoride. { 'sil·ə·kən ¦te·trə'flür,īd }

silicon transistor [ELECTR] A transistor in which silicon is used as the semiconducting material. { 'sil·ə·kən tran'zis·tər }

silicosiderosis [MED] Pneumoconiosis caused by inhalation of silicate-and iron-containing dust. { ¦sil·ə·kō,sid·ə'rō·səs }

silicosis [MED] Pneumoconiosis due to the inhalation of silica (SiO_2) particles. { ,sil·ə'kō·səs }

silicospiegel [MET] A spiegeleisen pig iron containing 15–20% manganese and 8–15% silicon and up to 4% carbon with the balance iron; used in making steel. { ¦sil·ə·kō'spē·gəl }

silicothermic process See Pidgeon process. { ¦sil·ə·kō'thər·mik 'präs·əs }

silification See silicification. { ,sil·ə·fə'kā·shən }

silique [BOT] A silicle-like capsule, but usually at least four times as long as it is wide, which opens by sutures at either margin and has parietal placentation. { si'lēk }

silk [GEOL] Microscopic needle-shaped crystalline inclusions of rutile in a natural gem from which subsurface reflections produce a whitish sheen resembling that of a silk fabric. [INV ZOO] A continuous protein fiber consisting principally of fibroin and secreted by various insects and arachnids, especially the silkworm, for use in spinning cocoons, webs, egg cases, and other structures. [TEXT] A thread or fabric made from silk secretions of the silkworm. { silk }

silk cotton See kapok. { 'silk ¦kät·ən }

silk cotton tree See kapok tree. { 'silk ¦kät·ən ,trē }

silk gland [INV ZOO] A gland in certain insects which secretes a viscous fluid in the form of filaments known as silk; it is a salivary gland in insects and an abdominal gland in spiders. { 'silk ,gland }

silk paper [MATER] **1.** A paper containing a small amount of silk fibers which give a mottled appearance. **2.** A safety paper sometimes used for postage and revenue stamps. { 'silk ,pā·pər }

silk-screen printing See screen printing. { 'silk ¦skrēn 'print·iŋ }

silkworm [INV ZOO] The larva of various moths, especially *Bombyx mori*, that produces a large amount of silk for building its cocoon. { 'silk,wərm }

silky fracture [MET] A metal fracture in which the broken surface is fine in texture and dull in appearance; characteristic of tough, strong metals. { 'sil·kē 'frak·chər }

sill [BUILD] The lowest horizontal member of a framed partition or of a window or door frame. [CIV ENG] **1.** A timber laid across the foot of a trench or a heading under the side truss. **2.** The horizontal overflow line of a dam spillway or other weir structure. **3.** A horizontal member on which a lift gate rests when closed. **4.** A low concrete or masonry dam in a small stream to retard bottom erosion. [CONT SYS] A type of robot articulation that has three degrees of freedom. [GEOL] **1.** Submarine ridge in relatively shallow water that separates a partly closed basin from another basin or from an adjacent sea. **2.** A tabular igneous intrusion that is oriented parallel to the planar structure of surrounding rock. [MIN ENG] **1.** A piece of wood laid across a drift to constitute a frame to support uprights of timber sets and to carry the track of the tramway. **2.** The floor of a gallery or passage in a mine. { sil }

sill anchor [BUILD] A fastener projecting from a foundation wall or foundation slab to secure the sill to the foundation. { 'sil ,aŋ·kər }

sill depth [OCEANOGR] The maximum depth at which there is horizontal communication between an ocean basin and the open ocean. Also known as threshold depth. { 'sil ,depth }

silled basin See restricted basin. { 'sild 'bās·ən }

sillenite [MINERAL] Bi_2O_3 A mineral composed of native bismuth oxide, is polymorphous with bismite, and occurs as earthy masses. { 'sil·ə,nīt }

sillimanite [MINERAL] Al_2SiO_5 A brown, pale-green, or white neosilicate mineral with vitreous luster crystallizing in the orthorhombic system; commonly occurs in slender crystals, often in fibrous aggregates; hardness is 6–7 on Mohs scale, and specific gravity is 3.23. Also known as fibrolite. { 'sil·ə·mə nīt }

silo [AERO ENG] A missile shelter that consists of a hardened vertical hole in the ground with facilities either for lifting the missile to a launch position, or for direct launch from the shelter. [CIV ENG] A large vertical, cylindrical structure, made of reinforced concrete, steel, or timber, and used for storing grain, cement, or other materials. { 'sī·lō }

siloxane [ORG CHEM] R_2SiO Any of a family of silica-based polymers in which R is an alkyl group, usually methyl; these polymers exist as oily liquids, greases, rubbers, resins, or plastics. Also known as oxosilane. { si'läk,sān }

Silphidae [INV ZOO] The carrion beetles, a family of coleopteran insects in the superfamily Staphylinoidea. { 'sil·fə,dē }

Silsbee effect [CRYO] The ability of an electric current to destroy superconductivity by means of the magnetic field that it generates, without raising the cryogenic temperature. { silz·bē i,fekt }

silt [GEOL] **1.** A rock fragment or a mineral or detrital particle in the soil having a diameter of 0.002–0.05 millimeter that is,

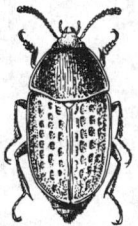

SILPHIDAE

Drawing of a carrion beetle.

smaller than fine sand and larger than coarse clay. **2.** Sediment carried or deposited by water. **3.** Soil containing at least 80% silt and less than 12% clay. { silt }

silt discharge rating *See* sediment discharge rating. { 'silt 'dis‚chärj ‚rād·iŋ }

silting [CIV ENG] The filling up or raising of the bed of a body of water by depositing silt. [GEOL] The deposition or accumulation of stream-deposited silt that is suspended in a body of standing water. { 'silt·iŋ }

silting index [ENG] The measurement of the tendency of a solids- or gel-carrying fluid to cause silting in close-tolerance devices, such as valves or other process-line flow constrictions. { 'silt·iŋ ‚in‚deks }

siltite *See* siltstone. { 'sil‚tīt }

silt loam [GEOL] A soil containing 50–88% silt, 0–27% clay, and 0–50% sand. { 'silt ‚lōm }

silt shale [PETR] A consolidated sediment consisting of no more than 10% sand and having a silt/clay ratio greater than 2:1. { 'silt ‚shāl }

silt soil [GEOL] A soil containing 80% or more of silt, and not more than 12% of clay and 20% of sand. { 'silt ‚sȯil }

siltstone [GEOL] Indurated silt having a shalelike texture and composition. Also known as siltite. { 'silt‚stōn }

silttil [GEOL] A chemically decomposed and eluviated till consisting of a friable, brownish, open-textured silt that contains a few small siliceous pebbles. { 'sil‚til }

Silurian [GEOL] **1.** A period of geologic time of the Paleozoic era, covering a time span of between 430–440 and 395 million years ago. **2.** The rock system of this period. { si'lu̇r·ē·ən }

Siluridae [VERT ZOO] A family of European catfish in the suborder Siluroidei in which the adipose dorsal fin is rudimentary or lacking. { si'lu̇r·ə‚dē }

Siluriformes [VERT ZOO] The catfishes, a distinctive order of actinopterygian fishes in the superorder Ostariophysi, distinguished by a complex Weberian apparatus that involves the fifth vertebrae and one to four pair of barbels. { si‚lu̇r·ə'fȯr‚mēz }

Siluroidei [VERT ZOO] A suborder of the Siluriformes. { ‚sil·yə'rȯid·ē‚ī }

Silvanidae [INV ZOO] An equivalent name for Cucujidae. { sil'van·ə‚dē }

silver [CHEM] A white metallic transition element, symbol Ag, with atomic number 47; soluble in acids and alkalies, insoluble in water; melts at 961°C, boils at 2212°C; used in photographic chemicals, alloys, conductors, and plating. [MET] A sonorous, ductile, malleable metal that is capable of a high degree of polish and that has high thermal and electric conductivity. { 'sil·vər }

silver acetate [ORG CHEM] CH_3COOAg A white powder, moderately soluble in water and nitric acid; used in medicine. { 'sil·vər 'as·ə‚tāt }

silver acetylide [INORG CHEM] Ag_2C_2 A white explosive powder used in detonators. { 'sil·vər ə'sed·əl‚īd }

silver alloy [MET] A metal consisting of silver and one or more additional metallic components. { 'sil·vər 'al‚ȯi }

silver arsenite [INORG CHEM] Ag_3AsO_3 A poisonous, light-sensitive, yellow powder; soluble in acids and alkalies, insoluble in water and alcohol; decomposes at 150°C; used in medicine. { 'sil·vər 'ärs·ən‚īt }

silver arsphenamine [PHARM] A brownish-black powder soluble in water; used in medicine. { 'sil·vər är'sfen·ə‚mēn }

silver battery [ELEC] A solid-state battery based on an Ag_4RbI_5 electrolyte that conducts positive silver ions. { 'sil·vər 'bad·ə·rē }

silver brazing [MET] Brazing in which silver-base alloys are used as the filler metal. { 'sil·vər 'brāz·iŋ }

silver brazing alloy *See* silver solder. { 'sil·vər 'brāz·iŋ 'al‚ȯi }

silver bromate [INORG CHEM] $AgBrO_3$ A poisonous, light- and heat-sensitive, white powder; soluble in ammonium hydroxide, slightly soluble in hot water; decomposed by heat. { 'sil·vər 'brō‚māt }

silver bromide [INORG CHEM] AgBr Yellowish, light-sensitive crystals; soluble in potassium bromide and potassium cyanide, slightly soluble in water; melts at 432°C; used in photographic films and plates. { 'sil·vər 'brō‚mīd }

silver-cadmium storage battery [ELEC] A storage battery that combines the excellent space and weight characteristics of silver-zinc batteries with long shelf life and other desirable

properties of nickel-cadmium batteries. { 'sil·vər 'kad·mē·əm 'stȯr·ij ‚bad·ə·rē }

silver carbonate [INORG CHEM] Ag_2CO_3 Yellowish, light-sensitive crystals; insoluble in water and alcohol, soluble in alkalies and acids; decomposes at 220°C; used as a reagent. { 'sil·vər 'kär·bə‚nət }

silver chloride [INORG CHEM] AgCl A white, poisonous, light-sensitive powder; slightly soluble in water, soluble in alkalies and acids; melts at 445°C; used in photography, photometry, silver plating, and medicine. { 'sil·vər 'klȯr‚īd }

silver chromate [INORG CHEM] Ag_2CrO_4 Dark-colored crystals insoluble in water, soluble in acids and in solutions of alkali chromates; used as an analytical reagent. { 'sil·vər 'krō‚māt }

silver coating *See* silver plating. { 'sil·vər 'kōd·iŋ }

silver cyanide [INORG CHEM] AgCN A poisonous, white, light-sensitive powder; insoluble in water, soluble in alkalies and acids; decomposes at 320°C; used in medicine and in silver plating. { 'sil·vər 'sī·ə‚nīd }

silver-disk pyrheliometer [ENG] An instrument used for the measurement of direct solar radiation; it consists of a silver disk located at the lower end of a diaphragmed tube which serves as the radiation receiver for a calorimeter; radiation falling on the silver disk is periodically intercepted by means of a shutter located in the tube, causing temperature fluctuations of the calorimeter which are proportional to the intensity of the radiation. { 'sil·vər 'disk ‚pīr‚hē·lē'äm·əd·ər }

silvered mica capacitor [ELECTR] A mica capacitor in which a coating of silver is deposited directly on the mica sheets to serve in place of conducting metal foil. { 'sil·vərd ‚mī·kə kə'pas·əd·ər }

silverfish [INV ZOO] Any of over 350 species of insects of the order Thysanura; they are small, wingless forms with biting mouthparts. { 'sil·vər‚fish }

silver fluoride [INORG CHEM] $AgF·H_2O$ A light-sensitive, yellow or brownish solid, soluble in water; dehydrated form melts at 435°C; used in medicine. Also known as tachiol. { 'sil·vər 'flu̇r‚īd }

silver foil [MET] Silver or a silver-colored metal in very thin sheets. { 'sil·vər ‚fȯil }

silver frost [METEOROL] A deposit of glaze built up on trees, shrubs, and other exposed objects during a fall of freezing precipitation; the product of an ice storm. Also known as silver thaw. { 'sil·vər 'frȯst }

silver glance *See* argentite. { 'sil·vər 'glans }

silver halide [INORG CHEM] A compound of silver and a halogen; for example, silver bromide (AgBr), silver chloride (AgCl), silver fluoride (AgF), and silver iodide (AgI). { 'sil·vər 'ha‚līd }

silver iodate [INORG CHEM] $AgIO_3$ A white powder, soluble in ammonium hydroxide and nitric acid, slightly soluble in water; melts above 200°C; used in medicine. { 'sil·vər 'ī·ə‚dāt }

silver iodide [INORG CHEM] AgI A pale-yellow powder, insoluble in water, soluble in potassium iodide-sodium chloride solutions and ammonium hydroxide; melts at 556°C; used in medicine, photography, and artificial rainmaking. { 'sil·vər 'ī·ə‚dīd }

silver lactate [ORG CHEM] $CH_3CHOHCOOAg·H_2O$ Gray-to-white, light-sensitive crystals; slightly soluble in water and in alcohol; used in medicine. Also known as actol. { 'sil·vər 'lak‚tāt }

silverline system [INV ZOO] A series of superficial argentophilic lines in many protozoans, especially ciliates. { 'sil·vər‚līn ‚sis·təm }

silver metallurgy [MET] The art and science of extracting silver metal economically from ores, and the reclamation of silver from industrial processes or from scrap metal. { 'sil·vər 'med·əl‚ər·jē }

silver migration [ELEC] A process, causing reduction in insulation resistance and dielectric failure; silver, in contact with an insulator, at high humidity, and subjected to an electrical potential, is transported ionically from one location to another. { 'sil·vər mī'grā·shən }

silver nitrate [INORG CHEM] $AgNO_3$ Poisonous, corrosive, colorless crystals; soluble in glycerol, water, and hot alcohol; melts at 212°C; used in external medicine, photography, hair dyeing, silver plating, ink manufacture, and mirror silvering, and as a chemical reagent. { 'sil·vər 'nī‚trāt }

SILURIAN

CENOZOIC	QUATERNARY	
	TERTIARY	
MESOZOIC	CRETACEOUS	
	JURASSIC	
	TRIASSIC	
PALEOZOIC	PERMIAN	
	CARBONIFEROUS	PENNSYLVANIAN
		MISSISSIPPIAN
	DEVONIAN	
	SILURIAN	
	ORDOVICIAN	
	CAMBRIAN	
PRECAMBRIAN		

Chart showing the position of the Silurian in relation to the various periods of geologic time.

silver nitrite [INORG CHEM] $AgNO_2$ Yellow or grayish-yellow needles which decompose at 140°C; soluble in hot water; used in organic synthesis and in testing for alcohols. { 'sil·vər 'nī,trīt }

silver orthophosphate See silver phosphate. { 'sil·vər ,or·thō'fäs,fāt }

silver oxide [INORG CHEM] Ag_2O An odorless, dark-brown powder with a metallic taste; soluble in nitric acid and ammonium hydroxide, insoluble in alcohol; decomposes above 300°C; used in medicine and in glass polishing and coloring, as a catalyst, and to purify drinking water. { 'sil·vər 'äk,sīd }

silver oxide cell [ELEC] A primary cell in which depolarization is accomplished by an oxide of silver. { 'sil·vər 'äk,sīd ,sel }

silver permanganate [INORG CHEM] $AgMnO_4$ Water-soluble, violet crystals that decompose in alcohol; used in medicine and in gas masks. { 'sil·vər pər'maŋ·gə,nāt }

silver phosphate [INORG CHEM] Ag_3PO_4 A poisonous, yellow powder; darkens when heated or exposed to light; soluble in acids and in ammonium carbonate, very slightly soluble in water; melts at 849°C; used in photographic emulsions and in pharmaceuticals, and as a catalyst. Also known as silver orthophosphate. { 'sil·vər 'fä,sfāt }

silver picrate [ORG CHEM] $C_6H_2O(NO_2)_3Ag·H_2O$ Yellow crystals, soluble in water, insoluble in ether and chloroform; used in medicine. { 'sil·vər 'pi,krāt }

silver plating [MET] Electrolytically depositing a coating of metallic silver on a base metal. Also known as silver coating. { 'sil·vər ,plād·iŋ }

silver potassium cyanide [ORG CHEM] $KAg(CN)_2$ Toxic, white crystals soluble in water and alcohol; used in silver plating and as a bactericide and antiseptic. Also known as potassium argentocyanide; potassium cyanoargentate. { 'sil·vər pə'tas·ē·əm 'sī·ə,nīd }

silverprint [GRAPHICS] A print made from photographic paper treated with silver halides. { 'sil·vər,print }

silver protein [ORG CHEM] A brown, hygroscopic powder containing 7.5–8.5% silver; made by reaction of a silver compound with gelatin in the presence of an alkali; used as an antibacterial. { 'sil·vər 'prō,tēn }

silver selenide [INORG CHEM] Ag_2Se A gray powder, insoluble in water, soluble in ammonium hydroxide; melts at 880°C. { 'sil·vər 'sel·ə,nīd }

silver solder [MET] A solder composed of silver, copper, and zinc, having a melting point lower than silver but higher than lead-tin solder. Also known as silver-brazing alloy. { 'sil·vər ,säd·ər }

silverstat regulator [ELEC] Multitapped resistor, the taps of which are connected to single-leaf silver contacts; variation of voltage causes a solenoid to open or close these contacts, shorting out more or less of the resistance in the exciter circuit as a means of regulating the output voltage to the desired value. { 'sil·vər,stat 'reg·yə,lād·ər }

silver storm See ice storm. { 'sil·vər 'storm }

silver suboxide [INORG CHEM] AgO A charcoal-gray powder that crystallizes in the cubic or orthorhombic system, and has diamagnetic properties; used in making silver oxide-zinc alkali batteries. Also known as argentic oxide. { 'sil·vər səb'äk,sīd }

silver sulfate [INORG CHEM] Ag_2SO_4 Light-sensitive, colorless, lustrous crystals; soluble in alkalies and acids, insoluble in alcohol; melts at 652°C; used as an analytical reagent. Also known as normal silver sulfate. { 'sil·vər 'səl,fāt }

silver sulfide [INORG CHEM] Ag_2S A dark, heavy powder, insoluble in water, soluble in concentrated sulfuric and nitric acids; melts at 825°C; used in ceramics and in inlay metalwork. { 'sil·vər 'səl,fīd }

silver thaw See silver frost. { 'sil·vər 'tho }

silvery iron [MET] A variety of cast iron with a high silicon content, a light-gray color, and a fine grain. { 'sil·və·rē 'ī·ərn }

silver-zinc storage battery [ELEC] A storage battery that gives higher current output and greater watt-hour capacity per unit of weight and volume than most other types, even at high discharge rates; used in missiles and torpedoes, where its high cost can be tolerated. { 'sil·vər ,ziŋk 'stor·ij ,bad·ə·rē }

silviculture [FOR] The theory and practice of controlling the establishment, composition, and growth of stands of trees for any of the goods and benefits that they may be called upon to produce. { 'sil·və,kəl·chər }

silylene [CHEM] A divalent silicon species (R_2Si, with two nonbonding electrons, where R = alkyl, aryl, or hydrogen); analogous to a carbene in carbon chemistry. { sə'li,lēn }

sima [PETR] A petrologic term for the lower layer of the earth's crust, composed of silica- and magnesia-rich rocks; source of basaltic magma; sima is equivalent to the lower part of the continental crust and the bulk of the oceanic crust. Also known as intermediate layer. { 'sī·mə }

SIMD [COMPUT SCI] A type of multiprocessor architecture in which there is a single instruction cycle, but multiple sets of operands may be fetched to multiple processing units and may be operated upon simultaneously within a single instruction cycle. Acronym for single-instruction-stream, multiple-data-stream.

similar decimals [MATH] Decimals that have the same number of decimal places. { 'sim·ə·lər 'des·məlz }

similar figures [MATH] Two figures or bodies that are identical except for size; similar figures can be placed in perspective, so that straight lines joining corresponding parts of the two figures will pass through a common point. { 'sim·ə·lər 'fig·yərz }

similar fold [GEOL] A fold in deformed beds in which the successive folds resemble each other. { 'sim·ə·lər 'fold }

similar fractions [MATH] Two or more common fractions that have the same denominator. { 'sim·ə·lər 'frak·shənz }

similarity coefficient [SYST] In numerical taxonomy, a factor S used to calculate the similarity between organisms, according to the formula $S = n_s/(n_s + n_d)$, where n_s represents the number of positive features shared by two strains, and n_d represents the number of features positive for one strain and negative for the other. { ,sim·ə'lar·əd·ē ,kō·i,fish·ənt }

similarity principle See principle of dynamical similarity. { ,sim·ə'lar·əd·ē ,prin·sə·pəl }

similarity transformation [MATH] **1.** A transformation of a euclidean space obtained from such transformations as translations, rotations, and those which either shrink or expand the length of vectors. **2.** A mapping that associates with each linear transformation P on a vector space the linear transformation $R^{-1}PR$ that results when the coordinates of the space are subjected to a nonsingular linear transformation R. **3.** A mapping that associates with each square matrix P the matrix $Q = R^{-1}PR$, where R is a nonsingular matrix and R^{-1} is the inverse matrix of R; if P is the matrix representation of a linear transformation, then this definition is equivalent to the second definition. { ,sim·ə'lar·əd·ē ,tranz·fər,mā·shən }

similarly placed conics [MATH] Conics of the same type (both ellipses, both parabolas, or both hyperbolas) whose corresponding axes are parallel. { 'sim·ə·lər·lē 'plāst 'kän·iks }

similar matrices [MATH] Two square matrices A and B related by the transformation $B = SAT$, where S and T are nonsingular matrices and T is the inverse matrix of S. { 'sim·i·lər 'mā·tri,sēz }

similar terms [MATH] Terms that contain the same unknown factors and the same powers of these factors. Also known as like terms. { 'sim·ə·lə· 'tərms }

similar triangles [MATH] Triangles whose corresponding angles are equal; the corresponding sides are then proportional in length. { 'sim·ə·lər 'trī,aŋ·gəlz }

similitude [ENG] A likeness or resemblance; for example, the scale-up of a chemical process from a laboratory or pilot-plant scale to a commercial scale. [PHYS] The use in scientific studies and engineering designs of the corresponding behavior between large and small objects or systems which are of similar nature and, more precisely, have geometrical, kinematic, and dynamical similarity. { si'mil·ə,tüd }

SIMM [COMPUT SCI] A printed circuit board that holds several semiconductor memory chips and is used to add memory to a computer. Acronym for single in-line memory module. { sim·

simmer [ENG] The detectable leakage of fluid in a safety valve below the popping pressure. { 'sim·ər }

Simmonds' disease [MED] Hypopituitarism with marked insufficiency of the target glands and profound cachexia. Also known as hypophyseal cachexia; hypopituitary cachexia. { 'sim·ənz di,zēz }

simo chart [IND ENG] A basic motion-time chart used to show the simultaneous nature of motions; commonly a therblig chart for two-hand work with motion symbols plotted vertically with respect to time, showing the therblig abbreviation and a brief description for each activity, and individual times values and body-member detail. Also known as simultaneous motion-cycle chart. { 'sī·mō ˌchärt }

Simon liquefier [CRYO] A device for liquefying helium in which helium is first cooled at high pressure by liquid or solid hydrogen and is then liquefied by a single adiabatic expansion. { 'sī·mən 'lik·wəˌfī·ər }

Simonsiellaceae [MICROBIO] A family of bacteria in the order Cytophagales; cells are arranged to form flat filaments capable of gliding motility when the flat surface is in contact with the substrate. { səˌmän·sē·ə'lās·ēˌē }

Simon's theory [ENG] A theory of drilling which includes the effects of drilling by percussion and by vibration with a rotary (oil well) bit, cable tool, and pneumatic hammer; the rate of penetration of a chisel-shaped bit into brittle rock may be defined as follows: $R = NAf_v/\pi D$, where R equals the rate of advance of bit, N equals the number of wings of bit, f_v equals the number of impacts per unit time, D equals the diameter of the bit, and A equals the cross-sectional area of the crater at the periphery of the drill hole. { 'sī·mənz ˌthē·ə·rē }

simoom [METEOROL] A strong, dry, dust-laden desert wind which blows in the Sahara, Israel, Syria, and the desert of Arabia; its temperature may exceed 130°F (54°C), and the humidity may fall below 10. { sə'müm }

simple [BIOL] **1.** Made up of one piece. **2.** Unbranched. **3.** Consisting of identical units, as a simple tissue. { 'sim·pəl }

simple aggregative index [STAT] A statistic computed for a collection of items by taking the ratio of the sum of their given-year values or amounts to the sum of their base-year values or amounts and usually multiplying by 100 to express the figure as a percentage. { ¦sim·pəl ˌag·rəˌgād·iv 'inˌdeks }

simple algebra [MATH] An algebra over a field that is also a simple ring. { 'sim·pəl ¦al·jə·brə }

simple alternative [STAT] An alternative to the null hypothesis which completely specifies the distribution of the observed random variables. { 'sim·pəl ȯl'tər·nəd·iv }

simple arc [MATH] The image of a closed interval under a continuous, injective mapping from the interval into a plane. Also known as Jordan arc. { 'sim·pəl 'ärk }

simple balance [ENG] An instrument for measuring weight in which a beam can rotate about a knife-edge or other point of support, the unknown weight is placed in one of two pans suspended from the ends of the beam and the known weights are placed in the other pan, and a small weight is slid along the beam until the beam is horizontal. { 'sim·pəl 'bal·əns }

simple branched tubular gland [ANAT] A structure consisting of two or more unbranched, tubular secreting units joining a common outlet duct. { 'sim·pəl 'brancht 'tüb·yə·lər 'gland }

simple buffering [COMPUT SCI] A technique for obtaining simultaneous performance of input/output operations and computing; it involves associating a buffer with only one input or output file (or data set) for the entire duration of the activity on that file (or data set). { 'sim·pəl 'bəf·ə·riŋ }

simple character [MATH] The character of an irreducible representation of a group. { 'sim·pəl 'kar·ik·tər }

simple closed curve [MATH] A closed curve which never crosses itself. { 'sim·pəl 'klōzd 'kərv }

simple compression [MATH] A transformation that compresses a configuration in a given direction, given by $x' = kx$, $y' = y$, $z' = z$, with $k < 1$, when the direction is that of the x axis. { ¦sim·pəl kəm¦presh·ən }

simple conic chart [MAP] A chart on a simple conic projection. { 'sim·pəl 'kän·ik 'chärt }

simple conic projection [MAP] A conic map projection in which the surface of a sphere or spheroid, such as the earth, is developed on a tangent cone which is then spread out to form a plane. { 'sim·pəl 'kän·ik prə'jek·shən }

simple continuous distillation See equilibrium flash vaporization. { 'sim·pəl kən'tin·yə·wəs ˌdis·tə'lā·shən }

simple crater [GEOL] A meteorite impact crater of relatively small diameter, characterized by a uniformly concave-upward shape and a maximum depth in the center, and lacking a central uplift. { 'sim·pəl 'krād·ər }

simple cross-bedding [GEOL] Cross-bedding in which the

lower bounding surfaces are nonerosional surfaces. { 'sim·pəl 'krȯs ˌbed·iŋ }

simple cubic lattice [CRYSTAL] A crystal lattice whose unit cell is a cube, and whose lattice points are located at the vertices of the cube. { 'sim·pəl 'kyü·bik 'lad·əs }

simple curve [MATH] A curve that does not cross itself or touch itself. { 'sim·pəl ¦kərv }

simple cusp See cusp of the first kind. { 'sim·pəl 'kəsp }

simple data structure [COMPUT SCI] An arrangement of data in a data base or file in which each grouping of data, such as a record, is of equal importance or significance. { 'sim·pəl 'dad·ə ˌstrək·chər }

simple dike [PETR] An igneous dike emplaced in a single episode. { 'sim·pəl 'dīk }

simple dipath [MATH] A directed path in which no two vertices are the same (except that the initial and final vertices may be the same). { 'sim·pəl 'dīˌpath }

simple electrostatic lens [ELECTR] An electrostatic lens that consists of a circular hole in a conducting plate with different electrostatic fields on the two sides. { ¦sim·pəl i¦lek·trəˌstad·ik 'lenz }

simple elongation [MATH] A transformation that elongates a configuration in a given direction, given by $x' = kx$, $y' = y$, $z' = z$, with $k > 1$, when the direction is that of the x axis. { 'sim·pəl ˌē̦loŋ'gā·shən }

simple engine [MECH ENG] An engine (such as a steam engine) in which expansion occurs in a single phase, after which the working fluid is exhausted. { 'sim·pəl 'en·jən }

simple event See elementary event. { 'sim·pəl i¦vent }

simple extension [MATH] An extension of a field that has an element c such that the extension field consists of the set of all quotients (with nonzero denominator) of polynomials in c with coefficients in the original field. { 'sim·pəl ik¦sten·shən }

simple fraction See common fraction. { 'sim·pəl 'frak·shən }

simple fruit [BOT] A fruit that has developed from a single carpel or several united carpels. { 'sim·pəl 'früt }

simple function [MATH] **1.** For a region D of the complex plane, an analytic, injective function on D. Also known as schlicht function. **2.** Any measurable function whose range is a finite set. **3.** See step function. { 'sim·pəl 'fəŋk·shən }

simple gland [ANAT] A gland having a single duct. { 'sim·pəl 'gland }

simple goiter [MED] Diffuse enlargement of the thyroid gland, usually not associated with constitutional features. { 'sim·pəl 'gȯid·ər }

simple graph [MATH] A graph with no loops and no parallel edges. { 'sim·pəl ¦graf }

simple group [MATH] A group G that is nontrivial and contains no normal subgroups other than the identity element and G itself. { ¦sim·pəl 'grüp }

simple harmonic current [ELEC] Alternating current, the instantaneous value of which is equal to the product of a constant, and the cosine of an angle varying linearly with time. Also known as sinusoidal current. { 'sim·pəl här'män·ik 'kə·rənt }

simple harmonic electromotive force [ELEC] An alternating electromotive force which is equal to the product of a constant and the cosine or sine of an angle which varies linearly with time. { 'sim·pəl här'män·ik i¦lek·trə̦mōd·iv 'fȯrs }

simple harmonic motion See harmonic motion. { 'sim·pəl här'män·ik 'mō·shən }

simple hypothesis [STAT] A hypothesis which completely specifies the distribution of the observed random variables. { 'sim·pəl hī'päth·ə·səs }

simple integral [MATH] An integral over only one variable. { 'sim·pəl 'int·ə·grəl }

simple lattice See primitive lattice. { 'sim·pəl 'lad·əs }

simple layering [BOT] A plant propagation technique in which one portion of the stem, still attached to the plant, is buried. { ˌsim·pəl 'lā·ə·riŋ }

simple leaf [BOT] A leaf having one blade, or a lobed leaf in which the separate parts do not reach down to the midrib. { 'sim·pəl 'lēf }

simple lens [OPTICS] A lens consisting of a single element. Also known as single lens. { 'sim·pəl 'lenz }

simple machine [MECH ENG] Any of several elementary machines, one or more being incorporated in every mechanical machine; usually, only the lever, wheel and axle, pulley (or

SIMPLE LEAF

The one-blade type of simple leaf.

block and tackle), inclined plane, and screw are included, although the gear drive and hydraulic press may also be considered simple machines. { 'sim·pəl mə'shēn }

Simple Mail Transfer Protocol [COMPUT SCI] An Internet standard for sending e-mail messages. Abbreviated SMTP. { ¦sim·pəl 'māl ‚tranz·fər ‚prōd·ə‚kȯl }

simple metal [SOLID STATE] A metal in which the electrons are basically free to move throughout the volume. { 'sim·pəl 'med·əl }

simple microscope [OPTICS] A diverging lens system, which can form an enlarged image of a small object. Also known as hand lens; magnifier; magnifying glass. { 'sim·pəl 'mī·krə‚skōp }

simple order See linear order. { 'sim·pəl 'ȯr·dər }

simple ore [GEOL] An ore of a single metal. { 'sim·pəl 'ȯr }

simple oscillator See harmonic oscillator. { 'sim·pəl 'äs·ə‚lād·ər }

simple path See path. { 'sim·pəl 'path }

simple pendulum [MECH] A device consisting of a small, massive body suspended by an inextensible object of negligible mass from a fixed horizontal axis about which the body and suspension are free to rotate. { 'sim·pəl 'pen·jə·ləm }

simple pistil [BOT] A pistil that consists of a single carpel. { 'sim·pəl 'pis·təl }

simple pit [BOT] A pit that lacks a border. { 'sim·pəl 'pit }

simple point See ordinary point. { 'sim·pəl ¦pȯint }

simple polyhedron [MATH] A polyhedron with no holes inside it; technically, a polyhedron that is topologically equivalent to a solid sphere. { 'sim·pəl ‚päl·ə'hē·drən }

simple protein [BIOCHEM] One of a group of proteins which, upon hydrolysis, yield exclusively amino acids; included are globulins, glutelins, histones, prolamines, and protamines. { 'sim·pəl 'prō‚tēn }

simple random samples [STAT] Samples in which every possible sample of size n, that is, every combination of n items from the number in the population, is equally likely to be part of the sample. { ¦sim·pəl 'ran·dəm 'sam·pəlz }

simple results [STAT] Results of observations such that on each trial of an experiment one and only one of these results will occur. { 'sim·pəl ri'zəls }

simple ring [ASTRON] See elementary ring structure. [MATH] A semisimple ring R such that for any two left ideals in R there is an isomorphism of R which maps one onto the other. { 'sim·pəl 'riŋ }

simple root [MATH] A polynomial $f(x)$ has c as a simple root if $(x - c)$ is a factor but $(x - c)^2$ is not.

simple salt [CHEM] One of four classes of salts in a classification system that depends on the character of completeness of the ionization; examples are $NaCl$, $NaHCO_3$, and $Pb(OH)Cl$. { 'sim·pəl 'sȯlt }

simple shear [GEOPHYS] Strain caused by differential movements on one set of parallel planes which results in internal rotation of fabric elements. [MATH] A transformation that corresponds to a shearing motion in which a coordinate axis in the plane or a coordinate plane in space does not move, having the form $x' = x$, $y' = ax + y$, $z' = z$, where a is a constant, for a suitable choice of axes. { 'sim·pəl 'shir }

simple sound source [ACOUS] Under free-field conditions, a source that emits sound uniformly in every direction. { 'sim·pəl 'saund ‚sȯrs }

simple spit [GEOGR] A spit, either straight or recurved, without the development of minor spits at its end or along its inner side. { 'sim·pəl 'spit }

simple stomach [ANAT] A stomach consisting of a single dilation of the alimentary canal, as found in humans, dogs, and many higher and lower vertebrates. { 'sim·pəl 'stəm·ək }

simple strain [MATH] A one-dimensional strain or a simple shear. { ¦sim·pəl 'strān }

simple tone [ACOUS] Also known as pure tone. **1.** A sound wave whose instantaneous sound pressure is a simple sinusoidal function of time. **2.** A sound sensation characterized by singleness of pitch. { 'sim·pəl 'tōn }

simple tubular gland [ANAT] A gland consisting of a single, tubular secreting unit. { 'sim·pəl 'tü·byə·lər 'gland }

simple twin [CRYSTAL] A twinned crystal composed of only two individuals in twin relation. { 'sim·pəl 'twin }

simple valley [GEOL] A valley that maintains a constant

relation to the general structure of the underlying strata. { 'sim·pəl 'val·ē }

simplex [MATH] An n-dimensional simplex in a euclidean space consists of $n + 1$ linearly independent points $p_0, p_1, \ldots, p_n$ together with all line segments $a_0 p_0 + a_1 p_1 + \cdots + a_n p_n$ where the $a_i \geq 0$ and $a_0 + a_1 + \cdots + a_n = 1$; a triangle with its interior and a tetrahedron with its interior are examples. [QUANT MECH] The eigenvalue of a nucleus or other object with an octupole (pear) shape under an operation consisting of rotation through 180° about an axis perpendicular to the symmetry axis, followed by inversion. { 'sim‚pleks }

simplex channel [COMMUN] A channel which permits transmission in one direction only. { 'sim‚pleks ¦chan·əl }

simplex concrete pile [CIV ENG] A molded-in-place pile made by using a hollow cylindrical mandrel which is filled with concrete after having been driven to the desired depth and raised a few feet at a time, the concrete flowing out at the bottom and filling the hole in the earth. { 'sim‚pleks 'kän‚krēt 'pīl }

simplex machine [TEXT] A warp knitting machine that is similar to a tricot machine, but employs two sets of needles and knits a double fabric. { 'sim‚pleks mə'shēn }

simplex method [MATH] A finite iterative algorithm used in linear programming whereby successive solutions are obtained and tested for optimality. { 'sim‚pleks ¦meth·əd }

simplex pump [MECH ENG] A pump with only one steam cylinder and one water cylinder. { 'sim‚pleks ¦pəmp }

simplex structure [COMPUT SCI] The structure of an information processing system designed in such a way that only the minimum amount of hardware is utilized to implement its function. { 'sim‚pleks ¦strək·chər }

simplex transmission [COMMUN] A mode of radio transmission in which communication takes place between two stations in only one direction at a time. { 'sim‚pleks tranz¦mish·ən }

simplex uterus [ANAT] A uterus consisting of a single cavity, representing the greatest degree of fusion of the Müllerian ducts; found in humans and apes. { 'sim‚pleks ¦yüd·ə·rəs }

simplicial complex [MATH] A set consisting of finitely many simplices where either two simplices are disjoint or intersect in a simplex which is a face common to each. Also known as geometric complex. { sim'plish·əl 'käm‚pleks }

simplicial graph [MATH] A graph in which no line starts and ends at the same point, and in which no two lines have the same pair of end points. { sim'plish·əl 'graf }

simplicial homology [MATH] A homology for a topological space where the nth group reflects how the space may be filled out by n-dimensional simplicial complexes and detects the presence of analogs of n-dimensional holes. { sim'plish·əl hə mäl·ə·jē }

simplicial mapping [MATH] A mapping of one simplicial complex into another in which the images of the simplexes of one complex are simplexes of the other complex. { sim'plish·əl 'map·iŋ }

simplicial subdivision [MATH] A decomposition of the simplices composing a simplicial complex which results in a simplicial complex with a larger number of simplices. { sim'plish·əl 'səb·di‚vizh·ən }

simply connected region [MATH] A region having no holes; all closed curves can be shrunk to a point without passing through points in the complement of the region. { 'sim‚plē kə¦nek·təd 'rē·jən }

simply connected space [MATH] A topological space whose fundamental group consists of only one element; equivalently, all closed curves can be shrunk to a point. { 'sim‚plē kə¦nek·təd 'spās }

simply normal number [MATH] A number whose expansion with respect to a given base (not necessarily 10) is such that all the digits occur with equal frequency. { ¦sim‚plē ¦nȯr·məl 'nəm·bər }

simply ordered set See linearly ordered set. { ¦sim‚plē ¦ȯrd·ərd 'set }

simply periodic function [MATH] A periodic function $f(x)$ for which there is a period a such that every period of $f(x)$ is an integral multiple of a. Also known as singly periodic function. { ¦sim‚plē ¦pir·ē‚äd·ik 'fəŋk·shən }

simpsonite [MINERAL] $AlTaO_4$ A hexagonal mineral composed of aluminum tantalum oxide and occurring in short crystals. { 'sim·sə‚nīt }

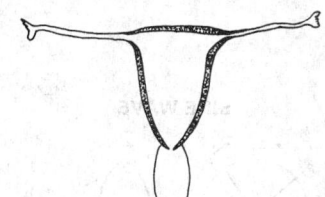

SIMPLEX UTERUS

The simplex uterus of the human. (From C. K. Weichert, *Elements of Chordate Anatomy*, 3d ed., McGraw-Hill, 1967)

Simpson's rule [MATH] Also known as parabolic rule. **1.** A basic approximation formula for definite integrals which states that the integral of a real-valued function f on an interval $[a,b]$ is approximated by $h[f(a) + 4f(a + h) + f(b)]/3$, where $h = (b - a)/2$; this is the area under a parabola which coincides with the graph of f at the abscissas a, $a + h$, and b. **2.** A method of approximating a definite integral over an interval which is equivalent to dividing the interval into equal subintervals and applying the formula in the first definition to each subinterval. [PETRO ENG] A mathematical relationship for calculating the oil- or gas-bearing net-pay volume of a reservoir; uses the contour lines from a subsurface geological map of the reservoir, including gas-oil and gas-water contacts. { 'simsənz ‚rül }

SIMS *See* secondary ion mass spectrometer. { simz }

SIMSCRIPT [COMPUT SCI] A high-level programming language used in simulation, in which systems are described in terms of sets, entities, which are groups of sets, and attributes, which are properties associated with entities. { 'sim‚skript }

simson *See* Simson line. { 'sim·sən }

Simson line [MATH] The Simson line of a point P on the circumcircle of a triangle ABC is the line passing through the collinear points L, M, and N, where L, M, and N are the projections of P upon the sides BC, CA, and AB, respectively. Also known as simson. { 'sim·sən ‚līn }

simulate [ENG] To mimic some or all of the behavior of one system with a different, dissimilar system, particularly with computers, models, or other equipment. { 'sim·yə‚lāt }

simulation [COMPUT SCI] The development and use of computer models for the study of actual or postulated dynamic systems. { ‚sim·yə'lā·shən }

simulation language [COMPUT SCI] A computer language used to write programs for the simulation of the behavior through time of such things as transportation and manufacturing systems; SIMSCRIPT is an example. { ‚sim·yə'lā·shən ‚laŋ·gwij }

simulator [COMPUT SCI] A routine which is executed by one computer but which imitates the operations of another computer. [ENG] A computer or other piece of equipment that simulates a desired system or condition and shows the effects of various applied changes, such as a flight simulator. { 'sim·yə‚lād·ər }

Simuliidae [INV ZOO] The black flies, a family of orthorrhaphous dipteran insects in the series Nematocera. { ‚sim·yə'lī·ə‚dē }

simultaneity [MECH] Two events have simultaneity, relative to an observer, if they take place at the same time according to a clock which is fixed relative to the observer. { ‚sī·məl·tə'nē·əd·ē }

simultaneous access *See* parallel access. { ‚sī·məl'tā·nē·əs 'ak‚ses }

simultaneous altitudes [ASTRON] Altitudes of two or more celestial bodies observed at the same time. { ‚sī·məl'tā·nē·əs 'al·tə‚tüdz }

simultaneous color television [ELECTR] A color television system in which the phosphors for the three primary colors are excited at the same time, not one after another; the shadow-mask color picture tube gives a simultaneous display. { ‚sī·məl'tā·nē·əs 'kəl·ər 'tel·ə‚vizh·ən }

simultaneous computer [COMPUT SCI] **1.** A computer, usually of the analog or hybrid type, in which separate units of hardware are used to carry out the various parts of a computation, the execution of different parts usually overlap in time, and the various hardware units are interconnected in a manner determined by the computation. **2.** A computer that serves to back up another computer and can replace it when it is not operating effectively. { ‚sī·məl'tā·nē·əs kəm'pyüd·ər }

simultaneous equations [MATH] A collection of equations considered to be a set of joint conditions imposed on the variables involved. { ‚sī·məl'tā·nē·əs i'kwā·zhənz }

simultaneous lobing [ELECTR] A radar direction-finding technique in which the signals received by two partly overlapping antenna lobes are compared in phase or power to obtain a measure of the angular displacement of a target from the equisignal direction. { ‚sī·məl'tā·nē·əs 'lōb·iŋ }

simultaneous motion-cycle chart *See* simo chart. { ‚sī·məl'tā·nē·əs 'mō·shən ‚sī·kəl ‚chärt }

simultaneous peripheral operations on line *See* spooling. { ‚sī·məl'tā·nē·əs pə'rif·ə·rəl ‚äp·ə·rā·shənz ön 'līn }

SINE WAVE

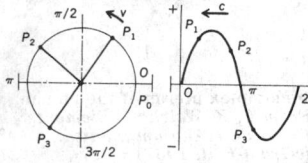

Relation of sine wave, simple harmonic motion, and uniform circular motion. Point P moves around the circle at constant tangential speed v; projection of P on vertical line moves up and down, executing simple harmonic motion. Plane on which the projection is plotted travels to the left at speed c, generating a sine wave.

sin A *See* sine. { 'sīn 'ā }

sincosite [MINERAL] $Ca(VO)_2(PO_4)_2 \cdot 5H_2O$ A leek-green mineral composed of hydrous calcium vanadyl phosphate and occurring in tetragonal scales or plates. { 'siŋ·kə‚sīt }

sine [MATH] The sine of an angle A in a right triangle with hypotenuse of length c given by the ratio a/c, where a is the length of the side opposite A; more generally, the sine function assigns to any real number A the ordinate of the point on the unit circle obtained by moving from $(1,0)$ counterclockwise A units along the circle, or clockwise $|A|$ units if A is less than 0. Denoted sin A. { 'sīn }

sine bar [DES ENG] A device consisting of a steel straight edge with two cylinders of equal diameter attached near the ends with their centers equidistant from the straightedge; used to measure angles accurately and to lay out work at a desired angle in relationship to a surface. { 'sīn ‚bär }

sine-cosine encoder [ELECTR] A shaft-position encoder having a special type of angle-reading code disk that gives an output which is a binary representation of the sine of the shaft angle. { 'sīn 'kō‚sīn in'kōd·ər }

sine-cosine generator *See* resolver. { 'sīn 'kō‚sīn 'jen·ə‚rād·ər }

sine curve [MATH] The graph of $y = \sin x$, where x and y are Cartesian coordinates. Also known as sinusoid. { 'sīn ‚kərv }

sine galvanometer [ENG] A type of magnetometer in which a small magnet is suspended in the center of a pair of Helmholtz coils, and the rest position of the magnet is measured when various known currents are sent through the coils. { 'sīn ‚gal·və'näm·əd·ər }

Sinemurian [GEOL] A European stage of geologic time; Lower Jurassic, above Hattangian and below Pliensbachian. { sin·ə'myùr·ē·ən }

sine potentiometer [ELECTR] A potentiometer whose direct-current output voltage is proportional to the sine of the shaft angle; used as a resolver in computer and radar systems. { 'sīn pə‚ten·chē'äm·əd·ər }

sine series [MATH] A Fourier series containing only terms that are odd in the independent variable, that is, terms involving the sine function. { 'sīn ‚sir‚ēz }

sine wave [PHYS] A wave whose amplitude varies as the sine of a linear function of time. Also known as sinusoidal wave. { 'sīn ‚wāv }

sine-wave modulated jamming [ELECTR] Jamming signal produced by modulating a continuous wave signal with one or more sine waves. { 'sīn ‚wāv ‚mäj·ə‚lād·əd 'jam·iŋ }

sine-wave oscillator *See* sinusoidal oscillator. { 'sīn ¦wāv 'as·ə‚lād·ər }

sine-wave response *See* frequency response. { 'sīn ¦wāv ri'späns }

singeing [TEXT] Passing a fabric over heated plates or gas flames during finishing to remove lint or loose threads from the surface. Also known as gassing. { 'sinj·iŋ }

singing [CONT SYS] An undesired, self-sustained oscillation in a system or component, at a frequency in or above the passband of the system or component; generally due to excessive positive feedback. { 'siŋ·iŋ }

singing margin [CONT SYS] The difference in level, usually expressed in decibels, between the singing point and the operating gain of a system or component. { 'siŋ·iŋ ‚mär·jən }

singing point [CONT SYS] The minimum value of gain of a system or component that will result in singing. { 'siŋ·iŋ ‚pöint }

singing sand *See* sounding sand. { 'siŋ·iŋ ¦sand }

singing-stovepipe effect [ELEC] Reception and reproduction of radio signals by ordinary pieces of metal in contact with each other, such as sections of stovepipe; it occurs when rusty bolts, faulty welds, or mechanically loose connections within strong radiated fields near transmitters produce intermodulation interference; the mechanically poor connections serve as nonlinear diodes. { 'siŋ·iŋ ‚stōv‚pīp i‚fekt }

single acting [MECH ENG] Acting in one direction only, as a single-acting plunger, or a single-acting engine (admitting the working fluid on one side of the piston only). { 'siŋ·gəl 'akt·iŋ }

single action [ORD] A method of fire in some revolvers and shoulder arms in which the hammer must be cocked by hand, in contrast to double action in which a single pull of the trigger both cocks and fires the weapon. { 'siŋ·gəl 'ak·shən }

single-action press [MECH ENG] A press having a single slide. { 'siŋ·gəl ¦ak·shən 'pres }

single-address instruction See one-address instruction. { 'siŋ·gəl ¦ad,res in'strək·shən }

single-atom laser [ATOM PHYS] A device in which atoms emit visible-wavelength photons at an increased rate as they pass through a resonant cavity one by one, consistent with a theory of quantized Rabi oscillations. Also known as microlaser. { ¦siŋ·gəl ¦ad·əm 'lā·zər }

single-atom maser [ATOM PHYS] A device in which atoms emit microwave photons at an increased rate as they pass through a resonant cavity one by one, consistent with a theory of quantized Rabi oscillations. Also known as micromaser. { ¦siŋ·gəl ¦ad·əm 'mā·zər }

single-axis gyroscope [ENG] A gyroscope suspended in just one gimbal whose bearings form its output axis; an example is a rate gyroscope. { 'siŋ·gəl ¦ak·səs 'jī·rə,skōp }

single-base powder [MATER] An explosive or propellant powder in which nitrocellulose is the only active ingredient. { 'siŋ·gəl ¦bās 'paúd·ər }

single-bevel groove weld [MET] A groove weld in which one member has a joint edge beveled from one side. { 'siŋ·gəl ¦bev·əl 'grüv ,weld }

single-blind technique [STAT] An experimental procedure in which the experimenters but not the subjects know the makeup of the test and control groups during the actual course of the experiments. { 'siŋ·gəl ¦blīnd tek'nēk }

single-block brake [MECH ENG] A friction brake consisting of a short block fitted to the contour of a wheel or drum and pressed up against the surface by means of a lever on a fulcrum; used on railroad cars. { 'siŋ·gəl ¦blák 'brāk }

single-board computer [COMPUT SCI] A computer consisting of a processor and memory on a single printed circuit board. { 'siŋ·gəl ¦bórd kəm'pyüd·ər }

single bus [ELEC] A substation switching arrangement that involves one common bus for all connections and one breaker per connection. { 'siŋ·gəl 'bəs }

single-button carbon microphone [ENG ACOUS] Microphone having a carbon-filled buttonlike container on only one side of its flexible diaphragm. { 'siŋ·gəl ¦bət·ən 'kär·bən 'mī·krə,fōn }

single-carrier theory [SOLID STATE] A theory of the behavior of a rectifying barrier which assumes that conduction is due to the motion of carriers of only one type; it can be applied to the contact between a metal and a semiconductor. { 'siŋ·gəl ¦kar·ē·ər 'thē·ə·rē }

single-channel multiplier [ELECTR] A type of photomultiplier tube in which electrons travel down a cylindrical channel coated on the inside with a resistive secondary-emitting layer, and gain is achieved by multiple electron impacts on the inner surface as the electrons are directed down the channel by an applied voltage over the length of the channel. { 'siŋ·gəl ¦chan·əl 'məl·tə,plī·ər }

single-channel simplex [COMMUN] Simplex operation that provides nonsimultaneous radio communications between stations using the same frequency channel. { 'siŋ·gəl ¦chan·əl 'sim,pleks }

single-chip computer [COMPUT SCI] A computer whose processor consists of a single integrated circuit. { 'siŋ·gəl ¦chip kəm'pyüd·ər }

single completion [PETRO ENG] An oil or gas well drilled to produce fluids from a single reservoir level or zone, and using a single tubing string. { 'siŋ·gəl kəm'plē·shən }

single-compound explosive [MATER] Explosive composed of a single chemical compound. { 'siŋ·gəl ¦käm,paúnd ik'splō·siv }

single crystal [CRYSTAL] A crystal, usually grown artificially, in which all parts have the same crystallographic orientation. { 'siŋ·gəl 'krist·əl }

single-current transmission [COMMUN] Telegraph transmission in which a current flows, in only one direction, during marking intervals, and no current flows during spacing intervals. { 'siŋ·gəl ¦kə·rənt tranz'mish·ən }

single cusp of the first kind See keratoid cusp. { ¦siŋ·gəl ,kəsp əv thə 'fərst ,kīnd }

single cusp of the second kind See ramphoid cusp. { ¦siŋ·gəl ,kəsp əv thə 'sek·ənd ,kīnd }

single-cut file [DES ENG] A file with one set of parallel teeth, extending diagonally across the face of the file. { 'siŋ·gəl ¦kət 'fīl }

single-cycle mountain [GEOL] A fold mountain that has been destroyed without reelevation of any of its important parts. { 'siŋ·gəl ¦sī·kəl 'maúnt·ən }

single-degree-of-freedom gyro [MECH] A gyro the spin reference axis of which is free to rotate about only one of the orthogonal axes, such as the input or output axis. { 'siŋ·gəl di¦grē əv ¦frē·dəm 'jī·rō }

single density [COMPUT SCI] Property of computer storage which holds the standard amount of data per unit of storage space. { ¦siŋ·gəl 'den·səd·ē }

single-drift flight [NAV] A flight made by using a single-drift correction angle based upon net drift between the point of departure and the point of destination. { 'siŋ·gəl ¦drift 'flīt }

single-edged push-pull amplifier circuit [ELECTR] Amplifier circuit having two transmission paths designed to operate in a complementary manner and connected to provide a single unbalanced output without the use of an output transformer. { 'siŋ·gəl ¦ejd ¦push ¦púl 'am·plə,fī·ər ,sər·kət }

single-effect evaporation [CHEM ENG] An evaporation process completed entirely in one vessel or by means of a single heating unit. { 'siŋ·gəl i¦fekt i,vap·ə'rā·shon }

single-electron transistor [ELECTR] A transistor whose dimensions are extremely small, in the nanometer range, causing it to exhibit characteristics that are sensitive to the transport and storage of single electrons. { ¦siŋ·gəl i,lek·trän tran'zis·tər }

single-end amplifier [ELECTR] Amplifier stage which normally employs only one tube or semiconductor or, if more than one tube or semiconductor is used, they are connected in parallel so that operation is asymmetric with respect to ground. Also known as single-sided amplifier. { 'siŋ·gəl end 'am·plə,fī·ər }

single-ended [ELEC] Unbalanced, as when one side of a transmission line or circuit is grounded. { 'siŋ·gəl 'end·əd }

single-ended ferry [NAV ARCH] A ferry vessel that has a normal hull form with propellers and rudders at the stern, so that it can proceed in only one direction without turning. { ¦siŋ·gəl ¦end·əd 'fer·ē }

single-ended Q machine [PL PHYS] A Q machine in which the plasma column is generated at a hot tungsten plate at one end and terminated by a cold metal plate at the other. { ¦siŋ·gəl ¦en·dəd 'kyü mə,shēn }

single-ended signal [ELECTR] A circuit signal that is the voltage difference between two nodes, one of which can be defined as being at ground or reference voltage. { ¦siŋ·gəl ¦en·dəd 'sig·nəl }

single-ended spread [ENG] A spread of geophones in which the shot point is located at one end of the arrangement. { ¦siŋ·gəl ¦end·əd 'spred }

single-event upset [ELECTR] A change in the state of a logic device from 0 to 1 or vice versa, as the result of the passage of a single cosmic ray. { ¦siŋ·gəl i¦vent 'əp,set }

single-frequency duplex [COMMUN] Duplex carrier communications that provide communications in opposite directions, but not simultaneously, over a single-frequency carrier channel, the transfer between transmitting and receiving conditions being automatically controlled by the voices or other signals of the communicating parties. { 'siŋ·gəl 'frē·kwən·sē 'dü,pleks }

single-frequency simplex [COMMUN] Single-frequency carrier communications in which manual rather than automatic switching is used to change over from transmission to reception. { 'siŋ·gə. 'frē·kwən·sē 'sim,pleks }

single-gun color tube [ELECTR] A color television picture tube having only one electron gun and one electron beam; the beam is sequentially deflected across phosphors for the three primary colors to form each color picture element, as in the chromatron. { 'siŋ·gəl ¦gən 'kəl·ər ,tüb }

single-hand drilling [ENG] A method of rock drilling in which the drill steel, which is held in the hand, is struck with a 4-pound (1.8-kilogram) hammer, the drill being turned between the blows. { 'siŋ·gəl ,han 'dril·iŋ }

single-heading flight See aerologation. { 'siŋ·gəl 'hed·iŋ 'flīt }

single-hop transmission [COMMUN] Radio transmission in which radio waves are reflected from the ionosphere only once

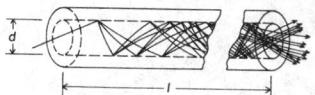

SINGLE-CHANNEL MULTIPLIER

Continuous-channel multiplier structure.

along their path from the transmitter to the receiver. { 'siŋ·gəl ¦häp trans'mish·ən }

single-impulse welding [MET] Spot, projection, or upset welding by means of a single current impulse. { ¦siŋ·gəl 'im,puls 'weld·iŋ }

single in-line memory module See SIMM. { ¦siŋ·gəl ¦in ¦līn 'mem·rē ,mä·jəl }

single in-line package [ELECTR] A packaged resistor network or other assembly that has a single row of terminals or lead wires along one edge of the package. Abbreviated SIP. { 'siŋ·gəl 'in,līn 'pak·ij }

single-instruction-stream, multiple-data-stream See SIMD. { ¦siŋ·gəl in¦strək·shən ,strēm ¦məl·tə·pəl 'dad·ə ,strēm }

single-instruction-stream, single-data-stream See SISD. { ¦siŋ·gəl in¦strək·shən ,strēm ¦siŋ·gəl 'dad·ə ,strēm }

single-J groove weld [MET] A groove weld in which one member has a joint edge in the form of a J from one side. { ¦siŋ·gəl ¦jā ¦grüv ,weld }

single-keyboard point-of-sale system [COMPUT SCI] A point-of-sale system based upon electronic cash registers as stand-alone units, each equipped with a few internal registers and some programming capability. { 'siŋ·gəl ¦kē,bord ¦point əv 'sāl ,sis·təm }

single knock-on [SOLID STATE] A sputtering event in which target atoms are ejected either directly by the bombarding projectiles or after a small number of collisions. { 'siŋ·gəl 'näk·on }

single-layer bit See surface-set bit. { 'siŋ·gəl ¦lā·ər 'bit }

single-layer solenoid [ELECTROMAG] A solenoid which has only one layer of wire, wound in a cylindrical helix. { 'siŋ·gəl ¦lā·ər 'so·lə,noid }

single-length [COMPUT SCI] Pertaining to the expression of numbers in binary form in such a way that they can be included in a single computer word. { 'siŋ·gəl 'leŋkth }

single lens See simple lens. { 'siŋ·gəl 'lenz }

single-loop feedback [CONT SYS] A system in which feedback may occur through only one electrical path. { 'siŋ·gəl ¦lüp 'fēd,bak }

single-loop servomechanism [CONT SYS] A servomechanism which has only one feedback loop. Also known as servo loop. { 'siŋ·gəl ¦lüp 'sər·vō,mek·ə,niz·əm }

single nucleotide polymorphism [GEN] A single base-pair difference between two copies of a deoxyribonucleic acid sequence from two individuals. Abbreviated SNP. { ¦siŋ·gəl ¦nü·klē·ə,tīd ,päl·ē·'mor·fiz·əm }

single packing [MIN ENG] Strip packing on a longwall face in which the widest pack is along the roadside. { 'siŋ·gəl 'pak·iŋ }

single-pass weld [MET] A weld made by depositing the filler metal with a single pass. { 'siŋ·gəl ¦pas weld }

single-path system [NAV] An electronic navigation system in which a measurement is made on the delay incurred in transmission of electromagnetic energy from a radio transmitter to a radio receiver; the single-path system measures absolute transmission time and produces a circular line of position. { 'siŋ·gəl ¦path ,sis·təm }

single-perforated grain [MATER] A cylindrical propellant grain with a single perforation located in its axis; this type of granulation is used in propelling charges for several calibers of guns, and in rockets. { 'siŋ·gəl ¦pər·fə,rād·əd 'grān }

single-phase [ELEC] Energized by a single alternating voltage. { 'siŋ·gəl 'fāz }

single-phase circuit [ELEC] Either an alternating-current circuit which has only two points of entry, or one which, having more than two points of entry, is intended to be so energized that the potential differences between all pairs of points of entry are either in phase or differ in phase by 180°. { 'siŋ·gəl ¦fāz 'sər·kət }

single-phase flow [CHEM ENG] The flow of a material, as a gas, single-phase liquid, or a solid, but not in any combination of the three. { 'siŋ·gəl ¦fāz ¦flō }

single-phase meter [ENG] A type of power-factor meter that contains a fixed coil that carries the load current, and crossed coils that are connected to the load voltage; there is no spring to restrain the moving system, which takes a position to indicate the angle between the current and voltage. { 'siŋ·gəl ¦fāz 'mēd·ər }

single-phase motor [ELEC] A motor energized by a single alternating voltage. { 'siŋ·gəl ¦fāz 'mōd·ər }

single-phase rectifier [ELECTR] A rectifier whose input voltage is a single sinusoidal voltage, in contrast to a polyphase rectifier. { 'siŋ·gəl ¦fāz 'rek·tə,fī·ər }

single-photon-emission computed tomography [MED] A technique which measures the emission of single photons of a given energy from radioactive tracers to construct images of the distribution of the tracers in the human body. Abbreviated SPECT. { 'siŋ·gəl ¦fō,tän i¦mish·ən kəm,pyüd·əd tō'mäg·rə·fē }

single-piece milling [MECH ENG] A milling method whereby one part is held and milled in one machine cycle. { 'siŋ·gəl ¦pēs 'mil·iŋ }

single-point grounding [ELEC] Grounding system that attempts to confine all return currents to a network that serves as the circuit reference; to be effective, no appreciable current is allowed to flow in the circuit reference, that is, the sum of the return currents is zero. { 'siŋ·gəl ¦point 'graund·iŋ }

single-point tool [ENG] A cutting tool having one face and one continuous cutting edge. { 'siŋ·gəl ¦point 'tül }

single-polarity pulse [ELEC] Pulse in which the sense of the departure from normal is in one direction only. { 'siŋ·gəl pə¦lar·əd·ē 'pəls }

single-polarity pulse-amplitude modulation See unidirectional pulse-amplitude modulation. { 'siŋ·gəl pə¦lar·əd·ē 'pəls 'am·plə,tüd ,mäj·ə'lā·shən }

single-pole double-throw [ELEC] A three-terminal switch or relay contact arrangement that connects one terminal to either of two other terminals. Abbreviated SPDT. { 'siŋ·gəl 'pōl ¦dəb·əl 'thrō }

single-pole single-throw [ELEC] A two-terminal switch or relay contact arrangement that opens or closes one circuit. Abbreviated SPST. { 'siŋ·gəl 'pōl ¦siŋ·gəl 'thrō }

single-precision number [COMPUT SCI] A number having as many digits as are ordinarily used in a given computer, in contrast to a double-precision number. Also known as short-precision number. { 'siŋ·gəl prə¦sizh·ən 'nəm·bər }

single-program, multiple-data See SPMD. { ¦siŋ·gəl ¦prō·grəm ¦məl·tə·pəl 'dad·ə }

single radial immunodiffusion [IMMUNOL] A technique for quantitating soluble proteins that involves placing the solution to be measured into a well cut into an agar or agarose gel containing antiserum specific for the protein. As the solution to be measured diffuses out of the well, it complexes with the antiserum and forms a ring, the size of which is proportional to the quantity of soluble protein in the well. Abbreviated SRID. Also known as Mancini method. { ¦siŋ·gəl ¦rād·ē·əl ,im·yə·nō·də'fyü·zhən }

single reference See random access. { 'siŋ·gəl 'ref·rəns }

single refraction [OPTICS] Any refraction that occurs in an isotropic crystal. { 'siŋ·gəl ri'frak·shən }

single-replacement reaction [CHEM] A chemical reaction in which an element replaces one element in a compound. { ¦siŋ·gəl ri'plās·mənt rē,ak·shən }

single sampling [IND ENG] A sampling inspection in which the lot is accepted or rejected on the basis of one sample. { 'siŋ·gəl 'sam·pliŋ }

single scattering [PHYS] A change in direction of a particle or photon because of a single collision. { 'siŋ·gəl 'skad·ər·iŋ }

singlesheet feed [COMPUT SCI] Equipment for feeding one sheet of paper to a computer printer at a time. { 'siŋ·gəl,shēt 'fēd }

single shot [ORD] Semiautomatic operation of an automatic gun, in which the trigger must be pulled for each shot fired. { 'siŋ·gəl 'shät }

single-shot blocking oscillator [ELECTR] Blocking oscillator modified to operate as a single-shot trigger circuit. { 'siŋ·gəl ¦shät 'bläk·iŋ 'äs·ə,lād·ər }

single-shot camera [OPTICS] An underwater camera that takes one picture on each lowering when the camera shutter is tripped by contact with the bottom. { 'siŋ·gəl ¦shät 'kam·rə }

single-shot exploder [ENG] A magneto exploder operated by the twist action given by a half turn of the firing key. { 'siŋ·gəl ¦shät ik'splōd·ər }

single-shot multivibrator See monostable multivibrator. { 'siŋ·gəl ¦shät ¦məl·ti'vī,brād·ər }

single-shot operation See single-step operation. { 'siŋ·gəl ¦shät ,äp·ə'rā·shən }

single-shot probability [ORD] Probability that a single projectile fired against a target will hit that target under a given set of conditions. { 'siŋ·gəl ¦shät präb·ə'bil·əd·ē }

single-shot survey [PETRO ENG] An oil well directional log or record with a single-reading device that is either run down into the drill pipe or positioned in a nonmagnetic drill collar. { 'siŋ·gəl ¦shät 'sər,vā }

single-shot trigger circuit [ELECTR] Trigger circuit in which one triggering pulse initiates one complete cycle of conditions ending with a stable condition. Also known as single-trip trigger circuit. { 'siŋ·gəl ¦shät 'trig·ər ¦sər·kət }

single-sideband [COMMUN] Pertaining to single-sideband communication. Abbreviated SSB. { 'siŋ·gəl 'sīd,band }

single-sideband communication [COMMUN] A communication system in which one of the two sidebands used in amplitude-modulation is suppressed; the carrier wave may be either transmitted, suppressed, or partially suppressed. { 'siŋ·gəl ¦sīd,band kə,myü·nə'kā·shən }

single-sideband modulation [COMMUN] Modulation resulting from elimination of all components of one sideband from an amplitude-modulated wave. { 'siŋ·gəl ¦sīd,band ,mäj·ə'lā·shən }

single-sideband transmission [COMMUN] Transmission of a carrier and substantially only one sideband of modulation frequencies, as in television where only the upper sideband is transmitted completely for the picture signal; the carrier wave may be either transmitted or suppressed, partially or totally. { 'siŋ·gəl ¦sīd,band tranz'mish·ən }

single-sided [COMPUT SCI] Pertaining to storage media that use only one of two sides for recording data. { 'siŋ·gəl 'sīd·əd }

single-sided amplifier See single-end amplifier. { 'siŋ·gəl ¦sīd·əd 'am·plə,fī·ər }

single-sided board [ELECTR] A printed wiring board that contains all of the interconnect material on one of the external layers. { ¦siŋ·gəl ¦sīd·əd 'bórd }

single-signal receiver [ELECTR] A highly selective superheterodyne receiver for code reception, having a crystal filter in the intermediate-frequency amplifier. { 'siŋ·gəl ¦sig·nəl ri'sē·vər }

single-sized aggregate [MATER] Aggregate in which most of the particles lie between narrow size limits. { 'siŋ·gəl ¦sīzd 'ag·rə·gət }

single-stage compressor [MECH ENG] A machine that effects overall compression of a gas or vapor from suction to discharge conditions without any sequential multiplicity of elements, such as cylinders or rotors. { 'siŋ·gəl ¦stāj kəm'pres·ər }

single-stage pump [MECH ENG] A pump in which the head is developed by a single impeller. { 'siŋ·gəl ¦stāj 'pəmp }

single-stage rocket [AERO ENG] A rocket or rocket missile to which the total thrust is imparted in a single phase, by either a single or multiple thrust unit. { 'siŋ·gəl ¦stāj 'räk·ət }

single-stand mill [MET] A rolling mill in which the product is in contact with only two rolls at a time. { 'siŋ·gəl ¦stand 'mil }

single-station analysis [METEOROL] The analysis or reconstruction of the weather pattern from more or less continuous meteorological observations made at a single geographic location, or the body of techniques employed in such an analysis. { 'siŋ·gəl ¦stā·shən ə'nal·ə·səs }

single-step operation [COMPUT SCI] A method of computer operation, used in debugging or detecting computer malfunctions, in which a program is carried out one instruction at a time, each instruction being performed in response to a manual control device such as a switch or button. Also known as one-shot operation; one-step operation; single-shot operation; step-by-step operation. { 'siŋ·gəl ¦step ,äp·ə'rā·shən }

single-stub transformer [ELECTROMAG] Shorted section of a coaxial line that is connected to a main coaxial line near a discontinuity to provide impedance matching at the discontinuity. { 'siŋ·gəl ¦stəb tranz'fór·mər }

single-stub tuner [ELECTROMAG] Section of transmission line terminated by a movable short-circuiting plunger or bar, attached to a main transmission line for impedance-matching purposes. { 'siŋ·gəl ¦stəb 'tün·ər }

singlet [QUANT MECH] An energy level that is not split by a relatively weak interaction, and thus is not a multiplet or a member of a multiplet. [SPECT] A spectral line that cannot be resolved into components at even the highest resolution. { 'siŋ·glət }

single-theodolite observation [METEOROL] The usual type of pilot-balloon observation, that is, using one theodolite. { 'siŋ·gəl thē¦äd·əl,īt äb·zər'vā·shən }

single thread [DES ENG] A screw thread having a single helix in which the lead and pitch are equal. { 'siŋ·gəl 'thred }

single threading [COMPUT SCI] Transaction processing in which one transaction is completed before another is begun. { 'siŋ·gəl 'thred·iŋ }

single-throw switch [ELEC] A switch in which the same pair of contacts is always opened or closed. { 'siŋ·gəl ¦thrō 'swich }

singleton [MATH] A set that has only one element. { 'siŋ·gəl·tən }

single-tone keying [COMMUN] Form of keying in which the modulating function causes the carrier to be modulated with a single tone for one condition, which may be either marking or spacing, and the carrier is unmodulated for the other condition. { 'siŋ·gəl ¦tōn 'kē·iŋ }

single-trip trigger circuit See single-shot trigger circuit. { 'siŋ·gəl ¦trip 'trig·ər ¦sər·kət }

single-tuned amplifier [ELECTR] An amplifier characterized by resonance at a single frequency. { 'siŋ·gəl ¦tünd 'am·plə,fī·ər }

single-tuned circuit [ELEC] A circuit whose behavior is the same as that of a circuit with a single inductance and a single capacitance, together with associated resistances. { 'siŋ·gəl ¦tünd 'sər·kət }

single-tuned interstage [ELECTR] An interstage circuit which is resonant at a single frequency. { 'siŋ·gəl ¦tünd 'in·tər,stāj }

single-U groove weld [MET] A groove weld in which the joint edge of both members is prepared in the form of a J from one side, giving a final U form to the completed weld. { 'siŋ·gəl ¦yü ¦grüv ,weld }

single-unit semiconductor device [ELECTR] Semiconductor device having one set of electrodes associated with a single carrier stream. { 'siŋ·gəl ¦yü·nət 'sem·i·kən,dək·tər di,vīs }

single-valued function [MATH] A function for which exactly one point in the range corresponds to each point in the domain; a function that associates to each value of the independent variable exactly one value of the dependent variable. Also known as one-valued function. { ¦siŋ·gəl ,val·yüd 'fəŋk·shən }

single-V groove weld [MET] A groove weld in which the joint edge of each member is beveled from the same side. { 'siŋ·gəl ¦vē ¦grüv ,weld }

single welded joint [MET] A joint welded from one side only. { 'siŋ·gəl 'wel·dəd 'jóint }

single-wire line [ELEC] **1.** Transmission line that uses the ground as one side of the circuit. **2.** A surface-wave transmission line that consists of a single conductor which has a dielectric coating or other treatment that confines the propagated energy close to the wire. { 'siŋ·gəl wīr 'līn }

singly linked ring [COMPUT SCI] A cyclic arrangement of data elements in which searches may be performed in either a clockwise or a counterclockwise direction, but not both. { 'siŋ·glē ¦liŋkt 'riŋ }

singly periodic function See simply periodic function. { ¦siŋ·glē ,pir·ē,äd·ik 'fəŋk·shən }

singular arc [CONT SYS] In an optimal control problem, that portion of the optimal trajectory in which the Hamiltonian is not an explicit function of the control inputs, requiring higher-order necessary conditions to be applied in the process of solution. { 'siŋ·gyə·lər 'ärk }

singular corresponding point [METEOROL] A center of elevation or depression on a constant-pressure chart (or a center of high or low pressure on a constant-height chart) considered as a reappearing characteristic of successive charts. { 'siŋ·gyə·lər ¦kär·ə¦spänd·iŋ 'póint }

singular integral See singular solution. { 'siŋ·gyə·lər ¦int·ə·grəl əv ə ,dif·ə'ren·chəl 'kwā·zhən }

singular integral equation [MATH] An integral equation where the integral appearing either has infinite limits of integration or the kernel function has points where it is infinite. { siŋ·gyə·lər ¦int·ə·grəl i'kwā·zhən }

singularity [MATH] A point where a function of real or complex variables is not differentiable or analytic. Also known as singular point of a function. [METEOROL] A characteristic meteorological condition which tends to occur on or near a specific calendar date more frequently than chance would indicate; an example is the January thaw. [RELAT] A region of space-time where one or more components of the Riemann curvature tensor becomes infinite. { ˌsiŋ·gyə'lar·əd·ē }

singularity theorems [RELAT] Theorems proving that singularities must develop in certain space-times, such as the universe, given only broad conditions, such as causality, and the existence of a trapped surface. { ˌsiŋ·gyə'lar·əd·ē ˌthir·əmz }

singular matrix [MATH] A matrix which has no inverse; equivalently, its determinant is zero. { 'siŋ·gyə·lər 'mā·triks }

singular point [MATH] **1.** For a differential equation, a point that is a singularity for at least one of the known functions appearing in the equation. **2.** A point on a curve at which the curve possesses no smoothly turning tangent, or crosses or touches itself, or has a cusp or isolated point. **3.** A point on a surface whose coordinates, x, y, and z, depend on the parameters u and v, at which the Jacobians $D(x,y)/D(u,v)$, $D(y,z)/D(u,v)$, and $D(z,x)/D(u,v)$ all vanish. **4.** See singularity. { 'siŋ·gyə·lər 'pȯint }

singular solution [MATH] For a differential equation, a solution that is not generic, that is, not obtainable from the general solution. Also known as singular integral. { 'siŋ·gyə·lər sə'lü·shən }

singular transformation [MATH] A linear transformation which has no corresponding inverse transformation. { 'siŋ·gyə·lər tranz·fər'mā·shən }

singular values [MATH] For a matrix A these are the positive square roots of the eigenvalues of A^*A, where A^* denotes the adjoint matrix of A. { 'siŋ·gyə·lər 'val·yüz }

singultus [MED] A repeated involuntary spasmodic contraction of the diaphragm followed by sudden closures of the glottis. Also known as hiccup. { 'siŋ·gəl·təs }

sinh See hyperbolic sine. { 'sīn'āch }

sinhalite [MINERAL] MgAl(BO$_4$) A mineral composed of magnesium aluminum borate; sometimes used as a gem. { 'sin·ə,līt }

sinistral fault See left lateral fault. { 'sin·əs·trəl 'fȯlt }

sinistral fold [GEOL] An asymmetric fold whose long limb, when viewed along its dip, appears to have a leftward offset. { 'sin·əs·trəl 'fōld }

sinistrorse [BIOL] Twisting or coiling counterclockwise. { ˌsin·ə¦strȯrs }

sinistrorse curve See left-handed curve. { ˌsin·ə¦strȯrs 'kərv }

sinistrorsum See left-handed curve. { ˌsin·ə'strȯrs·əm }

sink [COMMUN] Equipment at the end of a communications channel that receives signals and may perform other functions such as error detection. [ELECTROMAG] The region of a Rieke diagram where the rate of change of frequency with respect to phase of the reflection coefficient is maximum for an oscillator; operation in this region may lead to unsatisfactory performance by reason of cessation or instability of oscillations. [GEOL] **1.** A circular or ellipsoidal depression formed by collapse on the flank of or near to a volcano. **2.** A slight, low-lying desert depression containing a central playa or saline lake with no outlet, as where a desert stream comes to an end or disappears by evaporation. [MIN ENG] **1.** To excavate strata downward in a vertical line for the purpose of winning and working minerals. **2.** To drill or put down a shaft or borehole. [PHYS] A device or system where some extensive entity is absorbed, such as a heat sink, a sink flow, a load in an electrical circuit, or a region in a nuclear reactor where neutrons are strongly absorbed. { siŋk }

sinker [MIN ENG] **1.** A person who sinks mine shafts and puts in framing. **2.** A special movable pump used in shaft sinking. **3.** See sinker drill. { 'siŋ·kər }

sinker bar [MIN ENG] A short, heavy rod placed above the drill jars to increase the effect of the upward sliding jars in well-drilling with cable tools. { 'siŋ·kər ˌbär }

sinker drill [MIN ENG] A jackhammer type of rock drill used in shaft sinkings. Also known as sinker. { 'siŋ·kər ˌdril }

sink-float separation process [ENG] A simple gravity process used in ore dressing that separates particles of different sizes or composition on the basis of differences in specific gravity. { 'siŋk 'flōt ˌsep·ə'rā·shən ˌprä·səs }

sink flow [FL MECH] **1.** In three-dimensional flow, a point into which fluid flows uniformly from all directions. **2.** In two-dimensional flow, a straight line into which fluid flows uniformly from all directions at right angles to the line. { 'siŋk ˌflō }

sinkhead See feedhead. { 'siŋk,hed }

sinkhole [GEOL] Closed surface depressions in regions of karst topography produced by solution of surface limestone or the collapse of cavern roofs. { 'siŋk,hōl }

sinkhole plain [GEOL] A regionally extensive plain or plateau characterized by well-developed karst features. { 'siŋk ˌhōl ˌplān }

sinking [OCEANOGR] The downward movement of surface water generally caused by converging currents or when a water mass becomes denser than the surrounding water. Also known as downwelling. [OPTICS] In atmospheric optics, a refraction phenomenon, the opposite of looming, in which an object on, or slightly above, the geographic horizon apparently sinks below it. { 'siŋk·iŋ }

sinking-and-walling scaffold [MIN ENG] A platform designed for use in shaft sinking to enable sinking and walling to be performed simultaneously. Also known as Galloway sinking and walling stage. { 'siŋk·iŋ ən ¦wȯl·iŋ 'ska,fōld }

sinking bucket See hoppit. { 'siŋk·iŋ ,bək·ət }

sinking fund [IND ENG] A fund established by periodically depositing funds at compound interest in order to accumulate a given sum at a given future time for some specific purpose. { 'siŋk·iŋ ,fənd }

sinking pump [MIN ENG] A long, narrow, electrically driven centrifugal-type pump designed for keeping a shaft dry during sinking operations. { 'siŋk·iŋ ,pəmp }

sinking tubing [MET] Drawing tubing through a die or passing it through rolls without the use of a tool in the bore to control the inside diameter. { 'siŋk·iŋ ,tüb·iŋ }

sink mark [ENG] A shallow depression or dimple on the surface of an injection-molded plastic part due to collapsing of the surface following local internal shrinkage after the gate seals. { 'siŋk ,märk }

sinoatrial node [ANAT] A bundle of Purkinje fibers located near the junction of the superior vena cava with the right atrium which acts as a pacemaker for cardiac excitation. Abbreviated SA node. Also known as sinoauricular node. { ¦sī·nō'ā·trē·əl 'nōd }

sinoauricular node See sinoatrial node. { ¦sī·nō·ȯ'rik·yə·lər 'nōd }

sinoite [MINERAL] Si$_2$N$_2$O A nitride mineral known only in meteorites. { 'sīn·ə,wīt }

Sinope [ASTRON] A small satellite of Jupiter with a diameter of about 17 miles (27 kilometers), orbiting with retrograde motion at a mean distance of about 1.47×10^7 miles (2.37×10^7 kilometers). Also known as Jupiter IX. { 'sin·ə·pē }

sinople [MINERAL] A blood-red or brownish-red (with a tinge of yellow) variety of quartz containing inclusions of hematite. { 'sin·ə·pəl }

sinter [MET] **1.** The product of a sintering operation. **2.** A shaped body composed of metal powders and produced by sintering with or without previous compacting. [MINERAL] See siliceous sinter. [PETR] A chemical sedimentary rock deposited by precipitation from mineral waters, especially siliceous sinter and calcareous sinter. { 'sin·tər }

sintered copper [MET] Copper prepared by heating a compressed powder of the metal to form a solid mass. { 'sin·tərd 'käp·ər }

sintered steel [MET] Steel prepared by heating compressed iron powder and graphite to form a solid. { 'sin·tərd 'stēl }

sintering [MET] Forming a coherent bonded mass by heating metal powders without melting; used mostly in powder metallurgy. { 'sin·tə·riŋ }

sintering furnace [MET] A furnace in which presintering and sintering operations are carried out. { 'sin·tə·riŋ ,fər·nəs }

sinter setting See mechanical setting. { 'sin·tər ,sed·iŋ }

sinuate [BOT] Having a wavy margin with strong indentations. { 'sin·yə,wət }

sinus [BIOL] A cavity, recess, or depression in an organ, tissue, or other part of an animal body. { 'sī·nəs }

sinus gland [INV ZOO] An endocrine gland in higher crustaceans, lying in the eyestalk in most stalk-eyed species, which is the site of storage and release of a molt-inhibiting hormone. { 'sī·nəs ,gland }

sinus hairs See vibrissae. { 'sī·nəs ,herz }

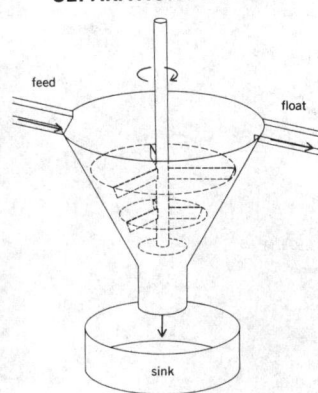

SINK-FLOAT SEPARATION PROCESS

Feed particles are introduced into suspension, whose specific gravity is between that of mineral and gangue particles; particles of higher specific gravity sink while those of lower specific gravity float; stirrer prevents suspension from setting out on the walls.

sinusitis [MED] Inflammation of a paranasal sinus. { ˌsī·nəˈsīd·əs }

sinus of Morgagni [ANAT] The space between the upper border of the levator veli palatini muscle and the base of the skull. { 'sī·nəs əv mȯrˈgän·yē }

sinusoid [ANAT] Any of the relatively large spaces comprising part of the venous circulation in certain organs, such as the liver. [MATH] *See* sine curve. { 'sī·nəˌsȯid }

sinusoidal angular modulation *See* angle modulation. { ˌsī·nəˈsȯid·əl 'aŋ·gyə·lər ˌmaj·əˈlā·shən }

sinusoidal current *See* simple harmonic current. { ˌsī·nəˈsȯid·əl 'kə·rənt }

sinusoidal function [MATH] The real or complex function sin(*u*) or any function with analogous continuous periodic behavior. { ˌsī·nəˈsȯid·əl 'fəŋk·shən }

sinusoidal oscillator [ELECTR] An oscillator circuit whose output voltage is a sine-wave function of time. Also known as harmonic oscillator; sine-wave oscillator. { ˌsī·nəˈsȯid·əl 'as·əˌlād·ər }

sinusoidal projection [MAP] An equal-area map projection in which the parallels are equally spaced, straight lines drawn to scale, the central meridian is a straight line perpendicular to the equator along which the scale is true, and the other meridians are sine curves, concaving toward the central meridian. { ˌsī·nəˈsȯid·əl prəˈjek·shən }

sinusoidal spiral [MATH] A plane curve whose equation in polar coordinates (r, θ) is $r^n = a^n \cos n\theta$, where a is a constant and n is a rational number. { ˌsī·nəˈsȯid·əl 'spī·rəl }

sinusoidal wave *See* sine wave. { ˌsī·nəˈsȯid·əl 'wāv }

sinus venosus [EMBRYO] The vessel in the transverse septum of the embryonic mammalian heart into which open the vitelline, allantoic, and common cardinal veins. [VERT ZOO] The chamber of the lower vertebrate heart to which the veins return blood from the body. { 'sī·nəs vəˈnō·səs }

SIP *See* single in-line package. { sip }

Siphinodentallidae [INV ZOO] A family of mollusks in the class Scaphopoda characterized by a subterminal epipodial ridge which is not slit dorsally and which terminates with a crenulated disk. { ˌsī·fə·nō·denˈtal·əˌdē }

siphon [BOT] A tubular element in various algae. [ENG] A tube, pipe, or hose through which a liquid can be moved from a higher to a lower level by atmospheric pressure forcing it up the shorter leg while the weight of the liquid in the longer leg causes continuous downward flow. [GEOL] A passage in a cave system that connects with a water trap. [INV ZOO] **1.** A tubular structure for intake or output of water in bivalves and other mollusks. **2.** The sucking-type of proboscis in many arthropods. { 'sī·fən }

Siphonales [BOT] A large order of green algae (Chlorophyta) which are coenocytic, nonseptate, and mostly marine. Also known as Bryopsidales; Caulerpales. { ˌsī·fəˈnā·lez }

Siphonaptera [INV ZOO] The fleas, an order of insects characterized by a small, laterally compressed, oval body armed with spines and setae, three pairs of legs modified for jumping, and sucking mouthparts. { ˌsī·fəˈnäp·trə }

siphon barograph [ENG] A recording siphon barometer. { 'sī·fən 'bar·əˌgraf }

siphon barometer [ENG] A J-shaped mercury barometer in which the stem of the J is capped and the cusp is open to the atmosphere. { 'sī·fən bəˈräm·əd·ər }

Siphonocladaceae [BOT] A family of green algae in the order Siphonocladales. { ˌsī·fə·nōˈkləˈdās·ēˌē }

Siphonocladales [BOT] An order of green algae in the division Chlorophyta including marine, mostly tropical forms. { ˌsī·fə·nō·kləˈdā·lēz }

siphonogamous [BOT] In plants, especially seed plants, the accomplishment of fertilization by means of a pollen tube. { ˌsī·fəˈnäg·ə·məs }

siphonoglyph [INV ZOO] A ciliated groove leading from the mouth to the gullet in certain anthozoans. { sīˈfän·əˌglif }

Siphonolaimidae [INV ZOO] A family of nematodes in the superfamily Monhysteroidea in which the stoma is modified into a narrow, elongate, hollow, spearlike apparatus. { ˌsī·fə·nōˌlāim·əˌdē }

Siphonolaimoidea [INV ZOO] A superfamily of marine nematodes in the order Monhysterida, having a stoma in the form of a very narrow tube or a spear, and very large amphids. { ˌsī·fə·nō·ləˈmȯid·ē·ə }

Siphonophora [INV ZOO] An order of the cnidarian class Hydrozoa characterized by the complex organization of components which may be connected by a stemlike region or may be more closely united into a compact organism. { ˌsī·fəˈnäf·rə }

siphonosome [INV ZOO] The lower part of a siphonophore colony, bearing the nutritive and reproductive zooids. { sīˈfän·əˌsōm }

siphonostele [BOT] A type of stele consisting of pith surrounded by xylem and phloem. { sīˈfän·əˌstēl }

Siphonotretacea [PALEON] A superfamily of extinct, inarticulate brachiopods in the suborder Acrotretidina of the order Acrotretida having an enlarged, tear-shaped, apical pedicle valve. { ˌsī·fə·nō·trəˈtās·ē·ə }

siphonozooid [INV ZOO] A zooid of certain alcyonarians that lacks tentacles and gonads. { ˌsī·fə·nəˈzō·ȯid }

siphon recorder [ENG] A recorder in which a small siphon discharges ink to make the record; used in submarine telegraphy. { 'sī·fən riˈkȯrd·ər }

siphon spillway [CIV ENG] An enclosed spillway passing over the crest of a dam in which flow is maintained by atmospheric pressure. { 'sī·fən 'spilˌwā }

Siphoviridae [VIROL] A family of linear double-stranded deoxyribonucleic acid-containing bacterial viruses (bacteriophages) characterized by a long noncontractile tail. Formerly known as Styloviridae. { ˌsī·fəˈvir·əˌdī }

siphuncle [INV ZOO] **1.** A honeydew-secreting tube (cornicle) in aphids. **2.** A tubular extension of the mantle extending through all the chambers to the apex of a shelled cephalopod. { 'sī·fəŋ·kəl }

Siphunculata [INV ZOO] The equivalent name for Anoplura. { si·fəŋ·kyəˈläd·ə }

siporex [MATER] A building material composed of sand, lime or cement, and aluminum powder which are mixed and cast into molds to be made into roof slabs, door lintels, and wall blocks which give excellent heat and sound insulation. { 'sī·pəˌreks }

Sipunculida [INV ZOO] A phylum of marine worms which dwell in burrows, secreted tubes, or adopted shells; the mouth and anus occur close together at one end of the elongated body, and the jawless mouth, surrounded by tentacles, is situated in an eversible proboscis. { sī·pəŋˈkyü·lə·də }

Sipunculoidea [INV ZOO] An equivalent name for Sipunculida. { sī·pəŋ·kyəˈlȯid·ē·ə }

siren [ENG ACOUS] An apparatus for generating sound by the mechanical interruption of the flow of fluid (usually air) by a perforated disk or cylinder. { 'sī·rən }

Sirenia [VERT ZOO] An order of aquatic placental mammals which include the living manatees and dugongs; these are nearly hairless, thick-skinned mammals without hindlimbs and with paddlelike forelimbs. { sīˈrē·nē·ə }

Siricicae [INV ZOO] The horntails, a family of the Hymenoptera in the superfamily Siricoidea; females use a stout, hornlike ovipositor to deposit eggs in wood. { səˈris·əˌdē }

Siricoidea [INV ZOO] A superfamily of wasps of the suborder Symphala in the order Hymenoptera. { ˌsir·əˈkȯid·ē·ə }

siriometer [ASTRON] A unit of length, formerly used in astronomical measurement, equal to 10^6 astronomical units, or 1.496×10^{17} meters. { ˌsir·ēˈäm·əd·ər }

Sirius [ASTRON] The brightest-appearing star in the sky; 8.7 light-years from the sun, spectral class A1V; it has a white dwarf companion. Also known as Dog Star. { 'sir·ē·əs }

sirocco [METEOROL] A warm south or southeast wind in advance of a depression moving eastward across the southern Mediterranean Sea or North Africa. { səˈrä·kō }

SIRS *See* satellite infrared spectrometer. { sərz }

sisal [BOT] *Agave sisalina.* An agave of the family Amaryllidaceae indigenous to Mexico and Central America; a coarse, stiff yellow fiber produced from the leaves is used for making twine and brush bristles. { 'sī·səl }

sisal-hemp wax [MATER] A hard wax derived from sisal waste; melts at 63°C, decomposes at 95°C. Also known as sisal wax. { 'sī·səl ˌhemp ˌwaks }

sisal wax *See* sisal-hemp wax. { 'sī·səl ˌwaks }

SISD [COMPUT SCI] A type of computer architecture in which there is a single instruction cycle, and operands are fetched in serial fashion into a single processing unit before execution. Acronym for single-instruction-stream, single-data-stream. { 'esˌīˌesˈdē }

siserskite [MINERAL] A light steel gray mineral consisting

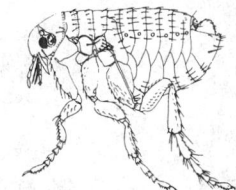

SIPHONAPTERA

Human flea, *Pulex irritans.* (From E. O. Essig, *College Entomology*, Macmillan, 1942)

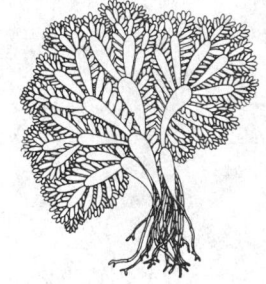

SIPHONOCLADALES

Anadyomene, habit of thallus, showing expanded blades.

SIRICIDAE

Drawing of a horntail. (From T. I. Storer and R. L. Usinger, *General Zoology*, 3d ed., McGraw-Hill, 1957)

of an alloy of osmium and iridium; occurs in tabular form. { 'sis·ər‚kīt }

sister chromatids [CYTOL] The two daughter strands of a chromosome after it has duplicated. { 'sis·tər 'krō·mə‚tədz }

sister hook [DES ENG] **1.** Either of a pair of hooks which can be fitted together to form a closed ring. **2.** A pair of such hooks. { 'sis·tər ‚hůk }

SIT See self-induced transparency; static induction transistor.

sitaparite See bixbyite. { sə'tap·ə‚rīt }

site [COMPUT SCI] **1.** A position available for the symbols of an inscription, for example, a digital place. **2.** A location on a tally that can bear either a mark or a blank; for example, a location that can be punched or left unpunched on a card. [ENG] Position of anything; for example, the position of a gun emplacement. { sīt }

Sitka cypress See Alaska cedar. { 'sit·kə 'sī·prəs }

sitting height [ANTHRO] A measure of the vertical distance (taken along the back) from the table surface to the crest of the head as the subject sits erectly on the table, knees pressed against the table edge and head in the eye-ear horizontal plane. { 'sid·iŋ ‚hīt }

situation-display tube [ELECTR] Large cathode-ray tube used to display tabular and vector messages pertinent to the various functions of an air defense mission. { ‚sich·ə'wā·shən di'splā ‚tüb }

situation map [MAP] A map showing the tactical or the administrative situation at a particular time. { ‚sich·ə'wā·shən ‚map }

situation therapy See milieu therapy. { ‚sich·ə'wā·shən ‚ther·ə·pē }

situs inversus [ANAT] Complete mirror-image inversion of the internal organs. { 'sīd·əs in'vər·səs }

situs solitus [ANAT] The normal pattern of internal organ placement. { ‚sīd·əs 'säl·ə·təs }

six-axis system [MECH ENG] A robot that has six degrees of freedom, three rectangular and three rotational. { 'siks ‚ak·səs 'sis·təm }

six-j-symbol [QUANT MECH] A coefficient that appears in the transformation between various modes of coupling eigenfunctions of three angular momenta; it is equal to the Racah coefficient, except perhaps in sign, and has greater symmetry than the Racah coefficient. { 'siks ‚jā 'sim·bəl }

six-phase circuit [ELEC] Combination of circuits energized by alternating electromotive forces which differ in phase by one-sixth of a cycle (60°). { 'siks ‚fāz 'sər·kət }

six-phase rectifier [ELECTR] A rectifier in which transformers are used to produce six alternating electromotive forces which differ in phase by one-sixth of a cycle, and which feed six diodes. { 'siks ‚fāz 'rek·tə‚fī·ər }

Six's thermometer [ENG] A combination maximum thermometer and minimum thermometer; the tube is shaped in the form of a U with a bulb at either end; one bulb is filled with creosote which expands or contracts with temperature variation, forcing before it a short column of mercury having iron indexes at either end; the indexes remain at the extreme positions reached by the mercury column, thus indicating the maximum and minimum temperatures; the indexes can be reset with the aid of a magnet. { 'sik·səz thər‚mäm·əd·ər }

six-tenths factor [IND ENG] An empirical relationship between the cost and the size of a manufacturing facility; as size increases, cost increases by an exponent of six-tenths, that is $cost_1/cost_2 = (size_1/size_2)^{0.6}$. { 'siks ‚tenths ‚fak·tər }

sixth-power law [FL MECH] A law stating that the size of particles that can be carried by a stream is proportional to the sixth power of its velocity. { 'siksth ‚paů·ər ‚lō }

sixty degrees Fahrenheit British thermal unit See British thermal unit. { 'siks·tē di‚grēz 'far·ən‚hīt 'brid·ish 'thər·məl ‚yü·nət }

six-vector [RELAT] An antisymmetrical, second-rank tensor in Minkowski space; that is, a tensor whose components, $T_{\mu\nu}$, with $\mu,\nu = 1,2,3,4$, satisfy $T_{\mu\nu} = -T_{\nu\mu}$; it has six independent components. { 'siks ‚vek·tər }

size [MATER] Materials used to surface-treat textiles, papers, and leathers; examples are starch, gelatins, casein, water-soluble gums, and waxes. Also known as sizing. [MATH] **1.** The number of edges of a graph. **2.** For a test of a hypothesis, the probability of a type I error. { sīz }

size analysis See particle-size analysis. { 'sīz ə‚nal·ə·səs }

size block See gage block. { 'sīz ‚bläk }

size classification See sizing. { 'sīz ‚klas·ə·fə‚kā·shən }

size constancy [PSYCH] The tendency of objects to appear to be about the same size whether seen nearby or farther away, except at very great distances. { 'sīz ‚kän·stən·sē }

size control [ELECTR] A control provided on a television receiver for changing the size of a picture either horizontally or vertically. { 'sīz kən‚trōl }

size dimension [DES ENG] In dimensioning, a specified value of a diameter, width, length, or other geometrical characteristic directly related to the size of an object. { 'sīz di‚men·shən }

size effect [MET] The effect of the size of a piece of metal on its properties and manufacturing variables; in general, mechanical properties are lower for a larger size. { 'sīz i‚fekt }

size enlargement [CHEM ENG] Making large particles out of small ones by crystallization, particle cementation, tableting, briquetting, agglomeration, flocculation, melting, casting, compaction and extrusion, and sintering or nodulizing. { 'sīz in‚lärj·mənt }

size-frequency analysis See particle-size analysis. { 'sīz 'frē·kwən·sē ə‚nal·ə·səs }

size of a critical region [STAT] For statistical hypotheses, the probability of committing a type I error, that is, rejecting the hypothesis tested when it is true. { ‚sīz əv ə ‚krid·ə·kəl 'rē·jən }

size of weld [MET] **1.** The joint penetration in a groove weld. **2.** The lengths of the nominal legs of a fillet weld. { 'sīz əv 'weld }

size reduction [MECH ENG] The breaking of large pieces of coal, ore, or stone by a primary breaker, or of small pieces by grinding equipment. { 'sīz ri‚dək·shən }

sizing [ENG] **1.** Separating an aggregate of mixed particles into groups according to size, using a series of screens. Also known as size classification. **2.** See sizing treatment. [MATER] See size. [MECH ENG] A finishing operation to correct surfaces and shapes to meet specified dimensions and tolerances. [MET] **1.** Final pressing of a metal powder compact after sintering. **2.** A cold-working operation in which a part is re-pressed in a die to improve surface hardness, smoothness, and dimensional accuracy. { 'sīz·iŋ }

sizing screen [DES ENG] A mesh sheet with standard-size apertures used to separate granular material into classes according to size; the Tyler standard screen is an example. { 'sīz·iŋ ‚skrēn }

sizing treatment [ENG] Also known as sizing; surface sizing. **1.** Application of material to a surface to fill pores and thus reduce the absorption of subsequently applied adhesive or coating; used for textiles, paper, and other porous materials. **2.** Surface-treatment applied to glass fibers used in reinforced plastics. { 'sīz·iŋ ‚trēt·mənt }

sjogrenite [MINERAL] $Mg_6Fe_2(OH)_{16}(CO_3)\cdot4H_2O$ A hexagonal mineral composed of hydrous basic magnesium iron carbonate. { 'shō·grə‚nīt }

Sjögren's syndrome [IMMUNOL] An autoimmune disease characterized by the destruction of salivary and lacrimal glands; damage by T lymphocytes within the glands may be accompanied by damage by immune complexes throughout the body. { 'shō·grənz ‚sin‚drōm }

Sk See Stefan number.

skarn [GEOL] A lime-bearing silicate derived from nearly pure limestone and dolomite with the introduction of large amounts of silicon, aluminum, iron, and magnesium. { skärn }

skate [VERT ZOO] Any of various batoid elasmobranchs in the family Rajidae which have flat bodies with winglike pectoral fins and a slender tail with two small dorsal fins. { skāt }

skatole [ORG CHEM] C_9H_9N A white, crystalline compound that melts at 93–95°C, dissolves in hot water, and has an unpleasant feceslike odor. Also known as 3-methylindole. { 'ska‚tōl }

skauk [HYD] An extensive field of crevasses in a glacier. { skók }

skavl See sastruga. { 'skav·əl }

skeet gun [ORD] A classification of shotguns which includes those with 26-inch (66-centimeter) improved cylinder barrels. { 'skēt ‚gən }

skeg [NAV ARCH] An extension of the stern end of the keel, which in some cases passes under the propeller and supports the rudder post. { skeg }

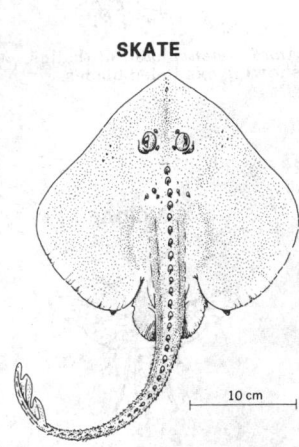

SKATE

Skate (Raja). (From H. B. Bigelow and W. C. Schroeder, Fishes of the Western North Atlantic, pt. 2, 1953)

10 cm

skeletal coding [COMPUT SCI] A set of incomplete instructions in symbolic form, intended to be completed and specialized by a processing program written for that purpose. { 'skel·əd·əl 'kōd·iŋ }

skeletal muscle [ANAT] A striated, voluntary muscle attached to a bone and concerned with body movements. { 'skel·əd·əl ,məs·əl }

skeletal system [ANAT] Structures composed of bone or cartilage or a combination of both which provide a framework for the vertebrate body and serve as attachment for muscles. Also known as skeleton. { 'skel·əd·əl ,sis·təm }

skeleton [ANAT] See skeletal system. [MATH] **1.** For a simplex, the set of all the vertices. **2.** For a simplicial complex, the class of all simplexes which belong to the simplicial complex and have dimension less than that of the simplicial complex. { 'skel·ət·ən }

skeleton crystal [CRYSTAL] A crystal formed in microscopic outline with incomplete filling in of the faces. { 'skel·ət·ən ,krist·əl }

skeleton framing [BUILD] Framing in which steel framework supports all the gravity loading of the structure; this system is used for skyscrapers. { 'skel·ət·ən ,frām·iŋ }

skeleton grain [GEOL] A relatively stable and not readily translocated grain of soil material, concentrated or reorganized by soil-forming processes. { 'skel·ət·ən ,grān }

skeleton layer [OCEANOGR] The structure that is formed at the bottom of sea ice while freezing, and consists of vertically oriented platelets of ice separated by layers of brine. { 'skel·ət·ən ,lā·ər }

skeleton texture [PETR] Descriptive of the texture of limestone that consists of an in-place accumulation of skeletal material, that is, the hard parts secreted by organisms. { 'skel·ət·ən ,teks·chər }

skelic index [ANTHRO] The ratio, multiplied by 100, of the length of the leg to that of the trunk. { 'skel·ik ,in,deks }

skelp [MET] A strip or sheet of steel which will be rolled and welded to form a tube. { skelp }

skerry [GEOL] A low, small, rugged and rocky island or reef. { 'sker·ē }

skew [COMPUT SCI] In character recognition, a condition arising at the read station whereby a character or a line of characters appears in a "twisted" manner in relation to a real or imaginary horizontal baseline. [ELECTR] **1.** The deviation of a received facsimile frame from rectangularity due to lack of synchronism between scanner and recorder; expressed numerically as the tangent of the angle of this deviation. **2.** The degree of nonsynchronism of supposedly parallel bits when bit-coded characters are read from magnetic tape. [MECH ENG] Gearing whose shafts are neither interesecting nor parallel. [SCI TECH] Deviating from rectangularity or a straight line. { skyü }

skewback [CIV ENG] The beveled or inclined support at each end of a segmental arch. { 'skyü ,bak }

skew bridge [CIV ENG] A bridge which spans a gap obliquely and is therefore longer than the width of the gap. { 'skyü ,brij }

skew chisel [ENG] A tool used for wood turning that has a straight cutting edge sharpened at an angle to the shank. { 'skyü ,chiz·əl }

skewed bridge [CIV ENG] A bridge for which the deck in plan is a parallelogram. { 'skyüd 'brij }

skewed density function [STAT] A density function which is not symmetrical, and which depends not only on the magnitude of the difference between the average value and the variate, but also on the sign of this difference. { 'skyüd 'den·səd·ē ,fəŋk·shən }

Skewes number [MATH] The first integer n for which the number of primes not greater than n is greater than the Cauchy principal value of the integral over x from 0 to n of the reciprocal of the natural logarithm of x. { 'skyüz ,nəm·bər }

skew failure [COMPUT SCI] In character recognition, the condition that exists during document alignment whereby the document reference edge is not parallel to that of the read station. { 'skyü ,fāl·yər }

skew field [MATH] A ring whose nonzero elements form a non-Abelian group with respect to the multiplicative operation. { 'skyü ,fēld }

skew Hermitian matrix [MATH] A square matrix which

equals the negative of its adjoint. { 'skyü hər'mish·ən 'mā·triks }

skew level gear [DES ENG] A level gear whose axes are not in the same place. { 'skyü 'lev·əl ,gir }

skew lines [MATH] Lines which do not lie in the same plane in euclidean three-dimensional space. { 'skyü ,līnz }

skew matrix See antisymmetric matrix. { 'skyü 'mā,triks }

skewness [STAT] The degree to which a distribution departs from symmetry about its mean value. { 'skyü·nəs }

skew product [MATH] A multiplicative operation or structure induced upon a Cartesian product of sets, where each has some algebraic structure. { 'skyü 'präd·əkt }

skew quadrilateral [MATH] A quadrilateral all four of whose vertices do not lie in a single plane. { ¦skyü ,kwäd·rə'lad·ə·rəl }

skew surface [MATH] A ruled surface that is not a developable surface. { 'skyü ,sər·fəs }

skew-symmetric determinant See antisymmetric determinant. { ¦skyü sə¦me·trik də'tər·mə·nənt }

skew-symmetric matrix See antisymmetric matrix. { 'skyü si¦me·trik 'mā·triks }

skew-symmetric tensor [MATH] A tensor where interchanging two indices will only change the sign of the corresponding component. { 'skyü si¦me·trik 'ten·sər }

skialith [PETR] A vague remnant of country rock assimilated in granite. { 'skī·ə,lith }

skiascope [OPTICS] An instrument used to study optical refraction within the eye. { 'skī·ə,skōp }

skiatron See dark-trace tube. { 'skī·ə,trän }

skid [AERO ENG] The metal bar or runner used as part of the landing gear of helicopters and planes. [ENG] **1.** A device attached to a chain and placed under a wheel to prevent its turning when descending a steep hill. **2.** A timber, bar, rail, or log placed under a heavy object when it is being moved over bare ground. **3.** A wood or metal platform support on wheels, legs, or runners used for handling and moving material. Also known as skid platform. [MECH ENG] A brake for a power machine. [MIN ENG] An arrangement upon which certain coal-cutting machines travel along the working faces. { skid }

skid boulder [GEOL] An isolated angular block of stone resting on the floor of a playa, derived from an outcrop near the playa margin, and associated with a trail or mark indicating that the boulder has recently slid across the mud surface. { 'skid ,bōl·dər }

Skiddavin See Arenigian. { skə'dav·ən }

skidding [FOR] The short distance movement of tree lengths or segments over unimproved terrain to loading points on transportation routes. { 'skid·iŋ }

skid-mounted [ENG] Equipment or processing systems mounted on a portable platform. { 'skid,maunt·əd }

skid platform See skid. { 'skid ,plat,fórm }

skidway [FOR] A platform, usually inclined, for mounting logs for loading and sawing. { 'skid wā }

skill score [METEOROL] In synoptic meteorology, an index of the degree of skill of a set of forecasts, expressed with reference to some standard such as forecasts based upon chance, persistence, or climatology. { 'skil ,skór }

skim coat [BUILD] A finish coat of plaster composed of lime putty and fine white sand. { 'skim ,kōt }

skim gate [MET] A gate used to prevent slag and other undesirable materials from passing into the casting. { 'skim ,gāt }

skim ice [HYD] First formation of a thin layer of ice on the water surface. { 'skim ,īs }

skimmer [VERT ZOO] Any of various ternlike birds, members of the Rhynchopidae, that inhabit tropical waters around the world and are unique in having the knifelike lower mandible substantially longer than the wider upper mandible. { 'skim·ər }

skimming [HYD] **1.** Diversion of water from a stream or conduit by shallow overflow in order to avoid diverting sand, silt, or other debris carried as bottom load. **2.** Withdrawal of fresh groundwater from a thin body or lens floating on salt water by means of shallow wells or infiltration galleries. { 'skim·iŋ }

skimming plant [CHEM ENG] A petroleum refinery designed to remove and finish only the lighter constituents of crude oil, such as gasoline and kerosine; the heavy ends are sold as fuel oil or for further processing elsewhere. { 'skim·iŋ ,plant }

SKIN FRICTION

The thin boundary layer, in which fluid flow is distorted by the passage of the body, fills the region between the body skin and the dashed line. $\delta(x)$ is the thickness of the boundary layer; u_1 is the velocity of fluid relative to the body outside the boundary layer; $u(y)$, the fluid velocity within the boundary at a distance y from the body skin, is graphed by the solid curved line. The friction force per unit surface area of the body, τ_w, equals the rate of deformation of an adjacent fluid element, $(\partial u/\partial y)_w$, times the dynamic viscosity μ.

SKINK

The blue-tailed skink (*Eumeces fasciatus*).

skin [AERO ENG] The covering of a body, such as the covering of a fuselage, a wing, a hull, or an entire aircraft. [ANAT] The external covering of the vertebrate body, consisting of two layers, the outer epidermis and the inner dermis. [BUILD] The exterior wall of a building. [ENG] In flexible bag molding, a protective covering for the mold; it may consist of a thin piece of plywood or a thin hardwood. [MET] A thin outside layer of metal differing in composition, structure, or other characteristics from the main mass of metal but not formed by bonding or electroplating. { skin }

skin antenna [ELECTROMAG] Flush-mounted aircraft antenna made by using insulating material to isolate a portion of the metal skin of the aircraft. { 'skin an,ten·ə }

skin depth [ELECTROMAG] The depth beneath the surface of a conductor, which is carrying current at a given frequency due to electromagnetic waves incident on its surface, at which the current density drops to one neper below the current density at the surface. { 'skin ,depth }

skin diving [ENG] Diving without breathing apparatus, using fins and faceplate only. { 'skin ,dīv·iŋ }

skin effect [ELEC] The tendency of alternating currents to flow near the surface of a conductor thus being restricted to a small part of the total sectional area and producing the effect of increasing the resistance. Also known as conductor skin effect; Kelvin skin effect. [PETRO ENG] The restriction to fluid flow through a reservoir adjacent to the borehole; calculated as a factor of reservoir pressure, product rate, formation volume and thickness, porosity, and other related parameters. { 'skin i,fekt }

skin erythema dose [NUCLEO] A unit of radioactive dose resulting from exposure to electromagnetic radiation, equal to the dose that slightly reddens or browns the skin of 80% of all persons within 3 weeks after exposure; it is approximately 1000 roentgens for gamma rays, 600 roentgens for x-rays. Abbreviated SED. { 'skin ,er·ə'thē·mə ,dōs }

skin friction [FL MECH] A type of friction force which exists at the surface of a solid body immersed in a much larger volume of fluid which is in motion relative to the body. { 'skin ,frik·shən }

skin-friction coefficient [METEOROL] A dimensionless drag coefficient expressing the proportionality between the frictional force per unit area, or the shearing stress exerted by the wind at the earth's surface, and the square of the surface wind speed. { 'skin ¦frik·shən ,kō·i'fish·ənt }

skink [VERT ZOO] Any of numerous small- to medium-sized lizards comprising the family Scincidae with a cylindrical body; short, sometimes vestigial, legs; cores of bone in the body scales; and pleurodont dentition. { skiŋk }

skin lamination [MET] Surface rupture in flat-rolled metals due to exposure of a subsurface lamination. { 'skin ,lam·ə,nā·shən }

skin resistance [ELEC] For alternating current of a given frequency, the direct-current resistance of a layer at the surface of a conductor whose thickness equals the skin depth. [NAV ARCH] The frictional resistance existing between the shell or skin of a ship and the water flowing over it. { 'skin ri,zis·təns }

skin test [IMMUNOL] A procedure for evaluating immunity status involving the introduction of a reagent into or under the skin. { 'skin ,test }

skintle [CIV ENG] To set bricks in an irregular fashion so that they are out of alignment with the face by 1/4 inch (6 millimeters) or more. { 'skint·əl }

skin tracking [ELECTROMAG] Tracking of an object by means of radar without using a beacon or other signal device on board the object being tracked. { 'skin ,trak·iŋ }

skiograph [ELECTR] An instrument used to measure the intensity of x-rays. { 'skī·ə,graf }

skiou See morvan. { skyō }

skip [COMPUT SCI] **1.** In fixed-instruction-length digital computers, to bypass or ignore one or more instructions in an otherwise sequential process. **2.** A device on a card punch that causes columns on a punch in fields where no punching is desired to move rapidly past the punching station. **3.** Action of a computer printer that moves rapidly over a line so that a blank line appears in the printout. [MECH ENG] *See* skip hoist. { skip }

skip bombing [ORD] Releasing one or more bombs from a plane flying at a low altitude, so that the bomb or bombs glance off the surface of the water or ground and strike the target. { 'skip ,bäm·iŋ }

skip cast [GEOL] The cast of a skip mark. { 'skip ,kast }

skip chain [COMPUT SCI] A programming technique which matches a word against a set of test words; if there is a match, control is transferred (skipped) to a routine, otherwise the word is matched with the next test word in sequence. { 'skip ,chān }

skip-dent [TEXT] A type of weaving in which holes are produced as a design in a fabric, usually massed in a patterned effect such as stripes. { 'skip ,dent }

skip distance [ELECTROMAG] The minimum distance that radio waves can be transmitted between two points on the earth by reflection from the ionosphere, at a specified time and frequency. [ENG] In angle-beam ultrasonic testing, the distance between the point of entry on the workpiece and the point of first reflection. { 'skip ,dis·təns }

skip effect [COMMUN] The existence of an oval-shaped area around a radio transmitter within which no radio signals are received, because ground signals are received only inside the oval and sky-wave signals are received only outside the oval. { 'skip i,fekt }

skip fading [ELECTROMAG] Fading due to fluctuations of ionization density at the place in the ionosphere where the wave is reflected which causes the skip distance to increase or decrease. { 'skip ,fād·iŋ }

skip flag [COMPUT SCI] The thirty-fifth bit of a channel command word which suppresses the transfer of data to main storage. { 'skip ,flag }

skip hoist [MECH ENG] A basket, bucket, or open car mounted vertically or on an incline on wheels, rails, or shafts and hoisted by a cable; used to raise materials. Also known as skip. { 'skip ,hoist }

skipjack See bluefish. { 'skip,jak }

skip keying [ELECTR] Reduction of radar pulse repetition frequency to submultiple of that normally used, to reduce mutual interference between radar or to increase the length of radar time base. { 'skip ,kē·iŋ }

skip logging [ENG] A phenomenon during acoustical (sonic) logging in which the acoustical energy is attenuated by low-elasticity formations and lacks the energy to trip the second sonic receiver (skips a cycle). Also known as cycle skip. { 'skip ,läg·iŋ }

skip mark [GEOL] A crescent-shaped mark that is one of a linear pattern of regularly spaced marks made by an object that skipped along the bottom of a stream. { 'skip ,märk }

skip-searched chain [COMPUT SCI] A chain which has pointers and can therefore be searched without examining each link. { 'skip ,sərcht ,chān }

skip shaft [MIN ENG] A mine shaft prepared for hauling a skip. { 'skip ,shaft }

skip trajectory [MECH] A trajectory made up of ballistic phases alternating with skipping phases; one of the basic trajectories for the unpowered portion of the flight of a reentry vehicle or spacecraft reentering earth's atmosphere. { 'skip trə,jek·trē }

skip vehicle [AERO ENG] A reentry body which climbs after striking the sensible atmosphere in order to cool the body and to increase its range. { 'skip ,vē·ə·kəl }

skip zone [ACOUS] A region in the air surrounding a source of sound in which no sound is heard, although the sound becomes audible at greater distances. Also known as zone of silence. [COMMUN] The area between the outer limit of reception of radio high-frequency ground waves and the inner limit of reception of sky waves, where no signal is received. { 'skip ,zōn }

skiron [METEOROL] The Greek name for the northwest wind, which is cold in winter but hot and dry in summer. { 'skē·rón }

skirt See apron; baseboard. { skərt }

skirting See apron; baseboard. { 'skərd·iŋ }

skirting block [BUILD] Also known as base block; plinth block. **1.** A corner block where a base strip and vertical enframement meet. **2.** A concealed block to which a baseboard is anchored. { 'skərd·iŋ ,bläk }

skirting plate [ORD] A thin plate which is placed a considerable distance in front of the main armor plate and which acts as a passive form of resistance to the jet of shaped-charge ammunition. { 'skərd·iŋ ,plāt }

skirt roof [BUILD] A false band of roofing projecting from between the stories of a building. { 'skərt ,rüf }

skiver [MATER] Thin, soft leather made from the grain side of a split sheepskin. { 'skī·vər }

skiving [MECH ENG] **1.** Removal of material in thin layers or chips with a high degree of shear or slippage of the cutting tool. **2.** A machining operation in which the cut is made with a form tool with its face at an angle allowing the cutting edge to progress from one end of the work to the other as the tool feeds tangentially past ten rotating workpieces. { 'skīv·iŋ }

skleropelite [PETR] An argillaceous or allied rock which has been indurated by low-grade metamorphism, is more massive and dense than shale, and differs from slate by the absence of cleavage. { sklə'räp·ə,līt }

skolite [MINERAL] A scaly, dark-green variety of glauconite rich in aluminum and calcium and deficient in ferric iron. { 'skō,līt }

skomerite [PETR] A fine-grained, compact extrusive rock containing microscopic grains and crystals of augite, olivine, and phenocrysts of decomposed plagioclase (probably albite) in a groundmass of plagioclase, thought to be more calcic than the phenocrysts. { 'skäm·ə,rīt }

skot [OPTICS] A unit of luminance, used particularly to measure low-level luminance, equal to 10^{-3} apostilb, or $10^{-3}/\pi$ nit. { skät }

Skraup synthesis [ORG CHEM] A method for the preparation of commercial synthetic quinoline by heating aniline and glycerol in the presence of sulfuric acid and an oxidizing agent to form pyridine unsubstituted quinolines. { skraup 'sin·thə,səs }

skull [ANAT] The bones and cartilages of the vertebrate head which form the cranium and the face. [MET] A layer of solidified metal or dross left in the pouring vessel after the molten metal has been poured. { skəl }

skull cracker [ENG] A heavy iron or steel ball that can be swung freely or dropped by a derrick to raze buildings or to compress bulky scrap. Also known as wrecking ball. { 'skəl ,krak·ər }

skull crucible [MET] A consumable-electrode vacuum arc melting and casting furnace; used in the production of turbine buckets for aircraft jet engines using a nickel-base high-temperature alloy. { 'skəl ,krü·sə·bəl }

skunk [VERT ZOO] Any one of a group of carnivores in the family Mustelidae characterized by a glossy black and white coat and two musk glands at the base of the tail. { skəŋk }

skutterudite [MINERAL] $(Co,Ni)As_3$ A tin-white mineral with metallic luster composed of cobalt and nickel arsenides; crystallizes in the isometric system but commonly is massive; hardness is 5.5–6 on Mohs scale, and specific gravity is 6.6; it is a minor ore of cobalt and nickel. { 'skəd·ə,rə,dīt }

sky [ASTRON] In the daytime the apparent blue dome resting on the earth along the horizon circle; at night the blue becomes nearly black. { skī }

sky compass [NAV] A type of astro compass, designed for use in the Arctic during long periods of twilight, which utilizes the polarization of sunlight in the sky; it operates whenever the zenith is clear whether or not the sun is visible; when the sun is more than 6.5° below or 10° above the horizon, readings are uncertain. { 'skī ,käm·pəs }

sky cover [METEOROL] In surface weather observations, the amount of sky covered but not necessarily concealed by clouds or by obscuring phenomena aloft, the amount of sky concealed by obscuring phenomena that reach the ground, or the amount of sky covered or concealed by a combination of the two phenomena. { 'skī ,kəv·ər }

sky diagram [ASTRON] A diagram of the heavens, indicating the apparent positions of various celestial bodies with reference to the horizon system of coordinates. { 'skī ,dī·ə,gram }

skyhook [MIN ENG] To drive bolts into the overhead rock of a mine in order to reinforce the ceiling. { 'skī,huk }

skyhook balloon [AERO ENG] A large plastic constant-level balloon for duration flying at very high altitudes, used for determining wind fields and measuring upper-atmospheric parameters. { 'skī,huk bə,lün }

skylight [ASTROPHYS] *See* diffuse sky radiation. [ENG] An opening in a roof or ship deck that is covered with glass or plastic and designed to admit daylight. { 'skī,līt }

skyline yarding [FOR] A technique that uses a skyline cable system to transport logs from a forest to a clear space along a road for trucking to a sawmill; it requires fewer roads than older cable systems, suspends logs off the ground to reduce erosion, and reduces damage to young trees that are left. { 'skī,līn ,yärd·iŋ }

sky map [ASTRON] A planar representation of areas of the sky showing positions of celestial bodies. [METEOROL] A pattern of variable brightness observable on the underside of a cloud layer, and caused by the different reflectivities of material on the earth's surface immediately beneath the clouds; this term is used mainly in polar regions. { 'skī ,map }

sky noise [ELECTROMAG] Noise produced by radio energy from stars. { 'skī ,nȯiz }

sky radiation *See* diffuse sky radiation. { 'skī ,rād·ē,ā·shən }

skyscraper [BUILD] A very tall, multistory building. { 'skī,skrāp·ər }

sky wave [ELECTROMAG] A radio wave that travels upward into space and may or may not be returned to earth by reflection from the ionosphere. Also known as ionospheric wave. { 'skī ,wāv }

sky-wave correction [ELECTR] The correction to be applied to the time difference readings of received sky waves to convert them to an equivalent ground-wave reading. { 'skī ,wāv kə'rek·shən }

sky-wave-synchronized loran [NAV] A type of loran in which the transmitting stations are synchronized by signals reflected from the ionosphere; used to obtain greater range and more accurate nighttime navigation. Abbreviated ss loran. { 'skī ,wāv ,siŋ·krə,nīzd 'lȯr,an }

sky-wave transmission delay [ELECTROMAG] Amount by which the time of transit from transmitter to receiver of a pulse carried by sky waves reflected once from the E layer exceeds the time of transit of the same pulse carried by ground waves. { 'skī ,wāv tranz'mish·ən di,lā }

slab [CIV ENG] That part of a reinforced concrete floor, roof, or platform which spans beams, columns, walls, or piers. [ELECTR] A relatively thick-cut crystal from which blanks are obtained by subsequent transverse cutting. [ENG] The outside piece cut from a log when sawing it into boards. [GEOL] A cleaved or finely parallel jointed rock, which splits into tabular plates from 1 to 4 inches (2.5 to 10 centimeters) thick. Also known as slabstone. [HYD] A layer in, or the whole-thickness of, a snowpack that is very hard and has the ability to sustain elastic deformation under stress. [MATER] A thin piece of concrete or stone. [MET] A piece of metal, intermediate between ingot and plate, with the width at least twice the thickness. [MIN ENG] A slice taken off the rib of an entry or room in a mine. { slab }

slabbing cutter [MECH ENG] A face-milling cutter used to make wide, rough cuts. { 'slab·iŋ ,kəd·ər }

slabbing machine [MIN ENG] A coal-cutting machine designed to make cuts in the side of a room or entry pillar preparatory to slabbing. { 'slab·iŋ mə,shēn }

slabbing method [MIN ENG] A method of mining pillars in which successive slabs are cut from one side or rib of the pillar after a room is finished, until as much of the pillar is removed as can safely be recovered. { 'slab·iŋ ,meth·əd }

slabbing mill [MET] A steel rolling mill for making slabs. { 'slab·iŋ ,mil }

slab cutter *See* plain milling cutter. { 'slab ,kəd·ər }

slab entry [MIN ENG] An entry which is widened or slabbed to provide a working place for a second miner. { 'slab ,en·trē }

slab jointing [GEOL] Jointing produced in rock by the formation of numerous cleaved or closely spaced parallel fissures dividing the rock into thin slabs. { 'slab ,jȯint·iŋ }

slab oil [MATER] White petroleum-based oil used by candymakers and bakers to oil the slab or surface on which the candy or pastry is worked. { 'slab ,ȯil }

slab pahoehoe [GEOL] A pahoehoe whose surface consists of a jumbled arrangement of slabs of flow crust. { 'slab pə'hō·ē,hō·ē }

slabstone *See* slab. { 'slab,stōn }

slack [ENG] Looseness or play in a mechanism, as the play in the trigger of a small-arms weapon. [GEOL] A hollow or depression between lines of shore dunes or in a sandbank or mudbank on a shore. { slak }

slack barrel [PETRO ENG] A petroleum-industry container used for shipment of petroleum paraffin; generally contains 235 to 245 pounds (107 to 111 kilograms) net, and is of lighter construction than the ordinary oil barrel but of the same general shape. { slak 'bar·əl }

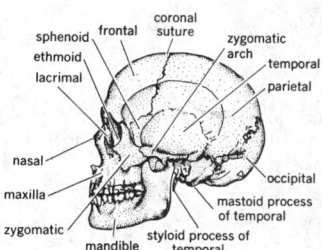

SKULL

sphenoid · frontal · coronal suture · zygomatic arch · temporal · parietal · ethmoid · lacrimal · nasal · maxilla · zygomatic · mandible · occipital · mastoid process of temporal · styloid process of temporal

Lateral view of human skull. *(From W. T. Foster, Anatomy, Foster Art Service)*

SKUNK

The spotted skunk (*Spilogate putorius*).

slack ice [HYD] Ice fragments on still or slow-moving water. { 'slak 'īs }

slackline cableway [MECH ENG] A machine, widely used in sand-and-gravel plants, employing an open-ended dragline bucket suspended from a carrier that runs upon a track cable, which can dig, elevate, and convey materials in one continuous operation. { 'slak,līn 'kā·bəl,wā }

slack quenching [MET] Formation of transformation products other than martensite as a result of quenching at a rate slower than the critical cooling rate. { 'slak ,kwench·iŋ }

slack time [ENG] For an activity in a PERT or critical-path-method network, the difference between the latest possible completion time of each activity which will not delay the completion of the overall project, and the earliest possible completion time, based on all predecessor activities. { 'slak ,tīm }

slack water [OCEANOGR] The interval when the speed of the tidal current is very weak or zero; usually refers to the period of reversal between ebb and flood currents. { 'slak 'wòd·ər }

slack wax [MATER] A soft, oily, crude wax obtained from the pressing of petroleum paraffin distillate or wax distillate. { 'slak 'waks }

slag [MET] A nonmetallic product resulting from the interaction of flux and impurities in the smelting and refining of metals. Also known as bloom. { slag }

slag cement [MATER] Cement produced by grinding blast-furnace slag and mixing it with lime, portland cement, or dehydrated gypsum. { 'slag si,ment }

slagging [MET] Freeing from or converting into slag. { 'slag·iŋ }

slag inclusion [MET] Slag entrapped in solidified metal. { 'slag in'klü·zhən }

slag sand [MATER] Slag that has been finely crushed for use in mortar and concrete. { 'slag ,sand }

slag wool See mineral wool. { 'slag ,wùl }

slaked lime See calcium hydroxide. { 'slākt 'līm }

slaking [GEOL] 1. Crumbling and disintegration of earth materials when exposed to air or moisture. 2. The breaking up of dried clay when saturated with water. { 'slāk·iŋ }

slamming [NAV ARCH] The impact of the bottom of a ship's bow hitting the water during a severe downward pitch. { 'slam·iŋ }

slamming stile [BUILD] The vertical strip that a closed door abuts; it receives the bolt when the lock engages. { 'slam·iŋ ,stīl }

slant [MIN ENG] 1. Any short, inclined crosscut connecting the entry with its air course to facilitate the hauling of coal. Also known as shoofly. 2. A heading driven diagonally between the dip and the strike of a coal seam. Also known as run. { slant }

slant culture [MICROBIO] A method for maintaining bacteria in which the inoculum is streaked on the surface of agar that has solidified in inclined glass tubes. { 'slant ,kəl·chər }

slant depth [DES ENG] The distance between the crest and root of a screw thread measured along the angle forming the flank of the thread. { 'slant ,depth }

slant distance [NAV] The straight-line distance between two points not at the same elevation as contrasted with ground distance, ground range, or the great-circle distance between the two positions. Also known as slant range. { 'slant ,dis·təns }

slant drilling [ENG] The drilling of a borehole or well at an angle to the vertical. { 'slant ,dril·iŋ }

slant height [MATH] 1. The common length of the elements of a right circular cone. 2. The common altitude of the lateral faces of a regular pyramid. { 'slant hīt }

slant plane [ORD] In antiaircraft artillery, the plane containing the target course line and the pintle center of the gun. { 'slant ,plān }

slant range See slant distance. { 'slant ,rānj }

slant visibility See oblique visual range. { 'slant ,viz·ə'bil·əd·ē }

slash [FOR] Debris, such as logs, chunks of wood, bark, and branches, in an open forest tract. { slash }

slashing [TEXT] The stiffening of warp yarn before the spinning operation, using a solution of oil and starch. { 'slash·iŋ }

slash-sawing See backsawing. { 'slash ,sò·iŋ }

slat [AERO ENG] A movable auxiliary airfoil running along the leading edge of a wing, remaining against the leading edge in normal flight conditions, but lifting away from the wing to form a slot at certain angles of attack. { slat }

slat conveyor [MECH ENG] A conveyor consisting of horizontal slats on an endless chain. { slat kən,vā·ər }

slate [PETR] A group name for various very-fine-grained rocks derived from mudstone, siltstone, and other clayey sediment as a result of low-degree regional metamorphism; characterized by perfect fissility or slaty cleavage which is a regular or perfect planar schistosity. { slāt }

Slater determinant [QUANT MECH] A quantum-mechanical wave function for n fermions, which is an $n \times n$ determinant whose entries are n different one-particle wave functions depending on the coordinates of each of the particles in the system. { 'slād·ər di,tər·mə·nənt }

slate ribbon [GEOL] A relict ribbon sstructure on the cleavage surface of slate, in which varicolored and straight, wavy, or crumpled stripes cross the cleavage surface. { 'slāt ,rib·ən }

Slater's rule [ELECTR] The ratio of the cathode radius to the anode radius of a magnetron is approximately equal to $(N-4)/(N+4)$, where N is the number of resonators. { 'slād·ərz ,rül }

slaty cleavage See flow cleavage. { 'slād·ē 'klē·vij }

slave [COMPUT SCI] A terminal or computer that is controlled by another computer. [CONT SYS] A device whose motions are governed by instructions from another machine. { slāv }

slave antenna [ELECTROMAG] A directional antenna positioned in azimuth and elevation by a servo system; the information controlling the servo system is supplied by a tracking or positioning system. { 'slāv an,ten·ə }

slave arm [ENG] A component of a remote manipulator that automatically duplicates the motions of a master arm, sometimes with changes of scale in displacement or force. { 'slāv ,ärm }

slaved gyro magnetic compass [NAV] A directional gyro compass with an input from a flux valve to keep the gyro oriented to magnetic north. { 'slāvd 'jī·rō mag'ned·ik 'käm·pəs }

slave mode See user mode. { 'slāv ,mōd }

slavery [INV ZOO] An interspecific association among ants in which members of one species bring pupae of another species to their nest, which, when adult, become slave workers in the colony. { 'slav·ə·rē }

slave station [NAV] In a radio navigation system such as loran, the transmitting station controlled or triggered by the signal received from the master station. { 'slāv 'stā·shən }

slave tube [ELECTR] A display monitor that is connected to another monitor and provides an identical display. { 'slāv ,tüb }

slavikite [MINERAL] MgFe$_2^{3+}$(SO$_4$)$_4$(OH)$_3$·18H$_2$O A greenish-yellow mineral composed of hydrous basic magnesium ferric sulfate and occurring as rhombohedral crystals. { 'slav·ə,kīt }

sled [ENG] An item equipped with runners and a suitable body designed to transport loads over ice and snow. { sled }

sledgehammer [DES ENG] A large heavy hammer that is usually wielded with two hands; used for driving stakes or breaking stone. { 'slej,ham·ər }

sleep [COMPUT SCI] State of a computer system that halts, or a program that appears to be doing nothing because the program is caught in an endless loop. [PHYSIO] A state of rest in which consciousness and activity are diminished. { slēp }

sleeper [CIV ENG] A timber, steel, or precast concrete beam placed under rails to hold them at the correct gage. { 'slēp·ər }

sleeping sickness See African sleeping sickness; encephalitis lethargica. { 'slēp·iŋ ,sik·nəs }

sleep paralysis [MED] Transient paralysis with spontaneous recovery occurring on falling asleep or on awakening. { 'slēp pə,ral·ə·səs }

sleep therapy See narcosis therapy. { 'slēp ,ther·ə·pē }

sleet [METEOROL] Colloquially in some parts of the United States, precipitation in the form of a mixture of snow and rain. { slēt }

sleeve [ELEC] 1. The cylindrical contact that is farthest from the tip of a phone plug. 2. Insulating tubing used over wires or components. Also known as bushing; sleeving. [ENG] A cylindrical part designed to fit over another part. { slēv }

sleeve antenna [ELECTROMAG] A single vertical half-wave

radiator, the lower half of which is a metallic sleeve through which the concentric feed line runs; the upper radiating portion, one quarter-wavelength long, connects to the center of the line. { 'slēv an,ten·ə }

sleeve bearing [MECH ENG] A machine bearing in which the shaft turns and is lubricated by a sleeve. { 'slēv ,ber·iŋ }

sleeve brick [MATER] Tube-shaped firebrick for lining slag vents. { 'slēv ,brik }

sleeve burner [ENG] A type of oil burner for domestic heating. { 'slēv ,bər·nər }

sleeve coupling [DES ENG] A hollow cylinder which fits over the ends of two shafts or pipes, thereby joining them. { 'slēv ,kəp·liŋ }

sleeve dipole antenna [ELECTROMAG] Dipole antenna surrounded in its central portion by a coaxial cable. { 'slēv 'dī,pōl an'ten·ə }

sleeve joint [DES ENG] A device for joining the ends of two wires or cables together, constructed by forcing the ends of the wires or cables into both ends of a hollow sleeve. { 'slēv ,jȯint }

sleeve valve [MECH ENG] An admission and exhaust valve on an internal-combustion engine consisting of one or two hollow sleeves that fit around the inside of the cylinder and move with the piston so that their openings align with the inlet and exhaust ports in the cylinder at proper stages in the cycle. { 'slēv ,valv }

sleeving See sleeve. { 'slēv·iŋ }

sleigh [ORD] Part of a gun carriage which supports the recoil mechanism and barrel of the gun and slides with the gun on recoil, guiding it in runways in the cradle. { slā }

slender-body theory [FL MECH] The theory of compressible inviscid fluid flow past bodies which have pointed noses and bases, or flat bases in supersonic flow only, and which satisfy the following conditions: (1) the ratio r of the maximum thickness to the length of the body must be small compared with unity, (2) the angle between the tangent plane to the body and the direction of motion must be small and of order r, and (3) the smoothness conditions. { ¦slen·dər 'bäd·ē ,thē·ə·rē }

slenderness ratio [AERO ENG] A configuration factor expressing the ratio of a missile's length to its diameter. [CIV ENG] The ratio of the length of a column L to the radius of gyration r about the principal axes of the section. { 'slen·dər·nəs ,rā·shō }

slewing [ENG] Moving a radar antenna or a sonar transducer rapidly in a horizontal or vertical direction, or both. [NAV] In sea ice navigation, the act of forcing a ship through ice by pushing apart adjoining ice floes. { 'slü·iŋ }

slewing mechanism [ENG] Device which permits rapid traverse or change in elevation of a weapon or instrument. { 'slü·iŋ ,mek·ə,niz·əm }

slewing motor [ELEC] A motor used to drive a radar antenna at high speed for slewing to pick up or track a target. { 'slü·iŋ ,mōd·ər }

slew rate [COMPUT SCI] The speed at which a logic-seeking print head advances to the succeeding line and finds the position where it is to start printing. [CONT SYS] The maximum rate at which a system can follow a command. [ELECTR] The maximum rate at which the output voltage of an operational amplifier changes for a square-wave or step-signal input; usually specified in volts per microsecond. { 'slü ,rāt }

slice [GEOL] An arbitrary section of some uniform standard, such as thickness of a stratigraphic unit that is otherwise indivisible for purposes of analytic study. [MIN ENG] **1.** A thin broad piece cut off, as a portion of ore cut from a pillar or face. **2.** To remove ore by successive slices. { slīs }

slice bar [ENG] A broad, flat steel blade used for chipping and scraping. { 'slīs ,bär }

slice drift [MIN ENG] In sublevel caving, the crosscuts driven between every other slice from 18 to 36 feet (5.5 to 11 meters) apart. { 'slīs ,drift }

slice method [METEOROL] A method of evaluating the static stability over a limited area at any reference level in the atmosphere; unlike the parcel method, the slice method takes into account continuity of mass by considering both upward and downward motion. { 'slīs ,meth·əd }

slicer See amplitude gate; slitting mill. { 'slīs·ər }

slicer amplifier See amplitude gate. { 'slīs·ər ,am·plə,fī·ər }

slicing [ELECTR] Transmission of only those portions of a waveform lying between two amplitude values. { 'slīs·iŋ }

slicing method [MIN ENG] Removal of a horizontal layer from a massive ore body. { 'slīs·iŋ ,meth·əd }

slick [OCEANOGR] Area in which capillary waves are absent or suppressed. { slik }

slickens [GEOL] A layer of fine silt deposited by a flooding stream. [MIN ENG] The light earth removed by sluicing in hydraulic mining. { 'slik·ənz }

slickenside [GEOL] A surface that is polished and smoothly striated and results from slippage along a fault plane. { 'slik·ən,sīd }

slickolite [GEOL] A vertically discontinuous slip-scratch surface made by slippage and shearing and developed on sharply dipping bedding planes of limestone that shapes the wall of a solution cavity. { 'slik·ə,līt }

slick paper proof See reproduction proof. { 'slik ¦pā·pər 'prüf }

slide [ENG] **1.** A sloping chute with a flat bed. **2.** A sliding mechanism. [GEOL] **1.** A vein of clay intersecting and dislocating a vein vertically, or the vertical dislocation itself. **2.** A rotational or planar mass movement of earth, snow, or rock resulting from failure under shear stress along one or more surfaces. [MECH ENG] The main reciprocating member of a mechanical press, guided in a press frame, to which the punch or upper die is fastened. [MIN ENG] **1.** An upright rail fixed in a shaft with corresponding grooves for steadying the cages. **2.** A trough used to guide and to support rods in a tripod when drilling an angle hole. Also known as rod slide. [ORD] **1.** Sliding part of the receiver of certain automatic weapons. **2.** Sliding catch on the breech mechanism of certain weapons. { slīd }

slide-back voltmeter [ELECTR] An electronic voltmeter in which an unknown voltage is measured indirectly by adjusting a calibrated voltage source until its voltage equals the unknown voltage. { 'slīd ,bak 'vōlt,mēd·ər }

slide conveyor [ENG] A slanted gravity slide for the forward downward movement of flowable solids, slurries, liquids, or small objects. { 'slīd kən,vā·ər }

slide gate [CIV ENG] A crest gate which has high frictional resistance to opening because it slides on its bearings in opening and closing. { 'slīd ,gāt }

slide projector See optical lantern. { 'slīd prə,jek·tər }

slider [ELEC] Sliding type of movable contact. [ORD] A fuse or exploder component that interrupts the explosive train when the device is in the unarmed condition, and that moves during arming in such a way as to render the explosive train operative. { 'slīd·ər }

slide rail See guardrail. { 'slīd ,rāl }

slider coupling [MECH ENG] A device for connecting shafts that are laterally misaligned. Also known as double-slider coupling; Oldham coupling. { 'slīd·ər ,kəp·liŋ }

slide rest [MECH ENG] An adjustable slide for holding a cutting tool, as on an engine lathe. { 'slīd ,rest }

slider support [ENG] A support designed to allow longitudinal movement of pipework in a horizontal plane. { 'slīd·ər sə'pȯrt }

slide rule [MATH] A mechanical device, composed of a ruler with sliding insert, marked with various number scales, which facilitates such calculations as division, multiplication, finding roots, and finding logarithms. { 'slīd ,rül }

slide-rule dial [ENG] A dial in which a pointer moves in a straight line over long straight scales resembling the scales of a slide rule. { 'slīd ,rül ,dīl }

slide valve [MECH ENG] A sliding mechanism to cover and uncover ports for the admission of fluid, as in some steam engines. { 'slīd ,valv }

slide-wire bridge [ELEC] A bridge circuit in which the resistance in one or more branches is controlled by the position of a sliding contact on a length of resistance wire stretched along a linear scale. { 'slīd ¦wīr ,brij }

slide-wire potentiometer [ELEC] A potentiometer (variable resistor) which employs a movable sliding connection on a length of resistance wire. { 'slīd ¦wīr pə,ten·chē'äm·əd·ər }

sliding See gravitational sliding. { 'slīd·iŋ }

sliding-block linkage [MECH ENG] A mechanism in which a crank and sliding block serve to convert rotary motion into translation, or vice versa. { 'slīd·iŋ ¦bläk 'liŋ·kij }

sliding-chain conveyor [MECH ENG] A conveying machine to handle cases, cans, pipes, or similar products on the plain

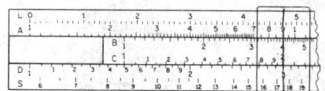

or modified links of a set of parallel chains. { 'slīd·iŋ ¦chān kən'vā·ər }

sliding contact *See* wiping contact. { 'slīd·iŋ 'kän,takt }

sliding fit [DES ENG] A fit between two parts that slide together. { 'slīd·iŋ 'fit }

sliding form *See* slip form. { 'slīd·iŋ ,fòrm }

sliding friction [MECH] Rubbing of bodies in sliding contact. { 'slīd·iŋ ,frik·shən }

sliding gear [DES ENG] A change gear in which speed changes are made by sliding gears along their axes, so as to place them in or out of mesh. { 'slīd·iŋ ,gir }

sliding-gear transmission [MECH ENG] A transmission system utilizing a pair of sliding gears. { 'slīd·iŋ ,gir tranz 'mish·ən }

sliding pair [MECH ENG] Two adjacent links, one of which is constrained to move in a particular path with respect to the other; the lower, or closed, pair is completely constrained by the design of the links of the pair. { 'slīd·iŋ 'per }

sliding scale [METEOROL] A set of combinations of ceilings and visibilities which constitute the operational weather limits at an airport; as the observed value of one element increases, the limiting value of the other element decreases, and vice versa. { 'slīd·iŋ 'skāl }

sliding-vane compressor [CHEM ENG] A rotary-element gas compressor in which spring-loaded sliding vanes (evenly spaced around a cylinder off-center in a surrounding chamber) pick up, compress, and discharge gas as the cylinder revolves. { 'slīd·iŋ ¦vān kəm'pres·ər }

sliding vector [MECH] A vector whose direction and line of application are prescribed, but whose point of application is not prescribed. { 'slīd·iŋ 'vek·tər }

sliding way [CIV ENG] One of the timbers which form the upper part of the cradle supporting a ship during its construction, and which slide over the ground ways with the ship when it is launched. { 'slīd·iŋ 'wā }

slime [ENG] Liquid slurry of very fine solids with slime- or mudlike appearance. Also known as mud; pulp; sludge. { slīm }

slime bacteria [MICROBIO] The common name for bacteria in the order Myxobacterales, so named for the layer of slime deposited behind cells as they glide on a surface. { 'slīm bak,tir·ē·ə }

slime disease [PL PATH] Any of several diseases of plants characterized by slimy rot of the parts. { 'slīm di,zēz }

slime flux [PL PATH] The fluid or viscous outflow from the bark or wood of a deciduous tree that is indicative of injury or disease. { 'slīm ,fləks }

slime fungus *See* slime mold. { 'slīm ,fəŋ·gəs }

slime gland [ZOO] A glandular structure in many animals producing a mucous material. { 'slīm ,gland }

slime mold [MYCOL] The common name for members of the Myxomycetes. Also known as slime fungus. { 'slīm ,mōld }

slim hole [ENG] A drill hole of the smallest practicable size, drilled with less-than normal-diameter tools, used primarily as a seismic shothole and for structure tests and sometimes for stratigraphic tests. [PETRO ENG] A diamond-drill borehole having a diameter of 5 inches (12.7 centimeters) or less. { 'slīm ,hōl }

sliming [OCEANOGR] The formation of films of algae on submerged structures. { 'slīm·iŋ }

sling [ENG] A length of rope, wire rope, or chain used for attaching a load to a crane hook. { sliŋ }

sling psychrometer [ENG] A psychrometer in which the wet- and dry-bulb thermometers are mounted upon a frame connected to a handle at one end by means of a bearing or a length of chain; the psychrometer may be whirled in the air for the simultaneous measurement of wet- and dry-bulb temperatures. { 'sliŋ si'kräm·əd·ər }

sling thermometer [ENG] A thermometer mounted upon a frame connected to a handle at one end by means of a bearing or length of chain, so that the thermometer may be whirled by hand. { 'sliŋ thər'mäm·əd·ər }

slip [CIV ENG] A narrow body of water between two piers. [CRYSTAL] The movement of one atomic plane over another in a crystal; it is one of the ways that plastic deformation occurs in a solid. Also known as glide. [ELEC] **1.** The difference between synchronous and operating speeds of an induction machine. Also known as slip speed. **2.** Method of interconnecting multiple wiring between switching units by which trunk

number 1 becomes the first choice for the first switch, trunk number 2 first choice for the second switch, trunk number 3 first choice for the third switch, and so on. [ELECTR] Distortion produced in the recorded facsimile image which is similar to that produced by skew but is caused by slippage in the mechanical drive system. [FL MECH] The difference between the velocity of a solid surface and the mean velocity of a fluid at a point just outside the surface. [GEOL] The actual relative displacement along a fault plane of two points which were formerly adjacent on either side of the fault. Also known as actual relative movement; total displacement. [MATER] A suspension of fine clay in water with a creamy consistency, used in the casting process and in decorating ceramic ware. Also known as slurry. [NAV ARCH] **1.** To part from an anchor by releasing the shackles from the anchor chain. **2.** The reduction in the distance a propeller advances, per unit time, due to yielding of the fluid. { slip }

slip additive [MATER] A plastics modifier that acts as an internal lubricant by exuding to the surface of the plastic during and immediately after processing to reduce friction and improve slip. { 'slip ,ad·əd·iv }

slipband [CRYSTAL] One of the microscopic parallel lines (Lüders' lines) on the surface of a crystalline material stretched beyond its elastic limit, located at the intersection of the surface with intracrystalline slip planes in the grains of the material. Also known as slip line. { 'slip,band }

slip bedding [GEOL] Convolute bedding formed as the result of subaqueous sliding. { 'slip ,bed·iŋ }

slip block [GEOL] A separate rock mass that has slid away from its original position and come to rest down the slope without undergoing much deformation. { 'slip ,bläk }

slip casting [ENG] A process in the manufacture of shaped refractories, cermets, and other materials in which the slip is poured into porous plaster molds. { 'slip ,kast·iŋ }

slip clay [MATER] Clay containing a high percentage of fluxing impurities; easily fusible and used in clayware to produce a natural glaze. { 'slip ,klā }

slip cleavage [GEOL] Cleavage that is superposed on slaty cleavage or schistosity, characterized by spaced cleavage with thin tabular bodies of rock between the cleavage planes. Also known as close-joints cleavage; crenulation cleavage; shear cleavage; strain-slip cleavage. { 'slip ,klē·vij }

slip direction [CRYSTAL] The crystallographic direction in which the translation of slip occurs. { 'slip di,rek·shən }

slip face [GEOL] The steeply sloping leeward surface of a sand dune. Also known as sandfall. { 'slip ,fās }

slip flow [FL MECH] A situation in which the mean free path of a gas is between 1 and 65% of the channel diameter; the gas layer next to the channel wall assumes a velocity of slip past the liquid, known as slip flow. { 'slip 'flō }

slip fold *See* shear fold. { 'slip ,fōld }

slip form [CIV ENG] A narrow section of formwork that can be easily removed as concrete placing progresses. { 'slip ,fòrm }

slip forming [ENG] A plastics-sheet forming technique in which some of the sheet is allowed to slip through the mechanically operated clamping rings during stretch-forming operations. { 'slip ,fòrm·iŋ }

slip friction clutch [MECH ENG] A friction clutch designed to slip when too much power is applied to it. { 'slip 'frik·shən ,kləch }

slip joint [CIV ENG] **1.** Contraction joint between two adjoining wall sections, or at the horizontal bearing of beams, slabs, or precast units, consisting of a vertical tongue fitted into a groove which allows independent movement of the two sections. **2.** A telescoping joint between two parts. [ENG] **1.** A method of laying-up plastic veneers in flexible-bag molding, wherein edges are beveled and allowed to overlap part or all of the scarfed area. **2.** A mechanical union that allows limited endwise movement of two solid items for example, pipe, rod, or duct with relation to each other. [GEOL] *See* shear joint. { 'slip ,jòint }

slip line *See* slipband. { 'slip ,līn }

slip-off slope [GEOL] The long, low, gentle slope on the inside of the downstream face of a stream meander. { 'slip ,òf ,slōp }

slippage [ENG] The leakage of fluid between the plunger and the bore of a pump piston. Also known as slippage loss.

SLIDING PAIR

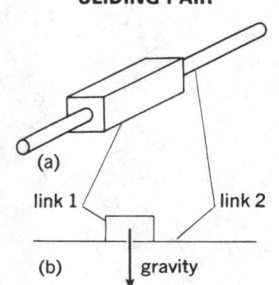

(a)

link 1   link 2

(b)   ↓ gravity

Two types of sliding pair.
(*a*) Motion of link 1 is constrained by the design of the links.
(*b*) Motion of link 1 is constrained by gravity.

[PETRO ENG] The movement of gas past or through a liquid-phase reservoir front; this movement occurs instead of driving the liquid forward; it can exist in gas-drive reservoirs or in gas-lift oil-well bores. [TEXT] A fabric defect on which warp threads slip over filling threads or vice versa as a result of loose weaving or of unevenly matched yarns. { 'slip·ij }

slippage loss [ENG] **1.** Unintentional movement between the faces of two solid objects. **2.** *See* slippage. { 'slip·ij ‚lòs }

slipped disk *See* herniated disk. { 'slipt 'disk }

slipper brake [MECH ENG] **1.** A plate placed against a moving part to slow or stop it. **2.** A plate applied to the wheel of a vehicle or to the track roadway to slow or stop the vehicle. { 'slip·ər ‚brāk }

slipping [MIN ENG] Enlarging an excavation by breaking one or more walls. { 'slip·iŋ }

slipping cut [MIN ENG] A drill-hole pattern used in a wide tunnel face, in which each successive vertical line of shots breaks to the face made by the previous round, so that the relieving cut moves across the end being blasted. Also known as slabbing cut; swing cut. { 'slip·iŋ ‚kət }

slip plane [CRYSTAL] *See* glide plane. [GEOL] A planar slip surface. [ENG] A plane visible by reflected light in a transparent material; caused by poor welding and shrinkage during cooling. { 'slip ‚plān }

slip ratio [MECH ENG] For a screw propeller, relates the actual advance to the theoretic advance determined by pitch and spin. { 'slip ‚rā·shō }

slip ring [ELEC] A conductive rotating ring which, in combination with a stationary brush, provides a continuous electrical connection between rotating and stationary conductors; used in electric rotating machinery, synchros, gyroscopes, and scanning radar antennas. { 'slip ‚riŋ }

slips [ENG] A wedge-shaped steel collar fabricated in two sections, designed to hold a string of casing between various portions of the drilling operation. { 'slips }

slip sheet [GEOL] A stratum or rock on the limb of an anticline that has slid down and away from the anticline; a gravity collapse structure. { 'slip ‚shēt }

slip speed *See* slip. { 'slip ‚spēd }

slip surface [GEOL] The displacement surface of a landslide. { 'slip ‚sər·fəs }

slip tongue [ENG] A pole on a horse-drawn wagon that is fastened by slipping it between two plates connected to the forecarriage. { 'slip ‚təŋ }

slip velocity [FL MECH] The difference in velocities between liquids and solids (or gases and liquids) in the vertical flow of two-phase mixtures through a pipe because of the slip between the two phases. { 'slip və‚läs·əd·ē }

slipway [CIV ENG] The space in a shipyard where a foundation for launching ways and keel blocks exists and which is occupied by a ship while under construction. [NAV ARCH] **1.** One of the timbers over which the cradle of a marine railway travels. **2.** A sloping passage in a whaling ship's stern, through which whales are hauled in. { 'slip‚wā }

slit [DES ENG] A long, narrow opening through which radiation or streams of particles enter or leave certain instruments. { slit }

slitless spectrograph [OPTICS] A type of astronomical spectrograph that does not use a slit, sufficient resolution being obtained from the small image sizes of individual stars, and through the use of an objective prism in front of the telescope. { 'slit·ləs 'spek·trə‚graf }

slitlet mask [SPECT] A metal plate used in astronomical spectroscopy, with several small slits at locations corresponding to astronomical objects of interest. { 'slit·lət ‚mask }

slit scan [COMPUT SCI] In character recognition, a magnetic or photoelectric device that obtains the horizontal structure of an inputted character by vertically projecting its component elements at given intervals. { 'slit ‚skan }

slit spectrograph [OPTICS] A type of astronomical spectrograph that uses a slit to provide resolution. { 'slit 'spek·trə‚graf }

slitter [MECH ENG] A synchronized feeder-knife variation of a rotary cutter; used for precision cutting of sheet material, such as metal, rubber, plastics, or paper, into strips. { 'slid·ər }

slitting [MECH ENG] The passing of sheet or strip material (metal, plastic, paper, or cloth) through rotary knives. { 'slid·iŋ }

slitting mill [LAP] A rotating disk used by gem cutters in slitting. Also known as slicer. { 'slid·iŋ ‚mil }

sliver [MATER] A piece of propellant grain of triangular cross section which remains unburned when the web of multiperforated grains has been burned through. [MET] A thin, elongated fragment of metal that has been rolled onto the surface of the parent metal and is attached by only one end. [TEXT] A round, untwisted strand of fiber removed from the carding or combing machine and used to spin yarn. { 'sliv·ər }

slop [CHEM ENG] A petroleum-refinery term for odds and ends of oil in the refinery; the slop must be rerun or further processed to make it suitable for use. Also known as slop oil. { släp }

slop culture [BOT] A method of growing plants in which surplus nutrient fluid is allowed to run through the sand or other medium in which the plants are growing. { 'släp ‚kəl·chər }

slope [GEOL] The inclined surface of any part of the earth's surface. [MATH] **1.** The slope of a line through the points (x_1, y_1) and (x_2, y_2) is the number $(y_2 - y_1)/(x_2 - x_1)$. **2.** The slope of a curve at a point p is the slope of the tangent line to the curve at p. [NAV] The projection of a flight path in the vertical plane. { slōp }

slope angle [MATH] The angle of inclination of a line in the plane, where this angle is measured from the positive x axis to the line in the counterclockwise direction. { 'slōp ‚aŋ·gəl }

slope control [MET] Electronic production of a change in the welding current within set limits and a selected interval of time. { 'slōp kən‚trōl }

slope conveyor [MECH ENG] A troughed belt conveyor used for transporting material on steep grades. { 'slōp kən‚vā·ər }

slope correction [GEOL] A tape correction applied to a distance measured on a slope in order to reduce it to a horizontal distance, between the vertical lines through its end points. Also known as grade correction. { 'slōp kə‚rek·shən }

slope course [ENG] A proving ground facility consisting of a large mound of earth with various sloping sides on which are roads having different grades; this slope course is used to measure the slope performance of military and other vehicles, including maximum speed on various grades, the most suitable gear for best performance, traction, and the holding ability of brakes { 'slōp ‚kòrs }

slope current *See* gradient current. { 'slōp ‚kə·rənt }

slope deviation [AERO ENG] The difference between planned and actual slopes of aircraft travel, expressed in either angular or linear measurement. { 'slōp ‚dē·vē‚ā·shən }

slope engineer [MIN ENG] In anthracite and bituminous coal mining, one who operates a hoisting engine to haul loaded and empty mine cars along a haulage road in a mine. { 'slōp ‚en·jə‚nir }

slope failure [GEOL] The downward and outward movement of a mass of soil beneath a natural slope or other inclined surface; four types of slope failure are rockfall, rock flow, plane shear, and rotational shear. { 'slōp ‚fāl·yər }

slope gully [GEOL] A small, discontinuous submarine valley, usually formed by slumping along a fault scarp or the slope of a river delta. Also known as sea gully. { 'slōp ‚gəl·ē }

slope-intercept form [MATH] In a cartesian coordinate system, the equation of a straight line in the form $y = mx + b$, where m is the slope of the line and b is its intercept on the y axis. { 'slōp ‚in·tər‚sept ‚fòrm }

slope mine [MIN ENG] A mine opened by a slope or an incline. { 'slōp ‚mīn }

slope of fall [MECH] Ratio between the drop of a projectile and its horizontal movement; tangent of the angle of fall. { 'slōp əv 'fòl }

slope stability [GEOL] The resistance of an inclined surface to failure by sliding or collapsing. { 'slōp stə‚bil·əd·ē }

slope stake [MIN ENG] Stake set at the point where the finished side slope of an excavation or embankment meets the original grade. { 'slōp ‚stāk }

slope wash [GEOL] **1.** The mass-wasting process, assisted by nonchanneled running water, by which rock and soil is transported down a slope, specifically by sheet erosion. **2.** The material that is or has been transported. { 'slōp ‚wäsh }

slop oil *See* slop. { 'släp ‚òil }

slosh test [ENG] A test to determine the ability of the control system of a liquid-propelled missile to withstand or overcome the dynamic movement of the liquid within its fuel tanks. { 'släsh ‚test }

slot [AERO ENG] **1.** An air gap between a wing and the length of a slat or certain other auxiliary airfoils, the gap providing space for airflow or room for the auxiliary airfoil to be depressed in such a manner as to make for smooth air passage on the surface. **2.** Any of certain narrow apertures made through a wing to improve aerodynamic characteristics. [COMPUT SCI] A punched-out area of a hand-sorted card to connect two or more guide holes; slots can be extended to the outside edge with notches. [DES ENG] A narrow, vertical opening. [ELEC] One of the conductor-holding grooves in the face of the rotor or stator of an electric rotating machine. [MIN ENG] To hole; to undercut or channel. { slät }

slot antenna [ELECTROMAG] An antenna formed by cutting one or more narrow slots in a large metal surface fed by a coaxial line or waveguide. { 'slät an,ten·ə }

slot-bound [COMPUT SCI] Condition of a computer when all the slots in the machine's bus are filled with printed circuit boards, so that it is not possible to expand the machine's capacity by plugging in additional boards. { 'slät ,baȯnd }

slot coupling [ELECTROMAG] Coupling between a coaxial cable and a waveguide by means of two coincident narrow slots, one in a waveguide wall and the other in the sheath of the coaxial cable. { 'slät ,kəp·liŋ }

slot distributor [ENG] A long, narrow discharge opening (slot) in a pipe or conduit; used for the extrusion of sheet material, such as plastics. { 'slät di'strib·yəd·ər }

slot dozing [ENG] A method of moving large quantities of material with a bulldozer using the same path for each trip so that the spillage from the sides of the blade builds up along each side; afterward all material pushed into the slot is retained in front of the blade. { 'slät ,dōz·iŋ }

slot extrusion [ENG] A method of extruding plastics-film sheet in which the molten thermoplastic compound is forced through a straight slot. { 'slät ik,strü·zhən }

sloth [VERT ZOO] Any of several edentate mammals in the family Bradypodidae found exclusively in Central and South America; all are slow-moving, arboreal species that cling to branches upside down by means of long, curved claws. { släth }

slot-mask picture tube [ELECTR] An in-line gun-type picture tube in which the shadow mask is perforated by short, vertical slots, and the screen is painted with vertical phosphor stripes. { 'slät ,mask 'pik·chər ,tüb }

slot radiator [ELECTROMAG] Primary radiating element in the form of a slot cut in the walls of a metal waveguide or cavity resonator or in a metal plate. { 'slät ,rād·ē,ād·ər }

slotted-head screw [DES ENG] A screw fastener with a single groove across the diameter of the head. { 'släd·əd 'hed 'skrü }

slotted line See slotted section. { 'släd·əd 'līn }

slotted nut [DES ENG] A regular hexagon nut with slots cut across the flats of the hexagon so that a cotter pin or safety wire can hold it in place. { 'släd·əd 'nət }

slotted section [ELECTROMAG] A section of waveguide or shielded transmission line in which the shield is slotted to permit the use of a traveling probe for examination of standing waves. Also known as slotted line; slotted waveguide. { 'släd·əd ,sek·shən }

slotted waveguide See slotted section. { 'släd·əd 'wāv,gīd }

slotter [MECH ENG] A machine tool used for making a mortise or shaping the sides of an aperture. { 'släd·ər }

slotting [MECH ENG] Cutting a mortise or a similar narrow aperture in a material using a machine with a vertically reciprocating tool. { 'släd·iŋ }

slotting machine [MECH ENG] A vertically reciprocating planing machine, used for making mortises and for shaping the sides of openings. { 'släd·iŋ mə,shēn }

slot washer [DES ENG] **1.** A lock washer with an indentation on its edge through which a nail or screw can be driven to hold it in place. **2.** A washer with a slot extending from its edge to the center hole to allow the washer to be removed without first removing the bolt. { 'slät ,wäsh·ər }

slot wedge [ELEC] The wedge that holds the windings in a slot in the rotor or stator core of an electrical machine. { 'slät ,wej }

slot weld [MET] Similar to plug weld, but the hole is elongated and may extend to the edge of a member without closing. { 'slät ,weld }

slough [ENG] The fragments of rocky material from the wall of a borehole. Also known as cavings. [HYD] A minor marshland or tidal waterway which usually connects other tidal areas; often more or less equivalent to a bayou. [MED] A necrotic mass of tissue in, or separating from, healthy tissue. { slaȯ }

slough ice [HYD] Slushy ice or snow. { 'slaȯ ,īs }

sloughing See caving. { 'sləf·iŋ }

slow-acting relay [ELECTROMAG] A time-delay relay in which an interval of several seconds may exist between energizing of the coil and pulling up of the armature; the delay can be obtained electrically by placing a solid copper ring on the core of the relay. Also known as slow-operate relay. { 'slō ,ak·tiŋ 'rē,lā }

slow-blow fuse [ELEC] A fuse that can withstand up to 10 times its normal operating current for a brief period, as required for circuits and devices which draw a very heavy starting current. { 'slō ,blō 'fyüz }

slow-curing liquid asphaltic material [MATER] An asphalt cement blended with slow-volatilizing gas oil. Also known as SC asphalt. { 'slō ,kyȯr·iŋ 'lik·wəd as'fȯl·tik mə'tir·ē·əl }

slow death [ELECTR] The gradual change of transistor characteristics with time; this change is attributed to ions which collect on the surface of the transistor. { 'slō 'deth }

slowed-down video [ELECTR] Technique or method of transmitting radar data over narrow-bandwidth circuits; the procedure involves storing the radar video over the time required for the antenna to move through the beam width, and the subsequent sampling of this stored video at some periodic rate at which all of the range intervals of interest are sampled at least once each beam width or per azimuth quantum; the radar returns are quantized at the gap-filler radar site. { 'slōd ,daȯn 'vid·ē·ō }

slow fire [ORD] Type of firing used in instructing beginners and in record practice, in which no time limit for completing a score is set. { 'slō ,fīr }

slow igniter cord [ENG] An igniter cord made with a central copper wire around which is extruded a plastic incendiary material with an iron wire embedded to give greater strength; the whole is enclosed in a thin extruded plastic coating. { 'slō ig'nīd·ər ,kȯrd }

slowing-down [NUCLEO] A decrease in the energy of a particle as a result of collisions with nuclei. { 'slō·iŋ 'daȯn }

slowing-down area [NUCLEO] One-sixth of the mean square distance from the source of a neutron in an infinite, homogeneous medium to the point at which the neutron reaches a given energy. { 'slō·iŋ 'daȯn 'er·ē·ə }

slowing-down density [NUCLEO] A measure of the rate at which neutrons lose energy in a nuclear reactor through collisions; equal to the number of neutrons that fall below a given energy per unit volume per unit time. { 'slō·iŋ 'daȯn 'den·səd·ē }

slowing-down kernel [NUCLEO] The probability per unit volume that a neutron will travel from one position in a homogeneous medium to another while slowing down through a specified energy range. { 'slō·iŋ 'daȯn 'kər·nəl }

slowing of clocks [RELAT] According to the special theory of relativity, a clock appears to tick less rapidly to an observer moving relative to the clock than to an observer who is at rest with respect to the clock. Also known as time dilation effect. { 'slō·iŋ əv 'kläks }

slow ion See large ion. { 'slō 'ī,än }

slow match [ENG] A match or fuse that burns at a known slow rate; used for igniting explosive charges. { 'slō 'mach }

slow memory See slow storage. { 'slō 'mem·rē }

slow-motion video disk recorder [ELECTR] A magnetic disk recorder that stores one field of video information per revolution, for instant replay at normal speed or any degree of slow motion down to complete stopping of action. { 'slō ,mō·shən 'vid·ē·ō ,disk ri'kȯrd·ər }

slow neutron [NUC PHYS] **1.** A neutron having low kinetic energy, up to about 100 electronvolts. **2.** See thermal neutron. { 'slō 'nü,trän }

slow-neutron spectroscopy [PHYS] The use of beams of slow neutrons, from nuclear reactors or nuclear accelerators, in studies of the structure or structural dynamics of solid, liquid, or gaseous matter, particularly the atomic and magnetic dynamics. Also known as inelastic neutron scattering; neutron spectroscopy. { 'slō 'nü,trän spek'träs·kə·pē }

slow nova [ASTRON] A nova whose brightness takes a

SLOTH

The three-toed sloth (*Bradypus tridactylus*).

SLOTTED NUT

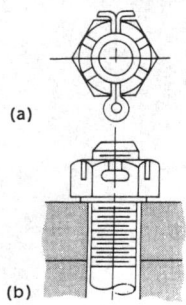

(a)

(b)

Diagrams of a slotted nut. (*a*) End view with cotter pin inserted. (*b*) Side view.

month or more to reach a maximum, and many years to decrease to the original value. { 'slō 'nŏ·və }

slow-operate relay See slow-acting relay. { 'slō ¦äp·ə,rāt 'rē,lā }

slow ray [OPTICS] In crystal optics, that component of light in any birefringent crystal section which travels with the lesser velocity and has the higher index of refraction. { 'slō 'rā }

slow-reacting substance of anaphylaxis [BIOCHEM] A group of leukotrienes released by mast cells during anaphylaxis that induces prolonged bronchoconstriction. { ¦slō rē,ak·tiŋ ¦səb·stəns əv ,an·ə·fə'lak·səs }

slow reactor [NUCLEO] A nuclear reactor in which fission is induced primarily by slow neutrons, as in a thermal reactor. { 'slō rē'ak·tər }

slow-release relay [ELECTROMAG] A time-delay relay in which there is an appreciable delay between deenergizing of the coil and release of the armature. { 'slō ri¦lēs 'rē,lā }

slow sand filter [CIV ENG] A bed of fine sand 20–48 inches (151–122 centimeters) deep through which water, being made suitable for human consumption and other purposes, is passed at a fairly low rate, 2,500,000 to 10,000,000 gallons per acre (23,000 to 94,000 cubic meters per hectare); an underdrain system of graded gravel and perforated pipes carries the water from the filters to the point of discharge. { 'slō 'sand ,fil·tər }

slow-scan television [COMMUN] Television system that uses a slow rate of horizontal scanning, requiring typically 8 seconds for each complete scan of the scene; suitable for transmitting printed matter, photographs, and illustrations. Abbreviated SSTV. { 'slō ¦skan 'tel·ə,vizh·ən }

slow-spiral drill See low-helix drill. { 'slō ¦spī·rəl 'dril }

slow storage [COMPUT SCI] In computers, storage with a relatively long access time. Also known as slow memory. { 'slō 'stȯr·ij }

slow time scale [COMPUT SCI] In simulation by an analog computer, a time scale in which the time duration of a simulated event is greater than the actual time duration of the event in the physical system under study. Also known as extended time scale. { 'slō 'tīm ,skāl }

slow-vibration direction [OPTICS] The direction of the electric field vector of the ray of light that travels with the smallest velocity in an anisotropic crystal and therefore corresponds to the maximum refractive index. { 'slō vī'brā·shən di,rek·shən }

slow virus [VIROL] Any member of a group of animal viruses characterized by prolonged periods of incubation and an extended clinical course lasting months or years. { 'slō 'vī·rəs }

slow virus infection [VIROL] A persistent viral infection characterized by a long preclinical period extending for months or years from the time of exposure. { 'slō 'vir·əs in,fek·shən }

slow wave [ELECTROMAG] A wave having a phase velocity less than the velocity of light, as in a ridge wave guide. { 'slō 'wāv }

slow-wave sleep See deep sleep. { 'slō ¦wāv 'slēp }

SLSI circuit See super-large-scale integrated circuit. { ,es,el,es'ī ,sər·kət }

slub [TEXT] A planned irregularity in yarn produced by alternately tightening and relaxing tensions during spinning. { 'sləb }

slud [GEOL] **1.** Muddy material which has moved downslope by solifluction. **2.** Ground that behaves as a viscous fluid, including material moved by solifluction and by mechanisms not limited to gravitational flow. { 'sləd }

sludge [CHEM ENG] **1.** Residue left after acid treatment of petroleum oils. **2.** Any semisolid waste from a chemical process. [CIV ENG] See sewage sludge. [ENG] **1.** Mud from a drill hole in boring. **2.** Sediment in a steam boiler. **3.** A precipitate from petroleum oils or liquid fuels, for example, the insoluble degradation products formed during the operation of an internal combustion engine. **4.** An amorphous deposit that has accumulated on the surface of a tube in a heat exchanger or of an evaporating device, but is not bonded to the fouled surface. **5.** See slime. [GEOL] A soft or muddy bottom deposit as on tideland or in a stream bed. [OCEANOGR] A dense, soupy accumulation of new sea ice consisting of incoherent floating frazil crystals. Also known as cream ice; sludge; slush. { sləj }

sludge acid [MATER] The residue from the sulfuric-acid treatment of petroleum lubricants. Also known as spent acid; waste acid. { 'sləj ,as·əd }

sludge assay [MIN ENG] The chemical assaying of drill cuttings for a specific metal or group of metals. { 'sləj 'as,ā }

sludge barrel See calyx. { 'sləj ,bar·əl }

sludge box [MIN ENG] A wooden box in which sludge settles from the mud flush. { 'sləj ,bäks }

sludge bucket See calyx. { 'sləj ,bək·ət }

sludge cake [OCEANOGR] An accumulation of sludge hardened into a cake strong enough to bear the weight of a man. { 'sləj ,kāk }

sludge coking [CHEM ENG] The recovery of sulfuric acid from dry acid sludge. { 'sləj ,kōk·iŋ }

sludged blood [MED] The intracapillary aggregation of erythrocytes associated with decreased blood flow in the involved capillary bed. { 'sləjd 'bləd }

sludge floe [OCEANOGR] Sludge that is hardened into a floe strong enough to bear the weight of a person. { 'sləj ,flō }

sludge ice See sludge. { 'sləj ,īs }

sludge lump [OCEANOGR] An irregular mass of sludge formed as a result of strong winds. { 'sləj ,ləmp }

sludge pit See slushpit. { 'sləj ,pit }

sludge pond See slushpit. { 'sləj ,pänd }

sludge pump [MECH ENG] See sand pump. [MIN ENG] A short iron pipe or tube fitted with a valve at the lower end, with which the sludge is extracted from a borehole. { 'sləj ,pəmp }

sludging See solifluction. { 'sləj·iŋ }

sluff [ENG] The mud cake detached from the wall of a borehole. [MIN ENG] The falling of decomposed, soft rocks from the roof or walls of mine openings. { sləf }

slug [ELECTROMAG] **1.** A heavy copper ring placed on the core of a relay to delay operation of the relay. **2.** A movable iron core for a coil. **3.** A movable piece of metal or dielectric material used in a wave guide for tuning or impedance-matching purposes. [GRAPHICS] A strip of metal used to space between lines of type. [INV ZOO] Any of a number of pulmonate gastropods which have a rudimentary shell and the body elevated toward the middle and front end where the mantle covers the lung region. [MECH] A unit of mass in the British gravitational system of units, equal to the mass which experiences an acceleration of 1 foot per second per second when a force of 1 pound acts on it; equal to approximately 32.1740 pound mass or 14.5939 kilograms. Also known as geepound. [MET] **1.** A small, roughly shaped piece of metal for subsequent processing, as by forging or extruding. **2.** The piece of material produced by piercing a hole in a sheet. [MIN ENG] To inject a borehole with cement, slurry, or various liquids containing shredded materials in an attempt to restore lost circulation by sealing off the openings in the borehole-wall rocks. [NUCLEO] A short fuel rod inserted in a hole or channel in the active lattice of a nuclear reactor. [ORD] **1.** As pertains to shaped charge ammunition, massive and relatively slow-moving remnant of the collapsed metal liner, as distinguished from the jet. **2.** A solid cast iron projectile used in test firing. { sləg }

slug bit See insert bit. { 'sləg ,bit }

slug flow See piston flow. { 'sləg ,flō }

slugging [MET] Adding a separate piece of material to a weld joint, resulting in a joint which does not meet specifications. { 'sləg·iŋ }

SLUG junction [CRYO] A Josephson junction consisting of a drop of lead-tin solder solidified around a niobium wire. { 'sləg ,jəŋk·shən }

slug tuner [ELECTROMAG] Waveguide tuner containing one or more longitudinally adjustable pieces of metal or dielectric. { 'sləg ,tün·ər }

slug tuning [ELECTROMAG] Means of varying the frequency of a resonant circuit by introducing a slug of material into either the electric field or magnetic field, or both. { 'sləg ,tün·iŋ }

sluice [CIV ENG] A passage fitted with a vertical sliding gate or valve to regulate the flow of water in a channel or lock. **2.** A body of water retained by a floodgate. **3.** A channel serving to drain surplus water. { 'slüs }

sluice box [MIN ENG] A long, inclined trough or launder with riffles in the bottom that provide a lodging place for heavy minerals in ore concentration. { 'slüs ,bäks }

sluice gate [CIV ENG] The vertical slide gate of a sluice. { 'slüs ,gāt }

SLUG

Field gray slug (*Deroceras agreste*).

sluice tender [MIN ENG] In metal mining, a laborer who tends sluice boxes. { 'slüs ¦ten·dər }

sluicing [MIN ENG] **1.** Washing auriferous earth through sluices provided with riffles and other gold-saving appliances. **2.** Separation of minerals in a flowing stream of water. **3.** Moving earth, sand, gravel, or other rock or mineral materials by flowing water. { 'slüs·iŋ }

sluicing pond *See* scouring basin. { 'slüs·iŋ ¦pänd }

slump [GEOL] A type of landslide characterized by the downward slipping of a mass of rock or unconsolidated debris, moving as a unit or several subsidiary units, characteristically with backward rotation on a horizontal axis parallel to the slope; common on natural cliffs and banks and on the sides of artificial cuts and fills. { sləmp }

slump ball [GEOL] A relatively flattened mass of sandstone resembling a large concretion, measuring from 0.8 inch to 10 feet (2 centimeters to 3 meters) across, commonly thinly laminated with internal contortions and a smooth or lumpy external form, and formed by subaqueous slumping. { 'sləmp ¦bȯl }

slump basin [GEOL] A shallow basin near the base of a canyon wall and on a shale hill or ridge, formed by small, irregular slumps. { 'sləmp ¦bās·ən }

slump bedding [GEOL] Also known as slurry bedding. **1.** Any disturbed bedding. **2.** Convolute bedding produced by subaqueous slumping or lateral movement of newly deposited sediment. { 'sləmp ¦bed·iŋ }

slump fault *See* normal fault. { 'sləmp ¦fȯlt }

slump fold [GEOL] An intraformational fold produced by slumping of soft sediments, as at the edge of the continental shelf. { 'sləmp ¦fōld }

slump overfold [GEOL] A fold consisting of hook-shaped masses of sandstone produced during slumping. { 'sləmp 'ō·vər¦fōld }

slump scarp [GEOL] A low cliff or rim of thin solidified lava occurring along the margins of a lava flow and against the valley walls or around steptoes after the central part of the lava crust collapsed due to outflow of still-molten underlying layers. { 'sləmp ¦skärp }

slump sheet [GEOL] A well-defined bed of limited thickness and wide horizontal extent, containing slump structures. { 'sləmp ¦shēt }

slump structure [GEOL] Any sedimentary structure produced by subaqueous slumping. { 'sləmp ¦strək·chər }

slump test [ENG] Determining the consistency of concrete by filling a conical mold with a sample of concrete, then inverting it over a flat plate and removing the mold; the amount by which the concrete drops below the mold height is measured and this represents the slump. { 'sləmp ¦test }

slurry [MATER] **1.** A semiliquid refractory material, such as clay, used to repair furnace refractories. **2.** A free-flowing, pumpable suspension of fine solid material in liquid. **3.** An emulsion of a sulfonated soluble oil in water used to cool and lubricate metal during cutting operations. **4.** A plastic mixture of portland cement and water pumped into an oil well; after hardening, it provides support for the casing and a seal for the well bore. **5.** *See* slip. { 'slər·ē }

slurry bedding *See* slump bedding. { 'slər·ē ¦bed·iŋ }

slurry bed reactor *See* ebullating-bed reactor. { 'slər·ē ¦bed rē¦ak·tər }

slurry blasting agent [MATER] A dense, insensitive, high-velocity explosive of great power and very high water resistance used principally for blasting hard rock or where blastholes are wet. { 'slər·ē 'blast·iŋ ¦ā·jənt }

slurry carrier *See* mineral tanker. { 'slər·ē ¦kar·ē·ər }

slurrying [ENG] The formation of a mud or a suspension from a liquid and nonsoluble solid particles. { 'slər·ē·iŋ }

slurry mining [MIN ENG] The hydraulic breakdown of a subsurface ore matrix with drill-hole equipment, and the eduction of the resulting slurry to the surface for processing. { 'slər·ē ¦mīn·iŋ }

slurry preforming [ENG] The preparation of reinforced plastics preforms by wet-processing techniques; similar to pulp molding. { 'slər·ē prē'fȯrm·iŋ }

slurry slump [GEOL] A slump in which the incoherent sliding mass is mixed with water and disintegrates into a quasiliquid slurry. { 'slər·ē ¦sləmp }

slurry truck [ENG] A mobile unit that transports dry blasting

ingredients, and mixes them in required proportions for introduction as explosive slurry into blastholes. { 'slər·ē ¦trək }

slush [HYD] Snow or ice on the ground that has been reduced to a soft, watery mixture by rain, warm temperature, or chemical treatment. [OCEANOGR] *See* sludge. { sləsh }

slush avalanche [GEOL] A rapid and far-reaching downslope transport of rock debris released by snow supersaturated with meltwater and marking the catastrophic opening of ice- and snow-dammed brooks to the spring flood. { 'sləsh 'av·ə¦lanch }

slush ball [HYD] An extremely compact accretion of snow, frazil, and ice particles. { 'sləsh ¦bȯl }

slush casting [MET] Producing a hollow casting without a core in the mold by rotating a liquid alloy in a hollow metal mold until a solid layer chills onto the mold, and then pouring off the remaining liquid. { 'sləsh ¦kast·iŋ }

slusher [ENG] A method for the application of vitreous enamel slip to ware by dashing it on the ware to cover all its parts, excess then being removed by shaking the ware. { 'sləsh·ər }

slush field [HYD] An area of water-saturated snow having a soupy consistency. Also known as snow swamp. { 'sləsh ¦fēld }

slushflow [HYD] **1.** A mudflow-like outburst of water-saturated snow along a stream course, commonly occurring in the Arctic Zone after intense thawing has produced more meltwater than can drain through the snow, and having a width generally several times greater than that of the stream channel. **2.** A flow of clear slush on a glacier, as in Greenland. { 'shəsh¦flō }

slush grouting [CIV ENG] Spreading a portland cement slurry over a surface that will subsequently be covered by concrete. { 'sləsh ¦graủd·iŋ }

slush icing [METEOROL] The accumulation of ice and water on exposed surfaces of aircraft when the craft is flown through wet snow or snow and liquid drops at temperatures near 0°C. { 'sləsh ¦īs·iŋ }

slushing compound [MATER] A temporary, corrosion-protective coating for metals; made of nondrying oil, grease, or other similar material. { 'sləsh·iŋ ¦käm¦paủnd }

slushing grease [MATER] A special grade of grease used as a metal coating to prevent corrosion. { 'sləsh·iŋ ¦grēs }

slushing oil [MATER] A nondrying oil which is strongly adhesive to metal and is applied to metal surfaces to minimize corrosion. { 'sləsh·iŋ ¦ȯil }

slush molding [ENG] A thermoplastic casting in which a liquid resin is poured into a hot, hollow mold where a viscous skin forms; excess slush is drained off, the mold is cooled, and the molded product is stripped out. { 'sləsh ¦mōld·iŋ }

slushpit [ENG] An excavation or diked area to hold water, mud, sludge, and other discharged matter from an oil well. Also known as mud pit; sludge pit; sludge pond. { 'shəsh¦pit }

slush pond [HYD] A pool or lake containing slush, on the ablation surface of a glacier. { 'sləsh ¦pänd }

slush pump [PETRO ENG] A pump used to circulate the drilling fluid during rotary drilling. { 'sləsh ¦pəmp }

Sm *See* samarium.

Smalian's formula [FOR] A cubic volume formula used in log scaling, expressed as cubic volume = $[(B + b)/2]L$, where B = the cross-sectional area at the large end of the log, b = the cross-sectional area at the small end of the log, and L = log length. { 'smȯl·yəns ¦fȯr·myə·lə }

small-angle scattering [PHYS] Scattering of a beam of electromagnetic or acoustic radiation, or particles, at small angles by particles or cavities whose dimensions are many times as large as the wavelength of the radiation or the de Broglie wavelength of the scattered particles. Also known as low-angle scattering. { ¦smȯl ¦aŋ·gəl 'skad·ər·iŋ }

small arm [ORD] A gun of small character; includes such hand and shoulder weapons as pistols, carbines, rifles, and shotguns. { 'smȯl ¦ärm }

small-arms ammunition [ORD] Ammunition for use in small arms. { 'smȯl ¦ärmz ¦am·yə'nish·ən }

small-arms sling [ORD] An item made of leather, webbing, or other material, designed to be attached to a carbine, mortar, rifle, rocket launcher, shotgun, or submachine gun; used as a means of carrying a small-arms weapon or to steady the weapon for firing. { 'smȯl ¦ärmz 'sliŋ }

small-bore practice [ORD] Practice in firing with small

arms using caliber-.22 ammunition instead of the standard service rounds. { 'smȯl ˌbȯr ˌprak·təs }

small calorie *See* calorie. { 'smȯl 'kal·ə·rē }

small circle [GEOD] A circle on the surface of the earth, the plane of which does not pass through the earth's center. [MATH] The intersection of a sphere with a plane that does not pass through the center of the sphere. [NAV] A planned or accomplished route of travel which is generated by the intersection of a sphere and a plane which does not pass through the center of the sphere. { 'smȯl 'sər·kəl }

small computer system interface [COMPUT SCI] An interface standard or format for personal computers that allows the connection of up to seven peripheral devices. Abbreviated SCSI (scuzzy). { 'smȯl kəmˌpyüd·ər ˌsis·təm 'in·tər·fās }

small-craft warning [METEOROL] A warning, for marine interests, of impending winds up to 28 knots (32 miles per hour or 52 kilometers per hour). { 'smȯl ˌkraft 'wȯrn·iŋ }

small-diameter blasthole [ENG] A blast hole $1\frac{1}{2}$ to 3 inches (3.8 to 7.6 centimeters) in diameter, in low-face quarries. { 'smȯl dīˌam·əd·ər 'blast·hōl }

small diurnal range [OCEANOGR] The difference in height between mean lower high water and mean higher low water. { 'smȯl dīˌərn·əl 'rānj }

small hail [METEOROL] Frozen precipitation consisting of small, semitransparent, roundish grains, each grain consisting of a snow pellet surrounded by a very thin ice covering, giving it a glazed appearance. { 'smȯl 'hāl }

small ice floe [OCEANOGR] An ice floe of sea ice 30 to 600 feet (9 to 180 meters) across. { 'smȯl 'īs ˌflō }

small intestine [ANAT] The anterior portion of the intestine in humans and other mammals; it is divided into three parts, the duodenum, the jejunum, and the ileum. { 'smȯl in'tes·tən }

small ion [METEOROL] An atmospheric ion of the type that has the greatest mobility; and hence, collectively, it is the principal agent of atmospheric conduction; evidence indicates that each ion is a singly charged atmospheric molecule (or, rarely, an atom) about which a few other neutral molecules are held by the electrical attraction of the central ionized molecule; estimates of the number of satellite molecules range as high as 12. Also known as fast ion; light ion. { 'smȯl 'īˌän }

small-ion combination [METEOROL] Either of two processes by which small ions disappear: the union of a small ion and a neutral Aitken nucleus to form a new large ion, or the neutralization of a large ion by the small ion. { 'smȯl 'īˌän ˌkäm·bə'nā·shən }

small-lot storage [IND ENG] Generally, a quantity of less than one pallet stack, stacked to maximum storage height; thus, the term refers to a lot consisting of from one container to two or more pallet loads, but is not of sufficient quantity to form a complete pallet column. { 'smȯl ˌlät 'stȯr·ij }

Small Magellanic Cloud [ASTRON] The smaller of the two star clouds near the south celestial pole; it is about 170,000 light-years away and contains a wide assortment of giant and variable stars, star clusters, and nebulae. Also known as Nubecula Minor. { 'smȯl maj·əˌlan·ik 'klaúd }

small nucleolar ribonucleic acid [MOL BIO] A type of antisense ribonucleic acid (RNA) that mediates site-specific cleavage and modification of ribosomal RNA precursors. Abbreviated snoRNA. { ˌsmȯl nüˌklē·ə·lər ˌrī·bō·nü·klē·ik 'as·əd }

small of the stock [ORD] The part of the stock of a small-arms weapon ordinarily gripped by the right hand, and located immediately behind the receiver and trigger assembly. { 'smȯl əv thə 'stäk }

small perturbation [PHYS] A disturbance imposed on a system in steady state, with amplitude assumed small of the first order; that is, the square of the amplitude is negligible in comparison with the amplitude, and the derivatives of the perturbation are assumed to be of the same order of magnitude as the perturbation. { 'smȯl ˌpərd·ər'bā·shən }

small polaron [SOLID STATE] A quasiparticle comprising a self-trapped electronic charge localized within a small region of a solid of spatial extent comparable to an interatomic dimension, and the atomic displacement pattern which produces the potential well within which the charge is bound. { 'smȯl 'pō·ləˌrän }

smallpox [MED] An acute, infectious, viral disease characterized by severe systemic involvement and a single crop of skin lesions which proceeds through macular, papular, vesicular, and pustular stages. Also known as variola. { 'smȯlˌpäks }

smallpox vaccine [IMMUNOL] A vaccine prepared from a glycerinated suspension of the exudate from cowpox vesicles obtained from healthy vaccinated calves or sheep. Also known as antismallpox vaccine; glycerinated vaccine virus; Jennerian vaccine; virus vaccinium. { 'smȯlˌpäks vakˌsēn }

small scale [GRAPHICS] A scale involving a large reduction in size, for example, 1:1,000,000; small-scale chart covers a large area. { 'smȯl 'skāl }

small-scale hydropower [MECH ENG] The generation of electricity by using hydraulic turbines in which the installed capacity of the plant lies within the range from 5 kilowatts to 5 megawatts. { 'smȯl ˌskāl 'hī·drəˌpaú·ər }

small-scale integration [ELECTR] Integration in which a complete major subsystem or system is fabricated on a single integrated-circuit chip that contains integrated circuits which have appreciably less complexity than for medium-scale integration. Abbreviated SSI. { 'smȯl ˌskāl ˌint·ə'grā·shən }

small-signal parameter [ELECTR] One of the parameters characterizing the behavior of an electronic device at small values of input, for which the device can be represented by an equivalent linear circuit. { 'smȯl ˌsig·nəl pə'ram·əd·ər }

small talk [COMPUT SCI] A high-level, user-friendly programming language that incorporates the functions of an operating system. { 'smȯl ˌtȯk }

small tropic range [OCEANOGR] The difference in height between tropic lower high water and tropic higher low water. { 'smȯl 'träp·ik 'rānj }

small-waterplane-area twin-hull ship *See* SWATH ship. { 'smȯl ˌwȯd·ərˌplān ˌer·ē·ə ˌtwin ˌhȯl 'ship }

smalt [MATER] A blue glass made by fusing silica and potash with cobalt oxides; used as a pigment for glass, ceramics, paints, and dyes. { smȯlt }

smaltite [MINERAL] $(Co,Ni)As_{3-x}$ A metallic-gray isometric mineral composed of nickel cobalt arsenide. { 'smȯlˌtīt }

smaragd *See* emerald. { 'smaˌragd }

smaragdite [MINERAL] A green amphibole mineral that is pseudomorphous after pyroxene in rocks such as eclogite. { smə'ragˌdīt }

Smarandache function [MATH] A function η defined on the integers with the property that $\eta(r)$ is the smallest integer m such that $m!$ is divisible by n. { ˌsmär·ən'dä·chē ˌfəŋk·shən }

smart bomb [ORD] A bomb equipped with sensors that can be programmed to detect or recognize distinctive features of its targets, and data transfer systems that allow information to flow between the bomb and a guidance unit for course direction and changes in flight. { 'smärt ˌbäm }

smart card [COMPUT SCI] A plastic card in which is embedded a microprocessor that is usually programmed to hold information about the card holder or user. Also known as chip card. { 'smärt ˌkärd }

smart materials [MATER] **1.** Materials that can significantly change their mechanical properties (such as shape, stiffness, and viscosity), or their thermal, optical, or electromagnetic properties, in a predictable or controllable manner in response to their environment. **2.** Materials that perform sensing and actuating functions, including piezoelectrics, electrostrictors, magnetostrictors, and shape-memory alloys. { ˌsmärt mə'tir·ē·əlz }

smart sensor [ENG] A microsensor integrated with signal-conditioning electronics such as analog-to-digital converters on a single silicon chip to form an integrated microelectromechanical component that can process information itself or communicate with an embedded microprocessor. Also known as intelligent sensor. { 'smärt 'sen·sər }

smart structures [ENG] Structures that are capable of sensing and reacting to their environment in a predictable and desired manner, through the integration of various elements, such as sensors, actuators, power sources, signal processors, and communications network. In addition to carrying mechanical loads, smart structures may alleviate vibration, reduce acoustic noise, monitor their own condition and environment, automatically perform precision alignments, or change their shape or mechanical properties on command. { ˌsmärt 'strək·chərz }

smart terminal *See* intelligent terminal. { 'smärt 'tər·mən·əl }

smart tool [CONT SYS] A robot end effector or fixed tool that uses sensors to measure the tool's position relative to reference markers or a workpiece or jig, and an actuator to

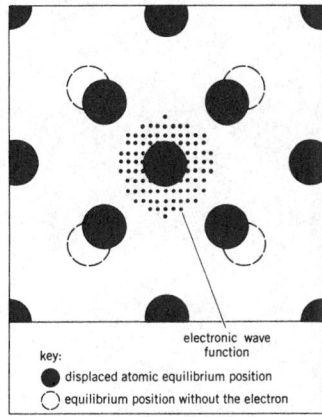

SMALL POLARON

key:
● displaced atomic equilibrium position
○ equilibrium position without the electron

electronic wave function

An electron self-trapped by the equilibrium atomic displacement around it to form a small polaron.

adjust the tool's position with respect to the workpiece. { 'smärt ,tül }

S matrix *See* scattering matrix. { 'es ,mā·triks }

S-matrix theory [PART PHYS] A theory of elementary particles based on the scattering matrix, and on its properties such as unitarity and analyticity. Also known as scattering-matrix theory. { 'es ,mā·triks 'thē·ə·rē }

SMATV system *See* satellite master antenna television system. { ¦es,em,ā,tē'vē ,sis·təm }

SMC X-1 [ASTRON] The most luminous x-ray pulsar known, located in the small Magellanic Cloud.

smear [BIOL] A preparation for microscopic examination made by spreading a drop of fluid, such as blood, across a slide and using the edge of another slide to leave a uniform film. [ELECTR] A television-picture defect in which objects appear to be extended horizontally beyond their normal boundaries in a blurred or smeared manner; one cause is excessive attenuation of high video frequencies in the television receiver. { smir }

smectic-A [PHYS CHEM] A subclass of smectic liquid crystals in which molecules are free to move within layers and are oriented perpendicular to the layers. { 'smek·dik 'ā }

smectic-B [PHYS CHEM] A subclass of smectic liquid crystals in which molecules in each layer are arranged in a close-packed lattice and are oriented perpendicular to the layers. { 'smek·dik 'bē }

smectic-C [PHYS CHEM] A subclass of smectic liquid crystals in which molecules are free to move within layers and are oriented with their axes tilted with respect to the normal to the layers. { 'smek·dik 'sē }

smectic phase [PHYS CHEM] A form of the liquid crystal (mesomorphic) state in which molecules are arranged in layers that are free to glide over each other with relatively small viscosity. { 'smek·dik ,fāz }

smectite [MINERAL] Dioctahedral (montmorillonite) and trioctahedral (saponite) clay minerals, and their chemical varieties characterized by swelling properties and high cation-exchange capacities. { 'smek,tīt }

smectogenic solid [PHYS CHEM] A solid which will form a smectic liquid crystal when heated. { ¦smek·tə¦jen·ik 'säl·əd }

smegma [PHYSIO] The sebaceous secretion that accumulates around the glans penis and the clitoris. { 'smeg·mə }

smell [PHYSIO] To perceive by olfaction. { smel }

smelling salts *See* aromatic spirits of ammonia. { 'smel·iŋ ,sóls }

smell prism [PSYCH] A diagram that attempts to systematize an odor classification system that has six main odor qualities: fruity, flowery, resinous, spicy, foul, and burnt. { 'smel ,priz·əm }

smell the bottom *See* feel the bottom. { smel thə 'bäd·əm }

smelter [MET] A furnace used for smelting. { 'smel·tər }

smelting [MET] The heating of ore mixtures accompanied by a chemical change resulting in liquid metal. { 'smelt·iŋ }

S meter *See* signal-strength meter. { 'es ,mēd·ər }

Smilacaceae [BOT] A family of monocotyledonous plants in the order Liliales; members are usually climbing, leafy-stemmed plants with tendrils, trimerous flowers, and a superior ovary. { ,smī·lə'kās·ē,ē }

smiley *See* emoticon. { 'smīl·ē }

Sminthuridae [INV ZOO] A family of insects in the order Collembola which have simple tracheal systems. { smin'-thyùr·ə,dē }

Smith-Baker microscope [OPTICS] A type of interference microscope in which a beam of polarized light is split by a birefringent calcite plate cemented to the front lens of the condenser, and reunited by another such plate cemented to the objective. { 'smith 'bā·kər 'mī·krə,skōp }

Smith chart [ELECTROMAG] A special polar diagram containing constant-resistance circles, constant-reactance circles, circles of constant standing-wave ratio, and radius lines representing constant line-angle loci; used in solving transmission-line and waveguide problems. { 'smith ,chärt }

Smithell's burner [ENG] Two concentric tubes that can be added to a bunsen burner to separate the inner and outer flame cones. { 'smith·əlz ,bər·nər }

Smitherm process [FOOD ENG] A direct heating process employing fat as the heating medium and used for deep-fat cooking of potato chips. { 'smi,thərm ,prä·səs }

smith forging [MET] Manual forging of small, hot metal

parts with flat or simple-shaped dies, as with a hammer and anvil. { 'smith ,fòrj·iŋ }

Smith-Helmholtz law [OPTICS] For a single refracting surface of sufficiently small aperture, the product of the index of refraction, distance from the optical axis, and the angle which a light ray makes with the optical axis at the object point is equal to the corresponding product at the image point. { 'smith 'helm,hōlts ,lò }

smithite [MINERAL] $AgAsS_2$ A red monoclinic mineral composed of silver arsenic sulfide and occurring as small crystals. { 'smi,thīt }

Smith-McIntyre sampler [MECH ENG] A device for taking samples of sediment from the ocean bottom; the digging and hoisting mechanisms are independent: the digging bucket is forced into the sediment before the hoisting action occurs. { 'smith 'mak·ən,tīr ,sam·plər }

Smith-Petersen nail [MED] A three-flanged nail used to fix fractures of the neck of the femur; it is inserted from just below the greater trochanter, through the neck, and into the head of the femur. { 'smith 'pēd·ər·sən ,nāl }

Smith-Purcell-Salisbury effect [PHYS] The emission of nearly coherent radiation when a beam of relativistic electrons strikes a metallic diffraction grating at grazing incidence. { ¦smith ¦pər·səl 'sólz·brē i,fekt }

Smith-Recklinghausen's disease *See* neurofibromatosis. { 'smith 'rek·liŋ,haúz·ənz di,zēz }

smithsonite [MINERAL] $ZnCO_3$ White, yellow, gray, brown, or green secondary carbonate mineral associated with sphalerite and commonly reniform, botryoidal, stalactitic, or granular; hardness is 5 on Mohs scale, and specific gravity is 4.30–4.45; it is an ore of zinc. Also known as calamine; dry-bone ore; zsaskaite; zinc spar. { 'smith·sə,nīt }

smog [METEOROL] Air pollution consisting of smoke and fog. { smäg }

smoke [ENG] Dispersions of finely divided (0.01–5.0 micrometers) solids or liquids in a gaseous medium. { smōk }

smokebox [MECH ENG] A chamber external to a boiler for trapping the unburned products of combustion. { 'smōk,bäks }

smoke candle [ORD] Munition which produces smoke by vaporizing an oil. { 'smōk ,kand·əl }

smoke chamber [ENG] That area in a fireplace directly above the smoke shelf. { 'smōk ,chām·bər }

smoke detector [ENG] A photoelectric system for an alarm when smoke in a chimney or other location exceeds a predetermined density. { 'smōk di,tek·tər }

smoke grenade [ORD] Hand grenade or rifle grenade containing a smoke-producing mixture, used for screening or signaling; sometimes charged with colored smoke such as red, green, yellow, or violet. { 'smōk gri,nād }

smoke horizon [METEOROL] The top of a smoke layer which is confined by a low-level temperature inversion in such a way as to give the appearance of the horizon when viewed from above against the sky; in such instances the true horizon is usually obscured by the smoke layer. { 'smōk hə,rīz·ən }

smokehouse [FOOD ENG] A building with cabinets generally built of sanitary stainless steel with excellent temperature and humidity control, in which volatilization and redeposition on food of certain components of hardwoods from burning of hardwood sawdust are accomplished. { 'smōk,haús }

smokeless [ORD] Indicating that ammunition is relatively smokeless when used in the weapon for which it is intended. { 'smōk·ləs }

smokeless powder [MATER] Nitrocellulose containing 13.1 percent nitrogen with small amounts of stabilizers (amines) and plasticizers usually present, as well as various modifying agents (nitrotoluene and nitroglycerin salts); used in ammunition. { 'smōk·ləs 'paúd·ər }

smoke point [ENG] The maximum flame height in millimeters at which kerosine will burn without smoking, tested under standard conditions; used as a measure of the burning cleanliness of jet fuel and kerosine. { 'smōk ,póint }

smoke pot [ORD] A cylindrical metal munition designed to produce smoke for screening or signaling purposes, either by combustion of a smoke-producing mixture, or by combustion of a fuel mixture to vaporize a smoke-producing oil; it may be with or without an igniting device and a filling, and it is not intended for throwing or for firing from weapons. { 'smōk ,pät }

smoke printing [GRAPHICS] A photocopy process in which electrically charged particles are deposited on paper or other material; the paper is held behind a sheet of glass which is backed with a thin metallic coating; the mist of particles is dispensed from behind the paper by an electrode which gives it a charge; a positive or negative print can be made. { 'smōk ˌprint·iŋ }

smoke projectile [ORD] Any projectile containing a smoke-producing chemical agent with means for properly dispersing the agent. { 'smōk prə¸jek·təl }

smoke rocket [ORD] A rocket with a warhead which contains material to produce a smoke cloud on functioning. { 'smōk ˌräk·ət }

smokes [METEOROL] Dense white haze and dust clouds common in the dry season on the Guinea coast of Africa, particularly at the approach of the harmattan. { smōks }

smoke screen [ORD] A screen of smoke used to hide a maneuver, force, place, or activity. { 'smōk ˌskrēn }

smoke shelf [ENG] A horizontal surface directly behind the throat of a fireplace to prevent downdrafts. { 'smōk ˌshelf }

smoke signal [ORD] A pyrotechnic item designed to produce a sign by means of smoke and used, for example, to provide identification, location, or warning. { 'smōk ˌsig·nəl }

smokestack [ENG] A chimney for the discharge of flue gases from a furnace operation such as in a steam boiler, powerhouse, heating plant, ship, locomotive, or foundry. { 'smōk¸stak }

smokestone See smoky quartz. { 'smōk¸stōn }

smoke technique [FL MECH] A technique used to measure only very-low-speed air velocity; smoke enables the fluid motion to be observed with the eye, and the smoke is timed over a measured distance along an airway of constant cross section to determine the velocity of flow. { 'smōk tek¸nēk }

smoke test [ENG] A test used on kerosine to determine the highest point to which the flame can be turned before smoking occurs. { 'smōk ˌtest }

smoke washer [ENG] A device for removing particles from smoke by forcing it through a spray of water. { 'smōk ˌwäsh·ər }

smoke-wire method [FL MECH] A method of flow visualization in which a thin stainless-steel wire coated with a thin film of kerosine is heated electrically to generate smoke, resulting in a series of time lines. { ¦smōk ¦wīr 'meth·əd }

smoky quartz [MINERAL] A smoky-yellow, smoky-brown, or brownish-gray, often transparent variety of crystalline quartz containing inclusions of carbon dioxide; may be used as a semiprecious stone. Also known as cairngorm; smokestone. { 'smōk·ē 'kwȯrts }

smoldering [CHEM] Combustion of a solid without a flame, often with emission of smoke. { 'smōl·driŋ }

S Monel [MET] An alloy similar to H Monel but containing 4% silicon. { 'es mō'nel }

smooth [OCEANOGR] Comparatively calm water between heavy seas. [STAT] To modify a sequential set of numerical data items in a manner designed to reduce the differences in value between adjacent items. { smüth }

smooth blasting [ENG] Blasting to ensure even faces without cracks in the rock. { 'smüth 'blast·iŋ }

smooth bore [ORD] A bore that is smooth and without rifling, for example, in shotguns and mortars. { 'smüth 'bȯr }

smooth chert [GEOL] A hard, dense, homogeneous chert (insoluble residue) characterized by a conchoidal-to-even fracture surface that is devoid of roughness and by a lack of crystallinity, granularity, or other distinctive structure. { 'smüth 'chərt }

smooth curve [MATH] The range of a function from a closed interval to a Euclidean space such that each of the Cartesian coordinates of the image point is a continuously differentiable function on the closed interval. { 'smüth 'kərv }

smooth drilling [ENG] Drilling in a rock formation in which a fast rotation of the drill stem, a fast rate of penetration, and a high recovery of core can be achieved with vibration-free rotation of the drill stem. { 'smüth 'dril·iŋ }

smoothed data [STAT] Information that has been averaged or processed with a curve-fitting algorithm so that curves that are free from singularities result when the data are plotted on a graph. { 'smüthd 'dad·ə }

smoothing [ENG] Making a level, or continuously even, surface. [MATH] Approximating or perturbing a function by one which has a higher degree of differentiability. [STAT] A process that uses either freehand methods, moving averages, or fitting a curve by least squares method to remove fluctuations in the data in a time series. { 'smüth·iŋ }

smoothing choke [ELECTR] Iron-core choke coil employed as a filter to remove fluctuations in the output current of a vacuum-tube rectifier or direct-current generator. { 'smüth·iŋ ˌchōk }

smoothing circuit See ripple filter. { 'smüth·iŋ ˌsər·kət }

smoothing filter See ripple filter. { 'smüth·iŋ ˌfil·tər }

smoothing mill [MECH ENG] A revolving stone wheel used to cut and bevel glass or stone. { 'smüth·iŋ ˌmil }

smoothing plane [DES ENG] A finely set hand tool, usually 5.5–10 inches (14–25.4 centimeters) long, for finishing small areas on wood. { 'smüth·iŋ ˌplān }

smooth manifold [MATH] A differentiable manifold whose local coordinate systems depend upon those of Euclidean space in an infinitely differentiable manner. { 'smüth 'man·ə¸fōld }

smooth map [MATH] An infinitely differentiable function. { 'smüth 'map }

smooth muscle [ANAT] The involuntary muscle tissue found in the walls of viscera and blood vessels, consisting of smooth muscle fibers. { 'smüth 'məs·əl }

smooth muscle fiber [HISTOL] Any of the elongated, nucleated, spindle-shaped cells comprising smooth muscles. Also known as involuntary fiber; nonstriated fiber; unstriated fiber. { 'smüth 'məs·əl ˌfī·bər }

smoothness conditions [FL MECH] Two conditions that must be satisfied by bodies studied in slender body theory: (1) the rate of change of the angle between the tangent plane to the body and the direction of the motion, evaluated along this direction, must be small and of the same order as the ratio of the maximum thickness of the body to its length; (2) the curvature of any section of the body in a plane normal to the direction of motion must be of the same order as the reciprocal of the maximum diameter of the section, at all points where the section is convex outward. { 'smüth·nəs kən¸dish·ənz }

smooth phase [GEOL] The part of stream traction whereby a mass of sediment travels as a sheet with gradually increasing density from the surface downward. { 'smüth ˌfāz }

smooth sea [OCEANOGR] Sea with waves no higher than ripples or small wavelets. { 'smüth 'sē }

smooth surface [MATH] A surface which has a tangent plane at each point, and for which the direction of the normal to this plane is a continuous function of the point of tangency. { ¦smüth 'sər·fəs }

smothered bottom [GEOL] A sedimentary surface on which complete, well-preserved, and commonly very fragile and delicate fossils were saved by an influx of mud that buried them instantly. { 'sməth·ərd 'bäd·əm }

smother kiln [ENG] A kiln into which smoke can be introduced for blackening pottery. { 'sməth·ər ˌkil }

SMOW See standard mean ocean water. { smaúw or ¦es¦em ¦ō'dät·əl¸yü }

SMPT See Simple Mail Transfer Protocol.

smudge [PL PATH] Any of several fungus diseases of cereals and other plants characterized by dark, sooty discolorations. { sməj }

smudge oil [MATER] A dark petroleum distillate or gas oil that is burned in citrus fruit-growing areas to prevent frost damage. { 'sməj ˌȯil }

smudging [ENG] A frost-preventive measure used in orchards; properly, it means the production of heavy smoke, supposed to prevent radiational cooling, but it is generally applied to both heating and smoke production. { 'sməj·iŋ }

smut [MET] A reaction product left on the surface of a metal after pickling. [PL PATH] Any of various destructive fungus diseases of cereals and other plants characterized by large dusty masses of dark spores on the plant organs. { smət }

smut fungus [MYCOL] The common name for members of the Ustilaginales. { 'smət ˌfəŋ·gəs }

Sn See tin.

S/N See signal-to-noise ratio.

snag [FOR] A standing dead tree. { snag }

snagging [MECH ENG] Removing surplus metal or large surface defects by using a grinding wheel. { 'snag·iŋ }

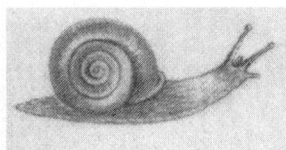

The white-lipped sand snail (*Triodopsis albalabris*).

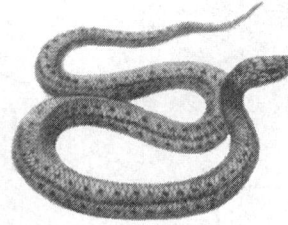

Garter snake (*Thomnophis ordinoides*).

snail [INV ZOO] Any of a large number of gastropod mollusks distinguished by a spiral shell that encloses the body, a head, a foot, and a mantle. { snāl }

snail fever *See* schistosomiasis. { 'snāl ,fē·vər }

snake [MET] **1.** A twisted and bent hod rod formed before subsequent rolling operations. **2.** A flexible mandrel used to prevent collapse of a shaped piece during bending operations. [VERT ZOO] Any of about 3000 species of reptiles which belong to the 13 living families composing the suborder Serpentes in the order Squamata. { snāk }

snake hole [ENG] **1.** A blasting hole bored directly under a boulder. **2.** A drill hole used in quarrying or bench blasting. { 'snāk ,hōl }

snaking [ENG] Towing a load with a long cable. { 'snāk·iŋ }

snaking stream *See* meandering stream. { 'snāk·iŋ ,strēm }

snap [METEOROL] A brief period of extreme (generally cold) weather setting in suddenly, as in a "cold snap." { snap }

SNAP [NUCLEO] A small nuclear power plant in which heat from radioisotope decay in a fuel such as strontium-90 is converted into electric energy, to provide power for spacecraft instrumentation, telemetry, and other applications. Derived from systems for nuclear auxiliary power. { snap }

snap-action switch [ELEC] A switch that responds to very small movements of its actuating button or lever and changes rapidly and positively from one contact position to the other; the trademark of one version is Micro Switch. Also known as sensitive switch. { 'snap ,ak·shən ,swich }

snap-back forming [ENG] A plastic-sheet-forming technique in which an extended, heated, plastic sheet is allowed to contract over a form shaped to the desired final contour. { 'snap ,bak ,form·iŋ }

snapback method *See* repetitive time method. { 'snap,bak ,meth·əd }

snap fastener [DES ENG] A fastener consisting of a ball on one edge of an article that fits in a socket on an opposed edge, and used to hold edges together, such as those of a garment. { 'snap ,fas·ən·ər }

snap flask [MET] A foundry flask having its sides latched on one corner to allow removal of the flask from around the sand mold. { 'snap ,flask }

snap gage [DES ENG] A device with two flat, parallel surfaces spaced to control one limit of tolerance of an outside diameter or a length. { 'snap ,gāj }

snap hook *See* spring hook. { 'snap ,huk }

snap-off diode [ELECTR] Planar epitaxial passivated silicon diode that is processed so a charge is stored close to the junction when the diode is conducting; when reverse voltage is applied, the stored charge then forces the diode to snap off or switch rapidly to its blocking state. { 'snap,of 'dī,ōd }

snap-on ammeter [ELEC] An ac ammeter having a magnetic core in the form of hinged jaws that can be snapped around the current-carrying wire. Also known as clamp-on ammeter. { 'snap,on 'am,ēd·ər }

snapper [ENG] A device for collecting samples from the ocean bottom, and which closes to prevent the sample from dropping out as it is raised to the surface. { 'snap·ər }

snapping finger *See* trigger finger. { 'snap·iŋ ,fiŋ·gər }

snap ring [DES ENG] A form of spring used as a fastener; the ring is elastically deformed, put in place, and allowed to snap back toward its unstressed position into a groove or recess. { 'snap ,riŋ }

snapshot [COMPUT SCI] The storing of the entire contents of the memory, including status indicators and hardware registers. { 'snap,shät }

snapshot dump [COMPUT SCI] An edited printout of selected parts of the contents of main memory, performed at one or more times during the execution of a program without materially affecting the operation of the program. { 'snap ,shät ,dəmp }

snapshot program [COMPUT SCI] A program that provides dumps of certain portions of memory when certain instructions are executed or when certain conditions are fulfilled. { 'snap ,shät ,prō·grəm }

snatch block [DES ENG] A pulley frame or sheave with an eye through which lashing can be passed to fasten it to a scaffold or pole. { 'snach ,bläk }

snatch plate [ENG] A thick steel plate through which a hole about one-sixteenth of an inch larger than the outside diameter

of the drill rod on which it is to be used is drilled; the plate is slipped over the drill rod and one edge is fastened to a securely anchored chain, and if rods must be pulled because high-pressure water is encountered, the eccentric pull of the chain causes the outside of the rods to be gripped and held against the pressure of water; the rod is moved a short distance out of the hole each time the plate is tapped. { 'snach ,plāt }

SNC group [GEOL] A group of meteorites comprising the shergottites, nakhlites, and chassignites, which are all believed to have originated from Mars. { ¦es¦en'sē ,grüp }

S-N diagram [ENG] In fatigue testing, a graphic representation of the relationship of stress *S* and the number of cycles *N* before failure of the material. { ¦es¦en 'dī·ə,gram }

sneak path [COMPUT SCI] In computers, an undesired circuit through a series-parallel configuration. { 'snēk ,path }

sneeze [PHYSIO] A sudden, noisy, spasmodic expiration through the mouth and nose. { snēz }

Snellen test [MED] A test for visual acuity presenting letters, numbers, or letter E's in various positions, with the symbols varying in size; if the smallest are read at a distance of 20 feet (6 meters), vision is recorded as 20/20, or normal. { 'snel·ən ,test }

Snell laws of refraction [OPTICS] When light travels from one medium into another the incident and refracted rays lie in one plane with the normal to the surface; are on opposite sides of the normal; and make angles with the normal whose sines have a constant ratio to one another. Also known as Descartes laws of refraction; laws of refraction. { 'snel 'lóz əv ri'frak·shən }

snezhura *See* snow slush. { 'snezh·ə·rə }

snifter valve [ENG] A valve on a pump that allows air to enter or escape, and accumulated water to be released. { 'snif·tər ,valv }

sniperscope [ORD] A snooperscope which, when combined with a carbine or other firearm, enables the operator to see and shoot at targets in the dark. { 'snī·pər,skōp }

snivet [ELECTR] Straight, jagged, or broken vertical black line appearing near the right-hand edge of a television receiver screen. { 'sniv·ət }

sno *See* elvegust. { snó }

SNOBOL [COMPUT SCI] A computer programming language that has significant applications in program compilation and generation of symbolic equations. Derived from String-Oriented-Symbolic-Language. { 'snō,ból }

Snoek effect [SOLID STATE] The preferential occupation by carbon impurity atoms of sites on one of the three faces in the cubic lattice of iron. { 'snúk i,fekt }

SNOM *See* near-field scanning optical microscopy.

snooperscope [ELECTR] An infrared source, an infrared image converter, and a battery-operated high-voltage direct-current source constructed in portable form to permit a foot soldier or other user to see objects in total darkness; infrared radiation sent out by the infrared source is reflected back to the snooperscope and converted into a visible image on the fluorescent screen of the image tube. { 'snüp·ər,skōp }

snorkel [ENG] Any tube which supplies air for an underwater operation, whether it be for material or personnel. [NAV ARCH] A tube or pair of tubes for air intake and exhaust that can be extended above the surface of the water for operating submerged submarines. { 'snór·kəl }

snout [GEOGR] A promontory or protruding mass of rock. [HYD] The protruding lower extremity of a glacier. [VERT ZOO] The elongated nose of various mammals. { snaút }

snow [ELECTR] Small, random, white spots produced on a television or radar screen by inherent noise signals originating in the receiver. [METEOROL] The most common form of frozen precipitation, usually flakes of starlike crystals, matted ice needles, or combinations, and often rime-coated. { snō }

snow accumulation [METEOROL] The actual depth of snow on the ground at any instant during a storm, or after any single snowstorm or series of snowstorms. { 'snō ə,kyù·myə,lā·shən }

snow avalanche [HYD] An avalanche of relatively pure snow; some rock and earth material may also be carried downward. Also known as snowslide. { 'snō ,av·ə,lanch }

snowbank glacier *See* nivation glacier. { 'snō,baŋk ,glā·shər }

snow banner [METEOROL] Snow being blown from a mountain crest. Also known as snow plume; snow smoke. { 'snō ¦ban·ər }

snow barchan [HYD] A crescentic or horseshoe-shaped snow dune of windblown snow with the ends pointing downwind. Also known as snow medano. { 'snō bär'kän }

snow bin [ENG] A box for measuring the amount of snowfall; a type of snow gage. { 'snō ¦bin }

snow blindness [MED] A transient visual impairment and actinic keratoconjunctivitis caused by exposure of the eyes to ultraviolet rays reflected from snow. Also known as solar photophthalmia. { 'snō ¦blīnd·nəs }

snow blink [METEOROL] A bright, white glare on the underside of clouds, produced by the reflection of light from a snow-covered surface; this term is used in polar regions with reference to the sky map. Also known as snow sky. { 'snō ¦bliŋk }

snow blower [MECH ENG] A machine that removes snow from a road surface or pavement using a screw-type blade to push the snow into the machine and from which it is ejected at some distance. { 'snō ¦blō·ər }

snowbreak [CIV ENG] Any barrier designed to shelter an object or area from snow. { 'snō¦brāk }

snowbridge [HYD] Snow bridging a crevasse in a glacier. { 'snō¦brij }

snow cap [HYD] **1.** Snow covering a mountain peak when no snow exists at lower elevations. **2.** Snow on the surface of a frozen lake. { 'snō ¦kap }

snow climate See polar climate. { 'snō ¦klī·mət }

snow cloud [METEOROL] A popular term for any cloud from which snow falls. { 'snō ¦klaůd }

snow concrete [HYD] Snow that is compacted at low temperatures by heavy objects (as by a vehicle) and that sets into a tough substance of considerably greater strength than uncompressed snow. Also known as snowcrete. { 'snō ¦kän¦krēt }

snow course [HYD] An established line, usually from several hundred feet to as much as a mile long, traversing representative terrain in a mountainous region of appreciable snow accumulation; along this course, measurements of snow cover are made to determine its water equivalent. { 'snō ¦kȯrs }

snow cover [HYD] **1.** All accumulated snow on the ground, including that derived from snowfall, snowslides, and drifting snow. Also known as snow mantle. **2.** The extent, expressed as a percentage, of snow cover in a particular area. { 'snō ¦kəv·ər }

snow-cover chart [METEOROL] A synoptic chart showing areas covered by snow and contour lines of snow depth. { 'snō ¦kəv·ər ¦chärt }

snowcreep [HYD] The slow internal deformation of a snowpack resulting from the stress of its own weight and metamorphism of snow crystals. { 'snō¦krēp }

snowcrete See snow concrete. { 'snō¦krēt }

snow crust [HYD] A crisp, firm, outer surface upon snow. { 'snō ¦krəst }

snow crystal [METEOROL] Any of several types of ice crystal found in snow; a snow crystal is a single crystal, in contrast to a snowflake which is usually an aggregate of many single snow crystals. { 'snō ¦krist·əl }

snow cushion [HYD] An accumulation of snow, commonly deep, soft, and unstable, deposited in the lee of a cornice on a steep mountain slope. { 'snō ¦kůsh·ən }

snow density [HYD] The ratio of the volume of meltwater that can be derived from a sample of snow to the original volume of the sample; strictly speaking, this is the specific gravity of the snow sample. { 'snō ¦den·səd·ē }

snowdrift [HYD] Snow deposited on the lee of obstacles, lodged in irregularities of a surface, or collected in heaps by eddies in the wind. { 'snō¦drift }

snowdrift glacier [HYD] A semipermanent mass of firn, formed by drifted snow in depressions in the ground or behind obstructions. Also known as catchment glacier; drift glacier. { 'snō¦drift ¦glā·shər }

snowdrift ice [HYD] Permanent or semipermanent masses of ice, formed by the accumulation of drifted snow in the lee of projections, or in depressions of the ground. Also known as glacieret. { 'snō¦drift ¦īs }

snow dune [HYD] An accumulation of wind-transported snow resembling the forms of sand dunes. { 'snō ¦dün }

snow dust [METEOROL] Fine snow crystals fragmented or driven by the wind. { 'snō ¦dəst }

snow eater [METEOROL] **1.** Any warm wind blowing over a snow surface; usually applied to a foehn wind. **2.** A fog over a snow surface; so called because of the frequently observed rapidity with which a snow cover disappears after a fog sets in. { 'snō ¦ēd·ər }

snowfall [METEOROL] **1.** The rate at which snow falls; in surface weather observations, this is usually expressed as inches of snow depth per 6-hour period. **2.** A snow storm. { 'snō¦fȯl }

snow fence [CIV ENG] An open-slatted board fence usually 4 to 10 feet (1.2 to 3.0 meters) high, placed about 50 feet (15 meters) on the windward side of a railroad track or highway; the fence serves to disrupt the flow of the wind so that the snow is deposited close to the fence on the leeward side, leaving a comparatively clear, protected strip parallel to the fence and slightly farther downwind. { 'snō ¦fens }

snowfield [HYD] **1.** A broad, level, relatively smooth and uniform snow cover on ground or ice at high altitudes or in mountainous regions above the snow line. **2.** The accumulation area of a glacier. **3.** A small glacier or accumulation of perennial ice and snow too small to be designated a glacier. { 'snō ¦fēld }

snowflake [MET] See flake. [METEOROL] An ice crystal or, much more commonly, an aggregation of many crystals which falls from a cloud; simple snowflakes (single crystal) exhibit beautiful variety of form, but the symmetrical shapes reproduced so often in photomicrographs are not actually found frequently in snowfalls; broken single crystals, fragments, or clusters of such elements are much more typical of actual snows. { 'snō¦flāk }

snowflake obsidian [PETR] An obsidian that contains white, gray, or reddish spherulites ranging in size from microscopic to a meter or more in diameter. { 'snō¦flāk äb'sid·ē·ən }

snow flurry [METEOROL] Popular term for snow shower, particularly of a very light and brief nature. { 'snō ¦flər·ē }

snowflush [GEOL] An accumulation of drifted snow, wind-blown soil, and wind-transported seeds on a lee slope, characteristically marked during the winter by a dark patch of soil. { 'snō¦fləsh }

snow forest climate [CLIMATOL] A major category in W. Köppen's climatic classification, defined by a coldest-month mean temperature of less than 26.6°F (3°C) and a warmest-month mean temperature of greater than 50°F (10°C). { 'snō ¦fär·əst ¦klī·mət }

snow fungus See Tremella fuciformis. { 'snō ¦fəŋ·gəs }

snow gage [HYD] An instrument for measuring the amount of water equivalent in a snowpack. Also known as snow sampler. { 'snō ¦gāj }

snow garland [HYD] A rare phenomenon in which snow is festooned from trees, fences, and so on, in the form of a rope of snow, several feet long and several inches in diameter; produced by surface tension acting in thin films of water bonding individual crystals; such garlands form only when the surface temperature is close to the melting point, for only then will the requisite films of slightly supercooled water exist. { 'snō ¦gär·lənd }

snow geyser [METEOROL] Fine, powdery snow blown upward by a snow tremor. { 'snō ¦gī·zər }

snow glide [HYD] The slow slip of a snowpack over the ground surface caused by the stress of its own weight. { 'snō ¦glīd }

snow grains [METEOROL] Precipitation in the form of very small, white opaque particles of ice; the solid equivalent of drizzle; the grains resemble snow pellets in external appearance, but are more flattened and elongated, and generally have diameters of less than 1 millimeter; they neither shatter nor bounce when they hit a hard surface. Also known as granular snow. { 'snō ¦grānz }

snow ice [HYD] Ice crust formed from snow, either by compaction or by the refreezing of partially thawed snow. { 'snō ¦īs }

snow line [GEOGR] **1.** A transient line delineating a snow-covered area or altitude. **2.** An area with more than 50% snow cover. **3.** The altitude or geographic line separating areas in which snow melts in summer from areas having perennial ice and snow. { 'snō ¦līn }

snow load [CIV ENG] The unit weight factor considered in the design of a flat or pitched roof for the probable amount of snow lying upon it. { 'snō ,lōd }

snow mantle *See* snow cover. { 'snō ,mant·əl }

snow mat [ENG] A device used to mark the surface between old and new snow, consisting of a piece of white duck 28 inches (71 centimeters) square, having in each corner triangular pockets in which are inserted slats placed diagonally to keep the mat taut and flat. { 'snō ,mat }

snow medano *See* snow barchan. { 'snō mə'dä·nō }

snowmelt [HYD] The water resulting from the melting of snow; it may evaporate, seep into the ground, or become a part of runoff. { 'snō,melt }

snow-melting system [CIV ENG] A system of pipes containing a circulating nonfreezing liquid or electric-heating cables, embedded beneath the surface of a road, walkway, or other area to be protected from snow accumulation. { 'snō ,melt·iŋ ,sis·təm }

snow niche *See* nivation hollow. { 'snō ,nich }

snowpack [HYD] The amount of annual accumulation of snow at higher elevations in the western United States, usually expressed in terms of average water equivalent. { 'snō,pak }

snow patch erosion *See* nivation. { 'snō ,pach i,rō·zhən }

snow pellets [METEOROL] Precipitation consisting of white, opaque, approximately round (sometimes conical) ice particles which have a snowlike structure and are about 2 to 5 millimeters in diameter; snow pellets are crisp and easily crushed, differing in this respect from snow grains, and they rebound when they fall on a hard surface and often break up. Also known as graupel; soft hail; tapioca snow. { 'snō ,pel·əts }

snow pillow [ENG] A device used to record the changing weight of the snow cover at a point, consisting of a fluid-filled bladder lying on the ground with a pressure transducer or a vertical pipe and float connected to it. { 'snō ,pil·ō }

snowplow [MECH ENG] A device for clearing away snow, as from a road or railway track. { 'snō,plaů }

snow plume *See* snow banner. { 'snō ,plüm }

snow point [PHYS CHEM] Referring to a gas mixture, the temperature at which the vapor pressure of the sublimable component is equal to the actual partial pressure of that component in the gas mixture; analogous to dew point. { 'snō ,pȯint }

snowquake *See* snow tremor. { 'snō,kwāk }

snow resistograph [ENG] An instrument for recording a hardness profile of a snow cover by recording the force required to move a blade up through the snow. { 'snō ri'zis·tə,graf }

snow ripple *See* wind ripple. { 'snō ,rip·əl }

snow roller [HYD] A mass of snow, shaped somewhat like a lady's muff, rather common in mountainous or hilly regions; it occurs when snow, moist enough to be cohesive, is picked up by wind blowing down a slope and rolled onward and downward until either it becomes too large or the ground levels off too much for the wind to propel it further; snow rollers vary in size from very small cylinders to some as large as 4 feet (1.2 meters) long and 7 feet (2.1 meters) in circumference. { 'snō ,rō·lər }

snow sampler [ENG] A hollow tube for collecting a sample of snow in place. Also known as snow tube. [HYD] *See* snow gage. { 'snō ,sam·plər }

snow scale *See* snow stake. { 'snō ,skāl }

snowshed [CIV ENG] A structure to protect an exposed area as a road or rail line from snow. [HYD] A drainage basin primarily supplied by snowmelt. { 'snō,shed }

snow sky *See* snow blink. { 'snō ,skī }

snowslide *See* snow avalanche. { 'snō,slīd }

snow sludge [OCEANOGR] Sludge formed mainly from snow. { 'snō ,sləj }

snow slush [HYD] Slush formed from snow that has fallen into water that is at a temperature below that of the snow. Also known as snezhura. { 'snō ,sləsh }

snow smoke *See* snow banner. { 'snō ,smōk }

snow stage [METEOROL] The thermodynamic process of sublimation of water vapor into snow in an idealized saturation-adiabatic or pseudoadiabatic expansion (lifting) of moist air; the snow stage begins at the condensation level when it is higher than the freezing level. { 'snō ,stāj }

snow stake [ENG] A wood scale, calibrated in inches, used in regions of deep snow to measure its depth; it is bolted to a wood post or angle iron set in the ground. Also known as snow scale. { 'snō ,stāk }

snow static [ELECTROMAG] Precipitation static caused by falling snow. { 'snō ,stad·ik }

snowstorm [METEOROL] A storm in which snow falls. { 'snō,stȯrm }

snow survey [HYD] The process of determining depth and water content of snow at representative points, for example, along a snow course. { 'snō ,sər,vā }

snow swamp *See* slush field. { 'snō ,swämp }

snow tremor [HYD] A disturbance in a snowfield, caused by the simultaneous settling of a large area of thick snow crust or surface layer. Also known as snowquake. { 'snō ,trem·ər }

snow tube *See* snow sampler. { 'snō ,tüb }

SNP *See* single nucleotide polymorphism. { 'es'en'pē }

SNR *See* signal-to-noise ratio; supernova remnant.

SNU *See* solar neutrino unit.

snub [MIN ENG] **1.** To increase the height of an undercut by means of explosives or otherwise. **2.** To check the descent of a car by the turn of a rope around a post. { snəb }

snubber [MECH ENG] A mechanical device consisting essentially of a drum, spring, and friction band, connected between axle and frame, in order to slow the recoil of the spring and reduce jolting. { 'snəb·ər }

snuffles [MED] Discharge from the nasal mucosa in congenital syphilis in infants. { 'snəf·əlz }

Snyder sampler [ENG] A mechanical device for obtaining small representative quantities from a moving stream of pulverized or granulated solids; it consists of a cast-iron plate revolving in a vertical plane on a horizontal axis with an inclined sample spout; the material to be sampled comes to the sampler by way of an inclined chute whenever the sample spout comes in line with the moving stream. { 'snī·dər 'sam·plər }

soak cleaning [MET] Cleaning the surface of a metal by immersion in a cleaning solution without electrolysis. { 'sōk ,klēn·iŋ }

soaking [MET] Heating an alloy, usually an ingot, to a temperature not far below its melting temperature and holding it there for a long time to eliminate segregation that occurred on solidification. { 'sōk·iŋ }

soaking drum [CHEM ENG] A heated petroleum-refinery process vessel used in connection with petroleum thermal-cracking coils to furnish the residence time needed to complete the cracking reaction. { 'sōk·iŋ ,drəm }

soaking pit [MET] A high-temperature, gas-fired, tightly covered, refractory-lined hole or pit into which a hot metal ingot (with liquid interior) is held at a fixed temperature until needed for rolling into sheet or other forms. { 'sōk·iŋ ,pit }

soap [MATER] **1.** A particular type of detergent, in which the water-solubilizing group is a carboxylate, COO−, and the positive ion is usually sodium, Na⁺, or potassium, K⁺. **2.** A soap compound mixed with a fragrance and other ingredients and then cast into soap bars of different shapes. { sōp }

soap bubble test [ENG] A leak test in which a soap solution is applied to the surface of the vessel under internal pressure test; soap bubbles form if the tracer gas leaks from the vessel. { 'sōp ,bəb·əl ,test }

soap builder [MATER] A substance mixed with soap to modify the alkali content, to add water-softening ability, or to improve otherwise the cleaning properties; examples are rosin and sodium phosphate. { 'sōp ,bil·dər }

soaprock *See* soapstone. { 'sōp,räk }

soapstone [MINERAL] **1.** A mineral name applied to steatite or to massive talc. Also known as soaprock. **2.** *See* saponite. [PETR] A metamorphic rock characterized by massive, schistose, or interlaced fibrous texture and a soft unctuous feel. { 'sōp,stōn }

soar [AERO ENG] To fly without loss of altitude, using no motive power other than updrafts in the atmosphere. { sȯr }

Sobemovirus [VIROL] A genus of plant viruses with nonenveloped icosahedral virions containing one molecule of linear, single-stranded, positive-sense ribonucleic acid; the type species is Southern bean mosaic virus. Also known as Southern bean mosaic virus group. { sə'bē·mə,vī·rəs }

sobole [BOT] An underground creeping stem. { 'sä·bə,lē }

social animal [ZOO] An animal that exhibits social behavior. { 'sō·shəl 'an·ə·məl }

social anthropology [ANTHRO] The study of social organization among nonliterate peoples. { 'sō·shəl ,an·thrə'päl·ə·jē }

social behavior [ZOO] Any behavior on the part of an organism stimulated by, or acting upon, another member of the same species. { 'sō·shəl bi'hā·vyər }

social hierarchy [VERT ZOO] The establishment of a dominance-subordination relationship among higher animal societies. { 'sō·shəl 'hī·ər,är·kē }

social insects [ECOL] Insect species in which individuals share resources and reproduce cooperatively. { 'sō·shəl 'in,seks }

socialization [PSYCH] The process whereby a child learns to get along with and to behave similarly to other people in the group, largely through imitation as well as group pressure. { ,sō·shə·lə'zā·shən }

social parasitism [VERT ZOO] An aberrant type of parasitism occurring in some birds, in which the female of one species lays her eggs in the nests of other species and permits the foster parents to raise the young. { 'sō·shəl 'par·ə,sə,diz·əm }

social phobia [PSYCH] An anxiety disorder characterized by intense fear of social or performance situations. { 'sōsh·əl 'fō·be·ə }

social psychiatry [PSYCH] Psychiatry especially concerned with the study of social influences on the cause and dynamics of emotional and mental illness, the use of the social environment in treatment, and preventive community programs, as well as the application of psychiatry to social issues, industry, law, education, and other such activities and organizations. { 'sō·shəl sī'kī·ə·trē }

social psychology [PSYCH] The study of the manner in which the attitudes, personality, and motivations of the individual influence, and are influenced by, the structure, dynamics, and behavior of the social group with which the individual interacts. { 'sō·shəl sī'käl·ə·jē }

social releaser [ZOO] A releaser stimulus which an animal receives from a member of its species. { 'sō·shəl ri'lē·sər }

society [ECOL] A secondary or minor plant community forming part of a community. [ZOO] An organization of individuals of the same species in which there are divisions of resources and of labor as well as mutual dependence. { sə'sī·əd·ē }

sociobiology [ANTHRO] A discipline that applies evolutionary biology to the study of animal social behavior, including human behavior; considered a synthesis of ethology, ecology, and evolution, in which social behavior is viewed as the result of natural selection and other biological processes. { ,sō·sē·ō·bī'äl·ə·jē }

socked in [METEOROL] In the early days of aviation, pertaining to weather at an airport when ceiling or visibility were of such low values that the airport was effectively closed to aircraft operations. { 'säkt 'in }

socket [ELEC] A device designed to provide electric connections and mechanical support for an electronic or electric component requiring convenient replacement. [ENG] A device designed to receive and grip the end of a tubular object, such as a tool or pipe. { 'säk·ət }

socket-head screw [DES ENG] A screw fastener with a geometric recess in the head into which an appropriate wrench is inserted for driving and turning, with consequent improved nontamperability. { 'säk·ət ,hed ,skrü }

socket wrench [DES ENG] A wrench with a socket to fit the head of a bolt or a nut. { 'säk·ət ,rench }

sockeye salmon [VERT ZOO] The species *Oncorhynchus nerka*, which is generally smaller and is uniquely adapted to rearing in interior lakes rather than streams or rivers. Also known as red salmon. { ,säk,ī 'sam·ən }

soda *See* sodium carbonate. { 'sōd·ə }

soda-acid extinguisher [ENG] A fire-extinguisher from which water is expelled at a high rate by the generation of carbon dioxide, the result of mixing (when the extinguisher is tilted) of sulfuric acid and sodium bicarbonate. { 'sōd·ə 'as·əd ik'stiŋ·gwə·shər }

soda alum *See* aluminum sodium sulfate. { 'sōd·ə 'al·əm }

soda ash [INORG CHEM] Na_2CO_3 The commercial grade of sodium carbonate; a powder soluble in water, insoluble in alcohol; used in glass manufacture and petroleum refining, and for soaps and detergents. Also known as anhydrous sodium carbonate; calcined soda. { 'sōd·ə 'ash }

soda blasting powder *See* B blasting powder. { 'sōd·ə 'blast·iŋ ,paùd·ər }

sodaclase *See* albite. { 'sōd·ə,klās }

soda crystals *See* metahydrate sodium carbonate. { 'sōd·ə ,krist·əlz }

soda-granite [PETR] **1.** A granite in which soda is more abundant than potash. **2.** A granite that contains soda-plagioclase instead of the orthoclase found in normal granite. { 'sōd·ə ,gran·ət }

soda lake [HYD] An alkali lake rich in dissolved sodium salts, especially sodium carbonate, sodium chloride, and sodium sulfate. Also known as natron lake. { 'sōd·ə ,lāk }

soda lime [MATER] A mixture of sodium or potassium hydroxide with calcium oxide; granules are used to absorb water vapor and carbon dioxide gas. { 'sōd·ə ,līm }

soda-lime glass [MATER] Glass made by fusion of sand with sodium carbonate, or sodium sulfate and lime, or limestone; used for window glass. { 'sōd·ə ,līm ,glas }

sodalite [MINERAL] $Na_2Al_3Si_3O_{12}Cl$ A blue or sometimes white, gray, or green mineral tectosilicate of the feldspathoid group, crystallizing in the isometric system, with vitreous luster, hardness of 5 on Mohs scale, and specific gravity of 2.2–2.4; used as an ornamental stone. { 'sōd·əl,īt }

soda mica *See* paragonite. { 'sōd·ə 'mī·kə }

soda microcline *See* anorthoclase. { 'sōd·ə 'mī·krə,klīn }

sodamide *See* sodium amide. { 'sōd·ə,mīd }

soda niter [MINERAL] $NaNO_3$ A colorless to white mineral composed of sodium nitrate, crystallizing in the rhombohedral division of the hexagonal system; hardness is $1^1/_2$ to 2 on Mohs scale and specific gravity is 2.266. Also known as nitratine; Peru saltpeter. { 'sōd·ə 'nīd·ər }

soda pulping process [CHEM ENG] The digestion of wood chips by caustic soda; used to manufacture pulp for paper products. { 'sōd·ə 'pəl·piŋ ,präs·əs }

sodar [ENG] Sound-wave transmitting and receiving equipment that is used to remotely measure the vertical turbulence structure and wind profile of the lower layer of the atmosphere by analyzing sound reflected in scattering by atmospheric turbulence. Derived from sonic detection and ranging. { 'sō,där }

soddyite [MINERAL] $(UO_2)_{12}Si_5O_{22}\cdot14H_2O$ A pale-yellow orthorhombic mineral composed of hydrous uranium silicate and occurring in fine-grained aggregates or crystals. { 'säd·ē,īt }

Soddy's displacement law [NUC PHYS] The atomic number of a nuclide decreases by 2 in alpha decay, increases by 1 in beta negatron decay, and decreases by 1 in beta positron decay and electron capture. { 'säd·ēz di'splās·mənt ,lö }

sodide [INORG CHEM] A member of a class of alkalides in which the metal anion is sodium (Na^+). { 'sä,dīd }

sodium [CHEM] A metallic element of group 1, symbol Na, with atomic number 11, atomic weight 22.98977; silver-white, soft, and malleable; oxidizes in air; melts at 97.6°C; used as a chemical intermediate and in pharmaceuticals, petroleum refining, and metallurgy; the source of the symbol Na is natrium. { 'sōd·ē·əm }

sodium-24 [NUC PHYS] A radioactive isotope of sodium, mass 24, half-life 15.5 hours; formed by deuteron bombardment of sodium; decomposes to magnesium with emission of beta rays. { 'sōd·ē·əm ,twen·tē'för }

sodium acetate [ORG CHEM] $NaC_2H_3O_2$ Colorless, efflorescent crystals, soluble in water and ether; melts at 324°C; used as a chemical intermediate and for pharmaceuticals, dyes, and dry colors. { 'sōd·ē·əm 'as·ə,tāt }

sodium acid carbonate *See* sodium bicarbonate. { 'sōd·ē·əm 'as·əd 'kär·bə·nət }

sodium acid chromate *See* sodium dichromate. { 'sōd·ē·əm 'as·əd 'krō,māt }

sodium acid fluoride *See* sodium bifluoride. { 'sōd·ē·əm 'as·əd 'flùr,īd }

sodium acid sulfate *See* sodium bisulfate. { 'sōd·ē·əm 'as·əd 'səl,fāt }

sodium acid sulfite *See* sodium bisulfite. { 'sōd·ē·əm 'as·əd 'səl,fīt }

sodium alginate [ORG CHEM] $C_6H_7O_6Na$ Colorless or light yellow filaments, granules, or powder which forms a viscous colloid in water; used in food thickeners and stabilizers, in medicine and textile printing, and for paper coating and water-base paint. Also known as algin; alginic acid sodium salt; sodium polymannuronate. { 'sōd·ē·əm 'al·jə,nāt }

sodium aluminate [INORG CHEM] $Na_2Al_2O_4$ A white powder soluble in water, insoluble in alcohol; melts at 1800°C; used as a zeolite-type of material and a mordant, and in water purification, milkglass manufacture, and cleaning compounds. { 'sōd·ē·əm ə'lü·mə,nāt }

sodium aluminosilicate [INORG CHEM] White, amorphous powder or beads of variable stoichiometry, partially soluble in strong acids and alkali hydroxide solutions between 80 and 100°C; used in food as an anticaking agent. Also known as sodium silicoaluminate. { 'sōd·ē·əm ə¦lü·mə·nō'sil·ə,kāt }

sodium aluminum phosphate [INORG CHEM] NaAl₃-H₁₄(PO₄)₈·4H₂O or Na₃Al₂H₁₅(PO₄)₈ White powder, soluble in hydrochloric acid; used as a food additive for baked products. { 'sōd·ē·əm ə'lü·mə·nəm 'fä,sfāt }

sodium aluminum silicofluoride [INORG CHEM] Na₅Al-(SiF₆)₄ A toxic, white powder, used for mothproofing and in insecticides. { 'sōd·ē·əm ə'lü·mə·nəm ,sil·ə·kō'flur,īd }

sodium aluminum sulfate See aluminum sodium sulfate. { 'sōd·ē·əm ə'lü·mə·nəm 'səl,fāt }

sodium amalgam [INORG CHEM] Na$_x$Hg$_x$ A fire-hazardous, silver-white crystal mass that decomposes in water; used to make hydrogen and as an analytical reagent. { 'sōd·ē·əm ə'mal·gəm }

sodium amalgam-oxygen cell [ELEC] Fuel cell system in which materials functioning in the dual capacity of fuel and anode are consumed continuously; low operating temperatures and high power-to-weight ratios are significant characteristics of the system. { 'sōd·ē·əm ə'mal·gəm 'äk·sə·jən ,sel }

sodium amide [INORG CHEM] NaNH₂ White crystals that decompose in water; melts at 210°C; a fire hazard; used to make sodium cyanide. Also known as sodamide. { 'sōd·ē·əm 'am,īd }

sodium *para*-aminobenzoate [PHARM] NH₂C₆H₄COONa Water-soluble crystals, used in medicine. Also known as PABA sodium. { 'sōd·ē·əm ¦par·ə ¦am·ə·nō'ben·zə·wāt }

sodium antimonate [INORG CHEM] NaSbO₃ A white, granular powder, used as an enamel opacifier and high-temperature oxidizing agent. Also known as antimony sodiate. { 'sōd·ē·əm an'tim·ə,nāt }

sodium arsanilate [ORG CHEM] C₆H₄NH₂(AsO·OH·ONa) A white, water-soluble, poisonous powder with a faint saline taste, used in medicine and as a chemical intermediate. { 'sōd·ē·əm är'san·əl,āt }

sodium arsenate [INORG CHEM] Na₃AsO₄·12H₂O Water-soluble, poisonous, clear, colorless crystals with a mild alkaline taste; melts at 86°C; used in medicine, insecticides, dry colors, and textiles, and as a germicide and a chemical intermediate. { 'sōd·ē·əm 'ärs·ən,āt }

sodium arsenite [INORG CHEM] NaAsO₂ A poisonous, water-soluble, grayish powder; used in antiseptics, dyeing, insecticides, and soaps for taxidermy. { 'sōd·ē·əm 'ärs·ən,īt }

sodium ascorbate [ORG CHEM] CH₂OH(CHOH)₂COH-COHCOONa White, odorless crystals; soluble in water, insoluble in alcohol; decomposes at 218°C; used in therapy for vitamin C deficiency. { 'sōd·ē·əm ə'skór,bāt }

sodium aurothiomalate See gold sodium thiomalate. { 'sōd·ē·əm ,ór·ə,thī·ō'ma,lāt }

sodium azide [INORG CHEM] NaN₃ Poisonous, colorless crystals; soluble in water and liquid ammonia; decomposes at 300°C; used in medicine and to make lead azide explosives. { 'sōd·ē·əm 'ā,zīd }

sodium barbital [PHARM] C₈H₁₁N₂NaO₃ A bitter, white powder, soluble in water; used in medicine. Also known as soluble barbital. { 'sōd·ē·əm 'bär·bə·təl }

sodium barbiturate [ORG CHEM] C₄H₃N₂O₃Na White to slightly yellow powder, soluble in water and dilute mineral acid; used in wood-impregnating solutions. { 'sōd·ē·əm bär'-bich·ə·rət }

sodium benzoate [ORG CHEM] NaC₇H₅O₂ Water- and alcohol-soluble, white, amorphous crystals with a sweetish taste; used as a food preservative and an antiseptic and in tobacco, pharmaceuticals, and medicine. { 'sōd·ē·əm 'ben·zə,wāt }

sodium benzoylacetone dihydrate [ORG CHEM] A metal chelate with low melting point (115°C) and slight solubility in acetone. { 'sōd·ē·əm ¦ben·zə,wil'as·ə,tōn dī'hī,drāt }

sodium bicarbonate [INORG CHEM] NaHCO₃ White, water-soluble crystals with an alkaline taste; loses carbon dioxide at 270°C; used as a medicine and a butter preservative, in food preparation, in effervescent salts and beverages, in ceramics, and to prevent timber mold. Also known as baking soda; bicarbonate of soda; sodium acid carbonate. { 'sōd·ē·əm bī'kär·bə,nət }

sodium bichromate See sodium dichromate. { 'sōd·ē·əm bī'krō,māt }

sodium bifluoride [INORG CHEM] NaHF₂ Poisonous, water-soluble, white crystals; decomposes when heated; used as a laundry-rinse neutralizer, preservative, and antiseptic, and in glass etching and tinplating. Also known as sodium acid fluoride. { 'sōd·ē·əm bī'flur,īd }

sodium bismuthate [INORG CHEM] NaBiO₃ A yellow to brown amorphous powder; used as an analytical reagent and in pharmaceuticals. { 'sōd·ē·əm 'biz·mə,thāt }

sodium bisulfate [INORG CHEM] NaHSO₄ Colorless crystals, soluble in water; the aqueous solution is strongly acidic; decomposes at 315°C; used for flux to decompose minerals, as a disinfectant, and in dyeing and manufacture of magnesia, cements, perfumes, brick, and glue. Also known as niter cake; sodium acid sulfate. { 'sōd·ē·əm bī'səl,fāt }

sodium bisulfide See sodium hydrosulfide. { 'sōd·ē·əm bī'səl,fīd }

sodium bisulfite [INORG CHEM] NaHSO₃ A colorless, water-soluble solid; decomposes when heated. Also known as sodium acid sulfite. { 'sōd·ē·əm bī'səl,fīt }

sodium bisulfite test [ANALY CHEM] A test for aldehydes in which aldehydes form a crystalline salt upon addition of a 40% aqueous solution of sodium bisulfite. { 'sōd·ē·əm bī'səl,fīt ,test }

sodium bitartrate [ORG CHEM] NaHC₄H₅O₆·H₂O A white, combustible, water-soluble powder that loses water at 100°C, decomposes at 219°C; used in effervescing mixtures and as an analytical reagent. Also known as acid sodium tartrate. { 'sōd·ē·əm bī'tär,trāt }

sodium borate [INORG CHEM] Na₂B₄O₇·10H₂O A water-soluble, odorless, white powder; melts between 75 and 200°C; used in glass, ceramics, starch and adhesives, detergents, agricultural chemicals, pharmaceuticals, and photography; the impure form is known as borax. Also known as sodium pyroborate; sodium tetraborate. { 'sōd·ē·əm 'bó,rāt }

sodium boroformate [ORG CHEM] NaH₂BO₃·2HCOOH·2H₂O Water-soluble, white crystals, melting at 110°C; used in textile treating and in tanning, and as a buffering agent. { 'sōd·ē·əm ,bór·ō'fór,māt }

sodium borohydride [INORG CHEM] NaBH₄ A flammable, hygroscopic, white to gray powder; soluble in water, insoluble in ether and hydrocarbons; decomposes in damp air; used as a hydrogen source, a chemical reagent, and a rubber foaming agent. { 'sōd·ē·əm ,bór·ō'hī,drīd }

sodium bromate [INORG CHEM] NaBrO₃ Odorless, white crystals; soluble in water, insoluble in alcohol; decomposes at 381°C; a fire hazard, used as an analytical reagent. { 'sōd·ē·əm 'brō,māt }

sodium bromide [INORG CHEM] NaBr White, water-soluble, crystals with a bitter, saline taste; absorbs moisture from air; melts at 758°C; used in photography and medicine, as a chemical intermediate, and to make bromides. { 'sōd·ē·əm 'brō,mīd }

sodium cacodylate [ORG CHEM] C₂H₆AsNaO₂ A herbicide used as a harvest aid. Also known as bollseye (trade name). { ,sōd·ē·əm ka'käd·əl,āt }

sodium-calcium feldspar See plagioclase. { 'sōd·ē·əm 'kal·sē·əm 'fel,spär }

sodium carbolate See sodium phenate. { 'sōd·ē·əm 'kär·bə,lāt }

sodium carbonate [INORG CHEM] Na₂CO₃ A white, water-soluble powder that decomposes when heated to about 852°C; used as a reagent; forms a monohydrate compound, Na₂CO₃·H₂O, and a decahydrate compound, Na₂CO₃·10H₂O. Also known as soda. { 'sōd·ē·əm 'kär·bə,nət }

sodium carbonate decahydrate See sal soda. { 'sōd·ē·əm 'kär·bə·nət ,dek·ə'hī,drāt }

sodium carbonate peroxide [INORG CHEM] 2Na₂CO₃·3H₂O A white, crystalline powder; used in household detergents, in dental cleansers, and for bleaching and dyeing. { 'sōd·ē·əm 'kär·bə·nət pə'räk,sīd }

sodium caseinate [ORG CHEM] A tasteless, odorless, water-soluble, white powder; used in medicine, foods, emulsification, and stabilization; formed by dissolving casein in sodium hydroxide and then evaporating. Also known as casein sodium; nutrose. { 'sōd·ē·əm 'kā·sē·ə,nāt }

sodium chlorate [INORG CHEM] $NaClO_3$ Water- and alcohol-soluble, colorless crystals with a saline taste; melts at 255°C; used as a medicine, weed killer, defoliant, and oxidizing agent, in matches, explosives, and bleaching. { 'sōd·ē·əm 'klȯr,āt }

sodium chloride [INORG CHEM] $NaCl$ Colorless or white crystals; soluble in water and glycerol, slightly soluble in alcohol; melts at 804°C; used in foods and as a chemical intermediate and an analytical reagent. Also known as common salt; table salt. { 'sōd·ē·əm 'klȯr,īd }

sodium chloride solution *See* normal saline. { 'sōd·ē·əm 'klȯr,īd sə,lü·shən }

sodium chlorite [INORG CHEM] $NaClO_2$ An explosive, white, mildly hygroscopic, water-soluble powder; decomposes at 175°C; used as an analytical reagent and oxidizing agent. { 'sōd·ē·əm 'klȯr,īt }

sodium chloroacetate [ORG CHEM] $ClCH_2COONa$ A white, water-soluble powder; used as a defoliant and in the manufacture of weed killers, dyes, and pharmaceuticals. { 'sōd·ē·əm ,klȯr·ō'as·ə,tāt }

sodium chloroplatinate [INORG CHEM] $Na_2PtCl_6\cdot4H_2O$ A yellow powder, soluble in alcohol and water; used for zinc etching, indelible ink, plating, and mirrors, and in photography and medicine. Also known as platinic sodium chloride; platinum sodium chloride; sodium platinichloride. { 'sōd·ē·əm ,klȯr·ō'plat·ən,āt }

sodium chromate [INORG CHEM] $Na_2CrO_4\cdot10H_2O$ Water-soluble, translucent, yellow, efflorescent crystals that melt at 20°C; used as a rust preventive and in inks, dyeing, and leather tanning. { 'sōd·ē·əm 'krō,māt }

sodium citrate [ORG CHEM] $C_6H_5Na_3O_7\cdot2H_2O$ A white powder with the taste of salt; soluble in water, slightly soluble in alcohol; has an acid taste; loses water at 150°C; decomposes at red heat; used in medicine as an anticoagulant, in soft drinks, cheesemaking, and electroplating. Also known as trisodium citrate. { 'sōd·ē·əm 'sī,trāt }

sodium cobaltinitrite [INORG CHEM] $Na_3Co(NO_2)_6\cdot{}^1\!/_2H_2O$ Purple, water-soluble, hygroscopic crystals; used as a reagent for analysis of potassium. { 'sōd·ē·əm ,kō,bȯl·tə'nī,trīt }

sodium-cooled reactor [NUCLEO] A nuclear reactor in which sodium metal in liquid form is used as the coolant. { 'sōd·ē·əm ,küld rē'ak·tər }

sodium cyanate [INORG CHEM] $NaOCN$ A poisonous, white powder; soluble in water, insoluble in alcohol and ether; used as a chemical intermediate and for the manufacture of medicine and the heat-treating of steels. { 'sōd·ē·əm 'sī·ə,nāt }

sodium cyanide [INORG CHEM] $NaCN$ A poisonous, water-soluble, white powder melting at 563°C; decomposes rapidly when standing; used to manufacture pigments, in heat treatment of metals, and as a silver- and gold-ore extractant. { 'sōd·ē·əm 'sī·ə,nīd }

sodium cyclamate [ORG CHEM] $C_6H_{11}NHSO_3Na$ White, water-soluble crystals; sweetness 30 times that of sucrose; formerly used as an artificial sweetener for foods, but now prohibited. { 'sōd·ē·əm 'sī·klə,māt }

sodium dehydroacetate [ORG CHEM] $C_8H_7NaO_4\cdot H_2O$ A tasteless, white powder, soluble in water and propylene glycol; used as a fungicide and plasticizer, in toothpaste, and for pharmaceuticals. { 'sōd·ē·əm dē,hī·drō'as·ə,tāt }

sodium diacetate [ORG CHEM] $CH_3COONa\cdot x(CH_3COOH)$ Combustible, white, water-soluble crystals with an acetic acid aroma; decomposes above 150°C; used to inhibit mold, and as a buffer, varnish hardener, sequestrant, and food preservative, and in mordants. { 'sōd·ē·əm dī'as·ə,tāt }

sodium diatrizoate [ORG CHEM] $C_{11}H_8NO_4I_3Na$ White, water-soluble crystals which give a radiopaque solution; used in medicine as a radiopaque medium. { 'sōd·ē·əm ,dī·ə'trī·zə,wāt }

sodium dichloroisocyanate [ORG CHEM] $HC_3N_3O_3NaCl$ A white, crystalline compound, soluble in water; used as a bactericide and algicide in swimming pools. { 'sōd·ē·əm dī'klȯr·ō,ī·sō'sī·ə,nāt }

sodium dichloroisocyanurate [ORG CHEM] $C_3N_3O_3Cl_2Na$ A white, crystalline powder; used in dry bleaches, detergents, and cleaning compounds, and for water and sewage treatment. { 'sōd·ē·əm dī'klȯr·ō,ī·sō,sī·ə'nȯr,āt }

sodium dichromate [INORG CHEM] $Na_2Cr_2O_7\cdot2H_2O$ Poisonous, red to orange deliquescent crystals; soluble in water, insoluble in alcohol; melts at 320°C; loses water of hydration upon prolonged heating at 105°C; used as a chemical intermediate and corrosion inhibitor and in the manufacture of pigments, leather tanning, and electroplating. Also known as bichromate of soda; sodium acid chromate; sodium bichromate. { 'sōd·ē·əm dī'krō,mat }

sodium diethyldithiocarbamate [ORG CHEM] $(C_2H_5)_2$-NCS_2Na A solid that is soluble in water and in alcohol; the trihydrate is used to determine small amounts of copper and to separate copper from other metals. { 'sōd·ē·əm dī'eth·əl·dī,thī·ō'kär·bə,mat }

sodium dimethyldithiocarbamate [ORG CHEM] $(CH_3)_2$-NCS_2Na Amber to light green liquid; used as a fungicide, corrosion inhibitor, and rubber accelerator. Abbreviated SDDC. { 'sōd·ē·əm dī'meth·əl·dī,thī·ō'kär·bə,mat }

sodium dinitro-*ortho*-cresylate [ORG CHEM] $CH_3C_6H_2$-$(NO_2)_2ONa$ A toxic, orange-yellow dye, used as a herbicide and fungicide. { 'sōd·ē·əm dī'nī trō ,ȯr·thō 'kres·ə,lat }

sodium dithionite *See* sodium hydrosulfite. { 'sōd·ē·əm dī'thī·ə,nīt }

sodium diuranate [ORG CHEM] $Na_2U_2O_7\cdot6H_2O$ A yellow-orange solid, soluble in dilute acids; used for colored glazes on ceramics and in the manufacture of fluorescent uranium glass. Also known as uranium yellow. { 'sōd·ē·əm dī'yür·ə,nāt }

sodium dodecylbenzenesulfanate [ORG CHEM] $C_{18}H_{29}$-SO_3Na Biodegradable, white to yellow flakes, granules, or powder, used as a synthetic detergent. { 'sōd·ē·əm ,dō·də,sil·ben zēn'səl·fə,nat }

sodium ethoxide *See* sodium ethylate. { 'sōd·ē·əm e'thák,sīd }

sodium ethylate [ORG CHEM] C_2H_5ONa A white powder formed from ethanol by replacement of the hydroxyl groups' hydrogen by monovalent sodium; used in organic synthesis. Also known as caustic alcohol; sodium ethoxide. { 'sōd·ē·əm eth·ə,lāt }

sodium 2-ethylhexyl sulfoacetate [ORG CHEM] $C_{10}H_{19}O_2$-SO_3Na Cream-colored, water-soluble flakes, used as a stabilizing agent in soapless shampoos. { 'sōd·ē·əm ,tü ,eth·əl'hek·səl ,səl·fō'as·ə,tāt }

sodium ethylxanthate [ORG CHEM] $C_2H_5OC(S)SNa$ A yellowish powder, soluble in water and alcohol; used as an ore flotation agent. Also known as sodium xanthate; sodium xanthogenate. { 'sōd·ē·əm ,eth·əl'zan,thāt }

sodium feldspar *See* albite. { 'sōd·ē·əm 'fel,spär }

sodium ferricyanide [INORG CHEM] $Na_3Fe(CN)_6\cdot H_2O$ A poisonous, deliquescent, red powder, soluble in water, insoluble in alcohol; used in printing and for the manufacture of pigments. Also known as red prussiate of soda. { 'sōd·ē·əm ,fer·ə'sī·ə,nīd }

sodium ferrocyanide [INORG CHEM] $Na_4Fe(CN)_6\cdot10H_2O$ Semitransparent crystals, soluble in water; insoluble in alcohol; used in photography, dyes, tanning, and blueprint paper. Also known as yellow prussiate of soda. { 'sōd·ē·əm ,fer·ə'sī·ə,nīd }

sodium fluoborate [INORG CHEM] $NaBF_4$ A white powder with a bitter taste; soluble in water, slightly soluble in alcohol; decomposes when heated, fuses below 500°C; used in electrochemical processes, as flux for nonferrous metals refining, and as an oxidation inhibitor. { 'sōd·ē·əm ,flü·ə'bȯr,āt }

sodium fluorescein *See* uranine. { 'sōd·ē·əm flü'res·ē·ən }

sodium fluoride [INORG CHEM] NaF A poisonous, water-soluble, white powder, melting at 988°C; used as an insecticide and a wood and adhesive preservative, and in fungicides, vitreous enamels, and dentistry. { 'sōd·ē·əm 'flür,īd }

sodium fluoroacetate [ORG CHEM] $C_2H_2FO_2Na$ A white powder, hygroscopic and nonvolatile; decomposes at 200°C; very soluble in water; used as a repellent for rodents and predatory animals. { 'sōd·ē·əm ,flür·ō'as·ə,tāt }

sodium fluosilicate [INORG CHEM] Na_2SiF_6 A poisonous, white, amorphous powder; slightly soluble in water; decomposes at red heat; used to fluoridate drinking water and to kill rodents and insects. Also known as sodium silicofluoride. { 'sōd·ē·əm ,flü·ə'sil·ə·kət }

sodium folate [ORG CHEM] $C_{19}H_{18}N_7NaO_6$ A yellow to yellow-orange liquid; used in medicine for folic acid deficiency. Also known as folic acid sodium salt. { 'sōd·ē·əm 'fō,lāt }

sodium formaldehyde sulfoxylate [ORG CHEM] $HCHO\cdot HSO_2Na\cdot2H_2O$ A white solid with a melting point of 64°C;

soluble in water and alcohol; used as a textile stripping agent and a bleaching agent for soap and molasses. Abbreviated SFS. { 'sōd·ē·əm fȯr'mal·də,hīd səl'fäk·sə,lāt }

sodium formate [ORG CHEM] HCOONa A mildly hygroscopic, white powder, soluble in water; has a formic acid aroma; melts at 245°C; used in medicine and as a chemical intermediate and reducing agent. { 'sōd·ē·əm 'fȯr,māt }

sodium glucoheptonate [ORG CHEM] $C_7H_{13}O_8Na$ A light tan, crystalline powder; used for cleaning metal, mercerizing, paint stripping, and aluminum etching. { 'sōd·ē·əm ¦glü·kō'hep·tə,nāt }

sodium gluconate [ORG CHEM] $C_6H_{11}NaO_7$ A water-soluble, yellow to white, crystalline powder, produced by fermentation; used in food and pharmaceutical industries, and as a metal cleaner. Also known as gluconic acid sodium salt. { 'sōd·ē·əm 'glü·kə,nāt }

sodium glucosulfone [PHARM] $C_{24}H_{34}N_2Na_2O_{18}S_3$ A leprostatic drug, and suppressant for dermatitis herpetiformis. Also known as glucosulfone sodium. { 'sōd·ē·əm ¦glü·kō'səl,fōn }

sodium glutamate [ORG CHEM] $COOH(CH_2)_2CH(NH_2)$- COONa A salt of an amino acid; a white powder, soluble in water and alcohol; used as a taste enhancer. Also known as monosodium glutamate (MSG). { 'sōd·ē·əm 'glüd·ə,māt }

sodium gold chloride [INORG CHEM] $NaAuCl_4·2H_2O$ Yellow crystals, soluble in water and alcohol; used in photography, fine glass staining, porcelain decorating, and medicine. Also known as gold salt; gold sodium chloride. { 'sōd·ē·əm 'gōld 'klȯr,īd }

sodium gold cyanide [ORG CHEM] $NaAu(CN)_2$ A yellow, water-soluble powder; used for gold plating radar and electric parts, jewelry, and tableware. Also known as gold sodium cyanide. { 'sōd·ē·əm 'gōld 'sī·ə,nīd }

sodium-graphite reactor [NUCLEO] A nuclear reactor that uses slightly enriched uranium as fuel, graphite as moderator, and liquid sodium as the coolant. { 'sōd·ē·əm 'gra,fīt rē,ak·tər }

sodium halide [INORG CHEM] A compound of sodium with a halogen; for example, sodium bromide (NaBr), sodium chloride (NaCl), sodium iodide (NaI), and sodium fluoride (NaF). { 'sōd·ē·əm 'ha,līd }

sodium halometallate [INORG CHEM] A compound of sodium with halogen and a metal; for example, sodium platinichloride, $Na_2PtCl_6·6H_2O$. { 'sōd·ē·əm ¦hal·ō'med·əl,āt }

sodium hexylene glycol monoborate [ORG CHEM] $C_6H_{12}O_3BNa$ An amorphous, white solid with a melting point of 426°C; used as a corrosion inhibitor, flame retardant, and lubricating-oil additive. { 'sōd·ē·əm 'hek·sə,lēn 'glī,kȯl ¦män·ə'bȯr,āt }

sodium hydrate See sodium hydroxide. { 'sōd·ē·əm 'hī,drāt }

sodium hydride [INORG CHEM] NaH A white powder, decomposed by water, and igniting in moist air; used to make sodium borohydride and as a drying agent and a reagent. { 'sōd·ē·əm 'hī,drīd }

sodium hydrogen phosphate [INORG CHEM] $NaH_2PO_4·H_2O$ Hygroscopic, transparent, water-soluble crystals; used as a purgative, reagent, and buffer. { 'sōd·ē·əm 'hī·drə·jən 'fä,sfāt }

sodium hydrogen sulfide See sodium hydrosulfide. { 'sōd·ē·əm 'hī·drə·jən 'səl,fīd }

sodium hydrosulfide [INORG CHEM] $NaSH·2H_2O$ Toxic, colorless, water-soluble needles, melting at 55°C; used in pulping of paper, processing dyestuffs, hide dehairing, and bleaching. Also known as sodium bisulfide; sodium hydrogen sulfide; sodium sulfhydrate. { 'sōd·ē·əm ¦hī·drə'səl,fīd }

sodium hydrosulfite [INORG CHEM] $Na_2S_2O_4$ A fire-hazardous, lemon to whitish-gray powder; soluble in water, insoluble in alcohol; melts at 55°C; used as a chemical intermediate and catalyst and in ore flotation. Also known as sodium dithionite. { 'sōd·ē·əm ¦hī·drə'səl,fīt }

sodium hydroxide [INORG CHEM] NaOH White, deliquescent crystals; absorbs carbon dioxide and water from air; soluble in water, alcohol, and glycerol; melts at 318°C; used as an analytical reagent and chemical intermediate, in rubber reclaiming and petroleum refining, and in detergents. Also known as sodium hydrate. { 'sōd·ē·əm hī'dräk,sīd }

sodium hypochlorite [INORG CHEM] NaOCl Air-unstable, pale-green crystals with sweet aroma; soluble in cold water, decomposes in hot water; used as a bleaching agent for paper

pulp and textiles, as a chemical intermediate, and in medicine. { 'sōd·ē·əm ¦hī·pō'klȯr,īt }

sodium hypophosphite [INORG CHEM] $NaH_2PO_2·H_2O$ Colorless, pearly, water-soluble crystalline plates or a white, granular powder; used in medicine and electroless nickel plating of plastic and metal. { 'sōd·ē·əm ¦hī·pō'fä,sfīt }

sodium hyposulfite See sodium thiosulfate. { 'sōd·ē·əm ¦hī·pō'səl,fīt }

sodium illite See brammalite. { 'sōd·ē·əm 'i,līt }

sodium iodate [INORG CHEM] $NaIO_3$ A white, water- and acetone-soluble powder; used as a disinfectant and in medicine. { 'sōd·ē·əm 'ī·ə,dāt }

sodium iodide [INORG CHEM] NaI A white, air-sensitive powder, deliquescent, with bitter taste; soluble in water, alcohol, and glycerin; melts at 653°C; used in photography and in medicine and as an analytical reagent. { 'sōd·ē·əm 'ī·ə,dīd }

sodium iodide scintillator [NUCLEO] A sodium iodide crystal activated with thallium; used especially in the measurement of the energy of gamma rays, by measuring the amplitude of pulses of light generated by electrons in the crystal which are excited by the gamma rays. { 'sōd·ē·əm 'ī·ə,dīd 'sint·əl,ād·ər }

sodium isopropylxanthate [ORG CHEM] $C_5H_7ONaS_2$ Light yellow, crystalline compound that decomposes at 150°C; soluble in water; used as a postemergence herbicide and as a flotation agent for ores. { 'sōd·ē·əm ¦ī·sə¦prō·pəl'zan,thāt }

sodium lactate [ORG CHEM] $CH_3CHOHCOONa$ A water-soluble, hygroscopic, yellow to colorless, syrupy liquid; solidifies at 17°C; used in medicine, as a corrosion inhibitor in antifreeze, and a hygroscopic agent. { 'sōd·ē·əm 'lak,tāt }

sodium lauryl sulfate [ORG CHEM] $CH_3(CH_2)_{10}CH_2$- OSO_3Na A water-soluble salt, produced as a white or cream powder, crystals, or flakes; used in the textile industry as a wetting agent and detergent. Also known as dodecyl sodium sulfate. { 'sōd·ē·əm 'lȯr·əl 'səl,fāt }

sodium lead alloy [MATER] A highly toxic, explosion-prone alloy of lead and sodium; contains 10% sodium and 90% lead when used to make tetraethyllead, and 2% sodium and 98% lead when used as a deoxidizer and homogenizer in lead-containing nonferrous alloys; reacts with moisture, acids, and oxidizing agents. { 'sōd·ē·əm 'led 'al,ȯi }

sodium lead hyposulfate See lead sodium thiosulfate. { 'sōd·ē·əm 'led ¦hī·pō'səl,fāt }

sodium lead thiosulfate See lead sodium thiosulfate. { 'sōd·ē·əm 'led ¦thī·ə'səl,fāt }

sodium-line reversal temperature measurement [PHYS] A method of measuring the temperature of a gas containing sodium vapor, in which the gas is placed in the path of a radiator of known temperature, and the temperature of the gas or the radiator is adjusted until the sodium D line disappears against the background of light from the radiator. { ¦sōd·ē·əm ,līn ri¦vər·səl 'tem·prə·chər ,mezh·ər·mənt }

sodium metaborate [INORG CHEM] $NaBO_2$ Water-soluble, white crystals, melting at 966°C; the aqueous solution is alkaline; made by fusing sodium carbonate with borax; used as an herbicide. { 'sōd·ē·əm ¦med·ə'bȯr,āt }

sodium metaphosphate [INORG CHEM] $(NaPO_3)_x$ Sodium phosphate groupings; cyclic forms range from $x = 3$ for the trimetaphosphate, to $x = 10$ for the decametaphosphate; sodium hexametaphosphate with $x = 10$ to 20 is probably a polymer; used for dental polishing, building detergents, and water softening, and as a sequestrant, emulsifier, and food additive. { 'sōd·ē·əm ¦med·ə'fä,sfāt }

sodium metasilicate See sodium silicate. { 'sōd·ē·əm ¦med·ə'sil·ə,kāt }

sodium metavanadate [INORG CHEM] $NaVO_3$ Colorless crystals or a pale green, crystalline powder with a melting point of 630°C; soluble in water; used in inks, fur dyeing, and photography, and as a corrosion inhibitor in gas scrubbers. { 'sōd·ē·əm ¦med·ə'van·ə,dāt }

sodium methiodal [ORG CHEM] ICH_2SO_3Na A white, crystalline powder, soluble in water and methanol; used in medicine as a radiopaque medium. { 'sōd·ē·əm me'thī·ə,dal }

sodium methoxide [ORG CHEM] CH_3ONa A salt produced as a free-flowing powder, soluble in methanol and ethanol; used as an intermediate in organic synthesis. Also known as sodium methylate. { 'sōd·ē·əm me'thäk,sīd }

sodium methylate See sodium methoxide. { 'sōd·ē·əm 'meth·ə,lāt }

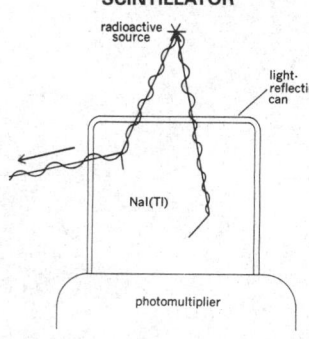

SODIUM IODIDE SCINTILLATOR

radioactive source

light-reflecting can

NaI(Tl)

photomultiplier

Illustration of γ-ray absorption in sodium iodide crystal used as scintillator. In event at left, photon coming from source undergoes Compton scattering in crystal and secondary quantum escapes unabsorbed. At right, photon is fully absorbed because of photoelectric effect.

sodium *N*-methyldithiocarbamate dihydrate [ORG CHEM] $CH_3NHC(S)SNa\cdot2H_2O$ A white, water-soluble, crystalline solid; used as a fungicide, insecticide, nematicide, and weed killer. { 'sōd·ē·əm ¦en ¦meth·əl·dī¦thī·ə'kär·bə,mät dī'hī ,drāt }

sodium molybdate [INORG CHEM] Na_2MoO_4 Water-soluble crystals, melting at 687°C; used as an analytical reagent, corrosion inhibitor, catalyst, and zinc-plating brightening agent, and in medicine. { 'sōd·ē·əm mə'lib,dāt }

sodium 12-molybdophosphate [INORG CHEM] $Na_3PMo_{12}O_{40}$ Yellow, water-soluble crystals; used in neuromicroscopy and photography, and as a water-resisting agent in plastic adhesives and cements. { 'sōd·ē·əm ¦twelv mə,lib·dō'fä,sfāt }

sodium monoxide [INORG CHEM] Na_2O A strong basic white powder soluble in molten caustic soda; forms sodium hydroxide in water; used as a dehydrating and polymerization agent. Also known as sodium oxide. { 'sōd·ē·əm mə'näk,sīd }

sodium naphthalenesulfonate [ORG CHEM] $C_{10}H_7SO_3Na$ Yellow, water-soluble crystalline plates or white scales; used as a liquefying agent in animal glue. { 'sōd·ē·əm ¦naf·thə,lēn'səl·fə,nāt }

sodium naphthionate [ORG CHEM] $NaC_{10}H_6(NH_2)SO_3\cdot4H_2O$ White, light-sensitive crystals, soluble in water and insoluble in ether; used in analysis (Riegler's reagent) for nitrous acid. { 'sōd·ē·əm naf'thī·ə,nāt }

sodium nitrate [INORG CHEM] $NaNO_3$ Fire-hazardous, transparent, colorless crystals with bitter taste; soluble in glycerol and water; melts at 308°C; decomposes when heated; used in manufacture of glass and pottery enamel and as a fertilizer and food preservative. { 'sōd·ē·əm 'nī,trāt }

sodium nitrate gelignites [MATER] A group of explosives, modifications of blasting gelatin in which varying percentages of nitroglycerin are replaced by sodium nitrate and combustible material; characterized by plastic consistency, high densities, medium velocity of detonation, good resistance to the effects of water, and fume characteristics which are suitable for underground workings. { 'sōd·ē·əm 'nī,trāt jə'lig,nīts }

sodium nitrite [INORG CHEM] $NaNO_2$ A fire-hazardous, air-sensitive, yellowish powder, soluble in water; decomposes above 320°C; used as an intermediate for dyestuffs and for pickling meat, textiles dyeing, and rust-proofing, and in medicine. { 'sōd·ē·əm 'nī,trīt }

sodium nitroferricyanide [INORG CHEM] $Na_2Fe(CN)_5NO\cdot2H_2O$ Water-soluble, transparent, reddish crystals; slowly decomposes in water; used as an analytical reagent. { 'sōd·ē·əm ¦nī·trō,fer·ə'sī·ə,nīd }

sodium oleate [ORG CHEM] $C_{17}H_{33}COONa$ A white powder with a tallow aroma; soluble in alcohol and water, with partial decomposition; used in medicine and textile waterproofing. { 'sōd·ē·əm 'ō·lē,āt }

sodium oxalate [ORG CHEM] $Na_2C_2O_4$ A poisonous, white powder; soluble in water, insoluble in alcohol; used for leather tanning and as an analytical reagent. { 'sōd·ē·əm 'äk·sə,lāt }

sodium oxide See sodium monoxide. { 'sōd·ē·əm 'äk,sīd }

sodium paraperiodate [INORG CHEM] $Na_3H_2IO_6$ White, crystalline solid, soluble in concentrated sodium hydroxide solutions; used to wet-strengthen paper and to aid in tobacco combustion. { 'sōd·ē·əm ¦par·ə·pər'ī·ə,dāt }

sodium pentaborate [INORG CHEM] $Na_2B_{10}O_{16}\cdot10H_2O$ A white, water-soluble powder; used in glassmaking, weed killers, and fireproofing compositions. { 'sōd·ē·əm ,pen·tə'bȯr,āt }

sodium pentachlorophenate [ORG CHEM] C_6Cl_5ONa A white or tan powder, soluble in water, ethanol, and acetone; used as a fungicide and herbicide. { 'sōd·ē·əm ¦pen·tə,klȯr·ə'fe,nāt }

sodium pentobarbitone See pentobarbital sodium. { 'sōd·ē·əm ,pen·tə'bär·bə,tōn }

sodium perborate [INORG CHEM] $NaBO_2\cdot H_2O_2\cdot3H_2O$ A white powder with a saline taste; slightly soluble in water, decomposes in moist air; used in deodorants, in dental compositions, and as a germicide. Also known as peroxydol. { 'sōd·ē·əm pər'bȯr,āt }

sodium perchlorate [INORG CHEM] $NaClO_4$ Fire-hazardous, white, deliquescent crystals; soluble in water and alcohol; melts at 482°C; explosive when in contact with concentrated sulfuric acid; used in jet fuel, as an analytical reagent, and for explosives. { 'sōd·ē·əm pər'klȯr,āt }

sodium permanganate [INORG CHEM] $NaMnO_4\cdot3H_2O$ A fire-hazardous, water-soluble, purple powder; decomposes when heated; used to make saccharin, as a disinfectant, and as an oxidizing agent. { 'sōd·ē·əm pər'maŋ·gə,nāt }

sodium peroxide [INORG CHEM] Na_2O_2 A fire-hazardous, white powder that yellows with heating; decomposes when heated; causes ignition when in contact with water; used as an oxidizing agent and a bleach, and in medicinal soap. { 'sōd·ē·əm pə'räk,sīd }

sodium persulfate [INORG CHEM] $Na_2S_2O_8$ A white, water-soluble, crystalline powder; used as a bleaching agent and in medicine. { 'sōd·ē·əm pər'səl,fāt }

sodium phenate [ORG CHEM] C_6H_5ONa White, deliquescent crystals, soluble in water and alcohol; decomposed by carbon dioxide in air; used as a chemical intermediate, antiseptic, and military gas absorbent. Also known as sodium carbolate; sodium phenolate. { 'sōd·ē·əm 'fe,nāt }

sodium phenolate See sodium phenate. { 'sōd·ē·əm 'fen·əl,āt }

sodium phenylacetate [ORG CHEM] $C_6H_5CH_2\cdot COONa$ Pale yellow, 50% aqueous solution which crystallizes at 15°C; used in the manufacture of penicillin G. { 'sōd·ē·əm ,fen·əl'as·ə,tāt }

sodium phenylphosphinate [ORG CHEM] $C_6H_5PH(O)(ONa)$ Crystals with a melting point of 355°C; used as an antioxidant and as a heat and light stabilizer. { 'sōd·ē·əm ,fen·əl'fä·sfə,nāt }

sodium phosphate [INORG CHEM] A general term encompassing the following compounds: sodium hexametaphosphate, sodium metaphosphate, dibasic sodium phosphate, hemibasic sodium phosphate, monobasic sodium phosphate, tribasic sodium phosphate, sodium pyrophosphate, and acid sodium pyrophosphate. { 'sōd·ē·əm 'fä,sfāt }

sodium phosphite [INORG CHEM] $Na_2HPO_3\cdot5H_2O$ White, hygroscopic crystals, melting at 53°C; soluble in water, insoluble in alcohol; used in medicine. { 'sōd·ē·əm 'fä,sfīt }

sodium phosphotungstate See sodium tungstophosphate. { 'sōd·ē·əm ,fä·sfō'təŋ,stāt }

sodium phytate [ORG CHEM] $C_6H_9O_2P_6Na_9$ A hygroscopic, water-soluble powder; used as a chelating agent for trace metals and in medicine. { 'sōd·ē·əm 'fī,tāt }

sodium picramate [ORG CHEM] $NaOC_6H_2(NO_2)_2NH_2$ A yellow salt, soluble in water; used in the manufacture of dye intermediates. { 'sōd·ē·əm 'pik·rə,māt }

sodium platinichloride See sodium chloroplatinate. { 'sōd·ē·əm ¦plat·ən·ə'klȯr,īd }

sodium plumbite [INORG CHEM] $Na_2PbO_2\cdot3H_2O$ A toxic, corrosive solution of lead oxide (litharge) in sodium hydroxide; used (as doctor solution) to sweeten gasoline. { 'sōd·ē·əm 'pləm,bīt }

sodium polymannuronate See sodium alginate. { 'sōd·ē·əm ¦päl·i·mə'nyur·ə,nāt }

sodium polysulfide [INORG CHEM] Na_2S_x Yellow-brown granules, used to make dyes and colors, and insecticides, as a petroleum additive, and in electroplating. { 'sōd·ē·əm ¦päl·i'səl,fīd }

sodium/potassium pump [CELL MOL] The cell membrane protein channel through which active transport of sodium ions out of the cell and potassium ions into the cell against their electrochemical gradients takes place. { ¦sōd·ē·əm pə'tas·ē·əm ,pəmp }

sodium propionate [ORG CHEM] CH_3CH_2COONa Deliquescent, transparent crystals; soluble in water, slightly soluble in alcohol; used as a fungicide, and mold preventive. { 'sōd·ē·əm 'prō·pē·ə,nāt }

sodium pyroborate See sodium borate. { 'sōd·ē·əm 'pī·rō'bȯr,āt }

sodium pyrophosphate [INORG CHEM] $Na_4P_2O_7$ A white powder; soluble in water, insoluble in alcohol and ammonia; melts at 880°C; used as a water softener and newsprint deinker, and to control drilling-mud viscosity. Also known as normal sodium pyrophosphate; tetrasodium pyrophosphate (TSPP). { 'sōd·ē·əm 'pī·rō'fä,sfāt }

sodium saccharin [ORG CHEM] $C_7H_4NNaO_3S\cdot2H_2O$ White crystals or a crystalline powder, soluble in water and slightly soluble in alcohol; used in medicine and as a nonnutritive food sweetener. { 'sōd·ē·əm 'sak·ə·rən }

sodium salicylate [ORG CHEM] HOC_6H_4COONa A shiny, white powder with sweetish taste and mild aromatic aroma;

soluble in water, glycerol, and alcohol; used in medicine and as a preservative. { 'sōd·ē·əm sə'lis·ə,lāt }

sodium selenate [INORG CHEM] Na$_2$SeO$_4$·10H$_2$O White, poisonous, water-soluble crystals; used as an insecticide. { 'sōd·ē·əm 'sel·ə,nāt }

sodium selenite [INORG CHEM] Na$_2$SeO$_3$·5H$_2$O White, water-soluble crystals; used in glass manufacture, as a bacteriological reagent, and for decorating porcelain. { 'sōd·ē·əm 'sel·ə,nīt }

sodium sesquicarbonate [INORG CHEM] Na$_2$CO$_3$· NaHCO$_3$·2H$_2$O White, water-soluble, needle-shaped crystals; used as a detergent, an alkaline agent for water softening and leather tanning, and a food additive. { 'sōd·ē·əm 'ses·kwē'kär·bə,nāt }

sodium sesquisilicate [INORG CHEM] Na$_6$Si$_2$O$_7$ A white, water-soluble powder; used for metals cleaning and textile processing. { 'sōd·ē·əm 'ses·kwē'sil·ə,kāt }

sodium silicate [INORG CHEM] Na$_2$SiO$_3$ A gray-white powder; soluble in alkalies and water, insoluble in alcohol and acids; used to fireproof textiles, in petroleum refining and corrugated paperboard manufacture, and as an egg preservative. Also known as liquid glass; silicate of soda; sodium metasilicate; soluble glass; water glass. { 'sōd·ē·əm 'sil·ə,kāt }

sodium silicoaluminate See sodium aluminosilicate. { 'sōd· ē·əm 'sil·ə·kō·ə'lü·mə,nāt }

sodium silicofluoride See sodium fluosilicate. { 'sōd·ē·əm 'sil·ə·kō'flur,īd }

sodium stannate [INORG CHEM] Na$_2$SnO$_3$·3H$_2$O Water- and alcohol-insoluble, whitish crystals; used in ceramics, dyeing, and textile fireproofing, and as a mordant. Also known as preparing salt. { 'sōd·ē·əm 'sta,nāt }

sodium stearate [ORG CHEM] NaC$_{18}$H$_{35}$O$_2$ A white powder with a fatty aroma; soluble in hot water and alcohol; used in medicine and toothpaste and as a waterproofing agent. { 'sōd·ē·əm 'stir,āt }

sodium subsulfite See sodium thiosulfate. { 'sōd·ē·əm 'səb'səl,fīt }

sodium succinate [ORG CHEM] Na$_2$C$_4$H$_4$O$_4$·6H$_2$O Water-soluble, white crystals; loses water at 120°C; used in medicine. { 'sōd·ē·əm 'sək·sə,nāt }

sodium sulfate [INORG CHEM] Na$_2$SO$_4$ Crystalline compound, melts at 888°C, soluble in water; used to make paperboard, kraft paper, glass, and freezing mixtures. { 'sōd·ē·əm 'səl,fāt }

sodium sulfhydrate See sodium hydrosulfide. { 'sōd·ē·əm 'səlf'hī,drāt }

sodium sulfide [INORG CHEM] Na$_2$S An irritating, water-soluble, yellow to red, deliquescent powder; melts at 1180°C; used as a chemical intermediate, solvent, photographic reagent, and analytical reagent. Also known as sodium sulfuret. { 'sōd·ē·əm 'səl,fīd }

sodium sulfite [INORG CHEM] Na$_2$SO$_3$ White, water-soluble, crystals with a sulfurous, salty taste; decomposes when heated; used as a chemical intermediate and food preservative, in medicine and paper manufacturing, and for dyes and photographic developing. { 'sōd·ē·əm 'səl,fīt }

sodium sulfite process [CHEM ENG] A process for the digestion of wood chips in a solution of magnesium, ammonium, or calcium disulfite containing free sulfur dioxide; used in papermaking. { 'sōd·ē·əm 'səl,fīt ,präˈsəs }

sodium sulfocyanate See sodium thiocyanate. { 'sōd·ē·əm 'səl·fō'sī·ə,nāt }

sodium/sulfur battery [ELEC] A storage battery that operates at temperatures of 300–350°C (570–660°F) and has a liquid sodium anode and liquid sulfur cathode separated by a solid ceramic electrolyte that conducts sodium ions. { 'sōd· ē·əm 'səl·fər 'bad·ə·rē }

sodium sulfuret See sodium sulfide. { 'sōd·ē·əm 'səl· fyə,ret }

sodium tartrate [ORG CHEM] Na$_2$C$_4$H$_4$O$_6$·2H$_2$O White, water-soluble crystals or granules; loses water at 150°C; used in medicine and as a food stabilizer and sequestrant. Also known as disodium tartrate; sal tartar. { 'sōd·ē·əm 'tär,trāt }

sodium TCA See sodium trichloroacetate. { 'sōd·ē·əm 'tē'sē'ā }

sodium tetraborate See sodium borate. { 'sōd·ē·əm 'te·tra'bor,āt }

sodium tetrafluorescein See easin. { 'sōd·ē·əm 'te·tra'flur· ə,sēn }

sodium tetraphenylborate [ORG CHEM] [(C$_6$H$_5$)$_4$B]Na A snow-white, crystalline compound, soluble in water and acetone; used as a reagent in the determination of the following ions: potassium, ammonium, rubidium, and cesium. { 'sōd· ē·əm 'tre·tra'fen·əl'bór,āt }

sodium tetrasulfide [INORG CHEM] Na$_2$S$_4$ Hygroscopic, yellow or dark-red crystals, melting at 275°C; used for insecticides and fungicides, ore flotation, and dye manufacture, and as a reducing agent. { 'sōd·ē·əm 'tet·rə'səl,fīd }

sodium thiocyanate [INORG CHEM] NaSCN A poisonous, water- and alcohol-soluble, deliquescent, white powder; melts at 287°C; used as an analytical reagent, solvent, and chemical intermediate, and for rubber treatment and textile dyeing and printing. Also known as sodium sulfocyanate. { 'sōd·ē·əm ,thī·ə'sī·ə,nāt }

sodium thioglycolate [ORG CHEM] C$_2$H$_3$NaO$_3$S A water-soluble compound produced as hygroscopic crystals; used as an ingredient in bacteriology media, and in hair-waving solutions. { 'sōd·ē·əm ,thī·ə'glī·kə,lāt }

sodium thiopental [PHARM] C$_{11}$H$_{17}$N$_2$O$_2$SNa A yellow-ish-white powder, soluble in water and alcohol; used in medicine as a barbiturate. { 'sōd·ē·əm ,thī·ə'pent·əl }

sodium thiosulfate [INORG CHEM] Na$_2$S$_2$O$_3$·5H$_2$O White, translucent crystals or powder with a melting point of 48°C; soluble in water and oil of turpentine; used as a fixing agent in photography, for extracting silver from ore, in medicine, and as a sequestrant in food. Also known as sodium hyposulfite; sodium subsulfite. { 'sōd·ē·əm ,thī·ə'səl,fāt }

sodium trichloroacetate [ORG CHEM] CCl$_3$COONa A toxic material, used in herbicides and pesticides. Abbreviated sodium TCA. { 'sōd·ē·əm trī'klór·ō'as·ə,tāt }

sodium 2,4,5-trichlorophenate [ORG CHEM] C$_6$H$_2$Cl$_3$ONa· 1$^{1}/_{2}$H$_2$O Buff to light brown flakes, soluble in water, methanol, and acetone; used as a bactericide and fungicide. { 'sōd· ē·əm 'tü 'fōr 'fīv trī'klórō'fe,nāt }

sodium tripolyphosphate [INORG CHEM] Na$_5$P$_3$O$_{10}$ A white powder with a melting point of 622°C; used for water softening and as a food additive and texturizer. Abbreviated STPP. { 'sōd·ē·əm trī'päl·i'fä,sfāt }

sodium tungstate [INORG CHEM] Na$_2$WO$_4$·2H$_2$O Water-soluble, colorless crystals; lose water at 100°C, melts at 692°C; used as a chemical intermediate analytical reagent, and for fireproofing. Also known as sodium wolframate. { 'sōd·ē·əm 'təŋ,stāt }

sodium tungstophosphate [INORG CHEM] Approximately 2Na$_2$O·P$_2$O$_5$·12WO$_3$·18H$_2$O A yellowish-white powder, soluble in water and alcohols; used to manufacture organic pigments, as an antistatic agent for textiles, in leather tanning, and as a water-resistant agent in plastic films, adhesives, and cements. Also known as sodium phosphotungstate. { 'sōd· ē·əm 'twelv 'təŋ·stō'fä,sfāt }

sodium undecylenate [ORG CHEM] C$_{11}$H$_{19}$O$_2$Na A white, water-soluble powder that decomposes above 200°C; used in cosmetics and pharmaceuticals as a bacteriostat and fungistat. { 'sōd·ē·əm ,ən,de·sə'le,nāt }

sodium-vapor lamp [ELECTR] A discharge lamp containing sodium vapor, used chiefly for outdoor illumination. { 'sōd· ē·əm 'vā·pər 'lamp }

sodium wolframate See sodium tungstate. { 'sōd·ē·əm 'wúl·frə,mīt }

sodium xanthate See sodium ethylxanthate. { 'sōd·ē·əm 'zan,thāt }

sodium xanthogenate See sodium ethylxanthate. { 'sōd·ē· əm zan'thä·jə,nāt }

sofar [NAV] A system of fixing a position at sea by exploding a charge under water, measuring the time for the shock waves to travel through water to three widely separated shore stations, and calculating the position of the explosive by triangulation; the explosive can be dropped from a lifeboat by survivors of air or sea disasters. Derived from sound fixing and ranging. { 'sō,fär }

soffione [GEOL] A jet of steam and other vapors issuing from the ground in a volcanic area. { ,sä·fē'ō·nē }

soffit [CIV ENG] The underside of a horizontal structural member, such as a beam or a slab. { 'säf·ət }

soffosian knob See frost mound. { sə'fō·zhən 'näb }

soft automation [ENG] Automatic control, chiefly through the use of computer processing, with relatively little reliance on computer hardware. { 'sóft ,ȯd·ə'mā·shən }

SODIUM/SULFUR BATTERY

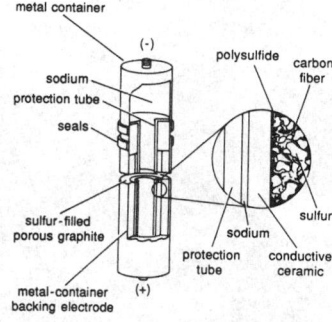

Schematic of a sodium/sulfur cell. *(Ford Aerospace and Communications Corp.)*

soft cataract [MED] A cataract, affecting the cortex of the lens of the eye, which is of soft consistency and has a milky appearance. { 'sȯft 'kad·ə,rakt }

soft chancre *See* chancroid. { 'sȯft 'shaŋ·kər }

soft coal *See* bituminous coal. { 'sȯft 'kōl }

soft computing [COMPUT SCI] A family of methods that imitate human intelligence with the goal of creating tools provided with some human-like capabilities (such as learning, reasoning, and decision making), and are based on fuzzy logic, neural networks, and probabilistic reasoning techniques such as genetic algorithms. { ,sȯft kəm'pyüd·iŋ }

soft copy [COMPUT SCI] Information that is displayed on a screen, given by voice, or stored in a form that cannot be read directly by a person, as on magnetic tape, disk, or microfilm. { 'sȯft 'käp·ē }

soft-copy terminal [COMPUT SCI] A computer terminal that presents its output through an electronic display, rather than printing it on paper. { 'sȯft ¦käp·ē 'tər·mən·əl }

soft coral [INV ZOO] The common name for cnidarians composing the order Alcyonacea; the colony is supple and leathery. { 'sȯft 'kär·əl }

soft crash [COMPUT SCI] A halt in computer operations in which the computer operator has enough warning time to take action to minimize the effects of the stoppage. { 'sȯft 'krash }

soft dot [GRAPHICS] A dot on a screened halftone with an excessive fringe having an area almost as large as the dot itself. { 'sȯft 'dät }

soft edit [COMPUT SCI] A checking and correction process that allows data in which problems have been identified to be accepted by a computer system. { 'sȯft 'ed·it }

soft electrophile [PHYS CHEM] A molecule that readily accepts electrons during a primary reaction step. { 'sȯft i'lek·trə,fīl }

softening agent [MATER] **1.** A substance that is added to another substance to increase softness; for example, stearic acid added to plastics, fat-liquoring agents to leather, and fatty alcohol to fabrics. **2.** A chemical that softens hard water by removing or trapping calcium and magnesium ions. { 'sȯf·ən·iŋ ,ā·jənt }

softening point [PHYS] For a substance which does not have a definite melting point, the temperature at which viscous flow changes to plastic flow. { 'sȯf·ən·iŋ ,pȯint }

softening range [PHYS] The temperature range in which material without a melting point goes from a rigid to a soft condition. { 'sȯf·ən·iŋ ,rānj }

soft error [COMPUT SCI] An error that occurs in automatic operations but does not recur when the operation is attempted a second time. { 'sȯft 'er·ər }

soft failure [COMPUT SCI] A failure that can be overcome without the assistance of a person with specialized knowledge to repair the device. { 'sȯft 'fāl·yər }

soft flow [ENG] The free-flowing characteristics of a plastic material under conventional molding conditions. { 'sȯft 'flō }

soft font [COMPUT SCI] A typeface or set of typefaces that is contained in the software of a computer system and is transmitted to the printer before printing. Also known as downloadable font. { 'sȯft 'fänt }

soft ground [MIN ENG] **1.** A mineral deposit which can be mined without drilling and shooting hard rock. **2.** The rock about underground openings that does not stand well and requires heavy timbering. { 'sȯft 'graùnd }

soft hail *See* snow pellets. { 'sȯft 'hāl }

soft hammer [ENG] A hammer having a head made of a soft material, such as copper, lead, rawhide, or plastic; used to prevent damage to a finished surface. { 'sȯft 'ham·ər }

soft-iron ammeter [ENG] An ammeter in which current in a coil causes two pieces of magnetic material within the coil, one fixed and one attached to a pointer, to become similarly magnetized and to repel each other, moving the pointer; used for alternating-current measurement. { 'sȯft 'ī·ərn 'am,ēd·ər }

soft landing [AERO ENG] The act of landing on the surface of a planet or moon without damage to any portion of the vehicle or payload, except possibly the landing gear. { 'sȯft 'land·iŋ }

soft limiting [ELECTR] Limiting in which there is still an appreciable increase in output for increases in input signal strength up into the range at which limiting action occurs. { 'sȯft 'lim·əd·iŋ }

soft magnetic material [ELECTROMAG] A magnetic material which is relatively easily magnetized or demagnetized. { ,sȯft mag'ned·ik mə'tir·ē·əl }

soft missile base [CIV ENG] A missile-launching base that is not protected against a nuclear explosion. { 'sȯft 'mis·əl ,bās }

soft page break [COMPUT SCI] A page break that is inserted in a document by a word-processing program, and can move if text is added, deleted, or reformatted above it. { ¦sȯft 'pāj ,brāk }

soft palate [ANAT] The posterior part of the palate which consists of an aggregation of muscles, the tensor veli palatini, levator veli palatini, azygos uvulae, palatoglossus, and palatopharyngeus, and their covering mucous membrane. { 'sȯft 'pal·ət }

soft patch [COMPUT SCI] A temporary change in a computer program's machine language that is carried out while the program is in memory, and thus prevails only for the duration of a single run of the program. [ENG] A patch in a crack in a vessel such as a steam boiler consisting of a soft material inserted in the crack and covered by a metal plate bolted or riveted to the vessel. { 'sȯft 'pach }

soft phosphate [MATER] Powdery, impure tricalcium phosphate separated in fertilizer manufacture from rock and pebble phosphates. { 'sȯft 'fä,sfāt }

soft point [ORD] A bullet with a soft point, intended to spread upon striking a target with some resistance, such as the flesh of game; not permitted in combat operations. { 'sȯft 'pȯint }

soft proof [GRAPHICS] An image shown on a calibrated monitor to evaluate graphic design. { ,sȯft 'prüf }

soft radiation [PHYS] Radiation whose particles or photons have a low energy, and, as a result, do not penetrate any type of material readily. { 'sȯft ,rād·ē'ā·shən }

soft return [COMPUT SCI] A control code that is automatically entered into a text document by the word-processing program to mark the end of a line, based on the current right margin. { 'sȯft ri'tərn }

soft rime [HYD] A white, opaque coating of fine rime deposited chiefly on vertical surfaces, especially on points and edges of objects, generally in supercooled fog. { 'sȯft 'rīm }

soft rock [MIN ENG] Rock that can be removed by air-operated hammers, but cannot be handled economically by a pick. [PETR] **1.** A broad designation for sedimentary rock. **2.** A rock that is relatively nonresistant to erosion. { 'sȯft 'räk }

soft rot [PL PATH] A mushy, watery, or slimy disintegration of plant parts caused by either fungi or bacteria. { 'sȯft 'rät }

soft rubber [MATER] A type of rubber that has been cured by adding 0.5 to 8% sulfur, without prolonged vulcanization. { 'sȯft 'rəb·ər }

soft sector [COMPUT SCI] A disk or drum format in which the locations of sectors are determined by control information written on the storage medium rather than by some physical means. { 'sȯft 'sek·tər }

soft-shell disease [INV ZOO] A disease of lobsters caused by a chitinous bacterium which extracts chitin from the exoskeleton. { 'sȯft ¦shel di,zēz }

soft shower [NUC PHYS] A cosmic-ray shower that cannot penetrate 6 to 8 inches (15 to 20 centimeters) of lead; consists mainly of electrons and positrons. { 'sȯft 'shaù·ər }

soft solder [MET] Solder composed of an alloy of lead and tin. Also known as low melting solder. { 'sȯft 'säd·ər }

soft soldering [MET] Soldering with a soft solder. { 'sȯft 'säd·ə·riŋ }

soft tube [ELECTR] **1.** An x-ray tube having a vacuum of about 0.000002 atmosphere (0.2 pascal), the remaining gas being left in intentionally to give less-penetrating rays than those of a more completely evacuated tube. **2.** *See* gassy tube. { 'sȯft ,tüb }

software [COMPUT SCI] The totality of programs usable on a particular kind of computer, together with the documentation associated with a computer or program, such as manuals, diagrams, and operating instructions. { 'sȯf,wer }

software compatibility [COMPUT SCI] Property of two computers, with respect to a particular programming language, in which a source program from one machine in that language will compile and execute to produce acceptably similar results in the other. { 'sȯf,wer kəm,pad·ə'bil·əd·ē }

software driver [COMPUT SCI] Software that is designed to

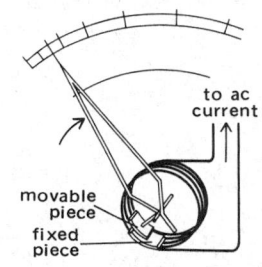

to ac current

movable piece

fixed piece

Schematic of the repulsion-type soft-iron ammeter showing the fixed and movable pieces of magnetic material.

handle the interaction between a computer and its peripheral equipment, changing the format of data as necessary. { 'sȯf,wer 'drīv·ər }

software engineering [COMPUT SCI] The systematic application of scientific and technological knowledge, through the medium of sound engineering principles, to the production of computer programs, and to the requirements definition, functional specification, design description, program implementation, and test methods that lead up to this code. { 'sȯf,wer ,en·jə'nir·iŋ }

software flexibility [COMPUT SCI] The ability of software to change easily in response to different user and system requirements. { 'sȯf,wer ,flek·sə'bil·əd·ē }

software floating point [COMPUT SCI] Special routines that allow high-level programming languages to perform floating-point arithmetic on computer hardware designed for integer arithmetic. { 'sȯf,wer 'flōd·iŋ 'pȯint }

software interface [COMPUT SCI] A computer language whereby computer programs can communicate with each other, and one language can call upon another for assistance. { 'sȯf,wer 'in·tər·fās }

software maintenance [COMPUT SCI] The correction of errors in software systems and the remedying of inadequacies in running the software. { 'sȯf,wer ,mānt·ən,əns }

software metric [COMPUT SCI] **1.** A rule for quantifying some characteristic or attribute of a computer software entity. **2.** One of a set of techniques whose aim is to measure the quality of a computer program. { 'sȯf,wer 'me·trik }

software monitor [COMPUT SCI] A system, used to evaluate the performance of computer software, that is similar to accounting packages, but can collect more data concerning usage of various components of a computer system and is usually part of the control program. { 'sȯf,wer ,män·əd·ər }

software multiplexing [COMPUT SCI] A procedure used in a time-sharing or multiprogrammed system in which the central processing unit, acting under control of a software algorithm, interleaves its attention between a family of programs waiting for service, in such a way that the programs appear to be processed in parallel. { 'sȯf,wer 'məl·ti,pleks·iŋ }

software package [COMPUT SCI] A program for performing some specific function or calculation which is useful to more than one computer user and is sufficiently well documented to be used without modification on a defined configuration of some computer system. { 'sȯf,wer ,pak·ij }

software path length [COMPUT SCI] The number of machine-language instructions required to carry out some specified task. Also known as path length. { 'sȯf,wer 'path ,leŋkth }

software piracy [COMPUT SCI] The process of copying commercial software without the permission of the originator. { 'sȯf,wer 'pir·ə·sē }

software protection [COMPUT SCI] The use of various techniques to prevent the unauthorized duplication of software. Also known as copy protection. { 'sȯf,wer prə,tek·shən }

soft waste [TEXT] The waste from yarn manufacturing prior to spinning, including some spinning waste; usually reprocessed in the mill. { 'sȯft 'wāst }

soft water [CHEM] Water that is free of magnesium or calcium salts. { 'sȯft 'wȯd·ər }

soft-wired numerical control See computer numerical control. { 'sȯf ,wīrd nù'mer·ə·kəl kən'trōl }

soft wood [MATER] Wood from a coniferous tree. { 'sȯft 'wùd }

soft x-ray [ELECTROMAG] An x-ray having a comparatively long wavelength and poor penetrating power. { 'sȯft 'eks,rā }

soft-x-ray absorption spectroscopy [SPECT] A spectroscopic technique which is used to get information about unoccupied states above the Fermi level in a metal or about empty conduction bands in an inoculator. { 'sȯft 'eks,ra əb'sȯrp·shən spek'träs·kə·pē }

soft-x-ray appearance potential spectroscopy [SPECT] A branch of electron spectroscopy in which a solid surface is bombarded with monochromatic electrons, and small but abrupt changes in the resulting total x-ray emission intensity are detected as the energy of the electrons is varied. Abbreviated SXAPS. { 'sȯft 'eks,ra ə'pir·əns pə'ten·chəl spek'träs·kə·pē }

sogasoid [PHYS] A system of solid particles dispersed in a gas. { 'säg·ə,sȯid }

Sohm Abyssal Plain [GEOL] A basin in the North Atlantic,

about 2400 fathoms (4390 meters) deep, between Newfoundland and the Mid-Atlantic Ridge. { 'sōm ə'bis·əl 'plān }

Sohncke's law [PHYS] The law that the stress per unit area normal to a crystallographic plane needed to produce a fracture in a crystal is a constant characteristic of a crystalline substance. { 'zōŋ·kəz ,lȯ }

soil [GEOL] **1.** Unconsolidated rock material over bedrock. **2.** Freely divided rock-derived material containing an admixture of organic matter and capable of supporting vegetation. { sȯil }

soil air [GEOL] The air and other gases in spaces in the soil; specifically, that which is found within the zone of aeration. Also known as soil atmosphere. { 'sȯil 'er }

soil atmosphere See soil air. { 'sȯil ,at·mə,sfir }

soil blister See frost mound. { 'sȯil ,blis·tər }

soil-cement [MATER] A compacted mixture of soil, cement, and water used as a base course or surface for roads and airport paving. { 'sȯil si,ment }

soil chemistry [GEOCHEM] The study and analysis of the inorganic and organic components and the life cycles within soils. { 'sȯil 'kem·ə·strē }

soil colloid [GEOL] Colloidal complex of soils composed principally of clay and humus. { 'sȯil 'kä,lȯid }

soil complex [GEOL] A mapping unit used in detailed soil surveys; consists of two or more recognized classifications. { 'sȯil 'käm,pleks }

soil conservation [ECOL] Management of soil to prevent or reduce soil erosion and depletion by wind and water. { 'sȯil ,kän·sər,vā·shən }

soil creep [GEOL] The slow, steady downhill movement of soil and loose rock on a slope. Also known as surficial creep. { 'sȯil ,krēp }

soil ecology [ECOL] The study of interactions among soil organisms and interactions between biotic and abiotic aspects of the soil environment. { 'sȯil i,käl·ə·jē }

soil element [GEOL] A unit that represents an arbitrarily small volume of soil within a soil mass. { 'sȯil ,el·ə·mənt }

soil erosion [GEOL] The detachment and movement of topsoil by the action of wind and flowing water. { 'sȯil i,rōzh·ən }

soil fertility [AGR] The ability of a soil to supply plant nutrients. { 'sȯil fər,til·əd·ē }

soil flow See solifluction. { 'sȯil ,flō }

soil fluction See solifluction. { 'sȯil ,flək·shən }

soil formation See soil genesis. { 'sȯil ,fȯr·mā·shən }

soil genesis [GEOL] The mode by which soil originates, with particular reference to processes of soil-forming factors responsible for the development of true soil from unconsolidated parent material. Also known as pedogenesis; soil formation. { 'sȯil ,jen·ə·səs }

soil line See soil pipe. { 'sȯil ,līn }

soil mechanics [ENG] The application of the laws of solid and fluid mechanics to soils and similar granular materials as a basis for design, construction, and maintenance of stable foundations and earth structures. { 'sȯil mi,kan·iks }

soil microbiology [MICROBIO] A study of the microorganisms in soil, their functions, and the effect of their activities on the character of the soil and the growth and health of plant life. { 'sȯil 'mī·krə·bī'äl·ə·jē }

soil moisture See soil water. { 'sȯil 'mȯis·chər }

soil physics [GEOPHYS] The study of the physical characteristics of soils; concerned also with the methods and instruments used to determine these characteristics. { 'sȯil 'fiz·iks }

soil pipe [CIV ENG] A cast-iron or plastic pipe for carrying discharges from toilet fixtures from a building into the soil drain. Also known as soil line. { 'sȯil ,pīp }

soil profile [GEOL] A vertical section of a soil, showing horizons and parent material. { 'sȯil 'prō,fīl }

soil remediation [AGR] The removal of harmful contaminants in soil. { 'sȯil rə,mē·dē,ā·shən }

soil rot [PL PATH] Plant rot caused by soil microorganisms. { 'sȯil ,rät }

soil science [GEOL] The study of the formation, properties, and classification of soil; includes mapping. Also known as pedology. { 'sȯil 'sī·əns }

soil separate [GEOL] Any of a group of rock or mineral particles, separated from a soil sample, having diameters less than 0.8 inch (2 millimeters) and ranging within the limits of one of the standard classifications of soil particle size. Also known as separate. { 'sȯil ˌsep·rət }

soil series [GEOL] A family of soils having similar profiles, and developing from similar original materials under the influence of similar climate and vegetation. { 'sȯil ˌsir·ēz }

soil shear strength [GEOL] The maximum resistance of a soil to shearing stresses. { 'sȯil 'shir ˌstreŋkth }

soil stabilizer [MATER] A chemical that alters the engineering property of a natural soil; used to stabilize soil slopes, to prepare for building foundations, and to prevent erosion. { 'sȯil 'stab·ə̣ˌlīz·ər }

soil stack [BUILD] The main vertical pipe into which flows the waste water from the soil pipes in a structure. { 'sȯil ˌstak }

soil stripes [GEOL] Alternating bands of fine and coarse material in a soil structure. { 'sȯil ˌstrīps }

soil structure [GEOL] Arrangement of soil into various aggregates, each differing in the characteristics of its particles. { 'sȯil ˌstrək·chər }

soil survey [GEOL] The systematic examination of soils, their description and classification, mapping of soil types, and the assessment of soils for various agricultural and engineering uses. { 'sȯil ˌsər‚vā }

soil thermograph [ENG] A remote-recording thermograph whose sensing element may be buried at various depths in the earth. { 'sȯil 'thər·mə̣ˌgraf }

soil thermometer [ENG] A thermometer used to measure the temperature of the soil, usually the mercury-in-glass thermometer. Also known as earth thermometer. { 'sȯil thər‚mäm·əd·ər }

soil vent *See* stack vent. { 'sȯil ˌvent }

soil water [HYD] Water in the belt of soil water. Also known as rhizic water; soil moisture. { 'sȯil ˌwȯd·ər }

soil-water belt *See* belt of soil water. { 'sȯil ˌwȯd·ər ˌbelt }

soil-water zone *See* belt of soil water. { 'sȯil ˌwȯd·ər ˌzōn }

sol [CHEM] A colloidal solution consisting of a suitable dispersion medium, which may be gas, liquid, or solid, and the colloidal substance, the disperse phase, which is distributed throughout the dispersion medium. { säl }

Sol *See* sun. { säl }

solaire [METEOROL] A name generally applied to winds from an easterly direction (that is, from the rising sun) in central and southern France. { sō'ler }

sol-air temperature [METEOROL] The temperature which, under conditions of no direct solar radiation and no air motion, would cause the same heat transfer into a house as that caused by the interplay of all existing atmospheric conditions. { 'säl'er ˌtem·prə·chər }

Solanaceae [BOT] A family of dicotyledonous plants in the order Polemoniales having internal phloem, mostly numerous ovules and seeds on axile placentae, and mostly cellular endosperm. { ˌsō·lə'nās·ē‚ē }

solanine [BIOCHEM] A bitter poisonous alkaloid derived from potato sprouts (*Solanum tuberosum*), tomatoes, and nightshade. { 'sō·lə‚nēn }

solano [METEOROL] A southeasterly or easterly wind on the southeast coast of Spain in summer; usually an extension of the sirocco; it is hot and humid and sometimes brings rain; when dry, it is dusty. { sō'lä·nō }

solar absorption index [GEOPHYS] A relation of the sun's angle at various latitudes and local times with the ionospheric absorption. { 'sō·lər əp'sȯrp·shən ˌin‚deks }

solar activity [ASTRON] Disturbances on the surface of the sun; examples are sunspots, prominences, and solar flares. { 'sō·lər ak'tiv·əd·ē }

solar air mass [METEOROL] The optical air mass penetrated by light from the sun for any given position of the sun. { 'sō·lər 'er ˌmas }

solar antapex [ASTRON] The point on the celestial sphere away from which the solar system is moving; it lies in the constellation Columba. { 'sō·lər ant'ā‚peks }

solar apex [ASTRON] A point toward which the solar system is moving; it is about 10° southwest of the star Vega. { 'sō·lər 'ā‚peks }

solar atmospheric tide [GEOPHYS] An atmospheric tide due to the thermal or gravitational action of the sun. { 'sō·lər 'at·mə̣ˌsfir·ik 'tīd }

solar attachment [ENG] A device for determining the true meridian directly from the sun; used an an attachment on a surveyor's transit or compass. { 'sō·lər ə'tach·mənt }

solar battery [ELECTR] An array of solar cells, usually connected in parallel and series. { 'sō·lər 'bad·ə·rē }

solar bridge [ASTRON] A bright, narrow, streak-shaped region which is sometimes observed across a large sunspot, dividing the umbra into two or more parts. { 'sō·lər 'brij }

solar burst [ASTROPHYS] A sudden increase in the radio-frequency energy radiated by the sun, generally associated with visible solar flares. { 'sō·lər 'bərst }

solar calendar [ASTRON] A calendar based on the time period known as the tropical year, which has 365.24220 days. { 'sō·lər 'kal·ən·dər }

solar cavity *See* heliosphere. { 'sō·lər 'kav·əd·ē }

solar cell [ELECTR] A *pn*-junction device which converts the radiant energy of sunlight directly and efficiently into electrical energy. { 'sō·lər 'sel }

solar chimney [ENG] A natural-draft drive device that uses solar radiation to provide upward momentum to a mass of air, thereby converting the thermal energy to kinetic energy, which can be extracted from the air with suitable wind machines. { 'sō·lər 'chim·nē }

solar climate [CLIMATOL] The hypothetical climate which would prevail on a uniform solid earth with no atmosphere; thus, it is a climate of temperature alone and is determined only by the amount of solar radiation received. { 'sō·lər 'klī·mət }

solar collector [ENG] An installation designed to gather and accumulate energy in the form of solar radiation. { 'sō·lər kə'lek·tər }

solar constant [METEOROL] The rate at which energy from the sun is received just outside the earth's atmosphere on a surface normal to the incident radiation and at the earth's mean distance from the sun; it is approximately 1367 watts per square meter. { 'sō·lər 'kän·stənt }

solar cooking [FOOD ENG] The preparation of food by concentrating solar radiation on a heater plate. { 'sō·lər 'kūk·iŋ }

solar corona [ASTRON] The upper, rarefied solar atmosphere which becomes visible around the darkened sun during a total solar eclipse. Also known as corona. { 'sō·lər kə'rō·nə }

solar cosmic rays *See* energetic solar particles. { 'sō·lər 'käz·mik 'rāz }

solar cycle [ASTRON] The periodic change in the number of sunspots; the cycle is taken as the interval between successive minima and is about 11.1 years. { 'sō·lər 'sī·kəl }

solar day [ASTRON] A time measurement, the duration of one rotation of the earth on its axis with respect to the sun; this may be a mean solar day or an apparent solar day as the reference is to the mean sun or apparent sun. { 'sō·lər 'dā }

solar dermatitis [MED] Any of various skin eruptions caused by exposure to the sun, excluding sunburn. { 'sō·lər ˌdər·mə'tīd·əs }

solar distillation [CHEM ENG] A procedure in which the sun's heat is used to evaporate seawater in order to produce sodium chloride and other salts or potable water. { 'sō·lər ˌdis·tə'lā·shən }

solar eclipse [ASTRON] An eclipse that takes place when the new moon passes between the earth and the sun and the shadow formed reaches the earth; may be classified as total, partial, or annular. { 'sō·lər i'klips }

solar energy [ASTROPHYS] The energy transmitted from the sun in the form of electromagnetic radiation. { 'sō·lər 'en·ər·jē }

solar engine [MECH ENG] An engine which converts thermal energy from the sun into electrical, mechanical, or refrigeration energy; may be used as a method of spacecraft propulsion, either directly by photon pressure on huge solar sails, or indirectly from solar cells or from a reflector-boiler combination used to heat a fluid. { 'sō·lər 'en·jən }

solar evaporation [HYD] The evaporation of water due to the sun's heat. { 'sō·lər iˌvap·ə'rā·shən }

solar-excited laser *See* sun-pumped laser. { 'sō·lər ik'sīd·əd 'lā·zər }

solar faculae [ASTRON] Bright streaks or regions on the surface of the sun, especially near solar sunspots. { 'sō·lər 'fak·yə‚lē }

solar flare [ASTROPHYS] An abrupt increase in the intensity of the H-α and other emission near a sunspot region; the brightness may be many times that of the associated plage. { 'sō·lər fler }

solar flux unit [ASTRON] A unit of solar radio emission per

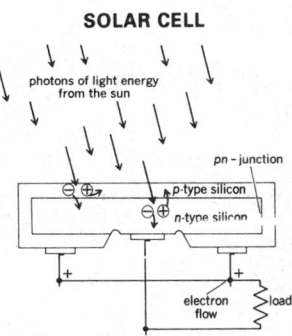

SOLAR CELL

Cross-sectional view of a silicon *pn*-junction solar cell, illustrating the creation of electron pairs by photons of light energy from the sun.

unit frequency interval, equal to 10^{-22} watt per square meter per hertz at the earth. { 'sō·lər ¦fläks ¦yü·nət }

solar furnace [ENG] An image furnace in which high temperatures are produced by focusing solar radiation. { 'sō·lər 'fər·nəs }

solar generator [ELEC] An electric generator powered by radiation from the sun and used in some satellites. { 'sō·lər 'jen·ə‚rād·ər }

solar heat exchanger drive [AERO ENG] A proposed method of spacecraft propulsion in which solar radiation is focused on an area occupied by a boiler to heat a working fluid that is expelled to produce thrust directly. Abbreviated SHED. { 'sō·lər 'hēt iks¦chān·jər ‚drīv }

solar heating [MECH ENG] The conversion of solar radiation into heat for technological, comfort-heating, and cooking purposes. { 'sō·lər 'hēd·iŋ }

solar heat storage [ENG] The storage of solar energy for later use; usually accomplished by the heating of water or fusing a salt, although sand and gravel have been used as storage media. { 'sō·lər 'hēt ‚stȯr·ij }

solar house [BUILD] A house with large expanses of glass designed to catch solar radiation for heating. { ¦sō·lər ¦haus }

solarimeter [ENG] 1. A type of pyranometer consisting of a Moll thermopile shielded from the wind by a bell glass. 2. *See* pyranometer. { ‚sō·lə'rim·əd·ər }

solarization [GRAPHICS] 1. Reversal of a photographic image due to great overexposure. 2. *See* Sabattier effect. [PHYS] Loss of transparency or coloration of glass exposed to sunlight or ultraviolet radiation. { ‚sō·lə·rə'zā·shən }

solar magnetic field [ASTROPHYS] The magnetic field that pervades the ionized and highly conducting gas composing the sun. { 'sō·lər mag'ned·ik 'fēld }

solar magnetograph [ENG] An instrument that utilizes the Zeeman effect to directly measure the strength and polarity of the complex patterns of magnetic fields at the sun's surface; comprises a telescope, a differential analyzer, a spectrograph, and a photoelectric or photographic means of differencing and recording. { 'sō·lər mag'ned·ə‚graf }

solar month [ASTRON] A time interval equal to one-twelfth of the solar year. { 'sō·lər 'mənth }

solar motion [ASTRON] The two main motions of the sun: relative motion with respect to the neighboring stars, or motion due to the rotation of the Milky Way of which the sun is a part. { 'sō·lər 'mō·shən }

solar nebula [ASTRON] The rotating flattened cloud of gas and dust from which the sun and the rest of the bodies in the solar system formed, about 4.56×10^9 years ago. { ¦sō·lər 'neb·yə·lə }

solar neutrino [ASTROPHYS] A neutrino produced in a nuclear reaction inside the sun. { 'sō·lər nü'trē·nō }

solar neutrino problem [ASTROPHYS] The difficulty of understanding why the observed flux of neutrinos from the sun is significantly lower than predicted by standard solar models; this discrepancy suggests the existence of some type of neutrino oscillation. { ¦sō·lər nü'trē·nō ‚präb·ləm }

solar neutrino unit [ASTROPHYS] A unit for measuring the capture rate of neutrinos emanating from the sun, equal to 10^{-36} per second per atom. Abbreviated SNU. { 'sō·lər nü'trē·nō ‚yü·nət }

solar noise *See* solar radio noise. { 'sō·lər 'nȯiz }

solar nutation [ASTRON] Nutation caused by the change in declination of the sun. { 'sō·lər nü'tā·shən }

solar orbit [ASTRON] An orbit of a planet or other celestial body or satellite about the sun. { 'sō·lər 'ȯr·bət }

solar parallax [ASTRON] The sun's mean equatorial horizontal parallax p, which is the angle subtended by the equatorial radius r of the earth at mean distance a of the sun. { 'sō·lər 'par·ə‚laks }

solar phase angle [ASTRON] The angular distance between the earth and the sun at a specified planet. { 'sō·lər 'fāz ‚aŋ·gəl }

solar photophthalmia *See* snow blindness. { 'sō·lər 'fäd·ō'thal·mē·ə }

solar physics [ASTROPHYS] The scientific study of all physical phenomena connected with the sun; it overlaps with geophysics in the consideration of solar-terrestrial relationships, such as the connection between solar activity and auroras. { 'sō·lər 'fiz·iks }

solar pond [MECH ENG] A type of nonfocusing solar collector consisting of a pool of salt water heated by the sun; used either directly as a source of heat or as a power source for an electric generator. Also known as salt-gradient solar pond. { 'sō·lər 'pänd }

solar power [MECH ENG] The conversion of the energy of the sun's radiation to useful work. { 'sō·lər 'pau·ər }

solar-powered aircraft [AERO ENG] An aircraft in which the energy required for propulsion is collected by arrays of solar photovoltaic cells mounted on the wings. { 'sō·lər ¦pau·ərd 'er‚kraft }

solar power satellite [ENG] A proposed collector of solar energy that would be placed in geostationary orbit where sunlight striking the satellite would be converted to electricity and then to microwaves, which would be beamed to earth. { ‚sō·lər ‚pau·ər 'sad·əl‚īt }

solar probe [AERO ENG] A space probe whose trajectory passes near the sun so that instruments on board may detect and transmit back to earth data about the sun. { 'sō·lər 'prōb }

solar prominence [ASTRON] Sheets of luminous gas emanating from the sun's surface; they appear dark against the sun's disk but bright against the dark sky, and occur only in regions of horizontal magnetic fields. { 'sō·lər 'präm·ə·nəns }

solar propagation [BOT] A method of rooting plant cuttings involving the use of a modified hotbed; bottom heat is provided by radiation of stored solar heat from bricks or stones in the bottom of the hotbed frame. { 'sō·lər ‚präp·ə'gā·shən }

solar propulsion [AERO ENG] Spacecraft propulsion with a system composed of a type of solar engine. { 'sō·lər prə'pəl·shən }

solar pumping [OPTICS] The use of sunlight focused directly into a laser rod for pumping to induce lasing action. { 'sō·lər 'pəmp·iŋ }

solar radiation [ASTROPHYS] The electromagnetic radiation and particles (electrons, protons, and rarer heavy atomic nuclei) emitted by the sun. { 'sō·lər ‚rād·ē'ā·shən }

solar-radiation observation [GEOPHYS] An evaluation of the radiation from the sun that reaches an observation point; the observing instrument is usually a pyrheliometer or pyranometer. { 'sō·lər ‚rād·ē·ā·shən ‚äb·zər‚vā·shən }

solar radio emission [ASTROPHYS] Radio-frequency electromagnetic radiation emitted from the sun, and increasing greatly in intensity during sunspots and flares. { 'sō·lər 'rād·ē·ō i‚mish·ən }

solar radio noise [ELECTROMAG] Radio noise originating at the sun, and increasing greatly in intensity during sunspots and flares; it is heard as a hissing noise on shortwave radio receivers. Also known as solar noise. { 'sō·lər 'rād·ē·ō ‚nȯiz }

solar rocket [AERO ENG] A rocket designed to carry instruments to measure and transmit parameters of the sun. { 'sō·lər 'räk·ət }

solar sail [AERO ENG] A surface of a highly polished material upon which solar light radiation exerts a pressure. Also known as photon sail. { 'sō·lər 'sāl }

solar satellite [AERO ENG] A space vehicle designed to enter into orbit about the sun. Also known as sun satellite. { 'sō·lər 'sad·əl‚īt }

solar sector [ASTRON] A region of the solar wind in which one magnetic polarity predominates. { 'sō·lər 'sek·tər }

solar sensor [ELECTR] A light-sensitive diode that sends a signal to the attitude-control system of a spacecraft when it senses the sun. Also known as sun sensor. { 'sō·lər 'sen·sər }

solar spectrum [ASTROPHYS] The spectrum of the sun's electromagnetic radiation extending over the whole electromagnetic spectrum, from wavelengths of 10^{-9} centimeter to 30 kilometers. { 'sō·lər 'spek·trəm }

solar spicule *See* spicule. { 'sō·lər 'spik·yül }

solar still [CHEM ENG] A device for evaporating seawater, in which water is confined in one or more shallow pools, over which is placed a roof-shaped transparent cover made of glass or plastic film; the sun's heat evaporates the water, leaving behind a residue of salt; the vapor from the evaporated water condenses on the surface of the cover and trickles down into gutters, which thus collect fresh water. { 'sō·lər 'stil }

solar system [ASTRON] The sun and the celestial bodies

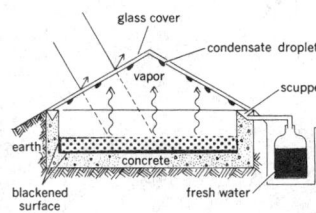

SOLAR STILL

Roof-type solar still. Dotted area above the concrete is sea water.

moving about it; the bodies are planets, satellites of the planets, asteroids, comets, and meteor swarms. { 'sō·lər ,sis·təm }

solar telescope [OPTICS] An observational instrument of the solar astronomer; it is designed so that heating effects produced by the sun do not distort the images; the two classes consist of those designed for observations of the brilliant solar disk, and those designed for the study of the much fainter prominences and the still fainter corona through the relatively bright, scattered light of the sky. { 'sō·lər 'tel·ə,skōp }

solar-terrestrial phenomena [GEOPHYS] All observed physical effects that are caused by solar activity; the phenomena may be in the atmosphere or on the earth's surface; an example is the aurora borealis. { 'sō·lər tə'res·trē·əl fə'näm·ə,nä }

solar tide [OCEANOGR] The tide caused solely by the tide-producing forces of the sun. { 'sō·lər 'tīd }

solar time [ASTRON] Time based on the rotation of the earth relative to the sun. { 'sō·lər 'tīm }

solar-topographic theory [CLIMATOL] The theory that the changes of climate through geologic time (the paleoclimates) have been due to changes of land and sea distribution and orography, combined with fluctuations of solar radiation of the order of 10–20% on either side of the mean. { 'sō·lər ¦täp·ə¦graf·ik 'thē·ə·rē }

solar tower [ASTRON] A tower which has a coelostat mounted on top to reflect the sun's light vertically downward so that it may be studied with a spectroheliograph or other astronomical instrument at the bottom of the tower. { 'sō·lər ,tau̇·ər }

solar turboelectric drive [AERO ENG] A proposed method of spacecraft propulsion in which solar radiation is focused on an area occupied by a boiler to heat a working fluid that drives a turbine generator system, producing electrical energy. Abbreviated STED. { 'sō·lər ¦tər·bō·i¦lek·trik 'drīv }

solar-type star [ASTRON] Any of the stars (yellow stars) of spectral type G, so called because the sun is in this class. { 'sō·lər ¦tīp 'stär }

solar ultraviolet radiation [ASTROPHYS] That portion of the sun's electromagnetic radiation that has wavelengths from about 400 to about 4 nanometers; this radiation may sufficiently ionize the earth's atmosphere so that propagation of radio waves is affected. { 'sō·lər ¦əl·trə¦vī·lət ,rād·ē'ā·shən }

solar units [ASTROPHYS] A set of units for measuring properties of stars, in which properties of the sun such as mass, diameter, density, and luminosity are set equal to unity. { 'sō·lər ,yü·nəts }

solar velocity [ASTRON] The Sun's velocity with respect to the local standard of rest. { 'sō·lər və'läs·əd·ē }

solar wind [GEOPHYS] The supersonic flow of gas, composed of ionized hydrogen and helium, which continuously flows from the sun out through the solar system with velocities of 180 to 600 miles (300 to 1000 kilometers) per second; it carries magnetic fields from the sun. { 'sō·lər 'wind }

Solasteridae [INV ZOO] The sun stars, a family of asteroid echinoderms in the order Spinulosida. { ,säl·ə'ster·ə,dē }

solation [PHYS CHEM] The change of a substance from a gel to a sol. { sə'lā·shən }

solder [MET] **1.** To join by means of solder. **2.** An alloy, such as of zinc and copper, or of tin and lead, used when melted to join metallic surfaces. { 'säd·ər }

solder-ball flip chip See flip chip. { 'säd·ər ,bȯl 'flip ,chip }

solder brazing [MET] Brazing by means of a relatively high-melting solder. { 'säd·ər ¦brāz·iŋ }

solder glass [MATER] A special glass having a relatively low softening point (below 500°C); used to join two pieces of higher-melting glass without softening and deforming them. { 'säd·ər ,glas }

soldering embrittlement [MET] The reduction in mechanical properties of a metal due to the local penetration of solder along grain boundaries. { 'säd·ə·riŋ em'brid·əl·mənt }

soldering flux [MET] A chemical substance which aids the flow of solder and serves to remove and prevent the formation of oxides on the pieces to be united. { 'säd·ə·riŋ ,fləks }

soldering gun [ENG] A soldering iron shaped like a gun. { 'säd·ə·riŋ ,gən }

soldering iron [ENG] A rod of copper with a handle on one end and pointed or wedge-shaped at the other end, and used for applying heat in soldering. { 'säd·ə·riŋ ,ī·ərn }

soldering lug [ELEC] A stamped metal strip used as a terminal to which wires can be soldered. { 'säd·ə·riŋ ,ləg }

soldering pencil [ENG] A small soldering iron, about the size and weight of a standard lead pencil, used for soldering or unsoldering joints on printed wiring boards. { 'säd·ə·riŋ ,pen·səl }

solderless contact See crimp contact. { 'säd·ər·ləs 'kän ,takt }

solderless wrapped connection See wire-wrap connection. { 'säd·ər·ləs 'rapt kə'nek·shən }

solder track [ELECTR] A conducting path on a printed circuit board that is formed by applying molten solder to the board. { 'säd·ər ,trak }

soldier course [CIV ENG] A course of bricks laid on their ends so that only their long sides are visible. { 'sōl·jər ,kȯrs }

sole [BUILD] The horizontal member beneath the studs in a framed building. [ELECTR] Electrode used in magnetrons and backward-wave oscillators to carry a current that generates a magnetic field in the direction wanted. [GEOGR] The lowest part of a valley. [GEOL] **1.** The bottom of a sedimentary stratum. **2.** The middle and lower portion of the shear surface of a landslide. **3.** The underlying fault plane of a thrust nappe. Also known as sole plane. [HYD] The basal ice of a glacier, often dirty in appearance due to contained rock fragments. { sōl }

soleil compensator [OPTICS] A compensator which resembles the Babinet compensator but is constructed so that the phase change is constant over the entire field. { sō'lā 'käm·pən,sād·ər }

sole injection [GEOL] An igneous intrusion that was put in place along a thrust plane. { 'sōl in,jek·shən }

sole mark [GEOL] An irregularity or penetration on the undersurface of a sedimentary stratum. { 'sōl ,märk }

Solemyidae [INV ZOO] A family of bivalve mollusks in the order Protobranchia. { ,säl·ə'mī·ə,dē }

Solenichthyes [VERT ZOO] An equivalent name for Gasterosteiformes. { ,säl·ə'nik·thē,ēz }

solenium [INV ZOO] A diverticulum of the enteron in certain hydroids. { sō'lē·nē·əm }

solenocyte [INV ZOO] Any of various hollow, flagellated cells in the nephridia of the larvae of certain annelids, mollusks, rotifers, and lancelets. { sō'lē·nə,sīt }

solenodon [VERT ZOO] Either of two species of insectivorous mammals comprising the family Solenodontidae; the almique (Atopogale cubana) is found only in Cuba, while the white agouta (Solenodon paradoxus) is confined to Haiti. { sō'lē·nə,dän }

Solenodontidae [VERT ZOO] The solenodons, a family of insectivores belonging to the group Lipotyphla. { sō,lē·nə'dänt·ə,dē }

Solenogastres [INV ZOO] The equivalent name for Aplacophora. { sō,lē·nə'ga,strēz }

solenoid [ELECTROMAG] Also known as electric solenoid. **1.** An electrically energized coil of insulated wire which produces a magnetic field within the coil. **2.** In particular, a coil that surrounds a movable iron core which is pulled to a central position with respect to the coil when the coil is energized by sending current through it. [METEOROL] See meteorological solenoid. { 'säl·ə,nȯid }

solenoidal [MATH] A vector field has this property in a region if its divergence vanishes at every point of the region. { ¦säl·ə¦nȯid·əl }

solenoidal index [METEOROL] The difference between the mean virtual temperature from the surface to some specified upper level averaged around the earth at 55° latitude, and the mean virtual temperature for the corresponding layer averaged at 35° latitude. { ¦säl·ə¦nȯid·əl 'in,deks }

solenoid brake [MECH ENG] A device that retards or arrests rotational motion by means of the magnetic resistance of a solenoid. { 'säl·ə,nȯid ,brāk }

solenoid group [MATH] A compact Abelian, topological group that is one-dimensional and connected. { 'sä·lə,nȯid ,grüp }

solenoid valve [MECH ENG] A valve actuated by a solenoid, for controlling the flow of gases or liquids in pipes. { 'säl·ə,nȯid ,valv }

Solenopora [PALEOBOT] A genus of extinct calcareous red algae in the family Solenoporaceae that appeared in the Late Cambrian and lasted until the Early Tertiary. { ,säl·ə'näp·rə }

Solenoporaceae [PALEOBOT] A family of extinct red algae

SOLENODON

The solenodon uses its elongated snout to grub for food.

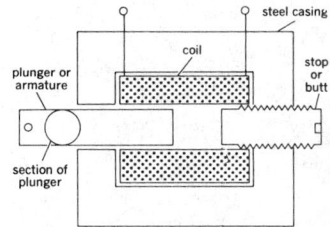

SOLENOID

A cross-sectional view of a steel-clad solenoid showing coil and movable core known as plunger or armature.

having compact tissue and the ability to deposit calcium carbonate within and between the cell walls. { sō͵lē·nə·pə'rās·ē͵ē }

solepiece [CIV ENG] One of two steel plates, port and starboard, whose forward parts are bolted to the ground ways supporting a ship about to be launched, while their aft parts are attached to the sliding ways; at the start of the launch, they are cut simultaneously with burning torches to release the ship. Also known as soleplate. [NAV ARCH] **1.** A member attached to the bottom of the rudder of a wooden ship, which brings it down to the false keel. **2.** A member which joins the sternpost and the rudderpost of a wooden ship. **3.** A projection from the keel of a ship. { 'sōl͵pēs }

sole plane *See* sole. { 'sōl ͵plān }

soleplate [BUILD] The plate on which stud bases butt in a stud partition. [CIV ENG] *See* solepiece. [ENG] **1.** The supporting base of a machine. **2.** A plate on which a bearing can be attached and, if necessary, adjusted slightly. { 'sōl͵plāt }

soleus [ANAT] A flat muscle of the calf; origin is the fibula, popliteal fascia, and tibia, and insertion is the calcaneus; plantarflexes the foot. { 'sō·lē·əs }

solfatara [GEOL] A fumarole from which sulfurous gases are emitted. { ͵säl·fä'tär·ə }

solfataric [ECOL] Relating to a volcanic area. { ͵sōl·fə'tär·ik }

sol-gel glass [PHYS CHEM] An optically transparent amorphous silica or silicate material produced by forming interconnections in a network of colloidal, submicrometer particles under increasing viscosity until the network becomes completely rigid, with about one-half the density of glass. { 'säl ͵jel 'glas }

sol-gel monolith process [MATER] A method for fabricating advanced ceramics in which a suspension of colloidal ceramic particles (sol) is converted to a gel by chemical treatment, and then dried and sintered to form a ceramic product. Many ceramic fibers are manufactured by the sol-gel monolith process. { ͵säl ͵jel 'män·ə͵lith ͵prä͵ses }

solid [PHYS] **1.** A substance that has a definite volume and shape and resists forces that tend to alter its volume or shape. **2.** A crystalline material, that is, one in which the constituent atoms are arranged in a three-dimensional lattice, periodic in three independent directions. { 'säl·əd }

solid angle [MATH] A surface formed by all rays joining a point to a closed curve. { 'säl·əd 'aŋ·gəl }

solid asphalt [MATER] Asphalt with a penetration number of less than 10 under specified test conditions. { 'säl·əd 'as͵fólt }

solid box [MECH ENG] A solid, unadjustable ring bearing lined with babbitt metal, used on light machinery. { 'säl·əd 'bäks }

solid car [MIN ENG] A mine car equipped with a swivel coupling and generally used with a rotary dump. { 'säl·əd 'kär }

solid coupling [MECH ENG] A flanged-face or a compression-type coupling used to connect two shafts to make a permanent joint and usually designed to be capable of transmitting the full load capacity of the shaft; a solid coupling has no flexibility. { 'säl·əd 'kəp·liŋ }

solid crib timbering [MIN ENG] Shaft timbering with cribs laid solidly upon one another. { 'säl·əd ͵krib 'tim·bə·riŋ }

solid cutter [DES ENG] A cutter made of a single piece of material. { 'säl·əd 'kəd·ər }

solid die [DES ENG] A one-piece screw-cutting tool with internal threads. { 'säl·əd 'dī }

solid-dielectric capacitor [ELEC] A capacitor whose dielectric is one of several solid materials such as ceramic, mica, glass, plastic film, or paper. { 'säl·əd ͵dī·ə͵lek·trik kə'pas·əd·ər }

solid drilling [ENG] In diamond drilling, using a bit that grinds the whole face, without preserving a core for sampling. { 'säl·əd 'dril·iŋ }

solid-electrolyte battery [ELEC] A primary battery whose electrolyte is either a solid crystalline salt, such as silver iodide or lead chloride, or an ion-exchange membrane; in either case, conductivity is almost entirely ionic. { 'säl·əd i'lek·trə͵līt 'bad·ə·rē }

solid-electrolyte fuel cell [ELEC] Self-contained fuel cell in which oxygen is the oxidant and hydrogen is the fuel; the oxidant and fuel are kept separated by a solid electrolyte which

has a crystalline structure and a low conductivity. { 'säl·əd i'lek·trə͵līt 'fyül ͵sel }

solid-electrolyte gas transducer [ENG] A device in which the concentration of a particular gas in a mixture is determined from the diffusion voltage across a heated solid electrolyte placed between this mixture and a reference gas. { 'säl·əd i'lek·trə͵līt 'gas tranz͵düs·ər }

solid electrolytic capacitor [ELEC] An electrolytic capacitor in which the dielectric is an anodized coating on one electrode, with a solid semiconductor material filling the rest of the space between the electrodes. { 'säl·əd i͵lek·trə͵lid·ik kə'pas·əd·ər }

solid explosive [MATER] An explosive employed in the form of a powder, a light-running granulated mass, or as solid sticks. { 'säl·əd ik'splō·siv }

solid geometry [MATH] The geometric study of objects in three-dimensional space, such as spheres and polyhedrons. { ͵säl·əd jē'äm·ə·trē }

solid helium [CRYO] A certain state which is not attained by helium under its own vapor pressure down to absolute zero, but which requires an external pressure of 25 atmospheres at absolute zero. { 'säl·əd 'hē·lē·əm }

solidification [PHYS] The change of a fluid (liquid or gas) into the solid state. { sə'lid·ə·fə'kā·shən }

solidification shrinkage [MET] Volume contraction of a metal during solidification. { sə'lid·ə·fə'kā·shən ͵shriŋ·kij }

solid injection system [MECH ENG] A fuel injection system for a diesel engine in which a pump forces fuel through a fuel line and an atomizing nozzle into the combustion chamber. { 'säl·əd in'jek·shən ͵sis·təm }

solid insulator [ELEC] An electric insulator made of a solid substance, such as sulfur, polystyrene, rubber, or porcelain. { 'säl·əd 'in·sə͵lād·ər }

solid ionization chamber [NUCLEO] A particle detector in which the gas filling a conventional ionization chamber is replaced by a large single crystal of suitably chosen material. { 'säl·əd ͵ī·ə·nə'zā·shən ͵chäm·bər }

solid laser [OPTICS] A laser in which either a crystalline or amorphous solid material, usually in the form of a rod, is excited by optical pumping; the most common crystalline materials are ruby, neodymium-doped ruby, and neodymium-doped yttrium aluminum garnet. { 'säl·əd 'lā·zər }

solid-liquid equilibrium [PHYS CHEM] **1.** The interrelation of a solid material and its melt at constant vapor pressure. **2.** The concentration relationship of a solid with a solvent liquid other than its melt. Also known as liquid-solid equilibrium. { 'säl·əd 'lik·wəd ͵ē·kwə'lib·rē·əm }

solid logic technology [ELECTR] A method of computer construction that makes use of miniaturized modules, resulting in faster circuitry because of the reduced distances that current must travel. { 'säl·əd ͵läj·ik tek'näl·ə·jē }

solid moment of inertia [PHYS] The integral of the products of the mass of each of the infinitesimal elements of the solid with the square of their distance from a given axis. { 'säl·əd 'mō·mənt əv i'nər·shə }

solid of revolution [MATH] A solid that can be generated by rotating a plane area about a line. { ͵säl·əd əv ͵rev·ə'lü·shən }

solid-phase welding [MET] A welding method in which the weld is consummated by pressure or by heat and pressure without fusion. { 'säl·əd ͵fāz 'weld·iŋ }

solid-piled [MATER] Pertaining to plywood which is fresh from clamps or a hot press, and which is piled onto a solid flat base without stickers and weighted down until it reaches its normal temperature and moisture content. Also known as bulked-down; dead-piled. { 'säl·əd 'pīld }

solid propellant [MATER] A rocket propellant in solid form, usually containing both fuel and oxidizer combined or mixed, and formed into a monolithic (not powdered or granulated) grain. Also known as solid rocket fuel; solid rocket propellant. { 'säl·əd prə'pel·ənt }

solid-propellant binder [MATER] The ingredient component of a propellant that is the agent for holding all the other ingredients together; contributes most to the physical or mechanical properties of the grain. { 'säl·əd prə͵pel·ənt 'bin·dər }

solid-propellant rocket engine [AERO ENG] A rocket engine fueled with a solid propellant; such motors consist essentially of a combustion chamber containing the propellant, and

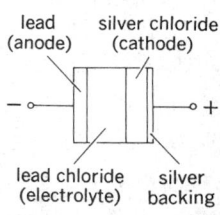

SOLID-ELECTROLYTE BATTERY

lead (anode)　　silver chloride (cathode)

lead chloride (electrolyte)　　silver backing

Schematic diagram of typical solid-electrolyte cell with solid crystalline salt electrolyte.

a nozzle for the exhaust jet. { 'säl·əd prə¦pel·ənt 'räk·ət ‚en·jən }

solid rocket [AERO ENG] A rocket that is propelled by a solid-propellant rocket engine. { 'säl·əd 'räk·ət }

solid rocket fuel See solid propellant. { 'säl·əd 'räk·ət ‚fyül }

solid rocket propellant See solid propellant. { 'säl·əd 'räk·ət prə‚pel·ənt }

solid Schmidt telescope [OPTICS] A type of Schmidt system which is constructed from a single block of glass, designed to operate at very small aperture ratios. { 'säl·əd 'shmit 'tel·ə‚skōp }

solid shafting [MECH ENG] A solid round bar that supports a roller and wheel of a machine. { 'säl·əd 'shaft·iŋ }

solid shank tool [ENG] A cutting tool in which the shank and cutting edges are machined from one piece. { 'säl·əd ‚shaŋk 'tül }

solid solution [PHYS] A homogeneous crystalline phase composed of several distinct chemical species, occupying the lattice points at random and existing in a range of concentrations. { 'säl·əd sə'lü·shən }

solid sphere [MATH] The union of a sphere with its interior. { 'säl·əd ‚sfir }

solid state [ENG] Pertaining to a circuit, device, or system that depends on some combination of electrical, magnetic, and optical phenomena within a solid that is usually a crystalline semiconductor material. [PHYS] The condition of a substance in which it is a solid. { 'säl·əd ‚stāt }

solid-state amorphizing reaction [MET] An interdiffusion reaction that takes place at constant temperature over long periods of time at a clean, oxide-free boundary between two crystalline metals that have a large chemical affinity, and results in the formation of an amorphous alloy of the two metals (a metallic glass). { 'säl·əd ‚stāt ə'mór‚fīz·iŋ rē'ak·shən }

solid-state battery [ELEC] A battery in which both the electrodes and the electrolyte are solid-state materials. { 'säl·əd ‚stāt 'bad·ə·rē }

solid-state circuit [ELECTR] Complete circuit formed from a single block of semiconductor material. { 'säl·əd ‚stāt 'sər·kət }

solid-state circuit breaker [ELECTR] A circuit breaker in which a Zener diode, silicon controlled rectifier, or solid-state device is connected to sense when load terminal voltage exceeds a safe value. { 'säl·əd ‚stāt 'sər·kət ‚brāk·ər }

solid-state component [ELECTR] A component whose operation depends on the control of electrical or magnetic phenomena in solids, such as a transistor, crystal diode, or ferrite device. { 'säl·əd ‚stāt kəm'pō·nənt }

solid-state counter [NUCLEO] A radiation counter whose sensitive material is a crystalline solid; for example, a crystal counter or a scintillation counter. { 'säl·əd ‚stāt 'kaunt·ər }

solid-state device [ELECTR] A device, other than a conductor, which uses magnetic, electrical, and other properties of solid materials, as opposed to vacuum or gaseous devices. { 'säl·əd ‚stāt di'vīs }

solid-state image sensor See charge-coupled image sensor. { 'säl·əd ‚stāt 'im·ij ‚sen·sər }

solid-state lamp See light-emitting diode. { 'säl·əd ‚stāt 'lamp }

solid-state laser [OPTICS] A laser in which a semiconductor material produces the coherent output beam. { 'säl·əd ‚stāt 'lā·zər }

solid-state maser [PHYS] A maser in which a semiconductor material produces the coherent output beam; two input waves are required: one wave, called the pumping source, induces upward energy transitions in the active material, and the second wave, of lower frequency, causes downward transitions and undergoes amplification as it absorbs photons from the active material. { 'säl·əd ‚stāt 'mā·zər }

solid-state memory [COMPUT SCI] A computer memory whose elements consist of integrated-circuit bistable multivibrators in which bits of information are stored as one of two states. { 'säl·əd ‚stāt 'mem·rē }

solid-state physics [PHYS] The branch of physics centering about the physical properties of solid materials. { 'säl·əd ‚stāt 'fiz·iks }

solid-state power amplifier [ELECTR] An amplifier that uses field-effect transistors to provide useful amplification at gigahertz frequencies. { ‚säl·əd ‚stāt 'pau·ər ‚am·plə‚fī·ər }

solid-state relay [ELECTR] A relay that uses only solid-state components, with no moving parts. Abbreviated SSR. { 'säl·əd ‚stāt 'rē‚lā }

solid-state switch [ELECTR] A microwave switch in which a semiconductor material serves as the switching element; a zero or negative potential applied to the control electrode will reverse-bias the switch and turn it off, and a slight positive voltage will turn it on. { 'säl·əd ‚stāt 'swich }

solid-state thyratron [ELECTR] A semiconductor device, such as a silicon controlled rectifier, that approximates the extremely fast switching speed and power-handling capability of a gaseous thyratron tube. { 'säl·əd ‚stāt 'thī·rə‚trän }

solid-state uninterruptible power system [ELEC] An uninterruptible power system in which the load operates continuously from the output of a dc-to-ac static inverter powered by a battery. { 'säl·əd ‚stāt ‚ən‚int·ə'rəp·tə·bəl 'pau·ər ‚sis·təm }

solid-state welding [MET] Welding processes which coalesce materials at temperatures below the melting point of the base metal by methods such as welding or diffusion welding without the use of filler metal. { 'säl·əd ‚stāt 'weld·iŋ }

solid stowing [MIN ENG] The complete filling of the waste area behind a longwall face with stone and dirt. { 'säl·əd 'stō·iŋ }

solid tantalum capacitor [ELEC] An electrolytic capacitor in which the anode is a porous pellet of tantalum; the dielectric is an extremely thin layer of tantalum pentoxide formed by anodization of the exterior and interior surfaces of the pellet; the cathode is a layer of semiconducting manganese dioxide that fills the pores of the anode over the dielectric. { 'säl·əd ‚tan·əl·əm kə'pas·əd·ər }

solidus [MATH] A sloping line that indicates division in a fraction. [PHYS CHEM] In a constitution or equilibrium diagram, the locus of points representing the temperature below which the various compositions finish freezing on cooling, or begin to melt on heating. { 'säl·əd·əs }

solidus curve [PHYS CHEM] A curve on the phase diagram of a system with two components which represents the equilibrium between the liquid phase and the solid phase. { 'säl·əd·əs ‚kərv }

solid-web girder [CIV ENG] A beam, such as a box girder, having a web consisting of a plate or other solid section but not a lattice. { 'säl·əd ¦web 'gər·dər }

solifluction [GEOL] A rapid soil creep, especially referring to downslope soil movement in periglacial areas. Also known as sludging; soil flow; soil fluction. { 'säl·ə'flək·shən }

solifluction lobe [GEOL] An isolated, tongue-shaped feature of the land surface with a steep front and a smooth upper surface formed by more rapid solifluction on certain sections of the slope. Also known as solifluction tongue. { ¦säl·ə'flək·shən ¦lōb }

solifluction mantle [GEOL] The locally derived, unsorted material moved downslope by solifluction. Also known as flow earth. { ¦säl·ə'flək·shən ¦mant·əl }

solifluction sheet [GEOL] A broad deposit of a solifluction mantle. { ¦säl·ə'flək·shən ¦shēt }

solifluction stream [GEOL] A narrow, streamlike deposit of a solifluction mantle. { ¦säl·ə'flək·shən ¦strēm }

solifluction tongue See solifluction lobe. { ¦säl·ə'flək·shən ¦təŋ }

solion [ELEC] An electrochemical device in which amplification is obtained by controlling and monitoring a reversible electrochemical reaction. { ¦säl'ī‚än }

soliquid [PHYS CHEM] A system in which solid particles are dispersed in a liquid. { ¦sä'lik·wəd }

solitary wave [PHYS] A traveling wave in which a single disturbance is neither preceded by nor followed by other such disturbances, but which does not involve unusually large amplitudes or rapid changes in variables, in contrast to a shock wave. { 'säl·ə‚ter·ē 'wāv }

soliton [MATH] A solution of a nonlinear differential equation that propogates with a characteristic constant shape. [PHYS] An isolated wave that propagates without dispersing its energy over larger and larger regions of space, and whose nature is such that two such objects emerge unchanged from a collision. { 'säl·ə‚tän }

solodize [GEOL] To improve a soil by removing alkalies from it. { 'sō·lə‚dīz }

Solod soil See Soloth soil. { 'sō·ləd ‚soil }

Solo man [PALEON] A relatively late but primitive form of fossil man from Java; this form had a small brain, heavy horizontal browridges, and a massive cranial base. { 'sō·lō 'man }

Solomon R unit [NUCLEO] A unit of radiation dose rate due to x-rays, equal to 2100 roentgens per hour. Also known as R unit. { 'säl·ə·mən 'är ˌyü·nət }

Solonchak soil [GEOL] One of an intrazonal, balamorphic group of light-colored soils rich in soluble salts. { ¦säl·ən¦chäk ˌsȯil }

Solonetz soil [GEOL] One of an intrazonal group of black alkali soils having a columnar structure. { ¦säl·ə¦nets ˌsȯil }

solore [METEOROL] A cold, night wind of the mountains following the course of the Drome River in southeastern France. { sə'lȯr }

Soloth soil [GEOL] One of an intrazonal halomorphic group of soils formed from saline material; the surface layer is soft and friable, and overlies a light-colored leached horizon which, in turn, overlies a dark horizon. Also known as Solod soil. { 'sō·lət ˌsȯil }

Solpugida [INV ZOO] The sun spiders, an order of nonvenomous, spiderlike, predatory arachnids having large chelicerae for holding and crushing prey. { säl'pyü·jəd·ə }

solstice [ASTRON] The two days (actually, instants) during the year when the earth is so located in its orbit that the inclination (about $23^1/_2°$) of the polar axis is toward the sun; the days are June 21 for the North Pole and December 22 for the South Pole; because of leap years, the dates vary a little. { 'sälz·təs }

solstitial colure [ASTRON] That great circle of the celestial sphere through the celestial poles and the solstices. { sälz'tish·əl kə'lúr }

solstitial points [ASTRON] Those points of the ecliptic that are 90° from the equinoxes north or south at which the greatest declination of the sun is reached. { sälz'tish·əl 'pȯins }

solstitial tidal currents [OCEANOGR] Tidal currents of especially large tropic diurnal inequality occurring at the time of solstitial tides. { sälz'tish·əl 'tīd·əl ˌkə·rəns }

solstitial tides [OCEANOGR] Tides occurring near the times of the solstices, when the tropic range is especially large. { sälz'tish·əl 'tīdz }

solubility [PHYS CHEM] The ability of a substance to form a solution with another substance. { ˌsäl·yə'bil·əd·ē }

solubility coefficient [PHYS CHEM] The volume of a gas that can be dissolved by a unit volume of solvent at a specified pressure and temperature. { ˌsäl·yə'bil·əd·ē ˌkō·iˌfish·ənt }

solubility curve [PHYS CHEM] A graph showing the concentration of a substance in its saturated solution in a solvent as a function of temperature. { ˌsäl·yə'bil·əd·ē ˌkərv }

solubility product constant [PHYS CHEM] A type of simplified equilibrium constant, K_{sp}, defined for and useful for equilibria between solids and their respective ions in solution; for example, the equilibrium

$$AgCl(s) \rightleftarrows Ag^+ + Cl^-, [Ag^+][Cl^-] \cong K_{sp}$$

where $[Ag^+]$ and $[Cl^-]$ are molar concentrations of silver ions and chloride ions. { ˌsäl·yə'bil·əd·ē ¦präd·əkt ˌkän·stənt }

solubility test [ANALY CHEM] **1.** A test for the degree of solubility of asphalts and other bituminous materials in solvents, such as carbon tetrachloride, carbon disulfide, or petroleum ether. **2.** Any test made to show the solubility of one material in another (such as liquid-liquid, solid-liquid, gas-liquid, or solid-solid). { ˌsäl·yə'bil·əd·ē ˌtest }

soluble [CHEM] Capable of being dissolved. { 'säl·yə·bəl }

soluble barbital *See* sodium barbital. { 'säl·yə·bəl 'bär·bəˌtal }

soluble castor oil *See* Turkey red oil. { 'säl·yə·bəl 'kas·tər ˌȯil }

soluble cutting oil [MATER] A petroleum oil containing an emulsifying agent to make it mix easily with water; used as a coolant for metal-cutting tools. { 'säl·yə·bəl 'kəd·iŋ ˌȯil }

soluble glass *See* sodium silicate. { 'säl·yə·bəl 'glas }

soluble guncotton *See* pyroxylin. { 'säl·yə·bəl 'gənˌkat·ən }

soluble indigo blue *See* indigo carmine. { 'säl·yə·bəl 'in·dəˌgō 'blü }

soluble nitrocellulose *See* pyroxylin. { 'säl·yə·bəl ¦nī·trō 'sel·yəˌlōs }

soluble oil [MATER] An oil that readily forms a stable emulsion or colloidal suspension in water. Also known as emulsifying oil. { 'säl·yə·bəl 'ȯil }

soluble starch [MATER] A group of water-soluble polymers formed from starch, such as the starches derived from corn or potato, by acetylation, acid hydrolysis, chlorination, or by action of enzymes to form starch acetates, ethers, and esters; used as textile sizing agents, emulsifying agents, and paper coatings. { 'säl·yə·bəl 'stärch }

solum [GEOL] The upper part of a soil profile, composed of A and B horizons in mature soil. Also known as true soil. { 'sō·ləm }

solute [CHEM] The substance dissolved in a solvent. { 'säl·yüt }

solute compartmentation [BOT] The sequestering of a plant cell's salt in a vacuole so that the salt does not poison the cell. { 'säl·yüt kəmˌpärt·mən'tā·shən }

solution [CHEM] A single, homogeneous liquid, solid, or gas phase that is a mixture in which the components (liquid, gas, solid, or combinations thereof) are uniformly distributed throughout the mixture. { sə'lü·shən }

solution ceramic [ELEC] A nonbrittle, inorganic ceramic insulating coating that can be applied to wires at a low temperature; examples include ceria, chromia, titania, and zirconia. { sə'lü·shən sə'ram·ik }

solution dyeing [TEXT] Adding dye to the chemical compound in the spinneret before extrusion. Also known as dope dyeing. { sə'lü·shən ˌdī·iŋ }

solution gas [PETRO ENG] Gaseous reservoir hydrocarbons dissolved in liquid reservoir hydrocarbons because of the prevailing pressures in the reservoir. Also known as dissolved gas. { sə'lü·shən ˌgas }

solution gas drive *See* internal gas drive. { sə'lü·shən ˌgas 'drīv }

solution-gas reservoir [PETRO ENG] Oil reservoir initially at or above the bubble-point pressure of the gas-oil mixture, and produced primarily by the expansion of the oil and its dissolved gas. Also known as dissolved-gas reservoir. { sə'lü·shən ˌgas 'rez·əvˌwär }

solution groove [GEOL] One of a series of continuous, subparallel furrows developed on an inclined or vertical surface of a soluble and homogeneous rock (such as the limestone walls of a cave) by the slow corroding action of trickling water. { sə'lü·shən ˌgrüv }

solution heat treatment [MET] Heating and holding an alloy at a temperature at which one (or more) constituent enters into solid solution, then cooling the alloy rapidly to prevent the constituent from precipitating. { sə'lü·shən 'hēt ˌtrēt·mənt }

solution mining [MIN ENG] The extraction of soluble minerals from subsurface strata by injection of fluids, and the controlled removal of mineral-laden solutions. { sə'lü·shən ˌmīn·iŋ }

solution poison [NUCLEO] A soluble nuclear poison, such as boric acid, added to the coolant of a nuclear reactor for purposes of reactivity control; generally used only during shutdown periods, and chemically removed from the coolant prior to resuming operation. { sə'lü·shən ˌpȯiz·ən }

solution polymerization [CHEM ENG] A process for producing an addition polymer by heating the monomer, solvent, initiator, and catalyst together, with polymerization continuing as the solvent is removed. { səˌlü·shən pəˌlim·ə·rə'zā·shən }

solution pool [GEOL] A pool in a rock that is formed by the dissolution of the rock in ocean water. { sə'lü·shən ˌpül }

solution porosity [PETRO ENG] A generic designation for reservoir-rock porosity created by solution action; some examples are crystalline limestone and dolomite, porous cap rock, and honeycombed anhydrite. { sə'lü·shən pəˌräs·əd·ē }

solution potholes [GEOL] Potholes produced in carbonate rocks by dissolution. { sə'lü·shən ˌpät·hōlz }

solution pressure [PHYS CHEM] **1.** A measure of the tendency of molecules or atoms to cross a bounding surface between phases and to enter into a solution. **2.** A measure of the tendency of hydrogen, metals, and certain nonmetals to pass into solution as ions. { sə'lü·shən ˌpresh·ər }

solution process [CHEM ENG] An oil-refining process for separating mercaptans from gasoline by washing with a caustic solution containing organic compounds in which the mercaptans are soluble. { sə'lü·shən ˌprä·səs }

SOLPUGIDA

Sun spider. *(From T. I. Storer and R. L. Usinger, General Zoology, 3d ed., McGraw-Hill, 1957)*

solution set [MATH] The set of values that satisfy a given equation. { sə'lü·shən ˌset }

solution transfer [GEOL] A process whereby pressure solution of detrital mineral grains at contact areas is followed by recrystallization on the less strained parts of the grain surfaces. { sə'lü·shən ˌtranz·fər }

solutizer-air regenerative process [CHEM ENG] A petroleum refinery process that is identical to the solutizer-steam regeneration process, except for the regeneration step; the newer units use uncatalyzed air regeneration. { sə'lü·ˌtīz·ər 'er rē'jen·ə·rəd·iv ˌprä·səs }

solutizer-steam regenerative process [CHEM ENG] A petroleum refinery process used to extract mercaptans from gasoline or naphtha; uses solutizers (potassium isobutyrate or potassium alkyl phenolate) in strong potassium hydroxide solution as the selective solvent. { sə'lü·ˌtīz·ər 'stēm rē'jen·ə·rəd·iv ˌprä·səs }

solutizer-tannin process [CHEM ENG] A petroleum refinery process that is an early variation of the solutizer-air regenerative process for extraction of mercaptans from gasoline; uses tannin-catalyzed oxidation for the regeneration step. { sə'lü·ˌtīz·ər 'tan·ən ˌprä·səs }

solutrope [CHEM] A ternary mixture with two liquid phases and a third component distributed between the phases, or selectively dissolved in one or the other of the phases; analogous to an azeotrope. { 'säl·yə·ˌtrōp }

solvable extension [MATH] A finite extension E of a field F such that the Galois group of the smallest Galois extension of F containing E is a solvable group. { 'säl·və·bəl ik'sten·chən }

solvable group [MATH] A group G which has subgroups $G_0, G_1, \ldots, G_n$, where $G_0 = G$, G_n = the identity element alone, and each G_i is a normal subgroup of G_{i-1} with the quotient group G_{i-1}/G_i Abelian. { 'säl·və·bəl 'grüp }

solvation [CHEM] The process of swelling, gelling, or dissolving of a material by a solvent; for resins, the solvent can be a plasticizer. { säl'vā·shən }

Solvay process [CHEM ENG] The process to make sodium carbonate and calcium chloride by treating sodium chloride with ammonia and carbon dioxide. { 'säl·vā ˌprä·səs }

solvent [CHEM] That part of a solution that is present in the largest amount, or the compound that is normally liquid in the pure state (as for solutions of solids or gases in liquids). { 'säl·vənt }

solvent deasphalting [CHEM ENG] A petroleum refinery process used to remove asphaltic and resinous materials from reduced crude oils, lubricating oil stocks, gas oils, or middle distillates through the extractive or precipitant action of solvents. Also known as solvent deresining. { 'säl·vənt dē'as·ˌfȯlt·iŋ }

solvent deresining See solvent deasphalting. { 'säl·vənt di·rez·ən·iŋ }

solvent dewaxing [CHEM ENG] A petroleum refinery process for solvent removal of wax from oils; the mixture of waxy oil and solvent is chilled, then filtered or centrifuged to remove the precipitated oil; the solvent is recovered for reuse. { 'säl·vənt di·ˌwaks·iŋ }

solvent dyeing [TEXT] The dyeing of synthetic textiles by using chlorinated hydrocarbon solvents (such as trichloroethylene or perchloroethylene) instead of water; used for nylons, polyesters, and acrylics. { 'säl·vənt ˌdī·iŋ }

solvent extraction [CHEM ENG] The separation of materials of different chemical types and solubilities by selective solvent action; that is, some materials are more soluble in one solvent than in another, hence there is a preferential extractive action; used to refine petroleum products, chemicals, vegetable oils, and vitamins. [NUCLEO] A process for removing uranium fuel residue from used fuel elements of a reactor; it generally involves decay cooling under water for up to 6 months, removal of cladding, dissolution, separation of reusable fuel, decontamination, and disposal of radioactive wastes. Also known as liquid extraction. { 'säl·vənt ik·ˌstrak·shən }

solvent front [ANALY CHEM] In paper chromatography, the wet moving edge of the solvent that progresses along the surface where the separation of the mixture is occurring. { 'säl·vənt ˌfrənt }

solvent molding [ENG] A process to form thermoplastic articles by dipping a mold into a solution or dispersion of the resin and drawing off (evaporating) the solvent to leave a plastic film adhering to the mold. { 'säl·vənt ˌmōld·iŋ }

solvent naphtha [MATER] Refined petroleum naphtha of restricted boiling range; used as a solvent and paint thinner and in dry cleaning; Stoddard solvent is a special grade of solvent naphtha. { 'säl·vənt ˌnaf·thə }

solvent recovery [CHEM ENG] For reuse purposes, the catching and recovery of solvent vapors from vent lines, process vessels, or other sources of evaporative loss, usually with a solid adsorbent material. { 'säl·vənt ri·ˌkəv·ə·rē }

solvent-refined [CHEM ENG] Pertaining to any product material whose final quality and condition is in part the result of a solvent treatment during processing of the feedstock material. { 'säl·vənt ri·ˌfīnd }

solvent-refined oil [MATER] A lubricating oil that has been solvent-treated during refining, such as most motor, aircraft, diesel-engine, steam-turbine, and other high-quality oils. { 'säl·vənt ri·ˌfīnd 'ȯil }

solvent refining [CHEM ENG] The process of treating a mixed material with a solvent that preferentially dissolves and removes certain minor constituents (usually the undesired ones); common in the petroleum refining industry. { 'säl·vənt ri·ˌfīn·iŋ }

solvent welding [ENG] A technique for joining plastic pipework in which a mixture of solvent and cement is applied to the pipe end and to the socket, with the parts then being joined and allowed to set. { 'säl·vənt ˌweld·iŋ }

solvmanifold [MATH] A homogeneous space obtained by factoring a connected solvable Lie group by a closed subgroup. { 'säl·v'man·ə·ˌfōld }

solvolysis [CHEM] A reaction in which a solvent reacts with the solute to form a new substance. { säl'väl·ə·səs }

solvus [PHYS CHEM] In a phase or equilibrium diagram, the locus of points representing the solid-solubility temperatures of various compositions of the solid phase. { 'säl·vəs }

soma [BIOL] The whole of the body of an individual, excluding the germ tract. { 'sō·mə }

Somali Current See East Africa Coast Current. { sə'mäl·ē 'kə·rənt }

Somasteroidea [INV ZOO] A subclass of Asterozoa comprising sea stars of generalized structure, the jaws often only partly developed, and the skeletal elements of the arm arranged in a double series of transverse rows termed metapinnules. { ˌsō·mə·stə'rȯid·ē·ə }

somatic aneuploidy [CELL MOL] An irregular variation in number of one or more individual chromosomes in the cells of a tissue. { sō'mad·ik 'a·nyü·ˌplȯid·ē }

somatic cell [BIOL] Any cell of the body of an organism except the germ cells. { sō'mad·ik 'sel }

somatic copulation [MYCOL] A form of reproduction in ascomycetes and basidiomycetes involving sexual fusion of undifferentiated vegetative cells. { sō'mad·ik ˌkäp·yə'lā·shən }

somatic crossing-over [CELL MOL] Crossing-over during mitosis in somatic or vegetative cells. { sō'mad·ik ˌkrȯs·iŋ 'ō·vər }

somatic death [BIOL] The cessation of characteristic life functions. { sō'mad·ik 'deth }

somatic embryogenesis [BOT] The production of embryoids from sporophytic or somatic plant cells. { sə'mad·ik ˌem·brē·ə'jen·ə·səs }

somatic mesoderm [EMBRYO] The external layer of the lateral mesoderm associated with the ectoderm after formation of the coelom. { sō'mad·ik 'mez·ə·ˌdərm }

somatic mutation [GEN] A genetic change limited to a somatic cell lineage; a major cause of cancer in humans. { sō'mad·ik myü'tā·shən }

somatic nervous system [PHYSIO] The portion of the nervous system concerned with the control of voluntary muscle and relating the organism with its environment. { sō'mad·ik 'nər·vəs ˌsis·təm }

somatic pairing [CELL MOL] The pairing of homologous chromosomes at mitosis in somatic cells; occurs in Diptera. { sō'mad·ik 'per·iŋ }

somatic reflex system [PHYSIO] An involuntary control system characterized by a control loop which includes skeletal muscles. { sō'mad·ik 'rē·ˌfleks ˌsis·təm }

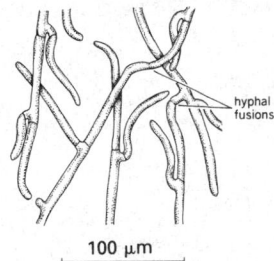

SOMATIC COPULATION

hyphal fusions

100 μm

Somatic copulation by sexual fusion of vegetative cells of hyphae.

somatization [PSYCH] A type of neurosis manifested in neurasthenias, hypochondriacal symptoms, and conversion hysterias. { ˌsō·məd·ə'zā·shən }

somatization disorder [PSYCH] A chronic fluctuating somatoform disorder in which the patient recurrently complains of multiple somatic symptoms that are referable to practically every organ system in the body but which, upon medical investigation, turn out not to be a diagnosable physical disease. { ˌsō·mə·tə'zā·shən dis‚órd·ər }

somatoblast [INV ZOO] **1.** An undifferentiated cleavage cell that gives rise to somatic cells in annelids. **2.** The outer cell layer of the nematogen in Dicyemida. { 'sō·məd·ə‚blast }

somatochrome [NEUROSCI] A nerve cell possessing a well-defined body completely surrounding the nucleus on all sides, the cytoplasm having a distinct contour, and readily taking a stain. { 'sō·məd·ə‚krōm }

somatoclonal variation [GEN] The appearance of new traits in plants that regenerate from a callus in tissue culture. { sə‚mad·ə¦klōn·əl ‚ver·ē'ā·shən }

somatocyst [INV ZOO] A cavity filled with air in the float of certain Siphonophora. { 'sō·məd·ə‚sist }

somatoform disorder [PSYCH] A psychiatric disorder in which the patient experiences physical symptoms although no actual physical disorder can be found to explain them. { sō'mad·ə‚fórm dis‚órd·ər }

somatogamy See pseudomixis. { ˌsō·mə'täg·ə·mē }

somatomedin [BIOCHEM] A growth factor similar to insulin that may be produced by a variety of cell types. Also known as insulin-like growth factor. { sə‚mad·ə'med·ən }

somatometry [ANAT] Measurement of the human body with the soft parts intact. { ˌsō·mə'täm·ə·trē }

somatophyte [BOT] A plant composed of distinct somatic cells that develop especially into mature or adult tissue. { 'sō·məd·ə‚fīt }

somatopleure [EMBRYO] A complex layer of tissue consisting of the somatic layer of the mesoblast together with the epiblast, forming the body wall in craniate vertebrates and the amnion and chorion in amniotes. { 'sō·məd·ə‚plùr }

somatopsychic [MED] Pertaining to both the body and mind. { ¦sō·məd·ə'sī·kik }

somatostatin [BIOCHEM] A peptide secreted by the hypothalamus which acts primarily to inhibit the release of growth hormone from the anterior pituitary. { ¦sō·məd·ə'stat·ən }

somatotonia [PSYCH] The temperamental trait associated with a mesomorphic somatotype, characterized by an active, aggressive, and risk-taking approach to life. { ˌsō·məd·ə'tōn·yə }

somatotropin [BIOCHEM] The growth hormone of the pituitary gland. { ˌsō·mə'tä·trə·pən }

somatotype [PSYCH] A basic body type; three primary components are ectomorph, mesomorph, and endomorph. { 'sō·məd·ə‚tīp }

Sombrero galaxy [ASTRON] A spiral galaxy in the constellation Virgo that is seen nearly edge-on, having a recession velocity of approximately 910 kilometers per second (565 miles per second). { səm'brer·ō 'gal·ik·sē }

somesthesis [PHYSIO] The general name for all systems of sensitivity present in the skin, muscles and their attachments, visceral organs, and nonauditory labyrinth of the ear. { ¦sōm·es'thē·səs }

somite See metamere. { 'sō‚mīt }

somma [GEOL] The rim of a volcano. { 'säm·ə }

Sommelet process [ORG CHEM] The preparation of thiophene aldehydes by treatment of thiophene with hexamethylenetetramine. { ˌsó·məl'yā ‚prä·səs }

Sommerfeld equation See Sommerfeld formula. { 'zóm·ər‚felt i‚kwā·zhən }

Sommerfeld fine-structure constant See fine-structure constant. { 'zóm·ər‚felt 'fīn ¦strək·chər ‚kän·stənt }

Sommerfeld formula [ELECTROMAG] An approximate formula for the field strength of electromagnetic radiation generated by an antenna at distances small enough so that the curvature of the earth may be neglected, in terms of radiated power, distance from the antenna, and various constants and parameters. Also known as Sommerfeld equation. { 'zóm·ər‚felt ‚fór·myə·lə }

Sommerfeld law for doublets [ATOM PHYS] According to the Bohr-Sommerfeld theory, the splitting in frequency of regular or relativistic doublets is $\alpha^2 R(Z - \sigma)^4/n^3(l + 1)$, where α is the fine structure constant, R is the Rydberg constant of the atom, Z is the atomic number, σ is a screening constant, n is the principal quantum number, and l is the orbital angular momentum quantum number. { 'zóm·ər‚felt 'ló fər 'dəb·ləts }

Sommerfeld model See free-electron theory of metals. { 'zóm·ər‚felt ‚mäd·əl }

Sommerfeld theory See free-electron theory of metals. { 'zóm·ər‚felt ‚thē·ə·rē }

Sommerfeld-Watson transformation See Watson-Sommerfeld transformation. { 'zóm·ər‚felt 'wät·sən ‚tranz·fər‚mā·shən }

somnambulism [PHYSIO] **1.** Sleepwalking. **2.** The performance of any fairly complex act while in a sleeplike state or trance. { säm'nam·byə‚liz·əm }

somnificant See hypnotic. { säm'nif·ə·kənt }

sonar [ENG] **1.** A system that uses underwater sound, at sonic or ultrasonic frequencies, to detect and locate objects in the sea, or for communication; the commonest type is echo-ranging sonar; other versions are passive sonar, scanning sonar, and searchlight sonar. Derived from sound navigation and ranging. **2.** See sonar set. { 'sō‚när }

sonar array [ELECTR] An arrangement of several sonar transducers or sonar projectors, appropriately spaced and energized to give proper directional characteristics. { 'sō‚när ə‚rā }

sonar attack plotter [ORD] A system that coordinates information from a sonar installation, a ship's gyrocompass, and related devices, and presents graphically the information needed to plan an antisubmarine attack. { 'sō‚när ə'tak ‚pläd·ər }

sonar beacon [ENG ACOUS] An underwater beacon that transmits sonic or ultrasonic signals for the purpose of providing bearing information; it may have receiving facilities that permit triggering an external source. { 'sō‚när ‚bē·kən }

sonar boomer transducer [ENG ACOUS] A sonar transducer that generates a large pressure wave in the surrounding water when a capacitor bank discharges into a flat, epoxy-encapsulated coil, creating opposed magnetic fields from the coil and from eddy currents in an adjacent aluminum disk, which cause the disk to be driven away from the coils with great force. { 'sō‚när 'büm·ər trans‚dü·sər }

sonar capsule [ENG ACOUS] A capsule that reflects high-frequency sound waves; the sonar capsule, if attached to a reentry body, may be used to locate the reentry body. { 'sō‚när ‚kap·səl }

sonar countermeasures [ORD] Actions taken to prevent or reduce an enemy's effective use of sonar. { 'sō‚när 'kaùnt·ər‚mezh·ərz }

sonar detector See sonar receiver. { 'sō‚när di‚tek·tər }

sonar dome [ENG] A streamlined, watertight enclosure that provides protection for a sonar transducer, sonar projector, or hydrophone and associated equipment, while offering minimum interference to sound transmission and reception. { 'sō‚när ‚dōm }

sonar navigation See sonic navigation. { 'sō‚när ‚nav·ə'gā·shən }

sonar projector [ENG ACOUS] An electromechanical device used under water to convert electrical energy to sound energy; a crystal or magnetostriction transducer is usually used for this purpose. { 'sō‚när prə‚jek·tər }

sonar receiver [ELECTR] A receiver designed to intercept and amplify the sound signals reflected by an underwater target and display the accompanying intelligence in useful form; it may also pick up other underwater sounds. Also known as sonar detector. { 'sō‚när ri'sē·vər }

sonar resolver [ELECTR] A resolver used with echo-ranging and depth-determining sonar to calculate and record the horizontal range of a sonar target, as required for depth-bombing. { 'sō‚när ri‚zäl·vər }

sonar self-noise [ELECTR] Unwanted sonar signals generated in the sonar equipment itself. { 'sō‚när 'self'nóiz }

sonar set [ENG] A complete assembly of sonar equipment for detecting and ranging or for communication. Also known as sonar. { 'sō‚när ‚set }

sonar target [ENG ACOUS] An object which reflects a sufficient amount of a sonar signal to produce a detectable echo signal at the sonar equipment. { 'sō‚när ‚tär·gət }

sonar transducer [ENG ACOUS] A transducer used under water to convert electrical energy to sound energy and sound energy to electrical energy. { 'sō‚när tranz‚dü·sər }

sonar transmission [ENG ACOUS] The process by which

underwater sound signals generated by a sonar set travel through the water. { 'sō,när tranz,mish·ən }

sonar transmitter [ELECTR] A transmitter that generates electrical signals of the proper frequency and form for application to a sonar transducer or sonar projector, to produce sound waves of the same frequency in water; the sound waves may carry intelligence. { 'sō,när tranz,mid·ər }

sonar window [ENG ACOUS] The portion of a sonar dome or sonar transducer that passes sound waves at sonar frequencies with little attenuation while providing mechanical protection for the transducer. { 'sō,när ,win·dō }

sondage [ARCHEO] A trial excavation made prior to an archeological excavation. { sän¦däzh }

sonde [ENG] An instrument used to obtain weather data during ascent and descent through the atmosphere, in a form suitable for telemetering to a ground station by radio, as in a radiosonde. [PETRO ENG] A downhole probe that measures the physical characteristics of the surrounding rocks, recording the measurements as a continuous depth record (log). { sänd }

Sondhauss tube [ACOUS] A device that converts heat to acoustic energy by heating a small glass bulb that is attached to a cool glass stem whose tip radiates sound. { 'zänd,häus ,tüb }

sone [ACOUS] A unit of loudness, equal to the loudness of a simple 1000-hertz tone with a sound pressure level 40 decibels above 0.0002 microbar; a sound that is judged by listeners to be *n* times as loud as this tone has a loudness of *n* sones. { sōn }

S-100 bus [ELECTR] A bus assembly with 100 conductors; widely used in microcomputer-based systems. { ,es¦wən'hən·drəd 'bəs }

son file [COMPUT SCI] The master file that is currently being updated. { 'sän ,fīl }

sonic [ACOUS] **1.** Of or pertaining to the speed of sound. **2.** Pertaining to that which moves at acoustic velocity, as in sonic flow. **3.** Designed to operate or perform at the speed of sound, as in sonic leading edge. { 'sän·ik }

sonic altimeter [ENG] An instrument for determining the height of an aircraft above the earth by measuring the time taken for sound waves to travel from the aircraft to the surface of the earth and back to the aircraft again. { 'sän·ik al'tim·əd·ər }

sonic anemometer [ENG] An anemometer which measures wind speed by means of the properties of wind-borne sound waves; it operates on the principle that the propagation velocity of a sound wave in a moving medium is equal to the velocity of sound with respect to the medium plus the velocity of the medium. { 'sän·ik ,an·ə'mäm·əd·ər }

sonicate [ENG] To apply high-frequency sound waves to matter. { 'sän·ə,kāt }

sonicator [ENG ACOUS] An instrument for producing high-intensity ultrasound, consisting of a converter that transforms electrical energy into mechanical energy in the form of oscillation of piezoelectric transducers at a frequency of 20 kilohertz, and a titanium horn that focuses this oscillation and radiates energy into the liquid being treated through a tip. { 'sän·ə,kād·ər }

sonic barrier [AERO ENG] A popular term for the large increase in drag that acts upon an aircraft approaching acoustic velocity; the point at which the speed of sound is attained and existing subsonic and supersonic flow theories are rather indefinite. Also known as sound barrier. { 'sän·ik 'bar·ē·ər }

sonic bearing [NAV] A bearing determined by measuring the direction from which a sound wave is coming. Also known as acoustic bearing. { 'sän·ik 'ber·iŋ }

sonic boom [ACOUS] A noise caused by a shock wave that emanates from an aircraft or other object traveling at or above sonic velocity. { 'sän·ik 'büm }

sonic chemical analyzer [ENG] A device to characterize the composition of a gas, liquid, or solid by the attenuation or change in the velocity of sound waves through a sample; the effect is related to molecular structure and intermolecular interactions. { 'sän·ik 'kem·ə·kəl 'an·ə,līz·ər }

sonic cleaning [ENG] Cleaning of contaminated materials by the action of intense sound in the liquid in which the material is immersed. { 'sän·ik 'klēn·iŋ }

sonic delay line See acoustic delay line. { 'sän·ik di'lā ,līn }

sonic depth finder [ENG] A sonar-type instrument used to measure ocean depth and to locate underwater objects; a sound pulse is transmitted vertically downward by a piezoelectric or

magnetostriction transducer mounted on the hull of the ship; the time required for the pulse to return after reflection is measured electronically. Also known as echo sounder. { 'sän·ik 'depth ,fīn·dər }

sonic detection and ranging See sodar. { ¦sän·ik di,tek·shən an 'rānj·iŋ }

sonic drilling [MECH ENG] The process of cutting or shaping materials with an abrasive slurry driven by a reciprocating tool attached to an audio-frequency electromechanical transducer and vibrating at sonic frequency. { 'sän·ik 'dril·iŋ }

sonic fix [NAV] A fix established by means of sound waves. Also known as acoustic fix. { 'sän·ik 'fiks }

sonic flaw detection [ENG] The process of locating imperfections in solid materials by observing internal reflections or a variation in transmission through the materials as a function of sound-path location. { 'sän·ik 'flo di,tek·shən }

sonic line of position [NAV] A line of position determined by means of sound waves. Also known as acoustic line of position. { 'sän·ik 'līn əv pə'zish·ən }

sonic liquid-level meter [ENG] A meter that detects a liquid level by sonic-reflection techniques. { 'sän·ik 'lik·wəd ¦lev·əl ,mēd·ər }

sonic logging [PETRO ENG] A logging method which records the time required for a sound wave to travel a specific distance through a formation. { 'sän·ik 'läg·iŋ }

sonic navigation [NAV] Navigation by means of sound waves whether or not they are within the audible range. Also known as acoustic navigation; sonar navigation. { 'sän·ik nav·ə'gā·shən }

sonic nucleation [CHEM ENG] In supersaturated solutions, the use of sonic or ultrasonic radiation to help bring about nucleation and corresponding crystallization of substances otherwise difficult to crystallize. { 'sän·ik ,nü·klē'ā·shən }

sonic pump [PETRO ENG] A type of lifting pump used in a shallow oil well to pump out the crude; consists of a string of tubing equipped with a check valve at each point, and mechanical means on the surface to vibrate the tubing string vertically; creates a harmonic condition that results in several hundred strokes per minute, with the strokes being a small fraction of an inch. { 'sän·ik 'pəmp }

sonic radiation [ACOUS] Acoustic radiation with a frequency between about 15 hertz and about 20,000 hertz. { 'sän·ik ,rād·ē'ā·shən }

sonics [ACOUS] The technology of sound, or elastic wave motion, as applied to problems of measurement, control, and processing. { 'sän·iks }

sonic sifter [MECH ENG] A high-speed vibrating apparatus used in particle size analysis. { 'sän·ik 'sif·tər }

sonic sounding [ENG] Determining the depth of the ocean bottom by measuring the time for an echo to return to a shipboard sound source. { 'sän·ik 'saurd·iŋ }

sonic spark chamber [NUCLEO] A spark chamber in which the position of a spark is determined by measuring the time it takes for sound from the spark to arrive at each of two microphones. { 'sän·ik 'spärk ,chām·bər }

sonic speed See speed of sound. { 'sän·ik 'spēd }

sonic thermometer [ENG] A thermometer based upon the principle that the velocity of a sound wave is a function of the temperature of the medium through which it passes. { 'sän·ik thər'mäm·əd·ər }

sonic velocity See speed of sound. { 'sän·ik və'läs·əd·ē }

sonic well logging [ENG] A well logging technique that uses a pulse-echo system to measure the distance between the instrument and a sound-reflecting surface; used to measure the size of cavities around brine wells, and capacities of underground liquefied petroleum gas storage chambers. { 'sän·ik 'wel ,läg·iŋ }

Sonnar lens [OPTICS] A modified triplet lens used as a photographic objective. { 'sō,när ,lenz }

Sonne dysentery [MED] An intestinal bacterial infection caused by *Shigella sonnei*. { 'zon·ə 'dis·ən,ter·ē }

Sonnenschein's reagent [ANALY CHEM] A solution of phosphomolybdic acid that forms a yellow precipitate with alkaloid sulfates. { 'zon·ən,shīn·z rē,ā·jənt }

sonobuoy [ENG] An acoustic receiver and radio transmitter mounted in a buoy that can be dropped from an aircraft by parachute to pick up underwater sounds of a submarine and transmit them to the aircraft; to track a submarine, several buoys are dropped in a pattern that includes the known or

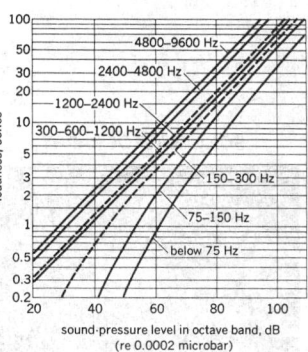
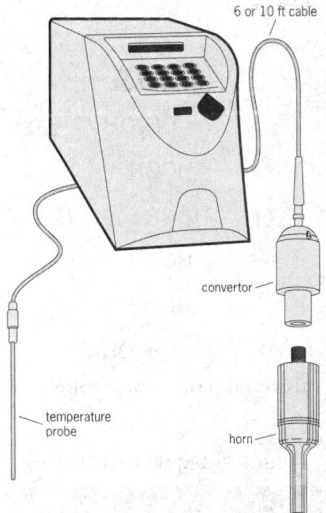
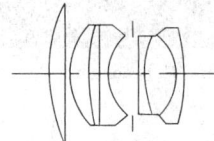

suspected location of the submarine, with each buoy transmitting an identifiable signal; an electronic computer then determines the location of the submarine by comparison of the received signals and triangulation of the resulting time-delay data. Also known as radio sonobuoy. { 'sän·ə¦bȯi }

Sonobuoy Reference System [NAV] An automatic electronic system for locating sonobuoys, employing angle-measuring equipment, distance-measuring equipment, or both. Abbreviated SRS. { 'sän·ə¦bȯi 'ref·rəns ¸sis·təm }

sonocatalysis [CHEM] **1.** Initiation of a catalytic reaction by irradiation with sound or ultrasound. **2.** Use of sound to impart catalytic activity to a chemical compound. { ¦sän·ə·kə'tal·ə·səs }

sonochemistry [CHEM] Any chemical change, such as in reaction type or rate, that occurs in response to sound or ultrasound. { ¦sän·ə'kem·ə·strē }

sonoelastography [ACOUS] An ultrasound technique for imaging the relative elastic properties of soft tissue and, in particular, for imaging hard tumors within the human body, in which vibrations (shear waves) with low frequencies (less than 1000 hertz) are propagated through tissue while real-time Doppler techniques are used to image the resulting vibration pattern on an ultrasound scanner. { ¸sō·nō·i·las'täg·rə·fē }

sonoencephalograph See echoencephalograph. { ¦sän·ō·in'sef·ə·lə¸graf }

sonogram [ACOUS] The image produced by ultrasonic imaging. { 'sän·ə¸gram }

sonograph [ENG] **1.** An instrument for recording sound or seismic vibrations. **2.** An instrument for converting sounds into seismic vibrations. { 'sän·ə¸graf }

sonography See acoustic imaging. { sə'näg·rə·fē }

sonoluminescence [PHYS] Luminescence produced by high-frequency sound waves or by phonons. { ¦sän·ō·ə¸lü·mə'nes·əns }

sonolysis [PHYS CHEM] The breaking of chemical bonds or formation of radicals using ultrasound. { sō'näl·ə·səs }

sonometer [ENG] **1.** In general, any device which consists of a thin metallic wire stretched over two bridges that are usually mounted on a soundboard and which is used to measure the vibration frequency, tension, density, or diameter of the wire, or to verify relations between these quantities. Also known as monochord. **2.** In particular, an instrument for measuring rock stress by means of a piano wire stretched between two bolts in the rock; any change of pitch after destressing is observed and used to indicate stress. { sə'näm·əd·ər }

sonora [METEOROL] A summer thunderstorm in the mountains and deserts of southern California and Baja California. { sə'nȯr·ə }

sonoscan [ENG] A type of acoustic microscope in which an unfocused acoustic beam passes through the object and produces deformations in a liquid-solid interface that are sensed by a laser beam reflected from the surface. { 'sän·ə¸skan }

soot [MATER] Impure black carbon with oily compounds obtained from the incomplete combustion of resinous materials, oils, wood, or coal. { süt }

soot blower [ENG] A system of steam or air jets used to maintain cleanliness, efficiency, and capacity of heat-transfer surfaces by the periodic removal of ash and slag from the heat-absorbing surfaces. { 'süt ¸blō·ər }

soot luminosity [OPTICS] The portion of the luminosity of a flame attributable to soot particles in the flame. { 'süt ¸lü·mə'näs·əd·ē }

sooty mold [MYCOL] Ascomycetous fungi of the family Capnodiaceae, with dark mycelium and conidia. [PL PATH] A plant disease, common on *Citrus* species, characterized by a dense velvety layer of a sooty mold on exposed parts of the plant. { 'süd·ē 'mōld }

sophisticated robot [CONT SYS] A robot that can be programmed and is controlled by a microprocessor. { sə'fis·tə¸kād·əd 'rō¸bät }

sophisticated vocabulary [COMPUT SCI] An advanced and elaborate set of instructions; a computer with a sophisticated vocabulary can go beyond the more common mathematical calculations such as addition, multiplication, and subtraction, and perform operations such as linearize, extract square root, and select highest number. { sə'fis·tə¸kād·əd və'kab·yə¸ler·ē }

soporific See hypnotic. { ¸säp·ə¦rif·ik }

sorbate [CHEM] A substance that has been either adsorbed

or absorbed. [ORG CHEM] A salt or an ester of sorbic acid. { 'sȯr¸bāt }

sorbent [MATER] A material, compound, or system that can provide a sorption function, such as adsorption, absorption, or desorption. { 'sȯr·bənt }

sorbic acid [ORG CHEM] $CH_3CH=CHCH=CHCOOH$ A white, crystalline compound; soluble in most organic solvents, slightly soluble in water; melts at 135°C; used as a fungicide and food preservative, and in the manufacture of plasticizers and lubricants. { 'sȯr·bik 'as·əd }

sorbide [CHEM] The generic term for anhydrides derived from sorbitol. { 'sȯr¸bīd }

sorbin See sorbose. { 'sȯr·bən }

sorbitol [ORG CHEM] $C_6H_8(OH)_6$ Combustible, white, water-soluble, hygroscopic crystals with a sweet taste; melt at 93 to 97.5°C (depending on the form); used in cosmetic creams and lotions, toothpaste, and resins; as a food additive; and for ascorbic acid fermentation. { 'sȯr·bə¸tȯl }

sorbose [BIOCHEM] $C_6H_{12}O_6$ A carbohydrate prepared by fermentation; produced as water-soluble crystals that melt at 165°C; used in the production of vitamin C. Also known as sorbin. { 'sȯr¸bōs }

sordawalite See tachylite. { 'sȯr'dä·wə¸līt }

soredium [BOT] A structure comprising algal cells wrapped in the hyphal tissue of lichens, as in certain Lecanorales; when separated from the thallus, it grows into a new thallus. { sȯ'rē·dē·əm }

Sorel cement See oxychloride cement. { sȯ'rel si¸ment }

Sörensen titration [ANALY CHEM] Titration with one of the Sörensen hydrogen-ion-concentration indicators. { 'sȯr·ən·sən tī¸trā·shən }

sore shin [PL PATH] A fungus disease of cowpea, cotton, tobacco, and other plants, beyond the seedling stage, marked by annular growth of the pathogen on the stem at the groundline. { 'sȯr 'shin }

Soret coefficient [PHYS] A tabulated value used in binary thermal diffusion calculations in gaseous systems; expressed as $D'/D = \alpha\, X_1X_2$, where D' is the coefficient of thermal diffusion, D is the coefficient of ordinary diffusion, α is the thermal diffusion constant, X_1 is the mole fraction of the lower-molecular-weight component, and X_2 is the mole fraction of the higher-molecular-weight component. { sȯ'rā ¸kō·i¸fish·ənt }

Soret effect [PHYS] Thermal diffusion in liquids. { sȯ'rā i¸fekt }

sorghum [BOT] Any of a variety of widely cultivated grasses, especially *Sorghum bicolor* in the United States, grown for grain and herbage; growth habit and stem form are similar to Indian corn, but leaf margins are serrate and spikelets occur in pairs on a hairy rachis. { 'sȯr·gəm }

Sorghum bicolor See sweet sorghum. { ¸sȯr·gəm 'bī¸kəl·ər }

sorghum downy mildew [AGR] A serious fungus disease that systematically invades sorghum plants, causing stripped leaves and barren stalks. { ¸sȯr·gəm ¸daù·nē 'mil¸dü }

sorghum head smut [AGR] A disease of sorghum plants that completely destroys the normal head and replaces it with masses of smut spores. { ¸sȯr·gəm 'hed ¸smət }

Soricidae [VERT ZOO] The shrews, a family of insectivorous mammals belonging to the Lipotyphla. { sə'ris·ə¸dē }

sorosilicate [MINERAL] A structural type of silicate whose crystal lattice has two SiO_4 tetrahedra sharing one oxygen atom. { ¸sȯr·ō'sil·ə·kət }

sorotiite [GEOL] A type of meteorite similar to the pallasites, with troilite substituting for olivine. { sə'räd·ē¸īt }

sorption [PHYS CHEM] A general term used to encompass the processes of adsorption, absorption, desorption, ion exchange, ion exclusion, ion retardation, chemisorption, and dialysis. { 'sȯrp·shən }

sorption pumping [ENG] A technique used to reduce the pressure of gas in an atmosphere; the gas is adsorbed on a granular sorbent material such as a molecular sieve in a metal container; when this sorbent-filled container is immersed in liquid nitrogen, the gas is sorbed. { 'sȯrp·shən ¸pəmp·iŋ }

sorrel tree See sourwood. { 'sär·əl ¸trē }

sort [COMPUT SCI] **1.** To rearrange a set of data items into a new sequence, governed by specific rules of precedence. **2.** The program designed to perform this activity. { sȯrt }

sort algorithm [COMPUT SCI] The methods followed in arranging a set of data items into a sequence according to precise rules. { 'sȯrt ¦al·gə¸rith·əm }

SONOELASTOGRAPHY

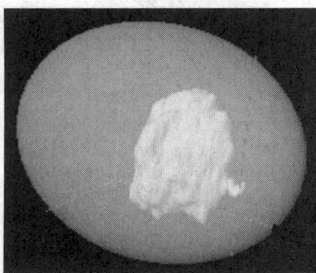

Three-dimensional sonoelastography image of a cancerous lesion in an excised prostate. The lesion is the light-gray solid mass within the gland boundary (transparent ellipsoid).

SORBITOL

$$\begin{array}{c} CH_2OH \\ | \\ HCOH \\ | \\ HOCH \\ | \\ HCOH \\ | \\ HCOH \\ | \\ CH_2OH \end{array}$$

Structural formula of sorbitol.

SORGHUM HEAD SMUT

Sorghum with head smut.

sorted [GEOL] **1.** Pertaining to a nongenetic group of patterned-ground features displaying a border of stones, including boulders, commonly alternating with very small particles, including silt, sand, and clay. **2.** Pertaining to an unconsolidated sediment or a cemented detrital rock consisting of particles of essentially uniform size or of particles lying within the limits of a single grade. { 'sȯrd·əd }

sorted polygon [GEOL] A patterned ground having a sorted appearance due to a border of stones and characterized by a polygonal mesh. Also known as stone polygon. { 'sȯrd·əd 'päl·i‚gän }

sorter See sequencer. { 'sȯrd·ər }

sort field [COMPUT SCI] A field in a record that is used in determining the final sorted sequence of the records. { 'sȯrt ‚fēld }

sort generator [COMPUT SCI] A computer program that produces other programs which arrange collections of items into sequences as specified by parameters in the original program. { 'sȯrt ‚jen·ə‚rād·ər }

sortie [AERO ENG] An operational flight by one aircraft. [ORD] A sudden attack made from a defensive position. { 'sȯrd·ē }

sortie number [ENG] A reference used to identify the images taken by all the sensors during one air reconnaissance sortie. { 'sȯrd·ē ‚nəm·bər }

sortie plot [MAP] An overlay representing the area on a map covered by imagery taken during one sortie. { 'sȯrd·ē ‚plät }

sorting [GEOL] The process by which similar in size, shape, or specific gravity sedimentary particles are selected and separated from associated but dissimilar particles by the agent of transportation. { 'sȯrd·iŋ }

sorting coefficient [GEOL] A sorting index equal to the square root of the ratio of the larger quartile (the diameter having 25% of the cumulative size-frequency distribution larger than itself) to the smaller quartile (the diameter having 75% of the cumulative size-frequency distribution larger than itself). { 'sȯrd·iŋ ‚kō·i‚fish·ənt }

sorting index [GEOL] A measure of the degree of sorting in a sediment based on the statistical spread of the frequency curve of particle sizes. { 'sȯrd·iŋ ‚in‚deks }

sorting table [ENG] Any horizontal conveyor where operators, along its side, sort bulk material, packages, or objects from the conveyor. { 'sȯrd·iŋ ‚tā·bəl }

sort key [COMPUT SCI] A key used as a basis for determining the sequence of items in a set. { 'sȯrt ‚kē }

sort/merge [COMPUT SCI] To combine two or more similar files, with the records arranged in the appropriate order, according to precise rules. { 'sȯrt 'mərj }

sort/merge package [COMPUT SCI] A set of programs capable of sorting and merging data files. { 'sȯrt 'mərj ‚pak·ij }

sort order [COMPUT SCI] The sequence into which a collection of records are arranged after they have been sorted. { 'sȯrt ‚ȯr·dər }

sort pass [COMPUT SCI] Any one of a collection of similar procedures carried out during a sort operation in which a part of the sort is completed. { 'sȯrt ‚pas }

sortworker [COMPUT SCI] A file created temporarily by a computer program to hold intermediate results when the amount of data to be sorted exceeds the available storage space. { 'sȯrt‚wər·kər }

sorus [BOT] **1.** A cluster of sporangia on the lower surface of a fertile fern leaf. **2.** A clump of reproductive bodies or spores in lower plants. { 'sȯr·əs }

SOS [COMMUN] The distress signal in radiotelegraphy, consisting of the letters S, O, and S of the international Morse code.

sosoloid [PHYS CHEM] A system consisting of particles of a solid dispersed in another solid. { 'säs·ə‚lȯid }

SOS response [GEN] A bacterial deoxyribonucleic acid (DNA) repair system in which cell division and DNA replication are blocked, and DNA repair, recombination, and mutation genes are induced. { ‚es‚ō'es ri‚späns }

SOTAS See standoff target acquisition system. { 'sō‚tas }

Sothic cycle [ASTRON] A time period of about 1460 years; this cycle is such that the New Year of the calendar used in ancient Egypt was in error by a whole year because the adopted year of 365 days is about a quarter of a day shorter than the mean solar year. { 'sä·thik ‚sī·kəl }

sou'easter See southeaster. { 'saủ‚ēs·tər }

souma [VET MED] A disease caused by *Tryparosoma vivax* in domestic and wild animals; the insect vectors are the tsetse fly and the stable fly. { 'sü·mə }

sound [ACOUS] An alteration of properties of an elastic medium, such as pressure, particle displacement, or density, that propagates through the medium, or a superposition of such alterations; sound waves having frequencies above the audible (sonic) range are termed ultrasonic waves; those with frequencies below the sonic range are called infrasonic waves. Also known as acoustic wave; sound wave. [PHYSIOL] The auditory sensation which is produced by these alterations. Also known as sound sensation. { saủnd }

sound absorption [ACOUS] A process in which sound energy is reduced when sound waves pass through a medium or strike a surface. Also known as acoustic absorption. { 'saủnd əb‚sȯrp·shən }

sound absorption coefficient [ACOUS] The ratio of sound energy absorbed to that arriving at a surface or medium. Also known as acoustic absorption coefficient; acoustic absorptivity. { 'saủnd ab‚sȯrp·shən ‚kō·i fish·ənt }

sound analyzer [ENG] An instrument which measures the amount of sound energy in various frequency bands; it generally consists of a set of fixed electrical filters or a tunable electrical filter, along with associated amplifiers and a meter which indicates the filter output. { 'saủnd ‚an·ə‚līz·ər }

sound and flash ranging [ORD] Two distinct and separate but supplementary systems of locating enemy weapons and, secondarily, adjusting friendly counterfire: by observation by sonic devices on the sound produced by the enemy weapon in firing or by the friendly projectile in exploding; or by visual observation of the flash produced or of the point of burst of the enemy weapon or friendly projectile. { ¦saủnd ən ¦flash 'rānj·iŋ }

sound attenuation [ACOUS] Diminution of the intensity of sound energy propagating in a medium; caused by absorption, spreading, and scattering. { 'saủnd ə‚ten·yə‚wā·shən }

sound band pressure level [ACOUS] The effective sound pressure for the sound energy in a given frequency band. { 'saủnd ¦band 'presh·ər ‚lev·əl }

sound barrier See sonic barrier. { 'saủnd ‚bar·ē·ər }

sound board [COMPUT SCI] An adapter which provides a computer with the capability of reproducing and recording digitally encoded sound. Also known as audio adapter; sound card. { 'saủn ‚bȯrd }

sound buoy [NAV] A buoy equipped with a characteristic sound signal, such as a bell or a whistle. { 'saủnd ‚bȯi }

sound card See sound board. { 'saủn kärd }

sound carrier [COMMUN] The television carrier that is frequency-modulated by the sound portion of a television program; the unmodulated center frequency of the sound carrier is 4.5 megahertz higher than the video carrier frequency for the same television channel. { 'saủnd ‚kar·ē·ər }

sound channel [ACOUS] A layer of seawater extending from about 2300 feet (700 meters) down to about 4950 feet (1500 meters), in which sound travels at about 1485 feet (450 meters) per second, the slowest it can travel in seawater; below 4950 feet (1500 meters) the speed of sound increases as a result of pressure. [ELECTR] The series of stages that handles only the sound signal in a television receiver. { 'saủnd ‚chan·əl }

sound detection [ACOUS] The discrimination of a sound from background noise, either by the ear or by an electronic instrument such as a volume indicator. { 'saủnd di‚tek·shən }

sound effects [ENG ACOUS] Mechanical devices or recordings used to provide lifelike imitations of various sounds. { 'saủnd i‚feks }

sound energy [ACOUS] The difference between the total energy and the energy which would exist if no sound waves were present. Also known as acoustic energy. { 'saủnd ‚en·ər·jē }

sound-energy density [ACOUS] Sound energy per unit volume; the commonly used unit is the erg per cubic centimeter. { 'saủnd ‚en·ər·jē ‚den·səd·ē }

sound-energy flux [ACOUS] Average over one period of the rate of flow of sound energy through any specified area; the unit is the erg per second. { 'saủnd ‚en·ər·jē ‚fləks }

sound exclusion [PETRO ENG] Several techniques used to prevent a borehole from sloughing in drilling a well through a reservoir rock that has an unconsolidated nature, similar to beach sand; the borehole can be lined with a screen, the sand

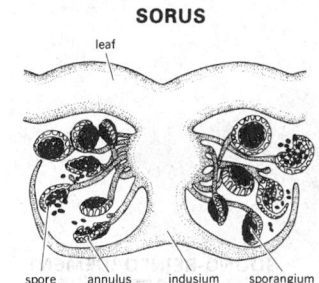

SORUS

leaf

spore　annulus　indusium　sporangium

Diagram of section through a fern leaf, showing details of a sorus. *(From W. W. Robbins, T. E. Weier, and C. R. Stocking, Botany: An Introduction to Plant Science, 3d ed., copyright © 1964 by John Wiley & Sons, Inc.; reprinted by permission)*

consolidated with a binding material, or prepack gravel liner can be used. { 'saund ik,sklü·zhən }

sound-field enhancement [ENG ACOUS] A system for enhancing the acoustical properties of both indoor and outdoor spaces, particularly for unamplified speech, song, and music; may consist of one or more microphones, systems for amplification and electronic signal processing, and one or more loudspeakers. { ¦saun ¦fēld in'hans·mənt }

sound film [ENG ACOUS] Motion picture film having a sound track along one side for reproduction of the sounds that are to accompany the film. { 'saund ,film }

sound filmstrip [ENG ACOUS] A filmstrip that has accompanying sound on a separate disk or tape, which is manually or automatically synchronized with projection of the pictures in the strip. { 'saund 'film,strip }

sound fixing and ranging See sofar. { 'saund 'fiks·iŋ ən 'rānj·iŋ }

sound gate [ENG ACOUS] The gate through which film passes in a sound-film projector for conversion of the sound track into audio-frequency signals that can be amplified and reproduced. { 'saund ,gāt }

sound head [ENG ACOUS] **1.** The section of a sound motion picture projector that converts the photographic or magnetic sound track to audible sound signals. **2.** In a sonar system, the cylindrical container for the transmitting projector and the receiving hydrophone. { 'saund ,hed }

sound image [ACOUS] The photographic image of a sound, as on a film sound track. { 'saund ,im·ij }

sounding [ENG] **1.** Determining the depth of a body of water by an echo sounder or sounding line. **2.** Measuring the depth of bedrock by driving a steel rod into the soil. **3.** Any penetration of the natural environment for scientific observation. [METEOROL] See upper-air observation. [MIN ENG] **1.** Knocking on a mine roof to see whether it is sound or safe to work under. **2.** Subsurface investigation by observing the penetration resistance of the subsurface material without drilling holes, by driving a rod into the ground or by using a penetrometer. { 'saund·iŋ }

sounding balloon [ENG] A small free balloon used for carrying radiosonde equipment aloft. { 'saund·iŋ bə,lün }

sounding device [PETRO ENG] An acoustical device used to measure the liquid level in a wellbore; for example, a Sonolog or an Echometer. { 'saund·iŋ di,vīs }

sounding lead [ENG] A lead used for determining the depth of water. { 'saund·iŋ ,led }

sounding line [ENG] The line attached to a sounding lead. Also known as lead line. { 'saund·iŋ ,līn }

sounding machine [ENG] An instrument for measuring the depth of water, consisting essentially of a reel of wire; to one end of this wire there is attached a weight which carries a device for measuring and recording the depth; a crank or motor reels in the wire. { 'saund·iŋ mə,shēn }

sounding pipe [NAV ARCH] A pipe through which the depth of liquid in a water or oil tank on board a ship can be measured or sounded. { 'saund·iŋ ,pīp }

sounding pole [ENG] A pole or rod used for sounding in shallow water, and usually marked to indicate various depths. { 'saund·iŋ ,pōl }

sounding rocket [AERO ENG] A rocket that carries aloft equipment for making observations of or from the upper atmosphere. { 'saund·iŋ ,räk·ət }

sounding sand [GEOL] Sand that emits musical, humming, or crunching sounds when disturbed. Also known as singing sand. { 'saund·iŋ ,sand }

sounding sextant See hydrographic sextant. { 'saund·iŋ ,sek·stənt }

sounding velocity [ACOUS] The vertical velocity of sound in water, usually assumed to be constant at 800 to 820 fathoms (1464 to 1501 meters) per second for sounding measurements. { 'saund·iŋ və,läs·əd·ē }

sounding wire [ENG] A wire used with a sounding machine in determining depth of water. { 'saund·iŋ ,wīr }

sound intensity [ACOUS] For a specified direction and point in space, the average rate at which sound energy is transmitted through a unit area perpendicular to the specified direction. { 'saund in,ten·səd·ē }

sound irradiator [ACOUS] A device for focusing sound waves so that sound of high intensity is produced at the focus. { 'saund i,rād·ē,ād·ər }

sound knot [MATER] A knot in a piece of lumber that is firmly fixed, undecayed, and as strong as the surrounding wood. Also known as tight knot. { 'saund ,nät }

sound lag [ACOUS] Time necessary for a sound wave to travel from its source to the point of reception. { 'saund ,lag }

sound level [ACOUS] The sound pressure level (in decibels) at a point in a sound field, averaged over the audible frequency range and over a time interval, with a frequency weighting and the time interval as specified by the American National Standards Association. { 'saund ,lev·əl }

sound-level meter [ENG] An instrument used to measure noise and sound levels in a specified manner; the meter may be calibrated in decibels or volume units and includes a microphone, an amplifier, an output meter, and frequency-weighting networks. { 'saund ¦lev·əl ,mēd·ər }

sound locator [ENG ACOUS] A device formerly used to detect aircraft in flight by sound, consisting of four horns, or sound collectors (two for azimuth detection and two for elevation), together with their associated mechanisms and controls, which enabled the listening operator to determine the position and angular velocity of an aircraft. { 'saund ,lō,kād·ər }

sound masking [ACOUS] The ability of one sound to make the ear incapable of perceiving another sound. { 'saund ,mask·iŋ }

sound navigation and ranging See sonar. { 'saund ,nav·ə'gā·shən ən 'rānj·iŋ }

sound power [ACOUS] The total sound energy radiated by a source per unit time, generally expressed in ergs per second or watts. Also known as acoustic power. { 'saund ,pau·ər }

sound-powered telephone [ENG ACOUS] A telephone operating entirely on current generated by the speaker's voice, with no external power supply; sound waves cause a diaphragm to move a coil back and forth between the poles of a powerful but small permanent magnet, generating the required audio-frequency voltage in the coil. { 'saund ,pau·ərd 'tel·ə,fōn }

sound pressure See effective sound pressure. { 'saund ,presh·ər }

sound pressure level [ACOUS] A value in decibels equal to 20 times the logarithm to the base 10 of the ratio of the pressure of the sound under consideration to a reference pressure; reference pressures in common use are 0.0002 microbar and 1 microbar. Abbreviated SPL. { 'saund ¦presh·ər ,lev·əl }

sound pressure spectrum level [ACOUS] Ten times the logarithm to base 10 of the ratio of the mean square pressure of the portion of sound within a specified frequency band to the mean square pressure of the portion of a reference sound within the same frequency band. Abbreviated SPSL. { ¦saund ,presh·ər ,spek·trəm ,lev·əl }

sound production [ENG ACOUS] Conversion of energy from mechanical or electrical into acoustical form, as in a siren or loudspeaker. { 'saund prə,dək·shən }

soundproofing See damping. { 'saund,prüf·iŋ }

sound ranging [ENG ACOUS] Determining the location of a gun or other sound source by measuring the travel time of the sound wave to microphones at three or more different known positions. { 'saund ,rānj·iŋ }

sound-ray diagram [ACOUS] A plot of the paths taken by sound rays in an acoustical system; analogous to a light-ray diagram in optics. { 'saund ¦rā ,dī·ə,gram }

sound reception [ENG ACOUS] Conversion of acoustical energy into another form, usually electrical, as in a microphone. { 'saund ri,sep·shən }

sound recording [ENG ACOUS] The process of recording sound signals so they may be reproduced at any subsequent time, as on a phonograph disk, motion picture sound track, or magnetic tape. { 'saund ri,kord·iŋ }

sound reduction factor [ACOUS] A measure of the reduction in the intensity of sound when it crosses an interface, equal to 10 times the common logarithm of the reciprocal of the sound transmission coefficient of the surface. { 'saund ri,dək·shən ,fak·tər }

sound reflection coefficient See acoustic reflectivity. { 'saund ri,flek·shən ,kō·i,fish·ənt }

sound-reinforcement system [ENG ACOUS] An electronic means for augmenting the sound output of a speaker, singer, or musical instrument in cases where it is either too weak to be heard above the general noise or too reverberant; basic

SOUND-REINFORCEMENT SYSTEM

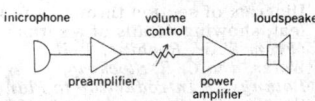

Basic sound-reinforcement system.

elements of such a system are microphones, amplifiers, volume controls, and loudspeakers. Also known as public address system. { 'saund ˌrē·in'fòrs·mənt ˌsis·təm }

sound-reproducing system [ENG ACOUS] A combination of transducing devices and associated equipment for picking up sound at one location and time and reproducing it at the same or some other location and at the same or some later time. Also known as audio system; reproducing system; sound system. { 'saund ˌrē·prə'düs·iŋ ˌsis·təm }

sound reproduction [ENG ACOUS] The use of a combination of transducing devices and associated equipment to pick up sound at one point and reproduce it either at the same point or at some other point, at the same time or at some subsequent time. { 'saund ˌrē·prə₁dək·shən }

sound sensation *See* sound. { 'saund sen₁sā·shən }

sound spectrograph [ENG ACOUS] An instrument that records and analyzes the spectral composition of audible sound. { 'saund 'spek·trə₁graf }

sound spectrum [ACOUS] A plot of the strength of a sound at various frequencies. { 'saund ₁spek·trəm }

sound speed [ENG] The speed of sound motion picture film, standardized at 24 frames per second (silent film speed is 18 frames per second). { 'saund ₁spēd }

soundstripe [ENG ACOUS] A longitudinal stripe of magnetic material placed on some motion picture films for recording a magnetic sound track. { 'saund₁strīp }

sound system *See* sound-reproducing system. { 'saund ₁sis·təm }

sound track [ENG ACOUS] A narrow band, usually along the margin of a sound film, that carries the sound record; it may be a variable-width or variable-density optical track or a magnetic track. { 'saund ₁trak }

sound transducer *See* electroacoustic transducer. { 'saund tranz₁düs·ər }

sound transmission [ACOUS] Passage of a sound wave through a medium or series of media. { 'saund tranz₁mish·ən }

sound transmission coefficient [ACOUS] The ratio of transmitted to incident sound energy at an interface in a sound medium; the value depends on the angle of incidence of the sound. Also known as acoustic transmission coefficient; acoustic transmissivity. { 'saund tranz₁mish·ən ₁kō·i₁fish·ənt }

sound trap [ELECTR] A wave trap in a television receiver circuit that prevents sound signals from entering the picture channels. [ENG ACOUS] A pit between adjoining instrument sections in a sound-recording studio, generally filled with fiberglass panels, to absorb sound that would otherwise propagate from instruments in one section to microphones in adjacent sections. { 'saund ₁trap }

sound velocity *See* speed of sound. { 'saund və₁läs·əd·ē }

sound volume velocity [ACOUS] The rate at which a substance flows through a specified area as a result of a sound wave. { 'saund ₁väl·yəm və₁läs·əd·ē }

sound wave *See* sound. { 'saund ₁wāv }

sound-wave photography [PHYS] A method of studying propagation, reflection, and refraction of sound waves, in which a sound wave is generated by a spark and is illuminated a fraction of a second later by a second spark, causing the wave to cast a shadow on a photographic plate. { 'saund ₁wāv fə'tag·rə₁fē }

sour [CHEM] Containing large amounts of malodorous sulfur compounds (such as mercaptans or hydrogen sulfide), as in crude oils, naphthas, or gasoline. { saur }

source [ELEC] The circuit or device that supplies signal power or electric energy or charge to a transducer or load circuit. [ELECTR] The terminal in a field-effect transistor from which majority carriers flow into the conducting channel in the semiconductor material. [MATH] The vertex with indegree 0 that is specified in the definition of an *s-t* network. [NUCLEO] A radioactive material packaged so as to produce radiation for experimental or industrial use. [PHYS] **1.** In general, a device that supplies some extensive entity, such as energy, matter, particles, or electric charge. **2.** A point, line, or area at which mass or energy is added to a system, either instantaneously or continuously. **3.** A point at which lines of force in a vector field originate, such as a point in an electrostatic field where there is positive charge. [SPECT] The arc

or spark that supplies light for a spectroscope. [THERMO] A device that supplies heat. { sòrs }

source address [COMPUT SCI] The first address of a two-address instruction (the sound address is known as the destination address). { 'sòrs 'ad₁res }

source area *See* provenance. { 'sòrs ₁er·ē·ə }

source bed [GEOL] The original stratigraphic horizon from which secondary sulfide minerals were derived. { 'sòrs ₁bed }

source code [COMPUT SCI] The statements in which a computer program is initially written before translation into machine language. { 'sòrs ₁kōd }

source data [SCI TECH] Data generated in the course of research. { 'sòrs ₁dad·ə }

source data automation *See* automation source data. { 'sòrs ₁dad·ə ₁òd·ə'mā·shən }

source data automation equipment [COMPUT SCI] Equipment (except paper tape and magnetic tape cartridge typewriters acquired separately and not operated in support of a computer) which, as a by-product of its operation, produces a record in a medium which is acceptable by automatic data-processing equipment. { 'sòrs ₁dad·ə ₁òd·ə'mā·shən i₁kwip·mənt }

source data capture [COMPUT SCI] The procedures for entering source data into a computer system. { 'sòrs ₁dad·ə ₁kap·chər }

source data entry [COMPUT SCI] Entry of data into a computer system directly from its source, without transcription. { 'sòrs ₁dad·ə ₁en·trē }

source degeneration [ELECTR] The addition of a circuit element between a transistor source and ground, with several effects, including a reduction in gain. { ₁sòrs di₁jen·ə'rā·shən }

source document [COMPUT SCI] The original medium containing the basic data to be used by a data-processing system, from which the data are converted into a form which can be read into a computer. Also known as original document. { 'sòrs ₁däk·yə·mənt }

source flow [FL MECH] **1.** In three-dimensional flow, a point from which fluid issues at a uniform rate in all directions. **2.** In two-dimensional flow, a line normal to the planes of flow, from which fluid flows uniformly in all directions at right angles to the line. { 'sòrs ₁flō }

source-follower amplifier *See* common-drain amplifier. { 'sòrs 'fäl·ə·wər 'am·plə₁fī·ər }

source function [ASTROPHYS] The emissivity of a stellar or other radiating material divided by its opacity. { 'sòrs ₁faŋk·shən }

source impedance [ELEC] Impedance presented by a source of energy to the input terminals of a device. { 'sòrs im₁pēd·əns }

sourceland *See* provenance. { 'sòrs₁land }

source language [COMPUT SCI] The language in which a program (or other text) is originally expressed. { 'sòrs ₁laŋ·gwij }

source level [ACOUS] The sound intensity, in decibels above a reference level, at a point which is a unit distance from a source and on an axis of the source. { 'sòrs ₁lev·əl }

source library [COMPUT SCI] A collection of computer programs in compiler language or assembler language. { 'sòrs ₁lī₁brer·ē }

source listing [COMPUT SCI] A printout of a source program. { 'sòrs ₁list·iŋ }

source material [NUCLEO] Material from which fissionable material can be extracted. { 'sòrs mə₁tir·ē·əl }

source module [COMPUT SCI] An organized set of statements in any source language recorded in machine-readable form and suitable for input to an assembler or compiler. { 'sòrs ₁mäj·ül }

source program [COMPUT SCI] The form of a program just as the programmer has written it, often on coding forms or machine-readable media; a program expressed in a source-language form. { 'sòrs ₁prō₁gram }

source program optimizer [COMPUT SCI] A routine for examining the source code of a program under development and providing information about use of the various portions of the code, enabling the programmer to modify those sections of the target program that are most heavily used in order to improve performance of the final, operational program. { 'sòrs ₁prō₁gram ₁äp·tə₁mīz·ər }

source region [METEOROL] An extensive area of the earth's

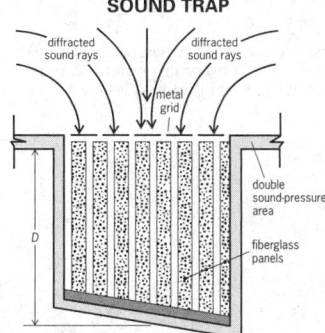

SOUND TRAP

diffracted sound rays

diffracted sound rays

metal grid

double sound-pressure area

fiberglass panels

D

Sound trap used to minimize reflections from overhead instrument to adjacent instrument microphones. Diffracted sound rays are absorbed into the pit where sound pressure is low.

surface characterized by essentially uniform surface conditions and so situated with respect to the general atmospheric circulation that an air mass may remain over it long enough to acquire its characteristic properties.   { 'sòrs ˌrē·jən }

source rock [GEOL]   **1.** Rock from which fragments have been derived which form a later, usually sedimentary rock. Also known as mother rock; parent rock.   **2.** Sedimentary rock, usually shale and limestone, deposited together with organic matter which was subsequently transformed to liquid or gaseous hydrocarbons.   { 'sòrs ˌräk }

source time [COMPUT SCI]   The time involved in fetching the contents of the register specified by the first address of a two-address instruction.   { 'sòrs ˌtīm }

source transition loss [ELECTR]   The transmission loss at the junction between an energy source and a transducer connecting that source to an energy load; measured by the ratio of the source power to the input power.   { 'sòrs tran'zish·ən ˌlòs }

sourcing [ELECTR]   Redesign or the modification of existing equipment to eliminate a source of radio-frequency interference.   { 'sòrs·iŋ }

sour corrosion [PETRO ENG]   Corrosion occurring in oil or gas wells where there is an iron sulfide corrosion product, and hydrogen sulfide is present in the produced reservoir fluid.   { 'saùr kə'rō·zhən }

sour crude [MATER]   Crude oil containing an abnormally large amount of sulfur compounds that, upon refining, liberate corrosive sulfur compounds; opposite to sweet crude.   { 'saùr ˌkrüd }

sour dirt [PETRO ENG]   Sulfate-impregnated soil or soil characterized by escaping sulfur dioxide or hydrogen sulfide; considered an indicator of oil in the area.   Also known as copper dirt.   { 'saùr ˌdərt }

sour gas [MATER]   Natural gas that contains corrosive, sulfur-bearing compounds, such as hydrogen sulfide and mercaptans.   { 'saùr ˌgas }

sourwood [BOT]   *Oxydendrum arboreum.* A deciduous tree of the heath family (Ericaceae) indigenous along the Alleghenies and having long, simple, finely toothed, long-pointed leaves that have an acid taste, and white, urn-shaped flowers.   Also known as sorrel tree.   { 'saùr,wùd }

Souslin's conjecture [MATH]   The conjecture that a topological space is homeomorphic to the real line if (1) it is totally ordered with no first or last element, (2) the open intervals are a base for its topology, (3) it is connected, and (4) any collection of disjoint open intervals.   { 'sü,slanz kən,jek·chər }

Souslin's line [MATH]   A topological space that satisfies the four conditions of Souslin's conjecture but is not separable, and thus not homeomorphic to the real line, in contradiction of Souslin's conjecture.   { 'sü,slanz ˌlīn }

Souslin's theorem [MATH]   The theorem that, if both a subset of a separable, complete metric space and its complement in this space are continuous images of Borel sets in this space, then the subset is itself a Borel set.   { 'sü,slanz thir·əm }

south [GEOD]   The direction 180° from north.   { saùth }

South African jade *See* Transvaal jade.   { 'saùth 'af·ri·kən 'jād }

South African tick-bite fever [MED]   An infectious tick-borne rickettsial disease of humans which is similar to fièvre boutonneuse.   { 'saùth 'af·ri·kən 'tik ˌbīt ˌfē·vər }

South America [GEOGR]   The southernmost of the Western Hemisphere continents, three-fourths of which lies within the tropics.   { 'saùth ə'mer·ə·kə }

South American blastomycosis [MED]   An infectious, yeastlike fungus disease of humans seen primarily in Brazil; caused by *Blastomyces brasiliensis* and characterized by massive enlargement of the cervical lymph nodes.   Also known as paracoccidioidomycosis.   { 'saùth ə'mer·ə·kən ˌblas-tō,mī'kō·səs }

South American leishmaniasis *See* American mucocutaneous leishmaniasis.   { 'saùth ə'mer·ə·kən ˌlēsh·mə'nī·ə·səs }

South American trypanosomiasis *See* Chagas' disease.   { 'saùth ə'mer·ə·kən trip,an·ə·sə'mī·ə·səs }

South Atlantic Current [OCEANOGR]   An eastward-flowing current of the South Atlantic Ocean that is continuous with the northern edge of the West Wind Drift.   { 'saùth at'lan·tik 'kə·rənt }

South Australian faunal region [ECOL]   A marine littoral region along the southwestern coast of Australia.   { 'saùth ò'strāl·yən 'fòn·əl ˌrē·jən }

SOURWOOD

Sourwood (*Oxydendrum arboreum*) leaf, twig, and enlargement of twig showing axial bud and leaf scar.

southbound node *See* descending node.   { 'saùth,baùnd 'nōd }

Southeast Drift Current [OCEANOGR]   A North Atlantic Ocean current flowing southeastward and southward from a point west of the Bay of Biscay toward southwestern Europe and the Canary Islands, where it continues as the Canary Current.   { saù'thēst 'drift 'kə·rənt }

southeaster [METEOROL]   A southeasterly wind, particularly a strong wind or gale; for example, the winter southeast storms of the Bay of San Francisco.   Also spelled sou'easter.   { saù'thēs·tər }

South Equatorial Current [OCEANOGR]   Any of several ocean currents, flowing westward, driven by the southeast trade winds blowing over the tropical oceans of the Southern Hemisphere and extending slightly north of the equator.   Also known as Equatorial Current.   { 'saùth ˌek·wə'tòr·ē·əl 'kə·rənt }

souther [METEOROL]   A south wind, especially a strong wind or gale.   { 'saùth·ər }

southerly burster [METEOROL]   A cold wind from the south in Australia.   { 'səth·ər·lē 'bər·stər }

southern bean mosaic virus [VIROL]   The type species of the plant-virus genus *Sobemovirus*. It is transmitted mechanically or via seed or the bean leaf beetle. Symptoms include crinkled leaves expressing a mild mosaic. Abbreviated SBMV.   { ˌsəth·ərn ˌbēn mō'zā·ik ˌvī·rəs }

southern bean mosaic virus group *See* Sobemovirus.   { ˌsəth·ərn ˌbēn mō'zā·ik ˌvī·rəs ˌgrüp }

Southern blotting [CELL MOL]   A technique for the detection of specific sequences among deoxyribonucleic acid (DNA) fragments whereby the fragments are separated by gel electrophoresis and then blotted onto a sheet of nitrocellulose for detection with radioactively labeled nucleic acid probes. It was first described by E. M. Southern in 1975.   { 'sə·th·ərn ˌbläd·iŋ }

Southern Cross *See* Crux.   { 'səth·ərn 'kròs }

Southern Crown *See* Corona Australis.   { 'səth·ərn 'kraùn }

Southern Fish *See* Piscis Australis.   { 'səth·ərn 'fish }

southern lights *See* aurora australis.   { 'səth·ərn 'līts }

Southern Polar Front *See* Antarctic Convergence.   { 'səth·ərn 'pō·lər 'frənt }

Southern Triangle *See* Triangulum Australe.   { 'səth·ərn 'trī,aŋ·gəl }

south foehn [METEOROL]   A foehn condition sustained by a strong south-to-north airflow across a transverse mountain barrier; the south foehn of the Alps may well be the most striking foehn in the world.   { 'saùth 'fān }

south frigid zone [GEOGR]   That part of the earth south of the Antarctic Circle.   { 'saùth 'frij·əd ˌzōn }

south geographical pole [GEOGR]   The geographical pole in the Southern Hemisphere, at latitude 90°S.   Also known as South Pole.   { 'saùth ˌjē·ə¦graf·ə·kəl ˌpōl }

south geomagnetic pole [GEOPHYS]   The geomagnetic pole in the Southern Hemisphere at approximately 78.5°S, longitude 111°E, 180° from the north geomagnetic pole.   Also known as south pole.   { 'saùth ˌjē·ō·mag'ned·ik ˌpōl }

South Indian Current [OCEANOGR]   An eastward-flowing current of the southern Indian Ocean that is continuous with the northern edge of the West Wind Drift.   { 'saùth 'in·dē·ən 'kə·rənt }

southing [NAV]   The distance a craft makes good to the south.   { 'saùth·iŋ }

South Pacific Current [OCEANOGR]   An eastward-flowing current of the South Pacific Ocean that is continuous with the northern edge of the West Wind Drift.   { 'saùth pə'sif·ik 'kə·rənt }

south point [ASTRON]   That imaginary point on the celestial sphere at which the meridian intersects the horizon; it is due south of the observer.   { 'saùth 'pòint }

south polar distance [ASTRON]   The angular distance between a celestial object and the south celestial pole.   { 'saùth 'pō·lər 'dis·təns }

south pole [ELECTROMAG]   The pole of a magnet at which magnetic lines of force are assumed to enter.   Also known as negative pole.   [GEOPHYS]   *See* south geomagnetic pole.   { 'saùth 'pōl }

South Pole *See* south geographical pole.   { 'saùth 'pōl }

south temperate zone [GEOGR]   That part of the earth

between the Tropic of Capricorn and the Antarctic Circle. { 'saùth 'tem·prət ˌzōn }

south tropical disturbance [ASTRON] An elongated dark band seen on the surface of Jupiter at about the latitude of the Great Red Spot; it has at times exceeded 180° of longitude in length and, like the Red Spot, appears and disappears intermittently. { 'saùth 'trap·ə·kəl di'stər·bəns }

southwester [METEOROL] A southwest wind, particularly a strong wind or gale. Also spelled sou'wester. { saùth'wes·tər }

sou'wester See southwester. { saù'wes·tər }

souzalite [MINERAL] $(Mg,Fe)_3(Al,Fe)_4(PO_4)_4(OH)_6 \cdot 2H_2O$ A green mineral composed of hydrous basic phosphate of magnesium, iron, and aluminum. { 'sō·zəˌlīt }

sow [MET] **1.** A mold of larger size than a pig. **2.** A channel that conducts molten metal to molds in a pig bed. [VERT ZOO] An adult female swine. { saù }

sow block [MET] In forging, a removable block set into the hammer anvil to reduce wear on the anvil. { 'saù ˌbläk }

Soxhlet extractor [CHEM] A flask and condenser device for the continuous extraction of alcohol- or ether-soluble materials. { 'säks·lət ik,strak·tər }

soya bean oil See soybean oil. { 'sȯi·ə ˌbēn ˌȯil }

soybean [BOT] Glycine max. An erect annual legume native to China and Manchuria and widely cultivated for forage and for its seed. { 'sȯiˌbēn }

soybean lecithin See lecithin. { 'sȯiˌbēn 'les·ə·thən }

soybean oil [MATER] A pale yellow, fixed drying oil produced by solvent extraction from soybeans; soluble in alcohol, chloroform, and ether; used for soap manufacture, cattle feeds, and printing inks, and in margarine, salad dressing, and high-protein foods. Also known as Chinese bean oil; soya bean oil; soy oil. { 'sȯiˌbēn ˌȯil }

soy lecithin See lecithin. { 'sȯi 'les·ə·thən }

soymilk [FOOD ENG] A milklike product derived from soybeans. { 'sȯiˌmilk }

soy oil See soybean oil. { 'sȯi ˌȯil }

space [ASTRON] **1.** Specifically, the part of the universe lying outside the limits of the earth's atmosphere. **2.** More generally, the volume in which all celestial bodies, including the earth, move. [COMMUN] The open-circuit condition or the signal causing the open-circuit condition in telegraphic communication; the closed-circuit condition is called the mark. [MATH] In context, usually a set with a topology on it or some other type of structure. { spās }

space attenuation [ACOUS] Loss of energy, expressed in decibels, of a signal in free air; caused by such factors as absorption, reflection, scattering, and dispersion. { 'spās ə,ten·yəˌwā·shən }

space biology [BIOL] A term for the various biological sciences and disciplines that are concerned with the study of living things in the space environment. { 'spās bī,äl·ə·jē }

space capsule [AERO ENG] A container, manned or unmanned, used for carrying out an experiment or operation in space. { 'spās ˌkap·səl }

space centrode [MECH] The path traced by the instantaneous center of a rotating body relative to an inertial frame of reference. { ¦spās 'sen,trōd }

space character See blank character. { 'spās ˌkar·ik·tər }

space charge [ELEC] The net electric charge within a given volume. [GEOPHYS] In atmospheric electricity, space charge refers to a preponderance of either negative or positive ions within any given portion of the atmosphere. { 'spās ˌchärj }

space-charge balanced flow [ELECTR] A method of focusing an electron beam in the interaction region of a traveling-wave tube; there is an axial magnetic field in the interaction region which is stronger than that in the gun region; at the transition between the two values of magnetic field strength, the beam is given a rotation in such a direction as to produce an inward force that counterbalances the outward forces from space charge and from the centrifugal forces set up by rotation. { 'spās ¦chärj 'bal·ənst 'flō }

space-charge debunching [ELECTR] A process in which the mutual interactions between electrons in a stream spread out the electrons of a bunch. { 'spās ¦chärj di'bənch·iŋ }

space-charge effect [ELECTR] Repulsion of electrons emitted from the cathode of a thermionic vacuum tube by electrons accumulated in the space charge near the cathode. { 'spās ¦chärj i,fekt }

space-charge grid [ELECTR] Grid operated at a low positive potential and placed between the cathode and control grid of a vacuum tube to reduce the limiting effect of space charge on the current through the tube. { 'spās ¦chärj ˌgrid }

space-charge layer See depletion layer. { 'spās ¦charj ˌlā·ər }

space-charge limitation [ELECTR] The current flowing through a vacuum between a cathode and an anode cannot exceed a certain maximum value, as a result of modification of the electric field near the cathode due to space charge in this region. { 'spās ¦charj ˌlim·ə'tā·shən }

space-charge polarization [ELEC] Polarization of a dielectric which occurs when charge carriers are present which can migrate an appreciable distance through the dielectric but which become trapped or cannot discharge at an electrode. Also known as interfacial polarization. { 'spās ¦chärj ˌpō·lə·rə'zā·shən }

space-charge region [ELECTR] Of a semiconductor device, a region in which the net charge density is significantly different from zero. { 'spās ¦chärj rē·jən }

space cloth [CHEM ENG] Woven cloth or wire used for solids screening, and for which the openings between the fibers or strands are designated in terms of space or clear opening. { 'spās ˌklȯth }

space communication [COMMUN] Communication between a vehicle in outer space and the earth or between two vehicles in space, using electromagnetic radiation. { 'spās kə,myü·nəˌkā·shən }

space cone [MECH] The cone in space that is swept out by the instantaneous axis of a rigid body during Poinsot motion. Also known as herpolhode cone. { 'spās ˌkōn }

space coordinates [MATH] A three-dimensional system of cartesian coordinates by which a point is located by three magnitudes indicating distance from three planes which intersect at a point. { 'spās kō'ȯrd·ən·əts }

spacecraft [AERO ENG] Devices, crewed and uncrewed, which are designed to be placed into an orbit about the earth or into a trajectory to another celestial body. Also known as space ship; space vehicle. { 'spāsˌkraft }

spacecraft ground instrumentation [ENG] Instrumentation located on the earth for monitoring, tracking, and communicating with manned spacecraft, satellites, and space probes. Also known as ground instrumentation. { 'spāsˌkraft 'graùnd ˌin·strəˌmən'tā·shən }

spacecraft launching [AERO ENG] The setting into motion of a space vehicle with sufficient force to cause it to leave the earth's atmosphere. { 'spāsˌkraft ˌlȯnch·iŋ }

spacecraft propulsion [AERO ENG] The use of rocket engines to accelerate space vehicles. { 'spāsˌkraft prə,pəl·shən }

spacecraft tracking [ENG] The determination of the positions and velocities of spacecraft through radio and optical means. { 'spāsˌkraft ˌtrak·iŋ }

space current [ELECTR] Total current flowing between the cathode and all other electrodes in a tube; this includes the plate current, grid current, screen grid current, and any other electrode current which may be present. { 'spās ˌkə·rənt }

space curve [MATH] A curve in three-dimensional Euclidean space; it may be a twisted curve or a plane curve. { 'spās ˌkərv }

spaced antenna [ELECTROMAG] Antenna system consisting of a number of separate antennas spaced a considerable distance apart, used to minimize local effects of fading at short-wave receiving stations. { 'spāst an'ten·ə }

spaced armor [ORD] An arrangement of armor plate, using two or more thicknesses, each thickness spaced from the adjoining one; used as a protective device, particularly against shaped-charge ammunition. { 'spāst 'är·mər }

space defense [ORD] All measures designed to reduce or nullify the effectiveness of hostile acts by vehicles (including missiles) while in space. { 'spās di,fens }

space detection and tracking system [ENG] System capable of detecting and tracking space vehicles from the earth, and reporting the orbital characteristics of these vehicles to a central control facility. Abbreviated SPADATS. { 'spās di¦tek·shən ən ¦trak·iŋ ˌsis·təm }

space diversity reception [ELECTROMAG] Radio reception involving the use of two or more antennas located several wavelengths apart, feeding individual receivers whose outputs are combined; the system gives an essentially constant output signal despite fading due to variable propagation characteristics, because fading affects the spaced-out antennas at different instants of time. { 'spās di·vər·səd·ē ri'sep·shən }

space-division multiple access [COMMUN] The use of the same portion of the electromagnetic spectrum over two or more transmission paths; in most applications, the paths are formed by multibeam antennas, and each beam is directed toward a different geographic area. Abbreviated SDMA. { 'spās də,vizh·ən ,məl·tə·pəl 'ak,ses }

spaced loading [ENG] Loading shot holes so that cartridges are separated by open spacers which do not prevent the concussion from one charge from reaching the next. { 'spāst 'lōd·iŋ }

space-dyed yarn [TEXT] Yarn dyed in one color for a specified length and in other colors for other lengths, the sequence being repeated. { 'spās¦dīd 'yärn }

space environment [ASTRON] The environment encountered by vehicles and living creatures in space, characterized by absence of atmosphere. { 'spās in,vi·ərn·mənt }

space factor [ELECTROMAG] **1.** The ratio of the space occupied by the conductors in a winding to the total cubic content or volume of the winding, or the similar ratio of cross sections. **2.** The ratio of the space occupied by iron to the total cubic content of an iron core. { 'spās ,fak·tər }

space-filling curve *See* Peano curve. { 'spās ,fil·iŋ ,kərv }

space fixed reference [NAV] An oriented reference system in space independent of earth phenomena for positioning. { 'spās ¦fikst 'ref·rəns }

space flight [AERO ENG] Travel beyond the earth's sensible atmosphere; space flight may be an orbital flight about the earth or it may be a more extended flight beyond the earth into space. { 'spās ,flīt }

space-flight trajectory [AERO ENG] The track or path taken by a spacecraft. { 'spās ¦flīt trə'jek·trē }

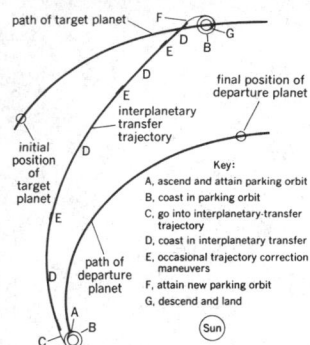

Key:
A, ascend and attain parking orbit
B, coast in parking orbit
C, go into interplanetary-transfer trajectory
D, coast in interplanetary transfer
E, occasional trajectory correction maneuvers
F, attain new parking orbit
G, descend and land

Simplified trajectory, a schematic diagram of an interplanetary mission.

space frame [BUILD] A three-dimensional steel building frame which is stable against wind loads. { 'spās ,frām }

space group [CRYSTAL] A group of operations which leave the infinitely extended, regularly repeating pattern of a crystal unchanged; there are 230 such groups. { 'spās ,grüp }

space-group extinction [CRYSTAL] The absence of certain classes of reflections in the x-ray diffraction pattern of a crystal due to the existence of symmetry elements in the space group of the crystal which are not present in its point group. { 'spās ¦grüp ik'stiŋk·shən }

space guidance [NAV] Guidance operations considered as consisting of the following three phases: ascent from earth to an orbit or space trajectory, operations in space requiring navigation or guidance, and descent to the surface of the earth or the moon. { 'spās ,gīd·əns }

space-hold [COMMUN] The transmission of a steady space signal over a transmission line which is carrying no traffic. { 'spās ,hōld }

space inversion *See* inversion. { 'spās in,vər·zhən }

SPACE SHUTTLE

space lattice [BUILD] A space frame built of lattice girders. [CRYSTAL] *See* lattice. { 'spās ,lad·əs }

spacelike path [RELAT] A trajectory in space-time such that a vector tangent to any point on the path is a spacelike vector. { 'spās,līk 'path }

spacelike surface [RELAT] A three-dimensional surface in a four-dimensional space-time which has the property that no event on the surface lies in the past or the future of any other event on the surface. { 'spās¦līk 'sər·fəs }

spacelike vector [RELAT] A four vector in Minkowski space whose space component has a magnitude which is greater than the magnitude of its time component multiplied by the speed of light. { 'spās¦līk 'vek·tər }

space medicine [MED] A branch of medicine that deals with the physiologic disturbances and diseases produced in man by high-velocity projection through and beyond the earth's atmosphere, flight through interplanetary space, and return to earth. { 'spās ,med·ə·sən }

space mission [AERO ENG] A journey by a vehicle, manned or unmanned, beyond the earth's atmosphere, usually for the purpose of collecting scientific data. { 'spās ,mish·ən }

space motion [ASTRON] Motion of a celestial body through space. { 'spās ,mō·shən }

space navigation [NAV] Determination of the three-dimensional position and velocity vector of a space vehicle relative to a selected frame of reference. { 'spās ,nav·ə,gā·shən }

space perception [PHYSIO] The awareness of the spatial

Launch of Columbia on its first orbital flight, April 2, 1981. (*NASA*)

properties and relations of an object, or of one's own body, in space; especially, the sensory appreciation of position, size, form, distance, and direction of an object, or of the observer himself, in space. { 'spās pər,sep·shən }

space permeability [ELECTROMAG] Factor that expresses the ratio of magnetic induction to magnetizing force in a vacuum; in the centimeter-gram-second electromagnetic system of units, the permeability of a vacuum is arbitrarily taken as unity; in the meter-kilogram-second-ampere system, it is $4\pi \times 10^{-7}$. { 'spās ,pər·mē·ə'bil·əd·ē }

space polar coordinates [MATH] A system of coordinates by which a point is located in space by its distance from a fixed point called the pole, the colatitude or angle between the polar axis (a reference line through the pole) and the radius vector (a straight line connecting the pole and the point), and the longitude or angle between a reference plane containing the polar axis and a plane through the radius vector and polar axis. { 'spās 'pō·lər kō'ōrd·ən·əts }

spaceport [AERO ENG] An installation used to test and launch spacecraft. { 'spās,pórt }

space power system [AERO ENG] An on-board assemblage of equipment to generate and distribute electrical energy on satellites and spacecraft. { 'spās ,paù·ər ,sis·təm }

space probe [AERO ENG] An instrumented vehicle, the payload of a rocket-launching system designed specifically for flight missions to other planets or the moon and into deep space, as distinguished from earth-orbiting satellites. { 'spās ,prōb }

space processing [ENG] The carrying out of various processes aboard orbiting spacecraft, utilizing the low-gravity, high-vacuum environment associated with these vehicles. { 'spās ,prä,ses·iŋ }

space quadrature [PHYS] A difference of a quarter-wavelength in the position of corresponding points of a wave in space. { 'spās ,kwäd·rə·chər }

space quantization [QUANT MECH] The quantization of the component of the angular momentum of a system in some specified direction. { 'spās ,kwän·tə'zā·shən }

spacer [ENG] **1.** A piece of metal wire twisted at one end to form a guard to keep the explosive in a shothole in place and twisted at the other end to form a guard to hold the tamping in its place. **2.** A piece of wood doweling interposed between charges to extend the column of explosive. **3.** A device for holding two members at a given distance from each other. Also known as spacer block. **4.** The tapered section of a pug joining the barrel to the die; clay is compressed in this section before it issues through the die. { 'spās·ər }

spacer block *See* spacer. { 'spās·ər ,bläk }

spacer deoxyribonucleic acid [MOL BIO] Untranscribed deoxyribonucleic acid (DNA) segments, usually containing repetitious DNA, of eukaryotic and some viral genomes flanking functional genetic regions (cistrons). { 'spās·ər dē¦äk·sē,rī·bō·nü,klē·ik 'as·əd }

space reconnaissance [AERO ENG] Reconnaissance of the surface of a planet from a space ship or satellite. { 'spās ri,kän·ə·səns }

space reddening [ASTRON] Reddening of light from distant stars caused by selective absorption of blue light by interstellar dust clouds. { 'spās ,red·ən·iŋ }

space reflection symmetry *See* parity. { 'spās ri¦flek·shən 'sim·ə·trē }

space request [COMPUT SCI] A parameter that specifies the amount of storage space required by a new file at the time the file is created. { 'spās ri,kwest }

space research [AERO ENG] Research involving studies of all aspects of environmental conditions beyond the atmosphere of the earth. { 'spās ri,sərch }

spacer strip [MET] A strip or bar of metal placed in the root of a weld joint, prepared for a groove weld, to serve as backing and maintain root opening during welding. { 'spās·ər ,strip }

space satellite [AERO ENG] A vehicle, crewed or uncrewed, for orbiting the earth. { 'spās ,sad·əl,īt }

space ship *See* spacecraft. { 'spās ,ship }

space shuttle [AERO ENG] A reusable orbital spacecraft, designed to travel from the earth to an orbital trajectory and then to return. { 'spās ,shəd·əl }

space simulator [AERO ENG] **1.** Any device which simulates one or more parameters of the space environment and which is used to test space systems or components. **2.** Specifically, a closed chamber capable of reproducing approximately the vacuum and normal environments of space. { 'spās ‚sim·yə‚lād·ər }

space station [AERO ENG] An autonomous, permanent facility in space for the conduct of scientific and technological research, earth-oriented applications, and astronomical observations. { 'spās ‚stā·shən }

space suit [ENG] A pressure suit for wear in space or at very low ambient pressures within the atmosphere, designed to permit the wearer to leave the protection of a pressurized cabin. { 'spās ‚süt }

space suppression [COMPUT SCI] Prevention of the normal movement of paper in a computer printer after the printing of a line of characters. { 'spās sə‚presh·ən }

space technology [AERO ENG] The systematic application of engineering and scientific disciplines to the exploration and utilization of outer space. { 'spās tek‚näl·ə·jē }

space-time [RELAT] A four-dimensional space used to represent the universe in the theory of relativity, with three dimensions corresponding to ordinary space and the fourth to time. Also known as space-time continuum. { 'spās 'tīm }

space-time continuum See space-time. { 'spās 'tīm kən‚tin·yə·wəm }

space-to-mark transition [COMMUN] The transition from the space condition to the mark condition in telegraphic communication. { 'spās tə ¦märk tran'zish·ən }

Space Tracking and Data Acquisition Network [ENG] A network of ground stations operated by the National Aeronautics and Space Administration, which tracks, commands, and receives telemetry for United States and foreign unmanned satellites. Abbreviated STADAN. { 'spās 'trak·iŋ ən ¦dad·ə ‚ak·wə'zīsh·ən ‚net‚wərk }

space vehicle See spacecraft. { 'spās ‚vē·ə·kəl }

space velocity [ASTRON] A star's true velocity with reference to the sun. [CHEM ENG] The relationship between feed rate and reactor volume in a flow process; defined as the volume or weight of feed (measured at standard conditions) per unit time per unit volume of reactor (or per unit weight of catalyst). { 'spās və‚läs·əd·ē }

space walk [AERO ENG] The movement of an astronaut outside the protected environment of a spacecraft during a space flight; the astronaut wears a spacesuit. { 'spās ‚wȯk }

space wave [ELECTROMAG] The component of a ground wave that travels more or less directly through space from the transmitting antenna to the receiving antenna; one part of the space wave goes directly from one antenna to the other; another part is reflected off the earth between the antennas. { 'spās ‚wāv }

space weapon [ORD] A weapon that travels through space and is directed against an enemy target whether on the ground, in the air, or in space. { 'spās ‚wep·ən }

space weather [GEOPHYS] The conditions on the sun and in the solar wind, magnetosphere, ionosphere, and thermosphere that can influence the performance and reliability of space-borne and ground-based technological systems and endanger human life or health. { 'spās ‚weth·ər }

spacing [GRAPHICS] The arrangement of characters, words, lines, and other elements to give the most pleasing effect on a printed page. { 'spās·iŋ }

spacing bias See bias telegraph distortion. { 'spās·iŋ ‚bī·əs }

spacing clamp [PETRO ENG] A clamp for maintaining the rod string in the correct pumping position while the well is in the final stages of being fitted to the pump. { 'spās·iŋ ‚klamp }

spacing pulse [COMMUN] In teletypewriter operation, the signal interval during which the selector unit is not operated. { 'spās·iŋ ‚pəls }

spacistor [ELECTR] A multiple-terminal solid-state device, similar to a transistor, that generates frequencies up to about 10,000 megahertz by injecting electrons or holes into a space-charge layer which rapidly forces these carriers to a collecting electrode. { spā'sis·tər }

spackling [ENG] The process of repairing a part of a plaster wall or mural by cleaning out the defective spot and then patching it with a plastering material. { 'spak·liŋ }

SPADATS See space detection and tracking system. { 'spā‚dats }

spade [DES ENG] A shovellike implement with a flat oblong blade; used for turning soil by pushing against the blade with the foot. { 'spād }

spade bolt [DES ENG] A bolt having a spade-shaped flattened head with a transverse hole, used to fasten shielded coils, capacitors, and other components to a chassis. { 'spād ‚bōlt }

spade drill [DES ENG] A drill consisting of three main parts: a cutting blade, a blade holder or shank, and a device, such as a screw, which fastens the blade to the holder; used for cutting holes over 1 inch (2.54 centimeters) in diameter. { 'spād ‚dril }

spade grip [ORD] D-shaped handle for pointing a gun, fastened on the rear of the receiver of certain flexible automatic weapons. { 'spād ‚grip }

spade lug [DES ENG] An open-ended flat termination for a wire lead, easily slipped under a terminal nut. { 'spād ‚ləg }

spadix [BOT] A fleshy spike that is enclosed in a leaflike spathe and is the characteristic inflorescence of palms and arums. [INV ZOO] A cone-shaped structure in male Nautiloidea formed of four modified tentacles, and believed to be homologous with the hectocotylus in male squids. { 'spā‚diks }

spaghetti [ELEC] Insulating tubing used over bare wires or as a sleeve for holding two or more insulated wires together; the tubing is usually made of a varnished cloth or a plastic. { spə'ged·ē }

spaghetti code [COMPUT SCI] Computer program code that lacks a coherent structure, and in which the sequence of program execution frequently jumps to a distant instruction in the program listing, making the program very difficult to follow. { spə'ged·ē ‚kōd }

spall [ENG] **1.** To reduce irregular stone blocks to an approximate size by chipping with a hammer. **2.** To break off thin chips from, and parallel to, the surface of a material, such as a metal or rock. [GEOL] **1.** A fragment removed from the surface of a rock by weathering. **2.** A relatively thin, sharp-edged fragment produced by exfoliation. **3.** A rock fragment produced by chipping with a hammer. [MIN ENG] To break ore. [ORD] A fragment torn from the surface of an armor plate. { spȯl }

spallation [NUC PHYS] A nuclear reaction in which the energy of the incident particle is so high that more than two or three particles are ejected from the target nucleus and both its mass number and atomic number are changed. Also known as nuclear spallation; spallation reaction. { spȯ'lā·shən }

spallation reaction See spallation. { spȯ'lā·shən rē‚ak·shən }

spalling [GEOL] The chipping or fracturing with an upward heaving, of rock caused by a compressional wave at a free surface. { 'spȯl·iŋ }

spalling hammer [ENG] A heavy axlike hammer with chisel edge, used for breaking and rough-dressing stone. { 'spȯl·iŋ ‚ham·ər }

spam [COMPUT SCI] Unsolicited commercial e-mail. { spam }

span [AERO ENG] **1.** The dimension of a craft measured between lateral extremities; the measure of this dimension. **2.** Specifically, the dimension of an airfoil from tip to tip measured in a straight line. [ENG] A structural dimension measured between certain extremities. [MATH] **1.** For a set A, the intersection of all sets that contain A and have some specified property. Also known as hull. **2.** For a set of vectors, the set of all possible linear combinations of those vectors. Also known as linear span. [STAT] The difference between the highest value and the lowest value in a range of values. { span }

spandex [TEXT] An elastic synthetic fiber; usually provides a core around which other fibers are wound. { 'span‚deks }

spandrel [BUILD] The part of a wall between the sill of a window and the head of the window below it. { 'span·drəl }

spandrel beam [BUILD] In steel or concrete construction, the exterior beam that extends from column to column and marks the floor level between stories. { 'span·drəl ‚bēm }

spandrel frame [BUILD] A triangular framing, as below a stair. { 'span·drəl ‚frām }

spandrel wall [BUILD] A wall on the outer surface of a vault to fill the spandrels. { 'span·drəl ‚wȯl }

spangolite [MINERAL] $Cu_6Al(SO_4)(OH)_{12}Cl \cdot 3H_2O$ A dark-green hexagonal mineral composed of hydrous basic sulfate and chloride of aluminum and copper and occurring as crystals. { 'spaŋ·gə‚līt }

Spanish collar See paraphimosis. { 'span·ish 'käl·ər }

spanishing [GRAPHICS] The depositing of ink in the valleys of a grained or embossed substrate. { 'span·ə·shiŋ }

Spanish lavender oil See spike oil. { 'span·ish ¦lav·ən·dər ¸òil }

Spanish spike oil See spike oil. { 'span·ish ¦spīk ¸òil }

Spanish white See bismuth subnitrate. { 'span·ish 'wīt }

spanned record [COMPUT SCI] A logical record which covers more than one block, used when the size of a data buffer is fixed or limited. { 'spand 'rek·ərd }

spanner [DES ENG] A wrench with a semicircular head having a projection or hole at one end. [ENG] **1.** A horizontal brace. **2.** An artificial horizon attachment for a sextant. { 'span·ər }

spanning subgraph [MATH] With reference to a graph G, a subgraph of G that contains all the vertices of G. { 'span·iŋ 'səb¸graf }

spanning tree [MATH] A spanning tree of a graph G is a subgraph of G which is a tree and which includes all the vertices in G. { 'span·iŋ ¸trē }

spar [AERO ENG] A principal spanwise member of the structural framework of an airplane wing, aileron, stabilizer, and such; it may be of one-piece design or a fabricated section. [MIN ENG] A small clay vein in a coal seam. [MINERAL] Any transparent or translucent, nonmetallic, light-colored, readily cleavable, crystalline mineral; examples are calespar and fluorspar. [NAV ARCH] A long, round stick of steel or wood, often tapered at one or both ends, and usually a part of a ship's masts or rigging. { spär }

sparagmite [GEOL] Late Precambrian fragmental rocks of Scandinavia, characterized by high proportions of microcline. { spə'rag¸mīt }

spar buoy [NAV] A long, thin, typically cylindrical buoy, ballasted at one end so that it floats in an approximately vertical position; used to mark the port side of a channel. { 'spar ¸bòi }

spar deck [NAV ARCH] A deck fitted from bow to stern on a superstructure having heavier scantlings than those under an awning deck. { 'spar ¸dek }

spare part [ENG] In supply usage, any part, component, or subassembly kept in reserve for the maintenance and repair of major items of equipment. { spär 'pärt }

spare-parts list [ENG] List approved by designated authorities, indicating the total quantities of spare parts, tools, and equipment necessary for the maintenance of a specified number of major items for a definite period of time. { 'spar ¦pärts ¸list }

Sparganiaceae [BOT] A family of monocotyledonous plants in the order Typhales distinguished by the inflorescence of globose heads, a vestigial perianth, and achenes that are sessile or nearly sessile. { spär¸gā·nē'ās·ē¸ē }

sparganosis [VET MED] An infection by the plerocercoid larva, or sparganum, of certain species of *Spirometra;* the adult form normally occurs in the intestine of dogs and cats. { ¸spär·gə'nō·səs }

sparganum [INV ZOO] The plerocercoid larva of a tapeworm. { 'spär·gə·nəm }

sparger See perforated-pipe distributor. { 'spär·jər }

sparging [CHEM ENG] The process of forcing air through water to remove undesirable gases. { 'spärj·iŋ }

Sparidae [VERT ZOO] A family of perciform fishes in the suborder Percoidei, including the porgies. { 'spar·ə¸dē }

sparite See sparry calcite. { 'spä¸rīt }

spark [ELEC] A short-duration electric discharge due to a sudden breakdown of air or some other dielectric material separating two terminals, accompanied by a momentary flash of light. Also known as electric spark; spark discharge; sparkover. { spärk }

spark arrester [ENG] **1.** An apparatus that prevents sparks from escaping from a chimney. **2.** A device that reduces or eliminates electric sparks at a point where a circuit is opened and closed. { 'spärk ə¸res·tər }

spark capacitor [ELEC] Capacitor connected across a pair of contact points, or across the inductance which causes the spark, for the purpose of diminishing sparking at these points. { 'spärk kə¸pas·əd·ər }

spark chamber [NUCLEO] A particle-detecting device in which the trajectory of a charged particle is made visible by a series of sparks that are triggered by the particle as it passes through an array of spark gaps. { 'spärk ¸chäm·bər }

spark coil [ELECTROMAG] An induction coil for producing spark discharges, as to initiate combustion in an internal combustion engine. { 'spärk ¸kòil }

spark-coil leak detector [ENG] A coil similar to a Tesla coil which detects leaks in a vacuum system by jumping a spark between the leak hole and the core of the coil. { 'spärk ¦kòil 'lēk di¸tek·tər }

spark counter [NUCLEO] A particle detector in which high-speed charged particles ionize a gas, consisting of argon mixed with an organic gas, triggering a spark between two plane parallel metal electrodes. { 'spärk ¸kaùnt·ər }

spark discharge See spark. { 'spärk 'dis¸chärj }

spark excitation [SPECT] The use of an electric spark (10,000 to 30,000 volts) to excite spectral line emissions from otherwise hard-to-excite samples; used in emission spectroscopy. { 'spärk ¸ek¸sī'tā·shən }

spark explosion method [ANALY CHEM] A technique for the analysis of hydrogen; the sample is mixed with an oxidant and exploded by a spark or hot wire, and the combustion products are then analyzed. { 'spärk ik'splō·zhən ¸meth·əd }

spark gap [ELEC] An arrangement of two electrodes between which a spark may occur; the insulation (usually air) between the electrodes is self-restoring after passage of the spark; used as a switching device, for example, to protect equipment against lightning or to switch a radar antenna from receiver to transmitter and vice versa. { 'spärk ¸gap }

spark-gap generator [ELEC] A high-frequency generator in which a capacitor is repeatedly charged to a high voltage and allowed to discharge through a spark gap into an oscillatory circuit, generating successive trains of damped high-frequency oscillations. { 'spärk ¦gap ¸jen·ə¸rād·ər }

spark-ignition combustion cycle See Otto cycle. { 'spärk ig¦nish·ən kəm'bəs·chən ¸sī·kəl }

spark-ignition engine [MECH ENG] An internal combustion engine in which an electrical discharge ignites the explosive mixture of fuel and air. { 'spärk ig¦nish·ən ¸en·jən }

sparking potential See breakdown voltage. { 'spärk·iŋ pə¸ten·chəl }

sparking voltage See breakdown voltage. { 'spärk·iŋ ¸vōl·tij }

spark killer See spark suppressor. { 'spärk ¸kil·ər }

spark knock [MECH ENG] The knock produced in an internal combustion engine precedes the arrival of the piston at the top dead-center position. { 'spärk ¸näk }

spark lead [MECH ENG] The amount by which the spark precedes the arrival of the piston at its top (compression) dead-center position in the cylinder of an internal combustion engine. { 'spärk ¸lēd }

sparkle metal [MET] A crude mixture of sulfides containing 74% copper produced by the smelting of copper ore. { 'spär·kəl ¸med·əl }

spark machining [MET] Cutting metal by repetitive sparking between a tool (the cathode) and the workpiece (the anode). { 'spärk mə¸shēn·iŋ }

sparkover See spark. { 'spärk¸ō·vər }

sparkover-initiated discharge machining [MECH ENG] An electromachining process in which a potential is impressed between the tool (cathode) and workpiece (anode) which are separated by a dielectric material; a heavy discharge current flows through the ionized path when the applied potential is sufficient to cause rupture of the dielectric. { 'spärk¸ō·vər i¦nish·ē¸ād·əd 'dis¸chärj mə¸shēn·iŋ }

sparkover voltage See flashover voltage. { 'spärk¸ō·vər ¸vōl·tij }

spark photography [OPTICS] Any type of photography in which a spark provides illumination and determines the length of exposure. { 'spärk fə¸tag·rə·fē }

spark plate [ELEC] A metal plate insulated from the chassis of an auto radio by a thin sheet of mica, and connected to the battery lead to bypass noise signals picked up by battery wiring in the engine compartment. { 'spärk ¸plāt }

spark plug [ELEC] A device that screws into the cylinder of an internal combustion engine to provide a pair of electrodes between which an electrical discharge is passed to ignite the explosive mixture. { 'spärk ¸pləg }

sparkproof [ENG] **1.** Treated with a material to prevent ignition or damage by sparks. **2.** Generating no sparks. { 'spärk¸prüf }

spark range [ORD] A firing range in which missiles in free flight can be photographed by the light from an electric spark which is triggered by passage of the projectile. { 'spärk ¸rānj }

spark recorder [ENG] Recorder in which the recording paper passes through a spark gap formed by a metal plate

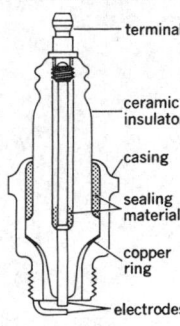

SPARK PLUG

— terminal
— ceramic insulator
— casing
— sealing material
— copper ring
— electrodes

Diagram of a typical spark plug.

underneath and a moving metal pointer above the paper; sparks from an induction coil pass through the paper periodically, burning small holes that form the record trace. { 'spärk ri'kòrd·ər }

spark spectrum [SPECT] The spectrum produced by a spark discharging through a gas or vapor; with metal electrodes, a spectrum of the metallic vapor is obtained. { 'spärk ˌspek·trəm }

spark suppressor [ELEC] A device used to prevent sparking between a pair of contacts when the contacts open, such as a resistor and capacitor in series between the contacts, or, in the case of an inductive circuit, a rectifier in parallel with the inductor. Also known as spark killer. { 'spärk sə,pres·ər }

spark test [MET] A technique for estimating the composition of a steel by observing the sparks it produces on a grinding wheel. { 'spärk ˌtest }

spark-tracing method [FL MECH] A method of flow visualization in which a series of spark discharges produced between two electrodes is photographed with an open shutter to record a system of time lines. { 'spärk ˌträs·iŋ ˌmeth·əd }

spark transmitter [ELECTR] A radio transmitter that utilizes the oscillatory discharge of a capacitor through an inductor and a spark gap as the source of radio-frequency power. { 'spärk tranz'mid·ər }

spark voltage [ELEC] The voltage required to create an arc across the gap of a spark plug. { 'spärk ˌvōl·tij }

Sparnacean [GEOL] A European stage of geologic time; upper upper Paleocene, above Thanetian, below Ypresian of Eocene. { spär'näsh·ən }

Sparrow's criterion [OPTICS] A criterion for the resolution of two light sources, according to which the light sources are resolved if there is some central dip in their combined diffraction pattern. { 'spar·ōz krī,tir·ē·ən }

sparry calcite [MINERAL] A clean, coarse-grained calcite crystal. Also known as calcsparite; sparite. { 'spär·ē 'kal,sīt }

sparry cement [GEOL] Clear, relatively coarse-grained calcite in the interstices of any sedimentary rock. { 'spär·ē si'ment }

sparry iron See siderite. { 'spär·ē 'ī·ərn }

sparse matrix [MATH] A matrix most of whose entries are zeros. { 'spärs 'mā·triks }

sparseness [MATH] The property of a nonlinear programming problem which has many variables, but whose objective and constraint functions each involve only relatively few variables. { 'spärs·nəs }

spartalite See zincite. { 'spärd·əl,īt }

sparteine [ORG CHEM] $C_{15}H_{26}N_2$ A poisonous, colorless, oily alkaloid; soluble in alcohol and ether, slightly soluble in water; boils at 173°C; used in medicine. Also known as lupinidine. { 'spärd·ē,ēn }

spar varnish [MATER] A flammable varnish made of drying oils, resins, thinners, and driers to provide a durable, water-resistant coating for outside or other severe service. { 'spär ˌvär·nish }

spasm [MED] An involuntary and abnormal contraction of isolated bundles of muscle or groups of muscles resulting from a chemical imbalance due to fatigue, ischemia, or trauma. { 'spaz·əm }

spasmodic turbidity current [GEOPHYS] A single, rapidly developed turbidity current. { spaz'mäd·ik tər'bid·əd·ē ,kə·rənt }

spasmophilia [MED] A morbid tendency to convulsions, and to tonic spasms, such as those observed in tetany, infantile spasms, or spasmus nutans. { ,spaz·mə'fil·ē·ə }

spastic colon See irritable colon. { 'spas·tik 'kō·lən }

spastic diplegia [MED] 1. Spastic paralysis of the arms and legs caused by diffuse lesions of the cerebral cortex. 2. A form of cerebral palsy, possibly due to prenatal or perinatal hypoxia or other injuries resulting in atrophic lobar sclerosis, or to congenital or developmental abnormalities. { 'spas·tik dī'plē·jə }

spastic ileus [MED] A form of ileus in which temporary obstruction is due to segmental intestinal spasm. Also known as dynamic ileus; hyperdynamic ileus. { 'spas·tik 'il·ē·əs }

spasticity See spastic paralysis. { spas'tis·əd·ē }

spastic paralysis [MED] A condition in which a group of muscles manifest increased tone, exaggerated tendon reflexes, depressed or absent superficial reflexes, and sometimes clonus,

due to an upper motor neuron lesion. Also known as spasticity. { 'spas·tik pə'ral·ə·səs }

spastic paraplegia [MED] Paralysis of the lower limbs with increased muscular tone and hyperactive tendon reflexes; commonly seen in diseases and injuries involving pyramidal tracts of the spinal cord. { 'spas·tik ,par·ə'plē·jə }

spastic strabismus [MED] A squint resulting from the contraction of an ocular muscle. { 'spas·tik strə'biz·məs }

Spatangoida [INV ZOO] An order of exocyclic Euechinoidea in which the posterior ambulacral plates form a shield-shaped area behind the mouth. { ,spat·ən'gòid·ə }

spathe [BOT] A large, usually colored bract or pair of bracts enclosing an inflorescence, especially a spadix, on the same axis. { spāth }

spathic iron See siderite. { 'spath·ik 'ī·ərn }

spatial [PHYS] Of or pertaining to space; occupying space; occurring in, or conditioned by, space; considered with relation to space. { 'spā·shəl }

spatial autocorrelation [GEOGR] In mathematical geography, the degree of interdependence among data arranged on a three-dimensional grid. { 'spā·shəl ˌòd·ō,kä·rə'lā·shən }

spatial coherence [PHYS] The existence of a correlation between the phases of waves at points separated in space at a given time. { 'spā·shəl kō'hir·əns }

spatial data management [COMPUT SCI] A technique whereby users retrieve information in data bases, document files, or other sources by making contact with picture symbols displayed on the screen of a video terminal through the use of such devices as light pens, joy sticks, and heat-sensitive screens for finger-touch activation. { 'spā·shəl 'dad·ə ,man·ij·mənt }

spatial dendrite [METEOROL] A complex ice crystal with fernlike arms that extend in many directions (spatially) from a central nucleus; its form is roughly spherical. Also known as spatial dendritic crystal. { 'spā·shəl 'den,drīt }

spatial dendritic crystal See spatial dendrite. { 'spā·shəl den·'drid·ik 'krist·əl }

spatial filter [OPTICS] An optical filter that consists of a very small aperture, such as a pinhole. { 'spā·shəl 'fil·tər }

spatial linkage [MECH ENG] A linkage that involves motion in all three dimensions. { 'spā·shəl 'liŋ·kij }

spatiotemporal [PHYS] Of or pertaining to space time; having extent and duration. { spā·shē·ō'tem·pə·rəl }

spatter [MET] Particles of metal expelled during arc or gas welding. { 'spad·ər }

spatter cone [GEOL] A low, steep-sided cone of small pyroclastic fragments built up on a fissure or vent. Also known as agglutinate cone; volcanello. { 'spad·ər ,kōn }

spatter dash [CIV ENG] 1. A finish put on stucco by dashing a mortar and sand mixture against it. 2. Paint spattered on a different-colored ground coat. { 'spad·ər ,dash }

spatter rampart [GEOL] A low, circular ridge of pyroclastics built up around the margins of small volcanoes. { 'spad·ər ,ram,pärt }

spatulate [BIOL] Shaped like a spoon. { 'spach·ə·lət }

spawn [BOT] Mycelium used for initiating mushroom propagation. [ZOO] 1. The collection of eggs deposited by aquatic animals, such as fish. 2. To produce or deposit eggs or discharge sperm; applied to aquatic animals. { spòn }

spay [VET MED] To remove the ovaries. { spā }

SPC See stored-program control.

SPDT See single-pole double-throw.

speaker See loudspeaker. { 'spēk·ər }

speaker identification [ENG ACOUS] The use of automated equipment to find the identity of a talker, in a known population of talkers, using the speech input. { ,spēk·ər ī,dent·ə·fə'kā·shən }

speaker verification [ENG ACOUS] The use of automated equipment to authenticate a claimed speaker identity from a voice signal based on speaker-specific characteristics reflected in spoken words or sentences. Abbreviated SV. { ,spēk·ər ,ver·i·fə'kā·shən }

spear [DES ENG] A rodlike fishing tool having a barbed-hook end, used to recover rope, wire line, and other materials from a borehole. { spir }

Spearman-Brown formula [STAT] A formula to estimate the reliability of a test n times as long as one for which reliability is known; the tests must be comparable in all aspects other than size. { ,spir·mən 'braun ,fòr·myə·lə }

Spearman's rank correlation coefficient [STAT] A statistic used as a measure of correlation in nonparametric statistics when the data are in ordinal form; a product moment correlation coefficient. Also known as Spearman's rho. { ¦spir·mənz ¦raŋk ¦kär·ə'lā·shən ¸kō·ə¸fish·ənt }

Spearman's rho *See* Spearman's rank correlation coefficient. { ¦spir·mənz 'rō }

spearmint [BOT] *Mentha spicata.* An aromatic plant of the mint family, Labiatae; the leaves are used as a flavoring in foods. { 'spir¸mint }

spearmint oil [MATER] A colorless to yellowish essential oil obtained from spearmint with characteristic taste and scent; soluble in alcohol, ether, and chloroform; used as a flavor and a source of carvone. { 'spir¸mint ¸óil }

special ammunition supply point [ORD] A mobile supply point where special ammunition is stored and issued to delivery units. { 'spesh·əl ¸am·yə¦nish·ən sə'plī ¸póint }

special atomic demolition munition [ORD] A very-low-yield, man-portable, atomic demolition munition which is detonated by a timer device. { 'spesh·əl ə'täm·ik ¸dem·ə'lish·ən myü¸nish·ən }

special cargo [IND ENG] Cargo which requires special handling or protection, such as pyrotechnics, detonators, watches, and precision instruments. { 'spesh·əl 'kär·gō }

special cause [ANALY CHEM] A cause of variance or bias in a measurement process that is external to the system. { ¦spesh·əl ¦kóz }

special character [COMPUT SCI] A computer-representable character that is not alphabetic, numeric, or blank. { 'spesh·əl 'kar·ik·tər }

special flight [AERO ENG] An air transport flight, other than a scheduled service, set up to move a specific load. { 'spesh·əl 'flīt }

special functions [MATH] The various families of solution functions corresponding to cases of the hypergeometric equation or functions used in the equation's study, such as the gamma function. { 'spesh·əl 'faŋk·shənz }

special gelatin [MATER] The brand name of a series of ammonia-gelatin-type dynamites used in open-pit mining, underground metal mining, quarrying, and construction. { 'spesh·əl 'jel·ət·ən }

special job cover map [MAP] A small-scale map used to record progress on photographic reconnaissance tasks covering very large areas; as each portion of the task is completed, the area covered is outlined on the map. { 'spesh·əl ¦jäb 'kəv·ər ¸map }

special Jordan algebra [MATH] A Jordan algebra that can be written as a symmetrized product over a matrix algebra. { 'spesh·əl ¦jórd·ən ¸al·jə·brə }

special nuclear material [NUCLEO] Fissionable and related material controlled directly by the Atomic Energy Commission, such as uranium enriched in the isotopes ^{235}U and ^{233}U, and plutonium. { 'spesh·əl 'nü·klē·ər mə'tir·ē·əl }

special observation [METEOROL] A category of aviation weather observation taken to report significant changes in one or more of the observed elements since the last previous record observation. { 'spesh·əl ¸äb·zər'vā·shən }

special orthogonal group of dimension n [MATH] The Lie group of special orthogonal transformations on an *n*-dimensional real inner product space. Symbolized SO$_n$; SO(n). { ¦spesh·əl ór¦thäg·ən·əl ¸grüp əv di¦men·chən 'en }

special orthogonal transformation [MATH] An orthogonal transformation whose matrix representation has determinant equal to 1. { ¦spesh·əl ór¦thäg·ən·əl ¸tranz·fər'mā·shən }

special-purpose buoy [NAV] A buoy for a special purpose which has no lateral significance, such as those buoys used to mark quarantine and anchorage areas, or dredging and survey operations. { 'spesh·əl ¦pər·pəs 'bói }

special-purpose computer [COMPUT SCI] A digital or analog computer designed to be especially efficient in a certain class of applications. { 'spesh·əl ¦pər·pəs kəm'pyüd·ər }

special-purpose item [ENG] In supply usage, any item designed to fill a special requirement, and having a limited application; for example, a wrench or other tool designed to be used for one particular model of a piece of machinery. { 'spesh·əl ¦pər·pəs 'īd·əm }

special-purpose language [COMPUT SCI] A programming language designed to solve a particular type of problem. { 'spesh·əl ¦pər·pəs 'laŋ·gwij }

special-purpose vehicle [ENG] A vehicle having a special chassis, or a general-purpose chassis incorporating major modifications, designed to fill a specialized requirement; all tractors (except truck tractors) and tracklaying vehicles, regardless of design, size, or intended purpose, are classified as special-purpose vehicles. { 'spesh·əl ¦pər·pəs 'vē·ə·kəl }

special relativity [RELAT] The division of relativity theory which relates the observations of observers moving with constant relative velocities and postulates that natural laws are the same for all such observers. { 'spesh·əl rel·ə'tiv·əd·ē }

special unitary group of dimension n [MATH] The Lie group of special unitary transformations on an *n*-dimensional inner product space over the complex numbers. Symbolized SU(n). { 'spesh·əl ¦yü·nə¸ter·ē ¸grüp əv di¦men·chən 'en }

special unitary transformation [MATH] A unitary transformation whose matrix representation has determinant equal to 1. { ¦spesh·əl ¦yü·nə¸ter·ē ¸tranz·fər'mā·shən }

special weapon [ORD] Any extraordinary modern weapon, such as an atomic, radiological, or biological weapon. { 'spesh·əl 'wep·ən }

special weather report [METEOROL] The encoded and transmitted weather report of a special observation. { 'spesh·əl 'weth·ər ri¸pórt }

speciation [EVOL] The evolution of species. { ¸spē·sē'ā·shən }

species [CHEM] A chemical entity or molecular particle, such as a radical, ion, molecule, or atom. Also known as chemical species. [NUC PHYS] *See* nuclide. [SYST] A taxonomic category ranking immediately below a genus and including closely related, morphologically similar individuals which actually or potentially interbreed. { 'spē·shēz }

species concept [EVOL] The idea that the diversity of nature is divisible into a finite number of definable species. { 'spē·shēz ¸kän¸sept }

species number [OCEANOGR] The first number in the argument number in a Doodson tide schedule; indicates approximately the period of a component of tidal potential. { 'spē·shēz ¸nəm·bər }

species population [ECOL] A group of similar organisms residing in a defined space at a certain time. { 'spē·shēz ¸päp·yə'lā·shən }

specific [PHYS] Indicating the amount of a physical quantity per unit mass, weight, volume, or area, or the ratio of the quantity for the substance under consideration to the same quantity for a standard substance, such as water. { spə'sif·ik }

specific acoustical impedance [ACOUS] The ratio of the pressure phasor associated with a sound wave at any given point in a medium to the velocity phasor at that point. { spə'sif·ik ə'küs·tə·kəl im'pēd·əns }

specific acoustical ohm *See* rayl. { spə'sif·ik ə'küs·tə·kəl 'ōm }

specific acoustical reactance [ACOUS] The magnitude of the imaginary part of the specific acoustical impedance. { spə'sif·ik ə'küs·tə·kəl rē'ak·təns }

specific acoustical resistance [ACOUS] The real part of the specific acoustical impedance. { spə'sif·ik ə'küs·tə·kəl ri'zis·təns }

specific active immunotherapy [IMMUNOL] Active immunotherapy that attempts to stimulate specific antitumor responses with tumor-associated antigens as the immunizing materials. { spə¦sif·ik ¦ak·tiv ¸im·yə·nō'ther·ə·pē }

specific activity [NUCLEO] **1.** The activity of a radioisotope of an element per unit weight of element present in the sample. **2.** The activity per unit mass of a pure radionuclide. **3.** The activity per unit weight of a sample of radioactive material. In these three cases, specific activity can be expressed in such units as millicuries per gram, disintegrations per second per milligram, or counts per minute per milligram. { spə'sif·ik ak'tiv·əd·ē }

specifications [ENG] An organized listing of basic requirements for materials of construction, product compositions, dimensions, or test conditions; a number of organizations publish standards (for example, American Society of Mechanical Engineers, American Petroleum Institute, and American Society for Testing and Materials), and many companies have their own specifications. Also known as specs. [IND ENG] A quantitative description of the required characteristics of a device, machine, structure, product, or process. { ¸spes·ə·fə'kā·shənz }

specific catalysis [CHEM] The acceleration of a given

chemical reaction by a unique catalyst rather than by a family of related substances. { spə¦sif·ik kə'tal·ə·səs }

specific charge [ELEC] The ratio of a particle's charge to its mass. { spə'sif·ik 'chärj }

specific conductance See conductivity. { spə'sif·ik kən'dək·təns }

specific cryptosystem [COMMUN] A general cryptosystem and a key or set of keys for controlling the cryptographic process. { spə'sif·ik 'krip·tō͵sis·təm }

specific energy [HYD] The energy at any cross section of an open channel, measured above the channel bottom as datum; numerically the specific energy is the sum of the water depth plus the velocity head, $v^2/2g$, where v is the velocity of flow and g the acceleration of gravity. [THERMO] The internal energy of a substance per unit mass. { spə'sif·ik 'en·ər·jē }

specific fuel consumption [MECH ENG] The weight flow rate of fuel required to produce a unit of power or thrust, for example, pounds per horsepower-hour. Abbreviated SFC. Also known as specific propellant consumption. { spə'sif·ik 'fyül kən͵səm·shən }

specific gravity [MECH] The ratio of the density of a material to the density of some standard material, such as water at a specified temperature, for example, 4°C or 60°F, or (for gases) air at standard conditions of pressure and temperature. Abbreviated sp gr. Also known as relative density. { spə'sif·ik 'grav·əd·ē }

specific-gravity bottle [ENG] A small bottle or flask used to measure the specific gravities of liquids; the bottle is weighed when it is filled with the liquid whose specific gravity is to be determined, when filled with a reference liquid, and when empty. Also known as density bottle; relative-density bottle. { spə'sif·ik ¦grav·əd·ē ¦bäd·əl }

specific-gravity hydrometer [ENG] A hydrometer which indicates the specific gravity of a liquid, with reference to water at a particular temperature. { spə'sif·ik ¦grav·əd·ē hī'dräm·əd·ər }

specific heat [THERMO] **1.** The ratio of the amount of heat required to raise a mass of material 1 degree in temperature to the amount of heat required to raise an equal mass of a reference substance, usually water, 1 degree in temperature; both measurements are made at a reference temperature, usually at constant pressure or constant volume. **2.** The quantity of heat required to raise a unit mass of homogeneous material one degree in temperature in a specified way; it is assumed that during the process no phase or chemical change occurs. { spə'sif·ik 'hēt }

specific humidity [METEOROL] In a system of moist air, the (dimensionless) ratio of the mass of water vapor to the total mass of the system. { spə'sif·ik hyü'mid·əd·ē }

specific impulse [AERO ENG] A performance parameter of a rocket propellant, expressed in seconds, equal to the thrust in pounds divided by the weight flow rate in pounds per second. Also known as specific thrust. { spə'sif·ik 'im͵pəls }

specific inductive capacity See dielectric constant. { spə'sif·ik in'dək·tiv kə'pas·əd·ē }

specific insulation resistance See volume resistivity. { spə'sif·ik ͵in·sə'lā·shən ri͵zis·təns }

specific ionization [NUCLEO] The number of ion pairs formed per unit distance along the track of an ion passing through matter. Also known as total specific ionization. { spə'sif·ik ͵ī·ə·nə'zā·shən }

specificity [CHEM] The selective reactivity that occurs between substances, such as between an antigen and its corresponding antibody. { ͵spes·ə'fis·əd·ē }

specific locus test [GEN] A technique used to detect recessive induced mutations in diploid organisms; a strain which carries several known recessive mutants in a homozygous condition is crossed with a nonmutant strain treated to induce mutations in its germ cells; induced recessive mutations allelic with those of the test strain will be expressed in the progeny. { spə'sif·ik 'lō·kəs ͵test }

specific mass shift [NUC PHYS] The portion of the mass shift that is produced by the correlated motion of different pairs of atomic electrons and is therefore absent in one-electron systems. { spə'sif·ik 'mas ͵shift }

specific power [NUCLEO] The power produced per unit mass of fuel present in a nuclear reactor. { spə'sif·ik 'pau·ər }

specific productivity index [PETRO ENG] Barrels per day of oil produced per pound decline in bottom-hole pressure per foot of effective reservoir thickness. { spə'sif·ik ͵präd·ək'tiv·əd·ē ͵in͵deks }

specific propellant consumption See specific fuel consumption. { spə'sif·ik prə'pel·ənt kən͵səm·shən }

specific reluctance See reluctivity. { spə'sif·ik ri'lək·təns }

specific repetition rate [ELECTR] The pulse repetition rate of a pair of transmitting stations of an electronic navigation system using various rates differing slightly from each other, as in loran. { spə'sif·ik ͵rep·ə'tish·ən ͵rāt }

specific resistance See electrical resistivity. { spə'sif·ik ri'zis·təns }

specific retention [GEOL] The ratio of the volume of water that a given body of rock or soil will retain after saturation, and the pull of gravity to the volume of the body itself. { spə'sif·ik ri'ten·chən }

specific retention volume [ANALY CHEM] The relationship among retention volume, void volume, and adsorbent weight, used to standardize gas chromatography adsorbents by the elution of a standard solute by a standard eluent from the adsorbent under test. { spə'sif·ik ri'ten·chən ͵väl·yəm }

specific rotation [OPTICS] The calculated rotation of light passing through a solution as related to the solution volume and depth, the amount of solute, and the observed optical rotation at a given wavelength and temperature. { spə'sif·ik rō'tā·shən }

specific routine [COMPUT SCI] Computer routine to solve a particular data-handling problem in which each address refers to explicitly stated registers and locations. { spə'sif·ik rü'tēn }

specific speed [MECH ENG] A number, N_s, used to predict the performance of centrifugal and axial pumps or hydraulic turbines: for pumps, $N_s = N \sqrt{Q}/H^{3/4}$; for turbines, $N_s = N \sqrt{F}/H^{5/4}$, where N_s is specific speed, N is the rotational speed in revolutions per minute, Q is the rate of flow in gallons per minute, H is head in feet, and P is shaft horsepower. { spə'sif·ik 'spēd }

specific surface [CHEM ENG] The surface area per unit weight or volume of a particulate solid; used in size-reduction (crushing and grinding) calculations. { spə'sif·ik 'sər·fəs }

specific susceptibility See mass susceptibility. { spə'sif·ik sə͵sep·tə'bil·əd·ē }

specific thrust See specific impulse. { spə'sif·ik 'thrəst }

specific viscosity [FL MECH] The specific viscosity of a polymer is the relative viscosity of a polymer solution of known concentration minus 1; usually determined at low concentration of the polymer; for example, 0.5 gram per 100 milliliters of solution, or less. { spə'sif·ik vi'skäs·əd·ē }

specific volume [MECH] The volume of a substance per unit mass; it is the reciprocal of the density. Abbreviated sp vol. { spə'sif·ik 'väl·yəm }

specific-volume anomaly [OCEANOGR] The excess of the actual specific volume of the sea water at any point in the ocean over the specific volume of sea water of salinity 35 parts per thousand (‰) and temperature 0°C at the same pressure. Also known as steric anomaly. { spə'sif·ik ¦väl·yəm ə'näm·ə·lē }

specific weight [MECH] The weight per unit volume of a substance. { spə'sif·ik 'wāt }

specific yield [HYD] The quantity of water which a unit volume of aquifer, after being saturated, will yield by gravity; it is expressed either as a ratio or as a percentage of the volume of the aquifer; specific yield is a measure of the water available to wells. { spə'sif·ik 'yēld }

specimen [SCI TECH] **1.** An item representative of others in the same class or group. **2.** A sample selected for testing, examination, or display. { 'spes·ə·mən }

speck [PL PATH] A fungus or bacterial disease of rice characterized by speckled grains. { spek }

speckle [OPTICS] A phenomenon in which the scattering of light from a highly coherent source, such as a laser, by a rough surface or inhomogeneous medium generates a random-intensity distribution of light that gives the surface or medium a granular appearance. { 'spek·əl }

speckle interferometry [OPTICS] The use of speckle patterns in the study of object displacements, vibration, and distortion, and in obtaining diffraction-limited images of stellar objects. { 'spek·əl ͵in·tər·fə'räm·ə·trē }

specs See specifications. { speks }

SPECT See single-photon-emission computed tomography. { spekt }

spectacle [ZOO] A colored marking in the form of rings

around the eyes, as in certain birds, reptiles, and mammals (as the raccoon). { 'spek·tə·kəl }

spectacle frame [NAV ARCH] A frame at or close to the sternposts of a twin-screw ship, through which pass propeller shafts. { 'spek·tə·kəl ‚frām }

spectacle stone See selenite. { 'spek·tə·kəl ‚stōn }

spectator ion [CHEM] An ion that serves to balance the electrical charges in a reaction environment without participating in product formation. { 'spek‚tād·ər 'ī‚än }

spectral approximation [MATH] A numerical approximation of a function of two or more variables that involves the expansion of the function into a generalized Fourier series, followed by computation of the Fourier coefficients. { ‚spek·trəl ə‚präk·sə'mā·shən }

spectral bandwidth [SPECT] The minimum radiant-energy bandwidth to which a spectrophotometer is accurate; that is, 1–5 nanometers for better models. { 'spek·trəl 'band‚width }

spectral centroid [OPTICS] An average wavelength; specifically, for a light filter or other light-transmitting device, a weighted average of the spectral energy distribution of the incident light, the transmittance of the device, and the luminosity function. { 'spek·trəl 'sen‚tróid }

spectral characteristic [OPTICS] The relation between wavelength and some other variable, such as between wavelength and emitted radiant power of a luminescent screen per unit wavelength interval. { 'spek·trəl ‚kar·ik·tə'ris·tik }

spectral classification [ASTRON] A classification of stars by characteristics revealed by study of their spectra; the six classes B, A, F, G, K, and M include 99% of all known stars. { 'spek·trəl ‚klas·ə·fə'kā·shən }

spectral color [OPTICS] **1.** A color corresponding to light of a pure frequency; the basic spectral colors are violet, blue-green, yellow, orange, and red. **2.** A color that is represented by a point on the chromaticity diagram that lies on a straight line between some point on the spectral color (first definition) locus and the achromatic points; purple, for example, is not a spectral color. { 'spek·trəl 'kəl·ər }

spectral density [ELECTROMAG] See spectral energy distribution. [MATH] The density function for the spectral measure of a linear transformation on a Hilbert space. [SYS ENG] See frequency spectrum. { 'spek·trəl 'den·səd·ē }

spectral dimension See fracton dimension. { 'spek·trəl di'men·shən }

spectral directional reflectance factor [ANALY CHEM] In spectrophotometric colorimetry, the ratio of the energy diffused in any desired direction by the object under analysis to that energy diffused in the same direction by an ideal perfect (energy) diffuser. { 'spek·trəl di'rek·shən·əl ri'flek·təns ‚fak·tər }

spectral emissivity [THERMO] The ratio of the radiation emitted by a surface at a specified wavelength to the radiation emitted by a perfect blackbody radiator at the same wavelength and temperature. { 'spek·trəl ‚ē‚mi'siv·əd·ē }

spectral energy distribution [ELECTROMAG] The power carried by electromagnetic radiation within some small interval of wavelength (of frequency) of fixed amount as a function of wavelength (of frequency). Also known as spectral density. { 'spek·trəl 'en·ər·jē ‚dis·trə‚byü·shən }

spectral extinction [OPTICS] The selective absorption of different wavelengths of light as a function of depth in water. { 'spek·trəl ik'stiŋk·shən }

spectral factorization [MATH] A process sometimes used in the study of control systems, in which a given rational function of the complex variable s is factored into the product of two functions, $F_R(s)$ and $F_L(s)$, each of which has all of its poles and zeros in the right and left half of the complex plane, respectively. { 'spek·trəl ‚fak·tə·rə'zā·shən }

spectral function [MATH] In the theory of stationary stochastic processes, the function

$$F(y) = (2/\pi) \int_0^\infty \rho(x)(\sin xy/x)(dx), \quad 0 \le y \le \infty$$

where $\rho(x)$ is the autocorrelation function of a stationary time series. { 'spek·trəl 'fəŋk·shən }

spectral hygrometer [ENG] A hygrometer which determines the amount of precipitable moisture in a given region of the atmosphere by measuring the attenuation of radiant energy

caused by the absorption bands of water vapor; the instrument consists of a collimated energy source, separated by the region under investigation and a detector which is sensitive to those frequencies that correspond to the absorption bands of water vapor. { 'spek·trəl hī'gräm·əd·ər }

spectral irradiance [OPTICS] The density of the radiant flux that is incident on a surface per unit of wavelength. { 'spek·trəl i'rād·ē·əns }

spectral line [SPECT] A discrete value of a quantity, such as frequency, wavelength, energy, or mass, whose spectrum is being investigated; one may observe a finite spread of values resulting from such factors as level width, Doppler broadening, and instrument imperfections. Also known as spectrum line. { 'spek·trəl ‚līn }

spectral locus See spectrum locus. { 'spek·trəl 'lō·kəs }

spectral luminosity classification See MK system. { 'spek·trəl ‚lü·mə'näs·əd·ē ‚klas·ə·fə‚kā·shən }

spectral luminous efficacy [OPTICS] The ratio of the luminous flux emitted by a monochromatic light source in lumens to its radiant flux in watts, as a function of the wavelength of the emitted light. { 'spek·trəl 'lü·mə·nəs 'ef·i·kə·sē }

spectral luminous efficiency See luminosity function. { 'spek·trəl 'lü·mə·nəs ə'fish·ən·sē }

spectral measure [MATH] A measure on the spectrum of an operator on a Hilbert space whose values are projection operators there; spectral theorems concerning linear operators often give an integral representation of the operator in terms of these projection valued measures. { 'spek·trəl 'mezh·ər }

spectral photography [OPTICS] A technique used in airborne surveys for mineral deposits; narrow-band-pass filters and special film are used to accentuate minor color effects caused by mineralization and alteration which would be undetectable by broad-band photography. { 'spek·trəl fə'täg·rə·fē }

spectral pyrometer See narrow-band pyrometer. { 'spek·trəl pī'räm·əd·ər }

spectral radiance [OPTICS] The radiant flux per unit wavelength or frequency interval per unit solid angle per unit of projected area of the source; the usual unit is watt per nanometer per steradian per square meter. { 'spek·trəl 'rād·ē·əns }

spectral radiance factor [ANALY CHEM] A situation when the desired directions for analysis of energy diffused from (reflected from) an object under spectrophotometric colorimetric analysis are all substantially the same (a solid angle of nearly zero steradians). { 'spek·trəl 'rād·ē·əns ‚fak·tər }

spectral radius [MATH] For the spectrum of an operator, this is the least upper bound of the set of all $|\lambda|$, where λ is in the spectrum. { 'spek·trəl 'rād·ē·əs }

spectral reflectance [ANALY CHEM] Situation when the desired directions for analysis of energy from (reflected from) an object under spectrophotometric colorimetric analysis is diffused in all directions (not directed as a single beam). { 'spek·trəl ri'flek·təns }

spectral regions [SPECT] Arbitrary ranges of wavelength, some of them overlapping, into which the electromagnetic spectrum is divided, according to the types of sources that are required to produce and detect the various wavelengths, such as x-ray, ultraviolet, visible, infrared, or radio-frequency. { 'spek·trəl ‚rē·jənz }

spectral response See spectral sensitivity. { 'spek·trəl ri'späns }

spectral selective photoelectric effect See selective photoelectric effect. { 'spek·trəl si‚lek·tiv ‚fōd·ō·i'lek·trik i‚fekt }

spectral sensitivity [ELECTR] Radiant sensitivity, considered as a function of wavelength. [PHYS] The response of a device or material to monochromatic light as a function of wavelength. Also known as spectral response. { 'spek·trəl ‚sen·sə'tiv·əd·ē }

spectral series [SPECT] Spectral lines or groups of lines that occur in sequence. { 'spek·trəl 'sir·ēz }

spectral-shift reactor [NUCLEO] A reactor in which, for control or other purposes, the neutron spectrum may be adjusted by varying the properties or amount of moderator. { 'spek·trəl ‚shift rē‚ak·tər }

spectral temperature [OPTICS] The temperature of a blackbody that produces the same spectral radiance as a given radiation field at a given wavelength or frequency and in a given direction. { 'spek·trəl 'tem·prə·chər }

spectral theorems [MATH] Spectral theorems enable

detailed study of various types of operators on Banach spaces by giving an integral or series representation of the operator in terms of its spectrum, eigenspaces, and simple projectionlike operators. { 'spek·trəl 'thir·əmz }

spectral transmission [OPTICS] The radiant flux which passes through a filter divided by the radiant flux incident upon it, for monochromatic light of a specified wavelength. { 'spek·trəl tranz'mish·ən }

spectral type [ASTRON] A label used to indicate the physical and chemical characteristics of a star as indicated by study of the star's spectra; for example, the stars in the spectral type known as class B are blue-white, and are referred to as helium stars because the dominant lines in their spectra are the lines in helium spectra. { 'spek·trəl ˌtīp }

Spectra Pritchard photometer [OPTICS] A photoelectric instrument for measuring the luminance of surfaces; it has a telescopic viewing system for imaging the bright surface to be measured on the cathode of a photoemissive tube, and a separate unit that combines the power supply with the controls and readout meter. { 'spek·trə 'prich·ərd fə'täm·əd·ər }

spectrobolometer [SPECT] An instrument that measures radiation from stars; measurement can be made in a narrow band of wavelengths in the electromagnetic spectrum; the instrument itself is a combination spectrometer and bolometer. { ¦spek·trō·bō'läm·əd·ər }

spectrofluorometer [SPECT] A device used in fluorescence spectroscopy to increase the selectivity of fluorometry by passing emitted fluorescent light through a monochromator to record the fluorescence emission spectrum. { ¦spek·trō·flù'räm·əd·ər }

spectrogram [SPECT] The record of a spectrum produced by a spectrograph. Also known as measured spectrum. { 'spek·trə·ˌgram }

spectrograph [SPECT] A spectroscope provided with a photographic camera or other device for recording the spectrum. { 'spek·trə·ˌgraf }

spectrography [SPECT] The use of photography to record the electromagnetic spectrum displayed in a spectroscope. { spek'träg·rə·fē }

spectroheliocinematograph [OPTICS] A camera used to make motion pictures of, for example, prominences of the sun; the camera utilizes monochromatic light; it is composed of a camera and a spectrohelioscope. { ¦spek·trō¦hē·lē·ō,sin·ə'm ad·ə,graf }

spectroheliogram [ASTRON] A photograph of the sun obtained by means of a spectroheliograph. { ¦spek·trō'hē·lē· ə,gram }

spectroheliograph [OPTICS] An instrument used to photograph the sun in one spectral band. { ¦spek·trō'hē·lē·ə,graf }

spectrohelioscope [OPTICS] An instrument based on the principle of the spectroheliograph but used for visual observation, and not for photography. { ¦spek·trō'hē·lē·ə,skōp }

spectrometer [SPECT] **1.** A spectroscope that is provided with a calibrated scale either for measurement of wavelength or for measurement of refractive indices of transparent prism materials. **2.** A spectroscope equipped with a photoelectric photometer to measure radiant intensities at various wavelengths. { spek'träm·əd·ər }

spectrometry [SPECT] The use of spectrographic techniques for deriving the physical constants of materials. { spek'träm· ə·trē }

spectrophone [ANALY CHEM] A cell containing the sample in the optoacoustic detection method; equipped with windows through which the laser beam enters the cell and a microphone for detecting sound. { 'spek·trə,fōn }

spectrophotometer [SPECT] An instrument that consists of a radiant-energy source, monochromator, sample holder, and detector; used for measurement of radiant flux as a function of wavelength and for measurement of absorption spectra. { ¦spek·trō·fə'täm·əd·ər }

spectrophotometric titration [ANALY CHEM] An analytical method in which the radiant-energy absorption of a solution is measured spectrophotometrically after each increment of titrant is added. { ¦spek·trō,fōd·ō'me·trik tī'trā·shən }

spectrophotometry [ANALY CHEM] A method of chemical analysis based on the absorption or attenuation by matter of electromagnetic radiation of a specified wavelength or frequency. The radiation interacts with specific features of the molecular species being determined, such as the vibrational or

rotational motions of the chemical bonds. The radiation can also interact with specific atoms or the whole molecule, for example, by causing the molecule to change its electronic energy state. { ¦spek·trō·fə'täm·ə·trē }

spectropolarimeter [OPTICS] A device used to measure optical rotation in solutions for different light wavelengths. { ¦spek·trō,pō·lə'rim·əd·ər }

spectropolarimetry [SPECT] The measurement of the polarization of light that has been dispersed into a continuum or line spectrum as a function of wavelength. { ¦spek·trə,pō· lə'rim·ə·trē }

spectropyrheliometer [SPECT] An astronomical instrument used to measure distribution of radiant energy from the sun in the ultraviolet and visible wavelengths. { ¦spek·trō¦pīr,hē· lē'äm·əd·ər }

spectroscope [SPECT] An optical instrument consisting of a slit, collimator lens, prism or grating, and a telescope or objective lens which produces a spectrum for visual observation. { 'spek·trə,skōp }

spectroscopic binary star [ASTRON] A binary star that may be distinguished from a single star only by noting the Doppler shift of the spectral lines of one or both stars as they revolve about their common center of mass. { ¦spek·trə¦skäp·ik 'bī,ner·ē 'stär }

spectroscopic displacement law [SPECT] The spectrum of an un-ionized atom resembles that of a singly ionized atom of the element one place higher in the periodic table, and that of a doubly ionized atom two places higher in the table, and so forth. { ¦spek·trə¦skäp·ik di splās·mənt ,lö }

spectroscopic parallax [ASTRON] Parallax as determined from examination of a stellar spectrum; critical spectral lines indicate the star's absolute magnitude, from which the star's distance, or parallax, can be deduced. { ¦spek·trə¦skäp·ik 'par·ə,laks }

spectroscopic splitting factor See Landé g factor. { ¦spek· trə¦skäp·ik 'splid·iŋ ,fak·tər }

spectroscopy [PHYS] The branch of physics concerned with the production, measurement, and interpretation of electromagnetic spectra arising from either emission or absorption of radiant energy by various substances. { spek'träs·kə·pē }

spectrum [MATH] If T is a linear operator of a normed space X to itself and I is the identity transformation ($I(x) \equiv x$), the spectrum of T consists of all scalars λ for which either $T - \lambda I$ has no inverse or the range of $T - \lambda I$ is not dense in X. [PHYS] **1.** A display or plot of intensity of radiation (particles, photons, or acoustic radiation) as a function of mass, momentum, wavelength, frequency, or some related quantity. **2.** The set of frequencies, wavelengths, or related quantities, involved in some process; for example, each element has a characteristic discrete spectrum for emission and absorption of light. **3.** A range of frequencies within which radiation has some specified characteristic, such as audio-frequency spectrum, ultraviolet spectrum, or radio spectrum. { 'spek·trəm }

spectrum analysis [PHYS] The measurement of the amplitude of the components of a complex waveform throughout the frequency range of the waveform. { 'spek·trəm ə,nal·ə·səs }

spectrum analyzer [ENG] Test instrument used to show the distribution of energy contained in the frequencies emitted by a pulse magnetron; also used to measure the Q of resonant cavities and lines, and to measure the cold impedance of a magnetron. { 'spek·trəm 'an·ə,līz·ər }

spectrum level [COMMUN] The level of the part of a specified signal at a specified frequency that is contained within a specified frequency bandwidth centered at the particular frequency. { 'spek·trəm ,lev·əl }

spectrum line See spectral line. { 'spek·trəm ,līn }

spectrum locus [OPTICS] The locus of points representing the chromaticities of spectrally pure stimuli in a chromaticity diagram. Also known as spectral locus. { 'spek·trəm ,lō· kəs }

spectrum of turbulence [ASTROPHYS] A relationship between the size of turbulent eddies in the sun's atmosphere and their average speed. { 'spek·trəm əv 'tər·byə·ləns }

spectrum-selectivity characteristic [ELECTR] Measure of the increase in the minimum input signal power over the minimum detectable signal required to produce an indication on a radar indicator, if the received signal has a spectrum different from that of the normally received signal. { 'spek·trəm ,si,lek· 'tiv·əd·ē ,kar·ik·tə,ris·tik }

spectrum signature [ELECTR] The spectral characteristics of the transmitter, receiver, and antenna of an electronic system, including emission spectra, antenna patterns, and other characteristics. { 'spek·trəm ,sig·nə·chər }

spectrum signature analysis [ELECTR] The evaluation of electromagnetic interference from transmitting and receiving equipment to determine operational and environment compatibility. { 'spek·trəm ,sig·nə·chər ə,nal·ə·səs }

spectrum variable [ASTRON] A main-sequence star of spectral class A whose spectrum displays anomalously strong lines of metals and rare earths whose intensity varies by about 0.1 magnitude over periods from 1 to 25 days. { 'spek·trəm 'ver·ē·ə·bəl }

specular hematite [MINERAL] A variety of hematite with a blue-gray color and bright metallic luster. { 'spek·yə·lər 'hē·mə,tīt }

specular iron *See* specularite. { 'spek·yə·lər 'ī·ərn }

specularite [MINERAL] A black or gray variety of hematite with brilliant metallic luster, occurring in micaceous or foliated masses, or in tabular or disklike crystals. Also known as gray hematite; iron glance; specular iron. { 'spek·yə·lə,rīt }

specular reflection [PHYS] Reflection of electromagnetic, acoustic, or water waves in which the reflected waves travel in a definite direction, and the directions of the incident and reflected waves make equal angles with a line perpendicular to the reflecting surface, and lie in the same plane with it. Also known as direct reflection; mirror reflection; regular reflection. { 'spek·yə·lər ri'flek·shən }

specular reflection factor [OPTICS] The ratio of the specularly reflected light to the incident light. { 'spek·yə·lər ri'flek·shən ,fak·tər }

specular reflection model [PHYS] A model for the behavior of gas molecules striking the surface of a solid body, in which the molecules are reflected so that the component of velocity tangent to the surface is unchanged while the component of velocity perpendicular to the surface is reversed. { 'spek·yə·lər ri'flek·shən ,mäd·əl }

specular reflector [OPTICS] A reflecting surface (polished metal or silvered glass) that gives a direct image of the source, with the angle of reflection equal to the angle of incidence. Also known as regular reflector; specular surface. { 'spek·yə·lər ri'flek·tər }

specular surface *See* specular reflector. { 'spek·yə·lər 'sər·fəs }

specular transmittance [ELECTROMAG] The ratio of the power carried by electromagnetic radiation which emerges from a body and is parallel to a beam entering the body, to the power carried by the beam entering the body. { 'spek·yə·lər tranz'mit·əns }

speculum [MED] A tubular instrument for inserting into a passage or cavity of the body to facilitate visual inspection or medication. [OPTICS] An optical instrument reflector of polished metal or of glass with a film of metal. { 'spek·yə·ləm }

speculum alloy [MET] A brilliant white, hard, brittle alloy composed of copper and tin in a 2:1 proportion and sometimes with additions of other elements. { 'spek·yə·ləm 'al,ȯi }

speech [PHYSIO] A complex process in which the eating and breathing mechanisms are used to generate patterns of sounds that form words and sentences to express thoughts. { spēch }

speech amplifier [ENG ACOUS] An audio-frequency amplifier designed specifically for amplification of speech frequencies, as for public-address equipment and radiotelephone systems. { 'spēch ,am·plə,fī·ər }

speech bandwidth [COMMUN] The range of speech frequencies that can be transmitted by a carrier telephone system. { 'spēch 'band,width }

speech clipper [ENG ACOUS] A clipper used to limit the peaks of speech-frequency signals, as required for increasing the average modulation percentage of a radiotelephone or amateur radio transmitter. { 'spēch ,klip·ər }

speech clipping [ACOUS] In tests of the intelligibility of speech signals, the limiting of peak signals to a maximum value, or the reduction of signals of less than a certain value to zero. { 'spēch ,klip·iŋ }

speech coder [COMMUN] A device that uses data-compression techniques to convert a high-bit-rate signal resulting from digital pulse-code modulation of speech to a low-rate digital signal that can be transmitted or stored. { 'spēch ,kōd·ər }

speech coil *See* voice coil. { 'spēch ,kȯil }

speech compression [COMMUN] Techniques that take advantage of certain properties of the speech signal to permit adequate information quality, characteristics, and the sequential pattern of a speaker's voice to be transmitted over a narrower frequency band than would otherwise be necessary. { 'spēch kəm,presh·ən }

speech frequency *See* voice frequency. { 'spēch ,frē·kwən·sē }

speech intelligibility *See* intelligibility. { 'spēch in,tel·ə·jə'bil·əd·ē }

speech interference level [ACOUS] The average sound pressure, in decibels above 0.0002 microbar, in the frequency range from 600 to 4800 hertz. Abbreviated SIL. { 'spēch ,in·tər¦fir·əns ,lev·əl }

speech interpolation [COMMUN] Method of obtaining more than one voice channel per voice circuit by giving each subscriber a speech path in the proper direction only at times when the subscriber's speech requires it. { 'spēch ,in·tər·pəl¦ā·shən }

speech inverter *See* scrambler. { 'spēch in,vərd·ər }

speech recognition [ENG ACOUS] The process of analyzing an acoustic speech signal to identify the linguistic message that was intended, so that a machine can correctly respond to spoken commands. { 'spēch ,rek·ig'nish·ən }

speech scrambler *See* scrambler. { 'spēch ,skram·blər }

speech synthesis *See* voice response. { 'spēch 'sin·thə·səs }

speed [GRAPHICS] The sensitivity of a photographic film, expressed according to one of several scales. [MECH] The time rate of change of position of a body without regard to direction; in other words, the magnitude of the velocity vector. [OPTICS] **1.** The light-gathering power of a lens, expressed as the reciprocal of the *f* number. **2.** The time that a camera shutter is open. [PHYS] In general, the rapidity with which a process takes place. { spēd }

speed circle [NAV] A circle having a radius equal to a given speed and drawn about a specified center; the term is used chiefly in connection with relative movement problems. { 'spēd ,sər·kəl }

speed cone [MECH ENG] A cone-shaped pulley, or a pulley composed of a series of pulleys of increasing diameter forming a stepped cone. { 'spēd ,kōn }

speed control [ELEC] A control that changes the speed of a motor or other drive mechanism, as for a phonograph or magnetic tape recorder. { 'spēd kən,trōl }

speed-course-latitude error [NAV] An error in both pendulous and nonpendulous types of gyro compasses resulting from movement of the gyro compass in other than an east-west direction. Also known as speed error. { 'spēd 'kȯrs 'lad·ə,tüd ,er·ər }

speed density metering [AERO ENG] A type of aircraft carburetion in which the fuel feed is regulated by the parameters of engine feed and intake manifold pressure. { 'spēd ¦den·səd·ē ,mēd·ə·riŋ }

speed error [NAV] **1.** Acceleration error due to a change in the speed of a craft. **2.** *See* speed-course-latitude error. { 'spēd ,er·ər }

speed-in [PETRO ENG] To start drilling by making a hole. { 'spēd¦in }

speed lathe [MECH ENG] A light, pulley-driven lathe, usually without a carriage or back gears, used for work in which the tool is controlled by hand. { 'spēd ,lāth }

speed-length ratio [NAV ARCH] The speed of a ship divided by the square root of its length. { 'spēd 'leŋkth 'rā·shō }

speed line [NAV] A line of position approximately perpendicular to the course, which can be used to determine the speed made good. { 'spēd ,līn }

speed made good [NAV] The actual average speed in knots which was maintained in proceeding along the intended track to the ultimate destination or an intermediate point. { 'spēd ,mād 'gu̇d }

speed-matching buffer [COMPUT SCI] A small computer storage unit that connects two devices operating at different data transfer rates; each device writes into and reads from the buffer at its own rate. { 'spēd ¦mach·iŋ 'bəf·ər }

speed of advance [NAV] The average speed in knots which

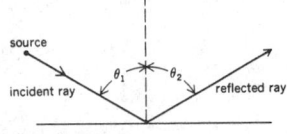

SPECULAR REFLECTION

Specular reflection, such as from a polished surface. Angle θ_1 between incident ray and perpendicular to surface equals angle θ_2 between reflected ray and perpendicular.

must be maintained during a passage to arrive at a destination at an appointed time. { 'spēd əv ad'vans }

speed of light [ELECTROMAG] The speed of propagation of electromagnetic waves in a vacuum, which is a physical constant equal to exactly 299,792.458 kilometers per second. Also known as electromagnetic constant; velocity of light. { 'spēd əv 'līt }

speed of response [PHYS] The time required for a system to react to some signal; for example, the delay time for a photon detector to react to a radiation pulse, or the time needed for a current or voltage in a circuit to reach a definite fraction of its final value as a result of an abrupt change in the electromotive force. { 'spēd əv ri'späns }

speed of sound [ACOUS] The phase velocity of a sound wave. Also known as sonic speed; sonic velocity; sound velocity; velocity of sound. { 'spēd əv 'saund }

speed of travel [MET] The speed at which a weld is made along its longitudinal axis; measured in inches or spots per minute. { 'spēd əv 'trav·əl }

speedometer [ENG] An instrument that indicates the speed of travel of a vehicle in miles per hour, kilometers per hour, or knots. { spi'däm·əd·ər }

speed over the ground [NAV] The speed of travel of a craft over the ground; usually called ground speed. { 'spēd ,ō·vər thə 'graund }

speed-payload tradeoff [MECH ENG] The relationship between the maximum speed with which a machine can move a workpiece and the maximum weight of the workpiece. { 'spēd 'pā,lōd 'trād,óf }

speed-power product [ELECTR] The product of the gate speed or propagation delay of an electronic circuit and its power dissipation. { 'spēd'pau·ər ,präd·əkt }

speed reducer [MECH ENG] A train of gears placed between a motor and the machinery which it will drive, to reduce the speed with which power is transmitted. { 'spēd ri,dü·sər }

speed regulator [ELEC] A device that maintains the speed of a motor or other device at a predetermined value or varies it in accordance with a predetermined plan. { 'spēd ,reg·yə,lād·ər }

speed-reliability tradeoff [MECH ENG] The relationship between the maximum speed at which a machine can move a workpiece and the reliability with which the machine's operations can be achieved to some degree of satisfaction. { 'spēd ri,lī·ə'bil·əd·ē 'trād,óf }

speed triangle [NAV] A vector diagram composed of vectors representing the actual courses and speeds of two craft and the relative course and speed; the third vector is the sum and represents them. { 'spēd ,trī,aŋ·gəl }

speed-up theorem [MATH] There is a computable function *f* with the property that for any algorithm *A* there is another algorithm *B* which computes *f* much faster than *A*. { 'spēd'əp ,thir·əm }

speiss [MET] A mixture of impure metal arsenides and antimonides resulting from the smelting of certain ores such as cobalt and lead. { spīs }

Spelaeogriphacea [INV ZOO] A peracaridan order of the Malacostraca comprised of the single species *Spelaeogriphus lepidops*, a small, blind, transparent, shrimplike crustacean with a short carapace that coalesces dorsally with the first thoracic somite. { ,spē·lē·ō·gri'fās·ē·ə }

spelean [GEOL] Of or pertaining to a feature in a cave. { spə'lē·ən }

speleology [GEOL] The study and exploration of caves. { ,spē·lē·äl·ə·jē }

speleothem [GEOL] A secondary mineral deposited in a cave by the action of water. Also known as cave formation. { 'spē·lē·ə,them }

spelling checker [COMPUT SCI] A program, used in conjunction with word-processing software, which automatically checks words in a text against a dictionary of commonly used words and identifies words that appear to be misspelled. { 'spel·iŋ ,chek·ər }

spelter [MET] A commercially pure grade of zinc used in galvanizing; contains lead or iron as impurities. { 'spel·tər }

spelter shakes See metal fume fever. { 'spel·tər ,shāks }

spelter solder [MET] Brass composed of equal parts of copper and zinc; used in brazing as a filler metal. Also known as brazing brass. { 'spel·tər ,säd·ər }

spencerite [MINERAL] $Zn_4(PO_4)_2(OH)_2 \cdot 3H_2O$ A pearly white monoclinic mineral composed of hydrous basic zinc phosphate and occurring in scaly masses and small crystals. { 'spen·sə,rīt }

spending beach [GEOL] In a wave basin, the beach on which the entering waves spend themselves, except for the small remainder entering the inner harbor. { 'spend·iŋ ,bēch }

spent acid See sludge acid. { 'spent 'as·əd }

spent fuel [NUCLEO] Nuclear reactor fuel that has been irradiated to the extent that it can no longer effectively sustain a chain reaction because its fissionable isotopes have been partially consumed and fission-product poisons have accumulated in it. { 'spent 'fyül }

spent iron sponge [MATER] Iron sponge saturated with sulfur; prone to spontaneous heating. Also known as spent oxide. { 'spent 'ī·ərn ,spənj }

spent liquor [MATER] The liquid effluent from the digestion of wood during pulping; contains wood chemicals (for example, lignin) and spent digestant (caustic, sulfite, or sulfate, depending on the process used). { 'spent 'lik·ər }

spent oxide See spent iron sponge. { 'spent 'äk,sīd }

spergenite [GEOL] A biocalcarenite containing ooliths and fossil debris and having a maximum quartz content of 10%. Also known as Bedford limestone; Indiana limestone. { 'spər·jə,nīt }

sperm See spermatozoon. { spərm }

spermaceti [MATER] A white, crystalline, oily (waxy) solid that separates from sperm oil; soluble in ether, chloroform, and carbon disulfide, insoluble in water; melts at 42 to 50°C; used in ointments, emulsions, candles, soaps, and cosmetics; and for linen finishing. Also known as spermaceti wax. { ,spər·mə'sed·ē }

spermaceti wax See spermaceti. { ,spər·mə'sed·ē ,waks }

spermatheca [ZOO] A sac in the female for receiving and storing sperm until fertilization; found in many invertebrates and certain vertebrates. Also known as seminal receptacle. { ,spər·mə'thē·kə }

spermatic cord [ANAT] The cord consisting of the ductus deferens, epididymal and testicular nerves and blood vessels, and connective tissue that extends from the testis to the deep inguinal ring. { spər'mad·ik 'kórd }

spermatid [HISTOL] A male germ cell immediately before assuming its final typical form. { 'spər·məd·əd }

spermatin [BIOCHEM] An albuminoid material occurring in semen. { 'spər·məd·ən }

spermatocele [MED] A cystic dilation of a duct in the head of the epididymis or in the rete testis. { spər'mad·ə,sēl }

spermatocyte [HISTOL] A cell of the last or next to the last generation of male germ cells which differentiates to form spermatozoa. { spər'mad·ə,sīt }

spermatogenesis [PHYSIO] The process by which spermatogonia undergo meiosis and transform into spermatozoa. { spər,mad·ə'jen·ə·səs }

spermatogonium [HISTOL] A primitive male germ cell, the last generation of which gives rise to spermatocytes. { spər,mad·ə'gō·nē·əm }

spermatophore [ZOO] A bundle or packet of sperm produced by certain animals, such as annelids, arthropods, and some vertebrates. { spər'mad·ə,fór }

spermatophyte [BOT] Any one of the seed-bearing vascular plants. { spər'mad·ə,fīt }

spermatorrhea [MED] Involuntary discharge of semen without orgasm. { spər,mad·ə'rē·ə }

spermatozoon [HISTOL] A mature male germ cell. Also known as sperm. { spər,mad·ə'zō·ən }

spermaturia [MED] The presence of sperm in the urine. { ,spər·mə'tür·ē·ə }

spermidine [BIOCHEM] $H_2N(CH_2)_3NH(CH_2)_4NH_2$ The triamine found in semen and other animal tissues. { 'spər·mə,dēn }

spermine [BIOCHEM] $C_{10}H_{26}N_4$ A tetramine found in semen, blood serum, and other body tissues. { 'spər,mēn }

spermiogenesis [CYTOL] Nuclear and cytoplasmic transformation of spermatids into spermatozoa. { ,spər·mē·ō'jen·ə·səs }

sperm nucleus [BOT] One of the two nuclei in a pollen grain that function in double fertilization in seed plants. { 'spərm ,nü·klē·əs }

sperm oil [MATER] A combustible, yellowish oil found in the head cavities and blubber of the sperm whale; soluble in

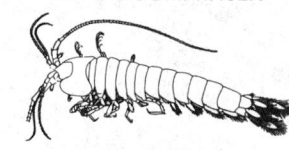

SPELAEOGRIPHACEA

Spelaeogriphus lepidops.

SPERM WHALE

The sperm whale (*Physeter catodon*).

SPHAEROPLEINEAE

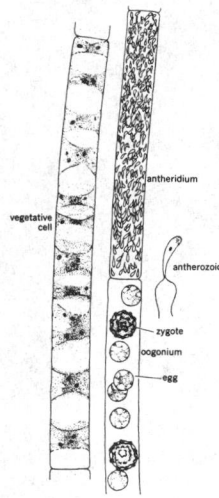

antheridium

vegetative cell

antherozoid

zygote

oogonium

egg

Representative cells of *Sphaeroplea* showing bandlike chloroplasts and heterogametes.

SPHECIDAE

Typical member of the family Sphecidae.

ether, chloroform, and benzene; used as a lubricant for precision machinery, for rustproofing metals, and in transmission fluids. { 'spərm ,oil }

sperm whale [VERT ZOO] *Physeter catodon.* An aggressive toothed whale belonging to the group Odontoceti of the order Cetacea; it produces ambergris and contains a mixture of spermaceti and oil in a cavity of the nasal passage. { 'spərm ,wāl }

Sperner set [MATH] A set *S* of subsets of a given set such that if *A* and *B* are in *S*, and *A* does not equal *B*, then neither *A* nor *B* is a subset of the other. Also known as antichain; clutter. { 'spər·nər ,set }

Sperner's theorem [MATH] A theorem which gives the maximum possible cardinality of an antichain in a finite set. { 'spər·nərz ,thir·əm }

speromagnetic state [SOLID STATE] The condition of a rare-earth glass in which the spins are oriented in fixed directions which are more or less random because of electric fields which exist in the glass. { ¦spir·ə·mag¦ned·ik 'stāt }

sperrylite [MINERAL] PtAs$_2$ A tin-white isometric mineral composed of platinum arsenide; the only platinum compound known to occur in nature; hardness is 6–7 on Mohs scale, and specific gravity is 10.60. { 'sper·ē,līt }

Sperry process [CHEM ENG] The electrolytic manufacture of basic lead carbonate (white lead) from desilvered lead that contains some bismuth; impure lead collects at the anode, and carbon dioxide is passed into the solution to convert the lead to carbonate. { 'sper·ē ,prä·səs }

spessartite [MINERAL] Mn$_3$Al$_2$(SiO$_4$)$_3$ A mineral composed of manganese aluminum silicate with small amounts of iron, magnesium, or other elements. [PETR] A lamprophyre composed of a sodic plagioclase groundmass in which green hornblende phenocrysts are embedded; also contains accessory olivine, biotite, apatite, and opaque oxides. { 'spes·ər,tīt }

sp gr *See* specific gravity.

Sphaeractinoidea [PALEON] An extinct group of fossil marine hydrozoans distinguished in part by the relative prominence of either vertical or horizontal trabeculae and by the presence of long, tabulate tubes called autotubes. { sfir,ak·tə'nóid·ē·ə }

Sphaeriales [MYCOL] An order of fungi in the subclass Euascomycetes characterized by hard, dark perithecia with definite ostioles. { ,sfir·ē'ā·lēz }

Sphaeriidae [INV ZOO] The minute bog beetles, a small family of coleopteran insects in the suborder Myxophaga. { sfə'rī·ə,dē }

Sphaerioidaceae [MYCOL] A family of fungi of the order Sphaeropsidales in which the pycnidia are black or dark-colored and are flask-, cone-, or lens-shaped with thin walls and a round, relatively small pore. { ,sfir·ē,ói'dās·ē,ē }

sphaerite [MINERAL] Light-gray or bluish mineral composed of hydrous aluminum phosphate and occurring in global concretions. { 'sfir,īt }

Sphaerocarpales [BOT] An order of liverworts in the subclass Marchantiidae, characterized by an envelope surrounding each antheridium and archegonium, absence of elaters, poor development of seta, and absence of thickenings in the unilayered wall of an indehiscent capsule. { ,sfir·ō·kär'pā·lēz }

Sphaeroceridae [INV ZOO] A family of myodarian cyclorrhaphous dipteran insects in the subsection Acalypteratae. { ,sfir·ō'ser·ə,dē }

Sphaerodoridae [INV ZOO] A family of polychaete annelids belonging to the Errantia in which species are characterized by small bodies, and are usually papillated. { ,sfir·ō'dór·ə,dē }

Sphaerolaimidae [INV ZOO] A family of free-living nematodes in the superfamily Monhysteroidea characterized by a spacious and deep stoma. { ,sfir·ō'lī·mə,dē }

sphaerolitic *See* spherulitic. { ,sfir·ə¦lid·ik }

Sphaeromatidae [INV ZOO] A family of isopod crustaceans in the suborder Flabellifera in which the body is broad and oval and the inner branch of the uropod is immovable. { ,sfir·ō'mad·ə,dē }

Sphaerophoraceae [BOT] A family of the Ascolichenes in the order Caliciales which are fruticose with a solid thallus. { sfə,räf·ə'rās·ē,ē }

Sphaeropleineae [BOT] A suborder of green algae in the order Ulotrichales distinguished by long, coenocytic cells, numerous bandlike chloroplasts, and heterogametes produced in undifferentiated vegetative cells. { ,sfir·ō'plīn·ē,ē }

Sphaeropsidaceae [MYCOL] An equivalent name for Sphaerioidaceae. { sfə,räp·sə'dās·ē,ē }

Sphaeropsidales [MYCOL] An order of fungi of the class Fungi Imperfecti in which asexual spores are formed in pycnidia, which may be separate or joined to vegetative hyphae, conidiophores are short or absent, and conidia are always slime spores. { sfə,räp·sə'dā·lēz }

Sphagnaceae [BOT] The single monogeneric family of the order Sphagnales. { sfag'nās·ē,ē }

Sphagnales [BOT] The single order of mosses in the subclass Sphagnobrya containing the single family Sphagnaceae. { sfag'nā·lēz }

Sphagnobrya [BOT] A subclass of the Bryopsida; plants are grayish-green with numerous, spirally arranged branches and grow in deep tufts or mats, commonly in bogs and in other wet habitats. { sfag'näb·rē·ə }

sphagnum bog [ECOL] A bog composed principally of mosses of the genus *Sphagnum* (Sphagnales) but also of other plants, especially acid-tolerant species, which tend to form peat. { 'sfag·nəm 'bäg }

sphalerite [MINERAL] (Zn,Fe)S The low-temperature form and common polymorph of zinc sulfide; a usually brown or black mineral that crystallizes in the hextetrahedral class of the isometric system, occurs most commonly in coarse to fine, granular, cleanable masses, has resinous luster, hardness of 3.5 on Mohs scale, and specific gravity of 4.1. Also known as blende; false galena; jack; lead marcasite; mock lead; mock ore; pseudogalena; steel jack. { 'sfal·ə,rīt }

Sphecidae [INV ZOO] A large family of hymenopteran insects in the superfamily Sphecoidea. { 'sfes·ə,dē }

Sphecoidea [INV ZOO] A superfamily of wasps belonging to the suborder Apocrita. { sfə'kóid·ē·ə }

Sphenacodontia [PALEON] A suborder of extinct reptiles in the order Pelycosauria which were advanced, active carnivores. { sfə,näk·ə'dän·chə }

sphene [MINERAL] CaTiSiO$_5$ A brown, green, yellow, gray, or black neosilicate mineral common as an accessory mineral in igneous rocks; it is monoclinic and has resinous luster; hardness is 5–5.5 on Mohs scale; specific gravity is 3.4–3.5. Also known as grothite; titanite. { sfēn }

sphenethmoid [VERT ZOO] A bone that surrounds the anterior portion of the brain in many amphibians. { sfēn'eth,móid }

Spheniscidae [VERT ZOO] The single family of the avian order Sphenisciformes. { sfə'nis·ə,dē }

Sphenisciformes [VERT ZOO] The penguins, an order of aquatic birds found only in the Southern Hemisphere and characterized by paddlelike wings, erect posture, and scalelike feathers. { sfə,nis·ə'fór,mēz }

sphenochasm [GEOL] A triangular gap of oceanic crust separating two continental blocks and converging to a point. { ,sfē·nə'kaz·əm }

Sphenodontidae [VERT ZOO] A family of lepidosaurian reptiles in the order Rhynchocephalia represented by a single living species, *Sphenodon punctatus*, a lizardlike form distinguished by lack of a penis. { ,sfē·nə'dänt·ə,dē }

sphenoid [CRYSTAL] An open crystal, occurring in monoclinic crystals of the sphenoidal class, and characterized by two nonparallel faces symmetrical with an axis of twofold symmetry. [SCI TECH] Wedge-shaped. { 'sfē,nóid }

sphenoid bone [ANAT] The butterfly-shaped bone forming the anterior part of the base of the skull and portions of the cranial, orbital, and nasal cavities. { 'sfē,nóid ¦bōn }

sphenoid sinus [ANAT] Either of a pair of paranasal sinuses located centrally between and behind the eyes, below the ethymoid sinus. { 'sfē,nóid ¦sī·nəs }

sphenolith [GEOL] A wedgelike igneous intrusion that is partly concordant and partly discordant. { 'sfēn·əl,ith }

sphenopalatine [ANAT] Of or pertaining to the region of or surrounding the sphenoid and palatine bones. { ,sfē·nō'pal·ə,tēn }

sphenopalatine foramen [ANAT] The space between the sphenoid and orbital processes of the palatine bone; it opens into the nasal cavity and gives passage to branches from the pterygopalatine ganglion and the sphenopalatine branch of the maxillary artery. { ,sfē·nō'pal·ə,tēn fə'rā·mən }

sphenoparietal index [ANTHRO] The ratio, multiplied by 100, of the breadth of the skull from stenion to stenion to its greatest breadth. { ¦sfē·nō·pə'rī·əd·əl 'in,deks }

Sphenopsida [BOT] A group of vascular cryptogams characterized by whorled, often very small leaves and by the absence of true leaf gaps in the stele; essentially equivalent to the division Equisetophyta. { sfə'näp·səd·ə }

Sphenyllopsida [PALEOBOT] An extinct class of embryophytes in the division Equisetophyta. { ˌsfēn·əl'äp·səd·ə }

spherand [ORG CHEM] A macrocyclic compound capable of completely enveloping a cation, having donor atoms (O, N, S) arranged such that they provide a solvation sphere to the encapsulated cation. { 'sfir·ənd }

spherator [PL PHYS] One of the class of low-β, low-density, quasi-steady-state closed devices (like Tokamak) used in studying production of electric power by fusion. { 'sfe,rād·ər }

sphere [MATH] **1.** The set of all points in a euclidean space which are a fixed common distance from some given point; in Euclidean three-dimensional space the Riemann sphere consists of all points (x,y,z) which satisfy the equation $x^2 + y^2 + z^2 = 1$. **2.** The set of points in a metric space whose distance from a fixed point is constant. { 'sfir }

sphere gap [ELEC] A spark gap between two equal-diameter spherical electrodes. { 'sfir ,gap }

sphere of attraction [PHYS CHEM] The distance within which the potential energy arising from mutual attraction of two molecules is not negligible with respect to the molecules' average thermal energy at room temperature. { 'sfir əv ə'trak·shən }

sphere photometer See integrating-sphere photometer. { 'sfir fə'täm·əd·ər }

spheres of Eudoxus [ASTRON] A theory of Eudoxus from about 400 B.C.; the planets, sun, and moon were on a series of concentric spheres rotating inside one another on different axes. { 'sfirz əv yü'däk·səs }

spherical aberration [OPTICS] Aberration arising from the fact that rays which are initially at different distances from the optical axis come to a focus at different distances along the axis when they are reflected from a spherical mirror or refracted by a lens with spherical surfaces. { 'sfir·ə·kəl ˌab·ə'rā·shən }

spherical angle [MATH] The figure formed by the intersection of two great circles on a sphere, and equal in size to the angle formed by the tangents to the great circles at the point of intersection. { ˈsfir·ə·kəl 'aŋ·gəl }

spherical antenna [ELECTROMAG] An antenna having the shape of a sphere, used chiefly in theoretical studies. { 'sfir·ə·kəl an'ten·ə }

spherical Bessel functions [MATH] Bessel functions whose order is half of an odd integer; they arise as the radial functions that result from solving Pockel's equation (or, equivalently, the time-independent Schrödinger equation for a free particle) by separation of variables in spherical coordinates. { ˈsfir·i·kəl 'bes·əl ˌfəŋk·shənz }

spherical capacitor [ELEC] A capacitor made of two concentric metal spheres with a dielectric filling the space between the spheres. { 'sfir·ə·kəl kə'pas·əd·ər }

spherical cone [MATH] **1.** A solid consisting of the cap and cone formed by the intersection of a plane with a sphere, the cone extending from the plane to the center of the sphere and the cap extending from the plane to the surface of the sphere. **2.** The surface of this solid. { ˈsfir·ə·kəl 'kōn }

spherical-coordinate robot [CONT SYS] A robot in which the degrees of freedom of the manipulator arm are defined primarily by spherical coordinates. { ˈsfir·ə·kəl kō'ȯrd·ən·ət 'rō,bät }

spherical coordinates [MATH] A system of curvilinear coordinates in which the position of a point in space is designated by its distance r from the origin or pole, called the radius vector, the angle ϕ between the radius vector and a vertically directed polar axis, called the cone angle or colatitude, and the angle θ between the plane of ϕ and a fixed meridian plane through the polar axis, called the polar angle or longitude. Also known as spherical polar coordinates. { 'sfir·ə·kəl kō'ȯrd·ən·əts }

spherical curve [MATH] A curve that lies entirely on the surface of a sphere. { 'sfir·ə·kəl 'kərv }

spherical cyclic curve See cyclic curve. { 'sfir·ə·kəl 'sī·klik 'kərv }

spherical degree [MATH] A solid angle equal to one-ninetieth of a spherical right angle. { 'sfir·ə·kəl di'grē }

spherical distance [MATH] The length of a great circle arc between two points on a sphere. { 'sfir·i·kəl 'dis·təns }

spherical-earth attenuation [ELECTROMAG] Attenuation over an imperfectly conducting spherical earth in excess of that over a perfectly conducting plane. { 'sfir·ə·kəl ˈərth ə,ten·yə,wā·shən }

spherical-earth factor [ELECTROMAG] The ratio of the electric field strength that would result from propagation over an imperfectly conducting spherical earth to that which would result from propagation over a perfectly conducting plane. { 'sfir·ə·kəl ˈərth ,fak·tər }

spherical excess [MATH] The sum of the angles of a spherical triangle, minus 180°. { 'sfir·ə·kəl ek'ses }

spherical geometry [MATH] The geometry of points on a sphere. { 'sfir·ə·kəl je'äm·ə·trē }

spherical harmonics [MATH] Solutions of Laplace's equation in spherical coordinates. { 'sfir·ə·kəl här'män·iks }

spherical image [MATH] Also known as spherical representation. **1.** For a point on a surface, the end point of the radius of a unit sphere parallel to the positive direction of the normal to the surface at the point. **2.** For a surface, a portion of a unit sphere consisting of all the end points of those radii of the sphere that are parallel to the positive directions of normals to the surface. Also known as Gaussian representation. **3.** See spherical indicatrix. { 'sfir·ə·kəl 'im·ij }

spherical indicatrix [MATH] For a space curve, those points on the unit sphere traced out by a radius moving from point to point always parallel with the tangent to the curve. Also known as spherical image; spherical indicatrix of the tangent; spherical representation; tangent indicatrix. { 'sfir·ə·kəl in'dik·ə,triks }

spherical indicatrix of the binormal See binormal indicatrix. { 'sfir·ə·kəl in'dik·ə,triks əv thə 'bī,nȯr·məl }

spherical indicatrix of the principal normal See principal normal indicatrix. { 'sfir·ə·kəl in'dik·ə,triks əv thə 'prin·sə·pəl ,nȯr·məl }

spherical indicatrix of the tangent See spherical indicatrix. { 'sfir·ə·kəl in'dik·ə,triks əv thə 'tan·jənt }

spherical lens [OPTICS] A lens whose surfaces form portions of spheres. { 'sfir·ə·kəl 'lenz }

spherical mirror [OPTICS] A mirror, either convex or concave, whose surface forms part of a sphere. { 'sfir·ə·kəl 'mir·ər }

spherical pendulum [MECH] A simple pendulum mounted on a pivot so that its motion is not confined to a plane; the bob moves over a spherical surface. { 'sfir·ə·kəl 'pen·jə·ləm }

spherical polar coordinates See spherical coordinates. { 'sfir·ə·kəl 'pō·lər kō'ȯrd·ən·əts }

spherical polygon [MATH] A part of a sphere that is bounded by arcs of great circles. { 'sfir·ə·kəl 'päl·ə,gän }

spherical powder [MATER] A powder consisting of globular-shaped particles. { 'sfir·ə·kəl 'paȯd·ər }

spherical pyramid [MATH] A solid bounded by a spherical polygon and portions of planes passing through the sides of the polygon and the center of the sphere. { 'sfir·ə·kəl 'pir·ə,mid }

spherical radius [MATH] For a circle on a sphere, the smaller of the spherical distances from one of the two poles of the circle to any point on the circle. { 'sfir·ə·kəl 'rād·ē·əs }

spherical representation See spherical image; spherical indicatrix. { 'sfir·ə·kəl ,rep·ri·zen'tā·shən }

spherical right angle [MATH] The solid angle subtended at the center of a sphere by a portion of the surface of the sphere bounded by a trirectangular spherical triangle; equal to π/2 steradians. { 'sfir·ə·kəl 'rīt 'aŋ·gəl }

spherical sailing [NAV] Any of the sailing computation methods which are used to solve the problems of course, distance, difference of latitude, difference of longitude, and departure which take into account the spherical or spheroidal shape of the earth. { 'sfir·ə·kəl 'sāl·iŋ }

spherical sector [MATH] A solid formed by rotating a sector of a circle about a diameter of the circle; the diameter may contain one of the radii bounding the circular sector or it may lie outside the circular sector. { 'sfir·ə·kəl 'sek·tər }

spherical segment [MATH] A solid that is bounded by a sphere and two parallel planes which intersect the sphere or are tangent to it. { 'sfir·ə·kəl 'seg·mənt }

spherical separator [PETRO ENG] A gas-oil separator in the form of a spherical vessel. { 'sfir·ə·kəl 'sep·ə,rād·ər }

spherical stress [MECH] The portion of the total stress that corresponds to an isotropic hydrostatic pressure; its stress tensor

SPHERAND

All-aromatic spherand.

is the unit tensor multiplied by one-third the trace of the total stress tensor. { 'sfir·ə·kəl 'stres }

spherical surface [MATH] A surface whose total curvature has a constant positive value but that is not necessarily a sphere. { 'sfir·ə·kəl 'sər·fəs }

spherical surface harmonics [MATH] Functions of the two angular coordinates of a spherical coordinate system which are solutions of the partial differential equation obtained by separation of variables of Laplace's equation in spherical coordinates. Also known as surface harmonics. { ¦sfir·ə·kəl ¦sər·fəs här'män· iks }

spherical triangle [MATH] A three-sided surface on a sphere the sides of which are arcs of great circles. { 'sfir·ə·kəl 'trī,aŋ·gəl }

spherical trigonometry [MATH] The study of spherical triangles from the viewpoint of angle, length, and area. { 'sfir·ə·kəl ,trig·ə'näm·ə·trē }

spherical wave [PHYS] A wave whose equiphase surfaces form a family of concentric spheres; the direction of travel is always perpendicular to the surfaces of the spheres. { 'sfir·ə·kəl 'wāv }

spherical weathering See spheroidal weathering. { 'sfir·ə·kəl 'weth·ə·riŋ }

spherical wedge [MATH] The portion of a sphere bounded by two semicircles and a lune (the surface of the sphere between the semicircles). { 'sfir·ə·kəl 'wej }

sphericity [SCI TECH] The degree to which a shape approaches that of a sphere. { sfə'ris·əd·ē }

spherochromatism [OPTICS] The variation of chromatic aberration with color of light. { ¦sfir·ō'krō·mə,tiz·əm }

spherocylindrical lens [OPTICS] A lens having one surface that is a portion of a sphere, while the other is a portion of a cylinder. { ¦sfir·ō·si¦lin·dri·kəl 'lenz }

spherocyte [PATH] A spherical red blood cell. { 'sfir·ə,sīt }

spherocytosis [MED] Preponderance of spherocytes in the blood. { ,sfir·ə,sī'tō·səs }

spheroid See ellipsoid of revolution. { 'sfir,ȯid }

spheroidal excess [MATH] The amount by which the sum of the three angles of a triangle on the surface of a spheroid exceeds 180°. { sfir'ȯid·əl ek,ses }

spheroidal galaxy See elliptical galaxy. { sfir'ȯid·əl 'gal·ik·sē }

spheroidal graphite cast iron See nodular cast iron. { sfir 'ȯid·əl 'gra,fīt 'kast 'ī·ərn }

spheroidal group [CRYSTAL] A group in the tetragonal symmetry system; the sphenoid is the typical form. { sfir'ȯid·əl 'grüp }

spheroidal harmonics [MATH] Solutions to Laplace's equation when phrased in ellipsoidal coordinates. { sfir'ȯid·əl här 'män·iks }

spheroidal recovery [GEOPHYS] The hypothetical return of the earth to spheroid form after it has been distorted. { sfir 'ȯid·əl ri'kəv·ə·rē }

spheroidal triangle [MATH] The figure formed by three geodesic lines joining three points on a spheroid. Also known as geodetic triangle. { sfir'ȯid·əl 'trī,aŋ·gəl }

spheroidal weathering [GEOL] Chemical weathering in which concentric or spherical shells of decayed rock are successively separated from a block of rock; commonly results in the formation of a rounded boulder of decomposition. Also known as concentric weathering; spherical weathering. { sfir 'ȯid·əl 'weth·ə·riŋ }

spheroidized carbides [MET] Globular forms of carbide, as formed in spheroidized steel. { 'sfir·ə,dīzd 'kär,bīdz }

spheroidized steel [MET] Steel that has been heat-treated to produce a spheroidized carbide structure. { 'sfir·ə,dīzd 'stēl }

spheroidizing [MET] Heating steels just below Ae₁ until the shape of cementite particles becomes relatively spherical. { 'sfir·ə,dīz·iŋ }

spherometer [ENG] A device used to measure the curvature of a spherical surface. { sfə'räm·əd·ər }

spheroplast [CYTOL] A plant cell which possesses only a partial or modified cell wall. [MICROBIO] A bacterial cell that assumes a spherical shape due to partial or complete absence of the wall. { 'sfir·ə,plast }

spherotoric lens [OPTICS] A lens having one surface that is a portion of a sphere, while the other is a portion of a torus. { ¦sfir·ə¦tȯr·ik 'lenz }

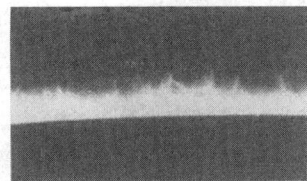

SPICULE

Large-scale photograph of the chromosphere in H-α light, with the disk of the sun artificially eclipsed and the hairy spicules projecting about the continuous chromosphere.

spherulite [GEOL] A spherical body or coarsely crystalline aggregate having a radial internal structure arranged about one or more centers. { 'sfir·ə,līt }

spherulitic [PETR] Relating to the texture of a rock composed of numerous spherulites. Also known as globular; sphaerolitic. { ¦sfir·ə¦lid·ik }

sphincter [ANAT] A muscle that surrounds and functions to close an orifice. { 'sfiŋk·tər }

sphincter of Oddi [ANAT] Sphincter of the hepatopancreatic ampulla. { 'sfiŋk·tər əv 'äd·ē }

Sphinctozoa [PALEON] A group of fossil sponges in the class Calcarea which have a skeleton of massive calcium carbonate organized in the form of hollow chambers. { ,sfiŋk·tə'zō·ə }

Sphindidae [INV ZOO] The dry fungus beetles, a family of coleopteran insects in the superfamily Cucujoidea. { 'sfin·də,dē }

Sphingidae [INV ZOO] The single family of the lepidopteran superfamily Sphingoidea. { 'sfin·jə,dē }

Sphingoidea [INV ZOO] A superfamily of Lepidoptera in the suborder Heteroneura consisting of the sphinx, hawk, or hummingbird moths; these are heavy-bodied forms with antennae that are thickened with a pointed apex, a well-developed proboscis, and narrow wings. { sfiŋ'gȯid·ē·ə }

sphingolipid [BIOCHEM] Any lipid, such as a sphingomyelin, that yields sphingosine or one of its derivatives as a product of hydrolysis. { ¦sfiŋ·gō'lip·əd }

sphingolipidosis [MED] Any of a group of hereditary metabolic disorders characterized by excessive accumulations of certain glycolipids and phospholipids in various tissues of the body. { ¦sfiŋ·gō,lip·ə'dō·səs }

sphingomyelin [BIOCHEM] A phospholipid consisting of choline, sphingosine, phosphoric acid, and a fatty acid. { ¦sfiŋ·gō'mī·ə·lən }

sphingosine [BIOCHEM] $C_{18}H_{37}O_2N$ A moiety of sphingomyelin, cerebrosides, and certain other phosphatides. { 'sfiŋ·gə,sēn }

sphygmomanometer [MED] An instrument for measuring the arterial blood pressure. { ¦sfig·mō·mə'näm·əd·ər }

sphygmophone [MED] A microphone attached to the wrist to pick up the sounds of the pulse. { 'sfig·mə,fōn }

Sphyraenidae [VERT ZOO] A family of shore fishes in the suborder Mugiloidei of the order Perciformes comprising the barracudas. { sfə'rē·nə,dē }

Sphyriidae [INV ZOO] A family of ectoparasitic Crustacea belonging to the group Lernaeopodoida; the parasite embeds its head and part of its thorax into the host. { sfə'rī·ə,dē }

Spica [ASTRON] A blue-white dwarf star of stellar magnitude 1.0, 160 light-years from the sun, spectral classification B1-V, in the constellation Virgo; the star α Virginis. { 'spī·kə }

spice [FOOD ENG] An aromatic vegetable material used for food seasoning. { spīs }

spicule [ASTRON] One of an irregular distribution of jets shooting up from the sun's chromosphere. Also known as solar spicule. [BOT] An empty diatom shell. [INV ZOO] A calcareous or siliceous, usually spikelike supporting structure in many invertebrates, particularly in sponges and alcyonarians. { 'spik·yül }

spiculin [BIOCHEM] An organic material making up a portion of a spicule. { 'spik·yə·lən }

spiculite [PETR] A spindle-shaped belonite thought to have formed by the coalescence of globulites. { 'spik·yə,līt }

spiculum [INV ZOO] A bristlelike copulatory organ in certain nematodes. Also known as copulatory spicule. { 'spik·yə·ləm }

spider [AGR] An attachment to a cultivator that pulverizes the soil. [COMPUT SCI] A program that searches the Internet for new, publicly accessible resources and transmits its findings to a database that is accessible to search engines. [ELEC] A structure on the shaft of an electric rotating machine that supports the core or poles of the rotor, consisting of a hub, spokes, and rim, or some similar arrangement. [ENG] **1.** The part of an ejector mechanism which operates ejector pins in a molding press. **2.** In extrusion, the membranes which support a mandrel within the head-die assembly. [ENG ACOUS] A highly flexible perforated or corrugated disk used to center the voice coil of a dynamic loudspeaker with respect to the pole piece without appreciably hindering in-and-out motion of the voice coil and its attached diaphragm.　[INV ZOO]　The common name for arachnids comprising the order Araneida.　[MECH ENG]　In a universal joint, a part with four projections that is pivoted between the forked ends of two shafts and transmits motion between the shafts.　Also known as cross.　[MET]　In founding, a device that consists of a frame with radiating arms or members and is used for strengthening a core or mold.　[PETRO ENG]　A steel block with a tapered opening which permits passage of pipe during movement into or from a well; designed to hold pipe suspended in the well when the slips are placed in the tapered opening and in contact with the pipe. { 'spīd·ər }

spider nevus　[MED]　A type of telangiectasis characterized by a central, elevated, tiny red dot, pinhead in size, from which blood vessels radiate like strands of a spider's web.　Also known as stellar nevus.　{ 'spīd·ər 'nē·vəs }

spiderweb antenna　[ELECTROMAG]　All-wave receiving antenna having several different lengths of doublets connected somewhat like the web of a spider to give favorable pickup characteristics over a wide range of frequencies.　{ 'spīd·ər,web an,ten·ə }

spiegeleisen　[MET]　An iron alloy containing 15–30% manganese and 5% carbon used in steelmaking.　{ 'spē·gə,līz·ən }

Spiegler's test　[PATH]　A test for the presence of protein in urine performed by overlaying clear acidulated urine on Spiegler's reagent (mercuric chloride, tartaric acid, glycerin, distilled water); opalescence at the fluid junction indicates protein.　{ 'spēg·lərz ,test }

spigot mortar　[ORD]　A mortar which propels a warhead larger than the bore of the mortar by means of a closed tube (spigot) attached to the warhead and extending into the mortar; the force of the propellant within the mortar acts upon the tube, thus propelling the warhead toward the target.　{ 'spik·ət ,mȯrd·ər }

spike　[BOT]　An indeterminate inflorescence with sessile flowers.　[DES ENG]　A large nail, especially one longer than 3 inches (7.6 centimeters), and often of square section.　[PHYS]　A short-duration transient whose amplitude considerably exceeds the average amplitude of the associated pulse or signal.　[SOLID STATE]　A sputtering event in which the process from impact of a bombarding projectile to the ejection of target atoms involves motion of a large number of particles in the target, so that collisions between particles become significant.　{ spīk }

spike antenna　See monopole antenna.　{ 'spīk an,ten·ə }

spiked core　See seed core.　{ 'spīkt 'kȯr }

spikelet　[BOT]　The compound inflorescence of a grass consisting of one or several bracteate spikes.　{ 'spīk·lət }

spike microphone　[ENG ACOUS]　A device for clandestine aural surveillance in which the sensor is a spike driven into the wall of the target area and mechanically coupled to the diaphragm of a microphone on the other side of the wall. { 'spīk ,mī·krə,fōn }

spike nozzle　[AERO ENG]　A nozzle in which gas is initially directed radially inward toward the nozzle axis, and expansion occurs only outside the nozzle when the gas is directly exposed to the external environment.　{ 'spīk ,näz·əl }

spike oil　[MATER]　A pale yellow essential oil extracted by distillation from the flowers of *Lavandula latifolia*; soluble in fixed oils and propylene glycol; used in soap and as an alcohol denaturant and flavoring agent.　Also known as lavender spike oil; Spanish lavender oil; Spanish spike oil.　{ 'spīk ,ȯil }

spike-tooth harrow　[AGR]　An implement with steel spikes extending downward from a frame and pulled by a tractor to pulverize and smooth plowed soil.　{ 'spīk,tüth 'har·ō }

spile　[MIN ENG]　**1.** A temporary lagging driven ahead on levels in loose ground.　**2.** A short piece of plank sharpened flatwise and used for driving into watery strata as sheet piling to assist in checking the flow of water.　{ spīl }

spiling　See forepoling.　{ 'spīl·iŋ }

spilite　[PETR]　An altered basalt containing albitized feldspar accompanied by low-temperature, hydrous crystallization products such as chlorite, calcite, and epidote.　{ 'spī,līt }

spill　[ENG]　The accidental release of some material, such as nuclear material or oil, from a container.　[NUCLEO]　The accidental release of radioactive material.　{ spil }

spill box　[CIV ENG]　A device such as a flume that maintains a constant head on a measuring weir or orifice.　{ 'spil ,bäks }

spilling　[OCEANOG]　The process by which steep waves break on approaching the shore; white water appears on the crest and the wave top gradually rolls over, without a crash. { 'spil·iŋ }

spilling breaker　See plunging breaker.　{ 'spil·iŋ ,brāk·ər }

spillover　[COMMUN]　The receiving of a radio signal of a different frequency from that to which the receiver is tuned, due to broad tuning characteristics.　[METEOROL]　That part of orographic precipitation which is carried along by the wind so that it reaches the ground in the nominal rain shadow on the lee side of the barrier.　{ 'spil ō·vər }

spillover positions　[COMMUN]　When a send channel is unusually busy or inoperative, the resulting backlogged traffic can be switched to spillover (storage) positions where it is held for immediate transmission when a channel becomes available. { 'spil,ō·vər pə,zish·ənz }

spill pit　See runoff pit.　{ 'spil ,pit }

spill stream　See overflow stream.　{ 'spil ,strēm }

spillway　[CIV ENG]　A passage in or about a dam or other hydraulic structure for escape of surplus water.　{ 'spil,wā }

spillway apron　[CIV ENG]　A concrete or timber floor at the bottom of a spillway to prevent soil erosion from heavy or turbulent flow.　{ 'spil,wā ,ā·prən }

spillway channel　[CIV ENG]　An outlet channel from a spillway.　{ 'spil,wā ,chan·əl }

spillway dam　See overflow dam.　{ 'spil,wā ,dam }

spillway gate　[CIV ENG]　A gate for regulating the flow from a reservoir.　{ 'spil,wā ,gāt }

spin　[MECH]　Rotation of a body about its axis.　[QUANT MECH]　The intrinsic angular momentum of an elementary particle or nucleus, which exists even when the particle is at rest, as distinguished from orbital angular momentum. { spin }

spina bifida　[MED]　A congenital anomaly characterized by defective closure of the spinal canal with herniation of the spinal cord meninges.　{ ¦spī·nə 'bī·fəd·ə }

spina bifida occulta　[MED]　An asymptomatic congenital anomaly consisting of incomplete fusion of the posterior arch of the vertebral canal without hernial protrusion of the meninges. { ¦spī·rə 'bī·fəd·ə ə'kəl·tə }

spinacene　See squalene.　{ 'spin·ə,sēn }

spinach　[BOT]　*Spinacia oleracea.* An annual potherb of Asiatic origin belonging to the order Caryophyllales and grown for its edible foliage.　{ 'spin·ich }

spinal anesthesia　[MED]　**1.** Anesthesia due to a lesion of the spinal cord.　**2.** Anesthesia produced by the injection of an anesthetic into the spinal subarachnoid space.　{ 'spīn·əl ,an·əs'thē·zhə }

spinal column　See spine.　{ 'spīn·əl ,käl·əm }

spinal cord　[NEUROSCI]　The cordlike posterior portion of the central nervous system contained within the spinal canal of the vertebral column of all vertebrates.　{ 'spīn·əl ,kȯrd }

spinal foramen　[ANAT]　Central canal of the spinal cord. { 'spīn·əl fə'rā·mən }

spinal ganglion　[NEUROSCI]　Any one of the sensory ganglions, each associated with the dorsal root of a spinal nerve. { 'spīn·əl 'gaŋ·glē·ən }

spinal nerve　[NEUROSCI]　Any of the paired nerves arising from the spinal cord.　{ 'spīn·əl ¦nərv }

spinal reflex　[NEUROSCI]　A reflex mediated through the spinal cord without the participation of the more cephalad structures of the brain or spinal cord.　{ 'spīn·əl 'rē,fleks }

spin axis　[PHYS]　The axis of rotation of a gyroscope. { 'spin ,ak·səs }

spincasting　[ENG]　A technique for manufacturing telescope mirrors in which molten glass is poured into a rotating mold and, as the glass cools and solidifies, the surface of the relatively thin mirror takes on a shape that is relatively close to the desired one, reducing substantially the need for grinding away excess glass.　{ 'spin kast·iŋ }

spin compensation　[MECH]　Overcoming or reducing the effect of projectile rotation in decreasing the penetrating capacity of the jet in shaped-charge ammunition.　{ 'spin ,käm·pən,sā·shən }

spin-decelerating moment　[MECH]　A couple about the axis of the projectile, which diminishes spin.　{ 'spin di¦sel·ə,rād·iŋ 'mō·mənt }

spin-density wave　[SOLID STATE]　The ground state of a metal in which the conduction-electron spin density has a sinusoidal variation in space.　{ 'spin ¦den·səd·ē 'wāv }

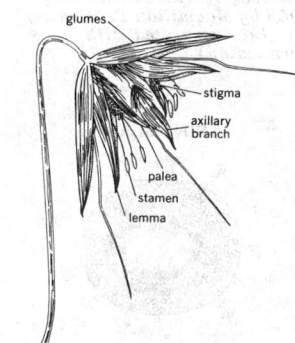

SPIKELET

glumes
stigma
axillary
branch
palea
stamen
lemma

Spikelet of wild oat (*Avena fatua*).

spin-dependent force [PHYS] A force between two particles which depends in some way on the spin, possibly on the angle between their spin directions, or on the angles between their spin directions and a line joining the particles. { 'spin di¦pen·dənt 'förs }

spindle [CELL MOL] A structure formed of fiberlike elements just before metaphase that extends between the poles of the achromatic figure and is attached to the centromeric regions of the chromatid pairs. [DES ENG] A short, slender or tapered shaft. [NAV] A spar serving as a beacon. { 'spin·dəl }

spindle fiber [CELL MOL] One of the fiberlike elements of the spindle; an aggregation of microtubules resulting from the polymerization of a series of small protein fibrils by primary −S−S− linkages. { 'spin·dəl ¡fī·bər }

spindle tuber [PL PATH] A virus disease of the potato characterized by spindliness of the tops and tubers. { 'spin·dəl ¡tü·bər }

spine [ANAT] An articulated series of vertebrae forming the axial skeleton of the trunk and tail, and being a characteristic structure of vertebrates. Also known as backbone; spinal column; vertebral column. [BOT] A rigid sharp-pointed process in plants; many are modified leaves. [GRAPHICS] The edge of a book that connects the covers and along which the sections are secured together in the binding. Also known as backbone. [INV ZOO] One of the processes covering the surface of a sea urchin. [VERT ZOO] **1.** One of the spiny rays supporting the fins of most fishes. **2.** A sharp-pointed modified hair on certain mammals, such as the porcupine. { spīn }

spin echo technique [NUC PHYS] A variation of the nuclear magnetic resonance technique in which the radio frequency field is applied in two pulses, separated by a time interval t, and a strong nuclear induction signal is observed at a time t after the second pulse. { 'spin ¡ek·ō tek,nēk }

spinel [MINERAL] **1.** $MgAl_2O_4$ A colorless, purplish-red, greenish, yellow, or black mineral, usually forming octahedral crystals, and characterized by great hardness; used as a gemstone. **2.** A group of minerals of general formula AB_2O_4, where A is magnesium, ferrous iron, zinc, or manganese, or a combination of them, and B is aluminum, ferric iron, or chromium. { spə'nel }

spin electronics See magnetoelectronics. { 'spin ¡i·lek,trän·iks }

spin filter [ELECTR] A device used in a Lamb-shift polarized ion source to cause those atoms having an undesired nuclear spin orientation to decay from their metastable state to the ground state, while those with the desired spin orientation are allowed to pass through without decay. { 'spin ,fil·tər }

spin-flip laser [OPTICS] A semiconductor laser in which the output wavelength is continuously tunable by a magnetic field; operation is based on exciting conduction-band electrons to a higher energy level by reversing the direction of the electrons as they spin about an axis in the direction of the magnetic field. { 'spin ¡flip 'lā·zər }

spin-flip scattering [QUANT MECH] Scattering of a particle with spin $1/2$ in which the direction of the particle's spin is reversed. { 'spin ¡flip 'skad·ə·riŋ }

spin glass [SOLID STATE] A substance in which the atomic spins are oriented in random but fixed directions. { 'spin ,glas }

spiniger [INV ZOO] Seta that tapers to a fine point, most frequently used in connection with compound seta (thus, compound spiniger). { 'spin·ə·jər }

spin isomer [NUC PHYS] An excited nuclear state which has an unusually long lifetime because of the large difference between the spin of the state and the spins of the states of lower energy into which it is permitted to decay. { 'spin,ī·sə·mər }

spin label [PHYS CHEM] A molecule which contains an atom or group of atoms exhibiting an unpaired electron spin that can be detected by electron spin resonance (ESR) spectroscopy and can be bonded to another molecule. { 'spin ,lā·bəl }

spin-lattice interaction [SOLID STATE] The state of a solid when the energy of electron spins is being shared with the thermal-vibration energy of the solid as a whole. { 'spin ¦lad·əs ,in·tə,rak·shən }

spin-lattice relaxation [SOLID STATE] Magnetic relaxation in which the excess potential energy associated with electron spins in a magnetic field is transferred to the lattice. { 'spin ¦lad·əs ,rē,lak,sā·shən }

spin magnetism [SOLID STATE] Paramagnetism or ferromagnetism that arises from polarization of electron spins in a substance. { 'spin ,mag·nə,tiz·əm }

Spinnbarkeit relaxation [FL MECH] A rheological effect illustrated by the pulling away of liquid threads when an object that has been immersed in a viscoelastic fluid is pulled out. { 'spin,bär,kīt ,rē,lak,sā·shən }

spinner [ENG] **1.** Automatically rotatable radar antenna, together with directly associated equipment. **2.** Part of a mechanical scanner which rotates about an axis, generally restricted to cases where the speed of rotation is relatively high. { 'spin·ər }

spinneret [ENG] An extrusion die with many holes through which plastic melt is forced to form filaments. [INV ZOO] An organ that spins fiber from the secretion of silk glands. [TEXT] A metal device with tiny holes through which a solution is forced at high speeds to make fine textile filaments. { ,spin·ə'ret }

spinney [ECOL] A small grove of trees or a thicket with undergrowth. { 'spin·ē }

spinning [ENG] The extrusion of a spinning solution (such as molten plastic) through a spinneret. [MECH ENG] Shaping and finishing sheet metal by rotating the workpiece over a mandrel and working it with a round-ended tool. Also known as metal spinning. [TEXT] Converting fibers or filaments into thread or yarn by drawing and twisting. { 'spin·iŋ }

spinning acoustic modes [ACOUS] Natural acoustic waveforms in a circular duct, consisting of pressure disturbances with wavefronts that follow cylindrical paths. { 'spin·iŋ ə,kü·stik 'mäd·əlz }

spinning-band column [ANALY CHEM] An analytical distillation column inside of which is a series of driven, spinning bands; centrifugal action of the bands throws a layer of liquid onto the inner surface of the column; used as an aid in liquid-vapor contact. { 'spin·iŋ ¦band ,käl·əm }

spinning machine [MECH ENG] **1.** A machine that winds insulation on electric wire. **2.** A machine that shapes metal hollow ware. [TEXT] A machine that spins yarn from staple fiber or continuous filament. { 'spin·iŋ mə,shēn }

spinoblast [INV ZOO] A statoblast having a float of air cells and barbs or hooks on the surface. { 'spī·nə,blast }

spinochrome [BIOCHEM] A type of echinochrome; an organic pigment that is known only from sea urchins and certain homopteran insects. { 'spī·nə,krōm }

spinodal decomposition [MINERAL] An unmixing process in which crystals with bulk composition in the central region of the phase diagram undergo exsolution. { spī'nōd·əl dē,käm·pə'zish·ən }

spinode See cusp. { 'spi,nōd }

spinor [MATH] **1.** A vector with two complex components, which undergoes a unitary unimodular transformation when the three-dimensional coordinate system is rotated; it can represent the spin state of a particle of spin $1/2$. **2.** More generally, a spinor of order (or rank) n is an object with 2^n components which transform as products of components of n spinors of rank one. **3.** A quantity with four complex components which transforms linearly under a Lorentz transformation in such a way that if it is a solution of the Dirac equation in the original Lorentz frame it remains a solution of the Dirac equation in the transformed frame; it is formed from two spinors (definition 1). Also known as Dirac spinor. { 'spin·ər }

spin-orbit coupling [QUANT MECH] The interaction between a particle's spin and its orbital angular momentum. { 'spin ¦ör·bət ,kəp·liŋ }

spin-orbit multiplet [PHYS] A collection of atomic or nuclear states which differ in energy only on account of spin-orbit coupling; the total spin angular momentum quantum number S and total orbital angular momentum quantum number L are the same for all states; the energy levels are labeled by the total angular momentum quantum number J. { 'spin ¦ör·bət 'məl·tə·plət }

spinous process [ANAT] Any slender, sharp-pointed projection on a bone. { 'spī·nəs ,prä·səs }

spin paramagnetism [SOLID STATE] Paramagnetism that arises from the electron spins in a substance. { 'spin ,par·ə'mag·nə,tiz·əm }

spin-parity [PART PHYS] A combined symbol J^P for an elementary particle's spin J, and its intrinsic parity P. { 'spin ,par·əd·ē }

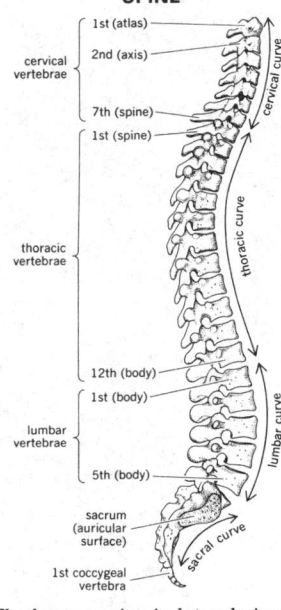

SPINE

1st (atlas)
cervical vertebrae
2nd (axis)
7th (spine)
1st (spine)
cervical curve
thoracic curve
thoracic vertebrae
12th (body)
1st (body)
lumbar vertebrae
lumbar curve
5th (body)
sacrum (auricular surface)
sacral curve
1st coccygeal vertebra

The human spine in lateral view. *(From W. J. Hamilton et al., Textbook of Human Anatomy, © 1956 by Macmillan Publishing Co., Inc.; reprinted with permission)*

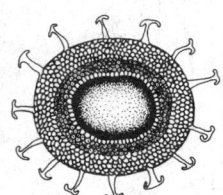

SPINOBLAST

Spinoblast of *Pectinatella magnifica*.

spin-polarized atomic hydrogen [PHYS CHEM] A system of hydrogen atoms cooled to a very low temperature in a very high magnetic field so that electron spins in almost all the atoms are antiparallel to the magnetic field, with the result that the atoms interact only through the weak triplet-state interaction so that no hydrogen molecules are formed. { 'spin ¦pō·lə‚rīzd ə'täm·ik 'hī·drə·jən }

spin-polarized low-energy electron diffraction [SOLID STATE] A version of low-energy electron diffraction in which electrons in the incident beam have their spins aligned in one direction; used in studies of the magnetic properties of atoms near the surface of a material. { 'spin ¦pō·lə‚rīzd ¦lō 'en·ər·jē i'lek‚trän di‚frak·shən }

spin quantum number [QUANT MECH] The ratio of the maximum observable component of a system's spin to Planck's constant divided by 2π; it is an integer or a half-integer. { 'spin 'kwän·təm ‚nəm·bər }

spin resonance *See* magnetic resonance. { 'spin ‚rez·ən·əns }

spin rocket [AERO ENG] A small rocket that imparts spin to a larger rocket vehicle or spacecraft. { 'spin ‚räk·ət }

spin safe [ORD] Referring to a fuse that is safe when experiencing a rotation equivalent to that attained during flight; that is, other arming forces are necessary to arm the fuse. { 'spin ‚sāf }

spin space [MATH] The two-dimensional vector space over the complex numbers, whose unitary unimodular transformations are a two-dimensional double-valued representation of the three-dimensional rotation group; its vectors can represent the various spin states of a particle with spin $\frac{1}{2}$, and its unitary unimodular transformations can represent rotations of this particle. { 'spin ‚spās }

spin-spin energy [PHYS] An interaction energy proportional to the dot product of the spin angular momenta of two systems. { 'spin 'spin ‚en·ər·jē }

spin-spin relaxation [SOLID STATE] Magnetic relaxation, observed after application of weak magnetic fields, in which the excess potential energy associated with electron spins in a magnetic field is redistributed among the spins, resulting in heating of the spin system. { 'spin 'spin ‚rē‚lak‚sā·shən }

spin stabilization [AERO ENG] Directional stability of a spacecraft obtained by the action of gyroscopic forces which result from spinning the body about its axis of symmetry. { 'spin ‚stā·bə·lə‚zā·shən }

spin state [QUANT MECH] Condition of a particle in which its total spin, and the component of its spin along some specified axis, have definite values; more precisely, the particle's wave function is an eigenfunction of the operators corresponding to these quantities. { 'spin ‚stāt }

spin temperature [SOLID STATE] For a system of electron spins in a lattice, a temperature such that the population of the energy levels of the spin system is given by the Boltzmann distribution with this temperature. { 'spin ‚tem·prə·chər }

spinthariscope [ELECTR] An instrument for viewing the scintillations of alpha particles on a luminescent screen, usually with the aid of a microscope. { spin'thar·ə‚skōp }

Spintheridae [INV ZOO] An amphinomorphan family of small polychaete annelids included in the Errantia. { spin'ther·ə‚dē }

spin transistor *See* magnetic switch. { 'spin tran‚zis·tər }

spintronics *See* magnetoelectronics. { spin'trän·iks }

Spinulosida [INV ZOO] An order of Asteroidea in which pedicellariae rarely occur, marginal plates bounding the arms and disk are small and inconspicuous, and spines occur in groups on the upper surface. { ‚spin·yə'läs·əd·ē }

spin-up [ASTRON] A sudden increase in the pulse frequency of a pulsar. { 'spin‚əp }

spin valve *See* magnetic switch. { 'spin ‚valv }

spin wave [SOLID STATE] A sinusoidal variation, propagating through a crystal lattice, of that angular momentum which is associated with magnetism (mostly spin angular momentum of the electrons). { 'spin ‚wāv }

spin welding [ENG] Fusion of two objects (for example, plastics) by forcing them together while one of the pair is spinning; frictional heat melts the interface, spinning is stopped, and the bodies are held together until they are frozen in place (welded). { 'spin ‚weld·iŋ }

spiny-rayed fish [VERT ZOO] The common designation for actinopterygian fishes, so named for the presence of stiff, unbranched, pointed fin rays, known as spiny rays. { 'spī·nē 'rād 'fish }

Spionidae [INV ZOO] A family of spioniform annelid worms belonging to the Sedentaria. { spī'än·ə‚dē }

spioniform worm [INV ZOO] A polychaete annelid characterized by the presence of a pair of short to long, grooved palpi near the mouth. { spī'än·ə‚fòrm 'wərm }

spiracle [INV ZOO] An external breathing orifice of the tracheal system in insects and certain arachnids. [VERT ZOO] **1.** The external respiratory orifice in cetaceous and amphibian larvae. **2.** The first visceral cleft in fishes. { 'spir·ə·kəl }

spiral [MATH] A simple curve in the plane which continuously winds about itself either into some point or out from some point. { 'spī·rəl }

spiral arms [ASTRON] The shape of sections of certain galaxies called spirals; these sections are two so-called arms composed of stars, dust, and gas extending from the center of the galaxy and coiled about it. { 'spī·rəl 'ärmz }

spiral band [METEOROL] Spiral-shaped radar echoes received from precipitation areas within intense tropical cyclones (hurricanes or typhoons); they curve cyclonically in toward the center of the storm and appear to merge to form the wall around the eye of the storm. Also known as hurricane band; hurricane radar band. { 'spī·rəl 'band }

spiral bevel gear [DES ENG] Bevel gear with curved, oblique teeth to provide gradual engagement and bring more teeth together at a given time than an equivalent straight bevel gear. { 'spī·rəl 'bev·əl ‚gir }

spiral binding [GRAPHICS] A book binding in which a spiral of wire or plastic strip is wound through holes at the edge of the book. { 'spī·rəl 'bīnd·iŋ }

spiral chute [DES ENG] A gravity chute in the form of a continuous helical trough spiraled around a column for conveying materials to a lower level. { 'spī·rəl 'shüt }

spiral cleavage [EMBRYO] A cleavage pattern characterized by formation of a cell mass showing spiral symmetry; occurs in mollusks. { 'spī·rəl 'klē·vij }

spiral conveyor *See* screw conveyor. { 'spī·rəl kən'vā·ər }

spiral cutterhead [MIN ENG] A rotary digging device which dislodges and feeds alluvial sand or gravel to the intake of a suction dredge. { 'spī·rəl 'kəd·ər‚hed }

spiral delay line [ELECTROMAG] A transmission line which has a helical inner conductor. { 'spī·rəl di'lā ‚līn }

spiral flow tank [CIV ENG] An aeration tank of the activated sludge process into which air is diffused in a spiral helical movement guided by baffles and proper location of diffusers. { 'spī·rəl 'flō ‚taŋk }

spiral flow test [ENG] The determination of the flow properties of a thermoplastic resin by measuring the length and weight of resin flowing along the path of a spiral cavity. { 'spī·rəl 'flō ‚test }

spiral four cable [ELEC] A quad cable in which the four conductors are twisted about a common axis, the two sets of opposite conductors being used as pairs. { 'spī·rəl 'fòr ‚kā·bəl }

spiral gage *See* spiral pressure gage. { 'spī·rəl 'gāj }

spiral galaxy [ASTRON] A type of galaxy classified on the basis of appearance of its photographic image; this type includes two main groups: normal spirals with circular symmetry of the nucleus and of the spiral arms, and barred spirals in which the dominant form is a luminous bar crossing the nucleus with spiral arms starting at the ends of the bar or tangent to a luminous rim on which the bar terminates. { 'spī·rəl 'gal·ik·sē }

spiral ganglion [NEUROSCI] The ganglion of the cochlear part of the vestibulocochlear nerve embedded in the spiral canal of the modiolus. { 'spī·rəl 'gaŋ·glē·ən }

spiral gear [MECH ENG] A helical gear that transmits power from one shaft to another, nonparallel shaft. { 'spī·rəl ‚gir }

spiral-jaw clutch [MECH ENG] A modification of the square-jaw clutch permitting gradual meshing of the mating faces, which have a helical section. { 'spī·rəl 'jò ‚kləch }

spiral layer *See* Ekman layer. { 'spī·rəl 'lā·ər }

spiral ligament [ANAT] The reticular connective tissue connecting the basilar membrane to the outer cochlear wall in the ear of mammals. { 'spī·rəl 'lig·ə·mənt }

spiral mold cooling [ENG] Cooling an injection mold by passing a liquid through a spiral cavity in the body of the mold. { 'spī·rəl ¦mōld 'kül·iŋ }

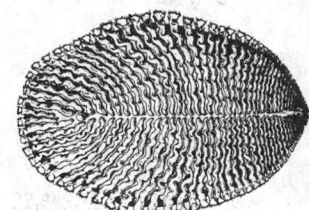

SPINTHERIDAE

Spinther, dorsal view.

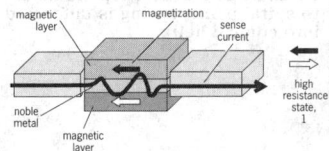

SPIN VALVE

Magnetic components used in the spin valve.

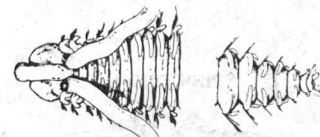

SPIONIDAE

Nerinides showing the anterior portion (left) and the tail section (right), in dorsal view.

spiral of Archimedes [MATH] The curve spiraling into the origin which in polar coordinates is given by the equation $r = a\theta$. Also known as Archimedes' spiral. { 'spī·rəl əv ,är·kə'mē·dēz }

spiral organ *See* organ of Corti. { 'spī·rəl 'ȯr·gən }

spiral pipe [DES ENG] Strong, lightweight steel pipe with a single continuous welded helical seam from end to end. { 'spī·rəl 'pīp }

spiral plate exchanger [CHEM ENG] A heat-transfer device made from a pair of plates rolled in a spiral to provide two relatively long, rectangular passages for heat-transfer between fluids in countercurrent flow. { 'spī·rəl ,plāt iks'chān·jər }

spiral pressure gage [ENG] A device for measurement of pressures; a hollow tube spiral receives the system pressure which deforms (unwinds) the spiral in direct relation to the pressure in the tube. Also known as spiral gage. { 'spī·rəl 'presh·ər ,gāj }

spiral ramp system [MIN ENG] Development of a mine by driving moderately inclined haulageways from the surface to underground ore horizons. { 'spī·rəl 'ramp ,sis·təm }

spiral ring structure [ASTRON] A lunar crater in which ridges spiral inward from the main wall across the floor. { 'spī·rəl ,riŋ ,strək·chər }

spiral scanning [ENG] Scanning in which the direction of maximum radiation describes a portion of a spiral; the rotation is always in one direction; used with some types of radar antennas. { 'spī·rəl 'skan·iŋ }

spiral spring [DES ENG] A spring bar or wire wound in an Archimedes spiral in a plane; each end is fastened to the force-applying link of the mechanism. { 'spī·rəl 'spriŋ }

spiral thermometer [ENG] A temperature-measurement device consisting of a bimetal spiral that winds tighter or opens with changes in temperature. { 'spī·rəl thər'mäm·əd·ər }

spiral-tube heat exchanger [ENG] A countercurrent heat-exchange device made of a group of concentric spirally wound coils, generally connected by manifolds; used for cryogenic exchange in air-separation plants. { 'spī·rəl,tüb 'hēt iks,chān·jər }

spiral valve [VERT ZOO] A spiral fold of mucous membrane in the small intestine of elasmobranchs and some primitive fishes which increases the surface area for absorption. { 'spī·rəl ,valv }

spiral welded pipe [DES ENG] A steel pipe made of long strips of steel plate fitted together to form helical seams, which are welded. { 'spī·rəl ,weld·əd 'pīp }

spiral wire column [ANALY CHEM] An analytical rectification (distillation) column with a wire spiral the length of the inside of the column to serve as a liquid-vapor contact surface. { 'spī·rəl ,wīr 'käl·əm }

spiramycin [MICROBIO] A complex of related antibiotics, which resemble erythromycin structurally and in antibacterial spectrum, produced by *Streptomyces ambofaciens*. { ,spī·rə'mīs·ən }

spiran [ORG CHEM] A polycyclic compound containing a carbon atom which is a member of two rings. { 'spī,ran }

spiraster [INV ZOO] A spiral spicule bearing rays in Porifera. { spī'ras·tər }

spire [ARCH] As a landmark, a prominent, slender, pointed structure surmounting a building; a spire is seldom less than two-thirds of the entire height, and its lines are rarely broken by stages or other features. [BOT] A narrow, tapering blade or stalk. { spīr }

spiricle [BOT] Any of the coiled threads in certain seed coats which uncoil when moistened. { 'spir·ə·kəl }

Spiriferida [PALEON] An order of fossil articulate brachiopods distinguished by the spiralium, a pair of spirally coiled ribbons of calcite supported by the crura. { ,spī·rə'fer·əd·ə }

Spiriferidina [PALEON] A suborder of the extinct brachiopod order Spiriferida including mainly ribbed forms having laterally or ventrally directed spires, well-developed interareas, and a straight hinge line. { spī,rif·ə·rə'dī·nə }

Spirillaceae [MICROBIO] A family of bacteria; motile, helically curved rods that move with a characteristic corkscrewlike motion. { 'spī·rə'lās·ē,ē }

Spirillinacea [INV ZOO] A superfamily of foraminiferan protozoans in the suborder Rotaliina characterized by a planispiral or low conical test with a wall composed of radial calcite. { spə,ril·ə'nās·ē·ə }

spirit [FOOD ENG] A flammable liquid mixture of water and ethyl alcohol that is separated from an alcoholic liquid or mash by distillation during the manufacture of whiskey. [ORG CHEM] A solution of alcohol and a volatile substance, such as an essential oil. { 'spir·ət }

spirit compass [NAV] A liquid compass using a mixture of alcohol and water. { 'spir·ət ,käm·pəs }

spirit duplicating [GRAPHICS] A contact process used in reproducing maps in which a map with several colors in accurate position relative to each other can be produced in one run through the machine; the map is drawn on a paper master having a carbon backing; different-colored carbons can be substituted to prepare a multicolor image. { 'spir·ət 'dü·plə,kād·iŋ }

spirit level *See* level. { 'spir·ət ,lev·əl }

spirit stain [MATER] A dye dissolved in methylated spirits; used to stain wood surfaces. { 'spir·ət ,stān }

spirit thermometer [ENG] A temperature-measurement device consisting of a closed capillary tube with a liquid (for example, alcohol) reservoir bulb at the bottom; as the bulb is heated, the liquid expands up into the capillary tubing, indicating the temperature of the bulb. { 'spir·ət thər'mäm·əd·ər }

spirit varnish [MATER] An artificial varnish consisting of resin, asphalt, or a cellulose ester dissolved in a volatile solvent. { 'spir·ət ,vär·nish }

spiro atom [ORG CHEM] A single atom that is the only common member of two ring structures. { 'spir·ō ,ad·əm }

Spirobrachiidae [INV ZOO] A family of the Brachiata in the order Thecanephria. { ,spī·rə·brə'kī·ə,dē }

Spirochaetaceae [MICROBIO] The single family of the order Spirochaetales. { ,spī·rə·kē'tās·ē,ē }

Spirochaetales [MICROBIO] An order of bacteria characterized by slender, helically coiled cells sometimes occurring in chains. { ,spī·rə·kē'tā·lēz }

spirochetal jaundice *See* Weil's disease. { ,spī·rə,kēd·əl 'jȯn·dəs }

spirochete [MICROBIO] The common name for any member of the order Spirochaetales. { 'spī·rə,kēt }

spirochetemia [MED] The presence of spirochetes in the blood. { ,spī·rə,kēd'ē·mē·ə }

spirocyst [INV ZOO] A thin-walled capsule that contains a long, unarmed, eversible, spirally coiled thread of uniform diameter; found in cnidarians. { 'spī·rə,sist }

spirograph [MED] An instrument for registering respiration. { 'spī·rə,graf }

spirometry [PHYSIO] The measurement, by a form of gas meter (spirometer), of volumes of air that can be moved in or out of the lungs. { spī'räm·ə·trē }

spironolactone [PHARM] $C_{24}H_{32}O_4S$ A steroid having a lactone ring attached at carbon-17; used as a diuretic. { 'spī·rə·nō'lak,tōn }

spiro ring system [ORG CHEM] A molecular structure with two ring structures having one atom in common; for example, spiropentane. { 'spī·rō 'riŋ ,sis·təm }

Spirotrichia [INV ZOO] A subclass of the protozoan class Ciliatea which contains those ciliates characterized by conspicuous, compound ciliary structures, known as cirri, and buccal organelles. { ,spī·rō'trik·ē·ə }

spirule [GRAPHICS] A plastic device consisting of a protractor and a pivoted arm with a logarithmic spiral, used to add angles between vectors and multiply magnitudes of vectors. { 'spī,rül }

Spirulidae [INV ZOO] A family of cephalopod mollusks containing several species of squids. { spī'rül·ə,dē }

Spiruria [INV ZOO] A subclass of nematodes in the class Secernentea. { spī'rür·ē·ə }

Spirurida [INV ZOO] An order of phasmid nematodes in the subclass Spiruria. { spī'rür·əd·ə }

Spiruroidea [INV ZOO] A superfamily of spirurid nematodes which are parasitic in the respiratory and digestive systems of vertebrates. { ,spī·rə'rȯid·ē·ə }

spit [ENG] To light a fuse. [GEOGR] A small point of land commonly consisting of sand or gravel and which terminates in open water. { spit }

Spitsbergen Current [OCEANOGR] An ocean current flowing northward and westward from a point south of Spitsbergen, and gradually merging with the East Greenland Current in the Greenland Sea; the Spitsbergen Current is the continuation of the northwestern branch of the Norwegian Current. { 'spits,bər·gən 'kə·rənt }

SPIRAL SPRING

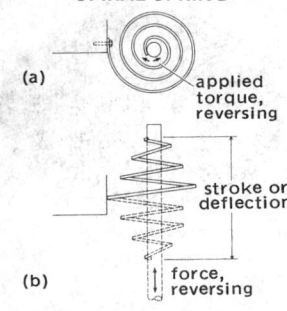

Spiral spring is unique in being able to respond to two types of forces. *(a)* In response to torsional force, spring remains flat spiral. *(b)* In response to translational force, perpendicular to spiral plane, spring is deformed into conical helix.

SPIRIFERIDINA

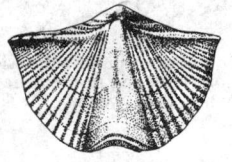

Dorsal view of *Spinocyrtia* valve.

spitted fuse [ENG] A slow-burning fuse which has been cut open at the lighting end for ease of ignition. { 'spid·əd 'fyüz }

spitting rock [ENG] A rock mass under stress that breaks and ejects small fragments with considerable velocity. { 'spid·iŋ ,räk }

SPL *See* sound pressure level.

splanchnic mesoderm [EMBRYO] The internal layer of the lateral mesoderm that is associated with the entoderm after the formation of the coelom. { 'splaŋk·nik 'mez·ə,dərm }

splanchnic nerve [NEUROSCI] A nerve carrying nerve fibers from the lower thoracic paravertebral ganglions to the collateral ganglions. { 'splaŋk·nik ¦nərv }

splanchnocranium [ANAT] Portions of the skull derived from the primitive skeleton of the gill apparatus. { ¦splaŋk·nə'krā·nē·əm }

splanchnopleure [EMBRYO] The inner layer of the mesoblast from which part of the wall of the alimentary canal and portions of the visceral organs are derived in coelomates. { 'splaŋk·nə,plür }

splash block [BUILD] A small masonry block with a concave surface placed on the ground below a downspout at a sloping angle to carry roof drainage water away from a building and to prevent erosion of the soil. { 'splash ,bläk }

splashdown [AERO ENG] 1. The landing of a spacecraft or missile on water. 2. The moment of impact of a spacecraft on water. { 'splash,daùn }

splash erosion [GEOL] Erosion resulting from the impact of falling raindrops. { 'splash i,rōzh·ən }

splash lubrication [ENG] An engine-lubrication system in which the connecting-rod bearings dip into troughs of oil, splashing the oil onto the cylinder and piston rods. { 'splash ,lü·brə,kā·shən }

splatter [COMMUN] Distortion due to overmodulation of a transmitter by peak signals of short duration, particularly sounds containing high-frequency harmonics; it is a form of adjacent-channel interference. { 'splad·ər }

splay [ENG] A slanted or beveled surface making an oblique angle with another surface. { splā }

splayed arch [CIV ENG] An arch whose opening has a larger radius in front than at the back. { 'splād 'ärch }

spleen [ANAT] A blood-forming lymphoid organ of the circulatory system, present in most vertebrates. { splēn }

splenectomy [MED] Surgical removal of the spleen. { splə'nek·tə·mē }

splenic fever *See* anthrax. { 'splen·ik 'fē·vər }

splenic flexure [ANAT] An abrupt turn of the colon beneath the lower end of the spleen, connecting the descending with the transverse colon. { 'splen·ik 'flek·shər }

splenium [ANAT] The rounded posterior extremity of the corpus callosum. [MED] A bandage. { 'splē·nē·əm }

splenomegaly [MED] Enlargement of the spleen. { ,splē·nə'meg·ə·lē }

splent coal *See* splint coal. { 'splent ¦kōl }

splice [ELEC] A joint used to connect two lengths of conductor with good mechanical strength and good conductivity. [ENG] To unite two parts, such as rope or wire, to form a continuous length. [GRAPHICS] To join two pieces of film together. { splīs }

spliced [GEOL] Relating to veins that pinch out and are overlapped at that point by another parallel vein. { splīst }

splice plate [CIV ENG] A plate for joining the web plates or the flanges of girders. { 'splīs ,plāt }

splicing tape [MATER] A pressure-sensitive nonmagnetic tape used for splicing magnetic tape and motion picture film; it has a hard adhesive that will not ooze and gum up the equipment or cause adjacent layers of tape or film on the reel to stick together. { 'splīs·iŋ ,tāp }

spline [DES ENG] One of a number of equally spaced keys cut integral with a shaft, or similarly, keyways in a hubbed part; the mated pair permits the transmission of rotation or translatory motion along the axis of the shaft. [ENG] A strip of wood, metal, or plastic. [GRAPHICS] A flexible strip used in drawing curves. [MATH] A function used to approximate a specified function on an interval, consisting of pieces which are defined uniquely on a set of subintervals, usually as polynomials or some other simple form, and which match up with each other and the prescribed function at the end points of the subintervals with a sufficiently high degree of accuracy. { splīn }

spline broach [MECH ENG] A broach for cutting straight-sided splines, or multiple keyways in holes. { 'splīn ,brōch }

splined shaft [DES ENG] A shaft with longitudinal gearlike ridges along its interior or exterior surface. { 'splīnd 'shaft }

splint [GEOL] *See* splint coal. [MED] A stiff or flexible material applied to an anatomical part in order to protect it, immobilize it, or restrict its motion. { splint }

splint coal [GEOL] A hard, dull, blocky, grayish-black, banded bituminous coal characterized by an uneven fracture and a granular texture; burns with intense heat. Also known as splent coal; splint. { 'splint ,kōl }

split [COMPUT SCI] To divide a data base, file, or other data set into two or more separate parts. [GEOL] A coal seam that cannot be mined as a single unit because it is separated by a parting of other sedimentary rock. Also known as coal split; split coal. [MIN ENG] 1. To divide the air current into separate circuits to ventilate more than one section of the mine. 2. Any division or branch of the ventilating current. { split }

split-altitude profile [AERO ENG] Flight profile at two separate altitudes. { 'split ¦al·tə,tüd prō,fīl }

split-anode magnetron [ELECTR] A magnetron in which the cylindrical anode is divided longitudinally into halves, between which extremely high-frequency oscillations are produced. { 'split ¦an,ōd 'mag·nə,trän }

split barrel [DES ENG] A core barrel that is split lengthwise so that it can be taken apart and the sample removed. { 'split 'bar·əl }

split-barrel sampler [DES ENG] A drive-type soil sampler with a split barrel. { 'split ¦bar·əl 'sam·plər }

split-base concept [ORD] A principle which applies to a deployed tactical combat unit that divides its resources between two separate operating bases. { 'split ¦bās 'kän,sept }

split bearing [DES ENG] A shaft bearing composed of two pieces bolted together. { 'split 'ber·iŋ }

split cameras [OPTICS] An assembly of two cameras disposed at a fixed overlapping angle relative to each other. { 'split 'kam·rəz }

split cavity [ENG] A cavity, such as in a mold, made in sections. { 'split 'kav·əd·ē }

split coal *See* split. { 'split ¦kōl }

split die [MET] 1. A screw-thread die made in one piece with a longitudinal slit connecting the outside to the central hole which allows size adjustment. 2. *See* segment die. { 'split 'dī }

split-field microtron [NUCLEO] A microtron whose magnetic field is divided into two separate sections in order to provide room for a large accelerating section. { 'split ¦fēld 'mī·krə,trän }

split fix [NAV] A fix by horizontal sextant angles obtained by measuring two angles between four objects, or suitable charted features, with no common center object being observed. { split 'fiks }

split flap [AERO ENG] A hinged plate forming the rear upper or lower portion of an airfoil; the lower portion may be deflected downward to give increased lift and drag; the upper portion may be raised over a portion of the wing for the purpose of lateral control. { 'split 'flap }

split gene [GEN] A eukaryotic gene in which the coding sequence is divided into two or more exons that are interrupted by a number of noncoding intervening sequences (introns). Also known as interrupted gene. { 'split 'jēn }

split-half method [STAT] A method used to gage the reliability of a test; two sets of scores are obtained from the same test, one set from odd items and one set from even items, and the scores of the two sets are correlated. { 'split ¦haf ,meth·əd }

split interstitial [CRYSTAL] A crystal defect in which a displaced atom forms a bond with a normal atom in such a way that neither atom is on the normal site but the two are symmetrically displaced from it. { 'split ¦in·tər'stish·əl }

split-lens interference [OPTICS] Interference produced by a Billet split lens. { 'split ¦lenz ,in·tər'fir·əns }

split link [DES ENG] A metal link in the shape of a two-turn helix pressed together. { 'split 'liŋk }

splitnut [ENG] A nut cut axially into halves to allow for rapid engagement (closed) or disengagement (open). { 'split¦nət }

split personality [PSYCH] The condition in which there is a separation of various components of the normal personality

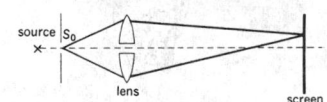

Billet split-lens interference. Light from source S_0 is transmitted through two separated parts of lens to screen, producing interference fringes on screen.

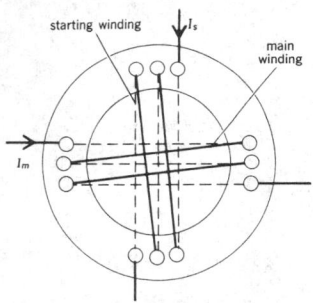

Windings of a split-phase motor; I_m is current of the main winding, and I_s is current of the starting winding.

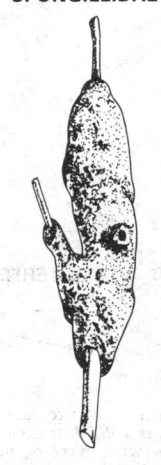

An encrusting spongillid sponge growing on a twig.

unit, and each component functions as an independent entity. { 'split ,pər·sə'nal·əd·ē }

split-phase motor [ELEC] A single-phase induction motor having an auxiliary winding connected in parallel with the main winding, but displaced in magnetic position from the main winding so as to produce the required rotating magnetic field for starting; the auxiliary circuit is generally opened when the motor has attained a predetermined speed. { 'split |fāz 'mōd·ər }

split pin [DES ENG] A pin with a split at one end so that it can be spread to hold it in place. { 'split 'pin }

split-plot design [STAT] An experimental design that enables an additional factor or treatment to be included at more than one level; each plot is split into two or more parts. { 'split ,plät di,zīn }

split-ring core lifter [DES ENG] A hardened steel ring having an open slit, an outside taper, and an inside or outside serrated surface; in its expanded state it allows the core to pass through it freely, but when the drill string is lifted, the outside taper surface slides downward into the bevel of the bit or reaming shell, causing the ring to contract and grip tightly the core which it surrounds. Also known as core catcher; core gripper; core lifter; ring lifter; split-ring lifter; spring lifter. { 'split |riŋ 'kȯr ,lif·tər }

split-ring lifter See split-ring core lifter. { 'split |riŋ 'lif·tər }

split-ring mold [ENG] A plastics mold in which a split-cavity block is assembled in a chase to permit the forming of undercuts in a molded piece. { 'split |riŋ 'mōld }

split-ring piston packing [MECH ENG] A metal ring mounted on a piston to prevent leakage along the cylinder wall. { 'split |riŋ 'pis·tən ,pak·iŋ }

split screen See partitioned display. { 'split 'skrēn }

split shovel [DES ENG] A shovel containing parallel troughs separated by slots; used for sampling ground ore. { 'split 'shəv·əl }

split-stator variable capacitor [ELECTR] Variable capacitor having a rotor section that is common to two separate stator sections; used in grid and plate tank circuits of transmitters for balancing purposes. { 'split |stād·ər 'ver·ē·ə·bəl kə'pas·əd·ər }

splitter [CHEM ENG] A petroleum-refinery term for a fractionating tower that produces only an overhead and bottom stream. [SYST] A taxonomist who divides taxa very finely. { 'splid·ər }

splitter vanes [ENG] A group of curved, parallel vanes located in a sharp (for example, miter) bend of a gas conduit; the vane shape and its location help guide the moving gas around the bend. { 'splid·ər ,vānz }

splitting [ELECTR] In the scope presentation of the standard loran (2000 kilohertz), signals the slow diminution of the leading or lagging edge of the pulse so that it resembles two pulses and eventually a single pulse, which appears to be normal but which may be displaced in time by as much as 10,000 microseconds; this phenomenon is caused by shifting of the E_1 reflections from the ionosphere, and if the deformation is that of the leading edge and is not detected, it will cause serious errors in the reading of the navigational parameter. [MIN ENG] **1.** Lamina of mica with a maximum thickness of 0.0012 inch (30 micrometers), split from blocks and thins. **2.** One of a pair of horizontal level headings driven through a pillar, in pillar workings, in order to mine the pillar coal. { 'splid·iŋ }

splitting field See Galois field. { 'splid·iŋ ,fēld }

split transducer [ENG] A directional transducer with electroacoustic transducing elements which are divided and arranged so that there is an electrical separation of each division. { 'split tranz'dü·sər }

split vertical photography [GRAPHICS] Photographs taken simultaneously by two cameras mounted at an angle from the vertical, one tilted to the left and one to the right, to obtain a small side lap. { 'split 'vərd·ə·kəl phə'täg·rə·fē }

split-word operation [COMPUT SCI] A computer operation performed with portions of computer words rather than whole words as is normally done. { 'split |wərd ,äp·ə'rā·shən }

SP logging See spontaneous-potential well logging. { |es|pē 'läg·iŋ }

SPMD [COMPUT SCI] A type of programming on a multiprocessor in which parallel programs all run the same subroutine but operate on different data. Acronym for single-program, multiple-data.

spodic horizon [GEOL] A soil horizon characterized by illuviation of amorphous substances. { 'späd·ik hə'rīz·ən }

Spodosol [GEOL] A soil order characterized by accumulations of amorphous materials in subsurface horizons. { 'späd·ə,sȯl }

spodumene [MINERAL] $LiAlSi_2O_6$ A white to yellowish-, purplish-, or emerald-green clinopyroxene mineral occurring in prismatic crystals; hardness is 6.5–7 on Mohs scale, and specific gravity 3.13–3.20; an ore of lithium. Also known as triphane. { 'spä·jə,mēn }

spoil [MIN ENG] **1.** The overburden or nonore material from a coal mine. **2.** A stratum of coal and dirt mixed. { spȯil }

spoil bank [MIN ENG] **1.** In surface mining, the accumulation of overburden. **2.** The place where spoil is deposited. Also known as spoil heap. { 'spȯil ,baŋk }

spoil dam [MIN ENG] An earthen dike forming a depression, in which returns from a borehole can be collected and retained. { 'spȯil ,dam }

spoiler [AERO ENG] A plate, series of plates, comb, tube, bar, or other device that projects into the airstream about a body to break up or spoil the smoothness of the flow, especially such a device that projects from the upper surface of an airfoil, giving an increased drag and a decreased lift. [ELECTROMAG] Rod grating mounted on a parabolic reflector to change the pencil-beam pattern of the reflector to a cosecant-squared pattern; rotating the reflector and grating 90° with respect to the feed antenna changes one pattern to the other. { 'spȯi·lər }

spoil heap See spoil bank. { 'spȯil ,hēp }

spoke [DES ENG] A bar or rod radiating from the center of a wheel. { spōk }

spokeshave [ENG] A small tool for planing convex or concave surfaces. { 'spōk,shāv }

spondylitis [MED] Inflammation of the vertebrae. Also known as ankylosing spondylitis. { ,spän·də'līd·əs }

spondyloarthropathy [MED] Any of a group of arthritis disorders characterized by involvement of the axial (central) skeleton and association with the histocompatibility system. { ,spän·də·lō·är'thräp·ə·thē }

spondylolisthesis [MED] Forward displacement of a vertebra upon the one below as a result of a bilateral defect in the vertebral arch, or erosion of the articular surface of the posterior facets due to degenerative joint disease. { |spän·də·lō,lis 'thē·səs }

sponge [CHEM ENG] Wood shavings coated with iron oxide and used as a catalyst in processes for removing hydrogen sulfide from industrial gases. [INV ZOO] The common name for members of the phylum Porifera. { spənj }

sponge gold See cake of gold. { 'spənj |gōld }

sponge grease [MATER] Fibrous, spongy, soda-base grease. { 'spənj |grēs }

sponge iron [MET] Iron in porous or powder form made without fusion by heating iron ore in a reducing gas or with charcoal. { 'spənj |ī·ərn }

sponge metal [MET] Any porous metal made by decomposition or reduction of a compound without melting. { 'spənj |med·əl }

sponge rubber See rubber sponge. { 'spənj |rəb·ər }

spongework [GEOL] A pattern of small irregular interconnecting cavities on walls of limestone caves. { 'spənj,wərk }

spongiform encephalopathy See prion disease. { ,spän·jə,fȯrm in,sef·ə'läp·ə·thē }

Spongiidae [INV ZOO] A family of sponges of the order Dictyoceratida; members are encrusting, massive, or branching in form and have small spherical flagellated chambers which characteristically join the exhalant canals by way of narrow channels. { spən'jī·ə,dē }

Spongillidae [INV ZOO] A family of fresh- and brackish water sponges in the order Haplosclerida which are chiefly gray, brown, or white in color, and encrusting, massive, or branching in form. { spən'jil·ə,dē }

spongin [BIOCHEM] A scleroprotein, occurring as the principal component of skeletal fibers in many sponges. { 'spən·jən }

spongioblast [EMBRYO] A primordial cell arising from the ectoderm of the embryonic neural tube which differentiates to form the neuroglia, the ependymal cells, the neurolemma sheath cells, the satellite cells of ganglions, and Müller's fibers of the retina. { 'spən·jē·ō,blast }

spongiocyte [HISTOL] **1.** A neuroglia cell. **2.** A cell of the adrenal cortex which has a spongy appearance due to the solution of lipids during tissue preparation for microscopical examination. { 'spən·jē·ō₁sīt }

Spongiomorphida [PALEON] A small, extinct Mesozoic order of fossil colonial Hydrozoa in which the skeleton is a reticulum composed of perforate lamellae parallel to the upper surface and of regularly spaced vertical elements in the form of pillars. { ₁spən·jē·ō'mȯr·fə·də }

Spongiomorphidae [PALEON] The single family of extinct hydrozoans comprising the order Spongiomorphida. { ₁spən·jē·ō'mȯr·fə₁dē }

spongocoel [INV ZOO] The branching, internal cavity of a sponge, connected to the outside by way of the osculum. { 'spän₁gə₁sēl }

spongolite [GEOL] A rock or sediment composed chiefly of the remains of sponges. Also known as spongolith. { 'spän₁gə₁līt }

spongolith See spongolite. { 'spaŋ₁gə₁lith }

spongy [MECH ENG] Property of a robot whose end effector has high compliance, so that a small force applied to it results in a large motion. { 'spən₁jē }

spongy mesophyll [BOT] A system of loosely and irregularly arranged parenchymal cells with numerous intercellular spaces found near the lower surface in well-differentiated broad leaves. Also known as spongy parenchyma. { 'spən·jē 'mē·zō₁fil }

spongy parenchyma See spongy mesophyll. { 'spən·jē pə'reŋ·kə·mə }

sponson mount [ORD] A gun mount positioned on the sponson of a tank or combat vehicle; practically abandoned because of vulnerability and limited field of fire, although widely used in earlier tanks. { 'spän·sən ₁maùnt }

sponsor [COMMUN] The advertiser who pays part or all of the cost of a television or radio program. { 'spän·sər }

spontaneous [PHYS] Occurring without application of an external agency, because of the inherent properties of an object. { spän'tā·nē·əs }

spontaneous abortion [MED] An unexpected, premature expulsion of the fetus. Also known as miscarriage. { spän'tā·nē·əs ə'bȯr·shən }

spontaneous amputation [MED] **1.** Congenital amputation. **2.** Amputation not caused by external trauma or injury, as in ainhum. { spän'tā·nē·əs ₁am·pyə'tā·shən }

spontaneous combustion [CHEM] Ignition that can occur when certain materials such as tung oil are stored in bulk, resulting from the generation of heat, which cannot be readily dissipated; often heat is generated by microbial action. Also known as spontaneous ignition. [MECH ENG] See autoignition. { spän'tā·nē·əs kəm'bəs·chən }

spontaneous emission [ELECTROMAG] The emission of radiation from a system in an excited state at a rate that does not depend on the presence of external fields. { spän'tan·ē·əs i'mish·ən }

spontaneous fission [NUC PHYS] Nuclear fission in which no particles or photons enter the nucleus from the outside. { spän'tā·nē·əs 'fi·shən }

spontaneous generation See abiogenesis. { spän'tā·nē·əs ₁jen·ə'rā·shən }

spontaneous heating [CHEM] The slow reaction of material with atmospheric oxygen at ambient temperatures; liberated heat, if undissipated, accumulates so that in the presence of combustible substances a fire will result. { spän'tā·nē·əs 'hēd·iŋ }

spontaneous ignition See spontaneous combustion. { spän'tā·nē·əs ig'nish·ən }

spontaneous magnetization [ELECTROMAG] Magnetization which a substance possesses in the absence of an applied magnetic field. { spän'tā·nē·əs ₁mag·nə·də'zā·shən }

spontaneous mutation [GEN] A mutation that occurs spontaneously, that is, in an individual not specifically exposed to a known mutagen. { spän'tā·nē·əs myü'tā·shən }

spontaneous nucleation [METEOROL] The nucleation of a phase change of a substance without the benefit of any seeding nuclei within or otherwise in contact with that substance; examples of such systems are a pure vapor condensing to its pure liquid state, a pure liquid freezing to its pure solid state, and a pure solution crystallizing to yield pure solute crystals. { spän'tā·nē·əs ₁nü·klē'ā·shən }

spontaneous pneumothorax [MED] Air that has entered the chest cavity and caused lung collapse without being the result of injury or deliberate introduction of air; caused by an overexpansion of air in the lungs. { spän'tā·nē·əs ₁nü·mō 'thȯr₁aks }

spontaneous polarization [ELEC] Electric polarization that a substance possesses in the absence of an external electric field. { spän'tā·nē·əs ₁pō·lə·rə'zā·shən }

spontaneous-potential well logging [ENG] The recording of the natural electrochemical and electrokinetic potential between two electrodes, one above the other, lowered into a drill hole; used to detect permeable beds and their boundaries. Also known as SP logging. { spän'tā·nē·əs pə'ten·chəl ₁wel ₁läg·iŋ }

spontaneous process [THERMO] A thermodynamic process which takes place without the application of an external agency, because of the inherent properties of a system. { spän'tā·nē·əs 'prä·səs }

spontaneous symmetry breaking [PHYS] A situation in which the solution of a set of physical equations fails to exhibit a symmetry possessed by the equations themselves; an example is a magnet, in which the underlying equations describing the metal do not distinguish any direction of space from any other, but the magnet certainly does, since it points in some definite direction. { spän'tā·nē·əs 'sim·ə·trē ₁brāk·iŋ }

spoofing [ELECTR] Deceiving or misleading the enemy in electronic operations, as by continuing transmission on a frequency after it has been effectively jammed by the enemy, using decoy radar transmitters to lead the enemy into a useless jamming effort, or transmitting radio messages containing false information for intentional interception by the enemy. { 'spüf·iŋ }

spool [GRAPHICS] A flanged cylinder in a camera on which unprocessed roll film is wound. [MECH ENG] **1.** The drum of a hoist. **2.** The movable part of a slide-type hydraulic valve. [TEXT] A cylinder made of wood or other material on which yarn is held during spinning or weaving. { spül }

spooling [COMPUT SCI] The temporary storage of input and output on high-speed input-output devices, typically magnetic disks and drums, in order to increase throughput. Acronym for simultaneous peripheral operations on line. { 'spül·iŋ }

spool-type roller conveyor [MECH ENG] A type of roller conveyor in which the rolls are of conical or tapered shape with the diameter at the ends of the roll larger than that at the center. { 'spül ₁tīp 'rō·lər kən₁vā·ər }

spoon [DES ENG] A slender rod with a cup-shaped projection at right angles to the rod, used for scraping drillings out of a borehole. [MIN ENG] An instrument in which earth or pulp may be delicately tested by washing to detect gold or amalgam. { spün }

spoon nail See koilonychia. { 'spün ₁nāl }

sporadic E layer [GEOPHYS] A layer of intense ionization that occurs sporadically within the E layer; it is variable in time of occurrence, height, geographical distribution, penetration frequency, and ionization density. { spə'rad·ik 'ē ₁lā·ər }

sporadic fault [COMPUT SCI] A hardware malfunction that occurs intermittently and at unpredictable times. { spə'rad·ik 'fȯlt }

sporadic meteor [ASTRON] A meteor which is not associated with one of the regularly recurring meteor showers or streams. { spə'rad·ik 'mēd·ē·ər }

sporadic reflections [ELECTROMAG] Sharply defined reflections of substantial intensity from the sporadic E layer at frequencies greater than the critical frequency of the layer; they are variable with respect to time of occurrence, geographic location, and range of frequencies at which they are observed. { spə'rad·ik ri'flek·shənz }

sporadic simple group [MATH] A simple group which cannot be classified in any known infinite family of simple groups. { spə'rad·ik ₁sim·pəl ₁grüp }

sporangiolum [MYCOL] A small sporangium that has few sporangiospores. { spə₁ran·jē'ōl·əm }

sporangiophore [BOT] A stalk or filament on which sporangia are borne. { spə'ran·jē·ə₁fȯr }

sporangiospore [BOT] A spore that forms in a sporangium. { spə'ran·jē·ə₁spȯr }

sporangium [BOT] A case in which asexual spores are formed and borne. { spə'ran·jē·əm }

spore [BIOL] A uni- or multicellular, asexual, reproductive or resting body that is resistant to unfavorable environmental

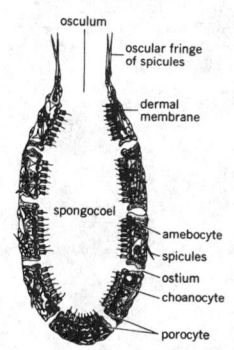

SPONGOCOEL

osculum
oscular fringe of spicules
dermal membrane
spongocoel
amebocyte
spicules
ostium
choanocyte
porocyte

Morphology of asconoid calcareous sponge—longitudinal section.

conditions and produces a new vegetative individual when the environment is favorable. { 'spòr }

spore mother cell [BOT] One of the cells of the archespore of a spore-bearing plant from which a spore, but usually a tetrad of spores, is produced. Also known as sporocyte. { 'spòr 'məth·ər ,sel }

Spörer minimum [ASTRON] A period of low sunspot activity that occurred between 1420 and 1570. { 'spər·ər 'min·ə·məm }

Spörer's law [ASTRON] A relationship to indicate the frequency of occurrence of sunspots and their progressive movement to lower latitudes on the sun. { 'spər·ərz ,lò }

spore stain [MICROBIO] A type of staining technique that utilizes the fact that the spore does not take up dyes readily but, once stained, it resists decolorization. The differentiating agent used may be a dilute solution of an organic acid, an acid dye, or another basic dye. { 'spòr ,stān }

sporidium [MYCOL] A small spore, especially one formed on a promycelium. { spə'rid·ē·əm }

sporinite [GEOL] A variety of exinite composed of spore exines which have been compressed parallel to the stratification. { 'spòr·ə,nīt }

sporoblast [INV ZOO] A sporozoan cell from which sporozoites arise. { 'spòr·ə,blast }

Sporobolomycetaceae [MYCOL] The single family of the order Sporobolomycetales. { spə,räb·ə·lō,mī·sə'tās·ē,ē }

Sporobolomycetales [MYCOL] An order of yeastlike and moldlike fungi assigned to the class Basidiomycetes characterized by the formation of sterigmata, upon which the asexual ballistospores are formed. { spə,räb·ə·lō,mī·sə'tā·lēz }

sporocarp [BOT] Any multicellular structure in or on which spores are formed. { 'spòr·ə,kärp }

sporocyst [BOT] A unicellular resting body from which asexual spores arise. [INV ZOO] **1.** A resistant envelope containing an encysted sporozoan. **2.** An encysted sporozoan. **3.** The first reproductive form of a digenetic trematode in which rediae develop. { 'spòr·ə,sist }

sporocyte *See* spore mother cell. { 'spòr·ə,sīt }

sporogenesis [BIOL] **1.** Reproduction by means of spores. **2.** Formation of spores. { ¦spòr·ə'jen·ə·səs }

sporogony [BIOL] Reproduction by means of spores. [INV ZOO] Propagative reproduction involving formation, by sexual processes, and subsequent division of a zygote. { spə'räg·ə·nē }

sporont [INV ZOO] A stage in the life history of sporozoans which gives rise to spores. { 'spòr,änt }

sporophore [MYCOL] A structure on the thallus of fungi which produces spores. { 'spòr·ə,fòr }

sporophyll [BOT] A modified leaf that develops sporangia. { 'spòr·ə,fil }

sporophyte [BOT] **1.** An individual of the spore-bearing generation in plants exhibiting alternation of generation. **2.** The spore-producing generation. **3.** The diplophase in a plant life cycle. { 'spòr·ə,fīt }

sporopollenin [BIOCHEM] A substance related to suberin and cutin but more resistant to decay that is found in the exine of pollen grains. { ¦spòr·ō'päl·ə·nən }

sporosac [INV ZOO] A degenerate gonophore in certain hydroid cnidarians. { 'spòr·ə,sak }

sporotrichosis [MED] A granulomatous fungus disease caused by *Sporotrichum schenckii,* with cutaneous lesions along the lymph channels and occasionally involving the internal organs. Also known as de Beurmann-Gougerot disease; Schenk's disease. { ,spòr·ə·tri'kō·səs }

Sporozoa [INV ZOO] A subphylum of parasitic Protozoa, typically producing spores during the asexual stages of the life cycle. { ,spòr·ə'zō·ə }

sporozoite [INV ZOO] A motile, infective stage of certain sporozoans, which is the result of sexual reproduction and which gives rise to an asexual cycle in the new host. { ,spòr·ə'zō,īt }

sporting gun [ORD] A classification of shotguns which includes those with 30-inch (762-millimeter) full choke barrels. { 'spòrd·iŋ ,gən }

sports medicine [MED] A branch of medicine concerned with the effects of exercise and sports on the human body, including treatment of injuries. { 'spòrts ,med·ə·sən }

sporulation [BIOL] The act and process of spore formation. { ,spòr·yə'lā·shən }

spot [ELECTR] In a cathode-ray tube, the area instantaneously affected by the impact of an electron beam. [ORD] **1.** To determine, by observation, deviations of ordnance from the target for the purpose of supplying necessary information for adjustment of fire. **2.** To place ordnance in a proper location. **3.** To locate or espy something, as an aircraft or troop concentration. { spät }

spot beam [COMMUN] A beam generated by a communications satellite antenna of sufficient size that the angular spread of energy in the beam is small, always smaller than the earth's angular beamwidth as seen from the satellite. { 'spät ,bēm }

spot blight *See* grease spot. { 'spät ,blīt }

spot blotch [PL PATH] A fungus disease of barley caused by *Helminthosporium sativum* and characterized by the appearance of dark, elongated spots on the foliage. { 'spät ,bläch }

spot carbon [GRAPHICS] Carbon paper that is carbonized only in certain areas so that only the corresponding areas of the original will be reproduced. { 'spät ,kär·bən }

spot check [IND ENG] A check or inspection of certain steps in an operation, process, or the like, of certain parts of a piece of equipment or of a representative lot of completed parts or articles; the steps or parts inspected would normally be only a small percentage of the total. { 'spät ,chek }

spot drilling [MECH ENG] Drilling a small hole or indentation in the surface of a material to serve as a centering guide in later machining operations. { 'spät ,dril·iŋ }

spot elevation [MAP] Elevation of a point on a map or chart, usually indicated by a dot accompanied by a number indicating the vertical distance of the point from the reference datum; spot elevations are used principally to indicate points higher than their surroundings. { 'spät ,el·ə,vā·shən }

spot facing [MECH ENG] A finished circular surface around the top of a hole to seat a bolthead or washer, or to allow flush mounting of mating parts. { 'spät ,fās·iŋ }

spot film [MED] A small, highly collimated radiograph of an anatomic part, usually obtained in conjunction with fluoroscopy. { 'spät ,film }

spot gluing [ENG] Applying heat to a glued assembly by dielectric heating to make the glue set in spots that are more or less regularly distributed. { 'spät ,glü·iŋ }

spot group [ASTROPHYS] A complex formation of the sun's surface consisting of a sunspot with several umbrae surrounded by a single penumbra, or of several sunspots which are close together and clearly associated. { 'spät ,grüp }

spot hover [AERO ENG] To remain stationary relative to a point on the ground while airborne. { 'spät ,hòv·ər }

spot jammer [ELECTR] A jammer that interferes with reception of a specific channel or frequency. { 'spät ,jam·ər }

spotlight [ELEC] **1.** A strong beam of light that illuminates only a small area about an object. **2.** A lamp that has a strongly focused beam. { 'spät,līt }

spot noise figure [ELECTR] Of a transducer at a selected frequency, the ratio of the output noise power per unit bandwidth to a portion thereof attributable to the thermal noise in the input termination per unit bandwidth, the noise temperature of the input termination being standard (290 K). { 'spät 'nòiz ,fig·yər }

spot punch [COMPUT SCI] A hand-operated device resembling a pair of pliers, for selectively punching holes in punch cards. { 'spät ,pənch }

spot-size error [ELECTR] The distortion of the radar returns on the radarscope presentation caused by the diameter of the electron beam which displays the returns of the scope and the lateral radiation across the scope of part of the glow produced when the electron beam strikes the phosphorescent coating of the cathode-ray tube. { 'spät ¦sīz ,er·ər }

spot speed [COMMUN] **1.** In television, the product of the length (in units of elemental area, that is, in spots) of scanning line by the number of scanning lines per second. **2.** In facsimile transmission, the speed of the scanning or recording spot within the available line. Also known as scanning speed. { 'spät ,spēd }

spotted phyllite [PETR] A phyllite rock containing dark spots that represent the beginning of porphyroblast development. { 'späd·əd 'fī,līt }

spotted slate [PETR] A type of slate containing dark spots that represent the beginning of porphyroblast development. { 'späd·əd 'slāt }

spotted wilt [PL PATH] A virus disease of various crop and wild plants, especially tomato, characterized by bronzing and downward curling of the leaves. { 'späd·əd 'wilt }

spotter *See* dotter. [ORD] An observer stationed for the purpose of observing and reporting results of naval gunfire to the firing agency, or in designating targets. { 'späd·ər }

spotter-tracer [ORD] In cartridge nomenclature, indicating that the bullet or projectile is equipped with a tracer and contains a filler suitable for spotting purposes. { 'späd·ər ¦trā·sər }

spot test [ANALY CHEM] The addition of a drop of reagent to a drop or two of sample solution to obtain distinctive colors or precipitates; used in qualitative analysis. { 'spät ‚test }

spottiness [ELECTR] Bright spots scattered irregularly over the reproduced image in a television receiver, due to man-made or static interference entering the television system at some point. { 'späd·ē·nəs }

spotting [ENG] Fitting one part of a die to another part by applying an oil color to the surface of the finished part and bringing this against the surface of the intended mating part, the high spots being marked by the transferred color. [MIN ENG] Bringing mine cars or surface wagons to the correct spot for loading, discharging, or any other purpose. { 'späd·iŋ }

spotting board [ORD] Device for determining the direction and size of deviations from the target; it converts the readings of spotters into usable form for firing data. { 'späd·iŋ ‚bȯrd }

spotting hoist [MIN ENG] A small haulage engine used for bringing mine cars into the correct position under a loading chute or feeder or some other point. { 'späd·iŋ ‚hȯist }

spotting line [ORD] Either the gun-target line, the observer-target line, or a reference line used by the observer or spotter in making spot corrections. { 'späd·iŋ ‚līn }

spotting pistol [ORD] A short automatic or semiautomatic firearm mounted coaxially on a larger-caliber gun, designed to conserve and to increase first-round hit probability of the ammunition used in the larger weapon; it employs a magazine and fires a spotter-tracer projectile which is ballistically matched with the trajectory of the projectile of the gun on which it is mounted. { 'späd·iŋ ‚pis·təl }

spotting rifle [ORD] An auxiliary item mounted coaxially on a larger-caliber gun used to assist a gunner in determining range; it is usually a magazine-fed, semiautomatic gun with a rifled barrel that utilizes ammunition incorporating a tracer element providing a smoke puff on impact. { 'späd·iŋ ‚rī·fəl }

spotty ore [MIN ENG] Ore in which the valuable material is concentrated irregularly as small particles. { 'späd·ē 'ȯr }

spot welding [MET] Resistance welding in which fusion is localized in small circular areas; sometimes also accomplished by various arc-welding processes. { 'spät ‚weld·iŋ }

spot wind [METEOROL] In air navigation, wind direction and speed, either observed or forecast if so specified, at a designated altitude over a fixed location. { 'spät ‚wind }

spot wobble [COMMUN] The process of giving the scanning spot of a television screen a small periodic motion transverse to the scanning lines at a frequency above the picture signal spectrum. { 'spät ‚wäb·əl }

spout hole [VERT ZOO] **1.** A blowhole of a cetacean mammal. **2.** A nostril of a walrus or seal. { 'spaut ‚hōl }

spouting [ENG] A term used in the feeding or ejection of powdered or granulated solids by means of vertical or slanted discharge spouts. { 'spaud·iŋ }

spouting horn [GEOL] A sea cave with a rearward or upward opening through which water spurts or sprays after waves enter the cave. Also known as chimney; oven. { 'spaud·iŋ ‚hȯrn }

sprag [ENG] A stake used as a brake for a vehicle by inserting it through the spokes of a wheel or digging it into the ground at an angle. [MIN ENG] A prop supporting the roof or ore in a mine. { sprag }

sprag clutch [MECH ENG] A clutch designed to transmit power in one direction only. { 'sprag ‚kləch }

spragger [MIN ENG] In coal mining, a laborer who rides trains of cars and controls their free movement down gently sloping inclines by throwing switches and by poking sprags between the wheel spokes to stop them. { 'sprag·ər }

sprag road [MIN ENG] A road so steep that sprags must be used on the wheels of ore cars during descent. { 'sprag ‚rōd }

sprain [MED] A wrenching of a joint, producing a stretching or laceration of the ligaments. { sprān }

sprain fracture [MED] An injury in which a tendon or ligament, together with a shell of bone, is torn from its attachment; occurs most commonly at the ankle. { 'sprān ‚frak·chər }

spray [ASTROPHYS] An explosive release of gas in all directions from the sun's chromosphere, with velocities as high as 930 miles (1500 kilometers) per second, which normally occurs in the first minutes of a flare. [ENG] A mechanically produced dispersion of liquid into a gas stream; as drops are large, the spray is unstable and the liquid will fall free of the gas stream when velocity decreases. { 'sprā }

spray angle [FL MECH] The angle formed by the cone of liquid leaving a nozzle orifice. { 'sprā ‚aŋ·gəl }

spray chamber [MECH ENG] A compartment in an air conditioner where humidification is conducted. { 'sprā ‚chām·bər }

spray dryer [MECH ENG] A machine for drying an atomized mist by direct contact with hot gases. { 'sprā ‚drī·ər }

sprayed metal mold [ENG] A plastics mold made by spraying molten metal onto a master form until a shell of predetermined thickness is achieved; the shell is then removed and backed up with plaster, cement, or casting resin; used primarily in plastic sheet forming. { 'sprād ‚med·əl ‚mōld }

sprayer plate [ENG] A rotating flat-faced or dished metal plate used in an oil burner to enhance atomization. { 'sprā·ər ‚plāt }

spray flow [FL MECH] A two-phase flow in which the liquid phase is the dispersed phase and exists in the form of many droplets, while the gas phase is the continuous phase. { 'sprā ‚flō }

spray gun [MECH ENG] An apparatus shaped like a gun which delivers an atomized mist of liquid. { 'sprā ‚gən }

spray nozzle [MECH ENG] A device in which a liquid is subdivided to form a stream (mist) of small drops. { 'sprā ‚näz·əl }

spray oil [MATER] A low-viscosity petroleum oil similar to lubricating oil; used to combat pests that attack trees and shrubbery. { 'sprā ‚ȯil }

spray painting [ENG] Applying a fine, even coat of paint by means of a spray nozzle. { 'sprā ‚pānt·iŋ }

spray point [ELEC] One of the sharp points arranged in a row and charged to a high direct-current potential, used to charge and discharge the conveyor belt in a Van de Graaff generator. { 'sprā ‚pȯint }

spray pond [ENG] An arrangement for cooling large quantities of water in open reservoirs or ponds; nozzles spray a portion of the water into the air for the evaporative cooling effect. { 'sprā ‚pänd }

spray probe [ENG] A device which detects a jet spray of tracer gas in vacuum testing for leaks. { 'sprā ‚prōb }

spray quenching [MET] Rapid cooling in a spray of water or oil. { 'sprā ‚kwench·iŋ }

spray region *See* fringe region. { 'sprā ‚rē·jən }

spray torch [ENG] In thermal spraying, a device used for the application of self-fluxing alloys; molten metal is propelled against the substrate by a stream of air and gas. { 'sprā ‚tȯrch }

spray tower [CHEM ENG] A vertical column, at the top of which is a liquid spray device; used to contact liquids with gas streams for absorption, humidification, or drying. { 'sprā ‚tau·ər }

spray transfer [MET] In arc welding, transfer of filler metal across the arc to the workpiece in the form of droplets. { 'sprā ‚tranz·fər }

spray-up [ENG] A term for a number of techniques in which a spray gun is used as the processing tool; for example, in reinforced plastics manufacture, fibrous glass and resin can simultaneously be spray-deposited into a mold or onto a form. { 'sprā‚əp }

spread [ENG] The layout of geophone groups from which data from a single shot are recorded simultaneously. [STAT] **1.** The range within which the values of a variable quantity occur. **2.** *See* sensitivity. { spred }

spreadable life *See* pot life. { 'spred·ə·bəl ‚līf }

spreader [CIV ENG] A wood or steel member inserted temporarily between form walls to keep them apart. [ELEC] An insulating crossarm used to hold apart the wires of a transmission line or multiple-wire antenna. [MECH ENG] **1.** A tool used in sharpening machine drill bits. **2.** A machine which spreads dumped material with its blades. [MIN ENG] **1.** A horizontal timber below the cap of a set, used to stiffen the legs, and to support the brattice when there are two air courses in the same gangway. **2.** A piece of timber stretched across a shaft as a temporary support of the walls. { 'spred·ər }

spreader beam [ENG] A rigid beam hanging from a crane

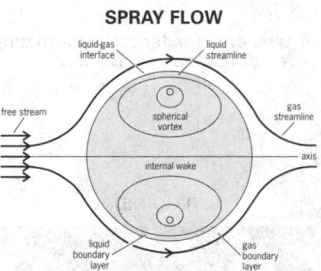

SPRAY FLOW

Cross section of a vaporizing droplet, showing the relative gas-droplet motion and internal circulation.

SPREADER STOKER

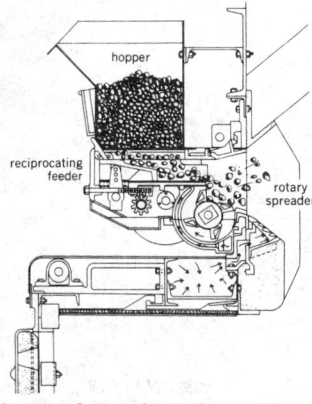

A type of spreader stoker utilizing a reciprocating coal feeder.

SPRIGGINIDAE

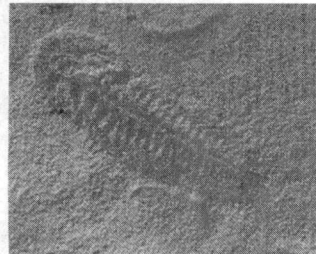

Spriggina floundersi fossil showing characteristic horseshoe-shaped head.

SPRING GRAVIMETER

Principle of a spring gravimeter. The variable length of the spring is *s*.

hook and fitted with a number of ropes at different points along its length; employed for such purposes as lifting reinforced concrete piles or large sheets of glass. { 'spred·ər ‚bēm }

spreader stoker [MECH ENG] A coal-burning system where mechanical feeders and distributing devices form a thin fuel bed on a traveling grate, intermittent-cleaning dump grate, or reciprocating continuous-cleaning grate. { 'spred·ər ‚stōk·ər }

spread footing [CIV ENG] A wide, shallow footing usually made of reinforced concrete. { 'spred ‚fùd·iŋ }

spreading anomaly [PHYS] That part of the propagation anomaly which may be identified with the geometry of the ray paths. { 'spred·iŋ ə‚näm·ə·lē }

spreading coefficient [THERMO] The work done in spreading one liquid over a unit area of another, equal to the surface tension of the stationary liquid, minus the surface tension of the spreading liquid, minus the interfacial tension between the liquids. { 'spred·iŋ ‚kō·i‚fish·ənt }

spreading concept See sea-floor spreading. { 'spred·iŋ ‚kän‚sept }

spreading factor See hyaluronidase. { 'spred·iŋ ‚fak·tər }

spreading fire [ORD] A notification by the spotter or the naval gunfire ship, depending on who is controlling the fire, to indicate that fire is about to be distributed over an area. { 'spred·iŋ ‚fīr }

spreading-floor hypothesis See sea-floor spreading. { 'spred·iŋ ‚flór hī‚päth·ə·səs }

spreading method [ELEC] A method of calculating the potential due to a set of point charges by replacing them with a continuous distribution of charge or a distribution of charge and polarization. { 'spred·iŋ ‚meth·əd }

spreading of inactivation [GEN] A type of position effect, as in the inactivation of nearby autosomal genes in an X-autosome translocation chromosome in mammalian females heterozygous for the translocation. { ‚spred·iŋ əv in‚ak·tə'vā·shən }

spread reflection [ELECTROMAG] Reflection of electromagnetic radiation from a rough surface with large irregularities. Also known as mixed reflection. { 'spred ri‚flek·shən }

spread spectrum transmission [ELECTR] Communications technique in which many different signal waveforms are transmitted in a wide band; power is spread thinly over the band so narrow-band radios can operate within the wide-band without interference; used to achieve security and privacy, prevent jamming, and utilize signals buried in noise. { 'spred ‚spek·trəm tranz‚mish·ən }

Sprengel pump [MECH ENG] An air pump that exhausts by trapping gases between drops of mercury in a tube. { 'spreŋ·gəl ‚pəmp }

sprig [DES ENG] A small brad having no head. [ENG] See glazier's point. { sprig }

Sprigginidae [INV ZOO] An extinct family of annelid worms distinguished by a horseshoe-shaped head. { spri'gin·ə‚dē }

spring [ASTRON] The period extending from the vernal equinox to the summer solstice; comprises the transition period from winter to summer. [ENG] To enlarge the bottom of a drill hole by small charges of a high explosive in order to make room for the full charge; to chamber a drill hole. [HYD] A general name for any discharge of deep-seated, hot or cold, pure or mineralized water. [MECH ENG] An elastic, stressed, stored-energy machine element that, when released, will recover its basic form or position. Also known as mechanical spring. { spriŋ }

springback [MET] **1.** Return of a metal part to its original shape after release of stress. **2.** The degree to which a metal returns to its original shape after forming operations. **3.** In flash, upset or pressure welding, the deflection in the welding machine caused by the upset pressure. { 'spriŋ‚bak }

spring balance [ENG] An instrument which measures force by determining the extension of a helical spring. { 'spriŋ ‚bal·əns }

spring bearings [NAV ARCH] Bearings designed to take the weight of the propeller shaft inside a vessel. { 'spriŋ ‚ber·iŋz }

spring bolt [DES ENG] A bolt which must be retracted by pressure and which is shot into place by a spring when the pressure is released. { 'spriŋ ‚bōlt }

spring box mold [ENG] A compression mold with a spacing fork that is removed after partial compression. { 'spriŋ ‚bäks ‚mōld }

spring brass [MET] Common brass containing 70–72% copper which has been cold-worked to make it stiffer. { 'spriŋ ‚bras }

spring buffer [ENG] A buffer in the form of a spring that stores and dissipates the kinetic energy of an impact. { 'spriŋ ‚bəf·ər }

spring calipers [ENG] Calipers in which tension against the adjusting nut is maintained by a circular spring. { 'spriŋ ‚kal·ə‚pərz }

spring clip [DES ENG] **1.** A U-shaped fastener used to attach a leaf spring to the axle of a vehicle. **2.** A clip that grips an inserted part under spring pressure; used for electrical connections. { 'spriŋ ‚klip }

spring clock [HOROL] A clock containing a spring that supplies the energy to operate the clock mechanism. { 'spriŋ ‚kläk }

spring collet [DES ENG] A bushing that surrounds and holds the end of the work in a machine tool; the bushing is slotted and tapered, and when the collet is slipped over it, the slot tends to close and the bushing thereby grips the work. { 'spriŋ ‚käl·ət }

spring contact [ELEC] A relay or switch contact mounted on a flat spring, usually of phosphor bronze. { 'spriŋ ‚kän‚takt }

spring cotter [DES ENG] A cotter made of an elastic metal that has been bent double to form a split pin. { 'spriŋ ‚käd·ər }

spring coupling [MECH ENG] A flexible coupling with resilient parts. { 'spriŋ ‚kəp·liŋ }

spring crust [HYD] A type of snow crust, formed when loose firn is recemented by a decrease in temperature; it is most common in late winter and spring. { 'spriŋ ‚krəst }

spring die [DES ENG] An adjustable die consisting of a hollow cylinder with internal cutting teeth, used for cutting screw threads. { 'spriŋ ‚dī }

spring equinox See vernal equinox. { 'spriŋ 'ē·kwə‚näks }

spring faucet [ENG] A faucet that is kept closed by a spring; force must be exerted to open it, and it closes when the force is removed. { 'spriŋ ‚fós·ət }

spring gravimeter [ENG] An instrument for making relative measurements of gravity; the elongation s of the spring may be considered proportional to gravity g, $s = (1/k)g$, and the basic formula for relative measurements is $g_2 - g_1 = k(s_2 - s_1)$. { 'spriŋ grə'vim·əd·ər }

spring hammer [MECH ENG] A machine-driven hammer actuated by a compressed spring or by compressed air. { 'spriŋ ‚ham·ər }

spring high water See mean high-water springs. { 'spriŋ 'hī ‚wód·ər }

spring hinge [DES ENG] A hinge fitted with one or more springs. { 'spriŋ ‚hinj }

spring hook [DES ENG] A hook closed at the end by a spring snap. Also known as snap hook. { 'spriŋ ‚hùk }

spring-joint caliper [DES ENG] An outside or inside caliper having a heavy spring joining the legs together at the top; legs are opened and closed by a knurled nut. { 'spriŋ ‚jóint ‚kal·ə·pər }

spring lifter See split-ring core lifter. { 'spriŋ ‚lif·tər }

spring-lifter case See lifter case. { 'spriŋ ‚lif·tər ‚kās }

spring line [NAV ARCH] A hawser run out from any part of a vessel to a point on shore, as a dock, to prevent the vessel from going ahead or astern. { 'spriŋ ‚līn }

spring-load [ENG] To load or exert a force on an object by means of tension from a spring or by compression. { 'spriŋ ‚lōd }

spring-loaded meter [ENG] A variable-area flowmeter in which the force on an obstruction in a tapered tube created by the fluid flowing past the obstruction is balanced by the force of a spring to which the obstruction is attached, and the resulting differential pressure is used to determine the flow rate. { 'spriŋ ‚lōd·əd ‚mēd·ər }

spring-loaded regulator [MECH ENG] A pressure-regulator valve for pressure vessels or flow systems; the regulator is preloaded by a calibrated spring to open (or close) at the upper (or lower) limit of a preset pressure range. { 'spriŋ ‚lōd·əd 'reg·yə‚lād·ər }

spring low water See mean low-water springs. { 'spriŋ 'lō ‚wód·ər }

spring modulus [MECH] The additional force necessary to deflect a spring an additional unit distance; if a certain spring has a modulus of 100 newtons per centimeter, a 100-newton weight will compress it 1 centimeter, a 200-newton weight 2 centimeters, and so on. { 'spriŋ 'mäj·ə·ləs }

spring needle *See* bearded needle. { 'spriŋ ‚nēd·əl }

spring pin [MECH ENG] An iron rod which is mounted between spring and axle on a locomotive, and which maintains a regulated pressure on the axle. { 'spriŋ ‚pin }

spring range [OCEANOGR] The mean semidiurnal range of tide when spring tides are occurring; the mean difference in height between spring high water and spring low water. Also known as mean spring range. { 'spriŋ 'rānj }

spring rise [OCEANOGR] The height of mean high-water springs above the chart datum. { 'spriŋ 'rīz }

spring scale [ENG] A scale that utilizes the deflection of a spring to measure the load. { 'spriŋ ‚skāl }

spring seepage [HYD] A spring of small discharge. Also known as weeping spring. { 'spriŋ ‚sēp·ij }

spring shackle [ENG] A shackle for supporting the end of a spring, permitting the spring to vary in length as it deflects. { 'spriŋ ‚shak·əl }

spring sludge *See* rotten ice. { 'spriŋ 'sləj }

spring snow [HYD] A coarse, granular snow formed during spring by alternate freezing and thawing. Also known as corn snow. { 'spriŋ 'snō }

spring steel [MET] Carbon or low-alloy steel which can be processed to give it the hardness and yield strength needed in springs. { 'spriŋ ‚stēl }

spring stop-nut locking fastener [DES ENG] A locking fastener that functions by a spring action clamping down on the bolt. { 'spriŋ ‚stäp‚nət 'läk·iŋ ‚fas·nər }

spring switch [CIV ENG] A railroad switch that contains a spring to return it to the running position after it has been thrown over by trailing wheels moving on the diverging route. { 'spriŋ ‚swich }

spring temper [MET] **1.** A steel temper characterized by an increased upper limit of elasticity; obtained by hardening and tempering in the usual way, then reheating until the steel turns blue. **2.** A similar temper in brass obtained by cold rolling. { 'spriŋ ‚tem·pər }

spring tidal currents [OCEANOGR] Tidal currents of increased speed occurring at the time of spring tides. { 'spriŋ 'tīd·əl ‚kə·rəns }

spring tide [OCEANOGR] Tide of increased range which occurs about every 2 weeks when the moon is new or full. { 'spriŋ 'tīd }

spring-tooth harrow [AGR] An implement pulled over plowed soil that has long curved teeth of spring steel, used to break clods, level the surface, and destroy weeds. { ‚spriŋ ‚tüth 'har·ō }

spring velocity [OCEANOGR] The average speed of the maximum flood and maximum ebb of a tidal current at the time of spring tides. { 'spriŋ və'läs·əd·ē }

springwood [BOT] The portion of an annual ring that is formed principally during the growing season; it is softer, more porous, and lighter than summerwood because of its higher proportion of large, thin-walled cells. { 'spriŋ‚wu̇d }

sprinkle [METEOROL] A very light shower of rain. { 'spriŋ·kəl }

sprinkler irrigation [AGR] A method of providing water to plants by pipelines which carry water under pressure from a pump or elevated source to lines, with sprinkler heads spaced at appropriate intervals. { 'spriŋ·klər ‚ir·i'gā·shən }

sprinkler system [ENG] A fire-protection system of pipes and outlets in a building, mine, or other enclosure for delivering a fire extinguishing liquid or gas, usually automatically by the action of heat on the sprinkler head. Also known as fire sprinkling system. { 'spriŋk·lər ‚sis·təm }

Sprint [ORD] A United States guided, surface-to-air, high-acceleration, antimissile missile with nuclear warhead capability employed in the Nike X system. { sprint }

sprite [METEOROL] A transient illumination that can appear over a laterally extensive thunderstorm, with a red body about 20 kilometers (12 miles) in diameter, extending up to an altitude of 85–90 kilometers (51–54 miles), and blue tendrils extending down to an altitude of about 45 kilometers (27 miles). { sprīt }

s-process [NUC PHYS] The synthesis of elements, predominantly in the iron group, over long periods of time through the capture of slow neutrons which are produced mainly by the reactions of α-particles with carbon-13 and neon-21. { 'es ‚prä·səs }

sprocket [DES ENG] A tooth on the periphery of a wheel or cylinder to engage in the links of a chain, the perforations of a motion picture film, or other similar device. { 'spräk·ət }

sprocket chain [MECH ENG] A continuous chain which meshes with the teeth of a sprocket and thus can transmit mechanical power from one sprocket to another. { 'spräk·ət ‚chān }

sprocket hole [ENG] One of a series of perforations at the edge of a motion picture film, paper tape, or roll of continuous stationery, which are engaged by the teeth of a sprocket wheel to drive the material through some device. { 'spräk·ət ‚hōl }

sprocket pulse [COMPUT SCI] **1.** A pulse generated by a magnetized spot which accompanies every character recorded on magnetic tape; this pulse is used during read operations to regulate the timing of the read circuits, and also to provide a count on the number of characters read from the tape. **2.** A pulse generated by the sprocket or driving hole in paper tape which serves as the timing pulse for reading or punching the paper tape. { 'spräk·ət ‚pəls }

sprocket wheel [DES ENG] A wheel with teeth or cogs, used for a chain drive or to engage the blocks on a cable. { 'spräk·ət ‚wēl }

spruce [BOT] An evergreen tree belonging to the genus *Picea* characterized by single, four-sided needles borne on small peglike projections, pendulous cones, and resinous wood. { sprüs }

spruce budworm [INV ZOO] The larva of a common moth, *Choristoneura fumiferana*, that is a destructive pest primarily of spruce and balsam fir. { ‚sprüs 'bəd‚wərm }

spruce oil [MATER] A combustible, colorless to yellow essential oil, soluble in fixed and mineral oils; derived from spruce needles and branches; the main components are bornyl acetate, cadinene, and pinene; used in flavors and veterinary liniments, and as an odorant for soaps and cosmetics. Also known as hemlock oil. { 'sprüs ‚oil }

spruce sulfite extract [MATER] A paper-manufacture byproduct from the sulfite-pulping process; used as a foundry core binder and road binder, and in tanning. { 'sprüs 'səl‚fīt 'ek‚strakt }

sprue [ENG] **1.** A feed opening or vertical channel through which molten material, such as metal or plastic, is poured in an injection or transfer mold. **2.** A slug of material that solidifies in the channel. [MED] A syndrome characterized by impaired absorption of food, water, and minerals by the small intestine; symptoms are the result of nutritional deficiencies. { sprü }

sprue bushing [ENG] A steel insert in an injection mold which contains the sprue hole and has a seat for the injection cylinder nozzle. { 'sprü ‚bu̇sh·iŋ }

sprue gate [ENG] A passageway for the flow of molten resin from the nozzle to the mold cavity. { 'sprü ‚gāt }

sprue puller [ENG] A pin with a Z-shaped slot to pull the sprue out of the sprue bushing in an injection mold. { 'sprü ‚pu̇l·ər }

sprung axle [MECH ENG] A supporting member for carrying the rear wheels of an automobile. { 'sprəŋ ‚ak·səl }

sprung weight [MECH ENG] The weight of a vehicle which is carried by the springs, including the frame, radiator, engine, clutch, transmission, body, load, and so forth. { 'sprəŋ ‚wāt }

SPSL *See* sound pressure spectrum level.

SPST *See* single-pole single-throw.

spud [DES ENG] **1.** A diamond-point drill bit. **2.** An offset type of fishing tool used to clear a space around tools stuck in a borehole. **3.** Any of various spade- or chisel-shaped tools or mechanical devices. **4.** *See* grouser. [MIN ENG] A nail, resembling a horseshoe nail, with a hole in the head, driven into mine timbering or into a wooden plug inserted in the rock to mark a surveying station. [NAV ARCH] A foot piece to provide support for the legs of the A frame of a floating dipper dredge. { spəd }

spudded-in [MIN ENG] A borehole that has been started and has reached bedrock or in which the standpipe has been set. { 'spəd·əd‚in }

Spumellina [INV ZOO] The equivalent name for Peripylina. { spyü'mel·ə·nə }

spun glass [MATER] Blown glass made of fine threads of glass. { 'spən 'glas }

spun rayon [TEXT] A yarn or fabric spun from rayon staple. { 'spən 'rā‚än }

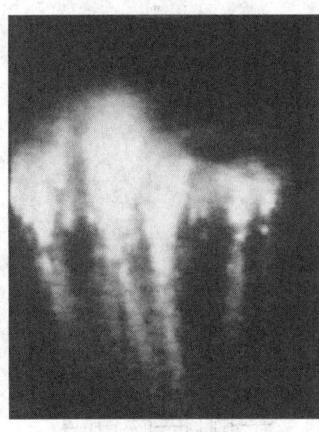

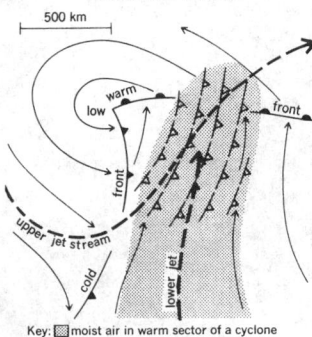

SQUALL LINE

500 km

Key: ▨ moist air in warm sector of a cyclone

Map showing successive locations of squall line moving eastward through unstable northern portion of tongue of moist air in warm sector of a cyclone. Solid arrows show general flow in low levels; broken arrows, axes of strongest wind at about 1 kilometer (lower jet) aboveground and at 10–12 kilometers (upper jet stream).

SQUAMOUS EPITHELIUM

Cellular arrangement in squamous epithelial tissue.

SQUARE-JAW CLUTCH

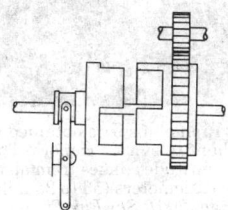

Square-jaw positive clutch.

spun silk [TEXT] A yarn made from short lengths of inferior filaments such as silk broken by moths as they emerged from their cocoons. { 'spən 'silk }

spun yarn [TEXT] A textile yarn spun and twisted from staple-length fiber, either natural or synthetic. { 'spən 'yärn }

spur [BOT] **1.** A hollow process at the base of a petal or sepal. **2.** A short fruit-bearing tree branch. **3.** A short projecting root. [GEOL] A ridge or rise projecting from a larger elevational feature. [HYD] *See* ram. [MATH] *See* trace. [PHYS] A cluster of ionized molecules near the path of an energetic charged particle, consisting of the molecule ionized directly by the charged particle, and secondary ionizations produced by electrons released in the primary ionization; it usually forms a side track from the path of the particle. [ZOO] A stiff, sharp outgrowth, as on the legs of certain birds and insects. { spər }

spur blight [PL PATH] A fungus disease of raspberries and blackberries caused by *Didymella applanata* which kills the fruit spurs and causes dark spotting of the cane. { 'spər ,blīt }

spur dike *See* groin. { 'spər ,dīk }

spur gear [DES ENG] A toothed wheel with radial teeth parallel to the axis. { 'spər ,gir }

spurious correlation [STAT] The value of the coefficient of correlation when it is computed correctly but its relationship implications are nonsensical or unreasonable. { 'spyùr·ē·əs ,kä·rə'lā·shən }

spurious disk [OPTICS] The nearly round image of perceptible diameter of a star as seen through a telescope, due to diffraction of light in the telescope. { 'spyùr·ē·əs 'disk }

spurious emission *See* spurious radiation. { 'spyùr·ē·əs i'mish·ən }

spurious modulation [ELECTR] Undesired modulation occurring in an oscillator, such as frequency modulation caused by mechanical vibration. { 'spyùr·ē·əs ,mäj·ə'lā·shən }

spurious radiation [ELECTROMAG] Any emission from a radio transmitter at frequencies outside its frequency band. Also known as spurious emission. { 'spyùr·ē·əs ,rād·ē'ā·shən }

spurious response [ELECTR] **1.** Response of a radio receiver to a frequency different from that to which the receiver is tuned. **2.** In electronic warfare, the undesirable signal images in the intercept receiver resulting from the mixing of the intercepted signal with harmonics of the local oscillators in the receiver. { 'spyùr·ē·əs ri'späns }

spur pile *See* batter pile. { 'spər ,pīl }

spurrite [MINERAL] $Ca_5(SiO_4)_2(CO_3)$ A light-gray mineral occurring in granular masses. { 'spər,īt }

spurt tone [COMMUN] Short audio-frequency tone used for signaling or dialing selection. { 'spərt ,tōn }

sputtering [ELECTR] Also known as cathode sputtering. **1.** The ejection of atoms or groups of atoms from the surface of the cathode of a vacuum tube as the result of heavy-ion impact. **2.** The use of this process to deposit a thin layer of metal on a glass, plastic, metal, or other surface in vacuum. { 'spəd·ə·riŋ }

sputter-ion pump *See* getter-ion pump. { 'spəd·ər ,ī,än ,pəmp }

sputum [PHYSIO] Material discharged from the surface of the respiratory passages, mouth, or throat; may contain saliva, mucus, pus, microorganisms, blood, or inhaled particulate matter in any combination. { 'spyüd·əm }

sp vol *See* specific volume.

sq *See* square.

SQL *See* Structural Query Language.

squadron [ORD] **1.** An organization consisting of two or more divisions of ships, or two or more divisions (U.S. Navy) or flights of aircraft; it is normally, but not necessarily, composed of ships or aircraft of the same type. **2.** The basic administrative aviation unit of the U.S. Army, Navy, Marine Corps, and Air Force. { 'skwäd·rən }

squalene [BIOCHEM] $C_{30}H_{50}$ A liquid triterpene which is found in large quantities in shark liver oil, and which appears to play a role in the biosynthesis of sterols and polycyclic terpenes; used as a bactericide and as an intermediate in the synthesis of pharmaceuticals. Also known as spinacene. { 'skwā,lēn }

Squalidae [VERT ZOO] The spiny dogfishes, a family of squaloid elasmobranchs recognized by their well-developed fin spines. { 'skwā·lə,dē }

squall [METEOROL] A strong wind with sudden onset and more gradual decline, lasting for several minutes; in the United States observational practice, a squall is reported only if a wind speed of 16 knots (8.23 meters per second) or higher is sustained for at least 2 minutes. { skwól }

squall cloud [METEOROL] A small eddy cloud sometimes formed below the leading edge of a thunderstorm cloud, between the upward and downward currents. { 'skwól ,klaùd }

squall line [METEOROL] A line of thunderstorms near whose advancing edge squalls occur along an extensive front; the region of thunderstorms is typically 12 to 30 miles (20 to 50 kilometers) wide and a few hundred to 1200 miles (2000 kilometers) long. { 'skwól ,līn }

Squamata [VERT ZOO] An order of reptiles, composed of the lizards and snakes, distinguished by a highly modified skull that has only a single temporal opening, or none, by the lack of shells or secondary palates, and by possession of paired penes on the males. { skwə'mäd·ə }

squamodisk [INV ZOO] In monogenetic trematodes, a disk bearing concentric circles of spines, scales, or ridges, and located on the opisthaptor. { 'skwä·mō,disk }

squamosal bone [ANAT] The part of the temporal bone in man corresponding with the squamosal bone in lower vertebrates. [VERT ZOO] A membrane bone lying external and dorsal to the auditory capsule of many vertebrate skulls. { skwə'mō·səl ,bōn }

squamous [BIOL] Covered with or composed of scales. { 'skwä·məs }

squamous-cell carcinoma [MED] A carcinoma composed principally of anaplastic, squamous epithelial cells. Also known as epidermoid carcinoma. { 'skwä·məs ,sel ,kärs·ən'ō·mə }

squamous epithelium [HISTOL] A single-layered epithelium consisting of thin, flat cells. { 'skwä·məs ,ep·ə'thē·lē·əm }

squamulose [BIOL] Covered with or composed of minute scales. { 'skwäm·yə,lōs }

square [MATH] **1.** The square of a number r is the number r^2, that is, r times r. **2.** The plane figure with four equal sides and four interior right angles. Also known as monomino. [MECH] Denotes a unit of area; if x is a unit of length, a square x is the area of a square whose sides have a length of $1x$; for example, a square meter, or a meter squared, is the area of a square whose sides have a length of 1 meter. Abbreviated sq. { skwer }

square degree [MATH] A unit of a solid angle equal to $(\pi/180)^2$ steradian, or approximately 3.04617×10^{-4} steradian. { 'skwer di'grē }

square dome [ARCH] A spherical dome with a square base. { 'skwer 'dōm }

square-edged orifice [ENG] An orifice plate with straight-through edges for the hole through which fluid flows; used to measure fluid flow in fluid conduits by means of differential pressure drop across the orifice. { 'skwer ,ejd 'ór·ə·fəs }

square engine [MECH ENG] An engine in which the stroke is equal to the cylinder bore. { 'skwer 'en·jən }

square-foot unit of absorption *See* sabin. { 'skwer 'fùt 'yü·nət əv əb'sórp·shən }

squarefree number [MATH] A positive integer which is not a multiple of the square of any integer other than 1. { ¦skwer ,frē 'nəm·bər }

square grade [MATH] A unit of solid angle equal to $(\pi/200)^2$ steradian, or approximately 2.46740×10^{-4} steradian. { 'skwer 'grād }

square groove weld [MET] A groove weld in which the joint edges are square. { 'skwer ¦grüv ,weld }

square-head bolt [DES ENG] A cylindrical threaded fastener with a square head. { 'skwer ¦hed ,bōlt }

square-jaw clutch [MECH ENG] A type of positive clutch consisting of two or more jaws of square section which mesh together when they are aligned. { 'skwer ¦jó ,kləch }

square joint *See* straight joint. { ¦skwer ¦jóint }

square key [DES ENG] A machine key of square, usually uniform, but sometimes tapered, cross section. { 'skwer ¦kē }

square-law demodulator *See* square-law detector. { 'skwer ¦ló dē'māj·ə,lād·ər }

square-law detector [ELECTR] A demodulator whose output voltage is proportional to the square of the amplitude-modulated input voltage. Also known as square-law demodulator. { 'skwer ¦lȯ di₁tek·tər }

square-loop ferrite [ELECTROMAG] A ferrite that has an approximately rectangular hysteresis loop. { 'skwer ¦lüp 'fe₁rīt }

square matrix [MATH] A matrix with the same number of rows and columns. { 'skwer 'mā·triks }

square mesh [DES ENG] A wire-cloth textile mesh count that is the same in both directions. { 'skwer 'mesh }

squareness ratio [ELECTROMAG] **1.** The magnetic induction at zero magnetizing force divided by the maximum magnetic induction, in a symmetric cyclic magnetization of a material. **2.** The magnetic induction when the magnetizing force has changed half-way from zero toward its negative limiting value divided by the maximum magnetic induction in a symmetric cyclic magnetization of a material. { 'skwer·nəs ₁rā·shō }

square-nose bit See flat-face bit. { 'skwer ₁nōz ₁bit }

square number [MATH] A number that is derived by squaring an integer. { 'skwer 'nəm·bər }

square-on See center. { ¦skwer 'ȯn }

square planar molecule [CHEM] A molecule in which a central atom possesses four valence bonds directed to the corners of a square, with all atoms lying in the same plane. { 'skwer ¦plā·nər ₁mäl·ə₁kyül }

square root [MATH] A square root of a real or complex number s is a number t for which $t^2 = s$. { 'skwer 'rüt }

square-root law [STAT] The standard deviation of the ratio of the number of successes to number of trials is inversely proportional to the square root of the number of trials. { 'skwer ¦rüt ₁lȯ }

square-root transformation [STAT] A conversion or transformation of data having a Poisson distribution where sample means are approximately proportional to the variances of the respective samples; replacing each measurement by its square root will often result in homogeneous variances. { 'skwer ₁rüt ₁tranz·fər'mā·shən }

square search [NAV] Search by a series of course lines such as to produce an expanding square pattern relative to either a fixed point or a moving point. Also known as expanding square search. { 'skwer 'sərch }

square-serif type [GRAPHICS] A contemporary type face; the serifs do not have any curvature and appear boxlike. { 'skwer ¦ser·əf ₁tīp }

square set [MIN ENG] A set of timbers composed of a cap, girt, and post which meet so as to form a solid 90° angle; they are so framed at the intersection as to form a compression joint, and join with three other similar sets. { 'skwer ₁set }

square-set block caving [MIN ENG] Block caving in which the ore is extracted through drifts supported by square sets. { 'skwer ¦set 'bläk ₁kāv·iŋ }

square-set stopes [MIN ENG] The use of square-set timbering to support the ground as ore is extracted. { 'skwer ¦set ₁stōps }

square thread [DES ENG] A screw thread having a square cross section; the width of the thread is equal to the pitch or distance between threads. { 'skwer 'thred }

square wave [ELEC] An oscillation the amplitude of which shows periodic discontinuities between two values, remaining constant between jumps. { 'skwer 'wāv }

square-wave amplifier [ELECTR] Resistance-coupled amplifier, the circuit constants of which are to amplify a square wave with the minimum amount of distortion. { 'skwer ¦wāv 'am·plə₁fī·ər }

square-wave generator [ELECTR] A signal generator that generates a square-wave output voltage. { 'skwer ¦wāv 'jen·ə₁rād·ər }

square-wave response [ELECTR] The response of a circuit or device when a square wave is applied to the input. { 'skwer ¦wāv ri₁späns }

square wheel [DES ENG] A wheel with a flat spot on its rim. { 'skwer ¦wēl }

squaring circuit [ELECTR] **1.** A circuit that reshapes a sine or other wave into a square wave. **2.** A circuit that contains nonlinear elements proportional to the square of the input voltage. { 'skwer·iŋ ₁sər·kət }

squaring shear [MECH ENG] A machine tool consisting of one fixed cutting blade and another mounted on a reciprocating crosshead; used for cutting sheet metal or plate. { 'skwer·iŋ ₁shir }

squaring the circle [MATH] For a circle with a specified radius, the problem of constructing a square that has the same area as the circle. { 'skwer·iŋ thə 'sər·kəl }

squarrose [BOT] Having stiff divergent bracts, or other processes. { 'skwä₁rōs }

squarrulose [BOT] Mildly squarrose. { 'skwä·rə₁lōs }

squash [BOT] Either of two plants of the genus *Cucurbita*, order Violales, cultivated for its fruit; some types are known as pumpkins. { skwäsh }

Squatinidae [VERT ZOO] A group of squaloid elasmobranchs of uncertain affinity characterized by a greatly extended rostrum with enlarged denticles along the margins; maximum length is under 4 feet (1.2 meters). { skwə'tin·ə₁dē }

squatting [NAV ARCH] The lowering of the stern of a ship when it is traveling at high speed. { 'skwäd·iŋ }

squawker See midrange. { 'skwȯk·ər }

squealing [ELECTR] A condition in which a radio receiver produces a high-pitched note or squeal along with the desired radio program, due to interference between stations or to oscillation in some receiver circuit. { 'skwēl·iŋ }

squeegee [DES ENG] A device consisting of a handle with a blade of rubber or leather set transversely at one end and used for spreading, pushing, or wiping liquids off or across a surface. { 'skwē₁jē }

squeezable waveguide [ELECTROMAG] A waveguide whose dimensions can be altered periodically; used in rapid scanning. { 'skwēz·ə·bəl 'wāv₁gīd }

squeeze [ENG] **1.** To inject a grout into a borehole under high pressure. **2.** The plastic movement of a soft rock in the walls of a borehole or mine working that reduced the diameter of the opening. [MIN ENG] **1.** The settling, without breaking, of the roof over a considerable area of working. Also known as creep; nip; pinch. **2.** The gradual upheaval of the floor of a mine due to the weight of the overlying strata. **3.** The sections in coal seams that have become constricted by the squeezing in of the overlying or underlying rock. [PHYS] Increasing external pressure upon the ears and sinuses in diving. { skwēz }

squeezed state [QUANT MECH] **1.** A quantum state for which one of a pair of conjugate variables, which cannot simultaneously possess definite values according to the Heisenberg uncertainty principle, is specified more accurately than in the vacuum state, at the expense of increasing the uncertainty in the value of the other variable. **2.** A quantum state in which one of a pair of conjugate variables is specified more accurately than in the vacuum state, and the product of the uncertainties of the conjugate variables is the minimum allowed by the Heisenberg uncertainty principle. { 'skwēzd ₁stāt }

squeeze roll [MECH ENG] A roller designed to exert pressure on material passing between it and a similar roller. { 'skwēz ₁rōl }

squeeze section [ELECTROMAG] Length of waveguide constructed so that alteration of the critical dimension is possible with a corresponding alteration in the electrical length. { 'skwēz ₁sek·shən }

squeeze time [MET] In resistance welding, the time between the initial application of the current and of the pressure. { 'skwēz ₁tīm }

squegger See blocking oscillator. { 'skweg·ər }

squegging [ELECTR] Condition of self-blocking in an electron-tube-oscillator circuit. { 'skweg·iŋ }

squegging oscillator See blocking oscillator. { 'skweg·iŋ ₁äs·ə₁läd·ər }

squelch [ELECTR] To automatically quiet a receiver by reducing its gain in response to a specified characteristic of the input. { skwelch }

squelch circuit See noise suppressor. { 'skwelch ₁sər·kət }

squib [ENG] A small tube filled with fine-grained black powder; upon the lighting and burning of the ignition match, the squib assumes a rocket effect and darts back into the hole to ignite the powder charge. [ORD] A small explosive device, similar in appearance to a detonator, but loaded with low explosive, so that its output is primarily heat (flash); usually electrically initiated, and provided to initiate action of pyrotechnic devices and rocket propellants. { skwib }

SQUID

Dorsal view of a squid (*Loligo*).

squid [INV ZOO] Any of a number of marine cephalopod mollusks characterized by a reduced internal shell, ten tentacles, an ink sac, and chromatophores. { skwid }

SQUID See superconducting quantum interference device. { skwid }

Squillidae [INV ZOO] The single family of the eumalacostracan order Stomatopoda, the mantis shrimp. { 'skwil·ə‚dē }

squinch [ARCH] A small arch across the interior corner of a structure to support a superimposed mass such as a dome or spire. Also known as squinch arch. { skwinch }

squinch arch See squinch. { 'skwinch ‚ärch }

squint [ELECTROMAG] **1.** The angle between the two major lobe axes in a radar lobe-switching antenna. **2.** The angular difference between the axis of radar antenna radiation and a selected geometric axis, such as the axis of the reflector. **3.** The angle between the full-right and full-left positions of the beam of a conical-scan radar antenna. [MED] See strabismus. { skwint }

squirrel [VERT ZOO] Any of over 200 species of arboreal rodents of the families Sciuridae and Anomaluridae having a bushy tail and long, strong hind limbs. { 'skwərl }

squirrel-cage motor [ELEC] An induction motor in which the secondary circuit consists of a squirrel-cage winding arranged in slots in the iron core. { 'skwərl ‚kāj ‚mōd·ər }

squirrel-cage rotor See squirrel-cage winding. { 'skwərl ‚kāj ‚rōd·ər }

squirrel-cage winding [ELEC] A permanently short-circuited winding, usually uninsulated, around the periphery of the rotor and joined by continuous end rings. Also known as squirrel-cage rotor. { 'skwərl ‚kāj ‚wīnd·iŋ }

squirt can [ENG] An oil can with a flexible bottom and a tapered spout; pressure applied to the bottom forces oil out the spout. { 'skwərt ‚kan }

squirt gun [ENG] A device with a bulb and nozzle; when the bulb is pressed, liquid squirts from the nozzle. { 'skwərt ‚gən }

squishing See compaction. { 'skwish·iŋ }

squitter [ELECTR] Random firing, intentional or otherwise, of the transponder transmitter in the absence of interrogation. { 'skwid·ər }

sr See steradian.

Sr See strontium.

SRAM See short-range attack missile; static random-access memory. { 'es‚ram }

SRA-size [ENG] One of a series of sizes to which untrimmed paper is manufactured; for reels of paper the standard sizes are 450, 640, 900, and 1280 millimeters; for sheets of paper the sizes are SRA0, 900 × 1280 millimeters; SRA1, 640 × 900 millimeters; and SRA2, 450 × 640 millimeters; SRA sizes correspond to A sizes when trimmed. { ¦es¦är'ā ‚sīz }

SRC See stored response chain.

SR cylinder oil [MATER] A viscous, unfiltered lubrication oil, generally made from reduced petroleum crudes that have had the lighter lubricant fractions removed by direct steam heating; used to lubricate steam engine cylinders and valves. Also known as steam-refined cylinder oil. { ¦es¦är 'sil·ən·dər ‚oil }

SRID See single radial immunodiffusion. { ¦es¦är¦ī'dē or 'es‚rid }

SRMS See structure resonance modulation spectroscopy.

SRP See signal-recognition particle. { ¦es¦är'pē }

SRS See Sonobuoy Reference System.

SRS-A See slow-reacting substance of anaphylaxis.

SRY [GEN] The male sex-determining gene on the Y chromosome in mammals. { ¦es¦är'wī }

SS See stainless steel.

SSA service See S-band single-access service. { ¦es¦es'ā ‚sər·vəs }

SSB See single-sideband.

SS Cygni stars See U Geminorum stars. { ¦es¦es 'sig·nē ‚stärz }

SSD See steady-state distribution.

SS 433 [ASTRON] A stellar object that shows evidence of ejection of two narrow streams of cool gas travelling in opposite direction from a cool object at a velocity of almost one-quarter the speed of light; the beams execute a repeating, rotating pattern about the central object once every 164 days. { ‚es‚for‚thərd·ē'thrē }

SSI See small-scale integration.

ss loran See sky-wave-synchronized loran. { ¦es¦es 'lôr‚an }

SSM See surface-to-surface missile.

SSP See static spontaneous potential.

SSR See solid-state relay.

SST See supersonic transport.

S star [ASTRON] A spectral classification of stars, comprising red stars with surface temperature of about 2200 K; prominent in the spectra is zirconium oxide. { 'es ‚stär }

s-state [QUANT MECH] A single-particle state whose orbital angular momentum quantum number is zero. { 'es ‚stät }

SSTV See slow-scan television.

SSU See Saybolt Seconds Universal.

St See stoke.

stab [ENG] In a drilling operation, to insert the threaded end of a pipe joint into the collar of the joint already placed in the hole and to rotate it slowly to engage the threads before screwing up. { stab }

stab culture [MICROBIO] A culture of anaerobic bacteria made by piercing a solid agar medium in a test tube with an inoculating needle covered with the bacterial inoculum. { 'stab ‚kəl·chər }

stabilator [AERO ENG] A one-piece horizontal tail that is swept back and movable; movement is controlled by motion of the pilot's control stick; usually used in supersonic aircraft. { 'stā·bə‚lād·ər }

stability [CHEM] The property of a chemical compound which is not readily decomposed and does not react with other compounds. [CONT SYS] The property of a system for which any bounded input signal results in a bounded output signal. [ENG] The property of a body, as an aircraft, rocket, or ship, to maintain its attitude or to resist displacement, and, if displaced, to develop forces and moments tending to restore the original condition. [FL MECH] The resistance to overturning or mixing in the water column, resulting from the presence of a positive (increasing downward) density gradient. [GEOL] **1.** The resistance of a structure, spoil heap, or clay bank to sliding, overturning, or collapsing. **2.** Chemical durability, resistance to weathering. [MATER] Of a fuel, the capability to retain its characteristics in an adverse environment, for example, extreme temperature. [MATH] Stability theory of systems of differential equations deals with those solution functions possessing some particular property that still maintain the property after a perturbation. [MECH] See dynamic stability. [PHYS] **1.** The property of a system which does not undergo any change without the application of an external agency. **2.** The property of a system in which any departure from an equilibrium state gives rise to forces or influences which tend to return the system to equilibrium. Also known as static stability. [PL PHYS] The property of a plasma which maintains its shape against externally applied forces (usually pressure of magnetic fields) and whose constituents can pass through confining fields only by diffusion of individual particles. { stə'bil·əd·ē }

stability augmentation system [AERO ENG] Automatic control devices which supplement a pilot's manipulation of the aircraft controls and are used to modify inherent aircraft handling qualities. Abbreviated SAS. Also known as stability augmentors. { stə'bil·əd·ē ȯg·mən'tā·shən ‚sis·təm }

stability augmentors See stability augmentation system. { stə'bil·əd·ē 'ȯg‚men·tərz }

stability chart [METEOROL] A synoptic chart that shows the distribution of a stability index. { stə'bil·əd·ē ‚chärt }

stability constant [CHEM] Refers to the equilibrium reaction of a metal cation and a ligand to form a chelating mononuclear complex; the absolute-stability constant is expressed by the product of the concentration of products divided by the product of the concentrations of the reactants; the apparent-stability constant (also known as the conditional- or effective-stability constant) allows for the nonideality of the system because of the combination of the ligand with other complexing agents present in the solution. { stə'bil·əd·ē ‚kän·stənt }

stability criterion [CONT SYS] A condition which is necessary and sufficient for a system to be stable, such as the Nyquist criterion, or the condition that poles of the system's overall transmittance lie in the left half of the complex-frequency plane. { stə'bil·əd·ē krī‚tir·ē·ən }

stability exchange principle [CONT SYS] In a linear system, which is either dynamically stable or unstable depending on the value of a parameter, the complex frequency varies with the parameter in such a way that its real and imaginary parts pass through zero simultaneously; the principle is often violated. { stə'bil·əd·ē iks'chānj ,prin·sə·pəl }

stability factor [ELECTR] A measure of a transistor amplifier's bias stability, equal to the rate of change of collector current with respect to reverse saturation current. { stə'bil·əd·ē ,fak·tər }

stability index [METEOROL] An indication of the local static stability of a layer of air. { stə'bil·əd·ē ,in,deks }

stability matrix See stiffness matrix. { stə'bil·əd·ē ,mā·triks }

stability subgroup See stabilizer. { stə'bil·əd·ē 'səb,grüp }

stability test [ENG] Accelerated test to determine the probable suitability of an explosive material for long-term storage. { stə'bil·əd·ē ,test }

stabilivolt [ELECTR] Gas tube that maintains a constant voltage drop across its terminals, essentially independent of current, over a relatively wide range. { stə'bil·ə,vōlt }

stabilization [CHEM ENG] A petroleum-refinery process for separating light gases from petroleum or gasoline, thus leaving a stable (less volatile) liquid so that it can be handled or stored with less change in composition. [CONT SYS] See compensation. [ELECTR] Feedback introduced into vacuum tube or transistor amplifier stages to reduce distortion by making the amplification substantially independent of electrode voltages and tube constants. [ELECTROMAG] Treatment of a magnetic material to improve the stability of its magnetic properties. [ENG] Maintenance of a desired orientation independent of the roll and pitch of a ship or aircraft. { ,stā·bə·lə'zā·shən }

stabilize [TEXT] To treat a fabric to prevent it from shrinking and stretching. { 'stā·bə,līz }

stabilized feedback See negative feedback. { 'stā·bə,līzd 'fēd,bak }

stabilized flight [AERO ENG] Maintenance of desired orientation in flight. { 'stā·bə,līzd 'flīt }

stabilized gasoline [MATER] Gasoline from which the "wild" or low-boiling (high vapor pressure, volatile) hydrocarbons have been removed by stabilization. { 'stā·bə,līzd 'gas·ə,lēn }

stabilized platform See stable platform. { 'stā·bə,līzd 'plat,fòrm }

stabilized winding [ELEC] Auxiliary winding used particularly in star-connected transformers to stabilize the neutral point of the fundamental frequency voltages, to protect the transformer and the system from excessive third-harmonic voltages; and to prevent telephone interference caused by third-harmonic currents and voltages in the lines and earth. Also known as tertiary winding. { 'stā·bə,līzd 'wīnd·iŋ }

stabilizer [AERO ENG] Any airfoil or any combination of airfoils considered as a single unit, the primary function of which is to give stability to an aircraft or missile. [CHEM ENG] The fractionation column in a petroleum refinery used to stabilize (remove fractions from) hydrocarbon mixtures. [ENG] **1.** A hardened, splined bushing, sometimes freely rotating, slightly larger than the outer diameter of a core barrel and mounted directly above the core barrel back head. Also known as ferrule; fluted coupling. **2.** A tool located near the bit in the drilling assembly to modify the deviation angle in a well by controlling the location of the contact point between the hole and the drill collars. [MATER] **1.** Any powdered or liquid additive used as an agent in soil stabilization. **2.** Any substance that tends to maintain the physical and chemical properties of a material. [MATH] The stabilizer of a point x in a Riemann surface X, relative to a group G of conformal mappings of X onto itself, is the subgroup G_x of G consisting of elements g such that $g(x) = x$. Also known as stability subgroup. [NAV ARCH] Any of the submerged fins used on ships to prevent rolling. { 'stā·bə,līz·ər }

stabilizer bar [MECH ENG] In an automotive vehicle, a shaft that interconnects the two lower suspension arms in order to reduce body roll when the vehicle is turning. Also known as sway bar. { 'stā·bə,līz·ər ,bär }

stabilizing magnetic field [PL PHYS] A magnetic field which is added to a device confining a plasma in order to increase the plasma's stability. { 'stā·bə,līz·iŋ mag'ned·ik 'fēld }

stabilizing treatment [MET] Any of various treatments intended to promote dimensional stability of a metal part or stabilize the structure of an alloy. { 'stā·bə,līz·iŋ ,trēt·mənt }

stabistor [ELECTR] A diode component having closely controlled conductance, controlled storage charge, and low leakage, as required for clippers, clamping circuits, bias regulators, and other logic circuits that require tight voltage-level tolerances. { stā'bis·tər }

stable [PHYS] Not subject to any change without the application of an external agency, such as radiation; said of a molecule, atom, nucleus, or elementary particle. { 'stā·bəl }

stable-base film [MATER] A particular type of film having high stability in regard to shrinkage and stretching. { 'stā·bəl 'bās 'film }

stable bundle [MATH] The bundle E^s of a hyperbolic structure. { 'stā·bəl 'bən·dəl }

stable element [ENG] Any instrument or device, such as a gyroscope, used to stabilize a radar antenna, turret, or other piece of equipment mounted on an aircraft or ship. { 'stā·bəl 'el·ə·mənt }

stable equilibrium [SCI TECH] Equilibrium in which any departure from the equilibrium state gives rise to forces or influences which tend to return the system to equilibrium. { 'stā·bəl ,ē·kwə'lib·rē·əm }

stable factor See factor VII. { 'stā·bəl 'fak·tər }

stable graph [MATH] A graph from which an edge can be deleted to produce a subgraph whose group of automorphisms is a subgroup of the group of automorphisms of the original graph. { 'stā·bəl 'graf }

stable homeomorphism conjecture [MATH] For dimension n, the assertion that each orientation-preserving homeomorphism of the real n space, R^n, into itself can be expressed as a composition of homeomorphisms, each of which is the identity on some nonempty open set in R^n. { 'stā·bəl 'hō·mē·ō'mòr,fiz·əm kən,jek·chər }

stable isobar [NUC PHYS] One of two or more stable nuclides which have the same mass number but differ in atomic number. { 'stā·bəl 'ī·sə,bär }

stable isotope [NUC PHYS] An isotope which does not spontaneously undergo radioactive decay. { 'stā·bəl 'ī·sə,tōp }

stable local oscillator See stalo. { 'stā·bəl 'lō·kəl 'äs·ə,lād·ər }

stable nucleus [NUC PHYS] A nucleus which does not spontaneously undergo radioactive decay. { 'stā·bəl 'nü·klē·əs }

stable orbit See equilibrium orbit. { 'stā·bəl 'òr·bət }

stable platform [AERO ENG] A gimbal-mounted platform, usually containing gyros and accelerometers, the purpose of which is to maintain a desired orientation in inertial space independent of craft motion. Also known as stabilized platform. { 'stā·bəl 'plat,fòrm }

stable strobe [ELECTR] Series of strobes which behaves as if caused by a single jammer. { 'stā·bəl 'strōb }

stable system [PHYS] A system that returns to a stationary state following sufficiently small perturbations. { 'stā·bəl 'sis·təm }

stable vertical [ENG] Vertical alignment of any device or instrument maintained during motion of the mount. { 'stā·bəl 'vərd·ə·kəl }

stack [BUILD] The portion of a chimney rising above the roof. [CHEM ENG] In gas works, a row of benches containing retorts. [COMPUT SCI] A portion of a computer memory used to temporarily hold information, organized as a linear list for which all insertions and deletions, and usually all accesses, are made at one end of the list. [ELECTR] See pileup. [ENG] **1.** To stand and rack drill rods in a drill tripod or derrick. **2.** Any structure or part thereof that contains a flue or flues for the discharge of gases. **3.** One or more filter cartridges mounted on a single column. **4.** Tall, vertical conduit (such as smokestack, flue) for venting of combustion or evaporation products or gaseous process wastes. **5.** The exhaust pipe of an internal combustion engine. [GEOL] An erosional, coastal landform that is a steep-sided, pillarlike rocky island or mass that has been detached by wave action from a shore made up of cliffs; applies particularly to a stack that is columnar in structure and has horizontal stratifications. Also known as marine stack; rank. [MET] The cone-shaped section of a blast furnace or cupola above the hearth and melting zone and extending to the throat. [NAV] To assign different altitudes by radio to aircraft awaiting their turns to land at an airport. { stak }

stack automaton [COMPUT SCI] A variation of a pushdown automaton in which the read-only head of the input tape is allowed to move both ways, and the read-write head on the pushdown storage is allowed to scan the entire pushdown list in a read-only mode. { 'stak ȯ'täm·ə,tän }

stack cutting [MET] Cutting a stack of metal plates with a single cut using a stream of oxygen. { 'stak ,kəd·iŋ }

stacked antennas [ELECTROMAG] Two or more identical antennas arranged above each other on a vertical supporting structure and connected in phase to increase the gain. { 'stakt an'ten·əz }

stacked array [ELECTROMAG] An array in which the antenna elements are stacked one above the other and connected in phase to increase the gain. { 'stakt ə'rā }

stacked-beam radar [ENG] Three-dimensional radar system that derives elevation by emitting narrow beams stacked vertically to cover a vertical segment, azimuth information from horizontal scanning of the beam, and range information from echo-return time. { 'stakt ,bēm 'rā,där }

stacked-dipole antenna [ELECTROMAG] Antenna in which directivity is increased by providing a number of identical dipole elements, excited either directly or parasitically; the resultant radiation pattern depends on the number of dipole elements used, the spacing and phase difference between the elements, and the relative magnitudes of the currents. { 'stakt 'dī,pōl an,ten·ə }

stacked-job processing [COMPUT SCI] A technique of automatic job-to-job transition, with little or no operator intervention. { 'stakt 'jäb ,prä,ses·iŋ }

stacked loops [ELECTROMAG] Two or more loop antennas arranged above each other on a vertical supporting structure and connected in phase to increase the gain. Also known as vertically stacked loops. { 'stakt 'lüps }

stack effect [MECH ENG] The pressure difference between the confined hot gas in a chimney or stack and the cool outside air surrounding the outlet. { 'stak i,fekt }

stacker [MECH ENG] A machine for lifting merchandise on a platform or fork and arranging it in tiers; operated by hand, or electric or hydraulic mechanisms. { 'stak·ər }

stacker-reclaimer [MECH ENG] Equipment which transports and builds up material stockpiles, and recovers and transports material to processing plants. { 'stak·ər rē'klām·ər }

stack gas [ENG] Gas passed through a chimney. { 'stak ,gas }

stacking [AERO ENG] The holding pattern of aircraft awaiting their turn to approach and land at an airport. [ELECTROMAG] The placing of antennas one above the other, connecting them in phase to increase the gain. { 'stak·iŋ }

stacking fault [CRYSTAL] A defect in a face-centered cubic or hexagonal close-packed crystal in which there is a change from the regular sequence of positions of atomic planes. { 'stak·iŋ ,fȯlt }

stack model [COMPUT SCI] A model for describing the runtime execution of programs written in block-structured languages, consisting of a program component, which remains unchanged throughout the execution of the program; a control component, consisting of an instruction pointer and an environment pointer; and a stack of records containing all the data the program operates on. { 'stak ,mäd·əl }

stack operation [COMPUT SCI] A computer system in which flags, return address, and all temporary addresses are saved in the core in sequential order for any interrupted routine so that a new routine (including the interrupted routine) may be called in. { 'stak ,äp·ə,rā·shən }

stack pointer [COMPUT SCI] A register which contains the last address of a stack of addresses. { 'stak ,pȯint·ər }

stack pollutants [ENG] Smokestack emissions subject to Environmental Protection Agency standards regulations, including sulfur oxides, particulates, nitrogen oxides, hydrocarbons, carbon monoxide, and photochemical oxidants. { 'stak pə,lüt·əns }

stack vent [ENG] An extension to the atmosphere of a waste stack or a soil stack above the highest horizontal branch drain or fixture branch that is connected to the stack. Also known as soil vent; waste vent. { 'stak ,vent }

stack welding [MET] Simultaneous spot welding of stacked plates. { 'stak ,weld·iŋ }

stactometer See stalagmometer. { stak'täm·əd·ər }

STADAN See Space Tracking and Data Acquisition Network. { 'stā,dan }

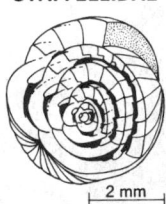

STAFFELLIDAE

↦ 2 mm ↤

A representative of the fusulinacean family Staffellidae.

stade [GEOL] A substage of a glacial stage marked by a secondary advance of glaciers. { stād }

Stader's splint [MED] A metal bar with pins affixed at right angles; the pins are driven into the fragments of a fracture, and the bar maintains the alignment. { 'stād·ərz ,splint }

stadia [ENG] A surveying instrument consisting of a telescope with special horizontal parallel lines or wires, used in connection with a vertical graduated rod. { 'stād·ē·ə }

stadia hairs [ENG] Two horizontal lines in the reticle of a theodolite arranged symmetrically above and below the line of sight. Also known as stadia wires. { 'stād·ē·ə ,herz }

stadial moraine See recessional moraine. { 'stād·ē·əl mə'rān }

stadia rod [ENG] A graduated rod used with a stadia to measure the distance from the observation point to the rod by observation of the length of rod subtended by the distance between the stadia hairs. { 'stād·ē·ə ,räd }

stadia tables [ENG] Mathematical tables from which may be found, without computation, the horizontal and vertical components of a reading made with a transit and stadia rod. { 'stād·ē·ə ,tā·bəlz }

stadia wires See stadia hairs. { 'stād·ē·ə ,wīrz }

stadimeter [ENG] An instrument for determining the distance to an object, but its height must be known; the angle subtended by the object's bottom and top as measured at the observer's position is proportional to the object's height; the instrument is graduated directly in distance. { sta'dim·əd·ər }

stadiometry [COMPUT SCI] In computer vision, the determination of the distance to an object based on the size of its image. { ,stād·ē'äm·ə·trē }

Staebler-Wronski effect [SOLID STATE] A reversible change (usually a reduction) in the dark conductivity and photoconductivity of hydrogenated amorphous silicon during and following illumination by light with sufficient energy to produce electron-hole pairs. { ¦stāb·lər 'rän·skē i,fekt }

staff bead [BUILD] **1.** A bead between a wooden frame and adjacent masonry. **2.** A molded or beaded angle of wood or metal set into the corner of plaster walls. { 'staf ,bēd }

Staffellidae [PALEON] An extinct family of marine protozoans (superfamily Fusulinacea) that persisted during the Pennsylvanian and Early Permian. { sta'fel·ə,dē }

staff gage [ENG] A graduated scale placed in a position so that the stage of a stream may be read directly therefrom; a type of river gage. { 'staf ,gāj }

Staffordian [GEOL] A European stage of geologic time forming the middle Upper Carboniferous, above Yorkian and below Radstockian, equivalent to part of the upper Westphalian. { sta'fȯrd·ē·ən }

stage [AERO ENG] A self-propelled separable element of a rocket vehicle or spacecraft. [ELECTR] A circuit containing a single section of an electron tube or equivalent device or two or more similar sections connected in parallel, push-pull, or push-push; it includes all parts connected between the control-grid input terminal of the device and the input terminal of the next adjacent stage. [GEOL] **1.** A developmental phase of an erosion cycle in which landscape features have distinctive characteristic forms. **2.** A phase in the historical development of a geologic feature. **3.** A major subdivision of a glacial epoch. **4.** A time-stratigraphic unit ranking below series and above chronozone, composed of rocks formed during an age of geologic time. [HYD] The elevation of the water surface in a stream as measured by a river gage with reference to some arbitrarily selected zero datum. Also known as stream stage. [MIN ENG] **1.** A certain length of underground roadway worked by one horse. **2.** A narrow thin dike, especially one where the material of which the dike is composed is soft. **3.** A platform on which mine cars stand. { stāj }

stage acidizing [PETRO ENG] An oil-reservoir acid treatment (acidizing) in which the formation is treated with two or more separate stages of acid, instead of a single large treatment. { 'stāj 'as·ə,dīz·iŋ }

staged crew [AERO ENG] An aircrew specifically positioned at an intermediate airfield to take over aircraft operating on an air route, thus relieving a complementary crew of flying fatigue and speeding up the traffic rate of the aircraft concerned. { 'stājd 'krü }

stage gain [ELECTR] The ratio of the output power of an amplifier stage to the input power, usually expressed in decibels. { 'stāj ,gān }

stage loader See feeder conveyor. { 'stāj ‚lōd·ər }

stage separation [PETRO ENG] A system for gas-oil separation of well fluids by a series of stages instead of a single operation. { 'stāj ‚sep·ə‚rā·shən }

stage theory [NEUROSCI] A theory of color vision which proposes that there are three or more types of cone receptors whose responses are conducted to higher visual centers, and that interactions occur at some stage along the conducting paths so that strong activity in one type of response inhibits that of other response paths. Also known as zone theory. { 'stāj ‚thē·ə·rē }

stagger [COMMUN] Periodic error in the position of the recorded spot along a recorded facsimile line. { 'stag·ər }

staggered blastholes [MIN ENG] Two rows of holes staggered to a triangular pattern to distribute the burden when shot-firing in thick coal seams. { 'stag·ərd 'blast‚hōlz }

staggered-intermittent fillet welding [MET] Making a line of intermittent fillet welds on each side of a joint in a manner such that the increments on one side are not opposite to those on the other side. { 'stag·ərd ‚in·tər'mit·ənt 'fil·ət ‚weld·iŋ }

staggered-line drive [PETRO ENG] In the placement and drilling of water-injection wells for water flood recovery of oil, the staggered (versus in-line) areal arrangement of the injection wells. { 'stag·ərd ‚līn 'drīv }

staggered tuning [ELECTR] Alignment of successive tuned circuits to slightly different frequencies in order to widen the overall amplitude-frequency response curve. { 'stag·ərd 'tün·iŋ }

staggering [COMMUN] Offsetting of two channels of different carrier systems from exact side-band frequency coincidence to avoid mutual interference. { 'stag·ə·riŋ }

staggering advantage [COMMUN] Effective reduction of interference between carrier channels, due to staggering. { 'stag·ə·riŋ ad‚van·tij }

staggers [VET MED] Any of various diseases of livestock, sheep, and horses manifested by lack of coordination in movement and a staggering gait. { 'stag·ərz }

stagger-tooth cutter [MECH ENG] Side-milling cutter with successive teeth having alternating helix angles. { 'stag·ər ‚tüth ‚kəd·ər }

stagger-tuned amplifier [ELECTR] An amplifier that uses staggered tuning to give a wide bandwidth. { 'stag·ər ‚tünd 'am·plə‚fī·ər }

stagger-tuned filter [ELECTR] A filter consisting of a cascade of amplifier stages with tuned coupling networks whose resonant frequencies and bandwidths may be easily adjusted to achieve an overall transmission function of desired shape (maximally flat or equal ripple). { 'stag·ər ‚tünd 'fil·tər }

staghead See witches'-broom disease. { 'stag‚hed }

staging [AERO ENG] The process or operation during the flight of a rocket vehicle whereby a full stage or half stage is disengaged from the remaining body and made free to decelerate or be propelled along its own flightpath. [COMPUT SCI] Moving blocks of data from one storage device to another. [GRAPHICS] See stopping out. { 'stāj·iŋ }

staging area [ORD] A general locality, containing accommodations for troops, established for the concentration of troop units and transient personnel between movements over the lines of communication. { 'stāj·iŋ ‚er·ē·ə }

staging base [ORD] 1. An advanced naval base for the anchoring, fueling, and refitting of transports and cargo ships, and for replenishing mobile service squadrons. 2. A landing and takeoff area with minimum servicing, supply, and shelter provided for the temporary occupancy of military aircraft during the course of movement from one location to another. { 'stāj·iŋ ‚bās }

stagnant glacier [HYD] A glacier which has ceased to move. { 'stag·nənt 'glā·shər }

stagnant water [HYD] Motionless water, not flowing in a stream or current. Also known as standing water. { 'stag·nənt 'wòd·ər }

stagnation [HYD] 1. The condition of a body of water unstirred by a current or wave. 2. The condition of a glacier that has stopped flowing. { stag'nā·shən }

stagnation point [FL MECH] A point in a field of flow about a body where the fluid particles have zero velocity with respect to the body. { stag'nā·shən ‚pòint }

stagnation pressure See dynamic pressure. { stag'nā·shən ‚presh·ər }

stagnation temperature See adiabatic recovery temperature. { stag'nā·shən ‚tem·prə·chər }

stagnum [HYD] A pool of water with no outlet. { 'stag·nəm }

stain [MATER] 1. A nonprotective coloring matter used on wood surfaces; imparts color without obscuring the wood grains. 2. Any colored, organic compound used to stain tissues, cells, cell components, cell contents, or other biological substrates for microscopic examination. { stān }

stained glass [ENG] Glass colored by any of several means and assembled to produce a varicolored mosaic or representation. { 'stānd 'glas }

stainierite See heterogenite. { 'stī·nē·ə‚rīt }

stainless alloy [MET] Any of a large and complex group of corrosion-resistant iron-chromium alloys (containing 10% or more chromium), sometimes containing other elements, such as nickel, silicon, molybdenum, tungsten, and niobium. Commonly known as stainless steel (SS). { 'stān·ləs 'al‚ói }

stainless-clad steel [MET] Steel clad on one or two sides with a stainless steel to provide a surface that is corrosion-resistant and attractive. { 'stān·ləs ‚klad 'stēl }

stainless iron See ferritic stainless steel. { 'stān·ləs 'ī·ərn }

stainless steel See stainless alloy. { 'stān·ləs 'stēl }

stair [CIV ENG] A series of steps between levels or from floor to floor in a building. { ster }

staircase signal [COMMUN] In television transmissions, a waveform that consists of a series of discrete steps resembling a staircase. { 'ster‚kās ‚sig·nəl }

stairway [CIV ENG] One or more flights of stairs connected by landings. { 'ster‚wā }

stairwell [BUILD] A vertical compartment that extends through a building to hold a stairway. { 'ster‚wel }

stake [ELEC] An iron peg used as a power electrode to transfer current into the ground in electrical prospecting. [ENG] 1. To fasten back or prop open with a piece of chain or otherwise the valves or clacks of a water barrel in order that the water may run back into the sump when necessary. 2. A pointed piece of wood driven into the ground to mark a boundary, survey station, or elevation. { stāk }

staking [ENG] Joining two parts together by fitting a projection on one part against a mating feature in the other part and then causing plastic flow at the joint. { 'stāk·iŋ }

staking out [ENG] Driving stakes into the earth to indicate the foundation location of a structure to be built. { 'stāk·iŋ 'aùt }

stalactite [GEOL] A conical or roughly cylindrical speleothem formed by dripping water and hanging from the roof of a cave; usually composed of calcium carbonate. { stə'lak‚tīt }

stalacto-stalagmite [GEOL] A columnar deposit formed by the union of a stalactite with its complementary stalagmite. Also known as column; pillar. { stə'lak·tō stə'lag‚mīt }

stalagmite [GEOL] A conical speleothem formed upward from the floor of a cave by the action of dripping water; usually composed of calcium carbonate. { stə'lag‚mīt }

stalagmometer [ENG] An instrument for measuring the size of drops suspended from a capillary tube, used in the drop-weight method. Also known as stactometer; stalogometer. { ‚stal·ig'mäm·əd·ər }

stale link [COMPUT SCI] A hyperlink to a document that has been erased or removed from the World Wide Web. Also known as black hole. { 'stāl 'liŋk }

stalked barnacle [INV ZOO] The common name for crustaceans composing the suborder Lepadomorpha. { 'stòkt 'bär·nə·kəl }

stall [AERO ENG] 1. The action or behavior of an airplane (or one of its airfoils) when by the separation of the airflow, as in the case of insufficient airspeed or of an excessive angle of attack, the airplane or airfoil tends to drop; the condition existing during this behavior. 2. A flight performance in which an airplane is made to lose flying speed and to drop by pointing the nose more steeply upward. 3. An act or instance of stalling. { stòl }

stall flutter [AERO ENG] A type of dynamic instability that takes place when the separation of flow around an airfoil occurs during the whole or part of each cycle of a flutter motion. { 'stòl ‚fləd·ər }

stalling angle See angle of stall. { 'stòl·iŋ ‚aŋ·gəl }

stalling angle of attack See critical angle of attack. { 'stòl·iŋ 'aŋ·gəl əv ə'tak }

STALACTITE

Diagram of stalactite and stalagmite formation. (After E. B. Branson et al., Introduction to Geology, McGraw-Hill, 3d ed., 1952)

stalling Mach number [AERO ENG] The Mach number of an aircraft when the coefficient of lift of the aerodynamic surfaces is the maximum obtainable for the pressure altitude, true airspeed, and angle of attack under which the craft is operated. { 'stȯl·iŋ 'mäk ,nəm·bər }

stallion [VERT ZOO] **1.** A mature male equine mammal. **2.** A male horse not castrated. { 'stal·yən }

stall torque [MECH ENG] The amount of torque provided by a motor at close to zero speed. { 'stȯl ,tȯrk }

stall warning indicator [AERO ENG] A device that determines the critical angle of attack for a given aircraft; usually operates from vane sensors, airflow sensors, tabs on leading edges of wings, and computing devices such as accelerometers or airspeed indicators. { 'stȯl ¦wȯrn·iŋ ,in·də,kād·ər }

stalo [ELECTR] A highly stable local radio-frequency oscillator used for heterodyning signals to produce an intermediate frequency in radar moving-target indicators; only echoes that have changed slightly in frequency due to reflection from a moving target produce an output signal. Derived from stable local oscillator. { 'stā,lō }

stalogometer See stalagmometer. { ,stal·ə'gäm·əd·ər }

stamen [BOT] The male reproductive structure of a flower, consisting of an anther and a filament. { 'stā·mən }

Stamey test [MED] A test of differential urinary excretion designed to detect unilateral renovascular disease. { 'stām·ē ,test }

staminate flower [BOT] A flower having stamens but lacking functional carpels. { 'stam·ə·nət 'flau̇·ər }

staminode [BOT] A stamen with no functional anther. { 'stā·mə,nōd }

stamp battery [MIN ENG] A machine for crushing very strong ores or rocks; consists essentially of a crushing member (gravity stamp) which is dropped on a die, the ore being crushed in water between the shoe and the die. { 'stamp ,bad·ə·rē }

stamp copper [MIN ENG] A copper-bearing rock that is stamped and washed before it is smelted. { 'stamp ,käp·ər }

stamper [ENG ACOUS] A negative, generally made of metal by electroforming, used for molding phonograph records. { 'stam·pər }

stamp head [MIN ENG] A heavy and nearly cylindrical cast-iron head fixed on the lower end of the stamp rod, shank, or lifter to give weight in stamping the ore. { 'stamp ,hed }

Stampian See Rupelian. { 'stam·pē·ən }

stamping [ELECTR] A transformer lamination that has been cut out of a strip or sheet of metal by a punch press. [MECH ENG] Almost any press operation including blanking, shearing, hot or cold forming, drawing, bending, and coining. [MIN ENG] Reducing to the desired fineness in a stamp mill; the grain is usually not so fine as that produced by grinding in pans. { 'stam·piŋ }

stamping mill [MIN ENG] A machine in which ore is finely crushed by descending pestles (stamps), usually operated by hydraulic power. Also known as crushing mill. { 'stam·piŋ ,mil }

stamukha [OCEANOGR] An individual piece of stranded ice. { ,sta,mü·kə }

stanchion [ENG] A structural steel member, usually larger than a strut, whose main function is to withstand axial compressive stresses. [NAV ARCH] An upright metal post on a ship. { 'stan·chən }

stand [ECOL] A group of plants, distinguishable from adjacent vegetation, which is generally uniform in species composition, age, and condition. [FOR] The amount of standing timber per unit area; usually expressed in terms of volume. [MET] A set of rolls used in a metal-rolling process. [OCEANOGR] The interval at high or low water when there is no appreciable change in the height of the tide. Also known as tidal stand. { stand }

stand-alone fuel [MATER] A fuel that is burned directly rather than being blended with another fuel. { 'stand ə¦lōn 'fyül }

stand-alone machine [COMPUT SCI] A machine capable of functioning independently of a master computer, either part of the time or all of the time. { 'stand ə¦lōn mə'shēn }

standard [PHYS] An accepted reference sample used for establishing a unit for the measurement of a physical quantity. { 'stan·dərd }

standard advanced base units [ORD] Personnel and materiel organized to function as advanced base units, including the functional components which are employed in the stops of naval advanced bases. { 'stan·dərd ad'vanst ¦bās 'yü·nəts }

standard air munitions package [ORD] A package of air-to-ground conventional munitions required for the 30-day support of a specific type of aircraft; designed to be air transportable in three equal increments. { 'stan·dərd 'er myü,nish·ənz ,pak·ij }

standard antenna [ELECTROMAG] An open single-wire antenna (including the lead-in wire) having an effective height of 4 meters. { 'stan·dərd an'ten·ə }

standard arm [ORD] A passive homing air-to-surface antiradiation missile designed as a follow-on to the Shrike. { 'stan·dərd 'ärm }

standard artillery atmosphere [METEOROL] A set of values describing atmospheric conditions on which ballistic computations are based, namely: no wind, a surface temperature of 15°C, a surface pressure of 1000 millibars, a surface relative humidity of 78%, and a lapse rate which yields a prescribed density-altitude relation. { 'stan·dərd är'til·ə·rē 'at·mə,sfir }

standard artillery zone [METEOROL] A vertical subdivision of the standard artillery atmosphere; it may be considered a layer of air of prescribed thickness and altitude. { 'stan·dərd är'til·ə·rē ,zōn }

standard atmosphere [METEOROL] A hypothetical vertical distribution of atmospheric temperature, pressure, and density which is taken to be representative of the atmosphere for purposes of pressure altimeter calibrations, aircraft performance calculations, aircraft and missile design, and ballistic tables; the air is assumed to obey the perfect gas law and hydrostatic equation, which, taken together, relate temperature, pressure, and density variations in the vertical; it is further assumed that the air contains no water vapor, and that the acceleration of gravity does not change with height. [PHYS] See atmosphere. { 'stan·dərd 'at·mə,sfir }

standard ballistic conditions [MECH] A set of ballistic conditions arbitrarily assumed as standard for the computation of firing tables. { 'stan·dərd bə'lis·tik kən'dish·ənz }

standard blocked F-format data set See FBS data set. { 'stan·dərd 'bläkt 'ef'fȯr,mat 'dad·ə ,set }

standard broadcast band See broadcast band. { 'stan·dərd 'brȯd,kast ,band }

standard broadcast channel [COMMUN] Band of frequencies occupied by the carrier and two side bands of a radio broadcast signal, with the carrier frequency at the center. { 'stan·dərd 'brȯd,kast ,chan·əl }

standard broadcasting [COMMUN] Radio broadcasting using amplitude modulation in the band of frequencies from 535 to 1605 kilohertz; carrier frequencies are placed 10 kilohertz apart. { 'stan·dərd 'brȯd,kast·iŋ }

standard calomel electrode [PHYS CHEM] A mercury-mercurous chloride electrode used as a reference (standard) measurement in polarographic determinations. { 'stan·dərd 'kal·ə·məl i'lek,trōd }

standard candle See international candle. { 'stan·dərd 'kand·əl }

standard capacitor [ELEC] A capacitor constructed in such a manner that its capacitance value is not likely to vary with temperature and is known to a high degree of accuracy. Also known as capacitance standard. { 'stan·dərd kə'pas·əd·ər }

standard cell [ELEC] A primary cell whose voltage is accurately known and remains sufficiently constant for instrument calibration purposes; the Weston standard cell has a voltage of 1.018636 volts at 20°C. { 'stan·dərd 'sel }

standard compass [NAV] A compass designated as the standard for a vessel; it is located in a favorable position, that is, a position with a minimum magnetic aberration, and is accurately calibrated. { 'stan·dərd 'käm·pəs }

standard conditions [PHYS] **1.** A temperature of 0°C and a pressure of 1 atmosphere (760 torr). Also known as normal temperature and pressure (NTP); standard temperature and pressure (STP). **2.** According to the American Gas Association, a temperature of 60°F (155/9°C) and a pressure of 762 millimeters (30 inches) of mercury. **3.** According to the Compressed Gas Institute, a temperature of 20°C (68°F) and a pressure of 1 atmosphere. [SOLID STATE] The allotropic form in which a substance most commonly occurs. { 'stan·dərd kən'dish·ənz }

standard coordinates [ASTRON] A coordinate system used to locate stars on a photographic plate, in which the coordinates

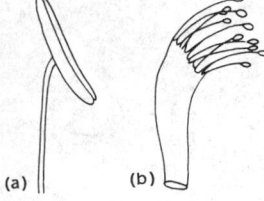

STAMEN

Two types of stamens:
(a) with versatile anther;
(b) monodelphous, with filaments united into one set.

are the differences in right ascension and declination between the position of each star and the assumed position of the plate center. { 'stan·dərd kō'órd·ən·əts }

standard depth-pressure recorder [PETRO ENG] A device for the measurement of pressures at the bottom of a well bore (that is, bottom-hole pressure); a spring-restrained piston moves a recording stylus on a pressure-sealed chart. { 'stan·dərd 'depth ¦presh·ər ri,kórd·ər }

standard deviate [STAT] For a variable x, the quantity $(x − \bar{x})/\sigma$, where $\bar{x}$ is the mean value of x and σ is the standard deviation of x. { ¦stan·dərd 'dē·vē·ət }

standard deviation [STAT] The positive square root of the expected value of the square of the difference between a random variable and its mean. { 'stan·dərd ,dē·vē·ā·shən }

standard electrode potential [PHYS CHEM] The reversible or equilibrium potential of an electrode in an environment where reactants and products are at unit activity. { 'stan·dərd i'lek,trōd pə,ten·chəl }

standard elemental time [IND ENG] A standard time for individual work elements. { 'stan·dərd 'el·ə,ment·əl 'tīm }

standard error [STAT] A measure of the variability any statistical constant would be expected to show in taking repeated random samples of a given size from the same universe of observations. { 'stan·dərd 'er·ər }

standard error of the estimate [STAT] Standard deviation of observed values about the regression line; computed by dividing the unexplained variation or the error sum of squares by its degrees of freedom. { ¦stan·dərd ¦er·ər əv thē 'es·tə·mət }

standard error of the forecast [STAT] Standard deviation of the estimate (point or interval) of a dependent variable for a given value of an independent variable. { ¦stan·dərd ¦er·ər əv thē 'fòr,kast }

standard error of the regression coefficient [STAT] The standard deviation of an estimated regression coefficient; depends on sample size and model assumptions. { ¦stan·dərd ¦er·ər əv thē ri'gresh·ən ,kō·ə,fish·ənt }

standard evaporator See short-tube vertical evaporator. { 'stan·dərd i'vap·ə,rād·ər }

standard fit [DES ENG] A fit whose allowance and tolerance are standardized. { 'stan·dərd 'fit }

standard form [COMPUT SCI] The form of a floating point number whose mantissa lies within a standard specified range of values. { 'stan·dərd 'fòrm }

standard free-energy increase [THERMO] The increase in Gibbs free energy in a chemical reaction, when both the reactants and the products of the reaction are in their standard states. { 'stan·dərd 'frē ¦en·ər·jē 'in,krēs }

standard-frequency signal [COMMUN] One of the highly accurate signals broadcast by government radio stations and used for testing and calibrating radio equipment all over the world; in the United States signals are broadcast by the National Bureau of Standards' radio stations WWV, WWVH, WWVB, and WWVL. { 'stan·dərd ¦frē·kwən·sē ,sig·nəl }

standard function See built-in function. { 'stan·dərd 'fəŋk·shən }

standard gage [CIV ENG] A railroad gage measuring 4 feet $8\frac{1}{2}$ inches (1.4351 meters). [DES ENG] A highly accurate gage used only as a standard for working gages. { 'stan·dərd 'gāj }

Standard Generalized Markup Language [COMPUT SCI] A system that encodes the logical structure and content of a document rather than its display formatting, or even the medium in which the document will be displayed; widely used in the publishing business and for producing technical documentation. Abbreviated SGML. { ¦stan·dərd ,jen·rə,līzd 'märk,əp ,laŋ·gwij }

standard gold [MET] A gold alloy containing 10% copper; at one time used for legal coinage in the United States. { 'stan·dərd 'gōld }

standard gravity [MECH] A value of the acceleration of gravity equal to 9.80665 meters per second per second. { 'stan·dərd 'grav·əd·ē }

standard heat of formation [THERMO] The heat needed to produce one mole of a compound from its elements in their standard state. { 'stan·dərd 'hēt əv fòr'mā·shən }

standard hole [DES ENG] A hole with zero allowance plus a specified tolerance; fit allowance is provided for by the shaft in the hole. { 'stan·dərd 'hōl }

standard hour [IND ENG] The quantity of output required of an operator to meet an hourly production quota. Also known as allowed hour. { 'stan·dərd 'aùr }

standard-hour plan [IND ENG] A wage incentive plan in which standard work times are expressed as standard hours and the worker is paid for standard hours instead of the actual work hours. { 'stan·dərd 'aùr 'plan }

standard illuminants [OPTICS] Three standard sources of light, designated A, B, and C, used in specifying the light used when colors are matched; A is light from a filament at a color temperature of 2575°C, and B and C, representing noon sunlight and normal daylight respectively, are obtained by modifying A with rigorously specified filters. { 'stan·dərd i'lü·mə·nəns }

standard inductor [ELECTROMAG] An inductor (coil) having high stability of inductance value, with little variation of inductance with current or frequency and with a low temperature coefficient; it may have an air core or an iron core; used as a primary standard in laboratories and as a precise working standard for impedance measurements. { 'stan·dərd in'dək·tər }

standard interface [COMPUT SCI] **1.** A joining place of two systems or subsystems that has a previously agreed-upon form, so that two systems may be readily connected together. **2.** In particular, a system of uniform circuits and input/output channels connecting the central processing unit of a computer with various units of peripheral equipment. { 'stan·dərd 'in·tər,fās }

standardization [ANALY CHEM] A process in which the value of a potential standard is fixed by a measurement made with respect to a standard whose value is known. [DES ENG] The adoption of generally accepted uniform procedures, dimensions, materials, or parts that directly affect the design of a product or a facility. [ENG] The process of establishing by common agreement engineering criteria, terms, principles, practices, materials, items, processes, and equipment parts and components. { ,stan·dər·də'zā·shən }

standardize [COMPUT SCI] To replace any given floating point representation of a number with its representation in standard form; that is, to adjust the exponent and fixed-point part so that the new fixed-point part lies within a prescribed standard range. { 'stan·dər,dīz }

standardized product [DES ENG] A product that conforms to specifications resulting from the same technical requirements. { 'stan·dər,dīzd 'präd·əkt }

standardized test statistic [STAT] A test statistic which has been reduced to standardized units. { 'stan·dər,dīzd 'test stə,tis·tik }

standardized units [STAT] A random variable Z has been reduced to standardized units when it has zero expected value and standard deviation 1; this is accomplished by dividing the difference of Z and the expected value of Z by the standard deviation of Z. { 'stan·dər,dīzd 'yü·nəts }

standard leak [ENG] Tracer gas allowed to enter a leak detector at a controlled rate in order to facilitate calibration and adjustment of the detector. { 'stan·dərd 'lēk }

standard lens [OPTICS] Usually the lens provided with a camera as standard equipment: in still cameras, the standard lens is one whose focal length is about equal to the length of the diagonal of the negative area normally provided by the camera; the normal field of view of a standard lens is about 53°. { 'stan·dərd 'lenz }

standard load [DES ENG] A load which has been preplanned as to dimensions, weight, and balance, and designated by a number or some classification. { 'stan·dərd 'lōd }

standard loran See loran A. { 'stan·dərd 'lò,ran }

standard mean ocean water [GEOL] An international reference standard used to determine oxygen and hydrogen isotopic content. Abbreviated SMOW. { ¦stan·dərd ,mēn 'ō·shən ,wòd·ər }

standard measure See standard score. { 'stan·dərd 'mezh·ər }

standard meridian [GEOD] The meridian used for reckoning standard time; throughout most of the world the standard meridians are those whose longitudes are exactly divisible by 15°. [MAP] A meridian of a map projection along which the scale is as stated. { 'stan·dərd mə'rid·ē·ən }

standard mineral [MINERAL] A mineral that, on the basis of chemical analyses, is theoretically capable of being present in a rock. Also known as normative mineral. { 'stan·dərd 'min·rəl }

standard model [PART PHYS] The modern theory of the interactions of elementary particles, comprising the Weinberg-Salam theory and quantum chromodynamics. { 'stan·dərd 'mäd·əl }

standard noise temperature [ELECTR] The standard reference temperature for noise measurements, equal to 290 K. { 'stan·dərd 'nȯiz ,tem·prə·chər }

standard noon [ASTRON] Twelve o'clock standard time, or the instant the mean sun is over the upper branch of the standard meridian. { 'stan·dərd 'nün }

standard output [IND ENG] The reciprocal of standard time. { 'stan·dərd 'aůt,půt }

standard parallel [MAP] A parallel on a map or chart along which the scale is as stated for that map or chart. [GEOD] The parallel or parallels of latitude used as control lines in the computation of a map projection. { 'stan·dərd 'par·ə,lel }

standard parallel port [COMPUT SCI] A parallel port that can transfer data in only one direction. { 'stan·dərd ,par·ə,lel 'pȯrt }

standard performance [IND ENG] The performance of an individual or of a group on meeting standard output. { 'stan·dərd pər'fȯr·məns }

standard pitch [ACOUS] A musical pitch based on 440 hertz for tone A; with this standard, the frequency of middle C is 261 hertz. { 'stan·dərd 'pich }

standard plane [CRYSTAL] The crystal plane whose Miller indices are (111), that is, whose intercepts on the crystal axes are proportional to the corresponding sides of a unit cell. { 'stan·dərd 'plān }

standard potential [PHYS CHEM] The potential of an electrode composed of a substance in its standard state, in equilibrium with ions in their standard states compared to a hydrogen electrode. { 'stan·dərd pə'ten·chəl }

standard preemphasis [COMMUN] Preemphasis in frequency-modulation and television aural broadcasting whose level lies between upper and lower limits specified by the Federal Communications Commission. { 'stan·dərd prē'em·fə·səs }

standard pressure [METEOROL] The arbitrarily selected atmospheric pressure of 1000 millibars to which adiabatic processes are referred for definitions of potential temperature, equivalent potential temperature, and so on. [PHYS] A pressure of 1 atmosphere (101,325 newtons per square meter), to which measurements of quantities dependent on pressure, such as the volume of a gas, are often referred. Also known as normal pressure. { 'stan·dərd 'presh·ər }

standard project flood [HYD] The volume of streamflow expected to result from the most severe combination of meteorological and hydrologic conditions which are reasonably characteristic of the geographic region involved, excluding extremely rare combinations. { 'stan·dərd 'präj,ekt ,fləd }

standard propagation [ELECTROMAG] Propagation of radio waves over a smooth spherical earth of specified dielectric constant and conductivity, under conditions of standard refraction in the atmosphere. { 'stan·dərd ,präp·ə'gā·shən }

standard rate turn [AERO ENG] A turn in an aircraft in which heading changes at the rate of 3° per second. { 'stan·dərd 'rāt 'tərn }

standard reference material [ANALY CHEM] A reference material distributed and certified by the appropriate national institute for standardization. { 'stan·dərd 'ref·rəns mə,tir·ē·əl }

standard refraction [ELECTROMAG] Refraction which would occur in an idealized atmosphere in which the index of refraction decreases uniformly with height at a rate of 39×10^{-6} per kilometer; standard refraction may be included in ground wave calculations by use of an effective earth radius of 8.5×10^6 meters, or $4/3$ the geometrical radius of the earth. { 'stan·dərd ri'frak·shən }

standard sand [MATER] A silica sand that is free of organic matter and is used in making test samples of concrete and cement. { 'stan·dərd 'sand }

standard score [STAT] A test score converted or transformed into a common scale, such as standard units, to effect a more reasonable scale of measurement in order to make comparisons between different tests. Also known as standard measure. { 'stan·dərd 'skȯr }

standard seawater See normal water. { 'stan·dərd 'sē,wȯd·ər }

standard shaft [DES ENG] A shaft with zero allowance minus a specified tolerance. { 'stan·dərd 'shaft }

standard solution See titrant. { 'stan·dərd sə'lü·shən }

standard star [ASTRON] A star whose position or other data are precisely known so that it is used as a reference to calculate positions of other celestial bodies, or of objects on earth. { 'stan·dərd 'stär }

standard state [PHYS] The stable and pure form of a substance at standard pressure and a specified temperature, usually 298 K (77°F). { 'stan·dərd 'stāt }

standard station See reference station. { 'stan·dərd 'stā·shən }

standard subroutine [COMPUT SCI] In computers, a subroutine which is applicable to a class of problems. { 'stan·dərd 'səb·rü,tēn }

standard target [ELECTROMAG] A radar target which will produce an echo of known power under various conditions; smooth metal spheres or corner reflectors of known dimensions are such targets, and they may be used to calibrate a radar or check its performance. { 'stan·dərd 'tär·gət }

standard temperature and pressure See standard conditions. { 'stan·dərd 'tem·prə·chər ən 'presh·ər }

standard terminal arrival route [NAV] A route established for the purpose of reducing pilot and controller workload and reducing communications congestion in air-traffic control arrival operations. Abbreviated STAR. { 'stan·dərd 'ter·mən·əl ə'rīv·əl ,rüt }

standard test-tone power [ELECTR] One milliwatt (0 decibels above one milliwatt) at 1000 hertz. { 'stan·dərd 'test ,tōn ,paů·ər }

standard time [ASTRON] The mean solar time, based on the transit of the sun over a specified meridian, called the time meridian, and adopted for use over an area that is called a time zone. [IND ENG] A unit time value for completion of a work task as determined by the proper application of the appropriate work-measurement techniques. Also known as direct labor standard; output standard; production standard; time standard. { 'stan·dərd 'tīm }

standard ton See ton. { 'stan·dərd 'tən }

standard trajectory [MECH] Path through the air that it is calculated a projectile will follow under given conditions of weather, position, and material, including the particular fuse, projectile, and propelling charge that are used; firing tables are based on standard trajectories. { 'stan·dərd trə'jek·trē }

standard visibility See meteorological range. { 'stan·dərd ,viz·ə'bil·əd·ē }

standard visual range See meteorological range. { 'stan·dərd 'vizh·ə·wəl 'ranj }

standard volume [PHYS] The volume of 1 mole of a gas at a pressure of 1 atmosphere and a temperature of 0°C. Also known as normal volume. { 'stan·dərd 'väl·yəm }

standard waveguide [ELECTROMAG] Any one of several rectangular waveguides whose dimensions have been specified by various agencies and which are in general use. { 'stan·dərd 'wāv,gīd }

standard wiggler [NUCLEO] A wiggler that has a relatively high magnetic field and produces radiation with a smooth spectral distribution similar to that of normal synchrotron radiation. { 'stan·dərd 'wig·lər }

standard wire rope [DES ENG] Wire rope made of six wire strands laid around a sisal core. Also known as hemp-core cable. { 'stan·dərd 'wīr 'rōp }

standby battery [ELEC] A storage battery held in reserve as an emergency power source in event of failure of regular power facilities at a radio station or other location. { 'stand,bī ,bad·ə,rē }

standby computer [COMPUT SCI] A computer in a duplex system that takes over when the need arises. { 'stand,bī kəm,pyüd·ər }

standby mode [ELEC] The operation of a circuit or device with unused portions of the circuit disconnected to reduce power consumption. { 'stan,bī mōd }

standby power source [ELEC] An uninterruptible power system in which the load normally operated from the commercial power line is switched to the output of a dc-to-ac static inverter powered by a battery in the event of a power failure. { 'stand,bī 'paů·ər ,sȯrs }

standby register [COMPUT SCI] In computers, a register into which information can be copied to be available in case the original information is lost or mutilated in processing. { 'stand¦bī ,rej·ə·stər }

standby replacement redundancy [COMPUT SCI] A form of redundancy in which there is a single active unit and a reserve of spare units, one of which replaces the active unit if it fails. { 'stand¦bī ri'plās·mənt ri,dən·dən·sē }

standby time [COMPUT SCI] **1.** The time during which two or more computers are tied together and available to answer inquiries or process intermittent actions on stored data. **2.** The elapsed time between inquiries when the equipment is operating on an inquiry application. { 'stand¦bī ,tīm }

stand fire [FOR] A forest fire igniting in the trunks of trees. { 'stand ,fīr }

standing cloud [METEOROL] Any stationary cloud maintaining its position with respect to a mountain peak or ridge. { 'stand·iŋ 'klaud }

standing crop [ECOL] The number of individuals or total biomass present in a community at one particular time. { 'stand·iŋ 'kräp }

standing ground [MIN ENG] Ground that will stand firm without timbering. { 'stand·iŋ 'graund }

standing knee height [ANTHRO] The vertical distance taken from the top of the kneecap to the floor as the subject stands. { 'stand·iŋ 'nē ,hīt }

standing-on-nines carry [COMPUT SCI] In high-speed parallel addition of decimal numbers, an arrangement that causes carry digits to pass through one or more nine digits, while signaling that the skipped nines are to be reset to zero. { ¦stand·iŋ òn ¦nīnz 'kar·ē }

standing operating procedure [AERO ENG] A set of instructions covering those features of operations which lend themselves to a definite or standardized procedure without loss of effectiveness; the procedure is applicable unless prescribed otherwise in a particular case; thus, the flexibility necessary in special situations is retained. { 'stand·iŋ 'äp·ə,rād·iŋ prə,sē·jər }

standing rigging [NAV ARCH] Rigging that is permanently secured such as shrouds, stays, bob-stays, martingales, and mast pendants. { 'stand·iŋ 'rig·iŋ }

standing valve [PETRO ENG] A sucker-rod-pump (oil well) discharge valve that remains stationary during the pumping cycle, in contrast to a traveling valve. { 'stand·iŋ 'valv }

standing water See stagnant water. { ¦stand·iŋ 'wód·ər }

standing wave [PHYS] A wave in which the ratio of an instantaneous value at one point to that at any other point does not vary with time. Also known as stationary wave. { 'stand·iŋ 'wāv }

standing-wave detector [ELECTROMAG] An electric indicating instrument used for detecting a standing electromagnetic wave along a transmission line or in a waveguide and measuring the resulting standing-wave ratio; it can also be used to measure the wavelength, and hence the frequency, of the wave. Also known as standing-wave indicator; standing-wave meter; standing-wave-ratio meter. { 'stand·iŋ ¦wāv di'tek·tər }

standing-wave indicator See standing-wave detector. { 'stand·iŋ ¦wāv 'in·də,kād·ər }

standing-wave loss factor [ELECTROMAG] The ratio of the transmission loss in an unmatched waveguide to that in the same waveguide when matched. { 'stand·iŋ ¦wāv ¦lòs ,fak·tər }

standing-wave meter See standing-wave detector. { 'stand·iŋ ¦wāv 'mēd·ər }

standing-wave method [ELECTROMAG] Any method of measuring the wavelength of electromagnetic waves that involves measuring the distance between successive nodes or antinodes of standing waves. { 'stand·iŋ ¦wāv 'meth·əd }

standing-wave producer [ELECTROMAG] A movable probe inserted in a slotted waveguide to produce a desired standing-wave pattern, generally for test purposes. { 'stand·iŋ ¦wāv prə'dü·sər }

standing-wave ratio [PHYS] **1.** The ratio of the maximum amplitude to the minimum amplitude of corresponding components of a wave in a transmission line or waveguide. **2.** The reciprocal of this ratio. { 'stand·iŋ ¦wāv 'rā·shō }

standing-wave-ratio meter See standing-wave detector. { 'stand·iŋ ¦wāv 'rā·shō ,mēd·ər }

standing ways See ground ways. { 'stand·iŋ 'wāz }

stand method [FOR] The practice of successively cutting trees of different ages so that ultimately the trees in the stand are new growth of a uniform age. { 'stand ,meth·əd }

standoff insulator [ELEC] An insulator used to support a conductor at a distance from the surface on which the insulator is mounted. { 'stan,dòf 'in·sə,lād·ər }

standoff jammer [ELECTR] An aircraft that patrols the target air space and engages in high-power jamming of both the acquisition or tracking devices and the closing vehicles, by using powerful transmitters excited by travelling-wave tubes. { 'stan,dòf 'jam·ər }

standoff target acquisition system [ORD] An airborne radar system designed to locate with extreme accuracy moving targets within enemy territory while ensuring maximum survivability. Abbreviated SOTAS. { 'stan,dòf 'tar·gət ,ak·wə'zish·ən ,sis·təm }

standoff weapon [ORD] A weapon which may be launched at a distance sufficient to allow attacking personnel to evade defensive fire from the target area. { 'stan,dòf ,wep·ən }

stand on [NAV] To proceed on the same course. { 'stand 'òn }

standpipe [ENG] **1.** A vertical pipe for holding a water supply for fire protection. **2.** A high tank or reservoir for holding water that is used to maintain a uniform pressure in a water-supply system. { 'stand,pīp }

standpipe system [ENG] A system that contains standpipes, pumps, siamese connections, piping, and equipment with hose outlets and is provided with an adequate supply of water for fire fighting. { 'stand,pīp sis·təm }

standstill [ASTRON] An interval in the cycle of a variable star during which its brightness remains nearly constant. { 'stan,stil }

standstill feature [CONT SYS] A device which insures that false signals such as fluctuations in the power supply do not cause a controller to be altered. { 'stan,stil ,fē·chər }

stanfieldite [MINERAL] $Ca_4(Mg,Fe,Mn)_5(PO_4)_6$ A phosphate mineral found only in meteorites. { 'stan,fēl,dīt }

Stanford achievement test [PSYCH] A group test employing a primary (grades 2 and 3) and an advanced (grades 4 through 9) examination to measure achievement at various grade levels. { 'stan·fərd ə'chēv·mənt ,test }

Stanhope lens [OPTICS] A thick biconvex lens with front and back surfaces having radii of curvature of two-thirds and one-third the lens thickness; used as a magnifier by placing the object to be viewed in contact with the front surface. { 'stan,hōp ,lenz }

stannane See tin hydride. { 'sta,nān }

stannic acid See stannic oxide. { 'stan·ik 'as·əd }

stannic anhydride See stannic oxide. { 'stan·ik an'hī,drīd }

stannic bromide [INORG CHEM] $SnBr_4$ Water- and alcohol-soluble, white crystals that fume when exposed to air, and melt at 31°C; used in mineral separations. Also known as tin bromide; tin tetrabromide. { 'stan·ik 'brō,mīd }

stannic chloride [INORG CHEM] $SnCl_4$ A colorless, fuming liquid; soluble in cold water, alcohol, carbon disulfide, and oil of turpentine; decomposed by hot water; boils at 114°C; used as a conductive coating and a sugar bleach, and in drugs, ceramics, soaps, and blueprinting. Also known as tin chloride; tin tetrachloride. { 'stan·ik 'klòr,īd }

stannic chromate [INORG CHEM] $Sn(CrO_4)_2$ Toxic, brownish-yellow crystals, slightly soluble in water; used to decorate porcelain and china. Also known as tin chromate. { 'stan·ik 'krō,māt }

stannic iodide [INORG CHEM] SnI_4 Yellow-reddish crystals; insoluble in water, soluble in alcohol, ether, chloroform, carbon disulfide, and benzene; decomposed by water, melt at 144°C, sublime at 180°C. Also known as tin iodide; tin tetraiodide. { 'stan·ik 'ī·ə,dīd }

stannic oxide [INORG CHEM] SnO_2 A white powder; insoluble in water, soluble in concentrated sulfuric acid; melts at 1127°C; used in ceramic glazes and colors, special glasses, putty, and cosmetics, and as a catalyst. Also known as flowers of tin; stannic acid; stannic anhydride; tin dioxide; tin oxide; tin peroxide. { 'stan·ik 'äk,sīd }

stannic sulfide [INORG CHEM] SnS_2 A yellow-brown powder; insoluble in water, soluble in alkaline sulfides; decomposes at red heat; used as a pigment and for imitation gilding. Also known as artificial gold; mosaic gold; tin bisulfide. { 'stan·ik 'səl,fīd }

stannite [MINERAL] Cu_2FeSnS_4 A steel-gray or iron-black mineral crystallizing in the tetragonal system and occurring in granular masses; luster is metallic, hardness is 4 on Mohs scale, and specific gravity is 4.3–4.53. Also known as bell-metal ore; tin pyrites. { 'sta,nīt }

stannous bromide [INORG CHEM] $SnBr_2$ A yellow powder; soluble in water, alcohol, acetone, ether, and dilute hydrochloric acid; browns in air; melts at 215°C. Also known as tin bromide. { 'stan·əs 'brō,mīd }

stannous chloride [INORG CHEM] $SnCl_2$ White crystals; soluble in water, alcohol, and alkalies; oxidized in air to the oxychloride; melt at 247°C; used as a chemical intermediate, reducing agent, and ink-stain remover, and for silvering mirrors. Also known as tin chloride; tin crystals; tin dichloride; tin salts. { 'stan·əs 'klȯr,īd }

stannous chromate [INORG CHEM] $SnCrO_4$ A brown powder; very slightly soluble in water; used to decorate porcelain. Also known as tin chromate. { 'stan·əs 'krō,māt }

stannous 2-ethylhexoate [ORG CHEM] $Sn(C_8H_{15}O_2)_2$ A light yellow liquid, soluble in benzene, toluene, and petroleum ether; used as a lubricant, a vulcanizing agent, and a stabilizer for transformer oil. { 'stan·əs ¦tü ¦eth·əl'hek·sə,wāt }

stannous fluoride [INORG CHEM] SnF_2 A white, lustrous powder; slightly soluble in water; used to fluoridate toothpaste and as a medicine. { 'stan·əs 'flu̇r,īd }

stannous oxalate [ORG CHEM] SnC_2O_4 A white, crystalline powder that decomposes at about 280°C; soluble in acids; used in textile dyeing and printing. Also known as tin oxalate. { 'stan·əs 'äk·sə,lāt }

stannous oxide [INORG CHEM] SnO An air-unstable, brown to black powder; insoluble in water, soluble in acids and strong bases; decomposes when heated; used as a reducing agent and chemical intermediate, and for glass plating. Also known as tin oxide; tin protoxide. { 'stan·əs 'äk,sīd }

stannous sulfate [INORG CHEM] $SnSO_4$ Heavy light-colored crystals; decomposes rapidly in water, loses SO_2 at 360°C; used for dyeing and tin plating. Also known as tin sulfate. { 'stan·əs 'səl,fāt }

stannous sulfide [INORG CHEM] SnS Dark crystals; insoluble in water, soluble (with decomposition) in concentrated hydrochloric acid; melts at 880°C; used as an analytical reagent and catalyst, and in bearing material. Also known as tin monosulfide; tin protosulfide; tin sulfide. { 'stan·əs 'səl,fīd }

stannum [CHEM] The Latin name for tin, thus the symbol Sn for the element. { 'stan·əm }

Stanton diagram [FL MECH] The plot of the airflow friction coefficient against the Reynolds number. { 'stant·ən ,dī·ə,gram }

Stanton number [THERMO] A dimensionless number used in the study of forced convection, equal to the heat-transfer coefficient of a fluid divided by the product of the specific heat at constant pressure, the fluid density, and the fluid velocity. Symbolized N_{St}. Also known as Margoulis number (M). { 'stant·ən ,nəm·bər }

stapedectomy [MED] Surgical reconstruction of the junction of the ossicular chain of the middle ear with the oval window of the inner ear, fixed in place by otosclerosis. { ,stā·pə'dek·tə·mē }

stapedius muscle [ANAT] The muscle which attaches to and controls the stapes in the inner ear. { stə'pēd·ē·əs ,məs·əl }

stapes [ANAT] The stirrup-shaped middle-ear ossicle, articulating with the incus and the oval window. Also known as columella. { 'stā·pēz }

STAPHYLINIDAE

A representative member of the coeleopteran superfamily Staphylinidae.

Staphylinidae [INV ZOO] The rove beetles, a very large family of coleopteran insects in the superfamily Staphylinoidea. { ,staf·ə'lin·ə,dē }

Staphylinoidea [INV ZOO] A superfamily of Coleoptera in the suborder Polyphaga. { ,staf·ə·lə'nȯid·ē·ə }

staphylococcal pneumonia [MED] A severe form of pneumonia caused by *Staphylococcus aureus*. { ,staf·ə·lō'käk·əl nu̇'mō·nē·ə }

staphylomycin [MICROBIO] An antibiotic composed of three active components produced by a strain of *Actinomyces* that inhibits growth of gram-positive microorganisms and acid-fast bacilli. { ,staf·ə·lō'mīs·ən }

staphylotoxin [BIOCHEM] Any of the various toxins elaborated by strains of *Staphylococcus aureus*, including hemolysins, enterotoxins, and leukocidin. { ,staf·ə·lō'täk·sən }

staple [DES ENG] A U-shaped loop of wire with points at both ends; used as a fastener. [TEXT] The average fiber length to be used in spinning a yarn. { 'stā·pəl }

stapler [ENG] **1.** A device for inserting wire staples into paper or wood. **2.** A hammer for inserting staples. { 'stā·plər }

star [ASTRON] A celestial body consisting of a large, self-luminous mass of hot gas held together by its own gravity; the sun is a typical star. [MATH] For a member S of a family of sets, the collection of all sets in the family that contain S as a subset. [NUCLEO] A star-shaped group of tracks made by ionizing particles originating at a common point in a nuclear emulsion, cloud chamber, or bubble chamber; some stars are produced by successive disintegrations of an atom in a radioactive series, and others by nuclear reactions of the spallation type, such as those due to cosmic-ray particles. Also known as nuclear star. [OPTICS] A light source that subtends a very small angle at the entrance pupil of an optical instrument and is used to test the instrument. { stär }

STAR *See* ship-tended acoustic relay; standard terminal arrival route. { stär }

star algebra [MATH] A real or complex algebra on which an involution is defined. { 'stär ,al·jə·brə }

star atlas [ASTRON] A series of star maps for different times, for example, for each month; the maps are generally drawn to a small scale and in book form. { 'stär ,at·ləs }

starboard [NAV ARCH] The side of a ship or airplane which is on one's right when one faces forward. { 'stär·bərd }

starburst galaxy [ASTRON] A galaxy that is presently undergoing a period of intense star formation. { 'stär,bərst ,gal·ik·sē }

starburst polymer *See* dendrimer. { ,stär,bərst 'päl·ə·mər }

star catalog [ASTRON] A comprehensive tabulation of data concerning the stars listed; the data may include, for example, apparent positions, brightness, motions, parallaxes, and other properties of stars. { 'stär ¦kad·əl,äg }

starch [BIOCHEM] Any one of a group of carbohydrates or polysaccharides, of the general composition $(C_6H_{10}O_5)_n$, occurring as organized or structural granules of varying size and markings in many plant cells; it hydrolyzes to several forms of dextrin and glucose; its chemical structure is not completely known, but the granules consist of concentric shells containing at least two fractions: an inner portion called amylose, and an outer portion called amylopectin. { stärch }

star chain [NAV] A group of synchronized, transmitting, navigation stations having the master station located at the center of a roughly circular area and surrounded by three or more slave stations located at some distance from the master. { 'stär ,chān }

star chart *See* star map. { 'stär ,chärt }

starch nitrate *See* nitrostarch. { 'stärch 'nī,trāt }

star cloud [ASTRON] An aggregation of thousands or of millions of stars spread over hundreds or thousands of light-years. { 'stär ,klau̇d }

star cluster [ASTRON] A group of stars held together by gravitational attraction; the two chief types are open clusters (composed of from 12 to hundreds of stars) and globular clusters (composed of thousands to hundreds of thousands of stars). Also known as cluster. { 'stär ,kləs·tər }

star color [ASTRON] The color of a star as a function of its radiation and related to its temperature; colors range from blue-white to deep red. { 'stär ,kəl·ər }

star-connected circuit [ELEC] Polyphase circuit in which all the current paths within the region that delimits the circuit extend from each of the points of entry of the phase conductors to a common conductor (which may be the neutral conductor). { 'stär kə¦nek·təd 'sər·kət }

star connection *See* star network. { 'stär kə¦nek·shən }

star count [ASTRON] A count of stars on a photographic plate. { 'stär ,kau̇nt }

star cut [LAP] A gem cut characterized by a hexagonal table surrounded by six facets in the form of an equilateral triangle. { 'stär ,kət }

star day [ASTRON] The time period between two successive passages of a star across the meridian. { 'stär 'dā }

star-delta switching starter [ELEC] A type of motor starter, used with three-phase induction motors, that switches the stator windings from a star connection to a delta connection. { 'stär 'del·tə ¦swich·iŋ ,stärd·ər }

star density [ASTRON] The average number of stars in a unit volume of space. { 'stär ,den·səd·ē }

star diagram [MATH] A graphical way of representing a partition of a positive integer using asterisks that are arranged in rows corresponding to the parts. { 'stär ,dī·ə,gram }

star drift [ASTRON] A description of two star groups in the Milky Way traveling through each other in opposite directions; individual stars have movements that are relative to each other. Also known as star stream. { 'stär ,drift }

star drill [DES ENG] A tool with a star-shaped point, used for drilling in stone or masonry. { 'stär ,dril }

star finder [ASTRON] A device such as a star map or celestial globe to facilitate the identification of stars. Also known as a star identifier. { 'stär ,fin·dər }

starfish [INV ZOO] The common name for echinoderms belonging to the subclass Asteroidea. { 'stär,fish }

star-free expression [COMPUT SCI] An expression containing only Boolean operations and concatenation, used to define the language corresponding to a counter-free machine. { 'stär ,frē ik'spresh·ən }

star globe See celestial globe. { 'stär ,glōb }

star grain [MATER] A rocket propellant grain with a cavity of star-shaped cross section. { 'stär ,grān }

star group [ASTRON] A number of stars that move in the same general direction at the same time. { 'stär ,grüp }

star identifier See star finder. { 'stär ī,den·tə,fī·ər }

Stark effect [SPECT] The effect on spectrum lines of an electric field which is either externally applied or is an internal field caused by the presence of neighboring ions or atoms in a gas, liquid, or solid. Also known as electric field effect. { 'stärk i,fekt }

Stark-Einstein law See Einstein photochemical equivalence law. { 'stärk 'īn,stīn ,lo }

Stark-Lunelund effect [ELECTROMAG] The polarization of light emitted from a beam of moving atoms in a region where there are no electric or magnetic fields. { 'stärk 'lün·ə,lənd i,fekt }

Stark number See Stefan number. { 'stärk ,nəm·bər }

star lamp [ELEC] A high-pressure xenon arc, used in a planetarium, which produces a tiny, intense point of light focused through thousands of individual lenses and pinholes, and projected to the planetarium's dome. { 'stär ,lamp }

starlike region [MATH] A region in the complex number plane such that the line segment joining any of its points to the origin lies entirely in the region. { 'stär,līk ,rē·jən }

starling [CIV ENG] A protective enclosure around the pier of a bridge that consists of piles driven close together and is often filled with gravel or stone to protect the pier by serving as a break to water, ice, or drift. { 'stär·liŋ }

Starling's law of the heart [PHYSIO] The energy associated with cardiac contraction is proportional to the length of the myocardial fibers in diastole. { 'stär·liŋz 'lo əv thə 'härt }

star map [ASTRON] A map indicating the relative apparent positions of the stars. Also known as star chart. { 'stär ,map }

star model See stellar model. { 'stär ,mäd·əl }

star motions [ASTRON] For the Milky Way, this includes rotation within the galaxy, motion which is described with respect to an external frame of reference; superposed on this systematic rotation are the individual motions of a star; each star moves in a somewhat elliptical orbit, with respect to the local standard of rest, the standard moving in a circular orbit around the galactic center. { 'stär ,mō·shənz }

star names [ASTRON] Nomenclature for the identification of stars; hundreds of stars have proper names that are traditional, for example, Betelgeuse; this star may be also identified as α Orionis (Alpha Orionis), α for its being the brightest visual star in the constellation Orion. { 'stär ,nāmz }

star network [COMMUN] A communications network in which all communications between any two points must pass through a central node. Also known as centralized configuration. [ELEC] A set of three or more branches with one terminal of each connected at a common node to give the form of a star. Also known as star connection. { 'stär ,net,wərk }

star place [ASTRON] The position of a star on the celestial sphere, usually measured by its right ascension and declination. { 'stär ,plās }

star polymer [ORG CHEM] A macromolecule having a small core of molecules (branch point) with branches radiating from the core. { 'stär ,päl·ə·mər }

star ruby [MINERAL] An asteriated variety of ruby with normally six chatoyant rays. { 'stär 'rü·bē }

star sapphire [MINERAL] A variety of sapphire exhibiting a six-pointed star resulting from the presence of microscopic crystals in various orientations within the gemstone. { 'stär 'sa,fīr }

star-shaped set [MATH] With respect to a point P of a euclidean space or vector space, a set such that if Q is a member of the set, then so is any point on the line segment PQ. { 'stär ,shāpt ,set }

star shell See illuminating projectile. { 'stär ,shel }

star spot [ASTRON] A region of reduced brightness of the surface of a star comparable to a sunspot on the Sun's surface. { 'stär,spät }

star stream See star drift. { 'stär 'strēm }

star streaming [ASTRON] A phenomenon that results from the mean random speeds of stars being different in different directions. { 'stär ,strēm·iŋ }

star subalgebra [MATH] A subalgebra of a star algebra which is mapped onto itself by the involution operation. { 'stär ,səb,al·jə·brə }

start bit [COMPUT SCI] The first bit transmitted in asynchronous data transmission to unequivocally indicate the start of the word { 'stärt ,bit }

start codon See initiation codon. { 'stärt 'kō,dän }

start dialing signal [COMMUN] Signal transmitted from the incoming end of a circuit, following the receipt of a seizing signal, to indicate that the necessary circuit conditions have been established for receiving the numerical routine information. { 'stärt 'dīl·iŋ ,sig·nəl }

started task [COMPUT SCI] A computer program that is kept permanently in main storage and, though not a part of the operating system, is treated as though it were. { 'stärd·əd 'task }

start element [COMMUN] The first element of a character in certain serial transmissions, used to permit synchronization. { 'stärt ,el·ə·mənt }

star telescope [OPTICS] An accessory of the marine navigational sextant designed primarily for star observations; it has a large objective to give a greater field of view and increased illumination; it is an erect telescope, that is, the object viewed is seen erect as opposed to the inverting telescope in which the object viewed is inverted. { 'stär ,tel·ə,skōp }

starter [ELEC] **1.** A device used to start an electric motor and to accelerate the motor to normal speed. **2.** See engine starter. [ELECTR] An auxiliary control electrode used in a gas tube to establish sufficient ionization to reduce the anode breakdown voltage. Also known as trigger electrode. [ENG] A drill used for making the upper part of a hole, the remainder of the hole being made with a drill of smaller gage, known as a follower. [MICROBIO] A culture of microorganisms, either pure or mixed, used to commence a process, for example, cheese manufacture. { 'stär·dər }

star test [OPTICS] A procedure in which a telescope is directed at a bright star and the in-focus and out-of-focus images and diffraction patterns of the star are examined to detect aberrations and abnormalities in the optical system. { 'stär ,test }

starting barrel [ENG] A short (12 to 24 inches or 30 to 60 centimeters) core barrel used to begin coring operations when the distance between the drill chuck and the bottom of the hole or to the rock surface in which a borehole is to be collared is too short to permit use of a full 5- or 10-foot-long (1.5- or 3.0-meter) core barrel. { 'stärd·iŋ ,bar·əl }

starting box [ELEC] A device for providing extra resistance in the armature of a motor while it is being started. { 'stärd·iŋ ,bäks }

starting friction See static friction. { 'stärd·iŋ ,frik·shən }

starting mix [MATER] In pyrotechnic devices, an easily ignited mixture which transmits flame from an initiating device to a less readily ignitable composition. { 'stärd·iŋ ,miks }

starting motor See engine starter. { 'stärd·iŋ ,mōd·ər }

starting reactor [ELEC] A reactor that is used to limit the starting current of electric motors, and usually consists of an iron-core inductor connected in series with the machine stator winding. { 'stärd·iŋ rē,ak·tər }

STARK EFFECT

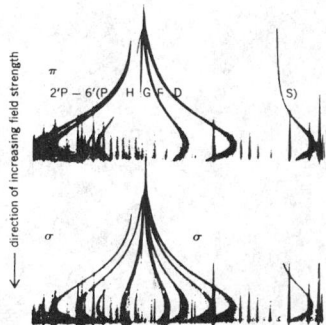

An example of the linear Stark effect in the 4144- and 4169-angstrom lines of helium showing large symmetrical pattern. Electric field strengths range continuously from 0 to 85,000 volts per centimeter. Symbols π and σ refer respectively to light polarized parallel and perpendicular to the electric field.

starting resistance [MECH ENG] The force needed to produce an oil film on the journal bearings of a train when it is at a standstill. { 'stärd·iŋ ri,zis·təns }

starting sheet [MET] A thin sheet of metal used as the initial cathode in electrowinning or electrorefining. { 'stärd·iŋ ,shēt }

starting taper [DES ENG] A slight end taper on a reamer to aid in starting. { 'stärd·iŋ ,tā·pər }

startle response [PHYSIO] The complex, involuntary, usually spasmodic psychophysiological response movement of an organism to a sudden unexpected stimulus. { 'stärd·əl ri,späns }

startover [COMPUT SCI] Program function that causes a computer that is not active to become active. { 'stär,dō·vər }

startover data transfer and processing program [COMPUT SCI] Program which controls the transfer of startover data from the active to the standby machine and their subsequent processing by the standby machine. { 'stär,dō·vər 'dad·ə ,tranz·fər ən 'prä,ses·iŋ ,prō,gram }

startpoint [MOL BIO] The deoxyribonucleic acid base pair that corresponds to the first nucleotide incorporated into the primary ribonucleic acid (RNA) transcript by RNA polymerase. { 'stärt,póint }

star tracker *See* astrotracker. { 'stär ,trak·ər }

start-stop multivibrator *See* monostable multivibrator. { 'stärt 'stäp ,məl·ti'vī,brād·ər }

start-stop printing telegraph [COMMUN] Form of printing telegraph in which the signal-receiving mechanisms, normally at rest, are started in operation at the beginning and stopped at the end of each character transmitted over the channel. { 'stärt 'stäp 'print·iŋ 'tel·ə·graf }

start-stop system [COMMUN] A telegraph system in which each group of code elements corresponding to a character is preceded by a start signal that prepares the receiving mechanism to receive and register a character, and is followed by a stop signal that brings the receiving mechanism to rest in preparation for the reception of the next character. { 'stärt 'stäp ,sis·təm }

start time [IND ENG] The calendar time at which the manufacturing work for a specific job begins on a machine or in a facility. { 'stärt ,tīm }

start-to-leak pressure [MECH ENG] The amount of inlet pressure at which the first bubble occurs at the outlet of a safety relief valve with a resilient disk when the valve is subjected to an air test under a water seal. { 'stärt tə 'lēk ,presh·ər }

start-up curve [IND ENG] A learning curve applied to a job for the purpose of adjusting work times that are longer than the standard because of the introduction of new jobs or new workers. { 'stärd,əp ,kərv }

start-up time *See* setup time. { 'stärd,əp ,tīm }

star valve [FOOD ENG] In a machine for preparing puffed foods, a thick wheel with vanes that rotates in a tight-fitting cylindrical casing and transfers the product from a higher pressure to a lower pressure, or vice versa. Also known as rotary lock. { 'stär ,valv }

starved basin [GEOL] A sedimentary basin in which rate of subsidence exceeds rate of sedimentation. { 'stärvd 'bās·ən }

starved joint [ENG] A glued joint containing insufficient or inadequate adhesive. Also known as hungry joint. { 'stärvd 'jóint }

star wheel [COMPUT SCI] The sensing device of a card punch, which is held in contact with the card under spring tension and which, detecting a hole, closes a contact point. { 'stär ,wēl }

stasis [MED] A cessation of the normal flow of blood or other body fluids. { 'stā·səs }

stasis dermatitis [MED] Chronic inflammation of the skin of the legs, resulting from poor circulation. { 'stā·səs ,dər·mə'tīd·əs }

stat [NUCLEO] A unit of radioactive disintegration rate equal to the disintegration rate of that quantity of radon that gives rise to a charge of 1 statcoulomb in 1 second when situated in air. { stat }

stat- [ELEC] A prefix indicating an electrical unit in the electrostatic centimeter-gram-second system of units; it is attached to the corresponding SI unit. { stat }

stat℧ *See* statmho.

statΩ *See* statohm.

statA *See* statampere. { 'stat¦ā }

statampere [ELEC] The unit of electric current in the electrostatic centimeter-gram-second system of units, equal to a flow of charge of 1 statcoulomb per second; equal to approximately 3.3356×10^{-10} ampere. Abbreviated statA. { stad'am,pir }

statC *See* statcoulomb. { 'stat,sē }

statcoulomb [ELEC] The unit of charge in the electrostatic centimeter-gram-second system of units, equal to the charge which exerts a force of 1 dyne on an equal charge at a distance of 1 centimeter in a vacuum; equal to approximately 3.3356×10^{-10} coulomb. Abbreviated statC. Also known as franklin (Fr); unit charge. { 'stat¦kü,läm }

state [CONT SYS] A minimum set of numbers which contain enough information about a system's history to enable its future behavior to be computed. [PHYS] The condition of a system which is specified as completely as possible by observations of a specified nature, for example, thermodynamic state, energy state. [QUANT MECH] The condition in which a system exists; the state may be pure and describable by a wave function or mixed and describable by a density matrix. { stāt }

state equations [CONT SYS] Equations which express the state of a system and the output of a system at any time as a single valued function of the system's input at the same time and the state of the system at some fixed initial time. { 'stāt i,kwā·zhənz }

state estimator *See* observer. { 'stāt ,es·tə,mād·ər }

state feedback [CONT SYS] A class of feedback control laws in which the control inputs are explicit memoryless functions of the dynamical system state, that is, the control inputs at a given time t_a are determined by the values of the state variables at t_a and do not depend on the values of these variables at earlier times $t \geq t_a$. { 'stāt 'fēd,bak }

state graph [COMPUT SCI] A directed graph whose nodes correspond to internal states of a sequential machine and whose edges correspond to transitions among these states. { 'stāt ,graf }

statement [COMPUT SCI] An elementary specification of a computer action or process, complete and not divisible into smaller meaningful units; it is analogous to the simple sentence of a natural language. { 'stāt·mənt }

statement editor [COMPUT SCI] A text editor in which the text is divided into superlines, that is, units greater than ordinary lines, resulting in easier editing and freedom from truncation problems. { 'stāt·mənt ,ed·əd·ər }

statement function *See* propositional function. { 'stāt·mənt ,fəŋk·shən }

state observer *See* observer. { 'stāt əb,zər·vər }

state of matter [PHYS] One of three fundamental conditions of matter: the solid, liquid, and gaseous states. { 'stāt əv 'mad·ər }

state of strain [MECH] A complete description, including the six components of strain, of the deformation within a homogeneously deformed volume. { 'stāt əv 'strān }

state of stress [MECH] A complete description, including the six components of stress, of a homogeneously stressed volume. { 'stāt əv 'stres }

state of the sea [OCEANOGR] A description of the properties of the wind-generated waves on the surface of the sea. { 'stāt əv th̲ə 'sē }

state of the sky [METEOROL] The aspect of the sky in reference to the cloud cover; the state of the sky is fully described when the amounts, kinds, directions of movement, and heights of all clouds are given. { 'stāt əv th̲ə 'skī }

state parameter *See* thermodynamic function of state. { 'stāt pə,ram·əd·ər }

state space [CONT SYS] The set of all possible values of the state vector of a system. { 'stāt ,spās }

state table [COMPUT SCI] A table that represents a sequential machine, in which the rows correspond to the internal states, the columns correspond to the input combinations, and the entries to the next state. { 'stāt ,tā·bəl }

state transition equation [CONT SYS] The equation satisfied by the $n \times n$ state transition matrix $\Phi(t,t_0)$: $\partial \Phi(t,t_0)/\partial t = A(t)\,\Phi(t,t_0)$, $\Phi(t_0,t_0) = I$; here I is the unit $n \times n$ matrix, and $A(t)$ is the $n \times n$ matrix which appears in the vector differential equation $dx(t)/dt = A(t)x(t)$ for the n-component state vector $x(t)$. { 'stāt tran'zish·ən i,kwā·zhən }

state transition matrix [CONT SYS] A matrix $\Phi(t,t_0)$ whose product with the state vector x at an initial time t_0 gives the state vector at a later time t; that is, $x(t) = \Phi(t,t_0)x(t_0)$. { 'stāt tran'zish·ən ˌmā·triks }

state variable [CONT SYS] One of a minimum set of numbers which contain enough information about a system's history to enable computation of its future behavior. [THERMO] *See* thermodynamic function of state. { 'stāt ˌver·ē·ə·bəl }

state-variable filter [ELECTR] A multiple-amplifier active filter that has three outputs for high-pass, band-pass, and low-pass transfer functions respectively. Also known as KHN filter. { 'stāt ˌver·ē·ə·bəl ˌfil·tər }

state vector [COMPUT SCI] *See* task descriptor. [CONT SYS] A column vector whose components are the state variables of a system. [QUANT MECH] A vector in Hilbert space which corresponds to the state of a quantum-mechanical system. { 'stāt ˌvek·tər }

statF *See* statfarad. { 'stad‚ef }

statfarad [ELEC] Unit of capacitance in the electrostatic centimeter-gram-second system of units, equal to the capacitance of a capacitor having a charge of 1 statcoulomb, across the plates of which the charge is 1 statvolt; equal to approximately 1.1126×10^{-12} farad. Abbreviated statF. { 'stat‚fa‚rad }

statH *See* stathenry. { 'stad‚āch }

stathenry [ELECTROMAG] The unit of inductance in the electrostatic centimeter-gram-second system of units, equal to the self-inductance of a circuit or the mutual inductance between two circuits if there is an induced electromotive force of 1 statvolt when the current is changing at a rate of 1 statampere per second; equal to approximately 8.9876×10^{11} henry. Abbreviated statH. { 'stat‚hen·rē }

static [COMMUN] A hissing, crackling, or other sudden sharp sound that tends to interfere with the reception, utilization, or enjoyment of desired signals or sounds. [PHYS] Without motion or change. { 'stad·ik }

static algorithm [COMPUT SCI] An algorithm whose operation is known in advance. Also known as deterministic algorithm. { 'stad·ik 'al·gə‚rith·əm }

statically admissible loads [MECH] Any set of external loads and internal forces which fulfills conditions necessary to maintain the equilibrium of a mechanical system. { 'stad·ik· əl·ē əd'mis·ə·bəl 'lōdz }

static bed [CHEM ENG] A layer of solids in a process vessel (absorber, catalytic reactor, packed distillation column, or granular filter bed) in which the particles rest upon one another at essentially the settled bulk density of the solids phase; contrasted to moving-solids or fluidized-solids beds. { 'stad·ik 'bed }

static breeze *See* convective discharge. { 'stad·ik 'brēz }

static characteristic [ELECTR] A relation between a pair of variables, such as electrode voltage and electrode current, with all other operating voltages for an electron tube, transistor, or other amplifying device maintained constant. { 'stad·ik ˌkar· ik·tə'ris·tik }

static charge [ELEC] An electric charge accumulated on an object. { 'stad·ik 'chärj }

static check [COMPUT SCI] Of a computer, one or more tests of computing elements, their interconnections, or both, performed under static conditions. { 'stad·ik 'chek }

static contraction *See* isometric contraction. { 'stad·ik kən'trak·shən }

static debugging routine [COMPUT SCI] A debugging routine which is used after the program being checked has been run and has stopped. { 'stad·ik dē'bəg·iŋ rü‚tēn }

static discharger [ELEC] A rubber-covered cloth wick about 6 inches (15 centimeters) long, sometimes attached to the trailing edges of the surfaces of an aircraft to discharge static electricity in flight. { 'stad·ik 'dis‚chär·jər }

static dump [COMPUT SCI] An edited printout of the contents of main memory or of the auxiliary storage, performed in a fixed way; it is usually taken at the end of a program run either automatically or by operator intervention. { 'stad·ik 'dəmp }

static electricity [ELEC] **1.** The study of the effects of macroscopic charges, including the transfer of a static charge from one object to another by actual contact or by means of a spark that bridges an air gap between the objects. **2.** *See* electrostatics. { 'stad·ik ˌi‚lek'tris·əd·ē }

static eliminator [ELECTR] Device intended to reduce the effect of atmospheric static interference in a radio receiver. { 'stad·ik i‚lim·ə‚nād·ər }

static equilibrium *See* equilibrium. { 'stad·ik ˌē·kwə'lib·rē· əm }

static error [STAT] Error independent of the time-varying nature of a variable. { 'stad·ik 'er·ər }

static firing [AERO ENG] The firing of a rocket engine in a hold-down position to measure thrust and to accomplish other tests. { 'stad·ik 'fīr·iŋ }

static fluid column [FL MECH] An unchanging height of fluid in a vertical pipe, well bore, process vessel, or tank. { 'stad·ik 'flü·əd ˌkäl·əm }

static friction [MECH] **1.** The force that resists the initiation of sliding motion of one body over the other with which it is in contact. **2.** The force required to move one of the bodies when they are at rest. Also known as limiting friction; starting friction. { 'stad·ik 'frik·shən }

static gearing ratio [AERO ENG] The ratio of the control-surface deflection in degrees to angular displacement of the missile which caused the deflection of the control surface. { 'stad·ik 'gir·iŋ ˌrā·shō }

static gel buildup [FL MECH] A method used to infer the degree of thixotropy of a fluid by viscometric measurement of its gel strength. { 'stad·ik 'jel ˌbild‚əp }

static granitization [PETR] The formation of a granitic rock by a metasomatic process in the absence of compressive forces or strains. { 'stad·ik ˌgran·əd·ə zā·shən }

static grizzly [MIN ENG] A grizzly in the form of a stationary bar screen which either allows suitable pieces of rock or ore to pass over and unwanted small sizes to drop through, or rejects oversize pieces while allowing suitable material to drop through. { 'stad·ik 'griz·lē }

static head [FL MECH] Pressure of a fluid due to the head of fluid above some reference point. { 'stad·ik 'hed }

static induction transistor [ELECTR] A type of transistor capable of operating at high current and voltage, whose current-voltage characteristics do not saturate, and are similar in form to those of a vacuum triode. Abbreviated SIT. { 'stad·ik in'dək·shən tran‚zis·tər }

static inverter [ELEC] A device that converts a dc voltage to a stable ac voltage for use in an uninterruptible power system. { 'stad·ik in'vərd·ər }

staticize [COMPUT SCI] **1.** To capture transient data in stable form, thus converting fleeting events into examinable information. **2.** To extract an instruction from the main computer memory and store the various component parts of it in the appropriate registers, preparatory to interpreting and executing it. { 'stad·ə‚sīz }

static level [FL MECH] Elevation of the water level or a pressure surface at rest. [HYD] The height to which water will rise in an artesian well; the static level of a flowing well is above the ground surface. { 'stad·ik 'lev·əl }

static limit *See* stationary limit. { 'stad·ik 'lim·ət }

static line [AERO ENG] A line attached to a parachute pack and to a strop or anchor cable in an aircraft so that when the load is dropped the parachute is deployed automatically. { 'stad·ik ˌlīn }

static load [MECH] A nonvarying load; the basal pressure exerted by the weight of a mass at rest, such as the load imposed on a drill bit by the weight of the drill-stem equipment or the pressure exerted on the rocks around an underground opening by the weight of the superimposed rocks. Also known as dead load. { 'stad·ik 'lōd }

static machine [ELEC] A machine for generating electric charges, usually by electric induction, sometimes used to build up high voltages for research purposes. { 'stad·ik mə‚shēn }

static magnetism *See* magnetostatics. { 'stad·ik 'mag· nə‚tiz·əm }

static metamorphism [GEOL] Regional metamorphism caused by heat and solvents at high lithostatic pressures. Also known as load metamorphism. { 'stad·ik ˌmed·ə'mór‚fiz· əm }

static moment [MECH] **1.** A scalar quantity (such as area or mass) multiplied by the perpendicular distance from a point connected with the quantity (such as the centroid of the area or the center of mass) to a reference axis. **2.** The magnitude of some vector (such as force, momentum, or a directed line segment) multiplied by the length of a perpendicular dropped from the line of action of the vector to a reference point. { 'stad ik 'mō·mənt }

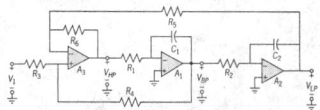

STATE-VARIABLE FILTER

State-variable (KHN) filter, with outputs V_{HP}, V_{BP}, and V_{LP} for high-pass, band-pass, and low-pass transfer funtions.

static oceanography [OCEANOGR] Branch of oceanography that deals with the physical and chemical nature of water in the ocean and with the shape and composition of the ocean bottom. { 'stad·ik ,ō·shə'näg·rə·fē }

static pressure [ACOUS] The pressure that would exist at a point in a medium if no sound waves were present. [FL MECH] **1.** The normal component of stress, the force per unit area, exerted across a surface moving with a fluid, especially across a surface which lies in the direction of fluid flow. **2.** The average of the normal components of stress exerted across three mutually perpendicular surfaces moving with a fluid. { 'stad·ik 'presh·ər }

static-pressure tap See pressure tap. { 'stad·ik ¦presh·ər ,tap }

static-pressure tube [ENG] A smooth tube with a rounded nose that has radial holes in the portion behind the nose and is used to measure the static pressure within the flow of a fluid. { 'stad·ik ¦presh·ər ,tüb }

static random-access memory [COMPUT SCI] A read-write random-access memory that uses either four transistors and two resistors to form a passive-load flip-flop, or six transistors to form a flip-flop with dynamic loads, for each cell in an array. Once data are loaded into the flip-flop storage elements, the flip-flop will indefinitely remain in that state until the information is intentionally changed or the power to the memory circuit is shut off. Abbreviated SRAM. { 'stad·ik 'rand·əm ¦ak,ses 'mem·rē }

static reaction [MECH] The force exerted on a body by other bodies which are keeping it in equilibrium. { 'stad·ik rē'ak·shən }

static reactive compensator [ELEC] A thyristor-controlled generator of reactive power that is used to compensate for reactive power in an electric power system in order to limit voltage variations. Also known as static var compensator. { 'stad·ik rē¦ak·tiv ,käm·pən'sād·ər }

static reflex [PHYSIO] Any one of a series of reflexes which are involved in the establishment of muscular tone for postural purpose. { 'stad·ik 'rē,fleks }

static regulator [ELECTR] Transmission regulator in which the adjusting mechanism is in self-equilibrium at any setting and requires control power to change the setting. { 'stad·ik ,reg·yə¦lād·ər }

statics [MECH] The branch of mechanics which treats of force and force systems abstracted from matter, and of forces which act on bodies in equilibrium. { 'stad·iks }

static seal See gasket. { 'stad·ik 'sēl }

static sensitivity [ELECTR] In phototubes, quotient of the direct anode current divided by the incident radiant flux of constant value. { 'stad·ik ,sen·sə'tiv·əd·ē }

static SP See static spontaneous potential. { 'stad·ik ¦es'pē }

static spontaneous potential [PETRO ENG] Theoretical maximum spontaneous potential current that can be measured in a down-hole, mud log in clean sand. Abbreviated SSP; static SP. { 'stad·ik spän'tā·nē·əs pə'ten·chəl }

static stability [METEOROL] The stability of an atmosphere in hydrostatic equilibrium with respect to vertical displacements, usually considered by the parcel method. Also known as convectional stability; convection stability; hydrostatic stability; vertical stability. [PHYS] See stability. { 'stad·ik stə'bil·əd·ē }

static storage [COMPUT SCI] Computer storage such that information is fixed in space and available at any time, as in flip-flop circuits, electrostatic memories, and coincident-current magnetic-core storage. { 'stad·ik 'stor·ij }

static subroutine [COMPUT SCI] In computers, a subroutine which involves no parameters other than the addresses of the operands. { 'stad·ik 'səb·rü,tēn }

static switching [ELEC] Switching of circuits by means of magnetic amplifiers, semiconductors, and other devices that have no moving parts. { 'stad·ik 'swich·iŋ }

static test [AERO ENG] In particular, a test of a rocket or other device in a stationary or hold-down position, either to verify structural design criteria, structural integrity, and the effects of limit loads, or to measure the thrust of a rocket engine. [ENG] A measurement taken under conditions where neither the stimulus nor the environmental conditions fluctuate. { 'stad·ik 'test }

static trees [GRAPHICS] A pattern created on film by undesirable emulsion exposures to static electricity. { 'stad·ik ,trēz }

static tube [ENG] A device used to measure the static (not kinetic or total) pressure in a stream of fluid; consists of a perforated, tapered tube that is placed parallel to the flow, and has a branch tube that is connected to a manometer. { 'stad·ik ,tüb }

static universe [ASTRON] A postulated universe that has a finite static volume and is closed. { 'stad·ik 'yü·nə,vərs }

static var compensator See static reactive compensator. { 'stad·ik 'vär ,käm·pən,sād·ər }

static variable [COMPUT SCI] A local variable that does not cease to exist upon termination of the block in which it can be accessed, but instead retains its most recent value until the next execution of this block. { 'stad·ik 'ver·ē·ə·bəl }

static weapon [ORD] A weapon which is used in place. { 'stad·ik 'wep·ən }

static work See isometric work. { 'stad·ik 'wərk }

station [COMMUN] See broadcast station. [COMPUT SCI] One of a series of essentially similar positions or facilities occurring in a data-processing system. [ELEC] An assembly line or assembly machine location at which a wiring board or chassis is stopped for insertion of one or more parts. [ELECTR] A location at which radio, television, radar, or other electric equipment is installed. [ENG] Any predetermined point or area on the seas or oceans which is patrolled by naval vessels. [MIN ENG] **1.** An enlargement in a mining shaft or gallery on any level used for a landing at any desired place and also for receiving loaded mine cars that are to be sent to the surface. **2.** An opening into a level which heads out of the side of an inclined plane; the point at which a surveying instrument is planted or observations are made. [SCI TECH] A geographic location at which scientific observations and measurements are made. { 'stā·shən }

stationary cone classifier [MECH ENG] In a pulverizer directly feeding a coal furnace, a device which returns oversize coal to the pulverizing zone. { 'stā·shə,ner·ē ¦kōn 'klas·ə,fī·ər }

stationary distribution [PHYS] A time-independent distribution of a scalar quantity. { 'stā·shə,ner·ē ,dis·trə'byü·shən }

stationary engine [MECH ENG] A permanently placed engine, as in a power house, factory, or mine. { 'stā·shə,ner·ē 'en·jən }

stationary ergodic noise [ELECTR] A stationary noise for which the probability that the noise voltage lies within any given interval at any time is nearly equal to the fraction of time that the noise voltage lies within this interval if a sufficiently long observation interval is recorded. { 'stā·shə,ner·ē ər'gäd·ik 'nóiz }

stationary field [PHYS] Field which does not change during the time interval under consideration. Also known as constant field. { 'stā·shə,ner·ē 'fēld }

stationary front See quasi-stationary front. { 'stā·shə,ner·ē 'frənt }

stationary limit [RELAT] In the Kerr solution to Einstein's equations, a surface on which a particle would have to move with the local light velocity in order to appear stationary to a distant observer, and inside which no particle can appear stationary to such an observer. Also known as static limit. { 'stā·shə,ner·ē 'lim·ət }

stationary noise [ELECTR] A random noise for which the probability that the noise voltage lies within any given interval does not change with time. { 'stā·shə,ner·ē 'nóiz }

stationary orbit [AERO ENG] A circular, equatorial orbit in which the satellite revolves about the primary body at the angular rate at which the primary body rotates on its axis; from the primary body, the satellite thus appears to be stationary over a point on the primary body; a stationary orbit must be synchronous, but the reverse need not be true. { 'stā·shə,ner·ē 'ór·bət }

stationary phase [ANALY CHEM] In chromatography, the nonmobile phase contained in the chromatographic bed. [MATH] A method used to find approximations to the integral of a rapidly oscillating function, based on the principle that this integral depends chiefly on that part of the range of integration near points at which the derivative of the trigonometric function involved vanishes. [MICROBIO] The period following termination of exponential growth in a bacterial culture when the number of viable microorganisms remains relatively constant for a time. { 'stā·shə₁ner·ē 'fāz }

stationary point [ASTRON] A point at which a planet's apparent motion changes from direct to retrograde motion, or vice versa. [MATH] **1.** A point on a curve at which the tangent is horizontal. **2.** For a function of several variables, a point at which all partial derivatives are 0. { 'stā·shə₁ner·ē 'póint }

stationary population [ECOL] A population containing a basically even distribution of age groups. { 'stā·shə₁ner·ē ₁päp·yə'lā·shən }

stationary satellite [AERO ENG] A satellite in a stationary orbit. { 'stā·shə₁ner·ē 'sad·əl₁īt }

stationary state *See* energy state. { 'stā·shə₁ner·ē 'stāt }

stationary stochastic process [MATH] A stochastic process $x(t)$ is stationary if each of the joint probability distributions is unaffected by a change in the time parameter t. { 'stā·shə₁ner·ē stō'kas·tik 'prä·səs }

stationary time principle *See* Fermat's principle. { 'stā·shə₁ner·ē 'tīm ₁prin·sə·pəl }

stationary time series [STAT] A time series which as a stochastic process is unchanged by a uniform increment in the time parameter defining it. { 'stā·shə₁ner·ē 'tīm ₁sir·ēz }

stationary wave *See* standing wave. { 'stā·shə₁ner·ē 'wāv }

station authentication [COMMUN] Security measure designed to establish the authenticity of a transmitting or receiving station. { 'stā·shən ó₁then·tə'kā·shən }

station buoy [NAV] A buoy used to mark the approximate station (position) of an important buoy or lightship should it be carried away or temporarily removed. Also known as marker buoy; watch buoy. { 'stā·shən ₁bói }

station continuity chart [METEOROL] A chart or graph on which time is one coordinate, and one or more of the observed meteorological elements at that station is the other coordinate. { 'stā·shən ₁kan·tə'nü·əd·ē ₁chärt }

station elevation [METEOROL] The vertical distance above mean sea level that is adopted as the reference datum level for all current measurements of atmospheric pressure at the station. { 'stā·shən ₁el·ə₁vā·shən }

stationkeeping [AERO ENG] Keeping a satellite in geosynchronous orbit within assigned boundaries, typically within a few tenths of a degree of longitude, generally with the assistance of jet thrusters. { 'stā·shən₁keep·iŋ }

station model [METEOROL] A specified pattern for entering, on a weather map, the meteorological symbols that represent the state of the weather at a particular observation station. { 'stā·shən ₁mäd·əl }

station pointer *See* three-arm protractor. { 'stā·shən ₁póint·ər }

station pole [CIV ENG] One of various rods used in surveying to mark stations, to sight points and lines; or to measure elevation with respect to the transit. { 'stā·shən ₁pōl }

station pressure [METEOROL] The atmospheric pressure computed for the level of the station elevation. { 'stā·shən ₁presh·ər }

station roof [BUILD] **1.** A roof supported by a single central post and having a shape that resembles an umbrella. Also known as umbrella roof. **2.** A long roof supported by a single row of posts and by cantilevers on one or both sides; typically used for railroad platforms. { 'stā·shən ₁rüf }

station stock level [ORD] Maximum quantity of supplies expressed in days of supply, permitted to be on hand or due in at any time at a military installation; this level is based on actual past issues and anticipated demands; it represents the requisitioning objective. { 'stā·shən ₁stäk ₁lev·əl }

station time [AERO ENG] Time at which crews, passengers, and cargo are to be on board air transport and ready for the flight. { 'stā·shən ₁tīm }

statistic [STAT] An estimate or piece of data, concerning some parameter, obtained from a sampling. { stə'tis·tik }

statistical analysis [STAT] The body of techniques used in statistical inference concerning a population. { stə'tis·tə·kəl ə'nal·ə·səs }

statistical computing *See* computational statistics. { stə'tis·tə·kəl kəm'pyüd·iŋ }

statistical control [ANALY CHEM] In an analytical procedure, a state that exists when the means of a large number of

individual values in the output of a measurement process tend to approach a limiting value known as the limiting mean. { stə'tis·tə·kəl kən't₁rōl }

statistical distribution *See* distribution. { stə'tis·tə·kəl ₁dis·trə'byü·shən }

statistical forecast [METEOROL] A weather forecast based upon a systematic statistical examination of the past behavior of the atmosphere, as distinguished from a forecast based upon thermodynamic and hydrodynamic considerations. { stə'tis·tə·kəl 'fór₁kast }

statistical hypothesis [STAT] A statement about the way a random variable is distributed. { stə'tis·tə·kəl hī'päth·ə·səs }

statistical independence [STAT] Two events are statistically independent if the probability of their occurring jointly equals the product of their respective probabilities. Also known as stochastic independence. { stə'tis·tə·kəl ₁in·də'pen·dəns }

statistical inference [STAT] The process of reaching conclusions concerning a population upon the basis of random samplings. { stə'tis·tə·kəl 'in·frəns }

statistical map [MAP] A special type of map in which the variation in quantity of a factor such as rainfall, population, or crops in a geographic area is indicated; a dot map is one type. { stə'tis·tə·kəl 'map }

statistical mechanics [PHYS] That branch of physics which endeavors to explain and predict the macroscopic properties and behavior of a system on the basis of the known characteristics and interactions of the microscopic constituents of the system, usually when the number of such constituents is very large. Also known as statistical thermodynamics. { stə'tis·tə·kəl mi'kan·iks }

statistical monitor [COMPUT SCI] A software monitor that collects information by periodically sampling activity in the system. { stə'tis·tə·kəl 'män·əd·ər }

statistical multiplexer [ELECTR] A device which combines several low-speed communications channels into a single high-speed channel, and which can manage more communications traffic than a standard multiplexer by analyzing traffic and choosing different transmission patterns. { stə'tis·tə·kəl 'məl·tə₁plek·sər }

statistical multiplexing [COMMUN] Time-division multiplexing in which time on a communications channel is assigned to multiple users on a demand basis, rather than periodically to each user. { stə₁tis·ti·kəl 'məl·tə₁pleks·iŋ }

statistical parallax [ASTRON] The mean parallax of a collection of stars that are all at approximately the same distance, as determined from their radial velocities and proper motions. { stə'tis·tə·kəl 'par·ə₁laks }

statistical quality control [IND ENG] The use of statistical techniques as a means of controlling the quality of a product or process. { stə'tis·tə·kəl 'kwäl·əd·ē kən₁trōl }

statistical thermodynamics *See* statistical mechanics. { stə'tis·tə·kəl ₁thər·mō·dī'nam·iks }

statistical weight [STAT] A number assigned to each value or range of values of a given quantity, giving the number of times this value or range of values is found to be observed. [STAT MECH] **1.** The number of microscopic states that correspond to a given macroscopic state. **2.** A multiplicative factor in the expression for the probability of finding a system in a given quantum state, usually equal to the number of degenerate substates contained in the state. { stə'tis·tə·kəl 'wāt }

statistics [MATH] A discipline dealing with methods of obtaining data, analyzing and summarizing it, and drawing inferences from data samples by the use of probability theory. { stə'tis·tiks }

statmho [ELEC] The unit of conductance, admittance, and susceptance in the electrostatic centimeter-gram-second system of units, equal to the conductance between two points of a conductor when a constant potential difference of 1 statvolt applied between the points produces in this conductor a current of 1 statampere, the conductor not being the source of any electromotive force; equal to approximately 1.1126×10^{-12} mho. Abbreviated stat℧. Also known as statsiemens (statS). { 'stat₁mō }

statoblast [INV ZOO] A chitin-encapsulated body which serves as a special means of asexual reproduction in the Phylactolaemata. { 'stad·ə₁blast }

statocone [INV ZOO] One of the minute calcareous granules found in the statocyst of certain animals. { 'stad·ə₁kōn }

statocyst [BOT] A cell containing statoliths in a fluid medium. Also known as statocyte. [INV ZOO] A sensory vesicle containing statoliths and which functions in the perception of the position of the body in space. { 'stad·ə,sist }

statocyte *See* statocyst. { 'stad·ə,sīt }

statohm [ELEC] The unit of resistance, reactance, and impedance in the electrostatic centimeter-gram-second system of units, equal to the resistance between two points of a conductor when a constant potential difference of 1 statvolt between these points produces a current of 1 statampere; it is equal to approximately 8.9876 × 10¹¹ ohms. Abbreviated statΩ. { 'stad,ōm }

statokinetic [PHYSIO] Pertaining to the balance and posture of the body or its parts during movement, as in walking. { ¦stad·ō·ki'ned·ik }

statolith [BOT] A sand grain or other solid inclusion which moves readily in the fluid contents of a statocyst, comes to rest on the lower surface of the cell, and is believed to function in gravity perception. [INV ZOO] A secreted calcareous body, a sand grain, or other solid inclusion contained in a statocyst. { 'stad·ə,lith }

stator [ELEC] The portion of a rotating machine that contains the stationary parts of the magnetic circuit and their associated windings. [MECH ENG] A stationary machine part in or about which a rotor turns. { 'stād·ər }

stator armature [ELEC] A stator which includes the main current-carrying winding in which electromotive force produced by magnetic flux rotation is induced; it is found in most alternating-current machines. { 'stād·ər 'är·mə·chər }

statoreceptor [PHYSIO] A sense organ concerned primarily with equilibrium. { ¦stād·ō·ri'sep·tər }

stator plate [ELEC] One of the fixed plates in a variable capacitor; stator plates are generally insulated from the frame of the capacitor. { 'stād·ər ,plāt }

statoscope [ENG] **1.** A barometer that records small variations in atmospheric pressure. **2.** An instrument that indicates small changes in an aircraft's altitude. { 'stad·ə,skōp }

statospore [BOT] In certain algae, an internally formed spore in its resting stage. { 'stad·ə,spȯr }

statS *See* statmho. { 'stat'es }

statsiemens *See* statmho. { 'stat'sē·mənz }

statT *See* stattesla. { 'stat'tē }

stattesla [ELECTROMAG] The unit of magnetic flux density in the electrostatic centimeter-gram-second system of units, equal to one statweber per square centimeter; equal to approximately 2.9979 × 10⁶ tesla. Abbreviated statT. { 'stat'tes·lə }

statuary bronze [MET] Special bronze alloys used for casting statues and other ornamental objects; a typical bronze for statuary work contains 90% copper, 6% tin, 3% zinc, and 1% lead. { 'stach·ə,wer·ē ¦bränz }

stature [ANTHRO] A measure of the distance from the floor to the vertex of the head, taken either front or back as the subject stands erectly with heels together. { 'stach·ər }

status asthmaticus [MED] Intractable asthma lasting from a few days to a week or longer. { 'stad·əs az'mad·ə·kəs }

status byte [COMPUT SCI] A byte of storage whose contents indicate the activities currently taking place in some part of the computer or various conditions governing the execution of a computer program; often, each bit is assigned a particular meaning. { 'stad·əs ,bīt }

status check [COMPUT SCI] The detection of software failures and verification of programs through the use of redundant computers. { 'stad·əs ,chek }

status epilepticus [MED] Occurrence of prolonged, generalized epileptic seizures in rapid succession with brief intervals of coma. { 'stad·əs ,ep·ə'lep·tə·kəs }

status line [COMPUT SCI] A conductor on the bus of a computer over which an addressed storage location or component transmits its status to the central processing unit. { 'stad·əs ,līn }

status register [COMPUT SCI] A register maintained by the central processing unit that contains a status byte with information about activities currently taking place there. { 'stad·əs ,rej·ə·stər }

status word [COMPUT SCI] A word indicating the state of the system or the diagnosis of a state into which the system has entered. { 'stad·əs ,wȯrd }

statute mile *See* mile. { 'stach·üt 'mīl }

statV *See* statvolt.

statvolt [ELEC] The unit of electric potential and electromotive force in the electrostatic centimeter-gram-second system of units, equal to the potential difference between two points such that the work required to transport 1 statcoulomb of electric charge from one to the other is equal to 1 erg; equal to approximately 299.79 volts. Abbreviated statV. { 'stat,vōlt }

statWb *See* statweber.

statweber [ELECTROMAG] The unit of magnetic flux in the electrostatic centimeter-gram-second system of units, equal to the magnetic flux which, linking a circuit of one turn, produces in it an electromotive force of 1 statvolt as it is reduced to zero at a uniform rate in 1 second; equal to approximately 299.79 webers. Abbreviated statWb. { 'stat,web·ər }

stauractine [INV ZOO] A sponge spicule in which the four rays lie in one plane. { stȯ'rak,tēn }

staurolite [MINERAL] FeAl₄(SiO₄)₂(OH)₂ A reddish-brown to black neosilicate mineral that crystallizes in the orthorhombic system, has resinous to vitreous luster, hardness is 7–7.5 on Mohs scale, and specific gravity is 3.7. Also known as cross-stone; fairy stone; grenatite; staurotide. { 'stȯr·ə,līt }

Stauromedusae [INV ZOO] An order of the class Scyphozoa in which the medusa is composed of a cuplike bell called a calyx and a stem that terminates in a pedal disk. { ¦stȯ·rō·mi'dü·sē }

Staurosporae [MYCOL] A spore group of the Fungi Imperfecti characterized by star-shaped or forked spores. { stȯ'räs·pə,rē }

staurotide *See* staurolite. { 'stȯr·ə,tīd }

stave [DES ENG] **1.** A rung of a ladder. **2.** Any of the narrow wooden strips or metal plates placed edge to edge to form the sides, top, or lining of a vessel or structure, such as a barrel. { stāv }

stay [ENG] In a structure, a tensile member which holds other members of the structure rigidly in position. { stā }

staybolt [DES ENG] A bolt with a thread along the entire length of the shaft; used to attach machine parts that are under pressure to separate. { 'stā,bōlt }

stayed-cable bridge [CIV ENG] A modified cantilever bridge consisting of girders or trusses cantilevered both ways from a central tower and supported by inclined cables attached to the tower at the top or sometimes at several levels. { 'stād ¦kā·bəl ,brij }

stay time [AERO ENG] In rocket engine usage, the average value of the time spent by each gas molecule or atom within the chamber volume. { 'stā ,tīm }

s-t cut [MATH] The set of all the arcs in an *s-t* network that originates in *X* and terminate in the complement of *X*, where *X* is a set of vertices in the *s-t* network that contains the source but not the terminal. { ¦es ¦tē ¦kət }

STD *See* sexually transmitted disease.

STDM *See* synchronous time-division multiplexing.

STD recorder *See* salinity-temperature-depth recorder. { ¦es,tē'dē ri,kȯrd·ər }

steadiness [CONT SYS] Freedom of a robot arm or end effector from high-frequency vibrations and jerks. [METEOROL] *See* persistence. { 'sted·ē·nəs }

steadite [MET] A hard structural constituent of cast iron consisting of the eutectic of ferrite and iron phosphide (Fe₃P); composition of the eutectic is 10.2% phosphorus and 89.8% iron; melts at 1049°C (1920°F). { 'ste,dīt }

steady bearing [NAV] An approaching or closing craft is said to be on a steady bearing if the compass bearing does not appreciably change. { 'sted·ē 'ber·iŋ }

steady flow [FL MECH] Fluid flow in which all the conditions at any one point are constant with respect to time. { 'sted·ē 'flō }

steady pin [ENG] **1.** A retaining device such as a dowel, pin, or key that prevents a pulley from turning on its axis. **2.** A guide pin used to lift a cope or pattern. { 'sted·ē ,pin }

steady rest [MECH ENG] A device that is used to support long, slender workpieces during turning or grinding and permits them to rotate without eccentric movement. { 'sted·ē ,rest }

steady state [PHYS] The condition of a body or system in which the conditions at each point do not change with time, that is after initial transients or fluctuations have disappeared. { 'sted·ē 'stāt }

steady-state cavitation *See* sheet cavitation. { 'sted·ē ¦stāt ,kav·ə'tā·shən }

steady-state conduction [THERMO] Heat conduction in

STATOR ARMATURE

Photograph of a stator armature of an alternating-current induction motor. (*Allis-Chalmers*)

STAUROMEDUSAE

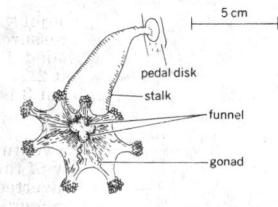

Medusa of *Lucernaria*; calyx is eight-sided and has eight groups of short, capped tentacles on its margin.

which the temperature and heat flow at each point does not change with time. { 'sted·ē ¦stāt kən'dək·shən }

steady-state creep *See* secondary creep. { 'sted·ē ¦stāt 'krēp }

steady-state current [ELEC] An electric current that does not change with time. { 'sted·ē ¦stāt 'kə·rənt }

steady-state distribution [ANALY CHEM] The equilibrium condition between phases in each step of a multistage, countercurrent liquid-liquid extraction. Abbreviated SSD. { 'sted·ē ¦stāt ¦dis·trə'byü·shən }

steady-state error [CONT SYS] The error that remains after transient conditions have disappeared in a control system. { 'sted·ē ¦stāt 'er·ər }

steady-state flow [CHEM ENG] Fluid flow without any change in composition or phase equilibria relationships. { 'sted·ē ¦stāt 'flō }

steady-state model [PETRO ENG] Electric or electrolytic analogs of a reservoir formation used to study the steady-state flow of fluids through porous media; includes gel, blotter, liquid, potentiometric, and similar models. { 'sted·ē ¦stāt 'mäd·əl }

steady-state reactor [NUCLEO] A reactor in which conditions such as temperature, reaction rate, and neutron flux do not change appreciably with time. { 'sted·ē ¦stāt rē'ak·tər }

steady-state theory [ASTRON] A cosmological theory which holds that the average density of matter does not vary with space or time in spite of the expansion of the universe; this requires that matter be continuously created. { 'sted·ē ¦stāt 'thē·ə·rē }

steady-state vibration [MECH] Vibration in which the velocity of each particle in the system is a continuous periodic quantity. { 'sted·ē ¦stāt vī'brā·shən }

steady-state wave motion [PHYS] Wave motion in which the wave quantities at each point in the region through which the wave is passing repeat themselves periodically. { 'sted·ē ¦stāt 'wāv ¦mō·shən }

steam [PHYS] Water vapor, or water in its gaseous state; the most widely used working fluid in external combustion engine cycles. { stēm }

steam accumulator [MECH ENG] A pressure vessel in which water is heated by steam during off-peak demand periods and regenerated as steam when needed. { 'stēm ə'kyü·mə‚lād·ər }

steam atomizing oil burner [ENG] A burner which has two supply lines, one for oil and the other for a jet of steam which assists in the atomization process. { 'stēm ¦ad·ə‚mīz·iŋ 'oil ‚bər·nər }

steam attemperation [MECH ENG] The control of the maximum temperature of superheated steam by water injection or submerged cooling. { 'stēm ə‚tem·pə'rā·shən }

steam bending [MECH ENG] Forming wooden members to a desired shape by pressure after first softening by heat and moisture. { 'stēm ‚bend·iŋ }

steam blow *See* blister. { 'stēm ‚blō }

steam boiler [MECH ENG] A pressurized system in which water is vaporized to steam by heat transferred from a source of higher temperature, usually the products of combustion from burning fuels. Also known as steam generator. { 'stēm ‚boi·lər }

steam bronze [MET] A leaded tin-bronze containing 88% copper, 6% tin, 4.5% zinc, 1.5% lead; used for steam valve bodies, gears, and bearings. { 'stēm ‚bränz }

steam calorimeter [ENG] **1.** A calorimeter, such as the Joly or differential steam calorimeter, in which the mass of steam condensed on a body is used to calculate the amount of heat supplied. **2.** *See* throttling calorimeter. { 'stēm 'kal·ə'rim·əd·ər }

steam capstan [NAV ARCH] A capstan operated by a steam engine. { 'stēm 'kap‚stan }

steam cock [ENG] A valve for the passage of steam. { 'stēm ‚käk }

steam condenser [MECH ENG] A device to maintain vacuum conditions on the exhaust of a steam prime mover by transfer of heat to circulating water or air at the lowest ambient temperature. { 'stēm kən‚den·sər }

steam cracking [CHEM ENG] High-temperature cracking of petroleum hydrocarbons in the presence of steam. { 'stēm 'krak·iŋ }

steam cure [ENG] To cure concrete or mortar in water vapor at an elevated temperature, at either atmospheric or high pressure. { 'stēm 'kyür }

steam cycle *See* Rankine cycle. { 'stēm ‚sī·kəl }

steam distillation [CHEM ENG] A distillation in which vaporization of the volatile constituents of a liquid mixture takes place at a lower temperature by the introduction of steam directly into the charge; steam used in this manner is known as open steam. Also known as steam stripping. { 'stēm ‚dis·tə'lā·shən }

steam drive [MECH ENG] Any device which uses power generated by the pressure of expanding steam to move a machine or a machine part. { 'stēm ‚drīv }

steam dryer [MECH ENG] A device for separating liquid from vapor in a steam supply system. { 'stēm ‚drī·ər }

steam-electric generator [ELEC] An electric generator driven by a steam turbine. { 'stēm i‚lek·trik 'jen·ə‚rād·ər }

steam emulsion test [ENG] A test used for measuring the ability of oil and water to separate, especially for steam-turbine oil; after emulsification and separation, the time required for the emulsion to be reduced to 3 milliliters or less is recorded at 5-minute intervals. { 'stēm i'məl·shən ‚test }

steam engine [MECH ENG] A thermodynamic device for the conversion of heat in steam into work, generally in the form of a positive displacement piston and cylinder mechanism. { 'stēm ‚en·jən }

steam engine indicator [ENG] An instrument that plots the steam pressure in an engine cylinder as a function of piston displacement. { 'stēm ‚en·jən 'in·cə‚kād·ər }

steam fog [METEOROL] Fog formed when water vapor is added to air which is much colder than the vapor's source; most commonly, when very cold air drifts across relatively warm water. Also known as frost smoke; sea mist; sea smoke; steam mist; water smoke. { 'stēm ‚fäg }

steam gage [ENG] A device for measuring steam pressure. { 'stēm ‚gāj }

steam-generating furnace *See* boiler furnace. { 'stēm ‚jen·ə‚rād·iŋ ‚fər·nəs }

steam generator *See* steam boiler. { 'stēm ‚jen·ə‚rād·ər }

steam hammer [MECH ENG] A forging hammer in which the ram is raised, lowered, and operated by a steam cylinder. { 'stēm ‚ham·ər }

steam-hammer oil *See* tempering oil.

steam-heated evaporator [MECH ENG] A structure using condensing steam as a heat source on one side of a heat-exchange surface to evaporate liquid from the other side. { 'stēm ¦hēd·əd i'vap·ə‚rād·ər }

steam heating [MECH ENG] A system that used steam as the medium for a comfort or process heating operation. { 'stēm ‚hēd·iŋ }

steam jacket [MECH ENG] A casing applied to the cylinders and heads of a steam engine, or other space, to keep the surfaces hot and dry. { 'stēm ‚jak·ət }

steam jet [ENG] A blast of steam issuing from a nozzle. [MIN ENG] A system of ventilating a mine by means of a number of jets of steam at high pressure kept constantly blowing off from a series of pipes in the bottom of the upcast shaft. { 'stēm ‚jet }

steam-jet cycle [MECH ENG] A refrigeration cycle in which water is used as the refrigerant; high-velocity steam jets provide a high vacuum in the evaporator, causing the water to boil at low temperature and at the same time compressing the flashed vapor up to the condenser pressure level. { 'stēm ‚jet ‚sī·kəl }

steam-jet ejector [MECH ENG] A fluid acceleration vacuum pump or compressor using the high velocity of a steam jet for entrainment. { 'stēm ‚jet i'jek·tər }

steam line [THERMO] A graph of the boiling point of water as a function of pressure. { 'stēm ‚līn }

steam locomotive [MECH ENG] A railway propulsion power plant using steam, generally in a reciprocating, noncondensing engine. { 'stēm ‚lō·kə‚mōd·iv }

steam loop [ENG] Two vertical pipes connected by a horizontal one, used to condense boiler steam so that it can be returned to the boiler without a pump or injector. { 'stēm ‚lüp }

steam mist *See* steam fog. { 'stēm ‚mist }

steam molding [ENG] The use of steam, either directly on the material or indirectly on the mold surfaces, as a heat source to mold parts from preexpanded polystyrene beads. { 'stēm ‚mōld·iŋ }

steam nozzle [MECH ENG] A streamlined flow structure in

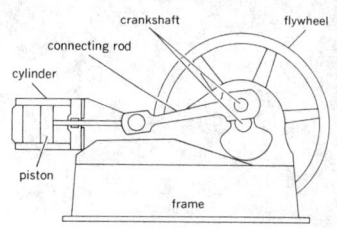

STEAM ENGINE

crankshaft   flywheel
connecting rod
cylinder
piston
frame

Principal parts of horizontal steam engine.

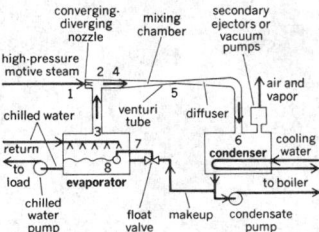

STEAM-JET CYCLE

converging-diverging nozzle   mixing chamber   secondary ejectors or vacuum pumps
high-pressure motive steam   2   4   air and vapor
1   5
chilled water   venturi tube   diffuser
3   6   cooling water
return to load   7   condenser   to boiler
8   evaporator
chilled water pump   float valve   makeup   condensate pump

Diagram of the basic steam-jet water-vapor cycle. High-pressure motive steam at 1 is expanded to low absolute pressure at 2. Water vapor in evaporator at 3 is entrained by motive steam at 4. Motive steam plus entrained moisture is forced through venturi tube at 5, in which velocity of the mixture is reduced and converted into pressure head at 6. Makeup water is added from 7 to 8.

STEAM LOCOMOTIVE

Steam locomotive uses high-pressure steam to move pistons which are connected to rods that force the driving wheels to rotate.

which heat energy of steam is converted to the kinetic form. { 'stēm ˌnäz·əl }

steam point [THERMO] The boiling point of pure water whose isotopic composition is the same as that of sea water at standard atmospheric pressure; it is assigned a value of 100°C on the International Practical Temperature Scale of 1968. { 'stēm ˌpȯint }

steam pump [MECH ENG] A pump driven by steam acting on the coupled piston rod and plunger. { 'stēm ˌpəmp }

steam purifier See steam separator. { 'stēm 'pyùr·ə‚fī·ər }

steam-refined asphalt [MATER] Petroleum-derived asphalt that has been refined in the presence of steam during the distillation of crude oil. { 'stēm ri‚fīnd 'as‚fȯlt }

steam-refined cylinder oil See SR cylinder oil. { 'stēm ri‚fīnd 'sil·ən·dər ‚ȯil }

steam-refined stock [MATER] Unfiltered petroleum products distilled with heat applied in the form of steam, for example, for gear oils, lubricating oils, and cylinder oils. { 'stēm ri‚fīnd 'stäk }

steam refining [CHEM ENG] A petroleum refinery distillation process, in which the only heat used comes from steam in open and closed coils near the bottom of the still; used to produce gasoline and naphthas where odor and color are of prime importance; where open steam is used, it is known as steam distillation. { 'stēm ri‚fīn·iŋ }

steam reheater [MECH ENG] A steam boiler component in which heat is added to intermediate-pressure steam, which has given up some of its energy in expansion through the high-pressure turbine. { 'stēm rē‚hēd·ər }

steam rig [PETRO ENG] A drilling rig which is powered by a number of portable boilers. { 'stēm ‚rig }

steam roller [MECH ENG] A road roller driven by a steam engine. { 'stēm ‚rō·lər }

steam separator [MECH ENG] A device for separating a mixture of the liquid and vapor phases of water. Also known as steam purifier. { 'stēm 'sep·ə‚rād·ər }

steamship [NAV ARCH] A ship propelled by a steam engine. { 'stēm‚ship }

steam shovel [MECH ENG] A power shovel operated by steam. { 'stēm ‚shəv·əl }

steam still [CHEM ENG] A still in which steam provides most of the heat; distillation requires a lower temperature than in standard equipment (except for a vacuum distillation unit). { 'stēm ‚stil }

steam stripping See steam distillation. { 'stēm 'strip·iŋ }

steam superheater [MECH ENG] A boiler component in which sensible heat is added to the steam after it has been evaporated from the liquid phase. { 'stēm 'sü·pər‚hēd·ər }

steam tracing [ENG] A steam-carrying heater (such as tubing or piping) next to or twisted around a process-fluid or instrument-air line; used to keep liquids from solidifying or condensing. { 'stēm ‚trās·iŋ }

steam trap [MECH ENG] A device which drains and removes condensate automatically from steam lines. { 'stēm ‚trap }

steam-tube dryer [MECH ENG] Rotary dryer with steam-heated tubes running the full length of the cylinder and rotating with the dryer shell. { 'stēm ‚tüb ‚drī·ər }

steam turbine [MECH ENG] A prime mover for the conversion of heat energy of steam into work on a rotating shaft, utilizing fluid acceleration principles in jet and vane machinery. { 'stēm ‚tər·bən }

steam valve [ENG] A valve used to regulate the flow of steam. { 'stēm ‚valv }

steam washer [ENG] A device for removing contaminants, such as silica, from the steam produced in a boiler. { 'stēm ‚wäsh·ər }

steapsin [BIOCHEM] An enzyme in pancreatic juice that catalyzes the hydrolysis of fats. Also known as pancreatic lipase. { stē'ap·sən }

stearamide [ORG CHEM] $CH_3(CH_2)_{16}CONH_2$ Colorless leaflets with a melting point of 109°C; used as a corrosion inhibitor in oil wells. { 'stir·ə·məd }

stearate [ORG CHEM] $C_{17}H_{35}COOM$ A salt or ester of stearic acid where M is a monovalent radical, for example, sodium stearate, $C_{17}H_{35}COONa$. { 'stir‚āt }

stearic acid [ORG CHEM] $CH_3(CH_2)_{16}COOH$ Nature's most common fatty acid, derived from natural animal and vegetable fats; colorless, waxlike solid, insoluble in water, soluble in alcohol, ether, and chloroform; melts at 70°C; used as a lubricant

and in pharmaceuticals, cosmetics, and food packaging. { 'stir·ik 'as·əd }

stearin [ORG CHEM] $C_3H_5(C_{18}H_{35}O_2)_3$ A colorless combustible powder; insoluble in water, soluble in alcohol, chloroform, and carbon disulfide; melts at 72°C; used in metal polishes, pastes, candies, candles, and soap, and to waterproof paper. Also known as glyceryl tristearate; tristearin. { 'stir·ən }

stearyl alcohol [ORG CHEM] $CH_3(CH_2)_{16}CH_2OH$ Oily white, combustible flakes; insoluble in water, soluble in alcohol, acetone, and ether; melt at 59°C; used in lubricants, resins, perfumes, and cosmetics, and as a surface-active agent. { 'sti‚ril 'al·kə‚hȯl }

steatite [PETR] A compact, massive, fine-ground rock composed principally of talc, but with much other material. { 'stē·ə‚tīt }

steatization [GEOL] Introduction of or replacement by talc or steatite. { stē‚ad·ə'zā·shən }

Steatornithidae [VERT ZOO] A family of birds in the order Caprimulgiformes which contains a single, South American species, the oilbird or guacharo (*Steatornis caripensis*). { ‚stē·ə‚tȯr'nith·ə‚dē }

steatorrhea [MED] **1.** Fatty stool. **2.** Increased flow of sebum. { ‚stē‚ad·ə'rē·ə }

Steckel rolling [MET] Cold metal rolling in which the strip is pulled through idler rolls by front tension only; the direction is reversed until the desired thickness is attained. { 'stek·əl ‚rōl·iŋ }

S tectonite [PETR] A tectonite whose fabric is dominated by planar surfaces of formation or deformation, such as slate. { 'es 'tek·tə‚nīt }

STED See solar turboelectric drive. { sted }

steel [MET] An iron base alloy, malleable under proper conditions, containing up to about 2% carbon. { stēl }

steel bronze [MET] A hardened bronze consisting of 92% copper and 8% tin; used as a substitute for steel in guns. { 'stēl 'bränz }

steel-cable conveyor belt [DES ENG] A rubber conveyor belt in which the carcass is composed of a single plane of steel cables. { 'stēl ‚kā·bəl kən'vā·ər ‚belt }

steel-case [ORD] In cartridge nomenclature, indicating that the cartridge case is made of steel. { 'stēl ‚kās }

steel-clad rope [DES ENG] A wire rope made from flat strips of steel wound helically around each of the six strands composing the rope. { 'stēl ‚klad 'rōp }

steel converter [MET] A retort in which cast iron is converted to steel; an example is the Bessemer converter. { 'stēl kən‚vərd·ər }

steel emery [MET] An abrasive material composed of chilled iron produced by forcing iron through a steam jet; used in tumbling barrels and for grinding stones. { 'stēl 'em·ə·rē }

steel engraving [GRAPHICS] Engraving on a steel plate, as used in printing. { 'stēl in'grāv·iŋ }

Steelflex coupling [MECH ENG] A flexible coupling made with two grooved steel hubs keyed to their respective shafts and connected by a specially tempered alloy-steel member called the grid. { 'stēl‚fleks 'kəp·liŋ }

steel foil [MET] A very thin sheet of steel, the thickness of which is measured in thousandths of an inch. { 'stēl 'fȯil }

steel jack [MIN ENG] A screw jack suitable in mechanical mining; used for legs or upright timbers. [MINERAL] See sphalerite. { 'stēl 'jak }

steel-jacket [ORD] In small-arms ammunition nomenclature, indicating that the bullet has a steel shell and core. { 'stēl 'jak·ət }

steelmaking [MET] Any of various processes for making steel from pig iron. { 'stel‚māk·iŋ }

steel sets [MIN ENG] Steel beam used in main entries of coal mines and in shafts of metal mines; I beams for caps and H beams for posts or wall plates. { 'stēl 'sets }

steel tunnel support [MIN ENG] One of the tunnel support systems made of steel; five types are continuous rib; rib and post; rib and wall plate; rib, wall plate, and post; and full circle rib. { 'stēl 'tən·əl sə‚pȯrt }

steel wool [MET] Fine steel threads matted into a mass. { 'stēl 'wùl }

steelyard [ENG] A weighing device with a counterbalanced arm supporting the load to be weighed on the short end. { 'stil·yərd }

steen [CIV ENG] To line an excavation such as a cellar or well with stone, cement, or similar material without the use of mortar. { stēn }

Steenrod algebra [MATH] The cohomology groups of a topological space have additive operations on them, which can be added and multiplied so as to form the Steenrod algebra. { 'sten‚räd 'al·jə·brə }

Steenrod squares [MATH] Operations which associate elements from different cohomology groups of a topological space and produce an element in another of the groups; these operations can be so added and multiplied as to produce the Steenrod algebra. { 'sten‚räd 'skwerz }

steepest descent method [MATH] **1.** Certain functions can be approximated for large values by an asymptotic formula derived from a Taylor series expansion about a saddle point. Also known as saddle point method. **2.** A method of approximating extreme values of the functions of two or more variables, in which the gradient of the function is used to obtain a sequence of approximations of the point at which the extreme value occurs. { 'stēp·əst di'sent ‚meth·əd }

steeple [ARCH] A tall, tapering structure atop a church tower. { 'stē·pəl }

steerable antenna [ELECTROMAG] A directional antenna whose major lobe can be readily shifted in direction. { 'stir·ə·bəl an'ten·ə }

steerage [NAV ARCH] The least desirable portions of a vessel used for accommodations for passengers who pay the lowest fare. { 'stir·ij }

steerageway [NAV ARCH] The least speed (forward or reverse) to allow a vessel to be steered. { 'stir·ij‚wā }

steering [METEOROL] Loosely used for any influence upon the direction of movement of an atmospheric disturbance exerted by another aspect of the state of the atmosphere; for example, a surface pressure system tends to be steered by isotherms, contour lines, or streamlines aloft, or by warm-sector isobars or the orientation of a warm front. { 'stir·iŋ }

steering arm [MECH ENG] An arm that transmits turning motion from the steering wheel of an automotive vehicle to the drag link. { 'stir·iŋ ‚ärm }

steering brake [MECH ENG] Means of turning, stopping, or holding a tracked vehicle by braking the tracks individually. { 'stir·iŋ ‚brāk }

steering compass [NAV] A compass by which a craft is steered; it sometimes refers to a gyro repeater, which is used for the same purpose as the steering compass; the term steering repeater is preferable. { 'stir·iŋ ‚kam·pəs }

steering engine [NAV ARCH] A steam-, electric-, or hydraulic-power machine used for turning the rudder, and having its valves or operating gear actuated by leads from the pilot house. { 'stir·iŋ ‚en·jən }

steering gear [MECH ENG] The mechanism, including gear train and linkage, for the directional control of a vehicle or ship. { 'stir·iŋ ‚gir }

steering level [METEOROL] A hypothetical level, in the atmosphere, where the velocity of the basic flow bears a direct relationship to the velocity of movement of an atmospheric disturbance embedded in the flow. { 'stir·iŋ ‚lev·əl }

steering repeater [NAV] A device which repeats at a distance the reading of the gyro compass; a craft is steered by reference to the device; sometimes called a steering compass. { 'stir·iŋ ri'pēd·ər }

steering wheel [MECH ENG] A hand-operated wheel for controlling the direction of the wheels of an automotive vehicle or of the rudder of a ship. { 'stir·iŋ ‚wēl }

Stefan-Boltzmann constant [STAT MECH] The energy radiated by a blackbody per unit area per unit time divided by the fourth power of the body's temperature; equal to (5.6696 ± 0.0010) × 10⁻⁸ (watt)(meter)⁻²(kelvin)⁻⁴. { 'shte‚fän 'bōlts ‚män ‚kän·stənt }

Stefan-Boltzmann law [STAT MECH] The total energy radiated from a blackbody is proportional to the fourth power of the temperature of the body. Also known as fourth-power law; Stefan's law of radiation. { 'shte‚fän 'bōlts‚män ‚lö }

Stefan number [THERMO] A dimensionless number used in the study of radiant heat transfer, equal to the Stefan-Boltzmann constant times the cube of the temperature times the thickness of a layer divided by the layer's thermal conductivity. Symbolized St. Also known as Stark number (Sk). { 'shte‚fän ‚nəm·bər }

Stefan's formula [OCEANOGR] A formula for the growth of thickness h of an ice cover on the ocean at various freezing temperatures, expressed as

$$h \approx \sqrt{\left(\frac{2l}{\lambda_i \rho_i}\right)} \psi$$

where l is the coefficient of thermal conductivity, λ_i is the latent heat of fusion, ρ_i is the density of ice, and ψ is the cold sum (in degree days below 0°C). { 'shte‚fänz ‚för·myə·lə }

Stefan's law of radiation See Stefan-Boltzmann law. { 'shte‚fänz 'lö əv 'rād·ē‚ā·shən }

Steffen process [FOOD ENG] A process used in sugar factories to concentrate the dilute waste liquor from beet molasses and recover sugar for use in a fermentation process to make monosodium glutamate. { 'stef·ən ‚prä·səs }

steganography [COMPUT SCI] The art and science of hiding a message in a medium, such as a digital picture or audio file, so as to defy detection. { ‚steg·ə'näg·rə·fē }

Steganopodes [VERT ZOO] Formerly, an order of birds that included the totipalmate swimming birds. { ‚steg·ə'näp· ə‚dēz }

Stegodontinae [PALEON] An extinct subfamily of elephantoid proboscideans in the family Elephantidae. { ‚steg· ə'dänt·ə‚nē }

Stegosauria [PALEON] A suborder of extinct reptiles of the order Ornithischia comprising the plated dinosaurs of the Jurassic which had tiny heads, great triangular plates arranged on the back in two alternating rows, and long spikes near the end of the tail. { ‚steg·ə'sör·ē·ə }

steigerite [MINERAL] $4AlVO_4 \cdot 13H_2O$ A canary-yellow mineral composed of hydrous aluminum vanadate and occurring in masses. { 'stī·gə‚rīt }

Steiner's hypocycloid See deltoid. { 'stīn·ərz ‚hī· pə'sī‚klöid }

Steiner's theorem See parallel axis theorem. { 'shtīn·ərz ‚thir·əm }

Steiner triple system [MATH] A balanced incomplete block design in which the number k of distinct elements in each block equals 3, and the number λ of blocks in which each combination of elements occurs together equals 1. { ‚stīn·ər ‚trip·əl 'sis·təm }

Steinheil lens [OPTICS] A type of magnifier lens in which a biconvex crown lens is cemented between a pair of flint lenses. { 'shtīn‚hīl ‚lenz }

Steinheim man [PALEON] A prehistoric man represented by a skull, without mandible, found near Stuttgart, Germany; the browridges are massive, the face is relatively small, and the braincase is similar in shape to that of *Homo sapiens*. { 'shtīn‚hīm ‚man }

Steinitz theorem [MATH] The theorem that an interior point of the convex span of a set in n-dimensional Euclidean space is also an interior point of the convex span of a subset of that set which has at most two n points. { 'shtī‚nits ‚thir·əm }

steinkern [GEOL] **1.** Rock material formed from consolidated mud or sediment that filled a hollow organic structure, such as a fossil shell. **2.** The fossil formed after dissolution of the mold. Also known as endocast; internal cast. { 'shtīn‚kərn }

Stein-Leventhal syndrome [MED] A complex of symptoms characterized by amenorrhea or abnormal uterine bleeding or both, enlarged polycystic ovaries, frequently hirsutism, and occasionally retarded breast development. { 'stīn 'lev· ən‚thöl ‚sin‚dröm }

Steinmann pin [MED] A surgical nail inserted in distal portions of such bones as the femur or tibia for skeletal tractions. { 'stīn·mən ‚pin }

Steinmetz coefficient [ELECTROMAG] The constant of proportionality in Steinmetz's law. { 'stīn‚mets ‚kō·i‚fish·ənt }

Steinmetz's law [ELECTROMAG] The energy converted into heat per unit volume per cycle during a cyclic change of magnetization is proportional to the maximum magnetic induction raised to the 1.6 power, the constant of proportionality depending only on the material. { 'stīn‚mets·əz ‚lö }

STEL See short-term exposure limit. { stel or ‚es‚tē‚ē'el }

stele [BOT] The part of a plant stem including all tissues and regions of plants from the cortex inward, including the pericycle, phloem, cambium, xylem, and pith. { 'stēl }

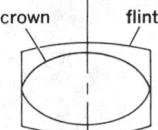

STEINHEIL LENS

crown | flint

Schematic of Steinheil aplanatic magnifier showing the crown lens within the flint lenses.

Stelenchopidae [INV ZOO] A family of polychaete annelids belonging to the Myzostomaria, represented by a single species from Crozet Island in the Antarctic Ocean. { ˌstel·ən'käp·ə‚dē }

stellar [ASTRON] Relating to or consisting of stars. { 'stel·ər }

stellar association [ASTRON] A loose grouping of stars which may have had a common origin. { 'stel·ər ə‚sō·sē'ā·shən }

stellar atmosphere [ASTRON] The envelope of gas and plasma surrounding a star; consists of about 90% hydrogen atoms and 9% helium atoms, by number of atoms. { 'stel·ər 'at·mə‚sfir }

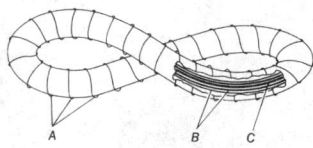

STELLARATOR

Diagram of a stellarator. External, closely spaced, current-carrying windings *A* produce a magnetic field whose lines of force run parallel to the walls, confining plasma *C* in tube's center.

stellarator [PL PHYS] A device for confining a high-temperature plasma, consisting of a tube, which closes in on itself in a figure-eight or race-track configuration, and external coils which generate magnetic fields whose lines of force run parallel to the walls of the tube and prevent the plasma from touching the walls. { 'stel·ə‚rād·ər }

stellar crystal *See* plane-dendritic crystal. { 'stel·ər 'krist·əl }

stellar evolution [ASTROPHYS] The changes in spectrum and luminosity that take place in the life of a star. { 'stel·ər ‚ev·ə'lü·shən }

stellar flare [ASTRON] Ejection of material from a star in an eruption that may last from a few minutes to an hour or more. { 'stel·ər 'fler }

stellar guidance [NAV] Guidance by means of celestial bodies, particularly the stars. { 'stel·ər 'gīd·əns }

stellar inertial guidance [NAV] The guidance of a flight-borne vehicle by a combination of celestial and inertial devices and techniques. { 'stel·ər i'nər·shəl 'gīd·əns }

stellar interferometer [OPTICS] An optical interferometer for measuring angular diameters of stars; it is attached to a telescope and measures interference rings at the telescope's focus. { 'stel·ər ‚in·tər·fə'räm·əd·ər }

stellar light [ASTRON] The part of the background illumination of the night sky that results from direct light from stars too faint to be visible to the unaided eye. { 'stel·ər 'līt }

stellar lightning [GEOPHYS] Lightning consisting of several flashes seeming to radiate from a single point. { 'stel·ər 'līt·niŋ }

stellar luminosity [ASTRON] A star's brightness; it is measured either in ergs per second or in units of solar luminosity or in absolute magnitude. { 'stel·ər lü·mə'näs·əd·ē }

stellar magnetic field [ASTROPHYS] A magnetic field, generally stronger than the earth's magnetic field, possessed by many stars. { 'stel·ər mag'ned·ik 'fēld }

stellar magnitude *See* magnitude. { 'stel·ər 'mag·nə‚tüd }

stellar mass [ASTROPHYS] The mass of a star, usually expressed in terms of the sun's mass. { 'stel·ər 'mas }

stellar model [ASTROPHYS] A mathematical characterization of the internal properties of a star. Also known as star model. { 'stel·ər 'mäd·əl }

stellar nevus *See* spider nevus. { 'stel·ər 'nē·vəs }

stellar parallax [ASTRON] The subtended angle at a star formed by the mean radius of the earth's orbit; it indicates distance to the star. { 'stel·ər 'par·ə‚laks }

stellar photometry [ASTRON] The measurement of the brightness of stars. { 'stel·ər fə'täm·ə·trē }

stellar population [ASTRON] Either of two classes of stars, termed population I and population II; population I are relatively young stars, found in the arms of spiral galaxies, especially the blue stars of high luminosity; population II stars are the much older, more evolved stars of lower metallic content; many high luminosity red giants and many variable stars are members of population II. { 'stel·ər ‚päp·yə'lā·shən }

stellar pulsation [ASTROPHYS] Expansion of a star followed by contraction so that its surface temperature and intrinsic brightness undergo periodic variation. { 'stel·ər pəl'sā·shən }

stellar rotation [ASTRON] Axial rotation of stars; surface rotational equatorial velocities of stars range from a few to 500 kilometers per second. { 'stel·ər rō'tā·shən }

stellar scintillation *See* astronomical scintillation. { 'stel·ər ‚sint·əl'ā·shən }

stellar spectroscopy [ASTRON] The techniques of obtaining spectra of stars and their study. { 'stel·ər spek'träs·kə·pē }

stellar spectrum [ASTRON] The spectrum of a star normally obtained with a slit spectrograph by black-and-white photography; the spectrum of a star in a large majority of cases shows absorption lines superposed on a continuous background. { 'stel·ər 'spek·trəm }

stellar structure [ASTROPHYS] The mathematical study of a rotating, chemically homogeneous mass of gas held together by its own gravitation; a representative model of the observable properties of a star; thermonuclear reactions are postulated to be the main source of stellar energy. { 'stel·ər ‚strək·chər }

stellar system [ASTRON] A gravitational system of stars. { 'stel·ər ‚sis·təm }

stellar temperature [ASTROPHYS] Any temperature above several million degrees, such as occurs naturally in the interior of the sun and other stars. { 'stel·ər 'tem·prə·chər }

stellar wind [ASTRON] The flow of ionized gas from the surface of a star into interstellar space. { 'stel·ər 'wind }

stellate ganglion [NEUROSCI] The ganglion formed by the fusion of the inferior cervical and the first thoracic sympathetic ganglions. { 'ste‚lāt 'gaŋ·glē·ən }

stellate reticulum [HISTOL] The part of the epithelial dental organ of a developing tooth which lies between the inner and the outer dental epithelium; composed of stellate cells with long, anastomosing processes in a mucoid fluid in the interstitial spaces. { 'ste‚lāt rə'tik·yə·ləm }

Stelleroidea [INV ZOO] The single class of echinoderms in the subphylum Asterozoa; characters coincide with those of the subphylum. { ‚stel·ə'roid·ē·ə }

stellite [MET] A hard, wear- and corrosion-resistant family of nonferrous alloys of cobalt (20–65), chromium (11–32), and tungsten (2–5); resistance to softening is exceptionally high at high temperature. { 'ste‚līt }

stem [BOT] The organ of vascular plants that usually develops branches and bears leaves and flowers. [ENG] **1.** The heavy iron rod acting as the connecting link between the bit and the balance of the string of tools on a churn rod. **2.** To insert packing or tamping material in a shothole. [NAV] To make headway against an obstacle, as a current. [NAV ARCH] The foremost part of a ship's hull. { stem }

STEM *See* scanning transmission electron microscope. { stem }

stem bag [MIN ENG] A fire-resisting paper bag filled with dry sand for stemming shotholes. { 'stem ‚bag }

stem blight [PL PATH] Any of various fungus blights that affect the plant stem. { 'stem ‚blīt }

stem break *See* browning. { 'stem ‚brāk }

stem canker [PL PATH] Any canker disease affecting the stem. { 'stem ‚kaŋ·kər }

stem cell [EMBRYO] A formative cell. [HISTOL] *See* hemocytoblast. { 'stem ‚sel }

stem-cell leukemia [MED] A form of leukemia in which the predominant cell type is so poorly differentiated that its series cannot be identified. { 'stem ‚sel lü'kē·mē·ə }

stem correction [THERMO] A correction which must be made in reading a thermometer in which part of the stem, and the thermometric fluid within it, is at a temperature which differs from the temperature being measured. { 'stem kə‚rek·shən }

stemming rod [ENG] A nonmetallic rod used to push explosive cartridges into position in a shothole and to ram tight the stemming. { 'stem·iŋ ‚räd }

Stemonitaceae [MYCOL] The single family of the order Stemonitales. { stē‚män·ə'tās·ē‚ē }

Stemonitales [MYCOL] An order of fungi in the subclass Myxogastromycetidae of the class Myxomycetes. { stē‚män·ə'tā·lēz }

stem rust [PL PATH] Any of several fungus diseases, especially of grasses, affecting the stem and marked by black or reddish-brown lesions. { 'stem ‚rəst }

stem-winding [MECH ENG] Pertaining to a timepiece that is wound by an internal mechanism turned by an external knob and stem (the winding button of a watch). { 'stem ‚wīnd·iŋ }

stench *See* odorant. { stench }

stencil [GRAPHICS] **1.** A template with either mechanically or hand cut openings. **2.** A metal foil with openings made by chemically etching, laser cutting, or electroforming processes. { 'sten·səl }

stencil printing [GRAPHICS] A method of transferring a pattern by brushing, spraying, or squeeging ink or paint through the open areas of a stencil cut from thin metal or cardboard. Also known as pochoir. { 'sten·səl ˌprint·iŋ }

Stenetrioidea [INV ZOO] A group of isopod crustaceans in the suborder Asellota consisting mostly of tropical marine forms in which the first pleopods are fused. { stəˌne·trēˈoid·ē·ə }

Stenocephalidae [INV ZOO] A family of Old World, neotropical Hemiptera included in the Pentatomorpha. { ˌsten·ə·səˈfal·ə,dē }

stenocephaly [MED] Unusual narrowness of the head. { ˌsten·əˈsef·ə·lē }

stenode circuit [ELECTR] Superheterodyne receiving circuit in which a piezoelectric unit is used in the intermediate-frequency amplifier to balance out all frequencies except signals at the crystal frequency, thereby giving very high selectivity. { 'steˌnōd ˌsər·kət }

Stenoglossa [INV ZOO] The equivalent name for Neogastropoda. { ˌsten·əˈgläs·ə }

stenohaline [ECOL] In marine organisms, indicating the ability to tolerate only a narrow range of salinities. { ¦sten·ə¦ha,līn }

Stenolaemata [INV ZOO] A class of marine ectoproct bryozoans having lophophores which are circular in basal outline and zooecia which are long, slender, tubular or prismatic, and gradually tapering to their proximal ends. { ˌsten·ə·lə'mäd·ə }

Stenomasteridae [PALEON] An extinct family of Euechinoidea, order Holasteroida, comprising oval and heart-shaped forms with fully developed pore pairs. { ˌsten·ə·mas'ter·ə,dē }

stenometer [ENG] An instrument for measuring distances; employs a telescope in which two target images a known distance apart are superimposed by turning a micrometer screw. { stə'näm·əd·ər }

stenoplastic [BIOL] Relating to an organism which exhibits a limited capacity for modification or adaptation to a new environment. { ¦sten·ə¦plas·tik }

Stenopodidea [INV ZOO] A section of decapod crustaceans in the suborder Natantia which includes shrimps having the third pereiopods chelate and much longer and stouter than the first pair. { ˌsten·ə·pə'did·ē·ə }

stenosis [MED] Constriction or narrowing, as of the heart or blood vessels. { stə'nō·səs }

Stenostomata [INV ZOO] The equivalent name for Cyclostomata. { ˌsten·ə·stə'mäd·ə }

stenotherm [BIOL] An organism able to tolerate only a small variation of temperature in the environment. { 'sten·əˌthərm }

stenothermic [BIOL] Indicating the ability to tolerate only a limited range of temperatures. { ¦sten·ə¦thər·mik }

Stenothoidae [INV ZOO] A family of amphipod crustaceans in the suborder Gammaridea containing semiparasitic and commensal species. { ˌsten·ə'thói,dē }

stenotopic [ECOL] Referring to an organism with a restricted distribution. { ¦sten·ə¦täp·ik }

Stensen's duct See parotid duct. { 'sten·sənz ˌdəkt }

Stensioellidae [PALEON] A family of Lower Devonian placoderms of the order Petalichthyida having large pectoral fins and a broad subterminal mouth. { ˌsten·shō'el·ə,dē }

Stenurida [PALEON] An order of Ophiuroidea, comprising the most primitive brittlestars, known only from Paleozoic sediments. { stə'nùr·əd·ə }

step [COMPUT SCI] A single computer instruction or operation. [ENG] A small offset on a piece of core or in a drill hole resulting from a sudden sidewise deviation of the bit as it enters a hard, tilted stratum or rock underlying a softer rock. [GEOL] A hitch or dislocation of the strata. [MIN ENG] The portion of a longwall face at right angles to the line of the face formed when a place is worked in front of or behind an adjoining place. [ORG CHEM] See elementary reaction. { step }

step aeration [CIV ENG] An activated sludge process in which the settled sewage is introduced into the aeration tank at more than one point. { 'step eˌrā·shən }

step-and-repeat camera [OPTICS] A type of camera providing a gridlike pattern of latent image frames in a given sequence. { 'step ən ri'pēt ˌkam·rə }

step angle [ELEC] The angle between two successive positions of a stepping motor. { 'step ˌaŋ·gəl }

step attenuator [ELECTR] An attenuator in which the attenuation can be varied in precisely known steps by means of switches. { 'step əˌten·yəˌwäd·ər }

step bearing [MECH ENG] A device which supports the bottom end of a vertical shaft. Also known as pivot bearing. { 'step ˌber·iŋ }

step block [ENG] A metal block, usually of steel or cast iron, with integral stepped sections to allow application of clamps when securing a workpiece to a machine tool table. { 'step ˌbläk }

step brazing [MET] Brazing consecutive joints at sequentially lower temperatures to maintain the integrity of preceding joints. { 'step ˌbrāz·iŋ }

step-by-step operation See single-step operation. { ¦step bī ¦step ˌäp·ə'rā·shən }

step-by-step switch [ELEC] A bank-and-wiper switch in which the wipers are moved by electromagnet ratchet mechanisms individual to each switch. { ¦step bī ¦step 'swich }

step-by-step system [COMMUN] See Strowger system. [CONT SYS] A control system in which the drive motor moves in discrete steps when the input element is moved continuously. { ¦step bī ¦step 'sis·təm }

step change [ELECTR] The change of a variable from one value to another in a single process, taking a negligible amount of time. { 'step ˌchānj }

step-climb profile [AERO ENG] The aircraft climbs a specified number of feet whenever its weight reaches a predetermined amount, thus stepping to an optimum altitude as gross weight decreases. { 'step ˌklīm 'prō,fīl }

step counter [COMPUT SCI] In computers, a counter in the arithmetic unit used to count the steps in multiplication, division, and shift operations. { 'step ˌkaùnt·ər }

step cut [LAP] A cut for gems in which the facets decrease in length the farther they are from the girdle, giving the appearance of steps; an example is the emerald cut. { 'step ˌkət }

step-down photophobic response [PHYSIO] A photophobic response elicited by a sudden decrease in light intensity. { 'step ˌdaùn ˌfōd·ə,fō·bik ri'späns }

step-down transformer [ELEC] A transformer in which the alternating-current voltages of the secondary windings are lower than those applied to the primary winding. { 'step ¦daùn tranz'fòr·mər }

step fault [GEOL] One of a set of closely spaced, parallel faults. Also known as distributive fault; multiple fault. { 'step ˌfòlt }

step function [MATH] **1.** A function f defined on an interval $[a,b]$ so that $[a,b]$ can be partitioned into a finite number of subintervals on each of which f is a constant. Also known as simple function. **2.** More generally, a real function with finite range. { 'step ˌfəŋk·shən }

step-function generator [ELECTR] A function generator whose output waveform increases and decreases suddenly in steps that may or may not be equal in amplitude. { 'step ¦fəŋk·shən 'jen·əˌrād·ər }

step gage [DES ENG] **1.** A plug gage containing several cylindrical gages of increasing diameter mounted on the same axis. **2.** A gage consisting of a body in which a blade slides perpendicularly; used to measure steps and shoulders. { 'step ˌgāj }

Stephanian [GEOL] A European stage of Upper Carboniferous geologic time, forming the Upper Pennsylvanian, above the Westphalian and below the Sakmarian of the Permian. { stə'fän·ē·ən }

Stephanidae [INV ZOO] A small family of the Hymenoptera in the superfamily Ichneumonoidea characterized by many-segmented filamentous antennae. { stə'fan·ə,dē }

stephanite [MINERAL] Ag_5SbS_4 An iron-black mineral crystallizing in the orthorhombic system and having a metallic luster; an ore of silver. Also known as black silver; brittle silver ore; goldschmidtine. { 'stef·ə,nīt }

Stephano [ASTRON] A small satellite of Uranus in a retrograde orbit with a mean distance of 4,940,000 miles (7,950,000 kilometers), eccentricity of 0.053, and sidereal period of 1.84 years. { 'stef·əˌnō or stə'fä·nō }

Stephan's Quintet [ASTRON] A group of five galaxies which lie close together, one of which has widely divergent red shifts. { 'stef·ənz kwin'tet }

step lake See paternoster lake. { 'step ˌlāk }

step-out time [GEOPHYS] In seismic prospecting, the time differentials in arrivals of a given peak or trough of a reflected or refracted event for successive detector positions on the earth's surface. { 'step ¦aut ¸tīm }

step-out well [PETRO ENG] A well drilled at a later time over remote, undeveloped portions of a partially developed continuous reservoir rock. Also known as delayed development well. { 'step ¦aut ¸wel }

steppe [GEOGR] An extensive grassland in the semiarid climates of southeastern Europe and Asia; it is similar to but more arid than the prairie of the United States. { step }

steppe climate [CLIMATOL] The type of climate in which precipitation though very slight, is sufficient for growth of short, sparse grass; typical of the steppe regions of south-central Eurasia. Also known as semiarid climate. { 'step ¸klī·mət }

stepped cone pulley [DES ENG] A one-piece pulley with several diameters to engage transmission belts and thereby provide different speed ratios. { 'stept ¦kōn ¸pül·ē }

stepped footing [CIV ENG] A widening at the bottom of a wall consisting of a series of steps in the proportion of one horizontal to two vertical units. { 'stept 'fud·iŋ }

stepped gear wheel [DES ENG] A gear wheel containing two or more sets of teeth on the same rim, with adjacent sets slightly displaced to form a series of steps. { 'stept 'gir ¸wēl }

stepped leader [GEOPHYS] The initial streamer of a lightning discharge; an intermittently advancing column of high ion density which established the channel for subsequent return streamers and dart leaders. { 'stept 'lēd·ər }

stepped screw [DES ENG] A screw from which sectors have been removed, the remaining screw surfaces forming steps. { 'stept 'skrü }

stepped-wave static inverter [ELEC] A static inverter that generates several pulses in each half cycle and combines them to achieve an output voltage which needs very little filtering. { 'stept ¦wāv 'stad·ik in'vərd·ər }

stepper motor [ELEC] A motor that rotates in short and essentially uniform angular movements rather than continuously; typical steps are 30, 45, and 90°; the angular steps are obtained electromagnetically rather than by the ratchet and pawl mechanisms of stepping relays. Also known as magnetic stepping motor; stepping motor; step-servo motor. { 'step·ər ¸mōd·ər }

stepping See zoning. { 'step·iŋ }

stepping motor See stepper motor. { 'step·iŋ ¸mōd·ər }

stepping reflex [PHYSIO] A reflex response of the newborn and young infant, characterized by alternating stepping movements with the legs, as in walking, elicited when the infant is held upright so that both soles touch a flat surface while the infant is moved forward to accompany any step taken. { 'step·iŋ ¸rē¸fleks }

stepping relay [ELEC] A relay whose contact arm may rotate through 360° but not in one operation. Also known as rotary stepping relay; rotary stepping switch; stepping switch. { 'step·iŋ ¸rē¸lā }

stepping switch See stepping relay. { 'step·iŋ ¸swich }

step pulley [MECH ENG] A series of pulleys of various diameters combined in a single concentric unit and used to vary the velocity ratio of shafts. Also known as cone pulley. { 'step ¸pul·ē }

step-recovery diode [ELECTR] A varactor in which forward voltage injects carriers across the junction, but before the carriers can combine, voltage reverses and carriers return to their origin in a group; the result is abrupt cessation of reverse current and a harmonic-rich waveform. { 'step ri¦kəv·rē 'dī¸ōd }

step response [CONT SYS] The behavior of a system when its input signal is zero before a certain time and is equal to a constant nonzero value after this time. { 'step ri¸späns }

step rocket See multistage rocket. { 'step ¸räk·ət }

step-servo motor See stepper motor. { 'step ¸sər·vō ¸mōd·ər }

step soldering [MET] Soldering consecutive joints at sequentially lower temperatures to maintain the integrity of preceding joints. { 'step ¸säd·ə·riŋ }

step strobe marker [ELECTR] Form of strobe marker in which the discontinuity is in the form of a step in the time base. { 'step 'strōb ¸mär·kər }

step tablet See density step tablet. { 'step ¸tab·lət }

step test [GRAPHICS] A series of exposures made to determine the optimum exposures of either film or paper prints. { 'step ¸test }

STEPPER MOTOR

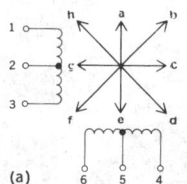

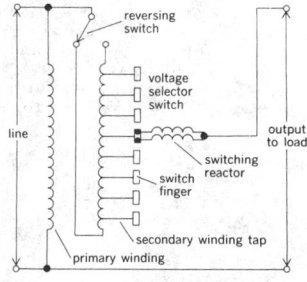

Rotor position	Windings to which voltage is applied
a	V_{1-2}
b	V_{1-2} and V_{4-5}
c	V_{4-5}
d	V_{4-5} and V_{3-2}
e	V_{3-2}
f	V_{3-2} and V_{6-5}
g	V_{6-5}
h	V_{6-5} and V_{1-2}

Permanent-magnet stepper motor. (*a*) Circuit diagram. (*b*) Excitation sequence for 45°-per-step clockwise rotation.

STEP VOLTAGE REGULATOR

Circuit diagram of a step voltage regulator. When switch moves from full-tap position shown, one finger contacts next secondary winding tap before other finger leaves first tap; switching reactors limit current circulating between the bridged taps in resulting position.

---

steptoe [GEOL] An isolated protrusion of bedrock, such as the summit of a hill or mountain, in a lava flow. { 'step¸tō }

step trench [ARCHEO] A trench cut in a mound in a series of steps. { 'step ¸trench }

step-up photophobic response [PHYSIO] A photophobic response elicited by a sudden increase in light intensity. { ¦step ¸əp ¸fōd·ə¸fō·bik ri'späns }

step-up transformer [ELEC] Transformer in which the energy transfer is from a low-voltage winding to a high-voltage winding or windings. { 'step¦əp tranz¸fór·mər }

step voltage regulator [ELEC] A type of voltage regulator used on distribution feeder lines; it provides increments or steps of voltage change. { 'step 'vōl·tij ¸reg·yə¸lād·ər }

step wedge [GRAPHICS] A series of tones in steps from white to black usually on film or a glass plate for testing purposes. { 'step ¸wej }

stepwise reaction [CHEM] A chemical reaction in which at least one reactive intermediate is produced and at least two elementary reactions are involved. { ¦step¸wīz rē¦ak·shən }

sterad See steradian. { 'sti¸rad }

steradian [MATH] The unit of measurement for solid angles; it is equal to the solid angle subtended at the center of a sphere by a portion of the surface of the sphere whose area equals the square of the sphere's radius. Abbreviated sr; sterad. { stə'rād·ē·ən }

steradiancy See radiance. { stə'rād·ē·ən·sē }

sterane [ORG CHEM] A cycloalkane derived from a sterol. { 'sti¸rān }

sterba curtain [ELECTROMAG] Type of stacked dipole antenna array consisting of one or more phased half-wave sections with a quarter-wave section at each end; the array can be oriented for either vertical or horizontal radiation, and can be either center or end fed. { 'stər·bə ¸kərt·ən }

stercobilin [BIOCHEM] Urobilin as a component of the brown fecal pigment. { ¸stər·kō'bī·lən }

stercobilinogen [BIOCHEM] A colorless reduction product of stercobilin found in feces. { ¸stər·kō·bī'lin·ə·jən }

Stercorariidae [VERT ZOO] A family of predatory birds of the order Charadriiformes including the skuas and jaegers. { ¸stər·kə·rə'rī·ə¸dē }

stercorite [MINERAL] $Na(NH_4)H(PO_4)\cdot4H_2O$ A white to yellowish and brown, triclinic mineral consisting of a hydrated acid phosphate of sodium and ammonium. { 'stər·kə¸rīt }

Sterculiaceae [BOT] A family of dicotyledonous trees and shrubs of the order Malvales distinguished by imbricate or contorted petals, bilocular anthers, and ten to numerous stamens arranged in two or more whorls. { stər¸kyü·lē'ās·ē }

stère [MECH] A unit of volume equal to 1 cubic meter; it is used mainly in France, and in measuring timber volumes. { stir }

steregon [MATH] The entire solid angle bounded by a sphere; equal to 4π steradians. { 'ster·ə¸gän }

stereo See stereophonic; stereo sound system. { 'ste·rē·ō }

stereo- [PHYS] A prefix used to designate a three-dimensional characteristic. { 'ste·rē·ō }

stereo amplifier [ENG ACOUS] An audio-frequency amplifier having two or more channels, as required for use in a stereo sound system. { 'ste·rē·ō 'am·plə¸fī·ər }

stereoblastula [EMBRYO] A blastula that lacks a cavity, making it unable to gastrulate. { ¦ster·ē·ə'blas·chə·lə }

stereo broadcasting [COMMUN] Broadcasting two sound channels for reproduction by a stereo sound system having a stereo tuner at its input, to afford a listener a sense of the spatial distribution of the sound sources. { 'ster·ē·ō 'bród¸kast·iŋ }

stereo camera See stereoscopic camera. { 'ster·ē·ō 'kam·rə }

stereocenter [ORG CHEM] A (chiral) carbon atom that has four different substituents bonded to it. Also known as a stereogenic atom. { 'ster·ē·ō¸sen·tər }

stereochemistry [PHYS CHEM] The study of the spatial arrangement of atoms in molecules and the chemical and physical consequences of such arrangement. { ¦ster·ē·ə'kem·ə¸strē }

stereocilia [CELL MOL] **1.** Nonmotile tufts of secretory microvilli on the free surface of cells of the male reproductive tract. **2.** Homogeneous cilia within simple membrane coverings; found on the free-surface hair cells. { ¦ster·ē·ə'sil·ē·ə }

stereocomparagraph [OPTICS] A projection device in which two-dimensional aerial photographs taken at slightly different angles are combined so as to give the appearance of tridimensionality. { ¦ster·ē·ō·kəm¦par·ə‚graf }

stereo comparator [OPTICS] An instrument that may be used to view two photographs taken of the stars in the same section of sky at different times; viewing the images stereoscopically may reveal stars that have moved between exposures or stars of varying brightness. { ¦ster·ē·ō kəm¦par·əd·ər }

stereo effect [ACOUS] Reproduction of sound in such a manner that the listener receives the sensation that individual sounds are coming from different locations, just as did the original sounds reaching the stereo microphone system. { ¦ster·ē·ō i‚fekt }

stereofluoroscopy [ELECTR] A fluoroscopic technique that gives three-dimensional images. { ¦ster·ē·ə·flü¦räs·kə·pē }

stereogastrula [EMBRYO] A gastrula that lacks a cavity. { ¦ster·ē·ə¦gas·trə·lə }

stereogenic atom See stereocenter. { ‚ster·ē·ə‚jen·ik ¦ad·əm }

stereogenic center See asymmetric carbon atom. { ‚ster·ē·ə¦jen·ik ¦sen·tər }

stereognomogram [CRYSTAL] The projection resulting from the superposition of the projection planes of a stereogram and a gnomogram. { ¦ster·ē·ə¦nō·mə‚gram }

stereognosis [PSYCH] The recognition or identification of objects by the sense of touch. { ¦ster·ē¦äg·nə·səs }

stereogram [GRAPHICS] A stereoscopic set of photographs or drawings correctly oriented and mounted for stereoscopic viewing. { ¦ster·ē·ə‚gram }

stereographic chart [MAP] A chart on the stereographic projection. { ¦ster·ē·ə¦graf·ik ¦chärt }

stereographic coverage [GRAPHICS] Photographic coverage with overlapping air photographs to provide a three-dimensional presentation of the picture; 60% overlap is considered normal, and 53% is generally regarded as the minimum. { ¦ster·ē·ə¦graf·ik ¦kəv·rij }

stereographic net See net. { ¦ster·ē·ə¦graf·ik ¦net }

stereographic projection [CRYSTAL] A method of displaying the positions of the poles of a crystal in which poles are projected through the equatorial plane of the reference sphere by lines joining them with the south pole for poles in the upper hemisphere, and with the north pole for poles in the lower hemisphere. [MAP] A perspective conformal, azimuthal map projection in which points on the surface of a sphere or spheroid, such as the earth, are conceived as projected by radial lines from any point on the surface to a plane tangent to the antipode of the point of projection; circles project as circles through the point of tangency, except for great circles which project as straight lines; the principal navigational use of the projection is for charts of the polar regions. Also known as azimuthal orthomorphic projection. [MATH] The projection of the Riemann sphere onto the euclidean plane performed by emanating rays from the north pole of the sphere through a point on the sphere. { ¦ster·ē·ə¦graf·ik prə¦jek·shən }

stereoisomers [ORG CHEM] Compounds whose molecules have the same number and kind of atoms and the same atomic arrangement, but differ in their spatial relationship. { ¦ster·ē·ō¦ī·sə·mərz }

stereolithography [IND ENG] A three-dimensional printing process whereby a CAD drawing of a part is processed to create a file of the part in slices and the part is constructed one slice (or layer) at a time (from bottom to top) by depositing layer upon layer of material (usually a liquid resin that can be hardened using a scanning laser); used for rapid prototyping. { ‚ster·ē·ō·li¦thäg·rə·fē }

stereomicrography [OPTICS] The taking of two microphotographs of the same field at different angles (a stereo pair), then viewing them simultaneously with a stereo viewer. { ¦ster·ē·ə·mī¦krag·rə·fē }

stereomicrometer [ENG] An instrument attached to an optical instrument (such as a telescope) to measure small angles. { ¦ster·ē·ə·mī¦kräm·əd·ər }

stereo multiplex [COMMUN] Stereo broadcasting by a frequency-modulation station, in which the outputs of two channels are transmitted on the same carrier by frequency-division multiplexing. { ¦ster·ē·ō ¦məl·tə‚pleks }

stereophonic [ENG ACOUS] Pertaining to three-dimensional pickup or reproduction of sound, as achieved by using two or more separate audio channels. Also known as stereo. { ¦ster·ē·ə¦fän·ik }

stereophonics [ENG ACOUS] The study of reproducing or reinforcing sound in such a way as to produce the sensation that the sound is coming from sources whose spatial distribution is similar to that of the original sound sources. { ¦ster·ē·ə¦fän·iks }

stereophonic sound system See stereo sound system. { ¦ster·ē·ə¦fan·ik ¦saund ‚sis·təm }

stereo pickup [ENG ACOUS] A phonograph pickup designed for use with standard single-groove two-channel stereo records; the pickup cartridge has a single stylus that actuates two elements, one responding to stylus motion at 45° to the right of vertical and the other responding to stylus motion at 45° to the left of vertical. { ¦ster·ē·ō ¦pik‚əp }

stereoplanigraph [ENG] An instrument for drawing topographic maps from observations of stereoscopic aerial photographs with a stereocomparator. { ¦ster·ē·ə¦plan·ə‚graf }

stereo preamplifier [ENG ACOUS] An audio-frequency preamplifier having two channels, used in a stereo sound system. { ¦ster·ē·ō ¦prē¦am·plə‚fī·ər }

stereopsis See stereoscopy. { ‚ster·ē¦äp·səs }

stereo rangefinder See stereoscopic rangefinder. { ¦ster·ē·ō ¦rārj‚fīn·dər }

stereo record [ENG ACOUS] A single-groove disk record having V-shaped grooves at 45° to the vertical; each groove wall has one of the two recorded channels. { ¦ster·ē·ō ¦rek·ərd }

stereo recorded tape [ENG ACOUS] Recorded magnetic tape having two separate recordings, one for each channel of a stereo sound system. { ster·ē·ō ri¦kòrd·əd ¦tāp }

stereoregular polymer See stereospecific polymer. { ¦ster·ē·ə¦reg·yə·lər ¦päl·i·mər }

stereorubber [ORG CHEM] Synthetic rubber, *cis*-polyisoprene, a polymer with stereospecificity. { ¦ster·ō·ə‚rəb·ər }

stereoscope [OPTICS] An optical instrument in which each eye views one of two photographs taken with the camera or object of study displaced, or simultaneously with two cameras, or with a stereoscopic camera, so that a sensation of depth is produced. { ¦ster·ē·ə‚skōp }

stereoscopic camera [OPTICS] A camera which takes photographs simultaneously with two similar lenses a few inches apart, for use in a stereoscope or other optical system which gives a sensation of depth to the viewer. Also known as stereo camera. { ¦ster·ē·ə¦skäp·ik ¦kam·rə }

stereoscopic heightfinder See stereoscopic rangefinder. { ¦ster·ē·ə¦skäp·ik ¦hīt‚fīn·dər }

stereoscopic microscope [OPTICS] A microscope with two eyepieces and two objectives, giving the viewer a sensation of depth. { ¦ste·rē·ə¦skäp·ik ¦mī·krə‚skōp }

stereoscopic model [GRAPHICS] A mental impression of an area or object seen as being in three dimensions when viewed stereoscopically on photographs. { ¦ster·ē·ə¦skäp·ik ¦mäd·əl }

stereoscopic pair [GRAPHICS] Two photographs with sufficient overlap of detail to make possible stereoscopic examination of an object or an area common to both. { ¦ster·ē·ə¦skäp·ik ¦per }

stereoscopic parallax See absolute stereoscopic parallax. { ¦ster·ē·ə¦skäp·ik ¦par·ə‚laks }

stereoscopic photography [OPTICS] A technique that simulates stereoscopic vision, in which two photographs are made with the camera or object of study displaced, or simultaneously with two cameras, or with a stereoscopic camera, and each of the photographs is viewed by one eye, using a stereoscope or other optical system. { ¦ster·ē·ə¦skäp·ik fə¦täg·rə·fē }

stereoscopic power [OPTICS] The magnifying power of a binoculars or other stereo system multiplied by the ratio of the distance between the objective axes to the distance between the eyepiece axes; it is a measure of the stereoscopic radius. Also known as total relief. { ¦ster·ē·ə¦skäp·ik ¦pau·ər }

stereoscopic radius [PHYSIO] The greatest distance at which there is a sensation of depth in vision due to the fact that the two eyes do not perceive exactly the same view. { ¦ster·ē·ə¦skäp·ik ¦rād·ē·əs }

stereoscopic rangefinder [OPTICS] An optical rangefinder which utilizes stereoscopic vision; it is essentially a large binoculars fitted with special reticles which allow a skilled user to superimpose the stereoscopic image formed by the pair of reticles over the images of the target seen in the eyepieces, so that the correct range is obtained when the reticle marks appear

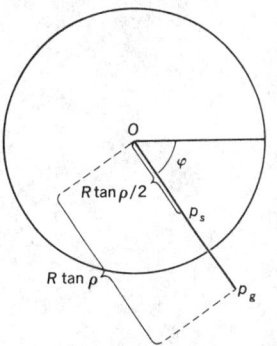

STEREOGNOMOGRAM

Stereognomogram of pole defined by azimuthal angle φ and colatitude ρ. Both projection planes are superimposed; p_s and p_g are on a radius making an angle φ with the origin, and at distances respectively equal to $R \tan \rho/2$ and $R \tan \rho$.

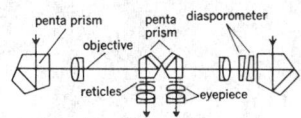

STEREOSCOPIC RANGEFINDER

Lens system of a typical stereoscopic rangefinder. Objective forms images in plane of respective reticles. Diasporometer is used to vary angle between beams reaching eyes from target until it equals angle between beams reaching eyes from two reticles.

to be suspended over the target and at the same apparent distance. Also known as stereo rangefinder; stereoscopic heightfinder. { ¦ster·ē·ə¦skäp·ik 'ranj,fīn·dər }

stereoscopic system [OPTICS] An optical system such as a binoculars or stereoscope that produces two images of the same object viewed from slightly different positions, so that a sensation of depth is created when one image is presented to each eye. { ‚ste·rē·ə¦skäp·ik 'sis·təm }

stereoscopic vision See stereoscopy. { ¦ster·ē·ə¦skäp·ik 'vizh·ən }

stereoscopy [PHYSIO] The phenomenon of simultaneous vision with two eyes in which there is a vivid perception of the distances of objects from the viewer; it is present because the two eyes view objects in space from two points, so that the retinal image patterns of the same object are slightly different in the two eyes. Also known as stereopsis; stereoscopic vision. { ‚ster·ē'äs·kə·pē }

stereoselective reaction [ORG CHEM] A chemical reaction in which one stereoisomer is produced or decomposed more rapidly than another. Also known as enantioselective reaction. { ¦ster·ē·ə·si'lek·tiv rē'ak·shən }

stereoselectivity [BIOCHEM] The selectivity of a reaction for forming one stereoisomer of a product in preference to another stereoisomer. { ¦ster·ē·ō·si·lek'tiv·əd·ē }

stereo sound system [ENG ACOUS] A sound reproducing system in which a stereo pickup, stereo tape recorder, stereo tuner, or stereo microphone system feeds two independent audio channels, each of which terminates in one or more loudspeakers arranged to give listeners the same audio perspective that they would get at the original sound source. Also known as stereo; stereophonic sound system. { 'ster·ē·ō 'saund ‚sis·təm }

stereospecificity [ORG CHEM] The condition of a polymer whose molecular structure has a fixed spatial (geometric) arrangement of its constituent atoms, thus having crystalline properties; for example, synthetic natural rubber, cis-polyisoprene. { ¦ster·ē·ō·spə·sə'fis·ə d·ē }

stereospecific polymer [ORG CHEM] A polymer with specific or definite order of arrangement of molecules in space, as in isotactic polypropylene; permits close packing of molecules and leads to a high degree of polymer crystallinity. Also known as stereoregular polymer. { ¦ster·ē·ō·spə'sif·ik 'päl·i·mər }

stereospecific synthesis [ORG CHEM] Catalytic polymerization of monomer molecules to produce stereospecific polymers, as with Ziegler or Natta catalysts (derived from a transition metal halide and a metal alkyl). { ¦ster·ē·ō·spə'sif·ik 'sin·thə·səs }

Stereospondyli [PALEON] A group of labyrinthodont amphibians from the Triassic characterized by a flat body without pleurocentra and with highly developed intercentra. { ‚ster·ē·ə'spän·də,lī }

stereo subcarrier [COMMUN] A subcarrier whose frequency is the second harmonic of the pilot subcarrier frequency used in frequency-modulation stereo broadcasting. { 'ster·ē·ō ‚səb'kar·ē·ər }

stereo tape recorder [ENG ACOUS] A magnetic-tape recorder having two stacked playback heads, used for reproduction of stereo recorded tape. { 'ster·ē·ō 'tāp ri‚kórd·ər }

stereotaxis [BIOL] An orientation movement in response to stimulation by contact with a solid body. Also known as thigmotaxis. { ‚ster·ē·ə'tak·səs }

stereotropism [BIOL] Growth or orientation of a sessile organism or part of an organism in response to the stimulus of a solid body. Also known as thigmotropism. { ‚ster·ē'ä·trə,piz·əm }

stereo tuner [ENG ACOUS] A tuner having provisions for receiving both channels of a stereo broadcast. { 'ster·ē·ō 'tün·ər }

stereotype [GRAPHICS] A duplicate printing plate made from type and cuts; a paper matrix, or mat, is forced down over the type and cuts to form a mold, into which molten metal is poured, resulting in a new metal printing surface that exactly duplicates the original. { 'ster·ē·ə,tīp }

sterhydraulic [MECH ENG] Pertaining to a hydraulic press in which motion or pressure is produced by the introduction of a solid body into a cylinder filled with liquid. { ¦ster·hī'dról·ik }

steric anomaly See specific-volume anomaly. { 'ster·ik ə'näm·ə·lē }

steric effect [PHYS CHEM] The influence of the spatial configuration of reacting substances upon the rate, nature, and extent of reaction. { 'ster·ik i‚fekt }

steric hindrance [ORG CHEM] The prevention or retardation of chemical reaction because of neighboring groups on the same molecule; for example, ortho-substituted aromatic acids are more difficult to esterify than are the meta and para substitutions. { 'ster·ik 'hin·drəns }

stericooling [FOOD ENG] A method of cooling or precooling fruit and vegetables prior to shipping; uses a cooling spray of water containing, in addition to salt in solution to lower freezing point, a fungicide or bactericide. { 'ster·ə,kül·iŋ }

sterigma [BOT] A peg-shaped structure to which needles are attached in certain conifers. [MYCOL] A slender stalk arising from the basidium of some fungi, from the top of which basidiospores are formed by abstriction. { stə'rig·mə }

sterile distribution [ECOL] A range of areas in which marine animals may live and spawn, but in which eggs do not hatch and larvae do not survive. { 'ster·əl ‚dis·trə'byü·shən }

sterile field See field. { 'ster·əl 'fēld }

sterility [PHYSIO] The inability to reproduce because of congenital or acquired reproductive system disorders involving lack of gamete production or production of abnormal gametes. { stə'ril·əd·ē }

sterilization [MICROBIO] An act or process of destroying all forms of microbial life on and in an object. { ‚ster·ə·lə'zā·shən }

sterilizer [ENG] An apparatus for sterilizing by dry heat, steam, or water. { 'ster·ə,līz·ər }

sterling silver [MET] A silver alloy having a defined standard of purity of 92.5% silver and the remaining 7.5% usually of copper. { 'stər·liŋ 'sil·vər }

stern [NAV ARCH] The aftermost part of a ship. { stərn }

Sternaspidae [INV ZOO] A monogeneric family of polychaete annelids belonging to the Sedentaria. { stər'nas·pə,dē }

stern attack [AERO ENG] In air intercept, an attack by an interceptor aircraft which terminates with a heading crossing angle of 45° or less. { 'stərn ə,tak }

sternbergite [MINERAL] AgFe₂S₃ A dark-brown or black mineral composed of silver iron sulfide and occurring as tabular crystals or flexible laminae. { 'stərn,bər,gīt }

sternebra [VERT ZOO] A segment of the sternum in vertebrates. { 'stər·nə·brə }

stern end bulb [NAV ARCH] A bulb-shaped structure which can be installed at the stern end of a ship to reduce both wave-making and eddy-making resistance. { 'stərn 'end ‚bəlb }

stern frame [NAV ARCH] 1. The timbers making up the upper part of the stern of a wooden ship. 2. A casting or forging, including the propeller post, sternpost, arch, and solepiece of a steel ship. { 'stərn ‚frām }

Stern-Gerlach effect [ATOM PHYS] The splitting of a beam of atoms passing through a strong, inhomogeneous magnetic field into several beams. { 'stərn 'ger,läk i‚fekt }

Sterninae [VERT ZOO] A subfamily of birds in the family Laridae, including the Arctic tern. { 'stər·nə,nē }

sternite [INV ZOO] 1. The ventral part of an arthropod somite. 2. The chitinous plate on the ventral surface of an abdominal segment of an insect. { 'stər,nīt }

stern layer [PHYS CHEM] One of two electrically charged layers of electrolyte ions, the layer of ions immediately adjacent to the surface, in the neighborhood of a negatively charged surface. { 'stərn ‚lā·ər }

sternocleidomastoid [ANAT] A muscle of the neck that flexes the head; origin is the manubrium of the sternum and clavicle, and insertion is the mastoid process. { ¦stər·nō¦klīd·ə'ma,stóid }

sternocostal [ANAT] Pertaining to the sternum and the ribs. { ¦stər·nə'käst·əl }

sternohyoid [ANAT] A muscle arising from the manubrium of the sternum and inserted into the hyoid bone. { ‚stər·nō'hī,óid }

Sternorrhyncha [INV ZOO] A series of the insect order Homoptera in which the beak appears to arise either between or behind the fore coxae, and the antennae are long and filamentous with no well-differentiated terminal setae. { ‚stər·nə'riŋ·kə }

sternothyroid [ANAT] Pertaining to the sternum and thyroid cartilage. { ¦stər·nə'thī,róid }

Sternoxia [INV ZOO] The equivalent name for Elateroidea. { stər′näk·sē·ə }

stern post [NAV ARCH] The main member of the after end of a ship, usually upright, running from the keel up to the bottom of the hull. { ′stərn ‚pōst }

stern tube [NAV ARCH] **1.** A long, circular bearing or bushing which supports the propeller shaft where it emerges from the stern of a ship. **2.** A torpedo tube at the stern of a ship. { ′stərn ‚tüb }

sternum [ANAT] The bone, cartilage, or series of bony or cartilaginous segments in the median line of the anteroventral part of the body of vertebrates above fishes, connecting with the ribs or pectoral girdle. { ′stər·nəm }

sternum height [ANTHRO] Vertical distance taken from the lower tip of the sternum to the floor as the subject stands. { ′stər·nəm ‚hīt }

sternway [NAV] Making way through the water in a direction opposite to the heading. { ′stərn‚wā }

stern wheel [NAV ARCH] A paddle wheel located at the vessel's stern and used to propel the vessel. { ′stərn ‚wēl }

Stern-Zartman experiment [STAT MECH] An experiment in which the distribution in speed of atoms or molecules in a beam emitted from an opening in an oven is measured by having the beam impinge on a rotating cylindrical drum, with a slit cut parallel to the drum axis, and measuring the density of atoms or molecules deposited on the inner surface of the drum, roughly opposite the slit, as a function of distance from a point directly opposite the slit; it is used to test the Maxwell-Boltzmann distribution law. { ′stərn ′zärt·mən ik‚sper·ə·mənt }

steroid [BIOCHEM] A member of a group of compounds, occurring in plants and animals, that are considered to be derivatives of a fused, reduced ring system, cyclopenta[α]-phenanthrene, which consists of three fused cyclohexane rings in a nonlinear or phenanthrene arrangement. { ′sti‚ròid }

sterol [BIOCHEM] Any of the natural products derived from the steroid nucleus; all are waxy, colorless solids soluble in most organic solvents but not in water, and contain one alcohol functional group. { ′sti‚ròl }

sterol regulatory element binding protein [BIOCHEM] A transcription factor required for the active transcription of genes that encode the low-density lipoprotein receptor and enzymes in cholesterol synthesis. { ste‚ròl ′reg·yə·lə tòr·ē ‚el·ə·mənt ′bīnd·iŋ ‚prō‚tēn }

sterone [BIOCHEM] A ketone derived from a steroid. { ′sti‚rōn }

sterrettite *See* kolbeckite. { ′ster·ə‚tīt }

sterrometal [MET] Hard brass containing a small amount of iron and manganese; used for hydraulic cylinders and marine castings. { ′ste·rō‚med·əl }

Stetefeldt furnace [MET] A furnace for desulfurizing and chloridizing silver ores; the ores are powdered, mixed with salt, and dropped through a hot atmosphere. { ′sted·ə‚felt ‚fər·nəs }

stethoscope [MED] An instrument for indirect auscultation for the detection and study of sounds arising within the body; sounds are conveyed to the ears of the examiner through rubber tubing connected to a funnel or disk-shaped endpiece. { ′steth·ə‚skōp }

stewartite [GEOL] A steel-gray, iron-containing variety of bort that has magnetic properties. [MINERAL] $Mn_3(DO)_2$· $4H_2O$ A brownish-yellow mineral composed of hydrous manganese phosphate occurring in minute crystals or fibrous tufts in pegmatites. { ′stü·ər‚tīt }

sthène [MECH] The force which, when applied to a body whose mass is 1 metric ton, results in an acceleration of 1 meter per second per second; equal to 1000 newtons. Formerly known as funal. { sthēn }

Sthenurinae [PALEON] An extinct subfamily of marsupials of the family Diprotodontidae, including the giant kangaroos. { sthə′nùr·ə‚nē }

stibiconite [MINERAL] $Sb_3O_6(OH)$ A pale yellow to yellowish- or reddish-white mineral consisting of a basic or hydrated oxide of antimony; occurs in massive form, as a powder, and in crusts. { ′stib·ə·kə‚nīt }

stibide *See* antimonide. { ′sti‚bīd }

stibiocolumbite [MINERAL] $Sb(Nb,Ta,Cb)O_4$ A dark brown to light yellowish- or reddish-brown, orthorhombic mineral consisting of an oxide of antimony and tantalum-columbium. { ‚stib·ē·ō′käl·əm‚bīt }

stibium [CHEM] The Latin name for antimony, thus the symbol Sb for the element. [MINERAL] *See* antimonite. { ′stib·ē·əm }

stibnate *See* potassium antimonate. { ′stib‚nāt }

stibnite *See* antimonite. { ′stib‚nīt }

stibophen [PHARM] $C_{12}H_4Na_5O_{16}S_4Sb·7H_2O$ A crystalline compound that is soluble in water, insoluble in organic solvents; used in medicine for protozoan infections. { ′stib·ə‚fen }

Stichaeidae [VERT ZOO] The pricklebacks, a family of perciform fishes in the suborder Blennioidei. { stə′kē·ə‚dē }

Stichocotylidae [INV ZOO] A family of trematodes in the subclass Aspidogastrea in which adults are elongate and have a single row of alveoli. { ‚stik·ə·kə′til·ə‚dē }

Stichopodidae [INV ZOO] A family of the echinoderm order Aspidochirotida characterized by tentacle ampullae and by left and right gonads. { ‚stik·ə′päd·ə‚dē }

stichtite [MINERAL] $Mg_6Cr_2(CO_3)(OH)_{16}·4H_2O$ A lilac-colored rhombohedral mineral composed of hydrous basic carbonate of magnesium and chromium. { ′sti‚kīt }

stick [ENG] **1.** A rigid bar hinged to the boom of a dipper or pull shovel and fastened to the bucket. **2.** A long slender tool bonded with an abrasive for honing or sharpening tools and for dressing of wheels. [ORD] A succession of missiles fired or released separately at predetermined intervals from a single aircraft. { stik }

stick gage [ENG] A suitably divided vertical rod, or stick, anchored in an open vessel so that the magnitude of rise and fall of the liquid level may be observed directly. { ′stik ‚gāj }

sticking [COMPUT SCI] In computers, the tendency of a flip-flop to remain in, or to spontaneously switch to, one of its two stable states. { ′stik·iŋ }

sticking coefficient [PHYS CHEM] The fraction of all atoms incident on a surface that are adsorbed on the surface. { ′stik·iŋ ‚kō·i‚fish·ənt }

Stickland reaction [BIOCHEM] An amino acid fermentation involving the coupled decomposition of two or more substrates. { ′stik·lənd rē‚ak·shən }

stickleback [VERT ZOO] Any fish which is a member of the family Gasterosteidae, so named for the variable number of free spines in front of the dorsal fin. { ′stik·əl‚bak }

stick shellac [MATER] Shellac in the form of a solid stick and in a variety of colors; used for filling imperfections in wood. { ′stik shə‚lak }

stick-slip friction [MECH] Friction between two surfaces that are alternately at rest and in motion with respect to each other. { ′stik ‚slip ‚frik·shən }

sticky charge *See* sticky grenade. { ′stik·ē ′chärj }

sticky end [BIOCHEM] Any of the single-stranded complementary ends of a deoxyribonucleic acid molecule. Also known as cohesive end. { ′stik·ē ‚end }

sticky grenade [ORD] A small explosive charge covered with an adhesive, intended to be thrown or placed by hand where the adhesive will hold the charge in place until detonated by a time fuse. Also known as sticky charge. { ′stik·ē grə′nād }

stiction [MECH] Friction that tends to prevent relative motion between two movable parts at their null position. { ′stik shən }

Stieltjes integral [MATH] The Stieltjes integral of a real function $f(x)$ relative to a real function $g(x)$ of bounded variation on an interval $[a,b]$ is defined, analogously to the Riemann integral, as a limit of a sum of terms $f(a_i) [g(x_i) − g(x_{i−1})]$ taken as partitions of the interval shrink. Denoted

$$\int_a^b f(x)dg(x)$$

Also known as Riemann-Stieltjes integral. { ′stēlt·yəs ‚int·ə·grəl }

Stieltjes transform [MATH] A form of the Laplace transform of a function where the usual Riemann integral is replaced by a Stieltjes integral. { ′stēlt·yəs ‚tranz·fòrm }

stiffened-shell fuselage *See* semimonocoque. { ′stif·ənd ‚shel ′fyü·sə‚läzh }

stiffener [CIV ENG] A steel angle or plate attached to a slender beam to prevent its buckling by increasing its stiffness. { 'stif·nər }

stiffleg derrick [MECH ENG] A derrick consisting of a mast held in the vertical position by a fixed tripod of steel or timber legs. Also known as derrick crane; Scotch derrick. { 'stif¦leg 'der·ik }

stiffness [ACOUS] *See* acoustic stiffness. [MECH] The ratio of a steady force acting on a deformable elastic medium to the resulting displacement. { 'stif·nəs }

stiffness coefficient [MECH] The ratio of the force acting on a linear mechanical system, such as a spring, to its displacement from equilibrium. { 'stif·nəs ¦kō·i¦fish·ənt }

stiffness constant [MECH] Any one of the coefficients of the relations in the generalized Hooke's law used to express stress components as linear functions of the strain components. Also known as elastic constant. { 'stif·nəs ¦kän·stənt }

stiffness matrix [MECH] A matrix **K** used to express the potential energy V of a mechanical system during small displacements from an equilibrium position, by means of the equation $V = 1/2\mathbf{q}^T\mathbf{Kq}$, where **q** is the vector whose components are the generalized components of the system with respect to time and $\mathbf{q}^T$ is the transpose of **q**. Also known as stability matrix. { 'stif·nəs ¦mā·triks }

stiffness reactance *See* acoustic stiffness reactance. { 'stif·nəs rē¦ak·təns }

stigma [BOT] The rough or sticky apical surface of the pistil for reception of the pollen. [INV ZOO] **1.** The eyespot of certain protozoans, such as *Euglena*. **2.** The spiracle of an insect or arthropod. **3.** A colored spot on many lepidopteran wings. [MECH] A unit of length used mainly in nuclear measurements, equal to 10^{-12} meter. Also known as bicron. { 'stig·mə }

stigmatic [OPTICS] **1.** Property of an optical system whose focal power is the same in all meridians. **2.** *See* homocentric. { stig'mad·ik }

stigmatic concave grating [OPTICS] An optical element with many parallel grooves on a concave optical surface; combines the two functions of light dispersion and focusing in one dispersive element; used in space optics, in food and metal analysis, and as a dispersive element of spectrophotometers and spectrographs. { stig'mad·ik 'kän¦kāv 'grād·iŋ }

stigmatism [PHYSIO] A condition of the refractive media of the eye in which rays of light from a point are accurately brought to a focus on the retina. { 'stig·mə,tiz·əm }

stigmator [ELECTR] A device that corrects asymmetries in an electron lens by superposing on the field of the lens a second adjustable field. { stig'mäd·ər }

stilb [OPTICS] A unit of luminance, equal to 1 candela per square centimeter. Abbreviated sb. { stilb }

Stilbaceae [MYCOL] The equivalent name for Stilbellaceae. { stil'bās·ē,ē }

Stilbellaceae [MYCOL] A family of fungi of the order Moniliales in which conidiophores are aggregated in long bundles or fascicles, forming synnemata or coremia, generally having the conidia in a head at the top. { ,stil·bə'lās·ē,ē }

stilbene [ORG CHEM] $C_6H_5CH:CHC_6H_5$ Colorless crystals soluble in ether and benzene, insoluble in water; melts at 124°C; used to make dyes and bleaches and as phosphors. Also known as diphenylethylene; toluylene. { 'stil,bēn }

stilbesterol *See* diethylstilbesterol. { stil'bes·tə,ról }

stilbite [MINERAL] $Ca(Al_2Si_7O_{18})·7H_2O$ A white, brown, or yellow mineral belonging to the zeolite family of silicates; crystallizes in the monoclinic system, occurs in sheaflike aggregates of tabular crystals, and has pearly luster; hardness is 3.5–4 on Mohs scale, and specific gravity is 2.1–2.2. Also known as desmine. { 'stil,bīt }

stile [BUILD] The upright outside framing piece of a window or door. { stīl }

Stiles method [PETRO ENG] A technique for computing oil recovery by waterflood methods, taking into account the distribution of varying permeability stratums throughout the reservoir. { 'stīlz ,meth·əd }

stiletto [ELECTR] An advanced electronic subsystem contained in United States strike aircraft type F-4D for detection, identification, and location of ground-based radars; the location of radar targets is determined by direction finding and passive ranging techniques; it is used for the delivery of guided and unguided weapons against the target radars under all weather conditions. { stə'led·ō }

still [CHEM ENG] A device used to evaporate liquids; heat is applied to the liquid, and the resulting vapor is condensed to a liquid state. { stil }

stillage [FOOD ENG] The residue grain from the manufacture of alcohol from grain; used as a feed supplement. { 'stil·ij }

stillbirth [MED] Birth of a dead infant. { 'stil,bərth }

stilling basin [ENG] A depressed area in a channel or reservoir that is deep enough to reduce the velocity of the flow. Also known as stilling box. { 'stil·iŋ ,bas·ən }

stilling box *See* stilling basin. { 'stil·iŋ ,bäks }

stillingia oil [MATER] A combustible, toxic, pale-yellow drying oil with a linseed-oil scent and a mustard taste, derived from tallow tree seeds; used in lubricants, candles, textile dressing, and soap. Also known as tallow-seed oil. { stə'lin·jē·ə ,oil }

Still's disease [MED] Juvenile rheumatoid arthritis in which involvement of the viscera is prominent. Also known as Chauffard-Still disease. { stilz di,zēz }

stillstand [GEOL] A period during which a land area, a continent, or an island remains stationary with respect to the interior of the earth or to sea level. { 'stil,stand }

still water [HYD] A portion of a stream having a very slight gradient and no visible current. { 'stil ¦wȯd·ər }

still-water level [OCEANOGR] The level that the sea surface would assume in the absence of wind waves. { 'stil ¦wȯd·ər ,lev·əl }

still wax *See* wax tailings. { 'stil ,waks }

still well [METEOROL] A device, used in evaporation pan measurements, which provides an undisturbed water surface and support for the hook gage; the U.S. Weather Bureau model consists of a brass cylinder, 8 inches (20.32 centimeters) high and 3.5 inches (8.89 centimeters) in diameter, mounted over a hole in a triangular galvanized iron base which is provided with leveling screws. { 'stil ,wel }

stilpnomelane [MINERAL] $K(Fe,Mg,Al)_3Si_4O_{10}(OH)_2·H_2O$ A black or greenish-black mineral composed of basic hydrous potassium iron magnesium aluminum silicate; occurs as fibers, incrustations, and foliated plates. { ,stilp·nō'me,lān }

stilt root [BOT] A prop root of a mangrove tree. { 'stilt ,rüt }

Stimson anchor [NAV ARCH] A cast-iron anchor in the shape of an airfoil and bridled in such a way that the slightest horizontal force, combined with the weight of the anchor, tends to plow it into the soft sediments of the sea floor. { 'stim·sən ,aŋ·kər }

stimulant [PHARM] A drug or agent that temporarily acts on muscles, nerves, or a sensory end organ, producing an increase in its state of activity. { 'stim·yə·lənt }

stimulated emission [ATOM PHYS] Emission of electromagnetic radiation by an atom or molecule as a result of its interaction with incident radiation of the same frequency. Also known as induced emission. [ELECTROMAG] The emission of radiation at a given frequency that results from the presence of an external radiation field of the same frequency. { 'stim·yə,läd·əd i'mish·ən }

stimulated-emission device [ELECTR] A device that uses the principle of amplification of electromagnetic waves by stimulated emission, namely, a maser or a laser. { 'stim·yə,läd·əd i'mish·ən di,vīs }

stimulated scattering [OPTICS] The amplification, by stimulated emission, of intense radiation from a pulsed laser that has been inelastically scattered; results in exponential growth of the scattered power and, in some cases, almost complete scattering of the incident light. { 'stim·yə,läd·əd 'skad·ə·riŋ }

stimulation deafness [MED] Deafness induced by noise; involves changes in the chemical interchange between the canals of the cochlea, as well as nerve destruction. { ,stim·yə'lā·shən 'def·nəs }

stimulation treatment [PETRO ENG] One of the techniques to increase (stimulate) oil- or gas-reservoir production, such as acidizing, fracturing, controlled underground explosions, or various cleaning techniques. Also known as well stimulation. { 'stim·yə'lā·shən ,trēt·mənt }

stimulator [MED] A neurosurgical device that supplies a controlled alternating-current voltage to two electrodes that are applied to a patient. { 'stim·yə,läd·ər }

STILBELLACEAE

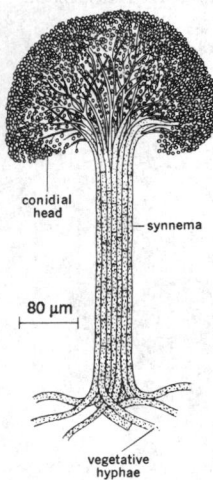

conidial head

synnema

80 μm

vegetative hyphae

Graphium ulmi, cause of Dutch elm disease.

stimulatory G protein [CELL MOL] A G protein (heterotrimetric GTP-binding protein) that activates adenylyl cyclase to produce cAMP in intracellular signaling pathways. Symbolized G_s. { |stim·yə·lə,tȯr·ē 'jē ,prō,tēn }

stimulon [GEN] A system of genes not physically linked together. { 'stim·yə,län }

stimulus [CONT SYS] A signal that affects the controlled variable in a control system. [PHYSIO] An agent that produces a temporary change in physiological activity in an organism or in any of its parts. { 'stim·yə·ləs }

stimulus filtering [PSYCH] The apparent awareness of only a few of the great number of stimuli bombarding an animal. { 'stim·yə·ləs ,fil·tər·iŋ }

stinger [PETRO ENG] **1.** A support which is attached to the stern of a pipe-laying barge and which controls the bending of the pipe as it leaves the barge to enter the water. **2.** A hollow tube, narrower at the bottom, that is forced into a gushing stream of oil from the wellhead to carry out well-control operations. [ZOO] A sharp piercing organ, as of a bee, stingray, or wasp, usually connected with a poison gland. { 'stiŋ·ər }

stinging cell See cnidoblast. { 'stiŋ·iŋ ,sel }

stingray [VERT ZOO] Any of various rays having a whiplike tail armed with a long serrated spine, at the base of which is a poison gland. { 'stiŋ,rā }

stink See gob stink. { 'stiŋk }

stinkdamp [MIN ENG] The hydrogen sulfide that occurs in mines. { 'stiŋk,damp }

stinkstone [GEOL] A stone containing decomposing organic matter that gives off an offensive odor when rubbed or struck. { 'stiŋk,stōn }

stipe [BOT] **1.** The petiole of a fern frond. **2.** The stemlike portion of the thallus in certain algae. [MYCOL] The short stalk or stem of the fruit body of a fungus, such as a mushroom. { stīp }

stipoverite See stishovite. { stə'päv·ə,rīt }

stippling [GRAPHICS] Graduation of shading by numerous separate touches; shadow areas on charts, for instance, are sometimes indicated by numerous dots decreasing in density as the depth increases. { 'stip·liŋ }

stipule [BOT] Either of a pair of appendages that are often present at the base of the petiole of a leaf. { 'stip·yül }

stir-in resin [MATER] A vinyl resin that does not require grinding in order to disperse in a plastisol or organisol. { 'stir,in 'rez·ən }

Stirling cycle [THERMO] A regenerative thermodynamic power cycle using two isothermal and two constant volume phases. { 'stir·liŋ ,sī·kəl }

Stirling engine [MECH ENG] An engine in which work is performed by the expansion of a gas at high temperature; heat for the expansion is supplied through the wall of the piston cylinder. { 'stir·liŋ ,en·jən }

Stirling numbers [MATH] The coefficients which occur in the Stirling interpolation formula for a difference operator. { 'stir·liŋ ,nəm·bərz }

Stirling numbers of the first kind [MATH] The numbers $s(n,r)$ giving the coefficient of x^r in the falling factorial polynomial $x(x - 1)(x - 2)\cdots(x - n + 1)$. { 'stər·liŋ ,num·bərz əv thə 'fərst ,kind }

Stirling numbers of the second kind [MATH] The numbers $S(n,r)$ giving the numbers of ways that n elements can be distributed among r indistinguishable cells so that no cell remains empty. { 'stər·liŋ ,nəm·bərz əv thə 'sek·ənd ,kīnd }

Stirling's formula [MATH] The expression $(n/e)^n \sqrt{2\pi n}$ is asymptotic to factorial n; that is, the limit as n goes to ∞ of their ratio is 1. { 'stir·liŋz ,fȯr·myə·lə }

Stirling's series [MATH] An asymptotic expansion for the logarithm of the gamma function, or an equivalent asymptotic expansion for the gamma function itself, from which Stirling's formula may be derived. { 'stər·liŋz ,sir,ēz }

Stirodonta [INV ZOO] Formerly, an order of Euechinoidea that included forms with stirodont dentition. { ,stir·ə'dän·tə }

stirodont dentition [INV ZOO] In Echinoidea, the condition in which the teeth are keeled within and the foramen magnum is open. { 'stir·ə,dänt den'tish·ən }

stirred-flow reactor [CHEM ENG] A reactor in which there is a device for achieving effective mixing, frequently in the form of a rapidly rotating basket holding the catalyst. { |stird 'flō rē,ak·tər }

stirring [PHYS] A turbulent process in which molecular diffusion and molecular heat conduction are speeded up. { 'stər·iŋ }

stirring effect [ELECTROMAG] The circulation in a molten metal carrying electric current as a result of the combined forces of the pinch and motor effects. { 'stər·iŋ i,fekt }

stirrup [CIV ENG] In concrete construction, a U-shaped bar which is anchored perpendicular to the longitudinal steel as reinforcement to resist shear. [MIN ENG] **1.** A piece of steel hung from a gallows frame to engage the endgate hooks when a mine car is tilted over; used at dumps. **2.** A screw joint suspended from the brakestaff of a spring pole, by which the boring rods are adjusted to the depth of the borehole. Also known as temper screw. { 'stər·əp }

stishovite [MINERAL] SiO_2 A polymorph of quartz, a dense, fine-grained mineral formed under very high pressure (about 1×10^6 pounds per square inch or 7×10^9 pascals); it is the only mineral in which the silicon atom has a coordination number of six; specific gravity is 4.28. Also known as stipoverite. { 'stish·ə,vīt }

stitch bonding [ENG] A method of making wire connections between two or more points on an integrated circuit by using impulse welding or heat and pressure while feeding the connecting wire through a hole in the center of the welding electrode. { 'stich ,bänd·iŋ }

stitching [ENG] Progressive welding of thermoplastic materials (resins) by successive applications of two small, mechanically operated, radio-frequency-heated electrodes; the mechanism is similar to that of a normal sewing machine. { 'stich·iŋ }

stitch rivet [ENG] One of a series of rivets joining the parallel elements of a structural member so that they act as a unit. { 'stich ,riv·ət }

stitch welding [MET] A series of spaced spot resistance welds. { 'stich ,weld·iŋ }

STN LCD See supertwisted nematic liquid-crystal display.

Stobbe reaction [ORG CHEM] A type of aldol condensation reaction represented by the reaction of benzophenone with dimethyl succinate and sodium methoxide to form monoesters of an α-alkylidene (or arylidene) succinic acid. { 'shtȯb·ə rē,ak·shən }

stochastic [MATH] Pertaining to random variables. { stō'kas·tik }

stochastic automaton See probabilistic automaton. { stō'kas·tik ȯ'täm·ə,tän }

stochastic calculus [MATH] The mathematical theory of stochastic integrals and differentials, and its application to the study of stochastic processes. { stō'kas·tik 'kal·kyə·ləs }

stochastic chain rule [MATH] A generalization of the ordinary chain rule to stochastic processes; it states that the process $U_t = u(X_t^1, X_t^2, \ldots, X_t^n)$ satisfies

$$dU = \sum_i \partial_i u\, dX^i + \tfrac{1}{2} \sum_{ij} \partial_i \partial_j u\, dX^i dX^j$$

with the conventions $(dt)^2 = 0$ and $dW^\alpha dW^\beta = \partial_{\alpha\beta} dt$, where the X^i are processes satisfying

$$dX_t^i = a_t^i dt + \sum_{\alpha=1}^m b_t^{i\alpha}\, dW_t^\alpha, \quad i = 1, 2, \ldots, n;$$

$\{W_t^\alpha,\ t \geq 0\}$, $\alpha = 1, 2, \ldots, m$, are independent Wiener processes; the dW_t^α are the corresponding random disturbances occurring in the infinitesimal time interval dt; the a_t^i and $b_t^{i\alpha}$ are independent of future disturbances, and $u(x_1, x_2, \ldots, x_n)$ is a function whose derivatives $\partial_i u$ and $\partial_i \partial_j u$ are continuous. Also known as Itô's formula. { stō'kas·tik 'chān ,rül }

stochastic control theory [CONT SYS] A branch of control theory that aims at predicting and minimizing the magnitudes and limits of the random deviations of a control system through optimizing the design of the controller. { stō'kas·tik kən'trōl ,thē·ə·rē }

stochastic cooling [NUCLEO] A method of reducing the energy spread, angular divergence, and geometric size of a charged-particle beam in which the displacement of an assembly of particles is measured and is then corrected by fast electric field pulses. { stō'kas·tik 'kül·iŋ }

stochastic differential [MATH] An expression representing

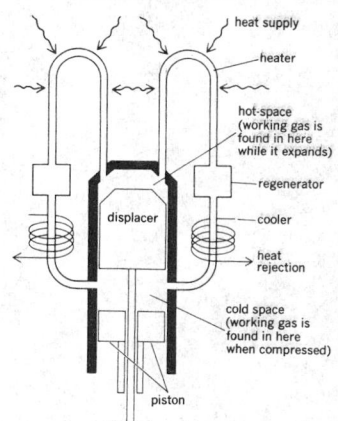

STIRLING ENGINE

heat supply
heater
hot-space (working gas is found in here while it expands)
regenerator
displacer
cooler
heat rejection
cold space (working gas is found in here when compressed)
piston

Diagram of Stirling engine, displacer type. Second piston, called displacer, transfers gas back and forth between hot space, at a fixed high temperature, and cold space, at a fixed low temperature.

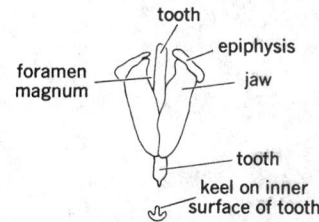

STIRODONT DENTITION

tooth
epiphysis
foramen magnum
jaw
tooth
keel on inner surface of tooth

Tooth and lantern structure in stirodont dentition.

the random disturbances occurring in an infinitesimal time interval; it has the form dW_t, where $\{W_t, t \geq 0\}$ is a Wiener process. { stō′kas·tik ˌdif·ə′ren·chəl }

stochastic independence *See* statistical independence. { stō′kas·tik in·də′pen·dəns }

stochastic integral [MATH] An integral used to construct the sample functions of a general diffusion process from those of a Wiener process; it has the form

$$\int_{W_0}^{W_s} a_t \, dW_t$$

where $\{W_t, t \geq 0\}$ is a Wiener process, dW_t represents the random disturbances occurring in an infinitesimal time interval dt, and a_t is independent of future disturbances. Also known as Itô's integral. { stō′kas·tik ′int·ə·grəl }

stochastic matrix [MATH] A square matrix with nonnegative real entries such that the sum of the entries of each row is equal to 1. { stō′kas·tik ′mā·triks }

stochastic process [MATH] A family of random variables, dependent upon a parameter which usually denotes time. Also known as random process. { stō′kas·tik ′prä·səs }

stochastic sequential machine *See* probabilistic sequential machine. { stō′kas·tik si′kwen·chəl mə′shēn }

stochastic variable *See* random variable. { stō′kas·tik ′ver·ē·ə·bəl }

stock [GEOL] *See* pipe. [IND ENG] **1.** A product or material kept in storage until needed for use or transferred to some ultimate point for use, for example, crude oil tankage or paper-pulp feed. **2.** Designation of a particular material, such as bright stock or naphtha stock. [PETR] A usually discordant, batholithlike body of intrusive igneous rock not exceeding 40 square miles (103.6 square kilometers) in surface exposure and usually discordant. { stäk }

stock accounting [IND ENG] The establishment and maintenance of formal records of material in stock reflecting such information as quantities, values, or condition. { ′stäk ə,kau̇nt·iŋ }

stockage objective [AERO ENG] The maximum quantities of material to be maintained on hand to sustain current operations; it will consist of the sum of stocks represented by the operating level and the safety level. { ′stä·kij əb,jek·tiv }

stock control [IND ENG] Process of maintaining inventory data on the quantity, location, and condition of supplies and equipment due in, on hand, and due out, to determine quantities of material and equipment available or required for issue and to facilitate distribution and management of material. { ′stäk kən,trōl }

stock coordination [IND ENG] A supply management function exercised usually at department level which controls the assignment of material cognizance for items or categories of material to inventory managers. { ′stäk kō,ȯrd·ən,ā·shən }

stock-dye [TEXT] Dyeing fibers before they are spun into yarn. { ′stäk ,dī }

stocking cutter [MECH ENG] **1.** A gear cutter having side rake or curved edges to rough out the gear-tooth spaces before they are formed by the regular gear cutter. **2.** A concave gear cutter ganged beside a regular gear cutter and used to finish the periphery of a gear blank by milling ahead of the regular cutter. { ′stäk·iŋ ′kəd·ər }

stock number [IND ENG] Number assigned to an item, principally to identify that item for storage and issue purposes. { ′stäk ,nəm·bər }

stockpile [ENG] A reserve stock of material, equipment, raw material, or other supplies. { ′stäk,pīl }

stock rail [CIV ENG] The fixed rail in a track, against which the switch rail operates. { ′stäk ,rāl }

stock record account [IND ENG] A basic record showing by item the receipt and issuance of property, the balances on hand, and such other identifying or stock control data as may be required by proper authority. { ′stäk ′rek·ərd ə,kau̇nt }

stockwork [GEOL] A mineral deposit in the form of a network of veinlets diffused in the country rock. { ′stäk,wərk }

Stoddard solvent [MATER] A petroleum naphtha product with a comparatively narrow boiling range; used mostly for dry cleaning. { ′städ·ərd ′säl·vənt }

Stodola method [MECH] A method of calculating the deflection of a uniform or nonuniform beam in free transverse vibration at a specified frequency, as a function of distance along the beam, in which one calculates a sequence of deflection curves each of which is the deflection resulting from the loading corresponding to the previous deflection, and these deflections converge to the solution. { ′stō·də·lə ,meth·əd }

stoichiometry [PHYS CHEM] The numerical relationship of elements and compounds as reactants and products in chemical reactions. { ˌstȯi·kē′äm·ə·trē }

stoke [FL MECH] A unit of kinematic viscosity, equal to the kinematic viscosity of a fluid with a dynamic viscosity of 1 poise and a density of 1 gram per cubic centimeter. Symbol St (formerly S). Also known as lentor (deprecated usage); stokes. { stōk }

stoker [MECH ENG] A mechanical means, as used in a furnace, for feeding coal, removing refuse, controlling air supply, and mixing with combustibles for efficient burning. { ′stō·kər }

stokes *See* stoke. { stōks }

Stokes-Adams syndrome [MED] Syncopic or convulsive attacks occurring in patients with complete heart block. { ′stōks ′ad·əmz ,sin,drōm }

Stokes drift [FL MECH] The drift of particles in a gravity wave, which arises from the fact that particle velocities are periodic with a mean which is not zero. { ′stōks ,drift }

Stokes flow [FL MECH] Fluid flow in which the Reynolds number is very small, so that the nonlinear terms in the Navier-Stokes equations can be neglected. { ′stōks ,flō }

Stokes frequencies [OPTICS] Scattered (secondary) light in the Raman effect (when a high-intensity light beam passes through a transparent medium) that occurs at frequencies smaller than the frequency of the primary beam. { ′stōks ,frē·kwən·sēz }

Stokes' integral theorem [MATH] The analog of Green's theorem in n-dimensional euclidean space; that is, a line integral of $F_1(x_1,x_2,\ldots,x_n)dx_1 + \cdots + F_n(x_1,x_2,\ldots,x_n)dx_n$ over a closed curve equals an integral of an expression containing various partial derivatives of $F_1,\ldots,F_n$ over a surface bounded by the curve. { ′stōks ′int·ə·grəl ,thir·əm }

stokesite [MINERAL] $CaSnSi_3O_9 \cdot 2H_2O$ A colorless orthorhombic mineral composed of hydrous calcium tin silicate occurring in crystals. { ′stōk,sīt }

Stokes' law [FL MECH] At low velocities, the frictional force on a spherical body moving through a fluid at constant velocity is equal to 6π times the product of the velocity, the fluid viscosity, and the radius of the sphere. [SPECT] The wavelength of luminescence excited by radiation is always greater than that of the exciting radiation. { ′stōks ,lȯ }

Stokes' lens [OPTICS] A variable-power compound lens made up of cylindrical lenses mounted so that the angle between their axes can be varied. { ′stōks ,lenz }

Stokes line [SPECT] A spectrum line in luminescent radiation whose wavelength is greater than that of the radiation which excited the luminescence, and thus obeys Stokes' law. { ′stōks ,līn }

Stokes number 1 [FL MECH] A dimensionless number used in the study of the dynamics of a particle in a fluid, equal to the product of the dynamic viscosity of the fluid and the particle's vibration time, divided by the product of the fluid density and a characteristic length. Symbol St. { ′stōks ′nəm·bər ′wən }

Stokes number 2 [ENG] A dimensionless number used in the calibration of rotameters, equal to $1.042 \, m_f g \rho (1 - \rho/\rho_f) R^3/\mu^2$, where ρ and μ are the density and dynamic viscosity of the fluid respectively, m_f and ρ_f are the mass and density of the float respectively, and R is the ratio of the radius of the tube to the radius of the float. Symbol St_2. { ′stōks ′nəm·bər ′tü }

Stokes parameters [OPTICS] Four quantities that fully describe the polarization of a beam of light. { ′stōks pə,ram·əd·ərz }

Stokes phenomenon [MATH] A change in the asymptotic representation of certain analytic functions that occurs in passing from one section of the complex plane to another. { ′stōks fə,näm·ə,nän }

Stokes shift [SPECT] The displacement of spectral lines or bands of luminescent radiation toward longer wavelengths than those of the absorption lines or bands. { ′stōks ,shift }

Stokes stream function [FL MECH] A one-component vector potential function used in analyzing and describing a steady, axially symmetric fluid flow; at any point it is equal to $\frac{1}{2\pi}$

times the mass rate of flow inside the surface generated by rotating the streamline on which the point is located about the axis of symmetry. { 'stōks 'strem ,faŋk·shən }

Stokes stretcher [MED] A basket-type stretcher constructed of tubular steel and strong wire mesh, and which acts as a splint for the entire body. { 'stōks 'strech·ər }

stoking [MET] Presintering or sintering a metal powder in such a way as to advance the compacts through the furnace at a fixed rate. Also known as continuous sintering. { 'stōk·iŋ }

STOL See short takeoff and landing. { 'stōl or |es|tē|ō el }

STOL aircraft [AERO ENG] Heavier-than-air craft that cannot take off and land vertically, but can operate within areas substantially more confined than those normally required by aircraft of the same size. Derived from short takeoff and landing aircraft. { 'stōl ,er·kraft }

stolon [BOT] See runner. [INV ZOO] An elongated projection of the body wall from which buds are formed giving rise to new zooids in Anthozoa, Hydrozoa, Bryozoa, and Ascidiacea. [MYCOL] A hypha produced above the surface and connecting a group of conidiophores. { 'stō·lən }

Stolonifera [INV ZOO] An order of the Alcyonaria, lacking a coenenchyme; they form either simple (*Clavularia*) or rather complex colonies (*Tubipora*). { ,stäl·ə'nif·rə }

stolzite [MINERAL] PbWO₄ A tetragonal mineral composed of native lead tungstate; it is isomorphous with wulfenite and dimorphous with raspite. { 'stōl,zīt }

stoma [BIOL] A small opening or pore in a surface. [BOT] One of the minute openings in the epidermis of higher plants which are regulated by guard cells and through which gases and water vapor are exchanged between internal spaces and the external atmosphere. { 'stō·mə }

stomach [ANAT] The tubular or saccular organ of the vertebrate digestive system located between the esophagus and the intestine and adapted for temporary food storage and for the preliminary stages of food breakdown. { 'stəm·ək }

stomatitis [MED] Inflammation of the soft tissues in the mouth. { ,stō·mə'tīd·əs }

stomatoblastula [INV ZOO] A blastula stage in some sponges capable of engulfing maternal amebocytes for nutrition. { |stō·məd·ō'blas·chə·lə }

stomatology [MED] The branch of medical science that concerns the anatomy, physiology, pathology, therapeutics, and hygiene of the oral cavity, of the tongue, teeth, and adjacent structures and tissues, and of the relationship of that field to the entire body. { ,stō·mə'täl·ə·jē }

Stomatopoda [INV ZOO] The single order of the Eumalacostraca in the superorder Hoplocarida distinguished by raptorial arms, especially the second pair of maxillipeds. { ,stō·mə'täp·əd·ə }

Stomiatoidei [VERT ZOO] A suborder of fishes of the order Salmoniformes including the lightfishes and allies, which are of small size and often grotesque form and are equipped with photophores. { ,stō·mē·ə'tȯid·ē,ī }

stomium [BOT] A region along a sporangium or pollen sac where dehiscence takes place. { 'stō·mē·əm }

stomocnidae nematocyst [INV ZOO] A nematocyst which has an open-ended thread. { stə'mäk·nə,dē nə'mad·ə,sist }

stomodaeum [EMBRYO] The anterior part of the embryonic alimentary tract formed as an invagination of the ectoderm. { ,stō·mə'dē·əm }

stone [GEOL] **1.** A small fragment of rock or mineral. **2.** See stony meteorite. [LAP] A cut and polished natural gemstone. [MECH] A unit of mass in common use in the United Kingdom, equal to 14 pounds or 6.35029318 kilograms. { stōn }

stoneboat [MIN ENG] A flat runnerless sled for transporting heavy material. { 'stōn|bōt }

stone bubble See lithophysa. { 'stōn |bəb·əl }

stone canal [INV ZOO] A canal in many echinoderms that has a more or less calcified wall and that leads from the madreporite to the ring vessel. { 'stōn kə,nal }

Stone-Čech compactification [MATH] The Stone-Čech compactification of a completely regular space is a Hausdorff space in which the original space forms a dense subset, such that any continuous function from the original space to a compact space has a unique continuous extension to the Hausdorff space. { 'stōn |chek kəm,pak·tə·fə'kā·shən }

stone cell See brachysclereid. { 'stōn ,sel }

stone coal See anthracite. { 'stōn ,kōl }

stone dust [MIN ENG] Inert dust spread on roadways in coal mines as a defense against the danger of coal-dust explosions; effective because the stone dust absorbs heat. { 'stōn ,dəst }

stone-dust barrier [MIN ENG] A device erected in mine roadways to arrest explosions; consists of trays or vee troughs loaded with stone dust, which are upset or overturned by the pressure wave in front of an explosion and the flame, producing a dense cloud of inert dust which blankets the flame and stops further propagation of the explosion. { 'stōn |dəst 'bar·ē·ər }

stone fruit See drupe. { 'stōn ,früt }

stone gobber [MIN ENG] In bituminous coal mining, one who removes stone and other refuse from coal mine floors and dumps the refuse into mine cars for disposal. { 'stōn ,gäb·ər }

stone ice See ground ice. { 'stōn ,īs }

stone polygon See sorted polygon. { 'stōn 'päl·i,gän }

stoner [FOOD ENG] A machine that removes stones and other undesirable material from coffee beans and other crops before the crops are processed further. { 'stō·nər }

stone ring [GEOL] A ring of stones surrounding a central area of finer material; characteristic of sorted circle and sorted polygon. { 'stōn 'riŋ }

Stone's representation theorem [MATH] This theorem determines the nature of all unitary representations of locally compact Abelian groups. { 'stōnz ,rep·rə·zən'tā·shən ,thir·əm }

Stone's theorem [MATH] Every Boolean ring is isomorphic to a ring of subsets of some set. { 'stōnz ,thir·əm }

stoneware [MATER] Vitrified ware with impermeable surface; used for corrosive materials in the laboratory and for some industrial operations. { 'stōn,wer }

Stone-Weierstrass theorem [MATH] If S is a collection of continuous real-valued functions on a compact space E, which contains the constant functions, and if for any pair of distinct points x and y in E there is a function f in S such that $f(x)$ is not equal to $f(y)$, then for any continuous real-valued function g on E there is a sequence of functions, each of which can be expressed as a polynomial in the functions of S with real coefficients, that converges uniformly to g. { 'stōn 'vī·ər,sträs ,thir·əm }

stonework [CIV ENG] A structure or the part of a structure built of stone. { 'stōn,wərk }

stonewort [BOT] The common name for algae comprising the class Charophyceae, so named because most species are lime-encrusted. { 'stōn,wȯrt }

Stoney gate [CIV ENG] A crest gate which moves along a series of rollers traveling vertically in grooves in masonry piers, independently of the gate and piers. { 'stō·nē ,gāt }

Stonnomida [INV ZOO] An order of Xenophyophorea distinguished by the presence of linellae in the test and a flexible body. { stə'näm·əd·ə }

Stonnomidae [INV ZOO] A family coextensive with the order Stonnomida. { stə'näm·ə,dē }

stony coral [INV ZOO] Any coral characterized by a calcareous skeleton. { 'stō·nē 'kär·əl }

stony-iron meteorite [GEOL] Any of the rare meteorites containing at least 25% of both nickel-iron and heavy basic silicates. Also known as iron-stony meteorite; lithosiderite; sideraerolite; siderolite; syssiderite. { 'stō·nē |ī·ərn 'mēd·ē·ə,rīt }

stony meteorite [GEOL] Any meteorite composed principally of silicate minerals, especially olivine, pyroxene, and plagioclase. Also known as aerolite; asiderite; meteoric stone; meteorolite; stone. { 'stō·nē 'mēd·ē·ə,rīt }

stooping [METEOROL] An atmospheric refraction phenomenon; a special case of sinking in which the curvature of light rays due to atmospheric refraction decreases with elevation so that the visual image of a distant object is foreshortened in the vertical. { 'stüp·iŋ }

stop [CONT SYS] A bound or final position of a robot's movement. [OPTICS] The aperture or useful opening of a lens, usually adjustable by means of a diaphragm. { stäp }

stop and stay See absolute stop. { 'stäp ən 'stā }

stop band See rejection band. { 'stäp ,band }

stop bath [GRAPHICS] When a negative or print is removed from the developer, it is usually placed in a stop bath to halt the action of the developer immediately; a common stop bath is a solution of 2 to 5% acetic or citric acid, or potassium metabisulfite. { 'stäp ,bath }

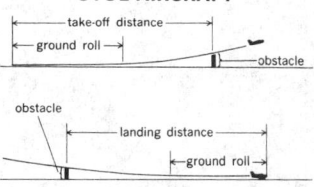

STOL AIRCRAFT

Takeoff and landing profile of STOL aircraft.

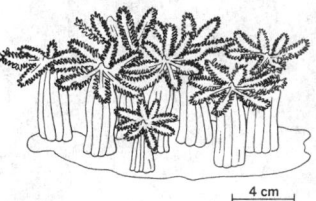

STOLONIFERA

Clavularia garciae.

stop bead [BUILD] A molding on the pulley stile of a window frame; forms one side of the groove for the inner sash. { 'stäp ‚bēd }

stop bits [COMPUT SCI] The last two bits transmitted in asynchronous data transmission to unequivocally indicate the end of a word. { 'stäp ‚bits }

stop cock [ENG] A small valve for stopping or regulating the flow of a fluid through a pipe. { 'stäp ‚käk }

stop code [COMPUT SCI] A character that is placed in a storage medium and, when encountered, causes the computer system to cease processing until it is directed to continue. { 'stäp ‚kōd }

stop codon See terminator codon. { 'stäp 'kō‚dän }

stop down [OPTICS] To reduce the size of a lens aperture. { 'stäp'daùn }

stope [MIN ENG] **1.** To excavate ore in a vein by driving horizontally upon it a series of workings, one immediately over the other, or vice versa; each horizontal working is called a stope because when a number of them are in progress, each working face under attack assumes the shape of a flight of stairs. **2.** Any subterranean extraction of ore except that which is incidentally performed in sinking shafts or driving levels for the purpose of opening the mine. { stōp }

stope assay plan [MIN ENG] A plan that details assay value of ore exposures in a stope. { 'stōp 'as‚ā ‚plan }

stope board [MIN ENG] A timber staging on the floor of a stope for setting a rock drill; the stage is tilted so that the bottom holes can be drilled in the same inclined direction. { 'stōp ‚bórd }

stope fillings [MIN ENG] Broken waste material or low-grade matter from a lode or vein used to fill stopes on abandonment. { 'stōp ‚fil·iŋz }

stope hoist [MIN ENG] A small, portable, compressed-air hoist for operating a scraper-loader or for pulling heavy timbers into position, often used in narrow stopes. { 'stōp ‚hóist }

stop element [COMMUN] The last element of a character in certain serial transmissions, used to ensure the recognition of the next start element. { 'stäp ‚el·ə·mənt }

stope pillar [MIN ENG] An ore column left in place to support the stope. { 'stōp ‚pil·ər }

stoper See stoping drill. { 'stō·pər }

stoping See magnetic stoping. { 'stō·pər }

stoping drill [MIN ENG] A small air or electric drill, usually mounted on an extensible column, for working stopes, raises, and narrow workings. Also known as stoper. { 'stōp·iŋ ‚dril }

stoping ground [MIN ENG] Part of an ore body opened by drifts and raises, and ready for breaking down. { 'stōp·iŋ ‚graùnd }

stop instruction [COMPUT SCI] An instruction in a computer program that causes execution of the program to stop. { 'stäp in‚strək·shən }

stoplight [ELEC] One of the lights that are installed at the rear of an automotive vehicle and are automatically turned on when the driver applies the brakes. { 'stäp‚līt }

stoplog [CIV ENG] A log, plank, or steel or concrete beam that fits into a groove or rack between walls or piers to prevent the flow of water through an opening in a dam, conduit, or other channel. { 'stäp‚läg }

stop loop See loop stop. { 'stäp ‚lüp }

stop number See f number. { 'stäp ‚nəm·bər }

stop nut [DES ENG] **1.** An adjustable nut that restricts the travel of an adjusting screw. **2.** A nut with a compressible insert that binds it so that a lock washer is not needed. { 'stäp ‚nət }

stopoff See resist. { 'stäp‚óf }

stoppage [ORD] A jam in an automatic weapon; the condition of a weapon being jammed. { 'stäp·ij }

stopped-flow method [CHEM] A method for studying chemical reactions in which the reactants are rapidly mixed, then abruptly stopped after a very short time. { 'stäpt ¦flō ‚meth·əd }

stopping [MIN ENG] A brattice, or more commonly, a masonry or brick wall built across old headings, chutes, or airways to confine the ventilating current to certain passages, or to lock up the gas in old workings, or to smother a mine fire. [NUCLEO] The decrease in kinetic energy of an ionizing particle as a result of energy losses along its path through matter. { 'stäp·iŋ }

stopping capacitor See coupling capacitor. { 'stäp·iŋ kə‚pas·əd·ər }

stopping off [MET] **1.** Local deposition of a protective coating, such as copper or fireclay, to prevent carburization, decarburization, or nitriding during heat treatment. **2.** Filling up a portion of the mold cavity to keep out molten metal. **3.** Applying a nonconducting layer to avoid electrodeposition in certain areas. { 'stäp·iŋ 'óf }

stopping out [GRAPHICS] A photomechanical process in which opaque material is applied to photographic negatives to cover the material that is not to be reproduced. Also known as staging. { 'stäp·iŋ 'aùt }

stopping potential [ELECTR] Voltage required to stop the outward movement of electrons emitted by photoelectric or thermionic action. { 'stäp·iŋ pə‚ten·chəl }

stopping power [NUCLEO] The energy lost by a charged particle passing through a substance per unit length of path; related concepts are mass, atomic, molecular, and relative stopping power. Also known as linear energy transfer (LET); linear stopping power. { 'stäp·iŋ ‚paù·ər }

stopping range [NUC PHYS] A postulated energy range for extremely high-energy heavy-ion collisions in which the target stops the projectile and the protons and neutrons of target and projectile are mixed together in a fireball forming a quark-gluon plasma. { 'stäp·iŋ ‚rānj }

stopping rule [STAT] A rule which specifies when observation is to be discontinued in sequential trials. { 'stäp·iŋ ‚rül }

stop signal [COMMUN] Signal that initiates the transfer of facsimile equipment from active to standby conditions. { 'stäp ‚sig·nəl }

stop valve [ENG] A valve that can be opened or closed to regulate or stop the flow of fluid in a pipe. { 'stäp ‚valv }

stopwatch [HOROL] A watch that can be started or stopped at will by pressing a small button; used to measure elapsed time. { 'stäp‚wäch }

stopwater [NAV ARCH] A canvas, backed with red lead or some other substance, fitted between metal parts of a ship to make it watertight. { 'stäp‚wód·ər }

storable propellant [MATER] A propellant capable of being placed and kept in a tank without benefit of special measures for temperature or pressure control. { 'stór·ə·bəl prə‚pel·ənt }

storage [COMPUT SCI] Any device that can accept, retain, and read back one or more times; the means of storing data may be chemical, electrical, magnetic, mechanical, or sonic. { 'stór·ij }

storage address register [COMPUT SCI] A register used to hold the address of a location in storage containing data that is being processed. { 'stór·ij ¦ad‚res ‚rej·ə·stər }

storage allocation [COMPUT SCI] The process of assigning storage locations to data or instructions in a digital computer. { 'stór·ij ‚al·ə'kā·shən }

storage and retrieval system [COMPUT SCI] An organized method of putting items away in a manner which permits their recall or retrieval from storage. Also known as storetrieval system. { 'stór·ij ən ri'trē·vəl ‚sis·təm }

storage area [COMPUT SCI] A specified set of locations in a storage unit. Also known as zone. { 'stór·ij ‚er·ē·ə }

storage battery [ELEC] A connected group of two or more storage cells or a single storage cell. Also known as accumulator; accumulator battery; rechargeable battery; secondary battery. { 'stór·ij ‚bad·ə·rē }

storage battery locomotive [MIN ENG] An underground locomotive powered by storage batteries. { 'stór·ij ‚bad·ə·rē ‚lō·kə‚mōd·iv }

storage block [COMPUT SCI] A contiguous area of storage whose contents can be handled in a single operation. { 'stór·ij ‚bläk }

storage buffer register [COMPUT SCI] A register used in some microcomputers during input or output operations to temporarily hold a copy of the contents of a storage location. { 'stór·ij ¦bəf·ər ‚rej·ə·stər }

storage calorifier See cylinder. { 'stór·ij kə'lór·ə‚fī·ər }

storage camera See iconoscope. { 'stór·ij ‚kam·rə }

storage capacity [COMPUT SCI] The quantity of data that can be retained simultaneously in a storage device; usually measured in bits, digits, characters, bytes, or words. Also known as capacity; memory capacity. { 'stór·ij kə‚pas·əd·ē }

storage cell [COMPUT SCI] An elementary (logically indivisible) unit of storage; the storage cell can contain one bit, character, byte, digit (or sometimes word) of data. [ELEC] An electrolytic cell for generating electric energy, in which the cell after being discharged may be restored to a charged condition by sending a current through it in a direction opposite to that of the discharging current. Also known as secondary cell. { 'stȯr·ij ˌsel }

storage compacting [COMPUT SCI] The practice, followed on multiprogramming computers which use dynamic allocation, of assigning and reassigning programs so that the largest possible area of adjacent locations remains available for new programs. { 'stȯr·ij kəmˌpakt·iŋ }

storage cycle [COMPUT SCI] **1.** Periodic sequence of events occurring when information is transferred to or from the storage device of a computer. **2.** Storing, sensing, and regeneration from parts of the storage sequence. { 'stȯr·ij ˌsī·kəl }

storage cycle time [COMPUT SCI] The time required to read and restore one word from a computer storage, or to write one word in computer storage. { 'stȯr·ij ¦sī·kəl ˌtīm }

storage density [COMPUT SCI] The number of characters stored per unit-length of area of storage medium (for example, number of characters per inch of magnetic tape). { stȯr·ij ˌden·səd·ē }

storage device [COMPUT SCI] A mechanism for performing the function of data storage: accepting, retaining, and emitting (unchanged) data items. Also known as computer storage device. { 'stȯr·ij diˌvīs }

storage disease [MED] Metabolic abnormality in which some substance (such as fats, proteins, or carbohydrates) accumulates in abnormal amounts in certain body tissues. { 'stȯr·ij diˌzēz }

storage dump [COMPUT SCI] A printout of the contents of all or part of a computer storage. Also known as memory dump; memory print. { 'stȯr·ij ˌdəmp }

storage element [COMPUT SCI] Smallest part of a digital computer storage used for storing a single bit. { 'stȯr·ij ˌel·ə·mənt }

storage equation [HYD] The equation of continuity applied to unsteady flow; it states that the fluid inflow to a given space during an interval of time minus the outflow during the same interval is equal to the change in storage; it is applied in hydrology to the routing of floods through a reservoir or a reach of a stream; the moisture continuity equation applied to the atmosphere is a modification of this. { 'stȯr·ij iˌkwā·zhən }

storage factor See Q. { 'stȯr·ij ˌfak·tər }

storage fill [COMPUT SCI] Storing a pattern of characters in areas of a computer storage that are not intended for use in a particular machine run; these characters cause the machine to stop if one of these areas is erroneously referred to. Also known as memory fill. { 'stȯr·ij ˌfil }

storage hierarchy [COMPUT SCI] The sequence of storage devices, characterized by speed, type of access, and size for the various functions of a computer; for example, core storage from programs and data, disks or drums for temporary storage of massive amounts of data, tapes and cards for back up storage. { 'stȯr·ij 'hī·ərˌär·kē }

storage integrator [COMPUT SCI] In an analog computer, an integrator used to store a voltage in the hold condition for future use while the rest of the computer assumes another computer control state. { 'stȯr·ij ˌint·əˌgrād·ər }

storage key [COMPUT SCI] A special set of bits associated with every word or character in some block of storage, which allows tasks having a matching set of protection key bits to use that block of storage. { 'stȯr·ij ˌkē }

storage location [COMPUT SCI] A digital-computer storage position holding one machine word and usually having a specific address. { 'stȯr·ij lōˌkā·shən }

storage mark [COMPUT SCI] The name given to a point location which defines the character space immediately to the left of the most significant character in accumulator storage. { 'stȯr·ij ˌmärk }

storage medium [COMPUT SCI] Any device or recording medium into which data can be copied and held until some later time, and from which the entire original data can be obtained. { 'stȯr·ij ˌmēd·ē·əm }

storage oscilloscope [ELECTR] An oscilloscope that can retain an image for a period of time ranging from minutes to days, or until deliberately erased to make room for a new image. { 'stȯr·ij əˌsil·əˌskōp }

storage pool [COMPUT SCI] A collection of similar data storage devices. { 'stȯr·ij ˌpül }

storage print [COMPUT SCI] In computers, a utility program that records the requested core image, core memory, or drum locations in absolute or symbolic form either on the line-printer or on the delayed-printer tape. { 'stȯr·ij ˌprint }

storage protection [COMPUT SCI] Any restriction on access to storage blocks, with respect to reading, writing, or both. Also known as memory protection. { 'stȯr·ij prəˌtek·shən }

storage register [COMPUT SCI] A register in the main internal memory of a digital computer storing one computer word. Also known as memory register. { 'stȯr·ij rej·ə·stər }

storage reservoir See impounding reservoir. { 'stȯr·ij ˌrez·əvˌwär }

storage-retrieval machine [CONT SYS] A computer-controlled machine for an automated storage and retrieval system that operates on rails and moves material either vertically or horizontally between a storage compartment and a transfer station. { ¦stȯr·ij ri'trēv·əl məˌshēn }

storage rings [NUCLEO] Annular vacuum chambers in which charged particles can be stored, without acceleration, by a magnetic field of suitable focusing properties; they are used to stretch effectively the duty cycle of a particle accelerator or to produce colliding beams of particles, resulting in a greater possible center of mass energy. { 'stȯr·ij ˌriŋz }

storage ripple [COMPUT SCI] A hardware function, used during maintenance periods, which reads or writes zeros or ones through available storage locations to detect a malfunctioning storage unit. { 'stȯr·ij ˌrip·əl }

storage routing See flood routing. { 'stȯr·ij ˌrüd·iŋ }

storage surface [COMPUT SCI] In computers, the surface (screen), in an electrostatic storage tube, on which information is stored. { 'stȯr·ij ˌsər·fəs }

storage tank See tank. { 'stȯr·ij ˌtaŋk }

storage time [ELECTR] **1.** The time required for excess minority carriers stored in a forward-biased *pn* junction to be removed after the junction is switched to reverse bias, and hence the time interval between the application of reverse bias and the cessation of forward current. **2.** The time required for excess charge carriers in the collector region of a saturated transistor to be removed when the base signal is changed to cut-off level, and hence for the collector current to cease. [PHYS] See decay time. { 'stȯr·ij ˌtīm }

storage-to-register instruction [COMPUT SCI] A machine-language instruction to move a word of data from a location in main storage to a register. { 'stȯr·ij tə 'rej·ə·stər inˌstrək·shən }

storage-to-storage instruction [COMPUT SCI] A machine-language instruction to move a word of data from one location in main storage to another. { 'stȯr·ij tə 'stȯr·ij inˌstrək·shən }

storage tube [ELECTR] An electron tube employing cathode-ray beam scanning and charge storage for the introduction, storage, and removal of information. Also known as electrostatic storage tube; memory tube (deprecated usage). { 'stȯr·ij ˌtüb }

storage-type camera tube See iconoscope. { 'stȯr·ij ¦tīp 'kam·rə ˌtüb }

store [COMPUT SCI] **1.** To record data into a (static) data storage device. **2.** To preserve data in a storage device. { stȯr }

store and forward [COMMUN] A procedure in data communications in which data are stored at some point between the sender and the receiver and are later forwarded to the receiver. { 'stȯr ən 'fȯr·wərd }

stored-energy welding [MET] Welding by means of energy accumulated electrostatically, electrochemically, or electromagnetically at a low rate. { 'stȯrd ¦en·ər·jē ˌweld·iŋ }

stored program [COMPUT SCI] A computer program that is held in a computer's main storage and carried out by a central processing unit that reads and acts on its instructions. { 'stȯrd ¦prō·gram }

stored-program computer [COMPUT SCI] A digital computer which executes instructions that are stored in main memory as patterns of data. { 'stȯrd ¦prō·gram kəm'pyüd·ər }

stored-program control [COMMUN] Electronic control of a telecommunications switching system by means of a program of instructions stored in bulk electronic memory. Abbreviated SPC. { 'stȯrd 'prō‚gram kən'trōl }

stored-program logic [COMPUT SCI] Program that is stored in a memory unit containing logical commands in order to perform the same processes on all problems. { 'stȯrd 'prō‚gram ‚läj·ik }

stored-program numerical control See computer numerical control. { 'stȯrd 'prō‚gram nü'mer·ə·kəl kən‚trōl }

stored response chain [COMPUT SCI] A fixed sequence of instructions that are stored in a file and acted on by an interactive computer program at a point where it would normally request instructions from the user, in order to save the user the trouble of repeatedly keying the same commands for a frequently used function. Abbreviated SRC. { 'stȯrd ri'späns ‚chān }

stored routine [COMPUT SCI] In computers, a series of instructions in storage to direct the step-by-step operation of the machine. { 'stȯrd rü'tēn }

stored word [COMPUT SCI] The actual linear combination of letters (or their machine equivalents) to be placed in the machine memory; this may be physically quite different from a dictionary word. { 'stȯrd 'wȯrd }

storethrough [COMPUT SCI] The process of updating data in main memory each time the central processing unit writes into a cache. { 'stȯr‚thrü }

store transmission bridge [ELEC] Transmission bridge, which consists of four identical impedance coils (the two windings of the back-bridge relay and live relay of a connector, respectively) separated by two capacitors, which couples the calling and called telephones together electrostatically for the transmission of voice-frequency (alternating) currents, but separates the two lines for the transmission of direct current for talking purposes (talking current). { 'stȯr tranz'mish·ən ‚brij }

storetrieval system See storage and retrieval system. { 'stȯ·ri‚trē·vəl ‚sis·təm }

stork [VERT ZOO] Any of several species of long-legged wading birds in the family Ciconiidae. { stȯrk }

storm [METEOROL] An atmospheric disturbance involving perturbations of the prevailing pressure and wind fields on scales ranging from tornadoes (0.6 mile or 1 kilometer across) to extratropical cyclones (up to 1800 miles or 3000 kilometers across); also the associated weather (rain storm or blizzard) and the like. { stȯrm }

storm beach [GEOL] A ridge composed of gravel or shingle built up by storm waves at the inner margin of a beach. { 'stȯrm ‚bēch }

storm cellar See cyclone cellar. { 'stȯrm ‚sel·ər }

storm center [METEOROL] The area of lowest atmospheric pressure of a cyclone; this is a more general expression than eye of the storm, which refers only to the center of a well-developed tropical cyclone, in which there is a tendency of the skies to clear. { 'stȯrm ‚sen·tər }

storm choke [PETRO ENG] A device installed in an oil-well tubing string below the surface to shut in the well when the flow reaches a predetermined rate; provides an automatic shutoff in case Christmas-tree or control valves are damaged. Also known as tubing safety valve. { 'stȯrm ‚chōk }

storm delta See washover. { 'stȯrm ‚del·tə }

storm detection [METEOROL] Any of the methods and techniques used to ascertain the formation of storms, including procedures for locating, tracking, and forecasting; special tools adapted to this purpose are radar and satellites to supplement meteorological charts and visual observations. { 'stȯrm di‚tek·shən }

storm drain [CIV ENG] A drain which conducts storm surface, or wash water, or drainage after a heavy rain from a building to a storm or a combined sewer. Also known as storm sewer. { 'stȯrm ‚drān }

storm ice foot [OCEANOGR] An ice foot produced by the breaking of a heavy sea or the freezing of wind-driven spray. { 'stȯrm 'īs ‚fut }

storm microseism [GEOPHYS] A microseism lasting 25 or more seconds, caused by ocean waves. { 'stȯrm 'mī·krə‚sīz·əm }

storm model [METEOROL] A physical, three-dimensional representation of the inflow, outflow, and vertical motion of air and water vapor in a storm. { 'stȯrm ‚mäd·əl }

storm sash See storm window. { 'stȯrm ‚sash }

stormscope [NAV] An airborne device that uses data from electrical signals generated by atmospheric electrical discharge

activity to generate a map of severe weather areas, and that displays, in real time, weather conditions in all directions. { 'stȯrm‚skōp }

storm sewage [CIV ENG] Refuse liquids and waste carried by sewers during or following a period of heavy rainfall. { 'stȯrm ‚sü·ij }

storm sewer See storm drain. { 'stȯrm ‚sü·ər }

storm surge [OCEANOGR] A rise above normal water level on the open coast due only to the action of wind stress on the water surface; includes the rise in level due to atmospheric pressure reduction as well as that due to wind stress. Also known as storm wave; surge. { 'stȯrm ‚sərj }

storm tide [OCEANOGR] Height of a storm surge or hurricane wave above the astronomically predicted sea level. { 'stȯrm ‚tīd }

storm track [METEOROL] The path followed by a center of low atmospheric pressure. { 'stȯrm ‚trak }

storm transposition [METEOROL] The transfer of precipitation patterns or DDA (depth-duration-area) values from the areas where they actually occurred to areas where they could occur; if necessary, the precipitation values are modified to account for differences in elevation or intervening barriers, and restrictions on change in shape or orientation of the storm may be imposed. { 'stȯrm ‚tranz·pə'zish·ən }

storm warning [METEOROL] A specially worded forecast of severe weather conditions, designed to alert the public to impending dangers; usually, this refers to a warning of potentially dangerous wind conditions for marine interests. { 'stȯrm ‚wȯrn·iŋ }

storm-warning signal [METEOROL] An arrangement of flags or pennants (by day) and lanterns (by night) displayed on a coastal storm-warning tower. { 'stȯrm ¦wȯrn·iŋ ‚sig·nəl }

storm-warning tower [METEOROL] A tower, generally constructed of steel, for displaying coastal storm-warning signals. { 'stȯrm ¦wȯrn·iŋ ‚tau̇·ər }

storm wave See storm surge. { 'stȯrm ‚wāv }

storm wind [METEOROL] In the Beaufort wind scale, a wind whose speed is from 56 to 63 knots (64 to 72 miles per hour or 104 to 117 kilometers per hour). { 'stȯrm ‚wind }

storm window [BUILD] A sash placed on the outside of an ordinary window to give added protection from the weather. Also known as storm sash. { 'stȯrm ‚win·dō }

Storrow whirling hygrometer [ENG] A hygrometer in which the two thermometers are mounted side by side on a brass frame and fitted with a loose handle so that it can be whirled in the atmosphere to be tested; the instrument is whirled at some 200 revolutions per minute for about 1 minute and the readings on the wet- and dry-bulb thermometers are recorded; used in conjunction with Glaisher's or Marvin's hygrometrical tables. { 'stä·rō 'wərl·iŋ hī'gräm·əd·ər }

story [BUILD] The space between two floors or between a floor and the roof. { 'stȯr·ē }

storyboard [GRAPHICS] A series of small drawings intended to show the sequence and continuity of a proposed motion picture, television production, or slide presentation; only key portions of the action or story are shown, which help to visualize the total idea. { 'stȯr·ē‚bȯrd }

story pole See story rod. { 'stȯr·ē ‚pōl }

story rod [DES ENG] A pole cut to the exact specified height from finished floor to ceiling and used as a measuring device in the course of construction. Also known as story pole. { 'stȯr·ē ‚räd }

stoss [GEOL] Of the side of a hill, knob, or prominent rock, facing the upstream side of a glacier. { stäs }

stoss-and-lee topography [GEOL] A type of glaciated landscape in which small hills or other landforms exhibit gentle eroded slopes on the up-glacier or upstream side and less eroded, steeper slopes on the lee side. { ¦stäs ənd ¦lē tə'päg·rə‚fē }

stove [ENG] A chamber within which a fuel-air mixture is burned to provide heat, the heat itself being radiated outward from the chamber; used for space heating, process-fluid heating, and steel blast furnaces. { stōv }

stove bolt [DES ENG] A coarsely threaded bolt with a slotted head, which with a square nut is used to join metal parts. { 'stōv ‚bōlt }

stovepipe [ENG] Large-diameter pipe made of sheet steel. { 'stōv‚pīp }

STORM

Key:

→ hurricanes

– – – extratropical cyclones (W-winter only)

----- summer position of intertropical convergence

▮ subtropical high

L_s semipermanent summer-heat lows

H_w winter continental anticyclones

Principal tracks of extratropical cyclones and hurricanes with significantly associated features in the Northern Hemisphere.

STORM DETECTION

Hurricane Donna as observed at 0840 EST, September 10, 1960, on the WSR-57, 10-centimeter radar set at Key West, Florida. A spiral overlay of crossover angle α = 15° has been fitted to the precipitation bands in order to indicate the location of the storm center. (*ESSA photograph*)

stover [AGR] Stalks and leaves, not including grain, of such forages as corn and sorghum. { 'stō·vər }

stovewood [MATER] Firewood sawed into short lengths for use in a stove. { 'stōv,wúd }

stoving *See* baking. { 'stōv·iŋ }

stowage diagram [NAV ARCH] A scaled drawing included in the loading plan of a ship for each deck or platform showing the exact location of all cargo, and specifying for each location such data as overall dimensions, location of obstructions, dimensions of the overhead hatch opening, dimensions of bow door or stern gate opening, minimum clearances to the overhead, bale cubic capacity, square feet of deck area, and the capacity of booms. { 'stō·ij ,dī·ə,gram }

stowage factor [NAV ARCH] The number which expresses the space, in cubic feet, occupied by a long ton of any commodity as prepared for shipment, including all crating or packaging. { 'stō·ij ,fak·tər }

stowage plan [NAV ARCH] A completed stowage diagram showing each type of material that has been loaded and its exact stowage location; each port of discharge is indicated by a particular color or other symbol; deck and between-deck cargo is shown in perspective, cargo stowed in the lower hold is shown in profile, but vehicles are shown in perspective, regardless of stowage. { 'stō·ij ,plan }

stowboard [MIN ENG] A mine heading used for storing waste. { 'stō,bórd }

STP *See* standard conditions.

STPP *See* sodium tripolyphosphate.

STR *See* self-tuning regulator.

strabismus [MED] Incoordinate action of the extrinsic ocular muscles resulting in failure of the visual axes to meet at the desired objective point. Also known as cast; heterotropia; squint. { strə'biz·məs }

straddle milling [MECH ENG] Face milling of two parallel vertical surfaces of a workpiece simultaneously by using two side-milling cutters. { 'strad·əl ,mil·iŋ }

straddle truck [MECH ENG] A self-loading outrigger type of industrial truck that straddles the load before lifting it between the outrigger arms. { 'strad·əl ,trək }

strafe [ORD] To rake a body of troops or other persons with gunfire or rocket fire at close range and from a flying aircraft, or to attack a roadway, railyard, factory, or other installation with bullets, projectiles, or rockets fired from a low firing airplane. { strāf }

straggling [PHYS] Random variations in some property associated with the passage of ions through matter. { 'strag·liŋ }

straight angle [MATH] An angle of measure one-half revolution or 180°, whose sides lie on the same straight line but extend in opposite directions from the vertex. { 'strāt ,aŋ·gəl }

straight beam [ENG] In ultrasonic testing, a longitudinal wave emitted from an ultrasonic search unit in a wavetrain which travels perpendicularly to the test surface. { 'strāt 'bēm }

straight bevel gear [DES ENG] A simple form of bevel gear having straight teeth which, if extended inward, would come together at the intersection of the shaft axes. { 'strāt 'bev·əl ,gir }

straight dynamite [MATER] Any of the powerful, quick-acting dynamites composed of nitroglycerin, a combustible such as wood meal, sodium nitrate, and an antacid such as calcium or magnesium carbonate; made in 15 to 60% strength, the percentage representing the proportion of nitroglycerin in the dynamite. { 'strāt 'dī·nə,mīt }

straightedge [DES ENG] A strip of wood, plastic, or metal with one or more long edges made straight with a desired degree of accuracy. { 'strād,ej }

straightening vanes [ENG] Horizontal vanes mounted on the inside of fluid conduits to reduce the swirling or turbulent flow ahead of the orifice or the venturi meters. { 'strāt·ən·iŋ ,vānz }

straight filing [ENG] Filing by pushing a file in a straight line across the work. { 'strāt 'fīl·iŋ }

straight-flow turbine [MECH ENG] A horizontal-axis, low-head hydraulic turbine in which the upstream and downstream reservoirs are connected by a straight tube into which the runners are integrated, with the generator placed directly on the periphery of these runners. { 'strāt ¦flō 'tər·bən }

straightforward circuit [COMMUN] Circuit in which signaling is automatic and in one direction. { 'strāt¦fór·wərd 'sər·kət }

straightforward trunking [COMMUN] In a manual telephone switchboard system, that method of operation in which the A operator gives the order to the B operator over the trunk on which talking later takes place. { 'strāt¦fór·wərd 'trəŋk·iŋ }

straight joint [BUILD] **1.** A continuous joint formed by the ends of parallel floor boards or masonry units and oriented perpendicularly to their length. **2.** A joint between two pieces of wood that are set edge to edge without tongues and grooves, dowels, or overlap to bind them. Also known as square joint. { 'strāt 'jóint }

straight line *See* line. { 'strāt ¦līn }

straight-line coding [COMPUT SCI] A digital computer program or routine (section of program) in which instructions are executed sequentially, without branching, looping, or testing. { 'strāt ¦līn 'kōd·iŋ }

straight-line mechanism [MECH ENG] A linkage so proportioned and constrained that some point on it describes over part of its motion a straight or nearly straight line. { 'strāt ¦līn 'mek·ə,niz·əm }

straight-line motion [CONT SYS] A method of moving a robot between via or way points in which the end effector moves only along segments of straight lines, stopping momentarily on any change in direction. { 'strāt ¦līn 'mō·shən }

straight piecework system *See* one-hundred-percent premium plan. { 'strāt 'pēs,wərk ,sis·təm }

straight polarity [MET] Arrangement of an arc welding circuit in which the electrode is connected to the negative terminal. { 'strāt pə'lar·əd·ē }

straight proportional system *See* one-hundred-percent premium plan. { 'strāt prə'pór·shən·əl ,sis·təm }

straight-run [CHEM ENG] Petroleum fractions derived from the straight distillation of crude oil without chemical reaction or molecular modification. Also known as virgin. { 'strāt 'rən }

straight-run distillation [CHEM ENG] Continuous nonreactive distillation of petroleum oil to separate it into products in the order of their boiling points. { 'strāt ,rən ,dis·tə'lā·shən }

straight-run gasoline [MATER] Gasoline comprised of only natural ingredients from crude oil or natural-gas liquids; for example, no cracked, polymerized, alkylated, reformed, or visbroken stock. { 'strāt ¦rən 'gas·ə,lēn }

straight-run pitch [MATER] A petroleum pitch that is run directly during distillation to the desired consistency without any compounding or aftertreatment. { 'strāt ¦rən 'pich }

straight-run stock *See* virgin stock. { 'strāt ¦rən 'stäk }

straight sinus [ANAT] A sinus of the dura mater running from the inferior sagittal sinus along the junction of the falx cerebri and tentorium to the transverse sinus. { 'strāt 'sī·nəs }

straight strap clamp [DES ENG] A clamp made of flat stock with an elongated slot for convenient positioning; held in place by a T bolt and nut. { 'strāt ¦strap 'klamp }

straight-tube boiler [MECH ENG] A water-tube boiler in which all the tubes are devoid of curvature and therefore require suitable connecting devices to complete the circulatory system. Also known as header-type boiler. { 'strāt ¦tüb 'bói·lər }

straight turning [MECH ENG] Work turned in a lathe so that the diameter is constant over the length of the workpiece. { 'strāt 'tərn·iŋ }

straight vertical antenna [ELECTROMAG] An antenna consisting of a straight vertical wire. { 'strāt 'vərd·ə·kəl an'ten·ə }

straightway pump [MECH ENG] A pump with suction and discharge valves arranged to give a direct flow of fluid. { 'strāt,wā ,pəmp }

straight wheel [DES ENG] A grinding wheel whose sides or face are straight and not in any way changed from a cylindrical form. { 'strāt ¦wēl }

strain [BIOL] An intraspecific group of organisms that possess only one or a few distinctive traits and are maintained as an artificial breeding group. [CYTOL] A population of cells derived either from a primary culture or from a cell line by the selection or cloning of cells having specific properties or markers. [MECH] Change in length of an object in some direction per unit undistorted length in some direction, not necessarily the same; the nine possible strains form a second-rank tensor. { strān }

STRAIGHT-TUBE BOILER

A straight-tube-type marine boiler, which has been fitted with a superheater and an air heater and arranged for oil firing. (*Babcock and Wilcox Co.*)

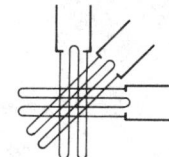

STRAIN ROSETTE

(a)

(b)

Strain rosettes. (a) 45° type; (b) 60° type.

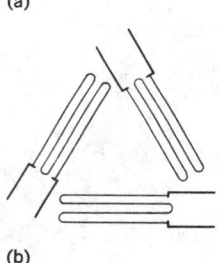

STRAKE

strake

vortex

Strake and resulting vortex flow of a combat aircraft.

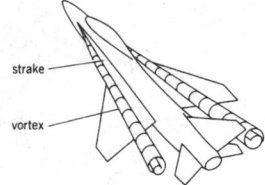

STRAND BURNER

electrical leads (ignition)

timing leads

pressurizing gas (usually nitrogen)

burn-through of lead no. 1 starts timer and burn-through of lead no. 2 stops timer

experimental strand (burns from top down)

strand support

Strand burner apparatus.

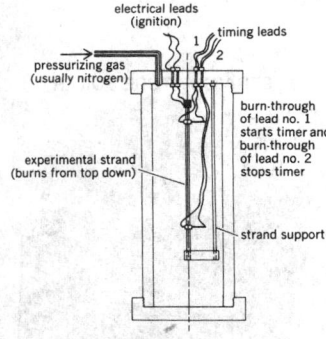

strain aging [MET] Change of mechanical properties of a metal by aging induced by plastic deformation. { 'strān ‚āj·iŋ }

strain axis See principal axis of strain. { 'strān ‚ak·səs }

strain bursts [MIN ENG] Rock bursts in which there is spitting, flaking, and sudden fracturing at the face, indicating increased pressure at the site. { 'strān ‚bərsts }

strained-layer superlattice [ELECTR] A structure consisting of alternating layers of two different semiconducting materials, each several nanometers thick, in which a mismatch between the lattice spacings of the two materials of up to several percent is accommodated by elastic strains in the thin layers without the generation of mismatch defects. { 'strānd ¦lā·ər ¦sü·pər'lad·əs }

strain ellipsoid [MECH] A mathematical representation of the strain of a homogeneous body by a strain that is the same at all points or of unequal stress at a particular point. Also known as deformation ellipsoid. { 'strān i'lip‚sóid }

strain energy [MECH] The potential energy stored in a body by virtue of an elastic deformation, equal to the work that must be done to produce this deformation. { 'strān ‚en·ər·jē }

strainer [ENG] A porous or screen medium used ahead of equipment to filter out harmful solid objects and particles from a fluid stream; used for example, in river-water intakes for process plants or to remove decomposition products from the circulating fluid in a hydraulic system. { 'strān·ər }

strain figure [PHYS] A series of markings, such as Lüder's lines, that may appear on the surface of a body subjected to stress, indicating its state of deformation. Also known as flow figure. { 'strān ‚fig·yər }

strain foil [ENG] A strain gage produced from thin foil by photoetching techniques; may be applied to curved surfaces, has low transverse sensitivity, exhibits negligible hysteresis under cycling loads, and creeps little under sustained loads. { 'strān ‚fóil }

strain gage [ENG] A device which uses the change of electrical resistance of a wire under strain to measure pressure. { 'strān ‚gāj }

strain-gage accelerometer [ENG] Any accelerometer whose operation depends on the fact that the resistance in a wire changes when it is strained; these devices are classified as bonded or unbonded. { 'strān ¦gāj ak‚sel·ə'räm·əd·ər }

strain-gage bridge [ENG] A bridge arrangement of four strain gages, cemented to a stressed part in such a way that two gages show increases in resistance and two show decreases when the part is stressed; the change in output voltage under stress is thus much higher than that for a single gage. { 'strān ¦gāj ‚brij }

strain hardening [MET] Increasing the hardness and tensile strength of a metal by cold plastic deformation. { 'strān ‚härd·ən·iŋ }

straining beam [CIV ENG] A short piece of timber in a truss that holds the ends of struts or rafters. Also known as straining piece. { 'strān·iŋ ‚bēm }

straining piece See straining beam. { 'strān·iŋ ‚pēs }

strain insulator [ELEC] An insulator used between sections of a stretched wire or antenna to break up the wire into insulated sections while withstanding the total pull of the wire. { 'strān ‚in·sə‚lād·ər }

strain propagation [PHYSIO] Transmission of a response to external stress within the body by mechanical or biological processes. { 'strān ‚präp·ə‚gā·shən }

strain rate [MECH] The time rate for the usual tensile test. { 'strān ‚rāt }

strain relief method [MIN ENG] A method for determining absolute strain and stress within rock in place by boring a smooth hole in the rock and inserting a gage capable of measuring diametral deformation, and overcoring the hole with a large coring bit; the change in the diameter of the hole when the rock cylinder is free to expand is a function of the original stress in the rock and its elastic modulus. { 'strān ri¦lēf ‚meth·əd }

strain restoration method [MIN ENG] A method for determining absolute strain and stress within rock in place by the installation of strain gages on the rock surface, cutting of a slot in the rock between the strain gages so that the surface rock is free to expand, installation of a flat jack in the slot, and application of hydraulic pressure to the flat jack until the rock is restored to its original state of strain; the original stress in the rock is presumed to be equal to the final pressure in the flat jack. { 'strān ‚res·tə'rā·shən ‚meth·əd }

strain rosette [MECH] A pattern of intersecting lines on a surface along which linear strains are measured to find stresses at a point. { 'strān rō‚zet }

strain seismograph [ENG] A seismograph that detects secular strains related to tectonic processes and tidal yielding of the solid earth; also detects strains associated with propagating seismic waves. { 'strān 'sīz·mə‚graf }

strain seismometer [ENG] A seismometer that measures relative displacement of two points in order to detect deformation of the ground. { 'strān sīz'mäm·əd·ər }

strain shadow See pressure shadow; undulatory extinction. { 'strān ‚shad·ō }

strain-slip [GEOL] A rock fracture resulting in a slight displacement. { 'strān ‚slip }

strain-slip cleavage See slip cleavage. { 'strān ¦slip ‚klē·vij }

strain tensor [MECH] A second-rank tensor whose components are the nine possible strains. { 'strān ‚ten·sər }

strait [GEOGR] **1.** A neck of land. **2.** A narrow waterway connecting two larger bodies of water. { strāt }

strake [AERO ENG] The slender forward extension of the inboard region of the wing of a combat aircraft, used to provide increased lift in the high angle-of-attack maneuvering condition. [BUILD] A course of clapboarding on a house. [CIV ENG] A row of steel plates installed on a tall steel chimney. [MIN ENG] A relatively wide trough set at a slope and covered with a blanket or corduroy for catching comparatively coarse gold and any valuable mineral. [NAV ARCH] A continuous band of planking or plating running fore and aft along the hull of a ship. { strāk }

strand [ENG] **1.** One of a number of steel wires twisted together to form a wire rope or cable or an electrical conductor. **2.** A thread, yarn, string, rope, wire, or cable of specified length. **3.** One of the fibers or filaments twisted or laid together into yarn, thread, rope, or cordage. [GEOL] A beach bordering a sea or an arm of an ocean. [NAV] To run aground; term strand usually refers to a serious grounding, while the term "ground" refers to any grounding, however slight. [TEXT] An element of a woven material. { strand }

strand burner [ENG] A device that determines the rate at which a propellant burns at various pressures by using a propellant strand. { 'strand ‚bər·nər }

stranded caisson See box caisson. { 'stran·dəd 'kā‚sän }

stranded conductor See stranded wire. { 'stran·dəd kən'dək·tər }

stranded-floe ice foot See stranded ice foot. { 'stran·dəd ¦flō 'īs ‚fút }

stranded ice [OCEANOGR] Ice held in place by virtue of being grounded. Also known as grounded ice. { 'stran·dəd 'īs }

stranded ice foot [OCEANOGR] An ice foot formed by the stranding of floes or small icebergs along a shore; it may be built up by freezing spray or breaking seas. Also known as stranded-floe ice foot. { 'stran·dəd 'īs ‚fút }

stranded wire [ELEC] A conductor composed of a group of wires or a combination of groups of wires, usually twisted together. Also known as stranded conductor. { 'stran·dəd 'wīr }

strand flat See wave-cut platform. { 'strand ‚flat }

stranding machine See closing machine. { 'strand·iŋ mə‚shēn }

strandline [GEOL] **1.** A beach raised above the present sea level. **2.** The level at which a body of standing water meets the land. **3.** See shoreline. { 'strand‚līn }

strange attractor [PHYS] A geometrical object in phase space with a fractal structure toward which the trajectory followed by a chaotic system converges in the course of time. { 'strānj ə'trak·tər }

strangeness conservation [PART PHYS] The principle that the sum of the strangeness numbers of the hadrons in an isolated system is constant; it is violated by the weak interactions. { 'strānj·nəs ‚kän·sər'vā·shən }

strangeness number [PART PHYS] A quantum number carried by hadrons, equal to the hypercharge minus the baryon number. Symbol S. { 'strānj·nəs ‚nəm·bər }

strange particle [PART PHYS] A hadron whose strangeness number is not zero, for example, a K-meson or a Σ-hyperon. { 'strānj 'pard·ə·kəl }

strange quark [PART PHYS] A quark with an electric charge of $-1/3$, baryon number of $1/3$, strangeness of -1, and 0 charm. Symbolized s. { 'stränj 'kwärk }

stranger [AERO ENG] In air intercept, an unidentified aircraft, bearing, distance, and altitude as indicated relative to an aircraft. { 'strän·jər }

strangulated hernia [MED] A hernia involving the intestine in which circulation of the blood and fecal current are blocked. { 'straŋ·gyə‚lād·əd 'hər·nē·ə }

strangulation [MED] **1.** Asphyxiation due to obstruction of the air passages, as by external pressure on the neck. **2.** Constriction of a part producing arrest of the circulation, as strangulation of a hernia. { ‚straŋ·gyə'lā·shən }

strap bolt [DES ENG] **1.** A bolt with a hook or flat extension instead of a head. **2.** A bolt with a flat center portion and which can be bent into a U shape. { 'strap ‚bōlt }

strap hammer [MECH ENG] A heavy hammer controlled and operated by a belt drive in which the head is slung from a strap, usually of leather. { 'strap ‚ham·ər }

strap hinge [DES ENG] A hinge fastened to a door and the adjacent wall by a long hinge. { 'strap ‚hinj }

strapped-down inertial navigation equipment [NAV] Inertial navigation equipment in which a stable platform and gimbal system are not used: the inertial devices are attached or strapped directly to the carrier; a computer utilizing gyro information resolves accelerations sensed along the carrier axes and refers these accelerations to an inertial frame of reference. Also known as gimballess inertial navigation equipment. { 'strapt ‚daún i'nər·shəl ‚nav·ə'gā·shən i‚kwip·mənt }

strapped magnetron [ELECTR] A multicavity magnetron in which resonator segments having the same polarity are connected together by small conducting strips to suppress undesired modes of oscillation. { 'strapt 'mag·nə‚trän }

strapped wall *See* battened wall. { 'strapt ‚wól }

strapping [ELEC] Connecting two or more points in a circuit or device with a short piece of wire or metal. [ELECTR] Connecting together resonator segments having the same polarity in a multicavity magnetron to suppress undesired modes of oscillation. [PETRO ENG] A petroleum industry procedure in which storage tanks are strapped (measured) on their outside with steel measuring tapes to calculate the volumetric capacity of the tank for increments of height. { 'strap·iŋ }

strapping option [COMPUT SCI] The rearrangement of jumpers on a printed circuit board to render a hardware feature operative or inoperative. { 'strap·iŋ ‚äp·shən }

strapping table [PETRO ENG] A tabular record of tank volume versus height so that taped (strapped) measurements of liquid depth can be converted into liquid volumes. Also known as gaging table. { 'strap·iŋ ‚tā·bəl }

Strasbourg turpentine [MATER] A balsam from the European white fir; a heavy, thick material, it is sometimes used in painting mediums and glazes, but an excessive amount causes smearing and slow drying of the paint. { 'stras‚bərg 'tər·pən‚tīn }

strategic [ORD] Of or pertaining to military measures or actions taken against the enemy's war-making effort. { strə'tē·jik }

strategic airlift [AERO ENG] The continuous, sustained air movement of units, personnel, and materiel in support of all U.S. Department of Defense agencies between area commands. { strə'tē·jik 'er‚lift }

strategic air warfare [ORD] Air combat and supporting operations designed to effect, through the systematic application of force to a selected series of vital targets, the progressive destruction and disintegration of the enemy's war-making capacity to a point where the ability or the will to wage war is no longer retained; vital targets may include key manufacturing systems, sources of raw material, critical material, stockpiles, power systems, transportation systems, communication facilities, concentrations of uncommitted elements of enemy armed forces, key agricultural areas, and other such target systems. { strə'tē·jik 'er ‚wór‚fär }

strategic attack [ORD] An attack by means of aerospace forces directed at selected vital targets of an enemy nation so as to destroy its war-making capacity or will to fight. { strə'tē·jik ə'tak }

strategic concentration [ORD] The assembly of designated forces in areas from which it is intended that operations of the assembled forces shall begin so that they are best disposed to initiate the plan of campaign. { strə'tē·jik ‚kän·sən'trā·shən }

strategic map [MAP] A map of medium scale, or smaller, used for planning of operations, including the movement, concentration, and supply of troops. { strə'tē·jik ‚map }

strategic material [IND ENG] A material needed for the industrial support of a war effort. { strə'tē·jik mə'tir·ē·əl }

strategic mobility [ORD] The capability of a unit, command, force, or thing that enables it to be readily moved in advance of engagement with hostile forces. { strə'tē·jik mō'bil·əd·ē }

strategic nuclear weapon [ORD] A nuclear weapon which is programmed primarily for use against strategic targets in strategic nuclear war. { strə'tē·jik 'nü·klē·ər 'wep·ən }

strategic research [SCI TECH] Research conducted to produce specific applied programs. { strə'tē·jik ri'sərch }

strategic reserve [ORD] That quantity of material which is placed in a particular geographic location due to strategic considerations or in anticipation of major interruptions in the supply distribution system; it is over and above the stockage objective. { strə'tē·jik ri'zərv }

strategic target [ORD] Any installation, network, group of buildings, or the like, considered vital to a country's war-making capacity and singled out for attack. { strə'tē·jik 'tär·gət }

strategic transport aircraft [AERO ENG] Aircraft designed primarily for the carriage of personnel or cargo over long distances. { strə'tē·jik 'tranz‚pórt 'er‚kraft }

strategy [ECOL] A group of related traits that evolved under the influence of natural selection and solve particular problems encountered by organisms. [MATH] In game theory a strategy is a specified collection of moves, which cover all possible situations, for the complete play of a given game. { 'strad·ə·jē }

strategy vector [MATH] A vector characterizing a mixed strategy, whose components are the probability weights of the strategy. { 'strad·ə·jē ‚vek·tər }

strath [GEOL] **1.** A broad, elongate depression with steep sides on the continental shelf. **2.** An extensive remnant of a broad, flat valley floor that has undergone degradation following uplift. { strath }

strath terrace [GEOL] An extensive remnant of a strath from a former erosion cycle. { 'strath ‚ter·əs }

stratification [GEOL] An arrangement or deposition of sedimentary material in layers, or of sedimentary rock in strata. [HYD] **1.** The arrangement of a body of water, as a lake, into two or more horizontal layers of differing characteristics, especially densities. **2.** The formation of layers in a mass of snow, ice, or firn. { ‚strad·ə·fə'kā·shən }

stratification index [GEOL] A measure of the beddedness of a stratigraphic unit, expressed as the number of beds in the unit per 100 feet (30 meters) of section. { ‚strad·ə·fə'kā·shən ‚in‚deks }

stratification plane [GEOL] A demarcation between two layers of sedimentary rock, often signifying that the layers were deposited under different conditions. { ‚strad·ə·fə'kā·shən ‚plān }

stratified charge engine [MECH ENG] An internal combustion engine that uses a fuel charge consisting of two layers; a rich mixture is close to the spark plug, and combustion promotes ignition of a lean mixture in the remainder of the cylinder. { 'strad·ə‚fīd 'chärj ‚en·jən }

stratified drift [GEOL] Fluvioglacial drift composed of material deposited by a meltwater stream or settled from suspension. { 'strad·ə‚fīd 'drift }

stratified film [PHYS CHEM] A film in which two thicknesses are present in a fixed configuration for a significant period of time. { 'strad·ə‚fīd 'film }

stratified flow [FL MECH] A two-phase flow in which a liquid flows along the bottom of a pipe and gas flows separately above it. { 'strad·ə‚fīd 'flō }

stratified fluid [FL MECH] A fluid having density variation along the axis of gravity, usually implying upward decrease of density, that is, a stratification characterized by static stability. { 'strad·ə‚fīd 'flü·əd }

stratified ocean [OCEANOGR] An ocean where there is a vertical gradient of density. { 'strad·ə‚fīd 'ō·shən }

stratified rock *See* sedimentary rock. { 'strad·ə‚fīd 'räk }

stratified sampling [STAT] A random sample of specified size is drawn from each stratum of a population. { 'strad·ə‚fīd 'sam·pliŋ }

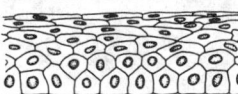

STRATIFIED SQUAMOUS EPITHELIUM

Drawing of a section through stratified squamous epithelium.

stratified squamous epithelium [HISTOL] A multiple-layered epithelium composed of thin, flat superficial cells and cuboidal and columnar deeper cells. { 'strad·ə‚fīd 'skwä·məs ‚ep·ə'thē·lē·əm }

stratiform [GEOL] **1.** Descriptive of a layered mineral deposit of either igneous or sedimentary origin. **2.** Consisting of parallel bands, layers, or sheets. [METEOROL] Description of clouds of extensive horizontal development, as contrasted to the vertically developed cumuliform types. { 'strad·ə‚fórm }

stratiformis [METEOROL] A cloud species consisting of a very extensive horizontal layer or layers which need not be continuous; this species is the most common form of the genera altocumulus and stratocumulus and is occasionally found in cirrocumulus. { ‚strad·ə'fór·məs }

stratigrapher [GEOL] A geologist who deals with stratified rocks, for example, the classification, nomenclature, correlation, and interpretation of rocks. { strə'tig·rə·fər }

stratigraphic geology See stratigraphy. { ‚strad·ə‚graf·ik jē'äl·ə·jē }

stratigraphic map [GEOL] A map showing the areal distribution, configuration, or aspect of a stratigraphic unit or surface, such as an isopach map or a lithofacies map. { ‚strad·ə‚graf·ik 'map }

stratigraphic oil fields [GEOL] Hydrocarbon reserves in stratigraphic (sedimentary) traps formed by the positioning of clastic materials through chemical deposition. { ‚strad·ə‚graf·ik 'óil ‚fēlz }

stratigraphic separation See stratigraphic throw. { ‚strad·ə‚graf·ik ‚sep·ə'rā·shən }

stratigraphic sequence See sequence. { ‚strad·ə‚graf·ik 'sē·kwəns }

stratigraphic throw [GEOL] The thickness of the strata which originally separated two beds brought into contact at a fault. Also known as stratigraphic separation. { ‚strad·ə‚graf·ik 'thrō }

stratigraphic trap [GEOL] Sealing of a reservoir bed due to lithologic changes rather than geologic structure. Also known as porosity trap; secondary stratigraphic trap. { ‚strad·ə‚graf·ik 'trap }

stratigraphic unit [GEOL] A stratum of rock or a body of strata classified as a unit on the basis of character, property, or attribute. { ‚strad·ə‚graf·ik 'yü·nət }

stratigraphy [GEOL] A branch of geology concerned with the form, arrangement, geographic distribution, chronologic succession, classification, correlation, and mutual relationships of rock strata, especially sedimentary. Also known as stratigraphic geology. { strə'tig·rə·fē }

Stratiomyidae [INV ZOO] The soldier flies, a family of orthorrhaphous dipteran insects in the series Brachycera. { ‚strad·ē·ō'mī·ə‚dē }

stratocumulus [METEOROL] A principal cloud type predominantly stratiform, in the form of a gray or whitish layer of patch, which nearly always has dark parts. { ‚strad·ō'kyü·myə·ləs }

stratopause [METEOROL] The boundary or zone of transition separating the stratosphere and the mesosphere; it marks a reversal of temperature change with altitude. { 'strad·ə‚póz }

stratoscope [OPTICS] A balloon-borne astronomical telescope for taking solar or other celestial photographs at high altitudes; subsequently, the photos are transmitted to a ground receiving station. { 'strad·ə‚skōp }

stratosphere [METEOROL] The atmospheric shell above the troposphere and below the mesosphere; it extends, therefore, from the tropopause to about 33 miles (55 kilometers), where the temperature begins again to increase with altitude. { 'strad·ə‚sfir }

stratosphere radiation [GEOPHYS] Any infrared radiation involved in the complex infrared exchange continually proceeding within the stratosphere. { 'strad·ə‚sfir ‚rād·ē'ā·shən }

stratospheric coupling [METEOROL] The interaction between disturbances in the stratosphere and those in the troposphere. { ‚strad·ə‚sfir·ik 'kəp·liŋ }

stratospheric ozone [METEOROL] Atmospheric ozone that is relatively concentrated in the lower stratosphere in a layer between 9 and 18 miles (15 and 30 kilometers) above the earth's surface, and plays a critical role for the biosphere by absorbing the damaging ultraviolet radiation with wavelengths 320 nanometers and lower. Also known as ozone layer. { ‚strad·ə‚sfir·ik 'ō‚zōn }

stratospheric steering [METEOROL] The steering of lower-level atmospheric disturbances along the contour lines of the tropopause, which lines are presumably roughly parallel to the direction of the wind at the tropopause level. { ‚strad·ə‚sfir·ik 'stir·iŋ }

stratotype [GEOL] A specifically bounded type section of rock strata to which a time-stratigraphic unit is ascribed, ideally consisting of a complete and continuously exposed and deposited sequence of correlatable strata, and extending from a readily identifiable basal boundary to a readily identifiable top boundary. { 'strad·ə‚tīp }

stratovolcano [GEOL] A volcano constructed of lava and pyroclastics, deposited in alternating layers. Also known as composite volcano. { ‚strad·ō·väl'kā·nō }

Stratton pseudoscope [OPTICS] A type of Wheatstone stereoscope in which the mirrors transpose the right- and left-eye views, producing a reversed stereoscopic effect. { ‚strat·ən 'süd·ə‚skōp }

stratum [GEOL] A mass of homogeneous or gradational sedimentary material, either consolidated rock or unconsolidated soil, occurring in a distinct layer and visually separable from other layers above and below. [SCI TECH] One in a sequence of distinct layers. [STAT] See subpopulation. { 'strad·əm }

stratum corneum [HISTOL] The outer layer of flattened keratinized cells of the epidermis. { 'strad·əm 'kór·nē·əm }

stratum disjunctum [HISTOL] The outermost layer of desquamating keratinized cells of the stratum corneum. { 'strad·əm dis'jəŋk·təm }

stratum germinativum [HISTOL] The innermost germinative layer of the epidermis. { 'strad·əm ‚jər·mə·nə'tī·vəm }

stratum granulosum [HISTOL] A layer of granular cells interposed between the stratum corneum and the stratum germinativum in the thick skin of the palms and soles. { 'strad·əm ‚gran·yə'lō·səm }

stratum lucidum [HISTOL] A layer of irregular transparent epidermal cells with traces of nuclei interposed between the stratum corneum and stratum germinativum in the thick skin of the palms and soles. { 'strad·əm 'lü·səd·əm }

stratus [METEOROL] A principal cloud type in the form of a gray layer with a rather uniform base; a stratus does not usually produce precipitation, but when it does occur it is in the form of minute particles, such as drizzle, ice crystals, or snow grains. { 'strad·əs }

Strauss reaction [IMMUNOL] The exudative swelling of the scrotum in male hamsters and guinea pigs upon subcutaneous or intraperitoneal inoculation of *Pseudomonas mallei*, the causative agent of glanders. { 'straùs rē‚ak·shən }

straw [AGR] Grain stalks after threshing and usually mixed with leaves and chaff. [BOT] A stem of grain, such as wheat or oats. { strò }

strawberry [BOT] A low-growing perennial of the genus *Fragaria*, order Rosales, that spreads by stolons; the juicy, usually red, edible fruit consists of a fleshy receptacle with numerous seeds in pits or nearly superficial on the receptacle. { 'strò‚ber·ē }

strawberry hemangioma [MED] A vascular birthmark characterized by a soft, raised, bright-red, lobular appearance. { 'strò‚ber·ē hi‚mān·jē'ō·mə }

strawboard [MATER] A type of pasteboard in which straw and a bonding material are incorporated with other paper materials having a low cellulose content. Also known as compressed straw slab. { 'strò‚bórd }

straw oil [MATER] A straw-colored petroleum paraffin oil; used for many process applications. { 'strò ‚óil }

strawwalker [AGR] A set of reciprocating notched bars inside a thresher or combine that push the straw to the rear. { 'strò‚wók·ər }

stray [GEOL] A lenticular rock formation encountered unexpectedly in drilling an oil or a gas well; it differs from an adjacent persistent formation in lithology and hardness. { strā }

stray capacitance [ELECTR] Undesirable capacitance between circuit wires, between wires and the chassis, or between components and the chassis of electronic equipment. { 'strā kə'pas·əd·əns }

stray current [ELEC] **1.** A portion of a current that flows over a path other than the intended path, and may cause electrochemical corrosion of metals in contact with electrolytes.

2. An undesirable current generated by discharge of static electricity; it commonly arises in loading and unloading petroleum fuels and some chemicals, and can initiate explosions. { 'strā ,kə·rənt }

stray current corrosion [MET] Corrosion of metals caused by a stray current. { 'strā ¦kə·rənt kə'rō·zhən }

stray emission [PHYS] Emission of radiation that serves no useful purpose. { 'strā i'mish·ən }

stray field [ELECTROMAG] Leakage of magnetic flux that spreads outward from a coil and does no useful work. { 'strā ¦fēld }

stray line [ENG] An ungraduated portion of the line connected to a current pole, used so that the pole will acquire the speed of the current before a measurement is begun. { 'strā ¦līn }

strays *See* atmospheric interference. { strāz }

stray sand [GEOL] A stray composed of sandstone. { 'strā 'sand }

streak [MINERAL] The color of a powdered mineral, obtained by rubbing the mineral on a streak plate. { strēk }

streak camera [OPTICS] A special type of high-speed motion picture camera that records an image as a continuous spreadout picture rather than as a sequence of separate frames; special viewing equipment must be used to analyze the image and reconstitute individual pictures. { strēk ,kam·rə }

streak lightning [GEOPHYS] Ordinary lightning, of a cloud-to-ground discharge, that appears to be entirely concentrated in a single, relatively straight lightning channel. { strēk ,līt·niŋ }

streak line [FL MECH] A line within a fluid which, at a given instant, is formed by those fluid particles which at some previous instant have passed through a specified fixed point in the fluid; an example is the line of color in a flow into which a dye is continuously introduced through a small tube, all dyed fluid particles having passed the tube's end. { strēk ,līn }

streak plate [MICROBIO] A method of culturing aerobic bacteria by streaking the surface of a solid medium in a petri dish with an inoculating wire or glass rod in a continuous movement so that most of the surface is covered; used to isolate majority members of a mixed population. { strēk ,plāt }

stream [COMPUT SCI] A collection of binary digits that are transmitted in a continuous sequence, and from which extraneous data such as control information or parity bits are excluded. [HYD] A body of running water moving under the influence of gravity to lower levels in a narrow, clearly defined natural channel. { strēm }

stream anchor [NAV ARCH] An anchor used in narrow channels to prevent the stern of the vessel moving with the tide. { 'strēm ,aŋ·kər }

stream-built terrace *See* alluvial terrace. { 'strēm ¦bilt 'ter·əs }

stream capacity [GEOL] The ability of a stream to carry detritus, measured at a given point per unit of time. { 'strēm kə,pas·əd·ē }

stream capture *See* capture. { 'strēm ,kap·chər }

stream channel [GEOL] A long, narrow, sloping troughlike depression where a natural stream flows or may flow. Also known as streamway. { 'strēm ,chan·əl }

stream-channel form ratio [GEOL] The mathematical relationship between a stream channel width, depth, and channel perimeter. { 'strēm ,chan·əl 'form ,rā·shō }

stream cipher [COMMUN] A cipher that makes use of an algorithmic procedure to produce an unending sequence of binary digits which is then combined either with plaintext to produce ciphertext or with ciphertext to recover plaintext. { 'strēm ,sī·fər }

stream current [HYD] A steady current in a stream or river. [OCEANOGR] A deep, narrow, well-defined fast-moving ocean current. { 'strēm ,kə·rənt }

stream day [CHEM ENG] Denoting a 24-hour actual operation of a processing unit, in contrast to the hours actually operated during a calendar (24-hour) day. { 'strēm 'dā }

stream editor [COMPUT SCI] A modification of a statement editor to allow superlines that expand and contract as necessary; the most powerful type of text editor. Also known as string editor. { 'strēm ,ed·əd·ər }

streamer [GEOPHYS] A sinuous channel of very high ion-density which propagates itself through a gas by continual establishment of an electron avalanche just ahead of its advancing tip; in lightning discharges, the stepped leader, and return streamer all constitute special types of streamers. { 'strē·mər }

streamer chamber [NUCLEO] A particle detector in which application of a pulse of about 500,000 volts for about 10 nanoseconds between the electrodes bounding the chamber accelerates the ions formed by charged particles moving through the gas filling the chamber, resulting in avalanches of similar ions and then streamers about 0.4 inch (1 centimeter) long which are photographed. { 'strēm·ər ,chām·bər }

stream erosion [GEOL] The progressive removal of exposed matter from the surface of a stream channel by a stream. { 'strēm i,rō·zhən }

stream feeder [GRAPHICS] A printing press device that feeds several overlapping sheets toward the grippers. { 'strēm ,fēd·ər }

streamflow [HYD] A type of channel flow, applied to surface runoff moving in a stream. { 'strēm,flō }

streamflow routing *See* flood routing. { 'strēm,flō 'rüd·iŋ }

stream frequency [GEOL] A measure of topographic texture expressed as the ratio of the number of streams in a drainage basin to the area of the basin. Also known as channel frequency. { 'strēm ,frē·kwən·sē }

stream function *See* Lagrange stream function. { 'strēm ,fəŋk·shən }

stream gage *See* river gage. { 'strēm ,gāj }

stream gradient [GEOL] The angle, measured in the direction of flow, between the water surface (for large streams) or the channel flow (for small streams) and the horizontal. Also known as stream slope. { 'strēm ,grād·ē·ənt }

stream-gradient ratio [GEOL] Ratio of the stream gradient of a stream channel of one order to the stream gradient of the next higher order channel in the same drainage basin. Also known as channel gradient ratio. { 'strēm ,grād·ē·ənt ,rā·shō }

streaming [COMPUT SCI] A malfunction in which a communicating device constantly transmits worthless data and thereby locks out all other devices on the line. { 'strēm·iŋ }

streaming current [ELEC] The electric current which is produced when a liquid is forced to flow through a diaphragm, capillary, or porous solid. { 'strēm·iŋ ,kə·rənt }

streaming media [COMPUT SCI] Audio or video files that can begin playing as they are being downloaded to a computer. { ¦strēm·iŋ ¦mēd·ē·ə }

streaming potential [ELEC] The difference in electric potential between a diaphragm, capillary, or porous solid and a liquid that is forced to flow through it. { 'strēm·iŋ pə,ten·chəl }

streaming tape [COMPUT SCI] A type of high-speed magnetic tape that is used as a backup storage for disks, particularly hard disks in microcomputer systems. { 'strēm·iŋ 'tāp }

stream-length ratio [HYD] Ratio of the mean length of a stream of a given order to the mean length of the next lower order stream in the same basin. { 'strēm ¦leŋkth ,rā·shō }

streamline [FL MECH] A line which is everywhere parallel to the direction of fluid flow at a given instant. { 'strēm,līn }

streamline flow [FL MECH] Flow of a fluid in which there is no turbulence; particles of the fluid follow well-defined continuous paths, and the flow velocity at a fixed point either remains constant or varies in a regular fashion with time. { 'strēm,līn ,flō }

streamlining [DES ENG] The contouring of a body to reduce its resistance to motion through a fluid. { 'strēm,līn·iŋ }

stream load [GEOL] Solid material transported by a stream. { 'strēm ,lōd }

stream morphology *See* river morphology. { 'strēm mor'fäl·ə·jē }

stream order [HYD] The designation by a dimensionless integer series (1, 2, 3, . . .) of the relative position of stream segments in the network of a drainage basin. Also known as channel order. { 'strēm ,or·dər }

stream piracy *See* capture. { 'strēm ,pī·rə·sē }

stream profile [HYD] The longitudinal profile of a stream. { 'strēm ,prō,fīl }

stream robbery *See* capture. { 'strēm ,räb·ə·rē }

stream segment [HYD] The part of a stream extending between designated tributary junctions. Also known as channel segment. { 'strēm ,seg·mənt }

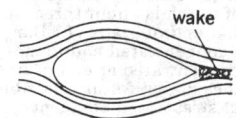

STREAMLINE FLOW

Flow indicated about a streamlined body which is traveling at subsonic speed.

streamsink [GEOL] An opening in the surface of the earth down which a stream disappears underground. { 'strēm,siŋk }

stream slope *See* stream gradient. { 'strēm ,slōp }

stream stage *See* stage. { 'strēm ,stāj }

stream takeoff [AERO ENG] Aircraft taking off in tail/column formation. { 'strēm 'tāk,óf }

stream terrace [GEOL] One of a series of level surfaces on a stream valley flanking and parallel to a stream channel and above the stream level, representing the uneroded remnant of an abandoned floodplain or stream bed. Also known as river terrace. { 'strēm ,ter·əs }

stream tin [GEOL] The mineral cassiterite occurring as pebbles in alluvial deposits. { 'strēm ,tin }

stream transport [GEOL] Movement of rock material in and by a stream. { 'strēm 'tranz,pórt }

stream tube [FL MECH] In fluid flow, an imaginary tube whose wall is generated by streamlines passing through a closed curve. { 'strēm ,tüb }

streamway *See* stream channel. { 'strēm,wā }

Streblidae [INV ZOO] The bat flies, a family of cyclorrhaphous dipteran insects in the section Pupipara; adults are ectoparasites on bats. { 'streb·lə,dē }

street [CIV ENG] A paved road for vehicular traffic in an urban area. { strēt }

street elbow [DES ENG] A pipe elbow with an internal thread at one end and an external thread at the other. { strēt ,el·bō }

Strehl ratio [OPTICS] The ratio of the peak field amplitude in the focus of an optical element to the diffraction-limited amplitude. { 'strāl ,rā·shō }

Strelitziaceae [BOT] A family of monocotyledonous plants in the order Zingiberales distinguished by perfect flowers with five functional stamens and without an evident hypanthium, penniveined leaves, and symmetrical guard cells. { strə,lit·sē'ās·ē,ē }

stremmatograph [ENG] An instrument for measuring longitudinal stress in rails as trains pass over. { strə'mad·ə,graf }

strengite [MINERAL] $FePO_4 \cdot 2H_2O$ A pale-red mineral crystallizing in the orthorhombic system, isomorphous with variscite and dimorphous with phosphosiderite, and specific gravity 2.87. { 'streŋ,īt }

strength [ACOUS] The maximum instantaneous rate of volume displacement produced by a sound source when emitting a wave with sinusoidal time variation. [MECH] The stress at which material ruptures or fails. { streŋkth }

strength of current [OCEANOGR] **1.** The phase of a tidal current at which the speed is a maximum. **2.** The velocity of the current at this time. { 'streŋkth əv 'kə·rənt }

strength of ebb [OCEANOGR] **1.** The ebb current at the time of maximum speed. **2.** The speed of the current at this time. { 'streŋkth əv 'eb }

strength-of-ebb interval [OCEANOGR] The time interval between the transit (upper or lower) of the moon and the next maximum ebb current at a place. { 'streŋkth əv 'eb 'in·tər·vəl }

strength of enemy forces [ORD] The description of an enemy unit of force in terms of men, weapons, and equipment. { 'streŋkth əv 'en·ə·mē 'fór·səz }

strength of flood [OCEANOGR] **1.** The flood current at the time of maximum speed. **2.** The speed of the current at this time. { 'streŋkth əv 'fləd }

strength-of-flood interval [OCEANOGR] The time interval between the transit (upper or lower) of the moon and the next maximum flood current at a place. { 'streŋkth əv 'fləd 'in·tər·vəl }

strepaster [INV ZOO] A short, spiny microscleric, monaxonic spicule. { stre'pas·tər }

strepogenin [BIOCHEM] A factor, possibly a peptide derivative of glutamic acid, reported to exist in certain proteins, acting as a growth stimulant in bacteria and mice in the presence of completely hydrolyzed protein. Also known as streptogenin. { strə'päj·ə·nən }

Strepsiptera [INV ZOO] An order of the Coleoptera that is coextensive with the family Stylopidae. { strep'sip·tə·rə }

streptobacillary fever *See* Haverhill fever. { ¦strep·tō·bə'sil·ə·rē 'fē·vər }

streptobiosamine [BIOCHEM] $C_{13}H_{23}NO_9$ A nitrogen-containing disaccharide, obtained when streptomycin undergoes acid hydrolysis; in the streptomycin molecule it is glycosidally linked to streptidine. { ¦strep·tō·bī'ä·sə,mēn }

Streptococcaceae [MICROBIO] A family of gram-positive cocci; chemoorganotrophs with fermentative metabolism. { ¦strep·tə·käk'sās·ē,ē }

Streptococceae [MICROBIO] Formerly a tribe of the family Lactobacillaceae including cocci that occur in pairs, short chains, or tetrads and which generally obtain energy by fermentation of carbohydrates or related compounds. { ¦strep·tə'käk·sē,ē }

streptogenin *See* strepogenin. { strep'tä·jə·nən }

streptokinase [BIOCHEM] An enzyme occurring as a component of fibrinolysin in cultures of certain hemolytic streptococci. { ¦strep·tō'kī,nās }

streptolysin [BIOCHEM] Any of a group of hemolysins elaborated by *Streptococcus pyogenes*. { ¦strep·tō'līs·ən }

Streptomycetaceae [MICROBIO] A family of soil-inhabiting bacteria in the order Actinomycetales; branched mycelia are produced by vegetative hyphae; spores are produced on aerial hyphae. { ¦strep·tō,mī·sə'tās·ē,ē }

streptomycin [MICROBIO] $C_{21}H_{39}O_{12}N_7$ A water-soluble antibiotic obtained from *Streptomyces griseus* that is used principally in the treatment of tuberculosis. { ¦strep·tə'mīs·ən }

streptothricin [MICROBIO] $C_{19}H_{34}O_7N_8$ An antibiotic produced by *Streptomyces lavendulae*; active against various gram-negative and gram-positive microorganisms. { ¦strep·tə'thrīs·ən }

stress [BIOL] A stimulus or succession of stimuli of such magnitude as to tend to disrupt the homeostasis of the organism. [MECH] The force acting across a unit area in a solid material resisting the separation, compacting, or sliding that tends to be induced by external forces. { stres }

STRESS [COMPUT SCI] A problem-oriented programming language used to solve structural engineering problems. Derived from structural engineering system solver. { stres }

stress amplitude [MECH ENG] One half the algebraic difference between the maximum and minimum stress in one fatigue test cycle. { 'stres ,am·plə,tüd }

stress analysis [PHYS] The determination of the stresses produced in a solid body when subjected to various external forces. { 'stres ə,nal·ə·səs }

stress axis *See* principal axis of stress. { 'stres ,ak·səs }

stress birefringence *See* mechanical birefringence. { 'stres ¦bī·ri·frin·jəns }

stress concentration [MECH] A condition in which a stress distribution has high localized stresses; usually induced by an abrupt change in the shape of a member; in the vicinity of notches, holes, changes in diameter of a shaft, and so forth, maximum stress is several times greater than where there is no geometrical discontinuity. { 'stres ,kän·sən,trā·shən }

stress concentration factor [MECH] A theoretical factor K_t expressing the ratio of the greatest stress in the region of stress concentration to the corresponding nominal stress. { 'stres ,kän·sən¦trā·shən ,fak·tər }

stress corrosion [MET] Corrosion that is accelerated by stress, applied or residual, in a metal. { 'stres kə,rō·zhən }

stress-corrosion cracking [MET] Failure by cracking under the conjoint action of a constant tensile stress, which is applied to residual, in certain chemical environments specific to the metal. { 'stres kə¦rō·zhən ,krak·iŋ }

stress crack [MECH] An external or internal crack in a solid body (metal or plastic) caused by tensile, compressive, or shear forces. { 'stres ,krak }

stress difference [MECH] The difference between the greatest and the least of the three principal stresses. { 'stres ,dif·rəns }

stressed skin construction [CIV ENG] A type of construction in which the outer skin and the framework interact, thus contributing to the flexural strength of the unit. { 'strest ¦skin kən'strak·shən }

stress ellipsoid [MECH] A mathematical representation of the state of stress at a point that is defined by the minimum, intermediate, and maximum stresses and their intensities. { 'stres i'lip,sóid }

stress equivalent [IND ENG] A quantitative expression that can be used to compare the physiological outputs generated by different types of work stress. { 'stres i,kwiv·ə·lənt }

stress function [MECH] A single function, such as the Airy stress function, or one of two or more functions, such as Maxwell's or Morera's stress functions, that uniquely define the

STRESS CONCENTRATION

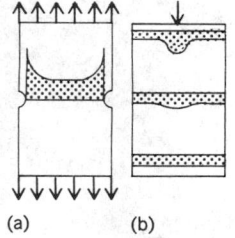

(a)   (b)

Stress concentrations. (*a*) Tensile stress distribution in a plate reduced by circular notches is shown qualitatively; stress at root of notch is about three times stress at end of plate. (*b*) Bar under concentrated end load; load concentrated at end of bar produces nonuniformly distributed normal stresses on adjacent sections with variation decreasing at more remote sections.

stresses in an elastic body as a function of position. { 'stres ,faŋk·shən }

stress intensity [MECH] Stress at a point in a structure due to pressure resulting from combined tension (positive) stresses and compression (negative) stresses. { 'stres in,ten·səd·ē }

stress lines See isostatics. { 'stres ,līnz }

stress mineral [MINERAL] Any mineral whose formation in metamorphosed rock is favored by shearing stress. { 'stres ,min·rəl }

stress-optic law [OPTICS] In a transparent, isotropic plate subjected to a biaxial stress field, the relative retardation R_t between the two components produced by temporary double refraction is equal to $Ct(p - q)$, which in turn is equal to $n\lambda$; C is the stress-optic coefficient, t the plate thickness, p and q the principal stresses, n the number of fringes which have passed the point during application of the load, and λ the wavelength of the light. { 'stres 'äp·tik ,lö }

stress raiser [MET] A notch, hole, or other discontinuity in contour or structure which causes localized stress concentration. { 'stres ,rā·zər }

stress range [MECH] The algebraic difference between the maximum and minimum stress in one fatigue test cycle. { 'stres ,rānj }

stress ratio [MECH] The ratio of minimum to maximum stress in fatigue testing, considering tensile stresses as positive and compressive stresses as negative. { 'stres ,rā·shō }

stress relief cracking [MET] Cracking between metal grains in the heat-affected zone of a weldment during exposure to high temperatures. { 'stres ri¦lēf ,krak·iŋ }

stress relieving [MET] Low-temperature heating to reduce residual stress. { 'stres ri,lēv·iŋ }

stress sensor [CONT SYS] A contact sensor that responds to the forces produced by mechanical contact. { 'stres ,sen·sər }

stress-strain curve See deformation curve. { 'stres 'strān ,kərv }

stress tensor [MECH] A second-rank tensor whose components are stresses exerted across surfaces perpendicular to the coordinate directions. { 'stres ,ten·sər }

stress test [COMPUT SCI] A test of new software or hardware under unusually heavy work loads. [ENG] A test of equipment under extreme conditions, outside the range anticipated in normal operation. [MED] A procedure involving continuous electrocardiographic monitoring during exercise, as running a treadmill, to test the circulatory response to physical stress. { 'stres ,test }

stress trajectories See isostatics. { 'stres trə,jek·trēz }

stress transmittal [IND ENG] Transfer of external force from a human-equipment interface to various points of the body. { 'stres tranz,mid·əl }

stress-wave emission See acoustic emission. { 'stres ¦wāv i,mish·ən }

stretch [GEOGR] See reach. [PETRO ENG] The increase in length of oil-well casing or tubing when freely suspended in fluid mediums. { strech }

stretched pebbles [GEOL] Pebbles in a sedimentary rock which have been elongated from their original shape by deformation. { 'strecht 'peb·əlz }

stretcher [CIV ENG] A brick or block that is laid with its length paralleling the wall. [MED] A litter usually made of canvas stretched on a frame for carrying injured, disabled, or dead persons. [MIN ENG] A bar used for roof support on roadways and which is either wedged against or pocketed into the sides of the roadway without support of legs or struts. { 'strech·ər }

stretcher bar [MIN ENG] A single screw column, capable of holding one machine drill; used in small drifts. { 'strech·ər ,bär }

stretcher bond [CIV ENG] A bond that consists entirely of stretchers, with each vertical joint lying between the centers of the stretchers above and below. { 'strech·ər ,bänd }

stretcher leveling [MET] Removing warp and distortion from a piece of metal by gripping it at each end and subjecting it to stress beyond the yield strength. Also known as patent leveling. { 'strech·ər ,lev·əl·iŋ }

stretcher strains See Lüders' lines. { 'strech·ər ,strānz }

stretch fault See stretch thrust. { 'strech ,fölt }

stretch former [MECH ENG] A machine used to form materials, such as metals and plastics, by stretching over a mold. { 'strech ,för·mər }

stretch forming [MECH ENG] Shaping metals and plastics by applying tension to stretch the heated sheet or part, wrapping it around a die, and then cooling it. Also known as wrap forming. { 'strech ,förm·iŋ }

stretching transformation [MATH] A homothetic transformation in which the ratio of similitude is less than 1. { 'strech·iŋ ,tranz·fər,mā·shən }

stretch out [IND ENG] A reduction in the delivery rate specified for a program without a reduction in the total quantity to be delivered. { 'strech ¦aut }

stretch reflex [PHYSIO] Contraction of a muscle in response to a sudden, brisk, longitudinal stretching of the same muscle. Also known as myostatic reflex. { 'strech ,rē,fleks }

stretch thrust [GEOL] A reverse fault developed as a result of shear in the middle limb of an overturned fold. Also known as stretch fault. { 'strech ,thrəst }

stria [BIOL] A minute line, band, groove, or channel. { 'strī·ə }

striated ground See striped ground. { 'strī,ād·əd 'graund }

striated muscle [HISTOL] Muscle tissue consisting of muscle fibers having cross striations. { 'strī,ād·əd 'məs·əl }

striation [ELECTR] A succession of alternately luminous and dark regions sometimes observed in the positive column of a glow-discharge tube near the anode. [GEOL] One of a series of parallel or subparallel scratches, small furrows, or lines on the surface of a rock or rock fragment; usually inscribed by rock fragments embedded at the base of a moving glacier. [MINERAL] One of a series of parallel, shallow depressions or narrow bands on the cleavage face of a mineral caused either by growth twinning or oscillatory growth of different crystal faces. { strī'ā·shən }

striation technique [ACOUS] A technique for making sound waves visible by using their ability to refract light waves. { strī'ā·shən tek,nēk }

strich See millimeter. { strich }

strictly convex space [MATH] A normal linear space such that, for any two vectors x and y, if $\|x + y\| = \|x\| + \|y\|$, then either $y = 0$ or $x = cy$, where c is a number. Also known as rotund space. { ¦strik·lē ¦kän,veks 'spās }

strictly decreasing function See decreasing function. { strik·lē di¦krēs·iŋ 'fəŋk·shən }

strictly dominant strategy [MATH] Relative to a given pure strategy for one player of a game, a second pure strategy for that player that has a greater payoff than the given strategy for any pure strategy of the opposing player. { ¦strik·lē ¦däm·ə·nənt 'strad·ə·jē }

strictly Hurwitz polynomial [MATH] A polynomial whose roots all have strictly negative real parts. { 'strik·lē 'hər,vits ,päl·i'nō·mē·əl }

strictly increasing function See increasing function. { ¦strik·lē in,krēs·iŋ 'fəŋk·shən }

stricture [MED] Abnormal narrowing of a passage, such as a vessel, duct, or the intestine. Also known as constriction. { 'strik·chər }

striding compass [ENG] A compass mounted on a theodolite for orientation. { 'strīd·iŋ ,käm·pəs }

stridor [MED] A peculiar, harsh, vibrating sound produced during respiration. { 'strīd·ər }

stridulation [INV ZOO] Creaking and other audible sounds made by certain insects, produced by rubbing various parts of the body together. { ,strij·ə'lā·shən }

Strigidae [VERT ZOO] A family of birds of the order Strigiformes containing the true owls. { 'strij·ə,dē }

Strigiformes [VERT ZOO] The order of birds containing the owls. { ,strij·ə'för,mēz }

strigose [BIOL] Covered with stiff, pointed, hairlike scales or bristles. { 'strī,gōs }

strigovite [MINERAL] $Fe_3(Al,Fe)_3Si_3O_{11}(OH)_7$ A dark-green mineral of the chlorite group, composed of basic aluminum iron silicate; occurs as crystalline incrustations. { 'strig·ə,vīt }

Strigulaceae [BOT] A family of Ascolichenes in the order Pyrenulales comprising crustose species confined to tropical evergreens, and which form extensive crusts on or under the cuticle of leaves. { ,strig·yə'lās·ē,ē }

strike [GEOL] The direction taken by a structural surface, such as a fault plane, as it intersects the horizontal. Also known as line of strike. [MET] **1.** A very thin, initially electroplated film or the plating solution with which to deposit such a film. **2.** A local crater in a metal surface due to accidental contact with the welding electrode. [ORD] Concerted air attack on a single objective. { strīk }

strike board [MIN ENG] A board at the top of a shaft from which the bucket is tipped; used in shaft sinking; formerly, the beam or plank at the shaft top on which the baskets were landed. Also known as strike tree. { 'strīk ‚bȯrd }

strike fault [GEOL] A fault whose strike is parallel with that of the strata involved. { 'strīk ‚fȯlt }

strike force [ORD] A force composed of appropriate units necessary to conduct strikes, attack, or assault operations. { 'strīk ‚fȯrs }

strike joint [GEOL] A joint that strikes parallel to the bedding or cleavage of the constituent rock. { 'strīk ‚jȯint }

strike note [ACOUS] The note which is the loudest heard when a bell is struck, and whose pitch is generally assigned to the bell. { 'strīk ‚nōt }

strike-off board [ENG] A straight-edge board used to remove excess, freshly placed plaster, concrete, or mortar from a surface. { 'strīk ‚ȯf ‚bȯrd }

strike photography [GRAPHICS] Air photographs taken during an air strike. { 'strīk fə‚täg·rə·fē }

strike plate [DES ENG] A metal plate or box which is set in a door jamb and is either pierced or recessed to receive the bolt or latch of a lock. { 'strīk ‚plāt }

strike plating [MET] Applying a thin electroplated film prior to depositing the principal electroplate. { 'strīk ‚plād·iŋ }

striker [ORD] A firing pin or a projection on the hammer of a firearm, which strikes the primer to initiate a propelling charge explosive train or a fuse explosive train. { ‚strī·kər }

striker plate [ORD] A plate in the breech of a firearm or gun, which supports the base of the cartridge and which is pierced with a hole through which the striker or firing pin hits the primer. { 'strī·kər ‚plāt }

strike separation [GEOL] The distance of separation on either side of a fault surface of two formerly adjacent beds. { 'strīk ‚sep·ə‚rā·shən }

strike-separation fault See lateral fault. { 'strīk ‚sep·ə¦rā·shən ‚fȯlt }

strike-shift fault See strike-slip fault. { 'strīk ¦shift ‚fȯlt }

strike slip [GEOL] The component of the slip of a fault that is parallel to the strike of the fault. Also known as horizontal displacement; horizontal separation. { 'strīk ‚slip }

strike-slip fault [GEOL] A fault whose direction of movement is parallel to the strike of the fault. Also known as strike-shift fault. { 'strīk ¦slip ‚fȯlt }

strike stream See subsequent stream. { 'strīk ‚strēm }

strike-through See bleed-through; show-through. { 'strīk ‚thrü }

strike tree See strike board. { 'strīk ‚trē }

striking hammer [ENG] A hammer used to strike a rock drill. { 'strīk·iŋ ‚ham·ər }

striking potential [ELECTR] **1.** Voltage required to start an electric arc. **2.** Smallest grid-cathode potential value at which plate current begins flowing in a gas-filled triode. { 'strīk·iŋ pə‚ten·chəl }

striking velocity See impact velocity. { 'strīk·iŋ və‚läs·əd·ē }

string [COMPUT SCI] A set of consecutive, adjacent items of similar type; normally a bit string or a character string. [ENG] A piece of pipe, casing, or other down-hole drilling equipment coupled together and lowered into a borehole. [GEOL] A very small vein, either independent or occurring as a branch of a larger vein. Also known as stringer. [MATH] One of the space curves that form a braid. [MECH] A solid body whose length is many times as large as any of its cross-sectional dimensions, and which has no stiffness. [PART PHYS] A proposed structure for elementary particles, consisting of a one-dimensional curve with zero thickness and length typically of the order of the Planck length, 10^{-35} m. { striŋ }

string bead [MET] A continuous weld bead made without appreciable transverse oscillation. { 'striŋ ‚bēd }

string break [COMPUT SCI] In the sorting of records, the situation that arises when there are no records having keys with values greater than the highest key already written in the sequence of records currently being processed. { 'striŋ ‚brāk }

string constant [COMPUT SCI] An arbitrary combination of letters, digits, and other symbols that is treated in a manner completely analogous to numeric constants. { 'striŋ ‚kän·stənt }

stringcourse [BUILD] A horizontal band of masonry, generally narrower than other courses and sometimes projecting, extending across the facade of a structure and in some instances encircling pillars or engaged columns. Also known as belt course. { 'striŋ‚kȯrs }

string duality [PART PHYS] The property of superstring theories that the strongly coupled behavior of each theory appears to be equivalent to that of some other weakly coupled system, and all the seemingly different superstring theories, as well as M-theory, are just different weakly coupled limits of a single theory. { ¦striŋ dü'al·əd·ē }

string editor See stream editor. { 'striŋ ‚ed·əd·ər }

string electrometer [ENG] An electrometer in which a conducting fiber is stretched midway between two oppositely charged metal plates; the electrostatic field between the plates displaces the fiber laterally in proportion to the voltage between the plates. { 'striŋ ‚i‚lek'träm·əd·ər }

stringer [CIV ENG] **1.** A long horizontal member used to support a floor or to connect uprights in a frame. **2.** An inclined member supporting the treads and risers of a staircase. [GEOL] See string. [MET] An elongated mass of microconstituents or foreign material in wrought metal oriented in the direction of working. { 'striŋ·ər }

stringer lode [GEOL] A lode that consists of many narrow veins in a mass of country rock. { 'striŋ·ər ‚lōd }

stringer plate [NAV ARCH] One of the plates that make up the outer strake of the deck of a ship, and which are usually heavier than those making up the rest of the deck. { 'striŋ·ər ‚plāt }

string galvanometer [ENG] A galvanometer consisting of a silver-plated quartz fiber under tension in a magnetic field, used to measure oscillating currents. Also known as Einthoven galvanometer. { 'striŋ ‚gal·və'näm·əd·ər }

stringing [PETRO ENG] The connecting of lengths of pipe end to end (tubing or casing) to make a string long enough to reach to the desired depth in a well bore. { 'striŋ·iŋ }

string manipulation [COMPUT SCI] The handling of strings of characters in a computer storage as though they were single units of data. { 'striŋ mə‚nip·yə‚lā·shən }

string manipulation language See string processing language. { 'striŋ mə‚nip·yə‚lā·shən ‚laŋ·gwij }

string milling [MECH ENG] A milling method in which parts are placed in a row and milled consecutively. { 'striŋ ‚mil·iŋ }

String-Oriented-Symbolic Language See SNOBOL. { 'striŋ ¦ȯr·ē‚ent·əd sim'bäl·ik 'laŋ‚gwij }

string processing language [COMPUT SCI] A higher-level programming language equipped with facilities to synthesize and decompose character strings, search them in response to arbitrarily complex criteria, and perform a variety of other manipulations. Also known as string manipulation language. { 'striŋ 'prä‚ses·iŋ ‚laŋ·gwij }

string shot [PETRO ENG] An oil-well stimulation technique in which a string of explosive (for example, Prima Cord) is hung opposite to the producing zone down a wellbore and detonated; used to remove deposits (gypsum, mud, or paraffin) from the formation face. { 'striŋ ‚shät }

stringy floppy [COMPUT SCI] A peripheral storage device for microcomputers that uses a removable magnetic tape cartridge with a 1/16-inch-wide (1.5875-millimeter) loop of magnetic tape. { 'striŋ·ē 'fläp·ē }

strip [ENG] **1.** To remove insulation from a wire. **2.** To break or otherwise damage the threads of a nut or bolt. [MATER] A long, narrow piece of rigid material of uniform width. [MIN ENG] To remove coal, stone, or other material from a quarry or from a working that is near the surface of the earth. [ORD] To dissassemble a piece of equipment, such as a gun, in order to clean, repair, or transport it. { strip }

strip-borer drill [MECH ENG] An electric or diesel skid- or caterpillar-mounted drill used at quarry or opencast sites to drill 3- to 6-inch-diameter (8- to 15-centimeter), horizontal blast holes up to 100 feet (30 meters) in length, without the use of flush water. { 'strip ‚bȯr·ər ‚dril }

strip-chart recorder [ENG] A recorder in which one or more writing pens or other recording devices trace changes in a measured variable on the surface of a strip chart that is moved at constant speed by a time-clock motor. { 'strip ‚chärt ri‚kȯrd·ər }

strip-cropping [AGR] Growing separate crops in adjacent strips that follow the contour of the land as a method of reducing soil erosion. { 'strip ,krap·iŋ }

stripe *See* stripe phase. { strīp }

striped ground [GEOL] A pattern of alternating stripes formed by frost action on a sloping surface. Also known as striated ground; striped soil. { 'strīpt 'graund }

striped soil *See* striped ground. { 'strīpt 'soil }

stripe phase [SOLID STATE] A phase exhibited by the electrons in certain solid systems, such as doped antiferomagnets and quantum Hall systems, in which the behavior of the electrons is similar to that of the molecules in a liquid crystal, exhibiting properties of both a crystal (orientational order and anisotropy) and a liquid (absence of space periodicity). Also known as stripe. { 'strīp ,fāz }

strip lights [NAV] Lights marking the edge of a landing strip. { 'strip ,līts }

strip line [ELECTROMAG] A strip transmission line that consists of a flat metal-strip center conductor which is separated from flat metal-strip outer conductors by dielectric strips. { 'strip ,līn }

strip-line circuit [ELECTROMAG] A circuit in which one or more strip transmission lines serve as filters or other circuit components. { 'strip ¦līn ,sər·kət }

strip method [FOR] A lumbering method in which timbers are cleared from a forest in strips; new growth in the strip results from seeds sown in the adjoining forest. { 'strip ,meth·əd }

strip mine [MIN ENG] An opencut mine in which the overburden is removed from a coal bed before the coal is taken out. { 'strip ,mīn }

strip mining [MIN ENG] The mining of coal by surface mining methods. { 'strip ,mīn·iŋ }

stripped atom [ATOM PHYS] An ionized atom which has appreciably fewer electrons than it has protons in the nucleus. { 'stript 'ad·əm }

stripped illite *See* degraded illite. { 'stript 'il,īt }

stripped plain [GEOL] The upper, exposed surface of a resistant stratum that forms a stripped structural surface when extended over a considerable area. { 'stript 'plān }

stripped structural surface [GEOL] An erosion surface formed in an area underlain by horizontal or gently sloping strata of unequal resistance where the overlying softer beds have been removed by erosion. Also known as stripped surface. { 'stript ¦strək·chə·rəl 'sər·fəs }

stripped surface *See* stripped structural surface. { 'stript 'sər·fəs }

stripper [CHEM ENG] An evaporative device for the removal of vapors from liquids; can be in a bubble-tray distillation tower, a vacuum vessel, or an evaporator; if it is a part of a distillation column below the feed tray, it is called the stripping section. [ENG] A hand or motorized tool used to remove insulation from wires. [PETRO ENG] A well from which oil production is quite small. Also known as stripper well. { 'strip·ər }

stripper plate [ENG] In plastics molding, a plate that strips a molded article free of core pins or force plugs. { 'strip·ər ,plāt }

stripper punch [MET] In powder metallurgy, a device used as the bottom or top of the die cavity which can be pushed into the die to eject the formed compact. { 'strip·ər ,pənch }

stripper rubber [PETRO ENG] A pressure-actuated seal used to control gas pressure in the casing-tubing annulus of low-pressure wells while inserting (running) or withdrawing (pulling) tubing. { 'strip·ər ,rəb·ər }

stripper well *See* stripper. { 'strip·ər ,wel }

stripping [CHEM ENG] In petroleum refining, the removal (by flash evaporation or steam-induced vaporation) of the more volatile components from a cut or fraction; used to raise the flash point of kerosine, gas oil, or lubricating oil. [GRAPHICS] In offset lithography, the process of positioning negatives or positives on a flat before making plates. [MET] Removing a coating from the surface of a metal. { 'strip·iŋ }

stripping agent [TEXT] A substance used to remove dyes from fabrics so that they can be redyed, for example, sodium hydrosulfite, titanous sulfate, and sodium and zinc formaldehyde sulfoxylates. { 'strip·iŋ ,ā·jənt }

stripping analysis [ANALY CHEM] An analytic process of solutions or concentrations containing ions, in which the ions are electrodeposited onto an electrode, stripped (dissolved)

from the material from the electrode, and weighed. { 'strip· iŋ ə,nal·ə·səs }

stripping area [MIN ENG] In stripping operations, an area encompassing the pay material, its bottom depth, the thickness of the layer of waste, the slope of the natural ground surface, and the steepness of the safe slope of cuts. { 'strip·iŋ ,er·ē·ə }

stripping a shaft [MIN ENG] **1.** Removing the timber from an abandoned shaft. **2.** Trimming or squaring the sides of a shaft. { 'strip·iŋ ə 'shaft }

stripping film [GRAPHICS] **1.** The process of assembling photographic negatives or positives to make printing plates. **2.** A film with an emulsion that can be removed from its base and transferred to another one. { 'strip·iŋ ,film }

stripping ratio [MIN ENG] The unit amount of spoil or waste that must be removed to gain access to a similar unit amount of ore or mineral material. { 'strip·iŋ ,rā·shō }

stripping reaction [NUC PHYS] A nuclear reaction in which part of the incident nucleus combines with the target nucleus, and the other part proceeds with most of its original momentum in practically its original direction; especially the reaction in which the incident nucleus is a deuteron and only a proton emerges from the target. { 'strip·iŋ rē,ak·shən }

stripping shovel [MIN ENG] A shovel with an especially long boom and stick, enabling it to reach further and pile higher. { 'strip·iŋ ,shəv·əl }

strip pit [MIN ENG] **1.** A coal or other mine worked by stripping. **2.** An open-pit mine. { 'strip ,pit }

strip plot [MAP] A portion of a map or overlay on which a number of photographs taken along a flight line is delineated without defining the outlines of individual prints. { 'strip ,plät }

strip printer [ENG] A device that prints computer, telegraph, or industrial output information along a narrow paper tape which resembles a ticker tape. { 'strip ,print·ər }

strip survey [FOR] A survey of the value of a strip of forest; used to estimate the value of a larger area of the forest. { 'strip ¦sər,vā }

strip transmission line [ELECTROMAG] A microwave transmission line consisting of a thin, narrow, rectangular metal strip that is supported above a ground-plane conductor or between two wide ground-plane conductors and is usually separated from them by a dielectric material. { 'strip tranz'mish·ən ,līn }

strobe [ELECTR] **1.** Intensified spot in the sweep of a deflection-type indicator, used as a reference mark for ranging or expanding the presentation. **2.** Intensified sweep on a plan-position indicator or B-scope; such a strobe may result from certain types of interference, or it may be purposely applied as a bearing or heading marker. **3.** Line on a console oscilloscope representing the azimuth data generated by a jammed radar site. **4.** A signaling pulse of very short duration. { strōb }

strobe circuit [ELECTR] A circuit that produces an output pulse only at certain times or under certain conditions, such as a gating circuit or a coincidence circuit. { 'strōb ,sər·kət }

strobe marker [ELECTR] A small bright spot, or a short gap, or other discontinuity produced on the trace of a radar display to indicate that part of the time base which is receiving attention. { 'strōb 'mär·kər }

strobe photography *See* stroboscopic photography. { 'strōb fə'täg·rə·fē }

strobe pulse [ELECTR] Pulse of duration less than the time period of a recurrent phenomenon used for making a close investigation of that phenomenon; the frequency of the strobe pulse bears a simple relation to that of the phenomenon, and the relative timing is usually adjustable. { 'strōb ,pəls }

strobilation [INV ZOO] Asexual reproduction by segmentation of the body into zooids, proglottids, or separate individuals. { ,sträb·ə'lā·shən }

strobilocercus [INV ZOO] A larval tapeworm that has undergone strobilation. { ,sträb·ə·lō'sər·kəs }

strobilus [BOT] **1.** A conelike structure made up of sporophylls, or spore-bearing leaves, as in Equisetales. **2.** The cone of members of the Pinophyta. { 'sträb·ə·ləs }

strobing [COMPUT SCI] The technique required to time-synchronize data appearing as pulses at the output of a computer memory. { 'strōb·iŋ }

stroboscope [ENG] An instrument for making moving bodies visible intermittently, either by illuminating the object with brilliant flashes of light or by imposing an intermittent shutter

STRIPE PHASE

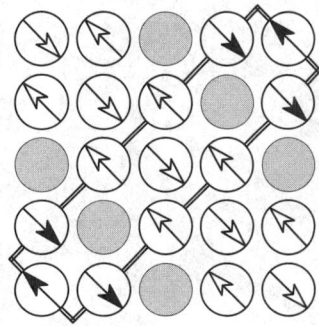

Idealized diagram of the spin and charge stripe pattern within an NiO_2 plane observed in the doped antiferromagnet La_2NiO_4. Arrows indicate the orientation of magnetic moments of metal atoms, locally antiparallel. The double line outlines the magnetic unit cell.

STRIP TRANSMISSION LINE

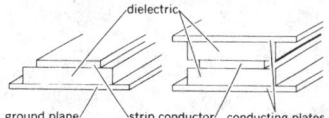

Components of strip transmission lines.

between the viewer and the object; a high-speed vibration can be made visible by adjusting the strobe frequency close to the vibration frequency. { 'strō·bə¦skōp }

stroboscopic direction finder [NAV] A radio direction finder employing a continuously rotating antenna or other directional element such as a goniometer; associated with the rotating element, either directly or remotely, is a mechanism which rotates a thin metal disk in synchronism with the antenna or directional element; the disk has a single thin radial slot which covers a neon tube so that its light can be seen only through the slot; the output of the receiver is connected to the neon tube so that illumination occurs at a reading corresponding to the bearing of the received signal. { ¦sträb·ə¦skäp·ik di'rek·shən ‚fīn·dər }

stroboscopic disk [ENG] A printed disk having a number of concentric rings each containing a different number of dark and light segments; when the disk is placed on a phonograph turntable or rotating shaft and illuminated at a known frequency by a flashing discharge tube, speed can be determined by noting which pattern appears to stand still or to rotate slowly. { ¦sträb·ə¦skäp·ik 'disk }

stroboscopic lamp See flash lamp. { ¦sträb·ə¦skäp·ik 'lamp }

stroboscopic motion [PSYCH] The illusion of motion that occurs when a stationary object is first seen briefly in one location and, following a short interval, is seen in another location. { ‚strō·bə‚skäp·ik 'mō·shən }

stroboscopic photography [GRAPHICS] The technique of producing pictures of both single and multiple exposure taken by flashes of light from electrical discharges. Also known as flash photography; strobe photography. { ¦sträb·ə¦skäp·ik fə'täg·rə·fē }

stroboscopic tachometer [ENG] A stroboscope having a scale that reads in flashes per minute or in revolutions per minute; the speed of a rotating device is measured by directing the stroboscopic lamp on the device, adjusting the flashing rate until the device appears to be stationary, then reading the speed directly on the scale of the instrument. { ¦sträb·ə¦skäp·ik tə'käm·əd·ər }

stroboscopic tube See strobotron. { ¦sträb·ə¦skäp·ik 'tüb }

strobotron [ELECTR] A cold-cathode gas-filled arc-discharge tube having one or more internal or external grids to initiate current flow and produce intensely bright flashes of light for a stroboscope. Also known as stroboscopic tube. { 'strō·bə‚trän }

stroke [COMPUT SCI] **1.** In optical character recognition, straight or curved portion of a letter, such as is commonly made with one smooth motion of a pen. Also known as character stroke. **2.** That segment of a printed or handwritten character which has been temporarily isolated from other segments for the purpose of analyzing it, particularly with regard to its dimensions and relative reflectance. Also known as character stroke. [ELECTR] The penlike motion of a focused electron beam in cathode-ray-tube diplays. [MECH ENG] The linear movement, in either direction, of a reciprocating mechanical part. Also known as throw. [MED] A sudden cerebrovascular accident. { strōk }

stroke analysis [COMPUT SCI] In character recognition, a method employed in character property detection in which an input specimen is dissected into certain prescribed elements; the sequence, relative positions, and number of detected elements are then used to identify the characters. { 'strōk ə‚nal·ə·səs }

stroke-bore ratio [MECH ENG] The ratio of the distance traveled by a piston in a cylinder to the diameter of the cylinder. { 'strōk 'bȯr ‚rā·shō }

stroke center line [COMPUT SCI] In character recognition, a line midway between the two average-edge lines; the center line describes the stroke's direction of travel. Also known as center line. { 'strōk 'sen·tər ‚līn }

stroke density [GEOPHYS] The areal density of lightning discharges over a given region during some specified period of time, as number per square mile per year. { 'strōk ‚den·səd·ē }

stroke edge [COMPUT SCI] In character recognition, a continuous line, straight or otherwise, which traces the outermost part of intersection of the stroke along the two sides of its greatest dimension. { 'strōk ‚ej }

stroke speed [COMMUN] Number of times per minute that a fixed line, perpendicular to the direction of scanning, is

crossed in one direction by a scanning or recording spot in a facsimile system. Also known as scanning frequency; scanning line frequency. { 'strōk ‚spēd }

stroke volume [PHYSIO] The amount of blood pumped during each cardiac contraction; quantitatively, the diastolic volume of the left ventricle minus the volume of blood in the ventricle at the end of systole. { 'strōk ‚väl·yəm }

stroke width [COMPUT SCI] In character recognition, the distance that obtains, at a given location, between the points of intersection of the stroke edges and a line drawn perpendicular to the stroke center line. { 'strōk ‚width }

stroma [ANAT] The supporting tissues of an organ, including connective and nervous tissues and blood vessels. { 'strō·mə }

stromal endometriosis See interstitial endometriosis. { 'strō·məl ¦en·dō‚mē·trē'ō·səs }

stromal myosis See interstitial endometriosis. { 'strō·məl mī'ō·səs }

Stromateidae [INV ZOO] A family of perciform fishes in the suborder Stromateoidei containing the butterfishes. { 'strō·mə'tē·ə‚dē }

Stromateoidei [VERT ZOO] A suborder of fishes of the order Perciformes in which most species have teeth in pockets behind the pharyngeal bone. { 'strō·mə·tē'ȯid·ē‚ī }

stromatite [GEOL] Chorismite having flat or folded parallel layers of two or more textural elements. Also known as stromatolith. { 'strō·mə‚tīt }

stromatolite [GEOL] A structure in calcareous rocks consisting of concentrically laminated masses of calcium carbonate and calcium-magnesium carbonate which are believed to be of calcareous algal origin; these structures are irregular to columnar and hemispheroidal in shape, and range from 1 millimeter to many meters in thickness. Also known as callenia. { strə'mad·əl‚īt }

stromatolith [GEOL] **1.** A complex sill-like igneous intrusion interfingered with sedimentary strata. **2.** See stromatite. { strə'mad·əl‚ith }

Stromatoporoidea [PALEON] An extinct order of fossil colonial organisms thought to belong to the class Hydrozoa; the skeleton is a coenosteum. { strə‚mad·ə·pə‚rȯid·ē·ə }

stromatosis See interstitial endometrosis. { ‚strō·mə'tō·səs }

Strombacea [PALEON] An extinct superfamily of gastropod mollusks in the order Prosobranchia. { sträm'bās·ē·ə }

Strombidae [INV ZOO] A family of gastopod mollusks comprising tropical conchs. { 'sträm·bə‚dē }

strombolian [GEOL] A type of volcanic eruption characterized by fire fountains of lava from a central crater. { sträm'bō·lē·ən }

stromeyerite [MINERAL] CuAgS A metallic-gray orthorhombic mineral with a blue tarnish composed of silver copper sulfide occurring in compact masses. { 'strō‚mī·ə‚rīt }

Strömgren four-color index See uvby system. { 'strəm·grən ¦fȯr ¦kəl·ər 'in‚deks }

Strömgren radius [ASTRON] The radius of a sphere surrounding the central star of an emission nebula within which the hydrogen is nearly completely ionized. { 'strəm·grən ‚rād·ē·əs }

Strömgren sphere [ASTRON] An approximately spherical region of ionized hydrogen that surrounds a hot star. { 'strəm·grən ‚sfir }

strong acid [CHEM] An acid with a high degree of dissociation in solution, for example, mineral acids, such as hydrochloric acid, HCl, sulfuric acid, H_2SO_4, or nitric acid, HNO_3. { 'strȯŋ 'as·əd }

strong algorithm [COMMUN] A cryptographic algorithm for which the cost or time required to obtain the message or key is prohibitively great in practice even though the message may be obtainable in theory. { 'strȯŋ 'al·gə‚rith·əm }

strong base [CHEM] A base with a high degree of dissociation in solution, for example, sodium hydroxide, NaOH, potassium hydroxide, KOH. { 'strȯŋ 'bās }

strong breeze [METEOROL] In the Beaufort wind scale, a wind whose speed is from 22 to 27 knots (25 to 31 miles per hour or 41 to 50 kilometers per hour). { 'strȯŋ 'brēz }

strong causality condition [RELAT] The strong causality condition is said to hold at a point p in a space-time if every neighborhood of p contains a neighborhood of p which no timelike or null curve intersects more than once. { ¦strȯŋ kȯ'zal·əd·ē kən‚dish·ən }

strong fix [NAV] A fix determined from horizontal sextant angles between objects well placed. { 'strȯŋ 'fiks }

strong gale [METEOROL] In the Beaufort wind scale, a wind whose speed is from 41 to 47 knots (47 to 54 miles per hour or 76 to 87 kilometers per hour). { 'strȯŋ gāl }

strong interaction [PART PHYS] One of the fundamental interactions of elementary particles, primarily responsible for nuclear forces and other interactions among hadrons. { 'strȯŋ ‚in·tər'ak·shən }

strongly connected digraph [MATH] A directed graph in which there is a directed path from every vertex to every other vertex. { ¦strȯŋ·lē kə¦nek·təd 'dī‚graf }

strongly continuous semigroup [MATH] A semigroup of bounded linear operators on a Banach space B, together with a bijective mapping T from the positive real numbers onto the semigroup, such that $T(0)$ is the identity operator on B, $T(s + t) = T(s)T(t)$ for any two positive numbers s and t and, for each element x of B, $T(t)x$ is a continuous function of t. { 'strȯŋ·lē kən¦tin·yə·wəs 'sem·i‚grüp }

strongly damped collision See deep inelastic collision. { 'strȯŋ·lē ‚dampt kə'lizh·ən }

strongly future asymptotically predictable space-time [RELAT] A future asymptotically predictable space-time such that a neighborhood of the event horizon is also predictable. { 'strȯŋ·lē ¦fyü·chər ‚ā‚sim¦täd·ə·klē prə¦dik·tə·bəl 'spās ‚tīm }

strongly typed language [CONT SYS] A high-level programming language in which the type of each variable must be declared at the beginning of the program, and the language itself then enforces rules concerning the manipulation of variables according to their types. { 'strȯŋ·lē ¦tīpt 'laŋ·gwij }

strong point [ORD] A strongly fortified and heavily armed point in a defense system, usually supported by auxiliary armed positions. { 'strȯŋ ‚pȯint }

strong topology [MATH] The topology on a normed space obtained from the given norm; the basic open neighborhoods of a vector x are sets consisting of all those vectors y where the norm of $x − y$ is less than some number. { 'strȯŋ tə'päl·ə·jē }

strongyle [INV ZOO] A monaxonic spicule rounded at each end. { 'strän‚jīl }

Strongyloidea [INV ZOO] The hookworms, an order or superfamily of roundworms which, as adults, are endoparasites of most vertebrates, including humans. { ‚strän·jə'lȯid·ē·ə }

strongyloidiasis [MED] An infestation of humans with one of the roundworms of the genus *Strongyloides*. { ‚strän·jə‚lȯi'dī·ə·səs }

strongylote [INV ZOO] Rounded at one end, referring to sponge spicules. { 'strän·jə‚lōt }

strontia See strontium oxide. { 'strän·chə }

strontianite [MINERAL] $SrCO_3$ A pale-green, white, gray, or yellowish mineral of the aragonite group having orthorhombic symmetry and occurring in veins or as masses; hardness is 3.5 on Mohs scale, and specific gravity is 3.76. { 'strän·chē·ə‚nīt }

strontium [CHEM] A metallic element in group II, symbol Sr, with atomic number 38, atomic weight 87.62; flammable, soft, pale-yellow solid; soluble in alcohol and acids, decomposes in water; melts at 770°C, boils at 1380°C; chemistry is similar to that of calcium; used as electron-tube getter. { 'strän·tē·əm }

strontium-90 [NUC PHYS] A poisonous, radioactive isotope of strontium; 28-year half life with β radiation; derived from reactor-fuel fission products; used in thickness gages, medical treatment, phosphor activation, and atomic batteries. { 'strän·tē·əm 'nīn·tē }

strontium acetate [ORG CHEM] $Sr(C_2H_3O_2)_2 \cdot \frac{1}{2}H_2O$ White, water-soluble crystals, loses water at 150°C; used for catalysts, as a chemical intermediate, and in medicine. { 'strän·tē·əm 'as·ə‚tāt }

strontium bromide [INORG CHEM] $SrBr_2 \cdot 6H_2O$ A white, hygroscopic powder soluble in water and alcohol; loses water at 180°C, melts at 643°C; used in medicine and as an analytical reagent. { 'strän·tē·əm 'brō‚mīd }

strontium carbonate [INORG CHEM] $SrCO_3$ A white powder slightly soluble in water, decomposes at 1340°C; used to make TV-tube glass, strontium salts, and ceramic ferrites, and in pyrotechnics. { 'strän·tē·əm 'kär·bə‚nāt }

strontium chlorate [INORG CHEM] $Sr(ClO_3)_2$ Shock-sensitive, highly combustible, white, water-soluble crystals that decompose at 120°C; used in pyrotechnics and tracer bullets. { 'strän·tē·əm 'klȯr‚āt }

strontium chloride [INORG CHEM] $SrCl_2$ Water- and alcohol-soluble white crystals, melts at 872°C; used in medicine and pyrotechnics and to make strontium salts. { 'strän·tē·əm 'klȯr‚īd }

strontium chromate [INORG CHEM] $SrCrO_4$ A light yellow, rust- and corrosion-resistant pigment used in metal coatings and for pyrotechnics. { 'strän·tē·əm 'krō‚māt }

strontium dioxide See strontium peroxide. { 'strän·tē·əm dī'äk‚sīd }

strontium fluoride [INORG CHEM] SrF_2 A white powder, soluble in hydrochloric acid and hydrofluoric acid; used in medicine and for single crystals for lasers. { 'strän·tē·əm 'flür‚īd }

strontium hydrate See strontium hydroxide. { 'strän·tē·əm hī‚drāt }

strontium hydroxide [INORG CHEM] $Sr(OH)_2$ Colorless deliquescent crystals that absorb carbon dioxide from air, soluble in hot water and acids, melts at 375°C; used by the sugar industry, in lubricants and soaps, and as a plastic stabilizer. Also known as strontium hydrate. { 'strän·tē·əm hī'dräk‚sīd }

strontium iodide [INORG CHEM] SrI_2 Air-yellowing, white crystals that decompose in moist air, melts at 515°C; used in medicine and as a chemicals intermediate. { 'strän·tē·əm 'ī·ə‚dīd }

strontium monosulfide See strontium sulfide. { 'strän·tē·əm ¦män·ə'səl‚fīd }

strontium nitrate [INORG CHEM] $Sr(NO_3)_2$ A white, water-soluble powder melting at 570°C; used in pyrotechnics, signals and flares, medicine, and matches, and as a chemicals intermediate. { 'strän·tē·əm 'nī‚trāt }

strontium oxalate [INORG CHEM] $SrC_2O_4 \cdot H_2O$ A white powder that loses water at 150°C; used in pyrotechnics and tanning. { 'strän·tē·əm 'äk·sə‚lāt }

strontium oxide [INORG CHEM] SrO A grayish powder, melts at 2430°C, becomes the hydroxide in water; used in medicine, pyrotechnics, pigments, greases, soaps, and as a chemicals intermediate. Also known as strontia. { 'strän·tē·əm 'äk‚sīd }

strontium peroxide [INORG CHEM] SrO_2 A strongly oxidizing, fire-hazardous, white, alcohol-soluble powder that decomposes in hot water; used in medicine, bleaching, and fireworks. Also known as strontium dioxide. { 'strän·tē·əm pər'äk‚sīd }

strontium salicylate [ORG CHEM] $Sr(C_7H_5O_3)_2 \cdot 2H_2O$ White crystals or powder with a sweet saline taste; soluble in water and alcohol; used in medicine and manufacture of pharmaceuticals. { 'strän·tē·əm sə'lis·ə‚lāt }

strontium sulfate [INORG CHEM] $SrSO_4$ White crystals insoluble in alcohol, slightly soluble in water and concentrated acids, melts at 1605°C; used in paper manufacture, pyrotechnics, ceramics, and glass. { 'strän·tē·əm 'səl‚fāt }

strontium sulfide [INORG CHEM] SrS A gray powder with a hydrogen sulfide aroma in moist air, slightly soluble in water, soluble (with decomposition) in acids, melts above 2000°C; used in depilatories and luminous paints and as a chemicals intermediate. Also known as strontium monosulfide. { 'strän·tē·əm 'səl‚fīd }

strontium titanate [INORG CHEM] $SrTiO_3$ A solid material, insoluble in water and melting at 2060°C; used in electronics and electrical insulation. { 'strän·tē·əm 'tīt·ən‚āt }

strontium unit [NUCLEO] A unit of concentration of strontium-90 in a medium relative to the concentration of calcium, equal to 10^{-12} curie of strontium per gram of calcium. Abbreviated SU. Also known as sunshine unit (deprecated usage). { 'strän·tē·əm ‚yü·nət }

strophanthin [PHARM] A glycoside or mixture of glycosides extracted from the plant *Strophanthus kombe*; used as a cardioactive drug in the treatment of various heart ailments. { strə'fan·thən }

strophiole [BOT] A crestlike excrescence around the hilum in certain seeds. { 'strä·fē‚ōl }

strophoid [MATH] **1.** A curve derived from a given curve C and two points, called the pole and the fixed point, consisting of the locus of points on a rotating line L passing through the pole whose distance from the intersection of L and C is equal to the distance of this intersection from the fixed point.

2. The special case of the first definition in which *C* is a straight line and the fixed point lies on *C*. { 'strä,fȯid }

Strophomenida [PALEON] A large diverse order of articulate brachiopods which first appeared in Lower Ordovician times and became extinct in the Late Triassic. { ,strä·fə'men·əd·ə }

Strophomenidina [PALEON] A suborder of extinct, articulate brachiopods in the order Strophomenida characterized by a concavo-convex shell, the pseudodeltidium and socket plates disposed subparallel to the hinge. { ,strä·fə,men·ə'dī·nə }

Strouhal frequency [FL MECH] The frequency of vortex shedding from a structure in a uniform flow. { 'strü·əl ,frē·kwən·sē }

Strouhal number [MECH] A dimensionless number used in studying the vibrations of a body past which a fluid is flowing; it is equal to a characteristic dimension of the body times the frequency of vibrations divided by the fluid velocity relative to the body; for a taut wire perpendicular to the fluid flow, with the characteristic dimension taken as the diameter of the wire, it has a value between 0.185 and 0.2 Symbolized S_r. Also known as reduced frequency. { 'strü·əl ,nəm·bər }

Strowger system [COMMUN] An automatic telephone switching system that uses successive step-by-step selector switches actuated by current pulses produced by rotation of a telephone dial. Also known as step-by-step system. { 'strō·gər ,sis·təm }

struck capacity [MIN ENG] The volume of water a mine car, tram, hoppet, or wagon would hold if the conveyance were of watertight construction. { 'strək kə,pas·əd·ē }

struck joint [CIV ENG] A mortar joint in brickwork formed by pressing the trowel in at the lower edge, so that a recess is formed at the bottom of the joint; suitable only for interior work. { 'strək ,jȯint }

structural adhesive [MATER] An adhesive capable of bearing loads of considerable magnitude; a structural adhesive will not fail when a bonded joint prepared from the thickness of metal, or other material typical for that industry, is stressed to its yield point. { 'strək·chə·rəl ad'hē·siv }

structural analysis [ENG] The determination of stresses and strains in a given structure. [PETR] *See* structural petrology. { 'strək·chə·rəl ə'nal·ə·səs }

structural bench [GEOL] A bench typifying the resistant edge of a terrace that is being reduced by erosion. Also known as rock bench. { 'strək·chə·rəl 'bench }

structural bulkhead [NAV ARCH] A bulkhead designed to function as a strength member of the ship's structure. { 'strək·chə·rəl 'bəlk,hed }

structural clay tile [MATER] Hollow burned-clay masonry unit with parallel cells used as facing tile, load-bearing tile, partition tile, fireproofing tile, furring tile, floor tile, and header tile. { 'strək·chə·rəl 'klā 'tīl }

structural concrete [MATER] A special type of concrete that is capable of carrying a structural load or forming an integral part of a structure. { 'strək·chə·rəl ,kän'krēt }

structural connection [CIV ENG] A means of joining the individual members of a structure to form a complete assembly. { 'strək·chə·rəl kə'nek·shən }

structural contour map [GEOL] A map representation of a subsurface stratigraphic unit; depicts the configuration of a rock surface by means of elevation contour lines. { 'strək·chə·rəl 'kän,tùr ,map }

structural deflections [MECH] The deformations or movements of a structure and its flexural members from their original positions. { 'strək·chə·rəl di'flek·shənz }

structural drawings [GRAPHICS] The design and working drawings for structures such as buildings, bridges, dams, tanks, and highways. { 'strək·chə·rəl 'drȯ·iŋz }

structural drill [MECH ENG] A highly mobile diamond- or rotary-drill rig complete with hydraulically controlled derrick mounted on a truck, designed primarily for rapidly drilling holes to determine the structure in subsurface strata or for use as a shallow, slim-hole producer or seismograph drill. { 'strək·chə·rəl 'dril }

structural drilling [ENG] Drilling done specifically to obtain detailed information delineating the location of folds, domes, faults, and other subsurface structural features indiscernible by studying strata exposed at the surface. { 'strək·chə·rəl 'dril·iŋ }

structural engineering [CIV ENG] A branch of civil engineering dealing with the design of structures such as buildings, dams, and bridges. { 'strək·chə·rəl ,en·jə'nir·iŋ }

structural engineering system solver *See* STRESS. { 'strək·chə·rəl ,en·jə'nir·iŋ 'sis·təm ,säl·vər }

structural fabric *See* fabric. { 'strək·chə·rəl |fab·rik }

structural foam [MATER] A type of cellular plastic with a dense outer skin surrounding a foamed core. { 'strək·chə·rəl |fōm }

structural formula [CHEM] A system of notation used for organic compounds in which the exact structure, if it is known, is given in schematic representation. { 'strək·chə·rəl 'fȯr·myə·lə }

structural frame [BUILD] The entire set of members of a building or structure required to transmit loads to the ground. { 'strək·chə·rəl 'frām }

structural gene *See* cistron. { 'strək·chə·rəl 'gēn }

structural geology [GEOL] A branch of geology concerned with the form, arrangement, and internal structure of the rocks. { 'strək·chə·rəl jē'äl·ə·jē }

structural glass [MATER] A vitreous material cast as cubes, rectangular blocks, and rectangular plates to be used as a decorative covering for masonry walls. { 'strək·chə·rəl 'glas }

structural high [GEOL] Any of various structural features such as a crest, culmination, anticline, or dome. { 'strək·chə·rəl 'hī }

structural information [COMPUT SCI] Information specifying the number of independently variable features or degrees of freedom of a pattern. { 'strək·chə·rəl ,in·fər'mā·shən }

structural isomers *See* constitutional isomers. { 'strək·chə·rəl 'ī·sə·mərz }

structural low [GEOL] Any of various structural features such as a basin, a syncline, a saddle, or a sag. { 'strək·chə·rəl 'lō }

structural petrology [PETR] The study of the internal structure of a rock to determine its deformational history. Also known as fabric analysis; microtectonics; petrofabric analysis; petrofabrics; petrogeometry; petromorphology; structural analysis. { 'strək·chə·rəl pi'träl·ə·jē }

structural riveting [ENG] Riveting structural members by using punched holes. { 'strək·chə·rəl 'riv·əd·iŋ }

structural shape [MET] A piece of metal of a standard design used in construction. { 'strək·chə·rəl 'shāp }

structural stability [MATH] Property of a differentiable flow on a compact manifold whose orbit structure is insensitive to small perturbations in the equations governing the flow or in the vector field generating the flow. { 'strək·chər·əl stə'bil·əd·ē }

structural steel [MET] Steel used in engineering structures, usually manufactured by either open-hearth or the electricfurnace process. { 'strək·chə·rəl |stēl }

structural terrace [GEOL] A terracelike landform developed where generally steeply inclined and otherwise uniformly dipping strata locally flatten. { 'strək·chə·rəl 'ter·əs }

structural tile [MATER] A hollow clay product which may be load-bearing or non-load-bearing; used for facing, flooring, or partitions. { 'strək·chə·rəl |tīl }

structural trap [GEOL] Containment in a reservoir bed of oil or gas due to flexure or fracture of the bed. { 'strək·chə·rəl 'trap }

structural valley [GEOL] A valley whose form and origin is attributable to the underlying geologic structure. { 'strək·chə·rəl 'val·ē }

structural wall *See* bearing wall. { 'strək·chə·rəl 'wȯl }

structural weight *See* construction weight. { 'strək·chə·rəl 'wāt }

structure [AERO ENG] The construction or makeup of an airplane, spacecraft, or missile, including that of the fuselage, wings, empennage, nacelles, and landing gear, but not that of the power plant, furnishings, or equipment. [CIV ENG] Something, as a bridge or a building, that is built or constructed and designed to sustain a load. [COMPUT SCI] For a dataprocessing system, the nature of the chain of command, the origin and type of data collected, the form and destination of results, and the procedures used to control operations. [GEOL] **1.** An assemblage of rocks upon which erosive agents have been or are acting. **2.** The sum total of the structural features of an area. [MINERAL] The form taken by a mineral, such as tabular or fibrous. [PETR] A macroscopic feature of a rock mass or rock unit, best seen in an outcrop. [SCI TECH] The arrangement and interrelation of the parts of an object. { 'strək·chər }

structure amplitude [SOLID STATE] The absolute value of a structure factor. { 'strək·chər ,am·plə,tüd }

structure cell *See* unit cell. { 'strək·chər ,sel }

structure constants [MATH] A set of numbers that serve as coefficients in expressing the commutators of the elements of a Lie algebra. { 'strək·chər ,kän·stəns }

structure contour [GEOL] A contour that portrays a structural surface, such as a fault. Also known as subsurface contour. { 'strək·chər ,kän,tur }

structure-contour map [GEOL] A map that uses structure contour lines to portray subsurface configuration. Also known as structure map. { 'strək·chər ,kän,tur ,map }

structured analysis [SYS ENG] A method of breaking a large problem or process into smaller components to aid in understanding, and then identifying the components and their interrelationships and reassembling them. { 'strək·chərd ə'nal·ə·səs }

structured data type [COMPUT SCI] The manner in which a collection of data items, which may have the same or different scalar data types, is represented in a computer program. { 'strək·chərd 'dad·ə ,tīp }

structured food *See* food analog. { 'strək·chərd 'füd }

structured grid [MATH] In the discretization of partial differential equations, an organized set of points formed by the intersections of the lines of a boundary-conforming curvilinear coordinate system, at which the equations are expressed in discrete form. { 'strək·chərd 'grid }

structured light [OPTICS] Light that is projected in a particular geometrical pattern that is used to aid in computer vision. { 'strək·chərd 'līt }

structured programming [COMPUT SCI] The use of program design and documentation techniques that impose a uniform structure on all computer programs. { 'strək·chərd 'prō,gram·iŋ }

Structured Query Language [COMPUT SCI] The standard language for accessing relational databases. Abbreviated SQL. { ,strək·chərd 'kwir·ē ,laŋ·gwij }

structured variable *See* record variable. { 'strək·chərd 'ver·ē·ə·bəl }

structured walkthrough [COMPUT SCI] A formal method of debugging a computer system or program, involving a systematic review to search for errors and inefficiencies. { 'strək·chərd 'wok,thrü }

structure factor [SOLID STATE] A factor which determines the amplitude of the beam reflected from a given atomic plane in the diffraction of an x-ray beam by a crystal, and is equal to the sum of the atomic scattering factors of the atoms in a unit cell, each multiplied by an appropriate phase factor. { 'strək·chər ,fak·tər }

structure map *See* structure-contour map. { 'strək·chər ,map }

structure number [DES ENG] A number, generally from 0 to 15, indicating the spacing of abrasive grains in a grinding wheel relative to their grit size. { 'strək·chər ,nəm·bər }

structure resonance [SPECT] An extremely narrow resonance exhibited by a small aerosol particle at a natural electromagnetic frequency at which the dielectric sphere oscillates, observed in the particle's scattered light excitation spectrum. { 'strək·chər ,rez·ən·əns }

structure resonance modulation spectroscopy [SPECT] The infrared modulation of visible scattered light near a structure resonance to determine the absorption spectrum of an aerosol particle. Abbreviated SRMS. { 'strək·chər ,rez·ən·əns ,mäj·ə¦lā·shən spek'träs·kə·pē }

structure section [GEOL] A vertical section showing the observed or inferred geologic structure on a vertical surface or plane. { 'strək·chər ,sek·shən }

structure-sensitive property [SOLID STATE] A property of a substance that depends on impurities and the imperfections of the crystal structure. { 'strək·chər ¦sen·sə·tiv 'präp·ərd·ē }

structure type [CRYSTAL] The structural arrangement of a crystal, regardless of the atomic elements present; it corresponds to the crystal's space group. { 'strək·chər ,tīp }

strut [AERO ENG] A bar supporting the wing or landing gear of an airplane. [CIV ENG] A long structural member of timber or metal, or a bar designed to resist pressure in the direction of its length. [ENG] **1.** A brace or supporting piece. **2.** A diagonal brace between two legs of a drill tripod or derrick. [MIN ENG] A vertical-compression member in a structure or in an underground timber set. [NAV ARCH] A bracket outside the hull of a ship, supporting the propeller shaft. Also known as propeller strut. { strət }

Struthionidae [VERT ZOO] The single family of the avian order Struthioniformes. { ,strü·thē'än·ə,dē }

Struthioniformes [VERT ZOO] A monofamilial order of ratite birds containing the single living species of ostrich (*Struthio camelus*). { ,strü·thē,än·ə'fór,mēz }

struvite [MINERAL] $Mg(NH_4)PO_4 \cdot 6H_2O$ A colorless to yellow or pale-brown mineral consisting of a hydrous ammonium magnesium phosphate, and occurring in orthorhombic crystals; hardness is 2 on Mohs scale, and specific gravity is 1.7. { 'strü,vīt }

strychnine [ORG CHEM] $C_{21}H_{22}O_2N_2$ An alkaloid obtained primarily from the plant nux vomica, formerly used for therapeutic stimulation of the central nervous system. { 'strik,nīn }

strychninization [MED] The condition resulting from large doses of strychnine. { ,strik·nə·nə'zā·shən }

Strychnos [BOT] A genus of tropical trees and shrubs of the order Loganiaceae. { 'strik,nōs }

Stuart factor [BIOCHEM] A procoagulant in normal plasma but deficient in the blood of patients with a hereditary bleeding disorder; may be closely related to prothrombin since both are formed in the liver by action of vitamin K. Also known as factor X; Stuart-Power factor. { 'stü·ərt ,fak·tər }

Stuart-Power factor *See* Stuart factor. { 'stü·ərt 'pau·ər ,fak·tər }

Stuart windmill *See* Fales-Stuart windmill. { 'stü·ərt 'win,mil }

stub [CIV ENG] A projection on a sewer pipe that provides an opening to accept a connection to another pipe or house sewer. [COMPUT SCI] **1.** The left-hand portion of a decision table, consisting of a single column, and comprising the condition stub and the action stub. **2.** A program module that is only partly completed, to the extent needed to fulfill the requirements of other modules in the computer system. [ELECTROMAG] **1.** A short section of transmission line, open or shorted at the far end, connected in parallel with a transmission line to match the impedance of the line to that of an antenna or transmitter. **2.** A solid projection one-quarter-wavelength long, used as an insulating support in a waveguide or cavity. { stəb }

stub angle [ELECTROMAG] Right-angle elbow for a coaxial radio-frequency transmission line which has the inner conductor supported by a quarter-wave stub. { 'stəb ,aŋ·gəl }

stub axle [MECH ENG] An axle carrying only one wheel. { 'stəb ¦ak·səl }

stubborn disease [PL PATH] A virus disease of citrus trees characterized by short internodes resulting in stiff brushy growth and chlorotic leaves. { 'stəb·ərn di'zēz }

stub cable [ELEC] Short branch off a principal cable; the end is often sealed until it is used at a later date; pairs in the stub are referred to as stubbed-out pairs. { 'stəb ¦kā·bəl }

stub entry [MIN ENG] A short, narrow entry turned from another entry and driven into the solid coal, but not connected with other mine workings. { 'stəb ,en·trē }

stub matching [ELECTROMAG] Use of a stub to match a transmission line to an antenna or load; matching depends on the spacing between the two wires of the stub, the position of the shorting bar, and the point at which the transmission line is connected to the stub. { 'stəb ,mach·iŋ }

stub mortise [ENG] A mortise which passes through only part of a timber. { 'stəb ,mord·əs }

Stubs gage [DES ENG] A number system for denoting the thickness of steel wire and drills. { 'stəbz ,gāj }

stub-supported coaxial [ELECTROMAG] Coaxial whose inner conductor is supported by means of short-circuited coaxial stubs. { 'stəb sə¦pórd·əd kō'ak·sē·əl }

stub-supported line [ELECTROMAG] A transmission line that is supported by short-circuited quarter-wave sections of coaxial line; a stub exactly a quarter-wavelength long acts as an insulator because it has infinite reactance. { 'stəb sə¦pórd·əd 'līn }

STRUT

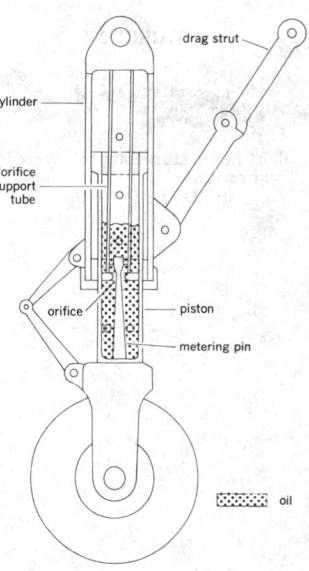

Diagram of oleopneumatic shock strut. As airplane sinks toward ground, piston forces oil through orifice, causing force which changes path of airplane.

STURGEON

Short-nosed sturgeon (*Acipenser brevirostrus*).

stub switch [ENG] A pair of short switch rails, held only at or near one end and free to move at the other end; used in mining and to some extent on narrow-gage industrial tramways. { 'stəb ,swich }

stub tenon [ENG] A tenon that fits into a stub mortise. { 'stəb ¦ten·ən }

stub tube [MECH ENG] A short tube welded to a boiler or pressure vessel to provide for the attachment of additional parts. { 'stəb ,tüb }

stub tuner [ELECTROMAG] Stub which is terminated by movable short-circuiting means and used for matching impedance in the line to which it is joined as a branch. { 'stəb ,tün·ər }

stucco [MATER] A smooth plasterlike material applied to the outside wall or other exterior surface of a building or structure. { 'stək·ō }

stud [BUILD] One of the vertical members in the walls of a framed building to which wallboards, lathing, or paneling is nailed or fastened. [DES ENG] **1.** A rivet, boss, or nail with a large, ornamental head. **2.** A short rod or bolt threaded at both ends without a head. { stəd }

stud driver [MECH ENG] A device, such as an impact wrench, for driving a hardened steel nail (stud) into concrete or other hard materials. { 'stəd ,drī·vər }

Student's distribution [STAT] The probability distribution used to test the hypothesis that a random sample of n observations comes from a normal population with a given mean. { 'stüd·əns ,dis·trə'byü·shən }

Student's t-statistic [STAT] A one-sample test statistic computed by $T = \sqrt{n}(\bar{X} - \mu_H)/S$, where $\bar{X}$ is the mean of a collection of n observations, S is the square root of the mean square deviation, and μ_H is the hypothesized mean. { 'stüd·əns 'tē stə,tis·tik }

Student's t-test [STAT] A test in a one-sample problem which uses Student's t-statistic. { 'stüd·əns 'tē ,test }

studio [COMMUN] A room in which television or radio programs are produced. { 'stüd·ē·ō }

stud link chain [NAV ARCH] Chain in which each link has a stud at its midlength perpendicular to the major axis to maintain the shape of the link. { 'stəb ¦liŋk ,chān }

stud wall [BUILD] A wall formed with timbers; studs are usually spaced 12–16 inches (30–41 centimeters) on center. { 'stəb ,wòl }

stud welding [MET] Arc-welding using the heat of an electric arc produced between a metal stud and another part, and then bringing the parts together under pressure. { 'stəd ,weld·iŋ }

stuffed mineral [MINERAL] A mineral having extra ions of a foreign element within its larger interstices. { 'stəft 'min·rəl }

stuffing [ENG] A method of sealing the mechanical joint between two metal surfaces; packing (stuffing) material is inserted within the seal area container (the stuffing or packing box), and compressed to a liquid-proof seal by a threaded packing ring follower. Also known as packing. { 'stəf·iŋ }

stuffing box [ENG] A packed, pressure-tight joint for a rod that moves through a hole, to reduce or eliminate fluid leakage. { 'stəf·iŋ ,bäks }

stuffing nut [ENG] A nut for adjusting a stuffing box. { 'stəf·iŋ ,nət }

stull [MIN ENG] A platform laid on timbers, braced across a working from side to side, to support workers or to carry ore or waste. { stəl }

stull piece [MIN ENG] **1.** A piece of timber placed slanting over the back of a level to prevent rock falling into the level from the stopes above. **2.** Timbers bracing the platform of a stull. { 'stəl ,pēs }

STYLOLITE

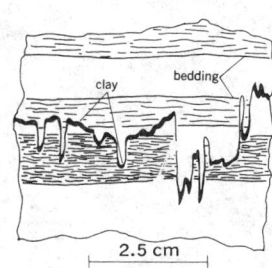

Stylolite in limestone.

stull stoping [MIN ENG] Stull timbers placed between the foot and hanging walls, which constitute the only artificial support provided during the excavation of a stope. { 'stəl ,stōp·iŋ }

stump [MIN ENG] A small pillar of coal left between the gangway or airway and the breasts to protect these passages; any small pillar. { stəmp }

stunt [PL PATH] Any of several plant diseases marked by reduction in size of the plant. { stənt }

stunt box [ELEC] A device to control the nonprinting functions of a teletypewriter terminal. { 'stənt ,bäks }

stupp [MIN ENG] A black residue from distilled mercury ore, consisting of soot, hydrocarbons, mercury and mercury compounds, and ore dust. { stəp }

sturgeon [VERT ZOO] Any of 10 species of large bottom-living fish which comprise the family Acipenseridae; the body has five rows of bony plates, and the snout is elongate with four barbels on its lower surface. { 'stər·jən }

Sturges rule [STAT] A rule for determining the desirable number of groups into which a distribution of observations should be classified; the number of groups or classes is $1 + 3.3 \log n$, where n is the number of observations. { 'stər·jəs ,rül }

Sturm-Liouville problem [MATH] The general problem of solving a given linear differential equation of order $2n$ together with $2n$-boundary conditions. Also known as eigenvalue problem. { 'stərm lyü'vil ,präb·ləm }

Sturm-Liouville system [MATH] A given differential equation together with its boundary conditions having Sturm-Liouville problem form. { 'stərm lyü'vil ,sis·təm }

Sturm separation theorem [MATH] The theorem that if u and v are real, linearly independent solutions of a second-order linear homogeneous differential equation in which the coefficient of the second derivative is unity and the other two coefficients are continuous functions, then there is exactly one zero of u between any two zeros of v. { ¦stərm ,sep·ə'rā·shən ,thir·əm }

Sturm sequence [MATH] For a polynomial $p(x)$, this is the sequence of functions $f_0(x), f_1(x), \ldots$, where $f_0(x) = p(x)$, $f_1(x) = p'(x)$, and $f_n(x)$ is the negative remainder that occurs by finding the greatest common divisor of $f_{-2}(x)$ and $f_{-1}(x)$ via the euclidean algorithm. { 'stərm ,sēkwəns }

Sturm's theorem [MATH] This gives a method to determine the number of real roots of a polynomial $p(x)$ which lie between two given values of x; the Sturm sequence of $p(x)$ provides the necessary information. { 'stərmz ,thir·əm }

sturtite [MINERAL] A black mineral composed of hydrous silicate of iron, manganese, calcium, and magnesium; occurs in compact masses. { 'stərd,īt }

stutter [COMMUN] Series of undesired black and white lines sometimes produced when a facsimile signal undergoes a sharp amplitude change. [MED] A speech disorder marked by repetition of words, syllables, or sounds, or by hesitations in manner by the speaker. { 'stəd·ər }

Stuve chart [METEOROL] A thermodynamic diagram with atmospheric temperature as the x axis and atmospheric pressure to the power 0.286 as the y ordinate, increasing downward; named after G. Stuve. Also known as adiabatic chart; pseudo-adiabatic chart. { 'stüv·ə ,chärt }

S twist [TEXT] A left-handed yarn twist in which the spirals resemble the letter S. { 'es ,twist }

sty See hordeolum. { stī }

Styginae [INV ZOO] A subfamily of butterflies in the family Lycaenidae in which the prothoracic legs in the male are nonfunctional. { 'stij·ə·nē }

Stygocaridacea [INV ZOO] An order of crustaceans in the superorder Syncarida characterized by having a furca. { ,stig·ə,kar·ə'dās·ē·ə }

Stylasterina [INV ZOO] An order of the class Hydrozoa, including several brightly colored branching or encrusting corallike cnidarians of warm seas. { stə,las·tə'rī·nə }

style [BOT] The portion of a pistil connecting the stigma and ovary. [ENG] See gnomon. [ZOO] A slender elongated process on an animal. { stīl }

stylet [GRAPHICS] A slender, pointed marking tool, as one used in graving. [INV ZOO] A slender, rigid, elongated appendage. [MED] **1.** A slender probe used for surgery. **2.** A thin wire inserted in a catheter to provide support or in a hollow needle to clear the passage. { 'stī·lət }

styloglossus [ANAT] A muscle arising from the styloid process of the temporal bone, and inserted into the tongue. { ,stī·lō'gläs·əs }

stylohyoid [ANAT] Pertaining to the styloid process of the temporal bone and the hyoid bone. { ¦stī·lō¦hī,oid }

styloid [ZOO] Resembling a style. { 'stī,loid }

stylolite [GEOL] An irregular surface, generally parallel to a bedding plane, in which small toothlike projections on one side of the surface fit into cavities of complementary shape on the other surface; interpreted to result diagenetically by pressure solution. { 'stī·lə,līt }

stylomastoid [ANAT] Relating to the styloid and the mastoid processes of the temporal bone. { ¦stī·lō¦ma,stóid }

Stylommatophora [INV ZOO] A large order of the molluscan subclass Pulmonata characterized by having two pairs of retractile tentacles with eyes located on the tips of the large tentacles. { ˌsti̇ˌlam·əˈtäf·ə·rə }

stylopodium [BOT] A conical or disk-shaped enlargement at the base of the style in plants of the family Umbelliferae. { ˌsti̇·ləˈpōd·ē·əm }

stylotypite *See* tetrahedrite. { ˈsti̇·ləˌti̇ˌpi̇t }

Styloviridae *See* Siphoviridae. { ˌsti̇l·əˈvir·ə‚di̇ }

stylus [COMPUT SCI] The pointed device used to draw images on a graphics tablet. [ENG ACOUS] The portion of a phonograph pickup that follows the modulations of a record groove and transmits the resulting mechanical motions to the transducer element of the pickup for conversion to corresponding audio-frequency signals. Also known as needle; phonograph needle; reproducing stylus. [GRAPHICS] A rather blunt metal point sometimes used in painting to make lightly ruled lines. { ˈsti̇·ləs }

stylus printing *See* matrix printing. { ˈsti̇·ləs ‚print·iŋ }

S-type asteroid [ASTRON] A type of asteroid whose surface is reddish and of moderate albedo, containing pyroxene and olivine silicates, probably mixed with metallic iron, similar to stony iron meteorites. { ˈes ˌti̇p ˈas·təˌrȯid }

S-type magma [GEOL] Magma formed from sedimentary source material. { ˈes ˌti̇p ˈmag·mə }

S-type symbiotic star [ASTRON] A member of a class of symbiotic stars that emit infrared radiation typical of red-giant atmospheres; they are relatively short and have binary periods shorter than 20 years. { ˈes ˌti̇p ˌsim·bēˌäd·ik ˈstär }

styphnic acid [ORG CHEM] $C_6H(OH)_2(HO_2)_3$ An explosive, yellow, crystalline compound, melting at 179–180°C, slightly soluble in water; used in explosives as a priming agent. { ˈstif·nik ˈas·əd }

Stypocapitellidae [INV ZOO] A family of polychaete annelids belonging to the Sedentaria and consisting of a monotypic genus found in western Germany. { ˌsti̇·pōˌkap·əˈtel·əˌdē }

styramate [PHARM] $C_9H_{11}NO_3$ A compound that crystallizes from chloroform solution, and melts at 111–112°C; used in medicine as a muscle relaxant. { ˈsti̇·rəˌmāt }

styrene [ORG CHEM] $C_6H_5CH:CH_2$ A colorless, toxic liquid with a strong aroma; insoluble in water, soluble in alcohol and ether; polymerizes rapidly, can become explosive; boils at 145°C; used to make polymers and copolymers, polystyrene plastics, and rubbers. Also known as phenylethylene; styrene monomer; vinylbenzene. { ˈsti̇·rēn }

styrene-acrylonitrile resin [ORG CHEM] A thermoplastic copolymer of styrene and acrylonitrile with good stiffness and resistance to scratching, chemicals, and stress. Also known as SAN. { ˈsti̇ˌrēn ˌak·rəˈlän·ə·trəl ˈrez·ən }

styrene-butadiene rubber [MATER] The most common type of synthetic rubber, made by the copolymerization of styrene and butadiene monomers; used in tires, footwear, adhesives, and sealants. Also known as SBR. { ˈsti̇ˌrēn ‚byüd·əˈdi̇ˌēn ˈrəb·ər }

styrene monomer *See* styrene. { ˈsti̇ˌrēn ˈmän·ə·mər }

styrene oxide [ORG CHEM] C_8H_8O A moderately toxic, combustible, colorless or straw-colored liquid miscible in acetone, ether, and benzene, and melts at 195°C; used as a chemical intermediate. { ˈsti̇ˌrēn ˈäkˌsi̇d }

styrene plastic [ORG CHEM] A plastic made by the polymerization of styrene or the copolymerization of styrene with other unsaturated compounds. { ˈsti̇ˌrēn ˈplas·tik }

styrene-rubber plastic [MATER] A plastic-rubber mixture consisting of at least 50% of a styrene plastic combined with rubber and various compounding ingredients. { ˈsti̇ˌrēn ‚rəb·ər ˈplas·tik }

styroflex cable [COMMUN] A radio-frequency cable whose protective covering is a special type of styrene tape. { ˈsti̇·rəˌfleks ‚kā·bəl }

SU *See* strontium unit.

sub [PETRO ENG] A short, threaded section of pipe used to adapt portions of the drilling string that otherwise cannot be screwed together because of variations in thread or design. { ˈsəb }

subacute bacterial endocarditis *See* bacterial endocarditis. { ‚səb·əˈkyüt bakˈtir·ē·əl ‚en·dōˌkärˈdi̇d·əs }

subadditive function [MATH] A function f such that $f(x + y)$ is less than or equal to $f(x) + f(y)$ for all x and y in its domain. { ‚səbˈad·əd·iv ‚fəŋk·shən }

subadditive set function [MATH] A set function with the property that the union of any finite or countable collection of sets in the range of the function is also in this range, and the value of the function at this union is equal to or less than the sum of its values at each set of the collection. { ‚səbˌad·əd·iv ‚set ‚fəŋk·shən }

subaerial [GEOL] Pertaining to conditions and processes occurring beneath the atmosphere or in the open air, that is, on or adjacent to the land surface. { ‚səbˈer·ē·əl }

subage [GEOL] A subdivision of a geologic age. { ˈsəbˌāj }

subalgebra [MATH] **1.** A subset of an algebra which itself forms an algebra relative to the same operations. **2.** A subalgebra (of sets) is any algebra (of sets) contained in some given algebra. { ‚səbˈal·jə·brə }

subalkaline [GEOCHEM] Pertaining to a soil in which the pH is 8.0 to 8.5, usually in a limestone or salt-marsh region. { ‚səbˈal·kəˌli̇n }

subalphabet [COMPUT SCI] A subset of an alphabet. { ‚səbˈal·fə‚bet }

subalpine *See* alpestrine. { ‚səbˈalˌpi̇n }

subangular [SCI TECH] Somewhat angular but free from sharp edges and corners. { ‚səbˈaŋ·gyə·lər }

Subantarctic Intermediate Water [OCEANOGR] A layer of water above the deep-water layer in the South Atlantic. { ‚səbˌantˈärd·ik ‚in·tərˌmēd·ē·ət ˈwȯd·ər }

subaperture [ENG] Any subset of an array of transmitters of acoustic or electromagnetic radiation. { ‚səbˈap·ə·chər }

subaqueous [HYD] Pertaining to conditions and processes occurring in, under, or beneath the surface of water, especially fresh water. { ‚səbˈā·kwē·əs }

subaqueous dune [GEOL] A dune resulting from entrainment of grains by the flow of moving water. { ‚səbˈā·kwē·əs ˈdün }

subaqueous mining [MIN ENG] Surface mining in which the mined material is removed from the bed of a natural body of water. { ‚səbˈā·kwē·əs ˈmi̇n·iŋ }

subarachnoid hemorrhage [MED] Bleeding between the pia mater and the arachnoid of the brain. { ‚səb·əˈrakˌnȯid ˈhem·rij }

subarachnoid space [ANAT] The space between the pia mater and the arachnoid of the brain. { ‚səb·əˈrakˌnȯid ˈspās }

subarctic [GEOGR] Pertaining to regions adjacent to the Arctic Circle or having characteristics somewhat similar to those of these regions. { ‚səbˈärd·ik }

subarctic climate *See* taiga climate. { ‚səbˈärd·ik ˈkli̇·mət }

subarid [CLIMATOL] Pertaining to regions that are moderately or slightly arid. { ‚səbˈar·əd }

subarkose [GEOL] Sandstone that is intermediate in composition between arkose and pure quartz sandstone; it contains less feldspar than arkose. { ‚səbˈärˌkōs }

subartesian well [HYD] A well that requires artificial pumping to raise water to the surface because confining pressure forces the water only part of the distance up the well shaft. { ‚səbˌärˈtē·zhən ˈwel }

subassembly [ELECTR] Two or more components combined into a unit for convenience in assembling or servicing equipment; an intermediate-frequency strip for a receiver is an example. [ENG] A structural unit, which, though manufactured separately, was designed for incorporation with other parts in the final assembly of a finished product. { ‚səb·əˈsem·blē }

subastral point *See* substellar point. { ‚səbˈas·trəl ˈpȯint }

subatmospheric heating system [MECH ENG] A system which regulates steam flow into the main throttle valve under automatic thermostatic control and maintains a fixed vacuum differential between supply and return by means of a differential controller and a vacuum pump. { ‚səbˌat·məˈsfir·ik ˈhēd·iŋ ˌsis·təm }

subatomic particle [PHYS] A particle which is smaller than an atom, namely, an elementary particle or an atomic nucleus. { ‚səb·əˈtäm·ik ˈpärd·ə·kəl }

subbase for a topology [MATH] A family S of subsets of a topological space X where by taking all finite intersections of sets from S and all unions of such intersections the entire topology of open sets of X is obtained. { ˈsəbˌbās fȯr ə təˈpäl·ə·ē }

subbituminous coal [GEOL] Black coal intermediate in rank between lignite and bituminous coal; has more carbon and less moisture than lignite. { ‚səb·bəˈtü·mə·nəs ˈkōl }

subboreal [ECOL] A biogeographic zone whose climatic condition approaches that of the boreal. { ¦səb'bȯr·ē·əl }

subbottom depth recorder [ENG] A compact seismic instrument which can provide continuous soundings of strata beneath the ocean bottom utilizing the low-frequency output of an intense electrical spark discharge source in water. { ¦səb'bäd·əm 'depth ri,kȯrd·ər }

subbottom reflection [GEOPHYS] The return of sound energy from a discontinuity in material below the surface of the sea bottom. { ¦səb'bäd·əm ri'flek·shən }

subboundary structure [MET] A network of low-angle grain boundaries of less than one degree within the main crystals of a metal. { ¦səb'baun·drē 'strək·chər }

subcaliber [ORD] Smaller than the standard caliber for a gun, generally used in practice firing. { ¦səb'kal·ə·bər }

subcaliber ammunition [ORD] Ammunition used with a gun or launching tube, usually in practice firing, of a caliber smaller than the standard. { ¦səb'kal·ə·bər ,am·yə'nish·ən }

subcaliber equipment [ORD] Any item of equipment, such as small guns, adapters, tubes, and accessories, used for firing subcaliber ammunition in practice drills with larger guns. { ¦səb'kal·ə·bər i'kwip·mənt }

subcaliber firing [ORD] Practice firing of subcaliber ammunition, in connection with drills in elevating, traversing, or aiming guns of larger caliber. { ¦səb'kal·ə·bər 'fīr·iŋ }

subcaliber gun [ORD] A gun mounted on the outside and above the tube of a larger gun; it is used in practice firing of subcaliber ammunition, in connection with aiming drills with the larger gun. { ¦səb'kal·ə·bər 'gən }

subcaliber mount [ORD] Special mount in or on the tube of a gun, upon which a gun of smaller caliber can be attached for practice firing. { ¦səb'kal·ə·bər 'maunt }

subcaliber rocket [ORD] A rocket designed especially to be fired from launching tubes of larger caliber than the rocket itself. { ¦səb'kal·ə·bər 'räk·ət }

subcapillary interstice [GEOL] An interstice in which the molecular attraction of its walls extends across the entire opening; it is smaller than a capillary interstice. { ¦səb'kap·ə,ler·ē in'tər·stəs }

subcardinal vein [VERT ZOO] Either of a pair of longitudinal veins of the mammalian embryo or the adult of some lower vertebrates which partly replace the postcardinals in the abdominal region, ventromedial to the mesonephros. { ¦səb'kärd·nəl 'vān }

subcarrier [COMMUN] 1. A carrier that is applied as a modulating wave to modulate another carrier. 2. *See* chrominance subcarrier. { ¦səb'kar·ē·ər }

subcarrier oscillator [ELECTR] 1. The crystal oscillator that operates at the chrominance subcarrier or burst frequency of 3.579545 megahertz in a color television receiver; this oscillator, synchronized in frequency and phase with the transmitter master oscillator, furnishes the continuous subcarrier frequency required for demodulators in the receiver. 2. An oscillator used in a telemetering system to translate variations in an electrical quantity into variations of a frequency-modulated signal at a subcarrier frequency. { ¦səb'kar·ē·ər 'äs·ə,lād·ər }

subcerebral plane [ANTHRO] The plane passing through a line traversing the lower angles of the parietal bones and the juncture of the superciliary ridge and the cheek bone. { ¦səb·sə're·brəl 'plān }

subchannel [COMPUT SCI] The portion of an input/output channel associated with a specific input/output operation. { ¦səb'chan·əl }

subclavian artery [ANAT] The proximal part of the principal artery in the arm or forelimb. { səb¦klā·vē·ən 'ärd·ə·rē }

subclavian vein [ANAT] The proximal part of the principal vein in the arm or forelimb. { səb¦klā·vē·ən 'vān }

subclavius [ANAT] A small muscle attached to the clavicle and the first rib. { ,səb'klā·vē·əs }

subclimax [ECOL] A community immediately preceding a climax in an ecological succession. { ¦səb'klī,maks }

subcluster [ASTRON] One of the several distinct clumps of galaxies that often compose an irregular cluster. { ¦səb,kləs·tər }

subclutter visibility [ELECTR] A measure of the effectiveness of moving-target indicator radar, equal to the ratio of the signal from a fixed target that can be canceled to the signal from a just visible moving target. { ¦səb¦kləd·ər ,viz·ə'bil·əd·ē }

subcollateral [ANAT] Ventrad of the collateral sulcus of the brain. { ¦səb·kə'lad·ə·rəl }

subcommutation [COMMUN] In telemetry, commutation of additional channels with output applied to individual channels of the primary commutator. { ¦səb,käm·yə'tā·shən }

subcomponent [DES ENG] A part of a component having characteristics of the component. { ¦səb·kəm,pō·nənt }

subcompound [CHEM] A compound, generally in the vapor phase, in which an element exhibits a valency lower than that exhibited in its ordinary compounds. { ¦səb'käm,paund }

subconchoidal [GEOL] Pertaining to a fracture that is partly or vaguely conchoidal in shape. { ¦səb·kən'kȯid·əl }

subconscious [PSYCH] Pertaining to mental activity beyond the level of consciousness, including the preconscious and the unconscious. { ¦səb'kän·shəs }

subconsequent stream *See* secondary consequent stream. { ¦səb'kän·sə·kwənt 'strēm }

subcontinent [GEOGR] 1. A landmass such as Greenland that is large but not as large as the generally recognized continents. 2. A large subdivision of a continent (for example, the Indian subcontinent) distinguished geologically or geomorphically from the rest of the continent. { ¦səb'känt·ən·ənt }

subcontract [ENG] A contract made with a third party by one who has contracted to perform work or service for whole or part performance of that work or service. { ¦səb'kän,trakt }

subcontractor [ENG] A manufacturer or organization that receives a contract from a prime contractor for a portion of the work on a project. { ¦səb'kän,trak·tər }

subcritical [NUCLEO] Having an effective multiplication constant less than one, so that a self-supporting chain reaction cannot be maintained in a nuclear reactor. { ¦səb'krid·ə·kəl }

subcritical assembly *See* subcritical reactor. { ¦səb'krid·ə·kəl ə'sem·blē }

subcritical flow *See* subsonic flow. { ¦səb'krid·ə·kəl 'flō }

subcritical mass [NUCLEO] A piece of fissionable material having an effective multiplication constant of less than one, so that it does not give rise to a self-supporting chain reaction. { ¦səb'krid·ə·kəl 'mas }

subcritical reactor [NUCLEO] A reactor having an effective multiplication constant of less than one, so that a self-supporting chain reaction cannot be maintained. Also known as subcritical assembly; teaching reactor. { ¦səb'krid·ə·kəl rē'ak·tər }

subcrop [GEOL] An occurrence of strata beneath the subsurface of an inclusive stratigraphic unit that succeeds an unconformity on which there is marked overstep. { 'səb,kräp }

subcutaneous connective tissue [HISTOL] The layer of loose connective tissue beneath the dermis. { ¦səb·kyü'tā·nē·əs kə'nek·tiv ,tish·ü }

subcutaneous emphysema [MED] The presence of air in the tissues just under the skin; when seen in diving it usually involves the skin of the neck and nearby areas. { ¦səb·kyü'tā·nē·əs ,em·fə'sē·mə }

subcutaneous mycosis [MED] Any of a wide spectrum of infections caused by a heterogeneous group of fungi characterized by the development of lesions (that usually remain localized or spread slowly by direct extension via the lymphatics) at sites of inoculation and initially involve the deeper layers of the dermis and subcutaneous tissues but eventually extend into the epidermis. { ,səb·kyü,tān·ē·əs mī'kō·səs }

subcycle generator [ELECTR] Frequency-reducing device used in telephone equipment which furnishes ringing power at a submultiple of the power supply frequency. { 'səb,sī·kəl 'jen·ə,rād·ər }

subdevice bit *See* select bit. { 'səb·di,vīs ,bit }

subdivided capacitor [ELEC] Capacitor in which several capacitors known as sections are mounted so that they may be used individually or in combination. { ¦səb·di'vīd·əd kə'pas·əd·ər }

subdivision graph [MATH] A graph which can be obtained from a given graph by breaking up each edge into one or more segments by inserting intermediate vertices between its two ends. { 'səb·di,vizh·ən ,graf }

subdominant [ECOL] A species which may appear more abundant at particular times of the year than the true dominant in a climax; for example, in a savannah trees and shrubs are more conspicuous than the grasses, which are the true dominants. { ¦səb'däm·ə·nənt }

subdrainage [CIV ENG] Natural or artificial removal of water from beneath a lined conduit. { ¦səb'drā·nij }

subdrilling [ENG] Refers to the breaking of the base in which boreholes are drilled 1 foot (0.3 meter) or several feet below the level of the quarry floor. { ¦səb'dril·iŋ }

subduction [GEOL] The process by which one crustal block descends beneath another, such as the descent of the Pacific plate beneath the Andean plate along the Andean Trench. { səb'dək·shən }

subduction zones [GEOL] Regions where portions of the earth's tectonic plates are diving beneath other plates, into the earth's interior. They are defined by deep oceanic trenches, lines of volcanoes parallel to the trenches, and zones of large earthquakes that extend from the trenches landward. { səb'dək·shən ‚zōnz }

subdural hematoma [MED] A mass of blood between the arachnoid and the dura mater. { səb'dùr·əl ‚hē·mə'tō·mə }

subdural hemorrhage [MED] Bleeding between the dura mater and the arachnoid. { səb'dùr·əl 'hem·rij }

subdwarf star [ASTRON] An intermediate star type; luminosity is between that of main sequence stars and the white dwarf stars on the Hertzsprung-Russell diagram; spectral classes F, G, and K are most numerous. { səb‚dwȯrf 'stär }

subdwarf symbiotic [ASTRON] A type of symbiotic star consisting of a combination of a cool red giant with a small hot subdwarf star, the latter probably the inner core of a former giant or super giant which has shed its outer envelope and is now contracting to become a white dwarf. Also known as planetary nebula symbiotic. { 'səb‚dwȯrf ‚sim·bē'äd·ik }

subendothelial layer [HISTOL] The middle layer of the tunica intima of veins and of medium and larger arteries, consisting of collagenous and elastic fibers and a few fibroblasts. { ¦səb‚en·də'thē·lē·əl 'lā·ər }

suberic acid [ORG CHEM] HOOC(CH₂)₆COOH A colorless, crystalline compound that melts at 143°C, and dissolves slightly in cold water; used in organic synthesis. Also known as octanedioic acid. { sü'ber·ik 'as·əd }

suberin [BIOCHEM] A fatty substance found in many plant cell walls, especially cork. { 'sü·bə·rən }

suberinite [GEOL] A variety of provitrinite composed of corky tissue. { sü'ber·ə‚nīt }

suberization [BOT] Infiltration of plant cell walls by suberin resulting in the formation of corky tissue that is impervious to water. { ‚sü·bə·rə'zā·shən }

suberose [BOT] Having a texture like cork due to or resembling that due to suberization. { 'sü·bə‚rōs }

subfactorial [MATH] For an integer n, the number that is expressed as $n!\{(1/2!) - (1/3!) + (1/4!) - \cdots + [(-1)^n/n!]\}$. { ‚səb‚fak'tȯr·ē·əl }

subfeldspathic [GEOL] Referring to mature lithic wacke or arenite containing an abundance of quartz grains with less than 10% feldspar grains. { ¦səb·fel'spath·ik }

subfield [MATH] **1.** A subset of a field which itself forms a field relative to the same operations. **2.** A subfield (of sets) is any field (of sets) contained in some given field of sets. { 'səb‚fēld }

subfloor [BUILD] The rough floor which rests on the floor joists and on which the finished floor is laid. Also known as blind floor; counterfloor. { 'səb‚flȯr }

subframe [COMMUN] In telemetry, a complete sequence of frames during which all subchannels of a specific channel are sampled once. { 'səb‚frām }

subgelisol [GEOL] Unfrozen ground beneath permafrost. { 'səb‚jel·ə‚sȯl }

subgeostrophic wind [METEOROL] Any wind of lower speed than the geostrophic wind required by the existing pressure gradient. { ¦səb¦jē·ō'sträf·ik 'wind }

subgiant CH star [ASTRON] A type of star that resembles the CH stars and barium stars but is less luminous and somewhat hotter, with some members of the class lying on or near the main sequence. { səb¦jī·ənt ‚sē'āch ‚stär }

subgiant star [ASTRON] A member of the family of stars whose luminosity is intermediate between giants and the main sequence in the Hertzsprung-Russell diagram; spectral classes G and K are most frequent. { ‚səb'jī·ənt 'stär }

subglacial [GEOL] Pertaining to the area in or at the bottom of, or immediately beneath, a glacier. { ‚səb'glā·shəl }

subglacial moraine *See* ground moraine. { ¦səb'glā·shəl mə'rān }

subgrade [CIV ENG] The soil or rock leveled off to support the foundation of a structure. { 'səb grād }

subgradient wind [METEOROL] A wind of lower speed than the gradient wind required by the existing pressure gradient and centrifugal force. { ¦səb'grād·ē·ənt 'wind }

subgrain [MET] The portion of a metal crystal or grain with an orientation that differs slightly from the orientation of neighboring portions of the same crystal. { 'səb‚grān }

subgraph [MATH] A graph contained in a given graph which has as its vertices some subset of the vertices of the original. { 'səb‚graf }

subgraywacke [PETR] An argillaceous sandstone with a composition intermediate between graywacke and orthoquartzite; a clay matrix is usually present but it amounts to less than 15%. { ¦səb'grā‚wak·ə }

subgroup [MATH] A subset N of a group G which is itself a group relative to the same operation. { 'səb‚grüp }

subharmonic [PHYS] A sinusoidal quantity having a frequency that is an integral submultiple of the frequency of some other sinusoidal quantity to which it is referred; a third subharmonic would be one-third the fundamental or reference frequency. { ¦səb·här'män·ik }

subharmonic function [MATH] A continuous function is subharmonic in a region R of the plane if its value at any point z_0 of R is less than or equal to its integral along a circle centered at z_0. { ¦səb·här'män·ik 'fəŋk·shən }

subharmonic triggering [ELECTR] A method of frequency division which makes use of a triggered multivibrator having a period of one cycle which allows triggering only by a pulse that is an exact integral number of input pulses from the last effective trigger. { ¦səb·här'män·ik 'trig·ə·riŋ }

subhedral [MINERAL] **1.** Pertaining to an individual mineral crystal that is partly bounded by its own crystal faces and partly bounded by surfaces formed against preexisting crystals. **2.** Descriptive of a crystal having partially developed crystal faces. { ¦səb'hē·drəl }

subhumid climate [CLIMATOL] A humidity province based on its typical vegetation. Also known as grassland climate; prairie climate. { ¦səb'hyü·məd 'klī·mat }

subidiomorphic *See* hypidiomorphic. { ¦səb‚id·ē·ə'mȯr·fik }

subirrigation *See* subsurface irrigation. { ¦səb‚ir·ə'gā·shən }

subjacent [GEOL] Being lower than but not directly underneath. { ‚səb'jās·ənt }

subjacent igneous body [GEOL] An igneous intrusion without a known floor, and which presumably enlarges downward. { ‚səb'jās·ənt 'ig·nē·əs 'bäd·ē }

subjamming visibility [ORD] Relates to how well a particular radar antijam technique can see through jamming signals. { ¦səb'jam·iŋ ‚viz·ə'bil·əd·ē }

subject copy [GRAPHICS] Material that is to be transmitted for facsimile reproduction. Also known as copy. { 'səb·jəkt ‚käp·ē }

subjective probability *See* personal probability. { səb'jek·tiv ‚präb·ə'bil·əd·ē }

subkiloton weapon [ORD] A nuclear weapon producing a yield below 1 kiloton. { ¦səb'kil·ə‚tän 'wep·ən }

sublacustrine [GEOL] Existing or formed on the bottom of a lake. { ¦səb·lə'kəs·trən }

sublacustrine channel [GEOL] A channel eroded in a lake bed either before the lake existed or by a strong current in the lake. { ¦səb·lə'kəs·trən ‚chan·əl }

sublethal gene *See* semilethal gene. { ¦səb'lē·thəl 'jēn }

subleukemic [MED] Less than leukemic; usually applied to states in which the peripheral blood manifestations of leukemia are temporarily suppressed. { ¦səb·ü'kē·mik }

sublevel [ATOM PHYS] *See* subshell. [MIN ENG] An intermediate level opened a short distance below the main level; or in the caving system of mining, a 15–20-foot (4.6–6.1-meter) level below the top of the ore body, preliminary to caving the ore between it and the level above. { ¦səb'lev·əl }

sublevel caving [MIN ENG] A stoping method in which relatively thin blocks of ore are caused to cave by successively undermining small panels. { ¦səb'lev·əl 'kāv·iŋ }

sublevel drive [MIN ENG] A drive often made in a section which divides the deposit into narrower panels and zones. { ¦səb'lev·əl 'drīv }

sublevel stoping [MIN ENG] A mining method involving overhand, underhand, and shrinkage stoping; the characteristic feature is the use of sublevels which are worked simultaneously, the lowest on a given block being farthest advanced and the sublevels above following one another at short intervals. { ¦səb'lev·əl ˈstōp·iŋ }

sublimation [PSYCH] A defense mechanism whereby the energies of undesirable instinctual cravings and impulses are converted into socially acceptable activities. [THERMO] The process by which solids are transformed directly to the vapor state or vice versa without passing through the liquid phase. { ˌsəb·lə'mā·shən }

sublimation cooling [THERMO] Cooling caused by the extraction of energy to produce sublimation. { ˌsəb·lə'mā·shən ¦kül·iŋ }

sublimation curve [THERMO] A graph of the vapor pressure of a solid as a function of temperature. { ˌsəb·lə'mā·shən ¦kərv }

sublimation energy [THERMO] The increase in internal energy when a unit mass, or 1 mole, of a solid is converted into a gas, at constant pressure and temperature. { ˌsəb·lə'mā·shən ¦en·ər·jē }

sublimation nucleus [METEOROL] Any particle upon which an ice crystal may grow by the process of sublimation. { ˌsəb·lə'mā·shən ¦nü·klē·əs }

sublimation point [THERMO] The temperature at which the vapor pressure of the solid phase of a compound is equal to the total pressure of the gas phase in contact with it; analogous to the boiling point of a liquid. { ˌsəb·lə'mā·shən ¦pȯint }

sublimation pressure [THERMO] The vapor pressure of a solid. { ˌsəb·lə'mā·shən ¦presh·ər }

sublimation vein [GEOL] A vein of mineral that has condensed from a vapor. { ˌsəb·lə'mā·shən ¦vān }

sublimatography [ANALY CHEM] A procedure of fractional sublimation in which a solid mixture is separated into bands along a condensing tube with a temperature gradient. { ˌsəb·lə·mə'täg·rə·fē }

sublimator [CHEM] Device used for the heating of solids (usually under vacuum) to the temperature at which the solid sublimes. { ˈsəb·lə¸mād·ər }

sublime [THERMO] To change from the solid to the gaseous state without passing through the liquid phase. { sə'blīm }

sublimed sulfur See flowers of sulfur. { sə'blīmd 'səl·fər }

subliminal [PHYSIO] Below the threshold of responsiveness, consciousness, or sensation to a stimulus. { sə'blim·ə·nəl }

subliming rocket propellant [MATER] A propellant characterized by sublimation of the material at the heated surface. { sə'blīm·iŋ 'räk·ət prə¸pel·ənt }

sublingual gland [ANAT] A complex of salivary glands located in the sublingual fold on each side of the floor of the mouth. { ˌsəb'liŋ·gwəl 'gland }

sublitharenite [PETR] A sandstone which contains between 5 and 25% rock fragments and in which the rock fragments are more abundant than feldspar grains. { ˌsəb'li·thar·ə¸nīt }

sublittoral zone [OCEANOGR] The benthic region extending from mean low water (2–3 fathoms or 40–60 meters, according to some authorities) to a depth of about 110 fathoms (200 meters), or the edge of a continental shelf, beyond which most abundant attached plants do not grow. { ˌsəb'lid·ə·rəl ¸zōn }

subluminous star [ASTRON] A star that is fainter than those of the same color on the main sequence. { səb'lüm·ə·nəs 'stär }

sublunar point [ASTRON] The moon's geographic zenith position at any particular moment in time. { ˌsəb'lü·nər 'pȯint }

subluxation [MED] An incomplete dislocation. { ˌsəb·lək'sā·shən }

submachine gun [ORD] A type of machine gun; a short-barreled shoulder firearm using pistol-type ammunition and capable of full automatic fire. { ˌsəb·mə'shēn ¸gən }

submandibular duct [ANAT] The duct of the submandibular gland which empties into the mouth on the side of the frenulum of the tongue. { ˌsəb·man'dib·yə·lər 'dəkt }

submandibular gland [ANAT] A large seromucous or mixed salivary gland located below the mandible on each side of the jaw. Also known as mandibular gland; submaxillary gland. { ˌsəb·man'dib·yə·lər 'gland }

submarine [NAV ARCH] A ship that can operate both on the surface of the water and completely submerged. [OCEANOGR] Being or functioning in the sea. { ˌsəb·mə'rēn }

submarine base [NAV ARCH] A base providing logistic support for submarines. { ˌsəb·mə'rēn ¦bās }

submarine bell [NAV] A bell whose signal is transmitted through water. { ¦səb·mə'rēn 'bel }

submarine blast [ENG] A charge of high explosives fired in boreholes drilled in the rock underwater for dislodging dangerous projections and for deepening channels. { ¦səb·mə'rēn 'blast }

submarine cable [ELEC] A cable designed for service under water; usually a lead-covered cable with steel armor applied between layers of jute. { ¦səb·mə'rēn 'kā·bəl }

submarine canyon [GEOL] Steep-sided valleys winding across the continental shelf or continental slope, probably originally produced by Pleistocene stream erosion, but presently the site of turbidity flows. { ¦səb·mə'rēn 'kan·yən }

submarine cave See submarine fan. { ¦səb·mə'rēn 'kāv }

submarine delta See submarine fan. { ¦səb·mə'rēn 'del·tə }

submarine earthquake See seaquake. { ¦səb·mə'rēn 'ərth¸kwāk }

submarine false target [ORD] A pyrotechnic item designed to be ejected from a submarine to confuse and disrupt underwater echo-ranging equipment and create a bubble wake which can be seen by aircraft and surface vessels. { ¦səb·mə'rēn 'fȯls ¸tär·gət }

submarine fan [GEOL] A shallow marine sediment that is fan- or cone-shaped and lies off the seaward opening of large rivers and submarine canyons. Also known as abyssal cave; abyssal fan; sea fan; submarine cave; submarine delta; subsea apron. { ¦səb·mə'rēn 'fan }

submarine gate [ENG] An edge gate with the opening from the runner into the mold positioned below the printing line or mold surface. { ¦səb·mə'rēn 'gāt }

submarine geology See geological oceanography. { ¦səb·mə'rēn jē'äl·ə·jē }

submarine isthmus [GEOL] A submarine elevation joining two land areas and separating two basins or depressions by a depth less than that of the basins. { ¦səb·mə'rēn 'is·məs }

submarine mine [MIN ENG] A mine for the extraction of minerals or ores under the sea. { ¦səb·mə'rēn 'mīn }

submarine navigation [NAV] **1.** Navigation of a submarine, whether submerged or surfaced. **2.** See underwater navigation. { ¦səb·mə'rēn ¸nav·ə'gā·shən }

submarine oscillator [ENG ACOUS] A large, electrically operated diaphragm horn which produces a powerful sound for signaling through water. { ¦səb·mə'rēn 'äs·ə¸läd·ər }

submarine peninsula [GEOL] An elevated portion of the submarine relief resembling a peninsula. { ¦səb·mə'rēn pə'nin·sə·lə }

submarine pipeline [ENG] A pipeline installed under water, resting on the bed of the waterway; frequently used for petroleum or natural gas transport across rivers, lakes, or bays. { ¦səb·mə'rēn 'pīp¸līn }

submarine pit [GEOL] A cavity on the bottom of the sea. Also known as submarine well. { ¦səb·mə'rēn 'pit }

submarine plain See plain. { ¦səb·mə'rēn 'plān }

submarine relief [GEOL] Relative elevations of the ocean bed, or the representation of them on a chart. { ¦səb·mə'rēn ri'lēf }

submarine sanctuary [NAV ARCH] Any of the restricted areas established for the conduct of noncombat submarine or antisubmarine exercises; they may be either stationary or moving and are normally designated only in rear areas. { ¦səb·mə'rēn 'saŋk·chə¸wer·ē }

submarine sentry [ENG] A form of underwater kite towed at a predetermined constant depth in search of elevations of the bottom; the kite rises to the surface upon encountering an obstruction. { ¦səb·mə'rēn 'sen·trē }

submarine sound signal [ACOUS] A sound signal transmitted through water. { ¦səb·mə'rēn 'saủn ¸sig·nəl }

submarine spring [HYD] A spring of water issuing from the bottom of the sea. { ¦səb·mə'rēn 'spriŋ }

submarine station [OCEANOGR] **1.** One of the places for which tide or tidal current predictions are determined by applying a correction to the predictions of a reference station. **2.** A tide or tidal current station at which a short series of observations have been made; these observations are reduced by comparison with simultaneous observations at a reference station. { ¦səb·mə'rēn 'stā·shən }

submarine striking forces [NAV ARCH] Submarines having guided or ballistic missile launching or guidance capabilities formed to launch offensive nuclear strikes. { ¦səb·mə'rēn 'strīk·iŋ ˌför·səz }

submarine topography [GEOL] Configuration of a surface such as the sea bottom or of a surface of given characteristics within the water mass. { ¦səb·mə'rēn tə'päg·rə·fē }

submarine trench *See* trench. { ¦səb·mə'rēn 'trench }

submarine trough *See* trough. { ¦səb·mə'rēn 'tróf }

submarine valley *See* valley. { ¦səb·mə'rēn 'val·ē }

submarine wave recorder [ENG] An instrument for measuring the changing water height above a hovering submarine by measuring the time required for sound emitted by an inverted echo sounder on the submarine to travel to the surface and return. { ¦səb·mə'rēn 'wāv riˌkórd·ər }

submarine weathering [GEOL] A slow alteration of the form, texture, and composition of the sea floor from chemical, thermal, and biological causes. { ¦səb·mə'rēn 'weth·ə·riŋ }

submarine well *See* submarine pit. { ¦səb·mə'rēn 'wel }

submaxillary gland *See* submandibular gland. { ¦səb'mak·səˌler·ē ˌgland }

submerged-arc furnace [MET] An arc-heating furnace in which the arcs may be completely submerged under the charge or in the molten bath under the charge. { səb'mərjd ¦ärk 'fər·nəs }

submerged-arc welding [MET] Arc welding with a bare metal electrode, the arc and tip of the electrode being shielded by a blanket of granular, fusible material. { səb'mərjd ¦ärk 'weld·iŋ }

submerged breakwater [OCEANOGR] A breakwater with its top below the still water level; when struck by a wave, part of the wave energy is reflected seaward and the remaining energy is largely dissipated in a breaker, transmitted shoreward as a multiple crest system, or transmitted shoreward as a simple wave system. { səb'mərjd 'brāk·wòd·ər }

submerged coastal plain [GEOL] The continental shelf as the seaward extension of a coastal plain on the land. Also known as coast shelf. { səb'mərjd 'kóst·əl 'plān }

submerged-combustion evaporator [ENG] A liquid-evaporation device in which heat is provided by combustion gases bubbling up through the liquid; the burner is submerged in the body of the liquid. { səb'mərjd kəmˌbəs·chən i'vap·əˌrād·ər }

submerged-combustion heater [ENG] A combustion device in which fuel and combustion air are mixed and ignited below the surface of a liquid; used in heaters and evaporators where absorption of the combustion products will not be detrimental. { səb'mərjd kəmˌbəs·chən hēd·ər }

submerged culture [MICROBIO] A method for growing pure cultures of aerobic bacteria in which microorganisms are incubated in a liquid medium subjected to continuous, vigorous agitation. { səb'mərjd 'kəl·chər }

submerged fermentation [MICROBIO] Industrial production of antibiotics, enzymes, and other substances by growing the microorganisms that produce the product in a submerged culture. { səb'mərjd ˌfər·mən'tā·shən }

submerged-foil hydrofoil [NAV ARCH] A hydrofoil craft with foils which are completely submerged and whose angle of attack is controlled by an autopilot and sensors to maintain the height and attitude of the craft. { səb'mərjd 'fóil 'hīdrəˌfóil }

submerged lands [GEOL] Lands covered by water at any stage of the tide, as distinguished from tidelands which are attached to the mainland or an island and are covered or uncovered with the tide; tidelands presuppose a high-water line as the upper boundary, submerged lands do not. { səb'mərjd 'lanz }

submerged screw log [NAV ARCH] A type of electric log which is actuated by the flow of water past a propeller. { səb'mərjd 'skrü ˌläg }

submerged shoreline *See* shoreline of submergence. { səb'mərjd 'shór·līn }

submerged weir [CIV ENG] A dam which, when in use, has the downstream water level at an elevation equal to or higher than the crest of the dam. { səb'mərjd 'wer }

submergence [GEOL] A change in the relative levels of water and land from either a sinking of the land or a rise of the water level. { səb'mər·jəns }

submersible pump [MECH ENG] A pump and its electric motor together in a protective housing which permits the unit to operate under water. { səb'mər·sə·bəl 'pəmp }

submetallic [OPTICS] Referring to a luster intermediate between metallic and nonmetallic, such as exhibited by the mineral chromite. { ¦səb·mə'tal·ik }

submillimeter astronomy [ASTRON] Astronomical observations carried out in the region of the electromagnetic spectrum with wavelengths from approximately 0.3 to 1.0 millimeter. { ¦səb¦mil·əˌmēd·ər ə'strän·ə·mē }

submillimeter wave [ELECTROMAG] An electromagnetic wave whose wavelength is less than 1 millimeter, corresponding to frequencies above 300 gigahertz. { ¦səb¦mil·əˌmēd·ər ˌwāv }

subminiature tube [ELECTR] An extremely small electron tube designed for use in hearing aids and other miniaturized equipment; a typical subminiature tube is about $1\frac{1}{2}$ inches (4 centimeters) long and 0.4 inch (1 centimeter) in diameter, with the pins emerging through the glass base. { ¦səb'min·yə·chər 'tüb }

submissile [ORD] One of several smaller missiles carried and released by a larger missile, especially in a warhead. { ¦səb'mis·əl }

submodule [MATH] A subset N of a module M over a ring R such that, if x and y are in N and a is in R, then $x + y$ and ax are in N, so that N is also a module over R. { ¦səb'mä·jəl }

submucosa [HISTOL] The layer of fibrous connective tissue that attaches a mucous membrane to its subadjacent parts. { ¦səb·myü'kō·sə }

submucous plexus [NEUROSCI] A visceral nerve network lying in the submucosa of the digestive tube. Also known as Meissner's plexus. { ¦səb'myü·kəs 'plek·səs }

submultiple [MATH] A number or quantity divided by an integer. { səb'məl·tə·pəl }

submultiple resonance [PHYS] Resonance at a frequency that is a submultiple of the frequency of the exciting impulses. { ¦səb'məl·tə·pəl 'rez·ən·əns }

subnormal [MATH] For a given point on a plane curve, the projection on the x axis of a rectangular coordinate system of the segment of the normal between the given point and the intersection of the normal with the x axis. { səb'nór·məl }

subnormal operator [MATH] An operator A on a Hilbert space **H** is said to be subnormal if there exists a normal operator B on a Hilbert space **K** such that **H** is a subspace of **K**, the subspace **H** is invariant under the operator B, and the restriction of B to **H** coincides with A. { ˌsəb·nór·məl 'äp·əˌrād·ər }

subnuclear particle *See* elementary particle. { səb'nü·klē·ər 'pärd·i·kəl }

sub-Nyquist sampling [COMMUN] **1.** Any technique of sampling an analog signal at a rate lower than the Nyquist rate in such a way as to preserve signal content without aliasing distortion. **2.** In particular, the sampling of television signals at a rate lower than the Nyquist rate and at an odd multiple of the frame rate, so that the aliasing components are placed into periodically spaced voids in the television spectrum where they can be removed by a comb filter at the receiver. { ¦səb 'nīˌkwist ˌsam·pliŋ }

suboptimization [SYS ENG] The process of fulfilling or optimizing some chosen objective which is an integral part of a broader objective; usually the broad objective and lower-level objective are different. { səbˌäp·tə·mə'zā·shən }

Suboscines [VERT ZOO] A major division of the order Passeriformes, usually divided into the suborders Eurylaimi, Tyranni, and Memirae. { ¦səb'äs·əˌnēz }

subpolar anticyclone *See* subpolar high. { ¦səb'pō·lər ¦ant·i'sīˌklōn }

subpolar glacier [HYD] A polar glacier with 30 to 60 feet (10 to 20 meters) of firn in the accumulation area where some melting occurs. { ¦səb'pō·lər 'glā·shər }

subpolar high [METEOROL] A high that forms over the cold continental surfaces of subpolar latitudes, principally in Northern Hemisphere winters; these highs typically migrate eastward and southward. Also known as polar anticyclone; polar high; subpolar anticyclone. { ¦səb'pō·lər 'hī }

subpolar icebreaker [NAV ARCH] An icebreaker that is capable of operations in the ice-covered waters of coastal seas and lakes outside the polar regions. { ˌsəb'pō·lər 'īsˌbrāk·ər }

subpolar low-pressure belt [METEOROL] A belt of low pressure located, in the mean, between 50 and 70° latitude; in the Northern Hemisphere, this belt consists of the Aleutian low and the Icelandic low; in the Southern Hemisphere, it is supposed to exist around the periphery of the Antarctic continent. { ¦səb'pō·lər ¦lō 'presh·ər ˌbelt }

SUBMERGED-FOIL HYDROFOIL

U.S. Navy Patrol Hydrofoil, Missile (PHM) class ship, with fully submerged foils. *(U.S. Navy)*

subpolar westerlies See westerlies. { ¦səb'pō·lər 'wes·tər,lēz }

subpopulation [STAT] A subset of population. Also known as stratum. { ¦səb,päp·yə'lā·shən }

subprogram [COMPUT SCI] A part of a larger program which can be converted independently into machine language. { ¦səb'prō,gram }

subpulse [ASTRON] The weaker component of a pulsar's periodic emission. { 'səb,pəls }

subrange [MATH] A subset of the range of values that a function may assume. { 'səb,rānj }

subrefraction [ELECTROMAG] Atmospheric refraction which is less than standard refraction. { ¦səb·ri'frak·shən }

subring [MATH] **1.** A subset I of a ring R where I is also a ring relative to the operations of R. **2.** A subring (of sets) is any ring (of sets) contained in some given ring (of sets). { 'səb,riŋ }

subroutine [COMPUT SCI] **1.** A body of computer instruction (and the associated constants and working-storage areas, if any) designed to be used by other routines to accomplish some particular purpose. **2.** A statement in FORTRAN used to define the beginning of a closed subroutine (first definition). { 'səb·rü,tēn }

subroutine library [COMPUT SCI] A collection of subroutines that is stored on a disk or other direct-access storage device and can be used by a programmer through facilities of the computer's operating system. { 'səb·rü,tēn lī,brēr·ē }

subsample [ANALY CHEM] A portion taken from a sample of material for which a chemical analysis has been specified. { ¦səb'sam·pəl }

subsampling [STAT] Taking samples from a sample of a population. { 'səb,sam·pliŋ }

subsatellite [AERO ENG] An object that is carried into orbit by, and subsequently released from, an artificial satellite. { 'səb,sad·əl,īt }

subscale [MET] Oxidation occurring within a metal instead of on the surface. { 'səb,skāl }

subscapularis [ANAT] A muscle arising from the costal surface of the scapula and inserted on the lesser tubercle of the humerus. { ,səb,skap·yə'lar·əs }

subschema [COMPUT SCI] An individual user's partial view of a data base. { 'səb,skē·mə }

subscriber line [ELEC] A telephone line between a central office and a telephone station, private branch exchange, or other end equipment. Also known as central office line; subscriber loop. { səb'skrīb·ər ,līn }

subscriber loop See subscriber line. { səb'skrīb·ər ,lüp }

subscriber multiple [ELEC] Bank of jacks in a manual switchboard providing outgoing access to subscriber lines, and usually having more than one appearance across the face of the switchboard. { səb'skrīb·ər 'məl·tə·pəl }

subscriber set See subset. { səb'skrīb·ər ,set }

subscriber station [COMMUN] The connection between a central office and an outside location, including the circuit, some circuit termination equipment, and possibly some associated input/output equipment. { səb'skrīb·ər ,stā·shən }

subscript [SCI TECH] A letter or symbol written below, and usually to the right, of another symbol for any of various purposes, such as to identify a particular element or elements of a set, to denote a constant value of a variable, or, in a chemical formula, to indicate the number of atoms of a particular kind in a molecule. { 'səb,skript }

subscripted variable [MATH] A symbolic name for an array of variables whose elements are identified by subscripts. { səb'skrip·təd 'ver·ē·ə·bəl }

subscription data base See information network. { səb'skrip·shən 'dad·ə ,bās }

subscription television [COMMUN] A television system in which programs are broadcast in coded or scrambled form, for reception only by subscribers who make payments for use of the decoding or unscrambling devices required to obtain a clear program. Also known as pay television. { səb'skrip·shən 'tel·ə,vizh·ən }

subsea apron See submarine fan. { 'səb,sē 'ā·prən }

subsequence [MATH] A subsequence of a given sequence is any sequence all of whose entries appear in the original sequence and in the same manner of succession. { 'səb·sə·kwəns }

subsequent [GEOL] Referring to a geologic feature that followed in time the development of a consequent feature of which it is a part. { 'səb·sə·kwənt }

subsequent drainage [HYD] Drainage by a stream developed subsequent to the system of which it is a part; drainage follows belts of weak rocks. { 'səb·sə·kwənt 'drā·nij }

subsequent fold See cross fold. { 'səb·sə·kwənt 'fōld }

subsequent stream [HYD] A stream that flows in the general direction of the strike of the underlying strata and is subsequent to the formation of the consequent stream of which it is a tributary. Also known as longitudinal stream; strike stream. { 'səb·sə·kwənt 'strēm }

subsequent valley [GEOL] A valley eroded by a stream developed subsequent to the system of which it is a part. { 'səb·sə·kwənt 'val·ē }

subsere [ECOL] A secondary community that succeeds an interrupted climax. { 'səb,sir }

subset [COMMUN] A telephone or other subscriber equipment connected to a communication system, such as a modem. Derived from subscriber set. [MATH] **1.** A subset A of a set B is a set all of whose elements are included in B. **2.** A fuzzy set A is a subset of a fuzzy set B if, for every element x, the value of the membership function of A at x is equal to or less than the value of the membership function of B at x. { 'səb,set }

subshell [ATOM PHYS] Electrons of an atom within the same shell (energy level) and having the same azimuthal quantum numbers. Also known as sublevel. { 'səb,shel }

subsidence [METEOROL] A descending motion of air in the atmosphere, usually with the implication that the condition extends over a rather broad area. [MIN ENG] A sinking down of a part of the earth's crust due to underground excavations. { səb'sīd·əns }

subsidence break [MIN ENG] A fracture in the rocks overlying a coal seam or mineral deposit resulting from mining operations. { səb'sīd·əns ,brāk }

subsidence inversion [METEOROL] A temperature inversion produced by the adiabatic warming of a layer of subsiding air; this inversion is enhanced by vertical mixing in the air layer below the inversion. { səb'sīd·əns in,vər·zhən }

subsidiary conduit [CIV ENG] Terminating branch of an underground conduit run extending from a manhole or handhole to a nearby building, handhole, or pole. { səb'sid·ē,er·ē 'kän·dü·ət }

subsidiary fracture See tension fracture. { səb'sid·ē,er·ē 'frak·chər }

subsidiary transport [MIN ENG] The conveying or haulage of coal or mineral from the working faces to a junction or loading point. { səb'sid·ē,er·ē 'tranz,pórt }

subsieve analysis [MET] Analysis by size distribution of metal powder particles all of which pass through a standard 44-micrometer sieve. { 'səb,siv ə'nal·ə·səs }

subsieve fraction [MET] The fraction of particles of a metal powder which pass through a standard 44-micrometer sieve. { 'səb,siv 'frak·shən }

subsine function [MATH] A function that is dominated by functions of the form $f(x) = A \sin (x + c)$, where A and c are constants, in the same way that convex functions are dominated by linear functions. { 'səb,sīn ,faŋk·shən }

subsistence farming [AGR] Growth of crops predominantly for consumption by the farm family rather than for sale. { səb¦sis·təns ¦färm·iŋ }

subsoil [GEOL] **1.** Soil underlying surface soil. **2.** See B horizon. { 'səb,sóil }

subsoil ice See ground ice. { 'səb,sóil 'īs }

subsolar point [ASTRON] The sun's zenith geographic position at any particular moment in time. { ¦səb'sō·lər 'póint }

subsolvus [PHYS CHEM] A range of conditions in which two or more solid phases can form by exsolution from an original homogeneous phase. { ¦səb'säl·vəs }

subsonic [ACOUS] See infrasonic. [PHYS] Of, pertaining to, or dealing with speeds less than acoustic velocity, as in subsonic aerodynamics. { ¦səb'sän·ik }

subsonic flight [AERO ENG] Movement of a vehicle through the atmosphere at a speed appreciably below that of sound waves; extends from zero (hovering) to a speed about 85% of sonic speed corresponding to ambient temperature. { ¦səb'sän·ik 'flīt }

subsonic flow [FL MECH] Flow of a fluid at a speed less

than that of the speed of sound in the fluid. Also known as subcritical flow. { ¦səb′sän·ik ′flō }

subsonic inlet [ENG] An entrance or orifice for the admission of fluid flowing at speeds less than the speed of sound in the fluid. { ¦səb′sän·ik ′in,let }

subsonic nozzle [ENG] A nozzle through which a fluid flows at speed less than the speed of sound in the fluid. { ¦səb′sän·ik ′näz·əl }

subsonic speed [FL MECH] A speed relative to surrounding fluid less than that of the speed of sound in the same fluid. { ¦səb′sän·ik ′spēd }

subspace [MATH] A subset of a space which, in the appropriate context, is a space in its own right. { ′səb,spās }

subspecies [SYST] A geographically defined grouping of local populations which differs taxonomically from similar subdivisions of species. { ′səb′spē·shēz }

substance [PHYS] Tangible material, occurring in macroscopic amounts. { ′səb·stəns }

substance P [BIOCHEM] An undecapeptide widely distributed in the central nervous system and found in highest concentrations in superficial layers of the dorsal horn of the spinal cord, in the trigeminal nerve nucleus, and in the substantia nigra; acts as a neurotransmitter. { ′səb·stəns ′pē }

substandard propagation [ELECTROMAG] The propagation of radio energy under conditions of substandard refraction in the atmosphere; that is, refraction by an atmosphere or section of the atmosphere in which the index of refraction decreases with height at a rate of less than 12 N units (unit of index of refraction) per 1000 feet (304.8 meters). { ¦səb′stan·dərd ,präp·ə′gā·shən }

substantive dye See direct dye. { ′səb·stən·tiv ′dī }

substation [ELEC] See electric power substation. [ENG] An intermediate compression station to repressure a fluid being transported by pipeline over a long distance. [MIN ENG] A subsidiary station for the conversion of power to the type, usually direct current, and voltage needed for mining equipment and fed into the mine power system. { ′səb,stā·shən }

substellar object See brown dwarf. { ,səb,stel·ər ′äb,jekt }

substellar point [ASTRON] The geographical position of a star; that point on the earth at which the star is in the zenith at a specified time. Also known as substastral point. { ¦səb,stel·ər ′pöint }

substituent [ORG CHEM] An atom or functional group substituted for another in a chemical structure. { səb′stich·ə·wənt }

substitute mode [COMPUT SCI] One method of exchange buffering, in which segments of storage function alternately as buffer and as program work area. { ′səb·stə,tüt ,mōd }

substitution [PSYCH] A defense mechanism whereby an unattainable or unacceptable goal, emotion, or object is replaced by one that is more attainable or acceptable. { ,səb·stə′tü·shən }

substitutional impurity [SOLID STATE] An atom or ion which is not normally found in a solid, but which resides at the position where an atom or ion would ordinarily be located in the lattice structure, and replaces it. { ,səb·stə′tü·shən·əl im′pyūr·əd·ē }

substitution alphabet [COMMUN] An alphabet used in a coded message in which each letter in the original message is replaced by another letter in the coded message, according to a set of rules. { ,səb·stə′tü·shən ′al·fə,bet }

substitution cipher [COMMUN] A cipher in which the characters of the original message are replaced by other characters according to a key. { ,səb·stə′tü·shən ,sī·fər }

substitution group See permutation group. { ,səb·stə′tü·shən ,grüp }

substitution method [PHYS] Any method of measurement, such as substitution weighing, in which a quantity is determined by substituting for it a known quantity which produces the same effect. { ,səb·stə′tü·shən ,meth·əd }

substitution reaction [CHEM] Replacement of an atom or radical by another one in a chemical compound. { ,səb·stə′tü·shən rē,ak·shən }

substitution solid solution [MET] A solid alloy having the atoms of the solute located at some lattice of points of the solvent. { ,səb·stə′tü·shən ′säl·əd sə′lü·shən }

substitution weighing [MECH] A method of weighing to allow for differences in lengths of the balance arms, in which the object to be weighed is first balanced against a counterpoise,

and the known weights needed to balance the same counterpoise are then determined. Also known as counterpoise method. { ,səb·stə′tü·shən ,wā·iŋ }

substitutive nomenclature [ORG CHEM] A system in which the name of a compound is derived by using the functional group (the substituent) as a prefix or suffix to the name of the parent compound to which it is attached; for example, in 2-chloropropane a chlorine atom has replaced a hydrogen atom on the central carbon of the propane chain. { ′səb·stə,tüd·iv ′nō·mən,klā·chər }

substrain [CELL MOL] A strain derived by isolation of a single cell or group of cells having properties or markers not shared by the other cells of the cell strain. { ′səb·,strān }

substrate [BIOCHEM] The substance with which an enzyme reacts. [ECOL] The foundation to which a sessile organism is attached. [ELECTR] The physical material on which a microcircuit is fabricated; used primarily for mechanical support and insulating purposes, as with ceramic, plastic, and glass substrates; however, semiconductor and ferrite substrates may also provide useful electrical functions. [ENG] Basic surface on which a material adheres, for example, paint or laminate. [ORG CHEM] A compound with which a reagent reacts. { ′səb,strāt }

substratosphere [METEOROL] A region of indefinite lower limit just below the stratosphere. { ′səb′strad·ə,sfir }

substratum [GEOL] Any layer underlying the true soil. { ′səb′strad·əm }

substring [COMPUT SCI] A sequence of successive characters within a string. { ′səb,striŋ }

substructure [CIV ENG] The part of a structure which is below ground. { ′səb′strək·chər }

subsurface contour See structure contour. { ¦səb′sər·fəs ′kän,tur }

subsurface current [OCEANOGR] An underwater current which is not present at the surface or whose core (region of maximum velocity) is below the surface. { ¦səb′sər·fəs ′kə·rənt }

subsurface flow [HYD] Interflow plus groundwater flow. { ¦səb′sər·fəs ′flō }

subsurface geology [GEOL] The study of geologic features beneath the land or sea-floor surface. Also known as underground geology. { ¦səb′sər·fəs jē′äl·ə·jē }

subsurface irrigation [AGR] A method of providing water to plants by raising the water table to the root zone of the crop or by carrying moisture to the root zone by perforated underground pipe. Also known as subirrigation. { ¦səb,sər·fəs ,ir·ə′gā·shən }

subsurface radar See ground-probing radar. { ¦səb,sər·fəs ′rā·dar }

subsurface tillage [AGR] A method of stirring the soil with blades that leaves stubble on or just below the surface. { ¦səb′sər·fəs ′til·ij }

subsurface waste disposal [ENG] A waste disposal method for manufacturing wastes in porous underground rock formations. { ¦səb′sər·fəs ′wāst di,spōz·əl }

subsurface wave [ELECTROMAG] Electromagnetic wave propagated through water or land; operating frequencies for communications may be limited to approximately 35 kilohertz due to attenuation of high frequencies. { ¦səb′sər·fəs ′wāv }

subsynchronous [ELEC] Operating at a frequency or speed that is related to a submultiple of the source frequency. { ′səb′siŋ·krə·nəs }

subsynchronous resonance [ELEC] An electrical resonant frequency on an alternating-current transmission line that is less than the line frequency, and results from the insertion of series capacitors to cancel out part of the line and system reactance. { səb¦siŋ·krə·nəs ′rez·ən·əns }

subsystem [ENG] A major part of a system which itself has the characteristics of a system, usually consisting of several components. { ′səb,sis·təm }

subtangent [MATH] For a given point on a plane curve, the projection on the x axis of a rectangular coordinate system of the segment of the tangent between the point of tangency and the intersection of the tangent with the x axis. { ′səb′tan·jənt }

subtend [BOT] To lie adjacent to and below another structure, often enclosing it. [MATH] A line segment or an arc of a circle subtends an angle with vertex at a specified point if the end points of the line segment or arc lie on the sides of the angle. { səb′tend }

The subtense bar used in the subtense technique of distance measurement. (*Lockwood, Kessler, and Bartlett Inc.*)

subtense bar [ENG] The horizontal bar of fixed length in the subtense technique of distance measurement method. { ¦səb¦tens ¦bär }

subtense technique [CIV ENG] A distance measuring technique in which the transit angle subtended by the subtense bar enables the computation of the transit-to-bar distance. { ¦səb¦tens tek′nēk }

subterranean ice See ground ice. { ¦səb·tə′rā·nē·ən ′īs }

subterranean stream [HYD] A subsurface stream that flows through a cave or a group of communicating caves. { ¦səb·tə′rā·nē·ən ′strēm }

subtilin [MICROBIO] An antibiotic substance obtained from *Bacillus subtilis*, active against gram-positive bacteria. { ′səb·tə·lən }

subtracted time [IND ENG] In a continuous timing technique, the difference between two successive readings of a stopwatch. { səb¦trak·təd ′tīm }

subtracter [COMPUT SCI] A computer device that can form the difference of two numbers or quantities. { səb′trak·tər }

subtraction [MATH] The addition of one quantity with the negative of another; in a system with an additive operation this is formally the sum of one element with the additive inverse of another. { səb′trak·shən }

subtraction formula [MATH] An equation expressing a function of the difference of two quantities in terms of functions of the quantities themselves. { səb′trak·shən ‚fòr·myə·lə }

subtraction sign [MATH] The symbol −, used to indicate subtraction. Also known as minus sign. { səb′trak·shən ‚sīn }

subtractive primaries [OPTICS] The three colors, usually yellow, magenta, and cyan (greenish-blue), which are mixed together in a subtractive process. { səb′trak·tiv ′prī‚mer·ēz }

subtractive process [OPTICS] The process of producing colors by mixing absorbing media or filters of subtractive primary colors. { səb′trak·tiv ′prä·səs }

subtractive synthesis [ENG ACOUS] A method of synthesizing musical tones, in which an electronic circuit produces a standard waveform (such as a sawtooth wave), which contains a very large number of harmonics at known relative amplitudes, and this circuit is followed by a variety of electric or electronic filters to convert the basic tone signals into the desired musical waveforms. { səb‚trak·tiv ′sin·thə·səs }

subtractor [ELECTR] A circuit whose output is determined by the differences in analog or digital input signals. { səb′trak·tər }

subtrahend [MATH] A quantity which is to be subtracted from another given quantity. { ′səb·trə‚hend }

subtree [MATH] A subgraph of a tree which is itself a tree. { ′səb‚trē }

Subtriquetridae [INV ZOO] A family of arthropods in the suborder Porocephaloidea. { ‚səb·trə′ke·trə‚dē }

subtropic [METEOROL] An indefinite belt in each hemisphere between the tropic and temperate regions; the polar boundaries are considered to be roughly 35–40° northern and southern latitudes, but vary greatly according to continental influence, being farther poleward on the western coasts of continents and farther equatorward on the eastern coasts. { səb′träp·ik }

subtropical anticyclone See subtropical high. { ‚səb′träp·ə·kəl ¦ant·i′sī‚klōn }

Subtropical Convergence [OCEANOGR] The zone of converging currents, generally located in midlatitudes. { ‚səb′träp·ə·kəl kən′vər·jəns }

subtropical cyclone [METEOROL] The low-level (surface chart) manifestation of a cutoff low. { ‚səb′träp·ə·kəl ′sī‚klōn }

subtropical easterlies See tropical easterlies. { ‚səb′träp·ə·kəl ′ēs·tər‚lēz }

subtropical easterlies index [METEOROL] A measure of the strength of the easterly wind between the latitudes of 20° and 35°N; the index is computed from the average sea-level pressure difference between these latitudes and is expressed as the east to west component of the corresponding geostrophic wind in meters and tenths of meters per second. { ‚səb′träp·ə·kəl ′ēs·tər‚lēz ‚in‚deks }

subtropical forest See temperate rainforest. { ‚səb′träp·ə·kəl ′fär·əst }

subtropical high [METEOROL] One of the semipermanent highs of the subtropical high-pressure belt; these highs appear as centers of action on mean charts of surface pressure; they lie over oceans and are best developed in the summer season. Also known as oceanic anticyclone; oceanic high; subtropical anticyclone. { ‚səb′träp·ə·kəl ′hī }

subtropical high-pressure belt [METEOROL] One of the two belts of high atmospheric pressure that are centered, in the mean, near 30°N and 30°S latitudes; these belts are formed by the subtropical highs. { ‚səb′träp·ə·kəl ¦hī ′presh·ər ‚belt }

subtropical westerlies See westerlies. { ‚səb′träp·ə·kəl ′wes·tər‚lēz }

subulate [BOT] Linear, delicate, and tapering to a sharp point. { ′səb·yə·lət }

Subulitacea [PALEON] An extinct superfamily of gastropod mollusks in the order Prosobranchia which possessed a basal fold but lacked an apertural sinus. { ‚səb·yə·lə′tās·ē·ə }

Subuluridae [INV ZOO] The equivalent name for Heterakidae. { ¦səb·yə′lur·ə‚dē }

Subuluroidea [INV ZOO] A superfamily of parasitic nematodes in the order Ascaridida characterized by weakly developed lips with sensilla and a thick-walled stoma that is armed with three teeth. { ‚səb·yə·lə′róid·ē·ə }

subumbrella [INV ZOO] The concave undersurface of the body of a jellyfish. { ¦səb·əm′brel·ə }

subunit See protomer. { ′səb‚yü·nət }

subunit vaccine [IMMUNOL] A type of noninfectious vaccine that consists of immunogenic viral proteins stripped free from whole virus particles, then purified from other irrelevant components, thereby reducing the risk of adverse reactions and residual infectious virus. { ‚səb‚yü·nit vak′sēn }

subvoice-grade channel [COMMUN] A channel whose bandwidth is smaller than the bandwidth of a voice-grade channel; it is usually a subchannel of a voice-grade line. { ¦səb′vòis ‚grād ‚chan·əl }

subway [CIV ENG] An underground passage. { ′səb‚wā }

subway-type transformer [ELEC] Transformer of submersible construction. { ′səb‚wā ¦tīp tranz′fòr·mər }

subwoofer [ENG ACOUS] A loudspeaker designed to reproduce extremely low audio frequencies, extending into the infrasonic range, generally used in conjunction with a crossover network, a woofer, and a tweeter. { ′səb‚wüf·ər }

succession [ECOL] A gradual process brought about by the change in the number of individuals of each species of a community and by the establishment of new species populations which may gradually replace the original inhabitants. [GEOL] A group of rock units or strata that succeed one another in chronological order. { sək′sesh·ən }

succession of crops [AGR] **1.** Growing a crop over a long season by either repeated sowings or a single sowing of varieties of the crop that mature at different rates. **2.** Growing two or more crops in a season on the same land by planting them in succession. { sək′sesh·ən əv ′kräps }

successive approximation converter [COMPUT SCI] An analog-to-digital converter which operates by successively considering each bit position in the digital output and setting that bit equal to 0 or 1 on the basis of the output of a comparator. { sək′ses·iv ə‚präk·sə′mā·shən kən‚vərd·ər }

successive approximations [MATH] Any method of solving a problem in which an approximate solution is first calculated, this solution is then used in computing an improved approximation, and the process is repeated as many times as desired. { sək′ses·iv ə‚präk·sə′mā·shənz }

successive fracture treatment [PETRO ENG] A second or third fracturing operation of an oil well in an oil reservoir to fracture a new part or zone. { sək′ses·iv ′frak·chər ‚trēt·mənt }

successor [MATH] **1.** For a vertex *a* in a directed graph, any vertex *b* for which there is an arc between *a* and *b* directed from *a* to *b*. **2.** For a positive integer, *n*, the next integer, *n* + 1. Also known as consequent. { sək′ses·ər }

successor job [COMPUT SCI] A job that uses the output of another job (predecessor) as its input, so that it cannot start until the other job has been successfully completed. { sək′ses·ər ‚jäb }

succinamide [BIOCHEM] $H_2NCOCH_2CONH_2$ The amide of succinic acid. { sək′sin·ə·mīd }

succinate [ORG CHEM] A salt or ester of succinic acid; for example, sodium succinate, $Na_2C_4H_4O_4 \cdot 6H_2O$, the reaction product of succinic acid and sodium hydroxide. { ′sək·sə‚nāt }

succinate dehydrogenase [BIOCHEM] A key enzyme in the citric acid cycle; it oxidizes succinate to fumarate. { ¦sək·sə‚nāt ‚dē·hī'dräj·ə‚nās }

succinic acid [ORG CHEM] $CO_2H(CH_2)_2CO_2H$ Water-soluble, colorless crystals with an acid taste; melts at 185°C; used as a chemical intermediate, in medicine, and to make perfume esters. { sək'sin·ik 'as·əd }

succinic acid dehydrogenase [BIOCHEM] An enzyme that catalyzes the dehydrogenation of succinic acid to fumaric acid in the presence of a hydrogen acceptor. Also known as succinic dehydrogenase. { sək'sin·ik 'as·əd dē'hī·drə·jə‚nās }

succinic acid 2,2-dimethylhydrazide [ORG CHEM] $C_6H_{12}O_3N_2$ White crystals with a melting point of 154–156°C; soluble in water; used as a growth regulator for many crops and ornamentals. Also known as aminocide. { sək'sin·ik 'as·əd ¦tü ¦tü dī‚meth·əl'hī·drə‚zīd }

succinic anhydride [ORG CHEM] $C_4H_4O_3$ Colorless or pale needles soluble in alcohol and chloroform; converts to succinic acid in water; melts at 120°C; used as a chemical and pharmaceutical intermediate and a resin hardener. { sək'sin·ik an'hī‚drīd }

succinic dehydrogenase See succinic acid dehydrogenase. { sək'sin·ik dē'hī·drə·jə‚nās }

succinimide [ORG CHEM] $C_4H_5O_2N \cdot H_2O$ Colorless or tannish water-soluble crystals with a sweet taste; melts at 126°C; used to make plant growth stimulants and as a chemical intermediate. { sək'sin·ə‚mīd }

succinite [MINERAL] An amber-colored variety of grossularite. { 'sək·sə‚nīt }

succinonitrite See ethylene cyanide. { ¦sək·sə·nō'nī‚trāt }

succinoxidase [BIOCHEM] A complex enzyme system containing succinic dehydrogenase and cytochromes that catalyzes the conversion of succinate ion and molecular oxygen to fumarate ion. { ‚sək·sən'äk·sə‚dās }

succinylcholine chloride [ORG CHEM] $[Cl(CH_3)_3N-(CH_2)_2OOCH_2]_2 \cdot 2H_2O$ Water-soluble white crystals with a bitter taste, melts at 162°C; used in medicine. { sək·sən·əl'kō‚lēn 'klór‚īd }

succinyldicholine [PHARM] A drug that is used to produce relaxation of muscle during surgery; it is hydrolyzed by serum cholinesterase rather than by acetylcholinesterase, limiting the duration of paralysis. { sək‚sin·əl·dī'kō‚lēn }

succinylsulfathiazole [PHARM] $C_{13}H_{13}N_3O_5S_2$ A poorly absorbed sulfonamide used as an intestinal antibacterial agent in preoperative preparation of patients for abdominal surgery, and also postoperatively to maintain a low bacterial count. { ¦sək·sən·əl‚səl·fə'thī·ə‚zōl }

succulent [BOT] Describing a plant having juicy, fleshy tissue. { 'sək·yə·lənt }

succus entericus [PHYSIO] The intestinal juice secreted by the glands of the intestinal mucous membrane; it is thin, opalescent, alkaline, and has a specific gravity of 1.011. { 'sək·əs in'ter·ə·kəs }

sucker [BOT] A shoot that develops rapidly from the lower portion of a plant, and usually at the expense of the plant. [ZOO] A disk-shaped organ in various animals for adhering to or holding onto an individual, usually of another species. { 'sək·ər }

sucker rod [PETRO ENG] A connecting rod between a downhole oil-well pump and the lifting or pumping device on the surface. { 'sək·ər ‚räd }

sucker-rod pump [PETRO ENG] A cylinder-piston-type pump used to displace oil into the oil-well tubing string, and to the surface. { 'sək·ər ‚räd ‚pəmp }

sucking louse [INV ZOO] The common name for insects of the order Anoplura, so named for the slender, tubular mouthparts. { 'sək·iŋ ‚laús }

Sucksmith ring balance [ENG] A magnetic balance in which the specimen is rigidly suspended from a phosphor bronze ring carrying two mirrors that convert small deflections of the specimen in a nonuniform magnetic field into large deflections of a light beam; used chiefly to measure paramagnetic susceptibility. { 'sək‚smith 'riŋ ‚bal·əns }

sucrase See saccharase. { 'sü‚krās }

sucrochemical [ORG CHEM] A chemical made from a feedstock derived from sucrose extracted from sugarcane or sugarbeet. { 'sü·krō‚kem·i·kəl }

sucrochemistry [ORG CHEM] A type of chemistry based on sucrose as a starting point. { ¦sü·krō'kem·i·strē }

sucrose [ORG CHEM] $C_{12}H_{22}O_{11}$ Combustible, white crystals soluble in water, decomposes at 160 to 186°C; derived from sugarcane or sugarbeet; used as a sweetener in drinks and foods and to make syrups, preserves, and jams. Also known as saccharose; table sugar. { 'sü‚krōs }

sucrose octoacetate [ORG CHEM] $C_{28}H_{38}O_{19}$ A bitter crystalline compound that forms needles from alcohol solution, melts at 89°C, and breaks down at 286°C or above; used as an adhesive, to impregnate and insulate paper, and in lacquers and plastics. { 'sü‚krōs ¦äk·tō as·ə‚tāt }

sucrosic See saccharoidal. { sü'krō·sik }

suction anemometer [ENG] An anemometer consisting of an inverted tube which is half-filled with water that measures the change in water level caused by the wind's force. { 'sək·shən ‚an·ə'mäm·əd·ər }

suction boundary layer control [AERO ENG] A technique that is used in addition to purely geometric means to control boundary layer flow; it consists of sucking away the retarded flow in the lower regions of the boundary through slots or perforations in the surface. { 'sək·shən 'baún·drē ‚lā·ər kən‚trōl }

suction cup [ENG] A cup, often of flexible material such as rubber, in which a partial vacuum is created when it is inverted on a surface; the vacuum tends to hold the cup in place. { 'sək·shər ‚kəp }

suction-cutter dredger [MECH ENG] A dredger in which rotary blades dislodge the material to be excavated, which is then removed by suction as in a sand-pump dredger. { 'sək·shən ¦kəd·ər ‚drej·ər }

suction dredge [NAV ARCH] A vessel equipped with a centrifugal pump to excavate under water. { 'sək·shən ‚drej }

suction head See suction lift. { 'sək·shən ‚hed }

suction lift [MECH ENG] The head, in feet, that a pump must provide on the inlet side to raise the liquid from the supply well to the level of the pump. Also known as suction head. { 'sək·shən ‚lift }

suction line [ENG] A pipe or tubing feeding into the inlet of a fluid impelling device (for example, pump, compressor, or blower), consequently under suction. { 'sək·shən ‚līn }

suction pump [MECH ENG] A pump that raises water by the force of atmospheric pressure pushing it into a partial vacuum under the valved piston, which retreats on the upstroke. { 'sək·shən ‚pəmp }

suction stroke [MECH ENG] The piston stroke that draws a fresh charge into the cylinder of a pump, compressor, or internal combustion engine. { 'sək·shən ‚strōk }

suction wave See rarefaction wave. { 'sək·shən ‚wāv }

Suctoria [INV ZOO] A small subclass of the protozoan class Ciliatea, distinguished by having tentacles which serve as mouths. { sək'tōr·ē·ə }

Suctorida [INV ZOO] The single order of the protozoan subclass Suctoria. { sək'tōr·ə·də }

sudamen [MED] A skin disease in which sweat accumulates under the superficial horny layers of the epidermis to form small, clear, transparent vesicles. { sü'dā·mən }

sudatoria See hyperhidrosis. { ‚süd·ə'tōr·ē·ə }

sudburite [GEOL] A basic basalt composed of hypersthene, augite, and magnetite, among other minerals. { 'səd·bə‚rīt }

sudden commencement [GEOPHYS] Magnetic storms which start suddenly (within a few seconds) and simultaneously all over the earth. { 'səd·ən kə'mens·mənt }

sudden death syndrome See sudden infant death syndrome. { 'səd·ən 'deth ‚sin‚drōm }

sudden infant death syndrome [MED] The sudden and unexpected death of an apparently normal infant that remains unexplained after the performance of an adequate autopsy. Abbreviated SIDS. Also known as crib death; sudden death syndrome. { 'səd·ən 'in·fənt ¦deth 'sin‚drōm }

sudden ionospheric disturbance [GEOPHYS] A complex combination of sudden changes in the condition of the ionosphere following the appearance of solar flares, and the effects of these changes. Abbreviated SID. { 'səd·ən ī‚än·ə‚sfir·ik di'stər·bəns }

sudomotor [NEUROSCI] Pertaining to the efferent nerves that control the activity of sweat glands. { ¦süd·ə'mōd·ər }

suede [MATER] Leather with a velvet finish on the flesh side of the skin; calfskin is the commonest suede leather. Also known as napped leather. { swād }

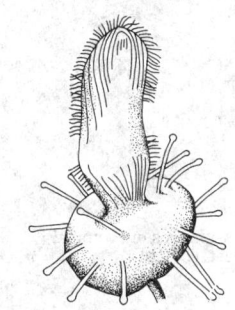

suede cloth [TEXT] Woven or knitted fabric with a surface which is finished to resemble suede leather. { 'swād ,klȯth }

suestada [METEOROL] Strong southeast winds occurring in winter along the coast of Argentina, Uruguay, and southern Brazil; they cause heavy seas and are accompanied by fog and rain; the counterpart of the northeast storm in North America. { swā'städ·ə }

suevite [GEOL] A grayish or yellowish fragmental rock associated with meteorite impact craters; resembles tuff breccia or pumiceous tuff but is of nonvolcanic origin. { 'swā,vīt }

sufficiency [STAT] Condition of an estimator that uses all the information about the population parameter contained in the sample observations. { sə'fish·ən·sē }

sufficient condition [MATH] A mathematical statement whose truth suffices to assure the truth of a given statement. { sə¦fish·ənt kən'dish·ən }

sufficient statistic [MATH] A statistic that contains all the information that can possibly be obtained from a sample to estimate a specified parameter of the sampled population. { sə¦fish·ənt stə'tis·tik }

suffix notation See reverse Polish notation. { 'səf,iks nō,tā·shən }

suffrutescent [BOT] Of or pertaining to a stem intermediate between herbaceous and shrubby, becoming partly woody and perennial at the base. { ¦sə,frü'tes·ənt }

suffruticose [BOT] Low stems which are woody, grading into herbaceous at the top. { sə'früd·ə,kōs }

sugar [BIOCHEM] A generic term for a class of carbohydrates usually crystalline, sweet, and water soluble; examples are glucose and fructose. { 'shug·ər }

sugar alcohol [ORG CHEM] Any of the acyclic linear polyhydric alcohols; may be considered sugars in which the aldehydic group of the first carbon atom is reduced to a primary alcohol; classified according to the number of hydroxyl groups in the molecule; sorbitol (D-glucitol, sorbite) is one of the most widespread of all the naturally occurring sugar alcohols. { 'shug·ər 'al·kə,hȯl }

sugarbeet [BOT] *Beta vulgaris.* A beet characterized by a white root and cultivated for the high sugar content of the roots. { 'shug·ər,bēt }

sugar berg [OCEANOGR] An iceberg of porous glacier ice. { 'shug·ər ,bərg }

sugarcane [BOT] *Saccharum officinarum.* A stout, perennial grass plant characterized by two-ranked leaves, and a many-jointed stalk with a terminal inflorescence in the form of a silky panicle; the source of more than 50% of the world's annual sugar production. { 'shug·ər,kān }

sugarcane gummosis See Cobb's disease. { 'shug·ər,kān gə'mō·səs }

sugarcane wax [MATER] Hard, tan to dark-green wax extracted from sugarcane; melts at 77°C; used in polishes, lubricants, and food wrappers. { 'shug·ər,kān ,waks }

sugarloaf sea [OCEANOGR] A sea characterized by waves that rise into sugarloaf shapes, with little wind, possibly resulting from intersecting waves. { 'shug·ər,lōf ,sē }

sugar maple [BOT] *Acer saccharum.* A commercially important species of maple tree recognized by its gray furrowed bark, sharp-pointed scaly winter buds, and symmetrical oval outline of the crown. { 'shug·ər 'mā·pəl }

sugar of lead See lead acetate. { 'shug·ər əv 'led }

sugar refining [FOOD ENG] The purification of sugar from sugarcane or sugarbeet by the removal of impurities (vegetable proteins and salts), followed by decolorization and crystallization of the pure sucrose. { 'shug·ər ri,fīn·iŋ }

sugar snow See depth hoar. { 'shug·ər ,snō }

sugary See saccharoidal. { 'shug·ə·rē }

Suhl amplifier [SOLID STATE] A parametric microwave amplifier which utilizes the instability of certain spin waves in a ferromagnetic material subjected to intense microwave fields. { 'sül ,am·plə,fī·ər }

Suhl effect [ELECTR] When a strong transverse magnetic field is applied to an *n*-type semiconducting filament, holes injected into the filament are deflected to the surface, where they may recombine rapidly with electrons or be withdrawn by a probe. { 'sül i,fekt }

suicide [IMMUNOL] Death of cells that have selectively taken up heavily radioactively labeled antigen. { 'sü·ə,sīd }

suicide inhibitor [BIOCHEM] A compound which resembles the normal substrate for an enzyme, but which interacts with

the enzyme to form a covalent bond and thus inactivates the enzyme. { 'sü·ə,sīd in'hib·əd·ər }

Suidae [VERT ZOO] A family of paleodont artiodactyls in the superfamily Suoidea including wild and domestic pigs. { 'sü·ə,dē }

suite [COMPUT SCI] A collection of related computer programs run one after another. { swēt }

sukhovei [METEOROL] Literally dry wind; a dry, hot, dusty wind in the south Russian steppes, which blows principally from the east and frequently brings a prolonged drought and crop damage. { sü·kō·vā }

sulcate [ZOO] Having furrows or grooves on the surface. { 'səl,kāt }

sulculus [ZOO] A small sulcus. { 'səl·kyə·ləs }

sulcus [ZOO] A furrow or groove, especially one on the surface of the cerebrum. { 'səl·kəs }

sulfadiazine [PHARM] $C_{10}H_{10}O_2N_4S$ An antibacterial sulfonamide used in the treatment of a variety of infections. { ¦səl·fə'dī·ə,zēn }

sulfa drug [PHARM] Any of a family of drugs of the sulfonamide type with marked bacteriostatic properties. { 'səl·fə ,drəg }

sulfaguanidine [PHARM] $C_7H_{10}N_4O_2S$ An intestinal antibacterial sulfonamide proposed for treatment of dysentery and for sterilization of the colon prior to gastrointestinal tract surgery. { ¦səl·fə'gwän·ə,dēn }

sulfallate [ORG CHEM] $C_8H_{14}NS_2Cl$ An oily liquid, used as a preemergence herbicide for vegetable crops and ornamentals. Also known as 2-chloroallyl diethyldithiocarbamate (CDEC). { səl'fa,lāt }

sulfamate [CHEM] A salt of sulfamic acid; for example, calcium sulfamate, $Ca(SO_3NH_2)_2 \cdot 4H_2O$. { 'səl·fə,māt }

sulfamerazine [PHARM] $C_{11}H_{12}N_4O_2S$ An antibacterial agent with uses similar to those of sulfadiazine, but generally used in combination with sulfadiazine and with sulfamethazine. { ¦səl·fə'mer·ə,zēn }

sulfamic acid [INORG CHEM] HSO_3NH_2 White, nonvolatile crystals slightly soluble in water and organic solvents, decomposes at 205°C; used to clean metals and ceramics, and as a plasticizer, fire retardant, chemical intermediate, and textile and paper bleach. { ¦səl'fam·ik 'as·əd }

sulfanilamide [PHARM] $C_6H_8O_2N_2S$ White crystals slightly soluble in water, soluble in alcohol and most sulfa drugs, but less effective and more toxic than its derivatives. { ,səl·fə'nil·ə,mīd }

sulfanilic acid [ORG CHEM] $C_6H_4NH_2 \cdot SO_3H \cdot H_2O$ Combustible, grayish-white crystals slightly soluble in water, alcohol, and ether, soluble in fuming hydrochloric acid; chars at 280–300°C; used in medicine and dyestuffs and as a chemical intermediate. { ¦səl·fə'nil·ik 'as·əd }

sulfapyridine [PHARM] $C_{11}H_{11}N_3O_2S$ A sulfonamide formerly used for the treatment of various infections but found to be too toxic for general use; now employed only as a suppressant for dermatitis herpetiformis. { ¦səl·fə'pir·ə,dēn }

sulfaquinoxaline See N'-2-quinoxalysulfanilimide. { ¦səl·fə·kwə'näk· sə,lēn }

sulfatase [BIOCHEM] Any of a group of esterases that catalyze the hydrolysis of sulfuric esters. { 'səl·fə,tās }

sulfate [CHEM] **1.** A compound containing the $-SO_4$ group, as in sodium sulfate, Na_2SO_4. **2.** A salt of sulfuric acid. { 'səl,fāt }

sulfate mineral [MINERAL] A mineral compound characterized by the sulfate radical SO_4. { 'səl,fāt 'min·rəl }

sulfate paper [MATER] Paper made by the sulfate process, and which cannot be bleached as white as soda or sulfite paper; strong papers, such as kraft paper, are made from unbleached sulfate materials. { 'səl,fāt ,pā·pər }

sulfate pulping [CHEM ENG] A wood-pulping process in which sodium sulfate is used in the caustic soda pulp-digestion liquor. Also known as kraft process; kraft pulping. { 'səl,fāt 'pəlp·iŋ }

sulfathiazole [PHARM] $C_9H_9N_3O_2S_2$ A sulfa drug formerly widely used in the treatment of pneumococcal, staphylococcal, and urinary tract infections; it has been replaced by less toxic sulfonamides. { ¦səl·fə'thī·ə,zōl }

sulfatide lipidosis See metachromatic leukodystrophy. { 'səl·fə,tīd ,lip·ə'dō·səs }

sulfating [ELEC] The formation of lead sulfate on the plates

SUGARBEET

Root with leaves of a typical sugarbeet. (*USDA*)

of lead-acid storage batteries reducing the energy-storing ability of the battery and eventually causing failure. { 'səl,fād·iŋ }

sulfation [CHEM] The conversion of a compound into a sulfate by the oxidation of sulfur, as in sodium sulfide, Na_2S, oxidized to sodium sulfate, Na_2SO_4; or the addition of a sulfate group, as in the reaction of sodium and sulfuric acid to form Na_2SO_4. { səl'fā·shən }

sulfenic acid [ORG CHEM] An oxy acid of sulfur with the general formula RSOH, where R is an alkyl or aryl group such as CH_3; known as the esters and halides. { |səl|fen·ik 'as·əd }

sulfenyl chloride [ORG CHEM] Any of a group of well-known organosulfur compounds with the general formula RSCl; although highly reactive compounds, they can generally be synthesized and isolated; examples are trichloromethanesulfenyl chloride and 2,4-dinitrobenzenesulfenyl chloride. { ,səl'fen·əl 'klór,īd }

sulf-heme protein [BIOCHEM] A heme protein that has reacted with sulfur to yield a new structure. { 'səlf ,hēm 'prō,tēn }

sulfhemoglobin [BIOCHEM] A greenish substance derived from hemoglobin by the action of hydrogen sulfide; it may appear in the blood following the ingestion of sulfanilamide and other substances. { |səlf'hē·mə,glō·bən }

sulfhydryl compound [CHEM] A compound with a —SH group. Also known as a mercapto compound. { |səlf'hī·drəl 'käm,paúnd }

sulfidation [CHEM] The chemical insertion of a sulfur atom into a compound. { ,səl·fə'dā·shən }

sulfide [CHEM] Any compound with one or more sulfur atoms in which the sulfur is connected directly to a carbon, metal, or other nonoxygen atom; for example, sodium sulfide, Na_2S. { 'səl,fīd }

sulfide dye [ORG CHEM] A dye containing sulfur and soluble in a 0.25–0.50% sodium sulfide solution, and used to dye cotton; the dyes are manufactured from aromatic polyamines or hydroxy amines; the amine group is primary, secondary, or tertiary, or may be an equivalent nitro, nitroso, or imino group; an example is the dye sulfur blue. Also known as sulfur dye. { 'səl,fīd ,dī }

sulfide mineral [MINERAL] A mineral compound characterized by the linkage of sulfur with a metal or semimetal. { 'səl,fīd ,min·rəl }

sulfidogen [MICROBIO] A strict anaerobe that reduces sulfur to hydrogen sulfide. { səl'fīd·ə,jen }

sulfinate [ORG CHEM] **1.** A compound containing the R_2SX_2 grouping, where X is a halide. **2.** A salt of sulfinic acid having the general formula R·OH·S:O. { səl·fə'nāt }

sulfinic acid [ORG CHEM] Any of the monobasic organic acids of sulfur with the general formula RS:O(OH); for example, ethanesulfinic acid, $C_2H_5SO_2H$. { |səl|fin·ik 'as·əd }

sulfinyl bromide See thionyl bromide. { səl·fə,nil 'brō,mīd }

sulfisoxazole [PHARM] $C_{11}H_{13}N_3O_3S$ A sulfonamide of general therapeutic utility; for parenteral administration the soluble salt sulfisoxazole diethanolamine is used; for pediatric use the tasteless derivative acetyl sulfisoxazole is given. { ,səl·fə'säk·sə,zōl }

sulfite [INORG CHEM] M_2SO_3 A salt of sulfurous acid, for example, sodium sulfite, Na_2SO_3. { 'səl,fīt }

sulfite paper [MATER] Paper made from sulfite pulp. { 'səl,fīt ,pā·pər }

sulfite pulp [MATER] Wood chips digested with a solution of magnesium, ammonium, or calcium disulfite, with free sulfur dioxide present; used to make paper and paper products from spruce and other coniferous woods. { 'səl,fīt ,pəlp }

sulfite waste liquor [MATER] Waste reactants and other impurities from the sulfite pulping of wood; used as a foaming and emulsifying agent, in adhesives and tanning, and for road construction. { 'səl,fīt ,wäst ,lik·ər }

sulfo- [CHEM] Prefix for a compound with either a divalent sulfur atom, or the presence of —SO_3H, the sulfo group in a compound. Also spelled sulpho-. { 'səl·fō or 'səl·fə }

sulfoborite [MINERAL] $Mg_6H_4(BO_3)_4(SO_4)_2·7 H_2O$ A mineral composed of hydrous acid sulfate and borate of magnesium. { ,səl·fə'bór,īt }

sulfobromophthalein sodium [PHARM] $C_{20}H_8Br_4Na_2$-$O_{10}S_2$ A hygroscopic, crystalline compound that has a bitter taste; soluble in water; used in humans and animals as a diagnostic and in liver function tests. { |səl·fō,brō·mō'thā,lēn 'sōd·ē·əm }

sulfocarbanilide See thiocarbanilide. { |səl·fō·kär'ban·ə·ləc }

sulfocarbimide See isothiocyanate. { |səl·fō'kär·bə,mīd }

sulfocyanate See thiocyanate. { |səl·fō'sī·ə,nāt }

sulfocyanic acid See thiocyanic acid. { |səl·fō·sī'an·ik 'as·əd }

sulfocyanide See thiocyanate. { |səl·fō'sī·ə,nīd }

sulfofication [GEOCHEM] Oxidation of sulfur and sulfur compounds into sulfates, occurring in soils by the agency of bacteria. { ,səl·fə·kə'kā·shən }

sulfohalite [MINERAL] $Na_6(SO_4)_2FCl$ A mineral composed of sulfate, chloride, and fluoride of sodium. { |səl·fō'ha,līt }

sulfolane [ORG CHEM] $C_4H_8SO_2$ A liquid with a boiling point of 285°C and outstanding solvent properties; used for extraction of aromatic hydrocarbons, fractionation of fatty acids, and textile finishing, and as a solvent and plasticizer. { 'səl·fə,lān }

Sulfolobus [MICROBIO] A genus of bacteria that is gram-negative, coccoid, chemolithotrophic, and thermoacidophilic. It is found worldwide in sulfur-rich hot springs and oxidizes sulfur for energy production. Its cells are highly irregular in shape, often lobed, but occasionally spherical. { səl'fäl·ə·bəs }

sulfonamide [ORG CHEM] One of a group of organosulfur compounds, RSO_2NH_2, prepared by the reaction of sulfonyl chloride and ammonia; used for sulfa drugs. { ,səl'fän·ə,mīd }

sulfonate [CHEM] A sulfuric acid derivative or a sulfonic acid ester containing a —SO_3— group. [ORG CHEM] Any of a group of petroleum hydrocarbons derived from sulfuric-acid treatment of oils, used as synthetic detergents, emulsifying and wetting agents, and chemical intermediates. { 'səl·fə,nāt }

sulfonated castor oil See Turkey red oil. { 'səl·fə,nād·əd 'kas·tər ,óil }

sulfonated oil [MATER] Mineral or vegetable oil treated with sulfuric acid to make a water-soluble (emulsifiable) form; used as lubricants, emulsifiers, defoamers, and softeners. { 'səl·fə,nād·əd 'óil }

sulfonation [CHEM] Substitution of —SO_3H groups (from sulfuric acid) for hydrogen atoms, for example, conversion of benzene, C_6H_6, into benzenesulfonic acid, $C_6H_5SO_3H$. { ,səl·fə'nā·shən }

sulfone [ORG CHEM] R_2SO_2 (or RSOOR) A compound formed by the oxidation of sulfides, for example, ethyl sulfone, $C_4H_{10}SO_2$, from ethyl sulfide, $C_4H_{10}S$; the use of sulfones, particularly 4,4'-sulfonyldianiline (dapsone) in the treatment of leprosy leads to apparent improvement; relapses associated with sulfone-resistant strains have been encountered. { 'səl fōn }

sulfonic acid [ORG CHEM] A compound with the radical —SO_2OH, derived by the sulfuric acid replacement of a hydrogen atom; for example, conversion of benzene, C_6H_6, to the water-soluble benzenesulfonic acid, $C_6H_5SO_3H$, by treatment with sulfuric acid; used to make dyes and drugs. { |səl|fän·ik 'as·əd }

sulfonyl [CHEM] Also known as sulfuryl. **1.** A compound containing the radical —SO_2—. **2.** A prefix denoting the presence of a sulfone group. { 'səl·fə,nil }

sulfonyl chloride See sulfuryl chloride. { 'səl·fə,nil 'klór,īd }

4,4'-sulfonyldianiline [PHARM] $C_{12}H_{12}N_2O_2S$ A sulfone that precipitates as crystals from alcohol; melting point 175–176°C; used in the treatment of leprosy. Also known as dapsone. { |fór ,fór,prīm |səl·fə nil·dī'an·ə·lən }

sulfophile element [GEOCHEM] An element occurring preferentially in an oxygen-free mineral. Also known as thiophile element. { 'səl·fə'fīl ,el·ə·mənt }

sulfosalicylic acid [ORG CHEM] $C_7H_6O_6S$ A trifunctional aromatic compound whose dihydrate is in the form of white crystals or crystalline powder; soluble in water and alcohol; melting point is 120°C; used as an indicator for albumin in urine and as a reagent for the determination of ferric ion; it also has industrial uses. { |səl·fō|sal·ə|sil·ik 'as·əd }

sulfoxide [ORG CHEM] R_2SO A compound with the group =SO; derived from oxidation of sulfides, the proportion of oxidant, such as hydrogen peroxide, and temperature being set to avoid excessive oxidation; an example is dimethyl sulfoxide, $(CH_3)_2SO$. { ,səl'fäk,sīd }

sulfur [CHEM] A nonmetallic element in group 16, symbol

S, atomic number 16, atomic weight 32.06, existing in a crystalline or amorphous form and in four stable isotopes; used as a chemical intermediate and fungicide, and in rubber vulcanization. [MINERAL] A yellow orthorhombic mineral occurring in crystals, masses, or layers, and existing in several allotropic forms; the native form of the element. { 'səl·fər }

sulfur-35 [NUC PHYS] Radioactive sulfur with mass number 35; radiotoxic, with 87.1-day half-life, β radiation; derived from pile irradiation; used as a tracer to study chemical reactions, engine wear, and protein metabolism. { 'səl·fər ¦thərd·ē'fīv }

sulfurated lime See calcium sulfide. { ¦səl·fə'rād·əd 'līm }

sulfuration [CHEM] The chemical act of combining an element or compound with sulfur. { ¦səl·fə'rā·shən }

sulfur bacteria [MICROBIO] Any of various bacteria having the ability to oxidize sulfur compounds. { 'səl·fər bak,tir·ē·ə }

sulfur ball [GEOL] A bubble of hot volcanic gas encased in a sulfurous mud skin that solidified on contact with air. { 'səl·fər ,bȯl }

sulfur bichloride See sulfur dichloride. { 'səl·fər bī'klȯr,īd }

sulfur bromide [INORG CHEM] S_2Br_2 A toxic, irritating, yellow liquid that reddens in air, soluble in carbon disulfide, decomposes in water, boils at 54°C. Also known as sulfur monobromide. { 'səl·fər 'brō,mīd }

sulfur cement [MATER] Cement used for connecting iron parts; made of equal parts of sulfur and pitch. { 'səl·fər si,ment }

sulfur chloride [INORG CHEM] S_2Cl_2 A combustible, water-soluble, oily, fuming, amber to yellow-red liquid with an irritating effect on the eyes and lungs, boils at 138°C; used to make military gas and insecticides, in rubber substitutes and cements, to purify sugar juices, and as a chemical intermediate. Also known as sulfur subchloride. { 'səl·fər 'klȯr,īd }

sulfur dichloride [INORG CHEM] SCl_2 A red-brown liquid boiling (when heated rapidly) at 60°C, decomposes in water; used to make insecticides, for rubber vulcanization, and as a chemical intermediate and a solvent. Also known as sulfur bichloride. { 'səl·fər dī'klȯr,īd }

sulfur dioxide [INORG CHEM] SO_2 A toxic, irritating, colorless gas soluble in water, alcohol, and ether; boils at −10°C; used as a chemical intermediate, in artificial ice, paper pulping, and ore refining, and as a solvent. Also known as sulfurous acid anhydride. { 'səl·fər dī'äk,sīd }

sulfur dome [MET] An inverted container containing a high concentration of sulfur dioxide gas, used in die casting to cover a pot of molten magnesium to prevent burning. { 'səl·fər ,dōm }

sulfur dye See sulfide dye. { 'səl·fər ,dī }

sulfur hexafluoride [INORG CHEM] SF_6 A colorless gas soluble in alcohol and ether, slightly soluble in water, sublimes at −64°C; used as a dielectric in electronics. { 'səl·fər ¦hek·sə'flůr,īd }

sulfur hexameter [ENG] An instrument used to measure or to continuously monitor the amount of sulfur hexafluoride present in a waveguide or other device in which this gas is used as a dielectric. { 'səl·fər hek'sam·əd·ər }

sulfuric acid [INORG CHEM] H_2SO_4 A toxic, corrosive, strongly acid, colorless liquid that is miscible with water and dissolves most metals, and melts at 10°C; used in industry in the manufacture of chemicals, fertilizers, and explosives, and in petroleum refining. Also known as dipping acid; oil of vitriol, vitriolic acid. { səl¦fyůr·ik 'as·əd }

sulfuric acid alkylation [CHEM ENG] A petroleum refinery alkylation process in which three-carbon, four-carbon, and five-carbon olefins combine with isobutane in the presence of a sulfuric acid catalyst to form high-octane, branched-chain hydrocarbons; used in motor gasoline. { səl¦fyůr·ik 'as·əd ,al·kə'lā·shən }

sulfuric chloride See sulfuryl chloride. { səl¦fyůr·ik 'klȯr,īd }

sulfur iodide See sulfur iodine. { səl¦fyůr·ik 'ī·ə,dīd }

sulfur iodine [INORG CHEM] I_2S_2 A gray-black brittle mass with an iodine aroma and a metallic luster, insoluble in water, soluble in carbon disulfide; used in medicine. Also known as iodine bisulfide; iodine disulfide; sulfur iodide. { səl¦fyůr·ik 'ī·ə,dīn }

sulfurized oil [MATER] Any of various mineral oils and fatty oils containing active sulfur to increase film strength and load-carrying ability; used generally as cutting fluids. { 'səl·fə,rīzd 'ȯil }

sulfur monobromide See sulfur bromide. { 'səl·fər ¦män·ə'brō,mīd }

sulfur monoxide [INORG CHEM] SO A gas at ordinary temperatures; produces an orange-red deposit when cooled to temperatures of liquid air; prepared by passing an electric discharge through a mixture of sulfur vapor and sulfur dioxide at low temperature. { 'səl·fər mə'näk,sīd }

sulfur-mud pool See mud pot. { 'səl·fər ¦məd ,pül }

sulfur number [ANALY CHEM] The number of milligrams of sulfur per 100 milliliters of sample, determined by electrometric titration; used in the petroleum industry for oils. { 'səl·fər ,nəm·bər }

sulfurous acid [INORG CHEM] H_2SO_3 An unstable, water-soluble, colorless liquid with a strong sulfur aroma; derived from absorption of sulfur dioxide in water; used in the synthesis of medicine and chemicals, manufacture of paper and wine, brewing, metallurgy, and ore flotation, as a bleach and analytic reagent, and to refine petroleum products. { 'səl·fə·rəs 'as·əd }

sulfurous acid anhydride See sulfur dioxide. { 'səl·fə·rəs 'as·əd an'hī,drīd }

sulfurous oxychloride See thionyl chloride. { 'səl·fə·rəs ¦äk·sē'klȯr,īd }

sulfur oxide [INORG CHEM] An oxide of sulfur, such as sulfur dioxide, SO_2, and sulfur trioxide, SO_3. { 'səl·fər 'äk,sīd }

sulfur oxychloride See thionyl chloride. { 'səl·fər ¦äk·sē'klȯr,īd }

sulfur spring [HYD] A spring containing sulfur compounds such as hydrogen sulfide. { 'səl·fər ,spriŋ }

sulfur subchloride See sulfur chloride. { 'səl·fər ¦səb'klȯr,īd }

sulfur test [ANALY CHEM] **1.** Method to determine the sulfur content of a petroleum material by combustion in a bomb. **2.** Analysis of sulfur in petroleum products by lamp combustion in which combustion of the sample is controlled by varying the flow of carbon dioxide and oxygen to the burner. { 'səl·fər ,test }

sulfur trioxide [INORG CHEM] SO_3 A toxic, irritating liquid in three forms, α, β, γ, with respective melting points of 62°C, 33°C, and 17°C; a strong oxidizing agent and fire hazard; used for sulfonation of organic chemicals. { 'səl·fər trī'äk,sīd }

sulfuryl See sulfonyl. { 'səl·fə,ril }

sulfuryl chloride [INORG CHEM] SO_2Cl_2 A colorless liquid with a pungent aroma, boils at 69°C, decomposed by hot water and alkalies; used as a chlorinating agent and solvent and for pharmaceuticals, dyestuffs, rayon, and poison gas. Also known as sulfonyl chloride; sulfuric chloride. { 'səl·fə,ril 'klȯr,īd }

sulfuryl fluoride [INORG CHEM] SO_2F_2 A colorless gas with a melting point of -136.7°C and a boiling point of 55.4°C; used as an insecticide and fumigant. { 'səl·fə,ril 'flůr,īd }

Sulidae [VERT ZOO] A family of aquatic birds in the order Pelecaniformes including the gannets and boobies. { 'sü·lə,dē }

sullage [CIV ENG] Drainage or wastewater from a building, farmyard, or street. [GEOL] Mud, silt, or other sediments carried and deposited by flowing water. [MET] Scoria flowing on molten metal carried in a ladle. { 'səl·ij }

Sullivan angle compressor [MECH ENG] A two-stage compressor in which the low-pressure cylinder is horizontal and the high-pressure cylinder is vertical; a compact compressor driven by a belt, or directly connected to an electric motor and diesel engine. { 'səl·ə·vən 'aŋ·gəl kəm,pres·ər }

Sullivan reaction [ORG CHEM] The formation of a red-brown color when cysteine is reacted with 1,2-naphthoquinone-4-sodium sulfate in a highly alkaline reducing medium. { 'səl·ə·vən rē,ak·shən }

sulpho- See sulfo-. { 'səl·fō }

sultriness [METEOROL] An oppressively uncomfortable state of the weather which results from the simultaneous occurrence of high temperature and high humidity, and often enhanced by calm air and cloudiness. { 'səl·trē·nəs }

sulvanite [MINERAL] Cu_3VS_4 A bronze-yellow mineral composed of copper vanadium sulfide occurring in masses. { 'səl·və,nīt }

Sulzer two-cycle engine [MECH ENG] An internal combustion engine utilizing the Sulzer Company system for the effective scavenging and charging of the two-cycle diesel engine. { 'səlt·sər 'tü ˌsī·kəl 'en·jən }

sum [MATH] **1.** The addition of numbers or mathematical objects in context. **2.** The sum of an infinite series is the limit of the sequence consisting of all partial sums of the series. **3.** The sum $A + B$ of two matrices A and B, with the same number of rows and columns, is the matrix whose element c_{ij} in row i and column j is the sum of corresponding elements a_{ij} in A and b_{ij} in B. { səm }

sumac wax See Japan wax. { 'sü·mak ˌwaks }

sumatra [METEOROL] A squall, with wind speeds occasionally exceeding 30 miles (48 kilometers) per hour, in the Malacca Strait between Malay and Sumatra during the southwest monsoon (April through November). { sü'mä·trə }

sumbul oil [MATER] Oil from the root of a muskroot (*Ferula sumbul*); formerly used in medicine as an antispasmodic. { səm'bùl ˌoil }

sumi ink See india ink. { 'sü·mē ˌiŋk }

sumi reed [GRAPHICS] A reed about a foot long with a sharp point at one end, used in ink drawings. { 'sü·mē ˌrēd }

summability method [MATH] A method, such as Hölder summation or Cesaro summation, of attributing a sum to a divergent series by using some process to average the terms in the series. { ˌsəm·əˈbil·əd·ē ˌmeth·əd }

summable function [MATH] A function whose Lebesgue integral exists. { ˌsəm·ə·bəl 'fəŋk·shən }

summary plotter [ORD] A polar coordinate chart showing the tactical distribution of ships and aircraft. { 'səm·ə·rē ˌpläd·ər }

summary recorder [COMPUT SCI] In computers, output equipment which records a summary of the information handled. { 'səm·ə·rē ri'kòrd·ər }

summation check [COMPUT SCI] An error-detecting procedure involving adding together all the digits of some number and comparing this sum to a previously computed value of the same sum. { səˈmā·shən ˌchek }

summation convention [MATH] An abbreviated notation used particularly in tensor analysis and relativity theory, in which a product of tensors is to be summed over all possible values of any index which appears twice in the expression. { səˈmā·shən kənˌven·chən }

summation network See summing network. { səˈmā·shən ˌnet·wərk }

summation principle [METEOROL] In United States weather observing practice, the rule which governs the assignment of sky cover amount to any layer of cloud or obscuring phenomenon, and to the total sky cover; in essence, this principle states that the sky cover at any level is equal to the summation of the sky cover of the lowest layer plus the additional sky cover provided at all successively higher layers up to and including the layer in question; thus, no layer can be assigned a sky cover less than a lower layer, and no sky cover can be greater than 1.0 (10/10). { səˈmā·shən ˌprin·sə·pəl }

summation sign [MATH] A capital Greek sigma (Σ) that indicates the members of a set are to be added together, and has numbers below and above it indicating the range of values of an index that are to be included in the summation. { səˈmā·shən ˌsīn }

summation tone [ACOUS] Combination tone, heard under certain circumstances, whose pitch corresponds to a frequency equal to the sum of the frequencies of the two components. { səˈmā·shən ˌtōn }

summation wave [PHYSIO] A sustained contraction of muscles, caused by the rapid firing of nerve impulses. { səˈmā·shən ˌwāv }

summer [ASTRON] The period from the summer solstice to the autumnal equinox; popularly and for most meteorological purposes, it is taken to include June through August in the Northern Hemisphere, and December through February in the Southern Hemisphere. { 'səm·ər }

summer black oil See tempering oil. { 'səm·ər 'blak ˌoil }

summer noon See daylight saving noon. { 'səm·ər 'nün }

summer savory oil See savory oil. { 'səm·ər sav·ə·rē ˌoil }

summer solstice [ASTRON] **1.** The sun's position on the ecliptic when it reaches its greatest northern declination. Also known as first point of Cancer. **2.** The date, about June 21, on which the sun has its greatest northern declination. { 'səm·ər säl·stəs }

summer time See daylight saving time. { 'səm·ər 'tīm }

summerwood [BOT] The less porous, usually harder portion of an annual ring that forms in the latter part of the growing season. { 'səm·ər·ˌwùd }

summing amplifier [ELECTR] An amplifier that delivers an output voltage which is proportional to the sum of two or more input voltages or currents. { 'səm·iŋ 'am·plə·ˌfī·ər }

summing network [ELEC] A passive electric network whose output voltage is proportional to the sum of two or more input voltages. Also known as summation network. { 'səm·iŋ ˌnet·wərk }

summit [SCI TECH] **1.** The apex of a pyramid. **2.** The highest point of, for example, a mountain, road, trajectory, or tower. { 'səm·ət }

summit plain See peak plain. { 'səm·ət ˌplān }

Sumner line [NAV] **1.** A line of position established by the Sumner method. **2.** Any celestial line of position (deprecated usage). { 'səm·nər ˌlīn }

Sumner method [NAV] The establishing of a line of position from the observation of the altitude of a celestial body by assuming two latitudes (or longitudes) and calculating the longitudes (or latitudes) through which the line of position passes; the line of position is the straight line connecting these two points (extended if necessary). { 'səm·nər ˌmeth·əd }

sum of states See partition function. { 'səm əv 'stāts }

sum over states See partition function. { 'səm ˌō·vər 'stāts }

sump [ENG] A pit or tank which receives and temporarily stores drainage at the lowest point of a circulating or drainage system. Also known as sump pit. { səmp }

sump fuse [ENG] A fuse used for underwater blasting. { 'səmp ˌfyüz }

sump pit See sump. { 'səmp ˌpit }

sump pump [MECH ENG] A small, single-stage vertical pump used to drain shallow pits or sumps. { 'səmp ˌpəmp }

sum rule [QUANT MECH] A formula for a transition between energy levels in which the sum of the transition strengths is expressed in a simple form. { 'səm ˌrül }

sun [ASTRON] The star about which the earth revolves; it is a globe of gas 8.65×10^5 miles (1.392×10^6 kilometers) in diameter, held together by its own gravity; thermonuclear reactions take place in the deep interior of the sun converting hydrogen into helium releasing energy which streams out. Also known as Sol. { sən }

sun-and-planet motion [MECH ENG] A train of two wheels moving epicyclically with a small wheel rotating a wheel on the central axis. { 'sər ən ˌplan·ət 'mō·shən }

sunburn [MED] Skin inflammation due to overexposure to sunlight. { 'sən·ˌbərn }

sun compass [NAV] A device which utilizes the shadow of a shadow pin, or gnomon, and a suitable dial to facilitate use of the sun for determination of direction. { 'sən ˌkäm·pəs }

sun crack See mud crack. { 'sən ˌkrak }

sun cross [METEOROL] A rare halo phenomenon in which bands of white light intersect over the sun at right angles; it appears probable that most of such observed crosses appear merely as a result of the superposition of a parhelic circle and a sun pillar. { 'sən ˌkrós }

sun crust [HYD] A type of snow crust, formed by refreezing of surface snow crystals after having been melted by the sun. { 'sən ˌkrəst }

sundew [BOT] Any plant of the genus *Drosera* of the family Droseraceae; the genus comprises small, herbaceous, insectivorous plants that grow on all continents, especially Australia. { 'sən·ˌdü }

sundial [HOROL] An instrument for telling time by the sun; it is composed of a stylus that casts a shadow and a dial plate, on which hour lines are marked and upon which the shadow falls. { 'sən·ˌdīl }

sun dog See parhelion. { 'sən ˌdòg }

sun drawing water [METEOROL] Popular designation for a phenomenon of the sun showing through scattered openings in a layer of clouds into a layer of turbid air that is hazy or dusty; bright bands are seen where the several beams of sunlight pass down through the subcloud layer; sailors called the phenomenon the backstays of the sun. { 'sən 'dró·iŋ 'wòd·ər }

sundtite See andorite. { 'sən·ˌtīt }

SUNFISH

Black crapple (*Pomoxis nigromaculatus*).

SUNFLOWER

Maturing sunflower. (*USDA*)

sunfish [VERT ZOO] Any of several species of marine and freshwater fishes in the families Centrarchidae and Molidae characterized by brilliant metallic skin coloration. { 'sən‚fish }

sunflower [BOT] *Helianthus annuus.* An annual plant native to the United States characterized by broad, ovate leaves growing from a single, usually long (3–20 feet or 1–6 meters) stem, and large, composite flowers with yellow petals. { 'sən‚flaú·ər }

sunflower oil [MATER] A combustible, pale-yellow, semidrying oil with a pleasant scent, expressed from the seeds of the common sunflower; soluble in alcohol, ether, and carbon disulfide; consists mostly of mixed triglycerides of fatty acids; used for resins, soaps, edible oils, and margarines. { 'sən‚flaú·ər ‚óil }

sun follower [ELECTR] A photoelectric pickup and an associated servomechanism used to maintain a sun-facing orientation, as for a space vehicle. Also known as sun seeker. { 'sən ‚fäl·ə·wər }

sun gear *See* central gear. { 'sən ‚gir }

sun-grazing comet [ASTRON] A comet whose orbit causes it to either collide with the sun or completely disintegrate in the outer solar atmosphere. { 'sən ‚gräz·iŋ 'käm·ət }

S-unit meter *See* signal-strength meter. { 'es ‚yü·nət ‚mēd·ər }

sunk [MIN ENG] Drilled downward, as a shaft. { səŋk }

sunk draft [BUILD] A recessed margin around a building stone that imparts a raised appearance to the stone. { 'səŋk ¦draft }

sunken rock [NAV] A rock that is a potential danger to navigation; for cartographic purposes, it is a rock whose summit is below the lower limit of the zone for a rock awash. { 'səŋ·kən 'räk }

sunk face [BUILD] A building stone from whose face some material has been removed in order to impart the appearance of a sunk panel. { 'səŋk ¦fās }

sunk panel [BUILD] A panel that is recessed below the face of its framing or other surrounding surface. { 'səŋk ¦pan·əl }

sunlamp [ELEC] A mercury-vapor gas-discharge tube used to produce ultraviolet radiation for therapeutic or cosmetic purposes. { 'sən‚lamp }

sun line [NAV] A line of position determined from a sextant observation of the sun. { 'sən ‚līn }

sunlit aurora [GEOPHYS] An aurora which occurs in the part of the upper atmosphere which is in the sunlight, above the earth's shadow. { 'sən‚lit ə'rór·ə }

sun opal *See* fire opal. { 'sən ‚ō·pəl }

sun pillar [METEOROL] A luminous streak of light, white or slightly reddened, extending above and below the sun, most frequently observed near sunrise or sunset; it may extend to about 20° above the sun, and generally ends in a point. Also known as light pillar. { 'sən ‚pil·ər }

sun-pumped laser [OPTICS] A continuous-wave laser in which pumping is achieved by concentrating the energy of the sun on the laser crystal with a parabolic mirror. Also known as solar-excited laser. { 'sən ‚pəmt 'lā·zər }

sunrise [ASTRON] The exact moment the upper limb of the sun appears above the horizon. { 'sən‚rīz }

sun satellite *See* solar satellite. { 'sən ‚sad·əl‚īt }

sunscald [PL PATH] An injury to woody plants which results in local death of the plant tissues; in summer it is caused by excessive action of the sun's rays, in winter, by the great variation of temperature on the side of trees that is exposed to the sun in cold weather. { 'sən‚skóld }

sun seeker *See* sun follower. { 'sən ‚sēk·ər }

sun sensor *See* solar sensor. { 'sən ‚sen·sər }

sunset [ASTRON] The exact moment the upper limb of the sun disappears below the horizon. { 'sən‚set }

sunshine [ASTRON] Direct radiation from the sun, as opposed to the shading of a location by clouds or by other obstructions. { 'sən‚shīn }

sunshine integrator [ENG] An instrument for determining the duration of sunshine (daylight) in any locality. { 'sən‚shīn ‚int·ə‚gräd·ər }

sunshine recorder [ENG] An instrument designed to record the duration of sunshine without regard to intensity at a given location; sunshine recorders may be classified in two groups according to the method by which the time scale is obtained: in one group the time scale is obtained from the motion of the

sun in the manner of a sun dial, in the second group the time scale is supplied by a chronograph. { 'sən‚shīn ri‚kórd·ər }

sunshine unit *See* strontium unit. { 'sən‚shīn ‚yü·nət }

sunspot [ASTRON] A dark area in the photosphere of the sun caused by a lowered surface temperature. { 'sən‚spät }

sunspot cycle [ASTRON] Variation of the size and number of sunspots in an 11-year cycle which is shared by all other forms of solar activity. { 'sən‚spät ‚sī·kəl }

sunspot maximum [ASTRON] The time in the solar cycle when the number of sunspots reaches a maximum value. { 'sən‚spät 'mak·sə·məm }

sunspot number *See* relative sunspot number. { 'sən‚spät ‚nəm·bər }

sunspot relative number *See* relative sunspot number. { 'sən‚spät 'rel·əd·iv ‚nəm·bər }

sunstone [MINERAL] An aventurine feldspar containing minute flakes of hematite; usually brilliant and translucent, it emits reddish or golden billowy reflection. Also known as heliolite. { 'sən ‚stōn }

sun strobe [ELECTR] The signal display seen on a radar plan-position-indicator screen when the radar antenna is aimed at the sun; the pattern resembles that produced by continuous-wave interference, and is due to radio-frequency energy radiated by the sun. { 'sən ‚strōb }

sunstroke [MED] Heat stroke resulting from prolonged exposure to the sun, characterized by extreme pyrexia, prostration, convulsion, and coma. Also known as thermic fever. { 'sən‚strōk }

sun's way [ASTRON] The path of the solar system through space. { 'sənz 'wā }

SU$_n$ symmetry [PART PHYS] A unitary symmetry based on a fundamental multiplet of *n* equivalent particles-in particular, *n* quarks. { ¦es 'yü 'səb¦en 'sim·ə·trē }

sun-synchronous orbit [AERO ENG] An earth orbit of a spacecraft so that the craft is always in the same direction relative to that of the sun; as a result, the spacecraft passes over the equator at the same spots at the same times. { 'sən ¦siŋ·krə·nəs 'ór·bət }

Sunyaev-Zeldovich effect [ASTRON] A change in the spectrum of the cosmic background radiation in the direction of galaxy clusters, whereby the radiation is less intense at wavelengths longer than 1.4 millimeters and more intense at shorter wavelengths as the result of hot gas in the clusters. { sùn¦yä‚ef zel'dō‚vich i‚fekt }

Suoidea [VERT ZOO] A superfamily of artiodactyls of the suborder Paleodonta which comprises the pigs and peccaries. { sü'óid·ē·ə }

sup *See* least upper bound.

superabrasive [MECH ENG] A material having characteristically long life and high grinding productivity such as cubic boron nitride or polycrystalline diamond. { 'sü·pər·ə‚brā·siv }

superacid [CHEM] **1.** An acidic medium that has a proton-donating ability equal to or greater than 100% sulfuric acid. **2.** A solution of acetic or phosphoric acid. { ¦sü·pər'as·əd }

superadditive function [MATH] A function *f* such that $f(x + y)$ is greater than or equal to $f(x) + f(y)$ for all *x* and *y* in its domain. { ¦sü·pər‚ad·əd·iv 'fəŋk·shən }

superadiabatic lapse rate [METEOROL] An environmental lapse rate greater than the dry-adiabatic lapse rate, such that potential temperature decreases with height. { ¦sü·pər‚ad·ē·ə'bad·ik 'laps ‚rāt }

superaerodynamics [FL MECH] That branch of gas dynamics dealing with the flow of gases at such low density that the molecular mean free path is not negligibly small; under these conditions the gas no longer behaves as a continuous fluid. { ¦sü·pər‚er·ō·dī'nam·iks }

superalloy [MET] A thermally resistant alloy for use at elevated temperatures where high stresses and oxidation are encountered. { ¦sü·pər'al‚ói }

supercalendered finish [MATER] A shiny, smooth-surface paper obtained by passing the paper between alternating fiber-filled and steel rolls with the application of steam and pressure. { ¦sü·pər'kal·ən·dərd 'fin·ish }

supercalendering [ENG] A calendering process that uses both steam and high pressure to give calendered material, for example, paper, a high-density finish. { ¦sü·pər'kal·ən·driŋ }

supercapillary interstice [GEOL] An interstice that is too large to hold water above the free water surface by surface tension; it is larger than a capillary interstice. { ¦sü·pər'kap·ə₁ler·ē in'tər·stəs }

supercardioid microphone [ENG ACOUS] A microphone whose response pattern resembles a cardioid but is exaggerated along the axis of maximum response, so that it is highly sensitive in one direction and insensitive in all others. Also known as superdirectional microphone. { ¦sü·pər₁kärd·ē₁óid 'mī·krə₁fōn }

supercavitating propeller [NAV ARCH] A marine propeller which has special blade sections so that at sufficiently high speed the whole back of each blade becomes enveloped by a smooth sheet of cavitation. { ¦sü·pər'kav·ə₁tād·iŋ prə'pel·ər }

supercavitation [FL MECH] An extreme form of cavitation in which a single bubble of gas forms around an object moving rapidly through water, enveloping it almost completely so that the water wets very little of the object's surface, thereby drastically reducing viscous drag. { ¦sü·pər₁kav·ə'ā·shən }

supercell [METEOROL] A thunderstorm with a persistent rotating updraft. While rare, it produces the most severe weather such as tornadoes, strong winds, and hail. { 'sü·pər₁sel }

supercentrifuge [MECH ENG] A centrifuge built to operate at faster speeds than an ordinary centrifuge. { ¦sü·pər'sen·trə₁fyüj }

supercharge method [ENG] A method for measuring the knock-limited power, under supercharge rich-mixture conditions, of fuels for use in spark-ignition aircraft engines. { ¦sü·pər₁chärj ₁meth·əd }

supercharger [MECH ENG] An air pump or blower in the intake system of an internal combustion engine used to increase the weight of air charge and consequent power output from a given engine size. { 'sü·pər₁chär·jər }

supercharging [MECH ENG] A method of introducing air for combustion into the cylinder of an internal combustion engine at a pressure in excess of that which can be obtained by natural aspiration. { 'sü·pər₁chärj·iŋ }

superchip See super-large-scale integrated circuit. { 'sü·pər₁chip }

supercilium [ANAT] The eyebrow. { ₁sü·pər'sil·ē·əm }

supercluster [ASTRON] An association of galaxy clusters and groups, typically composed of a few rich clusters and many poorer groups and isolated galaxies. { 'sü·pər₁kləs·tər }

supercobalt drill [DES ENG] A drill made of 8% cobalt highspeed steel; used for drilling work-hardened stainless steels, silicon chrome, and certain chrome-nickel alloy steels. { ¦sü·pər'kō₁bólt ₁dril }

supercoiling [MOL BIO] Winding of the deoxyribonucleic acid duplex on itself so that it crosses its own axis; may be in the same (positive) direction as, or opposite (negative) direction to, the turns of the double helix. { 'sü·pər₁kóil·iŋ }

supercompressibility factor See compressibility factor. { ¦sü·pər·kəm₁pres·ə'bil·əd·ē ₁fak·tər }

supercomputer [COMPUT SCI] A computer which is among those with the highest speed, largest functional size, biggest physical dimensions, or greatest monetary cost in any given period of time. { 'sü·pər₁kəm₁pyüd·ər }

superconducting accelerator [NUCLEO] A particle accelerator whose magnetic fields are generated by superconducting magnets. { ¦sü·pər·kən¦dək·tiŋ ik'sel·ə₁rād·ər }

superconducting alloy [MET] An alloy capable of exhibiting superconductivity, such as an alloy of niobium and zirconium or an alloy of lead and bismuth. { ¦sü·pər·kən¦dəkt·iŋ 'al₁ói }

superconducting ball [CRYO] A ball, typically with a radius of about 0.25 millimeter (0.01 inch), that is formed from the aggregation of several million microscopic superconducting particles in a strong electric field. { ¦sü·pər·kən₁duk·tiŋ 'ból }

superconducting circuit [CRYO] An electric circuit having elements which are in a superconducting state at least part of the time, such as a cryotron. { ¦sü·pər·kən¦dəkt·iŋ 'sər·kət }

superconducting computer [COMPUT SCI] A high-performance computer whose circuits employ superconductivity and the Josephson effect to reduce computer cycle time. { ¦sü·pər·kən¦dəkt·iŋ kəm'pyüd·ər }

superconducting cyclotron [NUCLEO] A cyclotron whose main coil is superconducting, allowing the magnetic field to be greatly increased. { ¦sü·pər·kən¦dək·tiŋ 'sī·klə₁trän }

superconducting device See cryogenic device. { ¦sü·pər·kən¦dəkt·iŋ di'vīs }

superconducting fault-current limiter [ELEC] A device which uses the transition of superconductors from zero to finite resistance to limit the fault current that results from a short circuit in an electric power system to a value that is not much higher than the nominal current. { ¦sü·pər·kən₁duk·tiŋ ¦fólt ₁kər·ənt ₁lim·əd·ər }

superconducting gyroscope See cryogenic gyroscope. { ¦sü·pər·kən¦dəkt·iŋ 'jī·rə₁skōp }

superconducting magnet [CRYO] An electromagnet whose coils are made of a type II superconductor with a high transition temperature and extremely high critical field, such as niobium tin, Nb₃Sn; it is capable of generating magnetic fields of 100,000 oersteds (8,000,000 amperes per meter) and more with no steady power dissipation. { ¦sü·pər·kən¦dəkt·iŋ 'mag·nət }

superconducting magnetic energy storage [ELEC] The storing of electrical energy, generally for use by an electrical utility during peak load period, as a circulating current in a large superconducting coil or magnet. { ¦sü·pər·kən¦dəkt·iŋ mag¦ned·ik ¦en·ər·jē 'stór·ij }

superconducting material See superconductor. { ¦sü·pər·kən¦dəkt·iŋ mə'tir·ē·əl }

superconducting memory [CRYO] A computer memory made up of a number of cryotrons, thin-film cryotrons, superconducting thin films, or other superconducting storage devices; these operate only under cryogenic conditions and dissipate power only during the read or write operation, which permits construction of large, dense memories. { ¦sü·pər·kən¦dəkt·iŋ 'mem·rē }

superconducting metal [MET] A metal capable of exhibiting superconductivity. { ¦sü·pər·kən¦dəkt·iŋ 'med·əl }

superconducting quantum interference device [ELECTR] A superconducting ring that couples with one or two Josephson junctions; applications include high-sensitivity magnetometers, near-magnetic-field antennas, and measurement of very small currents or voltages. Abbreviated SQUID. { ¦sü·pər·kən¦dəkt·iŋ ¦kwän·təm ₁in·tər¦fir·əns di₁vīs }

superconducting thin film [CRYO] A thin film of indium, tin, or other superconducting element, used as a cryogenic switching or storage device, as in a thin-film cryotron. { ¦sü·pər·kən¦dəkt·iŋ 'thin 'film }

superconductivity [SOLID STATE] A property of many metals, alloys, and chemical compounds at temperatures near absolute zero by virtue of which their electrical resistivity vanishes and they become strongly diamagnetic. { ¦sü·pər₁kän₁dək'tiv·əd·ē }

superconductor [SOLID STATE] Any material capable of exhibiting superconductivity; examples include iridium, lead, mercury, niobium, tin, tantalum, vanadium, and many alloys. Also known as cryogenic conductor; superconducting material. { ¦sü·pər·kən¦dək·tər }

supercontinent [GEOL] A large continental mass, such as Pangea, that existed early in geologic time and from which smaller continents formed and separated by fragmentation and drifting. { ¦sü·pər₁känt·ən·ənt }

supercooled cloud [METEOROL] A cloud composed of supercooled liquid waterdrops. { ¦sü·pər'küld 'klaud }

supercooling [THERMO] Cooling of a substance below the temperature at which a change of state would ordinarily take place without such a change of state occurring, for example, the cooling of a liquid below its freezing point without freezing taking place; this results in a metastable state. { ¦sü·pər'kül·iŋ }

supercritical [NUCLEO] Having an effective multiplication constant greater than 1, so that the rate of reaction rises rapidly in a nuclear reactor. [THERMO] Property of a gas which is above its critical pressure and temperature. { ¦sü·pər'krid·ə·kəl }

supercritical field [PHYS] A static field that is strong enough to cause the normal vacuum, which is devoid of real particles, to break down into a new vacuum in which real particles exist. { ¦sü·pər¦krid·ə·kəl 'fēld }

supercritical flow See rapid flow; supersonic flow. { ¦sü·pər¦krid·ə·kəl 'flō }

supercritical fluid [THERMO] A fluid at a temperature and pressure above its critical point; also, a fluid above its critical temperature regardless of pressure. { ¦sü·pər¦krid·ə·kəl 'flü·əd }

supercritical-fluid chromatography [ANALY CHEM] Any chemical separation technique using chromatography in which

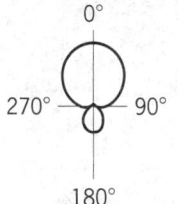

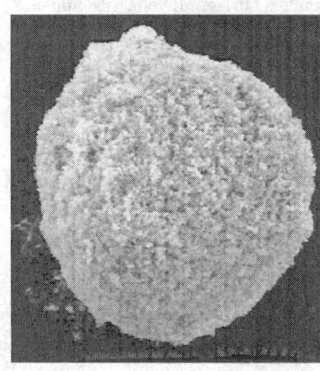

a supercritical fluid is used as the mobile phase. { ¦sü·pər‚krid·ə·kəl ¦flü·əd ‚krō·mə'täg·rə·fē }

supercritical-fluid extraction [CHEM ENG] A separation process that uses a supercritical fluid as the solvent. { ¦sü·pər¦krid·ə·kəl ¦flü·əd ik'strak·shən }

supercritical mass [NUCLEO] A mass of nuclear-reactor fuel whose effective multiplication factor is greater than 1. { ¦sü·pər'krid·ə·kəl 'mas }

supercritical reactor [NUCLEO] A nuclear reactor in which the effective multiplication constant is greater than 1 and consequently a reactor that is increasing its power level; if uncontrolled, a supercritical reactor will undergo a sudden and dangerous rise in power level. { ¦sü·pər'krid·ə·kəl rē'ak·tər }

supercritical wing [AERO ENG] A wing developed to permit subsonic aircraft to maintain an efficient cruise speed very close to the speed of sound; the middle portion of the wing is relatively flat with substantial downward curvature near the trailing edge; in addition, the forward region of the lower surface is more curved than that of the upper surface with a substantial cusp of the rearward portion of the lower surface. { ¦sü·pər'krid·ə·kəl 'wiŋ }

supercurrent [SOLID STATE] In the two-fluid model of superconductivity, the current arising from motion of superconducting electrons, in contrast to the normal current. { ¦sü·pər'kə·rənt }

superdeformed nuclear state [NUC PHYS] A nucleus in a highly excited state, formed in the collision of heavy nuclei, whose shape corresponds to an ellipsoid with an axis ratio approaching 2:1. { ¦sü·pər·di'fòrmd 'nü·klē·ər ¦stāt }

superdense state [ASTROPHYS] An extremely compact state of matter in which protons and electrons are pressed together to form neutrons, as in a neutron star. { ¦sü·pər¦dens 'stāt }

superdense theory See big bang theory. { ¦sü·pər¦dens ‚thē·ə·rē }

superdirectional microphone See supercardioid microphone. { ‚sü·pər·di‚rek·shən·əl 'mī·krə‚fōn }

superego [PSYCH] The subdivision of the psyche that acts as the conscience of the unconscious; the components, derived from both the id and the ego, are associated with standards of behavior and self-criticism. { ¦sü·pər'ē·gō }

superelastic collision [PHYS] A collision in which potential energy is converted into kinetic energy so that the total kinetic energy of the colliding objects is greater after the collision than before. { ¦sü·pər·i¦las·tik kə'lizh·ən }

superelevation [ORD] **1.** An added positive angle in antiaircraft gunnery that compensates for the fall of the projectile during the time of flight due to the pull of gravity. **2.** The angle the gun or launcher much be elevated above the gun-target line. { ¦sü·pər‚el·ə'vā·shən }

superemitron camera See image iconoscope. { ¦sü·pər'em·ə‚trän ‚kam·rə }

superexchange [SOLID STATE] A phenomenon in which two electrons from a double negative ion (such as oxygen) in a solid go to different positive ions and couple with their spins, giving rise to a strong antiferromagnetic coupling between the positive ions, which are too far apart to have a direct exchange interaction. { ¦sü·pər·iks'chānj }

superfecundation [PHYSIO] Multiple, simultaneous fertilization by a number of sperm of many eggs released at ovulation. { ¦sü·pər‚fē‚kən'dā·shən }

superfemale [GEN] In *Drosophila*, a female with three X chromosomes and two sets of autosomes resulting in sterility and generally early death. In humans, XXX females are fertile. { ¦sü·pər'fē‚māl }

superfiche [GRAPHICS] A sheet of film, usually 4 by 6 inches (10 by 15 centimeters) in size, containing either negative or positive images, or frames, of printed material reduced from 40 to 100 times by photographic reduction. { ¦sü·pər‚fēsh }

superficial cleavage [EMBRYO] Meroblastic cleavage restricted to the peripheral cytoplasm, as in the centrolecithal insect ovum. { ¦sü·pər¦fish·əl 'klē·vij }

superficial deposit See surficial deposit. { ¦sü·pər¦fish·əl di'päz·ət }

superficial expansivity See coefficient of superficial expansion. { ¦sü·pər¦fish·əl ‚ik‚span'siv·əd·ē }

superficial palmar arch [ANAT] The arterial anastomosis formed by the ulnar artery in the palm with a branch from the radial artery. Also known as palmar arch. { ¦sü·pər¦fish·əl 'päm·ər 'ärch }

superficial Rockwell hardness test [MET] A test to determine surface hardness of thin sheet material which applies relatively light loads producing minimal penetration and damage. { ¦sü·pər¦fish·əl 'räk‚wel 'härd·nəs ‚test }

superfines [MET] The portion of a metal powder composed of particles smaller than 10 micrometers. { 'sü·pər‚fīnz }

superfinishing [MET] Fine honing of a metal surface with abrasive stones. { ¦sü·pər¦fin·ə·shiŋ }

superfluid [PHYS] A collection of particles which obey Bose-Einstein statistics and are all in the lowest energy state allowed by quantum mechanics, having zero entropy and zero resistance to motion; examples are a fraction of the atoms in liquid helium II and a fraction of the pairs of electrons in a superconductor. { ¦sü·pər'flü·əd }

superfluidity [CRYO] The frictionless flow of liquid helium at temperatures very close to absolute zero through holes as small as 10^{-7} centimeter in diameter, and for particle velocities below a few centimeters per second. { ¦sü·pər'flü'id·əd·ē }

superfluorescence [ATOM PHYS] The process of spontaneous emission of electromagnetic radiation from a collection of excited atoms. { ¦sü·pər·flü'res·əns }

supergalaxy [ASTRON] A hypothetical very large group of galaxies which together fill an ellipsoidal space. { ¦sü·pər'gal·ik·sē }

supergene [GEN] A chromosomal segment protected from crossing over and therefore transmitted from generation to generation as if it were a single recon. [MINERAL] Referring to mineral deposits or enrichments formed by descending solutions. Also known as hypergene. { 'sü·pər‚jēn }

supergeostrophic wind [METEOROL] Any wind of greater speed than the geostrophic wind required by the pressure gradient. { ¦sü·pər‚jē·ō¦sträf·ik 'wind }

supergiant star [ASTRON] A member of the family containing the intrinsically brightest stars, populating the top of the Hertzsprung-Russell diagram; supergiant stars occur at all temperatures from 30,000 to 3000 K and have luminosities from 10^4 to 10^6 times that of the sun; the star Betelgeuse is an example. { ¦sü·pər'gī·ənt 'stär }

superglacial [HYD] Of or pertaining to the upper surface of a glacier or ice sheet. { ¦sü·pər'glā·shəl }

supergradient wind [METEOROL] A wind of greater speed than the gradient wind required by the existing pressure gradient and centrifugal force. { ¦sü·pər'grād·ē·ənt 'wind }

supergranulation [ASTRON] A system of convective cells, with typical diameters of 12,000 miles (20,000 kilometers), that cover the sun's surface. { ‚sü·pər‚gran·yə'lā·shən }

supergranulation cells [ASTRON] Convective cells in the solar photosphere with primarily horizontal flow and diameters of about 20,000 miles (30,000 kilometers). Also known as supergranules. { ¦sü·pər‚gran·yə'lā·shən ‚selz }

supergranules See supergranulation cells. { ¦sü·pər'gran·yülz }

supergravity [PHYS] A supersymmetry which is used to unify general relativity and quantum theory; it is formed by adding to the Poincaré group, as a symmetry of space-time, four new generators that behave as spinors and vary as the square root of the translations. { ¦sü·pər'grav·əd·ē }

supergroup [COMMUN] In carrier telephony, five groups (60 voice channels) multiplexed together and treated as a unit; a basic supergroup occupies the band between 312 and 552 kilohertz. [GEOL] A lithostratigraphic material unit of the highest order. { 'sü·pər‚grüp }

superharmonic function [MATH] A continuous complex function f whose value at a point z_0 exceeds its average values computed by the integral of f around a circle centered at z_0. { ¦sü·pər·här¦män·ik 'fəŋk·shən }

superheat [THERMO] Sensible heat in a gas above the amount needed to maintain the gas phase. { 'sü·pər‚hēt }

superheated vapor [THERMO] A vapor that has been heated above its boiling point. { ¦sü·pər'hēd·əd 'vā·pər }

superheater [MECH ENG] A component of a steam-generating unit in which steam, after it has left the boiler drum, is heated above its saturation temperature. { ¦sü·pər'hēd·ər }

superheating [THERMO] Heating of a substance above the temperature at which a change of state would ordinarily take place without such a change of state occurring, for example, the heating of a liquid above its boiling point without boiling

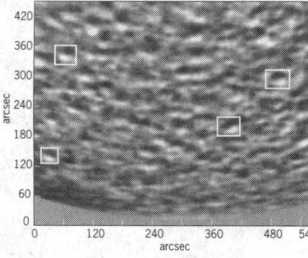

SUPERGRANULATION

Dopplergram showing supergranules near the solar limb. The slant view makes horizontal velocities appear dark (approaching) and bright (receding). Supergranules appear elongated due to foreshortening. Four supergranules are boxed. *(G. Simon, Air Force Research Lab, National Solar Observatory, Sunspot, NM, and R. Shine, Lockheed Martin Solar and Astrophysics Lab, Palo Alto, CA)*

taking place; this results in a metastable state. { ¦sü·pər'hēd·iŋ }

superheavy boson [PART PHYS] A hypothetical particle, postulated in grand unified theories, which would be responsible for interactions between quarks and leptons in the early universe, and also responsible for proton decay. Also known as X boson. { ¦sü·pər'hev·ē 'bō,sän }

superheavy element [INORG CHEM] A chemical element with an atomic number of 110 or greater. { ¦sü·pər'hev·ē 'el·ə·mənt }

superhelix [BIOCHEM] A macromolecular structure consisting of a number of alpha-helical polypeptide strands which are twisted together. { ¦sü·pər'hē,liks }

superhet See superheterodyne receiver. { ¦sü·pər,het }

superheterodyne receiver [ELECTR] A receiver in which all incoming modulated radio-frequency carrier signals are converted to a common intermediate-frequency carrier value for additional amplification and selectivity prior to demodulation, using heterodyne action; the output of the intermediate-frequency amplifier is then demodulated in the second detector to give the desired audio-frequency signal. Also known as superhet. { ¦sü·pər'he·trə,dīn ri'sē·vər }

superhigh frequency [COMMUN] A frequency band from 3000 to 30,000 megahertz, corresponding to wavelengths from 1 to 10 centimeters. Abbreviated SHF. { ¦sü·pər'hī 'frē·kwən·sē }

superhighway [CIV ENG] A broad highway, such as an expressway, freeway, turnpike, for high-speed traffic. { ¦sü·pər'hī,wā }

superimposed [GEOL] Pertaining to layered or stratified rocks. { ¦sü·pər·im'pōzd }

superimposed back pressure [MECH ENG] The static pressure at the outlet of an operating pressure relief device, resulting from pressure in the discharge system. { ¦sü·pər·im'pōzd 'bak ,presh·ər }

superimposed coding [COMPUT SCI] A means of placing many keywords in a single card area, where they can be scanned simultaneously. { ¦sü·pər·im'pōzd 'kōd·iŋ }

superimposed drainage [HYD] A naturally evolved drainage system that became established on a preexisting surface, now eroded, and whose course is unrelated to the present underlying geological structure. { ¦sü·pər·im'pōzd 'drā·nij }

superimposed fan [GEOL] An alluvial fan developed on, and having a steeper gradient than, an older fan. { ¦sü·pər·im'pōzd 'fan }

superimposed fold See cross fold. { ¦sü·pər·im'pōzd 'fōld }

superimposed glacier [GEOL] A glacier whose course is maintained despite different preexisting structures and lithologies as the glacier erodes downward. { ¦sü·pər·im'pōzd 'glā·shər }

superimposed stream [HYD] A stream, started on a new surface, that kept its course through the different preexisting lithologies and structures encountered as it eroded downward into the underlying rock. Also known as superinduced stream. { ¦sü·pər·im'pōzd 'strēm }

superimposed valley [GEOL] A valley eroded by or containing a superimposed stream. { ¦sü·pər·im'pōzd 'val·ē }

superincumbent [GEOL] Pertaining to a superjacent layer, especially one that is situated so as to exert pressure. { ¦sü·pər·in'kəm·bənt }

superinduced stream See superimposed stream. { ¦sü·pər·in'düst 'strēm }

superinfection [VIROL] An attack on a bacterial cell by several phages due to the introduction of large numbers of viruses into the bacterial culture. { ¦sü·pər·in'fek·shən }

superinsulation [CHEM ENG] A multilayer insulation for cryogenic systems, composed of many floating radiation shields in an evacuated double-wall annulus, closely spaced but thermally separated by a poor-conducting fiber. { ¦sü·pər,in·sə'lā·shən }

superionic conduction [SOLID STATE] Extremely fast conduction of ions in certain inorganic crystalline solids, approaching the ionic conductivity of aqueous sodium chloride. { ¦sü·pər,ī'än·ik kən'dək·shən }

superionic conductor [SOLID STATE] An ionic solid whose ionic conductivity is extremely high, on the order of 100 times that normally observed. { ¦sü·pər,ī'än·ik kən'dək·tər }

superior [BOT] **1.** Positioned above another organ or structure. **2.** Referring to a calyx that is attached to the ovary. **3.**

Referring to an ovary that is above the insertion of the floral parts. { sə'pir·ē·ər }

superior air [METEOROL] An exceptionally dry mass of air formed by subsidence and usually found aloft but occasionally reaching the earth's surface during extreme subsidence processes. { sə'pir·ē·ər 'er }

superior alveolar canals [ANAT] The alveolar canals of the maxilla. { sə'pir·ē·ər al've·ə·lər kə'nalz }

superior conjunction [ASTRON] A conjunction when an astronomical body is opposite the earth on the other side of the sun. { sə'pir·ē·ər kən'jəŋk·shən }

superior fruit See true fruit. { sə,pir·ē·ər 'früt }

superior ganglion [NEUROSCI] **1.** The upper sensory ganglion of the glossopharyngeal nerve, located in the upper part of the jugular foramen; it is inconstant. **2.** The upper sensory ganglion of the vagus nerve, located in the jugular foramen. { sə'pir·ē·ər 'gaŋ·glē·ən }

superior mesenteric artery [ANAT] A major branch of the abdominal aorta with branches to the pancreas and intestine. { sə'pir·ē·ər ,mez·ən'ter·ik 'ärd·ə·rē }

superior mirage [OPTICS] A spurious image of an object formed above the object's position by abnormal refraction conditions; opposite to an inferior mirage. { sə'pir·ē·ər mi'räzh }

superior planet [ASTRON] Any of the planets that are farther than the earth from the sun; includes Mars, Jupiter, Saturn, Uranus, Neptune, and Pluto. { sə'pir·ē·ər 'plan·ət }

superior tide [OCEANOGR] The tide in the hemisphere in which the moon is above the horizon. { sə'pir·ē·ər 'tīd }

superior transit See upper transit. { sə'pir·ē·ər 'tran·zət }

superior vena cava [ANAT] The principal vein collecting blood from the head, chest wall, and upper extremities and draining into the right atrium. { sə'pir·ē·ər ¦vē·nə 'kä·və }

superjacent [GEOL] Pertaining to a stratum situated immediately upon or over a particular lower stratum or above an unconformity. { ¦sü·pər'jā·sənt }

superjacent roadway system See Hirschback method. { ¦sü·pər·jā·sənt 'rōd,wā ,sis·təm }

superjacent waters [OCEANOGR] The waters above the continental shelf. { ¦sü·pər'jā·sənt 'wòd·ərz }

super-Jupiter See brown dwarf. { ¦sü·pər 'jü·pəd·ər }

super-large-scale integrated circuit [ELECTR] A very complex integrated circuit that has a high density of transistors and other components, for a total of 10^6 or more components. Also known as superchip. Abbreviated SLSI circuit. { ¦sü·pər ¦lärj ¦skāl ,in·tə'grād·əd 'sər·kət }

superlattice [ELECTR] A structure consisting of alternating layers of two different semiconductor materials, each several nanometers thick. [SOLID STATE] An ordered arrangement of atoms in a solid solution which forms a lattice superimposed on the normal solid solution lattice. Also known as artificial crystal; artificially layered structure; superstructure. { ¦sü·pər'lad·əs }

superlayer [FL MECH] A very thin, highly convoluted interface that separates the turbulent from the nonturbulent regions in a flow of high Reynolds number. { 'sü·pər,lā·ər }

superleak See lambda leak. { 'sü·pər,lēk }

superline [COMPUT SCI] A unit of text longer than an ordinary line, used in some of the more powerful text editors. { 'sü·pər,līn }

superluminal motion [ASTRON] Apparent proper motion exceeding the velocity of light in an astronomical object. { ,sü·pər,lim·ə·nəl 'mō·shən }

superluminal radio source [ASTRON] A radio source whose velocity appears to exceed that of light. { ¦sü·pər¦lüm·ən·əl 'rād·ē·ō ,sórs }

supermale [GEN] In *Drosophila*, a male with one X chromosome and three or more sets of autosomes, resulting in sterility and generally early death. { 'sü·pər,māl }

supermassive star [ASTRON] A star with a mass exceeding about 50 times that of the sun. { ¦sü·pər'mas·iv 'stär }

supermature [GEOL] Pertaining to a texturally mature clastic sediment whose grains have become rounded. { ¦sü·pər·mə'chùr }

super-metal-rich star [ASTRON] **1.** A low-luminosity giant star of spectral class K, strongly enhanced cyanogen radical (CN) bands, and apparently strong metal lines. **2.** A star that is significantly richer in metals than those of the Hyades. { ¦sü·pər ¦med·əl ,rich 'stär }

supermicro [COMPUT SCI] A computer resembling a

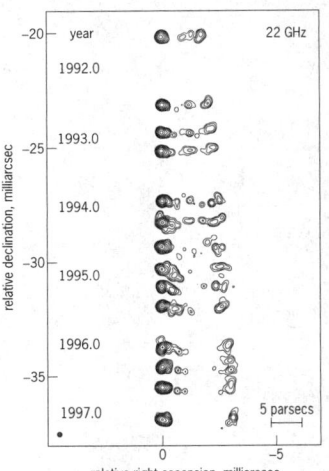

SUPERLUMINAL MOTION

A series of 14 very long baseline interferometry images of the quasar 3C 279 over a period of 6 years. All images show a bright feature on the left (the nucleus of the quasar), and a feature on the right, which moves away steadily with time. Superluminal speed is 4.3c, where c is the speed of light. (*Astronomical Society of the Pacific Conference Proceedings Series*)

supermini in design but scaled down to the size of a microcomputer, usually capable of working with a small number of users at once. { ¦sü·pər'mī·krō }

supermini [COMPUT SCI] A large minicomputer that has sufficient capability to compete with mainframe computers. { 'sü·pər,min·ē }

supermode laser [OPTICS] Frequency-modulated laser, the output of which is passed through a second phase modulator driven 180° out of phase and with the same modulation index as the first modulator; all of the energy of the previously existing laser modes is compressed into a single frequency with nearly the full power of the laser concentrated in that signal. { ¦sü·pər'mōd 'lā·zər }

supermolecule [PHYS CHEM] A single quantum-mechanical entity presumably formed by two reacting molecules and in existence only during the collision process; a concept in the hard-sphere collision theory of chemical kinetics. { ¦sü·pər'mäl·ə,kyül }

supermultiplet [QUANT MECH] A set of quantum-mechanical states each of which has the same value of some fundamental quantum numbers and differs from the other members of the set by other quantum numbers, which take values from a range of numbers dictated by the fundamental quantum numbers. { ¦sü·pər'məl·tə·plət }

supernatant [BIOCHEM] The overlying soluble liquid fraction of a sample that remains after precipitation of the insoluble solid component by centrifugation. { ,sü·pər'nāt·ənt }

supernatant liquor [ENG] The liquid above settled solids, as in a gravity separator. { ¦sü·pər'nāt·ənt 'lik·ər }

supernova [ASTRON] A star that suddenly bursts into very great brilliance as a result of its blowing up; it is orders of magnitude brighter than a nova. { ¦sü·pər'nō·və }

supernova remnant [ASTRON] A nebula consisting of an expanding shell of gas that has been ejected by a supernova. Abbreviated SNR. { ¦sü·pər'nō·və 'rem·nənt }

supernumerary bud See accessory bud. { ¦sü·pər'nü·mə,rer·ē 'bəd }

supernumerary chromosome [CELL MOL] A chromosome present in addition to the normal chromosome complement. Also known as accessory chromosome. { ¦sü·pər'nü·mə,rer·ē 'krō·mə,sōm }

supernumerary rainbow [OPTICS] One of a set of weakly colored rainbow arcs sometimes discernible inside a primary rainbow; they are of smaller angular width, and fade toward the common center. { ¦sü·pər'nü·mə,rer·ē 'rān,bō }

superoxide dismutase [ORG CHEM] An enzymatic antioxidant that removes the potentially toxic superoxide ion (O^{2-}) by disproportionating it to O_2 and hydrogen peroxide (H_2O_2). { ¦sü·pər,äk,sīd 'diz·myü,tās }

superoxide ion [CHEM] O_2^- An ion formed by the combination of one molecule of dioxygen (O_2) and one electron (e^-). { ,sü·pər'äk,sīd ,ī·ən }

superparamagnetic particle [SOLID STATE] A crystalline grain in a magnetic medium that is so small that its magnetic properties decrease with time due to thermal fluctuations. { ¦sü·pər,par·ə·mag,ned·ik 'pärd·i·kəl }

superparamagnetism See collective paramagnetism. { ¦sü·pər,par·ə'mag·nə,tiz·əm }

superphosphate [MATER] The most important phosphorus fertilizer, derived by action of sulfuric acid on phosphate rock (mostly tribasic calcium phosphate) to produce a mix of gypsum and monobasic calcium phosphate. { ¦sü·pər'fä,sfāt }

superplasticity [MET] The unusual ability of some metals and alloys to elongate uniformly by thousands of percent at elevated temperatures, much like hot polymers or glasses. { ¦sü·pər·pla'stis·əd·ē }

superposability See congruence. { ,sü·pər,pōz·ə'bil·əd·ē }

superposed [BOT] **1.** Growing vertically over another part. **2.** Of or pertaining to floral parts that are opposite each other. { ¦sü·pər'pōzd }

superposed circuit [COMMUN] Additional channel obtained from one or more circuits, normally provided for other channels, in a way that all channels can be used simultaneously without mutual interference. { ¦sü·pər'pōzd 'sər·kət }

superposed stream See consequent stream. { ¦sü·pər'pōzd 'strēm }

superposition [GEOL] **1.** The order in which sedimentary layers are deposited, the highest being the youngest. **2.** The process by which the layering occurs. [MATH] The principle

of superposition states that any given geometric figure in a euclidean space can be so moved about as not to change its size or shape. [PHYS] Addition of phenomena when the sum of two physically realizable disturbances is also physically realizable; for example, sound waves are superposable in this sense, but shock waves are not. { ,sü·pər·pə'zish·ən }

superposition eye [INV ZOO] A compound eye in which a given rhabdome receives light from a number of facets; visual acuity is reduced in this type of eye. { ,sü·pər·pə'zish·ən 'ī }

superposition integral [CONT SYS] An integral which expresses the response of a linear system to some input in terms of the impulse response or step response of the system; it may be thought of as the summation of the responses to impulses or step functions occurring at various times. { ,sü·pər·pə'zish·ən 'int·ə·grəl }

superposition principle See principle of superposition. { ,sü·pər·pə'zish·ən 'prin·sə·pəl }

superposition theorem See principle of superposition. { ,sü·pər·pə'zish·ən 'thir·əm }

superprint See overprint. { 'sü·pər,print }

superradiant scattering [RELAT] The scattering of radiation from a black hole in such a way that the scattered radiation carries more energy than the incident radiation. { ¦sü·pər'rād·ē·ənt 'skad·ə·riŋ }

superreflexive Banach space [MATH] A Banach space, B, such that any Banach space that is finitely representable in B is a reflexive Banach space. { ,sü·pər·ri¦flek·siv 'bä,näk ,spās }

superrefraction See ducting. { ,sü·pər·ri'frak·shən }

superregeneration [ELECTR] Regeneration in which the oscillation is broken up or quenched at a frequency slightly above the upper audibility limit of the human ear by a separate oscillator circuit connected between the grid and anode of the amplifier tube, to prevent regeneration from exceeding the maximum useful amount. { ¦sü·pər·ri,jen·ə'rā·shən }

superresolution [OCEANOGR] Separation of tides into components of different frequencies, without taking measurements for the full extent of the longest-period component. { ¦sü·pər,rez·ə'lü·shən }

supersaturation [METEOROL] The condition existing in a given portion of the atmosphere when the relative humidity is greater than 100%, in respect to a plane surface of pure water or pure ice. [PHYS CHEM] The condition existing in a solution when it contains more solute than is needed to cause saturation. Also known as supersolubility. { ¦sü·pər,sach·ə'rā·shən }

superscalar architecture [COMPUT SCI] A design that enables a central processing unit to send several instructions to different execution units simultaneously, allowing it to execute several instructions in each clock cycle. { ¦sü·pər,skā·lər 'är·kə,tek·chər }

super-Schmidt telescope [OPTICS] A type of Schmidt system that has a compound corrector plate consisting of a pair of opposing meniscus lenses and an achromatic doublet. { 'sü·pər 'shmit 'tel·ə,skōp }

superscript [SCI TECH] A letter or symbol written above, and usually to the right, of another symbol for any of various purposes, such as to denote a power or derivative, to identify a particular element of a set, or to indicate the mass number of an isotope. { 'sü·pər,skript }

supersensitive relay [ELEC] A relay that operates on extremely small currents, generally below 250 microamperes. { ¦sü·pər'sen·səd·iv 'rē,lā }

superset [COMPUT SCI] A programming language that contains all the features of a given language and has been expanded or enhanced to include other features as well. [MATH] A set whose elements include all the elements of a given set. { 'sü·pər,set }

supersolubility See supersaturation. { ¦sü·pər,säl·yə'bil·əd·ē }

supersonic See ultrasonic. [PHYS] Of, pertaining to, or dealing with speeds greater than the speed of sound. { ¦sü·pər¦sän·ik }

supersonic aerodynamics [FL MECH] The study of aerodynamics of supersonic speeds. { ¦sü·pər¦sän·ik ¦er·ō·dī'nam·iks }

supersonic aircraft [AERO ENG] Aircraft capable of supersonic speeds. { ¦sü·pər¦sän·ik 'er,kraft }

supersonic airfoil [AERO ENG] An airfoil designed to produce lift at supersonic speeds. { ¦sü·pər¦sän·ik 'er,fȯil }

supersonic combustion ramjet *See* scramjet. { ¦sü·pər¦sän·ik kəm¦bəs·chən 'ram,jet }

supersonic compressor [MECH ENG] A compressor in which a supersonic velocity is imparted to the fluid relative to the rotor blades, the stator blades, or both, producing oblique shock waves over the blades to obtain a high-pressure rise. { ¦sü·pər¦sän·ik kəm¦pres·ər }

supersonic diffuser [MECH ENG] A diffuser designed to reduce the velocity and to increase the pressure of fluid moving at supersonic velocities. { ¦sü·pər¦sän·ik di'fyü·zər }

supersonic flow [FL MECH] Flow of a fluid over a body at speeds greater than the speed of sound in the fluid, and in which the shock waves start at the surface of the body. Also known as supercritical flow. { ¦sü·pər¦sän·ik 'flō }

supersonic inlet [AERO ENG] An inlet of a jet engine at which single, double, or multiple shock waves form. { ¦sü·pər¦sän·ik 'in,let }

supersonic nozzle *See* convergent-divergent nozzle. { ¦sü·pər¦sän·ik 'näz·əl }

supersonic transport [AERO ENG] A transport plane capable of flying at speeds higher than the speed of sound. Abbreviated SST. { ¦sü·pər¦sän·ik 'tranz,pórt }

superspace [PHYS] The space of all three-geometries on a three-manifold, used in discussions of quantum gravity. { 'sü·pər,spās }

superstandard propagation [ELECTROMAG] The propagation of radio waves under conditions of superstandard refraction in the atmosphere, that is, refraction by an atmosphere or section of the atmosphere in which the index of refraction decreases with height at a rate of greater than 12 N units (unit of index of refraction) per 1000 feet (304.8 meters). { ¦sü·pər'stan·dərd ,präp·ə'gā·shən }

superstring theory [PART PHYS] A theory of elementary particles which obeys supersymmetry and in which the particles are one-dimensional, closed curves with zero thickness and length of the order of the Planck length, 10^{-35} m. { 'sü·pər,striŋ ,thē·ə·rē }

superstructure [CIV ENG] The part of a structure that is raised on the foundation. [NAV ARCH] The entire structure of a ship above the main deck. [SOLID STATE] *See* superlattice. { 'sü·pər,strək·chər }

supersymmetry [PART PHYS] A generalization of previously known symmetries of elementary particles to new kinds of supermultiplets that include both bosons and fermions; it is based on graded Lie algebras rather than on Lie algebras. { ¦sü·pər'sim·ə·trē }

supersystem [SCI TECH] A whole comprising systems as the major functions at the first level; conceptualized during the process of synthesis rather than analysis, it results from an awareness that a system under consideration has a relationship with one or more other systems. { 'sü·pər,sis·təm }

supertanker [NAV ARCH] An obsolete term for an unusually large tanker, about 30,000 to 50,000 tons deadweight; such a vessel is now referred to as a very large or ultra-large crude carrier. { 'sü·pər,taŋ·kər }

superthermal particles [ASTRON] Particles in the solar corona that have been accelerated by magnetic energy dissipation to very high energies, from 10^2 to 10^6 times the mean thermal energy of particles in the coronal gas. { ¦sü·pər,thər·məl 'pärd·ə·kəlz }

superthermal source [NUCLEO] A source of ultracold neutrons consisting of a container of liquid helium, cooled to about 1 K or less, which has walls that are good reflectors of ultracold neutrons but are transparent to electrons with energies of about 1 millielectronvolt, and which is placed in a beam of such electrons. { ¦sü·pər'thər·məl 'sórs }

supertransuranics [INORG CHEM] A group of relatively stable elements, with atomic numbers around 114 and mass numbers around 298, that are predicted to exist beyond the present periodic table of known elements. { ¦sü·pər,tranz·yü'ran·iks }

superturbulent flow [FL MECH] The flow of water in which the energy loss by friction is so great that Reynolds criterion for the transition of laminar to turbulent flow does not apply. { ¦sü·pər¦tər·byə·lənt 'flō }

supertweeter [ENG ACOUS] A loudspeaker designed to reproduce extremely high audio frequencies, extending into the ultrasonic range, generally used in conjunction with a crossover network, a tweeter, and a woofer. { 'süp·ər,twēd·ər }

supertwisted nematic liquid-crystal display [ELECTR] A display in which nematic liquid-crystal molecules are twisted more than 90°, and the picture elements respond to the average (root-mean-square) voltage applied by transistors connected to each row and column to switch the liquid. Abbreviated STN LCD. Also known as passive-matrix liquid-crystal display (PM LCD). { ¦sü·pər,twis·təd nə¦mad·ik ,lik·wəd 'krist·əl di,splā }

supervisor [COMPUT SCI] A collection of programs, forming part of the operating system, that provides services for and controls the running of user programs. { 'sü·pər,vī·zər }

supervisor call [COMPUT SCI] A mechanism whereby a computer program can interrupt the normal flow of processing and ask the supervisor to perform a function for the program that the program cannot or is not permitted to perform for itself. Also known as system call. { 'sü·pər,vī·zər ,kól }

supervisor interrupt [COMPUT SCI] An interruption caused by the program being executed which issues an instruction to the master control program. { 'sü·pər,vī·zər 'int·ə,rəpt }

supervisor mode [COMPUT SCI] A method of computer operation in which the computer can execute all its own instructions, including the privileged instruction not normally allowed to the programmer, in contrast to problem mode. { 'sü·pər,vī·zər ,mōd }

supervisory computer [COMPUT SCI] A minicomputer which accepts test results from satellite minicomputers, transmits new programs to the satellite minicomputers, and may further communicate with a larger computer. { 'sü·pər,vīz·ə·rē kəm'pyüd·ər }

supervisory control [ENG] A control panel or room showing key readings or indicators (temperature, pressure, or flow rate) from an entire operating area allowing visual supervision and control of the overall operation. { 'sü·pər,vīz·ə·rē kər'trōl }

supervisory control and data acquisition [ENG] A version of telemetry commonly used in wide-area industrial applications, such as electrical power generation and distribution and water distribution, which includes supervisory control of remote stations as well as data acquisition from those stations over a bidirectional communications link. Abbreviated SCADA. { ,sü·pər,vīz·ə·rē kən,trōl ən 'dad·ə ,ak·wə zish·ən }

supervisory controlled manipulation [ENG] A form of remote manipulation in which a computer enables the operator to teach the manipulator motion patterns to be remembered and repeated later. { 'sü·pər,vīz·ə·rē kən'trōld mə,nip·yə'lā·shən }

supervisory expert control system [CONT SYS] A control system in which an expert system is used to supervise a set of control, identification, and monitoring algorithms. { ,sü·pər,vīz·ə·rē ,ek,spərt kən'trōl ,sis·təm }

supervisory program [COMPUT SCI] A program that organizes and regulates the flow of work in a computer system, for example, it may automatically change over from one run to another and record the time of the run. { 'sü·pər,vīz·ə·rē 'prō,gram }

supervisory routine [COMPUT SCI] A program or routine that initiates and guides the execution of several (or all) other routines and programs; it usually forms part of (or is) the operating system. { 'sü·pər,vīz·ə·rē rü'tēn }

supervisory signal [ELEC] A signal which indicates the operating condition of a circuit or a combination of circuits in a switching apparatus or other electrical equipment to an attendant. { 'sü·pər,vīz·ə·rē 'sig·nəl }

supervisory system [ELEC] A system of control, indicating, and telemetry devices which operates between the stations of an electric power distribution system, using a single common channel to transmit signals. { 'sü·pər,vīz·ə·rē 'sis·təm }

supervoltage [ELEC] A voltage in the range of 500 to 2000 kilovolts, used for some x-ray tubes. { 'sü·pər'vōl·tij }

superwind [ASTRON] An energetic outflow of hot material along the minor axis of some starburst galaxies, detected in x-rays and the emission lines of hydrogen. { 'sü·pər,wind }

supination [ANAT] **1.** Turning the palm upward. **2.** Inversion of the foot. [CONT SYS] The orientation and motion of a robot component with its front or unprotected side facing upward and exposed. { ,sü·pə'nā·shən }

supplemental chords [MATH] Two chords joining a point on the circumference of a circle to the ends of a diameter of the circle.

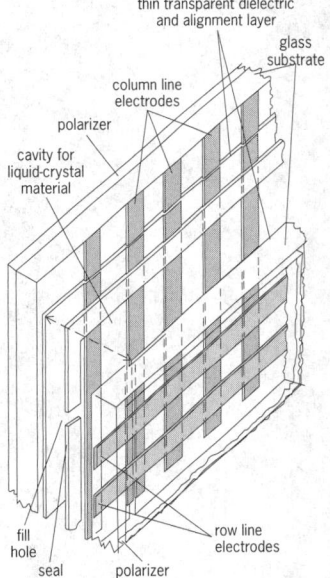

SUPERTWISTED NEMATIC LIQUID-CRYSTAL DISPLAY

thin transparent dielectric and alignment layer

glass substrate

column line electrodes

polarizer

cavity for liquid-crystal material

fill hole

seal

polarizer

row line electrodes

Column-line and row-line electrodes are made of transparent indium-tin oxide. During filling steps, the liquid-crystal material aligns to the alignment layer, which has been mechanically rubbed to import microgrooves on the surface. Alignment rub directions are rotated 180–270°.

supplementary angle [MATH] One angle is supplementary to another angle if their sum is 180°. { ¦səp·lə¦men·trē ′aŋ·gəl }

supplementary condition [QUANT MECH] In a quantized field theory, an auxiliary condition required of a state vector to make it correspond to an actual state. { ¦səp·lə¦men·trē kən′dish·ən }

supplementary group [ELEC] In wire communications, a group of trunks that directly connects local or trunk switching centers over other than a fundamental (or backbone) route. { ¦səp·lə¦men·trē ′grüp }

supplementary units [PHYS] Dimensionless units that are used along with base units to form derived units in the International System; this class contains only two, purely geometrical units: the radian and the steradian. { ¦səp·lə‚men·trē ′yü·nəts }

supplied-air respirator [ENG] An atmospheric-supplying device which provides the wearer with respirable air from a source outside the contaminated area; only those with manual or motor-operated blowers are approved for immediately harmful or oxygen-deficient atmospheres. { sə′plīd ¦er ′res·pə‚rād·ər }

supply chain management [IND ENG] An inventory process involving planning and processing orders; handling; transporting and storing all materials purchased, processed, or distributed; and managing inventories in a coordinated manner among all the players on the chain to fulfill customer orders as they arise rather than to build up stock level to fulfill anticipated future demand. { sə′plī ‚chān ‚man·ij·mənt }

supply control [IND ENG] The process by which an item of supply is controlled within the supply system, including requisitioning receipt, storage, stock control, shipment, disposition, identification, and accounting. [ORD] The process by which anticipated demands for supplies are balanced against the projected availability of supplies in order that qualitative acquisition or disposal action can be initiated. { sə′plī kən‚trōl }

supply current [GEOPHYS] The electrical current in the atmosphere which is required to balance the observed air-earth current of fair-weather regions by transporting positive charge upward or negative charge downward. { sə′plī ‚kə·rənt }

supply point [ORD] Any point where supplies are issued in detail. { sə′plī ‚póint }

supply voltage [ELEC] The voltage obtained from a power source for operation of a circuit or device. { sə′plī ‚vōl·tij }

support [MATH] The support of a real-valued function f on a topological space is the closure of the set of points where f is not zero. { sə′pórt }

support base [ENG] A place from which logistic support is provided for a group of launch complexes and their control center. { sə′pórt ‚bās }

support-coated capillary column [ANALY CHEM] A capillary column that utilizes a fine-granular solid support to disperse the stationary liquid. { sə′pórt ¦kōd·əd ′kap·ə‚ler·ē ‚käl·əm }

supported end [MECH] An end of a structure, such as a beam, whose position is fixed but whose orientation may vary; for example, an end supported on a knife-edge. { sə′pórd·əd ‚end }

support function [MATH] Relative to a convex body in a real inner product space, a function whose value at a point P is the maximum of the inner product of P and Q for Q in the convex body. { sə′pórt ‚fəŋk·shən }

supporting artillery [ORD] Artillery which executes fire missions in support of a specific unit, usually infantry, but remains under the command of the next higher artillery commander. { sə′pórd·iŋ är′til·ə·rē }

supporting fire [ORD] Fire delivered by supporting units to assist or protect a unit in combat. { sə′pórd·iŋ ′fīr }

supporting weapon [ORD] Any weapon used to assist or protect a unit of which it is not an organic part. { sə′pórd·iŋ ′wep·ən }

suppository [PHARM] A medicated solid body of varying weight and shape, intended for introduction into different orifices of the body, as the rectum, urethra, or vagina; usually suppositories melt or are softened at body temperature, though in some instances release of medication is effected through use of a hydrophilic vehicle; typical vehicles or bases are theobroma oil (cocoa butter), glycerinated gelatin, sodium stearate, and propylene glycol monostearate. { sə′päz·ə‚tór·ē }

suppressed carrier [COMMUN] A carrier in a modulated signal that is suppressed at the transmitter; the chrominance subcarrier in a color television transmitter is an example. { sə′prest ′kar·ē·ər }

suppressed-carrier modulation [COMMUN] Modulation resulting from elimination or partial suppression of the carrier component from an amplitude modulated wave. { sə′prest ′kar·ē·ər ‚mäj·ə′lā·shən }

suppressed-carrier transmission [COMMUN] Transmission in which the carrier component of the modulated wave is eliminated or partially suppressed, leaving the side bands to be transmitted. { sə′prest ′kar·ē·ər tranz′mish·ən }

suppressed-zero instrument [ENG] An indicating or recording instrument in which the zero position is below the lower end of the scale markings. { sə′prest ¦zir·ō ‚in·strə·mənt }

suppression [COMPUT SCI] **1.** Removal or deletion usually of insignificant digits in a number, especially zero suppression. **2.** Optional function in either on-line or off-line printing devices that permits them to ignore certain characters or groups of characters which may be transmitted through them. [ELECTR] Elimination of any component of an emission, as a particular frequency or group of frequencies in an audio-frequency of a radio-frequency signal. { sə′presh·ən }

suppression shield [NUCLEO] An array of high-efficiency, low-resolution gamma-ray detectors that surround a high-resolution gamma-ray detector and detect the background of gamma rays which is Compton scattered out of the germanium crystal, in order that these gamma rays may be excluded from the measurement. { sə′presh·ən ‚shēld }

suppressor [ELEC] **1.** In general, a device used to reduce or eliminate noise or other signals that interfere with the operation of a communication system, usually at the noise source. **2.** Specifically, a resistor used in series with a spark plug or distributor of an automobile engine or other internal combustion engine to suppress spark noise that might otherwise interfere with radio reception. [ELECTR] *See* suppressor grid. [SPECT] In an analytical procedure, a substance added to the analyte to reduce the extraneous emission, absorption, or light scattering caused by the presence of an impurity. { sə′pres·ər }

suppressor gene [GEN] A gene that reverses the effect of a mutation in another gene. { sə′pres·ər ‚jēn }

suppressor grid [ELECTR] A grid placed between two positive electrodes in an electron tube primarily to reduce the flow of secondary electrons from one electrode to the other; it is usually used between the screen grid and the anode. Also known as suppressor. { sə′pres·ər ‚grid }

suppressor mutation [GEN] A mutation that restores the function lost after a primary mutation at a different locus. { sə′pres·ər myü‚tā·shən }

suppressor pulse [ELECTR] Pulse used to disable an ionized flow field or beacon transponder during intervals when interference would be encountered. { sə′pres·ər ‚pəls }

suppuration [MED] Pus formation. { ‚səp·yə′rā·shən }

supraaural cushion [ACOUS] An earphone cushion that fits against the auricle. { ‚sü·prə‚ór·əl ′kush·ən }

supracardinal veins [VERT ZOO] Paired longitudinal veins in the mammalian embryo and various adult lower vertebrates in the thoracic and abdominal regions, dorsolateral to and on the sides of the descending aorta; they replace the postcardinal and subcardinal veins. { ¦sü·prə′kärd·nəl ‚vānz }

supracrustal rocks [GEOL] Rocks that overlie basement rock. { ¦sü·prə′krəst·əl ′räks }

suprafacial [ORG CHEM] The stereochemistry when, simultaneously, two sigma bonds are formed or broken on the same face of the component pi systems, such as in a cycloaddition reaction. { ¦sü·prə′fā·shəl }

supragelisol *See* suprapermafrost layer. { ¦sü·prə′jel·ə‚sól }

suprahyoid muscles [ANAT] The muscles attached to the upper margin of the hyoid bone. { ¦sü·prə′hī‚óid ‚məs·əlz }

supralateral tangent arcs [METEOROL] Two oblique luminous arcs, concave to the sun and tangent to the halo of 46° at points above the altitude of the sun. { ¦sü·prə′lad·ə·rəl ′tan·jənt ‚ärks }

supraliminal [PHYSIO] Above, or in excess of, a threshold. { ¦sü·prə′lim·ə·nəl }

supramolecular chemistry [CHEM] A highly interdisciplinary field covering the chemical, physical, and biological features of complex chemical species held together and organized by means of intermolecular (noncovalent) bonding interactions such as hydrogen bonds, van der Waals forces, and hydrophobic interactions. { ‚sü·prə·mə‚lek·yə·lər 'kem·ə‚strē }

supramolecule [PHYS CHEM] A stable system formed by two or more molecules held together and organized by intermolecular (noncovalent bonding) interactions. { ‚sü·prə 'mäl·ə‚kyül }

supranuclear [ANAT] In the nervous system, central to a nucleus. { ¦sü·prə'nü·klē·ər }

supraoccipital [ANAT] Situated above the occipital bone. { ¦sü·prə·äk'sip·əd·əl }

supraoptic [ANAT] Situated above the optic tract. { ¦sü·prə¦äp·tik }

suprapermafrost layer [HYD] The layer of ground above permafrost; it includes the active layer and possibly occurrences of talik and perelotok. Also known as supragelisol. { ¦sü·prə'pər·mə‚fròst ‚lā·ər }

suprarenal gland See adrenal gland. { ¦sü·prə'rēn·əl ‚gland }

suprascapula [ANAT] An anomalous bone sometimes found between the superior border of the scapula and the spines of the lower cervical or first thoracic vertebrae. { ¦sü·prə'skap·yə·lə }

suprasegmental reflex [PHYSIO] A reflex employing complex multineuronal channels to integrate the body and limb musculature with fixed positions or movements of the head. { ¦sü·prə·seg'ment·əl 'rē‚fleks }

suprasternal notch [ANAT] Jugular notch of the sternum. { ¦sü·prə'stərn·əl 'näch }

suprasternal space [ANAT] The triangular space above the manubrium, enclosed by the layers of the deep cervical fascia which are attached to the front and back of this bone. { ¦sü·prə'stərn·əl 'spās }

supratidal sediment [GEOL] The sediment deposited immediately above the high-tide level. { ¦sü·prə'tīd·əl 'sed·ə·mənt }

supratidal zone [GEOL] Pertaining to the shore area immediately marginal to and above the high-tide level. { ¦sü·prə'tīd·əl zōn }

supravital [BIOL] Pertaining to the staining of living cells after removal from a living animal or of still living cells from a recently killed animal. { ¦sü·prə'vīd·əl }

supremum See least upper bound. { su'prē·məm }

surbase [ARCH] **1.** A molding above the base of a wall. **2.** A molding or series of moldings above the base of a pedestal. { 'sər‚bās }

surcharge [CIV ENG] The load supported above the level of the top of a retaining wall. { 'sər‚chärj }

surcharged wall [CIV ENG] A retaining wall with an embankment on the top. { 'sər‚chärjd 'wòl }

surd [MATH] A sum of one or more roots of rational numbers, some or all of which are themselves irrational numbers. { sərd }

surf [OCEANOGR] Wave activity in the area between the shoreline and the outermost limit of breakers, that is, in the surf zone. { sərf }

surface [ENG] The outer part (skin with a thickness of zero) of a body; can apply to structures, to micrometer-sized particles, or to extended-surface zeolites. [MATH] A subset of three-space consisting of those points whose cartesian coordinates x, y, and z satisfy equations of the form $x = f(u,v)$, $y = g(u,v)$, $z = h(u,v)$, where f, g, and h are differentiable real-valued functions of two parameters u and v which take real values and vary freely in some domain. { 'sər·fəs }

surface acoustic wave [ACOUS] A sound wave that propagates along and is bound to the surface of a solid; ordinarily it contains both compressional and shear components. Abbreviated SAW. { 'sər·fəs ə'kü·stik 'wāv }

surface-acoustic-wave device [ELECTR] Any device, such as a filter, resonator, or oscillator, which employs surface acoustic waves with frequencies in the range $10^7 – 10^9$ hertz, traveling on the optically polished surface of a piezoelectric substrate, to process electronic signals. { 'sər·fəs ə'kü·stik 'wāv di‚vīs }

surface-acoustic-wave filter [ELECTR] An electric filter consisting of a piezoelectric bar with a polished surface along which surface acoustic waves can propagate, and on which are deposited metallic transducers, one of which is connected, via

thermocompression-bonded leads, to the electric source, while the other drives the load. { 'sər·fəs ə'kü·stik 'wāv ‚filtər }

surface-active agent [MATER] A soluble compound that reduces the surface tension of liquids, or reduces interfacial tension between two liquids or a liquid and a solid. Also known as surfactant. { 'sər·fəs ‚ak·tiv 'ā·jənt }

surface air leakage [MIN ENG] The amount of surface air entering the fan through the casing at the top of the upcast shaft, the air-lock doors, and fan-drift walls. { 'sər·fəs 'er ‚lē·kij }

surface alloying [MET] Deposition and metallurgical bonding of additional metals or alloys on the surfaces of ferrous or nonferrous metals; such additional materials become an integral part of the total mass, as distinguished from coatings which are bonded mechanically. { 'sər·fəs 'al‚ci·iŋ }

surface analysis [COMPUT SCI] A procedure in which a computer program writes a series of test characters onto a magnetic data storage medium and then reads them back to determine the location of any flaws in the medium. { 'sər·fəs ə‚nal·ə·səs }

surface analyzer [ENG] An instrument that measures or records irregularities in a surface by moving the stylus of a crystal pickup or similar device over the surface, amplifying the resulting voltage, and feeding the output voltage to an indicator or recorder that shows the surface irregularities magnified as much as 50,000 times. { 'sər·fəs ‚an·ə‚līz·ər }

surface area [ENG] Measurement of the extent of the area (without allowance for thickness) covered by a surface. { 'sər·fəs ‚er·ē·ə }

surface barrier [ELECTR] A potential barrier formed at a surface of a semiconductor by the trapping of carriers at the surface. { 'sər·fəs ‚bar·ē·ər }

surface-barrier diode [ELECTR] A diode utilizing thin-surface layers, formed either by deposition of metal films or by surface diffusion, to serve as a rectifying junction. { 'sər·fəs ‚bar·ē·ər 'dī‚ōd }

surface-barrier transistor [ELECTR] A transistor in which the emitter and collector are formed on opposite sides of a semiconductor wafer, usually made of n-type germanium, by training two jets of electrolyte against its opposite surfaces to etch and then electroplate the surfaces. { 'sər·fəs ‚bar·ē·ər tran'zis·tər }

surface boundary layer [METEOROL] That thin layer of air adjacent to the earth's surface, extending up to the so-called anemometer level (the base of the Ekman layer); within this layer the wind distribution is determined largely by the vertical temperature gradient and the nature and contours of the underlying surface, and shearing stresses are approximately constant. Also known as atmospheric boundary layer; friction layer; ground layer; surface layer. { 'sər·fəs 'baůn·drē ‚lā·ər }

surface burning See glowing combustion. { 'sər·fəs ‚bərn·iŋ }

surface carburetor [MECH ENG] A carburetor in which air is passed over the surface of gasoline to charge it with fuel. { 'sər·fəs 'kär·bə‚rād·ər }

surface-charge transistor [ELECTR] An integrated-circuit transistor element based on controlling the transfer of stored electric charges along the surface of a semiconductor. { 'sər·fəs ¦chärj tran'zis·tər }

surface chart [METEOROL] An analyzed synoptic chart of surface weather observations; essentially, a surface chart shows the distribution of sea-level pressure (therefore, the positions of highs, lows, ridges, and troughs) and the location and nature of fronts and air masses, plus the symbols of occurring weather phenomena, analysis of pressure tendency (isallobars), and indications of the movement of pressure systems and fronts. Also known as sea-level chart; sea-level-pressure chart; surface map. { 'sər·fəs chärt }

surface chemistry [PHYS CHEM] The study and measurement of the forces and processes that act on the surfaces of fluids (gases and liquids) and solids, or at an interface separating two phases; for example, surface tension. { 'sər·fəs ‚kem·ə·strē }

surface-coated mirror [OPTICS] A mirror produced by depositing a thin film of highly reflective material on a glass surface. { 'sər·fəs ¦kōd·əd 'mir·ər }

surface color [OPTICS] The color of light reflected from the surface of a body; in contrast to the color of light that is reflected

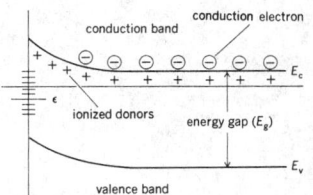

SURFACE BARRIER

Energy diagram of a surface barrier as employed in an n-type semiconductor. W = Fermi level; ϵ = highest energy level of surface state filled by electrons when surface is electrically neutral; E_c = energy of bottom of conduction band; E_v = energy of top of valence band.

after penetrating some distance into the body. { 'sər·fəs ˌkəl·ər }

surface combustion [ENG] Combustion brought about near the surface of a heated refractory material by forcing a mixture of air and combustible gases through it or through a hole in it, or having the gas impinge directly upon it; used in muffles, crucibles, and certain types of boiler furnaces. { 'sər·fəs kəmˌbəs·chən }

surface condenser [MECH ENG] A heat-transfer device used to condense a vapor, usually steam under vacuum, by absorbing its latent heat in cooling fluid, ordinarily water. { 'sər·fəs kənˌden·sər }

surface contamination [NUCLEO] The deposition and attachment of radioactive materials to a surface. { 'sər·fəs kənˌtam·ə·nā·shən }

surface-controlled avalanche transistor [ELECTR] Transistor in which avalanche breakdown voltage is controlled by an external field applied through surface-insulating layers, and which permits operation at frequencies up to the 10-gigahertz range. { 'sər·fəs kənˌtrōld 'av·əˌlanch tranˌzis·tər }

surface creep [GEOL] A stage of the wind erosion process in which grains of sand move each other along the surface. { 'sər·fəs ˌkrēp }

surface current [OCEANOGR] **1.** Water movement which extends to depths of 3–10 feet (1–3 meters) below the surface in nearshore areas, and to about 33 feet (10 meters) in deep-ocean areas. **2.** Any current whose maximum velocity core is at or near the surface. { 'sər·fəs ˌkə·rənt }

surface density [PHYS] The quantity of anything distributed over a surface per unit area of surface. { 'sər·fəs ˌden·səd·ē }

surface deposit *See* surficial deposit. { 'sər·fəs diˌpäz·ət }

surface detention [HYD] Water in temporary storage as a thin sheet over the soil surface during the occurrence of overland flow. { 'sər·fəs diˌten·chən }

surface drag [FL MECH] That portion of drag which is caused by skin friction. { 'sər·fəs ˌdrag }

surface drainage [HYD] Natural or artificial removal of excess groundwater. { 'sər·fəs ˌdrā·nij }

surface drilling [MIN ENG] Boreholes collared at the surface of the earth, as opposed to boreholes collared in mine workings or underwater. { 'sər·fəs ˌdril·iŋ }

surface duct [GEOPHYS] Atmospheric duct for which the lower boundary is the surface of the earth. { 'sər·fəs ˌdəkt }

surface-effect ship [MECH ENG] A transportation device with fixed side walls, which is supported by low-pressure, low-velocity air and operates on water only. { 'sər·fəs iˌfekt ˌship }

surface energy [FL MECH] The energy per unit area of an exposed surface of a liquid; generally greater than the surface tension, which equals the free energy per unit surface. { 'sər·fəs ˌen·ər·jē }

surface finish [ENG] The surface roughness of a component after final treatment, measured by a surface profile. { 'sər·fəs ˌfin·ish }

surface fire [FOR] A forest fire in which only surface litter and undergrowth burn. { 'sər·fəs ˌfīr }

surface flow *See* overland flow. { 'sər·fəs ˌflō }

surface force [MECH] An external force which acts only on the surface of a body; an example is the force exerted by another object with which the body is in contact. { 'sər·fəs ˌfôrs }

surface friction [GEOPHYS] The drag or skin friction of the earth on the atmosphere, usually expressed in terms of the shearing stress of the wind on the earth's surface. { 'sər·fəs ˌfrik·shən }

surface gage [DES ENG] **1.** A scribing tool in an adjustable stand, used to mark off castings and to test the flatness of surfaces. **2.** A gage for determining the distances of points on a surface from a reference plane. { 'sər·fəs ˌgāj }

surface geology [GEOL] The scientific study of the features at the surface of the earth. { 'sər·fəs jēˌäl·ə·jē }

surface grinder [MECH ENG] A grinding machine that produces a plane surface. { 'sər·fəs ˌgrīn·dər }

surface hardening [MET] Hardening the surface of steel by one of several processes, such as carburizing, carbonitriding, nitriding, flame or induction hardening, and surface working. { 'sər·fəs ˌhärd·ən·iŋ }

surface harmonics *See* spherical surface harmonics. { 'sər·fəs härˌmän·iks }

surface hoar [HYD] **1.** Fernlike ice crystals formed directly on a snow surface by sublimation; a type of hoarfrost. **2.**

Hoarfrost that has grown primarily in two dimensions, as on a window or other smooth surface. { 'sər·fəs ˌhôr }

surface ignition [ENG] The initiation of a flame in the combustion chamber of an automobile engine by any hot surface other than the spark discharge. { 'sər·fəs igˌnish·ən }

surface integral [MATH] The integral of a function of several variables with respect to surface area over a surface in the domain of the function. { 'sər·fəs ˌint·ə·grəl }

surface inversion [METEOROL] A temperature inversion based at the earth's surface; that is, an increase of temperature with height beginning at ground level. Also known as ground inversion. { 'sər·fəs inˌvər·zhən }

surface irrigation [AGR] Application of water to the soil by means of pipes or furrows along the surface. { 'sər·fəs ˌir·əˌgā·shən }

surface layer *See* surface boundary layer. { 'sər·fəs ˌlā·ər }

surface leakage [ELEC] The passage of current over the surface of an insulator. { 'sər·fəs ˌlē·kij }

surface lift [MIN ENG] In the freezing method of shaft sinking, freezing and heaving of the surface around the shaft due to the formation of ice and the variation of temperature. { 'sər·fəs ˌlift }

surface machining [MET] The cutting of three-dimensional shapes in a piece of work, by using numerical control equipment to transmit predetermined paths and designs. { 'sər·fəs məˌshēn·iŋ }

surface magnetic wave [ELECTROMAG] A magnetostatic wave that can be propagated on the surface of a magnetic material, as on a slab of yttrium iron garnet. { 'sər·fəs magˌned·ik 'wāv }

surface map *See* surface chart. { 'sər·fəs ˌmap }

surface micromachining [ENG] A set of processes based upon deposition, patterning, and selective etching of thin films to form a free-standing microsensor on the surface of a silicon wafer. { ˌsər·fəs ˌmī·krə·məˈshēn·iŋ }

surface mining [MIN ENG] Mining at or near the surface; includes placer mining, mining in open glory-hole or milling pits, mining and removing ore from opencuts by hand or with mechanical excavating and transportation equipment, and the removal of capping or overburden to uncover the ores. { 'sər·fəs ˌmīn·iŋ }

surface-mount technology [ELECTR] The technique of mounting electronic circuit components and their electrical connections on the surface of a printed board, rather than through holes. { 'sər·fəs ˌmaùnt tekˈnäl·ə·jē }

surface navigation [NAV] Navigation of a craft on the surface of the earth; in particular, navigation of vessels on the surface of water. { 'sər·fəs ˌnav·əˌgā·shən }

surface noise [ELECTR] The noise component in the electric output of a phonograph pickup due to irregularities in the contact surface of the groove. Also known as needle scratch. { 'sər·fəs ˌnôiz }

surface of center [MATH] The locus of points that are one of the two centers of principal curvature at some point on a given surface. { 'sər·fəs əv 'sen·tər }

surface of discontinuity [FL MECH] A surface within a fluid across which there is a discontinuity in fluid velocity; often generated in the wake of a body moving relative to the fluid. [METEOROL] An interface, applied to the atmosphere; for example, an atmospheric front is represented ideally by a surface of discontinuity of velocity, density, temperature, and pressure gradient. { ˌsər·fəs əv ˌdisˌkänt·ənˈü·əd·ē }

surface of Joachimsthal [MATH] A surface such that all the members of one of its two families of lines of curvature are plane curves and their planes all pass through a common axis. { ˌsər·fəs əv yōˈäk·əmzˌtäl }

surface of Liouville [MATH] A surface that can be assigned parameters u and v such that a linear element ds on the surface is given by $ds^2 = [f(u) + g(v)][du^2 + dv^2]$, where f and g are functions of u and v. { ˌsər·fəs əv 'lyü·vel }

surface of Monge [MATH] A surface generated by a plane curve as its plane rolls without slipping over a developable surface. { ˌsər·fəs əv 'mônzh }

surface of negative curvature [MATH] A surface whose Gaussian curvature is negative at every point. { ˌsər·fəs əv ˌneg·əd·iv ˈkər·və·chər }

surface of positive curvature [MATH] A surface whose Gaussian curvature is positive at every point. { ˌsər·fəs əv ˈpäs·əd·iv ˌkər·və·chər }

surface of revolution [MATH] A surface realized by rotating a planar curve about some axis in its plane. { ¦sər·fəs əv ˌrev·ə'lü·shən }

surface of section *See* Poincaré surface of section. { ¦sər·fəs əv 'sek·shən }

surface of translation [MATH] A surface that can be generated from two curves by translating either one of them parallel to itself in such a way that each of its points describes a curve that is a translation of the other curve. Also known as translation surface. { ¦sər·fəs əv tranz'lā·shən }

surface of Voss [MATH] A surface that has a conjugate system of geodesics. { ¦sər·fəs əv 'vos }

surface of zero curvature [MATH] A surface whose Gaussian curvature is zero at every point. { ¦sər·fəs əv 'zir·ō ˌkər·və·chər }

surface oil-film technique [FL MECH] A method of flow visualization in which a solid surface is coated with a mixture of oil and powdered pigment, and the airstream carries the oil away, leaving a streaky deposit of pigment that provides information on airflow. { ¦sər·fəs 'oil ˌfilm tek,nēk }

surface orientation [PHYS CHEM] Arrangement of molecules on the surface of a liquid with one part of the molecule turned toward the liquid. { 'sər·fəs ˌór·ē·ən'tā·shən }

surface passivation [ELECTR] A method of coating the surface of a *p*-type wafer for a diffused junction transistor with an oxide compound, such as silicon oxide, to prevent penetration of the impurity in undesired regions. { 'sər·fəs ˌpas·ə'vā·shən }

surface patch [MATH] A surface or a portion of a surface that is bounded by a closed curve. { 'sər·fəs ˌpach }

surface-penetrating radar *See* ground-probing radar. { ˌsər·fəs ˌpen·ə,trād·iŋ 'rā,där }

surface phase [GEOCHEM] A thin rock layer differing in geochemical properties from those of the volume phases on either side. Also known as volume phase. { 'sər·fəs ˌfāz }

surface physics [SOLID STATE] The study of the structure and dynamics of atoms and their associated electron clouds in the vicinity of a surface, usually at the boundary between a solid and a low-density gas. { 'sər·fəs 'fiz·iks }

surface-piercing hydrofoil [NAV ARCH] A hydrofoil craft which attains its stability by hydrodynamically balancing the amount of foil area above and below the water surface. { 'sər·fəs ¦pirs·iŋ 'hī·drəˌfoil }

surface pipe [PETRO ENG] The string of casing first set in a well, usually to shut off shallow, fresh-water sands from contamination by deeper, saline waters. { 'sər·fəs ˌpīp }

surface planer *See* surfacer. { 'sər·fəs ˌplā·nər }

surface plasmon [SOLID STATE] A quantum of a collective oscillation of charges on the surface of a solid induced by a time-varying electric field. { 'sər·fəs 'plaz,män }

surface plate [DES ENG] A plate having a very accurate plane surface used for testing other surfaces or to provide a true surface for accurately measuring and locating testing fixtures. { 'sər·fəs ˌplāt }

surface pressure [METEOROL] The atmospheric pressure at a given location on the earth's surface; the expression is applied loosely and about equally to the more specific terms: station pressure and sea-level pressure. [PHYS] *See* film pressure. { 'sər·fəs ˌpresh·ər }

surfacer [DES ENG] A machine that is used to dress or plane the surface of a material such as stone, metal, or wood. Also known as surface planer. { 'sər·fəs·ər }

surface reaction [CHEM] A chemical reaction carried out on a surface as on an adsorbent or solid catalyst. { 'sər·fəs rē,ak·shən }

surface recombination rate [SOLID STATE] The rate at which free electrons and holes at the surface of a semiconductor recombine, thus neutralizing each other. { 'sər·fəs rē,käm·bə'nā·shən ˌrāt }

surface recombination velocity [SOLID STATE] A measure of the rate of recombination between electrons and holes at the surface of a semiconductor, equal to the component of the electron or hole current density normal to the surface divided by the excess electron or hole volume charge density close to the surface. { 'sər·fəs rē,käm·bə·nā·shən və,läs·əd·ē }

surface resistivity [ELEC] The electric resistance of the surface of an insulator, measured between the opposite sides of a square on the surface; the value in ohms is independent of the size of the square and the thickness of the surface film. { 'sər·fəs rē,zis'tiv·əd·ē }

surface retention *See* surface storage. { 'sər·fəs ri,ten·chən }

surface rights [MIN ENG] **1.** Ownership of the surface land only, mineral rights being reserved. **2.** Ownership of the surface land plus mineral rights. **3.** The right of a mineral owner or an oil and gas lessee to use as much surface land as may be reasonably necessary for the conduct of operations under the lease. { 'sər·fəs ˌrīts }

surface rolling [MET] A cold-rolling process for hardening the surface of a metal. { 'sər·fəs ˌrōl·iŋ }

surface roughness [ENG] The closely spaced unevenness of a solid surface (pits and projections) that results in friction for solid-solid movement or for fluid flow across the solid surface. { 'sər·fəs ˌrəf·nəs }

surface runoff [HYD] Runoff that moves over the soil surface to the nearest surface stream. { 'sər·fəs 'rən,óf }

surface-set bit [DES ENG] A bit containing a single layer of diamonds set so that the diamonds protrude on the surface of the crown. Also known as single-layer bit. { 'sər·fəs ¦set ˌbit }

surface sizing *See* sizing treatment. { 'sər·fəs ˌsīz·iŋ }

surface soil [GEOL] The soil extending 5 to 8 inches (13 to 20 centimeters) below the surface. { 'sər·fəs ˌsoil }

surface state [SOLID STATE] An electron state in a semiconductor whose wave function is restricted to a layer near the surface. { 'sər·fəs ˌstāt }

surface storage [HYD] The part of precipitation retained temporarily at the ground surface as interception or depression storage so that it does not appear as infiltration or surface runoff either during the rainfall period or shortly thereafter. Also known as initial detention, surface retention. { 'sər·fəs ˌstór·ij }

surface temperature [METEOROL] Temperature of the air near the surface of the earth. [OCEANOGR] Temperature of the layer of seawater nearest the atmosphere. { 'sər·fəs ˌtem·prə·chər }

surface tension [FL MECH] The force acting on the surface of a liquid, tending to minimize the area of the surface; quantitatively, the force that appears to act across a line of unit length on the surface. Also known as interfacial force; interfacial tension; surface tensity. { 'sər·fəs ˌten·chən }

surface tension number [FL MECH] A dimensionless number used in studying mass transfer in packed columns equal to the square of the dynamic viscosity of a liquid times the length of the perimeter of a packing element, divided by the product of the surface area of the packing element, the surface tension, and the density of the liquid. Symbol T_s. { 'sər·fəs ˌten·chən ˌnəm·bər }

surface tensity *See* surface tension. { 'sər·fəs ˌten·səd·ē }

surface thermometer [ENG] A thermometer, mounted in a bucket, used to measure the temperature of the sea surface. { 'sər·fəs thər'mäm·əd·ər }

Rodriquez ferry in the RHS-160 series.

surface-to-air missile [ORD] A guided missile designed to be fired at an airborne target from the ground or from the deck of a surface ship; examples include Bomarc, Hawk, Nike, Talos, Tartar, Terrier, and Wizard. Abbreviated SAM. { 'sər·fəs tü 'er ˌmis·əl }

surface-to-surface missile [ORD] A guided missile designed to be fired at a surface target from a surface position on land or water; examples include Atlas, Corporal, Dart, Jupiter, Lacrosse, Mace, Matador, Navaho, Pershing, Polaris, Redstone, Regulus, Sergeant, Snark, Thor, and Titan. Abbreviated SSM. { 'sər·fəs tü 'sər·fəs ˌmis·əl }

surface treating [ENG] Any method of treating a material (metal, polymer, or wood) so as to alter the surface, rendering it receptive to inks, paints, lacquers, adhesives, and various other treatments, or resistant to weather or chemical attack. { 'sər·fəs ˌtrēd·iŋ }

surface vibrator [MECH ENG] A vibrating device used on the surface of a pavement or flat slab to consolidate the concrete. { 'sər·fəs ˌvī,brād·ər }

surface visibility [METEOROL] The visibility determined from a point on the ground, as opposed to control-tower visibility. { 'sər·fəs ˌviz·ə'bil·əd·ē }

surface wash *See* sheet erosion. { 'sər·fəs ˌwäsh }

surface water [HYD] All bodies of water on the surface of the earth. [OCEANOGR] *See* mixed layer. { 'sər·fəs ˌwód·ər }

surface waterproofing [ENG] Waterproofing concrete by painting a waterproofing liquid on the surface. { 'sər·fəs 'wód·ər ˌprüf·iŋ }

Patriot system firing an MIM-104 missile. (*Raytheon*)

surface wave [COMMUN] *See* ground wave. [ELECTROMAG] A wave that can travel along an interface between two different mediums without radiation; the interface must be essentially straight in the direction of propagation; the commonest interface used is that between air and the surface of a circular wire. [FL MECH] A wave that distorts the free surface that separates two fluid phases, usually a liquid and a gas. [MECH] *See* Rayleigh wave. [OCEANOGR] A progressive gravity wave in which the disturbance is of greatest amplitude at the air-water interface. { 'sər·fəs ,wāv }

surface-wave transmission line [ELECTROMAG] A single conductor transmission line energized in such a way that a surface wave is propagated along the line with satisfactorily low attenuation. { 'sər·fəs ¦wāv tranz'mish·ən ,līn }

surface weather observation [METEOROL] An evaluation of the state of the atmosphere as observed from a point at the surface of the earth, as opposed to an upper-air observation, and applied mainly to observations which are taken for the primary purpose of preparing surface synoptic charts. { 'sər·fəs 'weth·ər ,äb·zər,vā·shən }

surface wind [METEOROL] The wind measured at a surface observing station; customarily, it is measured at some distance above the ground itself to minimize the distorting effects of local obstacles and terrain. { 'sər·fəs ,wind }

surface zero *See* ground zero. { 'sər·fəs 'zir·ō }

surfacing [MET] Depositing filler metal on a metal surface by welding or spraying. { 'sər·fə·siŋ }

surfacing mat *See* overlay. { 'sər·fə·siŋ ,mat }

surfactant *See* surface-active agent. { sər'fek·tənt }

surfactant flooding *See* micellar flooding. { sər'fek·tənt ,fləd·iŋ }

surf beat [OCEANOGR] Oscillations of water level near shore, associated with groups of high breakers. { 'sərf ,bēt }

surficial creep *See* soil creep. { sər'fish·əl ,krēp }

surficial deposit [GEOL] Unconsolidated alluvial, residual, or glacial deposits overlying bedrock or occurring on or near the surface of the earth. Also known as superficial deposit; surface deposit. { sər'fish·əl di'päz·ət }

surficial geology [GEOL] The scientific study of surficial deposits, including soils. { sər'fish·əl jē'äl·ə·jē }

surf ripple [GEOL] A ripple mark formed on a sandy beach by wave-generated currents. { 'sərf ,rip·əl }

surf zone [OCEANOGR] The area between the landward limit of wave uprush and the farthest seaward breaker. { 'sərf ,zōn }

surge [ASTROPHYS] An unusually violent solar prominence that usually accompanies a smaller flare, consisting of a brilliant jet of gas which shoots out into the solar corona with a speed on the order of 180 miles (300 kilometers) per second and reaches a height on the order of 60,000 miles (100,000 kilometers). [ELEC] A momentary large increase in the current or voltage in an electric circuit. [ENG] **1.** An upheaval of fluid in a processing system, frequently causing a carryover (puking) of liquid through the vapor lines. **2.** The peak system pressure. **3.** An unstable pressure buildup in a plastic extruder leading to variable throughput and waviness of the hollow plastic tube. [FL MECH] A wave at the free surface of a liquid generated by the motion of a vertical wall, having a change in the height of the surface across the wavefront and violent eddy motion at the wavefront. [OCEANOGR] **1.** Wave motion of low height and short period, from about 1/2 to 60 minutes. **2.** *See* storm surge. { sərj }

surge admittance [ELEC] Reciprocal of surge impedance. { 'sərj ad,mit·əns }

surge arrester [ELEC] A protective device designed primarily for connection between a conductor of an electrical system and ground to limit the magnitude of transient overvoltages on equipment. Also known as arrester; lightning arrester. { 'sərj ə,res·tər }

surge bunker [MIN ENG] A large-capacity storage hopper, installed near the pit bottom or at the input end of a processing plant to provide uniform bulk deliveries. { 'sərj ,bən·kər }

surge column [PETRO ENG] A large-sized pipe of sufficient height to provide a static head able to absorb the surging liquid discharge of the process tank to which it is connected. Also known as boot. { 'sərj ,käl·əm }

surge current [ELEC] A short-duration, high-amperage electric current wave that may sweep through an electrical network, as a power transmission network, when some portion

of it is strongly influenced by the electrical activity of a thunderstorm. { 'sərj ,kə·rənt }

surge drum *See* accumulator. { 'sərj ,drəm }

surge electrode current *See* fault electrode current. { 'sərj i'lek,trōd ,kə·rənt }

surge generator [ELEC] A device for producing high-voltage pulses, usually by charging capacitors in parallel and discharging them in series. { 'sərj ,jen·ə,rād·ər }

surge header *See* accumulator. { 'sərj ,hed·ər }

surge impedance *See* characteristic impedance. { 'sərj im,pēd·əns }

surge line [METEOROL] A line along which a discontinuity in the wind speed occurs. { 'sərj ,līn }

surge protector [ELEC] A device placed in an electrical circuit to prevent the passage of surges and spikes that could damage electronic equipment. { 'sərj prə,tek·tər }

surgery [MED] The branch of medicine that deals with conditions requiring operative procedures. { 'sər·jə·rē }

surge stress [MECH] The physical stress on process equipment or systems resulting from a sudden surge in fluid (gas or liquid) flow rate or pressure. { 'sərj ,stres }

surge suppressor [ELECTR] A circuit that responds to the rate of change of a current or voltage to prevent a rise above a predetermined value; it may include resistors, capacitors, coils, gas tubes, and semiconducting disks. Also known as transient suppressor. { 'sərj sə,pres·ər }

surge tank [ENG] **1.** A standpipe or storage reservoir at the downstream end of a closed aqueduct or feeder pipe, as for a water wheel, to absorb sudden rises of pressure and to furnish water quickly during a drop in pressure. Also known as surge drum. **2.** An open tank to which the top of a surge pipe is connected so as to avoid loss of water during a pressure surge. [MIN ENG] In pumping of ore pulps, a relatively small tank which maintains a steady loading of the pump. { 'sərj ,taŋk }

surgical debridement [MED] The removal of foreign material and devitalized tissue using a scalpel or other sharp instrument. { ,sər·ji·kəl də'brēd·mənt }

surgical needle [MED] Any sewing needle used in a surgical operation. { 'sər·jə·kəl 'nēd·əl }

surging [ENG] Motion of a ship that alternately moves forward and aft, usually when moored. { 'sərj·iŋ }

surging breaker *See* plunging breaker. { 'sərj·iŋ 'brā·kər }

surging glacier [HYD] A glacier that alternates periodically between surges (brief periods of rapid flow) and stagnation. { 'sərj·iŋ 'glā·shər }

surjection [MATH] A mapping *f* from a set *A* to a set *B* such that for every element *b* of *B* there is an element *a* of *A* such that $f(a) = b$. Also known as surjective mapping. { sər'jek·shən }

surjective mapping *See* surjection. { sər'jek·tiv 'map·iŋ }

suroet [METEOROL] A persistent, rain-bearing southwest wind on the west coast of France. { sər·ə'wä }

surprint [GRAPHICS] **1.** To superimpose an image upon a previously printed image. **2.** An image that has been surprinted. { 'sər,print }

sursassite [MINERAL] $Mn_5Al_4Si_5O_{21}\cdot3H_2O$ A mineral which is composed of hydrous manganese aluminum silicate. { ,sər'sa,sīt }

surveillance [ENG] Systematic observation of air, surface, or subsurface areas or volumes by visual, electronic, photographic, or other means, for intelligence or other purposes. { sər'vā·ləns }

surveillance approach [NAV] An instrument approach conducted according to directions issued by a controller using information which appears on the surveillance radar display. { sər'vā·ləns ə,prōch }

surveillance radar [NAV] Ground radar used for traffic-control purposes in the approach and landing zone; it is used to assist controllers in converting random arrivals to regular landings and in positioning such aircraft so that they may make low approaches by the use of a fixed-beam, low-approach system or by a precision radar low-approach system. { sər'vā·ləns ,rā,där }

surveillance satellite [AERO ENG] A satellite whose function is to make systematic observations of the earth, usually by photographic means, for military intelligence or for other purposes. { sər'vā·ləns ,sad·əl,īt }

survey [ENG] **1.** The process of determining accurately the position, extent, contour, and so on, of an area, usually for the

SURGE

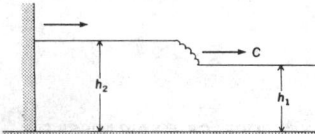

Sketch of a surge wave, in fluid mechanics. C = speed of propagation; h_1 = height of water before passage of surge; h_2 = height of water after passage of surge.

purpose of preparing a chart. **2.** The information so obtained. [NUCLEO] Measurement of radiation in the vicinity of a nuclear reactor or other source. { 'sər‚vā }

survey design *See* sample design. { 'sər‚vā di‚zīn }

survey foot [MECH] A unit of length, used by the U.S. Coast and Geodetic Survey, equal to 12/39.37 meter, or approximately 1.000002 feet. { 'sər‚vā 'fut }

surveying altimeter [ENG] A barometric-type instrument consisting of a pressure-sensitive element which contracts or expands in proportion to atmospheric pressure, connected through a linkage to a pointer; its dial is graduated in units of linear measurement (feet or meters) to indicate differences of elevation only. { sər'vā·iŋ al'tim·əd·ər }

surveying sextant *See* hydrographic sextant. { sər'vā·iŋ ‚seks·tənt }

survey instrument [NUCLEO] A portable instrument used to detect and measure radiation. Also known as survey meter. { 'sər‚vā ‚in·strə·mənt }

survey meter *See* survey instrument. { 'sər‚vā ‚mēd·ər }

surveyor's compass [ENG] An instrument used to measure horizontal angles in surveying. { sər'vā·ərz ‚käm·pəs }

surveyor's cross [ENG] An instrument for setting out right angles in surveying; consists of two bars at right angles with sights at each end. { sər'vā·ərz ‚krós }

surveyor's level [ENG] A telescope and spirit level mounted on a tripod, rotating vertically and having leveling screws for adjustment. { sər'vā·ərz ‚lev·əl }

surveyor's measure [ENG] A system of measurement used in surveying having the engineer's, or Gunter's chain, as a unit. { sər'vā·ərz ‚mezh·ər }

survey traverse *See* traverse. { 'sər‚vā trə'vərs }

survivability [ORD] The capability of a system to withstand a man-made hostile environment without suffering an abortive impairment of its ability to accomplish its designated mission. { sər‚vī·və'bil·əd·ē }

survivable route [COMMUN] A communication cable system begun in 1960 in which the cable, main stations, amplifiers, and power feed stations are placed underground; it incorporates the latest techniques of protection against natural disasters and nuclear blasts, and avoids possible target areas. { sər'vī·və·bəl 'rüt }

survival curve [NUCLEO] The curve obtained by plotting the number or percentage of organisms surviving at a given time against the dose of radiation, or the number surviving at different intervals after a particular dose of radiation. { sər'vī·vəl ‚kərv }

survival probability [ORD] The chance that a target will survive a given operation. { sər'vī·vəl ‚präb·ə‚bil·əd·ē }

survival ratio [BIOL] The number of organisms surviving irradiation by ionizing radiation divided by the number of organisms before irradiation. { sər'vī·vəl ‚rā·shō }

survivor curve [IND ENG] A curve showing the percentage of a group of machines or facilities surviving at a given age. { sər'vī·vər ‚kərv }

Surwell clinograph [ENG] A directional surveying instrument which records photographically the direction and magnitude of well deviations from the vertical; powered by batteries, it contains a box level gage (indicating vertical deviation), a gyroscopic compass (indicating azimuth direction) and a watch and a dial thermometer, so that a simultaneous record of amount and direction of deviation, temperature, and time can be made on 16-millimeter film. { 'sər‚wel 'klīn·ə‚graf }

susannite [MINERAL] $Pb_4(SO_4)(CO_3)_2(OH)_2$ A greenish or yellowish, rhombohedral mineral that is dimorphous with leadhillite. { sü'za‚nīt }

suscept [PL PATH] A plant that is susceptible to disease caused by either parasitic or nonparasitic plant pathogens. { sə'sept }

susceptance [ELEC] The imaginary component of admittance. { sə'sep·təns }

susceptance standard [ELEC] Standard that introduces calibrated small values of shunt capacitance into 50-ohm coaxial transmission arrays. { sə'sep·təns ‚stan·dərd }

susceptibility [ELEC] *See* electric susceptibility. [ELECTROMAG] *See* magnetic susceptibility. [ORD] The degree to which a device, equipment, or weapons system is open to effective attack due to one or more inherent weaknesses. { sə‚sep·tə'bil·əd·ē }

susceptometer [ENG] An instrument that measures paramagnetic, diamagnetic, or ferromagnetic susceptibility. { ‚sə'sep'täm·əd·ər }

suspended acoustical ceiling [BUILD] An acoustical ceiling which is suspended from either the roof or a higher ceiling. { sə'spen·dəd ə'kü·stə·kəl 'sē·liŋ }

suspended ceiling [BUILD] The suspension of the furring members beneath the structural members of a ceiling. { sə'spen·dəd 'sē·liŋ }

suspended formwork [CIV ENG] Formwork suspended from supports for the floor being cast. { sə'spen·dəd 'fórm‚wərk }

suspended load [GEOL] The part of the stream load that is carried for a long time in suspension. Also known as suspension load. { sə'spen·dəd 'lōd }

suspended solids *See* suspension. { sə'spen·dəd 'säl·ədz }

suspended span [CIV ENG] A simple span supported from the free ends of cantilevers. { sə'spen·dəd 'span }

suspended transformation [THERMO] The cessation of change before true equilibrium is reached, or the failure of a system to change immediately after a change in conditions, such as in supercooling and other forms of metastable equilibrium. { sə'spen·dəd ‚tranz·fər'mā·shən }

suspended tray conveyor [MECH ENG] A vertical conveyor having pendant trays or other carriers on one or more endless chains. { sə'spen·dəd ‚trā kən'vā·ər }

suspended tubbing [MIN ENG] A permanent method of lining a circular shaft, in which the tubbing (German type) is temporarily suspended from the next wedging curb above and for which no temporary supports are required; slurry is run in behind the tubbing by means of a funnel passing through the holes provided in the segments. { sə'spen·dəd 'təb·iŋ }

suspended water *See* vadose water. { sə'spen·dəd 'wód·ər }

suspension [CHEM] A mixture of fine, nonsettling particles of any solid within a liquid or gas, the particles being the dispersed phase, while the suspending medium is the continuous phase. Also known as suspended solids. [ENG] A fine wire or coil spring that supports the moving element of a meter. [MIN ENG] The bolting of rock to secure fragments or sections, such as small slabs barred down after blasting blocks of rock broken by fracture or joint patterns, which may subsequently loosen and fall. { sə'spen·shən }

suspension bridge [CIV ENG] A fixed bridge consisting of either a roadway or a truss suspended from two cables which pass over two towers and are anchored by backstays to a firm foundation. { sə'spen·shən ‚brij }

suspension cable [ENG] A freely hanging cable; may carry mainly its own weight or a uniformly distributed load. { sə'spen·shən ‚kā·bəl }

suspension current *See* turbidity current. { sə'spen·shən ‚kə·rənt }

suspension feeder [ZOO] An animal that feeds on small particles suspended in water; particles may be minute living plants or animals, or products of excretion or decay from these or larger organisms. { sə'spen·shən ‚fēd·ər }

suspension insulator [ELEC] A type of insulator used to support a conductor of an overhead transmission line, consisting of one or a string of insulating units suspended from a pole or tower, with the conductor attached to the end. { sə'spen·shən ‚in·sə‚lād·ər }

suspension load *See* suspended load. { sə'spen·shən ‚lōd }

suspension roast *See* flash roast. { sə'spen·shən ‚rōst }

suspension roof [BUILD] A roof that is supported by steel cables. { sə'spen·shən ‚rüf }

suspension system [MECH ENG] A system of springs, shock absorbers, and other devices supporting the upper part of a motor vehicle on its running gear. { sə'spen·shən ‚sis·təm }

suspensor [BOT] A mass of cells in higher plants that pushes the embryo down into the embryo sac and into contact with the nutritive tissue. [MYCOL] A hypha which bears an apical gametangium in fungi of the Mucorales. { sə'spen·sər }

sussexite [MINERAL] $MnBO_2OH$ A white mineral composed of basic manganese borate occurring in fibrous veins. { 'səs·iks‚ī }

sustainable agriculture [AGR] An integration of traditional, conservation-minded farming techniques with modern scientific advances that maximizes use of on-farm renewable resources instead of imported and nonrenewable resources,

SUSPENSION BRIDGE

Verrazano-Narrows Bridge, New York City, an example of a suspension bridge. (*Triborough Bridge and Tunnel Authority*)

while earning a return that is large enough for the farmer to continue in an ecologically harmless, regenerative way. { sə¦stan·ə·bəl 'ag·rə¸kəl·chər }

sustainable development [ENG] Development of industrial and natural resources that meets the energy needs of the present without compromising the ability of future generations to meet their needs in a similar manner. { sə¸stān·ə·bəl di'vel·əp·mənt }

sustainable forest management [FOR] Integrated management of the full range of environmental, social, and economic values of the forest to ensure future health and usefulness of the forest. { sə¦stan·ə·bəl 'fär·əst ¸man·ij·mənt }

sustained oscillation [CONT SYS] Continued oscillation due to insufficient attenuation in the feedback path. [PHYS] Oscillation in which forces outside the system, but controlled by the system, maintain a periodic oscillation of the system at a period or frequency that is nearly the natural period of the system. { sə'stānd ¸äs·ə'lā·shən }

sustained rate of fire [ORD] Actual rate of fire that a weapon can continue to deliver for an indefinite length of time without seriously overheating. { sə'stānd 'rāt əv 'fīr }

sustained yield [BIOL] In a biological resource such as timber or grain, the replacement of a harvest yield by growth or reproduction before another harvest occurs. { sə'stānd 'yēld }

sustainer rocket engine [AERO ENG] A rocket engine that maintains the velocity of a rocket vehicle once it has achieved its programmed velocity by use of a booster or other engine. { sə'stān·ər 'räk·ət ¸en·jən }

sustentacular cell [HISTOL] One of the supporting cells of an epithelium as contrasted with other cells with special function, as the nonnervous cells of the olfactory epithelium or the Sertoli cells of the seminiferous tubules. { ¦səs·tən¦tak·yə·lər ¸sel }

SU₃ symmetry [PARTIC PHYS] A unitary symmetry based on a fundamental multiplet of three equivalent particles—in particular, the approximate symmetry based on the up, down, and strange quarks, and the exact symmetry based on the three differently colored quarks of a given flavor. { ¦es ¦yü ¸səb¦thrē 'sim·ə·trē }

Sutherland's formula [STAT MECH] **1.** The absolute viscosity of a gas is proportional to $T^{3/2}/(C + T)$, where T is the absolute temperature and C is a constant for a given gas. **2.** The mean free path of a molecule in a gas is proportional to $1/[nd\sqrt{1 + (C/T)}]$, where n is the number of molecules per unit volume, d is the diameter of a molecule, T is the absolute temperature, and C is a constant. { 'səth·ər·lənz ¸fȯr·myə·lə }

Sutro weir [CIV ENG] A dam with at least one curved side and horizontal crest, so formed that the head above the crest is directly proportional to the discharge. { 'sü·trō ¸wer }

suture [BIOL] A distinguishable line of union between two closely united parts. [MED] A fine thread used to close a wound or surgical incision. { 'sü·chər }

sutured [PETR] Referring to rock texture in which mineral grains or irregularly shaped crystals interfere with their neighbors, producing interlocking, irregular contacts without interstitial spaces. { 'sü·chərd }

Sv See sievert.

SV See speaker verification.

svabite [MINERAL] $Ca_5(AsO_4)_3F$ A colorless, yellow, rose, or reddish-brown mineral composed of fluoride-arsenate of calcium. { 'sfä¸bīt }

svanbergite [MINERAL] $SrAl_3(PO_4)(SO_4)(OH)_6$ A colorless to yellow mineral composed of basic phosphate and sulfate of strontium and aluminum; it is isomorphous with corkite, hinsdalite, and woodhouseite. { 'sfän¸bər¸gīt }

svedberg [PHYS CHEM] A unit of sedimentation coefficient, equal to 10^{-13} second. { 'sfed¸bərg }

Svedberg equation [STAT MECH] An equation which states that the amplitude of vibration of a particle which exhibits Brownian motion is proportional to its period. { 'sfed¸bərg i¸kwā·zhən }

sverdrup [FL MECH] A unit of volume transport equal to 1,000,000 cubic meters per second. { 'sfər·drəp }

SW See switch.

swab [MIN ENG] **1.** A pistonlike device provided with a rubber cap ring that is used to clean out debris inside a borehole or casing. **2.** The act of cleaning the inside of a tubular object with a swab. [PETRO ENG] In petroleum drilling, to pull the

drill string so rapidly that the drill mud is sucked up and overflows the collar of the borehole, thus leaving an undesirably empty borehole. { swäb }

swage bolt [DES ENG] A bolt having indentations with which it can be gripped in masonry. { 'swāj ¸bōlt }

swaging [MET] Tapering a rod or tube or reducing its diameter by any of several methods, such as forging, squeezing, or hammering. Also known as cressing. { 'swāj·iŋ }

swale [GEOL] **1.** A slight depression, sometimes swampy, in the midst of generally level land. **2.** A shallow depression in an undulating ground moraine due to uneven glacial deposition. **3.** A long, narrow, generally shallow, troughlike depression which lies between two beach ridges and is aligned roughly parallel to the coastline. { swāl }

swallow buoy See swallow float. { 'swä·lō ¸bȯi }

swallow float [ENG] A tubular buoy used to measure current velocities; it can be adjusted to be neutrally buoyant and to drift at a selected density level while being tracked by shipboard listening devices. Also known as neutrally buoyant float; swallow buoy. { 'swä·lō ¸flōt }

swallow hole [GEOL] An opening that occurs occasionally at the bottom of a sinkhole which permits direct drainage from the surface into an underground channel. { 'swäl·ō ¸hōl }

swamp [ECOL] A waterlogged land supporting a natural vegetation predominantly of shrubs and trees. { swämp }

swamp buggy [MECH ENG] A wheeled vehicle that runs on sand, on mud, or through shallow water; used especially in swamps. { 'swämp ¸bəg·ē }

swamper [MIN ENG] **1.** A rear brakeman in a metal mine. **2.** A laborer who assists in hauling ore and rock, coupling and uncoupling cars, throwing switches, and loading and unloading carriers. { 'swäm·pər }

swamp fever See infectious anemia. { 'swämp ¸fē·vər }

swamping resistor [ELECTR] Resistor placed in the emitter lead of a transistor circuit to minimize the effects of temperature on the emitter-base junction resistance. { 'swämp·iŋ ri¸zis·tər }

swamp itch See schistosome dermatitis. { 'swämp ¸ich }

swan [VERT ZOO] Any of several species of large waterfowl comprising the subfamily Anatinae; they are herbivorous stout-bodied forms with long necks and spatulate bills. { swän }

Swan See Cygnus. { swän }

Swann bands [ASTROPHYS] Particular bands seen in the visible spectra of comets; they arise from the presence of dimeric carbon (C_2) in the comet's tail. { swän ¸banz }

Swan Nebula See Omega Nebula. { 'swän 'neb·yə·lə }

Swanscombe man [PALEON] A partial skull recovered in Swanscombe, Kent, England, which represents an early stage of *Homo sapiens* but differing in having a vertical temporal region and a rounded occipital profile. { 'swanz·kəm ¸man }

swap out [COMPUT SCI] The action of an operating system on a process wherein it blocks the process and writes the contents of its memory onto a disk in order to make available more memory for other current processes. { 'swäp ¸aut }

swapping [COMPUT SCI] A procedure in which a running program is temporarily suspended and moved onto secondary storage, and primary storage is reassigned to a more pressing job, in order to maximize the efficient use of primary storage. { 'swäp·iŋ }

sward [AGR] A portion of ground covered with grass. { sword }

swarf [ENG] Chips, shavings, and other fine particles removed from the workpiece by grinding tools. { swȯrf }

swarf cut [MET] The removal and cutting of metal in which the axis of the cutting tool is varied with respect to the part being machined. { 'swȯrf ¸kət }

swarmer cell [MICROBIO] The daughter cell which separates from the stalked mother cell in bacteria in the genus *Caulobacter*. { 'swȯr·mər ¸sel }

Swarts reaction [ORG CHEM] The reaction of chlorinated hydrocarbons with metallic fluorides to form chlorofluorohydrocarbons, such as CCl_2F_2, which is quite inert and nontoxic. { 'svärts rē¸ak·shən }

swartzite [MINERAL] $CaMg(UO_2)(CO_3)_3·12H_2O$ A green monoclinic mineral composed of hydrous carbonate of calcium, magnesium, and uranium. { 'swȯrt¸sīt }

swash [GEOL] **1.** A narrow channel or ground within a sand bank, or between a sand bank and the shore. **2.** A bar over which the sea washes. [OCEANOGR] The rush of water up onto the beach following the breaking of a wave. Also known as run-up; uprush. { 'swäsh }

swash bulkhead *See* swash plate. { 'swäsh 'bəlk‚hed }

swash mark [GEOL] A fine, wavy or arcuate line or minute ridge consisting of fine sand, seaweed, and other debris on a beach; marks the farthest advance of wave uprush. Also known as debris line; wave line; wavemark. { 'swäsh ‚märk }

swash plate [NAV ARCH] A partial bulkhead in the tank of an oil tanker; used to decrease the back-and-forth surge of liquid as the ship rocks or rolls. Also known as swash bulkhead. { 'swäsh ‚plāt }

swash-plate pump [MECH ENG] A rotary pump in which the angle between the drive shaft and the plunger-carrying body is varied. { 'swäsh¦plāt ‚pəmp }

swastika [MATH] A plane curve whose equation in Cartesian coordinates x and y is $y^4 - x^4 = xy$. { 'swäs·tə·kə }

SWATH ship [NAV ARCH] A ship consisting of a component below the surface that has most of the buoyant volume, a component above the surface that contains most of the usable volume, and struts that connect these two volumes and pierce the surface. Acronym for small-waterplane-area twin-hull ship. { 'swath ‚ship }

S wave [GEOPHYS] A seismic body wave propagated in the crust or mantle of the earth by a shearing motion of material; speed is 1.9–2.5 miles (3–4 kilometers) per second in the crust and 2.7–2.9 miles (4.4–4.6 kilometers) in the mantle. Also known as distortional wave; equivoluminal wave; rotational wave; secondary wave; shake wave; shear wave; tangential wave; transverse wave. { 'es ‚wāv }

sway [NAV ARCH] Lateral movement of the center of gravity of a ship. { swā }

swayback [MED] Increased lumbar lordosis with compensatory increased thoracic kyphosis. [VET MED] Sinking of the back, or lordosis. { 'swā‚bak }

sway bar *See* stabilizer bar. { 'swā ‚bär }

sway brace [CIV ENG] One or a pair of diagonal members designed to resist horizontal forces, such as wind. { 'swā ‚brās }

sway frame [CIV ENG] A unit in the system of members of a bridge that provides bracing against side sway; consists of two diagonals, the verticals, the floor beam, and the bottom strut. { 'swā ‚frām }

sweat [CHEM] Exudation of nitroglycerin from dynamite due to separation of nitroglycerin from its adsorbent. [MET] Exudate of low-melting-point constituents from a metal on solidification. [PHYSIO] The secretion of the sweat glands. Also known as perspiration. [SCI TECH] Formation of moisture beads on a surface as a result of concentration. { swet }

sweat cooling [AERO ENG] A technique for cooling combustion chambers or aerodynamically heated surfaces by forcing a coolant through a porous wall, resulting in film cooling at the interface. Also known as transpiration cooling. { 'swet ‚kül·iŋ }

sweated wax [MATER] A white, moisture-free petroleum wax with the oil removed by a sweating process, in which the unrefined wax is heated in shallow pans. { 'swed·əd ‚waks }

sweat gland [PHYSIO] A coiled tubular gland of the skin which secretes sweat. { 'swet ‚gland }

sweating [CHEM ENG] Separation of paraffin oil from low-melting petroleum wax obtained from paraffin wax in a chamber (sweater) by first cooling the mixture until it is a solid cake, then warming gradually to cause partial fusion of the mixture to allow drainage of liquid from the cake. Also known as exudation. { 'swed·iŋ }

sweating out [MET] Bringing small globules of low-melting constituents to the surface of an alloy during heat treatment (as lead out of bronze). { 'swed·iŋ ‚aút }

sweat soldering [MET] Soldering two parts by precoating with solder and merging them by application of heat. { 'swet ‚säd·ə·riŋ }

swedenborgite [MINERAL] $NaBe_4SbO_7$ A colorless to wine-yellow mineral composed of sodium beryllium antimony oxide. { 'swēd·ən‚bór‚gīt }

sweep [ELECTR] **1.** The steady movement of the electron beam across the screen of a cathode-ray tube, producing a steady bright line when no signal is present; the line is straight for a linear sweep and circular for a circular sweep. **2.** The steady change in the output frequency of a signal generator from one limit of its range to the other. [MET] A profile pattern used to form molds for symmetrical articles made by sweep casting. [ORD] **1.** Swift flight of a formation of combat airplanes over enemy territory. **2.** To cover a wide area with gunfire. { swēp }

sweep amplifier [ELECTR] An amplifier used with a cathode-ray tube, such as in a television receiver or cathode-ray oscilloscope, to amplify the sawtooth output voltage of the sweep oscillator, to shape the waveform for the deflection circuits of a television picture tube, or to provide balanced signals to the deflection plates. { 'swēp ‚am·plə‚fī·ər }

sweepback [AERO ENG] **1.** The backward slant of a leading or trailing edge of an airfoil. **2.** The amount of this slant, expressed as the angle between a line perpendicular to the plane of symmetry and a reference line in the airfoil. { 'swēp‚bak }

sweep circuit [ELECTR] The sweep oscillator, sweep amplifier, and any other stage used to produce the deflection voltage or current for a cathode-ray tube. Also known as scanning circuit. { 'swēp ‚sər·kət }

sweep-frequency reflectometer [ELECTROMAG] A reflectometer that measures standing-wave ratio and insertion loss in decibels over a wide range of frequencies, in either single- or sweep-frequency operation. { 'swēp ¦frē·kwən·sē ‚rē‚flek 'täm·əd·ər }

sweep generator Also known as sweep oscillator. [ELECTR] **1.** An electronic circuit that generates a voltage or current, usually recurrent, as a prescribed function of time; the resulting waveform is used as a time base to be applied to the deflection system of an electron-beam device, such as a cathode-ray tube. Also known as time-base generator; timing-axis oscillator. **2.** A test instrument that generates a radio-frequency voltage whose frequency varies back and forth through a given frequency range at a rapid constant rate; used to produce an input signal for circuits or devices whose frequency response is to be observed on an oscilloscope. { 'swēp ‚jen·ə‚rād·ər }

sweeping [NAV] **1.** The process of towing a line or object below the surface in order to determine whether an area is free from isolated submerged hazards to vessels and to determine the position of any such hazards that exist, or to determine the least depth of an area. **2.** The process of clearing an area or channel of mines or other hazards to navigation. { 'swēp·iŋ }

sweeping receivers [ELECTR] Automatically and continuously tuned receivers designed to stop and lock on when a signal is found, or to continually plot band occupancy. { 'swēp·iŋ ri‚sē·vərz }

sweep jamming [ELECTR] Jamming an enemy radarscope by sweeping the region of radar-beam coverage with electromagnetic waves having the same frequency as those received by the radarscope. { 'swēp ‚jam·iŋ }

sweep oscillator *See* sweep generator. { 'swēp ‚äs·ə‚lad·ər }

sweep-out pattern [PETRO ENG] The areal pattern of water advance in a petroleum reservoir, as for a waterflood operation. { 'swēp¦aút ‚pad·ərn }

sweep rate [ELECTR] The number of times a radar radiation pattern rotates during 1 minute; sometimes expressed as the duration of one complete rotation in seconds. { 'swēp ‚rāt }

sweepstakes route [ECOL] A means that allows chance migration across a sea on natural rafts, so that oceanic islands can be colonized. { 'swēp‚stāks ‚rüt }

sweep test [ELECTR] Test given coaxial cable with an oscilloscope to check attenuation. { 'swēp ‚test }

sweep-through jammer [ELECTR] A jamming transmitter which is swept through a radio-frequency band in short steps to jam each frequency briefly, producing a sound like that of an aircraft engine. { 'swēp¦thrü 'jam·ər }

sweep voltage [ELECTR] Periodically varying voltage applied to the deflection plates of a cathode-ray tube to give a beam displacement that is a function of time, frequency, or other data base. { 'swēp ‚vōl·tij }

sweet basil oil *See* basil oil. { 'swēt 'bā·zəl ‚óil }

sweet bay oil *See* volatile laurel oil. { 'swēt 'bā ‚óil }

sweet corrosion [PETRO ENG] Corrosion occurring in oil or gas wells where there is no iron-sulfide corrosion product, and there is no odor of hydrogen sulfide in the produced reservoir fluid. { 'swēt kə'rō·zhən }

sweet crudes [MATER] Crude petroleum oil containing little sulfur. { 'swēt 'krüdz }

sweetening [CHEM ENG] Improvement of a petroleum-product color and odor by converting sulfur compounds into disulfides with sodium plumbite (doctor treating), or by removing

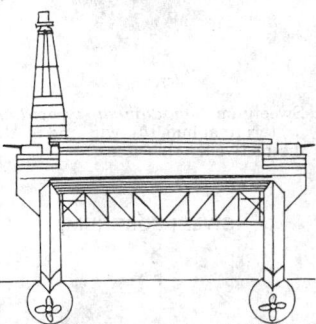

SWATH SHIP

Transverse view of Creed and Lewis seadrome (1943), the original concept of a SWATH ship.

SWEETGUM

Sweetgum (*Liquidambar styraciflua*);
(*a*) terminal bud, (*b*) twig, and (*c*) the
distinctive five-lobed leaf.

SWEET SORGHUM

Sweet sorghum in Oklahoma. The
scale indicates feet. 1 ft = 0.3 meter.
(*USDA*)

SWING BRIDGE

The vertical-lift span and the swing
bridge across Arthur Kill, between
Staten Island and New Jersey. The
vertical-lift replaced the swing, in
use since 1888. (*Baltimore and Ohio
Railroad*)

them by contacting the petroleum stream with alkalies or other
sweetening agents.   { 'swēt·ən·iŋ }

sweet gas [MATER] A petroleum natural gas containing no
corrosive components, such as hydrogen sulfide and mercaptans.   { 'swēt 'gas }

sweetgum [BOT] *Liquidambar styraciflua.* A deciduous tree
of the order Hamamelidales found in the southeastern United
States, and distinguished by its five-lobed, or star-shaped,
leaves, and by the corky ridges developed on the twigs.
{ 'swēt,gəm }

sweet oil *See* olive oil.   { 'swēt ,òil }

sweet orange oil [MATER] A sweet, yellow, mild essential
oil expressed from the peel of an orange, *Citrus aurantium;*
soluble in glacial acetic acid, somewhat in alcohol; used in
flavors, perfumes, and medicine.   Also known as orange oil.
{ 'swēt 'är·inj ,òil }

sweet potato [BOT] *Ipomoea batatas.* A tropical vine having variously shaped leaves, purplish flowers, and a sweet,
fleshy, edible tuberous root.   { 'swēt pə,tād·ō }

sweet roast *See* dead-roast.   { 'swēt ,rōst }

sweet sorghum [AGR] *Sorghum bicolor.* A crop plant grown
primarily for syrup production and for forage.   { ,swēt 'sòr·
gəm }

sweet spirits of niter *See* ethyl nitrite.   { 'swēt 'spir·əts əv
'nī·tər }

swell [GEOL] **1.** The volumetric increase of soils on being
removed from their compacted beds due to an increase in void
ratio.   **2.** A local enlargement or thickening in a vein or ore
deposit.   **3.** A low dome or quaquaversal anticline of considerable areal extent; long and generally symmetrical waves contribute to the mixing processes in the surface layer and thus to
its sound transmission properties.   **4.** Gently rising ground, or
a rounded hill above the surrounding ground or ocean floor.
[MIN ENG] *See* horseback. [OCEANOGR] Ocean waves
which have traveled away from their generating area; these
waves are of relatively long length and period, and regular in
character.   { swel }

swell-and-swale topography [GEOGR] A low-relief, undulating landscape characterized by gentle slopes and rounded
hills interspersed with shallow depressions.   { ¦swel ən ¦swāl
tə¦päg·rə·fē }

swell diameter [AERO ENG] In a body of revolution having
an ogival portion, such as a projectile, the swell diameter is in
the diameter of the maximum transverse section of the geometrical ogive.   { 'swel dī,am·əd·ər }

swell direction [OCEANOGR] The direction from which
swell is moving.   { 'swel di,rek·shən }

swelled ground [GEOL] A soil or rock that expands when
wetted.   { 'sweld 'graùnd }

swell forecast [OCEANOGR] Prediction of the frequency and
height of swell waves in a remote area from the characteristics
of the waves at their origin.   { 'swel ,fòr,kast }

swelling clay [GEOL] Clay that can absorb large amounts
of water, such as bentonite.   { 'swel·iŋ ,klā }

swep [ORG CHEM] $C_8H_7Cl_2NO_2$ A white, crystalline compound with a melting point of 112–114°C; insoluble in water;
used as a pre- and postemergence herbicide for rice, carrots,
potatoes, and cotton.   Also known as methyl-*N*-(3,4-dichloro-
phenyl)carbamate.   { swep }

sweptback wing [AERO ENG] An airplane wing on which
both the leading and trailing edges have sweepback, the trailing
edge forming an acute angle with the longitudinal axis of the
airplane aft of the root.   Also known as swept wing.   { 'swep
¦bak ,wiŋ }

swept-frequency analyzer [ELECTR] A spectrum analyzer
in which a ramp generator simultaneously moves a spot horizontally across an electronic display and increases the frequency
of a local oscillator; and any signal at the input, at a frequency
such that the difference between its frequency and the local
oscillator is within the bandwidth of an intermediate-frequency
filter, vertically deflects the spot on the display by an amount
proportional to the amplitude of the input signal being analyzed.
{ ¦swept ,frē·kwən·sē 'an·ə,līz·ər }

swept-frequency reflectometer [ELECTROMAG] A microwave reflectometer with a swept-signal source and an oscilloscope for displaying the output as a function of frequency.
{ ¦swept ,frē·kwən·sē ,rē,flek'täm·əd·ər }

swept wing *See* sweptback wing.   { 'swept ,wiŋ }

swim bladder [VERT ZOO] A gas-filled cavity found in the

body cavities of most bony fishes; has various functions in
different fishes, acting as a float, a lung, a hearing aid, and a
sound-producing organ.   { 'swim ,blad·ər }

swimmeret [INV ZOO] Any of a series of paired biramous
appendages under the abdomen of many crustaceans, used for
swimming and egg carrying.   { ,swim·ə,ret }

swimmer's conjunctivitis *See* inclusion conjunctivitis.
{ 'swim·ərz kən,jəŋk·tə'vīd·əs }

swimmer's itch *See* schistosome dermatitis.   { 'swim·ərz
'ich }

swimming bird [VERT ZOO] Any bird belonging to the orders
Charadriiformes and Pelecaniformes.   { 'swim·iŋ ,bird }

swimming-pool conjunctivitis *See* inclusion conjunctivitis.
{ 'swim·iŋ ¦pül kən,jəŋk·tə'vīd·əs }

swimming-pool reactor *See* pool reactor.   { 'swim·iŋ ¦pül
rē,ak·tər }

swine [AGR] A domesticated member of the family Suidae.
[VERT ZOO] Any of various species comprising the Suidae.
{ swīn }

swine erysipelas [VET MED] An infectious bacterial disease
of swine caused by *Erysipelothrix insidiosa* in which involvement of the skin is predominant.   { 'swin ,er·ə'sip·ə·ləs }

swine influenza [VET MED] A disease of swine caused by
the associated effects of a filterable virus and *Hemophilus suis*,
characterized by inflammation of the upper respiratory tract.
{ 'swin ,in·flü'en·zə }

swine plague [VET MED] Hemorrhagic septicemia of swine
caused by *Pasteurella suiseptica*, characterized by pleuropneumonia.   { 'swin ¦plāg }

swine pox [VET MED] A benign infection of young hogs
characterized by pox lesions on the body and inner surfaces of
the legs.   { 'swin ¦päks }

swinestone [PETR] Limestone containing black bituminous
matter, which gives off an objectionable odor when rubbed.
{ 'swīn,stōn }

swing [ELEC] Variation in frequency or amplitude of an electrical quantity.   [ENG] **1.** The arc or curve described by the
point of a pick or mandril when being used.   **2.** Rotation of
the superstructure of a power shovel on the vertical shaft in
the mounting.   **3.** To rotate a revolving shovel on its base.
{ swiŋ }

swing-around trajectory [AERO ENG] A planetary round-
trip trajectory which requires minimal propulsion at the destination planet, but instead uses the planet's gravitational field to
effect the bulk of the necessary orbit change to return to earth.
{ 'swiŋ ə¦raùnd trə'jek·trē }

swing bridge [CIV ENG] A movable bridge that pivots in a
horizontal plane about a center pier.   { 'swiŋ ,brij }

swing-by *See* flyby.   { 'swiŋ,bī }

swing cut *See* slipping cut.   { 'swiŋ ,kət }

swinger *See* revolver.   { 'swiŋ·ər }

swing-frame grinder [MECH ENG] A grinding machine
hanging by a chain so that it may swing in all directions for
surface grinding heavy work.   { 'swiŋ ,frām ,grīn·dər }

swing-hammer crusher [MIN ENG] A rotary crusher with
rotating hammers that break up ore by impelling it against
breaker plates.   { 'swiŋ ¦ham·ər ,krəsh·ər }

swinging [NAV] The process of placing a craft on various
headings and comparing magnetic compass readings with the
corresponding magnetic directions to determine deviation; this
usually follows compass adjustment or compass compensation,
and is done to obtain information for making a deviation table,
deviation card, or compass correction card.   { 'swiŋ·iŋ }

swinging a claim [MIN ENG] The adjustment of the boundaries of a mining claim to more nearly conform to the strike
of the vein.   { 'swiŋ·iŋ ə 'klām }

swinging buoy [NAV] A buoy placed at a favorable location
to assist a vessel to adjust its compass or swing ship; the bow
of the vessel is made fast to one such buoy, and the vessel is
swung by means of lines to a tug or to additional buoys.
{ 'swiŋ·iŋ 'bòi }

swinging choke [ELEC] An iron-core choke having a core
that can be operated almost at magnetic saturation; the inductance is then a maximum for small currents, and swings to a
lower value as current increases.   Also known as swinging
reactor.   { 'swiŋ·iŋ 'chōk }

swinging compass [NAV] An accurate portable magnetic
compass used to indicate magnetic headings during aircraft
magnetic compass calibration.   { 'swiŋ·iŋ ¦käm·pəs }

swinging load [ENG] The load in pressure equipment which changes at frequent intervals. { 'swiŋ·iŋ 'lōd }

swinging reactor *See* swinging choke. { 'swiŋ·iŋ rē'ak·tər }

swinging ship [NAV] The process of determining the deviation of the ship's magnetic compass by placing a vessel or an aircraft on various headings and comparing magnetic compass readings with the corresponding but previously determined magnetic directions; this procedure usually follows compass adjustment or compass compensation, and is done to obtain information for making a deviation table; usually called swinging when referred to an aircraft compass. Also known as compass calibration. { 'swiŋ·iŋ 'ship }

swinging the arc [NAV] The process of rotating a sextant about the line of sight to the horizon to determine the foot of the vertical circle through a body being observed. Also known as rocking the sextant. { 'swiŋ·iŋ thə 'ärk }

swinging traverse [ORD] Type of fire used against dense troop formations moving toward a machine gun position or rapidly moving targets; the traversing clamp is loosened so that a gunner makes rapid changes by exerting pressure against the pistol-grip. { 'swiŋ·iŋ trə'vərs }

swing joint [DES ENG] A pipe joint in which the parts may be rotated relative to each other. { 'swiŋ ‚jȯint }

swing pipe [ENG] A discharge pipe whose intake end can be raised or lowered on a tank. { 'swiŋ ‚pīp }

swing shift [IND ENG] Working arrangement in a three-shift, continuously run plant with working hours changed at regular intervals; during a swing shift the morning shift becomes the afternoon shift, while the afternoon shift becomes the morning shift of the next day, with only an 8-hour break on the first day of change. { 'swiŋ ‚shift }

swing-wing aircraft [AERO ENG] An aircraft whose wings fold back along the fuselage at high flight speeds. { 'swiŋ ‚wiŋ 'er‚kraft }

swirl error [NAV] The additional error in the reading of a magnetic compass during a turn due to friction in the compass liquid. { 'swərl ‚er·ər }

swirl flowmeter *See* vortex precession flowmeter. { 'swərl 'flō‚mēd·ər }

Swiss pattern file [DES ENG] A type of fine file used for precision filing of jewelry, instrument parts, and dies. { 'swis ‚pad·ərn 'fīl }

switch [COMPUT SCI] **1.** A hardware or programmed device for indicating that one of several alternative states or conditions have been chosen, or to interchange or exchange two data items. **2.** A symbol used to indicate a branch point, or a set of instructions to condition a branch. [CIV ENG] **1.** A device for enabling a railway car to pass from one track to another. **2.** The junction of two tracks. [ELEC] A manual or mechanically actuated device for making, breaking, or changing the connections in an electric circuit. Also known as electric switch. Symbolized SW. { swich }

switch angle [CIV ENG] The angle between the switch and stock rails of a railroad track, measured at the point of juncture between the gage lines. { 'swich ‚aŋ·gəl }

switchback [MIN ENG] A zigzag arrangement of railroad tracks by means of which a train can reach a higher or lower level by a succession of easy grades. { 'swich‚bak }

switchblade knife [DES ENG] A knife in which the blade is spring-loaded and swings open when released by a pushbutton. { 'swich‚blād 'nīf }

switchboard [COMMUN] A manually operated apparatus at a telephone exchange, on which the various circuits from subscribers and other exchanges are terminated to enable operators to establish communications either between two subscribers on the same exchange, or between subscribers on different exchanges. Also known as telephone switchboard. [ELEC] A single large panel or assembly of panels on which are mounted switches, circuit breakers, meters, fuses, and terminals essential to the operation of electric equipment. Also known as electric switchboard. { 'swich‚bȯrd }

switched capacitor [ELECTR] An integrated circuit element, consisting of a capacitor with two metal oxide semiconductor (MOS) switches, whose function is approximately equivalent to that of a resistor. { 'swicht kə'pas·əd·ər }

switched-capacitor filter [ELECTR] An integrated-circuit filter in which a resistor is simulated by a combination of a capacitor and metal oxide semiconductor switches that are turned on and off periodically at a high frequency. Also known as switched-C filter. { ‚swicht kə'pas·əd·ər ‚fil·tər }

switched-C filter *See* switched-capacitor filter. { ‚swicht 'sē ‚fil·tər }

switched circuit [COMMUN] A communications circuit or channel that can be turned on and off and made to serve various users. { 'swicht 'sər·kət }

switched line [COMMUN] A communications line, such as a dial telephone line, whose path can vary each time the line is used. { 'swicht 'līn }

switched-message network [COMPUT SCI] A data transmission system in which a user can communicate with any other user of the network. { 'swicht ‚mes·ij 'net‚wərk }

switched network [COMMUN] A communications network, such as the dial telephone network, in which any station may be connected with any other through the use of switching and control devices. { 'swicht 'net‚wərk }

switch function [ELECTR] A circuit having a fixed number of inputs and outputs designed such that the output information is a function of the input information, each expressed in a certain code or signal configuration or pattern. { 'swich ‚fəŋk·shən }

switchgear [ELEC] The aggregate of switching devices for a power or transforming station, or for electric motor control. { 'swich‚gir }

switch gene [GEN] A gene that causes the epigenotype to switch to a different developmental pathway. { 'swich ‚jēn }

switch hook [ELEC] A switch on a telephone set that operates when the receiver is placed on the hook or removed from it. { 'swich ‚hùk }

switching [ELEC] Making, breaking, or changing the connections in an electrical circuit. { 'swich·iŋ }

switching center [COMMUN] The equipment in a relay station for automatically or semiautomatically relaying communications traffic. { 'swich·iŋ ‚sen·tər }

switching circuit [ELEC] A constituent electric circuit of a switching or digital processing system which receives, stores, or manipulates information in coded form to accomplish the specified objectives of the system. { 'swich·iŋ ‚sər·kət }

switching control [COMMUN] Installation in a wire system where telephone or teletypewriter switchboards are installed to interconnect circuits. { 'swich·iŋ kən‚trōl }

switching device [ENG] An electrical or mechanical device or mechanism, which can bring another device or circuit into an operating or nonoperating state. Also known as switching mechanism. { 'swich·iŋ di‚vīs }

switching diode [ELECTR] A crystal diode that provides essentially the same function as a switch; below a specified applied voltage it has high resistance corresponding to an open switch, while above that voltage it suddenly changes to the low resistance of a closed switch. { 'swich·iŋ ‚dī‚ōd }

switching function [MATH] A switching function of n variables is a function that assigns to each binary sequence of length n the number 0 or the number 1. { 'swich·iŋ ‚fəŋk·shən }

switching gate [ELECTR] An electronic circuit in which an output having constant amplitude is registered if a particular combination of input signals exists; examples are the OR, AND, NOT, and INHIBIT circuits. Also known as logical gate. { 'swich·iŋ ‚gāt }

switching key *See* key. { 'swich·iŋ ‚kē }

switching mechanism *See* switching device. { 'swich·iŋ ‚mek·ə‚niz·əm }

switching node [COMMUN] A location in a communications network where messages or lines are routed. { 'swich·iŋ ‚nōd }

switching pad [ELECTR] Transmission-loss pad automatically cut in and out of a toll circuit for different desired operating conditions. { 'swich·iŋ ‚pad }

switching reactor [ELECTROMAG] A saturable-core reactor that has several input control windings and one or more output windings that essentially duplicate the functions of a relay. { 'swich·iŋ rē'ak·tər }

switching substation [ELEC] An electric power substation whose equipment is mainly for connections and interconnections, and does not include transformers. { 'swich·iŋ 'səb‚stā·shən }

switching surface [CONT SYS] In feedback control systems employing bang-bang control laws, the surface in state space which separates a region of maximum control effort from one of minimum control effort. { 'swich·iŋ ‚sər·fəs }

SYCAMORE

Terminal bud, leaf with seed capsule, and twig of American sycamore (*Platanus offidentalis*).

switching system [COMMUN] An assembly of switching and control devices provided so that any station in a communications system may be connected as desired with any other station. { 'swich·iŋ ,sis·təm }

switching theory [ELECTR] The theory of circuits made up of ideal digital devices; included are the theory of circuits and networks for telephone switching, digital computing, digital control, and data processing. { 'swich·iŋ ,thē·ə·rē }

switching-through relay [ELEC] Control relay of a line-finder selector, connector, or other stepping switch, which extends the loop of a calling telephone through to the succeeding switch in a switch train. { 'swich·iŋ ¦thrü 'rē,lā }

switching time [ELECTR] **1.** The time interval between the reference time and the last instant at which the instantaneous voltage response of a magnetic cell reaches a stated fraction of its peak value. **2.** The time interval between the reference time and the first instant at which the instantaneous integrated voltage response of a magnetic cell reaches a stated fraction of its peak value. { 'swich·iŋ ,tīm }

switching transistor [ELECTR] A transistor designed for on/off switching operation. { 'swich·iŋ tran'zis·tər }

switching trunk [ELEC] Trunk from a long-distance office to a local exchange office used for completing a long-distance call. { 'swich·iŋ ,trəŋk }

switching tube [ELECTR] A gas tube used for switching high-power radio-frequency energy in the antenna circuits of radar and other pulsed radio-frequency systems; examples are ATR tube; pre-TR tube; TR tube. { 'swich·iŋ ,tüb }

switch jack [ELEC] Any of the devices that provide terminals for the control circuits of the switch. { 'swich ,jak }

switchman [MIN ENG] A laborer who throws switches of mine tracks in a coal mine. { 'swich·mən }

switch-over travel [ELEC] That movement of a switch-operating lever which takes place after the switch has been actuated either to close or open its contacts. { 'swich ,ō·vər 'trav·əl }

switch plate [MIN ENG] An iron plate on tramroads in mines to change the direction of movement. { 'swich ,plāt }

switch pretravel [ELEC] That movement of a switch-operating level that takes place before the switch is actuated either to close or to open its contacts. { 'swich 'prē,trav·əl }

switch register [COMPUT SCI] A manual switch on the control panel by means of which a bit may be entered in a processor register. { 'swich ,rej·ə·stər }

switch room [COMMUN] Part of a central office building that houses switching mechanisms and associated apparatus. { 'swich ,rüm }

switch selectable addressing [COMPUT SCI] The setting of DIP switches in a peripheral or terminal device to determine the address that identifies the device to the computer system. { 'swich si¦lek·tə·bəl 'ad,res·iŋ }

switch train [ELEC] A series of switches in tandem. { 'swich ,trān }

swivel [DES ENG] A part that oscillates freely on a headed bolt or pin. [PETRO ENG] A short piece of casing having one end belled over a heavy ring, and having a large hole through both walls, the other end being threaded. { 'swiv·əl }

swivel block [DES ENG] A block with a swivel attached to its hook or shackle permitting it to revolve. { 'swiv·əl ,bläk }

swivel coupling [MECH ENG] A coupling that gives complete rotary freedom to a deflecting wedge-setting assembly. { 'swiv·əl ,kəp·liŋ }

swivel gun [ORD] A gun mounted on a pedestal so that it can be turned from side to side or up and down. { 'swiv·əl ,gən }

swivel head [MECH ENG] The assembly of a spindle, chuck, feed nut, and feed gears on a diamond-drill machine that surrounds, rotates, and advances the drill rods and drilling stem; on a hydraulic-feed drill the feed gears are replaced by a hydraulically actuated piston assembly. { 'swiv·əl ,hed }

swivel hook [DES ENG] A hook with a swivel connection to its base or eye. { 'swiv·əl ,huk }

swivel joint [DES ENG] A joint with a packed swivel that allows one part to move relative to the other. { 'swiv·əl ,joint }

swivel neck See water swivel. { 'swiv·əl ,nek }

swivel pin See kingpin. { 'swiv·əl ,pin }

swivel spindle [BUILD] A shaft in a door handle assembly

SYLLIDAE

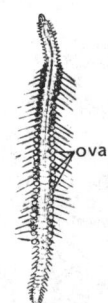

Exogone, of the Syllidae (Exogoninae), dorsal view.

designed with a center joint that permits one knob to remain fixed while the other is being turned. { 'swiv·əl ,spin·dəl }

Swordfish See Dorado. { 'sord,fish }

SWS See deep sleep.

SXAPS See soft-x-ray appearance potential spectroscopy.

syblet [COMMUN] One of the elements of speech; others are words, syllables, and phonemes. { 'sib·lət }

sycamore [BOT] **1.** Any of several species of deciduous trees of the genus *Platanus*, especially *P. occidentalis* of eastern and central North America, distinguished by simple, large, three-to five-lobed leaves and spherical fruit heads. **2.** The Eurasian maple (*Acer pseudoplatanus*). { 'sik·ə,mor }

Sycettida [INV ZOO] An order of calcareous sponges of the subclass Calcaronea in which choanocytes occur in flagellated chambers, and the spongocoel is not lined with these cells. { sə'sed·əd·ə }

Sycettidae [INV ZOO] A family of sponges in the order Sycettida. { sə'sed·ə,dē }

Sycidales [PALEON] A group of fossil aquatic plants assigned to the Charophyta, characterized by vertically ribbed gyrogonites. { ,sis·ə'dā·lēz }

sycon [INV ZOO] A canal system in sponges in which the flagellated layer is confined to outpocketings of the paragaster that are indirectly connected to the incurrent canals. { 'sī,kän }

syconium [BOT] A fleshy fruit, as a fig, with an enlarged pulpy receptacle internally lined with minute flowers. { sī'kō·nē·əm }

Sycorax [ASTRON] A small satellite of Uranus in a retrograde orbit with a mean distance of 7,550,000 miles (12,150,000 kilometers), eccentricity of 0.504, and sidereal period of 3.53 years. { 'sī'kor,aks }

sycosis [MED] An inflammatory disease affecting the hair follicles, particularly of the beard, and characterized by papules, pustules, and tubercles, perforated by hairs, together with infiltration of the skin and crusting. { sī'kō·səs }

Sydenham's chorea See Saint Vitus dance. { 'sīd·ən,hamz kə'rē·ə }

syenite [PETR] A visibly crystalline plutonic rock with granular texture composed largely of alkali feldspar, with subordinate plagioclose and mafic minerals; the intrusive equivalent of trachyte. { 'sī·ə,nīt }

syenodiorite [PETR] Plutonic rock consisting of acid plagioclase, orthoclase, and a ferromagnesian mineral. { ¦sī·ə·nō'dī·ə,rīt }

syenogabbro [PETR] Plutonic rock consisting of basic plagioclase, orthoclase, and a dark mineral such as augite. { ¦sī·ə·nō'ga,bro }

syllabic compandor [ELECTR] A compandor in which the effective gain variations are made at speeds allowing response to the syllables of speech but not to individual cycles of the signal wave. { si'lab·ik kəm'pan·dər }

Syllidae [INV ZOO] A large family of polychaete annelids belonging to the Errantia; identified by their long, linear, translucent bodies with articulated cirri; size ranges from minute *Exogone* to *Trypanosyllis*, which may be 4 inches (100 millimeters) long. { 'sil·ə,dē }

Syllinae [INV ZOO] A subfamily of polychaete annelids of the family Syllidae. { 'sil·ə,nē }

syllogism [MATH] A statement together with a conclusion; this usually has the form "if p then q." { 'sil·ə,jiz·əm }

Sylonidae [INV ZOO] A family of parasitic crustaceans in the order Rhizocephala. { sə'län·ə,dē }

Sylopidae [INV ZOO] A family of coleopteran insects in the superfamily Meloidea in which the elytra in males are reduced to small leathery flaps while the hindwings are large and fan-shaped. { sə'läp·ə,dē }

Sylow subgroup [MATH] A subgroup H of a given group G such that the order of H is p^n, where p is a prime and n is an integer, and p^n is the highest power of p dividing the order of G. { ¦sī,lō 'səb,grüp }

sylvanite [MINERAL] $(Au,Ag)Te_2$ A steel-gray, silver-white, or brass-yellow mineral that crystallizes in the monoclinic system and often occurs in implanted crystals. Also known as goldschmidtite; graphic tellurium; white tellurium; yellow tellurium. { 'sil·və,nīt }

sylvatic plague [VET MED] Plague occurring in rodents; may be transmitted to humans. Also known as endemic rural plague. { sil'vad·ik 'plāg }

sylvester [MIN ENG] A hand-operated device for withdrawing supports from the waste or old workings in a mine by means of a long chain which allows the device to be positioned at a safe distance from the support to be extracted. { sil'ves·tər }

Sylvester's theorem [MATH] If A is a matrix with distinct eigenvalues $\lambda_1, \ldots, \lambda_n$, then any analytic function $f(A)$ can be realized from the $\lambda_i, f(\lambda_i)$, and the matrices $A - \lambda_i I$, where I is the identity matrix. { sil'ves·tərz ,thir·əm }

Sylvicolidae [INV ZOO] A family of orthorrhaphous dipteran insects in the series Nematocera. { ,sil·və'käl·ə,dē }

sylvine See sylvite. { 'sil,vīn }

sylvite [MINERAL] KCl A salty-tasting, white or colorless isometric mineral, occurring in cubes or crystalline masses or as a saline residue; the chief ore of potassium. Also known as leopoldite; sylvine. { 'sil,vīt }

sym- [ORG CHEM] A chemical prefix; denotes structure of a compound in which substituents are symmetrical with respect to a functional group or to the carbon skeleton. { sim }

symballophone [ENG] A double stethoscope for the comparison and lateralization of sounds; permits the use of the acute function of the two ears to compare intensity and varying quality of sounds arising in the body or mechanical devices. { sim'bôl·ə,fōn }

symbiont [ECOL] A member of a symbiotic pair. { 'sim·bē,änt }

symbiosis [ECOL] **1.** An interrelationship between two different species. **2.** An interrelationship between two different organisms in which the effects of that relationship is expressed as being harmful or beneficial. Also known as consortism. { ,sim·bē'ō·səs }

symbiotic nova See novalike symbiotic. { ,sim·bē'äd·ik 'nō·və }

symbiotic objects [ASTRON] Stars whose spectra have characteristics of two disparate spectral classes { ,sim·bē'äd·ik 'äb,jeks }

symbiotic star [ASTRON] A stellar object whose optical spectrum displays features indicative of two very different stellar regimes: a stellar spectrum whose flux distribution and absorption lines suggest the presence of a cool star, and emission lines which can be formed only in a much hotter medium. { ,sim·bē'äd·ik 'stär }

symblepharon [MED] Adhesion of the eyelids to the eyeball. { sim'blef·ə,rän }

symbol [CHEM] Letter or combination of letters and numbers that represent various conditions or properties of an element, for example, a normal atom, O (oxygen); with its atomic weight, ^{16}O; its atomic number, $_8$^{16}O; as a molecule, O_2; as an ion, O^{2+}; in excited state, O^*; or as an isotope, ^{8}O. [SCI TECH] **1.** A design used on a diagram to represent a component or to identify specific characteristics, quantities, or objects. **2.** A sign letter or abbreviation used on a diagram or in an equation to represent a quantity or to identify an object. { 'sim·bəl }

symbolic address [COMPUT SCI] In coding, a programmer-defined symbol that represents the location of a particular datum item, instruction, or routine. Also known as symbolic number. { sim'bäl·ik 'ad,res }

symbolic age of neutrons See Fermi age. { sim'bäl·ik 'āj əv 'nü,tränz }

symbolic algebraic manipulation language [COMPUT SCI] An algebraic manipulation language which admits the most general species of mathematical expressions, usually representing them as general tree structures, but which lacks certain special algorithms found in seminumerical and ghost languages. { sim'bäl·ik ,al·jə'brā·ik mə,nip·yə'lā·shən ,laŋ·gwij }

symbolic assembly language listing [COMPUT SCI] A list that may be produced by a computer during the compilation of a program showing the source language statements together with the corresponding machine language instructions generated by them. { sim'bäl·ik ə'sem·blē ,laŋ·gwij ,list·iŋ }

symbolic assembly system [COMPUT SCI] A system for forming programs that can be run on a computer, consisting of an assembly language and an assembler. { sim'bäl·ik ə'sem·blē ,sis·təm }

symbolic coding [COMPUT SCI] Instruction written in an assembly language, using symbols for operations and addresses. Also known as symbolic programming. { sim'bäl·ik 'kōd·iŋ }

symbolic computation system See symbolic system. { sim,bäl·ik kəm'pyü'tā·shən ,sis·təm }

symbolic computing [COMPUT SCI] The development and use of symbolic systems. { sim,bäl·ik kəm'pyüd·iŋ }

symbolic debugging [COMPUT SCI] A method of correcting known errors in a computer program written in a source language, in which certain statements are compiled together with the program. { sim'bäl·ik dē'bəg·iŋ }

symbolic language [COMPUT SCI] A language which expresses addresses and operation codes of instructions in symbols convenient to humans rather than in machine language. { sim'bäl·ik 'laŋ·gwij }

symbolic logic [MATH] The formal study of symbolism and its use in the foundations of mathematical logic. { sim'bäl·ik 'läj·ik }

symbolic mathematical computation [COMPUT SCI] The manipulation of symbols, representing variables, functions, and other mathematical objects, and combinations of these symbols, representing formulas, equations, and expressions, according to mathematical rules, for example, the rules of algebra or calculus. { sim'bäl·ik 'math·ə,mad·ə·kəl ,käm·pyə'tā·shən }

symbolic number See symbolic address. { sim'bäl·ik 'nəm·bər }

symbolic name [COMPUT SCI] A name given to some entity that is actually something else; for example, the name of a table in a computer program actually represents the physical storage locations used to hold the data stored in that table, as well as the values stored in those locations. { sim'bäl·ik 'nām }

symbolic programming See symbolic coding. { sim'bäl·ik 'prō,gram·iŋ }

symbolic system [COMPUT SCI] A computer program that performs computations with constants and variables according to the rules of algebra, calculus, and other branches of mathematics. Also known as algebraic computation system; computer algebra system; symbolic computation system. { sim'bäl·ik 'sis·təm }

symbol input [COMPUT SCI] Includes all contextual symbols that may appear in a source text. { 'sim·bəl 'in,put }

symbolization [PSYCH] A general mechanism by which some mental representation comes to stand for some other thing, class of things, or attribute of something. { ,sim·bə·lə'zā·shən }

symbol sequence [COMPUT SCI] A sequence of contextual symbols not interrupted by space. { 'sim·bəl 'sē·kwəns }

symbol table [COMPUT SCI] A mapping for a set of symbols to another set of symbols or numbers. { 'sim·bəl 'tā·bəl }

symclosene See trichloroisocyanuric acid. { 'sim·klə,zēn }

symmetrical achromat lens [OPTICS] An older type of camera lens in which two positive achromatic meniscus lenses are symmetrically arranged about the stop. { sə'me·trə·kəl 'ak·rə,mat 'lenz }

symmetrical alternating quantity [PHYS] Alternating quantity of which all values separated by a half period have the same magnitude but opposite sign. { sə'me·trə·kəl 'ôl·tər,nād·iŋ 'kwän·əd·ē }

symmetrical architecture [COMPUT SCI] A type of computer design that allows any type of data to be used with any type of instruction. { si'me·trə·kəl 'ärk·ə,tek·chər }

symmetrical avalanche rectifier [ELECTR] Avalanche rectifier that can be triggered in either direction, after which it has a low impedance in the triggered direction. { sə'me·trə·kəl 'av·ə,lanch ,rek·tə,fī·ər }

symmetrical band-pass filter [ELECTR] A band-pass filter whose attenuation as a function of frequency is symmetrical about a frequency at the center of the pass band. { sə'me·trə·kəl 'band,pas ,fil·tər }

symmetrical band-reject filter [ELECTR] A band-rejection filter whose attenuation as a function of frequency is symmetrical about a frequency at the center of the rejection band. { sə'me·trə·kəl 'band ri,jekt ,fil·tər }

symmetrical clipper [ELECTR] A clipper in which the upper and lower limits on the amplitude of the output signal are positive and negative values of equal magnitude. { sə'me·trə·kəl 'klip·ər }

symmetrical deflection [ELECTR] A type of electrostatic deflection in which voltages that are equal in magnitude and opposite in sign are applied to the two deflector plates. { sə'me·trə·kəl di'flek·shən }

symmetrical distribution [STAT] A distribution in which observations equidistant from the central maximum have the

SYMMETRICAL ACHROMAT LENS

Geometry of symmetrical achromat lens.

same frequency. Also known as symmetric distribution. { si¦met·rə·kəl ‚dis·trə'byü·shən }

symmetrical fold [GEOL] A fold whose limbs have approximately the same angle of dip relative to the axial surface. Also known as normal fold. { sə'me·trə·kəl 'fōld }

symmetrical H attenuator [ELECTR] An H attenuator in which the impedance near the input terminals equals the corresponding impedance near the output terminals. { sə'me·trə·kəl 'āch ə‚ten·yə'wād·ər }

symmetrical inductive diaphragm [ELECTROMAG] A waveguide diaphragm which consists of two plates that leave a space at the center of the waveguide, and which introduces an inductance in the waveguide. { sə'me·trə·kəl in'dək·tiv 'dī·ə‚fram }

symmetrical lens [OPTICS] A lens system consisting of two parts, each of which is the mirror image of the other. { sə'me·trə·kəl 'lenz }

symmetrical O attenuator [ELECTR] An O attenuator in which the impedance near the input terminals equals the corresponding impedance near the output terminals. { sə'me·trə·kəl 'ō ə‚ten·yə‚wād·ər }

symmetrical pi attenuator [ELECTR] A pi attenuator in which the impedance near the input terminals equals the corresponding impedance near the output terminals. { sə'me·trə·kəl 'pī ə‚ten·yə‚wād·ər }

symmetrical T attenuator [ELECTR] A T attenuator in which the impedance near the input terminals equals the corresponding impedance near the output terminals. { sə'me·trə·kəl 'tē ə‚ten·yə‚wād·ər }

symmetrical transducer [ELECTR] A transducer is symmetrical with respect to a specified pair of terminations when the interchange of that pair of terminations will not affect the transmission. { sə'me·trə·kəl tranz'dü·sər }

symmetric chain [MATH] A sequence of subsets of a set of n elements such that each member of the sequence is a subset of the next one, each member of the sequence has a cardinality one greater than that of the previous member, and the sum of the cardinalities of the first and last members of the sequence equals n. { sə'me·trik 'chān }

symmetric chain decomposition [MATH] A partition of the set of all subsets of a finite set, X, into symmetric chains in X. { sə'me·trik ¦chān dē‚käm·pə'zish·ən }

symmetric design [MATH] A balanced incomplete block design in which the number b of blocks equals the number v of elements arranged among the blocks. { si¦me·trik di'zīn }

symmetric difference [MATH] The symmetric difference of two sets consists of all points in one or the other of the sets but not in both. { sə'me·trik 'dif·rəns }

symmetric distribution See symmetrical distribution. { sə¦me·trik ‚di·strə'byü·shən }

symmetric form [MATH] A bilinear form f which is unchanged under interchange of its independent variables; that is, $f(x,y) = f(y,x)$ for all values of the independent variables x and y. { si¦me·trik 'fòrm }

symmetric function [MATH] A function whose value is unchanged for any permutation of its variables. { sə'me·trik 'fəŋk·shən }

symmetric group [MATH] The group consisting of all permutations of a finite set of symbols. { sə'me·trik 'grüp }

symmetric list [COMPUT SCI] A list with sequencing pointers to previous as well as subsequent items. { sə'me·trik 'list }

symmetric matrix [MATH] A matrix which equals its transpose. { sə'me·trik 'mā·triks }

symmetric relation [MATH] The property of a relation on a set that requires y to be related to x whenever x is related to y. { sə'me·trik ri'lā·shən }

symmetric ripple mark [GEOL] A ripple mark whose cross-section profile is symmetric. { sə'me·trik 'rip·əl ‚märk }

symmetric space [MATH] A differentiable manifold which has a differentiable multiplication operation that behaves similarly to the multiplication of a complex number and its conjugate. { sə'me·trik 'spās }

symmetric spherical triangles [MATH] Spherical triangles whose corresponding angles and corresponding sides are equal but appear in opposite order as viewed from the center of the sphere. { sə'me·trik ¦sfir·ə·kəl 'trī‚aŋ·gəlz }

symmetric tensor [MATH] A tensor that is left unchanged by the interchange of two contravariant (or covariant) indices. { sə'me·trik 'ten·sər }

symmetric top molecule [PHYS CHEM] A nonlinear molecule which has one and only one axis of threefold or higher symmetry. { sə'me·trik ¦täp 'mäl·ə‚kyül }

symmetric transformation [MATH] A transformation T defined on a Hilbert space such that the inner products (x,Ty) and (Tx,y) are equal for any vectors x and y in the domain of T. { sə¦me·trik ‚tranz·fər'mā·shən }

Symmetrodonta [PALEON] An order of the extinct mammalian infraclass Pantotheria distinguished by the central high cusp, flanked by two smaller cusps and several low minor cusps, on the upper and lower molars. { ‚sim·ə·trə'dänt·ə }

symmetry [BIOL] The disposition of organs and other constituent parts of the body of living organisms with respect to imaginary axes. [MATH] **1.** A geometric object G has this property relative to some configuration S of its points if S determines two pieces of G which can be reflected onto each other through S. **2.** A rigid motion of a geometric figure that maps the figure onto itself. [PHYS] See invariance. { 'sim·ə‚trē }

symmetry axis See axis of symmetry; rotation axis. { 'sim·ə‚trē ‚ak·səs }

symmetry axis of rotary inversion See rotoinversion axis. { 'sim·ə‚trē ‚ak·səs əv 'rōd·ə·rē in'vər·zhən }

symmetry axis of rotoinversion See rotoinversion axis. { 'sim·ə‚trē ‚ak·səs əv ¦rōd·ō·in'vər·zhən }

symmetry breaking [PHYS] The deviation from exact symmetry exhibited by many physical systems; it encompasses explicit symmetry breaking and spontaneous symmetry breaking. { 'sim·ə·trē ‚brāk·iŋ }

symmetry center See center of symmetry. { 'sim·ə‚trē ‚sen·tər }

symmetry class See crystal class. { 'sim·ə‚trē ‚klas }

symmetry element [CRYSTAL] **1.** Some combination of rotations and reflections and translations which brings a crystal into a position that cannot be distinguished from its original position. Also known as symmetry operation; symmetry transformation. **2.** The rotational axes, mirror planes, and center of symmetry characteristic of a given crystal. { 'sim·ə‚trē ‚el·ə·mənt }

symmetry function See symmetry transformation. { 'sim·ə‚trē ‚fəŋk·shən }

symmetry group [MATH] A group composed of all rigid motions or similarity transformations of some geometric object onto itself. { 'sim·ə‚trē ‚grüp }

symmetry law See invariance principle. { 'sim·ə‚trē ‚lò }

symmetry number [PHYS CHEM] The number of indistinguishable orientations that a molecule can exhibit by being rotated around symmetry axes. { 'sim·ə‚trē ‚nəm·bər }

symmetry operation See symmetry element. { 'sim·ə‚trē ‚äp·ə‚rā·shən }

symmetry operation of the second kind [CRYSTAL] A combination of rotations, reflections, and translations that brings a crystal into a position that is a mirror image of its original position. { ¦sim·ə·trē ‚äp·ə‚rā·shən əv t͟hə 'sek·ənd ‚kīnd }

symmetry plane See plane of mirror symmetry. { 'sim·ə‚trē ‚plān }

symmetry principle [MATH] The centroid of a geometrical figure (line, area, or volume) is at a point on a line or plane of symmetry of the figure. See invariance principle. { 'sim·ə‚trē ‚prin·sə·pəl }

symmetry transformation [CRYSTAL] See symmetry element. [MATH] A rigid motion sending a geometric object onto itself; examples are rotations and, for the case of a polygon, permutations of the vertices. Also known as symmetry function. { 'sim·ə‚trē ‚tranz·fər'mā·shən }

symmict [GEOL] Referring to a sedimentation unit that is structureless and in which coarse- and fine-grained particles are mixed more extensively in the lower part. { 'sim·ikt }

symmictite [PETR] An eruptive breccia that is homogenized and is made up of a mixture of country rock and intrusive rock. { sə'mik‚tīt }

symmicton See diamicton. { sə'mikt·ən }

symon fault See horseback. { 'sī·mən ‚fòlt }

Symon's cone crusher [MIN ENG] A modified gyratory crusher used in secondary ore crushing that consists of a downward-flaring bowl within which is gyrated a conical crushing head; the main shaft is gyrated by means of a long eccentric which is driven by bevel gears. { 'sī·mənz 'kōn ‚krəsh·ər }

Symon's disk crusher [MIN ENG] A mill in which the crushing is done between two cup-shaped plates that revolve on shafts set at a small angle to each other. { 'sī·mənz 'disk ˌkrəsh·ər }

sympathetic detonation [ENG] Explosion caused by the transmission of a detonation wave through any medium from another explosion. { ˌsim·pə'thed·ik ˌdet·ən'ā·shən }

sympathetic nervous system [NEUROSCI] The portion of the autonomic nervous system, innervating smooth muscle and glands of the body, which upon stimulation produces a functional state of preparation for flight or combat. { ˌsim·pə'thed·ik 'nər·vəs ˌsis·təm }

sympathetic ophthalmia [MED] A granulomatous inflammation of the uveal tract following ocular injury or intraocular surgery. { ˌsim·pə'thed·ik äf'thal·mē·ə }

sympathetic vibration [PHYS] The driving of a mechanical or acoustical system at its resonant frequency by energy from an adjacent system vibrating at the same frequency. { ˌsim·pə'thed·ik vī'brā·shən }

sympathicotropic cell [HISTOL] Any of various cells possessing special affinity for the sympathetic nervous system. { sim'path·ə·kō'träp·ik 'sel }

sympathochromaffin cell [NEUROSCI] One of the precursors of sympathetic and medullary cells in the adrenal medulla. { sim·pə·thō·krō'maf·ən ˌsel }

sympatholitic [PHARM] Of or pertaining to an effect antagonistic to that of the sympathetic nervous system. { ˌsim·pə·thō'lid·ik }

sympathomimetic [PHARM] Having the ability to produce physiologic changes similar to those caused by action of the sympathetic nervous system. { ˌsim·pə·thō·mə'med·ik }

sympatric [ECOL] Of a species, occupying the same range as another species but maintaining identity by not interbreeding. { sim'pa·trik }

sympatric speciation [EVOL] Speciation that occurs without geographic isolation of a population. { sim'pa·trik ˌspē·shē'ā·shən }

sympetalous See gamopetalous. { sim'ped·əl·əs }

symphile [ECOL] An organism, usually a beetle, living as a guest in the nest of a social insect, such as an ant, where it is reared and bred in exchange for its exudates. { 'sim·fīl }

Symphyla [INV ZOO] A class of the Myriapoda comprising tiny, pale, centipedelike creatures which inhabit humus or soil. { 'sim·fə·lə }

symphysis [ANAT] An immovable articulation of bones connected by fibrocartilaginous pads. { 'sim·fə·səs }

symphysis pubis See pubic symphysis. { 'sim·fə·səs 'pyü·bəs }

Symphyta [INV ZOO] A suborder of the Hymenoptera including the sawflies and horntails characterized by a broad base attaching the abdomen to the thorax. { 'sim·fəd·ə }

symplectic group of dimension n [MATH] The Lie group of symplectic transformations on an *n*-dimensional vector space over the quaternions. Symbolized Sp(*n*). { sim'plek·tik 'grüp əv di'men·chən 'en }

symplectic transformation [MATH] A linear transformation of a vector space over the quaternions that leaves the lengths of vectors unchanged. { sim'plek·tik ˌtranz·fər'mā·shən }

symplectite See symplektite. { sim'plek·tīt }

symplektite [MINERAL] An intimate intergrowth of two different minerals. Also spelled symplectite. { sim'plek·tīt }

symplesite [MINERAL] $Fe_2(AsO_4)_3·8H_2O$ A blue to bluish-green triclinic mineral composed of hydrous iron arsenate. { 'sim·plə·sīt }

sympodium [BOT] A branching system in trees in which the main axis is composed of successive secondary branches, each representing the dominant fork of a dichotomy. { sim'pōd·ē·əm }

symporter [CELL MOL] A channel protein that simultaneously transports two different types of substrates (for example, sodium ion plus glucose) across a cell membrane, both in the same direction. { sim'pórd·ər }

symptom [MED] A phenomenon of physical or mental disorder or disturbance which leads to complaints on the part of the patient. { 'sim·təm }

symptomatology [MED] **1.** The science of symptoms. **2.** In common usage, the symptoms of disease taken together as a whole. { ˌsim·tə·mə'täl·ə·jē }

syn [ORG CHEM] In stereochemistry. on the same side of a reference plane; for example, the stereochemical outcome of an addition reaction where the new bonds are on the same side of the original pi bond is called syn addition. { sin }

synadelphite [MINERAL] $(Mn,Mg,Ca,Pb)(AsO_4)(OH)_5$ A black mineral composed of basic arsenate of manganese, often with magnesium, calcium, lead, or other metals. { ˌsin·ə'del·fīt }

Synallactidae [INV ZOO] A family of echinoderms of the order Aspidochirotida comprising mainly deep-sea forms which lack tentacle ampullae. { ˌsin·ə'lak·tə,dē }

synandrous [BOT] Having several united stamens. { sə'nan·drəs }

synangium [BOT] A compound sorus made up of united sporangia. [VERT ZOO] In lower vertebrates, a peripheral arterial trunk from which branches arise. { sə'nan·jē·əm }

synantectic [MINERAL] Refers to a mineral that was formed by the reaction of two other minerals. { ˌsin·ən'tek·tik }

synantexis [GEOL] Deuteric alteration. { ˌsin·ən'tek·səs }

Synanthae [BOT] An equivalent name for Cyclanthales. { sə'nan,thē }

Synanthales [BOT] An equivalent name for Cyclanthales. { ˌsin·ən'thā·lēz }

synapomorphy [SYST] A derived trait shared by two or more taxa that is believed to reflect their shared ancestry. { si'nap·ə,mór·fē }

synapse [NEUROSCI] A site where the axon of one neuron comes into contact with and influences the dendrites of another neuron or a cell body. { 'si,naps }

synapsis [CYTOL] Pairing of homologous chromosomes during the zygotene stage of meiosis. { sə'nap·səs }

synaptic transmission [NEUROSCI] The mechanisms by which a presynaptic neuron influences the activity of an anatomically adjacent postsynaptic neuron. { sə'nap·tik ˌtranz'mish·ən }

synapticulum [INV ZOO] A conical or cylindrical supporting process, as those extending between septa in some corals, or connecting gill bars in *Branchiostoma*. { ˌsin·ap'tik·yə·ləm }

synaptic vesicle [NEUROSCI] A small membrane-bound structure in the axon terminals of nerve cells that contains neurotransmitters and releases them by exocytosis when an action potential reaches the terminal. { si'nap·tik 'ves·ə·kəl }

Synaptidae [INV ZOO] A family of large sea cucumbers of the order Apodida lacking a respiratory tree and having a reduced water-vascular system. { sə'nap·tə,dē }

synaptinemal complex [CYTOL] Ribbonlike structures that extend the length of synapsing chromosomes and are believed to function in exchange pairing. { sə,nap·tə'nē·məl 'käm pleks }

synarthrosis [ANAT] An articulation in which the connecting material (fibrous connective tissue) is continuous, immovably binding the bones. { ˌsin·är'thrō·səs }

Synbranchiformes [VERT ZOO] An order of eellike actinopterygian fishes that, unlike true eels, have the premaxillae present as distinct bones. { sin,bran·kə'fór,mēz }

Synbranchii [VERT ZOO] The equivalent name for Synbranchiformes. { sin'bran·kē,ī }

sync See synchronization. { siŋk }

Syncarida [INV ZOO] A superorder of crustaceans of the subclass Malacostraca lacking a carapace and oostegites and having exopodites on all thoracic limbs. { siŋ'kar·əd·ə }

syncarp [BOT] A compound fleshy fruit. { 'sin,kärp }

syncarpous [BOT] Descriptive of a gynoecium having the carpels united in a compound ovary. { sin'kär·pəs }

sync generator See synchronizing generator. { 'siŋk ˌjen·ə,rād·ər }

synchisite See synchysite. { 'siŋ·kə,sīt }

synchondrosis [ANAT] A type of synarthrosis in which the bone surfaces are connected by cartilage. { ˌsin·kän'drō·səs }

synchorology [ECOL] A study which involves the distribution ranges of plant communities, phytosociological zones, vegetation and geographical complexes, dissemination spectra, and current plant migration patterns. { ˌsin·kə'räl·ə·jē }

synchro [ELEC] Any of several devices which are used for transmitting and receiving angular position or angular motion over wires, such as a synchro transmitter or synchro receiver. Also known as mag-slip (British usage); self-synchronous device; self-synchronous repeater; selsyn. [SCI TECH] Occurring at the same time or made to occur at the same time. { 'siŋ·krō }

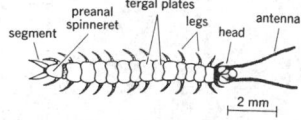

SYMPHYLA

A symphylan, *Scutigerella immaculata. (From R. E. Snodgrass, A Textbook of Arthropod Anatomy, copyright 1952 by Cornell University Press; used by permission)*

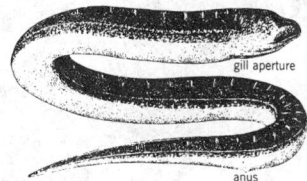

SYNBRANCHIFORMES

Rice eel *(Monopterus albus).*

synchro control transformer [ELEC] A transformer having its secondary winding on a rotor; when its three input leads are excited by angle-defining voltages, the two output leads deliver an alternating-current voltage that is proportional to the sine of the difference between the electrical input angle and the mechanical rotor angle. { 'siŋ·krō kən¦trōl tranz,fȯr·mər }

synchro control transmitter [ELEC] A high-accuracy synchro transmitter, having high-impedance windings. { 'siŋ·krō kən¦trōl tranz,mid·ər }

synchrocyclotron [NUCLEO] A circular particle accelerator for accelerating protons, deuterons, or alpha particles, in which the frequency of the accelerating voltage is modulated to maintain synchronism with the frequency of the particles which spiral outward and attain energies at which the relativistic mass increase becomes significant. Also known as frequency-modulated cyclotron; synchrophasotron. { ¦siŋ·krō'sī·klə,trän }

synchro differential motor [ELEC] Motor which is electrically similar to the synchro differential generator except that a damping device is added to prevent oscillations; both its rotor and stator are connected to synchro generators, and its function is to indicate the sum or difference between the two signals transmitted by the generators. { 'siŋ·krō ¦dif·ə'ren·chəl ,mōd·ər }

synchro differential receiver [ELEC] A synchro receiver that subtracts one electrical angle from another and delivers the difference as a mechanical angle. Also known as differential synchro. { 'siŋ·krō ,dif·ə'ren·chəl ri'sē·vər }

synchro differential transmitter [ELEC] A synchro transmitter that adds a mechanical angle to an electrical angle and delivers the sum as an electrical angle. Also known as differential synchro. { 'siŋ·krō ,dif·ə'ren·chəl tranz'mid·ər }

synchro generator See synchro transmitter. { 'siŋ·krō 'jen·ə,rād·ər }

synchromesh [MECH ENG] An automobile transmission device that minimizes clashing; acts as a friction clutch, bringing gears approximately to correct speed just before meshing. { 'siŋ·krō,mesh }

synchro motor See synchro receiver. { 'siŋ·krō ,mōd·ər }

synchrone [ASTRON] The geometrical locus of the dust grains ejected from the nucleus of a comet at the same time and having any value of beta. { 'siŋ,krōn }

synchronism [ELEC] Of a synchronous motor, the condition under which the motor runs at a speed which is directly related to the frequency of the power applied to the motor and is not dependent upon variables. [PHYS] Condition of two periodic quantities which have the same frequency, and whose phase difference is either constant or varies around a constant average value. { 'siŋ·krə,niz·əm }

synchronization [ENG] The maintenance of one operation in step with another, as in keeping the electron beam of a television picture tube in step with the electron beam of the television camera tube at the transmitter. Also known as sync. { ,siŋ·krə·nə'zā·shən }

synchronization indicator [ENG] An indicator that presents visually the relationship between two varying quantities or moving objects. { ,siŋ·krə·nə'zā·shən ,in·də,kād·ər }

synchronized blocking oscillator [ELECTR] A blocking oscillator which is synchronized with pulses occurring at a rate slightly faster than its own natural frequency. { 'siŋ·krə,nīzd 'bläk·iŋ 'äs·ə,lād·ər }

synchronized shifting [MECH ENG] Changing speed gears, with the gears being brought to the same speed before the change can be made. { 'siŋ·krə,nīzd 'shift·iŋ }

synchronizer [COMPUT SCI] A computer storage device used to compensate for a difference in rate of flow of information or time of occurrence of events when transmitting information from one device to another. [ELECTR] The component of a radar set which generates the timing voltage for the complete set. { 'siŋ·krə,nīz·ər }

synchronizing generator [ELECTR] An electronic generator that supplies synchronizing pulses to television studio and transmitter equipment. Also known as sync generator; sync-signal generator. { 'sin·krə,nīz·iŋ 'jen·ə,rād·ər }

synchronizing pulse [COMMUN] In pulse modulation, a pulse which is transmitted to synchronize the transmitter and the receiver; it is usually distinguished from signal-carrying pulses by some special characteristic. { 'sin·krə,nīz·iŋ ,pəls }

synchronizing reactor [ELEC] Current-limiting reactor for connecting momentarily across the open contacts of a circuit-interrupting device for synchronizing purposes. { 'sin·krə,nīz·iŋ rē'ak·tər }

synchronizing relay [ELEC] Relay which functions when two alternating-current sources are in agreement within predetermined limits of phase angle and frequency. { 'sin·krə,nīz·iŋ 'rē,lā }

synchronizing signal See sync signal. { 'sin·krə,nīz·iŋ ,sig·nəl }

synchronous [ENG] In step or in phase, as applied to two or more circuits, devices, or machines. [GEOL] Geological rock units or features formed at the same time. { 'siŋ·krə·nəs }

synchronous belt See timing belt. { 'siŋ·krə·nəs 'belt }

synchronous bombing [ORD] Bombing done with certain bombsights, such as the Norden bombsight, in which the travel of the telescope, focused upon the target, is synchronized with the ground speed of the airplane, and the course flown is determined by manual adjustment of the bombsight, the two together determining the dropping angle and correcting for drift so that the release occurs at the right instant. { 'siŋ·krə·nəs 'bäm·iŋ }

synchronous booster converter [ELEC] Synchronous converter having an alternating-current generator mounted on the same shaft and connected in series with it to adjust the voltage at the commutator of the converter. { 'siŋ·krə·nəs 'büs·tər kən'vərd·ər }

synchronous capacitor [ELEC] A synchronous motor running without mechanical load and drawing a large leading current, like a capacitor; used to improve the power factor and voltage regulation of an alternating-current power system. { 'siŋ·krə·nəs kə'pas·əd·ər }

synchronous clamp circuit See keyed clamp circuit. { 'siŋ·krə·nəs 'klamp ,sər·kət }

synchronous clock [HOROL] An electric clock driven by a synchronous motor, for operation from an alternating-current power system in which the frequency is accurately controlled. { 'siŋ·krə·nəs 'kläk }

synchronous communications [COMPUT SCI] The high-speed transmission and reception of long groups of characters at a time, requiring synchronization of the sending and receiving devices. { 'siŋ·krə·nəs kə,myü·nə'kā·shənz }

synchronous computer [COMPUT SCI] A digital computer designed to operate in sequential elementary steps, each step requiring a constant amount of time to complete, and being initiated by a timing pulse from a uniformly running clock. { 'siŋ·krə·nəs kəm'pyüd·ər }

synchronous converter [ELEC] A converter in which motor and generator windings are combined on one armature and excited by one magnetic field; normally used to change alternating to direct current. Also known as converter; electric converter. { 'siŋ·krə·nəs kən'vərd·ər }

synchronous data-link control [COMMUN] A bit-oriented protocol for managing the flow of information in a data-communications system, in full, half-duplex, or multipoint modes, that uses an error-check algorithm. { 'siŋ·krə·nəs 'dad·ə ,liŋk kən,trōl }

synchronous data transmission [COMMUN] Data transmission in which a clock defines transmission times for data; since start and stop bits for each character are not needed, more of the transmission bandwidth is available for message bits. { 'siŋ·krə·nəs 'dad·ə tranz,mish·ən }

synchronous demodulator See synchronous detector. { 'siŋ·krə·nəs dē'mäj·ə,lād·ər }

synchronous detector [ELECTR] **1.** A detector that inserts a missing carrier signal in exact synchronism with the original carrier at the transmitter; when the input to the detector consists of two suppressed-carrier signals in phase quadrature, as in the chrominance signal of a color television receiver, the phase of the reinserted carrier can be adjusted to recover either one of the signals. Also known as synchronous demodulator. **2.** See cross-correlator. { 'siŋ·krə·nəs di'tek·tər }

synchronous dynamic random access memory [COMPUT SCI] High-speed memory that is controlled by the system clock and can run at bus speeds up to 100 megahertz. Abbreviated SDRAM. { ¦siŋ·krə·nəs dī,nam·ik ,ran·dəm 'ak,ses ,mem·rē }

synchronous gate [ELECTR] A time gate in which the output intervals are synchronized with an incoming signal. { 'siŋ·krə·nəs 'gāt }

synchronous generator [ELEC] A machine that generates an alternating voltage when its armature or field is rotated by a motor, an engine, or other means. The output frequency is exactly proportional to the speed at which the generator is driven. { 'siŋ·krə·nəs 'jen·ə,rād·ər }

synchronous growth [MICROBIO] A population of bacteria in which all cells divide at approximately the same time. { 'siŋ·krə·nəs 'grōth }

synchronous inverter See dynamotor. { 'siŋ·krə·nəs in'vərd·ər }

synchronous machine [ELEC] An alternating-current machine whose average speed is proportional to the frequency of the applied or generated voltage. { 'siŋ·krə·nəs mə'shēn }

synchronous motor [ELEC] A synchronous machine that transforms alternating-current electric power into mechanical power, using field magnets excited with direct current. { 'siŋ·krə·nəs 'mōd·ər }

synchronous operation [ELECTR] **1.** An operation that takes place regularly or predictably with respect to the occurrence of a particular event in another process. **2.** In particular, an operation whose timing is controlled by pulses generated by an electronic clock. { 'siŋ·krə·nəs ,äp·ə'rā·shən }

synchronous orbit [AERO ENG] **1.** An orbit in which a satellite makes a limited number of equatorial crossing points which are then repeated in synchronism with some defined reference (usually earth or sun). **2.** Commonly, the equatorial, circular, 24-hour case in which the satellite appears to hover over a specific point of the earth. { 'siŋ·krə·nəs 'ȯr·bət }

synchronous phase modifier [ELEC] A synchronous motor that runs without mechanical load, and is provided with means for varying its power factor to simulate a capacitive or inductive reactor; used in voltage regulation of alternating-current power systems. { 'siŋ·krə·nəs 'fāz ,shif·tər }

synchronous pluton [GEOL] Any pluton whose time of emplacement coincides with a major orogeny. { 'siŋ·krə·nəs 'plü,tän }

synchronous radar bombing [ORD] A kind of radar bombing in which special airborne radar equipment containing rate and steering mechanisms is used to control the direction of flight of the bombing aircraft, solve the bombing problem, and automatically drop bombs at the proper release point. { 'siŋ·krə·nəs 'rā,där 'bäm·iŋ }

synchronous rectifier [ELECTR] A rectifier in which contacts are opened and closed at correct instants of time for rectification by a synchronous vibrator or by a commutator driven by a synchronous motor. { 'siŋ·krə·nəs 'rek·tə,fī·ər }

synchronous rotation [ASTRON] The rotation of a planet or satellite whose period is equal to its orbital period. { 'siŋ·krə·nəs rō'tā·shən }

synchronous satellite See geosynchronous satellite. { 'siŋ·krə·nəs 'sad·əl,īt }

synchronous speed [ELECTROMAG] The speed of rotation of a magnetic field in a synchronous machine; in revolutions per second, it is equal to twice the frequency of the alternating current in hertz, divided by the number of poles in the machine. { 'siŋ·krə·nəs 'spēd }

synchronous switch [ELECTR] A thyratron circuit used to control the operation of ignitrons in such applications as resistance welding. { 'siŋ·krə·nəs 'swich }

synchronous system [COMMUN] A telecommunication system in which transmitting and receiving apparatus operate continuously at substantially the same rate, and correction devices are used, if necessary, to maintain them in a fixed time relationship. { 'siŋ·krə·nəs 'sis·təm }

synchronous time-division multiplexing [COMMUN] A data transmission technique in which several users make use of a single channel by means of a system in which time slots are allotted on a fixed basis, usually in round-robin fashion. Abbreviated STDM. { 'siŋ·krə·nəs 'tīm də,vizh·ən 'məl·tə,pleks·iŋ }

synchronous timing [MET] Regulating the welding-transformer primary current in spot, seam, or projection welding so that the following conditions prevail: the first half-cycle is initiated at the proper time in relation to the voltage to ensure

a balanced current wave, each succeeding half-cycle is essentially the same as the first, and the last half-cycle is of the opposite polarity to the first. { 'siŋ·krə·nəs 'tīm·iŋ }

synchronous vibrator [ELECTROMAG] An electromagnetic vibrator that simultaneously converts a low direct voltage to a low alternating voltage and rectifies a high alternating voltage obtained from a power transformer to which the low alternating voltage is applied; in power packs, it eliminates the need for a rectifier tube. { 'siŋ·krə·nəs 'vī,brād·ər }

synchronous working [COMPUT SCI] The mode of operation of a synchronous computer, in which the starting of each operation is clock-controlled. { 'siŋ·krə·nəs 'wərk·iŋ }

synchrophasotron See synchrocyclotron. { ¦siŋ·krō'fāz·ə,trän }

synchro receiver [ELEC] A synchro that provides an angular position related to the applied angle-defining voltages; when two of its input leads are excited by an alternating-current voltage and the other three input leads are excited by the angle-defining voltages, the rotor rotates to the corresponding angular position; the torque of rotation is proportional to the sine of the difference between the mechanical and electrical angles. Also known as receiver synchro; selsyn motor; selsyn receiver; synchro motor. { 'siŋ·krō ri'sē·vər }

synchro resolver See resolver. { 'siŋ·krō ri'zäl·vər }

synchroscope [ELECTR] A cathode-ray oscilloscope designed to show a short-duration pulse by using a fast sweep that is synchronized with the pulse signal to be observed. [ENG] An instrument for indicating whether two periodic quantities are synchronous; the indicator may be a rotating-pointer device or a cathode-ray oscilloscope providing a rotating pattern; the position of the rotating pointer is a measure of the instantaneous phase difference between the quantities. { 'siŋ·krə,skōp }

synchro-shutter [ENG] A camera shutter with a circuit that flashes a light the instant the shutter opens. { 'siŋ·krō ,shəd·ər }

synchro system [ELEC] An electric system for transmitting angular position or motion; in the simplest form it consists of a synchro transmitter connected by wires to a synchro receiver; more complex systems include synchro control transformers and synchro differential transmitters and receivers. Also known as selsyn system. { 'siŋ·krō ,sis·təm }

synchro transmitter [ELEC] A synchro that provides voltages related to the angular position of its rotor; when its two input leads are excited by an alternating-current voltage, the magnitudes and polarities of the voltages at the three output leads define the rotor position. Also known as selsyn generator; selsyn transmitter; synchro generator; transmitter; transmitter synchro. { 'siŋ·krō tranz'mid·ər }

synchrotron [NUCLEO] A device for accelerating electrons or protons in closed orbits in which the frequency of the accelerating voltage is varied (or held constant in the case of electrons) and the strength of the magnetic field is varied so as to keep the orbit radius constant. { 'siŋ·krə,trän }

synchrotron process [ELECTROMAG] The emission of electromagnetic radiation by relativistic electrons orbiting in a magnetic field. { 'siŋ·krə,trän ,prä·səs }

synchrotron radiation [ELECTROMAG] Electromagnetic radiation generated by the acceleration of charged relativistic particles, usually electrons, in a magnetic field. { 'siŋ·krə,trän ,rād·ē,ā·shən }

synchysite [MINERAL] $(Ce,La)Ca(CO_3)_2F$ A mineral composed of fluoride and carbonate of calcium, cerium, and lanthanum. Also spelled synchisite. { 'siŋ·kə,sīt }

synclastic [MATH] Property of a surface or portion of a surface for which the centers of curvature of the principal sections at each point lie on the same side of the surface. { sin'klas·tik }

synclinal axis See trough surface. { sin'klīn·əl 'ak·səs }

synclinal valley [GEOL] Pertaining to a topographic valley whose sides coincide with a synclinal fold. { sin'klīn·əl 'val·ē }

syncline [GEOL] A fold having stratigraphically younger rock material in its core; it is concave upward. { 'sin,klīn }

synclinorium [GEOL] A composite synclinal structure in a region of lesser folds. { ,sin·klə'nȯr·ē·əm }

syncope [MED] Swooning or fainting; temporary suspension of consciousness. { 'siŋ·kə,pē }

syncrude See synthetic crude oil. { 'sin,krüd }

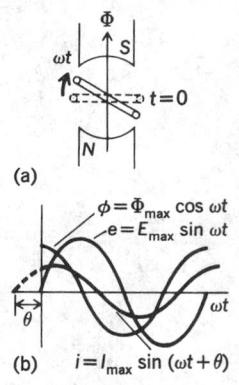

SYNCHRONOUS MACHINE

$\Phi = \Phi_{max} \cos \omega t$
$e = E_{max} \sin \omega t$
$i = I_{max} \sin (\omega t + \theta)$

Single-phase, two-pole synchronous machine. *(a)* Schematic diagram. Φ is direction of flux. Coil is perpendicular to pole axis at time $t = 0$. ω = angular velocity of coil; ωt = angular displacement at time t. Angular displacement is proportional to ωt. *(b)* Graphs of flux linking coil ϕ, voltage induced in coil e, and current in coil i as functions of time. θ is phase angle.

sync separator [ELECTR] A circuit that separates synchronizing pulses from the video signal in a television receiver. { 'siŋk ,sep·ə,rād·ər }

sync signal [COMMUN] A signal transmitted after each line and field to synchronize the scanning process in a television or facsimile receiver with that of the receiver. Also known as synchronizing signal. { 'siŋk ,sig·nəl }

sync-signal generator *See* synchronizing generator. { 'siŋk ¦sig·nəl 'jen·ə,rād·ər }

syncytial trophoblast *See* syncytiotrophoblast. { sin'sish·əl 'träf·ə,blast }

syncytiotrophoblast [CYTOL] An irregular sheet or net of deeply staining cytoplasm in which nuclei are irregularly scattered. Also known as plasmoditrophoblast; syncytial trophoblast. { sin¦sish·ē·ō'träf·ə,blast }

syncytium [CYTOL] A mass of multinucleated cytoplasm without division into separate cells. Also known as polykaryocyte. [INV ZOO] Multinucleated cell or gland. { sin'sish·ē·əm }

syndactyly [ANAT] The condition characterized by union of two or more digits, as in certain birds and mammals; it is a familial anomaly in humans. { sin'dakt·əl·ē }

syndesmosis [ANAT] An articulation in which the bones are joined by collagen fibers. { ,sin,dez'mō·səs }

syndet *See* synthetic detergent. { 'sin,det }

syndiotactic polymer [ORG CHEM] A vinyl polymer in which the side chains alternate regularly above and below the plane of the backbone. { ¦sin·dē·ə¦tak·tik 'päl·i·mər }

syndrome [MED] A group of signs and symptoms which together characterize a disease. Also known as complex. { 'sin,drōm }

syndromic hearing loss [MED] Hearing loss that occurs in the presence of one or more other symptoms. { sin,drōm·ik 'hir·iŋ ,lós }

syndynamics [ECOL] The study of the causes of and trends in successional changes within a plant community. { ¦sin·dī'nam·iks }

syndyne [ASTRON] The geometrical locus of the dust grains ejected from the nucleus of a comet continuously and having a particular value of beta. { 'sin,dīn }

synecology [ECOL] The study of environmental relations of groups of organisms, such as communities. { ¦sin·i'käl·ə·jē }

Synentognathi [VERT ZOO] The equivalent name for Beloniformes. { ,sin·ən'täg·nə·thē }

syneresis [CHEM] Spontaneous separation of a liquid from a gel or colloidal suspension due to contraction of the gel. { sə'ner·ə·səs }

synergetics [SCI TECH] The study of complex systems that involve the cooperative, nonlinear behavior of numerous subsystems and exhibit ordered structures far from thermodynamic equilibrium. { ,sin·ər'jed·iks }

synergic curve [AERO ENG] A curve plotted for the ascent of a rocket, space-air vehicle, or space vehicle, calculated to give optimum fuel economy and optimum velocity. { sə'nər·jik 'kərv }

synergid [BOT] Either of two small cells lying in the embryo sac in seed plants adjacent to the egg cell toward the micropylar end. { sə'nər·jəd }

synergism [ECOL] An ecological association in which the physiological processes or behavior of an individual are enhanced by the nearby presence of another organism. [MATER] An action where the total effect of two active components in a mixture is greater than the sum of their individual effects, for example, a mixture volume that is greater than the sum of the individual volumes, or in resin formulation, the use of two or more stabilizers, where the combination improves polymer stability more than expected from the additive effect of the stabilizers. { 'sin·ər,jiz·əm }

synergist [ANAT] A muscle that assists a prime mover muscle in performing a specific action. [MATER] A material that enhances the effect of another material so that when they are combined the total effect is greater than the sum of their individual effects. { 'sin·ər,jist }

synergy [PHARM] Suppression of a strain of infectious microbes by concentrations of two or more drugs which are not active singly. { 'sin·ər·jē }

synesthesia [PSYCH] The condition in which a sensory experience normally associated with one sensory system occurs when another sensory system is stimulated. { ,sin·əs'thēzh·ə }

synfuel *See* synthetic fuel. { 'sin,fyül }

Syngamidae [INV ZOO] A family of roundworms belonging to the Strongyloidea and including parasites of birds and mammals. { siŋ'gam·ə,dē }

syngamy [BIOL] Sexual reproduction involving union of gametes. { 'siŋ·gə·mē }

syngas *See* synthesis gas. { 'sin,gas }

syngeneic *See* isogeneic. { ,sin·jə'nē·ik }

syngenesious [BOT] Pertaining to an aggregate of stamens fused at the anthers. { ¦sin·jə¦nē·zhəs }

syngenesis [GEOL] In place formation of unconsolidated sediments. { sin'jen·ə·səs }

syngenetic [GEOL] **1.** Pertaining to a primary sedimentary structure formed contemporaneously with sediment deposition. **2.** Pertaining to a mineral deposit formed contemporaneously with the enclosing rock. Also known as ideogenous. { ¦sin·jə¦ned·ik }

syngenite [MINERAL] $K_2Ca(SO_4)_2 \cdot H_2O$ A colorless or white mineral composed of hydrous potassium calcium sulfate occurring in tabular crystals. { 'sin·jə,nīt }

Syngnathidae [VERT ZOO] A family of fishes in the order Gasterosteiformes including the seahorses and pipefishes. { siŋ'nath·ə,dē }

synkinematic *See* syntectonic. { ¦sin,kin·ə'mad·ik }

synkinesia [PHYSIO] Involuntary movement coincident with purposeful movements carried out by a distant part of the body, such as swinging the arms while walking. Also known as accessory movement; associated automatic movement. { ¦sin,kī¦nē·zhə }

synodic [ASTRON] Referring to conjunction of celestial bodies. { sə'näd·ik }

synodic month [ASTRON] A month based on the moon's phases. { sə'näd·ik 'mənth }

synodic period [ASTRON] The time period between two successive astronomical conjunctions of the same celestial objects. { sə'näd·ik 'pir·ē·əd }

synonym [SYST] A taxonomic name that is rejected as being incorrectly applied, or incorrect in form, or not representative of a natural genetic grouping. { 'sin·ə,nim }

synonymous substitution [GEN] *See* isonymous substitution. { si'nän·ə·məs ,səb·stə'tü·shən }

synopsis [SYST] In taxonomy, a brief summary of current knowledge about a taxon. { sə'näp·səs }

synoptic [METEOROL] Refers to the use of meteorological data obtained simultaneously over a wide area for the purpose of presenting a comprehensive and nearly instantaneous picture of the state of the atmosphere. { sə'näp·tik }

synoptic chart [METEOROL] Any chart or map on which data and analyses are presented that describe the state of the atmosphere over a large area at a given moment in time. { sə'näp·tik 'chärt }

synoptic climatology [CLIMATOL] The study and analysis of climate in terms of synoptic weather information, principally in the form of synoptic charts; the information thus obtained gives the climate (that is, average weather) of a given locality in a given synoptic situation rather than the usual climatic parameters which represent averages over all synoptic conditions. { sə'näp·tik ,klī·mə'täl·ə·jē }

synoptic code [METEOROL] In general, any code by which synoptic weather observations are communicated; among the synoptic codes in use are the international synoptic code, ship synoptic code, U.S. Airways code, and RECCO code. { sə'näp·tik 'kōd }

synoptic correlation *See* Eulerian correlation. { sə'näp·tik ,kär·ə'lā·shən }

synoptic meteorology [METEOROL] The study and analysis of synoptic weather information. { sə'näp·tik ,mēd·ē·ə'räl·ə·jē }

synoptic model [METEOROL] Any model specifying a space distribution of some meteorological elements; the distribution of clouds, precipitation, wind, temperature, and pressure in the vicinity of a front is an example of a synoptic model. { sə'näp·tik 'mäd·əl }

synoptic oceanography [OCEANOGR] The study of the physical spatial parameters of the ocean through analysis of simultaneous observations from many stations. { sə'näp·tik ,ō·shə'näg·rə·fē }

synoptic report [METEOROL] An encoded and transmitted synoptic weather observation. { sə'näp·tik ri'pórt }

synoptic scale *See* cyclonic scale. { sə'näp·tik 'skāl }

synoptic wave chart [OCEANOGR] A chart of an ocean area on which is plotted synoptic wave reports from vessels, along with computed wave heights for areas where reports are lacking; atmospheric fronts, highs, and lows are also shown; isolines of wave height and the boundaries of areas having the same dominant wave direction are drawn. { sə'näp·tik 'wāv ,chärt }

synoptic weather observation [METEOROL] A surface weather observation, made at periodic times (usually at 3- and 6-hourly intervals specified by the World Meteorological Organization), of sky cover, state of the sky, cloud height, atmospheric pressure reduced to sea level, temperature, dew point, wind speed and direction, amount of precipitation, hydrometeors and lithometeors, and special phenomena that prevail at the time of the observation or have been observed since the previous specified observation. { sə'näp·tik 'weth·ər ,äb·zər,vā·shən }

synorchidism [MED] Partial or complete fusion of the two testes within the abdomen or scrotum. { sə'nór·kə,diz·əm }

synorogenic [GEOL] Referring to a geologic process occurring at the same time as orogenic activity. { ¦sin,ór·ə'jen·ik }

synostosis [ANAT] A type of synarthrosis in which the bones are continuous. [MED] Union of originally separate bones into a single bone structure. { si,nä'stō·səs }

synovia *See* synovial fluid. { sə'nō·vē·ə }

synovial fluid [PHYSIO] A transparent viscid fluid secreted by synovial membranes. Also known as synovia. { sə'nō·vē·əl 'flü·əd }

synovial membrane [HISTOL] A layer of connective tissue which lines sheaths of tendons at freely moving articulations, ligamentous surfaces of articular capsules, and bursae. { sə'nō·vē·əl 'mem,brān }

synovitis [MED] Inflammation of a synovial membrane. { ,sin·ə'vīd·əs }

synpelmous [VERT ZOO] Having the two main flexor tendons of the toes united beyond the branches that go to each digit, as in certain birds. { sin'pel·məs }

synphylogeny [ECOL] The study of the trends and changes in plant communities through historical and evolutionary perspectives. { ¦sin·fə'läj·ə·nē }

synphysiology [ECOL] The study of the metabolic processes of plant communities or species which constantly compete with each other, by investigating water needs, transpiration, assimilation and production or organic matter, physiological effects of light, temperature, root exudates, and various other ecological factors. { ¦sin,fiz·ē'äl·ə·jē }

synsepalous *See* gamosepalous. { sin'sep·ə·ləs }

syntactic analysis [COMPUT SCI] The problem of associating a given string of symbols through a grammar to a programming language, so that the question of whether the string belongs to the language may be answered. { sin'tak·tik ə'nal·ə·səs }

syntactic error *See* syntax error. { sin'tak·tik 'er·ər }

syntactic extension [COMPUT SCI] An extension mechanism which creates new notations for existing or user-defined mechanisms in an extensible language. { sin'tak·tik ik'sten·shən }

syntactic foam [MATER] A cellular polymer made by dispersing rigid, microscopic particles in a fluid polymer and then curing it. { sin'tak·tik 'fōm }

syntactic model *See* linguistic model. { sin'tak·tik 'mäd·əl }

syntactics [COMMUN] The branch of semiotics that treats relations between symbols themselves, and defines valid relationships between the elements of a language. { sin'tak·tiks }

syntactic semigroup [SYS ENG] For a sequential machine, the set of all transformations performed by all input sequences. { sin'tak·tik 'sem·i,grüp }

syntax [COMPUT SCI] The set of rules needed to construct valid expressions or sentences in a language. { 'sin,taks }

syntax checker *See* syntax scanner. { 'sin,taks ,chek·ər }

syntax diagram [COMPUT SCI] A pictorial diagram showing the rules for forming an instruction in a computer programming language, and how the components of the statement are related. { 'sin,taks 'dī·ə,gram }

syntax-directed compiler [COMPUT SCI] A general-purpose compiler that can service a family of languages by providing the syntactic rules for language analysis in the form of data, typically in tabular form, rather than using a specific parsing algorithm for a particular language. Also known as syntax-oriented compiler. { 'sin,taks di¦rek·təd kəm'pīl·ər }

syntax error [COMPUT SCI] An error in the format of a statement in a computer program that violates the rules of the programming language employed. Also known as syntactic error. { 'sin,taks ,er·ər }

syntaxial overgrowth [MINERAL] A crystallographically oriented overgrowth of two alternating, chemically identical substances. { sin'tak·sē·əl 'ō·vər,grōth }

syntaxis [MAP] On a map, a sheaflike pattern of mountains converging on a common center { sin'tak·səs }

syntax-oriented compiler *See* syntax-directed compiler. { 'sin,taks ¦ór·ē,ent·əd kəm'pīl·ər }

syntax scanner [COMPUT SCI] A subprogram of a compiler or interpreter that checks the source program for syntax errors, and reports any such errors by printing the erroneous statement together with a diagnostic message. Also known as syntax checker. { 'sin,taks ,skan·ər }

syntectic [GEOL] *See* syntexis. [MET] Isothermal, reversible conversion of a solid phase into two conjugate liquid phases on applying heat. { sin'tek·tik }

syntectonic [GEOL] Refers to a geologic process or event occurring during tectonic activity. Also known as synkinematic. { ¦sin·tek'tän·ik }

Synteliidae [INV ZOO] The sap-flow beetles, a small family of coleopteran insects in the superfamily Histeroidea. { ,sint·əl'ī·ə,dē }

syntenic group [GEN] The loci on one chromosome in the complement, whether or not they show linkage in family studies (pedigree analysis). { sin'ten·ik ,grüp }

Syntexidae [INV ZOO] A family of the Hymenoptera in the superfamily Siricoidea. { sin'tek·sə,dē }

syntexis [GEOL] Magma made by the melting of two or more rock types and the assimilation of country rock. Also known as syntectic. { sin'tek·səs }

synthem [GEOL] A chronostratigraphic unit that defines an unconformity-bounded regional body of sediments and represents a cycle of sedimentation in response to changes in relative sea level or tectonics. { 'sin,them }

synthesis [CHEM] Any process or reaction for building up a complex compound by the union of simpler compounds or elements. [CONT SYS] *See* system design. { 'sin·thə·səs }

synthesis gas [CHEM ENG] A mixture of gases prepared as feedstock for a chemical reaction, for example, carbon monoxide and hydrogen to make hydrocarbons or organic chemicals, or hydrogen and nitrogen to make ammonia. Also known as syngas. { 'sin·thə·səs ,gas }

synthesizer [ELECTR] An electronic instrument which combines simple elements to generate more complex entities; examples are frequency synthesizer and sound synthesizer. { 'sin·thə,sīz·ər }

synthetase *See* ligase. { 'sin·thə,tās }

synthetic address *See* generated address. { sin'thed·ik 'ad,res }

synthetic aperture [ENG] A method of increasing the ability of an imaging system, such as radar or acoustical holography, to resolve small details of an object, in which a receiver of large size (or aperture) is in effect synthesized by the motion of a smaller receiver and the proper correlation of the detected signals. { sin'thed·ik 'ap·ə·chər }

synthetic-aperture radar [ENG] A radar system in which an aircraft moving along a very straight path emits microwave pulses continuously at a frequency constant enough to be coherent for a period during which the aircraft may have traveled about 1 kilometer; all echoes returned during this period can then be processed as if a single antenna as long as the flight path had been used. { sin'thed·ik ¦ap·ə·chər 'rā,där }

synthetic chromosome *See* artificial chromosome. { sin'thed·ik 'krō·mə,sōm }

synthetic crude [MATER] The total liquid, multicomponent hydrocarbon mixture resulting from a process involving molecular rearrangement of charge stock, as from oil shale or synthesis gas. Also known as synthetic oil. { sin'thed·ik 'krüd }

synthetic crude oil [MATER] A complex mixture of hydrocarbons that are somewhat similar to petroleum and are obtained from coal, synthesis gas, or oil shale and tar sands. Also known as syncrude. { sin'thed·ik ¦krüd 'oil }

synthetic data [IND ENG] Any production data applicable to a given situation that are not obtained by direct measurement. { sin'thed·ik 'dad·ə }

synthetic detergent [MATER] A liquid or solid material able to dissolve oily materials and disperse them (or emulsify them) in water. Also known as syndet. { sin'thed·ik di'tər·jənt }

synthetic division [MATH] A long division process for dividing a polynomial $p(x)$ by a polynomial $(x - a)$ where only the coefficients of these polynomials are used. { sin'thed·ik də'vizh·ən }

synthetic fiber *See* artificial fiber. { sin'thed·ik 'fī·bər }

synthetic fuel [MATER] A fuel that is artificially formulated and manufactured; frequently derived from fossil fuels that are less convenient or environmentally undesirable for direct use. Also known as synfuel. { sin'thed·ik 'fyül }

synthetic gem [MATER] A precious or semiprecious stone made by artificial processes, for example, synthetic diamonds made by extreme heat and pressure on carbon, used industrially; and synthetic rubies made by high-temperature crystallization of aluminum oxide, used in laser equipment. { sin'thed·ik 'jem }

synthetic graphite [MATER] Graphitic crystalline material made by the high-temperature and pressure processing of carbon. { sin'thed·ik 'gra,frīt }

synthetic language [COMPUT SCI] A pseudocode or symbolic language; fabricated language. { sin'thed·ik 'laŋ·gwij }

synthetic lubricant [MATER] Any of a group of products, some of them based on petroleum, used as lubricants where heat, chemical resistance, and other requirements can be better met than with straight petroleum products. { sin'thed·ik 'lü·brə·kənt }

synthetic mica [MATER] A fluorphlogopite mica made artificially by heating a large batch of raw material in an electric resistance furnace and allowing the mica to crystallize from the melt during controlled slow cooling. { sin'thed·ik 'mī·kə }

synthetic oil *See* synthetic crude. { sin'thed·ik 'ȯil }

synthetic quartz [MATER] A quartz crystal grown commercially at high temperature and pressure around a seed of quartz suspended in a solution which contains scraps of natural quartz crystals. { sin'thed·ik 'kwȯrts }

synthetic resin [ORG CHEM] Amorphous, organic, semisolid, or solid material derived from the polymerization of unsaturated monomers such as ethylene, butylene, propylene, and styrene. { sin'thed·ik 'rez·ən }

synthetic rubber [MATER] Synthetic products whose properties are similar to those of natural rubber, including elasticity and ability to be vulcanized; usually produced by the polymerization or copolymerization of petroleum-derived olefinic or other unsaturated compounds. { sin'thed·ik 'rəb·ər }

synthol process [CHEM ENG] A reaction of carbon monoxide and hydrogen with an iron and sodium carbonate catalyst; produces a mixture of higher alcohols, aldehydes, ketones, higher fatty acids, and aliphatic hydrocarbons, usable as a synthetic gasoline. { 'sin,thȯl ,prä·səs }

syntony [ELEC] Condition in which two oscillating circuits have the same resonant frequency. { 'sin·tə·nē }

Syntrophiidina [PALEON] A suborder of extinct articulate brachiopods of the order Pentamerida characterized by a strong dorsal median fold. { sin,träf·ē·ə'dī·nə }

syntrophism [BIOL] Mutual dependence of cells for nutritional needs, especially between strains of bacteria. { 'sin·trə,fiz·əm }

syntrophoblast [EMBRYO] The outer synctial layer of the trophoblast that forms the outermost fetal element of the placenta. { sin'träf·ə,blast }

syntype [SYST] Any specimen of a series when no specimen is designated as the holotype. Also known as cotype. { 'sin,tīp }

synusia [ECOL] A structural unit of a community characterized by uniformity of life-form or of height. { sə'nü·zhə }

Synxiphosura [PALEON] An extinct heterogeneous order of arthropods in the subclass Merostomata possibly representing an explosive proliferation of aberrant, terminal, and apparently blind forms. { ,sin,zif·ə'sùr·ə }

syphilis [MED] An infectious disease caused by the spirochete *Treponema pallidum,* transmitted principally by sexual intercourse. { 'sif·ə·ləs }

syphilitic meningoencephalitis *See* general paresis. { ¦sif·ə¦lid·ik mə¦niŋ·gō·in,sef·ə'līd·əs }

(a)   (b)

SYNTROPHIIDINA

Pedicle valve of *Imbricata* in the Syntrophiidina; *(a)* exterior view, *(b)* interior view.

syphilophobia [PSYCH] An abnormal fear of syphilis. { ,sif·ə·lə'fō·bē·ə }

Syringamminidae [INV ZOO] A family of Psamminida, with a fragile test constructed of tubes of xenophyae tightly cemented together. { ,sir·iŋ·gə'min·ə,dē }

syringe [MED] **1.** An apparatus commonly made of glass or plastic, fitted snugly onto a hollow needle, used to aspirate or inject fluids for diagnostic or therapeutic purposes. Also known as hypodermic syringe. **2.** A large glass barrel with a fitted rubber bulb at one end and a nozzle at the other, used primarily for irrigation purposes. { sə'rinj }

syringobulbia [MED] The presence of cavities in the medulla oblongata similar to those found in syringomyelia. { sə,riŋ·gō'bəl,bē·ə }

syringocystadenoma *See* syringoma. { sə,riŋ·gō¦sist,ad·ən'ō·mə }

syringocystoma *See* syringoma. { sə,riŋ·gō·si'stō·mə }

syringoma [MED] A multiple nevoid tumor of sweat glands. Also known as syringocystadenoma; syringocystoma. { ,sir·əŋ'gō·mə }

syringomyelia [MED] A chronic disease characterized by the presence of cavities surrounded by gliosis near the canal of the spinal cord and often extending to the medulla. { sə,riŋ·gō,mī'ē·lē·ə }

Syringophyllidae [PALEON] A family of extinct corals in the order Tabulata. { sə,riŋ·gō'fil·ə,dē }

syrinx [PALEON] A tube surrounding the pedicle in certain fossil brachiopods. [VERT ZOO] The vocal organ in birds. { 'sir·iŋks }

Syrphidae [INV ZOO] The flower flies, a family of cyclorrhaphous dipteran insects in the series Aschiza. { 'sər·fə,dē }

syserskite [MINERAL] Mineral composed of an alloy of osmium (50–80%) and iridium (20–50%). { 'sis·ər,skīt }

sysgen *See* system generation. { 'sis,jen }

SYSIN [COMPUT SCI] The principal input stream of an operating system. Derived from system input. { 'sis,in }

syssiderite *See* stony-iron meteorite. { sə'sid·ə,rīt }

Systellomatophora [INV ZOO] An order of the subclass Pulmonata in which the eyes are contractile but stalks are not retractile, the body is sluglike, oval, or lengthened, and the lung is posterior. { ¦sis·tə·lō·mə'täf·ə·rə }

system [ELECTR] A combination of two or more sets generally physically separated when in operation, and such other assemblies, subassemblies, and parts necessary to perform an operational function or functions. [ENG] A combination of several pieces of equipment integrated to perform a specific function; thus a fire control system may include a tracking radar, computer, and gun. [GEOL] **1.** A major time-stratigraphic unit of worldwide significance, representing the basic unit of Phanerozoic rocks. **2.** A group of related structures, such as joints. **3.** A chronostratigraphic unit, below erathem and above series. [PHYS] A region in space or a portion of matter that has a certain amount of one or more substances, ordered in one or more phases. [SCI TECH] A method of organizing entities or terms; in particular, organizing such entities into a larger aggregate. { 'sis·təm }

system analysis [CONT SYS] The use of mathematics to determine how a set of interconnected components whose individual characteristics are known will behave in response to a given input or set of inputs. { 'sis·təm ə,nal·ə·səs }

systematic analog network testing approach [ELECTR] An on-line minicomputer-based system with an integrated database and optimal human intervention, which provides computer printouts used in automatic testing of electronic systems; aimed at maximizing cost effectivity. Abbreviated SANTA. { ,sis·tə'mad·ik 'an·ə,läg 'net,wərk ,test·iŋ ə,prōch }

systematic desensitization [PSYCH] A behavior therapy technique that is used to modify phobic behaviors by constructing a hierarchy of anxiety-producing stimuli and gradually presenting them to the individual until they no longer produce anxiety. { ,sis·tə'mad·ik dē,sen·sə·tə'zā·shən }

systematic distortion [ELEC] Periodic or constant distortion, such as bias or characteristic distortion; the direct opposite of fortuitous distortion. { ,sis·tə'mad·ik di'stȯr·shən }

systematic error [ENG] An error due to some known physical law by which it might be predicted; these errors produced by the same cause affect the mean in the same sense, and do not tend to balance each other but rather give a definite bias to the mean. [STAT] An error which results from some bias in the measurement process and is not due to chance, in contrast to random error. { ,sis·tə'mad·ik 'er·ər }

systematic error-checking code [COMPUT SCI] A type of self-checking code in which a valid character consists of the minimum number of digits needed to identify the character and distinguish it from any other valid character, and a set of check digits which maintain a minimum specified signal distance between any two valid characters. Also known as group code. { ,sis·tə'mad·ik 'er·ər ¦chek·iŋ ,kōd }

systematic joints [GEOL] Joints occurring in patterns or sets and oriented perpendicular to the boundaries of the constituent rock unit. { ,sis·tə'mad·ik 'jóins }

systematic nomenclature [CHEM] A system for naming chemical compounds according to a specific set of rules, usually those developed by the International Union of Pure and Applied Chemistry. { ¦sis·tə,mad·ik 'nō·mən,klā·chər }

systematics [BIOL] The science of animal and plant classification. { ,sis·tə'mad·iks }

systematic sampling [MIN ENG] Extracting samples at evenly spaced periods or in fixed quantities from a unit of coal. { ,sis·tə'mad·ik 'sam·pliŋ }

systematic support [MIN ENG] The regular setting of timber or steel supports at fixed intervals irrespective of the condition of the roof and sides. { ,sis·tə'mad·ik sə'pórt }

system bandwidth [CONT SYS] The difference between the frequencies at which the gain of a system is $\sqrt{2}/2$ (that is, 0.707) times its peak value. { 'sis·təm 'band,width }

system calendar [COMPUT SCI] A register in a computer system that holds the date and year and provides them in response to supervisor calls to the operating system. { 'sis·təm 'kal·ən·dər }

system call *See* supervisor call. { 'sis·təm ,kól }

system catalog [COMPUT SCI] An index of all files controlled by the operating system of a large computer. { 'sis·təm 'kad·əl,äg }

system chart [COMPUT SCI] A flowchart that emphasizes the component operations which make up a system. { 'sis·təm ,chärt }

system check [COMPUT SCI] A check on the overall performance of the system, usually not made by built-in computer check circuits; for example, control total, hash totals, and record counts. { 'sis·təm ,chek }

system command [COMPUT SCI] A special instruction to a computer system to carry out a particular processing function, such as allowing a user to gain access to the system, running a program, activating a translator, or issuing a status report. { 'sis·təm kə,mand }

system design [COMPUT SCI] Determination in detail of the exact operational requirements of a system, resolution of these into file structures and input/output formats, and relation of each to management tasks and information requirements. [CONT SYS] A technique of constructing a system that performs in a specified manner, making use of available components. Also known as synthesis. { 'sis·təm di,zīn }

system designer [COMPUT SCI] A person who prepares final system documentation, analyzes findings, and synthesizes new system design. { 'sis·təm di,zīn·ər }

system documentation [COMPUT SCI] Detailed information, in either written or computerized form, about a computer system, including its architecture, design, data flow, and programming logic. { 'sis·təm ,däk·yə·mən'tā·shən }

system effectiveness [ENG] A measure of the extent to which a system may be expected to achieve a set of specific mission requirements expressed as a function of availability, dependability, and capability. { 'sis·təm i'fek·tiv·nəs }

Système International d'Unités *See* International System of Units. { si'stem ,in·tər,näs·ē·ə'näl dyùn·i'tāz }

system engineering *See* systems engineering. { 'sis·təm ,en·jə'nir·iŋ }

system evaluation [COMPUT SCI] A periodic evaluation of the system to assess its status in terms of original or current expectations and to chart its future direction. { 'sis·təm i,val·yə'wā·shən }

system flowchart *See* data flow diagram. { 'sis·təm 'flō ,chärt }

system generation [COMPUT SCI] A process that creates a particular and uniquely specified operating system; it combines user-specified options and parameters with manufacturer-supplied general-purpose or nonspecialized program subsections

to produce an operating system (or other complex software) of the desired form and capacity. Abbreviated sysgen. { 'sis·təm ,jen·ə'rā·shən }

systemic circulation [PHYSIO] The general circulation, as distinct from the pulmonary circulation. { si'stem·ik ,sər·kyə'lā·shən }

systemic inflammatory response syndrome [MED] The spectrum of elicited pathophysiologic changes (including blood clotting and changes in metabolism, heart rate, and respiration) resulting from excess production of inflammatory mediators (for example, histamines and leukotrienes), which orchestrate the process of inflammation through various processes. { si¦stem·ik in,flam·ə,tór·ē ri'späns ,sin,drōm }

systemic lupus erythematosus [MED] An inflammatory, multisystem, usually chronic disorder in which tissue injury is mediated by immune complexes. { sis¦tem·ik ,lü·pəs ,er·ə,the·mə'tō·səs }

system improvement time [COMPUT SCI] The machine downtime needed for the installation and testing of new components, large or small, and machine downtime necessary for modification of existing components; this includes all programming tests following the above actions to prove the machine is operating properly. { 'sis·təm im'prüv·mənt ,tīm }

system input *See* SYSIN. { 'sis·təm 'in,pùt }

system integration [COMPUT SCI] The procedures involved in combing separately developed modules of components so that they work together as a complete computer system. { 'sis·təm ,in·tə'grā·shən }

system-level timer [COMPUT SCI] A hardware device that is set by the operating system to interrupt it after a specified time interval, either to set deadlines for events or to remind the operating system to take some action. { 'sis·təm ¦lev·əl 'tīm·ər }

system library [COMPUT SCI] An organized collection of computer programs that is maintained on-line with a computer system by being held on a secondary storage device and is managed by the operating system. { 'sis·təm 'lī,brer·ē }

system life cycle [ENG] The continuum of phases through which a system passes from conception through disposition. { 'sis·təm 'līf ,sī·kəl }

system loader [COMPUT SCI] A computer program that loads all the other programs, including the operating system, into a computer's main storage. { 'sis·təm ,lōd·ər }

system master tapes [COMPUT SCI] Magnetic tapes containing programmed instructions necessary for preparing a computer prior to running programs. { 'sis·təm 'mas·tər 'tāps }

system of distinct representatives [MATH] A family of subsets S_i of a given finite set S such that the family has as many members as there are elements in S, and such that it is possible to assign each element x_i of S to a distinct subset S_i with x_i in S_i. { ¦sis·təm əv di¦stinkt ,rep·rə'zen·tə·tivz }

system of stages [MATH] A collection of nonempty sets that includes the intersection of any two sets that belong to the collection. { ¦sis·təm əv 'stāj·əz }

system operation [COMPUT SCI] The administration and operation of an automatic data-processing equipment-oriented system, including staffing, scheduling, equipment and service contract administration, equipment utilization practices, and time-sharing. { 'sis·təm ,äp·ə'rā·shən }

system optimization *See* optimization. { 'sis·təm ,äp·tə·mə'zā·shən }

system reliability [ENG] The probability that a system will accurately perform its specified task under stated environmental conditions. { 'sis·təm ri,lī·ə'bil·əd·ē }

system response *See* response. { 'sis·təm ri'späns }

system safety [ENG] The optimum degree of safety within the constraints of operational effectiveness, time, and cost, attained through specific application of system safety engineering throughout all phases of a system. { 'sis·təm 'sāf·tē }

system safety engineering [ENG] An element of systems management involving the application of scientific and engineering principles for the timely identification of hazards, and initiation of those actions necessary to prevent or control hazards within the system. { 'sis·təm 'sāf·tē ,en·jə,nir·iŋ }

systems analysis [ENG] The analysis of an activity, procedure, method, technique, or business to determine what must be accomplished and how the necessary operations may best be accomplished. { 'sis·təmz ə,nal·ə·səs }

systems architecting [SYS ENG] The discipline that combines elements which, working together, create unique structural and behavioral capabilities in a system that none could produce alone. Also known as systems architecture. { ¦sis·təmz 'är·kə,tek·tiŋ }

systems architecture *See* systems architecting. { 'sis·təmz ,är·kə,tek·chər }

systems definition [COMPUT SCI] A document describing a computer-based system for processing data or solving a problem, including a general description of the aims and benefits of the system and clerical procedures employed, and detailed program specification. Also known as systems specification. { 'sis·təmz ,def·ə,nish·ən }

systems ecology [ECOL] The combined approaches of systems analysis and the ecology of whole ecosystems and subsystems. { 'sis·təmz i'käl·ə·jē }

systems engineering [ENG] The design of a complex interrelation of many elements (a system) to maximize an agreed-upon measure of system performance, taking into consideration all of the elements related in any way to the system, including utilization of worker power as well as the characteristics of each of the system's components. Also known as system engineering. { 'sis·təmz ,en·jə,nir·iŋ }

systems for nuclear auxiliary power *See* SNAP. { 'sis·təmz fȯr 'nü·klē·ər ȯg'zil·ə·rē 'pau̇·ər }

systems implementation test [ENG] The test program that exercises the complete system in its actual environment to determine its capabilities and limitations; this test also demonstrates that the system is functionally operative, and is compatible with the other subsystems and supporting elements required for its operational employment. { 'sis·təmz ,im·plə·mən'tā·shən ,test }

systems integration [SYS ENG] A discipline that combines processes and procedures from systems engineering, systems management, and product development for the purpose of developing large-scale complex systems that involve hardware and software and may be based on existing or legacy systems coupled with totally new requirements to add significant functionality. { ¦sis·təmz ,in·tə'grā·shən }

systems-management reengineering *See* organizational reengineering. { ¦sis·təmz ,man·ij·mənt ,rē,en·jə'nir·iŋ }

system software [COMPUT SCI] Computer software involved with data and program management, including operating systems, control programs, and database management systems. { 'sis·təm 'sȯft,wer }

systems programming [COMPUT SCI] The development and production of programs that have to do with translation, loading, supervision, maintenance, control, and running of computers and computer programs. { 'sis·təmz ,prō,gram·iŋ }

systems specification *See* systems definition. { 'sis·təmz ,spes·ə·fə,kā·shən }

systems test [COMPUT SCI] The running of the whole system against test data, a complete simulation of the actual running system for purposes of testing out the adequacy of the

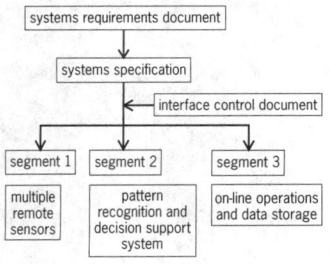

SYSTEMS INTEGRATION

Architecture of a typical system.

system. [ENG] A test of an entire interconnected set of components for the purpose of determining proper functions and interconnections. { 'sis·təmz ,test }

systems tract [GEOL] A discrete package of distinctive sediment types (facies) that are laid down during different phases of a cycle of sea-level change. { 'sis·təmz ,trakt }

system study [COMPUT SCI] A detailed study to determine whether, to what extent, and how automatic data-processing equipment should be used; it usually includes an analysis of the existing system and the design of the new system, including the development of system specifications which provide a basis for the selection of equipment. { 'sis·təm ,stəd·ē }

system supervisor [COMPUT SCI] A control program which ensures an efficient transition in running program after program and accomplishing setups and control functions. { 'sis·təm 'sü·pər,vīz·ər }

system unit [COMPUT SCI] **1.** An individual card, section of tape, or the like, which is manipulated during operation of the system; class 1 systems have one unit per document; class 2 systems have one unit per vocabulary term or concept. **2.** *See* case. { 'sis·təm ,yü·nət }

systole [PHYSIO] The contraction phase of the heart cycle. { 'sis·tə·lē }

systolic array [COMPUT SCI] An array of processing elements of cells connected to a memory which pulses data through the array in such a way that each data item can be used effectively at each cell it passes while being pumped from cell to cell along the array. { si'stäl·ik ə'rā }

syzygy [ASTRON] **1.** One of the two points in a celestial object's orbit where it is in conjunction with or opposition to the sun. **2.** Those points in the moon's orbit where the moon, earth, and sun are in a straight line. **3.** The alignment of any three objects within the solar system, or within any other system of objects in orbit about a star. [INV ZOO] End-to-end union of the sporonts of certain gregarine protozoans. { 'siz·ə·jē }

szaibelyite [MINERAL] $(Mn,Mg)(BO_2)(OH)$ A white to buff or straw yellow, orthorhombic mineral consisting of a basic borate of manganese and magnesium; occurs as veinlets, masses, or embedded nodules. { sä'bel,yīt }

szaskaite *See* smithsonite. { sə'skā,īt }

Szechtman cell [CHEM ENG] An electrolytic process for manufacture of chlorine that is a variation of both the mercury cell and molten salt cell. { 'sekt·mən ,sel }

Szilard-Chalmers effect [NUCLEO] The breaking of a chemical bond between a radioactive atom formed in a nuclear reaction and the molecule to which the atom belonged; used in the separation of isotopes by chemical means. { 'zi,lürd 'chä·mərz i,fekt }

szmikite [MINERAL] $MnSO_4·H_2O$ A monoclinic mineral composed of hydrous manganese sulfate. { 'smi,kīt }

szomolnokite [MINERAL] $FeSO_4·H_2O$ A yellow or brown monoclinic mineral composed of hydrous ferrous sulfate. { sə'mäl·nə,kīt }

T

t *See* troy system.

T *See* tera-; tesla.

TΩ *See* teraohm.

2,4,5-T *See* 2,4,5-trichlorophenoxyacetic acid.

2,4,6-T *See* trichlorophenol.

T_c *See* critical temperature.

Ta *See* tantalum.

Tabanidae [INV ZOO] The deer and horse flies, a family of orthorrhaphous dipteran insects in the series Brachycera. { tə'ban·ə‚dē }

tabbyite [MINERAL] A variety of solid asphalt found in the western United States; used as rubber filler and with roofing materials. { 'ta·bē‚īt }

tab-card cutter [DES ENG] A device for die-cutting card stock to uniform tabulating-card size. { 'tab ¦kärd ‚kəd·ər }

Taber ice *See* segregated ice. { 'tā·bər ‚īs }

tabes dorsalis [MED] A form of parenchymatous neurosyphilis in which there is demyelination and sclerosis of the posterior columns of the spinal cord. Also known as locomotor ataxia. { 'tā·bēz dȯr'sal·əs }

tabetisol *See* talik. { tə'bed·ə‚sȯl }

table [BUILD] A horizontal projection or molding on the exterior or interior face of a wall. [COMPUT SCI] A set of contiguous, related items, each uniquely identified either by its relative position in the set or by some label. [LAP] The flat face forming the top of a brilliant-cut stone. [MATH] An array or listing of computed quantities. [MECH ENG] That part of a grinding machine which directly or indirectly supports the work being ground. [MIN ENG] **1.** In placer mining, a wide, shallow sluice box designed to recover gold or other valuable material from screened gravel. **2.** A platform or plate on which coal is screened and picked. { 'tā·bəl }

tabled joint [CIV ENG] In cut stonework, a bed joint formed by a broad, shallow channel in the surface of one stone that fits a corresponding projection of the stone above or below. { 'tā·bəld ‚jȯint }

table-driven compiler [COMPUT SCI] A compiler in which the source language is described by a set of syntax rules. { 'tā·bəl ¦driv·ən kəm'pī·lər }

table-driven program [COMPUT SCI] A computer program that relies on tables stored outside of the program in the computer's memory to furnish data. { 'tā·bəl ¦driv·ən 'prō‚gram }

tabled whelk [INV ZOO] *Neptunea tabulata.* A marine gastropod mollusk about 5 inches (13 centimeters) in length and 2 inches (5 centimeters) in diameter, found at depths of 150–200 feet (45–60 meters), off the west coast of Canada and the United States. { 'tā·bəld 'welk }

table flotation [MIN ENG] A flotation process in which a slurry of ore is fed to a shaking table where floatable particles become glomerules, held together by minute air bubbles and edge adhesion; the glomerules roll across the table and are discharged nearly opposite the feed end; the process is helped by jets of low-pressure air. { 'tā·bəl flō‚tā·shən }

table iceberg *See* tabular iceberg. { 'tā·bəl 'īs‚bərg }

table knoll [GEOGR] A knoll with a comparatively smooth, flat top. { 'tā·bəl 'nōl }

tableland [GEOGR] A broad, elevated, nearly level, and extensive region of land that has been deeply cut at intervals by valleys or broken by escarpments. Also known as continental plateau. { 'tā·bəl‚and }

table look-up [COMPUT SCI] A procedure for calculating the location of an item in a table by means of an algorithm, rather than by conducting a search for the item. { 'tā·bəl 'lùk ‚əp }

table look-up device [ELECTR] A logic circuit in which the input signals are grouped as address digits to a memory device, and, in response to any particular combination of inputs, the memory device location that is addressed becomes the output. { 'tā·bəl ¦lùk ‚əp di‚vīs }

table management program [COMPUT SCI] A computer program that handles the creation and maintenance of tables and access to data stored in them. { 'tā·bəl ¦man·ij·mənt prō‚gram }

tablemount *See* guyot. { 'tā·bəl‚maùnt }

table mountain [GEOGR] A flat-topped mountain. { 'tā·bəl ‚maùnt·ən }

table reef [GEOL] A small, isolated organic reef which has a flat top and does not enclose a lagoon. { 'tā·bəl ‚rēf }

table salt *See* sodium chloride. { 'tā·bəl ‚sȯlt }

tablespoonful [MECH] A unit of volume used particularly in cookery, equal to 4 fluid drams or $1/2$ fluid ounce; in the United States this is equal to approximately 14.7868 cubic centimeters, in the United Kingdom to approximately 14.2065 cubic centimeters. Abbreviated tbsp. { 'tā·bəl‚spün‚fùl }

table sugar *See* sucrose. { 'tā·bəl ‚shùg·ər }

tableting [ENG] A punch-and-die procedure for the compaction of powdered or granular solids; used for pharmaceuticals, food products, fireworks, vitamins, and dyes. { 'tab·ləd·iŋ }

tabling [BUILD] Formation of a horizontal masonry joint by arranging building stones in a course so that they extend into the next course and thus prevent slippage. [MIN ENG] Separation of two materials of different densities by passing a dilute suspension over a slightly inclined table having a reciprocal horizontal motion or shake with a slow forward motion and a fast return. { 'tāb·liŋ }

tab stop [DES ENG] A column position to which the printing mechanism of a typewriter or computer printer advances upon receipt of a command. { 'tab ‚stäp }

tabula [PALEON] A transverse septum that closes off the lower part of the polyp cavity in certain extinct corals and hydroics. { 'tab·yə·lə }

tabular [GEOL] Referring to a sedimentary particle whose length is two to three times its thickness. { 'tab·yə·lər }

tabular berg *See* tabular iceberg. { 'tab·yə·lər 'bərg }

tabular crystal [CRYSTAL] A crystal that appears broad and flat due to two prominent parallel faces. { 'tab·yə·lər 'krist·əl }

tabular iceberg [OCEANOGR] An iceberg with clifflike sides and a flat top; usually arises by detachment from an ice shelf. Also known as table iceberg; tabular berg. { 'tab·yə·lər 'īs‚bərg }

tabular interpolation [MATH] Method of finding from a table the values of the dependent variable for intermediate values of the independent variable. { 'tab·yə·lər in‚tər·pə'lā·shən }

tabular language [COMPUT SCI] A part of a program which represents the composition of a decision table required by the problem considered. { 'tab·yə·lər 'laŋ·gwij }

tabular spar *See* wollastonite. { 'tab·yə·lər 'spär }

TABLELAND

View of an ideal tableland: Canyon de Chelly National Monument, northeastern Arizona. (*Spence Air Photos*)

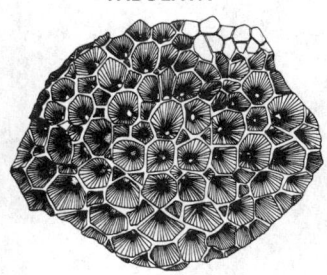

TABULATA

Specimen of *Michelinia convexa.*

Tabulata [PALEON] An extinct Paleozoic order of corals of the subclass Zoantharia characterized by an exclusively colonial mode of growth and by secretion of a calcareous exoskeleton of slender tubes. { ˌtab·yəˈläd·ə }

tabulate [COMPUT SCI] To order a set of data into a table form, or to print a set of data as a table, usually indicating differences and totals, or just totals. { ˈtab·yəˌlāt }

tabulated altitude [NAV] In navigational sight reduction tables, the altitude taken directly from a table for the entering arguments. { ˈtab·yəˌlād·əd ˈal·təˌtüd }

tabulated azimuth [NAV] Azimuth taken directly from a table, before interpolation. { ˈtab·yəˌlād·əd ˈaz·ə·məth }

tabulated azimuth angle [NAV] Azimuth angle taken directly from a table, before interpolation. { ˈtab·yəˌlād·əd ˈaz·ə·məth ˌaŋ·gəl }

tabulation character [COMPUT SCI] A character that controls the action of a computer printer and is not itself printed, although it forms part of the data to be printed. { ˌtab·yəˈlā·shən ˌkar·ik·tər }

tabulator [COMPUT SCI] A machine that reads information from punched cards and produces lists, tables, and totals on separate forms or continuous paper. { ˈtab·yəˌlād·ər }

tabun [ORG CHEM] $(CH_3)_2NP(O)(C_2H_5O)(CN)$ A toxic liquid with a boiling point of 240°C; soluble in organic solvents; used as a nerve gas. { ˈtäˌbùn }

TAB vaccine [IMMUNOL] A vaccine containing killed typhoid bacilli and the paratyphoid organisms (*Salmonella paratyphi* A and B) most frequently involved in paratyphoid fever. { ˈtab vakˈsēn }

Tacan *See* tactical air navigation system. { ˈtakˌan }

tache noire [MED] The primary painless lesion of the tick-borne typhus fevers of Africa, manifested by a raised red area with a typical black necrotic center which appears at the site of the tick bite. { täsh ˈnwär }

Tachinidae [INV ZOO] The tachina flies, a family of bristly, grayish or black Diptera whose larvae are parasitic in caterpillars and other insects. { təˈkin·əˌdē }

tachiol *See* silver fluoride. { ˈtak·ēˌól }

tachocline [ASTRON] A zone of radial shear in the interior of the sun that is located at the base of the convection zone at a distance from the sun's center equal to about 0.7 of the sun's radius, and represents an abrupt transition from the rotation rates in the convection zone to those in the underlying layers. { ˈtak·əˌklīn }

tachometer [ENG] An instrument that measures the revolutions per minute or the angular speed of a rotating shaft. { təˈkäm·əd·ər }

tachycardia [MED] Excessive rapidity of the heart's action. { ˌtak·əˈkärd·ē·ə }

tachy-electromagnetic pulse [ELECTROMAG] A pulse of electromagnetic radiation generated within a microsecond of a high-altitude nuclear explosion by high-energy electrons, produced by gamma rays from the explosion, spiralling in the earth's magnetic field. { ˈtak·ē iˌlek·trō·magˌned·ik ˈpəls }

Tachyglossidae [VERT ZOO] A family of monotreme mammals having relatively large brains with convoluted cerebral hemispheres; comprises the echidnas or spiny anteaters. { ˌtak·əˈgläs·əˌdē }

tachyhydrite [MINERAL] $CaMg_2Cl_6·12H_2O$ A honey yellow, hexagonal mineral consisting of a hydrated chloride of calcium and magnesium; occurs in massive form. { ˌtak·əˈhīˌdrīt }

tachylite [GEOL] A black, green, or brown volcanic glass formed from basaltic magma. Also known as basalt glass; basalt obsidian; hyalobasalt; jaspoid; sordawalite; wichtisite. { ˈtak·əˌlīt }

Tachyniscidae [INV ZOO] A family of myodarian cyclorrhaphous dipteran insects in the subsection Acalypteratae. { ˌtak·əˈnis·əˌdē }

tachyon [PART PHYS] A hypothetical particle that travels faster than light, consistent with the theory of relativity. { ˈtak·ēˌän }

tachyphylaxis [IMMUNOL] Rapid desensitization against doses of organ extracts or serum by the previous inoculation of small subtoxic doses of the same preparation. { ˌtak·ə·fəˈlak·səs }

tachypnea [MED] An abnormally rapid rate of respiration. { təˈkip·nē·ə }

tachysterol [BIOCHEM] The precursor of calciferol in the irradiation of ergosterol; an isomer of ergosterol. { təˈkisˌtəˌról }

tachytely [EVOL] Evolution at a rapid rate resulting in differential selection and fixation of new types. { ˈtak·əˌtel·ē }

tack [DES ENG] A small, sharp-pointed nail with a broad flat head. [MATER] Adhesive stickiness, such as occurs on the surface of a varnish or ink that has not completely dried. Also known as tackiness. [NAV] To change the course of a sailing vessel by coming about so as to take the wind from over the opposite bow (starboard or port). { tak }

tack coat [CIV ENG] A thin layer of bitumen, road tar, or emulsion laid on a road to enhance adhesion of the course above it. { ˈtak ˌkōt }

tackifier [MATER] A type of resinous material added to an elastomer to produce adhesives that will adhere on contact. { ˈtak·əˌfī·ər }

tackiness *See* tack. { ˈtak·ē·nəs }

tackiness agent [MATER] An additive used to impart adhesive properties to otherwise nonadhesive materials, such as oils and greases. { ˈtak·ē·nəs ˌā·jənt }

tacking [MET] Making small, isolated tack welds. { ˈtak·iŋ }

tackle [MECH ENG] Any arrangement of ropes and pulleys to gain a mechanical advantage. [NAV ARCH] An assemblage of lines and blocks in which the line passes through more than one block. { ˈtak·əl *or* ˈtāk·əl (naval usage) }

tack range [ENG] The length of time during which an adhesive will remain in the tacky-dry condition after application to an adherent. { ˈtak ˌrānj }

tack weld [MET] A joint between two pieces of metal made by welding at isolated points. { ˈtak ˌweld }

tacky *See* tacky dry. { ˈtak·ē }

tacky dry [MATER] Also known as tacky. **1.** In drying of adhesive, that stage at which the volatile constituents have evaporated or have been absorbed sufficiently to give the adhesive the desired degree of tack. **2.** In drying of paint, that stage at which the paint surface feels sticky when lightly touched. { ˈtak·ē ˈdrī }

tacnode *See* double cusp. { ˈtakˌnōd }

Taconian orogeny [GEOL] A process of formation of mountains in the latter part of the Ordovician period, particularly in the northern Appalachians. Also known as Taconic orogeny. { təˈkō·nē·ən óˈräj·ə·nē }

Taconic orogeny *See* Taconian orogeny. { təˈkän·ik óˈräj·ə·nē }

taconite [GEOL] The siliceous iron formation from which high-grade iron ores of the Lake Superior district have been derived; consists chiefly of fine-grained silica mixed with magnetite and hematite. { ˈtak·əˌnīt }

tactical [ORD] Of or pertaining to the arranging, positioning, or maneuvering of forces in contact or near contact with the enemy, so as to achieve an objective or objectives in a campaign or battle. { ˈtak·tə·kəl }

tactical air control center [COMMUN] The principal air-operations installation (land- or ship-based) from which all aircraft and air warning functions of tactical air operations are controlled. { ˈtak·tə·kəl ˈer kənˌtrōl ˌsen·tər }

tactical air controller [COMMUN] The officer who is in charge of all operations of the tactical air-control center, and who is responsible to the tactical air officer for the control of all aircraft and air warning facilities within the area of responsibility. { ˈtak·tə·kəl ˈer kənˌtrō·lər }

tactical air-control system [COMMUN] The organization and equipment necessary to plan, direct, and control tactical air operations and to coordinate air operations with other services; it is composed of communications-electronics facilities which provide the means for centralized control and decentralized execution of missions. { ˈtak·tə·kəl ˈer kənˌtrōl ˌsis·təm }

tactical aircraft shelter [CIV ENG] A shelter to house fighter-type aircraft and to provide protection to the aircraft from attack by conventional weapons, or damage from high winds or other elemental hazards. { ˈtak·tə·kəl ˈerˌkraft ˌshel·tər }

tactical air-direction center [AERO ENG] An air operations installation under the overall control of the tactical air-control center, from which aircraft and air warning service functions of tactical air operations in an area of responsibility are directed. { ˈtak·tə·kəl ˈer diˌrek·shən ˌsen·tər }

tactical air force [AERO ENG] An air force charged with carrying out tactical air operations in coordination with ground or naval forces. { 'tak·tə·kəl 'er ˌfòrs }

tactical airlift [AERO ENG] That airlift which provides the immediate and responsive air movement and delivery of combat troops and supplies directly into objective areas through air landing, extraction, airdrop, or other delivery techniques; and the air logistic support of all theater forces, including those engaged in combat operations, to meet specific theater objectives and requirements. { 'tak·tə·kəl 'er ˌlift }

tactical air navigation system [NAV] Short-range ultra-high-frequency air navigation system that provides accurate slant-range distance and bearing information; this information is presented to the pilot in two dimensions, that is, distance and bearing from a selected ground station. Also known as Tacan. { 'tak·tə·kəl 'er ˌnav·ə'gā·shən ˌsis·təm }

tactical air observer [AERO ENG] An officer trained as an air observer whose function is to observe from airborne aircraft and report on movement and disposition of friendly and enemy forces, on terrain and weather and hydrography, and to execute other missions as directed. { 'tak·tə·kəl 'er əb,zər·vər }

tactical air operations [AERO ENG] An air operation involving the employment of air power in coordination with ground or naval forces to gain and maintain air superiority, to prevent movement of enemy forces into and within the objective area and to seek out and destroy these forces and their supporting installations, and to join with ground or naval forces in operations within the objective area in order to assist directly in attainment of their immediate objective. { 'tak·tə·kəl 'er ˌäp·ə,rā·shənz }

tactical air reconnaissance [AERO ENG] The use of air vehicles to obtain information concerning terrain, weather, and the disposition, composition, movement, installations, lines of communications, and electronic and communication emissions of enemy forces. { 'tak·tə·kəl 'er ri,kän·ə·səns }

tactical air support [AERO ENG] Air operations carried out in coordination with surface forces which directly assist the land or naval battle. { 'tak·tə·kəl 'er sə,pòrt }

tactical air transport [AERO ENG] The use of air transport in direct support of airborne assaults, carriage of air-transported forces, tactical air supply, evacuation of casualties from forward airdromes, and clandestine operations. { 'tak·tə·kəl 'er 'tranz,pòrt }

tactical call sign [COMMUN] Call sign which identifies a tactical communications facility. { 'tak·tə·kəl 'kòl ˌsīn }

tactical command ship [NAV ARCH] A warship designed to serve as a command ship for a fleet or force commander; it is equipped with extensive communication equipment. { 'tak·tə·kəl kə'mand ˌship }

tactical communications system [COMMUN] A system which provides internal communications within tactical air elements, composed of transportable and mobile equipment assigned as unit equipment to the supporting tactical unit. { 'tak·tə·kəl kə,myü·nə'kā·shənz ˌsis·təm }

tactical control radar [ENG] Antiaircraft artillery radar which has essentially the same inherent capabilities as the target acquisition radar (physically it may be the same type of set) but whose function is chiefly that of providing tactical information for the control of elements of the antiaircraft artillery defenses in battle. { 'tak·tə·kəl kən'trol ˌrā,där }

tactical diameter [NAV] In marine operations, the distance gained to the right or left of the original course when a turn of 180° with a constant rudder angle has been completed. { 'tak·tə·kəl dī'am·əd·ər }

tactical electronic warfare [ELECTR] The application of electronic warfare to tactical air operations; tactical electronic warfare encompasses the three major subdivisions of electronic warfare: electronic warfare support measures, electronic countermeasures, and electronic counter-countermeasures. { 'tak·tə·kəl ˌi,lek'trän·ik 'wòr,fer }

tactical fire control [ORD] The manner in which fire power is employed with regard to selection of targets, to opening, suspending, or ceasing fire, and to classes of fire. { 'tak·tə·kəl 'fīr kən,trōl }

tactical frequency [COMMUN] Radio frequency assigned to a military unit to be used in the accomplishment of a tactical mission. { 'tak·tə·kəl 'frē·kwən·sē }

tactical map [ORD] A large-scale map used for tactical and administrative purposes. { 'tak·tə·kəl 'map }

tactical missile [ORD] A missile for use in tactical operations. { 'tak·tə·kəl 'mis·əl }

tactical mobility [ORD] The capability of a unit, command, task force, or the like that enables it to be readily moved while engaged in combat. { 'tak·tə·kəl mō'bil·əd·ē }

tactical nuclear weapon [ORD] A nuclear weapon which is programmed primarily for employment against tactical targets in tactical military operations. { 'tak·tə·kəl 'nü·klē·ər 'wep·ən }

tactical nuclear weapon employment [ORD] The use of nuclear weapons by land, sea, or air forces against opposing forces and supporting installations or facilities, in support of operations which contribute to the accomplishment of a military mission of limited scope, or in support of the military commander's scheme of maneuver, usually limited to the area of military operations. { 'tak·tə·kəl 'nü·klē·ər 'wep·ən im,plòi·mənt }

tactical range recorder [ENG] A sonar device in surface ships used to plot the time-range coordinates of submarines and determine firing of depth charges. { 'tak·tə·kəl 'rānj ri,kòrd·ər }

tactical reserve [ORD] A part of battalion, regiment, or similar force, held initially under the control of the commander as a maneuvering force to influence future action. { 'tak·tə·kəl ri'zərv }

tactical target [ORD] Any physical object, person, group of persons, or position singled out for attack during the course of battle or tactical operations in order to reduce or destroy the enemy's ability to sustain combat operations. { 'tak·tə·kəl 'tär·gət }

tactical transport aircraft [AERO ENG] Aircraft designed primarily for carrying personnel or cargo over short or medium distances. { 'tak·tə·kəl 'tranz,pòrt 'er,kraft }

tactical unit [ORD] An organization of troops, aircraft, or ships which is intended to serve as a single unit in combat, and may include service units required for its direct support. { 'tak·tə·kəl 'yü·nət }

tactical vehicle [ORD] Any vehicle designed for field requirements in combat and tactical operations, or for training personnel for such operations. { 'tak·tə·kəl 'vē·ə·kəl }

tactic polymer [ORG CHEM] A polymer with regularity or symmetry in the structural arrangement of its molecules, as in a stereospecific polymer such as some types of polypropylene. { 'tak·tə·kəl 'päl·i·mər }

tactile [PHYSIO] Pertaining to the sense of touch. { 'tak·təl }

tactile agnosia See astereognosis. { 'tak·təl ag'nō·zhə }

tactile feedback [COMPUT SCI] In haptics, devices that provide a user with the sensations of heat, pressure, and texture. { ˌtak·təl 'fēd,bak }

tactile hairs See vibrissae. { 'tak·təl 'herz }

tactile receptor See tactoreceptor. { 'tak·təl ri'sep·tər }

tactile sensor [CONT SYS] A transducer, usually associated with a robot end effector, that is sensitive to touch; comprises stress and touch sensors. { 'tak·təl 'sen·sər }

tactite [PETR] A rock with a complex mineralogical composition formed by contact metamorphism and metasomatism of carbonate rocks. { 'tak,tīt }

tactoid [BIOCHEM] A particle that appears as a spindle-shaped body under the polarizing microscope and occurs in mosaic virus, fibrin, and myosin. { 'tak,tòid }

tactoreceptor [PHYSIO] A sense organ that responds to touch. Also known as tactile receptor. { ˌtak·tō·ri'sep·tər }

tadpole [VERT ZOO] The larva of a frog or toad; at hatching it has a rounded body with a long fin-bordered tail, and the gills are external but shortly become enclosed. { 'tad,pōl }

tadpole shrimp [INV ZOO] Any of the phyllopod crustaceans that are members of the genus Lepidurus. { 'tad,pōl ˌshrimp }

taele See frozen ground. { 'tā·lə }

taenia [ANAT] A ribbon-shaped band of nerve fibers or muscle. { 'tē·nē·ə }

Taeniocidea [INV ZOO] An equivalent name for Cyclophyllidea. { ˌtē·nē·ə'did·ē·ə }

Taeniodonta [PALEON] An order of extinct quadrupedal land mammals, known from early Cenozoic deposits in North America. { ˌtē·nē·ə'dänt·ə }

Taenioidea [INV ZOO] An equivalent name for Cyclophyllidea. { ˌtē·nē·'òid·ē·ə }

Taeniolabidoidea [PALEON] An advanced suborder of the extinct mammalian order Multituberculata having incisors that

TADPOLE

Tadpole of the common frog (*Rana temporaria*).

TADPOLE SHRIMP

Lepidurus (Notostraca). *(From T. I. Storer and R. L. Usinger, General Zoology, 4th ed., McGraw-Hill, 1965)*

were self-sharpening in a limited way. { ˌtē·nē·ō‚lab·ə'dȯid·ē·ə }

taeniolite [MINERAL] $KLiMg_2Si_4O_{10}F_2$ A white or colorless mica mineral. { 'tē·nē·ə‚līt }

taenite [MINERAL] A meteoritic mineral consisting of a nickel-iron alloy, with a nickel content varying from about 27 to 65%. { 'tē‚nīt }

taenoglossate radula [INV ZOO] A long, narrow radula with seven teeth in each transverse row, found in certain pectinibranch bivalves. { ˌtē·nə'glä‚sāt 'raj·ə·lə }

Tafel slope [ELEC] The slope of a curve of overpotential or electrolytic polarization in volts versus the logarithm of current density. { 'tä·fəl ‚slōp }

taffeta [TEXT] A plain-woven, usually silk fabric that has a smooth finish and sheen on both sides. { 'taf·əd·ə }

taffrail log [ENG] A log consisting essentially of a rotator towed through the water by a braided log line attached to a distance-registering device usually secured at the taffrail, the railing at the stern. Also known as patent log. { 'taf‚rāl ‚läg }

tag [COMPUT SCI] **1.** A unit of information used as a label or marker. **2.** The symbol written in the location field of an assembly-language coding form, and used to define the symbolic address of the data or instruction written on that line. [NUCLEO] See isotopic tracer. { 'tag }

Tag closed-cup tester [ANALY CHEM] A laboratory device used to determine the flash point of mobile petroleum liquids flashing below 175°F (79.4°C). Also known as Tagliabue closed tester. { 'tag ‚klōzd ‚kəp 'tes·tər }

tag converting unit [COMPUT SCI] A device capable of reading the perforations of a price tag as input data. { 'tag kən'vərd·iŋ ‚yü·nət }

tag field [COMPUT SCI] A data item within a variant record that identifies the format to be used in the record. { 'tag ‚fēld }

tag format [COMPUT SCI] The arrangement of data in a short record inserted in a direct-access storage to indicate the location of an overflow record. { 'tag ‚fȯr‚mat }

tagged molecule [CHEM] A molecule having one or more atoms which are either radioactive or have a mass which differs from that of the atoms which normally make up the molecule. { 'tagd 'mäl·ə‚kyül }

tagilite See pseudomalachite. { 'tag·ə‚līt }

tag image file format [COMPUT SCI] File format used for storing bitmap images at any resolution. Abbreviated TIFF. { ‚tag ‚im·ij 'fīl ‚fȯr‚mat }

Tagliabue closed tester See Tag closed-cup tester. { ‚täl·yə'bü·ē 'klōsd 'tes·tər }

tagma [INV ZOO] A compound body section of a metameric animal that results from embryonic fusion of two or more somites. { 'tag·mə }

tagmosis [INV ZOO] The formation of groups of metameres into body regions with functional differences. { tag'mō·səs }

Tag-Robinson colorimeter [ENG] A laboratory device used to determine the color shades of lubricating and other oils; the color, reported as a number, is determined by varying the thickness of a column of oil until its color matches that of a standard color glass. { 'tag 'räb·ən·sən ‚kə·lə'rim·əd·ər }

tag sort [COMPUT SCI] A method of sorting data in which the addresses of records rather than the records themselves are used to determine the sequence. { 'tag ‚sȯrt }

tagua palm [BOT] *Phytelephas macrocarpa.* A palm tree of tropical America; the endosperm of the seed is used as an ivory substitute. { 'täg·wə ‚päm }

Tahuian [GEOL] A local Eocene time subdivision in Australia whose identification is based on foraminiferans. { tə'wī·ən }

taiga [ECOL] A zone of forest vegetation encircling the Northern Hemisphere between the arctic-subarctic tundras in the north and the steppes, hardwood forests, and prairies in the south. Also known as boreal forest. { 'tī·gə }

taiga climate [CLIMATOL] In general, a climate which produces taiga vegetation, that is, too cold for prolific tree growth but milder than the tundra climate and moist enough to promote appreciable vegetation. Also known as subarctic climate. { 'tī·gə ‚klī·mət }

tail [AERO ENG] **1.** The rear part of a body, as of an aircraft or a rocket. **2.** The tail surfaces of an aircraft or a rocket. [ASTRON] The part of a comet that extends from the coma in a direction opposite to the sun; it consists of dust and gas that have been blown away from the coma by the solar wind and

the sun's radiation pressure. [ELECTR] **1.** A small pulse that follows the main pulse of a radar set and rises in the same direction. **2.** The trailing edge of a pulse. [MATH] For a stochastic process represented by $x(t_1), x(t_2), \ldots$, the process obtained by deleting the first n terms, for some n. [VERT ZOO] **1.** The usually slender appendage that arises immediately above the anus in many vertebrates and contains the caudal vertebrae. **2.** The uropygium, and its feathers, of a bird. **3.** The caudal fin of a fish or aquatic mammal. { tāl }

tail assembly See empennage. { 'tāl ə‚sem·blē }

tailboard See tailgate. { 'tāl‚bȯrd }

tail clipping [ELECTR] Method of sharpening the trailing edge of a pulse. { 'tāl ‚klip·iŋ }

tail fin [AERO ENG] A fin at the rear of a rocket or other body. { 'tāl ‚fin }

tailgate [CIV ENG] The downstream gate of a canal lock. [ENG] A hinged gate at the rear of a vehicle that can be let down for convenience in loading. Also known as tailboard. { 'tāl‚gāt }

tail house [CHEM ENG] An installation in a refinery containing a look box, facilities for sampling, and controls for diverting the products to storage tanks or to other locations in the refinery for further processing. { 'tāl ‚haùs }

tailing [BUILD] The projecting portion of a stone or brick that has been set into a wall, for example, a cornice. { 'tāl·iŋ }

tailings [ENG] The lighter particles which pass over a sieve in milling, crushing, or purifying operations. [MIN ENG] **1.** The parts, or a part, of any incoherent or fluid material separated as refuse, or separately treated as inferior in quality or value. **2.** The decomposed outcrop of a vein or bed. **3.** The refuse material resulting from processing ground ore. { 'tāl·iŋz }

tailings settling tank [MIN ENG] A vessel in which solids are removed from the tailings effluent in mineral processing plants. { 'tāl·iŋz 'set·liŋ ‚taŋk }

tail out [PETRO ENG] To place sucker rods on a rack as they are pulled from a well during oil production. { 'tāl ‚aùt }

tail pulley [MECH ENG] A pulley at the tail of the belt conveyor opposite the normal discharge end; may be a drive pulley or an idler pulley. { 'tāl ‚pùl·ē }

tailrace [ENG] A channel for carrying water away from a turbine, waterwheel, or other industrial application. [MIN ENG] A channel for conveying mine trailings. { 'tāl‚rās }

tail rope [MIN ENG] **1.** The rope which passes around the return sheave in main-and-tail haulage or a scraper loader layout. **2.** The rope that is used to draw the empty cars back into a mine in a tail-rope system. **3.** A counterbalance rope attached beneath the cage when the cages are hoisted in balance. **4.** A hemp rope used for moving pumps in shafts. { 'tāl ‚rōp }

tail-rope system [MIN ENG] Haulage by a hoisting engine and two separate drums in which the main rope is attached to the front end of a trip of cars, and the tail rope is attached to the rear end of the trip. { 'tāl ‚rōp ‚sis·təm }

tailshaft [NAV ARCH] That part of the shaft of a ship's propeller extending through the stern tube. { 'tāl‚shaft }

tail sheave [MIN ENG] The return sheave for an endless rope or the tail rope of the main-and-tail-rope system, placed at the far end of a haulageway. { 'tāl ‚shēv }

tailstock [MECH ENG] A part of a lathe that holds the end of the work not being shaped, allowing it to rotate freely. { 'tāl‚stäk }

tail surface [AERO ENG] A stabilizing or control surface in the tail of an aircraft or missile. { 'tāl ‚sər·fəs }

tail track system [MIN ENG] A form of track layout for car or trip loading in which the track can be extended down the heading, turned right or left, or turned back, U-fashion, in an adjacent heading. { 'tāl ‚trak ‚sis·təm }

tail warning radar [ENG] Radar installed in the tail of an aircraft to warn the pilot that an aircraft is approaching from the rear. { 'tāl 'wȯrn·iŋ ‚rā‚där }

tailwater ditch [AGR] A channel made along the lower end of a field to carry surface runoff from irrigation furrows off the field. { 'tāl‚wȯd·ər ‚dich }

tailwind [METEOROL] A wind which assists the intended progress of an exposed, moving object, for example, rendering an airborne object's ground speed greater than its airspeed; the opposite of a headwind. Also known as following wind. { 'tāl‚wind }

taino [METEOROL] A tropical cyclone (hurricane) in parts of the Greater Antilles. { 'tī·nō }

TAENOGLOSSATE RADULA

Taenoglossate radula from *Lanistes*, a fresh-water gastropod. *(From R. R. Shrock and W. H. Twenhofel, Principles of Invertebrate Paleontology, 2d ed., McGraw-Hill, 1953)*

Tainter gate [CIV ENG] A spillway gate whose face is a section of a cylinder; rotates about a horizontal axis on the downstream end of the gate and can be closed under its own weight. Also known as radial gate. { 'tān·tər ,gāt }

takedown [COMPUT SCI] The actions performed at the end of an equipment operating cycle to prepare the equipment for the next setup; for example, to remove the tapes from the tape handlers at the end of a computer run is a takedown operation. { 'tāk,daùn }

takedown time [COMPUT SCI] The time required to take down a piece of equipment. { 'tāk,daùn ,tīm }

takeoff [AERO ENG] Ascent of an aircraft or rocket at any angle, as the action of a rocket vehicle departing from its launch pad or the action of an aircraft as it becomes airborne. { 'tāk,óf }

takeoff assist [AERO ENG] **1.** The extra thrust given to an airplane or missile during takeoff through the use of a rocket motor or other device. **2.** The device used in such a takeoff. { 'tāk,óf ə,sist }

takeoff weight [AERO ENG] The weight of an aircraft or rocket vehicle ready for takeoff, including the weight of the vehicle, the fuel, and the payload. { 'tāk,óf ,wāt }

take the ground [NAV] A ship takes the ground when the tide leaves it aground for want of sufficient depth of water. { 'tāk thə 'graùnd }

takeup [MECH ENG] A tensioning device in a belt-conveyor system for taking up slack of loose parts. { 'tāk,əp }

takeup pulley [MECH ENG] An adjustable idler pulley to accommodate changes in the length of a conveyor belt to maintain proper belt tension. { 'tāk,əp ,pùl·ē }

takeup reel [ENG] The reel that accumulates magnetic tape after it is recorded or played by a tape recorder. { 'tāk,əp ,rēl }

takt time [IND ENG] **1.** The rate of customer demand, calculated by dividing the available production time by the quantity the customer requires in that time. **2.** The reciprocal of the production rate. { 'tak ,tīm }

taku wind [METEOROL] A strong, gusty, east-northeast wind, occurring in the vicinity of Juneau, Alaska, between October and March; it sometimes attains hurricane force at the mouth of the Taku River, after which it is named. { 'tä·kü ,wind }

TALAR [NAV] A modular step-scan microwave landing system. Derived from tactical landing approach radar. { 'ta,lär }

talbot [OPTICS] A unit of luminous energy equal to the luminous energy carried by a luminous flux of 1 lumen during a period of 1 second. { 'tal·bət }

Talbot's bands [OPTICS] A series of dark bands that appear in the spectrum of white light when a glass plate of the proper thickness is placed across one half of the aperture of a spectroscope on the side of the blue end of the spectrum. { 'tal·bəts ,banz }

Talbot's curve [MATH] The negative pedal of an ellipse, with eccentricity greater than $\sqrt{2}/2$, with respect to its center. { 'tal·bəts ,kərv }

Talbot's law [OPTICS] The law that apparent brightness of an object flashing at a frequency greater than about 10 hertz is equal to its actual brightness times the ratio of the exposure time to the total time. { 'tal·bəts ,lò }

talbutal [PHARM] $C_{11}H_{16}N_2O_3$ A crystalline compound that melts at 108–110°C, and is soluble in alcohol, chloroform, acetone, and ether; used in medicine as a short-acting hypnotic and sedative. { 'tal·byə,tal }

talc [MINERAL] $Mg_3Si_4O_{10}(OH)_2$ A whitish, greenish, or grayish hydrated magnesium silicate mineral crystallizing in the monoclinic system; it is extremely soft (hardness is 1 on Mohs scale) and has a characteristic soapy or greasy feel. { talk }

talcose rock [PETR] A rock having a soft and soapy feel, that is, resembling talc. { 'tal,kōs ,räk }

talcosis [MED] A lung disease caused by inhalation of talc dust and characterized by chronic induration and fibrosis. { tal'kō·səs }

talc schist [PETR] A schist in which talc is the dominant schistose material. { 'talk ,shist }

talik [GEOL] A Russian term applied to permanently unfrozen ground in regions of permafrost; usually applies to a layer which lies above the permafrost but below the active layer, that is, when the permafrost table is deeper than the depth reached by winter freezing from the surface. Also known as tabetisol. { 'tä·lik }

talipes [MED] Any of several foot deformities, especially of congenital origin. { 'tal·ə,pēz }

talipes cavus [MED] A deformity of the foot marked by exaggeration of the longitudinal arch and by dorsal contractures of the toes. { 'tal·ə,pēz 'kā·vəs }

talipes equinovarus [MED] The most common form of clubfoot, characterized by an extreme turning down and under of the foot; it is seen more often in boys and tends to affect one foot only. { 'tal·ə,pēz ,ek·wi·no'va·rəs }

Talitridae [INV ZOO] A family of terrestrial amphipod crustaceans in the suborder Gammaridea. { tə'li·trə,dē }

talk-back circuit See interphone. { 'tòk ,bak ,sər·kət }

talking [MIN ENG] A series of small bumps or cracking noises within the mine walls, indicating that the rock is beginning to yield to stresses. { 'tòk·iŋ }

talking battery See quiet battery. { 'tòk·iŋ ,bad·ə·rē }

talk-listen switch [ENG ACOUS] A switch provided on intercommunication units to permit using the loudspeaker as a microphone when desired. { 'tòk 'lis·ən ,swich }

tall building [CIV ENG] A structure that, because of its height, is affected by lateral forces due to wind or earthquake to the extent that the forces constitute an important element in structural design. Also known as high-rise building. { 'tòl ,bil·diŋ }

tall fescue [AGR] *Festuca arundinacea.* A perennial cool-season plant that is used primarily as pasture and hay for beef cattle. { ,tòl 'fes·kyü }

tall fescue toxicosis [VET MED] A group of several animal disorders caused by grazing on tall fescue infected with the endophytic symbiotic fungus *Acremonium coenophialum.* { ,tòl ,fes·kyü ,tak·sə'kō·səs }

tall oil [MATER] A yellow-black, malodorous, resinous admixture of rosin, fatty acids, sterols, high-molecular-weight alcohols, and other materials, derived from wood-pulping waste liquors; used in paint drying oils, alkyd resins, linoleum, soaps, lubricants, and greases. Also known as liquid rosin; tallol. { 'tàl ,óil }

tallol See tall oil. { 'tä,lòl }

tallow [MATER] Animal fat with carbon chains containing 16–18 carbons, derived from cattle, sheep, and horses; used for soaps, leather dressings, candles, food, and greases, and as a chemical intermediate. { 'ta·lō }

tallow-seed oil See stillingia oil. { 'ta·lō ¦sēd ,óil }

tally [MIN ENG] A mark, number, or tin ticket placed by the miner on each car of coal or ore that is sent from the work place, thus facilitating a count or tally of all the filled cars. { 'tal·ē }

talon [VERT ZOO] A sharply hooked claw on the foot of a bird of prey. { 'tal·ən }

Talpidae [VERT ZOO] The moles, a family of insectivoran mammals; distinguished by the forelimbs which are adapted for digging, having powerful muscles and a spadelike bony structure. { 'tal·pə,dē }

talus [ANAT] See astragalus. [GEOL] Also known as rubble; scree. **1.** Coarse and angular rock fragments derived from and accumulated at the base of a cliff or steep, rocky slope. **2.** The accumulated heap of such fragments. { 'tal·əs }

talus creep [GEOL] The slow, downslope movement of talus. { 'tal·əs ,krēp }

talus glacier See rock glacier. { 'tal·əs ,glā·shər }

talus slope [GEOL] A steep, concave slope consisting of an accumulation of talus. Also known as debris slope. { 'tal·əs ,slōp }

tamarack [BOT] *Larix laricina.* A larch and a member of the pine family; it has an erect narrowly pyramidal habit, and grows in wet and moist soils in the northeastern United States, west to the Lake States and across Canada to Alaska; used for railroad ties, posts, sills, and boats. Also known as hackmarack. { 'tam·ə,rak }

tamarugite [MINERAL] $NaAl(SO_4)_2 \cdot 6H_2O$ A colorless, monoclinic mineral consisting of a hydrated sulfate of sodium and aluminum; occurs as crystals and masses. { ,tam·ə'rü,gīt }

Tamm-Dancoff method [QUANT MECH] A method of forming an approximate wave function of a system of interacting particles, particularly nucleons and mesons, by describing it as an algebraic sum of several possible states, the number of such states determining the order of the approximation. { 'tam 'dan,kòf ,meth·əd }

tamp [ENG] To tightly pack a drilled hole with clay or other stemming material after the charge has been placed. { tamp }

tamper [CIV ENG] A ramming device for compacting a granular material such as soil, backfill, or unformed concrete; usually powered by a motor. [NUCLEO] *See* reflector. [ORD] In a weapon, any substance that resists movement for a split microsecond, used so that the active materials may build up greater pressure behind the substance. { 'tam·pər }

tamping bag [ENG] A bag filled with stemming material such as sand for use in horizontal and upward sloping shotholes. { 'tamp·iŋ ,bag }

tamping bar [ENG] A piece of wood for pushing explosive cartridges or forcing the stemming into shotholes. { 'tamp·iŋ ,bär }

tamping plug [ENG] A plug of iron or wood used instead of tamping material to close up a loaded blasthole. { 'tamp·iŋ ,pləg }

tamping roller *See* sheepsfoot roller. { 'tamp·iŋ 'rō·lər }

tampion [ENG] A cone-shaped hand tool usually fashioned of hardwood that is forced into a lead pipe to increase its diameter. { 'tam·pē·ən }

tampon [MED] A plug of absorbent material, such as cotton or sponge, inserted into a cavity as packing. { 'tam,pän }

tan *See* tangent. { tan }

Tanaidacea [INV ZOO] An order of eumalacostracans of the crustacean superorder Peracarida; the body is linear, more or less cylindrical or dorsoventrally depressed, and the first and second thoracic segments are fused with the head, forming a carapace. { ,tan·ē·ə'dā·shə }

tan alt *See* shadow factor. { tan 'ȯlt }

Tanaostigmatidae [INV ZOO] A small family of hymenopteran insects in the superfamily Chalcidoidea. { tə,nā·ō·stig 'mad·ə,dē }

tanbark [MATER] The fibrous portion of ground oak or hemlock bark; it is burned in a mixture with other fuels to maintain combustion; also used on a running track for horses. { 'tan,bärk }

tandem [AERO ENG] The fore and aft configuration used in boosted missiles, long-range ballistic missiles, and satellite vehicles; stages are stacked together in series and are discarded at burnout of the propellant for each stage. [ELEC] Two-terminal pair networks are in tandem when the output terminals of one network are directly connected to the input terminals of the other network. { 'tan·dəm }

tandem accelerator [NUCLEO] An electrostatic accelerator in which negative hydrogen ions generated in a special ion source are accelerated as they pass from ground potential up to a high-voltage terminal, both electrons are then stripped from the negative ion by passage through a thin foil or gas cell, and the proton is again accelerated as it passes to ground potential. { 'tan·dəm ak'sel·ə,rād·ər }

tandem central office [COMMUN] A telephone office that makes connections between local offices in an area where there is such a high density of local offices that it would be uneconomical to make direct connections between them. Also known as tandem office. { 'tan·dəm 'sen·trəl 'ȯf·əs }

tandem compensation *See* cascade compensation. { 'tan·dəm ,käm·pən'sā·shən }

tandem connection *See* cascade connection. { 'tan·dəm kə'nek·shən }

tandem distributed numerical control [CONT SYS] A form of distributed numerical control involving a series of machines connected by a conveyor and automatic loading and unloading devices that are under control of the central computers. { 'tan·dəm di'strib·yəd·əd nü,mer·ə·kəl kən'trōl }

tandem-drive conveyor [MECH ENG] A conveyor having the conveyor belt in contact with two drive pulleys, both driven with the same motor. { 'tan·dəm ¦drīv kən'vā·ər }

tandem duplication [CYTOL] The occurrence of two identical sequences, one following the other, in a chromosome segment. { 'tan·dəm ,dü·plə'kā·shən }

tandem hoisting [MIN ENG] Hoisting in a deep shaft with two skips running in one shaft; the lower skip is suspended from the tail rope of the upper skip. { 'tan·dəm 'hȯist·iŋ }

tandem mill [MET] A rolling mill consisting of two or more stands in succession, synchronized so that the metal passes directly from one to another. { 'tan·dəm 'mil }

tandem office *See* tandem central office. { 'tan·dəm 'ȯf·əs }

tandem propellers [NAV ARCH] Two propellers on the same shaft, rotating in the same direction. { 'tan·dəm prə'pel·ərz }

tandem roller [MECH ENG] A steam- or gasoline-driven road roller in which the weight is divided between heavy metal rolls, of dissimilar diameter, one behind the other. { 'tan·dəm 'rō·lər }

tandem switching [COMMUN] System of routing telephone calls in which calls do not travel directly between local offices, but rather through a central office. { 'tan·dəm 'swich·iŋ }

tandem system [COMPUT SCI] A computing system in which there are two central processing units, usually with one controlling the other, and with data proceeding from one processing unit into the other. { 'tan·dəm ,sis·təm }

tang [ENG] **1.** The part of a file that fits into a handle. **2.** The end of a drill shank which allows transmission of torque from the drill press spindle to the body of the drill. { taŋ }

tangeite *See* calciovolborthite. { 'tan·jē,īt }

tangelo [BOT] A tree that is hybrid between a tangerine or other mandarin and a grapefruit or shaddock; produces an edible fruit. { 'tan·jə·lō }

tangent [MATH] **1.** A line is tangent to a curve at a fixed point *P* if it is the limiting position of a line passing through *P* and a variable point on the curve *Q*, as *Q* approaches *P*. **2.** The function which is the quotient of the sine function by the cosine function. Abbreviated tan. **3.** The tangent of an angle is the ratio of its sine and cosine. Abbreviated tan. { 'tan·jənt }

tangent arc [METEOROL] Generic name for several types of halo arcs that form as loci tangent to other halos; the halo of 22° occasionally exhibits the horizontal and vertical tangent arcs, and the halo of 46° exhibits the infralateral tangent arcs and the supralateral tangent arcs. { 'tan·jənt ¦ärk }

tangent bending [MET] In a single piece of metal, forming a series of identical bends with parallel axes. { 'tan·jənt ,bend·iŋ }

tangent bundle [MATH] The fiber bundle *T(M)* associated to a differentiable manifold *M* which is composed of the points of *M* together with all their tangent vectors. Also known as tangent space. { 'tan·jənt ,bənd·əl }

tangent circles [MATH] Two circles that have a single point in common. { ¦tan·jənt 'sər·kəlz }

tangent cone [MATH] A cone each of whose elements is tangent to a given quadric surface. { 'tan·jənt ,kōn }

tangent galvanometer [ENG] A galvanometer in which a small compass is mounted horizontally in the center of a large vertical coil of wire; the current through the coil is proportional to the tangent of the angle of deflection of the compass needle from its normal position parallel to the magnetic field of the earth. { 'tan·jənt ,gal·və'näm·əd·ər }

tangential acceleration [MECH] The component of linear acceleration tangent to the path of a particle moving in a circular path. { tan'jen·chəl ak,sel·ə'rā·shən }

tangential coma [OPTICS] For a lens that displays coma, the length of a tangent from the vertex of the patch of light formed in the focal plane by rays from an off-axis point to the comatic circle of rays from this point that pass near the edge of the lens. { tan'jen·chəl 'kō·mə }

tangential component [MATH] A component of a given vector acting at right angles to a given radius of a given circle. { tan'jen·chəl kəm'pō·nənt }

tangential coordinates [MATH] For a surface, a set of four coordinates, three of which are the direction cosines of the normal to the surface, while the fourth is the algebraic distance from the origin to the plane tangent to the surface. { tan'jen·chəl kō'ȯrd·ən·əts }

tangential curvature *See* geodesic curvature. { tan'jen·chəl 'kər·və·chər }

tangential developable *See* tangent surface. { tan'jen·chəl di'vel·əp·ə·bəl }

tangential focus *See* primary focus. { tan'jen·chəl 'fō·kəs }

tangential helical-flow turbine *See* helical-flow turbine. { tan'jen·chəl ¦hel·ə·kəl ,flō 'tər·bən }

tangential plane *See* meridional plane. { tan'jen·chəl 'plān }

tangential polar equation [MATH] An equation of a curve exprepssed in terms of the distance of a point *P* on the curve from a reference point *O* and the perpendicular distance from *O* to the tangent to the curve at *P*. { tan¦jen·chəl ¦pō·lər i'kwā·zhən }

tangential stress *See* shearing stress. { tan¦jen·chəl 'stres }

TANAIDACEA

Apseudes spinosus.

tangential surface [OPTICS] A surface containing the primary foci of points in a plane perpendicular to the optical axis of an astigmatic system. { tan¦jen·chəl 'sər·fəs }

tangential velocity [MECH] **1.** The instantaneous linear velocity of a body moving in a circular path; its direction is tangential to the circular path at the point in question. **2.** The component of the velocity of a body that is perpendicular to a line from an observer or reference point to the body. { tan¦jen·chəl və'läs·əd·ē }

tangential wave *See* S wave. { tan¦jen·chəl 'wāv }

tangential wave path [ELECTROMAG] In radio propagation over the earth, a path of propagation of a direct wave which is tangential to the surface of the earth; the tangential wave path is curved by atmospheric refraction. { tan¦jen·chəl 'wāv ,path }

tangent indicatrix *See* spherical indicatrix. { ¦tan·jənt in'dik·ə,triks }

tangent latitude error [NAV] The angle between the local meridian and the settling position or spin axis in a nonpendulous gyrocompass where damping is accomplished by offsetting the point of application of the force of a mercury ballistic. { 'tan·jənt 'lad·ə,tüd ,er·ər }

tangent offset [ENG] In surveying, a method of plotting traverse lines; angles are laid out by linear measurement, using a constant times the natural tangent of the angle. { 'tan·jənt 'óf,set }

tangent plane [MATH] The tangent plane to a surface at a point is the plane having every line in it tangent to some curve on the surface at that point. { 'tan·jənt 'plān }

tangent point *See* point of tangency. { 'tan·jənt ,póint }

tangent screw [ENG] A screw providing tangential movement along an arc, such as the screw which provides the final angular adjustment of a marine sextant during an observation. { 'tan·jənt ,skrü }

tangent space [MATH] **1.** The vector space of all tangent vectors at a given point of a differentiable manifold. **2.** *See* tangent bundle. { 'tan·jənt ,spās }

tangent surface [MATH] The ruled surface generated by the tangents to a specified space curve. Also known as tangential developable. { 'tan·jənt ,sər·fəs }

tangent vector [MATH] A tangent vector at a point of a differentiable manifold is any vector tangent to a differentiable curve in the manifold at this point; alternatively, a member of the tangent plane to the manifold at the point. { 'tan·jənt ,vek·tər }

tangent welding [MET] Arc welding in which two or more electrodes are in a plane parallel to the line of travel. { 'tan·jənt ¦weld·iŋ }

tangerine [BOT] Any of several trees of the species *Citrus reticulata;* the fruit is a loose-skinned mandarin with a deep-orange or scarlet rind. { 'tan·jə,rēn }

tangerine oil *See* mandarin oil. { 'tan·jə,rēn ,óil }

tangling [SOLID STATE] The reduction of motion of dislocations in a substance by increasing the total number of dislocations, so that they tangle and interfere with each other's motions. { 'taŋ·gliŋ }

tangoreceptor [PHYSIO] A sense organ in the skin that responds to touch and pressure. { ¦taŋ·gō·ri'sep·tər }

tanh *See* hyperbolic tangent. { 'tan¦äch }

tank [ELECTR] **1.** A unit of acoustic delay-line storage containing a set of channels, each forming a separate recirculation path. **2.** The heavy metal envelope of a large mercury-arc rectifier or other gas tube having a mercury-pool cathode. **3.** *See* tank circuit. [ENG] A large container for holding, storing, or transporting a liquid. { taŋk }

tankage [ENG] Contents of a storage tank. [MATER] Slaughter-house entrails and scraps used as fertilizer. { 'taŋ·kij }

tank balloon [ENG] An air- and vapor-tight flexible container fitted to the breather pipe of a gasoline storage tank to receive gasoline vapors; as the tank cools, the vapors return to the tank. { 'taŋk bə,lün }

tank barge [NAV ARCH] A barge equipped with tanks that may carry any one of a great variety of liquid commodities, such as petroleum and petroleum products, chemicals, and fertilizers. { 'taŋk ,bärj }

tank battery [PETRO ENG] A grouping of interconnected storage tanks situated to receive the output of one or more oil wells. { 'taŋk ,bad·ə·rē }

tank bottom [CHEM ENG] The liquid material in a tank below the level of the outlet pipe; often a mixture of the stored liquid with rust and other sediment. { 'taŋk ,bäd·əm }

tank car [ENG] Railroad car onto which is mounted a cylindrical, horizontal tank designed for the transport of liquids, chemicals, gases, meltable solids, slurries, emulsions, or fluidizable solids. { 'taŋk ,kär }

tank circuit [ELECTR] A circuit which exhibits resonance at one or more frequencies, and which is capable of storing electric energy over a band of frequencies continuously distributed about the resonant frequency, such as a coil and capacitor in parallel. Also known as electrical resonator; tank. { 'taŋk ,sər·kət }

tank ditch *See* antitank ditch. { 'taŋk ,dich }

tank dozer [ORD] Standard tank equipped with a detachable bulldozer blade. { 'taŋk ,dōz·ər }

tanker [NAV ARCH] A steel ship in which the hull is subdivided into tanks for carrying petroleum products or other liquid cargo in bulk. Also known as tankship. { 'taŋ·kər }

tank farm [PETRO ENG] An area in which a number of large-capacity storage tanks are located, generally used for crude oil or petroleum products. { 'taŋk ,färm }

tank gage [ENG] A device used to measure the contents of a liquid storage tank; can be manual or automatic. { 'taŋk ,gāj }

tank obstacle *See* antitank obstacle. { 'taŋk ,äb·stə·kəl }

tank periscope [OPTICS] A periscope permitting a tank occupant to observe without being exposed to bullet fire; employs a pair of plane, parallel, reflecting surfaces (either mirrors or prisms) so arranged in a mount that the path of light through the instrument forms a letter Z. { 'taŋk ,per·ə,skōp }

tank reactor [NUCLEO] A nuclear reactor in which the core is suspended in a closed tank, as distinct from an open-pool reactor. { 'taŋk rē¦ak·tər }

tank scale [ENG] A counterweighted suspension or platform weighing mechanism for tanks, hoppers, and similar solids or liquids containers. { 'taŋk ,skāl }

tankship *See* tanker. { 'taŋk,ship }

tank switch [PETRO ENG] An automatic control of lease tanks in oil fields, including controls of lines to fill tanks and for pipeline runs; can be electrically or pneumatically actuated. { 'taŋk ,swich }

tank truck [ENG] A truck body onto which is mounted a cylindrical, horizontal tank, designed for the transport of liquids, chemicals, gases, meltable solids, slurries, emulsions, or fluidizable solids. { 'taŋk ,trək }

tannase [BIOCHEM] An enzyme that catalyzes the hydrolysis of tannic acid to gallic acid; found in cultures of *Aspergillus* and *Penicillium.* { 'ta,nās }

tannic acid [ORG CHEM] **1.** $C_{14}H_{10}O_9$ A yellowish powder with an astringent taste; soluble in water and alcohol, insoluble in acetone and ether; derived from nutgalls; decomposes at 210°C; used as an alcohol denaturant and a chemical intermediate, and in tanning and textiles. Also known as digallic acid; gallotannic acid; gallotannin; tannin. **2.** $C_{76}H_{52}O_{46}$ Yellowish-white to light-brown amorphous powder or flakes; decomposes at 210–215°C; very soluble in alcohol and acetone; used as a mordant in dyeing, in photography, as a reagent, and in clarifying wine or beer. Also known as pentadigalloylglucose. { 'tan·ik 'as·əd }

tannin *See* tannic acid. { 'tan·ən }

tanning [ENG] A process of preserving animal hides by chemical treatment (using vegetable tannins, metallic sulfates, and sulfurized phenol compounds, or syntans) to make them immune to bacterial attack, and subsequent treatment with fats and greases to make them pliable. { 'tan·iŋ }

tanning agent [MATER] Any one of the tannins used to treat skins and hides to preserve the hide substance and to protect it from decay. { 'tan·iŋ ,ā·jənt }

tanning extract [MATER] Tannin-rich liquor extracted from woods and pulps; used in leather tanning. { 'tan·iŋ ,ek,strakt }

tan rot [PL PATH] A fungus disease of strawberries caused by *Pezizella lythri* and characterized by the appearance of tan depressions on the fruit. { 'tan ,rät }

tantalic acid anhydride *See* tantalum oxide. { tan'tal·ik 'as·əd an'hī,drīd }

tantalic chloride *See* tantalum chloride. { tan'tal·ik 'klór,īd }

tantalite [MINERAL] $(Fe,Mn)Ta_2O_6$ An iron-black mineral that crystallizes in the orthorhombic system and commonly occurs in short prismatic crystals; luster is submetallic, hardness

Tank barges carrying refrigerated ammonia.

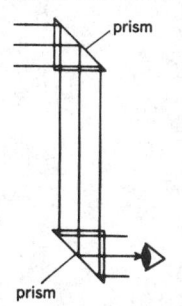

prism

prism

Diagram of a simple tank periscope with parallel reflecting surfaces.

is 6 on Mohs scale, and specific gravity is 7.95; principal ore of tantalum. { 'tant·əl‚līt }

tantalum [CHEM] A metallic transition element, symbol Ta, atomic number 73, atomic weight 180.9479; black powder or steel-blue solid soluble in fused alkalies, insoluble in acids (except hydrofluoric and fuming sulfuric); melts about 3000°C. [MET] A lustrous, platinum-gray ductile metal used in making dental and surgical tools, pen points, and electronic equipment. { 'tant·əl·əm }

tantalum capacitor [ELEC] An electrolytic capacitor in which the anode is some form of tantalum; examples include solid tantalum, tantalum-foil electrolytic, and tantalum-slug electrolytic capacitors. { 'tant·əl·əm kə'pas·əd·ər }

tantalum carbide [INORG CHEM] TaC Hard, chemical-resistant crystals melting at 3875°C; used in cutting tools and dies. { 'tant·əl·əm 'kär‚bīd }

tantalum chloride [INORG CHEM] TaCl₅ A highly reactive, pale-yellow powder decomposing in moist air; soluble in alcohol and potassium hydroxide; melts at 221°C; used to produce tantalum and as a chemical intermediate. Also known as tantalic chloride; tantalum pentachloride. { 'tant·əl·əm 'klȯr‚īd }

tantalum-foil electrolytic capacitor [ELEC] An electrolytic capacitor that uses plain or etched tantalum foil for both electrodes, with a weak acid electrolyte. { 'tant·əl·əm ¦fȯil i¦lek·trə¦lid·ik kə'pas·əd·ər }

tantalum nitride [INORG CHEM] TaN A very hard, black, water-insoluble solid, melting at 3360°C. { 'tant·əl·əm 'nī‚trīd }

tantalum nitride resistor [ELECTR] A thin-film resistor consisting of tantalum nitride deposited on a substrate, such as industrial sapphire. { 'tant·əl·əm 'nī‚trīd ri'zis·tər }

tantalum oxide [INORG CHEM] Ta₂O₅ Prisms insoluble in water and acids (except for hydrofluoric); melts at 1800°C; used to make tantalum, in optical glass and electronic equipment, and as a chemical intermediate. Also known as tantalic acid anhydride; tantalum pentoxide. { 'tant·əl·əm 'äk‚sīd }

tantalum pentachloride See tantalum chloride. { 'tant·əl·əm ¦pen·tə'klȯr‚īd }

tantalum pentoxide See tantalum oxide. { 'tant·əl·əm pen'täk‚sīd }

tantalum-slug electrolytic capacitor [ELEC] An electrolytic capacitor that uses a sintered slug of tantalum as the anode, in a highly conductive acid electrolyte. { 'tant·əl·əm ¦sləg i¦lek·trə¦lid·ik kə'pas·əd·ər }

T antenna [ELECTROMAG] An antenna consisting of one or more horizontal wires, with a lead-in connection being made at the approximate center of each wire. { 'tē an‚ten·ə }

tanteuxenite [MINERAL] (Y,Ce,Ca)(Ta,Nb,O)₂(O,OH)₆ A brown or black variety of euxenite with tantalum substituting for niobium. Also known as delorenzite; eschwegeite. { tan'tyük·sə‚nīt }

tantiron [MET] An iron alloy containing silicon, carbon, manganese, phosphorus, and sulfur; used for chemical equipment where resistance to acids is needed. { 'tan‚tī·ərn }

Tanyderidae [INV ZOO] The primitive crane flies, a family of orthorrhaphous dipteran insects in the series Nematocera. { ‚tan·ə'der·ə‚dē }

Tanypezidae [INV ZOO] A family of myodarian cyclorrhaphous dipteran insects in the subsection Acalyptreatae. { ‚tan·ə'pez·ə‚dē }

Taos hum [ACOUS] An irritating low-frequency sound of mysterious origin perceived by some individuals in and around Taos, New Mexico. { ¦tä‚ōs 'həm }

tap [DES ENG] **1.** A plug of accurate thread, form, and dimensions on which cutting edges are formed; it is screwed into a hole to cut an internal thread. **2.** A threaded cone-shaped fishing tool. [ELEC] A connection made at some point other than the ends of a resistor or coil. [ENG] A small, threaded hole drilled into a pipe or process vessel; used as connection points for sampling devices, instruments, or controls. [MET] **1.** A quantity of molten metal run out from a furnace at one time. **2.** To remove excess slag from the floor of a pot furnace. [MIN ENG] To intersect with a borehole and withdraw or drain the contained liquid, as water from a water-bearing formation or from underground workings. { tap }

tap bolt [DES ENG] A bolt with a head that can be screwed into a hole and held in place without a nut. Also known as tap screw. { 'tap ‚bōlt }

tap changer [ELEC] A device which is used to change the

ratio of the input and output voltages of a transformer over any one of a definite number of steps. { 'tap ‚chān·jər }

tap crystal [ELECTR] Compound semiconductor that stores current when stimulated by light and then gives up energy as flashes of light when it is physically tapped. { 'tap ‚krist·əl }

tap density [MET] The apparent density of a volume of metal powder obtained when its receptacle is tapped or vibrated. { 'tap ‚den·səd·ē }

tap drill [MECH ENG] A drill used to make a hole of a precise size for tapping. { 'tap ‚dril }

tape [COMPUT SCI] A ribbonlike material used to store data in lengthwise sequential position. [ENG] A graduated steel ribbon used, instead of a chain, in surveying. { tāp }

tape alternation [COMPUT SCI] The switching of a computer program back and forth between two tape units in order to avoid interruption of the program during mounting and removal of tape reels. { 'tāp ‚ȯl·tər‚nā·shən }

tape-automated bonding [ELECTR] A semiconductor chip (die) assembly method, where the chips are connected to polyimide (tape) carriers, complete with circuitry for attachment to a printed circuit board. The chip-bonded tape carriers typically are supplied on a reel (like a roll of film) for automated circuit assembly processes. { ¦tāp ‚ȯd·ə‚mād·əd 'bän·diŋ }

tape bootstrap routine [COMPUT SCI] A computer routine stored in the first block of a magnetic tape that instructs the computer to read certain programs from the tape. { 'tāp 'büt‚strap rü‚tēn }

tape cartridge [ENG ACOUS] A cartridge that holds a length of magnetic tape in such a way that the cartridge can be slipped into a tape recorder and played without threading the tape; in stereophonic usage, usually refers to an eight-track continuous-loop cartridge, which is larger than a cassette. Also known as cartridge. { 'tāp ‚kär·trij }

tape cluster See magnetic tape group. { 'tāp ‚kləs·tər }

tape-controlled carriage [COMPUT SCI] A device which uses a loop of punched-paper or plastic mylar tape to control the motion of paper through a computer printer or typewriter. { 'tāp kən¦trōld ‚kar·ij }

tape-controlled machine [MECH ENG] A machine tool whose movements are automatically controlled by means of a magnetic or punched tape. { 'tāp kən¦trōld mə‚shēn }

tape control unit [COMPUT SCI] A device which senses which tape unit is to be accessed for read or write purpose and opens up the necessary electronic paths. Formerly known as hypertape control unit. { 'tāp kən‚trōl ‚yü·nət }

tape correction [ENG] A quantity applied to a taped distance to eliminate or reduce errors due to the physical condition of the tape and the manner in which it is used. { 'tāp kə‚rek·shən }

tape crease [COMPUT SCI] A fold or wrinkle in a magnetic tape that results in an error in the reading or writing of data at that point. { 'tāp ‚krēs }

tape deck [ENG ACOUS] A tape-recording mechanism that is mounted on a motor board, including the tape transport, electronics, and controls, but no power amplifier or loudspeaker. { 'tāp ‚dek }

tape drive [COMPUT SCI] A tape reading or writing device consisting of a tape transport, electronics, and controls; it usually refers to magnetic tape exclusively. [ENG ACOUS] See tape transport. [MECH ENG] A device that transmits power from an actuator to a remote mechanism by flexible tapes and pulleys. { 'tāp ‚drīv }

tape editor [COMPUT SCI] A routine designed to help edit, revise, and correct a routine contained on a tape. { 'tāp ‚ed·ər }

tape-float liquid-level gage [ENG] A liquid-level measurement by a float connected by a flexible tape to a rotating member, in turn connected to an indicator mechanism. { 'tāp ¦flōt 'lik·wəd ¦lev·əl ‚gāj }

tape gage [ENG] A box- or float-type tide gage which consists essentially of a float attached to a tape and counterpoise; the float operates in a vertical box or pipe which dampens out short-period wind waves while admitting the slower tidal movement; for the standard installation, the tape is graduated with numbers increasing toward the float and is arranged with pulleys and counterpoise to pass up and down over a fixed reading mark as the tide rises and falls. { 'tāp ‚gāj }

tape grass [BOT] Vallisnerida spiralis. An aquatic flowering plant belonging to the family Hydrocharitaceae. Also known as eel grass. { 'tāp ‚gras }

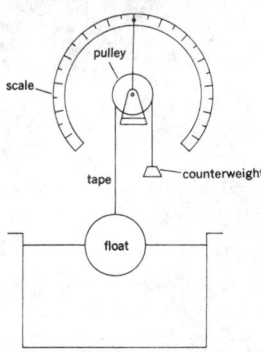

TAPE-FLOAT LIQUID-LEVEL GAGE

Diagram of tape-float liquid-level gage. (After D. M. Considine, ed., *Process Instruments and Controls Handbook*, McGraw-Hill, 1957)

tape group *See* magnetic tape group.   { 'tāp ‚grüp }

tape label [COMPUT SCI] A record appearing at the beginning or at the end of a magnetic tape to uniquely identify the tape as the one required by the system.   { 'tāp ‚lā·bəl }

tape library [COMPUT SCI] A special area, most often a room within a computer installation, used to store magnetic tapes.   { 'tāp ‚lī‚brer·ē }

tape-limited [COMPUT SCI] Pertaining to a computer operation in which the time required to read and write tapes exceeds the time required for computation.   { 'tāp ‚lim·əd·əd }

tape loop [ENG ACOUS] A length of magnetic tape having the ends spliced together to form an endless loop; used in message repeater units and in some types of tape cartridges to eliminate the need for rewinding the tape.   { 'tāp ‚lüp }

tape mark [COMPUT SCI] **1.** A special character or coding, an attached piece of reflective material, or other device that indicates the physical end of recording on a magnetic tape. Also known as destination warning mark; end-of-tape mark. **2.** A special character that divides a file of magnetic tape into sections, usually followed by a record with data describing the particular section of the file. Also known as control mark.   { 'tāp ‚märk }

tape operating system [COMPUT SCI] A computer operating system in which source programs and sometimes incoming data are stored on magnetic tape, rather than in the computer memory. Abbreviated TOS.   { 'tāp ‚äp·ə‚rād·iŋ ‚sis·təm }

tape player [ENG ACOUS] A machine designed only for playback of recorded magnetic tapes.   { 'tāp ‚plā·ər }

tape plotting system [COMPUT SCI] A digital incremental plotter in which the digital data are supplied from a magnetic or paper tape.   { 'tāp ‚pläd·iŋ ‚sis·təm }

tape pool [COMPUT SCI] A collection of tape drives.   { 'tāp ‚pül }

tape-processing simultaneity [COMPUT SCI] A feature of some computer systems whereby reading or writing of data can be carried out on all the tape units at the same time, while the central processing unit continues to process data.   { 'tāp ‚prä‚ses·iŋ ‚sī·məl·tə'nē·əd·ē }

tape punch [COMPUT SCI] A machine that punches code holes and feed holes in paper tape.   { 'tāp ‚pənch }

taper [AERO ENG] An airfoil feature in which either the thickness or the chord length or both decrease from the root to the tip. [ELEC] Continuous or gradual change in electrical properties with mechanical position such as rotation or length; for example, continuous change of cross section of a waveguide, or distribution of resistance in a potentiometer.   { 'tā·pər }

taper bit [DES ENG] A long, cone-shaped noncoring bit used in drilling blastholes and in wedging and reaming operations.   { 'tā·pər ‚bit }

tape reader *See* paper-tape reader.   { 'tāp ‚rēd·ər }

tape reading [COMPUT SCI] A process of feeding coded tapes through a tape-to-card punch to convert the coded information into punched cards; tapes can be prepared on the typewriter tape punch or on the card-controlled tape punch; the latter is capable of punching tape that can be transmitted by telegraph.   { 'tāp ‚rēd·iŋ }

tape recorder [ENG ACOUS] A device that records audio signals and other information on magnetic tape by selective magnetization of iron oxide particles that form a thin film on the tape; a recorder usually also includes provisions for playing back the recorded material.   { 'tāp ri‚kȯrd·ər }

tape recording [ENG ACOUS] The record made on a magnetic tape by a tape recorder.   { 'tāp ri‚kȯrd·iŋ }

tapered-bore [ORD] **1.** Referring to a gun with a tapered bore; the bore may be tapered throughout its length or only in the muzzle section. **2.** Pertaining to the ammunition for such a gun.   { 'tā·pərd 'bȯr }

tapered core bit [DES ENG] A core bit having a conical diamond-inset crown surface tapering from a borehole size at the bit face to the next larger borehole size at its upper, shank, or reaming-shell end.   { 'tā·pərd 'kȯr ‚bit }

tapered joint [DES ENG] A firm, leakproof connection between two pieces of pipe having the thread formed with a slightly tapering diameter.   { 'tā·pərd 'jȯint }

tapered pipeline [PETRO ENG] A changing of the pressure grade, either by change of wall thickness or material, of pipeline sections as working pressure is lessened.   { 'tā·pərd 'pīp‚līn }

tapered thread [DES ENG] A screw thread cut on the surface

of a tapered part; it may be either a pine or box thread, or a V-, Acme, or square-screw thread.   { 'tā·pərd 'thred }

tapered transmission line *See* tapered waveguide.   { 'tā·pərd tranz'mish·ən ‚līn }

tapered waveguide [ELECTROMAG] A waveguide in which a physical or electrical characteristic changes continuously with distance along the axis of the waveguide. Also known as tapered transmission line.   { 'tā·pərd 'wāv‚gīd }

tapered wheel [DES ENG] A flat-face grinding wheel with greater thickness at the hub than at the face.   { 'tā·pərd 'wēl }

taper gage [ENG] A precision gage that is used to check the accuracy of a standard taper.   { 'tā·pər ‚gāj }

taper-in-thickness ratio [AERO ENG] A gradual change in the thickness ratio along the wing span with the chord remaining constant.   { 'tā·pər in ‚thik·nəs ‚rā·shō }

taper key [DES ENG] A rectangular machine key that is slightly tapered along its length.   { 'tā·pər ‚kē }

taper pin [DES ENG] A small, tapered self-holding peg or nail used to connect parts together.   { 'tā·pər ‚pin }

taper pipe thread *See* pipe thread.   { 'tā·pər ‚pīp ‚thred }

taper plug gage [DES ENG] An internal gage in the shape of a frustrum of a cone used to measure internal tapers.   { 'tā·pər 'pləg ‚gāj }

taper reamer [DES ENG] A reamer whose fluted portion tapers toward the front end.   { 'tā·pər ‚rē·mər }

taper ring gage [DES ENG] An external gage having a conical internal contour; used to measure external tapers.   { 'tā·pər 'riŋ ‚gāj }

taper-rolling bearing [MECH ENG] A roller bearing capable of sustaining end thrust by means of tapered rollers and coned races.   { 'tā·pər 'rō·liŋ ‚ber·iŋ }

taper shank [DES ENG] A cone-shaped part on a tool that fits into a tapered sleeve on a driving member.   { 'tā·pər ‚shaŋk }

taper tap [DES ENG] A threaded cone-shaped tool for cutting internal screw threads.   { 'tā·pər ‚tap }

taper washer [DES ENG] A type of washer designed to be used underneath nuts with tapered flanges to enable the bolt assembly to fit properly when tightened.   { 'tā·pər ‚wäsh·ər }

tape search unit [COMPUT SCI] Small, fully transistorized, special-purpose, digital data-processing system using a stored program to perform logical functions necessary to search a magnetic tape in off-line mode, in response to a specific request.   { 'tā·pər 'sərch ‚yü·nət }

tape serial number [COMPUT SCI] A number identifying a magnetic tape which remains unchanged throughout the time the tape is used, even though all other information about the tape may change.   { 'tā·pər 'sir·ē·əl ‚nəm·bər }

tape skip [COMPUT SCI] A machine instruction to space forward and erase a portion of tape when a defect on the tape surface causes a write error to persist.   { 'tāp ‚skip }

tape speed [ENG ACOUS] The speed at which magnetic tape moves past the recording head in a tape recorder; standard speeds are $^{15}/_{16}$, $1^{7}/_{8}$, $3^{3}/_{4}$, $7^{1}/_{2}$, 15, and 30 inches per second (2.38125, 4.7625, 9.525, 19.05, 38.1, and 76.2 centimeters per second); faster speeds give improved high-frequency response under given conditions.   { 'tāp ‚spēd }

tape station [COMPUT SCI] A tape reading or writing device consisting of a tape transport, electronics, and controls; it may use either magnetic tape or paper tape.   { 'tāp ‚stā·shən }

tape-to-tape conversion [COMPUT SCI] A routine which directs a computer to copy information from one tape to another tape of a different kind; for example, from a seven-track onto a nine-track tape.   { 'tāp tə 'tāp kən'vər·zhən }

tape transport [COMPUT SCI] The mechanism that physically moves a tape past a stationary head. Also known as transport. [ENG ACOUS] The mechanism of a tape recorder that holds the tape reels, drives the tape past the heads, and controls various modes of operation. Also known as tape drive.   { 'tāp ‚tranz‚pȯrt }

tapetum [BOT] A layer of nutritive cells surrounding the spore mother cells in the sporangium in higher plants; it is broken down to provide nourishment for developing spores. [NEUROSCI] **1.** A reflecting layer in the choroid coat behind the neural retina, chiefly in the eyes of nocturnal mammals. **2.** A tract of nerve fibers forming part of the roof of each lateral ventricle in the vertebrate brain.   { tə'pēd·əm }

tape unit [COMPUT SCI] A tape reading or writing device consisting of a tape transport, electronics, controls, and possibly

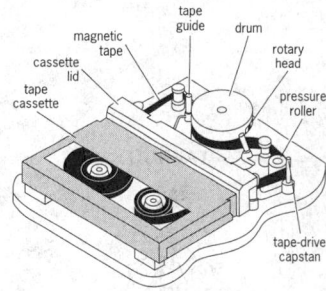

TAPE RECORDER

Rotary digital audio tape (R-DAT) recorder. (*After J. Watkinson, An Introduction to Digital Audio, Focal Press, Butterworth-Heinemann, Ltd., 1994*)

a cabinet; the cabinet may contain one or more magnetic tape stations. { 'tāp ˌyü·nət }

tape verifier [COMPUT SCI] A verifier for checking the accuracy of a punched paper tape by comparing it with a second manual punch of the same data; the machine stops whenever a character being punched the second time differs from that on the first tape. { tāp ˌver·ə‚fī·ər }

tapeworm [INV ZOO] Any member of the class Cestoidea; all are vertebrate endoparasites, characterized by a ribbonlike body divided into proglottids, and the anterior end modified into a holdfast organ. { 'tāp‚wərm }

tape-wound core [ELECTROMAG] A length of ferromagnetic material in tape form, wound in such a way that each turn falls directly over the preceding turn. { 'tāp‚waùnd 'kór }

taphocoenosis See thanatocoenosis. { ‚taf·ō·sē'nō·səs }

taphole [MET] A hole in a furnace or ladle through which molten metal is tapped. { 'tap‚hōl }

taphonomy [PALEON] The study of fossil preservation, including all events during the transition of organisms from the biosphere to the lithosphere. { tə'fän·ə·mē }

Taphrina caerulescens [MYCOL] A fungal pathogen that is the cause of leaf blister of oaks. { ta‚frī·nə ‚kī·rə'les·ənz }

Taphrina deformans [MYCOL] A fungal pathogen that is the cause of leaf curl of peach and almond trees. { ta‚frī·nə di'fòr·mənz }

taphrogenesis See taphrogeny. { ‚taf·rə'jen·ə·səs }

taphrogeny [GEOL] The formation of rift or trench phenomena, characterized by block faulting and associated subsidence. Also known as taphrogenesis. { tə'fräj·ə·nē }

taphrogeosyncline [GEOL] A geosyncline formed as a rift basin between faults. { ‚taf·rō‚jē·ō'sin‚klīn }

taping [ENG] The process of measuring distances with a surveyor's tape. { 'tāp·iŋ }

tapioca [FOOD ENG] A food, high in starch, that is made from the cassava plant. { ‚tap·ē'ō·kə }

tapioca snow See snow pellets. { ‚tap·ē'ō·kə 'snō }

tapiolite [MINERAL] Fe(Ta,Nb)$_2$O$_6$ A mineral that is isomorphous with mossite; occurs in pegmatites or detrital deposits; an ore of tantalum. { 'tap·ē·ə‚līt }

tapir [VERT ZOO] Any of several large odd-toed ungulates of the family Tapiridae that have a heavy, sparsely hairy body, stout legs, a prehensile muzzle, a short tail, and small eyes. { 'tā·pər }

Tapiridae [VERT ZOO] The tapirs, a family of perissodactyl mammals in the superfamily Tapiroidea. { tə'pir·ə‚dē }

Tapiroidea [VERT ZOO] A superfamily of the mammalian order Perissodactyla in the suborder Ceratomorpha. { ‚tap·ə'róid·ē·ə }

tapped control [ELECTR] A rheostat or potentiometer having one or more fixed taps along the resistance element, usually to provide a fixed grid bias or for automatic bass compensation. { 'tapt kən'trōl }

tapped-potentiometer function generator [ELECTR] A device used in analog computers for representing a function of one variable, consisting of a potentiometer with a number of taps held at voltages determined by a table of values of the variable; the input variable sets the angular position of a shaft that moves a slide contact, and the output voltage is taken from the slide contact. { 'tapt pə‚ten·chē‚äm·əd·ər 'fəŋk·shən jen·ə‚rād·ər }

tapped resistor [ELEC] A wire-wound fixed resistor having one or more additional terminals along its length, generally for voltage-divider applications. { 'tapt ri'zis·tər }

tappet [MECH ENG] A lever or oscillating member moved by a cam and intended to tap or touch another part, such as a push rod or valve system. { 'tap·ət }

tappet rod [MECH ENG] A rod carrying a tappet or tappets, as one for opening or closing the valves in a steam or an internal combustion engine. { 'tap·ət ‚räd }

tapping [MECH ENG] Forming an internal screw thread in a hole or other part by means of a tap. [MET] Opening the pouring hole of a melting furnace to remove molten metal. { 'tap·iŋ }

tapping screw See self-tapping screw. { 'tap·iŋ ‚skrü }

taproot [BOT] A root system in which the primary root forms a dominant central axis that penetrates vertically and rather deeply into the soil; it is generally larger in diameter than its branches. { 'tap‚rüt }

tap screw See tap bolt. { 'tap ‚skrü }

tap switch [ELEC] Multicontact switch used chiefly for connecting a load to any one of a number of taps on a resistor or coil. { 'tap ‚swich }

tap wrench [ENG] A tool used to clamp taps during tapping operations. { 'tap ‚rench }

tar [MATER] A viscous material composed of complex, high-molecular-weight compounds derived from the distillation of petroleum or the destructive distillation of wood or coal. { tär }

tar acid [MATER] A mixture of phenols (phenols, cresols, and xylenols) found in tars and tar distillates; toxic, combustible, and soluble in alcohol and coal-tar hydrocarbons; used as a wood preservative and an insecticide for farm animals and also to make disinfectants. { 'tär ‚as·əd }

taranakite [MINERAL] KAl$_3$(PO$_4$)$_3$(OH)·9H$_2$O A white, gray, or yellowish-white mineral consisting of a hydrated basic phosphate of potassium and aluminum. { ‚tar·ə'nä‚kīt }

tarantata [METEOROL] A strong breeze from the northwest in the Mediterranean region. { ‚tär·ən'täd·ə }

tarantula [INV ZOO] **1.** Any of various large hairy spiders of the araneid suborder Mygalomorphae. **2.** Any of the wolf spiders comprising the family Lycosidae. { tə'ran·chə·lə }

Tarantula See Loop Nebula. { tə'ran·chə·lə }

tarapacaite [MINERAL] K$_2$CrO$_4$ A bright canary yellow, orthorhombic mineral consisting of potassium chromate; occurs in tabular form. { ‚tär·ə·pə'kä‚īt }

tar base [CHEM] A basic nitrogen compound found in coal tar, for example, pyridine and quinoline. { 'tär ‚bās }

tarbuttite [MINERAL] Zn$_2$(PO$_4$)(OH) A triclinic mineral of varying color, consisting of basic zinc phosphate. { 'tär·bə‚tīt }

tar camphor See naphthalene. { 'tär ‚kam·fər }

Tardigrada [INV ZOO] A class of microscopic, bilaterally symmetrical invertebrates in the subphylum Malacopoda; the body consists of an anterior prostomium and five segments surrounded by a soft, nonchitinous cuticle, with four pairs of ventrolateral legs. { tär'dig·rə·də }

tar distillate [MATER] A petroleum product produced by a tar still to which is charged the tarlike bottoms from continuous crude stills, pressure stills, cracking coils, or other petroleum refinery equipment. { 'tär ‚dis·tə·lət }

tardive dyskinesia [MED] A movement disorder marked by involuntary twitching of the mouth, lips, tongue, arms, legs, or trunk; frequently associated with the use of neuroleptic drugs. { 'tär·div ‚dis·kə'nē·zhə }

tare [MECH] The weight of an empty vehicle or container; subtracted from gross weight to ascertain net weight. { ter }

tare effect [FL MECH] In wind tunnel testing, the forces and moments due to support assembly and mutual interference between support assembly and model. { 'ter i‚fekt }

target [ATOM PHYS] The atom or nucleus in an atomic or nuclear reaction which is initially stationary. [COMPUT SCI] An index card or test document used to assist, reference, or calibrate equipment. [ELECTR] **1.** In an x-ray tube, the anode or anticathode which emits x-rays when bombarded with electrons. **2.** In a television camera tube, the storage surface that is scanned by an electron beam to generate an output signal current corresponding to the charge-density pattern stored there. **3.** In a cathode-ray tuning indicator tube, one of the electrodes that is coated with a material that fluoresces under electron bombardment. [ENG] **1.** The sliding weight on a leveling rod used in surveying to enable the staffman to read the line of collimation. **2.** The point that a borehole or an exploratory work is intended to reach. **3.** In radar and sonar, any object capable of reflecting the transmitted beam. [ORD] **1.** A geographical area, complex, or installation planned for capture or destruction by military forces. **2.** A paper or pasteboard item of square or rectangular shape designed to be fired upon from a specified range during practice or while testing an automatic firearm such as an automatic rifle, machine gun, or submachine gun; it is used to establish a degree of accuracy; it usually consists of a series of geometric patterns of various shapes on a common background. [PHYS] An object or substance subjected to bombardment or irradiation by particles or electromagnetic radiation. { 'tär·gət }

target acquisition [AERO ENG] The process of optically, manually, mechanically, or electronically orienting a tracking system in direction and range to lock on a target. [ELECTR] **1.** The first appearance of a recognizable and useful echo signal

TAPIR

Brazilian tapir (Tapirus terrestris).

TAPROOT

Taproot system of a dandelion.

from a new target in radar and sonar. **2.** *See* acquire. { 'tär·gət ˌak·wə'zish·ən }

target acquisition radar [ENG] An antiaircraft artillery radar, normally of lesser range capabilities but of greater inherent accuracy than that of surveillance radar, whose normal function is to acquire aerial targets either by independent search or on direction of the surveillance radar, and to transfer these targets to tracking radars. { 'tär·gət ˌak·wə¦zish·ən 'rā,där }

target analysis [ORD] Examination of potential targets to determine their military importance, their relative priority for attack, and the capabilities of available weapons for such attack. { 'tär·gət əˌnal·ə·səs }

target angle [NAV] The relative bearing of one craft from another craft, measured clockwise through 360°. [ORD] The angle at the target subtended by the observing base line. { 'tär·gət ˌaŋ·gəl }

target approach point [AERO ENG] In air transport operations, a navigational checkpoint over which the final turn-in to the drop zone-landing zone is made. { 'tär·gət ə'prōch ˌpóint }

target array [ORD] A graphic representation of enemy forces, personnel, and facilities in a specific situation, accompanied by a target analysis. { 'tär·gət əˌrā }

target bearing [ORD] **1.** The true compass bearing of a target from a firing ship. **2.** The bearing of a target measured in the horizontal from the bow of one's own ship clockwise from 0 to 360°, or from the nose of one's own aircraft in hours of the clock. { 'tär·gət ˌber·iŋ }

target cell [PHYSIO] A cell that has receptors for the product of a signaling cell. { 'tär·gət ˌsel }

target central processing unit [COMPUT SCI] The type of central processing unit for which a language processor (assembler, compiler, or interpreter) generates machine language output. { 'tär·gət ¦sen·trəl 'prä,ses·iŋ ˌyü·nət }

target compound [ORG CHEM] In chemical synthesis, the molecule of interest. { 'tär·gət ˌkäm,paúnd }

target concentration [ORD] A grouping of geographically proximate targets. { 'tär·gət ˌkäns·ən,trā·shən }

target configuration [COMPUT SCI] The combination of input, output, and storage units and the amount of computer memory required to carry out an object program. { 'tär·gət kən,fig·yə,rā·shən }

target cross section *See* echo area. { 'tär·gət 'krós ˌsek·shən }

target-designating system [ELECTR] A system for designating to one instrument a target which has already been located by a second instrument; it employs electrical data transmitters and receivers which indicate on one instrument the pointing of another. { 'tär·gət ¦dez·ig,nād·iŋ ˌsis·təm }

target deviation [ORD] Distance from point of impact or point of burst to the target. { 'tär·gət ˌdē·vē'a·shən }

target discrimination [ELECTR] The ability of a detection or guidance system to distinguish a target from its background or to discriminate between two or more targets that are close together. { 'tär·gət di,skrim·ə,nā·shən }

target drone [AERO ENG] A pilotless aircraft controlled by radio from the ground or from a mother ship and used exclusively as a target for antiaircraft weapons. { 'tär·gət ˌdrōn }

target echo [ELECTROMAG] A radio signal reflected by an airborne or other target and received by the radar station which transmitted the original signal. { 'tär·gət ˌek·ō }

target glint *See* scintillation. { 'tär·gət ˌglint }

target identification [ORD] The act of determining the nature of a target, including whether it is a friend or foe. { 'tär·gət īˌden·tə·fə'kā·shən }

target indicating system [ORD] A system which indicates to the tracker of an antiaircraft automatic weapon the direction of approach of a suitable target, or the approach of a new target after engagement with one target has been broken off. { 'tär·gət 'in·də,kād·iŋ ˌsis·təm }

target information center [ORD] An intelligence center set up afloat or ashore for assembly, evaluation, interpretation, dissemination, and coordination of target information for supporting weapons, that is, artillery, naval gunfire, and air strike. { 'tär·gət ˌin·fər¦mā·shən ˌsen·tər }

target language [COMPUT SCI] The language into which a program (or text) is to be converted. { 'tär·gət ˌlaŋ·gwij }

target length [ORD] Length of a target as it appears to an observer or gunner at the moment the gun is fired. { 'tär·gət ˌleŋkth }

target noise [ELECTROMAG] Statistical variations in a radar echo signal due to the presence on the target of a number of reflecting elements randomly oriented in space; target noise can cause scintillation. { 'tär·gət ˌnóiz }

target offset [ORD] Horizontal angle at the target between a line from the target to the piece and a line from the target to the observation post. { 'tär·gət ¦óf,set }

target of opportunity [ORD] **1.** A target visible to a surface or air sensor or observer, which is within range of available weapons and against which fire has not been scheduled or requested. **2.** A nuclear target observed or detected after an operation begins that has not been previously considered, analyzed, or planned for a nuclear strike. { 'tär·gət əv ˌäp·ər'tü·nəd·ē }

target pack [COMPUT SCI] A disk pack that is used to maintain systems software and, in particular, to hold a copy of a system control program on which modifications are made and tested. { 'tär·gət ˌpak }

target pattern [AERO ENG] The flight path of aircraft during the attack phase. { 'tär·gət ˌpad·ərn }

target phase [COMPUT SCI] The stage of handling a computer program at which the object program is first carried out after it has been compiled. { 'tär·gət ˌfāz }

target program *See* object program. { 'tär·gət ¦prō,gram }

target range [ORD] Area equipped for practice in shooting at targets. { 'tär·gət ˌrānj }

target response [ORD] The effect on men, material, and equipment of blast, heat, light, and nuclear radiation resulting from the explosion of a nuclear weapon. { 'tär·gət ri,späns }

target routine *See* object program. { 'tär·gət rü,tēn }

target scintillation *See* scintillation. { 'tär·gət ˌsint·əl'ā·shən }

target seeker [ORD] **1.** A missile having a self-contained system that provides homing guidance to the target. Also known as homer. **2.** The device within such a missile that directs it to the target. { 'tär·gət ˌsēk·ər }

target selector [ORD] Component of both a target-designating system and a target-indicating system; it is an off-carriage observing instrument provided for the purpose of selecting an initial or new target, and it is electrically connected to the gun mount in such a manner as to slew the gun to the approximate azimuth and elevation of a selected target (when the selector is a component of a target-designating system), and to give the tracker an indication of the direction of approach of selected target (when the selector is a component of a target-indicating system). { 'tär·gət si,lek·tər }

target signal [ELECTROMAG] The radio energy returned to a radar by a target. Also known as echo signal; video signal. { 'tär·gət ˌsig·nəl }

target signature [ELECTR] Characteristic pattern of the target displayed by detection and classification equipment. { 'tär·gət ˌsig·nə·chər }

target spot [PL PATH] Any plant disease characterized by lesions in the form of concentric markings. { 'tär·gət ˌspät }

target spotter [ORD] Small, black metal disk attached to a target in practice shooting to show the shooter exactly where the bullet has hit. { 'tär·gət ˌspäd·ər }

target strength [ACOUS] A measure of the reflecting power of a sonar target, which is expressed in decibels by the equation $E + 2L - S$, where E is the echo level, L is the total transmission loss, and S is the source level. { 'tär·gət ˌstreŋkth }

target system [ORD] **1.** All the targets situated in a particular geographic area and functionally related. **2.** A group of targets which are so related that their destruction will produce some particular effect desired by the attacker. { 'tär·gət ˌsis·təm }

target timing [NAV] The timing of successive positions of a radar target, as plotted on a polar coordinate diagram, for the purpose of determining ground speed and track of a craft. { 'tär·gət ˌtīm·iŋ }

target-type flowmeter [ENG] A fluid-flow measurement device with a small circular target suspended centrally in the flow conduit; the target transmits force to a force-balance transmitter by means of a pivoted bar. { 'tär·gət ¦tīp 'flō,mēd·ər }

target volume [ELECTROMAG] The volume of that part of a precipitation-type radar target from which a target signal is received; if the precipitation completely fills the radar beam, the target volume is identical with the radar volume. { 'tär·gət ,väl·yəm }

target vulnerability [ORD] A factor considered in target selection that relates each potential target to a standard scale in terms of the degree to which it is considered vulnerable; each target is given a scale number. { 'tär·gət ,vəln·rə'bil·əd·ē }

tariff [COMMUN] The rate charged by a communications common carrier for the use of a specified service or facility. [IND ENG] A government-imposed duty on imported or exported goods. { 'tar·əf }

tarlatan [TEXT] A thin, open-mesh, transparent muslin used for stiffening or decorating. { 'tär·lət·ən }

tarn [GEOGR] A landlocked pool or small lake that may occur in a marsh or swamp, or that may occupy a basin amid mountain ranges. { tärn }

tarnish [MET] Discoloration of a metal surface due to the formation of a thin film of oxide, sulfide, or some other corrosion product. [MINERAL] The altered color and luster of a mineral surface; characteristic of copper-bearing minerals. { 'tär·nish }

tar paper [MATER] Heavy construction paper coated with and impregnated with tar. { 'tär ,pā·pər }

tarpaulin [MATER] A sheet of waterproof canvas or other material; used to cover and protect construction materials and equipment, athletic fields, vehicles, or other exposed objects. { 'tär·pə·lən }

TARPON

The tarpon (*Megalops atlantica*).

tarpon [VERT ZOO] *Megalops atlantica.* A herringlike fish of the family Elopidae weighing up to 300 pounds (136 kilograms) and reaching a length of 8 feet (2.4 meters); it has a single soft, rayed dorsal fin, strong jaws, a bony plate under the mouth, numerous small teeth, and coarse, bony flesh covered with large scales. { 'tär·pən }

tarragon [FOOD ENG] A herb prepared from the pungent leaves of the tarragon tree (*Artemisia dracunculus*). { 'tar·ə,gän }

tarragon oil *See* estragon oil. { 'tar·ə,gän ,óil }

tarring [ENG] The coating of piles for permanent underground work with prepared acid-free tar. { 'tär·iŋ }

tarsal gland [ANAT] Any of the sebaceous glands in the tarsal plates of the eyelids. Also known as Meibomian gland. { 'tär·səl ,gland }

tarsal tunnel syndrome [MED] A neurological foot disorder in which the posterior tibial nerve becomes compressed and damaged within the tarsal canal; symptoms include pain and burning that arises from behind the inside of the ankle and that may travel to the bottom of the foot. { ,tärs·əl 'tən·əl ,sin,drōm }

tar sand [GEOL] A type of oil sand; a sand whose interstices are filled with asphalt that remained after the escape of the lighter fractions of crude oil. { 'tär ,sand }

tar seep [GEOL] Natural tar that, because of its close proximity to the ground surface, seeps from cracks in the earth or from between rocks, often forming pits or pools. { 'tär ,sēp }

tarsia *See* intarsia. { 'tär·sē·ə }

tarsier [VERT ZOO] Any of several species of primates comprising the genus *Tarsius* of the family Tarsiidae characterized by a round skull, a flattened face, and large eyes that are separated from the temporal fossae in the orbital depression, and by adhesive pads on the expanded ends of the fingers and toes. { 'tär·sē,ā }

Tarsiidae [VERT ZOO] The tarsiers, a family of prosimian primates distinguished by incomplete postorbital closure and a greatly elongated ankle region. { tär'sī·ə,dē }

Tarsonemidae [INV ZOO] A small family of phytophagous mites in the suborder Trombidiformes. { ,tär·sə'nem·ə,dē }

tarsoplasty *See* blepharoplasty. { 'tär·sə,plas·tē }

tarsus [ANAT] **1.** The instep of the foot consisting of the calcaneus, talus, cuboid, navicular, medial, intermediate, and lateral cuneiform bones. **2.** The dense connective tissues supporting an eyelid. { 'tär·səs }

tartar *See* dental calculi. { 'tärd·ər }

tartar emetic [ORG CHEM] K(SbO)C$_4$H$_4$O$_6$·^{1}/$_2$H$_2$O A transparent crystalline compound, soluble in water; used to attract and kill moths, wasps, and yellow jackets. Also known as antimony potassium tartrate; potassium antimonyl tartrate. { 'tärd·ər i'med·ik }

tartaric acid [ORG CHEM] HOOC(CHOH)$_2$COOH Water- and alcohol-soluble colorless crystals with an acid taste, melts

TARSIER

The tarsier has adhesive pads on the expanded ends of the fingers and toes.

at 170°C; used as a chemical intermediate and a sequestrant and in tanning, effervescent beverages, baking power, ceramics, photography, textile processing, mirror silvering, and metal coloring. { tär'tar·ik 'as·əd }

tartary buckwheat [BOT] One of three buckwheat species grown commercially; the leaves are narrower than the other two species and arrow-shaped, and the flowers are smaller with inconspicuous greenish-white sepals. Also known as duck wheat; hulless buckwheat; rye buckwheat. { 'tärd·ə·rē 'bək,wēt }

tartrate [ORG CHEM] A salt or ester of tartaric acid, for example, sodium tartrate, Na$_2$C$_4$H$_4$O$_6$. { 'tär,trāt }

tartrazine [ORG CHEM] C$_{16}$H$_9$N$_4$O$_9$S$_2$ A bright orange-yellow, water-soluble powder, used as a food, drug, and cosmetic dye. { 'tär·trə,zēn }

task [COMPUT SCI] A set of instructions, data, and control information capable of being executed by the central processing unit of a digital computer in order to accomplish some purpose; in a multiprogramming environment, tasks compete with one another for control of the central processing unit, but in a nonmultiprogramming environment a task is simply the current work to be done. { task }

task analysis [IND ENG] A process for determining in detail the specific behaviors required of the personnel involved in a human-machine system. { 'task ə,nal·ə·səs }

task descriptor [COMPUT SCI] The vital information about a task in a multitask system which must be saved when the task is interrupted. Also known as state vector. { 'task di,skrip·tər }

task element [IND ENG] The smallest logically definable set of perceptions, decisions, and responses required of a human being in the performance of a task. { 'task ,el·ə·mənt }

task fleet [ORD] A mobile command consisting of ships and aircraft necessary for the accomplishment of a specific major task which may be of a continuing nature. { 'task ,flēt }

task management [COMPUT SCI] The functions, assumed by the operating system, of switching the processor among tasks, scheduling, sending messages or timing signals between tasks, and creating or removing tasks. { 'task ,man·ij·mənt }

task programmer [COMPUT SCI] A person who writes applications programs for controlling a robotic system. { 'task ,prō,gram·ər }

task switching [COMPUT SCI] Switching back and forth between two or more active programs without having to close or open any of them. Also known as context switching. { 'task ,swich·iŋ }

tasmanite [GEOL] An impure coal, transitional between cannel coal and oil shale. Also known as combustible shale; Mersey yellow coal; white coal; yellow coal. { 'taz·mə,nīt }

tassel [BOT] The male inflorescence of corn and certain other plants. { 'tas·əl }

T association [ASTRON] An association that includes many T Tauri stars. { 'tē ə,sō·sē,ā·shən }

taste [PHYSIO] A chemical sense by which flavors are perceived depending on taste, tactile, and warm and cold receptors in the mouth, as well as smell receptors in the nose. { tāst }

taste bud [ANAT] An end organ consisting of goblet-shaped clusters of elongate cells with microvilli on the distal end to mediate the sense of taste. { 'tāst ,bəd }

TAT *See* thematic apperception test.

TATA box [GEN] In eukaryotes, a short sequence of base pairs that is rich in adenine (A) and thymidine (T) residues and located about 25–30 nucleotides upstream of the transcriptional initiation site. { 'tä,tä *or* 'tē'ā',tē'ā ,bäks }

T attenuator [ELEC] **1.** A resistive attenuator with three resistors forming a T network. **2.** A power-tap type of attenuator which removes part of the power from a main line through a T connection and dissipates the power, without reflection into the main line. { 'tē ə,ten·yə,wād·ər }

tatting [TEXT] **1.** A knotted lace made by hand in various designs which are produced with the fingers and a shuttle. **2.** The process of making tatting lace by looping and knotting a single cotton thread with the fingers and a shuttle. { 'tad·iŋ }

Tauber test [ANALY CHEM] A color test for identification of pentose sugars; the sugars produce a cherry-red color when heated with a solution of benzidine in glacial acetic acid. { 'taú·bər ,test }

Taub NUT space [RELAT] An exact homogeneous, anisotropic solution of Einstein's equations of general relativity that contains many of the pathologies (such as closed timelike lines) possible in a space-time. { ¦taᵫb 'nət ˌspās }

tau meson [PART PHYS] Former name for the *K* meson, especially one which decays into three pions. { 'taᵫ 'mā̇ˌsän }

tauon *See* tau particle. { 'taᵫˌän }

tau particle [PART PHYS] A heavy, charged lepton with a mass of approximately 1777 megaelectronvolts, observed as a resonance in electron-positron collisions. Also known as tauon. { 'taᵫ ˌpärd·ə·kəl }

tau phenomenon *See* Geld-Benussi phenomenon. { 'taᵫ fə'näm·əˌnän }

tau protein [NEUROSCI] A protein found in the axons of healthy neurons, where it binds to other proteins called microtubules to form the cytoskeleton of the neuron and provide the tracks over which material can be transported from one part of the neuron to another. { 'taᵫ ˌprō̇ˌtēn }

Taurid meteor [ASTRON] A meteor shower occurring from about October 26 to November 16 in the Northern Hemisphere, with the maximum occurring about the first week in November; the radiant lies in the constellation Taurus. { 'tȯr·əd 'mēd·ē·ər }

taurine [ORG CHEM] $NH_2CH_2CH_2SO_3H$ A crystalline compound that decomposes at about 300°C; present in bile combined with cholic acid. { 'tȯˌrēn }

taurocholic acid [BIOCHEM] $C_{26}H_{45}NO_7S$ A common bile acid with a five-carbon chain; it is the product of the conjugation of taurine with cholic acid; crystallizes from an alcohol ether solution, and decomposes at about 125°C. Also known as cholaic acid; cholytaurine. { ˌtȯr·ə'kȯl·ik 'as·əd }

taurodont [ANAT] Of teeth, having a large pulp cavity and reduced roots. { 'tȯr·əˌdänt }

Taurus [ASTRON] A northern constellation; right ascension 4 hours, declination 15° north; it includes the star Aldebaran, useful in navigation. Also known as Bull. { 'tȯr·əs }

Taurus A [ASTRON] A strong, discrete radio source in the constellation Taurus, associated with the Crab Nebula. { 'tȯr·əs 'ā }

Taurus cluster [ASTRON] A cluster of stars observed in the region of the constellation Taurus; it is about 130 light-years (1.23×10^{18} meters) from the sun, and 58 light-years (5.49×10^{17} meters) in diameter. { 'tȯr·əs 'kləs·tər }

Taurus dark cloud [ASTRON] A large, relative nearby aggregate of dust and gas in which star formation is taking place. { 'tȯr·əs 'därk 'klaᵫd }

taut-band ammeter [ENG] A modification of the permanent-magnet movable-coil ammeter in which the jeweled bearings and control springs are replaced by a taut metallic band rigidly held at the ends; the coil is firmly attached to the band, and restoring torque is supplied by twisting of the band. { 'tȯt ¦band 'amˌēd·ər }

taut-line cableway [MECH ENG] A cableway whose operation is limited to the distance between two towers, usually 3000 feet (914 meters) apart, has only one carrier, and the traction cable is reeved at the carrier so that loads can be raised and lowered; the towers can be mounted on trucks or crawlers, and the machine shifted across a wide area. { 'tȯt ¦līn 'kā·bəlˌwā }

tautomerism [CHEM] The reversible interconversion of structural isomers of organic chemical compounds; such interconversions usually involve transfer of a proton. { tȯ'täm·əˌriz·əm }

tau-value [METEOROL] The time rate of change of *D* value at a fixed point defined by the relation $\tau = (\Delta_t D)/(\Delta_t t)$, where Δt is the change in time and $\Delta_t D$ is the change in *D* value during this time interval; tau-values are expressed in terms of feet per hour; tau-value lines are drawn on 4-D charts and constitute the time dimension of these charts. { 'taᵫ ˌval·yü }

tavistockite [MINERAL] $Ca_3Al_2(PO_4)_2(OH)_6$ A white, orthorhombic mineral consisting of a basic phosphate of calcium and aluminum. { 'tav·əˌstäˌkīt }

tawing [ENG] A tanning process in which alum is used as a partial tannage, supplementing or replacing chrome. { 'tȯ·iŋ }

Taxales [BOT] A small order of gymnosperms in the class Pinatae; members are trees or shrubs with evergreen, often needlelike leaves, with a well-developed fleshy covering surrounding the individual seeds, which are terminal or subterminal on short shoots. { 'sä·lēz }

taxi channel [CIV ENG] A defined path, on a water airport, intended for the use of taxiing aircraft. { 'tak·sē ˌchan·əl }

taxi-channel lights [NAV] Aeronautical ground lights arranged along a taxi channel to indicate the route to be followed by taxiing aircraft. { 'tak·sē ¦chan·əl ˌlīts }

taxis [PHYSIO] A mechanism of orientation by means of which an animal moves in a direction related to a source of stimulation. { 'tak·səs }

taxiway [CIV ENG] A specially prepared or designated path on an airport for taxiing aircraft. { 'tak·sēˌwā }

Taxocrinida [PALEON] An order of flexible crinoids distributed from Ordovician to Mississippian. { ˌtak·sə'krī·nəd·ə }

Taxodonta [INV ZOO] A subclass of pelecypod mollusks in which the hinge is of the taxodont type, that is, the dentition is a series of similar alternating teeth and sockets along the hinge margin. { ˌtak·sə'dänt·ə }

taxol [PHARM] $C_{47}H_{51}NO_{14}$ An alkaloid derived from the Pacific yew tree (*Taxus brevifolia*) that is used in the treatment of ovarian and breast cancer. { 'takˌsȯl }

taxon [SYST] A taxonomic group or entity. { 'takˌsän }

taxonomic category [SYST] One of a hierarchy of levels in the biological classification of organisms; the seven major categories are kingdom, phylum, class, order, family, genus, species. { ˌtak·sə¦näm·ik 'kad·əˌgȯr·ē }

taxonomy [SYST] A study aimed at producing a hierarchical system of classification of organisms which best reflects the totality of similarities and differences. { tak'sän·ə·mē }

Taxus brevifolia [BOT] The Pacific yew tree, from which the anticancer compound taxol is derived. { ˌtak·səs ˌbrev·i'fō·lē·ə }

Tayassuidae [VERT ZOO] The peccaries, a family of artiodactyl mammals in the superfamily Suoidea. { ˌtā·yə'sü·əˌdē }

Taylor connection [ELEC] A transformer connection for converting three-phase power to two-phase power, or vice versa. { 'tā·lər kəˌnek·shən }

Taylor effect [FL MECH] A phenomenon in which the relative motion of a homogeneous rotating liquid tends to be the same in all planes perpendicular to the axis of rotation. { 'tā·lər iˌfekt }

Taylor instability [FL MECH] An instability in Couette flow between rotating cylinders that arises at a critical angular velocity, in which periodic eddies along the cylinder axis form a regular pattern. { 'tā·lər ˌin·stə'bil·əd·ē }

taylorite *See* bentonite. { 'tā·ləˌrīt }

Taylor number [FL MECH] A nondimensional number arising in problems of a rotating viscous fluid, written as $T = (f^2 l^4)/\nu^2$, where *f* is the Coriolis parameter (or, for a cylindrical system, twice the rate of rotation of the system), *h* the depth of the fluid, and *ν* the kinematic viscosity; the square root of the Taylor number is a rotating Reynolds number, and the fourth root is proportional to the ratio of the depth *h* to the depth of the Ekman layer. { 'tā·lər ˌnəm·bər }

Taylor-Orowan dislocation *See* edge dislocation. { 'tā·lər ȯ'rō·wən ˌdis·lōˌkā·shən }

Taylor process [MET] A process for making extremely fine wire by stretching wire in a glass tube at elevated temperatures, or drawing wire through a bead of molten glass and then through dies. { 'tā·lər ˌprä·səs }

Taylor series [MATH] The Taylor series corresponding to a function $f(x)$ at a point x_0 is the infinite series whose *n*th term is $(1/n!) \cdot f^{(n)}(x_0)(x - x_0)^r$, where $f^{(n)}(x)$ denotes the *n*th derivative of $f(x)$. [NAV ARCH] Resistance charts based upon model tests of a series of ships derived by altering the proportions of a single parent form; used to study the effects of these alterations on resistance to the ship's motion, and to predict the powering requirements for new ships. { 'tā·lər ˌsir·ēz }

Taylor's theorem [MATH] The theorem that under certain conditions a real or complex function can be represented, in a neighborhood of a point where it is infinitely differentiable, as a power series whose coefficients involve the various order derivatives evaluated at that point. { 'tā·lərz ˌthir·əm }

Tay-Sachs disease [MED] A form of sphingolipidosis, transmitted as an autosomal recessive, in which there is an accumulation in neuronal cells of the neuraminic fraction of gangliosides; manifested clinically within the first few months of life by hypotonia progressing to spasticity, convulsions, and visual loss accompanied by the appearance of a cherry-red spot at the macula lutea. Also known as infantile amaurotic familial idiocy. { ¦tā ¦saks dˌzēz }

Tb *See* terbium.

TBE *See* binding energy.

TAURINE

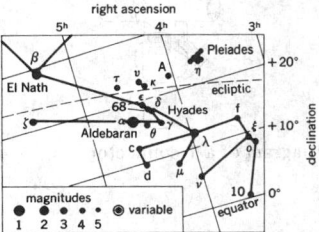

Structural formula of taurine.

TAURUS

Line pattern of the constellation Taurus. The grid lines represent the coordinates of the sky. The apparent brightness, or magnitude, of the stars is shown by the sizes of the dots, which are graded by appropriate numbers as indicated.

T beam [CIV ENG] A metal beam or bar with a T-shaped cross section. { 'tē ,bēm }

TBH *See* 1,2,3,4,5,6-hexachlorocyclohexane.

T bolt [DES ENG] A bolt with a T-shaped head, made to fit into a T-shaped slot in a drill swivel head or in the bed of a machine. { 'tē ,bōlt }

TBP *See* tributyl phosphate.

tbsp *See* tablespoonful.

TBSV *See* tomato bushy stunt virus.

TBT *See* tetrabutyl titanate.

Tc *See* technetium.

TCA *See* trichloroacetic acid.

T cell [IMMUNOL] One of a heterogeneous population of thymus-derived lymphocytes which participates in the immune responses. Also known as T lymphocyte. { 'tē ,sel }

T-cell receptor [IMMUNOL] Protein on the surface of T lymphocytes that specifically recognizes molecules of the major histocompatibility complex, either alone or in association with foreign antigens. Abbreviated TCR. { 'tē ,sel ri,sep·tər }

Tchebycheff *See* Chebyshev. { chəb·ə'shȯf }

Tchernozem *See* Chernozem. { 'chər·nə,zem }

Tchuprow-Neymann allocation [STAT] A technique of stratified sampling in which the size of each strata sample is proportional to the size of the strata population and the variance of the strata. { 'chü,prəf 'nā·mən ,al·ə,kā·shən }

T circulator [ELECTROMAG] A circulator in which three identical rectangular waveguides are joined asymmetrically to form a T-shaped structure, with a ferrite post or wedge at its center; power entering any waveguide emerges from only one adjacent waveguide. { 'tē ,sər·kyə,lād·ər }

T connector [ELEC] A type of electric connector that joins a through conductor to another conductor at right angles to it. { 'tē kə,nek·tər }

TCP *See* Transmission Control Protocol; tricresyl phosphate.

TCP/IP *See* Transmission Control Protocol/Internet Protocol.

TCR *See* T-cell receptor.

TD *See* transmitter-distributor.

TDE *See* 2,2-bis(*para*-chlorophenyl)-1,1-dichloroethane.

t distribution [STAT] A distribution used to test a hypothesis about a population mean when the population standard deviation is not known, the sample size is small, and the normal distribution is assumed for the sample mean. { 'tē ,dis·trə,byü·shən }

TDM *See* time-division multiplexing.

TDMA *See* time-division multiple access.

TDR *See* time-domain reflectometer.

TDRSS *See* Tracking and Data Relay Satellite System.

TDZ *See* touch-down zone.

Te *See* tellurium.

tea [BOT] *Thea sinensis.* A small tree of the family Theaceae having lanceolate leaves and fragrant white flowers; a caffeine beverage is made from the leaves of the plant. { tē }

TEA *See* transferred-electron amplifier.

teach [CONT SYS] To program a robot by guiding it through its motions, which are then recorded and stored in its computer. { tēch }

teach box *See* teach pendant. { 'tēch ,bäks }

teach-by-doing [CONT SYS] A method of programming a robot in which the operator guides the robot through its intended motions by holding it and performing the work. { ¦tēch ·bī 'dü·iŋ }

teach-by-driving [CONT SYS] Programming a robot by using a teach pendant. { ¦tēch ·bī 'drīv·iŋ }

teach gun *See* teach pendant. { 'tēch ,gən }

teaching interface [CONT SYS] The devices and hardware that are used to instruct robots and other machinery how to operate, and to specify their motions. { 'tēch·iŋ 'in·tər,fās }

teaching reactor *See* research reactor; subcritical reactor. { 'tēch·iŋ rē,ak·tər }

TEA chloride *See* tetraethylammonium chloride. { ,tē,ē'ā 'klȯr,īd }

teach mode [CONT SYS] The mode of operation in which a robot is instructed in its motions, usually by guiding it through these motions using a teach pendant. { 'tēch ,mōd }

teach pendant [CONT SYS] A hand-held device used to instruct a robot, specifying the character and types of motions it is to undertake. Also known as teach box; teach gun. { 'tēch ,pen·dənt }

teakwood [MATER] The strong, durable, yellowish-brown wood that is obtained from the teak tree, *Tectona grandis*. { 'tēk,wu̇d }

TEA laser [OPTICS] A gas laser in which a glow discharge is maintained without arc formation at atmospheric pressure (which is relatively high for a gas laser) by using a discharge which is transverse rather than parallel to the optic axis. Derived from transversely excited atmospheric pressure laser. { ,tē,ē'ā 'lā·zər }

teallite [MINERAL] PbSnS$_2$ A grayish-black, orthorhombic mineral consisting of lead tin sulfide. { 'tē,līt }

tear down [ENG] **1.** To disassemble a drilling rig preparatory to moving it to another drill site. **2.** To disassemble a machine or change the jigs and fixtures. { 'ter 'dau̇n }

tear-down time [IND ENG] The downtime of a machine following a given work order which usually involves removing parts such as jigs and fixtures and which must be completely finished before setting up for the next order. { 'ter¦dau̇n ,tīm }

teardrop balloon [AERO ENG] A sounding balloon which, when operationally inflated, resembles an inverted teardrop; this shape was determined primarily by aerodynamic considerations of the problem of obtaining maximum stable rates of a balloon ascension. { 'tir,dräp bə'lün }

tear fault [GEOL] A very steep to vertical fault associated with and perpendicular to the strike of an overthrust fault. { 'tar ,fȯlt }

tear gas [MATER] A substance (usually liquid) which, when atomized and of a certain concentration, causes temporary but intense eye irritation and a blinding flow of tears; chloroacetophenone is a common tear gas. Also known as lacrimator. { 'tir ,gas }

tear gland *See* lacrimal gland. { 'tir ,gland }

tears [COMMUN] In a television picture, a horizontal disturbance caused by noise, in which the picture appears to be torn apart. { tirz }

tear strength [MECH] The force needed to initiate or to continue tearing a sheet or fabric. { 'ter ,streŋkth }

teaser transformer [ELEC] Transformer, of two T-connected, single-phase units for three-phase to two-phase or two-phase to three-phase operation, which is connected between the midpoint of the main transformer and the third wire of the three-phase system. { 'tēz·ər tranz,fȯr·mər }

teaspoonful [MECH] A unit of volume used particularly in cookery and pharmacy, equal to 1$\frac{1}{3}$ fluid drams, or $\frac{1}{3}$ tablespoonful; in the United States this is equal to approximately 4.9289 cubic centimeters, in the United Kingdom to approximately 4.7355 cubic centimeters. Abbreviated tsp; tspn. { 'tē,spün,fu̇l }

technetium [CHEM] A transition element, symbol Tc, atomic number 43; derived from uranium and plutonium fission products; chemically similar to rhenium and manganese; isotope ^{99}Tc has a half-life of 200,000 years; used to absorb slow neutrons in reactor technology. [MET] Silver-gray metal with a high melting point, slightly magnetic. { tek'nē·shē·əm }

technetron [ELECTR] High-power multichannel field-effect transistor. { 'tek·nə,trän }

technical atmosphere [MECH] A unit of pressure in the metric technical system equal to one kilogram-force per square centimeter. Abbreviated at. { 'tek·nə·kəl 'at·mə,sfir }

technical characteristics [ENG] Those characteristics of equipment which pertain primarily to the engineering principles involved in producing equipment possessing desired characteristics, for example, for electronic equipment; technical characteristics include such items as circuitry, and types and arrangement of components. { 'tek·nə·kəl ,kar·ik·tə'ris·tiks }

technical cohesive strength [MET] Fracture stress in a notched tensile test. { 'tek·nə·kəl kō'he·siv 'streŋkth }

technical control board [ELEC] Testing position in a switch center or relay station with provisions for testing switches and associated access lines and trunks. { 'tek·nə·kəl kən'trōl ,bȯrd }

technical evaluation [ENG] The study and investigation to determine the technical suitability of material, equipment, or a system. { 'tek·nə·kəl i,val·yə'wā·shən }

technical information [ENG] Information, including scientific information, which relates to research, development, engineering, testing, evaluation, production, operation, use, and maintenance of equipment. { 'tek·nə·kəl ,in·fər'mā·shən }

technical inspection [ENG] Inspection of equipment to

T CONNECTOR

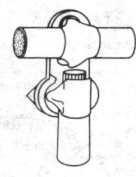

Diagram of a T connector.

determine whether it is serviceable for continued use or needs repairs. { 'tek·nə·kəl in'spek·shən }

technical intelligence [ORD] Intelligence pertaining to foreign or enemy technological developments capable of, or having, a practical application to warfare. { 'tek·nə·kəl in'tel·ə·jəns }

technical load [ELEC] Portion of a communications-electronics facility operational power load required for primary and ancillary equipment, including necessary lighting and air conditioning or ventilation required for full continuity of operation. { 'tek·nə·kəl 'lōd }

technical maintenance [ENG] A category of maintenance that includes the replacement of unserviceable major parts, assemblies, or subassemblies, and the precision adjustment, testing, and alignment of internal components. { 'tek·nə·kəl 'mānt·ən·əns }

technical manual [ENG] A publication containing detailed information on technical procedures, including instructions on the operation, handling, maintenance, and repair of equipment. { 'tek·nə·kəl ‚man·yə·wəl }

technical representative [IND ENG] A person who represents one or more manufacturers in an area and who gives technical advice on the application, installation, operation, and maintenance of their products, in addition to selling the products. { 'tek·nə·kəl ‚rep·ri¦zent·əd·iv }

technical specifications [ENG] A detailed description of technical requirements stated in terms suitable to form the basis for the actual design, development, and production processes of an item having the qualities specified in the operational characteristics. { 'tek·nə·kəl ‚spes·ə·fə'kā·shənz }

technical white oil See white oil. { 'tek·nə·kəl ‚wīt ‚öil }

technicolor [PART PHYS] A hypothetical fifth force proposed to explain the breakdown of symmetry in the unified electroweak interaction. { 'tek·nə‚kəl·ər }

technihadron [PART PHYS] Any of a class of hypothetical particles that would be bound together by technicolor forces in the same manner that the ordinary color force binds quarks, antiquarks, and gluons into hadrons. { ‚tek·ni'had‚rän }

technipion [PART PHYS] A technihadron that would be anomalously light compared to the characteristic technicolor energy scale, analogous to pions which are light compared to the strong interaction scale. { ‚tek·ni'pī‚än }

technologist [SCI TECH] A specialist who carries out a technique for the purpose of accomplishing a specified function, and extends a knowledge and skill of the technique for use on various products in different ways. { tek'näl·ə·jəst }

technology [SCI TECH] Systematic knowledge of and its application to industrial processes; closely related to engineering and science. { tek'näl·ə·jē }

technosphere [ECOL] The part of the physical environment affected through building or modification by humans. { 'tek·nə‚sfir }

Tectibranchia [INV ZOO] An order of mollusks in the subclass Opisthobranchia containing the sea hares and the bubble shells; the shell may be present, rudimentary, or absent. { ‚tek·tə'braŋ·kē·ə }

tectite See tektite. { 'tek‚tīt }

Tectiviridae [VIROL] A family of nontailed bacterial viruses (bacteriophages) characterized by a nonenveloped icosahedral particle containing a linear double-stranded deoxyribonucleic acid genome, includes the PRDI phage. { ‚tek·t·ə'vir·ə‚dī }

tectofacies [GEOL] A lithofacies that is interpreted tectonically. { ‚tek·tə'fā·shēz }

tectogene [GEOL] A long, relatively narrow downward fold of sialic crust considered to be an early phase in mountain-building processes. Also known as geotectogene. { 'tek·tə‚jēn }

tectogenesis See orogeny. { ‚tek·tə'jen·ə·səs }

tectonic analysis See petrotectonics. { tek'tän·ik ə'nal·ə·səs }

tectonic breccia [PETR] A breccia developed from brittle rocks, formed as a result of crustal movements and produced by lateral or vertical pressure. Also known as dynamic breccia; pressure breccia. { tek'tän·ik 'brech·ə }

tectonic conglomerate See crush conglomerate. { tek'tän·ik kən'gläm·ə·rət }

tectonic cycle [GEOL] The orogenic cycle which relates larger crustal features, such as mountain belts, to a series of stages of development. Also known as geosynclinal cycle. { tek'tän·ik 'sī·kəl }

tectonic framework [GEOL] The relationship in space and time of subsiding, stable, and rising tectonic elements in a sedimentary source area. { tek'tän·ik 'frām‚wərk }

tectonic land [GEOL] Linear fold ridges and volcanic islands which existed for a short time in the interior sections of an orogenic belt during the geosynclinal phase. { tek'tän·ik 'land }

tectonic lens [GEOL] An elongate, sausage-shaped body of rock formed by distortion of a continuous incompetent layer enclosed between competent layers, similar to a boudin, but genetically distinct. { tek'tän·ik 'lenz }

tectonic map [GEOL] A map which shows the architecture of the upper portion of the earth's crust. { tek'tän·ik 'map }

tectonic moraine [GEOL] An aggregation of boulders incorporated in the base of an overthrust mass. { tek'tän·ik mə'rān }

tectonic patterns [GEOL] The arrangement of the large structural units of the earth's crust, such as mountain systems, shields or stable areas, basins, arches, and volcanic archipelagoes. { tek'tän·ik 'pad·ərnz }

tectonic plate [GEOL] Any one of the internally rigid crustal blocks of the lithosphere which move horizontally across the earth's surface relative to one another. Also known as crustal plate. { tek'tän·ik 'plāt }

tectonic rotation [GEOL] Internal rotation of a tectonite in the direction of transport. { tek'tän·ik rō'tā·shən }

tectonics [CIV ENG] **1.** The science and art of construction with regard to use and design. **2.** Design relating to crustal deformations of the earth. [GEOL] A branch of geology that deals with regional structural and deformational features of the earth's crust, including the mutual relations, origin, and historical evolution of the features. Also known as geotectonics. { tek'tän·iks }

tectonite [PETR] A rock in which the history of its deformation is reflected in its fabric. { 'tek·tə‚nīt }

tectonoeustatism [OCEANOGR] Fluctuations of sea level due to changes in the capacities of the ocean basins resulting from earth movements. { ‚tek·tə·nō'yü·stə‚tiz·əm }

tectonomagnetism [GEOPHYS] Study of magnetic anomalies due to tectonic stress. { ‚tek·tə·nō'mag·nə‚tiz·əm }

tectonometer [ENG] An apparatus, including a microammeter, used on the surface to obtain knowledge of the structure of the underlying rocks. { ‚tek·tə'näm·əd·ər }

tectonophysicist [GEOPHYS] One who studies elastic deformation of flow and rupture of constituent materials of the earth's crust and makes deductions concerning the forces that cause these deformations. { ‚tek·tə·nō'fiz·ə‚sist }

tectonophysics [GEOPHYS] A branch of geophysics dealing with the physical processes involved in forming geological structures. { ‚tek·tə·nō'fiz·iks }

tectoquinone [ORG CHEM] $C_{15}H_{10}O_2$ A white compound with needlelike crystals; sublimes at 177°C; insoluble in water; used as an insecticide to treat wood. Also known as 2-methyl anthraquinone. { ‚tek·tō·kwə'nōn }

tectorial membrane [ANAT] **1.** A jellylike membrane covering the organ of Corti in the ear. **2.** A strong sheet of connective tissue running from the basilar part of the occipital bone to the dorsal surface of the bodies of the axis and third cervical vertebra. { tek'tor·ē·əl 'mem‚brān }

tectosilicate [MINERAL] A structural type of silicate in which all four oxygen atoms of the silicate tetrahedra are shared with neighboring tetrahedra; tectosilicates include quartz, the feldspars, the feldspathoids, and zeolites. Also known as framework silicate. { ‚tek·tō'sil·ə‚kāt }

tectosome [GEOL] A body of strata representing a tectotope. { 'tek·tə‚sōm }

tectosphere [GEOL] The region of the earth's crust occupied by the tectonic plates. { 'tek·tə‚sfir }

tectum [ANAT] A rooflike structure of the body, especially the roof of the midbrain including the corpora quadrigemina. { 'tek·təm }

tee [ENG] Shaped like the letter T. { tē }

tee joint [ENG] A joint in which members meet at right angles, forming a T. { 'tē ‚joint }

teel oil See sesame oil. { 'tēl ‚öil }

teeming [MET] Pouring molten metal, usually a ferrous metal, into an ingot mold from a furnace or ladle. { 'tēm·iŋ }

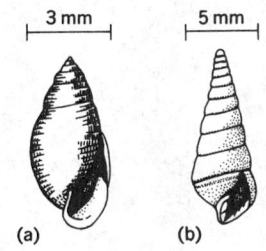

TECTIBRANCHIA

3 mm   5 mm

(a)   (b)

Shells of two genera of Tectibranchia. *(a) Acteon. (b) Pyramidella. (From A. M. Keen and J. C. Pearson, Illustrated Key to West North American Gastropod Genera, Stanford University Press, 1958)*

teepleite [MINERAL] $Na_2BO_2Cl \cdot 2H_2O$ A mineral composed of hydrous chloride and borate of sodium. { 'tē·pə‚līt }

teeth of the gale [METEOROL] An old nautical term for the direction from which the wind is blowing (upwind, windward); to sail into the teeth of the gale is to sail to windward. { 'tēth əv thə 'gāl }

Teflon shakes See metal fume fever. { 'tef‚län 'shāks }

TEG See tetraethylene glycol; triethylene glycol.

TEGFET See high-electron-mobility transistor. { 'teg‚fet }

tegmen [BIOL] An integument or covering. [BOT] The inner layer of a seed coat. [INV ZOO] A thickened forewing of Orthoptera, Coleoptera, and certain other insects. { 'teg‚mən }

tegmentum [ANAT] A mass of white fibers with gray matter in the cerebral peduncles of higher vertebrates. [BOT] The outer layer, or scales, of a leaf bud. [INV ZOO] The upper layer of a shell plate in Amphineura. { teg'men·təm }

tehuantepecer [METEOROL] A violent squally wind from north or north-northeast in the Gulf of Tehuantepec in winter; it originates in the Gulf of Mexico, as a norther which crosses the isthmus and blows through the gap between the Mexican and Guatemalan mountains. { tə'wän·tə‚pek·ər }

teichoic acid [BIOCHEM] A polymer of ribitol or glycerol phosphate with additional compounds such as glucose linked to the backbone of the polymer; found in the cell walls of some bacteria. { tā'kō·ik 'as·əd }

Teiidae [VERT ZOO] The tegus lizards, a diverse family of the suborder Sauria that is especially abundant and widespread in South America. { 'tē·ə‚dē }

T-E index See temperature-efficiency index. { ¦tē'ē ‚in‚deks }

teineite [MINERAL] $CuTeO_3 \cdot 2H_2$ A greenish to yellowish, probably triclinic mineral consisting of a hydrated sulfate-tellurate of copper; occurs as crystals. { 'tā‚nīt }

tektite [GEOL] A collective term applied to certain objects of natural glass of debatable origin that are widely strewn over the land and in sediments under the oceans; composition and size vary, and overall shapes resemble splash forms; most tektites are believed to be of extraterrestrial origin. Also known as obsidianite; tectite. { 'tek‚tīt }

TEL See tetraethyllead.

telain See provitrain. { 'te‚lān }

telamon [INV ZOO] A curved chitinous outgrowth of the cloacal wall in various male nematodes. { 'tel·ə‚män }

telangiectasis [MED] Localized dilation of capillaries forming dark-red, wartlike elevations varying in size from about 1 to 7 millimeters. { tə¦lan·jē'ek·tə·səs }

telautograph [COMMUN] A writing telegraph instrument in which manual movement of a pen at the transmitting position varies the current in two circuits in such a way as to cause corresponding movements of a pen at the remote receiving instrument; ordinary handwriting can thus be transmitted over wires. { te'lód·ə‚graf }

tele- [SCI TECH] Prefix meaning from a distance. { 'tel·ə }

telecast [COMMUN] A television broadcast intended for reception by the general public, involving the transmission of the picture and sound portions of the program. { 'tel·ə‚kast }

telecentric system [OPTICS] A telescopic system whose aperture stop is located at one of the foci of the objective lens. { ¦tel·ə¦sen·trik 'sis·təm }

teleceptor [PHYSIO] A sense receptor which transmits information about portions of the external environment which are not necessarily in direct contact with the organism, such as the receptors of the ear, eye, and nose. { 'tel·ə‚sep·tər }

telechir [CONT SYS] A handlike remote manipulator. { 'tel·ə‚kir }

telechirics [CONT SYS] The use of teleoperators or remote manipulators. { ¦tel·ə¦kir·iks }

telecine camera [ELECTR] A television camera used in conjunction with film or slide projectors to televise motion pictures and still images. { ¦tel·ə¦sin·ē 'kam·rə }

telecommunications [COMMUN] Communication over long distances. { ¦tel·ə·kə‚myü·nə'kā·shənz }

telecommunications coordinating committee [COMMUN] Committee organized by the U.S. State Department and composed of major government departments, agencies, and industrial organizations; makes recommendations on telecommunications matters affecting international telecommunications. { ¦tel·ə·kə‚myü·nə'kā·shənz kō'órd·ən‚ād·iŋ kə‚mid·ē }

teleconference [COMMUN] **1.** A two-way interactive meeting between relatively small groups of people remote from one another but linked by telecommunication facilities involving audio communication, and possibly also video, graphics, or facsimile. **2.** More broadly, any of various facilities allowing people to communicate among each other over some distance, encompassing teleseminars and telemeetings. { ¦tel·ə'kän‚frəns }

Telegeusidae [INV ZOO] The long-lipped beetles, a small family of colepteran insects in the superfamily Cantharoidea confined to the western United States. { ‚tel·ə'gyüs·ə‚dē }

telegraph alphabet See telegraph code. { 'tel·ə‚graf 'al·fə‚bet }

telegraph bandwidth [COMMUN] The difference between the limiting frequencies of a channel used to transmit telegraph signals. { 'tel·ə‚graf 'band‚width }

telegraph buoy [ENG] A buoy used to mark the position of a submarine telegraph cable. { 'tel·ə‚graf ‚bói }

telegraph cable [ELEC] A uniform conductive circuit consisting of twisted pairs of insulated wires or coaxially shielded wires or combinations of each, used to carry telegraph signals. { 'tel·ə‚graf ‚kā·bəl }

telegraph carrier [COMMUN] The single-frequency wave which is modulated by transmitting apparatus in carrier telegraphy. { 'tel·ə‚graf ‚kar·ē·ər }

telegraph circuit [COMMUN] The complete wire or radio circuit over which signal currents flow between transmitting and receiving apparatus in a telegraph system. { 'tel·ə‚graf ‚sər·kət }

telegraph code [COMMUN] A system of symbols for transmitting telegraph messages in which each letter or other character is represented by a set of long and short electrical pulses, or by pulses of opposite polarity, or by time intervals of equal length in which a signal is present or absent. Also known as telegraph alphabet. { 'tel·ə‚graf ‚kōd }

telegraph concentrator [ELEC] Switching arrangement by means of which a number of branch or subscriber lines or station sets may be connected to a lesser number of trunklines, operating positions, or instruments through the medium of manual or automatic switching devices to obtain more efficient use of facilities. { 'tel·ə‚graf 'käns·ən‚träd·ər }

telegraph distributor [ELEC] Device which effectively associates one direct-current or carrier-telegraph channel in rapid succession with the elements of one or more sending or receiving devices. { 'tel·ə‚graf di‚strib·yəd·ər }

telegraph emission [COMMUN] The signal transmitted by a telegraph system, classified by type of transmission, type of modulation, bandwidth, and supplementary characteristics. { 'tel·ə‚graf i‚mish·ən }

telegrapher's equation [MATH] The partial differential equation $(\partial^2 f/\partial x^2) = a^2(\partial^2 f/\partial y^2) + b(\partial f/\partial y) + cf$, where a, b, and c are constants; appears in the study of atomic phenomena. { tə'leg·rə·fərz i‚kwā·zhən }

telegraph grade [COMMUN] The class of communication circuits that can transmit only telegraphic signals, comprising the lowest types of circuits in regard to speed, accuracy, and cost. { 'tel·ə‚graf ‚grād }

telegraph interference [COMMUN] Any undesired electrical energy that tends to interfere with the reception of telegraph signals. { 'tel·ə‚graf ‚in·tər'fir·əns }

telegraph receiver [ELEC] A tape reperforator, teletypewriter, or other equipment which converts telegraph signals into a pattern of holes on a tape, printed letters, or other forms of information. { 'tel·ə‚graf ri‚sē·vər }

telegraph repeater [ELEC] A repeater inserted at intervals in long telegraph lines to amplify weak code signals, with or without reshaping of pulses, and to retransmit them automatically over the next section of the line. { 'tel·ə‚graf ri‚pēd·ər }

telegraph signal distortion [COMMUN] Time displacement of transitions between conditions, such as marking and spacing, with respect to their proper relative positions in perfectly timed signals; the total distortion is the algebraic sum of the bias and the characteristic and fortuitous distortions. { 'tel·ə‚graf ¦sig·nəl di‚stór·shən }

telegraph transmitter [ELEC] A device that controls an electric power source in order to form telegraph signals. { 'tel·ə‚graf tranz‚mid·ər }

telegraphy [COMMUN] Communication at a distance by means of code signals consisting of current pulses sent over wires or by radio. { tə'leg·rə·fē }

telemagmatic [GEOL] Pertaining to a hydrothermal mineral deposit that is distant from its magmatic source. { ¦tel·ə·mag 'mad·ik }

telemedicine [MED] The use of teleconferencing in medical diagnosis and treatment, allowing rural health-care facilities to perform diagnosis and treatment that would otherwise be available only in metropolitan areas. { ˌtel·ə'med·ə·sən }

telemeeting [COMMUN] A meeting between people remote from one another, but linked by audio and video telecommunications facilities that provide primarily one-way communication from a few people at one location to large numbers of people at other locations, and use temporary equipment or circuits. { ˈtel·ə,mēd·iŋ }

telemeteorograph [ENG] Any meteorological instrument, such as a radiosonde, in which the recording instrument is located at some distance from the measuring apparatus; for example, a meteorological telemeter. { ¦tel·ə,mēd·ē'ór·ə,graf }

telemeteorography [ENG] The science of the design, construction, and operation of various types of telemeteorographs. { ¦tel·ə,mēd·ē·ə'räg·rə·fe }

telemeteorometry [METEOROL] The study of making meteorological observations at a distance. { ¦tel·ə,mēd·ē·ə'räm·ə·trē }

telemeter [ENG] **1.** The complete measuring, transmitting, and receiving apparatus for indicating or recording the value of a quantity at a distance. Also known as telemetering system. **2.** To transmit the value of a measured quantity to a remote point. { 'tel·ə,mēd·ər }

telemetering [ENG] Transmitting the readings of instruments to a remote location by means of wires, radio waves, or other means. Also known as remote metering; telemetry. { ˌtel·ə'mēd·ə·riŋ }

telemetering antenna [ELECTROMAG] A highly directional antenna, generally mounted on a servo-controlled mount for tracking purposes, used at ground stations to receive telemetering signals from a guided missile or spacecraft. { ˌtel·ə'mēd·ə·riŋ an'ten·ə }

telemetering receiver [ELECTR] A device in a telemetering system which converts electrical signals into an indication or recording of the value of the quantity being measured at a distance. { ˌtel·ə'mēd·ə·riŋ ri'sē·vər }

telemetering system See telemeter. { ˌtel·ə'mēd·ə·riŋ ˌsis·təm }

telemetering transmitter [ELECTR] A device which converts the readings of instruments into electrical signals for transmission to a remote location by means of wires, radio waves, or other means. { ˌtel·ə'mēd·ə·riŋ tranz'mid·ər }

telemetering wave buoy [ENG] A buoy assembly that transmits a radio signal that varies in frequency proportional to the vertical acceleration experienced by the buoy, thereby conveying information about the buoy's vertical motion as it rides the waves. { ˌtel·ə'mēd·ə·riŋ ˈwāv ˌbói }

telemetry See telemetering. { tə'lem·ə·trē }

telencephalon [EMBRYO] The anterior subdivision of the forebrain in a vertebrate embryo; gives rise to the olfactory lobes, cerebral cortex, and corpora striata. { ¦tel·en'sef·ə,län }

teleology [SCI TECH] The doctrine that explanations of phenomena are to be sought in terms of final causes, purpose, or design in nature. { ˌtē·lē'äl·ə·jē }

teleoperation [ENG] **1.** The real-time control of remotely located machines that act as the eyes and hands of a person located elsewhere; it has been used in undersea and lunar exploration, mining, and microsurgery. **2.** Operation from a remote location. Also known as remote manipulation. { ˌtel·ē,äp· ə'rā·shən }

teleoperator See remote manipulator. { ˌtel·ē,äp·ə,rād·ər }

Teleosauridae [PALEON] A family of Jurassic reptiles in the order Crocodilia characterized by a long snout and heavy armor. { ˌtel·ē·ə'sór·ə,dē }

Teleostei [VERT ZOO] An infraclass of the subclass Actinopterygii, or rayfin fishes; distinguished by paired bracing bones in the supporting skeleton of the caudal fin, a homocercal caudal fin, thin cycloid scales, and a swim bladder with a hydrostatic function. { ˌtel·ē'äs·tē,ī }

telephone [COMMUN] A system of converting sound waves into variations in electric current that can be sent over wires and reconverted into sound waves at a distant point, used primarily for voice communication; it consists essentially of a telephone transmitter and receiver at each station, interconnecting wires or radio transmission systems, signaling devices, a central power supply, and switching facilities. Also known as telephone system. [ENG ACOUS] See telephone set. { 'tel·ə,fōn }

telephone-answering system [COMMUN] A special type of private branch exchange system used by a telephone-answering service bureau to provide secretarial service for its customers. { ˈtel·ə,fōn ¦an·sə·riŋ ˌsis·təm }

telephone carrier current [ELEC] A carrier current used for telephone communication over power lines or to obtain more than one channel on a single pair of wires. { 'tel·ə,fōn 'kar· ē·ər ,kə·rənt }

telephone central office See central office. { 'tel·ə,fōn 'sen· trəl 'óf·əs }

telephone channel [COMMUN] A one-way or two-way path suitable for the transmission of audio signals between two stations. { 'tel·ə,fōn ,chan·əl }

telephone circuit [ELEC] The complete circuit over which audio and signaling currents travel in a telephone system between the two telephone subscribers in communication with each other; the circuit usually consists of insulated conductors, as ground returns are now rarely used in telephony. { 'tel· ə,fōn ,sər·kət }

telephone data set [COMPUT SCI] Equipment interfacing a data terminal with a telephone circuit. { 'tel·ə,fōn 'dad·ə ,set }

telephone dial [ENG] A switch operated by a finger wheel, used to make and break a pair of contacts the required number of times for setting up a telephone circuit to the party being called. { 'tel·ə,fōn ,dīl }

telephone emission See telephone signal. { 'tel·ə,fōn i,mish·ən }

telephone induction coil [ELEC] A coil used in a telephone circuit to match the impedance of the line to that of a telephone transmitter or receiver. { 'tel·ə,fōn in'dək·shən ,kóil }

telephone influence factor [COMMUN] A measure of the interference of power-line harmonics with telephone lines, which is derived by weighting the terms in the mathematical expression for the total harmonic distortion of the power-line voltage. { 'tel·ə,fōn 'in·flü·əns ,fak·tər }

telephone line [ELEC] The conductors extending between telephone subscriber stations and central offices. { 'tel· ə,fōn ,līn }

telephone loading coil See loading coil. { 'tel·ə,fōn 'lōd· iŋ ,koil }

telephone modem [ELECTR] A piece of equipment that modulates and demodulates one or more separate telephone circuits, each containing one or more telephone channels; it may include multiplexing and demultiplexing circuits, individual amplifiers, and carrier-frequency sources. { 'tel·ə,fōn 'mō ,dem }

telephone pickup [ELEC] A large flat coil placed under a telephone set to pick up both voices during a telephone conversation for recording purposes. { 'tel·ə,fōn ,pik·əp }

telephone plug See phone plug. { 'tel·ə,fōn ,pləg }

telephone receiver [ENG ACOUS] The portion of a telephone set that converts the audio-frequency current variations of a telephone line into sound waves, by the motion of a diaphragm activated by a magnet whose field is varied by the electrical impulses that come over the telephone wire. { 'tel·ə,fōn ri,sē·vər }

telephone relay [ELEC] A relay having a multiplicity of contacts on long spring strips mounted parallel to the coil, actuated by a lever arm or other projection of the hinged armature; used chiefly for switching in telephone circuits. { 'tel· ə,fōn ,rē,lā }

telephone repeater [ELECTR] A repeater inserted at one or more intermediate points in a long telephone line to amplify telephone signals so as to maintain the required current strength. { 'tel·ə,fōn ri,pēd·ər }

telephone repeating coil [ELEC] A coil used in a telephone circuit for inductively coupling two sections of a line when a direct connection is undesirable. { 'tel·ə,fōn ri'pēd·iŋ ,kóil }

telephone ringer [ELECTROMAG] An electromagnetic device that actuates a clapper which strikes one or more gongs to produce a ringing sound; used with a telephone set to signal a called party. { 'tel·ə,fōn ,riŋ·ər }

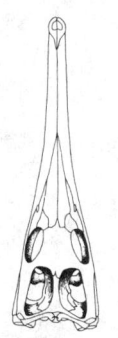

TELEOSAURIDAE

Skull of typical Teleosauridae, with long snout.

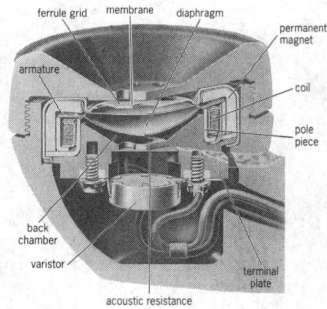

TELEPHONE RECEIVER

Cutaway view of a ring armature telephone receiver.

telephone set [ENG ACOUS] An assembly including a telephone transmitter, a telephone receiver, and associated switching and signaling devices. Also known as telephone. { 'tel·ə,fōn ,set }

telephone signal [COMMUN] The electrical signal transmitted by a telephone system, classified by type of transmission, type of modulation, bandwidth, and supplementary characteristics. Also known as telephone emission. { 'tel·ə,fōn ,sig·nəl }

telephone switchboard *See* switchboard. { 'tel·ə,fōn 'swich,bord }

telephone system *See* telephone. { 'tel·ə,fōn ,sis·təm }

telephone theory *See* frequency theory. { 'tel·ə,fōn ,thē·ə·rē }

telephone transmitter [ENG ACOUS] The microphone used in a telephone set to convert speech into audio-frequency electric signals. { 'tel·ə,fōn tranz,mid·ər }

telephony [COMMUN] The transmission of speech to a distant point by means of electric signals. { tə'lef·ə·nē }

telephoto *See* facsimile. { ¦tel·ə¦fōd·ō }

telephotography *See* facsimile. { ¦tel·ə·fə'täg·rə·fē }

telephoto lens [OPTICS] A lens for photographing distant objects; it is designed in a compact manner so that the distance from the front of the lens to the film plane is less than the focal length of the lens. { ¦tel·ə¦fōd·ō 'lenz }

telephotometer [ENG] A photometer that measures the received intensity of a distant light source. { ¦tel·ə·fə'täm·əd·ər }

telephotometry [OPTICS] The body of principles and techniques concerned with measuring atmospheric extinction by using various types of telephotometers. { ¦tel·ə·fə'täm·ə·trē }

teleport [COMMUN] A planned business development area that features direct and economic access to a large number of domestic and international satellites for users in the surrounding region, with the aid of a regional distribution network. { 'tel·ə,pórt }

telepresence [CONT SYS] The quality of sensory feedback from a teleoperator or telerobot to a human operator such that the operator feels present at the remote site. { ¦tel·ə'prez·əns }

teleprinter [COMPUT SCI] Any typewriter-type device capable of being connected to a computer and of printing out a set of messages under computer control. { 'tel·ə,print·ər }

teleprinting [COMMUN] Telegraphy in which the transmitter and receiver are teletypewriters. { 'tel·ə,print·iŋ }

teleprocessing [COMPUT SCI] **1.** The use of telecommunications equipment and systems by a computer. **2.** A computer service involving input/output at locations remote from the computer itself. { 'tel·ə,prä,ses·iŋ }

teleprocessing monitor [COMPUT SCI] A computer program that manages the transfer of information between local and remote terminals. Abbreviated TP monitor. { 'tel·ə,prä,ses·iŋ 'män·əd·ər }

telepsychrometer [ENG] A psychrometer in which the wet- and dry-bulb thermal elements are located at a distance from the indicating elements. { ¦tel·ə·sī'kräm·əd·ər }

Teleran *See* television and radar navigation system. { 'tel·ə,ran }

telerecording bathythermometer [ENG] A device which transmits measurements of sea water depth and temperature over a wire to a ship, where a graph of temperature versus depth is recorded. { 'tel·ə·ri,kórd·iŋ ¦bath·i·thər'mäm·əd·ər }

telering [ELECTR] In telephony, a frequency-selector device for the production of ringing power. { 'tel·ə,riŋ }

telerobot [CONT SYS] A type of teleoperator that embodies features of a robot and is programmed for communication with a human operator in a high-level language but can revert to direct control in the event of unplanned contingencies. { ,tel·ə'rō,bät }

telescope [ENG] Any device that collects radiation, which may be in the form of electromagnetic or particle radiation, from a limited direction in space. [OPTICS] An assemblage of lenses or mirrors, or both, that enhances the ability of the eye either to see objects with greater resolution or to see fainter objects. { 'tel·ə,skōp }

Telescope *See* Telescopium. { 'tel·ə,skōp }

telescope effect [PHYS] An effect in which stretching a high-polymer fiber increases hardness and reduces cross-sectional areas of the fibers. { ¦tel·ə,skäp·ik i¦fekt }

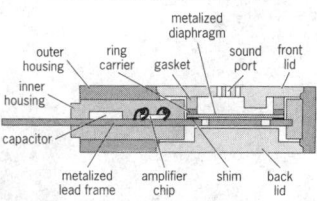

TELEPHONE TRANSMITTER

outer housing · ring carrier · gasket · metalized diaphragm · sound port · front lid · inner housing · capacitor · metalized lead frame · amplifier chip · shim · back lid

Cross section of an electret telephone transmitter.

telescope structure [GEOL] An alluvial fan structure characterized by younger fans with flatter gradients spreading out between older fans with steeper gradients. { 'tel·ə,skōp ,strək·chər }

telescopic alidade [ENG] An alidade used with a plane table, consisting of a telescope mounted on a straightedge ruler, fitted with a level bubble, scale, and vernier to measure angles, and calibrated to measure distances. { ¦tel·ə¦skäp·ik 'al·ə,dād }

telescopic comet [ASTRON] A comet in which only the coma is observed. { ¦tel·ə¦skäp·ik 'käm·ət }

telescopic derrick [ENG] A drill derrick divided into two or more sections, with the uppermost sections nesting successively into the lower sections. { ¦tel·ə¦skäp·ik 'de,rik }

telescopic loading trough [MIN ENG] A shaker conveyor trough of two sections, one nested in the other, used near the face for advancing the trough line without the necessity of adding either a standard or a short length of pan after each cut. { ¦tel·ə¦skäp·ik 'lōd·iŋ ,tróf }

telescopic series [MATH] The series whose nth term is $1/[(k + n - 1) (k + n)] = [1/(k + n - 1)] - [1/(k + n)]$, where k is not zero or a negative integer, and whose sum is $1/k$. { ¦tel·ə,skäp·ik 'sir,ēz }

telescopic sight [ORD] Gunsight equipped with a telescope. { ¦tel·ə¦skäp·ik 'sīt }

telescopic tripod [ENG] A drill or surveyor's tripod each leg of which is a series of two or more closely fitted nesting tubes, which can be locked rigidly together in an extended position to form a long leg or nested one within the other for easy transport. { ¦tel·ə¦skäp·ik 'trī,päd }

telescoping gage [DES ENG] An adjustable internal gage with a telescoping plunger that expands under spring tension in the hole to be measured; it is locked into position to allow measurement after being withdrawn from the hole. { ¦tel·ə¦skōp·iŋ 'gāj }

telescoping valve [MECH ENG] A valve, with sliding, telescoping members, to regulate water flow in a pipe line with minimum disturbance to stream lines. { ¦tel·ə¦skōp·iŋ 'valv }

Telescopium [ASTRON] A constellation, right ascension 19 hours, declination 50° south. Also known as Telescope. { ,tel·ə'skō·pē·əm }

teleseism [GEOPHYS] An earthquake that is far from the recording station. { 'tel·ə,sīz·əm }

teleseismology [GEOPHYS] The aspect of seismology dealing with records made at a distance from the source of the impulse. { ,tel·ə·sīz'mäl·ə·jē }

teleseminar [COMMUN] A form of long-distance, electronic communication, primarily one-way, to many destinations from one source, for educational purposes, involving audio communication, and possibly also video and some form of graphics. { ¦tel·ə'sem·ə,när }

Telestacea [INV ZOO] An order of the subclass Alcyonaria comprising individuals which form erect branching colonies by lateral budding from the body wall of an elongated axial polyp. { ,tel·ə'stā·shē·ə }

Telesto [ASTRON] A small, irregularly shaped satellite of Saturn that librates about the trailing Lagrangian point of Tethys's orbit. { te'les,tō }

telesynd [ELECTR] Telemeter or remote-control equipment which is synchronous in both speed and position. { 'tel·ə,sind }

teleterminal [COMPUT SCI] An instrument that integrates the functions of a telephone set and a computer terminal with keyboard and video screen. { ¦tel·ə'tər·mən·əl }

teletext [COMMUN] A data broadcasting service in which preprogrammed sequences of frames of data are broadcast cyclically, and a user equipped with a standard television receiver and a special decoder selects the desired frames of information for viewing. { 'tel·ə,tekst }

teletherapy [MED] Radiation treatment administered by using a source that is at a distance from the body, usually employing gamma-ray beams from radioisotope sources. { ¦tel·ə'ther·ə·pē }

telethermal [GEOL] Pertaining to a hydrothermal mineral deposit precipitated at a shallow depth and at a mild temperature. { ¦tel·ə'thər·məl }

telethermometer [ENG] A temperature-measuring system in which the heat-sensitive element is located at a distance from the indicating element. { ¦tel·ə·thər'mäm·əd·ər }

telethermoscope [ENG] A temperature telemeter, frequently used in a weather station to indicate the temperature at the instrument shelter located outside. { ¦tel·ə'thər·mə₁skōp }

telethesis [ENG] A robotic manipulation aid for the physically disabled that may be located remote from the body. There are two forms, operated by voice command, or operated through a body-powered prosthesis or a joystick. { tə'leth·ə·səs }

teletypesetter [GRAPHICS] A system automatically operating a linecasting machine (Linotype or Intertype) to produce lines of type at high speed under the control of perforated tape or equivalent electrical signals. { ¦tel·ə'tīp₁sed·ər }

teletypewriter [COMMUN] A special electric typewriter that produces coded electric signals corresponding to manually typed characters, and automatically types messages when fed with similarly coded signals produced by another machine. Also known as TWX machine. { ¦tel·ə'tīp₁rīd·ər }

teletypewriter code [COMMUN] Special code in which each code group is made up of five units, or elements, of equal length which are known as marking or spacing impulses; the five-unit start-stop code consists of five signal impulses preceded by a start impulse and followed by a stop impulse. { ¦tel·ə'tīp₁rīd·ər ₁kōd }

teletypewriter exchange service [COMMUN] A service furnished by telephone companies to subscribers in the United States, whereby any of the subscribers can communicate directly with any other subscriber via teletypewriter. Also known as TWX service. { ¦tel·ə tīp₁rīd·ər iks¦chānj ₁sər·vəs }

teletypewriter signal distortion [COMMUN] Of a start-stop teletypewriter signal, the shifting of the transition points of the signal pulses from their proper positions relative to the beginning of the start pulse; the magnitude of the distortion is expressed in percent of a perfect unit pulse length. { ¦tel·ə'tīp₁rīd·ər 'sig·nəl di₁stor·shən }

teleutospore *See* teliospore. { tə'lüd·ə₁spor }

televiewer [ENG] An acoustic camera that provides an ultrasonic image of the borehole wall during borehole logging. { 'tel·ə₁vyü·ər }

televise [COMMUN] To pick up a scene with a television camera and convert it into corresponding electric signals for transmission by a television station. { 'tel·ə₁vīz }

television [COMMUN] A system for converting a succession of visual images into corresponding electric signals and transmitting these signals by radio or over wires to distant receivers at which the signals can be used to reproduce the original images. Abbreviated TV. { 'tel·ə₁vizh·ən }

television and radar navigation system [NAV] Navigational system which employs ground-based search radar equipment along an airway to locate aircraft flying near that airway; transmits, by television, information pertaining to these aircraft and other information to the pilots of properly equipped aircraft, and provides information to the pilots appropriate for use in the landing approach. Also known as Teleran; television-radar air navigation. { 'tel·ə₁vizh·ən ən 'rā₁där ₁nav·ə'gā·shən ₁sis·təm }

television antenna [ELECTROMAG] An antenna suitable for transmitting or receiving television broadcasts; since television transmissions in the United States are horizontally polarized, the most basic type of receiving antenna is a horizontally mounted half-wave dipole. { 'tel·ə₁vizh·ən an₁ten·ə }

television bandwidth [COMMUN] The difference between the limiting frequencies of a television channel; in the United States, this is 6 megahertz. { 'tel·ə₁vizh·ən 'band₁width }

television broadcast band [COMMUN] Several groups of channels, each containing a number of 6-megahertz channels, that are available for assignment to television broadcast stations. { 'tel·ə₁vizh·ən 'brod₁kast ₁band }

television broadcasting [COMMUN] Transmission of television programs by means of radio waves for reception by the public. { 'tel·ə₁vizh·ən 'brod₁kast·iŋ }

television camera [ELECTR] The pickup unit used to convert a scene into corresponding electric signals; optical lenses focus the scene to be televised on the photosensitive surface of a camera tube, and the tube breaks down the visual image into small picture elements and converts the light intensity of each element in turn into a corresponding electric signal. Also known as camera. { 'tel·ə₁vizh·ən ₁kam·rə }

television camera tube *See* camera tube. { 'tel·ə₁vizh·ən 'kam·rə ₁tüb }

television channel [COMMUN] A band of frequencies 6 megahertz wide in the television broadcast band, available for assignment to a television broadcast station. { 'tel·ə₁vizh·ən ₁chan·əl }

television emission *See* television signal. { 'tel·ə₁vizh·ən i₁mish·ən }

television film scanner [ENG] A motion picture projector adapted for use with a television camera tube to televise 24-frame-per-second motion picture film at the 30-frame-per-second rate required for television. { 'tel·ə₁vizh·ən 'film ₁skan·ər }

television interference [COMMUN] Interference produced in television receivers by amateur radio and other local transmitters. Abbreviated TVI. { 'tel·ə₁vizh·ən ₁in·tər'fir·əns }

television monitor [ELECTR] **1.** A television set connected to the transmitter at a television station, used to continuously check the image picked up by a television camera and the sound picked up by the microphones. **2.** A closed-circuit television system used to provide continuous observation of such things as hazardous or remote locations, the readings of gages for process control, or microscopic or telescopic images, for greater convenience of viewing. { 'tel·ə₁vizh·ən ₁man·əd·ər }

television network [COMMUN] An arrangement of communication channels, suitable for transmission of video and accompanying audio signals, which link together groups of television broadcasting stations or closed-circuit television users in different cities so that programs originating at one point can be fed simultaneously to all others. { 'tel·ə₁vizh·ən ₁net₁wərk }

television pickup station [COMMUN] A land mobile station used for the transmission of television program material and related communications from the scene of an event occurring at a point remote to a television broadcast station. { 'tel·ə₁vizh·ən 'pik·əp ₁stā·shən }

television picture tube *See* picture tube. { 'tel·ə₁vizh·ən 'pik·chər ₁tüb }

television-radar air navigation *See* television and radar navigation system. { 'tel·ə₁vizh·ən 'rā₁där 'er ₁nav·ə₁gā·shən }

television receive only antenna [COMMUN] A parabolic reflector or dish large enough to receive signals intended for cable television systems from geostationary satellites, together with a feed horn that collects the signals reflected by the dish, a low-noise amplifier for preamplification, and a tunable satellite receiver. Abbreviated TVRO. { 'tel·ə₁vizh·ən ri'sēv ¦ōn·lē an'ten·ə }

television receiver [ELECTR] A receiver that converts incoming television signals into the original scenes along with the associated sounds. Also known as television set. { 'tel·ə₁vizh·ən ri₁sē·vər }

television reconnaissance [COMMUN] Reconnaissance in which television is used to transmit a scene from the reconnoitering point to another location on the surface or in the air. { 'tel·ə₁vizh·ən ri₁kän·ə·səns }

television recording [COMMUN] Permanent record of video signals that is recorded photographically, electronically, or by other means, and may be displayed through a television system or projected as a motion picture film. { 'tel·ə₁vizh·ən ri₁kord·iŋ }

television relay system *See* television repeater. { 'tel·ə₁vizh·ən 'rē₁lā ₁sis·təm }

television repeater [ELECTR] A repeater that transmits television signals from point to point by using radio waves in free space as a medium, such transmission not being intended for direct reception by the public. Also known as television relay system. { 'tel·ə₁vizh·ən ri₁pēd·ər }

television scanning [COMMUN] The process of scrutinizing the brightness of each element of detail contained in the image of a scene to be transmitted by television. { 'tel·ə₁vizh·ən ₁skan·iŋ }

television screen [ELECTR] The fluorescent screen of the picture tube in a television receiver. { 'tel·ə₁vizh·ən ₁skrēn }

television set *See* television receiver. { 'tel·ə₁vizh·ən ₁set }

television signal [COMMUN] A general term for the aural and visual signals that are broadcast together to provide the sound and picture portions of a television program. Also known as television emission. { 'tel·ə₁vizh·ən ₁sig·nəl }

television station [COMMUN] The installation, assemblage of equipment, and location where radio transmissions are sent or received. { 'tel·ə₁vizh·ən ₁stā·shən }

TELEVISION CAMERA

Television studio camera. (*Thomson Multimedia*)

television studio [COMMUN] A complex of rooms specifically designed for the origination of live television programs; in addition to the studio room in which the program takes place, there are three support rooms: the control room, equipment room, and prop room. { 'tel·ə,vizh·ən ,stüd·ē·ō }

television tower [ENG] A tall metal structure used as a television transmitting antenna, or used with another such structure to support a television transmitting antenna wire. { 'tel·ə,vizh·ən ,taů·ər }

television transmitter [ELECTR] An electronic device that converts the audio and video signals of a television program into modulated radio-frequency energy that can be radiated from an antenna and received on a television receiver. { 'tel·ə,vizh·ən tranz,mid·ər }

television tuner [ELECTR] A component in a television receiver that selects the desired channel and converts the frequencies received to lower frequencies within the passband of the intermediate-frequency amplifier; for very-high-frequency reception there are 12 discrete positions (channels 2–13); for ultra-high-frequency reception continuous tuning is usually employed. { 'tel·ə,vizh·ən ,tü·nər }

telewriter [COMMUN] System in which writing movement at the transmitting end causes corresponding movement of a writing instrument at the receiving end. { 'tel·ə,rīd·ər }

Telex [COMMUN] A worldwide teleprinter exchange service providing direct send and receive teleprinter connections between subscribers. Abbreviated TEX. { 'te,leks }

telford pavement [CIV ENG] A road pavement having a firm foundation of large stones and stone fragments, and a smooth hard-rolled surface of small stones. { 'tel·fərd ,pāv·mənt }

telinite [GEOL] A variety of provitrinite composed of plant cell-wall material. { 'tē·lə,nīt }

teliospore [MYCOL] A thick-walled spore of the terminal stage of Uredinales and Ustilaginales which is a probosidium or a group of probosidia. Also known as teleutospore. { 'tē·lē·ə,spór }

telium [MYCOL] In rust fungi, a specialized fruiting structure that produces teliospores. { 'tēl·ē·əm }

Tellerette [CHEM ENG] A type of inert packing with the appearance of a circular-wound spiral, used to create a large surface area to increase contact between falling liquid and rising vapor; used in gas-absorption operations. { 'tel·ə,rīt }

Teller-Redlich rule [PHYS CHEM] For two isotopic molecules, the product of the frequency ratio values of all vibrations of a given symmetry type depends only on the geometrical structure of the molecule and the masses of the atoms, and not on the potential constants. { 'tel·ər 'red·lik ,rül }

telltale [ENG] A marker on the outside of a tank that indicates on an exterior scale the amount of fluid inside the tank. { 'tel,tāl }

telltale compass [NAV] A marine magnetic compass, usually of the inverted type, frequently installed in the master's cabin. { 'tel,tāl ¦käm·pəs }

telltale float [CIV ENG] A water-level indicator in a reservoir. { 'tel,tāl ¦flōt }

telluric acid [INORG CHEM] H_6TeO_6 Toxic white crystals, slightly soluble in cold water, soluble in hot water and alkalies; melts at 136°C; used as an analytical reagent. Also known as hydrogen tellurate. { tə'lůr·ik 'as·əd }

telluric current See earth current. { tə'lůr·ik ,kə·rənt }

telluric line [SPECT] Any of the spectral bands and lines in the spectrum of the sun and stars produced by the absorption of their light in the atmosphere of the earth. { tə'lůr·ik 'līn }

tellurics [GEOPHYS] A geophysical exploration technique that measures variations in the conductivity (or resistivity) of rocks; often used for metallic mineral prospecting. { tə'lůr·iks }

tellurinic acid [ORG CHEM] A compound of tellurium with the general formula R₂TeOOH; an example is methanetellurinic acid, C_6H_5TeOOH. { ¦tel·yə¦rin·ik 'as·əd }

tellurite [MINERAL] TeO_2 A white or yellowish orthorhombic mineral consisting of tellurium dioxide, and occurring in crystals; it is dimorphous with paratellurite. { 'tel·yə,rīt }

tellurium [CHEM] A member of group 16, symbol Te, atomic number 52, atomic weight 127.60; dark-gray crystals, insoluble in water, soluble in nitric and sulfuric acids and potassium hydroxide; melts at 452°C, boils at 1390°C; used in alloys (with lead or steel), glass, and ceramics. { tə'lůr·ē·əm }

tellurium dibromide [INORG CHEM] TeBr₂ Toxic, hygroscopic, green- or gray-black crystals with violet vapor, soluble in ether, decomposes in water, and melts at 210°C. { tə'lůr·ē·əm dī'brō,mīd }

tellurium dichloride [INORG CHEM] TeCl₂ A toxic, amorphous, black or green-yellow powder decomposing in water, melting at 209°C. { tə'lůr·ē·əm dī'klór,īd }

tellurium dioxide [INORG CHEM] TeO_2 The most stable oxide of tellurium, formed when tellurium is burned in oxygen or air or by oxidation of tellurium with cold nitric acid; crystallizes as colorless, tetragonal, hexagonlike crystals that melt at 452°C. { tə'lůr·ē·əm dī'äk,sīd }

tellurium disulfide [INORG CHEM] TeS₂ A toxic, red powder, insoluble in water and acids. Also known as tellurium sulfide. { tə'lůr·ē·əm dī'səl,fīd }

tellurium glance See nagyagite. { tə'lůr·ē·əm 'glans }

tellurium hexafluoride [INORG CHEM] TeF₆ A colorless gas which is formed from the elements tellurium and fluorine; it is slowly hydrolyzed by water. { tə'lůr·ē·əm ¦hek·sə'flůr,īd }

tellurium method [FL MECH] A method of flow visualization in which a cloud of tellurium is generated in a liquid through an irreversible electrolytic reaction, forming a tracer. { tə'lůr·ē·əm ,meth·əd }

tellurium monoxide [INORG CHEM] TeO A black, amorphous powder, stable in cold dry air; formed by heating the mixed oxide TeSO₃. { tə'lůr·ē·əm mə'näk,sīd }

tellurium sulfide See tellurium disulfide. { tə'lůr·ē·əm 'səl,fīd }

tellurobismuthite [MINERAL] Bi₂Te₃ A pale lead gray, hexagonal mineral consisting of a bismuth and tellurium compound; occurs as irregular plates or foliated masses. { ¦tel·yə·rō'biz·mə,thīt }

telluroketone [ORG CHEM] One of a group of compounds with the general formula R₂CTe. { ¦tel·yə·rō'kē,tōn }

telluromercaptan [ORG CHEM] One of a group of compounds with the general formula RTeH. { ¦tel·yə·rō·mər'kap,tan }

tellurometer [ENG] A microwave instrument used in surveying to measure distance; the time for a radio wave to travel from one observation point to the other and return is measured and converted into distance by phase comparison, much as in radar. { ,tel·yə'räm·əd·ər }

tellurous acid [INORG CHEM] H_2TeO_3 Toxic, white crystals, soluble in alkalies and acids, slightly soluble in water and alcohol; decomposes at 40°C. { 'tel·yə·rəs 'as·əd }

TELNET See network terminal protocol. { 'tel,net }

teloblast [INV ZOO] A large cell that produces many smaller cells at the growing end of many embryos, especially in annelids and mollusks. { 'tel·ə,blast }

telocentric [CYTOL] Pertaining to a chromosome with a terminal centromere. { ¦tel·ə¦sen·trik }

telocoel [EMBRYO] A cavity of the telencephalon. { 'tel·ə,sēl }

telodendrion [NEUROSCI] The terminal branching of an axon. Also known as telodendron. { ¦tel·ə'den·drē,än }

telodendron See telodendrion. { ¦tel·ə'den·drən }

telogen [PHYSIO] A quiescent phase in the cycle of hair growth when the hair is retained in the hair follicle as a dead or "club" hair. { 'tel·ə·jən }

telogen effluvium [MED] Hair loss that is due to increased shedding occurring after a metabolic disturbance. { ,tel·ə·jən e'flü·vē·əm }

telolecithal [CYTOL] Of an ovum, having a large, evenly dispersed volume of yolk and a small amount of cytoplasm concentrated at one pole. { ¦tel·ō'les·ə·thəl }

telomerase [BIOCHEM] A deoxyribonucleic acid polymerase that elongates telomeres. { tə'läm·ə,rās }

telomere [CYTOL] A centromere in the terminal position on a chromosome. { 'tel·ə,mir }

telophase [CYTOL] The phase of meiosis or mitosis at which the chromosomes, having reached the poles, reorganize into interphase nuclei with the disappearance of the spindle and the reappearance of the nuclear membrane; in many organisms telophase does not occur at the end of the first meiotic division. { 'tel·ə,fāz }

Telosporea [INV ZOO] A class of the protozoan subphylum Sporozoa in which the spores lack a polar capsule and develop from an oocyst. { ¦tel·ə'spór·ē·ə }

telotaxis [BIOL] Tactic movement of an organism by the

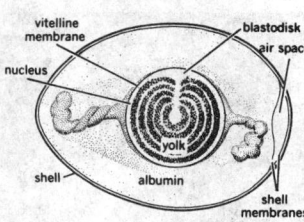

TELOLECITHAL

vitelline membrane
nucleus
blastodisk
air space
yolk
shell
albumin
shell membranes

Telolecithal ovum of a hen.

orientation of one or the other of two bilaterally symmetrical receptors toward the stimulus source. { 'tel·ō'tak·səs }

telotroch [INV ZOO] A preanal tuft of cilia in a trochophore larva. { 'tel·ə,träk }

telpher [MECH ENG] An electric hoist hanging from and driven by a wheeled cab rolling on a single overhead rail or a rope. { 'tel·fər }

Telsmith breaker [MECH ENG] A type of gyratory crusher, often used for primary crushing; consists of a spindle mounted in a long eccentric sleeve which rotates to impart a gyratory motion to the crushing head, but gives a parallel stroke, that is, the axis of the spindle describes a cylinder rather than a cone, as in the suspended spindle gyratory. { 'tel,smith ,brā·kər }

telson [INV ZOO] The postabdominal segment in lobsters, amphipods, and certain other invertebrates. { 'tel·sən }

telvar [ORG CHEM] The common name for the herbicide 3-(*para*-chlorophenyl)-1,1-dimethylurea; used as a soil sterilant. { 'tel,vär }

TEM *See* triethylenemelamine.

TEMA standard [CHEM ENG] Shell-and-tube heat-exchange standard designed to supplement the American Society of Mechanical Engineers code for unfired pressure vessels. { 'tē·mə ,stan·dərd }

TEM mode *See* transverse electromagnetic mode. { ,tē,ē'em ,mōd }

Temnocephalida [INV ZOO] A group of rhabdocoeles sometimes considered a distinct order but usually classified under the Neorhabdocoela; members are characterized by the possession of tentacles and adhesive organs. { ,tem·nō·sə'fal·əd·ə }

Temnochilidae [INV ZOO] The equivalent name for Ostomidae. { ,tem·nō'kil·ə,dē }

Temnopleuridae [INV ZOO] A family of echinoderms in the order Temnopleuroida whose tubercles are imperforate, though usually crenulate. { ,tem·nō'plür·ə,dē }

Temnopleuroida [INV ZOO] An order of echinoderms in the superorder Echinacea with a camarodont lantern, smooth or sculptured test, imperforate or perforate tubercles, and bronchial slits which are usually shallow. { ,tem·nō·plə'roid·ē·ə }

Temnospondyli [PALEON] An order of extinct amphibians in the subclass Labyrinthodontia having vertebrae with reduced pleurocentra and large intercentra. { ,tem·nō'spän·də,lī }

TE mode *See* transverse electric mode. { 'tē'ē ,mōd }

temper [ENG] 1. To moisten and mix clay, plaster, or mortar to the proper consistency for use. 2. *See* anneal. [MET] 1. The hardness and strength of a rolled metal. 2. The nominal carbon content of steel. 3. To soften hardened steel or cast iron by reheating to some temperature below the eutectoid temperature. 4. An alloy added to pure tin to make the finest pewter. { 'tem·pər }

tempera [MATER] 1. An opaque watercolor paint consisting of pigment ground in water and mixed with egg yolk. 2. A poster paint that uses glue or gum as a binder. { 'tem·pə·rə }

temperament [ACOUS] The adjustment of the pitch of the notes of a keyboard instrument so that the diatonic scale in all keys approximates just tuning; this permits modulation to any key. { 'tem·prə·mənt }

temperate and cold savanna [ECOL] A regional vegetation zone, very extensively represented in North America and in Eurasia at high altitudes; consists of scattered or clumped trees (very often conifers and mostly needle-leaved evergreens) and a shrub layer of varying coverage; mosses and, even more abundantly, lichens form an almost continuous carpet. { ,tem·prət ən ,kōld sə'van·ə }

temperate and cold scrub [ECOL] Regional vegetation zone whose density and periodicity vary a good deal; requires a considerable amount of moisture in the soil, whether from mist, seasonal downpour, or snowmelt; shrubs may be evergreen or deciduous; and undergrowth of ferns and other large-leaved herbs are quite frequent, especially at subalpine level; wind shearing and very cold winters prevent tree growth. Also known as bosque; fourré; heath. { ,tem·prət ən ,kōld 'skrəb }

temperate belt [CLIMATOL] A belt around the earth within which the annual mean temperature is less than 20°C (68°F) and the mean temperature of the warmest month is higher than 10°C (50°F). { 'tem·prət ,belt }

temperate climate [CLIMATOL] The climate of the middle latitudes; the climate between the extremes of tropical climate and polar climate. { 'tem·prət 'klī·mət }

temperate glacier [HYD] A glacier which, at the end of the melting season, is composed of firn and ice at the melting point. { 'tem·prət 'glā·shər }

temperate mixed forest [ECOL] A forest of the North Temperate Zone containing a high proportion of conifers with a few broad-leafed species. { 'tem·prət 'mikst ,fär·əst }

temperate phage [VIROL] A deoxyribonucleic acid phage, the genome of which can under certain circumstances become integrated with the genome of the host. { 'tem·prət 'fāj }

temperate rainforest [ECOL] A vegetation class in temperate areas of high and evenly distributed rainfall characterized by comparatively few species with large populations of each species; evergreens are somewhat short with small leaves, and there is an abundance of large tree ferns. Also known as cloud forest; laurel forest; laurisilva; moss forest; subtropical forest. { 'tem·prət 'rān,fär·əst }

temperate rainy climate [CLIMATOL] One of the major categories in W. Kippen's climatic classification; the coldest-month mean temperature is less than 64.4°F (18°C) and greater than 26.6°F (−3°C), and the warmest-month mean temperature is more than 50°F (10°C). { 'tem·prət 'rān·ē ,klī·mət }

temperate westerlies *See* westerlies. { 'tem·prət 'wes·tər·lēz }

temperate-westerlies index [METEOROL] A measure of the strength of the westerly wind between latitudes 35°N and 55°N; the index is computed from the average sea-level pressure difference between these latitudes and is expressed as the west to east component of geostrophic wind in meters and tenths of meters per second. { 'tem·prət 'wes·tər·lez 'in,deks }

temperate woodland [ECOL] A vegetation class similar to tropical woodland in spacing, height, and stratification, but it can be either deciduous or evergreen, broad-leaved or needle-leaved. Also known as parkland; woodland. { 'tem·prət 'wüd·lənd }

Temperate Zone [CLIMATOL] Either of the two latitudinal zones on the earth's surface which lie between 23°27' and 66°32' N and S (the North Temperate Zone and South Temperate Zone, respectively). { 'tem·prət ,zōn }

temperature [THERMO] A property of an object which determines the direction of heat flow when the object is placed in thermal contact with another object: heat flows from a region of higher temperature to one of lower temperature; it is measured either by an empirical temperature scale, based on some convenient property of a material or instrument, or by a scale of absolute temperature, for example, the Kelvin scale. { 'tem·prə·chər }

13.0 temperature *See* annealing point. { ',thər,tēn 'tem·prə·chər }

temperature-actuated pressure relief valve [MECH ENG] A pressure relief valve which operates when subjected to increased external or internal temperature. { 'tem·prə·chər ,ak·chə,wād·əd 'presh·ər ri,lēf ,valv }

temperature bath [THERMO] A relatively large volume of a homogeneous substance held at constant temperature, so that an object placed in thermal contact with it is maintained at the same temperature. { 'tem·prə·chər ,bath }

temperature belt [METEOROL] The belt which may be drawn on a thermograph trace or other temperature graph by connecting the daily maxima with one line and the daily minima with another. { 'tem·prə·chər ,belt }

temperature-chlorinity-depth recorder [ENG] An instrument in which an underwater unit suspended from a cable records temperature, chlorinity, and depth sequentially on a single-pen strip recorder, each quantity being recorded for several seconds at a time. { 'tem·prə·chər klō'rin·əd·ē 'depth ri,kord·ər }

temperature coefficient [PHYS] The rate of change of some physical quantity (such as resistance of a conductor or voltage drop across a vacuum tube) with respect to temperature. { 'tem·prə·chər ,kō·i,fish·ənt }

temperature coefficient of reactivity [NUCLEO] The change in reactor reactivity (per degree of temperature) occurring when the operating temperature of the reactor changes. { 'tem·prə·chər ,kō·i,fish·ənt əv ,rē ak'tiv·əd·ē }

temperature color scale [THERMO] The relation between an incandescent substance's temperature and the color of the light it emits. { 'tem·prə·chər 'kəl·ər ,skāl }

temperature-compensated Zener diode [ELECTR] Positive-temperature-coefficient reversed-bias Zener diode (*pn* junction) connected in series with one or more negative-temperature forward-biased diodes within a single package. { 'tem·prə·chər ¦käm·pən‚säd·əd 'zē·nər 'dī‚ōd }

temperature-compensating capacitor [ELEC] Capacitor whose capacitance varies with temperature in a known and predictable manner; used extensively in oscillator circuits to compensate for changes in the values of other parts with temperatures. { 'tem·prə·chər ¦käm·pən‚säd·iŋ kə'pas·əd·ər }

temperature compensation [ELECTR] The process of making some characteristic of a circuit or device independent of changes in ambient temperature. { 'tem·prə·chər ‚käm·pən‚sā·shən }

temperature control [ENG] A control used to maintain the temperature of an oven, furnace, or other enclosed space within desired limits. { 'tem·prə·chər kən‚trōl }

temperature-efficiency index [ECOL] For a given location, a measure of the long-range effectiveness of temperature (thermal efficiency) in promoting plant growth. Abbreviated T-E index. Also known as thermal-efficiency index. { 'tem·prə·chər i'fish·ən·sē ‚in‚deks }

temperature-efficiency ratio [ECOL] For a given location and month, a measure of thermal efficiency; it is equal to the departure, in degrees Fahrenheit, of the normal monthly temperature above 32°F (0°C) divided by 4: $(T - 32)/4$. Abbreviated T-E ratio. Also known as thermal-efficiency ratio. { 'tem·prə·chər i'fish·ən·sē ‚rā·shō }

temperature error [ENG] That instrument error due to nonstandard temperature of the instrument. { 'tem·prə·chər ‚er·ər }

temperature gradient [THERMO] For a given point, a vector whose direction is perpendicular to an isothermal surface at the point, and whose magnitude equals the rate of change of temperature in this direction. { 'tem·prə·chər ‚grād·ē·ənt }

temperature-humidity index [METEOROL] An index which gives a numerical value, in the general range of 70–80, reflecting outdoor atmospheric conditions of temperature and humidity as a measure of comfort (or discomfort) during the warm season of the year; equal to 15 plus 0.4 times the sum of the dry-bulb and wet-bulb temperatures in degrees Fahrenheit. Also known as comfort index; discomfort index. Abbreviated CI; DI; THI. { 'tem·prə·chər hyü'mid·ədē ‚in‚deks }

temperature-indicating compound [MATER] A temperature-sensitive material with a predetermined melting point; used to indicate when a predesignated temperature is reached in such processes as heat treating, welding, and molding. { 'tem·prə·chər ¦in·də‚kād·iŋ 'käm‚paund }

temperature inversion [METEOROL] A layer in the atmosphere in which temperature increases with altitude; the principal characteristic of an inversion layer is its marked static stability, so that very little turbulent exchange can occur within it; strong wind shears often occur across inversion layers, and abrupt changes in concentrations of atmospheric particulates and atmospheric water vapor may be encountered on ascending through the inversion layer. Also known as thermal inversion. [OCEANOGR] A layer of a large body of water in which temperature increases with depth. { 'tem·prə·chər in‚vər·zhən }

temperature log [PETRO ENG] A continuous record of temperature versus depth in an oil-well borehole. { 'tem·prə·chər ‚läg }

temperature profile recorder [ENG] A portable instrument for measuring temperature as a function of depth in shallow water, particularly in lakes, in which a thermistor element transmits data over an electrical cable to a recording drum and depth is measured by the amount of wire paid out. { 'tem·prə·chər ¦prō‚fīl ri‚kórd·ər }

temperature province [CLIMATOL] A major division of C.W. Thornthwaite's schemes of climatic classification, determined as a function of the temperature-efficiency index or the potential evapotranspiration. { 'tem·prə·chər ‚präv·əns }

temperature resistance coefficient [ELEC] The ratio of the change of electrical resistance in a wire caused by a change in its temperature of 1°C as related to its resistance at 0°C. { 'tem·prə·chər ri'zis·təns ‚kō·i‚fish·ənt }

temperature-salinity diagram [OCEANOGR] The plot of temperature versus salinity data of a water column; the resulting diagram identifies the water masses within the column, the column's stability, indicates the σ_T value via lines of constant σ_T printed on paper, and allows an estimate of the accuracy of the temperature and salinity measurements. Also known as

T-S curve; T-S diagram; T-S relation. { 'tem·prə·chər sə'lin·əd·ē ‚dī·ə‚gram }

temperature saturation [ELECTR] The condition in which the anode current of a thermionic vacuum tube cannot be further increased by increasing the cathode temperature at a given value of anode voltage; the effect is due to the space charge formed near the cathode. Also known as filament saturation; saturation. { 'tem·prə·chər ‚sach·ə‚rā·shən }

temperature scale [THERMO] An assignment of numbers to temperatures in a continuous manner, such that the resulting function is single valued; it is either an empirical temperature scale, based on some convenient property of a substance or object, or it measures the absolute temperature. { 'tem·prə·chər ‚skāl }

temperature-sensitive mutant [GEN] A mutant gene that is functional at high (low) temperature but is inactivated by lowering (elevating) the temperature. { 'tem·prə·chər ¦sen·səd·iv 'myüt·ənt }

temperature sensor [ENG] A device designed to respond to temperature stimulation. { 'tem·prə·chər ‚sen·sər }

temperature survey [PETRO ENG] An analysis of temperature changes or differences down an oil-well borehole, based on a temperature log; used to locate the top of casing cement, lost circulation zones, or gas entry zones. { 'tem·prə·chər ‚sər‚vā }

temperature transducer [ENG] A device in an automatic temperature-control system that converts the temperature into some other quantity such as mechanical movement, pressure, or electric voltage; this signal is processed in a controller, and is applied to an actuator which controls the heat of the system. { 'tem·prə·chər tranz‚dü·sər }

temperature wave [CRYO] A disturbance in which a variation in temperature propagates through a medium; the chief example of this is second sound. Also known as thermal wave. { 'tem·prə·chər ‚wāv }

temperature zone [CLIMATOL] A portion of the earth's surface defined by relatively uniform temperature characteristics, and usually bounded by selected values of some measure of temperature or temperature effect. { 'tem·prə·chər ‚zōn }

temper brittleness [MET] A brittle state resulting when certain low-alloy steels are slowly cooled in a range of 600–300°C or reheated in this range after quenching from 600°C. { 'tem·pər ‚brid·əl·nəs }

tempered glass [MATER] Glass that has been prestressed by heating followed by sudden quenching to give it two to four times the strength of ordinary glass. Also known as toughened glass. { 'tem·pərd 'glas }

tempering [MATER] Impregnating wood fibers or composition board with an oxidizing resin or drying oil followed by heat treatment, to improve strength, durability, and water resistance. [MET] Heat treatment of hardened steels to temperatures below the transformation temperature range, usually to improve toughness. { 'tem·pə·riŋ }

tempering air [ENG] Low-temperature air added to a heated airstream to regulate the stream temperature. { 'tem·pə·riŋ ‚er }

tempering oil [MATER] A high-viscosity neutral petroleum oil, such as a steam cylinder stock, used for the drawing or tempering of steel. Also known as steam-hammer oil; summer black oil. { 'tem·pə·riŋ ‚óil }

temper screw *See* stirrup. { 'tem·pər ‚skrü }

temper time [MET] In resistance welding, that part of the postweld interval during which the current is suitable for tempering or heat treatment. { 'tem·pər ‚tīm }

tempilstick [MATER] A crayon that, when applied to a surface, indicates when the surface temperature exceeds a given value by changing color. { 'tem·pəl‚stik }

template [COMPUT SCI] **1.** A prototype pattern against which observed patterns are matched in a pattern recognition system. **2.** A computer program that is used in conjunction with an electronic spreadsheet to solve a particular type of problem. [ENG] **1.** A two-dimensional representation of a machine or other equipment used for building layout design. **2.** A guide or a pattern used in manufacturing items. Also spelled templet. [MOL BIO] The macromolecular model for the synthesis of another macromolecule. { 'tem‚plət }

template matching [COMPUT SCI] The comparison of a picture or other data with a stored program or template, for purposes of identification or inspection. { 'tem‚plət ‚mach·iŋ }

temples [INV ZOO] The posterolateral angles of the head, in lice. { 'tem·pəlz }

templet See template. { 'tem·plət }

tempon [PHYS] A unit of time equal to the time required for light to traverse the classical radius of an electron. { 'tem,pän }

temporal [SCI TECH] Pertaining to or limited by time. { 'tem·prəl }

temporal arteritis See giant-cell arteritis. { 'tem·prəl ,ärd·ə'rīd·əs }

temporal bone [ANAT] The bone forming a portion of the lateral aspect of the skull and part of the base of the cranium in vertebrates. { 'tem·prəl ,bōn }

temporal coherence [PHYS] The existence of a correlation in time between the phases of waves at a point in space. { 'tem·prəl kō'hir·əns }

temporal decomposition [CONT SYS] The partitioning of the control or decision-making problem associated with a large-scale control system into subproblems based on the different time scales relevant to the associated action functions. { 'tem·prəl ,dē,käm·pə'zish·ən }

temporale [METEOROL] A rainy wind from the southwest to west resulting from a deflection of the southeast trades onto the Pacific coast of Central America. { ,tem·pō'rä·lē }

temporal-lobe epilepsy [MED] Recurrent seizures originating in lesions of the temporal lobe of the brain. { 'tem·prəl ¦lōb 'ep·ə,lep·sē }

temporal unit [GEOL] A stratigraphic unit defined in terms of time-related characteristics. { 'tem·prəl ,yü·nət }

temporamandibular joint disease [MED] A condition characterized by a faulty bite or misalignment of the teeth. Abbreviated TMJ. { ,tem·pə·rə·man,dib·yə·lər 'jóint di,zēz }

temporary base level [GEOL] Any base level, other than sea level, below which a land area temporarily cannot be reduced by erosion. Also known as local base level. { 'tem·pə,rer·ē 'bās ,lev·əl }

temporary file [COMPUT SCI] A file that is created during the execution of a computer program to hold interim results and is erased before the program is completed. { 'tem·pə,rer·ē 'fīl }

temporary geographic grid [ORD] A particular grid specified by military authorities for temporary and local use. { 'tem·pə,rer·ē ¦jē·ə¦graf·ik 'grid }

temporary hardness [CHEM] The portion of the total hardness of water that can be removed by boiling whereby the soluble calcium and magnesium bicarbonate are precipitated as insoluble carbonates. { 'tem·pə,rer·ē 'härd·nəs }

temporary plankton See meroplankton. { 'tem·pə,rer·ē 'plaŋk·tən }

temporary storage [COMPUT SCI] The storage capacity reserved or used for retention of temporary or transient data. { 'tem·pə,rer·ē 'stor·ij }

temporary structures [CIV ENG] Structures used to facilitate the construction of buildings, bridges, tunnels, and other above- and below-ground facilities by providing access, support, and protection for the facility as well as assuring the safety of the workers and the public. { ¦tem·pə,rer·ē 'strək·chərz }

TEM wave See transverse electromagnetic wave. { ,tē,ē'em ,wāv }

tenacity [TEXT] In yarn manufacture and textile engineering, the tensile strength of a yarn or a filament for its given size. { tə'nas·əd·ē }

Ten Broecke chart [THERMO] A graphical plot of heat transfer and temperature differences used to calculate the thermal efficiency of a countercurrent cool-fluid-warm-fluid heat-exchange system. { 'ten ,brü·kə ,chärt }

tendency [METEOROL] The local rate of change of a vector or scalar quality with time at a given point in space. { 'ten·dən·sē }

tendency chart See change chart. { 'ten·dən·sē ,chärt }

tendency equation [METEOROL] An equation for the local change of pressure at any point in the atmosphere, derived by combining the equation of continuity and an integrated form of the hydrostatic equation. { 'ten·dən·sē i,kwā·zhən }

tendency interval [METEOROL] The finite increment of time over which a change of the value of a meteorological element is measured in order to estimate its tendency; the most familiar example is the three-hour time interval over which local pressure differences are measured in determining pressure tendency. { 'ten·dən·sē ,in·tər·vəl }

tender [MECH ENG] A vehicle that is attached to a locomotive and carries supplies of fuel and water. [NAV ARCH] A naval auxiliary ship that serves as a mobile base for repair and limited resupply of other ships. { 'ten·dər }

tender plant [BOT] A plant that is incapable of resisting cold. { 'ten·dər ,plant }

Tendipedidae [INV ZOO] The midges, a family of orthorrhaphous dipteran insects in the series Nematocera whose larvae occupy intertidal wave-swept rocks on the seacoasts. { ,ten·də'ped·ə,dē }

tendon [ANAT] A white, glistening, fibrous cord which joins a muscle to some movable structure such as a bone or cartilage; tendons permit concentration of muscle force into a small area and allow the muscle to act at a distance. [CIV ENG] A steel bar or wire that is tensioned, anchored to formed concrete, and allowed to regain its initial length to induce compressive stress in the concrete before use. { 'ten·dən }

tendonitis [MED] Inflammation of a tendon. { ,ten·də'nīd·əs }

tendon sheath [ANAT] The synovial membrane surrounding a tendon. { 'ten·dər ,shēth }

tendril [BOT] A stem modification in the form of a slender coiling structure capable of twining about a support to which the plant is then attached. { 'ten·drəl }

tenebrescence [PHYS] Darkening and bleaching under suitable irradiation; materials having this property are called scotophors; darkening may be produced by primary x-rays or cathode rays, while bleaching may be produced by heat or by photons of appropriate wavelength. { ,ten·ə'bres·əns }

Tenebrionidae [INV ZOO] The darkling beetles, a large cosmopolitan family of coleopteran insects in the superfamily Tenebrionoidea; members are common pests of grains, dried fruits, beans, and other food products. { tə,neb·rē'än·ə,dē }

Tenebrionoidea [INV ZOO] A superfamily of the Coleoptera in the suborder Polyphaga. { tə,neb·rē·ə'nóid·ē·ə }

tenggara [METEOROL] A strong, dry, hazy, east or southeast wind during the east monsoon in the Spermunde Archipelago. { teŋ'gär·ə }

teniae coli [HISTOL] The three bands comprising the longitudinal layer of the tunica muscularis of the colon: the tenia libera, tenia mesocolica, and tenia omentalis. { 'tē·nē,ē 'kō·lī }

tennantite [MINERAL] $(Cu,Fe)_{12}As_4S_{13}$ A lead-gray mineral crystallizing in the isometric system; it is isomorphous with tetrahedrite; an important ore of copper. { 'ten·ən,tīt }

tenon [ENG] A tonguelike projection from the end of a framing member which is made to fit into a mortise. { 'ten·ən }

tenon saw [ENG] A precision saw that has a metal strip for stiffening along its back. { 'ten·ən ,só }

tenorite [MINERAL] CuO A triclinic mineral that occurs in small, shining, steel-gray scales, in black powder, or in black earthy masses; an ore of copper. { 'ten·ə,rīt }

tenosynovitis [MED] Inflammation of a tendon and its sheath. { ¦ten·ō,sī·nə'vīd·əs }

tenrec [VERT ZOO] Any of about 30 species of unspecialized, insectivorous mammals which are indigenous to Madagascar and have poor vision and clawed digits. { 'ten,rek }

Tenrecidae [VERT ZOO] The tenrecs, a family of insectivores in the group Lipotyphla. { ten'res·ə,dē }

ten's complement [MATH] In decimal arithmetic, the unique numeral that can be added to a given N-digit numeral to form a sum equal to 10^N (that is, a one followed by N zeros). { 'tenz 'käm·plə·mənt }

tensile bar [ENG] A molded, cast, or machined specimen of specified cross-sectional dimensions used to determine the tensile properties of a material by use of a calibrated pull test. Also known as tensile specimen; test specimen. { 'ten·səl ,bär }

tensile modulus [MECH] The tangent or secant modulus of elasticity of a material in tension. { 'ten·səl ,mäj·ə·ləs }

tensile specimen See tensile bar. { 'ten·səl ,spes·ə·mən }

tensile strength [MECH] The maximum stress a material subjected to a stretching load can withstand without tearing. Also known as hot strength. { 'ten·səl ,streŋkth }

tensile stress [MECH] Stress developed by a material bearing a tensile load. { 'ten·səl ,stres }

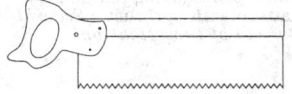

tensile test [ENG] A test in which a specimen is subjected to increasing longitudinal pulling stress until fracture occurs. { 'ten·səl ˌtest }

tensimeter [ENG] A device for measuring differences in the vapor pressures of two liquids in which the liquids are placed in sealed, evacuated bulbs connected by a differential manometer. { ten'sim·əd·ər }

tensiometer method [FL MECH] A method of determining the surface tension of a liquid that involves measuring the force necessary to remove a ring of known radius from the liquid surface, usually by means of a torsion balance. { ˌten·sē'äm·əd·ər ˌmeth·əd }

tensiometry [ENG] A discipline concerned with the measurement of tension or tensile strength. { ˌten·sē'äm·ə·trē }

tension [MECH] **1.** The condition of a string, wire, or rod that is stretched between two points. **2.** The force exerted by the stretched object on a support. [MECH ENG] A device on a textile manufacturing machine or a sewing machine that regulates the tautness and the movement of the thread or the fabric. Also known as tension device. { 'ten·chən }

tension crack [GEOL] An extension fracture caused by tensile stress. { 'ten·chən ˌkrak }

tension device See tension. { 'ten·chən di,vīs }

tension fault [GEOL] A fault in which crustal tension is a factor, such as a normal fault. Also known as extensional fault. { 'ten·chən ˌfȯlt }

tension fracture [GEOL] A minor rock fracture developed at right angles to the direction of maximum tension. Also known as subsidiary fracture. { 'ten·chən ˌfrak·chər }

tension headache See muscle-contraction headache. { 'ten·chən ˌhed,āch }

tension jack [MIN ENG] A type of jack with a jackscrew for wedging against the mine roof and a ratchet device for applying tension on a chain that is attached to the tail or foot section of a belt conveyor, and used to restore the proper tension to the belt. { 'ten·chən ˌjak }

tension joint [GEOL] A joint that is a tension fracture. { 'ten·chən ˌjȯint }

tension member [CIV ENG] A structural member subject to tensile stress. { 'ten·chən ˌmem·bər }

tension packer [PETRO ENG] A device to pressure-seal the annular space between an oil-well casing and tubing, held in place by tension against an upward push; a type of production packer. { 'ten·chən 'pak·ər }

tension pulley [MECH ENG] A pulley around which an endless rope passes mounted on a trolley or other movable bearing so that the slack of the rope can be readily taken up by the pull of the weights. { 'ten·chən ˌpu̇l·ē }

tension rod [DES ENG] A rod held in place by tension devices at the ends, such as a rod for a clothes closet. [ENG] A rod in a truss or other structure that connects opposite parts in order to prevent their spreading. { 'ten·chən ˌräd }

tension-type hanger [PETRO ENG] A type of tubing hanger for multiple- completion oil wells to allow for the varying lengths of tubing strings. { 'ten·chən ˌtīp 'haŋ·er }

tension wood [BOT] In some hardwood trees, wood characterized by the presence of gelatinous fibers and excessive longitudinal shrinkage; causes trees to lean. { 'ten·chən ˌwu̇d }

tensometer [ENG] A portable machine that is used to measure the tensile strength and other mechanical properties of materials. { ten'säm·əd·ər }

tensor [MATH] **1.** An object relative to a locally euclidean space which possesses a specified system of components for every coordinate system and which changes under a transformation of coordinates. **2.** A multilinear function on the cartesian product of several copies of a vector space and the dual of the vector space to the field of scalars on the vector space. { 'ten·sər }

tensor analysis [MATH] The abstract study of mathematical objects having components which express properties similar to those of a geometric tensor; this study is fundamental to Riemannian geometry and the structure of Euclidean spaces. Also known as tensor calculus. { 'ten·sər ə,nal·ə·səs }

tensor calculus See tensor analysis. { 'ten·sər ˌkal·kyə·ləs }

tensor contraction [MATH] For a tensor having an upper and a lower index, summation over the components in which these indexes have the same value, in order to obtain a new tensor two lower in rank. { 'ten·sər kən'trak·shən }

tensor differentiation [MATH] An operation on a tensor in which a term involving a Christoffel symbol is subtracted from the ordinary derivative, to obtain another tensor of one higher rank. { 'ten·sər ˌdif·ə,ren·chē'ā·shən }

tensor field [MATH] A tensor or collection of tensors defined in some open subset of a Riemann space. { 'ten·sər ˌfēld }

tensor force [NUC PHYS] A spin-dependent force between nucleons, having the same form as the interaction between magnetic dipoles; it is introduced to account for the observed values of the magnetic dipole moment and electric quadrupole moment of the deuteron. { 'ten·sər ˌfȯrs }

tensorial set [MATH] Any collection of quantities that are associated with a system of spatial coordinates and which undergo a linear transformation when this system rotates; examples are the components of a tensor and the eigenfunctions of a quantum mechanical operator. { ten'sȯr·ē·əl 'set }

tensor muscle [PHYSIO] A muscle that stretches a part or makes it tense. { 'ten·sər ˌməs·əl }

tensor product [MATH] **1.** The product of two tensors is the tensor whose components are obtained by multiplying those of the given tensors. **2.** In algebra, a multiplicative operation performed between modules. { 'ten·sər ˌpräd·əkt }

tensor quantity [MATH] A quantity mathematically represented by a tensor or possessing properties analogous to a tensor. { 'ten·sər ˌkwän·əd·ē }

tensor space [MATH] A fiber bundle composed of the points of a Riemannian manifold and tensor fields. { 'ten·sər ˌspās }

tentacle [INV ZOO] Any of various elongate, flexible processes with tactile, prehensile, and sometimes other functions, and which are borne on the head or about the mouth of many animals. { 'ten·tə·kəl }

Tentaculata [INV ZOO] A class of the phylum Ctenophora whose members are characterized by having variously modified tentacles. { ten,tak·yə'läd·ə }

tentaculocyst [INV ZOO] A sense organ located at the margin of the umbrella in some cnidarian medusoids, consisting of a modified tentacle with a cavity that often contains lithites. { ten'tak·yə·lō,sist }

tentaculozoid [INV ZOO] A slender tentacular individual of a hydrozoan colony. { ten'tak·yə·lō'zō,ȯid }

tented arch [FOREN SCI] A fingerprint pattern which possesses an angle, an upthrust, or two of the three basic characteristics of a loop. { 'ten·təd 'ärch }

tented ice [OCEANOGR] A type of pressure ice formed when ice is pushed up vertically, producing a flat-sided arch with a cavity between the raised ice and the water beneath. { 'ten·təd 'īs }

tenterhook [TEXT] One of the nails used to hold fabric on a frame (tenter) so that alterations in the original measurements can be corrected. { 'ten·tər,hu̇k }

tenthmeter See angstrom. { 'tenth,mēd·ər }

Tenthredinidae [INV ZOO] A family of hymenopteran insects in the superfamily Tenthredinoidea including economically important species whose larvae are plant pests. { ˌten·thrə'din·ə,dē }

Tenthredinoidea [INV ZOO] A superfamily of Hymenoptera in the suborder Symphyla. { ˌten·thrə·də'nȯid·ē·ə }

tentillum [INV ZOO] A contractile branch of a tentacle containing many nematocysts in certain siphonophores. { ten'til·əm }

tenting [OCEANOGR] The vertical displacement upward of ice under pressure to form a flat-sided arch with a cavity beneath. { 'tent·iŋ }

tentoxin [PL PATH] A species-selective pathotoxin produced by the fungus *Alternaria alternata* that causes variegated chlorosis in cucumber, cotton, lettuce, and many other sensitive plants. { 'ten,täk·sən }

Tenuipalpidae [INV ZOO] A small family of mites in the suborder Trombidiformes. { ˌten·yə·wə'pal·pə,dē }

TEP See triethyl phosphate.

tepary bean [BOT] *Phaseolus acutifolius* var. *latifolius*. One of the four species of beans of greatest economic importance in the United States. { 'tep·ə·rē ˌbēn }

tepee butte [GEOGR] A tepeelike hill or knoll, especially one comprising soft material capped by more resistant rock. { 'tē·pē ˌbyüt }

tepee structure [GEOL] A disharmonic sedimentary structure consisting of a fold that resembles an inverted depressed V in cross section. { 'tē·pē ˌstrək·chər }

tepetate See caliche. { ˌtep·ə'täd·ē }

TENTED ARCH

Fingerprint having tented arch pattern with upthrust. *(Federal Bureau of Investigation)*

TENTHREDINIDAE

Representative species of family Tenthredinidae. *(From T. I. Storer and R. L. Usinger, General Zoology, 3d ed., McGraw-Hill, 1957)*

tephigram [METEOROL] A thermodynamic diagram designed by Napier Shaw with temperature and logarithm of potential temperature as coordinates; isobars are gently curved lines and the chart is rotated so that pressure increases downward; vapor lines and saturation adiabats are curved; on this chart, energy is proportional to the area enclosed by the curve representing the process. { 'tef·ə,gram }

tephra [GEOL] All pyroclastics of a volcano. { 'tef·rə }

tephrite [PETR] A group of basaltic extrusive rocks composed chiefly of calcic plagioclase, augite, and nepheline or leucite, with some sodic sanidine. { 'te,frīt }

Tephritidae [INV ZOO] The fruit flies, a family of myodarian cyclorrhaphous dipteran insects in the subsection Acalyptratae. { tə'frid·ə,dē }

tephrochronology [GEOL] The dating of different layers of volcanic ash for the establishment of a sequence of geologic and archeologic occurrences. { ,tef·rō·krə'näl·ə·jē }

tephroite [MINERAL] Mn_2SiO_4 An olivine mineral that occurs with zinc and manganese minerals. { 'tef·rō,īt }

tephrostratigraphy [GEOL] The use of pyroclastic layers, in particular volcanic ash, as a correlational tool in the study of stratigraphic sequences. { ,tef·rō·strə'tig·rə·fē }

tequila [FOOD ENG] A distilled spirit that is Mexican in origin; obtained by the natural fermentation of the sweet sap of the agave plant and the distillation of the fermentation product. { tə'kē·lə }

tera- [MATH] A prefix representing 10^{12}, which is equivalent to 1,000,000,000,000 or a million million. Abbreviated T. { 'ter·ə }

terahertz [PHYS] A unit of frequency, equal to 10^{12} hertz, or 1,000,000 megahertz. Abbreviated THz. { 'ter·ə,hərts }

terahertz gap [ELECTROMAG] A region of the electromagnetic spectrum, roughly in the frequency range from 0.05 to 20 terahertz, that lies at the boundary between the microwave region where electronic devices such as high-speed transistors operate and the infrared and visible regions where photonic devices such as lasers operate, and whose exploitation has therefore suffered from a lack of bright sources or coherent sensitive detectors. { 'ter·ə,hərts ,gap }

terahertz technology [ENG] The generation, detection, and application (such as in communications and imaging) of electromagnetic radiation roughly in the frequency range from 0.05 to 20 terahertz, corresponding to wavelengths from 6 millimeters down to 15 micrometers. { ,ter·ə,hərts tek'näl·ə·jē }

teraohm [ELEC] A unit of electrical resistance, equal to 10^{12} ohms, or 1,000,000 megohms. Abbreviated TΩ { 'ter·ə,ōm }

teraohmmeter [ENG] An ohmmeter having a teraohm range for measuring extremely high insulation resistance values. { ,ter·ə'ōm,mēd·ər }

T-E ratio See temperature-efficiency ratio. { ,tē'ē 'rā·shō }

teratocarcinoma [MED] A teratoma with carcinomatous elements. { ,ter·ə·tō,kärs·ən'ō·mə }

teratogen [MED] An agent causing formation of a congenital anomaly or monstrosity. { tə'rad·ə·jən }

teratogenesis [MED] The formation of a fetal monstrosity. { ,te·rə·tō'jen·ə·səs }

teratology [MED] The science of fetal malformations and monstrosities. { ,ter·ə'täl·ə·jē }

teratoma [MED] A true neoplasm composed of bizarre and chaotically arranged tissues that are foreign embryologically as well as histologically to the area in which the tumor is found. { ,ter·ə'tō·mə }

teratophobia [PSYCH] An abnormal fear of deformed people and pregnant women. { ,ter·əd·ə'fō·bē·ə }

Teratornithidae [PALEON] An extinct family of vulturelike birds of the Pleistocene of western North America included in the order Falconiformes. { ,ter·ə·tόr'nith·ə,dē }

terawatt [PHYS] A unit of power, equal to 10^{12} watts, or 1,000,000 megawatts. Abbreviated TW. { 'ter·ə,wät }

terbacil [ORG CHEM] $C_9H_{13}ClN_2O_2$ A colorless, crystalline compound with a melting point of 175–177°C; used as an herbicide to control weeds in sugarcane, apples, peaches, citrus, and mints. { 'tər·bə,sil }

terbia See terbium oxide. { 'tər·bē·ə }

terbium [CHEM] A rare-earth element, symbol Tb, in the yttrium subgroup of the transition elements, atomic number 65, atomic weight 158.9254. { 'tər·bē·əm }

terbium chloride [INORG CHEM] $TbCl_3·6H_2O$ Water- and

alcohol-soluble, hygroscopic, colorless, transparent prisms; anhydrous form melts at 588°C. { 'tər·bē·əm 'klόr,īd }

terbium nitrate [INORG CHEM] $Tb(NO_3)_3·6H_2O$ A colorless, fire-hazardous (strong oxidant) powder, soluble in water; melts at 89°C. { 'tər·bē·əm 'nī,trāt }

terbium oxide [INORG CHEM] Tb_2O_3 A slightly hygroscopic, dark-brown powder soluble in dilute acids, absorbs carbon dioxide from air. Also known as terbia. { 'tər·bē·əm 'äk,sīd }

terbutol [ORG CHEM] The common name for the herbicide 2,6-di-*tert*-butyl-*p*-tolylmethylcarbamate; used as a selective preemergence crabgrass herbicide for turf. { 'tər·byə,tόl }

terbutryn [ORG CHEM] $C_{13}H_{19}N_5S$ A colorless powder with a melting point of 104–105°C; used for weed control for wheat, barley, and grain sorghum. { tər'byü·trən }

terbutylazine [ORG CHEM] $C_9H_{16}N_5Cl$ A white solid with a melting point of 177–179°C; used as a preemergence herbicide. { ,tər,byüd·əl'hī·lə,zēn }

tercentesimal thermometric scale See approximate absolute temperature. { ,tər·sen'tes·ə·məl ,thər·mə'me·trik 'skāl }

TERCOM See terrain contour matching. { 'tər,käm }

terdiurnal [METEOROL] Pertaining to a meteorological event that occurs three times a day. { ,tər·dī'ərn·əl }

Terebellidae [INV ZOO] A family of polychaete annelids belonging to the Sedentaria which are chiefly large, thick-bodied, tubicolous forms with the anterior end covered by a matted mass of tentacular cirri. { ,ter·ə'bel·ə,dē }

Terebratellidina [PALEON] An extinct suborder of articulate brachiopods in the order Terebratulida in which the loop is long and offers substantial support to the side arms of the lophophore. { ,ter·ə·brə,tel·ə'dīn·ə }

Terebratulida [INV ZOO] An order of articulate brachiopods that has a punctate shell structure and is characterized by the possession of a loop extending anteriorly from the crural bases, providing some degree of support for the lophophore. { ,ter·ə·brə'tül·əd·ə }

Terebratulidina [INV ZOO] A suborder of articulate brachiopods in the order Terebratulida distinguished by a short V- or W-shaped loop. { ,ter·ə·brach·ə·lə'dīn·ə }

Teredinidae [INV ZOO] The pileworms or shipworms, a family of bivalve mollusks in the subclass Eulamellibranchia distinguished by having the two valves reduced to a pair of small plates at the anterior end of the animal. { ,ter·ə'din·ə,dē }

terephthalic acid [ORG CHEM] $C_6H_4(COOH)_2$ A combustible white powder, insoluble in water, soluble in alkalies, sublimes above 300°C; used to make polyester resins for fibers and films and as an analytical reagent and poultry-feed additive. Also known as *para*-phthalic acid; TPA. { ,ter·əf'thal·ik 'as·əd }

terephthaloyl chloride [ORG CHEM] $C_6H_4(COCl)_2$ Colorless needles with a melting point of 82–84°C; soluble in ether; used in the manufacture of dyes, synthetic fibers, resins, and pharmaceuticals. { ,ter·əf'thal·ə,wil 'klόr,īd }

terete [BOT] Of a stem, cylindrical in section, but tapering at both ends. { tə'rēt }

tergite [INV ZOO] The dorsal plate covering a somite in arthropods and certain other articulate animals. { 'tər,jīt }

tergum [INV ZOO] A dorsal plate of the operculum in barnacles. { 'tər·gəm }

terlinguaite [MINERAL] Hg_2OCl A sulfur yellow to greenish-yellow, monoclinic mineral consisting of an oxychloride of mercury. { tər'liŋ·gwə,īt }

term [MATH] **1.** For an expression, any one of several quantities whose sum is the expression. **2.** For a fraction, either the numerator or the denominator. [SPECT] A set of $(2S-1)(2L+1)$ atomic states belonging to a definite configuration and to definite spin and orbital angular momentum quantum numbers S and L. { tərm }

terminal [ARCH] The ornamental finish, decorative element, or termination of an object, item of construction, or structural part. [COMPUT SCI] A site or location at which data can leave or enter a system. [ELEC] **1.** A screw, soldering lug, or other point to which electric connections can be made. Also known as electric terminal. **2.** The equipment at the end of a microwave relay system or other communication channel. **3.** One of the electric input or output points of a circuit or component. [MATH] The vertex with outdegree 0 that is specified in the definition of an *s-t* network. { 'tər·mən·əl }

terminal air-traffic control [NAV] A separation service

offered to aircraft operating in the vicinity of and at airports, involving a system of crewed control towers, radio communications, radar and other position location and guidance facilities, and information processing to make for the safe approach and landing of these aircraft. { 'tər·mən·əl 'er ˌtraf·ik kən,trōl }

terminal area [ELECTR] The enlarged portion of conductor material surrounding a hole for a lead on a printed circuit. Also known as land; pad. { 'tər·mən·əl ˌer·ē·ə }

terminal bar [CYTOL] One of the structures formed in certain epithelial cells by the combination of local modifications of contiguous surfaces and intervening intercellular substances; they become visible with the light microscope after suitable staining and appear to close the spaces between the epithelial cells of the intestine at their free surfaces. { 'tər·mən·əl ˌbär }

terminal block [COMMUN] A cluster of five captive screw terminals at which a telephone pair terminates; the center terminal is for the ground wire, and two other terminals are used for the tip and ring wires. { 'tər·mən·əl ˌbläk }

terminal board [ELEC] An insulating mounting for terminal connections. Also known as terminal strip. { 'tər·mən·əl ˌbȯrd }

terminal box [ELEC] An enclosure which includes, mounts, and protects one or more terminals or terminal boards; it may include a cover and such accessories as mounting hardware, brackets, locks, and conduit fittings. { 'tər·mən·əl ˌbäks }

terminal bud [BOT] A bud that develops at the apex of a stem. Also known as apical bud. { 'tər·mən·əl ˌbəd }

terminal clearance capacity [ENG] The amount of cargo or personnel that can be moved through and out of a terminal on a daily basis. { 'tər·mən·əl 'klir·əns kə,pas·əd·ē }

terminal control area [NAV] A control area or a portion thereof normally situated at the confluence of air-traffic control routes in the vicinity of one or more major airfields. { 'tər·mən·əl kən'trōl ˌer·ē·ə }

terminal cutout pairs [ELEC] Numbered, designated pairs brought out of a cable at a terminal. { 'tər·mən·əl 'kəd,aut ˌperz }

terminal disinfection [MICROBIO] The disinfection of sickrooms occupied by patients suffering from contagious disease. { ˌtərm·ən·əl ˌdis·in'fek·shən }

terminal endocarditis See verrucous endocarditis. { 'tər·mən·əl ˌen·dō·kär'dīd·əs }

terminal equipment [COMMUN] **1.** Assemblage of communications-type equipment required to transmit or receive a signal on a channel or circuit, whether it be for delivery or relay. **2.** In radio relay systems, equipment used at points where intelligence is inserted or derived, as distinct from equipment used to relay a reconstituted signal. **3.** Telephone and teletypewriter switchboards and other centrally located equipment at which wire circuits are terminated. { 'tər·mən·əl i,kwip·mənt }

terminal forecast [METEOROL] An aviation weather forecast for one or more specified air terminals. { 'tər·mən·əl ˌfȯr,kast }

terminal guidance [NAV] Guidance of a craft from an arbitrary point, at which midcourse guidance ends, to the destination. { 'tər·mən·əl ˌgīd·əns }

terminal hair [ANAT] One of three types of hair in man based on hair size, time of appearance, and structural variations; the larger, coarser hair in the adult that replaces the vellus hair. { 'tər·mən·əl ˌher }

terminal indecomposable future [RELAT] A means of attaching a causal boundary to a space-time; it is the future of a past inextendible timelike curve. Abbreviated TIF. { 'tər·mən·əl ˌin·di·kəmˌpōz·ə·bəl 'fyü·chər }

terminal indecomposable past [RELAT] A means of attaching a causal boundary to a space-time; it is the past of a future inextendible timelike curve. Abbreviated TIP. { 'tər·mən·əl ˌin·di·kəmˌpōz·ə·bəl 'past }

terminal leg See terminal stub. { 'tər·mən·əl ˌleg }

terminal line [MATH] One of the two rays that form an angle and may be regarded as having been rotated about a fixed point on another line (the initial line) to form the angle. { 'tərm·ən·əl ˌlīn }

terminal moraine [GEOL] An end moraine that extends as an arcuate or crescentic ridge across a glacial valley; marks the farthest advance of a glacier. Also known as marginal moraine. { 'tər·mən·əl mə'rān }

terminal nerve [NEUROSCI] Either of a pair of small cranial nerves that run from the nasal area to the forebrain, present in most vertebrates; the function is not known. { 'tər·mən·əl ˌnərv }

terminal network [COMPUT SCI] A system that links intelligent terminals through a communications channel. { 'tər·mən·əl 'net,wərk }

terminal operations [ENG] The reception, processing, and staging of passengers; the receipt, transit storage, and marshaling of cargo; the loading and unloading of ships or aircraft; and the manifesting and forwarding of cargo and passengers to destination. { 'tər·mən·əl ˌäp·ə'rā·shənz }

terminal pair [ELEC] An associated pair of accessible terminals, such as the input or output terminals of a device or network. { 'tər·mən·əl 'per }

terminal phase [ORD] The period of flight of a missile between the end of midcourse guidance and impact. { 'tər·mən·əl ˌfāz }

terminal pressure [ENG] A pressure drop across a unit when the maximum allowable pressure drop is reached, as for a filter press. { 'tər·mən·əl ˌpresh·ər }

terminal radar cab [NAV] An area in the main control tower of an airport where radar control and air-traffic control facilities are located together. { 'tər·mən·əl 'rā,där ˌkab }

terminal repeater [COMMUN] **1.** Assemblage of equipment designed specifically for use at the end of a communications circuit, as contrasted with the repeater designed for an intermediate point. **2.** Two microwave terminals arranged to provide for the interconnection of separate systems, or separate sections of a system. { 'tər·mən·əl ri'pēd·ər }

terminal room [COMMUN] In telephone practice, a room associated with a central office, private branch exchange, or private exchange, which contains distributing frames, relays, and similar apparatus, except that mounted in the switchboard section. { 'tər·mən·əl ˌrüm }

terminal sinus [EMBRYO] The vascular sinus bounding the area vasculosa of the blastoderm of a meroblastic ovum. Also known as marginal sinus. { 'tər·mən·əl 'sī·nəs }

terminal speed See terminal velocity. { 'tər·mən·əl ˌspēd }

terminal station [COMMUN] Receiving equipment and associated multiplex equipment used at the ends of a radio-relay system. { 'tər·mən·əl ˌstā·shən }

terminal strip See terminal board. { 'tər·mən·əl ˌstrip }

terminal stub [ELEC] Piece of cable that comes with a cable terminal for splicing into the main cable. Also known as terminal leg. { 'tər·mən·əl ˌstəb }

terminal throw velocity [ENG] The velocity at which a stream of air exiting a diffuser impinges on an object or surface. { 'tər·mən·əl 'thrō vəˌläs·əd·ē }

terminal unit [MECH ENG] In an air-conditioning system, a unit at the end of a branch duct through which air is transferred or delivered to the conditioned space. { 'tər·mən·əl ˌyü·nət }

terminal velocity [FL MECH] The velocity with which a body moves relative to a fluid when the resultant force acting on it (due to friction, gravity, and so forth) is zero. [PHYS] The maximum velocity attainable, especially by a freely falling body, under given conditions. Also known as terminal speed. { 'tər·mən·əl və'läs·əd·ē }

terminal vertex [MATH] A vertex in a rooted tree that has no successor. Also known as leaf. { 'tər·mən·əl 'vər,teks }

terminal very-high-frequency omnirange [NAV] A low-power very-high-frequency omnidirectional radio range intended to furnish service in the local area (about 30-mile, or 48-kilometer, radius). { 'tər·mən·əl 'ver·ē 'hī 'frē·kwən·sē 'äm·ni,ränj }

terminal voltage [ELEC] The voltage at the terminals connected to the source of electricity for an electric machine. { 'tər·mən·əl ˌvōl·tij }

terminate and stay resident See RAM resident. { ˌtər·məˌnāt ən ˌstā 'rez·ə·dənt }

terminated line [ELEC] Transmission line terminated in a resistance equal to the characteristic impedance of the line, so there is no reflection and no standing waves. { 'tər·məˌnād·əd 'līn }

terminating [ELEC] Closing of the circuit at either end of a line or transducer by connecting some device thereto; terminating does not imply any special condition such as the elimination of reflection. { 'tər·məˌnād·iŋ }

terminating continued fraction [MATH] A continued fraction that has a finite number of terms. { ¦tər·mə‚nād·iŋ kən¦tin·yüd 'frak·shən }

terminating decimal [MATH] A decimal that has only a finite number of nonzero digits to the right of the decimal point. Also known as finite decimal. { ‚tər·mə¦nād·iŋ 'des·məl }

termination [CHEM] The steps that end a chain reaction by destroying or rendering inactive the reactive intermediates. [ELECTROMAG] **1.** Load connected to a transmission line or other device; to avoid wave reflections, it must match the characteristic of the line or device. **2.** In waveguide technique, the point at which energy flowing along a waveguide continues in a nonwaveguide mode of propagation. { ‚tər·mə'nā·shən }

termination step [CHEM] In a chain reaction, the mechanism that halts the reaction. { ‚tər·mə'nā·shən ‚steps }

terminator [ASTRON] The line of demarcation between the dark and light portions of the moon or planets. { 'tər·mə‚nād·ər }

terminator codon [GEN] A UAA, UAG, or UGA trinucleotide in messenger ribonucleic acid (mRNA) that specifies termination of synthesis of the polypeptide (protein) product of the gene. Also known as stop codon. { 'tər·mə‚nād·ər 'kō‚dän }

Termitaphididae [INV ZOO] The termite bugs, a small family of Hemiptera in the superfamily Aradoidea. { ter‚mīd·ə'fid·ə‚dē }

termitarium [INV ZOO] A termites' nest. { ‚tər·mə'ter·ē·əm }

termite [INV ZOO] A soft-bodied insect of the order Isoptera; individuals feed on cellulose and live in colonies with a caste system comprising three types of functional individuals: sterile workers and soldiers, and the reproductives. Also known as white ant. { 'tər‚mīt }

termite shield [BUILD] A strip of metal, usually galvanized iron, bent down at the edges and placed between the foundation of a house and a timber floor, around pipes, and other places where termites can pass. { 'tər‚mīt ‚shēld }

termiticole [ECOL] An organism that lives in a termites' nest. { tər'mīd·ə‚kōl }

Termitidae [INV ZOO] A large family of the order Isoptera which contains the higher termites, representing 80% of the species. { tər'mid·ə‚dē }

termitophile [ECOL] An organism that lives in a termites' nest in a symbiotic association with the termites. { tər'mīd·ə‚fīl }

Termopsidae [INV ZOO] A family of insects in the order Isoptera composed of damp wood-dwelling forms. { tər'mäp·sə‚dē }

term rank [MATH] For a matrix in which each entry is either 0 or 1, the largest number of 1's that can be chosen from the matrix so that no two of them lie in the same row or in the same column. { 'tərm ‚raŋk }

term splitting [QUANT MECH] The separation of the energies of the states in a term; in the Russell-Saunders case this is produced by the spin-orbit interaction. { 'tərm ‚splid·iŋ }

ternary [SCI TECH] Consisting of three, as in a three-phase (that is, ternary) liquid system. { 'tər·nə·rē }

ternary alloy [MET] An alloy composed of three principal elements. { 'tər·nə·rē 'al‚ói }

ternary code [COMMUN] Code in which each code element may be any one of three distinct kinds or values. { 'tər·nə·rē ‚kōd }

ternary compound [CHEM] A molecule consisting of three different types of atoms; for example, sulfuric acid, H_2SO_4. { 'tər·nə·rē 'käm‚paùnd }

ternary diagram [PETR] A triangular diagram that graphically depicts the composition of a three-component mixture or ternary system. { 'tər·nə·rē 'dī·ə‚gram }

ternary expansion [MATH] The numerical representation of a real number relative to the base 3, the digits determined by how the given number can be written in terms of powers of 3. { 'tər·nə·rē ik'span·chən }

ternary incremental representation [COMPUT SCI] A type of incremental representation in which the value of the change in a variable is defined as +1, −1, or 0. { 'tər·nə·rē ‚in·krə'ment·əl ‚rep·ri·zən'tā·shən }

ternary notation [MATH] A system of notation using the base of 3 and the characters 0, 1, and 2. { 'tər·nə·rē nō'tā·shən }

ternary pulse code modulation [COMMUN] Pulse code

modulation in which each code element may be any one of three distinct kinds or values. { 'tər·nə·rē 'pəls ‚kōd ‚mäj·ə'lā·shən }

ternary quantic [MATH] A quantic that contains three variables. { 'tər·nə·rē 'kwän·tik }

ternary system [CHEM] Any system with three nonreactive components; in liquid systems, the components may or may not be partially soluble. { 'tər·nə·rē 'sis·təm }

ternate [BOT] Composed of three subdivisions, as a leaf with three leaflets. { 'tər‚nāt }

terne [MET] A lead alloy having a composition of 10–20% tin and 80–90% lead; used to coat iron or steel surfaces. { tərn }

terneplate [MATER] A sheet of iron or steel coated with a lead-tin alloy; used chiefly for roofing. { 'tərn‚plāt }

Ternifine man [PALEON] The name for a fossil human type, represented by three lower jaws and a parietal bone discovered in France and thought to be from the upper part of the middle Pleistocene. { 'tər·nə‚fēn 'man }

terpene [ORG CHEM] **1.** $C_{10}H_{16}$ A moderately toxic, flammable, unsaturated hydrocarbon liquid found in essential oils and plant oleoresins; used as an intermediate for camphor, menthol, and terpineol. **2.** A class of naturally occurring compounds whose carbon skeletons are composed exclusively of isopentyl (isoprene) C_5 units. Also known as isoprenoid. { 'tər‚pēn }

terpene alcohol [ORG CHEM] A generic name for an alcohol related to or derived from a terpene hydrocarbon, such as terpineol or borneol. { 'tər‚pēn 'al·kə‚hól }

terpene hydrochloride [ORG CHEM] $C_{10}H_{16}$·HCl A solid, water-insoluble material melting at 125°C; used as an antiseptic. Also known as artificial camphor; dipentene hydrochloride; pinene hydrochloride; turpentine camphor. { 'tər‚pēn ¦hī·drə'klór‚īd }

terpenoid [ORG CHEM] Any compound with an isoprenoid structure similar to that of the terpene hydrocarbons. { 'tər·pə‚nóid }

***para*-terphenyl** [ORG CHEM] $(C_6H_5)_2C_6H_4$ A combustible, toxic liquid boiling at 405°C; crystals are used for scintillation counters; polymerized with styrene to make plastic phosphor. { ¦par·ə ¦tər'fen·əl }

terpineol [ORG CHEM] $C_{10}H_{17}OH$ A combustible, colorless liquid with a lilac scent, derived from pine oil, soluble in alcohol, slightly soluble in water, boils at 214–224°C; used in medicine, perfumes, soaps, and disinfectants, and as an antioxidant, a flavoring agent, and a solvent; isomeric forms are alpha-, beta- and gamma-. { tər'pin·ē‚ól }

terpin hydrate [ORG CHEM] $CH_3(OH)C_6H_9C(CH_3)_2OH$·$H_2O$ Combustible, efflorescent, lustrous white prisms soluble in alcohol and ether, slightly soluble in water; melts at 116°C; used for pharmaceuticals and to make terpineol. Also known as dipentene glycol. { 'tər·pən 'hī‚drāt }

terpinolene [ORG CHEM] $C_{10}H_{16}$ A flammable, water-white liquid insoluble in water, soluble in alcohol, ether, and glycols, boils at 184°C; used as a solvent and as a chemical intermediate for resins and essential oils. { tər'pin·ə‚lēn }

terpinyl acetate [ORG CHEM] $C_{10}H_{17}OOCCH_3$ A combustible, colorless, liquid slightly soluble in water and glycerol, soluble in water, boils at 220°C; used in perfumes and flavors. { 'tər·pən·əl 'as·ə‚tāt }

terpolymer [ORG CHEM] A polymer that contains three distinct monomers; for example, acrylonitrile-butadiene-styrene terpolymer, ABS. { 'tər‚päl·i·mər }

terra [ASTRON] A bright upland or mountainous region on the surface of the moon, characterized by a lighter color than that of a mare, a relatively high albedo, and a rough texture formed by large intersecting or overlapping craters. { 'ter·ə }

terra alba [MATER] Pure white uncalcined gypsum, used as a filler in paper and paints and as a nutrient in growing yeast. { 'ter·ə 'al·bə }

terrace [BUILD] **1.** A flat roof. **2.** A colonnaded promenade. **3.** An open platform extending from a building, usually at ground level. [GEOL] **1.** A horizontal or gently sloping embankment of earth along the contours of a slope to reduce erosion, control runoff, or conserve moisture. **2.** A narrow coastal strip sloping gently toward the water. **3.** A long, narrow, nearly level surface bounded by a steeper descending slope on one side and by a steeper ascending slope on the other

TERMITE

Winged female termite of the reproductive caste.

side. **4.** A benchlike structure bordering an undersea feature. { 'ter·əs }

terraced pool [GEOGR] A shallow, rimmed pool on the surface of a reef. { 'ter·əst 'pül }

terracette [GEOL] A small steplike form developed on the surface of a slumped soil mass along a steep grassy incline. { ˌter·ə¦set }

terrachlor See pentachloronitrobenzene. { 'ter·ə,klȯr }

terracing See contour plowing. { 'ter·əs·iŋ }

terra-cotta [MATER] A brownish-orange clay used in the production of high-quality earthenware, vases, and statuettes, and for tile floors and roofing. { ¦ter·ə¦käd·ə }

terrain-avoidance radar [NAV] Airborne radar which provides a display of terrain ahead of a low-flying aircraft to permit horizontal avoidance of obstacles. { tə'rān ə¦vȯid·əns 'rā,där }

terrain-clearance indicator See absolute altimeter. { tə'rān ¦klir·əns ,in·də,kād·ər }

terrain contour matching [NAV] A method of correcting errors in the inertial navigation system of a cruise missile, in which predicted terrain profile data stored in the missile's computer are compared with the terrain profile below the missile; calculated by subtracting the radar-altimeter-derived height of the missile above the terrain from the missile's altitude as derived from inertial navigator and barometric measurements. Abbreviated TERCOM. { tə'rān 'kan,tür ,mach·iŋ }

terrain echoes See ground clutter. { tə'rān 'ek·ōz }

terrain-following radar [NAV] Airborne radar which provides a display of terrain ahead of a low-flying aircraft to permit manual control, or signals for automatic control, to maintain constant altitude above the ground. { tə'rān ¦fäl·ə·wiŋ 'rā,där }

terrain profile recorder See airborne profile recorder. { tə'rān ¦prō,fīl ri,kȯrd·ər }

terrain sensing [ENG] The gathering and recording of information about terrain surfaces without actual contact with the object or area being investigated; in particular, the use of photography, radar, and infrared sensing in airplanes and artificial satellites. { tə'rān ,sens·iŋ }

terra japonica See gambir. { 'ter·ə jə'pän·ə·kə }

terral levante [METEOROL] A land breeze of Spain and Brazil, sometimes a northwest squall of foehn character. { tə'räl lə'vän·tä }

terra miraculosa See bole. { 'ter·ə mi,rak·yə'lō·sə }

terrane [GEOL] A rock formation, a cluster of rock formations, or the general area of outcrops. { tə'rān }

terrapin [VERT ZOO] Any of several North American tortoises in the family Testudinidae, especially the diamondback terrapin. { 'ter·ə·pən }

terra rossa [GEOL] A reddish-brown soil overlying limestone bedrock. { 'ter·ə 'rȯs·ə }

terrazzo [MATER] A mosaic flooring surface made by embedding marble or granite chips in mortar, allowing the mortar to harden, and then grinding and polishing the surface. { tə'rät·sō }

terrestrial [SCI TECH] Of or pertaining to the earth. { tə'res·trē·əl }

terrestrial coordinates See geographical coordinates. { tə'res·trē·əl kō'ȯrd·ən·əts }

terrestrial ecosystem [ECOL] A community of organisms and their environment that occurs on the landmasses of continents and islands. { tə,res·trē·əl 'ek·ō,sis·təm }

terrestrial electricity [GEOPHYS] Electric phenomena and properties of the earth; used in a broad sense to include atmospheric electricity. Also known as geoelectricity. { tə'res·trē·əl ,i,lek'tris·əd·ē }

terrestrial environment [GEOGR] The earth's land area, including its human-made and natural surface and subsurface features, and its interfaces and interactions with the atmosphere and the oceans. { tə'res·trē·əl in'vī·rən·mənt }

terrestrial equator See astronomical equator. { tə'res·trē·əl i'kwād·ər }

terrestrial frozen water [HYD] Seasonally or perennially frozen waters of the earth, exclusive of the atmosphere. { tə'res·trē·əl 'frō·zən 'wȯd·ər }

terrestrial gravitation [GEOPHYS] The effect of gravitational attraction of the earth. { tə'res·trē·əl ,grav·ə'tā·shən }

terrestrial guidance See terrestrial-reference guidance. { tə'res·trē·əl 'gīd·əns }

terrestrial magnetism See geomagnetism. { tə'res·trē·əl 'mag·nə,tiz·əm }

terrestrial meridian See astronomical meridian. { tə'res·trē·əl mə'rid·ē·ən }

terrestrial planet [ASTRON] One of the four small planets near the sun (Earth, Mercury, Venus, and Mars). { tə'res·trē·əl 'plan·ət }

terrestrial radiation [GEOPHYS] Electromagnetic radiation originating from the earth and its atmosphere at wavelengths determined by their temperature. Also known as earth radiation; eradiation. { tə'res·trē·əl ,rād·ē'ā·shən }

terrestrial-reference flight [NAV] A flight with guidance by means of some terrestrial phenomena such as the magnetic or gravitational field. { tə'res·trē·əl 'ref·rəns ,flīt }

terrestrial-reference guidance [NAV] Long-range missile guidance in which the control system of the missile reacts to magnetic, gravitational, or other properties of the earth. Also known as terrestrial guidance. { tə'res·trē·əl 'ref·rəns ,gīd·əns }

terrestrial refraction [OPTICS] Any refraction phenomenon observed in the light originating from a source lying within the earth's atmosphere; this is applied only to refraction caused by inhomogeneities of the atmosphere itself, not, for example, to that caused by ice crystals suspended in the atmosphere. { tə'res·trē·əl ri'frak·shən }

terrestrial scintillation [OPTICS] A generic term for scintillation phenomena observed in light that reaches the eye from sources lying within the earth's atmosphere. Also known as atmospheric boil; atmospheric shimmer; optical haze. { tə'res·trē·əl ,sint·əl'ā·shən }

terrestrial sediment [GEOL] A sedimentary deposit on land above tidal reach. { tə'res·trē·əl 'sed·ə·mənt }

terrestrial telescope [OPTICS] Any telescope which produces an erect image. { tə'res·trē·əl 'tel·ə,skōp }

terrestrial triangle [NAV] A triangle on the surface of the earth, especially the navigational triangle. { tə'res·trē·əl 'trī,aŋ·gəl }

terrigenous sediment [GEOL] Shallow marine sedimentary deposits composed of eroded terrestrial material. { tə'rij·ə·nəs 'sed·ə·mənt }

territoriality [ZOO] A pattern of behavior in which one or more animals occupy and defend a definite area or territory. { ter·ə,tȯr·ē'al·əd·ē }

terry cloth [TEXT] An extremely absorbent woven or knitted cotton fabric, covered with loops on one or both sides. { 'ter·ē ,klȯth }

tert- [ORG CHEM] Abbreviation for tertiary; trisubstituted methyl radical with the central carbon attached to three other carbons ($R_1R_2R_3C-$). { tərt }

tertian malaria See vivax malaria. { 'tər·shən mə'ler·ē·ə }

Tertiary [GEOL] The older major subdivision (period) of the Cenozoic era, extending from the end of the Cretaceous to the beginning of the Quaternary, from 70,000,000 to 2,000,000 years ago. { 'tər·shē,er·ē }

tertiary air [MECH ENG] Combustion air added to primary and secondary air. { 'tər·shē,er·ē 'er }

tertiary alcohol [ORG CHEM] A trisubstituted alcohol in which the hydroxyl group is attached to a carbon that is joined to three carbons; for example, tert-butyl alcohol. { 'tər·shē,er·ē 'al·kə,hȯl }

tertiary amine [ORG CHEM] R_3N A trisubstituted amine in which the hydroxyl group is attached to a carbon that is joined to three carbons; for example, trimethylamine, $(CH_3)_3N$. { 'tər·shē,er·ē 'am,ēn }

tertiary carbon atom [ORG CHEM] A carbon atom bonded to three other carbon atoms with single bonds. { 'tər·shē,er·ē 'kär·bən 'ad·əm }

tertiary circulation [METEOROL] The generally small, localized atmospheric circulations, represented by such phenomena as local winds, thunderstorms, and tornadoes. { 'tər·shē,er·ē ,sər·kyə'lā·shən }

tertiary creep [MET] Creep strain occurring at an accelerating rate leading to fracture. { 'tər·shē,er·ē 'krēp }

tertiary crushing [MIN ENG] **1.** The preliminary breaking down of run-of-mine ore and sometimes coal. **2.** The third stage in crushing, following primary and secondary crushing. { 'tər·shē,er·ē 'krəsh·iŋ }

tertiary grinding [MIN ENG] The third-stage grinding in a

TERRAPIN

Diamondback terrapin
(*Malaclemys terrapin*).

TERTIARY

CENOZOIC	QUATERNARY	
	TERTIARY	
	CRETACEOUS	
MESOZOIC	JURASSIC	
	TRIASSIC	
PALEOZOIC	PERMIAN	
	CARBONIFEROUS	PENNSYLVANIAN
		MISSISSIPPIAN
	DEVONIAN	
	SILURIAN	
	ORDOVICIAN	
	CAMBRIAN	
	PRECAMBRIAN	

Chart showing the position of the Tertiary period in relation to the other periods and the eras of geologic time.

ball mill when a particularly fine grinding of ore is needed. { 'tər·shē,er·ē 'grīnd·iŋ }

tertiary hydrogen atom [ORG CHEM] A hydrogen atom that is bonded to a tertiary carbon atom. { 'tər·shē,er·ē 'hī·drə·jən 'ad·əm }

tertiary pyroelectricity [SOLID STATE] The polarization due to temperature and gradients and corresponding nonuniform stresses and strains when the crystal is heated nonuniformly; found in pyroelectric and nonpyroelectric crystals, that is, crystals which have no polar directions. Also known as false pyroelectricity. { 'tər·shē,er·ē |pī·rō,i,lek'tris·əd·ē }

tertiary recovery [PETRO ENG] A technique used to enhance the amount of oil recovered by secondary recovery methods. { 'tər·shē,er·ē ri'kəv·ə·rē }

tertiary sewage treatment [CIV ENG] A process for purification of wastewater in which nitrates and phosphates, as well as fine particles, are removed; the process follows removal of raw sludge and biological treatment. Also known as advanced sewage treatment. { |tər·shē,er·ē |sü·ij ,trēt·mənt }

tertiary sodium phosphate See trisodium phosphate. { 'tər·shē,er·ē 'sōd·ē·əm 'fä,sfāt }

tertiary storage [COMPUT SCI] Any of several types of computer storage devices, usually consisting of magnetic tape transports and mass storage tape systems, which have slower access times, larger capacity, and lower cost than main storage or secondary storage. { 'tər·shē,er·ē 'stòr·ij }

tertiary structure [BIOCHEM] The characteristic three-dimensional folding of the polypeptide chains in a protein molecule. { 'tər·shē,er·ē 'strək·chər }

tertiary winding See stabilized winding. { 'tər·shē,er·ē 'wīnd·iŋ }

teschemacherite [MINERAL] $(NH_4)HCO_3$ A colorless to white or yellowish, orthorhombic mineral consisting of ammonium bicarbonate; occurs as compact, crystalline masses. { 'tesh·ə,mäk·ə,rīt }

teschenite [PETR] A granular hypabyssal rock composed principally of calcic plagioclase, augite, and sometimes hornblende, with some brotite and analcime. { 'tesh·ə,nīt }

tesla [ELECTROMAG] The International System unit of magnetic flux density, equal to one weber per square meter. Symbolized T. { 'tes·lə }

Tesla coil [ELECTROMAG] An air-core transformer used with a spark gap and capacitor to produce a high voltage at a high frequency. { 'tes·lə ,kȯil }

Tessaratomidae [INV ZOO] A family of large tropical Hemiptera in the superfamily Pentatomoidea. { ,tes·ə·rə'täm·ə,dē }

Tessar lens [OPTICS] An anastigmatic lens made up of a negative lens at the aperture stop with two positive lenses, one in front and the other in back; the last positive lens is a cemented doublet. { 'te,sär ,lenz }

tesselation [MATH] A covering of a plane without gaps or overlappings by polygons, all of which have the same size and shape. { ,tes·ə'lā·shən }

tessellate [BOT] Marked by a pattern of small squares resembling a tiled pavement. { 'tes·ə,lāt }

tessera [MATER] A small rectangular piece of ceramic tile, stone, or other material used in a mosaic design. { 'tes·ə·rə }

tesseral harmonic [MATH] A spherical harmonic which is 0 on both a set of equally spaced meridians and a set of parallels of latitude of a sphere with center at the origin of spherical coordinates, dividing the sphere into rectangular and triangular regions. { |tes·ə·rəl här'män·ik }

test [IND ENG] A procedure in which the performance of a product is measured under various conditions. [INV ZOO] A hard external covering or shell that is calcareous, siliceous, chitinous, fibrous, or membranous. [PETRO ENG] A procedure for the analysis of current, potential and ultimate product flow, and pressure-decline properties of various types of petroleum reservoirs. { test }

testa [BOT] A seed coat. Also known as episperm. { 'tes·tə }

Testacellidae [INV ZOO] A family of pulmonate gastropods that includes some species of slugs. { ,tes·tə'sel·ə,dē }

test ammunition [ORD] 1. In a general sense, any ammunition used, or intended to be used, for test purposes. 2. Specifically, ammunition prepared for testing firearms. { 'test ,am·yə,nish·ən }

test bed [AERO ENG] A base, mount, or frame within or

upon which a piece of equipment, especially an engine, is secured for testing. { 'test ,bed }

testboard [ELEC] Switchboard equipped with testing apparatus, arranged so that connections can be made from it to telephone lines or central-office equipment for testing purposes. { 'test,bȯrd }

test chamber [AERO ENG] The test section of a wind tunnel. [ENG] A place, section, or room having special characteristics where a person or object is subjected to experimental procedures, as an altitude chamber. { 'test ,chām·bər }

test clip [ELEC] A spring clip used at the end of an insulated wire lead to make a temporary connection quickly for test purposes. { 'test ,klip }

test cross [GEN] A cross between an individual homozygous at one or more loci and a test subject; the phenotype of the progeny will reveal the subject's genotype at these loci. { 'test ,krȯs }

test data [COMPUT SCI] A set of data developed specifically to test the adequacy of a computer run or system; the data may be actual data that has been taken from previous operations, or artificial data created for this purpose. { 'test ,dad·ə }

test file [COMPUT SCI] A file consisting of test data. { 'test ,fīl }

test firing [AERO ENG] The firing of a rocket engine, either live or static, with the purpose of making controlled observations of the engine or of an engine component. { 'test ,fīr·iŋ }

test flight [AERO ENG] A flight to make controlled observations of the operation or performance of an aircraft or a rocket, or of a component of an aircraft or rocket. { 'test ,flīt }

test function [MATH] An infinitely differentiable function of several real variables used in studying solutions of partial differential equations. { 'test ,fəŋk·shən }

test hole [MIN ENG] A drill hole or shallow excavation to assess an ore body or to obtain rock samples to determine their structural and physical characteristics. { 'test ,hōl }

testicular feminization [MED] A hereditary disorder in which affected individuals are chromosomally XY but have a feminine phenotype and are sterile. The X-linked testicular feminization (Tfm) gene results in absent or defective testosterone receptors, leaving cells unable to bind androgens and respond even though normal male levels of androgen are present. Also known as androgen insensitivity syndrome. { tes|tik·yə·lər ,fem·ə·nə'zā·shən }

testicular hormone [BIOCHEM] Any of various hormones secreted by the testes. { te'stik·yə·lər 'hȯr,mōn }

testing level [ELEC] Value of power used for reference represented by 0.001 watt working in 600 ohms. { 'test·iŋ ,lev·əl }

testis [ANAT] One of a pair of male reproductive glands in vertebrates; after sexual maturity, the source of sperm and hormones. { 'tes·təs }

test jack [ELEC] 1. Appearance of a circuit or circuit element in jacks for testing purposes. 2. In recent practice, a jack multipled with the switchboard operating jack. { 'test ,jak }

test lead [ELEC] A flexible insulated lead, usually with a test prod at one end, used for making tests, connecting instruments to a circuit temporarily, or making other temporary connections. { 'test ,lēd }

test of hypothesis [STAT] A rule for rejecting or accepting a hypothesis concerning a population which is based upon a given sample of data. { 'test əv hī'päth·ə·səs }

test of significance [STAT] A test of a hypothetical population property against a sample property where an acceptance interval is used as the rule for rejection. { 'test əv sig'nif·ə·kəns }

test oscillator See signal generator. { 'test ,äs·ə,lād·ər }

testosterone [BIOCHEM] $C_{19}H_{28}O_2$ The principal androgenic hormone released by the human testis; may be synthesized from cholesterol and certain other sterols. { tes'täs·tə,rōn }

test pack [COMPUT SCI] A deck of punch cards that contains both a computer program and test data for carrying out a test run of the program. { 'test ,pak }

test pattern [COMMUN] A chart having various combinations of lines, squares, circles, and graduated shading, transmitted from time to time by a television station to check definition, linearity, and contrast for the complete system from camera to receiver. Also known as resolution chart. { 'test ,pad·ərn }

test pile [CIV ENG] A pile equipped with a platform on which a load of sand or pig iron is placed in order to determine the

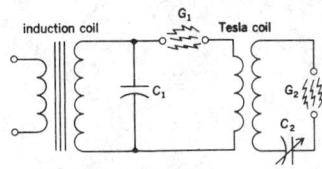

TESLA COIL

Circuit diagram of Tesla coil. G_1 and G_2 are spark gaps, C_1 is capacitor, C_2 is variable capacitor. The coils and C_1 act as a resonant circuit to produce high-frequency oscillation. Current in secondary (right-hand winding) of Tesla coil is both high-voltage and high-frequency.

load a pile can support (usually twice the working load) without settling. { 'test ¸pīl }

test pit [CIV ENG] An open excavation used to obtain soil samples in foundation studies. { 'test ¸pit }

test point [ELEC] A terminal or plug-in connector provided in a circuit to facilitate monitoring, calibration, or troubleshooting. { 'test ¸póint }

test prod [ELEC] A metal point attached to an insulating handle and connected to a test lead for convenience in making a temporary connection to a terminal while tests are being made. Also known as prod. { 'test ¸präd }

test program See check routine. { 'test ¸prō¸gram }

test reactor [NUCLEO] A nuclear reactor designed to test the behavior of materials and components under the neutron and gamma fluxes and temperature conditions of an operating reactor. { 'test rē¸ak·tər }

test record [COMPUT SCI] A record within a test file. { 'test ¸rek·ərd }

test routine See check routine. { 'test rü¸tēn }

test run [COMPUT SCI] The performance of a computer program to check that it is operating correctly, by using test data to generate results that can be compared with expected answers. { 'test ¦rən }

test section [AERO ENG] The section of a wind tunnel where objects are tested to determine their aerodynamic characteristics. { 'test ¸sek·shən }

test set [ELECTR] A combination of instruments needed for servicing a particular type of electronic equipment. { 'test ¸set }

test specimen See tensile bar. { 'test ¸spes·ə·mən }

test stand [AERO ENG] A stationary platform or table, together with any testing apparatus attached thereto, for testing or proving engines and instruments. { 'test ¸stand }

test statistic [STAT] A numerical value which summarizes the information contained in the sample data and which is a basis for testing a given hypothesis. { 'test stə¸tis·tik }

test system [COMPUT SCI] **1.** A computer system that is being tested before being used for production work. **2.** A version of a computer system that is retained, even after a live system is in use, chiefly to diagnose problems without interfering with the work of the live system. { 'test ¸sis·təm }

Testudinata [VERT ZOO] The equivalent name for Chelonia. { te¸styüd·ən'äd·ə }

Testudinellidae [INV ZOO] A family of free-swimming rotifers in the suborder Flosculariacea. { te¸styüd·ən'el·ə¸dē }

Testudinidae [VERT ZOO] A family of tortoises in the suborder Cryptodira; there are about 30 species found on all continents except Australia. { ¸test·yü'din·ə¸dē }

test under mask [COMPUT SCI] A procedure for checking the status of selected bits in a byte by comparing the byte with another byte in which these selected bits are set to one and the other bits are set to zero. { 'test ¸ən·dər 'mask }

test well [PETRO ENG] A well to determine the presence of petroleum oil and its potential commercial value in terms of abundance and accessibility. { 'test ¸wel }

tetanolysin [BIOCHEM] A hemolysin produced by *Clostridium tetani*. { ¦tet·ən·ō'līs·ən }

tetanophobia [PSYCH] An abnormal fear of tetanus. { ¸tet·ən·ō'fō·bē·ə }

tetanospasmin [BIOCHEM] A neurotoxin elaborated by the bacterium *Clostridium tetani* and which is responsible for the manifestations of tetanus. { ¸tet·ən·ō'spaz·mən }

tetanus [MED] An infectious disease of humans and animals caused by the toxin of *Clostridium tetani* and characterized by convulsive tonic contractions of voluntary muscles; infection commonly follows dirt contamination of deep wounds or other injured tissue. Also known as lockjaw. { 'tet·ən·əs }

tetanus antitoxin [IMMUNOL] A serum containing antibodies that neutralize tetanus toxin. { 'tet·ən·əs 'ant·i¸täk·sən }

tetanus toxoid [IMMUNOL] Detoxified tetanus toxin used to produce active immunity against tetanus. { 'tet·ən·əs 'täk¸sóid }

tetany [MED] A state of increased neuromuscular irritability caused by a decrease of serum calcium, manifested by intermittent numbness and cramps or twitchings of the extremities, laryngospasm, bizarre behavior, loss of consciousness, and convulsions. { 'tet·ən·ē }

tetany of the newborn [MED] A temporary increase of neuromuscular irritability during the first two months of life, especially in infants who are premature, are delivered by cesarean section, or receive an exchange transfusion, or in twins. { 'tet·ən·ē əv thə 'nü¸bórn }

Tethinidae [INV ZOO] A family of myodarian cyclorrhaphous dipteran insects in the subsection Acalyptratae. { tə'thin·ə¸dē }

Tethys [ASTRON] A satellite of the planet Saturn having a diameter of about 660 miles (1060 kilometers). [GEOL] **1.** A sea which existed for extensive periods of geologic time between the northern and southern continents of the Eastern Hemisphere. **2.** A composite geosyncline from which many structures of the present Alpine-Himalayan orogenic belt were formed. { 'tē·thəs }

tetraamylbenzene [ORG CHEM] $(C_5H_{11})_4C_6H_2$ A colorless liquid with a boiling range of 320–350°C; used as a solvent. { ¦te·trə¦am·əl'ben¸zēn }

Tetrabranchia [INV ZOO] A subclass of primitive mollusks of the class Cephalopoda; *Nautilus* is the only living form and is characterized by having four gills. { ¸te·trə'braŋ·kē·ə }

tetrabromobisphenol A [ORG CHEM] $(C_6H_2Br_2OH)_2C-(CH_3)_2$ An off-white powder with a melting point of 180–184°C; soluble in methanol and ether; used as a flame retardant for plastics, paper, and textiles. { ¦te·trə¦brō·mō¸bis'fē¸nól 'ā }

tetrabromophthalic anhydride [ORG CHEM] $C_6Br_4C_2O_3$ A pale yellow, crystalline solid with a melting point of 280°C; used as a flame retardant for paper, plastics, and textiles. { ¦te·trə¸brōm·¦af'thal·ik an'hī¸drīd }

tetrabutylthiuram monosulfide [ORG CHEM] $[(C_4H_9)_2-NCS]_2S$ A brown liquid, soluble in acetone, benzene, gasoline, and ethylene dichloride; used as a rubber accelerator. { ¦te·trə¦byüd·əl¦thī·yə¸ram ¦män·ə'səl¸fīd }

tetrabutyltin [ORG CHEM] $(C_4H_9)_4Sn$ A colorless or slightly yellow, oily liquid with a boiling point of 145°C; soluble in most organic solvents; used as a stabilizing agent and rust inhibitor for silicones, and as a lubricant and fuel additive. { ¸te·trə'byüd·əl·tən }

tetrabutyl titanate [ORG CHEM] $Ti(OC_4H_9)_4$ A combustible, colorless to yellowish liquid soluble in many solvents, boils at 312°C, decomposes in water; used in paints, surface coatings, and heat-resistant paints. Abbreviated TBT. { ¦te·trə¦byüd·əl 'tīt·ən¸āt }

tetrabutyl urea [ORG CHEM] $(C_4H_9)_4N_2CO$ A liquid with a boiling point of 305°C; used as a plasticizer. { ¦te·trə¦byüd·əl yù'rē·ə }

tetracaine hydrochloride [ORG CHEM] $C_{15}H_{24}O_2N_2·HCl$ Bitter-tasting, water-soluble crystals melting at 148°C; used as a local anesthetic. { 'te·trə¸kān ¸hī·drə'klór¸īd }

tetracene See naphthacene. { 'te·trə¸sēn }

Tetracentraceae [BOT] A family of dicotyledonous trees in the order Trochodendrales distinguished by possession of a perianth, four stamens, palmately veined leaves, and secretory idioblasts. { ¸te·trə¸sen'trās·ē¸ē }

tetrachlorobenzene [ORG CHEM] $C_6H_2Cl_4$ Water-insoluble, combustible white crystals that appear in two forms: 1,2,3,4-tetrachlorobenzene which melts at 47°C and is used in chemical synthesis and in dielectric fluids; and 1,2,4,5-tetrachlorobenzene which melts at 138°C and is used to make herbicides, defoliants, and electrical insulation. { ¦te·trə¦klór·ə'ben¸zēn }

***sym*-tetrachlorodifluoroethane** [ORG CHEM] CCl_2FCCl_2F A white, toxic liquid with a camphor aroma, soluble in alcohol, insoluble in water, boils at 93°C; used for metal degreasing. { ¦sim ¦te·trə¦klór·ə·dī¦flür·ō'eth¸ān }

***sym*-tetrachloroethane** [ORG CHEM] $CHCl_2CHCl_2$ A colorless, corrosive, toxic liquid with a chloroform scent, soluble in alcohol and ether, slightly soluble in water, boils at 147°C; used as a solvent, metal cleaner, paint remover, and weed killer. { ¦sim ¦te·trə¦klór·ō'eth¸ān }

tetrachloroethylene See perchloroethylene. { ¦te·trə¦klór·ō'eth·ə¸lēn }

tetrachlorophenol [ORG CHEM] C_6HCl_4OH Either of two toxic compounds: 2,3,4,6-tetrachlorophenol comprises brown flakes, soluble in common solvents, melting at 70°C, and is used as a fungicide; 2,4,5,6-tetrachlorophenol is a brown solid, insoluble in water, soluble in sodium hydroxide, has a phenol scent, melts at about 50°C, and is used as a fungicide and for wood preservatives. { ¦te·trə¦klór·ə'fē¸nól }

tetrachlorophthalic acid [ORG CHEM] $C_6Cl_4(CO_2H)_2$ Colorless plates, soluble in hot water; used in making dyes. { ¦te·trə¦klȯr·ə¦thal·ik 'as·əd }

tetrachlorophthalic anhydride [ORG CHEM] $C_6Cl_4(CO)_2O$ A white powder with a melting point of 254–255°C; slightly soluble in water; used in the manufacture of dyes and pharmaceuticals and as a flame retardant for epoxy resins. { ¦te·trə¦klȯr·ə¦thal·ik an'hī,drīd }

tetrachlorosilane See silicon tetrachloride. { ¦te·trə¦klȯr·ə'sī,lān }

tetrachord [ACOUS] The basis of a variety of ancient musical scales, consisting of four notes, with an interval of a perfect fourth between the highest and lowest notes. { 'te·trə,kȯrd }

Tetracorallia [PALEON] The equivalent name for Rugosa. { ,te·trə·kə'ral·yə }

tetracosane [ORG CHEM] $C_{24}H_{50}$ Combustible crystals insoluble in water, soluble in alcohol, melts at 52°C; used as a chemical intermediate. { ,te·trə'kō,sān }

Tetractinomorpha [INV ZOO] A heterogeneous subclass of Porifera in the class Demospongiae. { tə,trak·tə·nə'mȯr·fə }

tetracyanoethylene [ORG CHEM] $(CN)_2C:C(CN)_2$ A member of the cyanocarbon compounds; colorless crystals with a melting point of 198–200°C; used in dye manufacture. { ¦te·trə¦sī·ə·nō'eth·ə,lēn }

tetracycline [MICROBIO] **1.** Any of a group of broad-spectrum antibiotics produced biosynthetically by fermentation with a strain of *Streptomyces aureofaciens* and certain other species or chemically by hydrogenolysis of chlortetracycline. **2.** $C_{22}H_{24}O_8N_2$. A broad-spectrum antibiotic belonging to the tetracycline group of antibiotics; useful because of broad antimicrobial action, with low toxicity, in the therapy of infections caused by gram-positive and gram-negative bacteria as well as rickettsiae and large viruses such as psittacosis-lymphogranuloma viruses. { ,te·trə'sī,klēn }

tetrad [CYTOL] A group of four chromatids lying parallel to each other as a result of the longitudinal division of each of a pair of homologous chromosomes during the pachytene and later stages of the prophase of meiosis. { 'te,trad }

tetradactylous [VERT ZOO] Having four digits on a limb. { ¦te·trə¦dakt·əl·əs }

tetrad analysis [GEN] A method of genetic analysis possible in fungi, algae, bryophytes, and orchids in which the four products of an individual cell which has gone through meiosis are recovered as a group; it provides more direct and complete information regarding segregation and recombination mechanisms than is possible to obtain from meiotic products collected at random. { 'te,trad ə,nal·ə·səs }

tetrad axis [CRYSTAL] A rotation axis whose multiplicity is equal to 4. { 'te,trad ,ak·səs }

n-tetradecane [ORG CHEM] $C_{14}H_{30}$ A combustible, colorless, water-insoluble liquid boiling at 254°C; used as a solvent and distillation chaser and in organic synthesis. { ¦en ¦te·trə'de,kān }

1-tetradecene [ORG CHEM] $CH_2:CH(CH_2)_{11}CH_3$ A combustible, colorless, water-insoluble liquid boiling at 256°C; used as a solvent for perfumes and flavors and in medicine. { ¦wən ¦te·trə'de,sēn }

tetradecylamine [ORG CHEM] $C_{14}H_{29}NH_2$ A white solid with a melting point of 37°C; soluble in alcohol and ether; used in making germicides. { ¦te·trə·də'sil·ə,mēn }

tetradecyl mercaptan [ORG CHEM] $CH_3(CH_2)_{13}SH$ A combustible liquid with a boiling point of 176–180°C; used for processing synthetic rubber. Also known as myristyl mercaptan. { ¦te·trə'des·əl mər'kap,tan }

tetradentate ligand [INORG CHEM] A chelating agent which has four groups capable of attachment to a metal ion. Also known as quadridentate ligand. { ¦te·trə'den,tāt 'līg·ənd }

tetradic [MATH] An operator that transforms one dyadic into another. { tə'trad·ik }

tetrad of Fallot See tetralogy of Fallot. { 'te,trad əv fa'lō }

tetradymite [MINERAL] Bi_2Te_2S A pale steel-gray mineral that usually occurs in foliated masses in auriferous veins; has metallic luster, hardness of 1.5–2 on Mohs scale, and specific gravity of 7.2–7.6. { tə'trad·ə,mīt }

tetraethanolammonium hydroxide [ORG CHEM] $(HOCH_2CH_2)_4NOH$ A white, water-soluble, crystalline solid with a melting point of 123°C; used as a dye solvent and in metal-plating solutions. { ¦te·trə¦eth·ə,nȯl·ə'mō·nē·əm hī'dräk,sīd }

tetraethylammonium chloride [ORG CHEM] $(C_2H_5)_4NCl$ Colorless, hygroscopic crystals with a melting point of 37.5°C; soluble in water, alcohol, acetone, and chloroform; used in medicine. Abbreviated TEAC. Also known as TEA chloride. { ¦te·trə¦eth·əl·ə'mō·nē·əm 'klȯr,īd }

tetra-(2-ethylbutyl)silicate [ORG CHEM] $[(C_2H_5)C_4H_8O]_4Si$ A colorless liquid with a boiling point of 238°C at 50 mmHg (6660 pascals); used as a lubricant and hydraulic fluid. { ¦te·trə ¦tü ¦eth·əl¦byüd·əl 'sil·ə,kāt }

tetraethylene glycol [ORG CHEM] $HO(C_2H_4O)_3C_2H_4OH$ A combustible, hygroscopic, colorless, water-soluble liquid, boils at 327°C; used as a nitrocellulose solvent and plasticizer and in lacquers and coatings. Abbreviated TEG. { ¦te·trə¦eth·ə,lēn 'glī,kȯl }

tetraethylene glycol dimethacrylate [ORG CHEM] A colorless to pale straw-colored liquid with a boiling point of 200°C at 1 mmHg (133.32 pascals); soluble in styrene and some esters and aromatics; used as a plasticizer. { ¦te·trə¦eth·ə,lēn 'glī,kȯl ,dī·mə'thak·rə,lāt }

tetraethylenepentamine [ORG CHEM] $C_8H_{23}N_2$ A toxic, viscous liquid with a boiling point of 333°C and a freezing point of −30°C; soluble in water and organic solvents; used as a motor oil additive, in the manufacture of synthetic rubber, and as a solvent for dyes, acid gases, and sulfur. { ¦te·trə¦eth·ə,lēn'pent·ə,mēn }

tetraethyllead [ORG CHEM] $Pb(C_2H_5)_4$ A highly toxic lead compound that, when added in small proportions to gasoline, increases the fuel's antiknock quality. Abbreviated TEL. { ¦te·trə¦eth·əl'led }

tetraethylpyrophosphate [ORG CHEM] $C_8H_{20}O_7P_2$ A hygroscopic corrosive liquid miscible with although decomposed by water, and miscible with many organic solvents; inhibits the enzyme acetylcholinesterase; used as an insecticide in place of nicotine sulfate. { ¦te·trə¦eth·əl¦pī·rō'fä,sfāt }

tetrafluoroethylene [ORG CHEM] $F_2C:CF_2$ A flammable, colorless, heavy gas, insoluble in water, boils at 78°C; used as a monomer to make polytetrafluoroethylene polymers, for example, Teflon. Abbreviated TFE. { ¦te·trə¦flür·ō'eth·ə,lēn }

tetrafluorohydrazine [INORG CHEM] F_2NNF_2 A colorless liquid or gas with a calculated boiling point of −73°C; used as an oxidizer in rocket fuels. { ¦te·trə¦flür·ō'hī·drə,zēn }

tetrafluoromethane See carbon tetrafluoride. { ¦te·trə¦flür·ō'meth,ān }

tetrafunctional molecule [ORG CHEM] A chemical structure that possesses four highly reactive sites. { ,te·trə¦fəŋk·shən·əl 'mäl·ə,kyül }

tetragonal lattice [CRYSTAL] A crystal lattice in which the axes of a unit cell are perpendicular, and two of them are equal in length to each other, but not to the third axis. { te'trag·ən·əl 'lad·əs }

tetragonal trisoctahedron See trapezohedron. { te'trag·ən·əl ,tris¦äk·tə'hē·drən }

tetragonal tristetrahedron See deltohedron. { te'trag·ən·əl ,tris¦te·trə'hē·drən }

tetrahedral angle [MATH] A polyhedral angle with four faces. { ,te·trə¦hē·drəl 'aŋ·gəl }

tetrahedral group [MATH] The group of motions of three-dimensional space that transform a regular tetrahedron into itself. { ,te·trə¦hē·drəl 'grüp }

tetrahedral molecule [CHEM] A molecule whose structure forms a tetrahedron with a central atom possessing four valence bonds that are directed toward the four points of the tetrahedron. { ,te·trə¦hē·drəl 'mäl·ə,kyül }

tetrahedral symmetry [PHYS] Having the same rotation symmetries as a regular tetrahedron. { 'te·trə¦hē·drəl 'sim·ə·trē }

tetrahedrite [MINERAL] $(Cu,Fe,Zn,Ag)_{12}Sb_4S_{13}$ A grayish-black mineral crystallizing in the isometric system as tetrahedrons and occurring in massive or granular form; luster is metallic, hardness is 3.5–4 on Mohs scale, and specific gravity is 4.6–5.1; an important ore of copper. Also known as fahlore; gray copper ore; panabase; stylotypite. { ,te·trə'hē,drīt }

tetrahedron [CRYSTAL] An isometric crystal form in cubic crystals, in the shape of a four-faced polyhedron, each face of which is a triangle. [MATH] A four-sided polyhedron. { ,te·trə'hē·drən }

TETRACYCLINE

	R_1	R_2	R_3
Tetracycline	H	H	CH_3
Chlortetracycline	Cl	H	CH_3
Oxytetracycline	H	OH	CH_3
6-Demethyl tetracycline	H	H	H
7-Chloro-6-demethyl tetracycline	Cl	H	H

Chemical structural relationships of the tetracyclines (def. 1).

TETRAHEDRAL MOLECULE

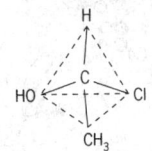

The tetrahedral molecule chloromethoxymethane.

TETRAODONTIFORMES

Gray triggerfish (*Balistes capriscus*).

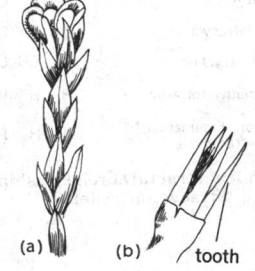

TETRAPHIDALES

(a)   (b)   tooth

Tetraphis pellucida, a representative species of the order Tetraphidales.
(a) Gemmiferous branch.
(b) Enlarged peristome.
(*From W. H. Welch, Mosses of Indiana, Ind. Dep. Conserv., 1957*)

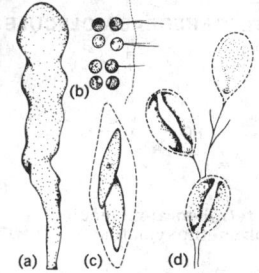

TETRASPORALES

(a)   (c)   (d)

Members of Tetrasporales.
(a) *Tetraspora*, habit of a gelatinous thallus;
(b) arrangement of cells with pseudocilia. (c) *Elakatothrix*, simple colony. (d) *Chlorangium*, an attached, dendroid colony.

tetrahexahedron [CRYSTAL] A form of regular crystal system with four triangular isosceles faces on each side of a cube; there are altogether 24 congruent faces. { ˌte·trə'hek·sə‚drän }

tetrahydrocannabinol [ORG CHEM] $C_{21}H_{30}O_2$ Any member of a group of isomers that are active components of marijuana. Abbreviated THC. { ˈte·trəˈhī·drə·kə'nab·ə‚nól }

tetrahydrofuran [ORG CHEM] C_4H_8O A clear, colorless liquid with a boiling point of 66°C; soluble in water and organic solvents; used as a solvent for resins and in adhesives, printing inks, and polymerizations. Abbreviated THF. { ˈte·trəˌhī·drə'fyù‚rän }

tetrahydrofurfuryl acetate [ORG CHEM] $C_7H_{12}O_3$ A colorless liquid with a boiling point of 194–195°C; soluble in water, alcohol, ether, and chloroform; used for flavoring. { ˈte·trəˈhī·drəˈfər·fə‚ril 'as·ə‚tāt }

tetrahydrofurfuryl alcohol [ORG CHEM] $C_4H_7OCH_2OH$ A hygroscopic, colorless liquid, miscible with water, boils at 178°C; used as a solvent for resins, in leather dyes, and in nylon. { ˈte·trəˈhī·drə'fər·fə‚ril 'al·kə‚hól }

tetrahydrofurfurylamine [ORG CHEM] $C_4H_7OCH_2NH_2$ A colorless to light yellow liquid with a distillation range of −150 to −156°C; used for fine-grain photographic development and to accelerate vulcanization. { ˈte·trəˈhī·drəˌfər·fə'ril·ə‚mēn }

tetrahydrofurfuryl oleate [ORG CHEM] $C_{23}H_{42}O_3$ A colorless liquid with a boiling point of 240°C at 2 mmHg (266.64 pascals); used as a plasticizer. { ˈte·trəˈhī·drəˌfər·fə'ril 'ōl·ē‚āt }

tetrahydrofurfuryl phthalate [ORG CHEM] $C_6H_4(COOCH_2-C_4H_7O)_2$ A colorless liquid with a melting point below 15°C; used as a plasticizer. { ˈte·trəˈhī·drəˌfər·fə'ril 'tha‚lāt }

tetrahydrolinalool [ORG CHEM] $C_{10}H_{21}OH$ A colorless liquid with a floral odor, used in perfumery and flavoring. { ˈte·trəˈhī·drə·lə'nal·ō‚ól }

tetrahydronaphthalene [ORG CHEM] $C_{10}H_{12}$ A colorless, oily liquid that boils at 206°C, and is miscible with organic solvents; used as an intermediate in chemical synthesis and as a solvent. { ˈte·trəˈhī·drə'naf·thə‚lēn }

tetrahydroxyadipic acid See mucic acid. { ˈte·trə·hīˈdräk·sē·ə‚dip·ik 'as·əd }

4,6,3′,4′-tetrahydroxyaurone See aureusidin. { ˈfòr siks ˈthrē‚prīm ˈfòrˌprīm ˈte·trə·hīˈdräk·sē'ó‚rōn }

Tetrahymenina [INV ZOO] A suborder of ciliated protozoans in the order Hymenostomatida. { ˌte·trə‚hī·mə'nīn·ə }

tetraiodoethylene [ORG CHEM] $I_2C:CI_2$ Light yellow crystals with a melting point of 187°C; soluble in organic solvents; used in surgical dusting powder and antiseptic ointments, and as a fungicide. Also known as iodoethylene. { ˈte·trəˌī·ə‚dō'eth·ə‚lēn }

tetraiodofluorescein [ORG CHEM] $C_{20}H_8O_5I_4$ A yellow, water-insoluble, crystalline compound; used as a dye. Also known as pyrosin. { ˈte·trəˌī·ə‚dō'flür·ə‚sēn }

tetraisopropylthiuram disulfide $[(CH_3CH_3CH)_2NCS]_2S_2$ A tan powder with a melting range of 95–99°C; soluble in benzene, chloroform, and gasoline; used as a rubber accelerator. { ˈte·trəˌī·sō‚prō·pəl'thī·yə‚ram dī'səl‚fīd }

tetrakis(hydroxymethyl)phosphonium chloride [ORG CHEM] $(HOCH_2)_4PCl$ A crystalline compound made from phosphine, formaldehyde, and hydrochloric acid; used as a flame retardant for cotton fabrics. Abbreviated THPC. { ˈte·trə·kəs·hīˈdräk·sē‚meth·əl·f aua'sfō·nē·əm 'klór‚īd }

tetralite See tetryl. { 'te·trə‚līt }

tetralogy of Fallot [MED] A congenital abnormality of the heart consisting of pulmonary stenosis, defect of the interventricular septum, hypertrophy of the right ventricle, and overriding or dextroposition of the aorta. Also known as tetrad of Fallot. { tə'träl·ə·jē əv fa'lō }

Tetralophodontinae [PALEON] An extinct subfamily of proboscidean mammals in the family Gomphotheridae. { ˈte·trə‚läf·ə'dänt·ə‚dē }

tetramer [ORG CHEM] A polymer that results from the union of four identical monomers; for example, the tetramer C_8H_8 forms from union of four molecules of C_2H_2. { 'te·trə·mər }

tetramerous [BIOL] Characterized by or having four parts. { te'tram·ə·rəs }

tetramethyldiaminobenzophenone [ORG CHEM] $[(CH_3)_2-NC_6H_4]_2CO$ White to greenish, crystalline leaflets with a melting point of 172°C; soluble in alcohol, ether, water, and warm benzene; used in the manufacture of dyes. Also known

as Michler's ketone. { ˈte·trəˌmethˈəlˈdīˌam·ə·nō·ben'zäf·ə‚nōn }

tetramethylene See cyclobutane. { ˈte·trəˈmeth·ə‚lēn }

tetramethylethylenediamine [ORG CHEM] $(CH_3)_2N_2(CH_2)_2$ A colorless liquid with a boiling point of 121–122°C; soluble in organic solvents and water; used in the formation of polyurethane, as a corrosion inhibitor, and for textile finishing agents. { ˈte·trəˈmethˈəlˈeth·ə‚lēnˈdī·ə‚mēn }

tetramethyllead [ORG CHEM] $Pb(CH_3)_4$ An organic compound of lead that, when added in small amounts to motor gasoline, increases the antiknock quality of the fuel; not widely used. { ˈte·trəˈmethˈəl'led }

tetramethylsilane [ORG CHEM] $(CH_3)_4Si$ A colorless, volatile, toxic liquid with a boiling point of 26.5°C; soluble in organic solvents; used as an aviation fuel. { ˈte·trəˈmethˈəl'sī‚lān }

tetramethylthiuram monosulfide [ORG CHEM] $[(CH_3)_2-NCS]_2S$ A yellow powder with a melting point of 104–107°C; soluble in acetone, benzene, and ethylene dichloride; used as a rubber accelerator, fungicide, and insecticide. { ˈte·trəˈmethˈəl'thī·yə‚ram ˈmän·ə·ə'səl‚fīd }

tetramethylurea [ORG CHEM] $C_5H_{12}N_2O$ A liquid that boils at 176.5°C, and is miscible in water and organic solvents; used as a reagent and solvent. { ˈte·trəˈmethˈəl·yú'rē·ə }

tetranitromethane [ORG CHEM] $C(NO_2)_4$ A powerful oxidant; toxic, colorless liquid with a pungent aroma, insoluble in water, soluble in alcohol and ether, boils at 126°C; used in rocket fuels and as an analytical reagent. { ˈte·trəˌnī·trō'meth‚ān }

Tetranychidae [INV ZOO] The spider mites, a family of phytophagous trombidiform mites. { ˌte·trə'nik·ə‚dē }

Tetraodontiformes [VERT ZOO] An order of specialized teleost fishes that includes the triggerfishes, puffers, trunkfishes, and ocean sunfishes. { ˌte·trə·ō‚dänt·ə'fòr‚mēz }

Tetraonidae [VERT ZOO] The ptarmigans and grouse, a family of upland game birds in the order Galliformes characterized by rounded tails and wings and feathered nostrils. { ˌte·trə'än·ə‚dē }

tetraphenyltin [ORG CHEM] $(C_6H_5)_4Sn$ A white powder with a melting point of 225–228°C; soluble in hot benzene, toluene, and xylene; used for mothproofing. { ˌte·trə'fen·əl‚tən }

Tetraphidaceae [BOT] The single family of the plant order Tetraphidales. { ˌte·trə·fə'dās·ē‚ē }

Tetraphidales [BOT] A monofamilial order of mosses distinguished by scalelike protonema and the peristomes of four rigid, nonsegmented teeth. { ˌte·trə·fə'dā‚lēz }

tetraphosphorus trisulfide See phosphorus sesquisulfide. { ˈte·trə'fä·sfō·rəs trī'səl‚fīd }

Tetraphyllidea [INV ZOO] An order of small tapeworms of the subclass Cestoda characterized by the variation in the structure of the scolex; all species are intestinal parasites of elasmobranch fishes. { ˌte·trə·fə'līd·ē·ə }

tetraploidy [CYTOL] The occurrence of related forms possessing in the somatic cells chromosome numbers four times the haploid number. { 'te·trə‚plòid·ē }

tetrapod [VERT ZOO] A four-footed animal. { 'te·trə‚päd }

Tetrapoda [VERT ZOO] The superclass of the subphylum Vertebrata whose members typically possess four limbs; includes all forms above fishes. { te'träp·əd·ə }

tetrapotassium pyrophosphate See potassium pyrophosphate. { ˈte·trə·pə'tas·ē·əm ˈpī·rō'fä‚sfāt }

tetrapropylene See dodecane. { ˈte·trə'prō·pə‚lēn }

tetrapyrrole [ORG CHEM] A chemical structure in which four pyrrole rings are joined in straight chains, as in a phycobilin, or as joined rings, as in a chlorophyll. { ˌte·trə'pī‚ról }

Tetrarhynchoidea [INV ZOO] The equivalent name for Trypanorhyncha. { ˌte·trə·riŋ'kòid·ē·ə }

tetrasaccharide [BIOCHEM] A carbohydrate which, on hydrolysis, yields four molecules of monosaccharides. { ˌte·trə'sak·ə‚rīd }

tetrasodium pyrophosphate See sodium pyrophosphate. { ˈte·trə'sōd·ē·əm ˈpī·rō'fä‚sfāt }

Tetrasporales [BOT] A heterogeneous and artificial assemblage of colonial fresh-water and marine algae in the division Chlorophyta. { ˌte·trə·spə'rā‚lēz }

tetraspore [BOT] One of the haploid asexual spores of the red algae formed in groups of four. { 'te·trə‚spór }

tetraterpene [ORG CHEM] A class of terpene compounds

that contain isoprene units; best known are the carotenoid pigments from plants and animals, such as lycopene, the red coloring matter in tomatoes. { ¦te·trə'tər¸pēn }

tetratohedral crystal [CRYSTAL] A crystal which has one quarter of the maximum number of faces allowed by the crystal system to which the crystal belongs. { 'te·trəd·ō'hē·drəl 'krist·əl }

tetraxon [INV ZOO] A type of sponge spicule with four axes. { tə'trak¸sän }

tetrazene [ORG CHEM] $H_2NC(NH)_3N_2C(NH)_3NO$ An explosive, colorless to yellowish solid practically insoluble in water and alcohol; used as an explosive initiator and in detonators. { 'te·trə¸zēn }

Tetrigidae [INV ZOO] The grouse locusts or pygmy grasshoppers in the order Orthoptera in which the front wings are reduced to small scalelike structures. { tə'trij·ə¸dē }

tetrode [ELECTR] A four-electrode electron tube containing an anode, a cathode, a control electrode, and one additional electrode that is ordinarily a grid. { 'te¸trōd }

tetrode junction transistor See double-base junction transistor. { 'te¸trōd 'jəŋk·shən tran¸zis·tər }

tetrode thyratron [ELECTR] A thyratron with two control electrodes. Also known as gas tetrode. { 'te¸trōd 'thī·rə¸trän }

tetrode transistor [ELECTR] A four-electrode transistor, such as a tetrode point-contact transistor or double-base junction transistor. { 'te¸trōd tran'zis·tər }

tetrodotoxin [BIOCHEM] $C_{11}H_{17}N_3O_8$ A toxin that blocks the action potential in the nerve impulse. { ¦te·trə·dō'täk·sən }

tetrol See furan. { 'te¸tról }

tetromino [MATH] One of the five plane figures that can be formed by joining four unit squares along their sides. { te'trä·mə·nō }

tetrose [BIOCHEM] Any of a group of monosaccharides that have a four-carbon chain; an example is erythrose, CH_2OH·$(CHOH)_2$·CHO. { 'te¸trōs }

tetryl [ORG CHEM] $(NO_2)_3C_6H_2N(NO_2)CH_3$ A yellow, water-insoluble, crystalline explosive material melting at 130°C; used in explosives and ammunition. Also known as tetralite. { 'te·trəl }

tetrytol [MATER] A high-explosive mixture of tetryl and trinitrotoluene (TNT) in any of several proportions which permit melt loading. { 'te·trə¸tól }

Tettigoniidae [INV ZOO] A family of insects in the order Orthoptera which have long antennae, hindlegs fitted for jumping, and an elongate, vertically flattened ovipositor; consists of the longhorn or green grasshopper. { ¸ted·ə·gə'nī·ə¸dē }

Teuthidae [VERT ZOO] The rabbitfishes, a family of perciform fishes in the suborder Acanthuroidei. { tü'thid·ə¸dē }

Teuthoidea [INV ZOO] An order of the molluscan subclass Coleoidea in which the rostrum is not developed, the proostracum is represented by the elongated pen or gladus, and ten arms are present. { tü'thóid·ē·ə }

TE wave See transverse electric wave. { ¦tē'ē ¸wāv }

Tex [TEXT] A unit of fiber fineness assessed by the weight in grams of 1000 meters of yarn; the lower the number the finer the yarn. { teks }

T$_E$X [GRAPHICS] A text processing system available for many types of computers and used in particular for high-quality typesetting of documents with mathematical content. { tek }

Texas fever [VET MED] A tick-borne infectious disease of cattle caused by the parasite *Babesia annulatus* which invades erythrocytes; characterized by fever, hemoglobinuria, and splenomegaly. { 'tek·səs ¸fē·vər }

Texas tower [ENG] A radar tower built in the sea offshore, to serve as part of an early-warning radar network. { 'tek·səs 'tau·ər }

text [COMMUN] The part of a message that conveys information, excluding bits or characters needed to facilitate transmission of the message. { tekst }

text-editing system [COMPUT SCI] A computer program, together with associated hardware, for the on-line creation and modification of computer programs and ordinary text. { 'tekst ¦ed·əd·iŋ ¸sis·təm }

textile [MATER] A material made of natural or artificial fibers and used for the manufacture of items such as clothing and furniture fittings. { 'tek·stəl }

textile microbiology [MICROBIO] That branch of microbiology concerned with textile materials; deals with microorganisms that are harmful either to the fibers or to the consumer, and microorganisms that are useful, as in the retting process. { 'tek·stəl ¦mī·krō·bī'äl·ə·jē }

textile oil [MATER] A specially compounded oil used to condition raw textile fibers, yarn, and fabric for manufacturing and finishing operations. { 'teks·təl ¸óil }

textile printing [TEXT] The specialized dyeing of restricted areas on fabrics; involves the preparation of printing paste, the development and fixing of color, and the addition of a thickening agent to confine paste applications to desired areas on the goods. { 'tek·stəl ¸print·iŋ }

textile softener [MATER] A chemical that attaches molecularly to textile fibers with the polar (charged) end of the cation oriented toward the fiber and the fatty tail exposed to give a feeling of softness to the fabric. { 'tek·stəl ¸sóf·nər }

text-to-speech synthesizer [ENG ACOUS] A voice response system that provides an automatic means to take a specification of any English text at the input and generate a natural and intelligible acoustic speech signal at the output by using complex sets of rules for predicting the needed phonemic states directly from the input message and dictionary pronunciations. { ¦tekst tə ¦spēch 'sin·thə¸sīz·ər }

Textulariina [INV ZOO] A suborder of foraminiferan protozoans characterized by an agglutinated wall. { ¸tek·styə·lə'rī·ən·ə }

texture [CRYSTAL] The nature of the orientation, shape, and size of the small crystals in a polycrystalline solid. [GEOL] The physical nature of the soil according to composition and particle size. [PETR] The physical appearance or character of a rock; applied to the megascopic or microscopic surface features of a homogeneous rock or mineral aggregate, such as grain size, shape, and arrangement. { 'teks·chər }

textured yarn [TEXT] A generic designation for continuous-filament manufactured yarns which are processed to give them a hand and appearance distinct from untreated yarn. { 'teks·chərd ¸yärn }

texturized food See food analog. { ¦teks·chə¸rīzd 'füd }

TFE See tetrafluoroethylene.

TFT See thin-film transistor.

th See thermie.

Th See thorium.

thalamus [ANAT] Either one of two masses of gray matter located on the sides of the third ventricle and forming part of the lateral wall of that cavity. { 'thal·ə·məs }

Thalassa [ASTRON] A satellite of Neptune orbiting at a mean distance of 31,000 miles (50,000 kilometers) with a period of 7.5 hours, and with a diameter of about 50 miles (80 kilometers). { thə'las·ə }

thalassemia [MED] A hereditary form of hemolytic anemia resulting from a defective synthesis of hemoglobin: thalassemia major is the homozygous state accompanied by clinical illness, and thalassemia minor is the heterozygous state and may not have evident clinical manifestations. Also known as Mediterranean anemia. { ¸thal·ə'sē·mē·ə }

thalassic [OCEANOGR] Of or pertaining to the smaller seas. { thə'las·ik }

Thalassinidea [INV ZOO] The mud shrimps, a group of thin-shelled, burrowing decapod crustaceans belonging to the Macrura; individuals have large chelate or subchelate first pereiopods, and no chelae on the third pereiopods. { ¸thə¸las·ə'nid·ē·ə }

thalassocratic [GEOL] **1.** Pertaining to a thalassocraton. **2.** Referring to a period of high sea level in the geologic past. { thə¦las·ə¦krad·ik }

thalassocraton [GEOL] A craton that is part of the oceanic crust. { ¸thal·ə'säk·rə¸tän }

thalassophile element [GEOCHEM] An element that is relatively more abundant in sea water than in normal continental waters, such as sodium and chlorine. { thə'las·ə¸fīl ¸el·ə·mənt }

thalassophobia [PSYCH] An abnormal fear of the sea. { thə¸las·ə'fō·bē·ə }

Thalattosauria [PALEON] A suborder of extinct reptiles in the order Eosuchia from the Middle Triassic. { thə¸lad·ə'sór·ē·ə }

Thaliacea [INV ZOO] A small class of pelagic Tunicata in which oral and atrial apertures occur at opposite ends of the body. { ‚thā·lē'ā·shē·ə }

thalidomide [PHARM] $C_{13}H_{10}N_2O_4$ A drug used as a sedative and hypnotic; may produce teratogenic effects when administered during pregnancy. { thə'lid·ə‚mīd }

thalline [ORG CHEM] $C_9H_6N(OCH_3)H_4$ Colorless rhomboids soluble in water and melting at 40°C. { 'tha‚lēn }

thallium [CHEM] A metallic element in group 13, symbol Tl, atomic number 81, atomic weight 204.383; insoluble in water, soluble in nitric and sulfuric acids, melts at 302°C, boils at 1457°C. [MET] Bluish-white metal with tinlike malleability, but a little softer; used in alloys. { 'thal·ē·əm }

thallium acetate [ORG CHEM] $TlOCOCH_3$ Toxic, white, deliquescent crystals, soluble in water and alcohol, melts at 131°C; used as an ore-flotation solvent and in medicine. { 'thal·ē·əm 'as·ə‚tāt }

thallium-beam clock [HOROL] A device similar to a cesium-beam clock, but using a beam of thallium atoms instead of cesium; advantages over cesium are reduced sensitivity to magnetic fields and a higher frequency; the beam, however, is much harder to detect and deflect. { 'thal·ē·əm ‚bēm 'kläk }

thallium bromide [INORG CHEM] TlBr A toxic, yellowish powder soluble in alcohol, slightly soluble in water, melts at 460°C; used in infrared radiation transmitters and detectors. Also known as thallous bromide. { 'thal·ē·əm 'brō‚mīd }

thallium carbonate [INORG CHEM] Tl_2CO_3 Toxic, shiny, colorless needles soluble in water, insoluble in alcohol, melts at 272°C; used as an analytical reagent and in artificial gems. Also known as thallous carbonate. { 'thal·ē·əm 'kär·bə‚nāt }

thallium chloride [INORG CHEM] TlCl A white, toxic, light-sensitive powder, slightly soluble in water, insoluble in alcohol, melts at 430°C; used as a chlorination catalyst and in medicine and suntan lamps. Also known as thallous chloride. { 'thal·ē·əm 'klȯr‚īd }

thallium hydroxide [INORG CHEM] $TlOH·H_2O$ Toxic yellow, water- and alcohol-soluble needles, decomposes at 139°C; used as an analytical reagent. Also known as thallous hydroxide. { 'thal·ē·əm hī'dräk‚sīd }

thallium iodide [INORG CHEM] TlI A toxic, yellow powder, insoluble in alcohol, slightly soluble in water, melts at 440°C; used in infrared radiation transmitters and in medicine. Also known as thallous iodide. { 'thal·ē·əm 'ī·ə‚dīd }

thallium monoxide [INORG CHEM] Tl_2O A black, toxic, water- and alcohol-soluble powder, melts at 300°C; used as an analytical reagent and in artificial gems and optical glass. Also known as thallium oxide; thallous oxide. { 'thal·ē·əm mə'näk‚sīd }

thallium nitrate [INORG CHEM] $TlNO_3$ Colorless, toxic, fire-hazardous crystals soluble in hot water, insoluble in alcohol, melts at 206°C, decomposes at 450°C; used as an analytical reagent and in pyrotechnics. Also known as thallous nitrate. { 'thal·ē·əm 'nī‚trāt }

thallium oxide See thallium monoxide. { 'thal·ē·əm 'äk‚sīd }

thallium sulfate [INORG CHEM] Tl_2SO_4 Toxic, water-soluble, colorless crystals melting at 632°C; used as an analytical reagent and in medicine, rodenticides, and pesticides. Also known as thallous sulfate. { 'thal·ē·əm 'səl‚fāt }

thallium sulfide [INORG CHEM] Tl_2S Lustrous, toxic, blue-black crystals insoluble in water, alcohol, and ether, soluble in mineral acids, melts at 448°C; used in infrared-sensitive devices. Also known as thallous sulfide. { 'thal·ē·əm 'səl‚fīd }

Thallobionta [BOT] One of the two subkingdoms of plants, characterized by the absence of specialized tissues or organs and multicellular sex organs. { ‚thal·ō·bī'änt·ə }

thallofide cell [ELECTR] A photoconductive cell in which the active light-sensitive material is thallium oxysulfide in a vacuum; it has maximum response at the red end of the visible spectrum and in the near infrared. { 'thal·ə‚fīd ‚sel }

Thallophyta [BOT] The equivalent name for Thallobionta. { thə'läf·əd·ə }

thallospore [BOT] A spore that develops by budding of hyphal cells. { 'thal·ə‚spȯr }

thallotoxicosis [MED] Poisoning due to ingestion of thallium or its derivatives. { ‚thal·ō‚täk·sə'kō·səs }

thallous bromide See thallium bromide. { 'thal·əs 'brō‚mīd }

thallous carbonate See thallium carbonate. { 'thal·əs 'kär·bə‚nāt }

thallous chloride See thallium chloride. { 'thal·əs 'klȯr‚īd }

thallous hydroxide See thallium hydroxide. { 'thal·əs hī'dräk‚sīd }

thallous iodide See thallium iodide. { 'thal·əs 'ī·ə‚dīd }

thallous nitrate See thallium nitrate. { 'thal·əs 'nī‚trāt }

thallous oxide See thallium monoxide. { 'thal·əs 'äk‚sīd }

thallous sulfate See thallium sulfate. { 'thal·əs 'səl‚fāt }

thallous sulfide See thallium sulfide. { 'thal·əs 'səl‚fīd }

thallus [BOT] A plant body that is not differentiated into special tissue systems or organs and may vary from a single cell to a complex, branching multicellular structure. { 'thal·əs }

thalweg [GEOGR] The middle of the principal navigable waterway which serves as a boundary between two states. [GEOL] **1.** A line connecting the lowest points along a stream bed or a valley. Also known as valley line. **2.** A line crossing all contour lines on a land surface perpendicularly. [HYD] Water seeping through the ground below the surface in the same direction as a surface stream course. { 'täl‚veg }

THAM See tromethamine.

thanatocoenosis [PALEON] The assemblage of dead organisms or fossils that occurred together in a given area at a given moment of geologic time. Also known as death assemblage; taphocoenosis. { ‚than·ə·tō·sə'nō·səs }

thanatology [MED] The study of the phenomenon of somatic death. { ‚than·ə'täl·ə·jē }

Thanetian [GEOL] A European stage of geologic time; uppermost Paleocene, above Montian, below Ypresian of Eocene. { thə'nē·shən }

Tharsis rise [ASTRON] An uplifted portion of the surface of Mars that stands several kilometers above the mean elevation of the planet and affects approximately one-quarter of the entire surface. { 'thär·səs ‚rīz }

Thaumaleidae [INV ZOO] A family of orthorrhaphous dipteran insects in the series Nematocera. { ‚thȯ·mə'lē·ə‚dē }

Thaumastellidae [INV ZOO] A monospecific family of the Hemiptera assigned to the Pentatomorpha found only in Ethiopia. { ‚thȯ·mə'stel·ə‚dē }

Thaumastocoridae [INV ZOO] The single family of the hemipteran superfamily Thaumastocoroidea. { thə‚mas·tə'kȯr·ə‚dē }

Thaumastocoroidea [INV ZOO] A monofamilial superfamily of the Hemiptera in the subdivision Geocorisae which occurs in Australia and the New World tropics. { thə‚mas·tə·kə'rȯid·ē·ə }

Thaumatoxenidae [INV ZOO] A family of cyclorrhaphous dipteran insects in the series Aschiza. { ‚thȯ·mə·täk'sen·ə‚dē }

thaw [CLIMATOL] A warm spell during which ice and snow melt, as a January thaw. { thȯ }

thaw house [ENG] A small building that is designed for thawing frozen dynamite and which is capacious enough for a supply of thawed dynamite for a day's work. { 'thȯ ‚haůs }

thawing [ENG] Warming dynamite, to reduce risk of premature explosion. [MIN ENG] Working permanently frozen ground by pumping water at a temperature of from 50 to 60°F (10 to 15.5°C) through pipes down into the frozen gravel. { 'thȯ·iŋ }

thawing index [CLIMATOL] The number of degree days, above and below 32°F, between the lowest and highest points on the cumulative degree-days time curve for one thawing season. { 'thȯ·iŋ ‚in‚deks }

thawing season [CLIMATOL] The period of time between the lowest point and the succeeding highest point on the time curve of cumulative degree days above and below 32°F. { 'thȯ·iŋ ‚sē·zən }

thaw pipe [MIN ENG] A string of pipe drilled into a string of drill rods that is frozen in a borehole in permafrost, through which water is circulated to thaw the ice and free the drill rods. { 'thȯ ‚pīp }

THC See tetrahydrocannabinol.

Theaceae [BOT] A family of dicotyledonous erect trees or shrubs in the order Theales characterized by alternate, exstipulate leaves, usually five petals, and mostly numerous stamens. { thē'ās·ē‚ē }

Theales [BOT] An order of dicotyledonous mostly woody plants in the subclass Dilleniidae with simple or occasionally compound leaves, petals usually separate, numerous stamens, and the calyx arranged in a tight spiral. { thē'ā·lēz }

theater of operations [ORD] Portion of a theater of war necessary for military operations, either offensive or defensive, pursuant to an assigned mission, and for the administration incident to such military operation. { 'thē·ə·dər əv ,äp·ə'rā·shənz }

theater of war [ORD] That area of land, sea, and air which is, or may become, involved directly in the operation of war. { 'thē·ə·dər əv 'wȯr }

theater stock level [ORD] Quantity of supplies authorized by the U.S. Army to be maintained in a theater of operations as stock on hand ready for issue. { 'thē·ə·dər 'stäk ,lev·əl }

theater television [ELECTR] A large projection-type television receiver used in theaters, generally for closed-circuit showing of important sport events. { 'thē·ə·dər 'tel·ə,vizh·ən }

thebaine [PHARM] $C_{19}N_{21}NO_3$ A white, crystalline alkaloid derived from opium; melting point is 193°C; soluble in alcohol and ether; used in medicine. Also known as *paramorphine*. { 'thē,bān }

Thebe [ASTRON] A small satellite of Jupiter, having an orbital radius of 137,900 miles (221,900 kilometers), and a radius of 28–34 miles (48–55 kilometers). Also known as Jupiter XIV. { 'thē·bē }

theca [ANAT] The sheath of dura mater which covers the spinal cord. [BOT] **1.** A moss capsule. **2.** A pollen sac. [HISTOL] The layer of stroma surrounding a Graafian follicle. [INV ZOO] The test of a testate protozoan or a rotifer. { 'thē·kə }

theca folliculi [HISTOL] The capsule surrounding a developing or mature Graafian follicle; consists of two layers, theca interna and theca externa. { 'thē·kə fə'lik·yə·lī }

Thecanephria [INV ZOO] An order of the phylum Brachiata containing a group of elongate, tube-dwelling tentaculate, deep-sea animals of bizarre structure. { ,thē·kə'nef·rē·ə }

thecate [BIOL] Having a theca. { 'thē,kāt }

Thecideidina [PALEON] An extinct suborder of articulate brachiopods doubtfully included in the order Terebratulida. { thə,sid·ē·ə'dīn·ə }

Thecodontia [PALEON] An order of archosaurian reptiles, confined to the Triassic and distinguished by the absence of a supratemporal bone, parietal foramen, and palatal teeth, and by the presence of an antorbital fenestra. { ,thek·ə'dän·chə }

Thelastomidae [INV ZOO] A family of nematode worms in the superfamily Oxyuroidea. { ,thel·ə'stäm·ə,dē }

thematic apperception test [PSYCH] A projective psychological test using a set of pictures suggesting life situations from which the subject constructs a story; designed to reveal to the trained interpreter some of the dominant drives, emotions, sentiments, complexes, and conflicts of personality. Abbreviated TAT. { thə'mad·ik ,ap·ər'sep·shən ,test }

Themis [ASTRON] An asteroid with a diameter of roughly 129 miles (207 kilometers), mean distance from the sun of 3.13 astronomical units, and C-type surface composition. { 'thē·məs }

thenar [ANAT] The ball of the thumb. { 'thē,när }

thenardite [MINERAL] Na_2SO_4 A colorless, grayish-white, yellowish, yellow-brown, or reddish, orthorhombic mineral consisting of sodium sulfate. { thə'när,dīt }

thenyl [ORG CHEM] $C_4H_3SCH_2-$ An organic radical based on methylthiophene; thus thenyl alcohol is also known as thiophenemethanol. { 'then·əl }

theobroma oil *See* cocoa butter. { ,thē·ə'brō·mə ,ȯil }

theobromine [ORG CHEM] $C_7H_8N_4O_2$ A toxic alkaloid found in cocoa, chocolate products, tea, and cola nuts; closely related to caffeine. { ,thē·ə'brō,mēn }

theodolite [OPTICS] An optical instrument used in surveying which consists of a sighting telescope mounted so that it is free to rotate around horizontal and vertical axes, and graduated scales so that the angles of rotation may be measured; the telescope is usually fitted with a right-angle prism so that the observer continues to look horizontally into the eyepiece, whatever the variation of the elevation angle; in meteorology, it is used principally to observe the motion of a pilot balloon. { thē'äd·əl,īt }

theophobia [PSYCH] Abnormal fear of a deity or of divine punishment. { ,thē·ə'fō·bē·ə }

Theophrastaceae [BOT] A family of tropical and subtropical dicotyledonous woody plants in the order Primulales characterized by flowers having staminodes alternate with the corolla lobes. { ,thē·ə·fras'tās·ē,ē }

theophylline [ORG CHEM] $C_7H_8N_4O_2·H_2O$ Alkaloid from tea leaves; bitter-tasting white crystals slightly soluble in water and alcohol, melts at 272°C; used in medicine. { ,thē·ə'fi,lēn }

theorem [MATH] A proven mathematical statement. { 'thir·əm }

theorem of corresponding states [STAT MECH] A theorem stating that two substances which have the same reduced temperature and the same reduced pressure have the same reduced volume. { 'thir·əm əv ,kär·ə'spänd·iŋ 'stāts }

theoretical air [ENG] The amount of air that is theoretically required for complete combustion. { ,thē·ə'red·ə·kəl 'er }

theoretical community ecology [ECOL] A branch of theoretical ecology that is concerned with factors determining the species composition and functional organization of communities, with a particular emphasis on interspecific interactions such as competition, predation, and mutualism. { ,thē·ə¦red·i·kəl kə,myün·əd·ē ē'käl·ə·jē }

theoretical cutoff frequency [ELEC] Of an electric structure, a frequency at which, disregarding the effects of dissipation, the attenuation constant changes from zero to a positive value or vice versa. { ,thē·ə'red·ə·kəl 'kəd,óf ,frē·kwən·sē }

theoretical draft [FL MECH] Draft in a ducted space, neglecting flow losses due to fluid friction. { ,thē·ə'red·ə·kəl 'draft }

theoretical ecology [ECOL] The use of mathematical models and verbal reasoning to provide a conceptual framework for the analysis of ecological systems. { ,thē·ə,red·ə·kəl ē'käl·ə·jē }

theoretical frequency [STAT] A distributional frequency that would result if the data followed a theoretical distribution law rather than the actual observed frequencies. { ,thē·ə'red·ə·kəl 'frē·kwən·sē }

theoretical linguistics [PSYCH] A branch of linguistics concerned with the form of language representation in the mind, that is, linguistic competence and the structure and components of mental grammar. { ,thē·ə,red·ə·kəl liŋ'gwis·tiks }

theoretical physics [PHYS] The description of natural phenomena in mathematical form. { thē·ə'red·ə·kəl 'fiz·iks }

theoretical plate [CHEM ENG] A distillation column plate or tray that produces perfect distillation (that is, produces the same difference in composition as that existing between a liquid mixture and the vapor in equilibrium with it); the packed-column equivalent of a theoretical plate is the HETP, or height (of packing) equivalent to a theoretical plate. { ,thē·ə'red·ə·kəl 'plāt }

theoretical relieving capacity [MECH ENG] The capacity of a theoretically perfect nozzle calculated in volumetric or gravimetric units. { ,thē·ə'red·ə·kəl ri'lēv·iŋ kə,pas·əd·ē }

theory [MATH] The collection of theorems and principles associated with some mathematical object or concept. [SCI TECH] An attempt to explain a certain class of phenomena by deducing them as necessary consequences of other phenomena regarded as more primitive and less in need of explanation. { 'thē·ə·rē }

theory of antecedent conflicts [PSYCH] A theory that the effects of an emotionally disturbing event in adult life may be greatly multiplied through the conditioning or sensitizing effects of the vicissitudes of early life. { 'thē·ə·rē əv ,ant·i'sēd·ənt 'kän,fliks }

theory of equations [MATH] The study of polynomial equations from the viewpoint of solution methods, relations among roots, and connections between coefficients and roots. { 'thē·ə·rē əv i'kwā·zhənz }

theory of games *See* game theory. { 'thē·ə·rē əv 'gāmz }

theralite [PETR] A dark-colored, visibly crystalline rock composed chiefly of pyroxene with smaller amounts of calcic plagioclase and nepheline. { 'ther·ə,līt }

therapeutic abortion [MED] Abortion performed when pregnancy jeopardizes the health or life of the mother. { ,ther·ə¦pyüd·ik ə'bȯr·shən }

therapeutic vaccination [IMMUNOL] Vaccination strategies that attempt to induce immune responses after infection with the target disease. { ,ther·ə,pyüd·ik ,vak·sə'nā·shən }

Therapsida [PALEON] An order of mammallike reptiles of the subclass Synapsida which first appeared in mid-Permian times and persisted until the end of the Triassic. { thə'rap·səd·ə }

Therberg system [IND ENG] A system of categorizing hand

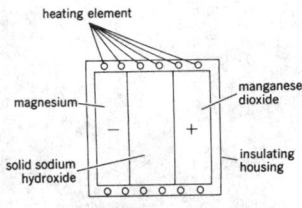

THERMAL CELL

Schematic of thermal cell. Cell is activated when heat from heating element melts sodium hydroxide.

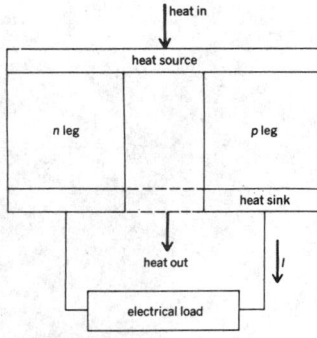

THERMAL CONVERTER

Simple thermal converter. One leg is *n*-type semiconductor material, the other is *p*-type material.

movements that is used in the standard motion-and-time analysis technique.   { 'thər,bərg ,sis·təm }

therblig See elemental motion.   { 'thər,blig }

therblig chart [IND ENG] An operation chart with the suboperations divided into basic motions, all designated with appropriate symbols.   { 'thər,blig ,chärt }

Therevidae [INV ZOO] The stiletto flies, a family of orthorrhaphous dipteran insects in the series Brachycera.   { thə 'rev·ə,dē }

Theria [VERT ZOO] A subclass of the class Mammalia including all living mammals except the monotremes.   { 'ther·ē·ə }

Theridiidae [INV ZOO] The comb-footed spiders, a family of the suborder Dipneumonomorphae.   { ,ther·ə'dī·ə,de }

therm [THERMO] A unit of heat energy, equal to 100,000 international table British thermal units, or approximately 1.055×10^8 joules.   { thərm }

thermactor See air-injection system.   { 'thər,mak·tər }

thermal [METEOROL] A relatively small-scale, rising current of air produced when the atmosphere is heated enough locally by the earth's surface to produce absolute instability in its lower layers. [THERMO] Of or concerning heat.   { 'thər·məl }

thermal agitation [SOLID STATE] Random movements of the free electrons in a conductor, producing noise signals that may become noticeable when they occur at the input of a high-gain amplifier. Also known as thermal effect.   { 'thər·məl ,aj·ə'tā·shən }

thermal ammeter See hot-wire ammeter.   { 'thər·məl 'am,ēd·ər }

thermal analysis [ANALY CHEM] Analytical techniques developed to continuously monitor physical or chemical changes of a sample which occur as the temperature of a sample is increased or decreased. Thermogravimetry, differential thermal analysis, and differential scanning calorimetry are the principal thermoanalytical methods. [MET] Determining transformations in a metal by observing the temperature-time relationship during uniform cooling or heating; phase transformations are indicated by irregularities in a smooth curve.   { 'thər·məl ə'nal·ə·səs }

thermal-arrest calorimeter [ENG] A vacuum device for measurement of heats of fusion; a sample is frozen under vacuum and allowed to melt as the calorimeter warms to room temperature.   { 'thər·məl ə¦rest ,kal·ə'rim·əd·ər }

thermal aureole See aureole.   { 'thər·məl 'òr·ē,ōl }

thermal barrier [AERO ENG] A limit to the speed of airplanes and rockets in the atmosphere imposed by heat from friction between the aircraft and the air, which weakens and eventually melts the surface of the aircraft. Also known as heat barrier. [BUILD] See thermal break.   { 'thər·məl 'bar·ē·ər }

thermal battery [ELEC] **1.** A combination of thermal cells. Also known as fused-electrolyte battery; heat-activated battery. **2.** A voltage source consisting of a number of bimetallic junctions connected to produce a voltage when heated by a flame.   { 'thər·məl 'bad·ə·rē }

thermal belt [ECOL] Any one of several possible horizontal belts of a vegetation type found in mountainous terrain, resulting primarily from vertical temperature variation. Also known as thermal zone.   { 'thər·məl ,belt }

thermal black [CHEM] A type of carbon black made by a thermal process using natural gas; used in the rubber industry.   { 'thər·məl 'blak }

thermal blooming [OPTICS] The phenomenon of self-defocusing in certain weakly absorbing media; it is prominent, for example, in the propagation of high-power infrared laser beams through the atmosphere.   { 'thər·məl 'blüm·iŋ }

thermal bond [NUCLEO] A thermally conductive bond, providing maximum transfer of heat, as between nuclear-reactor fuel and its protective cladding.   { 'thər·məl 'bänd }

thermal break [BUILD] A component that is a poor conductor of heat and is placed in an assembly containing highly conducting materials in order to reduce or prevent the flow of heat. Also known as thermal barrier.   { ¦thər·məl ¦brāk }

thermal breeder reactor [NUCLEO] A breeder reactor in which the fission chain reaction is sustained by thermal neutrons.   { 'thər·məl 'brēd·ər rē,ak·tər }

thermal bremsstrahlung [PL PHYS] Radiation emitted by electrons in a hot plasma when they are accelerated by positive ions.   { 'thər·məl 'brem,shträ·ləŋ }

thermal bulb [ENG] A device for measurement of temperature; the liquid in a bulb expands with increasing temperature,

pressuring a spiral Bourdon-type tube element and causing it to deform (unwind) in direct relation to the temperature in the bulb.   { 'thər·məl ¦bəlb }

thermal burn [MED] Tissue reaction to or injury resulting from application of heat.   { 'thər·məl ¦bərn }

thermal capacitance [THERMO] The ratio of the entropy added to a body to the resulting rise in temperature.   { 'thər·məl kə'pas·əd·əns }

thermal capacity See heat capacity.   { 'thər·məl kə'pas·əd·ē }

thermal cell [ELEC] A reserve cell that is activated by applying heat to melt a solidified electrolyte.   { 'thər·məl ¦sel }

thermal charge See entropy.   { 'thər·məl ¦chärj }

thermal climate [CLIMATOL] Climate as defined by temperature, and divided regionally into temperature zones.   { 'thər·məl 'klī·mət }

thermal column [NUCLEO] A column of moderating material extending through the shield into the reflector of a nuclear reactor, used to provide a source of thermal neutrons for research purposes.   { 'thər·məl ¦käl·əm }

thermal compressor [MECH ENG] A steam-jet ejector designed to compress steam at pressures above atmospheric.   { 'thər·məl kəm'pres·ər }

thermal conductance [THERMO] The amount of heat transmitted by a material divided by the difference in temperature of the surfaces of the material. Also known as conductance.   { 'thər·məl kən'dək·təns }

thermal conductimetry [THERMO] Measurement of thermal conductivities.   { 'thər·məl ,kän,dək'tim·ə·trē }

thermal conductivity [THERMO] The heat flow across a surface per unit area per unit time, divided by the negative of the rate of change of temperature with distance in a direction perpendicular to the surface. Also known as coefficient of conductivity; heat conductivity.   { 'thər·məl ,kan,dək'tiv·əd·ē }

thermal conductivity cell See katharometer.   { 'thər·məl ,kän ,dək'tiv·əd·ē ,sel }

thermal conductivity gage [ENG] A pressure measurement device for high-vacuum systems; an electrically heated wire is exposed to the gas under pressure, the thermal conductivity of which changes with changes in the system pressure.   { 'thər·məl ,kän,dək'tiv·əd·ē ,gāj }

thermal conductor [THERMO] A substance with a relatively high thermal conductivity.   { 'thər·məl kən'dək·tər }

thermal convection [METEOROL] Atmospheric currents, predominantly vertical, arising from the release of gravitational visibility; commonly produced by solar heating of the ground; the cause of convective (cumulus) clouds. Also known as free convection; gravitational convection. [THERMO] See heat convection.   { 'thər·məl kən'vek·shən }

thermal converter [ELECTR] A device that converts heat energy directly into electric energy by using the Seebeck effect; it is composed of at least two dissimilar materials, one junction of which is in contact with a heat source and the other junction of which is in contact with a heat sink. Also known as thermocouple converter; thermoelectric generator; thermoelectric power generator; thermoelement. [ENG] An instrument used with external resistors for ac current and voltage measurements over wide ranges, consisting of a conductor heated by an electric current, with one or more hot junctions of a thermocouple attached to it, so that the output emf responds to the temperature rise, and hence the current.   { 'thər·məl kən'vərd·ər }

thermal coulomb [THERMO] A unit of entropy equal to 1 joule per kelvin.   { 'thər·məl 'kü,läm }

thermal cracking [CHEM ENG] A petroleum refining process that decomposes, rearranges, or combines hydrocarbon molecules by the application of heat, without the aid of catalysts.   { 'thər·məl 'krak·iŋ }

thermal cross section [NUCLEO] The cross section of a thermal neutron.   { 'thər·məl 'kròs ,sek·shən }

thermal cutout [ELEC] A heat-sensitive switch that automatically opens the circuit of an electric motor or other device when the operating temperature exceeds a safe value.   { 'thər·məl 'kəd,aùt }

thermal cutting [MET] A group of processes to sever metals by melting or by chemical reaction of oxygen with the metal at elevated temperatures.   { 'thər·məl 'kəd·iŋ }

thermal degradation [CHEM] Molecular deterioration of materials (usually organics) because of overheat; can be avoided by low-temperature or vacuum processing, as for foods and pharmaceuticals. { 'thər·məl ˌdeg·rə'dā·shən }

thermal detector *See* bolometer. { 'thər·məl di'tek·tər }

thermal diffusion [PHYS CHEM] A phenomenon in which a temperature gradient in a mixture of fluids gives rise to a flow of one constituent relative to the mixture as a whole. Also known as thermodiffusion. { 'thər·məl di'fyü·zhən }

thermal diffusivity *See* diffusivity. { 'thər·məl ˌdi·fyü'siv·əd·ē }

thermal drift [ELECTR] Drift caused by internal heating of equipment during normal operation or by changes in external ambient temperature. { 'thər·məl 'drift }

thermal drilling [MECH ENG] A machining method in which holes are drilled in a workpiece by heat generated from the friction of a rotating tool. { 'thər·məl 'dril·iŋ }

thermal ecology [ECOL] Study of the independent and interactive biotic and abiotic components of naturally heated environments. { 'thər·məl i'käl·ə·jē }

thermal effect *See* thermal agitation. { 'thər·məl i,fekt }

thermal efficiency [CHEM ENG] In a tube-and-shell heat-exchange system, the ratio of the actual temperature range of the tube-side fluid (inlet versus outlet temperature) to the maximum possible temperature range. [THERMO] *See* efficiency. { 'thər·məl i'fish·ən·sē }

thermal-efficiency index *See* temperature-efficiency index. { 'thər·məl i¦fish·ən·sē ˌin,deks }

thermal-efficiency ratio *See* temperature-efficiency ratio. { 'thər·məl i¦fish·ən·sē ˌrā·shō }

thermal effusion *See* thermal transpiration. { 'thər·məl e'fyü·zhən }

thermal electromotive force [PHYS] An electromotive force arising from a difference in temperature at two points along a circuit, as in the Seebeck effect. { 'thər·məl i¦lek·trə¦mōd·iv 'förs }

thermal emissivity *See* emissivity. { 'thər·məl ˌē·mi'siv·əd·ē }

thermal energy [NUCLEO] Energy which is characteristic for thermal neutrons at room temperature, about 0.025 electronvolt. { 'thər·məl 'en·ər·jē }

thermal environment [IND ENG] Those aspects of the workplace that include local temperature, humidity, and air velocity as well as the presence of radiating surfaces. { 'thərm·əl in'vī·rən·mənt }

thermal equator *See* heat equator. { 'thər·məl i'kwād·ər }

thermal equilibrium [THERMO] Property of a system all parts of which have attained a uniform temperature which is the same as that of the system's surroundings. { 'thər·məl ˌē·kwə'lib·rē·əm }

thermal excitation [ATOM PHYS] The process in which atoms or molecules acquire internal energy in collisions with other particles. { 'thər·məl ˌek·sī'tā·shən }

thermal expansion [PHYS] The dimensional changes exhibited by solids, liquids, and gases for changes in temperature while pressure is held constant. { 'thər·məl ik'span·chən }

thermal expansion coefficient [PHYS] The fractional change in length or volume of a material for a unit change in temperature. { 'thər·məl ik¦span·chən ˌkō·i,fish·ənt }

thermal exposure [ORD] The total normal component of thermal radiation striking a given surface throughout the course of a detonation. { 'thər·məl ik'spō·zhər }

thermal farad [THERMO] A unit of thermal capacitance equal to the thermal capacitance of a body for which an increase in entropy of 1 joule per kelvin results in a temperature rise of 1 kelvin. { 'thər·məl 'far,ad }

thermal flame safeguard [MECH ENG] A thermocouple located in the pilot flame of a burner; if the pilot flame is extinguished, an elective circuit is interrupted and the fuel supply is shut off. { 'thər·məl 'flām ¦saf,gärd }

thermal flasher [ELEC] An electric device that opens and closes a circuit automatically at regular intervals because of alternate heating and cooling of a bimetallic strip that is heated by a resistance element in series with the circuit being controlled. { 'thər·məl 'flash·ər }

thermal flux *See* heat flux. { 'thər·məl 'fləks }

thermal gasoline [MATER] Gasoline produced in petroleum refineries by thermal processes, for example, thermal cracking and thermal reforming. { 'thər·məl 'gas·ə,lēn }

thermal gradient [GEOPHYS] The rate of temperature change with distance; for example, its increase with depth below the surface of the earth. { 'thər·məl 'grād·ē·ənt }

thermal henry [THERMO] A unit of thermal inductance equal to the product of a temperature difference of 1 kelvin and a time of 1 second divided by a rate of flow of entropy of 1 watt per kelvin. { 'thər·məl 'hen·rē }

thermal high [METEOROL] A high resulting from the cooling of air by a cold underlying surface, and remaining relatively stationary over the cold surface. { 'thər·məl 'hī }

thermal horsepower [ELEC] Electrical motor horsepower as determined by current readings from a thermal-type ammeter; will be higher than load horsepower determined from kilowatt-input methods. Also known as true motor load. { 'thər·məl 'hörs,pau·ər }

thermal hysteresis [THERMO] A phenomenon sometimes observed in the behavior of a temperature-dependent property of a body; it is said to occur if the behavior of such a property is different when the body is heated through a given temperature range from when it is cooled through the same temperature range. { 'thər·məl ˌhis·tə'rē·səs }

thermal imagery [ELECTR] Imagery produced by measuring and recording electronically the thermal radiation of objects. { 'thər·məl 'im·ij·rē }

thermal inductance [THERMO] The product of temperature difference and time divided by entropy flow. { 'thər·məl in'cək·təns }

thermal instability [FL MECH] The instability resulting in free convection in a fluid heated at a boundary. { 'thər·məl ˌin·stə'bil·əd·ē }

thermal instrument [ENG] An instrument that depends on the heating effect of an electric current, such as a thermocouple or hot-wire instrument. { 'thər·məl 'in·strə·mənt }

thermal inversion *See* temperature inversion. { 'thər·məl in'vər·zhən }

thermal ionization *See* Saha ionization. { 'thər·məl ˌī·ə·nə'zā·shən }

thermalize [NUC PHYS] To bring neutrons into thermal equilibrium with the surroundings. { 'thər·mə,līz }

thermal jet [METEOROL] A region in the atmosphere where isotherms or thickness lines are closely packed; therefore, a region of very strong thermal wind. { 'thər·məl 'jet }

thermal limit [ELEC] A limit on the power carried by an electric power system that results from the heating effects of the power carried by the devices. { 'thər·məl 'lim·ət }

thermal-liquid system [CHEM ENG] A system with a special liquid that acts as a heat sink or heat source (for example, steam, hot water, mercury, Dowtherm, molten salts, or mineral oils); used for process heating and cooling. { 'thər·məl ¦lik·wəd ˌsis·təm }

thermal-loss meter *See* heat-loss flowmeter. { 'thər·məl ¦lös ˌmēd·ər }

thermal low [METEOROL] An area of low atmospheric pressure due to high temperatures caused by intensive heating at the earth's surface; common to the continental subtropics in summer, thermal lows remain stationary over the area that produces them, their cyclonic circulation is generally weak and diffuse, and they are nonfrontal. Also known as heat low. { 'thər·məl 'lō }

thermal magnon [SOLID STATE] A magnon with a relatively short wavelength, on the order of 10^{-6} centimeter. { 'thər·məl 'mag,nän }

thermal mapper *See* line scanner. { 'thər·məl 'map·ər }

thermal metamorphism [PETR] Metamorphism that results from temperature-controlled and induced chemical reconstitution of preexisting rocks, with little influence of pressure. Also known as thermometamorphism. { 'thər·məl ¦med·ə¦mór,fiz·əm }

thermal microphone [ENG ACOUS] Microphone depending for its action on the variation in the resistance of an electrically heated conductor that is being alternately increased and decreased in temperature by sound waves. { 'thər·məl 'mī·krə,fōn }

thermal neutron [NUCLEO] One of a collection of neutrons whose energy distribution is identical with or similar to the Maxwellian distribution in the material in which they are found; the average kinetic energy of such neutrons at room temperature is about 0.025 electronvolt. Also known as slow neutron. { 'thər·məl 'nü,trän }

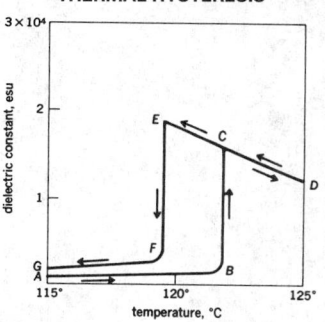

THERMAL HYSTERESIS

Plot of dielectric constant versus temperature for a single crystal of barium titanate, showing thermal hysteresis. On heating, dielectric constant was observed to follow path *ABCD*, and on cooling path *DCEFG*. (*After M. E. Drougard and D. R. Young, Phys. Rev., 95:1152-1153, 1954*)

thermal neutron analysis [ENG] A technique for detecting explosives, in which the object under inspection is conveyed through a cloud of thermal neutrons (generated by slowing down fast neutrons in multiple collisions in a moderator surrounding the source), and the characteristic high-energy gamma rays that are then emitted by the objects are used in analysis and imaging. { ˌthər·məl 'nü¦trän ə¦nal·ə·səs }

thermal noise [ELECTR] Electric noise produced by thermal agitation of electrons in conductors and semiconductors. Also known as Johnson noise; resistance noise. { 'thər·məl 'nȯiz }

thermal noise generator [ELECTR] A generator that uses the inherent thermal agitation of an electron tube to provide a calibrated noise source. { 'thər·məl ¦nȯiz ˌjen·ə¦rād·ər }

thermal ohm [THERMO] A unit of thermal resistance equal to the thermal resistance for which a temperature difference of 1 kelvin produces a flow of entropy of 1 watt per kelvin. Also known as fourier. { 'thər·məl 'ōm }

thermal phase *See* gradual phase. { 'thər·məl ˌfāz }

thermal photography *See* thermography. { 'thər·məl fə'täg·rə·fē }

thermal pollution [ECOL] The discharge of heated effluent into natural waters that causes a rise in temperature sufficient to upset the ecological balance of the waterway. { 'thər·məl pə'lü·shən }

thermal polymerization [CHEM ENG] A thermal, petroleum refining process used to convert light hydrocarbon gases into liquid fuels; paraffinic hydrocarbons are cracked to produce olefinic material which is concurrently polymerized by heat and pressure to form liquids, the product being known as polymer gasoline. { 'thər·məl pə¦lim·ə·rə'zā·shən }

thermal potential difference [THERMO] The difference between the thermodynamic temperatures of two points. { 'thər·məl pə¦ten·chəl 'dif·rəns }

thermal power plant [ENG] A facility to produce electric energy from thermal energy released by combustion of a fuel or consumption of a fissionable material. { 'thər·məl 'pau̇·ər ˌplant }

thermal printer [GRAPHICS] A nonimpact printer in which characters are formed by heating selected elements of a 5 × 7 or 7 × 9 dot matrix that is in contact with heat-sensitive paper. { 'thər·məl 'print·ər }

thermal probe [ENG] An instrument which measures the heat flow from ocean bottom sediment. [MECH ENG] A calorimeter in a boiler furnace which measures heat absorption rates. { 'thər·məl 'prōb }

thermal process [CHEM ENG] Any process that utilizes heat, without the aid of a catalyst, to accomplish chemical change; for example, thermal cracking, thermal reforming, or thermal polymerization. { 'thər·məl 'prä·səs }

thermal pulse [NUCLEO] The radiant power versus time pulse from a nuclear weapon detonation. { 'thər·məl 'pəls }

thermal pulse method [SOLID STATE] A method of measuring properties of insulating and conducting crystals, in which a heat pulse of known duration is measured after propagating through a crystal; the pulse can be generated by directing a laser pulse at an absorbing film evaporated onto one face of the crystal, and detected by a thin-film circuit on the other face. { 'thər·məl ¦pəls ˌmeth·əd }

thermal quality of snow *See* quality of snow. { 'thər·məl 'kwäl·əd·ē əv 'snō }

thermal radiation *See* heat radiation. { 'thər·məl ˌrād·ē'ā·shən }

thermal reactor [CHEM ENG] A device, system, or vessel in which chemical reactions take place because of heat (no catalysis); for example, thermal cracking, thermal reforming, or thermal polymerization. [NUCLEO] A nuclear reactor in which fission is induced primarily by neutrons of such low energy that they are in substantial thermal equilibrium with the material of the core. { 'thər·məl rē'ak·tər }

thermal reforming [CHEM ENG] A petroleum refining process using heat (but no catalyst) to effect molecular rearrangement of a low-octane naphtha to form high-octane motor gasoline. { 'thər·məl ri'fȯrm·iŋ }

thermal regenerative cell [ELEC] Fuel-cell system in which the reactants are regenerated continuously from the products formed during the cell reaction. { 'thər·məl rē'jen·rəd·iv 'sel }

thermal relay [ELEC] A relay operated by the heat produced by current flow. { 'thər·məl 'rē,lā }

thermal relief [ENG] A valve or other device that is preset to open when pressure becomes excessive due to increased temperature of the system. { 'thər·məl ri'lēf }

thermal resistance [ELECTR] *See* effective thermal resistance. [THERMO] A measure of a body's ability to prevent heat from flowing through it, equal to the difference between the temperatures of opposite faces of the body divided by the rate of heat flow. Also known as heat resistance. { 'thər·məl ri'zis·təns }

thermal resistivity [THERMO] The reciprocal of the thermal conductivity. { 'thər·məl rē¦zis¦tiv·əd·ē }

thermal resistor [ELEC] A resistor designed so its resistance varies in a known manner with changes in ambient temperature. { 'thər·məl ri'zis·tər }

thermal Rossby number [FL MECH] The nondimensional ratio of the inertial force due to the thermal wind and the Coriolis force in the flow of a fluid which is heated from below. { 'thər·məl 'räs·bē ˌnəm·bər }

thermal runaway [ELECTR] A condition that may occur in a power transistor when collector current increases collector junction temperature, reducing collector resistance and allowing a greater current to flow, which, in turn, increases the heating effect. { 'thər·məl 'rən·ə,wā }

thermal scattering [SOLID STATE] Scattering of electrons, neutrons, or x-rays passing through a solid due to thermal motion of the atoms in the crystal lattice. { 'thər·məl 'skad·ə·riŋ }

thermal shield [NUCLEO] A high-density heat-conducting portion of a shield placed close to the reflector of a nuclear reactor to absorb thermal neutrons, gamma rays, beta rays, and x-rays, whose absorption in the outer shield would generate excessive heat. { 'thər·məl 'shēld }

thermal shock [MECH] Stress produced in a body or in a material as a result of undergoing a sudden change in temperature. { 'thər·məl 'shäk }

thermal soakback [ENG] A phenomenon whereby, due to the lag in propagation of temperature changes through insulating materials, the maximum temperature of a thermally protected structure may be reached a certain time after the protective coating has reached its maximum temperature. { ˌthər·məl 'sōk,bak }

thermal spraying [MET] Spraying finely divided particles of powder or droplets of atomized metal wire or rod for coating a substrate. { 'thər·məl 'sprā·iŋ }

thermal spring [HYD] A spring whose water temperature is higher than the local mean annual temperature of the atmosphere. { 'thər·məl 'spriŋ }

thermal steering [METEOROL] The steering of an atmospheric disturbance in the direction of the thermal wind in its vicinity; equivalent to steering along thickness lines; for this purpose the thermal wind is usually taken from the earth's surface to a level in the middle troposphere. { 'thər·məl 'stir·iŋ }

thermal stratification [HYD] Horizontal layers of differing densities produced in a lake by temperature changes at different depths. { 'thər·məl ˌstrad·ə·fə'kā·shən }

thermal stress [MECH] Mechanical stress induced in a body when some or all of its parts are not free to expand or contract in response to changes in temperature. { 'thər·məl 'stres }

thermal stress cracking [MECH] Crazing or cracking of materials (plastics or metals) by overexposure to elevated temperatures and sudden temperature changes or large temperature differentials. { 'thər·məl ¦stres 'krak·iŋ }

thermal structure [PETR] A distinct structural pattern, such as a dome or anticline, defined by the arrangement of metamorphic zones of increasing grade. { 'thər·məl 'strək·chər }

thermal switch [ELEC] A temperature-controlled switch. Also known as thermoswitch. { 'thər·məl 'swich }

thermal telephone receiver [ENG ACOUS] A thermophone used as a telephone receiver. { 'thər·məl 'tel·ə,fōn ri,sē·vər }

thermal tide [METEOROL] A variation in atmospheric pressure due to the diurnal differential heating of the atmosphere by the sun; so-called in analogy to the conventional gravitational tide. { 'thər·məl 'tīd }

thermal time scale *See* Kelvin time scale. { 'thər·məl 'tīm ˌskāl }

thermal titration *See* thermometric titration. { 'thər·məl tī'trā·shən }

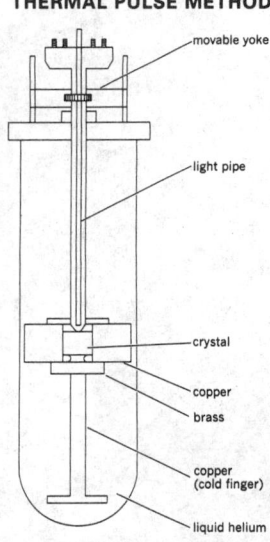

THERMAL PULSE METHOD

movable yoke

light pipe

crystal

copper

brass

copper (cold finger)

liquid helium

Apparatus for measuring heat pulses at various temperatures, showing components. Laser pulse propagates down light pipe onto absorbing film on face of crystal, and heat pulse propagates through crystal. *(From R. J. von Gutfeld and A. H. Nethercot, in W. P. Mason, ed., Physical Acoustics, vol. 5, 1969)*

thermal transducer [ENG] Any device which converts energy from some form other than heat energy into heat energy; an example is the absorbing film used in the thermal pulse method. { 'thər·məl tranz'dü·sər }

thermal transpiration [THERMO] The formation of a pressure gradient in gas inside a tube when there is a temperature gradient in the gas and when the mean free path of molecules in the gas is a significant fraction of the tube diameter. Also known as thermal effusion. { 'thər·məl ˌtranz·pə'rā·shən }

thermal tuning [ELEC] The process of changing the operating frequency of a system by using controlled thermal expansion to alter the geometry of the system. { 'thər·məl 'tün·iŋ }

thermal utilization factor [NUCLEO] The probability that a thermal neutron which is absorbed is absorbed usefully, as in a fissionable material. { 'thər·məl ˌyüd·əl·ə'zā·shən ˌfak·tər }

thermal value [THERMO] Heat produced by combustion, usually expressed in calories per gram or British thermal units per pound. { 'thər·məl ˌval·yü }

thermal valve [MECH ENG] A valve controlled by an element made of material that exhibits a significant change in properties in response to a change in temperature. { 'thər·məl 'valv }

thermal volt See kelvin. { 'thər·məl 'vōlt }

thermal vorticity [METEOROL] The vorticity of a thermal wind. { 'thər·məl vȯr'tis·əd·ē }

thermal vorticity advection [METEOROL] The advection or transport of the thermal vorticity by the thermal wind, in analogy to the advection of the vorticity by the wind. { 'thər·məl vȯr'tis·əd·ē ad'vek·shən }

thermal wattmeter [ENG] A wattmeter in which thermocouples are used to measure the heating produced when a current is passed through a resistance. { 'thər·məl 'wät·mēd·ər }

thermal wave [CRYO] See temperature wave. [SOLID STATE] A sound wave in a solid which has a short wavelength. { 'thər·məl 'wāv }

thermal wind [METEOROL] The mean wind-shear vector in geostrophic balance with the gradient of mean temperature of a layer bounded by two isobaric surfaces. { 'thər·məl 'wind }

thermal wind equation [METEOROL] An equation for the vertical variation of the geostrophic wind in hydrostatic equilibrium which may be written $-(\partial\mathbf{V}/\partial p) = (R/pf)\mathbf{k} \times \nabla_p T$, where $\mathbf{V}$ is the vector geostrophic wind, p the pressure (used here as the vertical coordinate), R the gas constant for air, f the Coriolis parameter, $\mathbf{k}$ a vertically directed unit vector, and ∇_p the isobaric del operator. { 'thər·məl ˌwind iˌkwā·zhən }

thermal x-rays [ELECTROMAG] The electromagnetic radiation, mainly in the soft (low-energy) x-ray region. { 'thər·məl 'eks·rāz }

thermal zone See thermal belt. { 'thər·məl 'zōn }

thermic boring [ENG] Boring holes into concrete by means of a high temperature, produced by a steel lance packed with steel wool which is ignited and kept burning by oxyacetylene or other gas. { 'thər·mik 'bȯr·iŋ }

thermic fever See sunstroke. { 'thər·mik 'fē·vər }

thermie [THERMO] A unit of heat energy equal to the heat energy needed to raise 1 tonne of water from 14.5°C to 15.5°C at a constant pressure of 1 standard atmosphere; equal to 10^6 fifteen-degrees calories or $(4.1855 \pm 0.0005) \times 10^6$ joules. Abbreviated th. { 'thər·mē }

thermion [ELECTR] A charged particle, either negative or positive, emitted by a heated body, as by the hot cathode of a thermionic tube. { 'thərm'ī,än }

thermionic [ELECTR] Pertaining to the emission of electrons as a result of heat. { ˌthər·mē'än·ik }

thermionic cathode See hot cathode. { ˌthər·mē'än·ik 'ka,thōd }

thermionic converter [ELECTR] A device in which heat energy is directly converted to electric energy; it has two electrodes, one of which is raised to a sufficiently high temperature to become a thermionic electron emitter, while the other, serving as an electron collector, is operated at a significantly lower temperature. Also known as thermionic generator; thermionic power generator; thermoelectric engine. { ˌthər·mē'än·ik kən'vərd·ər }

thermionic current [ELECTR] Current due to directed movements of thermions, such as the flow of emitted electrons from the cathode to the plate in a thermionic vacuum tube. { ˌthər·mē'än·ik 'kə·rənt }

thermionic detector [ELECTR] A detector using a hot-cathode tube. { ˌthər·mē'än·ik di'tek·tər }

thermionic diode [ELECTR] A diode electron tube having a heated cathode. { ˌthər·mē'än·ik 'dī,ōd }

thermionic emission [ELECTR] **1.** The outflow of electrons into vacuum from a heated electric conductor. Also known as Edison effect; Richardson effect. **2.** More broadly, the liberation of electrons or ions from a substance as a result of heat. { ˌthər·mē'än·ik i'mish·ən }

thermionic fuel cell [ELECTR] A thermionic converter in which the space between the electrodes is filled with cesium or other gas, which lowers the work functions of the electrodes, and creates an ionized atmosphere, controlling the electron space charge. { ˌthər·mē'än·ik 'fyül ˌsel }

thermionic generator See thermionic converter. { ˌthər·mē'än·ik 'jen·ə,rād·ər }

thermionic power generator See thermionic converter. { ˌthər·mē'än·ik 'pau̇·ər 'jen·ə,rād·ər }

thermionics [ELECTR] The study and applications of thermionic emission. { ˌthər·mē'än·iks }

thermionic triode [ELECTR] A three-electrode thermionic tube, containing an anode, a cathode, and a control electrode. { ˌthər·mē'än·ik 'trī,ōd }

thermionic tube [ELECTR] An electron tube that relies upon thermally emitted electrons from a heated cathode for tube current. Also known as hot-cathode tube. { ˌthər·mē'än·ik 'tüb }

thermionic work function [ELECTR] Energy required to transfer an electron from the fermi energy in a given metal through the surface to the vacuum just outside the metal. { ˌthər·mē'än·ik 'wərk ˌfəŋk·shən }

thermistor [ELECTR] A resistive circuit component, having a high negative temperature coefficient of resistance, so that its resistance decreases as the temperature increases; it is a stable, compact, and rugged two-terminal ceramiclike semiconductor bead, rod, or disk. Derived from thermal resistor. { thər'mis·tər }

thermit See thermite. { 'thər·mət }

thermite [MATER] A fire-hazardous mixture of ferric oxide and powdered aluminum; upon ignition by a magnesium ribbon, it reaches a temperature of 4000°F (2200°C), sufficient to soften steel; used for industrial purposes or as an incendiary bomb. Also spelled thermit. { 'thər,mīt }

thermit process [MET] An exothermic reaction when heating finely divided aluminum on a metal oxide causing reduction of the oxide. { 'thər·mət 'prä·səs }

thermit welding [MET] Welding with molten iron which is obtained by igniting aluminum and an iron oxide in a crucible, whereby the aluminum floats to the top of the molten metal and is poured off. { 'thər·mət ˌweld·iŋ }

thermoacidophile [BIOL] An organism that grows under extremely acidic conditions and at very high temperatures. { ˌthər·mō·a'sid·ə,fīl }

thermoacoustic array [ACOUS] A sound source consisting of a light beam (usually a laser beam) modulated at the frequency of the sound to be generated; the resulting sound has its maximum value in a direction perpendicular to the axis of the light beam, and its directivity pattern has no side lobes. { ˌthər·mō·əˈkü·stik ə'rā }

thermoacoustic effect [PHYS] **1.** Any effect that arises from the combination of the pressure oscillations of a sound wave with the accompanying adiabatic temperature oscillations. **2.** See optoacoustic effect. { ˌthər·mō·əˌkü·stik i'fekt }

thermoacoustic engine [ENG] A heat engine that harnesses the combination of the pressure oscillations of a sound wave with the accompanying adiabatic temperature oscillations. { ˌthər·mō·əˌkü·stik 'en·jən }

thermoacoustic refrigerator [ENG] A device that uses acoustic power to pump heat from a region of low temperature to a region of ambient temperature. { ˌthər·mō·ə,kü·stik ri'frij·ə,rād·ər }

thermoacoustics [PHYS] The study of phenomena that involve both thermodynamics and acoustics. { ˌthər·mō·ə'kü·stiks }

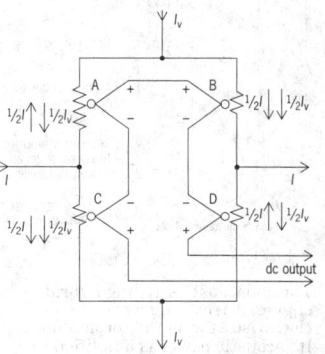

THERMAL WATTMETER

Thermal wattmeter circuit. I_V is the voltage signal; I is the current signal; B and C are thermal converters with current $\frac{1}{2}(I + I_V)$; A and D are thermal converters with current $\frac{1}{2}(I - I_V)$.

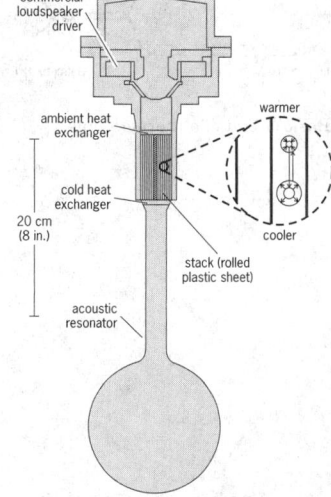

THERMOACOUSTIC REFRIGERATOR

An early model that cooled to −60°C (−76°F). Heat is carried up the temperature gradient in the stack. At right is a magnified view of the oscillating motion of at typical parcel of gas. The volume of the parcel depends on its pressure and temperature. (*After T.J. Hofler, Thermoacoustic Refrigerator Design and Performance, Ph.D. thesis, University of California at San Diego, 1996*)

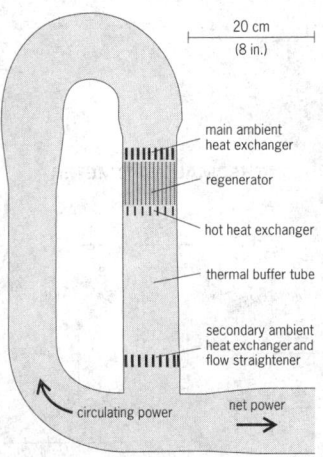

20 cm
(8 in.)

main ambient heat exchanger

regenerator

hot heat exchanger

thermal buffer tube

secondary ambient heat exchanger and flow straightener

circulating power

net power

Thermoacoustic-Stirling hybrid engine. A traveling wave runs clockwise around the toroidal loop. Its acoustic power is amplified in the regenerator so that new power up to 1 kilowatt (1.3 horsepower) can be extracted at the tee and delivered to a load. (*After S. Backhaus and G. Swift, Thermoacoustic-Stirling engine: Detailed study, J. Acous. Soc. Amer., 107:3148-3166, 2000*)

THERMOCOUPLE

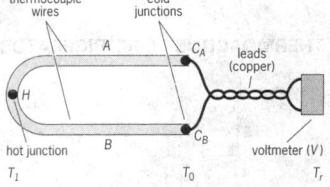

thermocouple wires

cold junctions

leads (copper)

voltmeter (*V*)

hot junction

T_1 T_0 T_r

Basic circuit of a thermocouple.

thermoacoustic-Stirling engine [ENG] A device in which the thermodynamic cycle of a Stirling engine is accomplished in a traveling-wave acoustic network, and acoustic power is produced from heat. { ¦thər·mō·ə¦kü·stik ¦stər·liŋ 'en·jən }

thermoammeter [ENG] An ammeter that is actuated by the voltage generated in a thermocouple through which is sent the current to be measured; used chiefly for measuring radio-frequency currents. Also known as electrothermal ammeter; thermocouple ammeter. { ¦thər·mō'am¸ēd·ər }

thermoanalysis *See* thermal analysis. { ¦thərmo·ə'nal·ə·səs }

thermobalance [ANALY CHEM] An analytical balance modified for thermogravimetric analysis, involving the measurement of weight changes associated with the transformations of matter when heated. { ¦thər·mō'bal·əns }

thermochemical calorie *See* calorie. { ¦thər·mō¦kem·ə·kəl 'kal·ə·rē }

thermochemistry [PHYS CHEM] The measurement, interpretation, and analysis of heat changes accompanying chemical reactions and changes in state. { ¦thər·mō¦kem·ə·strē }

thermochromism [PHYS] A reversible change in the color of a substance as its temperature is varied. { ¸thər·mə'krō¸miz·əm }

thermocline [GEOPHYS] **1.** A temperature gradient as in a layer of sea water, in which the temperature decrease with depth is greater than that of the overlying and underlying water. Also known as metalimnion. **2.** A layer in a thermally stratified body of water in which such a gradient occurs. { 'thər·mə¸klīn }

thermocoagulation [MED] Destruction of tissue by means of electrocautery or a high-frequency current. { ¦thər·mō·kō¸ag·yə'lā·shən }

thermocompression bonding [ENG] Use of a combination of heat and pressure to make connections, as when attaching beads to integrated-circuit chips; examples include wedge bonding and ball bonding. { ¦thər·mō·kəm'presh·ən 'bänd·iŋ }

thermocompression evaporator [MECH ENG] A system to reduce the energy requirements for evaporation by compressing the vapor from a single-effect evaporator so that the vapor can be used as the heating medium in the same evaporator. { ¦thər·mō·kəm'presh·ən i'vap·ə¸rād·ər }

thermocouple [ENG] A device consisting basically of two dissimilar conductors joined together at their ends; the thermoelectric voltage developed between the two junctions is proportional to the temperature difference between the junctions, so the device can be used to measure the temperature of one of the junctions when the other is held at a fixed, known temperature, or to convert radiant energy into electric energy. { 'thər·mə¸kəp·əl }

thermocouple ammeter *See* thermoammeter. { 'thər·mə¸kəp·əl 'am¸ēd·ər }

thermocouple converter *See* thermal converter. { 'thər·mə¸kəp·əl kən'vərd·ər }

thermocouple pyrometer *See* thermoelectric pyrometer. { 'thər·mə¸kəp·əl pī'räm·əd·ər }

thermocouple vacuum gage [ENG] A vacuum gage that depends for its operation on the thermal conduction of the gas present; pressure is measured as a function of the voltage of a thermocouple whose measuring junction is in thermal contact with a heater that carries a constant current; ordinarily, used over a pressure range of 10^{-1} to 10^{-3} millimeter of mercury. { 'thər·mə¸kəp·əl 'vak·yəm ¸gāj }

thermocyclogenesis [METEOROL] A theory of cyclogenesis by G. Stüve, in which the disturbance is initiated in the stratosphere and is reflected in the development of a disturbance in the lower troposphere. { ¦thər·mō¸sī·klə'jen·ə·səs }

thermodiffusion *See* thermal diffusion. { ¦thər·mō·di'fyü·zhən }

thermoduric bacteria [MICROBIO] Bacteria which survive pasteurization, but do not grow at temperatures used in a pasteurizing process. { ¦thər·mō'dúr·ik bak'tir·ē·ə }

thermodynamic cycle [THERMO] A procedure or arrangement in which some material goes through a cyclic process and one form of energy, such as heat at an elevated temperature from combustion of a fuel, is in part converted to another form, such as mechanical energy of a shaft, the remainder being rejected to a lower temperature sink. Also known as heat cycle. { ¦thər·mō·dī'nam·ik 'sī·kəl }

thermodynamic efficiency [IND ENG] An index for rating the effort required by a worker performing a task in terms of the ratio of work performed to the energy consumed. { ¦thər·mō·dī'nam·ik i'fish·ən·sē }

thermodynamic equation of state [THERMO] An equation that relates the reversible change in energy of a thermodynamic system to the pressure, volume, and temperature. { ¦thər·mō·dī'nam·ik i'kwā·zhən əv 'stāt }

thermodynamic equilibrium [THERMO] Property of a system which is in mechanical, chemical, and thermal equilibrium. { ¦thər·mō·dī'nam·ik ¸ē·kwə'lib·rē·əm }

thermodynamic function of state [THERMO] Any of the quantities defining the thermodynamic state of a substance in thermodynamic equilibrium; for a perfect gas, the pressure, temperature, and density are the fundamental thermodynamic variables, any two of which are, by the equation of state, sufficient to specify the state. Also known as state parameter; state variable; thermodynamic variable. { ¦thər·mō·dī'nam·ik 'fəŋk·shən əv 'stāt }

thermodynamic potential [THERMO] One of several extensive quantities which are determined by the instantaneous state of a thermodynamic system, independent of its previous history, and which are at a minimum when the system is in thermodynamic equilibrium under specified conditions. { ¦thər·mō·dī'nam·ik pə'ten·chəl }

thermodynamic potential at constant volume *See* free energy. { ¦thər·mō·dī'nam·ik pe¦ten·chəl at 'kän·stənt 'väl·yəm }

thermodynamic principles [THERMO] Laws governing the conversion of energy from one form to another. { ¦thər·mō·dī'nam·ik 'prin·sə·pəlz }

thermodynamic probability [THERMO] Under specified conditions, the number of equally likely states in which a substance may exist; the thermodynamic probability Ω is related to the entropy S by $S = k \ln \Omega$, where k is Boltzmann's constant. { ¦thər·mō·dī'nam·ik ¸präb·ə'bil·əd·ē }

thermodynamic process [THERMO] A change of any property of an aggregation of matter and energy, accompanied by thermal effects. { ¦thər·mō·dī'nam·ik 'prä·səs }

thermodynamic property [THERMO] A quantity which is either an attribute of an entire system or is a function of position which is continuous and does not vary rapidly over microscopic distances, except possibly for abrupt changes at boundaries between phases of the system; examples are temperature, pressure, volume, concentration, surface tension, and viscosity. Also known as macroscopic property. { ¦thər·mō·dī'nam·ik 'präp·ərd·ē }

thermodynamics [PHYS] The branch of physics which seeks to derive, from a few basic postulates, relationships between properties of matter, especially those which are affected by changes in temperature, and a description of the conversion of energy from one form to another. { ¦thər·mō·dī'nam·iks }

thermodynamic system [THERMO] A part of the physical world as described by its thermodynamic properties. { ¦thər·mō·dī'nam·ik 'sis·təm }

thermodynamic temperature scale [THERMO] Any temperature scale in which the ratio of the temperatures of two reservoirs is equal to the ratio of the amount of heat absorbed from one of them by a heat engine operating in a Carnot cycle to the amount of heat rejected by this engine to the other reservoir; the Kelvin scale and the Rankine scale are examples of this type. { ¦thər·mō·dī'nam·ik 'tem·prə·chər ¸skāl }

thermodynamic variable *See* thermodynamic function of state. { ¦thər·mō·dī'nam·ik 'ver·ē·ə·bəl }

thermoelasticity [PHYS] Dependence of the stress distribution of an elastic solid on its thermal state, or of its thermal conductivity on the stress distribution. { ¦thər·mō·i¸las'tis·əd·ē }

thermoelectric converter [ELECTR] A converter that changes solar or other heat energy to electric energy; used as a power source on spacecraft. { ¦thər·mō·i'lek·trik kən'vərd·ər }

thermoelectric cooler [ENG] An electronic heat pump based on the Peltier effect, involving the absorption of heat when current is sent through a junction of two dissimilar metals; it can be mounted within the housing of a device to prevent overheating or to maintain a constant temperature. { ¦thər·mō·i'lek·trik 'kü·lər }

thermoelectric cooling [ENG] Cooling of a chamber based

on the Peltier effect; an electric current is sent through a thermocouple whose cold junction is thermally coupled to the cooled chamber, while the hot junction dissipates heat to the surroundings. Also known as thermoelectric refrigeration. { ¦thər·mō·i'lek·trik 'kül·iŋ }

thermoelectric diffusion potential [PHYS CHEM] A potential difference across an electrolyte that results when a temperature gradient causes one constituent to attempt to flow relative to the other. { ¸thər·mō·i¦lek·trik də'fyü·zhən pə¸ten·chəl }

thermoelectric effect *See* thermoelectricity. { ¦thər·mō·i'lek·trik i¦fekt }

thermoelectric engine *See* thermionic converter. { ¦thər·mō·i'lek·trik 'en·jən }

thermoelectric generator *See* thermal converter. { ¦thər·mō·i'lek·trik 'jen·ə¸rād·ər }

thermoelectric heating [ENG] Heating based on the Peltier effect, involving a device which is in principle the same as that used in thermoelectric cooling except that the current is reversed. { ¦thər·mō·i'lek·trik 'hēd·iŋ }

thermoelectricity [PHYS] The direct conversion of heat into electrical energy, or the reverse; it encompasses the Seebeck, Peltier, and Thomson effects but, by convention, excludes other electrothermal phenomena, such as thermionic emission. Also known as thermoelectric effect. { ¦thər·mō¸i¸lek'tris·əd·ē }

thermoelectric junction *See* thermojunction. { ¦thər·mō·i'lek·trik 'jəŋk·shən }

thermoelectric laws [ENG] Basic relationships used in the design and application of thermocouples for temperature measurement; for example, the law of the homogeneous circuit, the law of intermediate metals, and the law of successive or intermediate temperatures. { ¦thər·mō·i'lek·trik 'lòz }

thermoelectric material [ELECTR] A material that can be used to convert thermal energy into electric energy or provide refrigeration directly from electric energy; good thermoelectric materials include lead telluride, germanium telluride, bismuth telluride, and cesium sulfide. { ¦thər·mō·i'lek·trik mə'tir·ē·əl }

thermoelectric nuclear battery [NUCLEO] A low-voltage battery in which a heat source, consisting of a radioactive isotope such as polonium-210, is hermetically sealed in a strong, dense capsule, and a series of thermocouples are alternately connected thermally, but not electrically, to the heat source and the outer surface of the capsule. { ¦thər·mō·i'lek·trik 'nü·klē·ər 'bad·ə·rē }

thermoelectric power generator *See* thermal converter. { ¦thər·mō·i'lek·trik 'pau·ər ¸jen·ə¸rād·ər }

thermoelectric properties [PHYS] Properties of materials associated with thermoelectricity, namely, the electromotive force generated in the Seebeck effect, the heat generated or absorbed in the Peltier and Thomson effects, and the influence of magnetic fields upon these quantities. { ¦thər·mō·i'lek·trik 'präp·ərd·ēz }

thermoelectric pyrometer [ENG] An instrument which uses one or more thermocouples to measure high temperatures, usually in the range between 800 and 2400°F (425 and 1315°C). Also known as thermocouple pyrometer. { ¦thər·mō·i'lek·trik pī'räm·əd·ər }

thermoelectric refrigeration *See* thermoelectric cooling. { ¦thər·mō·i'lek·trik ri¸frij·ə'rā·shən }

thermoelectric series [MET] A series of metals arranged in order of their thermoelectric voltage-generating ratings with respect to some reference metal, such as lead. { ¦thər·mō·i'lek·trik ¸sir·ēz }

thermoelectric solar cell [ELECTR] A solar cell in which the sun's energy is first converted into heat by a sheet of metal, and the heat is converted into electricity by a semiconductor material sandwiched between the first metal sheet and a metal collector sheet. { ¦thər·mō·i'lek·trik 'sō·lər 'sel }

thermoelectric thermometer [ENG] A type of electrical thermometer consisting of two thermocouples which are series-connected with a potentiometer and a constant-temperature bath; one couple, called the reference junction, is placed in a constant-temperature bath, while the other is used as the measuring junction. { ¦thər·mō·i'lek·trik thər'mäm·əd·ər }

thermoelectromotive force [ELEC] Voltage developed due to differences in temperature between parts of a circuit containing two or more different metals. { ¦thər·mō·i¦lek·trə¦mōd·iv 'fòrs }

thermoelectron [ELECTR] An electron liberated by heat, as from a heated filament. Also known as negative thermion. { ¦thər·mō·i'lek¸trän }

thermoelement *See* thermal converter. { ¦thər·mō'el·ə·mənt }

thermoforming [ENG] Forming of thermoplastic sheet by heating it and then pulling it down onto a mold surface to shape it. { 'thər·mə¸fòrm·iŋ }

thermogalvanic corrosion [MET] Corrosion associated with the passage of an electric current in which the anode and cathode are at different temperatures, the anode usually being the colder of the two. { ¦thər·mō·gal'van·ik kə'rō·zhən }

thermogalvanometer [ENG] Instrument for measuring small high-frequency currents by their heating effect, generally consisting of a direct-current galvanometer connected to a thermocouple that is heated by a filament carrying the current to be measured. { ¦thər·mō·gal·və'näm·əd·ər }

thermogenesis [PHYSIO] The production of heat. { ¸thər·mō'jen·ə·səs }

thermograd probe [ENG] An instrument that makes a record of temperature versus depth as it is lowered to the ocean floor, and measures heat flow through the ocean floor. { 'thər·mə¸grad 'prōb }

thermogram [ENG] The recording made by a thermograph. { 'thər·mə¸gram }

thermograph [ENG] An instrument that senses, measures, and records the temperature of the atmosphere. Also known as recording thermometer. [OPTICS] A far-infrared image-forming device that provides a thermal photograph by scanning a far-infrared image of an object or scene. { 'thər·mə¸graf }

thermograph correction card [ENG] A table for quick and accurate correction of the reading of a thermograph to that of the more accurate dry-bulb thermometer at the same time and place. { 'thər·mə¸graf kə'rek·shən ¸kärd }

thermography [ENG] A method of measuring surface temperature by using luminescent materials: the two main types are contact thermography and projection thermography. [GRAPHICS] **1.** A photocopying process in which the original copy is placed in contact with a transparent sheet and is exposed to infrared rays; heat from carbon or a metallic compound in the text ink then causes a chemical change in a substance laminated between the transparent sheet of paper and a white waxy back. **2.** Photography that uses radiation in the long-wavelength far-infrared region, emitted by objects at temperatures ranging from −170°F (−112°C) to over 300°F (149°C). Also known as thermal photography. [MED] A medical imaging technique based on detection of heat emitted by the body. { thər'mäg·rə·fē }

thermogravimetric analysis [ANALY CHEM] Chemical analysis by the measurement of weight changes of a system or compound as a function of increasing temperature. { ¦thər·mō¸grav·ə'me·trik ə'nal·ə·səs }

thermogravitational column [CHEM ENG] A device in which thermal diffusion results from the countercurrent flow of hot and cold material, thus increasing the separation of materials in a solution by the formation of a concentration gradient (difference). Also known as Clausius-Dickel column. { ¦thər·mō¸grav·ə'tā·shən·əl 'käl·əm }

thermohaline [OCEANOGR] Pertaining to the joint activity of salinity and temperature in the oceans. { ¸thər·mō'hā¸līn }

thermohaline convection [OCEANOGR] Vertical water movement observed when sea water, due to conditions of decreasing temperature or increasing salinity, becomes heavier than the water beneath it. { ¦thər·mō'hā¸līn kən'vek·shən }

thermointegrator [ENG] An apparatus, used in studying soil temperatures, for measuring the total supply of heat during a given period; it consists of a long nickel coil (inserted into the soil by an attached rod) forming a 100-ohm resistance thermometer and a 6-volt battery, the current used being recorded on a galvanometer; a mercury thermometer can be used. { ¦thər·mō'int·ə¸grād·ər }

thermoisopleth [CLIMATOL] An isopleth of temperature; specifically, a line on a climatic graph showing the variation of temperature in relation to two coordinates. { ¦thər·mō'īs·ə¸pleth }

thermojet [AERO ENG] Air-duct-type engine in which air is scooped up from the surrounding atmosphere, compressed, heated by combustion, and then expanded and discharged at high velocity. { 'thər·mə¸jet }

thermojunction [ELECTR] One of the surfaces of contact

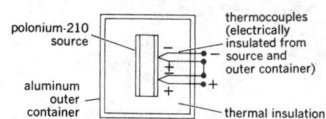

**THERMOELECTRIC
NUCLEAR BATTERY**

Diagram of thermoelectric nuclear battery.

between the two conductors of a thermocouple. Also known as thermoelectric junction. { ¦thər·mō'jəŋk·shən }

thermojunction battery [ELEC] Nuclear-type battery which converts heat into electrical energy directly by the thermoelectric or Seebeck effect. { ¦thər·mō'jəŋk·shən 'bad·ə·rē }

thermokarst topography [GEOL] An irregular land surface formed in a permafrost region by melting ground ice. { 'thər·mə,kärst tə'päg·rə·fē }

thermokinetic analysis [ANALY CHEM] A type of enthalpimetric analysis which uses kinetic titrimetry; involves rapid and continuous automatic delivery of a suitable titrant, under judiciously controlled experimental conditions with temperature measurement; the end points obtained are converted by mathematical procedures into valid stoichiometric equivalence points and used for determining reaction rate constants. { ¦thər·mō·ki¦ned·ik ə'nal·ə·səs }

thermoluminescence [ATOM PHYS] **1.** Broadly, any luminescence appearing in a material due to application of heat. **2.** Specifically, the luminescence appearing as the temperature of a material is steadily increased; it is usually caused by a process in which electrons receiving increasing amounts of thermal energy escape from a center in a solid where they have been trapped and go over to a luminescent center, giving it energy and causing it to luminesce. { ¦thər·mō,lü·mə'nes·əns }

thermoluminescent dating [ARCHEO] A method of dating the firing of pottery samples in which the amount of thermoluminescence from a heated sample is used to determine the number of trapped electrons resulting from absorption of alpha radiation. { ¦thər·mō,lü·mə¦nes·ənt 'dād·iŋ }

thermolysis [CHEM] *See* pyrolysis. [PHYSIO] The dissipation and dispersion of body heat by radiation or evaporation processes.

thermomagnetic effect [PHYS] An electrical or thermal phenomenon occurring when a conductor or semiconductor is placed simultaneously in a temperature gradient and a magnetic field; examples are the Ettingshausen-Nernst effect and the Righi-Leduc effect. { ¦thər·mō·mag'ned·ik i¦fekt }

thermomechanical effect *See* fountain effect. { ¦thər·mō·mi'kan·ə·kəl i¦fekt }

thermometal [MET] A bimetallic strip which, on temperature change, deflects because of differences in the coefficients of expansion of the two bonded metals. { 'thər·mō,med·əl }

thermometamorphism *See* thermal metamorphism. { ¦thər·mō¦med·ə¦mȯr,fiz·əm }

thermometer [ENG] An instrument that measures temperature. { thər'mäm·əd·ər }

thermometer anemometer [ENG] An anemometer consisting of two thermometers, one with an electric heating element connected to the bulb; the heated bulb cools in an airstream, and the difference in temperature as registered by the heated and unheated thermometers can be translated into air velocity by a conversion chart. { thər'mäm·əd·ər ,an·ə'mäm·əd·ər }

thermometer bird [VERT ZOO] The name applied to the brush turkey, native to Australia, because it lays its eggs in holes in mounds of earth and vegetation, with the heat from the decaying vegetation serving to incubate the eggs. { thər'mäm·əd·ər ,bərd }

thermometer-bulb liquid-level meter [ENG] Detection of liquid level by temperature measurement changes using an immersed bulb-type thermometer. { thər'mäm·əd·ər 'bəlb 'lik·wəd ¦lev·əl ,mēd·ər }

thermometer frame [ENG] A frame designed to hold two or more reversing thermometers; such a frame is often attached directly to a Nansen bottle. { thər'mäm·əd·ər ,frām }

thermometer screen *See* instrument shelter. { thər'mäm·əd·ər ,skrēn }

thermometer shelter *See* instrument shelter. { thər'mäm·əd·ər ,shel·tər }

thermometer support [ENG] A device used to hold liquid-in-glass maximum and minimum thermometers in the proper recording position inside an instrument shelter, and to permit them to be read and reset. { thər'mäm·əd·ər sə,pȯrt }

thermometric analysis [PHYS CHEM] A method for determination of the transformations a substance undergoes while being heated or cooled at an essentially constant rate, for example, freezing-point determinations. { ¦thər·mə¦me·trik ə'nal·ə·səs }

THERMOLUMINESCENCE

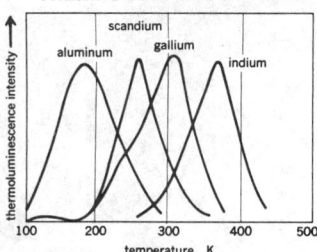

Plot of thermoluminescence intensity as a function of temperature in several zinc sulfide phosphors, each of which contains traces of copper, giving rise to luminescent centers, and different trivalent ions (as indicated), giving rise to traps. Curves reach a maximum and then decrease as traps are emptied.

thermometric conductivity *See* diffusivity. { ¦thər·mə¦me·trik ,kän,dək,tiv·əd·ē }

thermometric depth [OCEANOGR] The ocean depth, in meters, deduced from the difference between the paired protected and unprotected reversing thermometer readings; the unprotected reversing thermometer indicates higher temperature due to pressure effects on the instrument. { ¦thər·mə¦me·trik 'depth }

thermometric fluid [THERMO] A fluid that has properties, such as a large and uniform thermal expansion coefficient, good thermal conductivity, and chemical stability, that make it suitable for use in a thermometer. { ¦thər·mə¦me·trik 'flü·əd }

thermometric property [THERMO] A physical property that changes in a known way with temperature, and can therefore be used to measure temperature. { ¦thər·mə¦me·trik 'präp·ərd·ē }

thermometric titration [ANALY CHEM] A titration in an adiabatic system, yielding a plot of temperature versus volume of titrant; used for neutralization, precipitation, redox, organic condensation, and complex-formation reactions. Also known as calorimetric titration; enthalpy titration; thermal titration. { ¦thər·mə¦me·trik tī'trā·shən }

thermometry [THERMO] The science and technology of measuring temperature, and the establishment of standards of temperature measurement. { thər'mäm·ə·trē }

thermomigration [ELECTR] A technique for doping semiconductors in which exact amounts of known impurities are made to migrate from the cool side of a wafer of pure semiconductor material to the hotter side when the wafer is heated in an oven. { ¦thər·mō·mī¦grā·shən }

thermonatrite [MINERAL] $Na_2CO_3 \cdot H_2O$ A colorless to white, grayish, or yellowish, orthorhombic mineral consisting of sodium carbonate monohydrate; occurs as a crust or efflorescence. { ¦thər·mə'nā,trīt }

thermonuclear [NUCLEO] Referring to any process in which a very high temperature is used to bring about the fusion of light nuclei, with the accompanying liberation of energy. { ¦thər·mō'nü·klē·ər }

thermonuclear bomb *See* fusion bomb. { ¦thər·mō'nü·klē·ər 'bäm }

thermonuclear device [NUCLEO] A fusion bomb used for peaceful purposes, tests, or experiments. { ¦thər·mō'nü·klē·ər di'vīs }

thermonuclear reaction [NUC PHYS] A nuclear fusion reaction which occurs between various nuclei of light elements when they are constituents of a gas at very high temperature. { ¦thər·mō'nü·klē·ər rē'ak·shən }

thermonuclear rocket [NUCLEO] A type of thermal engine utilizing nuclear fusion to heat a working fluid: deuterium, tritium, and lithium are possible fuels. { ¦thər·mō'nü·klē·ər 'räk·ət }

thermonuclear weapon [NUCLEO] A fusion bomb in the form of a packaged unit ready for transportation or use by military forces. { ¦thər·mō'nü·klē·ər 'wep·ən }

thermooptic effect [PHYS] The change in optical properties of a material because of heat radiation. { ¦thər·mō'äp·tik ə'fekt }

thermoperiodicity [BOT] The totality of responses of a plant to appropriately fluctuating temperatures. { ¦thər·mō,pir·ē·ə'dis·əd·ē }

thermo-pervaporation *See* membrane distillation. { ¦thər·mō·pər,vap·ə'rā·shən }

thermophile [BIOL] An organism that thrives at high temperatures. { 'thər·mə,fīl }

thermophobia [PSYCH] An abnormal fear of heat. { ,thər·mə'fō·bē·ə }

thermophone [ENG ACOUS] An electroacoustic transducer in which sound waves having an accurately known strength are produced by the expansion and contraction of the air adjacent to a strip of conducting material, whose temperature varies in response to a current input that is the sum of a steady current and a sinusoidal current; used chiefly for calibrating microphones. { 'thər·mə,fōn }

thermophoresis [THERMO] The movement of particles in a thermal gradient from high to low temperatures. { ,thər·mə·fə'rē·səs }

thermopile [ENG] An array of thermocouples connected either in series to give higher voltage output or in parallel to give higher current output, used for measuring temperature or radiant energy or for converting radiant energy into electric power. { 'thər·mə,pīl }

thermopile generator [ELEC] An electricity source powered by the heating of an electrical resistor that can be connected to a thermopile to generate small amounts of electric current. { 'thər·mə,pīl 'jen·ə,rād·ər }

thermoplastic [MATER] A polymeric material with a linear macromolecular structure that will repeatedly soften when heated and harden when cooled; for example, styrene, acrylics, polyethylenes, vinyls, nylons, and fluorocarbons. Also known as thermoplastic resin. { 'thər·mə,plas·tik }

thermoplastic elastomer [ORG CHEM] A polymer that can be processed as a thermoplastic material but also possesses the properties of a conventional thermoset rubber. Abbreviated TPE. { ¦thər·mə¦pla·stik i'las·tə·mər }

thermoplastic insulation [MATER] Electrical insulation made of a thermoplastic material. { ¦thər·mə¦plas·tik ,in·sə'lā·shən }

thermoplastic recording [ELECTR] A recording process in which a modulated electron beam deposits charges on a thermoplastic film, and application of heat by radio-frequency heating electrodes softens the film enough to produce deformation that is proportional to the density of the stored electrostatic charges; an optical system is used for playback. { ¦thər·mə¦plas·tik ri'kórd·iŋ }

thermoplastic resin See thermoplastic. { ¦thər·mə¦plas·tik 'rez·ən }

thermopower [ELEC] A measure of the temperature-induced voltage in a conductor. { 'thər·mə,paú·ər }

thermoreception [PHYSIO] The process by which environmental temperature affects specialized sense organs (thermoreceptors). { ¦thər·mō·ri'sep·shən }

thermoreceptor [PHYSIO] A sense receptor that responds to stimulation by heat and cold. { ¦thər·mō·ri'sep·tər }

thermoregulation [PHYSIO] A mechanism by which mammals and birds attempt to balance heat gain and heat loss in order to maintain a constant body temperature when exposed to variations in cooling power of the external medium. { ¦thər·mō,reg·yə'lā·shən }

thermoregulator [ENG] A high-accuracy or high-sensitivity thermostat; one type consists of a mercury-in-glass thermometer with sealed-in electrodes, in which the rising and falling column of mercury makes and breaks an electric circuit. { ¦thər·mō'reg·yə,lād·ər }

thermorelay See thermostat. { ¦thər·mō'rē,lā }

thermoremanent magnetization [GEOPHYS] The permanent magnetization of igneous rocks, acquired at the time of cooling from the molten state. { ¦thər·mō'rem·ə·nənt ,mag·nəd·ə'zā·shən }

Thermosbaenacea [INV ZOO] An order of small crustaceans in the superorder Pancarida. { ,thər·məs·bə'nās·ē·ə }

Thermosbaenidae [INV ZOO] A family of the crustacean order Thermosbaenacea. { ,thər·məs'bē·nə,dē }

thermoscreen See instrument shelter. { 'thər·mə,skrēn }

thermosets [MATER] Polymeric materials that usually have a crosslinked network. They are formed into a permanent shape and are cured (set) by a chemical reaction that may require heat and pressure. Thermoset polymers are typically insoluble, and cannot be remelted or reformed into another shape after curing. Also known as thermosetting resin. { 'thər·mə,sets }

thermosetting resin See thermosets. { 'thər·mə,sed·iŋ 'rez·ən }

thermosiphon [MECH ENG] A closed system of tubes connected to a water-cooled engine which permit natural circulation and cooling of the liquid by utilizing the difference in density of the hot and cool portions. { ¦thər·mō'sī·fən }

thermosiphon reboiler [CHEM ENG] A liquid reheater (as for distillation-column bottoms) in which natural circulation of the boiling liquid is obtained by maintaining a sufficient liquid head. { ¦thər·mō'sī·fən ¦rē'bói·lər }

thermosphere [METEOROL] The atmospheric shell extending from the top of the mesosphere to outer space; it is a region of more or less steadily increasing temperature with height, starting at 40 to 50 miles (70 to 80 kilometers); the thermosphere includes, therefore, the exosphere and most or all of the ionosphere. { 'thər·mə,sfir }

thermostat [ENG] An instrument which measures changes in temperature and directly or indirectly controls sources of

heating and cooling to maintain a desired temperature. Also known as thermorelay. { 'thər·mə,stat }

thermostatic switch [ELEC] A temperature-operated switch that receives its operating energy by thermal conduction or convection from the device being controlled or operated. { ¦thər·mə¦stad·ik 'swich }

thermosteric anomaly [OCEANOGR] Component of the specific volume anomaly for a parcel of sea water at a pressure of 1 atmosphere due to its temperature being other than the standard temperature of C°C. { ¦thər·mə¦ster·ik ə'näm·ə·lē }

thermoswitch See thermal switch. { 'thər·mə,swich }

thermotaxis [BIOL] Orientation movement of a motile organism in response to the stimulus of a temperature gradient. { ¦thər·mō¦tak·səs }

thermotherapy [MED] The treatment of disease by heat of any kind; involves the local or general application of heat to the body. { ¦thər·mō'ther·ə·pē }

thermotropic liquid crystal [PHYS CHEM] A liquid crystal prepared by heating the substance. { ¦thər·mō¦träp·ik 'lik·wəd 'krist·əl }

thermotropic model [METEOROL] A model atmosphere used in numerical forecasting, in which the parameters are the height of one constant-pressure surface (usually 500 millibars) and one temperature (usually the mean temperature between 100 and 500 millibars). { ¦thər·mō¦träp·ik 'mad·əl }

thermovac evaporator [FOOD ENG] A type of vacuum evaporator used to remove moisture from foods for the purpose of preservation; temperature and velocities are minutely, accurately, and automatically controlled in all stages of the process. { 'thər·mə,vak i'vap·ə,rād·ər }

thermoviscous attenuation See classical attenuation. { ,thər·mō,vis·kəs ə,ten·yə wā·shən }

thermovoltmeter [ENG] A voltmeter in which a current from the voltage source is passed through a resistor and a fine vacuum-enclosed platinum heater wire; a thermocouple, attached to the midpoint of the heater, generates a voltage of a few millivolts, and this voltage is measured by a direct-current millivoltmeter. { ¦thər·mō'vōlt,mēd·ər }

therophyte [ECOL] An annual plant whose seed is the only overwintering structure. { 'ther·ə,fīt }

Theropoda [PALEON] A suborder of carnivorous bipedal saurischian reptiles which first appeared in the Upper Triassic and culminated in the uppermost Cretaceous. { thi'räp·əd·ə }

Theropsida [PALEON] An order of extinct mammallike reptiles in the subclass Synapsida. { thi'räp·səd·ə }

Thesium [BOT] The hymenium of the apothecium in lichens. { 'thē·sē·əm }

thesocyte [INV ZOO] An amebocyte in Porifera containing ergastic cytoplasmic inclusions. { 'thes·ə,sīt }

theta antigen [IMMUNOL] A cell membrane constituent which distinguishes T cells from other lymphocytes. { 'thād·ə 'ant·i·jən }

theta functions [MATH] Complex functions used in the study of Riemann surfaces and of elliptic functions and elliptic integrals; they are:

$$\theta_1(z) = 2 \sum_{n=0}^{\infty} (-1)^n q^{(n+1/2)^2} \sin(2n+1)z$$

$$\theta_2(z) = 2 \sum_{n=0}^{\infty} q^{(n-1/2)^2} \cos(2n+1)z$$

$$\theta_3(z) = 1 + 2 \sum_{n=1}^{\infty} q^{n^2} \cos 2nz$$

$$\theta_4(z) = 1 + 2 \sum_{n=1}^{\infty} (-1)^n q^{n^2} \cos 2nz$$

where $q = \exp \pi i \tau$, and τ is a constant complex number with positive imaginary part. { 'thād·ə ,faŋk·shənz }

thetagram [THERMO] A thermodynamic diagram with coordinates of pressure and temperature, both on a linear scale. { 'thād·ə,gram }

theta pinch [PL PHYS] A device for producing a controlled nuclear fusion reaction, in which plasma in a long torus or skinny tube is confined by a magnetic field produced by current-carrying coils, and is shock-heated and compressed by pulses in this field to produce the high temperatures at which fusion reactions take place; the magnetic field is then sustained in order to maintain the plasma confinement. { 'thäd·ə ,pinch }

theta polarization [ELECTROMAG] State of a wave in which the E vector is tangential to the meridian lines of some given spherical frame of reference. { 'thäd·ə ,pō·lə·rə'zā·shən }

theta rhythm [PSYCH] A brain rhythm having a frequency of about 4–7 hertz, and somewhat greater voltage than the alpha rhythm; thought to originate in the hippocampus. { 'thäd·ə ,rith·əm }

theta-theta [NAV] The generic term for electronic navigation systems in which position is derived by calculations based on the bearing from two or more emitters located at different but accurately known positions. { 'thäd·ə 'thäd·ə }

Thévenin equivalent circuit [ELEC] An equivalent circuit that consists of a series connection of a voltage source and a two-terminal circuit, where the voltage source is usually dependent on the electric signals applied to the input terminals. { tā·vō¦na i,kwiv·ə·lənt 'sər·kət }

Thévenin generator [ELEC] The voltage generator in the equivalent circuit of Thévenin's theorem. { tā·vō'na ,jen·ə,rād·ər }

Thévenin's theorem [ELEC] A theorem in network problems which allows calculation of the performance of a device from its terminal properties only: the theorem states that at any given frequency the current flowing in any impedance, connected to two terminals of a linear bilateral network containing generators of the same frequency, is equal to the current flowing in the same impedance when it is connected to a voltage generator whose generated voltage is the voltage at the terminals in question with the impedance removed, and whose series impedance is the impedance of the network looking back from the terminals into the network with all generators replaced by their internal impedances. Also known as Helmholtz's theorem. { tā·vō'naz ,thir·əm }

THF See tetrahydrofuran.

THI See temperature-humidity index.

thiabendazole [ORG CHEM] $C_{10}H_7N_3S$ A white powder with a melting point of 304–305°C; controls fungi on citrus fruits, sugarbeets, turf, and ornamentals, and roundworms of cattle and other animals. Also known as 2-(4-thiazolyl)benzimidazole. { ,thī·ə¦ben·də,zōl }

thiacetic acid See thioacetic acid. { thī·ə¦sēd·ik 'as·əd }

thiamine [BIOCHEM] $C_{12}H_{17}ClN_4OS$ A member of the vitamin B complex that occurs in many natural sources, frequently in the form of cocarboxylase. Also known as aneurine; vitamin B₁. { 'thī·ə·mən }

thiamine hydrochloride [ORG CHEM] $C_{12}H_{17}\cdot ON_4S\cdot HCl$ White, hygroscopic crystals soluble in water, insoluble in ether, with a yeasty aroma and a salty, nutlike taste, decomposes at 247°C; the form in which thiamine is generally employed. { 'thī·ə·mən ¦hī·drə'klȯr,īd }

thiamine pyrophosphate [BIOCHEM] The coenzyme or prosthetic component of carboxylase; catalyzes decarboxylation of various α-keto acids. Also known as cocarboxylase. { 'thī·ə·mən ¦pī·rō'fä,sfāt }

thianaphthene [ORG CHEM] C_8H_6S A crystalline compound with a melting point of 32°C; soluble in organic solvents; used in the production of pharmaceuticals. Also known as benzothiofuran. { ,thī·ə'naf,thēn }

Thiaridae [INV ZOO] A family of freshwater gastropod mollusks in the order Pectinibranchia. { ,thī'ar·ə,dē }

thiazole [ORG CHEM] C_3H_3NS A colorless to yellowish liquid with a pyridinelike aroma, slightly soluble in water, soluble in alcohol and ether; used as an intermediate for fungicides, dyes, and rubber accelerators. { 'thī·ə,zōl }

thiazole dye [ORG CHEM] One of a family of dyes in which the chromophore groups are $=C=N-$, $-S-C=$, and used mainly for cotton; an example is primuline. { 'thī·ə,zōl 'dī }

2-(4-thiazolyl)benzimidazole See thiabendazole. { ¦tü ¦fȯr thī¦az·ə,wil ,benz,im·ə'da,zōl }

thick-bedded [GEOL] Pertaining to a sedimentary bed that ranges in thickness from 60 to 120 centimeters (2–4 feet). { 'thik ,bed·əd }

thick disk [ASTRON] A component of the galactic stellar population that extends over distances up to 1500 parsecs from the galactic plane, and is approximately $9–10 \times 10^9$ years

---

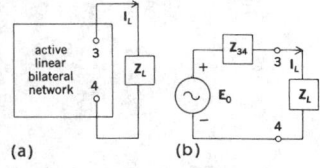

THÉVENIN'S THEOREM

(a)   (b)

The theorem states that circuit to left of terminals 3 and 4 in *(a)* may be replaced by that in *(b)*. Impedance Z_L and current I_L are same in both cases. E_0 is voltage of voltage generator, Z_{34} is series impedance.

---

old, intermediate in age between the thin disk and the halo. { ¦thik 'disk }

thickened fuel See gelatinized gasoline. { 'thik·ənd 'fyül }

thickened oil [MATER] Any oil to which a thickening agent has been added to increase viscosity or produce thixotropic properties; grease is a thickened oil. { 'thik·ənd 'ȯil }

thickener [ENG] A nonfilter device for the removal of liquid from a liquid-solids slurry to give a dewatered (thickened) solids product; can be by gravity settling or centrifugation. { 'thik·ə·nər }

thickening [CHEM ENG] The concentration of the solids in a suspension in order to recover a fraction with a higher concentration of solids than in the original suspension. [MIN ENG] Concentrating dilute slime pulp into a pulp containing a smaller percentage of moisture by rejecting the free liquid. { 'thik·ə·niŋ }

thicket See tropical scrub. { 'thik·ət }

thick-film capacitor [ELEC] A capacitor in a thick-film circuit, made by successive screen-printing and firing processes. { 'thik ¦film kə'pas·əd·ər }

thick-film circuit [ELECTR] A microcircuit in which passive components, of a ceramic-metal composition, are formed on a ceramic substrate by successive screen-printing and firing processes, and discrete active elements are attached separately. { 'thik ¦film 'sər·kət }

thick-film hybrid [ELECTR] An assembly consisting of a thick-film circuit pattern with mounting positions for the insertion of conventional silicon devices. { ,thik ,film 'hī·brəd }

thick-film resistor [ELEC] Fixed resistor whose resistance element is a film well over 0.001 inch (25 micrometers) thick. { 'thik ¦film ri'zis·tər }

thick-film sensor [ENG] A thick-film circuit that is fabricated from suitable materials to measure a physical quantity such as mechanical stress or temperature, or to perform a chemical sensing application such as the measurement of gas or liquid composition, acidity, or humidity. { ,thik ,film 'sen·sər }

thickhead See bluetongue. { 'thik,hed }

thick lens [OPTICS] A lens in which the separation between the two surfaces is too great to be ignored in calculations of such quantities as focal length and magnification. { ¦thik 'lenz }

thickness [METEOROL] The vertical depth, measured in geometric or geopotential units, of a layer in the atmosphere bounded by surfaces of two different values of the same physical quantity, usually constant-pressure surfaces. { 'thik·nəs }

thickness chart [METEOROL] A type of synoptic chart showing the thickness of a certain physically defined layer in the atmosphere; it almost always refers to an isobaric thickness chart, that is, a chart of vertical distance between two constant-pressure surfaces. { 'thik·nəs ,chärt }

thickness gage [ENG] A gage for measuring the thickness of a sheet of material, the thickness of an object, or the thickness of a coating; examples include penetration-type and backscattering radioactive thickness gages and ultrasonic thickness gages. { 'thik·nəs ,gāj }

thickness line [METEOROL] A line drawn through all geographic points at which the thickness of a given atmospheric layer is the same. Also known as relative contour; relative isohypse. { 'thik·nəs ,līn }

thickness noise [ACOUS] The component of propeller noise that is caused by the displacement of the air by the rotating propeller blade. { 'thik·nəs ,nȯiz }

thickness pattern [METEOROL] The general geometric distribution of thickness lines on a thickness chart. Also known as relative hypsography; relative topography. { 'thik·nəs ,pad·ərn }

thickness ratio [AERO ENG] The ratio of the maximum thickness of an airfoil section to the length of its chord. { 'thik·nəs ,rā·shō }

thick-skinned structure [GEOL] Any large-scale structure, such as a fold or fault, believed to have originated as a result of basement movement beneath overlying rocks. { 'thik ¦skind 'strək·chər }

thick-tailed bushbaby [VERT ZOO] *Galago crassicaudatus.* A primate animal in the family Lorisidae; one of the six species of bushbaby, the thick-tailed bushbaby is more aggressive and solitary than the other species, and grows to over 1 foot (30 centimeters) in length with an equally long tail. Also known as great galago. { 'thik ¦tāld 'bùsh,bā·bē }

thick-thin chart *See* isentropic thickness chart. { 'thik 'thin ‚chärt }

thief [PETRO ENG] In the petroleum industry, a device that permits the taking of samples from a predetermined location in the liquid body to be sampled. { 'thēf }

Thiele coordinates [CHEM ENG] A graphical method for calculating the solvent-free composition of two components being separated by solvent extraction. { 'tēl·ə kō‚ȯrd·ən·əts }

Thiele-Geddes method [CHEM ENG] A method for the prediction of the product distribution from a multicomponent distillation system. { 'tēl·ə 'ged·əs ‚meth·əd }

Thiele melting-point apparatus [ANALY CHEM] A stirred, specially shaped test-tube device used for the determination of the melting point of a crystalline chemical. { 'tēl·ə 'melt·iŋ ‚pȯint ‚ap·ə‚rad·əs }

Thiessen polygon method [METEOROL] A method of assigning areal significance to point rainfall values; perpendicular bisectors are constructed to the lines joining each measuring station with those immediately surrounding it; the bisectors form a series of polygons, each polygon containing one station; the value of precipitation measured at a station is assigned to the whole area covered by the enclosing polygon. { 'tē·sən 'päl·i‚gän ‚meth·əd }

thigh [ANAT] The upper part of the leg, from the pelvis to the knee. { thī }

thigh circumference [ANTHRO] The measurement around the thigh of the left leg midway between the crotch and the knee when the subject is in a standing position. { 'thī sər‚kəm·frəns }

thigmotaxis *See* stereotaxis. { ‚thig·mə¦tak·səs }

Thigmotrichida [INV ZOO] An order of ciliated protozoans in the subclass Holotrichia. { ‚thig·mō'trik·əd·ə }

thigmotropism *See* stereotropism. { thig'mä·trə‚piz·əm }

thill *See* underclay. { thil }

thimble [COMPUT SCI] A cone-shaped, rotating printing element on an impact printer having character slugs around the perimeter and a hammer that drives the appropriate slug forward to print the impression on paper. { 'thim·bəl }

thimble ionization chamber [NUCLEO] A small cylindrical or spherical ionization chamber, usually with walls made of organic material or air walls. { ‚thim·bəl ‚i·ə·nə'zā·shən ‚chām·bər }

thimerosal [ORG CHEM] $C_9H_9HgNaO_2S$ The sodium salt of ethyl mercury thiosalicylic acid used medicinally as an antimicrobial agent. Also known as merthiolate. { thi'mer·ə·sȯl }

thin [METEOROL] In aviation weather observations, the description of a sky cover that is predominantly transparent. { thin }

thin-bedded [GEOL] Pertaining to a sedimentary bed that ranges in thickness from 2 inches to 2 feet (5 to 60 centimeters). { 'thin ‚bed·əd }

thin disk [ASTRON] The youngest component of the galactic stellar population, still actively forming massive stars from molecular clouds and confined to within about 350 parsecs (1100 light-years) of the galactic plane. { ¦thin 'disk }

thin film [ELECTR] A film a few molecules thick deposited on a glass, ceramic, or semiconductor substrate to form a capacitor, resistor, coil, cryotron, or other circuit component. [MATER] A film of a material from one to several hundred molecules thick deposited on a solid substrate such as glass or ceramic or as a layer on a supporting liquid. { 'thin 'film }

thin-film capacitor [ELEC] A capacitor that can be constructed by evaporation of conductor and dielectric films in sequence on a substrate; silicon monoxide is generally used as the dielectric. { 'thin ¦film kə'pas·əd·ər }

thin-film circuit [ELECTR] A circuit in which the passive components and conductors are produced as films on a substrate by evaporation or sputtering; active components may be similarly produced or mounted separately. { 'thin ¦film 'sər·kət }

thin-film cryotron [ELECTR] A cryotron in which the transition from superconducting to normal resistivity of a thin film of tin or indium, serving as a gate, is controlled by current in a film of lead that crosses and is insulated from the gate. { 'thin ¦film 'krī·ə‚trän }

thin-film ferrite coil [ELECTROMAG] An inductor made by depositing a thin flat spiral of gold or other conducting metal on a ferrite substrate. { 'thin ¦film 'fe‚rīt 'kȯil }

thin-film field-emitter cathode [ELECTR] A sharply pointed microminiature electron field emitter with an integral low-voltage extraction gate. { 'thin ‚film ¦fēld i‚mid·ər 'kath‚ōd }

thin-film integrated circuit [ELECTR] An integrated circuit consisting entirely of thin films deposited in a patterned relationship on a substrate. { 'thin ¦film 'int·ə‚grād·əd 'sər·kət }

thin-film material [ELECTR] A material that can be deposited as a thin film in a desired pattern by a variety of chemical, mechanical, or high-vacuum evaporation techniques. { 'thin ¦film mə'tir·ē·əl }

thin-film memory *See* thin-film storage. { 'thin ¦film 'mem·rē }

thin-film resistor [ELEC] A fixed resistor whose resistance element is a metal, alloy, carbon, or other film having a thickness of about 0.000001 inch (25 nanometers). { 'thin ¦film ri'zis·tər }

thin-film semiconductor [ELECTR] Semiconductor produced by the deposition of an appropriate single-crystal layer on a suitable insulator. { 'thin ¦film 'sem·i·kən‚dək·tər }

thin-film solar cell [ELECTR] A solar cell in which a thin film of gallium arsenide, cadmium sulfide, or other semiconductor material is evaporated on a thin, flexible metal or plastic substrate; the rather low efficiency (about 2%) is compensated by the flexibility and light weight, making these cells attractive as power sources for spacecraft. { 'thin ¦film 'sō·lər 'sel }

thin-film storage [COMPUT SCI] A high-speed storage device that is fabricated by depositing layers, one molecule thick, of various materials which, after etching, provide microscopic circuits which can move and store data in small amounts of time. Also known as thin-film memory. { 'thin ¦film 'stȯr·ij }

thin-film transducer [SOLID STATE] A film a few molecules thick, usually consisting of cadmium sulfide, evaporated on a crystal substrate, used to convert microwave radiation into hypersonic sound waves in the crystal. { 'thin ¦film tran·z'dü·sər }

thin-film transistor [ELECTR] A field-effect transistor constructed entirely by thin-film techniques, for use in thin-film circuits. Abbreviated TFT. { 'thin ¦film tran'zis·tər }

think time [COMPUT SCI] Idle time between time intervals in which transmission takes place in a real-time system. { 'thiŋk ‚tīm }

thin-layer chromatography [ANALY CHEM] Chromatographing on thin layers of adsorbents rather than in columns; adsorbent can be alumina, silica gel, silicates, charcoals, or cellulose. { 'thin ¦lā·ər ‚krō·mə'täg·rə·fē }

thin lens [OPTICS] A lens whose thickness is small enough to be neglected in calculations of such quantities as object distance, image distance, and magnification. { 'thin 'lenz }

thin list *See* loose list. { 'thin 'list }

thinner [MATER] A liquid used to thin paint, varnish, cement, or other material to a desired consistency. { 'thin·ər }

Thinocoridae [VERT ZOO] The seed snipes, family of South American birds in the order Charadriiformes. { ‚thin·ə'kȯr·ə‚dē }

thin-out [GEOL] Gradual thinning of a stratum, vein, or other body of rock until the upper and lower surfaces meet and the rock disappears. { 'thin'aut }

thin-plate orifice [ENG] A thin-metal orifice sheet used in fluid-flow measurement in fluid conduits by means of differential pressure drop across the orifice. { 'thin ¦plāt 'ȯr·ə·fəs }

thin section [GEOL] A piece of rock or mineral specifically prepared to study its optical properties; the sample is ground to 0.03-millimeter thickness, then polished and placed between two microscope slides. Also known as section. { 'thin 'sek·shən }

thin-skinned structure [GEOL] Any large-scale structure, such as a fold or fault, confined to and originating within a thin layer of rocks above a surface of décollement. { 'thin ¦skind 'strək·chər }

thio- [CHEM] A chemical prefix derived from the Greek *theion*, meaning sulfur; indicates the replacement of an oxygen in an acid radical by sulfur with a negative valence of 2. { 'thī·ō }

thioacetamide [ORG CHEM] C_2H_5NS A crystalline compound with a melting point of 113–114°C; soluble in water and ethanol; used in laboratories in place of hydrogen sulfide. { ¦thī·ō·ə'sed·ə‚mīd }

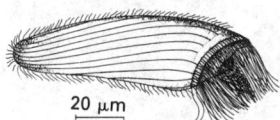

THIGMOTRICHIDA

20 μm

Drawing of *Boveria*, an example of a thigmotrichid, showing ciliature.

thioacetic acid [ORG CHEM] CH₃COSH A toxic, clear-yellow liquid with an unpleasant aroma, soluble in water, alcohol, and ether, boils at 82°C; used as an analytical reagent and a lacrimator. Also known as thiacetic acid. { ¦thī·ō·ə¦sēd·ik 'as·əd }

thioaldehyde [ORG CHEM] An organic compound that contains the −CHS radical and has the suffix -thial; for example, ethanethial, CH₃CHS. { ¦thī·ō'al·də₋hīd }

Thiobacillus ferrooxidans [MICROBIO] An aerobic rod-shaped microorganism that derives its energy from the oxidation of various sulfide minerals and soluble ferrous ion (Fe²⁺); it thrives in acidic environments of pH 1−3, conditions that would be fatal to most other life-forms. { ₋thī·ō·bə₋sil·əs fə'räk·sə₋danz }

Thiobacteriaceae [MICROBIO] Formerly a family of nonfilamentous, gram-negative bacteria of the suborder Pseudomonadineae characterized by the ability to oxidize hydrogen sulfide, free sulfur, and inorganic sulfur compounds to sulfuric acid. { ¦thī·ō·bak₋tir·ē'ās·ē₋ē }

thiobarbiturate [PHARM] A derivative of thiobarbituric acid that differs from the barbiturates in the replacement of one oxygen atom by sulfur but resembles the barbiturates in its effects. { ¦thī·ō·bär'bich·ə₋rāt }

thiobarbituric acid [ORG CHEM] C₆H₄N₂O₂S Malonyl thiourea, the parent compound of the thiobarbiturates; represents barbituric acid in which the oxygen atom of the urea component has been replaced by sulfur. { ¦thī·ō₋bär·bə'tùr·ik 'as·əd }

thiocarbamazine [PHARM] C₂₁H₁₇AsN₂O₅S₂ A white crystalline powder, freely soluble in dilute alkali; used in medicine as an amebicide. { ¦thī·ō·kär'bam·ə₋zēn }

thiocarbamide See thiourea. { ¦thī·ō'kär·bə₋mīd }

thiocarbanilide [ORG CHEM] CS(NHC₆H₅)₂ A gray powder with a melting point of 148°C; soluble in alcohol and ether; used for making dyes, and as a vulcanization accelerator and ore flotation agent. Also known as sulfocarbanilide. { ¦thī·ō₋kär·bə'ni₋līd }

thiocarbarsone [PHARM] C₁₁H₁₃AsN₂O₅S₂ A white crystalline powder, freely soluble in dilute alkali; used in medicine as an amebicide. { ¦thī·ō'kär·bə₋sōn }

thiocresol [PHARM] CH₃C₆H₄SH There are three isomers, of which only *meta*-thiocresol is a liquid; all three have a boiling point of 195°C and are soluble in alcohol or ether; used as an antiseptic. Also known as toluenethiol; tolylmercaptan. { ¦thī·ō'krē₋sōl }

thiocyanate [INORG CHEM] A salt of thiocyanic acid that contains the −SCN radical; for example, sodium thiocyanate, NaSCN. Also known as sulfocyanate; sulfocyanide; thiocyanide. { ¦thī·ō'sī·ə₋nāt }

thiocyanic acid [INORG CHEM] HSC:N A colorless, water-soluble liquid decomposing at 200°C; used to inhibit paper deterioration due to the action of light, and (in the form of organic esters) as an insecticide. Also known as rhodanic acid; sulfocyanic acid. { ¦thī·ō·sī¦an·ik 'as·əd }

thiocyanide See thiocyanate. { ¦thī·ō'sī·ə₋nīd }

thiocyanogen [INORG CHEM] NCSSCN White, light-unstable rhombic crystals melting at −2°C. { ¦thī·ō·sī·'an·ə·jən }

thiodiglycol [ORG CHEM] (CH₂CH₂OH)₂S A combustible, colorless, syrupy liquid soluble in water, alcohol, acetone, and chloroform, boils at 283°C; used as a chemical intermediate, textile-dyeing solvent, and antioxidant. { ¦thī·ō·dī'glī₋kòl }

thiodiglycolic acid [ORG CHEM] HOOCCH₂SCH₂COOH Combustible, colorless, water- and alcohol-soluble crystals melting at 128°C; used as an analytical reagent. { ¦thī·ō₋dī·glī'käl·ik 'as·əd }

3,3′-thiodiproprionic acid [ORG CHEM] (CH₂CH₂COOH)₂ A crystalline compound with a melting point of 134°C; soluble in hot water, acetone, and alcohol; used as an antioxidant for soap products and polymers of ethylene. { ¦thrē ¦thre₋prīm ¦thī·ō₋prō·prē'än·ik 'as·əd }

thioether [ORG CHEM] RSR A general formula for colorless, volatile organic compounds obtained from alkyl halides and alkali sulfides; the R groups can be the same, or different as in methylthioethane (CH₃SC₂H₅). { ¦thī·ō'ē·thər }

thioethyl alcohol See ethyl mercaptan. { ¦thī·ō'eth·əl 'al·kə₋hòl }

thioflavine T [ORG CHEM] C₁₆H₁₇N₂Cl A yellow basic dye, used for textile dyeing and fluorescent sign paints. { ¦thī·ō'flā·vən 'tē }

thiofuran See thiophene. { ¦thī·ō'fyù₋ran }

thioglycolic acid [ORG CHEM] HSCH₂COOH A liquid with a strong unpleasant odor; used as a reagent for metals such as iron, molybdenum, silver, and tin, and in bacteriology. Also known as mercaptoacetic acid. { ¦thī·ō·glī'käl·ik 'as·əd }

2-thiohydantoin [ORG CHEM] NHC(S)NHC(O)CH₂ Crystals or a tan powder with a melting point of 230°C; used in the manufacture of pharmaceuticals, rubber accelerators, and copper-plating brighteners. Also known as glycolythiourea. { ¦tü ¦thī·ō·hī'dänt·ə₋win }

thioindigo [MATER] A group of sulfur dyes made by treating the appropriate organic compound with sodium sulfide; colors are fast to washing and light. { ¦thī·ō'in·də₋gō }

thiol See mercaptan. { 'thī₋ òl }

thiolactic acid [ORG CHEM] CH₃CH(SH)COOH An oil with a disagreeable odor; used in toiletry preparation. Also known as 2-mercaptopropionic acid; 2-thiolpropionic acid. { ¦thī·ō¦lak·tik 'as·əd }

thiomalic acid [ORG CHEM] C₄H₆O₄S White crystals or powder with a melting point of 149−150°C; soluble in water, alcohol, and acetone; used as a sealer for fuel cells and machine and electrical parts, for caulking compounds, and as a propellant binder. Also known as mercaptosuccinic acid. { ¦thī·ō¦mal·ik 'as·əd }

thionic acid [INORG CHEM] H₂SₓO₆, where *x* varies from 2 to 6. [ORG CHEM] An organic acid with the radical −CSOH. { thī'än·ik 'as·əd }

thionyl bromide [ORG CHEM] SOBr₂ A red liquid boiling at 68°C (40 mmHg). Also known as sulfinyl bromide. { ¦thī·ən·əl 'brō₋mīd }

thionyl chloride [INORG CHEM] SOCl₂ A toxic, yellowish to red liquid with a pungent aroma, soluble in benzene, decomposes in water and at 140°C; boils at 79°C; used as a chemical intermediate and catalyst. Also known as sulfur oxychloride; sulfurous oxychloride. { ¦thī·ən·əl 'klòr₋īd }

thiopental sodium [ORG CHEM] C₁₁H₁₇O₂N₂NaS Yellow, water-soluble crystals with a characteristic aroma; used in medicine as a short-acting anesthetic. Also known as thiopentone sodium. { ¦thī·ō¦pen₋tal 'sōd·ē·əm }

thiopentone sodium See thiopental sodium. { ¦thī·ō¦pen₋tēn 'sōd·ē·əm }

thiophanate [ORG CHEM] C₁₄H₁₈N₄O₄S₂ A tan to colorless solid that decomposes at 195°C; slightly soluble in water; used to control fungus diseases of turf. { thī·ō'fa₋nāt }

thiophene [ORG CHEM] C₄H₄S A toxic, flammable, highly reactive, colorless liquid, insoluble in water, soluble in alcohol and ether, boils at 84°C; used as a chemical intermediate and to make condensation copolymers. Also known as thiofuran. { 'thī·ə₋fēn }

thiophenol [ORG CHEM] C₆H₅SH A toxic, fire-hazardous, water-white liquid with a disagreeable aroma, insoluble in water, soluble in alcohol and ether, boils at 168°C; used to make pharmaceuticals. Also known as phenyl mercaptan. { ¦thī·ō'fē₋nòl }

thiophile element See sulfophile element. { 'thī·ə₋fil ₋el·ə·mənt }

Thiorhodaceae [MICROBIO] Formerly a family of bacteria in the suborder Rhodobacteriineae composed of the purple, red, orange, and brown sulfur bacteria; characterized as strict anaerobes which oxidize hydrogen sulfide and store sulfur globules internally. { ₋thī·ə·rō'dās·ē₋ē }

thiosalicylic acid [ORG CHEM] HOOCC₆H₄SH A yellow solid with a melting point of 164−165°C; soluble in alcohol, ether, and acetic acid; used for making dyes. Also known as 2-mercaptobenzoic acid. { ¦thī·ō₋sal·ə¦sil·ik 'as·əd }

thiosemicarbazide [ORG CHEM] NH₂CSNHNH₂ A white, water- and alcohol-soluble powder melting at 182°C; used as an analytical reagent and in photography and rodenticides. { ¦thī·ō₋sem·i'kär·bə₋zīd }

thiosemicarbazone [PHARM] A class of chemical compounds used in treating tuberculosis; the most prominent member of the group is *para*-acetamidobenzaldehyde thiosemicarbazone. { ¦thī·ō₋sem·i'kär·bə₋zōn }

thiospinel [MINERAL] Any mineral with the spinel structure having the general formula AR₂S₄. { ¦thī·ō·spə'nel }

thiostreptone [MICROBIO] A polypeptide antibiotic produced by a species of *Streptomyces* that crystallizes from a chloroform methanol solution; used in veterinary medicine. { ¦thī·ō'strep,tōn }

thiosulfate [INORG CHEM] $M_2S_2O_3$ A salt of thiosulfuric acid and a base; for example, reaction of sodium hydroxide and thiosulfuric acid to produce sodium thiosulfate. { ¦thī·ə'səl,fāt }

thiosulfonic acid [ORG CHEM] Name for a group of oxy acids of sulfur, with the general formula RS_2O_2H; they are known as esters and salts. { ¦thī·ə,səl'fän·ik 'as·əd }

thiosulfuric acid [INORG CHEM] $H_2S_2O_3$ An unstable acid that decomposes readily to form sulfur and sulfurous acid. { ¦thī·ə,səl'fyùr·ik 'as·əd }

thiouracil [PHARM] $C_4H_4N_2OS$ A bitter-tasting white powder; used as an antithyroid drug that acts by interfering with thyroxine synthesis. { ¦thī·ō'yùr·ə,sil }

thiourea [ORG CHEM] $(NH_2)_2CS$ Bitter-tasting, white crystals with a melting point of 180–182°C; soluble in cold water and alcohol; used in photography and photocopying, as a rubber accelerator, and as an antithyroid drug in treating hyperthyroidism. Also known as thiocarbamide. { ¦thī·ō·yù'rē·ə }

thiram [ORG CHEM] $(CH_3)_2N-CS-S-S-CS-N(CH_3)_2$ (tetramethylthioperoxydicarbonic diamide) A fungicide, bacteriostat (in soap), antimicrobial agent (chemotherapeutic for plants), seed disinfectant, and vulcanizing agent. { 'thī,ram }

third-generation computer [COMPUT SCI] One of the general purpose digital computers introduced in the late 1960s; it is characterized by integrated circuits and has logical organization and software which permit the computer to handle many programs at the same time, allow one to add or remove units from the computer, permit some or all input/output operations to occur at sites remote from the main processor, and allow conversational programming techniques. { 'thərd ¦jen·ə¦rā·shən kəm'pyüd·ər }

third harmonic [PHYS] A sine-wave component having three times the fundamental frequency of a complex signal. { 'thərd här'män·ik }

third law of motion *See* Newton's third law. { 'thərd 'lò əv 'mō·shən }

third law of thermodynamics [THERMO] The entropy of all perfect crystalline solids is zero at absolute zero temperature. { 'thərd 'lò əv ¦thər·mō·də'nam·iks }

third-order climatological station [CLIMATOL] As defined by the World Meteorological Organization, a station, other than a precipitation station, at which the observations are of the same kind as those at a second-order climatological station, but are not so comprehensive, are made once a day only, and are made at other than the specified hours. { 'thərd ¦òr·dər ¦klī·mə·tə'läj·ə·kəl 'stā·shən }

third-order reaction [PHYS CHEM] A chemical reaction in which the rate of reaction is determined by the concentration of three reactants. { 'thərd ¦òr·dər rē'ak·shən }

third-order relief [GEOGR] Specific landform complexes that are smaller in extent and size than formations of subcontinental extent. { 'thərd ¦òr·dər ri'lēf }

third proportional [MATH] For numbers a and b, a number x such that $a/b = b/x$. { 'thərd prə'pòr·shən·əl }

third quadrant [MATH] **1.** The range of angles from 180 to 270°. **2.** In a plane with a system of Cartesian coordinates, the region in which the x and y coordinates are both negative. { 'thərd 'kwä·drənt }

third rail [CIV ENG] The electrified metal rail which carries current to the motor of an electric locomotive or other railway car. { 'thərd 'rāl }

third sound [CRYO] A type of wave propagated in thin films of superfluid helium (helium II), consisting of variations in film thickness and temperature. { 'thərd ,saùnd }

thirling *See* holing. { 'thərl·iŋ }

thirst [PHYSIO] A sensation, as of dryness in the mouth and throat, resulting from water deprivation. { 'thərst }

thirty-day forecast [METEOROL] A weather forecast for a period of 30 days; as issued by the U.S. Weather Bureau, the forecast concerns expected departures of temperature and precipitation from normal. { ¦thər·dē¦dā 'fòr,kast }

thirty-two nucleus [METEOROL] An unidentified type of freezing nucleus which first becomes active when a supercooled cloud is cooled to about −32°C. { ¦thər·dē¦tü 'nü·klē·əs }

Thisbe [ASTRON] An asteroid with a diameter of about 124 miles (200 kilometers). mean distance from the sun of 2.77 astronomical units, and C-type surface composition. { 'thiz·bē }

thistle [BOT] Any of the various prickly plants comprising the family Compositae. { 'this·əl }

thiuram [ORG CHEM] A chemical compound containing a R_2NCS radical; occurs mainly in disulfide compounds; the most common monosulfide compound is tetramethylthiuram monosulfide. { 'thī·yə ram }

thixotropic clay [GEOL] A clay that weakens when disturbed and increases in strength upon standing. { ¦thik·sə¦träp·ik 'klā }

thixotropy [PHYS CHEM] Property of certain gels which liquefy when subjected to vibratory forces, such as ultrasonic waves or even simple shaking, and then solidify again when left standing. { thik'sä·trə·pē }

Thlipsuridae [PALEON] A Paleozoic family of ostracod crustaceans in the suborder Platycopa. { thlip'sùr·ə,dē }

tholeiite [PETR] **1.** A group of basalts composed principally of plagioclase, pyroxene, and iron oxide minerals as phenocrysts in a glassy groundmass. **2.** Any rock in the group. { ¦hō'lē·ə,īt }

Thoma cavitation coefficient [MECH ENG] The equation for measuring cavitation in a hydraulic turbine installation, relating vapor pressure, barometric pressure, runner setting, tail water, and head. { 'tō·mə ,kav·ə'tā·shən ,kō·i,fish·ənt }

Thomas converter [MET] A basic Bessemer converter; that is, one in which air is forced upward through holes in the bottom of the steel container which has a basic lining, usually dolomite, and which employs a basic slag. { 'täm·əs kən,vərd·ər }

Thomas cyclotron [NUCLEO] A circular particle accelerator which operates like an ordinary cyclotron but employs a magnetic field that is variable in azimuth in such a way that cyclotron resonance at a fixed orbital frequency and radial and axial focusing can be maintained simultaneously. { 'täm·əs 'sī·klə,trän }

Thomas-Fermi atom model [ATOM PHYS] A method of approximating the electrostatic potential and the electron density in an atom in its ground state, in which these two quantities are related by the Poisson equation on the one hand, and on the other hand by a semiclassical formula for the density of quantum states in phase space. { 'täm·əs 'fer·mē 'ad·əm ,mäd·əl }

Thomas-Fermi equation [ATOM PHYS] The differential equation $x^{1/2}(d^2y/dx^2) = y^{3/2}$ that arises in calculating the potential in the Thomas-Fermi atom model; the physically meaningful solution satisfies the boundary conditions $y(0) = 1$ and $y(\infty) = 0$. { 'täm·əs 'fer·mē i,kwā·zhən }

Thomas meter [ENG] An instrument used to determine the rate of flow of a gas by measuring the rise in the gas temperature produced by a known amount of heat. { 'täm·əs ,mēd·ər }

Thomas precession [RELAT] The precession of a vector in an accelerated system, relative to an observer for whom the system has a given velocity and acceleration, when this vector appears to be constant to an observer attached to the system; this precession is the kinematical basis of one type of spin-orbit coupling. { 'täm·əs prē'sesh·ən }

Thomas-Reiche-Kuhn sum rule *See* f-sum rule. { 'täm·əs 'rīk·ə 'kyün 'səm ,rül }

Thompson submachine gun [ORD] Caliber-.45, air-cooled automatic weapon that can be carried and operated by one person. { 'täm·sən ¦səb·mə'shēn ,gən }

thomsenolite [MINERAL] $NaCaAlF_6 \cdot H_2O$ A colorless to white, monoclinic mineral consisting of a hydrated aluminofluoride of sodium and calcium; it is dimorphous with pachnolite. { 'täm·sə·nə,līt }

Thomson-Berthelot principle [PHYS CHEM] The assumption that the heat released in a chemical reaction is directly related to the chemical affinity, and that, in the absence of the application of external energy, that chemical reaction which releases the greatest heat is favored over others; the principle is in general incorrect, but applies in certain special cases. { 'täm·sən ber·tə'lō ,prin·sə·pəl }

Thomson bridge *See* Kelvin bridge. { 'täm·sən ,brij }

Thomson coefficient [PHYS] The ratio of the voltage existing between two points on a metallic conductor to the difference in temperature of those points. { 'täm·sən ,kō·i,fish·ənt }

Thomson cross section [ELECTROMAG] The total scattering cross section for Thomson scattering, equal to $(8/3)\pi(e^2/mc^2)^2$, where e and m are the charge (in electrostatic units) and mass of the scattering particle, and c is the speed of light. { 'täm·sən 'krós ‚sek·shən }

Thomson effect [PHYS] A thermoelectric effect in which heat flows into or out of a homogeneous conductor when an electric current flows between two points in the conductor at different temperatures, the direction of heat flow depending upon whether the current flows from colder to warmer metal or from warmer to colder. { 'täm·sən i‚fekt }

Thomson formula [ELECTROMAG] **1.** The formula for the intensity of scattered electromagnetic radiation in Thomson scattering as a function of the scattering angle ϕ; the intensity is proportional to $1 + \cos^2 \phi$. **2.** A formula for the period of oscillation of a current when a capacitor is discharged. Also known as Kelvin's formula. { 'täm·sən ‚fȯr·myə·lə }

Thomson heat [PHYS] The heat generated or absorbed in the Thomson effect in a reversible manner when a current passes through a conductor in which there is a temperature gradient; it is proportional to the product of the current and the temperature gradient. { 'täm·sən ‚hēt }

thomsonite [MINERAL] $NaCa_2Al_5Si_5O_{20}\cdot6H_2O$ Snow-white zeolite mineral forming orthorhombic crystals and occurring in masses of radiating crystals; hardness is 5–5.5 on Mohs scale. { 'täm·sə‚nīt }

Thomson parabolas [ELECTROMAG] A pattern of parabolas which appear on a photographic plate exposed to a beam of ions of an element which has passed through electric and magnetic fields applied in the same direction normal to the path of the ions; each parabola corresponds to a different charge-to-mass ratio, and thus to a different isotope. { 'täm·sən pə'rab·ə·ləz }

Thomson relations [PHYS] Equations in the study of thermoelectricity, relating the Peltier coefficient and the Thomson coefficient to the Seebeck voltage; they are derived by thermodynamics. Also known as Kelvin relations. { 'täm·sən ri‚lā·shənz }

Thomson scattering [ELECTROMAG] Scattering of electromagnetic radiation by free (or very loosely bound) charged particles, computed according to a classical nonrelativistic theory: energy is taken away from the primary radiation as the charged particles accelerated by the transverse electric field of the radiation, radiate in all directions. { 'täm·sən ‚skad·ə·riŋ }

Thomson voltage [PHYS] The voltage that exists between two points that are at different temperatures in a conductor. { 'täm·sən ‚vōl·tij }

thonzylamine hydrochloride [PHARM] $C_{16}H_{22}N_4O \cdot HCl$ A white, crystalline powder with a melting point of 173–176°C; soluble in water, alcohol, and chloroform; used in medicine as an antihistamine. { thän'zil·ə‚mēn ‚hī·drə'klȯr‚īd }

Thoracica [INV ZOO] An order of the subclass Cirripedia; individuals are permanently attached in the adult stage, the mantle is usually protected by calcareous plates, and six pairs of biramous thoracic appendages are present. { thə'ras·ə·kə }

thoracic cavity See thorax. { thə'ras·ik 'kav·əd·ē }

thoracic duct [ANAT] The common lymph trunk beginning in the crura of the diaphragm at the level of the last thoracic vertebra, passing upward, and emptying into the left subclavian vein at its junction with the left internal jugular vein. { thə'ras·ik 'dəkt }

thoracic vertebrae [ANAT] The vertebrae associated with the chest and ribs in vertebrates; there are 12 in humans. { thə'ras·ik 'vərd·ə‚brā }

thoracoabdominal breathing [PHYSIO] The process of air breathing in reptiles, birds, and mammals that depends upon aspiration or sucking inspiration, and involves trunk musculature to supply pulmonary ventilation. { ‚thȯr·ə·kō·ab'däm·ə·nəl 'brēth·iŋ }

Thoracostomopsidae [INV ZOO] A family of marine nematodes in the superfamily Enoploidea, which have the stomatal armature modified to form a hollow tube. { ‚thȯr·ə·kō·stō'mäp·sə‚dē }

Thoraeus filter [NUCLEO] A primary radiological filter of tin, combined with a secondary filter of copper to absorb the characteristic radiation of the tin and a third filter of aluminum to absorb the characteristic radiation of the copper; in the range of 200 to 400 kilovolts, such a filter hardens x-rays more efficiently than the usual combination of copper and aluminum. { 'thȯr·ē·əs 'fil·tər }

thorax [ANAT] The chest; the cavity of the mammalian body between the neck and the diaphragm, containing the heart, lungs, and mediastinal structures. Also known as thoracic cavity. [INV ZOO] The middle of three principal divisions of the body of certain classes of arthropods. { 'thȯr‚aks }

thoreaulite [MINERAL] $SnTa_2O_7$ A brown, monoclinic mineral consisting of an oxide of tin and tantalum; occurs as rough, prismatic crystals. { 'thȯ·rō‚līt }

thoria See thorium dioxide. { 'thȯr·ē·ə }

thorianite [MINERAL] ThO_2 A radioactive mineral that crystallizes in the isometric system, occurs in worn cubic crystals, is brownish black to reddish brown in color, and has resinous luster; hardness is 7 on the Mohs scale, and specific gravity is 9.7–9.8. { 'thȯr·ē·ə‚nīt }

thoriated emitter See thoriated tungsten filament. { 'thȯr·ē‚ād·əd i'mid·ər }

thoriated tungsten filament [ELECTR] A vacuum-tube filament consisting of tungsten mixed with a small quantity of thorium oxide to give improved electron emission. Also known as thoriated emitter. { 'thȯr·ē‚ād·əd ‚təŋ·stən 'fil·ə·mənt }

Thorictidae [INV ZOO] The ant blood beetles, a family of coleopteran insects in the superfamily Dermestoidea. { thə'rik·tə‚dē }

thorite [MINERAL] $ThSiO_4$ A brownish-yellow to brownish-black and black radioactive mineral that is tetragonal in crystallization; hardness is about 4.5 on Mohs scale, and specific gravity is 4.3–5.4. { 'thȯr‚īt }

thorium [CHEM] An element of the actinium series, symbol Th, atomic number 90, atomic weight 232.0381; soft, radioactive, insoluble in water and alkalies, soluble in acids, melts at 1750°C, boils at 4500°C. [MET] A heavy malleable metal that changes from silvery-white to dark gray or black in air; potential source of nuclear energy; used in manufacture of sunlamps. { 'thȯr·ē·əm }

thorium anhydride See thorium dioxide. { 'thȯr·ē·əm an'hī‚drīd }

thorium carbide [INORG CHEM] ThC_2 A yellow solid melting at above 2630°C, decomposes in water; used in nuclear fuel. { 'thȯr·ē·əm 'kär‚bīd }

thorium chloride [INORG CHEM] $ThCl_4$ Hygroscopic, toxic colorless crystal needles soluble in alcohol, melts at 820°C, decomposes at 928°C; used in incandescent lighting. Also known as thorium tetrachloride. { 'thȯr·ē·əm 'klȯr‚īd }

thorium dioxide [INORG CHEM] ThO_2 A heavy, white powder soluble in sulfuric acid, insoluble in water, melts at 3300°C; used in medicine, ceramics, flame spraying, and electrodes. Also known as thoria; thorium anhydride; thorium oxide. { 'thȯr·ē·əm dī'äk‚sīd }

thorium fluoride [INORG CHEM] ThF_4 A white, toxic powder, melts at 1111°C; used to make thorium metal and magnesium-thorium alloys and in high-temperature ceramics. { 'thȯr·ē·əm 'flür‚īd }

thorium nitrate [INORG CHEM] $Th(NO_3)_4\cdot4H_2O$ Explosive white crystals soluble in water and alcohol, strong oxidizer; the anhydrous form decomposes at 500°C; used in medicine and as an analytical reagent. { 'thȯr·ē·əm 'nī‚trāt }

thorium oxalate [ORG CHEM] $Th(C_2O_4)_2\cdot2H_2O$ A white, toxic powder soluble in alkalies and ammonium oxalate, insoluble in water and most acids, decomposes to thorium dioxide, ThO_2, above 300–400°C; used in ceramics. { 'thȯr·ē·əm 'äk·sə‚lāt }

thorium oxide See thorium dioxide. { 'thȯr·ē·əm 'äk‚sīd }

thorium reactor [NUCLEO] A nuclear reactor in which thorium surrounds the central enriched uranium core to give breeder operation. { 'thȯr·ē·əm rē‚ak·tər }

thorium series [NUCLEO] The series of nuclides resulting from the decay of thorium-232. { 'thȯr·ē·əm ‚sir·ēz }

thorium sulfate [INORG CHEM] $Th(SO_4)_2\cdot8H_2O$ A white powder soluble in ice water, loses water at 42° and 400°C. Also known as normal thorium sulfate. { 'thȯr·ē·əm 'səl‚fāt }

thorium tetrachloride See thorium chloride. { 'thȯr·ē·əm ‚te·trə'klȯr‚īd }

thorn [BOT] A short, sharp, rigid, leafless branch on a plant. [ZOO] Any of various sharp spinose structures on an animal. { thȯrn }

thornback [VERT ZOO] *Raja clavata.* A ray found in European waters and characterized by spines on its back. { 'thȯrn,bak }

thornbush [ECOL] A vegetation class that is dominated by tall succulents and profusely branching smooth-barked deciduous hardwoods which vary in density from mesquite bush in the Caribbean to the open spurge thicket in Central Africa; the climate is that of a warm desert, except for a rather short intense rainy season. Also known as Dorngeholz; Dorngestrauch; dornveld; savane armée; savane épineuse; thorn scrub. { 'thȯrn,bush }

Thorne-Żytkow object [ASTRON] A hypothetical object that results when a star in its helium-burning, or red-giant, phase swallows a companion neutron star that had resulted from the evolution of a very massive star. { ¦thȯrn 'zhit,kȯv ¦äb,jekt }

thorn forest [ECOL] A type of forest formation, mostly tropical and subtropical, intermediate between desert and steppe; dominated by small trees and shrubs, many armed with thorns and spines; leaves are absent, succulent, or deciduous during long dry periods, which may also be cool; an example is the caatinga of northeastern Brazil. { 'thȯrn ,fär·əst }

thorn scrub *See* thornbush. { 'thȯrn 'skrəb }

thorogummite [MINERAL] A silicate mineral and chemical variant of thorium silicate, with similar properties; isostructural with thorite and zircon; it is deficient in silica and contains small amounts of OH in substitution for oxygen. { ,thȯr·ə'gə,mīt }

thoron [NUC PHYS] The conventional name for radon-220. Symbolized Tn. { 'thȯr,än }

thoroughfare [CIV ENG] **1.** An important, unobstructed public street or highway. **2.** A street going through from one street to another. **3.** An inland waterway for passage of ships usually not between two bodies of water. { 'thər·ə,fer }

Thorpe reaction [ORG CHEM] The reaction by which, in presence of lithium amides, α,ω-dinitriles undergo base-catalyzed condensation to cyclic iminonitriles, which can be hydrolyzed and decarboxylated to cyclic ketones. { 'thȯrp rē,ak·shən }

thortveitite [MINERAL] $(Sc,Y)_2Si_2O_7$ A grayish-green mineral occurring in orthorhombic crystals; a source of scandium. { tȯrt'vī,tīt }

thou *See* mil. { thau̇ }

thought experiment *See* Gedanken experiment. { 'thȯt ik'sper·ə·mənt }

thousandth mass unit [PHYS] A unit of energy equal to the energy equivalent of a mass of 10^{-3} atomic mass unit according to the Einstein mass-energy relation, that is, to the product of 10^{-3} atomic mass unit and the square of the speed of light; equal to approximately 1.49176×10^{-13} joule. { 'thau̇z·ənth 'mas ¦yü·nət }

THPC *See* tetrakis(hydroxymethyl)phosphonium chloride.

thrashing [COMPUT SCI] An undesirable condition in a multiprogramming system, due to overcommitment of main memory, in which the various tasks compete for pages and none can operate efficiently. { 'thrash·iŋ }

thread [COMPUT SCI] A sequence of beads that are strung together. [DES ENG] A continuous helical rib, as on a screw or pipe. [GEOL] An extremely small vein, even thinner than a stringer. [MIN ENG] A more or less straight line of stall faces, having no cuttings, loose ends, fast ends, or steps. [TEXT] A continuous strand formed by spinning and twisting together short strands of textile fibers. { thred }

thread blight [PL PATH] A fungus disease of a number of tropical and semitropical woody plants, including cocoa and tea, caused by species of *Pellicularia* and *Marasmius* which form filamentous mycelia on the surface of twigs and leaves. { 'thred ,blīt }

thread contour [DES ENG] The shape of thread design as observed in a cross section along the major axis, for example, square or round. { 'thred ,kän,túr }

thread count [TEXT] An index of the compactness of a fabric determined by counting the number of warp yarns and filling yarns in 1 square inch (6.4516 square centimeters) of fabric. Also known as cloth count. { 'thred ,kau̇nt }

thread cutter [MECH ENG] A tool used to cut screw threads on a pipe, screw, or bolt. { 'thred ,kəd·ər }

threadfin [VERT ZOO] Common name for any of the fishes in the family Polynemidae. { 'thred,fin }

thread gage [DES ENG] A design gage used to measure screw threads. { 'thred ,gāj }

threading die [MECH ENG] A die which may be solid, adjustable, or spring adjustable, or a self-opening die head, used to produce an external thread on a part. { 'thred·iŋ ,dī }

threading machine [MECH ENG] A tool used to cut or form threads inside or outside a cylinder or cone. { 'thred·iŋ mə,shēn }

thread-lace scoria [GEOL] Scoria whose vesicle walls have collapsed and are represented only by a network of threads. { 'thred ¦lās 'skȯr·ē·ə }

thread plug [ENG] Mold part which shapes an internal thread onto a molded article; must be unscrewed from the finished piece. { 'thred ,pləg }

thread plug gage [DES ENG] A thread gage used to measure female screw threads. { 'thred ,pləg ,gāj }

thread protector [ENG] A short-threaded ring to screw onto a pipe or into a coupling to protect the threads while the pipe is being handled or transported. Also known as pipe-thread protector. { 'thred prə,tek·tər }

thread rating [ENG] The maximum internal working pressure allowable for threaded pipe or tubing joints; important for pressure systems, chemical processes, and oil-well systems. { 'thred ,rād·iŋ }

thread ring gage [DES ENG] A thread gage used to measure male screw threads. { 'thred 'riŋ ,gāj }

thread waste [TEXT] The hard, thready waste left on bobbins or collected during operations such as spinning, twisting, and weaving. { 'thred ,wāst }

threat [COMPUT SCI] An event that can cause harm to computers, to their data or programs, or to computations. { thret }

threat collision avoidance system [NAV] A system, based on air-traffic control transponders installed on aircraft, that issues an evasive maneuver command when it senses a collision threat. { 'thret kə¦lizh·ən ə'vȯid·əns ,sis·təm }

three-address code [COMPUT SCI] In computers, a multiple-address code which includes three addresses, usually two addresses from which data are taken and one address where the result is entered; location of the next instruction is not specified, and instructions are taken from storage in preassigned order. { 'thrē 'ad,res ,kōd }

three-address instruction [COMPUT SCI] In computers, an instruction which includes an operation and specifies the location of three registers. { 'thrē 'ad,res in'strək·shən }

three-alpha process [ASTROPHYS] A nuclear reaction in which three helium-4 nuclei (alpha particles) combine to form a carbon-12 nucleus, with the emission of a gamma ray; it converts helium into carbon in red giants. Also known as Salpeter process; triple-alpha process. { 'thrē ¦al·fə 'prä,sas }

three-arm protractor [NAV] An instrument consisting essentially of a circle graduated in degrees, to which is attached one fixed arm and two arms pivoted at the center and provided with clamps so that they can be set at any angle to the fixed arm, within the limits of the instrument; used for finding a ship's position when the angles between three fixed and known points are measured. Also known as station pointer. { 'thrē ¦ärm prō'trak·tər }

three-axis stabilization [AERO ENG] Directional stability of a spacecraft obtained without spin, usually with internal gyroscopes to maintain stability about each of three perpendicular axes. { ¦thrē ,ak·səs ,stā·bə·lə'zā·shən }

three-body problem [MECH] The problem of predicting the motions of three objects obeying Newton's laws of motion and attracting each other according to Newton's law of gravitation. { 'thrē ¦bäd·ē ,präb·ləm }

three-body recombination [ATOM PHYS] The combination of an electron with a positive ion in a gas in such a way that the incoming free electron transfers energy and momentum to another free electron in the neighborhood of the ion. { ¦thrē ,bäd·ē rē,käm·bə'nā·shən }

three-day fever *See* phlebotomus fever. { 'thrē ¦dā 'fē·vər }

three-decibel coupler [ELECTROMAG] Junction of two waveguides having a common H wall; the two guides are coupled together by H-type aperture coupling; the coupling is such that 50% of the power from either channel will be fed into the other. Also known as Riblet coupler; short-slot coupler. { 'thrē des·ə·bəl 'kəp·lər }

three-decision problem [STAT] A problem in which a choice must be made among three possible courses of action. { 'thrē di¦sizh·ən 'präb·ləm }

three-dimensional [SCI TECH] Giving the illusion of depth, in three dimensions. { ¦thrē di'men·chən·əl }

three-dimensional braiding *See* through-the-thickness braiding. { ¦thrē di¦men·chən·əl 'brād·iŋ }

three-dimensional display system [ELECTR] A radar display which shows range, azimuth, and elevation; for instance, a G display. { ¦thrē di¦men·chən·əl di'splā ¸sis·təm }

three-dimensional flow [FL MECH] Any fluid flow which is not a two-dimensional flow. { ¦thrē di¦men·chən·əl 'flō }

three-dimensional sound *See* virtual acoustics. { ¦thrē də¸men·shən·əl 'saúnd }

three-eighths rule [MATH] **1.** An approximation formula for definite integrals which states that the integral of a real-valued function f on an interval $[a, b]$ is approximated by $(3/8)h[f(a) + 3f(a + h) + 3f(a + 2h) + f(b)]$, where $h = (b − a)/3$; this is the integral of a third-degree polynomial whose value equals that of f at a, $a + h$, $a + 2h$, and b. **2.** A method of approximating a definite integral over an interval which is equivalent to dividing the interval into equal subintervals and applying the formula in the first definition to each subinterval. { ¦thrē 'āths ¸rül }

three-index symbols *See* Christoffel symbols. { ¦thrē 'in¸deks 'sim·bəlz }

three-input adder *See* full adder. { ¦thrē 'in¸pùt 'ad·ər }

three-input subtracter *See* full subtracter. { ¦thrē 'in¸pùt səb'trak·tər }

three-jaw chuck [DES ENG] A drill chuck having three serrated-face movable jaws that can grip and hold fast an inserted drill rod. { ¦thrē 'jó 'chək }

three-j number [QUANT MECH] A coefficient used in coupling eigenfunctions of two commuting angular momenta to form eigenfunctions of the total angular momentum; closely related to the Clebsch-Gordan coefficients. Also known as Wigner 3-*j* symbol. { ¦thrē 'jā 'nəm·bər }

three-junction transistor [ELECTR] A *pnpn* transistor having three junctions and four regions of alternating conductivity; the emitter connection may be made to the *p* region at the left, the base connection to the adjacent *n* region, and the collector connection to the *n* region at the right, while the remaining *p* region is allowed to float. { ¦thrē 'jəŋk·shən tran'zis·tər }

three-kiloparsec arm [ASTRON] A region approximately 3 kiloparsecs from the galactic center that displays strong absorption in the 21-centimeter line of atomic hydrogen. { ¦thrē ¦kil·ō¦pär¸sek 'ärm }

three-layer diode [ELECTR] A junction diode with three conductivity regions. { ¦thrē ¦lā·ər 'dī¸ōd }

three-level laser [OPTICS] A laser involving three energy levels, one of which is the ground state; laser action usually occurs between the intermediate and ground states. { ¦thrē ¦lev·əl 'lā·zər }

three-level maser [PHYS] A solid-state maser in which three energy levels are used; successful operation has been obtained with crystals of gadolinium ethyl sulfate and crystals of potassium chromecyanide at the temperature of liquid helium. { ¦thrē ¦lev·əl 'mā·zər }

three-level subroutine [COMPUT SCI] A subroutine in which a second subroutine is called, and a third subroutine is called by the second subroutine. { ¦thrē ¦lev·əl 'səb·rü¸tēn }

threeling *See* trilling. { 'thrēl·iŋ }

three-phase circuit [ELEC] A circuit energized by alternating-current voltages that differ in phase by one-third of a cycle or 120°. { ¦thrē ¦fāz 'sər·kət }

three-phase current [ELEC] Current delivered through three wires, with each wire serving as the return for the other two and with the three current components differing in phase successively by one-third cycle, or 120 electrical degrees. { ¦thrē ¦fāz 'kə·rənt }

three-phase four-wire system [ELEC] System of alternating-current supply comprising four conductors, three of which are connected as in a three-phase, three-wire system, the fourth being connected to the neutral point of the supply, which may be grounded. { ¦thrē ¦fāz 'fòr ¦wīr 'sis·təm }

three-phase magnetic amplifier [ELECTR] A magnetic amplifier whose input is the sum of three alternating-current voltages that differ in phase by 120°. { ¦thrē ¦fāz mag'ned·ik 'am·plə¸fī·ər }

three-phase motor [ELEC] An alternating-current motor operated from a three-phase circuit. { ¦thrē ¸fāz 'mōd·ər }

three-phase rectifier [ELEC] A rectifier supplied by three alternating-current voltages that differ in phase by one-third of a cycle or 120°. { ¦thrē ¦fāz 'rek·tə¸fī·ər }

three-phase seven-wire system [ELEC] System of alternating-current supply from groups of three single-phase transformers connected in Y to obtain a three-phase, four-wire grounded neutral system of higher voltage for power, the neutral wire being common to both systems. { ¦thrē ¦fāz 'sev·ən ¦wīr 'sis·təm }

three-phase system [PHYS] Any physical system in which three distinct phases coexist; phases can be liquid, solid, vapor (gas), or three mutually insoluble liquids, or any combination thereof. { ¦thrē ¦fāz 'sis·təm }

three-phase three-wire system [ELEC] System of alternating-current supply comprising three conductors between successive pairs of which are maintained alternating differences of potential successively displaced in phase by one-third cycle. { ¦thrē ¦fāz 'thrē ¦wīr 'sis·təm }

three-phase transformer [ELEC] A transformer used in a three-phase circuit, with three sets of primary and secondary windings on a single core. { ¦thrē ¸fāz tranz'fòr·mər }

three-piece set [MIN ENG] A set of timber consisting of a cap and its two supportive posts. { ¦thrē ¦pēs 'set }

three-plus-one address [COMPUT SCI] An instruction format containing an operation code, three operand address parts, and a control address. { ¦thrē ¸pləs ¦wən 'ad¸res }

three-ply [SCI TECH] Consisting of three distinct strands, layers, or veneers. { ¦thrē 'plī }

three-point bending [MET] Bending a piece of metal by placing the specimen on two supports and then applying a load on it between the supported ends. { ¦thrē ¦pòint 'bend·iŋ }

three-point method [GEOL] A method used to determine the dip and strike of a structural surface from three points of varying elevation along the surface. { ¦thrē ¦pòint 'meth·əd }

three-point problem [ENG] The problem of locating the horizontal position of a point of observation from the two observed horizontal angles subtended by three known sides of a triangle. { ¦thrē ¦pòint 'präb·ləm }

three-pulse cascaded canceler [ELECTR] A moving-target indicator technique in which two "two-pulse cancelers" are cascaded together; this improves the velocity response. { ¦thrē ¦pəls kas'kād·əd 'kan·slər }

three-quarters hard [MET] A temper designation for various nonferrous metals, such as aluminum, copper, and magnesium alloys, expressing degree of hardness achieved by mechanical working. { ¦thrē ¦kwòrd·ərz 'härd }

three-shift cyclic mining [MIN ENG] A system of cyclic mining on a longwall conveyor face, with coal cutting on one shift, hand filling and conveying on the next, and ripping, packing, and advancement of the face conveyor on the third shift. { ¦thrē ¦shift 'sī·klik 'mīn·iŋ }

three-space [MATH] A vector space over the real numbers whose basis has three vectors. { ¦thrē ¸spās }

three-way switch [ELEC] An electric switch with three terminals used to control a circuit from two different points. { ¦thrē ¦wā 'swich }

three-wire generator [ELEC] Electric generator with a balance coil connected across the armature, the midpoint of the coil providing the potential of the neutral wire in a three-wire system. { ¦thrē ¦wir 'jen·ə¸rād·ər }

three-wire system [ELEC] System of electric supply comprising three conductors, one of which (known as the neutral wire) is maintained at a potential midway between the potential of the other two (referred to as the outer conductors); part of the load may be connected directly between the outer conductors, the remainder being divided as evenly as possible into two parts, each of which is connected between the neutral and one outer conductor; there are thus two distinct supply voltages, one being twice the other. { ¦thrē ¦wir 'sis·təm }

threonine [BIOCHEM] $CH_3CHOHCH(NH_2)COOH$ A crystalline α-amino acid considered essential for normal growth of animals; it is biosynthesized from aspartic acid and is a precursor of isoleucine in microorganisms. { ¦thrē·ə¸nēn }

thresh [AGR] To extract grain or seed (as from wheat stalks) by treading, rubbing, striking with a flail, or using a threshing machine. { thresh }

thresher [AGR] A machine that separates grain or seeds from straw. Also known as threshing machine. { 'thresh·ər }

thresher shark [VERT ZOO] Common name for fishes in the family Alopiidae; pelagic predacious sharks of generally wide distribution that have an extremely long, whiplike tail with which they thrash the water, destroying schools of small fishes. { 'thresh·ər ,shärk }

threshing machine *See* thresher. { 'thresh·iŋ mə,shēn }

threshold [BUILD] A piece of stone, wood, or metal that lies under an outside door. [ELECTR] In a modulation system, the smallest value of carrier-to-noise ratio at the input to the demodulator for all values above which a small percentage change in the input carrier-to-noise ratio produces a substantially equal or smaller percentage change in the output signal-to-noise ratio. [ENG] The least value of a current, voltage, or other quantity that produces the minimum detectable response in an instrument or system. [GEOL] *See* riegel. [MATH] A logic operator such that, if *P, Q, R, S, . . .* are statements, then the threshold will be true if at least *N* statements are true, false otherwise. [PHYS] The minimum level of some input quantity needed for some process to take place, such as a threshold energy for a reaction, or the minimum level of pumping at which a laser can go into self-excited oscillation. [PHYSIO] The minimum level of a stimulus that will evoke a response in an irritable tissue. { 'thresh,hōld }

threshold contrast [OPTICS] The smallest contrast of luminance (or brightness) that is perceptible to the human eye under specified conditions of adaptation luminance and target visual angle. Also known as contrast sensitivity; contrast threshold; liminal contrast. { 'thresh,hōld ¦kän,trast }

threshold depth *See* sill depth. { 'thresh,hōld ,depth }

threshold detector [NUC PHYS] An element or isotope in which radioactivity is induced only by the capture of neutrons having energies in excess of a certain characteristic threshold value; used to determine the neutron spectrum from a nuclear explosion. { 'thresh,hōld di,tek·tər }

threshold dose [NUCLEO] The minimum radiation dose that will produce a detectable specified effect. { 'thresh,hōld ,dōs }

threshold element [COMPUT SCI] A logic circuit which has one output and several weighted inputs, and whose output is energized if and only if the sum of the weights of the energized inputs exceeds a prescribed threshold value. { 'thresh,hōld ,el·ə·mənt }

threshold frequency [ELECTR] The frequency of incident radiant energy below which there is no photoemissive effect. { 'thresh,hōld ,frē·kwən·sē }

threshold illuminance [OPTICS] The lowest value of illuminance which the eye is capable of detecting under specified conditions of background luminance and degree of dark adaptation of the eye. Also known as flux density threshold. { 'thresh,hōld i'lü·mə·nəns }

thresholding [COMPUT SCI] In machine vision, the comparison of an element's brightness or other characteristic with a set value or threshold. { 'thresh,hōld·iŋ }

threshold lights [NAV] Lights so placed as to indicate the longitudinal limits of that portion of a runway, channel, or landing path usable for landing. { 'thresh,hōld ,līts }

threshold limit value [MED] The average concentration of toxic gas to which the normal person can be exposed without injury for 8 hours per day, 5 days per week for an unlimited period; differs slightly from maximum allowable concentration in that threshold limit value is an average concentration. Abbreviated TLV. { 'thresh,hōld 'lim·ət ,val·yü }

threshold of audibility [PHYSIO] The minimum effective sound pressure of a specified signal that is capable of evoking an auditory sensation in a specified fraction of the trials; the threshold may be expressed in decibels relative to 0.0002 microbar (2 × 10⁻⁵ pascal) or 1 microbar (0.1 pascal). Also known as threshold of detectability; threshold of hearing. { 'thresh,hōld əv ,ód·ə'bil·əd·ē }

threshold of detectability *See* threshold of audibility. { 'thresh,hōld əv di,tek·tə'bil·əd·ē }

threshold of feelings [PHYSIO] The minimum effective sound pressure of a specified signal that, in a specified fraction of trials, will stimulate the ear to a point at which there is a sensation of feeling, discomfort, tickle, or pain; normally expressed in decibels relative to 0.0002 microbar (0.00002 pascal) or 1 microbar (0.1 pascal). { 'thresh,hōld əv 'fēl·iŋz }

threshold of hearing *See* threshold of audibility. { 'thresh,hōld əv 'hir·iŋ }

threshold of intelligibility for speech [PSYCH] The sound intensity level at which 50% of the words, nonsense syllables, or sentences used in the articulation test are correctly identified. { ¦thresh,hōld əv in,tel·ə·jə,bil·əd·ē fər 'spēch }

threshold of reaction [PHYS] The minimum energy, for an incident particle or photon, below which a particular reaction does not occur. { 'thresh,hōld əv rē'ak·shən }

threshold signal [ELECTROMAG] A received radio signal (or radar echo) whose power is just above the noise level of the receiver. Also known as minimum detectable signal. { 'thresh,hōld ,sig·nəl }

threshold speed [ENG] The minimum speed of current at which a particular current meter will measure at its rated reliability. { 'thresh,hōld ,spēd }

threshold switch [ELECTR] A voltage-sensitive alternating-current switch made from a semiconductor material deposited on a metal substrate; when the alternating-current voltage acting on the switch is increased above the threshold value, the number of free carriers present in the semiconductor material increases suddenly, and the switch changes from a high resistance of about 10 megohms to a low resistance of less than 1 ohm; in other versions of this switch, the threshold voltage is controlled by heat, pressure, light, or moisture. { 'thresh,hōld ,swich }

threshold treatment [CHEM ENG] The process of stopping a precipitation-type reaction at the threshold of precipitate formation; used in water-treatment reactions. { 'thresh,hōld ,trēt·mənt }

threshold value [COMPUT SCI] A point beyond which there is a change in the manner a program executes; in particular, an error rate above which the operating system shuts down the computer system on the assumption that a hardware failure has occurred. [CONT SYS] The minimum input that produces a corrective action in an automatic control system. { 'thresh,hōld ,val·yü }

threshold velocity [GEOPHYS] The minimum velocity at which wind or water begins to move particles of soil, sand, or other material at a given place under specified conditions. { 'thresh,hōld və,läs·əd·ē }

threshold voltage [ELECTR] **1.** In general, the voltage at which a particular characteristic of an electronic device first appears. **2.** The voltage at which conduction of current begins in a *pn* junction. **3.** The voltage at which channel formation occurs in a metal oxide semiconductor field-effect transistor. **4.** The voltage at which a solid-state lamp begins to emit light. [NUCLEO] The lowest voltage at which all pulses produced in a Geiger counter by any ionizing event are of the same size, regardless of the size of the initial ionizing event. { 'thresh,hōld ,vōl·tij }

Threskiornithidae [VERT ZOO] The ibises, a family of long-legged birds in the order Ciconiiformes. { ,thres·kē·ór'nith·ə,dē }

thribble [PETRO ENG] In drilling operations, a stand of pipe comprising three joints, each about 30 feet (9.4 meters) long. { 'thrib·əl }

thrip [INV ZOO] A small, slender-bodied phytophagous insect of the order Thysanoptera with suctorial mouthparts, a stout proboscis, a vestigial right mandible, and a fully developed left mandible, while wings may be present or absent. { thrip }

Thripidae [INV ZOO] A large family of thrips, order Thysanoptera, which includes the most common species. { 'thrip·ə,dē }

throat [ANAT] The region of the vertebrate body that includes the pharynx, the larynx, and related structures. [BOT] The upper, spreading part of the tube of a gamopetalous calyx or corolla. [DES ENG] The narrowest portion of a constricted duct, as in a diffuser or a venturi tube; specifically, a nozzle throat. [ENG] **1.** The smaller end of a horn or tapered waveguide. **2.** The area in a fireplace that forms the passageway from the firebox to the smoke chamber. { thrōt }

throatable [DES ENG] Of a nozzle, designed to allow a change in the velocity of the exhaust stream by changing the size and shape of the throat of the nozzle. { thrōd·ə·bəl }

throat balls *See* niter balls. { 'thrōt ,bólz }

throat depth [MET] The distance from the center line of the electrodes or platens of a resistance welding machine to the nearest point of interference for flat work. { 'thrōt ,depth }

throat microphone [ENG ACOUS] A contact microphone that is strapped to the throat of a speaker and reacts directly to throat vibrations rather than to the sound waves they produce. { 'thrōt 'mī·krə,fōn }

throat of fillet weld [MET] The thinnest part of a fillet weld,

THRIP

Flower thrip (*Frankliniella tritici*).

or the shortest distance from the root of a fillet weld to its face. { 'thrōt əv 'fil·ət 'weld }

throat velocity *See* critical velocity. { 'thrōt və,läs·əd·ē }

thrombin [BIOCHEM] An enzyme elaborated from prothrombin in shed blood which induces clotting by converting fibrinogen to fibrin. { 'thräm·bən }

thrombinogen *See* prothrombin. { thräm'bin·ə·jən }

thromboangitis obliterans [MED] Thrombosis with organization and a variable degree of associated inflammation in the arteries and veins of the extremities, occasionally of the viscera, progressing to fibrosis about these structures and associated nerves, and complicated by ischemic changes in the parts supplied. Also known as Buerger's disease. { ¦thräm·bō·an'jīd·əs ō'blid·ə,ranz }

thrombocyte [HISTOL] One of the minute protoplasmic disks found in vertebrate blood; thought to be fragments of megakaryocytes. Also known as blood platelet; platelet. { 'thräm·bə,sīt }

thrombocythemia *See* thrombocytosis. { ¦thräm·bō,sī'thē·mē·ə }

thrombocytopenia [MED] The condition of having an abnormally small number of platelets in the circulating blood. { ¦thräm·bō,sīd·ō'pē·nē·ə }

thrombocytopenic purpura [MED] Hemorrhages in the skin, mucous membranes, and elsewhere associated with a decreased number of thrombocytes per unit volume of blood. { ¦thräm·bō¦sīd·ō¦pē·nik 'pər·pə·rə }

thrombocytosis [MED] A condition characterized by an increase in the absolute number of thrombocytes in the circulation. Also known as piastrenemia; thrombocythemia. { ,thräm·bō,sī'tō·səs }

thromboembolectomy [MED] Surgical removal of an embolus that stems from a dislodged thrombus or part of a thrombus. { ¦thräm·bō,em·bə'lek·tə·mē }

thromboembolism [MED] An embolism resulting from a dislodged thrombus or part of a thrombus. { ¦thräm·bō'em·bə,liz·əm }

thrombokinase [BIOCHEM] A proteolytic enzyme in blood plasma that, together with thromboplastin, calcium, and factor V, converts prothrombin to thrombin. { ¦thräm·bō'kī,nās }

thrombophlebitis [MED] Inflammation of a vein associated with thrombosis. { ¦thräm·bō·fle'bīd·əs }

thromboplastin [BIOCHEM] Any of a group of lipid and protein complexes in blood that accelerate the conversion of prothrombin to thrombin. Also known as factor III; plasma thromboplastin component (PTC). { ,thräm·bō'plas·tən }

thromboplastinogen *See* antihemophilic factor. { ¦thräm·bō,plas'tin·ə·jən }

thrombosis [MED] Formation of a thrombus. { thräm 'bō·səs }

thrombotic thrombocytopenic purpura [MED] Thrombi in blood vessels associated with deposits of hyaline substances in the walls and with thrombocytopenia. Also known as Moschcowitz's disease. { thräm'bäd·ik ¦thräm·bō¦sīd·ō¦pen·ik 'pər·pə·rə }

thromboxane [BIOCHEM] Any member of a group of 20-carbon fatty acids related to the prostaglandins and derived mainly from arachidonic acid. { ,thräm'bäk,sān }

thrombus [MED] A blood clot occurring on the wall of a blood vessel where the endothelium is damaged. { 'thräm·bəs }

Throscidae [INV ZOO] The false metallic wood-boring beetles, a cosmopolitan family of the Coleoptera in the superfamily Elateroidea. { 'thräs·kə,dē }

throttle *See* throttle valve. { 'thräd·əl }

throttled flow [FL MECH] Flow which is forced to pass through a restricted area, where the velocity must increase. { 'thräd·əld 'flō }

throttle valve [MECH ENG] A choking device to regulate flow of a liquid, for example, in a pipeline, to an engine or turbine, from a pump or compressor. Also known as throttle. { 'thräd·əl 'valv }

throttling [AERO ENG] The varying of the thrust of a rocket engine during powered flight. [CONT SYS] Control by means of intermediate steps between full on and full off. [THERMO] An adiabatic, irreversible process in which a gas expands by passing from one chamber to another chamber which is at a lower pressure than the first chamber. { 'thräd·əl·iŋ }

throttling bar [ORD] A wedge-shaped bar attached to the interior wall of a recoil cyclinder of a gun; a rectangular notch in the recoil piston moves over the bar, forming an aperture through which the recoil oil must flow; the aperture decreases during recoil until at the end of recoil it is completely closed, thus stopping the flow of liquid and bringing the recoiling tube to rest. { 'thräd·əl·iŋ ,bär }

throttling calorimeter [ENG] An instrument utilizing the principle of constant enthalpy expansion for the measurement of the moisture content of steam; steam drawn from a steampipe through sampling nozzles enters the calorimeter through a throttling orifice and moves into a well-insulated expansion chamber in which its temperature is measured. Also known as steam calorimeter. { 'thräd·əl·iŋ ,kal·ə'rim·əd·ər }

throttling groove [ORD] A groove of varying width or depth cut on the interior wall of some recoil cyclinders of a gun to control the passage of oil, hence controlling the recoil resistance. { 'thräd·əl·iŋ ,grüv }

throttling rod [ORD] A tapered rod attached to a recoil cylinder of a gun; the piston rod is hollow to receive the throttling rod during recoil, and oil is forced through the orifice around the throttling rod; as recoil proceeds, the larger section of the throttling rod comes into the throttling orifice until, at the end of recoil, the throttling rod nearly seals the throttling orifice, thus bringing the recoiling tube to rest. { 'thräd·əl·iŋ ,räd }

throttling valve [ORD] Part of a variable recoil system of a gun, it is a spring-loaded valve through which recoil oil must flow during recoil; as the gun is elevated, the spring pressure is increased by means of a control arm; the valve thus offers greater resistance to the flow of recoil oil and the length of recoil is reduced; in a similar manner, depressing the gun decreases the spring pressure of the valve, with less resistance to oil flow and hence greater length of recoil. { 'thräd·əl·iŋ ,valv }

through arch [CIV ENG] An arch bridge from which the roadway is suspended as distinct from one which carries the roadway on top. { 'thrü ,ärch }

through bridge [CIV ENG] A bridge that carries the deck within the height of the superstructure. { 'thrü ,brij }

through-feed centerless grinding [MECH ENG] A metal cutting process by which the external surface of a cylindrical workpiece of uniform diameter is ground by passing the workpiece between a grinding and regulating wheel. { 'thrü ¦fēd 'sen·tər·ləs 'grīnd·iŋ }

through glacier [HYD] A two-ended glacier, consisting of two valley glaciers in a depression, flowing in opposite directions. { 'thrü ,glā·shər }

throughput [CHEM ENG] The volume of feedstock charged to a process equipment unit during a specified time. [COMMUN] A measure of the effective rate of transmission of data by a communications system. [COMPUT SCI] The productivity of a data-processing system, as expressed in computing work per minute or hour. [MIN ENG] The quantity of ore or other material passed through a mill or a section of a mill in a given time or at a given rate. { 'thrü,pùt }

through repeater [ELECTR] Microwave repeater that is not equipped to provide for connections to any local facilities other than the service channel. { 'thrü ri,pēd·ər }

throughstone *See* bond header. { 'thrü,stōn }

through street [CIV ENG] A street at which all cross traffic is required to stop before crossing or entering. Also known as throughway. { 'thrü ,strēt }

through-the-thickness braiding [ENG] A technique for preparing composite materials in which fibers are intertwined continuously, producing three-dimensional seamless patterns that resist growth of cracks and delamination in the finished parts. Also known as three-dimensional braiding. { ¦thrü thə ¦thik·nəs 'brād·iŋ }

through transmission [ENG] An ultrasonic testing method in which mechanical vibrations are transmitted into one end of the workpiece and received at the other end. { 'thrü tranz ,mish·ən }

through valley [GEOL] **1.** A depression eroded across a divide by glacier ice or meltwater streams. **2.** A valley excavated by a through glacier. { 'thrü ,val·ē }

throughway *See* expressway; through street. { 'thrü,wā }

through weld [MET] A long weld made through the unbroken surface of one member to the other member in a lap or tree joint. { 'thrü ,weld }

throw [ENG] The scattering of fragments in a blasting operation. [GEOL] The vertical component of dip separation on a fault, or generally the amount of vertical displacement on any fault. [MECH ENG] **1.** The maximum diameter of the circle moved by a rotary part. **2.** *See* stroke. { thrō }

throw-away device [ELECTR] An electronic component that is not serviced and is discarded and replaced upon failure. { 'thrō ə,wā di,vīs }

throwing [TEXT] The process of twisting and manipulating yarns to impart texture. { 'thrō·iŋ }

throwing power [MET] The ability of an electroplating solution to deposit metal uniformly on an irregularly shaped cathode. { 'thrō·iŋ ,paù·ər }

throwout [MECH ENG] In automotive vehicles, the mechanism or assemblage of mechanisms by which the driven and driving plates of a clutch are separated. { 'thrō,aùt }

throw-out spiral *See* lead-out groove. { 'thrō,aùt ,spī·rəl }

thrush [MED] A form of candidiasis due to infection by *Candida albicans* and characterized by small whitish spots on the tip and sides of the tongue and the mucous membranes of the buccal cavity. Also known as mycotic stomatitis; parasitic stomatitis. [VET MED] A disease of the frog of a horse's foot accompanied by a fetid discharge. { thrəsh }

thrust [GEOL] Overriding movement of one crystal unit over another. Also known as momentum thrust. [MECH] **1.** The force exerted in any direction by a fluid jet or by a powered screw. **2.** Force applied to an object to move it in a desired direction. [MECH ENG] The weight or pressure applied to a bit to make it cut. [MIN ENG] **1.** A crushing of coal pillars caused by excess weight of the superincumbent rocks, the floor being harder than the roof. **2.** The ruins of the fallen roof, after pillars and stalls have been removed. { thrəst }

thrust augmentation [AERO ENG] The increasing of the thrust of an engine or power plant, especially of a jet engine and usually for a short period of time, over the thrust normally developed. { 'thrəst ,óg·mən'tā·shən }

thrust augmenter [AERO ENG] Any contrivance used for thrust augmentation, as a venturi used in a rocket. { 'thrəst 'óg,men·tər }

thrust axis [AERO ENG] A line or axis through an aircraft or a rocket, along which the thrust acts; an axis through the longitudinal center of a jet or rocket engine, along which the thrust of the engine acts. Also known as axis of thrust; center of thrust. { 'thrəst ,ak·səs }

thrust bearing [MECH ENG] A bearing which sustains axial loads and prevents axial movement of a loaded shaft. { 'thrəst ,ber·iŋ }

thrust block *See* thrust nappe. { 'thrəst ,bläk }

thrust coefficient *See* nozzle thrust coefficient. { 'thrəst ,kō·i,fish·ənt }

thruster [AERO ENG] A control jet employed in spacecraft; an example would be one utilizing hydrogen peroxide. { 'thrəs·tər }

thrust fault [GEOL] A low-angle (less than a 45° dip) fault along which the hanging wall has moved up relative to the footwall. Also known as reverse fault; reverse slip fault; thrust slip fault. { 'thrəst ,fólt }

thrust horsepower [AERO ENG] **1.** The force-velocity equivalent of the thrust developed by a jet or rocket engine. **2.** The thrust of an engine-propeller combination expressed in horsepower; it differs from the shaft horsepower of the engine by the amount the propeller efficiency varies from 100%. [NAV ARCH] The product of the speed of advance of a marine propeller through the water, in feet per second, and the thrust delivered by the propeller, in pounds, divided by 550. { 'thrəst 'hórs,paù·ər }

thrust load [MECH ENG] A load or pressure parallel to or in the direction of the shaft of a vehicle. { 'thrəst ,lōd }

thrust meter [ENG] An instrument for measuring static thrust, especially of a jet engine or rocket. { 'thrəst ,mēd·ər }

thrust moraine *See* push moraine. { 'thrəst mə·rān }

thrust nappe [GEOL] The body of rock that makes up the hanging wall of a thrust fault. Also known as thrust block; thrust plate; thrust sheet; thrust slice. { 'thrəst ,nap }

thrust output [AERO ENG] The net thrust delivered by a jet engine, rocket engine, or rocket motor. { 'thrəst 'aùt,pùt }

thrust plate *See* thrust nappe. { 'thrəst ,plāt }

thrust-pound [AERO ENG] A unit of measurement for the thrust produced by a jet engine or rocket. { 'thrəst 'paùnd }

thrust power [AERO ENG] The power usefully expended on

thrust, equal to the thrust (or net thrust) times airspeed. { 'thrəst ,paù·ər }

thrust reverser [AERO ENG] A device or apparatus for reversing thrust, especially of a jet engine. { 'thrəst ri'vər·sər }

thrust section [AERO ENG] A section in a rocket vehicle that houses or incorporates the combustion chamber or chambers and nozzles. { 'thrəst ,sek·shən }

thrust sheet *See* thrust nappe. { 'thrəst ,shēt }

thrust slice *See* thrust nappe. { 'thrəst ,slīs }

thrust slip fault *See* thrust fault. { 'thrəst 'slip ,fólt }

thrust terminator [AERO ENG] A device for ending the thrust in a rocket engine, either through propellant cutoff (in the case of a liquid) or through diverting the flow of gases from the nozzle. { 'thrəst ,tər·mə,nād·ər }

thrust-weight ratio [AERO ENG] A quantity used to evaluate engine performance, obtained by dividing the thrust output by the engine weight less fuel. { 'thrəst 'wāt ,rā·shō }

thrust yoke [MECH ENG] The part connecting the piston rods of the feed mechanism on a hydraulically driven diamond-drill swivel head to the thrust block, which forms the connecting link between the yoke and the drive rod, by means of which link the longitudinal movements of the feed mechanism are transmitted to the swivel-head drive rod. Also known as back end. { 'thrəst ,yōk }

Thuban [ASTRON] A fourth-magnitude star of spectral class AO in the constellation Draco that was near the north celestial pole around 3000 B.C.; the star Alpha Draconis. { 'thü,ban }

thucolite [GEOL] Concentrations of carbonaceous matter in ancient sedimentary rocks. { 'thü·kə,līt }

thuja oil [MATER] An essential oil from white cedar leaves; pale-yellow, combustible oil with a camphor aroma, soluble in alcohol, ether, chloroform, carbon disulfide, and fixed oils; used in medicine, perfumery, and flavorings. Also known as arbor vitae oil. { 'thü·jə ,óil }

Thule group [ASTRON] An accumulation of asteroids whose sidereal period of revolution is in the ratio 3/4 with that of Jupiter. { 'tü·lē ,grüp }

thulia *See* thulium oxide. { 'thü·lē·ə }

thulite [MINERAL] A pink, rose-red, or purplish-red variety of epidote that contains manganese; used as an ornamental stone. { 'thü,līt }

thulium [CHEM] A rare-earth element, symbol Tm, of the lanthanide group, atomic number 69, atomic weight 168.9342; reacts slowly with water, soluble in dilute acids, melts at 1550°C, boils at 1727°C; the dust is a fire hazard; used as x-ray source and to make ferrites. { 'thü·lē·əm }

thulium-170 [NUC PHYS] The radioactive isotope of thulium, with mass number 170; used as a portable x-ray source. { 'thü·lē·əm ,wən'sev·ən·tē }

thulium chloride [INORG CHEM] $TmCl_3 \cdot 7H_2O$ Green, deliquescent crystals soluble in water and alcohol; melts at 824°C. { 'thü·lē·əm 'klór,īd }

thulium oxalate [ORG CHEM] $Tm_2(C_2O_4)_3 \cdot 6H_2O$ A toxic, greenish-white solid, soluble in aqueous alkali oxalates, loses one water at 50°C; used for analytical separation of thulium from common metals. { 'thü·lē·əm 'äk·sə,lāt }

thulium oxide [INORG CHEM] Tm_2O_3 A white, slightly hygroscopic powder that absorbs water and carbon dioxide from the air, and is slowly soluble in strong acids; used to make thulium metal. Also known as thulia. { 'thü·lē·əm 'äk,sīd }

thumbscrew [DES ENG] A screw with a head flattened in the same axis as the shaft so that it can be gripped and turned by the thumb and forefinger. { 'thəm,skrü }

thump [ENG ACOUS] Low-frequency transient disturbance in a system or transducer characterized audibly by the vocal imitation of the word. { thəmp }

Thunburg technique [BIOCHEM] A technique used to study oxidation of a substrate occurring by dehydrogenation reactions; methylene blue, a reversibly oxidizable indicator, substitutes for molecular oxygen as the ultimate hydrogen acceptor (oxidant), becoming reduced to the colorless leuco form. { 'thən,bərg tek,nēk }

thunder [GEOPHYS] The sound emitted by rapidly expanding gases along the channel of a lightning discharge. { 'thən·dər }

thunderbolt [GEOPHYS] In mythology, a lightning flash accompanied by a material bolt or dart and which causes great

THUMBSCREW

Drawing of thumbscrew showing flattened head. *(Reynolds Metals Co.)*

damage; it is still used as a popular term for a single lightning discharge accompanied by thunder. { 'thən·dər,bōlt }

thundercloud [METEOROL] A convenient and often used term for the cloud mass of a thunderstorm, that is, a cumulonimbus. { 'thən·dər,klaúd }

thunderhead See incus. { 'thən·dər,hed }

thundersquall See rainsquall. { 'thən·dər,skwòl }

thunderstorm [METEOROL] A convective storm accompanied by lightning and thunder and rain, rarely snow showers but often hail, and gusty squall winds at the onset of precipitation; the characteristic cloud is the cumulonimbus. { 'thən·dər,stòrm }

thunderstorm cell [METEOROL] The convection cell of a cumulonimbus cloud. { 'thən·dər,stòrm ,sel }

thunderstorm charge separation [GEOPHYS] **1.** The process by which the large electric field found within thunderclouds is generated. **2.** The processes by which particles bearing opposite electrical charges are given the charges and are transported to different regions of the active cloud. { 'thən·dər,stòrm 'chärj ,sep·ə,rā·shən }

thunderstorm day [METEOROL] An observational day during which thunder is heard at the station; precipitation need not occur. { 'thən·dər,stòrm 'dā }

thunk [COMPUT SCI] An additional subprogram created by the compiler to represent the evaluation of the argument of an expression in the call-by-name procedure. { thəŋk }

Thunnidae [VERT ZOO] The tunas, a family of perciform fishes; there are no scales on the posterior part of the body, and those on the anterior are fused to form an armored covering, the body is streamlined, and the tail is crescent-shaped. { 'thən·ə,dē }

thunniform motion [VERT ZOO] A type of locomotion in which a fish, such as a tuna, moves only the latter third of its body. { 'thən·ə,fòrm ,mō·shən }

Thuringian [GEOL] A European stage of Upper Permian geologic time, above the Saxonian and below the Triassic. { thə'rin·jē·ən }

thurm [ENG] To work wood across the grain with a saw and chisel in order to produce an effect similar to turning the piece on a lathe. { thərm }

Thurniaceae [BOT] A small family of monocotyledonous plants in the order Juncales distinguished by an inflorescence of one or more dense heads, vascular bundles of the leaf in vertical pairs, and silica bodies in the leaf epidermis. { ,thər·nē'ās·ē,ē }

Thylacinidae [VERT ZOO] A family of Australian carnivorous marsupials in the superfamily Dasyuroidea. { ,thī·lə'sīn·ə,dē }

Thylacoleonidae [PALEON] An extinct family of carnivorous marsupials in the superfamily Phalangeroidea. { ,thī·lə,kō·lē'än·ə,dē }

thylakoid [CYTOL] An internal membrane system which occupies the main body of a plastid; particularly well developed in chloroplasts. { 'thī·lə,kóid }

thyme [BOT] A perennial mint plant of the genus *Thymus*; pungent aromatic herb is made from the leaves. { tīm }

thyme camphor See thymol. { tīm ,kam·fər }

thymectomy [MED] Surgical removal of the thymus gland. { thī'mek·tə·mē }

Thymelaeaceae [BOT] A family of dicotyledonous woody plants in the order Myrtales characterized by a superior ovary with a solitary ovule, and petals, if present, are scalelike. { ,thī·mə·lē·ə'ās·ē,ē }

thyme oil [MATER] An essential oil found in the flowers of the thymes *Thymus vulgaris* or *T. zygis*, a colorless to reddish-brown liquid with a sharp taste and pleasant aroma, soluble in alcohol, slightly soluble in water; used in medicine, perfumery, cosmetics, flavoring, and soap. { tīm ,óil }

thymic aplasia [MED] Congenital absence of the thymus and of the parathyroids with deficient cellular immunity. Also known as Di George's syndrome. { 'thī·mik ə'plā·zhə }

thymic corpuscle [HISTOL] A characteristic, rounded, acidophil body in the medulla of the thymus; composed of hyalinized cells concentrically arranged about a core which is occasionally calcified. Also known as Hassal's body. { 'thī·mik 'kòr·pə·səl }

thymidine [BIOCHEM] $C_{10}H_{14}N_2O_5$ A nucleoside derived from deoxyribonucleic acid; essential growth factor for certain

microorganisms in mediums lacking vitamin B_{12} and folic acid. { 'thī·mə,dēn }

thymidylic acid [BIOCHEM] $C_{10}H_{15}N_2O_8P$ A mononucleotide component of deoxyribonucleic acid which yields thymine, D-ribose, and phosphoric acid on complete hydrolysis. { ¦thī¦mə¦dil·ik 'as·əd }

thymine [BIOCHEM] $C_5H_6N_2O_2$ A pyrimidine component of nucleic acid, first isolated from the thymus. { 'thī,mēn }

thymocyte [HISTOL] A lymphocyte formed in the thymus. { 'thī·mə,sīt }

thymol [ORG CHEM] $C_{10}H_{14}O$ A naturally occurring crystalline phenol obtained from thyme or thyme oil, melting at 515°C; used to kill parasites in herbaria, to preserve anatomical specimens, and in medicine as a topical antifungal agent. Also known as thyme camphor. { 'thī,mól }

thymol blue [ORG CHEM] $C_6H_4SO_2OC[C_6H_2(CH_3)(OH)CH-(CH_3)_2]_2$ Brown-green crystals soluble in alcohol and dilute alkalies, insoluble in water, decomposes at 223°C; used as acid-base pH indicator. { ¦thī,mól 'blü }

thymol blue method [FL MECH] A flow-visualization method in which the pH indicator thymol blue is added to the liquid under study and made to change color by electrolysis near a wire inserted in the liquid. { ¦thī,mól ¦blü ¦meth·əd }

thymol iodide [ORG CHEM] $[C_6H_2(CH_3)(OI)(C_3H_7)]_2$ A red-brown, light-sensitive powder with an aromatic aroma, soluble in ether and chloroform, insoluble in water; used in medicine and as a feed additive. { ¦thī,mól 'ī·ə,dīd }

thymolphthalein [ORG CHEM] $C_6H_4COOC[C_6H_2(CH_3)-(OH)CH(CH_3)_2]_2$ A white powder insoluble in water, soluble in alcohol and acetone, melts at 245°C; used in medicine and as an acid-base titration indicator. { ¦thī,mól¦thā,lēn }

thymoma [MED] A usually benign primary tumor of the thymus composed principally of lymphocytic and epithelial cells in varying proportions. { thī'mō·mə }

thymopharyngeal duct [EMBRYO] The third pharyngobranchial duct; it may elongate between the pharynx and thymus. { ¦thī·mō·fə'rin·jē·əl 'dəkt }

thymosin [IMMUNOL] Any of a group of hormones secreted by the thymus gland that stimulate lymphocyte production within the thymus and confer on lymphocytes elsewhere in the body the capacity to respond to antigenic stimulation. { 'thī·mə·sən }

thymulin [IMMUNOL] A zinc-dependent thymic hormone that regulates the differentiation of the immature thymocyte subpopulation and the function of mature T and natural killer cells and also functions as a transmitter between the neuroendocrine and immune systems. { 'thī·myü·lən }

thymus gland [ANAT] A lymphoid organ in the neck or upper thorax of all vertebrates; it is most prominent in early life and is essential for normal development of the circulating pool of lymphocytes. { 'thī·məs ,gland }

thyratron [ELECTR] A hot-cathode gas tube in which one or more control electrodes initiate but do not limit the anode current except under certain operating conditions. Also known as hot-cathode gas-filled tube. { 'thī·rə,trän }

thyratron gate [ELECTR] In computers, an AND gate consisting of a multielement gas-filled tube in which conduction is initiated by the coincident application of two or more signals; conduction may continue after one or more of the initiating signals are removed. { 'thī·rə,trän ,gāt }

thyratron inverter [ELECTR] An inverter circuit that uses thyratrons to convert direct-current power to alternating-current power. { 'thī·rə,trän in,vərd·ər }

thyrector [ELECTR] Silicon diode that acts as an insulator up to its rated voltage, and as a conductor above rated voltage; used for alternating-current surge voltage protection. { thī'rek·tər }

Thyrididae [INV ZOO] The window-winged moths, a small tropical family of lepidopteran insects in the suborder Heteroneura. { thī'rid·ə,dē }

thyristor [ELECTR] A transistor having a thyratronlike characteristic; as collector current is increased to a critical value, the alpha of the unit rises above unity to give high-speed triggering action. { thī'ris·tər }

thyrocalcitonin See calcitonin. { ¦thī·rō,kal·sə'tō·nən }

thyroglobulin [BIOCHEM] An iodinated protein found as the storage form of the iodinated hormones in the thyroid follicular lumen and epithelial cells. { ¦thī·rō'gläb·yə·lən }

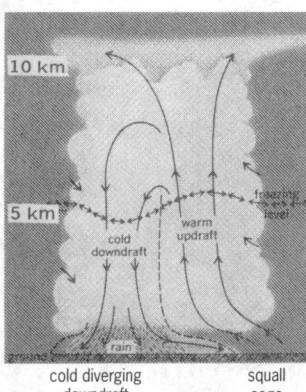

THUNDERSTORM

Diagram showing simplified circulation in a vertical section through a mature thunderstorm.

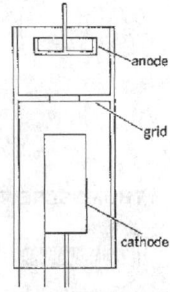

THYRATRON

Diagram of construction of a negative-grid thyratron.

thyroglossal cyst [MED] A cyst formed from the remnants of the thyroglossal duct. { ¦thī·rō¦gläs·əl 'sist }

thyroglossal duct [EMBRYO] A narrow temporary channel connecting the anlage of the thyroid with the surface of the tongue. { thī·rō¦gläs·əl 'dəkt }

thyroid [PHARM] Dried and powdered thyroid gland which contains about 0.2% iodine in combination, especially as thyroxine, and is used therapeutically in the treatment of thyroid deficiencies. { 'thī,ròid }

thyroid cartilage [ANAT] The largest of the laryngeal cartilages in humans and most other mammals, located anterior to the cricoid; in humans, it forms the Adam's apple. { 'thī,ròid ,kärt·lij }

thyroidectomy [MED] Surgical removal of the thyroid gland. { ,thī·ròi'dek·tə·mē }

thyroid gland [ANAT] An endocrine gland found in all vertebrates that produces, stores, and secretes the thyroid hormones. { 'thī,ròid ,gland }

thyroid hormone [BIOCHEM] Commonly, thyroxine or triiodothyronine, or both; a metabolically active compound formed and stored in the thyroid gland which functions to regulate the rate of metabolism. { 'thī,ròid 'hòr,mōn }

thyroiditis [MED] Inflammation of the thyroid gland. { ,thī,ròi'dīd·əs }

thyroid-stimulating hormone See thyrotropic hormone. { 'thī,ròid ¦stim·yə,lād·iŋ 'hòr,mōn }

thyroprotein [BIOCHEM] A protein secreted in the thyroid gland, such as thyroxine. { ¦thī·rō¦prō,tēn }

Thyropteridae [VERT ZOO] The New World disk-winged bats, a family of the Chiroptera found in Central and South America, characterized by a stalked sucking disk and a well-developed claw on the thumb. { ,thī,räp¦ter·ə,dē }

thyrotoxic myopathy [MED] A chronic disease associated with hyperthyroidism resulting in muscular atrophy. { ¦thī·rō¦täk·sik mī'äp·ə·thē }

thyrotoxicosis See hyperthyroidism. { ¦thī·rō¦täk·sə·kō·səs }

thyrotropic hormone [BIOCHEM] A hormone produced by the adenohypophysis which regulates thyroid gland function. Also known as thyroid-stimulating hormone (TSH). { ¦thī·rə¦träp·ik 'hòr,mōn }

thyrotropin [BIOCHEM] A thyroid-stimulating hormone produced by the adenohypophysis. { thī'rä·trə·pən }

thyroxine [BIOCHEM] $C_{15}H_{11}I_4NO_4$ The active physiologic principle of the thyroid gland; used in the form of the sodium salt for replacement therapy in states of hypothyroidism or absent thyroid function. { thī'räk,sēn }

thyrse [BOT] An inflorescence with a racemose primary axis and cymose secondary and later axes. { thərs }

Thysanidae [INV ZOO] A family of hymenopteran insects in the superfamily Chalcidoidea. { thī'san·ə,dē }

Thysanoptera [INV ZOO] The thrips, an order of small, slender insects having exopterygote development, sucking mouthparts, and exceptionally narrow wings with few or no veins and bordered by long hairs. { ,thī·sə'näp·tə·rə }

Thysanura [INV ZOO] The silverfish, machilids, and allies, an order of primarily wingless insects with soft, fusiform bodies. { ,thī·sə'nùr·ə }

THz See terahertz.

Ti See titanium.

tiba [ORG CHEM] $C_7H_3I_3O_2$ A colorless solid with a melting point of 226–228°C; insoluble in water; used as a growth regulator for fruit. { 'tī·bə }

tibia [ANAT] The larger of the two leg bones, articulating with the femur, fibula, and talus. { 'tib·ē·ə }

tibialis [ANAT] **1.** A muscle of the leg arising from the proximal end of the tibia and inserted into the first cuneiform and first metatarsal bones. **2.** A deep muscle of the leg arising proximally from the tibia and fibula and inserted into the navicular and first cuneiform bones. { ,tib·ē'al·əs }

tic douloureux See trigeminal neuralgia. { ,tik ,dü·lə'rü }

tick [COMMUN] A pulse broadcast at 1-second intervals by standard frequency and time broadcasting stations to indicate the exact time. [COMPUT SCI] A time interval equal to 1/60 second, used primarily in discussing computer operations. [INV ZOO] An arachnid comprising Ixodoidea; a bloodsucking parasite and important vector of various infectious diseases of humans and lower animals. { tik }

tick-bite paralysis [VET MED] A flaccid paralysis in animals, and occasionally in humans, caused by a feeding tick attached to the body. { 'tik ¦bīt pə'ral·ə·səs }

tick-borne typhus fever of Africa [MED] Any of several infections caused by *Rickettsia conori*, transmitted by ixodid ticks, and occurring in Africa and adjacent areas; includes boutonneuse fever, Marseilles fever, Kenya tick typhus fever, and South African tick-bite fever. { 'tik ¦bórn 'tī·fəs ,fē·vər əv 'af·ri·kə }

ticket converting [COMPUT SCI] The process of changing prepunched ticket stubs, 2.7 inches (6.9 centimeters) wide by 1 inch (2.5 centimeters) deep, into punched cards; the ticket is made up of a basic section and one or more stubs that are numerically prepunched and printed with identical information. { 'tik·ət kən,vərd·iŋ }

tick fever See Rocky Mountain spotted fever. { 'tik ,fē·vər }

tickle [PHYSIO] A tingling sensation of the skin or a mucous membrane following light, tactile stimulation. { 'tik·əl }

tickler coil [ELECTR] Small coil connected in series with the plate circuit of an electron tube and inductively coupled to a grid-circuit coil to establish feedback or regeneration in a radio circuit; used chiefly in regenerative detector circuits. { 'tik·lər ,kòil }

tick typhus See Rocky Mountain spotted fever. { 'tik ,tī·fəs }

tic polonga [VERT ZOO] *Vipera russellii*. A member of the Viperidae; one of the most deadly and most common snakes in India; it may reach a length of 5 feet (1.5 meters), is nocturnal in its habits, and pursues rodents into houses. Also known as Russell's viper. { ,tik pə'lòŋ·gə }

tidal air [PHYSIO] That air which is inspired and expired during normal breathing. { 'tīd·əl 'er }

tidal bore See bore. { 'tīd·əl 'bòr }

tidal channel [OCEANOGR] A major channel followed by tidal currents, extending from the ocean into a tidal marsh or tidal flat. { 'tīd·əl 'chan·əl }

tidal component See partial tide. { 'tīd·əl kəm'pō·nənt }

tidal constants [OCEANOGR] Tidal relations that remain essentially constant for any particular locality. { 'tīd·əl 'kän·stəns }

tidal constituent See partial tide. { 'tīd·əl kən'stich·ə·wənt }

tidal correction [GEOPHYS] A correction made in gravity observations to remove the effect of the earth's tides. { 'tīd·əl kə'rek·shən }

tidal current [OCEANOGR] The alternating horizontal movement of water associated with the rise and fall of the tide caused by the astronomical tide-producing forces. { 'tīd·əl 'kə·rənt }

tidal-current chart [OCEANOGR] A chart showing by arrows and numbers the average direction and speed of tidal currents at a particular part of the current cycle. { 'tīd·əl ¦kə·rənt ,chärt }

tidal-current tables [OCEANOGR] Tables issued annually which give daily predictions of the times of slack water and the times and velocities of the strength of flood and ebb currents for a number of reference stations, together with differences and constants for obtaining predictions at subordinate stations. { 'tīd·əl ¦kə·rənt ,tā·bəlz }

tidal cycle See tide cycle. { 'tīd·əl 'sī·kəl }

tidal datum [OCEANOGR] A level of the sea, defined by some phase of the tide, from which water depths and heights of tide are reckoned. Also known as tidal datum plane. { 'tīd·əl 'dad·əm }

tidal datum plane See tidal datum. { 'tīd·əl ¦dad·əm ,plān }

tidal day [OCEANOGR] The interval between two consecutive high waters of the tide at a given place, averaging 24 hours 51 minutes. { 'tīd·əl 'dā }

tidal delta [GEOL] A sand bar or shoal formed in the entrance of an inlet by the action of reversing tidal currents. { 'tīd·əl 'del·tə }

tidal difference [OCEANOGR] The difference in time or height of a high or low water at a subordinate station and at a reference station for which predictions are given in the tide tables; the difference applied to the prediction at the reference station gives the corresponding time or height for the subordinate station. { 'tīd·əl 'dif·rəns }

tidal energy [OCEANOGR] The energy in a tide flowing from a basin into an open sea. { 'tīd·əl 'en·ər·jē }

tidal epoch See phase lag. { 'tīd·əl 'ep·ik }

tidal excursion [OCEANOGR] The net horizontal distance over which a water particle moves during one tidal cycle of flood and ebb; the distances traversed during ebb and flood

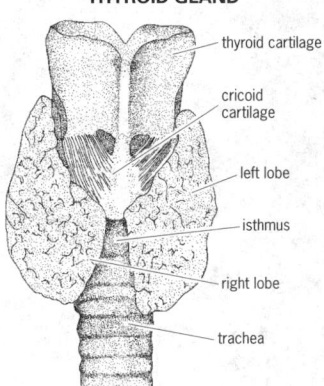

THYROID GLAND

Ventral view of a human thyroid gland.

THYSANURA

Firebrat, *Thermobia domestica*.

are rarely equal in nature, since there is usually a layered circulation in an estuary, with a net surface flow in one direction compensated by an opposite flow at depth. { 'tīd·əl ik'skər·zhən }

tidal flat [GEOL] A marshy, sandy, or muddy nearly horizontal coastal flatland which is alternately covered and exposed as the tide rises and falls. { 'tīd·əl 'flat }

tidal frequency [OCEANOGR] The rate of travel, in degrees per day, of a component of a tide, the component being created by a particular juxtaposition of forces in the sun-earth-moon system. { 'tīd·əl 'frē·kwən·sē }

tidal friction [OCEANOGR] The frictional effect of the tidal wave particularly in shallow waters that lengthens the tidal epoch and tends to slow the rotational velocity of the earth, thus increasing very slowly the length of the day. { 'tīd·əl 'frik·shən }

tidal glacier See tidewater glacier. { 'tīd·əl 'glā·shər }

tidal harbor [OCEANOGR] A harbor affected by the tides, in distinction to a harbor in which the water level is maintained by caissons or gates. { 'tīd·əl 'här·bər }

tidal inlet [GEOL] A natural inlet maintained by tidal currents. { 'tīd·əl 'in·lət }

tidalite [GEOL] Any sediment transported and deposited by tidal currents. { 'tīd·əl‚īt }

tidal lights [NAV] Lights at the entrance of a harbor to indicate tide and tidal current conditions within the harbor. { 'tīd·əl 'līts }

tidal lock See entrance lock. { 'tīd·əl 'läk }

tidal marsh [GEOGR] Any marsh whose surface is covered and uncovered by tidal flow. { 'tīd·əl 'märsh }

tidal platform ice foot [OCEANOGR] An ice foot between high and low water levels, produced by the the rise and fall of the tide. { 'tīd·əl ‚plat‚förm 'īs ‚fût }

tidal pool [OCEANOGR] An accumulation of sea water remaining in a depression on a beach or reef after the tide recedes. { 'tīd·əl 'pül }

tidal potential [OCEANOGR] Tidal forces expressed as components of a vector field. { 'tīd·əl pə‚ten·chəl }

tidal prism [OCEANOGR] The difference between the mean high-water volume and the mean low-water volume of an estuary. { 'tīd·əl 'priz·əm }

tidal quay [CIV ENG] A quay in an open harbor or basin with sufficient depth to enable ships lying alongside to remain afloat at any state of the tide. { 'tīd·əl 'kē }

tidal radius [ASTRON] The distance from the center of a planet in formation at which the planet's gravitational attraction for nearby gas equals that of the Sun. { 'tīd·əl 'rād·ē·əs }

tidal range See tide range. { 'tīd·əl 'rānj }

tidal scour [GEOL] Sea-floor erosion caused by strong tidal currents, resulting in removal of inshore sediments and formation of deep holes and channels. Also known as scour. { 'tīd·əl 'skaúr }

tidal stand See stand. { 'tīd·əl 'stand }

tidal volume [PHYSIO] The volume of air moved in and out of the lungs during a single normal respiratory cycle. { 'tīd·əl 'väl·yəm }

tidal water [OCEANOGR] Any water whose level changes periodically due to tidal action. { 'tīd·əl ‚wȯd·ər }

tidal wave [OCEANOGR] **1.** Any unusually high and generally destructive sea wave or water level along a shore. **2.** See tide wave. { 'tīd·əl ‚wāv }

tidal wind [METEOROL] A very light breeze which occurs in calm weather in inlets where the tide sets strongly; it blows onshore with rising tide and offshore with ebbing tide. { 'tīd·əl 'wind }

tide [OCEANOGR] The periodic rising and falling of the oceans resulting from lunar and solar tide-producing forces acting upon the rotating earth. { 'tīd }

tide amplitude [OCEANOGR] One-half of the difference in height between consecutive high water and low water; half the tide range. { 'tīd 'am·plə‚tüd }

tide-bound [NAV] Referring to a vessel unable to proceed because of insufficient depth of water due to tidal action. { 'tīd ‚baúnd }

tide bulge See tide wave. { 'tīd ‚bəlj }

tide crack [OCEANOGR] A crack in sea ice, parallel to the shore, caused by the vertical movement of the water due to tides; several such cracks often appear as a family. { 'tīd ‚krak }

tide curve [OCEANOGR] Any graphic representation of the

rise and fall of the tide; time is generally represented by the abscissas, and the height of the tide by the ordinates; for normal tides the curve so produced approximates a sine curve. { 'tīd ‚kərv }

tide cycle [OCEANOGR] A period which includes a complete set of tide conditions or characteristics, such as a tidal day or a lunar month. Also known as tidal cycle. { 'tīd ‚sī·kəl }

tide gage [ENG] A device for measuring the height of a tide; may be observed visually or may consist of an elaborate recording instrument. { 'tīd ‚gāj }

tide gate [CIV ENG] **1.** A restricted passage through which water runs with great speed due to tidal action. **2.** An opening through which water may flow freely when the tide sets in one direction, but which closes automatically and prevents the water from flowing in the other direction when the direction of flow is reversed. { 'tīd ‚gāt }

tidehead [OCEANOGR] The inland limit of water affected by a tide. { 'tīd‚hed }

tide hole [OCEANOGR] A hole made in ice to observe the height of the tide. { 'tīd ‚hōl }

tide indicator [ENG] That part of a tide gage which indicates the height of tide at any time; the indicator may be in the immediate vicinity of the tidal water or at some distance from it. { 'tīd 'in·də‚kād·ər }

tideland [GEOGR] Land which is under water at high tide and uncovered at low tide. { 'tīd·lənd }

tide lock See entrance lock. { 'tīd ‚läk }

tide machine [ENG] An instrument that computes, sometimes for years in advance, the times and heights of high and low waters at a reference station by mechanically summing the harmonic constituents of which the tide is composed. { 'tīd mə‚shēn }

tidemark [OCEANOGR] **1.** A high-water mark left by tidal water. **2.** The highest point reached by a high tide. { 'tīd‚märk }

tide notes [OCEANOGR] Notes included on nautical charts which give information on the mean range or the diurnal range of the tide, mean tide level, and extreme low water at key places on the chart. { 'tīd ‚nōts }

tide pole [ENG] A graduated spar used for measuring the rise and fall of the tide. Also known as tide staff. { 'tīd ‚pōl }

tide prediction [OCEANOGR] The mathematical process by which the times and heights of the tide are determined in advance from the harmonic constituents at a place. { 'tīd pri‚dik·shən }

tide-producing force [GEOPHYS] The slight local difference between the gravitational attraction of two astronomical bodies and the centrifugal force that holds them apart. { 'tīd prə‚düs·iŋ 'förs }

tide race [OCEANOGR] A strong tidal current or a channel in which such a current flows. { 'tīd ‚rās }

tide range [OCEANOGR] The difference in height between consecutive high and low waters. Also known as tidal range. { 'tīd ‚rānj }

tide rips See rips. { 'tīd ‚rips }

tide-rode [NAV] Referring to a ship riding at anchor and heading into the tidal current. { 'tīd ‚rōd }

tide signal [NAV] A visual signal displayed at the entrance of a harbor to indicate tidal conditions within the harbor. { 'tīd ‚sig·nəl }

tide staff See tide pole. { 'tīd ‚staf }

tide station [OCEANOGR] A place where observations of the tides are obtained. { 'tīd ‚stā·shən }

tide table [OCEANOGR] A table giving daily predictions, usually a year in advance, of the times and heights of the tide for a number of reference stations. { 'tīd ‚tā·bəl }

tidewater [OCEANOGR] **1.** A body of water, such as a river, affected by tides. **2.** Water inundating land at flood tide. { 'tīd‚wȯd·ər }

tidewater glacier [HYD] A glacier that descends into the sea and usually has a terminal ice cliff. Also known as tidal glacier. { 'tīd‚wȯd·ər 'glā·shər }

tide wave [OCEANOGR] A long-period wave associated with the tide-producing forces of the moon and sun, and identified with the rising and falling of the tide. Also known as tidal wave; tide bulge. { 'tīd ‚wāv }

tideway [OCEANOGR] A channel through which a tidal current runs. { 'tīd‚wā }

tidi-sound See time-division sound. { 'tī‚dē‚saúnd }

tie [CIV ENG] One of the transverse supports to which railroad rails are fastened to keep them to line, gage, and grade. [ELEC] **1.** Electrical connection or strap. **2.** *See* tie wire. [ENG] A beam, post, rod, or angle to hold two pieces together; a tension member in a construction. [MIN ENG] A support for the roof in coal mines. { tī }

tieback [MIN ENG] **1.** A beam serving a purpose similar to that of a fend-off beam, but fixed at the opposite side of the shaft or inclined road. **2.** The wire ropes or stay rods that are sometimes used on the side of the tower opposite the hoisting engine, either in place of or to reinforce the engine braces. { 'tī,bak }

tie bar [CIV ENG] **1.** A bar used as a tie rod. **2.** A rod connecting two switch rails on a railway to hold them to gage. [GEOL] *See* tombolo. { 'tī ,bär }

tie cable [ELEC] **1.** Cable between two distributing frames or distributing points. **2.** Cable between two private branch exchanges. **3.** Cable between a private branch exchange switchboard and main office. **4.** Cable connecting two other cables. { 'tī ,kā·bəl }

tied arch [CIV ENG] An arch having the horizontal reaction component provided by a tie between the skewbacks of the arch ends. { 'tīd 'ärch }

tied concrete column [CIV ENG] A concrete column reinforced with longitudinal bars and horizontal ties. { 'tīd 'kän ,krēt 'käl·əm }

Tiedenmann's body [INV ZOO] One of the small glands opening into the ring vessel in many echinoderms in which amebocytes are produced. { 'tēd·ən,mänz ,bäd·ē }

tie-down diagram [ENG] A drawing indicating the prescribed method of securing a particular item of cargo within a specific type of vehicle. { 'tī,daun ,dī·ə,gram }

tie-down point [ENG] An attachment point provided on or within a vehicle. { 'tī,daun ,pöint }

tie-down point pattern [ENG] The pattern of tie-down points within a vehicle. { 'tī,daun ,pöint 'pad·ərn }

tied rank [STAT] If two distinct observations have the same value, thus being given the same rank, they are said to be tied; this presents difficulties in the Wilcoxon two-sample test, the sign test, and the Fisher-Irwin test. { 'tīd 'raŋk }

tie line [COMMUN] **1.** A leased communication channel or circuit. **2.** *See* data link. [PHYS CHEM] A line on a phase diagram joining the two points which represent the composition of systems in equilibrium. Also known as conode. { 'tī ,līn }

tiemannite [MINERAL] HgSe A steel gray to blackish-lead gray mineral consisting of mercuric selenide; commonly occurs in massive form. { 'tē·mə,nīt }

tie plate [CIV ENG] A metal plate between a rail and a tie to hold the rail in place and reduce wear on the tie. [MECH ENG] A plate used in a furnace to connect tie rods. { 'tī ,plāt }

tie point [ELEC] Insulated terminal to which two or more wires may be connected. { 'tī ,pöint }

tier array [ELECTROMAG] Array of antenna elements, one above the other. { 'tir ə,rā }

tier building [CIV ENG] A multistory skeleton frame building. { 'tir ,bil·diŋ }

tie rod [CIV ENG] A structural member used as a brace to take tensile loads. [ENG] A round or square iron rod passing through or over a furnace and connected with buckstays to assist in binding the furnace together. [MECH ENG] A rod used as a mechanical or structural support between elements of a machine. [MIN ENG] Vertical rods mounted in overlying horizontal shaft timbers. { 'tī ,räd }

tie trunk [ELEC] Telephone line or channel directly connecting two private branch exchanges. { 'tī ,trəŋk }

Tietze extension theorem [MATH] A topological space X is normal if and only if every continuous function of a closed subset to [0,1] has a continuous extension to all of X. { 'tēt·sə ik'sten·chən ,thir·əm }

tie wire [ELEC] A short piece of wire used to tie an open-line wire to an insulator. Also known as tie. { 'tī ,wīr }

TIF *See* telephone influence factor; terminal indecomposable future.

TIFF *See* tag image file format. { tif }

TIGA *See* truncated icosahedral gravitational-wave antenna. { 'tē,ī'jē'ä *or* 'tī·gə }

tiger [VERT ZOO] *Felis tigris.* An Asiatic carnivorous mammal in the family Felidae characterized by a tawny coat with transverse black stripes and white underparts. { 'tī·gər }

tiger beetle [INV ZOO] The common name for any of the bright-colored beetles in the family Cicindelidae; there are about 1300 species distributed all over the world. { 'tī·gər ,bēd·əl }

tiger salamander [VERT ZOO] *Ambystoma tigrinum.* A salamander in the family Ambystomatidae, found in a variety of subspecific forms from Canada to Mexico and over most of the United States; lives in arid and humid regions, and is the only salamander in much of the Great Plains and Rocky Mountains. { 'tī·gər ,sal·ə,man·dər }

tiger's-eye [MINERAL] A yellowish-brown crystalline variety of quartz; a translucent, fibrous, broadly chatoyant gemstone that may be dyed other colors. { 'tī·gərz ,ī }

tiger shark *See* sand shark. { 'tī·gər ,shärk }

tight [ENG] **1.** Unbroken, crack-free, and solid rock in which a naked hole will stand without caving. **2.** A borehole made impermeable to water by cementation or casing. [MECH ENG] **1.** Inadequate clearance or the barest minimum of clearance between working parts. **2.** The absence of leaks in a pressure system. { tīt }

tight binding approximation [SOLID STATE] A method of calculating energy states and wave functions of electrons in a solid in which the wave function is assumed to be a sum of pure atomic wave functions centered about each of the atoms in the lattice, each multiplied by a phase factor; it is suitable for deep-lying energy levels. { 'tīt ,bīnd·iŋ ə präk·sə'mā·shən }

tight coupling *See* close coupling. { 'tīt 'kəp·liŋ }

tight fit [DES ENG] A fit between mating parts with slight negative allowance, requiring light to moderate force to assemble. { 'tīt 'fit }

tight fold *See* closed fold. { 'tīt 'fōld }

tight ion pair [ORG CHEM] An ion pair composed of individual ions which keep their stereochemical configuration; no solvent molecules separate the cation and anion. Also known as contact ion pair; intimate ion pair. { 'tīt 'ī,än ,per }

tight junction [CYTOL] An intercellular junction composed of a series of fusions of the junctional membrane, forming a continuous seal; serves as a selective barrier to small molecules and as a total barrier to large molecules. Also known as impermeable junction; occluding junction; zonula occludens. { 'tīt 'jəŋk·shən }

tight knot *See* sound knot. { 'tīt 'nät }

tightly coupled computer [COMPUT SCI] A computer linked to another computer in a manner that requires both computers to function as a single unit. { 'tīt·lē 'kəp·əld kəm'pyüd·ər }

tight sand [GEOL] A sand whose interstices are filled with finer grains of the matrix material, thus effectively destroying porosity and permeability. Also known as close sand. { 'tīt 'sand }

TIG welding *See* tungsten-inert gas welding. { 'tē'ī'jē ,weld·iŋ }

tilasite [MINERAL] CaMg(AsO₄)F A gray, gray-violet, olive green, or apple green, monoclinic mineral consisting of a fluorarsenate of calcium and magnesium. { 'til·ə,sīt }

tile [MATER] **1.** A piece of fired clay, stone, concrete, or other material used ornamentally to cover roofs, floors, or walls. **2.** A hollow building unit made of burned clay or other material. { tīl }

tileboard [MATER] A type of wallboard used for interior finishing in which the outer surface is a layer of hard glossy material, usually simulating tile. { 'tīl,bōrd }

tile painting [COMPUT SCI] **1.** The use of patterns to create shadings that fill shapes and areas on a monochrome display. **2.** The use of very small dots of two or more colors to make blends or shades that fill shapes and areas on a color display. { 'tīl ,pānt·iŋ }

tiling [COMPUT SCI] Dividing an electronic display into two or more nonoverlapping areas that display the outputs of different programs being run concurrently on a computer. { 'tīl·iŋ }

till [GEOL] Unsorted and unstratified drift consisting of a heterogeneous mixture of clay, sand, gravel, and boulders which is deposited by and underneath a glacier. Also known as boulder clay; glacial till; ice-laid drift. { til }

tillage [AGR] The operation or practice of cultivating soil in order to improve it for agricultural purposes. { 'til·ij }

till billow [GEOL] An undulating mass of glacial drift that is disposed in an irregular pattern with regard to the direction of movement of the ice. { 'til ,bil·ō }

tiller [BOT] A shoot that develops from an axillary or adventitious bud at the base of a stem. [NAV ARCH] A lever attached to the rudder of a boat or ship and used to turn the rudder from side to side, usually turned by hand in a boat and by mechanical devices in a ship. { 'til·ər }

Tilletiaceae [MYCOL] A family of fungi in the order Ustilaginales in which basidiospores form at the tip of the apibasidium. { tə‚lē·shē'ās·ē‚ē }

tilleyite [MINERAL] Ca$_5$(Si$_2$O$_7$)(CO$_3$)$_2$ A white mineral consisting of a carbonate and silicate of calcium. { 'til·ē‚īt }

tillite [PETR] A sedimentary rock formed by lithification of till, especially pre-Pleistocene till. { 'ti‚līt }

Tillodontia [PALEON] An order of extinct quadrupedal land mammals known from early Cenozoic deposits in the Northern Hemisphere and distinguished by large, rodentlike incisors, blunt-cuspid cheek teeth, and five clawed toes. { ‚til·ə'dän·chə }

tilloid [GEOL] A nonglacial till-like deposit. [PETR] A rock of uncertain origin which resembles tillite. { 'ti‚lȯid }

till plain [GEOL] An extensive, relatively flat area overlying a till. { 'til ‚plān }

till sheet [GEOL] A sheet, layer, or bed of till. { 'til ‚shēt }

tilt [AERO ENG] The inclination of an aircraft, winged missile, or the like from the horizontal, measured by reference to the lateral axis or to the longitudinal axis. [ELECTROMAG] **1.** Angle which an antenna forms with the horizontal. **2.** In radar, the angle between the axis of radiation in the vertical plane and a reference axis which is normally the horizontal. [METEOROL] The inclination to the vertical of a significant feature of the circulation (or pressure) pattern or of the field of temperature or moisture; for example, troughs in the westerlies usually display a westward tilt with altitude in the lower and middle troposphere. [OPTICS] The angle between the plane of a photograph from a downward-pointing camera and the horizontal plane. { tilt }

tilt angle [ELECTROMAG] The angle between the axis of radiation of a radar beam in the vertical plane and a reference axis (normally the horizontal). { 'tilt ‚aŋ·gəl }

tilt block [GEOL] A tilted fault block. { 'tilt ‚bläk }

tilt boundary [SOLID STATE] A boundary between two crystals that differ in orientation by only a few degrees, consisting of a series of edge dislocations; it is formed during polygonization. Also known as bend plane; polygon wall. { 'tilt ‚baún·drē }

tilted iceberg [OCEANOGR] A tabular iceberg that has become unbalanced, so that the flat, level top is inclined. { 'til·təd 'īs‚bərg }

tilted interface [GEOL] Oil-water interface in which water moves in a generally linear direction under an oil accumulation which is, for instance, in an anticline. { 'til·təd 'in‚tər‚fās }

tilt error [NAV] The error caused by the propagation of signals over the tilted ionosphere reflecting layer or, in systems requiring reception over two or more widely separated locations, by the different heights of the ionosphere reflecting layer for the various transmissions. { 'tilt ‚er·ər }

tilth [GEOL] The physical condition of a soil as expressed in terms of fitness for growth of specified plants or crops. { tilth }

tilting dozer [MECH ENG] A bulldozer whose blade can be pivoted on a horizontal center pin to cut low on either side. { 'tilt·iŋ 'dō·zər }

tilting idlers [MECH ENG] An arrangement of idler rollers in which the top set is mounted on vertical arms which pivot on spindles set low down on the frame of the roller stool. { 'tilt·iŋ 'īd·lərz }

tilting mixer [MECH ENG] A small-batch mixer consisting of a rotating drum which can be tilted to discharge the contents; used for concrete or mortar. { 'tilt·iŋ 'mik·sər }

tilting-type boxcar unloader [CIV ENG] A mechanism that is used to unload material such as grain from a boxcar; the car, with its door open, is held by end clamps on the specialized piece of track and tilted 15% from the vertical and then tilted endwise 40% to the horizontal to discharge the material at one end of the car, and 40% in the opposite direction to discharge the material from the opposite end. { 'tilt·iŋ ‚tīp 'bäks‚kär ən'lōd·ər }

tiltmeter [ENG] An instrument used to measure small changes in the tilt of the earth's surface, usually in relation to a liquid-level surface or to the rest position of a pendulum. { 'tilt‚mēd·ər }

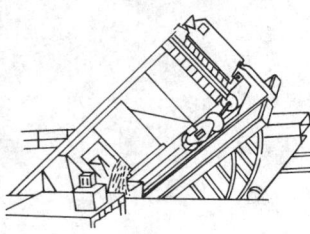

TILTING-TYPE BOXCAR UNLOADER

Drawing of a tilting-type unloader. *(Link-Belt Co.)*

tilt mold [MET] A mold that rotates from a horizontal to a vertical position during filling to reduce agitation and risk of dross entrapment. { 'tilt ‚mōld }

tilt/rotate code [ENG] A code that instructs a "golf ball" printing element which angle of tilt and rotation is needed to print a given character. { 'tilt'rō‚tāt ‚kōd }

tilt rotor [AERO ENG] An assembly of rapidly rotating blades on a vertical takeoff and landing aircraft, whose plane of rotation can be continuously varied from the horizontal to the vertical, permitting performance as helicopter blades or as propeller blades. { 'tilt ‚rōd·ər }

tilt slab construction See tilt-up construction. { 'tilt ‚slab kən‚strək·shən }

tilt-up construction [BUILD] A method for constructing concrete wall panels by casting them horizontally adjacent to their final positions and then tilting them into vertical positions after the concrete has cured. Also known as tilt slab construction. { 'tilt‚əp kən‚strək·shən }

timber [MATER] Wood used for building, carpentry, or joinery. { 'tim·bər }

timber connector [ENG] A metal fastener that has a series of sharp teeth digging into the wood and is tightened with bolts to join sections of timber in heavy construction. { 'tim·bər kə‚nek·tər }

timbered stope [MIN ENG] A stope made of square-set timbering or any of its variations. { 'tim·bərd 'stōp }

timbering [MIN ENG] The timber structure used for supporting the faces of an excavation during the progress of construction. { 'tim·bə·riŋ }

timbering machine [MIN ENG] An electrically driven machine to raise and hold timber in place while the supporting posts are being set, the posts having been cut to desired length previously by the machine's power-driven saw. { 'tim·bə·riŋ mə‚shēn }

timberline [ECOL] The elevation or latitudinal limits for arboreal growth. Also known as tree line. { 'tim·bər‚līn }

timber mat [MIN ENG] Broken timber forming the roof of an ore deposit that is being extracted by a caving method, such as top slicing. { 'tim·bər 'mat }

timber packer See pack builder. { 'tim·bər ‚pak·ər }

timber puller [MIN ENG] A machine used to remove the timber supports in a mine. { 'tim·bər ‚púl·ər }

timber trolley [MIN ENG] A carriage consisting of a timber or steel base, mounted on wheels, with U-shaped arms. { 'tim·bər ‚träl·ē }

timber truck [MIN ENG] Any truck or car used for hauling timber inside of a mine. { 'tim·bər ‚trək }

timbre [ACOUS] That attribute of auditory sensation in terms of which a listener can judge that two sounds similarly presented and having the same loudness and pitch are dissimilar. Also known as musical quality; quality of sound. { 'tam·bər }

time [PHYS] **1.** The dimension of the physical universe which, at a given place, orders the sequence of events. **2.** A designated instant in this sequence, as the time of day. Also known as epoch. { tīm }

time-and-altitude azimuth [NAV] In celestial navigation, the azimuth derived by a computation in which meridian angle, declination, and altitude are parameters, the values of which are either known or assumed. { ¦tīm ən 'al·tə‚tüd ¦az·ə·məth }

time and material contract [IND ENG] A contract providing for the procurement of supplies or services on the basis of direct labor hours at specified fixed hourly rates (which rates include direct and indirect labor, overhead, and profit), and material at cost. { ¦tīm ən mə'tir·ē·əl ‚kän‚trakt }

time and motion study [IND ENG] Observation, analysis, and measurement of the steps in the performance of a job to determine a standard time for each performance. Also known as time-motion study. { ¦tīm ən 'mō·shən ‚stəd·ē }

time assignment speech interpolation [COMMUN] Modulation technique based on the fact that speech is never a continuous stream of information, but consists of a large number of short signals; therefore, the period between the speech signals is used for transmitting other data including additional speech signals. { ¦tīm ə¦sīn·mənt 'spēch ‚in·tər·pə‚lā·shən }

time-average holographic interferometry [OPTICS] The study of holograms of a vibrating surface which have been averaged over time; illumination of such a hologram yields an image of the surface on which are superimposed interference fringes which are contour lines of equal displacement of the surface, enabling vibrational amplitudes of the surface to be measured with precision.   { 'tīm ¦av·rij ¦hāl·ə¦graf·ik ˌin·tər·fə'räm·ə·trē }

time azimuth [NAV] In celestial navigation, the azimuth derived by a calculation in which the meridian angle, the polar distance (or declination), and the latitude are parameters, the magnitudes of which are either known or assumed.   { 'tīm ˌaz·ə·math }

time base [ELECTR] A device which moves the fluorescent spot rhythmically across the screen of the cathode-ray tube.   { 'tīm ˌbās }

time-base generator *See* sweep generator.   { 'tīm ¦bās ˌjen·ə·ˌrād·ər }

time break [ENG] A distinctive mark shown on an exploration seismogram to indicate the exact detonation time of an explosive energy source.   { 'tīm ˌbrāk }

time-change component [ENG] A component which because of design limitations or safety is specified to be rebuilt or overhauled after a specified period of operation (for example, an engine or propeller of an airplane).   { 'tīm ¦chānj kəm‚pō·nənt }

time-code generator [ELECTR] A crystal-controlled pulse generator that produces a train of pulses with various predetermined widths and spacings, from which the time of day and sometimes also day of year can be determined; used in telemetry and other data-acquisition systems to provide the precise time of each event.   { 'tīm ¦kōd ˌjen·ə·ˌrā·dər }

time constant [PHYS] **1.** The time required for a physical quantity to rise from zero to $1 - 1/e$ (that is, 63.2%) of its final steady value when it varies with time t as $1 - e^{-kt}$. **2.** The time required for a physical quantity to fall to $1/e$ (that is, 36.8%) of its initial value when it varies with time t as e^{-kt}. **3.** Generally, the time required for an instrument to indicate a given percentage of the final reading resulting from an input signal. Also known as lag coefficient.   { 'tīm ˌkän·stənt }

time-controlled system *See* clock control system.   { 'tīm kən¦trōld ˌsis·təm }

time correlation [GEOL] A correlation of age or mutual time relations between stratigraphic units in separated areas.   { 'tīm ˌkär·ə'lā·shən }

time-current characteristics [ELEC] Of a fuse, the relation between the root-mean-square alternating current or direct current and the time for the fuse to perform the whole or some specified part of its interrupting function.   { 'tīm 'kə·rənt ˌkar·ik·tə‚ris·tiks }

time curve *See* time front.   { 'tīm ˌkərv }

time delay [PHYS] The time required for a signal to travel between two points in a circuit or for a wave to travel between two points in space.   { 'tīm di‚lā }

time-delay circuit [ELECTR] A circuit in which the output signal is delayed by a specified time interval with respect to the input signal. Also known as delay circuit.   { 'tīm di¦lā ˌsər·kət }

time-delay fuse [ELEC] A fuse in which the burnout action depends on the time it takes for the overcurrent heat to build up in the fuse and melt the fuse element.   { 'tīm di¦lā ˌfyüz }

time-delay relay [ELEC] A relay in which there is an appreciable interval of time between energizing or deenergizing of the coil and movement of the armature, such as a slow-acting relay and a slow-release relay.   { 'tīm di¦lā ˌrē‚lā }

time-derived channel [COMMUN] Any of the channels which result from time-division multiplexing of a channel.   { 'tīm di¦rīvd ˌchan·əl }

time diagram [ASTRON] A diagram in which the celestial equator appears as a circle, and celestial meridians and hour circles as radial lines; used to facilitate solution of time problems and other problems involving arcs of the celestial equator or angles at the pole, by indicating relations between various quantities involved; conventionally, the relationships are given as viewed from a point over the South Pole, in a westward direction or counterclockwise. Also known as diagram on the plane of the celestial equator; diagram on the plane of the equinoctial.   { 'tīm ˌdī·ə‚gram }

time dilation effect *See* slowing of clocks.   { 'tīm də‚lā·shən iˌfekt }

time-distance graph [AERO ENG] A graph used to determine the ground distance for air-route legs of a specified time interval; time-distance relationships are often simplified by

considering air, wind, and ground distances for flight legs of 1-hour duration.   { 'tīm 'dis·təns ˌgraf }

time-division data links [COMMUN] Radio communications which use time-division techniques for channel separation.   { 'tīm di‚vizh·ən 'dad·ə ˌliŋks }

time-division multiple access [COMMUN] A technique that allows multiple users who are geographically dispersed to gain access to a communications channel, by permitting each user access to the full pass-band of the channel for a limited time, after which the access right is assigned to another user. Abbreviated TDMA.   { ¦tīm də‚vizh·ən ¦məl·tə·pəl 'ak‚ses }

time-division multiplexing [COMPUT SCI] The interleaving of bits or characters in time to compensate for the slowness of input devices as compared to data transmission lines. [COMMUN] A process for transmitting two or more signals over a common path by using successive time intervals for different signals. Also known as time multiplexing. Abbreviated TDM.   { 'tīm di¦vizh·ən ˌməl·tə‚pleks·iŋ }

time-division multiplier *See* mark-space multiplier.   { 'tīm di¦vizh·ən ˌməl·tə‚plī·ər }

time-division sound [COMMUN] A method under development, employing time-division multiplexing, which enables the audio signal of a television broadcast to be transmitted over the same channel as the video signal. Also known as tidisound.   { 'tīm di¦vizh·ən ˌsaůnd }

time-division switching system [ELECTR] A type of electronic switching system in which input signals on lines and trunks are sampled periodically, and each active input is associated with the desired output for a specific phase of the period.   { 'tīm di‚vizh·ən 'swich·iŋ ˌsis·təm }

time-domain reflectometer [ELECTR] An instrument that measures the electrical characteristics of wideband transmission systems, subassemblies, components, and lines by feeding in a voltage step and displaying the superimposed reflected signals on an oscilloscope equipped with a suitable time-base sweep. Abbreviated TDR.   { 'tīm də‚mān ˌrē‚flek'täm·əd·ər }

time factor *See* time scale.   { 'tīm ˌfak·tər }

time fire [ORD] Fire in which fuses are set to act after a fixed time interval and before impact.   { 'tīm ˌfīr }

time formula [IND ENG] A formula to determine the standard time of an operation as a function of one or more variables in the operation.   { 'tīm ˌfór·mya·lə }

time front [AERO ENG] A locus of points representing the maximum ground distances from a departure point that can be covered by an aircraft in a prescribed time interval. Also known as hour-out line; time curve.   { 'tīm ˌfrənt }

time fuse [ENG] A fuse which contains a graduated time element to regulate the time interval after which the fuse will function.   { 'tīm ˌfyüz }

time gate [ELECTR] A circuit that gives an output only during chosen time intervals.   { 'tīm ˌgāt }

time-height section [ELECTR] A facsimile trace of a vertically directed radar; specifically, a cloud-detection radar.   { 'tīm 'hīt ˌsek·shən }

time hopping [COMMUN] A spread spectrum technique, usually used in combination with other methods, in which the transmitted pulse occurs in a manner determined by a pseudorandom code which places the pulse in one of several possible positions per frame.   { 'tīm ˌhäp·iŋ }

time-interval measurement [HOROL] A process that consists either in calculating the duration between two known epochs, or in counting the repetitions of a recurring phenomenon from an arbitrary starting point, as with an electronic digital-reading counter, which counts the cycles of an oscillator.   { 'tīm ˌin·tər·vəl ˌmezh·ər·mənt }

time-interval radiosonde *See* pulse-time-modulated radiosonde.   { 'tīm ˌin·tər·vəl 'rād·ē·ō‚sänd }

time-invariant system [CONT SYS] A system in which all quantities governing the system's behavior remain constant with time, so that the system's response to a given input does not depend on the time it is applied.   { 'tīm in‚ver·ē·ənt ˌsis·təm }

time lag [ORD] The amount by which the time of fall of a bomb, released under given conditions, would exceed that of an ideal bomb released under identical conditions. [PHYS] The time between a cause and a resultant effect, as between occurrence of a primary ionizing event and its count by a counter.   { 'tīm ˌlag }

time-lapse photography [GRAPHICS] Motion picture photography in which a single frame is exposed at regular intervals;

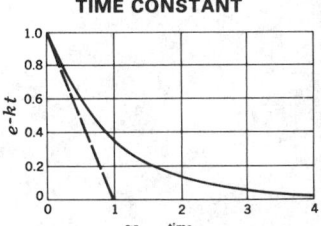

TIME CONSTANT

Graph of exponentially decreasing variable e^{-kt} (def. 2) as a function of time in time constants. Dotted line indicates variable that continues to decrease at same rate that exponential variable decreases at time $t = 0$.

when the film is projected at normal speed, the action appears to be speeded up.   { 'tīm ¦laps fə¦täg·rə·fē }

timelike path [RELAT] A trajectory in space-time such that a vector tangent to any point on the path is a timelike vector. { 'tīm‚līk 'path }

timelike surface [RELAT] A surface in space-time whose normal vector is everywhere spacelike. { 'tīm‚līk 'sər·fəs }

timelike vector [RELAT] A four vector in Minkowski space whose space component has a magnitude which is less than the magnitude of its time component multiplied by the speed of light. { 'tīm‚līk 'vek·tər }

time line [FL MECH] A line of tracers in a fluid, produced by a voltage or laser-beam pulse, whose deformation follows the contour of the local velocity profile. [GEOL] **1.** A line that indicates equal geologic age in a correlation diagram. **2.** A rock unit represented by a time line. { 'tīm ‚līn }

time-mark generator [ELECTR] A signal generator that produces highly accurate clock pulses which can be superimposed as pips on a cathode-ray screen for timing the events shown on the display.   { 'tīm ¦märk ‚jen·ə·rād·ər }

time measurement [HOROL] A process that consists in counting the repetitions of any recurring phenomenon and, if the interval between successive recurrences is sensible, in subdividing it.   { 'tīm ‚mezh·ər·mənt }

time meridian [ASTRON] Any meridian used as a reference for reckoning time, particularly a zone or standard meridian. { 'tīm mə‚rid·ē·ən }

time modulation [COMMUN] Modulation in which the time of occurrence of a definite portion of a waveform is varied in accordance with a modulating signal. { 'tīm ‚mäj·ə‚lā·shən }

time-motion study See time and motion study. { 'tīm 'mō·shən ‚stəd·ē }

time-multiplexing See multiprogramming; time-division multiplexing. { 'tīm ‚məl·tə‚pleks·iŋ }

time-of-day clock [COMPUT SCI] An electronic device that registers the actual time, generally accurate to 0.1 second, through a 24-hour cycle, and transmits its reading to the central processing unit of a computer upon demand. { ¦tīm əv ¦dā ‚kläk }

time of delivery [COMMUN] The time at which the addressee or responsible relay agency receipts for a message.   { 'tīm əv di'liv·ə·rē }

time of flight [MECH] Elapsed time in seconds from the instant a projectile or other missile leaves a gun or launcher until the instant it strikes or bursts. [PHYS] The elapsed time from the instant a particle leaves a source to the instant it reaches a detector. { 'tīm əv 'flīt }

time-of-flight mass spectrometer [SPECT] A mass spectrometer in which all the positive ions of the material being analyzed are ejected into the drift region of the spectrometer tube with essentially the same energies, and spread out in accordance with their masses as they reach the cathode of a magnetic electron multiplier at the other end of the tube. { ¦tīm əv ¦flīt 'mas spek'träm·əd·ər }

time-of-flight spectrometer [ENG] Any instrument in which the speed of a particle is determined directly by measuring the time it takes to travel a measured distance. { ¦tīm əv ¦flīt spek'träm·əd·ər }

time of origin [COMMUN] The time at which a message is released for transmission. { 'tīm əv 'är·ə·jən }

time of receipt [COMMUN] The time at which a receiving station completes reception of a message. { 'tīm əv ri'sēt }

time of set [MATER] The time required for freshly mixed concrete to stiffen (initial set, about 1 hour) or to attain a minimum specified hardness (final set, about 10 hours); actual times vary with the type of cement used. { 'tīm əv 'set }

time on target [ORD] **1.** A method of firing on a target in which various artillery units time their fire so that all projectiles reach the target simultaneously. **2.** A measure of the ability of a fire-control system or gunner to keep a weapon aimed at a moving target or to keep a weapon mounted on a moving vehicle aimed at a target; a useful evaluation of stabilizers. { 'tīm ȯn 'tär·gət }

timeout [CONT SYS] A test of the reliability of robotic software in which the robot is halted if a portion of software does not function properly until the problem is corrected. { 'tīm‚aut }

time over target [ORD] The time at which an aircraft or formation of aircraft arrives over a designated point for the purpose of conducting an air mission on a target. { 'tīm 'ō·vər 'tär·gət }

time phase [PHYS] Two disturbances are in time phase if they reach corresponding peak values at the same instants of time, though not necessarily at the same points in space. { 'tīm ‚fāz }

time phasing [IND ENG] Production scheduling of components for product assembly so that each component is available at the correct time. { 'tīm ‚fāz·iŋ }

time-projection chamber [NUCLEO] A particle detector consisting of a large cylinder filled with gas in which charged particles leave ionization trails, to which are applied strong, uniform electric and magnetic fields, both paralleed to the cylinder axis. { 'tīm prə¦jek·shən ‚chām·bər }

time-pulse distributor [ELECTR] A device or circuit for allocating timing pulses or clock pulses to one or more conducting paths or control lines in specified sequence. { 'tīm ‚pəls di‚strib·yəd·ər }

time quadrature [PHYS] **1.** Differing by a time interval corresponding to one-fourth the time of one cycle of the frequency in question. **2.** An integration over time. { 'tīm ‚kwäd·rə‚chər }

time quantum See time slice. { 'tīm ‚kwän·təm }

time quenching [MET] Interrupted quenching in which the time in the quenching medium is controlled. { 'tīm ‚kwench·iŋ }

timer [COMPUT SCI] A hardware device that can interrupt a computer program after a time interval specified by the program, generally to remind the program to take some action. [ELECTR] A circuit used in radar and in electronic navigation systems to start pulse transmission and synchronize it with other actions, such as the start of a cathode-ray sweep. [ENG] **1.** A device for automatically starting or stopping a machine or other device. **2.** See interval timer. [MECH ENG] A device that controls timing of the ignition spark of an internal combustion engine at the correct time. { 'tīm·ər }

timer clock [COMPUT SCI] An electronic device in the central processing unit of a computer which times events that occur during the operation of the system in order to carry out such functions as changing computer time, detecting looping and similar error conditions, and keeping a log of operations. { 'tī·mər ‚kläk }

time redundancy [COMPUT SCI] Performing a computation more than once and checking the results in order to increase reliability. { 'tīm ri‚dən·dən·sē }

time reference scanning beam [NAV] **1.** A radio or radar beam which is swept to and fro over a sector of space and in such a manner that the interval between reception of successive pulses at an aircraft indicates the aircraft's altitude or azimuth. **2.** A system of ground equipment that generates such beams at microwave frequencies to furnish guidance to aircraft making microwave landings. Abbreviated TRSB. { 'tīm ‚ref·rəns 'skan·iŋ ‚bēm }

time-resolved laser spectroscopy [SPECT] A method of studying transient phenomena in the interaction of light with matter through the exposure of samples to extremely short and intense pulses of laser light, down to subnanosecond or subpicosecond duration. { 'tīm ri¦zälvd 'lā·zər spek'träs·kə‚pē }

time reversal [PHYS] The replacement of the time coordinate t by its negative $-t$ in the equations of motion of a dynamical system; the time reversal operator, a symmetry operator for a quantum-mechanical system, contains also the complex conjugation operator and a matrix operating on the spin coordinate. { 'tīm ri‚vər·səl }

time-reversal invariance [PHYS] A symmetry of the fundamental (microscopic) equations of a system such that, if it holds, the time reversal of any motion of the system is also a motion of the system. { 'tīm ri¦vər·səl in'ver·ē·əns }

time-reversal reflection See optical phase conjugation. { 'tīm ri¦vər·səl ri‚flek·shən }

time-reversal test [STAT] A test used with index numbers that is satisfied when the new index is the reciprocal of the original index if the functions of the base period and given period are interchanged; the advantage of index numbers meeting the criteria of the test is that a symmetric comparison of the two periods is obtained and the results are consistent whether one or the other period is used as a base. { 'tīm ri‚vər·səl ‚test }

time-rock unit *See* time-stratigraphic unit. { 'tīm 'räk ‚yü‑nət }

time scale [COMPUT SCI] The ratio of the time duration of an event as simulated by an analog computer to the actual time duration of the event in the physical system under study. Also known as time factor. { 'tīm ‚skāl }

time separation [AERO ENG] The time interval between adjacent aircraft flying approximately the same path. { 'tīm ‚sep·ə‚rā·shən }

time series [STAT] A statistical process analogous to the taking of data at intervals of time. { 'tīm ‚sir·ēz }

time-series analysis [MATH] The general study of mathematical systems or processes analogous to that of data taken at time intervals. { 'tīm ‚sir·ēz ə‚nal·ə·səs }

time-share [COMPUT SCI] To perform several independent processes almost simultaneously by interleaving the operations of the processes on a single high-speed processor. { 'tīm ‚sher }

time-shared amplifier [ELECTR] An amplifier used with a synchronous switch to amplify signals from different sources one after another. { 'tīm ‚sherd ‚am·plə‚fī·ər }

time-sharing [COMPUT SCI] The simultaneous utilization of a computer system from multiple terminals. [IND ENG] Division of the time required for observation, decision making, and responding by an operator among the activities or tasks that must be performed almost simultaneously. { 'tīm ‚sher·iŋ }

time sight [NAV] A common method for determining longitude by celestial observations; these observations are reduced by solving the navigational triangle for meridian angle and require known or assumed values for altitude, latitude, and declination; the meridian angle is converted to local hour angle and compared with Greenwich hour angle, and the latter is read from astronomical tables. { 'tīm ‚sīt }

time signal [COMMUN] An accurate signal which is broadcast by radio and marks a specified time or time interval, used for setting timepieces and for determining their errors; in particular, a radio signal broadcast at accurately known times each day on a number of different frequencies by WWV and other stations. { 'tīm ‚sig·nəl }

time signal service [COMMUN] Radio communications service for the transmission of time signals of stated high precision, intended for general reception. { 'tīm 'sig·nəl ‚sər·vəs }

time slice [COMPUT SCI] A time interval during which a time-sharing system is processing one particular computer program. Also known as time quantum. { 'tīm ‚slīs }

times sign *See* multiplication sign. { 'tīmz ‚sīn }

time standard [HOROL] A recurring phenomenon, used as a reference for establishing a unit of time; the presently accepted standard is the second, defined to be 9,192,631,770 transitions between two specified hyperfine levels of the atom of cesium-133. [IND ENG] *See* standard time. { 'tīm ‚stan·dərd }

time-stratigraphic facies [GEOL] A stratigraphic facies based on the amount of geologic time during which deposition and nondeposition of sediment occurred. { 'tīm ¦strad·ə¦graf·ik ‚fā·shēz }

time-stratigraphic unit [GEOL] A stratigraphic unit based on geologic age or time of origin. Also known as chronolith; chronolithologic unit; chronostratic unit; chronostratigraphic unit; time-rock unit. { 'tīm ¦strad·ə¦graf·ik ‚yü·nət }

time study [IND ENG] A work measurement technique, generally using a stopwatch or other timing device, to record the actual elapsed time for performance of a task, adjusted for any observed variance from normal effort or pace, unavoidable or machine delays, rest periods, and personal needs. { 'tīm ‚stəd·ē }

time switch [ENG] A clock-controlled switch used to open or close a circuit at one or more predetermined times. { 'tīm ‚swich }

time system [CONT SYS] A system of clocks and control devices, with or without a master timepiece, to indicate time at various remote locations. { 'tīm ‚sis·təm }

time tick [COMMUN] A radio-broadcast time signal consisting of one or more short audible sounds or beats; in particular, one which is generated by an accurately controlled pulsed radio signal. { 'tīm ‚tik }

time transfer [HOROL] The transmission of information from time and frequency measurements between timing laboratories in different parts of the world for the purpose of comparison and synchronization, generally by two-way radio transmission via satellite. { 'tīm ‚tranz·fər }

time-transgressive *See* diachronous. { 'tīm tranz‚gres·iv }

time-varying system [CONT SYS] A system in which certain quantities governing the system's behavior change with time, so that the system will respond differently to the same input at different times. { 'tīm ¦ver·ē·iŋ ‚sis·təm }

time-weighted average [SCI TECH] The average exposure to a contaminant or condition (such as noise) to which workers may be exposed without adverse effect over a period such as in an 8-hour day or 40-hour week. Abbreviated TWA. { ¦tīm ‚wād·əd 'av·rij }

time zone [ASTRON] To avoid the inconvenience of the continuous change of mean solar time with longitude, the earth is divided into 24 time zones, each about 15° wide and centered on standard longitudes, 0°, 15°, 30°, and so on; within each zone the time kept is the mean solar time of the standard meridian. { 'tīm ‚zōn }

timing [MECH ENG] Adjustment in the relative position of the valves and crankshaft of an automobile engine in order to produce the largest effective output of power. [ORD] Adjustment of a small-arms weapon so that it will perform each function at a predetermined point in the cycle of operation. { 'tīm·iŋ }

timing-axis oscillator *See* sweep generator. { 'tīm·iŋ ‚ak·səs ‚äs·ə‚lād·ər }

timing belt [DES ENG] A power transmission belt with evenly spaced teeth on the bottom side which mesh with grooves cut on the periphery of the pulley to produce a positive, no-slip, constant-speed drive. Also known as cogged belt; synchronous belt. [MECH ENG] A positive drive belt that has axial cogs molded on the underside of the belt which fit into grooves on the pulley; prevents slip, and makes accurate timing possible; combines the advantages of belt drives with those of chains and gears. Also known as positive drive belt. { 'tīm·iŋ ‚belt }

timing-belt pulley [MECH ENG] A pulley that is similar to an uncrowned flat-belt pulley, except that the grooves for the belt's teeth are cut in the pulley's face parallel to the axis. { 'tīm·iŋ ‚belt ‚pul·ē }

timing circuit *See* clock. { 'tīm·iŋ ‚sər·kət }

timing error [COMPUT SCI] An error made in planning or writing a computer program, usually in underestimating the time that will be taken by input/output or other operations, which causes unnecessary delays in the execution of the program. { 'tīm·iŋ ‚er·ər }

timing gears [MECH ENG] The gear train of reciprocating engine mechanisms for relating camshaft speed to crankshaft speed. { 'tīm·iŋ ‚girz }

timing loop [COMPUT SCI] A set of instructions in a computer program whose execution time is known and whose only function is to cause a delay in processing by causing the loop to be executed an appropriate number of times. { 'tīm·iŋ ‚lüp }

timing motor [ELEC] A motor which operates from an alternating-current power system synchronously with the alternating-current frequency, used in timing and clock mechanisms. Also known as clock motor. { 'tīm·iŋ ‚mōd·ər }

timing relay [ELEC] Form of auxiliary relay used to introduce a definite time delay in the performance of a function. { 'tīm·iŋ ‚rē‚lā }

timing signal [COMPUT SCI] A pulse generated by the clock of a digital computer to provide synchronization of its activities. [ELECTR] Any signal recorded simultaneously with data on magnetic tape for use in identifying the exact time of each recorded event. { 'tīm·iŋ ‚sig·nəl }

Timken film strength [ENG] A test used on a gear lubricant to determine the amount of pressure the film of oil can withstand before rupturing. { 'tim·kən film ‚streŋkth }

Timken wear test [ENG] A test used on a gear lubricant to determine its abrasive effect on gear metals. { 'tim·kən 'wer ‚test }

TIMOTHY

Drawing of timothy (*Phleum pratense*) showing leafy stems and cylindrical inflorescence.

TINTINNIDA

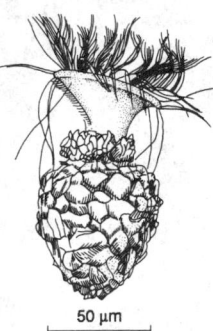

50 μm

A living specimen of a representative species of the order Tintinnida in the genus *Tintinnopsis*, shown protruding from its lorica, or shell.

timothy [BOT] *Phleum pratense.* A perennial hay grass of the order Cyperales characterized by moderately leafy stems and a dense cylindrical inflorescence. { 'tim·ə·thē }

tin [CHEM] Metallic element in group IV, symbol Sn, atomic number 50, atomic weight 118.69; insoluble in water, soluble in acids and hot potassium hydroxide solution; melts at 232°C, boils at 2260°C. [MET] A lustrous silver-white ductile, malleable metal used in alloys, for solder, terneplate, and tinplate. { tin }

Tinamidae [VERT ZOO] The single family of the avian order Tinamiformes. { ti'nam·ə,dē }

Tinamiformes [VERT ZOO] The tinamous, an order of South and Central American birds which are superficially fowllike but have fully developed wings and are weak fliers. { ti,nam·ə'fȯr,mēz }

tin bisulfide See stannic sulfide. { 'tin bī'səl,fīd }

tin bromide See stannic bromide; stannous bromide.

tin bronze [MET] A tin-copper alloy. { 'tin 'bränz }

tincal See borax. { 'tin,kal }

tincalconite [MINERAL] Na₂B₄O₇·5H₂O A colorless to dull-white mineral, crystallizing in the rhombohedral system; one of the principal ores of borax and boron compounds. Also known as mohavite; octahedral borax. { tin'kal·kə,nīt }

tin chloride See stannic chloride; stannous chloride. { 'tin 'klȯr,īd }

tin chromate See stannic chromate; stannous chromate. { 'tin 'krō,māt }

tin crystals See stannous chloride. { 'tin ,krist·əlz }

tincture [MATER] A dilute solution (aqueous or aqueous alcoholic) of a drug or chemical; more dilute than fluid extracts, less volatile than spirits. { 'tiŋk·chər }

tincture of iodine [PHARM] A medicinal preparation used as an anti-infective containing 20 grams iodine and 24 grams sodium iodide in 1000 milliliters of alcohol. Also known as iodine solution; iodine tincture. { 'tiŋk·chər əv 'ī·ə,dīn }

tin dichloride See stannous chloride. { 'tin dī'klȯr,īd }

tin difluoride See stannous fluoride. { 'tin dī'flur,īd }

tin dioxide See stannic oxide. { 'tin dī'äk,sīd }

tinea [MED] Group of skin diseases caused by various fungi, for example, tinea pedis (athlete's foot) and tinea capitis (ringworm infection of the scalp). { ,tin·ē·ə }

tinea favosa See favus. { 'tin·ē·ə fə'vō·sə }

Tineidae [INV ZOO] A family of small moths in the superfamily Tineoidea distinguished by an erect, bristling vestiture on the head. { ti'nē·ə,dē }

Tineoidea [INV ZOO] A superfamily of heteroneuran Lepidoptera which includes small moths that usually have well-developed maxillary palpi. { ,tin·ē'ȯid·ē·ə }

tin fluoride See stannous fluoride. { 'tin 'flur,īd }

tinfoil [MET] Foil made of tin or a tin alloy. { 'tin,fȯil }

Tingidae [INV ZOO] The lace bugs, the single family of the hemipteran superfamily Tingoidea. { 'tin·jə,dē }

tingle [BUILD] A support used in masonry to reduce sagging in a long layer of bricks. [DES ENG] **1.** A small nail. **2.** A flexible metal clip used to hold a sheet of material such as glass or metal. [ENG] A patch designed to cover a hole in a boat. { 'tiŋ·gəl }

Tingoidea [INV ZOO] A superfamily of the Hemiptera in the subdivision Geocorisae characterized by the wings with many lacelike areolae. { tiŋ'gȯid·ē·ə }

tin hydride [INORG CHEM] SnH₄ A gas boiling at −52°C. Also known as stannane. { 'tin 'hī,drīd }

tin iodide See stannic iodide. { 'tin 'ī·ə,dīd }

tin monosulfide See stannous sulfide. { 'tin 'män·ə'səl,fīd }

tinned wire [MET] Copper wire that has been coated during manufacture with a layer of tin or solder to prevent corrosion and simplify soldering of connections. { 'tind 'wīr }

tinner's rivet [DES ENG] A special-purpose rivet that has a flat head, used in sheet metal work. { 'tin·ərz ,riv·ət }

tinning [MET] **1.** Covering or preserving with tin. **2.** A protective coating of tin. { 'tin·iŋ }

tinnitus [MED] A ringing, roaring, or hissing sound in one or both ears. { 'tin·əd·əs }

tin oxalate See stannous oxalate. { 'tin 'äk·sə,lāt }

tin oxide See stannic oxide; stannous oxide. { 'tin 'äk,sīd }

tin peroxide See stannic oxide. { 'tin pə'räk,sīd }

tin pest [MET] Transformation of tin to a brittle, gray variety occurring spontaneously at temperatures below 0°C. { 'tin 'pest }

tinplate [MET] Thin sheet iron or steel coated with tin. { 'tin,plāt }

tin protosulfide See stannous sulfide. { 'tin ,prōd·ō'səl,fīd }

tin protoxide See stannous oxide. { 'tin prə'täk,sīd }

tin pyrites See stannite. { 'tin 'pī,rīts }

tin salts See stannous chloride. { 'tin ,sȯlts }

tinsel cord [ELEC] A highly flexible cord used for headphone leads and test leads, in which the conductors are strips of thin metal foil or tinsel wound around a strong but flexible central cord. { 'tin·səl ,kȯrd }

tin stone See cassiterite. { 'tin ,stōn }

tin sulfate See stannous sulfate. { 'tin 'səl,fāt }

tin sulfide See stannous sulfide. { 'tin 'səl,fīd }

tin sweat [MET] Exudation of tin-rich low-melting-point material from a tin-bronze surface as a result of inverse segregation in bronze casting, or overheating of the alloy. { 'tin ,swet }

tint [GRAPHICS] An even tone area on a printed page, often produced by screening black or a color to give a lighter shade. [OPTICS] The mixture of a pure color with white. { 'tint }

tin tetrabromide See stannic bromide. { 'tin ¦te·trə'brō,mīd }

tin tetrachloride See stannic chloride. { 'tin ¦te·trə'klȯr,īd }

tin tetraiodide See stannic iodide. { 'tin ¦te·trə'ī·ə,dīd }

tinticite [MINERAL] Fe₃(PO₄)₂(OH)₃·3¹/₂H₂O A creamy white mineral with a yellowish-green tint, consisting of a hydrated basic iron phosphate. { 'tin·ti,kīt }

Tintinnida [INV ZOO] An order of ciliated protozoans in the subclass Spirotrichia whose members are conical or trumpet-shaped pelagic forms bearing shells. { tin'tin·əd·ə }

tint of passage [OPTICS] The color produced when a plate which is colorless, but which rotates the plane of polarization of polarized light passing through it by an amount which depends on the wavelength of the light, is placed between crossed polarizers. { 'tint əv 'pas·ij }

tintometer [OPTICS] A device used to estimate the intensity of a colored solution by comparing it with standard solutions or colored glass slides, as with the Lovibond tintometer. { tin 'täm·əd·ər }

tip [DES ENG] A piece of material secured to and differing from a cutter tooth or blade. [ELEC] The contacting part at the end of a phone plug. [ELECTR] A small protuberance on the envelope of an electron tube, resulting from the closing of the envelope after evacuation. { tip }

TIP See terminal indecomposable past.

tipburn [PL PATH] A disease of certain cultivated plants, such as potato and lettuce, characterized by browning of the leaf margins due to excessive loss of water. { 'tip,bərn }

tip clearance [NAV ARCH] The clearance or distance between the circumference of the tip circle of a propeller and the hull of the vessel. { 'tip ,klir·əns }

Tiphiidae [INV ZOO] A family of the Hymenoptera in the superfamily Scolioidea. { tə'fī·ə,dē }

tip-in [GRAPHICS] **1.** A halftone illustration on smooth, coated paper that has been inserted on a page of a book printed on rough paper; a blank space is left on the rough page and the halftone is attached (tipped in) by a narrow coating of paste applied at the top edge. **2.** The insertion of a plate or full-page illustration on smoother or heavier stock than the rest of the book. { 'tip,in }

tip jack [ELEC] A small single-hole jack for a single-pin contact plug. Also known as pup jack. { 'tip ,jak }

tip layering [BOT] A plant propagation technique in which only the stem tip is buried; used to reproduce trailing blackberries and black raspberries. { 'tip ,lā·ə·riŋ }

tipped bit [DES ENG] A drill bit in which the cutting edge is made of especially hard material. { 'tipt 'bit }

tipped solid cutters [DES ENG] Cutters made of one material and having tips or cutting edges of another material bonded in place. { 'tipt 'säl·əd 'kəd·ərz }

tipper [MIN ENG] An apparatus for emptying coal or ore cars by turning them upside down and then righting them, with a minimum of manual labor. { 'tip·ər }

tipping-bucket rain gage [ENG] A type of recording rain gage; the precipitation collected by the receiver empties into one side of a chamber which is partitioned transversely at its center and is balanced bistably upon a horizontal axis; when a predetermined amount of water has been collected, the chamber tips, spilling out the water and placing the other half of the chamber under the receiver; each tip of the bucket is recorded on a chronograph, and the record obtained indicates the amount and rate of rainfall. { 'tip·iŋ ¸bək·ət 'rān ¸gāj }

tipple [MIN ENG] **1.** The place where the mine cars are tipped and emptied of their coal. **2.** The tracks, trestles, and screens at the entrance to a colliery, where coal is screened and loaded. { 'tip·əl }

tip side [ELEC] Conductor of a circuit which is associated with the tip of a plug or the top spring of a jack; by extension, it is common practice to designate by these terms the conductors having similar functions or arrangements in circuits where plugs or jacks may not be involved. { 'tip ¸sīd }

Tipulidae [INV ZOO] The crane flies, a family of orthorrhaphous dipteran insects in the series Nematocera. { tə'pyül·ə¸dē }

tire [ENG] A continuous metal ring, or pneumatic rubber and fabric cushion, encircling and fitting the rim of a wheel. { tīr }

tire iron [DES ENG] A single metal bar having bladelike ends of various shapes to insert between the rim and the bead of a pneumatic tire to remove or replace the tire. { 'tīr ¸ī·ərn }

Tiros satellite [AERO ENG] Television infrared observation satellite; a meteorological satellite that takes television pictures of cloud cover, using radiation sensors and cameras; it stores and transmits this information on ground command. { 'tī¸rōs 'sad·əl¸īt }

tirrill burner [ENG] A modification of the bunsen burner which allows greater flexibility in the adjustment of the air-gas mixture. { 'tir·əl ¸bər·nər }

Tirrill regulator [ELEC] A device for regulating the voltage of a generator, in which the field resistance of the exciter is short-circuited temporarily when the voltage drops. { 'tir·əl ¸reg·yə¸lād·ər }

Tischenko reaction [ORG CHEM] The formation of an ester by the condensation of two molecules of aldehyde utilizing a catalyst of aluminum alkoxides in the presence of a halide. { ti'shəŋ·kō rē¸ak·shən }

tissue [HISTOL] An aggregation of cells more or less similar morphologically and functionally. [TEXT] A sheer woven fabric or gauze, usually of fine quality. { 'tish·ü }

tissue culture [CYTOL] Growth of tissue cells in artificial media. { 'tish·ü ¸kəl·chər }

tissue dose [NUCLEO] The dose received by a tissue in the region of interest, expressed in roentgens for x-rays and gamma rays. { 'tish·ü ¸dōs }

tissue engineering [MED] The creation of tissues or organs to replace lost form or function. { 'tish·ü ¸en·jə¸nir·iŋ }

tissue paper [MATER] Extremely lightweight paper, available in many colors and used in craft projects and in collage painting. { 'tish·ü ¸pā·pər }

tissue plasminogen activator [MED] A proteolytic enzyme that can convert plasminogen to plasmin. { ¦tish·ü plaz¸min·ə·jən 'ak·tə¸vād·ər }

tissue roentgen See rep. { 'tish·ü 'rent·gən }

tissue typing [MED] A procedure involving a test or a series of tests to determine the compatibility of tissues from a prospective donor and a recipient prior to transplantation. { 'tish·ü ¸tīp·iŋ }

Titan [ASTRON] The largest satellite of Saturn, with a diameter estimated to be about 3440 miles (3550 kilometers). { 'tīt·ən }

titanate [INORG CHEM] A salt of titanic acid; titanates of the M_2TiO_3 type are called metatitanates, those of the M_4TiO_4 type are called orthotitanates; an example is sodium titanate, $(Na_2O)_2Ti_2O_5$. { 'tīt·ən¸āt }

titanaugite [MINERAL] $Ca(Mg,Fe,Ti)(Si,Al)_2O_6$ A variety of augite rich in titanium and occurring in basaltic rocks. { ¦tīt·ən'ȯ¸gīt }

titanellow See titanium trioxide. { ¦tīt·ən'el·ō }

titania See titanium dioxide. { tī'tā·nē·ə }

Titania [ASTRON] A satellite of Uranus, with a diameter estimated to be 990 miles (1600 kilometers). { tī'tā·nē·ə }

titanic acid [INORG CHEM] H_2TiO_3 A white, water-insoluble powder; used as a dyeing mordant. Also known as metatitanic acid; titanic hydroxide. { tī'tan·ik 'as·əd }

titanic anhydride See titanium dioxide. { tī'tan·ik an'hī¸drīd }

titanic chloride See titanium tetrachloride. { tī'tan·ik 'klȯr¸īd }

titanic hydroxide See titanic acid. { tī'tan·ik hī'dräk¸sīd }

titanic iron ore See ilmenite. { tī'tan·ik 'ī·ərn ¸ȯr }

titanic sulfate See titanium sulfate. { tī'tan·ik 'səl¸fāt }

titanite See sphene. { 'tīt·ən¸īt }

titanium [CHEM] A metallic transition element, symbol Ti, atomic number 22, atomic weight 47.90; ninth most abundant element in the earth's crust; insoluble in water, melts at 1660°C, boils above 3000°C. [MET] A lustrous, silvery-gray, strong, light metal that is hard and brittle when cold, malleable when heated, and ductile when pure; used in the pure state or in alloys for aircraft and chemical-plate metals, for surgical instruments, and in cermets, and metal-ceramic brazing. { tī'tā·nē·əm }

titanium boride [INORG CHEM] TiB_2 A hard solid that resists oxidation at elevated temperatures and melts at 2980°C; used as a refractory and in alloys, high-temperature electrical conductors, and cermets. { tī'tā·nē·əm 'bȯr¸īd }

titanium carbide [INORG CHEM] TiC Very hard gray crystals insoluble in water, soluble in nitric acid and aqua regia, melts at about 3140°C; used in cermets, arc-melting electrodes, and tungsten-carbide tools. { tī'tā·nē·əm 'kär¸bīd }

titanium chloride See titanium dichloride. { tī'tā·nē·əm 'klȯr¸īd }

titanium dichloride [INORG CHEM] $TiCl_2$ A flammable, alcohol-soluble, black powder that decomposes in water, and in vacuum at 475°C, and burns in air. Also known as titanium chloride. { tī'tā·nē·əm dī'klȯr¸īd }

titanium dioxide [INORG CHEM] TiO_2 A white, water-insoluble powder that melts at 1560°C, and which is produced commercially from the titanium dioxide minerals ilmenite and rutile; used in paints and cosmetics. Also known as titania; titanic anhydride; titanium oxide; titanium white. { tī'tā·nē·əm dī'äk¸sīd }

titanium hydride [INORG CHEM] TiH_2 A black metallic powder whose dust is an explosion hazard and which dissociates above 288°C; used in powder metallurgy, hydrogen production, foamed metals, glass solder, and refractories, and as an electronic gas getter. { tī'tā·nē·əm 'hī¸drīd }

titanium nitride [INORG CHEM] TiN Golden-brown brittle crystals melting at 2927°C; used in refractories, alloys, cermets, and semiconductors. { tī'tā·nē·əm 'nī¸trīd }

titanium oxalate [ORG CHEM] $Ti_2(C_2O_4)_3\cdot10H_2O$ Toxic, yellow prisms soluble in water, insoluble in alcohol; used to make titanic acid and titanium metal. Also known as titanous oxalate. { tī'tā·nē·əm 'äk·sə¸lāt }

titanium oxide See titanium dioxide; titanium trioxide. { tī'tā·nē·əm 'äk¸sīd }

titanium peroxide See titanium trioxide. { tī'tā·nē·əm pə'räk¸sīd }

titanium sesquisulfate See titanous sulfate. { tī'tā·nē·əm ¦ses·kwə'səl¸fāt }

titanium sulfate [INORG CHEM] $Ti(SO_4)_2\cdot9H_2O$ Caked solid, soluble in water, toxic, highly acidic; used as a dye stripper, reducing agent, laundry chemical, and in treatment of chrome yellow colors. Also known as titanic sulfate; titanyl sulfate. { tī'tā·nē·əm 'səl¸fāt }

titanium tetrachloride [INORG CHEM] $TiCl_4$ A colorless, toxic liquid soluble in water, fumes when exposed to moist air, boils at 136°C; used to make titanium and titanium salts, as a dye mordant and polymerization catalyst, and in smoke screens and pigments. Also known as titanic chloride. { tī'tā·nē·əm ¦te·trə'klȯr¸īd }

titanium trichloride [INORG CHEM] $TiCl_3$ Toxic, dark-violet, deliquescent crystals soluble in alcohol and some amines, decomposes in water with heat evolution, decomposes above 440°C; used as a reducing agent, chemical intermediate, polymerization catalyst, and laundry stripping agent. Also known as titanous chloride. { tī'tā·nē·əm trī'klȯr¸īd }

titanium trioxide [INORG CHEM] TiO_3 Yellow titanium oxide used to make ivory shades in ceramics. Also known as titanellow; titanium oxide; titanium peroxide. { tī'tā·nē·əm trī'äk¸sīd }

titanium white See titanium dioxide. { tī'tā·nē·əm 'wīt }

Titanoideidae [PALEON] A family of extinct land mammals in the order Pantodonta. { ¸tīt·ən·ȯi'dē·ə¸dē }

titanothere [PALEON] Any member of the family Brontotheriidae. { tī'tan·ə¸thir }

titanous chloride See titanium trichloride. { tī'tan·əs 'klȯr¸īd }

titanous oxalate See titanium oxalate. { tī'tan·əs 'äk·sə¸lāt }

titanous sulfate [INORG CHEM] $Ti_2(SO_4)_3$ Green crystals soluble in dilute hydrochloric and sulfuric acids, insoluble in water and alcohol; used as a textile reducing agent. Also known as titanium sesquisulfate. { tī'tan·əs 'səl,fāt }

titanyl sulfate *See* titanium sulfate. { 'tīt·ən·əl 'səl,fāt }

Titchmarsh's theorem [MATH] The proposition that, if $f(x)$ and $g(x)$ are continuous functions on the positive real numbers and are not identically equal to 0, then their convolution is not identically 0. { 'tich,märsh·əz ,thir·əm }

titer [CHEM] **1.** The concentration in a solution of a dissolved substance as shown by titration. **2.** The least amount or volume needed to give a desired result in titration. **3.** The solidification point of hydrolyzed fatty acids. [TEXT] **1.** The weight per unit length of yarn. **2.** The number of filaments in reeled silk thread. { 'tī·tər }

Tithonian [GEOL] Southern European equivalent of the Portlandian stage (uppermost Jurassic) of geologic time. { ti'thō·nē·ən }

Titius-Bode law *See* Bode's law. { 'tēt·sē·əs 'bōd·ə ,lȯ }

title bar [COMPUT SCI] An area at the top of a window that contains the name of the file or application in the window. { 'tīd·əl ,bär }

titrand [ANALY CHEM] The substance that is analyzed in a titration procedure. { 'tī,trand }

titrant [ANALY CHEM] A solution of known concentration and composition used for analytical titrations. Also known as standard solution. { 'tī'trənt }

titration [ANALY CHEM] A method of analyzing the composition of a solution by adding known amounts of a standardized solution until a given reaction (color change, precipitation, or conductivity change) is produced. { ti'trā·shən }

titrimetric analysis *See* volumetric analysis. { ¦tī·trə,me·trik ə'nal·ə·səs }

tivano [METEOROL] A night breeze blowing down the valley at Lake Como in Italy. { ti'vä·nō }

tjaele *See* frozen ground. { 'chā·lē }

tjenting needle [GRAPHICS] The traditional tool for applying wax in the batik process; it is used for fine-line work and for outlining areas to be filled in by brush; the melted wax flows out from the fine needle spout. { 'chän·tiŋ ,nēd·əl }

T junction [ELECTR] A network of waveguides with three waveguide terminals arranged in the form of a letter T; in a rectangular waveguide a symmetrical T junction is arranged by having either all three broadsides in one plane or two broadsides in one plane and the third in a perpendicular plane. { 'tē ,jəŋk·shən }

Tl *See* thallium.

T²L *See* transistor-transistor logic.

T1 line [COMMUN] High-speed digital connection that transmits data at 1.5 million bits per second through the telephone-switching network. { ,tē'wən ,līn }

T3 line [COMMUN] High-speed digital connection that transmits data at 45 million bits per second through the telephone-switching network. { ,tē'thrē ,līn }

TLP *See* transient lunar phenomena.

TLV *See* threshold limit value.

T lymphocyte *See* T cell. { 'tē 'lim·fə,sīt }

Tm *See* thulium.

TMA *See* trimethylamine.

TME *See* metric-technical unit of mass.

T method [BOT] A budding method in which a T-shaped cut is made through the bark at the internode of the stock, the bark of the scion is separated from the xylem along the cambium and removed, and the scion is forced into the incision on the stock. { 'tē ,meth·əd }

TMJ *See* temporamandibular joint disease.

TM mode *See* transverse magnetic mode. { ¦tē'em ,mōd }

TMV *See* tobacco mosaic virus.

TM wave *See* transverse magnetic wave. { ¦tē'em ,wāv }

Tn *See* thoron.

T network [ELEC] A network composed of three branches, with one end of each branch connected to a common junction point, and with the three remaining ends connected to an input terminal, an output terminal, and a common input and output terminal, respectively. { 'tē ,net,wərk }

TNF *See* 2,4,7-trinitrofluorenone.

tNOX *See* tumor nicotinamide adenine dinucleotide oxidase. { 'tē,näks }

TNT *See* 2,4,6-trinitrotoluene.

TNT-ammonium nitrate explosive [MATER] An explosive containing ammonium nitrate sensitized with trinitrotoluene; a

proportion of aluminum powder or calcium silicide may be added to increase power and sensitiveness. { ¦te,en¦tē ə'mō·nē·əm 'nī,trāt ig'splō·siv }

TNT equivalent [NUCLEO] A measure of the energy released in the detonation of a nuclear weapon, expressed in terms of the weight of TNT that would release the same amount of energy when exploded; usually expressed in kilotons or megatons of TNT; based on the release of 10^9 calories (approximately 4.18×10^9 joules) of energy by 1 ton of TNT. { ¦te,en¦tē i'kwiv·ə·lənt }

toad [VERT ZOO] Any of several species of the amphibian order Anura, especially in the family Bufonidae; glandular structures in the skin secrete acrid, irritating substances of varying toxicity. { tōd }

toadstool [MYCOL] Any of various fleshy, poisonous or inedible fungi with a large umbrella-shaped fruiting body. { 'tōd,stül }

to-and-fro ropeway *See* jig back. { ¦tü ən ¦frō 'rōp,wā }

Toarcian [GEOL] A European stage of geologic time; Lower Jurassic (above Pliensbachian, below Bajocian). { tō'är·shən }

tobacco [BOT] **1.** Any plant of the genus *Nicotinia* cultivated for its leaves, which contain 1–3% of the alkaloid nicotine. **2.** The dried leaves of the plant. { tə'bak·ō }

tobacco budworm [INV ZOO] The larva of a noctuid moth, *Heliothis virescens*, that damages the buds and young leaves of tobacco. { tə,bak·ō 'bəd,wərm }

tobacco jack *See* wolframite. { tə'bak·ō ,jak }

tobacco mosaic [PL PATH] Any of a complex of virus diseases of tobacco and other solanaceous plants in which the leaves are mottled with light- and dark-green patches, sometimes interspersed with yellow. { tə'bak·ō mō,zā·ik }

tobacco mosaic virus [VIROL] The type species of the genus *Tobamovirus*; it infects tobacco, tomato, and other solanaceous plants, causing defoliation and/or mosaic symptoms on leaves, stems, and fruit. Abbreviated TMV. { tə¦bak·ō mō¦zā·ik ,vī·rəs }

tobacco mosaic virus group *See* Tobamovirus. { tə¦bak·ō mō¦zā·ik ¦vī·rəs ,grüp }

tobacco rattle virus [VIROL] The type species of the genus *Tobravirus*; it infects a wide range of plants, usually via a nematode vector. Abbreviated TRV. { tə¦bak·ō 'rad·əl ,vī·rəs }

tobacco rattle virus group *See* Tobravirus. { tə¦bak·ō 'rad·əl ¦vī·rəs ,grüp }

tobacco ring spot virus [VIROL] The type species of the genus *Nepovirus*; it has a wide host range and is transmitted via seeds and nematodes. Abbreviated TRSV. { tə¦bak·ō 'riŋ ,spät ,vī·rəs }

tobacco ring spot virus group *See* Nepovirus. { tə¦bak·ō 'riŋ ,spät ¦vī·rəs ,grüp }

tobacco streak virus [VIROL] The type species of the genus *Ilarvirus*. { tə¦bak·ō 'stāk ,vī·rəs }

tobacco streak virus group *See* Ilarvirus. { tə¦bak·ō 'stāk ¦vī·rəs ,grüp }

Tobamovirus [VIROL] A genus of plant viruses characterized by particles with a rigid helical rod containing linear single-stranded ribonucleic acid. Tobacco mosaic virus is the type species. Also known as Tobacco mosaic virus group. { tə¦bam·ə,vī·rəs }

Tobravirus [VIROL] A genus of plant viruses characterized by two types of rigid helical rods containing linear single-stranded ribonucleic acid. The type species is tobacco rattle virus. Also known as tobacco rattle virus group. { 'tō·brə,vī·rəs }

tocopherol [ORG CHEM] Any of several substances having vitamin E activity that occur naturally in certain oils; α-tocopherol possesses the highest biological activity. { tə'käf·ə,röl }

tocophobia [PSYCH] Abnormal fear of children. { ,täk·ə'fō·bē·ə }

Todidae [VERT ZOO] The todies, a family of birds in the order Coraciiformes found in the West Indies. { 'tō·də,dē }

todorokite [GEOL] A hydrated manganese oxide mineral containing calcium, barium, potassium, sodium, and sometimes magnesium; a major constituent of manganese nodules, which occur in large quantities (>10^{12} tons) on the ocean floor. { tə'dȯr·ə,kīt }

toe [ANAT] One of the digits on the foot of humans and other vertebrates. [CIV ENG] The part of a base of a dam or

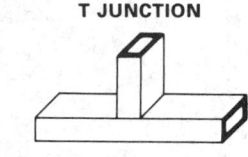

T JUNCTION

Drawing showing one type of waveguide T junction.

TOAD

A toad of the genus *Bufo*, covered with warty structures that secrete poisons for purposes of defense.

retaining wall on the side opposite to the retained material. [GEOL] The leading edge of a thrust nappe. [MET] The junction between the face of a weld and the base metal. [MIN ENG] **1.** The burden of material between the bottom of the borehole and the free face. **2.** The bottom of the borehole. **3.** A spurn, or small pillar of coal. **4.** The base of a bank in an open-pit mine. { tō }

toeboard [BUILD] A board placed around a platform or on a sloping roof to prevent personnel or materials from falling off. [ENG] A support or reinforcement that forms the lowest vertical face of a cabinet or similar installation, at toe level, and is frequently recessed. { 'tō₁bȯrd }

toe crack [MET] A crack in the base metal at the toe of a weld. { 'tō ₁krak }

toe cut [ENG] In underground blasting, the cut obtained by the use of toe holes. { 'tō ₁kət }

toe hole [ENG] A blasting hole, usually drilled horizontally or at a slight inclination into the base of a bank, bench, or slope of a quarry or open-pit mine. { 'tō ₁hōl }

toe-in [MECH ENG] The degree (usually expressed in fractions of an inch) to which the forward part of the front wheels of an automobile are closer together than the rear part, measured at hub height with the wheels in the normal "straight ahead" position of the steering gear. { 'tō ₁in }

toenailing [ENG] The technique of driving a nail at an angle to join two pieces of lumber. { 'tō¦nāl·iŋ }

toe-out [MECH ENG] The outward inclination of the wheels of an automobile at the front on turns due to setting the steering arms at an angle. { 'tō ₁au̇t }

toeplate See kickplate. { 'tō₁plāt }

Toepler-Holtz machine [ELEC] An early type of machine for continuously producing electrical charges at high voltage by electrostatic induction, superseded by the Wimhurst machine. Also known as Holtz machine. { 'tep·lər 'hōlts mə₁shēn }

toe-to-toe drilling [ENG] The drilling of vertical large-diameter blasting holes in quarries and opencast pits. { ¦tō tə ¦tō 'dril·iŋ }

toe wall [CIV ENG] A low wall constructed at the bottom of an embankment to prevent slippage or spreading of the soil. { 'tō ₁wȯl }

tofan [METEOROL] A violent spring storm common in the mountains of Indonesia. { tō'fän }

to-from indicator [NAV] An indicator that shows whether an aircraft is flying toward or away from an omnirange station. Also known as sense indicator. { tü 'frəm ₁in·də₁kād·ər }

Togaviridae [VIROL] A family of positive-strand ribonucleic acid (RNA)–containing viruses characterized by spherical enveloped particles with an icosahedral nucleocapsid containing linear single-stranded RNA; it contains the genera *Alphavirus* (arbovirus A; prototype Sindbis virus), *Flavivirus* (arbovirus B; prototype yellow fever), Rubivirus (rubella virus), and Pestivirus (mucosal disease virus). { ₁tō·gə'vir·ə₁dī }

toggle [COMPUT SCI] **1.** To switch back and forth between two stable states or modes of operation. **2.** A hardware or software device that carries out this switching action. [ELECTR] To switch over to an alternate state, as in a flip-flop. [MECH ENG] A form of jointed mechanism for the amplification of forces. { 'täg·əl }

toggle bolt [DES ENG] A bolt having a nut with a pair of pivotal wings that close against a spring; wings open after emergence through a hole or passage in a thin or hollow wall to fasten the unit securely. { 'täg·əl ₁bōlt }

toggle condition [ELECTR] Condition of a flip-flop circuit in which the internal state of the flip-flop changes from 0 to 1 or from 1 to 0. { 'täg·əl kən₁dish·ən }

toggle press [MECH ENG] A mechanical press in which a toggle mechanism actuates the slide. { 'täg·əl ₁pres }

toggle switch [ELEC] A small switch that is operated by manipulation of a projecting lever that is combined with a spring to provide a snap action for opening or closing a circuit quickly. [ELECTR] An electronically operated circuit that holds either of two states until changed. { 'täg·əl ₁swich }

toise [GEOD] A unit of length equal to about 6.4 feet (1.95 meters); used in early geodetic surveys. { 'tȯiz }

tokamak [PL PHYS] A device for confining a plasma within a toroidal chamber, which produces plasma temperatures, densities, and confinement times greater than that of any other such device; confinement is effected by a very strong externally applied toroidal field, plus a weaker poloidal field produced by a toroidally directed plasma current, and this current causes ohmic heating of the plasma. { 'täk·ə₁mak }

token [COMMUN] A unique grouping of bits that is transmitted as a unit in a communications network and used as a signal to notify stations in the network when they have control and are free to send information or take other specified actions. [COMPUT SCI] **1.** A distinguishable unit in a sequence of characters. **2.** A single byte that is used to represent a keyword in a programming language in order to conserve storage space. **3.** A physical object, such as a badge or identity card, issued to authorized users of a computing system, building, or area. { 'tō·kən }

tokenization [COMPUT SCI] The conversion of keywords of a programming language to tokens in order to conserve storage space. { ₁tō·kən·ə'zā·shən }

token-passing protocol [COMMUN] The assignment of data communications channels to units which communicate according to a fixed priority sequence. { 'tō·kən ¦pas·iŋ 'prōd·ə₁kȯl }

token-sharing network [COMMUN] A communications network in which all the stations are linked to a common bus and control is determined by a group of bits (token) that is passed along the bus from station to station. { 'tō·kən ¦sher·iŋ 'net₁wərk }

tolazoline hydrochloride [ORG CHEM] $C_{10}H_{12}N_2 \cdot HCl$ Water-soluble white crystals, melting at 173°C; used as a sympatholytic and vasodilator. Also known as priscol. { täl'az·ə₁lēn ¦hī·drə'klȯr₁īd }

tolbutamide [PHARM] $C_{12}H_{18}N_2O_3S$ A hypoglycemic drug effective when administered orally. { täl'byüd·ə₁mīd }

toleragen [IMMUNOL] A substance which, in appropriate dosages, produces a state of specific immunological tolerance in humans or animals. { 'täl·ə·rə·jən }

tolerance [DES ENG] The permissible variations in the dimensions of machine parts. [ENG] A permissible deviation from a specified value, expressed in actual values or more often as a percentage of the nominal value. [PHARM] **1.** The ability of enduring or being less responsive to the influence of a drug or poison, particularly when acquired by continued use of the substance. **2.** The allowable deviation from a standard, as the range of variation permitted for the content of a drug in one of its dosage forms. { 'täl·ə·rəns }

tolerance chart [DES ENG] A chart indicating graphically the sequence in which dimensions must be produced on a part so that the finished product will meet the prescribed tolerance limits. { 'täl·ə·rəns ₁chärt }

tolerance dose See permissible dose. { 'täl·ə·rəns ₁dōs }

tolerance interval [ANALY CHEM] That range of values within which it has been calculated that a specified percentage of individual values of measurements will lie with a stated confidence level. { 'täl·ə·rəns ₁in·tər·vəl }

tolerance limits [DES ENG] The extreme values (upper and lower) that are permitted by the tolerance. { 'täl·ə·rəns ₁lim·əts }

tolerance unit [DES ENG] A unit of length used to express the degree of tolerance allowed in fitting cylinders into cylindrical holes, equal, in micrometers, to $0.45 \, D^{1/3} + 0.001 \, D$, where D is the cylinder diameter in millimeters. { 'täl·ə·rəns ₁yü·nət }

***ortho*-tolidine** [ORG CHEM] $C_6H_3(CH_3)NH_2$ Light-sensitive, combustible white to reddish crystals soluble in alcohol and ether, slightly soluble in water, melts at 130°C; used as an anlytical reagent and a curing agent for urethane resins. { ¦ȯr·thō 'täl·ə₁dēn }

toll [COMMUN] **1.** Charge made for a connection beyond an exchange boundary. **2.** Any part of telephone plant, circuits, or services for which toll charges are made. { tōl }

toll call [COMMUN] Telephone call to points beyond the area within which telephone calls are covered by a flat monthly rate or are charged for on a message unit basis. { 'tōl ₁kȯl }

toll center [COMMUN] A telephone central office where trunks from end offices are joined to the long-distance system, and operators are present; it is a class-4 office. { 'tōl ₁sen·tər }

toll enrichment [NUCLEO] A proposed arrangement whereby privately owned uranium could be enriched in uranium-235 content in government facilities upon payment of a service charge by the owners. { 'tōl in'rich·mənt }

Tollen's aldehyde test [ANALY CHEM] A test that uses an ammoniacal solution of silver oxides to test for aldehydes and ketones. { 'täl·ənz 'al·də₁hīd ₁test }

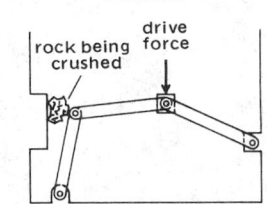

TOGGLE

Toggle mechanism used in a rock crusher; relatively small drive force causes large force to be applied to rock.

toll line [COMMUN] A telephone line or channel that connects different telephone exchanges. { 'tōl ‚līn }

toll office [COMMUN] A telephone central office which serves mainly to terminate and interconnect toll lines and various types of trunks. { 'tōl ‚óf·əs }

toll terminal loss [COMMUN] The part of the overall transmission loss on a toll connection that is attributable to the facilities from the toll center through the tributary office, to and including the subscriber's equipment. { 'tōl 'tər·mən·əl ‚lós }

Tolman and Stewart effect [ELEC] The development of negative charge at the forward end of a metal rod which is suddenly stopped after rapid longitudinal motion. { 'täl·mən ən 'stü·ərt i‚fekt }

tolnaftate [ORG CHEM] $C_{19}H_{17}NOS$ An agricultural fungicide; it is also used medically as an antifungal agent. { tōl'naf‚tāt }

toluene [ORG CHEM] $C_6H_5CH_3$ A colorless, aromatic liquid derived from coal tar or from the catalytic reforming of petroleum naphthas; insoluble in water, soluble in alcohol and ether, boils at 111°C; used as a chemical intermediate, for explosives, and in high-octane gasolines. Also known as methylbenzene; phenylmethane; toluol. { 'täl·yə‚wēn }

para-toluenesulfonic acid [ORG CHEM] $C_6H_4(SO_3H)(CH_3)$ Toxic, colorless, combustible crystals soluble in water, alcohol, and ether; melts at 107°C; used in dyes and as a chemical intermediate and organic catalyst. { ¦par·ə ¦täl·yə‚wēn¦səl¦fän·ik 'as·əd }

toluenethiol See thiocresol. { ¦täl·yə‚wēn'thī‚ól }

toluene 2,4-diisocyanate [ORG CHEM] $CH_3C_6H_3(NCO)_2$ A liquid (at room temperature) with a sharp, pungent odor; miscible with ether, acetone, and benzene; used to make polyurethane foams and other elastomers, and also as a protein cross-linking agent. { 'täl·yə‚wēn ¦tü ¦fór ‚dī¦ī·sō'sī·ə‚nāt }

α-toluic acid See phenylacetic acid. { ¦al·fə tə'lü·ik 'as·əd }

meta-toluic acid [ORG CHEM] $C_6H_4CH_3COOH$ White to yellow, combustible crystals soluble in alcohol and ether, slightly soluble in water, melts at 109°C; used as a chemical intermediate and base for insect repellants. Also known as meta-toluylic acid. { ¦med·ə tə'lü·ik 'as·əd }

ortho-toluic acid [ORG CHEM] $C_6H_4CH_3COOH$ White, combustible crystals soluble in alcohol and chloroform, slightly soluble in water, melts at 104°C; used as a bacteriostat. Also known as ortho-toluylic acid. { ¦ór·thō tə'lü·ik 'as·əd }

para-toluic acid [ORG CHEM] $C_6H_4CH_3COOH$ Transparent, combustible crystals soluble in alcohol and ether, slightly soluble in water, melts at 180°C; used in agricultural chemicals and as an animal feed supplement. Also known as para-toluylic acid. { ¦par·ə tə'lü·ik 'as·əd }

α-toluic aldehyde See phenylacetaldehyde. { ¦al·fə tə'lü·ik 'al·də‚hīd }

meta-toluidine [ORG CHEM] $CH_3C_6H_4NH_2$ A combustible, colorless, toxic liquid soluble in alcohol and ether, slightly soluble in water, boils at 203°C; used for dyes and as a chemical intermediate. { ¦med·ə tə'lü·ə‚dēn }

ortho-toluidine [ORG CHEM] $CH_3C_6H_4NH_2$ A light-green, light-sensitive, combustible, toxic liquid soluble in alcohol and ether, very slightly soluble in water, boils at 200°C; used for dyes and textile printing and as a chemical intermediate. { ¦ór·thō tə'lü·ə‚dēn }

para-toluidine [ORG CHEM] $CH_3C_6H_4NH_2$ Toxic, combustible, white leaflets soluble in alcohol and ether, very slightly soluble in water, boils at 200°C; used as an analytical reagent and in dyes. { ¦par·ə tə'lü·ə‚dēn }

toluol See toluene. { 'täl·yə‚wól }

toluylene See stilbene. { 'täl·yə·wə‚lēn }

toluylene red See neutral red. { 'täl·yə·wə‚lēn 'red }

meta-toluylic acid See meta-toluic acid. { ¦med·ə ¦täl·yə¦wil·ik 'as·əd }

ortho-toluylic acid See ortho-toluic acid. { ¦ór·thō ¦täl·yə¦wil·ik 'as·əd }

para-toluylic acid See para-toluic acid. { ¦par·ə ¦täl·yə¦wil·ik 'as·əd }

tolylmercaptan See thiocresol. { ¦tä‚lil·mər'kap‚tan }

para-tolylsulfonylmethylnitrosamide [ORG CHEM] $C_8H_{10}N_2O_3S$ Yellow crystals with a melting point of 62°C; soluble in ether, petroleum ether, benzene, carbon tetrachloride, and chloroform; a precursor to diazomethane; a useful reagent for the preparation of a wide range of biologically active compounds for gas chromatography analysis. { ¦par·ə ¦tä‚lil¦səl·fə‚nil¦meth·əl‚nī'trō·sə‚mīd }

tomatine [ORG CHEM] $C_{50}H_{83}NO_{21}$ A glycosidal alkaloid obtained from the leaves and stems from the tomato plant; the crude extract is known as tomatin: white, toxic crystals; used as a plant fungicide and as a precipitating agent for cholesterol. { 'täm·ə‚tēn }

tomato [BOT] A plant of the genus *Lycopersicon*, especially *L. esculentum*, in the family Solanaceae cultivated for its fleshy edible fruit, which is red, pink, orange, yellow, white, or green, with fleshy placentas containing many small, oval seeds with short hairs and covered with a gelatinous matrix. { tə'mäd·ō }

tomato bushy stunt virus [VIROL] The type species of the plant-virus genus *Tombusvirus*. Abbreviated TBSV. { tə‚mäd·ō 'bùsh·ē ‚stənt ‚vī·rəs }

tomato bushy stunt virus group See Tombusvirus. { tə‚mäd·ō 'bùsh·ē ‚stənt ‚vī·rəs ‚grüp }

tombolo [GEOL] A sand or gravel bar or spit that connects an island with another island or an island with the mainland. Also known as connecting bar; tie bar; tying bar. { 'täm·bə‚lō }

tombolo cluster See complex tombolo. { 'täm·bə‚lō ‚kləs·tər }

tombolo series See complex tombolo. { 'täm·bə‚lō ‚sir·ēz }

Tombusviridae [VIROL] A family of plant viruses characterized by nonenveloped icosahedral particles containing linear, positive-sense ribonucleic acid; genera include *Tombusvirus*, *Carmovirus*, *Necrovirus*, *Dianthovirus*, and *Machlomovirus*. { ‚täm·bəs'vir·ə‚dī }

Tombusvirus [VIROL] A genus of the family Tombusviridae; tomato bushy stunt virus is the type species. Also known as Tomato bushy stunt virus group. { 'tam·bəs‚vī·rəs }

tomentose [BOT] Covered with densely matted hairs. { tə'men‚tōs }

tomentum [ANAT] The deep layer of the pia mater composed principally of minute blood vessels. [BIOL] Pubescence consisting of densely matted wooly hairs. { tə'men·təm }

tomium [VERT ZOO] The cutting edge of a bird's beak. { 'tō·mē·əm }

tomography See sectional radiography. { tə'mäg·rə·fē }

Tomopteridae [INV ZOO] The glass worms, a family of pelagic polychaete annelids belonging to the group Errantia. { tə‚mäp'ter·ə‚dē }

ton [IND ENG] A unit of volume of sea freight, equal to 40 cubic feet or approximately 1.1327 cubic meters. Also known as freight ton; measurement ton; shipping ton. [MECH] **1.** A unit of weight in common use in the United States, equal to 2000 pounds or 907.18474 kilogram-force. Also known as just ton; net ton; short ton. **2.** A unit of mass in common use in the United Kingdom equal to 2240 pounds, or to 1016.0469088 kilogram-force. Also known as gross ton; long ton. **3.** A unit of weight in troy measure, equal to 2000 troy pounds, or to 746.4834432 kilogram-force. **4.** See tonne. [MECH ENG] A unit of refrigerating capacity, that is, of rate of heat flow, equal to the rate of extraction of latent heat when one short ton of ice of specific latent heat 144 international table British thermal units per pound is produced from water at the same temperature in 24 hours; equal to 200 British thermal units per minute, or to approximately 3516.85 watts. Also known as standard ton. [NAV ARCH] A unit of internal capacity of ships, equal to 100 cubic feet or approximately 2.8317 cubic meters. Also known as register ton. [NUCLEO] The energy released by one metric ton of chemical high explosives calculated at the rate of 1000 calories per gram; equal to 4.18×10^9 joules; used principally in expressing the energy released by a nuclear bomb. { tən }

tonalite See quartz diorite. { 'tōn·əl‚īt }

tondal [MECH] A unit of force equal to the force which will impart an acceleration of 1 foot per second to a mass of 1 long ton; equal to approximately 309.6911 newtons. { 'tónd·əl }

tone [ACOUS] **1.** A sound oscillation capable of exciting an auditory sensation having pitch. **2.** An auditory sensation having pitch. [GRAPHICS] Each distinguishable shade variation from black to white on photographs. { 'tōn }

tone-and-voice pager [COMMUN] A receiver in a radio paging system equipped with a speaker that broadcasts a 7- to 20-second message spoken by a caller over a telephone. { ¦tōn ən ¦vóis 'pāj·ər }

tone code ranging [NAV] In navigation by satellite, a method of transmitting a high-precision time signal using the reduced peak power available to a satellite; an audio frequency is used to phase an audio oscillator in the user equipment, and a train of short pulses is generated in synchronism with the zero crossings of the audio tone to provide high-resolution timing. { 'tōn ¦kōd ¦rānj·iŋ }

tone control [ELECTR] A control used in an audio-frequency amplifier to change the frequency response so as to secure the most pleasing proportion of bass to treble; individual bass and treble controls are provided in some amplifiers. { 'tōn kən¸trōl }

tone dialing See push-button dialing. { 'tōn ¸dīl·iŋ }

tone generator [ELECTR] A signal generator used to generate an audio-frequency signal suitable for signaling purposes or for testing audio-frequency equipment. { 'tōn ¸jen·ə¸rād·ər }

tone localizer See equisignal localizer. { 'tōn ¸lō·kə¸līz·ər }

tone-modulated waves [COMMUN] Waves obtained from continuous waves by amplitude-modulating them at audio frequency in a substantially periodic manner. { 'tōn ¦mäj·ə¸lād·əd ¸wāvz }

tone modulation [COMMUN] Type of code-signal transmission obtained by causing the radio-frequency carrier amplitude to vary at a fixed audio frequency. { 'tōn ¸mäj·ə¸lā·shən }

tone-only pager [COMMUN] A receiver in a radio paging system that alerts the user to call a specific telephone number. { 'tōn ¦on·lē 'pāj·ər }

tone-operated net-loss adjuster [COMMUN] System for stabilizing the net loss of a telephone circuit by a tone transmitted between conversations. { 'tōn ¦äp·ə¸rād·əd 'net ¦lȯs ə¸jəs·tər }

toner [GRAPHICS] The fine, black, resinous powder used in electrostatic imaging processes to make an electrostatic image readable; the toner is either deposited directly on coated paper or transferred from a charged surface to ordinary paper, then fused to the paper by heating. { 'tō·nər }

tone reversal [COMMUN] Distortion of the recorder copy in facsimile which causes the various shades of black and white not to be in the proper order. { 'tōn ri¸vər·səl }

tongara [METEOROL] A hazy, southeast wind in the Macassar Strait. { 'täŋ'gär·ə }

tong hold [MET] The end of a forging billet that is gripped by the operator's tongs; it is removed at the end of the forging operation. { 'täŋ ¸hōld }

Tongrian [GEOL] A European stage of geologic time; lower Oligocene (above Ludian of Eocene, below Rupelian). Also known as Lattorfian. { 'täŋ·grē·ən }

tongs [DES ENG] Any of various devices for holding, handling, or lifting materials and consisting of two legs joined eccentrically by a pivot or spring. { taŋz }

tongue [ANAT] A muscular organ located on the floor of the mouth in humans and most vertebrates which may serve various functions, such as taking and swallowing food or tasting or as a tactile organ or sometimes a prehensile organ. [GEOL] **1.** A minor rock-stratigraphic unit of limited geographic extent; it disappears laterally in one direction. **2.** A lava flow branching from a larger flow. [OCEANOGR] **1.** A protrusion of water into a region of different temperature, or salinity, or dissolved oxygen concentrating. **2.** A protrusion of one water mass into a region occupied by a different water mass. { təŋ }

tongue and groove [DES ENG] A joint in which a projecting rib on the edge of one board fits into a groove in the edge of another board. { 'təŋ ən 'grüv }

tongue worm See acorn worm. { 'təŋ ¸wərm }

tonic convulsion See tonic postural epilepsy; tonic spasm. { 'tän·ik kən'vəl·shən }

tonic labyrinthine reflexes [PHYSIO] Rotation or deviation of the head causes extension of the limbs on the same side as the chin, and flexion of the opposite extremities: dorsiflexion of the head produces increased extensor tonus of the upper extremities and relaxation of the lower limbs, while ventroflexion of the head produces the reverse; seen in the young infant and patients with a lesion at the midbrain level or above. { 'tän·ik ¸lab·ə'rin¸thēn 'rē¸flek·səz }

tonic neck reflexes [PHYSIO] Reflexes in which rotation or deviation of the head causes extension of the limbs on the same side as the chin, and flexion of the opposite extremities; dorsiflexion of the head produces increased extension tonus of the upper extremities and relaxation of the lower limbs, and ventroflexion of the head, the reverse; seen normally in incomplete forms in the very young infant, and thereafter in patients with a lesion at the midbrain level or above. { 'tän·ik 'nek 'rē¸flek·səz }

tonic postural epilepsy [MED] A form of epilepsy in which seizures are characterized by a rigid posture with the arms and legs extended, hands pronated, and feet held in plantar flexion. Also known as tonic convulsion. { 'tän·ik 'päs·chə·rəl 'ep·ə¸lep·sē }

tonic spasm [MED] A spasm which persists for some time without relaxation. Also known as tonic convulsion. { 'tän·ik 'spaz·əm }

ton-kilometer [MIN ENG] A unit of measurement equal to the weight in tons of material transported in a mine multiplied by the number of kilometers driven. { 'tən 'kil·ə¸mēd·ər }

ton-mile [CIV ENG] In railroading, a standard measure of traffic, based on the rate of carriage per mile of each passenger or ton of freight. { 'tən 'mīl }

tonnage [NAV ARCH] A measure of the size of a ship; it is usually taken to mean gross tonnage or net tonnage, but may also refer to deadweight or displacement tonnage. { 'tən·ij }

tonnage deck [NAV ARCH] In vessels having three or more decks to the hull, the tonnage deck is the second deck from the keel; in all other cases, it is the upper deck of the hull. { 'tən·ij ¸dek }

tonne [MECH] A unit of mass in the metric system, equal to 1000 kilograms or to approximately 2204.62 pound mass. Also known as metric ton; millier; ton; tonneau. { tən }

tonneau See tonne. { tə'nō }

tonofibril [CELL MOL] Any of the fibrils converging on desmosomes in epithelial cells. { ¦tä·nō'fī·brəl }

tonometer [MED] An electronic instrument that measures hydrostatic pressure within the eye: when placed in position, a tiny movable plate is pressed against the cornea, flattening a circular section of the cornea (no eyeball anesthesia is required); a current is then sent through a small electromagnet, of such value that it will just pull the plate away from the eye; the value of the current is proportional to eye pressure; a measurement can be made in about 1 second; used in diagnosis of glaucoma. Also known as electronic tonometer. { tō'näm·əd·ər }

tonoplast [BOT] The membrane surrounding a plant-cell vacuole. { 'tän·ə¸plast }

tonsil [ANAT] **1.** Localized aggregation of diffuse and nodular lymphoid tissue found in the throat where the nasal and oral cavities open into the pharynx. **2.** See palatine tonsil. { 'tän·səl }

tonsillectomy [MED] Surgical removal of the palatine tonsil. { ¸tan·sə'lek·tə·mē }

tonsillitis [MED] Inflammation of the tonsils. { ¸tan·sə'līd·əs }

tonstein [GEOL] Kaolinitic bands in certain coalfields which have characteristic fossil fauna from short-lived but widespread marine invasions. { 'tän¸shtīn }

tonus [PHYSIO] The degree of muscular contraction when not undergoing shortening. Also known as muscle tone. { 'tō·nəs }

tool [ENG] Any device, instrument, or machine for the performance of an operation, for example, a hammer, saw, lathe, twist drill, drill press, grinder, planer, or screwdriver. [IND ENG] To equip a factory or industry for production by designing, making, and integrating machines, machine tools, and special dies, jigs, and instruments, so as to achieve manufacture and assembly of products on a volume basis at minimum cost. { tül }

toolbar [COMPUT SCI] A row or column of on-screen push buttons containing icons that represent frequently accessed commands. { 'tül¸bär }

tool bit [ENG] A piece of high-strength metal, usually steel, ground to make single-point cutting tools for metal-cutting operations. { 'tül ¸bit }

toolbox [ENG] A box to hold tools. { 'tül¸bäks }

tool-center point [CONT SYS] The location on the end effector or tool of a robot manipulator whose position and orientation define the coordinates of the controlled object. { 'tül sen·tər ¸póint }

tool changer [MECH ENG] In program-controlled machines and robotics, a mechanism that allows the use of multiple tools. { 'tül ¸chānj·ər }

tool-check system [IND ENG] A system for temporary issue

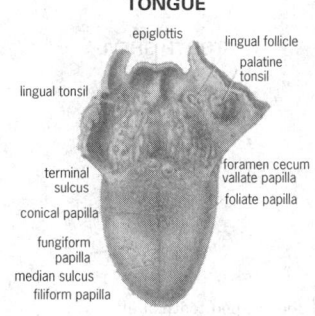

TONGUE

epiglottis
lingual follicle
palatine tonsil
lingual tonsil
terminal sulcus
foramen cecum
vallate papilla
conical papilla
foliate papilla
fungiform papilla
median sulcus
filiform papilla

The human tongue, dorsal view. (*From M. W. Woerdeman, Altas of Human Anatomy, vol. 2, McGraw-Hill, 1950*)

of tools in which the employee is issued a number of small metal checks stamped with the same number; a check is surrendered for each tool obtained from the crib. { 'tül ,chek ,sis·təm }

tool design [DES ENG] The division of mechanical design concerned with the design of tools. { 'tül di,zīn }

tool-dresser [MECH ENG] A tool-stone-grade diamond inset in a metal shank and used to trim or form the face of a grinding wheel. { 'tül ,dres·ər }

tool extractor [ENG] An implement for grasping and withdrawing drilling tools when broken, detached, or lost in a borehole. { 'tül ik,strak·tər }

tool-function controller [CONT SYS] A unit that selects and controls tools for machining operations; it may be internal or external to the main controller. { 'tül ¦fəŋk·shən kən'trōl·ər }

toolhead [MECH ENG] The adjustable tool-carrying part of a machine tool. { 'tül,hed }

tooling [MECH ENG] Tools or end effectors with which a robot performs the actual work on a workpiece. { 'tül·iŋ }

tool joint [ENG] A coupling element for a drill pipe; designed to support the weight of the drill stem and the strain of frequent use, and to provide a leakproof seal. { 'tül ,jóint }

tool-length compensation [CONT SYS] Programming of machining operations so that all tools are positioned correctly in advance for any tasks to be carried out. { 'tül ¦leŋkth ,käm·pən'sā·shən }

toolmaker's vise *See* universal vise. { 'tül,māk·ərz ,vīs }

tool mark [GEOL] Any of the wide variety of current marks, such as groove marks, prod marks, and skip marks, produced by the continuous contact or intermittent impact of solid, current-borne objects against a muddy bottom. { 'tül ,märk }

tool nipper [MIN ENG] A person whose duty it is to carry powder, drills, and tools to the various levels of the mine and to bring dull tools and drills to the surface. { 'tül ,nip·ər }

tool offset [MECH ENG] The adjustment of tool positions in machines to compensate for their wear, finishing, or displacement from an axis. { 'tül ,óf,set }

tool post [MECH ENG] A device to clamp and position a tool holder on a machine tool. { 'tül ,pōst }

tool steel [MET] Any of various steels capable of being hardened sufficiently so as to be a suitable material for making cutting tools. { 'tül ¦stēl }

tooth [ANAT] One of the hard bony structures supported by the jaws in mammals and by other bones of the mouth and pharynx in lower vertebrates serving principally for prehension and mastication. [DES ENG] **1.** One of the regular projections on the edge or face of a gear wheel. **2.** An angular projection on a tool or other implement, such as a rake, saw, or comb. [GRAPHICS] **1.** The coarse or abrasive quality of a paper or a painting ground that assists in the application of charcoal, pastels, or paint. **2.** A paper texture that holds ink more readily. [INV ZOO] Any of various sharp, horny, chitinous, or calcareous processes on or about any part of an invertebrate that functions like or resembles vertebrate jaws. { 'tüth }

toothache [MED] Pain in or about a tooth. Also known as odontalgia. { 'tü,thāk }

tooth decay [MED] Caries of the teeth. { 'tüth di,kā }

tooth point [DES ENG] The chamfered cutting edge of the blade of a face mill. { 'tüth ,póint }

tooth shell [INV ZOO] A mollusk of the class Scaphopoda characterized by the elongate, tube-shaped, or cylindrical shell which is open at both ends and slightly curved. { 'tüth ,shel }

top [GEOL] *See* overburden. [MECH] A rigid body, one point of which is held fixed in an inertial reference frame, and which usually has an axis of symmetry passing through this point; its motion is usually studied when it is spinning rapidly about the axis of symmetry. [PART PHYS] The new quantum number associated with the top quark. Also known as truth. [QUANT MECH] *See* rotator. { täp }

top and bottom process [MET] A process in which sodium sulfide is added to molten copper-nickel sulfide to form a two-layer melt, with the bulk of the nickel in the bottom layer. { ¦täp ən ¦bäd·əm ¦prä·səs }

topaz [MINERAL] $Al_2SiO_4(F,OH)$ A red, yellow, green, blue, or brown neosilicate mineral that crystallizes in the orthorhombic system and commonly occurs in prismatic crystals with pyramidal terminations; hardness is 8 on Mohs scale, and specific gravity is 3.4–3.6; used as a gemstone. { 'tō,paz }

topaz quartz *See* citrine. { 'tō,paz 'kwórtz }

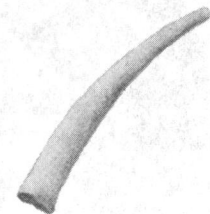

TOOTH SHELL

Tube-shaped tooth shell.

TOPOGON LENS

Components of a Topogon lens.

top-benching [MIN ENG] The method by which the bench is removed from above, as with a dragline. { ¦täp ,bench·iŋ }

top-blown rotary converter [MET] A rotary converter used for making nickel and steel; oxygen and other gases are fed to the furnace by a lance at the elevated end of the converter to permit formation of a metal product without oxidation. { ¦täp ¦blōn 'rōd·ə·rē kən'vərd·ər }

top cager [MIN ENG] A person at the top of a mine shaft who superintends the lowering and raising of the cage, and, at most mines, the removing of loaded cars from the placing of empty cars in the cage. { ¦täp ,kā·jər }

top cut [MIN ENG] A machine cut made in the coal at or near the top of the working face in a mine. { ¦täp ,kət }

top dead center [MECH ENG] The dead-center position of an engine piston and its crankshaft arm when at the top or outer end of its stroke. { ¦täp 'ded 'sen·tər }

top-down analysis [COMPUT SCI] A predictive method of syntactic analysis which, starting from the root symbol, attempts to predict the means by which a string was generated. { ¦täp ¦daún ə'nal·ə·səs }

top-down design [IND ENG] A design methodology that proceeds from the highest level to the lowest and from the general to the particular, and that provides a formal mechanism for breaking complex process designs into functional descriptions, reviewing progress, and allowing modifications. { 'täp ¦daún di'zīn }

top dyeing [TEXT] A method for dyeing combed wool yarn before spinning by placing it in large vats and circulating dye liquor through the yarn at increased temperatures. { ¦täp ,dī·iŋ }

top grafting [BOT] Grafting a scion of one variety of tree onto the main branch of another. { ¦täp ,graft·iŋ }

top hamper *See* rigging. { ¦täp ,ham·pər }

top hooker *See* lander. { ¦täp ,húk·ər }

tophus [MED] A localized swelling principally in cartilage and connective tissues in or adjacent to the small joints of the hands and feet; occurs specifically in gout. { 'tō·fəs }

top lander *See* lander. { ¦täp 'lan·dər }

top-loaded vertical antenna [ELECTROMAG] Vertical antenna constructed so that, because of its greater size at the top, there results modified current distribution, giving a more desirable radiation pattern in the vertical plane. { ¦täp ¦lōd·əd 'vərd·ə·kəl an'ten·ə }

topmark [NAV] A characteristic shape secured at the top of a buoy or beacon to aid in its identification. { 'täp,märk }

topochemical control [CHEM] In a chemical reaction, product formation that is determined by the orientation of molecules in the crystal. { ¦täp·ə¦kem·ə·kəl kən'trōl }

topocline [ECOL] A graded series of characters exhibited by a species or other closely related organisms along a geographical axis. { 'täp·ə,klīn }

topogenesis *See* morphogenesis. { ¦täp·ə'jen·ə·səs }

Topogon lens [OPTICS] A periscopic lens with supplementary thick menisci to permit the correction of aperture aberrations for a moderate aperture and a large field; one or two plane-parallel plates are sometimes added to correct distortion. { 'täp·ə,gän ¦lenz }

topographical latitude *See* geodetic latitude. { ¦täp·ə¦graf·ə·kəl 'lad·ə,tüd }

topographic anatomy [ANAT] The use of bony and soft tissue landmarks on the surface of the body to indicate the known location of deeper structures. { ¦täp·ə¦graf·ik ə'nad·ə·mē }

topographic climax [ECOL] A climax plant community under a uniform macroclimate over which minor topographic features such as hills, rivers, valleys, or undrained depressions exert a controlling influence. { ¦täp·ə¦graf·ik 'klī,maks }

topographic curl effect [OCEANOGR] A term in Ekman's differential equation for the effects of variable wind stress, variable depth, variable friction, and variable latitude on the deep current; tends to make the curl G (velocity of deep current) positive when the current flows over increasing depth and negative when the depth decreases in the direction of the current. { ¦täp·ə¦graf·ik 'kərl i,fəkt }

topographic infancy *See* infancy. { ¦täp·ə¦graf·ik 'in·fən·sē }

topographic map [MAP] A large-scale map showing relief and man-made features of a portion of a land surface distinguished by portrayal of position, relation, size, shape, and elevation of the features. { ¦täp·ə¦graf·ik 'map }

topographic maturity See maturity. { ¦täp·ə¦graf·ik mə'chür·əd·ē }

topographic old age See old age. { ¦täp·ə¦graf·ik 'ōld 'āj }

topographic passage [OCEANOGR] A pass or gap through a sea-floor feature that possesses high topography, such as a ridge or a plateau. { ¦täp·ə¦graf·ik 'pas·ij }

topographic profile See profile. { ¦täp·ə¦graf·ik 'prō,fīl }

topographic survey [ENG] A survey that determines ground relief and location of natural and man-made features thereon. { ¦täp·ə¦graf·ik 'sər,vā }

topographic unconformity [GEOGR] A lack of harmony or conformity between two parts of a landscape or two kinds of topography. { ¦täp·ə¦graf·ik ,ən·kən'fòr·məd·ē }

topographic youth See youth. { ¦täp·ə¦graf·ik 'yüth }

topography [GEOGR] **1.** The general configuration of a surface, including its relief; may be a land or water-bottom surface. **2.** The natural surface features of a region, treated collectively as to form. { tə'päg·rə·fē }

topoisomerases [BIOCHEM] Any of a group of enzymes capable of relaxing, unwinding, unpackaging, or changing the degree of supercoiling of deoxyribonucleic acid fiber. { ¦tä·pō·ī'säm·ə,rās·əz }

topological dynamics [MATH] The study and application of transformations, or groups of such transformations (particularly topological transformation groups), defined on a topological space (usually compact), with particular regard to properties of interest in the qualitative theory of differential equations. { ¦täp·ə¦läj·ə·kəl dī'nam·iks }

topological groups [MATH] Groups which also have a topology with the property that the group operation and the inverse operation determine continuous functions. { ¦täp·ə¦läj·ə·kəl 'grüps }

topological K theory See K theory. { ¦täp·ə¦läj·ə·kəl 'kā ,thē·ə·rē }

topological linear space See topological vector space. { ¦täp·ə¦läj·ə·kəl 'lin·ē·ər ,spās }

topologically closed set See closed set. { ¦täp·ə¦läj·ə·klē ¦klōzd 'set }

topologically complete space [MATH] A topological space that is homeomorphic to a complete metric space. { ¦täp·ə¦läj·ə·klē kəm¦plēt 'spās }

topological mapping See homeomorphism. { ¦täp·ə¦läj·ə·kəl 'map·iŋ }

topological order [CRYO] An internal order in a quantum Hall state which describes the quantum motions of the electrons with respect to one another; it is different from any other known order in not being associated with any symmetries or breaking of symmetries. { ¦täp·ə¦läj·ə·kəl 'órd·ər }

topological product [MATH] The topological space obtained from taking the cartesian product of topological spaces. { ¦täp·ə¦läj·ə·kəl 'präd·əkt }

topological property [MATH] A property that holds true for any topological space homeomorphic to one possessing the property. { ¦täp·ə¦läj·ə·kəl 'präp·ərd·ē }

topological shielding [ELEC] An optimal lightning protection system in which a series of shields (such as a building's sheet metal or a metal cabinet), each one surrounding the next, are connected so that deleterious voltage and power levels are reduced at each successive inner shield. { ¦täp·ə¦läj·ə·kəl 'shēld·iŋ }

topological simplex [MATH] A topological space that is homeomorphic to a simplex. { ¦täp·ə¦läj·ə·kəl 'sim,pleks }

topological simplicial complex See triangulable space. { ¦täp·ə¦läj·ə·kəl sim¦plish·əl 'käm,pleks }

topological soliton [PHYS] A soliton whose confinement can be related to the existence of multiple states of minimum energy whose arrangement around the wave has special geometrical properties. { ¦täp·ə¦läj·ə·kəl 'säl·ə,tän }

topological space [MATH] A set endowed with a topology. { ¦täp·ə¦läj·ə·kəl 'spās }

topological vector space [MATH] A vector space which has a topology with the property that vector addition and scalar multiplication are continuous functions. Also known as linear topological space; topological linear space. { ¦täp·ə¦läj·ə·kəl 'vek·tər ,spās }

topology [COMPUT SCI] The physical or logical arrangement of the stations (nodes) in a communications network. [MATH] **1.** A collection of subsets of a set *X*, which includes *X* and the empty set, and has the property that any union or finite intersection of its members is also a member. **2.** The generalized study of properties of spaces invariant under deformations and stretchings. { tə'päl·ə·jē }

topology of circuits [ELEC] The study of electric networks in terms of the geometry of their connections only; used in finding such properties of circuits as equivalence and duality, and in analyzing and synthesizing complex circuits. { tə'päl·ə·jē əv 'sər·kəts }

toponium [PART PHYS] A hypothetical meson that is made up of the top quark *t* and its antiquark *t̄*. { tä'pō·nē·əm }

topotaxis See tropism. { ¦täp·ə¦tak·səs }

topotype [SYST] A specimen of a species not of the original type series collected at the type locality. { 'täp·ə,tīp }

topped crude [MATER] A residual product remaining after the removal by distillation or other means of an appreciable quantity of the more volatile components of crude petroleum. { 'täpt 'krüd }

topping [CHEM ENG] The distillation of crude petroleum to remove the light fractions only; the unrefined distillate is called tops. [CIV ENG] A layer of mortar placed over concrete to form a finishing surface on a floor, driveway, sidewalk, or curb. [TEXT] A step in the dyeing process in which a dyed material is placed in a bath of another color. { 'täp·iŋ }

topping governor See limit governor. { 'täp·iŋ ,gəv·ə·nər }

topping joint [CIV ENG] In concrete finishing, a small space or break set at regular intervals, particularly over expansion joints, to allow for contraction and expansion of the topping layer. { 'täp·iŋ ,jòint }

topping lift [NAV ARCH] A rope or chain extending from the head of a boom or gaff to a mast or to the vessel's structure for the purpose of supporting the weight and permitting the boom or gaff end to be raised or lowered. { 'täp·iŋ ,lift }

top plate [BUILD] **1.** The top horizontal member of a building frame to which the rafters are fastened. **2.** The horizontal member of a building frame at the top of the partition studs. { 'täp ,plāt }

topple [MECH] In gyroscopes for marine or aeronautical use, the condition of a sudden upset gyroscope or a gyroscope platform evidenced by a sudden and rapid precession of the spin axis due to large torque disturbances such as the spin axis striking the mechanical stops. Also known as tumble. { 'täp·əl }

topple axis [MECH] Of a gyroscope, the horizontal axis, perpendicular to the horizontal spin axis, around which topple occurs. Also known as tumble axis. { 'täp·əl ,ak·səs }

top quark [PARTIC PHYS] A quark with a mass of approximately 175 GeV, electric charge of $+^2/_3$, baryon number of $^1/_3$, zero isotopic spin, strangeness, and charm, and a new quantum number associated with it. Also known as *t* quark; truth quark. Symbolized *t*. { 'täp ,kwärk }

top rail [BUILD] The uppermost horizontal member of a unit of framing, such as a door or a sash. { 'täp ,rāl }

topset bed [GEOL] One of the nearly horizontal sedimentary layers deposited on the top surface of an advancing delta. { 'täp,set ,bed }

top shell [INV ZOO] Any of the marine snails of the family Trochidae characterized by a spiral conical shell with a flat base. { 'täp ,shel }

topside sounder [AERO ENG] A satellite designed to measure ion concentration in the ionosphere from above the ionosphere. { 'täp,sīd ,saún·dər }

top slicing [MIN ENG] A method of stoping in which the ore is extracted by excavating a series of horizontal (sometimes inclined) timbered slices alongside each other, beginning at the top of the ore body and working progressively downward. { 'täp ,slīs·iŋ }

top slicing and cover caving [MIN ENG] A mining method that entails the working of the ore body from the top down in successive horizontal slices that may follow one another sequentially or simultaneously; the overburden or cover is caved after mining a unit. { 'täp ,slīs·iŋ ən 'kəv·ər ,kāv·iŋ }

topsoil [GEOL] **1.** Soil presumed to be fertile and used to cover areas of special planting. **2.** Surface soil, usually corresponding with the A horizon, as distinguished from subsoil. { 'täp,sòil }

top steam [CHEM ENG] Steam admitted near the top of a shell still to purge the still, and to prevent a vacuum from forming when pumping out the liquid contents. { 'täp ,stēm }

topwork [BOT] A procedure employed to propagate seedless varieties of fruit and hybrids, to change the variety of fruit, and to correct pollination problems, using any of three methods: root grafting, crown grafting, and top grafting. { 'täp,wərk }

tor [GEOGR] An isolated, rough pinnacle or rocky peak. [MECH] *See* pascal. { tòr }

TORAN [NAV] A radiolocation receiver used with the Omega long-range navigation system having a digitized output and a position location capability of better than 165 feet (50 meters). { 'tòr,an }

torbanite [GEOL] A variety of coal that resembles a carbonaceous shale in outward appearance; it is fine-grained, black to brown, and tough. Also known as bitumenite; kerosine shale. { 'tòr·bə,nīt }

torbernite [MINERAL] $Cu(UO_2)_2(PO_4)_2 \cdot 8-12H_2O$ A green radioactive mineral crystallizing in the tetragonal system and occurring in tabular crystals or in foliated form. Also known as chalcolite; copper uranite; cuprouranite; uran-mica. { 'tòr·bər,nīt }

torch [BUILD] To apply lime mortar under the top edges of roof tiles or slates. [ENG] A gas burner used for brazing, cutting, or welding. { tòrch }

torchon [TEXT] Bobbin lace made of coarse thread in simple fanlike or diamond-shaped design, with little background. { 'tòr,shän }

toric lens [OPTICS] A lens whose surfaces form portions of toric surfaces. Also known as toroidal lens. { 'tòr·ik ,lenz }

toric surface [MATH] A surface generated by rotating an arc of a circle about a line that lies in the plane of the circle but does not pass through its center. Also known as toroidal surface. { 'tòr·ik ,sər·fəs }

tornado [METEOROL] An intense rotary storm of small diameter, the most violent of weather phenomena; tornadoes always extend downward from the base of a convective-type cloud, generally in the vicinity of a severe thunderstorm. { tòr'nād·ō }

tornado belt [METEOROL] The district of the United States in which tornadoes are most frequent; it encompasses the great lowland areas of the central and upper Mississippi, the Ohio, and lower Missouri River valleys. { tòr'nād·ō ,belt }

tornado cellar *See* cyclone cellar. { tòr'nād·ō ,sel·ər }

tornado cloud *See* tuba. { tòr'nād·ō ,klaůd }

tornado echo [METEOROL] A type of radar precipitation echo which has been observed in connection with a number of tornadoes; it frequently appears, on plan-position-indicator scopes, in the form of the figure 6 in the southwest sector of the storm; this echo has not been noted with all radar-observed tornadoes. { tòr'nād·ō ,ek·ō }

tornadotron [ELECTR] Millimeter-wave device which generates radio-frequency power from an enclosed, orbiting electron cloud, excited by a radio-frequency field, when subjected to a strong, pulsed magnetic field. { tòr'nād·ə,trän }

tornaria [INV ZOO] The larva of some acorn worms (Enteropneusta) which is large and marked by complex bands of cilia. { tòr'nar·ē·ə }

tornote [INV ZOO] A monaxon spicule in certain Porifera having both ends terminating abruptly in points. { 'tòr,nōt }

Toro [ASTRON] A small asteroid with a diameter of about 3.6 miles (6 kilometers), whose orbit, with semimajor axis of 1.368 astronomical units and eccentricity of 0.44, oscillates with that of Venus; it is about 0.13 astronomical unit from Earth at closest approach. { 'tòr·ō }

toroid *See* doughnut; toroidal magnetic circuit. { 'tòr,ȯid }

toroidal [SCI TECH] Shaped like a doughnut. { tə'rȯid·əl }

toroidal coil *See* toroidal magnetic circuit. { tə'rȯid·əl 'kȯil }

toroidal coordinate system [MATH] A three-dimensional coordinate system whose coordinate surfaces are the toruses and spheres generated by rotating the families of circles defining a two-dimensional bipolar coordinate system about the perpendicular bisector of the line joining the common points of intersection of one of the families, together with the planes passing through the axis of rotation. { tə¦rȯid·əl kō'ȯrd·ən·ət ,sis·təm }

toroidal core [ELECTROMAG] The doughnut-shaped piece of magnetic material in a toroidal magnetic circuit. { tə'rȯid·əl 'kȯr }

toroidal discharge *See* ring discharge. { tə¦rȯid·əl 'dis,chärj }

toroidal lens *See* toric lens. { tə'rȯid·əl ,lenz }

toroidal magnetic circuit [ELECTROMAG] Doughnut-shaped piece of magnetic material, together with one or more coils of current-carrying wire wound about the doughnut, with the permeability of the magnetic material high enough so that the magnetic flux is almost completely confined within it. Also known as toroid; toroidal coil. { tə'rȯid·əl mag'ned·ik 'sər,kət }

toroidal surface *See* toric surface. { tə'rȯid·əl ,sər·fəs }

toromatic transmission [MECH ENG] A semiautomatic transmission; it contains a compound planetary gear train with a torque converter. { ¦tòr·ə¦mad·ik tranz'mish·ən }

torose [INV ZOO] **1.** Having knobby prominences on the surface. **2.** Cylindrical with alternate swellings and contractions. { 'tȯ,rōs }

torose load cast [GEOL] One of a group of elongate load casts with alternate contractions and swellings, which may terminate down current in bulbous, teardrop, or spiral forms. { 'tȯ,rōs 'lōd ,kast }

Torpedinidae [VERT ZOO] The electric rays or torpedoes, a family of batoid sharks. { ,tòr·pə'din·ə,dē }

torpedo [ENG] An encased explosive charge slid, lowered, or dropped into a borehole and exploded to clear the hole of obstructions or to open communications with an oil or water supply. Also known as bullet. [ORD] A missile designed to contain an explosive charge and to be launched into water, where it is self-propelling and usually directable; used against ships or other targets in the water. { tòr'pēd·ō }

torpedo air flask [ORD] A cylindrical item having various compartments for housing compressed air, fuel, water, and chemicals, which when combined form the propelling charge of aerial and underwater torpedoes. { tòr'pēd·ō 'er ,flask }

torpedo boat [NAV ARCH] Small, fast vessel that is equipped with torpedo tubes, carries very light guns, and uses its inconspicuousness and speed to get within torpedo range of a target. { tòr'pēd·ō ,bōt }

torpedo defense net [ORD] A net employed to close an inner harbor to torpedoes fired from seaward or to protect an individual ship at anchor or underway. { tòr'pēd·ō di¦fens ,net }

torpedo exercise head [ORD] An item designed for attachment to a torpedo main assemblage to complete a torpedo for a practice run; it may contain recording instruments. { tòr'pēd·ō 'ek·sər,sīz ,hed }

torpedo exploder mechanism [ORD] An electrical or mechanical device designed to actuate the explosive train of a torpedo warhead by means of a physical impact or an influence signal; it may contain a disarming device. { tòr'pēd·ō ik'splōd·ər ,mek·ə,niz·əm }

torpedo warhead extension [ORD] A metallic cylindrical item designed to change the center of gravity of a torpedo; it may be explosive-filled. { tòr'pēd·ō 'wȯr,hed ik,sten·chən }

torpor [PHYSIO] The condition in hibernating poikilotherms during winter when body temperature drops in a parallel relation to ambient environmental temperatures. { 'tòr·pər }

torque [MECH] **1.** For a single force, the cross product of a vector from some reference point to the point of application of the force with the force itself. Also known as moment of force; rotation moment. **2.** For several forces, the vector sum of the torques (first definition) associated with each of the forces. { tòrk }

torque amplifier [COMPUT SCI] An analog computer device having input and output shafts and supplying work to rotate the output shaft in positional correspondence with the input shaft without imposing any significant torque on the input shaft. { 'tòrk ,am·plə,fī·ər }

torque arm [MECH ENG] In automotive vehicles, an arm to take the torque of the rear axle. { 'tòrk ,ärm }

torque-coil magnetometer [ENG] A magnetometer that depends for its operation on the torque developed by a known current in a coil that can turn in the field to be measured. { 'tòrk ,kȯil ,mag·nə'täm·əd·ər }

torque constant [ELEC] The ratio of the torque delivered by a motor to the current supplied to it. { 'tòrk ,kän·stənt }

torque converter [MECH ENG] A device for changing the torque speed or mechanical advantage between an input shaft and an output shaft. { 'tòrk kən,vərd·ər }

torque-load characteristic [ENG] For electric motors, the armature torque developed versus the load on the motor at constant speed. { 'tòrk ,lōd ,kar·ik·tə,ris·tik }

torquemeter [ENG] An instrument to measure torque. { 'tòrk,mēd·ər }

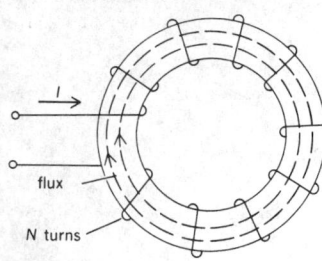

torque motor [ELECTROMAG] A motor designed primarily to exert torque while stalled or rotating slowly. { 'tȯrk ‚mōd·ər }

torque reaction [MECH ENG] On a shaft-driven vehicle, the reaction between the bevel pinion with its shaft (which is supported in the rear axle housing) and the bevel ring gear (which is fastened to the differential housing) that tends to rotate the axle housing around the axle instead of rotating the axle shafts alone. { 'tȯrk rē‚ak·shən }

torque ripple See cog. { 'tȯrk ‚rip·əl }

torque-speed characteristic [ELEC] For electric motors, the relationship of developed torque to armature speed. { 'tȯrk 'spēd ‚kar·ik·tə‚ris·tik }

torque-tube flowmeter [ENG] A liquid-flow measurement device in which a flexible torque tube transmits bellows motion (caused by differential pressure from the liquid flow through the pipe) to the recording pen arm. { 'tȯrk ‚tüb 'flō‚mēd·ər }

torque-type viscometer [ENG] A device that measures liquid viscosity by the torque needed to rotate a vertical paddle submerged in the liquid; used for both Newtonian and non-Newtonian liquids and for suspensions. { 'tȯrk ‚tīp vi'skäm·əd·ər }

torque-winding diagram [MECH ENG] A diagram showing how the winding load on a winch drum varies and is used to decide the method of balancing needed; made by plotting the turning moment in pounds per foot on the vertical axis against time, or revolutions or depth on the horizontal axis. { 'tȯrk ‚wīnd·iŋ ‚dī·ə‚gram }

torque wrench [ENG] **1.** A hand or power tool used to turn a nut on a bolt that can be adjusted to deliver a predetermined amount of force to the bolt when tightening the nut. **2.** A wrench that measures torque while being turned. { 'tȯrk ‚rench }

torque yarn [TEXT] A yarn that tends to rotate or twist when hung freely. { 'tȯrk ‚yärn }

torr [MECH] A unit of pressure, equal to 1/760 atmosphere; it differs from 1 millimeter of mercury by less than one part in seven million; approximately equal to 133.3224 pascals. { 'tȯr }

Torrert [GEOL] A suborder of the soil order Vertisol; it is the driest soil of the order and forms cracks that tend to remain open; occurs in arid regions. { 'tȯr·ərt }

torreyite [MINERAL] $(Mg,Mn,Zn)_7(SO_4)(OH)_{12}\cdot4H_2O$ A bluish-white mineral consisting of a hydrated basic sulfate of magnesium, manganese, and zinc; occurs in massive form. { 'tȯr·ē‚īt }

Torricellian barometer See mercury barometer. { ‚tȯr·ə‚chel·ē·ən bə'räm·əd·ər }

Torricellian vacuum [FL MECH] The space enclosed above a column of mercury when a tube, closed at one end, is filled with mercury and then placed, open end downward, in a well of mercury; this space is evacuated except for mercury vapor. { ‚tȯr·ə‚chel·ē·ən 'vak·yəm }

Torricelli's law of efflux [FL MECH] The velocity of efflux of liquid from an orifice in a container is equal to that which would be attained by a body falling freely from rest at the free surface of the liquid to the orifice. { ‚tȯr·ə‚chel·ēz 'lȯ əv 'e‚fləks }

Torridincolidae [INV ZOO] A small family of coleopteran insects in the suborder Myxophaga found only in Africa and Brazil. { tə‚rid·ən'käl·ə‚dē }

Torrid Zone [CLIMATOL] The zone of the earth's surface which lies between the Tropics of Cancer and Capricorn. { 'tär·əd ‚zōn }

Torrox [GEOL] A suborder of the soil order Oxisol that is low in organic matter, well drained, and dry most of the year; believed to have been formed under rainier climates of past eras. { 'tȯr‚äks }

torsel [BUILD] A section of wood, stone, or steel that supports one end of a beam or joist and distributes the load. { 'tȯr·səl }

torsiometer [MECH ENG] An instrument which measures power transmitted by a rotating shaft; consists of angular scales mounted around the shaft from which twist of the loaded shaft is determined. Also known as torsionmeter. { ‚tȯr·shē'äm·əd·ər }

torsion [MATH] The rate of change of the positive direction of the binormal of a space curve with respect to arc length along the curve; its sign is defined as positive if it is in the same direction as the principal normal, and negative if it is in the opposite direction. Also known as second curvature. [MECH] A twisting deformation of a solid body about an axis in which lines that were initially parallel to the axis become helices. { 'tȯr·shən }

torsional angle [MECH] The total relative rotation of the ends of a straight cylindrical bar when subjected to a torque. [PHYS CHEM] The angle between bonds on adjacent atoms. { 'tȯr·shən·əl 'aŋ·gəl }

torsional compliance [MECH] The reciprocal of the torsional rigidity. { ‚tȯr·shə‚nəl kəm'plī·əns }

torsional hysteresis [MECH] Dependence of the torques in a twisted wire or rod not only on the present torsion of the object but on its previous history of torsion. { ‚tȯr·shə·nəl ‚his·tə'rē·səs }

torsional mode delay line [COMPUT SCI] A device in which torsional vibrations are propagated through a solid material to make use of the propagation time of the vibrations to obtain a time delay for the signals. { 'tȯr·shən·əl ‚mōd di'lā‚līn }

torsional modulus [MECH] The ratio of the torsional rigidity of a bar to its length. Also known as modulus of torsion. { 'tȯr·shən·əl 'mäj·ə·ləs }

torsional pendulum [MECH] A device consisting of a disk or other body of large moment of inertia mounted on one end of a torsionally flexible elastic rod whose other end is held fixed; if the disk is twisted and released, it will undergo simple harmonic motion, provided the torque in the rod is proportional to the angle of twist. Also known as torsion pendulum. { 'tȯr·shən·əl 'pen·jə·ləm }

torsional rigidity [MECH] The ratio of the torque applied about the centroidal axis of a bar at one end of the bar to the resulting torsional angle, when the other end is held fixed. { 'tȯr·shən·əl ri'jid·əd·ē }

torsional vibration [MECH] A periodic motion of a shaft in which the shaft is twisted about its axis first in one direction and then in the other; this motion may be superimposed on rotational or other motion. { 'tȯr·shən·əl vī'brā·shən }

torsional wave [PHYS] A wave motion in which the vibrations of the medium are periodic rotational motions around the direction of propagation. { 'tȯr·shən·əl 'wāv }

torsion balance [ENG] An instrument, consisting essentially of a straight vertical torsion wire whose upper end is fixed while a horizontal beam is suspended from the lower end; used to measure minute gravitational, electrostatic, or magnetic forces. { 'tȯr·shən ‚bal·əns }

torsion bar [MECH ENG] A spring flexed by twisting about its axis; found in the spring suspension of truck and passenger car wheels, in production machines where space limitations are critical, and in high-speed mechanisms where inertia forces must be minimized. { 'tȯr·shən ‚bär }

torsion coefficients [MATH] For a finitely generated abelian group G, the orders of the finite cyclic groups such that G is the direct sum of these groups and infinite cyclic groups. { 'tȯr·shən ‚kō·ə‚fish·əns }

torsion damper [MECH ENG] A damper used on automobile internal combustion engines to reduce torsional vibration. { 'tȯr·shən ‚dam·pər }

torsion element [MATH] **1.** A torsion element of an Abelian group G is an element of G with finite period. **2.** A torsion element of a module M over an entire, principal ring R is an element x in M for which there exists an element a in R such that $a \neq 0$ and $ax = 0$. { 'tȯr·shən ‚el·ə·mənt }

torsion fault See wrench fault. { 'tȯr·shən ‚fȯlt }

torsion-free group [MATH] A group whose only torsion element is the unit element. { 'tȯr·shən ‚frē ‚grüp }

torsion function [MECH] A harmonic function, $\phi(x,y) = w/\tau$, expressing the warping of a cylinder undergoing torsion, where the x, y, and z coordinates are chosen so that the axis of torsion lies along the z axis, w is the z component of the displacement, and τ is the torsion angle. Also known as warping function. { 'tȯr·shən ‚fəŋk·shən }

torsion galvanometer [ENG] A galvanometer in which the force between the fixed and moving systems is measured by the angle through which the supporting head of the moving system must be rotated to bring the moving system back to its zero position. { 'tȯr·shən ‚gal·və'näm·əd·ər }

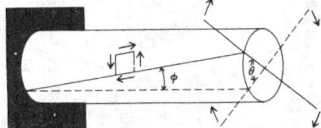

TORSION

Cylindrical bar in torsion, showing deformation of a small element of the bar; θ represents torsional angle; ϕ represents helical angle.

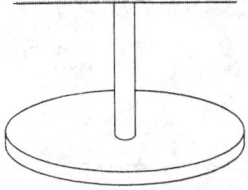

TORSIONAL PENDULUM

Diagram of a torsional pendulum.

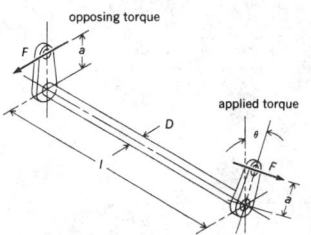

TORSION BAR

Diagram of torsion bar showing important dimensions involved in its design; θ is angle of twist, F is twisting force, a is radius of arm, l is length of bar, D is diameter of bar.

torsion group [MATH] **1.** A group whose elements all have finite period. **2.** For a topological space X, one of a sequence of finite groups $G_n(X)$ such that the homology group $H_n(X)$ is the direct sum of $G_n(X)$ and a number of infinite cyclic groups. { 'tȯr·shən ‚grüp }

torsion hygrometer [ENG] A hygrometer in which the rotation of the hygrometric element is a function of the humidity; such hygrometers are constructed by taking a substance whose length is a function of the humidity and twisting or spiraling it under tension in such a manner that a change in length will cause a further rotation of the element. { 'tȯr·shən hī'gräm·əd·ər }

torsionmeter *See* torsiometer. { 'tȯr·shən‚mēd·ər }

torsion module [MATH] A module M over an entire principal ring R is said to be a torsion module if for any element x in M there exists an element a in R such that $a \neq 0$ and $ax = 0$. { 'tȯr·shən ‚mä·jül }

torsion pendulum *See* torsional pendulum. { 'tȯr·shən 'pen·jə·ləm }

torsion-string galvanometer [ENG] A sensitive galvanometer in which the moving system is suspended by two parallel fibers that tend to twist around each other. { 'tȯr·shən ¦striŋ ‚gal·və'näm·əd·ər }

torsion subgroup [MATH] The torsion subgroup of an Abelian group G is the subset of all torsion elements of G. { ¦tȯr·shən 'səb‚grüp }

torsion submodule [MATH] The torsion submodule of a module E over an entire principal ring is the submodule consisting of all torsion elements of E. { ¦tȯr·shən 'səb‚mä·jül }

torso mountain *See* monadnock. { 'tȯr·sō ‚maůnt·ən }

torticollis [MED] A deformity of the neck resulting from contraction of the cervical muscles or fascia. Also known as wryneck. { ‚tȯrd·ə'käl·əs }

tortoise [VERT ZOO] Any of various large terrestrial reptiles in the order Chelonia, especially the family Testudinidae. { 'tȯrd·əs }

Tortonian [GEOL] A European stage of geologic time: Miocene (above Helvetian, below Sarmatian). { tȯr'tō·nē·ən }

Tortricidae [INV ZOO] A family of phytophagous moths in the superfamily Tortricoidea which have a stout body, lightly fringed wings, and threadlike antennae. { tȯr'tris·ə‚dē }

Tortricoidea [INV ZOO] A superfamily of small wide-winged moths in the suborder Heteroneura. { ‚tȯr·trə'kȯid·ē·ə }

Torulopsidales [MYCOL] The equivalent name for Cryptococcales. { ‚tȯr·ə‚läp·sə'dā·lēz }

torulosis *See* cryptococcosis. { ‚tȯr·ə'lō·səs }

torus [ANAT] A rounded protuberance occurring on a body part. [BOT] The thickened membrane closing a bordered pit. [MATH] **1.** The surface of a doughnut-shaped object. **2.** The topological space obtained by identifying the opposite sides of a rectangle. **3.** The group which is the product of two circles. { 'tȯr·əs }

Torymidae [INV ZOO] A family of hymenopteran insects in the superfamily Chalcidoidea. { tȯ'rim·ə‚dē }

TOS *See* tape operating system.

tosca [METEOROL] A southwest wind on Lake Garda in Italy. { 'tȯs·kə }

toss bombing [ORD] A method of bombing where an aircraft flies on a line toward the target, pulls up in a vertical plane, releasing the bomb at an angle that will compensate for the effect of gravity drop on the bomb. { 'tȯs ‚bäm·iŋ }

total air [ENG] The actual quantity of air supplied for combustion of fuel in a boiler, expressed as a percentage of theoretical air. { 'tōd·əl 'er }

total allowable catch [OCEANOGR] A fishery management approach to assign an annual quota that, if exceeded, will terminate the fishery for that year; the total allowable catch is set at a level to prevent a catch so large that the stock will be overfished. { ¦tōd·əl ə‚laů·ə·bəl 'kach }

total binding energy *See* binding energy. { 'tōd·əl 'bīnd·iŋ 'en·ər·jē }

total carbon [MET] The sum of free and combined carbon in a ferrous alloy, especially steel. { 'tōd·əl 'kär·bən }

total coincidence [MECH ENG] The condition in which all the joints of a robot become locked in position. { 'tōd·əl kō'in·səd·əns }

total conductivity [GEOPHYS] In atmospheric electricity, the sum of the electrical conductivities of the positive and negative

ions found in a given portion of the atmosphere. { 'tōd·əl ‚kän·dək'tiv·əd·ē }

total curvature [MATH] *See* Gaussian curvature. [OPTICS] The difference between the reciprocals of the radii of curvature of the two surfaces of a lens. { 'tōd·əl 'kər·və·chər }

total cyanide [MET] Total amount of cyanide contained in an electroplating bath, including both simple and complex ions. { 'tōd·əl 'sī·ə‚nīd }

total deadlock [COMPUT SCI] A deadlock that involves all the tasks in a multiprogramming system. { 'tōd·əl 'ded‚läk }

total differential [MATH] The total differential of a function of several variables, $f(x_1, x_2, \ldots, x_n)$, is the function given by the sum of terms $(\partial f / \partial x_i) dx_i$ as i runs from 1 to n. Also known as differential. { 'tōd·əl ‚dif·ə'ren·chəl }

total displacement *See* slip. { 'tōd·əl di'splās·mənt }

total drift [NAV] In marine gyroscopes, the algebraic sum of the apparent and real precession or drift; the apparent drift is that caused by the gyroscope attempting to maintain its orientation in inertial space while the vehicle on which the gyroscope is mounted moves to new locations. { 'tōd·əl 'drift }

total eclipse [ASTRON] An eclipse that obscures the entire surface of the moon or sun. { 'tōd·əl i'klips }

total energy ball [NUCLEO] An array of high-efficiency, low-resolution gamma-ray detectors that completely surrounds the target of a heavy-ion accelerator, and is used to measure the total energy deposited by the gamma-ray shower from a heavy-ion collision and the number of component detectors in which this energy is deposited. { 'tōd·əl 'en·ər·jē ‚bȯl }

total evaporation *See* evapotranspiration. { 'tōd·əl i‚vap·ə'rā·shən }

total harmonic distortion [ELECTR] Ratio of the power at the fundamental frequency, measured at the output of the transmission system considered, to the power of all harmonics observed at the output of the system because of its nonlinearity, when a single frequency signal of specified power is applied to the input of the system; it is expressed in decibels. { 'tōd·əl här'män·ik di'stȯr·shən }

total head [FL MECH] The sum of the velocity head and the pressure head corresponding to the static pressure. { ¦tōd·əl ¦hed }

total heat *See* enthalpy. { 'tōd·əl 'hēt }

total heat of dilution *See* heat of dilution. { 'tōd·əl 'hēt əv di'lü·shən }

total heat of solution *See* heat of solution. { 'tōd·əl 'hēt əv sə'lü·shən }

total impulse [AERO ENG] The product of the thrust and the time over which the thrust is produced, expressed in pounds (force)-seconds; used especially in reference to a rocket motor or a rocket engine. { 'tōd·əl 'im‚pəls }

total internal reflection [OPTICS] A phenomenon in which electromagnetic radiation in a given medium which is incident on the boundary with a less-dense medium (one having a lower index of refraction) at an angle less than the critical angle is completely reflected from the boundary. { 'tōd·əl in'tərn·əl ri'flek·shən }

totality [ASTRON] **1.** The portion of a total eclipse of the sun during which the sun is entirely covered by the moon at a specified location on the earth's surface. **2.** The portion of a total eclipse of the moon or other body during which the eclipsed body is entirely within the umbra of the eclipsing body. { tō'tal·əd·ē }

total lift [AERO ENG] The upward force produced by the gas in a balloon; it is equal to the sum of the free lift, the weight of the balloon, and the weight of auxiliary equipment carried by the balloon. { 'tōd·əl 'lift }

total lung capacity [PHYSIO] The volume of gas contained within the lungs at the end of a maximum inspiration. { 'tōd·əl 'ləŋ kə‚pas·əd·ē }

totally bounded set *See* precompact set. { 'tōd·əl·ē 'baůn·dəd 'set }

totally disconnected [MATH] A topological space has this property if the largest connected subset containing any given point is only the point itself. { 'tōd·əl·ē ‚dis·kə'nek·təd }

totally imaginary field [MATH] An extension field F of the field of rational numbers such that no embedding of F in the complex numbers is contained in the real numbers. { ¦tōd·əl·ē i‚maj·ə‚ner·ē 'fēld }

totally ordered set *See* linearly ordered set. { ¦tōd·əl·ē 'ȯrd·ərd 'set }

TORTOISE

Desert tortoise (*Gopherus agassizi*).

totally stable system [PHYS] A system that returns to a stationary state following arbitrarily large perturbations. { ¦tōd·əl·ē 'stā·bəl ¦sis·təm }

total order [MATH] **1.** The total order of an analytic function in a domain D is the algebraic sum of its orders at all poles and zeros in D. **2.** *See* linear order. { 'tōd·əl 'ȯr·dər }

total porosity [GEOL] The ratio of total void space in porous oil-reservoir rock to the bulk volume of the rock itself. { 'tōd·əl pə'räs·əd·ē }

total pressure [FL MECH] *See* dynamic pressure. [MECH] The gross load applied on a given surface. [MIN ENG] The total ventilating pressure in a mine, usually measured in the fan drift. { 'tōd·əl 'presh·ər }

total quality management [SYS ENG] A philosophy and set of guiding concepts that provides a comprehensive means of improving total organization performance and quality by examining each process through which work is done in a systematic, integrated, consistent, organization-wide manner. Abbreviated TQM. { 'tōd·əl 'kwäl·əd·ē ,man·ij·mənt }

total radiation pyrometer [ENG] A pyrometer which focuses heat radiation emitted by a hot object on a detector (usually a thermopile or other thermal type detector), and which responds to a broad band of radiation, limited only by absorption of the focusing lens, or window and mirror. { 'tōd·əl 'rād·ē¦ā·shən pī'räm·əd·ər }

total relief *See* stereoscopic power. { 'tōd·əl ri'lēf }

total slip *See* net slip. { 'tōd·əl 'slip }

total solids [CHEM] The total content of suspended and dissolved solids in water. { 'tōd·əl 'säl·ədz }

total space [MATH] The topological space E in the bundle (E, p, B). { 'tōd·əl 'spās }

total span [ANTHRO] An anthropometric determination of the distance between the tips of the middle fingers at maximum arm stretch without straining. { 'tōd·əl 'span }

total specific ionization *See* specific ionization. { 'tōd·əl spə'sif·ik ,ī·ə·nə'zā·shən }

total stability [PHYS] The property of a system that returns to an equilibrium configuration after the system is subjected to arbitrarily large perturbations. { ¦tōd·əl stə'bil·əd·ē }

total system error [NAV] The difference between the actual flight path of an aircraft and the assigned flight path, equal to the sum of the navigation system error and the flight technical error. { ¦tōd·əl 'sis·təm ,er·ər }

total transfer [COMPUT SCI] A method of calculating minor, intermediate, and major control totals during preparation of printed results on a punched-card tabulator, in which only minor totals are calculated by adding values in card fields, while intermediate totals are calculated by adding minor totals, and major totals are calculated by adding intermediate totals. { 'tōd·əl 'tranz·fər }

total variation [MATH] For a real function defined on an interval, the least upper bound of the function's variation relative to all possible partitions of the interval. { 'tōd·əl ,ver·ē'ā·shən }

total vorticity [FL MECH] Usually, the magnitude of the vorticity vector, all components included, as opposed to the vertical (component of the) vorticity. { 'tōd·əl vȯr'tis·əd·ē }

total wetting [FL MECH] The situation in which a liquid surface meets a solid surface with zero contact angle. { ,tōd·əl 'wed·iŋ }

totient *See* Euler's phi function. { 'tō·shənt }

totipalmate [VERT ZOO] Having all four toes connected by webs, as in the Pelecaniformes. { ,tōd·ə'pȧ,māt }

totipotence [EMBRYO] Capacity of a blastomere to develop into a fully formed embryo. { tō'tip·əd·əns }

totipotent cell [EMBRYO] A cell capable of differentiating into every type of cell found in an organism, and of forming the entire organism. { 'tōd·ə,pōt·ent ,sel }

totitive [MATH] An integer that is less than a given integer and relatively prime to it. { 'tōd·ə,tiv }

toucan [VERT ZOO] Any of numerous fruit-eating birds, of the family Ramphastidae, noted for their large and colorful bills. { 'tü,kan }

Toucan *See* Tucana. { 'tü,kan }

touch [PHYSIO] The array of sensations arising from pressure sensitivity of the skin. { təch }

touch call *See* push-button dialing. { 'təch ,kȯl }

touch control [ELEC] A circuit that closes a relay when two metal areas are bridged by a finger or hand. { 'təch kən,trōl }

touch-down dispersion [NAV] The statistical scatter of the points of contact between the aircraft and the landing surface during a large number of automatic landings. { 'təch¦daun di,spər·zhən }

touch-down zone [NAV] In air operations, the first 3000 feet (914 meters) of runway beginning at the threshold. Abbreviated TDZ. { 'təch¦daun ,zōn }

touch-down zone elevation [NAV] The highest runway centerline elevation in the touch-down zone. { 'təch¦daun ,zōn ,el·ə'vā·shən }

touch feedback [ENG] A type of force feedback in which servos provide the manipulator fingers with a sense of resistance when an object is grasped, so that the operator does not crush the object. { 'təch ,fēd,bak }

touch screen [COMPUT SCI] An electronic display that allows a user to send signals to a computer by touching an area on the display with a finger, pencil, or other object. { 'təch ,skrēn }

touch sensor [CONT SYS] A device such as a small, force-sensitive switch that uses contact to generate feedback in robotic systems. { 'təch ,sen·sər }

toughened glass *See* tempered glass. { 'təf·ənd ,glas }

toughness [MECH] A property of a material capable of absorbing energy by plastic deformation; intermediate between softness and brittleness. { 'təf·nəs }

tough pitch copper [MET] Copper refined in a reverberatory furnace to adjust the oxygen content to 0.2-0.5%. { 'təf ,pich ,käp·ər }

tour *See* shift. { tür }

Tourette's syndrome [MED] A syndrome characterized by repetitive tics, movement disorders, uncontrolled grunts, and occasionally verbal obscenities. Also known as Gilles de la Tourette syndrome. { tü'rets ,sin,drōm }

touriello [METEOROL] A south wind of foehn type descending from the Pyrenees in the Ariège valley, France; it is especially violent in February and March, when it melts the snow, flooding the rivers and sometimes causing avalanches. { ,tür·ē'el·ō }

tourmaline [MINERAL] $(Na, Ca)(Al, Fe, Li, Mg)_3Al_6(BO_3)_3\text{-}Si_6O_{18}(OH)_4$ Any of a group of cyclosilicate minerals with a complex chemical composition, vitreous to resinous luster, and variable color; crystallizes in the ditrigonal-pyramidal class of the hexagonal system, has piezoelectric properties, and is used as a gemstone. { 'tür·mə,lēn }

Tournaisian [GEOL] European stage of lowermost Carboniferous time. { tür'nā·zhən }

tournament [MATH] A graph in which there is one line between every pair of points and no loops, and in which a unique direction is assigned to every line. { 'tür·nə·mənt }

tourniquet [MED] An apparatus for controlling hemorrhage from, or circulation in, a limb or part of the body, where pressure can be brought upon the blood vessels by means of straps, cords, rubber tubes, or pads. { 'tür·nə·kət }

Touschek effect [PHYS] In electron storage rings, an effect in which the maximum particle concentration in the circulating electron bunches is restricted at low energies by the loss of electrons in Møller scattering. { 'tüs,chek i,fekt }

Toussaint's formula [METEOROL] A rule for the linear decrease of temperature with height in an atmosphere for which the temperature at mean sea level is 15°C, and given by the formula $t = 15 - 0.0065z$, where t is the temperature in degrees Celsius, and z is the geometric height in meters above mean sea level. { tü'sanz ,fȯr·myə·lə }

tow [ENG] **1.** To haul by a rope or chain, for example, to haul a disabled ship by another vessel or an automotive vehicle by another vehicle. **2.** To propel by pushing, as a tugboat piloting a ship. [TEXT] **1.** The broken, short, matted fiber that is removed during separation of long fibers of flax, hemp, or jute. **2.** A large number of continuous filaments collected in ropelike form without a definite twist. **3.** The coarsest linen yarns used to make crash. { tō }

towbar [ENG] An element which connects to a vehicle that is not equipped with an integral drawbar, for the purpose of towing or moving the vehicle. { 'tō,bär }

towboat [NAV ARCH] A relatively flat-bottomed vessel designed to push tows of barges on inland waterways; it has a square bow on which heavy, upright knees are fixed for the purpose of lashing the barges against the towing vessel. { 'tō,bōt }

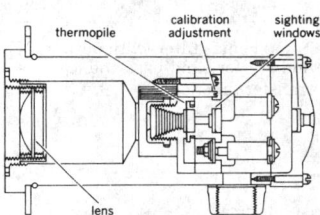

TOTAL RADIATION PYROMETER

thermopile   calibration adjustment   sighting windows

lens

Diagram of total radiation pyrometer. Lens focuses heat radiation onto thermopile. *(Honeywell Inc.)*

TOURNAMENT

An oriented complete graph.

towed artillery [ORD] Artillery weapons designed for movement as trailed loads behind prime movers or draft animals; some adjustment of the weapon is necessary to place it in firing position. { 'tōd är'til·ə·rē }

towed load [MECH] The weight of a carriage, trailer, or other equipment towed by a prime mover. { 'tōd 'lōd }

tower [CHEM ENG] A vertical, cylindrical vessel used in chemical and petroleum processing to increase the degree of separation of liquid mixtures by distillation or extraction. Also known as column. [ELECTROMAG] A tall metal structure used as a transmitting antenna, or used with another such structure to support a transmitting antenna wire. [ENG] A concrete, metal, or timber structure that is relatively high for its length and width, and used for various purposes, including the support of electric power transmission lines, radio and television antennas, and rockets and missiles prior to launching. [MATH] For a set S with a given algebraic structure, this is a set of subsets, $S_0 = S, S_1, S_2, \ldots, S_n$, such that S_{i+1} is a subset of S_i, $i = 1, 2, \ldots, n - 1$, and each S_i is closed under all possible operations in the algebraic structure of S. { 'taủ·ər }

tower bolt *See* barrel bolt. { 'taủ·ər ,bōlt }

tower case [COMPUT SCI] A system unit that stands in a vertical position. { 'taủ·ər ,kās }

tower crane [CIV ENG] A crane mounted on top of a tower which is sometimes incorporated in the frame of a building. { 'taủ·ər ,krān }

tower excavator [MIN ENG] A cableway excavator designed specifically for levee work but which is used extensively in the stripping of overburden, spoil, or waste in surface mining: basically, it is a Sauerman-type excavator with towers either fixed or movable, and when the head tower is located on the spoil pile and the tail tower on the unexcavated wall, pits of almost unlimited width can be dug. { 'taủ·ər 'eks·kə,vād·ər }

towering [METEOROL] A refraction phenomenon; a special case of looming in which the downward curvature of the light rays due to atmospheric refraction increases with elevation so that the visual image of a distant object appears to be stretched in the vertical direction. { 'taủ·ə·riŋ }

towering cumulus [METEOROL] A descriptive term, used mostly in weather observing, for the cloud type cumulus congestus. { 'taủ·ə·riŋ 'kyü·myə·ləs }

tower launcher [ORD] A missile launcher that is vertical (or nearly so) and high enough to give directional stability to the missile. Also known as vertical tower launcher. { 'taủ·ər ,lȯnch·ər }

tower loader [MIN ENG] A front-end loader whose bucket is lifted along tracks on a more or less vertical tower. { 'taủ·ər ,lōd·ər }

tower loading [ELEC] Load placed on a tower by its own weight, the weight of the wires with or without ice covering, the insulators, the wind pressure normal to the line acting both on the tower and the wires, and the pull from the wires. { 'taủ·ər ,lōd·iŋ }

tower radiator [ELECTROMAG] Metal structure used as a transmitting antenna. { 'taủ·ər 'rād·ē,ād·ər }

tower telescope [ASTRON] A telescope, usually of long focal length, that is situated underneath a solar tower to study the sun. { 'taủ·ər ,tel·ə,skōp }

towing spar *See* position buoy. { 'tō·iŋ ,spär }

towing tank *See* model basin. { 'tō·iŋ ,taŋk }

town plan *See* city plan. { 'taủn ,plan }

Townsend avalanche *See* avalanche. { 'taủn·zənd ,av·ə,lanch }

Townsend characteristic [ELECTR] Current-voltage characteristic curve for a phototube at constant illumination and at voltages below that at which a glow discharge occurs. { 'taủn·zənd ,kar·ik·tə,ris·tik }

Townsend coefficient [ELECTR] The number of ionizing collisions by an electron per centimeter of path length in the direction of the applied electric field in a radiation counter. { 'taủn·zənd ,kō·i,fish·ənt }

Townsend discharge [ELECTR] A discharge which occurs at voltages too low for it to be maintained by the electric field alone, and which must be initiated and sustained by ionization produced by other agents; it occurs at moderate pressures, above about 0.1 torr, and is free of space charges. { 'taủn·zənd ,dis,chärj }

Townsend ionization *See* avalanche. { 'taủn·zənd ,ī·ə·nə,zā·shən }

Townsend second ionization coefficient [NUCLEO] The number of electrons released from the cathode of an ionization chamber per initial ionizing collision in the gas. Also known as secondary ionization coefficient. { ¦taủn·zənd ,sek·ənd ,ī·ə·nə'zā·shən ,kō·ə,fish·ənt }

towrope [NAV ARCH] A hawser of either fiber or wire used for towing a vessel. { 'tō,rōp }

towrope horsepower *See* effective horsepower. { 'tō,rōp 'hȯrs,paủ·ər }

towrope resistance [NAV ARCH] The total resistance overcome in towing a ship or model; it equals the sum of the frictional resistance, eddy making, and wave making. { 'tō,rōp ri,ziz·təns }

tow target [ORD] Target for antiaircraft fire or aerial gunnery practice, towed behind an aircraft. { 'tō ,tär·gət }

Tow-Thomas filter [ELECTR] A multiple-amplifier active filter that has the advantage of ease of design but the disadvantage of lacking a high-pass output in its basic configuration. { ¦tō 'täm·əs ,fil·tər }

toxa [INV ZOO] A curved sponge spicule. { 'täk·sə }

Toxasteridae [PALEON] A family of Cretaceous echinoderms in the order Spatangoida which lacked fascioles and petals. { ,täk·sə'ster·ə,dē }

toxemia [MED] A condition in which the blood contains toxic substances, either of microbial origin or as by-products of abnormal protein metabolism. { täk'sē·mē·ə }

toxemia of pregnancy *See* preeclampsia. { täk'sē·mē·ə əv 'preg·nən·sē }

toxic [MED] Relating to a harmful effect by a poisonous substance on the human body by physical contact, ingestion, or inhalation. { 'täk·sik }

toxic amaurosis [MED] Blindness following the introduction of toxic substances into the body, such as ethyl and methyl alcohol, tobacco, lead, and metabolites of uremia and diabetes. { täk·sik ,a,mȯ'rō·səs }

toxic epidermal necrolysis [MED] Intraepidermal blistering and separation of the outer epidermis, giving the appearance and the management problems of a scald, caused by infection with *Staphylococcus aureus* strains producing one of the epidermolytic toxins, usually of phage group II. Also known as scalded skin syndrome. { ¦tak·sik ,ep·ə,dər·məl nə'kräl·ə·səs }

toxic goiter *See* hyperthyroidism. { 'täk·sik 'gȯid·ər }

toxic hepatitis [MED] Inflammation of the liver caused by chemical agents ingested or inhaled into the body, such as chlorinated hydrocarbons and some alkaloids. { 'täk·sik ,hep·ə'tīd·əs }

toxicity [PHARM] **1.** The quality of being toxic. **2.** The kind and amount of poison or toxin produced by a microorganism, or possessed by a chemical substance not of biological origin. { täk'sis·əd·ē }

toxicological study [MED] The study of how much poison must be present to produce an effect on animals or plant systems; may also include what type of effect is produced and how it is detected. { ,täk·sə·kə,läj·ə·kəl 'stəd·ē }

toxicology [PHARM] The study of poisons, including their nature, effects, and detection, and methods of treatment. { ,tak·sə'käl·ə·jē }

toxicophobia [PSYCH] Abnormal fear of being poisoned. { ,täk·sə·kō'fō·bē·ə }

toxic psychosis [MED] A brain disorder due to a toxic agent such as lead or alcohol. { 'täk·sik sī'kō·səs }

toxic shock syndrome [MED] A serious, sometimes life-threatening disease usually caused by a toxin produced by some strains of the bacterium *Staphylococcus aureus*. The signs and symptoms are fever, abnormally low blood pressure, nausea and vomiting, diarrhea, muscle tenderness, and a reddish rash, followed by peeling of the skin. { ,täk·sik 'shäk ,sin,drōm }

toxicyst [INV ZOO] A type of trichocyst in Protozoa which may, upon contact, induce paralysis or lysis of the prey. { 'täk·sə,sist }

toxigenicity [MICROBIO] A microorganism's capability for producing toxic substances. { ,täk·sə·jə'nis·əd·ē }

toxin [BIOCHEM] Any of various poisonous substances produced by certain plant and animal cells, including bacterial toxins, phytotoxins, and zootoxins. { 'täk·sən }

Toxodontia [PALEON] An extinct suborder of mammals representing a central stock of the order Notoungulata. { ,täk·sə'dän·chə }

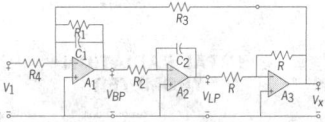

TOW-THOMAS FILTER

Tow-Thomas filter, with outputs V_{BP} and V_{LP} for band-pass and low-pass transfer functions.

Toxoglossa [INV ZOO] A group of carnivorous marine gastropod mollusks distinguished by a highly modified radula (toxoglossate). { ˌtäkˈsəˈgläsˈə }

toxoglossate radula [INV ZOO] A radula in certain carnivorous gastropods having elongated, spearlike teeth often perforated by the ducts of large poison glands. { ˌtäkˈsəˈglä¦sät ˈrajˈəˈlə }

toxoid [IMMUNOL] Detoxified toxin, but with antigenic properties intact; toxoids of tetanus and diphtheria are used for immunization. { ˈtäkˌsȯid }

Toxoplasmea [INV ZOO] A class of the protozoan subphylum Sporozoa composed of small, crescent-shaped organisms that move by body flexion or gliding and are characterized by a two-layered pellicle with underlying microtubules, micropyle, paired organelles, and micronemes. { ˌtäkˈsəˈplazˈmēˈə }

Toxoplasmida [INV ZOO] An order of the class Toxoplasmea; members are parasites of vertebrates. { ˌtäkˈsəˈplazˈmədˈə }

toxoplasmin [BIOCHEM] The *Toxoplasma* antigen; used in a skin test to demonstrate delayed hypersensitivity to toxoplasmosis. { ˌtäkˈsəˈplazˈmən }

toxoplasmosis [MED] Infection by the protozoan *Toxoplasma gondii*, manifested clinically in severe cases by jaundice, hepatomegaly, and splenomegaly. { ˌtakˈsəˈplazˈmōˈsəs }

Toxopneustidae [INV ZOO] A family of Tertiary and extant echinoderms of the order Temnopleuroida where the branchial slits are deep and the test tends to be absent. { ˌtäkˈsəˈnyüˈstəˌdē }

Toxotidae [VERT ZOO] The archerfishes, a family of small fresh-water forms in the order Perciformes. { täkˈsädˈəˌdē }

TPA See terephthalic acid.

T pad [ELEC] A pad made up of resistance elements arranged as a T network (two resistors inserted in one line, with a third between their junction and the other line). { ˈtē ˌpad }

TPE See thermoplastic elastomer.

T phage [VIROL] Any of a series (T1–T7) of deoxyribonucleic acid phages which lyse strains of the gram-negative bacterium *Escherichia coli* and its relatives. { ˈtē ˌfāj }

TP monitor See teleprocessing monitor. { ˈtēˈpē ˌmänˈədˈər }

TPN See triphosphopyridine nucleotide.

TPR See airborne profile recorder.

TQM See total quality management.

t quark See top quark. { ˈtē ˌkwärk }

Tr See trace.

trabeated [ARCH] Designed or constructed of beams and lintels, as distinguished from construction based on arches and vaults. { ˈtraˈbēˌādˈəd }

trabecula [ANAT] A band of fibrous or muscular tissue extending from the capsule or wall into the interior of an organ. { trəˈbekˈyəˈlə }

trace [COMPUT SCI] To provide a record of every step, or selected steps, executed by a computer program, and by extension, the record produced by this operation. [ELECTR] The visible path of a moving spot on the screen of a cathode-ray tube. Also known as line. [ENG] The record made by a recording device, such as a seismometer or electrocardiograph. [GEOL] The intersection of two geological surfaces. [MATH] **1.** The trace of a matrix is the sum of the entries along its principal diagonal. Designated Tr. Also known as spur. **2.** The trace of a linear transformation on a finite-dimensional vector space is the trace (in the sense of the first definition) of the matrix associated with it. **3.** One of the curves along which a given surface cuts a coordinate plane. **4.** A point at which a given straight line in space passes through a coordinate plane. Also known as piercing point. **5.** The projection of a given straight line in space on a coordinate plane. [METEOROL] A precipitation of less than 0.005 inch (0.127 millimeter). [SCI TECH] An extremely small but detectable quantity of a substance. { trās }

trace analysis [ANALY CHEM] Analysis of a very small quantity of material of a sample by such techniques as polarography or spectroscopy. { ˈtrās əˌnalˈəˈsəs }

trace element [ANALY CHEM] An element in a sample that has an average concentration of less than 100 parts per million atoms or less than 100 micrograms per gram. [BIOCHEM] A chemical element that is needed in minute quantities for the proper growth, development, and physiology of the organism. Also known as micronutrient. [GEOCHEM] An element found in small quantities (usually less than 1.0%) in a mineral. Also known as accessory element; guest element. { ˈtrās ˌelˈəˈmənt }

trace fossil [GEOL] A trail, track, or burrow made by an animal and found in ancient sediments such as sandstone, shale, or limestone. Also known as ichnofossil. { ˈtrās ˌfäsˈəl }

trace heating [ENG] Heating the layer between insulation and pipes in an insulated pipework system to reduce viscosity and thereby facilitate flow of the liquid. { ˈtrās ˌhēdˈiŋ }

trace interval [ELECTR] Interval corresponding to the direction of sweep used for delineation. { ˈtrās ˌinˈtərˈvəl }

tracer [CHEM] A foreign substance, usually radioactive, that is mixed with or attached to a given substance so the distribution or location of the latter can later be determined; used to trace chemical behavior of a natural element in an organism. Also known as tracer element. [ENG] A thread of contrasting color woven into the insulation of a wire for identification purposes. { ˈtrāˈsər }

tracer bullet [ORD] A bullet containing a pyrotechnic mixture to make the flight of the projectile visible by day and night. { ˈtrāˈsər ˌbulˈət }

tracer element See tracer. { ˈtrāˈsər ˌelˈəˈmənt }

tracer gas [ENG] In vacuum testing for leaks, a gas emitting through a leak in a pressure system and subsequently conducted into the detector. { ˈtrāˈsər ˌgas }

tracer milling [MECH ENG] Cutting a duplicate of a three-dimensional form by using a mastic form to direct the tracer-controlled cutter. { ˈtrāˈsər ˌmilˈiŋ }

tracer mixture [ORD] A pyrotechnic composition used for loading tracer bullets. { ˈtrāˈsər ˌmiksˈchər }

trace routine [COMPUT SCI] A routine which tracks the execution of a program, step by step, to locate a program malfunction. Also known as tracing routine. { ˈtrās ˌrüˌtēn }

trace sensitivity [ELECTR] The ability of an oscilloscope to produce a visible trace on the scope face for a specified input voltage. { ˈtrās ˌsenˈsəˌtīvˈədˈē }

trace slip [GEOL] That component of the net slip in a fault which is parallel to the trace of an index plane on a fault plane. { ˈtrās ˌslip }

trace-slip fault [GEOL] A fault whose net slip is trace slip. { ˈtrās ˌslip ˌfȯlt }

trace statement [COMPUT SCI] A statement, included in certain programming languages, that causes certain error-checking procedures to be carried out on specified segments of a source program. { ˈtrās ˌstātˈmənt }

trachea [ANAT] The cartilaginous and membranous tube by which air passes to and from the lungs in humans and many vertebrates. [BOT] A xylem vessel resembling the trachea of vertebrates. [INV ZOO] One of the anastomosing air-conveying tubules composing the respiratory system in most insects. { ˈtrāˈkēˈə }

tracheary [BOT] Water-conducting. { ˈtrāˈkēˌerˈē }

tracheid [BOT] An elongate, spindle-shaped xylem cell, lacking protoplasm at maturity, and having secondary walls laid in various thicknesses and patterns over the primary wall. { ˈtrāˈkēˈəd }

Tracheophyta [BOT] A large group of plants characterized by the presence of specialized conducting tissues (xylem and phloem) in the roots, stems, and leaves. { ˌtrāˈkēˈäfˈədˈē }

tracheophyte See Tracheophyta. { ˈtrāˈkēˈəˌfīt }

trachoma [MED] An infectious disease of the conjunctiva and cornea caused by *Chlamydia trachomatis* producing photophobia, pain, and excessive lacrimation. { trəˈkōˈmə }

trachomatous conjunctivitis [MED] Inflammation of the conjunctiva associated with trachoma, characterized by a subepithelial cellular infiltration with a follicular distribution. { trəˈkämˈədˈəs kənˌjəŋkˈtəˈvīdˈəs }

trachybasalt [PETR] An extrusive rock characterized by calcic plagioclase and sanidine, with augite, olivine, and possibly minor analcime or leucite. { ˌtraˈkēˈbəˈsȯlt }

Trachylina [INV ZOO] An order of moderate-sized jellyfish of the class Hydrozoa distinguished by having balancing organs and either a small polyp stage or none. { ˌtrakˈəˈlīnˈə }

Trachymedusae [INV ZOO] A group of marine jellyfish, recognized as a separate order or as belonging to the order Trachylina whose tentacles have a solid core consisting of a single row of endodermal cells. { ˌtrāˈkēˈməˈdüˌsē }

TOXOGLOSSATE RADULA

(a) (b)

Toxoglossate radula from marine gastropods: (*a*) *Conus* species; (*b*) *Terebra* species. (*From R. R. Shrock and W. H. Twenhoffel, Principles of Invertebrate Paleontology, 2d ed., McGraw-Hill, 1953*)

TRACE FOSSIL

Feeding trail, a trace fossil resulting from the grazing of animals along the surface of a sediment, in ichnogenus *Cosmorhaphe*, from Tertiary sediments at Pologne. (*From R. C. Moore, ed., Treatise on Invertebrate Paleontology, pt. W, University of Kansas Press, 1962*)

TRACHEID

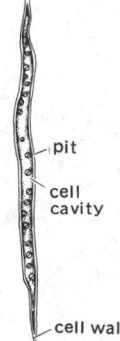

pit

cell cavity

cell wall

Drawing of a tracheid, showing characteristic features. (*From H. J. Fuller and O. Tippo, College Botany, rev. ed., Holt, 1954*)

Trachypsammiacea [INV ZOO] An order of colonial anthozoan cnidarians characterized by a dendroid skeleton. { ¦tra·kē¸sam·ē'ās·ē·ə }

Trachystomata [VERT ZOO] The name given to the Meantes when the group is considered to be an order. { ¦tra·kē'stō·məd·ē }

trachyte [PETR] The light-colored, aphanitic rock (the volcanic equivalent of syenite), composed largely of alkali feldspar with minor amounts of mafic minerals. { 'tra¸kīt }

trachytoid texture [GEOL] The texture of a phaneritic extrusive igneous rock in which the microlites of a mineral, not necessarily feldspar, in the groundmass have a subparallel or randomly divergent alignment. { 'trak·ə¸tȯid ¸teks·chər }

tracing distortion [ENG ACOUS] The nonlinear distortion introduced in the reproduction of a mechanical recording because the curve traced by the motion of the reproducing stylus is not an exact replica of the modulated groove. { 'trās·iŋ di¸stȯr·shən }

tracing paper [MATER] Thin paper used both for tracing and original drawings, and of various types and surfaces. { 'trās·iŋ ¸pā·pər }

tracing routine *See* trace routine. { 'trās·iŋ rü¸tēn }

track [AERO ENG] The actual line of movement of an aircraft or a rocket over the surface of the earth; it is the projection of the history of the flight path on the surface. Also known as flight track. [COMPUT SCI] The recording path on a rotating surface. [DES ENG] As applied to a pattern of setting diamonds in a bit crown, an arrangement of diamonds in concentric circular rows in the bit crown, with the diamonds in a specific row following in the track cut by a preceding diamond. [ELECTR] **1.** A path for recording one channel of information on a magnetic tape, drum, or other magnetic recording medium; the location of the track is determined by the recording equipment rather than by the medium. **2.** The trace of a moving target on a plan-position-indicator radar screen or an equivalent plot. [ENG] **1.** The groove cut in a rock by a diamond inset in the crown of a bit. **2.** A pair of parallel metal rails for a railway, railroad, tramway, or for any wheeled vehicle. [MECH ENG] **1.** The slide or rack on which a diamond-drill swivel head can be moved to positions above and clear of the collar of a borehole. **2.** A crawler mechanism for earth-moving equipment. Also known as crawler track. [NAV] **1.** To follow the movements of an object by keeping the reticle of an optical system or a radar beam on the object, by plotting its bearing and distance at frequent intervals, or by a combination of the two. **2.** To navigate by following the movements of a craft without regard for future positions; this is used when frequent changes of an unanticipated amount are expected in course or speed or both. **3.** A recommended route on a nautical chart, such as a North Atlantic Track. [NUCLEO] **1.** The visible path of an ionizing particle in a particle detector, such as a cloud chamber, bubble chamber, spark chamber, or nuclear photographic emulsion. **2.** *See* race track. { trak }

track adaptation effect [ACOUS] A drop in the sound insulation of a wall at the frequency at which bending oscillations excited by obliquely incident sound waves correspond to the free bending oscillations of the wall. Also known as coincidence effect. { ¦trak a¸dap'tā·shən i¸fekt }

track angle [NAV] Track measured from 0° at the reference direction clockwise or counterclockwise through 90 or 180°; it is labeled with the reference direction as a prefix and the direction of measurement from the reference direction as a suffix; thus, track angle N44°W is 44° west of north, or 316°. { 'trak ¸aŋ·gəl }

trackball [COMPUT SCI] A ball inset in the console of a video display terminal, which can be rotated by the operator, and whose motion is followed by a cursor on the display screen. { 'trak¸bȯl }

track cable [ENG] Steel wire rope, usually a locked-coil rope which supports the wheels of the carriers of a cableway. { 'trak ¸kā·bəl }

track cable scraper [MIN ENG] A type of excavator that uses a bottomless scraper bucket which conveys its load over the ground and is operated by a two-drum hoist which controls a track cable that spans the working area and a haulage cable that leads to the front of the bucket. { 'trak ¸kā·bəl 'skrāp·ər }

track chart [NAV] A chart showing recommended, required, or established tracks, and usually indicating turning points, courses, and distances. { 'trak ¸chärt }

track crawling [NAV] Keeping a craft close to a course line by frequent small changes of heading. { 'trak ¸krȯl·iŋ }

tracker [COMPUT SCI] An input device used in a virtual environment, which is capable of reporting its location in space and its orientation. { 'trak·ər }

track gage [CIV ENG] The width between the rails of a railroad track; in the United States the standard gage is 4 feet 8¹/₂ inches. { 'trak ¸gāj }

track haulage [MIN ENG] Movement or transportation of excavated or mined materials in cars or trucks that run on rails. { 'trak ¸hȯl·ij }

track homing [NAV] The process of following a line of position known to pass through an objective. { 'trak ¸hōm·iŋ }

track hopper [ENG] A hopper-shaped receiver mounted beside or below railroad tracks, into which railroad boxcars or bottom-dump cars are discharged; used for solid materials. { 'trak ¸häp·ər }

tracking [ELEC] A leakage or fault path created across the surface of an insulating material when a high-voltage current slowly but steadily forms a carbonized path. [ELECTR] The condition in which all tuned circuits in a receiver accurately follow the frequency indicated by the tuning dial over the entire tuning range. [ENG] **1.** A motion given to the major lobe of a radar or radio antenna such that some preassigned moving target in space is always within the major lobe. **2.** The process of following the movements of an object; may be accomplished by keeping the reticle of an optical system or a radar beam on the object, by plotting its bearing and distance at frequent intervals, or by a combination of techniques. [ENG ACOUS] **1.** The following of a groove by a phonograph needle. **2.** Maintaining the same ratio of loudness in the two channels of a stereophonic sound system at all settings of the ganged volume control. [NAV] Navigation which follows the movements of a craft but does not anticipate future positions. { 'trak·iŋ }

Tracking and Data Relay Satellite System [COMMUN] A system providing telecommunication services between low-earth-orbiting user spacecraft and user control centers; it consists of a series of geostationary spacecraft and an earth terminal located at White Sands, New Mexico. Abbreviated TDRSS. { ¦trak·iŋ an ¦dad·ə ¦rē¸lā 'sad·ə¸līt ¸sis·təm }

tracking beam [ORD] The beam that is aimed directly at the target at all times in antimissile warfare; data obtained from this beam are transmitted to the counterattacking guided missile over what is known as the guidance beam. { 'trak·iŋ ¸bēm }

tracking cross [COMPUT SCI] A cross displayed on the screen of a video terminal which automatically follows a light pen. Also known as tracking cursor. { 'trak·iŋ ¸krȯs }

tracking cursor *See* tracking cross. { 'trak·iŋ ¸kər·sər }

tracking error [ENG ACOUS] Deviation of the vibration axis of a phonograph pickup from tangency with a groove; true tangency is possible for only one groove when the pickup arm is pivoted; the longer the pickup arm, the less is the tracking error. { 'trak·iŋ ¸er·ər }

tracking filter [ELECTR] Electronic device for attenuating unwanted signals while passing desired signals, by phase-lock techniques that reduce the effective bandwidth of the circuit and eliminate amplitude variations. { 'trak·iŋ ¸fil·tər }

tracking jitter [ENG] Minor variations in the pointing of an automatic tracking radar. { 'trak·iŋ ¸jid·ər }

tracking network [ENG] A group of tracking stations whose operations are coordinated in tracking objects through the atmosphere or space. { 'trak·iŋ ¸net¸wərk }

tracking problem [CONT SYS] The problem of determining a control law which when applied to a dynamical system causes its output to track a given function; the performance index is in many cases taken to be of the integral square error variety. { 'trak·iŋ ¸präb·ləm }

tracking radar [ENG] Radar used to monitor the flight and obtain geophysical data from space probes, satellites, and high-altitude rockets. { 'trak·iŋ ¸rā¸där }

tracking station [ENG] A radio, radar, or other station set up to track an object moving through the atmosphere or space. { 'trak·iŋ ¸stā·shən }

tracking system [ENG] Apparatus, such as tracking radar, used in following and recording the position of objects in the sky. { 'trak·iŋ ¸sis·təm }

tracking telescope [OPTICS] A long-focal-length telescope mounted to track missiles in flight precisely while collecting missile performance data. { 'trak·iŋ ¸tel·ə¸skōp }

track in range [ELECTR] To adjust the gate of a radar set so that it opens at the correct instant to accept the signal from a target of changing range from the radar. { 'trak in 'rānj }

track-laying vehicle [ORD] Vehicle which travels upon two endless tracks, one on each side of the machine; it has high mobility and can maneuver, is usually armed and frequently armored, and is intended for tactical use; tanks are one type of track-laying vehicle. { 'trak ‚lā·iŋ ‚vē·ə·kəl }

trackless mine [MIN ENG] A mine in which rubber-tired vehicles are used for haulage and transport. { 'trak·ləs 'mīn }

trackless tunneling [MIN ENG] Tunneling by means of loaders mounted on caterpillars. { 'trak·ləs 'tən·əl·iŋ }

track made good [AERO ENG] The actual path of an aircraft over the surface of the earth, or its graphic representation. { 'trak ‚mād 'gúd }

track pitch [ELECTR] The physical distance between track centers. { 'trak ‚pich }

track-return power system [ELEC] A system for distributing electric power to trains or other vehicles, in which the track rails are used as an uninsulated return conductor. { 'trak ri‚tərn 'paú·ər ‚sis·təm }

trackshifter [ENG] A machine or appliance used to shift a railway track laterally. { 'trak ‚shif·tər }

track telling [COMMUN] The process of communicating air surveillance and tactical data information between command and control systems and facilities within the systems. { 'trak ‚tel·iŋ }

track-to-track access time [COMPUT SCI] The time required for a read-write head to move between the adjacent cylinders of a disk. { ‚trak tə ‚trak 'ak‚ses ‚tīm }

track-while-scan [ELECTR] Electronic system used to detect a radar target, compute its velocity, and predict its future position without interfering with continuous radar scanning. { ‚trak ‚wīl 'skan }

traction [GEOL] Transport of sedimentary particles along and parallel to a bottom surface of a stream channel by rolling, sliding, dragging, pushing, or saltation. [GRAPHICS] A defect in a paint coating in which the film cracks and wide fissures reveal the underlying surface. [MECH] Pulling friction of a moving body on the surface on which it moves. { 'trak·shən }

tractional force [FL MECH] The force exerted on particles under flowing water by the current; it is proportional to the square of the velocity. { 'trak·shən·əl 'fórs }

traction alopecia [MED] Hair loss that is the result of prolonged, tightly pulled hairstyles. { ‚trak·shən ‚al·ə'pē·shə }

traction-control system [MECH ENG] An acceleration sensor-control system which, when a driving tire has no traction, slows the wheel movement by braking or reduces the engine speed and torque if braking alone will not prevent wheel spin. { 'trak·shən kən'trōl ‚sis·təm }

traction diverticulum [MED] A circumscribed sacculation, usually of the esophagus, with bulging of the full thickness of the wall; caused by the pull of adhesions arising from adjacent organs. { 'trak·shən ‚dī·vər'tik·yə·ləm }

traction meter [ENG] A load-sensing device placed between a locomotive and the car immediately behind it to measure pulling force exerted by the locomotive. { 'trak·shən ‚mēd·ər }

traction tube [ENG] A device for measuring the minimum water velocities capable of moving various sizes of sand grains; it consists of a horizontal glass tube half-filled with sand. { 'trak·shən ‚tüb }

tractor [MECH ENG] **1.** An automotive vehicle having four wheels or a caterpillar tread used for pulling agricultural or construction implements. **2.** The front pulling section of a semitrailer. Also known as truck-tractor. { 'trak·tər }

tractor drill [MECH ENG] A drill having a crawler mounting to support the feed-guide bar on an extendable arm. { 'trak·tər ‚dril }

tractor-feed printer See pin-feed printer. { 'trak·tər ‚fēd 'print·ər }

tractor fuel See engine distillate. { 'trak·tər ‚fyül }

tractor gate [CIV ENG] A type of outlet control gate used to release water from a reservoir; there are two types, roller and wheel. { 'trak·tər ‚gāt }

tractor loader [MECH ENG] A tractor equipped with a tipping bucket which can be used to dig and elevate soil and rock fragments to dump at truck height. Also known as shovel dozer; tractor shovel. { 'trak·tər ‚lōd·ər }

tractor shovel See tractor loader. { 'trak·tər ‚shəv·əl }

tractrix [MATH] A curve in the plane where every tangent to it has the same length. Also known as equitangential curve. { 'trak‚triks }

trade air [METEOROL] The type of air of which the trade winds consist, and whose chief thermodynamic characteristic is the presence of the trade-wind inversion. { 'trād ‚er }

trade cumulus See trade-wind cumulus. { 'trād ‚kyü·myə·ləs }

trade-wind [METEOROL] The wind system, occupying most of the tropics, which blows from the subtropical highs toward the equatorial trough, a major component of the general circulation of the atmosphere; the winds are northeasterly in the Northern Hemisphere and southeasterly in the Southern Hemisphere; hence they are known as the northeast trades and southeast trades, respectively. { 'trād ‚wind }

trade-wind cumulus [METEOROL] The characteristic cumulus cloud of the trade winds over the oceans in average, undisturbed weather conditions; the individual cloud usually exhibits a blocklike appearance since its vertical growth ends abruptly in the lower stratum of the trade-wind inversion; a group of fully grown clouds shows considerable uniformity in size and shape. Also known as trade cumulus. { 'trād ‚wind ‚kyü·myə·ləs }

trade-wind desert [CLIMATOL] **1.** An area of very little rainfall and high temperature which occurs where the trade winds or their equivalent (such as the harmattan) blow over land; the best examples are the Sahara and Kalahari deserts. **2.** The arid cold-water coasts on the western shores of North and South America and Africa. { 'trād ‚wind ‚dez·ərt }

trade-wind inversion [METEOROL] A characteristic temperature inversion usually present in the trade-wind streams over the eastern portions of the tropical oceans; it is formed by broad-scale subsidence of air from high altitudes in the eastern extremities of the subtropical highs; while descending, the current meets the opposition of the low-level maritime air flowing equatorward; the inversion forms at the meeting point of these two strata which flow horizontally in the same direction. { 'trād ‚wind in‚vər·zhən }

traersu [METEOROL] A violent east wind of Lake Garda in Italy. { 'trer‚zü }

traffic [COMMUN] The messages transmitted and received over a communication channel. [ENG] The passage or flow of vehicles, pedestrians, ships, or planes along defined routes such as highways, sidewalks, sea lanes, or air lanes. { 'traf·ik }

trafficability [CIV ENG] Capability of terrain to bear traffic, or the extent to which the terrain will permit continued movement of any or all types of traffic. { ‚traf·ə·kə'bil·əd·ē }

traffic-circulation map [MAP] A map showing traffic routes and the measures for traffic regulation; indicates the roads for use of certain classes of traffic, the location of traffic-control stations, and the directions in which traffic may move. Also known as circulation map. { 'traf·ik ‚sər·kyə'lā·shən ‚map }

traffic control [ENG] Control of the movement of vehicles, such as airplanes, trains, and automobiles, and the regulatory mechanisms and systems used to exert or enforce control. { 'traf·ik kən‚trōl }

traffic cop [CONT SYS] The portion of a programmable controller's executive program concerned with input/output. { 'traf·ik ‚käp }

traffic density [CIV ENG] The average number of vehicles that occupy 1 mile or 1 kilometer of road space, expressed in vehicles per mile or per kilometer. { 'traf·ik ‚den·səd·ē }

traffic diagram [COMMUN] Chart or illustration used to show the movement and control of traffic over a communications system. { 'traf·ik dī·ə‚gram }

traffic distribution [COMMUN] Routing of communications traffic through a terminal to a switchboard or dialing center. { 'traf·ik ‚di·strə‚byü·shən }

traffic engineering [CIV ENG] The determination of the required capacity and layout of highway and street facilities

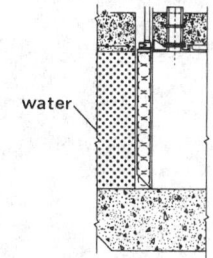

that can safely and economically serve vehicular movement between given points. { 'traf·ik ‚en·jə‚nir·iŋ }

traffic flow [CIV ENG] The total number of vehicles passing a given point in a given time, expressed as vehicles per hour. { 'traf·ik ‚flō }

traffic flow security [COMMUN] Transmission of an uninterrupted flow of random text on a wire or radio link between two stations with no indication to an interceptor of what portions of this steady stream constitute encrypted message text and what portions are merely random filler. { 'traf·ik ‚flō si‚kyùr·əd·ē }

traffic forecast [COMMUN] Traffic level prediction on which communications system management decisions and engineering effort are based. { 'traf·ik ‚fòr‚kast }

traffic noise [ENG] The general disturbance in sonar transmissions which is due to ships but is not associated with a specific vessel. { 'traf·ik ‚nòiz }

traffic pattern [AERO ENG] The traffic flow that is prescribed for aircraft landing at, taxiing on, and taking off from an airport; the usual components of a traffic pattern are upwind leg, crosswind leg, downwind leg, base leg, and final approach. { 'traf·ik ‚pad·ərn }

traffic recorder [ENG] A mechanical counter or recorder used to determine traffic movements (hourly variations and total daily volumes of traffic at a point) on an existing route; the air-impulse counter, magnetic detector, photoelectric counter, and radar detector are used. { 'traf·ik ri‚kórd·ər }

traffic signal [CIV ENG] With the exception of traffic signs, any power-operated device for regulating, directing, or warning motorists or pedestrians. { 'traf·ik ‚sig·nəl }

tragacanth [MATER] The gummy exudate produced by certain Asiatic species of *Astragalus*; consists of a soluble portion containing uronic acid and arabinose, and an insoluble portion that absorbs water and swells to make a stiff opalescent mucilage. { 'traj·ə‚kanth }

tragion-nasal root [ANTHRO] A measure of the distance from the tragion to the deepest concavity of the nasal root. { 'traj·ē‚än 'nā·zəl ‚rüt }

Tragulidae [VERT ZOO] The chevrotains, a family of pecoran ruminants in the superfamily Traguloidea. { trə'gyül·ə‚dē }

Traguloidea [VERT ZOO] A superfamily of pecoran ruminants, composed of the most primitive forms with large canines; the chevrotain is the only extant member. { ‚tra·gyə'lòid·ē·ə }

tragus [ANAT] **1.** The prominence in front of the opening of the external ear. **2.** One of the hairs in the external ear canal. { 'trā·gəs }

trail [ASTRON] A luminous trace left in the sky by the passage of a large meteor. [GEOL] A line of rock fragments that were picked up by glacial ice at a localized outcropping and left scattered along a fairly well-defined tract during the movement of a glacier. [GRAPHICS] One of the lines left on a photographic plate during prolonged exposure to starlight if the motion of the plate was not synchronized with the apparent motion of the sky. [ORD] **1.** In bombing, the line between the point of impact of the bomb and a point on the ground directly beneath the aircraft at the moment of impact, assuming that the aircraft stays on course after release of the bomb and maintains a constant speed. **2.** Rear part of a gun carriage which connects the piece with a limber or tractor. { trāl }

T rail [CIV ENG] A rail shaped like a T in cross section due to a wide head, web, and flanged base. { 'tē ‚rāl }

trail angle [AERO ENG] The angle at an aircraft between the vertical and the line of sight to an object over which the aircraft has passed. { 'trāl ‚aŋ·gəl }

trailer [ELECTR] A bright streak at the right of a dark area or dark line in a television picture, or a dark area or streak at the right of a bright part; usually due to insufficient gain at low video frequencies. [MECH ENG] The section of a semitrailer that is pulled by the tractor. { 'trā·lər }

trailer label [COMPUT SCI] A record appearing at the end of a magnetic tape that uniquely identifies the tape as one required by the system. { 'trā·lər ‚lā·bəl }

trailer record [COMPUT SCI] A record which contains data pertaining to an associated group of records immediately preceding it. { 'trā·lər ‚rek·ərd }

trail formation [AERO ENG] Aircraft flying singly or in elements in such manner that each aircraft or element is in line behind the preceding aircraft or element. [ENG] Vehicles proceeding one behind the other at designated intervals. Also known as column formation. { 'trāl fòr‚mā·shən }

trailing antenna [ELECTROMAG] An aircraft radio antenna having one end weighted and trailing free from the aircraft when in flight. { 'trāl·iŋ an‚ten·ə }

trailing edge [AERO ENG] The rear section of a multipiece airfoil, usually that portion aft of the rear spar. [ELECTR] The major portion of the decay of a pulse. { 'trāl·iŋ 'ej }

trailing-edge tab [AERO ENG] One of the devices on the aircraft elevator that reduce or eliminate hinge movements required to deflect the elevator during flight. { 'trāl·iŋ ‚ej ‚tab }

trailing pad [COMPUT SCI] Characters placed to the right of information in a field of data to fulfill length requirements or for cosmetic purposes. { 'trāl·iŋ ‚pad }

trailing zero [MATH] Any zero following the last nonzero integer of a number. { 'trāl·iŋ 'zir·ō }

trail pheromone [PHYSIO] A type of pheromone used by social insects and some lepidopterans to recruit others of its species to a food source. { 'trāl ‚fer·ə‚mōn }

train [ASTRON] The bright tail of a comet or meteor. [ENG] To aim or direct a radar antenna in azimuth. { trān }

train bombing [ORD] A method of bombing in which two or more bombs are released at a predetermined interval from one aircraft as the result of a single actuation of the bomb-release mechanism. { 'trān ‚bäm·iŋ }

trainer [ELECTR] A piece of equipment used for training operators of radar, sonar, and other electronic equipment by simulating signals received under operating conditions in the field. { 'trā·nər }

training aid [ENG] Any item which is developed or procured primarily to assist in training and the process of learning. { 'trān·iŋ ‚ād }

training ammunition [ORD] Ammunition used for training persons in marksmanship, handling weapons, and so forth. { 'trān·iŋ ‚am·yə‚nish·ən }

training data [CONT SYS] Data entered into a robot's computer at the beginning of an operation. { 'trān·iŋ ‚dad·ə }

training time [COMPUT SCI] The machine time expended in training employees in the use of the equipment, including such activities as mounting, console operation, converter operation, and printing operation, and time spent in conducting required demonstrations. { 'trān·iŋ ‚tīm }

training wall [CIV ENG] A wall built along the bank of a river or estuary parallel to the direction of flow to direct and confine the flow. { 'trān·iŋ ‚wòl }

train printer [COMPUT SCI] A computer printer in which the characters are carried in a track and a hammer strikes the proper character against the paper as it passes the print position. { 'trān ‚print·ər }

train shed [CIV ENG] **1.** A structure to protect trains from weather. **2.** The part of a railroad station that covers the tracks. { 'trān ‚shed }

trajectory [GEOPHYS] The path followed by a seismic wave. [MATH] A curve that intersects all the members of a given family of curves at the same angle. [MECH] The curve described by an object moving through space, as of a meteor through the atmosphere, a planet around the sun, a projectile fired from a gun, or a rocket in flight. { trə'jek·trē }

trajectory chart [ORD] Diagram of a side view of the paths of projectiles fired at various elevations under standard conditions; the trajectory chart is different for different guns, projectiles, and fuses. { trə'jek·trē ‚chärt }

trajectory control [CONT SYS] A type of continuous-path control in which a robot's path is calculated based on mathematical models of joint acceleration, arm loads, and actuating signals. { trə'jek·trē kən‚trōl }

trajectory-measuring system [ENG] A system used to provide information on the spatial position of an object at discrete time intervals throughout a portion of the trajectory or flight path. { trə'jek·trē ‚mezh·ə·riŋ ‚sis·təm }

trama [MYCOL] The loosely woven hyphal tissue between adjacent hymenia in basidiomycetes. { 'trä·mə }

Trametes versicolor [MYCOL] A brightly colored mushroom that appears to have antitumor properties; it is a common inhabitant of the woods worldwide. Also known as Coriolus versicolor; turkey tail mushroom. { trə‚mēd‚ēz 'vər·sə‚kəl·ər }

trammel [ENG] A device consisting of a bar, each of whose ends is constrained to move along one of two perpendicular lines; used in drawing ellipses and in the Rowland mounting. { 'tram·əl }

tramming [MIN ENG] Pushing tubs, mine cars, or trams by hand. { 'tram·iŋ }

tramontana [METEOROL] A cold wind from the northeast or north, particularly on the west coast of Italy and northern Corsica, but also in the Balearic Islands and the Ebro valley in Catalonia. { ˌträ·mōn'tä·nə }

tramp metal [MIN ENG] Unwanted metal which finds its way into the mill ore stream. { 'tramp ˌmed·əl }

tramp-metal detector [MIN ENG] A sensing device which detects presence of unwanted metal in an ore stream, and sounds an alarm or removes the metal. { 'tramp ˌmed·əl di̇ˌtek·tər }

tramway [MECH ENG] An overhead rail, rope, or cable on which wheeled cars run to convey a load. { 'tram·ˌwā }

tranexamic acid [PHARM] $C_8H_{15}NO_2$ Crystals which soften at 270°C and are soluble in water; used as a hemostatic agent. Abbreviated AMCHA. { ˌtran·ik'sam·ik 'as·əd }

tranquilizer [PHARM] **1.** Any agent that brings about a state of relief from anxiety, or peace of mind. **2.** Any agent that produces a calming or sedative effect without inducing sleep. **3.** Any drug, such as chlorpromazine, used primarily for its calming and antipsychotic effects, or such as meprobamate, used for symptomatic treatment of common psychoneuroses and as an adjunct in somatic disorders complicated by anxiety and tension. { 'traŋ·kwəˌlīz·ər }

transacter [COMPUT SCI] A system in which data from sources in a number of different locations, as in a factory, are transmitted to a data-processing center and immediately processed by a computer. { tran'sak·tər }

transactinide elements [CHEM] In the periodic table, elements with atomic numbers higher than 103. { tranz'ak·tə‚nīd ˌel·ə·məns }

transaction [COMPUT SCI] General description of updating data relevant to any item. { tran'sak·shən }

transaction data [COMPUT SCI] A set of data in a data-processing area in which the incidence of the data is essentially random and unpredictable; hours worked, quantities shipped, and amounts invoiced are examples from, respectively, the areas of payroll, accounts receivable, and accounts payable. { tran'sak·shən ˌdad·ə }

transaction file See detail file. { tran'sak·shən ˌfīl }

transaction processing system [COMPUT SCI] A system which processes predefined transactions, one at a time, with direct, on-site entry of the transactions into a terminal, and which produces predefined outputs and maintains the necessary data base. { tran'sak·shən 'prä·ses·iŋ ˌsis·təm }

transaction record See change record. { tran'sak·shən ˌrek·ərd }

transaction tape See change tape. { tran'sak·shən ˌtāp }

transadmittance [ELECTR] A specific measure of transfer admittance under a given set of conditions, as in forward transadmittance, interelectrode transadmittance, short-circuit transadmittance, small-signal forward transadmittance, and transadmittance compression ratio. { tranz'ad'mit·əns }

transaminase [BIOCHEM] One of a group of enzymes that catalyze the transfer of the amino group of an amino acid to a keto acid to form another amino acid. Also known as aminotransferase. { tranz'am·ə‚nās }

transamination [CHEM] **1.** The transfer of one or more amino groups from one compound to another. **2.** The transposition of an amino group within a single compound. { tranˌsam·ə'nā·shən }

transcapsidation [VIROL] Change in the capsid of PARA (particle aiding replication of adenovirus) from one type of adenovirus to another. { ˌtranzˌkap·sə'dā·shən }

transceiver [COMPUT SCI] A computer terminal that can transmit and receive information to and from an input/output channel. [ELECTR] A radio transmitter and receiver combined in one unit and having switching arrangements such as to permit both transmitting and receiving. Also known as transmitter-receiver. { tran'sē·vər }

transceiver data link [COMPUT SCI] Integrated data processing by means of punched cards, using transceivers as terminal equipment; the transmission path can be wire or radio. { tran'sē·vər 'dad·ə ˌliŋk }

transcendence base [MATH] A transcendence base of a field E over a subfield F is a subset S of E which is algebraically

independent over F and is not a proper subset of any other subset S′ which is algebraically independent over F. { tran'sen·dəns ˌbās }

transcendence degree [MATH] The transcendence degree of a field E of a subfield F is the number of elements in a transcendence base of E over F. Also known as transcendence dimension. { tran'sen·dəns diˌgrē }

transcendence dimension See transcendence degree. { tran'sen·dəns diˌmen·chən }

transcendental curve [MATH] The graph of a transcendental function. { ˌtran·sən'den·təl 'kərv }

transcendental element [MATH] An element of a field K is transcendental relative to a subfield F if it satisfies no polynomial whose coefficients come from F. { ˌtranˌsen'dent·əl 'el·ə·mənt }

transcendental field extension [MATH] A field extension K of F where the elements of K not in F are all transcendental relative to F. { ˌtranˌsen'dent·əl 'fēld ikˌsten·chən }

transcendental functions [MATH] Functions which cannot be given by any algebraic expression involving only their variables and constants. { ˌtranˌsen'dent·əl 'fəŋk·shənz }

transcendental number [MATH] An irrational number that is the root of no polynomial with rational-number coefficients. { ˌtranˌsen'dent·əl 'nəm·bər }

transcendental term [MATH] In an expression, a term that cannot be expressed solely by numbers and algebraic symbols. { ˌtran·sən'den·təl 'tərm }

transconductance [ELECTR] **1.** An electron-tube rating, equal to the change in plate current divided by the change in control-grid voltage that causes it, when the plate voltage and all other voltages are maintained constant. Also known as grid-anode transconductance; grid-plate transconductance; mutual conductance. Symbolized G_m; g_m. **2.** A field-effect-transistor rating, equal to the change in drain current divided by the change in gate-to-source voltage that causes it, when the drain voltage and all other voltages are maintained constant. Symbolized g_{fs}. **3.** An amplifier parameter, equal to the change in output current divided by the change in input voltage that causes it. Symbolized g_m. { ˌtranz·kən·dək·təns }

transconductance amplifier [ELECTR] An amplifier whose output current (rather than output voltage) is proportional to its input voltage. { ˌtranz·kənˌduk·təns 'am·pləˌfī·ər }

transconductance-C filter [ELECTR] An integrated-circuit filter that combines the functions of an amplifier and a simulated resistor into a transconductance amplifier. { 'tranz·kənˌduk·təns 'sē ˌfil·tər }

transconductor See transconductance amplifier. { ˌtranz·kən·ˌdək·tər }

transcontinental ballistic missile [ORD] A ballistic missile having a range of at least 12,500 miles (20,000 kilometers), so it can be fired from any point on the earth's surface and reach any surface target. { ˌtranzˌkänt·ən'ent·əl bə'lis·tik 'mis·əl }

transcranial magnetic stimulation [MED] A neurophysiologic technique that stimulates the human brain via magnetic pulses generated by a small electromagnet placed on the scalp. { ˌtranzˌkrā·nē·əl magˌned·ik ˌstim·yə'lā·shən }

transcribe [COMPUT SCI] To copy, with or without translating, from one external computer storage medium to another. [ELECTR] To record, as to record a radio program by means of electric transcriptions or magnetic tape for future rebroadcasting. { tranz'krīb }

transcriber [COMPUT SCI] The equipment used to convert information from one form to another, as for converting computer input data to the medium and language used by the computer. { tranz'krī·bər }

transcriptase See ribonucleic acid polymerase. { tran'skripˌtās }

transcription [ENG ACOUS] A recording of a complete radio program, made especially for broadcast purposes. Also known as electrical transcription. [MOL BIO] The process by which ribonucleic acid is formed from deoxyribonucleic acid. { tranz'krip·shən }

transcription attenuation [CELL MOL] A form of gene expression regulation in bacteria whereby transcription is terminated shortly after it begins by preventing its continuation beyond the attenuator site. { tranz'krip·shən aˌten·yə'wā·shən }

transcription unit [MOL BIO] The segment of deoxyribonucleic acid between the sites of initiation and termination of

TRANSCONDUCTANCE-C FILTER

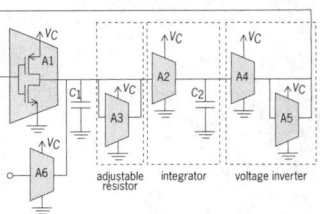

Circuitry of the filter, with amplifier A1 shown in detail.

transcription by ribonucleic acid polymerase. { tranz'krip· shən ,yü·nət }

transcrystalline [MET] Across the crystals of a metal; used of cracks in metals. Also known as intracrystalline; transgranular. { ¦tranz'krist·əl·ən }

transcurrent fault [GEOL] A strike-slip fault characterized by a steeply inclined surface. Also known as transverse thrust. { ¦tranz¦kə·rənt 'fȯlt }

transcytosis [CELL MOL] A form of intracellular vesicular traffic in which endocytosed macromolecules are transferred across the cell and released (via exocytosis) at the opposite plasma membrane domain. { ,tranz,sī'tō·səs }

transducer [ENG] Any device or element which converts an input signal into an output signal of a different form; examples include the microphone, phonograph pickup, loudspeaker, barometer, photoelectric cell, automobile horn, doorbell, and underwater sound transducer. { tranz'dü·sər }

transducer loss [ELECTR] The ratio of the power available to a transducer from a specified source to the power that the transducer delivers to a specified load; usually expressed in decibels. { tranz'dü·sər ,lȯs }

transduction [MICROBIO] Transfer of genetic material between bacterial cells by bacteriophages. { tranz'dək·shən }

transductor See magnetic amplifier; saturable reactor. { tranz'dək·tər }

transect [SCI TECH] To cut across, or to cut transversely. { tran'sekt }

transesterification [ORG CHEM] Conversion of an organic acid ester into another ester of that same acid. { ¦trans·e¦ster· ə·fə'kā·shən }

transfection [GEN] Infection of a cell with viral deoxyribonucleic acid or ribonucleic acid. { trans'fek·shən }

transfer [COMPUT SCI] See jump. [MIN ENG] A vertical or inclined connection between two or more levels, used as an ore pass. [NAV] **1.** The distance a vessel moves perpendicular to its initial direction in making a turn of 90° with a constant rudder angle. **2.** The distance a vessel moves perpendicular to its initial direction for turns of less than 90°. { 'tranz·fər }

transfer admittance [ELECTR] An admittance rating for electron tubes and other transducers or networks; it is equal to the complex alternating component of current flowing to one terminal from its external termination, divided by the complex alternating component of the voltage applied to the adjacent terminal on the cathode or reference side; all other terminals have arbitrary external terminations. { 'tranz·fər ad,mit·əns }

transferase [BIOCHEM] Any of various enzymes that catalyze the transfer of a chemical group from one molecule to another. { 'tranz·fə,rās }

transfer caliper [DES ENG] A caliper having one leg which can be opened (or closed) to remove the instrument from the piece being measured; used to measure inside recesses or over projections. { 'tranz·fər ,kal·ə·pər }

transfer car [MIN ENG] A quarry car provided with transverse tracks, on which the gang car may be conveyed to or from the saw gang. { 'tranz·fər ,kär }

transfer card See transition card. { 'tranz·fər ,kärd }

transfer case [MECH ENG] In a vehicle with more than one driving axle, a housing fitted with gears that distribute the driving power among the axles. { 'tranz·fər ,kās }

transfer chamber [ENG] In plastics processing, a vessel in which thermosetting plastic is softened by heat and pressure before being placed in a closed mold for final curing. { 'tranz· fər ,chām·bər }

transfer characteristic [ELECTR] **1.** Relation, usually shown by a graph, between the voltage of one electrode and the current to another electrode, with all other electrode voltages being maintained constant. **2.** Function which, multiplied by an input magnitude, will give a resulting output magnitude. **3.** Relation between the illumination on a camera tube and the corresponding output-signal current, under specified conditions of illumination. { 'tranz·fər ,kar·ik·tə,ris·tik }

transfer check [COMPUT SCI] Check (usually automatic) on the accuracy of the transfer of a word in a computer operation. { 'tranz·fər ,chek }

transfer chute [ENG] A chute used at a transfer point in a conveyor system; the chute is designed with a curved base or some other feature so that the load is discharged in a centralized stream and in the same direction as the receiving conveyor. { 'tranz·fər ,shüt }

transfer conditionally [COMPUT SCI] To copy, exchange, read, record, store, transmit, or write data or to change control or jump to another location according to a certain specified rule or in accordance with a certain criterion. { tranz'fər kən'dish·ən·ə·lē }

transfer constant [ENG] A transducer rating, equal to one-half the natural logarithm of the complex ratio of the product of the voltage and current entering a transducer to that leaving the transducer when the latter is terminated in its image impedance; alternatively, the product may be that of force and velocity or pressure and volume velocity; the real part of the transfer constant is the image attenuation constant, and the imaginary part is the image phase constant. Also known as transfer factor. { 'tranz·fər ,kän·stənt }

Transfer Control Protocol See Transmission Control Protocol. { ,tranz·fər kən'trōl ,prōd·ə,kȯl }

transfer ellipse See transfer orbit. { 'tranz·fər i,lips }

transference [PSYCH] The unconscious transfer of the patient's feelings and reactions originally associated with important persons in the patient's life, usually father, mother, or siblings, toward others and in the analytic situation, toward the analyst. { tranz'fər·əns }

transference number See transport number. { tranz'fər·əns ,nəm·bər }

transfer factor See transfer constant. { 'tranz·fər ,fak·tər }

transfer function [CONT SYS] The mathematical relationship between the output of a control system and its input: for a linear system, it is the Laplace transform of the output divided by the Laplace transform of the input under conditions of zero initial-energy storage. { 'tranz·fər ,fəŋk·shən }

transfer grille [ENG] In an air-conditioning system, a grille that permits air to flow from one space to another; may be one of a pair if installed on opposite sides of a wall or door. { 'tranz·fər ,gril }

transfer immunity See adoptive immunity. { 'tranz,fər i,myün·əd·ē }

transfer impedance [ELEC] The ratio of the voltage applied at one pair of terminals of a network to the resultant current at another pair of terminals, all terminals being terminated in a specified manner. { 'tranz·fər im,pēd·əns }

transfer-in-channel command [COMPUT SCI] A command used to direct channel control to a specified location in main storage when the next channel command word is not stored in the next location in sequence. { ¦tranz·fər in 'chan·əl kə,mand }

transfer instruction [COMPUT SCI] Step in computer operation specifying the next operation to be performed, which is not necessarily the next instruction in sequence. { 'tranz·fər in,strək·shən }

transfer interpreter [COMPUT SCI] A variation of a punched-card interpreter that senses a punched card and prints the punched information on the following card. Also known as posting interpreter. { 'tranz·fər in,tər·prəd·ər }

transfer machine [MECH ENG] **1.** Equipment that moves parts from one production location in a factory to another. **2.** A device that holds a workpiece and moves it automatically through the stages of a manufacturing process. { 'tranz·fər mə,shēn }

transfer matrix [CONT SYS] The generalization of the concept of a transfer function to a multivariable system; it is the matrix whose product with the vector representing the input variables yields the vector representing the output variables. { 'tranz·fər ,mā·triks }

transfer-matrix method [MECH] A method of analyzing vibrations of complex systems, in which the system is approximated by a finite number of elements connected in a chainlike manner, and matrices are constructed which can be used to determine the configuration and forces acting on one element in terms of those on another. { 'tranz·fər ,mā·triks ,meth·əd }

transfer molding [ENG] Molding of thermosetting materials in which the plastic is softened by heat and pressure in a transfer chamber, then forced at high pressure through suitable sprues, runners, and gates into a closed mold for final curing. { 'tranz· fər ,mōld·iŋ }

transfer of fire [ORD] Shifting of fire from one target to another, applying the corrections for the first target to the data for the second target. { 'tranz·fər əv 'fīr }

transfer operation [COMPUT SCI] An operation which moves information from one storage location or one storage medium to another (for example, read, record, copy, transmit, exchange). { 'tranz·fər ,ap·ə,rā·shən }

transfer orbit [AERO ENG] In interplanetary travel, an elliptical trajectory tangent to the orbits of both the departure planet and the target planet. Also known as transfer ellipse. { 'tranz·fər ,or·bət }

transfer rate [COMPUT SCI] The speed at which data are moved from a direct-access device to a central processing unit. { 'tranz·fər ,rāt }

transfer ratio [ENG] From one point to another in a transducer at a specified frequency, the complex ratio of the generalized force or velocity at the second point to the generalized force or velocity applied at the first point; the generalized force or velocity includes not only mechanical quantities, but also other analogous quantities such as acoustical and electrical; the electrical quantities are usually electromotive force and current. { 'tranz·fər ,rā·shō }

transfer reaction [NUC PHYS] A nuclear reaction in which one or more nucleons are exchanged between the target nucleus and an incident projectile. { 'tranz·fər rē,ak·shən }

transferred arc [MET] In plasma arc welding, an arc established between the electrode and the workpiece. { 'tranz'fərd 'ärk }

transferred-electron amplifier [ELECTR] A diode amplifier, which generally uses a transferred-electron diode made from doped n-type gallium arsenide, that provides amplification in the gigahertz range to well over 50 gigahertz at power outputs typically below 1 watt continuous-wave. Abbreviated TEA. { 'tranz'fərd i¦lek,trän 'am·plə,fī·ər }

transferred-electron device [ELECTR] A semiconductor device, usually a diode, that depends on internal negative resistance caused by transferred electrons in gallium arsenide or indium phosphide at high electric fields; transit time is minimized, permitting oscillation at frequencies up to several hundred megahertz. { 'tranz'fərd i¦lek,trän di'vīs }

transferred-electron effect [SOLID STATE] The variation in the effective drift mobility of charge carriers in a semiconductor when significant numbers of electrons are transferred from a low-mobility valley of the conduction band in a zone to a high-mobility valley, or vice versa. { 'tranz'fərd i¦lek,trän i,fekt }

transfer register [ENG] A transfer grille fitted with a mechanism for controlling the volume of airflow. { 'tranz·fər ,rej·ə·stər }

transfer ribonucleic acid [CELL MOL] The smallest ribonucleic acid molecule found in cells; its structure is complementary to messenger ribonucleic acid and it functions by transferring amino acids from the free state to the polymeric form of growing polypeptide chains. Abbreviated t-RNA. { 'tranz·fər ¦rī·bō·nü¦klē·ik 'as·əd }

transferrin [BIOCHEM] Any of various beta globulins in blood serum which bind and transport iron to the bone marrow and storage areas. { 'tranz'fer·ən }

transfer robot [CONT SYS] A fixed-sequence robot that moves parts from one location to another. { 'tranz·fər 'rō,bät }

transfer switch [ELEC] A switch for transferring one or more conductor connections from one circuit to another. { 'tranz·fər ,swich }

transfer test [COMMUN] Verification of transmitted information by temporary storing, retransmitting, and comparing. { 'tranz·fər ,test }

transfer unit [CHEM ENG] The relationship between the overall rate coefficient (for whatever transfer operation is being calculated), column volume, and fluid volumetric flow rate in fixed-bed sorption operations. { 'tranz·fər ,yü·nət }

transfinite induction [MATH] A reasoning process by which if a theorem holds true for the first element of a well-ordered set N and is true for an element n whenever it holds for all predecessors of n, then the theorem is true for all members of N. { tranz'fī,nīt in'dək·shən }

transfinite number [MATH] Any ordinal or cardinal number equal to or exceeding aleph null. { tranz'fī,nīt 'nəm·bər }

transfluxor [ELECTROMAG] A magnetic core having two or more apertures and three or more legs for flux; used as a computer memory element, crossbar switch, channel commutator, or control element. { ¦tranz'flək·sər }

transform [COMPUT SCI] To change the form of digital-computer information without significantly altering its meaning. [MATH] **1.** An expression, commonly used in harmonic analysis, formed from a given function f by taking an integral of

$f \cdot g$, where g is a member of an orthogonal family of functions. **2.** The value of a transformation at some point. **3.** A matrix B related to a given matrix A by $B = C^{-1}AC$, where C is a nonsingular matrix. **4.** *See* conjugate. { tranz'form (verb) *or* 'tranz,form (noun) }

transformation [CRYSTAL] *See* inversion. [ELEC] For two networks which are equivalent as far as conditions at the terminals are concerned. a set of equations giving the admittances or impedances of the branches of one circuit in terms of the admittances or impedances of the other. [GEN] **1.** Transfer and incorporation of foreign deoxyribonucleic acid (DNA) into a cell and subsequent recombination of part or all of that DNA into the cell's genome. Also known as bacterial transformation; genetic transformation. **2.** Conversion of a normal cell to a neoplastic cell by a cascade of events under the control of different classes of oncogenes. Also known as cellular transformation. [GRAPHICS] The process of projecting a photograph (mathematically, graphically, or photographically) from its plane onto another plane by translation, rotation, or scale change. [IMMUNOL] Change in a lymphocyte from a small, resting lymphocyte into a large lymphocyte following stimulation by antigens or lectin, or viral infection. Also known as lymphocyte transformation. [MATH] A function, usually between vector spaces. { ,tranz·fər'mā·shən }

transformation constant *See* decay constant. { ,tranz·fər'mā·shən ,kän·stənt }

transformation group [MATH] **1.** A collection of transformations which forms a group with composition as the operation. **2.** A dynamical system or, more generally, a topological group G together with a topological space X where each g in G gives rise to a homeomorphism of X in a continuous manner with respect to the algebraic structure of G. { ,tranz·fər'mā·shən ,grüp }

transformation matrix [ELECTROMAG] A two-by-two matrix which relates the amplitudes of the traveling waves on one side of a waveguide junction to those on the other. { ,tranz·fər'mā·shən ,mā·triks }

transformation methods [MATH] A category of numerical methods for finding the eigenvalues of a matrix, in which a series of orthogonal transformations are used to reduce the matrix to some simpler matrix, usually a triple-diagonal one, before an attempt is made to find the eigenvalues. { ,tranz·fər'mā·shən ,meth·ədz }

transformation of similitude *See* homothetic transformation. { ,tranz·fər'mā·shən əv si'mil·ə,tüd }

transformation series *See* radioactive series. { ,tranz·fər'mā·shən ,sir·ēz }

transformation temperature [MET] **1.** The temperature at which a change in phase occurs in a metal during heating or cooling. **2.** The maximum or minimum temperature of a transformation temperature range. { ,tranz·fər'mā·shən ,tem·prə·chər }

transformation-temperature ranges [MET] The ranges of temperatures within which austenite forms during heating and transforms during cooling. { ,tranz·fər'mā·shən 'tem·prə·chər ,rān·jəz }

transformation theory [QUANT MECH] The study of coordinate and other transformations in quantum mechanics, especially those which leave some properties of the system invariant. { ,tranz·fər'mā·shən ,thē·ə·rē }

transformation twin [CRYSTAL] A crystal twin developed by a growth transformation from a higher to a lower symmetry. { ,tranz·fər'mā·shən ,twin }

transformer [ELECTROMAG] An electrical component consisting of two or more multiturn coils of wire placed in close proximity to cause the magnetic field of one to link the other; used to transfer electric energy from one or more alternating-current circuits to one or more other circuits by magnetic induction. { tranz'for·mər }

transformer bridge [ELEC] A network consisting of a transformer and two impedances, in which the input signal is applied to the transformer primary and the output is taken between the secondary center-tap and the junction of the impedances that connect to the outer leads of the secondary. { tranz'for·mər ,brij }

transformer-coupled amplifier [ELECTR] Audio-frequency amplifier that uses untuned iron-core transformers to provide coupling between stages. { tranz'for·mər ¦kəp·əld 'am·plə,fī·ər }

TRANSFER RIBONUCLEIC ACID

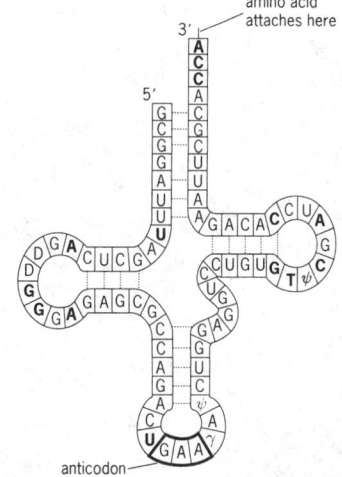

The structure of a t-RNA molecule. Bold letters indicate invariant bases.

transformer coupling [ELEC] *See* inductive coupling. [ELECTR] Interconnection between stages of an amplifier which employs a transformer for connecting the plate circuit of one stage to the grid circuit of the following stage; a special case of inductive coupling. { tranz'fȯr·mər ˌkəp·liŋ }

transformer hybrid *See* hybrid set. { tranz'fȯr·mər 'hī·brəd }

transformer load loss [ELEC] Losses in a transformer which are incident to the carrying of the load; load losses include resistance loss in the windings due to load current, stray loss due to stray fluxes in the windings, core clamps, and so on, and to circulating current, if any, in parallel windings. { tranz'fȯr·mər 'lōd ˌlȯs }

transformer loss [ELEC] Ratio of the signal power that an ideal transformer of the same impedance ratio would deliver to the load impedance, to the signal power that the actual transformer delivers to the load impedance; this ratio is usually expressed in decibels. { tranz'fȯr·mər ˌlȯs }

transformer oil [MATER] A high-quality insulating oil in which windings of large power transformers are sometimes immersed to provide high dielectric strength, high insulation resistance, high flash point, freedom from moisture, and freedom from oxidization. { tranz'fȯr·mər ˌȯil }

transformer read-only store [COMPUT SCI] In computers, read-only store in which the presence or absence of mutual inductance between two circuits determines whether a binary 1 or 0 is stored. { tranz'fȯr·mər 'rēd ˌōn·lē 'stȯr }

transformer rectifier [ELEC] A combination of a transformer and a rectifier that allows input alternating current to be varied and then rectified into direct current. { tranz'fȯr·mər 'rek·tə͵fī·ər }

transformer substation [ELEC] An electric power substation whose equipment includes transformers. { tranz'fȯr·mər 'səb͵stā·shən }

transformer voltage ratio [ELEC] Ratio of the root-mean-square primary terminal voltage to the root-mean-square secondary terminal voltage under specified conditions of load. { tranz'fȯr·mər 'vōl·tij ˌrā·shō }

transform fault [GEOL] A strike-slip fault with offset ridges characteristic of a midoceanic ridge. { 'tranz͵fȯrm ˌfȯlt }

transforming principle [MICROBIO] Deoxyribonucleic acid which effects transformation in bacterial cells. { tranz'fȯrm·iŋ ˌprin·sə·pəl }

transforming section [ELECTROMAG] Length of waveguide or transmission line of modified cross section, or with a metallic or dielectric insert, used for impedance transformation. { tranz'fȯrm·iŋ ˌsek·shən }

transfusion [MED] The administration of blood, or one of its components, as a part of treatment. { tranz'fyü·zhən }

transgene [GEN] Genetic material from one organism that has been experimentally transferred to another, so that the host acquires the genetic traits of the transferred genes in its chromosomal composition. { 'tranz͵jēn }

transgenic organism [GEN] An organism into whose genome a gene or genes from another organism have been experimentally transferred and can be expressed. { tranz'jen·ik 'ȯr·gə͵niz·əm }

transgranular *See* transcrystalline. { ͵tranz'gran·yə·lər }

transgression [GEOL] Geologic evidence of landward extension of the sea. Also known as invasion; marine transgression. [OCEANOGR] Extension of the sea over land areas. { tranz'gresh·ən }

transgressive deposit [GEOL] Sediment deposited during transgression of the sea or during subsidence of the land. { tranz'gres·iv di'päz·ət }

transgressive overlap *See* onlap. { tranz'gres·iv 'ō·vər͵lap }

transhybrid loss [ELEC] In a carrier telephone system, the transmission loss at a given frequency measured across a hybrid circuit joined to a given two-wire termination and balancing network. { ͵tranz'hī·brəd 'lȯs }

transient [PHYS] A pulse, damped oscillation, or other temporary phenomenon occurring in a system prior to reaching a steady-state condition. { 'tranch·ənt }

transient analyzer [ELECTR] An analyzer that generates transients in the form of a succession of equal electric surges of small amplitude and adjustable waveform, applies these transients to a circuit or device under test, and shows the resulting output waveforms on the screen of an oscilloscope. { 'tranch·ənt ˌan·ə͵līz·ər }

transient distortion [ELECTR] Distortion due to inability to amplify transients linearly. { 'tranch·ənt di͵stȯr·shən }

transient equilibrium [NUCLEO] Radioactive equilibrium in which the lifetime of the parent is sufficiently short that the quantity present decreases appreciably in the period under consideration. { ͵tran·zhənt ͵ē·kwə'lib·rē·əm }

transient grating photoacoustics *See* impulsive stimulated thermal scattering. { 'tranch·ənt 'grād·iŋ ˌfōd·ō·ə'kü·stiks }

transient ischemic attack [MED] A brief loss of nerve function caused by a temporary lack of adequate blood flow and oxygen to the brain due to a rupture in the carotid arteries leading to the brain. { ͵tranch·ənt i͵skēm·ik ə'tak }

transient liquid phase bonding *See* diffusion brazing. { ͵tranch·ənt ͵lik·wəd ͵fāz 'bän·diŋ }

transient lunar phenomena [ASTRON] Local obscurations and reddish glows that are sometimes observed in certain areas of the moon. Abbreviated TLP. { 'tranch·ənt 'lü·nər fə'näm·ə·nä }

transient motion [PHYS] An oscillatory or other irregular motion occurring while a quantity is changing to a new steady-state value. { 'tranch·ənt 'mō·shən }

transient overshoot [PHYS] The maximum value of the overshoot of a quantity as a result of a sudden change in conditions. { 'tranch·ənt 'ō·vər͵shüt }

transient phenomena [ELEC] Rapidly changing actions occurring in a circuit during the interval between closing of a switch and settling to a steady-state condition, or any other temporary actions occurring after some change in a circuit. { 'tranch·ənt fə͵näm·ə·nä }

transient problem *See* initial-value problem. { 'tranch·ənt 'präb·ləm }

transient program [COMPUT SCI] A computer program that is stored in a computer's main memory only while it is being executed. { 'tranch·ənt 'prō·grəm }

transient RANS modeling *See* TRANS modeling. { 'tranch·ənt 'ranz ˌmäd·əl·iŋ }

transient response [PHYS] The behavior of a system following a sudden change in its input. { 'tranch·ənt ri'späns }

transient situational disturbance [PSYCH] A form of personality disorder, more or less transient, and generally an acute symptom response to a specific situation, without persistent personality disturbance. { 'tranch·ənt ͵sich·ə'wā·shən·əl di'stər·bəns }

transient suppressor *See* surge suppressor. { 'tranch·ənt sə'pres·ər }

transient x-ray source *See* x-ray nova. { 'tranch·ənt 'eks͵rā ˌsȯrs }

transillumination [ENG] **1.** Indirect lighting on a console panel that uses edge and backlighting techniques on clear, fluorescent, or layered plastic materials. **2.** Transmission of light through sections of material in order to enhance inspection for deviations in quality. [OPTICS] Illumination of translucent material by light being transmitted through the material from the rear. { ͵tranz·ə͵lü·mə'nā·shən }

transistance [ELECTR] The characteristic that makes possible the control of voltages or currents so as to accomplish gain or switching action in a circuit; examples of transistance occur in transistors, diodes, and saturable reactors. { tran'zis·təns }

transistor [ELECTR] An active component of an electronic circuit consisting of a small block of semiconducting material to which at least three electrical contacts are made, usually two closely spaced rectifying contacts and one ohmic (nonrectifying) contact; it may be used as an amplifier, detector, or switch. { tran'zis·tər }

transistor amplifier [ELECTR] An amplifier in which one or more transistors provide amplification comparable to that of electron tubes. { tran'zis·tər ˌam·plə͵fī·ər }

transistor biasing [ELECTR] Maintaining a direct-current voltage between the base and some other element of a transistor. { tran'zis·tər ˌbī·əs·iŋ }

transistor characteristics [ELECTR] The values of the impedances and gains of a transistor. { tran'zis·tər ˌkar·ik·tə͵ris·tiks }

transistor chip [ELECTR] An unencapsulated transistor of very small size used in microcircuits. { tran'zis·tər ˌchip }

transistor circuit [ELECTR] An electric circuit in which a transistor is connected. { tran'zis·tər ˌsər·kət }

transistor clipping circuit [ELECTR] A circuit in which a transistor is used to achieve clipping action; the bias at the input is set at such a level that output current cannot flow during a portion of the amplitude excursion of the input voltage or current waveform. { tran'zis·tər 'klip·iŋ ˌsər·kət }

transistor gain [ELECTR] The increase in signal power produced by a transistor. { tran'zis·tər ˌgān }

transistor input resistance [ELECTR] The resistance across the input terminals of a transistor stage. Also known as input resistance. { tran'zis·tər 'inˌpu̇t riˌzis·təns }

transistor magnetic amplifier [ELECTR] A magnetic amplifier together with a transistor preamplifier, the latter used to make the signal strong enough to change the flux in the core of the magnetic amplifier completely during a half-cycle of the power supply voltage. { tran'zis·tər mag'ned·ik 'am·pləˌfī·ər }

transistor memory See semiconductor memory. { tran'zis·tər ˌmem·rē }

transistor radio [ELECTR] A radio receiver in which transistors are used in place of electron tubes. { tran'zis·tər ˌrād·ē·ō }

transistor-transistor logic [ELECTR] A logic circuit containing two transistors, for driving large output capacitances at high speed. Abbreviated T²L; TTL. { tran'zis·tər tran'zis·tər 'läj·ik }

transit [ASTRON] **1.** A celestial body's movement across the meridian of a place. Also known as meridian transit. **2.** Passage of a smaller celestial body across a larger one. **3.** Passage of a satellite's shadow across the disk of its primary. [ENG] **1.** A surveying instrument with the telescope mounted so that it can measure horizontal and vertical angles. Also known as transit theodolite. **2.** To reverse the direction of the telescope of a transit by rotating 180° about its horizontal axis. Also known as plunge. [NAV] A positive-fixing system employing low-orbit satellites which constantly emit continuous-wave signals; on the surface vehicle, the signals are received and the Doppler shift is recorded; position is determined by computation based on the shift. { 'trans·ət }

transit circle [ENG] A type of astronomical transit instrument having a micrometer eyepiece that has an extra pair of moving wires perpendicular to the vertical set to measure the zenith distance or declination of the celestial object in conjunction with readings taken from a large, accurately calibrated circle attached to the horizontal axis. Also known as meridian circle; meridian transit. { 'trans·ət ˌsər·kəl }

transit declinometer [ENG] A type of declinometer; a surveyor's transit, built to exacting specifications with respect to freedom from traces of magnetic impurities and quality of the compass needle, has a 17-power telescope for sighting on a mark and for making solar and stellar observations to determine true directions. { 'trans·ət ˌdek·lə'näm·əd·ər }

transit instrument See transit telescope. { 'trans·ət ˌin·strə·mənt }

transition [CELL MOL] A mutation resulting from the substitution in deoxyribonucleic acid or ribonucleic acid of one purine or pyrimidine for another. [COMMUN] Change from one circuit condition to the other; for example, the change from mark to space or from space to mark. [QUANT MECH] The change of a quantum-mechanical system from one energy state to another. [THERMO] A change of a substance from one of the three states of matter to another. { tran'zish·ən }

transitional epithelium [HISTOL] A form of stratified epithelium found in the urinary bladder; cells vary between squamous, when the tissue is stretched, and columnar, when not stretched. { tran'zish·ən·əl ˌep·ə'thē·lē·əm }

transitional fit [DES ENG] A fit with varying clearances due to specified tolerances on the shaft and sleeve or hole. { tran'zish·ən·əl 'fit }

transitional flow [FL MECH] A flow in which the viscous and Reynolds stresses are of approximately equal magnitude; it is transitional between laminar and turbulent flow. { tran'zish·ən·əl 'flō }

transition altitude [AERO ENG] The altitude in the vicinity of an aerodrome at or below which the vertical position of an aircraft is controlled by reference to true altitude. { tran zish·ən 'al·təˌtüd }

transition boiling [PHYS CHEM] A stage in the boiling process that follows fully developed nucleate boiling, precedes film boiling, and has features of both of those stages, in which a decrease in the heat flux accompanies an increase in wall superheat, making it highly unstable. { tranˌzish·ən 'bȯil·iŋ }

transition card [COMPUT SCI] In reading a deck of punched cards by a computer, a card that causes the computer to stop reading cards and begin executing a program. Also known as transfer card. { tran'zish·ən ˌkärd }

transition curve See easement curve. { tran'zish·ən ˌkərv }

transition element [CHEM] One of a group of metallic elements in which the members have the filling of the outermost shell to 8 electrons interrupted to bring the penultimate shell from 8 to 18 or 32 electrons; includes elements 21 through 29 (scandium through copper), 39 through 47 (yttrium through silver), 57 through 79 (lanthanum through gold), and all known elements from 89 (actinium) on. Also known as transition metal. [ELECTROMAG] An element used to couple one type of transmission system to another, as for coupling a coaxial line to a waveguide. { tran'zish·ən ˌel·ə·mənt }

transition factor See reflection factor. { tran'zish·ən ˌfak·tər }

transition flow [AERO ENG] A flow of fluid about an airfoil that is changing from laminar flow to turbulent flow. { tran'zish·ən ˌflō }

transition frequency [ENG ACOUS] The frequency corresponding to the intersection of the asymptotes to the constant-amplitude and constant-velocity portions of the frequency-response curve for a disk recording; this curve is plotted with output-voltage ratio in decibels as the ordinate, and the logarithm of the frequency as the abscissa. Also known as crossover frequency; turnover frequency. [QUANT MECH] The characteristic frequency of radiation emitted or absorbed by a quantum-mechanical system as it changes from one energy state to another; equal to the energy difference between the states divided by Planck's constant. { tran'zish·ən ˌfrē·kwən·sē }

transition function [COMPUT SCI] A function which determines the next state of a sequential machine from the present state and the present input. { tran'zish·ən ˌfəŋk·shən }

transition interval [ANALY CHEM] In a titrimetric analysis, the range in concentration of the species being determined over which a variation in a chemical indicator can be observed visually. { tran'zish·ən ˌin·tər·vəl }

transition lattice [MET] An unstable, intermediate configuration formed in a metal lattice during solid-state reactions such as precipitation or transformation. { tran'zish·ən ˌlad·əs }

transition level [NAV] The flight level below which heights are expressed in feet above mean sea level and are based on an approved station altimeter setting. { tran'zish·ən ˌlev·əl }

transition loss [ELEC] At a junction between a source and a load, the ratio of the available power to the power delivered to the load. { tran'zish·ən ˌlȯs }

transition metal See transition element. { tran'zish·ən ˌmed·əl }

transition moment [QUANT MECH] Any type of multipole moment which determines radiative transitions between states; it consists of an integral of the product of the conjugate of the final state wave function, a multipole moment operator, and the initial state wave function. { tran'zish·ən ˌmō·mənt }

transition point [ELECTROMAG] A point at which the constants of a circuit change in such a way as to cause reflection of a wave being propagated along the circuit. [THERMO] Either the temperature at which a substance changes from one state of aggregation to another (a first-order transition), or the temperature of culmination of a gradual change, such as the lambda point, or Curie point (a second-order transition). Also known as transition temperature. { tran'zish·ən ˌpȯint }

transition probability [MATH] Conditional probability concerning a discrete Markov chain giving the probabilities of change from one state to another. [QUANT MECH] The probability per unit time that a quantum-mechanical system will make a transition from a given initial state to a given final state. { tran'zish·ən ˌpräb·ə'bil·əd·ē }

transition radiation detector [PART PHYS] A detector of energetic charged particles that makes use of the radiation emitted as the particle crosses boundaries between regions with different indices of refraction. { tran'zish·ən ˌrād·ē·ā·shən diˌtek·tər }

transition region [ASTRON] A layer of the solar atmosphere only a few hundred miles thick between the chromosphere and the corona across which the temperature rises rapidly from a few times 10^4 K to the order of 10^6 K. [SOLID STATE] The region between two homogeneous semiconductors in which the impurity concentration changes. { tran'zish·ən ˌrē·jən }

TRANSIT CIRCLE

Six-inch (15-centimeter) transit circle at the U.S. Naval Observatory. (*Official U.S. Naval Observatory photograph*)

TRANSIT DECLINOMETER

Transit declinometer with compass needle, telescope, and microscope. (*U.S. Coast and Geodetic Survey*)

TRANSITIONAL EPITHELIUM

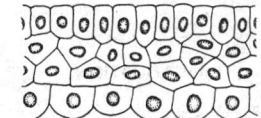

Cellular arrangement in transitional epithelium.

transition state See activated complex. { tran'zish·ən ,stāt }

transition temperature [CHEM] The temperature at which an enantiotropic polymorph is converted into a different form. [MET] The temperature at which a fracture changes from tough to brittle in various tests, such as notched-bar impact test. [THERMO] See transition point. { tran'zish·ən ,tem·prə·chər }

transition time [ANALY CHEM] The time interval needed for a working (nonreference) electrode to become polarized during chronopotentiometry (time-measurement electrolysis of a sample). { tran'zish·ən ,tīm }

transition to chaos [PHYS] The process by which a system evolves from periodic toward chaotic behavior as one or more parameters governing the behavior of the system are varied. { tran'zish·ən tə 'kā,äs }

transition zone [FL MECH] Those conditions of fluid flow in which the nature of the flow is changing from laminar to turbulent. [GEOL] **1.** A region within the upper mantle bordering the lower mantle, at a depth of 246–600 miles (410–1000 kilometers), characterized by a rapid increase in density of about 20% and an increase in seismic wave velocities. **2.** A region within the outer core, transitional to the inner core. { tran'zish·ən ,zōn }

transitive group [MATH] A group of permutations of a finite set such that for any two elements in the set there exists an element of the group which takes one into the other. { 'tran·səd·iv ,grüp }

transitive relation [MATH] A relation < on a set such that if $a < b$ and $b < c$, then $a < c$. { 'tran·səd·iv ri'lā·shən }

transit mix [MATER] Concrete or mortar mixed in a rotating cylinder en route to or at the construction site. { 'trans·ət ,miks }

transitory target [ORD] A target that obtains only for a limited period of time, as in the case of a troop concentration which may be dissipated in a short time. { 'trans·ə,tȯr·ē 'tär·gət }

transitron [ELECTR] Thermionic-tube circuit whose action depends on the negative transconductance of the suppressor grid of a pentode with respect to the screen grid. { 'tran·sə,trän }

transitron oscillator [ELECTR] A negative-resistance oscillator in which the screen grid is more positive than the anode, and a capacitor is connected between the screen grid and the suppressor grid; the suppressor grid periodically divides the current between the screen grid and the anode, thereby producing oscillation. { 'tran·sə,trän 'äs·ə,lād·ər }

transit survey [ENG] A ground surveying method in which a transit instrument is set up at a control point and oriented, and directions and distances to observed points are recorded. { 'trans·ət 'sər,vā }

transit telescope [OPTICS] A telescopic instrument adapted to the observation of the passage, or transit, of an astronomical object across the meridian of an observer; consists of a telescope mounted on a single fixed horizontal axis of rotation which has a central hollow cube (sometimes a sphere) and two conical semiaxes ending in cylindrical pivots; the objective and eyepiece halves of the instrument are also fastened to the cube of the instrument, perpendicular to the horizontal axis. Also known as transit instrument. { 'trans·ət 'tel·ə,skōp }

transit theodolite See transit. { 'trans·ət thē'äd·əl,īt }

transit time [ELECTR] The time required for an electron or other charge carrier to travel between two electrodes in an electron tube or transistor. { 'trans·ət ,tīm }

transit-time microwave diode [ELECTR] A solid-state microwave diode in which the transit time of charge carriers is short enough to permit operation in microwave bands. { 'trans·ət ,tīm 'mī·krə,wāv 'dī,ōd }

transit-time mode [ELECTR] A mode of operation of a Gunn diode in which a charge dipole, consisting of an electron accumulation and a depletion layer, travels through the semiconductor at a frequency dependent on the length of the semiconductor layer and the drift velocity. { 'trans·ət ,tīm ,mōd }

transketolase [BIOCHEM] An enzyme that cleaves a substrate at the position of the carbonyl carbon and transports a two-carbon fragment to an acceptor compound to form a new compound. { ¦tranz'kēd·ə,lās }

translate [COMPUT SCI] To convert computer information from one language to another, or to convert characters from one representation set to another, and by extension, the computer instruction which directs the latter conversion to be carried out. { tran'slāt }

TRANSLOCATION

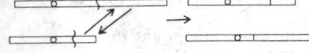

Translocation in which nonhomologous chromosomes become broken, switching their broken ends in the process of rejoining.

translating circuit See translator. { tran'slād·iŋ ,sər·kət }

translating roller [ORD] A double-thread screw by means of which a breechblock is drawn longitudinally from its position in the breech of a large-caliber gun. { tran'slād·iŋ ,rō·lər }

translation [MATH] **1.** A function changing the coordinates of a point in a euclidean space into new coordinates relative to axes parallel to the original. **2.** A function on a group to itself given by operating on each element by some one fixed element. **3.** Let E be a finitely generated extension of a field k, F be an extension of k, and both E and F be contained in a common field; the translation of E to F is the extension EF of F, where EF is the compositum of E and F. Also known as lifting. [MECH] The linear movement of a point in space without any rotation. [MOL BIO] The process by which the linear sequence of nucleotides in a molecule of messenger ribonucleic acid directs the specific linear sequence of amino acids, as during protein synthesis. { tran'slā·shən }

translational energy [PHYS CHEM] The kinetic energy of gaseous or liquid molecules that is associated with their motion within their particular chemical systems. { tran¦slā·shən·əl 'en·ər·jē }

translational fault [GEOL] A fault in which there has been uniform movement in one direction and no rotational component of movement. Also known as translatory fault. { tran'slā·shən·əl 'fȯlt }

translation algorithm [COMPUT SCI] A specific, effective, essentially computational method for obtaining a translation from one language to another. { tran'slā·shən 'al·gə,rith·əm }

translational motion [MECH] Motion of a rigid body in such a way that any line which is imagined rigidly attached to the body remains parallel to its original direction. { tran'slā·shən·əl 'mō·shən }

translational movement [GEOL] Movement, as of fault blocks, that is uniform, without rotation, so that parallel features maintain their orientation. { tran'slā·shən·əl 'müv·mənt }

translation gliding See crystal gliding. { tran'slā·shən ,glīd·iŋ }

translation group [CRYSTAL] The collection of all translation operations which carry a crystal lattice into itself. { tran'slā·shən ,grüp }

translation operation [PHYS] The process of moving an object along a straight line in such a way that any line which is fixed with respect to the object remains parallel to its original direction. { tran'slā·shən ,äp·ə,rā·shən }

translation surface See surface of translation. { tran'slā·shən ,sər·fəs }

translator [COMPUT SCI] A computer network or system having a number of inputs and outputs, so connected that when signals representing information expressed in a certain code are applied to the inputs, the output signals will represent the same information in a different code. Also known as translating circuit. [ELECTR] A combination television receiver and low-power television transmitter, used to pick up television signals on one frequency and retransmit them on another frequency to provide reception in areas not served directly by television stations. { tran'slād·ər }

translator routine [COMPUT SCI] A program which accepts statements in one language and outputs them as statements in another language. { tran'slād·ər rü,tēn }

translatory fault See translational fault. { 'tran·slə,tȯr·ē 'fȯlt }

transliterate [COMPUT SCI] To represent the characters or words of one language by corresponding characters or words of another language. { tran'slid·ə,rāt }

translocation [BOT] Movement of water, mineral salts, and organic substances from one part of a plant to another. [CELL MOL] The transfer of a chromosome segment from its usual position to a new position in the same or in a different chromosome. { ¦tranz·lō'kā·shən }

translucent attritus [GEOL] Attritus composed principally of transparent humic degradation matter. Also known as humodurite. { tran'slüs·əns ə'trīd·əs }

translucent medium [OPTICS] A medium which transmits rays of light so diffused that objects cannot be seen distinctly; examples are various forms of glass which admit considerable light but impede vision. { tran'slüs·əns 'mēd·ē·əm }

translucidus [METEOROL] A cloud variety occurring in a layer, patch, or extensive sheet, the greater part of which is sufficiently translucent to reveal the position of the sun, or through which higher clouds may be discerned; this variety is found in the general altocumulus, altostratus, stratocumulus, and stratus. { tran'slüs·əd·əs }

translunar [ASTRON] Beyond the orbit of the moon. { tran'slü·nər }

transmembrane distillation See membrane distillation. { ‚tranz¦mem‚brān ‚dis·tə'lā·shən }

transmethylase [BIOCHEM] A transferase enzyme involved in catalyzing chemical reactions in which methyl groups are transferred from a substrate to a new compound. { ‚tranz‑meth·ə‚lās }

transmethylation [BIOCHEM] A metabolic reaction in which a methyl group is transferred from one compound to another; methionine and choline are important donors of methyl groups. { ¦tranz‚meth·ə'lā·shən }

transmissibility [MECH] A measure of the ability of a system either to amplify or to suppress an input vibration, equal to the ratio of the response amplitude of the system in steady-state forced vibration to the excitation amplitude; the ratio may be in forces, displacements, velocities, or accelerations. { tranz‚mis·ə'bil·əd·ē }

transmission [ELECTR] **1.** The process of transferring a signal, message, picture, or other form of intelligence from one location to another location by means of wire lines, radio, light beams, infrared beams, or other communication systems. **2.** A message, signal, or other form of intelligence that is being transmitted. [ELECTROMAG] See transmittance. [MECH ENG] The gearing system by which power is transmitted from the engine to the live axle in an automobile. Also known as gearbox. { tranz'mish·ən }

transmission access [ELEC] The use of electric power lines and other power transmitting facilities by parties other than the owners of the lines. Also known as common carriage. { tranz'mish·ən 'ak‚ses }

transmission anomaly [ACOUS] The ratio of the transmission loss of underwater sound at a given distance from the source to the inverse square of this distance, usually expressed in decibels. { tranz'mish·ən ə‚näm·ə·lē }

transmission band [ELECTROMAG] Frequency range above the cutoff frequency in a waveguide, or the comparable useful frequency range for any other transmission line, system, or device. { tranz'mish·ən ‚band }

transmission coefficient [PHYS] **1.** The value of some quantity associated with the resultant field produced by incident and reflected waves at a given point in a transmission medium divided by the corresponding quantity in the incident wave. **2.** The ratio of transmitted to incident energy flux or flux of some other quantity at a discontinuity in a transmission medium; for sound waves, it is called the sound transmission coefficient. **3.** The ratio of the transmitted flux of some quantity to the incident flux for a substance of unit thickness. [QUANT MECH] See penetration probability. { tranz'mish·ən ‚kō·i‚fish·ənt }

transmission control character [COMMUN] A character included in a message to control its routing to the intended destination. { tranz'mish·ən kən'trōl ‚kar·ik·tər }

Transmission Control Protocol [COMMUN] The set of standards that is responsible for breaking down and reassembling the data packets transmitted on the Internet, for ensuring complete delivery of the packets, and for controlling data flow. Abbreviated TCP. { tranz‚mish·ən kən'trōl ‚prōd·ə‚kól }

Transmission Control Protocol/Internet Protocol [COMPUT SCI] The Internet's principal communication standard, dictating how packets of information are sent and received across multiple networks. TCP breaks down and reassembles packets, and IP ensures that the packets are sent to the correct destination. Abbreviated TCP/IP. { tranz‚mish·ən kən'trōl ‚prōd·ə‚kól 'in·tər‚net ‚prōd·ə‚kól }

transmission diffraction [ANALY CHEM] A type of electron diffraction analysis in which the electron beam is transmitted through a thin film or powder whose smallest dimension is no greater than a few tenths of a micrometer. { tranz'mish·ən di‚frak·shən }

transmission dynamometer [ENG] A device for measuring torque and power (without loss) between a propulsion power plant and the driven mechanism, for example, wheels or propellers. { tranz'mish·ən dī·nə'mäm·əd·ər }

transmission electron microscope [ELECTR] A type of electron microscope in which the specimen transmits an electron beam focused on it, image contrasts are formed by the scattering of electrons out of the beam, and various magnetic lenses perform functions analogous to those of ordinary lenses in a light microscope. { tranz'mish·ən i'lek‚trän 'mī·krə‚skōp }

transmission electron radiography [ELECTR] A technique used in microradiography to obtain radiographic images of very thin specimens; the photographic plate is in close contact with the specimen, over which is placed a lead foil and then a light-tight covering; hardened x-rays shoot through the light-tight covering. { tranz'mish·ən i‚lek‚tran ‚rād·ē'äg·rə·fē }

transmission facilities [COMMUN] All equipment and the medium required to transmit a message. { tranz'mish·ən fə‚sil·əd·ēz }

transmission factor [PHYS] The ratio of the flux of some quantity transmitted through a body to the incident flux. { tranz'mish·ən ‚fak·tər }

transmission function [GEOPHYS] A mathematical formulation of relationships between infrared transmission in the atmosphere, the path length, and the concentration of absorbing gases. { tranz'mish·ən ‚fəŋk·shən }

transmission gain See gain. { tranz'mish·ən ‚gān }

transmission gate [ELECTR] A gate circuit that delivers an output waveform that is a replica of a selected input during a specific time interval which is determined by a control signal. { tranz'mish·ən ‚gāt }

transmission grating [OPTICS] A diffraction grating produced on a transparent base so radiation is transmitted through the grating instead of being reflected from it. { tranz'mish·ən ‚grād·iŋ }

transmission interface converter [COMPUT SCI] A device that converts data to or from a form suitable for transfer over a channel connecting two computer systems or connecting a computer with its associated data terminals. { tranz'mish·ən 'in·tər‚fās kən‚vərd·ər }

transmission level [COMMUN] The ratio of the signal power at any point in a transmission system to the signal power at some point in the system chosen as a reference point; usually expressed in decibels. { tranz'mish·ən ‚lev·əl }

transmission line [ELEC] A system of conductors, such as wires, waveguides, or coaxial cables, suitable for conducting electric power or signals efficiently between two or more terminals. { tranz'mish·ən ‚līn }

transmission-line admittance [ELEC] The complex ratio of the current flowing in a transmission line to the voltage across the line, where the current and voltage are expressed in phasor notation. { tranz'mish·ən ¦līn ad‚mit·əns }

transmission-line attenuation [ELEC] The decrease in power of a transmission-line signal from one point to another, expressed as a ratio or in decibels. { tranz'mish·ən ¦līn ə‚ten·yə‚wā·shən }

transmission-line cable [ELEC] The coaxial cable, waveguide, or microstrip which forms a transmission line; a number of standard types have been designated, specified by size and materials. { tranz'mish·ən ¦līn ‚kā·bəl }

transmission-line constants See transmission-line parameters. { tranz'mish·ən ¦līn ‚kän·stəns }

transmission-line current [ELEC] The amount of electrical charge which passes a given point in a transmission line per unit time. { tranz'mish·ən ¦līn ‚kə·rənt }

transmission-line efficiency [ELEC] The ratio of the power of a transmission-line signal at one end of the line to that at the other end where the signal is generated. { tranz'mish·ən ¦līn i‚fish·ən sē }

transmission-line impedance [ELEC] The complex ratio of the voltage across a transmission line to the current flowing in the line, where voltage and current are expressed in phasor notation. { tranz'mish·ən ¦līn im‚pēd·əns }

transmission-line parameters [ELEC] The quantities which are necessary to specify the impedance per unit length of a transmission line, and the admittance per unit length between various conductors of the line. Also known as linear electrical parameters; line parameters; transmission line constants. { tranz'mish·ən ¦līn pə‚ram·əd·ərz }

transmission-line power [ELEC] The amount of energy carried past a point in a transmission line per unit time. { tranz'mish·ən ¦līn ‚paů·ər }

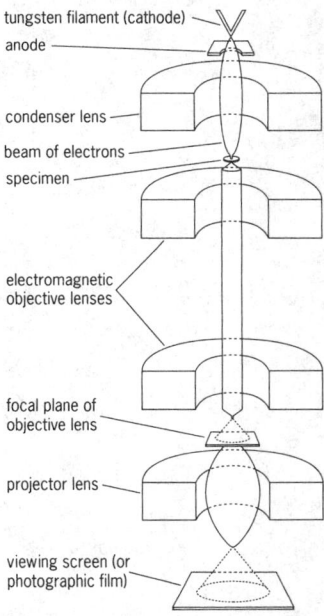

TRANSMISSION ELECTRON MICROSCOPE

tungsten filament (cathode)
anode
condenser lens
beam of electrons
specimen
electromagnetic objective lenses
focal plane of objective lens
projector lens
viewing screen (or photographic film)

The optical path in a transmission electron microscope.

transmission-line reflection coefficient [ELEC] The ratio of the voltage reflected from the load at the end of a transmission line to the direct voltage. { tranz'mish·ən ¦līn ri'flek·shən ‚kō·i‚fish·ənt }

transmission-line theory [ELEC] The application of electrical and electromagnetic theory to the behavior of transmission lines. { tranz'mish·ən ¦līn ‚thē·ə·rē }

transmission-line transducer loss [ELEC] The ratio of the power delivered by a transmission line to a load to that produced at the generator, expressed in decibels; equal to the sum of the attenuation of the line and the mismatch loss. { tranz'mish·ən ¦līn trans'dü·sər‚lós }

transmission-line voltage [ELEC] The work that would be required to transport a unit electrical charge between two specified conductors of a transmission line at a given instant. { tranz'mish·ən ¦līn ‚vōl·tij }

transmission loss [COMMUN] **1.** The ratio of the power at one point in a transmission system to the power at a point farther along the line; usually expressed in decibels. **2.** The actual power that is lost in transmitting a signal from one point to another through a medium or along a line. Also known as loss. { tranz'mish·ən ‚lós }

transmission mode See mode. { tranz'mish·ən ‚mōd }

transmission modulation [ELECTR] Amplitude modulation of the reading-beam current in a charge storage tube as the beam passes through apertures in the storage surface; the degree of modulation is controlled by the stored charge pattern. { tranz'mish·ən ‚mäj·ə'lā·shən }

transmission oil [MATER] A lubricant especially compounded for automobile transmissions. { tranz'mish·ən ‚óil }

transmission plane [OPTICS] The plane of vibration of polarized light that will pass through a Nicol prism or other polarizer. { tranz'mish·ən ‚plān }

transmission primaries [COMMUN] The set of three color primaries that correspond to the three independent signals contained in the color television picture signal. { tranz'mish·ən 'prī‚mer·ēz }

transmission range See night visual range. { tranz'mish·ən ‚rānj }

transmission regulator [ELECTR] In electrical communications, a device that maintains substantially constant transmission levels over a system. { tranz'mish·ən ‚reg·yə‚lād·ər }

transmission security [COMMUN] Component of communications security which results from all measures designed to protect transmissions from unauthorized interception, traffic analysis, and imitative deception. { tranz'mish·ən si‚kyúr·əd·ē }

transmission speed [COMMUN] The number of information elements sent per unit time; usually expressed as bits, characters, bands, word groups, or records per second or per minute. { tranz'mish·ən ‚spēd }

transmission substation [ELEC] An electric power substation associated with high voltage levels. { tranz'mish·ən 'səb‚stā·shən }

transmission time [COMMUN] Absolute time interval from transmission to reception of a signal. { tranz'mish·ən ‚tīm }

transmission tower [ENG] A concrete, metal, or timber structure used to carry a transmission line. { tranz'mish·ən ‚taú·ər }

transmissivity [ELECTROMAG] The ratio of the transmitted radiation to the radiation arriving perpendicular to the boundary between two mediums. { ‚tranz·mə'siv·əd·ē }

transmissometer [ENG] An instrument for measuring the extinction coefficient of the atmosphere and for the determination of visual range. Also known as hazemeter; transmittance meter. { ‚tranz·mə'säm·əd·ər }

transmissometry [OPTICS] The technique of determining the extinction characteristics of a medium by measuring the transmission of a light beam of known initial intensity directed into that medium. { ‚tranz·mə'säm·ə·trē }

transmit [COMMUN] To send a message, program, or other information to a person or place by wire, radio, or other means. [COMPUT SCI] To move data from one location to another. { tranz'mit }

transmittability [COMMUN] The ability of standard electronic and mechanical elements and automatic communications equipment to handle a code under various signal-to-noise ratios; for example, a code with a variable number of elements such as Morse presents technical problems in automatic interpretation not encountered in a fixed-length code. { tranz‚mid·ə'bil·əd·ē }

transmittance [ANALY CHEM] During absorption spectroscopy, the amount of radiant energy transmitted by the solution under analysis. [ELECTROMAG] The radiant power transmitted by a body divided by the total radiant power incident upon the body. Also known as transmission. { tranz'mid·əns }

transmittance meter See transmissometer. { tranz'mid·əns ‚mēd·ər }

transmittancy [ELECTROMAG] The transmittance of a solution divided by that of the pure solvent of the same thickness. { tranz'mid·ən·sē }

transmitted-carrier operation [COMMUN] Form of amplitude-modulated carrier transmission in which the carrier wave is transmitted. { tranz'mid·əd ¦kar·ē·ər ‚äp·ə‚rā·shən }

transmitted wave See refracted wave. { tranz'mid·əd 'wāv }

transmitter [COMMUN] **1.** In telephony, the carbon microphone that converts sound waves into audio-frequency signals. **2.** See radio transmitter. [ELEC] See synchro transmitter. { tranz'mid·ər }

transmitter-distributor [ELEC] In teletypewriter operations, a motor-driven device which translates teletypewriter code combinations form perforated tape into electrical impulses, and transmits these impulses to one or more receiving stations. Abbreviated TD. { tranz'mid·ər di'strib·yəd·ər }

transmitter noise See frying noise. { tranz'mid·ər ‚nóiz }

transmitter off [COMMUN] A signal sent by a receiving device to a transmitter, directing it to stop sending information if it is doing so, or not to send information if it is preparing to do so. Abbreviated XOFF. { tranz'mid·ər 'óf }

transmitter on [COMMUN] A signal sent by a receiving device to a transmitter, directing it to transmit any information it has to send. Abbreviated XON. { tranz'mid·ər 'ón }

transmitter-receiver See transceiver. { tranz'mid·ər ri'sē·vər }

transmitter synchro See synchro transmitter. { tranz'mid·ər ‚siŋ·krō }

transmitting loop loss [COMMUN] That part of the repetition equivalent assignable to the station set, subscriber line, and battery supply circuit which is on the transmitting end. { tranz'mid·iŋ ¦lüp ‚lós }

transmitting mode [COMPUT SCI] Condition of an input/output device, such as a magnetic tape when it is actually reading or writing. { tranz'mid·iŋ ‚mōd }

transmittivity [ELECTROMAG] The internal transmittance of a piece of nondiffusing substance of unit thickness. { ‚transmə'tiv·əd·ē }

TRANS modeling [FL MECH] A type of turbulence modeling which is based on solving the Reynolds-averaged Navier-Stokes equations in a time-dependent mode. Derived from transient RANS modeling. { 'tranz ‚mäd·əl·iŋ or 'tē‚ranz }

transmutation [NUC PHYS] A nuclear process in which one nuclide is transformed into the nuclide of a different element. Also known as nuclear transformation. { ‚trans·myü'tā·shən }

transobuoy [ENG] A free-floating or moored automatic weather station developed for the purpose of providing weather reports from the open oceans; it transmits barometric pressure, air temperature, sea-water temperature, and wind speed and direction. { 'tran·sə‚bói }

transolver [ELEC] A synchro having a two-phase cylindrical rotor within a three-phase stator, for use as a transmitter or a control transformer with no degradation of accuracy or nulls. { tran'säl·vər }

transom [BUILD] A window above a door. [NAV ARCH] The flat, vertical aft end of a ship or boat as distinguished from a canoe-shaped or cruiser stern. { 'tran·səm }

transonic [PHYS] That which occurs or is occurring within the range of speed in which flow patterns change from subsonic to supersonic (or vice versa), about Mach 0.8 to 1.2, as in transonic flight or transonic flutter. { tran'sän·ik }

transonic flight [AERO ENG] Flight of vehicles at speeds near the speed of sound (660 miles per hour or 1060 kilometers per hour, at 35,000 feet or 10,700 meters altitude), characterized by great increase in drag, decrease in lift at any altitude, and abrupt changes in the moments acting on the aircraft; the vehicle may shake or buffet. { tran'sän·ik 'flīt }

transonic flow [FL MECH] Flow of a fluid over a body in

the range just above and just below the acoustic velocity. { tran'sän·ik 'flō }

transonic range [FL MECH] The range of speeds between the speed at which one point on a body reaches supersonic speed, and the speed at which all points reach supersonic speed. { tran'sän·ik 'ränj }

transonic speed [FL MECH] The speed of a body relative to the surrounding fluid at which the flow is in some places on the body subsonic and in other places supersonic { tran'sän·ik 'spēd }

transonic wind tunnel [ENG] A type of high-speed wind tunnel capable of testing the effects of airflow past an object at speeds near the speed of sound, Mach 0.7 to 1.4; sonic speed occurs where the cross section of the tunnel is at a minimum, that is, where the test object is located. { tran'sän·ik 'wind ,tən·əl }

transorbital lobotomy [MED] A lobotomy performed through the roof of the orbit. { tranz'òr·bəd·əl lə'bäd·ə·mē }

transosonde [ENG] The flight of a constant-level balloon, whose trajectory is determined by tracking with radio-direction-finding equipment; thus, it is a form of upper-air, quasi-horizontal sounding. { 'tran·zə,sänd }

transparency [GRAPHICS] An image fixed on a clear base by means of a photographic, printing, chemical, or other process, especially adaptable for viewing by transmitted light. [OPTICS] The ability of a substance to transmit light of different wavelengths, sometimes measured in percent of radiation which penetrates a distance of 1 meter. { tranz'par·ən·sē }

transparency range [NUC PHYS] A postulated energy range for extremely high-energy heavy-ion collisions in which the projectile passes through the target and emerges with its temperature and density raised to the point at which a quark-gluon plasma forms. { tranz'par·ən·sē ränj }

transparent [COMPUT SCI] Pertaining to a device or system that processes data without the user being aware of or needing to understand its operation. [PHYS] Permitting passage of radiation or particles. { tranz'par·ənt }

transparent medium [OPTICS] **1.** A medium which has the property of transmitting rays of light in such a way that the human eye may see through the medium distinctly. **2.** A medium transparent to other regions of the electromagnetic spectrum, such as x-rays and microwaves. { tranz'par·ənt 'mēd·ē·əm }

transparent sky cover [METEOROL] In United States weather-observing practice, that portion of sky cover through which higher clouds and blue sky may be observed; opposed to opaque sky cover. { tranz'par·ənt 'skī ,kəv·ər }

transpassive region [PHYS CHEM] That portion of an anodic polarization curve in which metal dissolution increases as the potential becomes noble. { tranz'pas·iv ¦rē·jən }

transphasor [OPTICS] A nonlinear optical device that uses one light beam to modulate another, in a manner analogous to an electronic transistor, and that operates through the transference of a phase shift from one beam to the other. { ¦tranz 'fāz·ər }

transpiration [BIOL] The passage of a gas or liquid (in the form of vapor) through the skin, a membrane, or other tissue. { ,tranz·pə'rā·shən }

transpiration cooling See sweat cooling. { ,tranz·pə'rā·shən 'kül·iŋ }

transplantation [BIOL] **1.** The artificial removal of part of an organism and its replacement in the body of the same or of a different individual. **2.** To remove a plant from one location and replant it in another place. { ,tranz·plan'tā·shən }

transplantation antigen [IMMUNOL] An antigen in a cell which induces a histocompatibility reaction when the cell is transplanted into an organism not having that antigen. { ,tranz·plan'tā·shən 'ant·i·jən }

transplantation disease [MED] Disease ascribable to an immunological graft-versus-host reaction which occurs after transplantation of adult lymphoid cells to incompatible recipients who cannot reject them. { ,tranz·plan'tā·shən di,zēz }

transplanter [AGR] A special kind of equipment designed for the planting of cuttings or small plants; it transports one or more workers who assist the action of the machine in placing plants in a furrow and covering them; it commonly supplies a small quantity of water to each plant. { tranz'plan·tər }

transplutonium element [INORG CHEM] An element having

an atomic number greater than that of plutonium (94). { ¦tranz·plə'tō·nē·əm 'el·ə·mənt }

transpolarizer [ELEC] An electrostatically controlled circuit impedance that can have about 30 discrete and reproducible impedance values: two capacitors, each having a crystalline ferroelectric dielectric with a nearly rectangular hysteresis loop, are connected in series and act as a single low impedance to an alternating-current sensing signal when both capacitors are polarized in the same direction; application of 1-microsecond pulses of appropriate polarity increases the impedance in steps. { tranz'pō·lə,rīz·ər }

transponder [COMMUN] **1.** A transmitter-receiver capable of accepting the challenge of an interrogator and automatically transmitting an appropriate reply. **2.** A receiver-transmitter, such as on satellites, which receives a transmission and retransmits it at another radio frequency. { tranz'pän·dər }

transponder beacon See responder beacon. { tranz'pän·dər ,bē·kən }

transponder dead time [ELECTR] Time interval between the start of a pulse and the earliest instant at which a new pulse can be received or produced by a transponder. { tranz'pän·dər 'ded ,tīm }

transponder set [ELECTR] A complete electronic set which is designed to receive an interrogation signal, and which retransmits coded signals that can be interpreted by the interrogating station; it may also utilize the received signal for actuation of additional equipment such as local indicators or servo amplifiers { tranz'pän·dər ,set }

transponder suppressed time delay [ELECTR] Overall fixed time delay between reception of an interrogation and transmission of a reply to this interrogation. { tranz'pän·dər sə'prest 'tīm di,lā }

transport [COMPUT SCI] **1.** To convey as a whole from one storage device to another in a digital computer. **2.** See tape transport. [ENG] Conveyance equipment such as vehicular transport, hydraulic transport, and conveyor-belt setups. [NAV ARCH] A ship designed to carry military personnel from one place to another. Also known as troop ship. { trans'pòrt (verb), 'tranz,pòrt (noun) }

transportable computer [COMPUT SCI] A microcomputer that can be carried about conveniently but, in contrast to a portable computer, requires an external power source. { tranz'pòrd·ə·bəl kəm'pyüd·ər }

transportation [GEOL] A phase of sedimentation concerned with movement by natural agents of sediment or any loose or weathered material from one place to another. { ,tranz· pər'tā·shən }

transportation emergency [ENG] A situation which is created by a shortage of normal transportation capability and of a magnitude sufficient to frustrate movement requirements, and which requires extraordinary action by the designated authority to ensure continued movement. { ,tranz·pər'tā·shən i,mər· jən·sē }

transportation engineering [ENG] That branch of engineering relating to the movement of goods and people; major types of transportation are highway, water, rail, subway, air, and pipeline. { ,tranz·pər'tā·shən ,en·jə,nir·iŋ }

transportation lag See distance/velocity lag. { ,tranz·pər'tā· shən ,lag }

transportation priorities [ENG] Indicators assigned to eligible traffic which establish its movement precedence; appropriate priority systems apply to the movement of traffic by sea and air. { ,tranz·pər'tā·shən prī,är·əd·ēz }

transportation problem [IND ENG] A programming problem that is concerned with the optimal pattern of the distribution of goods from several points of origin to several different destinations, with the specified requirements at each destination. { ,tranz·pər'tā·shən ,präb·ləm }

transport capacity [ENG] The number of persons or the tonnage (or volume) of equipment which can be carried by a vehicle under given conditions. { 'tranz,pòrt kə,pas·əd·ē }

transport case [ENG] A moistureproof nonconductive wood, plastic, or fabric container used to transport safely small quantities of dynamite sticks to and from blasting sites. { 'tranz,pòrt ,kās }

transport cross section [PHYS] The product of the total scattering cross section and the average value of $1 - \cos\theta$, where θ is the laboratory scattering angle. { 'tranz,pòrt 'kròs ,sek·shən }

transport delay unit [COMPUT SCI] A device used in analog computers which produces an output signal as a delayed form of an input signal. Also known as delay unit; transport unit. { 'tranz¦pȯrt di'lā ¦yü·nət }

transport effect [PHYS] Any phenomenon, such as diffusion, thermal conductivity, or electrical conductivity, that involves the movement of some entity, such as matter, energy, or electric charge. { 'tranz¦pȯrt i¦fekt }

transporter crane [MECH ENG] A long lattice girder supported by two lattice towers which may be either fixed or moved along rails laid at right angles to the girder; a crab with a hoist suspended from it travels along the girder. { trans'pȯrd·ər ¦krān }

transport lag See distance/velocity lag. { 'tranz¦pȯrt ¦lag }

transport mean free path [NUCLEO] 1. A path length equal to three times the diffusion coefficient of neutron flux in a nuclear reactor when Fick's law is applicable. 2. A modification of the mean free path to take into account anisotropy of scattering and the persistence of velocities. { 'tranz¦pȯrt 'mēn 'frē ¦path }

transport network [ENG] The complete system of the routes pertaining to all means of transport available in a particular area, made up of the network particular to each means of transport. { 'tranz¦pȯrt ¦net¦wərk }

transport number [PHYS CHEM] The fraction of the total current carried by a given ion in an electrolyte. Also known as transference number. { 'tranz¦pȯrt ¦nəm·bər }

transport properties [PHYS] Properties of a compound or material associated with mass or heat transport; for example, viscosity and thermal conductivity of liquids, gases, or solids. { 'tranz¦pȯrt ¦präp·ərd·ēz }

transport unit See transport delay unit. { 'tranz¦pȯrt ¦yü·nət }

transport vehicle [MECH ENG] Vehicle primarily intended for personnel and cargo carrying. { 'tranz¦pȯrt ¦vē·ə·kəl }

transpose [MATH] The matrix obtained from a given matrix by interchanging its rows and columns. { 'tranz¦pōz }

transposition [COMMUN] Interchanging the relative positions of conductors at regular intervals along a transmission line to reduce cross talk. [MATH] A permutation of a set of symbols which exchanges exactly two while leaving all others unaffected. { ¦tranz·pə'zish·ən }

transposition cipher [COMMUN] A cipher in which the order of the characters in the original message is changed. { ¦tranz·pə'zish·ən ¦sī·fər }

transposon [GEN] A genetic element which comprises large discrete segments of deoxyribonucleic acid and is capable of moving from one chromosomal site to another in the same organism or in a different organism. { ¦tranz¦pō¦zän }

transradar [COMMUN] Bandwidth compression system developed for long-range narrow-band transmission of radio signals from a radar receiver to a remote location. { 'tränz¦rā¦där }

transrectification [ELEC] Rectification that occurs in one circuit when an alternating voltage is applied to another circuit. { tranz¦rek·tə·fə'kā·shən }

transrectification characteristic [ELECTR] Graph obtained by plotting the direct-voltage values for one electrode of a vacuum tube as abscissas against the average current values in the circuit of that electrode as ordinates, for various values of alternating voltage applied to another electrode as a parameter; the alternating voltage is held constant for each curve, and the voltages on other electrodes are maintained constant. { tranz¦rek·tə·fə'kā·shən ¦kar·ik·tə¦ris·tik }

transrectifier [ELECTR] Device, ordinarily a vacuum tube, in which rectification occurs in one electrode circuit when an alternating voltage is applied to another electrode. { tranz'rek·tə¦fī·ər }

transresistance [ELEC] The ratio of the voltage between any two connections of a four-terminal junction to the current passing between the other two connections. { ¦tranz·ri'zis·təns }

transresistance amplifier [ELECTR] An amplifier whose output voltage is proportional to its input current. { ¦tranz·ri¦zis·təns 'am·plə¦fī·ər }

transsexual [PSYCH] An individual whose chromosomes, gonads, and body habitus mark that individual as a member of one sex, but who feels psychically to be of the other sex, with an overwhelming desire for sex reassignment through surgical and hormonal intervention. { tran'sek·shə·wəl }

transuranic elements [CHEM] Elements that have atomic numbers greater than 92; all are radioactive and are products of artificial nuclear changes. Also known as transuranium elements. { ¦tranz·yü'ran·ik 'el·ə·məns }

transuranium elements See transuranic elements. { ¦tranz·yü'rā·nē·əm 'el·ə·məns }

Transvaal jade [MINERAL] A mineral that is not a true jade but a green grossularite garnet. Also known as South African jade. { trans'väl 'jād }

transversal [MATH] 1. A line intersecting a given family of lines. Also known as semisecant. 2. A curve orthogonal to a hypersurface. 3. If π is a given map of a set X onto a set Y, a transversal for π is a subset T of X with the property that T contains exactly one point of $\pi^{-1}(y)$ for each $y \in Y$. { trans'vər·səl }

transverse axis [MATH] The portion of a line passing through the foci of a hyperbola that lies between the two branches of the hyperbola. { trans¦vərs 'ak·səs }

transverse baffle See cross-flow baffle. { trans¦vərs ¦baf·əl }

transverse bar [GEOL] A slightly submerged sand bar extending perpendicular to the shoreline. { trans¦vərs 'bär }

transverse basin See exogeosyncline. { trans¦vərs 'bās·ən }

transverse chromatic aberration [OPTICS] A departure of an optical image-forming system from ideal behavior in which different colors have conjugate image planes which are separated transversely. { trans¦vərs krō'mad·ik ¦ab·ə'rā·shən }

transverse colon [ANAT] The portion of the colon between the right and left colic flexures, extending transversely across the upper abdomen. { trans¦vərs 'kō·lən }

transverse cyclindrical orthomorphic chart See transverse Mercator chart. { trans¦vərs sə'lin·drə·kəl ¦ȯr·thə¦mȯr·fik 'chärt }

transverse cylindrical orthomorphic projection See transverse Mercator projection. { trans¦vərs sə'lin·drə·kəl ¦ȯr·thə¦mȯr·fik prə'jek·shən }

transverse Doppler effect [ELECTROMAG] An aspect of the optical Doppler effect, occurring when the direction of motion of the source relative to an observer is perpendicular to the direction of the light received by the observer; the observed frequency is smaller than the source frequency by the factor $[1-(v/c)^2]^{1/2}$, where v is the speed of the source and c is the speed of light. { trans¦vərs 'däp·lər i¦fekt }

transverse dune [GEOL] A sand dune with a nearly straight ridge crest formed by the merger of crescentic dunes; elongated at right angles to the direction of prevailing winds, with a gentle windward slope and a steep leeward slope. { trans¦vərs 'dün }

transverse electric mode [ELECTROMAG] A mode in which a particular transverse electric wave is propagated in a waveguide or cavity. Abbreviated TE mode. Also known as H mode (British usage). { trans¦vərs i¦lek·trik ¦mōd }

transverse electric wave [ELECTROMAG] An electromagnetic wave in which the electric field vector is everywhere perpendicular to the direction of propagation. Abbreviated TE wave. Also known as H wave (British usage). { trans¦vərs i¦lek·trik 'wāv }

transverse electromagnetic mode [ELECTROMAG] A mode in which a particular transverse electromagnetic wave is propagated in a waveguide or cavity. Abbreviated TEM mode. { trans¦vərs i¦lek·trō·mag¦ned·ik ¦mōd }

transverse electromagnetic wave [ELECTROMAG] An electromagnetic wave in which both the electric and magnetic field vectors are everywhere perpendicular to the direction of propagation. Abbreviated TEM wave. { trans¦vərs i¦lek·trō·mag¦ned·ik 'wāv }

transverse equator [MAP] A meridian the plane of which is perpendicular to the axis of a transverse projection; it serves as the origin for measurement of transverse latitude. { trans¦vərs i'kwäd·ər }

transverse fault [GEOL] A fault whose strike is more or less perpendicular to the general structural trend of the region. { trans¦vərs 'fȯlt }

transverse fold See cross fold. { trans¦vərs 'fōld }

transverse frame [NAV ARCH] A ship frame consisting of a large number of relatively small, closely spaced, athwartship frames, reinforced in the bottom by vertical floor plates and working in conjunction with widely spaced, fore-and-aft, deep girders, such as the keel, longitudinals, and side stringers. { trans¦vərs 'frām }

transverse gallery [MIN ENG] An auxiliary crosscut made

in thick deposits across the ore body in order to divide it into sections along the strike. { trans¦vərs 'gal·rē }

transverse graticule [MAP] A fictitious graticule based upon a transverse projection. { trans¦vərs 'grad·ə,kyül }

transverse interference [ELEC] Interference occurring across terminals or between signal leads. { trans¦vərs ¸in·tər'fir·əns }

transverse joint *See* cross joint. { trans¦vərs 'jóint }

transverse latitude [MAP] Angular distance from a transverse equator. Also known as inverse latitude. { trans¦vərs 'lad·ə,tüd }

transversely excited atmospheric pressure laser *See* TEA laser. { trans¦vərs·lē ek¦sīd·əd ¦at·mə¦sfir·ik 'presh·ər 'lā·zər }

transverse magnetic mode [ELECTROMAG] A mode in which a particular transverse magnetic wave is propagated in a waveguide or cavity. Abbreviated TM mode. Also known as E mode (British usage). { trans¦vərs mag¦ned·ik 'mōd }

transverse magnetic wave [ELECTROMAG] An electromagnetic wave in which the magnetic field vector is everywhere perpendicular to the direction of propagation. Abbreviated TM wave. Also known as E wave (British usage). { trans¦vərs mag¦ned·ik 'mōd 'wāv }

transverse magnetization [ENG ACOUS] Magnetization of a magnetic recording medium in a direction perpendicular to the line of travel and parallel to the greatest cross-sectional dimension. { trans¦vərs ¸mag·nəd·ə'zā·shən }

transverse magnetoresistance [ELECTROMAG] One of the galvanomagnetic effects, in which a magnetic field perpendicular to an electric current gives rise to an electrical potential change in the direction of the current. { trans¦vərs ¦mag·ned·ō·ri'zis·təns }

transverse mass [RELAT] The ratio of a force acting on a relativistic particle in a direction perpendicular to its velocity to the resulting acceleration; equal to $m_0 (1 - v^2/c^2)^{-1/2}$, where m_0 is the particle's rest mass, v is its speed, and c is the speed of light. { trans¦vərs 'mas }

transverse Mercator chart [MAP] A chart on the transverse Mercator projection. Also known as inverse cylindrical orthomorphic chart; inverse Mercator chart; transverse cylindrical orthomorphic chart. { trans¦vərs mər¦kād·ər ¸chärt }

transverse Mercator projection [MAP] A conformal map projection in which the regular Mercator projection is rotated (transversed) 90° in azimuth, the central meridian corresponding to the line which represents the equator on the regular Mercator; the characteristics as to scale are identical to those of the regular Mercator, except that the scale is dependent on distances east or west of the meridian instead of north or south of the equator. Also known as inverse cylindrical orthomorphic projection; inverse Mercator projection; transverse cylindrical orthomorphic projection. { trans¦vərs mər¦kād·ər prə,jek·shən }

transverse meridian [MAP] A great circle perpendicular to a transverse equator. { trans¦vərs mə'rid·ē·ən }

transverse metacenter [NAV ARCH] The point of intersection of the vertical through the center of buoyancy of a ship in the position of equilibrium with the vertical through the new center of buoyancy when the ship is slightly heeled. { trans¦vərs 'med·ə,sen·tər }

transverse parallel [MAP] A circle or line parallel to a transverse equator, connecting all points of equal transverse latitude. Also known as inverse parallel. { trans¦vərs 'par·ə,lel }

transverse piezoelectric effect [SOLID STATE] The manifestation of the piezoelectric effect in which the applied stress is perpendicular to the direction of the resultant electric field, or in which the applied electric field is perpendicular to the direction of the resultant stress. { ¦tranz,vərs pē,āt·sō·i'lek·trik i,fekt }

transverse pole [MAP] One of the two points 90° from a transverse equator. { trans¦vərs 'pōl }

transverse ray aberration [OPTICS] The transverse displacement from the ideal image point to the ray intersection with the ideal image plane, a measure of monochromatic aberration. { 'tranz,vərs ¦rā ,ab·ə'rā·shən }

transverse recording [ELECTR] Technique for recording television signals on magnetic tape using a four-transducer rotating head. { trans¦vərs ri'kórd·iŋ }

transverse rhumb line [MAP] A line making the same oblique angle with all fictitious meridians of a transverse Mercator projection; transverse parallels and meridians may be

considered special cases of the transverse rhumb line. Also known as inverse rhumb line. { trans¦vərs 'rəm ¸līn }

transverse ripple mark [GEOL] A ripple mark formed nearly perpendicular to the direction of the current. { trans¦vərs 'rip·əl ¸märk }

transverse stability [ENG] The ability of a ship or aircraft to recover an upright position after waves or wind roll it to one side. { trans¦vərs stə'bil·əd·ē }

transverse thrust *See* transcurrent fault. { trans¦vərs 'thrəst }

transverse valley [GEOL] **1.** A valley perpendicular to the general strike of the underlying strata. **2.** A valley cutting perpendicularly across a ridge, range, or chain of mountains. Also known as cross valley. { trans¦vərs 'val·ē }

transverse vibration [MECH] Vibration of a rod in which elements of the rod move at right angles to the axis of the rod. { trans¦vərs vī'brā·shən }

transverse wave [GEOPHYS] *See* S wave. [PHYS] A wave in which the direction of the disturbance at each point of the medium is perpendicular to the wave vector and parallel to surfaces of constant phase. { trans¦vərs 'wāv }

transversion [MOL BIO] A mutation resulting from the substitution in deoxyribonucleic acid or ribonucleic acid of a purine for a pyrimidine or a pyrimidine for a purine. { trans'vər·zhən }

trap [AERO ENG] That part of a rocket motor that keeps the propellant grain in place. [CIV ENG] A bend or dip in a soil drain which is always full of water, providing a water seal to prevent odors from entering the building. [COMPUT SCI] An automatic transfer of control of a computer to a known location, this transfer occurring when a specified condition is detected by hardware. [ELECTR] **1.** A tuned circuit used in the radio-frequency or intermediate-frequency section of a receiver to reject undesired frequencies; traps in television receiver video circuits keep the sound signal out of the picture channel. Also known as rejector. **2.** *See* wave trap. [ENG] A sealed passage such as a U-shaped bend in a pipe or pump that prevents the return flow of liquid or gas. [GEOL] *See* oil trap. [MECH ENG] A device which reduces the effect of the vapor pressure of oil or mercury on the high-vacuum side of a diffusion pump. [PETR] Any dark-colored, fine-grained, nongranitic, hypabyssal or extrusive rock. Also known as trappide; trap rock. [SOLID STATE] Any irregularity, such as a vacancy, in a semiconductor at which an electron or hole in the conduction band can be caught and trapped until released by thermal agitation. Also known as semiconductor trap. { trap }

trap address [COMPUT SCI] The location at which control is transferred in case of an interrupt as soon as the current instruction is completed. { 'trap 'ad,res }

TRAPATT diode [ELECTR] A *pn* junction diode, similar to the IMPATT diode, but characterized by the formation of a trapped space-charge plasma within the junction region; used in the generation and amplification of microwave power. Derived from trapped plasma avalanche transit time diode. { 'tra,pat ,dī,ōd }

trapdoor [BUILD] A hinged, sliding, or lifting door to cover an opening in a roof, ceiling, or floor. An undocumented entry point into a computer program, which is generally inserted by a programmer to allow discreet access to the program. [COMPUT SCI] An undocumented entry point into a computer program, which is generally inserted by a programmer to allow discreet access to the program. { 'trap,dór }

trapdoor fault [GEOL] A circular fault that is hinged at one end. { 'trap¦dór ,fólt }

trapeziform [BIOL] Having the form of a trapezium. { trə'pē·zə,form }

trapezium [MATH] A quadrilateral where no sides are parallel. { trə'pē·zē·əm }

Trapezium [ASTRON] Four very hot stars that appear to the eye as a single star in the Great Nebula of Orion; the star symbol is M42. { trə'pē·zē·əm }

trapezium distortion [ELECTR] A defect in a cathode-ray tube in which the trace is confined within a trapezium rather than a rectangle, usually as a result of interaction between the two pairs of deflection plates. { trə'pē·zē·əm di,stór·shən }

trapezohedron [CRYSTAL] An isometric crystal form of 24 faces, each face of which is an irregular four-sided figure. Also known as icositetrahedron; leucitohedron; tetragonal trisoctahedron. { trə¦pē·zō¦hē·drən }

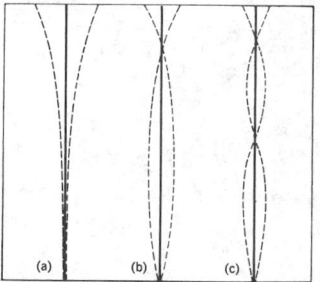

TRANSVERSE VIBRATION

Transverse vibrations of a long circular rod, rigidly clamped at one end and free at the other.
(a) Fundamental mode.
(b) First-overtone mode.
(c) Second-overtone mode.

trapezoid [MATH] A quadrilateral having two parallel sides. { 'trap·ə‚zȯid }

trapezoidal excavator [MECH ENG] A digging machine which removes earth in a trapezoidal cross-section pattern for canals and ditches. { ¦trap·ə¦zȯid·əl 'eks·kə‚vād·ər }

trapezoidal generator [ELECTR] Electronic stage designed to produce a trapezoidal voltage wave. { ¦trap·ə¦zȯid·əl 'jen·ə‚rād·ər }

trapezoidal integration [MATH] A numerical approximation of an integral by means of the trapezoidal rule. { ¦trap·ə¦zȯid·əl ‚int·ə'grā·shən }

trapezoidal pulse [ELECTR] An electrical pulse in which the voltage rises linearly to some value, remains constant at this value for some time, and then drops linearly to the original value. { ¦trap·ə¦zȯid·əl 'pəls }

trapezoidal rule [MATH] The rule that the integral from a to b of a real function $f(x)$ is approximated by

$$\frac{b-a}{2n}\left[f(a) + \sum_{j=1}^{n-1} 2f(x_j) + f(b)\right]$$

where $x_0 = a$, $x_j = x_{j-1} + (b-a)/n$ for $j = 1, 2, \ldots, n-1$. { ¦trap·ə¦zȯid·əl 'rül }

trapezoidal wave [ELECTR] A wave consisting of a series of trapezoidal pulses. { ¦trap·ə¦zȯid·əl 'wāv }

trap mine [ORD] Land mine designed to explode unexpectedly when enemy personnel attempt to move an object. { 'trap ‚mīn }

trapped-air process [ENG] A procedure for the blow-mold forming of closed plastic objects; the bottom pinch is conventional and, after blowing, sliding pinchers close off the top to form a sealed-air, inflated product. { 'trapt ¦er 'prä·səs }

trapped fuel [ENG] The fuel in an engine or fuel system that is not in the fuel tanks. { 'trapt 'fyül }

trapped plasma avalanche transit time diode See TRAPATT diode. { 'trapt 'plaz·mə 'av·ə‚lanch 'trans·ət ‚tīm 'dī‚ōd }

trapped radiation [GEOPHYS] Radiation from space that has become trapped in the magnetic field of the earth, as in the Van Allen belt. { 'trapt ‚rād·ē'ā·shən }

trappide See trap. { 'tra‚pīd }

trapping [CHEM] A method for intercepting a reactive intermediate or molecule and removing it from the system or converting it to a more stable form for further study and identification. [COMMUN] See guided propagation. [GRAPHICS] The process of an already printed ink film accepting a succeeding or overprinted ink film. { 'trap·iŋ }

trapping mode [COMPUT SCI] A procedure by means of which the computer, upon encountering a predetermined set of conditions, saves the program in its present status, executes a diagnostic procedure, and then resumes the processing of the program as of the moment of interruption. { 'trap·iŋ ‚mōd }

trap rock See trap. { 'trap ‚räk }

trap seal [CIV ENG] The vertical distance between the crown weir and the top of the dip of the trap in a plumbing system. { 'trap ‚sēl }

trash heap [COMPUT SCI] An area in a computer's memory that has been assigned to a program but contains data which are no longer useful and are therefore wasteful of storage space. { 'trash ‚hēp }

trash screen [CIV ENG] A screen placed in a waterway to prevent the passage of trash. { 'trash ‚skrēn }

Traube's rule [PHYS CHEM] In dilute solutions, the concentration of a member of a homologous series at which a given lowering of surface tension is observed decreases threefold for each additional methylene group in a given series. { 'traủ·bəz ‚rül }

trauma [MED] An injury caused by a mechanical or physical agent. [PSYCH] A severe psychic injury. { 'traủ·mə }

traumatic pneumonosis [MED] The acute, noninflammatory pathologic pulmonary changes produced by a large momentary deceleration. { trə'mad·ik ‚nü·mə'nō·səs }

traumatotropism [BIOL] Orientation response of an organ of a sessile organism in response to a wound. { ‚traủ·mə'tä·trə‚piz·əm }

Trauzl test [ENG] A test to determine the relative disruptive power of explosives, in which a standard quantity of explosive (10 grams) is placed in a cavity in a lead block and exploded; the resulting volume of cavity in the block is compared with the volume produced under the same conditions by a standard explosive, usually trinitrotoluene (TNT). { 'traủt·səl ‚test }

trave [BUILD] **1.** A division or bay (as in a ceiling) made by or appearing to be made by crossbeams. **2.** See crossbeam. { trāv }

travel [MECH ENG] The vertical distance of the path of an elevator or escalator as measured from the bottom terminal landing to the top terminal landing. { 'trav·əl }

travel chart [IND ENG] A tabulation of the various distances traveled by personnel or material between points in a manufacturing facility. { 'trav·əl ‚chärt }

travel envelope [IND ENG] The clearance in space required by an automated guided vehicle when the vehicle is carrying a load with the maximum permissible dimensions. { 'trav·əl ‚en·və‚lōp }

traveling block [MECH ENG] The movable unit, consisting of sheaves, frame, clevis, and hook, connected to, and hoisted or lowered with, the load in a block-and-tackle system. Also known as floating block; running block. { 'trav·əl·iŋ 'bläk }

traveling cable [ELEC] A cable that provides electrical contact between a fixed electrical outlet and an elevator or dumbwaiter car in the hoistway. { 'trav·əl·iŋ 'kā·bəl }

traveling charge [ORD] A propelling charge which travels along the bore with the projectile as burning takes place. Also known as Langweiler charge. { 'trav·əl·iŋ 'chärj }

traveling compartment [MIN ENG] The section of a mine shaft used for raising and lowering the miners. { 'trav·əl·iŋ kəm‚pärt·mənt }

traveling detector [ENG] Radio-frequency probe which incorporates a detector used to measure the standing-wave ratio in a slotted-line section. { 'trav·əl·iŋ di'tek·tər }

traveling dune See wandering dune. { 'trav·əl·iŋ 'dün }

traveling gantry crane [ENG] A type of hoisting machine with a bridgelike structure spanning the area over which it operates and running along tracks at ground level. { 'trav·əl·iŋ 'gan·trē ‚krān }

traveling-grate stoker [MECH ENG] A type of furnace stoker; coal feeds by gravity into a hopper located on top of one end of a moving (traveling) grate; as the grate passes under the hopper, it carries a bed of fresh coal toward the furnace. { 'trav·əl·iŋ ¦grāt 'stō·kər }

traveling microscope [OPTICS] A low-power microscope equipped with a graticule and rails enabling it to move horizontally or vertically, used to make accurate length determinations. { 'trav·əl·iŋ 'mī·krə‚skōp }

traveling position [ORD] Position of a weapon when ready for traveling, as opposed to firing position. { 'trav·əl·iŋ pə'zish·ən }

traveling road [MIN ENG] A roadway used by miners for walking to and from the face, that is, from the shaft bottom or main entry to the workings. { 'trav·əl·iŋ 'rōd }

traveling salesman problem [MATH] The problem of performing successively a number of tasks, represented by vertices of a graph, with the least expenditure on transitions from one task to another, represented by edges of the graph with journey costs attached. { 'trav·əl·iŋ 'sālz·mən ‚präb·ləm }

traveling-screen dryer [CHEM ENG] A moving screen belt on which damp material is conveyed through a heated drying zone. Also known as screen dryer. { 'trav·əl·iŋ ¦skrēn 'drī·ər }

traveling valve [PETRO ENG] A sucker-rod-pump (oil well) discharge valve that moves with the plunger of a stationary-barrel-type pump, and with the barrel of a traveling-barrel-type pump; contrasted with a standing valve. { 'trav·əl·iŋ 'valv }

traveling wave [PHYS] A wave in which energy is transported from one part of a medium to another, in contrast to a standing wave. { 'trav·əl·iŋ 'wāv }

traveling-wave amplifier [ELECTR] An amplifier that uses one or more traveling-wave tubes to provide useful amplification of signals at frequencies of the order of thousands of megahertz. Also known as traveling-wave-tube amplifier (TWTA). { 'trav·əl·iŋ ¦wāv 'am·plə‚fī·ər }

traveling-wave antenna [ELECTROMAG] An antenna in which the current distributions are produced by waves of charges propagated in only one direction in the conductors. Also known as progressive-wave antenna. { 'trav·əl·iŋ ¦wāv an'ten·ə }

traveling-wave magnetron [ELECTR] A traveling-wave tube in which the electrons move in crossed static electric and magnetic fields that are substantially normal to the direction of wave propagation, as in practically all modern magnetrons. { 'trav·əl·iŋ ¦wāv 'mag·nə,trän }

traveling-wave magnetron oscillations [ELECTR] Oscillations sustained by the interaction between the space-charge cloud of a magnetron and a traveling electromagnetic field whose phase velocity is approximately the same as the mean velocity of the cloud. { 'trav·əl·iŋ ¦wāv 'mag·nə,trän ,äs·ə,lā·shənz }

traveling-wave maser [PHYS] A ruby maser used with a comblike slow-wave structure and a number of yttrium iron garnet isolators to give L-band amplification (390 to 1550 megahertz); operation is at the temperature of liquid helium (4.2 K). { 'trav·əl·iŋ ¦wāv 'mā·zər }

traveling-wave parametric amplifier [ELECTR] Parametric amplifier which has a continuous or iterated structure incorporating nonlinear reactors and in which the signal, pump, and difference-frequency waves are propagated along the structure. { 'trav·əl·iŋ ¦wāv ¦par·ə¦me·trik 'am·plə,fī·ər }

traveling-wave phototube [ELECTR] A traveling-wave tube having a photocathode and an appropriate window to admit a modulated laser beam; the modulated laser beam causes emission of a current-modulated photoelectron beam, which in turn is accelerated by an electron gun and directed into the helical slow-wave structure of the tube. { 'trav·əl·iŋ ¦wāv 'fōd·ə,tüb }

traveling-wave tube [ELECTR] An electron tube in which a stream of electrons interacts continuously or repeatedly with a guided electromagnetic wave moving substantially in synchronism with it, in such a way that there is a net transfer of energy from the stream to the wave; the tube is used as an amplifier or oscillator at frequencies in the microwave region. { 'trav·əl·iŋ ¦wāv ,tüb }

traveling-wave-tube amplifier See traveling-wave amplifier. { ¦trav·əl·iŋ ,wāv ,tüb 'am·plə,fī·ər }

travel-time curve [GEOPHYS] A plot of P-, S-, and L-wave travel times used by seismologists to locate earthquakes. { 'trav·əl ,tīm ,kərv }

traverse [ENG] **1.** A survey consisting of a set of connecting lines of known length, meeting each other at measured angles. Also known as survey traverse. **2.** Movement to right or left on a pivot or mount, as of a gun, launcher, or radar antenna. [GEOL] A line of survey or sampling across a thin section of geological region. [METEOROL] A westerly wind in central France; it is moderate to strong, generally squally, humid and thundery in summer, especially on slopes facing west; it is cold in winter and spring and brings snow or hail showers. [NAV] A series of directions and distances, as those involved when a sailing vessel beats into the wind or a steam vessel zigzags. { tra'vərs }

traverse adjustment See balancing a survey. { tra'vərs ə,jəs·mənt }

traverse sailing [NAV] A method of determining the equivalent course and distance made good by a craft following a track consisting of a series of rhumb lines. { tra'vərs 'sāl·iŋ }

traverse table [NAV] A table giving relative values of various parts of plane right triangles, for use in solving such triangles, particularly in connection with various sailings. { tra'vərs ,tā·bəl }

traversia [METEOROL] A South American nautical term (especially Chile) for a west wind from the sea. { ,tra·vər'sē·ə }

traversier [METEOROL] In the Mediterranean, a dangerous wind blowing directly into port. { tra,ver·sē'ā }

traversing mechanism [ENG] Mechanism by which a gun or other device can be turned in a horizontal plane. { trə'vərs·iŋ ,mek·ə,niz·əm }

travertine [GEOL] Concretionary limestone deposited at the mouth of a hot spring. { 'trav·ər,tēn }

trawl [ENG] A baglike net whose mouth is kept open by boards or by a leading diving vane or depressor at the foot of the opening and a spreader bar at the top; towed by a ship at specified depths for catching forms of marine life. { trôl }

trawler [NAV ARCH] A ship designed for catching fish with a trawl. { 'trȯ·lər }

tray elevator [MECH ENG] A device for lifting drums, barrels, or boxes; a parallel pair of vertical-mounted continuous chains turn over upper and lower drive gears, and spaced trays

on the chains cradle and lift the objects to be moved. { 'trā ,el·ə,vād·ər }

tray tower [CHEM ENG] A vertical process tower for liquid-vapor contacting (as in distillation, absorption, stripping, evaporation, spray drying, dehumidification, humidification, flashing, rectification, dephlegmation), along the height of which is a series of trays designed to cause intimate contact between the falling liquid and the rising vapor. { 'trā ,taù·ər }

tread [CIV ENG] **1.** The horizontal part of a step in a staircase. **2.** The distance between two successive risers in a staircase. [ENG] The part of a wheel or tire that bears on the road or rail. { tred }

treanorite See allanite. { 'trā·nə,rīt }

treater [CHEM ENG] A vessel or system for the contacting of a process stream with reagent (treating) chemicals; for example, acid treating or caustic treating. { 'trēd·ər }

treating CHEM ENG] Usually, the contacting of a fluid stream (for example, water, sewage, petroleum products, or mixed gases) with chemicals to improve the fluid properties by removing, sequestering, or converting undesirable impurities. { 'trēd·iŋ }

Trebidae [INV ZOO] A family of copepod crustaceans of the order Caligoida which are external parasites on selachians. { 'treb·ə,dē }

treble [ACOUS] High audio frequencies, such as those handled by a tweeter in a sound system. { 'treb·əl }

tree [BOT] A perennial woody plant at least 20 feet (6 meters) in height at maturity, having an erect stem or trunk and a well-developed crown or leaf canopy. [COMPUT SCI] A data structure in which each element may be logically followed by two or more other elements, there is one element with no predecessor, every other element has a unique predecessor, and there are no circular lists. [ELECTR] A set of connected circuit branches that includes no meshes; responds uniquely to each of the possible combinations of a number of simultaneous inputs. Also known as decoder. [MATH] A connected graph contained in a given connected graph having all the vertices of the original but without any closed circuit. [MET] A projecting treelike aggregate of crystals formed at areas of high local current density in electroplating. { trē }

tree automaton [COMPUT SCI] An automaton that processes inputs in the form of trees, usually trees associated with parsing expressions in context-free languages. { 'trē ,ȯd·ə,mä·shən }

tree climate [CLIMATOL] Any type of climate which supports the growth of trees, including the tropical rainy climates, temperate rainy climates, and snow-forest climates. { 'trē ,klī·mət }

tree diagram [COMPUT SCI] A flow diagram which has no closed paths. { 'trē ,dī·ə,gram }

tree fern [BOT] The common name for plants belonging to the families Cyatheaceae and Dicksoniaceae; all are ferns that exhibit an arborescent habit. { 'trē ,fərn }

tree frog [VERT ZOO] Any of the arboreal frogs comprising the family Hylidae characterized by expanded digital adhesive disks. { 'trē ,fräg }

treeline [ECOL] The altitudinal or latitudinal limit beyond which conditions do not permit the growth of trees. { 'trē,līn }

tree pruning [COMPUT SCI] A strategy for eliminating branches of the complete game tree associated with a given position in a game such as chess or checkers, creating subtrees that explore a limited number of continuations for a limited number of moves. { 'trē ,prün·iŋ }

tree-ring hydrology See dendrohydrology. { ¦trē ¦riŋ hī'dräl·ə·jē }

trefoil [MATH] A multifoil consisting of three congruent arcs of a circle arranged around an equilateral triangle. { 'trē,fȯil }

trellis drainage [HYD] A drainage pattern characterized by parallel main streams and secondary tributaries intersected at right angles by tributaries. Also known as espalier drainage; grapevine drainage. { 'trel·əs ,drā·nij }

Trematoda [INV ZOO] A loose grouping of acoelomate, parasitic flatworms of the phylum Platyhelminthes; they exhibit cephalization, bilateral symmetry, and well-developed holdfast structures. { trem·ə'tōd·ə }

trematodiasis [MED] Infection caused by a member of the Trematoda (trematodes). { trem·ə·tə'dī·ə·səs }

Trematosauria [PALEON] A group of Triassic amphibians in the order Temnospondyli. { ,trem·əd·ə'sȯr·ē·ə }

trembling ill See louping ill. { 'trem·bliŋ ¦il }

TRAWL

Isaacs-Kidd midwater trawl, showing depressor at foot of opening and spreader bar at top. (*Scripps Institution of Oceanography*)

Tremella fuciformis [MYCOL] A mushroom that grows on deciduous trees in the southern United States and in warm climates worldwide. Once primarily grown for its medicinal properties (it boosts immunological function and stimulates leukocyte activity), it is now used mostly for food. Also known as snow fungus. { trə,mel·ə ,fyü·sə'för·məs }

Tremellales [MYCOL] An order of basidiomycetous fungi in the subclass Heterobasidiomycetidae in which basidia have longitudinal walls. { ,trem·ə'lā·lēz }

tremolo circuit [ENG ACOUS] A device which imparts a simple periodic amplitude modulation on the sound produced by an electronic instrument. { 'trem·ə·lō ,sər·kət }

tremie [ENG] An apparatus for placing concrete underwater, consisting of a large metal tube with a hopper at the top end and a valve arrangement at the bottom, submerged end. { 'trem·ē }

tremolite [MINERAL] $Ca_2Mg_5Si_8O_{22}(OH)_2$ Magnesium-rich monoclinic calcium amphibole that forms one end member of a group of solid-solution series with iron, sodium, and aluminum; occurs in long blade-shaped or short stout prismatic crystals and also in masses or compound aggregates. { 'trem·ə,līt }

tremor [GEOPHYS] A minor earthquake. Also known as earthquake tremor; earth tremor. [MED] Involuntary, rhythmic trembling of voluntary muscles resulting from alternate contraction and relaxation of opposing muscle groups. { 'trem·ər }

trench [GEOGR] **1.** A narrow, straight, elongate, U-shaped valley between two mountain ranges. **2.** A narrow stream-eroded canyon, gulley, or depression with steep sides. [GEOL] A long, narrow, deep depression of the sea floor, with relatively steep sides. Also known as submarine trench. { trench }

trench duct [CIV ENG] A metal-lined trough set into a concrete floor with removable cover plates that are level with the top of the floor; used to house electrical connections. { 'trench ,dəkt }

trencher See trench excavator. { 'trench·ər }

trench excavator [MECH ENG] A digging machine, usually on crawler tracks, and having either a movable wheel or a continuous chain on which buckets are mounted. Also known as bucket-ladder excavator; ditcher; trencher; trenching machine. { 'trench 'ek·skə,vād·ər }

trench fever [MED] A louse-borne infection that is caused by Rickettsia quintana and is characterized by headache, chills, rash, pain in the legs and back, and often by a relapsing fever. { 'trench ,fē·vər }

trenching machine See trench excavator. { 'trench·iŋ mə,shēn }

trench mouth See Vincent's infection. { 'trench ,maùth }

trench sampling [MIN ENG] A slight refinement of grab sampling in which the ore material to be sampled is spread out flat and channeled in one direction with a shovel, and the material for the sample is taken at regular intervals along the channel. { 'trench ,sam·pliŋ }

trench shield [CIV ENG] A movable shoring system consisting of steel plates and braces that are bolted or welded together; used to support the walls of a trench while work is in progress. { 'trench ,shēld }

trend [GEOL] The direction of an outcrop of a layer, vein, fold, or other kind of geologic feature. Also known as direction. [STAT] The general drift, tendency, or bent of a set of statistical data as related to time or another related set of statistical data. { trend }

trennschaukel apparatus [ENG] An instrument for determining the thermal diffusion factors of gases and gas mixtures, consisting of 20 suitably interconnected tubes whose top ends are maintained at the same temperature and whose bottom ends are maintained at the same temperature, with the temperature of the top ends greater than that of the bottom ends. { 'tren,shaù·kəl ,ap·ə,rad·əs }

Trentepohliaceae [BOT] A family of green algae belonging to the Ulotrichales having thick walls, bandlike or reticulate chloroplasts, and zoospores or isogametes produced in enlarged, specialized cells. { ,tren·tə,pō·lē'ās·ē,ē }

Trentonian [GEOL] A North American stage of geologic time; Middle Ordovician (above Wilderness, below Edenian); equivalent to the upper Mohawkian. { tren'tō·nē·ən }

trepanning tool [MECH ENG] A cutting tool in the form of a circular tube, having teeth on the end; the workpiece or tube, or both, are rotated and the tube is fed axially into the workpiece, leaving behind a narrow grooved surface in the workpiece. { trə'pan·iŋ ,tül }

Treponema pallidum [MICROBIO] A pathogenic spirochete that causes the sexually transmitted disease syphilis. { ,trep·ə,nē·mə 'pal·ə·dəm }

Treponemataceae [MICROBIO] Formerly a family of the bacterial order Spirochaetales including the spirochetes less than 20 micrometers long and less than 5 micrometers in diameter; most species are parasitic. { ,trep·ə,nē·mə'tās·ē,ē }

treponematosis [MED] Infection caused by any species of the genus Treponema. Also known as treponemiasis. { ,trep·ə,nē·mə'tō·səs }

treponemiasis See treponematosis. { ,trep·ə·nə'mī·ə·səs }

Trepostomata [PALEON] An extinct order of ectoproct bryozoans in the class Stenolaemata characterized by delicate to massive colonies composed of tightly packed zooecia with solid calcareous zooecial walls. { ,trep·ə'stō·məd·ə }

treptomorphism See isochemical metamorphism. { ¦trep·tə'mór,fiz·əm }

Treroninae [VERT ZOO] The fruit pigeons, a subfamily of the avian family Columbidae distinguished by the gaudy coloration of the feathers. { trə'rän·ə,nē }

Tresca criterion [MECH] The assumption that plastic deformation of a material begins when the difference between the maximum and minimum principal stresses equals twice the yield stress in shear. { 'tres·kə krī,tir·ē·ən }

trestle [CIV ENG] A series of short bridge spans supported by a braced tower. [ENG] **1.** A movable support usually with legs that spread diagonally. **2.** A braced structure of timber, reinforced concrete, or steel spanning a land depression to carry a road or railroad. { 'tres·əl }

trestle bent [CIV ENG] A transverse frame that supports the ends of the stringers in adjoining spans of a trestle. { 'tres·əl ,bent }

tretamine See triethylenemelamine. { 'tred·ə,mēn }

Tretothoracidae [INV ZOO] A family of the Coleoptera in the superfamily Tenebrionoidea which contains a single species found in Queensland, Australia. { ,tred·ə·thə'ras·ə,dē }

Trevelyan rocker [PHYS] A prismatic metal block having one edge grooved to form two ridges; it vibrates when heated and placed on the grooved edge, providing a simple example of heat-maintained vibrations. { trə'vel·yən ,räk·ər }

TRF receiver See tuned-radio-frequency receiver. { ,tē,är'ef ri,sē·vər }

triacetate [TEXT] A fiber manufactured from cellulose acetate in which 92% or more of the hydroxyl groups are acetylated. { trī'as·ə,tāt }

triacetin [ORG CHEM] $C_3H_5(CO_2CH_3)_3$ A colorless, combustible oil with a bitter taste and a fatty aroma; found in cod liver and butter; soluble in alcohol and ether, slightly soluble in water; boils at 259°C; used in plasticizers, perfumery, cosmetics, and external medicine and as a solvent and food additive. { trī'as·əd·ən }

triacetyloleandomycin [MICROBIO] An antibiotic produced by Streptomyces antibioticus and used clinically in the treatment of pneumonia, osteomyelitis, furuncles, and carbuncles. { trī·ə¦sēd·əl,ō·lē,an·də'mīs ·ən }

triad [COMPUT SCI] A group of three bits, pulses, or characters forming a unit of data. [ELECTR] A triangular group of three small phosphor dots, each emitting one of the three primary colors on the screen of a three-gun color picture tube. [NAV] See triplet. { 'trī,ad }

triad axis [CRYSTAL] A rotation axis whose multiplicity is equal to 3. { 'trī,ad ,ak·səs }

triaene [INV ZOO] An elongated spicule in certain Porifera with three rays diverging from one end. { 'trī,ēn }

triage [MED] The process of determining which casualties (as from an accident, disaster, military battle, or explosion of nuclear weapons) need urgent treatment, which ones are well enough to go untreated, and which ones are beyond hope of benefit from treatment. { trē'äzh }

Triakidae [VERT ZOO] A family of galeoid sharks in the carcharinid line. { trī'ak·ə,dē }

trial [STAT] One of a series of duplicate experiments. { trīl }

trial batch [ENG] A batch of concrete mixed to determine the water-cement ratio that will produce the required slump and compressive strength; from a trial batch, one can also compute the yield, cement factor, and required quantities of each material. { ¦trīl 'bach }

trial fire [ORD] Deliberate gunfire laid on a fixed point or target to determine the corrections for firing data. { 'trīl ‚fīr }

trial pit [MIN ENG] A shallow hole, 2 to 3 feet (60 to 90 centimeters) in diameter, put down to test shallow minerals or to establish the nature and thickness of superficial deposits and depth to bedrock. { 'trīl ‚pit }

trial shots [ENG] The experimental shots and rounds fired in a sinking pit, tunnel, opencast, or quarry to determine the best drill-hole pattern to use. { 'trīl ‚shäts }

triamcinolone [ORG CHEM] $C_{21}H_{27}FO_6$ White, toxic crystals; insoluble in water, soluble in dimethylformamide; melts at 266°C; used as an intermediate for ion-exchange resin, wetting and frothing agent, and photographic developer. { ‚trī·am'sin·əl‚än }

triamylamine [ORG CHEM] $(C_5H_{11})_3N$ A combustible, colorless, toxic liquid; soluble in gasoline, insoluble in water; used to inhibit corrosion and in insecticides. { ‚trī·ə'mil·ə‚mēn }

triamyl borate [ORG CHEM] $(C_5H_{11})_3BO_3$ A combustible, colorless liquid with an alcoholic aroma; soluble in alcohol and ether; boils at 220–280°C; used in varnishes. { trī'am·əl 'bór‚āt }

triandrous [BOT] Possessing three stamens. { trī'an·drəs }

triangle [MATH] The figure realized by connecting three noncollinear points by line segments. { 'trī‚aŋ·gəl }

Triangle See Triangulum. { 'trī‚aŋ·gəl }

triangle cut [MIN ENG] A zigzag arrangement of drill holes permitting larger openings to be obtained as the drill holes can break out between the preceding row of holes. { 'trī‚aŋ·gəl ‚kət }

triangle equation See angle equation. { 'trī‚aŋ·gəl i‚kwā·zhən }

triangle inequality [MATH] For real or complex numbers or vectors in a normed space x and y, the absolute value or norm of $x + y$ is less than or equal to the sum of the absolute values or norms of x and y. { 'trī‚aŋ·gəl ‚in·i'kwäl·ə·dē }

triangle of forces [MECH] A triangle, two of whose sides represent forces acting on a particle, while the third represents the combined effect of these forces. { 'trī‚aŋ·gəl əv 'fór·səs }

triangle of vectors [MATH] A triangle, two of whose sides represent vectors to be added, while the third represents the sum of these two vectors. { 'trī‚aŋ·gəl əv 'vek·tərz }

triangle of velocities [NAV] The fundamental triangle associated with dead-reckoning, composed of the following vectors: heading and true airspeed, track and groundspeed, and wind speed and wind direction. { 'trī‚aŋ·gəl əv və'läs·əd·ēz }

triangulable space [MATH] A topological space that is homeomorphic to the set of points that belong to the simplexes of a simplicial complex. Also known as polyhedron; topological simplicial complex. { trī¦aŋ·gyə·lə·bəl 'spās }

triangular facet [GEOL] A triangular-shaped steep-sloped hill or cliff formed usually by the erosion of a fault-truncated hill. { trī'aŋ·gyə·lər 'fas·ət }

triangular ligament See urogenital diaphragm. { trī'aŋ·gyə·lər 'lig·ə·mənt }

triangular matrix [MATH] A matrix where either all entries above or all entries below the principal diagonal are zero. { trī'aŋ·gyə·lər 'mā·triks }

triangular method [MIN ENG] A method of ore reserve estimation based on the assumption that a linear relationship exists between the grade difference and the distance between all drill holes. { trī'aŋ·gyə·lər 'meth·əd }

triangular-notch weir [CIV ENG] A measuring weir with a V-shaped notch for measuring small flows. Also known as V-notch weir. { trī'aŋ·gyə·lər ¦näch 'wer }

triangular numbers [MATH] The numbers 1, 3, 6, 10, . . . , which are the numbers of dots in successive triangular arrays, and are given by the expression $(n + 1)(n/2)$, where $n = 1, 2, 3, ...$ { trī¦aŋ·gyə·lər 'nəm·bərz }

triangular prism [MATH] A prism whose bases are triangles. { trī¦aŋ·gyə·lər 'priz·əm }

triangular pulse [ELECTR] An electrical pulse in which the voltage rises linearly to some value, and immediately falls linearly to the original value. { trī'aŋ·gyə·lər 'pəls }

triangular pyramid [MATH] A pyramid whose base is a triangle. { trī'aŋ·gyə·lər 'pir·ə‚mid }

triangular wave [ELECTR] A wave consisting of a series of triangular pulses. { trī'aŋ·gyə·lər 'wāv }

triangulation [ENG] A surveying method for measuring a large area of land by establishing a base line from which a network of triangles is built up; in a series, each triangle has at least one side common with each adjacent triangle. [MATH] A decomposition of a topological manifold into subsets homeomorphic with a polyhedron in some Euclidean space. [NAV] Determination of the position of a ship or aircraft by obtaining bearings of the moving object with reference to two fixed radio stations a known distance apart; this gives the values of one side and all angles of a triangle, from which the position can be computed. { trī‚aŋ·gyə'lā·shən }

triangulation mark [ENG] A bronze disk set in the ground to identify a point whose latitude and longitude have been determined by triangulation. { trī‚aŋ·gyə'lā·shən ‚märk }

triangulation problem [MATH] The problem of whether each topological n manifold admits a piecewise linear structure. { trī‚aŋ·gyə'lā·shən ‚präb·ləm }

Triangulum [ASTRON] A northern constellation, right ascension 2 hours, declination 30°N. Also known as Triangle. { trī'aŋ·gyə·ləm }

Triangulum Australe [ASTRON] A southern constellation, right ascension 16 hours declination 65°S. Also known as Southern Triangle. { trī'aŋ·gyə·ləm ó'strä·lē }

Triangulum Nebula [ASTRON] A nebula that is part of a small cluster of galaxies known as the local group; the nebula is labeled M 33. { trī'aŋ·gyə·ləm 'neb·yə·lə }

Triassic [GEOL] The first period of the Mesozoic era, lying above Permian and below Jurassic, 180–225 million years ago. { trī'a‚sik }

triatomic [CHEM] Consisting of three atoms. { ¦trī·ə'täm·ik }

Triatominae [INV ZOO] The kissing bugs, a subfamily of hemipteran insects in the family Reduviidae, distinguished by a long, slender rostrum. { trī·ə'täm·ə‚nē }

triaxial pinch [PL PHYS] A device for heating a confined plasma, in which a discharge in an annular space between two concentric cylindrical conductors forms a cylindrical sheet of plasma, and this plasma is then confined and compressed by magnetic fields produced by currents flowing in the axial direction in the discharge itself and in the two conductors. { trī'ak·sē·əl 'pinch }

triaxon [INV ZOO] A spicule in Porifera having three axes which cross each other at right angles. { trī'ak‚sän }

triazole [ORG CHEM] A five-membered chemical ring compound with three nitrogens in the ring; for example, $C_2H_3N_3$; proposed for use as a photoconductor and for copying systems. { 'trī·ə‚zōl }

tribasic calcium phosphate See calcium phosphate. { trī'bā·sik 'kal·sē·əm 'fä‚sfāt }

tribasic zinc phosphate See zinc phosphate. { trī'bā·sik 'ziŋk 'fä‚sfāt }

tribo- [PHYS] A prefix meaning pertaining to or resulting from friction. { 'trī·bo }

triboelectricity See frictional electricity. { ¦trī·bō‚i‚lek'tris·əd·ē }

triboelectric series [ELEC] A list of materials that produce an electrostatic charge when rubbed together, arranged in such an order that a material has a positive charge when rubbed with a material below it in the list, and has a negative charge when rubbed with a material above it in the list. { ¦trī·bō·i¦lek·trik 'sir·ēz }

triboelectrification [ELEC] The production of electrostatic charges by friction. { ¦trī·bō·i‚lek·trə·fə'kā·shən }

tribology [PHYS] The study of the phenomena and mechanisms of friction, lubrication, and wear of surfaces in relative motion. { trī'bäl·ə·jē }

triboluminescence [ATOM PHYS] Luminescence produced by friction between two materials. { ¦trī·bō·ə‚lü·mə'nes·əns }

tribometer [ENG] A device for measuring coefficients of friction, consisting of a loaded sled subject to a measurable force. { trī'bäm·əd·ər }

tribromoethanol [PHARM] $C_2H_3Br_3O$ A white crystalline compound, melting at 79–82°C; used in medicine in anesthesia. Also known as tribromoethyl alcohol. { trī¦brō·mō'eth·ə‚nól }

tribromoethyl alcohol See tribromoethanol. { trī¦brō·mō'eth·əl 'al·kə‚hól }

TRIASSIC

CENOZOIC	QUATERNARY	
	TERTIARY	
MESOZOIC	CRETACEOUS	
	JURASSIC	
	TRIASSIC	
PALEOZOIC	PERMIAN	
	CARBONIFEROUS	PENNSYLVANIAN
		MISSISSIPPIAN
	DEVONIAN	
	SILURIAN	
	ORDOVICIAN	
	CAMBRIAN	
PRECAMBRIAN		

Chart showing position of the Triassic period in relation to the other periods and to the eras of geologic time.

tributary [HYD] A stream that feeds or flows into or joins a larger stream or a lake. Also known as contributory; feeder; side stream; tributary stream. { 'trib·yə,ter·ē }

tributary area concept [MIN ENG] A theory that the weight to be supported by a square coal pillar should equal the weight of the rock strata above it plus the weight of rock equal in area to one-half the size of the opening adjacent to all four sides of the pillar. { 'trib·yə,ter·ē ¦er·ē·ə ,kän,sept }

tributary glacier [GEOL] A glacier that flows into a larger glacier. { 'trib·yə,ter·ē 'glā·shər }

tributary station [COMMUN] Communications terminal consisting of equipment compatible for the introduction of messages into or reception from its associated relay station. { 'trib·yə,ter·ē 'stā·shən }

tributary stream See tributary. { 'trib·yə,ter·ē 'strēm }

tributary waterway [HYD] Any body of water that flows into a larger body, that is, a creek in relation to a river, a river in relation to a bay, and a bay in relation to the open sea. { 'trib·yə,ter·ē 'wȯd·ər,wā }

tributoxyethyl phosphate [ORG CHEM] [CH₃(CH₂)₃-O(CH₂)₂O]PO A light yellow, oily liquid with a boiling range of 215–228°C; soluble in organic solvents; used as a plasticizer and flame retardant, and in floor waxes. { ,trī·byü¦täk·sē'eth·əl 'fä,sfāt }

tributyl borate [ORG CHEM] (C₄H₉)₃BO₃ A combustible, water-white liquid miscible with common organic liquids; boils at 232°C; used in welding fluxes and as a chemical intermediate and textile flame-retardant. { trī'byüd·əl 'bȯr,āt }

tributyl phosphate [ORG CHEM] (C₄H₉)₃PO₄ A combustible, toxic, stable liquid; soluble in most solvents, and very slightly soluble in water; boils at 292°C; used as a heat-exchange medium, pigment-grinding assistant, antifoam agent, and solvent. Abbreviated TBP. { trī'byüd·əl 'fä,sfāt }

tributyltin acetate [ORG CHEM] (C₄H₉)₃Sn−OOCCH₃ An organic compound of tin, used as an antimicrobial agent in the paper, wood, plastics, leather, and textile industries. { trī'byüd·əl·tən 'as·ə,tāt }

tributyltin chloride [ORG CHEM] (C₄H₉)₃SnCl A colorless liquid with a boiling point of 145–147°C; soluble in alcohol, benzene, and other organic solvents; used as a rodenticide. { trī'byüd·əl·tən 'klȯr,īd }

tricaine [ORG CHEM] C₁₀H₁₅NO₅S Fine, needlelike crystals, soluble in water; used as an anesthetic for fish. { 'trī,kān }

tricalcium phosphate See calcium phosphate. { trī'kal·sē·əm 'fä,sfāt }

tricamera photography [GRAPHICS] Photography obtained by simultaneous exposure of three cameras systematically disposed in the air vehicle at fixed overlapping angles relative to each other in order to cover a wide field. { ¦trī'kam·rə fə'täg·rə·fē }

tricarboxylic acid cycle See Krebs cycle. { trī,kär·bäk'sil·ik 'as·əd ,sī·kəl }

Triceratops [PALEON] Herbivorous dinosaur, 30 feet (9 meters) long and weighing 6 tons, from the Late Cretaceous Period that had long sharp horns over each eye and a short horn on its nose. { trī'ser·ə,täps }

trichalcite [MINERAL] Cu₅Ca(AsO₄)₂(CO₃)(OH)₄·6H₂O A verdigris green to blue-green, orthorhombic mineral consisting of hydrated copper arsenate. Also known as tyrolite. { trī'kal,sīt }

Trichechidae [VERT ZOO] The manatees, a family of nocturnal, solitary sirenian mammals in the suborder Trichechiformes. { trə'kek·ə,dē }

Trichechiformes [VERT ZOO] A suborder of mammals in the order Sirenia which contains the manatees and dugongids. { trə,kek·ə'fȯr,mēz }

trichesthesia [PHYSIO] A form of tactile sensibility in hair-covered regions of the body. { ,trik·əs'thēzh·ə }

Trichiaceae [MYCOL] A family of slime molds in the order Trichiales. { ,trik·ē'ās·ē,ē }

Trichiales [MYCOL] An order of Myxomycetes in the subclass Myxogastromycetidae. { ,trik·ē'ā·lēz }

trichiasis [MED] An eye disorder characterized by the misdirected inward growth of the eyelashes, causing trauma to the cornea. { tri'kī·ə·səs }

trichinosis [MED] Infection by the nematode *Trichinella spiralis* following ingestion of encysted larvae in raw or partially cooked pork; characterized by eosinophilia, nausea, fever, diarrhea, stiffness and painful swelling of muscles, and facial edema. { ,trik·ə'nō·səs }

trichite [PETR] A black, straight or curved, hairlike crystallite. { 'tri,kīt }

Trichiuridae [VERT ZOO] The cutlass-fishes, a family of the suborder Scombroidei. { ,trik·ē'yùr·ə,dē }

trichloroacetaldehyde See chloral. { trī,klȯr·ō,as·ə'tal·də,hīd }

trichloroacetic acid [ORG CHEM] CCl₃COOH Toxic, deliquescent, colorless crystals with a pungent aroma; soluble in water, alcohol, and ether; boils at 198°C; used as a chemical intermediate and laboratory reagent, and in medicine, pharmacy, and herbicides. Abbreviated TCA. { trī¦klȯr·ō·ə¦sed·ik 'as·əd }

trichloroacetic aldehyde See chloral. { trī¦klȯr·ō·ə¦sed·ik 'al·də,hīd }

trichlorobenzene [ORG CHEM] C₆H₃Cl₃ Either of two toxic compounds: 1,2,3-trichlorobenzene forms white crystals, soluble in ether, insoluble in water, boiling at 221°C, and is used as a chemical intermediate; 1,2,4-trichlorobenzene is a combustible, colorless liquid, soluble in most organic solvents and oils, insoluble in water, boiling at 213°C, and is used as a solvent and in dielectric fluids, synthetic transformer oils, lubricants, and insecticides. { trī¦klȯr·ō'ben,zēn }

trichloroethanal See chloral. { trī¦klȯr·ō'eth·ə,nal }

trichloroethane [ORG CHEM] C₂H₃Cl₃ Either of two nonflammable, irritating liquid isomeric compounds: 1,1,1-trichloroethane (CH₃CCl₃) is toxic, soluble in alcohol and ether, insoluble in water, and boils at 75°C; it is used as a solvent, aerosol propellant, and pesticide and for metal degreasing, and is also known as methyl chloroform; 1,1,2-trichloroethane (CHCl₂CH₂Cl) is clear and colorless, is soluble in alcohols, ethers, esters, and ketones, insoluble in water, has a sweet aroma, and boils at 114°C; it is used as a chemical intermediate and solvent, and is also known as vinyl trichloride. { trī¦klȯr·ō'eth,ān }

2,2,2-trichloroethanol [PHARM] CCl₃CH₂OH A hygroscopic liquid with a boiling point of 151–153°C; soluble in water and miscible with alcohol or ether; used in medicine as a hypnotic and anesthetic. Also known as trichloroethyl alcohol. { ¦tü ¦tü ¦tü trī¦klȯr·ō'eth·ə,nȯl }

trichloroethyl alcohol See 2,2,2-trichloroethanol. { trī¦klȯr·ō'eth·əl 'al·kə,hȯl }

trichloroethylene [ORG CHEM] CHCl:CCl₂ A heavy, stable, toxic liquid with a chloroform aroma; slightly soluble in water, soluble with greases and common organic solvents; boils at 87°C; used for metal degreasing, solvent extraction, and dry cleaning and as a fumigant and chemical intermediate. { trī¦klȯr·ō'eth·ə,lēn }

trichlorofluoromethane [ORG CHEM] CCl₃F A toxic, noncombustible, colorless liquid boiling at 24°C; used as a chemical intermediate, solvent, refrigerant, aerosol prepellant, and blowing agent (plastic foams) and in fire extinguishers. Also known as fluorocarbon-11; fluorotrichloromethane. { trī¦klȯr·ō¦flùr·ō'meth,ān }

trichloroiminocyanuric acid See trichloroisocyanuric acid. { trī¦klȯr·ō¦im·ə·nō¦sī·ə¦nùr·ik 'as·əd }

trichloroisocyanuric acid [ORG CHEM] C₃Cl₃N₃O₃ A crystalline substance that releases hypochlorous acid on contact with water; melting point is 246–247°C; soluble in chlorinated and highly polar solvents; used as a chlorinating agent, disinfectant, and industrial deodorant. Also known as symclosene; trichloroiminocyanuric acid. { trī¦klȯr·ō¦ī·sō¦sī·ə¦nùr·ik 'as·əd }

trichloromethane See chloroform. { trī¦klȯr·ō'meth,ān }

trichloromethyl chloroformate [ORG CHEM] ClCOOCCl₃ A toxic, colorless liquid with a boiling point of 127–128°C; soluble in alcohol, ether, and benzene; used in organic synthesis, and as a military poison gas during World War I. Also known as diphosgene. { trī¦klȯr·ō'meth·əl ¦klȯr·ə'fȯr,māt }

trichloromethylsilane See methyltrichlorosilane. { trī,klȯr·ə,meth·əl'sī,lān }

trichloronate See ortho-ethyl(O-2,4,5-trichlorophenyl)-ethyl-phosphonothioate. { trī'klȯr·ə,nāt }

trichloronitromethane See chloropicrin. { trī¦klȯr·ō¦nī·trō'meth,ān }

trichlorophenol [ORG CHEM] C₆H₂Cl₃OH Either of two

toxic nonflammable compounds with a phenol aroma: 2,4,5-trichlorophenol is a gray solid, is soluble in alcohol, acetone, and ether, melts at 69°C, and is used as a fungicide and bactericide; 2,4,6-trichlorophenol forms yellow flakes, is soluble in alcohol, acetone, and ether, boils at 248°C, and is used as a fungicide, defoliant, and herbicide; it is also known as 2,4,6-T. { trī̄klȯr·ō'fēˌnȯl }

2,4,5-trichlorophenoxyacetic acid [ORG CHEM] $C_6H_2Cl_3$-OCH_2CO_2H A toxic, light-tan solid; soluble in alcohol, insoluble in water; melts at 152°C; used as a defoliant, plant hormone, and herbicide. Also known as 2,4,5-T. { ˌtü ˈfȯr ˈfīv trī̄klȯr·ō·fəˌnäk·sē·əˌsēd·ik 'as·əd }

1,2,3-trichloropropane [ORG CHEM] $CH_2ClCHClCH_2Cl$ A toxic, colorless liquid with a boiling point of 156.17°C; used as a paint and varnish remover and degreasing agent. { ˌwən ˌtü ˈthrē trī̄klȯr·ō'prōˌpān }

1,1,2-trichloro-1,2,2-trifluoroethane [ORG CHEM] CCl_2-$FCClF_2$ A colorless, volatile liquid with a boiling point of 47.6°C; used as a solvent for dry cleaning, as a refrigerant, and in fire extinguishers. Also known as trifluorotrichloroethane. { ˌwən ˌwən ˌtü trī̄klȯr·ō ˌwən ˌtü ˌtü trī̄flür·ō'ethˌān }

trichobezoar [MED] A ball of hair or similar concretion in the stomach or intestine. { ˌtrik·ə'bē·zō·ər }

trichobothrium [INV ZOO] An erect, bristlelike sensory hair found on certain arthropods, insects, and other invertebrates. { ˌtrik·ə'bäth·rē·əm }

trichobranchiate gill [INV ZOO] A gill with filamentous branches arranged in several series around the axis; found in some decapod crustaceans. { ˌtrik·ə'braŋ·kēˌāt ˈgil }

Trichobranchidae [INV ZOO] A family of polychaete annelids belonging to the Sedentaria; most members are rare and live at great ocean depths. { ˌtrik·ə'braŋ·kəˌdē }

trichocercous cercaria [INV ZOO] A trematode larva distinguished by a spiny tail. { ˌtrik·ə'sər·kəs sər'kar·ē·ə }

Trichocomaceae [MYCOL] A small tropical family of ascomycetous fungi in the order Eurotiales with ascocarps from which a tuft of capillitial threads extrudes, releasing the ascospores after dissolution of the asci. { ˌtrik·ə·kə'mās·ēˌ̄ē }

trichocyst [INV ZOO] A minute structure in the cortex of certain protozoans that releases filamentous or fibrillar threads when discharged. { 'trik·əˌsist }

Trichodactylidae [INV ZOO] A family of fresh-water crabs in the section Brachyura, found mainly in tropical regions. { ˌtrik·ə·dak'til·əˌdē }

trichoepithelioma [MED] A benign tumor characterized by small, round, yellow, or flesh-colored papules, chiefly on the center of the face. { ˌtrik·ōˌep·əˌthē·lē'ō·mə }

Trichogrammatidae [INV ZOO] A family of the Hymenoptera in the superfamily Chalcidoidea whose larvae are parasitic in the eggs of other insects. { ˌtrik·ə·grə'mad·əˌdē }

trichogyne [BOT] A terminal portion of a procarp or archicarp which receives a spermatium. { 'trik·əˌjīn }

trichome [BOT] An appendage derived from the protoderm in plants, including hairs and scales. [INV ZOO] A brightly colored tuft of hairs on the body of a myrmecophile that releases an aromatic substance attractive to ants. { 'trī̄kōm }

Trichomonadida [INV ZOO] An order of the protozoan class Zoomastigophorea which contains four families of uninucleate species. { ˌtrik·ə·mə'näd·əd·ə }

Trichomonadidae [INV ZOO] A family of flagellate protozoans in the order Trichomonadida. { ˌtrik·ə·mə'näd·əˌdē }

trichomoniasis [MED] An infection caused by a species of the genus *Trichomonas.* { ˌtrik·ə·mə'nī·ə·səs }

Trichomycetes [MYCOL] A class of true fungi, division Fungi. { ˌtrik·əˌmī'sēd·ēs }

trichomycin [MICROBIO] An antibiotic produced by *Streptomyces hachijoensis* and *S. abikoensis;* a water-soluble yellow powder that inhibits yeasts and fungi. { ˌtrik·ə'mīs·ən }

Trichoniscidae [INV ZOO] A primitive family of isopod crustaceans in the suborder Oniscoidea found in damp littoral, halophilic, or riparian habitats. { ˌtrik·ə'nis·əˌdē }

trichopathophobia [PSYCH] Extreme anxiety and fear regarding growth, color, or diseases of hair. { ˌtrik·əˌpath·ə'fō·bē·ə }

Trichophilopteridae [INV ZOO] A family of lice in the order Mallophaga adapted to life upon the lemurs of Madagascar. { ˌtrik·əˌfil·əp'ter·əˌdē }

trichophytin [IMMUNOL] A group antigen obtained from filtrates of *Trichophyton mentagrophytes;* used in a skin test to

---

ascertain past or present infection with the dermatophytes. { ˌtrik·ə'fīt·ən }

Trichoptera [INV ZOO] The caddis flies, an aquatic order of the class Insecta; larvae are wormlike and adults have two pairs of well-veined hairy wings, long antennae, and mouthparts capable of lapping only liquids. { trə'käp·tə·rə }

Trichopterygidae [INV ZOO] The equivalent name for Ptiliidae. { trəˌkäp·tə'rij·əˌdē }

trichosclereids [BOT] Sclereid cells that are long and slender, resembling fibers, with which they intergrade. { ˌtrī̄·kō skler·ē·ədz }

Trichostomatida [INV ZOO] An order of ciliated protozoans in the subclass Holotrichia in which no true buccal ciliature is present but there is a vestibulum. { ˌtrik·ə·stō'mad·əd·ə }

Trichostrongylidae [INV ZOO] A family of parasitic roundworms belonging to the Strongyloidea; hosts are cattle, sheep, goats, swine, and cats. { ˌtrik·ə·strän'jil·əˌdē }

trichotillomania [MED] An obsessive-compulsive disorder characterized by artificial alopecia secondary to manipulation of the hair. { ˌtrik·ōˌtil·ə 'män·ē·ə }

trichotomy property [MATH] The property of a linear order < on a set S that for any two elements a and b in S exactly one of the statements $a < b$, $a = b$, or $b < a$ is true. Also known as comparison property. { trī'käd·ə·mē ˌpräp·ərd·ē }

trichroism [OPTICS] Phenomenon exhibited by certain optically anisotropic transparent crystals when subjected to white light, in which a cube of the material is found to transmit a different color through each of the three pairs of parallel faces. { 'trī̄krōˌiz·əm }

trichromatic theory [OPTICS] A theory of color vision which states that three primary colors may be chosen in such a way that, combined in various proportions, they can match any color. { ˌtrī̄·krō'mad·ik 'thē·ə·rē }

Trichuroidea [INV ZOO] A group of nematodes parasitic in various vertebrates and characterized by a slender body sometimes having a thickened posterior portion. { ˌtrik·yə'rȯid·ē·ə }

trickle charge [ELEC] A continuous charge of a storage battery at a low rate to maintain the battery in a fully charged condition. { 'trik·əl ˌchärj }

trickle cooler See cascade cooler. { 'trik·əl ˌkü·lər }

trickle drain [CIV ENG] A drain that is set vertically in water, such as a pond, with its top open and level with the normal water surface in order to carry off excess water. { 'trik·əl ˌdrān }

trickle hydrodesulfurization [CHEM ENG] A fixed-bed, petroleum refining process for desulfurization of middle distillates and gas oils; catalyst is cobalt molybdenum on alumina. { 'trik·əl ˌhī·drō·dēˌsəl·fə·rə'zā·shən }

trickling [COMPUT SCI] The temporary transfer of momentarily unneeded data from main storage to secondary storage devices. { 'trik·liŋ }

trickling filter [CIV ENG] A bed of broken rock or other coarse aggregate onto which sewage or industrial waste is sprayed intermittently and allowed to trickle through, leaving organic matter on the surface of the rocks, where it is oxidized and removed by biological growths. { 'trik·liŋ ˌfil·tər }

Tricladida [INV ZOO] The planarians, an order of the Turbellaria distinguished by diverticulated intestines with a single anterior branch and two posterior branches separated by the pharynx. { trī'klad·əd·ə }

triclinic crystal [CRYSTAL] A crystal whose unit cell has axes which are not at right angles, and are unequal. Also known as anorthic crystal. { trī'klin·ik 'krist·əl }

triclinic system [CRYSTAL] The most general and least symmetric crystal system, referred to by three axes of different length which are not at right angles to one another. { trī'klin·ik 'sis·təm }

tricolor picture tube See color picture tube. { 'trī̄kəl·ər 'pik·chər ˌtüb }

tricone bit [ENG] A rock bit with three toothed, conical cutters, each of which is mounted on friction-reducing bearings. { 'trī̄kōn 'bit }

triconodont [VERT ZOO] **1.** A tooth with three main conical cusps. **2.** Having such teeth. { trī'kän·əˌdänt }

Triconodonta [PALEON] An extinct mammalian order of small flesh-eating creatures of the Mesozoic era having no angle or a pseudoangle on the lower jaw and triconodont molars. { trī̄kän·ə'dänt·ə }

tricosane [ORG CHEM] $CH_3(CH_2)_{21}CH_3$ Combustible,

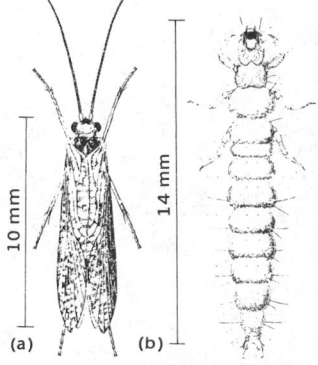

TRICHOPTERA

10 mm 14 mm

(a) (b)

Rhyacophila, a widespread genus of Trichoptera. (a) Adult, showing well-veined wings and long antennae. (b) Free-living larva.

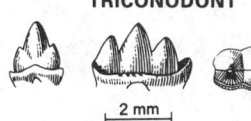

TRICONODONT

2 mm

(a) (b) (c)

Triconodont molars, showing three main cusps in longitudinal series. (a) External view of a lower molar of *Amphilestes,* (b) Internal and (c) occlusal views of a lower molar of *Priacodon.*

glittering crystals; soluble in alcohol, insoluble in water; melts at 48°C; used as a chemical intermediate. { 'trī·kə₊sān }

tricot [TEXT] A warp-knitted fabric, that is, flat-knitted with fine ribs on the face (lengthwise) and ribs on the back (widthwise). { 'trē·kō }

tricotine [TEXT] A fabric similar to gabardine but with a steep double-twill weave resembling a knitted surface. Also known as cavalry twill. { ¦trik·ə¦tēn }

tricresyl phosphate [ORG CHEM] (CH₃C₆H₄O)₃PO A combustible, colorless liquid; insoluble in water, soluble in common solvents and vegetable oils; boils at 420°C; used as a plasticizer, plastics fire retardant, air-filter medium, and gasoline and lubricant additive. Abbreviated TCP. { trī'kres·əl 'fä₊sfāt }

Trictenotomidae [INV ZOO] A small family of Indian and Malaysian beetles in the superfamily Tenebrionoidea. { ₊trik·tə·nō'täm·ə₊dē }

tricuspid valve [ANAT] A valve consisting of three flaps located between the right atrium and right ventricle of the heart. { trī'kəs·pəd 'valv }

tricycle landing gear [AERO ENG] A landing-gear arrangement that places the nose gear well forward of the center of gravity on the fuselage and the two main gears slightly aft of the center of gravity, with sufficient distance between them to provide stability against rolling over during a yawed landing in a crosswind, or during ground maneuvers. { 'trī·sik·əl 'land·iŋ ₊gir }

tricyclic dibenzopyran See xanthene. { trī'sī·klik dī¦ben·zō'pī·rən }

Tridacnidae [INV ZOO] A family of bivalve mollusks in the subclass Eulamellibranchia which contains the giant clams of the tropical Pacific. { trī'dak·nə₊dē }

Tridactylidae [INV ZOO] The pygmy mole crickets, a family of insects in the order Orthoptera, highly specialized for fossorial existence. { ₊trī·dak'til·ə₊dē }

n-tridecane [ORG CHEM] CH₃(CH₂)₁₁CH₃ A combustible liquid; soluble in alcohol, insoluble in water; boils at 226°C; used as a distillation chaser and chemical intermediate. { ¦en trī'de₊kān }

tridecanol See tridecyl alcohol. { trī'dek·ə₊nól }

tridecyl alcohol [ORG CHEM] C₁₂H₂₅CH₂OH An isomer mixture; a white, combustible solid with a pleasant aroma; melts at 31°C; used in detergents and perfumery and to make synthetic lubricants. Also known as tridecanol. { trī'des·əl 'al·kə₊hól }

tridentate ligand [INORG CHEM] A chelating agent having three groups capable of attachment to a metal ion. Also known as terdentate ligand. { trī'den₊tāt 'līg·ənd }

trident of Newton [MATH] The curve in the plane given by the equation $xy = ax^3 + bx^2 + cx + d$, where $a \neq 0$; this cuts the x axis in one or three points and is asymptotic to the y axis if $d \neq 0$. { 'trīd·ənt əv 'nüt·ən }

tridiagonal matrix [MATH] A square matrix in which all entries other than those on the principal diagonal and the two adjacent diagonals are zero. { ¦trī·dī'ag·ən·əl 'mā·triks }

triductor [ELEC] Arrangement of iron-core transformers and capacitors used to triple a power-line frequency. { trī'dək·tər }

tridymite [MINERAL] SiO₂ A white or colorless crystal occurring in minute, thin, tabular crystals or scales; a high-temperature polymorph of quartz. { 'trid·ə₊mīt }

triethanolamine [ORG CHEM] (HOCH₂CH₂)₃N A viscous, hygroscopic liquid with an ammonia aroma, soluble in chloroform, water, and alcohol, and boiling at 335°C; used in dry-cleaning soaps, cosmetics, household detergents, and textile processing, for wool scouring, and as a corrosion inhibitor. { trī₊eth·ə'näl·ə₊mēn }

triethanolamine stearate See trihydroxyethylamine stearate. { trī₊eth·ə'näl·ə₊mēn 'stir₊āt }

triethylamine [ORG CHEM] (C₂H₅)₃N A colorless, toxic, flammable liquid with an ammonia aroma; soluble in water and alcohol; boils at 90°C; used as a solvent, rubber-accelerator activator, corrosion inhibitor, and propellant, and in penetrating and waterproofing agents. { trī¦eth·ə·lo¦mēn }

triethylborane [ORG CHEM] (C₂H₅)₃B A colorless liquid with a boiling point of 95°C; used as a jet fuel or igniter for jet engines and as a fuel additive. Also known as boron triethyl; triethylborine. { trī₊eth·əl'bór₊ān }

triethylborine See triethylborane. { trī₊eth·əl'bór₊ēn }

---

triethylene glycol [ORG CHEM] HO(C₂H₄O)₃H A colorless, combustible, hygroscopic, water-soluble liquid; boils at 287°C; used as a chemical intermediate, solvent, bactericide, humectant, and fungicide. Abbreviated TEG. { trī'eth·ə₊lēn 'glī₊kól }

triethylenemelamine [ORG CHEM] NC[N(CH₂)₂]NC[N-(CH₂)₂]NC[N(CH₂)₂] White crystals, soluble in water, alcohol, acetone, chloroform, and methanol; polymerizes at 160°C; used in medicine and insecticides and as a chemosterilant. Abbreviated TEM. Also known as tretamine. { trī¦eth·ə₊lēn'mel·ə₊mēn }

triethylenetetramine [ORG CHEM] NH₂(C₂H₄NH)₂C₂H₄NH₂ A yellow, water-soluble liquid with a boiling point of 277.5°C; used in detergents and in the manufacture of dyes and pharmaceuticals. { trī¦eth·ə₊lēn'te·trə₊mēn }

triethylic borate See ethyl borate. { ₊trī·ə'thil·ik 'bór₊āt }

triethyl phosphate [ORG CHEM] (C₂H₅)₃PO₄ A toxic, colorless liquid that acts as a cholinesterase inhibitor; boiling point is 216°C; soluble in organic solvents; used as a solvent and plasticizer and for pesticides manufacture. Abbreviated TEP. { trī¦eth·əl 'fä₊sfāt }

trifid [BIOL] Divided into three lobes separated by narrow sinuses partway to the base. { 'trī₊fid }

Trifid nebula [ASTRON] An emission nebula in Sagittarius that consists mostly of hydrogen ionized by hot, young stars, and displays dark lanes formed by dust. { 'trī₊fid 'neb·yə·lə }

trifilter hydrophotometer [ENG] An instrument that uses red, green, and blue filters to measure the transparency of the water at three wavelengths. { 'trī₊fil·tər ¦hī·drō·fə'täm·əd·ər }

trifluorochlorethylene resin [ORG CHEM] A fluorocarbon used as a base for polychlorotrifluoroethylene resin. { trī¦flur·ō¦klór·ō'eth·ə₊lēn ₊rez·ən }

trifluoromethane See fluoroform. { trī¦flur·ō¦klór·ō'meth ₊ān }

trifluorotrichloroethane See 1,1,2-trichloro-1,2,2-trifluoroethane. { trī¦flur·ō·trī¦klór·ō'eth₊ān }

trifoliate [BOT] Having three leaves or leaflets. { trī'fō·lē₊āt }

trifoliosis [VET MED] An acute photosensitization characterized by superficial necrosis of white or light-skinned animals feeding on certain leguminous plants. { trī₊fō·lē'ō·səs }

Trifolium hybridium See alsike clover. { trī₊fō·lē·əm hī'brid·ē·əm }

triformal See sym-trioxane. { trī'fór·məl }

trigamma function [MATH] The derivative of the digamma function. { 'trī₊gam·ə ₊fəŋk·shən }

trigatron [ELECTR] Gas-filled, spark-gap switch used in line pulse modulators. { 'trig·ə₊trän }

trigeminal nerve [NEUROSCI] The fifth cranial nerve in vertebrates; either of a pair of composite nerves rising from the side of the medulla, and with three great branches: the ophthalmic, maxillary, and mandibular nerves. { trī'jem·ə·nəl 'nərv }

trigeminal neuralgia [MED] Sudden severe pains of unknown cause along the path of one or more branches of the trigeminal nerve. Also known as tic douloureux. { trī'jem·ə·nəl nù'ral·jə }

trigger [COMPUT SCI] To execute a jump to the first instruction of a program after the program has been loaded into the computer. Also known as initiate. [ELECTR] **1.** To initiate an action, which then continues for a period of time, as by applying a pulse to a trigger circuit. **2.** The pulse used to initiate the action of a trigger circuit. **3.** See trigger circuit. [ORD] A metallic item, part of the firing mechanism of a firearm, designed to release a firing pin by the application of pressure by the finger. { 'trig·ər }

trigger action [ELECTR] Use of a weak input pulse to initiate main current flow suddenly in a circuit or device. { 'trig·ər ₊ak·shən }

trigger bolt See auxiliary dead latch. { 'trig·ər ₊bōlt }

trigger circuit [ELECTR] **1.** A circuit or network in which the output changes abruptly with an infinitesimal change in input at a predetermined operating point. Also known as trigger. **2.** A circuit in which an action is initiated by an input pulse, as in a radar modulator. **3.** See bistable multivibrator. { 'trig·ər ₊sər·kət }

trigger control [ELECTR] Control of thyratrons, ignitrons, and other gas tubes in such a way that current flow may be started or stopped, but not regulated as to rate. { 'trig·ər kən₊trōl }

trigger diode [ELECTR] A symmetrical three-layer avalanche diode used in activating silicon-controlled rectifiers; it has a symmetrical switching mode, and hence fires whenever the breakover voltage is exceeded in either polarity. Also known as diode ac switch (diac). { 'trig·ər 'dī,ōd }

triggered spark gap [ELEC] A fixed spark gap in which the discharge passes between two electrodes but is initiated by an auxiliary trigger electrode to which low-power pulses are applied at regular intervals by a pulse amplifier. { 'trig·ərd 'spärk ,gap }

trigger electrode See starter. { 'trig·ər i,lek·trōd }

trigger extension [ORD] A metallic item attached to a trigger in order to extend its length. { 'trig·ər ik,sten·chən }

trigger finger [MED] A symptom of tenosynovitis manifested as a temporary partial obstruction in flexion or extension of a finger that is followed by a snapping into the final position; results from a thickening of a tendon or localized reduction in the tendon sheath. Also known as snapping finger. { 'trig·ər ,fiŋ·gər }

trigger housing [ORD] An item, usually of metal, designed to fit into the framework of a carbine, machine gun, pistol, or rifle; used to provide a mounting for a trigger. { 'trig·ər ,hauz·iŋ }

triggering [ELECTR] Phenomenon observed in some high-performance magnetic amplifiers with very low leakage rectifiers; as the input current is decreased in magnitude, the amplifier remains at cutoff for some time, and the output then suddenly shoots upward. { 'trig·ə·riŋ }

trigger level [ELECTR] In a transponder, the minimum input to the receiver which is capable of causing a transmitter to emit a reply. { 'trig·ər ,lev·əl }

trigger motor [ORD] An electric motor on certain types of automatic weapons that operates the sear mechanism for rapid fire. { 'trig·ər ,mōd·ər }

trigger pull [MECH] Resistance offered by the trigger of a rifle or other weapon; force which must be exerted to pull the trigger. { 'trig·ər ,pul }

trigger pulse [ELECTR] A pulse that starts a cycle of operation. Also known as tripping pulse. { 'trig·ər ,pəls }

trigger shaft [ORD] The shaft which passes transversely through the breechblock and firing lock of a gun and whose arm is actuated by the firing mechanism; movement of the arm rotates the shaft and thus imparts movement to the trigger fork of the firing lock. { 'trig·ər ,shaft }

trigger squeeze [ORD] Method of firing a rifle or similar weapon in which the trigger is not pulled, but squeezed gradually by an independent action of the forefinger. { 'trig·ər ,skwēz }

trigger switch [ELEC] A switch that is actuated by pulling a trigger, and is usually mounted in a pistol-grip handle. { 'trig·ər ,swich }

trigger tube [ELECTR] A cold-cathode gas-filled tube in which one or more auxiliary electrodes initiate the anode current but do not control it. { 'trig·ər ,tüb }

trigistor [ELECTR] A pnpn device with a gating control acting as a fast-acting switch similar in nature to a thyratron. { tri'gis·tər }

Triglidae [VERT ZOO] The searobins, a family of perciform fishes in the suborder Cottoidei. { 'trig·lə,dē }

triglyceride [ORG CHEM] $CH_2(COOCR_1)CH(OOCR_2)CH_2$-$(OOCR_3)$ A naturally occurring ester of normal, fatty acids and glycerol; used in the manufacture of edible oils, fats, and monoglycerides. { trī'glis·ə,rīd }

Trigonalidae [INV ZOO] A small family of hymenopteran insects in the superfamily Proctotrupoidea. { ,trig·ə'nal·ə,dē }

trigonal lattice See rhombohedral lattice. { trī'gōn·əl 'lad·əs }

trigonal planar molecule [CHEM] A molecule having a central atom that is bonded to three other atoms, with all four atoms lying in the same plane. { trī'gōn·əl 'plā,när 'mäl·ə,kyül }

trigonal system [CRYSTAL] A crystal system which is characterized by threefold symmetry, and which is usually considered as part of the hexagonal system since the lattice may be either hexagonal or rhombohedral. { trī'gōn·əl 'sis·təm }

trigone [ANAT] A triangular area inside the bladder limited by the openings of the ureters and urethra. [BOT] A thickening of plant cell walls formed when three or more cells adjoin. { 'trī,gōn }

trigonite [MINERAL] $MnPb_3H(AsO_3)_3$ A sulfur yellow to yellowish-brown or dark brown, monoclinic mineral consisting of an acid arsenite of lead and manganese; occurs in domatic form. { 'trī·gə,nīt }

trigonometric cofunctions [MATH] Trigonometric functions that are equal when their arguments are complementary angles, such as sine and cosine, tangent and cotangent, and secant and cosecant. { ,trig·ə·nə¦me·trik ,kō¦faŋk·shənz }

trigonometric functions [MATH] The real-valued functions such as sin(x), tan(x), and cos(x) obtained from studying certain ratios of the sides of a right triangle. Also known as circular functions. { ,trig·ə·nə¦me·trik 'faŋk·shənz }

trigonometric leveling [ENG] A method of determining the difference of elevation between two points, by using the principles of triangulation and trigonometric calculations. { ,trig·ə·nə¦me·trik 'lev·əl·iŋ }

trigonometric parallax [ASTRON] A parallax that may be determined for the nearest stars (less than 300 light-years or 2.84×10^{18} m) by a direct method utilizing trigonometry. { ,trig·ə·rə¦me·trik 'par·ə,laks }

trigonometric polynomial [MATH] A finite series of functions of the form $a_n \cos nx + b_n \sin nx$; occasionally used synonymously with trigonometric series. { ,trig·ə·nə¦me·trik ,päl·i'nō·mē·əl }

trigonometric series [MATH] An infinite series of functions with nth term of the form $a_n \cos nx + b_n \sin nx$. { ,trig·ə·nə¦me·trik 'sir·ēz }

trigonometric substitutions [MATH] The substitutions $x = a \sin u$, $x = a \tan u$ and $x = a \sec u$, which are used to rationalize expressions of the form $\sqrt{a^2 - x^2}$, $\sqrt{x^2 + a^2}$, and $\sqrt{x^2 - a^2}$, respectively, when they appear in integrals. { ,trig·ə·nə¦me·trik ,səb·stə'tü·shənz }

trigonometry [MATH] The study of triangles and the trigonometric functions. { ,trig·ə·näm·ə·trē }

Trigonostylopoidea [PALEON] A suborder of Paleocene-Eocene ungulate mammals in the order Astrapotheria. { ,trig·ə·nō,stil·ə'pòid·ē·ə }

trigonous [BIOL] **1.** Having three corners. **2.** Having a triangular cross section. { 'trig·ə·nəs }

trihedral [MATH] Any figure obtained from three noncoplanar lines intersecting in a common point. { trī'hē·drəl }

trihedral angle [MATH] A polyhedral angle with three faces. { trī¦hē·drəl 'aŋ·gəl }

trihydroxyethylamine stearate [ORG CHEM] $C_{24}H_{49}NO_5$ A cream-colored, waxy solid with a melting point of 42–44°C; soluble in methanol, ethanol, mineral oil, and vegetable oil; used as an emulsifier in cosmetics and pharmaceuticals. Also known as triethanolamine stearate. { ,trī·hī¦dräk·sē¦eth·ə·lə¦mēn 'stir,āt }

triiodomethane See iodoform. { t,rī,ī¦ō·də'meth,ān }

triisobutylene [ORG CHEM] $(C_4H_8)_3$ A mixture of isomers; combustible liquid boiling at 348–354°C; used as a chemical and resin intermediate, lubricating oil additive, and motor-fuel alkylation feedstock. { trī,ī·sō'byüd·əl,ēn }

trilateration [ENG] The measurement of a series of distances between points on the surface of the earth, for the purpose of establishing relative positions of the points in surveying. { trī,lad·ə'rā·shən }

trill See trilling. { tril }

trilling [CRYSTAL] A cyclic crystal twin consisting of three individual crystals. Also known as threeling; trill. { 'tril·iŋ }

trillion [MATH] **1.** The number 10^{12}. **2.** In British and German usage, the number 10^{18}. { 'tril·yən }

Trilobita [PALEON] The trilobites, a class of extinct Cambrian-Permian arthropods characterized by an exoskeleton covering the dorsal surface, delicate biramous appendages, body segments divided by furrows on the dorsal surface, and a pygidium composed of fused segments. { ,trī·lə'bīd·ə }

Trilobitoidea [PALEON] A class of Cambrian arthropods that are closely related to the Trilobita. { ,trī·lō·bə'tòid·ē·ə }

Trilobitomorpha [INV ZOO] A subphylum of the Arthropoda including Trilobita. { trī,läb·əd·ə'mòr·fə }

trilocular [BIOL] Having three cavities or cells. { trī'äk·yə·lər }

trim [AERO ENG] The orientation of an aircraft relative to the airstream, as indicated by the amount of control pressure required to maintain a given flight performance. [ELECTR] Fine adjustment of capacitance, inductance, or resistance of a component during manufacture or after installation in a circuit.

TRIGONAL PLANAR MOLECULE

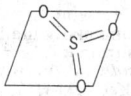

The trigonal planar molecule sulfur trioxide.

[NAV ARCH] **1.** The deviation of a ship from an even keel fore and aft. **2.** To add or remove water from the variable ballast tanks of a submarine to maintain neutral buoyancy. { trim }

trim by head [NAV ARCH] That condition of trim in which a vessel inclines forward so that its actual plane of flotation is not coincident with or parallel to its designed plane of flotation. { 'trim bī 'hed }

trim by stern [NAV ARCH] That condition of trim in which a vessel inclines aft so that its actual plane of flotation is not coincident with or parallel to its designed plane of flotation. { 'trim bī 'stərn }

Trimenoponidae [INV ZOO] A family of lice in the order Mallophaga occurring as parasites on South American rodents. { ˌtrī·mə·nəˈpän·ə·dē }

trimer [CHEM] An oligomer whose molecule is composed of three identical monomers. { 'trī·mər }

trimercuric orthophosphate *See* mercuric phosphate. { ˌtrī·mərˈkyùr·ik ˌȯr·thōˈfäˌsfāt }

trimercurous orthophosphate *See* mercurous phosphate. { ˌtrə·mərˈkyùr·əs ˌȯr·thōˈfäˌsfāt }

Trimerellacea [PALEON] A superfamily of extinct inarticulate brachiopods in the order Lingulida; they have valves, usually consisting of calcium carbonate. { trəˌmer·əˈläs·ē·ə }

Trimerophytatae *See* Trimerophytopsida. { trəˌmer·əˈfīd·əˌtē }

Trimerophytopsida [PALEOBOT] A group of extinct land vascular plants with leafless, dichotomously branched stems that bear terminal sporangia. { trəˌmer·ə·fəˈtäpˌsəd·ə }

trimerous [BOT] Having parts in sets of three. [INV ZOO] In insects, having the tarsus divided or apparently divided into three segments. { 'trim·ə·rəs }

trimethadione [PHARM] C_6H_9NO White, crystalline granules with a melting point of 45–47°C; soluble in water, alcohol, chloroform, and ether; used in medicine as an anticonvulsant. { trīˌmethˈəˈdīˌon }

trimethylamine [ORG CHEM] $(CH_3)_3N$ A colorless, liquefied gas with a fishy odor and a boiling point of −4°C; soluble in water, ether, and alcohol; used as a warning agent for natural gas, a flotation agent, and insect attractant. Abbreviated TMA. { trīˌmethˈəˈləˌmēn }

***uns*-trimethylbenzene** *See* pseudocumene. { ¦əns trīˌmethˈəlˈbenˌzēn }

trimethyl borate [ORG CHEM] $B(OCH_3)_3$ A water-white liquid, boiling at 67–68°C; used as a solvent for resins, waxes, and oils, and as a catalyst and a reagent in analysis of paint and varnish. Also known as methyl borate. { trīˈmethˈəl ˈbȯrˌāt }

trimethylchlorosilane [ORG CHEM] $(CH_3)_3SiCl$ A colorless liquid with a boiling point of 57°C; soluble in ether and benzene; used as a water-repelling agent. { trīˌmethˈəlˈklȯr·ōˈsiˌlān }

trimethylethylene *See* methyl butene. { trīˌmethˈəlˈethˈəˌlēn }

trimethylol aminomethane *See* tromethamine. { trīˈmethˈəˌlȯl əˌmē·nōˈmethˌān }

trimethylolethane [ORG CHEM] $CH_3C(CH_2OH)_3$ Colorless crystals, soluble in alcohol and water; used in the manufacture of varnishes and drying oils. Also known as methyltrimethylolmethane; pentoglycerine. { trīˌmethˈəˌlȯl ˈethˌān }

trimethylvinylammonium hydroxide *See* neurine. { trīˌmethˈəlˌvīnˈəlˈmō·nē·əm hīˈdräkˌsīd }

trimetric drawing [GRAPHICS] A form of nonperspective pictorial drawing in which the object being drawn is turned so that three mutually perpendicular edges are unequally foreshortened. { trīˈmeˈtrik ˈdrȯ·iŋ }

trimmer [BUILD] One of the single or double joists or rafters that go around an opening in the framing type of construction. [MIN ENG] **1.** A piece of bent wire used to regulate the size of the flame of a safety lamp without removing the top of the lamp. **2.** A worker who arranges coal in the hold of a vessel (miner, ship) as the coal is discharged into it from bins. **3.** A person who cleans miners' lamps. **4.** An apparatus for trimming a pile of coal into a regular form (as a cone or prism). **5.** *See* rib hole. { 'trim·ər }

trimmer capacitor [ELEC] A relatively small variable capacitor used in parallel with a larger variable or fixed capacitor to permit exact adjustment of the capacitance of the parallel combination. { 'tim·ər kəˌpas·əd·ər }

trimmer conveyor [MECH ENG] A self-contained, lightweight portable conveyor, usually of the belt type, for use in unloading and delivering bulk materials from trucks to domestic storage places, and for trimming bulk materials in bins or piles. { 'tim·ər kənˌvā·ər }

trimmer potentiometer [ELEC] A potentiometer which is used to provide a small-percentage adjustment and is often used with a coarse control. { 'tim·ər pəˌten·chēˈäm·əd·ər }

trimming [MET] Removing irregular edges from a drawn part, or parting-line flash from a forging, or gates, risers, and fins from a casting. { 'trim·iŋ }

trimorphous [BIOL] Characterized by occurring in three distinct forms, as an organ or whole organism. { trīˈmȯr·fəs }

trim size [GRAPHICS] The size of a map or page when the excess paper outside the margin has been trimmed off after printing. { trim ˌsīz }

trimuon event [PART PHYS] An inelastic collision of a neutrino or antineutrino with a nucleus in which there are three muons among the products of the collision. { trīˈmyüˌän iˌvent }

Trinidad asphalt [MATER] Natural asphaltic material found in Trinidad; contains about 47% bitumen and 28% clay, and the remainder is water. { 'trin·əˌdad 'asˌfȯlt }

trinitrobenzene [ORG CHEM] $C_6H_3(NO_2)_3$ A yellow crystalline compound, soluble in alcohol and ether; used as an explosive. { trīˌnī·trōˈbenˌzēn }

2,4,7-trinitrofluorenone [ORG CHEM] $C_{13}H_5N_3O_7$ A yellow, crystalline compound with a melting point of 175.2–176°C; forms crystalline complexes with indoles for identification by mass spectroscopy. Abbreviated TNF. { ¦tü ¦fȯr ¦sev·ən trīˌnī·trōˈflürˌəˌnōn }

trinitromethane [ORG CHEM] $CH(NO_2)_3$ A crystalline compound, melting at 150°C, decomposing above 25°C; used to make explosives. Also known as nitroform. { trīˌnī·trōˈmethˌān }

2,4,6-trinitrotoluene [ORG CHEM] $CH_3C_6H_2(NO_2)_3$ Toxic, flammable, explosive, yellow crystals; soluble in alcohol and ether, insoluble in water; melts at 81°C; used as an explosive and chemical intermediate and in photographic chemicals. Abbreviated TNT. { ¦tü ¦fȯr ¦siks trīˌnī·trōˈtälˈyəˌwen }

trinomial [MATH] A polynomial comprising three terms. [SYST] A nomenclatural designation for an organism composed of three terms: genus, species, and subspecies or variety. { trīˈnō·mē·əl }

trinomial distribution [STAT] A multinomial distribution in which there are three distinct outcomes. { trīˈnō·mē·əl ˌdi·strəˈbyü·shən }

trinomial nomenclature [SYST] The designation of subspecies by a three-word name. { trīˈnō·mē·əl 'nō·mənˌklā·chər }

trinomial surd [MATH] A sum of three roots of rational numbers, at least two of which are irrational numbers that cannot be combined without evaluating them. { trīˈnō·mē·əl 'sərd }

trinucleotide repeat expansion [GEN] An increase in the number of copies of a trinucleotide that is normally already present in multiple adjacent copies. For example, the X-linked mental retardation 1 (*XLMRI*) locus in humans usually contains 6–50 tandem repeats of CCG, but this is expanded to 200–2000 copies in the fragile-X syndrome. { ˌtrīˈnü·klē·əˌtīd riˈpēt ik ˌspanˈshən }

triode [ELECTR] A three-electrode electron tube containing an anode, a cathode, and a control electrode. { 'trīˌōd }

triode clamp [ELECTR] A keyed clamp circuit utilizing triodes, such as a circuit which contains a complementary pair of bipolar transistors. { 'trīˌōd 'klamp }

triode clipping circuit [ELECTR] A clipping circuit that utilizes a transistor or vacuum triode. { 'trīˌōd 'klip·iŋ ˌsər·kət }

triode laser [ELECTR] Gas laser whose light output may be modulated by signal voltages applied to an integral grid. { 'trīˌōd 'lā·zər }

triode transistor [ELECTR] A transistor that has three terminals. { 'trīˌōd tranˈzis·tər }

Trionychidae [VERT ZOO] The soft-shelled turtles, a family of reptiles in the order Chelonia. { trīˈəˈnik·əˌdē }

triose [BIOCHEM] A group of monosaccharide compounds that have a three-carbon chain length. { 'trīˌōs }

***sym*-trioxane** [ORG CHEM] $(CH_2O)_3$ White, flammable, explosive crystals; soluble in water, alcohol, and ether; melts at 62°C; used as a chemical intermediate, disinfectant, and

---

TRIMERELLACEA

Internal views of Silurian trimerellacean *Trimerella*: (a) pedicle valve; (b) brachial valve. (*From C. D. Walcott, Cambrian Brachiopoda, USGS Monogr. no. 51, 1912*)

(a)   (b)

TRIMETRIC DRAWING

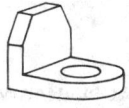

Trimetric drawing, in which three axes are unequally inclined to drawing surface. (*From T. E. French and C. J. Vierck, Graphic Science, McGraw-Hill, 1958*)

fuel. Also known as metaformaldehyde; triformol, trioxin. { ¦sim trī'äk₁sān }

trioxygen *See* ozone. { trī'äk·sə·jən }

trip [ENG] To release a lever or set free a mechanism. [MIN ENG] **1.** The line of cars hauled by mules or by motor, or run on a slope, plane, or sprag road. **2.** An automatic arrangement for dumping cars. [ORD] Part of the mechanism of some firearms, released by the action of the trigger. { trip }

tripalmitin [ORG CHEM] $C_3H_5(OOCC_{15}H_{31})_3$ A white, water-insoluble powder that melts at 65.5°C; used in the preparation of leather dressings and soaps. { trī'päm·ə·tən }

trip change [MIN ENG] The period during which the loaded cars are taken away and empties are brought back. { 'trip ₁chānj }

trip coil [ELECTROMAG] A type of solenoid in which the moving armature opens a circuit breaker or other protective device when the coil current exceeds a predetermined value. { 'trip ₁kȯil }

trip hammer [MECH ENG] A large power hammer whose head is tripped and falls by cam or lever action. { 'trip ₁ham·ər }

triphane *See* spodumene. { 'trī₁fān }

triphenylmethane dye [ORG CHEM] A family of dyes with a molecular structure derived from $(C_6H_5)_3CH$, usually by NH_2, OH, or HSO_3 substitution for one of the C_6H_5 hydrogens; includes many coal tar dyes, for example, rosaniline and fuchsin. { trī¦fen·əl'meth₁ān 'dī }

triphenylmethyl radical [ORG CHEM] A free radical in which three phenyl rings are bonded to a single carbon. Also known as trityl radical. { trī¦fen·əl¦meth·əl 'rad·ə·kəl }

triphenyl phosphate [ORG CHEM] $(C_6H_5O)_3PO$ A crystalline compound with a melting point of 49–50°C; soluble in benzene, chloroform, ether, and acetone; used as a substitute for camphor in celluloid, as a plasticizer in lacquers and varnishes, and to impregnate roofing paper. { trī'fen·əl 'fä₁sfāt }

triphenylphosphine [ORG CHEM] $(C_6H_5)_3P$ A crystalline compound with a melting point of 80.5°C; soluble in ether, benzene, chloroform, and glacial acetic acid; used as an initiator of polymerization and in organic synthesis. { trī¦fen·əl'fä₁sfēn }

triphenyltetrazolium chloride [ORG CHEM] $C_{19}H_{15}ClN_4$ A crystalline compound, soluble in water, alcohol, and acetone; used as a sensitive reagent for reducing sugars. Also known as red tetrazolium. { trī¦fen·əl¦te·trə'zō·lē·əm 'klȯr₁īd }

triphenyltinacetate *See* fentinacetate. { trī¦fen·əl·tə'nas·ə₁tāt }

triphosphopyridine dinucleotide *See* triphosphopyridine nucleotide. { trī¦fä·sfō'pir·ə₁dēn dī'nü·klē·ə₁tīd }

triphosphopyridine nucleotide [BIOCHEM] $C_{21}H_{28}N_7$-$O_{17}P_3$ A grayish-white powder, soluble in methanol and in water; a coenzyme and an important component of enzymatic systems concerned with biological oxidation-reduction systems. Abbreviated TPN. Also known as codehydrogenase II; coenzyme II; triphosphopyridine dinucleotide. { trī¦fä·sfō'pir·ə₁dēn 'nü·klē·ə₁tīd }

triphylite [MINERAL] $Li(Fe^{2+},Mn^{2+})PO_4$ A grayish-green or bluish-gray mineral crystallizing in the orthorhombic system; it is isomorphous with lithiophilite. { 'trif·ə₁līt }

tripinnate [BIOL] Being bipinnate and having each division pinnate. { trī'pi₁nāt }

trip lamp [MIN ENG] A removable self-contained mine lamp, designed for marking the rear end of a train (trip) of mine cars. { 'trip ₁lamp }

triple-alpha process *See* three-alpha process. { 'trip·əl ¦al·fə ₁präs·əs }

triple-base propellant [MATER] A propellant with three principal active ingredients, such as nitrocellulose, nitroglycerin, and nitroguanidine. { 'trip·əl bās prə'pel·ənt }

triple collision [PHYS] A process in which three particles collide simultaneously. { 'trip·əl kə'lizh·ən }

triple-conversion receiver [ELECTR] Communications receiver having three different intermediate frequencies to give higher adjacent-channel selectivity and greater image-frequency suppression. { 'trip·əl kən₁vər·zhən ri₁sē·vər }

triple detection *See* double-superheterodyne reception. { 'trip·əl di₁tek·shən }

triple-diagonal matrix *See* continuant matrix. { ¦trip·əl dī¦ag·ən·əl 'mā·triks }

triple ejection rack [ORD] A device designated to carry up to three bombs or other munitions on a single station of an aircraft. { 'trip·əl i'jek·shən ₁rak }

triple entry [MIN ENG] A system of opening a mine by driving three parallel entries as main entries. { 'trip·əl 'en·trē }

triple harmonic [PHYS] A harmonic whose frequency is three times the fundamental frequency. { 'trip·əl här'män·ik }

triple-length working [COMPUT SCI] Processing of data by a computer in which three machine words are used to represent each data item, in order to achieve the desired precision in the results. { 'trip·əl ¦leŋkth 'wərk·iŋ }

triple-mode Cepheid [ASTRON] A bent Cepheid that displays three nearly identical pulsation periods. { 'trip·əl ¦mōd 'sef·ē·əd }

triple modular redundancy [COMPUT SCI] A form of redundancy in which the original computer unit is triplicated and each of the three independent units feeds into a majority voter, which outputs the majority signal. { 'trip·əl 'mäj·ə·lər ri'dən·dən₁sē }

triple phosphate [CHEM] A phosphate containing magnesium, calcium, and ammonium ions. [MATER] A species of phosphate rock that contains three times as much phosphoric acid as superphosphate. { 'trip·əl 'fäs₁fāt }

triple point [PHYS CHEM] A particular temperature and pressure at which three different phases of one substance can coexist in equilibrium. { 'trip·əl 'pȯint }

triple response [PHYSIO] The three stages of vasomotor reaction consisting of reddening, flushing of adjacent skin, and development of wheals, when a pointed instrument is drawn heavily across the skin. { 'trip·əl ri'späns }

triple scalar product *See* scalar triple product. { 'trip·əl 'skā·lər ₁präd·əkt }

triple-stub transformer [ELECTROMAG] Microwave transformer in which three stubs are placed a quarter-wavelength apart on a coaxial line and adjusted in length to compensate for impedance mismatch. { 'trip·əl ¦stəb tranz'fȯr·mər }

triple superphosphate [MATER] A phosphatic fertilizer produced by the reaction of phosphate rock with phosphoric acid so as to give higher concentrations of calcium phosphate than for ordinary superphosphate. { 'trip·əl ¦sü·pər'fäs₁fāt }

triplet [GEN] A three-base unit in deoxyribonucleic or ribonucleic acid which codes for a particular amino acid in a protein chain. [NAV] A group of three synchronous transmitting stations operating as a system to provide signals for determination of position. Also known as triad. [OPTICS] A compound lens made up of three components. { 'trip·lət }

triple thread [DES ENG] A multiple screw thread having three threads or starts equally spaced around the periphery; the lead is three times the pitch. { 'trip·əl 'thred }

triplet state [ATOM PHYS] Electronic state of an atom or molecule whose total spin angular momentum quantum number is equal to 1. [QUANT MECH] Any multiplet having three states. { 'trip·lət ₁stāt }

triplet-state chlorophyll [BIOCHEM] The state of chlorophyll that occurs when two of the electrons in chlorophyll have their magnetic moments in parallel. { ₁trip·lət ₁stāt 'klȯr·ə₁fil }

triple vector product [MATH] The triple vector product of vectors **a**, **b**, and **c** is the cross product of **a** with the cross product of **b** and **c**; written $\mathbf{a} \times (\mathbf{b} \times \mathbf{c})$. { ¦trip·əl ¦vek·tər ₁präd·əkt }

triplex cable [ELEC] An electrical cable consisting of three individually insulated wires that are twisted together and covered by an outer layer of protective material. { 'trip₁leks ₁kā·bəl }

triplex chain block [MECH ENG] A geared hoist using an epicyclic train. { 'trip₁leks 'chān ₁bläk }

triplexer [ELECTR] Dual duplexer that permits the use of two receivers simultaneously and independently in a radar system by disconnecting the receivers during the transmitted pulse. { 'tri₁plek·sər }

triplex system [COMMUN] Telegraph system in which two messages in one direction and one message in the other direction can be sent simultaneously over a single circuit. { 'tri₁pleks ₁sis·təm }

triplite [MINERAL] $(Mn,Fe,Mg,Ca)_2(PO_4)(F,OH)$ A dark brown, chestnut brown, reddish-brown, or salmon pink, monoclinic mineral consisting of a fluophosphate of iron, manganese, magnesium, and calcium; occurs in massive form. { 'trip₁līt }

TRIPHOSPHOPYRIDINE NUCLEOTIDE

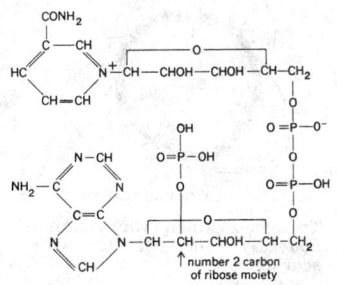

↑ number 2 carbon of ribose moiety

Structural formula of oxidized form of triphosphopyridine nucleotide.

triploblastic [EMBRYO] Having three embryonic germ layers: an ectoderm, mesoderm, and endoderm. { ˌtrip·lə'blas·tik }

Triploblastica [ZOO] Animals that develop from three germ layers. { ˌtrip·ō'blas·tə·kə }

triploidy [CYTOL] The occurrence of related forms possessing chromosome numbers three times the haploid number. { 'tri,plói·ē }

trip magnet See phase magnet. { 'trip ˌmag·nət }

tripod [DES ENG] An adjustable, collapsible three-legged support, as for a camera or surveying instrument. { 'tri,päd }

tripodal grasp [IND ENG] A basic grasp whereby an object is held by the thumb, index finger, and middle finger, to provide delicate rotational control. Also known as manipulative grasp. { ˌtrī'pōd·əl 'grasp }

tripod drill [MECH ENG] A reciprocating rock drill mounted on three legs and driven by steam or compressed air; the drill steel is removed and a longer drill inserted about every 2 feet (61 centimeters). { 'trī'päd ˌdril }

tripoli [GEOL] A lightweight, porous, friable, siliceous sedimentary rock that may have a white, gray, pink, red, or yellow color; used for polishing metals and stones. { 'trip·ə·lē }

tripolite See diatomaceous earth. { 'trip·ə,līt }

tripotassium orthophosphate See potassium phosphate. { ˌtrī·pə'tas·ē·əm ˌor·thō'fä,sfāt }

tripper [CIV ENG] A device activated by a passing train to work a signal or switch or to apply brakes. [MECH ENG] A device that snubs a conveyor belt causing the load to be discharged. { 'trip·ər }

tripping [MIN ENG] **1.** The process of pulling or lowering drill-string equipment in a borehole. **2.** To open a latch or locking device, thereby allowing a door or gate to open to empty the contents of a skip or bailer. { 'trip·iŋ }

tripping device [ELEC] Mechanical or electromagnetic device used to bring a circuit breaker or starter to its off or open position, either when certain abnormal electrical conditions occur or when a catch is actuated manually. { 'trip·iŋ di,vīs }

tripping pulse See trigger pulse. { 'trip·iŋ ,pəls }

trippkeite [MINERAL] $CuAs_2O_4$ A greenish-blue, tetragonal mineral consisting of copper arsenite. { 'trip·kē,īt }

triprene [ORG CHEM] $C_{18}H_{32}O_2S$ An amber liquid used as a growth regulator for crops. { 'trī,prēn }

trip rider [MIN ENG] A rider who throws switches, gives signals, and makes couplings. Also known as rope rider. { 'trip ,rīd·ər }

triprotic acid [CHEM] An acid that has three ionizable hydrogen atoms in each molecule. { trī'präd·ik 'as·əd }

trip spear [ENG] A fishing tool intended to recover lost casing; if the casing is found to be immovable, the hold is broken by operating the trip release. { 'trip ,spir }

triptane [ORG CHEM] C_7H_{16} A hydrocarbon compound made commercially in small quantities, but having one of the highest antiknock ratings known. { 'trip,tān }

tripuhyite [MINERAL] $FeSb_2O_6$ A greenish-yellow to dark brown mineral consisting of iron antimonate; occurs as microcrystalline aggregates. { ˌtrip·ə'wē,īt }

Tripylina [INV ZOO] A subdivision of the protozoan order Oculosida in which the major opening (astropyle) usually contains a perforated plate. { ˌtrip·ə'lī·nə }

trirectangular spherical triangle [MATH] A spherical triangle with three right angles. { ˌtrī·rek¦taŋ·gyə·lər ¦sfir·ə·kəl 'trī,aŋ·gəl }

TRIS See tromethamine. { tris }

trisaccate pollen [BOT] A three-pored pollen grain, often having a triangular outline in cross section. { trī'sa,kāt 'päl·ən }

trisaccharide [BIOCHEM] A carbohydrate which, on hydrolysis, yields three molecules of monosaccharides. { trī'sak·ə,rīd }

trisamine See tromethamine. { 'tris·ə,mēn }

tris buffer See tromethamine. { 'tris ,bəf·ər }

tris[2-(2,4-dichlorophenoxy)ethyl]phosphite [ORG CHEM] $C_{24}H_{21}Cl_6O_6P$ A dark liquid that boils above 200°C; used as a preemergence herbicide for corn, peanuts, and strawberries. Abbreviated 2,4-DEP. { ¦tris ¦tü ¦tü ¦for dī¦klor·ō·fə¦näk·sē ¦eth·əl 'fä,sfīt }

trisection [MATH] The problem of dividing an angle into three equal parts, which is impossible to do with straight edge and compass alone. { trī'sek·shən }

trisectrix [MATH] The planar curve given by $x^3 + xy^2 + ay^2 - 3ax^2 = 0$ which is symmetric about the x axis and asymptotic to the line $x = -a$; this is useful in studying the trisection of an angle problem. Also known as trisectrix of Maclaurin. { trī'sek·triks }

trisectrix of Catalan See Tschirnhausen's cubic. { trī'sek·triks əv 'kad·ə,lan }

trisectrix of Maclaurin See trisectrix. { trī'sek·triks əv mə'klor·ən }

trisistor [ELECTR] Fast-switching semiconductor consisting of an alloyed junction *pnp* device in which the collector is capable of electron injection into the base; characteristics resemble those of a thyratron electron tube, and switching time is in the nanosecond range. { trī'zis·tər }

triskaidekaphobia [PSYCH] Superstitious fear of the number thirteen. { ˌtri,skī,dek·ə'fō·bē·ə }

trisodium citrate See sodium citrate. { trī'sōd·ē·əm 'sī,trāt }

trisodium orthophosphate See trisodium phosphate. { trī'sōd·ē·əm ¦or·thō'fä,sfāt }

trisodium phosphate [INORG CHEM] Na_3PO_4 A water-soluble crystalline compound; used as a cleaning compound and as a water softener. Abbreviated TSP. Also known as tertiary sodium phosphate; trisodium orthophosphate. { trī'sōd·ē·əm 'fä,sfāt }

trisomic syndrome [MED] Any pathological condition characterized by the presence in triplicate of one of the chromosomes of a complement. { trī'sōm·ik 'sin,drōm }

trisomy [CYTOL] The presence in triplicate of one of the chromosomes of the complement. { 'trī,sō·mē }

trisomy 13–15 See D_1 trisomy. { 'trī,sō·mē 'thir¦tēn 'fif¦tēn }

trisomy 18 syndrome [MED] A congenital disorder due to trisomy of all or a large part of chromosome 18, characterized by severe mental deficiency, hypertonicity with clenched hands, and anomalies of the hands, sternum, pelvis, and facies; most infants so afflicted fail to thrive. Also known as Edwards' syndrome; E trisomy. { 'trī,sō·mē 'ā¦tēn ,sin,drōm }

trisomy 21 syndrome See Down's syndrome. { 'trī,sō·mē 'twen·tē¦wən ,sin,drōm }

tristate logic [ELECTR] A form of transistor-transistor logic in which the output stages or input and output stages can assume three states; two are the normal low-impedance 1 and 0 states, and the third is a high-impedance state that allows many tristate devices to time-share bus lines. { 'trī,stāt 'läj·ik }

tristearin See stearin. { ¦trī'stir·ən }

tristeza [PL PATH] A viral disease spread by three species of aphids that causes rapid decline or death of trees of sweet orange, grapefruit, and tangerine propagated on certain susceptible rootstock varieties. { tris'tā·zə }

tristimulus colorimeter [OPTICS] A colorimeter that measures a color stimulus in terms of tristimulus values. { trī'stim·yə·ləs ,kə·lə'rim·əd·ər }

tristimulus values [OPTICS] The magnitudes of three standard stimuli needed to match a given sample of light. { trī'stim·yə·ləs ,val·yüz }

trisulfide [CHEM] A binary chemical compound that contains three sulfur atoms in its molecule, for example, iron trisulfude, Fe_2S_3. { trī'səl,fīd }

trit [MATH] A digit in a balanced ternary system, that is, a balanced digit system with base 3. { trit }

tritanopia [MED] A defect in a third constituent essential for color vision, as in violet blindness. { ˌtrīt·ən'ō·pē·ə }

triterpene [ORG CHEM] One of a class of compounds having molecular skeletons containing 30 carbon atoms, and theoretically composed of six isoprene units; numerous and widely distributed in nature, occurring principally in plant resins and sap; an example is ambrein. { trī'tər,pēn }

tri-tet oscillator [ELECTR] Crystal-controlled, electron-coupled, vacuum-tube oscillator circuit which is isolated from the output circuit through use of the screen grid electrode as the oscillator anode; used for multiband operation because it generates strong harmonics of the crystal frequency. { 'trī,tet 'äs·ə,lād·ər }

trithioacetaldehyde [ORG CHEM] $(C_4H_4S_2)_3$ A colorless, water-insoluble, crystalline compound; used as a hypnotic. { trī¦thī·ō,as·ə'tal·də,hīd }

tritiated [CHEM] Pertaining to matter in which tritium atoms

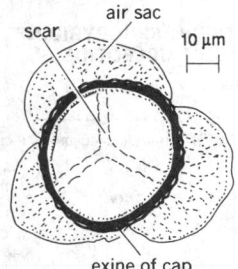
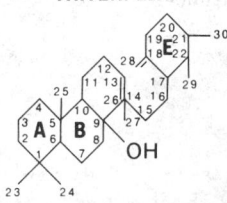

have replaced one or more atoms of ordinary hydrogen. { 'trish·ē‚ād·əd }

triticale [AGR] A cereal derived from crossing wheat and rye and then doubling the number of chromosomes in the hybrid. { 'trid·ə‚kāl }

tritium [NUC PHYS] The hydrogen isotope having mass number 3; it is one form of heavy hydrogen, the other being deuterium. Symbolized ³H; T. { 'trid·ē·əm }

triton [NUC PHYS] The nucleus of tritium. { 'trī‚tän }

Triton [ASTRON] The largest satellite of Neptune, with a diameter of about 1681 miles (2705 kilometers), orbiting at a mean distance of 220,500 miles (354,800 kilometers) with a period of 5 days 21.0 hours. { 'trīt·ən }

tritonal [MATER] An explosive composed of 80% trinitrotoluene (TNT) and 20% powdered aluminum; can be melt-loaded, and is used in bombs for its blast effect. { 'trīt·ən‚al }

triton block [MATER] Block of pressed trinitrotoluene (TNT) used for demolition purposes. { 'trī‚tän ‚bläk }

tritonymph [INV ZOO] The third stage of development in certain acarids. { 'trīd·ō‚nimf }

trityl radical See triphenylmethyl radical. { 'trīd·əl ‚rad·ə·kəl }

triuranium octoxide [INORG CHEM] U_3O_8 Olive green to black crystals or granules, soluble in nitric acid and sulfuric acid; decomposes at 1300°C; used in nuclear technology and in the preparation of other uranium compounds. Also known as uranous-uranic oxide; uranyl uranate. { ‚trī·yù'rā·nē·əm äk'täk‚sīd }

Triuridaceae [BOT] A family of monocotyledonous plants in the order Triuridales distinguished by unisexual flowers and several carpels with one seed per carpel. { trī‚yùr·ə'dās·ē‚ē }

Triuridales [BOT] A small order of terrestrial, mycotrophic monocots in the subclass Alismatidae without chlorophyll, and with separate carpels, trinucleate pollen, and a well-developed endosperm. { trī‚yùr·ə'dā·lēz }

trivial graph [MATH] A graph with one vertex and no edges. { 'triv·ē·əl ‚graf }

trivial name [ORG CHEM] A common name for a chemical compound derived from the names of the natural source of the compound at the time of its isolation and before anything is known about its molecular structure. { 'triv·ē·əl ‚nām }

trivial solution [MATH] A solution of a set of homogeneous linear equations in which all the variables have the value zero. { ‚triv·ē·əl sə'lü·shən }

trivial topology [MATH] For a set S, a topology whose only members are the set itself and the empty set. Also known as indiscreet topology. { ‚triv·ē·əl tə'päl·ə·jē }

trivium [INV ZOO] The three rays opposite the madreporite in starfish. { 'triv·ē·əm }

TRMM See Tropical Rainfall Measuring Mission.

t-RNA See transfer ribonucleic acid.

Trochacea [PALEON] A recent subfamily of primitive gastropod mollusks in the order Aspidobranchia. { trō'käsh·ē·ə }

trochal disk [INV ZOO] A flat or funnel-shaped ciliated disk at the anterior end of a rotifer that functions in locomotion and food ingestion. { 'trō·kəl ‚disk }

trochanter [ANAT] A process on the proximal end of the femur in many vertebrates, which serves for muscle attachment and, in birds, for articulation with the ilium. [INV ZOO] The second segment of an insect leg, counting from the base. { trō'kan·tər }

Trochidae [INV ZOO] A family of gastropod mollusks in the order Aspidobranchia, including many of the top shells. { 'träk·ə‚dē }

Trochili [VERT ZOO] A suborder of the avian order Apodiformes. { 'träk·ə‚lī }

Trochilidae [VERT ZOO] The hummingbirds, a tropical New World family of the suborder Trochili with tubular tongues modified for nectar feeding; slender bills and the ability to hover are further feeding adaptations. { trä'kil·ə‚dē }

Trochiliscales [PALEOBOT] A group of extinct plants belonging to the Charophyta in which the gyrogonites are dextrally spiraled. { trə‚kil·ə'skā·lēz }

trochlea [ANAT] A pulleylike anatomical structure. { 'träk·lē·ə }

trochlear nerve [NEUROSCI] The fourth cranial nerve; either of a pair of somatic motor nerves which innervate the superior oblique muscle of the eyeball. { 'träk·lē·ər ‚nərv }

trochoblast [INV ZOO] A cell bearing cilia on a trochophore. { 'träk·ə‚blast }

Trochodendraceae [BOT] A family of dicotyledonous trees in the order Trochodendrales distinguished by the absence of a perianth and stipules, numerous stamens, and pinnately veined leaves. { ‚träk·ō·den'drās·ē‚ē }

Trochodendrales [BOT] An order of dicotyledonous trees in the subclass Hamamelidae characterized by primitively vesselless wood and unique, elongate, often branched idioblasts in the leaves. { ‚träk·ō·den'drā·lēz }

trochoid [ANAT] See pivot joint. [MATH] The path in the plane obtained from a point on the radius of a circle or the extension of the radius as the circle rolls along a fixed straight line. { 'trō‚kóid }

trochoidal mass analyzer [PHYS] A mass spectrometer in which the ion beams traverse trochoidal paths within mutually perpendicular electric and magnetic fields. { trə'kóid·əl 'mas‚an·ə‚līz·ər }

trochoidal wave [FL MECH] A progressive oscillatory wave whose form is that of a prolate cycloid or trochoid; it is approximated by waves of small amplitude. { trə'kóid·əl 'wāv }

trochophore [INV ZOO] A generalized but distinct free-swimming larva found in several invertebrate groups, having a pear-shaped form with an external circlet of cilia, apical ciliary tufts, a complete functional digestive tract, and paired nephridia with excretory tubules. Also known as trochosphere. { 'träk·ə‚fór }

trochosphere See trochophore. { 'träk·ə‚sfir }

trochus [INV ZOO] The inner band of cilia on a trochal disk. { 'trō·kəs }

troctolite [PETR] A gabbro composed principally of calcic plagioclase and olivine. Also known as forellenstein. { 'träk·tə‚līt }

troegerite [MINERAL] $(UO_2)_3(AsO_4)_2 \cdot 12H_2O$ A lemon yellow, tetragonal mineral consisting of a hydrated uranium arsenate. { 'treg·ə‚rīt }

troffer [ELEC] A long, recessed lighting unit having its opening flush with the surface of the ceiling and serving as a support and reflector for lamps. { 'träf·ər }

Troglodytidae [VERT ZOO] The wrens, a family of songbirds in the order Passeriformes. { ‚träg·lə'did·ə‚dē }

Trogonidae [VERT ZOO] The trogons, the single, pantropical family of the avian order Trogoniformes. { trō'gän·ə‚dē }

Trogoniformes [VERT ZOO] An order of brightly colored, slow-moving birds characterized by a unique foot structure with the first and second toes directed backward. { trō‚gän·ə'fór‚mēz }

troilite [MINERAL] FeS A meteorite mineral crystallizing in the hexagonal system; a variety of pyrrhotite. { 'trói‚līt }

Trojan asteroid [ASTRON] **1.** One of a group of asteroids orbiting near the equilateral Lagrangian stability points of the sun-Jupiter system, which are located on Jupiter's orbit, 60° ahead of or 60° behind Jupiter. Also known as Jupiter Trojan. **2.** More generally, an object that is orbiting near one of the equilateral Lagrangian stability points of any pair of bodies. { ‚trō·jən 'as·tə‚róid }

Trojan horse [COMPUT SCI] A computer program that has an unannounced function in addition to a desirable apparent function. { ‚trō·jən 'hórs }

troland [OPTICS] A unit of retinal illuminance, equal to the retinal illuminance produced by a surface whose luminance is one nit when the apparent area of the entrance pupil of the eye is 1 square millimeter. Also known as luxon; photon. { 'trō·lənd }

Troland and Fletcher theories [PHYSIO] Theories of hearing according to which the time nature of a sound stimulation affects the sensation of pitch. { 'trō·lənd ən 'flech·ər ‚thē·ə·rēz }

trolley [GEOL] A basin-shaped depression in strata. Also known as lum. [MECH ENG] **1.** A wheeled car running on an overhead track, rail, or ropeway. **2.** An electric streetcar. { 'träl·ē }

trolley locomotive [MECH ENG] A locomotive operated by electricity drawn from overhead conductors by means of a trolley pole. { 'träl·ē ‚lōk·ə'mōd·iv }

trolley pole [ELEC] The pole which conducts electricity from the trolley wire to the trolley. { 'träl·ē ‚pōl }

trolley wire [ELEC] The means by which power is conveyed to an electric trolley locomotive; it is an overhead wire which

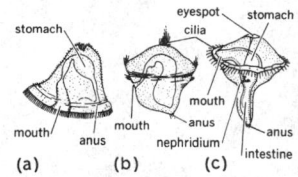

TROCHILISCALES

1 mm

Dextrally spiraled gyrogonite of Trochiliscales.

TROCHOPHORE

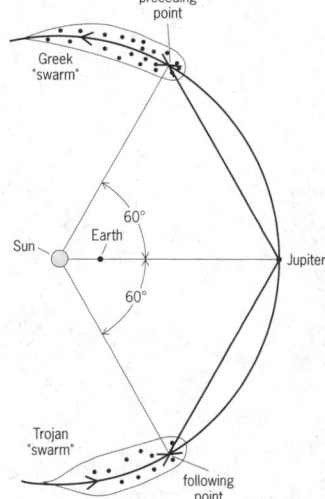

Some trochophore larvae, showing pear-shaped form, circlet of cilia, digestive tract and nephridium. *(a)* Bryozoan. *(b) Patella*, a mollusk. *(c) Polygardius*, an annelid. *(From T. I. Storer and R. L. Usinger, General Zoology, 3d ed., McGraw-Hill, 1957)*

TROJAN ASTEROID

Lagrangian stability points of the Sun-Jupiter system and the Trojan asteroids located near them.

conducts power to the locomotive by the trolley pole. { 'trāl·ē ‚wīr }

Trombiculidae [INV ZOO] The chiggers, or red bugs, a family of mites in the suborder Trombidiformes whose larvae are parasites of vertebrates. { ‚träm·bə'kyül·ə‚dē }

Trombidiformes [INV ZOO] The trombidiform mites, a suborder of the Acarina distinguished by the presence of a respiratory system opening at or near the base of the chelicerae. { ‚träm·bə·də'fȯr‚mēz }

trombone [ELECTROMAG] U-shaped, adjustable, coaxial-line matching assembly. { träm'bōn }

tromethamine [ORG CHEM] $C_4H_{11}NO_3$ A crystalline compound with a melting point of 171–172°C; soluble in water, ethylene glycol, methanol, and ethanol; used to make surface-active agents, vulcanization accelerators, and pharmaceuticals, and as a titrimetric standard.   Also known as THAM; trimethylol aminomethane; TRIS; trisamine; tris buffers. { trō'meth·ə‚mēn }

tromino [MATH] One of the two plane figures that can be formed by joining three unit squares along their sides. { 'träm·ə‚nō }

trommel [MIN ENG] **1.** A revolving cylindrical screen used to grade coarsely crushed ore: the ore is fed into the trommel at one end, the fine material drops through the holes, and the coarse is delivered at the other end.   Also known as trommel screen. **2.** To separate coal into various sizes by passing it through a revolving screen. { 'träm·əl }

trommel screen See trommel. { 'träm·əl ‚skrēn }

trona [MINERAL] $Na_2(CO_3)·Na(HCO_3)·2H_2O$ A gray-white or yellowish-white mineral that crystallizes in the monoclinic system and occurs in fibrous or columnar layers or masses.   Also known as urao. { 'trō·nə }

troop ship See transport. { 'trüp ‚ship }

Tropaeolaceae [BOT] A family of dicotyledonous plants in the order Geraniales characterized by strongly irregular flowers, simple peltate leaves, eight stamens, and schizocarpous fruit. { ‚trō·pē·ō'lās·ē‚ē }

tropeoline 00 [ORG CHEM] $NaSO_3C_6H_4NNC_6H_4NHC_6H_5$ An acid-base indicator with a pH range of 1.4–3.0, color change (from acid to base) red to yellow; used as a biological stain. { trō'pē·ə·lən 'zir·ō 'zir·ō }

Tropept [GEOL] A suborder of the order Inceptisol, characterized by moderately dark A horizons with modest additions of organic matter, B horizons with brown or reddish colors, and slightly pale C horizons; restricted to tropical regions with moderate or high rainfall. { 'trä‚pept }

trophallaxis [ECOL] Exchange of food between organisms, not only of the same species but between different species, especially among social insects. { ‚träf·ə'lak·səs }

trophic [BIOL] Pertaining to or functioning in nutrition. { 'träf·ik }

trophic ecology [ECOL] The study of the feeding relationships of organisms in communities and ecosystems. { ‚träf·ik ē'käl·ə·jē }

trophic level [ECOL] Any of the feeding levels through which the passage of energy through an ecosystem proceeds; examples are photosynthetic plants, herbivorous animals, and microorganisms of decay. { 'träf·ik ‚lev·əl }

trophobiosis [ECOL] A nutritional relationship associated only with certain species of ants in which alien insects supply food to the ants and are milked by the ants for their secretions. { ‚träf·ō‚bī'ō·səs }

trophoblast [EMBRYO] A layer of ectodermal epithelium covering the outer surface of the chorion and chorionic villi of many mammals. { 'träf·ə‚blast }

trophocyte [INV ZOO] A nutritive cell of the ovary or testis of an insect. { 'träf·ə‚sīt }

trophogenic [ECOL] Originating from nutritional differences rather than resulting from genetic determinants, such as various castes of social insects. { ¦träf·ə¦jen·ik }

tropholytic [ECOL] Pertaining to the deep zone in a lake where dissimilation of organic matter predominates. { ¦träf·ə¦lid·ik }

trophophase [MICROBIO] A period in culture production characterized by active microbial cell growth and the formation of primary metabolites. { 'träf·ə‚fāz }

trophosome [INV ZOO] The nutritional zooids of a hydroid colony. { 'träf·ə‚sōm }

trophotaeniae [VERT ZOO] Vascular rectal processes which establish placental relationships with the ovarian tissue in live-bearing fishes. { ‚träf·ō'tē·nē‚ī }

trophozoite [INV ZOO] A vegetative protozoan; used especially of a parasite. { ‚träf·ə'zō‚īt }

trophus [INV ZOO] Masticatory apparatus in Rotifera. { 'träf·əs }

tropical air [METEOROL] A type of air whose characteristics are developed over low latitudes. { 'träp·ə·kəl 'er }

tropical climate [CLIMATOL] A climate which is typical of equatorial and tropical regions, that is, one with continually high temperatures and with considerable precipitation, at least during part of the year. { 'träp·ə·kəl 'klī·mət }

tropical cyclone [METEOROL] The general term for a cyclone that originates over tropical oceans; at maturity, the tropical cyclone is one of the most intense storms of the world; winds exceeding 175 knots (324 kilometers per hour) have been measured, and the rain is torrential. { 'träp·ə·kəl 'sī‚klōn }

tropical disturbance [METEOROL] A cyclonic wind system of the tropics, of lesser intensity than a tropical cyclone. { 'träp·ə·kəl di'stər·bəns }

tropical easterlies [METEOROL] The trade winds when shallow and exhibiting a strong vertical shear; at about 500 feet (152 meters) the easterlies give way to the upper westerlies, which are sufficiently strong and deep to govern the course of cloudiness and weather.   Also known as subtropical easterlies. { 'träp·ə·kəl 'ēs·tər·lēz }

tropical finish [ENG] A finish that is applied to electronic equipment to resist the high relative humidity, fungus, and insects encountered in tropical climates. { 'träp·ə·kəl 'fin·ish }

tropical front See intertropical front. { 'träp·ə·kəl 'frənt }

tropicalize [ENG] To prepare electronic equipment for use in a tropical climate by applying a coating that resists moisture and fungi. { 'träp·ə·kə‚līz }

tropical life zone [ECOL] A subdivision of the eastern division of Merriam's life zones; an example is southern Florida, where the vegetation is the broadleaf evergreen forest; typical and important plants are palms and mangroves; typical and important animals are the armadillo and alligator; typical and important crops are citrus fruits, avocado, and banana. { 'träp·ə·kəl 'līf ‚zōn }

tropical meteorology [METEOROL] The study of the tropical atmosphere; the dividing lines, in each hemisphere, between the tropical easterlies and the mid-latitude westerlies in the middle troposphere roughly define the poleward boundaries of this region. { 'träp·ə·kəl ‚mēd·ē·ə'räl·ə·jē }

tropical monsoon climate [CLIMATOL] One of the tropical rainy climates; it is sufficiently warm and rainy to produce tropical rainforest vegetation, but it does exhibit the monsoon climate influences in that it has a winter dry season. { 'träp·ə·kəl män'sün ‚klī·mət }

tropical month [ASTRON] The average period of the revolution of the moon about the earth with respect to the vernal equinox, a period of 27 days 7 hours 43 minutes 4.7 seconds, or approximately $27^1/_3$ days. { 'träp·ə·kəl 'mənth }

tropical myositis [MED] A severe purulent infection of muscle or a muscle group, usually in the leg; decompression drainage is frequently necessary. { ‚träp·ə·kəl ‚mī·ə'sīd·əs }

Tropical Rainfall Measuring Mission [MET] A meteorological satellite used for mapping tropical precipitation in order to better understand the earth's climate system and to verify climate models.   Abbreviated TRMM. { ¦träp·ə·kəl 'rān‚fȯl ‚mezh·ər·iŋ 'mish·ən }

tropical rainforest [ECOL] A vegetation class consisting of tall, close-growing trees, their columnar trunks more or less unbranched in the lower two-thirds, and forming a spreading and frequently flat crown; occurs in areas of high temperature and high rainfall.   Also known as hylaea; selva. { 'träp·ə·kəl 'rān‚ər·əst }

tropical rainforest climate [CLIMATOL] In general, the climate which produces tropical rainforest vegetation, that is, a climate of unbroken warmth, high humidity, and heavy annual precipitation.   Also known as tropical wet climate. { 'träp·ə·kəl 'rān‚fär·əst ‚klī·mət }

tropical rainy climate [CLIMATOL] A major category in W. Köppen's climatic classification, characterized by a mean temperature of the coldest month of 64.4°F (18°C) or higher, and by a mean annual precipitation, in inches, greater than $0.44(t − a)$, where t is the mean annual temperature in degrees

Fahrenheit, and *a* equals 32 for precipitation chiefly in winter, 19.4 for evenly distributed precipitation, and 6.8 for precipitation chiefly in summer. { 'träp·ə·kəl ¦rän·ē ¸klī·mət }

tropical savanna *See* tropical woodland. { 'träp·ə· kəlsə'van·ə }

tropical savanna climate [CLIMATOL] In general, the type of climate which produces the vegetation of the tropical and subtropical savanna; thus, a climate with a winter dry season, a relatively short but heavy rainy summer season, and high year-round temperatures. Also known as savanna climate; tropical wet and dry climate. { 'träp·ə·kəl sə'van·ə ¸klī·mət }

tropical scrub [ECOL] A class of vegetation composed of low woody plants (shrubs), sometimes growing quite close together, but more often separated by large patches of bare ground, with clumps of herbs scattered throughout; an example is the Ghanaian evergreen coastal thicket. Also known as brush; bush; fourré; mallee; thicket. { 'träp·ə·kəl 'skrəb }

tropical sprue [MED] A malabsorption disease found in tropical and subtropical climates that resembles gluten enteropathy in its manifestations but does not improve with a gluten-free diet; it is thought that it may be an infectious diarrhea. { ¸träp·ə·kəl 'sprü }

tropical wet and dry climate *See* tropical savanna climate. { 'träp·ə·kəl ¦wet ən ¦drī ¸klī·mət }

tropical wet climate *See* tropical rainforest climate. { 'träp· ə·kəl 'wet ¸klī·mət }

tropical woodland [ECOL] A vegetation class similar to a forest but with wider spacing between trees and sparse lower strata characterized by evergreen shrubs and seasonal graminoids; the climate is warm and moist. Also known as parkland; savanna-woodland; tropical savanna. { 'träp·ə·kəl 'wùd·lənd }

tropical year [ASTRON] A unit of time equal to the period of one revolution of the earth about the sun measured between successive vernal equinoxes; it is 365.2422 mean solar days or 365 days 5 hours 48 minutes 46 seconds. Also known as astronomical year. { 'träp·ə·kəl 'yir }

tropic higher-high-water interval [OCEANOGR] The lunitidal interval pertaining to the higher high waters at the time of tropic tides. { 'träp·ik 'hī·ər ¦hī ¸wòd·ər ¸in·tər·vəl }

tropic high-water inequality [OCEANOGR] The average difference between the heights of the two high waters of the tidal day at the time of tropic tides. { 'träp·ik 'hī ¸wòd·ər ¸in· i'kwäl·əd·ē }

tropic lower-low-water interval [OCEANOGR] The lunitidal interval pertaining to the lower low waters at the time of tropic tides. { 'träp·ik 'lō·ər ¦lō ¸wòd·ər ¸in·tər·vəl }

tropic low-water inequality [OCEANOGR] The average difference between the heights of the two low waters of the tidal day at the time of tropic tides. { 'träp·ik 'lō ¸wòd·ər ¸in· i'kwäl·əd·ē }

Tropic of Cancer [ASTRON] A small circle on the celestial sphere connecting points with declination 23.45° north of the celestial equator, the northernmost declination of the sun. [GEOD] A parallel of latitude 23.45° north of the equator, marking the northernmost latitude at which the sun reaches its zenith. { 'träp·ik əv 'kan·sər }

Tropic of Capricorn [ASTRON] A small circle on the celestial sphere connecting points with declination 23.45° south of the celestial equator, the southernmost declination of the sun. [GEOD] A parallel of latitude 23.45° south of the equator, marking the southernmost latitude at which the sun reaches its zenith. { 'träp·ik əv 'kap·ri¸kòrn }

tropics [CLIMATOL] Any portion of the earth characterized by a tropical climate. { 'träp·iks }

tropic tidal currents [OCEANOGR] Tidal currents of increased diurnal inequality occurring at the time of tropic tides. { 'träp·ik 'tīd·əl 'kə·rəns }

tropic tide [OCEANOGR] A tide occurring when the moon is near maximum declination; the diurnal inequality is then at a maximum. { 'träp·ik 'tīd }

tropic velocity [OCEANOGR] The speed of the greater flood or greater ebb at the time of tropic currents. { 'träp·ik və'läs· əd·ē }

Tropiometridae [INV ZOO] A family of feather stars in the class Crinoidea which are bottom crawlers. { ¸träp·ē·ō'me· trə¸dē }

tropism [BIOL] Orientation movement of a sessile organism

in response to a stimulus. Also known as topotaxis. { 'trō¸piz·əm }

tropocollagen [BIOCHEM] The fundamental units of collagen fibrils. { ¸träp·ō'käl·ə·jən }

tropometer [ENG] An instrument for measuring the angle through which one end of a bar is twisted in determining the strength of a material in torsion. { trə'päm·əd·ər }

tropomyosin [BIOCHEM] A muscle protein similar to myosin and implicated as being part of the structure of the Z bands separating sarcomeres from each other. { ¸träp·ə'mī·ə·sən }

troponin [BIOCHEM] A protein species located at specific stations every 36.5 nanometers on the actin helix in muscle sarcomere. { 'trō·pə·nən }

tropopause [METEOROL] The boundary between the troposphere and stratosphere, usually characterized by an abrupt change of lapse rate; the change is in the direction of increased atmospheric stability from regions below to regions above the tropopause; its height varies from 9 to 12 miles (15 to 20 kilometers) in the tropics to about 6 miles (10 miles) in polar regions. { 'trōp·ə¸pòz }

tropopause chart [METEOROL] A synoptic chart showing the contour lines of the tropopause and tropopause break lines. { 'trōp·ə¸pòz ¸chärt }

tropopause fold [METEOROL] A phenomenon occurring in the stratosphere in which a tapering cone of dry, ozone-rich air intrudes into the troposphere. { 'trōp·ə¸pòz ¸fōld }

tropopause inversion [METEOROL] The decrease in the lapse rate of temperature encountered at the level of the tropopause. Also known as upper inversion. { 'trōp·ə¸pòz in'vər·zhən }

tropophytia [BOT] Plants that thrive in a climate that undergoes marked periodic changes. { ¸träp·ə'fī·shə }

troposcatter *See* tropospheric scatter. { 'trōp·ō¸skad·ər }

troposphere [METEOROL] That portion of the atmosphere from the earth's surface to the tropopause, that is, the lowest 10 to 20 kilometers of the atmosphere. { 'trōp·ə¸sfir }

tropospheric duct *See* duct. { ¦trōp·ə¦sfir·ik 'dəkt }

tropospheric ducting *See* ducting. { ¦trōp·ə¦sfir·ik 'dəkt·in }

tropospheric scatter [COMMUN] Scatter propagation of radio waves caused by irregularities in the refractive index of air in the troposphere; used for long-distance communications, with the aid of relay facilities, 180–300 miles (300–500 kilometers) apart. Also known as troposcatter. { ¦trōp·ə¦sfir·ik 'skad·ər }

tropospheric superrefraction [GEOPHYS] Phenomenon occurring in the troposphere whereby radio waves are bent sufficiently to be returned to the earth. { ¦trōp·ə¦sfir·ik ¦sü· pər·ri'frak·shən }

tropospheric wave [COMMUN] A radio wave that is propagated by reflection from a region of abrupt change in dielectric constant or its gradient in the troposphere. { ¦trōp·ə¦sfir·ik 'wāv }

trouble-location problem [COMPUT SCI] In computers, a test problem used in a diagnostic routine. { 'trəb·əl lō¦kā· shən ¸präb·ləm }

troubleshoot [COMPUT SCI] To find and correct errors and faults in a computer, usually in the hardware. { 'trəb·əl¸shüt }

trough [GEOL] **1.** A small, straight depression formed just offshore on the bottom of a sea or lake and on the landward side of a longshore bar. **2.** Any narrow, elongate depression in the surface of the earth. **3.** An elongate depression on the sea floor that is wider and shallower than a trench. Also known as submarine trench. **4.** The line connecting the lowest points of a fold. [METEOROL] An elongated area of relatively low atmospheric pressure; the opposite of a ridge. { tròf }

trough aloft *See* upper-level trough. { 'tròf ə'lòft }

trough crossbedding [GEOL] A variety of crossbedding in which the lower crossbedding surfaces are smoothly curved, rather than planar. { 'tròf 'kròs¸bed·in }

troughed belt conveyor [MECH ENG] A belt conveyor with the conveyor belt edges elevated on the carrying run to form a trough by conforming to the shape of the troughed carrying idlers or other supporting surface. { 'tròf 'belt kən¸vā·ər }

troughed roller conveyor [MECH ENG] A roller conveyor having two rows of rolls set at an angle to form a trough over which objects are conveyed. { 'tròf 'rō·lər kən¸vā·ər }

trough fault [GEOL] One of a set of two faults bounding a graben. { 'tròf ¸fòlt }

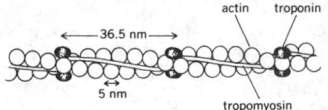

TROPOMYOSIN

Thin filament of muscle tissue, made up primarily of protein actin, showing position of tropomyosin, and also of troponin. *(From S. Ebashi, M. Endo, and I. Ohtsuki, Control of muscle contraction, Quart. Rev. Biophys., 2:351–384, 1969)*

troughing idler [MECH ENG] A belt idler having two or more rolls arranged to turn up the edges of the belt so as to form the belt into a trough. { 'tróf·iŋ ˌīd·lər }

troughing rolls [MECH ENG] The rolls of a troughing idler that are so mounted on an incline as to elevate each edge of the belt into a trough. { 'tróf·iŋ ˌrōlz }

trough plane See trough surface. { 'tróf ˌplān }

trough reef See reverse saddle. { 'tróf ˌrēf }

trough surface [GEOL] A surface or plane connecting the troughs of the bed of a syncline. Also known as synclinal axis; trough plane. { 'tróf ˌsər·fəs }

trough valley See U-shaped valley. { 'tróf ˌval·ē }

trough washer [MIN ENG] A sloping wooden trough, $1\frac{1}{2}$ to 2 feet wide, 8 to 12 feet long, and 1 foot deep (about 50 by 300 by 30 centimeters), open at the tail end but closed at the head end; it is used to float adhering clay or fine stuff from the coarser portions of an ore or coal. { 'tróf 'wäsh·ər }

trout [VERT ZOO] Any of various edible fresh-water fishes in the order Salmoniformes that are generally much smaller than the salmon. { 'traút }

Trouton-Noble experiment [ELECTROMAG] An experiment to detect ether drift by measuring the deflection of a charged parallel plate capacitor which is suspended so that it is free to turn. [RELAT] An experiment to measure the torque acting on a suspended, charged parallel-plate capacitor; the observed absence of the torque supports the special theory of relativity. { ˌtrüt·ən 'nō·bəl ik'sper·ə·mənt }

Trouton's rule [PHYS CHEM] An approximation rule for the derivation of molar heats of vaporization of normal liquids at their boiling points. [THERMO] The rule that, for a nonassociated liquid, the latent heat of vaporization in calories is equal to approximately 22 times the normal boiling point on the Kelvin scale. { 'traút·ənz ˌrül }

trowel [DES ENG] Any of various hand tools consisting of a wide, flat or curved blade with a short wooden handle; used by gardeners, plasterers, and bricklayers. { 'traúl }

troweling machine [MECH ENG] A motorized device used to spread concrete by operating orbiting steel trowels on radial arms rotated on a vertical shaft. { 'träwl·iŋ mə,shēn }

troy ounce See ounce. { 'trói 'aúns }

troy pound See pound. { 'trói 'paúnd }

troy system [MECH] A system of mass units used primarily to measure gold and silver; the ounce is the same as that in the apothecaries' system, being equal to 480 grains or 31.1034768 grams. Abbreviated t. Also known as troy weight. { 'trói ,sis·təm }

troy weight See troy system. { 'trói ,wāt }

TRSB See time reference scanning beam.

TRSV See tobacco ring spot virus.

Trube's correlation [PETRO ENG] An empirical correlation (based on pseudocritical properties) for compressibilities of undersaturated oil-reservoir fluids. { 'trü·bəz ˌkär·ə,lā·shən }

Trucherognathidae [PALEON] A family of conodonts in the order Conodontophorida in which the attachment scar permits the conodont to rest on the jaw ramus. { ˌtrü·chə·räg'nath·ə,dē }

truck [MECH ENG] A self-propelled wheeled vehicle, designed primarily to transport goods and heavy equipment; it may be used to tow trailers or other mobile equipment. [MIN ENG] See barney. { trək }

truck crane [MECH ENG] A crane carried on the bed of a motortruck. { 'trək ,krān }

truck-mounted drill rig [MECH ENG] A drilling rig mounted on a lorry or caterpillar tracks. { 'trək ˌmaúnt·əd 'dril ,rig }

truck-tractor See tractor. { 'trək ˌtrak·tər }

trudellite [MINERAL] $Al_{10}(SO_4)_3Cl_{12}(OH)_{12}\cdot 30H_2O$ An amber yellow, hexagonal mineral consisting of a hydrated basic sulfate-chloride of aluminum; occurs as compact masses. { 'trü·de,līt }

true [GEOD] Related to true north. [SCI TECH] **1.** Actual, as contrasted with fictitious, such as true sun. **2.** Related to a fixed point, either on the earth or in space, in contrast with relative, which is related to a moving point. { trü }

true airspeed [AERO ENG] The actual speed of an aircraft relative to the air through which it flies, that is, the calibrated airspeed corrected for temperature, density, or compressibility. { 'trü ˌer,spēd }

true-airspeed indicator [AERO ENG] An instrument for measuring true airspeed. Also known as true-airspeed meter. { 'trü 'er,spēd ,in·də,kād·ər }

true-airspeed meter See true-airspeed indicator. { 'trü 'er,spēd ,mēd·ər }

true air temperature [METEOROL] Basic air temperature corrected for heat of compression error due to high-speed motion of the thermometer through the air, as on an aircraft. { 'trü 'er ,tem·prə·chər }

true altitude See corrected altitude. { 'trü 'al·tə,tüd }

true amplitude [NAV] In celestial navigation, amplitude relative to true east or true west. { 'trü 'am·plə,tüd }

true anomaly See anomaly. { 'trü ə'näm·ə·lē }

true azimuth [NAV] Azimuth relative to true north. { 'trü 'az·ə·məth }

true bearing [NAV] Bearing relative to true north; compass bearing corrected for magnetic deviation. { 'trü 'ber·iŋ }

true-boiling-point analysis [CHEM ENG] A standard laboratory technique used to predict the refining qualities of crude petroleum; gives distillation cuts for gasoline, kerosine, distillate (diesel) fuel, cracking, and lube distillate stocks. Also known as true-boiling-point distillation. { 'trü 'bóil·iŋ ˌpóint ə,nal·ə·səs }

true-boiling-point distillation See true-boiling-point analysis. { 'trü 'bóil·iŋ ˌpóint ,dis·tə,lā·shən }

true complement See radix complement. { 'trü 'käm·plə·mənt }

true condensing point See critical condensation temperature. { 'trü kən'dens·iŋ ,póint }

true convergence [GEOD] The angle at which one meridian is inclined to another on the surface of the earth. { 'trü kən'vər·jəns }

true course [NAV] Course relative to true north. { 'trü 'kórs }

true crater [GEOL] The primary depression formed by impact or explosion before modification by slumping or by deposition of ejected material. Also known as primary crater. { 'trü 'krād·ər }

true dip See dip. { 'trü 'dip }

true direction [NAV] Horizontal direction expressed as angular distance from true north. { 'trü də'rek·shən }

true electrolyte [PHYS CHEM] A substance in the solid state that consists entirely of ions. { ˌtrü i'lek·trə,līt }

true formation resistivity [GEOPHYS] Electrical resistivity of a clean (nonshaly) porous reservoir formation containing hydrocarbons and formation water; value is greater than the resistivity when there is added water incursion. { 'trü fòr'mā·shən ,rē·zis'tiv·əd·ē }

true freezing point [PHYS CHEM] The temperature at which the liquid and solid forms of a substance exist in equilibrium at a given pressure (usually 1 standard atmosphere, or 101,325 pascals). { 'trü 'frēz·iŋ ,póint }

true fruit [BOT] A fruit that is derived from only the ovary and its contents; usually derived from a superior (inserted above the other floral parts) ovary. Also known as superior fruit. { 'trü 'früt }

true heading [NAV] Heading relative to true north. { 'trü 'hed·iŋ }

true homing [NAV] The process of following a course such that the true bearing of the craft from the objective is maintained constant. { 'trü 'hōm·iŋ }

true horizon [OPTICS] The boundary of a horizontal plane passing through a point of vision, or in photogrammetry, the perspective center of a lens system. { 'trü hə'rīz·ən }

true mean temperature [METEOROL] As adopted by the International Meteorological Organization, a monthly or annual mean air temperature based upon hourly observations at a given place, or on some combination of less frequent observations designed to represent this mean as nearly as possible. { 'trü 'mēn 'tem·prə·chər }

true-motion radar [ELECTR] A radar set which provides a true-motion radar presentation on the plan-position indicator, as opposed to the relative-motion, true-or-relative-bearing, presentation most commonly used. { 'trü 'mō·shən 'rä,där }

true-motion radar presentation [ELECTR] A radar plan-position indicator presentation in which the center of the scope represents the same geographic position, until reset, with all moving objects, including the user's own craft, moving on the scope. { 'trü ˌmō·shən 'rä,där ,prez·ən,tā·shən }

true motor load See thermal horsepower. { 'trü 'mōd·ər ,lōd }

TRUCHEROGNATHIDAE

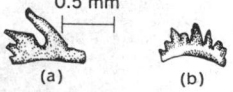

0.5 mm   1 mm

(a)   (b)

Typical examples, (a) *Polycaulodus* and (b) *Curtognathus*. (After illustration in R. R. Shrock and W. H. Twenhofel, *Principles of Invertebrate Paleontology*, McGraw-Hill, 2d ed., 1953)

TRUCK

Conventional truck chassis before installation of special body. (*White Motor Co.*)

true north [NAV] **1.** The direction of the north geographical pole. **2.** The reference direction for measurement of true directions. { 'trü 'north }

true place [ASTRON] The position of a star on the celestial sphere as it would be observed from the center of the sun, referred to the celestial equator and celestial equinox at the moment of observation. { 'trü 'plās }

true porcelain See porcelain. { 'trü 'pȯr·slən }

true rake [MECH ENG] The angle, measured in degrees, between a plane containing a tooth face and the axial plane through the tooth point in the direction of chip flow. { 'trü 'rāk }

true soil See solum. { 'trü 'sȯil }

true solar day See apparent solar day. { 'trü 'sō·lər 'dā }

true solar time See apparent solar time. { 'trü 'sō·lər 'tīm }

true sun See apparent sun. { 'trü 'sən }

true width [MIN ENG] The width of thickness of a vein or stratum as measured perpendicular to or normal to the dip and the strike; the true width is always the least width. { 'trü 'width }

true wind [METEOROL] Wind relative to a fixed point on the earth. { 'trü 'wind }

true wind direction [METEOROL] The direction, with respect to true north, from which the wind is blowing. { 'trü 'wind də,rek·shən }

truffle [BOT] The edible underground fruiting body of various European fungi in the family Tuberaceae, especially the genus *Tuber*. { 'trəf·əl }

truing [MECH ENG] **1.** Cutting a grinding wheel to make its surface run concentric with the axis. **2.** Aligning a wheel to be concentric and in one plane. { 'trü·iŋ }

trumpet buoy [NAV] A buoy provided with a trumpet having a distinctive tone. { 'trəm·pət ,bȯi }

trumpeter [VERT ZOO] A bird belonging to the Psophiidae, a family with three South American species; the common trumpeter (*Psophia crepitans*) is the size of a pheasant and resembles a long-legged guinea fowl. { 'trəm·pəd·ər }

truncate [BIOL] Abbreviated at an end, as if cut off. [CONT SYS] To stop a robotic process before it has been completed. [MATH] **1.** To drop digits at the end of a numerical value; the number 3.14159265 is truncated to five figures in 3.1415, whereas it would be 3.1416 if rounded off to five figures. **2.** To approximate the sum of an infinite series by the sum of a finite number of its terms. **3.** To terminate an infinite sequence of successively better approximations of a quantity after a finite number of such approximations. **4.** To construct from a geometric solid another solid consisting of those portions of the original solid that lie between two planes. { 'trəŋ,kāt }

truncated cone [MATH] The portion of a cone between two nonparallel planes whose line of intersection lies outside the cone. { ¦trəŋ·kād·əd 'kōn }

truncated distribution [STAT] A distribution fashioned from another distribution by deleting that part of the distribution to the right or left of a random variable value. { ¦trəŋ,kād·əd ,dis·trə'byü·shən }

truncated icosahedral gravitational-wave antenna [ENG] A resonant-mass antenna for detecting gravitational radiation in which the shape of the mass is a truncated icosahedron, which is much more efficient for this purpose than a cylinder. Abbreviated TIGA. { ¦trəŋ,kād·əd ī,käs·ə¦hē·drəl ,grav·ə¦tā·shən·əl 'wāv an,ten·ə }

truncated icosahedron [MATH] An Archimedean solid with 32 faces (20 regular hexagons and 12 regular pentagons) and 60 vertices, a shape used in the construction of soccer balls. { ¦trəŋ,kād·əd ī¦käs·ə'hē·drən }

truncated landform [GEOGR] A landform which has been cut off by erosion, creating a steep side or cliff. { ¦trəŋ,kād·əd 'land,fȯrm }

truncated paraboloid [ELECTROMAG] Paraboloid antenna in which a portion of the top and bottom have been cut away to broaden the main lobe in the vertical plane. { 'trəŋ,kād·əd pə'rab·ə,lȯid }

truncated prism [MATH] The part of a prism that lies between two nonparallel planes that cut the prism and intersect outside the prism. { ¦trəŋ·kād·əd 'priz·əm }

truncated pyramid [MATH] The part of a pyramid between the base and a plane that is not parallel to the base. { ¦trəŋ·kād·əd 'pir·ə,mid }

truncation [MATH] **1.** Approximating the sum of an infinite series by the sum of a finite number of its terms. **2.** See rounding. { trəŋ'kā·shən }

truncation error [ENG] The error resulting from the analysis of a partial set of data in place of a complete or infinite set. [MATH] **1.** The computation error resulting from use of only a finite number of terms of an infinite series. **2.** The error resulting from the approximation of a derivative or differential by a finite difference. { trəŋ'kā·shən ,er·ər }

truncus arteriosis [EMBRYO] The embryonic arterial trunk between the bulbous arteriosis and the ventral aorta in anamniotes and early stages of amniotes. { ¦trəŋ·kəs är,tir·ē'ō·səs }

trunk [ANAT] The main mass of the human body, exclusive of the head, neck, and extremities; it is divided into thorax, abdomen, and pelvis. [BOT] The main stem of a tree. [COMMUN] **1.** A path over which information is transferred in a computer. **2.** A telephone line connecting two central offices. Also known as trunk circuit. { trəŋk }

trunk buoy [ENG] A mooring buoy having a pendant extending through an opening in the buoy, with the ship's anchor chain or mooring line being secured to this pendant. { trəŋk ,bȯi }

trunk circuit See trunk. { 'trəŋk ,sər·kət }

trunk exchange [COMMUN] A telephone exchange whose main function is to interconnect trunks. { 'trəŋk iks'chānj }

trunk feeder [ELEC] An electric power transmission line that connects two generating stations, or a generating station and an important substation, or two electrical distribution networks. { 'trəŋk ,fēd·ər }

trunk group [COMMUN] The collection of trunks of a given type or characteristic that connect two switching points. { 'trəŋk ,grüp }

trunk height [ANTHRO] A vertical measurement (taken in front) of the distance from the table top to the upper edge of the sternum, when the individual is seated as in a sitting-height measurement. { 'trəŋk ,hīt }

trunk roadway [MIN ENG] The main developing heading from the pit bottom and is usually driven along the strike of the coal seam. { 'trəŋk 'rōd,wā }

trunk sewer [CIV ENG] A sewer receiving sewage from many tributaries serving a large territory. { 'trəŋk ,sü·ər }

trunk stream See main stream. { 'trəŋk ,strēm }

trunnion [DES ENG] **1.** Either of two opposite pivots, journals, or gudgeons, usually cylindrical and horizontal, projecting one from each side of a piece of ordnance, the cylinder of an oscillating engine, a molding flask, or a converter, and supported by bearings to provide a means of swiveling or turning. **2.** A pin or pivot usually mounted on bearings for rotating or tilting something. [ENG] A tubular section of steel welded to the side of a pipe in order to help support the pipe. { 'trən·yən }

truss [CIV ENG] A frame, generally of steel, timber, concrete, or a light alloy, built from members in tension and compression. { trəs }

truss bridge [CIV ENG] A fixed bridge consisting of members vertically arranged in a triangular pattern. { 'trəs ,brij }

trussed beam [CIV ENG] A beam stiffened by a steel tie rod to reduce its deflection. { 'trəst 'bēm }

trussed rafter [BUILD] A triangulated beam in a trussed roof. { 'trəst 'raf·tər }

truss rod [CIV ENG] A rod attached to the ends of a trussed beam which transmits the strain due to downward pressure. { 'trəs ,räd }

truth See top. { trüth }

truth quark [PART PHYS] See top quark. { 'trüth ,kwärk }

truth set [MATH] A set containing all the elements that make a given statement of relationships true when they are substituted in this statement. { 'trüth ,set }

truth table [MATH] A table listing statements concerning an event and their respective truth values. { 'trüth ,tā·bəl }

truth value [MATH] The result of a logical proposition; either "true" or "false" in classical logic. { 'trüth ,val·yü }

TRV See tobacco rattle virus.

Tryblidiidae [PALEON] An extinct family of Paleozoic mollusks. { ,trib·lə'dī·ə,dē }

trypan blue [MATER] An acid diazo dye of the benzopurpurine series used as a vital stain. { 'tri,pan 'blü }

Trypanorhyncha [INV ZOO] An order of tapeworms of the subclass Cestoda; all are parasites in the intestine of elasmobranch fishes. { trə,pan·ə'riŋ·kə }

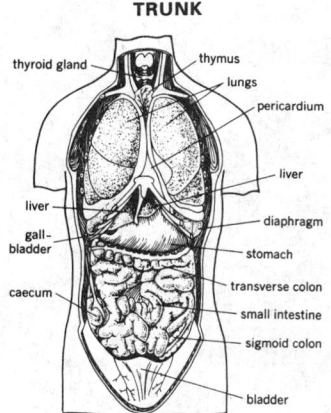

TRUNK

thyroid gland · thymus · lungs · pericardium · liver · diaphragm · stomach · transverse colon · small intestine · sigmoid colon · bladder · liver · gallbladder · caecum

Human trunk, showing principal organs. (*From Franz Frohse et al., Atlas of Human Anatomy, rev. ed., Barnes and Noble, 1957*)

Trypanosomatidae [INV ZOO] A family of Protozoa, order Kinetoplastida, containing flagellated parasites which exhibit polymorphism during their life cycle. { trə‚pan·ə·sō'mad·ə‚dē }

trypanosome [INV ZOO] A flagellated protozoan of the genus *Trypanosoma*. { trə'pan·ə‚sōm }

trypanosomiasis [MED] Any of many diseases of humans and animals caused by infection with species of *Trypanosoma* and transmitted by tsetse flies and other insects. { trə‚pan·ə·sō'mī·ə·səs }

trypsin [BIOCHEM] A proteolytic enzyme which catalyzes the hydrolysis of peptide linkages in proteins and partially hydrolyzed proteins; derived from trypsinogen by the action of enterokinase in intestinal juice. { 'trip·sən }

trypsinogen [BIOCHEM] The zymogen of trypsin, secreted in the pancreatic juice. Also known as protrypsin. { trip'sin·ə·jən }

tryptophan [BIOCHEM] $C_{11}H_{12}O_2N_2$ An amino acid obtained from casein, fibrin, and certain other proteins; it is a precursor of indoleacetic acid, serotonin, and nicotinic acid. { 'trip·tə‚fan }

try square [ENG] An instrument consisting of two straightedges secured at right angles to each other, used for laying off right angles and testing whether work is square. { 'trī ‚skwer }

tschermakite [MINERAL] $Ca_2Mg_3(Al,Fe^{3+})_2(Al_2Si_6)O_{22}(OH,F)_2$ An amphibole mineral. { 'chər·mə‚kīt }

Tschirnhausen's cubic [MATH] A plane curve consisting of the envelope of the line through a variable point *P* on a parabola which is perpendicular to the line from the focus of the parabola to *P*. Also known as l'Hôpital's cubic; trisectrix of Catalan. { 'chərn‚haúz·ənz 'kyü·bik }

Tschudi engine [MECH ENG] A cat-and-mouse engine in which the pistons, which are sections of a torus, travel around a toroidal cylinder; motion of the pistons is controlled by two cams which bear against rollers attached to the rotors. { 'chü·dē ‚en·jən }

T score [STAT] A score utilized in setting up norms for standardized tests; obtained by linearly transforming normalized standard scores. { 'tē ‚skór }

T-S curve *See* temperature-salinity diagram. { ¦tē¦es ‚kərv }

T-S diagram *See* temperature-salinity diagram. { ¦tē¦es ‚dī·ə‚gram }

T-section filter [ELEC] T network used as an electric filter. { 'tē ‚sek·shən ‚fil·tər }

tsetse fly [INV ZOO] Any of various South African muscoid flies of the genus *Glossina*; medically important as vectors of sleeping sickness or trypanosomiasis. { 'set‚sē ‚flī }

TSH *See* thyrotropic hormone.

tsi [MECH] A unit of force equal to 1 ton-force per square inch; equal to approximately 1.54444×10^7 pascals. { sī *or* ‚tē‚es'ī }

T slot [DES ENG] A recessed slot, in the form of an inverted T, in the table of a machine tool, to receive the square head of a T-slot bolt. { 'tē ‚slät }

tsp *See* teaspoonful.

TSP *See* trisodium phosphate.

T_0 space [MATH] A topological space where, for each pair of points, at least one has a neighborhood not containing the other. Also known as Kolmogorov space. { ‚tē 'zir·ō ‚spās }

T_1 space [MATH] A topological space where, for each pair of distinct points, each one has a neighborhood not containing the other. Also known as Fréchet space. { 'tē ‚wən ‚spās }

T_2 space *See* Hausdorff space. { 'tē ‚tü ‚spās }

T_3 space [MATH] A regular topological space that is also a T_1 space. { 'tē ‚thrē ‚spās }

$T_{3\,1/2}$ space *See* Tychonoff space. { ¦tē ‚thrē an ə 'haf ‚spās }

T_4 space [MATH] A normal space that is also a T_1 space. { 'tē ‚fór ‚spās }

tspn *See* teaspoonful.

TSPP *See* sodium pyrophosphate.

T square [GRAPHICS] A straightedge rule with a crosspiece at one end by which parallel lines are drawn perpendicular to the edge of the drawing board. { 'tē ‚skwer }

TSR *See* RAM resident.

T-S relation *See* temperature-salinity diagram. { ¦tē¦es ri‚lā·shən }

Tsuga canadensis *See* Eastern hemlock. { ‚tsü·gə ‚kan·ə'den·səs }

tsumebite [MINERAL] $Pb_2Cu(PO_4)(SO_4)(OH)$ An emerald green, monoclinic mineral consisting of a hydrated basic phosphate and sulfate of lead and copper. { 'tsü·mə‚bīt }

tsunami [OCEANOGR] An ocean wave or series of waves generated by any large, abrupt disturbance of the sea-surface by an earthquake in marine and coastal regions, as well as by a suboceanic landslide, volcanic eruption, or asteroid impact. { tsü'nä·mē }

tsunamiite [GEOL] **1.** A sedimentary deposit resulting from a tsunami generated by an asteroid or comet impact. **2.** Rock deposited by a tsunami. Also known as tsunamite. { tsü 'näm·ē‚īt }

tsunamite *See* tsunamiite. { 'tsü·nə‚mīt }

Tsushima Current [OCEANOGR] That part of the Kuroshio Current flowing northeastward through the Korea Strait and along the Japanese coast in the Sea of Japan. { 'tsü·shē‚mä 'kə·rənt }

tsutsugamushi disease [MED] A rickettsial disease of humans caused by *Rickettsia tsutsugamushi*, transmitted by larval mites, and characterized by headache, high fever, and a rash. Also known as scrub typhus. { ‚tsü·sə·gə'mü·shē di‚zēz }

T switch [ELECTR] An electrical switch that joins a machine to either of two other devices. { 'tē ‚swich }

Tsytovich effect [ELECTROMAG] An effect wherein the index of refraction of a medium is much less than unity so that the phase velocity of electromagnetic waves in the medium exceeds the speed of light. { 'sīd·ə‚vich i‚fekt }

T Tauri star [ASTRON] A star, with mass from 0.5 to 2.5 solar masses, in an early stage of formation at which interaction with its associated nebulosity, as well as possible internal instabilities, make it variable in luminosity and render its spectrum very peculiar. Also known as nebular variable. { 'tē 'tór·ē ‚stär }

t-test [STAT] A statistical test involving means of normal populations with unknown standard deviations; small samples are used, based on a variable *t* equal to the difference between the mean of the sample and the mean of the population divided by a result obtained by dividing the standard deviation of the sample by the square root of the number of individuals in the sample. { 'tē‚test }

TTL *See* transistor-transistor logic.

tuba [METEOROL] A cloud column or inverted cloud cone, pendant from a cloud base; this supplementary feature occurs mostly with cumulus and cumulonimbus; when it reaches the earth's surface it constitutes the cloudy manifestation of an intense vortex, namely, a tornado or waterspout. Also known as pendant cloud; tornado cloud. { 'tü·bə }

tubal bladder [VERT ZOO] A urine reservoir organ that is an enlargement of the mesonephric ducts in most fish; there are four types: duplex, bilobed, simplex with ureters tied, and simplex with separate ureters. { 'tü·bəl 'blad·ər }

tubal ligation [MED] Surgical tying of the uterine tubes to prevent conception. { 'tü·bəl lī'gā·shən }

tubatoxin *See* rotenone. { 'tü·bə‚täk·sən }

tubbing [MIN ENG] The watertight cast-iron lining of a circular shaft built up of segments with the space outside the tubbing grouted to add strength and to improve watertightness. { 'təb·iŋ }

tube [BIOL] A narrow channel within the body of an animal or plant. [ELECTR] *See* electron tube. [ENG] **1.** A long cylindrical body with a hollow center used especially to convey fluid. **2.** *See* inner tube. [GEOL] A passage in a cave having smooth sides and an elliptical to nearly circular cross section. [ORD] The main part of a gun, the cylindrical piece of metal surrounding the bore; tube is frequently used in referring to artillery weapons, and barrel is more frequently used in referring to small arms. { 'tüb }

tube bank [MECH ENG] An array of tubes designed to be used as a heat exchanger. { 'tüb ‚baŋk }

tube bundle [ENG] In a shell-and-tube heat exchanger, an assembly of parallel tubes that is tied together with tie rods. { 'tüb ‚bən·dəl }

tube cell [BOT] That nucleus of a pollen grain believed to influence the growth and development of the pollen tube. Also known as tube nucleus. { 'tüb ‚sel }

tube cleaner [MECH ENG] A device equipped with cutters or brushes used to clean tubes in heat transfer equipment. { 'tüb ‚klēn·ər }

tube coefficient [ELECTR] Any of the constants that

describe the characteristics of a thermionic vacuum tube, such as amplification factor, mutual conductance, or alternating-current plate resistance. { 'tüb ¦kō·i¦fish·ənt }

tube core [AERO ENG] One type of sandwich configuration used in structural materials in aircraft; aluminum, steel, and titanium have been used for face materials with cores of wood, rubber, plastics, steel, and aluminum in the form of tubes. { 'tüb ¦kȯr }

tube door [MECH ENG] A door in a boiler furnace wall which facilitates the removal or installation of tubes. { 'tüb ¦dȯr }

tube foot [INV ZOO] One of the tentaclelike outpushings of the radial vessels of the water-vascular system in echinoderms; may be suctorial, or serve as stiltlike limbs or tentacles. { 'tüb ¦fu̇t }

tube heating time [ELECTR] Time required for a tube to attain operating temperature. { 'tüb ¦hēd·iŋ ¦tīm }

tube hole [ENG] A hole in a tube sheet through which a tube is passed prior to sealing. { 'tüb ¦hōl }

tubeless tire [ENG] A tire that does not require an inner tube to hold air. { ¦tüb·ləs 'tīr }

tube mill [MECH ENG] A revolving cylinder used for fine pulverization of ore, rock, and other such materials; the material, mixed with water, is fed into the chamber from one end, and passes out the other end as slime. { 'tüb ¦mil }

tube noise [ELECTR] Noise originating in a vacuum tube, such as that due to shot effect and thermal agitation. { 'tüb ¦nȯiz }

tube nucleus *See* tube cell. { 'tüb 'nü·klē·əs }

tube of flux *See* tube of force. { 'tüb əv 'fləks }

tube of force [ELEC] A region of space bounded by a tubular surface consisting of the lines of force which pass through a given closed curve. Also known as tube of flux. { 'tüb əv 'fȯrs }

tube plug [ENG] A solid plug inserted into the end of a tube in a tube sheet. { 'tüb ¦pləg }

tuber [BOT] The enlarged end of a rhizome in which food accumulates, as in the potato. { 'tü·bər }

tuber cinereum [ANAT] An area of gray matter extending from the optic chiasma to the mammillary bodies and forming part of the floor of the third ventricle. { 'tü·bər si'ner·ē·əm }

tubercle [BIOL] A small knoblike prominence. [MET] A mound of corrosive products on the surface of a metal that is subjected to local corrosive attack. { 'tü·bər·kəl }

Tuberculariaceae [MYCOL] A family of fungi of the order Moniliales having short conidia that form cushion-shaped, often waxy or gelatinous aggregates (sporodochia). { tə¦bər·kyə¦la·rē'ās·ē¦ē }

tuberculate [BIOL] Having or characterized by knoblike processes. { tə'bər·kyə·lət }

tuberculation [MET] Corrosive attack with formation of tubercles. { tə¦bər·kyə'lā·shən }

tuberculin [IMMUNOL] A preparation containing tuberculoproteins derived from *Mycobacterium tuberculosis* used in the tuberculin test to determine sensitization to tubercle bacilli. { tə'bər·kyə·lən }

tuberculin test [IMMUNOL] A test for past or present infection with tubercle bacilli based on a delayed hypersensitivity reaction at the site where tuberculin or purified protein derivative was introduced. { tə'bər·kyə·lən ¦test }

tuberculosis [MED] A chronic infectious disease of humans and animals primarily involving the lungs caused by the tubercle bacillus, *Mycobacterium tuberculosis*, or by *M. bovis*. Also known as consumption. { tə¦bər·kyə'lō·səs }

tube reducing [MET] Reducing the diameter and wall thickness of tubing by means of a mandrel and rolls. { 'tüb ri¦düs·iŋ }

tuberosity [ANAT] A large or obtuse prominence, especially as on bone for muscle attachment. { ¦tü·bə'räs·əd·ē }

tuberous organ [PHYSIO] An electroreceptor most sensitive to high-frequency electric signals, and distributed over the body surface of electric fish. { 'tü·bə·rəs 'ȯr·gən }

tuberous sclerosis [MED] A familial neurocutaneous syndrome characterized in its complete form by epilepsy, adenoma sebaceum, and mental deficiency, and pathologically by nodular sclerosis of the cerebral cortex. Also known as Bourneville's disease. { 'tü·bə·rəs sklə'rō·səs }

tube seat [ENG] The surface of the tube hole in a tube sheet which contacts the tube. { 'tüb ¦sēt }

tube sheet [ENG] A mounting plate for elements of a larger

item of equipment; for example, filter cartridges, or tubes for heat exchangers, coolers, or boilers. { 'tüb ¦shēt }

tube shield [ENG] A shield designed to be placed around an electron tube. { 'tüb ¦shēld }

tube socket [ENG] A socket designed to accommodate electrically and mechanically the terminals of an electron tube. { 'tüb ¦säk·ət }

tube-still heater [CHEM ENG] A firebox containing a pipe coil through which oil for a tube still (pipe still) is pumped. { 'tüb ¦stil ¦hēd·ər }

tube stock [MET] Semifinished metal tubing. { 'tüb ¦stäk }

tube tester [ELECTR] A test instrument designed to measure and indicate the condition of electron tubes used in electronic equipment. { 'tüb ¦tes·tər }

tube turbining [MECH ENG] Cleaning tubes by passing a power-driven rotary device through them. { 'tüb ¦tər·bən·iŋ }

tube voltage drop [ELECTR] In a gas tube, the anode voltage during the conducting period. { 'tüb 'vōl·tij ¦dräp }

tube voltmeter *See* vacuum-tube voltmeter. { 'tüb 'vōlt¦mēd·ər }

tubeworms [INV ZOO] Marine polychaete worms (particularly many species in the family Serpulidae) which construct permanent calcareous tubes on rocks, seaweeds, dock pilings, and ship bottoms. The individual tubes with hard walls of calcite-aragonite, ranging from 0.04 to 0.4 inch (1 millimeter to 1 centimeter) in diameter and from 0.16 to 4 inches (4 millimeters to 10 centimeters) in length, are firmly cemented to a hard substrate and to each other. { 'tüb¦wərmz }

Tubicola [INV ZOO] An order of sedentary polychaete annelids that surround themselves with a calcareous tube or one which is composed of agglutinated foreign particles. { tü'bik·ə·lə }

tubing [ENG] Material in the form of a tube, most often seamless. { 'tüb·iŋ }

tubing hanger *See* hanger. { 'tüb·iŋ ¦haŋ·ər }

tubing head [PETRO ENG] A spool-type unit or housing attached to the top flange on the uppermost oil-well-casing head to support the tubing string and to seal the annulus between the tubing string and the production casing string. { 'tüb·iŋ ¦hed }

tubing-head adapter flange [PETRO ENG] An intermediate flange used in oil wells to connect the top tubing-head flange to the master valve (Christmas tree) and to provide support for the tubing. { 'tüb·iŋ ¦hed ə'dap·tər ¦flanj }

tubing pump [PETRO ENG] A type of oil-well, sucker-rod pump in which the pump barrel is attached to the tubing string, and lowered into the well bore with the tubing. { 'tüb·iŋ ¦pəmp }

tubing safety valve *See* storm choke. { 'tüb·iŋ 'sāf·tē ¦valv }

Tubulanidae [INV ZOO] A family of the order Palaeonemertini. { ¦tüb·yə'lan·ə¦dē }

tubular capacitor [ELEC] A paper or electrolytic capacitor having the form of a cylinder, with leads usually projecting axially from the ends; the capacitor plates are long strips of metal foil separated by insulating strips, rolled into a compact tubular shape. { 'tü·byə·lər kə'pas·əd·ər }

tubular exchanger *See* shell-and-tube exchanger. { 'tü·byə·lər iks'chānj·ər }

tubular gland [ANAT] A secreting structure whose secretory endpieces are tubelike or cylindrical in shape. { 'tü·byə·lər 'gland }

tubule [ANAT] A slender, elongated microscopic tube in an anatomical structure. { 'tü·byül }

Tubulidentata [VERT ZOO] An order of mammals which contains a single living genus, the aardvark (*Orycteropus*) of Africa. { ¦tü·byə·lə·den'täd·ə }

tubulin [BIOCHEM] A globular protein containing two subunits; 10–14 molecules are arranged to form a microtubule. { 'tü·byə·lən }

tubuloacinous gland *See* tubuloalveolar gland. { ¦tü·byə·lō'as·ə·nəs ¦gland }

tubuloalveolar gland [ANAT] A secreting structure having both tubular and alveolar secretory endpieces. Also known as acinotubular gland; tubuloacinous gland. { ¦tü·byə·lō·al'vē·ə·lər ¦gland }

Tucana [ASTRON] A constellation in the southern hemisphere; right ascension 23 hours, declination 60° south. Also known as Toucan. { ü'kä·nə }

TUBE CORE

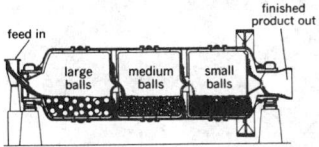

Tube core type of sandwich construction in a fuselage.

TUBE MILL

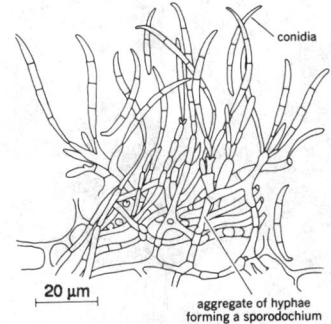

Three-compartment tube mill pulverizer, containing three sizes of balls. (*Hardinge Co., Inc.*)

TUBERCULARIACEAE

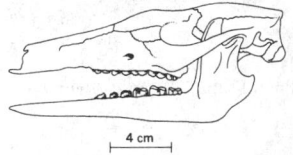

Sporodochium of *Fusarium lini*, a representative species of Tuberculariaceae, showing sickle-shaped, multicelled conidia.

TUBULIDENTATA

Skull and jaw of *Orycteropus gaudryi*, a Pliocene tubulidentate from Samos, Greece.

tuck-and-pat pointing See tuck pointing. { ¦tək ən ¦pat ˌpȯint·iŋ }

tuck joint pointing See tuck pointing. { ¦tək ˌjȯint ˌpȯint·iŋ }

tuck pointing [BUILD] The finishing of old masonry joints in which the joints are first cleaned out and then filled with fine mortar which projects slightly or has a fillet of putty or lime. Also known as tuck-and-pat pointing; tuck joint pointing. { ¦tək ˌpȯint·iŋ }

tufa [GEOL] A spongy, porous limestone formed by precipitation from evaporating spring and river waters, often onto leaves and stems of neighboring plants. Also known as calcareous sinter; calcareous tufa. { 'tü·fə }

tufaceous [GEOL] Pertaining to or similar to tufa. { tü'fā·shəs }

tuff [GEOL] Consolidated volcanic ash, composed largely of fragments (less than 4 millimeters) produced directly by volcanic eruption; much of the fragmented material represents finely comminuted crystals and rocks. { təf }

tuffaceous [GEOL] Pertaining to sediments which contain up to 50% tuff. { tə'fā·shəs }

tuff ball See mud ball. { 'təf ˌbȯl }

tuff lava See welded tuff. { 'təf ˌläv·ə }

tuft See mound. { təft }

tuft method [FL MECH] A technique of surface flow visualization in which an array of short pieces of flexible string or yarn are attached to a surface in such a way that they can move freely under the influence of a flow. { 'təft ˌset }

tugboat [NAV ARCH] A powerful, strongly built boat with shaped hull and bow, designed to tow or push other vessels or barges in harbors, on inland waterways, and at sea. { 'təg¦bōt }

tugger [MIN ENG] A small portable pneumatic or electric hoist mounted on a column and used in a mine. { 'təg·ər }

Tukey lemma [MATH] The proposition that any nonempty family of finite character has a maximal member. { 'tü·kē¦lem·ə }

Tukon tester [ENG] A device that uses a diamond (Knoop) indenter applying average loads of 1 to 2000 grams to determine microhardness of a metal. { 'tü¦kän ˌtes·tər }

tularemia [VET MED] A bacterial infection of wild rodents caused by *Pasteurella tularensis*; it may be generalized, or it may be localized in the eyes, skin, or lymph nodes, or in the respiratory tract or gastrointestinal tract; may be transmitted to humans and to some domesticated animals. { ˌtü·lə'rē·mē·ə }

tulip [BOT] Any of various plants with showy flowers constituting the genus *Tulipa* in the family Liliaceae; characterized by coated bulbs, lanceolate leaves, and a single flower with six equal perianth segments and six stamens. { 'tü·ləp }

tulip poplar See tulip tree. { 'tü·ləp ¦päp·lər }

tulip tree [BOT] *Liriodendron tulipifera*. A tree belonging to the magnolia family (Magnoliaceae) distinguished by leaves which are squarish at the tip, true terminal buds, cone-shaped fruit, and large greenish-yellow and orange-colored flowers. Also known as tulip poplar. { 'tü·ləp ˌtrē }

tulle See malines. { tül }

Tully-Fisher relation [ASTRON] A relation between the rotational velocity of a galaxy, as reflected in the width of the 21-centimeter line, and the intrinsic luminosity of the galaxy. { ¦təl·ē 'fish·ər ri¦lā·shən }

tumble See topple. { 'təm·bəl }

tumble axis See topple axis. { 'təm·bəl ˌak·səs }

tumble home [ARCH] An inclination inward from the greatest breadth of a structure. Also known as tumble in. [NAV ARCH] The curve of a boat or ship's upper side toward the centerline, causing the sides to be convex. { 'təm·bəl ¦hōm }

tumble in See tumble home. { 'təm·bəl ¦in }

tumble-plating process [MET] A method of zinc-coating small metal parts by first applying zinc powder with an adhesive, then tumbling with glass beads to roll out the powder into a continuous coat. { 'təm·bəl ¦plād·iŋ ˌprä·səs }

tumbler [ENG] **1.** A device in a lock cylinder that must be moved to a particular position, as by a key, before the bolt can be thrown. **2.** A device or mechanism in which objects are tumbled. { 'təm·blər }

tumbler feeder See drum feeder. { 'təm·blər ˌfēd·ər }

tumbler gears [MECH ENG] Idler gears interposed between spindle and stud gears in a lathe gear train; used to reverse rotation of lead screw or feed rod. { 'təm·blər ˌgirz }

tumbleweed [BOT] Any of various plants that break loose from their roots in autumn and are driven by the wind in rolling masses over the ground. { 'təm·bəlˌwēd }

tumbling [AERO ENG] An attitude situation in which the vehicle continues on its flight, but turns end over end about its center of mass. [ENG] A surface-finishing operation for small articles in which irregularities are removed or surfaces are polished by tumbling them together in a barrel, along with wooden pegs, sawdust, and polishing compounds. [MECH ENG] Loss of control in a two-frame free gyroscope, occurring when both frames of reference become coplanar. { 'təm·bliŋ }

tumbling mill [MECH ENG] A grinding and pulverizing machine consisting of a shell or drum rotating on a horizontal axis. { 'təm·bliŋ ˌmil }

tumid [BIOL] Marked by swelling or inflation. { 'tü·məd }

tumor [MED] Any abnormal mass of cells resulting from excessive cellular multiplication. { 'tü·mər }

tumorigenic [MED] Tumor-forming. { ¦tü·mə·rə'jen·ik }

tumor necrosis factor [IMMUNOL] A monokine that induces leukocytosis, fever, weight loss, the acute-phase reaction, and necrosis of some tumors. { ¦tü·mər nə'krō·səs ˌfak·tər }

tumor nicotinamide adenine dinucleotide oxidase [BIOCHEM] A cancer-specific growth protein (an unregulated nicotinamide adenine dinucleotide oxidase form) whose inhibition may be the underlying mechanism of green and black tea catechins' slowing of cancer cell growth. Abbreviated tNOX. { ¦tü·mər ˌnik·ə¦tin·ə¦mīd ¦ad·ən¦ēn dī¦nü·klē·ə¦tīd 'äk·səˌdās }

tumor suppressor gene [CELL MOL] A class of genes which, when mutated, predispose an individual to cancer by causing the loss of function of the particular tumor suppressor protein encoded by the gene. { ¦tüm·ər sə'pres·ər ˌjēn }

tumor suppressor protein [CELL MOL] A protein that helps protect a normal cell from becoming malignant; for example, p53. { ¦tüm·ər sə'pres·ər prō¦tēn }

tumuli lava [GEOL] A type of lava flow forming ovoid mounds, a few feet high and a few tens of feet long, caused by buckling up of the crust. { 'tü·myəˌlī 'lä·və }

tuna [VERT ZOO] Any of the large, pelagic, cosmopolitan marine fishes which form the family Thunnidae including species that rank among the most valuable of food and game fish. { 'tü·nə }

tunable echo box [ELECTROMAG] Echo box consisting of an adjustable cavity operating in a single mode; if calibrated, the setting of the plunger at resonance will indicate the wavelength. { 'tü·nə·bəl 'ek·ō ˌbäks }

tunable filter [ELECTR] An electric filter in which the frequency of the passband or rejection band can be varied by adjusting its components. { 'tü·nə·bəl 'fil·tər }

tunable laser [OPTICS] A laser in which the frequency of the output radiation can be tuned over part or all of the ultraviolet, visible, and infrared regions of the spectrum. { 'tü·nə·bəl 'lā·zər }

tunable magnetron [ELECTR] Magnetron which can be tuned mechanically or electronically by varying its capacitance or inductance. { 'tü·nə·bəl 'mag·nəˌträn }

tuna boat See tuna clipper. { 'tü·nə ˌbōt }

tuna clipper [NAV ARCH] A large craft used for tuna fishing on the Pacific coast; it is usually diesel-powered, and is equipped with refrigeration brine tanks. Also known as tuna boat. { 'tü·nə 'klip·ər }

tundish [MET] A funnel or pouring basin used for transferring a stream of molten metal. { 'tənˌdish }

tundra [ECOL] An area supporting some vegetation between the northern upper limit of trees and the lower limit of perennial snow on mountains, and on the fringes of the Antarctic continent and its neighboring islands. Also known as cold desert. { 'tən·drə }

tundra climate [CLIMATOL] The climate which produces tundra vegetation; it is too cold for the growth of trees but does not have a permanent snow-ice cover. { 'tən·drə ˌklī·mət }

Tundra orbit [AERO ENG] An inclined, elliptical, geosynchronous earth satellite orbit designed to provide communications satellite service coverage at high latitudes; it has an inclination of 63.4° degrees and an eccentricity of 0.2684, so that the satellite spends 16 hours each day over the hemisphere (northern or southern) where service coverage is intended and 8 hours over the opposite hemisphere. { 'tən·drə ˌȯr·bət }

TULIP TREE

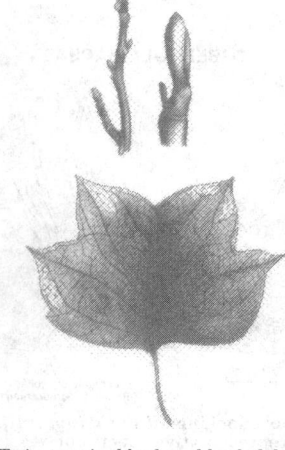

Twig, terminal bud, and leaf of the tulip tree (*Liriodendron tulipifera*).

TUNA

Bluefin tuna (*Thunnus thynnus*).

tune [ELECTR] To adjust for resonance at a desired frequency. { tün }

tuned amplifier [ELECTR] An amplifier in which the load is a tuned circuit; load impedance and amplifier gain then vary with frequency. { ¦tünd 'am·plə₁fī·ər }

tuned-anode oscillator [ELECTR] A vacuum-tube oscillator whose frequency is determined by a tank circuit in the anode circuit, coupled to the grid to provide the required feedback. Also known as tuned-plate oscillator. { ¦tünd 'an₁ōd ₁äs·ə₁läd·ər }

tuned-anode tuned-grid oscillator See tuned-grid tuned-anode oscillator. { ¦tünd 'an₁ōd ¦tünd 'grid ₁äs·ə₁läd·ər }

tuned-base oscillator [ELECTR] Transistor oscillator in which the frequency-determining resonant circuit is located in the base circuit; comparable to a tuned-grid oscillator. { ¦tünd 'bās ₁äs·ə₁läd·ər }

tuned cavity See cavity resonator. { ¦tünd 'kav·əd·ē }

tuned circuit [ELECTR] A circuit whose components can be adjusted to make the circuit responsive to a particular frequency in a tuning range. Also known as tuning circuit. { ¦tünd 'sər·kət }

tuned-collector oscillator [ELECTR] A transistor oscillator in which the frequency-determining resonant circuit is located in the collector circuit; this is comparable to a tuned-anode electron-tube oscillator. { ¦tünd kə'lek·tər ₁äs·ə₁läd·ər }

tuned filter [ELECTR] Filter that uses one or more tuned circuits to attenuate or pass signals at the resonant frequency. { ¦tünd 'fil·tər }

tuned-grid oscillator [ELECTR] Oscillator whose frequency is determined by a parallel-resonant circuit in the grid coupled to the plate to provide the required feedback. { ¦tünd 'grid ₁äs·ə₁läd·ər }

tuned-grid tuned-anode oscillator [ELECTR] A vacuum-tube oscillator whose frequency is determined by a tank circuit in the grid circuit, coupled to the anode to provide the required feedback. Also known as tuned-anode tuned-grid oscillator. { ¦tünd 'grid ¦tünd 'an₁ōd ₁äs·ə₁läd·ər }

tuned-plate oscillator See tuned-anode oscillator. { ¦tünd 'plāt ₁äs·ə₁läd·ər }

tuned-radio-frequency receiver [ELECTR] A radio receiver consisting of a number of amplifier stages that are tuned to resonance at the carrier frequency of the desired signal by a gang capacitor; the amplified signals at the original carrier frequency are fed directly into the detector for demodulation, and the resulting audio-frequency signals are amplified by an a-f amplifier and reproduced by a loudspeaker. Abbreviated TRF receiver. { ¦tünd 'rād·ē·ō ¦frē·kwən·sē ri₁sē·vər }

tuned-radio-frequency transformer [ELECTR] A transformer used for selective coupling in radio-frequency stages. { ¦tünd 'rād·ē·ō ¦frē·kwən·sē tranz₁fȯr·mər }

tuned-reed frequency meter See vibrating-reed frequency meter. { ¦tünd 'rēd 'frē·kwən·sē ₁mēd·ər }

tuned relay [ELEC] A relay having mechanical or other resonating arrangements that limit response to currents at one particular frequency. { ¦tünd 'rē₁lā }

tuned resonating cavity [ELECTROMAG] Resonating cavity half a wavelength long or some multiple of a half wavelength, used in connection with a waveguide to produce a resultant wave with the amplitude in the cavity greatly exceeding that of the wave in the waveguide. { ¦tünd 'rez·ən₁ād·iŋ ₁kav·əd·ē }

tuned transformer [ELEC] Transformer whose associated circuit elements are adjusted as a whole to be resonant at the frequency of the alternating current supplied to the primary, thereby causing the secondary voltage to build up to higher values than would otherwise be obtained. { ¦tünd tranz'fȯr·mər }

tuner [ELECTR] The portion of a receiver that contains circuits which can be tuned to accept the carrier frequency of the alternating current supplied to the primary, thereby causing the secondary voltage to build up to higher values than would otherwise be obtained. { 'tü·nər }

tungar tube [ELECTR] A gas tube having a heated thoriated tungsten filament serving as cathode and a graphite disk serving as anode in an argon-filled bulb at a low pressure; used chiefly as a rectifier in battery chargers. { 'təŋ₁är ₁tüb }

tung nut [BOT] The seed of the tung tree (Aleurites fordii), which is the source of tung oil. { 'təŋ ₁nət }

tung oil [MATER] A yellow, combustible drying oil extracted from the seed of the tung tree; soluble in ether, chloroform, carbon disulfide, and oils; used in formulations for paints, varnishes, varnish driers, paper waterproofing, and linoleum. Also known as China wood oil. { 'təŋ ₁ȯil }

tungstate [INORG CHEM] M_2WO_4 A salt of tungstic acid; for example, sodium tungstate, Na_2WO_4. [MINERAL] Any species of mineral containing the radical WO_4, such as wolframite. { 'təŋ₁stāt }

tungstate white See barium tungstate. { 'təŋ₁stāt 'wīt }

tungsten [CHEM] Also known as wolfram. A metallic transition element, symbol W, atomic number 74, atomic weight 183.85; soluble in mixed nitric and hydrofluoric acids; melts at 3400°C. [MET] A hard, brittle, ductile, heavy gray-white metal used in the pure form chiefly for electrical purposes and with other substances in dentistry, pen points, x-ray-tube targets, phonograph needles, and high-speed tool metal, and as a radioactive shield. { 'təŋ·stən }

tungsten boride [INORG CHEM] WB_2 A silvery solid; insoluble in water, soluble in aqua regia and concentrated acids; melts at 2900°C; used as a refractory for furnaces and chemical process equipment. { 'təŋ·stən 'bȯr₁īd }

tungsten carbide [INORG CHEM] WC A hard, gray powder; insoluble in water; readily attacked by nitric-hydrofluoric acid mixture; melts at 2780°C; used in tools, dies, ceramics, cermets, and wear-resistant mechanical parts, and as an abrasive. { 'təŋ·stən 'kär₁bīd }

tungsten carbonyl See tungsten hexacarbonyl. { 'təŋ·stən 'kär·bə₁nil }

tungsten disulfide [INORG CHEM] WS_2 A grayish-black solid with a melting point above 1480°C; used as a lubricant and aerosol. { 'təŋ·stən dī'səl₁fīd }

tungsten filament [ELEC] A filament used in incandescent lamps, and as an incandescent cathode in many types of electron tubes, such as thermionic vacuum tubes. { 'təŋ·stən 'fil·ə·mənt }

tungsten-halogen lamp [ELECTR] A lamp containing a halogen, usually iodine or bromine, which combines with tungsten evaporated from the filament. { 'təŋ·stən 'hal·ə·jən ₁lamp }

tungsten hexacarbonyl [INORG CHEM] $W(CO)_6$ A white, refractive, crystalline solid which decomposes at 150°C; used for tungsten coatings on base metals. Also known as tungsten carbonyl. { 'təŋ·stən ¦hek·sə'kär·bə₁nil }

tungsten hexachloride [INORG CHEM] WCl_6 Dark blue or violet crystals with a melting point of 275°C; soluble in organic solvents; used for tungsten coatings on base metals and as a catalyst for olefin polymers. { 'təŋ·stən ¦hek·sə'klȯr₁īd }

tungsten-inert gas welding [MET] Welding in which an arc plasma from a nonconsumable tungsten electrode radiates heat onto the work surface, to create a weld puddle in a protective atmosphere provided by a flow of inert shielding gas; heat must then travel by conduction from this puddle to melt the desired depth of weld. Abbreviated TIG welding. { 'təŋ·stən i'nərt ₁gas ₁weld·iŋ }

tungstenite [MINERAL] WS_2 A dark lead gray mineral consisting of tungsten disulfide; occurs in massive form, in scaly or feathery aggregates. { 'təŋ·stə₁nīt }

tungsten lake See phosphotungstic pigment. { 'təŋ·stən 'lāk }

tungsten oxychloride [INORG CHEM] $WOCl_4$ Dark red crystals with a melting point of approximately 211°C; soluble in carbon disulfide; used for incandescent lamps. { 'təŋ·stən ¦äk·sə'klȯr₁īd }

tungsten steel [MET] Steel containing tungsten with other alloys; formerly used for cutting and forging tools but replaced by high-speed steel. { 'təŋ·stən 'stēl }

tungstic acid [INORG CHEM] H_2WO_4 A yellow powder; insoluble in water, soluble in alkalies; used as a color-resist mordant for textiles, as an ingredient in plastics, and for the manufacture of tungsten metal products. Also known as orthotungstic acid; wolframic acid. { 'təŋ·stik 'as·əd }

tungstic acid anhydride See tungstic oxide. { 'təŋ·stik 'as·əd an'hī₁drīd }

tungstic anhydride See tungstic oxide. { 'təŋ·stik an'hī₁drīd }

tungstic oxide [INORG CHEM] WO_3 A heavy, canary-yellow powder; soluble in caustic, insoluble in water; melts at 1473°C; used in alloys, in fabric fireproofing, for ceramic pigments, and for the manufacture of tungsten metal. Also known as anhydrous wolframic acid; tungstic acid anhydride; tungstic anhydride; tungstic trioxide. { 'təŋ·stik 'äk₁sīd }

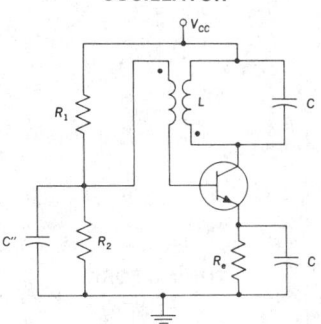

TUNED-COLLECTOR OSCILLATOR

Circuit diagram of tuned-collector oscillator. Frequency is determined by resonant circuit containing capacitance C and self-inductance L. Resistances R_1, R_2, and R_e determine quiescent bias. C' and C'' are capacitances. V_{CC} is collector supply voltage, with respect to emitter.

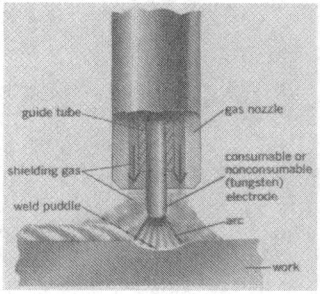

TUNGSTEN-INERT GAS WELDING

Components of tungsten-inert gas welding process.

TUNG TREE

Leaf of tung tree.

tungstic trioxide *See* tungstic oxide. { 'təŋ·stik trī'äk,sīd }

tungstite [MINERAL] $WO_3 \cdot H_2O$ A bright yellow, golden yellow, or yellowish-green mineral thought to consist of hydrated tungsten oxide; occurs in massive form and as platy crystals. { 'təŋ,stīt }

tung tree [BOT] *Aleurites fordii*. A plant of the spurge family in the order Euphorbiales, native to central and western China and grown in the southern United States. { 'təŋ ,trē }

tunica [BIOL] A membrane or layer of tissue that covers or envelops an organ or other anatomical structure. { 'tü·nə·kə }

tunica adventitia *See* adventitia. { 'tü·nə·kə ,ad·vən'tish·ə }

tunica intima *See* intima. { 'tü·nə·kə 'in·tə·mə }

tunica mucosa *See* mucous membrane. { 'tü·nə·kə myü'kō·zə }

Tunicata [INV ZOO] A subphylum of the Chordata characterized by restriction of the notochord to the tail and posterior body of the larva, absence of mesodermal segmentation, and secretion of an outer covering or tunic about the body. { ,tü·nə'käd·ə }

tuning [COMPUT SCI] The use of various techniques involving adjustments to both hardware and software to improve the operating efficiency of a computer system. [ELECTR] The process of adjusting the inductance or the capacitance or both in a tuned circuit, for example, in a radio, television, or radar receiver or transmitter, so as to obtain optimum performance at a selected frequency. { 'tün·iŋ }

tuning capacitor [ELEC] A variable capacitor used for tuning purposes. { 'tün·iŋ kə,pas·əd·ər }

tuning circuit *See* tuned circuit. { 'tün·iŋ ,sər·kət }

tuning coil [ELEC] A variable inductance coil for adjusting the frequency of an oscillator or tuned circuit. { 'tün·iŋ ,kóil }

tuning core [ELECTROMAG] A ferrite core that is designed to be moved in and out of a coil or transformer to vary the inductance. { 'tün·iŋ ,kòr }

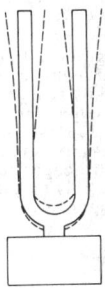

TUNING FORK

A tuning fork vibrating at its fundamental frequency.

tuning fork [ENG] A U-shaped bar for hard steel, fused quartz, or other elastic material that vibrates at a definite natural frequency when struck or when set in motion by electromagnetic means; used as a frequency standard. { 'tün·iŋ ,fórk }

tuning indicator [ELECTR] A device that indicates when a radio receiver is tuned accurately to a radio station, such as a meter or a cathode-ray tuning indicator; it is connected to a circuit having a direct-current voltage that varies with the strength of the incoming carrier signal. { 'tün·iŋ ,in·də,käd·ər }

tuning range [ELECTR] The frequency range over which a receiver or other piece of equipment can be adjusted by means of a tuning control. { 'tün·iŋ ,rānj }

tuning screw [ELECTROMAG] A screw that is inserted into the top or bottom wall of a waveguide and adjusted as to depth of penetration inside for tuning or impedance-matching purposes. { 'tün·iŋ ,skrü }

tuning stub [ELECTROMAG] Short length of transmission line, usually shorted at its free end, connected to a transmission line for impedance-matching purposes. { 'tün·iŋ ,stəb }

tuning susceptance [ELECTR] Normalized susceptance of an anti-transmit-receive tube in its mount due to the deviation of its resonant frequency from the desired resonant frequency. { 'tün·iŋ sə,sep·təns }

tuning wand [ELEC] Rod of insulating material having a brass plug at one end and a powered iron core at the other end; used for checking receiver alignment. { 'tün·iŋ ,wänd }

tunnel [ENG] A long, narrow, horizontal or nearly horizontal underground passage that is open to the atmosphere at both ends; used for aqueducts and sewers, carrying railroad and vehicular traffic, various underground installations, and mining. { 'tən·əl }

tunnel-bearing grease [MATER] Lubricating grease for the main engine and propeller shaft (in the shaft tunnel) of ships. { 'tən·əl 'ber·iŋ ,grēs }

tunnel blasting [ENG] A method of heavy blasting in which a heading is driven into the rock and afterward filled with explosives in large quantities, similar to a borehole, on a large scale, except that the heading is usually divided in two parts on the same level at right angles to the first heading, forming in plan a T, the ends of which are filled with explosives and the intermediate parts filled with inert material like an ordinary borehole. { 'tən·əl ,blast·iŋ }

tunnel borer [MECH ENG] Any boring machine for making

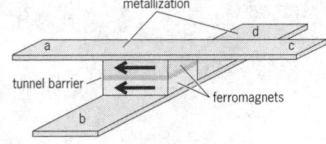

TUNNELING MAGNETORESISTANCE

Tunneling magnetoresistance memory cell. The magnetization of the bottom layer is pinned in a particular orientation (made susceptible to switching by an applied magnetic field).

a tunnel; often a ram armed with cutting faces operated by compressed air. { 'tən·əl ,bòr·ər }

tunnel carriage [MECH ENG] A machine used for rapid tunneling, consisting of a combined drill carriage and manifold for water and air so that immediately the carriage is at the face, drilling may commence with no lost time for connecting up or waiting for drill steels; the air is supplied at pressures of 95 to 100 pounds per square inch (655,000 to 689,000 pascals). { 'tən·əl ,kar·ij }

tunnel cave *See* natural tunnel. { 'tən·əl ,kāv }

tunnel diode [ELECTR] A heavily doped junction diode that has a negative resistance at very low voltage in the forward bias direction, due to quantum-mechanical tunneling, and a short circuit in the negative bias direction. Also known as Esaki tunnel diode. { 'tən·əl ,dī,ōd }

tunnel effect [QUANT MECH] The ability of a particle to pass through a region of finite extent in which the particle's potential energy is greater than its total energy; this is a quantum-mechanical phenomenon which would be impossible according to classical mechanics. Also known as tunneling. { 'tən·əl i,fekt }

tunnel gun [ORD] A gun mounted inside an airplane fuselage and firing through an aperture. { 'tən·əl ,gən }

tunneling *See* tunnel effect. { 'tən·əl·iŋ }

tunneling cryotron [ELECTR] A low-temperature current-controlled switching device that has two electrodes of superconducting material separated by an insulating film, forming a Josephson junction, and a control line whose currents generate magnetic fields that switch the device between two states characterized by the presence or absence of electrical resistance. { 'tən·əl·iŋ 'krī·ə,trän }

tunneling magnetoresistance [SOLID STATE] A type of magnetoresistance displayed by a trilayer thin-film structure consisting of two metallic ferromagnetic thin films sandwiching an insulating film that is thin enough (less than about 2 nanometers) that electrons can pass through it via quantum-mechanical tunneling. Also known as junction magnetoresistance. { ,tən·əl·iŋ mag,ned·ō·ri'zis·təns }

tunneling microscope *See* scanning tunneling microscope. { 'tən·əl·iŋ 'mī·krə,skōp }

tunnel junction [ELECTR] A two-terminal electronic device having an extremely thin potential barrier to electron flow, so that the transport characteristic (the current-voltage curve) is primarily governed by the quantum-mechanical tunneling process which permits electrons to penetrate the barrier. { 'tən·əl ,jəŋk·shən }

tunnel liner [CIV ENG] Any of various materials, especially timber, concrete, and cast iron, applied to the inner surface of a vehicular or railroad tunnel. [MIN ENG] The timber, brick, concrete, or steel supports erected in a mine tunnel to maintain dimensions and safe working conditions. { 'tən·əl ,līn·ər }

tunnel rectifier [ELECTR] Tunnel diode having a relatively low peak-current rating as compared with other tunnel diodes used in memory-circuit applications. { 'tən·əl ,rek·tə,fī·ər }

tunnel resistor [ELECTR] Resistor in which a thin layer of metal is plated across a tunneling junction, to give the combined characteristics of a tunnel diode and an ordinary resistor. { 'tən·əl ri,zis·tər }

tunnel set [MIN ENG] Timbers 6 to 8 inches (15 to 20 centimeters) in diameter and of sufficient height to support the roof of the tunnel. { 'tən·əl ,set }

tunnel system [MIN ENG] A method of mining in which tunnels or drifts are extended at regular intervals from the floor of the pit into the ore body. { 'tən·əl ,sis·təm }

tunnel triode [ELECTR] Transistorlike device in which the emitter-base junction is a tunnel diode and the collector-base junction is a conventional diode. { 'tən·əl ,trī,ōd }

tunnel vault *See* barrel vault. { 'tən·əl ,vòlt }

Tupaiidae [VERT ZOO] The tree shrews, a family of mammals in the order Insectivora. { tü'pī·ə,dē }

tupelo [BOT] Any of various trees belonging to the genus *Nyssa* of the sour gum family, Nyssaceae, distinguished by small, obovate, shiny leaves, a small blue-black drupaceous fruit, and branches growing at a wide angle from the axis. { 'tü·pə,lō }

tuple [COMPUT SCI] A horizontal row of data items in a relational data structure; corresponds to a record or segment in other types of data structures. { 'tü·pəl }

turanite [MINERAL] $Cu_5(VO_4)_2(OH)_4$ An olive green, orthorhombic mineral consisting of basic copper vanadate;

occurs as reniform crusts and spherical concretions. { 'tŭr·ə‚nīt }

Turbellaria [INV ZOO] A class of the phylum Platyhelminthes having bodies that are elongate and flat to oval or circular in cross section. { ‚tər·bə'lar·ē·ə }

turbidimeter [OPTICS] A device that measures the loss in intensity of a light beam as it passes through a solution with particles large enough to scatter the light. { ‚tər·bə'dim·əd·ər }

turbidimetric analysis [ANALY CHEM] A scattered-light procedure for the determination of the weight concentration of particles in cloudy, dull, or muddy solutions; uses a device that measures the loss in intensity of a light beam as it passes through the solution. Also known as turbidimetry. { ‚tər·bə·də‚me·trik ə'nal·ə·səs }

turbidimetric titration [ANALY CHEM] Titration in which the end point is indicated by the developing turbidity of the titrated solution. { ‚tər·bə·də‚me·trik 'tī‚trā·shən }

turbidimetry See turbidimetric analysis. { ‚tər·bə'dim·ə·trē }

turbidite [GEOL] Any sediment or rock transported and deposited by a turbidity current, generally characterized by graded bedding, large amounts of matrix, and commonly exhibiting a Bouma sequence. { 'tər·bə‚dīt }

turbidity [ANALY CHEM] **1.** Measure of the clarity of an otherwise clear liquid by using colorimetric scales. **2.** Cloudy or hazy appearance in a naturally clear liquid caused by a suspension of colloidal liquid droplets or fine solids. [ASTRON] The formation of disks centered on stars in long-exposure photographs as a result of light scattering by grains in the emulsion. [METEOROL] Any condition of the atmosphere which reduces its transparency to radiation, especially to visible radiation. { 'tər·bid·əd·ē }

turbidity coefficient [OPTICS] A factor in the absorption (light) law equation that describes the extinction of the incident light beam. { tər'bid·əd·ē ‚kō·i‚fish·ənt }

turbidity current [OCEANOGR] A highly turbid, relatively dense current carrying large quantities of clay, silt, and sand in suspension which flows down a submarine slope through less dense sea water. Also known as density current; suspension current. { tər'bid·əd·ē ‚kə·rənt }

turbidity factor [GEOPHYS] A measure of the atmospheric transmission of incident solar radiation; if I_0 is the flux density of the solar beam just outside the earth's atmosphere, I the flux density measured at the earth's surface with the sun at a zenith distance which implies an optical air mass m, and $I_{m,w}$ the intensity which would be observed at the earth's surface for a pure atmosphere containing 1 centimeter of precipitable water viewed through the given optical air mass, then turbidity factor θ is given by θ = (ln I_0 − ln I)/(ln I_0 − ln $I_{m,w}$). { tər'bid·əd·ē ‚fak·tər }

turbidostat [MICROBIO] A device in which a bacterial culture is maintained at a constant volume and cell density (turbidity) by adjusting the flow rate of fresh medium into the growth tube by means of a photocell and appropriate electrical connections. { tər'bid·ə‚stat }

turbinate [BOT] Shaped like an inverted cone. [INV ZOO] Spiral with rapidly decreasing whorls from base to apex. { 'tər·bə·nət }

turbine [MECH ENG] A fluid acceleration machine for generating rotary mechanical power from the energy in a stream of fluid. { 'tər·bən }

turbine generator [ELEC] An electric generator driven by a steam, hydraulic, or gas turbine. { 'tər·bən ‚jen·ə·rād·ər }

turbine propulsion [MECH ENG] Propulsion of a vehicle or vessel by means of a steam or gas turbine. { 'tər·bən prə‚pəl·shən }

turbine pump See regenerative pump. { 'tər·bən ‚pəmp }

Turbinidae [INV ZOO] A family of gastropod mollusks including species of top shells. { tər'bin·ə‚dē }

turbining [MECH ENG] The removal of scale or other foreign material from the internal surface of a metallic cylinder. { 'tər·bən·iŋ }

turboalternator [ELEC] An alternator, such as a synchronous generator, which is driven by a steam turbine. { ‚tər·bō'ól·tər‚nād·ər }

turboblower [MECH ENG] A centrifugal or axial-flow compressor. { 'tər·bō‚blō·ər }

turbodrill [PETRO ENG] A rotary tool used in drilling oil or gas wells in which the bit is rotated by a turbine motor inside the well. { 'tər·bō‚dril }

turbofan [AERO ENG] An air-breathing jet engine in which additional propulsive thrust is gained by extending a portion of the compressor or turbine blades outside the inner engine case. { 'tər·bō‚fan }

turbogrid plate [CHEM ENG] A tray for distillation columns that consists of a flat grid of parallel slots extending over the entire cross-sectional area of the column; the liquid level on each tray is maintained by a dynamic balance between down-flowing liquid and up-flowing vapor. { 'tər·bō‚grid 'plāt }

turbojet [AERO ENG] A jet engine incorporating a turbine-driven air compressor to take in and compress the air for the combustion of fuel (or for heating by a nuclear reactor), the gases of combustion (or the heated air) being used both to rotate the turbine and to create a thrust-producing jet. { 'tər·bō‚jet }

turbonada [METEOROL] A short thundersquall on the north Spanish coast, sometimes accompanied by waterspouts. { ‚tər·bə'näd·ə }

turbopause See homopause. { 'tər·bə‚póz }

turboprop [AERO ENG] A gas turbine power plant that produces shaft power to drive aircraft propellers. { 'tər·bō‚präp }

turbopump [MECH ENG] A pump that is powered by a turbine. { 'tər·bō‚pəmp }

turboramjet [AERO ENG] An aircraft engine that is a hybrid of a turbofan and a ramjet; operates as a ramjet for efficient propulsion at very high supersonic cruise speeds, or as a turbofan for relatively efficient propulsion at low flight speeds. { ‚tər·bō'ram‚jet }

turboshaft [MECH ENG] A gas turbine engine that is similar to a turboprop but operates through a transmission system to power a device such as a helicopter rotor or pump. { 'tər·bō‚shaft }

turbosphere [METEOROL] The region of the atmosphere in which turbulence frequently exists. { 'tər·bə‚sfir }

turbosupercharger [MECH ENG] A centrifugal air compressor, gas-turbine driven, usually used to increase induction system pressure in an internal combustion reciprocating engine. { ‚tər·bō'sü·pər‚chär·jər }

turbulence See turbulent flow. { 'tər·byə·ləns }

turbulence energy See eddy kinetic energy. { 'tər·byə·ləns ‚en·ər·jē }

turbulence modeling [FL MECH] The construction of models of the Reynolds stresses in turbulent flow. { 'tər·byə·ləns ‚mäd·əl·iŋ }

turbulent boundary layer [FL MECH] The layer in which the Reynolds stresses are much larger than the viscous stresses. { 'tər·byə·lənt 'baún·drē ‚lā·ər }

turbulent burner [ENG] An atomizing burner which mixes fuel and air to produce agitated flow. { 'tər·byə·lənt 'bər·nər }

turbulent diffusion See eddy diffusion. { 'tər·byə·lənt di'fyü·zhən }

turbulent flow [FL MECH] Motion of fluids in which local velocities and pressures fluctuate irregularly, in a random manner. Also known as turbulence. { 'tər·byə·lənt 'flō }

turbulent flux See eddy flux. { 'tər·byə·lənt 'fləks }

turbulent heat conduction [OCEANOGR] Conduction of heat in water by lateral and vertical eddy diffusion, with currents. { 'tər·byə·lənt 'hēt kən‚dək·shən }

turbulent Lewis number [PHYS] A dimensionless number used in the study of combined turbulent heat and mass transfer, equal to the ratio of the eddy mass diffusivity to the eddy thermal diffusivity. Symbolized Le_T. { 'tər·byə·lənt 'lü·əs ‚nəm·bər }

turbulent Prandtl number [PHYS] A dimensionless number used in the study of heat transfer in turbulent flow, equal to the ratio of the eddy viscosity to the eddy thermal diffusivity. Symbolized Pr_T. { 'tər·byə·lənt 'pränt·əl ‚nəm·bər }

turbulent Schmidt number [FL MECH] A dimensionless number used in the study of mass transfer in turbulent flow, equal to the ratio of the eddy viscosity to the eddy mass diffusivity. Symbolized Sc_T. { tər·byə·lənt 'shmit ‚nəm·bər }

turbulent shear force [FL MECH] A shear force in a fluid which arises from turbulent flow. { 'tər·byə·lənt 'shir‚fórs }

turbulization [ENG] In a heat-transfer process involving the interaction of a solid, heat-conducting, and impermeable surface with a surrounding fluid, destruction of the boundary layer

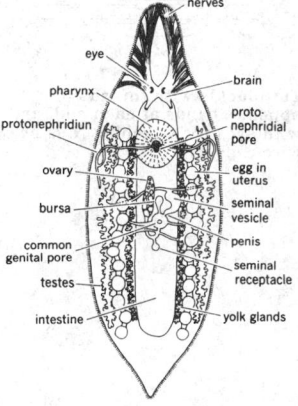

TURBELLARIA

A typical hermaphroditic turbellarian, *Mesostoma ehrenbergii wardii.*

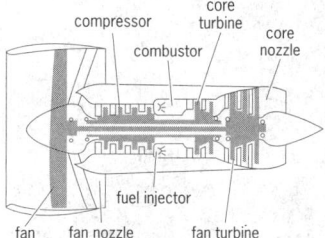

TURBOFAN

Configuration of a high-bypass, separate-flow turbofan.

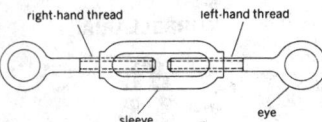

TURNBUCKLE

right-hand thread   left-hand thread

sleeve   eye

Turnbuckle with threads of opposite hands at each end, and with eyes for attaching rods or wire rope.

TURNIP

Turnip, showing enlarged root and foliage.

---

in order to intensify the convective heat transfer. { ,tər·bə·lə'zā·shən }

Turdidae [VERT ZOO] The thrushes, a family of passeriform birds in the suborder Oscines. { 'tər·də,dē }

turgid [MED] Swollen and congested. { 'tər·jəd }

turgor [BOT] Distension of a plant cell wall and membrane by the fluid contents. { 'tər·gər }

turgor movement [BOT] A reversible change in the position of plant parts due to a change in turgor pressure in certain specialized cells; movement of *Mimosa* leaves when touched is an example. { 'tər·gər ,müv·mənt }

turgor pressure [BOT] The actual pressure developed by the fluid content of a turgid plant cell. { 'tər·gər ,presh·ər }

Turing computable function [MATH] A function that can be computed on a Turing machine. { 'tur·iŋ kəm'pad·ə·bəl 'fəŋk·shən }

Turing machine [COMPUT SCI] A mathematical idealization of a computing automation similar in some ways to real computing machines; used by mathematicians to define the concept of computability. { 'tur·iŋ mə,shēn }

Turing's thesis *See* Church's thesis. { 'tur·iŋz ,thē·səs }

turion [BOT] A scaly shoot, such as asparagus, developed from an underground bud. { 'tur·ē,än }

turkey [VERT ZOO] Either of two species of wild birds, and any of various derived domestic breeds, in the family Meleagrididae characterized by a bare head and neck, and in the male a large pendant wattle which hangs on one side from the base of the bill. { 'tər·kē }

Turkey red oil [MATER] Sulfonated castor oil, soluble in water; autoignites at 833°F (445°C); used in textiles, leather, and paper coatings, for manufacture of soaps, and as an alizarin dye assistant. Also known as soluble castor oil; sulfonated castor oil. { 'tər·kē ¦red ,oil }

Turkey stone *See* turquoise. { 'tər·kē ,stōn }

turkey tail mushroom *See* Trametes versicolor. { 'tər·kē ,tāl ,məsh,rüm }

Turkish geranium oil *See* palmarosa oil. { 'tər·kish jə¦rān·ē·əm ,oil }

Turk's head rolls [MET] A group of four idler rolls, arranged in a square or rectangular pattern, through which strip metal can be drawn to form angled sections. { 'tərks ¦hed ,rōlz }

turmeric [BOT] *Curcuma longa.* An East Indian perennial of the ginger family (Zingiberaceae) with a short stem, tufted leaves, and short thick rhizomes; a spice with a pungent, bitter taste and a musky odor is derived from the rhizome. [MATER] An orange-red or reddish-brown dye obtained from the rhizome of turmeric. { 'tər·mər·ik }

turn [ELEC] One complete loop of wire. [MATH] *See* circle. { 'tərn }

turnaround [CHEM ENG] In petroleum refining, the shutdown of a unit after a normal run for maintenance and repair work, then putting the unit back into operation. [ENG] The length of time between arriving at a point and departing from that point; it is used in this sense for the turnaround of vehicles, ships in ports, and aircraft. { 'tərn·ə,raund }

turnaround cycle [ENG] A term used in conjunction with vehicles, ships, and aircraft, and comprising the following: loading time at home, time to and from destination, unloading and loading time at destination, unloading time at home, planned maintenance time, and, where applicable, time awaiting facilities. { 'tərn·ə,raund ,sī·kəl }

turnaround document [COMPUT SCI] A document, such as a punch card, that is produced by a computer, can be read by humans, and can be reread into the machine. { 'tərn·ə,raund ,däk·yə·mənt }

turnaround system [COMPUT SCI] In character recognition, a system in which the input data to be read have previously been printed by the computer with which the reader is associated; an application is invoice billing and the subsequent recording of payments. { 'tərn·ə,raund ,sis·təm }

turnaround time [COMPUT SCI] The delay between submission of a job for a data-processing system and its completion. { 'tərn·ə,raund ,tīm }

turnbuckle [DES ENG] A sleeve with a thread at one end and a swivel at the other, or with threads of opposite hands at each end so that by turning the sleeve connected rods or wire rope will be drawn together and tightened. { 'tərn,bək·əl }

Turnbull's blue [INORG CHEM] A blue pigment that precipitates from the reaction of potassium ferricyanide with a ferrous salt. { 'tərn,bülz 'blü }

Turner's syndrome [MED] A sex aberration in humans in which the chromosome complement includes only one sex chromosome, an X. { 'tər·nərz ,sin,drōm }

Turnicidae [VERT ZOO] The button quails, a family of Old World birds in the order Gruiformes. { tər'nis·ə,dē }

turning [MECH ENG] Shaping a member on a lathe. { 'tərn·iŋ }

turning area *See* enroute turning area. { 'tərn·iŋ ,er·ē·ə }

turning bar *See* chimney bar. { 'tərn·iŋ ,bär }

turning basin [CIV ENG] An open area at the end of a canal or narrow waterway to allow boats to turn around. { 'tərn·iŋ ,bās·ən }

turning-block linkage [MECH ENG] A variation of the sliding-block mechanical linkage in which the short link is fixed and the frame is free to rotate. Also known as the Wentworth quick-return motion. { 'tərn·iŋ ¦bläk ,liŋ·kij }

turning buoy [NAV] A buoy marking a turn, as in a channel. { 'tərn·iŋ ,boi }

turning center [MECH ENG] A numerically controlled lathe that sometimes functions together with a robot in boring and other machining work. { 'tərn·iŋ ,sen·tər }

turning circle [NAV] The path approximating a circle of 360° or more described by the pivot point of the ship as it makes a turn. { 'tərn·iŋ ,sər·kəl }

turning error *See* northerly turning error. { 'tərn·iŋ ,er·ər }

turning table [ENG] In plastics molding, a rotating table or wheel carrying various molds in a multimold, single-parison blow-molding operation. { 'tərn·iŋ ,tā·bəl }

turning value [MATH] A relative maximum or relative minimum of a function. { 'tərn·iŋ ,val·yü }

turnip [BOT] *Brassica rapa* or *B. campestris* var. *rapa.* An annual crucifer of Asiatic origin belonging to the family Brassicaceae in the order Capparales and grown for its foliage and edible root. { 'tər·nəp }

turnip yellow mosaic virus [VIROL] The type species of the genus *Tymovirus*; it is transmitted mechanically and via beetles, causing chloroplast clumping. Abbreviated TYMV. { ¦tər·nəp ,yel·ō mō'zā·ik ,vī·rəs }

turnip yellow mosaic virus group *See* Tymovirus. { ¦tər·nəp ,yel·ō mō'zā·ik ¦vī·rəs ,grüp }

turnkey [COMPUT SCI] A complete computer system delivered to a customer in running condition, with all necessary premises, hardware and software equipment, supplies, and operating personnel. { 'tərn,kē }

turnkey contract [ENG] A contract in which an independent agent undertakes to furnish for a fixed price all materials and labor, and to do all the work needed to complete a project. { 'tərn,kē 'kän,trakt }

turnoff mass [ASTRON] The mass of those stars in a cluster that are at the turnoff point. { 'tərn,of ,mas }

turnoff point [ASTRON] The point on a Hertzsprung-Russell diagram of a star cluster at which stars leave the main sequence and move toward the giant branch. { 'tərn,of ,point }

turn-off time [ELECTR] The time that is takes a gate circuit to shut off a current. { 'tərn,of ,tīm }

turn of the tide *See* change of tide. { 'tərn əv the̱ 'tīd }

turn-on time [ELECTR] The time that it takes a gate circuit to allow a current to reach its full value. { 'tərn,on ,tīm }

turnout [ENG] **1.** A contrivance consisting of a switch, a frog, and two guardrails for passing from one track to another. **2.** The branching off of one rail track from another. **3.** A siding. [MIN ENG] To shovel coal toward the track for more convenient loading. { 'tərn,aut }

turnover [MOL BIO] The number of substrate molecules transformed by a single molecule of enzyme per minute, when the enzyme is operating at maximum rate. { 'tərn,ō·vər }

turnover cartridge [ENG ACOUS] A phonograph pickup having two styli and a pivoted mounting that places in playing position the correct stylus for a particular record speed. { 'tərn,ō·vər ,kär·trij }

turnover frequency *See* transition frequency. { 'tərn,ō·vər ,frē·kwən·sē }

turnover number [BIOCHEM] The number of molecules of a substrate acted upon in a period of 1 minute by a single enzyme molecule, with the enzyme working at a maximum rate. [CHEM ENG] In an industrial catalytic process, a value that indicates the amount of feed or substrate converted per a measured amount of catalyst. { 'tərn·ō·vər ˌnəm·bər }

turnover rate [CHEM ENG] In an industrial catalytic process, a value corresponding to the turnover number per specified unit of time. { 'tərn·ō·vər ˌrāt }

turnpike [CIV ENG] A toll expressway. { 'tərn‚pīk }

turnplow See moldboard plow. { 'tərn‚plau }

turns ratio [ELEC] The ratio of the number of turns in a secondary winding of a transformer to the number of turns in the primary winding. { 'tərnz ˌrā·shō }

turnstile [ENG] A barrier that rotates about a vertical axis and usually is arranged to allow the passage of only one person at a time through an opening. { 'tərn‚stīl }

turnstile antenna [ELECTROMAG] An antenna consisting of one or more layers of crossed horizontal dipoles on a mast, usually energized so the currents in the two dipoles of a pair are equal and in quadrature; used with television, frequency modulation, and other very-high-frequency or ultra-high-frequency transmitters to obtain an essentially omnidirectional radiation pattern. { 'tərn‚stīl an‚ten·ə }

turntable [ENG ACOUS] The rotating platform on which a disk record is placed for recording or playback. { 'tərn‚tā·bəl }

turntable rumble [ENG ACOUS] Low-frequency vibration that is mechanically transmitted to a recording or reproducing turntable and superimposed on the reproduction. Also known as rumble. { 'tərn‚tā·bəl ˌrəm·bəl }

Turonian [GEOL] A European stage of geologic time: Upper or Middle Cretaceous (above Cenomanian, below Coniacian). { tü'rō·nē·ən }

turpentine [MATER] An essential oil produced by steam distillation of pine woods and from gum turpentine; used as a solvent and a thinner for paints and varnishes. { 'tər·pən‚tīn }

turpentine camphor See terpene hydrochloride. { 'tər·pən‚tīn 'kam·fər }

turquoise [MINERAL] $CuAl_6(PO_4)_4(OH)_8 \cdot 4H_2O$ A semitranslucent sky-blue, bluish-green, apple-green, or greenish-gray mineral that crystallizes in the triclinic system and occurs in veinlets or as crusts of massive, concretionary, and stalactite shapes; an important gem mineral. Also known as calaite; Turkey stone. { 'tər‚kwóiz }

turret [ORD] A dome-shaped or cylindrical armored structure containing one or more guns and located on forts, warships, airplanes, and tanks. { 'tə·rət }

turret coal cutter [MIN ENG] A coal cutter in which the horizontal jib can be adjusted vertically to cut at different levels in the seam; for example, an overcut. { 'tə·rət 'kōl ˌkəd·ər }

turret ice See ropak. { 'tə·rət ‚īs }

turret lathe [MECH ENG] A semiautomatic lathe differing from the engine lathe in having the tailstock replaced with a multisided, indexing tool holder or turret designed to hold several tools. { 'tə·rət ‚lāth }

turret mount [ORD] A gun mount positioned in the turret of a tank or combat vehicle; power-driven multiple gun turret mounts improve control in tracking aerial targets and also increase firepower. { 'tə·rət ‚maunt }

turret robot [CONT SYS] A tower-shaped robot whose manipulator makes circular motions about the robot's base. { 'tər·ət 'rō‚bät }

turret tuner [ELECTR] A television tuner having one set of pretuned circuits for each channel, mounted on a drum that is rotated by the channel selector; rotation of the drum connects each set of tuned circuits in turn to the receiver antenna circuit, radio-frequency amplifier, and r-f oscillator. { 'tə·rət ‚tü·nər }

turtle [COMPUT SCI] A cursor with the attributes of both position and direction; usually, an arrow that points in the direction it is about to move and generates a line along its path. [VERT ZOO] Any of about 240 species of reptiles which constitute the order Chelonia distinguished by the two bony shells enclosing the body. { 'tərd·əl }

turtle oil [MATER] The oil derived from the muscles and genital glands of the giant sea turtle; melts at 25°C; used in cosmetics. { 'tərd·əl ‚oil }

turtle stone See septarium. { 'tərd·əl ‚stōn }

tusche [MATER] A liquid lithographic ink that can be used with pen or brush on lithographic stones or metal plates; it is also used in the silk-screen process in a glue-resist system of pattern making; an extremely greasy material. { 'tush·ə }

tussock [ECOL] A small hummock of generally solid ground in a bog or marsh, usually covered with and bound together by the roots of low vegetation such as grasses, sedges, or ericaceous shrubs. { 'təs·ək }

tutorial [COMPUT SCI] A method of computer-assisted instruction that involves a collection of screen formats, generally arranged in sequences that can be selected from a menu, and presented in response to the terminal operator's request. { tü'tór·ē·əl }

tuyere [MET] An opening in the shell and refractory lining of a furnace through which air is forced. { twē'yer }

TV See television.

TV camera scanner [COMPUT SCI] In optical character recognition, a device that images an input character onto a sensitive photoconductive target of a camera tube, thereby developing an electric charge pattern on the inner surface of the target; this pattern is then explored by a scanning beam which traces out a rectangular pattern with the result that a waveform is produced which represents the character's most probable identity. { tē'vē ˌkam·rə ‚skan·ər }

TVI See television interference.

TVRO See television receive-only antenna.

TW See terawatt.

TWA See time-weighted average.

Twaddell scale [ENG] A scale for specific gravity of solutions that is the first two digits to the right of the decimal point multiplied by two; for example, a specific gravity of 1.4202 is equal to 84.04°Tw. { twə'del ‚skāl }

tweeter [ENG ACOUS] A loudspeaker designed to handle only the higher audio frequencies, usually those well above 3000 hertz, generally used in conjunction with a crossover network and a woofer. { 'twēd·ər }

twelve-color problem [MATH] The problem of showing that 12 colors are sufficient to color a map on which each country has at most one colony, and that there is such a map which requires 12 colors. { 'twelv ˌkəl·ər ‚präb·ləm }

twenty-nine feature [COMPUT SCI] A device used on some punched-card machines to represent values from 0 through 29 by a maximum of two punches on a single column; x and y punches represent 10 and 20, and these are added to punches in positions 0 through 9. { ‚twen·tē‚nīn 'fē·chər }

twilight [ASTRON] An intermediate period of illumination of the sky before sunrise and after sunset; the three forms are civil, nautical, and astronomical. { 'twī‚līt }

twilight arch See bright segment. { 'twī‚līt ‚ärch }

twilight compass [NAV] An instrument for indicating direction by use of the polarizing effect of the earth's atmosphere on sunlight. Also known as sky compass. { 'twī‚līt ‚kəm·pəs }

twilight correction [ASTRON] In the interpretation of the records of sunshine recorders, the difference between the time of sunrise and the time at which a record of sunshine first began to be made by the sunshine recorder; and conversely at sunset; this correction is added only when the horizon is clear during the period. { 'twī‚līt kə‚rek·shən }

twilight phenomena [OPTICS] Those meteorological optical phenomena which occur during twilight, including such effects as the antitwilight arch, dark segment, bright segment, green flash, purple light, and crepuscular rays. { 'twī‚līt fə‚näm·ə·nä }

twilight zone [ASTRON] That zone of the earth or other planet in twilight at any time. [ELECTROMAG] Anything resembling the twilight zone of the earth, as the narrow sector on each side of the equisignal zone of a four-course radio range station, in which one signal is barely heard above the monotone on-course signal. { 'twī‚līt ‚zōn }

twill weave [TEXT] A woven pattern of diagonal or twill lines that run upward to the right or left of the fabric face. { 'twil ‚wēv }

twin [BIOL] One of two individuals born at the same time. [CRYSTAL] See twin crystal. { 'twin }

twin arithmetic units [COMPUT SCI] A feature of some computers where the essential portions of the arithmetic section are virtually duplicated. { 'twin ə'rith·mə‚tik ‚yü·nəts }

twin axial cable [COMMUN] A transmission line consisting of two coaxial cables enclosed within a single sheath, each

TURTLE

Box turtle, *Terrapene carolina*. *(From J. J. Shomon, ed., Virginia Wildlife, 15(6):27, 1954)*

TWILL WEAVE

Three-shaft twill. Two warp yarns are interlaced with one filling yarn. *(From M. D. Potter and B. P. Corbman, Fiber to Fabric, 3d ed., McGraw-Hill, 1959)*

used to transmit signals in one direction. { 'twin 'ak·sē·əl 'kā·bəl }

twin axis [CRYSTAL] The crystal axis about which one individual of a twin crystal may be rotated (usually 180°) to bring it into coincidence with the other individual. { 'twin 'ak·səs }

twin band [MET] A line on a polished or etched surface representing the section through crystal twins. { 'twin 'band }

twin boundary [CRYSTAL] A grain boundary whose lattice structures are mirror images of each other in the plane of the boundary. { 'twin 'baún·drē }

twin-cable ropeway [MECH ENG] An aerial ropeway which has parallel track cables with carriers running in opposite directions; both rows of carriers are pulled by the same traction rope. { 'twin ¦kāb·əl 'rōp‚wā }

twin check [COMPUT SCI] Continuous check of computer operation, achieved by the duplication of equipment and automatic comparison of results. { 'twin 'chek }

twin crystal [CRYSTAL] A compound crystal which has one or more parts whose lattice structure is the mirror image of that in the other parts of the crystal. Also known as crystal twin; twin. { 'twin 'krist·əl }

twine [MATER] A strong string made up of two or more strands twisted together. { twīn }

twin entry [MIN ENG] A pair of parallel entries, one of which is an intake air course and the other the return air course; rooms can be worked from both entries. { 'twin 'en·trē }

twiner [BOT] A climbing stem that winds about its support, as pole beans or many tropical lianas. { 'twī·nər }

twin-geared press [MECH ENG] A crank press having the drive gears attached to both ends of the crankshaft. { 'twin ¦gird 'pres }

twinkling stars [ASTRON] Rapid fluctuations of the brightness and size of the images of stars caused by turbulence in the earth's atmosphere. { 'twink·liŋ 'stärz }

twin law [CRYSTAL] A statement relating two or more individuals of a twin to one another in terms of their crystallography (twin plane, twin axis, and so on). { 'twin ‚lò }

twinned double bonds *See* cumulative double bonds. { ¦twind ¦dəb·əl 'bänz }

twinning [CRYSTAL] The development of a twin crystal by growth, translation, or gliding. { 'twin·iŋ }

twinning plane *See* twin plane. { 'twin·iŋ ‚plān }

twin paradox *See* clock paradox. { 'twin ‚par·ə‚däks }

twin plane [CRYSTAL] The plane common to and across which the individual crystals or components of a crystal twin are symmetrically arranged or reflected. Also known as twinning plane. { 'twin ‚plān }

twin primes [MATH] A pair of prime numbers that differ by 2. { ¦twin 'prīmz }

twin ring structures [ASTRON] Consanguineous ring structures whose component craters are of the same size, as well as being similar in form. { 'twin 'riŋ ‚strək·chərz }

Twins *See* Gemini. { twinz }

twin-T filter [ELEC] An electric filter consisting of a parallel-T network with values of network elements chosen in such a way that the outputs due to each of the paths precisely cancel at a specified frequency. { 'twin 'tē ‚fil·tər }

twin-T network *See* parallel-T network. { 'twin 'tē ‚net‚wərk }

twist [DES ENG] In a fiber, rope, yarn, or cord, the turns about its axis per unit length; usually expressed as TPI (turns per inch). [ELECTROMAG] A waveguide section in which there is a progressive rotation of the cross section about the longitudinal axis of the waveguide. { twist }

twist boundary [SOLID STATE] A boundary between two crystals that differ in orientation by only a few degrees, consisting of a series of screw dislocations. { 'twist ‚baún·drē }

twist drill [DES ENG] A tool having one or more helical grooves, extending from the point to the smooth part of the shank, for ejecting cuttings and admitting a coolant. { 'twist ‚dril }

twisted curve [MATH] A curve that does not lie wholly in any one plane. { ¦twis·təd 'kərv }

twisted pair [ELEC] A cable composed of two small insulated conductors twisted together without a common covering. Also known as copper pair. { ¦twis·təd 'per }

twister [METEOROL] In the United States, a colloquial term for tornado. [SOLID STATE] A piezoelectric crystal that generates a voltage when twisted. { 'twis·tər }

twisting [TEXT] In a ply yarn or cord, the process of combining two or more ends with a twist to give greater strength and smoothness, increased uniformity, or novel effects. { 'twist·iŋ }

twist-lock connector [ELEC] A power plug and receptacle in which the plug must be twisted after insertion to lock it in place, to guard against the plug accidentally being knocked loose. { 'twist ‚läk kə'nek·tər }

twist of rifling [ORD] Inclination of the spiral grooves (rifling) to the axis of the bore of a weapon; it is expressed as the number of calibers of length in which the rifling makes one complete turn. { 'twist əv 'rīf·liŋ }

twistor [COMPUT SCI] Computer memory element consisting of a helix of a magnetic wire wound under tension at a 45° angle on a short piece of nonmagnetic wire, with a fine-wire solenoid wound over the helix. { 'twis·tər }

twist-yarn cloth [TEXT] Fabric made from two or more colored yarns twisted or plied together, giving a mottled effect. { 'twist ‚yärn ‚klòth }

Twitchell reagent [ORG CHEM] A catalyst for the acid hydrolysis of fats; a sulfonated addition product of naphthalene and oleic acid, that is, a naphthalenestearosulfonic acid. { 'twich·əl rē‚ā·jənt }

two-address code [COMPUT SCI] In computers, a code using two-address instructions. { 'tü 'ad‚res ‚kōd }

two-address instruction [COMPUT SCI] In computers, an instruction which includes an operation and specifies the location of two registers. { 'tü 'ad‚res in‚strək·shən }

two-beam interference [PHYS] Interference between two waves. { 'tü ¦bēm ‚in·tər'fir·əns }

two-body force [PHYS] A force between two particles which is not affected by the existence of other particles in the vicinity, such as a gravitational force or a Coulomb force between charged particles. { 'tü ¦bäd·ē 'fòrs }

two-body problem [MECH] The problem of predicting the motions of two objects obeying Newton's laws of motion and exerting forces on each other according to some specified law such as Newton's law of gravitation, given their masses and their positions and velocities at some initial time. { 'tü ¦bäd·ē 'präb·ləm }

two-carrier theory [SOLID STATE] A theory of the conduction properties of a material in bulk or in a rectifying barrier which takes into account the motion of both electrons and holes. { 'tü ¦kar·ē·ər 'thē·ə·rē }

two-component model [FL MECH] A dynamical model of a gas flow in which the gas has two components of differing temperature. { 'tü kəm¦pō·nənt 'mäd·əl }

two-component neutrino theory [PART PHYS] A theory according to which the neutrino and antineutrino have exactly zero rest mass, and the neutrino spin is always antiparallel to its motion, while the antineutrino spin is parallel to its motion. { 'tü ¦kəm¦pō·nənt nü'trē·nō ‚thē·ə·re }

two-cycle [MATH] The repetition of numbers generated by a mapping on every second iteration of the mapping. { ¦tü 'sī·kəl }

two-cycle engine [MECH ENG] A reciprocating internal combustion engine that requires two piston strokes or one revolution to complete a cycle. { 'tü ¦sī·kəl 'en·jən }

two-decision problem [STAT] The problem of deciding, using statistical information, between two actions or decisions. { 'tü ¦di¦sizh·ən 'präb·ləm }

two-degrees-of-freedom gyro [MECH] A gyro whose spin axis is free to rotate about two orthogonal axes, not counting the spin axis. { 'tü ¦di¦grēz əv ¦frē·dəm 'jī·rō }

two-dimensional chromatography [ANALY CHEM] A paper chromatography technique in which the sample is resolved by standard procedures (ascending, descending, or horizontal solvent movement) and then turned at right angles in a second solvent and re-resolved. { 'tü ¦di¦men·shən·əl ‚krō·mə'täg·rə·fē }

two-dimensional electron gas [SOLID STATE] A system of electrons that are confined by opposing forces to a thin planar region adjacent to an interface or within a thin layer of material, but are free to move along the plane scattering off each other. { ‚tü di'men·shən·əl i¦lek‚trän 'gas }

two-dimensional electron gas field-effect transistor *See* high-electron-mobility transistor. { ‚tü di'men·shən·əl i¦lek‚trän 'gas ¦fēld i¦fekt tran'zis·tər }

two-dimensional flow [FL MECH] Fluid flow in which all

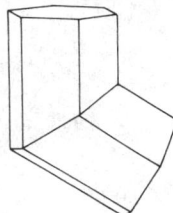

TWIN CRYSTAL

Example of twin crystal, consisting of hexagonal prism twinned on a pyramid face.

TWIST

A 90° twist for a rectangular waveguide.

flow occurs in a set of parallel planes with no flow normal to them, and the flow is identical in each of these parallel planes. { 'tü ¦di¦men·shən·əl 'flō }

two-dimensional gel electrophoresis [CELL MOL] A type of gel electrophoresis in which proteins are first separated by charge and then by molecular weight, enabling the analysis of complex protein mixtures. { ¦tü·də‚men·shən·əl ¦jel i‚lek·trō·fə'rē·səs }

two-dimensional storage [CCMPUT SCI] A direct-access storage device in which the storage locations assigned to a particular file do not have to be physically adjacent, but instead may be taken from one or more seek areas. { 'tü ¦di¦men·shən·əl 'stór·ij }

two-fluid cell [PHYS CHEM] Cell having different electrolytes at the positive and negative electrodes. { 'tü ¦flü·əd 'sel }

two-fluid model [CRYO] A theoretical model of helium II which assumes that it consists of two interpenetrating components, a normal fluid and a superfluid with zero entropy, viscosity, and thermal conductivity. { 'tü‚flü·əd ‚mäd·əl }

two-gap head [COMPUT SCI] One of two separate magnetic tape heads, one for reading and the other for recording data. { 'tü ¦gap 'hed }

two-hit model [GEN] The hypothesis that a cancer arises in an individual heterozygous for a mutant gene for a dominantly inherited form of cancer only if an additional mutation occurs in a somatic cell; for example, in the normal allele at the same locus, as in human retinoblastoma. { 'tü ‚hit ¦mäd·əl }

two-hop transmission [COMMUN] Propagation of radio waves in which the waves are reflected from the ionosphere, then reflected from the ground, and then reflected from the ionosphere again before reaching the receiver. { 'tü ¦häp tranz'mish·ən }

two-input subtracter See half-subtracter. { 'tü ¦in‚pút səb'trak·tər }

two-layer ocean [OCEANOGR] An idealized ocean in which a layer of uniform density near the surface overlays a deep layer of uniform but distinctly higher-density water. { 'tü ¦lā·ər 'ō·shən }

two-level mold [ENG] Placement of one cavity of a plastics mold above another instead of alongside it; reduces clamping force needed. { 'tü ¦lev·əl 'mōld }

two-level subroutine [COMPUT SCI] A subroutine in which entry is made to a second, lower-level subroutine. { 'tü ¦lev·əl 'səb·rü‚tēn }

two-lip end mill [MECH ENG] An end-milling cutter having two cutting edges and straight or helical flutes. { 'tü ¦lip 'end ‚mil }

two-out-of-five code [COMPUT SCI] An encoding of the decimal digits using five binary bits and having the property that every code element contains two 1s and three 0s. { 'tü aúd·əv ¦fīv 'kōd }

two-part adhesive [MATER] A glue supplied in two parts, a resin and an accelerator, which are mixed only just before application. { 'tü ¦pärt ad'hē·siv }

two-part code [COMMUN] Randomized code consisting of an encoding section in which the plain text groups are arranged in alphabetical or other significant order accompanied by their code groups in nonalphabetical or random order, and a decoding section in which the code groups are arranged in alphabetical or numerical order and are accompanied by their meanings given in the encoding section. { 'tü ¦pärt 'kōd }

two-part experiment [STAT] An experiment in which two operations or actions are performed; for example, throwing two dice, drawing two marbles from a box, throwing a die and then drawing a marble from a box. { 'tü ¦pärt ik'sper·ə·mənt }

two-pass compiler [COMPUT SCI] A language processor that goes through the program to be translated twice; on the first pass it checks the syntax of statements and constructs a table of symbols, while on the second pass it actually translates program statements into machine language. { 'tü ¦pas kəm 'pīl·ər }

two-person game [MATH] A game consisting of exactly two players with competing interests. { 'tü ¦pər·sən 'gām }

two-phase [PHYS] Having a phase difference of one quarter-cycle or 90°. Also known as quarter-phase. { 'tü ¦fāz }

two-phase alternating-current circuit [ELEC] A circuit in which there are two alternating currents on separate wires, the two currents being 90° out of phase. { 'tü ¦fāz 'ól·tər‚nād·iŋ kə·rənt ‚sər·kət }

two-phase current [ELEC] Current delivered through two pairs of wires or at a phase difference of one-quarter cycle (90°) between the current in the two pairs. { 'tü ¦fāz 'kə·rənt }

two-phase five-wire system [ELEC] System of alternating-current supply comprising five conductors, four of which are connected as in a two-phase four-wire system, the fifth being connected to the neutral points of each phase. { 'tü ¦fāz 'fīv ¦wīr 'sis·təm }

two-phase flow [CRYO] Flow of helium II, or of electrons in a superconductor thought of as consisting of two interpenetrating, noninteracting fluids, a superfluid component which exhibits no resistance to flow and is responsible for superconducting properties, and a normal component, which behaves as does an ordinary fluid or as conduction electrons in a nonsuperconducting metal. [FL MECH] Cocurrent movement of two phases (for example, gas and liquid) through a closed conduit or duct (for example, a pipe). { 'tü ¦fāz 'flō }

two-phase four-wire system [ELEC] System of alternating-current supply comprising two pairs of conductors, between one pair of which is maintained an alternating difference of potential displaced in phase by one-quarter of a period from an alternating difference of potential of the same frequency maintained between the other pair. { 'tü ¦fāz 'fór ¦wīr 'sis·təm }

two-phase three-wire system [ELEC] System of alternating-current supply comprising three conductors, between one of which (known as the common return) and each of the other two are maintained alternating difference of potential displaced in phase by one-quarter of a period with relation to each other. { 'tü ¦fāz 'thrē ¦wīr 's.s·təm }

two-photon absorption [PHYS CHEM] A relatively weak photon absorption and excitation process that occurs when a sufficiently intense light source, such as a laser beam, is used, making it possible for a molecule to absorb simultaneously two photons, each of approximately half the energy (twice the wavelength) normally required to reach an excited state. The probability of a molecule absorbing two photons simultaneously is proportional to the square of the intensity of the input beam. { ‚tü ¦fō‚tän əb'sórp·shən }

two-photon coherent state [QUANT MECH] A quantum state of an electromagnetic field in which the product of the uncertainties of two quadrature components of the field is the minimum allowed by the Heisenberg uncertainty principle. { 'tü ¦fō‚tän kō¦hir·ənt 'stāt }

two-piece set [MIN ENG] A set of timbers consisting of a cap and a single post. { 'tü ¦pēs 'set }

two-plus-one address instruction [COMPUT SCI] An instruction in a computer program which has two addresses specifying the locations of operands and one address specifying the location in which the result is to be entered. { 'tü ‚pləs ¦wən 'ad‚res in‚strək·shən }

two-point press [MECH ENG] A mechanical press in which the slide is actuated at two points. { 'tü ¦póint 'pres }

two-point threshold [PHYSIO] The distance on the skin separating two pointed stimulators that is required to experience two rather than one point of stimulation. { 'tü ‚póint 'thresh‚hōld }

two-port junction [ELECTROMAG] A waveguide junction with two openings; it can consist either of a discontinuity or obstacle in a waveguide, or of two essentially different waveguides connected together. { 'tü ¦pórt 'jəŋk·shən }

two-port system [CONT SYS] A system which has only one input or excitation and only one response or output. { 'tü ¦pórt 'sis·təm }

two-position propeller [AERO ENG] An airplane propeller whose blades are limited to two angles, one for take off and climb and the other for cruising. { 'tü ¦pə¦zish·ən prə'pel·ər }

two-pulse canceler [ELECTR] A moving-target indicator canceler which compares the phase variation of two successive pulses received from a target; discriminates against signals with radial velocities which produce a Doppler frequency equal to a multiple of the pulse repetition frequency. { 'tü ¦pəls 'kan·slər }

two-quadrant multiplier [COMPUT SCI] Of an analog computer, a multiplier in which operation is restricted to a single sign of one input variable only. { 'tü ¦kwäd·rənt 'məl·tə‚plī·ər }

two-range Decca [NAV] A Decca radio navigation system

modified to provide circular lines of position. { 'tü ¦ranj'-dek·ə }

two's complement [MATH] A number derived from a given *n*-bit number by requiring the two numbers to sum to a value of 2^n. { 'tüz 'käm·plə·mənt }

two-sheet detector [GRAPHICS] A device which stops the printing press when more than one sheet attempts to feed into the grippers. { 'tü ¦shēt di'tek·tər }

two-sided ideal [MATH] A two-sided ideal *I* is a sub-ring of a ring *R* where the products *xy* and *yx* are always in *I* for every *x* in *R* and *y* in *I*. { 'tü ¦sīd·əd ī'dēl }

two-sided sampling plans [IND ENG] Any sampling plan whereby the acceptability of material is determined against upper and lower limits. { 'tü ¦sīd·əd 'sam·pliŋ ¦planz }

two-sided test [STAT] A test which rejects the null hypothesis when the test statistic *T* is either less than or equal to *c* or greater than or equal to *d*, where *c* and *d* are critical values. { 'tü ¦sīd·əd 'test }

two-slit experiment [QUANT MECH] A thought experiment that demonstrates the essence of the wave-particle duality, in which radiation (either light or massive particles) passes a diaphragm with two openings, and interference fringes can be observed behind the diaphragm even when the intensity of radiation is so low that the photons or massive particles can be detected one by one. { 'tü ¦slit ik'sper·ə·mənt }

two-slit interference *See* Young's two-slit interference. { 'tü ¦slit ‚in·tər'fir·əns }

two-source frequency keying [COMMUN] Keying in which the modulating wave shifts the output frequency between predetermined values derived from independent sources. { 'tü ¦sörs 'frē·kwən·sē ‚kē·iŋ }

two-sphere [MATH] The surface of a ball; the two-dimensional sphere in euclidean three-dimensional space obtained from all points whose distance from the origin is one. { 'tü ‚sfir }

two-stage design [STAT] The design of an experiment which employs a pilot study in order to decide how to design the main experiment. { 'tü ¦stāj di'zīn }

two-stage experiment [STAT] An experiment in two parts, the outcome of the first part deciding the procedure for the second. { 'tü ¦stāj ik'sper·ə·mənt }

two-stage hoisting [MIN ENG] Deep shaft hoisting with two winders, one at the surface, and the other at mid-depth in the shaft. { 'tü ¦stāj 'hòist·iŋ }

two-stage sampling [STAT] Sampling from a population whose members are themselves sets of objects and then sampling from the sets selected in the first sampling; for example, to first draw a sample of states and then to draw a sample of representatives to Congress from each state selected. { 'tü ¦stāj 'sam·pliŋ }

two-state Turing machine [COMPUT SCI] A variation of a Turing machine in which only two states are allowed, although the number of symbols may be large. { 'tü ¦stāt 'tür·iŋ mə‚shēn }

two-step grooving system [ENG] A method of spooling a drum in which the wire rope, controlled by grooves, moves parallel to the drum flanges for one-half the circumference and then crosses over to start the next wrap. Also known as counterbalance system. { 'tü ¦step 'grüv·iŋ ‚sis·təm }

two-step test [MED] Repeated ascents over two 9-inch (23-centimeter) steps as a simple exercise test of cardiovascular function. Also known as Master's two-step test. { 'tü ¦step 'test }

two-stroke cycle [MECH ENG] An internal combustion engine cycle completed in two strokes of the piston. { 'tü ¦strōk 'sī·kəl }

two-symbol Turing machine [COMPUT SCI] A variation of a Turing machine in which only two symbols are permitted, although the number of states may be large. { 'tü ¦sim·bəl 'tür·iŋ mə‚shēn }

two-tailed test [STAT] A statistical test in which the critical region consists of those values of a test statistic less than a given value as well as those values greater than another given value. Also known as two-tail test. { 'tü ‚tāld ‚test }

two-tail test *See* two-tailed test. { 'tü ‚tāl ‚test }

two-tone diaphone [ENG ACOUS] A diaphone producing blasts of two tones, the second tone being of a lower pitch than the first tone. { 'tü ¦tōn 'dī·ə‚fōn }

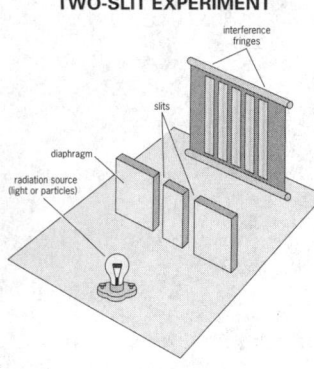

TWO-SLIT EXPERIMENT

Principle of the two-slit interference experiment, showing wave-particle duality.

two-tone keying [COMMUN] Keying in which the modulating wave causes the carrier to be modulated with one frequency for the marking condition and modulated with a different frequency for the spacing condition. { 'tü ¦tōn 'kē·iŋ }

two-tone modulation [COMMUN] In teletypewriter operation, a method of modulation in which two different carrier frequencies are employed for the two signaling conditions; the transition from one frequency to the other is abrupt, with resultant phase discontinuities. { 'tü ¦tōn ‚mäj·ə'lā·shən }

two-valued logic [MATH] A system of logic where each statement has two possible values or states, truth or falsehood. { 'tü ¦val·yüd 'läj·ik }

two-valued variable [MATH] A variable which assumes values in a set containing exactly two elements, often symbolized as 0 and 1. { 'tü ¦val·yüd 'ver·ē·ə·bəl }

two-way coupling [FL MECH] The property of a particle flow in which there is a mutual interaction between the particles and the fluid. { ‚tü ¦wā 'kəp·liŋ }

two-way series [MATH] An expression of the form $\cdots + x_{-2} + x_{-1} + x_0 + x_1 + x_2 + \cdots$, where the x_i are real or complex numbers. { ‚tü ¦wā 'sir·ēz }

two-way slab [CIV ENG] A concrete slab supported by beams along all four edges and reinforced with steel bars arranged perpendicularly. { 'tü ¦wā 'slab }

two-way time transfer [HOROL] A method of determining the time difference between atomic clocks at different locations in which each station transmits time signals to the other via a geostationary satellite. { 'tü ¦wā 'tīm ‚tranz·fər }

two-way valve [MECH ENG] A mechanical device that controls the flow of fluid by allowing flow in either of two directions. { 'tü ¦wā 'valv }

two-wire circuit [ELEC] A metallic circuit formed by two conductors insulated from each other; in contrast with a four-wire circuit, it uses only one line or channel for transmission of electric waves in both directions. { 'tü ¦wīr 'sər·kət }

two-wire repeater [ELECTR] Repeater that provides for transmission in both directions over a two-wire circuit; in carrier transmission, it usually operates on the principle of frequency separation for the two directions of transmission. { 'tü ¦wīr ri'pēd·ər }

two-year ice *See* second-year ice. { 'tü ¦yir 'īs }

TWTA [ELECTR] *See* traveling-wave amplifier.

TWX machine *See* teletypewriter. { ‚tē ‚dəb·əl·yü 'eks mə‚shēn }

TWX service *See* teletypewriter exchange service. { ‚tē ‚dəb·əl·yü 'eks ‚sər·vəs }

Twyman-Green interferometer [OPTICS] An interferometer similar to the Michelson interferometer except that it is illuminated with a point source of light instead of an extended source. { 'twī·mən 'grēn ‚in·tər·fə'räm·əd·ər }

Twystron [ELECTR] Very-high-power, hybrid microwave tube, combining the input section of a high-power klystron with the output section of a traveling wave tube, characterized by high operating efficiency and wide bandwidths. { 'twī‚strän }

tychite [MINERAL] $Na_6Mg_2(SO_4)(CO_3)_4$ A white, isometric mineral consisting of a sulfate-carbonate of sodium and magnesium. { 'tī‚kīt }

Tycho [ASTRON] A crater on the near side of the moon. { 'tī·kō }

Tychonic system [ASTRON] A theory of the planetary motion proposed by the astronomer Tycho Brahe in which the earth is stationary, with the sun and moon revolving about it but all the other planets revolving about the sun. { tī'kän·ik ‚sis·təm }

Tychonoff space [MATH] A completely regular space that is also a T_1 space. Also known as $T_{3\frac{1}{2}}$ space. { tī'kä‚nòf ‚spās }

Tychonoff theorem [MATH] A product of topological spaces is compact if and only if each individual space is compact. { tī'kä‚nòf 'thir·əm }

Tycho's Nova [ASTRON] A supernova that appeared in the constellation Cassiopeia in 1572; the star B Cassiopeiae. Also known as Tycho's star. { 'tī·kōz 'nō·və }

Tycho's star *See* Tycho's Nova. { 'tī·kōz 'stär }

tyfon *See* typhon. { 'tī‚fän }

tying bar *See* tombolo. { 'tī·iŋ ‚bär }

Tylenchida [INV ZOO] An order of soil-dwelling or phytoparasitic nematodes in the subclass Rhabdita. { tī'leŋ·kəd·ə }

Tylenchoidea [INV ZOO] A superfamily of mainly soil and

insect-associated nematodes in the order Tylenchida with a stylet for piercing live cells and sucking the juices. { tī·lən'kóid·ē·ə }

Tyler screen [CHEM ENG] A screen standard for the openings in screen-type mediums based on meshes per linear inch; convertible to the U.S. Sieve Series. { 'tī·lər ‚skrēn }

Tyler Standard screen scale [ENG] A scale for classifying particles in which the particle size in micrometers is correlated with the meshes per inch of a screen. { 'ti·lər 'stan·dərd 'skrēn ‚skāl }

Tylopoda [VERT ZOO] An infraorder of artiodactyls in the suborder Ruminantia that contains the camels and extinct related forms. { tī'läp·əd·ə }

tylose [BOT] A mass of parenchymal cells appearing somewhat frothlike in the pores of some hardwood trees. { 'tī‚lōs }

tylostyle [INV ZOO] A uniradiate spicule in Porifera with a point at one end and a knob at the other end. { 'tī·lə‚stīl }

tylote [INV ZOO] A slender sponge spicule with a knob at each end. { 'tī‚lōt }

Tymovirus [VIROL] A genus of plant viruses characterized by isometric particles containing one molecule of linear, positive-sense, single-stranded ribonucleic acid; turnip yellow mosaic virus is the type species. Also known as turnip yellow mosaic virus group. { 'tīm·ə‚vī·rəs }

tympan [GRAPHICS] One or more layers of paper placed on the impression surface of a printing press to improve presswork quality. { 'tim‚pən }

tympanic cavity [ANAT] The irregular, air-containing, mucous-membrane-lined space of the middle ear; contains the three auditory ossicles and communicates with the nasopharynx through the auditory tube. { tim'pan·ik 'kav·əd·ē }

tympanic membrane [ANAT] The membrane separating the external from the middle ear. Also known as eardrum; tympanum. { tim'pan·ik 'mem‚brān }

tympanoplasty [MED] A surgical procedure performed to eradicate disease in the middle ear cavity or to reconstruct the conductive mechanism. { ‚tim·pə·nō'plas·tē }

tympanum [ANAT] See tympanic membrane. [INV ZOO] A thin membrane covering an organ of hearing in insects. { 'tim·pə·nəm }

TYMV See turnip yellow mosaic virus.

Tyndall cone [OPTICS] The luminous path of a beam of light resulting from the Tyndall effect. { 'tind·əl ‚kōn }

Tyndall effect [OPTICS] Visible scattering of light along the path of a beam of light as it passes through a system containing discontinuities, such as the surfaces of colloidal particles in a colloidal solution. { 'tind·əl i‚fekt }

Tyndall flowers [HYD] Small water-filled cavities, often of basically hexagonal shape, which appear in the interior of ice masses upon which light is falling. { 'tind·əl ‚flaü·erz }

Tyndallization [ENG] Heat sterilization by steaming the food or medium for a few minutes at atmospheric pressure on three or four successive occasions, separated by 12- to 18-hour intervals of incubation at a temperature favorable for bacterial growth. { ‚tind·əl·ə'zā·shən }

type [GRAPHICS] The relief or plane characters used to generate printed characters of various styles and sizes. [SYST] A specimen on which a species or subspecies is based. { 'tīp }

typeahead buffer [COMPUT SCI] A temporary storage device in a keyboard or microcomputer that holds information typed on the keyboard before the central processing unit is ready to accept it. { 'tīp·ə‚hed 'bəf·ər }

type I assembly [ELECTR] An assembly consisting entirely of surface-mounted electronic components, on either one or both sides of a printed board. { 'tīp ¦wən ə'sem·blē }

type II assembly [ELECTR] An assembly of both surface-mounted and leaded electronic components, in which the surface-mounted components are on both sides of the printed board. { 'tīp ¦tü ə'sem·blē }

type III assembly [ELECTR] An assembly of both surface-mounted and leaded electronic components, in which the surface-mounted components are only on the bottom side of the printed board. { 'tīp ¦thrē ə'sem·blē }

type A wave See continuous wave. { 'tīp ¦ā ‚wāv }

type A1 wave [COMMUN] An unmodulated, keyed, continuous wave. { 'tīp ¦ā¦wən ‚wāv }

type A2 wave [COMMUN] A modulated, keyed, continuous wave. { 'tīp ¦ā¦tü ‚wāv }

type A3 wave [COMMUN] A continuous wave modulated by music, speech, or other sounds. { 'tīp ¦ā¦thrē ‚wāv }

type A4 wave [COMMUN] A superaudio frequency-modulated continuous wave, as used in facsimile systems. { 'tīp ¦ā¦fòr ‚wāv }

type A5 wave [COMMUN] A superaudio frequency-modulated continuous wave, as used in television. { 'tīp ¦ā¦fīv ‚wāv }

type A9 wave [COMMUN] A composite transmission and continuous wave that is not type A1, A2, A3, A4, or A5 wave. { 'tīp ¦ā¦nīn ‚wāv }

type bar [GRAPHICS] A long, narrow box or magazine from which projects the type used at a particular print position of a printer, and which contains the entire collection of characters available to that print position. { 'tīp ‚bär }

type-bar printer [GRAPHICS] A serial printer in which two characters are mounted on a type bar, as in some electric typewriters and early teletypewriters, and desired characters are printed one at a time. { 'tīp ‚bär 'print·ər }

type B wave [COMMUN] A keyed, damped wave. { 'tīp ¦bē ‚wāv }

type C1 carbonaceous chondrite [GEOL] A type of carbonaceous chondrite that is strongly magnetic, has a lower density than the other two types, contains sulfates, and has a carbon content of about 3.5%. { 'tīp ¦sē¦wən ‚kär·bə'nā·shəs 'kän‚drīt }

type C2 carbonaceous chondrite [GEOL] A type of carbonaceous chondrite that is weakly magnetic or nonmagnetic, has most of its sulfur present as free sulfur, and contains about 2.5% carbon. { 'tīp ¦sē¦tü ‚kär·bə'nā·shəs 'kän‚drīt }

type C3 carbonaceous chondrite [GEOL] A type of carbonaceous chondrite that has a lower percentage of water and a higher density than the other two types, and usually consists largely of olivine. { 'tīp ¦sē¦thrē ‚kär·bə'nā·shəs 'kän‚drīt }

Type II Cepheids See W Virginis stars. { 'tīp ¦tü 'sef·ē·ədz }

type drum [COMPUT SCI] A steel cylinder containing 128 to 144 lateral bands, each band containing the alphabet, the digits 0–9, and the standard set of punctuation marks such as commas and periods, and revolving at high speed; printing is achieved by a hammer facing each band and activated at the right time to cause a character to be printed on the paper flowing between hammers and drum. { 'tīp ‚drəm }

type I error [STAT] One of two types of errors in testing hypotheses: incorrectly rejecting the hypothesis tested when it is true. Also known as error of the first kind. { ‚tīp 'wən ‚er·ər }

type II error [STAT] One of two types of error in testing hypotheses: incorrectly accepting the hypothesis tested when an alternate hypothesis is true. Also known as error of the second kind. { ‚tīp 'tü ‚er·ər }

typeface See face. { 'tīp‚fās }

type gage [GRAPHICS] A ruler, calibrated in picas, used by a printer for making measurements during the composition process. Also known as line gage. { 'tīp ‚gāj }

type-α leader [GEOPHYS] A stepped leader of lightning which exhibits very little branching and whose individual steps are short and so weakly luminous as to be difficult to discern. { 'tīp 'al·fə ‚lēd·ər }

type-β leader [GEOPHYS] A stepped leader of lightning in which the upper portion of the channel is characterized by longer and brighter steps than those found in the lower portion of the channel, a consequence of excessive branching in the upper parts under the influence of strong fields set up by heavy space charges near and around the upper end of the channel. { 'tīp 'bād·ə ‚lēd·ər }

type locality [GEOL] **1.** The place at which a stratigraphic unit is typically displayed and from which it derives its name. **2.** The place where a geologic feature was first recognized and described. { 'tīp lō‚kal·əd·ē }

type-M carcinotron See M-type backward-wave oscillator. { 'tīp ¦em 'kärs·ən·ə‚trän }

type metal [MET] Any of various low-melting-point alloys, composed mainly of lead (50–90%), antimony (2–30%), and tin (2–20%), used for casting printers' type. { 'tīp ‚med·əl }

type-O carcinotron See O-type backward-wave oscillator. { 'tīp ¦ō 'kärs·ən·ə‚trän }

type I of Cori [MED] See von Gierke's disease. { 'tīp 'wən əv 'kòr·ē }

type section [GEOL] That sequence of strata identified as

TYNDALL CONE

The luminous light path known as the Tyndall cone. (*H. Steeves and R. G. Babcock*)

the original sequence for a location or area; the standard against which other stratigraphy of parts of the area are compared. Also known as section. { 'tīp ,sek·shən }

type size [GRAPHICS] For any font, the distance from the top of the tallest character to the bottom of the lowest character. { 'tīp ,sīz }

type I superconductor [CRYO] A superconductor for which there is a single critical magnetic field; magnetic flux is completely excluded from the interior of the material at field strengths below this critical field, while at field strengths above the critical field, magnetic flux penetrates the superconductor completely and it reverts to the normal state. { 'tīp ¦wən ¦sü·pər·kən'dək·tər }

type II superconductor [CRYO] A superconductor for which there are two critical magnetic fields; magnetic flux is completely excluded from the interior of the material only at field strengths below the smaller critical field, and at field strengths between the two critical fields the magnetic flux consists of flux vortices in the form of filaments embedded in the superconducting material. Also known as high-field superconductor (HFS). { 'tīp ¦tü ¦sü·pər·kən'dək·tər }

type I supernova [ASTRON] A member of a class of supernovae that lack hydrogen in their spectra and have relatively regular light curves. { 'tīp ¦wən ¦sü·pər'nō·və }

type Ia supernova [ASTRON] A member of a subclass of the type I supernovae that brighten relatively smoothly to a maximum whose brightness is relatively uniform among members of the subclass about 2 weeks after the explosion; decline in brightness quasi-exponentially thereafter; show strong lines of silicon, sulfur, calcium, and iron in their peak spectra; have late-time spectra dominated by strong emission lines of iron and cobalt; have line widths that imply velocities from 4000 to 12,000 miles (7000 to 20,000 kilometers) per second, with the highest velocities seen early on; and are believed to be exploding white dwarf stars. { ¦tīp ,wən¦ā ¦sü·pər'nō·və }

type Ib supernova [ASTRON] A member of a subclass of the type I supernovae that lack a strong line feature at 615.0 nanometers due to ionized silicon; are approximately a factor of 4 fainter than type Ia and have light curves that are a little broader and slower to decline; have, in addition to iron emission, broad lines of oxygen and calcium in their late-time spectra; and are believed to be explosions of massive stars that are devoid of hydrogen. { ¦tīp ,wən¦bē ¦sü·pər¦nō·və }

type Ic supernova [ASTRON] A member of a subclass of the type I supernovae that resemble type Ib supernovae closely but have a weak or absent line of neutral helium at 587.6 nanometers. { ¦tīp ,wən¦sē ¦sü·pər¦nō·və }

type II supernova [ASTRON] A member of a class of supernovae that display prominent lines of hydrogen in their spectra and have irregular light curves; they are believed be explosions of young massive stars, still in possession of their hydrogenic surface layers. { 'tīp ¦tü ¦sü·pər'nō·və }

type II-L supernova [ASTRON] A member of a subclass of the type II supernovae that begin to fade almost immediately in a quasi-linear way. { ¦tīp ,tü 'el ¦sü·pər¦nō·və }

type II-P supernova [ASTRON] A member of a subclass of the type II supernovae that display enduring emission at a nearly constant rate for several months. { ¦tīp ,tü 'pē ¦sü·pər¦nō·və }

typewriter [GRAPHICS] A machine that produces printed copy, character by character, as the typewriter is operated; essential parts are an input keyboard, a set of raised characters, inking means, a platen, and a mechanism for advancing the position at which successive characters are imprinted. { 'tīp,rīd·ər }

typewriter terminal [COMPUT SCI] An electric typewriter combined with an ASCII or other code generator that provides code output for feeding a computer, calculator, or other digital equipment; the terminal also produces hard copy when driven by incoming code signals. { 'tīp,rīd·ər ,tər·mən·əl }

Typhaceae [BOT] A family of monocotyledonous plants in the order Typhales characterized by an inflorescence of dense, cylindrical spikes and absence of a perianth. { tī'fās·ē,ē }

Typhales [BOT] An order of marsh or aquatic monocotyledons in the subclass Commelinidae with emergent or floating stems and leaves and reduced, unisexual flowers having a single ovule in an ovary composed of a single carpel. { tī'fā·lēz }

Typhlopidae [VERT ZOO] A family of small, burrowing circumtropical snakes, suborder Serpentes, with vestigial eyes and toothless jaws. { ti'fläp·ə,dē }

Typhloscolecidae [INV ZOO] A family of pelagic polychaete annelids belonging to the Errantia. { ,tif·lō·skō'les·ə,dē }

typhlosole [INV ZOO] A dorsal longitudinal invagination of the intestinal wall in certain invertebrates serving to increase the absorptive surface. { 'tif·lə,sōl }

typhoid fever [MED] A highly infectious, septicemic disease of humans caused by *Salmonella typhi* which enters the body by the oral route through ingestion of food or water contaminated by contact with fecal matter. { 'tī,fóid 'fē·vər }

typhoid vaccine [IMMUNOL] A type of killed vaccine used for active immunity production; made from killed typhoid bacillus (*Salmonella typhi*). { 'tī,fóid vak'sēn }

typhon [ENG ACOUS] A diaphragm horn which operates under the influence of compressed air or steam. Also spelled tyfon. { 'tī,fän }

typhoon [METEOROL] A severe tropical cyclone in the western Pacific. { tī'fün }

typhoon wind See hurricane wind. { tī'fün ,wind }

typhus fever [MED] Any of three louse-borne human diseases caused by *Rickettsia prowazakii* characterized by fever, stupor, headaches, and a dark-red rash. { 'tī·fəs 'fē·vər }

typography [GRAPHICS] The techniques involved in letterpress printing, including style, arrangements, and appearance of the printed matter. { tī'päg·rə·fē }

typography point [GEOGR] See point. { tī'päg·rə·fē ,póint }

Typotheria [PALEON] A suborder of extinct rodentlike herbivores in the order Notoungulata. { ,tī·pə'thir·ē·ə }

typp [TEXT] A unit of the reciprocal of line density used in the textile industry, equal to the reciprocal of the line density of a thread whose length is 1000 yards and whose length is 1 pound; equal to approximately 2015.91 meters per kilogram. { tip }

tyramine [PHARM] $HOC_6H_4CH_2CH_2NH_2$ A crystalline compound with a melting point of 164–165°C; soluble in water and boiling alcohol; used in medicine as an adrenergic drug. Also known as tyrosamine. { 'tī·rə,mēn }

Tyranni [VERT ZOO] A suborder of suboscine Passeriformes containing birds with limited song power and having the tendon of the hind toe separate and the intrinsic muscles of the syrinx reduced to one pair. { tə'ra,nī }

Tyrannidae [VERT ZOO] The tyrant flycatchers, a family of passeriform birds in the suborder Tyranni confined to the Americas. { tə'ran·ə,dē }

Tyrannoidea [VERT ZOO] The flycatchers, a superfamily of suboscine birds in the suborder Tyranni. { tir·ə'nóid·ē·ə }

Tyrannosaur [PALEON] A large carnivorous therapod dinosaur, 40 feet (12 meters) long and weighing 6 tons, from the Late Cretaceous Period that had powerful hindlimbs, short forelimbs, a large skull (4 feet long), and very powerful jaws. { tə'ran·ə,sór }

tyrocidine [MICROBIO] A peptide antibiotic produced by *Bacillus brevis*; used to control fungi, bacteria, and protozoa. { ,tir·ə'sīd·ən }

tyrolite See trichalcite. { 'tir·ə,līt }

tyrosamine See tyramine. { tə'räs·ə,mēn }

tyrosinase [BIOCHEM] An enzyme found in plants, molds, crustaceans, mollusks, and some bacteria which, in the presence of oxygen, catalyzes the oxidation of monophenols and polyphenols with the introduction of −OH groups and the formation of quinones. { 'tir·ə·sə,nās }

tyrosine [BIOCHEM] $C_9H_{11}NO_3$ A phenolic alpha amino acid found in many proteins; a precursor of the hormones epinephrine, norepinephrine, thyroxine, and triiodothyronine, and of the black pigment melanin. { 'tir·ə,sēn }

tyrosine hydroxylase [BIOCHEM] A specialized enzyme located only in catecholamine-containing nerve cells, where it serves as the primary regulatory or rate-limiting step in catecholamine biosynthesis. { ,tī·rə,sēn hī'dräk·sə,lās }

tyrosinemia [MED] An inborn metabolic disorder in which there is a deficiency of the enzyme p-hydroxyphenylpyruvic acid oxidase with abnormally high blood levels of tyrosine and sometimes methionine. { ,tir·ə'sē·mē·ə }

tyrosinosis [MED] Excretion of excessive amounts of tyrosine and its first oxidation products in the urine. { ,tir·ə·sə'nō·səs }

tyrothricin [MICROBIO] A polypeptide mixture produced by

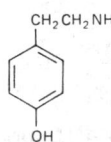

TYRAMINE

CH₂CH₂NH₂

OH

Structural formula.

TYROSINE

OH

CH₂

C

H₂N   H   COOH

Structural formula.

Bacillus brevis and consisting of the antibiotic substances gramicidin and tyrocidine; effective as an antibacterial applied locally in infections due to germ-positive organisms. { ‚tir·ə'thrīs·ən }

Tyson's gland [ANAT] A small scent gland in the human male which secretes the smegma. Also known as preputial gland. { 'tī·sənz ‚gland }

Tytonidae [VERT ZOO] The barn owls, a family of birds in the order Strigiformes distinguished by an unnotched sternum which is fused to large clavicles. { tī'tän·ə‚dē }

tyuyamunite [MINERAL] $Ca(UO_2)_2(VO_4)_2 \cdot 5\text{-}8H_2O$ A yellow orthorhombic mineral occurring in incrustations as a secondary mineral; an ore of uranium. Also known as calciocarnotite. { ‚yü·ə'mü‚nīt }

tyvelose [BIOCHEM] A dideoxy sugar found in bacterial lipopolysaccharides. { 'tī·və‚lōs }

U

u *See* up quark.

U *See* uranium.

UART *See* universal asynchronous receiver transmitter. { 'yü,ärt }

ubac [METEOROL] The shady (usually north) side of an Alpine mountain, characterized by a lower timberline and snow line than the sunny side. { ü,bäk }

u-band [OPTICS] The absorption band in the ultraviolet resulting from a U-center type of point lattice defect. { 'yü ,band }

U-bend die [MECH ENG] A die with a square or rectangular cross section which provides two edges over which metal can be drawn. { 'yü ,bend ,dī }

ubiquitin [BIOCHEM] A small, 76-amino-acid, highly conserved protein present in the cytoplasm and nucleus of all eukaryotes (but not eubacteria and archaea). The covalent, ATP-dependent linkage of multiple ubiquitin molecules to proteins serves as a signal for their degradation by the 26S proteasome. { yü'bik·wə,tin }

U blades [DES ENG] Curved bulldozer blades designed to increase moving capacity of tractor equipment. { 'yü ,blādz }

Übler effect [FL MECH] An effect in which secondary normal stresses cause bubbles or suspended bodies to lag behind the fluid in the accelerated flow of a viscoelastic fluid through a narrowing tube. { ē·blər i,fekt }

U bolt [DES ENG] A U-shaped bolt with threads at the ends of both arms to receive nuts. { 'yü ,bōlt }

UBV photometry [ASTRON] A system of three-color photometry used to obtain specific stellar magnitudes; the system is based on the comparison of stars' magnitudes with a standard sequence of about 400 stars. { 'yü,bē¦vē fə'täm·ə·trē }

UBVRI system [ASTRON] An extension of the UBV system through measurements of an object's apparent magnitude with red (R) and near-infrared (I) filters. { ¦yü¦bē¦vē¦är'ī ,sis·təm }

UBV system [ASTRON] A system of stellar magnitudes in which an object's apparent magnitude is measured at three wavelengths, labeled U, at 360 nanometers; B, at 420 nanometers; and V, at 540 nanometers; and is characterized by the color indices $B-V$ and $U-B$, which are both defined to be 0 for a star of spectral type A0. Also known as Johnson-Morgan system. { 'yü,bē¦vē 'sis·təm }

U center [CRYSTAL] The color-center type of point lattice defect in ionic crystals created by the incorporation of an impurity such as hydrogen into alkali halides. { 'yü ,sen·tər }

U Cephei [ASTRON] A binary star; in this double-star eclipsing system, one component has reached its Roche limit (a dynamical barrier beyond which the size of neither star can expand) while the other is distinctly smaller than this limit. { 'yü 'sef·ē,ī }

U coefficient [QUANT MECH] A coefficient that appears in the transformation between modes of coupling eigenfunctions of three angular momenta; it is equal to the product of the Racah coefficient and $[(2j_{12} + 1)(2j_{23} + 1)]^{1/2}$, where j_{12} and j_{23} are the intermediate angular momenta in the respective modes. { 'yü ,kō·i,fis'·ənt }

Uda antenna *See* Yagi-Uda antenna. { 'ü·də an,ten·ə }

Udalf [GEOL] A suborder of the soil order Alfisol; brown so formed in a udic moisture regime and in a mesic or warmer temperature regime. { 'ü,dälf }

Ide [VERT ZOO] A pendulous organ consisting of several an ary glands enclosed in a single envelope; each gland has its own nipple; found in some mammals, such as the cow and goat. { 'əd·ər }

Udert [GEOL] A suborder of the soil order Vertisol; formed in a humid region so that surface cracks remain open only for 2–3 months. { 'üd,ərt }

UDMH *See* uns-dimethylhydrazine.

Udoll [GEOL] A suborder of the Mollisol soil order; found in humid, temperate, and warm regions where maximum rainfall comes during growing season; has thick, very dark A horizons, brown B horizons, and paler C horizons. { 'üd,ól }

udometer *See* rain gage. { yü'däm·əd·ər }

UDP *See* uridine diphosphate; User Datagram Protocol.

UDPG *See* uridine diphosphoglucose.

Udult [GEOL] A suborder of the soil order Ultisol; organic-carbon content is low, argillic horizons are reddish or yellowish; formed in a udic moisture regime. { 'üd,əlt }

Uehling effect [ATOM PHYS] The departure of the electrostatic potential of a point charge from a Coulomb potential, due to vacuum polarization. { 'ēl·iŋ i,fekt }

U figure *See* U index. { 'yü ,fig·yər }

UFO *See* unidentified flying object.

U format [COMPUT SCI] A record format which the input/output control system treats as completely unknown and unpredictable. { 'yü ,fòr,mat }

U Geminorum stars [ASTRON] A class of variable stars known as dwarf novae; their light curves resemble those of novae, with range brightness variations of about 4 magnitudes; examples are U Gemini and SS Cygni. Also known as SS Cygni stars. { 'yü ,jem·ə'nòr·əm ,stärz }

Ugine Sejournet process *See* Sejournet process. { ü,zhēn se·zhər'nā ,prä·səs }

UHF *See* ultrahigh frequency.

Uhlbricht sphere [OPTICS] A sphere whose inside surface has a diffusely reflecting white finish, used in an integrating sphere photometer. { úl,brikt ,sfir }

uhligite [MINERAL] A black, pseudoisometric mineral consisting of an oxide of titanium and calcium, with zirconium and aluminum replacing titanium. { 'ü·lə,gīt }

U index [GEOPHYS] The difference between consecutive daily mean values of the horizontal component of the geomagnetic field. Also known as U figure. { 'yü ,in,deks }

Uintatheriidae [PALEON] The single family of the extinct mammalian order Dinocerata. { yù,win·tə·thə'rī·ə,dē }

Uintatheriinae [PALEON] A subfamily of extinct herbivores in the family Uintatheriidae including all horned forms. { yù,win·tə·thə'rī·ə,nē }

UJT *See* unijunction transistor.

Ulatisian [GEOL] A mammalian age in a local stage classification of the Eocene in use on the Pacific Coast based on foraminifers. { ,yü·lə'tē·zhən }

ULCC *See* ultralarge crude carrier.

ulcer [MED] Localized interruption of the continuity of an epithelial surface, with an inflamed base. { 'əl·sər }

ulcerative colitis [MED] An idiopathic inflammatory disease of the mucosa and submucosa of the colon manifested clinically by pain, diarrhea, and rectal bleeding. { 'əl·sə,rād·iv kə'līd·əs }

ulcerative endocarditis [MED] Acute bacterial endocarditis. { 'əl·sə,rād·iv ¦en·dō 'kär'dīd·əs }

ulcerative gingivitis *See* Vincent's infection. { 'əl·sə,rād·iv ,jin·jə'vīd·əs }

ulexite [MINERAL] $NaCaB_5O_9 \cdot 8H_2O$ A white mineral that

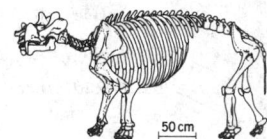

UINTATHERIINAE

Skeleton of *Uintatherium*.

50 cm

U
V

crystallizes in the triclinic system and forms rounded reniform masses of extremely fine acicular crystals. Also known as cotton ball. { 'ü·lek,sīt }

ULIRG *See* ultraluminous infrared galaxy.

ullage [ENG] The amount that a container, such as a fuel tank, lacks of being full. [NAV ARCH] The distance from the surface of the oil in a cargo tank of an oil tanker to the top of the hatch, or to the top of the inspection cover in the hatch. { 'əl·ij }

ullage rocket [AERO ENG] A small rocket used in space to impart an acceleration to a tank system to ensure that the liquid propellants collect in the tank in such a manner as to flow properly into the pumps or thrust chamber. { 'əl·ij ,räk·ət }

ullmannite [MINERAL] NiSbS A steel-gray to black mineral consisting of nickel antimonide and sulfide, usually with a little arsenic, occurring massive, and having a metallic luster. Also known as nickel-antimony glance. { 'əl·mə,nīt }

Ullmann reaction [ORG CHEM] A variation of the Fittig synthesis, using copper powder instead of sodium. { 'əl·mən rē,ak·shən }

Ulloa's ring *See* Bouguer's halo. { ü'yō·əz ,riŋ }

Ullrich-Turner syndrome [MED] A complex of symptoms including webbing of the neck, short stature, cubitus valgus, and hypogonadism in the male. Also known as male Turner's syndrome. { 'əl·rik 'tər·nər ,sin,drōm }

Ulmaceae [BOT] A family of dicotyledonous trees in the order Urticales distinguished by alternate stipulate leaves, two styles, a pendulous ovule, and lack of a latex system. { əl'mās·ē,ē }

ulmic acid *See* ulmin. { 'əl·mik 'as·əd }

ulmin [GEOL] Alkali-soluble organic substances derived from decaying vegetable matter; occurs as amorphous brown to black gel material. Also known as carbohumin; fundamental jelly; fundamental substance; gelose; humin; humogelite; jelly; ulmic acid; vegetable jelly. { 'əl·mən }

ulna [ANAT] The larger of the two bones of the forearm or forelimb in vertebrates; articulates proximally with the humerus and radius and distally with the radius. { 'əl·nə }

ulnar deviation [BIOPHYS] A position of the human hand in which the wrist is bent toward the end finger. { ¦əl·nər ,dē·vē'ā·shən }

ulnar loop [FOREN SCI] A loop fingerprint pattern which flows in the direction of the ulna bone, toward the little finger. { ,əl·nər 'lüp }

Ulotrichaceae [BOT] A family of green algae in the suborder Ulotrichineae; contains both attached and floating filamentous species with cells having parietal, platelike or bandlike chloroplasts. { yü,lä·trə'kās·ē,ē }

Ulotrichales [BOT] A large, artificial order of the Chlorophyta composed mostly of fresh-water, branched or unbranched filamentous species with mostly cylindrical, uninucleate cells having cellulose, but often mucilaginous walls. { yü,lä·trə'kā·lēz }

Ulotrichineae [BOT] A suborder of the Ulotrichales characterized by short cylindrical cells. { yü,lä·trə'kīn·ē,ē }

ulrichite *See* uraninite. { 'əl·rə,kīt }

ULSI circuit *See* ultralarge scale integrated circuit. { ¦yü¦el¦es¦ī 'sər·kət }

ultimate analysis [ANALY CHEM] The determination of the percentage of elements contained in a chemical substance. { 'əl·tə·mət ə'nal·ə·səs }

ultimate bearing capacity [CIV ENG] The average load per unit area that will cause failure by rupture of a supporting soil mass. { 'əl·tə·mət 'ber·iŋ kə,pas·əd·ē }

ultimate elongation [MET] The percentage of permanent deformation remaining after tensile rupture. { 'əl·tə·mət ,ē·lón'gā·shən }

ultimate lines [ASTRON] Special spectral lines that can be used to indicate the existence of an element in the sun or other star. { 'əl·tə·mət 'līnz }

ultimate load *See* breaking load. { 'əl·tə·mət ,lōd }

ultimate-load design [DES ENG] Design of a beam that is proportioned to carry at ultimate capacity the design load multiplied by a safety factor. Also known as limit-load design; plastic design; ultimate-strength design. { 'əl·tə·mət 'lōd di,zīn }

ultimate recovery [PETRO ENG] Estimated total (ultimate) recovery of hydrocarbon fluids expected from a reservoir during its productive lifetime. { 'əl·tə·mət ri'kəv·ə·rē }

ultimate set [ENG] The ratio of the length of a specimen plate or bar before testing to the length at the moment of fracture; usually expressed as a percentage. [MATER] The final degree of firmness achieved by a plastic compound as a result of curing, evaporation of the volatile materials, and polymerization at the surface. { 'əl·tə·mət 'set }

ultimate strength [MECH] The tensile stress, per unit of the original surface area, at which a body will fracture, or continue to deform under a decreasing load. { 'əl·tə·mət 'streŋkth }

ultimate-strength design *See* ultimate-load design. { ¦əl·tə·mət ¦streŋkth di,zīn }

ultimobranchial body [EMBRYO] One of the small, endocrine structures which originate as terminal outpocketings from each side of the embryonic vertebrate pharynx; can produce the hormone calcitonin. { ¦əl·tə·mō'braŋ·kē·əl 'bäd·ē }

Ultisol [GEOL] A soil order characterized by typically moist soils, with horizons of clay accumulation and a low supply of bases. { 'əl·tə,sól }

ultra-audion circuit [ELECTR] Regenerative detector circuit in which a parallel resonant circuit is connected between the grid and the plate of a vacuum tube, and a variable capacitor is connected between the plate and cathode to control the amount of regeneration. { 'əl·trə 'ód·ē·,än ,sər·kət }

ultra-audion oscillator [ELECTR] Variation of the Colpitts oscillator circuit; the resonant circuit employs a section of transmission line. { 'əl·trə 'ód·ē·,än ,äs·ə,lād·ər }

ultrabasic [PETR] Of igneous rock, having a low silica content, as opposed to the higher silica contents of acidic, basic, and intermediate rocks. { ¦əl·trə'bā·sik }

ultrabasite *See* diaphorite. { ¦əl·trə'bā,sīt }

ultracentrifuge [ENG] A laboratory instrument which develops centrifugal fields of more than 100,000 times gravity, used for the quantitative measurement of sedimentation velocity or sedimentation equilibrium, or for the separation of solutes in liquid solutions to study high polymers, particularly proteins, nucleic acids, viruses, and other macromolecules of biological origin. { ¦əl·trə'sen·trə,fyüj }

ultracold atom [PHYS] One of a collection of atoms cooled to a thermodynamic temperature below 1 millikelvin. { ¦əl·trə,kōld 'ad·əm }

ultracold molecules [PHYS] A collection of molecules cooled to a thermodynamic temperature below 1 millikelvin. { ¦əl·trə,kōld 'mäl·ə,kyülz }

ultracold neutron [PHYS] A neutron whose energy is of the order of 10^{-7} electronvolt or less, so that it is totally reflected from various materials and suitably constructed magnetic fields, regardless of the angle of incidence, and can be stored in suitably constructed bottles. { ¦əl·trə¦kōld 'nü,trän }

ultrafiche [GRAPHICS] A sheet of film, usually 4 by 6 inches (10 by 15 centimeters) in size, containing either negative or positive images, or frames, of printed material reduced more than 100 times by photographic reduction. { 'əl·trə,fēsh }

ultrafilter [MATH] A filter base which has no properly subordinated filter base. { ¦əl·trə'fil·tər }

ultrafiltration [CHEM ENG] Separation of colloidal or very fine solid materials by filtration through microporous or semipermeable mediums. { ¦əl·trə'fil'trā·shən }

ultragravity waves [FL MECH] Gravity waves which are characterized by periods in the 0.1–1.0 second range. { ¦əl·trə'grav·əd·ē ,wāvz }

ultrahigh frequency [COMMUN] The band of frequencies between 300 and 3000 megahertz in the radio spectrum, corresponding to wavelengths of 10 centimeters to 1 meter. Abbreviated UHF. { ¦əl·trə'hī 'frē·kwən·sē }

ultrahigh-frequency tuner [ELECTR] A tuner in a television receiver for reception of stations transmitting in the ultrahigh-frequency band (channels 14–83); it usually employs continuous tuning. { ¦əl·trə'hī 'frē·kwən·sē 'tü·nər }

ultrahigh vacuum [PHYS] A vacuum in which the pressure is of the order of 10^{-10} millimeter of mercury or less. { ¦əl·trə'hī 'vak·yəm }

ultralarge crude carrier [NAV ARCH] A liquid-cargo vessel over 250,000 tons. Abbreviated ULCC. { ¦əl·trə'lärj 'krüd ,kar·ē·ər }

ultralarge-scale integrated circuit [ELECTR] A complex integrated circuit that contains more than 1,000,000 elements. Abbreviated ULSI circuit. { ¦əl·trə'lärj 'skāl 'int·ə,grād·əd 'sər·kət }

ultralight aircraft [AERO ENG] An extremely lightweight,

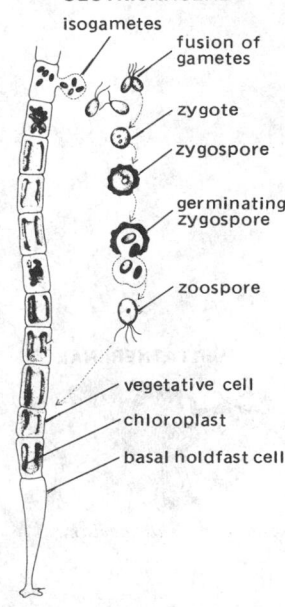

ULOTRICHACEAE

isogametes
fusion of gametes
zygote
zygospore
germinating zygospore
zoospore

vegetative cell
chloroplast
basal holdfast cell

Ulothrix, an attached form of Ulotrichaceae, showing cylindrical cells and other features.

single-seat aircraft with low flight speed, power, and fuel capacity, used for sport or recreation. { 'əl·trə‚līt 'er‚kraft }

ultralow-velocity zone [GEOPHYS] Thin, mushy layer detected in some places along the earth's core-mantle boundary where seismic waves slow down. { ‚əl·trə·lō və'läs·əd ē ‚zōn }

ultraluminous infrared galaxy [ASTRON] One of a class of galaxies that emit intense radiation at infrared wavelengths, 100 times or more as much as the Milky Way Galaxy. Abbreviated ULIRG. { ¦əl·trə¦lü·mə·nəs ‚in·frə‚rəd 'gal·ik·sē }

ultramafic [PETR] Referring to igneous rock composed principally of mafic minerals, such as olivine and pyroxene. { ¦əl·trə'maf·ik }

ultramarine blue [INORG CHEM] A blue pigment; a powder with heat resistance, used for enamels on toys and machinery, white baking enamels, printing inks, and cosmetics, and in textile printing. { ¦əl'trə·mə'rēn 'blü }

ultramicrobalance [ENG] A differential weighing device with accuracies better than 1 microgram; used for analytical weighings in microanalysis. { ¦əl·trə'mī·krō‚bal·əns }

ultramicroscope [OPTICS] An instrument for investigating particles of submicroscopic dimensions: it consists of a high-intensity illumination system for producing a Tyndall cone in a colloidal system, coupled with a compound microscope to examine the points of light scattered from the individual particles. { ¦əl·trə'mī·krə‚skōp }

ultramicrotome [ENG] A microtome which uses a glass or diamond knife, allowing sections of cells to be cut 300 nanometers in thickness. { ¦əl·trə'mī·krə‚tōm }

ultraperformance composite [MATER] A material that is made from a combination of polymers and can perform under extreme atmospheric conditions for extended periods. { ‚əl·trə·pər¦fȯr·məns kəm'päz·ət }

ultraphotic rays [ELECTROMAG] Rays outside the visible part of the spectrum, including infrared and ultraviolet rays. { ¦əl·trə¦fäd·ik 'rāz }

ultrarelativistic [RELAT] Having a speed that is nearly equal to the speed of light. { ¦əl·trə‚rel·ə·tə'vis·tik }

ultrasensitive mass spectrometry [ANALY CHEM] A form of mass spectrometry in which the ions to be detected are accelerated to megaelectronvolt energies in a particle accelerator and passed through a thin gas cell or foil, stripping away outer electrons, so that contaminating molecules dissociate into lower-mass fragments, and isobars can be distinguished by particle detectors that measure ionization rate and total energy. { ¦əl·trə'sen·səd·iv 'mas spek'träm·ə·trē }

ultrashort waves [COMMUN] Radio waves shorter than 10 meters in wavelength; corresponding to frequencies above 30 megahertz. { ¦əl·trə'shȯrt 'wāvz }

ultra-small-aperture terminal [COMMUN] An antenna less than 20 inches (0.5 meter) in diameter that is used for reception of direct broadcasts from geosynchronous satellites. Abbreviated USAT. { ¦əl·trə ¦smȯl ¦ap·ə·chər 'tərm·ə·nəl }

ultrasonic [ACOUS] Pertaining to signals, equipment, or phenomena involving frequencies just above the range of human hearing, hence above about 20,000 hertz. Also known as supersonic (deprecated usage). { ¦əl·trə'sän·ik }

ultrasonic atomizer [MECH ENG] An atomizer in which liquid is fed to, or caused to flow over, a surface which vibrates at an ultrasonic frequency; uniform drops may be produced at low feed rates. { ¦əl·trə'sän·ik 'ad·ə‚mīz·ər }

ultrasonic bonding [MET] Bonding of two identical or dissimilar metals by mechanical pressure combined with a wiping motion produced by ultrasonic vibration. { ¦əl·trə'sän·ik 'bänd·iŋ }

ultrasonic camera [ELECTR] A device which produces a picture display of ultrasonic waves sent through a sample to be inspected or through live tissue; a piezoelectric crystal is used to convert the ultrasonic waves to voltage differences, and the voltage pattern on the crystal modulates the intensity of an electronic beam scanning the crystal; this beam in turn controls the intensity of a beam in a television tube. { ¦əl·trə'sän·ik 'kam·rə }

ultrasonic cleaning [ENG] A method used to clean debris and swarf from surfaces by immersion in a solvent in which ultrasonic vibrations are excited. { ¦əl·trə'sän·ik 'klēn·iŋ }

ultrasonic coagulation [PHYS] The bonding of small particles into large aggregates by the action of ultrasonic waves. { ¦əl·trə'sän·ik kō‚ag·yə'lā·shən }

ultrasonic communication [COMMUN] Communication accomplished through water by keying the sound output of echo-ranging sonar on ships or submarines or by using other such devices. { ¦əl·trə'sän·ik kə‚myü·nə'kā·shən }

ultrasonic delay line [ENG ACOUS] A delay line in which use is made of the propagation time of sound through a medium such as fused quartz, barium titanate, or mercury to obtain a time delay of a signal. Also known as ultrasonic storage cell. { ¦əl·trə'sän·ik di'lā ‚līn }

ultrasonic depth finder [ENG] A direct-reading instrument which employs frequencies above the audible range to determine the depth of water; it measures the time interval between the emission of an ultrasonic signal and the return of its echo from the bottom. { ¦əl·trə'sän·ik 'depth ‚fīn·dər }

ultrasonic drill [MECH ENG] A drill in which a magnetostrictive transducer is attached to a tapered cone serving as a velocity transformer; with an appropriate tool at the end of the transformer, practically any shape of hole can be drilled in hard, brittle materials such as tungsten carbide and gems. { ¦əl·trə'sän·ik 'dril }

ultrasonic drilling [MECH ENG] A vibration drilling method in which ultrasonic vibrations are generated by the compression and extension of a core of electrostrictive or magnetostrictive material in a rapidly alternating electric or magnetic field. { ¦əl·trə'sän·ik 'dril·iŋ }

ultrasonic echo [ACOUS] An ultrasonic wave that has been reflected or otherwise returned with sufficient magnitude and time delay to be perceived in some manner as a wave distinct from that directly transmitted. { ¦əl·trə'sän·ik 'ek·ō }

ultrasonic flaw detector [ENG ACOUS] An ultrasonic generator and detector used together, much as in radar, to determine the distance to a wave-reflecting internal crack or other flaw in a solid object. { ¦əl·trə'sän·ik 'flȯ di‚tek·tər }

ultrasonic generator [ENG ACOUS] A generator consisting of an oscillator driving an electroacoustic transducer, used to produce acoustic waves above about 20 kilohertz. { ¦əl·trə'sän·ik 'jen·ə‚rād·ər }

ultrasonic imaging See acoustic imaging. { ¦əl·trə'sän·ik 'im·ij·iŋ }

ultrasonic imaging device [ENG ACOUS] An imaging device in which a wave is generated by a transducer external to the body; the reflected wave is detected by the same transducer. { ¦əl·trə'sän·ik 'im·ij·iŋ di‚vīs }

ultrasonic inspectoscope [ENG ACOUS] An instrument that transmits sound waves, at frequencies between 500 kilohertz and 15 megahertz, into a metal casting or other solid piece and determines the presence of flaws by reflections or by an interruption of the sound-wave transmission through the piece. { ¦əl·trə'sän·ik in'spek·tə‚skōp }

ultrasonic leak detector [ENG] An instrument which detects ultrasonic energy resulting from the transition from laminar to turbulent flow of a gas passing through an orifice. { ¦əl·trə'sän·ik 'lēk di‚tek·tər }

ultrasonic light modulator [OPTICS] Device containing a fluid which, by action of ultrasonic waves passing through the fluid, modulates a beam of light passed transversely through the fluid. { ¦əl·trə'sän·ik 'līt ‚mäj·ə‚lād·ər }

ultrasonic machining [MECH ENG] The removal of material by abrasive bombardment and crushing in which a flat-ended tool of soft alloy steel is made to vibrate at a frequency of about 20,000 hertz and an amplitude of 0.001–0.003 inch (0.0254–0.0762 millimeter) while a fine abrasive of silicon carbide, aluminum oxide, or boron carbide is carried by a liquid between tool and work. { ¦əl·trə'sän·ik mə'shēn·iŋ }

ultrasonic medical tomography [ACOUS] A form of acoustic tomography in which ultrasonic pulses are emitted by an acoustoelectric transducer and echoes are received from acoustic impedance discontinuities along the assumed line-of-sight propagation path. { ¦əl·trə'sän·ik 'med·ə·kəl tō'mäg·rə·fē }

ultrasonic metal inspection [MET] The application of ultrasonic vibrations to materials for detection of flaws, pits, or wall thickness. { ¦əl·trə'sän·ik 'med·əl in‚spek·shən }

ultrasonic microscope [OPTICS] A special type of microscope which employs ultrasonic radiation. { ¦əl·trə'sän·ik 'mī·krə‚skōp }

ultrasonic radiation [ACOUS] Ultrasonic waves propagating through a solid, liquid, or gaseous medium. { ¦əl·trə'sän·ik ‚rād·ē'ā·shən }

ultrasonics [ACOUS] The science of ultrasonic sound waves. { ¦əl·trə'sän·iks }

ultrasonic sealing [ENG] A method for sealing plastic film by using localized heat developed by vibratory mechanical pressure at ultrasonic frequencies. { ¦əl·trə'sän·ik 'sēl·iŋ }

ultrasonic soldering [MET] A method used for aluminum soldering by ultrasonically vibrating the soldering iron to disrupt the oxide film on the metal. { ¦əl·trə'sän·ik 'säd·ə·riŋ }

ultrasonic storage cell *See* ultrasonic delay line. { ¦əl·trə'sän·ik 'stȯr·ij ˌsel }

ultrasonic testing [ENG] A nondestructive test method that employs high-frequency mechanical vibration energy to detect and locate structural discontinuities or differences and to measure thickness of a variety of materials. { ¦əl·trə'sän·ik 'test·iŋ }

ultrasonic therapy *See* ultrasound diathermy. { ¦əl·trə'sän·ik 'ther·ə·pē }

ultrasonic thickness gage [ENG] A thickness gage in which the time of travel of an ultrasonic beam through a sheet of material is used as a measure of the thickness of the material. { ¦əl·trə'sän·ik 'thik·nəs ˌgāj }

ultrasonic transducer [ENG ACOUS] A transducer that converts alternating-current energy above 20 kilohertz to mechanical vibrations of the same frequency; it is generally either magnetostrictive or piezoelectric. { ¦əl·trə'sän·ik tranz'dü·sər }

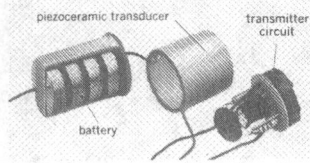
ultrasonic transmitter [ENG ACOUS] A device used to track seals, fish, and other aquatic animals: the device is fastened to the outside of the animal or fed to it, and has a loudspeaker which is made to vibrate at an ultrasonic frequency, propagating ultrasonic waves through the water to a special microphone or hydrophone. { ¦əl·trə'sän·ik tranz'mid·ər }

ultrasonic wave [ACOUS] A sound wave that has a frequency above about 20,000 hertz. { ¦əl·trə'sän·ik 'wāv }

ultrasonic welding [MET] A nonfusion welding process in which the atomic movement required for coalescence is stimulated by ultrasonic vibrations. { ¦əl·trə'sän·ik 'weld·iŋ }

ultrasonography *See* acoustic imaging. { }

ultrasonoscope [ENG] An instrument that displays an echosonogram on an oscilloscope; usually has auxiliary output to a chart-recording instrument. { ¦əl·trə'sän·ə,skōp }

ultrasound [ACOUS] Sound with a frequency above about 20,000 Hz, the upper limit of human hearing. { 'əl·trə,saund }

ultrasound diathermy [MED] The application of high-frequency sound waves (0.7 to 1.0 megahertz) and the conversion of this mechanical energy into heat for local thermotherapy. Also known as ultrasonic therapy. { ¦əl·trə,saund 'dī·ə,thər·mē }

ultraspeed welding *See* commutator-controlled welding. { 'əl·trə,spēd 'weld·iŋ }

ultraspherical polynomials *See* Gegenbauer polynomials. { ¦əl·trə'sfer·ə·kəl ˌpäl·i'nō·mē·əlz }

ultrastable material [MATER] A material having extremely low thermal expansivity, extremely high temporal stability, and high stiffness-to-density ratio. { ¦əl·trə¦stā·bəl mə'tir·ē·əl }

ultrastrip [GRAPHICS] A piece of film, usually 1.5 inches by 7 inches (3.5 by 18 centimeters) in size, containing either negative or positive images, or frames, of printed material reduced 150 times by photographic reduction. { 'əl·trə,strip }

ultrastructure [MOL BIO] The ultimate physiochemical organization of protoplasm. { ¦əl·trə'strək·chər }

ultraviolet [PHYS] Pertaining to ultraviolet radiation. Abbreviated UV. { ¦əl·trə'vī·lət }

ultraviolet absorber [OPTICS] Any substance that absorbs ultraviolet radiant energy, then dissipates the energy in a harmless form; used in plastics and rubbers to decrease light sensitivity. { ¦əl·trə'vī·lət əb'sȯr·bər }

ultraviolet absorber fixative [MATER] A protective fixative that includes a material to filter ultraviolet light from the sun and from artificial light; it helps to keep colors from fading. { ¦əl·trə'vī·lət əb¦sȯr·bər ,fik·səd·iv }

ultraviolet absorption [OPTICS] Absorption of specific ultraviolet radiation wavelengths by a material; for example, by a sample solution during spectroscopic analysis. { ¦əl·trə'vī·lət əb'sȯrp·shən }

ultraviolet absorption spectrophotometry [SPECT] The study of the spectra produced by the absorption of ultraviolet radiant energy during the transformation of an electron from the ground state to an excited state as a function of the wavelength

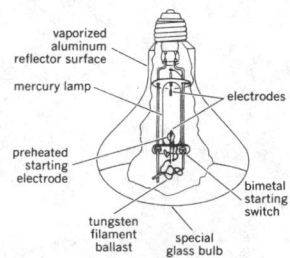
causing the transformation. { ¦əl·trə'vī·lət əb'sȯrp·shən ,spek·trō·fə'täm·ə·trē }

ultraviolet astronomy [ASTRON] Astronomical investigations utilizing observations carried out in the spectral region from approximately 350 to 90 nanometers. { ¦əl·trə'vī·lət ə'strän·ə·mē }

ultraviolet-bright star [ASTRON] A star that is brighter than stars on the horizontal branch and bluer than stars on the giant branch. { ¦əl·trə'vī·lət 'brīt 'stär }

ultraviolet catastrophe [STAT MECH] The prediction of the Rayleigh-Jeans law that the energy radiated by a blackbody at extremely short wavelengths is extremely large, and the total energy radiated is infinite, whereas in reality it must be finite. { ¦əl·trə'vī·lət kə'tas·trə·fē }

ultraviolet densitometry [SPECT] An ultraviolet-spectrophotometry technique for measurement of the colors on thin-layer chromatography absorbents following elution. { ¦əl·trə'vī·lət ,den·sə'täm·ə·trē }

ultraviolet-erasable programmable read-only memory [COMPUT SCI] An integrated-circuit memory chip in which the stored information can be erased only by ultraviolet light and the circuit can be reprogrammed with new information that can be stored indefinitely. Abbreviated UV EPROM; UVPROM. { ¦əl·trə'vī·lət i'rās·ə·bəl 'prō,gram·ə·bəl ¦rēd ,ȯn·lē 'mem·rē }

ultraviolet-erasable programmable read-only memory eraser [COMPUT SCI] A device that removes the contents of ultraviolet-erasable programmable read-only memory chips by exposing them to ultraviolet light. { ¦əl·trə'vī·lət i'rās·ə·bəl 'prō,gram·ə·bəl ¦rēd ,ȯn·lē 'mem·rē i'rā·sər }

ultraviolet imagery [ELECTROMAG] That imagery produced as a result of sensing ultraviolet radiations reflected from a given target surface. { ¦əl·trə'vī·lət 'im·ij·rē }

ultraviolet lamp [ELECTR] A lamp providing a high proportion of ultraviolet radiation, such as various forms of mercury-vapor lamps. { ¦əl·trə'vī·lət 'lamp }

ultraviolet light *See* ultraviolet radiation. { ¦əl·trə'vī·lət 'līt }

ultraviolet microscope [OPTICS] A special type of microscope which uses electromagnetic radiation in the range 180–400 nanometers; it requires reflecting optics or special quartz and crystal objectives. { ¦əl·trə'vī·lət 'mī·krə,skōp }

ultraviolet photoemission spectroscopy [SPECT] A spectroscopic technique in which photons in the energy range 10–200 electronvolts bombard a surface and the energy spectrum of the emitted electrons gives information about the states of electrons in atoms and chemical bonding. Abbreviated UPS. { ¦əl·trə'vī·lət ¦fōd·ō·i'mish·ən spek'träs·kə·pē }

ultraviolet photography [GRAPHICS] Photography in which the subject is illuminated with ultraviolet light, and either the resulting fluorescence (in the fluorescence method) or the reflected ultraviolet light (in the reflected ultraviolet method) is detected by the camera. { ¦əl·trə'vī·lət fə'täg·rə·fē }

ultraviolet radiation [ELECTROMAG] Electromagnetic radiation in the wavelength range 4–400 nanometers; this range begins at the short-wavelength limit of visible light and overlaps the wavelengths of long x-rays (some scientists place the lower limit at higher values, up to 40 nanometers. Also known as ultraviolet light. { ¦əl·trə'vī·lət ,rād·ē'ā·shən }

ultraviolet spectrometer [SPECT] A device which produces a spectrum of ultraviolet light and is provided with a calibrated scale for measurement of wavelength. { ¦əl·trə'vī·lət spek'träm·əd·ər }

ultraviolet spectrophotometry [SPECT] A method of chemical analysis based on the absorption of electromagnetic radiation between 200 and 400 nanometers; used extensively for quantitative analysis of simple inorganic ions and their complexes, as well as organic molecules that have at least one double bond. { ¦əl·trə'vī·lət ¦spek·trō·fə'täm·ə·trē }

ultraviolet spectroscopy [SPECT] Absorption spectroscopy involving electromagnetic wavelengths in the range 4–400 nanometers. { ¦əl·trə'vī·lət spek'träs·kə·pē }

ultraviolet spectrum [ELECTROMAG] **1.** The range of wavelengths of ultraviolet radiation, covering 4–400 nanometers. **2.** A display or graph of the intensity of ultraviolet radiation emitted or absorbed by a material as a function of wavelength or some related parameter. { ¦əl·trə'vī·lət 'spek·trəm }

ultraviolet stabilizer *See* UV stabilizer. { ¦əl·trə'vī·lət 'stā·bə,līz·ər }

ultraviolet star [ASTRON] A very hot star that is evolving toward the white dwarf stage; usually the central star of a planetary nebula. { ¦əl·trə'vī·lət ¦stär }

ultraviolet telescope [OPTICS] An assemblage of mirrors, with special coatings imparting high ultraviolet reflectivity, which forms magnified ultraviolet images of objects in the same manner as an optical telescope forms images in visible light. { ¦əl·trə'vī·lət 'tel·ə,skōp }

ultravulcanian [GEOL] A type of volcanic eruption characterized by periodic violent gaseous explosions of lithic dust and solid blocks, with little if any fiery scoria. { ¦əl·trə·vəl'kā·nē·ən }

Ulvaceae [BOT] A large family of green algae in the order Ulvales. { ,əl'vās·ē,ē }

Ulvales [BOT] An order of algae in the division Chlorophyta in which the thalli are macroscopic, attached tubes or sheets. { ,əl'vā·lēz }

umangite [MINERAL] Cu₃Se₂ A dark cherry red mineral consisting of copper selenide; occurs in massive form, in small grains or fine granular aggregates. { ü'maŋ,gīt }

umbel [BOT] An indeterminate inflorescence with the pedicels all arising at the top of the peduncle and radiating like umbrella ribs; there are two types, simple and compound. { 'əm·bəl }

Umbellales [BOT] An order of dicotyledonous herbs or woody plants in the subclass Rosidae with mostly compound or conspicuously lobed or dissected leaves, well-developed schizogenous secretory canals, separate petals, and an inferior ovary. { ,əm·bə'lā·lēz }

Umbelliferae [BOT] A large family of aromatic dicotyledonous herbs in the order Umbellales; flowers have an ovary of two carpels, ripening to form a dry fruit that splits into two halves, each containing a single seed. { ,əm·bə'lif·ə,rē }

umber [MATER] A naturally occurring brown siliceous earth, deriving its color from iron oxides and manganese oxide; used as a paint pigment. { 'əm·bər }

umbilic See umbilical point. { üm'bil·ik }

umbilical artery [EMBRYO] Either of a pair of arteries passing through the umbilical cord to carry impure blood from the mammalian fetus to the placenta. { əm'bil·ə·kəl 'ärd·ə·rē }

umbilical connections [AERO ENG] Electrical and mechanical connections to a launch vehicle prior to lift off; the umbilical tower adjacent to the vehicle on the launch pad supports these connections which supply electrical power, control signals, data links, propellant loading, high pressure gas transfer, and air conditioning. { əm'bil·ə·kəl kə'nek·shənz }

umbilical cord [AERO ENG] Any of the servicing electrical or fluid lines between the ground or a tower and an uprighted rocket vehicle before the launch. Also known as umbilical. [EMBRYO] The long, cylindrical structure containing the umbilical arteries and vein, and connecting the fetus with the placenta. { əm'bil·ə·kəl ,kord }

umbilical duct See vitelline duct. { əm'bil·ə·kəl ,dəkt }

umbilical hernia [MED] Herniation through the umbilical ring. Also known as annular hernia. { əm'bil·ə·kəl 'hər·nē·ə }

umbilical point [MATH] A point on a surface at which the normal curvature is the same in all directions. Also known as navel point; umbilic. { əm'bil·ə·kəl ,point }

umbilical tower [AERO ENG] A vertical structure supporting the umbilical cords running into a rocket in launching position. { əm'bil·ə·kəl 'tau̇·ər }

umbilical vein [EMBRYO] A vein passing through the umbilical cord and conveying purified, nutrient-rich blood from placenta to fetus. { əm'bil·ə·kəl 'vān }

Umbilicariaceae [BOT] The rock tripes, a family of Ascolichenes in the order Lecanorales having a large, circular, umbilicate thallus. { ,əm·bə·lə,kar·ē'ās·ē,ē }

umbilicus [ANAT] The navel; the round, depressed cicatrix in the median line of the abdomen, marking the site of the aperture through which passed the fetal umbilical vessels. { əm'bil·ə·kəs }

umbo [ANAT] A rounded elevation of the surface of the tympanic membrane. [INV ZOO] A prominence above the hinge of a bivalve mollusk shell. { 'əm·bō }

umbonate [BIOL] Having or forming an umbo. { 'əm·bə·nət }

umbra [ASTRON] The dark, central region of a sunspot. [OPTICS] That portion of a shadow which is screened from

light rays emanating from any part of an extended source. { 'əm·brə }

umbrella antenna [ELECTROMAG] Antenna in which the wires are guyed downward in all directions from a central pole or tower to the ground, somewhat like the ribs of an open umbrella. { əm'trel·ə an,ten·ə }

umbrella roof See station roof. { əm'brel·ə ,rüf }

Umbrept [GEOL] A suborder of the Inceptisol soil order; has dark A horizon more than 10 inches (25 centimeters) thick, brown B horizons, and slightly paler C horizons; soil is strongly acid, and clay minerals are crystalline; occurs in cool or temperate climates. { 'ən,brept }

Umbriel [ASTRON] A satellite of Uranus orbiting at a mean distance of 165,300 miles (266,000 kilometers) with a period of 4.144 days, and with a diameter of about 740 miles (1190 kilometers). { 'əm·brē·əl }

Umkehr effect [OPTICS] An anomaly of the relative zenith intensities of scattered sunlight at certain wavelengths in the ultraviolet as the sun approaches the horizon; it is due to the presence of the ozone layer. { 'üm,ker i,fekt }

Umklapp process [SOLID STATE] The interaction of three or more waves in a solid, such as lattice waves or electron waves, in which the sum of the wave vectors is not equal to zero but, rather, is equal to a vector in the reciprocal lattice. Also known as flip-over process. { 'üm,kläp ,prä·səs }

UMP See uridylic acid.

umpire [MIN ENG] An assay made by a third party to settle the difference in assays made by the purchaser and the seller of ore. { 'əm,pīr }

unaka [GEOL] A large residual mass rising above a peneplain that is less well developed than one having a monadnock. { ü'näk·ə }

unakite [PETR] An altered igneous rock composed principally of epidote, pink orthoclase, and quartz. { 'ü·nə,kīt }

unambiguous name [COMPUT SCI] The name of a file or other data item that completely specifies the item to a computer system. { ¦ən·am'big·yə·wəs 'nām }

unamplified back bias [ELECTR] Degenerative voltage developed across a fast time constant circuit within an amplifier stage itself. { ¦ən'am·plə,fīd 'bak ,bī·əs }

unarmed [ORD] The condition of a fuse (or other firing device) in which the necessary steps to put it in condition to function have not taken place; while in the unarmed state the fuse is safe to handle, store, and transport. { ¦ən'ärmd }

unary operation [MATH] An operation in which only a single operand is required to produce a unique result; some examples are negation, complementation, square root, transpose, inverse, and conjugate. { 'yü·nə,rē ,äp·ə,rā·shən }

unattended operation [COMPUT SCI] An operation in which components in the hardware of a communications terminal or data-processing system operate automatically, allowing handling of signals or data without human intervention. { ¦ən·ə'ten·dəd ,äp·ə'rā·shən }

unattended time [COMPUT SCI] Time during which a computer is turned off but is not undergoing maintenance. Also known as unused time. { ¦ən·ə'ten·dəd ,tīm }

unavailable energy [THERMO] That part of the energy which, when an irreversible process takes place, is initially in a form completely available for work and is converted to a form completely unavailable for work. { ¦ən·ə¦vāl·ə·bəl 'en·ər·jē }

unavoidable delay [IND ENG] Any delay in a task, the occurrence of which is outside the control or responsibility of the worker. { ¦ən·ə'vȯid·ə·bəl di lā }

unavoidable-delay allowance [IND ENG] An adjustment of standard time to allow for unavoidable delays in a task. { ¦ən·ə'vȯid·ə·bəl di'lā ə,lau̇·əns }

unavoidable set of configurations [MATH] A set of graphs such that any planar graph has at least one member of the set as a subgraph. { ¦ən·ə'vȯid·ə·bəl 'set əv kən,fig·yə'rā·shənz }

unbalanced cutter chain [MIN ENG] A cutter chain which carries more picks along the bottom line than along the top line. { ¦ən bal·ənst 'kəd·ər ,chān }

unbalanced hoisting [MIN ENG] The method of hoisting in small one-compartment shafts with only one cage in operation, as opposed to balanced winding. { ¦ən'bal·ənst 'hȯist·iŋ }

unbalanced line [ELEC] A transmission line in which the voltages on the two conductors are not equal with respect to ground; a coaxial line is an example. { ¦ən'bal·ənst 'līn }

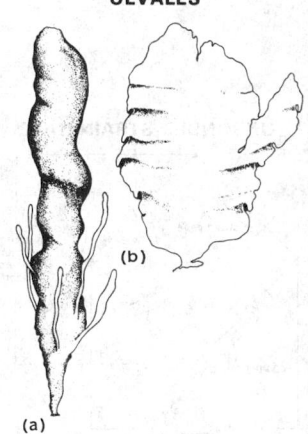

ULVALES

Ulvales. (a) *Enteromorpha*, a tubular form. (b) *Ulva*, the sea lettuce, an expanded thallus.

UMBEL

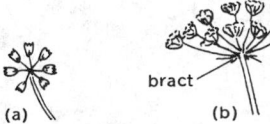

bract

Two types of umbel inflorescence: (a) simple; (b) compound.

unbalanced output [ELEC] An output in which one of the two input terminals is substantially at ground potential. { ¦ən'bal·ənst 'aut̩put̩ }

unbalanced shothole [MIN ENG] A shothole in which the explosive charge breaks down the coal at the back of the machine cut while leaving the front portion standing or in large blocks. { ¦ən'bal·ənst 'shät̩hōl }

unbalanced wire circuit [ELEC] A wire circuit whose two sides are inherently electrically unlike. { ¦ən'bal·ənst ¦wīr ̩sər·kət }

unbiased estimate [STAT] An estimate for a parameter θ whose expected value is θ. { ¦ən'bī·əst 'es·tə·mət }

unblanking pulse [ELECTR] Voltage applied to a cathode-ray tube to overcome bias and cause trace to be visible. { ¦ən'blaŋk·iŋ ̩pəls }

unbonded member [CIV ENG] A posttensioned member that is made of prestressed concrete and has the tensioning force applied only against the end anchorages. { ən¦bänd·əd 'mem·bər }

unbonded strain gage [ENG] A type of strain gage that consists of a grid of fine wires strung under slight tension between a stationary frame and a movable armature; pressure applied to the bellows or to the diaphragm sensing element moves the armature with respect to the frame, increasing tension in one half of the filaments and decreasing tension in the rest. { ¦ən'bän·dəd 'strān ̩gāj }

unbounded manifold [MATH] A manifold with no boundary. { ¦ən'baun·dəd 'man·ə̩fōld }

unbounded set of real numbers [MATH] A set with the property that, if R is any positive real number, there is a number in the set which is smaller than $-R$ or a number larger than R. { ¦ən'baun·dəd 'set əv 'rēl 'nəm·bərz }

unbounded wave [PHYS] A wave which propagates through a nondissipative, homogeneous medium which is infinite in extent, without any boundaries. { ¦ən'baun·dəd 'wāv }

unbreakable cipher [COMMUN] A cipher for which the message or key cannot be obtained through cryptanalysis, even with an unlimited amount of computational power, data storage, and calendar time. { ¦ən'brāk·ə·bəl 'sī·fər }

unbundling [COMPUT SCI] The separate pricing of software products and services from equipment charges. { ¦ən'bənd·liŋ }

uncage [ENG] To release the caging mechanism of a gyroscope, that is, the mechanism that erects the gyroscope or locks it in position. { ¦ən'kāj }

uncatalog [COMPUT SCI] To remove an entry from the system catalog so that the file named in the entry can no longer be accessed by the operating system. { ¦ən'kad·əl̩äg }

uncertainty [SCI TECH] The estimated amount by which an observed or calculated value may depart from the true value. { ¦ən'sərt·ən·tē }

uncertainty principle [QUANT MECH] The precept that the accurate measurement of an observable quantity necessarily produces uncertainties in one's knowledge of the values of other observables. Also known as Heisenberg uncertainty principle; indeterminacy principle. { ¦ən'sərt·ən·tē ̩prin·sə·pəl }

uncertainty relation [QUANT MECH] The relation whereby, if one simultaneously measures values of two canonically conjugate variables, such as position and momentum, the product of the uncertainties of their measured values cannot be less than approximately Planck's constant divided by 2π. Also known as Heisenberg uncertainty relation. { ¦ən'sərt·ən·tē ri̩lā·shən }

uncharged [ELEC] Having no electric charge. { ¦ən'chärjd }

uncharged demolition target [ENG] A demolition target which has been prepared to receive the demolition agent, the necessary quantities of which have been calculated, packaged, and stored in a safe place. { ¦ən'chärjd ̩dem·ə'lish·ən ̩tär·gət }

uncharged species [CHEM] A chemical entity with no net electric charge. Also known as neutral species. { ¦ən'chärjd 'spē·shēz }

uncinate trophus [INV ZOO] A trophus in rotifers characterized by a hooked or curved uncus. { 'ən·sə·nət 'träf·əs }

uncinus [INV ZOO] In annelids, a minute, pectiniform neuroseta. { ən'sī·nəs }

uncompetitive enzyme inhibition [BIOCHEM] The prevention of an enzymic process as a result of the interaction of an inhibitor with the enzyme-substrate complex or a subsequent intermediate form of the enzyme, but not with the free enzyme. { ¦ən·kəm'ped·əd·iv 'en̩zīm ̩in·ə'bish·ən }

unconcentrated wash See sheet erosion. { ¦ən'käns·ən̩trād·əd 'wäsh }

unconditional [COMPUT SCI] Not subject to conditions external to the specific instruction. { ¦ən·kən'dish·ən·əl }

unconditional convergence [MATH] A convergent series converges unconditionally if every series obtained by rearranging its terms also converges; equivalent to absolute convergence. { ¦ən·kən'dish·ən·əl kən'vər·jəns }

unconditional inequality [MATH] An inequality which holds true for all values of the variables involved, or which contains no variables; for example, $y + 2 > y$, or $4 > 3$. Also known as absolute inequality. { ¦ən·kən'dish·ən·əl ̩in·i'kwäl·əd·ē }

unconditional jump [COMPUT SCI] A digital-computer instruction that interrupts the normal process of obtaining instructions in an ordered sequence, and specifies the address from which the next instruction must be taken. Also known as unconditional transfer. { ¦ən·kən'dish·ən·əl 'jəmp }

unconditional transfer See unconditional jump. { ¦ən·kən'dish·ən·əl 'tranz·fər }

unconfined explosion [ENG] Explosion occurring in the open air where the (atmospheric) pressure is constant. { ¦ən·kən'fīnd ik'splō·zhən }

unconformable [GEOL] Pertaining to strata that do not conform in position, dip, or strike to the older underlying rocks. { ¦ən·kən'fór·mə·bəl }

unconformity [GEOL] The relation between adjacent rock strata whose time of deposition was separated by a period of nondeposition or of erosion; a break in a stratigraphic sequence. { ¦ən·kən'fór·məd·ē }

unconformity iceberg [OCEANOGR] An iceberg consisting of more than one kind of ice, such as blue water-formed ice and névé; such an iceberg often contains many crevasses and silt bands. { ¦ən·kən'fór·məd·ē 'īs̩bərg }

unconscious [MED] Insensible; in a state lacking conscious awareness, with reflexes abolished. [PSYCH] **1.** Pertaining to behavior or experience not controlled by the ego. **2.** The part of the mind, mental functioning, or personality not in the immediate field of awareness. { ¦ən'kän·shəs }

unconsolidated material [GEOL] Loosely arranged or unstratified sediment whose particles are not cemented together. { ¦ən·kən'säl·ə̩dād·əd mə'tir·ē·əl }

unconstrained optimization problem [MATH] A nonlinear programming problem in which there are no constraint functions. { ¦ən·kən'strānd ̩äp·tə·mə'zā·shən ̩präb·ləm }

uncontrolled fragments [ORD] The segments of the casing propelled outward at detonation of an explosive device in which no design provision has been made to control the size and shape of the segments. { ¦ən·kən'trōld 'frag·məns }

uncontrolled mosaic [GRAPHICS] A mosaic composed of uncorrected photographs, the details of which have been matched from print to print without ground control or other orientation; accurate measurement and direction cannot be accomplished. { ¦ən·kən'trōld mō'zā·ik }

unconventional warfare [ORD] A type of warfare that includes the interrelated fields of guerrilla warfare, evasion and escape, and subversion, in which operations are conducted within enemy or enemy-controlled territory by predominantly indigenous personnel, usually supported and directed in varying degrees by an external source. { ¦ən·kən'ven·chən·əl 'wór·fer }

uncorrecting [NAV] The process of converting true to magnetic, compass, or gyro direction, or magnetic to compass direction. { ¦ən·kə'rek·tiŋ }

uncorrelated random variables [STAT] Two random variables whose correlation coefficient is zero. { ¦ən'kär·ə̩läd·əd 'ran·dəm 'ver·ē·ə·bəlz }

uncountable set [MATH] An infinite set which cannot be put in one-to-one correspondence with the set of integers; for example, the set of real numbers. { ¦ən'kaunt·ə·bəl 'set }

uncouple [ENG] To unscrew or disengage. { ¦ən'kəp·əl }

uncoupling phenomena [SPECT] Deviations of observed spectra from those predicted in a diatomic molecule as the magnitude of the angular momentum increases, caused by interactions which could be neglected at low angular momenta. { ¦ən'kəp·liŋ fə̩näm·ə·nä }

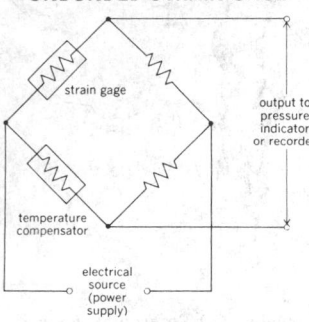

UNBONDED STRAIN GAGE

Circuit diagram of an unbonded strain gage.

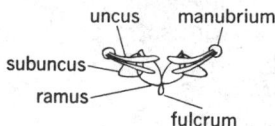

UNCINATE TROPHUS

Uncinate trophus of a predacious rotifer, *Stephanoceros*.

unctuous [MATER] Greasy, oily, or soapy to the touch. { 'əŋk·shəs }

undamped wave [PHYS] A continuous wave produced by oscillations having constant amplitude. { ¦ən'damt 'wāv }

undecagon [MATH] An 11-sided polygon. { ən'dek·ə₁gän }

undecahedron [MATH] A polyhedron with 11 faces. { ən₁dek·ə'hē·drən }

undecanal [ORG CHEM] $CH_3(CH_2)_9CHO$ A sweet-smelling, colorless liquid, soluble in oils and alcohol; used in perfumes and flavoring. Also known as hendecanal; *n*-undecyclic aldehyde. { ən'dek·ə₁nal }

undecane [ORG CHEM] $CH_3(CH_2)_9CH_3$ A colorless, combustible liquid, boiling at 385°F (196°C), flash point at 149°F (65°C); used as a chemical intermediate and in petroleum research. Also known as hendecane. { ₁ən'de₁kān }

undecanoic acid [ORG CHEM] $CH_3(CH_2)_9COOH$ Colorless crystals, soluble in alcohol and ether, insoluble in water; melts at 29°C; used as a chemical intermediate. { ¦ən¦dek·ə¦nō·ik 'as·əd }

2-undecanone See methyl nonyl ketone. { ¦tü ₁ən'dek·ə₁nōn }

undecomino [MATH] One of the 17,073 plane figures that can be formed by joining 11 unit squares along their sides. { ən·də'käm·ə·nō }

undecyl [ORG CHEM] $C_{11}H_{23}$ The radical of undecane. Also known as hendecyl. { ¦ən¦des·əl }

undecylenic acid [ORG CHEM] $CH_2\cdot CH(CH_2)_8COOH$ A light-colored, combustible liquid with a fruity aroma; soluble in alcohol, ether, chloroform, and benzene, almost insoluble in water; used in medicine, perfumes, flavors, and plastics. { ¦ən¦des·ə¦lin·ik 'as·əd }

undecylenic alcohol [ORG CHEM] $C_{11}H_{22}O$ A colorless liquid with a citrus odor, soluble in 70% alcohol; used in perfumes. { ¦ən¦des·ə¦lin·ik 'al·kə₁hol }

undecylenyl acetate [ORG CHEM] $C_{13}H_{24}O_2$ A colorless liquid with a floral-fruity odor, soluble in 80% alcohol; used in perfumes and for flavoring. { ¦ən¦des·əl¦en·əl 'as·ə₁tāt }

***n*-undecylic aldehyde** See undecanal. { ¦en ¦ən·də¦sil·ik 'al·də₁hīd }

underbead crack [MET] A crack in the heat-affected zone of a weldment which usually does not reach the base-metal surface. { 'ən·dər₁bēd ₁krak }

underbody [ENG] The lower portion or underside of the body of a vehicle or airplane. { 'ən·dər₁bäd·ē }

underbunching [ELECTR] In velocity-modulated electron streams, a condition representing less than the optimum bunching. { ¦ən·dər¦bənch·iŋ }

undercarriage [AERO ENG] The landing gear assembly for an aircraft. [ORD] Fixed or movable base on which the top carriage of a weapon moves. { 'ən·dər₁kar·ij }

undercast [METEOROL] A cloud layer of ten-tenths (1.0) coverage as viewed from an observation point above the layer; the term is used in pilot reporting of in-flight weather conditions. [MIN ENG] An air crossing in which one airway is deflected to pass under the other airway. { 'ən·dər₁kast }

underchain haulage [MIN ENG] Haulage in which the chains are placed beneath the mine car at certain intervals with suitable hooks that thrust against the car axle. { ¦ən·dər¦chān 'hòl·ij }

underclay [GEOL] A layer of clay or other fine-grained detrital material underlying a coal bed or comprising the floor of a coal seam. Also known as coal clay; root clay; seat clay; seat earth; thill; underearth; warrant. { 'ən·dər₁klā }

underclay limestone [GEOL] A thin, fresh-water limestone that is relatively free of fossils and is dense and nodular; found in underlying coal deposits. { 'ən·dər₁klā 'līm₁stōn }

undercliff [GEOL] A subordinate cliff or terrace formed by material which has fallen or slid from above. { 'ən·dər₁klif }

undercoat See undercoater; undercoating. { 'ən·dər₁kōt }

undercoater [MATER] A type of solvent-thinned paint that is formulated to have good adhesion to a substrate and furnish a good base for additional coats of paint, and having a relatively small amount of pigment. Also known as undercoat. { 'ən·dər¦kōd·ər }

undercoating [MATER] **1.** A waterproof protective coating applied to the underside of a vehicle to resist corrosion. **2.** A coat of paint over which another coat is applied. Also known as undercoat. { 'ən·dər₁kōd·iŋ }

underconsolidation [GEOL] Less than normal consolidation of sedimentary material for the existing overburden. { ¦ən·dər·kən₁säl·ə'dā·shən }

undercooling [MET] Cooling a metal below the transformation temperature without obtaining the transformation. { 'ən·dər₁kül·iŋ }

undercooling effect [SOLID STATE] The effect whereby a superconductor can be cooled below its critical temperature without the onset of superconductivity. { ¦ən·dər¦kül·iŋ i₁fekt }

undercurrent [OCEANOGR] A water current flowing beneath a surface current at a different speed or in a different direction. { 'ən·dər₁kə·rənt }

undercurrent relay [ELEC] A relay designed to operate when its coil current falls below a predetermined value. { 'ən·dər₁kə·rənt 'rē₁lā }

undercut [ELECTR] Undesirable lateral etching by chemicals in the fabrication of semiconductor devices. [ENG] Underside recess either cut or molded into an object so as to leave a topside lip or protuberance. [MET] **1.** An unfilled groove melted into the base metal at the toe of a weld. **2.** To fail to machine a part to a sufficient extent. [MIN ENG] To cut below or in the lower part of a coal bed by chipping away the coal with a pick or mining machine; cutting is usually done on the level of the floor of the mine, extending laterally the entire face and 5 or 6 feet (1.5 or 1.8 meters) into the material. { 'ən·dər₁kət }

undercutting [CHEM ENG] In distillation, the technique of taking the products coming off the distillation tower at a temperature below the desired ultimate boiling point range to prevent contaminating the products with the compound that would distill just beyond the ultimate boiling point range. [GEOL] Erosion of material at the base of a steep slope, cliff, or other exposed rock. { ¦ən·dər¦kəd·iŋ }

underdamping [PHYS] Condition of a system when the amount of damping is sufficiently small so that, when the system is subjected to a single disturbance, either constant or instantaneous, one or more oscillations are executed by the system. { ¦ən·dər¦dam·piŋ }

underdeck tonnage [NAV ARCH] The enclosed volume of a vessel below the tonnage deck, expressed in tons of 100 cubic feet (approximately 2.8317 cubic meters). { 'ən·dər₁dek ₁tən·ij }

underdevelopment [GRAPHICS] Insufficient development of a photographic print; processing to a degree lower than the optimum density. { ¦ən·dər·də¦vel·əp·mənt }

underdraft [MET] Downward curving of a metal part on leaving the rolls because of higher speed of the upper roll. { 'ən·dər₁draft }

underdrain [CIV ENG] A subsurface drain with holes into which water flows when the water table reaches the drain level. { 'ən·dər₁drān }

underdrive press [MECH ENG] A mechanical press having the driving mechanism located within or under the bed. { 'ən·dər₁drīv 'pres }

underearth See underclay. { 'ən·dər₁ərth }

underfeed stoker [ENG] A coal-burning system in which green coal is fed from beneath the burning fuel bed. { 'ən·dər₁fēd 'stō·kər }

underfill [MET] A depression on the face of the weld which falls below the surface of the adjacent base metal. { 'ən·dər₁fil }

underfit stream [HYD] A misfit stream that appears to be too small to have eroded the valley in which it flows. { 'ən·dər₁fit 'strēm }

underfloor raceway [BUILD] A raceway for electric wires which runs beneath the floor. { 'ən·dər₁flòr 'rās₁wā }

underflow [COMPUT SCI] The generation of a result whose value is smaller than the smallest quantity that can be represented or stored by a computer. { 'ən·dər₁flō }

underflow conduit [GEOL] A permeable deposit underlying a surface stream channel. { 'ən·dər₁flō 'kän₁dü·ət }

underground [ENG] Situated, done, or operating beneath the surface of the ground. { ¦ən·dər¦graùnd }

underground burst [ORD] The explosion of a nuclear weapon beneath the surface of the ground. { ¦ən·dər¦graùnd 'bərst }

underground gasification See gasification. { ¦ən·dər¦graùnd ₁gas·ə·fə¦kā·shən }

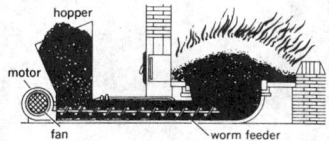

UNDERFEED STOKER

Section through an underfeed stoker.

underground geology See subsurface geology. { ¦ən·dər¦graůnd jē′äl·ə·jē }

underground glory-hole method [MIN ENG] A mining method used in large deposits with a very strong roof: the deposit is divided by levels and on every level chutes are raised to the next level; mining starts from the mouth of the chutes in such a way as to develop a funnel-shaped excavation (mill, or glory) with slopes so steep that the broken ore falls into the chutes and thus to the cars on the lower level; a sufficiently strong pillar is left for protection at the higher level. Also known as underground milling. { ¦ən·dər¦graůnd ′glȯr·ē ‚hōl ‚meth·əd }

underground ice See ground ice. { ¦ən·dər¦graůnd ′īs }

underground milling See underground glory-hole method. { ¦ən·dər¦graůnd ′mil·iŋ }

underground stem [BOT] Any of the stems that grow underground and are often mistaken for roots; principal kinds are rhizomes, tubers, corms, bulbs, and rhizomorphic droppers. { ¦ən·dər¦graůnd ′stem }

underground stream [HYD] A subsurface body of water flowing in a definite current in a distinct channel. { ¦ən·dər¦graůnd ′strēm }

underhand stoping [MIN ENG] Mining downward or from upper to lower level; the stope may start below the floor of a level and be extended by successive horizontal slices, either worked sequentially or simultaneously in a series of steps; the stope may be left as an open stope or supported by stulls or pillars. { ′ən·dər‚hand ′stōp·iŋ }

underhand work [MIN ENG] Picking or drilling downward. { ′ən·dər‚hand ′wərk }

underhead crack [MET] A subsurface crack in the heat-affected zone of the base metal near a weld. { ′ən·dər‚hed ′krak }

underhole [MIN ENG] To mine out a portion of the bottom of a seam, by pick or powder, thus leaving the top unsupported and ready to be blown down by shots, broken down by wedges, or mined with a pick or bar. { ′ən·dər‚hōl }

underhung crane [MECH ENG] An overhead traveling crane in which the end trucks carry the bridge suspended below the rails. { ′ən·dər‚həŋ ′krān }

underlap [COMMUN] **1.** In facsimile transmission, the space between the recorded elemental area in one recording line and the adjacent elemental area in the next recording line, when these areas are smaller than normal; or the space between the elemental areas in the direction of the recording line. **2.** The amount by which the effective height of the scanning spot falls short of the nominal width of the scanning line. { ′ən·dər‚lap }

underlay [GRAPHICS] The process of equalizing the impression of the press by inserting small pieces of paper beneath the type form. { ¦ən·dər¦lā }

underlay shaft [MIN ENG] A shaft sunk in the footwall and following the dip of a vein. Also known as footwall shaft; underlier. { ¦ən·dər¦lā ‚shaft }

underlie [GEOL] To lie or be situated under; to occupy a lower position, or to pass beneath. { ′ən·dər‚lī }

underlier See underlay shaft. { ′ən·dər‚lī·ər }

underlying graph [MATH] A directed graph, that results from replacing each directed arc with an undirected edge. { ¦ən·dər‚lī·iŋ ′graf }

undermelting [HYD] The melting from below of any floating ice. { ¦ən·dər¦melt·iŋ }

undermine [MIN ENG] To excavate the earth beneath, especially for the purpose of causing to fall; to form a mine under. { ¦ən·dər¦mīn }

undermining See sapping. { ′ən·dər‚mīn·iŋ }

underpainting [GRAPHICS] A layer of paint that is intended to be seen through a subsequent paint layer; an underpainted section of red, for example, can be altered to a violet or purple color with the application of a thin blue glaze. { ′ən·dər‚pānt·iŋ }

underpinning [CIV ENG] **1.** Permanent supports replacing or reinforcing the older supports beneath a wall or a column. **2.** Braced props temporarily supporting a structure. [MIN ENG] Building up the wall of a mine shaft to join that above it. { ′ən·dər‚pin·iŋ }

underplate [DES ENG] An unfinished plate which forms part of an armored front for a mortise lock, and which is fastened to the case. { ′ən·dər‚plāt }

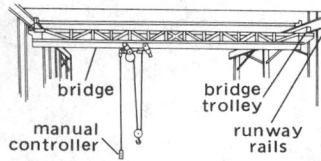

UNDERHUNG CRANE

bridge   bridge trolley   manual controller   runway rails

An underhung crane, a basic type of overhead-traveling crane.

underream [ENG] To enlarge a drill hole below the casing. { ¦ən·dər¦rēm }

undersaturated [PETR] Pertaining to igneous rock composed of unsaturated minerals, that is, without free silica. { ¦ən·dər¦sach·ə‚rād·əd }

undersaturated fluid [PHYS CHEM] Any fluid (liquid or gas) capable of holding additional vapor or liquid components in solution at specified conditions of pressure and temperature. { ¦ən·dər¦sach·ə‚rād·əd ′flü·əd }

undersea mining [MIN ENG] The working of economic deposits (usually coal) situated in strata or rocks below the seabed. { ¦ən·dər¦sē ′mīn·iŋ }

undershoot [CONT SYS] The amount by which a system's response to an abrupt change in input falls short of that desired. { ′ən·dər‚shüt }

undershot wheel [MECH ENG] A water wheel operated by the impact of flowing water against blades attached around the periphery of the wheel, the blades being partly or totally submerged in the moving stream of water. { ′ən·dər‚shät ‚wēl }

undersize [ENG] That part of a crushed material (for example, ore) which passes through a screen. { ′ən·dər‚sīz }

underspin [MECH] Property of a projectile having insufficient rate of spin to give proper stabilization. { ′ən·dər‚spin }

understory [FOR] A foliage layer occurring beneath and shaded by the main canopy of a forest. { ′ən·dər‚stȯr·ē }

understressing [MET] Repeated stressing below the fatigue limit or below the final applied stress; can improve fatigue properties as a result of strain-aging effects. { ¦ən·dər¦stres·iŋ }

underthrow distortion [COMMUN] Distortion occurring in facsimile when the maximum signal amplitude is too low. { ¦ən·dər¦thrō di′stȯr·shən }

underthrust [GEOL] A thrust fault in which the lower, active rock mass has been moved under the upper, passive rock mass. { ¦ən·dər¦thrəst }

undertow [OCEANOGR] A subsurface seaward movement by gravity flow of water carried up on a sloping beach by waves or breakers. { ′ən·dər‚tō }

undervoltage protection [ELEC] An undervoltage relay which removes a motor from service when a low-voltage condition develops, so that the motor will not draw excessive current, or which prevents a large induction or synchronous motor from starting under low-voltage conditions. { ′ən·dər‚vōl·tij prə′tek·shən }

undervoltage relay [ELEC] A relay designed to operate when its coil voltage falls below a predetermined value. { ¦ən·dər‚vōl·tij ′rē‚lā }

underwater acoustics [ACOUS] Study of the propagation of sound waves in water, especially in the oceans, and of phenomena produced by these sound waves. Also known as hydroacoustics. { ¦ən·dər¦wȯd·ər ə′küs·tiks }

underwater burst [ORD] The explosion of a nuclear weapon beneath the surface of the water. { ¦ən·dər¦wȯd·ər ′bərst }

underwater camera [OPTICS] A camera designed for use under the surface of the water; it is usually a conventional type enclosed in a casing to withstand water pressure, preferably with a correction lens to compensate for aberrations caused by the water. { ¦ən·dər¦wȯd·ər ′kam·rə }

underwater demolition [ORD] Destruction or fragmentation of underwater obstacles by use of explosive charges placed by diver personnel; primarily employed as a short-range emergency measure to accomplish a military objective with promptness and economy of material. { ¦ən·dər¦wȯd·ər ‚dem·ə′lish·ən }

underwater demolition team [ORD] Naval unit organized and equipped to perform beach reconnaissance and underwater demolition in an amphibious operation. { ¦ən·dər¦wȯd·ər ‚dem·ə′lish·ən ‚tēm }

underwater mine [ORD] A mine designed to be located underwater and exploded by means of propeller vibration, magnetic attraction, contact, or remote control. { ¦ən·dər¦wȯd·ər ′mīn }

underwater-mine extender mechanism [ORD] An item designed to extend the detonator into the booster for arming an underwater mine by means of hydrostatic pressure. { ¦ən·dər¦wȯd·ər ′mīn ik′sten·dər ‚mek·ə‚niz·əm }

underwater-mine fairing [ORD] An item designed to be mounted on an underwater mine; it is shaped to reduce or equally distribute air resistance when suspended and launched from an aircraft, and it may have collapsible fins. { ¦ən·dər¦wȯd·ər 'mīn ¦fer·iŋ }

underwater navigation [NAV] The navigation of a submerged vessel. Also known as submarine navigation. { ¦ən·dər¦wȯd·ər ¦nav·ə'gā·shən }

underwater obstacle [NAV] A natural or artificial obstacle which is located to seaward of the high-water line and wholly or partly submerged, and which acts as a barrier or obstruction to the passage of ships, landing ships, craft, vehicles, or torpedoes. { ¦ən·dər¦wȯd·ər 'äb·stə·kəl }

underwater ordnance [ORD] Munitions designed for use underwater; for example, torpedoes. { ¦ən·dər¦wȯd·ər 'ȯrd·nəns }

underwater sound [ACOUS] The production, transmission, and reception of sounds in the ocean; used for locating submarines and other submerged objects, and to determine the physical structure of the ocean and its bottom, and to study organisms found in the sea. { ¦ən·dər¦wȯd·ər 'saùnd }

underwater sound projector [ENG ACOUS] A transducer used to produce sound waves in water. { ¦ən·dər¦wȯd·ər 'saùnd prə¦jek·tər }

underwater telephone [COMMUN] A method of voice communication using underwater sound as a means of transmission; it functions similarly to a conventional telephone system except that the energy is carried by sound waves in the water rather than by electrical signals through a wire. { ¦ən·dər¦wȯd·ər 'tel·ə¦fōn }

underwater television [COMMUN] The technique of using remotely controlled television equipment under the surface of the water. { ¦ən·dər¦wȯd·ər 'tel·ə¦vizh·ən }

underwater transducer [ENG ACOUS] A device used for the generation or reception of underwater sounds. { ¦ən·dər¦wȯd·ər tranz'dü·sər }

underwater vehicle [OCEANOGR] A submersible work platform designed to be operated either remotely or directly. { ¦ən·dər¦wȯd·ər 'vē·ə·kəl }

underway [NAV] **1.** A ship without moorings; not secured in any way to the ground or a wharf. **2.** Craft in motion, particularly the start of such motion after a standstill. { 'ən·dər'wā }

underway bottom sampler See underway sampler. { ¦ən·dər¦wā 'bäd·əm ¦sam·plər }

underway sampler [ENG] A device for collecting samples of sediment on the ocean bottom, consisting of a cup in a hollow tube; on striking the bottom, the cup scoops up a small sample which is forced into the tube which is then closed with a lid, and the device is hoisted to the surface. Also known as scoopfish; underway bottom sampler. { ¦ən·dər¦wā 'sam·plər }

underwing [INV ZOO] Either of a pair of posterior wings on certain insects, as the moth. { 'ən·dər¦wiŋ }

Underwood chart [CHEM ENG] A graphical solution of mass balances for a single equilibrium stage in the calculation of a solvent-extraction operation. { 'ən·dər¦wùd ¦chärt }

Underwood distillation method [CHEM ENG] A method for calculation of liquid separations from binary distillation systems operated at partial reflux. { 'ən·dər¦wùd ¦dis·tə'lā·shən ¦meth·əd }

undetected error rate [COMMUN] The number of bits (or other units of information) which are received but are not detected or corrected by error-control equipment, divided by the total number of bits (or other units of information) transmitted. Also known as residual error rate. { ¦ən·di'tek·təd 'er·ər ¦rāt }

undetermined multipliers See Lagrangian multipliers. { ¦ən·di'tər·mənd 'məl·tə¦plī·ərz }

undirected graph [MATH] A graph whose edges are not assigned directions. { ¦ən·də¦rek·təd 'graf }

undistorted wave [COMMUN] Periodic wave in which both the attenuation and velocity of propagation are the same for all sinusoidal components, and in which no sinusoidal component is present at one point that is not present at all points. { ¦ən·di'stȯrd·əd 'wāv }

undisturbed [ENG] Pertaining to a sample of material, as of soil, subjected to so little disturbance that it is suitable for determinations of strength, consolidation, permeability characteristics, and other properties of the material in place. { ¦ən·di'stərbd }

undisturbed motion [PHYS] The steady state of a system before perturbations are introduced. { ¦ən·di'stərbd 'mō·shən }

undisturbed-one output [ELECTR] "One" output of a magnetic cell to which no partial-read pulses have been applied since that cell was last selected for writing. { ¦ən·di'stərbd 'wən ¦aút¦pút }

undisturbed-zero output [ELECTR] "Zero" output of a magnetic cell to which no partial-write pulses have been applied since that cell was last selected for reading. { ¦ən·di'stərbd 'zir·ō ¦aút¦pút }

undoing [PSYCH] A defense mechanism by which something unacceptable and already done is symbolically acted out in reverse, usually repetitiously, in the hope of relieving anxiety. { ən'dü·iŋ }

undulant fever See brucellosis. { 'ən·jə·lənt 'fē·vər }

undulating light [NAV] A continuously luminous light which alternately increases and decreases in brightness in cyclic sequence; the expression is applied primarily to aeronautical lights, the marine light equivalent being fixed and flashing light. { 'ən·jə¦lād·iŋ 'līt }

undulator [NUCLEO] A wiggler that has a relatively low magnetic field, resulting in coherent emission of radiation from all wiggles simultaneously, and has a relatively low-energy, directed or collimated beam of radiation that is comparable to a laser in many characteristics. Also known as interference wiggler. { 'ən·jə¦ād·ər }

undulatory extinction [OPTICS] Extinction that occurs successively in adjacent areas as the microscope stage is turned. Also known as oscillatory extinction; strain shadow; wavy extinction. { 'ən·jə·lə¦tȯr·ē ik'stiŋk·shən }

undulatus See billow cloud. { ən·jə'läd·əs }

unduloid [MATH] A surface formed by the rotation of a wavy line around a straight line parallel to its axis of symmetry. { 'ən·dyə¦lȯid }

unequal homologous recombination [GEN] Deoxyribonucleic acid exchange between identical chromosome regions that are not precisely paired. { ən¦ē·kwəl hə¦mäl·ə·gəs rē¦käm·bə'nā·shən }

unexploded ordnance [ORD] An object containing explosives which did not function as intended, or an object which contains some type of delay-action device. { ¦ən·ik'splōd·əd 'ȯrd·nəns }

unfavorable current [NAV] A current flowing in such a direction as to decrease the speed of a vessel. { ¦ən'fāv·rə·bəl 'kə·rənt }

unfavorable winds [NAV] Winds which delay the progress of a craft in a desired direction; used chiefly in connection with sailing vessels. { ¦ən'fāv·rə·bəl 'winz }

unfinished bolt [DES ENG] One of three degrees of finish in which standard hexagon wrench-head bolts and nuts are available; only the thread is finished. { ¦ən'fin·isht 'bōlt }

unfinished fabric [TEXT] A woolen fabric that is not fulled and sheared. { ¦ən'fin·isht 'fab·rik }

unfired pressure vessel [CHEM ENG] A pressure vessel that is not in direct contact with a heating flame. { ¦ən'fīrd 'presh·ər ¦ves·əl }

unfired tube [ELECTR] Condition of a TR (transmit-receive), ATR, or pre-TR tube in which there is no radio-frequency glow discharge at either the resonant gap or resonant window. { ¦ən'fīrd 'tüb }

unformatted file [COMPUT SCI] Any data file, such as a text file, that does not have various properties such as a consistent structure with regard to record length and order of data elements. { ¦ən'fȯr¦mad·əd 'fīl }

unfreezing [GEOL] The upward movement of stones to the surface as a result of repeated freezing and thawing of the containing soil. { ¦ən'frēz·iŋ }

ungemachite [MINERAL] $K_3Na_9Fe(SO_4)_6(OH)_3 \cdot 9H_2O$ A colorless to pale yellow, hexagonal mineral consisting of a hydrated basic sulfate of potassium, sodium, and iron; occurs in tabular form. { 'əŋ·gə·mä¦kīt }

unglazed tile [MATER] A hard, dense tile of homogeneous composition that derives its texture and color from the materials and method of manufacture. { ¦ən¦glāzd 'tīl }

ungula [MATH] A solid bounded by a portion of a circular cylindrical surface and portions of two planes, one of which is perpendicular to the generators of the cylindrical surface. [VERT ZOO] A nail, hoof, or claw. { 'əŋ·gyə·lə }

ungulate [VERT ZOO] Referring to an animal that has hoofs. { 'əŋ·gyə·lət }

ungulicutate [VERT ZOO] Having claws or nails. { ¦əŋ·gyə'lik·ə,tāt }

unguligrade [VERT ZOO] Walking on hoofs. { 'əŋ·gyə·lə,grād }

uniaxial crystal [OPTICS] A doubly refracting crystal which has a single axis along which light can propagate without exhibiting double refraction. { ¦yü·nē'ak·sē·əl 'krist·əl }

uniaxial stress [MECH] A state of stress in which two of the three principal stresses are zero. { ¦yü·nē'ak·sē·əl 'stres }

unicellular [BIOL] Composed of a single cell. { ¦yü·nə'sel·yə·lər }

unicellular gland [ANAT] A gland consisting of a single cell. { ¦yü·nə'sel·yə·lər ,gland }

unicuspid [ANAT] Having one cusp, as certain teeth. { ¦yü·nə'kəs·pəd }

unidentate ligand [CHEM] A ligand that donates one pair of electrons in a complexation reaction to form coordinate bonds. { ¦yü·nē¦den,tāt 'līg·ənd }

unidentified flying object [SCI TECH] Any reported flying object which cannot be identified or explained. Abbreviated UFO. Also known as flying saucer. { ¦ən·ī'den·tə,fīd 'flī·iŋ 'äb,jekt }

unidirectional [PHYS] **1.** Flowing in only one direction, such as direct current. **2.** Radiating in only one direction. { ¦yü·nə·də'rek·shən·əl }

unidirectional antenna [ELECTROMAG] An antenna that has a single well-defined direction of maximum gain. { ¦yü·nə·də'rek·shən·əl an'ten·ə }

unidirectional coupler [ELECTR] Directional coupler that samples only one direction of transmission. { ¦yü·nə·də'rek·shən·əl 'kəp·lər }

unidirectional hydrophone [ENG ACOUS] A hydrophone mainly sensitive to sound that is incident from a single solid angle of one hemisphere or less. { ¦yü·nə·də'rek·shən·əl 'hī·drə,fōn }

unidirectional log-periodic antenna [ELECTROMAG] A broad-band antenna in which the cut-out portions of a log-periodic antenna are mounted at an angle to each other, to give a unidirectional radiation pattern in which the major radiation is in the backward direction, off the apex of the antenna; impedance is essentially constant for all frequencies, as is the radiation pattern. { ¦yü·nə·də'rek·shən·əl 'läg ,pir·ē'äd·ik an'ten·ə }

unidirectional microphone [ENG ACOUS] A microphone that is responsive predominantly to sound incident from one hemisphere, without picking up sounds from the sides or rear. { ¦yü·nə·də'rek·shən·əl 'mī·krə,fōn }

unidirectional pulse-amplitude modulation [COMMUN] Modulation of pulse-amplitude type in which all pulses rise in the same direction. Also known as single-polarity pulse-amplitude modulation. { ¦yü·nə·də'rek·shən·əl ¦pəls 'am·plə,tüd ,maj·ə'lā·shən }

unidirectional pulses [ELECTR] Single polarity pulses which all rise in the same direction. { ¦yü·nə·də'rek·shən·əl 'pəl·səz }

unidirectional transducer [ELECTR] Transducer that measures stimuli in only one direction from a reference zero or rest position. Also known as unilateral transducer. { ¦yü·nə·də'rek·shən·əl tranz'dü·sər }

unified field theory [RELAT] Any theory which attempts to express gravitational theory and electromagnetic theory within a single unified framework; usually, an attempt to generalize Einstein's general theory of relativity from a theory of gravitation alone to a theory of gravitation and classical electromagnetism. { ¦yü·nə,fīd 'fēld ,thē·ə·rē }

unified screw thread [DES ENG] Three series of threads: coarse (UNC), fine (UNF), and extra fine (UNEF); a 1/4-inch-diameter (0.006-millimeter) thread in the UNC series has 20 threads per inch, while in the UNF series it has 28. { ¦yü·nə,fīd 'skrü ,thred }

unifilar suspension [ENG] The suspension of a body from a single thread, wire, or strip. { ¦yü·nə'fil·ər sə'spen·chən }

uniflow engine [MECH ENG] A steam engine in which steam enters the cylinder through valves at one end and escapes through openings uncovered by the piston as it completes its stroke. { ¦yü·nə,flō 'en·jən }

uniform bound [MATH] A number M such that $|f_n(x)| < M$

for every x and for every function in a given sequence of functions $\{f_n(x)\}$. { ¦yü·nə,fòrm 'baùnd }

uniform boundedness principle [MATH] A family of pointwise bounded, real-valued continuous functions on a complete metric space X is uniformly bounded on some open subset of X. { ¦yü·nə,fòrm 'baùn·dəd·nəs ,prin·sə·pəl }

uniform circular motion [MECH] Circular motion in which the angular velocity remains constant. { ¦yü·nə,fòrm 'sər·kyə·lər 'mō·shən }

uniform click track [ENG ACOUS] A click track with regularly spaced clicks. { ¦yü·nə,fòrm 'klik ,trak }

uniform continuity [MATH] A property of a function f on a set, namely: given any $\epsilon > 0$ there is a $\delta > 0$ such that $|f(x_1) - f(x_2)| < \epsilon$ provided $|x_1 - x_2| < \delta$ for any pair x_1, x_2 in the set. { ¦yü·nə,fòrm känt·ən'ü·əd·ē }

uniform convergence [MATH] A sequence of functions $\{f_n(x)\}$ converges uniformly on E to $f(x)$ if given $\epsilon > 0$ there is an N such that $|f_n(x) - f(x)| < \epsilon$ for all x in E provided $n > N$. { ¦yü·nə,fòrm kən'vər·jəns }

uniform corrosion [MET] Corrosion which takes place uniformly over the entire exposed surfaces. { ¦yü·nə,fòrm kə'rō·zhən }

uniform distribution [STAT] The distribution of a random variable in which each value has the same probability of occurrence. Also known as rectangular distribution. { ¦yü·nə,fòrm ,di·strə'byü·shən }

uniform field [PHYS] A field which, at the instant under consideration, has the same value at every point in the region under consideration. { ¦yü·nə,fòrm 'fēld }

uniform-field microtron [NUCLEO] A microtron with a small accelerating section which is placed in a uniform magnetic field that guides the electrons in circular orbits. { ¦yü·nə,fòrm 'fēld 'mī·krə,trän }

uniformitarianism [GEOL] Classically, the concept that the present is the key to the past; the principle that contemporary geologic processes have occurred in the same regular manner and with essentially the same intensity throughout geologic time, and that events of the geologic past can be explained by phenomena observable today. Also known as actualism; principle of uniformity. { ,yü·nə,fòr·mə'ter·ē·ə,niz·əm }

uniformity [MATH] A family of subsets of the direct product of a topological space with itself that is used to derive a uniform topology for the space. Also known as uniform structure. { ,yü·nə'fòr·məd·ē }

uniformity trial [STAT] The repetition of an experiment under exactly the same controlled conditions as in the original trial. { ,yü·nə'fòr·məd·ē ,trīl }

uniform line [ELEC] Line which has substantially identical electrical properties through its length. { ¦yü·nə,fòrm 'līn }

uniform load [MECH] A load distributed uniformly over a portion or over the entire length of a beam; measured in pounds per foot. { ¦yü·nə,fòrm 'lōd }

uniform luminance [OPTICS] Property of a surface for which the luminous intensity of any area of the surface is proportional to the area. { ¦yü·nə,fòrm 'lü·mə·nəns }

uniformly accessible storage See random-access memory. { ,yü·nə'fòrm·lē ak'ses·ə·bəl 'stòr·ij }

uniformly convex space [MATH] A normed vector space such that for any number $\epsilon > 0$ there is a number $\delta > 0$ such that, for any two vectors x and y, if $\|x\| \leq 1 + \delta$, $\|y\| \leq 1 + \delta$, and $\|x + y\| > 2$, then $\|x - y\| < \epsilon$. Also known as uniformly rotund space. { ¦yü·nə,fòrm·lē 'kän,veks 'spās }

uniformly equicontinuous family of functions See equicontinuous family of functions. { ,yü·nə¦fòrm·lē ¦ek·wē·kə¦tin·yə·wəs ¦fam·lē əv 'fəŋk·shənz }

uniformly most powerful test [STAT] A test which is simultaneously most powerful for all alternatives of interest in an experiment. { ,yü·nə¦fòrm·lē ,mōst 'paù·ər·fəl ,test }

uniformly rotund space See uniformly convex space. { ¦yü·nə,fòrm·lē rō¦tənd 'spās }

uniformly nonsquare Banach space [MATH] A Banach space for which there is a positive number ϵ such that there are no nonzero elements, x and y, that have the same norm and satisfy the inequalities $\|x + y\| > (2 - \epsilon) \|x\|$ and $\|x - y\| > (2 - \epsilon) \|x\|$. { ¦yü·nə,fòrm·lē ¦nän,skwer 'bä,nak ,spās }

uniformly summable series [MATH] For a given summability method and for a given interval, a series for which the sequence that defines the sum converges uniformly on the interval. { ¦yü·nə,fòrm·lē ¦səm·ə·bəl 'sir,ēz }

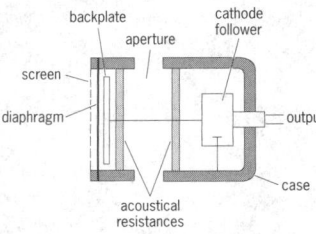

UNIDIRECTIONAL MICROPHONE

backplate
cathode follower
aperture
screen
diaphragm
output
case
acoustical resistances

Sectional view of a unidirectional microphone employing a condenser transducer.

uniform mat [CIV ENG] A type of foundation mat, consisting of a reinforced concrete slab of constant thickness, supporting walls, and columns; it is thick, rigid, and strong. { 'yü·nə,fȯrm 'mat }

uniform plane wave [ELECTROMAG] Plane wave in which the electric and magnetic intensities have constant amplitude over the equiphase surfaces; such a wave can only be found in free space at an infinite distance from the source. { 'yü·nə,fȯrm 'plān 'wāv }

uniform resource locator [COMPUT SCI] The unique Internet address assigned to a Web document or resource by which it can be accessed by all Web browsers. The first part of the address specifies the applicable Internet protocol, for example, http or ftp; the second part provides the IP address or domain name of the location. Abbreviated URL. { ,yü·nə,fȯrm ri,sȯrs 'lō·kād·ər }

uniform scale [MATH] A scale in which equal distances correspond to equal numerical values. Also known as linear scale. { 'yü·nə,fȯrm 'skāl }

uniform space [MATH] A topological space X whose topology is derived from a family of subsets of $X \times X$, called a uniformity; intuitively, this gives a notion of "nearness" which is uniform throughout the space. { 'yü·nə,fȯrm 'spās }

uniform structure See uniformity. { 'yü·nə,fȯrm 'strək·chər }

uniform system of maritime buoyage [NAV] The international system of buoyage in accordance with fixed rules as to shape, color, lights, topmarks, position, and so forth. { 'yü·nə,fȯrm 'sis·təm əv 'mar·ə,tīm 'bȯi·ij }

uniform topology [MATH] The topology of a uniform space. { 'yü·nə,fȯrm tə'päl·ə·jē }

uniform twist rifling [ORD] Rifling in which the degree of twist is constant from the origin of rifling to the muzzle, the path of the groove being a uniform spiral. { 'yü·nə,fȯrm ¦twist 'rīf·liŋ }

unijunction transistor [ELECTR] An n-type bar of semiconductor with a p-type alloy region on one side; connections are made to base contacts at either end of the bar and to the p-region. Abbreviated UJT. Formely known as double-base diode; double-base junction diode. { 'yü·nə,jəŋk·shən tran'zis·tər }

unilateral bearing [NAV] A bearing obtained with a radio direction finder which does not have a possible reciprocal ambiguity. { ¦yü·nə¦lad·ə·rəl 'ber·iŋ }

unilateral conductivity [ELECTR] Conductivity in only one direction, as in a perfect rectifier. { ¦yü·nə¦lad·ə·rəl ,kän·dək'tiv·əd·ē }

unilateral hermaphroditism [ZOO] A type of hermaphroditism in which there is a combination of ovatestis on one side of the body with an ovary or testis on the other side. { ¦yü·nə'lad·ə·rəl hər'maf·rə·dī,tiz·əm }

unilateralization [ELECTR] Use of an external feedback circuit in a high-frequency transistor amplifier to prevent undesired oscillation by canceling both the resistive and reactive changes produced in the input circuit by internal voltage feedback; with neutralization, only the reactive changes are canceled. { ,yü·nə,lad·ə·rə·lə'zā·shən }

unilateral shift [MATH] A bounded linear operator on the Hilbert space of infinite sequences of complex numbers whose squared moduli form convergent series; it takes the sequence $(x_1, x_2, \ldots)$ to the sequence $(0, x_1, x_2, \ldots)$. { ,yü·ni¦lad·ə·rəl 'shift }

unilateral surface [MATH] A one-sided surface; equivalently, any nonorientable two-dimensional manifold such as the Möbius strip and the Klein bottle. { ¦yü·nə¦lad·ə·rəl 'sər·fəs }

unilateral tolerance method [DES ENG] Method of dimensioning and tolerancing wherein the tolerance is taken as plus or minus from an explicitly stated dimension; the dimension represents the size or location which is nearest the critical condition (that is maximum material condition), and the tolerance is applied either in a plus or minus direction, but not in both directions, in such a way that the permissible variation in size or location is away from the critical condition. { ¦yü·nə'lad·ə·rəl 'täl·ə·rəns ,meth·əd }

unilateral transducer See unidirectional transducer. { ¦yü·nə'lad·ə·rəl tranz'dü·sər }

unilocular [BIOL] Having a single cavity. { ¦yü·nē'läk·yə·lər }

unimodal [STAT] Referring to a distribution with only one mode. { ,yü·nə'mōd·əl }

unimodal sequence [MATH] A finite sequence of n real numbers, $a_1, a_2, \ldots, a_n$, for which there is a positive integer, j, greater than 1 and less than n, such that a_i is greater than a_{i-1} for i greater than 1 and less than j, a_j is greater than or equal to a_{j-1}, and a_i is less than a_{i-1} for i greater than j and equal to or less than n. { ¦yü·nə,mȯd·əl 'sē·kwəns }

unimodular matrix [MATH] A unimodulus matrix with integer entries. { ¦yü·nə'mäj·ə·lər 'mā·triks }

unimodulus matrix [MATH] A square matrix whose determinant is 1. { ¦yü·nə'mäj·ə·ləs 'mā·triks }

unimolecular reaction [PHYS CHEM] A chemical reaction involving only one molecular species as a reactant; for example, $2H_2O \rightarrow 2H_2 + O_2$, as in the electrolytic dissociation of water. { ¦yü·nə·mə'lek·yə·lər rē'ak·shən }

unimorph [ELECTR] A piezoelectric microphone that consists of a single piezoelectric disk cemented to a thin metal plate. { 'yü·nə,mȯrf }

uninhibited bladder [MED] An abnormal urinary bladder that shows only a variable loss of cerebral inhibition over reflex bladder contractions, representing, of all neurogenic bladders, the least variance from normal. { ¦ən·in'hib·əd·əd 'blad·ər }

uninterruptible power system [ELEC] A system that provides protection against primary alternating-current power failure and variations in power-line frequency and voltage. Abbreviated UPS. { ¦ən¦in·tə'rəp·tə·bəl 'pau̇·ər ,sis·təm }

union [COMPUT SCI] A data structure that can store items of different types, but can store only one item at a time. [DES ENG] A screwed or flanged pipe coupling usually in the form of a ring fitting around the outside of the joint. [MATH] **1.** A union of a given family of sets is a set consisting of those elements that are members of at least one set in the family. Also known as join. **2.** For two fuzzy sets A and B, the fuzzy set whose membership function has a value at any element x that is the maximum of the values of the membership functions of A and B at x. **3.** The union of two Boolean matrices A and B, with the same number of rows and columns, is the Boolean matrix whose element c_{ij} in row i and column j is the union of corresponding elements a_{ij} in A and b_{ij} in B. **4.** The union of two graphs is the graph whose set of vertices is the union of the sets of vertices of the two graphs, and whose set of edges is the union of the sets of edges of the two graphs. { 'yün·yən }

union catalog [COMPUT SCI] A merged listing of the contents of two or more catalogs (of libraries, for example). { 'yün·yən 'kad·əl,äg }

union cloth [TEXT] Fabric made by using one kind of fiber for the warp and another kind for the filling. { 'yün·yən ,klȯth }

union dye [TEXT] A special dye used on union cloth. { 'yün·yən ,dī }

Unionidae [INV ZOO] The fresh-water mussels, a family of bivalve mollusks in the subclass Eulamellibranchia; the larvae, known as glochidia, are parasitic on fish. { ,yü·nē'än·ə,dē }

union joint [DES ENG] A threaded assembly used for the joining of ends of lengths of installed pipe or tubing where rotation of neither length is feasible. { 'yün·yən ,jȯint }

union rule of probability [STAT] The probability that the union of two events E_i and E_j equals the sum total of the probability of the sample points in either E_i or E_j minus the probability of being in both E_i and E_j. { 'yün·yən ¦rül əv ,präb·ə'bil·əd·ē }

union shop [IND ENG] An establishment in which union membership is not a requirement for original employment but becomes mandatory after a specified period of time. { 'yün·yən 'shäp }

uniparental disomy [GEN] Inheritance of both chromosomes or alleles of a homologous pair from one parent. { ¦yü·nə·pə¦rent·əl dī'sō·mē }

unipath See nonshared control unit. { 'yü·nə,path }

unipolar [ELEC] Having but one pole, polarity, or direction; when applied to amplifiers or power supplies, it means that the output can vary in only one polarity from zero and, therefore, must always contain a direct-current component. { ¦yü·nə'pō·lər }

Unipolarina [INV ZOO] A suborder of the protozoan order Myxosporida characterized by spores with one to six (never five) polar capsules located at the anterior end. { ,yü·nə,pō·lə'rīn·ə }

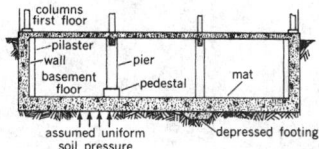

UNIFORM MAT

Foundation construction with a uniform mat.

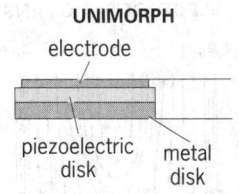

UNIMORPH

Unimorph piezoelectric transducer.

unipolar machine *See* homopolar generator. { ¦yü·nə'pō·lər mə'shēn }

unipolar transistor [ELECTR] A transistor that utilizes charge carriers of only one polarity, such as a field-effect transistor. { ¦yü·nə'pō·lər tran'zis·tər }

unipole [ELECTROMAG] A hypothetical antenna that radiates or receives signals equally well in all directions. Also known as isotropic antenna. { 'yü·nə,pōl }

uniporter [CELL MOL] A channel protein that transfers only one substrate at a time across the membrane. { 'yü·nə,pȯrd·ər }

unipotential cathode *See* indirectly heated cathode. { ¦yü·nə·pə'ten·chəl 'kath,ōd }

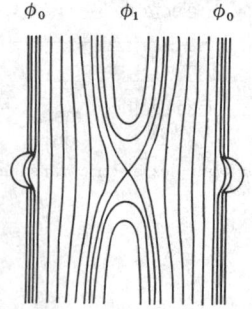

UNIPOTENTIAL ELECTROSTATIC LENS

ϕ_0 ϕ_1 ϕ_0

A unipotential electrostatic lens, a type of axially symmetric electrostatic lens. ϕ_0 = common potential for outer two apertures; ϕ_1 = lower potential for central aperture.

unipotential electrostatic lens [ELECTR] An electrostatic lens in which the focusing is produced by application of a single potential difference; in its simplest form it consists of three apertures of which the outer two are at a common potential, and the central aperture is at a different, generally lower, potential. { ¦yü·nə·pə'ten·chəl i¦lek·trə¦stad·ik 'lenz }

uniprocessor [COMPUT SCI] A computer that has a single central processing unit and works sequentially on only one program at a time. { 'yü·nə,prä,ses·ər }

unique-factorization domain [MATH] An integral domain in which every element that is neither a unit nor a prime has an expression as the product of a finite number of primes, and this expression is unique except for unit factors and the order of factors. Also known as factorial ring; unique-factorization ring. { yü¦nēk ,fak·tə·rə'zā·shə dō,mān }

unique-factorization ring *See* unique-factorization domain. { yü¦nēk ,fak·tə·rə'zā·shən ,riŋ }

unique factorization theorem [MATH] A positive integer may be expressed in precisely one way as a product of prime numbers. { yü¦nēk ,fak·tə·rə'zā·shən ,thir·əm }

unit [ENG] An assembly or device capable of independent operation, such as a radio receiver, cathode-ray oscilloscope, or computer subassembly that performs some inclusive operation or function. [MATH] An element of a ring with identity that has both a left inverse and a right inverse. [ORD] **1.** Any military element whose structure is prescribed by competent authority, such as a table of organization and equipment; specifically, part of an organization. **2.** A standard of basic quantity into which an item of supply is divided, issued, or used. [PHYS] A quantity adopted as a standard of measurement. { 'yü·nət }

unital left module [MATH] A left module over a ring with a unit element, 1, such that, for any element x of the module, $1 \cdot x = x$. { ¦yü·nat·əl ¦left 'mäj·əl }

unital module [MATH] A module over a ring with a unit element, 1, such that $1 \cdot x = x$ for any element x of the module. { ¦yü·nə·təl ,mäj·əl }

unit-area acoustical ohm *See* rayl. { 'yü·nət 'er·ē·ə ə'küs·tə·kəl 'ōm }

unitarity condition [PART PHYS] The condition that the scattering matrix for any process be unitary, as a result of the fact that the probability for the system to end in some final state must be unity. { ,yü·nə'tar·əd·ē kən,dish·ən }

unitary air conditioner [MECH ENG] A small self-contained electrical unit enclosing a motor-driven refrigeration compressor, evaporative cooling coil, air-cooled condenser, filters, fans, and controls. { 'yü·nə,ter·ē 'er kən,dish·ən·ər }

unitary decuplet [PART PHYS] A collection of 10 hadrons whose isospin and hypercharge values form a symmetrical pattern, and which are related by unitary symmetry operations. { 'yü·nə,ter·ē dē'kəp·lət }

unitary group [MATH] The group of unitary transformations on a k-dimensional complex vector space. Usually denoted $U(k)$. { 'yü·nə,ter·ē 'grüp }

unitary matrix [MATH] A matrix whose inverse is equal to the complex conjugate of its transpose. { 'yü·nə,ter·ē 'mā·triks }

unitary octet [PART PHYS] A collection of eight hadrons whose isospin and hypercharge values form a symmetrical pattern, and which are related by unitary symmetry operations. { 'yü·nə,ter·ē äk'tet }

unitary space [MATH] A finite-dimensional inner-product space over the field of complex numbers. { 'yü·nə,ter·ē 'spās }

unitary spin [PART PHYS] A quantum number associated with SU₃ symmetry and which determines the SU₃ supermultiplet to which a particle belongs, such as singlet, octet, or decuplet. { 'yü·nə,ter·ē 'spin }

unitary symmetry [PART PHYS] An approximate internal symmetry law obeyed by the strong interactions of elementary particles; a system of particles has such a symmetry if all the particles can be described as compounds of a fundamental multiplet of particles, and if all physical properties of the system are unchanged by an arbitrary unitary transformation of this fundamental multiplet. { 'yü·nə,ter·ē 'sim·ə·trē }

unitary transformation [MATH] A linear transformation on a vector space which preserves inner products and norms; alternatively, a linear operator whose adjoint is equal to its inverse. { 'yü·nə,ter·ē ,tranz·fər'mā·shən }

unit assembly [IND ENG] Assemblage of machine parts which constitutes a complete auxiliary part of an end item, and which performs a specific auxiliary function, and which may be removed from the parent item without itself being disassembled. { 'yü·nət ə'sem·blē }

unit ball [MATH] The set of all points in euclidean n-space whose distance from the origin is at most 1. { 'yü·nət 'bȯl }

unit binormal [MATH] A unit vector in the same direction as the binormal to a point on a surface or space curve. { 'yü·nət bī'nȯr·məl }

unit cell [CRYSTAL] A parallelepiped which will fill all space under the action of translations which leave the crystal lattice unchanged. Also known as structure cell. [MIN ENG] In flotation, a single cell. { 'yü·nət 'sel }

unit charge *See* statcoulomb. { 'yü·nət 'chärj }

unit circle [MATH] The locus of points in the plane which are precisely one unit from the origin. { 'yü·nət 'sər·kəl }

unit complex number [MATH] A complex number whose absolute value is 1. { ¦yü·nət 'käm,pleks 'nəm·bər }

unit construction [BUILD] An assembly comprising two or more walls, plus floor and ceiling construction, ready for shipping to a building site. { 'yü·nət kən'strək·shən }

unit conversion factor *See* conversion factor. { ¦yü·nət kən'vər·zhən ,fak·tər }

unit cost [IND ENG] Cost allocated to a specified unit of a product; computed as the cost over a period of time divided by the number of units produced. { 'yü·nət 'kȯst }

unit delay [ELECTR] A network whose output is equal to the input delayed by one unit of time. { 'yü·nət di'lā }

unit die [MET] A die block having more than one cavity insert and allowing several different castings to be made. { 'yü·nət 'dī }

United States airways code [METEOROL] A synoptic code for communicating aviation weather observations. Also known as airways code. { yə'nīd·əd 'stāts 'er,wāz ,kōd }

United States standard dry seal thread [DES ENG] A modified pipe thread used for pressure-tight connections that are to be assembled without lubricant or sealer in refrigeration pipes, automotive and aircraft fuel-line fittings, and gas and chemical shells. { yə'nīd·əd 'stāts 'stan·dərd 'drī ¦sēl ,thred }

United States Survey foot [GEOD] The foot used by the U.S. Coast and Geodetic Survey in which 1 inch is equal to 2.540005 centimeters. { ¦yü¦es 'sər,vā 'fut }

unitegmic [BOT] Referring to an ovule having a single integument. { ,yü·nə'teg·mik }

unit element [MATH] An element in a ring which acts as a multiplicative identity. { 'yü·nət 'el·ə·mənt }

uniterm [COMPUT SCI] A word, symbol, or number used as a description for retrieval of information from a collection; especially, such a description used in a coordinate indexing system. { 'yü·nə,tərm }

uniterm system [COMPUT SCI] An information retrieval system which uses uniterm cards; cards representing words of interest in a search are selected and compared visually; if identical members are found to appear on the uniterm card undergoing comparison, those numbers represent documents to be examined in connection with the search. { 'yü·nə,tərm ,sis·təm }

unit fraction [MATH] A common fraction whose numerator is unity. { 'yü·nət ,frak·shən }

unit heater [MECH ENG] A heater consisting of a fan for circulating air over a heat-exchange surface, all enclosed in a common casing. { 'yü·nət 'hēd·ər }

unit impulse *See* delta function. { 'yü·nət 'im,pəls }

unitized body [ENG] An automotive body that has the body and frame in one unit; side members are designed on the principle of a bridge truss to gain stiffness, and sheet metal of the body is stressed so that it carries some of the load. { 'yü·nə‚tīzd 'bäd·ē }

unitized cargo [IND ENG] Grouped cargo carried aboard a ship in pallets, containers, wheeled vehicles, and barges or lighters. { 'yü·nə‚tīzd 'kär·gō }

unitized film [GRAPHICS] Film which is filed or used by an individual frame or a group of frames under one classification. { 'yü·nə‚tīzd 'film }

unitized load [IND ENG] A single item or a number of items packaged, packed, or arranged in a specified manner and capable of being handled as a unit; unitization may be accomplished by placing the item or items in a container or by banding them securely together. Also known as unit load. { 'yü·nə‚tīzd 'lōd }

unitized tooling [DES ENG] A die having its upper and lower members incorporated into a self-contained unit arranged to maintain the die members in alignment. { 'yü·nə‚tīzd ‚tül·iŋ }

unit length [COMMUN] Basic element of time used in determining signaling speeds in message transmission. { 'yü·nət 'leŋkth }

unit load See unitized load. { 'yü·nət 'lōd }

unit loading [ORD] The loading of troop units with their equipment and supplies in the same ships, aircraft, or land vehicles. { 'yü·nət 'lōd·iŋ }

unit magnetic pole [ELECTROMAG] Two equal magnetic poles of the same sign have unit value when they repel each other with a force of 1 dyne if placed 1 centimeter apart in a vacuum. { 'yü·nət mag'ned·ik 'pōl }

unit mold [ENG] A simple plastics mold composed of a simple cavity without further mold devices; used to produce sample containers having shapes difficult to blow-mold. { 'yü·nət 'mōld }

unit normal [MATH] A unit vector in the direction of the principal normal to a surface or space curve. { 'yü·nət 'nȯr·məl }

unit of coal [MIN ENG] The quantity of coal from which the sample is taken and which the sample represents. { 'yü·nət əv 'kōl }

unit of fire [ORD] A basic load of ammunition. { 'yü·nət əv 'fīr }

unit of issue [IND ENG] In reference to special storage, the quantity of an item, such as each number, dozen, gallon, pair, pound, ream, set, or yard. [ORD] The standard or basic quantity in which an item of supply is issued. { 'yü·nət əv 'ish·ü }

unit operations [CHEM ENG] The basic physical operations of chemical engineering in a chemical process plant, that is, distillation, fluid transport, heat and mass transfer, evaporation, extraction, drying, crystallization, filtration, mixing, size separation, crushing and grinding, and conveying. { 'yü·nət ‚äp·ə'rā·shənz }

unit operator [MATH] The identity operator. { 'yü·nət 'äp·ə‚rād·ər }

unitor [COMPUT SCI] In computers, a device or circuit which performs a function corresponding to the Boolean operation of union. { 'yü·nə·tər }

unit power [MET] A unit describing machinability of a metal; the power needed to remove a unit volume in unit time, usually expressed as horsepower per cubic inch per minute. { 'yü·nət 'pau̇·ər }

unit process [CHEM ENG] In chemical manufacturing, a process that involves chemical conversion. { 'yü·nət ‚prä·ses }

unit procurement cost [IND ENG] The net basic cost paid or estimated to be paid for a unit of a particular item including, where applicable, the cost of government-furnished property and the cost of manufacturing operations performed at government-owned facilities. { 'yü·nət prə'kyu̇r·mənt ‚kȯst }

unit record [COMPUT SCI] Any of a collection of records, all of which have the same form and the same data elements. { 'yü·nət 'rek·ərd }

unit record device [COMPUT SCI] Any piece of equipment such as punch card readers, card punch, and line printers. { 'yü·nət 'rek·ərd di‚vīs }

unit replacement [ORD] Method of repair in which a defective, worn, or damaged group of parts of a weapon or other equipment is replaced by a complete new group of parts. { 'yü·nət ri'plās·mənt }

unit reserves [ORD] Prescribed quantities of supplies carried by a unit as a reserve to cover emergencies. { 'yü·nət ri'zərvz }

unit sphere [MATH] The set of points in three-space (more generally *n*-space) which are precisely one unit distance from the origin. { 'yü·nət 'sfir }

unit strain [MECH] **1.** For tensile strain, the elongation per unit length. **2.** For compressive strain, the shortening per unit length. **3.** For shear strain, the change in angle between two lines originally perpendicular to each other. { 'yü·nət 'strān }

unit strength [ORD] As applied to a friendly or enemy unit, relates to the number of personnel, amount of supplies, armament equipment and vehicles, and the total logistic capabilities. { 'yü·nət 'streŋkth }

unit stress [MECH] The load per unit of area. { 'yü·nət 'stres }

unit string [COMPUT SCI] A string that has only one element. { 'yü·nət 'striŋ }

unit systems [OPTICS] Optical systems that have a lateral magnification of +1 or −1. [PHYS] Groups of units suitable for use in measurement of physical quantities and in the convenient statement of physical laws relating physical quantities. { 'yü·nət ‚sis·təmz }

unit tangent [MATH] A unit vector in the tangent plane at a point of a surface. { 'yü·nət 'tan·jənt }

unit test [COMPUT SCI] The testing of a module within a computer system. { 'yü·nət ‚test }

unit train [MIN ENG] A system for delivering coal in which a string of cars, with distinctive markings and loaded to full visible capacity, is operated without service frills or stops along the way for cars to be cut in and out. { 'yü·nət 'trān }

unit vector [MATH] A vector whose length is one unit. { 'yü·nət 'vek·tər }

unity coupling [ELECTROMAG] Perfect magnetic coupling between two coils, so that all magnetic flux produced by the primary winding passes through the entire secondary winding. { 'yü·nəd·ē ‚kəp·liŋ }

unity gain bandwidth [ELECTR] Measure of the gain-frequency product of an amplifier; unity gain bandwidth is the frequency at which the open-loop gain becomes unity, based on 6 decibels per octave crossing. { 'yü·nəd·ē ‚gān 'band‚width }

unity power factor [ELEC] Power factor of 1.0, obtained when current and voltage are in phase, as in a circuit containing only resistance or in a reactive circuit at resonance. { 'yü·nəd·ē 'pau̇·ər ‚fak·tər }

univalent function See injection. { ‚yü·nə‚vā·lənt 'fəŋk·shən }

univariant system [THERMO] A system which has only one degree of freedom according to the phase rule. { ‚yü·nə‚ver·ē·ənt 'sis·təm }

univariate distribution [STAT] A frequency distribution of only one variate. { ‚yü·nə‚ver·ē·ət ‚dis·trə'byü·shən }

universal algebra [MATH] The study of algebraic systems such as groups, rings, modules, and fields and the examination of what families of theorems are analogous in each system. { ‚yü·nə‚vər·səl 'al·jə·brə }

universal asynchronous receiver transmitter [COMPUT SCI] An electronic circuit that converts bytes of data between the parallel format in which bits are stored side by side within a device and the serial format whereby bits are transmitted sequentially over a communications line. Abbreviated UART. { ‚yü·nə‚vər·səl ā'siŋ·krə·nəs ri'sē·vər tranz'mid·ər }

universal chuck [ENG] A self-centering chuck whose jaws move in unison when a scroll plate is rotated. { ‚yü·nə‚vər·səl 'chək }

universal conductance fluctuations [ELECTR] Fluctuations in the conductance of a quantum wire, as a function of applied voltage, whose root-mean-square value depends only on the geometry of the device, and which are reproducible in any one sample but vary in detail from one sample to another. { ‚yü·nə‚vər·səl kən'dək·təns ‚flək·chə‚wā·shənz }

universal constants See fundamental constants. { ‚yü·nə‚vər·səl 'kän·stəns }

universal dividing head [MECH ENG] An accessory fixture on a milling machine that rotates the workpiece to specified angles between machining steps. { ‚yü·nə‚vər·səl di'vīd·iŋ ‚hed }

universal donor [IMMUNOL] An individual of O blood

group; can give blood to persons of all blood types. { ¦yü·nə¦vər·səl 'dō·nər }

universal element [MATH] An element of a Boolean algebra that includes every element of the algebra. { ¦yü·nə¦vər·səl 'el·ə·mənt }

universal gas constant See gas constant. { ¦yü·nə¦vər·səl 'gas ¦kän·stənt }

universal grinding machine [MECH ENG] A grinding machine having a swivel table and headstock, and a wheel head that can be rotated on its base. { ¦yü·nə¦vər·səl 'grīnd·iŋ mə¦shēn }

universal gripper [CONT SYS] A versatile robot component that can grasp most kinds of objects. { ¦yü·nə¦vər·səl 'grip·ər }

universal instrument See altazimuth. { ¦yü·nə¦vər·səl 'inz·trə·mənt }

universality [STAT MECH] The hypothesis that the critical exponents of a substance are the same within broad classes of substances of widely varying characteristics, and depend only on the microscopic symmetry properties of the substance. { ¦yü·nə·vər'sal·əd·ē }

universality class [STAT MECH] A class of substances which have the same critical exponents according to the universality hypothesis. { ¦yü·nə·vər'sal·əd·ē ¦klas }

universal joint [MECH ENG] A linkage that transmits rotation between two shafts whose axes are coplanar but not coinciding. { ¦yü·nə¦vər·səl 'jöint }

universal language [COMPUT SCI] A programming language that is widely employed to write programs that can be run on a wide variety of computers. { ¦yü·nə¦vər·səl 'laŋ·gwij }

universally attracting object [MATH] An object O in a category C such that there exists a unique morphism of each object of C into O. { ¦yü·nə¦vər·sə·lē ə'trak·tiŋ ¦äb¦jekt }

universally repelling object [MATH] An object O of a category C such that there exists a unique morphism of O into each object of C. { ¦yü·nə¦vər·sə·lē ri'pel·iŋ ¦äb¦jekt }

universal mill [MET] A rolling mill having both horizontal and vertical sets of rolls. { ¦yü·nə¦vər·səl 'mil }

universal motor [ELEC] A motor that may be operated at approximately the same speed and output on either direct current or single-phase alternating current. Also known as ac/dc motor. { ¦yü·nə¦vər·səl 'mōd·ər }

universal object [MATH] An object which is universally attracting or universally repelling. { ¦yü·nə¦vər·səl 'äb¦jekt }

universal output transformer [ENG ACOUS] An output transformer having a number of taps on its winding, to permit its use between the audio-frequency output stage and the loudspeaker of practically any radio receiver by proper choice of connections. { ¦yü·nə¦vər·səl 'aùt¦pùt tranz¦för·mər }

universal plotting sheet [NAV] A plotting sheet on which either the latitude or longitude lines are omitted, to be drawn in by the user, making it possible to quickly construct a plotting sheet for any part of the earth's surface. { ¦yü·nə¦vər·səl 'pläd·iŋ ¦shēt }

Universal Polar Stereographic Grid [MAP] A particular grid based upon the polar stereographic projection according to specifications laid down by military authorities; it may be superimposed on any map. { ¦yü·nə¦vər·səl 'pō·lər ¦ster·ē·ō¦graf·ik 'grid }

universal product code [COMPUT SCI] **1.** A 10-digit bar code on the outside of a package for electronic scanning at supermarket checkout counters; each digit is represented by the ratio of the widths of adjacent stripes and white areas. **2.** The corresponding combinations of binary digits into which the scanned bars are converted for computer processing that provides continuously updated inventory data and printout of the register tape at the checkout counter. { ¦yü·nə¦vər·səl 'präd·əkt ¦kōd }

universal quantifier [MATH] A logical relation, often symbolized λ, that may be expressed by the phrase "for all" or "for every"; if P is a predicate, the statement (λx)P(x) is true if P(x) is true for all values of x in the domain of P, and is false otherwise. { ¦yü·nə¦vər·səl 'kwän·tə¦fī·ər }

universal receiver See ac/dc receiver. { ¦yü·nə¦vər·səl ri'sē·vər }

universal recipient [IMMUNOL] An individual of AB blood group; can receive a blood transfusion of all blood types, A, AB, B, or O. { ¦yü·nə¦vər·səl ri'sip·ē·ənt }

universal resonance curve [ELEC] A plot of Y/Y_0 against $Q_0\delta$ for a series-resonant circuit, or of Z/Z_0 against $Q_0\delta$ for a parallel-resonant circuit, where Y and Z are the admittance and impedance of a circuit, Y_0 and Z_0 are the values of these quantities at resonance, Q_0 is the Q value of the circuit at resonance, and δ is the deviation of the frequency from resonance divided by the resonant frequency; it can be applied to all resonant circuits. { ¦yü·nə¦vər·səl 'rez·ən·əns ¦kərv }

universal robot [CONT SYS] A robot whose end effector would be flexible enough to perform any desired task. { ¦yü·nə¦vər·səl 'rō¦bät }

universal sequence [PHYS] A sequence of periodic states which reappear as a parameter governing a mapping is varied after the mapping has undergone a transition to chaos by a sequence of period-doubling bifurcations, and which have a particular order that does not depend on the details of the mapping, within certain limits. { ¦yü·nə¦vər·səl 'sē·kwəns }

universal serial bus [COMPUT SCI] A serial interface that can transfer data at up to 480 million bits per second and connect up to 127 daisy-chained peripheral devices. Abbreviated USB. { ¦yü·nə¦vər·səl ¦sir·ē·əl 'bəs }

universal set [MATH] A set that contains all the elements of concern in the study of a particular problem. { ¦yü·nə¦vər·səl 'set }

universal shunt See Ayrton shunt. { ¦yü·nə¦vər·səl 'shənt }

universal stage [OPTICS] A stage attached to the rotating stage of a polarizing microscope that has three, four, or five axes and thin sections of low-symmetry minerals to be tilted about two mutually perpendicular horizontal axes. Also known as Fedorov stage; U stage. { ¦yü·nə¦vər·səl 'stāj }

universal time See Greenwich mean time. { ¦yü·nə¦vər·səl 'tīm }

universal time 0 [ASTRON] The uncorrected time of the earth's rotation as measured by the transit of stars across the observer's meridian. Abbreviated UT 0. { ¦yü·nə¦vər·səl ¦tīm 'zir·ō }

universal time 1 [ASTRON] Universal time 0 corrected for polar motion; it is the true angular rotation. Abbreviated UT 1. { ¦yü·nə¦vər·səl ¦tīm 'wən }

universal time 2 [ASTRON] Universal time 1 corrected for seasonal variations in the earth's rotation. Abbreviated UT 2. { ¦yü·nə¦vər·səl ¦tīm 'tü }

universal time coordinated [ASTRON] The coordinated time kept by a uniformly running clock, approximating the measure UT 2. Abbreviated UTC. { ¦yü·nə¦vər·səl ¦tīm kō'örd·ən¦ād·əd }

universal transmission function [GEOPHYS] A mathematical relationship that attempts to describe quantitatively the complex infrared propagation (including absorption and reradiation) in the atmosphere. { ¦yü·nə¦vər·səl tranz'mish·ən ¦fəŋk·shən }

universal transverse Mercator grid [MAP] A particular grid based upon a transverse Mercator projection, according to specifications laid down by military authorities; it may be superimposed on any map. { ¦yü·nə¦vər·səl tranz'vərs mər¦kād·ər ¦grid }

universal Turing machine [COMPUT SCI] A Turing machine that can simulate any Turing machine. { ¦yü·nə¦vər·səl 'tùr·iŋ mə¦shēn }

universal vise [ENG] A vise which has two or three swivel settings so that the workpiece can be set at a compound angle. Also known as toolmaker's vise. { ¦yü·nə¦vər·səl 'vīs }

universal wavelength function [OPTICS] One of four functions which enable one to compute easily, with reasonable accuracy, the refractive index of glass or other transparent material when this index is known for four standard wavelengths. { ¦yü·nə¦vər·səl 'wāv¦leŋkth ¦fəŋk·shən }

universe [ASTRON] The totality of astronomical things, events, relations, and energies capable of being described objectively. { 'yü·nə¦vərs }

univibrator See monostable multivibrator. { ¦yü·nə'vī¦brād·ər }

Unix [COMPUT SCI] An operating system that was designed for use with microprocessors and with the C programming language, and that has been adopted for use with several 16-bit-microprocessor microcomputers. { 'yü·niks }

unlimited ceiling [METEOROL] A ceiling that exists when the total sky cover is less than 0.6%, or when the total transparent sky cover is 0.5% or more, or when surface-based obscuring phenomena are classed as partial obscuration (that is, they

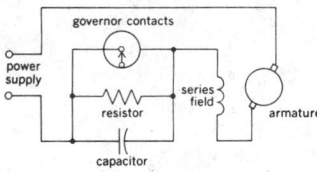

UNIVERSAL MOTOR

Universal motor circuit diagram. Centrifugal governor switches small resistor in or out, to obtain more constant speed with variations in load.

UNIVERSAL PRODUCT CODE

The first five digits of the code identify the manufacturer; the second five digits identify the specific product.

obscure 0.9% or less of the sky) and no layer aloft is reported as broken or overcast.　{ |ən'lim·əd·əd 'sē·liŋ }

unload [COMPUT SCI] To remove or copy data from a computer system.　{ ən'lōd }

unloaded Q [ELECTR] The Q of a system when there is no external coupling to it.　{ |ən'lōd·əd 'kyü }

unloader [MECH ENG] A power device for removing bulk materials from railway freight cars or highway trucks; in the case of railway cars, the car structure may aid the unloader; a transitional device between interplant transportation means and intraplant handling equipment.　{ ən'lōd·ər }

unloading [CHEM ENG] **1.** The release downstream of a trapped contaminant. **2.** A filter medium failure and release of system pressure. **3.** The depressuring or emptying of a process unit.　{ ən'lōd·iŋ }

unloading amplifer [ELECTR] Amplifier that is capable of reproducing or amplifying a given voltage signal while drawing negligible current from the voltage source.　{ |ən'lōd·iŋ 'am·plə,fī·ər }

unloading circuit [COMPUT SCI] In an analog computer, a computing element or combination of computing elements capable of reproducing or amplifying a given voltage signal while drawing negligible current from the voltage source.　{ |ən'lōd·iŋ ,sər·kət }

unloading conveyor [MECH ENG] Any of several types of portable conveyors adapted for unloading bulk materials, packages, or objects from conveyances.　{ |ən'lōd·iŋ kən'vā·ər }

unloading device [COMPUT SCI] Equipment that holds programs and other data that have been copied or removed from a computer system.　{ ən'lōd·iŋ di,vīs }

unmodified instruction See basic instruction.　{ |ən'mäd·ə,fīd in'strək·shən }

unmyelinated [HISTOL] Lacking myelin, either as a normal condition or as the result of a disease.　{ |ən'mī·ə·lə,nād·əd }

unpack [COMPUT SCI] **1.** To recover the individual data items contained in packed data. **2.** More specifically, to convert a packed decimal number into individual digits (and sometimes a sign).　{ |ən'pak }

unpaired electron [ATOM PHYS] An orbital electron for which there is no other electron in the same atom with the same energy but opposite spin.　{ |ən,perd i'lek,trän }

unpitched sound [ACOUS] Sound that includes a wide range of frequencies and thus does not have a definite pitch.　{ |ən,pichd 'saúnd }

unpolarized light [OPTICS] Light in which the electric vector is oriented in a random, unpredictable fashion.　{ |ən'pō·lə,rīzd 'līt }

unpolarized particle beam [PHYS] A beam of particles with spin in which the directions of the spins are random.　{ |ən'pō·lə,rīzd 'pard·ə·kəl ,bēm }

unproductive cough See dry cough.　{ ,ən·prə|dək·tiv 'kóf }

unproductive development [MIN ENG] The drifts, tunnels, and crosscuts driven in stone, preparatory to opening out production faces in a coal seam or ore body.　{ |ən·prə'dək·tiv də'vel·əp·mənt }

unprotect [COMPUT SCI] To remove restrictions on access to a file so that any computer program can read and alter the data contained in it.　{ |ən·prə'tekt }

unprotected reversing thermometer [ENG] A reversing thermometer for sea-water temperature which is not protected against hydrostatic pressure.　{ |ən·prə'tek·təd ri'vərs·iŋ thər'mäm·əd·ər }

unproven area [MIN ENG] An area in which it has not been established by drilling operations whether oil or gas may be found in commercial quantities.　{ |ən'prüv·ən 'er·ē·ə }

unrelated frequencies [STAT] The long run frequency of any result in one part of an experiment is approximately equal to the long run conditional frequency of that result, given that any specified result has occurred in the other part of the experiment.　{ |ən'lād·əd 'frē·kwən,sēz }

unreserved minerals [MIN ENG] Minerals which belong to the owner of the land on which or in which they are located.　{ |ən·ri'zərvd 'min·rəlz }

unresolved pneumonia See organizing pneumonia.　{ |ən·ri'zälvd nù'mōn·yə }

unrestricted element [IND ENG] An element of an operation that is entirely under the control of a worker.　{ |ən·ri'strik·təd 'el·ə·mənt }

unrestricted visibility [METEOROL] The visibility when no

obstruction to vision exists in sufficient quantity to reduce the visibility to less than 7 miles (11.3 kilometers).　{ |ən·ri'strik·təd ,viz·ə'bil·əd·ē }

uns-, unsym- [ORG CHEM] A chemical prefix denoting that the substituents of an organic compound are structurally unsymmetrical with respect to the carbon skeleton, or with respect to a function group (for example, double or triple bond).　{ əns, |ən·sim }

unsaturated [MINERAL] Referring to a mineral that will not form in the presence of free silica.　{ |ən'sach·ə,rād·əd }

unsaturated compound [CHEM] Any chemical compound with more than one bond between adjacent atoms, usually carbon, and thus reactive toward the addition of other atoms at that point; for example, olefins, diolefins, and unsaturated fatty acids.　{ |ən'sach·ə,rād·əd 'käm,paúnd }

unsaturated hydrocarbon [ORG CHEM] One of a class of hydrocarbons that have at least one double or triple carbon-to-carbon bond that is not in an aromatic ring; examples are ethylene, propadiene, and acetylene.　{ |ən'sach·ə,rād·əd |hī·drə'kär·bən }

unsaturated standard cell [ELEC] One of two types of Weston standard cells (batteries); used for voltage calibration work not requiring an accuracy greater than 0.01%.　{ |ən'sach·ə,rād·əd 'stan·dərd 'sel }

unsaturated zone See zone of aeration.　{ |ən'sach·ə,rād·əd 'zōn }

unsaturation [ORG CHEM] A state in which the atomic bonds of an organic compound's chain or ring are not completely satisfied (that is, not saturated); usually applies to carbon, but can include other ring or chain atoms; unsaturation usually results in a double bond (as for olefins) or a triple bond (as for the acetylenes).　{ |ən,sach·ə'rā·shən }

unscheduled maintenance [IND ENG] Those unpredictable maintenance requirements that had not been previously planned or programmed but require prompt attention and must be added to, integrated with, or substituted for previously scheduled workloads.　{ |ən'skej·əld 'mānt·ən·əns }

unscrambler [IND ENG] A part of a feeding and packaging line that aids in arranging cartons for the filling machines; there are rotary, straight-line, and walking-beam types.　{ |ən'skram·blər }

unseen fire [ORD] Fire which is continuously aimed at the future position of an aircraft, the aim being derived from radar sources.　{ |ən'sēn 'fīr }

unsettled [METEOROL] Pertaining to fair weather which may at any time become rainy, cloudy, or stormy.　{ |ən'sed·əld }

unsigned integer [MATH] A whole number that is equal to or greater than zero and does not carry a positive or negative sign.　{ ən'sīnd 'int·ə·jər }

unsigned real number [MATH] A number that does not carry a sign indicating whether it is positive or negative, and that is therefore assumed to be positive.　{ ən'sīnd 'rēl 'nəm·bər }

Unsin engine [MECH ENG] A type of rotary engine in which the trochoidal rotors of eccentric-rotor engines are replaced with two circular rotors, one of which has a single gear tooth upon which gas pressure acts, and the second rotor has a slot which accepts the gear tooth.　{ 'ən·sən ,en·jən }

unsolicited message [COMPUT SCI] A warning or error message that is automatically issued by a computer program when it detects a problem, and that does not depend on the operator making a query.　{ ,ən·sə'lis·əd·əd 'mes·ij }

unsprung axle [MECH ENG] A rear axle in an automobile in which the housing carries the right and left rear-axle shafts and the wheels are mounted at the outer end of each shaft.　{ |ən'sprəŋ 'ak·səl }

unsprung weight [MECH ENG] The weight of the various parts of a vehicle that are not carried on the springs, such as wheels, axles, and brakes.　{ |ən'sprəŋ 'wāt }

unstable [PHYS] Capable of undergoing spontaneous change, as in a radioactive nuclide or an excited nuclear system.　{ |ən'stā·bəl }

unstable colon See irritable colon.　{ |ən'stā·bəl 'kō·lən }

unstable equilibrium [PHYS] An equilibrium state of a system in which any departure of the system from equilibrium gives rise to forces or tendencies moving the system further away from equilibrium; for example, mechanical equilibrium in which the potential energy is a maximum, as a sphere sitting on top of a hill.　{ |ən'stā·bəl ē·kwə'lib·rē·əm }

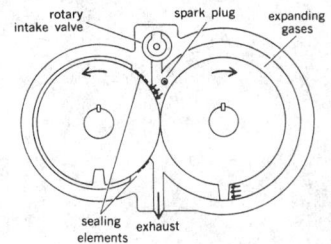

UNSIN ENGINE

rotary intake valve — spark plug — expanding gases

sealing elements — exhaust

Section through an Unsin engine showing two circular rotors.

unstable graph [MATH]　A graph from which it is not possible to delete an edge to produce a subgraph whose group of automorphisms is a subgroup of the group of automorphisms of the original graph. { ¦ən‚stā·bəl 'graf }

unstable isotope *See* radioisotope. { ¦ən'stā·bəl 'ī·sə‚tōp }

unstable particle [PARTIC PHYS]　**1.** Any elementary particle that spontaneously decays into other particles. **2.** An elementary particle that can decay through the strong interactions, as opposed to a semistable particle; it has a lifetime on the order of 10^{-23} second. { ¦ən'stā·bəl 'pärd·ə·kəl }

unstable wave [PHYS]　A wave motion whose amplitude increases with time or whose total energy increases at the expense of its environment. { ¦ən'stā·bəl 'wāv }

unsteady flow [FL MECH]　Fluid flow in which properties of the flow change with respect to time. { ¦ən'sted·ē 'flō }

unsteady-state flow [FL MECH]　A condition of fluid flow in which the volumetric ratios of two or more phases (liquid-gas, liquid-liquid, and so on) vary along the course of flow; can be the result of changes in temperature, pressure, or composition. { ¦ən'sted·ē ¦stāt 'flō }

unstriated fiber *See* smooth muscle fiber. { ¦ən'strī‚ād·əd 'fī·bərz }

unstructured grid [MATH]　In the discretization of partial differential equations, a collection of triangular elements or a random distribution of points at which the equations are expressed in discrete form. { ¦ən·strək·chərd 'grid }

unsurveyed area [MAP]　An area on a map or chart where both relief and planimetric data are unavailable, and which is usually labeled unsurveyed; or an area on a map or chart which shows little or no charted data because accurate information is limited or not available. { ¦ən·sər'vād 'er·ē·ə }

untuned [ELEC]　Not resonant at any of the frequencies being handled. { ¦ən'tünd }

unused time *See* unattended time. { ¦ən'yüzd 'tīm }

unvoiced sound [LING]　Speech sounds produced when the vocal folds are apart and are not vibrating. { ¦ən‚vȯist 'saund }

unwater [ENG]　To remove or draw off water; to drain. { ¦ən'wȯd·ər }

unwind [COMPUT SCI]　In computers, to rearrange and code a sequence of instructions to eliminate red-tape operations. [MECH ENG]　To reverse the direction of rotation of a threaded device. { ¦ən'wīnd }

up [ENG]　Fully in operation. [GRAPHICS]　In flat-bed terminology, indicating one or more documents being in position to photograph, for example, two up. { əp }

up-and-down method [STAT]　A technique which uses a unit sequential method of testing; the level of variable at each experiment is raised or lowered depending on the outcome of the previous test. { ¦əp ən ¦daun ‚meth·əd }

upcast [MIN ENG]　**1.** The opening through which the return air ascends and is removed from the mine; the opposite of downcast or intake. **2.** An upward current of air passing through a shaft. **3.** Material that has been thrown up, as by digging. { 'əp‚kast }

up-converter [ELECTR]　Type of parametric amplifier which is characterized by the frequency of the output signal being greater than the frequency of the input signal. { 'əp kən‚vərd·ər }

update [COMPUT SCI]　**1.** In computers, to modify an instruction so that the address numbers it contains are increased by a stated amount each time the instruction is performed. **2.** To change a record by entering current information; for example, to enter a new address or account number in the record pertaining to an employee or customer. Also known as posting. { 'əp¦dāt }

update service [COMPUT SCI]　A service that guarantees installation of updates to products within a certain period of time after they become available. { 'əp‚dāt ‚sər·vəs }

up-Doppler [ENG ACOUS]　The sonar situation wherein the target is moving toward the transducer, so the frequency of the echo is greater than the frequency of the reverberations received immediately after the end of the outgoing ping; opposite of down-Doppler. { 'əp ‚däp·lər }

updraft carburetor [MECH ENG]　For a gasoline engine, a fuel-air mixing device in which both the fuel jet and the airflow are upward. { 'əp‚draft 'kär·bə‚rād·ər }

updraft furnace [MECH ENG]　A furnace in which volumes of air are supplied from below the fuel bed or supply. { 'əp‚draft 'fər·nəs }

updrift [OCEANOGR]　The direction which is opposite that of the prevailing movement of littoral material. { 'əp‚drift }

upflow [CHEM]　In an ion-exchange unit, an operation in which solutions enter at the bottom of the unit and leave at the top. { 'əp‚flō }

upgrade [MIN ENG]　**1.** To increase the commercial value of a coal or mineral product by appropriate treatment. **2.** To increase the quality rating of diamonds beyond or above the rating implied by their particular classification. { 'əp¦grād }

uphole [MIN ENG]　A borehole collared in an underground working place and drilled in a direction pointed above the horizontal plane of the drill-machine swivel head. { 'əp‚hōl }

uphole time [GEOPHYS]　The time that a seismic pulse requires to travel from an explosion at some depth in a shot hole to the surface of the earth. { 'əp‚hōl ‚tīm }

upland [GEOGR]　**1.** An extensive region of high land. **2.** The higher ground of a region, in contrast to a valley, plain, or other low-lying land. **3.** The elevated land above the low areas along a stream or between hills. { 'əp·lənd }

uplift pressure [CIV ENG]　Pressure in an upward direction against the bottom of a structure, as a dam, a road slab, or a basement floor. { 'əp‚lift ‚presh·ər }

uplink [COMMUN]　The radio or optical transmission path upward from the earth to a communications satellite, or from the earth to aircraft. { 'əp‚liŋk }

upload [COMPUT SCI]　To transfer or copy data from a smaller computer, such as a microcomputer, to a larger computer. { 'əp‚lōd }

upmilling [MECH ENG]　Milling a workpiece by rotating the cutter against the direction of feed of the workpiece. { 'əp‚mil·iŋ }

upper [GEOL]　Pertaining to rocks or strata that normally overlie those of earlier formations of the same subdivision of rocks. { 'əp·ər }

upper air [METEOROL]　The region of the atmosphere which is above the lower troposphere; although no distinct lower limit is set, the term is generally applied to levels above that at which the pressure is 850 millibars. { 'əp·ər 'er }

upper-air chart *See* upper-level chart. { 'əp·ər 'er 'chärt }

upper-air disturbance [METEOROL]　A disturbance of the flow pattern in the upper air, particularly one which is more strongly developed aloft than near the ground. Also known as upper-level disturbance. { 'əp·ər 'er di‚stər·bəns }

upper-air observation [METEOROL]　A measurement of atmospheric conditions aloft, above the effective range of a surface weather observation. Also known as sounding; upper-air sounding. { 'əp·ər 'er ‚äb·zər‚vā·shən }

upper-air sounding *See* upper-air observation. { 'əp·ər 'er ‚saund·iŋ }

upper anticyclone *See* upper-level anticyclone. { 'əp·ər ‚ant·i'sī‚klōn }

upper-arm circumference [ANTHRO]　A measure of the horizontal circumference at the largest part of the biceps muscle. { 'əp·ər ¦ärm sər'kəm·frəns }

upper atmosphere [METEOROL]　The general term applied to the atmosphere above the troposphere. { 'əp·ər 'at·mə‚sfir }

upper-atmosphere dynamics [METEOROL]　Motion of the atmosphere above 300 miles (500 kilometers); predominant dynamical phenomena are internal gravity waves, tides, sound waves, turbulence, and large-scale circulation. { 'əp·ər ¦at·mə‚sfir dī'nam·iks }

upper band *See* upper bright band. { 'əp·ər 'band }

upper bound [MATH]　**1.** If S is a subset of an ordered set A, an upper bound b for S in A is an element b of A such that $x \leq b$ for all x belonging to A. **2.** An upper bound on a function f with values in a partially ordered set A is an element of A which is larger than every element in the range of f. { 'əp·ər 'baund }

upper branch [GEOD]　That half of a meridian or celestial meridian from pole to pole which passes through a place or its zenith. { 'əp·ər 'branch }

upper bright band [METEOROL]　A level of enhanced radar echo occasionally observed at a higher altitude than the bright band of the melting level; it is attributable to the growth of a layer of ice crystals in a supercooled cloud into snow pellets. Also known as radar upper band; upper band. { 'əp·ər 'brīt ‚band }

Upper Cambrian [GEOL]　The latest epoch of the Cambrian

period of geologic time, beginning approximately 510 million years ago. { 'əp·ər ˌkam·brē·ən }

Upper Carboniferous [GEOL] The European epoch of geologic time equivalent to the Pennsylvanian of North America. { 'əp·ər ˌkär·bə'nif·ə·rəs }

upper consolute temperature See consolute temperature. { 'əp·ər 'kän·sə,lüt 'tem·prə·chər }

upper control limit [IND ENG] A horizontal line on a control chart at a specified distance above the central line; if all the plotted points fall between the upper and lower control lines, the process is said to be in control. { 'əp·ər kən'trōl ˌlim·ət }

Upper Cretaceous [GEOL] The late epoch of the Cretaceous period of geologic time, beginning about 90 million years ago. { 'əp·ər kri'tā·shəs }

upper critical field [SOLID STATE] The magnetic field strength above which a type II superconductor is completely normal. Symbolized H_{c2}. { 'əp·ər ˌkrid·i·kəl 'fēld }

upper critical solution temperature See consolute temperature. { 'əp·ər ˌkrid·ə·kəl sə'lü·shən 'tem·prə·chər }

upper culmination See upper transit. { 'əp·ər ˌkəl·mə'nā·shən }

upper curtate [COMPUT SCI] The upper or top part of a punch card; on a standard card it contains the 11 (or X), 12 (or Y), and 0 punch positions, that is, the zone punch positions. { 'əp·ər 'kər,tāt }

upper cyclone See upper-level cyclone. { 'əp·ər 'sī,klōn }

Upper Devonian [GEOL] The latest epoch of the Devonian period of geologic time, beginning about 365 million years ago. { 'əp·ər də'vō·nē·ən }

upper explosive limit See upper flammable limit. { 'əp·ər ik'splō·siv ˌlim·ət }

upper face height [ANTHRO] A measure of the distance from the nasion to the lower gum edge between the two central upper teeth. { 'əp·ər 'fās ˌhīt }

upper flammable limit [CHEM] The maximum percentage of flammable gas or vapor in the air above which ignition cannot take place because the ratio of the gas to oxygen is too high. Also known as upper explosive limit. { 'əp·ər 'flam·ə·bəl ˌlim·ət }

upper front [METEOROL] A front which is present in the upper air but does not extend to the ground. { 'əp·ər 'frənt }

upper half-power frequency [ELECTR] The frequency on an amplifier response curve which is greater than the frequency for peak response and at which the output voltage is $1/\sqrt{2}$ (that is, 0.707) of its midband or other reference value. { 'əp·ər ˌhaf 'paů·ər 'frē·kwən·sē }

upper high See upper-level anticyclone. { 'əp·ər 'hī }

Upper Huronian See Animikean. { 'əp·ər hyü'rō·nē·ən }

upper integral [MATH] The upper Riemann integral for a real-valued function $f(x)$ on an interval is computed to be the infimum of all finite sums over all partitions of the interval, the sums having terms given by $(x_i - x_{i-1})y_i$, where the x_i are from a partition, and y_i is the largest value of $f(x)$ over the interval from x_{i-1} to x_i. { 'əp·ər 'int·ə·grəl }

upper inversion See tropopause inversion. { 'əp·ər in'vər·zhən }

Upper Jurassic [GEOL] The latest epoch of the Jurassic period of geologic time, beginning approximately 155 million years ago. { 'əp·ər jù'ras·ik }

upper-level anticyclone [METEOROL] An anticyclonic circulation existing in the upper air; this often refers to such anticyclones only when they are much more pronounced at upper levels than at and near the earth's surface. Also known as high aloft; high-level anticyclone; upper anticyclone; upper high; upper-level high. { 'əp·ər ˌlev·əl 'ant·i'sī,klōn }

upper-level chart [METEOROL] A synoptic chart of meteorological conditions in the upper air, almost invariably referring to a standard constant-pressure chart. Also known as upper-air chart. { 'əp·ər ˌlev·əl 'chärt }

upper-level cyclone [METEOROL] A cyclonic circulation existing in the upper air, and specifically, as seen on an upper-level constant-pressure chart; often restricted to describe cyclones associated with relatively little cyclonic circulation in the lower atmosphere. Also known as high-level cyclone; low aloft; upper cyclone; upper-level low; upper low. { 'əp·ər ˌlev·əl 'sī,klōn }

upper-level disturbance See upper-air disturbance. { 'əp·ər ˌlev·əl di'stər·bəns }

upper-level high See upper-level anticyclone. { 'əp·ər ˌlev·əl 'hī }

upper-level low See upper-level cyclone. { 'əp·ər ˌlev·əl 'lō }

upper-level ridge [METEOROL] A pressure ridge existing in the upper air, especially one that is stronger aloft than near the earth's surface. Also known as high-level ridge; ridge aloft; upper ridge. { 'əp·ər ˌlev·əl 'rij }

upper-level trough [METEOROL] A pressure trough existing in the upper air, but sometimes restricted to the troughs that are much more pronounced aloft than near the earth's surface. Also known as high-level trough; trough aloft; upper trough. { 'əp·ər ˌlev·əl 'trȯf }

upper-level winds See winds aloft. { 'əp·ər ˌlev·əl 'winz }

upper limb [ASTRON] That half of the outer edge of a celestial body having the greatest altitude. { 'əp·ər 'lim }

upper limit See limit superior. { 'əp·ər 'lim·ət }

upper low See upper-level cyclone.

upper mantle [GEOL] The portion of the mantle lying above a depth of about 600 miles (1000 kilometers). Also known as outer mantle; peridotite shell. { 'əp·ər 'mant·əl }

Upper Mississippian [GEOL] The latest epoch of the Mississippian period of geologic time. { 'əp·ər ˌmis·ə'sip·ē·ən }

upper mixing layer [METEOROL] The region of the upper mesosphere between about 30 and 50 miles (50 and 80 kilometers; that is, immediately above the mesopeak) through which there is a rapid decrease of temperature with height and where there appears to be considerable turbulence. { 'əp·ər 'miks·iŋ ˌlā·ər }

Upper Ordovician [GEOL] The latest epoch of the Ordovician period of geologic time, beginning approximately 440 million years ago. { 'əp·ər ˌȯr·də'vish·ən }

Upper Pennsylvanian [GEOL] The latest epoch of the Pennsylvanian period of geologic time. { 'əp·ər ˌpen·səl'vā·nyən }

Upper Permian [GEOL] The latest epoch of the Permian period of geologic time, beginning about 245 million years ago. { 'əp·ər 'pər·mē·ən }

upper punch [MET] In powder metallurgy, the member of the die assembly that moves downward into the die body to transmit pressure to the metal powder in the cavity. { 'əp·ər ˌpənch }

upper ridge See upper-level ridge. { 'əp·ər 'rij }

upper semicontinuous decomposition [MATH] A partition of a topological space with the property that for every member D of the partition and for every open set U containing D there is an open set V containing D which is contained in U and is the union of members of the partition. { 'əp·ər ˌsem·i·kən'tin·yə·wəs dē,käm·pə'zish·ən }

upper semicontinuous function [MATH] A real-valued function $f(x)$ is upper semicontinuous at a point x_0 if, for any small positive ϵ, $f(x)$ always is less than $f(x_0) + \epsilon$ for all x in some neighborhood of x_0. { 'əp·ər ˌsem·i·kən'tin·yə·wəs 'fəŋk·shən }

upper sideband [COMMUN] The higher of two frequencies or groups of frequencies produced by a modulation process. { 'əp·ər 'sīd,band }

Upper Silurian [GEOL] The latest epoch of the Silurian period of geologic time. { 'əp·ər sə'lùr·ē·ən }

upper transit [ASTRON] The movement of a celestial body across a celestial meridian's upper branch. Also known as superior transit; upper culmination. { 'əp·ər 'trans·ət }

Upper Triassic [GEOL] The latest epoch of the Triassic period of geologic time, beginning about 200 million years ago. { 'əp·ər trī'as·ik }

upper trough See upper-level trough. { 'əp·ər 'trȯf }

upper winds See winds aloft. { 'əp·ər 'winz }

up quark [PARTIC PHYS] A quark with an electric charge of $-2/3$, baryon number of $1/3$, and 0 strangeness and charm. Symbolized u. { 'əp ˌkwärk }

upright [CIV ENG] A vertical structural member, post, or stake. { 'əp,rīt }

uprush [METEOROL] The strong upward-flow air current in cumulus clouds during their stage of rapid development, often preceding a thunderstorm. Also known as vertical jet. [OCEANOGR] See swash. { 'əp,rəsh }

UPS See ultraviolet photoemission spectroscopy. See uninterruptible power system.

upset [ENG] To increase the diameter of a rock drill by blunting the end. [MATER] A defect occurring in timber as a result of a severe blow that breaks the fibers across the grain of the wood. [MET] A localized increase in the cross-sectional area of a metal during working, caused by the application of pressure; enables a head to be formed on fasteners such as bolts. [MIN ENG] **1.** A narrow heading connecting two levels in inclined coal. **2.** A capsized or broken skip. { 'əpˌset (noun); əp'set (verb) }

upsetted moraine See push moraine. { ¦əp¦sed·əd mə'rān }

upsetting test [MET] A test used to identify the role of variables in forging, which demonstrates that the force to forge is a function of the strength of the material, coefficient of friction, and ratio of the lateral to thickness dimensions of the workpiece. { ¦əp¦sed·iŋ ˌtest }

upset welding [MET] A resistance welding process in which coalescence is produced simultaneously over the entire area of abutting surfaces or progressively along a joint; pressure is applied before heating is started and is maintained throughout the heating period. { 'əˌset ˌweld·iŋ }

upsilon particle [PART PHYS] One of a family of elementary particles having about 10 times the mass of the proton and consisting of an atomlike combination of a bottom quark with its antiquark. { 'əp·sə·lən ˌpärd·ə·kəl }

upslope fog [METEOROL] A type of fog formed when air flows upward over rising terrain and, consequently, is adiabatically cooled to or below its dew point. { 'əpˌslōp 'fäg }

upslope time [MET] In resistance welding, the time associated with an increase in current when slope control is used. { 'əpˌslōp 'tīm }

upstand [BUILD] That section of a roof covering that turns up against a vertical surface. Also known as upturn. { 'əpˌstand }

upstream [CHEM ENG] That portion of a process stream that has not yet entered the system or unit under consideration; for example, upstream to a refinery or to a distillation column. [HYD] Toward the source of a stream. { 'əp¦strēm }

upstream face [CIV ENG] The side of a dam nearer the source of water. { 'əp¦strēm 'fās }

uptake [ENG] A large pipe for exhaust gases from a boiler furnace that runs upward to a chimney or smokestack. { 'əpˌtāk }

upthrow [GEOL] **1.** The fault side that has been thrown upward. **2.** The amount of vertical fault displacement. { 'əpˌthrō }

up time [COMPUT SCI] The time during which equipment is either producing work or is available for productive work. Also known as available time. [IND ENG] A period during which value is being added to a product by a machine or a process. { 'əp ˌtīm }

upturn See upstand. { 'əpˌtərn }

Upupidae [VERT ZOO] The Old World hoopoes, a family of birds in the order Coraciiformes whose young are hatched with sparse down. { yü'püp·əˌdē }

upward bias [STAT] The overestimation or overstatement by a statistical measure of the event it is attempting to describe. { ¦əp·wərd 'bī·əs }

upward compatibility [COMPUT SCI] The ability of a newer or larger computer to accept programs from an older or smaller one. { 'əp·wərd kəmˌpad·ə'bil·əd·ē }

upwarp [GEOL] A broad anticline with gently sloping limbs formed as a result of differential uplift. { 'əpˌwȯrp }

upwelling [OCEANOGR] The process by which water rises from a deeper to a shallower depth, usually as a result of divergence of offshore currents. { ¦əp¦wel·iŋ }

upwind [METEOROL] In the direction from which the wind is flowing. { 'əp¦wind }

upwind effect [METEOROL] The effect of an orographic barrier in producing orographic precipitation windward of the base of the barrier, because the airflow is forced upward before the barrier slope is actually reached. { 'əp¦wind iˌfekt }

urachus [EMBRYO] A cord or tube of epithelium connecting the apex of the urinary bladder with the allantois; its connective tissue forms the median umbilical ligament. { 'yur·ə·kəs }

uracil [BIOCHEM] $C_4H_4N_2O_2$ A pyrimidine base important as a component of ribonucleic acid. { 'yur·əˌsil }

Uralean [GEOL] A stage of geologic time in Russia: uppermost Carboniferous (above Gzhelian, below Sakmarian of Permian). { yu'räl·ē·ən }

uralite [MINERAL] A green variety of secondary amphibole;

URACIL

Structural formula of uracil.

it is usuallyy fibrous or acicular and is formed by alteration of pyroxene. { 'yur·əˌlīt }

uralitization [GEOL] **1.** A process of replacement whereby pyroxene undergoes alteration resulting in uralite. **2.** Development of amphibole from pyroxene. { yəˌral·əd·ə'zā·shən }

uranate [INORG CHEM] A salt of uranic acid; for example, sodium uranate, Na_2UO_4. { 'yur·əˌnāt }

urania See uranium dioxide. { yə'rā·nē·ə }

uranic chloride See uranium tetrachloride. { yu'ran·ik 'klȯrˌīd }

uranic oxide See uranium dioxide. { yu'ran·ik 'äkˌsīd }

Uraniidae [INV ZOO] A tropical family of moths in the superfamily Geometroidea including some slender-bodied, brilliantly colored diurnal insects which lack a frenulum and are often mistaken for butterflies. { ˌyur·ə'nī·əˌdē }

uranin See uranine. { 'yur·ə·nən }

uranine [ORG CHEM] $Na_2C_{20}H_{10}O_5$ A brown or orange-red hygroscopic powder soluble in water; used as a yellow dye for silk and wool, a marker in the ocean to facilitate air and sea rescues, and as an analytical reagent. Also known as sodium fluorescein; uranin; uranine yellow. { 'yur·əˌnēn }

uranine yellow See uranine. { 'yur·əˌnēn yel·ō }

uraninite [MINERAL] UO_2 A black, brownish-black, or dark-brown radioactive mineral that is isometric in crystallization; often contains impurities such as thorium, radium, cerium, and yttrium metals, and lead; the chief ore of uranium; hardness is 5.5–6 on Mohs scale, and specific gravity of pure UO_2 is 10.9, but that of most natural material is 9.7–7.5. Also known as coracite; ulrichite. { 'yur·ə·nəˌnīt }

uranium [CHEM] A metallic element in the actinide series, symbol U, atomic number 92, atomic weight 238.03; highly toxic and radioactive; ignites spontaneously in air and reacts with nearly all nonmetals; melts at 1132°C, boils at 3818°C; used in nuclear fuel and as the source of ^{235}U and plutonium. [MET] A dense, silvery, ductile, strongly electropositive metal. { yə'rā·nē·əm }

uranium acetate See uranyl acetate. { yə'rā·nē·əm 'as·əˌtāt }

uranium age [GEOL] The age of a mineral as calculated from the numbers of ionium atoms present originally, now, and when equilibrium is established with uranium. { yə'rā·nē·əm ˌāj }

uranium carbide [INORG CHEM] One of the carbides of uranium, such as uranium monocarbide; used chiefly as a nuclear fuel. { yə'rā·nē·əm 'kärˌbīd }

uranium decay series See uranium series. { yə'rā·nē·əm di'kā ˌsir·ēz }

uranium dioxide [INORG CHEM] UO_2 Black, highly toxic, spontaneously flammable, radioactive crystals; insoluble in water, soluble in nitric and sulfuric acids; melts at approximately 3000°C; used to pack nuclear fuel rods and in ceramics, pigments, and photographic chemicals. Also known as urania; uranic oxide; uranium oxide. { yə'rā·nē·əm dī'äkˌsīd }

uranium enrichment [NUCLEO] A process carried out on uranium, in which the ratio of the abundance of the isotope uranium-235 to that of the isotope uranium-238 is increased above that found in natural uranium. { yə'rā·nē·əm in'rich·mənt }

uranium hexafluoride [INORG CHEM] UF_6 Highly toxic, radioactive, corrosive, colorless crystals; soluble in carbon tetrachloride, fluorocarbons, and liquid halogens; it reacts vigorously with alcohol, water, ether, and most metals, and it sublimes; used to separate uranium isotopes in the gaseous-diffusion process. { yə'rā·nē·əm ˌhek·sə'flurˌīd }

uranium hydride [INORG CHEM] UH_3 A highly toxic, gray to black powder that ignites spontaneously in air, and that conducts electricity; used for making powdered uranium metal, for hydrogen-isotope separation, and as a reducing agent. { yə'rā·nē·əm 'hīˌdrīd }

uranium-lead dating [GEOL] A method for calculating the geologic age of a material in years based on the radioactive decay rate of uranium-238 to lead-206 and of uranium-235 to lead-207. { yə'rā·nē·əm 'led 'dād·iŋ }

uranium nitrate See uranyl nitrate. { yə'rā·nē·əm 'nīˌtrāt }

uranium ocher See gummite. { yə'rā·nē·əm 'ō·kər }

uranium oxide See uranium dioxide; uranium trioxide. { yə'rā·nē·əm 'äkˌsīd }

uranium-radium series See uranium series. { yə'rā·nē·əm 'rād·ē·əm ˌsir·ēz }

uranium reactor [NUCLEO] A nuclear reactor in which the

principal fuel is uranium; the uranium may be natural, with the naturally occurring ratio of 1 atom of uranium-235 to about 139 atoms of uranium-238, or may be enriched to have a higher proportion of fissile uranium-233 or uranium-235 atoms. { yə′rā·nē·əm rē′ak·tər }

uranium series [NUC PHYS] The series of nuclides resulting from the decay of uranium-238, including isotopes of uranium, thorium, protactinium, radium, radon, polonium, lead, bismuth, and thallium with mass number $4n+2$, where n is an integer. Also known as uranium decay series; uranium-radium series. { yə′rā·nē·əm ‚sir·ēz }

uranium sulfate *See* uranyl sulfate. { yə′rā·nē·əm ′səl‚fāt }

uranium tetrachloride [INORG CHEM] UCl_4 Poisonous, radioactive, hygroscopic, dark-green crystals; soluble in alcohol and water; melts at 590°C, boils at 792°C. Also known as uranic chloride. { yə′rā·nē·əm ‚te·trə′klȯr‚īd }

uranium tetrafluoride [INORG CHEM] UF_4 Toxic, radioactive, corrosive green crystals; insoluble in water; melts at 1036°C; used in the manufacture of uranium metal. Also known as green salt. { yə′rā·nē·əm ‚te·trə′flur‚īd }

uranium trioxide [INORG CHEM] UO_3 A poisonous, radioactive, red to yellow powder; soluble in nitric acid, insoluble in water; decomposes when heated; used in ceramics and pigments and for uranium refining. Also known as orange oxide; uranium oxide. { yə′rā·nē·əm trī′äk‚sīd }

uranium yellow *See* sodium diuranate. { yə′rā·nē·əm yel·ō }

uran-mica *See* torbernite. { ′yùr‚an ′mī·kə }

uranocircite [MINERAL] $Ba(UO_2)_2(PO_4)_2·8H_2O$ A yellow-green, tetragonal mineral consisting of a hydrated phosphate of barium and uranium; occurs as crystals. { ‚yùr·ə·nō′sər‚sīt }

uranography [ASTRON] The science of mapping stars, groups of stars, and star clusters. { ‚yùr·ə′näg·rə·fē }

uranometry [ASTRON] The science of the measurement of the celestial sphere and celestial bodies. { ‚yùr·ə′näm·ə·trē }

uranophane [MINERAL] $Ca(UO_2)_2Si_2O_7·6H_2O$ A yellow or orange-yellow radioactive secondary mineral; it is dimorphous with β-uranophane. Also known as uranotile. { yə′ran·ə‚fān }

uranopilite [MINERAL] $(UO_2)_6(SO_4)(OH)_{10}·12H_2O$ A bright yellow, lemon yellow, or golden yellow, monoclinic mineral consisting of a hydrated basic sulfate of uranium; occurs as encrustations and masses. { ‚yùr·ə·nō′pī‚līt }

uranosphaerite [MINERAL] $Bi_2O_3·2UO_3·3H_2O$ An orange-yellow or brick red, orthorhombic mineral consisting of a hydrated oxide of bismuth and uranium. { ‚yùr·ə·nō′sfi‚rīt }

uranospinite [MINERAL] $Ca(UO_2)_2(AsO_4)_2·8H_2O$ A lemon yellow to siskin green, tetragonal mineral consisting of a hydrated arsenate of calcium and uranium; occurs in tabular form. { ‚yùr·ə′näs·pə‚nīt }

uranotantalite *See* samarskite. { ‚yùr·ə·nō′tant·əl‚īt }

uranothorite [MINERAL] A uranium-bearing variety of thorite. { ‚yùr·ə·nō′thȯr‚īt }

uranotile *See* uranophane. { yə′ran·ə‚tīl }

uranous-uranic oxide *See* triuranium octoxide. { ′yùr·ə·nəs yə′ran·ik ′äk‚sīd }

Uranus [ASTRON] A planet, seventh in the order of distance from the sun; it has five known satellites, and its equatorial diameter is about four times that of the earth. { ′yùr·ə·nəs or yù′rā·nəs }

uranyl acetate [INORG CHEM] $UO_2(C_2H_3O_2)_2·2H_2O$ Poisonous, radioactive yellow crystals, decomposed by light; soluble in cold water, decomposes in hot water; loses water of crystallization at 110°C, decomposes at 275°C; used in medicine and as an analytical reagent and bacterial oxidant. Also known as uranium acetate. { ′yùr·ə‚nil ′as·ə‚tāt }

uranyl nitrate [INORG CHEM] $UO_2(NO_3)_2·6H_2O$ Toxic, explosive, unstable yellow crystals; soluble in water, alcohol, and ether; melts at 60°C and boils at 118°C; used in photography, in medicine, and for uranium extraction and uranium glaze. Also known as uranium nitrate; yellow salt. { ′yùr·ə‚nil ′nī‚trāt }

uranyl salts [INORG CHEM] Salts of UO_3 that ionize to form UO_2^{2+} and that are yellow in solution; for example, uranyl chloride, UO_2Cl_2. { ′yùr·ə‚nil ′sȯls }

uranyl sulfate [INORG CHEM] $UO_2SO_4·3^1/_2H_2O$ and $UO_2SO_4·3H_2O$ Poisonous, radioactive yellow crystals; soluble in water and concentrated hydrochloric acid; used as an

analytical reagent. Also known as uranium sulfate. { ′yùr·ə‚nil ′səl‚fāt }

uranyl uranate *See* triuranium octoxide. { ′yùr·ə‚nil ′yùr·ə‚nāt }

urao *See* trona. { ′yù·raù }

urate calculi [PATH] Kidney stones composed of uric acid salts and found particularly in people suffering from gout. { ′yù‚rāt ′kal·kyə‚lī }

urbacid [ORG CHEM] $C_7H_{15}AsN_2S_3$ A colorless, crystalline compound with a melting point of 144°C; insoluble in water; used to control apple scale and diseases of coffee trees. { ′ər·bə‚sid }

urban forestry [FOR] The management of tree resources in and around cities and towns. { ‚ər·bən ′fär·ə·strē }

urban geography [GEOGR] The study of the site, evolution, morphology, spatial patterns, and classification of densely populated areas. { ′ər·bən jē′äg·rə·fē }

urban geology [GEOL] The study of geological aspects of planning and managing high-density population centers and their surroundings. { ′ər·bən jē′äl·ə·jē }

urban heat island [METEOROL] Increased urban temperatures of 1–2°C higher for daily maxima and 1–9°C for daily minima compared to rural environs resulting from changes in moisture balance due to impermeable surfaces, decreased humidity, or alteration in heat balance. { ′ər·bən ′hēt ‚ī·lənd }

urbanization [CIV ENG] The state of being or becoming a community with urban characteristics. { ‚ər·bə·nə′zā·shən }

urban renewal [CIV ENG] Redevelopment and revitalization of a deteriorated urban community. { ′ər·bən ri′nü·əl }

urban typhus *See* murine typhus. { ′ər·bən ′tī·fəs }

Urca process [ASTROPHYS] A series of nuclear reactions, chiefly among the iron group of elements, that are postulated as a cause of stellar collapse, due to the energy lost to neutrinos that are rapidly formed in the reactions. { ′ər·kə ‚prä‚ses }

urceolate [BIOL] Shaped like an urn. { ′ər‚sē·ə·lət }

urea [ORG CHEM] $CO(HN_2)_2$ A natural product of protein metabolism found in urine; synthesized as white crystals or powder with a melting point of 132.7°C; soluble in water, alcohol, and benzene; used as a fertilizer, in plastics, adhesives, and flameproofing agents, and in medicine. Also known as carbamide. { yù′rē·ə }

urea anhydride *See* cyanamide. { yù′rē·ə an′hī‚drīd }

urea dewaxing [CHEM ENG] A continuous, petroleum refinery process used to produce low-pour-point oils; urea forms a filterable solid complex (adduct) with the straight-chain wax paraffins in the stock { yù′rē·ə dē′waks·iŋ }

urea-formaldehyde resin [ORG CHEM] A synthetic thermoset resin derived by the reaction of urea (carbamide) with formaldehyde or its polymers. Also known as urea resin. { yù′rē·ə fȯr′mal·də‚hīd ′rez·ən }

urea nitrate [ORG CHEM] $CO(NH_2)_2·HNO_3$ Colorless, explosive, fire-hazardous crystals; soluble in alcohol, slightly soluble in water; decomposes at 152°C; used in explosives and to make urethane. { yù′rē·ə ′nī‚trāt }

urea peroxide [ORG CHEM] $CO(NH_2)_2·H_2O_2$ An unstable, fire-hazardous white powder; soluble in water, alcohol, and ethylene glycol; decomposes at 75–85°C or by moisture; used as a source of water-free hydrogen peroxide, as a disinfectant, in cosmetics and pharmaceuticals, and for bleaching. { yù′rē·ə pə′räk‚sīd }

urea resin *See* urea-formaldehyde resin. { yù′rē·ə ′rez·ən }

urease [BIOCHEM] An enzyme that catalyzes the degradation of urea to ammonia and carbon dioxide; obtained from the seed of jack bean. { ′yùr·ē‚ās }

Urechinidae [INV ZOO] A family of echinoderms in the order Holasteroida which have an ovoid test lacking a marginal fasciole. { ‚yùr·ə′kin·ə‚dē }

Uredinales [MYCOL] An order of parasitic fungi of the subclass Heterobasidiomycetidae characterized by the teleutospore, a spore with one or more cells, each of which is a modified hypobasidium; members causes plant diseases known as rusts. { yə‚red·ən′ā·lēz }

Urediniomycetes [MYCOL] A class of fungi in the subdivision Basidiomycotina that causes plant rust diseases; members have thick-walled teliospores, produced in the terminal state in pustules. Also known as rust fungi. { ‚yùr·ə‚din·ē·ō·mī′sē‚dēz }

urediniospore [MYCOL] A spore produced by a uredinium.

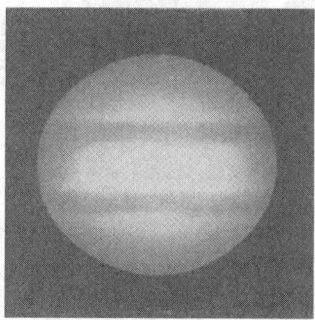

Telescopic appearance of Uranus, when the earth is in the equatorial plane of the globe.

Also known as urediospore; uredospore. { ˌyu̇r·ə'din·ē·ə̩spȯr }

uredinium [MYCOL] A reproductive tissue of the rust fungus which gives rise to urediniospores. { ˌyu̇r·ə'din·ē·əm }

urediospore See urediniospore. { yu̇'red·ē·əˌspȯr }

uredospore See urediniospore. { yu̇'red·əˌspȯr }

ureilite [GEOL] An achondritic stony meteorite consisting principally of olivine and clinobronzite, with some nickel-iron, troilite, diamond, and graphite. { yə'rē·əˌlīt }

uremia [MED] A condition resulting from kidney failure and characterized by azotemia, chronic acidosis, anemia, and a variety of systemic signs and symptoms. { yə'rē·mē·ə }

ureotelic [BIOL] Referring to animals that produce urea as their main nitrogenous excretion. { yə̩rē·ə'tel·ik }

ureter [ANAT] A long tube conveying urine from the renal pelvis to the urinary bladder or cloaca in vertebrates. { 'yu̇r·əd·ər }

urethane [ORG CHEM] $CO(NH_2)OC_2H_5$ A combustible, toxic, colorless powder; soluble in water and alcohol; melts at 49°C; used as a solvent and chemical intermediate and in biochemical research and veterinary medicine. Also known as ethyl carbamate; ethyl urethane. { 'yu̇r·əˌthān }

urethra [ANAT] The canal in most mammals through which urine is discharged from the urinary bladder to the outside. { yə'rē·thrə }

urethral gland [ANAT] One of the small, branched tubular mucous glands in the mucosa lining the urethra. { yə'rē·thrəl 'gland }

urethritis [MED] Inflammation of the urethra. { ˌyu̇r·ə'thrīd·əs }

ureyite [MINERAL] $NaCrSi_2O_6$ A meteoritic mineral of the pyroxene group. Also known as cosmochlore; kosmochlor. { 'yu̇r·ēˌīt }

uric acid [BIOCHEM] $C_5H_4N_4O_3$ A white, crystalline compound, the excretory end product in amino acid metabolism by uricotelic species. { 'yu̇r·ik 'as·əd }

uricase [BIOCHEM] An enzyme present in the liver, spleen, and kidney of most mammals except humans; converts uric acid to allantoin in the presence of gaseous oxygen. { 'yu̇r·əˌkās }

uricotelism [PHYSIO] An adaptation of terrestrial reptiles and birds which effectively provides for detoxification of ammonia and also for efficient conservation of water due to a relatively low rate of glomerular filtration and active secretion of uric acid by the tubules to form a urine practically saturated with urate. { ˌyu̇r·ə'käd·əlˌiz·əm }

uridine [BIOCHEM] $C_9H_{12}N_2O_6$ A crystalline nucleoside composed of one molecule of uracil and one molecule of D-ribose; a component of ribonucleic acid. { 'yu̇r·əˌdīn }

uridine diphosphate [BIOCHEM] The chief transferring coenzyme in carbohydrate metabolism. Abbreviated UDP. { 'yu̇r·əˌdīn dī̩fä̩sfāt }

uridine diphosphoglucose [BIOCHEM] A compound in which α-glucopyranose is esterified, at carbon atom 1, with the terminal phosphate group of uridine-5′-pyrophosphate (that is, uridine diphosphate); occurs in animal, plant, and microbial cells; functions as a key in the transformation of glucose to other sugars. Abbreviated UDPG. { 'yu̇r·əˌdīn dī̩fä'sfō'glü̩kōs }

uridine monophosphate See uridylic acid. { 'yu̇r·əˌdīn ̩män·ə'fäˌsfāt }

uridine phosphoric acid See uridylic acid. { 'yu̇r·əˌdīn fä'sfȯr·ik 'as·əd }

uridylic acid [BIOCHEM] $C_9H_{13}N_2O_9P$ Water-and alcohol-soluble crystals, melting at 202°C; used in biochemical research. Also known as uridine monophosphate (UMP); uridine phosphoric acid. { ̩yu̇r·ə'dil·ik 'as·əd }

urinalysis [PATH] Analysis of the urine, involving chemical, physical, and microscopic tests. { ˌyu̇r·ə'nal·ə·səs }

urinary bladder [ANAT] A hollow organ which serves as a reservoir for urine. { 'yu̇r·əˌner·ē 'blad·ər }

urinary system [ANAT] The system which functions in the elaboration and excretion of urine in vertebrates; in humans and most mammals, consists of the kidneys, ureters, urinary bladder, and urethra. { 'yu̇r·əˌner·ē ̩sis·təm }

urinary tract infection [MED] An inflammatory process occurring in the kidney, ureter, bladder, or adjacent structures that occurs when microorganisms (usually *Escherichia coli*) enter through the urethra. { ˌyu̇r·əˌner·ē 'trakt in̩fek·shən }

urination [PHYSIO] The discharge of urine from the bladder. Also known as micturition. { ˌyu̇r·ə'nā·shən }

URIC ACID

Structural formula of uric acid.

urine [PHYSIO] The fluid excreted by the kidneys. { 'yu̇r·ən }

uriniferous tubule [ANAT] One of the numerous winding tubules of the kidney. Also known as nephric tubule. { ̩yu̇r·ə̩nif·ə·rəs 'tü̩byül }

URL See uniform resource locator.

urn [BOT] The theca of a moss. { ərn }

urobilin [BIOCHEM] A bile pigment produced by reduction of bilirubin by intestinal bacteria and excreted by the kidneys or removed by the liver. { ˌyu̇r·ə'bī·lən }

urobilinogen [BIOCHEM] A chromogen, formed in feces and present in urine, from which urobilin is formed by oxidation. { ˌyu̇r·ə'bil·ə·jən }

urocanic acid [BIOCHEM] $C_6H_6N_2O_2$ A crystalline compound formed as an intermediate in the degradative pathway of histidine. { ̩yu̇r·ə̩kan·ik 'as·əd }

Urochordata [INV ZOO] The equivalent name for Tunicata. { ̩yu̇r·ə̩kȯr'däd·ə }

urochrome [BIOCHEM] $C_{43}H_{51}O_{26}N$ Yellow pigment found in normal urine. { 'yu̇r·əˌkrōm }

Urodela [VERT ZOO] The tailed amphibians or salamanders, an order of the class Amphibia distinguished superficially from frogs and toads by the possession of a tail, and from caecilians by the possession of limbs. { ˌyu̇r·ə'dē·lə }

urogenital diaphragm [ANAT] The sheet of tissue stretching across the pubic arch, formed by the deep transverse perineal and the sphincter urethrae muscles. Also known as triangular ligament. { ̩yu̇r·ə̩jen·əd·əl 'dī·ə̩fram }

urogenital system [ANAT] The combined urinary and genital system in vertebrates, which are intimately related embryologically and anatomically. Also known as genitourinary system. { ̩yu̇r·ə̩jen·əd·əl ̩sis·təm }

urogomphus [INV ZOO] In certain insect larvae, a process on the terminal segment. Also known as pseudocercus. { ˌyu̇r·ə'gäm·fəs }

urography [MED] Radiography of any portion of the urinary tract; most often follows the intravenous administration of iodinated contrast material. { yə'räg·rə·fē }

urokinase [BIOCHEM] An enzyme, present in human urine, that catalyzes the conversion of plasminogen to plasmin. { ˌyu̇r·ə'kī̩nās }

urolithiasis [MED] **1.** Condition associated with the presence of urinary calculi. **2.** Formation or presence of urinary calculi. { ˌyu̇r·ə·lə'thī·ə·səs }

urology [MED] The scientific study of urine and the diseases and abnormalities of the urinary and urogenital tracts. { yə'räl·ə·jē }

uronic acid [ORG CHEM] One of the compounds that are similar to sugars, except that the terminal carbon has been oxidized from the alcohol to a carboxyl group; for example, galacturonic acid and glucuronic acid. { yə'rän·ik 'as·əd }

uropepsin [BIOCHEM] The end product of the secretion of pepsinogen into the blood by gastric cells; occurs in urine. { ̩yu̇r·ə'pep·sən }

uropod [INV ZOO] One of the flattened abdominal appendages of various crustaceans that with the telson forms the tail fan. { 'yu̇r·əˌpäd }

uroporphyrin [BIOCHEM] Any of several isomeric, metal-free porphyrins, occurring in small quantities in normal urine and feces; molecule has four acetic acid ($-CH_2COOH$) and four propionic acid ($-CH_2CH_2COOH$) groups. { ̩yu̇r·ə'pȯr·fə·rən }

Uropygi [INV ZOO] The tailed whip scorpions, an order of arachnids characterized by an elongate, flattened body which bears in front a pair of thickened, raptorial pedipalps set with sharp spines and used to hold and crush insect prey. { ˌyu̇r·ə'pī̩jī }

uropygial gland [VERT ZOO] A relatively large, compact, bilobed, secretory organ located at the base of the tail (uropygium) of most birds having a keeled sternum. Also known as oil gland; preen gland. { ̩yu̇r·ə̩pij·ē·əl ̩gland }

urostyle [VERT ZOO] An unsegmented bone representing several fused vertebrae and forming the posterior part of the vertebral column in Anura. { 'yu̇r·əˌstīl }

Urostylidae [INV ZOO] A family of hemipteran insects in the superfamily Pentatomoidea. { ˌyu̇r·ə'stīl·ə̩dē }

urotropin See cystamine. { yə'rä·trə·pən }

Ursa Major [ASTRON] A northern constellation, right ascension 11 hours, declination 50°N; it contains a group of seven stars known as the Big Dipper. { 'ər·sə 'mā·jər }

Ursa Major cluster [ASTRON] **1.** An open cluster of stars, including 5 bright stars of the constellation Ursa Major, centered about 75 light-years from the sun and spread over a volume of 30 light-years length and 18 light-years width. **2.** A cluster of galaxies in the constellation Ursa Major having a redshift of about 0.051. { 'ər·sə 'mā·jər ‚kləs·tər }

Ursa Minor [ASTRON] A northern constellation, right ascension 15 hours, declination 70°N; its brightest star, Polaris, is almost at the north celestial pole; seven of the eight stars form a dipper ouline. Also known as Little Bear; Little Dipper. { 'ər·sə 'mī·nər }

Ursa Minor system [ASTRON] A dwarf spheroidal galaxy in the Local Group, about 2.1×10^5 light-years (1.2×10^{18} miles or 2.0×10^{18} kilometers) away. { 'ər·sə 'mī·nər ‚sis·təm }

Ursidae [VERT ZOO] A family of mammals in the order Carnivora including the bears and their allies. { 'ər·sə‚dē }

ursids [ASTRON] A shower of meteors occurring about December 22 from a radiant in the constellation Ursa Minor. { 'ər·sədz }

ursolic acid [BIOCHEM] $C_{30}H_{48}O_3$ A pentacyclic terpene that crystallizes from absolute alcohol solution, found in leaves and berries of plants; used in pharmaceutical and food industries as an emulsifying agent. { ər'säl·ik 'as·əd }

urstromthal [GEOL] A large channel cut by a stream of water from melting ice, flowing along the edge of an ice sheet. { 'ùr‚ström‚täl }

Urticaceae [BOT] A family of dicotyledonous herbs in the order Urticales characterized by a single unbranched style, a straight embryo, and the lack of milky juice (latex). { ‚ərd·ə'kās·ē‚ē }

Urticales [BOT] An order of dicotyledons in the subclass Hamamelidae; woody plants or herbs with simple, usually stipulate leaves, and reduced clustered flowers that usually have a vestigial perianth. { ‚ərd·ə'kā·lēz }

urticaria [MED] Hives or nettle rash; a skin condition characterized by the appearance of intensely itching wheals or welts with elevated, usually white centers and a surrounding area of erythema. Also known as hives { ‚ərd·ə'kar·ē·ə }

Urysohn lemma [MATH] If A and B are disjoint, closed sets in a normal space X, there is a real-valued function f such that $0 \leq f(x) \leq 1$ for all $x \in X$, and $f(A) = 0$ and $f(B) = 1$. { 'ùr·ē‚zōn ‚lem·ə }

Urysohn theorem [MATH] The theorem that a regular T_1 space whose topology has a countable base is metrizable. { 'ùr·ē‚zōn ‚thir·əm }

usability [IND ENG] The characteristics which enter into a product's design and are related to its quality and reliability that enable users to perform tasks quickly and error free, as well as reduce the time and mental effort to learn or operate the product. Also known as ease of use; user friendliness. { ‚yüz·ə'bil·əd·ē }

usable [COMPUT SCI] Pertaining to a computer system that is easy for all users to work with. { 'yüz·ə·bəl }

usable iron ore [MET] A steel industry term for high-grade iron ore, concentrates, or agglomerates which can be used in blast furnaces or other processing plants. { 'yüz·ə·bəl 'ī·ərn ‚ór }

usable life See pot life. { 'yüz·ə·bəl 'līf }

usable rate of fire [ORD] Normal rate of fire of a gun in actual use, measured in units of shots per minute; the usable rate of fire is considerably less than a gun's maximum rate of fire, which is a theoretical value based on the purely mechanical operation of a weapon. { 'yüz·ə·bəl 'rāt əv 'fīr }

U Sagittae [ASTRON] An eclipsing binary star in which one component has attained its Roche limit, while its mate is distinctly smaller than this limit. { 'yü 'saj·ə‚tē }

USAT See ultra-small-aperture terminal. { 'yü‚sat or 'yü‚es‚ā'tē }

USB See universal serial bus.

use-dilution test [MICROBIO] A bioassay method for testing disinfectants for use on surfaces where a substantial reduction of bacterial contamination is not achieved by prior cleaning; the test organisms *Salmonella choleraesuis* and *Staphylococcus aureus* are deposited in stainless steel cylinders which are then exposed to the action of the test disinfectant. { 'yüs də‚lü·shən ‚test }

UseNet [COMPUT SCI] A global network of newsgroups that is linked by the Internet and other wide-area networks. { 'yüz‚net }

user [COMMUN] An individual, installation, or activity having access to a switching center through a local private branch exchange, or by dialing an access code. [COMPUT SCI] Anyone who requires the use of services of a computing system or its products. { 'yü·zər }

user datagram protocol [COMMUN] A communications protocol providing a direct way to send and receive datagrams over an IP network but with few error recovery resources; used mainly for broadcasting over a network, for example, with streaming media. { ‚yüz·ər 'cad·ə‚gram ‚prōd·ə‚kól }

user-defined function [COMPUT SCI] A subroutine written by the programmer to calculate and return the value of a mathematical function. { 'yü·zər di fīnd 'fəŋk·shən }

user-defined type [COMPUT SCI] A data type that is not provided by a strongly typed language but is instead created by the programmer for a particular computer program. { 'yü·zər di'fīnd 'tīp }

user exit [COMPUT SCI] A point in a computer program at which a user can cause control to be transferred outside the program. { 'yü·zər 'eg·zət }

user friendliness See usability. { 'yü·zər 'frend·lē·nəs }

user friendly [COMPUT SCI] Property of a system that is easy for an untrained person to use and sets up an easily understood dialog between the user and the computer. { 'yü·zər 'frend·lē }

user group [COMPUT SCI] An organization of users of the computers of a particular vendor, which shares information and ideas, and may develop system software and influence vendors to change their products. { 'yü·zər ‚grüp }

userID [COMPUT SCI] The name used to log in to a network, remote server, and so on. { ‚yüz·ər‚ī'dē }

user interface [COMPUT SCI] **1.** The point at which a user or a user department or organization interacts with a computer system. **2.** The part of an interactive computer program that sends messages to and receives instructions from a terminal user. { 'yü·zər 'in·tər‚fās }

user mode [COMPUT SCI] The mode of operation exercised by the user programs of a computer system in which there is a class of privileged instructions that is not permitted, since these can be executed only by the operating system or executive system. Also known as slave mode. { 'yü·zər ‚mōd }

username See mailbox name. { 'yüz·ər‚nām }

user program [COMPUT SCI] A computer program written by the person who uses it or by personnel of the organization that will use it. Also known as roll your own; user-written code. { 'yü·zər 'prō‚gram }

user-programmable memory [COMPUT SCI] That portion of the internal storage of a microcomputer that is available for programs entered or loaded in by the user. { 'yü·zər prō'gram·ə·bəl 'mem·rē }

user-to-user service [COMMUN] Method of switching that enables direct user-to-user connection which does not include message store-and-forward service. { 'yü·zər tə 'yü·zər 'sər·vəs }

user-written code See user program. { 'yü·zər 'writ·ən 'kōd }

U-shaped abutment [CIV ENG] A bridge abutment with wings perpendicular to the face which act as counterforts; a very stable abutment, often used for architectural effect. { 'yü ‚shāpt ə'bət·mənt }

U-shaped distribution [STAT] A frequency distribution whose shape approximates that of the letter U. { 'yü ‚shāpt dis·trə'byü·shən }

U-shaped valley [GEOL] A type of valley with a broad floor and steep walls produced by glacial erosion. Also known as trough valley; U valley. { 'yü ‚shāpt 'val·ē }

using agency [ORD] Any element of the U.S. Armed Forces having command or service functions, and requiring materiel for use in performance of its mission. { 'yüz·iŋ ‚ā·jən‚sē }

Usneaceae [BOT] The beard lichens, a family of Ascolichenes in the order Lecanorales distinguished by their conspicuous fruticose growth form. { ‚əs·nē'ās·ē‚ē }

usnic acid [BIOCHEM] $C_{18}H_{16}O_7$ Yellow crystals, insoluble in water, slightly soluble in alcohol and ether, melts about

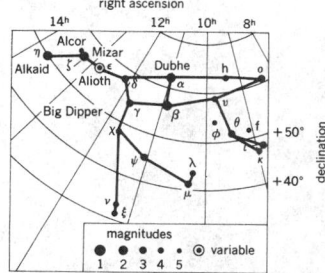

URSA MAJOR

Line pattern of the constellation Ursa Major. The grid lines represent the coordinates of the sky. The apparent brightness, or magnitude, of the stars is shown by the sizes of the dots, which are graded by appropriate numbers as indicated.

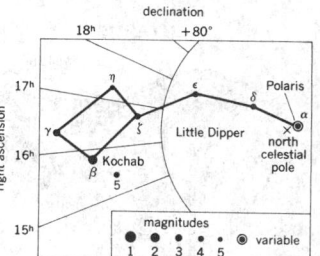

URSA MINOR

Line pattern of the constellation Ursa Minor. The grid lines represent the coordinates of the sky. The apparent brightness, or magnitude, of the stars is shown by the sizes of the dots, which are graded by appropriate numbers as indicated.

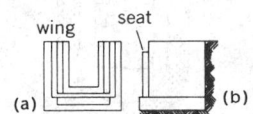

U-SHAPED ABUTMENT

U-shaped abutment design for a bridge. (*a*) Plan. (*b*) Side elevation.

198°C; found in lichens; used as an antibiotic. Also known as usninic acid. { 'əs·nik 'as·əd }

usninic acid *See* usnic acid. { əs'nin·ik 'as·əd }

USP acid test [ANALY CHEM] A United States Pharmacopoeia test to determine the carbonizable substances present in petroleum white oils. { ¦yü¦es'pē 'as·əd ‚test }

U stage *See* universal stage. { 'yü ‚stāj }

Ustalf [GEOL] A suborder of the soil order Alfisol; red or brown soil formed in a ustic moisture regime and in a mesic or warmer temperature regime. { 'üst‚älf }

Ustert [GEOL] A suborder of the Vertisol soil order; has a faint horizon and is dry for an appreciable period or more than one period of the year. { 'üst‚ərt }

Ustilaginaceae [MYCOL] A family of fungi in the order Ustilaginales in which basidiospores bud from the sides of the septate epibasidium. { ‚əs·tə‚laj·ə'nās·ē‚ē }

Ustilaginales [MYCOL] An order of the subclass Heterobasidiomycetidae comprising the smut fungi which parasitize plants and cause diseases known as smut or bunt. { ‚əs·tə‚laj·ə'nā·lēz }

Ustoll [GEOL] A suborder of the soil order Mollisol; formed in a ustic moisture regime and in a mesic or warmer temperature regime; may have a calcic, petrocalcic, or gypsic horizon. { 'üst‚ȯl }

Ustox [GEOL] A suborder of the soil order Oxisol that is low to moderate in organic matter, well drained, and dry for at least 90 cumulative days each year. { 'üst‚äks }

Ustult [GEOL] A suborder of the soil order Ultisol; brownish or reddish, with low to moderate organic-carbon content; a well-drained soil of warm-temperate and tropical climates with moderate or low rainfall. { 'üst‚əlt }

UT 0 *See* universal time 0.

UT 1 *See* universal time 1.

UT 2 *See* universal time 2.

utahite *See* jarosite. { 'yü·tȯ‚īt }

UTC *See* universal time coordinated.

uterus [ANAT] The organ of gestation in mammals which receives and retains the fertilized ovum, holds the fetus during development, and becomes the principal agent of its expulsion at term. { 'yüd·ə·rəs }

uterus bicornis [ANAT] A uterus divided into two horns or compartments; an abnormal condition in humans but normal in many mammals, such as carnivores. { 'yüd·ə·rəs bī'kȯr·nəs }

utilidor [CIV ENG] An insulated, heated conduit built below the ground surface or supported above the ground surface to protect the contained water, steam, sewage, and fire lines from freezing. { yü'til·ə‚dȯr }

utility [ENG] One of the nonprocess (support) facilities for a manufacturing plant; usually considered as facilities for steam, cooling water, deionized water, electric power, refrigeration, compressed and instrument air, and effluent treatment. { yü'til·əd·ē }

utility routine [COMPUT SCI] A program or routine of general usefulness, usually not very complicated, and applicable to many jobs or purposes. { yü'til·əd·ē rü‚tēn }

utilization factor [ELEC] In electric power distribution, the maximum demand of a system or part of a system divided by its rated capacity. { ‚yüd·əl·ə'zā·shən ‚fak·tər }

utilization rate [AERO ENG] The amount of flying time produced in a specific period expressed in hours per period per aircraft. Also known as flying hour rate. { ‚yüd·əl·ə'zā·shən ‚rāt }

utilization ratio [COMPUT SCI] The ratio of the effective time on a computer to the total up time. { ‚yüd·əl·ə'zā·shən ‚rā·shō }

utricle *See* utriculus. { 'yü·trə·kəl }

utriculus [ANAT] **1.** That part of the membranous labyrinth of the ear into which the semicircular canals open. **2.** A small, blind pouch extending from the urethra into the prostate. Also known as utricle. { yü'trik·yə·ləs }

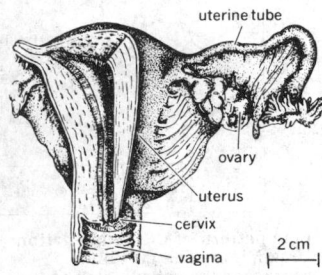

UTERUS

Human uterus and associated structures. *(From L. B. Arey, Developmental Anatomy, 7th ed., Saunders, 1965)*

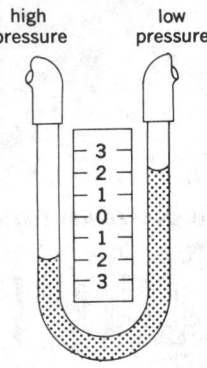

U-TUBE MANOMETER

high pressure   low pressure

Components of U-tube manometer for measuring liquid flow.

U-tube heat exchanger [CHEM ENG] A heat-exchanger system consisting of a bundle of U tubes (hairpin tubes) surrounded by a shell (outer vessel); one fluid flows through the tubes, and the other fluid flows through the shell, around the tubes. { 'yü ¦tüb 'hēt iks‚chän·jər }

U-tube manometer [ENG] A manometer consisting of a U-shaped glass tube partly filled with a liquid of known specific gravity; when the legs of the manometer are connected to separate sources of pressure, the liquid rises in one leg and drops in the other; the difference between the levels is proportional to the difference in pressures and inversely proportional to the liquid's specific gravity. Also known as liquid-column gage. { 'yü ¦tüb mə'näm·əd·ər }

uuencode [COMPUT SCI] A protocol for sending binary files in ASCII text format over the Internet, particularly e-mail attachments. { ‚yü'yü·in‚kōd }

Uukuvirus [VIROL] A genus of the viral family Bunyaviridae that is transmitted via ticks to a wide range of vertebrate hosts. { yü'yük·ə‚vī·rəs }

UV *See* ultraviolet.

UV-A [METEOROL] Ultraviolet radiation produced by the sun, ranging in wavelength from 320 to 400 nanometers; biologically, it is the least damaging of the sun's rays.

UV-B [METEOROL] Ultraviolet radiation produced by the sun, ranging in wavelength from 280 to 320 nanometers; it is biologically damaging. Stratospheric ozone absorbs much of it.

UV-C [METEOROL] Ultraviolet radiation produced by the sun, ranging in wavelength from 200 to 280 nanometers; biologically, it is the most damaging of the sun's rays. Stratospheric ozone strongly absorbs it and, as a result, the solar spectrum at the earth's surface contains only the UV-A and UV-B radiation.

uvala [GEOGR] Broad-bottomed lowlands. { 'ü·və·lə }

U valley *See* U-shaped valley. { 'yü ‚val·ē }

U-value [ENG] A measure of heat transmission through a building part or a given thickness of insulating material, expressed as the number of British thermal units that will flow in 1 hour through 1 square foot of the structure or material from air to air with a temperature differential of 1°F. { 'yü ‚väl·yü }

uvanite [MINERAL] $U_2V_6O_{21}\cdot15H_2O$ A brownish-yellow, orthorhombic mineral consisting of a hydrated uranium vanadate; occurs as crystalline masses and coatings. { 'yü·və‚nīt }

uvarovite [MINERAL] $Ca_3Cr_2(SiO_4)_3$ The emerald-green, calcium-chromium end member of the garnet group. Also known as ouvarovite; uwarowite. { 'ü'var·ə‚vīt }

uvby system [ASTRON] A four-color stellar magnitude system based on measurements in the ultraviolet, violet, blue, and yellow regions. Also known as Strömgren four-color index. { 'yüv·bē ‚sis·təm }

UV Ceti stars [ASTRON] A class of stars that have brief outbursts of energy over their surface areas; they may have an increase of about 1 magnitude for periods of 1 hour; the type star is UV Ceti. Also known as flare stars. { ¦yü¦vē 'sed·ē ‚stärz }

uvea [ANAT] The pigmented, vascular layer of the eye: the iris, ciliary body, and choroid. { 'yü·vē·ə }

uveitis *See* iridocyclochoroiditis. { ‚yü·vē'īd·əs }

UV EPROM *See* ultraviolet-erasable programmable read-only memory. { ‚yü‚vē 'ē‚präm }

uviol glass [MATER] A type of glass that is highly transparent to ultraviolet radiation. { 'yü·vē‚ȯl ‚glas }

UVPROM *See* ultraviolet-erasable programmable read-only memory. { ‚yü‚vē'präm }

UV stabilizer [CHEM] Any chemical compound that, admixed with a thermoplastic resin, selectively absorbs ultraviolet rays; used to prevent ultraviolet degradation of polymers. Also known as ultraviolet stabilizer. { ¦yü¦vē 'stā·bə‚līz·ər }

uvula [ANAT] **1.** A fingerlike projection in the midline of the posterior border of the soft palate. **2.** A lobe of the vermiform process of the lower surface of the cerebellum. { 'yü·vyə·lə }

uwarowite *See* uvarovite. { ü'var·ə‚vīt }

V

V *See* electric potential; vanadium; volt.

VA *See* volt-ampere.

vac *See* millibar.

vacancy [SOLID STATE] A defect in the form of an unoccupied lattice position in a crystal { 'vā·kən·sē }

vaccination [IMMUNOL] Inoculation of viral or bacterial organisms or antigens to produce immunity in the recipient. { ˌvak·sə'nā·shən }

vaccination encephalitis [MED] Encephalitis caused by vaccination with rabies vaccine. { ˌvak·sə'nā·shən in,sefə'līd·əs }

vaccine [IMMUNOL] A suspension of killed or attenuated bacteria or viruses or fractions thereof, injected to produce active immunity. { vak'sēn }

vaccinia [VET MED] A contagious disease of cows which is characterized by vesicopustular lesions of the skin that are prone to appear on the teats and udder, and which is transmissible to humans by handling infected cows and by vaccination; confers immunity against smallpox. Also known as cowpox. { vak 'sin·ē·ə }

vacuole [CELL MOL] A membrane-bound cavity within a cell; may function in digestion, storage, secretion, or excretion. [GEOL] *See* vesicle. { 'vak·yəˌwōl }

vacuum [PHYS] **1.** Theoretically, a space in which there is no matter. **2.** Practically, a space in which the pressure is far below normal atmospheric pressure so that the remaining gases do not affect processes being carried on in the space. [QUANT MECH] The lowest possible energy state of a system, conceived of as a polarizable gas of virtual particles, fluctuating randomly. { 'vak·yəm }

vacuum-arc casting [MET] A process for producing metal ingots that are low in oxygen and need no further treatment to remove it, whereby an ultrapure metal or metal alloy powder is pressed, sintered, and melted (all under vacuum). { ˌvak·yəmˌärk 'kast·iŋ }

vacuum-arc centrifuge [NUCLEC] A type of plasma centrifuge in which a fully ionized plasma consisting of ions of the cathode material is produced by a vacuum-arc discharge between a carbon or metal cathode and a grounded mesh anode, and this then propagates as a rotating plasma column. { 'vak·yəmˌärk 'sen·trəˌfyüj }

vacuum behavior [PSYCH] The carrying out of a series of model action patterns in apparent absence of any obviously appropriate behavior. { 'vak·yəm biˌhā·vyər }

vacuum brake [MECH ENG] A form of air brake which operates by maintaining low pressure in the actuating cylinder; braking action is produced by opening one side of the cylinder to the atmosphere so that atmospheric pressure, aided in some designs by gravity, applies the brake. { 'vak·yəm ˌbrāk }

vacuum brazing [MET] Brazing utilizing a chamber at subatmospheric pressure. { 'vak·yəm 'brāz·iŋ }

vacuum breaker [ENG] A device used to relieve a vacuum formed in a water supply line to prevent backflow. Also known as backflow preventer. { 'vak·yəm 'brāk·ər }

vacuum capacitor [ELEC] A capacitor with separated metal plates or cylinders mounted in an evacuated glass envelope to obtain a high breakdown voltage rating. { 'vak·yəm kə'pas·əd·ər }

vacuum casting [MET] Metal casting in a vacuum. { 'vak·yəm 'kast·iŋ }

vacuum circuit breaker [ELEC] A circuit breaker in which a pair of contacts is hermetically sealed in a vacuum envelope; the contacts are separated by using a bellows to move one of them; an arc is produced by metallic vapor boiled from the electrodes, and is extinguished when the vapor particles condense on solid surfaces. { 'vak·yəm 'sər·kət ˌbrā·kər }

vacuum cleaner [MECH ENG] An electrically powered mechanical appliance for the dry removal of dust and loose dirt from rugs, fabrics, and other surfaces. { 'vak·yəm ˌklē·nər }

vacuum concrete [CIV ENG] Concrete poured into a framework that is fitted with a vacuum mat to remove water not required for setting of the cement; in this framework, concrete attains its 28-day strength in 10 days and has a 25% higher crushing strength. { 'vak·yəm 'känˌkrēt }

vacuum condensing point [CHEM] Temperature at which the sublimate (vaporized solid) condenses in a vacuum. Abbreviated vcp. { 'vak·yəm kən'dens·iŋ ˌpȯint }

vacuum cooling [FOOD ENG] A system of cooling fruits and vegetables prior to shipping; water is caused to evaporate from the surfaces of the food by producing a vacuum around it. { 'vak·yəm ˌkül·iŋ }

vacuum correction [PHYS] The correction to the reading of a mercury barometer required by the imperfections in the vacuum above the mercury column, due to the presence of water vapor and air; this correction is a function of both temperature and pressure. { 'vak·yəm kə,rek·shən }

vacuum crystallizer [CHEM ENG] Crystallizer in which a warm saturated solution is fed to a lagged, closed vessel maintained under vacuum; the solution evaporates and cools adiabatically, resulting in crystallization. { 'vak·yəm 'krist·əlˌīz·ər }

vacuum degassing [MET] A process for removing gases from a metal either by melting or heating the solid metal in a vacuum. { 'vak·yəm dē'gas·iŋ }

vacuum deposition [MET] Deposition of a thin coating of metal by condensation on a cool work surface in vacuum. { 'vak·yəm ˌdep·ə'zish·ən }

vacuum diffusion [ELECTR] Diffusion of impurities into a semiconductor material in a continuously pumped hard vacuum. { 'vak·yəm di'fyü·zhən }

vacuum distillation [CHEM ENG] Liquid distillation under reduced (less than atmospheric) pressure; used to lower boiling temperatures and lessen the risk of thermal degradation during distillation. Also known as reduced-pressure distillation. { 'vak·yəm ˌdis·tə'lā·shən }

vacuum drying [ENG] The removal of liquid from a solid material in a vacuum system; used to lower temperatures needed for evaporation to avoid heat damage to sensitive material. { 'vak·yəm 'drī·iŋ }

vacuum evaporation [ENG] Deposition of thin films of metal or other materials on a substrate, usually through openings in a mask, by evaporation from a boiling source in a hard vacuum. { 'vak·yəm iˌvap·ə'rā·shən }

vacuum evaporator [ENG] A vacuum device used to evaporate metals and spectrographic carbon to coat (replicate) a specimen for electron spectroscopic analysis or for electron microscopy. { 'vak·yəm iˌvap·əˌrād·ər }

vacuum filter [ENG] A filter device into which a liquid-solid slurry is fed to the high-pressure side of a filter medium, with liquid pulled through to the low-pressure side of the medium and a cake of solids forming on the outside of the medium. { 'vak·yəm ˌfil·tər }

VACUOLE

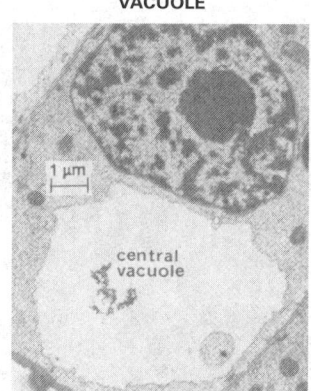

Electron micrograph of a partially mature cell from a developing root tip showing the central vacuole.

VACUUM CIRCUIT BREAKER

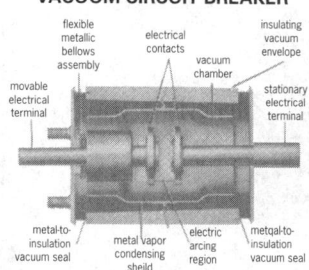

Cutaway view of vacuum circuit breaker showing component parts.

vacuum filtration [ENG] The separation of solids from liquids by passing the mixture through a vacuum filter. { 'vak·yəm fil'trā·shən }

vacuum flashing [CHEM ENG] The heating of a liquid that, upon release to a lower pressure (vacuum), undergoes considerable vaporization (flashing). Also known as flash vaporization. { 'vak·yəm 'flash·iŋ }

vacuum fluorescent lamp [ELECTR] An evacuated display tube in which the anodes are coated with a phosphor that glows when electrons from the cathode strike it, to create a display. { 'vak·yəm flù'res·ənt 'lamp }

vacuum forming [ENG] Plastic-sheet forming in which the sheet is clamped to a stationary frame, then heated and drawn down into a mold by vacuum. { 'vak·yəm 'fòrm·iŋ }

vacuum frame [GRAPHICS] In platemaking, a contact printing device that holds copy and reproduction material by means of a vacuum to attain a uniform pressure during exposure. { 'vak·yəm 'frām }

vacuum freeze dryer [ENG] A type of indirect batch dryer used to dry materials that would be destroyed by the loss of volatile ingredients or by drying temperatures above the freezing point. { 'vak·yəm 'frēz ,drī·ər }

vacuum fusion [MET] A technique for determining the oxygen, hydrogen, and sometimes nitrogen content of metals; can be applied to a wide variety of metals with the exception of alkali and alkaline earth metals. { 'vak·yəm 'fyü·zhən }

vacuum gage [ENG] A device that indicates the absolute gas pressure in a vacuum system. { 'vak·yəm ,gāj }

vacuum gripper [CONT SYS] A robot component that uses a suction cup connected to a vacuum source to lift and handle objects. { 'vak·yəm 'grip·ər }

vacuum heating [MECH ENG] A two-pipe steam heating system in which a vacuum pump is used to maintain a suction in the return piping, thus creating a positive return flow of air and condensate. { 'vak·yəm 'hēd·iŋ }

vacuum mat [CIV ENG] A rigid flat metal screen faced by a linen filter, the back of which is kept under partial vacuum; used to suck out surplus air and water from poured concrete to produce a dense, well-shrunk concrete. { 'vak·yəm ,mat }

vacuum measurement [ENG] The determination of a fluid pressure less in magnitude than the pressure of the atmosphere. { 'vak·yəm 'mezh·ər·mənt }

vacuum metallizing [MET] A special form of vapor deposition in which a metal coating is applied to the surface of a polymer. { 'vak·yüm 'med·əl,īz·iŋ }

vacuum metallurgy [MET] The melting, shaping, and treating of metals and alloys under reduced pressure that ranges from subatmospheric pressure to ultra-high vacuum. { 'vak·yəm 'med·əl,ər·jē }

vacuum microelectronics [ELECTR] The technology of the vacuum transistor and similar microminiature devices based on electron field emission into a vacuum. { 'vak·yüm ,mī·krō·i,lek'trän·iks }

vacuum pan salt [CHEM ENG] A salt made from salt brine boiled at reduced pressure in a triple-effect evaporator. { 'vak·yəm 'pan ,sòlt }

vacuum pencil [ENG] A pencillike length of tubing connected to a small vacuum pump, for picking up semiconductor slices or chips during fabrication of solid-state devices. { 'vak·yəm ,pen·səl }

vacuum phototube [ELECTR] A phototube that is evacuated to such a degree that its electrical characteristics are essentially unaffected by gaseous ionization; in a gas phototube, some gas is intentionally introduced. { 'vak·yəm 'fōd·ō,tüb }

vacuum plating See vapor deposition. { 'vak·yəm 'plād·iŋ }

vacuum polarization [QUANT MECH] A process in which an electromagnetic field gives rise to virtual electron-positron pairs that effectively alter the distribution of charges and currents that generated the original electromagnetic field. { 'vak·yəm ,pō·lə·rə'zā·shən }

vacuum printing frame [GRAPHICS] A frame employed in the graphic arts to keep materials flat or to keep them in absolute contact with one another; the frame is useful in making positives, contact prints, screened tints, and plates. { 'vak·yəm 'print·iŋ ,frām }

vacuum pump [MECH ENG] A compressor for exhausting air and noncondensable gases from a space that is to be maintained at subatmospheric pressure. { 'vak·yəm ,pəmp }

vacuum Rabi oscillation [ATOM PHYS] A process in which an excited atom placed in a very small cavity emits a photon much more quickly than it would in free space if the cavity is resonant with the radiation emitted by the photon. { 'vak·yəm ¦rä·bē ,äs·ə'lā·shən }

vacuum relay [ELEC] A sensitive relay having its contacts mounted in a highly evacuated glass housing, to permit handling radio-frequency voltages as high as 20,000 volts without flashover between contacts even though contact spacing is but a few hundredths of an inch when open. { 'vak·yəm 'rē,lā }

vacuum relief valve [ENG] A pressure relief device which is designed to allow fluid to enter a pressure vessel in order to avoid extreme internal vacuum. { 'vak·yəm ri'lēf ,valv }

vacuum shelf dryer [ENG] A type of indirect batch dryer which generally consists of a vacuum-tight cubical or cylindrical chamber of cast-iron or steel plate, heated supporting shelves inside the chamber, a vacuum source, and a condenser; used extensively for drying pharmaceuticals, temperature-sensitive or easily oxidizable materials, and small batches of high-cost products where any product loss must be avoided. { 'vak·yəm 'shelf,drī·ər }

vacuum support [MECH ENG] That portion of a rupture disk device which prevents deformation of the disk resulting from vacuum or rapid pressure change. { 'vak·yəm sə,pòrt }

vacuum switch [ELEC] A switch having its contacts in an evacuated envelope to minimize sparking. { 'vak·yəm ,swich }

vacuum thermobalance [ANALY CHEM] An instrument used in thermogravimetry consisting of a precision balance and furnace that have been adapted for continuously measuring or recording changes in weight of a substance as a function of temperature; used in many types of physicochemical reactions where rates of reaction and energies of activation for vaporization, sublimation, and chemical reaction can be obtained. { 'vak·yəm ¦thər·mō'bal·əns }

vacuum transistor [ELECTR] A microminiature electronic device based on the control of electron emission from a field-emitter array into a vacuum. { 'vak·yəm tran'zis·tər }

vacuum tube [ELECTR] An electron tube evacuated to such a degree that its electrical characteristics are essentially unaffected by the presence of residual gas or vapor. { 'vak·yəm ,tüb }

vacuum-tube amplifier [ELECTR] An amplifier employing one or more vacuum tubes to control the power obtained from a local source. { 'vak·yəm ¦tüb 'am·plə,fī·ər }

vacuum-tube circuit [ELECTR] An electric circuit in which a vacuum tube is connected. { 'vak·yəm ¦tüb 'sər·kət }

vacuum-tube clipping circuit [ELECTR] A circuit in which a vacuum tube is used to achieve clipping action; the bias at the input is set at such a level that output current cannot flow during a portion of the amplitude excursion of the input voltage or current waveform. { 'vak·yəm ¦tüb 'klip·iŋ 'sər·kət }

vacuum-tube electrometer [ELECTR] An electrometer in which the ionization current in an ionization chamber is amplified by a special vacuum triode having an input resistance above 10,000 megohms. { 'vak·yəm ¦tüb ,i,lek'träm·əd·ər }

vacuum-tube keying [ELECTR] Code-transmitter keying system in which a vacuum tube is connected in series with the plate supply lead of a frequency-controlling stage of the transmitter; when the key is open, the tube blocks, interrupting the plate supply to the output stage; closing the key allows the plate current to flow through the keying tube and the output tubes. { 'vak·yəm ¦tüb 'kē·iŋ }

vacuum-tube modulator [ELECTR] A modulator employing a vacuum tube as a modulating element for impressing an intelligence signal on a carrier. { 'vak·yəm ¦tüb 'mäj·ə,lād·ər }

vacuum-tube oscillator [ELECTR] A circuit utilizing a vacuum tube to convert direct-current power into alternating-current power at a desired frequency. { 'vak·yəm ¦tüb 'äs·ə,lād·ər }

vacuum-tube rectifier [ELECTR] A rectifier in which rectification is accomplished by the unidirectional passage of electrons from a heated electrode to one or more other electrodes within an evacuated space. { 'vak·yəm ¦tüb 'rek·tə,fī·ər }

vacuum-tube voltmeter [ENG] Any of several types of instrument in which vacuum tubes, acting as amplifiers or rectifiers, are used in circuits for the measurement of alternating-current or direct-current voltage. Abbreviated VTVM. Also known as tube voltmeter. { 'vak·yəm ¦tüb 'vōlt,mēd·ər }

vacuum-type insulation [CHEM ENG] Highly reflective double-wall structure with high vacuum between the walls; used as insulation for cryogenic systems; Dewar flasks have vacuum-type insulation. { 'vak·yəm ¦tīp in·sə'lā·shən }

vacuum ultraviolet radiation [ELECTROMAG] Ultraviolet radiation with a wavelength of less than 200 nanometers; absorption of radiation in this region by air and other gases requires the use of evacuated apparatus for transmission. Abbreviated VUV radiation. Also known as extreme ultraviolet radiation (EUV radiation). { 'vak·yəm ¦əl·trə'vī·lət ‚rād·ē'ā·shən }

vacuum ultraviolet spectroscopy [SPECT] Absorption spectroscopy involving electromagnetic wavelengths shorter than 200 nanometers; so called because the interference of the high absorption of most gases necessitates work with evacuated equipment. { 'vak·yəm ¦əl·trə'vī lət spek'träs·kə·pē }

VAD See vapor-phase axial deposition. { vad or ‚vē‚ā'dē }

vadose water [HYD] Water in the zone of aeration. Also known as kremastic water; suspended water; wandering water. { 'vā‚dōs ‚wȯd·ər }

vadose zone See zone of aeration. { 'vā‚dōs ‚zōn }

vaesite [MINERAL] NiS₂ An isometric mineral with pyrite structure composed of sulfide of nickel. { 'vä‚sīt }

vagility [ECOL] The ability of organisms to disseminate. { və'jil·əd·ē }

vagina [ANAT] The canal from the vulvar opening to the cervix uteri. { və'jī·nə }

vagina fibrosa tendinis [ANAT] A fibrous sheath surrounding the tendon of a muscle and usually confining the tendon in a bony groove. { və'jī·nə fī'brō·sə 'ten·də·nəs }

vaginal protocele See rectocele. { 'vaj·ən·əl 'prōd·ə‚sēl }

vaginate [BIOL] Invested in a sheath. { 'vaj·ə‚nāt }

vaginismus [MED] Painful vaginal spasm. { ‚vaj·ə'niz·məs }

vaginitis [MED] 1. Inflammation of the vagina. 2. Inflammation of a tendon sheath. { ‚vaj·ə'nīd·əs }

vagotonine [BIOCHEM] An endocrine substance which is thought to be elaborated by cells of the pancreas and which regulates autonomic tonus. { və'gäd·ə‚nēn }

vagus [ANAT] The tenth cranial nerve; either of a pair of sensory and motor nerves forming an important part of the parasympathetic system in vertebrates. { 'vā·gəs }

Vaisala comparator [OPTICS] An interferometer measuring distances on the order of 100 meters with accuracies on the order of 1 part in 10⁷ { ¦vī·sə·lə kəm'par·əd·ər }

Väisälä period [ACOUS] The natural period of oscillation of a small parcel of air which is displaced adiabatically in a vertical direction, in an isothermal atmosphere horizontally stratified by gravity; equal to 337 seconds for a sound velocity of 333 meters per second. { 'väl·sä·lä ‚pir·ē·əd }

valais wind [METEOROL] The notable valley wind that blows along the Rhone Valley from the upper end of Lake Geneva (Valais Canton); it is sufficiently strong and regular to distort the growth of trees. { va'lā ‚wind }

valence [BIOCHEM] The relative ability of a biological substance to react or combine. [CHEM] A positive number that characterizes the combining power of an element for other elements, as measured by the number of bonds to other atoms which one atom of the given element forms upon chemical combination; hydrogen is assigned valence 1, and the valence is the number of hydrogen atoms, or their equivalent, with which an atom of the given element combines. [MATH] The number of lines incident on a specified point of a graph. { 'vā·ləns }

valence angle See bond angle. { 'vā·ləns ‚aŋ·gəl }

valence band [SOLID STATE] The highest electronic energy band in a semiconductor or insulator which can be filled with electrons. { 'vā·ləns ‚band }

valence bond [PHYS CHEM] The bond formed between the electrons of two or more atoms. { 'vā·ləns ‚bänd }

valence-bond method [PHYS CHEM] A method of calculating binding energies and other parameters of molecules by taking linear combinations of electronic wave functions, some of which represent covalent structures, others ionic structures; the coefficients in the linear combination are calculated by the variational method. Also known as valence-bond resonance method. { 'vā·ləns ¦bänd ‚meth·əd }

valence-bond resonance method See valence-bond method. { 'vā·ləns ¦bänd 'rez·ən·əns ‚meth·əd }

valence-bond theory [CHEM] A theory of the structure of chemical compounds according to which the principal requirements for the formation of a covalent bond are a pair of electrons and suitably oriented electron orbitals on each of the atoms being bonded; the geometry of the atoms in the resulting coordination polyhedron is coordinated with the orientation of the orbitals on the central atom. { 'vā·ləns ¦bänd ‚thē·ə·rē }

valence crystal See covalent crystal. { 'vā·ləns ‚krist·əl }

valence electron [ATOM PHYS] An electron that belongs to the outermost shell of an atom. [SOLID STATE] See conduction electron. { 'vā·ləns i‚lek‚trän }

valence number [CHEM] A number that is equal to the valence of an atom or ion multiplied by +1 or −1, depending on whether the ion is positive or negative, or equivalently on whether the atom in the molecule under consideration has lost or gained electrons from its free state. { 'vā·ləns ‚nəm·bər }

valence shell [ATOM PHYS] The electrons that form the outermost shell of an atom. { 'vā·ləns ‚shel }

valence transition [PHYS CHEM] A change in the electronic occupation of the $4f$ or $5f$ orbitals of the rare-earth or actinide atoms in certain substances at a certain temperature, pressure, or composition. { 'vā·ləns tran‚zish·ən }

valencianite [MINERAL] A variety of potassium feldspar from Mexico. { və'len·chə‚nīt }

valency effect [SOLID STATE] The dependence of the transition temperature of a superconductor on the concentration of doping atoms. { 'vā·lən·sē i‚fekt }

valentinite [MINERAL] Sb₂O₃ A colorless to snow white mineral consisting of antimony trioxide. { 'val·ən‚tē‚nīt }

valeral See n-valeraldehyde. { 'val·ə·rəl }

n-valeraldehyde [ORG CHEM] CH₃(CH₂)₃CHO A flammable liquid, soluble in ether and alcohol, slightly soluble in water; boils at 102°C; used in flavors and as a rubber accelerator. { ¦en ¦val·ər'al·də‚hīd }

valeramide [ORG CHEM] CH₃(CH₂)₃CONH₂ Water-soluble, colorless crystals, melting at 127°C. Also known as pentanamide; valeric amide. { 'val·ər'am·əd }

valerianic acid See valeric acid. { və¦lir·ē¦an·ik 'as·əd }

valerian oil [MATER] A combustible, yellow to brown liquid with a penetrating aroma; soluble in alcohol, acetone, and other organic solvents; derived from the roots and rhizome of the garden heliotrope (Valeriana officinalis), the main components being pinene, camphene, borneol, and esters; used in medicine, flavors, and industrial odorants and to perfume tobacco. { və'lir·ē·ən ‚ȯil }

valeric acid [ORG CHEM] CH₃(CH₂)₃COOH A combustible, toxic, colorless liquid with a penetrating aroma; soluble in water, alcohol, and ether; boils at 185°C; used to make flavors, perfumes, lubricants, plasticizers, and pharmaceuticals. Also known as valerianic acid. { və'lir·ik 'as·əd }

γ-valerolactone [ORG CHEM] C₅H₈O₂ A combustible, mostly immiscible, colorless liquid, boiling at 205°C; used as a dye-bath coupling agent, in brake fluids and cutting oils, and as a solvent for adhesives, lacquers, and insecticides. { ¦gam·ə və¦lir·ō'lak‚tōn }

valid [SYST] Describing a taxon classified on the basis of distinctive characters of accepted importance. { 'val·əd }

validation [COMPUT SCI] The act of testing for compliance with a standard. { ‚val·ə'dā·shən }

validity [MATH] Correctness; especially the degree of closeness by which iterated results approach the correct result. { və'lid·əd·ē }

validity check [COMPUT SCI] Computer check of input data, based on known limits for variables in given fields. { və'lid·əd·ē ‚chek }

valid program [COMPUT SCI] A computer program whose statements, individually and together, follow the syntactical rules of the programming language in which it is written, so that they are capable of being translated into a machine language program. { 'val·əd 'prō‚gram }

valine [BIOCHEM] C₅H₁₁NO₂ An amino acid considered essential for normal growth of animals, and biosynthesized from pyruvic acid. { 'va‚lēn }

vallate papilla [ANAT] One of the large, flat papillae, each surrounded by a trench, in a group anterior to the sulcus terminalis of the tongue. Also known as circumvallate papilla. { 'va‚lāt pə'pil·ə }

vallerite [MINERAL] CuFeS₂ A sulfide mineral found in meteorites. { və'lir‚īt }

VALINE

H₃C   CH₃
   \   /
    CH
    |
    C
   / \
H₂N   COOH
    H

Structural formula of valine.

Valles Marineris [ASTRON] A system of canyons which extends for over 3000 miles (5000 kilometers) along the equatorial region of Mars; it is over 300 miles (500 kilometers) wide in places and drops to more than 4 miles (6 kilometers) below the surrounding surface. { ¦val·əs ¸mar·ə'ner·is }

valley [BUILD] An inside angle formed where two sloping sides intersect. [GEOGR] A generally broad area of flat, low-lying land bordered by higher ground. [GEOL] A relatively shallow, wide depression of the sea floor with gentle slopes. Also known as submarine valley. { 'val·ē }

valley attenuation [ELECTR] For an electric filter with an equal ripple characteristic, the maximum attenuation occurring at a frequency between two frequencies where the attenuation reaches a minimum value. { 'val·ē ¸a¸ten·yə¸wā·shən }

valley bottom See valley floor. { 'val·ē ¸bäd·əm }

valley breeze [METEOROL] A gentle wind blowing up a valley or mountain slope in the absence of cyclonic or anticyclonic winds, caused by the warming of the mountainside and valley floor by the sun. { 'val·ē ¸brēz }

valley fill [GEOL] Unconsolidated sedimentary deposit which fills or partly fills a valley. { 'val·ē ¸fil }

valley filling [ELEC] The addition of loads to an electric power system in off-peak periods. { 'val·ē ¸fil·iŋ }

valley flat [GEOL] The small plain at the bottom of a narrow valley with steep sides. { 'val·ē ¸flat }

valley floor [GEOL] The broad, flat bottom of a valley. Also known as valley bottom; valley plain. { 'val·ē ¸flór }

valley glacier [HYD] A glacier that flows down the walls of a mountain valley. { 'val·ē ¸glā·shər }

valley iceberg [OCEANOGR] An iceberg weathered in such a manner that a large U-shaped slot extends through the iceberg. Also known as drydock iceberg. { 'val·ē 'īs¸bərg }

valley line See thalweg. { 'val·ē ¸līn }

valley plain See valley floor. { 'val·ē ¸plān }

valley rafter [BUILD] A part of the roof frame that extends diagonally from an inside corner plate to the ridge board at the intersection of two roof surfaces. { 'val·ē ¸raf·tər }

valley roof [BUILD] A pitched roof with one or more valleys. { 'val·ē ¸rüf }

valley train [GEOL] A long, narrow body of outwash, deposited by meltwater far beyond the margin of an active glacier and extending along the floor of a valley. Also known as outwash train. { 'val·ē ¸trān }

valley wind [METEOROL] A wind which ascends a mountain valley (up-valley wind) during the day; the daytime component of a mountain and valley wind system. { 'val·ē ¸wind }

Valoniaceae [BOT] A family of green algae in the order Siphonocladales consisting of plants that are essentially unicellular, coenocytic vesicles, spherical or clavate, and up to 2.4 inches (6 centimeters) in diameter. { və¸lō·nē'ās·ē¸ē }

valuation [MATH] A scalar function of a field which has properties similar to those of absolute value. { ¸val·yə'wā·shən }

value [MATH] **1.** The value of a function f at an element x is the element y which f associates with x; that is, $y = f(x)$. **2.** The expected payoff of a matrix game when each player follows an optimal strategy. [MIN ENG] The economically valuable metals contained in ore or tailings. [SCI TECH] The magnitude of a quantity. { 'val·yü }

value-added network [COMMUN] A communications network that provides not only communications channels but also other services such as automatic error detection and correction, protocol conversions, and store-and-forward message services. { 'val·yü ¦ad·əd 'net¸wərk }

value analysis See value engineering. { 'val·yü ə¸nal·ə·səs }

value control See value engineering. { 'val·yü kən¸trōl }

value engineering [IND ENG] The systematic application of recognized techniques which identify the function of a product or service, and provide the necessary function reliably at lowest overall cost. Also known as value analysis; value control. { 'val·yü ¸en·jə¸nir·iŋ }

value group [MATH] For a discrete valuation v on a field K, this is the group formed by the elements $v(x)$ corresponding to nonzero elements x in K. { 'val·yü ¸grüp }

value index [STAT] An index member which is the ratio of the value of all items in a given period to the value of all items in the base period. { 'val·yü ¸in¸deks }

value of isotope mixture [CHEM] A measure of the effort required to prepare a quantity of an isotope mixture; it is proportional to the amount of the mixture, and also depends on the composition of the mixture to be prepared and the composition of the original mixture. { 'val·yü əv 'ī·sə¸tōp ¸miks·chər }

value parameter [COMPUT SCI] A parameter whose value is copied by a subprogram which can then alter its copy without affecting the original. { 'val·yü pə'ram·əd·ər }

value theory [SYS ENG] A concept normally associated with decision theory; it strives to evaluate relative utilities of simple and mixed parameters which can be used to describe outcomes. { 'val·yü ¸thē·ə·rē }

Valvatacea [PALEON] A superfamily of extinct gastropod mollusks in the order Prosobranchia. { ¸val·və'tā·shə }

valvate [BOT] Having valvelike parts, as those which meet edge to edge or which open as if by valves. { 'val¸vāt }

Valvatida [INV ZOO] An order of echinoderms in the subclass Asteroidea. { ¸val·və'tīd·ə }

Valvatina [INV ZOO] A suborder of echinoderms in the order Phanerozonida in which the upper marginals lie directly over, and not alternate with, the corresponding lower marginals. { ¸val·və'tīn·ə }

valve [ANAT] A flat of tissue, as in the veins or between the chambers in the heart, which permits movement of fluid in one direction only. [BOT] **1.** A segment of a dehiscing capsule or legume. **2.** The lidlike portion of certain anthers. [ELECTR] See electron tube. [INV ZOO] **1.** One of the distinct, articulated pieces composing the shell of certain animals, such as barnacles and brachiopods. **2.** One of two shells encasing the body of a bivalve mollusk or a diatom. [MECH ENG] A device used to regulate the flow of fluids in piping systems and machinery. { valv }

valve arrester [ELEC] A type of lightning arrester which consists of a single gap or multiple gaps in series with current-limiting elements; gaps between spaced electrodes prevent flow of current through the arrester except when the voltage across them exceeds the critical gap flashover. { 'valv ə¸res·tər }

valve follower [MECH ENG] A linkage between the cam and the push rod of a valve train. { 'valv ¸fäl·ə·wər }

valve guide [MECH ENG] A channel which supports the stem of a poppet valve for maintenance of alignment. { 'valv ¸gīd }

valve head [MECH ENG] The disk part of a poppet valve that gives a tight closure on the valve seat. { 'valv ¸hed }

valve-in-head engine See overhead-valve engine. { ¦valv in ¦hed 'en·jən }

valve lifter [MECH ENG] A device for opening the valve of a cylinder as in an internal combustion engine. { 'valv ¸lif·tər }

valve positioner [CONT SYS] A pneumatic servomechanism which is used as a component in process control systems to improve operating characteristics of valves by reducing hysteresis. Also known as pneumatic servo. { 'valv pə¸zish·ə·nər }

valve seat [DES ENG] The circular metal ring on which the valve head of a poppet valve rests when closed. { 'valv ¸sēt }

valve stem [MECH ENG] The rod by means of which the disk or plug is moved to open and close a valve. { 'valv ¸stem }

valve train [MECH ENG] The valves and valve-operating mechanism for the control of fluid flow to and from a piston-cylinder machine, for example, steam, diesel, or gasoline engine. { 'valv ¸trān }

Valvifera [INV ZOO] A suborder of isopod crustaceans distinguished by having a pair of flat, valvelike uropods which hinge laterally and fold inward beneath the rear part of the body. { val'vif·ə·rə }

valvula [BIOL] A small valve. [INV ZOO] One of the small processes forming a sheath for the ovipositor in certain insects. { 'val·vyə·lə }

valvulate [BIOL] Having valvules. { 'val·vyə·lət }

vamidothion [ORG CHEM] $C_7H_{16}NO_4PS_2$ A white wax with a melting point of 40°C; very soluble in water; used to control pests in orchards, vineyards, rice, cotton, and ornamentals. { ¦vam·əd·ō'thī¸än }

vampire [VERT ZOO] The common name for bats making up the family Desmodontidae which have teeth specialized for cutting and which subsist on a blood diet. { 'vam¸pīr }

Vampyrellidae [INV ZOO] A family of protozoans in the order Proteomyxida including species which invade filamentous algae and sometimes higher plants. { ¸vam·pə'rel·ə¸dē }

Vampyromorpha [INV ZOO] An order of dibranchiate cephalopod mollusks represented by Vampyroteuthis infernalis, an

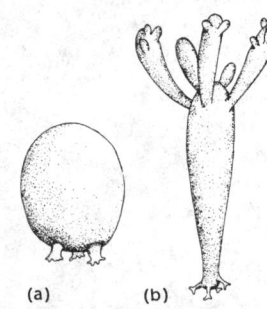

VALONIACEAE

Valonia. (a) Spherical form. (b) Clavate form.

(a)   (b)

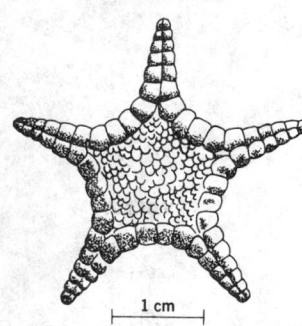

VALVATINA

1 cm

A representative valvate starfish, *Iconaster perierctus*.

inhabitant of the deeper waters of tropical and temperate seas. { ¦vam·pə·rō'mȯr·fə }

van [MIN ENG] **1.** A test of the value of an ore, made by washing (vanning) a small quantity, after powdering it, on the point of a shovel. **2.** To separate, as ore from veinstone, by washing it on the point of a shovel. **3.** A shovel used in ore dressing. { van }

vanadate [MINERAL] Any of several mineral compounds characterized by pentavalent vanadium and oxygen in the anion; an example is vanadinite. { 'van·ə‚dāt }

vanadic acid [INORG CHEM] Any of various acids that do not exist in a pure state and are found in various alkali and other metal vanadates; forms are meta- (HVO_3), ortho- (H_3VO_4), and pyro- ($H_4V_2O_7$). { və'nād·ik 'as·əd }

vanadic acid anhydride See vanadium pentoxide. { və'nād·ik 'as·əd an'hī‚drīd }

vanadic sulfate See vanadyl sulfate. { və'nād·ik 'səl‚fāt }

vanadic sulfide See vanadium sulfide. { və'nād·ik 'səl‚fīd }

vanadinite [MINERAL] $Pb_5(VO_4)_3Cl$ A red, yellow, or brown opatite mineral often occurring as globular masses encrusting other minerals in lead mines; an ore of vanadium and lead hardness is 2.75–3 on Mohs scale, and specific gravity is 6.66–7.10. { və'nād·ən‚īt }

vanadium [CHEM] A metallic transition element, symbol V, atomic number 23; soluble in strong acids and alkalies; melts at 1900°C, boils about 3000°C; used as a catalyst. [MET] A silvery-white, ductile metal resistant to corrosion; used in alloy steels and as an x-ray target. { və'nād·ē·əm }

vanadium carbide [INORG CHEM] VC Hard, black crystals, melting at 2800°C, boiling at 3900°C; insoluble in acids, except nitric acid; used in cutting-tool alloys and as a steel additive. { və'nād·ē·əm 'kär‚bīd }

vanadium dichloride [INORG CHEM] VCl_2 Toxic, green crystals, soluble in alcohol and ether; decomposes in hot water; used as a reducing agent. Also known as vanadous chloride. { və'nād·ē·əm dī'klȯr‚īd }

vanadium oxide [INORG CHEM] A compound of vanadium with oxygen, for example, vanadium tetroxide (V_2O_4), vanadium trioxide or sesquioxide (V_2O_3), vanadium oxide (VO), and vanadium pentoxide (V_2O_5). { və'nād·ē·əm 'äk‚sīd }

vanadium oxytrichloride [INORG CHEM] $VOCl_3$ A toxic, yellow liquid that dissolves or reacts with many organic substances; hydrolyzes in moisture; boils at 126°C; used as an olefin-polymerization catalyst and in organovanadium synthesis. { və'nād·ē·əm 'äk·sē·trī'klȯr‚īd }

vanadium pentasulfide See vanadium sulfide. { və'nād·ē·əm ¦pen·tə'səl‚fīd }

vanadium pentoxide [INORG CHEM] V_2O_5 A toxic, yellow to red powder, soluble in alkalies and acids, slightly soluble in water; melts at 690°C; used in medicine, as a catalyst, as a ceramics coloring, for ultraviolet-resistant glass, photographic developers, textiles dyeing, and nuclear reactors. Also known as vanadic acid anhydride. { və'nād·ē·əm ¦pen'täk‚sīd }

vanadium sesquioxide See vanadium trioxide. { və'nād·ē·əm ¦ses·kwē'äk‚sīd }

vanadium steel [MET] A low-alloy steel containing 0.10–0.15% vanadium. { və'nād·ē·əm 'stēl }

vanadium sulfate See vanadyl sulfate. { və'nād·ē·əm 'səl‚fāt }

vanadium sulfide [INORG CHEM] V_2S_5 A toxic, black-green powder; insoluble in water, soluble in alkalies and acids; decomposes when heated; used to make vanadium compounds. Also known as vanadic sulfide; vanadium pentasulfide. { və'nād·ē·əm 'səl‚fīd }

vanadium tetrachloride [INORG CHEM] VCl_4 A toxic, red liquid; soluble in ether and absolute alcohol; boils at 154°C; used in medicine and to manufacture vanadium and organovanadium compounds. { və'nād·ē·əm te·trə'klȯr‚īd }

vanadium tetraoxide [INORG CHEM] V_2O_4 A toxic blue-black powder; insoluble in water, soluble in alkalies and acids; melts at 1967°C; used as a catalyst. { və'nād·ē·əm te·trə'äk‚sīd }

vanadium trichloride [INORG CHEM] VCl_3 Toxic, deliquescent, pink crystals; soluble in ether and absolute alcohol; decomposes in water and when heated; used to prepare vanadium and organovanadium compounds. { və'nād·ē·əm ¦trī'klȯr‚īd }

vanadium trioxide [INORG CHEM] V_2O_3 Toxic, black crystals; soluble in alkalies and hydrofluoric acid, slightly soluble in water; melts at 1970°C; used as a catalyst. Also known as vanadium sesquioxide. { və'nād·ē·əm trī'äk‚sīd }

vanadous chloride See vanadium dichloride. { və'nād·əs 'klȯr‚īd }

vanadyl chloride [INORG CHEM] $V_2O_2Cl_4·5H_2O$ Toxic, deliquescent, water- and alcohol-soluble green crystals; used to mordant textiles. { və'nād·əl 'klȯr‚īd }

vanadyl sulfate [INORG CHEM] $VOSO_4·2H_2O$ Blue, toxic, water-soluble crystals; used as a reducing agent, catalyst, glass and ceramics colorant, and mordant. Also known as vanadic sulfate; vanadium sulfate. { və'nād·əl 'səl‚fāt }

Van Allen radiation belt [GEOPHYS] One of the belts of intense ionizing radiation in space about the earth formed by high-energy charged particles which are trapped by the geomagnetic field. { va'na·ən ‚rād·ē'a·shən ‚belt }

Van Atta array [ELECTROMAG] Antenna array in which pairs of corner reflectors or other elements equidistant from the center of the array are connected together by a low-loss transmission line in such a way that the received signal is reflected back to its source in a narrow beam to give signal enhancement without amplification. { va'nad·ə ə‚rā }

vancomycin [MICROBIO] A complex antibiotic substance produced by *Streptomyces orientalis;* useful for treatment of severe staphylococcic infections. { ‚van·kə'mīs·ən }

van Creveld-von Gierke's disease See von Gierke's disease. { van 'kre‚väld fən 'gir·kēz di‚zez }

Van Deemter rate theory [ANALY CHEM] A theory that the sample phase in gas chromatography flows continuously, not stepwise. { van 'dām·tər 'rāt ‚thē·ə·rē }

Van de Graaff accelerator [ELECTR] A Van de Graaff generator equipped with an evacuated tube through which charged particles may be accelerated. { 'van də ‚graf ak‚sel·ə‚rād·ər }

Van de Graaff generator [ELECTR] A high-voltage electrostatic generator in which electrical charge is carried from ground to a high-voltage terminal by means of an insulating belt and is discharged onto a large, hollow metal electrode. { 'van də ‚graf ‚jen·ə‚rād·ər }

Van den Bergh reaction [PATH] A liver function test in which diazotized serum or plasma is compared with a standard solution of diazotized bilirubin. { 'van dən ‚bərg rē‚ak·shən }

vandenbrandite [MINERAL] $CuO·UO_3·2H_2O$ A dark green to black mineral consisting of a hydrated oxide of copper and uranium; occurs in small crystals and massive form. { ‚van·dən'bran‚dīt }

Vandermonde determinant [MATH] The determinant of the $n \times n$ matrix whose ith row appears as $1, x_i, x_i^2, \ldots, x_i^{n-1}$ where the x_i^k appear as variables in a given polynomial equation; this provides information about the roots. { 'van·dər‚mȯnd di‚tər·mə·rənt }

Vandermonde's theorem [MATH] A theorem stating that a binomial $(x + y)^a$, where a is an exponent involving the variables x and y, can be stated in terms of a sum of expressions $x^c y^d$, where the exponents c and d involve the variables x and y also. { 'van·dər‚mȯrdz ‚thir·əm }

Van der Pol oscillator [ELECTR] A type of relaxation oscillator which has a single pentode tube and an external circuit with a capacitance that causes the device to switch between two values of the screen voltage. [PHYS] A vibrating system that is governed by an equation of the form $\ddot{x} + \epsilon(-x + 1/3x^3) + x = 0$. { 'van dər ‚pōl ‚äs·ə‚lād·ər }

van der Waals adsorption [PHYS CHEM] Adsorption in which the cohesion between gas and solid arises from van der Waals forces. { 'van dər ‚wȯlz ad‚sȯrp·shən }

van der Waals attraction See van der Waals force. { 'van dər ‚wȯlz ə‚trak·shən }

van der Waals covolume [PHYS CHEM] The constant b in the van der Waals equation, which is approximately four times the volume of an atom of the gas in question multiplied by Avogadro's number. { 'van dər ‚wȯlz ¦kō'väl·yəm }

van der Waals equation [PHYS CHEM] An empirical equation of state which takes into account the finite size of the molecules and the attractive forces between them: $p = [RT/(v - b)] - (a/v^2)$, where p is the pressure, v is the volume per mole, T is the absolute temperature, R is the gas constant, and a and b are constants. { 'van dər ‚wȯlz i‚kwā·zhən }

van der Waals force [PHYS CHEM] An attractive force between two atoms or nonpolar molecules, which arises because a fluctuating dipole moment in one molecule induces a dipole moment in the other, and the two dipole moments then interact.

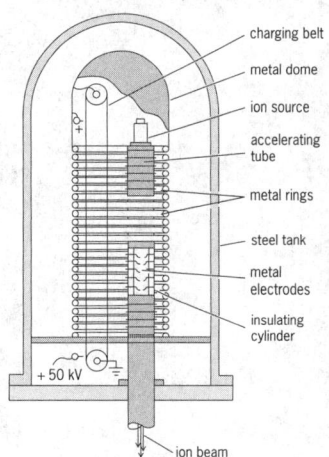

VAN DE GRAAFF GENERATOR

- charging belt
- metal dome
- ion source
- accelerating tube
- metal rings
- steel tank
- metal electrodes
- insulating cylinder
+ 50 kV
- ion beam

Generator for operation in high-pressure gas.

Also known as dispersion force; van der Waals attraction. { 'van dər ˌwȯlz ˌfȯrs }

van der Waals molecule [PHYS CHEM] A molecule that is held together by van der Waals forces. { 'van·dər ˌwälz ˌmäl·əˌkyül }

van der Waals radius [PHYS CHEM] One-half the distance between two atoms of an element that are as close to each other as possible without being formally bonded to each other except for van der Waals forces. { ˌvan·dərˌwälz 'rād·ē·əs }

van der Waals structure [CRYSTAL] The structure of a molecular crystal. { 'van dər ˌwȯlz ˌstrək·chər }

van der Waals surface tension formula [THERMO] An empirical formula for the dependence of the surface tension on temperature: $\gamma = Kp_c^{2/3}T_c^{1/3}(1 - T/T_c)^n$, where γ is the surface tension, T is the temperature, T_c and p_c are the critical temperature and pressure, K is a constant, and n is a constant equal to approximately 1.23. { 'van dər ˌwȯlz 'sər·fəs ˌten·chən ˌfȯr·myə·lə }

van der Waerden number [MATH] For two positive integers, k and r, the smallest positive integer, $n(k,r)$, that satisfies van der Waerden's theorem. { ˌvan·dər 'werd·ən ˌnəm·bər }

van der Waerden's theorem [MATH] The theorem that for any positive integers k and r there is a positive integer n such that, if the first n integers are divided into k classes, then there exists an arithmetic progression of r terms that all belong to the same class. { ˌvan·dər 'werd·ənz ˌthir·əm }

Vandermonde matrix [MATH] A matrix in which each entry in the first row is 1, and each entry in the ith row is the corresponding entry in the second row to the $(i − 1)$ power. { 'van·dərˌmônd ˌma·triks }

Van Dorn sampler [ENG] A sediment sampler that consists of a Plexiglas cylinder closed at both ends by rubber force cups; in the armed position the cups are pulled outside the cylinder and restrained by a releasing mechanism, and after the sample is taken, a length of surgical rubber tubing connecting the cups is sufficiently prestressed to permit the force cups to retain the sample in the cylinder. { van 'dȯrn ˌsam·plər }

vandyke [GRAPHICS] A process used for photocopying; the material is paper-sensitized with ferric iron and silver salts and then exposed to strong light; upon processing, a negative print of white lines on a brown background results; from this a brown-line print can be made. { van'dīk }

vane [AERO ENG] A device that projects ahead of an aircraft to sense gusts or other actions of the air so as to create impulses or signals that are transmitted to the control system to stabilize the aircraft. [MECH ENG] A flat or curved surface exposed to a flow of fluid so as to be forced to move or to rotate about an axis, to rechannel the flow, or to act as the impeller; for example, in a steam turbine, propeller fan, or hydraulic turbine. [NAV] A sight on an instrument used for observing bearings, such as on a pelorus or azimuth circle. [VERT ZOO] The expanded web part of a feather. { vān }

vane anemometer [ENG] A portable instrument used to measure low wind speeds and airspeeds in large ducts; consists of a number of vanes radiating from a common shaft and set to rotate when facing the wind. { 'vān an·əˈmäm·əd·ər }

vane-anode magnetron [ELECTR] Cavity magnetron in which the walls between adjacent cavities have parallel plane surfaces. { 'vān 'an·ˌōd 'mag·nəˌträn }

vane attenuator See flap attenuator. { 'vān əˈten·yəˌwäd·ər }

vane feather See contour feather. { 'vān ˌfeth·ər }

vane motor rotary actuator [MECH ENG] A type of rotary motor actuator which consists of a rotor with several spring-loaded sliding vanes in an elliptical chamber; hydraulic fluid enters the chamber and forces the vanes before it as it moves to the outlets. { 'vān ˌmōd·ər 'rōd·ə·rē 'ak·chəˌwäd·ər }

vane-type instrument [ENG] A measuring instrument utilizing the force of repulsion between fixed and movable magnetized iron vanes, or the force existing between a coil and a pivoted vane-shaped piece of soft iron, to move the indicating pointer. { 'vān ˌtīp ˌin·strə·mənt }

Vaneyellidae [INV ZOO] A family of holothurian echinoderms in the order Dactylochirotida. { ˌvā·nēˈel·əˌdē }

vang [NAV ARCH] **1.** A rope which supports or holds steady a boom or spar. **2.** In particular, one of the two ropes which run from the top of the gaff and steady it when the sail is not set. { vaŋ }

Vanhorniidae [INV ZOO] A monospecific family of the

Hymenoptera in the superfamily Proctotrupoidea. { ˌvan ˌhȯrˈnī·əˌdē }

vanilla [COMPUT SCI] Referring to a generalized system, usually software, that has not been subjected to special modifications, enhancements, or customization. Also known as plain vanilla; pure vanilla. { vəˈnil·ə }

vanilla extract [FOOD ENG] Flavoring prepared from vanilla beans with or without the addition of sugar, dextrose, or glycerol; contains soluble matter from not less than 10 grams of vanilla beans in 100 milliliters. { vəˈnil·ə 'ek·ˌstrakt }

vanillic aldehyde See vanillin. { vəˈnil·ik 'al·dəˌhīd }

vanillin [ORG CHEM] $C_8H_8O_3$ A combustible solid, soluble in water, alcohol, ether, and chloroform; melts at 82°C; used in pharmaceuticals, perfumes, and flavors, and as an analytical reagent. Also known as vanillic aldehyde. { vəˈnil·ən }

vanishing tide [OCEANOGR] When a high water and low water "melt" together into a period of several hours with a nearly constant water level. { 'van·ish·iŋ 'tīd }

vanner [MIN ENG] A machine for dressing ore; the name is given to various patented devices in which the peculiar motions of the shovel in the miner's hands in the operation of making a van are, or are supposed to be, successfully imitated. Also known as vanning machine. { 'van·ər }

vanning machine See vanner. { 'van·iŋ məˌshēn }

vanoxite [MINERAL] $(V_4)^{4+}(V_2)^{5+}O_{13}\cdot 8H_2O$ A black mineral consisting of a hydrous oxide of vanadium; occurs as microscopic crystals and in massive form. { vaˈnäkˌsīt }

V antenna [ELECTROMAG] An antenna having a V-shaped arrangement of conductors fed by a balanced line at the apex; the included angle, length, and elevation of the conductors are proportioned to give the desired directivity. Also spelled vee antenna. { 'vē anˌten·ə }

van't Hoff equation [PHYS CHEM] An equation for the variation with temperature T of the equilibrium constant K of a gaseous reaction in terms of the heat of reaction at constant pressure, ΔH: $d(\ln K)/dT = \Delta H/RT^2$, where R is the gas constant. Also known as van't Hoff isochore. { van'tȯf iˌkwā·zhən }

van't Hoff factor [PHYS] The ratio of the observed osmotic pressure of a solution to that predicted by van't Hoff's law. { van'tȯf ˌfak·tər }

van't Hoff formula [ORG CHEM] The expression that the number of stereoisomers of a sugar molecule is equal to 2^n, where n is the number of asymmetric carbon atoms. { van'tȯf ˌfȯr·myə·lə }

van't Hoff isochore See van't Hoff equation. { van'tȯf 'īˌsäˌkȯr }

van't Hoff isotherm [PHYS CHEM] An equation for the change in free energy during a chemical reaction in terms of the reaction, the temperature, and the concentration and number of molecules of the reactants. { van'tȯf 'īˌsəˌthərm }

vanthoffite [MINERAL] $Na_6Mg(SO_4)_4$ A colorless mineral consisting of a sulfate of sodium and magnesium; occurs in massive form. { van'tȯˌfīt }

van't Hoff's law [PHYS] The law that the osmotic pressure of a dissolved substance equals the gas pressure it would exert if it were an ideal gas that occupied the same volume as that of the solution. { van'tȯfs ˌlȯ }

Van Vleck equation [QUANT MECH] An equation based on quantum theory for the molar paramagnetism of a magnetically susceptible material from magnetic moment, absolute temperature, and various constants. { van 'vlek iˌkwā·zhən }

Van Vleck paramagnetism [QUANT MECH] The paramagnetism of a collection of atoms, ions, or molecules, as computed by quantum theory; the atoms, ions, or molecules in a magnetic field are distributed among the various allowed energy levels according to a Boltzmann distribution, and the magnetization of the system is computed by finding the average component of angular momentum parallel to the field. { van 'vlek ˌpar·əˈmag·nəˌtiz·əm }

vapor [THERMO] A gas at a temperature below the critical temperature, so that it can be liquefied by compression, without lowering the temperature. { 'vā·pər }

vapor barrier [CIV ENG] A layer of material applied to the inner (warm) surface of a concrete wall or floor to prevent absorption and condensation of moisture. { 'vā·pər ˌbar·ē·ər }

vapor blasting [MET] Cleaning the surface of a metal with a fine abrasive suspended in water and propelled at high speed by air or steam. Also known as liquid honing; vapor honing. { 'vā·pər 'blast·iŋ }

vapor-compression cycle [MECH ENG] A refrigeration cycle in which refrigerant is circulated through a machine which allows for successive boiling (or vaporization) of liquid refrigerant as it passes through an expansion valve, thereby producing a cooling effect in its surroundings, followed by compression of vapor to liquid. { 'vā·pər kəm'presh·ən ˌsī·kəl }

vapor cycle [THERMO] A thermodynamic cycle, operating as a heat engine or a heat pump, during which the working substance is in, or passes through, the vapor state. { 'vā·pər ˌsī·kəl }

vapor degreasing [ENG] A type of cleaning procedure for metals to remove grease, oils, and lightly attached solids; a solvent such as trichloroethylene is boiled, and its vapors are condensed on the metal surfaces. { 'vā·pər dē'grēs·iŋ }

vapor deposition [MET] Producing a film of metal on a heated surface, often in a vacuum, either by decomposition of the vapor of a compound at the work surface or by direct reaction between the work surface and the vapor. Also known as vacuum plating. { 'vā·pər ˌdep·ə'zish·ən }

vapor-dominated hydrothermal reservoir [GEOL] Any geothermal system mainly producing dry steam; the Geysers area of northern California and the Larderelle region of Italy are two examples. { 'vā·pər ˌdom·ə·ˌnād·əd ˌhī·drə'thər·məl 'rez·əv·ˌwär }

vapor-filled thermometer [ENG] A gas- or vapor-filled temperature measurement device that moves or distorts in response to temperature-induced pressure changes from the expansion or contraction of the sealed, vapor-containing chamber. { 'vā·pər ˌfild thər'mäm·əd·ər }

vapor honing See vapor blasting. { 'vā·pər 'hōn·iŋ }

vaporimeter [ENG] An instrument used to measure a substance's vapor pressure, especially that of an alcoholic liquid, in order to determine its alcohol content. { ˌvap·ə'rim·əd·ər }

vaporization See volatilization. { ˌvā·pə·rə'zā·shən }

vaporization coefficient [THERMO] The ratio of the rate of vaporization of a solid or liquid at a given temperature and corresponding vapor pressure to the rate of vaporization that would be necessary to produce the same vapor pressure at this temperature if every vapor molecule striking the solid or liquid were absorbed there. { ˌvā·pə·rə'zā·shən ˌkō·ə·fish·ənt }

vaporization cooling [ENG] Cooling by volatilization of a nonflammable liquid having a low boiling point and high dielectric strength; the liquid is flowed or sprayed on hot electronic equipment in an enclosure where it vaporizes, carrying the heat to the enclosure walls, radiators, or heat exchanger. Also known as evaporative cooling. { ˌvā·pə·rə'zā·shən ˌkül·iŋ }

vaporizer [CHEM ENG] A process vessel in which a liquid is heated until it vaporizes; heat can be indirect (steam or heat-transfer fluid) or direct (hot gases or submerged combustion). { 'vā·pəˌrīz·ər }

vapor lamp See discharge lamp. { vā·pər ˌlamp }

vapor-liquid equilibrium See liquid-vapor equilibrium. { 'vā·pər 'lik·wəd ˌe·kwə'lib·rē·əm }

vapor-liquid separation [CHEM ENG] The removal of liquid droplets from a flowing stream of gas or vapor; accomplished by impingement, cyclonic action, and absorption or adsorption operations. { 'vā·pər 'lik·wəd ˌsep·ə'rā·shən }

vapor lock [FL MECH] Interruption of the flow of fuel in a gasoline engine caused by formation of vapor or gas bubbles in the fuel-feeding system. { 'vā·pər ˌläk }

vapor-phase axial deposition [ENG] A method of fabricating graded-index optical fibers in which fine glass particles of silicon dioxide and germanium dioxide are synthesized and deposited on a rotating seed rod, and the synthesized porous preform is then pulled up and passes through a hot zone, undergoing dehydration and sintering, to become a porous preform. Abbreviated VAD. { 'vā·pər ˌfāz 'ak·sē·əl ˌdep·ə'zish·ən }

vapor-phase epitaxy [SOLID STATE] The use of chemical vapor deposition to grow epitaxial layers. Abbreviated VPE. { 'vā·pər ˌfāz 'ep·ə·ˌtak·sē }

vapor-phase reactor [CHEM ENG] A heavy steel vessel for carrying out chemical reactions on an industrial scale where efficient control over a vapor phase is needed, for example, in an oxidation process. { 'vā·pər ˌfāz rē'ak·tər }

vapor pressure [METEOROL] The partial pressure of water vapor in the atmosphere. [THERMO] For a liquid or solid,

the pressure of the vapor in equilibrium with the liquid or solid. { 'vā·pər ˌpresh·ər }

vapor-pressure deficit See saturation deficit. { 'vā·pər ˌpresh·ər 'def·ə·sət }

vapor-pressure osmometer [ANALY CHEM] A device for the determination of molecular weights by the decrease of vapor pressure of a solvent upon addition of a soluble sample. { 'vā·pər ˌpresh·ər äz'mäm·əd·ər }

vapor-pressure thermometer [ENG] A thermometer in which the vapor pressure of a homogeneous substance is measured and from which the temperature can be determined; used mostly for low-temperature measurements. { 'vā·pər ˌpresh·ər thər'mäm·əd·ər }

vapor rate [CHEM ENG] In distillation, the upward flow rate of vapor through a distillation column. { 'vā·pər ˌrāt }

vapor-recovery unit [ENG] **1.** A device or system to catch vaporized materials (usually fuels or solvents) as they are vented. **2.** In petroleum refining, a process unit to which gases and vaporized gasoline from various processing operations are charged, separated, and recovered for further use. { 'vā·pər ri'kəv·ə·rē ˌyü·nət }

vapor suppression [NUCLEO] A safety system that can be incorporated in the design of structures housing water-cooled nuclear reactors: the space surrounding the reactor is vented into pools of water open to the outside air; if surges of hot vapor are released from the reactor in an accident, their energy is dissipated in the pools of water; gases not condensed are scrubbed clean of radioactive particles by the bubbling. Also known as pressure suppression. { 'vā·pər sə,presh·ən }

vapor trail See condensation trail. { 'vā·pər ˌtrāl }

vapor volume equivalent [PETRO ENG] The volume of vapor to which a specified amount of liquid would be equivalent at designated standard conditions (for example, 14.65 psia and 60°F, or 15.5°C); used in the petroleum industry to calculate the specific gravity of fluids from gas-condensate wells. { 'vā·pər 'väl·yəm i'kwiv·ə·lənt }

var See volt-ampere reactive.

VAR See visual-aural range.

vara [CIV ENG] A surveyors' unit of length equal to $33\frac{1}{3}$ inches (84.7 centimeters). { 'vär·ə }

varactor [ELECTR] A semiconductor device characterized by a voltage-sensitive capacitance that resides in the space-charge region at the surface of a semiconductor bounded by an insulating layer. Also known as varactor diode; variable-capacitance diode; varicap; voltage-variable capacitor. { va'rak·tər }

varactor diode See varactor. { va'rak·tər 'dīˌōd }

varactor tuning [ELECTR] A method of tuning in which varactor diodes are used to vary the capacitance of a tuned circuit. { va'rak·tər 'tün·iŋ }

Varanidae [VERT ZOO] The monitors, a family of reptiles in the suborder Sauria found in the hot regions of Africa, Asia, Australia, and Malaya. { və'ran·ə,dē }

vardar [METEOROL] A cold fall wind blowing from the northwest down the Vardar valley in Greece to the Gulf of Salonica; it occurs when atmospheric pressure over eastern Europe is higher than over the Aegean Sea, as is often the case in winter. Also known as vardarac. { 'värˌdär }

vardarac See vardar. { 'vär·dəˌrak }

var hour [ELEC] A unit of the integral of reactive power over time, equal to a reactive power of 1 var integrated over 1 hour; equal in magnitude to 3600 joules. Also known as reactive volt-ampere hour; volt-ampere-hour reactive. { 'vär ˌau̇r }

var hour meter [ENG] An instrument that measures and registers the integral of reactive power over time in the circuit to which it is connected. { 'vär ˌau̇r ˌmēd·ər }

variable [COMPUT SCI] A data item, or specific area in main memory, that can assume any of a set of values. [MATH] A symbol which is used to represent some undetermined element from a given set, usually the domain of a function. { 'ver·ē·ə·bəl }

variable acceleration See dynamic resolution. { ˌver·ē·ə·bəl ik,sel·ə'rā·shən }

variable-area exhaust nozzle [AERO ENG] On a jet engine, an exhaust nozzle of which the exhaust exit opening can be varied in area by means of some mechanical device, permitting variation in the jet velocity. { ˌver·ē·ə·bəl ˌer·ē·ə ig'zȯst ˌnäz·əl }

variable-area meter [ENG] A flowmeter that works on the

principle of a variable restrictor in the flowing stream being forced by the fluid to a position to allow the required flow-through. { 'ver·ē·ə·bəl ¦er·ē·ə 'mēd·ər }

variable-area track [ENG ACOUS] A sound track divided laterally into opaque and transparent areas; a sharp line of demarcation between these areas corresponds to the waveform of the recorded signal. { 'ver·ē·ə·bəl ¦er·ē·ə 'trak }

variable attenuator [ELECTR] An attenuator for reducing the strength of an alternating-current signal either continuously or in steps, without causing appreciable signal distortion, by maintaining a substantially constant impedance match. { 'ver·ē·ə·bəl ə'ten·yə,wād·ər }

variable-bandwidth filter [ELECTR] An electric filter whose upper and lower cutoff frequencies may be independently selected, so that almost any bandwidth may be obtained; it usually consists of several stages of *RC* filters, each separated by buffer amplifiers; tuning is accomplished by varying the resistance and capacitance values. { 'ver·ē·ə·bəl ¦band,width ,fil·tər }

variable-block [COMPUT SCI] Pertaining to an arrangement of data in which the number of words or characters in a block can vary, as determined by the programmer. { 'ver·ē·ə·bəl 'bläk }

variable-capacitance diode See varactor. { 'ver·ē·ə·bəl kə¦pas·əd·əns 'dī,ōd }

variable capacitor [ELEC] A capacitor whose capacitance can be varied continuously by moving one set of metal plates with respect to another. { 'ver·ē·ə·bəl kə'pas·əd·ər }

variable carrier modulation See controlled carrier modulation. { 'ver·ē·ə·bəl 'kar·ē·ər ,mäj·ə'lā·shən }

variable ceiling [METEOROL] After United States weather-observing practice, a condition in which the ceiling rapidly increases and decreases while the ceiling observation is being made; the average of the observed values is used as the reported ceiling, and it is reported only for ceilings of less than 3000 feet (914 meters). { 'ver·ē·ə·bəl 'sēl·iŋ }

variable click track [ENG ACOUS] A click track with irregularly spaced clicks. { 'ver·ē·ə·bəl 'klik ,trak }

variable connector [COMPUT SCI] A flow chart symbol representing a sequence connection which is not fixed, but which can be varied by the flow-charted procedure itself; it corresponds to an assigned GO TO in a programming language such as FORTRAN. { 'ver·ē·ə·bəl kə,nek·tər }

variable costs [IND ENG] Costs which vary directly with the number of units produced; direct labor and material are examples. { 'ver·ē·ə·bəl 'kòsts }

variable coupling [ELEC] Inductive coupling that can be varied by moving one coil with respect to another. { 'ver·ē·ə·bəl 'kəp·liŋ }

variable-cycle engine [AERO ENG] A type of gas turbine jet engine whose cycle parameters, such as pressure ratio, temperature, gas flow paths, and air-handling characteristics, can be varied between those of a turbojet and a turbofan, enabling it to combine the advantages of both. { 'ver·ē·ə·bəl ¦sī·kəl 'en·jən }

variable-cycle operation [COMPUT SCI] An operation that requires a variable number of regularly timed execution cycles for its completion. { 'ver·ē·ə·bəl ¦sī·kəl ,äp·ə'rā·shən }

variable-density sound track [ENG ACOUS] A constant-width sound track in which the average light transmission varies along the longitudinal axis in proportion to some characteristic of the applied signal. { 'ver·ē·ə·bəl ¦den·səd·ē 'saun ,trak }

variable-depth sonar [ENG] Sonar in which the projector and receiving transducer are mounted in a watertight pod that can be lowered below a vessel to an optimum depth for minimizing thermal effects when detecting underwater targets. { 'ver·ē·ə·bəl ¦depth 'sō,när }

variable diode function generator [ELECTR] An improvement of a diode function generator in which fully adjustable potentiometers are used for breakpoint and slope resistances, permitting the programming of analytic, arbitrary, and empirical functions, including inflections. Abbreviated VDFG. { 'ver·ē·ə·bəl ¦dī,ōd 'fəŋk·shən ,jen·ə,rād·ər }

variable element [IND ENG] **1.** An element with a time that varies significantly from cycle to cycle as a function of one or more variables occurring within the job. **2.** An element that is common to two different jobs but whose time varies because of differences between the two jobs. { ¦var·ē·ə·bəl 'el·ə·mənt }

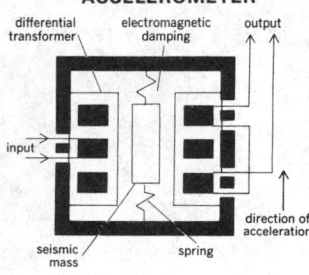
variable field [COMPUT SCI] A field of data whose length is allowed to vary within certain specified limits. [PHYS] Field which changes during the time under consideration. { 'ver·ē·ə·bəl 'fēld }

variable flow [FL MECH] Fluid flow in which the velocity changes both with time and from point to point. { 'ver·ē·ə·bəl 'flō }

variable-focal-length lens See zoom lens. { 'ver·ē·ə·bəl ¦fō·kəl ¦leŋkth 'lenz }

variable-focus condenser [OPTICS] A condenser that is used to obtain a large illuminated field area, and has two lenses, the first of which can be adjusted to bring light to a focus between the lenses. { 'ver·ē·ə·bəl ¦fō·kəs kən'den·sər }

variable force [MECH] A force whose direction or magnitude or both change with time. { 'ver·ē·ə·bəl 'fòrs }

variable geometry aircraft [AERO ENG] Aircraft with variable profile geometry, such as variable sweep wings. { 'ver·ē·ə·bəl jē¦äm·ə·trē 'er,kraft }

variable inductance See variable inductor. { 'ver·ē·ə·bəl in'dək·təns }

variable-inductance accelerometer [ENG] An accelerometer consisting of a differential transformer with three coils and a mass which passes through the coils and is suspended from springs; the center coil is excited from an external alternating-current power source, and two end coils connected in series opposition are used to produce an ac output which is proportional to the displacement of the mass. { 'ver·ē·ə·bəl in¦dək·təns ik,sel·ə'räm·əd·ər }

variable inductor [ELECTROMAG] A coil whose effective inductance can be changed. Also known as variable inductance. { 'ver·ē·ə·bəl in'dək·tər }

variable-length field [COMPUT SCI] A data field in which the number of characters varies, the length of the field being stored within the field itself. { 'ver·ē·ə·bəl ¦leŋkth 'fēld }

variable-length operation [COMPUT SCI] A computer operation whose operands are allowed to have a variable number of bits or characters. { 'ver·ē·ə·bəl ¦leŋkth ,äp·ə'rā·shən }

variable-length record [COMPUT SCI] A data or file format that allows each record to be exactly as long as needed. { 'ver·ē·ə·bəl ¦leŋkth 'rek·ərd }

variable-length word [COMPUT SCI] A computer word whose length is determined by the programmer. { 'ver·ē·ə·bəl ¦leŋkth 'wərd }

variable-mu tube [ELECTR] An electron tube in which the amplification factor varies in a predetermined manner with control-grid voltage; this characteristic is achieved by making the spacing of the grid wires vary regularly along the length of the grid, so that a very large negative grid bias is required to block anode current completely. Also known as remote-cutoff tube. { 'ver·ē·ə·bəl ¦myü 'tüb }

variable nebula [ASTRON] A nebula whose shape and brightness vary; an example is in the constellation Monoceros. { 'ver·ē·ə·bəl 'neb·yə·lə }

variable parameter [COMPUT SCI] A parameter whose storage address is passed to a subprogram so that the subprogram can alter its value. [PHYS] A parameter which may be varied to assume any value in some range. { 'ver·ē·ə·bəl pə'ram·əd·ər }

variable-pitch propeller [ENG] A controllable-pitch propeller whose blade angle may be adjusted to any angle between the low and high pitch limits. { 'ver·ē·ə·bəl ¦pich prə'pel·ər }

variable point [COMPUT SCI] A system of numeration in which the location of the decimal point is indicated by a special character at that position. { 'ver·ē·ə·bəl 'pòint }

variable radio-frequency radiosonde [ENG] A radiosonde whose carrier frequency is modulated by the magnitude of the meteorological variables being sensed. { 'ver·ē·ə·bəl 'rād·ē·ō ¦frē·kwən·sē 'rād·ē·ō,sänd }

variable recoil [ORD] In recoil systems, the variation of the length of recoil according to the elevation in such manner as to prevent the gun from striking the ground when fired at high angles. { 'ver·ē·ə·bəl 'rē,kòil }

variable reduction [GRAPHICS] A characteristic of microfilming cameras; the ability to produce various sized images of a single original. { 'ver·ē·ə·bəl ri'dək·shən }

variable-reluctance microphone See magnetic microphone. { 'ver·ē·ə·bəl ri¦lək·təns 'mī·krə,fōn }

variable-reluctance pickup [ENG ACOUS] A phonograph

pickup that depends for its operation on variations in the reluctance of a magnetic circuit due to the movements of an iron stylus assembly that is a part of the magnetic circuit. Also known as magnetic cartridge; magnetic pickup; reluctance pickup. { 'ver·ē·ə·bəl ri‚lək·təns 'pik‚əp }

variable-reluctance stepper motor [ELEC] A stepper motor having a soft iron rotor with teeth or poles so positioned that they cannot simultaneously align with all the stator poles. { 'ver·ē·ə·bəl ri‚lək·təns 'step·ər ‚mōd·ər }

variable-reluctance transducer [ELECTROMAG] A transducer in which a slug of magnetic material is moved between two coils by the displacement being monitored; this changes the reluctance of the coils, thereby changing their impedance. { 'ver·ē·ə·bəl ri‚lək·təns tranz'dü·sər }

variable-resistance accelerometer [ENG] Any accelerometer which operates on the principle that electrical resistance of any conductor is a function of its dimensions; when the dimensions of the conductor are varied mechanically, as constant current flows through it, the voltage across it varies as a function of this mechanical excitation; examples include the strain-gage accelerometer, and an accelerometer making use of a slide-wire potentiometer. { 'ver·ē·ə·bəl ri‚zis·təns ik‚sel·ə'räm·əd·ər }

variable resistor See rheostat. { 'ver·ē·ə·bəl ri'zis·tər }

variable-sequence robot [CONT SYS] A robot controlled by instructions that can be modified. { 'ver·ē·ə·bəl ‚sē·kwəns 'rō‚bät }

variable speech control [ELECTR] A method of removing small portions of speech from a tape recording at regular intervals and stretching the remaining sounds to fill the gaps, so that recorded speech can be played back at twice or even 2¹/₂ times the original speed without changing pitch and without significant loss of intelligibility. Abbreviated VSC. { 'ver·ē·ə·bəl 'spēch kən‚trōl }

variable-speed drive [MECH ENG] A mechanism transmitting motion from one shaft to another that allows the velocity ratio of the shafts to be varied continuously. { 'ver·ē·ə·bəl ‚spēd 'drīv }

variable-speed generator [ELEC] A generator whose speed can be adjusted within certain limits, with a method of regulation that causes it to deliver a constant voltage. { 'ver·ē·ə·bəl ‚spēd 'jen·ə‚rād·ər }

variable-speed motor [ELEC] A motor whose speed depends upon the load that it carries. { 'ver·ē·ə·bəl spēd 'mōd·ər }

variable-speed scanning [ELECTR] Scanning method whereby the speed of deflection of the scanning beam in the cathode-ray tube of a television camera is governed by the optical density of the film being scanned. { 'ver·ē·ə·bəl ‚spēd 'skan·iŋ }

variable star [ASTRON] A star that has a detectable change in its intensity which is often accompanied by other physical changes; changes in brightness may be a few thousandths of a magnitude to 20 magnitudes or even more. { 'ver·ē·ə·bəl 'stär }

variable-thickness microbridge [CRYO] A Josephson junction formed by a short, narrow constriction in a thin superconducting film, which is thinner than the rest of the film. { 'ver·ē·ə·bəl 'thik·nəs 'mī·krō‚brij }

variable-time fuse See proximity fuse. { 'ver·ē·ə·bəl 'tīm ‚fyüz }

variable-transconductance circuit [ELECTR] A circuit used in four-quadrant multipliers that employs a simple differential transistor pair in which one variable input to the base of one transistor controls the device's gain or transconductance, and one transistor amplifies the other's variable input, applied to the common emitter point, in proportion to the control input. { 'ver·ē·ə·bəl ‚tranz·kən‚dək·təns 'sər·kət }

variable transformer [ELEC] An iron-core transformer having provisions for varying its output voltage over a limited range or continuously from zero to maximum output voltage, generally by means of a contact arm moving along exposed turns of the secondary winding. Also known as adjustable transformer; continuously adjustable transformer. { 'ver·ē·ə·bəl tranz'fór·mər }

variable visibility [METEOROL] After United States weather observing practice, a condition in which the prevailing visibility fluctuates rapidly while the observation is being made; the average of the observed values is used as the reported visibility,

and it is reported only for visibilities of less than 3 miles (4.8 kilometers). { 'ver·ē·ə·bəl ‚viz·ə'bil·əd·ē }

variable-volume air system [MECH ENG] An air-conditioning system in which the volume of air delivered to each controlled zone is varied automatically from a preset minimum to a maximum value, depending on the load in each zone. { ‚ver·ē·ə·bəl 'väl·yəm 'er ‚sis·təm }

variable waveguide attenuator [ELECTROMAG] Device designed to introduce attenuation into a waveguide circuit by moving a lossy vane either sideways across the waveguide or into the waveguide through a longitudinal slot. { 'ver·ē·ə·bəl 'wāv‚gīd ə‚ten·yə‚wād·ər }

variable-word-length [COMPUT SCI] A phrase referring to a computer in which the number of characters addressed is not a fixed number but is varied by the data or instruction. { 'ver·ē·ə·bəl 'wərd ‚leŋkth }

variance [STAT] The square of the standard deviation. { 'ver·ē·əns }

variance ratio test [STAT] A technique for comparing the spreads or variabilities of two sets of figures to determine whether the two sets of figures were drawn from the same population. Also known as F test. { 'ver·ē·əns 'rā·shō ‚test }

variant [COMMUN] 1. One of two or more cipher or code symbols which have the same plain text equivalent. 2. One of several plain text meanings that are represented by a single code group. { 'ver·ē·ənt }

variant record [COMPUT SCI] A record variable whose format is made to depend on some circumstance; for example, a record dealing with rates of pay might contain information on hourly rates for some employees and weekly or monthly salaries for others. { 'ver·ē·ənt 'rek·ərd }

variate [MATH] See random variable. [STAT] The numerical value of a measurement to be used for statistical handling. { 'ver·ē·ət }

variate difference method [STAT] A technique for estimating the correlation between the random parts of two given time series. { 'ver·ē‚āt 'dif·rəns ‚meth·əd }

variation See declination. { ‚ver·ē'ā·shən }

variational inequality [ASTRON] An inequality in the moon's motion, due mainly to the tangential component of the sun's attraction. { ‚ver·ē'ā·shən·əl ‚in·i'kwäl·əd·ē }

variational method [QUANT MECH] A method of calculating an upper bound on the lowest energy level of a quantum-mechanical system and an approximation for the corresponding wave function; in the integral representing the expectation value of the Hamiltonian operator, one substitutes a trial function for the true wave function, and varies parameters in the trial function to minimize the integral. { ‚ver·ē'ā·shən·əl ‚meth·əd }

variational principle [MATH] A technique for solving boundary value problems that is applicable when the given problem can be rephrased as a minimization problem. { ‚ver·ē'ā·shən·əl ‚prin·sə·pəl }

variation diagram [PETR] A diagram constructed by plotting the chemical compositions of rocks in an igneous rock series in order to show the genetic relationships and the nature of the processes that have affected the series. Also known as Harker diagram. { ‚ver·ē'ā·shən ‚dī·ə‚gram }

variation of latitude [GEOPHYS] Change of the latitude of a place on earth because of the irregular movement of the north and south poles; the movement is caused by the earth's shifting on its axis. { ‚ver·ē'ā·shən əv 'lad·ə‚tüd }

variation per day [GEOPHYS] The change in the value of any geophysical quantity during 1 day. { ‚ver·ē'ā·shən pər 'dā }

variation per hour [GEOPHYS] The change in the value of any geophysical quantity during 1 hour. { ‚ver·ē'ā·shən pər 'aur }

variation per minute [GEOPHYS] The change in the value of any geophysical quantity during 1 minute. { ‚ver·ē'ā·shən pər 'min·ət }

varicap See varactor. { 'var·ə‚kap }

varicella See chickenpox. { ‚var·ə'sel·ə }

varicocele [MED] Dilatation of the veins of the pampiniform plexus of the spermatic cord, forming a soft, elastic, often uncomfortable swelling. { 'var·ək·ə‚sēl }

varicose [MED] Pertaining to blood vessels that are dilated, knotted, and tortuous. { 'var·ə‚kōs }

varicose vein [ANAT] An enlarged tortuous blood vessel that occurs chiefly in the superficial veins and their tributaries in

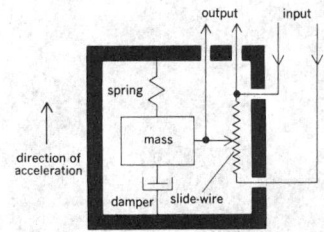

VARIOMETER

Variometer for measuring horizontal intensity or declination of terrestrial magnetic field, equipped with Helmholtz coil for calibration. (*U. S. Coast and Geodetic Survey*)

VARLEY LOOP TEST

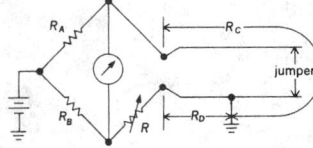

Diagram of circuit used in Varley loop test for location of leakage to ground. Resistance R is adjusted until no current flows through detector, whereupon $R_A/R_B = R_C/(R + R_D)$.

VASCULAR BUNDLE

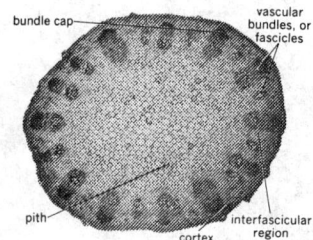

Cross section of sunflower (*Helianthus*) stem showing vascular bundles.

---

the lower extremities. Also known as varicosity. { 'var·ə‚kōs 'vān }

varicosity *See* varicose vein. { ‚var·ə'käs·əd·ē }

variegate [BIOL] Having irregular patches of diverse colors. { 'ver·ē·ə‚gāt }

variegated position effect [GEN] A phenomenon observed in some cases when a chromosome aberration causes a wild-type gene from euchromatin to be relocated adjacent to heterochromatin; the phenotypic expression of the wild-type allele will be unstable, producing patches of phenotypically mutant tissue that differ from the surrounding wild-type tissue. { 'ver·ē·ə‚gād·əd pə¦zish·ən i‚fekt }

variety [SYS ENG] The logarithm (usually to base 2) of the number of discriminations that an observer or a sensing system can make relative to a system. [SYST] A taxonomic group or category inferior in rank to a subspecies. { və'rī·əd·ē }

varifocal lens *See* zoom lens. { ¦ver·ə¦fō·kəl 'lenz }

Varignon's theorem [MECH] The theorem that the moment of a force is the algebraic sum of the moments of its vector components acting at a common point on the line of action of the force. { var·ən'yōnz ‚thir·əm }

varindor [ELECTROMAG] Inductor in which the inductance varies markedly with the current in the winding. { 'var·ən‚dȯr }

variocoupler [ELECTROMAG] In radio practice, a transformer in which the self-impedance of windings remains essentially constant while the mutual impedance between the windings is adjustable. { ‚ver·ē·ō 'kəp·lər }

variograph [ENG] A recording variometer. { 'ver·ē·ə ‚graf }

variola *See* smallpox. { ‚ver·ē'ō·lə }

variole [GEOL] A spherule the size of a pea, usually consisting of radiating plagioclase or pyroxene crystals. { 'ver·ē‚ōl }

variolitic [PETR] Referring to the texture of basic igneous rock composed of varioles in a finer-grained matrix. { ¦ver·ē·ə¦lid·ik }

variolosser [ELEC] Device in which loss can be controlled by a voltage or current. { 'ver·ē·ə‚lȯs·ər }

variometer [ELECTROMAG] A variable inductance having two coils in series, one mounted inside the other, with provisions for rotating the inner coil in order to vary the total inductance of the unit over a wide range. [ENG] A geomagnetic device for detecting and indicating changes in one of the components of the terrestrial magnetic field vector, usually magnetic declination, the horizontal intensity component, or the vertical intensity component. { ‚ver·ē'äm·əd·ər }

varioplex [ELEC] Telegraph switching system that establishes connections on a circuit-sharing basis between a multiplicity of telegraph transmitters in one locality and respective corresponding telegraph receivers in another locality over one or more intervening telegraph channels; maximum use of channel capacity is secured by momentarily storing the signals and allocating circuit time in rotation among the transmitters having information in storage. { 'ver·ē·ə‚pleks }

Variscan orogeny [GEOL] The late Paleozoic orogenic era in Europe, extending through the Carboniferous and Permian. Also known as Hercynian orogeny. { va'ris·kən ȯ'räj·ə·nē }

varistor [ELECTR] A two-electrode semiconductor device having a voltage-dependent nonlinear resistance; its resistance drops as the applied voltage is increased. Also known as voltage-dependent resistor. { və'ris·tər }

varix [INV ZOO] A conspicuous ridge across each whorl of certain univalves marking the ancestral position of the outer lip of the aperture. [MED] A dilated and tortuous vein, artery, or lymphatic vessel. { 'var·iks }

Varley loop test [ELEC] A method of using a Wheatstone bridge to determine the distance from the test point to a fault in a telephone or telegraph line or cable. { 'vär·lē 'lüp ‚test }

var measurement [ELEC] The measurement of reactive power in a circuit. { 'vär ‚mezh·ər·mənt }

varmeter [ENG] An instrument for measuring reactive power in vars. Also known as reactive volt-ampere meter. { 'vär‚mēd·ər }

varnish [MATER] A transparent surface coating which is applied as a liquid and then changes to a hard solid; all varnishes are solutions of resinous materials in a solvent. { 'vär·nish }

varnished cambric [TEXT] Linen or cotton fabric that has been impregnated with varnish or insulating oil and baked;

used for electrical insulating purposes, especially for between-layer insulation in transformers. { 'vär‚nisht 'kam·brik }

varnish makers' and painters' naphtha [MATER] A petroleum naphtha that has a narrow boiling range and is used mainly as a thinner in paint and varnish. Abbreviated VM & P naphtha. { 'vär·nish ¦māk·ərz ən ¦pān·tərz 'naf·thə }

varnish paper *See* insulating paper. { 'vär·nish ‚pā·pər }

varnish tree [BOT] *Rhus vernicifera*. A member of the sumac family (Anacardiaceae) cultivated in Japan; the cut bark exudes a juicy milk which darkens and thickens on exposure and is applied as a thin film to become a varnish of extreme hardness. Also known as lacquer tree. { 'vär·nish ‚trē }

varulite [MINERAL] $(Na,Ca)(Mn,Fe)_2(PO_4)_2$ An olive green, orthorhombic mineral consisting of a phosphate of sodium, calcium, manganese, and iron; occurs in massive form. { 'vär·ə‚līt }

varve [GEOL] A sedimentary bed, layer, or sequence of layers deposited in a body of still water within a year's time, and usually during a season. Also known as glacial varve. { 'värv }

varve clay *See* varved clay. { 'värv ‚klā }

varved clay [GEOL] A lacustrine sediment of distinct layers consisting of varves. Also known as varve clay. { 'värvd ‚klā }

vasa vasorum [ANAT] The blood vessels supplying the walls of arteries and veins. { 'vā·zə va'sȯr·əm }

vascular [ANAT] Pertaining to blood vessels or other channels for the conveyance of a body fluid. { 'vas·kyə·lər }

vascular bundle [BOT] A strandlike part of the plant vascular system containing xylem and phloem. { 'vas·kyə·lər 'bənd·əl }

vascular cambium [BOT] The lateral meristem which produces secondary xylem and phloem. { 'vas·kyə·lər 'kam·bē·əm }

vascular endothelial growth factor [MED] A soluble factor that acts through specific cell-surface receptors on endothelial cells to critically regulate vasculogenesis. { ¦vas·kyə·lər ‚en·dō¦thē·lē·əl 'grȯth ‚fak·tər }

vascularization [MED] Abnormal or excessive formation of blood vessels. [PHYSIO] The formation of new blood vessels within tissue. { ‚vas·kyə·lə·rə'zā·shən }

vascular nevus [MED] A birthmark arising either as a developmental abnormality or as a postnatal benign neoplasm of a blood vessel. { 'vas·kyə·lər 'nē·vəs }

vascular ray [BOT] A ray derived from cambium and found in the stele of some vascular plants, often separating vascular bundles. { 'vas·kyə·lər 'rā }

vascular retinopathy [MED] Pathological changes in the retina associated with diseases such as arterial hypertension, chronic nephritis, eclampsia, and advanced arteriosclerosis. Also known as retinal retinitis. { 'vas·kyə·lər ‚ret·ən'äp·ə·thē }

vascular tissue [BOT] The conducting tissue found in higher plants, consisting principally of xylem and phloem. { 'vas·kyə·lər 'tish·ü }

vasculitis [MED] Inflammation of a blood vessel or a lymph vessel. Also known as angiitis. { ‚vas·kyə'līd·əs }

vasculogenesis [PHYSIO] The formation and differentiation of the vascular system. { ‚vas·kyə·lə'jen·ə·səs }

vas deferens [ANAT] The portion of the excretory duct system of the testis which runs from the epididymal duct to the ejaculatory duct. Also known as ductus deferens. { 'vas 'def·ə·rənz }

vasectomy [MED] Cutting, or removing a section from, the ductus deferens. { va'sek·tə·mē }

vashegyite [MINERAL] $2Al_4(PO_4)_3(OH)_3 \cdot 27H_2O$ A white or pale green to yellow and brownish mineral consisting of a hydrous basic aluminum phosphate; occurs in massive and microcrystalline forms. { 'väsh‚he‚jīt }

vasoconstrictor [PHYSIO] A nerve or an agent that causes blood vessel constriction. { ¦vā·zō·kən'strik·tər }

vasodilator [PHYSIO] A nerve or an agent that causes blood vessel dilation. { ¦vā·zō'dī‚lād·ər }

vasogenic shock [MED] Failure of peripheral circulation due to vasodilation of arterioles and capillaries. { ¦vāz·ə¦jen·ik 'shäk }

vasography [MED] **1.** Radiography of blood vessels. **2.** Radiographic study of the vas deferens. { vā'zäg·rə·fē }

vasoinhibitor [PHARM] An agent that restricts or prevents the functioning of vasomotor nerves. { ‚vā·zō·in'hib·əd·ər }

vasoligation [MED] Surgical ligation of a vas deferens. { ‚vā·zō·lī'gā·shən }

vasomotion [PHYSIO] Change in the diameter of a blood vessel. Also known as angiokinesis. { 'vā·zə‚mō·shən }

vasomotor [PHYSIO] Pertaining to the regulation of the constriction or expansion of blood vessels. { 'vā·zə‚mōd·ər }

vasomotor center [PHYSIO] A large, diffuse area in the reticular formation of the lower brainstem; stimulation of different portions of this center causes either a rise in blood pressure and tachycardia (pressor area) or a fall in blood pressure and bradycardia (depressor area). { 'vāz·ə'mōd·ər ‚sen·tər }

vasoneuropathy [MED] Any disease involving both blood vessels and nerves. { ‚vā·zō·nü'räp·ə·thē }

vasopressin [BIOCHEM] A peptide hormone which is elaborated by the posterior pituitary and which has a pressor effect; used medicinally as an antidiuretic. Also known as antidiuretic hormone (ADH). { ‚vā·zō'pres·ən }

vasospasm [MED] A spasm of the blood vessels. Also known as angiospasm. { 'vā·zō‚spaz·əm }

vasotocin [BIOCHEM] A hormone from the neurosecretory cells of the posterior pituitary of lower vertebrates; increases permeability to water in amphibian skin and in bladder. { ‚vā·zə'tōs·ən }

vat dye [MATER] One of the dyes that are easily reduced to a soluble and colorless form in which they easily impregnate fibers; subsequent oxidation produces the final color; examples are indigo and indanthrene blue. { 'vat ‚dī }

vaterite [MINERAL] $CaCO_3$ A rare hexagonal mineral consisting of unstable calcium carbonate; it is trimorphous with calcite and aragonite. { 'väd·ə‚rīt }

vat printing assistant [MATER] The carrier for the dye in the printing of fabrics with vat dyes; a mixture of gums and reducing and wetting agents to assist in penetrating the fabric. { 'vat ‚print·iŋ ə‚sis·tənt }

vaudaire [METEOROL] A violent south wind; a foehn of Lake Geneva in Switzerland. Also known as vauderon. { vō'der }

vauderon See vaudaire. { vō·də'rōn }

vault [ARCH] An arched masonry structure usually forming a ceiling or a roof. [BIOL] An anatomical structure that is arched or dome-shaped. { vȯlt }

vauquelinite [MINERAL] $Pb_2Cu(CrO_4)PO_4(OH)$ A monoclinic mineral of varying color, consisting of a basic chromate-phosphate of lead and copper. { 'vȯk·lə‚nīt }

vauxite [MINERAL] $FeAl_2(PO_4)_2(OH)_2 \cdot 7H_2O$ A sky blue to Venetian blue, triclinic mineral consisting of a hydrated basic phosphate of iron and aluminum. { 'vȯk‚sīt }

V band [ELECTROMAG] A radio-frequency band of 46.0 to 56.0 gigahertz. [SPECT] Absorption bands that appear in the ultraviolet part of the spectrum due to color centers produced in potassium bromide by exposure of the crystal at temperature of liquid nitrogen (81 K) to intense penetrating x-rays. { 'vē ‚band }

V-beam radar [ELECTROMAG] A volumetric radar system that uses two fan beams to determine the distance, bearing, and height of a target: one beam is vertical and the other inclined; the beams intersect at ground level and rotate continuously about a vertical axis; the time difference between the arrivals of the echoes of the two beams is a measure of target elevation. { 'vē ‚bēm 'rā‚där }

V belt [DES ENG] An endless power-transmission belt with a trapezoidal cross section which runs in a pulley with a V-shaped groove; it transmits higher torque at less width and tension than a flat belt. [MECH ENG] A belt, usually endless, with a trapezoidal cross section which runs in a pulley with a V-shaped groove, with the top surface of the belt approximately flush with the top of the pulley. { 'vē ‚belt }

V-bend die [MECH ENG] A die with a triangular cross-sectional opening to provide two edges over which bending is accomplished. { 'vē ‚bend 'dī }

V block [ENG] A square or rectangular steel block having a 90° V groove through the center, and sometimes provided with clamps to secure round workpieces. { 'vē ‚bläk }

v-body See nucleosome. { 'vē ‚bäd·ē }

V-bucket carrier [MECH ENG] A conveyor consisting of two strands of roller chain separated by V-shaped steel buckets; used for elevating and conveying nonabrasive materials, such as coal. { 'vē ‚bək·ət ‚kar·ē·ər }

V-chip [COMMUN] An electronic device that can be programmed to prevent viewing of television programs. { 'vē ‚chip }

VCO See voltage-controlled oscillator.

V coefficient [QUANT MECH] Either of two coefficients used in the coupling of eigenfunctions of two angular momenta, differing from the Wigner 3-j symbol by at most a sign. Symbolized V and $\bar{V}$. { 'vē ‚kō·i‚fish·ənt }

V connection See open-delta connection. { 'vē kə‚nek·shən }

vcp See vacuum condensing point.

VCR See videocassette recorder.

VCSEL See vertical-cavity surface-emitting laser. { 'vē ‚sē‚es‚ē'el or 'vē'se‚sel }

V cut [ENG] In mining and tunneling, a cut where the material blasted out in plan is like the letter V; usually consists of six or eight holes drilled into the face, half of which form an acute angle with the other half. { 'vē ‚kət }

VD See venereal disease.

VDFG See variable diode function generator.

V(D)J recombination [GEN] A special type of chromatin diminution in which site-specific breakage within immunoglobulin heavy- and light-chain gene clusters and T-cell receptor gene clusters, elimination of part of each cluster, and rejoining of the ends of the remaining deoxyribonucleic acid fragments make possible an enormous number of different antibody and cell-surface antigen receptor specificities. { 'vē‚dē‚jā rē‚käm·bə'nā·shən }

VDT See display terminal.

veatchite [MINERAL] $Sr_2B_{11}O_{16}(OH)_5 \cdot H_2O$ A white mineral consisting of hydrous strontium borate. { 'vē‚chīt }

Vectian See Aptian. { 'vek·chən }

vectograph [GRAPHICS] A picture or drawing having self-contained light polarization; at each point of such a picture, one can control the direction and magnitude of polarization, and the image can be expressed as a nonuniform vector field. { 'vek·tə‚graf }

vectopluviometer [ENG] A rain gage or array of rain gages designed to measure the inclination and direction of falling rain; vectopluviometers may be constructed in the fashion of a wind vane so that the receiver always faces the wind, or they may consist of four or more receivers arranged to point in cardinal directions. { ‚vek·tō‚plü·vē'äm·əd·ər }

vector [COMPUT SCI] See jump vector. [MATH] **1.** An element of a vector space. **2.** A matrix consisting of a single row or a single column of entries. [MED] An agent, such as an insect, capable of mechanically or biologically transferring a pathogen from one organism to another. [NAV] To guide a pilot, navigator, aircraft, or missile from one point to another within a given time by means of a direction communicated to the craft. [PHYS] A quantity which has both magnitude and direction, and whose components transform from one coordinate system to another in the same manner as the components of a displacement. Also known as polar vector. { 'vek·tər }

vector analysis [MATH] The formal study of vectors. { 'vek·tər ə‚nal·ə·səs }

vector bundle [MATH] A locally trivial bundle whose fibers are isomorphic vector spaces. { 'vek·tər ‚bən·dəl }

vectorcardiogram [PHYSIO] The part of the pathway of instantaneous vectors during one cardiac cycle. Also known as monocardiogram. { ‚vek·tər'kärd·ē·ə‚gram }

vectorcardiography [PHYSIO] A method of recording the magnitude and direction of the instantaneous cardiac vectors. { ‚vek·tər‚kärd·ē'äg·rə·fē }

vector coupling coefficient [QUANT MECH] One of the coefficients used to express an eigenfunction of the sum of two angular momenta in terms of sums of products of eigenfunctions of the original two angular momenta. Also known as Clebsch-Gordan coefficient; Wigner coefficient. { 'vek·tər ‚kəp·liŋ ‚kō·i‚fish·ənt }

vector current [PART PHYS] A current which behaves as a vector under Lorentz transformations, rather than as an axial vector. { 'vek·tər ‚kə·rənt }

vectored attacks [ORD] Attacks in which a weapon carrier (air, surface, or subsurface) not holding contact on the target, is vectored to the weapon delivery point by a unit (air, surface, or subsurface) which holds contact on the target. { 'vek·tərd ə'taks }

vectored interrupt [COMPUT SCI] A signal that instructs a computer program to temporarily halt the processing it is doing

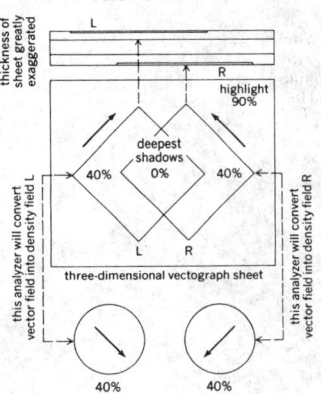

VECTOGRAPH

thickness of sheet greatly exaggerated

this analyzer will convert vector field into density field L

this analyzer will convert vector field into density field R

L

R

highlight 90%

deepest shadows
40% 0% 40%

L R

three-dimensional vectograph sheet

40% analyzer

40% analyzer

Diagram of superposed stereoscopic vectographs. The solid arrows show the axes of polarization; L indicates the left-eye image; R, the right-eye image. The percentages refer to light transmission.

and transfer control to a routine whose address is given by an entry in a jump vector specified by a value included in the signal. { 'vek·tərd 'in·tə·rəpt }

vector effect *See* selective photoelectric effect. { 'vek·tər i‚fekt }

vector equation [MATH] An equation involving vectors. { 'vek·tər i‚kwā·zhən }

vector field [MATH] **1.** The field of vectors arising from considering a system of differential equations on a differentiable manifold. **2.** A function whose range is in a vector space. [PHYS] A field which is characterized by a vector function. { 'vek·tər ‚fēld }

vector function [PHYS] A function of position and time whose value at each point is a vector. Also known as vector point function. { 'vek·tər ‚fəŋk·shən }

vector graphics [COMPUT SCI] A computer graphics image-coding technique which codes only the image itself as a series of lines, according to the cartesian coordinates of the lines' origins and terminations. Also known as object-oriented graphics. { 'vek·tər ‚graf·iks }

vector gunsight [ORD] A gunsight that contains a device that computes the vector required for the bullet to follow if it is to strike its target; used especially in firing at moving targets. { 'vek·tər 'gən‚sīt }

vectorial structure *See* directional structure. { vek‚tór·ē·əl 'strək·chər }

vector impedance meter [ENG] An instrument that not only determines the ratio between voltage and current, to give the magnitude of impedance, but also determines the phase difference between these quantities, to give the phase angle of impedance. { 'vek·tər im'pēd·əns ‚mēd·ər }

vector meson [PART PHYS] A meson which has spin quantum number 1 and negative parity, and may be described by a vector field; examples include the ω, ρ, φ, and K* mesons. { 'vek·tər ‚mā‚sän }

vector model of atomic structure [ATOM PHYS] A model of atomic structure in which spin and orbital angular momenta of the electrons are represented by vectors, with special rules for their addition imposed by underlying quantum-mechanical considerations. { 'vek·tər ‚mäd·əl əv ə'täm·ik 'strək·chər }

vector momentum *See* momentum. { 'vek·tər mə'men·təm }

vector multiplication [MATH] The operation which associates to each ordered pair of vectors the cross product of these two vectors. { ‚vek·tər ‚məl·tə·plə'kā·shən }

vector point function *See* vector function. { 'vek·tər 'póint ‚fəŋk·shən }

vector potential [ELECTROMAG] A vector function whose curl is equal to the magnetic induction. Symbolized **A**. Also known as magnetic vector potential. [PHYS] Any vector function whose curl is equal to some solenoidal vector field. { 'vek·tər pə‚ten·chəl }

vector power [ELEC] Vector quantity equal in magnitude to the square root of the sum of the squares of the active power and the reactive power. { 'vek·tər ‚pau̇·ər }

vector-power factor [ELEC] Ratio of the active power to the vector power; it is the same as power factor in the case of simple sinusoidal quantities. { 'vek·tər ‚pau̇·ər ‚fak·tər }

vector processing [COMPUT SCI] A procedure for speeding the processing of information by a computer, in which pipelined units perform arithmetic operations on uniform, linear arrays of data values, and a single instruction involves the execution of the same operation on every element of the array. { 'vek·tər 'prä‚ses·iŋ }

vector product *See* cross product. { 'vek·tər ‚präd·əkt }

vector quantization [COMPUT SCI] A data compression technique in which a finite sequence of values is presented as resembling the template (from among the choices available to a given codebook) that minimizes a distortion measure. { ‚vek·tər ‚kwän·tə'zā·shən }

vector random variable [MATH] A vector whose entries are random variables that are defined on the same sample space of an experiment. { ‚vek·tər ‚ran·dəm 'ver·ē·ə·bəl }

vector resolver *See* resolver. { 'vek·tər ri‚zäl·vər }

vector space [MATH] A system of mathematical objects which have an additive operation producing a group structure and which can be multiplied by elements from a field in a manner similar to contraction or magnification of directed line segments in euclidean space. Also known as linear space. { 'vek·tər ‚spās }

vector steering [AERO ENG] A steering method for rockets and spacecraft wherein one or more thrust chambers are gimbal-mounted so that the direction of the thrust force (thrust vector) may be tilted in relation to the center of gravity of the vehicle to produce a turning movement. { 'vek·tər ‚stir·iŋ }

vector sum [MATH] For a set of located vectors in euclidean space, v_1, v_2, . . . , v_n, this is the vector whose initial point is the initial point of v_1 and whose terminal point is the terminal point of v_n, when the vectors are laid end to end so that the terminal point of one vector v_i is the initial point of the next vector v_{i+1}. Also known as resultant. { 'vek·tər ‚səm }

vector table *See* jump vector. { 'vek·tər ‚tā·bəl }

vector voltmeter [ENG] A two-channel high-frequency sampling voltmeter that measures phase as well as voltage of two input signals of the same frequency. { 'vek·tər 'vōlt ‚mēd·ər }

vee antenna *See* V antenna. { 'vē an‚ten·ə }

vee path [ENG] In ultrasonic testing, the path of an angle beam from an ultrasonic search unit in which the waves are reflected off the opposite surface of the test piece and returned to the examination surface in a manner which has the appearance of the letter V. { 'vē ‚path }

veer [NAV ARCH] To pay or let out, as to veer anchor chain. { vir }

veering [METEOROL] **1.** In international usage, a change in wind direction in a clockwise sense (for example, south to southwest to west) in either hemisphere of the earth. **2.** According to widespread usage among United States meteorologists, a change in wind direction in a clockwise sense in the Northern Hemisphere, counterclockwise in the Southern Hemisphere. { 'vir·iŋ }

Vega [ASTRON] One of the brightest stars, apparent magnitude 0.1; it is a main sequence star of spectral type A0, distance is 8 parsecs, and it is 40 times brighter than the sun. Also known as α Lyrae. { 'vā·gə }

Vega-excess star [ASTRON] A star from which is detected far-infrared emission greatly in excess of what the star alone should produce, believed to originate from a cloud of dust surrounding the star that may be analogous to the Kuiper belt dust in the solar system but has hundreds or thousands of times as much dust. { 'vā·gə 'ek‚ses ‚stär }

Vegard's Law [MET] Linear relation between lattice parameters and composition of solid solution alloys expressed as atomic percentage. { ve'gärz ‚lȯ }

vegetable [AGR] The edible portion of a usually herbaceous plant; customarily served with the main course of a meal. [BOT] Resembling or relating to plants. { 'vej·tə·bəl }

vegetable black [MATER] Carbon made by the incomplete combustion or destructive distillation of vegetable matter, for example, wood. { 'vej·tə·bəl ‚blak }

vegetable diastase *See* diastase. { 'vej·tə·bəl 'dī·əs‚tās }

vegetable dye [MATER] Any colorant that is obtained from a vegetable source; for example, indigo or madder. { 'vej·tə·bəl ‚dī }

vegetable fat [MATER] A semisolid vegetable oil, used chiefly for food; for example, Suari fat, ucuhuba tallow, Mahuba fat, gamboge butter (gurgi, murga), Sierra Leone butter (kamga, lamy), and Mafura tallow. { 'vej·tə·bəl ‚fat }

vegetable glue [MATER] Mostly starch- or dextrine-based glues mixed with gums, resins, or antioxidants; tapioca paste is the most common; used on cheaper plywoods, postage stamps, envelopes, and labels. { 'vej·tə·bəl ‚glü }

vegetable ivory [MATER] A material from the ivory nut, a seed of the palm *Phytelephas macrocarpa*, which grows in tropical America; the nut has a white color and fine texture and is used to make buttons and similar small articles. { 'vej·tə·bəl ‚īv·rē }

vegetable jelly *See* ulmin. { 'vej·tə·bəl ‚jel·ē }

vegetable oil [MATER] An edible, mixed glyceride oil derived from plants (fruit, leaves, and seeds), including cottonseed, linseed, tung, and peanut; used in food oils, shortenings, soaps, and medicine, and as a paint drying oil. { 'vej·tə·bəl ‚ȯil }

vegetable parchment [MATER] A paperlike material made from a base of cotton rags or alpha cellulose called waterleaf, and containing no sizing or filling materials; used for documents and food packaging. { 'vej·tə·bəl 'pärch·mənt }

vegetable tanning [ENG] Leather tanning using plant extracts, such as tannic acid. { 'vej·tə·bəl 'tan·iŋ }

vegetable wax [MATER] A waxy substance of vegetable origin, composed of fatty acids in combination with higher alcohols (instead of glycerin, as in fats and oils); includes Japan wax, jojoba oil, candelilla, and carnauba wax. { 'vej·tə·bəl ,waks }

vegetation [BOT] The total mass of plant life that occupies a given area. { ,vej·ə'tā·shən }

vegetational plant geography [ECOL] A field of study concerned with the mapping of vegetation regions and the interpretation of these in terms of environmental or ecological influences. { ,vej·ə'tā·shən·əl 'plant jē,äg·rə·fē }

vegetation and ecosystem mapping [BOT] An art and a science concerned with the drawing of maps which locate different kinds of plant cover in a geographic area. { ,vej·ə'tā·shən ən 'ek·ō,sis·təm 'map·iŋ }

vegetation management [ECOL] The art and practice of manipulating vegetation such as timber, forage, crops, or wild life, so as to produce a desired part or aspect of that material in higher quantity or quality. { ,vej·ə'tā·shən ,man·ij·mənt }

vegetation zone [ECOL] **1.** An extensive, even transcontinental, band of physiognomically similar vegetation on the earth's surface. **2.** Plant communities assembled into regional patterns by the area's physiography, geological parent material, and history. { ,vej·ə'tā·shən ,zōn }

vegetative [BIOL] Having nutritive or growth functions, as opposed to reproductive. { 'vej·ə,tād·iv }

vegetative propagation [BOT] Production of a new plant from a portion of another plant, such as a stem or branch. { 'vej·ə,tād·iv ,präp·ə'gā·shən }

vegetative state [VIROL] The noninfective state during which the genome of a phage multiplies and directs host synthesis of substances needed for production of infective particles. { 'vej·ə,tād·iv ,stāt }

vehicle [AERO ENG] **1.** A structure, machine, or device, such as an aircraft or rocket, designed to carry a burden through air or space. **2.** More restrictively, a rocket vehicle. [MATER] The fluid component of a paint or printing ink; acts as a carrier for the pigment. [MECH ENG] A self-propelled wheeled machine that transports people or goods on or off roads; automobiles and trucks are examples. { 've·ə·kəl }

vehicle control system [AERO ENG] A system, incorporating control surfaces or other devices, which adjusts and maintains the altitude and heading, and sometimes speed, of a vehicle in accordance with signals received from a guidance system. Also known as flight control system. { 've·ə·kəl kən'trōl ,sis·təm }

vehicle mass ratio [AERO ENG] The ratio of the final mass of a vehicle after all propellant has been used, to the initial mass. { 've·ə·kəl 'mas 'rā·shō }

vehicular telephony [COMMUN] The transmission of speech signals to and from mobile radio stations installed in automotive vehicles; typically, each station is equipped with a transmitter and a receiver. { vē'hik·yə·lər tə'lef·ə·nē }

veil [BIOL] See velum. [METEOROL] A very thin cloud through which objects are visible. { vāl }

veiling glare [OPTICS] The reduction in contrast of an optical image caused by superposition of scattered light. { 'vāl·iŋ ,glär }

Veillonella [MICROBIO] The type genus of the family Veillonellaceae; small cells occurring in pairs, chains, and clusters. { vā·yō'nel·ə }

Veillonellaceae [MICROBIO] The single family of gram-negative, anaerobic cocci; characteristically occur in pairs with adjacent sides flattened; parasites of homotherms, including humans, rodents, and pigs. { ,vā·yō·nə'läs·ē,ē }

Veil Nebula See Cygnus loop. { 'vāl 'neb·yə·lə }

vein [ANAT] A relatively thin-walled blood vessel that carries blood from capillaries to the heart in vertebrates. [BOT] One of the vascular bundles in a leaf. [GEOL] A mineral deposit in tabular or shell-like form filling a fracture in a host rock. [INV ZOO] **1.** One of the thick, stiff ribs providing support for the wing of an insect. **2.** A venous sinus in invertebrates. { vān }

veined gneiss [PETR] A composite gneiss with irregular layering. { 'vānd 'nīs }

veining [MET] Lines in a polished and etched metal surface marking slight imperfections in structure of an otherwise single grain. { 'vān·iŋ }

veinite [GEOL] A genetic type of veined gneiss in which the vein material was secreted from the rock itself. { 'vā,nīt }

vein quartz [PETR] A rock composed chiefly of sutured quartz crystals of pegmatitic or hydrothermal origin of variable size. { 'vān ,kwôrts }

Vela [ASTRON] A southern constellation, right ascension 9 hours, declination 50°S. Also known as Sail. { 'vē·lə }

velamen [BOT] The corky epidermis covering the aerial roots of an epiphytic orchid. { və'lā·mən }

Vela pulsar [ASTRON] A pulsar with a period of 80 milliseconds, about 1500 light-years (1.4×10^{19} meters) away in the constellation Vela, whose variation has been detected at radio, gamma-ray, and optical wavelengths; probably associated with the Vela supernova remnant. { 'vē·lə 'pəl,sär }

velarium [INV ZOO] The velum of certain scyphozoans and cubomedusans distinguished by the presence of canals lined with endoderm. { və'lar·ē·əm }

Vela supernova remnant [ASTRON] A gaseous nebula that is the result of a supernova whose light reached earth about 10,000 years ago. { 'vē·lə ¦sü·pər'nō·və ,rem·nənt }

Vela X [ASTRON] A compact, nonthermal radio source associated with the Vela pulsar but displaced from it by about 0.7°. { 'vē·lə 'eks }

Vela X-1 [ASTRON] A pulsing, eclipsing x-ray source in the constellation Vela that is a particularly intense emitter of hard x-rays. { 'vē·lə ,eks'wən }

Vela X-2 [ASTRON] The pulsed x-ray emission associated with the Vela pulsar. { 'vē·lə ,eks'tü }

veld See veldt. { velt }

veldt [ECOL] Grasslands of eastern and southern Africa that are usually level and mixed with trees and shrubs. Also spelled veld. { velt }

veliger [INV ZOO] A mollusk larval stage following the trochophore, distinguished by an enlarged girdle of ciliated cells (velum). { 'vē·lə·jər }

Veliidae [INV ZOO] A family of the Hemiptera in the subdivision Amphibicorisae composed of small water striders which have short legs and a longitudinal groove between the eyes. { və'lī·ə,dē }

vellum [MATER] A high-grade paper made to resemble genuine parchment. { 'vel·əm }

vellus [ANAT] Fine body hair that is present until puberty. { 'vel·əs }

velocimeter [ENG] An instrument for measuring the speed of sound in water; two transducers transmit acoustic pulses back and forth over a path of fixed length, each transducer immediately initiating a pulse upon receiving the previous one; the number of pulses occurring in a unit time is measured. { ,vel·ə'sim·əd·ər }

Velocipedidae [INV ZOO] A tropical family of hemipteran insects in the superfamily Cimicoidea. { və,läs·ə'ped·ə,dē }

Velociraptor [PALEON] A carnivorous theropod dinosaur, 7 feet (2 meters) long, with birdlike features from the Late Cretaceous that had strong grasping hands with claws, powerful hindlimbs, and jaws containing sharp teeth. { və'läs·ə,rap·tər }

velocity [MECH] **1.** The time rate of change of position of a body; it is a vector quantity having direction as well as magnitude. Also known as linear velocity. **2.** The speed at which the detonating wave passes through a column of explosives, expressed in meters or feet per second. { və'läs·əd·ē }

velocity analysis [MECH] A graphical technique for the determination of the velocities of the parts of a mechanical device, especially those of a plane mechanism with rigid component links. { və'läs·əd·ē ə,nal·ə·səs }

velocity coefficient [FL MECH] The ratio of the actual velocity of gas emerging from a nozzle to the velocity calculated under ideal conditions; it is less than 1 because of friction losses. Also known as coefficient of velocity. { və'läs·əd·ē ,kō·i,fish·ənt }

velocity constant [CONT SYS] The ratio of the rate of change of the input command signal to the steady-state error, in a control system where these two quantities are proportional. { və'läs·əd·ē ,kän·stənt }

velocity control See rate control. { və'läs·əd·ē kən,trōl }

velocity curve [ASTRON] A graphical representation of the line-of-sight velocity (versus time) of a star or components of a spectroscopic binary system. { və'läs·əd·ē ,kərv }

velocity discontinuity See seismic discontinuity. { və'läs·əd·ē dis,känt·ən'ü·əd·ē }

velocity dispersion [PHYS] The root-mean-square value of

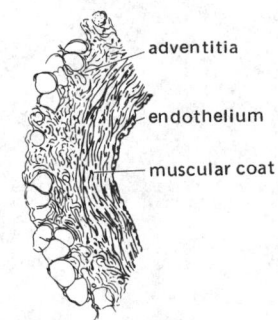

VEIN

Portion of cross section through common digital vein of a human. *(From A. A. Maximow and W. Bloom, A Textbook of Histology, 6th ed., Saunders, 1953)*

VELOCIMETER

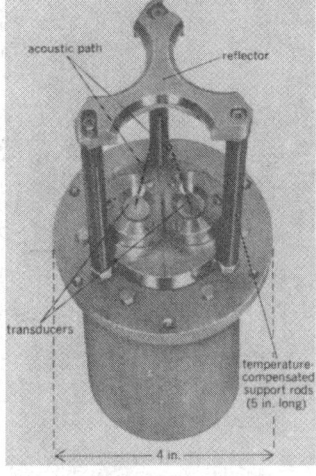

A typical commercial velocimeter with protecting cage removed (1 inch = 2.54 centimeters). *(NUS crop.)*

the magnitudes of the random velocities of particles about their mean velocity.　{ və'läs·əd·ē di'spər·zhən }

velocity-distance relation [ASTRON] The relation wherein all the exterior galaxies are moving away from the galaxy that the sun is part of, with velocities that are greater with increasing distance of the galaxy.　{ və'läs·əd·ē 'dis·təns ri,lā·shən }

velocity distribution [STAT MECH] For the molecules of a gas, a function of velocity whose value at any velocity v is proportional to the number of molecules with velocities in an infinitesimal range about v, per unit velocity range.　{ və'läs·əd·ē ,di·strə,byü·shən }

velocity error [CONT SYS] The difference between the rate of change of the actual position of a control system component and the rate of change of the desired position.　{ və'läs·əd·ē ,er·ər }

velocity filter [ELECTR] Storage tube device which blanks all targets that do not move more than one resolution cell in less than a predetermined number of antenna scans.　{ və'läs·əd·ē ,fil·tər }

velocity fire [ORD] Preparatory fire conducted to determine the velocity for a particular combination of weapon, propellant, projectile, and fuse.　{ və'läs·əd·ē ,fīr }

velocity-focusing mass spectrograph See velocity spectrograph.　{ və'läs·əd·ē ¦fō·kəs·iŋ 'mas 'spek·trə,graf }

velocity gradient [FL MECH] The rate of change of velocity of propagation with distance normal to the direction of flow. [GEOPHYS] See seismic gradient.　{ və'läs·əd·ē ,grād·ē·ənt }

velocity head [FL MECH] The square of the speed of flow of a fluid divided by twice the acceleration of gravity; it is equal to the static pressure head corresponding to a pressure equal to the kinetic energy of the fluid per unit volume.　{ və'läs·əd·ē ,hed }

velocity-head tachometer [ENG] A type of tachometer in which the device whose speed is to be measured drives a pump or blower, producing a fluid flow, which is converted to a pressure.　{ və'läs·əd·ē ¦hed tə'käm·əd·ər }

velocity hydrophone [ENG ACOUS] A hydrophone in which the electric output essentially matches the instantaneous particle velocity in the impressed sound wave.　{ və'läs·əd·ē 'hī·drə,fōn }

velocity level [ACOUS] A sound rating in decibels, equal to 20 times the logarithm to the base 10 of the ratio of the particle velocity of the sound to a specified reference particle velocity.　{ və'läs·əd·ē ,lev·əl }

velocity microphone [ENG ACOUS] A microphone whose electric output depends on the velocity of the air particles that form a sound wave; examples are a hot-wire microphone and a ribbon microphone.　{ və'läs·əd·ē 'mī·krə,fōn }

velocity-modulated oscillator [ELECTR] Oscillator which employs velocity modulation to produce radio-frequency power. Also known as klystron oscillator.　{ və'läs·əd·ē ,mäj·ə¦lād·əd 'äs·ə,lād·ər }

velocity modulation [ELECTR] **1.** Modulation in which a time variation in velocity is impressed on the electrons of a stream. **2.** A television system in which the intensity of the electron beam remains constant throughout a scan, and the velocity of the spot at the screen is varied to produce changes in picture brightness (not in general use).　{ və'läs·əd·ē ,mäj·ə'lā·shən }

velocity of light See speed of light.　{ və'läs·əd·ē əv 'līt }

velocity-of-light cylinder [ASTRON] A cylinder whose axis is the axis of rotation of a neutron star and whose radius is such that the velocity of a plasma rotating with the neutron star would equal the velocity of light at the surface of the cylinder. Also known as light cylinder.　{ və¦läs·əd·ē əv ¦līt 'sil·ən·dər }

velocity-of-light radius [ASTRON] The radius of the velocity-of-light cylinder. Also known as light radius.　{ və¦läs·əd·ē əv ¦līt 'rād·ē·əs }

velocity of sound See speed of sound.　{ və'läs·əd·ē əv 'saund }

velocity pickup [ELEC] A device that generates a voltage proportional to the relative velocity between two principal elements of the pickup, the two elements usually being a coil of wire and a source of magnetic field.　{ və'läs·əd·ē 'pik,əp }

velocity potential [FL MECH] For a fluid flow, a scalar function whose gradient is equal to the velocity of the fluid.　{ və'läs·əd·ē pə,ten·chəl }

velocity pressure See wind pressure.　{ və'läs·əd·ē ,presh·ər }

velocity profile [FL MECH] A graph of the speed of a fluid flow as a function of distance perpendicular to the direction of flow.　{ və'läs·əd·ē ,prō,fīl }

velocity ratio [MECH ENG] The ratio of the velocity given to the effort or input of a machine to the velocity acquired by the load or output. [OCEANOGR] The ratio of the speed of tidal current at a subordinate station to the speed of the corresponding current at the reference station.　{ və'läs·əd·ē ,rā·shō }

velocity resonance See phase resonance.　{ və'läs·əd·ē ,rez·ən·əns }

velocity servomechanism [CONT SYS] A servomechanism in which the feedback-measuring device generates a signal representing a measured value of the velocity of the output shaft. Also known as rate servomechanism.　{ və'läs·əd·ē 'sər·vō,mek·ə,niz·əm }

velocity shaped canceler See cascaded feedback canceler.　{ və'läs·əd·ē ¦shāpt 'kan·slər }

velocity spectrograph [PHYS] A mass spectrograph in which only positive ions having a certain velocity pass through all three slits and enter a chamber where they are deflected by a magnetic field in proportion to their charge-to-mass ratio. Also known as velocity-focusing mass spectrograph.　{ və'läs·əd·ē 'spek·trə,graf }

velocity-type flowmeter [ENG] A turbine-type fluid-flow measurement device in which the fluid flow actuates the movement of a wheel or turbine-type impeller, giving a volume-time reading. Also known as current meter; rotating meter.　{ və'läs·əd·ē ¦tīp 'flō,mēd·ər }

velodrome [ARCH] A stadium or arena having a banked track, used for bicycle or motorcycle racing.　{ 'vel·ə,drōm }

velour [TEXT] A fabric with a short pile on the surface.　{ və'lur }

velour paper [MATER] A paper with a velvetlike finish, produced by flocking the surface with fine bits of rayon, nylon, cotton, or wool; it is sometimes embossed in various patterns.　{ və'lur ,pā·pər }

velum [BIOL] A veil- or curtainlike membrane. [INV ZOO] A swimming organ on the larva of certain marine gastropod mollusks that develops as a contractile ciliated collar-shaped ridge. Also known as veil. [METEOROL] An accessory cloud veil of great horizontal extent draped over or penetrated by cumuliform clouds; velum occurs with cumulus and cumulonimbus.　{ 'vē·ləm }

velvet [TEXT] A fabric with a short, thick-set pile of silk, cotton, or other fiber on a back that is closely woven and of the same or different fibers.　{ 'vel·vət }

velveteen [TEXT] Cotton or rayon pile fabric with short, close-filling loops cut by sharp knives to create an erect, velvety pile; unlike velvet, which is woven face to face, velveteen is woven singly.　{ ,vel·və'tēn }

vena cava [ANAT] One of two large veins which in air-breathing vertebrates conveys blood from the systemic circulation to the right atrium.　{ ,vē·nə 'kā·və }

vena contracta [FL MECH] The contraction of a jet of liquid which comes out of an opening in a container to a cross section smaller than the opening.　{ ,vē·nə kən'trak·tə }

venation [BOT] The system or pattern of veins in the tissues of a leaf. [INV ZOO] The arrangement of veins in an insect wing.　{ ve'nā·shən }

vendaval [METEOROL] A stormy southwest wind on the southern Mediterranean coast of Spain and in the Straits of Gibraltar; it occurs with a low advancing from the west in late autumn, winter, or early spring, and is often accompanied by thunderstorms and violent squalls.　{ ,ven·də'väl }

veneer [MATER] **1.** A thin sheet of wood of uniform thickness used for facing furniture or, when bonded, used to make plywood. **2.** A facing, as of brick or marble, on the outside of a wall.　{ və'nir }

veneered construction [BUILD] A type of construction in which the framework is faced with a thin external layer of material, such as marble.　{ və¦nird kən'strək·shən }

venereal bubo See lymphogranuloma venereum.　{ və'nir·ē·əl 'bü·bō }

venereal disease [MED] Any of several contagious diseases generally acquired during sexual intercourse; includes gonorrhea, syphilis, chancroid, granuloma inguinale, and lymphogranuloma venereum. Abbreviated VD. { və'nir·ē·əl di'zēz }

venereal wart [MED] A warty growth of the penis, frequent in some parts of the world, and probably acquired during sexual intercourse. { və'nir·ē·əl 'wȯrt }

venetian cloth [TEXT] A wool or cotton fabric with a smooth texture and a warp face. { və'nēsh·ən 'klȯth }

Venetian red [INORG CHEM] A pigment with a true red hue; contains 15–40% ferric oxide and 60–80% calcium sulfate. { və'nēsh·ən 'red }

venetian rose point *See* rose point lace. { və'nēsh·ən 'rōz ˌpȯint }

venipuncture [MED] A surgical puncture of a vein, such as for withdrawing blood or injecting medication. { 'ven·ə,pəŋk·chər }

venite [PETR] Migmatite having mobile portions which were formed by exudation from the rock itself. { 'vē,nīt }

Venn diagram [MATH] A pictorial representation of set theoretic operations such as union, intersection, and complementation of sets. { 'ven ˌdī·ə,gram }

venom [PHYSIO] Any of various poisonous materials secreted by certain animals, such as snakes or bees. { 'ven·əm }

venous pressure [PHYSIO] Tension of the blood within the veins. { 'vē·nəs ˌpresh·ər }

vent [ENG] **1.** A small passage made with a needle through stemming, for admitting a squib to enable the charge to be lighted. **2.** A hole, extending up through the bearing at the top of the core-barrel inner tube, which allows the water and air in the upper part of the inner tube to escape into the borehole. **3.** A small hole in the upper end of a core-barrel inner tube that allows water and air in the inner tube to escape into the annular space between the inner and outer barrels. **4.** An opening provided for the discharge of pressure or the release of pressure from tanks, vessels, reactors, processing equipment, and so on. **5.** A pipe for providing airflow to or from a drainage system or for circulating air within the system to protect trap seals from siphonation and back pressure. [GEOL] The opening of a volcano on the surface of the earth. [MET] A small opening in a casting mold to allow for the escape of gases. [ZOO] The external opening of the cloaca or rectum, especially in fish, birds, and amphibians. { vent }

vent da Mùt [METEOROL] A strong, wet wind of Lake Garda in Italy. { ˌvent dä 'müt }

vent des dames [METEOROL] A daily sea breeze of about 15 miles (24 kilometers) per hour from the southwest in summer on the Mediterranean coast east of the Rhone delta, extending some 20 miles (32 kilometers) inland. { vȯn de 'däm }

vent du midi [METEOROL] A south wind in the center of the Massif Central and the southern Cevennes (France); it is warm, moist, and generally followed by a southwest wind with heavy rain. { vȯn dyü mē'dē }

vented baffle *See* reflex baffle. { 'ven·təd 'baf·əl }

vented battery [ELEC] A nickel-cadmium or other battery which lacks provisions for recombination of gases produced during normal operation, so that these gases must be vented to the atmosphere to avoid rupture of the cell case. { 'ven·təd 'bad·ə·rē }

vented-box system [ELECTR] A loudspeaker system in which the woofer is mounted in a box with a vent connecting the air inside the box to the outside. Also known as ported system. { 'ven·təd 'bäks ˌsis·təm }

venter [ANAT] The abdomen, or other body cavity containing organs. [BOT] The thickened basal portion of an archegonium. [INV ZOO] **1.** The undersurface of an arthropod's abdomen. **2.** The outer, convex part of a curved or coiled gastropod or cephalopod shell. { 'ven·tər }

ventifact [GEOL] A stone or pebble whose shape, wear, faceting, cut, or polish is the result of sandblasting. Also known as glyptolith; rillstone; wind-cut stone; wind-grooved stone; wind-polished stone; wind-scoured stone; wind-shaped stone. { 'ven·tə,fakt }

ventilation [ENG] Provision for the movement, circulation, and quality control of air in an enclosed space. [METEOROL] The process of causing representative air to be in contact with the sensing elements of observing instruments; especially

applied to producing a flow of air past the bulb of a wet-bulb thermometer. { ˌvent·əl'ā·shən }

ventilator [ENG] A device with an adjustable aperture for regulating the flow of fresh or stagnant air. [MECH ENG] A mechanical apparatus for producing a current of air, as a blowing or exhaust fan. { 'vent·əl,ād·ər }

vento di sotto [METEOROL] Breezes blowing up-lake on Lake Garda in Italy. { ˌven·tō dī 'sȯ·tō }

ventral [BOT] On the lower surface of a dorsiventral plant structure, such as a leaf. [ZOO] On or belonging to the lower or anterior surface of an animal, that is, on the side opposite the back. { 'ven·trəl }

ventral aorta [VERT ZOO] The arterial trunk or trunks between the heart and the first aortic arch in embryos or lower vertebrates. { 'ven·trəl ā'ȯrd·ə }

ventral hernia [MED] A hernia of the abdominal wall not involving the umbilical, femoral, or inguinal openings. Also known as abdominal hernia. { 'ven·trəl 'hər·nē·ə }

ventralia [INV ZOO] Paired sensory bristles on the ventral aspect of the head of gnathostomulids. { ven'tral·yə }

ventral light reflex [INV ZOO] A basic means of orientation in aquatic invertebrates, such as shrimp, which swim belly up toward the light. { 'ven·trəl 'līt ˌrē,fleks }

ventral rib [VERT ZOO] Any of the ribs which lie in the septa dividing the trunk musculature into segments in fish. Also known as pleural rib. { 'ven·trəl 'rib }

ventricle [ANAT] **1.** A chamber, or one of two chambers, in the vertebrate heart which receives blood from the atrium and forces it into the arteries by contraction of the muscular wall. **2.** One of the interconnecting, fluid-filled chambers of the vertebrate brain that are continuous with the canal of the spinal cord. [ZOO] A cavity in a body part or organ. { 'ven·trə·kəl }

ventricose [BIOL] Swollen or distended, especially on one side. { ˌven·trə·kōs }

ventricular depolarization complex *See* QRS complex. { ven'trik·yə·lər di,pō·lə·rə'zā·shən ˌkäm,pleks }

ventricular septum *See* interventricular septum. { ven'trik·yə·lər 'sep·təm }

ventriculus [ZOO] A ventricle that performs digestive functions, such as a stomach or a gizzard. { ven'trik·yə·ləs }

ventromedial nucleus [ANAT] A central nervous system nucleus in the hypothalamus that appears to be the satiation center; bilateral surgical damage to this nucleus results in overeating. { ˌven·trō'mēd·ē·əl 'nü·klē·əs }

vent stack [BUILD] The portion of a soil stack above the highest fixture. { 'vent ˌstak }

venture life [IND ENG] The period of time during which expenditures and reimbursements involving a given venture occur. Also known as financial life. { 'ven·chər ˌlīf }

Venturia inaequalis [MYCOL] A fungal pathogen that causes apple scab disease. { ven,tur·ē·ə ,in·ē'kwäl·əs }

Venturian [GEOL] A North American stage of middle Pliocene geologic time, above Repettian and below Wheelerian. { ven'chür·ē·ən }

venturi flume [ENG] An open flume with a constricted flow which causes a drop in the hydraulic grade line; used in flow measurement. { ven'tür·ē ˌflüm }

venturi meter [ENG] An instrument for efficiently measuring fluid flow rate in a piping system; a nozzle section increases velocity and is followed by an expanding section for recovery of kinetic energy. { ven'tür·ē ˌmēd·ər }

venturi scrubber [CHEM ENG] A gas-cleaning device in which liquid injected at the throat of a venturi is used to scrub dust and mist from the gas flowing through the venturi. { ven'tür·ē ˌskrəb·ər }

venturi tube [ENG] A constriction that is placed in a pipe and causes a drop in pressure as fluid flows through it, consisting essentially of a short straight pipe section or throat between two tapered sections; it can be used to measure fluid flow rate (a venturi meter), or to draw fuel into the main flow stream, as in a carburetor. { ven'tür·ē ˌtüb }

venule [ANAT] A small vein. { 'ven·yül }

Venus [ASTRON] The planet second in distance from the sun; the linear equatorial diameter of the solid globe is 7521 miles (12,104 kilometers); the mass is about 0.815 (earth = 1). { 'vē·nəs }

Venus' flytrap [BOT] *Dionaea muscipula.* An insectivorous plant (order Sarraceniales) of North and South Carolina; the

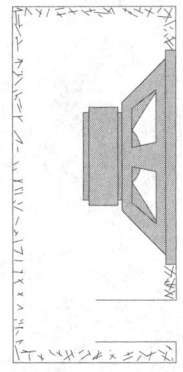
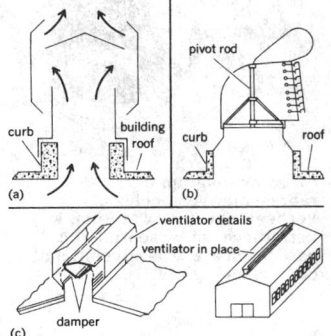

two halves of a leaf blade can swing upward and inward as though hinged, thus trapping insects between the closing halves of the leaf blade. { 'vē·nəs 'flī,trap }

Venus hairstone *See* rutilated quartz. { 'vē·nəs 'her,stōn }

Venus probe [AERO ENG] A probe for exploring and reporting on conditions on or about the planet Venus, such as Pioneer and Mariner probes of the United States, and Venera probes of the Soviet Union. { 'vē·nəs 'prōb }

veranillo [CLIMATOL] The lesser dry season, made up of a few weeks of hot dry weather, that breaks up the summer rainy season on the Pacific coast of Mexico and Central America. { ver·ə'nēl·yō }

verano [CLIMATOL] In Mexico and Central America, the main dry season, generally occurring from November through April. { ve'rä·nō }

veratria *See* veratrine. { və'ra·trē·ə }

veratrine [MATER] An alakaloid mixture from the seeds of sabadilla (*Schoenocaulon officinale*); which is toxic, colorless, soluble in alcohol and ether, very slightly soluble in water, and melts at about 150°C; used in medicine. Also known as veratria. { 'ver·ə,trēn }

verb [COMPUT SCI] In COBOL, the action indicating part of an unconditional statement. { vərb }

verbal information verification [ENG ACOUS] A method of talker authentication that involves checking the content of a spoken password or pass-phrase, such as a personal identification number, a social security number, or a mother's maiden name. Abbreviated VIV. { ˌvər·bəl ˌin·fərˌmā·shən ˌver·i·fəˈkā·shən }

verbal learning [PSYCH] A field of experimental psychology which studies the formation of certain verbal associations; deals with acquisition of the associations. { 'vər·bəl 'lərn·iŋ }

Verbeekinidae [PALEON] A family of extinct marine protozoans in the superfamily Fusulinacea. { ˌver,bā'kin·ə,dē }

Verbenaceae [BOT] A family of variously woody or herbaceous dicotyledons in the order Lamiales characterized by opposite or whorled leaves and regular or irregular flowers, usually with four or two functional stamens. { ˌvər·bə'nās·ē,ē }

verbena oil [MATER] A volatile oil from lemon verbena leaves; contains 30% citral; used to make perfumes. { vər'bē·nə ,óil }

verdant zone *See* frostless zone. { 'vərd·ənt ,zōn }

Verdet constant [OPTICS] A constant of proportionality in the equation of the Faraday effect; it is equal to the angle of rotation of plane-polarized light in a magnetized substance divided by the product of the length of the light path in the substance and the strength of the magnetic field. { ˌvər'dā ,kän·stənt }

verdigris *See* cupric acetate. { 'vərd·ə,grēs }

verge [BUILD] The edge of a sloping roof which projects over a gable. { vərj }

vergeboard [BUILD] One of the boards utilized as the finish of the eaves on the gable end of a structure. Also known as bargeboard; gableboard. { 'vərj,bórd }

vergence [GEOL] The direction of overturning or of inclination of a fold. { 'vər·jəns }

verglas *See* glaze. { vər'glä }

veridicality [PSYCH] The correct perception of an object, that is, in agreement with the object's real properties. { və,rid·ə'kal·əd·ē }

verification [COMPUT SCI] The process of checking the results of one data transcription against the results of another data transcription; both transcriptions usually involve manual operations. { ˌver·ə·fə'kā·shən }

verification fire [ORD] Preparatory fire to test the mechanical adjustment of guns and fire-control equipment, and to measure the accuracy of corrections determined by calibration and trial fire. { ˌver·ə·fə'kā·shən ,fīr }

verifier [COMPUT SCI] A device for checking card punching semimechanically; it mimics keypunch machine operation, but reads prepunched cards without punching any new holes, and signals if the card does not agree with data entered through the verifier keyboard in some column. { 'ver·ə,fī·ər }

verify [COMMUN] To ensure that the meaning and phraseology of the transmitted message convey the exact intention of the originator. [COMPUT SCI] To determine whether an operation has been completed correctly, and in particular, to

check the accuracy of keypunching by using a verifier. { 'ver·ə,fī }

Vermes [INV ZOO] An artificial taxon considered to be a phylum in some systems of classification, but variously defined as including all invertebrates except arthropods, or including all vermiform invertebrates. { 'vər·mēz }

vermiculite [MINERAL] $(Mg,Fe,Al)_3(Al,Si)_4O_{10}(OH)_2 \cdot 4H_2O$ A clay mineral constituent similar to chlorite and montmorillonite, and consisting of trioctahedral mica sheets separated by double water layers; sometimes used as a textural material in painting, or as an aggregate in certain plaster formulations used in sculpture. { vər'mik·yə,līt }

vermiform [BIOL] Wormlike; resembling a worm. { 'vər·mə,fórm }

vermiform appendix [ANAT] A small, blind sac projecting from the cecum. Also known as appendix. { 'vər·mə,fórm ə'pen,diks }

vermifuge [PHARM] An agent that expels worms or intestinal animal parasites. { 'vər·mə,fyüj }

Vermilingua [VERT ZOO] An infraorder of the mammalian order Edentata distinguished by lack of teeth and in having a vermiform tongue; includes the South American true anteaters. { ˌvər·mə'liŋ·gwə }

vermilion *See* mercuric sulfide. { vər'mil·yən }

vermiphobia [PSYCH] An abnormal fear of worms or of infection by worms. { ˌvər·mə'fō·bē·ə }

vermis [ANAT] The median lobe of the cerebellum. { 'vər·məs }

vernadskite *See* antlerite. { vər'nadz,kīt }

vernal [GEOPHYS] Pertaining to spring. { 'vərn·əl }

vernal equinox [ASTRON] The sun's position on the celestial sphere about March 21; at this time the sun's path on the ecliptic crosses the celestial equator. Also known as first point of Aries; March equinox; spring equinox. { 'vərn·əl 'ē·kwə,näks }

vernalization [BOT] The induction in plants of the competence or ripeness to flower by the influence of cold, that is, at temperatures below the optimal temperature for growth. { ˌvərn·əl·ə'zā·shən }

vernation [BOT] The characteristic arrangement of young leaves within the bud. Also known as prefoliation. { vər'nā·shən }

vernier [ENG] A short, auxiliary scale which slides along the main instrument scale to permit accurate fractional reading of the least main division of the main scale. { 'vər·nē·ər }

vernier caliper [ENG] A caliper rule with an attached vernier scale. { 'vər·nē·ər 'kal·ə·pər }

vernier capacitor [ELEC] Variable capacitor placed in parallel with a larger tuning capacitor to provide a finer adjustment after the larger unit has been set approximately to the desired position. { 'vər·nē·ər kə'pas·əd·ər }

vernier dial [ENG] A tuning dial in which each complete rotation of the control knob causes only a fraction of a revolution of the main shaft, permitting fine and accurate adjustment. { 'vər·nē·ər 'dīl }

vernier engine [AERO ENG] A rocket engine of small thrust used primarily to obtain a fine adjustment in the velocity and trajectory of a rocket vehicle just after the thrust cutoff of the last sustainer engine, and used secondarily to add thrust to a booster or sustainer engine. Also known as vernier rocket. { 'vər·nē·ər 'en·jən }

vernier rocket *See* vernier engine. { 'vər·nē·ər 'räk·ət }

vernier sextant [NAV] A marine sextant providing a precise reading by means of a vernier used directly with the arc, and having either a clamp screw or an endless tangent screw for controlling the position of the index arm. { 'vər·nē·ər 'sek·stənt }

vernine *See* guanosine. { 'vər,nēn }

vernitel [ELECTR] Precision device which makes possible the transmission of data with high accuracy over standard frequency modulated-frequency modulated telemetering systems. { 'vər·nə,tel }

vernix caseosa [EMBRYO] A cheesy deposit on the surface of the fetus derived from the stratum corneum, sebaceous secretion, and remnants of the epitrichium. { 'vər·niks ,kā·sē'ō·sə }

vernolate [ORG CHEM] $C_{10}H_{21}NOS$ An amber liquid, used to control weeds in sweet potatoes, peanuts, soybeans, and tobacco. { 'vərn·əl,āt }

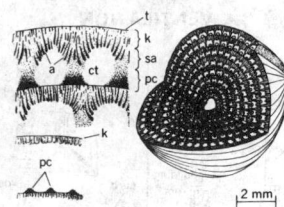

VERBEEKINIDAE

Cutaway diagram of representative species of Verbeekinidae; t = tectum, k = keriotheca, sa = septula, pc = parachromata, a = alveoli, and ct = chamberlets.

VERNIER

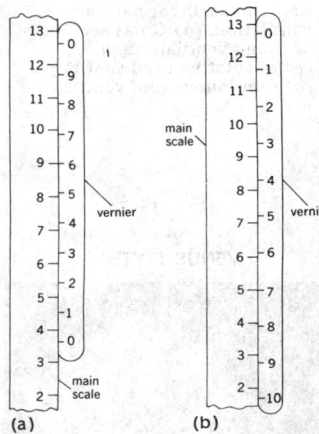

(a)   (b)

Two types of vernier scales. (*a*) Direct (reading 3.6). (*b*) Retrograde (reading 12.7).

veronal [PHARM] A white crystalline powder formerly used medicinally as a hypnotic drug to induce sleep. Also known as barbital. { 'ver·ə‚nól }

verrou *See* riegel. { və'rü }

verruca [BIOL] A wartlike elevation on the surface of a plant or animal. { və'rü·kə }

verruca peruana *See* verruca peruviana. { və'rü·kə ‚per·ə'wä·nə }

verruca peruviana [MED] A benign eruptive form of bartonellosis with chronic cutaneous lesions. Also known as verruca peruana. { və'rü·kə pə‚rü·vē'a·nə }

Verrucariaceae [BOT] A family of crustose lichens in the order Pyrenulales typically found on rocks, especially in intertidal or salt-spray zones along rocky coastlines. { və‚rü·kə·rē'ās·ē‚ē }

Verrucomorpha [INV ZOO] A suborder of the crustacean order Thoracica composed of sessile, asymmetrical barnacles. { və‚rü·kə'mór·fə }

verrucose [BIOL] Having the surface covered with wartlike protuberances. { və'rü‚kōs }

verrucous endocarditis [MED] Small thrombotic, nonbacterial, wartlike lesions on the heart valves and endocardium, occurring frequently in systemic lupus erythematosus. Also known as terminal endocarditis. { və'rü·kəs ‚en·dō·kär'dīd·əs }

vers *See* versed sine.

versatile anther [BOT] An anther whose attachment is near its middle, thus enabling it to swing freely. { 'vər·səd·əl 'an·thər }

versatile automatic test equipment [ELECTR] Computer-controlled tester, for missile electronic systems, that troubleshoots faults by deductive logic and isolates them to the plug-in module or component level. { 'vər·səd·əl ‚öd·ō‚mad·ik 'test i‚kwip·mənt }

versed cosine *See* coversed sine. { 'vərst 'kō‚sīn }

versed sine [MATH] The versed sine of *A* is 1 − cosine *A*. Denoted vers. Also known as versine. { 'vərst 'sīn }

versiera *See* witch of Agnesi. { vər'sē'er·ə }

versine *See* versed sine. { 'vər‚sīn }

verso [GRAPHICS] **1.** A left-hand page in a book, usually carrying an even page number. **2.** The side of a page that is to be read second. { 'vər·sō }

vertebra [ANAT] One of the bones that make up the spine in vertebrates. { 'vərd·ə·brə }

vertebral arch [ANAT] An arch formed by the paired pedicles and laminas of a vertebra; the posterior part of a vertebra which together with the anterior part, the body, encloses the vertebral foramen in which the spinal cord is lodged in vertebrates. Also known as neural arch { 'vərd·ə·brəl 'ärch }

vertebral column *See* spine. { 'vərd·ə·brəl 'käl·əm }

Vertebrata [VERT ZOO] The major subphylum of the phylum Chordata including all animals with backbones, from fish to human. { ‚vərd·ə'bräd·ə }

vertebrate zoology [ZOO] That branch of zoology concerned with the study of members of the Vertebrata. { 'vərd·ə·brət zō'äl·ə·jē }

vertebratus [METEOROL] A cloud variety (applied mainly to the genus cirrus), the elements of which are arranged in a manner suggestive of vertebrae, ribs, or a fish skeleton. { ‚vərd·ə'bräd·əs }

vertex [ASTRON] **1.** The highest point that a celestial body attains. **2.** On a great circle, that point that is closest to a pole. [MATH] **1.** For a polygon or polyhedron, any of those finitely many points which together with line segments or plane pieces determine the figure or solid. **2.** The common point at which the two sides of an angle intersect. **3.** The fixed point through which pass all the elements of a cone or conical surface. **4.** An intersection of a conic with one of its axes of symmetry. **5.** A member of the set of points that are connected by the edges. Also known as node. **6.** For a simplex, one of the finite set of points on which a simplex is based. [OPTICS] One of the points where the surface of a lens intersects the optical axis. { 'vər‚teks }

vertex angle [MATH] In a triangle, the angle opposite the base. { 'vər‚teks ‚aŋ·gəl }

vertex cover [MATH] A set of vertices in a graph such that every edge in the graph is incident to at least one vertex in this set. { 'vər‚teks ‚kəv·ər }

vertex-covering number [MATH] For a graph, the smallest possible number of vertices in a vertex cover. { 'vər‚teks ‚kəv·ər·iŋ ‚nəm·bər }

vertex detector [NUCLEO] A particle detector designed to provide high-precision (typically 5 to 20 micrometers) measurements of points along the trajectories of charged particles very close to the interaction point (typically 0.4 to 4 inches. or 1 to 10 centimeters) of a high-energy collider. { 'vər‚teks di‚tek·tər }

vertex-disjoint paths [MATH] In a graph, two paths with common end points that have no other points in common. { ‚vər‚teks 'dis‚jóint ‚paths }

vertex domination number [MATH] For a graph, the smallest possible number of vertices in a dominating vertex set. Also known as external stability number. { ‚vər‚teks ‚däm·ə'nā·shən ‚nəm·bər }

vertex-induced graph [MATH] A subgraph whose edges consist of all the edges in the original graph that join pairs of vertices in the subgraph. Also known as induced subgraph. { ‚vər‚teks in‚düst 'graf }

vertex power [OPTICS] The reciprocal of the back focal length of a lens. { 'vər‚teks ‚paú·ər }

vertical air photograph [GRAPHICS] An air photograph taken with the optical axis of the camera perpendicular to the earth's surface. { 'vərd·ə·kəl 'er 'fōd·ə‚graf }

vertical anemometer [METEOROL] An instrument which records the vertical component of the wind speed. { 'vərd·ə·kəl ‚an·ə'mäm·əd·ər }

vertical angles [MATH] The two angles produced by a pair of intersecting lines and lying on opposite sides of the point of intersection. { 'vərd·ə·kəl 'aŋ·gəlz }

vertical antenna [ELECTROMAG] A vertical metal tower, rod, or suspended wire used as an antenna. { 'vərd·ə·kəl an'ten·ə }

vertical axis [NAV ARCH] The vertical line near the center of gravity of a craft, perpendicular to both the longitudinal and lateral axes, around which it yaws. { 'vərd·ə·kəl 'ak·səs }

vertical-axis propeller [NAV ARCH] A type of propeller wheel that consists of a circular horizontal disk set flush into a vessel's bottom, rotates about a vertical axis, and carries near its periphery a number of spadelike vertical blades. { 'vərd·ə·kəl ‚ak·səs prə'pel·ər }

vertical ballistic transistor [ELECTR] A transistor in which, ideally, electrons traverse ballistically (that is without scattering) a very short region that separates a cathode from an anode, and whose effective cross section is modulated by the potential of metal gate contacts. { ‚vərd·i·kəl bə‚lis·tik tran'zis·tər }

vertical band saw [MECH ENG] A band saw whose blade operates in the vertical plane; ideal for contour cutting. { 'vərd·ə·kəl 'band ‚só }

vertical blanking [ELECTR] Blanking of a television picture tube during the vertical retrace. { 'vərd·ə·kəl 'blaŋk·iŋ }

vertical boiler [MECH ENG] A fire-tube boiler having vertical tubes between top head and tube sheet, connected to the top of an internal furnace. { 'vərd·ə·kəl 'bói·lər }

vertical boring mill [MECH ENG] A large type of boring machine in which a rotating workpiece is fastened to a horizontal table, which resembles a four-jaw independent chuck with extra radial T slots, and the tool has a traverse motion. { 'vərd·ə·kəl 'bór·iŋ ‚mil }

vertical broaching machine [MECH ENG] A broaching machine having the broach mounted in the vertical plane. { 'vərd·ə·kəl 'brōch·iŋ mə‚shēn }

vertical-cavity surface-emitting laser [OPTICS] A very small semiconductor laser in which stacks of dielectric mirrors above and below the optically active region form the optical cavity, the active gain medium consists of one or more semiconductor quantum wells placed parallel to the mirrors at an antinode of the cavity resonance, and lasing light emission is from the surface of the semiconductor substrate, normal to the plane of the gain medium. Abbreviated VCSEL. { 'vərd·i·kəl ‚kav·əd·ē ‚sər·fəs i‚mid·iŋ 'lā·zər }

vertical centering control [ELECTR] The centering control provided in a television receiver or cathode-ray oscilloscope to shift the position of the entire image vertically in either direction on the screen. { 'vərd·ə·kəl 'sen·tər·iŋ kən‚trōl }

vertical center keelson [NAV ARCH] The lower middle line girder, which, in conjunction with a flat plate keel on the bottom and a rider plate on top, forms the principal fore-and-aft strength member in the bottom of a ship. { 'vərd·ə·kəl 'sen·tər 'kēl·sən }

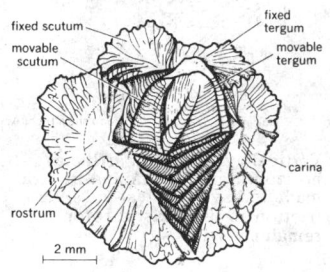

VERRUCOMORPHA

Verruca stroemia.

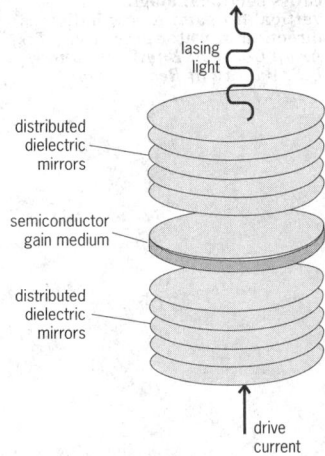

VERTICAL-CAVITY SURFACE-EMITTING LASER

Distributed dielectric mirrors form the optical cavity. Lasing light is normal to the plane of the gain medium.

VERTICAL GYRO

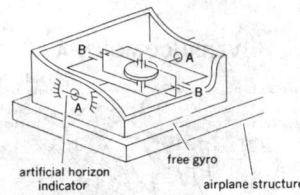

Vertical gyro used as artificial horizon. Bearings AA and BB are made to have smallest possible friction so that axis of rotor remains vertical.

VERTICAL-LIFT GATE

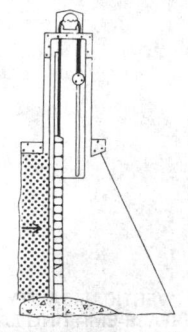

Cross-sectional diagram of vertical-lift gate. Arrow indicates direction of water pressure. (*U.S. Army Corps of Engineers and U.S. Bureau of Reclamation*)

vertical circle [ASTRON] A great circle of the celestial sphere, through the zenith and nadir of the celestial sphere; vertical circles are perpendicular to the horizon. { 'vərd·ə·kəl 'sər·kəl }

vertical compliance [ENG ACOUS] The ability of a stylus to move freely in a vertical direction while in the groove of a phonograph record. { 'vərd·ə·kəl kəm'plī·əns }

vertical component effect *See* antenna effect. { 'vərd·ə·kəl kəm'pō·nənt i‚fekt }

vertical conveyor [MECH ENG] A materials-handling machine designed to move or transport bulk materials or packages upward or downward. { 'vərd·ə·kəl kən'vā·ər }

vertical-current recorder [ENG] An instrument which records the vertical electric current in the atmosphere. { 'vərd·ə·kəl ‚kə·rənt ri‚kórd·ər }

vertical curve [CIV ENG] A curve inserted between two lengths of a road or railway which are at different slopes. { 'vərd·ə·kəl 'kərv }

vertical danger angle [NAV] The maximum or minimum angle between the top and bottom of an object of known height, as observed from a craft, indicating the limit of safe approach to an off-lying danger. { 'vərd·ə·kəl 'dān·jər ‚aŋ·gəl }

vertical definition *See* vertical resolution. { 'vərd·ə·kəl ‚def·ə'nish·ən }

vertical deflection oscillator [ELECTR] The oscillator that produces, under control of the vertical synchronizing signals, the sawtooth voltage waveform that is amplified to feed the vertical deflection coils on the picture tube of a television receiver. Also known as vertical oscillator. { 'vərd·ə·kəl di'flek·shən 'äs·ə‚lād·ər }

vertical deviation [ORD] In antiaircraft artillery, the distance between the target and the point of burst in the plane normal to the line of position along a line perpendicular to the lateral deviation. { 'vərd·ə·kəl ‚dē·vē'ā·shən }

vertical differential chart [METEOROL] A synoptic chart showing the difference in value of a meteorological element between two levels in the atmosphere; a common example is the thickness chart. { 'vərd·ə·kəl ‚dif·ə'ren·chəl ‚chärt }

vertical dip slip *See* vertical slip. { 'vərd·ə·kəl 'dip ‚slip }

vertical drop [MECH] The drop of an object in trajectory or along a plumb line, measured vertically from its line of departure to the object. { 'vərd·ə·kəl 'dräp }

vertical earth rate [NAV] To compensate for the effect of earth rate, the rate at which a gyroscope must be turned about its vertical axis to permit the spin axis to remain in the meridian; it is maximum at the poles, zero at the equator, and varies as the sine of the latitude. { 'vərd·ə·kəl 'ərth ‚rāt }

vertical-face breakwater [CIV ENG] A breakwater whose mound of rubble does not rise above the water, but is surmounted by a vertical-face superstructure of masonry or concrete; may be built without mound rubble, provided sea bed is firm. { 'vərd·ə·kəl ‚fās 'brāk‚wód·ər }

vertical field balance [ENG] An instrument that measures the vertical component of the magnetic field by means of the torque that the field component exerts on a horizontal permanent magnet. { 'vərd·ə·kəl 'fēld ‚bal·əns }

vertical field-strength diagram [ELECTROMAG] Representation of the field strength at a constant distance from an antenna and in a vertical plane passing through the antenna. { 'vərd·ə·kəl 'fēld ‚streŋkth ‚dī·ə‚gram }

vertical firing [MECH ENG] The discharge of fuel and air perpendicular to the burner in a furnace. { 'vərd·ə·kəl 'fīr·iŋ }

vertical force instrument *See* heeling adjuster. { 'vərd·ə·kəl ‚fórs 'in·strə·mənt }

vertical guide idlers [MECH ENG] Idler rollers about 3 inches (8 centimeters) in diameter so placed as to make contact with the edge of the belt conveyor should it run too much to one side. { 'vərd·ə·kəl ‚gīd 'īd·lərz }

vertical gyro [AERO ENG] A two-degree-of-freedom gyro with provision for maintaining its spin axis vertical; output signals are produced by gimbal angular displacements which correspond to components of the angular displacements of the base about two orthogonal axes; used in aircraft to measure both bank angle and pitch attitude. { 'vərd·ə·kəl 'jī·rō }

vertical hold control [ELECTR] The hold control that changes the free-running period of the vertical deflection oscillator in a television receiver, so the picture remains steady in the vertical direction. { 'vərd·ə·kəl ‚hōld kən‚trōl }

vertical illuminator [OPTICS] A microscope designed for observing surfaces of opaque substances such as metals, which has a mechanism for passing light down through the objective lens in order to illuminate the surface to be observed with a beam perpendicular to the surface. { 'vərd·ə·kəl i'lü·mə‚nād·ər }

vertical-incidence transmission [ELECTROMAG] Transmission of a radio wave vertically to the ionosphere and back. { 'vərd·ə·kəl ‚in·sə·dəns tranz'mish·ən }

vertical instruction [COMPUT SCI] An instruction in machine language to carry out a single operation or a time-ordered series of a fixed number and type of operation on a single set of operands. { 'vərd·ə·kəl in'strək·shən }

vertical intensity [GEOPHYS] The magnetic intensity of the vertical component of the earth's magnetic field, reckoned positive if downward, negative if upward. { 'vərd·ə·kəl in'ten·səd·ē }

vertical intensity variometer [ENG] A variometer employing a large permanent magnet and equipped with very fine steel knife-edges or pivots resting on agate planes or saddles and balanced so that its magnetic axis is horizontal. Also known as Z variometer. { 'vərd·ə·kəl in'ten·səd·ē ‚ver·ē'äm·əd·ər }

vertical interval reference [ELECTR] A reference signal inserted into a television program signal every 1/60 second, in line 19 of the vertical blanking period between television frames, to provide references for luminance amplitude, black-level amplitude, sync amplitude, chrominance amplitude, and color-burst amplitude and phase. Abbreviated VIR. { 'vərd·ə·kəl ‚in·tər·vəl 'ref·rəns }

vertical jet *See* uprush. { 'vərd·ə·kəl 'jet }

vertical launch [ORD] A launch in which the missile or vehicle starts from a vertical position. { 'vərd·ə·kəl 'lónch }

vertical lead [ORD] The vertical angle by which the gun must be moved from the line of position in order for the trajectory to pass through the target; it is the algebraic sum of the principal vertical deflection, the vertical pointing correction, and the superelevation. { 'vərd·ə·kəl 'lēd }

vertical-lift bridge [CIV ENG] A movable bridge with a span that rises on towers, lifted by steel ropes. { 'vərd·ə·kəl ‚lift 'brij }

vertical-lift gate [CIV ENG] A dam spillway gate of which the movable parts are raised and lowered vertically to regulate water flow. { 'vərd·ə·kəl ‚lift 'gāt }

vertical linearity control [ELECTR] A linearity control that permits narrowing or expanding the height of the image on the upper half of the screen of a television picture tube, to give linearity in the vertical direction so circular objects appear as true circles; usually mounted at the rear of the receiver. { 'vərd·ə·kəl ‚lin·ē‚ar·əd·ē kən‚trōl }

vertically stacked loops *See* stacked loops. { 'vərd·ə·klē ‚stakt 'lüps }

vertical metal oxide semiconductor technology [ELECTR] For semiconductor devices, a technology that involves essentially the formation of four diffused layers in silicon and etching of a V-shaped groove to a precisely controlled depth in the layers, followed by deposition of metal over silicon dioxide in the groove to form the gate electrode. Abbreviated VMOS technology. { 'vərd·ə·kəl ‚med·əl ‚äk‚sīd ‚sem·i·kən‚dək·tər tek'näl·ə·jē }

vertical obstacle sonar [ENG] An active sonar used to determine heights of objects in the path of a submersible vehicle; its beam sweeps along a vertical plane, about 30° above and below the direction of the vehicle's motion. Abbreviated VOS. { 'vərd·ə·kəl ‚äb·stə·kəl 'sō‚när }

vertical or short takeoff and landing aircraft *See* V/STOL aircraft. { 'vərd·ə·kəl ór ‚shórt ‚tāk‚óf and ‚land·iŋ 'er‚kraft }

vertical oscillator *See* vertical deflection oscillator. { 'vərd·ə·kəl 'äs·ə‚lād·ər }

vertical parity check *See* lateral parity check. { 'vərd·ə·kəl 'par·əd·ē ‚chek }

vertical pitch [GRAPHICS] The number of lines per inch on a printed page. { 'vərd·ə·kəl 'pich }

vertical polarization [COMMUN] Transmission of linear polarized radio waves whose electric field vector is perpendicular to the earth's surface. { 'vərd·ə·kəl ‚pō·lə·rə'zā·shən }

vertical-position welding [MET] Welding in which the weld axis is essentially vertical. { 'vərd·ə·kəl pə'zish·ən 'weld·iŋ }

vertical recording [ELECTR] Magnetic recording in which bits are magnetized in directions perpendicular to the surface of the recording medium, allowing the bits to be smaller. Also known as perpendicular recording. [ENG ACOUS] A type of disk recording in which the groove modulation is perpendicular to the surface of the recording medium, so the cutting stylus moves up and down rather than from side to side during recording. Also known as hill-and-dale recording. { 'vərd·ə·kəl ri'kȯrd·iŋ }

vertical redundancy check *See* lateral parity check. { 'vərd·ə·kəl ri'dən·dən·sē ,chek }

vertical resolution [ELECTR] The number of distinct horizontal lines, alternately black and white, that can be seen in the reproduced image of a television or facsimile test pattern; it is primarily fixed by the number of horizontal lines used in scanning. Also known as vertical definition. { 'vərd·ə·kəl ,rez·ə'lü·shən }

vertical retort process [MET] A zinc-smelting method using a vertical retort of silicon carbide brick. Also known as New Jersey retort process. { 'vərd·ə·kəl ri¦tȯrt ,präs·əs }

vertical retrace [ELECTR] The return of the electron beam to the top of the screen at the end of each field in television. { 'vərd·ə·kəl 'rē,trās }

vertical scale [DES ENG] The ratio of the vertical dimensions of a laboratory model to those of the natural prototype; usually exaggerated in relation to the horizontal scale. { 'vərd·ə·kəl 'skāl }

vertical seismograph [ENG] An instrument that records the vertical component of the ground motion during an earthquake. { 'vərd·ə·kəl 'sīz·mə,graf }

vertical separation [AERO ENG] A specified vertical distance measured in terms of space between aircraft in flight at different altitudes or flight levels. [GEOL] The vertical component of the dip slip in a fault. { 'vərd·ə·kəl ,sep·ə'rā·shən }

vertical separator [PETRO ENG] A gas-oil separator in the form of a vertical cylindrical tank. { 'vərd·ə·kəl 'sep·ə,rād·ər }

vertical slip [GEOL] The vertical component of the net slip in a fault. Also known as vertical dip slip. { 'vərd·ə·kəl 'slip }

vertical speed indicator *See* rate-of-climb indicator. { 'vərd·ə·kəl ¦spēd ,in·də,kād·ər }

vertical stability *See* static stability. { 'vərd·ə·kəl stə'bil·əd·ē }

vertical stretching [METEOROL] A process in which ascending vertical motion of air increases with altitude, or descending motion decreases with (increasing) altitude. { 'vərd·ə·kəl 'strech·iŋ }

vertical sweep [ELECTR] The downward movement of the scanning beam from top to bottom of the picture being televised. { 'vərd·ə·kəl 'swēp }

vertical synchronizing pulse [ELECTR] One of the six pulses that are transmitted at the end of each field in a television system to keep the receiver in field-by-field synchronism with the transmitter. Also known as picture synchronizing pulse. { 'vərd·ə·kəl 'siŋ·krə,nīz·iŋ ,pəls }

vertical tab [COMPUT SCI] A control character that causes a computer printer to jump from its current line to another preset line further down the page. { 'vərd·ə·kəl 'tab }

vertical tail [AERO ENG] A part of the tail assembly of an aircraft; consists of a fin (a symmetrical airfoil in line with the center line of the fuselage) fixed to the fuselage or body and a rudder which is movable by the pilot. { 'vərd·ə·kəl 'tāl }

vertical takeoff and landing [AERO ENG] A flight technique in which an aircraft rises directly into the air and settles vertically onto the ground. Abbreviated VTOL. { 'vərd·ə·kəl 'tāk,ȯf ən 'land·iŋ }

vertical tower launcher *See* tower launcher. { 'vərd·ə·kəl ¦taů·ər 'lȯnch·ər }

vertical transmission [GEN] Passage of genetic information from one cell or individual organism to its progeny by conventional heredity mechanisms. { 'vərd·ə·kəl tranz'mish·ən }

vertical traverse [MECH ENG] The angle through which a robot's arm can swing up and down, typically 30°. { 'vərd·ə·kəl trə'vərs }

vertical turbine pump *See* deep-well pump. { 'vərd·ə·kəl 'tər·bən ,pəmp }

vertical turret lathe [DES ENG] Similar in principle to the horizontal turret lathe but capable of handling heavier, bulkier workpieces; it is constructed with a rotary, horizontal worktable

whose diameter (30–74 inches, or 76–188 centimeters) normally designates the capacity of the machine; a crossrail mounted above the worktable carries a turret, which indexes in a vertical plane with tools that may be fed either across or downward. { 'vərd·ə·kəl 'tə·rət ,lāth }

vertical visibility [METEOROL] According to United States weather observing practice, the distance that an observer can see vertically into a surface-based obscuring phenomenon, such as fog, rain, or snow. { 'vərd·ə·kəl ,vis·ə'bil·əd·ē }

vertical vorticity [FL MECH] The vertical component of the vorticity vector. { 'vərd·ə·kəl vȯr'tis·əd·ē }

verticillate [BOT] Whorled, in an arrangement resembling the spokes of a wheel. { ¦vərd·ə¦si,lāt }

vertigo [MED] The sensation that the outer world is revolving about the patient (objective vertigo) or that the patient is moving in space (subjective vertigo). { 'vərd·ə,gō }

Vertisol [GEOL] A soil order formed in regoliths high in clay; subject to marked shrinking and swelling with changes in water content; low in organic content and high in bases. { 'vərd·ə,sȯl }

very close pack ice [OCEANOGR] Sea ice so concentrated that there is little if any open water. { ¦ver·ē ¦klȯs 'pak ,īs }

very high frequency [COMMUN] The band of frequencies from 30 to 300 megahertz in the radio spectrum, corresponding to wavelengths of 1 to 10 meters. Abbreviated VHF. { ¦ver·ē ¦hī 'frē·kwən·sē }

very high frequency omnidirectional radio range [NAV] A radio navigation aid operating at very high frequency and supplying bearing information for the entire 360° of azimuth. Abbreviated VOR. { ¦ver·ē ¦hī 'frē·kwən·sē ¦äm·nə·də'rek·shən·əl 'rād·ē·ō ,rānj }

very high frequency oscillator [ELECTR] An oscillator whose frequency lies in the range from a few to several hundred megahertz; it uses distributed, rather than lumped, impedances, such as parallel wire transmission lines or coaxial cables. { ¦ver·ē ¦hī 'frē·kwən·sē 'äs·ə,lad·ər }

very high frequency tuner [ELECTR] A tuner in a television receiver for reception of stations transmitting in the very high frequency band; it generally has 12 discrete positions corresponding to channels 2–13. { ¦ver·ē ¦hī 'frē·kwən·sē 'tün·ər }

Very Large Array [ASTRON] An array near Socorro, New Mexico, of 27 separate radio telescopes on movable platforms, arranged along the arms of a Y, designed to provide radio pictures which have an angular resolution comparable with that of the best optical telescopes. Abbreviated VLA. { ¦ver·ē ¦lärj ə'rā }

very large crude carrier [NAV ARCH] A liquid-cargo vessel in the 100,000- to 250,000-ton range. Abbreviated VLCC. { ¦ver·ē ¦lärj 'krüd ,kar·ē·ər }

very large scale integrated circuit [ELECTR] A complex integrated circuit that contains between 20,000 and 1,000,000 transistors. Abbreviated VLSI circuit. { ¦ver·ē ¦lärj ¦skāl 'int·ə,grād·əd 'sər·kət }

very long baseline interferometry [ELECTR] A method of improving angular resolution in the observation of radio sources; these are simultaneously observed by two radio telescopes which are very far apart, and the signals are recorded on magnetic tapes which are combined electronically or on a computer. Abbreviated VLBI. { ¦ver·ē ¦lȯŋ ¦bās·līn ,in·tər·fə'räm·ə·trē }

very long range material requirements [ORD] Items required by operational and organizational concepts established for a period 10 years hence and beyond. { ¦ver·ē ¦lȯŋ ¦rānj mə'tir·ē·əl ri,kwīr·məns }

very long range radar [ELECTR] Equipment whose maximum range on a reflecting target of 10.76 square feet (1 square meter) normal to the signal path exceeds 800 miles (1300 kilometers), provided line of sight exists between the target and the radar. { ¦ver·ē ¦lȯŋ ¦rānj 'rā,där }

very low frequency [COMMUN] The band of frequencies from 3 to 30 kilohertz in the radio spectrum, corresponding to wavelengths of 10 to 100 kilometers. Abbreviated VLF. { ¦ver·ē ¦lō 'frē·kwən·sē }

very open pack ice [OCEANOGR] Sea ice whose concentration ranges between one-tenth and three-tenths of the sea surface. { ¦ver·ē ¦ō·pən 'pak ,īs }

very short range radar [ELECTR] Equipment whose range on a reflecting target of 10.76 square feet (1 square meter) normal to the signal path is less than 50 miles (80 kilometers),

VERY LARGE ARRAY

Aerial view of the Very Large Array radio telescope near Socorro, New Mexico. *(National Radio Astronomy Observatory)*

provided line of sight exists between the target and the radar. { ¦ver·ē ¦shȯrt ¦ranj 'rā,där }

very small aperture terminal [COMMUN] An antenna approximately 6 feet (1.8 meters) or less in diameter, that is used for both broadcast reception and interactive communications via geosynchronous satellites. Abbreviated VSAT. { ¦ver·ē ¦smȯl ¦ap·ə·chər 'tərm·ə·nəl }

vesicant [PHARM] An agent that causes blistering. { 'ves·ə·kənt }

vesication [MED] 1. A blister. 2. Formation of a blister. { ˌves·ə'kā·shən }

vesicle [BIOL] A small, thin-walled bladderlike cavity, usually filled with fluid. [GEOL] A cavity in lava formed by entrapment of a gas bubble during solidification. Also known as air sac; bladder; saccus; vacuole; wing. { 'ves·ə·kəl }

vesicular [SCI TECH] Characterized by abundant vesicles. { və'sik·yə·lər }

vesicular-arbuscular mycorrhizal fungi [MYCOL] Mycorrhizal fungi that grow into the root cortex of the host plant and penetrate root cells to form two kinds of specialized structures, arbuscules and vesicles. Also known as arbuscular mycorrhizae. { və'sik·yə·lər är¦bəs·kyə·lər ˌmī·kə,rīz·əl 'fən,jī }

vesicular film [GRAPHICS] A film that is sensitive to ultraviolet light and is developed by heat, without chemicals. { və'sik·yə·lər 'film }

vesicular stomatitis [VET MED] A viral disease, most often of horses, cattle, and pigs, characterized by fever and by vesicular and erosive lesions on the tongue, gums, lips, feet, and teats. { ve,sik·yə·lər ˌstō·mə'tīd·əs }

vesicular stomatitis virus [VIROL] A virus in the genus *Vesiculovirus,* family Rhabdoviridae; the causative agent of vesicular stomatitis. { və'sik·yə·lər ˌstō·mə'tīd·əs ,vī·rəs }

vesicular structure [PETR] A structure that is common in many volcanic rocks and which forms when magma is brought to or near the earth's surface; may form a structure with small cavities, or produce a pumiceous structure or a scoriaceous structure. { və'sik·yə·lər 'strək·chər }

Vesiculovirus [VIROL] A genus of the viral family Rhabdoviridae; type species is vesicular stomatitis virus. { və'sik·yə·lō,vī·rəs }

Vespertilionidae [VERT ZOO] The common bats, a large cosmopolitan family of the Chiroptera characterized by a long tail, extending to the edge of the uropatagium; almost all members are insect-eating. { ˌves·pər,til·ē'än·ə,dē }

vespertine [VERT ZOO] Active in the evening. { 'ves·pər,tīn }

Vespidae [INV ZOO] A widely distributed family of Hymenoptera in the superfamily Vespoidea including hornets, yellow jackets, and potter wasps. { 'ves·pə,dē }

Vespoidea [INV ZOO] A superfamily of wasps in the suborder Apocrita. { ve'spȯid·ē·ə }

vessel [BOT] A water-conducting tube or duct in the xylem. [ENG] A container or structural envelope in which materials are processed, treated, or stored; for example, pressure vessels, reactor vessels, agitator vessels, and storage vessels (tanks). [NAV ARCH] Any craft that can carry people or cargo over the surface of the water. { 'ves·əl }

vessel segment [BOT] A single cell or unit of a plant vessel. { 'ves·əl ,seg·mənt }

vessel traffic service [NAV] A program that provides marine traffic management of an advisory nature, and occasional emergency control, to reduce collisions and strandings in heavily trafficked ports. { 'ves·əl 'traf·ik ,sər·vəs }

Vesta [ASTRON] The third-largest asteroid with a diameter of about 300 miles (500 kilometers), mean distance from the sun of 2.362 astronomical units, and a unique surface composition resembling basaltic, achondritic meteorites. { 'ves·tə }

vestibular apparatus [ANAT] The anatomical structures concerned with the vestibular portion of the eighth cranial nerve; includes the saccule, utricle, semicircular canals, vestibular nerve, and vestibular nuclei of the ear. { və'stib·yə·lər 'ap·ə,rad·əs }

vestibular membrane of Reissner See Reissner's membrane. { və'stib·yə·lər 'mem,brān əv 'rīs·nər }

vestibular nerve [NEUROSCI] A somatic sensory branch of the auditory nerve, which is distributed about the ampullae of the semicircular canals, macula sacculi, and macula utriculi. { və'stib·yə·lər 'nərv }

vestibular reflexes [PHYSIO] The responses of the vestibular apparatus to strong stimulation; responses include pallor, nausea, vomiting, and postural changes. { və'stib·yə·lər 'rē,flek·səz }

vestibule [ANAT] 1. The central cavity of the bony labyrinth of the ear. 2. The parts of the membranous labyrinth within the cavity of the bony labyrinth. 3. The space between the labia minora. 4. See buccal cavity. [BUILD] A hall or chamber between the outer door and the interior, or rooms, of a building. { 'ves·tə,byül }

vestibule school [IND ENG] A school organized by an industrial concern to train new employees in specific tasks or prepare employees for promotion. { 'ves·tə,byül ,skül }

vestibule training [IND ENG] A procedure used in operator training in which the training location is separate from the main productive areas of the plant; includes student carrels, lecture rooms, and in many instances the same type of equipment that the trainee will use in the work station. { 'ves·tə,byül ,trān·iŋ }

vestibulocerebellar [NEUROSCI] Pertaining to the vestibular fibers and the cerebellum. { və'stib·yə·lō,ser·ə'bel·ər }

vestibulocochlear nerve See auditory nerve. { və'stib·yə·lə'käk·lē·ər ,nərv }

vestibulospinal tract [NEUROSCI] A tract of nerve fibers that originates principally from the lateral vestibular nucleus and descends in the anterior funiculus of the spinal cord. { və'stib·yə·lə'spīn·əl ,trakt }

vestige [BIOL] A degenerate anatomical structure or organ that remains from one more fully developed and functional in an earlier phylogenetic form of the individual. { 'ves·tij }

vestigial [BIOL] Of, being, or resembling a vestige. { və'stij·ē·əl }

vestigial sideband [COMMUN] The transmitted portion of an amplitude-modulated sideband that has been largely suppressed by a filter having a gradual cutoff in the neighborhood of the carrier frequency; the other sideband is transmitted without much suppression. Abbreviated VSB. { və'stij·ē·əl 'sīd,band }

vestigial-sideband filter [ELECTR] A filter that is inserted between a transmitter and its antenna to suppress part of one of the sidebands. { və'stij·ē·əl 'sīd,band ,fil·tər }

vestigial-sideband transmission [COMMUN] A type of radio signal transmission for amplitude modulation in which the normal complete sideband on one side of the carrier is transmitted, but only a part of the other sideband is transmitted. Also known as asymmetrical-sideband transmission. { və'stij·ē·əl 'sīd,band tranz,mish·ən }

vesuvian See leucite; vesuvianite. { və'sü·vē·ən }

Vesuvian eruption See Vulcanian eruption. { və'sü·vē·ən i'rəp·shən }

Vesuvian garnet See leucite. { və'sü·vē·ən 'gär·nət }

vesuvianite [MINERAL] $Ca_{10}Mg_2Al_4(SiO_4)_5(Si_2O_7)_2(OH)_4$ A brown, yellow, or green mineral found in contact-metamorphosed limestones. Also known as idocrase; vesuvian. { və'sü·vē·ə,nīt }

veszelyite [MINERAL] $(Cu,Zn)_3(PO_4)(OH)_3·2H_2O$ A greenish-blue to dark blue, monoclinic mineral consisting of a hydrated basic phosphate of copper and zinc. { 'ves·əl,yīt }

vetch [BOT] Any of a group of mostly annual legumes, especially of the genus *Vicia,* with weak viny stems terminating in tendrils and having compound leaves; some varieties are grown for their edible seed. { vech }

veterinary medicine [MED] The branch of medical practice which treats of the diseases and injuries of animals. { 'vet·ən,er·ē 'med·ə·sən }

vetiver oil [MATER] A combustible, viscous essential oil from partially dried roots of *Vetiveria zizanioides* (East Indian grass), with a violet scent; soluble in fixed oils; used in perfumery. Also known as cuscus oil; vetivert. { 'ved·ə·vər ,ȯil }

vetivert See vetiver oil. { 'ved·ə·vərt }

VF See voice frequency.

V factor [MICROBIO] Phosphopyridine nucleotide, a growth factor required by the parasitic bacteria of the genus *Haemophilus.* { 'vē ,fak·tər }

V flume [MIN ENG] A V-shaped flume, supported by trestlework, and used by miners for bringing down timber and wood from the high mountains; the flume water is also used for mining purposes. { 'vē ,flüm }

V format [COMPUT SCI] A data record format in which the

logical records are of variable length and each record begins with a record length indication. { 'vē ˌfȯr₁mat }

V471 Tauri star [ASTRON] A binary star, displaying irregular variability, in which one component is a main-sequence star and the other is either a white dwarf star or a star that is evolving toward the white dwarf state and is surrounded by a planetary nebula. { ¦vē₁fȯr₁sev·ən·tē₁wən 'tȯr·ē ₁stär }

VFR *See* visual flight rules.

VFR between layers [AERO ENG] A flight condition wherein an aircraft is operated under modified visual flight rules while in flight between two layers of clouds or obscuring phenomena, each of which constitutes a ceiling. { ₁vē₁ef'är bi₁twēn 'lā·ərz }

VFR on top [AERO ENG] A flight condition wherein an aircraft is operated under modified visual flight rules while in flight above a layer of clouds or an obscuring phenomenon sufficient to constitute a ceiling. { ₁vē₁ef'är ȯn 'täp }

VFR terminal minimums [AERO ENG] A set of operational weather limits at an airport, that is, the minimum conditions of ceiling and visibility under which visual flight rules may be used. { ₁vē₁ef'är 'tər·mən·əl 'min·ə·məmz }

VFR weather [METEOROL] In aviation terminology, route or terminal weather conditions which allow operation of aircraft under visual flight rules. { ₁vē₁ef'är 'weth·ər }

VGC *See* viscosity-gravity constant.

V guide [MECH ENG] A V-shaped groove serving to guide a wedge-shaped sliding machine element. { 'vē ₁gīd }

VHF *See* very high frequency.

VI *See* viscosity index.

via [ELECTR] A pathway that is etched to allow electrical contact between different layers of a semiconductor device. { 'vē·ə *or* 'vī·ə }

viable [BIOL] Able to live and develop normally. { 'vī·ə·bəl }

viaduct [CIV ENG] A bridge structure supported on high towers with short masonry or reinforced concrete arched spans. { 'vī·ə₁dəkt }

Vianaidae [INV ZOO] A small family of South American Hemiptera in the superfamily Tingordea. { ₁vē·ə'nā·ə₁dē }

via point [CONT SYS] A point located midway between the starting and stopping positions of a robot tool tip, through which the tool tip passes without stopping. Also known as way point. { 've·ə ₁pȯint }

vibraculum [INV ZOO] A specially modified bryozoan zooid with a bristlelike seta that sweeps debris from the surface of the colony. { və'brak·yə·ləm }

vibrating capacitor [ELEC] A capacitor whose capacitance is varied in a cyclic manner to produce an alternating voltage proportional to the charge on the capacitor; used in a vibrating-reed electrometer. { 'vī₁brād·iŋ kə'pas·əd·ər }

vibrating conveyor *See* oscillating conveyor. { 'vī₁brād·iŋ kən'vā·ər }

vibrating coring tube [ENG] A sediment corer made to vibrate in order to eliminate the resistance of compacted ocean floor sediments, sands, and gravel. { 'vī₁brād·iŋ 'kȯr·iŋ ₁tüb }

vibrating feeder [MECH ENG] A feeder for bulk materials (pulverized or granulated solids), which are moved by the vibration of a slightly slanted, flat vibrating surface. { 'vī₁brād·iŋ 'fēd·ər }

vibrating grizzlies [MECH ENG] Bar grizzlies mounted on eccentrics so that the entire assembly is given a forward and backward movement at a speed of some 100 strokes a minute. { 'vī₁brād·iŋ 'griz·lēz }

vibrating needle [ENG] A magnetic needle used in compass adjustment to find the relative intensity of the horizontal components of the earth's magnetic field and the magnetic field at the compass location. { 'vī₁brād·iŋ 'nēd·əl }

vibrating pebble mill [MECH ENG] A size-reduction device in which feed is ground by the action of vibrating, moving pebbles. { 'vī₁brād·iŋ 'peb·əl ₁mil }

vibrating-reed electrometer [ENG] An instrument using a vibrating capacitor to measure a small charge, often in combination with an ionization chamber. { 'vī₁brād·iŋ ¦rēd ₁i₁lek'träm·əd·ər }

vibrating-reed frequency meter [ENG] A frequency meter consisting of steel reeds having different and known natural frequencies, all excited by an electromagnet carrying the alternating current whose frequency is to be measured. Also

known as Frahm frequency meter; reed frequency meter; tuned-reed frequency meter. { 'vī₁brād·iŋ ¦rēd 'frē·kwən·sē ₁mēd·ər }

vibrating-reed magnetometer [ENG] An instrument that measures magnetic fields by noting their effect on the vibration of reeds excited by an alternating magnetic field. { 'vī₁brād·iŋ ¦rēd ₁mag·nə'täm əd·ər }

vibrating-reed tachometer [ENG] A tachometer consisting of a group of reeds of different lengths, each having a specific natural frequency of vibration; observation of the vibrating reed when in contact with a moving mechanical device indicates the frequency of vibration for the device. { 'vī₁brād·iŋ ¦rēd tə'käm·əd·ər }

vibrating screen [MECH ENG] A sizing screen which is vibrated by solenoid or magnetostriction, or mechanically by eccentrics or unbalanced spinning weights. { 'vī₁brād·iŋ 'skrēn }

vibrating screen classifier [MECH ENG] A classifier whose screening surface is hung by rods and springs, and moves by means of electric vibrators. { 'vī₁brād·iŋ ¦skrēn 'klas·ə₁fī·ər }

vibrating wire transducer [ENG] A device for measuring ocean depth, consisting of a very fine tungsten wire stretched in a magnetic field so that it vibrates at a frequency that depends on the tension in the wire, and thereby on pressure and depth. { 'vī₁brād·iŋ ¦wīr tranz'dü·sər }

vibration [MECH] A continuing periodic change in a displacement with respect to a fixed reference. [PHYS CHEM] Oscillation of atoms about their equilibrium positions within a molecular system. { vī'brā·shən }

vibrational energy [PHYS CHEM] For a diatomic molecule, the difference between the energy of the molecule idealized by setting the rotational energy equal to zero, and that of a further idealized molecule which is obtained by gradually stopping the vibration of the nuclei without placing any new constraint on the motions of electrons. { vī'brā·shən·əl 'en·ər·jē }

vibrational level [PHYS CHEM] An energy level of a diatomic or polyatomic molecule characterized by a particular value of the vibrational energy. { vī'brā·shən·əl 'lev·əl }

vibrational quantum number [PHYS CHEM] A quantum number v characterizing the vibrational motion of nuclei in a molecule; in the approximation that the molecule behaves as a quantum-mechanical harmonic oscillator, the vibrational energy is $h(v + 1/2)f$, where h is Planck's constant and f is the vibration frequency. { vī'brā·shən·əl 'kwän·təm ₁nəm·bər }

vibrational spectrum [SPECT] The molecular spectrum resulting from transitions between vibrational levels of a molecule which behaves like the quantum-mechanical harmonic oscillator. { vī'brā·shən·əl 'spek·trəm }

vibrational sum rule [SPECT] **1.** The rule that the sums of the band strengths of all emission bands with the same upper state is proportional to the number of molecules in the upper state, where the band strength is the emission intensity divided by the fourth power of the frequency. **2.** The sums of the band strengths of all absorption bands with the same lower state is proportional to the number of molecules in the lower state, where the band strength is the absorption intensity divided by the frequency. { vī'brā·shən·əl 'səm ₁rül }

vibrational transition [PHYS CHEM] A transition between two quantized levels of a molecule that have different vibrational energies. Also known as vibronic transition. { vī'brā·shən·əl tran'zish·ən }

vibration damping [MECH ENG] The processes and techniques used for converting the mechanical vibrational energy of solids into heat energy. { vī'brā·shən 'damp·iŋ }

vibration drilling [MECH ENG] Drilling in which a frequency of vibration in the range of 100 to 20,000 hertz is used to fracture rock. { vī'brā·shən 'dril·iŋ }

vibration galvanometer [ENG] An alternating-current galvanometer in which the natural oscillation frequency of the moving element is equal to the frequency of the current being measured. { vī'brā·shən ₁gal·və näm·əd·ər }

vibration isolation [ENG] The isolation, in structures, of those vibrations or motions that are classified as mechanical vibration; involves the control of the supporting structure, the placement and arrangement of isolators, and control of the internal construction of the equipment to be protected. { vī'brā·shən ₁ī·sə'lā·shən }

vibration limit [CIV ENG] The amount of time during which

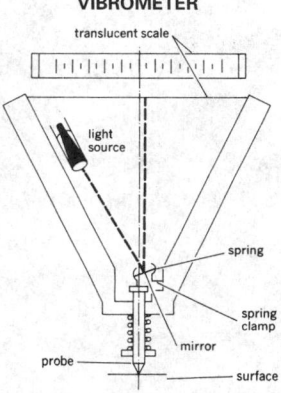

VIBROMETER

translucent scale

light source

spring

spring clamp

mirror

probe

surface

Mechanooptical vibrometer. The motion given to the probe by the vibrating surface is used to rock a mirror and thereby actuate an optical lever arm. A light beam reflected from the mirror and focused onto the scale provides an indication of the vibration amplitude. *(General Electric Co.)*

fresh concrete remains mobile when subjected to vibration. { vī'brā·shən ˌlim·əl }

vibration machine [MECH ENG] A device for subjecting a system to controlled and reproducible mechanical vibration. Also known as shake table. { vī'brā·shən mə,shēn }

vibration magnetometer [ENG] An instrument that measures the period of vibration of a magnetic needle to determine the horizontal magnetic field strength at the needle. { vī'brā·shən ˌmag·nə'täm·əd·ər }

vibration meter See vibrometer. { vī'brā·shən ˌmēd·ər }

vibration pickup [ELEC] An electromechanical transducer capable of converting mechanical vibrations into electrical voltages. { vī'brā·shən ˌpik,əp }

vibration puddling [CIV ENG] A technique used to achieve proper consolidation of concrete; vibrating machines may be drawn vertically through the cement, or used on the surface, or placed against the form holding the concrete in place. Also known as mechanical puddling. { vī'brā·shən 'pəd·liŋ }

vibration separation [MECH ENG] Classification or separation of grains of solids in which separation through a screen is expedited by vibration or oscillatory movement of the screening mediums. { vī'brā·shən ˌsep·ə'rā·shən }

vibration suppression [MECH ENG] The prevention of undesirable vibration, either through passive means such as damping or through active techniques involving feedback control. { vī'brā·shən sə,presh·ən }

vibrato [ACOUS] A musical embellishment that depends primarily on periodic variations of frequency which are often accompanied by variations in amplitude and waveform. { vi'bräd·ō }

vibrator [ELEC] An electromechanical device used primarily to convert direct current to alternating current but also used as a synchronous rectifier; it contains a vibrating reed which has a set of contacts that alternately hit stationary contacts attached to the frame, reversing the direction of current flow; the reed is activated when a soft-iron slug at its tip is attracted to the pole piece of a driving coil. [MECH ENG] An instrument which produces mechanical oscillations. { 'vī,brād·ər }

vibrator power supply [ELEC] A power supply using a vibrator to produce the varying current necessary to actuate a transformer, the output of which is then rectified and filtered. { 'vī,brād·ər 'paů·ər sə,plī }

vibrator-type inverter [ELEC] A device that uses a vibrator and an associated transformer or other inductive device to change direct-current input power to alternating-current output power. { 'vī,brād·ər ˌtīp in'vərd·ər }

vibratory centrifuge [MECH ENG] A high-speed rotating device to remove moisture from pulverized coal or other solids. { 'vī·brə,tür·ē 'sen·trə,fyüj }

vibratory equipment [MECH ENG] Reciprocating or oscillating devices which move, shake, dump, compact, settle, tamp, pack, screen, or feed solids or slurries in process. { 'vī·brə,tór·ē i'kwip·mənt }

vibratory hammer [MECH ENG] A type of pile hammer which uses electrically activated eccentric cams to vibrate piles into place. { 'vī·brə,tór·ē 'ham·ər }

vibratory saw [MED] A hand-operated saw, used in surgery, that can cut through hard materials such as bone or a cast, but leaves soft tissue such as skin or muscle untouched. { 'vī·brə,tór·ē 'só }

Vibrionaceae [MICROBIO] A family of gram-negative, facultatively anaerobic rods; cells are straight or curved and usually motile by polar flagella; generally found in water. { ˌvib·rē·ō'nās·ē,ē }

vibriosis [VET MED] An infectious bacterial disease, primarily of cattle, sheep, and goats, caused by *Vibrio fetus* and characterized by abortion, retained placenta, and metritis. { ˌvib·rē'ō·səs }

vibrissae [VERT ZOO] Hairs with specialized erectile tissue; found on all mammals except humans. Also known as sinus hairs; tactile hairs; whiskers. { vī'bri,sē }

vibroenergy separator [MECH ENG] A screen-type device for classification or separation of grains of solids by a combination of gyratory motion and auxiliary vibration caused by balls bouncing against the lower surface of the screen cloth. { ˌvī·brō'en·ər·jē 'sep·ə,rād·ər }

vibrograph [ENG] An instrument that provides a complete oscillographic record of a mechanical vibration; in one form a

moving stylus records the motion being measured on a moving paper or film. { 'vī·brə,graf }

vibrometer [ENG] An instrument designed to measure the amplitude of a vibration. Also known as vibration meter. { vī'bräm·əd·ər }

vibronic transition See vibrational transition. { vī,brän·ik tran'zish·ən }

vibrotron [ELECTR] A triode electron tube having an anode that can be moved or vibrated by an externally applied force. { 'vī·brə,trän }

vic- [ORG CHEM] A chemical prefix indicating vicinal (neighboring or adjoining) positions on a carbon structure (ring or chain); used to identify the location of substituting groups when naming derivatives. { vik }

vicariants [ECOL] Two or more closely related taxa, presumably derived from one another or from a common immediate ancestor, that inhabit geographically distinct areas. { vī'kar·ē·əns }

Vicat needle [ENG] An apparatus used to determine the setting time of cement by measuring the pressure of a special needle against the cement surface. { vē'kä ˌnēd·əl }

vicinal [ORG CHEM] Referring to neighboring or adjoining positions on a carbon structure (ring or chain). { 'vis·ən·əl }

vicinal faces [CRYSTAL] Macroscopic crystal faces which are inclined only a few minutes of arc to crystal faces with low Miller indices, and which therefore must have high Miller indices themselves. { 'vis·ən·əl ˌfās·əz }

Vickers hardness test See diamond-pyramid hardness test. { 'vik·ərz 'härd·nəs ˌtest }

Victaulic coupling [DES ENG] A development in which a groove is cut around each end of pipe instead of the usual threads; two ends of pipe are then lined up and a rubber ring is fitted around the joint; two semicircular bands, forming a sleeve, are placed around the ring and are drawn together with two bolts, which have a ridge on both edges to fit into the groove of the pipe; as the bolts are tightened, the rubber ring is compressed, making a watertight joint, while the ridges fitting in the grooves make it strong mechanically. { vik'tól·ik 'kəp·liŋ }

Victoria blue [ORG CHEM] $C_{33}H_{31}N_3 \cdot HCl$ Bronze crystals, soluble in hot water, alcohol, and ether; used as a dye for silk, wool, and cotton, as a biological stain, and to make pigment toners. { vik'tór·ē·ə 'blü }

vicuna [VERT ZOO] *Lama vicugna.* A rare, wild ruminant found in the Andes mountains; the fiber of the vicuna is strong, resilient, and elastic but is the softest and most delicate of animal fibers. { vī'kün·yə }

video [ELECTR] **1.** Pertaining to picture signals or to the sections of a television system that carry these signals in either unmodulated or modulated form. **2.** Pertaining to the demodulated radar receiver output that is applied to a radar indicator. { 'vid·ē·ō }

video adapter [COMPUT SCI] A printed circuit board that is plugged into a computer and generates the text and graphics images on a monitor. Also known as video board; video display board. { 'vid·ē·ō ə,dap·tər }

video amplifier [ELECTR] A low-pass amplifier having a band width on the order of 2–10 megahertz, used in television and radar transmission and reception; it is a modification of an *RC*-coupled amplifier, such that the high-frequency half-power limit is determined essentially by the load resistance, the internal transistor capacitances, and the shunt capacitance in the circuit. { 'vid·ē·ō 'am·plə,fī·ər }

video board See video adapter. { 'vid·ē·ō ,bórd }

videocassette [ELECTR] A compact plastic case containing a magnetic tape for video recording and playing. { ˌvid·ē·ō,kə'set }

videocassette recorder [ELECTR] A device for video recording and playing of magnetic tapes that are contained in plastic cases. Abbreviated VCR. { ˌvid·ē·ō,kə'set ri,kórd·ər }

videoconference [COMMUN] A teleconference that employs some type of television camera or other video equipment. { ˌvid·ē·ō 'kän·frəns }

videoconferencing [COMPUT SCI] Two-way interactive, digital communication through video streaming on the Internet, or by communications satellite, video telephone, and so forth. { ˌvid·ē·ō' kän·frəns·iŋ }

video correlator [ELECTR] Radar circuit that enhances automatic target detection capability, provides data for digital target plotting, and gives improved immunity to noise, interference, and jamming. { 'vid·ē·ȯ 'kär·ə lād·ər }

video data digital processing [COMMUN] Digital processing of video signals for pictures transmitted over a television link; the computer compares each scanned line with adjacent lines and eliminates extreme changes caused by electromagnetic interference. { 'vid·ē·ȯ ¦dad·ə 'dij·əd·əl 'prä,ses·iŋ }

video dial-tone [COMMUN] A service that enables a subscriber to select a video information provider from among many such providers offering service in a neighborhood, and to access on demand, from this provider, a movie or multimedia content, or similar services such as games and home shopping. { ¸vid·ē·ȯ 'dīl,tōn }

video discrimination [ELECTR] Radar circuit used to reduce the frequency band of the video amplifier stage in which it is used. { 'vid·ē·ȯ di,skrim·ə'nā·shən }

video disk recorder [ELECTR] A video recorder that records television visual signals and sometimes aural signals on a magnetic, optical, or other type of disk which is usually about the size of a long-playing phonograph record. { 'vid·ē·ȯ ¦disk ri,kȯrd·ər }

video disk storage See optical disk storage. { 'vid·ē·ȯ ¦disk ,stȯr·ij }

video display board See video adapter. { ¦vid·ē·ȯ di'splä ,bȯrd }

video display terminal See display terminal. { 'vid·ē·ȯ di'splä ,tər·mən·əl }

video frequency [COMMUN] One of the frequencies existing in the output of a television camera when an image is scanned; it may be any value from almost zero to well over 4 megahertz. { 'vid·ē·ȯ 'frē·kwən·sē }

video game [ELECTR] A form of interactive entertainment in which the player responds to electronically generated images that appear on a video display screen. { 'vid·ē·ȯ ,gām }

video integrator [ELECTR] **1.** Electric counter-countermeasures device that is used to reduce the response to nonsynchronous signals such as noise, and is useful against random pulse signals and noise. **2.** Device which uses the redundancy of repetitive signals to improve the output signal-to-noise ratio, by summing the successive video signals. { 'vid·ē·ȯ 'int·ə,grād·ər }

videomagnetograph [ENG] A sensitive and accurate device for measuring the strength and sign of solar magnetic fields, using the signal that results when successive images in right- and left-circularly polarized light are subtracted; the images are taken in the wing of a spectral line, using a birefringent filter. { ¸vid·ē·ō·mag'ned·ə,graf }

video masking [ELECTR] Method of removing chaff echoes and other extended clutter from radar displays. { 'vid·ē·ȯ 'mask·iŋ }

videomicroscopy [OPTICS] The use of television cameras to brighten magnified images that are otherwise too dark to be seen with the naked eye. { ¦vid·ē·ō·mī'kräs·kə·pē }

video monitor [COMPUT SCI] The cathode-ray-tube screen of a video display terminal. Also known as display screen; monitor. { 'vid·ē·ō 'män·əd·ər }

videophone See video telephone. { 'vid·ē·ə,fōn }

video player [ELECTR] A player that converts a video disk, videotape, or other type of recorded television program into signals suitable for driving a home television receiver. { 'vid·ē·ō ,plā·ər }

video RAM [COMPUT SCI] Dynamic random-access memory optimized for use with video displays. { ¸vid·ē·ō 'ram }

video recorder [ELECTR] A magnetic tape recorder capable of storing the video signals for a television program and feeding them back later to a television transmitter or directly to a receiver. { 'vid·ē·ō ri,kȯrd·ər }

video replay [ELECTR] Also known as videotape replay. **1.** A procedure in which the audio and video signals of a television program are recorded on magnetic tape and then the tape is run through equipment later to rebroadcast the live scene. **2.** A similar procedure in which the scene is rebroadcast almost immediately after it occurs. Also known as instant replay. { 'vid·ē·ō 'rē,plā }

video sensing [COMPUT SCI] In optical character recognition, a scanning technique in which the document is flooded with light from an ordinary light source, and the image of the character is reflected onto the face of a cathode-ray tube, where it is scanned by an electron beam. { 'vid·ē·ō 'sens·iŋ }

video signal [COMMUN] In television, the signal containing all of the visual information together with blanking and synchronizing pulses. [ELECTROMAG] See target signal. { 'vid·ē·ō ,sig·nəl }

videotape [ELECTR] A magnetic tape designed primarily for recording of television programs. { 'vid·ē·ō ,tāp }

videotape recorder [ELECTR] A device for video recording and playing of a magnetic tape either in a video cassette or on an open reel. { ¦vid·ē·ō'tāp ri,kȯrd·ər }

videotape recording [ELECTR] A method of recording television programs on magnetic tape for later rebroadcasting or replay. Abbreviated VTR. { 'vid·ē·ō ¦tāp ri,kȯrd·iŋ }

videotape replay See video replay. { 'vid·ē·ō ¦tap 'rē,plā }

video telephone [COMMUN] A communication instrument which transmits visual images along with the attendant speech. Also known as videophone. { 'vid·ē·ō 'tel·ə,fōn }

videotex [COMMUN] An electronic home information delivery system, either teletext or videotext. { 'vid·ē·ō,teks }

videotext [COMMUN] A computer communication service which uses information from a database, and which allows the user, equipped with a limited computer terminal, to interact with the service in selecting information to be displayed, so as to provide electronic mail, teleshopping, financial services, calculation services, and such. { ¦vid·ē·ō'tekst }

video transformer [ELECTR] A transformer designed to transfer, from one circuit to another, the signals containing picture information in television. { 'vid·ē·ō tranz'fȯr·mər }

video transmitter See visual transmitter. { 'vid·ē·ō tranz 'mid·ər }

vidicon [ELECTR] A camera tube in which a charge-density pattern is formed by photoconduction and stored on a photoconductor surface that is scanned by an electron beam, usually of low-velocity electrons; used chiefly in industrial television cameras. { 'vid·ə,kän }

Vienna definition language [COMPUT SCI] A language for defining the syntax and semantics of programming languages; consists of a syntactic metalanguage for defining the syntax of programming and data structures, and a semantic metalanguage which specifies programming language semantics operationally in terms of the computations to which programs give rise during execution. { vē'en·ə ,def·ə'nish·ən ,laŋ·gwij }

view camera [OPTICS] A camera that can be focused at both front and back, with adjustments for tilts, swings, shifts, and rise and fall, to control the shape of the subject in the image; it has a groundglass on the back which enables the photographer to view the image to be recorded. { 'vyü ,kam·rə }

viewfinder [ELECTR] An auxiliary optical or electronic device attached to a television camera so the operator can see the scene as the camera sees it. [OPTICS] A device which provides the user of a camera with the view of the subject that is focused by the lens. { 'vyü,fīn·dər }

viewing screen See screen. { 'vyü·iŋ ,skrēn }

viewing storage tube See direct-view storage tube. { 'vyü·iŋ 'stȯr·ij ,tüb }

viewing time [ELECTR] Time during which a storage tube is presenting a visible output corresponding to the stored information. { 'vyü·iŋ ,tīm }

viewport See window. { 'vyü,pȯrt }

view printer [GRAPHICS] In microfilm technology, a reader with built-in facilities to expose and process enlargements. { 'vyü ,print·ər }

vigia [NAV] **1.** A rock or shoal whose existence or position is doubtful. **2.** A warning note of a vigia condition on a chart. { və'hē·ə }

vignette [GRAPHICS] To lighten the edges of a photograph or a halftone illustration so it gradually disappears into the surrounding white of the paper, leaving no sharp edge. { vin'yet }

vignetting [OPTICS] Reduction in intensity of illumination near the edges of an optical instrument's field of view caused by obstruction of light rays by the edge of the aperture. { vin'yed·iŋ }

vigoureux printing [TEXT] Printing worsted fibers before spinning in order to achieve a mixture of light and dark shades in the finished fabric. { ¦vē·gə¦rü ,print·iŋ }

Vigreaux column [ANALY CHEM] An obsolete apparatus used in laboratory fractional distillation; it is a long glass tube

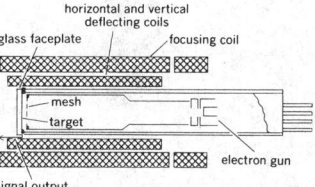

VIDICON

horizontal and vertical deflecting coils
glass faceplate
focusing coil
mesh
target
electron gun
signal output

Cross section of vidicon tube and its associated deflection and focusing coils. Target consists of transparent signal electrode deposited on faceplate of tube and thin layer of photoconductive material deposited over electrode. Mesh ensures uniform deceleration and collimation of electron beam.

with indentation in its walls; a thermometer is placed at the top of the tube and a side arm is attached to a condenser. { ve'grō ˌkäl·əm }

Villari effect [PHYS] A change of magnetic induction within a ferromagnetic substance in a magnetic field when the substance is subjected to mechanical stress. { və'lär·ē iˌfekt }

Villari reversal [PHYS] A change in the sign of the Villari effect which occurs with some ferromagnetic materials when the magnetic field strength reaches a certain value. { və'lär·ē riˌvər·səl }

villiaumite [MINERAL] NaF A carmine, isometric mineral consisting of sodium fluoride; occurs in masssive form. { vē'yō‚mīt }

villous [BOT] Having a surface covered with long, soft, shaggy hairs. { 'vil·əs }

villous adenoma [MED] A slow-growing, potentially malignant neoplasm of the mucosa of the rectum; manifested by bleeding and mucoid diarrhea. { 'vil·əs ˌad·ən'ō·mə }

villous placenta See epitheliochorial placenta. { 'vil·əs plə'sen·tə }

villous tenosynovitis [MED] A chronic inflammatory reaction of a tendon sheath producing hypertrophy of the lining, with the formation of redundant folds and villi. { 'vil·əs ˌten·ō‚sin·ə'vīd·əs }

villus [ANAT] A fingerlike projection from the surface of a membrane. { 'vil·əs }

vinal [TEXT] Any manufactured fiber made of a long-chain synthetic polymer composed of at least 50% by weight of vinyl alcohol units, with the total of the vinyl alcohol units and various acetate units being at least 85% by weight of the fiber. { 'vīn·əl }

vinasse [MATER] Residue from the fermentation of molasses or grapes; used as a fertilizer and a source of potassium salts. Also known as schlempe. { və'nas }

vinblastine [PHARM] $C_{46}H_{58}O_9N_4$ An alkaloid obtained from the periwinkle plant (Vinca rosea) and used, as the sulfate salt, as an antineoplastic drug. { vin'blaˌstēn }

Vinca rosea [BOT] A low, creeping evergreen perennial. Also known as periwinkle. { ˌviŋ·kə 'rō·zē·ə }

Vincent's angina See Vincent's infection. { 'vin·səns 'an·jə·nə }

Vincent's infection [MED] A noncontagious bacterial infection of the oral mucosa characterized by ulceration and formation of a gray pseudomembrane; caused by certain fusiform bacteria and spirochetes; formerly known as Vincent's angina. Also known as fusospirochetosis; Plaut-Vincent's infection; trench mouth; ulcerative gingivitis. { 'vin·səns inˌfek·shən }

vincristine [PHARM] $C_{46}H_{56}O_{10}N_4$ An alkaloid extracted from the periwinkle plant (Vinca rosea) and used, as the sulfate salt, as an antineoplastic drug. Also known as leurocristine. { vin'kriˌstēn }

Vindobonian [GEOL] A European stage of geologic time, middle Miocene. { ˌvin·də'bō·nē·ən }

vine [BOT] A plant having a stem that is too flexible or weak to support itself. { vīn }

vinegar [MATER] The product of the incomplete oxidation to acetic acid of ethyl alcohol produced by a primary fermentation of vegetable materials; contains not less than 4 grams of acetic acid per gallon; used in preparation of pickled fruits and vegetables and in salad dressing. { 'vin·ə·gər }

vinegar bacteria See Acetobacter. { 'vin·ə·gər bakˌtir·ē·ə }

vinegar eel [INV ZOO] Turbatrix aceti. A very small nematode often found in large numbers in vinegar fermentation. Also known as vinegar worm. { 'vin·ə·gər ˌēl }

vinegar worm See vinegar eel. { 'vin·ə·gər ˌwərm }

vinetine See oxyacanthine. { 'vin·əˌtēn }

viniculture See viticulture. { 'vin·iˌkəl·chər }

vinyl acetal resin [ORG CHEM] $[CH_2CH(OC_2H_5)]_x$ A colorless, odorless, light-stable thermoplastic that is unaffected by water, gasoline, or oils; soluble in lower alcohols, benzene, and chlorinated hydrocarbons; used in lacquers, coatings, and molded objects. Also known as polyvinyl acetal resin. { 'vīn·əl 'as·əˌtal 'rez·ən }

vinyl acetate [ORG CHEM] $CH_3COOCH:CH_2$ A colorless, water-insoluble, flammable liquid that boils at 73°C; used as a chemical intermediate and in the production of polymers and copolymers (for example, the polyvinyl resins). { 'vīn·əl 'as·əˌtāt }

vinyl acetate resin [ORG CHEM] $(CH_2:CHOOCCH_3)_x$ An odorless thermoplastic formed by the polymerization of vinyl acetate; resists attack by water, gasoline, and oils; soluble in lower alcohols, benzene, and chlorinated hydrocarbons; used in lacquers, coatings, and molded products. { 'vīn·əl 'as·əˌtāt 'rez·ən }

vinylacetylene [ORG CHEM] $H_2CCHCCH$ A combustible dimer of acetylene, boiling at 5°C; used for the manufacture of neoprene rubber and as a chemical intermediate. { ‚vīn·ələ'sed·əlˌēn }

vinyl alcohol [ORG CHEM] $CH_2:CHOH$ A flammable, unstable liquid found only in ester or polymer form. Also known as ethenol. { 'vīn·əl 'al·kəˌhòl }

vinylation [CHEM] Formation of a vinyl-derived product by reaction with acetylene; for example, vinylation of alcohols gives vinyl ethers, such as vinyl ethyl ether. { ‚vīn·əl'ā·shən }

vinylbenzene See styrene. { ‚vīn·əl'benˌzēn }

vinyl chloride [ORG CHEM] $CH_2:CHCl$ A flammable, explosive gas with an ethereal aroma; soluble in alcohol and ether, slightly soluble in water; boils at −14°C; an important monomer for polyvinyl chloride and its copolymers; used in organic synthesis and in adhesives. Also known as chloroethene; chloroethylene. { 'vīn·əl 'klòrˌīd }

vinyl chloride resin [ORG CHEM] $(CH_2CHCl)_x$ A white-power polymer made by the polymerization of vinyl chloride; used to make chemical-resistant pipe (when unplasticized) or bottles and parts (when plasticized). { 'vīn·əl 'klòrˌīd ˌrez·ən }

vinylcyanide See acrylonitrile. { ‚vīn·əl'sī·əˌnīd }

vinyl ether [ORG CHEM] $CH_2:CHOCH:CH_2$ A colorless, light-sensitive, flammable, explosive liquid; soluble in alcohol, acetone, ether, and chloroform, slightly soluble in water; boils at 39°C; used as an anesthetic and a comonomer in polyvinyl chloride polymers. Also known as divinyl ether; divinyl oxide. { 'vīn·əl 'ē·thər }

vinyl ether resin [ORG CHEM] Any of a group of vinyl ether polymers; for example, polyvinyl methyl ether, polyvinyl ethyl ether, and polyvinyl butyl ether. { 'vīn·əl 'ē·thər 'rez·ən }

vinyl group [ORG CHEM] $CH_2=CH-$ A group of atoms derived when one hydrogen atom is removed from ethylene. { 'vīn·əl ˌgrüp }

vinylidene chloride [ORG CHEM] $CH_2:CCl_2$ A colorless, flammable, explosive liquid, insoluble in water; boils at 37°C; used to make polymers copolymerized with vinyl chloride or acrylonitrile (Saran). { vī'nil·əˌdēn 'klòrˌīd }

vinylidene resin [ORG CHEM] A polymer made up of the $(-H_2CCX_2-)$ unit, with X usually a chloride, fluoride, or cyanide radical. Also known as polyvinylidene resin. { vī'nil·əˌdēn 'rez·ən }

vinylog [ORG CHEM] Any of the organic compounds that differ from each other by a vinylene linkage $(-CH=CH-)$; for example, ethyl crotonate is a vinylog of ethyl acetate and of the next higher vinylog, ethyl sorbate. { 'vīn·əlˌäg }

vinyl plastic See polyvinyl resin. { 'vīn·əl 'plas·tik }

vinyl polymerization [ORG CHEM] Addition polymerization where the unsaturated monomer contains a $CH_2=C$ group. { 'vīn·əl pəˌlim·ə·rə'zā·shən }

vinylpyridine [ORG CHEM] $C_5H_4NCH:CH_2$ A toxic, combustible liquid; soluble in water, alcohol, hydrocarbons, esters, ketones, and dilute acids; used to manufacture elastomers and pharmaceuticals. { ‚vīn·əl'pir·əˌdēn }

N-vinyl-2-pyrrolidone [ORG CHEM] C_6H_9ON A colorless, toxic, combustible liquid, boiling at 148°C (100 mmHg); used as a chemical intermediate and to make polyvinyl pyrrolidone. { en ‚vīn·əl ‚tü pə'räl·əˌdōn }

vinylstyrene See divinylbenzene. { ‚vīn·əl'stīˌrēn }

vinyltoluene [ORG CHEM] $CH_2:CHC_6H_4CH_3$ A colorless, flammable, moderately toxic liquid; soluble in ether and methanol, slightly soluble in water; boils at 170°C; used as a chemical intermediate and solvent. Also known as methyl styrene. { ‚vīn·əl'täl·yəˌwēn }

vinyl trichloride See trichloroethane. { 'vīn·əl trī'klòrˌīd }

vinyl trichlorosilone [ORG CHEM] $CH_2CH_3SiCl_3$ A liquid that boils at 90.6°C and is soluble in organic solvents; used in silicones and adhesives. { 'vīn·əl trī‚klòr·ō'siˌlōn }

vinyon [TEXT] Any manufactured fiber made of a long-chain synthetic polymer composed of at least 85% by weight of vinyl chloride units. { 'vinˌyän }

Violaceae [BOT] A family of dicolyledonous plants in the order Violales characterized by polypetalous, mostly perfect, hypogynous flowers with a single style and five stamens. { ˌvī·ə'lās·ēˌē }

Violales [BOT] A heterogeneous order of dicotyledons in the subclass Dilleniidae distinguished by a unilocular, compound ovary and mostly parietal placentation. { ˌvī·ə'lā·lēz }

violarite [MINERAL] Ni_2FeS_4 A violet-gray mineral of the linnaeite group consisting of a sulfide of nickel and iron; found in meteorites. { vī'ō·ləˌrīt }

violet [OPTICS] The hue evoked in an average observer by monochromatic radiation having a wavelength in the approximate range from 390 to 455 nanometers; however, the same sensation can be produced in a variety of other ways. { 'vī·ə·lət }

violet layer [ASTRON] A layer of particles in the upper Martian atmosphere that scatter and absorb electromagnetic radiation at shorter wavelengths, making the atmosphere opaque to blue, violet, and ultraviolet light. { 'vī·ə·lət ˌlā·ər }

violle [OPTICS] A unit of luminous intensity, equal to the luminous intensity of 1 square centimeter of platinum at its temperature of solidification; it is found experimentally to be equal to 20.17 candelas. { vyól }

viologen [CHEM] Any member of a group of chlorides of certain quaternary bases derived from γ,γ'-dipyridyl that are used as oxidation-reduction indicators; color is exhibited in the reduced form. { vī'äl·ə·jən }

viologen display [ELECTR] An electrochromic display based on an electrolyte consisting of an aqueous solution of a dipositively charged organic salt, containing a colorless cation that undergoes a one-electron reduction process to produce a purple radical cation, upon application of a negative potential to the electrode. { vē'äl·ə·jən diˌsplā }

viomycin [MICROBIO] A polypeptide antibiotic or mixture of antibiotic substances produced by strains of *Streptomyces griseus* var. *purpureus* (*Streptomyces puniceus*); the sulfate salt is administered intramuscularly for treatment of tuberculosis resistant to other therapy. { ˌvī·ə'mīs·ən }

viper [VERT ZOO] The common name for reptiles of the family Viperidae; thick-bodied poisonous snakes having a pair of long fangs, present on the anterior part of the upper jaw, which fold along the roof of the mouth when the jaws are closed. { 'vī·pər }

Viperidae [VERT ZOO] A family of reptiles in the suborder Serpentes found in Eurasia and Africa; all species are proglyphodont. { vī'per·əˌdē }

VIR See vertical interval reference.

viral encephalomyelitides [MED] A group of several encephalitis diseases caused by various viruses; includes epidemic encephalitis, equine encephalitides, and Japanese B encephalitis. { 'vī·rəl inˌsef·ə·lōˌmī·ə'lid·ə·dēz }

viral gastroenteritis [MED] An acute infectious gastroenteritis thought to be caused by various viruses and characterized by diarrhea, nausea, vomiting, and variable systemic symptoms. { 'vī·rəl ˌgas·trōˌent·ə'rīd·əs }

viral hepatitis [MED] A type of hepatitis caused by two distinct viruses, A and B; type A is also known as infectious hepatitis, type B as serum hepatitis. { 'vī·rəl ˌhep·ə'tīd·əs }

viral pneumonia [MED] A form of pneumonia caused by a virus of various types, in which the inflammatory reaction predominates in the septa, and the alveoli contain fibrin, edema fluid, and some inflammatory cells. { 'vī·rəl nù'mō·nyə }

viral shedding [VIROL] Excretion of virus from a specific site in the body or from a lesion. { 'vī·rəl 'shed·iŋ }

virazon [METEOROL] **1.** The very strong southwesterly sea breeze experienced where the coastal chains of the Andes Mountains descend steeply to the sea; it sets in about 10 a.m. and reaches its greatest strength at about 3 p.m. **2.** A westerly sea breeze of Spain and Portugal. { vir·ə'zón }

viremia [MED] Presence of viral particles in the blood. { vī'rē·mē·ə }

Vireonidae [VERT ZOO] The vireos, a family of New World passeriform birds in the suborder Oscines. { ˌvir·ē'än·əˌdē }

virga [METEOROL] Wisps or streaks of water or ice particles falling out of a cloud but evaporating before reaching the earth's surface as precipitation. Also known as fall streaks; Fallstreifen; precipitation trails. { 'vər·gə }

virgate [BOT] Banded. { 'vərˌgāt }

virgate trophus [INV ZOO] A piercing type of trophus in rotifers that is thin and slightly toothed. { 'vərˌgāt 'träf·əs }

virgin See straight-run. { 'vər·jən }

Virgin See Virgo. { 'vər·jən }

virgin medium [COMPUT SCI] A material designed to have data recorded on it which is as yet completely lacking any information, such as a paper tape without any punched holes, not even feed holes; in contrast to an empty medium. { 'vər·jən 'mēd·ē·əm }

virgin neutron [NUCLEO] A neutron from any source, before it makes a collision. { 'vər·jən 'nü,trän }

virgin stock [MATER] A petroleum-derived liquid stream processed from natural (virgin) crude oil; it contains no cracked or otherwise chemically modified material. Also known as straight-run stock. { 'vər·jən 'stäk }

virgin wool [TEXT] Wool used for the first time, directly after clipping from sheep. { 'vər·jən 'wúl }

Virgo [ASTRON] A constellation, right ascension 13 hours, declination 0°. Also known as Virgin. { 'vər·gō }

Virgo A [ASTRON] A radio galaxy; it is associated with the galaxy M 87 (NGC 4486). { 'vər·gō 'ā }

Virgo cluster [ASTRON] A cluster of galaxies which is the nearest to the galaxy that includes the sun; the cluster is centered in the constellation Virgo and is about 1.6×10^7 light-years (1.51×10^{23} m) from earth. { 'vər·gō ˌkləs·tər }

virgo-forcipate trophus [INV ZOO] A type of muscular chamber in rotifers containing jaws of a cuticular material intermediate between a piercing and grasping type of structure. { 'vər·gō 'fórˌsə·pət ˌträf·əs }

Virgo supercluster See Local supercluster. { 'vər·gō 'sü·pərˌkləs·tər }

Virgo X-1 [ASTRON] An x-ray source that is identical with Virgo A. { 'vər·gō 'eks·wən }

virial coefficients [THERMO] For a given temperature T, one of the coefficients in the expansion of P/RT in inverse powers of the molar volume, where P is the pressure and R is the gas constant. { 'vir·ē·əl ˌkō·i'fish·əns }

virial of a system [STAT MECH] The average over a long period of time of $-1/2$ the sum over the particles in the system of the scalar product of the total force acting on the particle and its radius vector. { 'vir·ē·əl əv ə 'sis·təm }

virial theorem See Clausius virial theorem. { 'vir·ē·əl ˌthir·əm }

virial-theorem mass [ASTRON] The mass of a cluster of stars or galaxies calculated from the observed mean-square velocity of the objects and application of the virial theorem. { vir·ē·əl 'thir·əm ˌmas }

viridans streptococci [MICROBIO] A group of pathogenic and saprophytic streptococci including strains not causing beta hemolysis, although many cause alpha hemolysis, and none which elaborate a C substance. { 'vir·əˌdanz ˌstrep·tə'käk·ē }

virilism [MED] See gynandry. [PSYCH] **1.** Masculinity. **2.** Manifestation of male behavioral patterns in the female. { 'vir·əˌliz·əm }

virion [VIROL] The complete, mature virus particle. { 'vir·ēˌän }

Virmel engine [MECH ENG] A cat-and-mouse engine that employs vanelike pistons whose motion is controlled by a gear-and-crank system; each set of pistons stops and restarts when a chamber reaches the spark plug. { 'vər'mel ˌen·jən }

viroid [MICROBIO] The smallest known agents of infectious disease, characterized by the absence of encapsidated proteins. { 'vīˌroid }

virology [MICROBIO] The study of submicroscopic organisms known as viruses. { vī'räl·ə·jē }

virotoxin [BIOCHEM] One of a group of toxins present in the mushroom *Amanita virosa*. { ˌvī·rəˌtäk·sən }

virtual acoustics [ENG ACOUS] Digitally processing sounds so that they appear to come from particular locations in three-dimensional space, with the goal of simulating the complex acoustic field experienced by a listener within a natural environment. Also known as auralization; three-dimensional sound. { ˌvər·chə·wəl ə'küs·tiks }

virtual address [COMPUT SCI] A symbol that can be used as a valid address part but does not necessarily designate an actual location. { 'vər·chə·wəl 'adˌres }

virtual cathode [ELECTR] The locus of a space-charge-potential minimum such that only some of the electrons approaching it are transmitted, the remainder being reflected

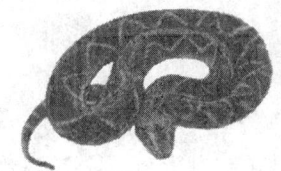

VIPER

The fer-de-lance (*Bothrops atrox*).

VIRGATE TROPHUS

Virgate trophus of *Notommata*. (*a*) Ventral view and (*b*) upper view, showing thin, slightly toothed structure.

VIRGO

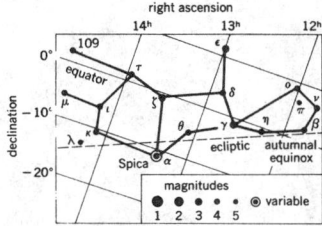

Line pattern of the constellation Virgo. The grid lines represent the coordinates of the sky. The apparent brightness, or magnitude, of the stars is shown by the size of the dots, graded by appropriate numbers as indicated.

back to the electron-emitting cathode. { 'vər·chə·wəl 'kath‚ōd }

virtual decimal point *See* assumed decimal point. { 'vər·chə·wəl 'des·məl ‚pȯint }

virtual direct-access storage [COMPUT SCI] A device used with mass-storage systems, whereby data are retrieved prior to usage by a batch-processing program and automatically transcribed onto disk storage. { 'vər·chə·wəl də‚rekt ‚ak‚ses 'stȯr‚ij }

virtual displacement [MECH] **1.** Any change in the positions of the particles forming a mechanical system. **2.** An infinitesimal change in the positions of the particles forming a mechanical system, which is consistent with the geometrical constraints on the system. { 'vər·chə·wəl di'splās·mənt }

virtual entropy [THERMO] The entropy of a system, excluding that due to nuclear spin. Also known as practical entropy. { 'vər·chə·wəl 'en·trə‚pē }

virtual environment *See* virtual reality. { 'vər·chə·wəl in'vī·rən·mənt }

virtual gravity [METEOROL] The force of gravity on a parcel of air, reduced by centrifugal force due to the motion of the parcel relative to the earth. { 'vər·chə·wəl 'grav·əd·ē }

virtual height [GEOPHYS] The apparent height of a layer in the ionosphere, determined from the time required for a radio pulse to travel to the layer and return, assuming that the pulse propagates at the speed of light. Also known as equivalent height. { 'vər·chə·wəl 'hīt }

virtual image [OPTICS] An optical image from which rays of light only appear to diverge, without actually being focused there. { 'vər·chə·wəl 'im·ij }

virtual leak [ENG] The semblance of the vacuum system leak caused by a gradual desorptive release of gas at a rate which cannot be accurately predicted. { 'vər·chə·wəl 'lēk }

virtual level [NUC PHYS] The energy of a virtual state. { 'vər·chə·wəl ‚lev·əl }

virtual machine [COMPUT SCI] A portion of a computer system or of a computer's time that is controlled by an operating system and functions as though it were a complete system, although in reality the computer is shared with other independent operating systems. { 'vər·chə·wəl mə'shēn }

virtual manufacturing [IND ENG] The modeling of manufacturing systems using audiovisual or other sensory features to simulate or design an actual manufacturing environment, or the prototyping and manufacture of a proposed product mainly through effective use of computers; used to predict potential problems and inefficiencies in product functionality and manufacturability before real manufacturing occurs. { ‚vər·chə·wəl ‚man·ə'fak·chər·iŋ }

virtual memory [COMPUT SCI] A combination of primary and secondary memories that can be treated as a single memory by programmers because the computer itself translates a program or virtual address to the actual hardware address. { 'vər·chə·wəl 'mem·rē }

virtual meridian [NAV] The meridian in which the spin axis of a gyro compass will settle as a result of speed-course-latitude error. { 'vər·chə·wəl mə'rid·ē·ən }

virtual object [OPTICS] A collection of points which may be regarded as a source of light rays for a portion of an optical system but which does not actually have this function. { 'vər·chə·wəl 'äb·jekt }

virtual orbital [PHYS CHEM] An orbital that is either empty or unoccupied while in the ground state. { 'vər·chə·wəl 'ȯr·bəd·əl }

virtual particle *See* virtual quantum. { 'vər·chə·wəl 'pard·ə·kəl }

virtual PPI reflectoscope [ENG] A device for superimposing a virtual image of a chart on a plan position indicator (PPI) pattern; the chart is usually prepared with white lines on a black background to the scale of the plan position indicator range scale. { 'vər·chə·wəl ‚pē‚pē'ī ri'flek·tə‚skōp }

virtual pressure [METEOROL] The pressure of a parcel of moist air when it has the same density as a parcel of dry air at the same temperature. { 'vər·chə·wəl 'presh·ər }

virtual private network [COMMUN] A wide-area network whose links are provided by a common carrier although they appear to the users to behave like dedicated lines, and whose computers use a common cryptographic key to send messages from one computer in the network to another. Abbreviated VPN. { ‚vər·chə·wəl ‚prī·vət 'net‚wərk }

virtual process [QUANT MECH] A process which contributes in a stage of a theoretical model but is not, by itself, physically realizable. { 'vər·chə·wəl ‚präs·əs }

virtual quantum [QUANT MECH] A photon or other particle in an intermediate state which appears in matrix elements connecting initial and final states in second-and higher-order perturbation theory; energy is not conserved in the transitions to or from the intermediate state. Also known as virtual particle. { 'vər·chə·wəl 'kwän·təm }

virtual reality [COMPUT SCI] A simulation of an environment that is experienced by a human operator provided with a combination of visual (computer-graphic), auditory, and tactile presentations generated by a computer program. Also known as artificial reality; immersive simulation; virtual environment; virtual world. { ‚vər·chə·wəl rē'al·əd·ē }

Virtual Reality Modeling Language [COMPUT SCI] The markup specification for three-dimentional (virtual reality) objects and environments on the Web. Abbreviated VRML. { ‚vər·chə·wəl rē‚al·əd·ē 'mäd·əl·iŋ ‚laŋ·gwij }

virtual source [ACOUS] A source of sound which is composed of the same material as that in which the sound propagates and which does not have sharply delineated boundaries; such sources include thermal sources (such as lightning and thermoacoustic arrays), turbulence (as in rocket and jet engines), and sound itself (as in a parametric acoustic array). { 'vər·chə·wəl 'sȯrs }

virtual state [NUC PHYS] An unstable state of a compound nucleus which has a lifetime many times longer than the time it takes a nucleon, with the same energy as it has in the virtual state, to cross the nucleus. { 'vər·chə·wəl 'stāt }

virtual temperature [METEOROL] In a system of moist air, the temperature of dry air having the same density and pressure as the moist air. { 'vər·chə·wəl 'tem·prə·chər }

virtual work [MECH] The work done on a system during any displacement which is consistent with the constraints on the system. { 'vər·chə·wəl 'wərk }

virtual work principle *See* principle of virtual work. { 'vər·chə·wəl ‚wərk ‚prin·sə·pəl }

virtual world [COMPUT SCI] A navigable visual digital environment. { ‚vər·chə·wəl 'wərld }

virulence [MICROBIO] The disease-producing power of a microorganism; infectiousness. { 'vir·ə·ləns }

virulence cassette *See* pathogenicity island. { 'vir·ə·ləns kə‚set }

virus [COMPUT SCI] A computer program that replicates itself and transfers itself to another computing system. [VIROL] A large group of infectious agents ranging from 10 to 250 nanometers in diameter, composed of a protein sheath surrounding a nucleic acid core and capable of infecting all animals, plants, and bacteria; characterized by total dependence on living cells for reproduction and by lack of independent metabolism. { 'vī·rəs }

virus hepatitis *See* infectious hepatitis. { 'vī·rəs ‚hep·ə'tīd·əs }

virus interference [MICROBIO] A phenomenon which may be defined as protection of host cells against one virus, conferred as a result of prior infection with a different virus. { 'vī·rəs ‚in·tər'fir·əns }

visbreaking *See* viscosity breaking. { 'vis‚brāk·iŋ }

viscera [ANAT] The organs within the cavities of the body of an organism. { 'vis·ə·rə }

visceral arch [ANAT] One of the series of mesodermal ridges covered by epithelium bounding the lateral wall of the oral and pharyngeal regions of vertebrates; embryonic in higher forms, they contribute to formation of the face and neck. [VERT ZOO] One of the first two arches of the series in gill-bearing forms. { 'vis·ə·rəl 'ärch }

visceral leishmaniasis [MED] A severe, generalized, and often fatal infection, caused by any of three pathogenic hemoflagellates of the genus *Leishmania*, affecting organs rich in endothelial cells; accompanied by fever, spleen and liver enlargement, anemia, leukopenia, skin pigmentation, and changes in plasma protein. { 'vis·ə·rəl ‚lēsh·mə'nī·ə·səs }

visceral peritoneum [ANAT] That portion of the peritoneum covering the organs of the abdominal cavity. { 'vis·ə·rəl ‚per·ə·tə'nē·əm }

visceral pouch *See* pharyngeal pouch. { 'vis·ə·rəl 'pau̇ch }

visceratonia [PSYCH] The behavioral type assigned to the endomorphic somatotype, manifested by a desire for assimilation and the conservation of energy through social interactions, relaxation, and love of food. { ˌvis·ə·rə'tō·nē·ə }

visceromegaly [MED] Enlargement of the organs in the abdomen, such as the liver, spleen, pancreas, stomach, or kidneys. { ˌvis·ə·rō'meg·ə·lē }

visceroptosis [MED] Prolapse of a viscus, especially the intestine; downward displacement of the intestine in the abdominal cavity. Also known as enteroptosis. { ˌvis·ə·räp'tō·səs }

viscid [BOT] Having a sticky surface, as certain leaves. { 'vis·əd }

viscoelastic fluid [FL MECH] A fluid that displays viscoelasticity. { ˌvis·kō·iˌlas·tik 'flü·əd }

viscoelasticity [MECH] Property of a material which is viscous but which also exhibits certain elastic properties such as the ability to store energy of deformation, and in which the application of a stress gives rise to a strain that approaches its equilibrium value slowly. { ˌvis·kō·iˌlas'tis· əd·ē }

viscoelastic theory [MECH] The theory which attempts to specify the relationship between stress and strain in a material displaying viscoelasticity. { ˌvis·kō·iˌlas·tik 'thē·ə·rē }

viscometer [ENG] An instrument designed to measure the viscosity of a fluid. { vi'skäm·əd·ər }

viscometer gage [ENG] A vacuum gage in which the gas pressure is determined from the viscosity of the gas. { vi'skäm·əd·ər ˌgāj }

viscometric analysis [FL MECH] Measurement of the flow properties of substances by viscometry. { ˌvis·kəˌme·trik ə'nal·ə·səs }

viscometry [ENG] A branch of rheology; the study of the behavior of fluids under conditions of internal shear; the technology of measuring viscosities of fluids. { vi'skäm·ə·trē }

viscose process [CHEM ENG] A process for the manufacture of rayon by treating cellulose with caustic soda, and with carbon disulfide to form cellulose xanthate, which is then dissolved in a weak caustic solution to form the viscose; fibers are used as silk substitutes. { 'visˌkōs 'prä·səs }

viscosity [FL MECH] The resistance that a gaseous or liquid system offers to flow when it is subjected to a shear stress. Also known as flow resistance; internal friction. { vi'skäs·əd·ē }

viscosity blending chart [CHEM ENG] A graphical means for estimating the viscosity at a given temperature of a blend of petroleum products. { vi'skäs·əd·ē 'blend·iŋ ˌchärt }

viscosity breaking [CHEM ENG] A petroleum refinery process used to lower or break the viscosity of high-viscosity residuum by thermal cracking of molecules at relatively low temperatures. Also known as visbreaking. { vi'skäs·əd·ē 'brāk·iŋ }

viscosity coefficient [FL MECH] An empirical number used in equations of fluid mechanics to account for the effects of viscosity. { vi'skäs·əd·ē ˌkō·iˌfish·ənt }

viscosity conversion table [CHEM ENG] A table or chart with which kinematic viscosity, in centistokes, can be converted to Saybolt viscosity, in seconds, at the same temperature. { vi'skäs·əd·ē kən'vər·zhən ˌtā·bəl }

viscosity curve [FL MECH] A graph showing the viscosity of a liquid or gaseous material as a function of temperature. { vi'skäs·əd·ē ˌkərv }

viscosity gage See molecular gage. { vi'skäs·əd·ē ˌgāj }

viscosity-gravity constant [CHEM ENG] An index of the chemical composition of crude oil; defined as the general relation between specific gravity and Saybolt Universal viscosity; the constant is low for paraffinic crude oils, high for naphthenic crude oils. Abbreviated VGC. { vi'skäs·əd·ē 'grav·əd·ē ˌkän·stənt }

viscosity index [CHEM ENG] An arbitrary scale used to show the magnitude of viscosity changes in lubricating oils with changes in temperature. Abbreviated VI. { vi'skäs·əd·ē ˌin·deks }

viscosity manometer See molecular gage. { viˌskäs·əd·ē mə'näm·əd·ər }

viscosity-temperature chart [CHEM ENG] A chart with which the kinematic or Saybolt viscosity of a petroleum oil at any temperature within a limited range may be ascertained, provided viscosities at two temperatures are known. { vi'skäs·əd·ē 'tem·prə·chər ˌchärt }

viscous damping [MECH ENG] A method of converting mechanical vibrational energy of a body into heat energy, in which a piston is attached to the body and is arranged to move through liquid or air in a cylinder or bellows that is attached to a support. { 'vis·kəs 'damp·iŋ }

viscous dissipation function [FL MECH] A quadratic function of spatial derivatives of components of fluid velocity which gives the rate at which mechanical energy is converted into heat in a viscous fluid per unit volume. Also known as dissipation function. { 'vis·kəs ˌdis·ə'pā·shən ˌfəŋk·shən }

viscous drag [FL MECH] That part of the rearward force on an aircraft that results from the airstream carrying air forward with it through viscous adherence. { 'vis·kəs 'drag }

viscous-drag gas-density meter [ENG] A device to measure gas-mixture densities; driven impellers in sample and standard chambers create measurable turbulences (drags) against respective nonrotating impellers. { 'vis·kəs ˌdrag ˌgas ˌden·səd·ē ˌmēd·ər }

viscous fillers [MECH ENG] A packaging machine that fills viscous product into cartons; there are two basic types, straight-line and rotary plunger; the former operates intermittently on a given number of containers, while the latter fills and discharges containers continuously. { 'vis·kəs 'fil·ərz }

viscous filter [ENG] An air-cleaning filter having a surface coated with a viscous liquid to trap particulates in the airstream. { ˌvis·kəs ˌfil·tər }

viscous flow [FL MECH] **1.** The flow of a viscous fluid. **2.** The flow of a fluid through a duct under conditions such that the mean free path is small in comparison with the smallest transverse section of the duct. { 'vis·kəs 'flō }

viscous fluid [FL MECH] A fluid whose viscosity is sufficiently large to make the viscous forces a significant part of the total force field in the fluid. { 'vis·kəs 'flü·əd }

viscous force [FL MECH] The force per unit volume or per unit mass arising from viscous effects in fluid flow. { 'vis·kəs 'fòrs }

viscous impingement filter [ENG] A filter made up of a relatively loosely arranged medium, such that the airstream is forced to change direction frequently as it passes through the filter medium; the medium usually consists of spun-glass fibers, metal screens, or layers of crimped expanded metal whose surfaces are coated with a tacky oil. { 'vis·kəs im'pinj·mənt ˌfil·tər }

viscous lubrication See complete lubrication. { 'vis·kəs ˌlü·brə'kā·shən }

viscous magnetization See viscous remanent magnetization. { 'vis·kəs ˌmag·nəd·ə'zā·shən }

viscous neutral oil [MATER] The bottoms from reducing, by distillation, of a petroleum neutral oil fraction; such oils are frequently blended with bright stock to make finished oils of various viscosities. { 'vis·kəs ˌnü·trəl 'òil }

viscous remanent magnetization [GEOPHYS] A process in which grains of magnetic minerals, which are either too small or too finely divided by undergrowths of different chemical composition to retain a permanent magnetization indefinitely, acquire a new direction of magnetization when the direction of the earth's magnetic field changes. Abbreviated VRM. Also known as viscous magnetization. { 'vis·kəs 'rem·ə·nənt ˌmag·nəd·ə'zā·shən }

viscous sublayer [FL MECH] In a turbulent flow, a very thin region next to a wall, typically only 1% of the boundary layer thickness, where turbulent mixing is impeded and transport occurs partly or, if the limit as the wall is approached, entirely by viscous diffusion. { ˌvis·kəs 'səbˌlā·ər }

viscus [ANAT] Singular of viscera. { 'vis·kəs }

vise [DES ENG] A tool consisting of two jaws for holding a workpiece; opened and closed by a screw, lever, or cam mechanism. { vīs }

Viséan [GEOL] A European stage of lower Carboniferous geologic time forming the lowermost Upper Mississippian, above Tournaisian and below lower Namurian. { vi'sā·ən }

visibility [METEOROL] In weather observing practice, the greatest distance in a given direction at which it is just possible to see and identify with the unaided eye, in the daytime, a prominent dark object against the sky at the horizon and, at nighttime, a known, preferably unfocused, moderately intense light source. { ˌviz·ə'bil·əd·ē }

visibility factor [ELECTR] The ratio of the minimum signal input detectable by ideal instruments connected to the output

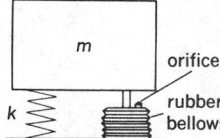

VISCOUS DAMPING

System employing viscous damping with air, used to reduce vibration of body with mass *m*, attached to spring with force constant *k*.

of a receiver, to the minimum signal power detectable by a human operator through a display connected to the same receiver. Also known as display loss. { ‚viz·ə'bil·əd·ē ‚fak·tər }

visibility function *See* luminosity function. { ‚viz·ə'bil·əd·ē ‚fəŋk·shən }

visibility meter [ENG] An instrument for making direct measurements of visual range in the atmosphere or of the physical characteristics of the atmosphere which determine the visual range. [OPTICS] A type of photometer that operates on the principle of artificially reducing the visibility of objects to threshold values (borderline of seeing and not seeing) and measuring the amount of the reduction on an appropriate scale. { ‚viz·ə'bil·əd·ē ‚mēd·ər }

visible absorption spectrophotometry [SPECT] Study of the spectra produced by the absorption of visible-light energy during the transformation of an electron from the ground state to an excited state as a function of the wavelength causing the transformation. { 'viz·ə·bəl əb'sórp·shən ‚spek·trō·fə'täm·ə·trē }

visible bearing [NAV] A bearing obtained by visual observation using a pelorus or azimuth circle. { 'viz·ə·bəl 'ber·iŋ }

visible horizon [ASTRON] That line where earth and sky appear to meet, and the projection of this line upon the celestial sphere. { 'viz·ə·bəl hə'rīz·ən }

visible radiation *See* light. { 'viz·ə·bəl ‚rād·ē·ā·shən }

visible spectrophotometry [SPECT] In spectrophotometric analysis, the use of a spectrophotometer with a tungsten lamp that has an electromagnetic spectrum of 380–780 nanometers as a light source, glass or quartz prisms or gratings in the monochromator, and a photomultiplier cell as a detector. { 'viz·ə·bəl ‚spek·trō·fə'täm·ə·trē }

visible spectrum [SPECT] **1.** The range of wavelengths of visible radiation. **2.** A display or graph of the intensity of visible radiation emitted or absorbed by a material as a function of wavelength or some related parameter. { 'viz·ə·bəl 'spek·trəm }

vision [PHYSIO] The sense which perceives the form, color, size, movement, and distance of objects. Also known as sight. { 'vizh·ən }

vision light [BUILD] A viewing window set in a fire door, usually glazed with wire glass. { 'vizh·ən ‚līt }

vision slit [ORD] A narrow opening or slit in armor to permit viewing, especially one in a tank or other armored vehicle. { 'vizh·ən ‚slit }

visor tin [MINERAL] Twin crystals of cassiterite characterized by a notch. { 'vī·zər 'tin }

visual achromatism [OPTICS] In an optical system, the removal of chromatic aberration or chromatic differences of magnification between light at the wavelength of the Fraunhofer C line at 656.3 nanometers and the F line at 486.1 nanometers in order to minimize these defects at wavelengths at which the human eye is most sensitive. Also known as optical achromatism. { 'vizh·ə·wəl ā'krō·mə‚tiz·əm }

visual acuity [PHYSIO] The ability to see fine details of an object; specifically, the ability to see an object whose angle subtended at the eye is 1 minute of arc. { 'vizh·ə·wəl ə'kyü·əd·ē }

visual aid to navigation [NAV] An object such as a tower or tripod target whose position is indicated on government-approved navigation charts and on which bearings may be taken by the naked eye or with optical assistance; the aid may or may not be illuminated. { 'vizh·ə·wəl 'ād tə ‚nav·ə'gā·shən }

visual angle [OPTICS] The angle which an object subtends at the nodal point of the eye of an observer. { 'vizh·ə·wəl ‚aŋ·gəl }

visual-aural range [NAV] A very-high-frequency radio range that provides one course for display to the pilot on a zero-center left-right indicator and another course, at right angles to the first, in the form of aural A-N radio range signals; the A-N aural signals provide a means for differentiating between the two directions of the visual course. Abbreviated VAR. { 'vizh·ə·wəl 'ór·əl 'rānj }

visual binaries [ASTRON] Binary stars that to the naked eye seem to be single stars, but when viewed through the telescope, are separated into pairs. Also known as visual doubles. { 'vizh·ə·wəl 'bī‚ner·ēz }

visual bombing [ORD] Bombing done by sighting on an

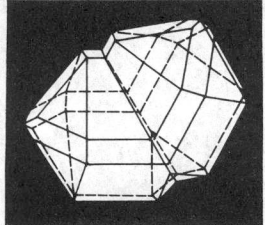

Visor tin with characteristic notch. (*From C. S. Hurlbut, Jr., Dana's Manual of Mineralogy, 17th ed., Wiley, 1959*)

aiming point, under conditions where the aiming point is visible from the bombing aircraft. { 'vizh·ə·wəl 'bäm·iŋ }

visual bombsight [ORD] A bombsight designed for aiming a bomb when the aiming point is visible. { 'vizh·ə·wəl 'bäm‚sīt }

visual capture [PSYCH] In perception, the dominance of vision over other sense modalities, such that what is felt or heard conforms to what is seen. { ‚vizh·ə·wəl 'kap·chər }

visual colorimetry [ANALY CHEM] A procedure for the determination of the color of an unknown solution by visual comparison to color standards (solutions or color-tinted disks). { 'vizh·ə·wəl ‚kal·ə'rim·ə·trē }

visual comparator *See* optical comparator. { 'vizh·ə·wəl kəm'par·əd·ər }

visual display unit *See* display tube. { 'vizh·ə·wəl di'splā ‚yü·nət }

visual doubles *See* visual binaries. { 'vizh·ə·wəl 'dəb·əlz }

visual fire control [ORD] Technical control of some artillery fire using an optical tracking instrument. { 'vizh·ə·wəl 'fīr kən‚trōl }

visual fix [NAV] A fix established by visual observation of fixed objects. { 'vizh·ə·wəl 'fiks }

visual flight [AERO ENG] An aircraft flight occurring under conditions which allow navigation by visual reference to the earth's surface at a safe altitude and with sufficient horizontal visibility, and operating under visual flight rules. Also known as VFR flight. { 'vizh·ə·wəl 'flīt }

visual flight rules [AERO ENG] A set of regulations set down by the U.S. Civil Aeronautics Board (in Civil Air Regulations) to govern the operational control of aircraft during visual flight. Abbreviated VFR. { 'vizh·ə·wəl 'flīt ‚rülz }

visual Herschel effect *See* Herschel effect. { 'vizh·ə·wəl 'hər·shəl i‚fekt }

visualization [COMPUT SCI] The process of converting data into a geometric or graphic representation. { ‚vizh·ə·lə'zā·shən }

visual learning [PSYCH] A type of sensory learning that is controlled by the cortical visual areas of the brain. { 'vizh·ə·wəl 'lərn·iŋ }

visual line of position [NAV] A line of position determined by visual observation of a landmark or aid to navigation. { 'vizh·ə·wəl 'līn əv pə'zish·ən }

visually coupled display *See* helmet-mounted display. { 'vizh·ə·lē 'kəp·əld di'splā }

visual magnitude [ASTRON] A celestial body's magnitude as seen by the eye of the observer. { 'vizh·ə·wəl 'mag·nə‚tüd }

visual object agnosia [PSYCH] The inability to name or give other evidence of recognizing visually presented objects. { ‚vizh·ə·wəl ‚äb‚jekt ag'nō·zhə }

visual photometer [OPTICS] A photometer in which the luminance of two surfaces is compared by human vision; it usually utilizes the Lummer-Brodhun sightbox or some adaptation of its principles. { 'vizh·ə·wəl fə'täm·əd·ər }

visual pigment [BIOCHEM] Any of various photosensitive pigments of vertebrate and invertebrate photoreceptors. { 'vizh·ə·wəl 'pig·mənt }

visual projection area [NEUROSCI] The receptive center for visual images in the cortex of the brain, located in the walls and margins of the calcarine sulcus of the occipital lobe. Also known as Brodmann's area 17. { 'vizh·ə·wəl prə'jek·shən ‚er·ē·ə }

visual purple *See* rhodopsin. { 'vizh·ə·wəl 'pər·pəl }

visual radio range [NAV] Any range facility whose course is flown by visual instrumentation not associated with aural reception. { 'vizh·ə·wəl 'rād·ē·ō ‚rānj }

visual range [METEOROL] The distance, under daylight conditions, at which the apparent contrast between a specified type of target and its background becomes just equal to the threshold contrast of an observer. { 'vizh·ə·wəl 'rānj }

visual receptive field [PHYSIO] That area of the retina within which stimulation with light or a light pattern causes a response in a particular receptor or neuron in the visual pathway. { ‚vizh·ə·wəl ri‚sep·tiv ‚fēld }

visual-righting reflex [PHYSIO] A reflex mechanism whereby righting of the head and body is caused by visual stimuli. Also known as optical-righting reflex. { 'vizh·ə·wəl 'rīd·iŋ ‚rē‚fleks }

visual scanner [ELECTR] Device that optically scans printed or written data and generates an analog or digital signal. { 'vizh·ə·wəl 'skan·ər }

visual seizure [MED] A form of epileptic seizure in which the patient experiences visual sensations in the form of light flashes, sometimes of varied colors. { 'vizh·ə·wəl 'sē·zhər }

visual servoing [CONT SYS] The use of a solid-state camera on the end effector of a robot to provide feedback. { 'vizh·ə·wəl 'sər·vō·iŋ }

visual storage tube [ELECTR] Any electrostatic storage tube that also provides a visual readout. { 'vizh·ə·wəl 'stȯr·ij ,tüb }

visual telephony [COMMUN] The transmission of picture information (television) over telephone lines. { 'vizh·ə·wəl tə'lef·ə·nē }

visual transmitter [ELECTR] Those parts of a television transmitter that act on picture signals, including parts that act on the audio signals as well. Also known as picture transmitter; video transmitter. { 'vizh·ə·wəl tranz'mid·ər }

visual yellow [BIOCHEM] An intermediary substance formed from rhodopsin in the retina after exposure to light; it is ultimately broken down to retinene and vitamin A. { 'vizh·ə·wəl 'yel·ō }

Vitaceae [BOT] A family of dicotyledonous plants in the order Rhamnales; mostly tendril-bearing climbers with compound or lobed leaves, as in grapes (*Vitis*). { vī'tās·ē,ē }

vital capacity [PHYSIO] The volume of air that can be forcibly expelled from the lungs after the deepest inspiration. { 'vīd·əl kə'pas·əd·ē }

Vitali set [MATH] A set of real numbers such that the difference of any two members of the set is an irrational number and any real number is the sum of a rational number and a member of the set. { və'täl·ē ,set }

vitalism [BIOL] The theory that the activities of a living organism are under the guidance of an agency which has none of the attributes of matter or energy. { 'vīd·əl,iz·əm }

vitamer [BIOCHEM] One of several very similar chemical compounds that can perform a specific vitamin function. { 'vīd·ə·mər }

vitamin [BIOCHEM] An organic compound present in variable, minute quantities in natural foodstuffs and essential for the normal processes of growth and maintenance of the body; vitamins do not furnish energy, but are essential for energy transformation and regulation of metabolism. { 'vīd·ə·mən }

vitamin A [BIOCHEM] $C_{20}H_{29}OH$ A pale-yellow alcohol that is soluble in fat and insoluble in water; found in liver oils and carotenoids, and produced synthetically; it is a component of visual pigments and is essential for normal growth and maintenance of epithelial tissue. Also known as antiinfective vitamin; antixerophthalmic vitamin; retinol. { 'vīd·ə·mən 'ā }

vitamin A aldehyde See retinal. { 'vīd·ə·mən 'ā 'al·də,hīd }

vitamin B₁ See thiamine. { 'vīd·ə·mən 'bē,wən }

vitamin B₂ See riboflavin. { 'vīd·ə·mən 'bē,tü }

vitamin B₃ See pantothenic acid. { 'vīd·ə·mən 'bē,thrē }

vitamin B₆ [BIOCHEM] A vitamin which exists as three chemically related and water-soluble forms found in food: pyridoxine, pyridoxal, and pyridoxamine; dietary requirements and physiological activities are uncertain. { 'vīd·ə·mən 'bē,siks }

vitamin B₁₂ [BIOCHEM] A group of closely related polypyrrole compounds containing trivalent cobalt; the antipernicious anemia factor, essential for normal hemopoiesis. Also known as cobalamin; cyanocobalamin; extrinsic factor. { 'vīd·ə·mən 'bē,twelv }

vitamin B complex [BIOCHEM] A group of water-soluble vitamins that include thiamine, riboflavin, nicotinic acid, pyridoxine, panthothenic acid, incsitol, *p*-aminobenzoic acid, biotin, folic acid, and vitamin B₁₂. { 'vīd·ə·mən 'bē 'käm,pleks }

vitamin B₆ hydrochloride See pyridoxine hydrochloride. { 'vīd·ə·mən 'bē,siks 'hī·drə'klȯr,īd }

vitamin B₂ phosphate See riboflavin 5′-phosphate. { 'vīd·ə·mən 'bē,tü 'fä,sfāt }

vitamin C See ascorbic acid. { 'vīd·ə·mən 'sē }

vitamin D [BIOCHEM] Either of two fat-soluble, sterol-like compounds, calciferol or ergocalciferol (vitamin D₂) and cholecalciferol (vitamin D₃); occurs in fish liver oils and is essential for normal calcium and phosphorus deposition in bones and teeth. Also known as antirachitic vitamin. { 'vīd·ə·mən 'dē }

vitamin D₃ See cholecalciferol. { 'vīd·ə·mən 'dē,thrē }

vitamin E [BIOCHEM] Any of a series of eight related compounds called tocopherols, α-tocopherol having the highest biological activity; occurs in wheat germ and other naturally occurring oils and is believed to be needed in certain human physiological processes. { 'vīd·ə·mən 'ē }

vitamin G See riboflavin. { 'vīd·ə·mən 'jē }

vitamin K [BIOCHEM] Any of three yellowish oils which are fat-soluble, nonsteroid, and nonsaponifiable; it is essential for formation of prothrombin. Also known as antihemorrhagic vitamin; prothrombin factor. { 'vīd·ə·mən 'kā }

vitamin K₁ See phytonadione. { 'vīd·ə·mən 'kā,wən }

vitamin P [BIOCHEM] A substance, such as citrin or one or more of its components, believed to be concerned with maintenance of the normal state of the walls of small blood vessels. { 'vīd·ə·mən 'pē }

vitamin P complex See bioflavonoid. { 'vīd·ə·mən 'pē ,käm pleks }

vitavite See moldavite. { 'vīd·ə,vīt }

vitellarium [INV ZOO] The part of the ovary in certain rotifers and flatworms that produces nutritive cells filled with yolk. Also known as yolk larva. { ,vid·əl'ar·ē·əm }

vitelline artery [EMBRYO] An artery passing from the yolk sac to the primitive aorta in young vertebrate embryos. Also known as omphalomesenteric artery. { vī'tel,ēn 'ärd·ə·rē }

vitelline duct [EMBRYO] The constricted part of the yolk sac opening into the midgut region of the future ileum. Also known as omphalomesenteric duct; umbilical duct. { vī'tel,ēn 'dəkt }

vitelline membrane [CYTOL] The cytoplasmic membrane on the surface of the mammalian ovum. { vī'tel,ēn 'mem,brān }

vitelline vein [EMBRYO] Any of the embryonic veins in vertebrates uniting the yolk sac and the sinus venosus; their proximal fused ends form the portal vein. Also known as omphalomesenteric vein. { vī'tel,ēn 'vān }

vitellogenesis [PHYSIO] The process by which yolk is formed in the ooplasm of an oocyte. { vī,tel·ə'jen·ə·səs }

Viterbi algorithm [COMMUN] A decoding procedure for convolutional codes that uses the maximum-likelihood method. { vi,ter·bē 'al·gə,rith·əm }

viticulture [AGR] That division of horticulture concerned with grape growing, studies of grape varieties, methods of culture, and insect and disease control. Also known as viniculture. { 'vid·ə,kəl·chər }

viticulturist [AGR] A grower of grapes. { 'vit·ə,kəl·chə·rəst }

vitiligo [MED] A skin disease characterized by an acquired ochromia in areas of various sizes and shapes. { ,vid·əl'ī,gō }

vitrain [GEOL] A brilliant black coal lithotype with vitreous luster and cubical cleavage. Also known as pure coal. { 'vi,trān }

Vitreoscillaceae [MICROBIO] Formerly a family of bacteria in the order Beggiatoales; included organisms which have a filamentous habit and move by gliding, but never store sulfur, and rely on organic nutrients in their metabolism. { ,vi·trē,äs·ə'lās·ē,ē }

vitreous body See vitreous humor. { 'vi·trē·əs 'bäd·ē }

vitreous chamber [ANAT] A cavity of the eye posterior to the crystalline lens and anterior to the retina, which is filled with vitreous humor. { 'vi·trē·əs 'chām·bər }

vitreous copper See chalcocite. { 'vi·trē·əs 'käp·ər }

vitreous enamel [MATER] A glass coating applied to a metal by covering the surface with a powdered glass frit and heating until fusion occurs. Also known as porcelain enamel. { 'vi·trē·əs i'nam·əl }

vitreous humor [PHYSIO] The transparent gel-like substance filling the greater part of the globe of the eye, the vitreous chamber. Also known as vitreous body. { 'vi·trē·əs 'hyü·mər }

vitreous luster [OPTICS] A type of luster resembling that of glass. { 'vi·trē·əs ,ləs·tər }

vitreous silica See silica glass. { 'vi·trē·əs 'sil·ə·kə }

vitreous silver See argentite. { 'vi·trē·əs 'sil·vər }

vitreous state [SOLID STATE] A solid state in which the atoms or molecules are not arranged in any regular order, as in a crystal, and which crystallizes only after an extremely long time. Also known as glassy state. { 'vi·trē·əs ,stāt }

vitric [GEOL] Referring to a pyroclastic material which is

VITAMIN A

Structural formula for vitamin A (retinol).

VITAMIN D

(a)

(b)

Structural formulas for the two vitamin D compounds. (*a*) Vitamin D₂ (calciferol). (*b*) Vitamin D₃ (cholecalciferol).

VOCAL SAC

Toad of the genus *Bufo* giving mating call with vocal sac expanded. (*American Museum of Natural History photograph*)

characteristically glassy, that is, contains more than 75% glass. { 'vi·trik }

vitric tuff [GEOL] Tuff composed principally of volcanic glass fragments. { 'vi·trik 'təf }

vitrification [ENG] Heat treatment of a material such as a ceramic to produce a glazed surface. [GEOL] Formation of a glassy or noncrystalline material. [MED] An experimental procedure for preserving human organs in which chemicals are added prior to cooling to prevent crystallization of water within and outside the cells, so that with coolings the molecules essentially become fixed in place. { ˌvi·trə·fəˈkā·shən }

vitrified brick [MATER] A type of brick that has been glazed to render it impervious to water and highly resistant to corrosion. { ˈvi·trəˌfīd 'brik }

vitrified-clay pipe [MATER] A pipe, made of clay treated in a kiln to induce vitrification, with the surface glazed for watertightness; used for drainage. { 'vi·trəˌfīd ˈklā 'pīp }

vitrified wheel [DES ENG] A grinding wheel with a glassy or porcelain bond. { 'vi·trəˌfīd 'wēl }

vitrinite [GEOL] A maceral group that is rich in oxygen and composed of humic material associated with peat formation; characteristic of vitrain. { 'vi·trəˌnīt }

vitrinoid [GEOL] Vitrinite occurring in bituminous coking coals; characterized by a reflectance of 0.5–2.0%. { 'vi·trəˌnȯid }

vitriolic acid *See* sulfuric acid. { ˌvi·trēˈäl·ik 'əsˌed }

vitriol stone [MINERAL] A hard, crystalline material, mainly a mixture of ferric sulfate and aluminum sulfate, that is extracted from weathered pyritic schist and used in the manufacture of sulfuric acid. { 'vi·trē·əl ˌstōn }

vitrophyre [PETR] Any porphyritic igneous rock whose groundmass is glassy. Also known as glass porphyry. { 'vi·trəˌfīr }

vittate [BOT] **1.** Having longitudinal stripes. **2.** Bearing specialized oil tubes (vittae). { 'viˌtāt }

viuga [METEOROL] A cold north or northeast storm of the Russian steppes, lasting about 3 days. { 'vyü·gə }

VIV *See* verbal information verification.

vivax malaria [MED] Malaria caused by *Plasmodium vivax* and characterized by typical paroxysms occurring every few days, commonly every 2 days. Also known as benign tertian malaria; tertian malaria. { 'vīˌvaks məˈler·ē·ə }

Viverridae [VERT ZOO] A family of carnivorous mammals in the superfamily Feloidea composed of the civets, genets, and mongooses. { vīˈver·əˌdē }

vivianite [MINERAL] Fe₃(PO₄)₂·8H₂O A colorless, blue, or green mineral in the unaltered state (darkens upon oxidation); crystallizes in the monoclinic system and occurs in earth form and as globular and encrusting fibrous masses. Also known as blue iron earth; blue ocher. { 'vi·vē·əˌnīt }

Viviparidae [INV ZOO] A family of fresh-water gastropod mollusks in the order Pectinibranchia. { ˌvī·vəˈpar·əˌdē }

viviparous [PHYSIO] Bringing forth live young. { vīˈvip·ə·rəs }

vixen file [DES ENG] A flat file with curved teeth; used for filing soft metals. { 'vik·sən ˌfīl }

V jewels [DES ENG] Jewel bearings used in conjunction with a conical pivot, the bearing surface being a small radius located at the apex of a conical recess; found primarily in electric measuring instruments. { 'vē ˌjülz }

VLA *See* Very Large Array.

Vlasov equation [PL PHYS] A modification of the Boltzmann transport equation for the study of a plasma, in which particles interact only through the mutually induced space-charge field, and collisions are assumed to be negligible. Also known as collisionless Boltzmann equation. { 'vla·sȯf iˌkwā·zhən }

Vlasov-Maxwell equations [PL PHYS] Equations for the propagation of electromagnetic radiation in a hot, collisionless plasma. { 'vlä·sȯf 'maksˌwel iˌkwā·zhənz }

VLBI *See* very long baseline interferometry.

VLCC *See* very large crude carrier.

VLF *See* very low frequency.

VLSI circuit *See* very large scale integrated circuit. { ˈvēˌelˌelˈsēˈī 'sər·kət }

VMOS technology *See* vertical metal oxide semiconductor technology. { 'vēˌmȯs tekˌnäl·ə·jē }

VM & P naphtha *See* varnish makers' and painters' naphtha. { ˈvēˈem ən ˈpē 'naf·thə }

V-notch weir *See* triangular-notch weir. { 'vē ˌnäch 'wer }

VOC *See* volatile organic compounds.

vocal cord *See* vocal fold. { 'vō·kəl ˌkȯrd }

vocal fold [ANAT] Either of a pair of folds of tissue covered by mucous membrane in the larynx. Also known as vocal cord. { 'vō·kəl ˌfōld }

vocal sac [VERT ZOO] An expansible pocket of skin beneath the chin or behind the jaws of certain frogs; may be inflated to a great volume and serves as a resonator. { 'vō·kəl ˌsak }

Vochysiaceae [BOT] A family of dicotyledonous plants in the order Polygalales characterized by mostly three carpels, usually stipulate leaves, one fertile stamen, and capsular fruit. { vōˌkizhˈē·āsˌēˌē }

vocoder [ELECTR] A system of electronic apparatus for synthesizing speech according to dynamic specifications derived from an analysis of that speech. { vōˈkōd·ər }

vodas [ELECTR] A voice-operated switching device used in transoceanic radiotelephone circuits to suppress echoes and singing sounds automatically; it connects a subscriber's line automatically to the transmitting station as soon as he starts speaking and simultaneously disconnects it from the receiving station, thereby permitting the use of one radio channel for both transmitting and receiving without appreciable switching delay as the parties alternately talk. Derived from voice-operated device anti-singing. { 'vōˌdas }

voder [ELECTR] An electronic system that uses electron tubes and filters, controlled through a keyboard, to produce voice sounds artificially. Derived from voice operation demonstrator. { 'vō·dər }

vodka [FOOD ENG] A colorless and unaged alcoholic beverage distilled from rye or wheat mash or sometimes from potatoes; it is highly rectified during distillation and thus is a very pure neutral spirit without a pronounced taste. { 'väd·kə }

vogad [ELECTR] An automatic gain control circuit used to maintain a constant speech output level in long-distance radiotelephony. Derived from voice-operated gain-adjusted device. { 'vōˌgad }

Vogel-Colson-Russell effect *See* Russell effect. { 'vō·gəl 'kōl·sən 'rəs·əl iˌfekt }

vogesite [PETR] A syenitic lamprophyre composed of phenocrysts of hornblende in a groundmass of orthoclase and hornblende. { 'vō·gəˌsīt }

Voges-Proskauer test [MICROBIO] One of the four tests of the IMVIC test; a qualitative test for the formation of acetyl methylcarbinol from glucose, in which solutions of α-naphthol, potassium hydroxide, and creatinine are added to an incubated culture of test organisms in a glucose broth, and a pink to rose color indicates a positive reaction. { 'fō·gə 'präsˌkaù·ər ˌtest }

voglite [MINERAL] An emerald green to grass green, triclinic mineral consisting of a hydrated carbonate of calcium, copper, and uranium; occurs as coatings of scales. { 'vōˌglīt }

Vogt-Russell theorem [ASTROPHYS] A theorem that states that if the pressure, opacity, and rate of energy generation in a star depend only on the local values of temperature, density, and chemical composition, then the star's structure is uniquely determined by its mass and chemical composition. Also known as Russell-Vogt theorem. { 'fōkt 'rəs·əl ˌthir·əm }

voice call sign [COMMUN] A call sign provided primarily for voice communications. { 'vȯis 'kȯl ˌsīn }

voice channel [COMMUN] A communication channel having sufficient bandwidth to carry voice frequencies intelligibly; the minimum bandwidth for an analog voice channel is about 3000 hertz for good intelligibility. { 'vȯis ˌchan·əl }

voice coder [ELECTR] Device that converts speech input into digital form prior to encipherment for secure transmission and converts the digital signals back to speech at the receiver. { 'vȯis ˌkō·dər }

voice coil [ENG ACOUS] The coil that is attached to the diaphragm of a moving-coil loudspeaker and moves through the air gap between the pole pieces due to interaction of the fixed magnetic field with that associated with the audio-frequency current flowing through the voice coil. Also known as loudspeaker voice coil; speech coil (British usage). { 'vȯis ˌkȯil }

voice/data system [COMMUN] Integrated communications system for transmitting both voice and digital data. { 'vȯis 'dad·ə ˌsisˌtəm }

voice digitization [ELECTR] The conversion of analog voice signals to digital signals. { 'vȯis ˌdij·əd·əˈzā·shən }

voiced sound [LING] A speech sound, such as "d," in which the auditory vibration of the vocal folds is included in the production of the sound. { ¦vȯist 'saůnd }

voice frequency [COMMUN] An audio frequency in the range essential for transmission of speech of commercial quality, from about 300 to 3400 hertz. Abbreviated VF. Also known as speech frequency. { 'vȯis ¦frē·kwən·sē }

voice-frequency carrier telegraphy [COMMUN] Carrier telegraphy in which the carrier currents have frequencies such that the modulated currents may be transmitted over a voice-frequency telephone channel. { 'vȯis ¦frē·kwən·sē ¦kar·ē·ər tə'leg·rə·fē }

voice-frequency dialing [ELECTR] Method of dialing by which the direct-current pulses from the dial are transformed into voice-frequency alternating-current pulses. { 'vȯis ¦frē·kwən·sē ˌdī·liŋ }

voice-frequency telegraph system [COMMUN] Telegraph system permitting the use of many channels on a single circuit; a different audio frequency is used for each channel, being keyed in the conventional manner; the various audio frequencies at the receiving end are separated by suitable filter circuits and fed to their respective receiving circuits. { 'vȯis ¦frē·kwən·sē 'tel·ə·graf ˌsis·təm }

voice-grade channel [COMMUN] A channel whose bandwidth is large enough to transmit voice-frequency signals. { 'vȯis ¦grād ˌchan·əl }

voice mail [COMMUN] A method of storing voice-recorded messages and delivering them electronically to an intended receiver. { 'vȯis ˌmāl }

voice-operated device [ELECTR] Any of several devices in a telephone system which are brought into operation by a sound signal, or some characteristic of such a signal. { 'vȯis ¦äp·ə,rād·əd di,vīs }

voice-operated device anti-singing See vodas. { 'vȯis ¦äp·ə,rād·əd di,vīs ¦an·tē'siŋ·iŋ }

voice-operated gain-adjusted device See vogad. { 'vȯis ¦äp·ə,rād·əd 'gān ə¦jəs·təd di,vīs }

voice-operated loss control and suppressor [ELECTR] Voice-operated device which switches loss out of the transmitting branch and inserts loss in the receiving branch under control of the subscriber's speech. { 'vȯis ¦äp·ə,rād·əd 'lȯs kən,trōl ən sə'pres·ər }

voice operation demonstrator See voder. { 'vȯis ˌäp·ə¦rā·shən 'dem·ən,strād·ər }

voice print [ENG ACOUS] A voice spectrograph that has individually distinctive patterns of voice characteristics that can be used to identify one person's voice from other voice patterns. { 'vȯis ,print }

voice recognition unit [COMMUN] A computer peripheral device that recognizes a limited number of spoken words and converts them into equivalent digital signals which can serve as computer input or initiate other desired actions. { 'vȯis ,rek·ig'nish·ən ,yü·nət }

voice response [COMPUT SCI] A computer-controlled recording system in which basic sounds, numerals, words, or phrases are individually stored for playback under computer control as the reply to a keyboarded query. [ENG ACOUS] The process of generating an acoustic speech signal that communicates an intended message, such that a machine can respond to a request for information by talking to a human user. Also known as speech synthesis. { 'vȯis ri,späns }

voice store and forward [COMMUN] A computer-supported system that converts spoken messages to digital format, stores them temporarily, and then transmits them to a receiver where they are converted back to sound. { 'vȯis ¦stȯr ən 'fȯr·wərd }

voice synthesizer [ELECTR] A synthesizer that simulates speech in any language by assembling a language's elements or phonemes under digital control, each with the correct inflection, duration, pause, and other speech characteristics. { 'vȯis ,sin·thə,sīz·ər }

void [COMPUT SCI] In optical character recognition, an island of insufficiently inked paper within the area of the intended character stroke. { vȯid }

voidage [MATER] The volume of the voids in a sample of powder divided by its overall volume (that is, the total volume occupied by the voids and the solid material). { 'vȯid·ij }

void channels [ENG] The open passages of a porous or packed medium through which liquid or gas can flow. { 'vȯid ,chan·əlz }

void coefficient [NUCLEO] A rate of change in the reactivity of a water reactor system resulting from a formation of steam bubbles as the power level and temperature increase. { 'vȯid ,kō·i,fish·ənt }

void ratio [SCI TECH] The ratio of the volume of void space to the volume of solid substance in any material consisting of void space and solid material, such as a soil sample, a sediment, or a powder. { vȯid ,rā·shō }

void swelling [NUCLEO] An increase in the external dimension of solid materials after irradiation. { 'vȯid ,swel·iŋ }

Voigt body See Kelvin body. { 'fȯit ,bäd·ē }

Voigt effect [OPTICS] Double refraction of light passing through a substance that is placed in a magnetic field perpendicular to the direction of light propagation. { 'fȯit i,fekt }

Voigt notation [MECH] A notation employed in the theory of elasticity in which elastic constants and elastic moduli are labeled by replacing the pairs of letters *xx, yy, zz, yz, zx,* and *xy* by the number 1, 2, 3, 4, 5, and 6 respectively. { 'fȯit nō,tā·shən }

voile [TEXT] A lightweight sheer fabric, made of cotton, worsted, silk, rayon, or acetate, with a crisp feel and airy appearance, woven from yarns having considerably more twist than warp yarns. { vȯil }

Voith-Schneider propulsion system [NAV ARCH] A system for ship propulsion, serving both as propeller and rudder, that consists of one or two vertical-axis rotors located underwater at the stern; rotor disks are flush with the shell plating which is approximately horizontal, and five to eight spadelike vertical-impeller blades are fitted near the periphery of the disks. { 'fȯit 'shnī·dər prə'pəl·shən ,sis·təm }

vol See volume.

Volans [ASTRON] A southern constellation, right ascension 8 hours, declination 70°S. Also known as Flying Fish; Piscis Volans. { 'vō,lanz }

volar [ANAT] Pertaining to, or on the same side as, the palm of the hand or the sole of the foot. { 'vō·lər }

volatile [CHEM] Readily passing off by evaporation. { 'väl·əd·əl }

volatile component [GEOL] A component of magma whose vapor pressures are high enough to allow them to be concentrated in any gaseous phase. Also known as volatile flux. { 'väl·əd·əl kəm'pō·nənt }

volatile file [COMPUT SCI] Any file in which data are rapidly added or deleted. { 'väl·əd·əl 'fīl }

volatile fluid [CHEM] A liquid with the tendency to become vapor at specified conditions of temperature and pressure. { 'väl·əd·əl 'flü·əd }

volatile flux See volatile component. { 'väl·əd·əl 'fləks }

volatile laurel oil [MATER] A bright-yellow liquid with an aromatic aroma, distilled from the leaves or berries of the laurel, *Laurus nobilis;* main components are cineole and pinene; soluble in alcohol, ether, benzene, and chloroform; used in perfumes, flavors, and medicine. Also known as sweet bay oil. { 'väl·əd·əl 'lȯr·əl ,ȯil }

volatile memory See volatile storage. { 'väl·əd·əl 'mem·rē }

volatile-oil reservoir [PETRO ENG] A type of bubble-point oil reservoir in which the temperature is high and the liquid density is low (leading to a volatilized oil situation), reducing the amount of producible liquids. { 'väl·əd·əl ¦ȯil 'rez·əv,wär }

volatile organic compounds [ENG] Organic chemicals that produce vapors readily at room temperature and normal atmospheric pressure, including gasoline and solvents such as toluene, xylene, and tetrachloroethylene. They form photochemical oxidants (including ground-level ozone) that affect health, damage materials, and cause crop and forest losses. Many are also hazardous air pollutants. Abbreviated VOC. { 'väl·ə,tȯl ȯr,gan·ik 'käm,paůnz }

volatile storage [COMPUT SCI] A storage device that must be continuously supplied with energy, or it will lose its retained data. Also known as volatile memory. { 'väl·əd·əl 'stȯr·ij }

volatility [THERMO] The quality of having a low boiling point or subliming temperature at ordinary pressure or, equivalently, of having a high vapor pressure at ordinary temperatures. { ,väl·ə'til·əd·ē }

volatility product [CHEM] The product of the concentrations of two or more molecules or ions that react to form a volatile substance. { ,väl·ə'til·əd·ē 'präd·əkt }

volatilization [THERMO] The conversion of a chemical substance from a liquid or solid state to a gaseous or vapor state by the application of heat, by reducing pressure, or by a combination of these processes. Also known as vaporization. { ˌväl·əd·əl·ə'zā·shən }

volborthite [MINERAL] $Cu_3(UO_4)_2 \cdot 3H_2O$ An olive green to green and yellowish-green, monoclinic mineral consisting of hydrated copper vanadate. { 'väl₁bȯr₁thīt }

volcanello *See* spatter cone. { ˌväl·kə'nel·ō }

volcanic arc *See* island arc. { väl₁kan·ik 'ärk }

volcanic ash [GEOL] Fine pyroclastic material; particle diameter is less than 4 millimeters. { väl'kan·ik 'ash }

volcanic bombs [GEOL] Pyroclastic ejecta; the lava fragments, liquid or plastic at the time of ejection, acquire rounded forms, markings, or internal structure during flight or upon landing. { väl'kan·ik 'bämz }

volcanic breccia [PETR] A pyroclastic rock that is composed of angular volcanic fragments having a diameter larger than 2 millimeters and that may or may not have a matrix. { väl'kan·ik 'brech·ə }

volcanic foam *See* pumice. { väl'kan·ik 'fōm }

volcanic gases [GEOL] Volatile matter composed principally of about 90% water vapor, and carbon dioxide, sulfur dioxide, hydrogen, carbon monoxide, and nitrogen, released during an eruption of a volcano. { väl'kan·ik 'gas·əz }

volcanic glass [GEOL] Natural glass formed by the cooling of molten lava, or one of its liquid fractions, too rapidly to allow crystallization. { väl'kan·ik 'glas }

volcanicity *See* volcanism. { ˌväl·kə'nis·əd·ē }

volcaniclastic rock [PETR] Clastic rock containing volcanic material in any proportion. { väl₁kan·ə₁klas·tik 'räk }

volcanic mud [GEOL] Sediment containing large quantities of ash from a volcanic eruption, mixed with water. { väl'kan·ik 'məd }

volcanic mudflow [GEOL] The flow of volcanic mud down the slope of a volcano. { väl'kan·ik 'məd₁flō }

volcanic neck [GEOL] A residual remnant of the pipe or throat of a volcano that was filled with solidified lava after its final eruption. { väl'kan·ik 'nek }

volcanic rift zone [GEOL] A zone comprising volcanic fissures with underlying dike assemblages; occurs in Hawaii. { väl'kan·ik 'rift ₁zōn }

volcanic rock [GEOL] Finely crystalline or glassy igneous rock resulting from volcanic activity at or near the surface of the earth. Also known as extrusive rock. { väl'kan·ik 'räk }

volcanics [PETR] Igneous rocks that solidified after reaching or nearing the earth's surface. { väl'kan·iks }

volcanic theory [ASTRON] A theory which holds that most features of the moon's surface were formed by volcanic eruptions, lava flows, and subsidences when lunar rocks were plastic. Also known as igneous theory; plutonic theory. { väl'kan·ik 'thē·ə·rē }

volcanic vent [GEOL] The channelway or opening of a volcano through which magma ascends to the surface; two general types are fissure and pipelike vents. { väl'kan·ik 'vent }

volcanism [GEOL] The movement of magma and its associated gases from the interior into the crust and to the surface of the earth. Also known as volcanicity. { 'väl·kə₁niz·əm }

volcano [GEOL] **1.** A mountain or hill, generally with steep sides, formed by the accumulation of magma erupted through openings or volcanic vents. **2.** The vent itself. { väl'kā·nō }

volcanology [GEOL] The branch of geology that deals with volcanism. { ˌväl·kə'näl·ə·jē }

vole [VERT ZOO] Any of about 79 species of rodent in the tribe Microtini of the family Cricetidae; individuals have a stout body with short legs, small ears, and a blunt nose. { vōl }

Volhard titration [ANALY CHEM] Determination of the halogen content of a solution by titration with a standard thiocyanate solution. { 'fȯl₁härt tī'trā·shən }

volley [ENG] A round of holes fired at any one time. [ORD] Burst of fire, especially a salute fired by a detachment of riflemen. { 'väl·ē }

volley bombing [ORD] Simultaneous or nearly simultaneous release of a number of bombs. { 'väl·ē ₁bäm·iŋ }

volley fire [ORD] Artillery fire in which each piece fires a specified number of rounds without regard to the other pieces and as fast as accuracy will permit. { 'väl·ē ₁fīr }

volt [ELEC] The unit of potential difference or electromotive force in the meter-kilogram-second system, equal to the potential difference between two points for which 1 coulomb of

electricity will do 1 joule of work in going from one point to the other. Symbolized V. { vōlt }

Volta effect *See* contact potential difference. { 'vōl·tə i₁fekt }

voltage [ELEC] Potential difference or electromotive force measured in volts. { 'vōl·tij }

voltage amplification [ELECTR] The ratio of the magnitude of the voltage across a specified load impedance to the magnitude of the input voltage of the amplifier or other transducer feeding that load; often expressed in decibels by multiplying the common logarithm of the ratio by 20. { 'vōl·tij ₁am·plə·fə'kā·shən }

voltage amplifier [ELECTR] An amplifier designed primarily to build up the voltage of a signal, without supplying appreciable power. { 'vōl·tij 'am·plə₁fī·ər }

voltage-amplitude-controlled clamp [ELECTR] A single diode clamp in which the diode functions as a clamp whenever the potential at point A rises above V_R; the diode is then in its forward-biased condition and acts as a very low resistance. { 'vōl·tij ₁am·plə₁tüd kən₁trōld 'klamp }

voltage coefficient [ELEC] For a resistor whose resistance varies with voltage, the ratio of the fractional change in resistance to the change in voltage. { 'vōl·tij ₁kō·i₁fish·ənt }

voltage-controlled oscillator [ELECTR] An oscillator whose frequency of oscillation can be varied by changing an applied voltage. Abbreviated VCO. { 'vōl·tij kən₁trōld 'äs·ə₁lād·ər }

voltage corrector [ELEC] Active source of regulated power placed in series with an unregulated supply to sense changes in the output voltage (or current), and to correct for these changes by automatically varying its own output in the opposite direction, thereby maintaining the total output voltage (or current) constant. { 'vōl·tij kə₁rek·tər }

voltage-current dual [ELEC] A pair of circuits in which the elements of one circuit are replaced by their dual elements in the other circuit according to the duality principle; for example, currents are replaced by voltages, capacitances by resistances. { 'vōl·tij 'kə·rənt 'dül }

voltage-dependent resistor *See* varistor. { 'vōl·tij di₁pen·dənt ri'zis·tər }

voltage derating [ELEC] The reduction of a voltage rating to extend the lifetime of an electric device or to permit operation at a high ambient temperature. { 'vōl·tij 'dē'rād·iŋ }

voltage divider [ELEC] A tapped resistor, adjustable resistor, potentiometer, or a series arrangement of two or more fixed resistors connected across a voltage source; a desired fraction of the total voltage is obtained from the intermediate tap, movable contact, or resistor junction. Also known as potential divider. { 'vōl·tij di₁vīd·ər }

voltage doubler [ELECTR] A transformerless rectifier circuit that gives approximately double the output voltage of a conventional half-wave vacuum-tube rectifier by charging a capacitor during the normally wasted half-cycle and discharging it in series with the output voltage during the next half-cycle. Also known as doubler. { 'vōl·tij ₁dəb·lər }

voltage drop [ELEC] The voltage developed across a component or conductor by the flow of current through the resistance or impedance of that component or conductor. { 'vōl·tij ₁dräp }

voltage feed [ELECTROMAG] Excitation of a transmitting antenna by applying voltage at a point of maximum potential (at a voltage loop or antinode). { 'vōl·tij ₁fēd }

voltage flare [ELEC] A higher than normal voltage purposely supplied to exposure lamps for a short period to produce full brilliance. { 'vōl·tij ₁fler }

voltage gain [ELECTR] The difference between the output signal voltage level in decibels and the input signal voltage level in decibels; this value is equal to 20 times the common logarithm of the ratio of the output voltage to the input voltage. { 'vōl·tij ₁gān }

voltage generator [ELECTR] A two-terminal circuit element in which the terminal voltage is independent of the current through the element. { 'vōl·tij jen·ə₁rād·ər }

voltage gradient [ELEC] The voltage per unit length along a resistor or other conductive path. { 'vōl·tij ₁grād·ē·ənt }

voltage level [ELEC] At any point in a transmission system, the ratio of the voltage existing at that point to an arbitrary value of voltage used as a reference. { 'vōl·tij ₁lev·əl }

VOLE

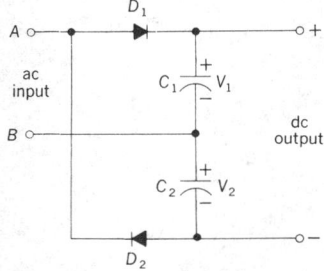

Vole showing characteristic features.

VOLTAGE DOUBLER

Circuit diagram of full-wave voltage doubler. When alternating-current input voltage is positive at terminal A, diode D_1 conducts, producing voltage V_1 across capacitor C_1. On other half cycle, diode D_2 conducts, producing voltage V_2 across capacitor C_2.

voltage measurement [ELEC] Determination of the difference in electrostatic potential between two points. { 'vōl·tij ,mezh·ər·mənt }

voltage multiplier [ELEC] *See* instrument multiplier. [ELECTR] A rectifier circuit capable of supplying a direct-current output voltage that is two or more times the peak value of the alternating-current voltage. { 'vōl·tij ,məl·tə,plī·ər }

voltage-multiplier circuit [ELEC] A rectifier circuit capable of supplying a direct-current output voltage that is two or more times the peak value of the alternating-current input voltage; useful for high-voltage, low-current supplies. { 'vōl·tij ¦məl·tə,plī·ər ,sər·kət }

voltage node [ELECTROMAG] Point having zero voltage in a stationary wave system, as in an antenna or transmission line; for example, a voltage node exists at the center of a half-wave antenna. { 'vōl·tij ,nōd }

voltage phasor [ELEC] A line whose length represents the magnitude of a sinusoidally varying voltage and whose angle with the positive *x*-axis represents its phase. { 'vōl·tij ,fā·zər }

voltage quadrupler [ELECTR] A rectifier circuit containing four diodes, which supplies a direct-current output voltage which is four times the peak value of the alternating-current input voltage. { 'vōl·tij kwə,drüp·lər }

voltage-range multiplier *See* instrument multiplier. { 'vōl·tij ¦rānj ,məl·tə,plī·ər }

voltage rating [ELEC] The maximum sustained voltage that can safely be applied to an electric device without risking the possibility of electric breakdown. Also known as working voltage. { 'vōl·tij ,rād·iŋ }

voltage ratio [ELEC] The root-mean-square primary terminal voltage of a transformer divided by the root-mean-square secondary terminal voltage under a specified load. { 'vōl·tij ,rā·shō }

voltage reflection coefficient [ELECTROMAG] The ratio of the phasor representing the magnitude and phase of the electric field of the backward-traveling wave at a specified cross section of a waveguide to the phasor representing the forward-traveling wave at the same cross section. { 'vōl·tij ri¦flek·shən ,kō·i,fish·ənt }

voltage-regulating transformer [ELECTROMAG] Saturated-core type of transformer which holds output voltage to within a few percent (5% above or below normal) with input variations up to 20% above or below normal; considerable harmonic distortion results unless extensive filters are employed. { 'vōl·tij ¦reg·yə,lād·iŋ tranz'fȯr·mər }

voltage regulation [ELEC] The ratio of the difference between no-load and full-load output voltage of a device to the full-load output voltage, expressed as a percentage. { 'vōl·tij ,reg·yə,lā·shən }

voltage regulator [ELECTR] A device that maintains the terminal voltage of a generator or other voltage source within required limits despite variations in input voltage or load. Also known as automatic voltage regulator; voltage stabilizer. { 'vōl·tij ,reg·yə,lād·ər }

voltage-regulator diode [ELECTR] A diode that maintains an essentially constant direct voltage in a circuit despite changes in line voltage or load. { 'vōl·tij ,reg·yə,lād·ər ,dī,ōd }

voltage-regulator tube [ELECTR] A glow-discharge tube in which the tube voltage drop is approximately constant over the operating range of current; used to maintain an essentially constant direct voltage in a circuit despite changes in line voltage or load. Also known as VR tube. { 'vōl·tij ,reg·yə,lād·ər ,tüb }

voltage saturation *See* anode saturation. { 'vōl·tij ,sach·ə,rā·shən }

voltage stabilizer *See* voltage regulator. { 'vōl·tij ,stā·bə,līz·ər }

voltage transformer [ELEC] An instrument transformer whose primary winding is connected in parallel with a circuit in which the voltage is to be measured or controlled. Also known as potential transformer. { 'vōl·tij tranz,fȯr·mər }

voltage-tunable tube [ELECTR] Oscillator tube whose operating frequency can be varied by changing one or more of the electrode voltages, as in a backward-wave magnetron. { 'vōl·tij ¦tün·ə·bəl 'tüb }

voltage-variable capacitor *See* varactor. { 'vōl·tij ¦ver·ē·əbəl kə'pas·əd·ər }

voltaic cell [ELEC] A primary cell consisting of two dissimilar metal electrodes in a solution that acts chemically on one or both of them to produce a voltage. { vōl'tā·ik 'sel }

voltaic pile [ELEC] An early form of primary battery, consisting of a pile of alternate pairs of dissimilar metal disks, with moistened pads between pairs. { vōl'tā·ik 'pīl }

voltaite [MINERAL] A greenish-black to black, isometric mineral consisting of a hydrated potassium iron sulfate. { 'väl·tə,īt }

voltameter *See* coulometer. { väl'tam·əd·ər }

voltammeter [ELEC] An instrument that may be used either as a voltmeter or ammeter. { väl'tam·əd·ər }

voltammetry [PHYS CHEM] Any electrochemical technique in which a faradaic current passing through the electrolysis solution is measured while an appropriate potential is applied to the polarizable or indicator electrode; for example, polarography. { väl'täm·ə·trē }

volt-ampere [ELEC] The unit of apparent power in the International System; it is equal to the apparent power in a circuit when the product of the root-mean-square value of the voltage, expressed in volts, and the root-mean-square value of the current, expressed in amperes, equals 1. Abbreviated VA. { 'vōlt 'am,pir }

volt-ampere hour [ELEC] A unit for expressing the integral of apparent power over time, equal to the product of 1 volt-ampere and 1 hour, or to 3600 joules. { 'vōlt 'am,pir 'aúr }

volt-ampere-hour reactive *See* var hour. { 'vōlt 'am,pir 'aúr rē'ak·tiv }

volt-ampere reactive [ELEC] The unit of reactive power in the International System; it is equal to the reactive power in a circuit carrying a sinusoidal current when the product of the root-mean-square value of the voltage, expressed in volts, by the root-mean-square value of the current, expressed in amperes, and by the sine of the phase angle between the voltage and the current, equals 1. Abbreviated var. Also known as reactive volt-ampere. { 'vōlt 'am,pir rē'ak·tiv }

Volta series *See* displacement series. { 'vōl·tə ,sir·ēz }

volt box [ELEC] A series of resistors arranged so that a desired fraction of a voltage can be measured, and the voltage thereby computed. { 'vōlt bäks }

Volterra dislocation [SOLID STATE] A model of a dislocation which is formed in a ring of crystalline material by cutting the ring, moving the cut surfaces over each other, and then rejoining them. { vȯl'ter·ə ,dis·lō'kā·shən }

Volterra equations [MATH] Given functions $f(x)$ and $K(x,y)$, these are two types of equations with unknown function y:

$$f(x) = \int_a^x K(x,t)\,y(t)\,dt$$

$$y(x) = f(x) + \lambda \int_a^x K(x,t)\,y(t)\,dt$$

{ vȯl'ter·ə i,kwā·shənz }

voltmeter [ENG] An instrument for the measurement of potential difference between two points, in volts or in related smaller or larger units. { 'vōlt,mēd·ər }

voltmeter-ammeter [ENG] A voltmeter and an ammeter combined in a single case but having separate terminals. { 'vōlt,mēd·ər 'am,ēd·ər }

voltmeter-ammeter method [ELEC] A method of measuring resistance in which simultaneous readings of the voltmeter and ammeter are taken, and the unknown resistance is calculated from Ohm's law. { 'vōlt,mēd·ər 'am,ēd·ər ,meth·əd }

voltmeter sensitivity [ELEC] Ratio of the total resistance of the voltmeter to its full scale reading in volts, expressed in ohms per volt. { 'vōlt,mēd·ər ,sen·sə'tiv·əd·ē }

volt-ohm-milliammeter [ENG] A test instrument having a number of different ranges for measuring voltage, current, and resistance. Also known as circuit analyzer; multimeter; multiple-purpose tester. { 'vōlt 'ōm ¦mil·ē'am,ēd·ər }

voltzite [MINERAL] Zn_5S_4O A rose red, yellowish, or brownish mineral consisting of an oxysulfide of zinc; occurs in implanted spherical globules and as a crust. { 'vält,sīt }

volume [ACOUS] The intensity of a sound. [COMPUT SCI] A single unit of external storage, all of which can be read or written by a single access mechanism or input/output device. [ENG ACOUS] The magnitude of a complex audio-frequency

VOLTAGE REGULATOR

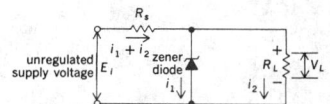

Circuit diagram of Zener-diode voltage regulator. Output load voltage V_L across load resistance R_L is maintained constant despite variation in input E_i. Current through series resistance R_s is sum of diode current i_1 and i_2.

current as measured in volume units on a standard volume indicator. [MATH] A measure of the size of a body or definite region in three-dimensional space; it is equal to the least upper bound of the sum of the volumes of nonoverlapping cubes that can be fitted inside the body or region, where the volume of a cube is the cube of the length of one of its sides. Abbreviated vol. { 'väl·yəm }

volume acoustic wave *See* bulk acoustic wave. { 'väl·yəm ə'küs·tik 'wāv }

volume by slicing [MATH] A method of computing the volume of a solid by integrating over the volumes of infinitesimal slices of the solid bounded by parallel planes. { ¦väl·yəm bī 'slīs·iŋ }

volume compressor [ENG ACOUS] An audio-frequency circuit that limits the volume range of a radio program at the transmitter, to permit using a higher average percent modulation without risk of overmodulation; also used when making disk recordings, to permit a closer groove spacing without overcutting. Also known as automatic volume compressor. { 'väl·yəm kəm,pres·ər }

volume control [ENG ACOUS] A potentiometer used to vary the loudness of a reproduced sound by varying the audio-frequency signal voltage at the input of the audio amplifier. { 'väl·yəm kən,trōl }

volume control system [ENG ACOUS] An electronic system that regulates the signal amplification or limits the output of a circuit, such as a volume compressor or a volume expander. { 'väl·yəm kən,trōl ,sis·təm }

volume dose *See* integral dose. { 'väl·yəm ,dōs }

volume expander [ENG ACOUS] An audio-frequency control circuit sometimes used to increase the volume range of a radio program or recording by making weak sounds weaker and loud sounds louder; the expander counteracts volume compression at the transmitter or recording studio. Also known as automatic volume expander. { 'väl·yəm ik,span·dər }

volume flow rate [FL MECH] The volume of the fluid that passes through a given surface in a unit time. { 'väl·yəm 'flō ,rāt }

volume indicator [ENG ACOUS] A standardized instrument for indicating the volume of a complex electric wave such as that corresponding to speech or music; the reading in volume units is equal to the number of decibels above a reference level which is realized when the instrument is connected across a 600-ohm resistor that is dissipating a power of 1 milliwatt at 100 hertz. Also known as volume unit meter. { 'väl·yəm ,in·də,kād·ər }

volume integral [MATH] An integral of a function of several variables with respect to volume measure taken over a three-dimensional subset of the domain of the function. { 'väl·yəm 'int·ə·grəl }

volume label [COMPUT SCI] A record that contains information about the contents of a particular storage device, usually a disk or magnetic tape, and is written somewhere on that device. { 'väl·yəm ,lā·bəl }

volume lifetime [SOLID STATE] Average time interval between the generation and recombination of minority carriers in a homogeneous semiconductor. { 'väl·yəm 'līf,tīm }

volume-limiting amplifier [ELECTR] Amplifier containing an automatic device that functions only when the input signal exceeds a predetermined level, and then reduces the gain so the output volume stays substantially constant despite further increases in input volume; the normal gain of the amplifier is restored when the input volume returns below the predetermined limiting level. { 'väl·yəm ¦lim·əd·iŋ 'am·plə,fī·ər }

volume meter [ENG] Any flowmeter in which the actual flow is determined by the measurement of a phenomenon associated with the flow. { 'väl·yəm ,mēd·ər }

volumenometer [ENG] An instrument for determining the volume of a body by measuring the pressure in a closed air space when the specimen is present and when it is absent. { väl,yü·mə'näm·əd·ər }

volume phase *See* surface phase. { 'väl·yəm ,fāz }

volume range [ELEC] In a transmission system, the difference, expressed in decibels, between the maximum and minimum volumes that can be satisfactorily handled by the system. [ENG ACOUS] The difference, expressed in decibels, between the maximum and minimum volumes of a complex audio-frequency signal occurring over a specified period of time. { 'väl·yəm ,rānj }

volume recombination rate [SOLID STATE] The rate at which free electrons and holes within the volume of a semiconductor recombine and thus neutralize each other.

volume resistivity [ELEC] Electrical resistance between opposite faces of a 1-centimeter cube of insulating material, commonly expressed in ohm-centimeters. Also known as specific insulation resistance. { 'väl·yəm ,rē,zis'tiv·əd·ē }

volume shift *See* field shift. { 'väl·yəm ,shift }

volume susceptibility [PHYS CHEM] The magnetic susceptibility of a specified volume (for example, 1 cubic centimeter) of a magnetically susceptible material. { 'väl·yəm sə,sep·tə'bil·əd·ē }

volume table of contents [COMPUT SCI] A list of all the files in a volume, usually with descriptions of their contents and locations. Abbreviated VTOC. { 'väl·yəm ¦tā·bəl əv 'kän,tens }

volume target [ELECTROMAG] A radar target composed of a large number of objects too close together to be resolved. { 'väl·yəm 'tär·gət }

volumeter [ENG] Any instrument for measuring volumes of gases, liquids, or solids. { 'väl·yə,mēd·ər }

volume test [COMPUT SCI] The processing of a volume of actual data to check for program malfunction. { 'väl·yəm ,test }

volume transport [OCEANOGR] The volume of moving water measured between two points of reference and expressed in cubic meters per second. { 'väl·yəm ,tranz,pórt }

volumetric analysis [ANALY CHEM] Quantitative analysis of solutions of known volume but unknown strength by adding reagents of known concentration until a reaction end point (color change or precipitation) is reached; the most common technique is by titration. Also known as titrimetric analysis. { ¦väl·yə¦me·trik ə'nal·ə·səs }

volumetric efficiency [MECH ENG] In describing an engine or gas compressor, the ratio of volume of working substance actually admitted, measured at a specified temperature and pressure, to the full piston displacement volume; for a liquid-fuel engine, such as a diesel engine, volumetric efficiency is the ratio of the volume of air drawn into a cylinder to the piston displacement. { ¦väl·yə¦me·trik i'fish·ən·sē }

volumetric flask [ANALY CHEM] A laboratory flask primarily intended for the preparation of definite, fixed volumes of solutions, and therefore calibrated for a single volume only. { ¦väl·yə¦me·trik 'flask }

volumetric performance [PETRO ENG] The volume production of gas and oil from a reservoir; usually expressed as gas-oil ratio. { ¦väl·yə¦me·trik pər'fór·məns }

volumetric pipet [ANALY CHEM] A graduated glass tubing used to measure quantities of a solution; the tube is open at the top and bottom, and a slight vacuum (suction) at the top pulls liquid into the calibrated section; breaking the vacuum allows liquid to leave the tube. { ¦väl·yə¦me·trik pī'pet }

volumetric radar [ENG] Radar capable of producing three-dimensional position data on a multiplicity of targets. { ¦väl·yə¦me·trik 'rā,där }

volumetric storage [COMMUN] Any data storage technology in which information is stored throughout a three-dimensional volume rather than merely on a surface. { ¦väl·yə¦me·trik 'stór·ij }

volumetric strain [MECH] One measure of deformation; the change of volume per unit of volume. { ¦väl·yə¦me·trik 'strān }

volume unit [ENG ACOUS] A unit for expressing the audio-frequency power level of a complex electric wave, such as that corresponding to speech or music; the power level in volume units is equal to the number of decibels above a reference level of 1 milliwatt as measured with a standard volume indicator. Abbreviated VU. { 'väl·yəm ,yü·nət }

volume unit meter *See* volume indicator. { 'väl·yəm ,yü·nət ,mēd·ər }

volume velocity [ACOUS] The rate of flow of a medium through a specified area due to a sound wave. { 'väl·yəm və,läs·əd·ē }

voluntary muscle [PHYSIO] A muscle directly under the control of the will of the organism. { 'väl·ən,ter·ē 'məs·əl }

volute [DES ENG] A spiral casing for a centrifugal pump or a fan designed so that speed will be converted to pressure without shock. { və'lüt }

volute pump [MECH ENG] A centrifugal pump housed in a spiral casing. { və'lüt 'pəmp }

Volutidae [INV ZOO] A family of gastropod mollusks in the order Neogastropoda. { və'lüd·ə,dē }

volutin [BIOCHEM] A basophilic substance, thought to be a nucleic acid, occurring as granules in the cytoplasm and vacuoles of algae and other microorganisms. { 'väl·yəd·ən }

volva [MYCOL] A cuplike membrane surrounding the base of the stipe in certain gill fungi. { 'väl·və }

volvent nematocyst [INV ZOO] A nematocyst in the form of an unarmed, coiled tube that is closed at the end. { 'väl·vənt nə'mad·ə,sist }

Volvocales [BOT] An order of one-celled or colonial green algae in the division Chlorophyta; individuals are motile with two, four, or rarely eight whiplike flagella. { ,väl·və'kā·lēz }

Volvocida [INV ZOO] An order of the protozoan class Phytamastigophorea; individuals are grass-green or colorless, have one, two, four, or eight flagella, and thick cell walls of cellulose. { väl'väs·əd·ə }

volvulus [MED] A twisting of the bowel upon itself so as to occlude the lumen and, in severe cases, compromise its circulation. { 'väl·vyə·ləs }

Vombatidae [VERT ZOO] A family of marsupial mammals in the order Diprotodonta in some classification systems. { väm'bad·ə,dē }

vomer [ANAT] A skull bone below the ethmoid region constituting part of the nasal septum in most vertebrates. { 'vō·mər }

vomeronasal cartilage [ANAT] A strip of hyaline cartilage extending from the anterior nasal spine upward and backward on either side of the septal cartilage of the nose and attached to the anterior margin of the vomer. Also known as Jacobson's cartilage. { ¦väm·ə·rō¦nāz·əl 'kärt·lij }

vomiting gas [MATER] Any one of a group of toxic gases, such as adamsite, that causes coughing, sneezing, sometimes vomiting, and other effects. { 'väm·əd·iŋ ,gas }

von Arx current meter [ENG] A type of current-measuring device using electromagnetic induction to determine speed and, in some models, direction of deep-sea currents. { fòn 'ärks 'kə·rənt ,mēd·ər }

von Gierke's disease [MED] A form of glycogenosis characterized by marked diminution in or absence of hepatic glucose-6-phosphatase, resulting in hepatic glycogenosis, hypoglycemia, and acidosis. Also known as glycogen storage disease; hepatic glycogenosis; type I of Cori; van Crevald-von Gierke's disease. { fòn 'gir·kēz di,zēz }

von Kármán See Kármán. { fòn 'kär·män }

von Klitzing constant [PHYS] 1. The quantity $RK = h/e^2$, where h is Planck's constant and e is the charge of the electron; materials that exhibit the quantum Hall effect have a Hall resistance equal to RK/n, where n is either an integer or a rational fraction. Also known as quantized Hall resistance. 2. The conventional value of this quantity adopted by international agreement on January 1, 1990, to establish a standard for the ohm, $RK - 90 = 25,812.807$ ohms. { fòn 'klit·siŋ ,kän·stənt }

von Klitzing effect See quantum Hall effect. { fòn 'klit·siŋ i,fekt }

von Mises yield criterion [MECH] The assumption that plastic deformation of a material begins when the sum of the squares of the principal components of the deviatoric stress reaches a certain critical value. { fòn ¦mēz·əz 'yēld ,krī,tir·ē·ən }

von Neumann algebra [MATH] A subalgebra A of the algebra $B(H)$ of bounded linear operators on a complex Hilbert space, such that the adjoint operator of any operator in A is also in A, and A is closed in the strong operator topology in $B(H)$. Also known as ring of operators; W* algebra. { fòn ¦nòi·män 'al·jə·brə }

von Neumann bottleneck [COMPUT SCI] An inefficiency inherent in the design of any von Neumann machine that arises from the fact that most computer time is spent in moving information between storage and the central processing unit rather than operating on it. { fòn 'nòi,män 'bäd·əl,nek }

von Neumann machine [COMPUT SCI] A stored-program computer equipped with a program counter. { fòn 'nòi,män mə,shēn }

VOR See very high frequency omnidirectional radio range.

Vorce diaphragm cell [CHEM ENG] A cylindrical cell with graphite anodes and asbestos-covered cathode, used in the electrolytic process for the manufacture of chlorine. { 'vòrs 'dī·ə,fram ,sel }

vorobievite See vorobyevite. { və'rō,bē·ə,vīt }

vorobyevite [MINERAL] A rose-red, purplish-red, or pinkish cesium-containing variety of beryl; used as a gem. Also known as morganite; rosterite; vorobievite; worobieffite. { və'rō,bē·ə,vīt }

Vortac [NAV] A ground radio station consisting of a collocated very-high-frequency omnidirectional radio range (VOR) and Tacan facility; this station permits obtaining polar coordinates by the use of VOR receiver and distance-measuring equipment, or by Tacan equipment alone. { 'vòr,tak }

vortex [FL MECH] 1. Any flow possessing vorticity; for example, an eddy, whirlpool, or other rotary motion. 2. A flow with closed streamlines, such as a free vortex or line vortex. 3. See vortex tube. [SOLID STATE] See fluxoid. { 'vòr,teks }

vortex amplifier [ENG] A fluidic device in which the supply flow is introduced at the circumference of a shallow cylindrical chamber; the vortex field developed can substantially reduce or throttle flow; used in fluidic diodes, throttles, pressure amplifiers, and a rate sensor. { 'vòr,teks 'am·plə,fī·ər }

vortex breakdown [FL MECH] An abrupt change in the structure of the core of a swirling flow. { ,vòr,teks 'brāk,daùn }

vortex burner [ENG] Combustion device in which the combustion air is fed tangentially into the burner, creating a spin (vortex) to mix it with the fuel as it is injected. { 'vòr,teks 'bər·nər }

vortex cage meter [ENG] In flow measurement, a type of quantity meter which exerts only a slight retardation on the flowing fluid; the elements rotate at a speed that is linear with fluid velocity; revolutions are counted either by coupling to a local mounted counter or by a proximity detector for remote transmission. { 'vòr,teks 'kāj ,mēd·ər }

vortex distribution method [FL MECH] An analytic method used in ideal aerodynamics which ignores the thickness of the profile of the aerodynamic figure being studied. { 'vòr,teks ,di·strə'byü·shən ,meth·əd }

vortex filament [FL MECH] The line of concentrated vorticity in a line vortex. Also known as vortex line. { 'vòr,teks ,fil·ə·mənt }

vortex generator [AERO ENG] Any of the small, upright vanes that are attached to aircraft surfaces to inhibit boundary-layer separation and thereby reduce drag. { 'vòr,teks ,jen·ə,rād·ər }

vortex line [FL MECH] 1. A line drawn through a fluid such that it is everywhere tangent to the vorticity. 2. See vortex filament. { 'vòr,teks ,līn }

vortex precession flowmeter [ENG] An instrument for measuring gas flows from the rate of precession of vortices generated by a fixed set of radial vanes placed in the flow. Also known as swirl flowmeter. { 'vòr,teks prē'sesh·ən 'flō ,mēd·ər }

vortex ring [FL MECH] A line vortex in which the line of concentrated vorticity is a closed curve. Also known as collar vortex; ring vortex. { 'vòr,teks ,riŋ }

vortex shedding [FL MECH] In the flow of fluids past objects, the shedding of fluid vortices periodically downstream from the restricting object (for example, smokestacks, pipelines, or orifices). { 'vòr,teks ,shed·iŋ }

vortex-shedding meter [ENG] A flowmeter in which fluid velocity is determined from the frequency at which vortices are generated by an obstruction in the flow. { 'vòr,teks ¦shed·iŋ ,mēd·ər }

vortex sheet [FL MECH] A surface across which there is a discontinuity in fluid velocity, such as in slippage of one layer of fluid over another; the surface may be regarded as being composed of vortex filaments. { 'vòr,teks ,shēt }

vortex street [FL MECH] A series of vortices which are systematically shed from the downstream side of a body around which fluid is flowing rapidly. Also known as vortex trail; vortex train. { 'vòr,teks ,strēt }

vortex thermometer [ENG] A thermometer, used in aircraft, which automatically corrects for adiabatic and frictional temperature rises by imparting a rotary motion to the air passing the thermal sensing element. { 'vòr,teks thər'mäm·əd·ər }

vortex trail See vortex street. { 'vòr,teks ,trāl }

Volvent nematocyst of *Hydra*. (*After Schulze, from T. I. Storer and R. L. Usinger, General Zoology, 3d ed., McGraw-Hill, 1957*)

VOLVOCALES

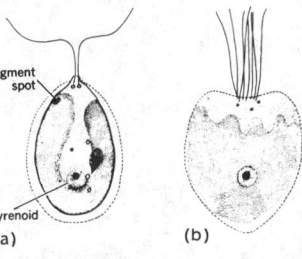

pigment spot

pyrenoid

(a)　　　　(b)

Unicellular algae of the order Volvocales. (*a*) *Chlamydomonas*, a unicellular organism with two flagella. (*b*) *Polyblepharides*, showing massive chloroplast and eight flagella.

VOLVOCIDA

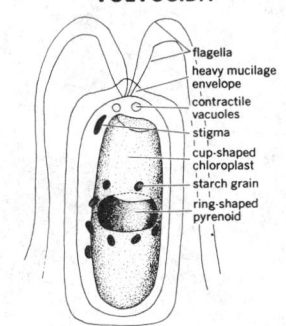

flagella
heavy mucilage envelope
contractile vacuoles
stigma
cup-shaped chloroplast
starch grain
ring-shaped pyrenoid

Representative species of Volvocida, in genus *Carteria*, with four equal flagella.

VORTEX CAGE METER

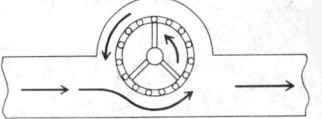

Diagram of vortex cage meter. Arrows indicate direction of fluid flow.

vortex train *See* vortex street. { 'vȯr‚teks ‚trān }

vortex tube [FL MECH] A tubular surface consisting of the collection of vortex lines which pass through a small closed curve. Also known as vortex. { 'vȯr‚teks ‚tüb }

vertical field *See* rotational field. { ¦vȯrd·ə·kəl 'fēld }

vorticity [FL MECH] For a fluid flow, a vector equal to the curl of the velocity of flow. { vȯr'tis·əd·ē }

vorticity equation [FL MECH] An equation of fluid mechanics describing horizontal circulation in the motion of particles around a vertical axis: $(d/dt) (S + f) = - (S + f) \text{div}_h c$, where $(S + f)$ is the absolute vorticity (S is the relative vorticity and f is the Coriolis parameter) and $\text{div}_h c$ is the horizontal divergence of the fluid velocity. { vȯr'tis·əd·ē i‚kwā·zhən }

vorticity-transport hypothesis [FL MECH] The hypothesis that, owing to the existence of pressure fluctuations, vorticity, and not momentum, is conservative in turbulent eddy flux. { vȯr'tis·əd·ē ¦tranz‚pȯrt hī‚päth·ə·səs }

VOS *See* vertical obstacle sonar.

vougesite [PETR] A lamprophyre having an orthoclase and hornblende groundmass in which are embedded hornblende phenocrysts. { 'vüzh‚sīt }

vowel [LING] A voiced "open sound," in which the quality of the sound is determined by its placement in the mouth. { 'vau̇·əl }

voxel [COMPUT SCI] The smallest box-shaped part of a three-dimensional image or scan. Derived from volume pixel. { 'väks·əl }

voyage [NAV] **1.** The outward and homeward passage of a trip by sea. **2.** A trip by sea. { 'vȯi·əj }

V particle [PART PHYS] The name first used for the unstable particles whose decay is responsible for the production of characteristic V-shaped tracks observed in cloud chambers exposed to cosmic radiation; they are neutral semistable particles such as neutral K mesons or lambda hyperons. { 'vē ‚pärd·ə·kəl }

VPE *See* vapor-phase epitaxy.

VPN *See* virtual private network.

VPR chart [NAV] A type of radar chart for use with VPR (virtual plan position indicator reflectoscope). { ‚vē‚pē'är ‚chärt }

v-process [NUC PHYS] The synthesis of certain elements and nuclides in type II supernovas; in this process, the inelastic scattering of neutrinos emitted from the core of the supernova excites states that then decay via single or multiple nucleon emission. { 'vē ‚prä·səs }

vrbaite [MINERAL] $Tl_4Hg_3Sb_2As_8S_{20}$ A dark gray-black, orthorhombic mineral that occurs in small crystals. { 'vər‚bə‚īt }

VRC *See* lateral parity check.

vriajem *See* friagem. { 'frē·ə‚jem }

VRM *See* viscous remanent magnetization.

VRML *See* Virtual Reality Modeling Language.

VR tube *See* voltage-regulator tube. { ¦vē'är ‚tüb }

VSAT *See* very small aperture terminal. { 'vē‚sat }

VSB *See* vestigial sideband.

VSC *See* variable speech control.

V-shaped depression [METEOROL] On a surface chart, a low or trough about which the isobars display a pronounced V shape, with the point of the V usually extending equatorward from the parent low. { 'vē ¦shāpt di'presh·ən }

V-shaped valley [GEOL] A valley having a cross-sectional profile in the form of the letter V, commonly produced by stream erosion. Also known as V valley. { 'vē ¦shāpt 'val·ē }

V/STOL aircraft [AERO ENG] An aircraft that can take off vertically or by a short running. Abbreviation for vertical or short takeoff and landing. { 'vē‚stȯl ‚er‚kraft }

vt fuse *See* proximity fuse. { ¦vē'tē ‚fyüz }

VTOC *See* volume table of contents.

VTOL *See* vertical takeoff and landing.

VTOL aircraft [AERO ENG] A heavier-than-air craft that can take off and land vertically. Abbreviation for vertical takeoff and landing. { 'vē‚tȯl ‚er‚kraft }

V-tool *See* parting tool. { 'vē‚tül }

VTR *See* videotape recording.

VTVM *See* vacuum-tube voltmeter.

v-type engine [MECH ENG] An engine in which the cylinders are arranged in two rows set at an angle to each other, with the crankshaft running through the point of a V. { 'vē ‚tīp ‚en·jən }

VU *See* volume unit. { vyü *or* ¦vē'yü }

vug [PETR] A small cavity in a vein or rock usually lined with minerals differing in composition from those of the enclosing rock. Also known as bughole. { vəg }

Vulcan [ASTRON] A hypothetical planet that was supposed to have an orbit within the orbit of Mercury; its existence was considered about 1859 and in the next few years, but it is generally considered by present-day astronomers to be nonexistent. { 'vəl·kən }

Vulcanian eruption [GEOL] A volcanic eruption characterized by periodic explosive events. Also known as paroxysmal eruption; Plinian eruption; Vesuvian eruption. { ¦vəl¦kā·nē·ən i'rəp·shən }

vulcanization [CHEM ENG] A chemical reaction of sulfur (or other vulcanizing agent) with rubber or plastic to cause cross-linking of the polymer chains; it increases strength and resiliency of the polymer. Also known as cure. { ‚vəl·kə·nə'zā·shən }

vulcanized fiber [MATER] A laminated plastic made by chemically treating layers of 100% rag-content paper to gelatinize the paper and fuse the layers into a solid mass; when dried under pressure, it forms a hard, tough material having good electrical properties along with mechanical strength and dimensional stability. { 'vəl·kə‚nīzd 'fī·bər }

vulgar establishment *See* high-water full and change. { 'vəl·gər i'stab·lish·mənt }

vulgar fraction *See* common fraction. { ¦vəl·gər 'frak·shən }

vulnerability [COMPUT SCI] A weakness in a computing system that can result in harm to the system or its operations, especially when this weakness is exploited by a hostile person or organization or when it is present in conjunction with particular events or circumstances. { ‚vəl·nə·rə'bil·əd·ē }

Vulpecula [ASTRON] A northern constellation, right ascension 20 hours, declination 25°N. Also known as Little Fox. { ‚vəl'pek·yə·lə }

vulture [VERT ZOO] The common name for any of various birds of prey in the families Cathartidae and Accipitridae of the order Falconiformes; the head of these birds is usually naked. { 'vəl·chər }

vulva [ANAT] The external genital organs of women. { 'vəl·və }

vulval gland [ANAT] A scent gland in the vulval tissues of the human female. { 'vəl·vəl ‚gland }

vulvovaginitis [MED] Simultaneous inflammation of the vulva and the vagina. { ¦vəl·vō‚vaj·ə'nīd·əs }

vuthan [METEOROL] In southern South America, an intense storm. { ¦vü‚tän }

VUV radiation *See* vacuum ultraviolet radiation. { ¦vē¦yü'vē ‚rād·ē‚ā·shən }

V valley *See* V-shaped valley. { 'vē ‚val·ē }

VV Cephei stars [ASTRON] A class of long-period eclipsing binary stars, with M supergiant primaries, a blue (usually B) supergiant or giant secondaries, and small variations in light. { ¦vē¦vē 'sef·ē‚ī ‚stärz }

W

W *See* tungsten; watt.

WAAS *See* Wide-Area Augmentation System. { wäs *or* ¦dəb·əl·yü¦ā¦ā'es }

wacke [PETR] Sandstone composed of a mixture of angular and unsorted or poorly sorted fragments of minerals and rocks and an abundant matrix of clay and fine silt. { 'wak·ə }

Wacker process [CHEM ENG] A process for the oxidation of ethylene to acetaldehyde by oxygen in the presence of palladium chloride and cupric chloride. { 'wak·ər ¸prä·səs }

wackestone [PETR] A limestone composed of mud (micrite) containing more than 10% particles (grains) with diameters greater than 20 micrometers scattered throughout. { 'wak·ə¸stōn }

wad [MINERAL] A massive, generally soft, amorphous, earthy, dark-brown or black mineral composed principally of manganese oxides with some other minerals, and formed by decomposition of manganese minerals. Also known as black ocher; bog manganese; earthy manganese. [ORD] A felt or cardboard pad used to secure the propellant in place in cartridges. Also known as wadding. { wäd }

Wadati-Benioff zone *See* Benioff zone. { ¦wä¦dä·tē 'ben·ē¸òf ¸zōn }

wad cutter [ORD] Bullet designed for target shooting, shaped to cut a clean hole in a paper target. { 'wäd ¸kəd·ər }

wadding *See* wad. { 'wäd·iŋ }

wader *See* shore bird. { 'wäd·ər }

wadi [GEOL] In the desert regions of southwestern Asia and northern Africa, a stream bed or channel, or a steep-sided ravine, gulley, or valley, which carries water only during the rainy season. Also spelled wady. { 'wäd·ē }

wading bird [VERT ZOO] Any of the long-legged, long-necked birds composing the order Ciconiiformes, including storks, herons, egrets, and ibises. { 'wäd·iŋ ¸bərd }

Wadsworth mounting [OPTICS] **1.** A device in which light passes through a prism and is then reflected from a plane mirror; it has the effect of a constant-deviation prism. **2.** A mounting for a diffraction grating in which the slit is placed at the principal focus of a concave mirror, so that the light falling on the grating is in a parallel beam; it greatly reduces astigmatism. { 'wädz¸wərth ¸maùnt·iŋ }

wady *See* wadi. { 'wäd·ē }

wafer [ELECTR] A thin semiconductor slice on which matrices of microcircuits can be fabricated, or which can be cut into individual dice for fabricating single transistors and diodes. [ENG] A flat element for a process unit, as in a series of stacked filter elements. { 'wā·fər }

wafer lever switch [ELECTR] A lever switch in which a number of contacts are arranged on one or both sides of one or more wafers, for engaging one or more contacts on a movable wafer segment actuated by the operating lever. { 'wā·fər lev·ər ¸swich }

wafer socket [ELECTR] An electron-tube socket consisting of one or two wafers of insulating material having holes in which are spring metal clips that grip the terminal pins of a tube. { 'wā·fər ¸säk·ət }

waffle weave [TEXT] Weave pattern of fabrics that resembles the cellular honeybee comb. Also known as honeycomb weave. { 'wäf·əl ¸wēv }

wage curve [IND ENG] A graphic representation of the relationship between wage rates and point values for key jobs. { 'wāj ¸kərv }

wage incentive plan [IND ENG] A wage system which provides additional pay for qualitative and quantitative performance which exceeds standard or normal levels. Also known as incentive wage system. { 'wāj in'sen·tiv ¸plan }

Wagner earth connection *See* Wagner ground. { 'wag·nər 'ərth kə¸nek·shən }

Wagner ground [ELEC] A ground connection used with an alternating-current bridge to minimize stray capacitance errors when measuring high impedances; a potentiometer is connected across the bridge supply oscillator, with its movable tap grounded. Also known as Wagner earth connection. { 'wag·nər graùnd }

wagnerite [MINERAL] $Mg_2(PO_4)F$ A yellow, grayish, flash-red, or greenish, monoclinic mineral consisting of magnesium fluophosphate. { 'väg·nə¸rīt }

Wagner's reagent [ANALY CHEM] An aqueous solution of iodine and potassium iodide, used for microchemical analysis of alkaloids. Also known as Wagner's solution. { 'väg·nərz rē¸ā·jənt }

Wagner's solution *See* Wagner's reagent. { 'väg·nərz sə¸lü·shən }

wagon drill [MECH ENG] **1.** A vertically mounted, pneumatic, percussive-type rock drill supported on a three- or four-wheeled wagon. **2.** A wheel-mounted diamond drill machine. { 'wag·ən ¸dril }

wagonhead vault *See* barrel vault. { 'wag·ən¸hed ¸vòlt }

wagon vault *See* barrel vault. { 'wag·ən ¸vòlt }

Wahl correlation [PETRO ENG] A pressure-volume-temperature (PVT) correlation used to estimate the total oil recovery from a solution-gas-drive oil reservoir; it is based on assumed PVT data, and may be in error. { 'väl ¸kär·ə¸lā·shən }

Waidner-Burgess standard [OPTICS] A unit of luminous intensity equal to the luminous intensity of 1 square centimeter of a blackbody at the melting point of platinum, or to 60 candelas. { 'wīd·nər 'bər·jəs ¸stan·dərd }

wainscot [BUILD] A decorative or protective panel installed over the lower portion of an interior partition or wall. { 'wānz·kət }

wairakite [MINERAL] $CaAl_2Si_4O_{12} \cdot 2H_2O$ A zeolite mineral that is isostructural with analcime. { 'wī·rə¸kīt }

waist [ENG] The center portion of a vessel or container that has a smaller cross section than the adjacent areas. { wāst }

wait [CONT SYS] Cessation of motion of a robot manipulator, under computer control, until further notice. { wāt }

waiting line [IND ENG] A line formed by units waiting for service. Also known as queue. { 'wād·iŋ ¸līn }

waiting time *See* idle time. { 'wād·iŋ ¸tīm }

wait state [COMPUT SCI] The state of a computer program in which it cannot use the central processing unit normally because the unit is waiting to complete an input/output operation. { 'wāt ¸stāt }

wake [FL MECH] The region behind a body moving relative to a fluid in which the effects of the body on the fluid's motion are concentrated. { wāk }

wake flow [FL MECH] Turbulent eddying flow that occurs downstream from bluff bodies. { 'wāk ¸flō }

wake gain [NAV ARCH] The increase in the effective thrust of a propeller, for a given power delivered thereto, because of the forward motion of the water dragged along behind a vessel's hull. { 'wāk ¸gān }

wake-induced flutter *See* buffeting flutter. { ¦wāk in¸düst 'fləd·ər }

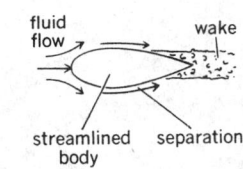

WAKE

Wake formed downstream from a streamlined body.

wake-induced galloping See buffeting flutter. { ¦wāk in¦düst 'gal·əp·iŋ }

wake stream theory [OCEANOGR] The theory that, in a stratified ocean, a compensation current must develop on the right side of a wake stream, flowing in the same direction, and a countercurrent in the opposite direction must appear to the left. { 'wāk ¦strēm ,thē·ə·rē }

Walden's rule [PHYS CHEM] A rule which states that the product of the viscosity and the equivalent ionic conductance at infinite dilution in electrolytic solutions is a constant, independent of the solvent; it is only approximately correct. { 'wȯl·dənz ,rül }

Waldeyer's ring [ANAT] A circular arrangement of the lymphatic tissues formed by the palatine and pharyngeal tonsils and the lymphatic follicles at the base of the tongue and behind the posterior pillars of the fauces. { 'väl,dī·ərz ,riŋ }

Wald-Wolfowitz run test [STAT] A procedure used in nonparametric statistics to determine whether the means of two independently drawn samples were taken from the same population. { ¦wȯld ¦wȯlf·ə,wits 'rən ,test }

wale [CIV ENG] See waler. [TEXT] 1. A rib or raised cord in a woven fabric. 2. A lengthwise row of loops in a knitted fabric. { wāl }

waler [CIV ENG] A horizontal reinforcement utilized to keep newly poured concrete forms from bulging outward. Also spelled whaler. Also known as wale. { 'wā·lər }

W* algebra See von Neumann algebra. { ¦dəb·əl,yü stär 'al·jə·brə }

walk [MATH] In graph theory, a set of vertices $(v_0, v_1, \ldots, v_n)$ in a graph, such that v_i and v_{i+1} are joined by a common edge for $i = 0, 1, \ldots, n-1$. Also known as path. { wȯk }

walk down [ELECTR] A malfunction in a magnetic core of a computer storage in which successive drive pulses or digit pulses cause changes in the magnetic flux in the core that persist after the magnetic fields associated with pulses have been removed. Also known as loss of information. { 'wȯk ,daùn }

walkie-talkie See pack unit. { ¦wȯ·kē ¦tȯ·kē }

walking beam [MECH ENG] A lever that oscillates on a pivot and transmits power in a manner producing a reciprocating or reversible motion; used in rock drilling and oil well pumping. { 'wȯk·iŋ ,bēm }

walking bird [VERT ZOO] Any bird of the order Columbiformes, including the pigeons, doves, and sandgrouse. { 'wȯk·iŋ ,bərd }

walking dragline [MECH ENG] A large-capacity dragline built with moving feet; disks 20 feet (6 meters) in diameter support the excavator while working. { 'wȯk·iŋ ,drag,līn }

walking machine [MECH ENG] A machine designed to carry its operator over various types of terrain; the operator sits on a platform carried on four mechanical legs, and movements of his arms control the front legs of the machine while movements of his legs control the rear legs of the machine. { 'wȯk·iŋ mə·shēn }

walking props See self-advancing supports. { 'wȯk·iŋ ,präps }

walk-off [GRAPHICS] In lithography, the failure of part of the image to stick to the metal plate while printing. { 'wȯk,ȯf }

walkthrough [COMPUT SCI] A step-by-step review of a computer program or system during its design to search for errors and problems. { 'wȯk,thrü }

walkthrough method [CONT SYS] The instruction of a robot by taking it through its sequences of motions, so that these actions are stored in its memory and recalled when necessary. { 'wȯk¦thrü ,meth·əd }

wall [ENG] A vertical structure or member forming an enclosure or defining a space. [GEOL] The side of a cave passage. [MIN ENG] 1. The side of a level or drift. 2. The country rock bounding a vein laterally. 3. The face of a longwall working or stall, commonly called coal wall. { wȯl }

Wallach transformation [ORG CHEM] By the use of concentrated sulfuric acid, an azoxybenzene is converted into a *para*-hydroxyazobenzene. { 'väl·ək ,tranz·fər,mā·shən }

wall anchor [BUILD] A steel strap fastened to the end of every second or third common joist and built into the brickwork of a wall to provide lateral support. Also known as joist anchor. { 'wȯl ,aŋ·kər }

wall-attachment amplifier [FL MECH] A bistable fluidic device utilizing two walls set back from the supply jet port, control ports, and channels to define two downstream outputs. Also known as flip-flop amplifier. { 'wȯl ə¦tach·mənt 'am·plə,fī·ər }

wallboard [MATER] Panels of various materials for surfacing ceilings and walls, including asbestos cement sheet, plywood, gypsum plasterboard, and laminated plastics. { 'wȯl,bȯrd }

wall box [BUILD] 1. A frame or box set into a wall to receive a beam or joist. Also known as beam box; wall frame. 2. A frame set into a wall to provide a sealed space for pipework to pass through. [ELEC] A metal box set into a wall to hold switches, receptacles, or similar electrical wiring components. { 'wȯl ,bäks }

wall cake [PETRO ENG] In drilling operations, the solid deposit along the hole wall due to filtration of the fluid part of the mud into the formation. { 'wȯl ¦kāk }

wall-coated capillary column [ANALY CHEM] A capillary column characterized by a layer of stationary liquid coated directly on the inner wall of a coiled capillary tube. { 'wȯl ¦kōd·əd 'kap·ə,ler·ē ,käl·əm }

wall coping [CIV ENG] The covering course on top of a brick or stone wall. { 'wȯl ,kōp·iŋ }

wall crane [MECH ENG] A jib crane mounted on a wall. { 'wȯl ,krān }

wall effect [ELECTR] The contribution to the ionization in an ionization chamber by electrons liberated from the walls. { 'wȯl i,fekt }

wall energy [SOLID STATE] The energy per unit area of the boundary between two ferromagnetic domains which are oriented in different directions. { 'wȯl ¦en·ər·jē }

waller See pack builder. { 'wȯl·ər }

Walley engine [MECH ENG] A multirotor engine employing four approximately elliptical rotors that turn in the same clockwise sense, leading to excessively high rubbing velocities. { 'wäl·ē ,en·jən }

wall frame See wall box. { 'wȯl ,frām }

wall friction [FL MECH] The drag created in the flow of a liquid or gas because of contact with the wall surfaces of its conductor, such as the inside surfaces of a pipe. { 'wȯl ,frik·shən }

wall furnace [MECH ENG] A self-contained vented furnace that is permanently attached to a wall and provides heated air directly to the surrounding space. { 'wȯl ,fər·nəs }

wall grille [BUILD] A perforated plate or a framed structure composed of rods or bars that is used to cover a wall opening to restrict vision but allow movement of air. { 'wȯl ,gril }

wall guard [BUILD] A protective strip of resilient material applied to the surface of a wall (especially along a corridor) several feet off the floor to prevent damage by vehicles used within a building. { 'wȯl ,gärd }

wall hanger [BUILD] A bracket installed in a masonry wall to support the end of a horizontal member. { 'wȯl ,haŋ·ər }

Wallis formulas [MATH] Formulas that determine the values of the definite integrals from 0 to $\pi/2$ of the functions $\sin^n (x)$, $\cos^n (x)$, and $\cos^m (x) \sin^n (x)$ for positive integers m and n. Also known as Wallis theorem. { 'wäl·əs ,fȯr·myə,ləz }

Wallis product [MATH] An infinite product representation of $\pi/2$, namely,

$$\frac{\pi}{2} = \frac{2}{1}\frac{2}{3}\frac{4}{3}\frac{4}{5} \cdots \frac{2n}{2n-1}\frac{2n}{2n+1} \cdots .$$

{ 'wäl·əs ,präd·əkt }

Wallis theorem See Wallis formulas. { 'wäl·əs ,thir·əm }

wall off [ENG] To seal cracks or crevices in the wall of a borehole with cement, mud cake, compacted cuttings, or casing. { 'wȯl ,ȯf }

wall outlet [ELEC] An outlet mounted on a wall, from which electric power can be obtained by inserting the plug of a line cord. { 'wȯl ,aùt·lət }

wallpaper [COMPUT SCI] The design or image used as a computer monitor background. { 'wȯl,pā·pər }

wall plate [BUILD] A piece of timber laid flat along the tip of the wall; it supports the rafters. Also known as raising plate. [MIN ENG] A horizontal timber supported by posts resting on sills and extending lengthwise on each side of the tunnel; roof supports rest on the wall plates. { 'wȯl ,plāt }

wall ratio [DES ENG] Ratio of the outside radius of a gun, a tube, or jacket to the inside radius; or ratio of the corresponding diameters. { 'wȯl ,rā·shō }

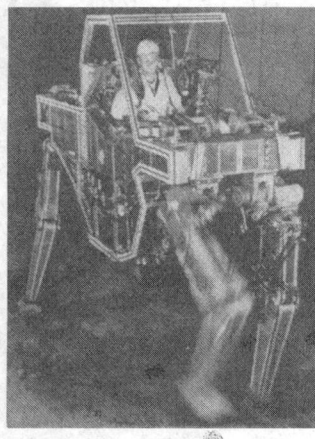

WALKING MACHINE

Research prototype of four-legged walking machine which was developed for use over various types of terrain. (*General Electric Co.*)

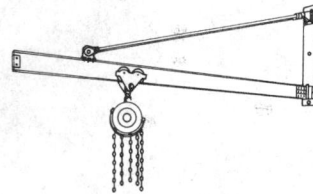

WALL CRANE

Wall crane, showing mounting on wall.

wall reef [GEOL] A linear, steep-sided coral reef constructed on a reef wall. { 'wȯl ¦rēf }

wall rock [GEOL] Rock that encloses a vein. { 'wȯl ¦räk }

wall-rock alteration [GEOL] Alteration of wall rock adjacent to hydrothermal veins by the fluid responsible for formation of the mineral deposit. { ¦wȯl ¦räk ˌȯl·tə'rā·shən }

wall-sided glacier [HYD] A glacier unconfined by a marked ravine or valley. { 'wȯl ¦sīd·əd 'glā·shər }

wall spacer [CIV ENG] A metal tie that holds a form for poured concrete in position until the concrete has set. { 'wȯl ˌspās·ər }

wall superheat [THERMO] The difference between the temperature of a surface and the saturation temperature (boiling point at the ambient pressure) of an adjacent liquid that is heated by the surface. { ¦wȯl 'sü·pər¸hēt }

wall tie [BUILD] A rigid, corrosion-resistant metal tie fitted into the bed joints across the cavity of a cavity wall. { 'wȯl ¸tī }

walnut [BOT] The common name for about a dozen species of deciduous trees in the genus *Juglans* characterized by pinnately compound, aromatic leaves and chambered or laminate pith; the edible nut of the tree is distinguished by a deeply furrowed or sculptured shell. { 'wȯl·nət }

walpurgite [MINERAL] $Bi_4(UO_2)(AsO_4)_2O_4 \cdot 3H_2O$ A wax yellow to straw yellow, triclinic mineral consisting of a hydrated arsenate of bismuth and uranium. Also known as waltherite. { wäl'pər¸jīt }

walrus [VERT ZOO] *Odobenus rosmarus*. The single species of the pinniped family Odobenidae distinguished by the upper canines in both sexes being prolonged as tusks. { 'wȯl·rəs }

Walter engine [MECH ENG] A multirotor rotary engine that uses two different-sized elliptical rotors. { 'wȯl·tər ¸en·jən }

waltherite *See* walpurgite. { 'väl·tə¸rīt }

WAN *See* wide-area network.

wand [COMPUT SCI] A hand-held device that contains an optical scanner to sense bar codes and other patterns and transmits the data to a computer. { wänd }

wander *See* apparent wander. *See* scintillation. { 'wän·dər }

wander correction [NAV] A correction to compensate for wander error in bubble sextant readings. { 'wän·dər kə¸rek·shən }

wandering dune [GEOL] A sand dune that has moved as a unit in the leeward direction of the prevailing winds, and that is characterized by the lack of vegetation to anchor it. Also known as migratory dune; traveling dune. { 'wän·də·riŋ 'dün }

wandering sequence [MET] A welding sequence in which the increments of weld bead are longitudinally deposited in a random fashion. { 'wän·də·riŋ ¸sē·kwəns }

wandering water *See* vadose water. { 'wän·də·riŋ 'wȯd·ər }

wane [MATER] A rounded edge of bark along an edge or at a corner of a section of lumber. { wān }

waning moon [ASTRON] The moon between full and new when its visible part is decreasing. { 'wän·iŋ 'mün }

Wankel engine [MECH ENG] An eccentric-rotor-type internal combustion engine with only two primary moving parts, the rotor and the eccentric shaft; the rotor moves in one direction around the trochoidal chamber containing peripheral intake and exhaust ports. Also known as rotary-combustion engine. { 'väŋ·kəl ¸en·jən }

Wanner optical pyrometer [ENG] A type of polarizing pyrometer in which beams from the source under investigation and a comparison lamp are polarized at right angles and then passed through a Nicol prism and a red filter; the source temperature is determined from the angle through which the Nicol prism must be rotated in order to equalize the intensities of the resulting patches of light. { ¦wän·ər ¦äp·tə·kəl pī'räm·əd·ər }

Wannier function [SOLID STATE] The Fourier transform of a Bloch function defined for an entire band, regarded as a function of the wave vector. { 'vän·yā ¸fəŋk·shən }

want *See* nip. { wänt }

warble tone [ACOUS] A tone whose frequency varies periodically several times per second over a small range; used to prevent standing-wave patterns from forming in reverberation chambers. { 'wȯr·bəl ¸tōn }

wardite [MINERAL] $Na_2CaAl_{12}(PO_4)_8(OH)_{18} \cdot 6H_2O$ A blue-green to pale green, tetragonal mineral consisting of a hydrated basic phosphate of sodium, calcium, and aluminum. { 'wȯr¸dīt }

Ward-Leonard speed-control system [CONT SYS] A system for controlling the speed of a direct-current motor in which the armature voltage of a separately excited direct-current motor is controlled by a motor-generator set. { 'wȯrd 'len·ərd 'spēd kən¸trōl ¸sis·təm }

warehouse [IND ENG] A building used for storing merchandise and commodities. { 'wer¸haus }

war game [ORD] A simulation, by whatever means, of a military operation involving two or more opposing forces, using rules, data, and procedures designed to depict an actual or assumed real-life situation. { 'wȯr ¸gām }

war gas [ORD] A toxic or irritant chemical agent, regardless of its physical state, whose properties may be effectively exploited in the field of war. { 'wȯr ¸gas }

warhead [ORD] An item which is designed to be mounted in or on a torpedo, guided missile, rocket, or bomb; it may contain high-explosive, nuclear, chemical, biological, or inert materials. { 'wȯr¸hed }

warhead installation [ORD] A warhead plus those additionally required items (contained in the adaptation kit) that are needed to mate the warhead with a specific carrier. { 'wȯr¸hed ¸in·stə'lā·shən }

warm-air drop *See* warm pool. { 'wȯrm ¸er 'dräp }

warm-air heating [MECH ENG] Heating by circulating warm air; system contains a direct-fired furnace surrounded by a bonnet through which air circulates to be heated. { 'wȯrm ¸er 'hēd·iŋ }

warm air mass [METEOROL] An air mass that is warmer than the surrounding air; an implication that the air mass is warmer than the surface over which it is moving. { 'wȯrm ¸er 'mas }

warm anticyclone *See* warm high. { 'wȯrm ¦ant·i'sī¸klōn }

warm-blooded *See* homoiothermal. { 'wȯrm ¦bləd·əd }

warm boot [COMPUT SCI] To boot a computer system after it has been running. { ¦wȯrm 'büt }

warm braw [METEOROL] A warm, dry, foehn wind which persists for up to 8 days during the east monsoon in the Schouten Islands off the north coast of New Guinea. { 'wȯrm 'brȯ }

warm-core anticyclone *See* warm high. { 'wȯrm ¦kȯr ¦ant·i'sī¸klōn }

warm-core cyclone *See* warm low. { 'wȯrm ¦kȯr 'sī¸klōn }

warm-core high *See* warm high. { 'wȯrm ¦kȯr 'hī }

warm-core low *See* warm low. { 'wȯrm ¦kȯr 'lō }

warm cyclone *See* warm low. { 'wȯrm 'sī¸klōn }

warm drop *See* warm pool. { 'wȯrm 'dräp }

warm front [METEOROL] Any nonoccluded front, or portion thereof, which moves in such a way that warmer air replaces colder air. { 'wȯrm ¸frənt }

warm high [METEOROL] At a given level in the atmosphere, any high that is warmer at its center than at its periphery. Also known as warm anticyclone; warm-core anticyclone; warm-core high. { 'wȯrm 'hī }

warm low [METEOROL] At a given level in the atmosphere, any low that is warmer at its center than at its periphery; the opposite of a cold low. Also known as warm-core cyclone; warm-core low; warm cyclone. { 'wȯrm 'lō }

warm pool [METEOROL] A region, or pool, of relatively warm air surrounded by colder air; the opposite of a cold pool; commonly applied to warm air of appreciable vertical extent isolated in high latitudes when a cutoff high is formed. Also known as warm-air drop; warm drop. { 'wȯrm 'pül }

warm sector [METEOROL] The area of warm air, within the circulation of a wave cyclone, which lies between the cold front and warm front of a storm. { 'wȯrm ¸sek·tər }

warm start [COMPUT SCI] A resumption of computer operation, following a problem-generated shutdown, in which programs running on the system can resume at the point they were at when the shutdown occurred and data is not lost. { 'wȯrm ¸stärt }

warm tongue [METEOROL] A pronounced poleward extension or protrusion of warm air. { 'wȯrm ¸təŋ }

warm-tongue steering [METEOROL] The steering influence apparently exerted upon a tropical cyclone by an upper-level warm tongue which often extends a considerable distance into regions adjacent to the cyclone. { 'wȯrm ¦təŋ ¸stir·iŋ }

warm-up time [ENG] A span of time between the first application of power to a system and the moment when the system can function fully. { 'wȯrm¸əp ¸tīm }

warm wave *See* heat wave. { 'wȯrm ¸wāv }

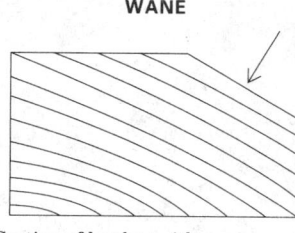

WANE

Section of lumber with a wane.

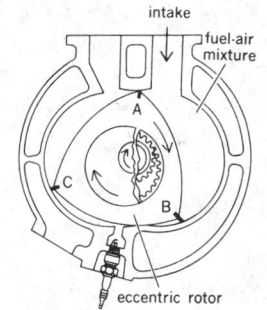

WANKEL ENGINE

intake

fuel-air mixture

A

C

B

eccentric rotor

Diagram of Wankel engine showing gears on rotor and shaft. Rotor divides inner volume into three chambers, each analogous to cyclinder in standard piston engine. Here, chamber AB is terminating the intake phase and commencing the compression phase, chamber BC is terminating the compression phase, and chamber CA is commencing the exhaust phase.

warning agent *See* odorant. { 'wȯrn·iŋ ˌā·jənt }

warning beacon *See* warning radio beacon. { 'wȯrn·iŋ ˌbē·kən }

warning device [COMPUT SCI] A visible or audible alarm to inform the operator of a machine condition. { 'wȯrn·iŋ di‚vīs }

warning message [COMPUT SCI] A diagnostic message that is issued when a computer program detects an error or potential problem but continues processing. { 'wȯrn·iŋ ˌmes·ij }

warning net [COMMUN] A communication system established for the purpose of disseminating warning information of enemy movement or action to all interested commands. { 'wȯrn·iŋ ˌnet }

warning pipe [ENG] An overflow pipe with a conspicuous outlet permitting prompt observation of discharge. { 'wȯrn·iŋ ˌpīp }

warning radio beacon [NAV] An auxiliary radio beacon that is located at a lightship to warn vessels of their proximity to the lightship; it is of short range, and sounds a warbling note for 1 minute immediately following the main radio beacon on the same frequency. Also known as warning beacon. { 'wȯrn·iŋ 'rād·ē·ō ˌbē·kən }

warning-receiver system [ELECTR] An electronic countermeasure system, carried on a tactical or transport aircraft, which is programmed to alert a pilot when his aircraft is being illuminated by a specific radar signal above predetermined power thresholds. { 'wȯrn·iŋ ri¦sē·vər ˌsis·təm }

warning stage [HYD] The stage, on a fixed river gage, at which it is necessary to begin issuing warnings or river forecasts if adequate precautionary measures are to be taken before flood stage is reached. { 'wȯrn·iŋ ˌstāj }

warp [GEOL] **1.** An upward or downward flexure of the earth's crust. **2.** A layer of sediment deposited by water. [NAV] To move a vessel or other waterborne object from one point to another by pulling on lines fastened to a fixed buoy, wharf, or such. [TEXT] Yarn extending lengthwise, under tension on a loom. Also known as end. { wȯrp }

warpage [MECH] The action, process, or result of twisting or turning out of shape. { 'wȯr·pij }

warp-faced fabric [TEXT] A fabric whose face has more ends than picks on it, the face being formed by warp yarns. { 'wȯrp ¦fāst 'fab·rik }

warping function *See* torsion function. { 'wȯrp·iŋ ˌfəŋk·shən }

warp knitting [TEXT] A knitting process in which a group of yarns form rows running lengthwise by an interlocking process. { 'wȯrp ˌnid·iŋ }

warp prints [TEXT] Blurred designs achieved in woven fabrics by printing the warp threads before fabrics are woven; subsequent weaving gives the soft blurred effect. { 'wȯrp ˌprins }

warrant *See* underclay. { 'wär·ənt }

Warren truss [CIV ENG] A truss having only sloping members between the top and bottom horizontal members. { 'wär·ən ˌtrəs }

war reserves [ORD] Stocks of material amassed in peacetime to meet the increase in military requirements consequent upon an outbreak of war, and intended to provide the interim support essential to sustain operations until resupply can be effected. { 'wȯr ri‚zərvz }

warringtonite *See* brochantite. { 'wär·iŋ·tə‚nīt }

Warrior *See* Orion. { 'wär·ē·ər }

war surplus [ORD] A military article of supply or piece of equipment that has been declared surplus because it is obsolete, unserviceable, or excess to current and reserve military requirements. { 'wȯr 'sər‚pləs }

wart [MED] A papillomatous growth which occurs singly or in groups on the skin surface; thought to be caused by a viral agent. { wȯrt }

warwickite [MINERAL] $(Mg,Fe)_3Ti(BO_4)_2$ A dark brown to dull black, orthorhombic mineral consisting of a titanoborate of magnesium and iron; occurs as prismatic crystals. { 'wȯr·i‚kīt }

Wasatch winds [METEOROL] Strong, easterly, jet-effect winds blowing out of the mouths of the canyons of the Wasatch Mountains onto the plains of Utah. { 'wä‚sach 'winz }

wash [AERO ENG] The stream of air or other fluid sent backward by a jet engine or a propeller. [BUILD] Any member that serves to carry water away from a section of a structure.

[ENG] **1.** To clean cuttings or other fragmental rock materials out of a borehole by the jetting and buoyant action of a copious flow of water or a mud-laden liquid. **2.** The erosion of core or drill string equipment by the action of a rapidly flowing stream of water or mud-laden drill-circulation liquid. [FL MECH] The surge of disturbed air or other fluid resulting from the passage of something through the fluid. [FOOD ENG] In the manufacture of whiskey, the fermented wort from which the spirit is distilled. [GEOL] **1.** An alluvial placer. **2.** A piece of land washed by a sea or river. **3.** *See* alluvial cone. [GRAPHICS] To dip negatives and prints in water after fixing to remove every remaining soluble silver halide-fixing agent complexes. [MET] **1.** A coating applied to the face of a mold prior to casting. **2.** A sand expansion defect on the surface finish of a casting due to radiation from the metal rising in the mold and causing increased volume and shear of the interface sand on the upper layers. { wäsh }

washability [MIN ENG] Coal properties determining the amenability of a coal to improvement in quality by cleaning. { ˌwäsh·ə'bil·əd·ē }

wash-and-strain ice foot [OCEANOGR] An ice foot formed from ice casts and slush and attached to a shelving beach, between the high and low waterlines; high waves and spray may cause it to build up above the high waterline. { ¦wäsh ən ¦strān 'īs ˌfút }

washboard course [ENG] A test course for vehicles consisting of a series of waves or convolutions having arbitrary amplitude and frequency; a common type is the so-called sine-wave course. { 'wäsh‚bȯrd ˌkȯrs }

wash boring *See* jet drilling. { 'wäsh ˌbȯr·iŋ }

wash-built terrace *See* alluvial terrace. { 'wäsh ¦bilt 'ter·əs }

wash coat [ENG] A sealer consisting of a very thin, semitransparent coat of paint. { 'wäsh ˌkōt }

washer [DES ENG] A flattened, ring-shaped device used to improve the tightness of a screw fastener. [ENG] **1.** A device for removing dirt and soluble impurities from pulp and paper stock. **2.** A system for washing photographic materials to remove soluble products of developing or fixing. **3.** A power-driven machine for washing clothes and household linens. Also known as washing machine. **4.** *See* scrubber. { 'wäsh·ər }

washer method [MATH] A method of computing the volume of a solid of revolution that is hollow about its axis, by integrating over the volumes of infinitesimal washer-shaped slices bounded by planes perpendicular to the axis of revolution. { 'wäsh·ər ˌmeth·əd }

washer thermistor [ELECTR] A thermistor in the shape of a washer, which may be as large as 0.75 inch (1.9 centimeters) in diameter and 0.50 inch (1.3 centimeters) thick; it is formed by pressing and sintering an oxide-binder mixture. { 'wäsh·ər thər‚mis·tər }

washing [ANALY CHEM] **1.** In the purification of a laboratory sample, the cleaning of residual liquid impurities from precipitates by adding washing solution to the precipitates, mixing, then decanting, and repeating the operation as often as needed. **2.** The removal of soluble components from a mixture of solids by using the effect of differential solubility. [CHEM ENG] In a process operation, cleaning of a solids bed (settler) or cake (filter) with a liquid in which the solid is not soluble. { 'wäsh·iŋ }

washing machine *See* washer. { 'wäsh·iŋ mə‚shēn }

washing plant [MIN ENG] A plant where slimes are removed from relatively coarse ore by washing, tumbling, or scrubbing. { 'wäsh·iŋ ˌplant }

washing soda *See* sal soda. { 'wäsh·iŋ ˌsōd·ə }

Washita stone [MATER] A relatively porous and not very dense oilstone, used chiefly for whetstones and for sharpening coarse tools. { 'wäsh·əd·ə ˌstōn }

wash load [GEOL] The finer part of the total sediment load of a stream which is supplied from bank erosion or an external upstream source, and which can be carried in large quantities. { 'wäsh ˌlōd }

wash metal [MET] Molten metal used to clean out a furnace, ladle, or other container. { 'wäsh ˌmed·əl }

Washoe zephyr [METEOROL] The chinook on the Nevada side of the Sierra Nevada Mountains of northern California. { 'wä‚shō 'zef·ər }

wash oil *See* absorption oil. { 'wäsh ˌoil }

washout [ENG] **1.** An overlarge well bore caused by the solvent and erosional action of drilling fluid. **2.** A fluid-cut opening resulting from leaking fluid. [MIN ENG] *See* horseback. { 'wäsh‚aút }

washover [GEOL] Material deposited by overwash, especially a small delta produced by storm waves and built on the landward side of a bar or barrier. Also known as storm delta; wave delta. { 'wäsh‚ō·vər }

wash plain *See* outwash plain. { 'wäsh ‚plān }

wash primer [MET] A synthetic vehicle primer containing phosphoric acid and zinc chromate; used as a corrosion-inhibiting first paint coat on metals { 'wäsh ‚prī·mər }

wash slope [GEOL] The gentle slope on a hillside occurring below the gravity slope and lying at the foot of an escarpment or steep rock face; usually covered by an accumulation of talus. Also known as haldenhang. { 'wäsh ‚slōp }

wash water [CHEM ENG] Water contacted with process streams (liquid or gas), packed beds, or filter cakes to flush or dissolve out impurities. { 'wäsh ‚wöd·ər }

wasp [INV ZOO] The common name for members of 67 families of the order Hymenoptera; all are important as parasites or predators of injurious pests { wäsp }

Wasserman test [IMMUNOL] A complement-fixation test for syphilis using sensitized lipid extracts of beef heart as antigen. { 'was·ər·mən ‚test }

waste [ENG] **1.** Rubbish from a building. **2.** Dirty water from mining, industrial, and domestic use. **3.** The amount of excavated material exceeding fill. [MIN ENG] **1.** The barren rock in a mine. **2.** The refuse from ore dressing and smelting plants. **3.** The fine coal made in mining and preparing coal for market. { wāst }

waste acid *See* sludge acid. { 'wāst ‚as·əd }

waste bank [MIN ENG] A bank made of earth excavated during the digging of a ditch and laid parallel to it. { 'wāst ‚baŋk }

waste filling [MIN ENG] Material used for support in heavy ground and in large stopes to prevent failure of rock walls and to minimize or control subsidence and to make it possible to extract pillars of ore left in the earlier stages of mining; material used for filling includes waste rock sorted out in the stopes or mined from rock walls, milltailing, sand and gravel, smelter slag, and rock from surface open cuts or quarries. { 'wāst ‚fil·iŋ }

waste heat [ENG] Sensible heat in gases not subject to combustion and used for processes downstream in a system. { 'wāst ‚hēt }

waste-heat boiler [CHEM ENG] A heat-retrieval unit using hot by-product gas or oil from chemical processes; used to produce steam in a boiler-type system. Also known as gas-tube boiler. { 'wāst ‚hēt 'bói·lər }

waste lubrication [ENG] A method in which a lubricant is delivered to a bearing surface by the wicking action of cloth waste or yarn. { 'wāst ‚lü·brə‚kā·shən }

waste pipe [CIV ENG] A pipe to carry waste water from a basin, bath, or sink in a building. { 'wāst ‚pīp }

waste plain *See* alluvial plain. { 'wāst ‚plān }

waste raise [MIN ENG] An excavation in the mine in which barren rock and other material is broken up for use as filling at a stope. { 'wāst ‚rāz }

waster dump [ARCHEO] A refuse deposit of vessels (or fragments of vessels) that were cracked, warped, or otherwise damaged and made unusable during firing. { 'wās·tər ‚dəmp }

waste rock [MIN ENG] Valueless rock that must be fractured and removed in order to gain access to or upgrade ore. Also known as muck; mullock. { 'wāst ‚räk }

waste silk [TEXT] Short silk filaments which are left on a reel after removing the long filaments; used for spun silk. { 'wāst ‚silk }

waste vent *See* stack vent. { 'wāst ‚vent }

watch [COMMUN] The service performed by a qualified operator when on duty in the radio room of a vessel. Also known as radio watch. [HOROL] A small timepiece of a size convenient to be carried on the person. { wäch }

watch buoy *See* station buoy. { 'wäch ‚bói }

watchdog timer [CONT SYS] In a flexible manufacturing system, a safety device in the form of a control interface on an automated guided vehicle that shuts down part or all of the system under certain conditions. { 'wäch‚dòg ‚tīm·ər }

watch error [HOROL] The amount by which watch time differs from the correct time; it is usually expressed as seconds per day and labeled fast (F) and slow (S). { 'wäch ‚er·ər }

watch rate [HOROL] The amount gained or lost by a watch or clock in unit time; it is usually expressed in seconds per 24 hours, to an accuracy of 0.1 second, and labeled gaining or losing. Also known as daily rate. { 'wäch ‚rāt }

watch time [HOROL] The hour of the day as indicated by a watch or clock. { 'wäch ‚tīm }

water [CHEM] H_2O Clear, odorless, tasteless liquid that is essential for most animal and plant life and is an excellent solvent for many substances; melting point 0°C (32°F), boiling point 100°C (212°F); the chemical compound may be termed hydrogen oxide. { 'wöd·ər }

water absorption tube [ANALY CHEM] A glass tube filled with a solid absorbent (calcium chloride or silica gel) to remove water from gaseous streams during or after chemical analyses. { 'wöd·ər əb'sörp·shən ‚tüb }

water-activated battery [ELEC] A primary battery that contains the electrolyte but requires the addition of or immersion in water before it is usable. { 'wöd·ər ‚ak·tə‚vād·əd 'bad·ə·rē }

water atmosphere [METEOROL] The concept of a separate atmosphere composed only of water vapor. { 'wöd·ər 'at·mə‚sfir }

water ballast [NAV ARCH] Water confined to double-bottom tanks, peak tanks, or other designated compartments, for use in obtaining satisfactory draft, trim, or stability. { 'wöd·ər ‚bal·əst }

water ballast tank [NAV ARCH] A tank in which sea water for ballast is confined. { 'wöd·ər 'bal·əst ‚taŋk }

water bar [BUILD] A strip of material attached to the sill of a window or external door to prevent penetration by water. Also known as weather bar. { 'wöd·ər ‚bär }

water-base mud [PETRO ENG] Oil-well drilling mud in which the liquid component is water, into which are mixed the thickeners and other additives. { 'wöd·ər ‚bās 'məd }

water-base paint [MATER] Paint in which the vehicle or binder is dissolved in water or in which the vehicle or binder is dispersed as an emulsion; an example of the dispersion type is latex paint. Also known as water-thinned paint. { 'wöd·ər ‚bās 'pānt }

Water Bearer *See* Aquarius. { 'wöd·ər 'ber·ər }

water-bearing strata [GEOL] Ground layers below the standing water level. { 'wöd·ər 'ber·iŋ 'strad·ə }

water block [PETRO ENG] The tendency of accumulated water-oil emulsion around the lower (producing) end of an oil well borehole to block the movement of formation fluids through the formation and toward the borehole. { 'wöd·ər ‚bläk }

water boiler *See* water-boiler reactor. { 'wöd·ər ‚bói·lər }

water-boiler reactor [NUCLEO] A homogeneous reactor that uses enriched uranium as fuel and ordinary water as moderator; the fuel is uranyl sulfate dissolved in water. Also known as water boiler. { 'wöd·ər 'bói·lər rē‚ak·tər }

water-borne [SCI TECH] Floating on or transported by water. { 'wöd·ər ‚bórn }

water-borne disease [MED] A disease transmitted by drinking water or by contact with potable or bathing water. { 'wöd·ər ‚bórn di'zēz }

water brake [ENG] An absorption dynamometer for measuring power output of an engine shaft; the mechanical energy is converted to heat in a centrifugal pump, with a free casing where turning moment is measured. { 'wöd·ər ‚brāk }

water brash [MED] Severe distress caused by regurgitation of gastric acid into the throat. { 'wöd·ər ‚brash }

water break [MET] A break in the continuity of the film of water on the surface of a metal withdrawn from an aqueous bath. { 'wöd·ər ‚brāk }

water budget *See* hydrologic accounting. { 'wöd·ər ‚bəj·ət }

water bug [INV ZOO] Any insect which lives in an aquatic habitat during all phases of its life history. { 'wöd·ər ‚bəg }

water calorimeter [ENG] A calorimeter that measures radiofrequency power in terms of the rise in temperature of water in which the r-f energy is absorbed. { 'wöd·ər ‚kal·ə'rim·əd·ər }

water clock [HOROL] An ancient device to estimate time; the operation depended upon the slow emptying of water from

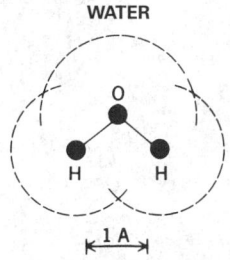

WATER

Water molecule, showing positions of atoms. Dotted circles show effective sizes of isolated atoms. Scale bar is 1 angstrom.

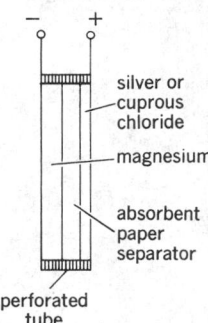

WATER-ACTIVATED BATTERY

silver or cuprous chloride

magnesium

absorbent paper separator

perforated tube

Schematic diagram of cell of water-activated battery.

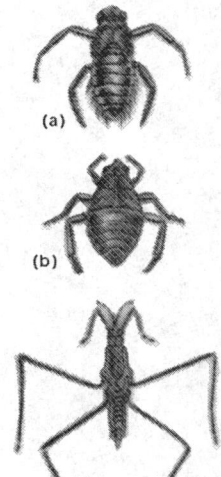

WATER BUG

(a)

(b)

(c)

Water bugs of order Hemiptera. (*a*) Water boatman. (*b*) Giant water bug, small species. (*c*) Water strider.

one graduated vessel into another, and the graduations marked the time periods. { 'wȯd·ər ‚kläk }

water cloud [METEOROL] Any cloud composed entirely of liquid water drops; to be distinguished from an ice-crystal cloud and from a mixed cloud. { 'wȯd·ər ‚klaúd }

watercolor [MATER] A pigment ground in a solution of gum arabic, water, and plasticizer, such as glycerin; the glycerin film retards drying in the tube and prevents brittleness in the paint film. { 'wȯd·ər‚kəl·ər }

watercolor paper [MATER] A special drawing paper with a surface texture suitable to accept watercolors; the better grades can withstand the harsh scraping that is sometimes necessary to produce highlights; for permanent painting, the paper should be 100% rag, not wood pulp. { 'wȯd·ər‚kəl·ər ‚pā·pər }

watercolor pigment [INORG CHEM] A permanent pigment used in watercolor painting, for example, titanium oxide (white). { 'wȯd·ər‚kəl·ər ‚pig·mənt }

water column [MECH ENG] A tubular column located at the steam and water space of a boiler to which protective devices such as gage cocks, water gage, and level alarms are attached. { 'wȯd·ər ‚käl·əm }

water conservation [ECOL] The protection, development, and efficient management of water resources for beneficial purposes. { 'wȯd·ər ‚kän·sər'vā·shən }

water content [HYD] The liquid water present within a sample of snow (or soil) usually expressed in percent by weight; the water content in percent of water equivalent is 100 minus the quality of snow. Also known as free-water content; liquid-water content. { 'wȯd·ər ‚kän‚tent }

water-cooled condenser [MECH ENG] A steam condenser which is for the maintenance of vacuum, and in which water is the heat-receiving fluid. { 'wȯd·ər ¦küld kən'den·sər }

water-cooled furnace [MECH ENG] A fuel-fired furnace containing tubes in which water is circulated to limit heat loss to the surroundings, control furnace temperature, and generate steam. { 'wȯd·ər ¦küld 'fər·nəs }

water-cooled reactor [NUCLEO] A nuclear reactor in which water is used as a primary coolant. { 'wȯd·ər ¦küld rē'ak·tər }

water-cooled tube [ELECTR] An electron tube that is cooled by circulating water through or around the anode structure. { 'wȯd·ər ¦küld 'tüb }

water cooling [ELECTR] Cooling the electrodes of an electron tube by circulating water through or around them. [ENG] Cooling in which the primary coolant is water. { 'wȯd·ər ‚kül·iŋ }

watercourse [HYD] **1.** A stream of water. **2.** A natural channel through which water may run or does run. { 'wȯd·ər‚kȯrs }

watercress [BOT] *Nasturtium officinale.* A perennial cress generally grown in flooded soil beds and used for salads and food garnishing. { 'wȯd·ər‚kres }

water curb *See* garland. { 'wȯd·ər ‚kərb }

water cycle *See* hydrologic cycle. { 'wȯd·ər ‚sī·kəl }

water demineralizing [CHEM ENG] The removal of minerals (for example, compounds of Ca, Mg, and Na) from water by chemical, ion-exchange, or distillation procedures. { 'wȯd·ər dē'min·rə‚līz·iŋ }

water-drive reservoir [PETRO ENG] An oil or gas reservoir in which pressure is maintained to a greater or lesser extent by an influx of water as the oil or gas is removed. { 'wȯd·ər ¦drīv 'rez·əv‚wär }

water dropper [ELEC] A simple electrostatic generator in which each of two series of water drops falls through cylindrical metal cans into lower cans with funnels, and the cans are electrically connected in such a way that charge accumulates on them, energy being supplied by the gravitational force on the water drops. { 'wȯd·ər ‚dräp·ər }

water equivalent [METEOROL] The depth of water that would result from the melting of the snowpack or of a snow sample; thus, the water equivalent of a new snowfall is the same as the amount of precipitation represented by that snowfall. { 'wȯd·ər i'kwiv·ə·lənt }

water exchange [OCEANOGR] The volume and rate of water exchange between air and a body of water in a specific location, or between several bodies of water, controlled by such factors as tides, winds, river discharge, and currents. { 'wȯd·ər iks‚chānj }

waterfall [HYD] A perpendicular or nearly perpendicular descent of water in a stream. { 'wȯd·ər‚fȯl }

waterfall lake *See* plunge pool. { 'wȯd·ər‚fȯl ‚lāk }

waterflooding *See* flooding. { 'wȯd·ər¦fläd·iŋ }

water-flow pyrheliometer [ENG] An absolute pyrheliometer, in which the radiation-sensing element is a blackened, water calorimeter; it consists of a cylinder, blackened on the interior, and surrounded by a special chamber through which water flows at a constant rate; the temperatures of the incoming and outgoing water, which are monitored continuously by thermometers, are used to compute the intensity of the radiation. { 'wȯd·ər ¦flō ‚pir‚hē·lē'äm·əd·ər }

waterfowl [VERT ZOO] Aquatic birds which constitute the order Anseriformes, including the swans, ducks, geese, and screamers. { 'wȯd·ər‚faúl }

water front [GEOGR] An area partly bounded by water. { 'wȯd·ər ‚frənt }

water gage [ENG] A gage glass with attached fittings which indicates water level in a vessel. { 'wȯd·ər ‚gāj }

water gap [GEOL] A deep and narrow pass that cuts to the base of a mountain ridge, and through which a stream flows; the Delaware Water Gap is an example. { 'wȯd·ər ‚gap }

water garland *See* garland. { 'wȯd·ər ‚gär·lənd }

water gas [MATER] A mixture of carbon monoxide and methane produced by passing steam through deep beds of incandescent coal; used for industrial heating and as a gas engine fuel. { 'wȯd·ər ‚gas }

water-gas coke [MATER] Coke which is used in the manufacture of water gas, and which should have a low ash content, a softening temperature of about 2500°F (1370°C), a low sulfur content, and a size larger than 2 inches (5 centimeters). { 'wȯd·ər ¦gas ‚kōk }

water-gas reaction [CHEM ENG] A method used to prepare carbon monoxide by passing steam over hot coke or coal at 600–1000°C. { 'wȯd·ər ¦gas rē‚ak·shən }

water glass *See* sodium silicate. { 'wȯd·ər ‚glas }

water hammer [FL MECH] Pressure rise in a pipeline caused by a sudden change in the rate of flow or stoppage of flow in the line. { 'wȯd·ər ‚ham·ər }

water heater [MECH ENG] A tank for heating and storing hot water for domestic use. { 'wȯd·ər ‚hēd·ər }

Waterhouse-Frederikson syndrome [MED] The association of bacteremia, particularly acute meningococcemia, massive skin hemorrhage, shock, and acute adrenal hemorrhage and insufficiency. { 'wȯd·ər‚haús ¦fred·rik·sən ‚sin‚drōm }

water influx [PETRO ENG] **1.** The incursion of water (natural or injected) into oil- or gas-bearing formations. **2.** One of the mechanisms of oil production in which the water movement (drive) displaces and moves the reservoir fluids toward the well borehole. { 'wȯd·ər 'in‚fləks }

water jacket [ENG] A casing for circulation of cooling water. { 'wȯd·ər ‚jak·ət }

waterjet [MIN ENG] *See* hydroexcavation. [NAV ARCH] A ship or boat propulsion device consisting of a pump, usually located within the hull, that receives water through an inlet duct and discharges it through a nozzle at increased velocity to produce propulsive thrust. Also known as marine jet. { 'wȯd·ər ‚jet }

water-jet cutting [ENG] A machining method that uses a jet of pressurized water containing abrasive powder for cutting steel and other dense materials. { 'wȯd·ər ‚jet ‚kəd·iŋ }

water joint [CIV ENG] A joint in a stone pavement containing stones that are set slightly higher to prevent water from settling in the joint. { 'wȯd·ər ‚jȯint }

water knockout drum [PETRO ENG] A device for removal of water from oil well fluids (gas, or gas with oil). Also known as water knockout trap; water knockout vessel. { 'wȯd·ər 'näk‚aút ‚drəm }

water knockout trap *See* water knockout drum. { 'wȯd·ər 'näk‚aút ‚trap }

water knockout vessel *See* water knockout drum. { 'wȯd·ər 'näk‚aút ‚ves·əl }

water lane [NAV] A designated lane or strip of water marked and maintained for the takeoff and landing of seaplanes. { 'wȯd·ər ‚lān }

water leg [ENG] The vertical area of a vessel or accessory to a vessel for the collection of water. Also known as sump. { 'wȯd·ər ‚leg }

waterless offset lithography [GRAPHICS] A method of printing in which the nonimage area of the plate is coated with silicone to repel the ink (in contrast to water in traditional lithographic printing) and the image is printed from ink that adheres to the image in the metal base. In practice, the image from a plate is offset onto the rubber blanket of an impression cylinder, and transferred to a sheet of paper. Also known as dryography. { ¦wȯd·ər‚les ¦óf‚set li'thäg·rə·fē }

waterless zone [HYD] The lowest hydrologic zone, generally beginning several miles beneath the land surface and characterized by the absence of water in the pore spaces due to the great pressure and density of the rock. { 'wȯd·ər·ləs 'zōn }

water level *See* water table. { 'wȯd·ər ‚lev·əl }

waterline [GEOL] *See* shoreline. [HYD] *See* water table. [NAV ARCH] **1.** The intersection of the surface of the water with the side of a ship. **2.** A line painted on the hull of a ship showing the level of the water when the ship is properly trimmed. { 'wȯd·ər‚līn }

water load [ELECTROMAG] A matched waveguide termination in which the electromagnetic energy is absorbed in water, the resulting rise in the temperature of the water is a measure of the output power. { 'wȯd·ər ‚lōd }

Waterloo Fortran IV *See* WATFIV. { 'wȯd·ər‚lü 'fȯr‚tran 'fȯr }

water loss *See* evapotranspiration. { 'wȯd·ər ‚lós }

water main [CIV ENG] The water pipe in a street from which water is delivered to individual service pipes supplying domestic property. { 'wȯd·ər ‚mān }

watermark [GRAPHICS] A localized modification of the structure and opacity of a sheet of paper so that a pattern or design can be seen when the sheet is held to the light. { 'wȯd·ər‚märk }

water mass [OCEANOGR] A body of water identified by its temperature-salinity curve or chemical composition, and normally consisting of a mixture of two or more water types. { 'wȯd·ər ‚mas }

watermelon [BOT] *Citrullus vulgaris.* An annual trailing vine with light-yellow flowers and leaves having five to seven deep lobes; the edible, oblong or roundish fruit has a smooth, hard, green rind filled with sweet, tender, juicy, pink to red tissue containing many seeds { 'wȯd·ər‚mel·ən }

water meter [ENG] An instrument for measuring the amount of water passing a specified point in a piping system. { 'wȯd·ər ‚mēd·ər }

water microbiology [MICROBIO] An aspect of microbiology that deals with the normal and adventitious microflora of natural and artificial water bodies. { 'wȯd·ər ¦mī·krō·bī'äl·ə·jē }

water moccasin [VERT ZOO] *Agkistrodon piscivorus.* A semiaquatic venomous pit viper; skin is brownish or olive on the dorsal aspect, paler on the sides, and has indistinct black bars. Also known as cottonmouth. { 'wȯd·ər ‚mäk·ə·sən }

water-moderated reactor [NUCLEO] A nuclear reactor in which water is the principal moderator. { 'wȯd·ər¦mäd·ə‚rād·əd rē'ak·tər }

Water Monster *See* Hydra. { 'wȯd·ər ‚män·stər }

water noise [ACOUS] Underwater acoustic energy resulting primarily from the movement of the water itself. { 'wȯd·ər ‚nóiz }

water of crystallization *See* water of hydration. { 'wȯd·ər əv ‚krist·əl·ə'zā·shən }

water of hydration [CHEM] Water present in a definite amount and attached to a compound to form a hydrate; can be removed, as by heating, without altering the composition of the compound. Also known as water of crystallization. { 'wȯd·ər əv hī'drā·shən }

water opal *See* hyalite. { 'wȯd·ər ‚ō·pəl }

water opening [OCEANOGR] A break in sea ice, revealing the sea surface. { 'wȯd·ər ‚ō·pə·niŋ }

water paint [MATER] A paint in which the vehicle or binder is dissolved in water; examples are calcimine in which the vehicle is glue, and casein paints in which the vehicle is casein. { 'wȯd·ər ‚pānt }

water path [ENG] In ultrasonic testing, distance from an ultrasonic search unit to the test piece in an immersion or water column examination. { 'wȯd·ər ‚path }

water plane [NAV ARCH] A plane coincident with or parallel to the surface of the water and limited by the line of its intersection with the vessel's hull. { 'wȯd·ər ‚plān }

water-plane area [NAV ARCH] The area of the water plane at which the ship floats. { 'wȯd·ər ‚plān ‚er·ē·ə }

water-plane coefficient [NAV ARCH] The ratio of the area of the water plane of a ship at the surface of the water to the

product of its beam and its length on the waterline. { 'wȯd·ər ‚plän ‚kō·i‚fish·ənt }

water pollution [ECOL] Contamination of water by materials such as sewage effluent, chemicals, detergents, and fertilizer runoff. { 'wȯd·ər pə‚lü·shən }

water potential [PHYSIO] The difference in free energy or chemical potential (per unit molal volume) between pure water and water in cells and solutions. { 'wȯd·ər pə‚ten·chəl }

waterpower [MECH] Power, usually electric, generated from an elevated water supply by the use of hydraulic turbines. { 'wȯd·ər‚paů·ər }

waterproof [ENG] Impervious to water. { 'wȯd·ər‚prüf }

waterproof grease [MATER] A viscous lubricating material that does not dissolve in water and that resists being washed out of bearings or gears; it usually has a low content of oil and metallic soaps of aluminum, barium, calcium, or strontium. { 'wȯd·ər‚prüf 'grēs }

waterproofing agent [MATER] A substance used to make textiles, paper, wood, and other porous or absorbent materials impervious to penetration by water. { ¦wȯd·ər¦prüf·iŋ ‚ā·jənt }

water-pump lubricant [MATER] A lubricating grease suitable for the types of automotive water pumps that require grease lubrication. { 'wȯd·ər ‚pəmp ‚lü·brə·kənt }

water purification [CIV ENG] Any of several processes in which undesirable impurities in water are removed or neutralized; for example, chlorination, filtration, primary treatment, ion exchange, and distillation. { 'wȯd·ər ‚pyùr·ə·fə'kā·shən }

water putty [MATER] A powder that forms a puttylike paste when mixed with water and is used to fill small holes or cracks in wood. { 'wȯd·ər ‚pəd·ē }

water-reducing agent [MATER] An additive for freshly mixed mortar or concrete for increasing workability without increasing water content, or for maintaining workability with a reduced amount of water. { 'wȯd·ər ri‚düs·iŋ ‚ā·jənt }

water repellent [MATER] Chemicals used to treat textiles, leather, paper, or wood to make them resistant (but not proof) to wetting by water; includes various types of resins, aluminum of zirconium acetates, or latexes. { 'wȯd·ər ri‚pel·ənt }

water requirement [HYD] The total quantity of water required to mature a specified crop under field conditions; includes applied irrigation, water precipitation, and groundwater available to the crop. { 'wȯd·ər ri‚kwīr·mənt }

water retting [MICROBIO] A type of retting process in which the stalks of fiber plants are immersed in cold or warm, slowly renewed water, for 4 days to several weeks. The active organism is *Clostridium felsineum* and related types, which break down the pectin to a mixture of organic acids (chiefly acetic and butyric), alcohols (butanol, ethanol, and methanol), carbon dioxide (CO_2), and hydrogen (H_2). { 'wȯd·ər ‚red·iŋ }

water rheostat *See* electrolytic rheostat. { 'wȯd·ər 'rē·ə‚stat }

water right [ENG] The right to use water for mining, agricultural, or other purposes. { 'wȯd·ər ‚rīt }

water ring *See* garland. { 'wȯd·ər ‚riŋ }

water sample [ENG] A portion of water brought up from a depth to determine its composition. { 'wȯd·ər ‚sam·pəl }

water saturation [CHEM] **1.** A solid adsorbent that holds the maximum possible amount of water under specified conditions. **2.** A liquid solution in which additional water will cause the appearance of a second liquid phase. **3.** A gas that is at or just under its dew point because of its water content. { 'wȯd·ər ‚sach·ə'rā·shən }

water scrubber [CHEM ENG] A device or system in which gases are contacted with water (either by spray or bubbling through) to wash out traces of water-soluble components of the gas stream. { 'wȯd·ər ‚skrəb·ər }

water seal [ENG] A seal formed by water to prevent the passage of gas. { 'wȯd·ər ‚sēl }

water-sealed holder [ENG] A low-pressure gas holder which consists of cylindrical sections or lifts telescoping into a pit or tank filled with water; the inside section is closed in on top. { 'wȯd·ər ‚sēld 'hōl·dər }

water seasoning [MATER] A wood treatment process in which lumber is soaked in water prior to air drying. { 'wȯd·ər ‚sēz·ən·iŋ }

watershed [HYD] The drainage area of a stream. { 'wȯd·ər‚shed }

water sky [METEOROL] The dark appearance of the underside of a cloud layer when it is over a surface of open water. { 'wȯd·ər ˌskī }

water smoke *See* steam fog. { 'wȯd·ər ˌsmōk }

Water Snake *See* Hydrus. { 'wȯd·ər ˌsnāk }

water snow [HYD] Snow that, when melted, yields a more than average amount of water; thus, any snow with a high water content. { 'wȯd·ər ˌsnō }

water softening [CHEM] Removal of scale-forming calcium and magnesium ions from hard water, or replacing them by the more soluble sodium ions; can be done by chemicals or ion exchange. { 'wȯd·ər ˌsȯf·ə·niŋ }

waterspout [ENG] A pipe or orifice through which water is discharged or by which it is conveyed. [METEOROL] A tornado occurring over water; rarely, a lesser whirlwind over water, comparable in intensity to a dust devil over land. { 'wȯd·ərˌspau̇t }

water-supply engineering [CIV ENG] A branch of civil engineering concerned with the development of sources of supply, transmission, distribution, and treatment of water. { 'wȯd·ər səˌplī ˌen·jə'nir·iŋ }

water swivel [DES ENG] A device connecting the water hose to the drill-rod string and designed to permit the drill string to be rotated in the borehole while water is pumped into it to create the circulation needed to cool the bit and remove the cuttings produced. Also known as gooseneck; swivel neck. { 'wȯd·ər ˌswiv·əl }

water table [BUILD] A ledge or slight projection of the masonry or wood construction on the exterior of a foundation wall, or just above it, to protect the foundation by directing rainwater away from the wall. Also known as canting strip. [HYD] The planar surface between the zone of saturation and the zone of aeration. Also known as free-water elevation; free-water surface; groundwater level; groundwater surface; groundwater table; level of saturation; phreatic surface; plane of saturation; saturated surface; water level; waterline. { 'wȯd·ər ˌtā·bəl }

water-thinned paint *See* water-base paint. { 'wȯd·ər ˌthind 'pānt }

watertight subdivision [NAV ARCH] A part of a ship that can be sealed off so that water cannot enter it. { 'wȯd·ərˌtīt 'səb·də,vizh·ən }

water tower [CIV ENG] A tower or standpipe for storing water in areas where ordinary water pressure is inadequate for distribution to consumers. { 'wȯd·ər ˌtau̇·ər }

water trap [GEOL] A chamber or part of a cave system that is filled with water, due to the dipping of the roof or ceiling below the water level. { 'wȯd·ər ˌtrap }

water treatment [CIV ENG] Purification of water to make it suitable for drinking or for any other use. { 'wȯd·ər ˌtrēt·mənt }

water-tube boiler [MECH ENG] A steam boiler in which water circulates within tubes and heat is applied from outside the tubes to generate steam. { 'wȯd·ər ˌtüb ˌbȯi·lər }

water tunnel [AERO ENG] A device similar to a wind tunnel, but using water as the working fluid instead of air or other gas. [CIV ENG] A tunnel to transport water in a water-supply system. { 'wȯd·ər ˌtən·əl }

water type [OCEANOGR] Ocean water of a specified temperature and salinity. { 'wȯd·ər ˌtīp }

water vapor [PHYS] Water in the form of a vapor, especially when below the boiling point and diffused. { 'wȯd·ər ˌvā·pər }

water-vapor absorption [METEOROL] The absorption of certain wavelengths of infrared radiation by atmospheric water vapor; a process of fundamental importance in the energy budget of the earth's atmosphere. { 'wȯd·ər ˌvā·pər əbˌsȯrp·shən }

water-vapor laser [OPTICS] A laser whose active substance is water vapor, and which emits infrared radiation at wavelengths of 27.97, 47.7, 78.46, and 118.6 micrometers. { 'wȯd·ər ˌvā·pər 'lā·zər }

water vascular system [INV ZOO] An internal closed system of reservoirs and ducts containing a watery fluid, found only in echinoderms. { 'wȯd·ər 'vas·kyə·lər ˌsis·təm }

waterwall [MECH ENG] The side of a boiler furnace consisting of water-carrying tubes which absorb radiant heat and thereby prevent excessively high furnace temperatures. { 'wȯd·ərˌwȯl }

waterway [CIV ENG] A channel for the escape or passage of water. [NAV] A navigable stream or canal. { 'wȯd·ərˌwā }

water well [CIV ENG] A well sunk to extract water from a zone of saturation. { 'wȯd·ər ˌwel }

water-wettable [CHEM] Denoting the capability of a material to accept water, or of being hydrophilic or hydrophoric. { 'wȯd·ər ˌwed·ə·bəl }

waterwheel [MECH ENG] A vertical wheel on a horizontal shaft that is made to revolve by the action or weight of water on or in containers attached to the rim. { 'wȯd·ərˌwēl }

water white [CHEM] A grade of color for liquids that has the appearance of clear water; for petroleum products, a plus 21 in the scale of the Saybolt chromometer. { 'wȯd·ər 'wīt }

water-white kerosene [MATER] Kerosine or refined oil from the crude still before it is treated or rerun; has the whitest (nearest to colorless) of the three standard kerosine colors, namely, water white, prime white, and standard white. { 'wȯd·ər ¦wīt 'ker·əˌsēn }

waterworks [CIV ENG] The whole system of supply and treatment utilized in acquisition and distribution of water to consumers. { 'wȯd·ərˌwərks }

water year [HYD] Any 12-month period, usually selected to begin and end during a relatively dry season, used as a basis for processing streamflow and other hydrologic data; the period from October 1 to September 30 is most widely used in the United States. { 'wȯd·ər 'yir }

WATFIV [COMPUT SCI] A programming language based on FORTRAN that is used in learning environments and is characterized by fast compilation and excellent diagnostic messages and debugging aids. Acronym for Waterloo Fortran IV. { 'watˌfīv }

WATS *See* Wide Area Telephone Service. { wäts }

Watson equation [PHYS CHEM] Calculation method to extend heat of vaporization data for organic compounds to within 10 or 15°C of the critical temperature; uses known latent heats of vaporization and reduced temperature data. { 'wät·sən iˌkwā·zhən }

Watson factor *See* characterization factor. { 'wät·sən ˌfak·tər }

Watson-Sommerfeld transformation [MATH] A procedure for transforming a series whose lth term is the product of the lth Legendre polynomial and a coefficient, a_l, having certain properties, into the sum of a contour integral of $a(l)$ and terms involving poles of $a(l)$, where $a(l)$ is a meromorphic function such that $a(l)$ equals a_l at integral values of l; used in studying rainbows, propagation of radio waves around the earth, scattering from various potentials, and scattering of elementary particles. Also known as Sommerfeld-Watson transformation. { 'wät·sən 'zȯm·ərˌfelt iˌkwā·zhən }

watt [PHYS] The unit of power in the meter-kilogram-second system of units, equal to 1 joule per second. Symbolized W. { wät }

wattage rating [ELEC] A rating expressing the maximum power that a device can safely handle continuously. { 'wäd·ij ˌrād·iŋ }

watt balance [PHYS] A device for making a highly accurate comparison of electrical and mechanical powers, in which the force on a current-carrying coil in a constant magnetic field is balanced by the gravitational force on an accurately measured mass, and then, with the current source removed, the coil is moved at constant speed in the same magnetic field and the induced potential difference is measured. { 'wät ˌbal·əns }

watt current *See* active current. { 'wät ˌkə·rənt }

wattevilleite [MINERAL] $Na_2Ca(SO_4)_2 \cdot 4H_2O$ A snow white mineral consisting of a hydrated sulfate of sodium and calcium; occurs as aggregates of acicular or hairlike crystals. { 'wät·viˌlīt }

watt-hour [ELEC] A unit of energy used in electrical measurements, equal to the energy converted or consumed at a rate of 1 watt during a period of 1 hour, or to 3600 joules. Abbreviated Wh. { 'wät ¦au̇r }

watt-hour capacity [ELEC] Number of watt-hours which can be delivered from a storage battery under specified conditions as to temperature, rate of discharge, and final voltage. { 'wät ¦au̇r kəˌpas·əd·ē }

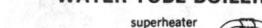

WATER-TUBE BOILER

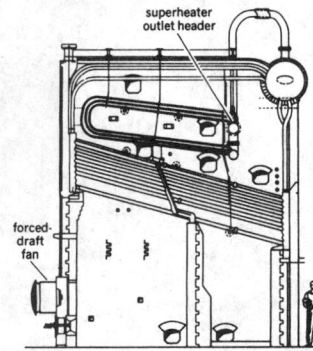

Diagram of straight-tube type of water-tube boiler.

watt-hour meter [ENG] A meter that measures and registers the integral, with respect to time, of the active power of the circuit in which it is connected; the unit of measurement is usually the kilowatt-hour. { 'wät ¦aur ˌmēd·ər }

wattle gum [MATER] A gum arabic extracted from the Australian and East African trees of the genus *Acacia*, as *A. dealbata*, and other species (called mimosa in Kenya); contains 65% tannin. { 'wad·əl ˌgəm }

wattless component *See* reactive component. { 'wät,ləs kəm'pō·nənt }

wattless current *See* reactive current. { 'wät,ləs 'kə·rənt }

wattless power *See* reactive power. { 'wät,ləs 'pau̇·ər }

wattmeter [ENG] An instrument that measures electric power in watts ordinarily. { 'wät,mēd·ər }

Watt's curve [MATH] The curve traced out by the midpoint of a line segment whose end points move along two circles of equal radius. { 'wäts ˌkərv }

watt-second [PHYS] Amount of energy corresponding to 1 watt acting for 1 second; 1 watt-second is equal to 1 joule. { 'wät ˌsek·ənd }

Watt's law [THERMO] A law which states that the sum of the latent heat of steam at any temperature of generation and the heat required to raise water from 0°C to that temperature is constant; it has been shown to be substantially in error. { 'wäts ˌlȯ }

wave [FL MECH] A disturbance which moves through or over the surface of a liquid, as of a sea. [PHYS] A disturbance which propagates from one point in a medium to other points without giving the medium as a whole any permanent displacement. { wāv }

wave aberration [OPTICS] The departure of the geometrical wavefront from a reference sphere with its vertex at the center of the exit pupil and its center of curvature located at the ideal image point; a measure of monochromatic aberration. { 'wāv ˌab·ə,rā·shən }

wave acoustics [ACOUS] The study of the propagation of sound based on its wave properties. { 'wāv ə,küs·tiks }

wave amplitude [PHYS] The magnitude of the greatest departure from equilibrium of the wave disturbance. { 'wāv ˌam·plə,tüd }

wave analyzer *See* harmonic analyzer. { 'wāv ,an·ə,līz·ər }

wave angle [ELECTROMAG] The angle, either in bearing or elevation, at which a radio wave leaves a transmitting antenna or arrives at a receiving antenna. { 'wāv ,aŋ·gəl }

wave antenna [ELECTROMAG] Directional antenna composed of a system of parallel, horizontal conductors, varying from a half to several wavelengths long, terminated to ground at the far end in its characteristic impedance. { 'wāv an,ten·ə }

wave base [HYD] The depth at which sediments are not stirred by wave action, usually about 33 feet (10 meters). Also known as wave depth. { 'wāv ˌbās }

wave basin [GEOGR] A basin close to the inner entrance of a harbor in which the waves from the outer entrance are absorbed, thus reducing the size of the waves entering the inner harbor. { 'wāv ˌbās·ən }

wave-built platform *See* alluvial terrace. { 'wāv ¦bilt 'plat,fȯrm }

wave-built terrace *See* alluvial terrace. { 'wāv ¦bilt 'ter·əs }

wave celerity *See* phase velocity. { 'wāv sə,ler·əd·ē }

wave clutter *See* sea clutter. { 'wāv ,kləd·ər }

wave converter [ELECTROMAG] Device for changing a wave of a given pattern into a wave of another pattern, for example, baffle-plate converters, grating converters, and sheath-reshaping converters for waveguides. { 'wāv kən,vərd·ər }

wave-corpuscle duality *See* wave-particle duality. { 'wāv ¦kȯr·pə·səl dü'al·əd·ē }

wave crest [PHYS] The position at which the disturbance of a progressive wave attains its maximum positive value. { 'wāv ,krest }

wave-cut bench [GEOL] A level or nearly level narrow platform produced by wave erosion and extending outward from the base of a wave-cut cliff. Also known as beach platform; high-water platform. { 'wāv ¦kət 'bench }

wave-cut cliff [GEOL] A cliff formed by the erosive action of waves on rock. { 'wāv ¦kət 'klif }

wave-cut notch [GEOL] An indentation cut into a sea cliff at water level by wave action. { 'wāv ¦kət 'näch }

wave-cut plain *See* wave-cut platform. { 'wāv ¦kət 'plān }

wave-cut platform [GEOL] A gently sloping surface which is produced by wave erosion and which extends into the sea for a considerable distance from the base of the wave-cut cliff. Also known as cut platform; erosion platform; strand flat; wave-cut plain; wave-cut terrace; wave platform. { 'wāv ¦kət 'plat,fȯrm }

wave-cut terrace *See* wave-cut platform. { 'wāv ¦kət 'ter·əs }

wave cyclone [METEOROL] A cyclone which forms and moves along a front; the circulation about the cyclone center tends to produce a wavelike deformation of the front. Also known as wave depression. { 'wāv ,sī,klōn }

wave delta *See* washover. { 'wāv ,del·tə }

wave depression *See* wave cyclone. { 'wāv di,presh·ən }

wave depth *See* wave base. { 'wāv ,depth }

wave disturbance [METEOROL] In synoptic meteorology, the same as wave cyclone, but usually denoting an early state in the development of a wave cyclone, or a poorly developed one. { 'wāv di,stər·bəns }

wave duct [ELECTROMAG] **1.** Waveguide, with tubular boundaries, capable of concentrating the propagation of waves within its boundaries. **2.** Natural duct, formed in air by atmospheric conditions, through which waves of certain frequencies travel with more than average efficiency. { 'wāv ,dəkt }

wave equation [PHYS] **1.** In classical physics, a special equation governing waves that suffer no dissipative attenuation; it states that the second partial derivative with respect to time of the function characterizing the wave is equal to the square of the wave velocity times the Laplacian of this function. Also known as classical wave equation; d'Alembert's wave equation. **2.** Any of several equations which relate the spatial and time dependence of a function characterizing some physical entity which can propagate as a wave, including quantum-wave equations for particles. { 'wāv i,kwā·zhən }

wave erosion *See* marine abrasion. { 'wāv i,rō·zhən }

wave filter [ELEC] A transducer for separating waves on the basis of their frequency; it introduces relatively small insertion loss to waves in one or more frequency bands and relatively large insertion loss to waves of other frequencies. { 'wāv ,fil·tər }

wave forecasting [OCEANOGR] The theoretical determination of future wave characteristics based on observed or forecasted meteorological phenomena. { 'wāv 'fȯr,kast·iŋ }

waveform [PHYS] The pictorial representation of the form or shape of a wave, obtained by plotting the displacement of the wave as a function of time, at a fixed point in space. { 'wāv,fȯrm }

waveform-amplitude distortion *See* frequency distortion. { 'wāv,fȯrm ¦am plə,tüd di,stȯr·shən }

waveform analysis [PHYS] The determination of the amplitude and phase of the components of a complex waveform, either mathematically or by means of electronic instruments. { 'wāv,fȯrm ə,nal·ə·səs }

waveform coder *See* waveform compression. { 'wāv,fȯrm ˌkōd·ər }

waveform compression [COMMUN] Any technique for compressing audio or video signals in which a facsimile of the source-signal waveform is replicated at the receiver with a level of distortion that is judged acceptable. Also known as waveform coder { 'wāv,fȯrm kəm,presh·ən }

wavefront [PHYS] **1.** A surface of constant phase. **2.** The portion of a wave envelope that is between the beginning zero point and the point at which the wave reaches its crest value, as measured either in time or distance. { 'wāv,frənt }

wavefront reversal *See* optical phase conjugation. { 'wāv,frənt ri'vər·səl }

wavefront splitting [OPTICS] Any method of producing interference in which light from a single source is split into two parts which can then be recombined; examples include Young's two-slit experiment, the Fresnel double mirror, and the Fresnel biprism. { 'wāv,frənt 'splid·iŋ }

wave function *See* Schrödinger wave function. { 'wāv ,fəŋk·shən }

wave gage [ENG] A device for measuring the height and period of waves. { 'wāv ,gāj }

wave gait [MECH ENG] A mode of motion of a mobile robot with several legs in which its components have a wavy motion. { 'wāv ,gāt }

wave group [PHYS] A series of waves in which the wave direction, length, and height vary only slightly. { 'wāv ,grüp }

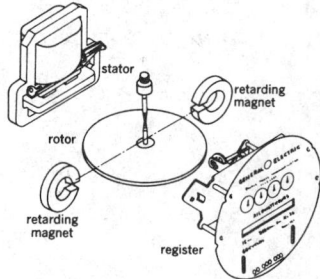
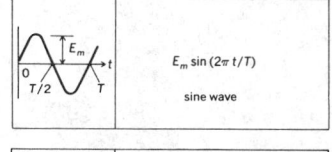
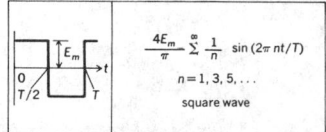

waveguide [ELECTROMAG] **1.** Broadly, a device which constrains or guides the propagation of electromagnetic waves along a path defined by the physical construction of the waveguide; includes ducts, a pair of parallel wires, and a coaxial cable. Also known as microwave waveguide. **2.** More specifically, a metallic tube which can confine and guide the propagation of electromagnetic waves in the lengthwise direction of the tube. { 'wāv‚gīd }

waveguide assembly [ELECTROMAG] An item consisting of one or more definite lengths of straight or formed, flexible or rigid, prefabricated hollow tubing of conductive material; the tubing has a predetermined cross-section, and is designed to guide or conduct high-frequency electromagnetic energy through its interior; one or more ends are terminated. { 'wāv‚gīd ə¦sem·blē }

waveguide attenuation [ELECTROMAG] The decrease from one point of a waveguide to another, in the power carried by an electromagnetic wave in the waveguide. { 'wāv‚gīd ə‚ten·yə'wā·shən }

waveguide bend [ELECTROMAG] A section of waveguide in which the direction of the longitudinal axis is changed; an **E**-plane bend in a rectangular waveguide is bent along the narrow dimension, while an **H**-plane bend is bent along the wide dimension. Also known as waveguide elbow. { 'wāv‚gīd ¦bend }

waveguide cavity [ELECTROMAG] A cavity resonator formed by enclosing a section of waveguide between a pair of waveguide windows which form shunt susceptances. { 'wāv‚gīd ¦kav·əd·ē }

waveguide connector [ELECTROMAG] A mechanical device for electrically joining and locking together separable mating parts of a waveguide system. Also known as waveguide coupler. { 'wāv‚gīd kə¦nek·tər }

waveguide coupler *See* waveguide connector. { 'wāv‚gīd ¦kəp·lər }

waveguide critical dimension [ELECTROMAG] Dimension of waveguide cross section which determines the cutoff frequency. { 'wāv‚gīd 'krid·ə·kəl də'men·shən }

waveguide cutoff frequency [ELECTROMAG] Frequency limit of propagation along a waveguide for waves of a given field configuration. { 'wāv‚gīd ¦kəd‚óf ¦frē·kwən‚sē }

waveguide discontinuity *See* discontinuity. { 'wāv‚gīd dis‚känt·ən'ü·əd·ē }

waveguide elbow *See* waveguide bend. { 'wāv‚gīd ¦el·bō }

waveguide filter [ELECTROMAG] A filter made up of waveguide components, used to change the amplitude-frequency response characteristic of a waveguide system. { 'wāv‚gīd ¦fil·tər }

waveguide hybrid [ELECTROMAG] A waveguide circuit that has four arms so arranged that a signal entering through one arm will divide and emerge from the two adjacent arms, but will be unable to reach the opposite arm. { 'wāv‚gīd ¦hī·brəd }

waveguide junction *See* junction. { 'wāv‚gīd ¦jəŋk·shən }

waveguide plunger *See* piston. { 'wāv‚gīd ¦plən·jər }

waveguide probe *See* probe. { 'wāv‚gīd ¦prōb }

waveguide propagation [COMMUN] Long-range communications in the 10-kilohertz frequency range by the waveguide characteristics of the atmospheric duct formed by the ionospheric D layer and the surface of the earth. { 'wāv‚gīd ‚präp·ə'gā·shən }

waveguide resonator *See* cavity resonator. { 'wāv‚gīd ¦rez·ən‚ād·ər }

waveguide shim [ELECTROMAG] Thin resilient metal sheet inserted between waveguide components to ensure electrical contact. { 'wāv‚gīd ¦shim }

waveguide slot [ELECTROMAG] A slot in a waveguide wall, either for coupling with a coaxial cable or another waveguide, or to permit the insertion of a traveling probe for examination of standing waves. { 'wāv‚gīd ‚slät }

waveguide switch [ELECTROMAG] A switch designed for mechanically positioning a waveguide section so as to couple it to one of several other sections in a waveguide system. { 'wāv‚gīd ‚swich }

waveguide synthesis [ENG ACOUS] A method of synthesizing the sounds of a string or wind instrument that simulates traveling waves on a string or inside a bore or horn using digital delay lines. { ‚wāv‚gīd 'sin·thə·səs }

waveguide window *See* iris. { 'wāv‚gīd ¦win·dō }

wave height [OCEANOGR] The height of a water-surface wave is generally taken as the height difference between the wave crest and the preceding trough. [PHYS] Twice the wave amplitude. { 'wāv ‚hīt }

wave-height correction [NAV] A correction to a sextant altitude required because of the elevation of parts of the sea surface by wave action. { 'wāv ¦hīt kə'rek·shən }

wave impedance [ELECTROMAG] The ratio, at every point in a specified plane of a waveguide, of the transverse component of the electric field to the transverse component of the magnetic field. { 'wāv im‚pēd·əns }

wave intensity [PHYS] The average amount of energy transported by a wave in the direction of wave propagation, per unit area per unit time. { 'wāv in‚ten·səd·ē }

wave interference *See* interference. { 'wāv ‚in·tər'fir·əns }

wavelength [PHYS] The distance between two points having the same phase in two consecutive cycles of a periodic wave, along a line in the direction of propagation. { 'wāv‚leŋkth }

wavelength constant *See* phase constant. { 'wāv‚leŋkth ‚kän·stənt }

wavelength-division multiplexing [COMMUN] The sharing of the total available pass-band of a transmission medium in the optical portion of the electromagnetic spectrum by assigning individual information streams to signals of different wavelengths. Abbreviated WDM. { ¦wāv‚leŋth də¦vizh·ən 'məl·tə‚pleks·iŋ }

wavelength shifter [ELECTR] A photofluorescent compound used with a scintillator material to increase the wavelengths of the optical photons emitted by the scintillator, thereby permitting more efficient use of the photons by the phototube or photocell. { 'wāv‚leŋkth ‚shif·tər }

wavelength standards [SPECT] Accurately measured lengths of waves emitted by specified light sources for the purpose of obtaining the wavelengths in other spectra by interpolating between the standards. { 'wāv‚leŋkth ‚stan·dərdz }

wavelet [MATH] One of a collection of mathematical functions that serve as the elementary building blocks of a mathematical tool for analyzing and synthesizing functions, and for forming representations of signals in both time and frequency. { 'wāv·lət }

wave line *See* swash mark. { 'wāv ‚līn }

wavellite [MINERAL] $Al_3(PO_4)_2(OH)_3 \cdot 5H_2O$ A white to yellow, green, or black mineral crystallizing in the orthorhombic system and occurring in small hemispherical aggregates. { 'wā·və‚līt }

wave-making resistance *See* wave resistance. { ¦wāv ¦māk·iŋ ri‚zis·təns }

wavemark *See* swash mark. { 'wāv‚märk }

wave mechanics *See* Schrödinger's wave mechanics. { 'wāv mi‚kan·iks }

wavemeter [ENG] A device for measuring the geometrical spacing between successive surfaces of equal phase in an electromagnetic wave. { 'wāv‚mēd·ər }

wave microphone [ENG ACOUS] Any microphone whose directivity depends upon some type of wave interference, such as a line microphone or a reflector microphone. { 'wāv 'mī·krə‚fōn }

wave motion [PHYS] The process by which a disturbance at one point is propagated to another point more remote from the source with no net transport of the material of the medium itself; examples include the motion of electromagnetic waves, sound waves, hydrodynamic waves in liquids, and vibration waves in solids. Also known as propagation; wave propagation. { 'wāv ‚mō·shən }

wave motor [MECH ENG] A motor that depends on the lifting power of sea waves to develop its usable energy. { 'wāv ‚mōd·ər }

wave noise [ELECTR] Noise in the electric current of a detector that results from fluctuations in the intensity of electromagnetic radiation falling on the detector. { 'wāv ‚nóiz }

wave normal [PHYS] **1.** A unit vector which is perpendicular to an equiphase surface of a wave, and has its positive direction on the same side of the surface as the direction of propagation. **2.** One of a family of curves which are everywhere perpendicular to the equiphase surfaces of a wave. { 'wāv 'nór·məl }

wave number [PHYS] The reciprocal of the wavelength of a wave, or sometimes 2π divided by the wavelength. Also known as reciprocal wavelength. { 'wāv ‚nəm·bər }

wave optics [OPTICS] The branch of optics which treats of

WAVE MOTION

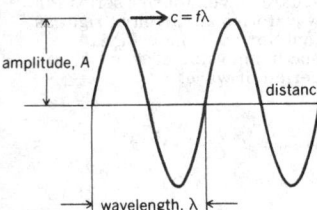

Relation between frequency *f*, wavelength λ, and velocity *c* in wave motion.

light (or electromagnetic radiation in general) with explicit recognition of its wave nature. { 'wāv ,äp·tiks }

wave packet [PHYS] In wave phenomena, a superposition of waves of differing lengths, so phased that the resultant amplitude is negligibly small except in a limited portion of space whose dimensions are the dimensions of the packet. Also known as packet. { 'wāv ,pak·ət }

wave-particle duality [QUANT MECH] The principle that both matter and electromagnetic radiation exhibit phenomena in which they behave as waves and other phenomena in which they behave as particles, the two aspects being associated by the de Broglie relations. Also known as duality principle; wave-corpuscle duality. { 'wāv ,pärd·ə·kəl dü'al·əd·ē }

wave period [PHYS] The time between the attainment of successive maxima, at a fixed point, of a quantity characterizing a wave. { 'wāv ,pir·ē·əd }

wave plate [OPTICS] A plate of material which is linearly birefringent. Also known as retardation plate; retardation sheet. { 'wāv ,plāt }

wave platform See wave-cut platform. { 'wāv 'plat,fòrm }

wave polarization See polarization. { 'wāv ,pō·lə·rə,zā·shən }

wave propagation See wave motion. { 'wāv ,präp·ə,gā·shən }

wave refraction [PHYS] The process by which the direction of a wave train moving in shallow water at an angle to the contours is changed. { 'wāv ri,frak·shən }

wave resistance [FL MECH] The portion of fluid resistance to a body moving on the surface of a liquid that results from energy dissipation in the formation of waves on the liquid surface. Also known as wave-making resistance. { 'wāv ri,zis·təns }

waverider [AERO ENG] A supersonic or hypersonic vehicle that has an attached shock wave along its entire leading edge, so that the vehicle appears to be riding on the top of its shock wave. { 'wāv,rīd·ər }

wave ripple mark See oscillation ripple mark. { 'wāv 'rip·əl ,märk }

wave setdown [OCEANOGR] A decrease in the mean water level in the region in which breakers form near the seashore, caused by the presence of a pressure field. { 'wāv 'set,daùn }

wave setup [OCEANOGR] An increase in the mean water level shoreward of the region in which breakers form at the seashore, caused by the onshore flux of momentum against the beach. { 'wāv 'sed,əp }

wave shaper [ENG] Of explosives, an insert or core of inert material or of explosives having different detonation rates, used for changing the shape of the detonation wave. { 'wāv ,shāp·ər }

wave-shaping circuit [ELECTR] An electronic circuit used to create or modify a specified time-varying electrical quantity, usually voltage or current, using combinations of electronic devices, such as vacuum tubes or transistors, and circuit elements, including resistors, capacitors, and inductors. { 'wāv ,shāp·iŋ ,sər·kət }

wave soldering See flow soldering. { 'wāv ,säd·ə·riŋ }

wave speed See phase velocity. { 'wāv ,spēd }

wave system [OCEANOGR] In ocean wave studies, a group of waves which have the same height, length, and direction of movement. { 'wāv ,sis·təm }

wave tail [ELECTR] Part of a signal-wave envelope (in time or distance) between the steady-state value (or crest) and the end of the envelope. { 'wāv ,tāl }

wave theory of cyclones [METEOROL] A theory of cyclone development based upon the principle of wave formation on an interface between two fluids; in the atmosphere, a front is taken as such an interface. { 'wāv 'thē·ə·rē əv 'sī,klōnz }

wave theory of light [OPTICS] A theory which assumes that light is a wave motion, rather than a stream of particles. { 'wāv 'thē·ə·rē əv 'līt }

wave tilt [ELECTROMAG] Forward inclination of a radio wave due to its proximity to ground. { 'wāv ,tilt }

wave train [PHYS] A series of waves produced by the same disturbance. { 'wāv ,trān }

wave trap [CIV ENG] A device used to reduce the size of waves from sea or swell entering a harbor before they penetrate as far as the quayage; usually in the form of diverging breakwaters, or small projecting breakwaters situated close within the entrance. [ELECTR] A resonant circuit connected to the

antenna system of a receiver to suppress signals at a particular frequency, such as that of a powerful local station that is interfering with reception of other stations. Also known as trap. { 'wāv ,trap }

wave trough [PHYS] The lowest part of a wave form between successive wave crests. { 'wāv ,tróf }

wave vector [PHYS] A vector whose direction is the direction of phase propagation of a wave at each point in space, and whose magnitude is sometimes set at $2\pi/\lambda$ and sometimes at $1/\lambda$, where λ is the wavelength. { 'wāv ,vek·tər }

wave-vector space [SOLID STATE] The space of the wave vectors of the state functions of some system; this would be used, for example, for electron wave functions in a crystal and thermal vibrations of a lattice. Also known as k-space; reciprocal space. { 'wāv ,vek·tər ,spās }

wave velocity See phase velocity. { 'wāv və,läs·əd·ē }

wavy extinction See undulatory extinction. { 'wāv·ē ik 'stiŋk·shən }

wax [MATER] Any of a group of substances resembling beeswax in appearance and character, and in general distinguished by their composition of esters and higher alcohols, and by their freedom from fatty acids. { waks }

wax-block photometer See Joly photometer. { 'waks 'bläk fō'tam·əd·ər }

wax-coating machine [GRAPHICS] A machine that applies a pressure-sensitive coating of wax to the backs of proofs, stats, photos, overlays, and other materials. { 'waks 'kōd·iŋ mə,shēn }

wax distillate [MATER] A neutral distillate from distillation of crude oil that contains a high percentage of crystallizable paraffin wax; used as a primary base for paraffin wax and neutral lubricating oils. { 'waks 'dis·tə·lət }

waxed paper [MATER] Paper that is treated or coated with wax to make it waterproof and greaseproof; used for wrapping. { 'wakst 'pā·pər }

wax engraving [GRAPHICS] A type of letterpress platemaking, used especially for ruled forms, in which lines of type are impressed in wax resulting in a mold suitable for electrotyping. { 'waks in'grāv·iŋ }

wax fractionation [CHEM ENG] A continuous solvent-recovery/crystallization petroleum-refinery process for the production of waxes with low oil content from wax concentrates; for example, MEK (methyl ethyl ketone) deoiling. { 'waks ,frak·shə'nā·shən }

wax gland See ceruminous gland. { 'waks ,gland }

waxing moon [ASTRON] The moon between new and full, when its visible part is increasing. { 'waks·iŋ 'mün }

wax manufacturing [CHEM ENG] A petroleum refinery process similar to wax fractionation for the manufacture of oil-free waxes by chilling and crystallization from a solvent. { 'waks ,man·ə'fak·chə·riŋ }

wax master See wax original. { 'waks 'mas·tər }

wax original [ENG ACOUS] An original recording made on a wax surface and used to make a master. Also known as wax master. { 'waks ə'rij·ən·əl }

wax stain [MATER] A semitransparent pigment mixed with beeswax, and thinned with turpentine. { 'waks 'stān }

wax tailings [MATER] A sticky, pitchlike substance with dark-brown color, the last volatile product distilling off an oil charge before it is coked; used as wood preservative and in manufacture of roofing paper. Also known as petroleum tailings; still wax. { 'waks 'tāl·iŋz }

waxy [MINERAL] A type of mineral luster that is soft like that of wax. { 'wak·sē }

waxy-electrolyte battery [ELEC] A primary battery in which the electrolyte is a waxy material, such as polyethylene glycol, in which is dissolved a small amount of a salt, such as zinc chloride; the electrodes are frequently made of zinc and manganese dioxide, and the electrolyte is melted and painted on a paper sheet to form the separator. { 'wak·sē i,lek·trə,līt 'bad·ə·rē }

way point [CONT SYS] See via point. [NAV] A reference point between the point of departure and the destination, particularly a point on a course line the coordinates of which are defined in relation to an electronic aid to navigation. { 'wā ,point }

ways [CIV ENG] **1.** The tracks and sliding timbers used in launching a vessel. **2.** The building slip or space upon which the sliding timbers or ways, supporting a vessel to be launched,

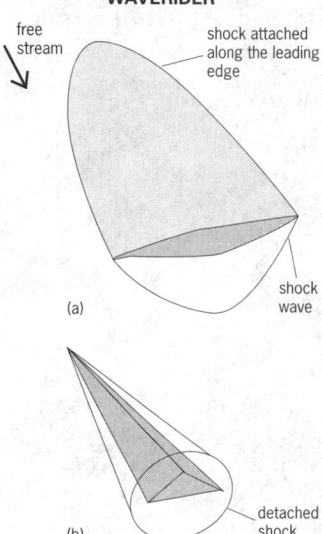

WAVERIDER

free stream

shock attached along the leading edge

(a)

shock wave

(b)

detached shock

Comparison of supersonic and hypersonic vehicles. (*a*) Waverider. (*b*) Generic hypersonic vehicle configuration.

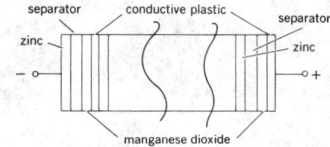

WAXY-ELECTROLYTE BATTERY

separator   conductive plastic   separator

zinc   zinc

manganese dioxide

Diagram of battery stack of cells using a waxy electrolyte.

travel. [MECH ENG] Bearing surfaces used to guide and support moving parts of machine tools; may be flat, V-shaped, or dovetailed. { wāz }

Wb *See* weber.

W boson [PART PHYS] An intermediate vector boson with positive or negative electric charge that mediates the charged-current weak interactions. Also known as W particle. { 'dəb·əl,yü ,bō,sän }

W chromosome [GEN] The sex chromosome present only in females in species with female heterogamety. { 'dəb·əl,yü ,krō·mə,sōm }

W coefficient *See* Racah coefficient. { 'dəb·əl,yü ,kō·i,fish·ənt }

WDM *See* wavelength-division multiplexing.

weak acid [CHEM] An acid that does not ionize greatly; for example, acetic acid or carbonic acid. { 'wēk 'as·əd }

weak convergence [MATH] A sequence of elements x_1, x_2, ... from a topological vector space X converges weakly if the sequence $f(x_1), f(x_2)$, ... converges for every continuous linear functional f on X. { 'wēk kən'vər·jəns }

weak coupling [PART PHYS] The coupling of four fermion fields in the weak interaction, having a strength many orders of magntiude weaker than that of the strong or electromagnetic interactions. { 'wēk 'kəp·liŋ }

weak energy condition [RELAT] The condition in general relativity theory that all observers see a nonnegative energy density. { ¦wēk 'en·ər·jē kən,dish·ən }

weak fix [NAV] A fix determined from horizontal sextant angles between objects poorly located. { 'wēk 'fiks }

weak ground [MIN ENG] Roof and walls of underground excavations which would be in danger of collapse unless suitably supported. { 'wēk 'graùnd }

weak interaction [PART PHYS] One of the fundamental interactions among elementary particles, responsible for beta decay of nuclei, and for the decay of elementary particles with lifetimes greater than about 10^{-10} second, such as muons, K mesons, and lambda hyperons; it is several orders of magnitude weaker than the strong and electromagnetic interactions, and fails to conserve strangeness or parity. Also known as beta interaction. { 'wēk ,in·tər'ak·shən }

weak-line T Tauri star [ASTRON] A T Tauri star that lacks strong emission lines in its optical spectrum, and lacks both strong stellar winds and a circumstellar accretion disk. Also known as naked T Tauri star; weak T Tauri star. { ¦wēk ,līn ¦tē 'tȯr·ē ,stär }

weakly complete space [MATH] A topological vector space in which an element x is associated with any weakly convergent sequence of elements x_n such that the limit of $f(x_n)$ equals $f(x)$ for any continuous linear functional f. { ¦wēk·lē kəm¦plēt 'spās }

weakly connected digraph [MATH] A directed graph whose underlying graph is a connected graph. { ¦wēk·lē kə,nek·təd 'dī,graf }

weakly interacting massive particle [PART PHYS] A hypothetical massive elementary particle, interacting only through gravity and the weak nuclear interaction. Abbreviated WIMP. { ¦wēk·lē ,in·tər¦ak·tiŋ ¦mas·iv 'pärd·ə·kəl }

weak-strong duality [PHYS] A property of some physical systems that have a dual description such that, when a coupling constant governing the strengths of interactions in the original system is large, the coupling constant in the dual description is small, so that the system can be accurately described by means of perturbation theory. { ¦wēk ¦strȯŋ dü'al·əd·ē }

weak topology [MATH] A topology on a topological vector space X whose open neighborhoods around a point x are obtained from those points y of X for which every $f_i(x)$ is close to $f_i(y)$, f_i appearing in a finite list of linear functionals. { 'wēk tə'päl·ə·jē }

weak T Tauri star *See* weak-line T Tauri star. { ¦wēk ¦tē 'tȯr·ē ,stär }

weapon [ORD] An instrument of combat, either offensive or defensive, used to destroy, injure, defeat, or threaten an enemy; for example, a gun, bayonet, bomb, or missile. { 'wep·ən }

weapon delivery [ORD] The total action required to locate the target, establish the necessary release conditions, and maintain guidance to the target if required; it includes the detection recognition and the acquisition of the target, and the weapon release and weapon guidance. { 'wep·ən di,liv·ə·rē }

WEASEL

The common or European weasel (*Mustela nivalis*).

weapon of mass destruction [ORD] Nuclear, bacteriological, or other weapon capable of causing widespread death or destruction. { 'wep·ən əv 'mas di'strək·shən }

weapon record book [ORD] Record book used to keep data on the performance, maintenance, and inspection of a gun or other weapon. { 'wep·ən 'rek·ərd ,bùk }

weapons system [ORD] Two or more instruments of combat operating as a single unit of striking power in military combat; specifically, a system in which two instruments of combat are required to perform a single mission. { 'wep·ənz ,sis·təm }

weapon target line [ORD] An imaginary straight line from a weapon to a target. { 'wep·ən 'tär·gət ,līn }

wear [ENG] Deterioration of a surface due to material removal caused by relative motion between it and another part. { wer }

wearing course [CIV ENG] The top layer of surfacing on a road. { 'wer·iŋ ,kȯrs }

wear tables [ORD] Tables indicating the decrease of muzzle velocity expected as the result of firing a certain number of equivalent rounds; although tubes may vary considerably from the wear rate indicated in such tables, the tables may be used to correct calibration data between periods of calibration. { 'wer ,tā·bəlz }

weasel [VERT ZOO] The common name for at least 12 species of small, slim carnivores which belong to the family Mustelidae and which have a reddish-brown coat with whitish underparts; species in the northern regions have white fur during the winter and are called ermine. { 'wē·zəl }

weather [METEOROL] **1.** The state of the atmosphere, mainly with respect to its effects upon life and human activities; as distinguished from climate, weather consists of the short-term (minutes to months) variations of the atmosphere. **2.** As used in the making of surface weather observations, a category of individual and combined atmospheric phenomena which must be drawn upon to describe the local atmospheric activity at the time of observation. { 'weth·ər }

weather bar *See* water bar. { 'weth·ər ,bär }

weather central [METEOROL] An organization which collects, collates, evaluates, and disseminates meteorological information in such a manner that it becomes a principal source of such information for a given area. { 'weth·ər 'sen·trəl }

weathercocking [AERO ENG] The aerodynamic action causing alignment of the longitudinal axis of a rocket with the relative wind after launch. Also known as weather vaning. { 'weth·ər,käk·iŋ }

weathercock stability *See* directional stability. { 'weth·ər,käk stə'bil·əd·ē }

weather deck [NAV ARCH] **1.** The uppermost deck of a ship. **2.** Any deck that does not have overhead protection from the weather. { 'weth·ər ,dek }

weathered crude [MATER] Crude petroleum that, owing to evaporation and other natural causes during storage and handling, has lost an appreciable quantity of its more volatile components. { 'weth·ərd 'krüd }

weathered iceberg [OCEANOGR] An iceberg which is irregular in shape, due to an advanced stage of ablation; it may have overturned. { 'weth·ərd 'īs,bərg }

weathered joint *See* weather-struck joint. { ¦weth·ərd ¦jȯint }

weathered layer [GEOPHYS] The zone of the earth which lies immediately below the surface and is characterized by low wave velocities. { 'weth·ərd 'lā·ər }

weather forecast [METEOROL] A forecast of the future state of the atmosphere with specific reference to one or more associated weather elements. { 'weth·ər ,fȯr,kast }

weathering [GEOL] Physical disintegration and chemical decomposition of earthy and rocky materials on exposure to atmospheric agents, producing an in-place mantle of waste. Also known as clastation; demorphism. { 'weth·ə,riŋ }

weathering correction [GEOPHYS] A velocity correction which is applied to seismic data, necessitated by the diminished velocity of seismic wave propagation in weathered rock. { 'weth·ə,riŋ kə,rek·shən }

weathering-potential index [GEOL] A measure of the susceptibility of a rock or mineral to weathering. { 'weth·ə,riŋ pə¦ten·chəl ,in,deks }

weathering rind [GEOL] The outer layer of a pebble, boulder, or other rock fragment that has formed as a result of chemical weathering. { 'weth·ər·iŋ ,rīnd }

weathering test [PETRO ENG] A standard test for liquid petroleum gas in which the heavy components in a given sample are determined by evaporation of the volatile components. Also known as boilaway test. { 'weth·ə,riŋ ,test }

weathering velocity [GEOPHYS] The velocity of propagation of seismic waves through weathered rock. { 'weth·ə,riŋ və,läs·əd·ē }

weather map [METEOROL] A chart portraying the state of the atmospheric circulation and weather at a particular time over a wide area; it is derived from a careful analysis of simultaneous weather observations made at many observing points in the area. { 'weth·ər ,map }

weather-map type See weather type. { 'weth·ər ,map ,tīp }

weather minimum [METEOROL] The worst weather conditions under which aviation operations may be conducted under either visual or instrument flight rules; usually prescribed by directives and standing operating procedures in terms of minimum ceiling, visibility, or specific hazards to flight. { 'weth·ər ,min·ə·məm }

weather modification [METEOROL] The changing of natural weather phenomena by technical means; so far, only on the microscale of condensation and freezing nuclei has it been possible to exert modifying influences. { 'weth·ər ,mäd·ə·fə'kā·shən }

weather observation [METEOROL] An evaluation of one or more meteorological elements that describe the state of the atmosphere either at the earth's surface or aloft. { 'weth·ər ,äb·zər,vā·shən }

weather observation radar See weather radar. { 'weth·ər ,äb·zər,vā·shən 'rā,där }

weatherometer [ENG] A device used to subject articles and finishes to accelerated weathering conditions; for example, a rich ultraviolet source, water spray, or salt water { ,weth·ə'räm·əd·ər }

weather pit [GEOL] A shallow depression (depth up to 6 inches or 15 centimeters) on the flat or gently sloping summit of large exposures of granite or granitic rocks, attributed to strongly localized solvent action of impounded water. { 'weth·ər ,pit }

weatherproof [ENG] Able to withstand exposure to weather without damage. { 'weth·ər,prüf }

weather radar [ENG] Generally, any radar which is suitable or can be used for the detection of precipitation or clouds. Also known as weather observation radar. { 'weth·ər 'rā,där }

weather resistance [ENG] The ability of a material, paint, film, or the like to withstand the effects of wind, rain, or sun and to retain its appearance and integrity. { 'weth·ər ri,zis·təns }

weather shore [METEOROL] As observed from a vessel, the shore lying in the direction from which the wind is blowing. { 'weth·ər ,shór }

weather side [METEOROL] The side of a ship exposed to the wind or weather. { 'weth·ər ,sīd }

weather signal [METEOROL] A visual signal displayed to indicate a weather forecast. { 'weth·ər ,sig·nəl }

weather station [METEOROL] A place and facility for the observation, measurement, and recording and transmission of data of the variable elements of weather; one of the most effective network facilities is that of the U.S. Weather Bureau. { 'weth·ər ,stā·shən }

weather strip [BUILD] A piece of material, such as wood or rubber, applied to the joints of a window or door to stop drafts. { 'weth·ər ,strip }

weather-struck joint [CIV ENG] A horizontal joint in a course of masonry in which the mortar at the upper edge has been pressed in, forming a convex surface that sheds water. Also known as weathered joint. { 'weth·ər ,strək ,jóint }

weather type [METEOROL] A series of generalized synoptic situations, usually presented in chart form; weather types are selected to represent typical pressure patterns, and were originally devised as a method for lengthening the effective time-range of forecasts. Also known as weather-map type. { 'weth·ər ,tīp }

weather vaning See weathercocking. { 'weth·ər ,vān·iŋ }

weather window [PETRO ENG] That part of the year when the weather is suitable for operations, such as pipelaying or platform installation, which cannot be undertaken in adverse sea conditions. { 'weth·ər ,win·dō }

weave [TEXT] **1.** To make cloth by interlacing strands of warp and filling threads. **2.** A cloth made by weaving. **3.** The pattern of a woven fabric. { wēv }

web [ARCH] The portion of a ribbed vault between ribs. [CIV ENG] The vertical strip connecting the upper and lower flanges of a rail or girder. [GRAPHICS] The continuous length of paper formed when paper pulp moves through a papermaking machine; the web is then cut into sheets or wrapped onto rolls. [MATER] In a grain of propellant, the minimum thickness of the grain between any two adjacent surfaces. [MECH ENG] For twist drills and reamers, the central portion of the tool body that joins the lands. [MET] In forging, the thin section of metal remaining at the bottom of a depression or at the location of the punches. [OPTICS] See wire. [TEXT] A fabric as it is being woven on a loom. [VERT ZOO] The membrane between digits in many birds and amphibians. { web }

Web See World Wide Web. { web }

web angle See chisel-edge angle. { 'web ,aŋ·gəl }

Web browser See browser. { 'web ,braủz·ər }

weber [ELECTROMAG] The unit of magnetic flux in the meter-kilogram-second system, equal to the magnetic flux which, linking a circuit of one turn, produces in it an electromotive force of 1 volt as it is reduced to zero at a uniform rate in 1 second. Symbolized Wb. { 'vā·bər }

Weber-Christian disease [MED] Febrile, relapsing, nodular nonsuppurative panniculitis. { 'vā·bər 'kris·chən di,zēz }

Weber differential equation [MATH] A special case of the confluent hypergeometric equation that has as solution a confluent hypergeometric series. Also known as Weber-Hermite equation. { 'vā·bər ,dif·ə'ren·chəl i'kwā·zhən }

Weber-Hermite equation See Weber differential equation. { 'vā·bər er'mēt i,kwā·zhən }

Weberian apparatus [VERT ZOO] A series of bony ossicles which form a chain connecting the swim bladder with the inner ear in fishes of the superorder Ostariophysi. { vā'bir·ē·ən ,ap·ə,rad·əs }

Weberian ossicle [VERT ZOO] One of a chain of three or four small bones that make up the Weberian apparatus. { vā'bir·ē·ən 'äs·ə·kəl }

weberite [MINERAL] Na_2MgAlF_7 A light gray, orthorhombic mineral consisting of an aluminofluoride of sodium and magnesium; occurs as grains and masses. { 'vā·bə,rīt }

Weber number 1 [FL MECH] A dimensionless number used in the study of surface tension waves and bubble formation, equal to the product of the square of the velocity of the wave or the fluid velocity, the density of the fluid, and a characteristic length, divided by the surface tension. Symbolized N_{We1}, We. { 'vā·bər ,nəm·bər 'wən }

Weber number 2 [FL MECH] A dimensionless number, equal to the square root of Weber number 1. Symbolized N_{We2}. { 'vā·bər ,nəm·bər 'tü }

Weber number 3 [CHEM ENG] A dimensionless number used in interfacial area determination in distillation equipment, equal to the surface tension divided by the product of the liquid density, the acceleration of gravity, and the depth of liquid on the tray under consideration. Symbolized N_{We3}. { 'vā·bər ,nəm·bər 'thrē }

Weber's law [PHYSIO] The law that the stimulus increment which can barely be detected (the just noticeable difference) is a constant fraction of the initial magnitude of the stimulus; this is only an approximate rule of thumb. { 'vā·bərz ,ló }

web-fed press [GRAPHICS] Printing press designed to accept paper from rolls instead of sheets; large web presses offer great speed and economy for long press runs, but high makeready and plate costs make them too costly for short runs. { 'web ,fed 'pres }

web frame [NAV ARCH] A built-up member consisting of a web plate with single or double bars riveted or welded to its edges; web frames are placed several frame spaces apart with smaller frames in between. { 'web ,frām }

Web page [COMPUT SCI] A document written in HTML and available for viewing on the World Wide Web; may contain images, sound, video, formatted text, and hyperlinks. { 'web ,pāj }

web plate [ENG] A steel plate that forms the web of a beam, girder, or truss. { 'web ,plāt }

Web server [COMPUT SCI] A program that processes document requests; it also has a database, which is a repository of data and content. { 'web ,sər·vər }

Web services [COMPUT SCI] A collection of SML-based

standards that enable electronic communication and interaction independently of the computer platforms or specific technologies used by the communication parties.   { 'web ‚serv·ə·səs }

Web site   [COMPUT SCI]   A collection of thematically related, hyperlinked World Wide Web services, mainly HTML documents, usually located on a specific Web server and reachable through a URL assigned to the site.   { 'web ‚sīt }

websterite   *See* aluminite.   { 'web·stə‚rīt }

Weddell Current   [OCEANOGR]   A surface current which flows in an easterly direction from the Weddell Sea outside the limit of the West Wind Drift.   { we'del 'kə·rənt }

weddellite   [MINERAL] $CaC_2O_4 \cdot 2H_2O$ A colorless to white or yellowish-brown to brown, tetragonal mineral consisting of calcium oxalate dihydrate.   { wə'de‚līt }

Wedener-Bergeron process   *See* Bergeron-Findeisen theory.   { 'vād·ən·ər ‚ber·zhə‚rōn ‚prä·səs }

wedge   [COMMUN]   A convergent pattern of equally spaced black and white lines, used in a television test pattern to indicate resolution.   [DES ENG]   A piece of resistant material whose two major surfaces make an acute angle.   [ELECTROMAG]   A waveguide termination consisting of a tapered length of dissipative material introduced into the guide, such as carbon.   [ENG] In ultrasonic testing, a device which directs waves of ultrasonic energy into the test piece at an angle.   [MATH]   A polyhedron whose base is a rectangle and whose lateral faces consist of two equilateral triangles and two trapezoids.   [METEOROL]   *See* ridge.   [OPTICS]   **1.** An optical filter in which the transmission decreases continuously or in steps from one end to the other. **2.** A refracting prism of very small angle, inserted in an optical train to introduce a bend in the ray path.   { wej }

wedge-base lamp   [ELEC]   A small indicator lamp that has wire leads folded back on opposite sides of a flat glass base.   { 'wej ‚bās 'lamp }

wedge bit   [DES ENG]   A tapered-nose noncoring bit, used to ream out the borehole alongside the steel deflecting wedge in hole-deflection operations.   Also known as bull-nose bit; wedge reaming bit; wedging bit.   { 'wej ‚bit }

wedge bonding   [ENG]   A type of thermocompression bonding in which a wedge-shaped tool is used to press a small section of the lead wire onto the bonding pad of an integrated circuit.   { 'wej ‚bänd·iŋ }

wedge core lifter   [MECH ENG]   A core-gripping device consisting of a series of three or more serrated-face, tapered wedges contained in slotted and tapered recesses cut into the inner surface of a lifter case or sleeve; the case is threaded to the inner tube of a core barrel, and as the core enters the inner tube, it lifts the wedges up along the case taper; when the barrel is raised, the wedges are pulled tight, gripping the core.   { 'wej ‚kȯr ‚lif·tər }

wedge filter   [NUCLEO]   A radiation filter so constructed that its thickness or transmission characteristics vary continuously or in steps from one edge to the other; used to increase the uniformity of radiation in certain types of treatment.   { 'wej ‚fil·tər }

wedge photometer   [ENG]   A photometer in which the luminous flux density of light from two sources is made equal by pushing into the beam from the brighter source a wedge of absorbing material; the wedge has a scale indicating how much it reduces the flux density, so that the luminous intensities of the sources may be compared.   { 'wej fə'täm·əd·ər }

wedge product   [MATH]   A product defined on forms such that a wedge product of a p-form and a q-form results in a $p + q$ form.   { 'wej ‚präd·əkt }

wedge reaming bit   *See* wedge bit.   { 'wej 'rēm·iŋ ‚bit }

wedge spectrograph   [SPECT]   A spectrograph in which the intensity of the radiation passing through the entrance slit is varied by moving an optical wedge.   { 'wej 'spek·trə‚graf }

wedging   [ENG]   **1.** A method used in quarrying to obtain large, regular blocks of building stones; a row of holes is drilled, either by hand or by pneumatic drills, close to each other so that a longitudinal crevice is formed into which a gently sloping steel wedge is driven, and the block of stone can be detached without shattering.   **2.** The act of changing the course of a borehole by using a deflecting wedge.   **3.** The lodging of two or more wedge-shaped pieces of core inside a core barrel, and therefore blocking it.   **4.** The material, moss, or wood used to render the shaft lining tight.   { 'wej·iŋ }

wedging bit   *See* wedge bit.   { 'wej·iŋ ‚bit }

weed   [BOT]   A plant that is useless or of low economic value,

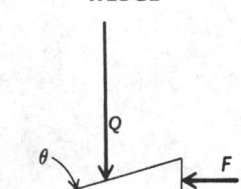

WEDGE

The shape of a wedge and a diagram of forces acting on it. F is smaller applied force, Q is larger force to be exerted, θ is angle between surfaces of wedge.

especially one growing on cultivated land to the detriment of the crop.   { wēd }

week   [ASTRON]   A time period of 7 days which has been accepted from ancient Babylon; the 7 days of the week were first given names of the seven celestial bodies: the sun, moon, and five visible planets.   { wēk }

weep hole   [CIV ENG]   A hole in a wood sill, retaining wall, or other structure to allow accumulated water to escape.   { 'wēp ‚hōl }

weeping core   [PETRO ENG]   A core cut that is covered with tearlike drops of fluid when it is brought to the surface; usually indicative of a deposit that will produce little oil.   { 'wēp·iŋ 'kȯr }

weeping spring   *See* spring seepage.   { 'wēp·iŋ 'spriŋ }

weevil   [INV ZOO]   Any of various snout beetles whose larvae destroy crops by eating the interior of the fruit or grain, or bore through the bark into the pith of many trees.   { 'wē·vəl }

weft   *See* filling.   { weft }

weft knitting   [TEXT]   A knitting process in which a continuous yarn is carried in crosswise rows.   { 'weft ‚nid·iŋ }

Wegener's granulomatosis   [MED]   A rare disease of unknown causation characterized by necrotizing granulomas in the air passages, necrotizing vasculitis, and glomerulitis.   { 'vā·gə·nərz ‚gran·yə·lə·mə'tō·səs }

Wehnelt cathode   *See* oxide-coated cathode.   { 'vān‚elt ‚kath‚ōd }

wehrlite   [MINERAL]   BiTe A mineral that is a native alloy of bismuth and tellurium.   Also known as mirror glance.   [PETR] A peridotite composed principally of olivine and clinopyroxene with accessory opaque oxides.   { 'wer‚līt }

Weibull distribution   [STAT]   A distribution that describes life-time characteristics of parts and components.   { 'wī‚bùl ‚dis·trə‚byü·shən }

weibullite   [MINERAL] $Pb_4Bi_6S_9Se_4$ A steel gray mineral consisting of lead bismuth sulfide with selenium replacing the sulfide; occurs in indistinct prismatic crystals in massive form.   { 'wī‚bù‚līt }

Weierstrass' approximation theorem   [MATH]   A continuous real-valued function on a closed interval can be uniformly approximated by polynomials.   { 'vī·ər‚shträs ə‚präk·sə'mā·shən ‚thir·əm }

Weierstrass functions   [MATH]   Used in the calculus of variations, these determine functions satisfying the Euler-Lagrange equation and Jacobi's condition while maximizing a given definite integral.   { 'vī·ər‚shträs ‚fəŋk·shənz }

Weierstrassian elliptic function   [MATH]   A function that plays a central role in the theory of elliptic functions; for z, g_2 and g_3 real or complex numbers, let y be that number such that

$$z = \int_y^\infty \frac{dt}{\sqrt{4t^3 - g_2 t - g_3}} \; ;$$

the Weierstrassian elliptic function of z with parameters g_2 and g_3 is $p(z; g_2, g_3) = y$.   { ‚vī·ər‚shträs·ē·ən i'lip·tik 'fəŋk·shən }

Weierstrass M test   [MATH]   An infinite series of numbers will converge or functions will converge uniformly if each term is dominated in absolute value by a nonnegative constant M_n, where these M_n form a convergent series.   Also known as Weierstrass' test for convergence.   { 'vī·ər‚shträs 'em ‚test }

Weierstrass' test for convergence   *See* Weierstrass M test.   { 'vī·ər‚shträs 'test fər kən'vər·jəns }

Weierstrass transform   [MATH]   This transform of a real function $f(y)$ is the function given by the integral from $-\infty$ to ∞ of $(4\pi t)^{-1/2} \exp[-(x-y)^2/4] f(y) dy$; this is used in studying the heat equation.   { 'vī·ər‚shträs 'tranz‚fȯrm }

Weigert effect   [OPTICS]   Dichroism introduced in a silver-silver chloride photographic emulsion by a beam of linearly polarized light.   { 'vī·gərt i‚fekt }

weighing rain gage   [ENG]   A type of recording rain gage, consisting of a receiver in the shape of a funnel which empties into a bucket mounted upon a weighing mechanism; the weight of the catch is recorded, on a clock-driven chart, as inches of precipitation; used at climatological stations.   { 'wā·iŋ 'rān gāj }

weight   [MATH]   **1.** The unique nonnegative integer assigned to an edge or arc in a network or directed network.   **2.** The sum of the weights (first definition) of all the arcs in an s-t cut.   **3.** The nonnegative integer assigned to a vertex in a generalized *s-t* network.   **4.** The sum of the weights of all the arcs and vertices in a generalized *s-t* cut.   [MECH]   **1.** The gravitational force with which the earth attracts a body.   **2.** By extension, the gravitational force with which a star, planet, or satellite attracts a nearby body.   { wāt }

weight and balance sheet   [AERO ENG]   A sheet which records the distribution of weight in an aircraft and shows the center of gravity of an aircraft at takeoff and landing.   { wat ən ¦bal·əns ˌshēt }

weight barometer   [ENG]   A mercury barometer which measures atmospheric pressure by weighing the mercury in the column or the cistern.   { 'wāt bə₁räm·əd·ər }

weight density   [PHYS]   The weight of a body or portion of a body divided by its volume.   { 'wāt ₁den·səd·ē }

weighted aggregative index   [STAT]   A statistic for a collection of items weighted so as to reflect the relative importance of the items with regard to the overall phenomenon which the index is designed to describe; a price index is an example.   { ¦wād·əd ₁a·grə₁gād·iv 'in₁deks }

weighted area masks   [COMPUT SCI]   In character recognition, a set of characters (each character residing in the character reader in the form of weighted points) which theoretically render all input specimens unique, regardless of the size or style.   { 'wād·əd 'er·ē·ə ₁masks }

weighted average   [STAT]   The number obtained by adding the product of α_i times the *i*th number in a set of *N* numbers for *i* = 1, 2, . . . , *N*, where α_i are numbers (weights) such that $\alpha_1 + \alpha_2 + \cdots + \alpha_N = 1$. Also known as weighted mean.   { 'wād·əd 'av·rij }

weighted code   [COMPUT SCI]   A method of representing a decimal digit by a combination of bits, in which each bit is assigned a weight, and the value of the decimal digit is found by multiplying each bit by its weight and then summing the results.   { 'wād·əd 'kōd }

weighted mean   *See* weighted average.   { ¦wād·əd 'mēn }

weighted moving average   [STAT]   A method used for smoothing data in a time series in which each observation being averaged is given a weight which reflects its relative importance in calculating the average.   { ¦wād·əd 'müv·iŋ 'av·rij }

weighted oscillator strength   *See* gf-value.   { 'wād·əd 'äs·ə₁lād·ər ₁streŋkth }

weight factor   [STAT MECH]   The number of microstates that correspond to a given macrostate.   { 'wāt ₁fak·tər }

weight function   [MATH]   **1.** Two real valued functions *f* and *g* are orthogonal relative to a weight function σ on an interval if the integral over the interval of *f·g·*σ vanishes.   **2.** A function defined on the edges of a network or the arcs of a directed network, whose value at each edge or arc is the unique nonnegative integer assigned to that edge or arc.   **3.** A function defined on the vertices of a generalized *s-t* network, whose value at each vertex is a nonnegative integer.   { 'wāt ₁fəŋk·shən }

weighting   [ENG]   The artificial adjustment of measurements to account for factors that, in the normal use of the device, would otherwise be different from conditions during the measurements.   [TEXT]   The chemical or mechanical process of adding weight or body to a fabric or yarn, especially silk, by the addition of various materials.   { 'wād·iŋ }

weighting network   [ENG ACOUS]   One of three or more circuits in a sound-level meter designed to adjust its response; the A and B weighting networks provide responses approximating the 40- and 70-phon equal loudness contours, respectively, and the C weighting network provides a flat response up to 8000 hertz.   { 'wād·iŋ ₁net₁wərk }

weightlessness   [MECH]   A condition in which no acceleration, whether of gravity or other force, can be detected by an observer within the system in question.   Also known as zero gravity.   { 'wāt·ləs·nəs }

weightlessness switch   *See* zero-gravity switch.   { 'wāt·ləs·nəs ₁swich }

weight-loaded regulator   [ENG]   A pressure-regulator valve for pressure vessels or flow systems; the regulator is preloaded by counterbalancing weights to open (or close) at the upper (or lower) limit of a preset pressure range.   { 'wāt ¦lōd·əd 'reg·yə₁lād·ər }

weight thermometer   [ENG]   A glass vessel for determining the thermal expansion coefficient of a liquid by measuring the mass of liquid needed to fill the vessel at two different temperatures.   { 'wāt ₁thər₁mäm·əd·ər }

weight titration   [ANALY CHEM]   A titration in which the

amount of titrant required is determined in terms of the weight that must be added to reach the end point.   { 'wāt tī'trā·shən }

weight zone   [ORD]   A weight range having specified minimum and maximum weights; artillery projectiles of 75-millimeter caliber and larger are sometimes grouped into weight zones and marked with appropriate symbols; the selection of projectiles of a single weight zone for a specific firing problem results in improved ballistic uniformity.   { 'wāt ₁zōn }

Weiland effect   [GRAPHICS]   A photographic effect in which a photographic material undergoes greater blackening when exposed at a very high intensity for a short time and then at a lower intensity for a long time than with the reverse sequence.   { 'wī·lənd i₁fekt }

Weil-Felix test   [IMMUNOL]   An agglutination test for various rickettsial infections based on production of nonspecific agglutinins in the blood of infected patients, and using various strains of *Proteus vulgaris* as antigen.   { 'vīl 'fā·liks ₁test }

Weil's disease   [MED]   A severe form of leptospirosis characterized by jaundice, oliguria, circulatory collapse, and tendency to hemorrhage.   Also known as icterohemorrhagic fever; leptospirosis icterohemorrhagia; spirochetal jaundice.   { 'vīlz di₁zēz }

Weinberg-Salam theory   [PART PHYS]   A gage theory in which the electromagnetic and weak nuclear interactions are described by a single unifying framework in which both have a characteristic coupling paramenter equal to the fine-structure constant; it predicts the existence of intermediate vector bosons and neutral current interactions.   Also known as Salam-Weinberg theory.   { 'wīn₁bərg sə'läm ₁thē·ə·rē }

Weingarten formulas   [MATH]   Equations concerning the normals to a surface at a point.   { 'wīn₁gart·ən ₁for·myə·ləs }

Weingarten surface   [MATH]   A surface such that either of the principal radii is uniquely determined by the other.   { 'wīn₁gärt·ən ₁sər·fəs }

weinschenkite   [MINERAL]   **1.** $YPO_4 \cdot 2H_2O$ A white mineral consisting of a hydrous yttrium phosphate.   Also known as churchite.   **2.** A dark-brown variety of hornblende high in ferric iron, aluminum, and water.   { 'vīn₁sheŋ₁kīt }

weir   [CIV ENG]   A dam in a waterway over which water flows, serving to regulate water level or measure flow.   { wer }

weird number   [MATH]   An abundant number that is not a semiperfect number.   { ¦wird 'nəm·bər }

weir tank   [PETRO ENG]   A type of oil-field storage tank with high- and low-level weir boxes and liquid-level controls for metering the liquid content of the tank.   { 'wer ₁taŋk }

Weissenberg effect   [FL MECH]   An alteration of the normal stresses in a non-Newtonian fluid on account of elasticity, so that such a fluid, when placed between two concentric, rotating cylinders, can rise on the inner cylinder in spite of centrifugal forces.   { 'vīs·ən₁bərg i₁fekt }

Weissenberg method   [SOLID STATE]   A method of studying crystal structure by x-ray diffraction in which the crystal is rotated in a beam of x-rays, and a photographic film is moved parallel to the axis of rotation; the crystal is surrounded by a sleeve which has a slot that passes only diffraction spots from a single layer of the reciprocal lattice, permitting positive identification of each spot in the pattern.   { 'vīs·ən₁berk ₁meth·əd }

weissite   [MINERAL] Cu_5Te_3 A dark bluish-black mineral consisting of copper telluride; occurs in massive form.   { 'wī₁sīt }

Weiss magneton   [ATOM PHYS]   A unit of magnetic moment, equal to 1.853×10^{-24} joule/tesla, about one-fifth of the Bohr magneton; it is experimentally derived, the magnetic moments of certain molecules being close to integral multiples of this quantity.   { 'ves 'mag·nə₁tän }

Weiss molecular field   [SOLID STATE]   The effective magnetic field postulated in the Weiss theory of ferromagnetism, which acts on atomic magnetic moments within a domain, tending to align them, and is in turn generated by these magnetic moments.   { 'ves mə'lek·yə·lər 'fēld }

Weiss theory   [SOLID STATE]   A theory of ferromagnetism based on the hypotheses that below the Curie point a ferromagnetic substance is composed of small, spontaneously magnetized regions called domains, and that each domain is spontaneously magnetized because a strong molecular magnetic field tends to align the individual atomic magnetic moments within the domain.   Also known as molecular field theory.   { 'ves ₁thē·ə·rē }

Weisz ring oven   [ANALY CHEM]   A device for vaporization

WEIGHTLESSNESS

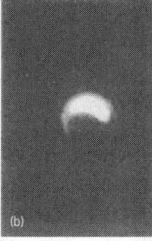

Effect of weightlessness on combustion. (a) Flame on Earth is elongated. (b) Flame in space forms a circle because the lack of convection does not allow waste material to flow away from the flame. (*NASA*)

of solvent from filter paper, leaving the solute in a ring (circular) shape; used for qualitative analysis of very small samples. { 'vīs 'riŋ ,əv·ən }

Weizäcker-Williams method [QUANT MECH] A method of calculating the bremsstrahlung emitted when two particles, whose relative kinetic energies are much larger than their rest energies, collide; in the rest frame of one of the particles, the field of the other is equivalent to a set of virtual photons, and Compton scattering of these photons by the particle at rest is computed. { 'vīt,sek·ər 'wil·yəms ,meth·əd }

Weizsaecker's theory [ASTRON] A theory of the origin of the solar system; it hypothesizes primeval turbulent eddies which become permanent and self-gravitating; Weizsaecker does not discuss the origin of the gas clouds. { 'vīt,sek·ərz ,thē·ə·rē }

welcome page See home page. { 'wel·kəm 'pāj }

weld [MET] A union made between two metals by welding. { weld }

weldability [MET] Suitability of a metal to be welded under specified conditions. { ,wel·də'bil·əd·ē }

weld bead [MET] A deposit of filler metal from a single welding pass. Also known as bead. { 'weld ,bēd }

weldbonding [MET] A process for joining metals in which adhesive, typically an epoxy paste, is applied to the parts, which are then clamped together, spot-welded, and put into an oven (250°F, or 121°C, for 1 hour) to cure the adhesive. { 'weld ,bänd·iŋ }

weld decay [MET] Intercrystalline corrosion of austenitic stainless steels near welded areas; caused by chromium carbide precipitation along grain boundaries of alloy subject to prolonged heating in the temperature range 400–850°C. { 'weld di,kā }

weld delay time [MET] Delay of the time current in spot, seam, or projection welding with respect to starting the forge delay timer used to synchronize pressure and heat. { 'weld di'lā ,tīm }

welded tuff [PETR] A pyroclastic deposit hardened by the action of heat, pressure from overlying material, and hot gases. Also known as tuff lava. { 'wel·dəd 'təf }

welder [MET] 1. A machine used in welding. Also known as welding machine. 2. A person who performs a welding operation. { 'wel·dər }

weld gage [ENG] A device used to check the shape and size of welds. { 'weld ,gāj }

welding [GEOL] Consolidation of sediments by pressure; water is squeezed out and cohering particles are brought within the limits of mutual molecular attraction. [MET] Joining two metals by applying heat to melt and fuse them, with or without filler metal. { 'weld·iŋ }

welding current [ELEC] The current that flows through a circuit while a weld is being made. { 'weld·iŋ ,kə·rənt }

welding cycle [MET] The complete sequence of events involved in making a resistance weld. { 'weld·iŋ ,sī·kəl }

welding electrode [MET] 1. In arc welding, the current-carrying rod or rods used to strike an arc between rod and work. 2. In resistance welding, the component of a machine through which current and pressure are applied to the work. { 'weld·iŋ i,lek,trōd }

welding force See electrode force. { 'weld·iŋ ,fórs }

welding generator [ELEC] A generator used for supplying the welding current. { 'weld·iŋ jen·ə,rād·ər }

welding ground See work lead. { 'weld·iŋ ,graùnd }

welding machine See welder. { 'weld·iŋ mə,shēn }

welding rod [MET] Filler metal in the form of a rod or heavy wire. { 'weld·iŋ ,räd }

welding schedule [MET] A record of all welding machine settings plus identification of the machine needed to produce a weld for a given material of a given size and finish. { 'weld·iŋ ,skej·əl }

welding sequence [MET] The order for welding component parts of a weldment or structure. { 'weld·iŋ ,sē·kwəns }

welding stress [MET] Residual stress resulting from localized heating and cooling during welding. { 'weld·iŋ ,stress }

welding tip [ENG] A replaceable nozzle for a gas torch used in welding. [MET] An electrode used in spot or projection welding. { 'weld·iŋ ,tip }

welding torch [ENG] A gas-mixing and burning tool for the welding of metal. { 'weld·iŋ ,tórch }

welding transformer [ELEC] A high-current, low-voltage

power transformer used to supply current for welding. { 'weld·iŋ tranz,fór·mər }

weld interval [MET] The total heat and cool times for making one multiple-impulse weld. { 'weld ,in·tər·vəl }

weld-interval timer [ENG] A device used to control weld interval. { 'weld ,in·tər·vəl ,tīm·ər }

weld line [ENG] See flow line. [MET] The junction of the weld metal and base metal, or the junction of base-metal parts when filler metal is not used. { 'weld ,līn }

weld mark See flow line. { 'weld ,märk }

weldment [ENG] An assembly or structure whose component parts are joined by welding. { 'weld·mənt }

weld metal [MET] The metal constituting the fused zone in spot, seam, or projection welding. { 'weld ,med·əl }

weld time [MET] The time that the welding current is applied to the work in single-impulse and flash welding. { 'weld ,tīm }

weld zone [MET] The region of a weld that includes both the weld metal and the heat-affected zone. { 'weld ,zōn }

Welge method [PETRO ENG] A method of calculation of the anticipated oil-recovery performance of a gas-cap-drive oil reservoir. { 'wel·gē ,meth·əd }

well [BUILD] An open shaft in a building, extending vertically through floors to accommodate stairs or an elevator. [ENG] A hole dug into the earth to reach a supply of water, oil, brine, or gas. { wel }

wellbore See borehole. { 'wel,bór }

wellbore hydraulics [PETRO ENG] A branch of oil production engineering that deals with the motion of fluids (oil, gas, or water) in wellbore tubing or casing, or the annulus between tubing and casing. { 'wel,bór hī'dró·liks }

welcome page See home page. { 'wel·kəm ,pāj }

well completion [PETRO ENG] The final sealing off of a drilled well (after drilling apparatus is removed from the borehole) with valving, safety, and flow-control devices. { 'wel kəm,ple·shən }

well conditioning [PETRO ENG] 1. Preparation of a well for sampling procedures by control of production rate and associated pressure drawdown. 2. Removal of accumulated scale, wax, mud, and sand from the inner surfaces of a wellbore, or breakage of water blocks to increase production of oil or gas. { 'wel kən,dish·ən·iŋ }

well core [ENG] A sample of rock penetrated in a well or other borehole obtained by use of a hollow bit that cuts a circular channel around a central column or core. { 'wel ,kór }

well-deck vessel [NAV ARCH] A merchant vessel having a sunken deck fitted between the forecastle and a long poop or continuous bridge house or raised quarterdeck. { 'wel ,dek ,ves·əl }

well drill [MECH ENG] A drill, usually a churn drill, used to drill water wells. { 'wel ,dril }

well-formed formula [MATH] A finite sequence or string of symbols that is grammatically or syntactically correct for a given set of grammatical or syntactical rules. { 'wel ¦fórmd ,for·myə·lə }

wellhead [CIV ENG] The top of a well. [HYD] The place where a stream emerges from the ground. { 'wel,hed }

wellhole [MIN ENG] 1. A large-diameter vertical hole used in quarries and opencast pits for taking heavy explosive charges in blasting. 2. The sump, or portion of a shaft below the place where skips are caged at the bottom of the shaft, in which water collects. { 'wel,hōl }

well injectivity [PETRO ENG] The ability of an injection well (water or gas) to receive injected fluid; can be negatively influenced by formation plugging, borehole scale, or liquid blocking around the lower end of the borehole. { 'wel ,in·jek'tiv·əd·ē }

well logging [ENG] The technique of analyzing and recording the character of a formation penetrated by a drill hole in petroleum exploration and exploitation work. { 'wel ,läg·iŋ }

well-ordered set [MATH] A linearly ordered set where every subset has a least element. { 'wel ¦ór·dərd 'set }

well-ordering principle [MATH] The proposition that every set can be endowed with an order so that it becomes a well-ordered set; this is equivalent to the axiom of choice. { 'wel ¦ór·dər·iŋ 'prin·sə·pəl }

well performance [PETRO ENG] The measurement of a well's production of oil or gas as related to the well's anticipated productive capacity, pressure drop, or flow rate. { 'wel pər,fȯr·məns }

wellpoint [CIV ENG] A component of a wellpoint system consisting of a perforated pipe about 4 feet (1.2 meters) long and about 2 inches (5 centimeters) in diameter, equipped with a ball valve, a screen, and a jetting tip. { 'wel,pȯin }

wellpoint system [CIV ENG] A method of keeping an excavated area dry by intercepting the flow of groundwater with pipe wells located around the excavation area. { 'wel,pȯint ,sis·təm }

well-posed problem [MATH] A problem that has a unique solution which depends continuously on the initial data. { 'wel ¦pōzd 'präb·ləm }

well-regulated system [CONT SYS] A system with a regulator whose action, together with that of the environment, prevents any disturbance from permanently driving the system from a state in which it is stable, that is, a state in which it retains its structure and survives. { 'wel ¦reg·yə,lād·əd ,sis·təm }

well shooting [ENG] The firing of a charge of nitroglycerin, or other high explosive, in the bottom of a well for the purpose of increasing the flow of water, oil, or gas. { 'wel ,shüd·iŋ }

well-sorted [GEOL] Referring to a sorted sediment that consists of particles of approximately the same size and has a sorting coefficient of less than 2.5. { 'wel ¦sȯrd·əd }

well spacing [PETRO ENG] Areal location and interrelationship between producing oil or gas wells in an oil field; calculated for the maximum ultimate production from a given reservoir. { 'wel ,spās·iŋ }

well stimulation See stimulation treatment. { 'wel ,stim·yə'lā·shən }

well-type manometer [ENG] A type of double-leg, glass-tube manometer; one leg has a relatively small diameter, and the second leg is a reservoir; the level of the liquid in the reservoir does not change appreciably with change of pressure; a mercury barometer is a common example. { 'wel ¦tīp mə'näm·əd·ər }

welt [BUILD] **1.** In sheet-metal roofing, a seam consisting of two joined sheets of metal whose edges have been folded over each other and fastened down flat. **2.** A strip of wood fastened over a flush seam or joint for added strength. [ENG] A strip that has been fastened to the edges of plates that form a butt joint in a steam boiler. { welt }

Welwitschiales [BOT] An order of gymnosperms in the subdivision Geneticae represented by the single species *Welwitschia mirabilis* of southwestern Africa; distinguished by only two leaves and short, unbranched, cushion- or saucer-shaped woody main stem which tapers to a very long taproot. { wel,wich·ē'ā·lēz }

Wenlockian [GEOL] A European stage of geologic time: Middle Silurian (above Tarannon below Ludlovian). { wen 'läk·ē·ən }

Wentworth classification [GEOL] A logarithmic grade for size classification of sediment particles starting at 1 millimeter and using the ratio of 1/2 in one direction (and 2 in the other), providing diameter limits to the size classes of 1, 1/2, 1/4, etc. and 1, 2, 4, etc. { 'went,wərth ,klas·ə·fə'kā·shən }

Wentworth quick-return motion See turning-block linkage. { 'went,wərth 'kwik ri¦tərn ,mō·shən }

Wentworth scale [GEOL] A geometric grade scale for sedimentary particles ranging from clay particles (diameter less than 1/250 millimeter) to boulders (diameters greater than 256 millimeters), in which the size classes are related to one another by a constant ratio of 1/2 (4, 2, 1, ¹/₂, etc.). { 'went,wərth ,skāl }

Wentzel-Kramers-Brillouin method [QUANT MECH] Method of approximating quantum-mechanical wave functions and energy levels, in which the logarithm of the wave function is expanded in powers of Planck's constant, and all except the first two terms are neglected. Also known as phase integral method; WKB method. { 'vent·səl 'krä·mərz brē'wan ,meth·əd }

Werdnig-Hoffmann disease [MED] Infantile spinal muscular atrophy. { 'ver,nik 'hȯf,män ci,zēz }

Werfenian stage See Scythian stage. { ver'fē·nē·ən ,stāj }

Werner band [SPECT] A band in the ultraviolet spectrum of molecular hydrogen extending from 116 to 125 nanometers. { 'ver·nər ,band }

Werner complex See coordination compound. { 'ver·nər ,käm,pleks }

wernerite See scapolite. { 'ver·nə,rīt }

Werner's syndrome [MED] A complex of symptoms, thought to be inherited as an autosomal recessive, including premature senescence, dwarfism, cataracts, scleroderma-like changes of the skin, osteoporosis, and multiglandular dysfunction. { 'ver·nərz ,sin,drōm }

Wernicke's aphasia [PSYCH] Fluent aphasia associated with injury to the area of the left temporal lobe just posterior to the primary auditory complex that is characterized by severely impaired auditory comprehension, and fluent but markedly paraphasic speech, sometimes at faster than normal rate. { ¦ver·ni·kēz ə'fāzh ə }

Wernicke's area [NEUROSCI] An area located in the left temporal lobe just posterior to the primary auditory complex that is involved with speech comprehension; injury to this area results in fluent aphasia. { 'ver·ni·kēz ,er·ē·ə }

Wernicke's encephalopathy [MED] A disease due to thiamine deficiency, characterized by vomiting, ophthalmoplegia, ptosis, nystagmus, ataxia, weakness, dementia, and hemorrhaging of neurons around the third ventricle, cerebral aqueduct, and mammillary bodies. { 'ver·nə,kēz in,sef·ə'läp·ə·thē }

Wertheim effect [ELECTROMAG] A potential difference that appears between the ends of a wire twisted in a longitudinal magnetic field. { 'vert,hīm i,fekt }

west [GEOGR] The direction 90° to the left or 270° to the right of north. { west }

West African cedar See Sapele mahogany. { 'west ¦af·rə·kən 'sēd·ər }

West Australia Current [OCEANOGR] The complex current flowing northward along the west coast of Australia; it is strongest from November to January, and weakest and variable from May to July; it curves toward the west to join the South Equatorial Current. { 'west ȯ'strāl·yə 'kə·rənt }

westerlies [METEOROL] The dominant west-to-east motion of the atmosphere, centered over the middle latitudes of both hemispheres; at the earth's surface, the westerly belt (or west-wind belt) extends, on the average, from about 35 to 65° latitude. Also known as circumpolar westerlies; middle-latitude westerlies; mid-latitude westerlies; polar westerlies; subpolar westerlies; subtropical westerlies; temperate westerlies; zonal westerlies; zonal winds. { 'wes·tər·lēz }

westerly wave [METEOROL] An atmospheric wave disturbance embedded in the mid-latitude westerlies. { 'wes·tər·lē 'wāv }

Western blotting [CELL MOL] A protein detection technique in which proteins are separated by one- or two-dimensional gel electrophoresis, transferred (blotted) to a nitrocellulose sheet, and then treated with radioactive antibodies (or antibodies coupled to a fluorescent dye or an easily detectable enzyme) that are specific to the protein of interest. { 'wes·tərn ,bläd·iŋ }

Western Equatorial Countercurrent [OCEANOGR] Weak, narrow bands of eastward-flowing water observed in some winter months in the western Atlantic near the equator. { 'wes·tərn ,ek·wə'tȯr·ē·əl kaunt·ər,kə·rənt }

Western equine encephalitis [MED] A type of equine encephalitis which occurs chiefly west of the Mississippi River; the chief vector is the culicine mosquito *Culex tarsalis*. { 'wes·tərn 'ē,kwīn in,sef·ə'līd·əs }

West Greenland Current [OCEANOGR] The current flowing northward along the west coast of Greenland into the Davis Strait; part of this current joins the Labrador Current, while the other part continues into Baffin Bay. { 'west ¦grēn·lənd 'kə·rənt }

westing [NAV] The distance a craft makes good to the west. { 'west·iŋ }

West Nile fever [MED] An acute, usually mild, mosquito-borne virus disease occurring in summer, chiefly in Egypt, Israel, Africa, India, and Korea; signs are fever and lymphadenopathy, sometimes with a rash. { 'west ¦nīl 'fē·vər }

Weston standard cell [ELEC] A standard cell used as a highly accurate voltage source for calibrating purposes; the positive electrode is mercury, the negative electrode is cadmium, and the electrolyte is a saturated cadmium sulfate solution; the Weston standard cell has a voltage of 1.018636 volts at 20°C. { 'wes·tən 'stan·dərd 'sel }

Westphal balance [ENG] A direct-reading instrument for determining the densities of solids and liquids; a plummet of

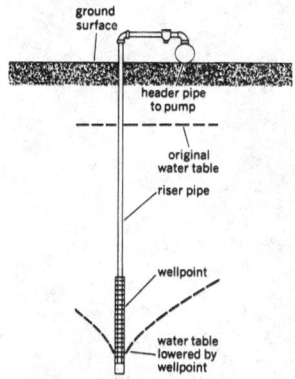

WELLPOINT SYSTEM

Diagram showing setup of components of wellpoint system.

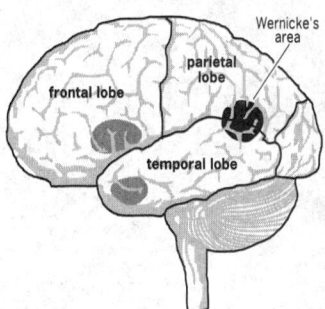

WERNICKE'S AREA

Lateral surface view of the cerebral hemisphere showing Wernicke's area. (*After C. R. Noback, The Human Nervous System, 4th ed., McGraw-Hill, 1991*)

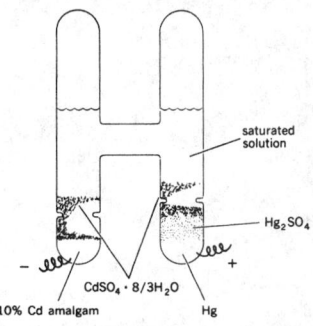

WESTON STANDARD CELL

Cross-sectional diagram of Weston standard cell.

known mass and volume is immersed in the liquid whose density is to be measured or, alternatively, a sample of the solid whose density is to be measured is immersed in a liquid of known density, and the loss in weight is measured, using a balance with movable weights. { 'west₁fȯl ₁bal·əns }

Westphalian [GEOL] A European stage of Upper Carboniferous geologic time, forming the Middle Pennsylvanian, above upper Namurian and below Stephanian. { west'fāl·yən }

Westphal-Pilcz reflex *See* pupillary reflex. { west₁fȯl 'pils ₁rē₁fleks }

Westphal's pupillary reflex *See* pupillary reflex. { 'west₁fȯlz ₁pyü·pə₁ler·ē 'rē₁fleks }

west point [ASTRON] That point on the celestial sphere that is due west of observer; at this point the celestial equator crosses the horizon. { 'west 'pȯint }

westward intensification [OCEANOGR] The intensification of ocean currents to the west, derived from a mathematical model that includes the effects of zonal wind stress at the sea surface and internal friction. { 'west·wərd in₁ten·sə·fə'kā·shən }

West Wind Drift *See* Antarctic Circumpolar Current. { 'west ₁wind 'drift }

wet [PHYS] A liquid is said to wet a solid if the contact angle between the solid and the liquid, measured through the liquid, lies between 0 and 90°, and not to wet the solid if the contact angle lies between 90 and 180°. { wet }

wet adiabat *See* saturation adiabat. { 'wet 'ad·ē·ə₁bat }

wet and dry bulb thermometer *See* psychrometer. { ₁wet ən ₁drī ₁bəlb thər₁mäm·əd·ər }

wet ashing [ORG CHEM] The conversion of an organic compound into ash (decomposition) by treating the compound with nitric or sulfuric acid. { 'wet 'ash·iŋ }

wet assay [MIN ENG] The determination of the quantity of a desired constituent in ores, metallurgical residues, and alloys by the use of the processes of solution, flotation, or other liquid means. { 'wet 'as₁ā }

wet blasting [ENG] Shot firing in wet holes. [MET] Liquid honing in which an impeller wheel drives the liquid suspension. { 'wet 'blast·iŋ }

wet-bulb depression [METEOROL] The difference in degrees between the dry-bulb temperature and the wet-bulb temperature. { 'wet ₁bəlb di'presh·ən }

wet-bulb temperature [METEOROL] **1.** Isobaric wet-bulb temperature, that is, the temperature an air parcel would have if cooled adiabatically to saturation at constant pressure by evaporation of water into it, all latent heat being supplied by the parcel. **2.** The temperature read from the wet-bulb thermometer; for practical purposes, the temperature so obtained is identified with the isobaric wet-bulb temperature. { 'wet ₁bəlb 'tem·prə·chər }

wet-bulb thermometer [ENG] A thermometer having the bulb covered with a cloth, usually muslin or cambric, saturated with water. { 'wet ₁bəlb thər'mäm·əd·ər }

wet cell [ELEC] A primary cell in which there is a substantial amount of free electrolyte in liquid form. { 'wet ₁sel }

wet-cell caplight [MIN ENG] A rechargeable head lamp; the batteries are worn on the belt. { 'wet ₁sel 'kap₁līt }

wet classifier [ENG] A device for the separation of solid particles in a mixture of solids and liquid into fractions, according to particle size or density by methods other than screening; operates by the difference in the settling rate between coarse and fine or heavy and light particles in a tank-confined liquid. { 'wet 'klas·ə₁fī·ər }

wet climate [CLIMATOL] A climate whose vegetation is of the rainforest type. Also known as rainforest climate. { 'wet 'klī·mət }

wet collector *See* scrubber. { 'wet kə'lek·tər }

wet contact [ELEC] Contact through which direct current flows. { 'wet 'kän₁takt }

wet cooling tower [MECH ENG] A structure in which water is cooled by atomization into a stream of air; heat is lost through evaporation. Also known as evaporative cooling tower. { 'wet 'kül·iŋ ₁taù·ər }

wet corrosion [MET] Corrosion caused by exposure to aqueous solutions. { 'wet kə'rō·zhən }

wet criticality [NUCLEO] Reactor criticality achieved with the coolant present. { 'wet ₁krid·ə'kal·əd·ē }

wet drawing [MET] Drawing in which the dies and blocks are completely immersed in the lubricant. { 'wet 'drȯ·iŋ }

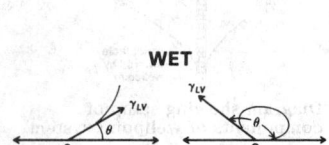

WET

Contact angle between the solid and the liquid is indicated by θ. *(a)* Liquid wets solid, *(b)* liquid does not wet solid. Arrows represent interfacial tensions γ_{SL}, γ_{SV}, and γ_{LV}, where S, L, and V refer to solid, liquid and vapor.

wet drill [MECH ENG] A percussive drill with a water feed either through the machine or by means of a water swivel, to suppress the dust produced when drilling. { 'wet ₁dril }

wet electrolytic capacitor [ELEC] An electrolytic capacitor employing a liquid electrolyte. { 'wet i₁lek·trə₁lid·ik kə'pas·əd·ər }

wet emplacement [AERO ENG] A launch emplacement that provides a deluge of water for cooling the flame bucket, the rocket engines, and other equipment during the launch of a missile. { 'wet im'plās·mənt }

wet engine [MECH ENG] An engine with its oil, liquid coolant (if any), and trapped fuel inside. { 'wet 'en·jən }

wet flashover voltage [ELECTR] The voltage at which an electric discharge occurs between two electrodes that are separated by an insulator whose surface has been sprayed with water to simulate rain. { 'wet 'flash₁ō·vər ₁vōl·tij }

wet gas [MATER] Natural gas produced along with crude petroleum in oil fields or from gas-condensate fields; in addition to methane, it contains ethane, propane, butanes, and some higher hydrocarbons, such as pentane and hexane. { 'wet 'gas }

wet grinding [MECH ENG] **1.** The milling of materials in water or other liquid. **2.** The practice of applying a coolant to the work and the wheel to facilitate the grinding process. { 'wet 'grīnd·iŋ }

wet hole [ENG] A borehole that traverses a water-bearing formation from which the flow of water is great enough to keep the hole almost full of water. { 'wet 'hōl }

wetlands [ECOL] An area characterized by a high content of soil moisture, such as a swamp or bog. { 'wet ₁lanz }

wet mill [MECH ENG] **1.** A grinder in which the solid material to be ground is mixed with liquid. **2.** A mill in which the grinding energy is developed by a fast-flowing liquid stream; for example, a jet pulverizer. { 'wet 'mil }

wet-reed relay [ELEC] Reed-type relay containing mercury at the relay contacts to reduce arcing and contact bounce. { 'wet ₁rēd 'rē₁lā }

wet rot [MATER] Fungal decay of wood with a high moisture content. { 'wet ₁rät }

wet scrubber [ENG] A device designed to clean a gas stream by bringing it into contact with a liquid. { 'wet 'skrəb·ər }

wet season *See* rainy season. { 'wet ₁sēz·ən }

wet sleeve [MECH ENG] A cylinder liner which is exposed to the coolant over 70% or more of its surface. { 'wet 'slēv }

wet slip [CIV ENG] An opening between two wharves or piers where dock trials are usually conducted, and the final fitting out is done. { 'wet 'slip }

wet snow [METEOROL] Deposited snow that contains a great deal of liquid water. { 'wet 'snō }

wet spinning [TEXT] Producing synthetic and man-made filaments by extruding the chemical solution through spinnerets into a chemical bath where they coagulate. { 'wet ₁spin·iŋ }

wet stowage [ORD] Method of stowing major caliber ammunition in combat vehicles by placing it in racks surrounded by nonflammable liquid, to reduce ammunition fire hazards. { 'wet 'stō·ij }

wet strength [MATER] **1.** The strength of a material saturated with water. **2.** The ability to withstand water (as for paper products) with a wet-strength additive or resin finish. { 'wet ₁streŋkth }

wet-strength paper [MATER] Paper with increased water resistance due to processing and interlocking of fibers, as well as impregnation with small amounts of resins (for example, melamine or urea formaldehyde). Also known as wet-strong paper. { 'wet ₁streŋkth 'pā·pər }

wet-strong paper *See* wet-strength paper. { 'wet ₁strȯŋ 'pā·pər }

wettability [CHEM] The ability of any solid surface to be wetted when in contact with a liquid; that is, the surface tension of the liquid is reduced so that the liquid spreads over the surface. { ₁wed·ə'bil·əd·ē }

wet tabling [MIN ENG] A tabling process in which a pulp of two or more minerals flows across an inclined, riffled plane surface, is shaken longwise, and is water-washed crosswise. { 'wet ₁tāb·liŋ }

wetted [CHEM] Pertaining to material that has accepted water or other liquid, either on its surface or within its pore structure. { 'wed·əd }

wetted perimeter [GEOL] The portion of the perimeter of a steam channel cross section which is in contact with the water. { 'wed·əd pə'rim·əd·ər }

wetted surface [NAV ARCH] The surface of a ship's hull in contact with the water under specified conditions. { 'wed·əd 'sər·fəs }

wetted-wall column [CHEM ENG] A vertical column that operates with the inner walls wetted by the liquid being processed; used in theoretical studies of mass transfer rates and in analytical distillations; an example is a spinning-band column. { 'wed·əd ¦wȯl 'käl·əm }

wet-test meter [ENG] A device to measure gas flow by counting the revolutions of a shaft upon which water-sealed, gas-carrying cups of fixed capacity are mounted. { 'wet ¦test ¦mēd·ər }

wetting [ELECTR] The coating of a contact surface with an adherent film of mercury. [MET] Spreading liquid filler metal or flux on a solid base metal. { 'wed·iŋ }

wetting agent [CHEM ENG] A substance that increases the rate at which a liquid spreads across a surface when it is added to the liquid in small amounts. [GRAPHICS] A substance that renders a surface nonrepellent to a wetting liquid. { 'wed·iŋ ¦ā·jənt }

wetting angle [FL MECH] A contact angle which lies between 0 and 90°. { 'wed·iŋ ¦aŋ·gəl }

wetting phase [PETRO ENG] In a two-phase oil reservoir system (oil and water), one phase (water) will wet the pore surfaces of the reservoir formation, the other (oil) will not. { 'wed·iŋ ¦fāz }

wet well [MECH ENG] A chamber which is used for collecting liquid, and to which the suction pipe of a pump is attached. { 'wet ¦wel }

wetwood [PL PATH] Wood having a water-soaked appearance because of a high water content; may be caused by bacteria or by physiological factors. { 'wet¦wu̇d }

Weyl equations [QUANT MECH] Two sets of relativistic wave equations into which the Dirac equation decomposes for a massless, spin-1/2 particle. { vīl i¦kwā·zhənz }

Weyl tensor [RELAT] A tensor with the symmetries of the curvature tensor such that all contractions on its indices vanish; the curvature tensor is decomposable in terms of the metric, the scalar curvature, and the Weyl tensor. { 'vīl ¦ten·sər }

whale [VERT ZOO] A large marine mammal of the order Cetacea; the body is streamlined, the broad flat tail is used for propulsion, and the limbs are balancing structures. { wāl }

Whale See Cetus. { wāl }

whaleback dune [GEOL] A smooth, elongated mound or hill of desert sand shaped generally like a whale's back; formed by passage of a succession of longitudinal dunes along the same path. Also known as sand levee. { 'wāl¦bak ¦dün }

whaleback roof See rainbow roof. { 'wāl¦bak ¦rüf }

whaleboat [NAV ARCH] **1.** A long, narrow rowboat that has a large sheer and both ends sharp and inclined to the perpendicular; formerly used to hunt whales. **2.** A long, narrow rowboat or motorboat that has both ends sharp and rounded in the manner of the original whaleboats, and that is equipped with buoyancy tanks; it is often carried on merchant ships and warships. { 'wāl¦bōt }

whalebone See baleen. { 'wāl¦bōn }

whale oil [MATER] A combustible, nontoxic, yellow-brown fixed oil obtained from whale blubber; soluble in alcohol, ether, chloroform, carbon disulfide, and benzene; used as a lubricant, illuminant, and leather dressing, and in soapmaking and fat manufacture. Also known as blubber oil. { 'wāl ¦ȯil }

whaler See waler. { 'wāl·ər }

wharf [CIV ENG] A structure of open construction built parallel to the shoreline; used by vessels to receive and discharge passengers and cargo. { 'wȯrf }

Wharton's duct See submandibular duct. { 'wȯrt·ənz ¦dəkt }

wheat [BOT] A food grain crop of the genus *Triticum;* plants are self-pollinating; the inflorescence is a spike bearing sessile spikelets arranged alternately on a zigzag rachis. { wēt }

wheat germ [BOT] The embryo of a wheat grain. { 'wēt ¦jərm }

wheat germ oil [MATER] A light-yellow oil extracted from wheat germ; used as a source for vitamin E, as a dietary supplement, and in medicine. { 'wēt ¦jərm ¦ȯil }

wheat middlings See sharps. { 'wēt ¦mid·liŋz }

Wheatstone bridge [ELEC] A four-arm bridge circuit, all arms of which are predominately resistive; used to measure the electrical resistance of an unknown resistor by comparing it with a known standard resistance. Also known as resistance bridge; Wheatstone network. { 'wēt¦stōn 'brij }

Wheatstone network See Wheatstone bridge. { 'wēt¦stōn 'net¦wərk }

Wheatstone stereoscope [OPTICS] A type of stereoscope that uses plane mirrors to enable the eyes to form a fused image of two pictures whose separation is greater than the interocular distance. { ¦wēt¦stōn 'ster·ē·ō¦skōp }

wheel [DES ENG] A circular frame with a hub at the center for attachment to an axle, about which it may revolve and bear a load. { wēl }

wheelbarrow [ENG] A small, hand-pushed vehicle with a single wheel and axle between the front ends of two shafts that support a boxlike body and serve as handles at the rear. Also known as barrow. { 'wēl¦bar·ō }

wheel base [DES ENG] The distance in the direction of travel from front to rear wheels of a vehicle, measured between centers of ground contact under each wheel. { 'wēl ¦bās }

wheel-bearing lubricant [MATER] A lubricating grease with the character, structure, and consistency needed to make it suitable for use in antifriction wheel bearings. { 'wēl ¦ber·iŋ ¦lü·brə·kənt }

wheel dresser [ENG] A tool for cleaning, resharpening, and restoring the mechanical accuracy of the cutting faces of grinding wheels. { 'wēl ¦dres·ər }

wheeled crane [MECH ENG] A self-propelled crane that rides on a rubber-tired chassis with power for transportation provided by the same engine that is used for hoisting. { 'wēld 'krān }

Wheeler-Feynman theory [RELAT] A relativistic action-at-a-distance theory in which it is assumed that there are enough absorbers in the universe to serve as sinks for all actions that emanate from any charged particle; radiation damping is a consequence of the theory. { 'wēl·ər 'fīn·mən ¦thē·ə·rē }

Wheelerian [GEOL] A North American stage of upper Pliocene geologic time, above the Venturian and below the Hallian. { wē'lir·ē·ən }

wheelhouse See pilothouse. { 'wēl¦hau̇s }

wheeling [ELEC] The transfer of one utility's energy over another utility's lines for delivery to a third utility. { 'wēl·iŋ }

wheel load capacity [CIV ENG] The capacity of airfield runways, taxiways, parking areas, or roadways to bear the pressures exerted by aircraft or vehicles in a gross weight static configuration. { 'wēl 'lōd kə¦pas·əd·ē }

wheel printer [COMPUT SCI] A line printer that prints its characters from the rim of a wheel around which is the type for the alphabet, numerals, and other characters. { 'wēl ¦print·ər }

wheel sleeve [DES ENG] A flange used as an adapter on precision grinding machines where the hole in the wheel is larger than the machine arbor. { 'wēl ¦slēv }

wheel static [ELECTR] Interference encountered in automobile-radio installations due to static electricity developed by friction between the tires and the street. { 'wēl ¦stad·ik }

whelk [INV ZOO] A gastropod mollusk belonging to the order Neogastropoda; species are carnivorous but also scavenge. { welk }

wherryite [MINERAL] A light green mineral consisting of a basic carbonate-sulfate of lead and copper; occurs in massive form. { 'wer·ē¦īt }

whetstone [MATER] Any hard, fine-grained, naturally occurring, usually siliceous rock suitable for sharpening cutting instruments. { 'wet¦stōn }

Wheweel equation [MATH] An equation which relates the arc length along a plane curve to the angle of inclination of the tangent to the curve. { 'wā¦wēl i¦kwā·zhən }

whewellite [MINERAL] $Ca(C_2O_4) \cdot H_2O$ A colorless or yellowish or brownish, monoclinic mineral consisting of calcium oxalate monohydrate; occurs as crystals. { 'hyü·ə¦līt }

whey [FOOD ENG] The watery part of milk separated from the curd in the process of making cheese. { wā }

whiffletree switch [ELECTR] In computers, a multiposition electronic switch composed of gate tubes and flip-flops, so named because its circuit diagram resembles a whiffletree. { 'wif·əl¦trē ¦swich }

WHILE statement [COMPUT SCI] A statement in a computer program that is executed repeatedly, as long as a specified condition holds true. { 'wīl ¦stāt·mənt }

whip antenna [ELECTROMAG] A flexible vertical rod

WHALE

Sperm whale (*Physeter catodon*).

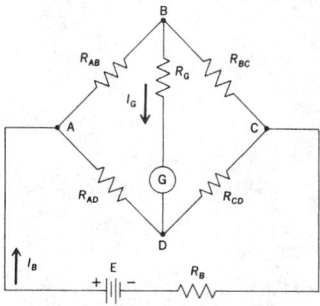

WHEATSTONE BRIDGE

Circuit diagram of Wheatstone bridge, used to measure resistance R_{CD} in terms of known resistances R_{AB}, R_{BC}, and R_{AD}; the latter are adjusted until current I_G through detector G equals zero. R_G = internal resistance of detector. Current I_B flows through battery with open-circuit voltage E and internal resistance R_B.

antenna, used chiefly on vehicles. Also known as fishpole antenna. { 'wip an,ten·ə }

whipcord [TEXT] A twill fabric with fine diagonal cords that is woven of hard-twisted cotton or worsted woolen yarns. { 'wip,kȯrd }

whip grafting [BOT] A method of grafting by fitting a small tongue and notch cut in the base of the scion into corresponding cuts in the stock. { 'wip ,graft·iŋ }

Whipple's disease [MED] A disease characterized by infiltration of the intestinal wall and lymphatics by macrophages filled with glycoprotein. Also known as intestinal lipodystrophy. { 'wip·əlz di,zēz }

whippoorwill storm See frog storm. { ¦wip·ər¦wil ,stȯrm }

whipstock [PETRO ENG] A long wedge dropped or placed in a petroleum well in order to deflect the drill from some obstruction. { 'wip,stäk }

whipworm disease [MED] A chronic, wasting diarrhea produced by heavy parasitization of the large intestine by the nematode *Trichuris trichiura*, particularly in undernourished children in the tropics. { 'wip,wərm di,zēz }

whirlpool [OCEANOGR] Water in rapid rotary motion. { 'wərl,pül }

Whirlpool galaxy [ASTRON] A spiral galaxy of type Sc (open spiral structure), seen face on, in the constellation Canes Venatici. { 'wərl,pül ,gal·ik·sē }

whirly [METEOROL] A small violent storm, a few yards (or meters) to 100 yards (91 meters) or more in diameter, frequent in Antarctica near the time of the equinoxes. { 'wər·lē }

whisker See crystal whisker. { 'wis·kər }

whiskers See vibrissae. { 'wis·kərz }

whiskey [FOOD ENG] A potable alcoholic beverage made by distilling fermented grain mashes and aging the distillate in wood, usually oak; principal sources of grain are barley, wheat, rye, oats, and corn. { 'wis·kē }

whispering gallery [ACOUS] A domed gallery in which weak sounds can be heard at great distances. { 'wis·pər·iŋ ,gal·rē }

whispering-gallery resonance [PHYS] A resonance that rises in the propagation of waves around the circumference of a circular structure when an integral number of wavelengths can fit into the circumference. { ¦wis·pər·iŋ ,gal·rē ¦rez·ən·əns }

whistle buoy [NAV] A buoy equipped with a whistle; in the United States it is usually a conical buoy with a whistle located on its top. { 'wis·əl ,bȯi }

whistler [GEOPHYS] An effect that occurs when a plasma disturbance, caused by a lightning discharge, travels out along lines of magnetic force of the earth's field and is reflected back to its origin from a magnetically conjugate point on the earth's surface; the disturbance may be picked up electromagnetically and converted directly to sound; the characteristic drawn-out descending pitch of the whistler is a dispersion effect due to the greater velocity of the higher-frequency components of the disturbance. { 'wis·lər }

whistler wave See electron cyclotron wave. { 'wis·lər ,wāv }

whistling meteor [ELECTROMAG] Name applied to a radio meteor when a special system for detection is used in which the presence of the meteor is indicated by a rapidly changing audio-frequency radio signal. { 'wis·liŋ ,mēd·ē·ər }

white adipose tissue [HISTOL] The most common type of adipose tissue, representing stored food reserves and thermal and physical insulation. { ¦wīt ,ad·ə,pōs 'tish·ü }

white ant See termite. { ¦wīt 'ant }

white band disease [INV ZOO] A coral reef disease that is typified by a loss of tissue that is visible as a band of bare white skeleton. { ,wīt 'band diz,ēz }

white blood cell See leukocyte. { ¦wīt 'bləd ,sel }

white body [PHYS] A hypothetical substance whose surface absorbs no electromagnetic radiation of any wavelength, that is, one which exhibits zero absorptivity for all wavelengths. { 'wīt ,bäd·ē }

whitecap [OCEANOGR] A cloud of bubbles at the sea surface caused by a breaking wave. { 'wīt,kap }

white carbon black [MATER] A white silica powder made from silicon tetrachloride; used as a replacement for carbon black in rubber compounding. { 'wīt 'kär·bən ,blak }

white cast iron [MET] An extremely hard cast iron, rapidly cooled from the melt; contains about 3% carbon in the form of cementite and fine pearlite. { 'wīt ,kast 'ī·ərn }

white cement [MATER] Pure white portland cement, made

WHIRLPOOL GALAXY

Whirlpool galaxy (NGC 5194), type Sc, and a companion irregular satellite (NGC 5195).

from raw materials with a low iron content, or by using a reducing flame to fire the clinker. { 'wīt si'ment }

white clay See kaolin. { 'wīt 'klā }

white coal See tasmanite. { 'wīt 'kōl }

white coat [BUILD] The finishing coat in plastering. { 'wīt ,kōt }

white cobalt See cobaltite. { 'wīt 'kō,bȯlt }

white compression [COMMUN] In facsimile or television the reduction in picture-signal gain at levels corresponding to light areas, with respect to the gain at the level for midrange light values; the overall effect of white compression is to reduce contrast in the highlights of the picture. { 'wīt kəm,presh·ən }

white copperas See zinc sulfate. { 'wīt 'käp·rəs }

white corpuscle See leukocyte. { 'wīt 'kȯr·pə·səl }

white cutch See gambir. { 'wīt 'kəch }

white damp [MIN ENG] In mining, carbon monoxide (CO); a gas that may be present in the afterdamp of a gas or coal-dust explosion, or in the gases given off by a mine fire; it is an important constituent of illuminating gas, supports combustion, and is very poisonous. { 'wīt ,damp }

white diarrhea See pullorum disease. { 'wīt ,dī·ə'rē·ə }

white dwarf star [ASTRON] An intrinsically faint star of very small radius and high density; the mass is about 0.6 that of the sun and the average radius is about 5000 miles (8000 kilometers); it is one final stage of stellar evolution with thermonuclear energy sources extinct. { 'wīt ¦dwȯrf 'stär }

white feldspar See albite. { 'wīt 'fel,spär }

whitefish [VERT ZOO] Any of various food fishes in the family Salmonidae, especially of the genus *Coregonus*, characterized by an adipose dorsal fin and nearly toothless mouth. { 'wīt,fish }

white frost See hoarfrost. { 'wīt 'frȯst }

white garnet See leucite. { 'wīt 'gär·nət }

white graphite See hexagonal boron nitride. { ,wīt 'graf,īt }

whiteheart malleable iron [MET] White cast iron malleableized and decarburized by heat treatment in an oxidizing material at 900°C for 100–150 hours; decarburization produces a light-colored fracture, in contrast to blackheart malleable iron, which is not decarburized. Also known as blackheart malleable iron. { 'wīt,härt 'mal·yə·bəl 'ī·ərn }

white infarct [MED] An infarct in which hemorrhage is slight, or that has been decolorized by removal of blood or its pigments. { 'wīt 'in,färkt }

white iron [MET] A brittle cast iron whose total carbon content is in the combined forms, and containing little or no graphite; a fresh fracture is white. { 'wīt 'ī·ərn }

white iron ore See siderite. { 'wīt 'ī·ərn ,ȯr }

white lead [INORG CHEM] Basic lead carbonate of variable composition, the oldest and most important lead paint pigment; also used in putty and ceramics. { 'wīt 'led }

white level [COMMUN] The carrier signal level corresponding to maximum picture brightness in television and facsimile. { 'wīt ,lev·əl }

white light [OPTICS] Any radiation producing the same color sensation as average noon sunlight. { 'wīt 'līt }

white light hologram [OPTICS] A reflection hologram which can be viewed with an ordinary light source. { 'wīt ¦līt 'häl·ə,gram }

white metal [MET] 1. Any of several white-colored metals and their alloys of relatively low melting points, such as lead, tin, antimony, and zinc. 2. A copper matte of about 77% copper, obtained from the smelting of sulfide copper ores. { 'wīt ,med·əl }

white-metal bearing alloy See lead-base babbitt. { 'wīt ,med·əl ¦ber·iŋ 'al,ȯi }

white mica See muscovite. { 'wīt 'mī·kə }

white mineral oil [MATER] A highly refined, colorless hydrocarbon oil with low volatility; used as a laxative and in medicine. Also known as liquid petrolatum; paraffinum liquidum. { 'wīt 'min·rəl ,ȯil }

white muscardine [INV ZOO] A disease of the silkworm caused by the fungus *Beauveria bassiana*. { ,wīt 'məs·kər,dēn }

white nickel See rammelsbergite. { 'wīt 'nik·əl }

whitening filter [ELECTR] An electrical filter which converts a given signal to white noise. Also known as prewhitening filter. { 'wīt·niŋ ,fil·tər }

white noise [PHYS] Random noise that has a constant energy per unit bandwidth at every frequency in the range of interest. { 'wīt ˌnȯiz }

white object [OPTICS] An object that reflects all wavelengths of light with substantially equal high efficiencies and with considerable diffusion. { 'wīt 'äb·jekt }

white oil [MATER] Any of various highly refined, colorless hydrocarbon oils of low volatility and a wide range of viscosities; used for lubrication of food and textile machinery and as medicinal and mineral oils. Also known as technical white oil. { 'wīt ˌȯil }

white olivine See forsterite. { 'wīt 'äl·ə̇ˌvēn }

whiteout [METEOROL] An atmospheric optical phenomenon of the polar regions in which the observer appears to be engulfed in a uniformly white glow; shadows, horizon, and clouds are not discernible; sense of depth and orientation are lost; dark objects in the field of view appear to float at an indeterminable distance. Also known as milky weather. { 'wīdˌau̇t }

white phosphorus [CHEM] The element phosphorus in its allotropic form, a soft, waxy, poisonous solid melting at 44.5°C; soluble in carbon disulfide, insoluble in water and alcohol; self-igniting in air. Also known as yellow phosphorus. { 'wīt 'fä·sfə·rəs }

white phosphorus grenade [ORD] Hand grenade or rifle grenade containing a main charge of white phosphorus and a small explosive burster charge for scattering the main charge; used for smoke and some incendiary effect. { 'wīt ˈfä·sfə·rəs grəˈnād }

white portland cement [MATER] Finely ground portland cement made from pure calcite limestone and white clay. { 'wīt 'pȯrt·lənd siˈment }

white potato See potato. { 'wīt pəˈtād·ō }

whiteprint [GRAPHICS] A print made by a plan copying process, produced on diazo paper or film, the emulsion containing a diazonium compound and a coupling or activating component; the process is based on sensitivity to ultraviolet light; development is by ammonia vapors, anhydrous ammonia gas, an alkaline solution which includes the coupler, or heat. { 'wītˌprint }

white radiation See continuous radiation. { ˈwīt ˌrād·ēˈā·shən }

white rainbow See fogbow. { 'wīt 'rānˌbō }

Whiterock [GEOL] A North American stage of lowermost Middle Ordovician geologic time, above lower Ordovician and below Marmor. { 'wītˌräk }

white schorl See albite. { 'wīt 'shȯrl }

white signal [COMMUN] Signal at any point in a facsimile system produced by the scanning of a minimum density area of the subject copy. { 'wīt 'sigˌnəl }

white spirit See petroleum ether. { ˈwīt 'spir·ət }

white squall [METEOROL] A sudden squall in tropical or subtropical waters, which lacks the usual squall cloud and whose approach is signaled only by the whiteness of a line of broken water or whitecaps. { 'wīt 'skwȯl }

white stochastic process [MATH] A stochastic process such that there is no correlation between any of its components at different times, including autocorrelations. { 'wīt stōˈkas·tik 'prä·səs }

white tellurium See sylvanite. { 'wīt təˈlu̇r·ē·əm }

white-to-black amplitude range [COMMUN] **1.** In a facsimile system employing positive amplitude modulation, the ratio of signal voltage (or current) for picture white to the signal voltage (or current) for picture black at any point in the system. **2.** In a facsimile system employing negative amplitude modulation, the ratio of the signal voltage (or current) for picture black to the signal voltage (or current) for picture white; this ratio is often expressed in decibels. { ˈwīt tə ˈblak 'am·pləˌtüd ˌrānj }

white-to-black frequency swing [COMMUN] In a facsimile system employing frequency modulation, the numerical difference between the signal frequencies corresponding to picture white and picture black at any point in the system. { ˈwīt tə ˈblak 'frē·kwən·sē ˌswiŋ }

white transmission [COMMUN] **1.** In an amplitude-modulated system, that form of transmission in which the maximum transmitted power corresponds to the minimum density of the subject copy. **2.** In a frequency-modulation system, that form of transmission in which the lowest transmitted frequency corresponds to the minimum density of the subject copy. { 'wīt tranzˈmish·ən }

white vitriol See zinc sulfate. { 'wīt 'vi·trēˌȯl }

whitewash [MATER] A simple mixture of hydrated lime and water, used mostly for painting fences and outbuildings; common whitewash is not water-resistant and rubs off easily. { 'wītˌwäsh }

white water [OCEANOGR] Frothy water, as in whitecaps or breakers. { 'wīt ˌwȯd·ər }

whiting [OCEANOGR] A patch of seawater that contains a substantial amount of calcium carbonate and therefore appears white relative to surrounding water. { 'wīd·iŋ }

whitleyite [GEOL] An achondritic stony meteorite consisting essentially of enstatite with fragments of black chondrite. { 'wit·lēˌīt }

whitlockite [MINERAL] Ca₉(Mg,Fe)H(PO₄)₇ A rare mineral that forms hexagonal crystals. { 'witˌläˌkīt }

Whitney number [MATH] The *k*th Whitney number of a ranked poset is the number of elements of rank *k*. { 'wit·nē ˌnəm·bər }

Whitney sum [MATH] A tangent bundle *TX* over a differentiable manifold *X* is a Whitney sum of continuous bundles *A* and *B* over *X* if for each *x* the fibers of *A* and *B* at *x* are complementary subspaces of the tangent space at *x*. { 'wit·nē ˌsəm }

Whittaker differential equation [MATH] A special form of Gauss' hypergeometric equation with solutions as special cases of the confluent hypergeometric series. { 'wid·ə·kər ˌdif·əˌren·chəl iˈkwā·zhən }

Whitworth screw thread [DES ENG] A British screw thread standardized to form and dimension. { 'witˌwərth 'skrü ˌthred }

whizzer mill See Jeffrey crusher. { 'wiz·ər ˌmil }

whole-body counter [NUCLEO] A radiation counter that directly measures radioactivity in the entire human body. { 'hōl ˌbäd·ē 'kau̇nt·ər }

whole gale [METEOROL] **1.** In storm-warning terminology, a wind of 48 to 63 knots (55 to 72 miles, or 89 to 133 kilometers, per hour). **2.** In the Beaufort wind scale, a wind whose speed is from 48 to 55 knots (55 to 63 miles, or 89 to 102 kilometers, per hour). { 'hōl 'gāl }

whole number [MATH] An integer equal to or greater than zero; one of the numbers 0, 1, 2, 3.... { 'hōl ˌnəm·bər }

whole range point [AERO ENG] The point vertically below an aircraft at the moment of impact of a bomb released from that aircraft, assuming that the aircraft's velocity has remained unchanged. { 'hōl 'rānj ˌpȯint }

whole step See whole tone. { 'hōl 'step }

whole tone [ACOUS] The interval between two sounds whose basic frequency ratio is approximately equal to the sixth root of 2. Also known as whole step. { 'hōl 'tōn }

whooping cough See pertussis. { 'hu̇p·iŋ ˌkȯf }

whooping crane [VERT ZOO] *Grus americana.* A member of a rare North American migratory species of wading birds; the entire species forms a single population. { 'hu̇p·iŋ ˌkrān }

whorl [BOT] An arrangement of several identical anatomical parts, such as petals, in a circle around the same point. [FOREN SCI] A fingerprint pattern in which at least two deltas are present with a recurve in front of each. { wərl }

whr See watt-hour.

Whytt's reflex See pupillary reflex. { 'wī·əts ˌrē·fleks }

wiborgite See rapakivi. { 'wīˌbȯrˌgīt }

wichtisite See tachylite. { 'wik·təˌsīt }

wicket dam [CIV ENG] A movable dam consisting of a number of rectangular panels of wood or iron hinged to a sill and propped vertically; the prop is hinged and can be tripped to drop the wickets flat on the sill. { 'wik·ət ˌdam }

wicking [ENG] The flow of solder under the insulation of covered wire. { 'wik·iŋ }

Widal test [IMMUNOL] A macroscopic or microscopic agglutination test for the diagnosis of typhoid fever and other *Salmonella* infections by using killed or preserved bacteria as the antigen. { we'däl ˌtest }

wide-angle lens [OPTICS] An optical lens having a large angular field, generally greater than 80°. { 'wīd ˌaŋ·gəl 'lenz }

wide-aperature digital VOR [NAV] A very high-frequency omnidirectional radio range in which azimuth angle is determined from a system of crossed wide-baseline interferometers that measure the angles between the line of sight and each of the two baselines. { 'wīd ˌap·ə·chər 'dij·əd·əl ˌvēˌōˈär }

Wide-Area Augmentation System [NAV] A satellite-based augmentation system developed by the Federal Aviation

WHOOPING CRANE

Whooping cranes (*Grus americana*). (*Bureau of Sport Fisheries and Wildlife*).

WHORL

delta

Plain whorl fingerprint pattern. (*Federal Bureau of Investigation*)

Administration in the United States. Abbreviated WAAS. { 'wīd ¦er·ē·ə óg·mən'tā·shən ‚sis·təm }

wide-area data service [COMMUN] Automatic wide-area teletypewriter data exchange service, using leased commercial lines. { 'wīd ¦er·ē·ə 'dad·ə ‚ser·vəs }

wide-area DGPS [NAV] A version of differential GPS which provides error corrections over a large geographic area, based on the measurements by a widely distributed network of monitor stations that are processed at a centrally located facility. { ¦wīd ¦er·ē·ə ¦de¦jē¦pē'es }

wide-area network [COMMUN] A system consisting of a set of nodes that are interconnected by a set of links, and generally covers a large geographic area, usually on the order of hundreds of miles. Abbreviated WAN. { 'wīd ¦er·ē·ə 'net‚wərk }

Wide Area Telephone Service [COMMUN] A special telephone service that allows a customer to call anyone in one or more of six regions into which the continental United States has been divided, on a direct dialing basis, for a flat monthly charge related to the number of regions to be called. Abbreviated WATS. { 'wīd ¦er·ē·ə 'tel·ə‚fōn ‚sər·vəs }

wide band [ELECTR] Property of a tuner, amplifier, or other device that can pass a broad range of frequencies. { 'wīd ¦band }

wide-band amplifier [ELECTR] An amplifier that will pass a wide range of frequencies with substantially uniform amplification. { 'wīd ¦band 'am·plə‚fī·ər }

wide-band communications system [COMMUN] Communications system which provides numerous channels of communications on a highly reliable and secure basis which are relatively invulnerable to interruption by natural phenomena or countermeasures; included are multichannel telephone cable, tropospheric scatter, multichannel line-of-sight radio system such as microwave, and satellites. { 'wīd ¦band kə‚myü·nə'kā·shənz ‚sis·təm }

wide-band ratio [COMMUN] Ratio in a system of the occupied frequency bandwidth to the intelligence bandwidth. { 'wīd ¦band 'rā·shō }

wide-band repeater [ELECTR] Airborne system that receives a radio-frequency signal for transmission; used in reconnaissance missions when low-altitude reconnaissance aircraft require an airborne relay platform for beyond line-of-sight data transmission to a readout station. { 'wīd ¦band ri'pēd·ər }

wide-band switching [ELECTR] Basically, four-wire circuits using correed matrices with electronic controls capable of switching wide-band facilities up to 50 kilohertz in bandwidth. { 'wīd ¦band 'swich·iŋ }

wide-band transformer [ELEC] A transformer that can transfer electric energy from one circuit to another at any of a broad range of frequencies. { 'wīd ¦band tranz'fȯr·mər }

wide berth [NAV] A vessel which keeps well away from another ship or navigational hazard is said to give the other ship or hazard a wide berth. { 'wīd 'bərth }

wide cross [GEN] A mating between individuals of different genera. { 'wīd 'krós }

wide-flange beam See H beam. { 'wīd ¦flanj 'bēm }

wide-open [ELECTR] Refers to the untuned characteristic or lack of frequency selectivity. { 'wīd 'ō·pən }

Widmanstatten patterns [GEOL] Characteristic figures that appear on the surface of an iron meteorite when the meteorite is cut, polished, and etched with acid. { 'vit·mən‚shtät·ən ‚pad·ərnz }

Widrow-Hoff least-mean-squares algorithm [COMMUN] An algorithm that is widely used in adaptive signal processing; for time-discrete analysis with a finite-response filter, it is represented by the first-order difference equation $\mathbf{W}_{k+1} = \mathbf{W}_k + 2\mu e_k \mathbf{X}_k$, where k is a time index that takes on integral values; $\mathbf{W}$ is a vector whose components are the coefficients of the filter; μ is the convergence coefficient; e_k is the residual signal or error, equal to $y_k - \hat{y}_k$, where y_k and $\hat{y}_k$ are the outputs of the plant (where the unprocessed signal is generated) and the filter, respectively; and $\mathbf{X}$ is a vector whose components are the present value of the input and $L - 1$ past values of the input, where L is the number of filter coefficients. { ¦wid·rō ¦hȯf ‚lēst ‚mēn ‚skwerz 'al·gə‚rith·əm }

width [COMMUN] **1.** The horizontal dimension of a television or facsimile picture. **2.** The time duration of a pulse. [MATH] For a plane convex, the greatest lower bound on the distance separating two parallel lines such that the set lies between them. { width }

width control [ELECTR] Control that adjusts the width of the pattern on the screen of a cathode-ray tube in a television receiver or oscilloscope. { 'width kən‚trōl }

Wiedemann effect [ELECTROMAG] The twist produced in a current-carrying wire when placed in a longitudinal magnetic field. Also known as circular magnetostriction. { 'vēd·ə‚män i‚fekt }

Wiedemann-Franz law [SOLID STATE] The law that the ratio of the thermal conductivity of a metal to its electrical conductivity is a constant, independent of the metal, times the absolute temperature. Also known as Lorentz relation. { 'vēd·ə‚män 'fränts ‚lȯ }

Wiedemann's additivity law [PHYS CHEM] The law that the mass (or specific) magnetic susceptibility of a mixture or solution of components is the sum of the proportionate (by weight fraction) susceptibilities of each component in the mixture. { 'vēd·ə‚mänz ‚ad·ə'tiv·əd·ē ‚lȯ }

Wiegand effect [ELEC] The generation of an electrical pulse in a coil wrapped around or located near a Wiegand wire subjected to a changing magnetic field. { 've·gänt i‚fekt }

Wiegand module [ELEC] The apparatus for generating an electrical pulse by means of the Wiegand effect, consisting of a Wiegand wire, two small magnets, and a pickup coil. { 've·gänt ‚mäj·əl }

Wiegand wire [ELEC] A work-hardened wire whose magnetic permeability is much greater near its surface than at its center. { 've·gänt ‚wīr }

Wien bridge oscillator [ELECTR] A phase-shift feedback oscillator that uses a Wien bridge as the frequency-determining element. { 'vēn ‚brij 'äs·ə‚lād·ər }

Wien capacitance bridge [ELEC] A four-arm alternating-current bridge used to measure capacitance in terms of resistance and frequency; two adjacent arms contain capacitors respectively in parallel and in series with resistors, while the other two arms are nonreactive resistors; bridge balance depends on frequency. { 'vēn kə'pas·əd·əns ‚brij }

Wien constant [STAT MECH] The product of the temperature and the wavelength at which the intensity of radiation from a blackbody reaches its maximum; it is equal to approximately 2898 micrometer-kelvins. { 'vēn ‚kän·stənt }

Wien-DeSauty bridge See DeSauty's bridge. { 'vēn də·sō'tē ‚brij }

Wien effect [PHYS CHEM] An increase in the conductance of an electrolyte at very high potential gradients. { 'vēn i‚fekt }

Wiener experiment [OPTICS] An experiment in which a front-faced mirror is covered with a thick photographic emulsion which is then exposed to light incident perpendicular to the surface; upon development, it is found that standing waves are set up in the emulsion whose nodes coincide with those of the electric vector, rather than those of the magnetic vector. { 'vē·nər ik‚sper·ə·mənt }

Wiener-Hopf equations [MATH] Integral equations arising in the study of random walks and harmonic analysis; they are

$$g(x) = \int_0^\infty K(|x - t|) f(t) \, dt$$

$$f(x) = \int_0^\infty K(|x + t|) f(t) \, dt + g(x)$$

where g and K are known functions on the positive real numbers and f is the unknown function. { 'vē·nər 'hȯpf i‚kwā·zhənz }

Wiener-Hopf technique [MATH] A method used in solving certain integral equations, boundary-value problems, and other problems, which involves writing a function that is holomorphic in a vertical strip of the complex z plane as the product of two functions, one of which is holomorphic both in the strip and everywhere to the right of the strip, while the other is holomorphic in the strip and everywhere to the left of the strip. { ¦vēn·ər 'hȯpf ‚tek‚nēk }

Wiener-Khintchine theorem [MATH] The theorem that determines the form of the correlation function of a given stationary stochastic process. { 've·nər kin'chēn ‚thir·əm }

Wiener process [MATH] A stochastic process with normal density at each stage, arising from the study of Brownian motion, which represents the limit of a sequence of experiments. Also known as Gaussian noise. { 've·nər ‚prä·səs }

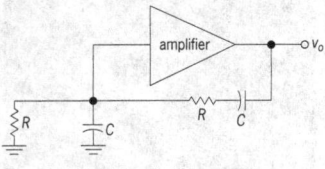

WIEN BRIDGE OSCILLATOR

Circuit of a Wien bridge oscillator.

Wien frequency bridge [ELEC] A modification of the Wien capacitance bridge, used to measure frequencies. { vēn 'frē·kwən·sē ,brij }

Wien inductance bridge [ELEC] A four-arm alternating-current bridge used to measure inductance in terms of resistance and frequency; two adjacent arms contain inductors respectively in parallel and in series with resistors, while the other two arms are nonreactive resistors; bridge balance depends on frequency. { vēn in'dək·təns ,brij }

Wien-Maxwell bridge See Maxwell bridge. { 'vēn 'maks·wel ,brij }

Wien's displacement law [STAT MECH] A law for blackbody radiation which states that the wavelength at which the maximum amount of radiation occurs is a constant equal to approximately 2898 times the product of 1 micrometer and 1 kelvin. Also known as displacement law; Wien's radiation law. { 'vēnz di'splās·mənt ,lò }

Wien's distribution law [STAT MECH] A formula for the spectral distribution of radiation from a blackbody, which is a good approximation to the Planck radiation formula at sufficiently low temperatures or wavelengths, for example, in the visible region of the spectrum below 3000 K. Also known as Wien's radiation law. { 'vēnz ,di·strə'byü·shən ,lò }

Wien's radiation law [STAT MECH] **1.** The law that the intensity of radiation emitted by a blackbody per unit wavelength, at that wavelength at which this intensity reaches a maximum, is proportional to the fifth power of the temperature. **2.** See Wien's displacement law. **3.** See Wien's distribution law. { 'vēnz ,rād·ē'ā·shən ,lò }

Wierl equation [ELECTR] A formula for the intensity of an electron beam scattered through a specified angle by diffraction from the molecules in a gas. { 'virl i'kwā·zhən }

Wiese formula [ENG] An empirical relationship for motor fuel antiknock values above 100 in relation to performance numbers; basis for the American Society for Testing and Materials scale, in which octane numbers above 100 are related to increments of tetraethyllead added to isooctane. { 'vē·zə ,fòr·myə·lə }

Wiesen See meadow. { 'vēz·ən }

WIG craft See wing-in-ground-effect craft. { wig ¦kraft }

wiggler [NUCLEO] A system of transverse, periodic electromagnetic fields, generally used in a storage ring or synchrotron to increase the intensity of synchrotron radiation, increase reaction rates in high-energy physics experiments, produce quasi-monochromatic photon beams, and amplify photon beams from other sources, as in the free-electron laser. { 'wig·lər }

wiggle stick See divining rod. { 'wig·əl ,stik }

Wigner coefficient See vector coupling coefficient. { 'wig·nər ,kō·i,fish·ənt }

Wigner-Eckart theorem [QUANT MECH] A theorem in the quantum theory of angular momentum which states that the matrix elements of a tensor operator can be factored into two quantities, the first of which is a vector-coupling coefficient, and the second of which contains the information about the physical properties of the particular states and operator, and is completely independent of the magnetic quantum numbers. { 'wig·nər 'ek·ərt ,thir·əm }

Wigner effect See discomposition effect. { 'wig·nər i,fekt }

Wigner energy [NUCLEO] Energy that is stored in graphite structures forming part of a nuclear reactor when carbon atoms in the graphite are displaced by bombarding neutrons. { 'wig·nər ,en·ər·jē }

Wigner force [NUCLEO] A short-range nonexchange force between nucleons, postulated to explain various phenomena. { 'wig·nər ,fòrs }

Wigner nuclides [NUC PHYS] The most important class of mirror nuclides, comprising pairs of odd-mass-number isobars for which the atomic number and the neutron number differ by 1. { 'wig·nər 'nü,klīdz }

Wigner release [NUCLEO] The removal of thermal and radiation-induced stress (Wigner energy) from graphite structures in a nuclear reactor by heating the graphite until displaced carbon atoms return to their normal positions. { 'wig·nər ri,lēs }

Wigner-Seitz cell [CRYSTAL] A polyhedron about an atom in a face-centered cubic structure, made by drawing planes which perpendicularly bisect the lines to the nearest neighbors; in a body-centered cubic structure, bisecting planes of lines to nearest neighbors and next-nearest neighbors are used; such polyhedra fill space. { 'wig·nər 'zīts ,sel }

Wigner-Seitz method [SOLID STATE] A method of approximating the band structure of a solid: Wigner-Seitz cells surrounding atoms in the solid are approximated by spheres, and band solutions of the Schrödinger equation for one electron are estimated by using the assumption that an electronic wave function is the product of a plane wave function and a function whose gradient has a vanishing radial component at the sphere's surface. { 'wig·nər 'zīts ,meth·əd }

Wigner's theorem [QUANT MECH] **1.** The theorem that, if ψ is an eigenfunction of the Hamiltonian operator and R is a symmetry element of the Hamiltonian, then $R\psi$ is an eigenfunction of the Hamiltonian having the same eigenvalue as ψ. **2.** Angular momentum of the electron spin is conserved in a collision of the second kind. { 'wig·nərz ,thir·əm }

Wigner supermultiplet [NUC PHYS] A set of quantum-mechanical states of a collection of nucleons which form the basis of a representation of SU(4), especially appropriate when spin and isospin dependence of the nuclear interaction may be disregarded; several combinations of spin and isospin multiplets may occur in a supermultiplet. { 'wig·nər ¦sü·pər'məl·tə·plət }

Wigner three-j symbol See three-j number. { 'wig·nər 'thrē ¦jā ,sim·bəl }

Wiik classification [GEOL] A classification of carbonaceous chondrites into three types, C_1, C_2, and C_3. { 'wik ,klas·ə·fə'kā·shən }

Wijs' iodine monochloride solution [ANALY CHEM] A solution in glacial acetic acid of iodine monochloride; used to determine iodine numbers. Also known as Wijs' special solution. { vīs 'ī·ə,dīn ¦män·ə'klòr,īd sə,lü·shən }

Wijs' special solution See Wijs' iodine monochloride solution. { vīs 'spesh·əl sə,lü·shən }

Wilcoxon one-sample test [STAT] A rank test for testing the hypothesis $\mu = \mu_H$ against the alternative $\mu > \mu_H$ under the assumption that observations are symmetrically distributed about μ_H; here μ_H is a given number and μ is the (unknown) mean of a random variable. { 'wil,käk·sən ¦wən ¦sam·pəl ,test }

Wilcoxon paired comparison distribution [STAT] The distribution of the rank sum V_- (or V_+) of the negative differences (or positive differences) of observations in paired comparisons. { 'wil,käk·sən ¦perd kəm¦par·ə·sən ,di·strə,byü·shən }

Wilcoxon paired comparison test [STAT] The test based upon the rank sum V_- (or V_+) of the negative differences (or positive differences) of observations in paired comparisons. { 'wil,käk·sən ¦perd kəm¦par·ə·sən ,test }

Wilcoxon two-sample distribution [STAT] The distribution of the Wilcoxon two-sample test statistic; it consists of the rank sums of treated subjects. { 'wil,käk·sən ¦tü ¦sam·pəl ,di·strə byü·shən }

Wilcoxon two-sample test [STAT] The test based upon the rank sum of treated (or untreated) subjects. { 'wil,käk·sən ¦tü ¦sam·pəl ,test }

wild boar [VERT ZOO] Sus scrofa. A wild hog with coarse, grizzled hair and enlarged tusks or canines on both jaws. Also known as boar. { 'wīld 'bòr }

wild card [COMPUT SCI] A symbolic character in a search argument such that any character will satisfy it. { 'wīld ,kärd }

wildcat [NAV ARCH] The drum of an anchor windlass, with projections on its rim that engage the anchor chain. { 'wīl,kat }

wildcat drilling [MIN ENG] The drilling of boreholes in unproved territory. Also known as cold nosing; wildcatting. { 'wīl,kat dril·iŋ }

wildcatting See wildcat drilling. { 'wīl,kad·iŋ }

wild cinnamon See bayberry. { 'wīld 'sin·ə·mən }

Wilderness [GEOL] A North American stage of Middle Ordovician geologic time, above Porterfield and below Trentonian. { 'wil·dər·nəs }

Wild fence [ENG] A wooden enclosure about 16 feet (4.8 meters) square and 8 feet (2.4 meters) high with a precipitation gage in its center; the function of the fence is to minimize eddies around the gage, and thus ensure a catch which will be representative of the actual rainfall or snowfall. { 'wīld ,fens }

wildfire [PL PATH] A bacterial disease of tobacco caused by Pseudomonas tabaci and characterized by the appearance of

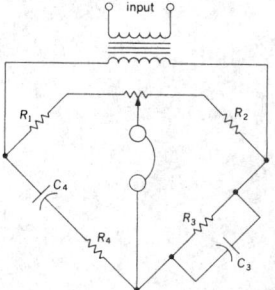

brown spots surrounded by yellow rings, which turn dark, rot, and fall out.　{ 'wīl,fīr }

wildflysch [GEOL] A type of flysch facies that represents a stratigraphic unit with irregularly sorted boulders resulting from fragmentation, and twisted, confused beds resulting from slumping or sliding due to the influence of gravity. { 'vilt,flish }

wild gasoline [MATER] Unstabilized casinghead gasoline. { 'wīld 'gas·ə,lēn }

wildness [MET] A condition that exists when molten metal, during cooling, evolves so much gas that it becomes violently agitated, forcibly ejecting metal from its container. { 'wīl·nəs }

wild shot [ORD] Artillery shot which is completely out of the normal pattern of dispersion; that is, shot whose impact is more than four probable errors, and more than six firing table probable errors from the center of impact. { 'wīld 'shät }

wild snow [METEOROL] Newly deposited snow which is very fluffy and unstable; in general, it falls only during a dead calm at very low air temperatures. { 'wīld 'snō }

wild type [GEN] The most prevalent allele or character in the wild organism. { 'wīld ,tīp }

Wilfley table [MIN ENG] A flat, rectangular surface that can be tilted and shaken about the long axis and has horizontal riffles for imposing restraint in removing minerals from classified sand. Also known as shaking table. { 'wil·flē ,tā·bəl }

wilkeite [MINERAL] Ca₅(SiO₄,PO₄,SO₄)₃(O,OH,F) A rose red or yellow, hexagonal mineral consisting of a basic sulfate-silicate-phosphate of calcium. { 'wil·kē,īt }

Willans line [MECH ENG] The line (nearly straight) on a graph showing steam consumption (pounds per hour) versus power output (kilowatt or horsepower) for a steam engine or turbine; frequently extended to show total fuel consumed (pounds per hour) for gas turbines, internal combustion engines, and complete power plants. { 'wil·ənz ,līn }

willemite [MINERAL] Zn₂SiO₄ A white, greenish-yellow, green, reddish, or brown mineral that forms rhombohedral crystals and exhibits intense bright-yellow fluorescence in ultraviolet light; a minor ore of zinc. { 'wil·ə,mīt }

Williams-Hazen formula [FL MECH] In a liquid-flow system, a method for calculation of head loss due to the friction in a pipeline. { 'wil·yəmz 'hāz·ən ,fòr·mya·lə }

Williamsoniaceae [PALEOBOT] A family of extinct plants in the order Cycadeoidales distinguished by profuse branching. { ,wil·yəm,sō·nē'ās·ē,ē }

Williamson synthesis [ORG CHEM] The synthesis of ethers utilizing an alkyl iodide and sodium alcoholate. { 'wil·yəm·sən 'sin·thə·səs }

Williams refractometer [OPTICS] A refractometer in which light from a single slit is divided into two beams by a pentagonal prism. { 'wil·yəmz ,rē,frak,täm·əd·ər }

Williams tube [ELECTR] A cathode-ray storage tube in which information is stored as a pattern of electric charges produced, maintained, read, and erased by suitably controlled scanning of the screen by the electron beam. { 'wil·yəmz ,tüb }

williwaw [METEOROL] A very violent squall in the Straits of Magellan; it may occur in any month but occurs most frequently in winter. { 'wil·ē,wò }

willow [BOT] A deciduous tree and shrub of the genus *Salix*, order Salicales; twigs are often yellow-green and bear alternate leaves which are characteristically long, narrow, and pointed, usually with fine teeth along the margins. { 'wil·ō }

willy-willy [METEOROL] In Australia, a severe tropical cyclone. { 'wil·ē'wil·ē }

Wilms' tumor [MED] A malignant renal tumor composed principally of mesodermal tissues. Also known as nephroblastoma. { 'vilmz ,tü·mər }

Wilson-Bappu effect [ASTRON] A linear relation between the absolute magnitudes of late-type stars and the width of the K₂ emission core in the resonance line of ionized calcium (CaII) at a wavelength of 3933 nanometers. { 'wil·sən 'bä·pü i,fekt }

Wilson cloud chamber [NUCLEO] A cloud chamber containing air supersaturated with water vapor by sudden expansion, in which rapidly moving nuclear particles such as alpha or beta rays produced ionization tracks by condensation of vapor on the ions produced by the rays. { 'wil·sən 'klaùd ,chäm·bər }

Wilson effect [ASTRON] An effect in which the penumbra

of a sunspot appears narrower in the direction toward the sun's center than in the direction toward the sun's limb. { 'wil·sən i,fekt }

Wilson electroscope [ELEC] An electroscope that has a single gold leaf which, when charged, is attracted to a grounded metal plate inclined at an angle that maximizes the instrument's sensitivity. { 'wil·sən i'lek·trə,skōp }

Wilson experiment [ELECTROMAG] An experiment that tests the validity of electromagnetic theory; a hollow cylinder of dielectric material, having layers of metal on its outer and inner cylindrical surfaces, is rotated about its axis in a magnetic field parallel to the axis; a sensitive electrometer, connected to the metal layers, indicates a charge that has the magnitude and sign predicted by theory. { 'wil·sən ik,sper·ə·mənt }

Wilson's disease [MED] A hereditary disease of ceruloplasmin formation transmitted as an autosomal recessive and characterized by decreased serum ceruloplasmin and copper values, and increased excretion of copper in the urine. Also known as hepatolenticular degeneration. { 'wil·sənz di,zēz }

Wilson's theorem [MATH] The number (n − 1)! + 1 is divisible by n if and only if n is a prime. { 'wil·sənz ,thir·əm }

wilt [PL PATH] Any of various plant diseases characterized by drooping and shriveling, following loss of turgidity. { wilt }

wilting point [BOT] A condition in which a plant begins to use water from its own tissues for transpiration because soil water has been exhausted. { 'wilt·iŋ ,pòint }

WIMP See weakly interacting massive particle. { wimp }

Wimshurst machine [ELEC] An electrostatic generator consisting of two glass disks rotating in opposite directions, having sectors of tinfoil and collecting combs so arranged that static electricity is produced for charging Leyden jars or discharging across a gap. { 'wimz,hərst mə,shēn }

winch [MECH ENG] A machine having a drum on which to coil a rope, cable, or chain for hauling, pulling, or hoisting. { winch }

Winchester disk [COMPUT SCI] A type of disk storage device characterized by nonremovable or sealed disk packs; extremely narrow tracks; a lubricated surface that allows the head to rest on the surface during start and stop operations; and servomechanisms which utilize a magnetic pattern, recorded on the medium itself, to position the head. { 'win·ches·tər ,disk }

Winchester technology [COMPUT SCI] Innovations designed to achieve disks with up to 6 × 10⁸ bytes per disk drive; the technology includes nonremovable or sealed disk packs, a read/write head that weighs only 0.25 gram and floats above the surface, magnetic orientation of iron oxide particles on the disk surface, and lubrication of the disk surface. { 'win·ches·tər tek'näl·ə·jē }

winch operator See hoistman. { 'winch ,äp·ə,rād·ər }

wind [ELECTR] The manner in which magnetic tape is wound onto a reel; in an A wind, the coated surface faces the hub; in a B wind, the coated surface faces away from the hub. [METEOROL] The motion of air relative to the earth's surface; usually means horizontal air motion, as distinguished from vertical motion, and air motion averaged over the response period of the particular anemometer. { wind }

windage [MECH] **1.** The deflection of a bullet or other projectile due to wind. **2.** The correction made for such deflection. { 'win·dij }

windage loss [ENG] In a ventilating or air-conditioning system, the decrease in the water content of the circulating air due to the loss of entrained droplets of water; expressed as a percentage of the rate of circulation. { 'win·dij ,lòs }

windage scale [ORD] Scale for adjusting a sight to allow for the effect of the wind on a bullet in flight. Also known as wind gage. { 'win·dij ,skāl }

windage yaw [ORD] In aerial gunnery, the yaw produced by the relative motion of the gun and the air; the tangent of the windage yaw is the crosswind, or the component of the airspeed perpendicular to the axis of the bore, divided by the muzzle velocity (the initial velocity relative to the gun). { 'win·dij ,yò }

wind box [ENG] A plenum chamber that supplies air for combustion to a stoker, gas burner, or oil burner. { 'wind ,bäks }

windbreak [ENG] Any device designed to obstruct wind flow and intended for protection against any ill effects of wind. { 'win,brāk }

windburn [BOT] Injury to plant foliage, caused by strong,

WILLOW

Twig, terminal bud, and leaf of the Babylon weeping willow (*Salix babylonia*).

hot, dry winds. [MED] A superficial inflammation of the skin, analogous to sunburn, caused by exposure to wind, especially a hot dry wind, inducing a dilation of the surface blood vessels. { 'win,bərn }

wind charger [ELEC] A wind-driven direct-current generator used for charging storage batteries. { 'win 'chär·jər }

wind chill [METEOROL] That part of the total cooling of a body caused by air motion. { 'win ,chil }

wind-chill index [METEOROL] The cooling effect of any combination of temperature and wind, expressed as the loss of body heat in kilogram calories per hour per square meter of skin surface; it is only an approximation because of individual body variations in shape, size, and metabolic rate. { 'win ,chil ,in,deks }

wind cone [ENG] A tapered fabric sleeve, shaped like a truncated cone and pivoted at its larger end on a standard, for the purpose of indicating wind direction; since the air enters the fixed end, the small end of the cone points away from the wind. Also known as wind sleeve; wind sock. { 'win ,kōn }

wind correction [ENG] Any adjustment which must be made to allow for the effect of wind; especially, the adjustments to correct for the effect on a projectile in flight, on sound received by sound ranging instruments, and on an aircraft flown by dead reckoning navigation. { 'win kə'rek·shən }

wind crust [HYD] A type of snow crust, formed by the packing action of wind on previously deposited snow; wind crust may break locally but, unlike wind slab, does not constitute an avalanche hazard. { 'win 'krəst }

wind current [METEOROL] Generally, any of the quasi-permanent, large-scale wind systems of the atmosphere, for example, the westerlies, trade winds, equatorial easterlies, or polar easterlies. { 'win 'kə,rənt }

wind-cut stone See ventifact. { 'win 'kət 'stōn }

wind deflection [MECH] Deflection caused by the influence of wind on the course of a projectile in flight. { 'win di,flek·shən }

wind direction [METEOROL] The direction from which wind blows. { 'win də,rek·shən }

wind-direction indicator [ENG] A device to indicate the direction from which the wind blows; an example is a weather vane. { 'win də'rek·shən ,in·də,kād·ər }

wind-direction shaft [METEOROL] A representational mark for wind direction on a synoptic chart, it is a straight line drawn directly upwind from the station circle; the wind arrow is completed by adding the wind-speed barbs and pennants to the outer end of the shaft. { 'win də'rek·shən ,shaft }

wind divide [METEOROL] A semipermanent feature of the atmospheric circulation (usually a high-pressure ridge) on opposite sides of which the prevailing wind directions differ greatly. { 'win də,vīd }

wind drift [ACOUS] Shift in the apparent position of a sound source or target observed by sound apparatus; it is caused by the effect of wind on sound waves, which changes their direction and increases or decreases sound lag. [OCEANOGR] See drift current. { 'win ,drift }

wind-driven current See drift current. { 'win ,driv·ən 'kə·rənt }

winder [BUILD] A step, generally wedge-shaped, with a tread that is wider at one end than the other; often used in spiral staircases. { 'wīn·dər }

wind erosion [GEOL] Detachment, transportation, and deposition of loose topsoil or sand by the action of wind. { 'wind i,rō·zhən }

wind factor [NAV] In air navigation, a measure of the net effect of wind on the ground speed of an aircraft; it is the magnitude of the wind vector component parallel to the heading of an aircraft, averaged over the entire flight positive if a tailwind, negative if a headwind { 'win ,fak·tər }

wind-fire angle [ORD] The horizontal angle measured clockwise from the plane of fire to the direction from which the ballistic wind is blowing. { 'win ,fīr ,aŋ·gəl }

wind gage See windage scale. { 'win ,gāj }

wind gap [GEOL] A shallow, relatively high-level notch in the upper part of a mountain ridge, usually an abandoned water gap. Also known as air gap; wind valley. { 'win ,gap }

wind-grooved stone See ventifact. { 'win ,grüvd 'stōn }

wind guard [CIV ENG] A building component that protects the building or some part of it against the wind, for example, a chimney cap. { 'win,gärd }

winding [ELEC] **1.** One or more turns of wire forming a continuous coil for a transformer, relay, rotating machine, or other electric device. **2.** A conductive path, usually of wire, that is inductively coupled to a magnetic storage core or cell. [MATER] A material that is wound or coiled around a cylindrical object such as a mandrel. { 'wīnd·iŋ }

winding engine See hoist. { 'wīnd·iŋ ,en·jən }

winding number [MATH] The number of times a given closed curve winds in the counterclockwise direction about a designated point in the plane. Also known as index. { 'wīnd·iŋ ,nəm·bər }

windlass See anchor windlass. { 'win·dləs }

wind measurement [METEOROL] The determination of three parameters: the size of an air sample, its speed, and its direction of motion. { 'win ,mezh·ər·mənt }

windmill [MECH ENG] Any of various mechanisms, such as a mill, pump, or electric generator, operated by the force of wind against vanes or sails radiating about a horizontal shaft. { 'win ,mil }

windmill anemometer [ENG] A rotation anemometer in which the axis of rotation is horizontal; the instrument has either flat vanes (as in the air meter) or helicoidal vanes (as in the propeller anemometer); the relation between wind speed and angular rotation is almost linear. { 'win,mil ,an·ə'mäm·əd·ər }

windmilling [MECH ENG] The rotation of a propeller from the force of the air when the engine is not operating. { 'win ,mil·iŋ }

wind noise [ACOUS] Noise caused by turbulent airflow over and around an object. { 'win ,nȯiz }

window [AERO ENG] An interval of time during which conditions are favorable for launching a spacecraft on a specific mission. [BUILD] An opening in the wall of a building or the body of a vehicle to admit light and usually to permit vision through a transparent or translucent material, usually glass. [COMPUT SCI] A separate viewing area on a display screen that is established by the computer software. Also known as viewport. [ELECTR] A material having minimum absorption and minimum reflection of radiant energy, sealed into the vacuum envelope of a microwave or other electron tube to permit passage of the desired radiation through the envelope to the output device. [ELECTROMAG] A hole in a partition between two cavities or waveguides, used for coupling. [GEOL] A break caused by erosion of a thrust sheet or a large recumbent anticline that exposes the rocks beneath the thrust sheet. Also known as fenster. [GEOPHYS] Any range of wavelengths in the electromagnetic spectrum to which the atmosphere is transparent. [HYD] The unfrozen part of a river surrounded by river ice during the winter. [MATER] A globular defect in a thermoplastic sheet or film caused by incomplete plasticization; similar to a fisheye. [NUCLEO] **1.** An aperture for the passage of particles or radiation in a nuclear reactor. **2.** An energy range of relatively high transparency in the total neutron cross section of a material; such windows arise from interference between potential and resonance scattering in elements of intermediate atomic weight, and can be of importance in neutron shielding. [ORD] A confusion reflector consisting of strips of chaff, wire, or bars cut to give resonance at expected enemy radar frequencies, and dropped in clusters from aircraft or expelled from shells or rockets as a radar countermeasure. { 'win·dō }

window bar [BUILD] **1.** A bar for securing a casement window or window shutters. **2.** A bar that prevents ingress or egress through a window. **3.** See sash bar. { 'win,dō ,bär }

window editor [COMPUT SCI] An interactive program that allows the user to view and alter stored information by using the video display as though it provides a view of a portion of storage that can be moved around. { 'win,dō ,ed·əd·ər }

window frost [HYD] A thin deposit of hoarfrost often found on interior surfaces of windows in winter, and frequently exhibiting beautiful fernlike patterns. { 'win·dō ,frȯst }

window ice [HYD] A thin deposit of ice which forms by the freezing of many tiny drops of water that have condensed on the indoors side of a cold window surface. { 'win·dō ,īs }

windowing [COMPUT SCI] **1.** The procedure of selecting a portion of a large drawing to be displayed on the screen of a computer graphics system, usually by placing a rectangular window over a compressed version of the entire drawing displayed on the screen. **2.** Dividing an electronic display into

Field winding (def. 1) on cylindrical rotor. (*National Electric Coil, Division of McGraw-Edison Company*)

areas that display the outputs of different programs and can overlap in the same manner as pieces of paper on a desk, partially concealing the contents of pages underneath. { 'win,dō·iŋ }

window rocket [ORD] A rocket filled with window that is to be expelled at a desired height. { 'win·dō ,räk·ət }

wind-polished stone *See* ventifact. { 'win ¦päl·əsht 'stōn }

wind power [MECH ENG] The extraction of kinetic energy from the wind and conversion of it into a useful type of energy: thermal, mechanical, or electrical. { 'win ,paú·ər }

wind pressure [MECH] The total force exerted upon a structure by wind. Also known as velocity pressure. { 'win ,presh·ər }

wind ripple [METEOROL] One of a series of wavelike formations on a snow surface, an inch or so in height, at right angles to the direction of wind. Also known as snow ripple. { 'win ,drip·əl }

wind-rode [NAV ARCH] A ship riding at anchor is said to be wind-rode when it is heading into the wind. { 'win ,drōd }

wind rose [METEOROL] A diagram in which statistical information concerning direction and speed of the wind at a location may be summarized; a line segment is drawn in each of perhaps eight compass directions from a common origin; the length of a particular segment is proportional to the frequency with which winds blow from that direction; thicknesses of a segment indicate frequencies of occurrence of various classes of wind speed. { 'win ,drōz }

windrow [GEOL] Any accumulation of material formed by wind or tide action. { 'win,drō }

winds aloft [METEOROL] Generally, the wind speeds and directions at various levels in the atmosphere above the domain of surface weather observations, as determined by any method of winds-aloft observation. Also known as upper-level winds; upper winds. { 'winz ə'lóft }

winds-aloft observation [METEOROL] The measurement and computation of wind speeds and directions at various levels above the surface of the earth. { 'winz ə'lóft ,äb·zər'vā·shən }

wind scoop [METEOROL] A saucerlike depression in the snow near obstructions such as trees, houses, and rocks, caused by the eddying action of the deflected wind. { 'win ,sküp }

wind-scoured stone *See* ventifact. { 'win ¦skaúrd 'stōn }

wind-shaped stone *See* ventifact. { 'win ¦shapt 'stōn }

wind shear [METEOROL] The local variation of the wind vector or any of its components in a given direction. { 'win ,shir }

windshield [ENG] A transparent glass screen that protects the passengers and compartment of a vehicle from wind and rain. { 'win,shēld }

wind shield *See* rain-gage shield. { 'win ,shēld }

wind-shift line [METEOROL] A line or narrow zone along which there is an abrupt change of wind direction. { 'win ¦shift ,līn }

wind ship [NAV ARCH] A ship equipped with sails so that it can be propelled by wind, generally to augment steam propulsion. { 'win ,ship }

wind slab [HYD] A type of snow crust; a patch of hard-packed snow, which is packed as it is deposited in favored spots by the wind, in contrast to wind crust. { 'wind ,slab }

wind sleeve *See* wind cone. { 'win ,slēv }

wind sock *See* wind cone. { 'win ,säk }

wind speed [METEOROL] The rate of motion of air. { 'win ,spēd }

wind star [NAV] A method of solution for the speed and direction of the wind by observing drift on three different headings which form an approximate equilateral triangle. { 'win ,stär }

windstorm [METEOROL] A storm in which strong wind is the most prominent characteristic. { 'win,stórm }

wind stress [METEOROL] The drag or tangential force per unit area exerted on the surface of the earth by the adjacent layer of moving air. { 'win ,stres }

wind tee [ENG] A weather vane shaped like the letter T or like an airplane, situated on an airport or landing field to indicate the wind direction. Also known as landing tee. { 'win ,tē }

wind tide [OCEANOGR] **1.** The vertical rise in the still-water level on the leeward side of a body of water, particularly the ocean or other large body, caused by wind stresses on the surface of the water. **2.** The difference in still-water level

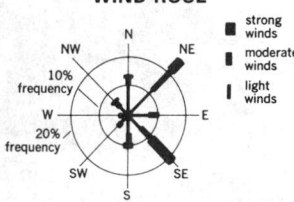

WIND ROSE

Example of standard wind rose.

WIND SHIP

Wind ship cruising with sails fully stretched. Hydraulic sail-furling components and mast and turntable connection are visible in the foreground.

between the windward and leeward sides of such a body caused by wind stresses. { 'win ,tīd }

wind triangle [AERO ENG] A vector diagram showing the effect of the wind on the flight of an aircraft; it is composed of the wind direction and wind speed vector, the true heading and true airspeed vector, and the resultant track and ground speed vector. { 'wīn 'trī,aŋ·gəl }

wind tunnel [ENG] A duct in which the effects of airflow past objects can be determined. { 'win ,tən·əl }

wind-tunnel balance [AERO ENG] A device or apparatus that measures the aerodynamic forces and moments acting upon a body tested in a wind tunnel. { 'win ,tən·əl ,bal·əns }

wind-tunnel instrumentation [ENG] Measuring devices used in wind-tunnel tests; in addition to conventional laboratory instruments for fluid flow, thermometry, and mechanical measurements, there are sensing devices capable of precision measurement in the small-scale environment of the test setup. { 'win ,tən·əl ,in·strə·mən'tā·shən }

wind turbine [ELEC] An advanced type of windmill designed to convert wind energy into electrical energy. { 'win 'tər,bīn }

windup [MECH ENG] The twisting of a shaft under a torsional load, usually resulting in vibration and other undesirable effects as the shaft relaxes. { 'wīn,dəp }

wind valley *See* wind gap. { 'win ,val·ē }

wind vane [ENG] An instrument used to indicate wind direction, consisting basically of an asymmetrically shaped object mounted at its center of gravity about a vertical axis; the end which offers the greater resistance to the motion of air moves to the downwind position; the direction of the wind is determined by reference to an attached oriented compass rose. { 'win ,vān }

wind velocity [METEOROL] The speed and direction of wind. { 'win və,läs·əd·ē }

windward [METEOROL] In the general direction from which the wind blows. { 'win·wərd }

wind wave [OCEANOGR] A wave resulting from the action of wind on a water surface. { 'win ,wāv }

wine [FOOD ENG] An alcoholic beverage made by fermentation of the juice of fruits or berries, especially grapes; classified on the basis of color, sweetness, alcoholic content, variety of grape, presence of carbon dioxide, and region where the grapes are grown. { wīn }

wine lees [FOOD ENG] Sediment or deposit that forms in the bottom of wine casks during the fermentation process; used as a source of tartaric acid and tartrates. { 'wīn ,lēz }

wing [AERO ENG] **1.** A major airfoil. **2.** An airfoil on the side of an airplane's fuselage or cockpit, paired off by one on the other side, the two providing the principal lift for the airplane. [GEOL] *See* vesicle. [ZOO] Any of the paired appendages serving as organs of flight on many animals. { win }

wing assembly [AERO ENG] An aeronautical structure designed to maintain a guided missile in stable flight; it consists of all panels, sections, fastening devices, chords, spars, plumbing accessories, and electrical components necessary for a complete wing assembly. { 'win ə,sem·blē }

wing axis [AERO ENG] The locus of the aerodynamic centers of all the wing sections of an airplane. { 'win ,ak·səs }

wing dam *See* groin. { 'win ,dam }

wing drop [AERO ENG] A phenomenon experienced by an air vehicle during maneuvers at moderate to high angles of attack, in which an abrupt reduction of lift from one side of the vehicle prior to the other side creates a rolling moment, which causes loss of control of the airplane's roll attitude. { 'win ,dräp }

winged headland [GEOGR] A seacliff with two bays or spits, one on either side. { 'wiŋd 'hed·lənd }

Winged Horse *See* Pegasus. { 'wiŋd 'hórs }

winged missile [ORD] A missile that has wings, distinguished from wingless missiles such as bullets, projectiles, and certain rockets. { 'wiŋd 'mis·əl }

wing gun [ORD] A fixed gun mounted on the wing of an airplane. { 'win ,gən }

wing-in-ground-effect craft [AERO ENG] A type of air-cushion vehicle that is similar in design to an aircraft and requires forward speed to provide lift; it flies very close to the surface, generally over water, to take advantage of the increased lift and reduction in induced drag that are characteristic of such flight. Also known as wing-in-surface-effect craft. Abbreviated WIG craft. { ¦wiŋ in ¦graünd i¸fekt ¸kraft }

wing-in-surface-effect craft See wing-in-ground-effect craft. { ¦wiŋ in ¦sər·fəs i¸fekt ¸kraft }

wingless abutment [CIV ENG] A straight-sided bridge abutment designed to resist pressure in back and provide a bridge seat. { ¦wiŋ·ləs ə'bət·mənt }

winglet [AERO ENG] A small, nearly vertical surface mounted at the tip of an aircraft wing to decrease drag resistance. { 'wiŋ·lət }

wing loading [AERO ENG] A measure of the load carried by an airplane wing per unit of wing area; commonly used units are pounds per square foot and kilograms per square meter. { 'wiŋ ¸lōd·iŋ }

wing nut [DES ENG] An internally threaded fastener with wings to permit it to be tightened or loosened by finger pressure only. Also known as butterfly nut. { 'wiŋ ¸nət }

wing panel [AERO ENG] That portion of a multipiece wing section that usually lies between the front and rear spars; it may be designed to include either the leading edge or the trailing edge as an integral part, but never both, and excludes control surfaces. { 'wiŋ ¸pan·əl }

wing profile [AERO ENG] The outline of a wing section. { 'wiŋ ¸prō·fīl }

wing rib [AERO ENG] A chordwise member of the wing structure of an airplane, used to give the wing section its form and to transmit the load from the fabric to the spars. { 'wiŋ ¸rib }

wing screw [DES ENG] A screw with a wing-shaped head that can be turned manually. { 'wiŋ ¸skrü }

wing section See airfoil profile. { 'wiŋ ¸sek·shən }

wing spot generator [ELECTR] Electronic circuit that grows wings on the video target signal of a type G indicator; these wings are inversely proportional in size to the range. { 'wiŋ ¸spät ¸jen·ə¸rād·ər }

wing structure [AERO ENG] In an aircraft, the combination of outside fairing panels that provide the aerodynamic lifting surfaces and the inside supporting members that transmit the lifting force to the fuselage; the primary load-carrying portion of a wing is a box beam (the prime box) made up usually of two or more vertical webs, plus a major portion of the upper and lower skins of the wing, which serve as chords of the beam. { 'wiŋ ¸strək·chər }

wing-tip rake [AERO ENG] The shape of the wing when the tip edge is straight in plan but not parallel to the plane of symmetry; the amount of rake is measured by the acute angle between the straight portion of the wing tip and the plane of symmetry; the rake is positive when the trailing edge is longer than the leading edge. { 'wiŋ ¸tip ¸rāk }

Winkler titration [ANALY CHEM] A chemical method for estimating the dissolved oxygen in seawater: manganous hydroxide is added to the sample and reacts with oxygen to produce a manganese compound which in the presence of acid potassium iodide liberates an equivalent quantity of iodine that can be titrated with standard sodium thiosulfate. { 'viŋ·klər tī¸trā·shən }

winning [MIN ENG] **1.** A new mine opening. **2.** The portion of a coal field laid out for working. **3.** Mining. { 'win·iŋ }

winnowing gold [MIN ENG] Tossing up dry powdered auriferous material in air, and catching the heavier particles not blown away. { 'win·ə·wiŋ 'gōld }

winter [ASTRON] The period from the winter solstice, about December 22, to the vernal equinox, about March 21; popularly and for most meteorological purposes, winter is taken to include December, January, and February in the Northern Hemisphere, and June, July, and August in the Southern Hemisphere. { 'win·tər }

Winteraceae [BOT] A family of dicotyledonous plants in the order Magnoliales distinguished by hypogynous flowers, exstipulate leaves, air vessels absent, and stamens usually laminar. { ¸win·tə'rās·ē¸ē }

winter buoy [NAV] An unlighted buoy without sound signal which is maintained in certain areas during winter months when other aids are temporarily removed or extinguished. { 'win·tər 'bȯi }

wintergreen oil See methyl salicylate. { 'win·tər¸grēn ¸ȯil }

winter ice [OCEANOGR] Level sea ice more than 8 inches (20 centimeters) thick, and less than 1 year old; the stage which follows young ice. { 'win·tər 'īs }

winterization [ENG] The preparation of equipment for operation in conditions of winter weather; this applies to preparation not only for cold temperatures, but also for snow, ice, and strong winds. { ¸win·tə·rə'zā·shən }

winter load line [NAV ARCH] The waterline to which a vessel is allowed to load when going to sea in the wintertime. { 'win·tər 'lōc ¸līn }

winter marker [NAV] An unlighted marker without sound signal which is maintained in certain areas during the winter months when other aids are temporarily removed or extinguished. { 'win·tər 'mär·kər }

winter solstice [ASTRON] **1.** The sun's position on the ecliptic (about December 22). Also known as first point of Capricorn. **2.** The date (December 22) when the greatest southern declination of the sun occurs. { 'win·tər 'säl·stəs }

winter-talus ridge [GEOL] A wall-like arcuate ridge on the floor of a cirque formed by freezing activity that dislodged boulders from a cirque wall covered with a snowbank. Also known as nivation ridge. { 'win·tər 'tā·ləs ¸rij }

winze [MIN ENG] A vertical or inclined opening or excavation connecting two levels in a mine, differing from a raise only in construction; a winze is sunk underhand, and a raise is put up overhand. { winz }

wiped joint [MET] A joint wherein filler metal is applied in liquid form, and the joint is wiped mechanically to distribute the metal. { 'wīpt 'jȯint }

wipe-on plate [GRAPHICS] In offset lithography, a plate with a light-sensitive coating. { 'wīp¸ȯn ¸plāt }

wiper [ELEC] That portion of the moving member of a selector, or other similar device, in communications practice, which makes contact with the terminals of a bank. { 'wī·pər }

wiping [NAV] The process of reducing the amount of permanent magnetism in a naval vessel by placing a single coil horizontally around the vessel and moving it, while energized, up and down along the sides of the vessel. { 'wīp·iŋ }

wiping contact [ELEC] A switch or relay contact designed to move laterally with a wiping motion after it touches a mating contact. Also known as self-cleaning contact; sliding contact. { 'wīp·iŋ ¸kän¸takt }

wiping effect [MET] Activation of a metal surface by mechanically rubbing or wiping to enhance the formation of a conversion coating. { 'wīp·iŋ i¸fekt }

wire [ELEC] A single bare or insulated metallic conductor having solid, stranded, or tinsel construction, designed to carry current in an electric circuit. Also known as electric wire. [MET] A thin, flexible, continuous length of metal, usually of circular cross section. [OPTICS] A filament, usually consisting of a stretched strand of spider's web or a fine metal wire, mounted in the field of view of a telescope eyepiece to serve as a reference or for measurements. Also known as web. { 'wīr }

wire bonding [ELEC] Lead-covered tie used to connect two cable sheaths until a splice is permanently closed and covered. [ELECTR] **1.** A method of connecting integrated-circuit chips to their substrate, using ultrasonic energy to weld very fine wires mechanically from metallized terminal pads along the periphery of the chip to corresponding bonding pads on the substrate. **2.** The attachment of very fine aluminum or gold wire (by thermal compression or ultrasonic welding) from metallized terminal pads along the periphery of an integrated circuit chip to corresponding bonding pads on the surface of the package leads. { 'wīr ¸bänd·iŋ }

wire cloth [DES ENG] Screen composed of wire crimped or woven into a pattern of squares or rectangles. { 'wīr ¸klȯth }

wire comb [ENG] A tool for roughening a base coat of plaster in order to improve bonding of the next coat. Also known as wire scratcher. { 'wīr ¸kōm }

wire-cut brick [MATER] A brick cut from clay shaped by extrusion before burning; the long bar of extruded clay is cut into bricks by a set of wires 9 inches (23 centimeters) apart. { 'wīr ¸kət 'brik }

wired-program computer [COMPUT SCI] A computer in which the sequence of instructions that form the operating program is created by interconnection of wires on a removable control panel. { 'wīrd ¸prō·gram kəm'pyüd·ər }

Craft used over water as a commuter vehicle. (*FlareCraft Corp.*)

WINGLESS ABUTMENT

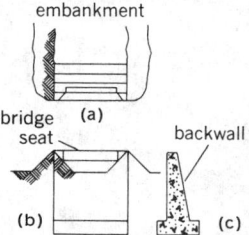

A wingless abutment of a bridge. (*a*) Plan. (*b*) Front elevation. (*c*) Section.

WING NUT

A wing nut. (*Reynolds Metals Co.*)

WIRE BONDING

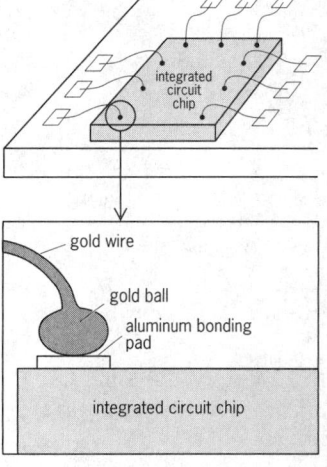

Wire bonding technique for connecting an integrated circuit chip to a substrate.

wire drag [ENG] An apparatus for surveying rocky underwater areas where normal sounding methods are insufficient to ensure the discovery of all existing submerged obstructions, small shoals, or rocks above a given depth or for determining the least depth of an area; it consists essentially of a buoyed wire towed at the desired depth by two launches. { 'wīr 'drag }

wire drawing [MET] The reduction of the diameter of a metal rod or wire by pulling it through a die or a series of dies. { 'wī·ər ˌdrȯ·iŋ }

wire-fabric reinforcing [CIV ENG] Reinforcing concrete or mortar with a welded wire fabric. { 'wīr ˈfab·rik ˌrē·ən'fȯrs·iŋ }

wire facsimile system [COMMUN] A facsimile system in which messages are sent over wires or cables, rather than by radio. { 'wīr fak'sim·ə·lē ˌsis·təm }

wireframe [COMPUT SCI] **1.** In computer-aided design, a line-drawn model. **2.** In computer graphics, an image-rendering technique in which only edges and vertices are shown. { ˌwīr'frām }

wire flame spray gun [ENG] A device which utilizes the heat from a gas flame and material in the form of wire or rod to perform a flame-spraying operation. { 'wīr ˈflām 'sprā ˌgən }

wire-frame model [COMPUT SCI] In computer-aided design, the representation of all surfaces of a three-dimensional object in outline form. { 'wī·ər ˌfrām ˌmäd·əl }

wire fusing current [ELEC] The electric current which will cause a wire to melt. { 'wīr ˈfyüz·iŋ ˌkə·rənt }

wire gage [DES ENG] **1.** A gage for measuring the diameter of wire or thickness of sheet metal. **2.** A standard series of sizes arbitrarily indicated by numbers, to which the diameter of wire or the thickness of sheet metal is usually made, and which is used in describing the size or thickness. { 'wīr ˌgāj }

wire glass [MATER] Sheet glass with woven wire mesh embedded in the center of the sheet; used in building construction for windows, doors, floors, and skylights. { 'wīr ˌglas }

wiregrating [ELECTROMAG] A series of wires placed in a waveguide that allow one or more types of waves to pass and block all others. { 'wīrˌgrād·iŋ }

wire holder [ELEC] A special type of electrical insulator fitted with a mounting screw or mounting bolt and having a hole for securing an electrical wire or cable. { 'wīr ˌhōl·dər }

wire insulation [MATER] A flexible insulation used to cover an electric wire. { 'wīr ˌin·sə'lā·shən }

wire lath [ENG] A netting formed of welded wire, usually with a paper backing, and used as a base for plaster. { ˈwīr 'lath }

wireless cable [COMMUN] A television broadcasting system in which signals are collected and transmitted to towers for net transmission to homes outfitted with special antennas. { ˌwīr·ləs 'kā·bəl }

wire line [DES ENG] **1.** Any cable or rope made of steel wires twisted together to form the strands. **2.** A steel wire rope 5/16 inch (7.94 millimeters) or less in diameter. [ELECTR] One or more current-conducting wires or cables, used for communication, control, or telemetry. [PETRO ENG] A line or cable used to lower and raise devices and gages in oil well boreholes; used for logging instruments and bottom-hole pressure gages. { 'wīr ˌlīn }

wire-line coring [PETRO ENG] A method for obtaining samples of reservoir rocks during the drilling phase of oil wells. { 'wīr ˈlīn 'kȯr·iŋ }

wire-link telemetry [COMMUN] Telemetry in which electric signals are sent over transmission lines, rather than by radio. Also known as hard-wire telemetry. { 'wīr ˈliŋk tə'lem·ə·trē }

wire mile [ELEC] Unit of measure of the length of two-conductor wire between two points; the length of the route multiplied by the number of circuits gives the number of wire miles. { 'wīr ˈmīl }

wire nail [DES ENG] A nail made of wire and having a circular cross section. { 'wīr ˌnāl }

wirephoto [COMMUN] **1.** A photograph transmitted over wires to a facsimile receiver. **2.** See facsimile. { 'wīrˌfōd·ō }

wire printing See matrix printing. { 'wīr ˌprint·iŋ }

wire recorder [ENG ACOUS] A magnetic recorder that utilizes a round stainless steel wire about 0.004 inch (0.01 centimeter) in diameter instead of magnetic tape. { 'wīr riˌkȯrd·ər }

wire recording [ENG ACOUS] Magnetic recording by use of a magnetized wire. { 'wīr riˌkȯrd·iŋ }

wire rod [MET] A metal rod used in wiredrawing. { 'wīr ˌräd }

wire rope [ENG] A rope formed of twisted strands of wire. { 'wīr ˌrōp }

wire saw [MECH ENG] A machine employing one- or three-strand wire cable, up to 16,000 feet (4900 meters) long, running over a pulley as a belt; used in quarries to cut rock by abrasion. { 'wīr 'sȯ }

wire scratcher See wire comb. { 'wīr ˌskrach·ər }

wiresonde [ENG] An atmospheric sounding instrument which is supported by a captive balloon and used to obtain temperature and humidity data from the ground level to a height of a few kilometers; height is determined by means of a sensitive altimeter, or from the amount of cable released and the angle which the cable makes with the ground, and the information is telemetered to the ground through a wire cable. { 'wīrˌsänd }

wire stripper [ENG] A hand-operated tool or special machine designed to cut and remove the insulation for a predetermined distance from the end of an insulated wire, without damaging the solid or stranded wire inside. { 'wīr ˌstrip·ər }

wire tack [DES ENG] A tack made from wire stock. { 'wīr ˌtak }

wiretap [COMMUN] A secretly made and concealed connection to a telephone line, office intercommunication line, or other wiring system, for the purpose of monitoring conversations and activities in a room from a remote location without knowledge of the participants, legally or illegally. { 'wīrˌtap }

wire telegraphy [COMMUN] Telegraphy in which messages are sent over wires or cables, rather than by radio. { 'wīr tə'leg·rə·fē }

wire train [ENG] An assembly that normally consists of an extruder, a crosshead and die, a means of cooling, and feed and take-up spools for the wire; used to coat wire with resin. { 'wīr ˌträn }

wireway [ENG] A trough which is lined with sheet metal and has hinged covers, designed to house electrical conductors or cables. { 'wīrˌwā }

wire weight gage [ENG] A river gage in which a weight suspended on a wire is lowered to the water surface from a bridge or other overhead structure to measure the distance from a point of known elevation on the bridge to the water surface; the distance is usually measured by counting the number of revolutions of a drum required to lower the weight, and a counter is provided which reads the water stage directly. { 'wīr ˈwāt ˌgāj }

wire-wound cryotron [CRYO] A cryotron that consists of a central insulated wire surrounded by a control coil; it is designed so that a relatively small current passed through the control coil produces a magnetic field which makes the gate resistive. { 'wīr ˈwau̇nd 'krī·ə,trän }

wire-wound potentiometer [ELEC] A potentiometer which is similar to a slide-wire potentiometer, except that the resistance wire is wound on a form and contact is made by a slider which moves along an edge from turn to turn. { 'wīr ˈwau̇nd pə,ten·chē'äm·əd·ər }

wire-wound resistor [ELEC] A resistor employing as the resistance element a length of high-resistance wire or ribbon, usually Nichrome, wound on an insulating form. { 'wīr ˈwau̇nd ri'zis·tər }

wire-wound rheostat [ELEC] A rheostat in which a sliding or rolling contact moves over resistance wire that has been wound on an insulating core. { 'wīr ˈwau̇nd 'rē·ə,stat }

wire-wrap connection [ELEC] A solderless connection made by wrapping several turns of bare wire around a sharp-corner rectangular terminal under tension, using either a power tool or hand tool. Also known as solderless wrapped connection; wrapped connection. { 'wīr ˈwrap kə,nek·shən }

wiring [ELEC] The installation and utilization of a system of wire for conduction of electricity. Also known as electric wiring. [ENG] A forming process in which the edge of a sheet-metal part is rolled over a wire to produce a tubular rim containing the wire. [SCI TECH] A system of wires. { 'wīr·iŋ }

wiring board See control panel. { 'wīr·iŋ ˌbȯrd }

wiring diagram See circuit diagram. { 'wīr·iŋ ˌdī·ə,gram }

wiring harness [ELEC] An array of insulated conductors bound together by lacing cord, metal bands, or other binding, in an arrangement suitable for use only in specific equipment for which the harness was designed; it may include terminations. { 'wir·iŋ ,här·nəs }

Wirsung's duct [ANAT] The adult pancreatic duct in man, sheep, ganoid fish, teleost fish, and frog. { 'vir,zùŋ ,dəkt }

Wisconsin [GEOL] Pertaining to the fourth, and last, glacial stage of the Pleistocene epoch in North America; followed the Sangamon interglacial, beginning about 85,000 ± 15,000 years ago and ending 7000 years ago. { wi'skän·sən }

Wisconsin false blossom See false blossom. { wi'skän·sən 'fòls 'bläs·əm }

Wiskott-Aldrich syndrome [MED] A hereditary immunodeficiency disease characterized by a low number of small platelets in the blood, eczema, and defective cell-mediated immunity. { ¦wis·kət 'òl¦drich ,sin,drōm }

wisper wind [METEOROL] A cold night wind, blowing out of the valley of the Wisper River in Germany during clear weather. { 'wis·pər ,wind }

witches'-broom disease [PL PATH] An abnormal cluster of small branches or twigs that grow on a tree or shrub as a result of attack by fungi, viruses, dwarf mistletoes, or insect injury. Also known as hexenbesen; staghead. { ¦wich·əz ¦brüm di,zēz }

witch hazel [MATER] A water extract from the dried leaves of the witch hazel shrub (Hamamelis virginiana); a solution of 14% alcohol with 1% witch hazel extract is commonly known as witch hazel; used as a tonic and sedative. { 'wich ,hā·zəl }

witch of Agnesi [MATH] The curve, symmetric about the y axis and asymptotic in both directions to the x axis, given by $x^2y = 4a^2(2a − y)$. Also known as versiera. { 'wich əv än'nyä·zē }

witherite [MINERAL] $BaCO_3$ A yellowish- or grayish-white mineral of the aragonite group that has orthorhombic symmetry, hardness of 31/4 on Mohs scale, and specific gravity 4.3. { 'with·ə,rīt }

withertip [PL PATH] A blighting of the terminal shoots or the tips of leaves associated with certain plant diseases, such as anthracnose of citrus plants. { 'with·ər,tip }

Witte-Margules equation [OCEANOGR] A formula expressing the slope of the boundary layer between two water masses of different densities and velocities, taking into account the rotation of the earth. Also known as Margules equation. { 'vid·ə 'mär·gyə·lēz i,kwā·zhən }

Witt-Grothendieck group [MATH] The Grothendieck group of the monoid consisting of isometry classes of nondegenerate symmetric forms on vector spaces over a given field, where the product of two such forms is given by their orthogonal sum. { ¦wit 'grōt·ən,dēk ,grüp }

Witt group [MATH] The group of isometry classes of symmetric forms on vector spaces over a given field, where the product of two such forms is given by their orthogonal sum. { 'wit ,grüp }

wittichenite [MINERAL] Cu_3BiS_3 A steel gray to tin white, orthorhombic mineral consisting of copper bismuth sulfide; occurs in tabular and massive form. { 'wid·ə·kə,nīt }

Wittig ether rearrangement [ORG CHEM] The rearrangement of benzyl and alkyl ethers when reacted with a methylating agent, producing secondary and tertiary alcohols. { 'vid·ik 'ē·thər ə,ränj·mənt }

wittite [MINERAL] $Pb_5Bi_6(S,Se)_{14}$ A light lead gray, orthorhombic or monoclinic mineral consisting of a sulfide of lead and bismuth. { 'wi,tīt }

Witt's theorem [MATH] If F and F′ are subspaces of a vector space E with a nondegenerate, symmetric form g, then an isometry of g from F onto F′ can be extended to an isometry of g from E onto itself. { 'wits ,thir·əm }

Witt theory [CHEM] A theory of the mechanism of dyeing stating that all colored organic compounds (called chromogens) contain certain unsaturated chromophoric groups which are responsible for the color, and if these compounds also contain certain auxochromic groups, they possess dyeing properties. { 'wit ,thē·ə·rē }

WKB method See Wentzel-Kramers-Brillouin method. { ,dəb·əl·yü,kā'bē ,meth·əd }

Wnt proteins [CELL MOL] A widely conserved family of secreted signaling molecules that regulate many processes during animal development but, when misregulated, can also contribute to several types of cancer. { ¦dəb·əl·yü¦en¦tē 'prō,tēnz }

Wobbe index [THERMO] A measure of the amount of heat released by a gas burner with a constant orifice, equal to the gross calorific value of the gas in British thermal units per cubic foot at standard temperature and pressure divided by the square root of the specific gravity of the gas. { 'wä·bə ,in,deks }

wobble friction [ENG] A force that occurs in prestressed concrete when the prestressing tendon deviates from its specified profile. { 'wäb·əl ,frik·shən }

wobble pairing [MOL BIO] The ability of a transfer ribonucleic acid molecule to recognize more than one codon. { 'wäb·əl ,per·iŋ }

wobble wheel roller [MECH ENG] A roller with freely suspended pneumatic tires used in soil stabilization. { 'wäb·əl ¦wēl ,rō·lər }

wobbulator [ELECTR] A signal generator in which a motor-driven variable capacitor is used to vary the output frequency periodically between two known limits, as required for displaying a frequency-response curve on the screen of a cathode-ray oscilloscope. { 'wäb·yə,lād·ər }

Wolbachieae [MICROBIO] A tribe of the family Rickettsiaceae; rickettsialike organisms found principally in arthropods. { ¦wōl'bak·ē·ē }

wolf [ACOUS] A dissonant interval which appears when the meantone scale is extended to include chromatic notes. [VERT ZOO] Any of several wild species of the genus Canis in the family Canidae which are fierce and rapacious, sometimes attacking humans; includes the red wolf, gray wolf, and coyote. { wùlf }

Wolf 359 [ASTRON] A star of absolute magnitude 16.6; it is 7.8 light-years from the sun and is a variable flare star, which may emit bursts of light and even radio noise. { wùlf ¦thrē¦fif·tē'nīn }

wolfachite [MINERAL] $Ni(As,Sb)S$ A silver white to tin white mineral consisting of nickel, arsenic, and antimony sulfide; occurs in small crystals and in aggregates. { 'vōl,fäk,īt }

Wolfcampian [GEOL] A North American provincial series of geologic time; lowermost Permian (below Leonardian, above Virgilian of Pennsylvania). { wùlf'kam·pē·ən }

wolfeite [MINERAL] $(Fe,Mn)_2(PO_4)(OH)$ A pinkish, wine yellow to yellowish-brown or reddish-brown, monoclinic mineral consisting of a basic phosphate of iron and manganese. { 'wùl,fīt }

Wolffian duct See mesonephric duct. { 'wùl·fē·ən 'dəkt }

Wolf-Kishner reduction [ORG CHEM] Conversion of aldehydes and ketones to corresponding hydrocarbons by heating their semicarbazones, phenylhydrazones, and hydrazones with sodium ethoxide or by heating the carbonyl compound with excess sodium ethoxide and hydrazine sulfate. { 'wùlf 'kish·nər ri'dək·shən }

Wolf-Lundmark-Melotte galaxy [ASTRON] A dwarf irregular galaxy that is about 1.3 to 1.8 megaparsecs distant and is probably a member of the Local Group. { 'vòlf 'lùnd,märk mə'lät ,gal·ik·sē }

Wolf number See relative sunspot number. { 'vòlf ,nəm·bər }

wolfram See tungsten. wolframite. { 'wùl·frəm }

wolframic acid See tungstic acid. { wùl'fram·ik 'as·əd }

wolframine See wolframite. { 'wùl·frə,mēn }

wolframite [MINERAL] $(Fe,Mn)WO_4$ A brownish- or grayish-black mineral occurring in short monoclinic, prismatic, bladed crystals; the most important ore of tungsten. Also known as tobacco jack; wolfram; wolframine. { 'wùl·frə,mīt }

wolfram white See barium tungstate. { 'wùl·frəm wīt }

Wolf-Rayet nebula [ASTRON] A bright ring-shaped nebula that is radiatively ionized by a central Wolf-Rayet star. Also known as Wolf-Rayet ring. { ¦vòlf rī¦ā ,neb·yə·lə }

Wolf-Rayet ring See Wolf-Rayet nebula. { ¦vòlf rī¦ā ,riŋ }

Wolf-Rayet star [ASTRON] A member of a class of very hot stars (100,000–35,000 K) which characteristically show broad bright emission lines in their spectra; luminosities are high, probably in the range 10^4–10^5 times that of the sun; these stars are probably very young and represent an early short-lived stage in stellar evolution. { ¦vòlf rī¦ā ,stär }

wolfsbane See aconite. { 'wùlfs,bān }

Wolf-Wolfer number See relative sunspot number. { 'vòlf 'vùl·fər ,rəm·bər }

wollastonite [MINERAL] $CaSiO_3$ A white to gray inosilicate mineral (a pyroxenoid) that crystallizes in the triclinic system in tabular crystals and has a pearly or silky luster on the cleavages;

WOLLASTON POLARIZING PRISM

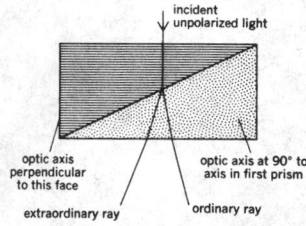

Diagram of Wollaston polarizing prism.

WOLTER TYPE I X-RAY TELESCOPE

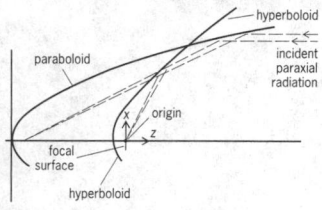

Two surfaces of revolution (paraboloid and hyperboloid) reflect the incident x-rays to a common focus. The horizontal scale is greatly compressed. The y axis is perpendicular to the plane of the figure.

hardness is 5–5.5 on Mohs scale, and specific gravity is 2.85. Also known as tabular spar. { 'wül·ə·stə,nīt }

Wollaston polarizing prism [OPTICS] A device for producing linearly polarized beams of light, consisting of two adjacent quartz wedges with their optic axes perpendicular to each other and to the direction of incident light. { 'wül·ə·stən 'pō·lə,rīz·iŋ ,priz·əm }

Wollaston wire [ENG] An extremely fine platinum wire, produced by enclosing a platinum wire in a silver sheath, drawing them together, and using acid to dissolve away the silver; used in electroscopes, microfuses, and hot-wire instruments. { 'wül·ə·stən ,wīr }

Wolter type I x-ray telescope [ASTRON] A telescope that uses mirrors that form two surfaces of revolution, a paraboloid and a hyperboloid, to reflect incident x-rays to a common focus. { ¦wól·tər ¦tīp ,wən ¦eks,rā 'tel·ə,skōp }

wolverine [VERT ZOO] *Gulo gulo.* A carnivorous mammal which is the largest and most vicious member of the family Mustelidae. { ¦wül·və¦rēn }

wood [BOT] The hard fibrous substance that makes up the trunks and large branches of trees beneath the bark. [ECOL] A dense growth of trees, more extensive than a grove and smaller than a forest. [MATER] Lumber or timber obtained from trees. { wüd }

wood alcohol *See* methyl alcohol. { 'wüd 'al·kə,hól }

wood block [GRAPHICS] A picture, design, or lettering carved on a block of wood with hand tools; the nonprinting areas of the block are carefully carved away, and only the raised areas of the design carry the ink; it differs from a wood engraving because the design is carved along the side of the block parallel with the grain, rather than perpendicular to it; it is therefore characterized by broad areas and strokes running with the grain. Also known as woodcut. { 'wüd ¦bläk }

wood-block printing [GRAPHICS] Making prints from wood blocks by use of a press. { 'wüd ¦bläk 'print·iŋ }

wood-carving tools [DES ENG] The tools normally used in wood carving; they consist of adzes, chisels, gouges, files, and rasps, all of which vary in size and shape. { 'wüd ¦kärv·iŋ ,tülz }

wood coal *See* bituminous wood. { 'wüd 'kōl }

wood copper *See* olivenite. { 'wüd 'käp·ər }

woodcut *See* wood block. { 'wüd,kət }

Wood effect [OPTICS] Transparence of alkali metals to ultraviolet light. { 'wüd i,fekt }

wooden bomb [ORD] A concept which pictures a weapon as being completely reliable and having an infinite shelf life while at the same time requiring no special handling, storage, or surveillance. { 'wüd·ən 'bäm }

wood engraving [GRAPHICS] An engraving carved on the edge grain of a piece of wood. { 'wüd in,grāv·iŋ }

wood ether *See* dimethyl ether. { 'wüd 'ē·thər }

wood filler [MATER] A paste designed to fill the pores of open-grained woods such as ash, chestnut, mahogany, and oak. { 'wüd 'fil·ər }

wood flour [MATER] Dried wood ground to a very fine powder and used in plastic wood, in molding of certain plastics, as an extender in some glues, and in metal-casting operations. { 'wüd 'flaür }

woodhouseite [MINERAL] $CaAl_3(PO_4)(SO_4)(OH)_6$ A colorless to flesh-colored or white, hexagonal mineral consisting of a basic sulfate-phosphate of calcium and aluminum; occurs in small crystals and tabular form. { 'wüd,haü,sīt }

woodland *See* forest; temperate woodland. { 'wüd·lənd }

woodpecker [VERT ZOO] A bird of the family Picidae characterized by stiff tail feathers and zygodactyl feet which enable them to cling to a tree trunk while drilling into the bark for insects. { 'wüd,pek·ər }

wood physics [MATER] The area of wood science concerned with the physical and mechanical properties of wood and the factors which affect them. { 'wüd 'fiz·iks }

wood preservative [MATER] A material used to coat wood to kill insects and fungi, but not usually classed as an insecticide; coal tar creosote and its derivatives are the most widely used wood preservatives. { 'wüd pri,zər·vəd·iv }

wood pulp *See* pulp. { 'wüd ,pəlp }

wood ray [BOT] A vascular ray consisting of a radial row of parenchyma cells in secondary xylem. Also known as xylem ray. { 'wüd ,rā }

Woodruff key [DES ENG] A self-aligning machine key made by a side-milling cutter in the form of a segment of a disk. { 'wü·drəf ,kē }

wood screw [DES ENG] A threaded fastener with a pointed shank, a slotted or recessed head, and a sharp tapered thread of relatively coarse pitch for use only in wood. { 'wüd ,skrü }

Wood's glass [MATER] A type of glass that has a high transmission factor for ultraviolet radiation but is relatively opaque to visible radiation. { 'wüdz ,glas }

Wood's metal [MET] A fusible alloy of the Cerro Corporation that contains 50% bismuth, 25% lead, 12.5% tin, and 12.5% cadmium, and melts at 158°F (70–72°C); used for automatic sprinkler plugs. { 'wüdz ,med·əl }

woodstave pipe [DES ENG] A pipe made of narrow strips of wood placed side by side and banded with wire, metal collars, and inserted joints, used largely for municipal water supply, outfall sewers, and mining irrigation. { 'wüd,stāv ,pīp }

woodstone *See* silicified wood. { 'wüd,stōn }

wood sugar *See* xylose. { 'wüd ,shüg·ər }

wood tin [MINERAL] A riniform, brownish variety of cassiterite with fibers radiating concentrically and resembling dry wood. Also known as dneprovskite. { 'wüd 'tin }

wood vinegar *See* pyroligneous acid. { 'wüd 'vin·ə·gər }

Woodward-Hoffmann rule [ORG CHEM] A concept which can predict or explain the stereochemistry of certain types of reactions in organic chemistry; it is also described as the conservation of orbital symmetry. { 'wüd·wərd 'häf·mən ,rül }

woodwardite [MINERAL] $Cu_4Al_2(SO_4)(OH)_{12}\cdot2\text{-}4H_2O$ A greenish-blue to turquoise blue mineral consisting of a hydrated basic sulfate of copper and aluminum; occurs as botryoidal concretions and in spherulitic form. { 'wüd·wər,dīt }

Woodward's Reagent K *See* N-ethyl-5-phenylisoxazolium-3'-sulfonate. { 'wüd·wərdz rē¦ā·jənt 'kā }

woody lignite *See* bituminous wood. { 'wüd·ē 'lig,nīt }

woody structure [MET] A fibrous appearance in a fracture, particularly found in wrought iron and extruded aluminum alloys, usually associated with elongated inclusions or grains. { 'wüd·ē 'strək·chər }

woof *See* filling. { wüf }

woofer [ENG ACOUS] A large loudspeaker designed to reproduce low audio frequencies at relatively high power levels; usually used in combination with a crossover network and a high-frequency loudspeaker called a tweeter. { 'wüf·ər }

wool [TEXT] A textile fiber made from raw wool characterized by absorbency, resiliency, and insulation. [VERT ZOO] The soft undercoat of various animals such as sheep, angora, goat, camel, alpaca, llama, and vicuna. { wül }

wool fat *See* wool grease. { 'wül ,fat }

wool grease [MATER] A highly complex mixture of wax ester, alcohols, and fatty acids coating the surface of sheep wool fibers and obtained by scouring the wool with soap or synthetic detergent; used in the manufacture of lanolin and its derivatives, for dressing leather, and in lubricating and slushing oils, soaps, and ointments. Formerly known as degras. Also known as wool fat; wool oil; wool wax. { 'wül ,grēs }

wool-sorter's disease *See* anthrax. { 'wül ¦sórd·ərz di,zēz }

wool wax *See* wool grease. { 'wül ,waks }

word [COMPUT SCI] The fundamental unit of storage capacity for a digital computer, almost always considered to be more than eight bits in length. Also known as computer word. { wórd }

word-addressable computer *See* word-oriented computer. { 'wórd ə¦dres·ə·bəl kəm¦pyüd·ər }

word boundary [COMPUT SCI] A storage address that is a multiple of the word length of a computer. { 'wórd ,baün·drē }

word concatenation system [ENG ACOUS] The simplest form of voice response system, which retrieves previously spoken versions of words or phrases and carefully forms them into a sequence without pauses, to approximate normally spoken word sequences. { 'wórd kən,kat·ən'ā·shən ,sis·təm }

word format [COMPUT SCI] Arrangement of characters in a word, with each position or group of positions in the word containing certain specified data. { 'wórd 'fór,mat }

word length [COMPUT SCI] The number of bits, digits, characters, or bytes in one word. { 'wórd ,leŋkth }

word mark [COMPUT SCI] A nondata punctuation bit used

to delimit a word in a variable-word-length computer. { 'wȯrd ˌmärk }

word-oriented computer [COMPUT SCI] A computer in which the locations of words are addressed, and the bits and characters within the words can be addressed only through use of special instructions. Also known as word-addressable computer. { 'wȯrd ˈȯr·ēn·tad kam'pyüd·ar }

word processing [COMPUT SCI] The use of computers or computerlike equipment to write, edit, and format text. { 'wȯrd ˈprä,ses·iŋ }

word processor [COMPUT SCI] **1.** A computer that is either dedicated to word processing or is used with a software package that supports word processing, together with a printer. **2.** A person who operates such a device. { 'wȯrd ˌprä,ses·ar }

word rate [COMPUT SCI] In computer operations, the frequency derived from the elapsed period between the beginning of the transmission of one word and the beginning of the transmission of the next word. { 'wȯrd ˌrāt }

words per minute [COMMUN] A measure of the speed with which messages can be transmitted by a telegraph system. Abbreviated WPM. { 'wərdz par 'min·at }

word time See minor cycle. { 'wərd ˌtīm }

word wrap [COMPUT SCI] A procedure whereby a word processor automatically ends each line when it is full and starts the next line with the next word, never breaking a word. Also known as wrap mode. { 'wərd ˌrap }

work See load. [IND ENG] The physical or mental effort expended in the performance of a task. [MECH] The transference of energy that occurs when a force is applied to a body that is moving in such a way that the force has a component in the direction of the body's motion; it is equal to the line integral of the force over the path taken by the body. { wərk }

workability [MATER] The ease with which concrete can be placed. { ˌwər·kə'bil·əd·ē }

work angle [MET] In arc welding, the angle in a plane normal to the weld axis between the electrode and one member of the joint. { 'wərk ˌaŋ·gəl }

work assembly [COMPUT SCI] The clerical activities related to organizing collections of data records and computer programs or series of related programs. { 'wərk ə,sem·blē }

work breakdown structure [IND ENG] A hierarchy designed to organize, define, and display all the work that must be performed in order to accomplish the objectives of a project. { ˈwərk ˈbrāk,daún ˌstrək·chər }

work cycle [IND ENG] A sequence of tasks, operations, and processes, or a pattern of manual motions, elements, and activities that is repeated for each unit of work. { 'wərk ˌsī·kəl }

work design See job design. { 'wərk di,zīn }

worked-out [MIN ENG] Exhausted, referring to a coal seam or ore deposit. { 'wərkt 'aút }

worked penetration [ENG] Penetration of a sample of lubricating grease immediately after it has been brought to a specified temperature and subjected to strokes in a standard grease worker. { 'wərkt ˌpen·ə'trā·shən }

work element [IND ENG] In planning a manufacturing process, a single task that cannot be subdivided. { 'wərk ˌel·ə·mənt }

worker [INV ZOO] One of the neuter, usually sterile individuals making up a caste of social insects, such as ants, termites, or bees, which labor for the colony. { 'wər·kər }

work file [COMPUT SCI] A file created to hold data temporarily during processing. { 'wərk ˌfīl }

work function [SOLID STATE] The minimum energy needed to remove an electron from the Fermi level of a metal to infinity; usually expressed in electronvolts. [THERMO] See free energy. { 'wərk ˌfəŋk·shən }

work hardening [MET] Increased hardness accompanying plastic deformation of a metal below the recrystallization temperature range. { 'wərk ˌhärd·ən·iŋ }

workhead See headstock. { 'wərk,hed }

working [COMMUN] Carrying on radio communication with a station by means of telegraphy, telephony, or facsimile for a purpose other than calling. [MIN ENG] **1.** The whole strata excavated in working a seam. **2.** Ground or rocks shifting under pressure and producing noise. [NAV] In sea ice navigation, making headway through an ice pack by boring, breaking, and slewing. { 'wərk·iŋ }

working area [IND ENG] A portion of the workplace in which a worker moves about while fulfilling work tasks. { 'wərk·iŋ ˌer·ē·ə }

working electrode [PHYS CHEM] The electrode used in corrosion testing by an electrochemical cell. { 'wərk·iŋ i'lek,rōd }

working envelope [MECH ENG] The surface bounding the maximum extent and reach of a robot's wrist, excluding the tool tip. Also known as working profile. { 'wərk·iŋ 'en·və,lōp }

working life See work life. { ˌwərk·iŋ ˌlīf }

working load [ENG] The maximum load that any structural member is designed to support. { 'wərk·iŋ ˌlōd }

working place [MIN ENG] The place in a mine at which coal or ore is being actually mined. { 'wərk·iŋ ˌplās }

working point [ARCH] A point that is designated on a construction drawing and is then used as reference for other points. { ˈwərk·iŋ ˈpóint }

working pressure [ENG] The allowable operating pressure in a pressurized vessel or conduit, usually calculated by ASME (American Society of Mechanical Engineers) or API (American Petroleum Institute) codes. { 'wərk·iŋ ˌpresh·ər }

working profile See working envelope. { 'wərk·iŋ 'prō,fīl }

working program [COMPUT SCI] A valid program which, when translated into machine language, can be executed on a computer. { 'wərk·iŋ ˌprō,gram }

working Q See loaded Q. { 'wərk·iŋ 'kyü }

working set [COMPUT SCI] The smallest collection of instruction and data words of a given computer program which should be loaded into the main storage of a computer system so that efficient processing is possible. { 'wərk·iŋ 'set }

working-set window [COMPUT SCI] A fixed time interval during which the working set is referenced. { 'wərk·iŋ ˌset 'win·dō }

working solution [GRAPHICS] A solution ready for use. { 'wərk·iŋ sə'lü·shən }

working space See working storage. { 'wərk·iŋ ˌspās }

working space-volume [MECH ENG] The volume enclosed by a robot's working envelope. { 'wərk·iŋ 'spās 'väl·yəm }

working storage [COMPUT SCI] **1.** An area of main memory that is reserved by the programmer for storing temporary or intermediate values. Also known as working space. **2.** In COBOL (computer language), a section in the data division used for describing the name, structure, usage, and initial value of program variables that are neither constants nor records of input/output files. { 'wərk·iŋ 'stȯr·ij }

working voltage See voltage rating. { 'wərk·iŋ ˌvōl·tij }

work-kinetic energy theorem [MECH] The theorem that the change in the kinetic energy of a particle during a displacement is equal to the work done by the resultant force on the particle during this displacement. { 'wərk ki'ned·ik ˌen·ər·jē ˌthir·əm }

work lead [MET] The electrical conductor connecting the source of current to the work in arc welding. Also known as ground lead; welding ground. { 'wərk ˌlēd }

work life [CHEM ENG] The period of time a resin or an adhesive will remain usable after it is mixed with a catalyst and other ingredients. Also known as pot life; working life. { 'wərk ˌlīf }

Workman-Reynolds effect [GEOPHYS] A mechanism for electric charge separation during freezing of slightly impure water; when a very dilute solution of certain salts freezes rapidly, a strong potential difference is established between the solid and liquid phases; for some salts, the ice attains negative charge, for others, positive; this mechanism has been suggested as one possible mode of thunderstorm charge separation in those portions of a thunderstorm downdraft where snow-pellet or hail particles sweep out supercooled waterdrops. { 'wərk·mən 'ren·əlz i,fekt }

work measurement [IND ENG] **1.** Determination of the difficulty of a given task by using both physiologic and biomechanical parameters to evaluate compatibility of available motions with motions required to perform the task. **2.** See ergonometrics. { 'wərk ˌmezh·ər·mənt }

work metabolism [PHYSIO] Metabolism in excess of resting metabolism that can be related to the performance of a specific task. { 'wərk mə,tab·ə,liz·əm }

work of adhesion See adhesional work. { 'wərk əv ad'hē·zhən }

work package [IND ENG] The amount of work required to complete a given job that falls within the responsibility of a single unit of the organization handling the project. { 'wərk ‚pak·ij }

work physiology [IND ENG] An aspect of industrial engineering that takes into account metabolic cost, measurement and prevention of work strain, and other ergonomic factors in the design of tasks and workplaces. { 'wərk ‚fiz·ē‚äl·ə·jē }

workpiece [IND ENG] An object that is being manufactured. { 'wərk‚pēs }

workpiece program [CONT SYS] A program that directs the machining of a component under numerical or computer control. { 'wərk‚pēs ‚prō‚gram }

work print [GRAPHICS] The first print of a motion picture; used for preliminary screening and editing. { 'wərk ‚print }

work sampling [IND ENG] A technique to measure work activity as related to delays consisting of intermittent observations of actual work and delays. Also known as activity sampling; frequency study; ratio delay study. { 'wərk ‚sam·pliŋ }

Workshop See Sculptor. { 'wərk‚shäp }

workspace [COMPUT SCI] In a string processing language, the portion of computer memory that contains the string currently being processed. { 'wərk‚spās }

work standardization [IND ENG] The establishment of uniformity of working conditions, tools, equipment, technical procedures, administrative procedures, workplace arrangements, motion sequences, materials, quality requirements, and similar factors which affect the performance of work. { 'wərk ‚stan·dər·də'zā·shən }

work station [COMPUT SCI] A workplace where a person can interact with a computer on a conversational basis, either a microcomputer and printer or a terminal connected to a remote computer. [IND ENG] A workplace that is included in a production system or on a piece of equipment at which an individual worker may spend only a portion of a working shift. { 'wərk ‚stā·shən }

work station independence [CONT SYS] Property of a numerical control or robot program which does not depend on the nature of the work station. { 'wərk‚stā·shən ‚in·də'pen·dəns }

work strain [PHYSIO] The response of the human body to work stress experienced in the performance of a task. { 'wərk ‚strān }

work stress [IND ENG] Any external force that acts on the body of a worker during the performance of a task. { 'wərk ‚stres }

work tape [COMPUT SCI] A magnetic tape that is available for general use during data processing. { 'wərk ‚tāp }

work task [IND ENG] A specified amount of work, set of responsibilities, or occupation assigned to an individual or to a group. { 'wərk ‚task }

work tolerance [IND ENG] A time period during which a worker can effectively perform a task without a rest period while maintaining acceptable levels of physiological and emotional well-being. { 'wərk ‚täl·ə·rəns }

work unit [IND ENG] An amount of work or the result of an amount of work that is treated as an integer (a single piece of information) when work is being characterized quantitatively. { 'wərk ‚yü·nət }

world [RELAT] Pertaining to Lorentz transformations and four-dimensional space-time, rather than rotations and three-dimensional space, as in world scalar, world vector, world line. { 'wərld }

world calendar [ASTRON] A proposed calendar in which the present 12 months are retained but the days are divided into four equal quarters; January, April, July, and October begin on Sunday and have 31 days, the other months have 30 days, so that there are 364 days with the 365th day following December 30 in no month; leap-year days would follow June 30. { 'wərld 'kal·ən·dər }

world coordinates [CONT SYS] A robotic coordinate system that is fixed with respect to the Earth. { 'wərld kō'órd·ən·əts }

world geographic reference system [NAV] A geographic reference system used by the U.S. Air Force for aircraft position reports, target designations, and other tactical air operations. Abbreviated georef. { 'wərld ‚jē·ə‚graf·ik 'ref·rəns ‚sis·təm }

world line [RELAT] A path in four-dimensional space-time that represents a continuous sequence of events relating to a given particle. { 'wərld ‚līn }

world modeling [CONT SYS] Robot programming that

allows the system to perform complex tasks, based on stored data. { 'wərld 'mäd·əl·iŋ }

world rift system [GEOL] The system of interconnected midocean ridges which is the locus of tensional splitting and magma upwelling believed responsible for sea-floor spreading. { 'wərld 'rift ‚sis·təm }

World Wide Web [COMPUT SCI] A part of the Internet that contains linked text, image, sound, and video documents. Abbreviated WWW. Also known as Web. { ‚wərld ‚wīd 'web }

worm [COMPUT SCI] A computer program that seeks to replicate itself and to spread, with the goal of consuming and exhausting computer resources, thereby causing computing systems to fail. [DES ENG] A shank having at least one complete tooth (thread) around the pitch surface; the driver of a worm gear. [INV ZOO] **1.** The common name for members of the Annelida. **2.** Any of various elongated, naked, soft-bodied animals resembling an earthworm. [MET] Sweat of molten metal which exudes through the crust of solidifying metal in a casting, and is caused by gas evolution. { wərm }

WORM [COMPUT SCI] Pertaining to a storage device, such as an optical disk, that allows the user to record data only once and to read back the data an unlimited number of times. Abbreviation for write-one, read-many. { wərm }

worm conveyor See screw conveyor. { 'wərm kən'vā·ər }

worm gear [DES ENG] A gear with teeth cut on an angle to be driven by a worm; used to connect nonparallel, nonintersecting shafts. { 'wərm ‚gir }

wormseed oil [MATER] An essential oil distilled from the seeds and leaf stems of the plant *Chenopodium anthelminticum*, which is grown in Maryland, and which contains the alkaloid ascoridole; used in worm treatment of animals. Also known as Baltimore oil. { 'wərm‚sēd ‚óil }

worm wheel [DES ENG] A gear wheel with curved teeth that meshes with a worm. { 'wərm ‚wēl }

wormwood oil See absinthe oil. { 'wərm‚wùd ‚óil }

worobieffite See vorobyevite. { wə'rō·bē·ə‚fīt }

worst case evaluation [COMPUT SCI] A testing situation in which the most unfavorable possible combination of circumstances is evaluated. { 'wərst ‚kās i‚val·yə'wā·shən }

worsted [TEXT] **1.** Yarn spun from combed long-staple wool fibers. **2.** Fabric made from this yarn. **3.** Yarns, with any fiber content, manufactured by using the worsted spinning system. { 'wùs·təd }

worsted system [TEXT] A manufacturing technique for yarns in which the short fibers are removed by combing, producing a compact yarn with parallel fibers. { 'wùs·təd ‚sis·təm }

wort [FOOD ENG] A clear infusion of malt grain extract, used by brewers for fermentation. { wórt }

wound botulism [MED] Botulism that involves production of toxin by the organisms infecting or colonizing a wound. { ‚wünd 'bäch·ə‚liz·əm }

wound periderm [BOT] A protective tissue that develops within injured plant organs beneath wound surfaces. { 'wünd ‚per·ə‚dərm }

wound shock See hypovolemic shock. { 'wünd ‚shäk }

woven-screen storage [COMPUT SCI] Digital storage plane made by weaving wires coated with thin magnetic films; when currents are sent through a selected pair of wires that are at right angles in the screen, storage and readout occur at the intersection of the two wires. { 'wō·vən ‚skrēn 'stór·ij }

wove paper [MATER] Paper characterized by a uniform, unlined surface and a soft, smooth finish. { 'wōv ‚pā·pər }

wow [ENG ACOUS] A low-frequency flutter; when caused by an off-center hole in a disk record, occurs once per revolution of the turntable. { waù }

W particle See W boson. { 'dəb·əl‚yü ‚pärd·ə·kəl }

WPM See words per minute.

wrap [GRAPHICS] In a book or magazine, an insert, usually on a different stock, which is not tipped in but wrapped around one of the signatures, and which usually consists of at least two leaves (four pages) securely stitched into the binding. { rap }

wrap-around grasp [IND ENG] A basic grasp whereby an object is held against the palm by the fingers wrapped around it, with the thumb opposing the index finger. { 'rap·ə‚raùnd ‚grasp }

wrap-around hanger [PETRO ENG] An oil-well tubing

hanger made up of two hinged halves with a resilient sealing element between two steel mandrels. { ′rap ə¦raùnd ′haŋ·ər }

wrap-around plate [GRAPHICS] In rotary letterpress, a thin, one-piece relief printing plate which is wrapped around the press cylinder. { ′rap ə¦raùnd ′plāt }

wrap-around type [GRAPHICS] A block of type that is contoured so that it surrounds a graphic object. { ′rap ə‚raùnd ‚tīp }

wrap forming *See* stretch forming. { ′rap ‚fòrm·iŋ }

wrap mode *See* word wrap. { ′rap ‚mōd }

wrapped connection *See* wire-wrap connection. { ′rapt kə·′nek·shən }

wrapper sheet [MECH ENG] **1.** The outer plate enclosing the firebox in a fire-tube boiler. **2.** The thinner sheet of a boiler drum having two sheets. { ′rap·ər ‚shēt }

wrecking ball *See* skull cracker { ′rek·iŋ ‚bòl }

wrecking bar *See* ripping bar. { ′rek·iŋ ‚bär }

wrecking strip [CIV ENG] A small section that is fitted into a form for poured concrete and is easily removed before the main panels to facilitate disassembly of the main components of the form. { ′rek·iŋ ‚strip }

wren [VERT ZOO] Any of the various small brown singing birds in the family Troglodytidae; they are insectivorous and tend to inhabit dense, low vegetation. { ren }

wrench [ENG] A manual or power tool with adapted or adjustable jaws or sockets either at the end or between the ends of a lever for holding or turning a bolt, pipe, or other object. [MECH] The combination of a couple and a force which is parallel to the torque exerted by the couple. { rench }

wrench fault [GEOL] A lateral fault with a more or less vertical fault surface. Also known as basculating fault; torsion fault. { ′rench ‚fòlt }

wrench-head bolt [DES ENG] A bolt with a square or hexagonal head designed to be gripped between the jaws of a wrench. { ′rench ¦hed ‚bōlt }

Wright's inbreeding coefficient *See* inbreeding coefficient. { ¦rīts ′in‚brēd·iŋ ‚kō·ə‚fish·ənt }

Wright's phenomenon [ASTRON] The phenomenon that the diameter of Mars appears greater when viewed from earth in ultraviolet light than when viewed in infrared light. { ′rīts fə‚nam·ə‚nän }

Wright system [PETRO ENG] A method for mining oil from partially drained sands that involves drilling a shaft through the productive strata, followed by long, slanting holes drilled radially in all directions from the shaft bottom into the oil sands. { ′rīt ‚sis·təm }

Wright telescope [OPTICS] A modification of the Schmidt system in which the spherical primary mirror is replaced by an ellipsoidal mirror, and the corrector plate is modified accordingly. { ′rīt ¦tel·ə‚skōp }

wringing fit [DES ENG] A fit of zero-to-negative allowance. { ′riŋ·iŋ ′fit }

wrinkle ridge [ASTRON] A prominent, well-defined, often sinuous ridge on a lunar mare, with gently sloping sides and a height of generally less than 500 feet (150 meters). { ′riŋ· kəl ‚rij }

wrinkling [MATER] **1.** Distortion occurring in a film of paint that gives the appearance of ripples. **2.** Development of a roughened surface on a film of sealant. Also known as crinkling; riveling. [MET] Waviness around the edges of a drawn metal product. { ′riŋk·liŋ }

wrist [ANAT] The part joining forearm and hand. [MECH ENG] A set of rotary joints to which the end effector of a robot is attached. Also known as wrist socket. { rist }

wrist breadth [ANTHRO] A measurement from the outside projection of the distal part of the ulna to the radius at the wrist joint. { ′rist ′bredth }

wrist pin *See* piston pin. { ′ris ‚pin }

wrist socket *See* wrist. { ′rist ‚säk·ət }

wrist thickness [ANTHRO] The measurement transverse to the wrist breadth. { ′rist ′thik·nəs }

writable control storage [COMPUT SCI] A section of the control storage holding microprograms which can be loaded from a console file or under microprogramming control. { ′rīd·ə·bəl kən′trōl ‚stòr·ij }

write [COMPUT SCI] **1.** To transmit data from any source onto an internal storage medium. **2.** A command directing that an output operation be performed. { rīt }

write enable ring [COMPUT SCI] A file protection ring that

must be attached to the hub of a reel of magnetic tape in order to physically allow data to be transcribed onto the reel. Also known as write ring. { ′rīt i¦nā·bəl ‚riŋ }

write error [COMPUT SCI] **1.** A condition in which information cannot be written onto or into a storage device, due to dust, dirt, damage to the recording surface, or damaged electronic components. **2.** A condition in which there is an inconsistency between the pattern of bits transmitted to the write head of a magnetic tape drive and the pattern sensed immediately afterward by the read head. { ′rīt ‚er·ər }

write head [ELECTR] Device that stores digital information as coded electrical pulses on a magnetic drum, disk, or tape. { ′rīt ‚hed }

write inhibit ring [COMPUT SCI] A file protection ring that physically prevents data from being written on a reel of magnetic tape when it is attached to the hub of the reel. { ′rīt in¦hib·ət ‚riŋ }

write-once, read-many *See* WORM. { ¦rīt ′wəns ¦rēd ′men·ē }

write protection [COMPUT SCI] **1.** Any procedure used to prevent writing on storage media. **2.** Any software technique that allows a computer program to read from any area in storage but not to write outside its own area. { ′rīt prə‚tek·shən }

writer [COMPUT SCI] The part of a job entry system that controls output, in particular, the printer and the spool file. { ′rīd·ər }

write ring *See* write enable ring. { ′rīt ‚riŋ }

write time [COMPUT SCI] The time required to transcribe a data item into a computer storage device. { ′rīt ‚tīm }

write to operator [COMPUT SCI] A message issued by a computer program and displayed on the system console that provides information or indicates the status of the program and requires no action by the operator. Abbreviated WTO. { ′rīt tü ′äp·ə‚rād·ər }

write to operator with reply [COMPUT SCI] A message issued by a computer program and displayed on the system console that requires action by the operator in order for execution of the program to continue. Abbreviated WTOR. { ′rīt tü ′äp·ə‚rād·ər with ri′plī }

writing speed [ELECTR] Lineal scanning rate of the electron beam across the storage surface in writing information on a cathode-ray storage tube. { ′rīd·iŋ ‚spēd }

Wronskian [MATH] An $n \times n$ matrix whose ith row is a list of the $(i - 1)$st derivatives of a set of functions $f_1, \ldots, f_n$; ordinarily used to determine linear independence of solutions of linear homogeneous differential equations. { ′vrän·skē·ən }

wrought alloy [MET] An alloy that has been mechanically worked after casting. { ′ròt ′al‚ói }

wrought iron [MET] A commercial iron consisting of slag fibers, primarily iron silicate, embedded in a ferrite matrix. { ′ròt ′ī·ərn }

wryneck *See* torticollis. { ′rī‚nek }

W stars [ASTRON] Stars of the W spectral class; their spectra contain an abundance of highly ionized elements such as He, C, N, and O, and they are intensely hot with surface temperatures of about 50,000 to 100,000 K. { ′dəb·əl·yü ‚stärz }

WTO *See* write to operator.

WTOR *See* write to operator with reply.

W-truss [CIV ENG] A truss having upper and lower chords joined by web members that form a shape resembling the letter W. { ′dəb·əl‚yü ‚trəs }

wuchereriasis [MED] Infection with worms of the genus *Wuchereria*. Also known as Bancroft's filariasis. { ‚wù·kə· rē′rī·ə·səs }

Wulf electrometer [ENG] **1.** A variant of the string electrometer in which charged metal plates are replaced by charged knife-edges. **2.** An electrometer in which two conducting fibers are placed side by side, and their separation upon charging is measured. { ¦wùlf i‚lek′träm·əd·ər }

wulfenite [MINERAL] PbMoO$_4$ A yellow, orange, orange-yellow, or orange-red tetragonal mineral occurring in tabular crystals or granular masses; an ore of molybdenum. Also known as yellow lead ore. { ′wùl·fə‚nīt }

Wulff process [CHEM ENG] A chemical process to make acetylene and ethylene by cracking a hydrocarbon gas (for example, butane) with high-temperature steam in a regenerative furnace. { ′wùlf ‚rä·səs }

Wullenweber antenna [ELECTROMAG] An antenna array consisting of two concentric circles of masts, connected to be

W-TRUSS

Side view of a W-truss.

electronically steerable; used for ground-to-air communication at Strategic Air Command bases. { 'wul·ən,web·ər an,ten·ə }

Würm [GEOL] **1.** A European stage of geologic time: uppermost Pleistocene (above Riss, below Holocene). **2.** Pertaining to the fourth glaciation of the Pleistocene epoch in the Alps, equivalent to the Wisconsin glaciation in North America, following the Riss-Würm interglacial. { vừrm }

W Ursae Majoris stars [ASTRON] Eclipsing variable stars whose brightness is continuously varying in periods of a few hours; they are composed of two close stars that have a common gaseous envelope. { ¦dəb·əl·yü 'ər,sī mə'jór·əs ,stärz }

Wurster process See air-suspension encapsulation. { 'wər·stər ,prä·səs }

Wurtz-Fittig reaction [ORG CHEM] A modified Wurtz reaction in which an aromatic halide reacts with an aklyl halide in the presence of sodium and an anhydrous solvent to form alkylated aromatic hydrocarbons. { 'wərtz 'fid·ig rē,ak·shən }

wurtzilite [GEOL] A black, massive, sectile, infusible, asphaltic pyrobitumen derived from the metamorphosis of petroleum. { 'wərt·sə,līt }

wurtzite [MINERAL] (Zn,Fe)S A brownish-black hexagonal mineral consisting of zinc sulfide and occurring in hemimorphic pyramidal crystals, or in radiating needles and bundles. { 'wərt,sīt }

Wurtz reaction [ORG CHEM] Synthesis of hydrocarbons by treating alkyl iodides in ethereal solution with sodium according to the reaction $2CH_3I + 2Na \rightarrow CH_3CH_3 + 2NaI$. { 'wərts rē,ak·shən }

wustite [MINERAL] FeO Iron oxide. { 'wùs,tīt }

W Virginis stars [ASTRON] Periodic variable stars with periods of about 10 to 30 days; they exhibit two surges of activity from the same star so that there is a doubling of their spectral lines. Also known as Type II Cepheids. { ¦dəb·əl·yü vər'jin·əs ,stärz }

WWV [COMMUN] The call letters of a radio station maintained by the National Institute of Standards and Technology to provide standard radio and audio frequencies and other technical services, such as precision time signals and radio propagation disturbance warnings; the station broadcasts on 2.5, 5, 10, 15, 20, 25, 30, and 35 megahertz at various times.

WWVH [COMMUN] The National Institute of Standards and Technology radio station at Maui, Hawaii, broadcasting services similar to those of WWV on 5, 10, and 15 megahertz.

WWW See World Wide Web.

wye [ELEC] Polyphase circuit whose phase differences are 120° and which when drawn resembles the letter Y. [ENG] A pipe branching off a straight main run at an angle of 45°. Also known as Y; yoke. { wī }

wye branch See Y branch. { 'wī ,branch }

wye connection See Y network. { 'wī kə,nek·shən }

wye fitting See Y fitting. { 'wī ,fid·iŋ }

wye level See Y level. { 'wī ,lev·əl }

Wynyardiidae [PALEON] An extinct family of herbivorous marsupial mammals in the order Diprotodonta. { ,win·yər'dī·ə,dē }

X *See* siegbahn.

XAFS *See* x-ray absorption fine structure.

XANES *See* x-ray absorption near-edge structure.

xanthan [BIOCHEM] A polysaccharide produced by *Xanthomonas campestris* that is used in oil recovery to help improve water flooding and oil displacement. { 'zan·thən }

xanthan gum [ORG CHEM] A high-molecular-weight (5–10 million) water-soluble natural gum; a heteropolysaccharide made up of building blocks of D-glucose, D-mannose, and D-glucuronic acid residues; produced by pure culture fermentation of glucose with *Xanthomonas campestris*. { 'zan·thən ,gəm }

xanthate [ORG CHEM] A water-soluble salt of xanthic acid, usually potassium or sodium; used as an ore-flotation collector. { 'zan,thāt }

xanthelasma [MED] Raised yellow plaques occurring around the eyelids, resulting from lipid-filled cells in the dermis. { ,zan·thə'laz·mə }

xanthene [ORG CHEM] CH$_2$(C$_6$H$_4$)$_2$O Yellowish crystals that are soluble in ether, slightly soluble in water and alcohol; melts at 100°C; used as a fungicide and chemical intermediate. Also known as tricyclic dibenzopyran. { 'zan,thēn }

xanthene dye [ORG CHEM] Any of a family of dyes related to the xanthenes; the chromophore groups are (C$_6$H$_4$). { 'zan,thēn 'dī }

9-xanthenone *See* genicide. { 'nīn 'zan·thə,nōn }

Xanthidae [INV ZOO] The mud crabs, a family of decapod crustaceans in the section Brachyura. { 'zan·thə,dē }

xanthine [ORG CHEM] C$_5$H$_4$N$_4$O$_2$ A toxic yellow-white purine base that is found in blood and urine, and occasionally in plants; it is a powder, insoluble in water and acids, soluble in caustic soda; sublimes when heated; used in medicine and as a chemical intermediate. Also known as dioxopurine. { 'zan,thēn }

xanthine oxidase [BIOCHEM] A flavoprotein enzyme catalyzing the oxidation of certain purines. { 'zan,thēn 'äk·sə,dās }

xanthism [BIOL] A color variation in which an animal's normal coloring is largely replaced by yellow pigments. Also known as xanthochroism. { 'zan,thiz·əm }

xanthochroism *See* xanthism. { ,zan·thrə'krō,iz·əm }

xanthochroite *See* greenockite. { ,zan·thrə'krō,īt }

xanthoconite [MINERAL] Ag$_3$AsS$_3$ A dark red to dull orange to clove brown mineral consisting of silver arsenic sulfide. { zan'thäk·ə,nīt }

xanthoma [MED] A yellowish mass of lipid-filled histocytes occurring in subcutaneous tissue, often around tendons. { zan'thō·mə }

xanthomatosis [MED] A condition marked by the deposit of a yellowish or orange lipoid material in the reticuloendothelial cells, the skin, and the internal organs. { ,zan,thō·mə'tō·səs }

Xanthomonas citri [MICROBIO] The bacterial pathogen that causes citrus canker. { ,zan·thə,mō·nəs 'si,trē }

xanthomycin [MICROBIO] An antibiotic produced by a strain of *Streptomyces* and composed of two varieties, A and B; active in low concentrations against a number of gram-positive microorganisms. { ,zan·thə'mīs·ən }

xanthone [ORG CHEM] CO(C$_6$H$_4$)$_2$O White needle crystals that are found in some plant pigments; insoluble in water, soluble in alcohol, chloroform, and benzene; melts at 173°C, sublimes at 350°C; used as a larvicide, as a dye intermediate, and in perfumes and pharmaceuticals. { 'zan,thōn }

xanthophore [CYTOL] A yellow chromatophore. { 'zan·thə,fôr }

Xanthophyceae [BOT] A class of yellow-green to green flagellate organisms of the division Chrysophyta; zoologists classify these organisms in the order Heterochlorida. { ,zan·thə'fīs·ē,ē }

xanthophyll [BIOCHEM] C$_{40}$H$_{56}$O$_2$ Any of a group of yellow, alcohol-soluble carotenoid pigments that are oxygen derivatives of the carotenes, and are found in certain flowers, fruits, and leaves. Also known as carotenol; lutein. { 'zan·thə,fil }

xanthophyllite *See* clintonite. { ,zan·thə'fi,līt }

xanthosiderite *See* goethite. { ,zan·thō'sī·də,rīt }

xanthoxenite [MINERAL] Ca$_2$Fe(PO$_4$)$_2$(OH)·1^1/$_2$H$_2$O A pale yellow to brownish-yellow, monoclinic or triclinic mineral consisting of a hydrated basic phosphate of calcium and iron; occurs as masses and crusts. { zan'thäk·sə,nīt }

xanthurenic acid [BIOCHEM] C$_{10}$H$_7$NO$_4$ Sulfur yellow crystals with a melting point of 286°C; soluble in aqueous alkali hydroxides and carbonates; excreted by pyridoxine-deficient animals after ingestion of tryptophan. { ,zan·thyə,ren·ik 'as·əd }

Xantusiidae [VERT ZOO] The night lizards, a family of reptiles in the suborder Sauria. { ,zan·tə'sī·ə,dē }

x axis [CRYSTAL] A reference axis within a quartz crystal. [MATH] **1.** A horizontal axis in a system of rectangular coordinates. **2.** That line on which distances to the right or left (east or west) of the reference line are marked, especially on a map, chart, or graph. { 'eks ,ak·səs }

X band [COMMUN] A radio-frequency band extending from 8 to 12 gigahertz. { 'eks ,band }

X boson *See* superheavy boson. { 'eks 'bō,sän }

X chromosome [GEN] The sex chromosome occurring in double dose in the homogametic female sex and in single dose in the heterogametic male sex in mammals, *Drosophila*, and many other organisms. { 'eks 'krō·mə,sōm }

X coefficient *See* nine-*j* symbol. { 'eks ,kō·i,fish·ənt }

x component [MATH] The projection of a vector quantity on the *x* axis of a coordinate system. { 'eks kəm,pō·nənt }

x coordinate [MATH] One of the coordinates of a point in a two- or three-dimensional cartesian coordinate system, equal to the directed distance of a point from the *y* axis in a two-dimensional system, or from the plane of the *y* and *z* axes in a three-dimensional system, measured along a line parallel to the *x* axis. { 'eks kō'ôrd·ən·ət }

X cut [CRYSTAL] A quartz-crystal cut made in such a manner that the *x* axis is perpendicular to the faces of the resulting slab. { 'eks ,kət }

Xe *See* xenon.

Xenarthra [VERT ZOO] A suborder of mammals in the order Edentata including sloths, anteaters, and related forms; posterior vertebrae have extra articular facets and vertebrae in the hip, and shoulder regions tend to be fused. { zə'när·thrə }

X engine [MECH ENG] An in-line engine with the cylinder banks so arranged around the crankshaft that they resemble the letter X when the engine is viewed from the end. { 'eks ,en·jən }

xenobiotic [BIOCHEM] A chemical that is not normally found in the body, such as a drug. { ,zēn·ə·bī'äd·ik }

xenoblast [MINERAL] A mineral which has grown during metamorphism without development of its characteristic crystal faces. Also known as allotrioblast. { 'zēn·ə,blast }

xenocryst [CRYSTAL] A crystal in igneous rock that resembles a phenocryst and is foreign to the enclosing body of rock. Also known as chadacryst. { 'zēn·ə‚krist }

xenogamy [BOT] Cross-fertilization between flowers on different plants. { zə'näg·ə·mē }

xenogeneic [IMMUNOL] Referring to cells, tissues, or organs used in transplantation that originate in a different species. { ¦zēn·ə·jə'nē·ik }

xenograft [IMMUNOL] A graft performed between members of different species. { 'zēn·ə‚graft }

xenolith [PETR] An inclusion in an igneous rock which is not genetically related, such as an unmelted fragment of country rock. Also known as accidental inclusion; exogenous inclusion. { 'zēn·ə‚lith }

xenomorphic See allotriomorphic. { ¦zēn·ə¦mȯr·fik }

xenon [CHEM] An element, symbol Xe, member of the noble gas family, group 0, atomic number 54, atomic weight 131.291; colorless, boiling point −108°C (1 atmosphere, or 101,325 pascals), noncombustible, nontoxic, and nonreactive; used in photographic flash lamps, luminescent tubes, and lasers, and as an anesthetic. { 'zē‚nän }

xenon-135 [NUC PHYS] A radioactive isotope of xenon produced in nuclear reactors; readily absorbs neutrons; half-life is 9.2 hours. { 'zē‚nän ¦wən¦thərd·ē'fīv }

xenon arc lamp [ELEC] An arc lamp filled with xenon giving a light intensity approaching that of the carbon arc; particularly valuable in projecting motion pictures. { 'zē‚nän 'ärk ‚lamp }

xenon flash lamp [ELEC] A flash tube containing xenon gas, which produces an intense peak of radiant energy at a wavelength of 566 nanometers when a high direct-current pulsed voltage is applied between electrodes at opposite ends of the tube. { 'zē‚nän 'flash ‚lamp }

xenon override [NUCLEO] In a nuclear reactor, the excess reactivity provided to compensate for the poisoning effect of xenon buildup. { 'zē‚nän 'ō·və‚rīd }

xenon poisoning [NUCLEO] The accumulation in a nuclear reactor of xenon-135, formed by beta decay of iodine-135; xenon-135 has the highest cross section for thermal neutron capture of any known reactor poison. { 'zē‚nän 'pȯiz·ən·iŋ }

xenophyae [INV ZOO] In xenophyophores, the inorganic portion of the test consisting of foreign elements, such as sponge spicules, foraminiferan or radiolarian tests, and mineral particles. { zə'näf·ē‚ē }

Xenophyophorea [INV ZOO] A class of giant marine benthic Rhizopoda. { zə¦näf·ē·ə'fȯr·ē·ə }

Xenophyophorida [INV ZOO] An order of Protozoa in the subclass Granuloreticulosia; includes deep-sea forms that develop as discoid to fan-shaped branching forms which are multinucleate at maturity. { ¦zē·nō‚fī·ə'fȯr·əd·ə }

Xenopneusta [INV ZOO] A small order of wormlike animals belonging to the Echiurida. { ‚zēn·əp'nùs·tə }

Xenopterygii [VERT ZOO] The equivalent name for Gobiesociformes. { zə‚näp·tə'rij·ē‚ī }

Xenosauridae [VERT ZOO] A family of four rare species of lizards in the suborder Sauria; composed of the Chinese lizard (*Shinisaurus crocodilurus*) and three Central American species of the genus *Xenosaurus*. { ‚zēn·ə'sȯr·ə‚dē }

xenothermal [MINERAL] Pertaining to a mineral deposit formed at high temperature but at shallow to moderate depth. { ¦zēn·ə¦thər·məl }

xenotime [MINERAL] Y(PO₄) A tetragonal mineral of varying color, consisting of yttrium phosphate. { 'zēn·ə‚tīm }

Xenungulata [PALEON] An order of large, digitigrade, extinct, tapirlike mammals with relatively short, slender limbs and five-toed feet with broad, flat phalanges; restricted to the Paleocene deposits of Brazil and Argentina. { zə‚nùŋ·gyə'läd·ə }

xenyl [ORG CHEM] The functional group C₆H₅C₆H₄−. { 'zen·əl }

Xeralf [GEOL] A suborder of the soil order Alfisol, having good drainage, and found in regions with rainy winters and dry summers in mediterranean climates; the surface horizons tend to become massive and hard during the dry seasons, with some soils having duripans that interfere with root growth. { 'zir‚älf }

xerarch succession [ECOL] A type of succession that originates in a dry habitat. { 'zer‚ärk sək‚sesh·ən }

Xerert [GEOL] A suborder of the soil order Vertisol, formed in a Mediterranean climate; wide surface cracks open and close once a year. { 'zir‚ərt }

xeric [ECOL] **1.** Of or pertaining to a habitat having a low or inadequate supply of moisture. **2.** Of or pertaining to an organism living in such an environment. { 'zer·ik }

xeroderma pigmentosum [MED] A genodermatosis characterized by premature degenerative changes in the form of keratoses, malignant epitheliomatosis, and hyper- and hypopigmentation. { ‚zir·ə'dər·mə ‚pig·mən'tō·səm }

xerodermosteosis See Sjögren's syndrome. { ‚zir·ō·dər‚mäs·tē'ō·səs }

xerogel [CHEM] **1.** A gel whose final form contains little or none of the dispersion medium used. **2.** An organic polymer capable of swelling in suitable solvents to yield particles possessing a three-dimensional network of polymer chains. { 'zer·ə‚jel }

xerography [GRAPHICS] A printing method developed by the Xerox Corporation; a negative image is formed by a resinous powder on an electrically charged plate, and this image is transferred and thermally fixed onto a paper as a positive. { zə'räg·rə·fē }

Xeroll [GEOL] A suborder of the soil order Mollisol, formed in a xeric moisture regime; may have a calcic, petrocalcic, or gypsic horizon, or a duripan. { 'zir‚ȯl }

xeromorphic [PLANT ECOL] Referring to a plant that is able to survive in dry environments. { ‚zir·ə'mȯr·fik }

xerophthalmia [MED] Dryness and thickening of the conjunctiva, sometimes following chronic conjunctivitis, disease of the lacrimal apparatus, or vitamin A deficiency. { ‚zi‚räf'thal·mē·ə }

xerophyte [ECOL] A plant adapted to life in areas where the water supply is limited. { 'zir·ə‚fīt }

xeroradiography [GRAPHICS] An electrostatic image-forming process in which x-rays or gamma rays form an electrostatic image on a photoconductive insulating medium; the charged image areas attract and hold a fine powder called a toner, and then the powder image is transferred to paper and fused there by heat. { ¦zi·rō‚rād·ē'äg·rə·fē }

xerosere [ECOL] A temporary community in an ecological succession on dry, sterile ground such as rock, sand, or clay. { 'zir·ə‚sir }

xerosis conjunctivae [MED] A condition marked by silver-gray, shiny, triangular spots on both sides of the cornea, within the region of the palpebral aperture, consisting of dried epithelium, flaky masses, and microorganisms. { zə'rō·səs kən‚jəŋk·tə‚vē }

xerothermal period See xerothermic period. { ¦zir·ə¦thər·məl 'pir·ē·əd }

xerothermic [CLIMATOL] Characterized by dryness and heat. { ¦zir·ə¦thər·mik }

xerothermic period [GEOL] A postglacial interval of a warmer, drier climate. Also known as xerothermal period. { ¦zir·ə¦thər·mik 'pir·ē·əd }

xerotolerance [PHYSIO] The ability to grow in extremely dry habitats. { 'zer·ə‚täl·ə·rəns }

Xerult [GEOL] A suborder of the soil order Ultisol, formed in a xeric moisture regime; brownish or reddish soil with a low to moderate organic-carbon content. { 'zir‚əlt }

X frame [DES ENG] An automotive frame which either has side rails bent in at the center of the vehicle, making the overall form that of an X, or has an X-shaped member which joins the side rails with diagonals for added strength and resistance to torsional stresses. { 'eks ‚frām }

x-height [GRAPHICS] The height of the lowercase letter x in a font. { 'eks ‚hīt }

xi hyperon [PART PHYS] Also known as xi particle. **1.** Collective name for the xi-minus and xi-zero particles, which form an isotopic-spin multiplet of quasi-stable baryons, designated Ξ, having a hypercharge of −1, a total isotopic spin of ¹/₂, a spin of ¹/₂, positive parity, and an average mass of approximately 1318 megaelectronvolts. Also known as cascade hyperon; cascade particle. **2.** A baryon belonging to any isotopic-spin multiplet having a hypercharge of −1 and a total isotopic spin of ¹/₂; designated by $\Xi_{JP}(m)$, where m is the mass of the baryon in MeV, and J and P are its spin and parity (if known); the $\Xi_{3/2+}(1530)$ is sometimes designated Ξ*. { 'zī 'hī·pə‚rän }

xi-minus particle [PART PHYS] A negatively charged xi

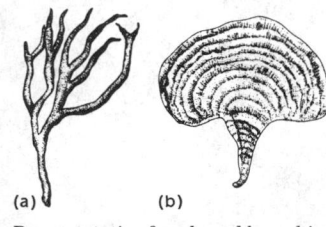

XENOPHYOPHORIDA

(a)   (b)

Representative fan-shaped branching forms of xenophyophorids.
(a) *Stanomma dendroides.*
(b) *Stannophyllum zonarium.*

hyperon, designated Ξ^-. Also known as cascade particle. { 'zī ‖mī·nəs ‚pärd·ə·kəl }

xi particle *See* xi hyperon. { 'zī ‚pärd·ə·kəl }

xiphidio cercaria [INV ZOO] A digenetic trematode larva having a stylet in the oral sucker. { zə'fid·ē·ō sər'kar·ē·ə }

Xiphiidae [VERT ZOO] The swordfishes, a family of perciform fishes in the suborder Scombroidei characterized by a tremendously pronounced bill. { zə'fī·ə‚dē }

xiphisternum [ANAT] The elongated posterior portion of the sternum. { ‖zif·ə'stər·nəm }

Xiphodontidae [PALEON] A family of primitive tylopod ruminants in the superfamily Anaplotherioidea from the late Eocene to the middle Oligocene of Europe. { ‚zif·ə'dänt·ə‚dē }

Xiphosura [INV ZOO] The equivalent name for Xiphosurida. { ‚zif·ə'sùr·ə }

Xiphosurida [INV ZOO] A subclass of primitive arthropods in the class Merostomata characterized by cephalothoracic appendages, ocelli, book lungs, a somewhat trilobed body, and freely articulating styliform telson. { ‚zif·ə'sùr·ə·də }

Xiphydriidae [INV ZOO] A family of the Hymenoptera in the superfamily Siricoidea. { ‚zi·ə'drī·ə‚dē }

XIST [GEN] The X-linked gene whose RNA transcript mediates X-chromosome inactivation in mammals.

xi-zero particle [PART PHYS] An uncharged xi hyperon, designated Ξ^0. { 'zī ‖zir·ō ‚pard·ə·kəl }

X-linked adrenoleukodystrophy [MED] An inherited peroxisomal disorder in which a defective peroxisomal membrane protein results in impaired β-oxidation of very long chain fatty acids, which subsequently accumulate in the adrenal glands, nervous system, and testes and disrupt normal activity. { ‖eks ‚liŋkt ə‚drē·nō‚lü·kə'dis·trə·fē }

XML *See* Extensible Markup Language.

XOFF *See* transmitter off. { eks'òf }

XON *See* transmitter on. { eks'òn }

XOR *See* exclusive or. { eks'òr }

X organ [INV ZOO] A cluster of neurosecretory cells of the medulla terminales, a portion of the brain lying in the eyestalk in stalk-eyed crustaceans. { 'eks ‚òr·gən }

x-parallax *See* absolute stereoscopic parallax. { 'eks 'par·ə‚laks }

Y-process [NUC PHYS] The synthesis of certain nuclides in stars through nuclear reactions in which gamma rays remove neutrons, protons, or alpha particles from the nucleus; it is one mode of the *p*-process. { 'wī ‚prä·səs }

XPS *See* x-ray photoelectron spectroscopy.

x-radiation *See* x-rays. { ‖eks ‚rād·ē'ā·shən }

x-ray absorption [ELECTROMAG] The taking up of energy from an x-ray beam by a medium through which the beam is passing. { 'eks ‚rā əb'sòrp·shən }

x-ray absorption fine structure [SPECT] The structure in the x-ray absorption spectrum of a substance at energies above the absorption edge, including both the x-ray absorption near-edge structure and the extended x-ray absorption fine structure. Abbreviated XAFS. { ‚eks ‚rā əb'sòrp·shən ‖fīn 'strək·chər }

x-ray absorption near-edge structure [PHYS] A ripplelike structure in the x-ray absorption spectrum of a substance, at energies just above the absorption edge associated with liberation of core electrons into the continuum, and much closer to the absorption edge than the energies associated with extended x-ray absorption fine structure. Abbreviated XANES. { 'eks ‚rā əb'sòrp·shən ‖nir ‚ej 'strək·chər }

x-ray analysis [PHYS] The use of x-ray radiations to detect heavy elements in the presence of lighter ones, to give critical-edge absorption to identify elemental composition, and to identify crystal structures by diffraction patterns. { 'eks ‚rā ə'nal·ə·səs }

x-ray astronomy [ASTRON] The study of x-rays mainly from sources outside the solar system; it includes the study of novae and supernovae in the Milky Way Galaxy, together with extra-galactic radio sources. { 'eks ‚rā ə'strän·ə·mē }

x-ray background [ASTRON] Diffuse, almost isotropic x-radiation from beyond the solar system, believed to be the summed contribution of many unresolved sources. { 'eks ‚rā 'bak‚graùnd }

x-ray binary [ASTRON] An x-ray source that is a member of a binary system. { 'eks ‚rā 'bī‚ner·ē }

x-ray burster [ASTRON] One of a class of celestial x-ray sources which produce bursts of x-rays in the 1–20-kiloelectronvolt range and which are characterized by rise times of less than a few seconds and decay times of a few seconds to a few minutes; the peak luminosity is of the order of 10^{38} ergs per second (10^{31} watts) and the sources have an average equivalent temperature of 10^8 K. { 'eks ‚rā 'bər·stər }

x-ray cluster [ASTRON] A cluster of galaxies that is pervaded by a diffuse medium that emits x-rays. { 'eks ‚rā ‚kləs·tər }

x-ray crystallography [CRYSTAL] The study of crystal structure by x-ray diffraction techniques. Also known as roentgen diffractometry. { 'eks ‚rā ‚krist·əl'äg·rə·fē }

x-ray crystal spectrometer [SPECT] An instrument designed to produce an x-ray spectrum and measure the wavelengths of its components, by diffracting x-rays from a crystal with known lattice spacing. { 'eks ‚rā 'krist·əl spek'träm·əd·ər }

x-ray diffraction [PHYS] The scattering of x-rays by matter, especially crystals, with accompanying variation in intensity due to interference effects. Also known as x-ray microdiffraction. { 'eks ‚rā di'frak·shən }

x-ray diffraction analysis [CRYSTAL] Analysis of the crystal structure of materials by passing x-rays through them and registering the diffraction (scattering) image of the rays. { 'eks ‚rā di'frak·shən ə‚nal·ə·səs }

x-ray diffractometer [ENG] An instrument used in x-ray analysis to measure the intensities of the diffracted beams at different angles. { 'eks ‚rā di‚frak'täm·əd·ər }

x-ray emission *See* x-ray fluorescence. { 'eks ‚rā i'mish·ən }

x-ray film [GRAPHICS] A film base coated, usually on both sides, with an emulsion designed for use with x-rays. { 'eks ‚rā ‚film }

x-ray fluorescence [ATOM PHYS] Emission by a substance of its characteristic x-ray line spectrum upon exposure to x-rays. Also known as x-ray emission. { 'eks ‚rā flü'res·əns }

x-ray fluorescence analysis [SPECT] A nondestructive physical method used for chemical elemental analysis in which a material is irradiated by photons or charged particles of sufficient energy to cause its elements to emit (fluoresce) characteristic x-ray line spectra. { 'eks ‚rā flü'res·əns ə‚nal·ə·səs }

x-ray fluorescent emission spectrometer [SPECT] An x-ray crystal spectrometer used to measure wavelengths of x-ray fluorescence; in order to concentrate beams of low intensity, it has bent reflecting or transmitting crystals arranged so that the theoretical curvature required can be varied with the diffraction angle of a spectrum line. { 'eks ‚rā flü'res·ənt i‖mish·ən spek'träm·əd·ər }

x-ray fluorimetry *See* x-ray fluorescence analysis. { 'eks ‚rā flü'räm·ə·trē }

x-ray generator [ELECTR] A metal from whose surface large amounts of x-rays are emitted when it is bombarded with high-velocity electrons; metals with high atomic weight are the most efficient generators. { 'eks ‚rā ‚jen·ə‚rād·ər }

x-ray goniometer [ENG] A scale designed to measure the angle between the incident and refracted beams in x-ray diffraction analysis. { 'eks ‚rā ‚gō·nē'äm·əd·ər }

x-ray hardness [ELECTROMAG] The penetrating ability of x-rays; it is an inverse function of the wavelength. { 'eks ‚rā 'härd·nəs }

x-ray holography [ELECTROMAG] The use of holographic techniques to image objects beyond the reach of optical microscopes by using high-intensity coherent sources of electromagnetic radiation with wavelengths between 0.1 and 10 nanometers. Also known as microholography. { 'eks ‚rā hō'läg·rə·fē }

x-ray image spectrography [SPECT] A modification of x-ray fluorescence analysis in which x-rays irradiate a cylindrically bent crystal, and Bragg diffraction of the resulting emission produces a slightly enlarged image with a resolution of about 50 micrometers. { 'eks ‚rā ‖im·ij spek'träg·rə·fē }

x-ray irradiation [PHYS] Subjection of a material, object, or patient to x-rays. { 'eks ‚rā i‚rād·ē'ā·shən }

x-ray laser [ELECTROMAG] A device that uses the principle of amplification by stimulated emission of radiation to produce an intense beam of coherent x-rays. { 'eks ‚rā 'lā·zər }

x-ray lithography [ELECTR] Lithography in which the resist is exposed to a well-collimated, high-intensity x-ray beam projected through a special mask in close proximity to the silicon slice. { 'eks ‚rā li'thäg·rə·fē }

x-ray machine [ENG] The x-ray tube, power supply, and

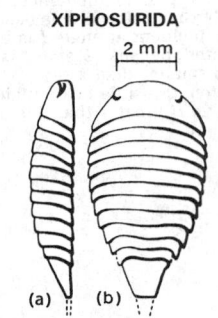

XIPHOSURIDA

Drawing of *Paleomerus* species. (*a*) Side view. (*b*) Dorsal view.

X-RAY IMAGE SPECTROGRAPHY

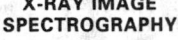
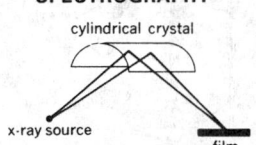

Arrangement of x-ray source, cylindrical crystal, and photographic film in x-ray image spectrography.

X-RAY TELESCOPE

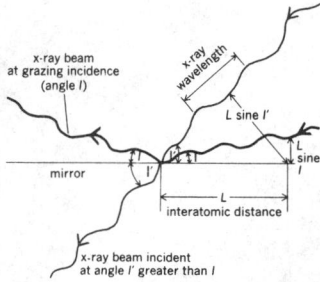

Diagram illustrating phenomenon of total external reflection of x-rays at grazing incidence, on which x-ray telescope is based. Beam incident at angle *I* is totally reflected, because *L* sine *I* is much smaller than x-ray wavelength, while beam incident at angle *I'* greater than *I* is not reflected.

associated equipment required for producing x-ray photographs. { 'eks ,rā mə,shēn }

x-ray microdiffraction *See* x-ray diffraction. { 'eks ,rā ¦mī·krō·di'frak·shən }

x-ray microprobe *See* microprobe. { 'eks ,rā 'mī·krə,prōb }

x-ray microscope [ENG] **1.** A device in which an ultra-fine-focus x-ray tube or electron gun produces an electron beam focused to an extremely small image on a transmission-type x-ray target that serves as a vacuum seal; the magnification is by projection; specimens being examined can thus be in air, as also can the photographic film that records the magnified image. **2.** Any of several instruments which utilize x-radiation for chemical analysis and for magnification of 100–1000 diameters; it is based on contact or projection microradiography, reflection x-ray microscopy, or x-ray image spectrography. { 'eks ,rā 'mī·krə,skōp }

x-ray monochromator [ENG] An instrument in which x-rays are diffracted from a crystal to produce a beam having a narrow range of wavelengths. { 'eks ,rā ¦män·ə¦krō‚mäd·ər }

x-ray nebulae [ASTRON] The remnant of an ancient supernova that has been identified as a source of x-rays; an example is the Crab Nebula. { 'eks ,rā 'neb·yə‚lī }

x-ray nova [ASTRON] An x-ray source which appears suddenly in the sky, dramatically increases in intensity over a few days, and then decays away with a lifetime of several months. Also known as transient x-ray source. { 'eks ,rā 'nō·və }

x-ray optics [ELECTROMAG] A title-by-analogy of those phases of x-ray physics in which x-rays demonstrate properties similar to those of light waves. Also known as roentgen optics. { 'eks ,rā 'äp·tiks }

x-ray photoelectron spectroscopy [SPECT] A form of electron spectroscopy in which a sample is irradiated with a beam of monochromatic x-rays and the energies of the resulting photoelectrons are measured. Abbreviated XPS. Also known as electron spectroscopy for chemical analysis (ESCA). { 'eks ,rā ¦fōd·ō·i¦lek‚trän spek'träs·kə·pē }

x-ray powder diffractometer *See* powder diffraction camera. { 'eks ,rā 'paùd·ər ‚di‚frak'täm·əd·ər }

x-ray powder method *See* powder method. { 'eks ,rā 'paùd·ər ‚meth·əd }

x-ray projection microscopy *See* projection microradiography. { 'eks ,rā prə¦jek·shən mī'krās·kə·pē }

x-rays [PHYS] A penetrating electromagnetic radiation, usually generated by accelerating electrons to high velocity and suddenly stopping them by collision with a solid body, or by inner-shell transitions of atoms with atomic number greater than 10; their wavelengths range from about 10^{-5} angstrom to 10^3 angstroms, the average wavelength used in research being about 1 angstrom. Also known as roentgen rays; x-radiation. { 'eks ,rāz }

x-ray spectrograph [SPECT] An x-ray spectrometer equipped with photographic or other recording apparatus; one application is fluorescence analysis. { 'eks ,rā 'spek·trə‚graf }

x-ray spectrometer [SPECT] An instrument for producing the x-ray spectrum of a material and measuring the wavelengths of the various components. { 'eks ,rā spek'träm·əd·ər }

x-ray spectrometry [SPECT] A technique for quantitative analysis of the elemental composition of specimens. Irradiation of a sample by high-energy electrons, protons, or photons ionizes some of the atoms, which then emit characteristic x-rays whose wavelength depends on the atomic number of the element and whose intensity is related to the concentration of that element. { 'eks ,rā spek'träm·ə·trē }

x-ray spectroscopy *See* x-ray spectrometry. { 'eks ,rā spek'träs·kə·pē }

x-ray spectrum [SPECT] A display or graph of the intensity of x-rays, produced when electrons strike a solid object, as a function of wavelengths or some related parameter; it consists of a continuous bremsstrahlung spectrum on which are superimposed groups of sharp lines characteristic of the elements in the target. { 'eks ,rā ‚spek·trəm }

x-ray star [ASTROPHYS] A source of x-rays from outside the solar system; examples are the point x-ray sources Scorpius X-1, Cygnus X-2, and the Crab x-ray source. { 'eks ,rā ‚stär }

x-ray target [ELECTR] The metal body with which high-velocity electrons collide, in a vacuum tube designed to produce x-rays. { 'eks ,rā ‚tär·gət }

x-ray telescope [ENG] An instrument designed to detect x-rays emanating from a source outside the earth's atmosphere and to resolve the x-rays into an image; they are carried to high altitudes by balloons, rockets, or space vehicles; although several types of x-ray detector, involving gas counters, scintillation counters, and collimators, have been used, only one, making use of the phenomenon of total external reflection of x-rays from a surface at grazing incidence, is strictly an x-ray telescope. { 'eks ,rā 'tel·ə‚skōp }

x-ray therapy [MED] Medical treatment by controlled application of x-rays; a type of radiotherapy. { 'eks ,rā ‚ther·ə·pē }

x-ray thickness gage [ENG] A thickness gage used for measuring and indicating the thickness of moving cold-rolled sheet steel during the rolling process without making contact with the sheet; an x-ray beam directed through the sheet is absorbed in proportion to the thickness of the material and its atomic number. { 'eks ,rā 'thik·nəs ‚gāj }

x-ray tube [ELECTR] A vacuum tube designed to produce x-rays by accelerating electrons to a high velocity by means of an electrostatic field, then suddenly stopping them by collision with a target. { 'eks ,rā ‚tüb }

x-ray unit *See* siegbahn. { 'eks ,rā ‚yü·nət }

XS-3 code *See* excess-three code. { 'ek‚ses 'thrē ‚kōd }

X server [COMPUT SCI] Software that draws the screen image and handles standard input in an X Windows System; in contrast to typical usage of the term server, an X server is located on the user's computer; the client is the application that is displayed, which may be located on a remote node of the network. { 'eks ‚ser·vər }

X test [STAT] A one-sample test which rejects the hypothesis $\mu = \mu_H$ in favor of the alternative $\mu > \mu_H$ if $X - \mu_H \geq c$ where c is an appropriate critical value, X is the arithmetic mean of observations, μ_H is a given number, and μ is the (unknown) expected value of the random variable X. { 'eks ‚test }

XU *See* siegbahn.

X unit *See* siegbahn. { 'eks ‚yü·nət }

X wave *See* extraordinary wave. { 'eks ‚wāv }

X Windows System [COMPUT SCI] A graphical environment providing window management for computer applications; originally developed to provide a graphical user interface for Unix systems, it has been ported to other platforms. { ¦eks 'win‚dōz ‚sis·təm }

x,y chromaticity diagram *See* Maxwell triangle. { ¦eks¦wī ‚krō·mə'tis·əd·ē ‚dī·ə‚gram }

XY coordinate plotter *See* coordinate plotter. { ¦eks¦wī kō'ȯrd·ən·ət ‚pläd·ər }

Xyelidae [INV ZOO] A family of hymenopteran insects in the superfamily Megalodontoidea. { zī'el·ə‚dē }

xylan [BIOCHEM] A polysaccharide composed of the pentose sugar D-xylose. { 'zī‚lan }

xylem [BOT] The principal water-conducting tissue and the chief supporting tissue of higher plants; composed of tracheids, vessel members, fibers, and parenchyma. { 'zī·ləm }

xylene [ORG CHEM] $C_6H_4(CH_3)_2$ Any one of the family of isomeric, colorless aromatic hydrocarbon liquids, produced by the destructive distillation of coal or by the catalytic reforming of petroleum naphthenic fractions; used for high-octane and aviation gasolines, solvents, chemical intermediates, and the manufacture of polyester resins. Also known as dimethylbenzene; xylol. { 'zī‚lēn }

meta-xylene [ORG CHEM] $1,3\text{-}C_6H_4(CH_3)_2$ A flammable, toxic liquid; insoluble in water, soluble in alcohol and ether; boils at 139°C; used as an intermediate for dyes, a chemical intermediate, and a solvent, and in insecticides and aviation fuel. { ¦med·ə 'zī‚lēn }

ortho-xylene [ORG CHEM] $1,2\text{-}C_6H_4(CH_3)_2$ A flammable, moderately toxic liquid; insoluble in water, soluble in alcohol and ether; boils at 144°C; used to make phthalic anhydride, vitamins, pharmaceuticals, and dyes, and in insecticides and motor fuels. { ¦ȯr·thō 'zī‚lēn }

para-xylene [ORG CHEM] $1,4\text{-}C_6H_4(CH_3)_2$ A toxic, combustible liquid; insoluble in water, soluble in alcohol and ether; boils at 139°C; used as a chemical intermediate, and to synthesize terephthalic acid, vitamins, and pharmaceuticals, and in insecticides. { ¦par·ə 'zī‚lēn }

xylenol [ORG CHEM] $(CH_3)_2C_6H_3OH$ Highly toxic, combustible crystals; slightly soluble in water, soluble in most

XYLEM

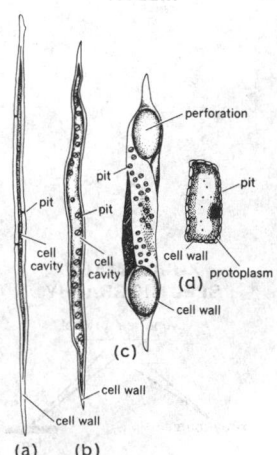

Xylem cell types. *(a)* Wood (xylem) fiber. *(b)* Tracheid. *(c)* Vessel member. *(d)* Xylem parenchyma cell. All can vary widely in structure. *(From H. J. Fuller and O. Tippo, College Botany, rev. ed., Holt, 1954)*

organic solvents; melts at 20–76°C; used as a chemical interme-diate, disinfectant, solvent, and fungicide, and for pharmaceuti-cals and dyestuffs. { 'zī·lə,nċl }

xylidine [ORG CHEM] $(CH_3)_2C_6H_3NH_2$ A toxic, combusti-ble liquid; soluble in alcohol and ether, slightly soluble in water; boils about 220°C; used as a chemical intermediate and to make dyes and pharmaceuticals. { zī·lə,dēn }

xylinite [GEOL] A variety of provitrinite consisting of xylem or lignified tissue. { 'zī·lə,nīt }

xylite See xylitol. { 'zī,līt }

xylitol [ORG CHEM] $CH_2OH(CHOH)_3CH_2OH$ Pentahydric alcohols derived from xylose. Also known as xylite. { 'zī·lə,tȯl }

Xylocopidae [INV ZOO] A family of hairy tropical bees in the superfamily Apoidea. { ,zīl·ə'käp·ə,dē }

xyloid coal See bituminous wood. { 'zī,lȯid 'kōl }

xyloid lignite See bituminous wood. { 'zī,lȯid 'lig nīt }

xylol See xylene. { 'zī,lȯl }

Xylomyiidae [INV ZOO] A family of orthorrhaphous dipteran insects in the series Brachycera. { ,zīl·ə'mī·ə,dē }

xylophagous [BIOL] Referring to an organism which feeds on wood. { zī'läf·ə·gəs }

xylose [BIOCHEM] $C_5H_{10}O_5$ A pentose sugar found in many woody materials; combustible, white crystals with a sweet taste; soluble in water and alcohol; melts about 148°C; used as a nonnutritive sweetener and in dyeing and tanning. Also known as wood sugar. { 'zī,lōs }

XY plotter See coordinate plotter. { ¦eks¦wī 'pläd·ər }

XY recorder [ENG] A recorder that traces on a chart the relation of two variables, neither of which is time. { ¦eks¦wī ri'kȯrd·ər }

Xyridaceae [BOT] A family of terrestrial monocotyledonous plants in the order Commelinales characterized by an open leaf sheath, three stamens, and a simple racemose head for the inflorescence. { ,zir·ə'dās·ē,ē }

XY switching system [ELECTR] A telephone switching sys-tem consisting of a series of flat bank and wiper switches in which the wipers move in a horizontal plane, first in one direc-tion and then in another under the control of pulses from a subscriber's dial; the switches are stacked on frames, and are operated one after another. { ¦eks¦wī 'swich·iŋ ,sis·təm }

Y

Y *See* wye; yttrium.

yacca gum *See* acaroid resin. { 'yak·ə ,gəm }

yacht [NAV ARCH] A sailing or power boat used for pleasure cruises or racing. { yät }

Yagi antenna *See* Yagi-Uda antenna. { 'yäg·ē an,ten·ə }

Yagi-Uda antenna [ELECTROMAG] An end-fire antenna array having maximum radiation in the direction of the array line; it has one dipole connected to the transmission line and a number of equally spaced unconnected dipoles mounted parallel to the first in the same horizontal plane to serve as directors and reflectors. Also known as Uda antenna; Yagi antenna. { 'yäg·ē 'üd·ə an,ten·ə }

yag laser *See* yttrium-aluminum-garnet laser. { 'yag ,lā·zər }

yak [VERT ZOO] *Poephagus grunniens.* A heavily built, long-haired mammal of the order Artiodactyla, with a shoulder hump; related to the bison, and resembles it in having 14 pairs of ribs. { yak }

yalca [METEOROL] A local name for a severe snowstorm with a strong squally wind which occurs in the Andes Mountain passes of northern Peru. { 'yäl·kə }

yam [BOT] **1.** A plant of the genus *Dioscorea* grown for its edible fleshy root. **2.** An erroneous name for the Puerto Rico variety of sweet potato; the edible, starchy tuberous root of the plant. { yam }

yamase [METEOROL] A cool, onshore, easterly wind in the Senriku district of Japan in summer. { yä'mä·sē }

Yang-Mills theory [PART PHYS] A theory of nuclear forces between nucleons based on the hypothesis that they can be derived by imposing local isospin invariance; this implies that the interaction must occur through the exchange of three massless vector bosons. { 'yaŋ 'milz ,thē·ə·rē }

yard [CIV ENG] A facility for building and repairing ships. [MECH] A unit of length in common use in the United States and United Kingdom, equal to 0.9144 meter, or 3 feet. Abbreviated yd. [NAV ARCH] A long spar, tapered at the ends, attached at its middle to a mast and running athwartships, and used to support a sail. { yärd }

yardage [MECH] An amount expressed in yards. [MIN ENG] The extra compensation a miner receives in addition to the mining price for working in a narrow place or in deficient coal, usually at a certain price per yard advanced. { 'yärd·ij }

yardang [GEOL] A long, irregular ridge with a sharp crest sited between two round-bottomed troughs that have been carved by wind erosion in a desert region. { 'yär,daŋ }

yardang trough [GEOL] A long, shallow, round-bottomed groove, furrow, or trough cut into a desert floor by wind erosion and separated by a yardang from the neighboring trough. { 'yär,daŋ 'trof }

yardarm [NAV ARCH] One of the ends of a yard. { 'yär,därm }

yard crane *See* crane truck. { 'yärd ,krān }

yard drain [CIV ENG] A drain for clearing an open area of surface water. { 'yärd ,drān }

yard lumber [BUILD] A category of lumber up to 5 inches (12.5 centimeters) thick. { 'yärd ,ləm·bər }

yard maintenance [ENG] A category of maintenance that includes the complete rebuilding of parts, subassemblies, or components. { 'yärd ,maint·ən·əns }

yardstick compass [GRAPHICS] A simple arrangement consisting of two fully adjustable clips that slide on an ordinary yardstick; one clip has a sharp point, the other a pencil holder; the compass is used to make large, accurate circles. { 'yärd ,stik ,käm·pəs }

Yarkovsky effect [ASTRON] The effect of a small particle's rotation on its orbit about the sun, due to anisotropic reradiation. { yär'käf·skē i,fekt }

Yarmouth interglacial [GEOL] The second interglacial stage of the Pleistocene epoch in North America, following the Kansan glacial stage and before the Illinoian. { 'yär·məth ,in·tər'glā·shəl }

yarn [TEXT] A continuous strand of two or more plies of carded or combed fibers twisted together, or a single filament of natural or synthetic fibers used for weaving or knitting. { 'yärn }

yarn-dyed [TEXT] Pertaining to a fabric made of yarns dyed before weaving or knitting. { 'yärn ,dīd }

yaw [MECH] **1.** The rotational or oscillatory movement of a ship, aircraft, rocket, or the like about a vertical axis. Also known as yawing. **2.** The amount of this movement, that is, the angle of yaw. **3.** To rotate or oscillate about a vertical axis. { yo }

yaw acceleration [MECH] The angular acceleration of an aircraft or missile about its normal or Z axis. { 'yo ak,sel·ə'rā·shən }

yaw angle *See* angle of yaw. { 'yo ,aŋ·gəl }

yaw axis [MECH] A vertical axis through an aircraft, rocket, or similar body, about which the body yaws; it may be a body, wind, or stability axis. Also known as yawing axis. { 'yo ,ak·səs }

yaw damper [AERO ENG] A control system or device that reduces the yaw of an aircraft, guided missile, or the like. { 'yo ,dam·pər }

yaw in bore [ORD] The maximum angle between the axis of the bore of a gun and the axis of the projectile which can occur due to the clearance between the bore diameter and the bourrelets. { 'yo in 'bor }

yaw indicator [AERO ENG] A device that measures the angular direction of the airflow relative to the longitudinal vertical plane of the aircraft; this may be accomplished by a balanced vane or by a differential pressure sensor that aligns the detector to the airflow, and in so doing transmits the measured angle between the normal axis and the detector as the yaw angle. { 'yo ,in·də,kād·ər }

yawing *See* yaw. { 'yo·iŋ }

yawing axis *See* yaw axis. { 'yo·iŋ ,ak·səs }

yaws [MED] An infectious tropical disease of humans caused by the spirochete *Treponema pertenue;* manifested by a primary cutaneous lesion followed by a granulomatous skin eruption. { 'yoz }

yaw simulator [CONT SYS] A test instrument used to derive and thereby permit study of probable aerodynamic behavior in controlled flight under specific initial conditions; certain components of the missile guidance system, such as the receiver or servo loop, are connected into the simulator circuitry; also, certain aerodynamic parameters of the specific missile must be known and set into the simulator; applicable to the yaw plane. { 'yo ,sim·yə,lāc·ər }

y axis [CRYSTAL] A line perpendicular to two opposite parallel faces of a quartz crystal. [MATH] **1.** A vertical axis in a system of rectangular coordinates. **2.** That line on which distances above or below (north or south) the reference line are marked, especially on a map, chart, or graph. { 'wī ,ak·səs }

Yb *See* ytterbium.

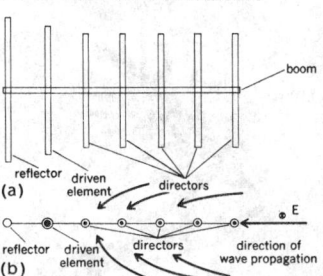

YAGI-UDA ANTENNA

Yagi-Uda antenna. (*a*) View from above. (*b*) Side view; arrows show energy flow close to parasitic elements.

YAK

Yak (*Poephagus grunniens*).

YAM

Puerto Rico sweet potato known as the yam. (*USDA*)

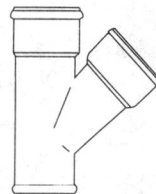

Y BRANCH

One type of Y branch.

YEW

Female branchlet and single leaf of Japanese yew (*Taxus cuspidata*).

Y block [MET] A Y-shaped test casting used to appraise low-shrinkage alloys. { 'wī ‚bläk }

Y branch [ENG] A Y-shaped branch in a piping system. Also known as wye branch. { 'wī ‚branch }

Y chromosome [GEN] The sex chromosome found only in the heterogametic male sex, as in mammals and *Drosophila*. { 'wī 'krō·mə‚sōm }

Y circulator [ELECTROMAG] Circulator in which three identical rectangular waveguides are joined to form a symmetrical Y-shaped configuration, with a ferrite post or wedge at its center; power entering any waveguide will emerge from only one adjacent waveguide. { 'wī 'sər‚kyə‚lād·ər }

y component [MATH] The projection of a vector quantity on the *y* axis of a coordinate system. { 'wī kəm‚pō·nənt }

Y connection See Y network. { 'wī kə‚nek·shən }

y coordinate [MATH] One of the coordinates of a point in a two- or three-dimensional coordinate system, equal to the directed distance of a point from the *x* axis in a two dimensional system, or from the plane of the *x* and *z* axes in a three-dimensional coordinate system, measured along a line parallel to the *y* axis. { 'wī kō‚órd·ən·ət }

Y cut [CRYSTAL] A quartz-crystal cut such that the *y* axis is perpendicular to the faces of the resulting slab. { 'wī ‚kət }

yd See yard.

Y-delta transformation [ELEC] One of two electrically equivalent networks with three terminals, one being connected internally by a Y configuration and the other being connected internally by a delta transformation. Also known as delta-Y transformation; pi-T transformation. { 'wī 'del·tə ‚tranz·fər‚mā·shən }

year [ASTRON] Any of several units of time based on the revolution of the earth about the sun; the tropical year to which the calendar is adjusted is the period required for the sun's longitude to increase 360°; it is about 365.24220 mean solar days. Abbreviated yr. { yir }

yeast [MYCOL] A collective name for those fungi which possess, under normal conditions of growth, a vegetative body (thallus) consisting, at least in part, of simple, individual cells. { yēst }

yellow [OPTICS] The hue evoked in an average observer by monochromatic radiation having a wavelength in the approximate range from 577 to 597 nanometers; however, the same sensation can be produced in a variety of other ways. { 'yel·ō }

yellow arsenic See orpiment. { yel·ō 'ärs·ən·ik }

yellow cake [MIN ENG] The final precipitate formed in the milling of uranium ores. { 'yel·ō 'kāk }

yellow cedar See Alaska cedar. { 'yel·ō 'sēd·ər }

yellow coal See tasmanite. { 'yel·ō 'kōl }

yellow copperas See copiapite. { 'yel·ō 'käp·rəs }

yellow cypress See Alaska cedar. { 'yel·ō 'sī·prəs }

yellow dwarf [PL PATH] Any of several plant viral diseases characterized by yellowing of the foliage and stunting of the plant. { 'yel·ō 'dwórf }

yellow fat cell [HISTOL] A large, generally spherical fat cell with a thin shell of protoplasm and a single enlarged fat droplet which appears yellowish. { 'yel·ō 'fat ‚sel }

yellow fever [MED] An acute, febrile, mosquito-borne viral disease characterized in severe cases by jaundice, albuminuria, and hemorrhage. { 'yel·ō 'fē·vər }

yellow-green algae [BOT] The common name for members of the class Xanthophyceae. { 'yel·ō ‚grēn 'al·jē }

yellow lead ore See wulfenite. { 'yel·ō 'led ‚ór }

yellow leaf blotch [PL PATH] A fungus disease of alfalfa caused by *Pyrenopeziza medicaginis* characterized by the appearance of yellow or orange blotches with small black dots on the foliage. { 'yel·ō 'lēf ‚bläch }

yellow metal See Muntz metal. { 'yel·ō ‚med·əl }

yellow mud [GEOL] Mud containing sediment having a characteristic yellow color, resulting from certain iron compounds. { 'yel·ō 'məd }

yellow ocher [MATER] A form of limonite used as a yellow pigment. { ‚yel·ō 'ō·kər }

yellow phosphorus See white phosphorus. { 'yel·ō 'fä·sfə·rəs }

yellow precipitate See mercuric oxide. { 'yel·ō pri'sip·ə‚tāt }

yellow prussiate of potash See potassium ferrocyanide. { 'yel·ō 'prəs·ē‚āt əv 'päd‚ash }

yellow prussiate of soda See sodium ferrocyanide. { 'yel·ō 'prəs·ē‚āt əv 'sōd·ə }

yellow pyoktanin See auramine hydrochloride. { 'yel·ō pī'äk·tə·nən }

yellow pyrite See chalcopyrite. { 'yel·ō 'pī‚rīt }

yellow quartz See citrine. { 'yel·ō 'kwórts }

yellows [PL PATH] Any of various fungus diseases of plants characterized by yellowing of the leaves which later turn brown, become brittle, and die; affects cabbage, lettuce, cauliflower, peach, sugarbeet, and other plants. { 'yel·ōz }

yellow salt See uranyl nitrate. { 'yel·ō 'sólt }

yellow scale [MATER] The commercial name for low-grade paraffin wax. { 'yel·ō 'skāl }

Yellow Sea [GEOGR] An inlet of the Pacific Ocean between northeastern China and Korea. { 'yel·ō 'sē }

yellow snow [HYD] Snow with a golden or yellow appearance because of the presence of pine or cypress pollen. { 'yel·ō 'snō }

yellow tellurium See sylvanite. { 'yel·ō tə'lúr·ē·əm }

Yerkes system See MK system. { 'yər·kēz ‚sis·təm }

Yersinia [MICROBIO] A genus of gram-negative, facultative, rod-shaped bacteria in the Enterobacteriaceae family that shares many physiological properties with related *Escherichia coli*, including metabolic processes and sensitivity to certain bacteriophages. { yər'sin·ē·ə }

yew [BOT] A genus of evergreen trees and shrubs, *Taxus*, with the fruit, an aril, containing a single seed surrounded by a scarlet, fleshy, cuplike envelope; the leaves are flat and acicular. { yü }

Y factor [PETRO ENG] An empirical relationship of bubble-point data (pressure and formation volume) used to smooth oil reservoir solution-gas/oil-ratio data for graphical presentation. { 'wī ‚fak·tər }

Y fitting [CIV ENG] A pipe fitting with one end subdivided to form two openings, usually at a 45° angle to the run of the pipe. Also known as wye fitting. { 'wī ‚fid·iŋ }

Y gun [ORD] Two-barreled, antisubmarine gun shaped like the letter Y, used to throw depth charges to either side of the stern of the vessel on which the gun is mounted. { 'wī ‚gən }

yield [ENG] Product of a reaction or process as in chemical reactions or food processing. [MECH] That stress in a material at which plastic deformation occurs. [ORD] The total effective energy released in a nuclear explosion; usually expressed in terms of the equivalent tonnage of trinitrotoluene (TNT) required to produce the same energy release. { yēld }

yield factor [IND ENG] The ratio of the amount of material that results from an industrial process to the amount of material that went into it. { 'yēld ‚fak·tər }

yielding arches [MIN ENG] Steel arches installed in underground openings as the ground is removed to support loads caused by changing ground movement or by faulted and fractured rock; when the ground load exceeds the design load of the arch as installed, yielding takes places in the joint of the arch, permitting the overburden to settle into a natural arch of its own and thus tending to bring all forces into equilibrium. { 'yēld·iŋ 'ärch·əz }

yielding floor [MIN ENG] A soft floor which heaves and flows into open spaces when subjected to heavy pressure from packs or pillars. { 'yēld·iŋ 'flór }

yielding prop [MIN ENG] A steel prop which is adjustable in length and incorporates a sliding or flexible joint which comes into operation when the roof pressure exceeds a set load or value. { 'yēld·iŋ 'präp }

yield monitor [AGR] An instrument used in precision agriculture to measure either the volume or the mass of the harvested portion of a crop using a variety of engineering principles, including light interception, radiation absorption, measurement of impact force, and directly weighing the crop. { 'yēld ‚man·əd·ər }

yield-pillar system [MIN ENG] A method of roof control whereby the natural strength of the roof strata is maintained by the relief of pressure in working areas and the controlled transference of load to abutments which are clear of the workings and roadways. { 'yēld ‚pil·ər ‚sis·təm }

yield point [MECH] The lowest stress at which strain increases without increase in stress. { 'yēld ‚póint }

yield rate [IND ENG] The amount of satisfactory material available after the completion of a given manufacturing process expressed as a percentage of the total amount produced. { 'yēld ‚rāt }

yield strength [MECH] The stress at which a material exhibits a specified deviation from proportionality of stress and strain. { 'yēld ,streŋkth }

yield stress [FL MECH] The minimum stress needed to cause a Bingham plastic to flow. [MECH] The lowest stress at which extension of the tensile test piece increases without increase in load. { 'yēld ,stres }

yield temperature [ENG] The temperature at which a fusible plug device melts and is dislodged by its holder and thus relieves pressure in a pressure vessel; it is caused by the melting of the fusible material, which is then forced from its holder. { 'yēld ,tem·prə·chər }

yig See yttrium iron garnet. { yig }

yig device [ELECTR] A filter, oscillator, parametric amplifier, or other device that uses an yttrium-iron-garnet crystal in combination with a variable magnetic field to achieve wideband tuning in microwave circuits. Derived from yttrium-iron-garnet device. { 'yig di,vīs }

yig filter [ELECTR] A filter consisting of an yttrium-iron-garnet crystal positioned in a magnetic field provided by a permanent magnet and a solenoid; tuning is achieved by varying the amount of direct current through the solenoid; the bias magnet serves to tune the filter to the center of the band, thus minimizing the solenoid power required to tune over wide bandwidths. { 'yig ,fil·tər }

yig-tuned parametric amplifier [ELECTR] A parametric amplifier in which tuning is achieved by varying the amount of direct current flowing through the solenoid of a yig filter. { 'yig ,tünd ¦par·ə¦me·trik 'am plə,fī·ər }

yig-tuned tunnel-diode oscillator [ELECTR] Microwave oscillator in which precisely controlled wide-band tuning is achieved by varying the current through a tuning solenoid that acts on a yig filter in the tunnel-diode oscillator circuit. { 'yig ¦tünd ¦tən·əl ¦dī,ōd ¦äs·ə,lād·ər }

YIQ [COMMUN] The method of representing colors in color television systems, where Y represents the luminosity, which is also the black-and-white signal, I represents red minus the luminosity, and Q represents blue minus the luminosity.

Y junction [ELECTROMAG] A waveguide in which the longitudinal axes of the waveguide form a Y. { 'wī ,jəŋk·shən }

ylang-ylang oil See ilang-ilang oil. { 'ē,läŋ 'ē,läŋ ,óil }

ylem [ASTROPHYS] The primordial matter which according to the big bang theory existed just prior to the formation of the chemical elements. { 'ī·ləm }

Y level [ENG] A surveyor's level with Y-shaped rests to support the telescope. Also known as wye level. { 'wī ,lev·əl }

ylide [ORG CHEM] An organic compound which contains two adjacent atoms bearing formal positive and negative charges, and in which both atoms have full octets of electrons. { 'i·līd }

Y ligament See iliofemoral ligament. { 'wī ,lig·ə·mənt }

ylium ion See enium ion. { 'ī·lē·əm 'ī,än }

Y network [ELEC] A star network having three branches. Also known as wye connection; Y connection. { 'wī ,net,wərk }

yogurt [FOOD ENG] A fermented milk food made by adding cultures of *Lactobacillus acidophilus* and *Streptococcus thermophilus* to skimmed cow's milk and milk solids. { 'yō·gərt }

yohimbine [PHARM] $C_{21}H_{26}N_2O_3$ An alkaloid derived from plants that is a weak blocker of alpha-adrenergic receptors; used as a mydriatic and in the treatment of impotence. { yō'him,bīn }

yoke [ARCH] A horizontal member forming the head of a window frame. [DES ENG] A clamp or similar device to embrace and hold two other parts. [ELECTR] See deflection yoke. [ELECTROMAG] Piece of ferromagnetic material without windings, which permanently connects two or more magnet cores. [ENG] **1.** A bar of wood used to join the necks of draft animals for working together. **2.** See wye. [MECH ENG] A slotted crosshead used instead of a connecting rod in some steam engines. [COMPUT SCI] Two or more read/write heads that are physically joined together and move as a unit over a disk, so that it is possible to read from or write to adjacent tracks without moving the head. { yōk }

yoked basin See zeugogeosyncline. { 'yōkt 'bās·ən }

yolk [BIOCHEM] **1.** Nutritive material stored in an ovum. **2.** The yellow spherical mass of food material that makes up the central portion of the egg of a bird or reptile. { yōk }

yolk larva See vitellarium. { 'yōk ,lär·və }

yolk sac [EMBRYO] A distended extraembryonic extension, heavy-laden with yolk, through the umbilicus of the midgut of the vertebrate embryo. { 'yōk ,sak }

Yonden square [STAT] An experimental design that is an incomplete block design derived from a Latin square by dropping one or more rows and by treating columns as blocks. Also known as incomplete Latin square. { 'yän·dən ,skwer }

Y organ [INV ZOO] Either of a pair of nonneural structures found in the anterior portion of the crustacean body; source of the molting hormone, ecdysone. { 'wī ,ór·gən }

Yorkian [GEOL] A European stage of geologic time forming part of the lower Upper Carboniferous, above Lanarkian and below Staffordian, equivalent to part of the lower Westphalian. { 'yór·kē·ən }

York-Scheibel column See Scheibel extractor. { 'yórk 'shī·bəl ,käl·əm }

youg [METEOROL] A hot wind during unsettled summer weather in the Mediterranean. { yóg }

Young construction [OPTICS] A graphical procedure for tracing a light ray through a boundary between two media having different refractive indices. { 'yəŋ kən,strək·shən }

Young-Helmholtz laws [MECH] Two laws describing the motion of bowed strings; the first states that no overtone with a node at the point of excitation can be present; the second states that when the string is bowed at a distance of $1/n$ times the string's length from one of the ends, where n is an integer, the string moves back and forth with two constant velocities, one of which has the same direction as that of the bow and is equal to it, while the other has the opposite direction and is $n-1$ times as large. { ¦yəŋ 'helm,hōlts ,lóz }

Young-Helmholtz theory [NEUROSCI] A theory of color vision according to which there are three types of color receptors that respond to short, medium, and long waves respectively; primary colors are those that stimulate most successfully the three types of receptors. Also known as Helmholtz theory. { ¦yəŋ 'helm,hōlts ,thē·ə·rē }

young ice [HYD] Newly formed ice in the transitional stage of development from ice crust to winter ice. { 'yəŋ ¦īs }

Younginiformes [PALEON] A suborder of extinct small lizardlike reptiles in the order Eosuchia, ranging from the Middle Permian to the Lower Triassic in South Africa. { ,yəŋ·gə·nə'fór,mēz }

Young's inequality [MATH] An inequality that applies to a function $y = f(x)$ that is continuous and strictly increasing for $x \geqq 0$ and satisfies $f(0) = 0$, with inverse function $x = g(y)$; it states that, for any positive numbers a and b in the ranges of x and y, respectively, the product ab is equal to or less than the sum of the integral from 0 to a of $f(x)dx$ and the integral from 0 to b of $g(y)dy$. { 'yəŋz ,in·ə'kwäl·əd·ē }

Young's modulus [MECH] The ratio of a simple tension stress applied to a material to the resulting strain parallel to the tension. Also known as modulus of elasticity { 'yəŋz ,mäj·ə·ləs }

Young's two-slit interference [OPTICS] Interference of light from two parallel slits which are illuminated by light from a single slit, which in turn is illuminated by a source; the interference can be seen by letting the light fall on a screen, which then shows a series of parallel fringes. Also known as double-slit interference; two-slit interference. { 'yəŋz ¦tü ¦slit ,in·tər'fir·əns }

youth [GEOL] The first stage of the cycle of erosion in which the original surface or structure is the dominant topographic feature; characterized by broad, flat-topped interstream divides, numerous swamps and shallow lakes, and progressive increase of local relief. Also known as topographic youth. { yüth }

y parameter [ELECTR] One of a set of four transistor equivalent-circuit parameters, used especially with field-effect transistors, that conveniently specify performance for small voltage and current in an equivalent circuit; the equivalent circuit is a current source with shunt impedance at both input and output. { 'wī pə,ram·əd·ər }

Yponomeutidae [INV ZOO] A heterogeneous family of small, often brightly colored moths in the superfamily Tineoidea; the head is usually smooth with reduced or absent ocelli. { ē,pän·ə'myüd·ə,dē }

y-process [NUC PHYS] The synthesis of certain nuclides in stars through nuclear reactions in which gamma rays remove neutrons, protons, or alpha particles from the nucleus; it is one mode of the *p*-process. { 'wī ,prä·sós }

Ypsilothuriidae [INV ZOO] A family of echinoderms in the

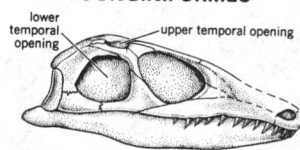

YOUNGINIFORMES

Lateral view of skull of *Youngina*, a representative of Younginiformes, Upper Permian to Lower Triassic. (*From A. S. Romer, Vertebrate Paleontology, 3d ed., University of Chicago Press, 1966*)

order Dactylochirotida having 8–10 tentacles, a permanent spire on the plates of the test, and the body fusiform or U-shaped. { ˌip·sə·lōˈthùr·ə,dē }

yr *See* year.

yrast state [NUC PHYS] An energy state of a nucleus whose energy is less than that of any other state with the same spin. { ēˈrast ˌstāt }

Y search [NAV] A sector search with one of the legs modified to provide for coverage of the interior of the sector. { ˈwī ˌsərch }

Y signal *See* luminance signal. { ˈwī ˌsig·nəl }

ytterbia *See* ytterbium oxide. { iˈtər·bē·ə }

ytterbium [CHEM] A rare-earth metal of the yttrium sub-group, symbol Yb, atomic number 70, atomic weight 173.04; lustrous, malleable, soluble in dilute acids and liquid ammonia, reacts slowly with water; melts at 824°C, boils at 1427°C; used in chemical research, lasers, garnet doping, and x-ray tubes. { iˈtər·bē·əm }

ytterbium oxide [INORG CHEM] Yb_2O_3 A colorless compound, melts at 2346°C, dissolves in hot dilute acids; used to prepare alloys, ceramics, and special glasses. Also known as ytterbia. { iˈtər·bē·əm ˈäk,sīd }

yttria *See* yttrium oxide. { ˈi·trē·ə }

yttrium [CHEM] A rare-earth metal, symbol Y, atomic number 39, atomic weight 88.9059; dark-gray, flammable (as powder), soluble in dilute acids and potassium hydroxide solution, and decomposes in water; melts at 1500°C, boils at 2927°C; used in alloys and nuclear technology and as a metal deoxidizer. { ˈi·trē·əm }

yttrium acetate [ORG CHEM] $Y(C_2H_3O_2)_3 \cdot 8H_2O$ Colorless, water-soluble crystals used as an analytical reagent. { ˈi·trē·əm ˈas·ə,tāt }

yttrium-aluminum-garnet laser [OPTICS] A four-level infrared laser in which the active material is neodymium ions in an yttrium-aluminum-garnet crystal; it can provide a continuous output power of several watts. Abbreviated yag laser. { ˈi·trē·əm əˈlü·mə·nəm ˈgär·nət ˈlā·zər }

yttrium chloride [INORG CHEM] $YCl_3 \cdot 6H_2O$ Reddish, transparent, water- and alcohol-soluble prisms; decomposes at 100°C; used as an analytical reagent. { ˈi·trē·əm ˈklȯr,īd }

yttrium iron garnet [MATER] $Y_3Fe_5O_{12}$ A synthetic ferri-magnetic material with the garnet crystal structure; used in microwave ferrite devices because of its very narrow ferromag-netic resonance absorption line. Abbreviated yig. { ˈi·trē·əm ˈī·ərn ˈgär·nət }

yttrium-iron-garnet device *See* yig device. { ˈi·trē·əm ˈī·ərn ˈgär·nət di,vīs }

yttrium oxide [INORG CHEM] Y_2O_3 A yellowish powder, insoluble in water, soluble in dilute acids; used as television tube phosphor and microwave filters. Also known as yttria. { ˈi·trē·əm ˈäk,sīd }

yttrium sulfate [INORG CHEM] $Y_2(SO_4)_3 \cdot 8H_2O$ Reddish crystals that are soluble in concentrated sulfuric acid, slightly soluble in water; decomposes at 700°C; used as an analytical reagent. { ˈi·trē·əm ˈsəl,fāt }

yttrocrasite [MINERAL] $(Y,Th,U,Ca)_2Ti_4O_{11}$ A black, orthorhombic mineral consisting of an oxide of rare earths and titanium. { ˌi·trəˈkrā,sīt }

yttrotantalite [MINERAL] $(Y,U,Fe)(Ta,Nb)O_4$ A black or brown, orthorhombic mineral consisting of an oxide of iron, yttrium, uranium, columbium, and tantalum; occurs in prismatic and tabular form. { ˌi·trəˈtant·əl,īt }

Yucatán Current [OCEANOGR] A rapid northward flowing current along the western side of the Yucatán Strait; generally loops to the north and exits as the Florida Current. { ˈyü·kəˌtän ˈkə·rənt }

yugawaralite [MINERAL] $CaAl_2Si_6O_{16} \cdot 4H_2O$ A zeolite mineral consisting of hydrous calcium aluminum silicate. { ˌyü·gəˈwär·ə,līt }

Yukawa force [NUC PHYS] The strong, short-range force between nucleons, as calculated on the assumption that this force is due to the exchange of a particle of finite mass (Yukawa meson), just as electrostatic forces are interpreted in quantum electrodynamics as being due to the exchange of photons. { yüˈkä·wä ˌfȯrs }

Yukawa meson [PARTIC PHYS] A particle, having a finite rest mass, whose exchange between nucleons is postulated to account for the strong, short-range forces between them; such a contributor is the pi meson. { yüˈkä·wä ˌmā,sän }

Yukawa potential [NUC PHYS] The potential function that is associated with the Yukawa force, with the form $V(r) = -V_0(b/r) \exp(-r/b)$, where r is the distance between the nucleons and V_0 and b are constants, giving measures of the strength and range of the force respectively. { yüˈkä·wä pə,ten·chəl }

Z

ZAA spectrometry See Zeeman-effect atomic absorption spectrometry. { ¦zē¦ā¦ā spek'träm·ə·trē }

Zalambdalestidae [PALEON] A family of extinct insectivorous mammals belonging to the group Proteutherea; they occur in the Late Cretaceous of Mongolia. { zə,lam·də'les·tə,dē }

ZAMS See zero-age main sequence.

Zanclidae [VERT ZOO] The Moorish idols, a family of Indo-Pacific perciform fishes in the suborder Acanthuroidei. { 'zaŋ·klə,dē }

Zanzibar gum [MATER] A combustible, hard, fossil-type copal, which is insoluble in most solvents and melts at about 245°C; used in varnishes. { 'zan·zə,bär ,gəm }

Zapoididae [VERT ZOO] The Northern Hemisphere jumping mice, a family of the order Rodentia with long legs and large feet adapted for jumping. { zə'pòid·ə,dē }

zaratite [MINERAL] $Ni_3(CO_3)(OH)_4·4H_2O$ An emerald-green mineral consisting of a hydrous basic nickel carbonate and occurring in incrustations or compact masses. { 'zär·ə,tīt }

zastruga See sastruga. { 'zas·trə·gə }

z axis [CRYSTAL] The optical axis of a quartz crystal, perpendicular to both the x and y axes. [MATH] One of the three axes in a three-dimensional Cartesian coordinate system; in a rectangular coordinate system it is perpendicular to the x and y axes. { 'zē ,ak·səs }

Z⁰ boson [PART PHYS] An intermediate vector boson which has zero electric charge and mediates the neutral current weak interactions. Also known as Z^0 particle. { 'zē¦zir·ō 'bō,sän }

Z cam [ASTRON] A representative type of variable star; it is eruptive with a cycle of about 10–600 days; magnitude ranges from 2 to 6. { 'zē ,kam }

Z Camelopardalis stars [ASTRON] A class of dwarf novae which exhibit unpredictable, and sometimes very protracted, pauses in the decline from maximum to minimum brightness. { 'zē kə,mel·ə'pärd·əl·əs ,stärz }

Z chromosome [GEN] The sex chromosome found in both sexes in species with female heterogamety. { 'zē ,krō·mə,sōm }

Z coefficient [QUANT MECH] A coefficient used in the transformation between modes of coupling eigenfunctions of three angular momenta, and especially in calculating matrix elements in beta decay and similar problems. { 'zē ,kō·i,fish·ənt }

z component [MATH] The projection of a vector quantity on the z axis of a coordinate system. { 'zē kəm,pō·nənt }

z coordinate [MATH] One of the coordinates of a point in a three-dimensional coordinate system, equal to the directed distance of a point from the plane of the x and y axes, measured along a line parallel to the z axis. { 'zē kō,órd·ən·ət }

zebra [VERT ZOO] Any of three species of African mammals belonging to the family Equidae distinguished by a coat of black and white stripes. { 'zē·brə }

zebu [VERT ZOO] A domestic breed of cattle, indigenous to India, belonging to the family Bovidae, distinguished by long drooping ears, a dorsal hump between the shoulders, and a dewlap under the neck; known as the Brahman in the United States. { 'zē·bü }

Zechstein [GEOL] A European series of geologic time, especially in Germany; Upper Permian (above Rothliegende). { 'zek,shtīn }

Zeckendorf's theorem [MATH] The theorem that any positive integer can be expressed as a sum of distinct Fibonacci numbers, no two of which are consecutive. { 'zek·ən,dórfs ,thir·əm }

zee [CIV ENG] A metal member whose cross section has a modified Z shape: the internal angles are slightly less than 90°. { zē }

Zeeman displacement [SPECT] The separation, in wave numbers, of adjacent spectral lines in the normal Zeeman effect in a unit magnetic field, equal (in centimeter-gram-second Gaussian units) to $e/4\pi mc^2$, where e and m are the charge and mass of the electron, or to approximately 4.67×10^{-5} (centimeter)$^{-1}$(gauss)$^{-1}$. { 'zā·mən di,splās·mənt }

Zeeman effect [SPECT] A splitting of spectral lines in the radiation emitted by atoms or molecules in a static magnetic field. { 'zā·mən ,fekt }

Zeeman-effect atomic absorption spectrometry [SPECT] A type of atomic absorption spectrometry in which either the light source or the sample is placed in a magnetic field, splitting the spectral lines under observation into polarized components, and a rotating polarizer is placed between the source and the sample, enabling the absorption caused by the element under analysis to be separated from background absorption. Abbreviated ZAA spectrometry. { 'zē·mən i,fekt ə'täm·ik əp'sórp·shən spek'träm·ə·trē }

Zeeman energy [ATOM PHYS] The energy of interaction between an atomic or molecular magnetic moment and an applied magnetic field. { 'zā·mən ,en·ər·jē }

Zeiformes [VERT ZOO] The dories, a small order of teleost fishes, distinguished by the absence of an orbitosphenoid bone, a spinous dorsal fin, and a pelvic fin with a spine and five to nine soft rays. { ,zē·ə'fór,mēz }

zein [MATER] A combustible, white to yellowish protein powder derived from corn; insoluble in water, soluble in dilute alcohol; used in inks, fibers, microencapsulation, and coatings for paper and food. { 'zē·ən }

Zeitgeber [PHYSIO] A periodic environmental condition or event that acts to set or reset an innate biological rhythm of an organism. { 'tsīt,gā·bər }

Zemorrian [GEOL] A North American stage of Oligocene and Miocene geologic time, above Refugian and below Saucesian. { zə'mór·ē·ən }

Zener breakdown [ELECTR] Nondestructive breakdown in a semiconductor, occurring when the electric field across the barrier region becomes high enough to produce a form of field emission that suddenly increases the number of carriers in this region. Also known as Zener effect. { 'zē·nər 'brāk,daùn }

Zener diode [ELECTR] A semiconductor breakdown diode, usually constructed of silicon, in which reverse-voltage breakdown is based on the Zener effect. { 'zē·nər 'dī,ōd }

Zener diode voltage regulator See diode voltage regulator. { 'zē·nər 'dī,ōd 'vōl·tij ,reg·yə,lād·ər }

Zener effect See Zener breakdown.

Zener voltage See breakdown voltage. { 'zē·nər ,vōl·tij }

zenith [ASTRON] That point of the celestial sphere vertically overhead. { 'zē·rəth }

zenithal chart See azimuthal chart. { 'zē·nə·thəl 'chärt }

zenithal hourly rate [ASTRON] The number of meteors in a meteor shower which would be observed per hour if the radiant of the meteor shower were overhead and there were no moonlight. { 'zē·nə·thəl 'aùr·lē 'rāt }

zenithal rain [METEOROL] In the tropics or subtropics, the rainy season which recurs annually or semiannually at about the time that the sun is most nearly overhead (at zenith). { 'zē·nə·thəl 'rān }

ZEBRA

Grevy's zebra (*Equus grevyi*).

ZEBU

Zebu, or Brahman, a domestic breed of cattle.

ZEIFORMES

John dory (*Zenopsis ocellata*).

zenith angle [ASTRON] The angle between the direction to the zenith and the direction of a light ray. { 'zē·nəth ˌaŋ·gəl }

zenith distance [ASTRON] Angular distance from the zenith; the arc of a vertical circle between the zenith and a point on the celestial sphere, measured from the zenith through 90°, for bodies above the horizon. Also known as co-altitude. { 'zē·nəth ˌdis·təns }

zenith telescope [OPTICS] A type of telescope that is fixed in the vertical or moves only a small amount from the vertical; it is used to get positional measurement of stars moving near the zenith. { 'zē·nəth ˌtel·ə,skōp }

zenocentric coordinates [ASTRON] Coordinates that indicate the position of a point on the surface of Jupiter, determined by the direction of a line joining the center of Jupiter to the point. { ¦zēn·ə¦sen·trik kō'órd·ən·əts }

zenodiagnosis [MED] A procedure of using a suitable arthropod to transfer an infectious agent from a patient to a susceptible laboratory animal. { ¦zē·nō,dī·əg'nō·səs }

zenographic coordinates [ASTRON] Coordinates that indicate the position of a point on the surface of Jupiter, determined by the direction of a line perpendicular to the mean surface at the point. { ¦zēn·ə¦graf·ik kō'órd·ən·əts }

Zeno's paradox [MATH] An erroneous group of paradoxes dealing with motion; the most famous one concerns two objects, one chasing the other which has a given head start, where the chasing one moves faster yet seemingly never catches the other. { 'zē·nōz 'par·ə,däks }

Zeoidea [VERT ZOO] An equivalent name for Zeiformes. { zē'óid·ē·ə }

zeolite [MINERAL] **1.** A group of white or colorless, sometimes red or yellow, hydrous tectosilicate minerals characterized by an aluminosilicate tetrahedral framework, ion-exchangeable large cations, and loosely held water molecules permitting reversible dehydration. **2.** Any mineral of the zeolite group, such as analcime, chabazite, natrolite, and stilbite. { 'zē·ə,līt }

zeolite catalyst [INORG CHEM] Hydrated aluminum and calcium (or sodium) silicates (for example, $CaO·2Al_2O_3·5SiO_2$ or $Na_2O·2Al_2O_3·5SiO_2$) made with controlled porosity; used as a catalytic cracking catalyst in petroleum refineries, or loaded with catalyst for other chemical reactions. { 'zē·ə,līt 'kad·əl·əst }

zeolite facies [PETR] Metamorphic rocks formed in the transitional period from diagenesis to metamorphism, at pressures of about 2000–3000 bars and temperatures of 200–300°C. { 'zē·ə,līt 'fā·shēz }

zeolitization [GEOL] Introduction of or replacement by a zeolite mineral. { zē,äl·əd·ə'zā·shən }

Zeomorphi [VERT ZOO] An equivalent name for Zeiformes. { ,zē·ə'mór·fī }

zeotrope [PHYS CHEM] A nonazeotropic liquid mixture which may be separated by distillation, and in which the components are miscible in all proportions (homogeneous zeotrope or homozeotrope) or not miscible in all proportions (heterogeneous zeotrope or heterozeotrope). { 'zē·ə,trōp }

zephyr [METEOROL] Any soft, gentle breeze. { 'zef·ər }

Zepp antenna [ELECTROMAG] Horizontal antenna which is a multiple of a half-wavelength long and is fed at one end by one lead of a two-wire transmission line that is some multiple of a quarter-wavelength long. { 'zep an,ten·ə }

Zerewitinoff reagent [ANALY CHEM] A light-colored methylmagnesium iodide-*n*-butyl ether solution that reacts rapidly with moisture and oxygen; used to determine water, alcohols, and amines in inert solvents. { ,zir·ə'wit·ən,óf rē,ā·jənt }

zero [MATH] **1.** The additive identity element of an algebraic system. **2.** Any point where a given function assumes the value zero. { 'zir·ō }

zero-access instruction [COMPUT SCI] An instruction consisting of an operation which does not require the designation of an address in the usual sense; for example, the instruction, "shift left 0003," has in its normal address position the amount of the shift desired. { 'zir·ō ¦ak,ses in'strək·shən }

zero-access storage [COMPUT SCI] Computer storage for which waiting time is negligible. { 'zir·ō ¦ak,ses 'stór·ij }

zero-address instruction format [COMPUT SCI] An instruction format in which the instruction contains no address, used when an address is not needed to specify the location of the operand, as in repetitive addressing. Also known as addressless instruction format. { 'zir·ō ¦ad,res in¦strək·shən ,fór,mat }

zero adjuster [ENG] A device for adjusting the pointer position of an instrument or meter to read zero when the measured quantity is zero. { 'zir·ō ə,jəs·tər }

zero-age main sequence [ASTRON] The position on the Hertzsprung-Russell diagram of a star that has just reached the main sequence, when it has reached hydrostatic equilibrium and thermonuclear reactions have begun in its core, but these reactions have not had time to produce an appreciable change in composition. Abbreviated ZAMS. { 'zir·ō ¦āj 'mān 'sē·kwəns }

zero beat [ELEC] The condition in which a circuit is oscillating at the exact frequency of an input signal, so no beat tone is produced or heard. { 'zir·ō ,bēt }

zero-beat reception See homodyne reception. { 'zir·ō ¦bēt ri'sep·shən }

zero bevel gear [DES ENG] A special form of bevel gear having curved teeth with a zero-degree spiral angle. { 'zir·ō ¦bev·əl 'gir }

zero bias [ELECTR] The condition in which the control grid and cathode of an electron tube are at the same direct-current voltage. { 'zir·ō 'bī·əs }

zero-bias tube [ELECTR] Vacuum tube which is designed so that it may be operated as a class B amplifier without applying a negative bias to its control grid. { 'zir·ō ¦bī·əs 'tüb }

zero branch [SPECT] A spectral band whose Fortrat parabola lies between two other Fortrat parabolas, with its vertex almost on the wave number axis. { 'zir·ō 'branch }

zero compression [COMPUT SCI] Any of a number of techniques used to eliminate the storage of nonsignificant leading zeros during data processing in a computer. { 'zir·ō kəm'presh·ən }

zero condition [COMPUT SCI] The state of a magnetic core or other computer memory element in which it represents the value 0. Also known as nought state; zero state. { 'zir·ō kən,dish·ən }

zero curtain [GEOL] The layer of ground between the active layer and permafrost where the temperature remains nearly constant at 0°C. { 'zir·ō ,kərt·ən }

zero defects [IND ENG] A program for improving product quality to the point of perfection, so there will be no failures due to defects in construction. { 'zir·ō 'dē,feks }

zero deflection [ORD] Adjustment of a sight exactly parallel to the axis of the bore of the gun to which it is attached. { 'zir·ō di'flek·shən }

zero divisor See divisor of zero. { 'zir·ō di'vīz·ər }

zero error [ELECTR] Delay time occurring within the transmitter and receiver circuits of a radar system; for accurate range data, this delay time must be compensated for in the calibration of the range unit. { 'zir·ō 'er·ər }

zero-field emission See field-free emission current. { 'zir·ō ¦fēld i'mish·ən }

zero fill [COMPUT SCI] To place leading zeros in the portion of a field to the left of a numeric value. { 'zir·ō ,fil }

zero flag [COMPUT SCI] A bit in a status register that is set to 1 to indicate that another register in the central processing unit contains all zeros or that two compared values are equal, and is set to 0 to indicate the contrary. { 'zir·ō ,flag }

zerogel [CHEM] A gel which has dried until apparently solid; sometimes it will swell or redisperse to form a sol when treated with a suitable solvent. { 'zir·ō,jel }

zero geodesic See null geodesic. { 'zir·ō jē·ə'des·ik }

zero gravity See weightlessness. { 'zir·ō 'grav·əd·ē }

zero-gravity switch [ELEC] A switch that closes as weightlessness or zero gravity is approached; in one version, a conductive sphere of mercury encompasses two contacts at zero gravity but flattens away from the upper contact under the influence of gravity. Also known as weightlessness switch. { 'zir·ō ¦grav·əd·ē ,swich }

zero height of burst [ORD] Condition obtained when rounds fired with the same fuse setting and the same quadrant elevation result in an equal number of airbursts and bursting of the projectile at the instant of ground impact. { 'zir·ō 'hīt əv 'bərst }

zero in [ORD] **1.** To adjust the sight settings of a weapon by calibrated results of firings. **2.** To adjust any device to another so that automatic synchronization results. { 'zir·ō 'in }

zero layer [OCEANOGR] A reference level in the ocean, at which horizontal motion is at a minimum. { 'zir·ō ,lā·ər }

zero length [AERO ENG] In rocket launchers, zero length indicates that the launcher is designed to hold the rocket in position for launching but not to give it guidance. { 'zir·ō 'leŋkth }

zero-length launcher [ORD] A launcher that holds a vehicle in position and releases the rocket simultaneously at two points so that the buildup of thrust, normally rocket thrust, is sufficient to take the missle or vehicle directly into the air without need of a takeoff run and without imposing a pitch rate release. { 'zir·ō ¦leŋkth 'lónch·ər }

zero-length rocket [ORD] A rocket with a sufficient thrust to launch a vehicle directly into the air. { 'zir·ō ¦leŋkth 'räk·ət }

zero level [ENG ACOUS] Reference level used for comparing sound or signal intensities; in audio-frequency work, a power of 0.006 watt is generally used as zero level; in sound, the threshold of hearing is generally assumed as the zero level. { 'zir·ō ¦lev·əl }

zero-level address [COMPUT SCI] The operand contained in an instruction so structured as to make immediate use of the operand. { 'zir·ō ¦lev·əl 'ad,res }

zero-lift angle [AERO ENG] The angle of attack of an airfoil when its lift is zero. { 'zir·ō ¦lift 'aŋ·gəl }

zero-lift chord [AERO ENG] A chord taken through the trailing edge of an airfoil in the direction of the relative wind when the airfoil is at zero-lift angle of attack. { 'zir·ō ¦lift 'kórd }

zero method See null method. { 'zir·ō ¦meth·əd }

zero-order hold [CONT SYS] A device which converts a sampled output into an output which is held constant between samples at the last sampled value. { 'zir·ō ¦órd·ər 'hōld }

zero-order reaction [PHYS CHEM] A reaction for which reaction rate is independent of the concentrations of the reactants; for example, a photochemical reaction in which the rate is determined by the intensity of light. { 'zir·ō ¦órd·ər rē'ak·shən }

zero output [ELECTR] **1.** Voltage response obtained from a magnetic cell in a zero state by a reading or resetting process. **2.** Integrated voltage response obtained from a magnetic cell in a zero state by a reading or resetting process; a ratio of a one output to a zero output is a one-to-zero ratio. { 'zir·ō 'aut,put }

zero phase-sequence relay [ELEC] Relay which functions in conformance with the zero phase-sequence component of the current, voltage, or power of the circuit. { 'zir·ō ¦fāz ¦sē·kwəns 'rē,lā }

zero point [MATH] A complex number at which an analytic function assumes the value zero. [ORD] The location of the center of a burst of an atomic missile at the instant of detonation; the zero point may be in the air, or on or beneath the surface of land or water, dependent upon the type of burst, and it is thus to be distinguished from ground zero. { 'zir·ō 'póint }

zero-point energy [STAT MECH] The kinetic energy retained by the molecules of a substance at a temperature of absolute zero. { 'zir·ō ¦póint 'en·ər,jē }

zero-point entropy [STAT MECH] The entropy that a substance such as glass, which is not in thermodynamic equilibrium, retains at a temperature of absolute zero. { 'zir·ō ¦póint 'en·trə·pē }

zero-point pressure [STAT MECH] The pressure exerted by a degenerate electron gas, whose electrons are in motion even at absolute zero. { 'zir·ō ¦póint 'presh·ər }

zero-point vibration [STAT MECH] The vibrational motion which molecules in a crystal lattice, or particles in any oscillator potential, retain at a temperature of absolute zero; it is quantum-mechanical in origin. Also known as residual vibration. { 'zir·ō ¦póint vī'brā·shən }

zero population growth [ECOL] A theory which advocates that there be no increase in population, that each person replace only oneself, and that birth control be practiced in all nations. { 'zir·ō 'päp·yə,lā·shən ,grōth }

zero potential [ELEC] Expression usually applied to the potential of the earth, as a convenient reference for comparison. { 'zir·ō pə'ten·chəl }

zero-power reactor [NUCLEO] An experimental nuclear reactor operated at low neutron flux and at a power level so low that no forced cooling is required; fission product activity in the fuel is then sufficiently low to permit handling the fuel after use. { 'zir·ō 'paů·ər rē'ak·tər }

zero state See zero condition. { 'zir·ō ,stāt }

zero subcarrier chromaticity [COMMUN] Chromaticity, in color television, which is intended to be displayed when the subcarrier amplitude is zero. { 'zir·ō səb¦kar·ē·ər ,krō·mə'tis·əd·ē }

zero-sum game [MATH] A two-person game where the sum of the payoffs to the two players is zero for each move. { 'zir·ō ¦səm ,gām }

zero suppression [COMPUT SCI] A process of replacing leading (nonsignificant) zeros in a numeral by blanks; it is an editing operation designed to make computable numerals easily readable to the human eye. { 'zir·ō sə'presh·ən }

zeroth law of thermodynamics [THERMO] A law that if two systems are separately found to be in thermal equilibrium with a third system, the first two systems are in thermal equilibrium with each other, that is, all three systems are at the same temperature. { ¦zir,ōth ¦ló əv ,thər·mō·dī'nam·iks }

zero time reference [ELECTR] Reference point in time from which the operations of various radar circuits are measured. { 'zir·ō 'tīm ,ref·rəns }

zero twist [ORD] Rifling with no twist; some designs have this condition at the origin of rifling for guns with increasing twist rifling. { 'zir·ō 'twist }

zero vector [MATH] The element 0 of a vector space such that, for any vector v in the space, the vector sum of 0 and v is v. { 'zir·ō ,vek·tər }

zeta [ARCH] A small, closed room. { 'zād·ə }

Zeta Aurigae star [ASTRON] A binary system with a supergiant primary in spectral class K and a main-sequence secondary. { 'zād·ə 'ór·ə,gē ,stär }

zeta function See Riemann zeta function. { 'zād·ə ,fəŋk·shən }

Zeta Geminorum stars [ASTRON] A subgroup of classical Cepheid variable stars whose variation of magnitude with time for one complete cycle produces a quasi-bell-shaped curve. { 'zād·ə ¦jem·ə'nór·əm ,stärz }

zeta potential [PHYS] The electrical potential that exists across the interface of all solids and liquids. Also known as electrokinetic potential. { 'zād·ə pə,ten·chəl }

zeugmatography [MED] A technique that combines nuclear magnetic resonance spectroscopy with methods of focusing and scanning radio waves to produce cross-sectional images of the human body that indicate proton density and relaxation times, and distinguish normal tissue from tumors. Also known as nuclear magnetic resonance tomography. { ,züg·mə'täg·rə·fē }

zeugogeosyncline [GEOL] A geosyncline in a craton or stable area, within which is also an uplifted area, receiving clastic sediments. Also known as yoked basin. { ¦zü·gō¦jē·ō'sin ,klīn }

zeunerite [MINERAL] $Cu(UO_2)_2(AsO_4)_2 \cdot 10\text{-}16H_2O$ A green secondary mineral of the autunite group consisting of a hydous copper uranium arsenate; it is isomorphous with uranospinite. { 'zói·nə,rīt }

zeylanite See ceylonite. { 'zā·lə,nīt }

zibeline [TEXT] A soft, lustrous woolen fabric with a long nap running in one direction, often containing mohair, alpaca, or camel's hair. { 'zib·ə,lēn }

Ziegler catalyst [MATER] A special catalyst developed to produce stereospecific polymers, and derived from a transition-metal halide and a metal hydride or metal alkyl. { 'zē·glər ,kad·əl·əst }

Ziegler process [CHEM ENG] A process for the low-pressure linear polymerization of ethylene and stereospecific polymerization of propylene; the product is a high-density polymer or elastomer. { 'zē·glər ,präs·əs }

Ziehl-Neelsen stain [MICROBIO] A procedure for acid-fast staining of tubercle bacilli with carbol fuchsin. { 'zēl 'nēl·sən ,stān }

zigzag lightning [GEOPHYS] Ordinary lightning of a cloud-to-ground discharge that appears to have a single, but very irregular, lightning channel; when viewed from a suitable angle, this may be observed as beaded lightning. { 'zig,zag ,līt·niŋ }

zigzag nanotube [PHYS CHEM] A carbon nanotube formed from a graphite sheet that is rolled up so that it has a zigzag edge. { ,zig,zag 'nan·ō,tüb }

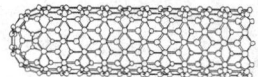

ZIGZAG NANOTUBE

Model of a zigzag nanotube, formed by rolling up a graphite sheet so that it has a zigzag edge.

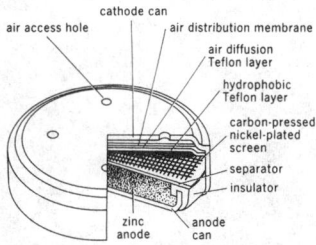

ZINC/AIR CELL

air access hole
cathode can
air distribution membrane
air diffusion Teflon layer
hydrophobic Teflon layer
carbon-pressed nickel-plated screen
separator
insulator
zinc anode
anode can

Major components of zinc/air cell; components are not drawn to scale. *(Duracell Inc.)*

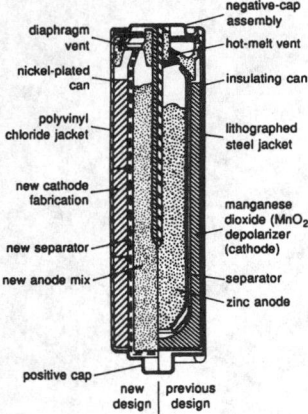

ZINC/ALKALINE/MANGANESE DIOXIDE BATTERY

negative-cap assembly
diaphragm vent
hot-melt vent
nickel-plated can
insulating can
polyvinyl chloride jacket
lithographed steel jacket
new cathode fabrication
manganese dioxide (MnO₂) depolarizer (cathode)
new separator
new anode mix
separator
zinc anode
positive cap
new design | previous design

Cross section of zinc/alkaline/ manganese dioxide cylindrical cell. *(Duracell Inc.)*

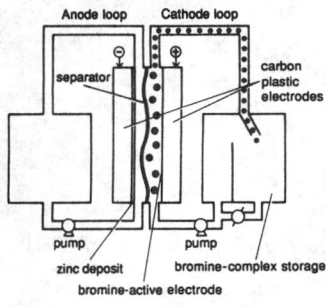

ZINC/BROMINE BATTERY

Anode loop   Cathode loop
separator
carbon plastic electrodes
pump   pump
zinc deposit
bromine-complex storage
bromine-active electrode

Schematic of a zinc/bromine cell. *(Exxon Research and Engineering Co.)*

zigzag reflections [ELECTROMAG] From a layer of the ionosphere, high-order multiple reflections which may be of abnormal intensity; they occur in waves which travel by multihop ionosphere reflections and finally turn back toward their starting point by repeated reflections from a slightly curved or sloping portion of an ionized layer. { 'zig,zag ri,flek·shənz }

zigzag rule [ENG] A folding ruler having pivoted sections that lock when the ruler is opened. { 'zig,zag ,rül }

Zimm plot [ANALY CHEM] A graphical determination of the root-square-mean end-to-end distances of coillike polymer molecules during scattered-light photometric analyses. { 'zim ,plät }

zinc [CHEM] A metallic transition element, symbol Zn, atomic number 30, atomic weight 65.38; explosive as powder; soluble in acids and alkalies, insoluble in water; strongly electropositive; melts at 419°C; boils at 907°C. [MET] A shiny, bluish-white, lustrous metal that is ductile when pure; used in alloys, metal coatings, electrical fuses, anodes, and dry cells. { ziŋk }

zinc-65 [NUC PHYS] A radioactive isotope of zinc, which has a 250-day half-life with beta and gamma radiation; used in alloy-wear tracer studies and body metabolism studies. { 'ziŋk ¦siks·tē'fīv }

zinc acetate [ORG CHEM] $Zn(C_2H_3O_2)_2 \cdot 2H_2O$ Pearly-white crystals with an astringent taste; soluble in water and alcohol; decomposes at 200°C; used to preserve wood in textile dyeing, and as an analytical reagent, a feed additive, and a polymer cross-linking agent. { 'ziŋk 'as·ə,tāt }

zinc/air cell [ELEC] A primary battery with very high energy density that uses atmospheric oxygen as the active cathode material and zinc as the anode; cans containing the anode and cathode are separated by an annular insulator. { ¦ziŋk 'er ,sel }

zinc/alkaline/manganese dioxide battery [ELEC] The most widely used primary battery, with an anode consisting of zinc powder, an electrolyte consisting of an aqueous solution of potassium hydroxide, and a cathode consisting of a blend of electrolytic manganese dioxide and graphite. { ¦ziŋk ¦al·kə,līn ¦maŋ·gə,nēs dī'äk,sīd ,bad·ə·rē }

zincaluminite [MINERAL] $Zn_6Al_6(SO_4)_2(OH)_{26} \cdot 5H_2O$ A white to bluish-white and pale blue mineral consisting of a basic hydrated sulfate of zinc and aluminum; occurs in tufts and crusts. { ¦ziŋk·ə¦lü·mə,nīt }

zinc arsenite [INORG CHEM] $Zn(AsO_2)_2$ A toxic white powder that is insoluble in water, soluble in alkalies; used as an insecticide and timber preservative. Also known as zinc meta-arsenite. { 'ziŋk 'ärs·ən,īt }

zincate [INORG CHEM] A reaction product of zinc with an alkali metal or with ammonia; for example, sodium zincate, Na_2ZnO_2. { 'ziŋ,kāt }

zinc baryta white See lithopone. { 'ziŋk bə¦rīd·ə 'wīt }

zinc borate [INORG CHEM] $3ZnO \cdot 2B_2O_3$ A white, amorphous powder that is soluble in dilute acids, slightly soluble in water; melts at 980°C; used in medicine, as a ceramics flux, as an inhibitor for mildew, and to fireproof textiles. { 'ziŋk 'bor,āt }

zinc bromide [INORG CHEM] $ZnBr_2$ Water- and alcohol-soluble, white crystals that melt at 294°C; used in medicine, manufacture of rayon, and photography, and in a radiation viewing screen. { 'ziŋk 'brō,mīd }

zinc/bromine battery [ELEC] A storage battery that is based on the reaction of zinc (Zn) and bromine (Br₂) to form zinc bromide (ZnBr₂), and that uses liquid electrolytes which circulate through stacks of bipolar cells. { ¦ziŋk ¦brō,mēn 'bad·ə·rē }

zinc carbonate [INORG CHEM] $ZnCO_3$ White crystals that are insoluble in water, soluble in alkalies and acids; used in ceramics and ointments, and as a fireproofing agent and feed additive. { 'ziŋk 'kär·bə,nāt }

zinc chills See metal fume fever. { 'ziŋk ,chilz }

zinc chloride [INORG CHEM] $ZnCl_2$ Water- and alcohol-soluble, white, fire-hazardous crystals that melt at 290°C, and are irritating to the skin; used as a catalyst and in electroplating, wood preservation, textile processing, petroleum refining, medicine, and feed additives. { 'ziŋk 'klor,īd }

zinc chromate [INORG CHEM] $ZnCrO_4$ A toxic, yellow powder that is insoluble in water, soluble in acids; used as a pigment in paints (artists', automotive, primer), varnishes, linoleum, and epoxy laminates. { 'ziŋk 'krō,māt }

zinc cyanide [INORG CHEM] $Zn(CN)_2$ A toxic, white powder that is insoluble in water and alcohol, soluble in alkalies and dilute acids; melts at 800°C; used as an analytical reagent and insecticide, and in medicine and metal plating. { 'ziŋk 'sī·ə,nīd }

zinc dust [MATER] Finely divided zinc metal, at least 97% pure, that is used as a pigment in paint primer for galvanized iron and other metal substrates. { 'ziŋk ,dəst }

zinc finger [BIOCHEM] A small structural domain that is organized around a zinc ion and is found in many gene-regulatory proteins. { 'ziŋk ¦fiŋ·gər }

zinc fluoride [INORG CHEM] ZnF_2 A toxic white powder that is slightly soluble in water and melts at 872°C; used in enamels, ceramic glazes, and galvanizing. { 'ziŋk 'flur,īd }

zinc formate [ORG CHEM] $Zn(CHO_2)_2 \cdot 2H_2O$ Toxic, white crystals that are soluble in water, insoluble in alcohol; used as a catalyst, weatherproofing agent, and wood preservative. { 'ziŋk 'for,māt }

zinc halide [INORG CHEM] A binary compound of zinc and a halogen; for example, ZnBr₂, ZnCl₂, ZnF₂, and ZnI₂. { 'ziŋk 'ha,līd }

zinc hydroxide [INORG CHEM] $Zn(OH)_2$ Colorless, water-soluble crystals that decompose at 125°C; used as a chemical intermediate and in rubber compounding and surgical dressings. { 'ziŋk hī'dräk,sīd }

zincite [MINERAL] (Zn,Mn)O A deep-red to orange-yellow brittle mineral; an ore of zinc. Also known as red oxide of zinc; red zinc ore; ruby zinc; spartalite. { 'ziŋ,kīt }

zinckenite See zinkenite. { 'ziŋ·kə,nīt }

zinc metaarsenite See zinc arsenite. { 'ziŋk ¦med·ə'ärs·ən,āt }

zinc naphthenate [ORG CHEM] $Zn(C_6H_5COO)_2$ A combustible, viscous, acetone-soluble solid; used in paints, varnishes, and resins, and as a drier and wetting agent, insecticide, fungicide, and mildewstat. { 'ziŋk 'naf·thə,nāt }

zinc orthoarsenate [INORG CHEM] $Zn_3(AsO_4)_2$ A toxic white powder that is insoluble in water, soluble in alkalies; used as an insecticide and wood preservative. { 'ziŋk ¦or·thō'ärs·ən,āt }

zinc orthophosphate See zinc phosphate. { 'ziŋk ¦or·thō'fä,sfāt }

zinc oxide [INORG CHEM] ZnO A bitter-tasting, white to gray powder that is insoluble in water, soluble in alkalies and acids; melts at 1978°C; used as a pigment, mold-growth inhibitor, and dietary supplement, and in cosmetics, electronics, and color photography. { 'ziŋk 'äk,sīd }

zinc phosphate [INORG CHEM] $Zn_3(PO_4)_2$ A white powder that is insoluble in water, soluble in acids and ammonium hydroxide; melts at 900°C; used in coatings for steel, aluminum, and other metals, and in dental cements and phosphors. Also known as tribasic zinc phosphate; zinc orthophosphate. { 'ziŋk 'fä,sfāt }

zinc phosphide [INORG CHEM] Zn_3P_2 A toxic, alcohol-insoluble, gray gritty powder that reacts violently with oxidizing agents; melts at over 420°C, decomposes in water; used as a rat poison and in medicine. { 'ziŋk 'fä,sfīd }

zinc selenide [INORG CHEM] ZnSe A water-insoluble, moderately toxic, yellow to reddish solid that is a fire hazard when in contact with water and acids; melts above 1100°C; used as infrared optical windows. { 'ziŋk 'sel·ə,nīd }

zinc-silver chloride primary cell [ELEC] A reserve primary cell that is activated by adding water; it can have a high capacity, up to 40 watt-hours per pound, and long life after activation. { 'ziŋk 'sil·vər ¦klor,īd 'prī,mer·ē 'sel }

zinc spar See smithsonite. { 'ziŋk 'spär }

zinc spinel See gahnite. { 'ziŋk spə'nel }

zinc sulfate [INORG CHEM] $ZnSO_4 \cdot 7H_2O$ Efflorescent, water-soluble, colorless crystals with an astringent taste; used to preserve skins and wood and as a paper bleach, analytical reagent, feed additive, and fungicide. Also known as white copperas; white vitriol; zinc vitriol. { 'ziŋk 'səl,fāt }

zinc sulfide [INORG CHEM] ZnS A yellowish powder that is insoluble in water, soluble in acids; exists in two crystalline forms (alpha, or wurtzite, and beta, or sphalerite); beta becomes alpha at 1020°C, and sublimes at 1180°C; used as a pigment for paints and linoleum, in opaque glass, rubber, and plastics, for hydrosulfite dyeing process, as x-ray and television screen phosphor, and as a fungicide. { 'ziŋk 'səl,fīd }

zinc sulfide white See lithopone. { 'ziŋk 'səl,fīd 'wīt }

zinc telluride [INORG CHEM] ZnTe Moderately toxic, reddish crystals that melt at 1238°C and decompose in water. { 'ziŋk 'tel·yə,rīd }

zinc vitriol See zinc sulfate. { 'ziŋk 'vi·trē,ȯl }

zinc white See Chinese white. { 'ziŋk 'wīt }

Zingiberaceae [BOT] A family of aromatic monocotyledonous plants in the order Zingiberales characterized by one functional stamen with two pollen sacs, distichously arranged leaves and bracts, and abundant oil cells. { ,zin·jə·bə'rās·ē,ē }

Zingiberales [BOT] An order of monocotyledonous herbs or scarcely branched shrubs in the subclass Commelinidae characterized by pinnately veined leaves and irregular flowers that have well-differentiated sepals and petals, an inferior ovary, and either one or five functional stamens. { ,zin·jə·bə'rā·lēz }

zinkenite [MINERAL] $Pb_6Sb_{14}S_{27}$ A steel-gray orthorhombic mineral consisting of a lead antimony sulfide and occurring in crystals and in masses; has metallic luster, hardness of 3–3.5 on Mohs scale, and specific gravity of 5.30–5.35. Also spelled zinckenite. { 'ziŋ·kə,nīt }

zinnwaldite [MINERAL] $K_2(Li,Fe,Al)_6(Si,Al)_8O_{20}(OH,F)_4$ A pale-violet, yellowish, brown, or dark-gray mica mineral; an iron-bearing variety of lepidolite; the characteristic mica of greisens. { 'tsin,väl,dīt }

zip [COMPUT SCI] Open standard for file compression and decompression used with personal computers. { zip }

zip mode [COMPUT SCI] Mode of operation of a plotter in which each input plot command represents a velocity increment and causes an increase or decrease in speed relative to either axis or both axes. { 'zip ,mōd }

zippeite [MINERAL] $(UO_2)_2(SO_4)(OH)_2·nH_2O$ An orange-yellow, orthorhombic mineral consisting of a hydrated basic sulfate of uranium. { 'tsip·ə,ī }

zipper [ENG] A generic name for slide fasteners in which two sets of interlocking teeth of the same design provide sturdy and continuous closure for adjacent pieces of textile, leather, and other materials. { 'zip·ər }

zipper conveyor [MECH ENG] A type of conveyor belt with zipperlike teeth that mesh to form a closed tube; used to handle fragile materials. { 'zip·ər kən,vā·ər }

zircaloy [MET] Any member of a group of alloys containing mainly zirconium that possess resistance to corrosion and stability over a wide range of temperatures and types of radiation. { 'zərk·ə,lȯi }

zircon [MINERAL] $ZrSiO_4$ A brown, green, pale-blue, red, orange, golden-yellow, grayish, or colorless neosilicate mineral occurring in tetragonal prisms; it is the chief source of zirconium; the colorless varieties provide brilliant gemstones. Also known as hyacinth; jacinth; zirconite. { 'zər,kän }

zirconia See zirconium oxide. { ,zər'kō·nē·ə }

zirconia brick [MATER] A type of brick containing zirconium oxide, used to line metallurgical furnaces. { ,zər'kō·nē·ə 'brik }

zirconic anhydride See zirconium oxide. { ,zər'kän·ik an'hī,drīd }

zirconite See zircon. { 'zər·kə,nīt }

zirconium [CHEM] A metallic transition element, symbol Zr, atomic number 40, atomic weight 91.22; occurs as crystals, flammable as powder; insoluble in water, soluble in hot, concentrated acids; melts at 1850°C, boils at 4377°C. [MET] A hard, lustrous, grayish metal that is strong and ductile; used in alloys, pyrotechnics, welding fluxes, and explosives. { ,zər'kō·nē·əm }

zirconium-95 [NUC PHYS] A radioactive isotope of zirconium; half-life of 63 days with beta and gamma radiation; used to trace petroleum-pipeline flows and in the circulation of a catalyst in a cracking plant. { ,zər'kō·nē·əm ,nīn·tē'fīv }

zirconium boride [INORG CHEM] ZrB_2 A hard, toxic, gray powder that melts at 3000°C; used as an aerospace refractory, in cutting tools, and to protect thermocouple tubes. Also known as zirconium diboride. { ,zər'kō·nē·əm 'bȯr,īd }

zirconium carbide [INORG CHEM] ZrC Hard, gray crystals that are soluble in water, soluble in acids; as powder, it ignites spontaneously in air; melts at 3400°C, boils at 5100°C; used as an abrasive, refractory, and metal cladding, and in cermets, incandescent filaments, and cutting tools. { ,zər'kō·nē·əm 'kär,bīd }

zirconium chloride See zirconium tetrachloride. { ,zər'kō·nē·əm 'klȯr,īd }

zirconium diboride See zirconium boride. { ,zər'kō·nē·əm dī'bȯr,īd }

zirconium dioxide See zirconium oxide. { ,zər'kō·nē·əm dī'äk,sīd }

zirconium halide [INORG CHEM] A compound of zirconium with a halogen; for example, $ZrBr_2$, $ZrCl_2$, $ZrCl_3$, $ZrCl_4$, $ZrBr_2$, $ZrBr_3$, ZrF_4, and ZrI_4. { ,zər'kō·nē·əm 'ha,līd }

zirconium hydride [INORG CHEM] ZrH_2 A flammable, gray-black powder; used in powder metallurgy and nuclear moderators, and as a reducing agent, vacuum-tube getter, and metal-foaming agent. { ,zər'kō·nē·əm 'hī,drīd }

zirconium hydroxide [INORG CHEM] $Zr(OH)_4$ A toxic, amorphous white powder; insoluble in water, soluble in dilute mineral acids; decomposes at 550°C; used in pigments, glass, and dyes, and to make zirconium compounds. { ,zər'kō·nē·əm hī'dräk,sīd }

zirconium lamp [ELECTR] A high-intensity point-source lamp having a zirconium oxide cathode in an argon-filled bulb, used because of its low emanation of long-wavelength light and its concentrated source. { ,zər'kō·nē·əm 'lamp }

zirconium nitride [INORG CHEM] ZrN A hard, brassy powder that is soluble in concentrated acids; melts at 2930°C; used in refractories, cermets, and laboratory crucibles. { ,zər'kō·nē·əm 'nī,trīd }

zirconium orthophosphate See zirconium phosphate. { ,zər'kō·nē·əm ,ȯr·thō'fä,sfāt }

zirconium oxide [INORG CHEM] ZrO_2 A toxic, heavy white powder that is insoluble in water, soluble in mineral acids; melts at 2700°C; used in ceramic glazes, special glasses, and medicine, and to make piezoelectric crystals. Also known as zirconia; zirconic anhydride; zirconium dioxide. { ,zər'kō·nē·əm 'äk,sīd }

zirconium oxide-based oxygen transducer [ENG] A device in which the concentration of oxygen in a mixture of gases is determined from the diffusion voltage across a heated, suitably doped zirconium oxide material placed between this mixture and a reference gas. { zər'kōn·ē·əm 'äk,sīd ,bāst ¦äks·ə·jən tranz'düs·ər }

zirconium oxychloride [INORG CHEM] $ZrOCl_2·8H_2O$ White crystals that are soluble in water, insoluble in organic solvents, and acidic in aqueous solution; used for textile dyeing and oil-field acidizing, in cosmetics and greases, and for antiperspirants and water repellents. Also known as zirconyl chloride. { ,zər'kō·nē·əm ¦äk·sē'klȯr,īd }

zirconium phosphate [INORG CHEM] $ZrO(H_2PO_4)_2·3H_2O$ A toxic, dense white powder that is insoluble in water, soluble in acids and organic solvents; decomposes on heating; used as an analytical reagent, coagulant, and radioactive-phosphor carrier. Also known as zirconium orthophosphate. { ,zər'kō·nē·əm 'fä,sfät }

zirconium tetrachloride [INORG CHEM] $ZrCl_4$ Toxic, alcohol-soluble, white lustrous crystals; sublimes above 300°C and decomposes in water; used to make pure zirconium and for water-repellent textiles and as an analytical reagent. Also known as zirconium chloride. { ,zər'kō·nē·əm ¦te·trə'klȯr,īd }

zircon sand [MATER] A refractory sand consisting principally of zirconium silicate and characterized by low thermal expansion and high thermal conductivity. { 'zər,kän 'sand }

zirconyl chloride See zirconium oxychloride. { 'zər·kən·əl 'klȯr,īd }

zirkelite [MINERAL] A black mineral consisting of an oxide of zirconium, titanium, calcium, ferrous iron, thorium, uranium, and rare earths. { 'zər·kə,līt }

zitterbewegung [QUANT MECH] An oscillatory motion of an electron suggested in some interpretations of the Dirac electron theory, having a frequency greater than $4\pi mc^2/h$, where m is the electron's mass, c is the speed of light, and h is Planck's constant, or approximately 1.5×10^{21} hertz. { ¦tsid·ər·be'vā,gùŋ }

Ziv-Lempel compression [COMPUT SCI] A data compression technique in which data is represented by a sequence of numbers standing for the positions of character strings in a dictionary; this dictionary initially contains every character in the alphabet and is continually enlarged by forming new strings from the string just compressed and the upcoming character in the text. { 'ziv 'lem·pəl kəm'presh·ən }

Z line [HISTOL] The line formed by attachment of the actin filaments between two sarcomeres. { 'zē ,līn }

ZOANTHIDEA

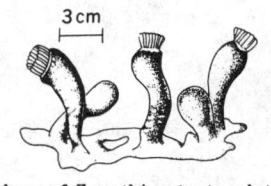

3 cm

Colony of *Zoanthina tentaculata*.

ZOEA

0.5 mm

Zoea larva of crab, showing characteristic features. (*Smithsonian Institution*)

Z-marker beacon [NAV] Transmitter equipment installed as part of a four-course radio range; it radiates vertically to indicate to aircraft when they pass directly over the range station; it is usually not keyed for identification. { 'zē 'märk·ər ‚bē·kən }

Zn *See* zinc.

Zoantharia [INV ZOO] A subclass of the class Anthozoa; individuals are monomorphic and most have retractile, simple, tubular tentacles. { ‚zō·ən'thar·ē·ə }

Zoanthidea [INV ZOO] An order of anthozoans in the subclass Zoantharia; these are mostly colonial, sedentary, skeletonless, anemonelike animals that live in warm, shallow waters and coral reefs. { ‚zō·ən'thid·ē·ə }

Zoarcidae [VERT ZOO] The eelpouts, a family of actinopterygian fishes in the order Gadiformes which inhabit cold northern and far southern seas. { zō'är·sə‚dē }

zobaa [METEOROL] In Egypt, a lofty whirlwind of sand resembling a pillar, moving with great velocity. { zō'bä }

Zodiac [ASTRON] A band of the sky extending 8° on each side of the ecliptic, within which the moon and principal planets remain. { 'zō·dē‚ak }

zodiacal cone *See* zodiacal pyramid. { zō'dī·ə·kəl 'kōn }

zodiacal constellations [ASTRON] The constellations Aries, Taurus, Gemini, Cancer, Leo, Virgo, Libra, Scorpio, Sagittarius, Capricorn, Aquarius, and Pisces which are assigned to 12 equal portions of the zodiac. { zō'dī·ə·kəl ‚kän·stə'lā·shənz }

zodiacal counterglow *See* gegenschein. { zō'dī·ə·kəl 'kaůnt·ər‚glō }

zodiacal dust [ASTRON] A cloud of dust that fills the plane of the solar system interior to the asteroid belt, and is responsible for zodiacal light. { zō'dī·ə·kəl 'dəst }

zodiacal light [GEOPHYS] A diffuse band of luminosity occasionally visible on the ecliptic; it is sunlight diffracted and reflected by dust particles in the solar system within and beyond the orbit of the earth. { zō'dī·ə·kəl 'līt }

zodiacal pyramid [GEOPHYS] The pattern formed by the zodiacal light. Also known as zodiacal cone. { zō'dī·ə·kəl 'pir·ə·mid }

zoea [INV ZOO] An early larval stage of decapod crustaceans distinguished by a relatively large cephalothorax, conspicuous eyes, and large, fringed antennae. { zō'ē·ə }

zoisite [MINERAL] Ca$_2$Al$_3$Si$_3$O$_{12}$(OH) A white, gray, brown, green, or rose-red orthorhombic mineral of the epidote group consisting of a basic calcium aluminum silicate and occurring massive or in prismatic crystals. { 'zói‚sīt }

Zollinger-Ellison syndrome [MED] Gastric hypersecretion and hyperacidity, fulminating intractable atypical peptic ulceration, and hyperplasia of the islet cells of the pancreas. { 'zäl·ən·jər 'el·ə·sən ‚sin‚drōm }

zona fasciculata [ANAT] The middle tissue layer of the adrenal cortex where glucocorticoids, mainly cortisol and corticosterone, are synthesized and secreted. { ‚zō·nə fə‚sik·yə'läd·ə }

zona glomerulosa [ANAT] The outer tissue layer of the adrenal cortex where mineralocorticoids, primarily aldosterone, are synthesized and secreted. { ‚zō·nə glə‚mer·yə'lō·sə }

zonal [METEOROL] Latitudinal, easterly or westerly, opposed to meridional. { 'zōn·əl }

zonal centrifuge [BIOL] A centrifuge that uses a rotating chamber of large capacity in which to separate cell organelles by density-gradient centrifugation. { 'zōn·əl 'sen·trə‚fyüj }

zonal circulation *See* zonal flow. { 'zōn·əl ‚sər·kyə'lā·shən }

zonal flow [METEOROL] The flow of air along a latitude circle; more specifically, the latitudinal (east or west) component of existing flow. Also known as zonal circulation. { 'zōn·əl 'flō }

zonal harmonics [MATH] Spherical harmonics which do not depend on the azimuthal angle; they are proportional to Legendre polynomials of cos θ, where θ is the colatitude. { 'zōn·əl här'män·iks }

zonal index [METEOROL] A measure of strength of the middle-latitude westerlies, expressed as the horizontal pressure difference between 35° and 55° latitude, or as the corresponding geostrophic wind. { 'zōn·əl 'in‚deks }

zonal kinetic energy [METEOROL] The kinetic energy of the mean zonal wind, obtained by averaging the zonal component of the wind along a fixed latitude circle. { 'zōn·əl ki'ned·ik 'en·ər·jē }

zonal soil [GEOL] In early classification systems in the United States, a soil order including soils with well-developed characteristics that reflect the influence of agents of soil genesis. Also known as mature soil. { 'zōn·əl 'sói̇l }

zonal theory [GEOL] A theory of the formation of mineral deposition and sequence patterns, based on the changes in a mineral-bearing fluid as it passes upward from a magmatic source. { 'zōn·əl 'thē·ə·rē }

zonal westerlies *See* westerlies. { 'zōn·əl 'wes·tər‚lēz }

zonal wind [METEOROL] The wind, or wind component, along the local parallel of latitude, as distinguished from the meridional wind. { 'zōn·əl 'wind }

zonal winds *See* westerlies. { 'zōn·əl 'winz }

zonal wind-speed profile [METEOROL] A diagram in which the speed of the zonal flow is one coordinate and latitude the other. { 'zōn·əl 'win ‚spēd ‚prō‚fīl }

zona pellucida [HISTOL] The thick, solid, elastic envelope of the ovum. Also known as oolemma. { 'zō·nə pə'lüs·əd·ə }

zona reticularis [ANAT] The inner tissue layer of the adrenal cortex that is involved in the synthesis and secretion of sex steroid precursors and, to less extent, glucocorticoids. { ‚zō·nə re‚tik·yə'lär·əs }

zonation [ECOL] Arrangement of organisms in biogeographic zones. [GEOL] The condition of being arranged in zones. { zō'nā·shən }

zonda [METEOROL] A hot wind in Argentina. { 'zän·də }

zone [ANALY CHEM] *See* band. [COMPUT SCI] **1.** One of the top three rows of a punched card, namely, the 11, 12, and zero rows. **2.** *See* storage area. [CRYSTAL] **2.** A set of crystal faces which intersect (or would intersect, if extended) along edges which are all parallel. [GEOGR] An area or region of latitudinal character. [GEOL] A belt, layer, band, or strip of earth material such as rock or soil. [MATH] The portion of a sphere lying between two parallel planes that intersect the sphere. [MECH ENG] **1.** In a heating or air-conditioning system, one or more spaces whose temperature is regulated by a single control. **2.** A subdivision of a sprinkler, water-supply, or standpipe system. [ORD] **1.** Any tactical area of importance, generally parallel to the front, such as a fortified area, a defensive position, a combat zone, or a traffic-control zone. **2.** An area in which projectiles will fall when a given propelling charge is used and the elevation is varied between the minimum and the maximum; in practice, generally limited to howitzer and mortar firings. { zōn }

zone axis [CRYSTAL] A line through the center of a crystal which is parallel to all the faces of a zone. { zōn 'ak·səs }

zone bit [COMPUT SCI] One of a set of bits used to indicate some grouping of characters. { zōn ‚bit }

zone blanking [ELECTR] Method of turning off the cathode-ray tube during part of the sweep of an antenna. { 'zōn 'blaŋk·iŋ }

zone charge [ORD] The number of increments of propellant in a propellant charge of semifixed rounds, corresponding to the intended zone of fire; for example, zone charge five consists of five increments of propellant. { 'zōn ‚chärj }

zone control [ENG] The zoning of a process or building, and the independent heating or temperature controls for each zone. { 'zōn kən‚trōl }

zoned decimal [COMPUT SCI] A format for use with EBCDIC input and output permitting a sign overpunch in the low order position of the field; thus, +1234 would be represented as: 1111/0001/1111/0010/1111/0011/1100/0100. { 'zōnd 'des·məl }

zone description [NAV] The number, with its sign, that must be added to or subtracted from the zone time to obtain the Greenwich mean time; the zone description is usually a whole number of hours. { 'zōn di‚skrip·shən }

zone fire [ORD] Artillery or mortar fire that is designed to cover an area in which a target is situated. { 'zōn ‚fīr }

zone heat [CIV ENG] A central heating system arranged to allow different temperatures to be maintained at the same time in two or more areas of a building. { 'zōn ‚hēt }

zone indices [CRYSTAL] Three integers identifying a zone of a crystal; they are the crystallographic coordinates of a point joined to the origin by a line parallel to the zone axis. { 'zōn 'in·də‚sēz }

zone law [CRYSTAL] A law which states that the Miller indices (*h, k, l*) of any crystal plane lying in a zone with zone indices (u, v, w) satisfy the equation $hu + lv + kw = 0$. { 'zōn ¦lō }

zone marker [NAV] Very-high-frequency radio station designed to radiate signals vertically in a cone-shaped pattern to define a zone above a radio range station. { 'zōn ¦mär·kər }

zone melting crystallization [CHEM ENG] A method for purification of crystalline solids; the sample, packed in a narrow column, is heated so that a molten zone passes down through the sample, carrying impurities with it. { 'zōn ¦mel·tiŋ ¦krist·əl·ə'zā·shən }

zone meridian [ASTRON] The meridian used for reckoning zone time; this is generally the nearest meridian whose longitude is exactly divisible by 15°. { 'zōn mə'rid·ē·ən }

zone noon [ASTRON] Twelve o'clock zone time, or the instant the mean sun is over the upper branch of the zone meridian. { 'zōn ¦nün }

zone of accumulation See B horizon. { 'zōn əv ə,kyü·mə'lā·shən }

zone of aeration [GEOL] The subsurface sediment above the water table containing air and water. Also known as unsaturated zone; vadose zone; zone of suspended water. { 'zōn əv e'rā·shən }

zone of avoidance [ASTRON] An irregularly shaped area in the Milky Way Galaxy in which no extragalactic nebulae are observed because of the presence of interstellar matter. { 'zōn əv ə'vóid·əns }

zone of cementation [GEOL] The layer of the earth's crust in which unconsolidated deposits are cemented by percolating water containing dissolved minerals from the overlying zone of weathering. Also known as belt of cementation. { 'zōn əv ,sē,men'tā·shən }

zone of illuviation See B horizon. { 'zōn əv i,lü·vē'ā·shən }

zone of intersection [NAV] That part of a civil airway which overlaps and lies within any part of another civil airway. { 'zōn əv ,in·tər'sek·shən }

zone of maximum precipitation [METEOROL] In a mountain region, the belt of elevation at which the annual precipitation is greatest. { 'zōn əv 'mak·sə·məm pri,sip·ə'tā·shən }

zone of optimal proportion [IMMUNOL] One of three zones considered to appear when antigen and antibody are mixed; it is that zone in which there is no uncombined antigen or antibody. Also known as equivalence zone. { 'zōn əv 'äp·tə·məl prə'pór·shən }

zone of physiological zero [PHYSIO] The band of skin temperatures between about 86 and 97°F (30 and 36°C). { ,zōn əv ,fiz·ē·ə,läj·ə·kəl 'zir·ō }

zone of saturation [HYD] A subsurface zone in which water fills the interstices and is under pressure greater than atmospheric pressure. Also known as phreatic zone; saturated zone. { 'zōn əv ,sach·ə'rā·shən }

zone of silence See skip zone. { 'zōn əv 'sī·ləns }

zone of soil water See belt of soil water. { 'zōn əv 'sóil ,wód·ər }

zone of suspended water See zone of aeration. { 'zōn əv sə¦spen·dəd ,wód·ər }

zone plate [OPTICS] A plate with alternate transparent and opaque rings, designed to block off every other Fresnel half-period zone; light from a point source passing through the plate produces an intense point image much like that produced by a lens. { 'zōn ,plāt }

zone-position indicator [ENG] Auxiliary radar set for indicating the general position of an object to another radar set with a narrower field. { 'zōn pə¦zish·ən 'in·də,kād·ər }

zone purification See zone refining. { 'zōn ,pyùr·ə·fə¦kā·shən }

zone refining [MET] A technique to purify materials in which a narrow molten zone is moved slowly along the complete length of the specimen to bring about impurity segregation, and which depends on differences in composition of the liquid and solid in equilibrium. Also known as zone purification. { 'zōn ri'fīn·iŋ }

zone theory See stage theory. { 'zōn ,thē·ə·rē }

zone time [ASTRON] The local mean time of a reference or zone meridian whose time is kept throughout a designated zone; the zone meridian is usually the nearest meridian whose longitude is exactly divisible by 15°. { 'zōn ,tīm }

zoning [CIV ENG] Designation and reservation under a master plan of land use for light and heavy industry, dwellings, offices, and other buildings; use is enforced by restrictions on types of buildings in each zone. [CRYSTAL] A variation in the composition of a crystal from core to margin due to a separation of the crystal phases during its growth by loss of equilibrium in a continuous reaction series. [ELECTROMAG] The displacement of various portions of the lens or surface of a microwave reflector so the resulting phase front in the near field remains unchanged. Also known as stepping. { 'zōn·iŋ }

zonite [INV ZOO] A body segment in Diplopoda. { 'zō,nīt }

zonochlorite See pumpellyite. { ,zō·nō'klór,īt }

zonula occludens See tight junction. { ¦zōn·yə·lə ə'klüd·ənz }

zoocecidium [PL PATH] A plant gall usually caused by an insect. { ¦zō·ə·sə¦sid·ē·əm }

zoochlorellae [BIOL] Unicellular green algae which live as symbionts in the cytoplasm of certain protozoans, sponges, and other invertebrates. { ¦zō·ə·klə'rel,ē }

zoochory [BOT] Dispersal of plant disseminules by animals. { 'zō·ə,klór·ē }

zooecium [INV ZOO] The exoskeleton of a feeding zooid in bryozoans. { zō-'ē·shē·əm }

zoogeographic region [ECOL] A major unit of the earth's surface characterized by faunal homogeneity. { ¦zō·ə¦jē·ə¦graf·ik 'rē·jən }

zoogeography [BIOL] The science that attempts to describe and explain the distribution of animals in space and time. { ¦zō·ə,jē'äg·rə·fē }

zoogloea [MICROBIO] A gelatinous or mucilaginous mass characteristic of certain bacteria grown in organic-rich fluid media. { ,zō·ə'glē·ə }

zooid [INV ZOO] A more or less independent individual of colonial animals such as bryozoans and coral. { 'zō,óid }

zoology [BIOL] The science that deals with knowledge of animal life. { zō'äl·ə·jē }

zoom [ENG] To enlarge or reduce the size of an image in an optical system or electronic display. { 'züm }

Zoomastigina [INV ZOO] The equivalent name for Zoomastigophorea. { ,zō·ə,mas·tə¦jīn·ə }

Zoomastigophorea [INV ZOO] A class of flagellate protozoans in the subphylum Sarcomastigophora; some are simple, some are specialized, and all are colorless. { ,zō·ə,mas·tə·gə'fór·ē·ə }

zoom lens [OPTICS] A system of lenses in which two or more parts are moved with respect to each other to obtain a continuously variable focal length and hence magnification, while the image is kept in the same image plane. Also known as variable-focal-length lens; varifocal lens. { 'züm ¦lenz }

zoonoses [BIOL] Diseases which are biologically adapted to and normally found in lower animals but which under some conditions also infect humans. { ,zō·ə'nō·sēz }

zoophobia [PSYCH] An abnormal fear of animals. { ,zō·ə'fō·bē·ə }

zooplankton [ECOL] Microscopic animals which move passively in aquatic ecosystems. { ¦zō·ə'plaŋk·tən }

zoosphere [ECOL] The world community of animals. { 'zō·ə,sfir }

zoosporangium [BOT] A spore case bearing zoospores. { ¦zō·ə·spə'ran·jē·əm }

zoospore [BIOL] An independently motile spore. { 'zō·ə,spór }

zooxanthellae [BIOL] Microscopic yellow-green algae which live symbiotically in certain radiolarians and marine invertebrates. { ,zō·ə·zan'thē,lē }

Zoraptera [INV ZOO] An order of insects, related to termites and psocids, which live in decaying wood, sheltered from light; most individuals are wingless, pale in color, and blind. { zə'rap·tə·rə }

Zorn's lemma [MATH] If every linearly ordered subset of a partially ordered set has a maximal element in the set, then the set has a maximal element. { 'zórnz 'lem·ə }

Zoroasteridae [INV ZOO] A family of deep-water asteroid echinoderms in the order Forcipulatida. { ,zór·ō·ə'ster·ə,dē }

Zorotypidae [INV ZOO] The single family, containing one genus, *Zorotypus*, in the order Zoraptera. { ,zór·ə'tīp·ə,dē }

zorsite [MINERAL] $Ca_2Al_3Si_3O_{12}(OH)$ White, gray, brown, green, or rose-red orthorhombic mineral of the epidote group; an essential constituent of saussurite. { 'zór,sīt }

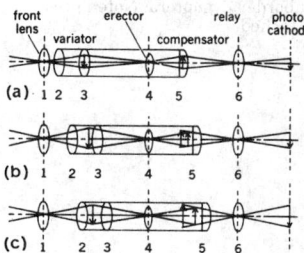

ZOOM LENS

Zoom lens in three operating positions. *(a)* Wide-angle. *(b)* Medium-angle. *(c)* Telephoto. Lens elements 2, 3, and 5 are mounted in movable barrel connected to zoom handle. Lens elements 1, 4, and 6 are stationary, mounted in lens housing. *(From D. G. Fink, ed., Television Engineering Handbook, McGraw-Hill, 1957)*

ZYGNEMATACEAE

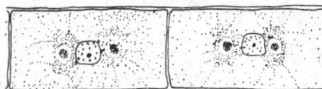

Vegetative cells of *Zygnema* species, type genus of the family Zygnemataceae, showing characteristic pair of stellate chloroplasts in each cell. (*From G. M. Smith, Cryptogamic Botany, 2d ed., McGraw-Hill, 1955*)

ZYGOPTERA

Two adult damsel flies of the suborder Zygoptera. Note mites attached.

zoster *See* herpes zoster. { 'zäs·tər }

Zosteraceae [BOT] A family of monocotyledonous plants in the order Najadales; the group is unique among flowering plants in that they grow submerged in shallow ocean waters near the shore. { ˌzäs·tə'rās·ē‚ē }

Zosterophyllatae *See* Zosterophyllopsida. { ¦zäs·tə·rō'fil·ə‚tē }

Zosterophyllopsida [PALEOBOT] A group of early land vascular plants ranging from the Lower to the Upper Devonian; individuals were leafless and rootless. { ¦zäs·tə·rō·fə'läp·səd·ə }

Z parameter [ELECTR] One of a set of four transistor equivalent-circuit parameters; they are the inverse of the *Y* parameters. { zē pə‚ram·əd·ər }

Z⁰ particle *See* Z⁰ boson. { 'zē ‚zir·ō ‚pärd·ə·kəl }

Zr *See* zirconium.

Z score [STAT] A measure of how many standard deviations a raw score is from the mean. { 'zē ‚skór }

Zsigmondy gold number [CHEM] The number of milligrams of protective colloid necessary to prevent 10 milliliters of gold sol from coagulating when 0.5 milliliter of 10% sodium chloride solution is added. { 'zig·món·dē 'gōld ‚nəm·bər }

Z time *See* Greenwich mean time. { 'zē ‚tīm }

z-transfer function *See* pulsed transfer function. { 'zē 'tranz·fər ‚fəŋk·shən }

z-transform [MATH] The *z*-transform of a sequence whose general term is f_n is the sum of a series whose general term is $f_n z^{-n}$, where z is a complex variable; n runs over the positive integers for a one-sided transform, over all the integers for a two-sided transform. { 'zē 'tranz‚fórm }

Z twist [TEXT] A right-handed yarn twist in which the spiral slants like the middle part of the letter Z. { 'zē ‚twist }

Zulu time *See* Greenwich mean time. { 'zü·lü ‚tīm }

Zuni [ORD] A United States air-to-surface unguided rocket with solid propellant; can be armed with various types of heads, including flares, fragmentation, and armor-piercing. { 'zü·nē }

Zurich number *See* relative sunspot number. { 'zür·ik ‚nəm·bər }

Z variometer *See* vertical intensity variometer. { 'zē ‚ver·ē'äm·əd·ər }

Zwicky dark matter [ASTRON] Matter of unknown nature that is postulated to exist outside the visible parts of galaxies or between galaxies in order to account for the radial velocity dispersion of galaxies in clusters. { ‚zwik·ē 'därk ‚mad·ər }

Zwischengebirge *See* median mass. { 'tsfish·ən‚gə'bir·gə }

zwitterion *See* dipolar ion. { 'tsfid·ər‚ī‚än }

Zygaenidae [INV ZOO] A diverse family of small, often brightly colored African moths in the superfamily Zygaenoidea. { zī'jēn·ə‚dē }

Zygaenoidea [INV ZOO] A superfamily of moths in the suborder Heteroneura characterized by complete venation, rudimentary palpi, and usually a rudimentary proboscis. { ˌzī·jə'nóid·ē·ə }

Zyglo method [ENG] A procedure for visualizing incipient cracks caused by fatigue failure, in which the part is immersed in a special activated penetrating oil and viewed under black light. { 'zī·glō ‚meth·əd }

Zygnemataceae [BOT] A family of filamentous plants in the order Conjugales; they are differentiated into genera by chloroplast morphology, which may be spiral, bandlike, or cushionlike. { zig‚nēm·ə'tās·ē‚ē }

zygodactyl [VERT ZOO] Of birds, having a toe arrangement of two in front and two behind. { ¦zī·gō¦dak·təl }

zygomatic bone [ANAT] A bone of the side of the face below the eye; forms part of the zygomatic arch and part of the orbit in mammals. Also known as malar bone. { ¦zī·gə¦mad·ik 'bōn }

zygomorphic [BIOL] Bilaterally symmetrical. { ¦zī·gə‚mór·fik }

Zygomycetes [MYCOL] A class of fungi in the division Eumycetes. { ‚zī·gō‚mī'sēd·ēz }

Zygomycotina [MYCOL] A subdivision of fungi characterized by distinctive sexual and asexual reproductive stages. { ‚zī·gə‚mī·kə'tē·nə }

zygopophysis [ANAT] One of the articular processes of the neural arch of a vertebra. { ‚zī·gə'päf·ə·səs }

Zygoptera [INV ZOO] The damsel flies, a suborder of insects in the order Odonata; individuals are slender, dainty creatures, often with bright-blue or orange coloring and usually with clear or transparent wings. { zī'gäp·tə·rə }

zygospore [BOT] A thick-walled cell or resting spore that results from the fusion of similar reproductive cells, especially in organisms that reproduce by conjugation. { 'zī·gə‚spór }

zygote [EMBRYO] **1.** An organism produced by the union of two gametes. **2.** The fertilized ovum before cleavage. { 'zī‚gōt }

zygotene [CYTOL] The stage of meiotic prophase during which homologous chromosomes synapse; visible bodies in the nucleus are now bivalents. Also known as amphitene. { 'zī·gə‚tēn }

zygotic induction [VIROL] Phage induction following conjugation of a lysogenic bacterium with a nonlysogenic one. { zī'gäd·ik in'dək·shən }

zymase [BIOCHEM] A complex of enzymes that catalyze glycosis. { 'zī‚mās }

zymogen [BIOCHEM] The inactive precursor of an enzyme; liberates an active enzyme on reaction with an appropriate kinose. Also known as proenzyme. { 'zī·mə·jən }

zymogen granules [BIOCHEM] Granules of zymogen in gland cells, particularly those of the pancreatic acini and of the gastric chief cells. { 'zī·mə·jən 'gran‚yülz }

zymogenic [MICROBIO] Obtaining energy by amylolitic processes. { ¦zī·mə¦jen·ik }

zymophore [BIOCHEM] The active portion of an enzyme. { 'zī·mə‚fór }

zymosis *See* fermentation. { zī'mō·səs }

zymosterol [BIOCHEM] $C_{27}H_{43}OH$ An unsaturated sterol obtained from yeast fat; yields cholesterol on hydrogenation. { zī'mäs·tə‚ról }

Zythiaceae [MYCOL] A family of fungi of the order Sphaeropsidales which contains many plant and insect pathogens. { ‚zith·ē·'ās·ē‚ē }

ZZ Ceti stars [ASTRON] A small class of variable-luminosity white dwarfs with small amplitude oscillations having periods of 10–100 seconds and effective temperatures of 10,000–14,000 K (18,000–25,000°F). { ¦zē¦zē 'sed·ē ‚stärz }

APPENDIX

Measurement systems

U.S. Customary System and the metric system

Over the past 200 years or so, scientists and engineers have used two major systems of units in measurement. These are commonly called the U.S. Customary System (inherited from the British Imperial System) and the metric system (developed at the time of the French Revolution).

In the U.S. Customary System the units yard and pound with their divisions, such as the inch, and multiples, such as the ton, are basic. The metric system has been adopted for general use by most countries. It is used nearly everywhere for precise measurements in science. The meter and kilogram with their multiples, such as the kilometer, and fractions, such as the gram, are basic to the metric system. Until the second half of the twentieth century, most of the base units in these systems were defined in terms of specific physical artifacts, such as the International Prototype Kilogram, a platinum-iridium cyclinder maintained in Sèvres, France.

In the U.S. Cutomary System, units of the same kind are related almost at random. For example, there are the units of length, the inch, yard, and mile. In the metric system the relationships between units of the same kind are strictly decimal (millimeter, meter, and kilometer).

However, in technical writing there is no uniformity within each of these two systems as to the choice of units for the same quantities. For example, the hour or the second, the foot or the inch, and the centimeter or the millimeter could be chosen as the unit of measurement for the quantities time and length.

International System

To simplify matters and to make communication more understandable, an internationally accepted system of units came into use in 1960. This is termed the International System of Units, which is abbreviated SI in all languages (from the French Système International d'Unités).

Fundamentally the system is metric, with base units whose definitions have been modified from time to time in order to allow for their more accurate realization as techniques of measurement have evolved. In several instances, artifact standards have been replaced by physically invariant quantities, such as atomic transition frequencies and fundamental physical constants. For example, the meter is the SI is defined as the length of the path traveled by light in vacuum during a time interval of 1/299 792 458 of a second. The effect of this definition, which was adopted in 1983, is to fix the speed of light at exactly 299 792 458 meters per second.

The second in the SI is defined as the duration of 9 192 631 770 periods of the radiation corresponding to the transition between two hyperfine levels of the ground state of the cesium-133 atom.

Interestingly, the kilogram, the SI unit of mass, is still the mass of the kilogram kept at Sèvres. However, it is probable that eventually the unit will be redefined in terms of atomic mass or through a definition that will fix the value of the Planck constant h.

Although the SI is increasingly used by scientists and engineers, there are some other units in everyday use which will probably remain, for example, minute, hour, day, dergree (angle), and liter. The point should be made, however, that these terms will not be employed in a scientific context if the SI is fully adopted.

Because of their extremely common use among scientists, several units are still permitted in conjunction with SI units, for example, the electronvolt, rad, roentgen, barn, and curie. In time their usage might be phased out.

One futher point is that in October 1967 the Thirteenth General Conference of Weights and Measures decided to name the SI unit of thermodynamic temperature "kelvin" (symbol K) instead of "degree Kelvin" (symbol °K). For example, the notation is 273 K and not 273°K.

The base units and derived units of the SI are shown in **Table 1** and **2.**

In the SI the prefixes differ from a unit in steps of 10^3. A list of prefix terms, symbols, and their factors is given in **Table 3.** Some examples of the use of these prefixes follow:

$$1000 \text{ m} = 1 \text{ kilometer} = 1 \text{ km}$$
$$1000 \text{ V} = 1 \text{ kilovolt} = 1 \text{ kV}$$
$$1\,000\,000 \,\Omega = 1 \text{ megohm} = 1 \text{ M}\Omega$$
$$0.000\,000\,001 \text{ s} = 1 \text{ nanosecond} = 1 \text{ ns}$$

Only one prefix is to be employed for a unit. For example:

$$1000 \text{ kg} = 1 \text{ Mg} \qquad \text{not } 1 \text{ kkg}$$
$$10^{-9} \text{ s} = 1 \text{ ns} \qquad \text{not } 1 \text{ m}\mu s$$
$$1\,000\,000 \text{ m} = 1 \text{ Mm} \qquad \text{not } 1 \text{ kkm}$$

Also, when a unit is raised to a power, the power applies to the whole unit including the prefix. For example:

$$\text{km}^2 = (\text{km})^2 = (1000 \text{ m})^2 = 10^6 \text{ m}^2$$
$$\text{not } 1000 \text{ m}^2$$

Some common units defined in terms of SI units are given in **Table 4** (the definitions in the fourth column are exact).

TABLE 1. Base units of the International System

Quantity	Name of unit	Unit symbol
length	meter	m
mass	kilogram	kg
time	second	s
electric current	ampere	A
temperature	kelvin	K
luminous intensity	candela	cd
amount of substance	mole	mol

Conversion factors for the measurement systems

This Dictionary has retained the U.S. Customary and metric systems, but has incorporated SI units in many cases. Conversion factors between the three measurement systems are given in **Table 5** for some prevalent units; in each of the subtables the user proceeds as follows:

To convert a quantity expressed in a unit in the

TABLE 2. Derived units of the International System*

Quantity	Name of unit	Unit symbol, or unit expressed in terms of other SI units	Unit expressed in terms of SI base units
plane angle	radian	rad	$m/m = 1$
solid angle	steradian	sr	$m^2/m^2 = 1$
area	square meter		m^2
volume	cubic meter		m^3
frequency	hertz	Hz	s^{-1}
density	kilogram per cubic meter		kg/m^3
velocity	meter per second		m/s
angular velocity	radian per second	rad/s	$m/(m \cdot s) = s^{-1}$
acceleration	meter per second squared		m/s^2
angular acceleration	radian per second squared	rad/s²	$m/(m \cdot s^2) = s^{-2}$
volumetric flow rate	cubic meter per second		m^3/s
force	newton	N	$kg \cdot m/s^2$
surface tension	newton per meter, joule per square meter	N/m, J/m²	kg/s^2
pressure	pascal, newton per square meter	Pa, N/m²	$kg/(m \cdot s^2)$
viscosity, dynamic	pascal-second, newton-second per square meter	Pa · s, N · s/m²	$kg/(m \cdot s)$
viscosity, kinematic	meter squared per second		m^2/s
work, torque, energy, quantity of heat	joule, newton-meter, watt-second	J, N · m, W · s	$kg \cdot m^2/s^3$
power, heat flux	watt, joule per second	W, J/s	$kg \cdot m^2/s^3$
heat flux density	watt per square meter	W/m²	kg/s^3
volumetric heat release rate	watt per cubic meter	W/m³	$kg/(m \cdot s^3)$
heat transfer coefficient	watt per square meter kelvin	W/(m² · K)	$kg/(s^3 \cdot K)$
heat capacity (specific)	joule per kilogram kelvin	J/kg · K)	$m^2/(s^2 \cdot K)$
capacity rate	watt per kelvin	W/K	$kg \cdot m^2/(s^3 \cdot K)$
thermal conductivity	watt per meter kelvin	W/(m · K), $\dfrac{J \cdot m}{s \cdot m^2 \cdot K}$	$kg \cdot m/(s^3 \cdot K)$
quantity of electricity	coulomb	C	$A \cdot s$
electromotive force	volt	V, W/A	$kg \cdot m^2/(A \cdot s^3)$
electric field strength	volt per meter	V/m	$kg \cdot m/(A \cdot s^3)$
electric resistance	ohm	Ω, V/A	$kg \cdot m^2/(A^2 \cdot s^3)$
electric conductance	siemens	S, A/V	$A^2 \cdot s^3/(kg \cdot m^2)$
electric conductivity	ampere per volt meter	A/(V · m)	$A^2 \cdot s^3/(kg \cdot m^3)$
electric capacitance	farad	F, A · s/V	$A^2 \cdot s^4/(kg \cdot m^2)$
magnetic flux	weber	Wb, V · s	$kg \cdot m^2/A \cdot s^2)$
inductance	henry	H, V · s/A	$kg \cdot m^2/(A^2 \cdot s^2)$
magnetic permeability	henry per meter	H/m	$kg \cdot m/(A^2 \cdot s^2)$
magnetic flux density	tesla, weber per square meter	T, Wb/m²	$kg/(A \cdot s^2)$
magnetic field strength	ampere per meter		A/m
magnetomotive force	ampere		A
luminous flux	lumen	lm, cd · sr	$cd \cdot m^2/m^2 = cd$
luminance	candela per square meter		cd/m^2
illumination	lux, lumen per square meter	lx, lm/m², cd · sr/m²	$cd \cdot m^2/m^4 = cd/m^2$
activity (of radionuclides)	becquerel	Bq	s^{-1}
absorbed dose	gray	Gy, J/kg	m^2/s^2
dose equivalent	sievert	Sv, J/Kg	m^2/s^2
catalytic activity	katal	kat	mol/s

*The degree Celsius (°C), listed in Table 4, is also a derived unit of the International System.

TABLE 3. Prefixes for units in the International System

Prefix	Symbol	Power	Example	Prefix	Symbol	Power	Example
yotta	Y	10^{24}		deci	d	10^{-1}	
zetta	Z	10^{21}		centi	c	10^{-2}	centimeter (cm)
exa	E	10^{18}		milli	m	10^{-3}	milligram (mg)
peta	P	10^{15}		micro	μ	10^{-6}	microgram (μg)
tera	T	10^{12}	terawatt (TW)	nano	n	10^{-9}	nanosecond (ns)
giga	G	10^{9}	gigawatt (GW)	pico	p	10^{-15}	picofarad (pF)
mega	M	10^{6}	megahertz (MHz)	femto	f	10^{-15}	femtosecond (fs)
kilo	k	10^{3}	kilometer (km)	atto	a	10^{-18}	
hecto	h	10^{2}		zepto	z	10^{-21}	
deka	da	10^{1}		yocto	y	10^{-24}	

TABLE 4. Some common units defined in terms of SI units

Quantity	Name of unit	Unit symbol	Definition of unit
length	inch	in.	2.54×10^{-2} m
mass	pound (avoirdupois)	lb	0.45359237 kg
force	kilogram-force	kgf	9.80665 N
pressure	atmosphere	atm	101325 Pa
pressure	torr	torr	(101325/760) Pa
pressure	conventional millimeter of mercury*	mmHg	$13.5951 \times 980.665 \times 10^{-2}$ Pa
energy	kilowatt-hour	kWh	3.6×10^{6} J
energy	thermochemical calorie	cal	4.184 J
energy	international steam table calorie	cal_{IT}	4.1868 J
thermodynamic temperature (T)	degree Rankine	°R	(5/9) K
customary temperature (t)	degree Celsius	°C	$t(°C) = T(K) - 273.15$
customary temperature (t)	degree Fahrenheit	°F	$t(°F) = [1.8 \times t(°C)] + 32 = T(°R) - 459.67$
radioactivity	curie	Ci	3.7×10^{10} Bq
energy†	electronvolt	eV	$eV = 1.60218 \times 10^{-19}$ J
mass†	unified atomic mass unit	u	$u = 1.66054 \times 10^{-27}$ kg

*The conventional millimeter of mercury, symbol mmHg (not mm Hg), is the pressure exerted by a column exactly 1 mm high of a fluid of density exactly 13.5951 g cm^{-3} in a place where the gravitational acceleration is exactly 980.665 cm · s^{-2}. The mmHg differs from the torr by less than 2×10^{-7} torr.
†These units defined in terms of the best available experimental values of certain physical constants may be converted to SI units. The factors for conversion of these units are subject of change in the light of new experimental measurements of the constants involved.

left-hand column to the equivalent in a unit in the top row of a subtable, multiply the quantity by the factor common to both units. For example, to convert 7 ft to the equivalent in meters, go to subtable A, "Units of lenght." and find 1 ft in the lefthand column and m in the top row. The conversion factor common to these units is 0.3048. Therefore, 7 ft = 7 × 0.3048 = 2.1336 m.

The conversion factors have been carried out to seven significant figures, as derived from the fundamental constants and the definitions of the units. However, this does not mean that the factors are always known to that accuracy. Numbers followed by ellipses are to be continued indefinitely with repetition of the same pattern of digits. Factors written with fewer than seven significant digits are exact values. Numbers followed by an asterisk are definitions of the relation between the two units.

Units of temperature in measurement systems

Temperature is a basic physical quantity. It is a measure of the thermal energy of random motion of particles in a system. As such it has been chosen as one of the base quantities in the SI. It is to be treated as are the units of length, mass, time, electric current, and luminous intensity. In the SI the unit of length is the meter, the unit of time the second, and so on. The question arises as to the choice of the unit of temperature in the SI.

In the past it was customary to refer to scales of temperature, for example, the Celsius and Fahrenheit scales. On the Celsius scale, 0 designates the freezing point (ice point) and 100 the boiling point (steam point) of water. Corresponding numbers on the Fahrenheit scale are 32 and 212. There are 100 units between the ice point and steam point on the Celsius scale, and 180 units between these points in the Fahrenheit system.

By measuring the volume changes of a gas within the 100-unit interval for the ice point and steam point of water on the Celsius scale, it was found that a numerical value could be assigned for a basic unit of temperature. Careful measurement of this ice-steam interval in a gas thermometer determined that the ice point of water should be assigned the value of 273.15

kelvins. The unit of temperature was thus called the kelvin with the symbol K. Further experiments led to the decision to define the kelvin in the SI along the same lines but in terms of the triple point of water. This is the temperature and pressure at which ice liquid water, and water vapor coexist at equilibrium The triple point was chosen because it was a more reproducible value than the ice point.

TABLE 5. Conversion factors for the U.S. Customary System, metric system, and International System

A. Units of length

Units	cm	m	in.	ft	yd	mi
1 cm	= 1	0.01*	0.3937008	0.03280840	0.01093613	6.213712×10^{-6}
1 m	= 100.	1	39.37008	3.280840	1.093613	6.213712×10^{-4}
1 in.	= 2.54*	0.0254	1	0.08333333...	0.02777777...	1.578283×10^{-5}
1 ft	= 30.48	0.3048	12.*	1	0.3333333...	$1.893939... \times 10^{-4}$
1 yd	= 91.44	0.9144	36.	3.*	1	$5.681818... \times 10^{-4}$
1 mi	= 1.609344×10^{5}	1.609344×10^{3}	6.336×10^{4}	5280.*	1760.	1

B. Units of area

Units	cm²	m²	in.²	ft²	yd²	mi²
1 cm²	= 1	10^{-4}*	0.1550003	1.076391×10^{-3}	1.195990×10^{-4}	3.861022×10^{-11}
1 m²	= 10^{4}	1	1550.003	10.76391	1.195990	3.861022×10^{-7}
1 in.²	= 6.4516*	6.4516×10^{-4}	1	$6.944444... \times 10^{-3}$	7.716049×10^{-4}	2.490977×10^{-10}
1 ft²	= 929.0304	0.09290304	144.*	1	0.7777777...	3.587007×10^{-8}
1 yd²	= 8361.273	0.8361273	1296.	9.*	1	3.228306×10^{-7}
1 mi²	= 2.589988×10^{10}	2.589988×10^{6}	4.014490×10^{9}	2.78784×10^{7}*	3.0976×10^{6}	1

C. Units of volume

Units	m³	cm³	liter	in.³	ft³	qt	gal
1 m³	= 1	10^{6}	10^{3}	6.102374×10^{4}	35.31467×10^{-3}	1.056688	264.1721
1 cm³	= 10^{-6}	1	10^{-3}	0.06102374	3.531467×10^{-5}	1.056688×10^{-3}	2.641721×10^{-4}
1 liter	= 10^{-3}	1000.*	1	61.02374	0.03531467	1.056688	0.2641721
1 in.³	= 1.638706×10^{-5}	16.38706*	0.01638706	1	5.787037×10^{-4}	0.01731602	4.329004×10^{-3}
1 ft³	= 2.831685×10^{-2}	28316.85	28.31685	1728.*	1	2.992208	7.480520
1 qt	= 9.46353×10^{-4}	946.353	0.946353	57.75	0.0342014	1	0.25
1 gal (U.S.)	= 3.785412×10^{-3}	3785.412	3.785412	231.*	0.1336806	4.*	1

D. Units of mass

Units	g	kg	oz	lb	metric ton	ton
1 g	= 1	10^{-3}	0.03527396	2.204623×10^{-3}	10^{-6}	1.102311×10^{-6}
1 kg	= 1000.	1	35.27396	2.204623	10^{-3}	1.102311×10^{-3}
1 oz (avdp)	= 28.34952	0.02834952	1	0.0625	2.834952×10^{-5}	3.125×10^{-5}
1 lb (avdp)	= 453.5924	0.4535924	16.*	1	4.535924×10^{-4}	$5. \times 10^{-4}$
1 metric ton	= 10^{8}	1000.*	35273.96	2204.623	1	1.102311
1 ton	= 907184.7	907.1847	32000.	2000.*	0.9071847	1

*Usage of the asterisk is defined in the accompanying text.

TABLE 5. Conversion factors for the U.S. Customary System, metric system, and International System (cont.)

E. Units of density

Units	$g \cdot cm^{-3}$	$g \cdot L^{-1}, kg \cdot m^{-3}$	$oz \cdot in.^{-3}$	$lb \cdot in.^{-3}$	$lb \cdot ft^{-3}$	$lb \cdot gal^{-1}$
$1\ g \cdot cm^{-3}$	= 1	1000.	0.5780365	0.03612728	62.42795	8.345403
$1\ g \cdot L^{-1}, kg \cdot m^{-3}$	$= 10^{-3}$	1	5.780365×10^{-4}	3.612728×10^{-5}	0.06242795	8.345403×10^{-3}
$1\ oz \cdot in.^{-3}$	= 1.729994	1729.994	1	0.0625	108.	14.4375
$1\ lb \cdot in.^{-3}$	= 27.67991	27679.91	16.	1	1728.	231.
$1\ lb \cdot ft^{-3}$	= 0.01601847	16.01847	9.259259×10^{-3}	5.7870370×10^{-4}		0.1336806
$1\ lb \cdot gal^{-1}$	= 0.1198264	119.8264	4.749536×10^{-3}	4.3290043×10^{-3}	7.480519	1

F. Units of pressure

Units	$Pa, N \cdot m^{-2}$	$dyn \cdot cm^{-2}$	bar	atm	$kgf \cdot cm^{-2}$	mmHg (torr)	in. Hg	$lbf \cdot in.^{-2}$
$1\ Pa, 1\ N \cdot m^{-2}$	= 1	10	10^{-5}	9.869233×10^{-6}	1.019716×10^{-5}	7.500617×10^{-3}	2.952999×10^{-4}	1.450377×10^{-4}
$1\ dyn \cdot cm^{-2}$	= 0.1	1	10^{-6}	9.869233×10^{-7}	1.019716×10^{-6}	7.500617×10^{-4}	2.952999×10^{-5}	1.450377×10^{-5}
1 bar	$= 10^{5*}$	10^{6}	1	0.9869233	1.019716	750.0617	29.52999	14.50377
1 atm	= 101325.0*	1013250	1.013250	1	1.033227	760.	29.92126	14.69595
$1\ kgf \cdot cm^{-2}$	= 98066.5	980665	0.980665	0.9678411	1	735.5592	28.95903	14.22334
1 mmHg (torr)	= 133.3224	1333.224	1.333224×10^{-3}	2.78784×10^{-3}	$1.35955099 \times 10^{-3}$	1	0.03937008	0.01933678
1 in. Hg	= 3386.388	33863.88	0.03386388	0.03342105	0.03453155	25.4	1	0.4911541
$1\ lbf \cdot in.^{-2}$	= 6894.757	68947.57	0.06894757	0.06804596	0.07030696	51.71493	2.036021	1

G. Units of energy

Units	g mass (energy equiv)	J	eV	cal	cal_{IT}	Btu_{IT}	kWh	hp-h	ft-lbf	$ft^3 \cdot lbf \cdot in.^{-2}$	liter-atm
1 g mass (energy equiv)	= 1	8.987552×10^{13}	5.609589×10^{32}	2.148076×10^{13}	2.146640×10^{13}	8.513555×10^{10}	2.496542×10^{7}	3.347918×10^{7}	6.628878×10^{13}	4.603388×10^{11}	8.870024×10^{11}
1 J	$= 1.112750 \times 10^{-14}$	1	6.241510×10^{18}	0.2390057	0.2388459	9.473172×10^{-4}	$2.777777... \times 10^{-7}$	3.725062×10^{-7}	0.7375622	5.121960×10^{-3}	9.869233×10^{-3}
1 eV	$= 1.782662 \times 10^{-33}$	1.602176×10^{-19}	1	3.829293×10^{-20}	3.826733×10^{-20}	1.519570×10^{-22}	4.450490×10^{-26}	5.968206×10^{-26}	1.181705×10^{-19}	8.206283×10^{-22}	1.581225×10^{-21}
1 cal	$= 4.655328 \times 10^{-14}$	4.184*	2.611448×10^{19}	1	0.9993312	3.965667×10^{-3}	$1.1622222... \times 10^{-6}$	1.558562×10^{-6}	3.085960	2.143028×10^{-2}	0.04129287
$1\ cal_{IT}$	$= 4.658443 \times 10^{-14}$	4.1868*	2.613195×10^{19}	1.000669	1	3.968321×10^{-3}	1.163000×10^{-6}	1.559609×10^{-6}	3.088025	2.144462×10^{-2}	0.04132050
$1\ Btu_{IT}$	$= 1.173908 \times 10^{-11}$	1055.056	6.585141×10^{21}	252.1644	251.9958	1	2.930711×10^{-4}	3.930148×10^{-4}	778.1693	5.403953	10.41259
1 kWh	$= 4.005540 \times 10^{-8}$	3600000.*	2.246944×10^{25}	860420.7	859845.2	3412.142	1	1.341022	2655224.	18349.06	35529.24
1 hp-h	$= 2.986931 \times 10^{-8}$	2684519.	1.675545×10^{25}	641615.6	641186.5	2544.33	0.7456998	1	1980000.*	13750.	26494.15
1 ft-lbf	$= 1.508551 \times 10^{-14}$	1.355818	8.462351×10^{18}	0.3240483	0.3238315	1.285067×10^{-3}	3.766161×10^{-7}	$5.050505... \times 10^{-7}$	1	$6.944444... \times 10^{-3}$	0.01338088
$1\ ft^3 \cdot lbf \cdot in.^{-2}$	$= 2.172313 \times 10^{-12}$	195.2378	1.218579×10^{21}	46.66295	46.63174	0.1850497	5.423272×10^{-5}	$7.272727... \times 10^{-5}$	144.*	1	1.926847
1 liter-atm	$= 1.127393 \times 10^{-12}$	101.3250	6.324210×10^{20}	24.21726	24.20106	0.09603757	2.814583×10^{-5}	3.774419×10^{-5}	74.73349	0.5189825	1

*Usage of the asterisk is defined in the accompanying text.

This change led to the SI definition of temperature in terms of the temperature of the triple point of water, which contains exactly 273.16 kelvins.

It follows that the Celsius temperture (°C) is an intermediate scale. It is useful in defining Kelvin temperature in the SI. Celsius temperature (t) is related to Kelvin temperature (K) as follows:

$$t_{\text{ice point}} = 0°C$$
$$t_{\text{steam point}} = 100°C$$
$$0 \text{ K} = -273.15°C$$

A summary of the conventions in the SI as proposed in the Thirteenth General Conference of Weights and Measures pertaining to temperature units is given below.

1. The unit of SI temperature is the kelvin, symbol K.

2. The word "scale" is not to be used except in terms of measurement of temperature between certain fixed points on the Celsius scale.

3. The terms "thermodynamic scale" or "absolute scale" are not to be used to describe temperature. The degree sign is to be eliminated with the symbol K.

4. When Celsius temperatures are used (°C), it is understood that the temperature unit is the kelvin.

Not all scientists and engineers have adopted the SI of temperature terminology. Furthermore, many engineers in the United States still use the Fahrenheit system in discussing practical engineering systems.

In converting Fahrenheit (°F) to Celsius (°C) the following formula applies.

$$°C = \frac{°F - 32°}{1.8}$$

In converting Celsius to Fahrenheit the following formula can be used.

$$°F = (°C \times 1.8) + 32°$$

In changing from Celsius terminology (t) to kelvin units (K) the following formula can be used.

$$K = t + 273.15$$

Chemistry

Symbols for the chemical elements

The mass number, atomic number, number of atoms, and ionic charge of an element are often indicated by means of four indices placed around the symbol. The positions occupied are left upper index, mass number; left lower index, atomic number; right upper index, ionic charge; and right lower index, number of atoms of an element in a molecule or formula unit of a given species: for example, $^{12}_{6}C$, Ca^{2+}, O_2, and Al_2O_3. The atomic number, which is redundant, is omitted in most cases: that is, $^{12}_{6}C$ can be written as ^{12}C.

Ionic charge is indicated by a plus or minus superscript following the symbol of the ion; for multiple charges an arabic superscript numeral precedes the plus or minus sign, for example, Na^+, NO_3^-, Ca^{2+}, PO_4^{3+}.

An alphabetical list of the elements, their symbols, and their atomic numbers is shown in **Table 6.** Elements 110–112 have been reported, but no official names have been assigned.

Chemical nomenclature

The International Union of Pure and Applied Chemistry (IUPAC) has established definitive rules for chemical nomenclature. Chemical species are identified in the Dictionary by a systematic name, frequently accompanied by a formula. Occasionally common names are used.

For inorganic compounds, systematic names of compounds are formed by identifying the constituents and their proportions in a specific order, for example, dinitrogen oxide (N_2O). Also accepted by IUPAC is Stock's system, in which the proportions of the constituents are indicated indirectly, and roman numerals are used to represent the oxidation number or stoichiometric valence of an element, for example, iron(II) chloride ($FeCl_2$). Complex compounds are also named according to rules specified by IUPAC; an example is potassium oxodichloroimidophosphate, $K[POCl_2(NH)]$. Examples of accepted trivial names are diborane (B_2H_6), silane (SiH_4), and ammonia (NH_3).

There also are definitive rules for naming organic compounds. Because of the infinite variety of disciplines and industrial applications involving organic compounds, the rules encompass different types of names. Sometimes a single compound can correctly be identified by a number of names; for example, chloral hydrate is also known as 2,2,2-trichloro-1,1-ethanediol and trichloroacetaldehyde monohydrate.

TABLE 6. The chemical elements

Name	Symbol	At. no.	Name	Symbol	At. no.	Name	Symbol	At. no.
Actinium	Ac	89	Germanium	Ge	32	Praseodymium	Pr	59
Aluminum	Al	13	Gold	Au	79	Promethium	Pm	61
Americium	Am	95	Hafnium	Hf	72	Protactinium	Pa	91
Antimony	Sb	51	Hassium	Hs	108	Radium	Ra	88
Argon	Ar	18	Helium	He	2	Radon	Rn	86
Arsenic	As	33	Holmium	Ho	67	Rhenium	Re	75
Astatine	At	85	Hydrogen	H	1	Rhodium	Rh	45
Barium	Ba	56	Indium	In	49	Rubidium	Rb	37
Berkelium	Bk	97	Iodine	I	53	Ruthenium	Ru	44
Beryllium	Be	4	Iridium	Ir	77	Rutherfordium	Rf	104
Bismuth	Bi	83	Iron	Fe	26	Samarium	Sm	62
Bohrium	Bh	107	Krypton	Kr	36	Scandium	Sc	21
Boron	B	5	Lanthanum	La	57	Seaborgium	Sg	106
Bromine	Br	35	Lawrencium	Lr	103	Selenium	Se	34
Cadmium	Cd	48	Lead	Pb	82	Silicon	Si	14
Calcium	Ca	20	Lithium	Li	3	Silver	Ag	47
Californium	Cf	98	Lutetium	Lu	71	Sodium	Na	11
Carbon	C	6	Magnesium	Mg	12	Strontium	Sr	38
Cerium	Ce	58	Manganese	Mn	25	Sulfur	S	16
Cesium	Cs	55	Meitnerium	Mt	109	Tantalum	Ta	73
Chlorine	Cl	17	Mendelevium	Md	101	Technetium	Tc	43
Chromium	Cr	24	Mercury	Hg	80	Tellurium	Te	52
Cobalt	Co	27	Molybdenum	Mo	42	Terbium	Tb	65
Copper	Cu	29	Neodymium	Nd	60	Thallium	Tl	81
Curium	Cm	96	Neon	Ne	10	Thorium	Th	90
Dubnium	Db	105	Neptunium	Np	93	Thulium	Tm	69
Dysprosium	Dy	66	Nickel	Ni	28	Tin	Sn	50
Einsteinium	Es	99	Niobium	Nb	41	Titanium	Ti	22
Element 110*		110	Nitrogen	N	7	Tungsten	W	74
Element 111*		111	Nobelium	No	102	Uranium	U	92
Element 112*		112	Osmium	Os	76	Vanadium	V	23
Erbium	Er	68	Oxygen	O	8	Xenon	Xe	54
Europium	Eu	63	Palladium	Pd	46	Ytterbium	Yb	70
Fermium	Fm	100	Phosphorus	P	15	Yttrium	Y	39
Fluorine	F	9	Platinum	Pt	78	Zinc	Zn	30
Francium	Fr	87	Plutonium	Pu	94	Zirconium	Zr	40
Gadolinium	Gd	64	Polonium	Po	84			
Gallium	Ga	31	Potassium	K	19			

*This element does not have an official name or symbol.

Symbols in scientific writing

Symbols commonly encountered in scientific writing are listed in **Table 7**. Symbols following the ellipses and separated by commas are alternatives that are used only when there is some reason for not using the symbol given first. The letters of the Greek alphabet, frequently used to represent terms, are shown in **Table 8**.

Some frequently encountered symbols for particles and quanta are as follows:

neutron	n	pion	π
proton	p	muon	μ
deuteron	d	electron	e
triton	t	neutrino	ν
alpha particle	α	photon	γ

The meaning of abbreviated notations for nuclear reactions should be the following:

initial nuclide (incoming particle(s) or quanta, outgoing particle(s) or quanta) final nuclide

Some examples are:

$$^{14}\mathrm{N}(\alpha, p)^{17}\mathrm{O} \qquad ^{59}\mathrm{Co}(n, \gamma)^{60}\mathrm{Co}$$
$$^{23}\mathrm{Na}(\gamma, 3n)^{20}\mathrm{Na} \qquad ^{31}\mathrm{P}(\gamma, pn)^{29}\mathrm{Si}$$

TABLE 7. Commonly used symbols in scientific writing

Space, time, mass, and related quantities

length l
height h
radius r
diameter d
path, length of arc s
plane angle $\alpha, \beta, \gamma, \theta, \phi, \psi$
solid angle ω
area A, S
volume $V \ldots v$
specific volume v
wavelength λ
wavenumber σ, ν
time t
period or other characteristic interval T, τ
frequency ν, f
angular frequency $(2\pi\nu)$ ω
velocity $\nu \ldots u, w$
angular velocity ω
acceleration a
acceleration of free fall g
mass m
moment of inertia I
density ρ
relative density d

Molecular and related quantities

molecular mass m
molar mass M
Avogadro's number N_0, L, N
number of molecules N
number of moles n
mole fraction $x \ldots X, y$
molality m
concentration c
molar concentration of substance B $c_B, [B], c(B)$
molecular concentration C
partition function Q
statistical weight $g \ldots p$
symmetry number σ
characteristic temperature θ
diameter of molecule $\sigma \ldots D$
mean free path l
diffusion coefficient D
osmotic pressure Π
surface concentration Γ

Mechanical and related quantities

force F
force due to gravity (weight) $G \ldots W$
moment of force M
power P
pressure p, P
traction σ
shear stress τ
modulus of elasticity E
shear modulus G
compressibility κ
compression modulus $(1/\kappa)$ K
viscosity η
fluidity ϕ
kinematic viscosity v
friction coefficient f
surface tension $\gamma \ldots \sigma$
angle of contact θ

Thermodynamic and related quantities

temperature $\theta \ldots t$
temperature, absolute T
gas constant R
Boltzmann constant k
heat q, Q
work w, A
energy (Gibbs ε) $E \ldots U$
entropy (Gibbs η) S
*Helmholtz free energy (Gibbs Ψ) A
enthalpy (Gibbs χ) H
*Gibbs function (ζ) $G \ldots F$
heat capacity C
specific heat at constant pressure c_p
specific heat at constant volume c_v
ratio of specific heats c_p/c_v γ, κ
chemical potential μ
activity, absolute λ
activity, relative a
activity coefficient f, γ
osmotic coefficient g, ϕ
thermal conductivity γ
Joule-Thomson coefficient μ

*The terms for the Helmholtz and Gibbs energies were modified by action of the IUPAC Council. Montreal, August 1961, as follows: Helmholtz energy (Gibbs $\Psi = E_2 TS$): A Gibbs energy (Gibbs $\zeta = H_2 TS$): G

TABLE 7. Commonly used symbols in scientific writing (cont.)

Chemical reactions

stoichiometric number of molecules (negative for reactants, positive for products) ν

standard equation of chemical reaction $\Sigma\nu_B B = 0$

affinity $(-\Sigma\nu_{B\mu B})$ of a reaction A

equilibrium constant K

equilibrium quotient or equilibrium product (of molalities) Q

extent of reaction $(dn_B = \nu_B d\xi)$ ξ

degree of reaction (e.g., degree of dissociation) α

rate constant k

collision number (collisions per unit volume and unit time) Z

rate constant corresponding to the rate Z z

rate of reaction $\nu \ldots r, s, J$

Light

Planck's constant h

Planck's constant divided by 2π $\hbar$

quantity of light q

radiant power, flux of light (dQ/dt) Φ

illumination $(d\Phi/dS)$ E

luminance L, B

luminous emittance H

absorption factor (fraction of incident radiant power which is absorbed) α

reflection factor (fraction of incident radiant power which is reflected) ρ

transmission factor (fraction of incident radiant power which is transmitted) τ

transmittance $(T = I/I_0)$ T

absorption (extinction) coefficient $[\kappa/c = \ln(1/T)]$ κ

absorbance (extinction) $[A = \log(1/T)]$ $A \ldots E$

absorptivity (specific absorbance) [decadic absorption or extinction coefficient] a

molar absorptivity (molar decadic absorption or extinction coefficient) $[\varepsilon/c = A]$ ε

refraction index n

refractivity r

angle of optical rotation α

Electricity and magnetism

elementary change e

quantity of electricity Q

change density ρ

surface change density σ

electric current $I \ldots i$

electric current density J

electric potential V

electric field strength E

electric displacement D

electrokinetic potential ζ

capacitance C

permittivity (dielectric constant) ϵ

dielectric polarization P

dipole moment μ

electric polarizability of a molecule α, γ

magnetic field strength H

magnetic induction B

magnetic permeability μ

magnetization M

magnetic susceptibility χ

resistance R

resistivity ρ

self inductance L

mutual inductance M, L_{12}

reactance X

impedance Z

admittance Y

Electrochemistry

Faraday's constant (the faraday) F

charge number of an ion, plus or minus z

degree of electrolytic dissociation α

ionic strength $I \ldots \mu$

electrolytic conductivity (specific conductance) κ

equivalent or molar conductance of electrolyte or ion A

transport number t, T

electromotive force E

overpotential η

TABLE 8. Greek alphabet

Upper and lower cases	Name	Upper and lower cases	Name
A α	Alpha	N ν	Nu
B β	Beta	Ξ ξ	Xi
Γ ξ	Gamma	O o	Omicron
Δ δ ∂	Delta	Π π	Pi
E ϵ ε	Epsilon	P ρ	Rho
Z ζ	Zeta	Σ σ	Sigma
H η	Eta	T τ	Tau
Θ θ	Theta	Υ υ	Upsilon
I ι	Iota	Φ ϕ φ	Phi
K κ	Kappa	X χ	Chi
Λ λ	Lambda	Ψ ψ	Psi
M μ	Mu	Ω ω	Omega

Periodic table of the elements

Filled Shells	**1**				KEY	

KEY

Oxidation States →
+1
+3
79 ← Atomic Num

Au ← Symbol

Gold ← Name

196.97

-32-18-1 ← Electron Configuration

Atomic Weight → 196.97

	1	2			Transition Element		
			3	**4**	**5**	**6**	**7**	**8**

Group/Period	1	2	3	4	5	6	7	8
Filled Shells **1**	+1 **1** H Hydrogen 1.0079 1							
	+1 **3** Li Lithium 6.941 2-1	+2 **4** Be Beryllium 9.012 2-2						
	+1 **11** Na Sodium 22.990 2-8-1	+2 **12** Mg Magnesium 24.305 2-8-2						
2	+1 **19** K Potassium 39.098 -8-8-1	+2 **20** Ca Calcium 40.078 -8-8-2	+3 **21** Sc Scandium 44.956 -8-9-2	+2 +3 +4 **22** Ti Titanium 47.867 -8-10-2	+2 +3 +4 +5 **23** V Vanadium 50.942 -8-11-2	+2 +3 +6 **24** Cr Chromium 51.996 -8-13-1	+2 +3 +4 +7 **25** Mn Manganese 54.938 -8-13-2	+2 +3 **26** Fe Iron 55.845 -8-14-2
2-8	+1 **37** Rb Rubidium 85.468 -18-8-1	+2 **38** Sr Strontium 87.62 -18-8-2	+3 **39** Y Yttrium 88.906 -18-9-2	+4 **40** Zr Zirconium 91.224 -18-10-2	+3 +5 **41** Nb Niobium 92.906 -18-12-2	+6 **42** Mo Molybdenum 95.94 -18-13-1	+4 +6 +7 **43** Tc Technetium (98) -18-14-1	+3 **44** Ru Ruthenium 101.07 -18-15-1
2-8-18	+1 **55** Cs Cesium 132.91 -18-8-1	+2 **56** Ba Barium 137.33 -18-8-2	57-71	+4 **72** Hf Hafnium 178.49 -32-10-2	+5 **73** Ta Tantalum 180.95 -32-11-2	+6 **74** W Tungsten 183.84 -32-12-2	+4 +6 +7 **75** Re Rhenium 186.21 -32-13-2	+3 +4 **76** Os Osmium 190.23 -32-14-2
2-8-18-32	+1 **87** Fr Francium (223) -18-8-1	+2 **88** Ra Radium (226) -18-8-2	89-103	+4 **104** Rf Rutherfordium (261) -32-10-2	**105** Db Dubnium (262) -32-11-2	**106** Sg Seaborgium (266) -32-12-2	**107** Bh Bohrium (264) -32-13-2	**108** Hs Hassium (277) -32-14-2

Lanthanides	+3 **57** La Lanthanum 138.91 -18-9-2	+3 +4 **58** Ce Cerium 140.12 -19-9-2	+3 **59** Pr Praseodymium 140.91 -21-8-2	+3 **60** Nd Neodymium 144.24 -22-8-2	+3 **61** Pm Promethium (145) -23-8-2	+2 +3 **62** Sm Samarium 150.36 -24-8-2	+2 +3 **63** Eu Europium 151.96 -25-8-2
Actinides	+3 **89** Ac Actinium (227) -18-9-2	+4 **90** Th Thorium 232.04 -18-10-2	+4 +5 **91** Pa Protactinium 231.04 -20-9-2	+3 +4 +5 +6 **92** U Uranium 238.03 -21-9-2	+3 +4 +5 +6 **93** Np Neptunium (237) -22-9-2	+3 +4 +5 +6 **94** Pu Plutonium (244) -24-8-2	+3 +4 +5 +6 **95** Am Americium (243) -25-8-2

*Atomic weights are those of the most commonly available long-lived isotopes on the 1999 IUPAC Atomic Weights of the Elements. A value in parentheses denotes the mass number of the longest-lived isotope.

18

					0 **2** **He** Helium 4.0026 2
13	**14**	**15**	**16**	**17**	

| +3   **5**
 B
 Boron
 10.811
 2-3 | +2 +4   **6**
 C
 Carbon
 12.011
 2-4 | ±1 ±2 ±3 +4 +5   **7**
 N
 Nitrogen
 14.007
 2-5 | −2   **8**
 O
 Oxgen
 15.999
 2-6 | −1   **9**
 F
 Fluorine
 18.998
 2-7 | 0   **10**
 Ne
 Neon
 20.180
 2-8 |
| +3   **13**
 Al
 Aluminum
 26.982
 2-8-3 | +2 +4   **14**
 Si
 Silicon
 28.086
 2-8-4 | ±3 +5   **15**
 P
 Phosphorus
 30.974
 2-8-5 | +4 +6 −2   **16**
 S
 Sulfur
 32.065
 2-8-6 | ±1 +5 +7   **17**
 Cl
 Chlorine
 35.453
 2-8-7 | 0   **18**
 Ar
 Argon
 39.948
 2-8-8 |

9	**10**	**11**	**12**						
+2 +3 **27** **Co** Cobalt 58.933 -8-15-2	+2 +3 **28** **Ni** Nickel 58.693 -8-16-2	+1 +2 **29** **Cu** Copper 63.546 -8-18-1	+2 **30** **Zn** Zinc 65.39 -8-18-2	+3 **31** **Ga** Gallium 69.723 -8-18-3	+2 +4 **32** **Ge** Germanium 72.64 -8-18-4	±3 +5 **33** **As** Arsenic 74.922 -8-18-5	+4 +6 −2 **34** **Se** Selenium 78.96 -8-18-6	−1 +5 **35** **Br** Bromine 79.904 -8-18-7	0 **36** **Kr** Krypton 83.80 -8-18-8
+3 **45** **Rh** Rhodium 102.91 -18-16-1	+2 +4 **46** **Pd** Palladium 106.42 -18-18-0	+1 **47** **Ag** Silver 107.87 -18-18-1	+2 **48** **Cd** Cadmium 112.41 -18-18-2	+3 **49** **In** Indium 114.82 -18-18-3	+2 +4 **50** **Sn** Tin 118.71 -18-18-4	±3 +5 **51** **Sb** Antimony 121.76 -18-18-5	+4 +6 −2 **52** **Te** Tellurium 127.60 -18-18-6	±1 +5 +7 **53** **I** Iodine 126.90 -18-18-7	0 **54** **Xe** Xenon 131.29 -18-18-8
+3 +4 **77** **Ir** Iridium 192.22 -32-15-2	+2 +4 **78** **Pt** Platinum 195.08 -32-17-1	+1 +3 **79** **Au** Gold 196.97 -32-18-1	+1 +2 **80** **Hg** Mercury 200.59 -32-18-2	−1 −3 **81** **Tl** Thallium 204.38 -32-18-3	+2 +4 **82** **Pb** Lead 207.2 -32-18-4	+3 +5 **83** **Bi** Bismuth 208.98 -32-18-5	+2 +4 **84** **Po** Polonium (209) -32-18-6	±1 +5 +7 **85** **At** Astatine (210) -32-18-7	0 **86** **Rn** Radon (222) -32-18-8
109 **Mt** Meitnerium (268) -32-15-2	**110**	**111**	**112**						

+3 **64** **Gd** Gadolinium 157.25 -25-9-2	+3 **65** **Tb** Terbium 158.92 -27-8-2	+3 **66** **Dy** Dysprosium 162.50 -28-8-2	+3 **67** **Ho** Holmium 164.93 -29-8-2	+3 **68** **Er** Erbium 167.26 -30-8-2	+3 **69** **Tm** Thulium 163.93 -31-8-2	+2 +3 **70** **Yb** Ytterbium 173.04 -32-8-2	+3 **71** **Lu** Lutetium 174.97 -32-9-2
+3 **96** **Cm** Curium (247) -25-9-2	+3 +4 **97** **Bk** Berkelium (247) -27-8-2	+3 **98** **Cf** Californium (251) -28-8-2	+3 **99** **Es** Einsteinium (252) -29-8-2	+3 **100** **Fm** Fermium (257) -30-8-2	+2 +3 **101** **Md** Mendelevium (258) -31-8-2	+2 +3 **102** **No** Nobelium (259) -32-8-2	+3 **103** **Lr** Lawrencium (262) -32-9-2

Mathematics

Mathematical signs and symbols

Symbol	Definition	Symbol	Definition	Symbol	Definition
+	plus (sign of addition)	$\neq$, $\ne$	not equal to	$\partial u/\partial x$	partial derivative of u with respect to x
+	positive	$\rightarrow \doteq$	approaches	$\int$	integral of
−	minus (sign of subtraction)	$\propto$	varies as	$\int_b^a$	integral of, between limits a and b
−	negative	∞	infinity		
$\pm$ ($\mp$)	plus or minus (minus or plus)	$\sqrt{}$	square root of	$\oint$	line integral around a closed path
$\times$	times, by (multiplication sign)	$\sqrt[3]{}$	cube root of	Σ	(sigma) summation of
$\cdot$	multiplied by	$\therefore$	therefore	$f(x)$, $F(x)$	functions of x
$\div$	sign of division	$\parallel$	parallel to	∇	del or nabla, vector differential operator
/	divided by	( ) [ ] { }	parentheses, brackets and braces;	∇^2	Laplacian operator
:	ratio sign, divided by, is to		quantities enclosed by them to be	£	Laplace operational symbol
::	equals, as (proportion)		taken together in multiplying	4!	factorial $4 = 1 \times 2 \times 3 \times 4$
<	less than		dividing, etc.	$\|x\|$	absolute value of x
>	greater than	$\overline{AB}$	length of line from A to B	$\dot{x}$	first derivative of x with respect to time
$\ll$	much less than	π	(pi) = 3.14159 +	$\ddot{x}$	second derivative of x with respect to time
$\gg$	much greater than	°	degrees	$\mathbf{A} \times \mathbf{B}$	vector product; magnitude of $\mathbf{A}$ times
=	equals	′	minutes		magnitude of $\mathbf{B}$ times sine of
$\equiv$	identical with	″	seconds		the angle from $\mathbf{A}$ to $\mathbf{B}$; $AB \sin \overline{AB}$
$\sim$	similar to	$\angle$	angle	$\mathbf{A} \cdot \mathbf{B}$	scalar product of $\mathbf{A}$ and $\mathbf{B}$;
$\approx$	approximately equals	dx	differential of x		magnitude of $\mathbf{A}$ times magnitude
$\cong$	approximately equals, congruent	Δ	(delta) difference		of $\mathbf{B}$ times cosine of the angle from
$\leq$	equal to or less than	Δx	increment of x		$\mathbf{A}$ to $\mathbf{B}$; $AB \cos \overline{AB}$
$\geq$	equal to or greater than				

Mathematical notation

Mathematical logic

p, q, $P(x)$	Sentences, propositional functions, propositions
$\neg p$, $\sim p$, non p, Np	Negation, read "not p" ($\neq$: read "not equal")
$p \vee q$, $p + q$, Apq	Disjunction, read "p or q," "p, q," or both
$p \wedge q$, $p \cdot q$, $p\&q$, Kpq	Conjunction, read "p and q"
$p \rightarrow q$, $p \supset q$, $p \Rightarrow q$, Cpq	Implication, read "p implies q" or "if p then q"
$p \leftrightarrow q$, $p \equiv q$, $p \Leftrightarrow q$, Epq, p iff q	Equivalence, read "p is equivalent to q" or "p if and only if q"
n.a.s.c.	Read "necessary and sufficient condition"
( ), [ ], { }, $\ldots$	Parentheses
V, $\forall$, Σ	Universal quantifier, read "for all" or "for every"
$\exists$, $\exists$, Π	Existential quantifier, read "there is a" or "there exists"
$\vdash$	Assertion sign ($p \vdash q$: read "q follows from p"; $\vdash p$: read "p is or follows from an axiom," or "p is a tautology"
0, 1	Truth, falsity (values)
=	Identity
$\overset{Df}{=}$, $\overset{df}{=}$, $\overset{}{=}_{df}$, $\equiv$	Definitional identity
∎	"End of proof"; "QED"

Set theory, relations, functions

X, Y	Sets
$x \in X$	x is a member of the set X
$x \notin X$	x is not a member of X
$A \subset X$, $A \subseteq X$	Set A is contained in set X
$A \not\subset X$, $A \not\subseteq X$	A is not contained in X
$X \cup Y$, $X + Y$	Union of sets X and Y
$X \cap Y$, $X \cdot Y$	Intersection of sets X and Y
$+$, $\div$, $\bigcirc$	Symmetric difference of sets
$\cup X_i$, ΣX	Union of all the sets X_i
$\cap X_i$, ΠX_i	Intersection of all the sets X_i
$\varnothing$, 0, Λ	Null set, empty set
X', $\complement X$, CX	Complement of the set X
$X - Y$, $X \backslash Y$	Difference of sets X and Y
$\hat{x}(P(x))$, $\{x \| P(x)\}$, $\{x{:}P(x)\}$	The set of all x with the property P
(x,y,z), (x, y, z)	Ordered set of elements x, y, and z; to be distinguished from (x,z,y), for example
$\{x,y,z\}$	Unordered set, the set whose elements are x, y, z, and no others
$\{a_1, a_2, \ldots, a_n\}$, $\{a_i\}_{i=1,2,\ldots,n}$, $\{a_i\}_{i=1}^n$	The set whose members are a_i, where i is any whole number from 1 to n
$\{a_1, a_2 \ldots\}$, $\{a_i\}_{i=1,2,\ldots}$, $\{a_i\}_{i=1}^{\infty}$	The set whose members are a_i, where i is any positive whole number
$X \times Y$	Cartesian product, set of all (x,y) such that $x \in X$, $y \in Y$

$\{a_i\}_{i\in I}$	The set whose elements are a_i, where $i \in I$
$Ry, R\{x,y\}$	Relation
$\equiv, \cong, \sim, \approx$	Equivalence relations, for example, congruence
$=, \geq, >, \varepsilon, \geqslant, \leqq, \leq, <$	Transitive relations, for example, numerical order
$X \to Y, X \xrightarrow{f} Y, X \to Y, f \in Y^X$	Function, mapping, transformation
$f^{-1}, \overset{-1}{f}, X \xleftarrow{f^{-1}} Y$	Inverse mapping
$g \circ f$	Composite functions: $(g \circ f)(x) = g(f(x))$
$f(X)$	Image of X by f
$f^{-1}(X)$	Inverse-image set, counter image
1-1, one-one	Read "one-to-one correspondence"
$\begin{array}{c} X \xrightarrow{f} Y \\ \phi\downarrow \quad \downarrow\psi \\ W \xrightarrow{g} Z \end{array}$	Diagram: the diagram is commutative in case $\psi \circ f = g \circ \phi$
$f|A$	Partial mapping, restriction of function f to set A
$\overline{X}, \text{card } X, |X|$	Cardinal of the set A
$\aleph_0, d$	Denumerable infinity
$c, 2^{\aleph_0}$	Power of continuum
ω	Order type of the set of positive integers
	Read "countably"

Number, numerical functions

1.4; 1,4; 1·4	Read "one and four-tenths"
1(1)2(10)100	Read "from 1 to 20 in intervals of 1, and from 20 to 100 in intervals of 10"
const	Constant
$A \geqq 0$	The number A is nonnegative, or the matrix A is positive definite, or the matrix A has nonnegative entries
$x|y$	Read "x divides y"
$x \equiv y \bmod p$	Read "x congruent to y modulo p"
$a_0 + \cfrac{1}{a_1} + \cfrac{1}{a_2} + \cdots,$ $a_0 + \cfrac{1|}{|a_1} + \cdots$	Continued fractions
$[a,b]$	Closed interval
$[a,b), [a,b[$	Half-open interval (open at the right)
$(a,b),]a,b[$	Open interval
$[a,\infty), [a,\to[$	Interval closed at the left, infinite to the right
$(-\infty, \infty), [\leftarrow,\to[$	Set of all real numbers
$\max_{x\in X} f(x), \max\{f(x)|x \in X\}$	Maximum of $f(x)$ when x is in the set X
min	Minimum
sup, l.u.b.	Supremum, least upper bound
inf, g.l.b.	Infimum, greatest lower bound
$\lim_{x\to a} f(x) = b, \lim_{x=a} f(x) = b, f(x) \to b$ as $x \to a$	b is the limit of $f(x)$ as x approaches a
$\lim_{x\to a^-} f(x), \lim_{x=a-0} f(x), f(a-)$	Limit of $f(x)$ as x approaches a from the left
$\limsup, \overline{\lim}$	Limit superior
$\liminf, \underline{\lim}$	Limit inferior
l.i.m.	Limit in the mean
$z = x + iy = re^{i\theta}, \zeta = \xi + i\eta, w = u + iv = \rho e^{i\phi}$	Complex variables
z^*	Complex conjugate
Re, $\Re$	Real part
Im, $\Im$	Imaginary part
arg	Argument
$\dfrac{\partial(u,v)}{\partial(x,y)}, \dfrac{D(u,v)}{D(x,y)}$	Jacobian, functional determinant
$\displaystyle\int_E f(x)\, d\mu(x)$	Integral (for example, Lebesgue integral) of function f over set E with respect to measure μ
$f(n) \sim \log n$ as $n \to \infty$	$f(n)/\log n$ approaches 1 as $n \to \infty$
$f(n) = O(\log n)$ as $n \to \infty$	$f(n)/\log n$ is bounded as $n \to \infty$
$f(n) = o(\log n)$	$f(n)/\log n$ approaches zero
$f(x) \nearrow b, f(x) \uparrow b$	$f(x)$ increases, approaching the limit b
$f(x) \downarrow b, f(x) \searrow b$	$f(x)$ decreases, approaching the limit b

a.e., p.p.	Almost everywhere
ess sup	Essential supremum
$C^0, C^0(X), C(X)$	Space of continuous functions
$C^k, C^k[a,b]$	The class of functions having continuous kth derivative (on $[a,b]$)
C'	Same as C^1
$\text{Lip}_\alpha, \text{Lip } \alpha$	Lipschitz class of functions
$L^p, L_p, L^p[a,b]$	Space of functions having integrable absolute pth power (on $[a,b]$)
L'	Same as L^1
$(C\,\alpha), (C,p)$	Cesàro summability

Special functions

$[x]$	The integral part of x
$\dbinom{n}{k}, {}^nC_k, {}_nC_k$	Binomial coefficient $n!/k!(n-k)!$
$\left(\dfrac{n}{p}\right)$	Legendre symbol
$e^x, \exp x$	Exponential function
$\sinh x, \cosh x, \tanh x$	Hyperbolic functions
$\text{sn } x, \text{cn } x, \text{dn } x$	Jacobi elliptic functions
$\wp(x)$	Weierstrass elliptic function
$\Gamma(x)$	Gamma function
$J_v(x)$	Bessel function
$\chi_X(x)$	Characteristic function of the set X: $\chi_X(x) = 1$ in case $x \in X$, otherwise $\chi_X(x) = 0$
$\text{sgn } x$	Signum: sgn $0 = 0$, while sgn $x = x/|x|$ for $x \neq 0$
$\delta(x)$	Dirac delta function

Algebra, tensors, operators

$+, \cdot, \times, \circ, \top, \tau$	Laws of composition in algebraic systems
$e,\ 0$	Identity, unit, neutral element (of an additive system)
$e,\ 1,\ I$	Identity, unit, neutral element (of a general algebraic system)
$e,\ e,\ E,\ P$	Idempotent
a^{-1}	Inverse of a
$\text{Hom}(M,N)$	Group of all homomorphisms of M into N
G/H	Factor group, group of cosets
$[K:k]$	Dimension of K over k
$\oplus, +$	Direct sum
$\otimes$	Tensor product, Kronecker product
$\wedge$	Exterior product, Grassmann product
$\vec{x}, \mathbf{x}, \mathfrak{x}, x$	Vector
$\vec{x} \cdot \vec{y}, \mathbf{x} \cdot \mathbf{y}, (\mathfrak{x}, \mathfrak{y})$	Inner product, scalar product, dot product
$\mathbf{x} \times \mathbf{y}, [\mathfrak{x},\mathfrak{y}], \mathbf{x} \wedge \mathbf{y}$	Outer product, vector product, cross product
$|\mathfrak{x}|, |x|, \|x\|, \|x\|_p$	Norm of the vector x
Ax, xA	The image of x under the transformation A
δ_{ij}	Kronecker delta: $\delta_{ij} = 1$, while $\delta_{ij} = 0$ for $i \neq j$
$A', {}^tA, A^t, {}^tA$	Transpose of the matrix A
$A^*, \bar{A}$	Adjoint, Hermitian conjugate of A
$\text{tr } A, \text{Sp } A$	Trace of the matrix A
$\det A, |A|$	Determinant of the matrix A
$\Delta^n f(x), \Delta_k^n f, \Delta^n f(x)$	Finite differences
$[x_0,x_1], [x_0,x_1,x_2], \underset{x_1}{\Delta u_{x_0}}, [x_0,x_1]_f$	Divided differences
$\nabla f, \text{grad } f$	Read "gradient of f"
$\nabla \cdot v, \text{div } v$	Read "divergence of $\mathbf{v}$"
$\nabla \times v, \text{curl } v, \text{rot } \mathbf{v}$	Read "curl of $\mathbf{v}$"
$\nabla^2, \Delta, \text{div grad}$	Laplacian
$[X,Y]$	Poisson bracket, or commutator, or Lie product
$GL(n,R)$	Full linear group of degree n over field R
$O(n,R)$	Full orthogonal group
$SO(n,R), O^+(n,R)$	Special orthogonal group

Topology

E^n	Euclidean n space
S^n	n sphere
$\rho(p,q), d(p,q)$	Metric, distance (between points p and q)
$\overline{X}, X^-$, cl X, X^c	Closure of the set X
FrX, fr$X$, ∂X, bdry X	Frontier, boundary of X
int X, $\overset{\circ}{X}$	Interior of X
T_2 space	Hausdorff space
F_σ	Union of countably many closed sets
G_δ	Intersection of countably many open sets
dim X	Dimensionality, dimension of X
$\pi_1(X)$	Fundamental group of the space X
$\pi_n(X)$, $\pi_n(X,A)$	Homotopy groups
$H_n(X)$, $H_n(X,A;G)$, $H_*(X)$	Homology groups
$H^n(X)$, $H^n(X,A;G)$, $H^*(X)$	Cohomology groups

Probability and statistics

X, Y	Random variables
$P(X \leq 2)$, $\mathrm{Pr}(X \leq 2)$	Probability that $X \leq 2$
$P(X \leq 2 | Y \geq 1)$	Conditional probability
$E(X), \mathscr{E}(X)$	Expectation of X
$E(X | Y \geq 1)$	Conditional expectation
c.d.f.	Cumulative distribution function
p.d.f.	Probability density function
c.f.	Characteristic function
$\bar{x}$	Mean (especially, sample mean)
σ, s.d.	Standard deviation
σ^2, Var, var	Variance
$\mu_1, \mu_2, \mu_3, \mu_i, \mu_{ij}$	Moments of a distribution
ρ	Coefficient of correlation
$\rho_{12\cdot34}$	Partial correlation coefficient

Fundamental constants

Recommended values (1998) of selected fundamental constants of physics and chemistry[a]

Quantity	Symbol	Numerical value[b]	Unit[c]	Relative uncertainty (standard deviation)
UNIVERSAL CONSTANTS				
Speed of light in vacuum	c, c_0	299 792 458	m s^{-1}	(exact)
Magnetic constant	μ_0	$4\pi \times 10^{-7}$	N A^{-2}	
		$= 12.566\ 370\ 614\ldots \times 10^{-7}$	N A^{-2}	(exact)
Electric constant, $1/\mu_0 c^2$	ε_0	$8.854\ 187\ 817\ldots \times 10^{-12}$	F m^{-1}	(exact)
Characteristic impedance of vacuum, $\sqrt{\mu_0/\varepsilon_0} = \mu_0 c$	Z_0	$376.730\ 313\ 461\ldots$	Ω	(exact)
Newtonian constant of gravitation	G	$6.673(10) \times 10^{-11}$	m^3 kg^{-1} s^{-2}	1.5×10^{-3}
	$G/\hbar c$	$6.707(10) \times 10^{-39}$	(GeV/c^2)$^{-2}$	1.5×10^{-3}
Planck constant	h	$6.626\ 068\ 76(52) \times 10^{-34}$	J s	7.8×10^{-8}
in eV s		$4.135\ 667\ 27(16) \times 10^{-15}$	eV s	3.9×10^{-8}
hc in eV m		$1.239\ 841\ 857(49) \times 10^{-6}$	eV m	3.9×10^{-8}
$h/2\pi$	$\hbar$	$1.054\ 571\ 596(82) \times 10^{-34}$	J s	7.8×10^{-8}
in eV s		$6.582\ 118\ 89(26) \times 10^{-16}$	eV s	3.9×10^{-8}
$\hbar c$ in eV m		$197.326\ 9602(77) \times 10^{-9}$	eV m	3.9×10^{-8}
Planck mass, $(\hbar c/G)^{1/2}$	m_p	$2.1767(16) \times 10^{-8}$	kg	7.5×10^{-4}
Planck length, $\hbar/m_p c = (\hbar G/c^3)^{1/2}$	l_p	$1.6160(12) \times 10^{-35}$	m	7.5×10^{-4}
Planck time, $l_p/c = (\hbar G/c^5)^{1/2}$	t_p	$5.3906(40) \times 10^{-44}$	s	7.5×10^{-4}
ELECTROMAGNETIC CONSTANTS				
Elementary charge	e	$1.602\ 176\ 462(63) \times 10^{-19}$	C	3.9×10^{-8}
Magnetic flux quantum, $h/2e$	Φ_0	$2.067\ 833\ 636(81) \times 10^{-15}$	Wb	3.9×10^{-8}
Josephson constant[d], $2e/h$	K_J	$483\ 597.898(19) \times 10^9$	Hz V^{-1}	3.9×10^{-8}
von Klitzing constant[e], $h/e^2 = \mu_0 c/2\alpha$	R_K	$25\ 812.807\ 572(95)$	Ω	3.7×10^{-9}
Bohr magneton, $e\hbar/2m_e$	μ_B	$927.400\ 899(37) \times 10^{-26}$	J T^{-1}	4.0×10^{-8}
in eV T^{-1}		$5.788\ 381\ 749(43) \times 10^{-5}$	eV T^{-1}	7.3×10^{-9}
	μ_B/h	$13.996\ 246\ 24(56) \times 10^9$	Hz T^{-1}	4.0×10^{-8}
	μ_B/hc	$46.686\ 4521(19)$	m^{-1}T^{-1}	4.0×10^{-8}
	$\mu_B k$	$0.671\ 7131(12)$	K T^{-1}	1.7×10^{-6}
Nuclear magneton, $e\hbar/2m_p$	μ_N	$5.050\ 783\ 17(20) \times 10^{-27}$	J T^{-1}	4.0×10^{-8}
in eV T^{-1}		$3.152\ 451\ 238(24) \times 10^{-8}$	eV T^{-1}	7.6×10^{-9}
	μ_N/h	$7.622\ 593\ 96(31)$	MHz T^{-1}	4.0×10^{-8}
	μ_N/hc	$2.542\ 623\ 66(10) \times 10^{-2}$	m^{-1}T^{-1}	4.0×10^{-8}
	μ_N/k	$3.658\ 2638(64) \times 10^{-4}$	K T^{-1}	1.7×10^{-6}
ATOMIC AND NUCLEAR CONSTANTS				
General				
Fine-structure constant, $e^2/4\pi\varepsilon_0 \hbar c$	α	$7.297\ 352\ 533(27) \times 10^{-3}$		3.7×10^{-9}
Inverse fine-structure constant	α^{-1}	$137.035\ 999\ 76(50)$		3.7×10^{-9}
Rydberg constant, $\alpha^2 m_e c/2h$	R_∞	$10\ 973\ 731.568\ 549(83)$	m^{-1}	7.6×10^{-12}
	$R_\infty c$	$3.289\ 841\ 960\ 368(25) \times 10^{15}$	Hz	7.6×10^{-12}
$R_\infty hc$ in eV		$13.605\ 691\ 72(53)$	eV	3.9×10^{-8}
Bohr radius, $\alpha/4\pi R_\infty = 4\pi\varepsilon_0 \hbar^2/m_e e^2$	a_0	$0.529\ 177\ 2083(19) \times 10^{-10}$	m	3.7×10^{-9}
Hartree energy, $e^2/4\pi\varepsilon_0 a_0 = 2R_\infty hc =$				
$\alpha^2 m_e c^2$	E_h	$4.359\ 743\ 81(34) \times 10^{-18}$	J	7.8×10^{-8}
in eV		$27.211\ 3834(11)$	eV	3.9×10^{-8}

[a]Footnotes are on page 2342.

Recommended values (1998) of selected fundamental constants of physics and chemistry[a] (*cont.*)

Quantity	Symbol	Numerical value[b]	Unit[c]	Relative uncertainty (standard deviation)
ATOMIC AND NUCLEAR CONSTANTS (cont.)				
Electroweak				
Fermi coupling constant[f]	$G_F/(\hbar c)^3$	$1.166\ 39(1) \times 10^{-5}$	GeV^{-2}	8.6×10^{-6}
Electron, e^-				
Electron mass	m_e	$9.109\ 381\ 88(72) \times 10^{-31}$	kg	7.9×10^{-8}
in u		$5.485\ 799\ 110(12) \times 10^{-4}$	u	2.1×10^{-9}
Energy equivalent in MeV	$m_e c^2$	$0.510\ 998\ 902(21)$	MeV	4.0×10^{-8}
Electron charge to mass quotient	$-e/m_e$	$-1.758\ 820\ 174(71) \times 10^{11}$	C kg^{-1}	4.0×10^{-8}
Compton wavelength, $h/m_e c$	λ_C	$2.426\ 310\ 215(18) \times 10^{-12}$	m	7.3×10^{-9}
$\lambda_C/2\pi = \alpha a_0 = \alpha^2/4\pi R_\infty$	λ_C	$386.159\ 2642(28) \times 10^{-15}$	m	7.3×10^{-9}
Classical electron radius, $\alpha^2 a_0$	r_e	$2.817\ 940\ 285(31) \times 10^{-15}$	m	1.1×10^{-8}
Thomson cross section, $(8\pi/3)r_e^2$	σ_e	$0.665\ 245\ 854(15) \times 10^{-28}$	m^2	2.2×10^{-8}
Electron magnetic moment	μ_e	$-928.476\ 362(37) \times 10^{-26}$	J T^{-1}	4.0×10^{-8}
Electron magnetic moment anomaly, $\|\mu_e\|/\mu_B - 1$	a_e	$1.159\ 652\ 1869(41) \times 10^{-3}$		3.5×10^{-9}
Electron gyromagnetic ratio, $2\|\mu_e\|/\hbar$	γ_e	$1.760\ 859\ 794(71) \times 10^{11}$	$s^{-1}\ T^{-1}$	4.0×10^{-8}
	$\gamma_e/2\pi$	$28\ 024.9540(11)$	MHz T^{-1}	4.0×10^{-8}
Muon, μ^-				
Muon mass	m_μ	$1.883\ 531\ 09(16) \times 10^{-28}$	kg	8.4×10^{-8}
in u		$0.113\ 428\ 9168(34)$	u	3.0×10^{-8}
Energy equivalent in MeV	$m_\mu c^2$	$105.658\ 3568(52)$	MeV	4.9×10^{-8}
Muon-electron mass ratio	m_μ/m_e	$206.768\ 2657(63)$		3.0×10^{-8}
Muon Compton wavelength, $h/m_\mu c$	$\lambda_{C,\mu}$	$11.734\ 441\ 97(35) \times 10^{-15}$	m	2.9×10^{-8}
$\lambda_{C,\mu}/2\pi$	$\lambda_{C,\mu}$	$1.867\ 594\ 444(55) \times 10^{-15}$	m	2.9×10^{-8}
Muon magnetic moment	μ_μ	$-4.490\ 448\ 13(22) \times 10^{-26}$	J T^{-1}	4.9×10^{-8}
Muon magnetic moment anomaly, $\|\mu_\mu\|/(e\hbar/2m_\mu) - 1$	a_μ	$1.165\ 916\ 02(64) \times 10^{-3}$		5.5×10^{-7}
Tau, τ^-				
Tau mass[g]	m_τ	$3.167\ 88(52) \times 10^{-27}$	kg	1.6×10^{-4}
in u		$1.907\ 74(31)$	u	1.6×10^{-4}
Energy equivalent in MeV	$m_\tau c^2$	$1\ 777.05(29)$	MeV	1.6×10^{-4}
Tau-electron mass ratio	m_τ/m_e	$3\ 477.60(57)$		1.6×10^{-4}
Proton, p				
Proton mass	m_p	$1.672\ 621\ 58(13) \times 10^{-27}$	kg	7.9×10^{-8}
in u		$1.007\ 276\ 466\ 88(13)$	u	1.3×10^{-10}
Energy equivalent in MeV	$m_p c^2$	$938.271\ 998(38)$	MeV	4.0×10^{-8}
Proton-electron mass ratio	m_p/m_e	$1\ 836.152\ 6675(39)$		2.1×10^{-9}
Proton charge to mass quotient	e/m_p	$9.578\ 834\ 08(38) \times 10^7$	C kg^{-1}	4.0×10^{-8}
Proton Compton wavelength, $h/m_p c$	$\lambda_{C,p}$	$1.321\ 409\ 847(10) \times 10^{-15}$	m	7.6×10^{-9}
$\lambda_{C,p}/2\pi$	$\lambda_{C,p}$	$0.210\ 308\ 9089(16) \times 10^{-15}$	m	7.6×10^{-9}
Proton magnetic moment	μ_p	$1.410\ 606\ 633(58) \times 10^{-26}$	J T^{-1}	4.1×10^{-8}
to nuclear magneton ratio	μ_p/μ_N	$2.792\ 847\ 337(29)$		1.0×10^{-8}
Shielded proton magnetic moment[h]	μ_p'	$1.410\ 570\ 399(59) \times 10^{-26}$	J T^{-1}	4.2×10^{-8}
to nuclear magneton ratio	μ_p'/μ_N	$2.792\ 775\ 597(31)$		1.1×10^{-8}
Proton gyromagnetic ratio, $2\mu_p/\hbar$	γ_p	$2.675\ 222\ 12(11) \times 10^8$	$s^{-1}\ T^{-1}$	4.1×10^{-8}
	$\gamma_p/2\pi$	$42.577\ 4825(18)$	MHz T^{-1}	4.1×10^{-8}
Shielded proton gyromagnetic ratio[h], $2\mu_p'/\hbar$	γ_p'	$2.675\ 153\ 41(11) \times 10^8$	$s^{-1}\ T^{-1}$	4.2×10^{-8}
	$\gamma_p'/2\pi$	$42.576\ 3888(18)$	MHz T^{-1}	4.2×10^{-8}
Neutron, n				
Neutron mass	m_n	$1.674\ 927\ 16(13) \times 10^{-27}$	kg	7.9×10^{-8}
in u		$1.008\ 664\ 915\ 78(55)$	u	5.4×10^{-10}
Energy equivalent in MeV	$m_n c^2$	$939.565\ 330(38)$	MeV	4.0×10^{-8}
Neutron magnetic moment	μ_n	$-0.966\ 236\ 40(23) \times 10^{-26}$	J T^{-1}	2.4×10^{-7}
to nuclear magneton ratio	μ_n/μ_N	$-1.913\ 042\ 72(45)$		2.4×10^{-7}
Deuteron, d				
Deuteron mass	m_d	$3.343\ 583\ 09(26) \times 10^{-27}$	kg	7.9×10^{-8}
In u		$2.013\ 553\ 212\ 71(35)$	u	1.7×10^{-10}
Energy equivalent in MeV	$m_d c^2$	$1\ 875.612\ 762(75)$	MeV	4.0×10^{-8}
Deuteron magnetic moment	μ_d	$0.433\ 073\ 457(18) \times 10^{-26}$	J T^{-1}	4.2×10^{-8}
to nuclear magneton ratio	μ_d/μ_N	$0.857\ 438\ 2284(94)$		1.1×10^{-8}

Recommended values (1998) of selected fundamental constants of physics and chemistry[a] *(cont.)*

Quantity	Symbol	Numerical value[b]	Unit[c]	Relative uncertainty (standard deviation)
PHYSICOCHEMICAL CONSTANTS				
Avogadro constant	N_A, L	$6.022\,141\,99(47) \times 10^{23}$	mol^{-1}	7.9×10^{-8}
Atomic mass constant, $m_u = \frac{1}{12} m(^{12}C) = 1$ u $= 10^{-3}$ kg mol$^{-1}/N_A$	m_u	$1.660\,538\,73(13) \times 10^{-27}$	kg	7.9×10^{-8}
Energy equivalent in MeV	$m_u c^2$	$931.494\,013(37)$	MeV	4.0×10^{-8}
Faraday constant[i], $N_A e$	F	$96\,485.3415(39)$	C mol^{-1}	4.0×10^{-8}
Molar Planck constant	$N_A h$	$3.990\,312\,689(30) \times 10^{-10}$	J s mol^{-1}	7.6×10^{-9}
	$N_A hc$	$0.119\,626\,564\,92(91)$	J m mol^{-1}	7.6×10^{-9}
Molar gas constant	R	$8.314\,472(15)$	J mol^{-1} K^{-1}	1.7×10^{-6}
Boltzmann constant, R/N_A	k	$1.380\,6503(24) \times 10^{-23}$	J K^{-1}	1.7×10^{-6}
in eV K^{-1}		$8.617\,342(15) \times 10^{-5}$	eV K^{-1}	1.7×10^{-6}
	k/h	$2.083\,6644(36) \times 10^{10}$	Hz K^{-1}	1.7×10^{-6}
	k/hc	$69.503\,56(12)$	m^{-1}K^{-1}	1.7×10^{-6}
k^{-1} in K eV^{-1}		$11\,604.506(20)$	K eV^{-1}	1.7×10^{-6}
Molar volume of ideal gas, RT/p, for $T = 273.15$ K, $p = 101.325$ kPa	V_m	$22.413\,996(39) \times 10^{-3}$	m^3 mol^{-1}	1.7×10^{-6}
Loschmidt constant, N_A/V_m	n_0	$2.686\,7775(47) \times 10^{25}$	m^{-3}	1.7×10^{-6}
Stefan-Boltzmann constant, $(\pi^2/60)k^4/\hbar^3 c^2$	σ	$5.670\,400(40) \times 10^{-8}$	W m^{-2} K^{-4}	7.0×10^{-6}
First radiation constant, $2\pi hc^2$	c_1	$3.741\,771\,07(29) \times 10^{-16}$	W m^2	7.8×10^{-8}
Second radiation constant, hc/k	c_2	$1.438\,7752(25) \times 10^{-2}$	m K	1.7×10^{-8}
Wien displacement law constant, $b = \lambda_{max}T = c_2/4.965\,114\,231\ldots$	b	$2.897\,7686(51) \times 10^{-3}$	m K	1.7×10^{-6}
NON-SI UNITS ACCEPTED FOR USE WITH THE SI				
Electronvolt: (e/C) J	eV	$1.602\,176\,462(63) \times 10^{-19}$	J	3.9×10^{-8}
(Unified) atomic mass unit: 1 u $= m_u = \frac{1}{12} m(^{12}C)$	u	$1.660\,538\,73(13) \times 10^{-27}$	kg	7.9×10^{-8}

[a]This table presents a selection of the values of the fundamental constants recommended by the Committee on Data for Science and Technology (CODATA). These "1998 CODATA recommended values" form a self-consistent set based on the data available through December 31, 1998, and are generally recognized for use in all fields of science and technology. There is a detailed description of the data and analysis that led to these results in P. J. Mohr and B. N. Taylor, CODATA recommended values of the fundamental constants: 1998, in the *Journal of Physical and Chemical Reference Data*, vol. 28, no. 6, pp. 1713-1852, 1999, and the *Reviews of Modern Physics*, vol. 72, no. 2, pp. 351-495, 2000, prepared under the auspices of the CODATA Task Group on Fundamental Constants. This table is a selection from Table XXIV of that paper, with a few additions based on Tables XXVIII and XXX. The recommended values are also available on the World Wide Web at http://physics.nist.gov/cuu/Constants/index.html.

[b]The digits in parentheses represent one-standard-deviation uncertainties in the final two digits of the quoted values.

[c]A = ampere, C = coulomb, F = farad, Hz = hertz, J = joule, K = kelvin, kg = kilogram, m = meter, mol = mole, N = newton, Pa = pascal, s = second, T = tesla, W = watt, Wb = weber, Ω = ohm, eV = electronvolt, u = (unified) atomic mass unit. Prefixes: k = 10^3, M = 10^6, G = 10^9.

[d]The conventional value of the Josephson constant, adopted internationally for realizing repesentations of the volt using the Josephson effect, is $K_{J-90} = 483\,597.9$ GHz V^{-1}.

[e]The conventional value of the von Klitzing constant, adopted internationally for realizing representations of the ohm using the quantum Hall effect, is $R_{K-90} = 25\,812.807\ \Omega$

[f]Value recommended by the Particle Data Group in the 1998 Review of Particle Physics (C. Caso et al., *The European Physical Journal*, vol. C3, pp. 1-794, 1998). The value recommended in the 2002 Review of Particle Physics [K. Hagiwara et al. (Particle Data Group), *Physical Review D*, vol. 66, Paper 010001, 2002 (http://pdg.lbl.gov)] is $G_F/(\hbar c)^3 = 1.166\,37(1) \times 10^{-5}$ GeV^{-2}.

[g]This and all other values involving m_τ are based on the value of $m_\tau c^2$ in MeV recommended by the Particle Data Group in the 1998 Review of Particle Physics (*ibid.*), but with a standard uncertainty of 0.29 MeV rather than the quoted uncertainty of -0.26 MeV, $+0.29$ MeV. The value of $m_\tau c^2$ recommended in the 2002 Review of Particle Physics (*ibid.*) is 1776.99 MeV with an uncertainty of -0.26 MeV, $+0.29$ MeV.

[h]Based on nuclear magnetic resonance (NMR) frequency of protons in a sphere of pure water (H_2O) at 25°C surrounded by vacuum.

[i]The numerical value of F to be used in coulometric chemical measurements is 96 485.3432(76) [relative uncertainty = 7.9×10^{-8}] when the relevant current is measured in terms of representations of the volt and ohm based on the Josephson and quantum Hall effects and the conventional values of the Josephson and von Klitzing constants, K_{J-90} and R_{K-90}.

Elementary particles

The elementary particles[a]

Gauge bosons $J_C^P = 1^-$ Self-conjugate except $\overline{W^+} = W^-$.

Name	Symbol	Charge[b]	Mass, GeV	Couplings
Photon	γ	0	0	$A \Rightarrow \gamma A$
Gluon[c]	g	0	0	$A \Rightarrow gA'$
Weak bosons[d]				
Charged	$W^\pm$	± 1	80	$U \Rightarrow W^+D$
Neutral	Z^0	0	91	$A \Rightarrow Z^0A$

Fermions $J = \frac{1}{2}$ All have distinct antiparticles, except perhaps the neutrinos.

Name	Charge[b]	Symbol	Mass, GeV	Symbol	Mass, GeV	Symbol	Mass, GeV
Leptons							
Neutrinos	0	ν_e	$<3 \times 10^{-9}$	ν_μ	<0.0002	ν_τ	<0.02
Charged leptons[e]	-1	e	0.0005	μ	0.106^f	τ	1.78^f
Quarks[c]							
Up type	$\frac{2}{3}$	u	0.001–0.005	c	1.15–1.35	t	170–180
Down type	$-\frac{1}{3}$	d	0.003–0.009	s	0.075–0.17	b	4.0–4.4

[a] The graviton, with $J_C^P = 2^+$, has been omitted, since t plays no role in high-energy particle physics.

[b] In units of the proton charge.

[c] The gluon is a color SU_3 octet {8}; each quark is a color triplet {3}. These colored particles are confined constituents of hadrons; they do not appear as free particles.

[d] The observation of weak bosons was reported in 1983.

[e] Any further charged leptons have mass greater than 92 GeV.

[f] The μ and τ leptons are unstable, with the following mean life and principal decay modes (branching ratios in %):

$$\mu \quad \tau_\mu = 2.2 \times 10^{-6}\ \text{s} \qquad e\overline{\nu}_e\nu_\mu\ 100$$

$$\tau \quad \tau_\tau = 2.9 \times 10^{-13}\ \text{s} \qquad \begin{array}{l} \mu\overline{\nu}_\mu\nu_\tau\ 17 \\ e\overline{\nu}_e\nu_\tau\ 18 \\ (\text{hadrons})^-\nu_\tau\ 65 \end{array}$$

Schematic electronic symbols*

Ammeter		Coaxial cable	
Amplifier, general		Crystal, piezoelectric	
Amplifier, inverting		Delay line	
Amplifier, operational		Diac	
and.gate		Diode, field-effect	
Antenna, balanced		Diode, general	
Antenna, general		Diode, Gunn	
Antenna, loop		Diode, light-emitting	
Antenna, loop, multiturn		Diode, photosensitive	
Battery		Diode, PIN	
Capacitor, feedthrough		Diode, Schottky	
Capacitor, fixed		Diode, tunnel	
Capacitor, variable		Diode, varactor	
Capacitor, variable, split-rotor		Diode, Zener	
Capacitor, variable, split-stator		Directional coupler	
Cathode, electron-tube, cold		Directional wattmeter	
Cathode, electron-tube, directly heated		Exclusive-OR gate	
Cathode, electron-tube indirectly heated		Female contact, general	
Cavity resonator		Ferrite bead	
Cell, electrochemical			
Circuit breaker			

*From S. Gibilisco, *The Illustrated Dictionary of Electronics*, 8th ed., McGraw-Hill, 2001.

Filament, electron-tube		Inductor, powdered-iron core	
Fuse		Inductor, powdered-iron core, bifilar	
Galvanometer		Inductor, powdered-iron core, tapped	
Grid, electron-tube		Inductor, powdered-iron core, variable	
Ground, chassis			or
Ground, earth		Integrated, circuit, general	
Headset		Jack, coaxial or photo	
Handset, double		Jack, phone, two-conductor	
Headset, single		Jack, phone, three-conductor	
Headset, stereo		Key, telegraph	
Inductor, air core		Lamp, incandescent	
Inductor, air core, bifilar		Lamp, neon	
Inductor, air core, tapped		Male contact, general	
Inductor, air core, variable		Meter, general	
Inductor, iron core		Microammeter	
Inductor, iron core, bifilar		Microphone	
Inductor, iron core, tapped		Microphone, directional	
		Milliammeter	
		NAND gate	
		Negative voltage connection	
Inductor iron core, variable		NOR gate	

NOT gate	
Optoisolator	
OR gate	
Outlet, two-wire, nonpolarized	
Outlet, two-wire, polarized	
Outlet, three-wire	
Outlet, 234-V	
Plate, electron-tube	
Plug, two-wire, nonpolarized	
Plug, two-wire, polarized	
Plug, three-wire	
Plug, 234-V	
Plug, coaxial or phono	
Plug, phone, two-conductor	
Plug, phone, three-conductor	
Positive voltage connection	
Potentiometer	
Probe, radio-frequency	

Rectifier, gas-filled	
Rectifier, high-vacuum	
Rectifier, semiconeductor	
Rectifier, silicon-controlled	
Relay, double-pole, double-throw	
Relay, double-pole, single-throw	
Relay, single-pole, double-throw	
Relay, single-pole, single-throw	
Resistor, fixed	
Resistor, preset	
Resistor, tapped	
Resonator	
Rheostat	
Saturable reactor	
Signal generator	
Solar battery	

Solar cell

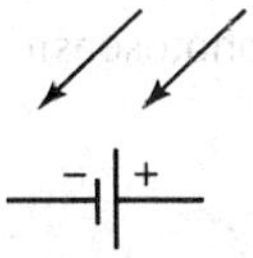

Source, constant-current

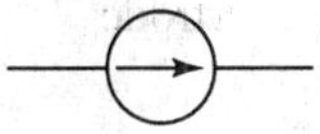

Source, constant-voltage

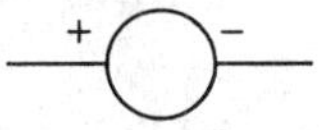

Speaker

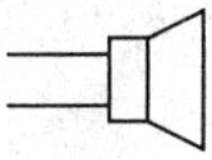

Switch, double-pole, double-throw

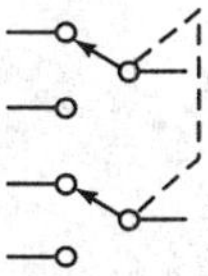

Switch, double-pole, rotary

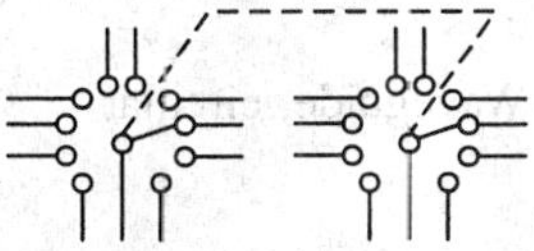

Switch, double-pole, single-throw

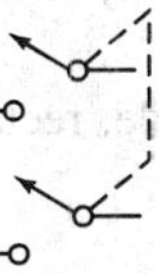

Switch, momentary-contact

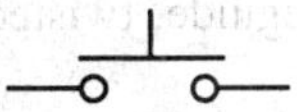

Switch, silicon-controlled

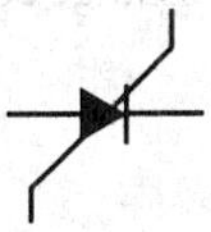

Switch, single-pole, rotary

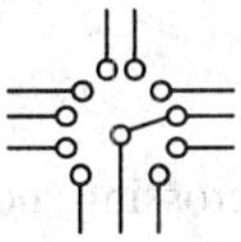

Switch, single-pole, double-throw

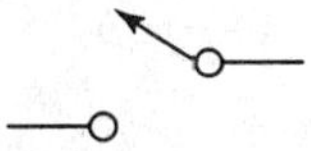

Switch, single-pole, single-throw

Terminals, general, balanced

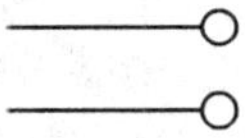

Terminals, general, unbalanced

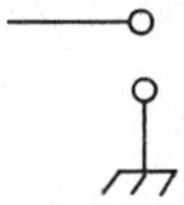

Test point

Thermocouple

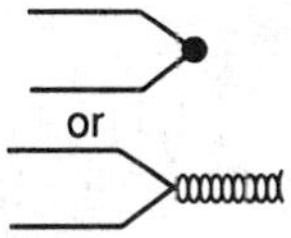

Transformer, air core

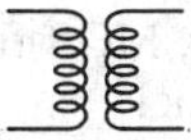

Transformer, air core, step-down

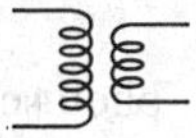

Transformer, air core, step-up

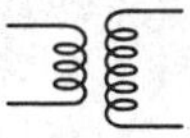

Transformer, air core, tapped primary

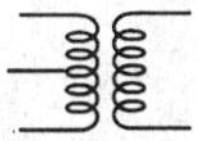

Transformer, air core, tapped secondary

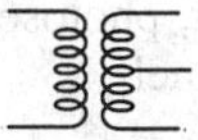

Transformer, iron core

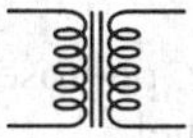

Transformer, iron core, step-down

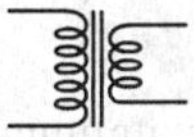

Transformer, iron core, step-up

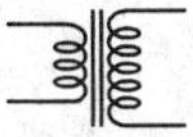

Transformer, iron core, tapped primary

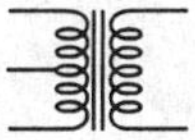

Transformer, iron core, tapped secondary

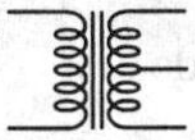

Transformer, powdered-iron core

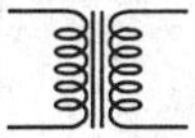

Transformer, powdered-iron core, step-down

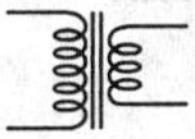

Transformer, powdered-iron core, step-up

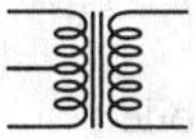

Transformer, powdered-iron core, tapped primary

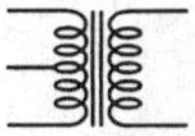

Transformer, powdered-iron core, tapped secondary

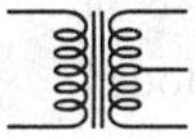

Transistor, bipolar, NPN

Transistor, bipolar, PNP

Transistor, field-effect, N-channel

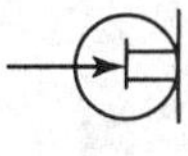

Transistor, field-effect, P-channel

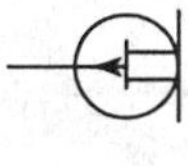

Transistor, MOS field-effect, N-channel

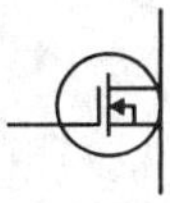

Transistor, MOS field-effect, *P*-channel

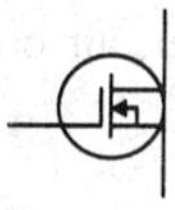

Transistor, photosensitive, *NPN*

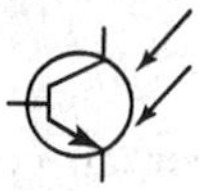

Transistor, photosensitive, *PNP*

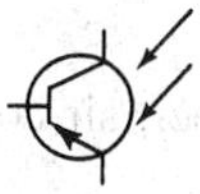

Transistor, photosensitive, field-effect, *N*-channel

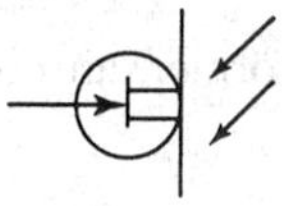

Transistor, photosensitive, field-effect, *P*-channel

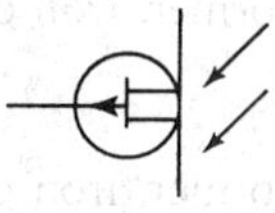

Transistor, unijunction

Triac

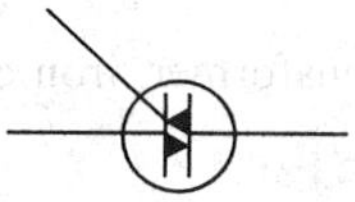

Tube, diode

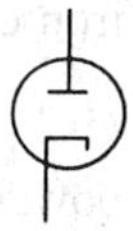

Tube, heptode

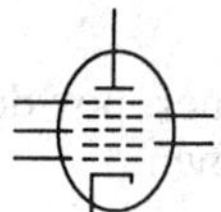

Tube, hexode

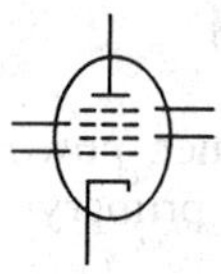

Tube, pentode

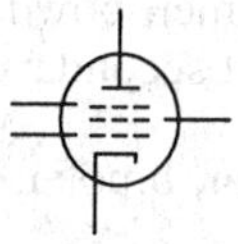

Tube, photosensitive

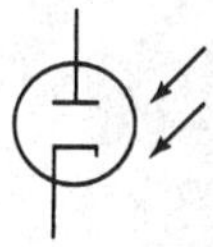

Tube, tertrode

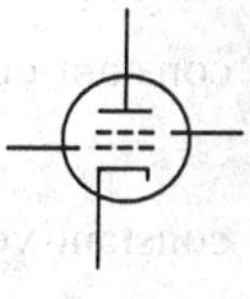

Tube, trirode

Voltmeter

Wattmeter

Waveguide, circular

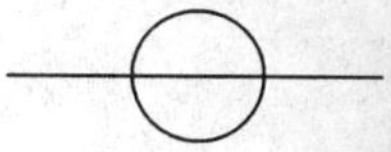

Waveguide, flexible

Waveguide, rectangular

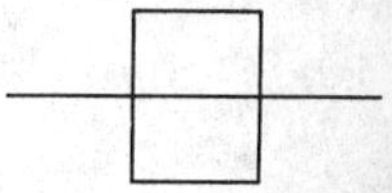

Waveguide, twisted

Wires, crossing, connected

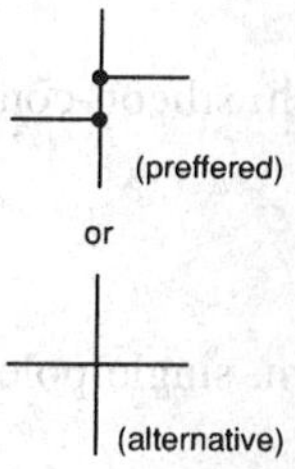

(preffered)

or

(alternative)

Wires, crossing, not connected

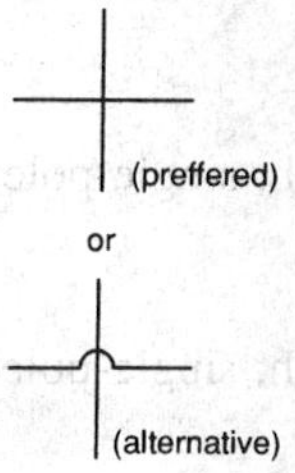

(preffered)

or

(alternative)

Geologic time scale and related aspects

Eon	Era	Period (Interval length, $\times 10^6$ years)		Epoch	Beginning of epoch, $\times 10^6$ years ago	Forms of life	Other features
Phanerozoic	Cenozoic "Age of Mammals"	Quaternary (1.8)		Recent	0.01	Modern humans	Extensive glaciation during portions of Quaternary
				Pleistocene	1.8	Early humans	
		Tertiary (63.2)		Pliocene	5.3	Large carnivores	
				Miocene	23.8	Whales, apes, grazing animals	Formation of most present mountain ranges
				Oligocene	33.7	Large browsing animals	
				Eocene	54.8	Rise of modern floras	
				Paleocene	65	First placental mammals	
	Mesozoic "Age of Reptiles"	Cretaceous (79)		Upper (Base Santonian)		Last of dinosaurs	Widespread seas over much of present-day Europe, North Africa, and North America
				Middle (Base Albian)		Last of ammonites	
				Lower	144	Rise of flowering plants	
		Jurassic (62)		Upper (Base Callovian)		Toothed birds	Beginning of formation of Atlantic Ocean
				Middle (Base Bajocian)		Flying reptiles	
				Lower	206	First primitive mammals	
		Triassic (44)		Upper		Rise of dinosaurs	
				Middle		Rise of ammonites	
				Lower	250	Rise of cycads	
	Paleozoic "Age of Invertebrates"	Permian (40)		Upper (Base Ochoan)		Primitive reptiles	Mountain building during late Paleozoic
				Middle (Base Guadalupian)		Last of trilobites	
				Lower	290	Glossopteris flora	
		Carboniferous[a]	Pennsylvanian (32)	Upper		Spread of amphibians	Widespread coal swamps
				Middle		Great coal forests	
				Lower	322	Climax of spore-forming plants	
			Mississippian (32)	Upper		Abundant sharks	
				Middle		Climax of crinoids and blastoids	
				Lower	354		
		Devonian (63)		Upper		First forests	
				Middle		Rise of ferns	
				Lower	417	Earliest known amphibians	
		Silurian (26)		Upper		Appearance of land plants First known scorpions	
				Middle		Expansion of brachiopods and corals	
				Lower	443		
		Ordovician (47)		Upper		Appearance of primitive fishes	
				Middle		Climax of trilobites	
				Lower	490	Rise of cephalopods	
		Cambrian (53)		Upper (Base Croixian)		Abundant trilobites	
				Middle		Many kinds of shelled invertebrates	
				Lower	543		
Proterozoic	Precambrian			Neoproterozoic (357[b])	900	Marine algae, wormlike organisms, other simple forms	Extensive glaciation Oldest dated rocks (Greenland)
				Mesoproterozoic (700[b])	1600		
				Paleoproterozoic (900[b])	~2500		
Archean[c]				Late (500[b])	3000	Blue-green algae (cyanobacteria)	Abundant dark sediments Formation of earliest known rocks
				Middle (400[b])	3400		
				Early (>400[b])	>3800	Bacteria	

[a] The period known as Carboniferous in Europe is divided into two periods in North American usage.
[b] Length of epoch, $\times 10^6$ years.
[c] The beginning of the Archean corresponds to the age of the oldest preserved continental crust.

Classification of living organisms

Domain Archaea[a]
　Phylum Crenarchaeota
　　Class Thermoprotei
　　　Order Thermoproteales
　　　Order Desulfurococcales
　　　Order Sulfolobales
　Phylum Euryarchaeota
　　Class Methanobacteria
　　　Order Methanobacteriales
　　Class Methanococci
　　　Order Methanococcales
　　　Order Methanomicrobiales
　　　Order Methanosarcinales
　　Class Halobacteria
　　　Order Halobacteriales
　　Class Thermoplasmata
　　　Order Thermoplasmatales
　　Class Thermococci
　　　Order Thermococcales
　　Class Archaeoglobi
　　Class Methanopyri
　　　Order Methanopyrales

Domain Bacteria
　Phylum Aquificae
　　Class Aquificae
　　　Order Aquificales
　Phylum Thermotogae
　　Class Thermotogae
　　　Order Thermotogales
　Phylum Thermodesulfobacteria
　　Class Thermodesulfobacteria
　　　Order Thermodesulfobacteriales
　Phylum Deinococcus-Thermus
　　Class Deinococci
　　　Order Deinococcales
　　　Order Thermales
　Phylum Chrysiogenetes
　　Class Chrysiogenetes
　　　Order Chrysiogenales
　Phylum Chloroflexi
　　Class Chloroflexi
　　　Order Chloroflexales
　　　Order Herpetosiphonales

Phylum Thermomicrobia
　Class Thermomicrobia
　　Order Thermomicrobiales
Phylum Nitrospira
　Class Nitrospira
　　Order Nitrospirales
Phylum Deferribacteres
　Class Deferribacteres
　　Order Deferribacterales
Phylum Cyanobacteria
　Class Cyanobacteria
Phylum Chlorobi
　Class Chlorobia
　　Order Chlorobiales
Phylum Proteobacteria
　Class Alphaproteobacteria
　　Order Rhodospirillales
　　Order Rickettsiales
　　Order Rhodobacterales
　　Order Sphingomonadales
　　Order Caulobacterales
　　Order Rhizobiales
　Class Betaproteobacteria
　　Order Burkholderiales
　　Order Hydrogenophilales
　　Order Methylophilales
　　Order Neisseriales
　　Order Nitrosomonadales
　　Order Rhodocyclales
　Class Gammaproteobacteria
　　Order Chromatiales
　　Order Acidithiobacillales
　　Order Xanthomonadales
　　Order Cardiobacteriales
　　Order Thiotrichales
　　Order Legionellales
　　Order Methylococcales
　　Order Oceanospirillales
　　Order Pseudomonadales
　　Order Alteromonadales
　　Order Vibrionales
　　Order Aeromonadales
　　Order Enterobacteriales
　　Order Pasteurellales

Class Deltaproteobacteria
　Order Desulfurellales
　Order Desulfovibrionales
　Order Desulfobacterales
　Order Desulfuromonadales
　Order Syntrophobacterales
　Order Bdellovibrionales
　Order Myxococcales
Class Epsilonproteobacteria
　Order Campylobacterales
Phylum Firmicutes
　Class Clostridia
　　Order Clostridiales
　　Order Thermoanaerobacteriales
　　Order Haloanaerobiales
　Class Mollicutes
　　Order Mycoplasmatales
　　Order Entomoplasmatales
　　Order Acholeplasmatales
　　Order Anaeroplasmatales
　Class Bacilli
　　Order Bacillales
　　Order Lactobacillales
Phylum Actinobacteria
　Class Actinobacteria
　　Subclass Acidimicrobidae
　　　Order Acidimicrobiales
　　　　Suborder Acidimicrobineae
　　Subclass Rubrobacteridae
　　　Order Rubrobacterales
　　　　Suborder Rubrobacterineae
　　Subclass Coriobacteridae
　　　Order Coriobacteriales
　　　　Suborder Coriobacterineae
　　Subclass Sphaerobacteridae
　　　Order Sphaeriobacterales
　　　　Suborder Sphaerobacterineae
　　Subclass Actinobacteridae
　　　Order Actinomycetales
　　　　Suborder Actinomycineae
　　　　Suborder Micrococcineae
　　　　Suborder Corynebacterineae
　　　　Suborder Micromonosporineae
　　　　Suborder Propionibacterineae

[a] Derived from G.M. Garrity, et. al., *Taxonomic Outline of the Prokaryotes*, Release 2, January 2002, Springer-Verlag, New York. 350 p. http://dx.doi.org/10.1007/bergeysoutline. Readers interested in determining taxonomic composition of lower taxa may obtain this document, free of charge.
[b] Condensed from Jan A. Pechenik, *Biology of the Invertebrates*, 4th ed., McGraw-Hill 2000.
[c] Condensed from Donald Linzey, *Vertebrate Biology*, Appendix I: Classification of Living Vertebrates, McGraw-Hill, 2001.
Note: The contributions of the following to the updating of this classification scheme are gratefully acknowledged: Dr. Craig Bailey; Dr. Mark Chase; Dr. George M. Garrity; Dr. S.C. Jong; Dr. Robert Knowlton; Dr. Donald Linzey.

Suborder Pseudonocardineae
Suborder Streptomycineae
Suborder Streptosporangineae
Suborder Frankineae
Suborder Glycomycineae
Order Bifidobacteriales
Phylum Planctomycetes
Class Planctomycetacia
Order Planctomycetales
Phylum Chlamydiae
Class Chlamydiae
Order Chlamydiales
Phylum Spirochaetes
Class Spirochaetes
Order Spirochaetales
Phylum Fibrobacteres
Class Fibrobacteres
Order Fibrobacterales
Phylum Acidobacteria
Class Acidobacteria
Order Acidobacteriales
Phylum Bacteroidetes
Class Bacteroidetes
Order Bacteroidales
Class Flavobacteria
Order Flavobacteriales
Class Sphingobacteria
Order Sphingobacteriales
Phylum Fusobacteria
Class Fusobacteria
Order Fusobacteriales
Phylum Verrucomicrobia
Class Verrucomicrobiae
Order Verrucomicrobiales
Phylum Dictyoglomus
Class Dictyoglomi
Order Dictyoglomales

Domain Eukarya[b]

Kingdom Protista
Phylum Metamonada
Phylum Trichozoa
Subphylum Parabasala
Class Trichomonadea
Class Hypermastigotea

Subkingdom Neozoa
Phylum Choanozoa
Phylum Amoebozoa
Subphylum Lobosa

Subphylum Conosa
Class Archamoebae
Class Mycetozoa
Phylum Foraminifera
Phylum Percolozoa
Phylum Euglenozoa
Class Euglenoidea
Class Saccostomae
Phylum Sporozoa
Subphylum Gregarinae
Subphylum Coccidiomorpha
Subphylum Perkinsida
Subphylum Manubrispora
Phylum Ciliophora
Phylum Radiozoa
Phylum Heliozoa
Phylum Rhodophyta
Class Rhodophyceae
Subclass Bangiophycidae
Order Bangiales
Order Compsopogonales
Order Porphyridiales
Order Rhodochaetales
Subclass Florideophycidae
Order Acrochaetiales
Order Ahnfeltiales
Order Balbianiales
Order Balliales
Order Batrachospermales
Order Bonnemaisoniales
Order Ceramiales
Order Colaconematales
Order Corallinales
Order Gelidiales
Order Gigartinales
Order Gracilariales
Order Halymeniales
Order Hildenbrandiales
Order Nemaliales
Order Palmariales
Order Plocamiales
Order Rhodogorgonales
Order Rhodymeniales
Order Thoreales
Phylum Chrysophyta
Class Bacillariophyceae
Subclass Bacillariophycidae
Order Achnanthales
Order Bacillariales
Order Cymbellales
Order Dictyoneidales

Order Lyrellales
Order Mastogloiales
Order Naviculales
Order Rhopalodiales
Order Surirellales
Order Thallassiophysales
Subclass Biddulphiophycidae
Order Anaulales
Order Biddulphiales
Order Hemiaulales
Order Triceratiales
Subclass Chaetocerotophycidae
Order Chaetocerotales
Order Leptocylindrales
Subclass Corethrophycidae
Order Cymatosirales
Subclass Coscinodiscophycidae
Order Arachnoidiscales
Order Asterolamprales
Order Aulacoseirales
Order Chrysaanthemodiscales
Order Coscinodiscales
Order Ethmodiscales
Order Melosirales
Order Orthoseirales
Order Paraliales
Order Stictocyclales
Order Stictodiscales
Subclass Cymatosirophycidae
Order Cymatosirales
Subclass Eunotiophycidae
Order Eunotiales
Subclass Fragilariophycidae
Order Ardissoneales
Order Cyclophorales
Order Climacospheniales
Order Fragilariales
Order Licmorphorales
Order Protoraphidales
Order Rhabdonematales
Order Rhaphoneidales
Order Striatellales
Order Tabellariales
Order Thalassionematales
Order Toxariales
Subclass Lithodesmiophycidae
Order Lithodesmiales
Subclass Rhizosoleniophycidae
Order Rhizosoleniales

Subclass Thalassiosirophycidae
  Order Thalassiosirales
Class Bolidophyceae
  Order Bolidomonadales
Class Chrysomerophyceae
  Order Chrysomeridales *nom. nud.*
Class Chrysophyceae
  Order Chromulinales
  Order Hibberdiales
Class Dictyochophyceae
  Order Dictyochales
  Order Pedinellales
  Order Rhizochromulinales
Class Eustigmatophyceae
  Order Eustigmatales
Class Pelagophyceae
  Order Pelagomonadales
  Order Sarcinochrysidales
Class Phaeophyceae
  Order Ascoseirales
  Order Chordariales
  Order Cutleriales
  Order Desmarestiales
  Order Dictyosiphonales
  Order Dictyotales
  Order Durvillaeales
  Order Ectocarpales
  Order Fucales
  Order Laminariales
  Order Scytosiphonales
  Order Sphacelariales
  Order Sporochnales
  Order Tilopteridiales
Class Phaeothamniophyceae
  Order Phaeothamniales
  Order Pleurochloridellales
Class Pinguiophyceae
  Order Pinguiochrysidales
Class Raphidophyceae
  Order Rhaphidomonadales
Class Synurophyceae
  Order Synurales
Class Xanthophyceae
  (=Tribophyceae)
  Order Botrydiales
  Order Chloramoebales
  Order Heterogloeales
  Order Mischococcales
  Order Rhizochloridales
  Order Tribonematales
  Order Vaucheriales

Phylum Cryptophyta
  Class Cryptophyceae
    Order Cryptomonadales
    Order Cryptococcales
Phylum Glaucocystophyta
  Class Glaucocystophyceae
    Order Cyanophorales
    Order Glaucocystales
    Order Gloeochaetales
Phylum Prymnesiophyta (=Haptophyta)
  Class Pavlovophyceae
    Order Pavlovales
  Class Prymnesiophyceae
    Order Coccolithales
    Order Isochrysidales
    Order Phaeocystales
    Order Prymnesiales
Phylum Dinophyta
  Class Dinophyceae
    Order Actiniscales
    Order Blastodiniales
    Order Chytriodiniales
    Order Desmocapsales
    Order Desmomonadales
    Order Dinophysales
    Order Gonyaulacales
    Order Gymnodiniales
    Order Kolkwitziellales
    Order Nannoceratopsiales
    Order Noctilucales
    Order Oxyrrhinales
    Order Peridiniales
    Order Phytodiniales
    Order Prorocentrales
    Order Ptychodiscales
    Order Pyrocysales
    Order Suessiales
    Order Syndiniales
    Order Thoracosphaerales
Phylum Chlorophyta
  Class Charophyceae
    Order Charales
    Order Chlorokybales
    Order Coleochaetales
    Order Klebsormidiales
    Order Zygnematales
  Class Chlorophyceae
    Order Chaetophorales
    Order Chlorococcales
    Order Cladophorales
    Order Odeogoniales

    Order Sphaeropleales
    Order Volvocales
    Order Pleurastrales
  Class Prasinophyceae
    Order Chlorodendrales
    Order Mamiellales
    Order Pseudoscourfeldiales
    Order Pyramimonidales
  Class Trebouxiophyceae
    Order Trebouxiales
  Class Ulvophyceae
    Order Bryopsidales
    Order Caulerpales
    Order Codiolales
    Order Dasycladales
    Order Halimedales
    Order Prasiolales
    Order Siphonocladales
    Order Trentepohliales
    Order Ulotrichales
    Order Ulvales
Phylum Euglenophyta
  Class Euglenophyceae
    Order Euglenales
    Order Euglenamorphales
    Order Eutreptiales
    Order Heteronematales
    Order Rhabdomonadales
    Order Sphenomonadales
Phylum Acrasiomycota
  Class Acrasiomycetes
    Order Acrasiales
Phylum Dictyosteliomycota
  Class Dictyosteliomycetes
    Order Dictyosteliales
Phylum Myxomycota
  Class Myxomycetes
    Order Liceales
    Order Echinosteliales
    Order Trichiales
    Order Physarales
    Order Stemonitales
    Order Ceratiomyxales
  Class Protosteliomycetes
    Order Protosteliales
Phylum Plasmodiophoromycota
  Class Plasmodiophoromycetes
    Order Plasmodiophorales
Phylum Oomycota
  Class Oomycetes
    Order Saprolegniales

Order Salilagenidiales

Order Lagenidiales

Order Leptomitales

Order Myzocytiopsidales

Order Rhipidiales

Order Pythiales

Order Peronosporales

Phylum Hyphochytriomycota

Class Hyphochytriomycetes

Order Hyphochytriales

Phylum Labyrinthulomycota

Class Labyrinthulomycetes

Order Labyrinthulales

Phylum Chytridiomycota

Class Chytridiomycetes

Order Blastocladiales

Order Chytridiales

Order Monoblepharidales

Order Neocallimastigales

Order Spizellomycetales

Phylum Zygomycota

Class Trichomycetes

Order Amoebidiales

Order Asellariales

Order Eccrinales

Order Harpellales

Class Zygomycetes

Order Mucorales

Order Dimargaritales

Order Kickxellales

Order Endogonales

Order Glomales

Order Entomophthorales

Order Zoopagales

Phylum Ascomycota

Class Archiascomycetes

Order Taphrinales

Order Schizosaccharomycetales

Class Saccharomycetes

Order Saccharomycetales

Class Plectomycetes

Order Eurotiales

Order Ascosphaerales

Order Onygenales

Class Laboulbeniomycetes

Order Laboulbeniales

Order Spathulosporales

Class Pyrenomycetes

Order Hypocreales

Order Melanosporales

Order Microascales

Order Phylachorales

Order Ophiostomatales

Order Diaporthales

Order Calosphaceriales

Order Xylariales

Order Sordariales

Order Meliolales

Order Halosphaeriales

Class Discomycetes

Order Medeolariales

Order Rhytismatales

Order Ostropales

Order Cyttariales

Order Helotiales

Order Neolectales

Order Gyalectales

Order Lecanorales

Order Lichinales

Order Peltigerales

Order Pertusariales

Order Teloschistales

Order Caliciales

Order Pezizales

Class Loculoascomycetes

Order Coryneliales

Order Dothideales

Order Myriangiales

Order Arthoniales

Order Pyrenulales

Order Asterinales

Order Capnodiales

Order Chaetothyriales

Order Patellariales

Order Pleosporales

Order Melanommatales

Order Trichotheliales

Order Verrucariales

Phylum Basidiomycota

Class Basidiomycetes

Subclass Heterobasidiomycetes

Order Agaricostilbales

Order Atractiellales

Order Auriculariales

Order Heterogastridiales

Order Tremellales

Subclass Homobasidiomycetes

Order Agaricales

Order Boletales

Order Bondarzewiales

Order Cantharellales

Order Ceratobasidiales

Order Cortinariales

Order Dacrymycetales

Order Fistulinales

Order Ganodermatales

Order Gautieriales

Order Gomphales

Order Hericiales

Order Hymenochaetales

Order Hymenogastrales

Order Lachnocladiales

Order Lycoperdales

Order Melanogastrales

Order Nidulariales

Order Phallales

Order Poriales

Order Russulales

Order Schizophyllales

Order Sclerodermatales

Order Stereales

Order Thelephorales

Order Tulasnellales

Order Tulostomatales

Class Ustomycetes

Order Cryptobasidiales

Order Cryptomycocolacales

Order Exobasidiales

Order Graphiolales

Order Platyglocales

Order Sporidiales

Order Ustilaginales

Class Teliomycetes

Order Septobasidiales

Order Uredinales

Phylum Deuteromycetes
(Asexual Ascomycetes and
Basidiomycetes)

Class Hyphomycetes

Order Hyphomycetales

Order Stilbellales

Order Tuberculariales

Class Agonomycetes

Order Agonomycetales

Class Coelomycetes

Order Melanconiales

Order Sphaeropsidales

Order Pycnothyriales

Kingdom Plantae

Subkingdom Embryobionta

Division Hepaticophyta

Class Junermanniopsida

Order Calobryales
Order Jungermanniales
Order Metzgeriales
Class Marchantiopsida
Order Sphaerocarpales
Order Monocleales
Order Marchantiales
Division Anthocerotophyta
Class Anthocerotopsida
Order Anthocerotales
Division Bryophyta
Class Sphagnicopsida
Order Sphagnicales
Class Andreaeopsida
Order Andreaeles
Class Bryopsida
Order Archidiales
Order Bryales
Order Buxbaumiales
Order Dicranales
Order Encalyptales
Order Fissidentales
Order Funariales
Order Grimmiales
Order Hookeriales
Order Hypnobryales
Order Isobryales
Order Orthotrichales
Order Pottiales
Order Orthotrichales
Order Seligerales
Order Splachnales
Division Lycophyta
Class Lycopsida
Order Isoetales
Order Lycopodiales
Order Selaginellales
Division Polypodiophyta
Class Polypodopsida
Order Equisetales
Order Marattiales
Order Ophioglossales
Order Polypodiales
Order Psilotales
Division Pinophyta
Class Ginkgopsida
Order Ginkgoales
Class Cycadopsida
Order Cycadales
Class Pinopsida
Order Pinales

Order Podocarpales
Order Gnetales
Division Magnoliophyta
[unplaced orders]
Order Ceratophyllales
Order Chloranthales
Class Amborellopsida
Order Amborellales
Class Austrobaileyales
Order Austrobaileyales
Class Liliopsida
Order Acorales
Order Alismatales
Order Arecales
Order Asparagales
Order Commelinales
Order Dioscoreales
Order Liliales
Order Pandanales
Order Poales
Order Zingiberales
Class Magnoliopsida
Order Magnoliales
Order Laurales
Order Piperales
Order Canellales
Class Nymphaeopsida
Order Nymphaeales
Class Rosopsida
[unplaced orders]
Order Berberidopsidales
Order Buxales
Order Gunnerales
Order Proteales
Order Saxifragales
Order Santalales
Order Trochodendrales
Subclass Caryophyllidae
Order Caryophyllales
Order Dilleniales
Subclass Ranunculidae
Order Ranunculales
Subclass Rosidae
[unplaced orders]
Order Crossosomatales
Order Geraniales
Order Myrtales
Order Vitales
Superorder Rosanae
Order Celastrales
Order Cucurbitales

Order Fabales
Order Fagales
Order Malpighiales
Order Oxalidales
Order Rosales
Order Zygophyllales
Superorder Malvanae
Order Brassicales
Order Malvales
Order Sapindales
Subclass Asteridae
[unplaced order]
Order Boraginales
Superorder Cornanae
Order Cornales
Superorder Ericanae
Order Ericales
Superorder Lamianae
Order Garryales
Order Gentianales
Order Lamiales
Order Solanales
Superorder Asteranae
Order Apiales
Order Aquifoliales
Order Asterales
Order Dipsacales

Kingdom Animalia

Subkingdom Parazoa
Phylum Porifera
Subphylum Cellularia
Class Demospongiae
Class Calcarea
Subphylum Symplasma
Class Hexactinellida
Phylum Placozoa

Subkingdom Eumetazoa
Phylum Cnidaria (=Coelenterata)
Class Scyphozoa
Order Stauromedusae
Order Coronatae
Order Semaeostomeae
Order Rhizostomeae
Class Cubozoa
Order Cubomedusae
Class Hydrozoa
Order Hydroida
Order Milleporina

Order Stylasterina
Order Trachylina
Order Siphonophora
Order Chondrophora
Order Actinulida
Class Anthozoa
Subclass Alcyonaria (=Octocorallia)
Order Stolonifera
Order Gorgonacea
Order Alcyonacea
Order Pennatulacea
Subclass Zoantharia (=Hexacorallia)
Order Actinaria
Order Corallimorpharia
Order Scleractinia
Order Zoanthinaria (=Zoanthidea)
Order Ceriantharia
Order Ptychodactiaria
Order Antipatharia
Phylum Ctenophora
Class Tentaculata
Order Cydippida
Order Platyctenida
Order Lobata
Order Cestida
Order Ganeshida
Order Thalassocalycida
Class Nuda
Order Beroida
Phylum Platyhelminthes
Class Turbellaria
Order Acoela
Order Rhabdocoela
Order Catenulida
Order Macrostomida
Order Nemertodermatida
Order Lecithoepitheliata
Order Polycladida
Order Prolecithophora
(=Holocoela)
Order Proseriata
Order Tricladida
Order Neorhabdocoela
Class Cestoda
Subclass Cestodaria
Subclass Eucestoda
Order Caryophyllidea
Order Spathebothriidea
Order Trypanorhyncha
Order Pseudophyllidea
Order Tetraphyllidea

Order Cyclophyllidea
Class Monogenea
Class Trematoda
Subclass Digenea
Order Strigeidida
Order Azygiida
Order Echinostomida
Order Plagiorchiida
Order Opisthorchiida
Subclass Aspidogastrea
(=Aspidobothrea)
Phylum Mesozoa
Class Orthonectida
Class Rhombozoa
Order Dicyemida
Order Heterocyemida
Phylum Myxozoa (=Myxospora)
Phylum Nemertea (=Rhynchocoela,
Nemertinea)
Class Anopla
Order Palaeonemertea
(=Palaeonemertini)
Order Heteronemertea
Class Enopla
Order Hoplonemertea
(=Hoplonemertini)
Order Bdellonemertea
Phylum Gnathostomuilda
Order Filospermoidea
Order Bursovaginoidea
Phylum Gastrotricha
Order Chaetonotida
Order Macrodasyida
Phylum Cycliophora
Phylum Rotifera
Class Monogononta
Order Ploima
Order Flosculariaceae
Order Collothecaceae
Class Bdelloidea
Class Seisonidea
Phylum Acanthocephala
Class Archiacanthocephala
Class Eoacanthocephala
Class Palaeacanthocephala
Phylum Nematoda (=Nemata)
Class Adenophorea
Subclass Enoplia
Order Enoplida
Order Dorylaimida
Order Trichocephalida

Order Mermithida
Subclass Chromadoria
Class Secernentea
Subclass Rhabditia
Order Rhabditida
Order Ascaridida
Order Strongylida
Subclass Spiruria
Order Spirurida
Order Camallanida
Subclass Diplogasteria
Phylum Nematomorpha
Class Nectonematoida
Class Gordioida
Phylum Priapulida
Phylum Kinorhyncha (=Echinoderida)
Class Cyclorhagida
Class Homalorhagida
Phylum Loricifera
Phylum Mollusca
Subphylum Aculifera
Class Polyplacophora
Class Aplacophora
Subclass Neomeniophora
(=Solenogastres)
Subclass Chaetodermomorpha
(=Caudofoveata)
Subphylum Conchifera
Class Monoplacophora
Class Gastropoda
Subclass Prosobranchia
Order Archaeogastropoda
Order Mesogastropoda
(=Taenioglossa)
Order Neogastropoda
Subclass Opisthobranchia
Order Cephalaspidea
Order Runcinoidea
Order Acochlidioidea
Order Sacoglossa (=Ascoglossa)
Order Anaspidea (=Aplysiacea)
Order Notaspidea
Order Thecosomata
Order Gymnosomata
Order Nudibranchia
Subclass Pulmonata
Order Archaeopulmonata
Order Basommatophora
Order Stylommatophora
Order Systellommatophora
Class Bivalvia (=Pelecypoda)

Subclass Protobranchia (=Palaeotaxodonta, Cryptodonta)
Subclass Pteriomorphia
Subclass Paleoheterodonta
Subclass Heterodonta
Subclass Anomalodesmata
Class Scaphopoda
Class Cephalopoda
Subclass Nautiloidea
Subclass Coleoidea (=Dibranchiata)
Order Sepioidea
Order Teuthoidea (=Decapoda)
Order Vampyromorpha
Order Octopoda
Phylum Annelida
Class Polychaeta
Order Phyllodocida
Order Spintherida
Order Eunicida
Order Spionida
Order Chaetopterida
Order Magelonida
Order Psammodrilida
Order Cirratulida
Order Flabelligerida
Order Opheliida
Order Capitellida
Order Oweniida
Order Terebellida
Order Sabellida
Order Protodrilida
Order Myzostomida
Class Clitellata
Subclass Oligochaeta
Order Lumbriculida
Order Haplotaxida
Subclass Hirudinea
Order Rhynchobdellae
Order Arhynchobdellae
Order Branchiobdellida
Order Acanthobdellida
Class Pogonophora (=Siboglinidae)
Subclass Perviata (=Frenulata)
Subclass Obturata (=Vestimentifera)
Class Echiura
Order Echiura
Order Xenopneusta
Order Heteromyota
Phylum Sipuncula
Phylum Arthropoda
Subphylum Chelicerata

Class Merostomata
Order Xiphosura
Class Arachnida
Order Scorpiones
Order Uropygi
Order Amblypygi
Order Araneae
Order Ricinulei
Order Pseudoscorpiones
Order Solifugae (=Solpugida)
Order Opiliones
Order Acari
Class Pycnogonida (=Pantopoda)
Subphylum Mandibulata
Class Myriapoda
Order Chilopoda
Order Diplopoda
Order Symphyla
Order Pauropoda
Class Insecta (=Hexapoda)
Subclass Apterygota
Order Thysanura
Order Collembola
Subclass Pterygota
Superorder Hemimetabola
Order Ephemeroptera
Order Odonata
Order Blattaria
Order Mantodea
Order Isoptera
Order Grylloblattaria
Order Orthoptera
Order Phasmida (=Phasmatoptera)
Order Dermaptera
Order Embiidina
Order Plecoptera
Order Psocoptera
Order Anoplura
Order Mallophaga
Order Thysanoptera
Order Hemiptera
Order Homoptera
Superorder Holometabola
Order Neuroptera
Order Coleoptera
Order Strepsiptera
Order Mecoptera
Order Siphonaptera
Order Diptera
Order Trichoptera
Order Lepidoptera

Order Hymenoptera
Class Crustacea
Subclass Cephalocarida
Subclass Malacostraca
Superorder Syncarida
Superorder Hoplocarida
Order Stomatopoda
Superorder Peracarida
Order Thermosbaenacea
Order Mysidacea
Order Cumacea
Order Tanaidacea
Order Isopoda
Order Amphipoda
Superorder Eucarida
Order Euphausiacea
Order Decapoda
Subclass Branchiopoda
Order Notostraca
Order Cladocera
Order Conchostraca
Order Anostraca
Subclass Ostracoda
Order Myodocopa
Order Podocopa
Subclass Mystacocarida
Subclass Copepoda
Order Calanoida
Order Harpacticoida
Order Cyclopoida
Order Monstrilloida
Order Siphonostomatoida
Order Poecilostomatoida
Subclass Branchiura
Subclass Pentastomida
Order Cephalobaenida
Order Porocephalida
Subclass Tantulocarida
Subclass Remipedia
Subclass Cirripedia
Order Acrothoracica
Order Ascothoracica
Order Thoracica
Order Rhizocephala
Phylum Tardigrada
Class Heterotardigrada
Class Mesotardigrada
Class Eutardigrada
Order Parachela
Order Apochela
Phylum Onychophora

Phylum Phoronida

Phylum Brachiopoda

  Class Inarticulata

    Order Lingulida

    Order Acrotretida

  Class Articulata

    Order Rhynchonellida

    Order Terebratulida

Phylum Bryozoa (=Ectoprocta, Polyzoa)

  Class Phylactolaemata

  Class Stenolaemata

  Class Gymnolaemata

    Order Ctenostomata

    Order Cheilostomata

Phylum Entoprocta (=Kamptozoa)

Phylum Chaetognatha

  Class Sagittoidea

    Order Phragmophora

    Order Aphragmophora

Phylum Echinodermata

Subphylum Crinozoa

  Class Crinoidea

    Order Millericrinida

    Order Cyrtocrinida

    Order Bourgueticrinida

    Order Isocrinida

    Order Comatulida

Subphylum Asterozoa

  Class Stelleroidea

  Subclass Somasteroidea

  Subclass Ophiuroidea

    Order Phrynophiurida

    Order Ophiurida

  Subclass Asteroidea

    Order Platyasterida

    Order Paxillosida

    Order Valvatida

    Order Spinulosida

    Order Forcipulata

    Order Brisingida

  Class Concentricycloidea

Subphylum Echinozoa

  Class Echinoidea

    Order Cidaroida

    Order Echinothuroida

    Order Diadematoida

    Order Arbacioida

    Order Temnopleuroida

    Order Echinoida

    Order Holectypoida

    Order Clypeasteroida

    Order Spatangoida

  Class Holothuroidea

    Order Dendrochirotida

    Order Aspidochirotida

    Order Elasipodida

    Order Apodida

    Order Molpadiida

Phylum Hemichordata

  Class Enteropneusta

  Class Pterobranchia

Phylum Chordata

Subphylum Urochordata (=Tunicata)

  Class Ascidiacea

    Order Asplousobranchia

    Order Phlebobranchia

    Order Stolidobranchia

  Class Larvacea (=Appendicularia)

  Class Thaliacea

    Order Pyrosomida

    Order Doliolida

    Order Salpida

Subphylum Cephalochordata

  (=Acrania)

Phylum Chordata[c]

Subphylum Vertebrata

  Superclass Agnatha

  Class Myxini

    Order Myxiniformes

  Class Cephalaspidomorphi

    Order Petromyzontiformes

  Superclass Gnathostomata

  Class Chondrichthyes

  Subclass Holocephali

    Order Chimaeriformes

  Subclass Elasmobranchii

    Order Hexanchiformes

    Order Squaliformes

    Order Pristiophoriformes

    Order Squatiniformes

    Order Pristiformes

    Order Rhinobatiformes

    Order Torpediniformes

    Order Myliobatiformes

    Order Heterodontiformes

    Order Orectolobiformes

    Order Lamniformes

    Order Carchiniformes

  Class Sarcopterygii

  Subclass Coelacanthimorpha

    Order Coelacanthiformes

  Subclass Porolepimorpha and Dipnoi

    Order Ceratodontiformes

    Order Lepidosireniformes

  Class Actinopterygii

  Subclass Chondrostei

    Order Polypteriformes

    Order Acipenseriformes

  Subclass Neopterygii

    Order Semionotiformes

    Order Amiiformes

Division Teleostei

Subdivision Osteoglossomorpha

    Order Osteoglossiformes

Subdivision Elopomorpha

    Order Elopiformes

    Order Albuliformes

    Order Anguilliformes

    Order Saccopharyngiformes

Subdivision Clupeomorpha

    Order Clupeiformes

Subdivision Euteleostei

  Superorder Ostariophysi

    Order Gonorhynchiformes

    Order Cypriniformes

    Order Characiformes

    Order Siluriformes

    Order Gymnotiformes

  Superorder Protacanthopterygii

    Order Esociformes

    Order Osmeriformes

    Order Salmoniformes

  Superorder Stenopterygii

    Order Stomiiformes

    Order Ateleopodiformes

  Superorder Cyclosquamata

    Order Aulopiformes

  Superorder Scopelomorpha

    Order Myctophiformes

  Superorder Lampridiomorpha

    Order Lampridiformes

  Superorder Polymixiomorpha

    Order Polymixiiformes

  Superorder Paracanthopterygii

    Order Percopsiformes

    Order Ophidiiformes

    Order Gadiformes

    Order Batrachoidiformes

    Order Lophiiformes

  Superorder Acanthopterygii

    Order Mugiliformes

    Order Atherinomorpha

Order Beloniformes
Order Cyprinodontiformes
Order Stephanoberyciformes
Order Beryciformes
Order Zeiformes
Order Gasterosteiformes
Order Synbranchiformes
Order Scorpaeniformes
Order Perciformes
Order Pleuronectiformes
Order Tetraodontiformes
Class Amphibia
Subclass Lissamphibia
Order Gymnophiona
Order Caudata (Urodela)
Order Anura—frogs and toads
Class Reptilia
Subclass Anapsida
Order Testudines
Subclass Diapsida
Order Sphenodonta
Order Squamata
Suborder Lacertilia
Suborder Serpentes
Order Crocodylia
Infraclass Eoaves
Order Struthioniformes

Order Tinamiformes
Infraclass Neoaves
Order Craciformes
Order Galliformes
Order Anseriformes
Order Turniciformes
Order Piciformes
Order Galbuliformes
Order Bucerotiformes
Order Upupiformes
Order Trogoniformes
Order Coraciiformes
Order Coliiformes
Order Cuculiformes
Order Psittaciformes
Order Apodiformes
Order Trochiliformes
Order Musophagiformes
Order Strigiformes
Order Columbiformes
Order Gruiformes
Order Ciconiiformes
Suborder Charadrii
Suborder Ciconii
Order Passeriformes
Class Mammalia (Synapsida)
Order Monotremata

Order Didelphimorphia
Order Paucituberculata
Order Microbiotheria
Order Dasyuromorphia
Order Peramelemorphia
Order Notoryctemorphia
Order Diprotodontia
Order Xenarthra
Order Insectivora
Order Scandentia
Order Dermoptera
Order Chiroptera
Order Primates
Order Carnivora
Order Cetacea
Order Sirenia
Order Proboscidea
Order Perissodactyla
Order Hyracoidea
Order Tubulidentata
Order Artiodactyla
Order Pholidota
Order Rodentia
Suborder Sciurognathi
Suborder Hystricognathi
Order Lagomorpha
Order Macroscelidea

Biographical listing*

Abbe, Ernst (1840–1905), German physicist. Developed optical instruments, such as an apochromatic objective and a crystal refractometer.

Abel, Frederick Augustus (1827–1902), English chemist. Expert on the chemistry of explosives; originated the Abel test for determination of the flash point of petroleum.

Abel, John Jacob (1857–1938), American pharmacologist and physiologist. Isolated epinephrine, and insulin in crystal form.

Abel, Niels Henrik (1802–1829), Norwegian mathematician. Contributed to the theory of elliptical functions.

Abell, George Ogden (1927–1983), American astronomer. Research in problems relating to organization, structure, and distribution of galaxies; observational cosmology; and planetary nebulae.

Abney, William de Wiveleslie (1843–1920), English photographic chemist and physicist. Photographed the infrared solar spectrum.

Adams, John Couch (1819–1892). English astronomer. Discovered, independently of U. J. J. Leverrier, Neptune.

Adanson, Michel (1727–1806), French naturalist. Classified plants in his *Les Familles Naturelles des Plantes*.

Addison, Thomas (1793–1860), English physician. Identified pernicious anemia and Addison's disease of the adrenal cortex.

Adler, Alfred (1870–1937), Austrian psychiatrist and psychologist. Founded the school of individual psychology.

Adrian of Cambridge, Edgar Douglas Adrian, Baron (1889–1977), English physiologist. Investigated physiology of nervous system; showed that change in electric potential in electroencephalograph is due to electrical activity of cortex; Nobel Prize, 1932.

Afzelius, Adam (1750–1837), Swedish botanist. Founded the Linnaean Institute.

Agassiz, Jean Louis Rodolphe (1807–1873), Swiss-born American naturalist. Wrote books on ichthyology, especially relating to classification.

Agnesi, Maria Gaetana (1718–1799), Italian mathematician. Author of *Instituzioni Analitiche*, a complete treatment of algebra and analysis; shared in the discovery of a cubic curve ("witch of Agnesi").

Agricola, Georgius, real name Georg Bauer (1494–1555), German physician and mineralogist. Known as the father of systematic mineralogy.

Ahlfors, Lars Valerian (1907–1996), Finnish-American mathematician. Did research on covering surfaces related to Riemann surfaces of inverse functions of entire and meromorphic functions; opened up new fields of analysis; Fields Medal, 1936.

Airy, George Biddell (1801–1892), English astronomer. Discovered inequality in the motions of Venus and the Earth; determined the mass of the Earth.

Aitken, John (1839–1919), Scottish physicist. Studied dust particles in the atmosphere, known as Aitken nuclei.

Aitken, Robert Grant (1864–1951), American astronomer. Discovered more than 3000 binary stars.

Alder, Kurt (1902–1958), German chemist. Codeveloper of the Diels-Alder reaction for diene synthesis; contributed to stereochemistry; Nobel Prize, 1950.

Alembert, Jean le Rond d' (1717–1783), French mathematician. Developed d'Alembert's principle and the calculus of partial differences.

Alferov, Zhores Ivanovich (1930–), Russian physicist and electronics engineer. Developed semiconductor heterostructures used in high-speed and opto-electronics, including fast transistors, laser diodes, and light-emitting diodes; Nobel Prize, 2000.

Alivén, Hannes Olof Gösta (1908–1995), Swedish physicist. Studies in magnetohydrodynamics, planetary physics, antiferromagnetism, and ferrimagnetism; Nobel Prize, 1970.

Alhazen (965–1038), Arab mathematician and astronomer. Provided the first accounts of atmospheric refraction and reflection from concave surfaces; constructed spherical and parabolic mirrors.

al-Khwarizmi (780–?850), Arab mathematician. Wrote treatises on arithmetic and algebra, which were important in the mathematical knowledge of medieval Europe.

Allen, Edgar (1892–1943), American biologist. Discovered estrogen; investigated hormonal mechanisms controlling female reproductive cycle.

Allen, Willard Myron (1904–1993), American physician. With G. W. Corner, discovered progesterone, and proved it necessary for development of embryo in early pregnancy; with O. Wintersteiner, synthesized crystalline progesterone.

Altman, Sidney (1939–), American chemist. Discovered an unusual enzyme that contains ribonucleic acid (RNA) in addition to a protein, leading to the discovery that RNA molecules have catalytic properties similar to those of enzymes; Nobel Prize, 1989.

Alvarez, Luis Walter (1911–1988), American physicist. Pioneer in building liquid hydrogen bubble chambers, and in developing measurement devices and computer systems to analyze data from these chambers; discovered large numbers of short-lived elementary particles; Nobel Prize, 1968.

Amagat, Émile (1841–1915), French physicist. Investigated relationship of pressure, density, and temperature in gases and liquids, particularly at high pressure.

Amici, Giovanni Battista (1786–1863), Italian astronomer, optician, and naturalist. Invented the Amici microscope; designed parabolic mirrors for reflecting telescopes.

Ampère, André Marie (1775–1836), French physicist and mathematician. Founder of electrodynamics; formulated Ampère's law; invented the astatic needle.

Anaximander (611–547 B.C.), Greek astronomer and mathematician. Reputed inventor of geographical maps; formulated the concept of the universe as infinite (apeiron).

Anderson, Carl David (1905–1991), American physicist. Discovered the meson in cosmic rays; discovered the positron; Nobel Prize, 1936.

Anderson, Philip Warren (1923–), American physicist. Demonstrated existence of electronic localization in disordered solids, and of localized magnetism in metals; Nobel Prize, 1977.

Andrade, Edward Neville da Costa (1887–1971), English physicist. Discovered Andrade's creep law and a law governing variation of viscosity of liquids with temperature.

Andrews, Roy Chapman (1884–1960), American naturalist. Discovered many plant and animal fossils.

Anfinsen, Christian Boehmer (1916–1995), American biochemist. Discovered how three-dimensional structures of ribonuclease and other proteins are formed; Nobel Prize, 1972.

Angström, Anders Jonas (1814–1874), Swedish physicist. Mapped the solar spectrum; discovered hydrogen in the solar atmosphere.

Apollonius of Perga (247–205 B.C.), Greek mathematician. Wrote about conic sections; coined the terms parabola, ellipse, and hyperbola.

Appleton, Edward Victor (1892–1965), English physicist. Demonstrated the existence of the ionosphere and discovered its region known as the Appleton layer; contributed to the development of radar; Nobel Prize, 1947.

Arago, Dominique François (1786–1853), French astronomer and physicist. Discovered the magnetic properties of nonferrous materials, and the production of magnetism by electricity.

Arber, Werner (1929–), Swiss molecular biologist. Determined the molecular mechanism of host-controlled restriction modification of bacterial viruses and discovered the restriction enzymes; Nobel Prize, 1978.

Archimedes (287–212 B.C.), Greek physicist and mathematician. Formulated Archimedes' principle; invented the compound pulley and Archimedes' screw.

Argand, Jean Robert (1768–1822), Swiss mathematician. Developed the Argand diagram.

Argelander, Friedrich Wilhelm August (1799–1875), German astronomer. Prepared a star catalog; introduced decimal division of stellar magnitudes.

Aristarchus of Samos (310–250 B.C.), Greek astronomer. Invented the hemispherical sundial; determined the movement of the Earth around the stationary Sun; added a correction factor of 1/1623 of a day to the length of the year.

Aristotle (384–322 B.C.), Greek philosopher. Exponent of the methodology and division of sciences; contributed to physics, astronomy, meteorology, psychology, and biology.

Arkwright, Richard (1732–1792), English inventor. Developed the first practical mechanized spinning frame, utilizing rollers.

Arrhenius, Svante August (1859–1927), Swedish physicist and chemist. Developed theory of electrolytic dissociation; investigated osmosis and viscosity of solutions; Nobel Prize, 1903.

Arsonval, Jacques Arsène d' (1851–1940), French physicist and physiologist. Pioneered in electrotherapy; invented d'Arsonval galvanometer.

Aston, Francis William (1877–1945), English physicist and chemist. Discovered isotopes in nonradioactive elements by using the mass spectrograph he invented; Nobel Prize, 1922.

Atiyah, Michael Francis (1929–), British mathematician. Work centered on the interaction between geometry and analysis; developed K theory in collaboration with F. Hirzebruch; with I. M. Singer, proved the index theorem concerning elliptic differential operators on compact differentiable manifolds, which was later seen to have applications to theoretical physics; Fields Medal, 1966.

Atwood, George (1746–1807), English mathematician. Invented the Atwood machine.

Audubon, John James (1785–1851), Haitian-born American ornithologist and artist. Made drawings and paintings of birds and animals.

Auger, Pierre Victor (1899–1943), French physicist. Discovered the Auger effect.

Avicenna (979–1037), Arab physician. Wrote the medical text *Canon Medicinae*.

Avogadro, Amedeo (1776–1856), Italian physicist. Formulated Avogadro's law.

Axelrod, Julius (1912–), American biochemist and pharmacologist. Showed that many drugs act by modifying storage of neurotransmitters at nerve terminals; made discoveries concerning metabolism, and mechanisms for formation and inactivation of norepinephrine; Nobel Prize, 1970.

Ayrton, William Edward (1847–1908), English physicist and electrical engineer. Invented the ammeter, voltmeter, and other electrical measuring instruments.

Baade, Walter (1893–1960), German-born American astronomer. Formulated concept of stellar populations; increased distance scale of universe by factor of 2.

Babbage, Charles (1792–1851), English mathematician. Devised a primitive computer to calculate and print mathematical and astronomical tables.

Babcock, Harold Delos (1882–1968) **and Horace Welcome** (1912–), American astronomers.

---

*This listing includes many of the individuals whose names appear in dictionary terms, as well as Nobel laureates.

Invented the Babcock magnetograph, observed weak solar magnetic fields, and discovered stellar magnetic fields.

Babcock, Stephen Moulton (1843–1931), American agricultural chemist. Pioneer in nutrition; devised the Babcock test to measure fat content in milk.

Babinet, Jacques (1794–1872), French physicist. Invented a polariscope and a goniometer.

Back, Ernst E. A. (1881–1959), German physicist. Developed improved spectrographs; made spectroscopic observations leading to Paschen-Back effect.

Badger, Richard McLean (1896–1974), American physical chemist and spectroscopist. Studied structures of polyatomic molecules; formulated Badger's rule concerning molecular bonds.

Baekeland, Leo Hendrik (1864–1944) Belgian-born, American chemist. Invented the phenolformaldehyde polymer, Bakelite, the first commercial synthetic polymer.

Baer, Karl Ernst von (1792–1876), Estonian embryologist. Discovered the mammalian ovum and the notochord; developed the theory of embryonic germ layers.

Baeyer, Johann Friedrich Wilhelm Adolf von (1835–1917), German chemist. Synthesized indigo and hydroaromatic compounds; Nobel Prize, 1905.

Baily, Francis (1774–1844), English astronomer. A founder of the Royal Astronomical Society; first observed phenomenon of Baily's beads.

Baire, René Louis (1874–1932), French mathematician. Contributed to theory of functions of real variables; introduced concept of Baire functions.

Baker, Alan (1939–), British mathematician. Proved a generalization of the Gelfond-Schneider theorem; from this work, generated transcendental numbers not previously identified and solved problems in the theory of Diophantine equations; Fields Medal, 1970.

Balfour, Francis Maitlant (1851–1882), English biologist. Founder of comparative embryology.

Balmer, Johann Jakob (1825–1898), Swiss physicist. Expressed the mathematical formula for frequencies of hydrogen lines in the visible spectrum.

Baltimore, David (1938–), American virologist. Investigated interaction between ribonucleic acid tumor viruses and genetic material; independently of H. M. Temin, discovered reverse transcriptase; Nobel Prize, 1975.

Banach, Stefan (1892–1945), Polish mathematician. Laid foundations of contemporary functional analysis; introduced concept of Banach space and discovered its fundamental properties.

Bang, Bernhard Laurits Frederik (1848–1932), Danish veterinarian. Discovered method of eradicating bovine tuberculosis; discovered *Brucella abortus*, the agent of contagious abortion (Bang's disease) and brucellosis.

Banting, Frederick Grant (1891–1941), Canadian physician. With J. J. R. Macleod and C. H. Best, discovered insulin and its role in diabetes; Nobel Prize, 1923.

Bárány, Robert (1876–1936), Austrian physician. Developed new methods of diagnosing ear diseases; Nobel Prize, 1914.

Bardeen, John (1908–1991), American physicist. With L. N. Cooper and J. R. Schrieffer, formulated a theory of superconductivity; invented the transistor; Nobel Prize, 1956 and 1972.

Barkhausen, Heinrich Georg (1881–1956), German electronic engineer and physicist. Contributed to theory and application of electron tubes; with K. Kurz, developed Barkhausen-Kurz oscillator; discovered Barkhausen effect.

Barkla, Charles Glover (1877–1944), English physicist. Described characteristics of x-rays and other short-wave emissions of elements; Nobel Prize, 1917.

Barnard, Edward Emerson (1857–1923), American astronomer. Discovered 16 comets, the fifth satellite of Jupiter, and dark nebulae; contributed to celestial photography.

Barnett, Samuel Jackson (1873–1956), American physicist. Discovered Barnett effect and used it to measure the gyromagnetic ratio of ferromagnetic

materials; gave experimental proof of existence of ionosphere.

Barr, Murray Llewellyn (1908–1995), Canadian anatomist. Discovered the Barr body on the X chromosome of the human female.

Bartholin, Kaspar (1655–1738), Danish physician. Discovered Bartholin's glands of the vagina and a sublingual duct.

Bartholin, Thomas (1616–1680), Danish physician. Discovered lymphatic glands; described the lymphatic system.

Bartlett, James Holly (1904–), American physicist. Introduced concept of Bartlett force; did research on nuclear shell model, electrochemical potentiostat, and restricted three-body problem.

Barton, Derek Harold Richard (1918–1998), British chemist. Developed and expanded concept of conformation to include large molecules with complex ring systems; Nobel Prize, 1969.

Basov, Nicolai Gennediyevich (1922–2001), Soviet physicist. Conducted fundamental studies in quantum electronics; with A. M. Prokhorov, developed quantum optical generators; Nobel Prize, 1964.

Bassham, James Alan (1922–), American chemist. Helped to elucidate basic photosynthetic carbon cycle.

Bates, Henry Walter (1825–1892), English naturalist. Discovered Batesian mimicry among butterflies and moths.

Baudot, Émile (1845–1903), French engineer. Invented an improved telegraph transmitter.

Baumé, Antoine (1728–1804), French chemist. Invented a graduated hydrometer which utilizes the Baumé scale.

Bautz, Laura Patricia (1940–), American astronomer. Collaborated with W. W. Morgan in developing Bautz-Morgan classification of galaxy clusters.

Bayer, Johann (1572–1625), German astronomer. Charted 12 constellations; first to use Greek letters to designate the order of brightness of stars in a constellation.

Bayes, Thomas (1702–1761), English mathematician. Formulated a basis for statistical inference.

Bayliss, William Maddock (1860–1924), English physiologist. Did research on electrophysiology of heart action; discovered the hormone secretin.

Beadle, George Wells (1903–1989), American geneticist. With E. L. Tatum, proved that genes affect heredity by controlling cell chemistry; Nobel Prize, 1958.

Beams, Jesse Wakefield (1898–1977), American physicist. Developed vacuum-type ultracentrifuges, used in purification and molecular weight determination of large-molecular-weight substances, isotope separation, and determination of the gravitational constant.

Beattle, James Alexander (1895–1981), American chemist and physicist. Studied ionic theory and thermodynamics; with P. W. Bridgman, proposed Beattie and Bridgman equation for gases.

Beaufort, Francis (1774–1857), English hydrographer. Devised scale of wind velocity.

Beaumont, William (1785–1853), American physician. Did pioneering studies of digestion and gastric juices.

Béchamp, Pierre Jacques Antoine (1816–1908), French chemist. Discovered a method of preparing aniline.

Beckmann, Ernst Otto (1853–1923), German chemist. Discovered Beckmann molecular transformation; invented the Beckmann thermometer.

Becquerel, Antoine César (1788–1878), French physicist. Pioneer in electrochemistry; first to extract metals from ore by electrolysis.

Becquerel, Antoine Henri (1852–1908), French physicist. A discoverer of radioactivity in uranium.

Bednorz, Johannes Georg (1950–), German physicist. With K. A. Müller, discovered high-temperature superconductivity in copper oxide ceramic materials; Nobel Prize, 1987.

Beebe, Charles William (1877–1962), American naturalist. Pioneer in deep-sea exploration; made ornithological collections.

Beer, August (1825–1863), German physicist. Discovered Beer's law of light absorption.

Behring, Emil Adolph von (1854–1917), German bacteriologist. Produced diphtheria and tetanus antitoxins; Nobel Prize, 1901.

Békésy, Georg von (1899–1972), Hungarian-born American physicist. Studied hearing processes, especially inner-ear mechanics; Nobel Prize, 1961.

Bell, Alexander Graham (1847–1922), Scottish-born American inventor. Invented the telephone, photophone, graphophone, and one of the earliest gramophones.

Bell, Charles (1774–1842), Scottish anatomist. Discovered that sensory and motor nerves are anatomically and functionally distinct.

Bellman, Richard Ernest (1920–1984), American mathematician. Research in analytic number theory, differential equations, stochastic processes, dynamic programming, and mathematical biosciences; discovered Bellman's principle of optimality.

Benacerraf, Baruj (1920–), American immunologist. Discovered immune-response (Ir) genes that control specific immune responses to thymus-dependent antigens; Nobel Prize, 1980.

Benioff, Hugo (1899–1968), American geophysicist. Investigated earthquakes, particularly through instrumental seismology.

Bentham, George (1800–1884), English botanist. With J. Hooker, wrote *Genera Plantarum*.

Berg, Paul (1926–), American biochemist. Investigated the biochemistry of deoxyribonucleic acid (DNA) and designed a technique for gene splicing; Nobel Prize, 1980.

Bergey, David Hendricks (1860–1937), American bacteriologist. Authority on classification of bacteria.

Bergius, Friedrich (1884–1949), Polish-born German chemist. Developed Bergius process for hydrogenation of coal to a petroleumlike oil; Nobel Prize, 1931.

Bergström, Sune Karl (1916–), Swedish biochemist and medical scientist. Studied the metabolism of unsaturated fatty acids and determined the chemical structure of prostaglandins; Nobel Prize, 1982.

Bernard, Claude (1813–1878), French physiologist. Studied digestion; discovered that glycogen is produced by the liver.

Berners-Lee, Tim (1955–), British-born physicist and software engineer. Proposed the World Wide Web, and then invented hypertext markup language (HTML), HyperText Transfer Protocol (HTTP), the Internet addressing scheme (Universal Resource Locator or URL), and the first Web browser.

Bernoulli, Daniel (1700–1782), Swiss mathematician born in the Netherlands. Founder of mathematical physics; worked on hydrodynamics and differential equations; formulated the Bernoulli equation.

Bernoulli, Jacques or Jacob (1654–1705), Swiss mathematician. Contributed to mathematics of curves, calculus, and probability; developed the Bernoulli number.

Bernoulli, Jean or Johann (1667–1748), Swiss mathematician. A founder of calculus of variations; contributed to exponential calculus, complex numbers, geodesics, and trigonometry.

Berthelot, Pierre Eugène Marcellin (1827–1907), French chemist. Founder of thermochemistry; first to synthesize organic compounds; demonstrated nitrogen fixation.

Bertrand, Joseph Louis François (1822–1900), French mathematician. Contributed to analysis, differential geometry, and probability theory.

Berzelius, Jons Jakob (1779–1848), Swedish chemist. Discovered the elements cerium, selenium, thorium, and silicon; developed a system for classification and nomenclature of compounds.

Bessel, Friedrich Wilhelm (1784–1846), German astronomer and mathematician. Developed Bessel functions; determined parallax of the star 61 Cygni; postulated existence of Neptune and dark stars.

Bessemer, Henry (1813–1898), English engineer. Invented the Bessemer process, the first method for manufacturing steel on a large scale.

Best, Charles Herbert (1899–1978), Canadian physiologist and medical researcher. Associated with F. G. Banting and J. J. R. Macleod in the discovery of insulin.

Bethe, Hans Albrecht (1906–), German-born American physicist. Formulated the theory of energy production in stars; research on nuclear physics; Nobel Prize, 1967.

Bhabha, Homi Jehangir (1909–1966), Indian physicist. With W. Heitler, developed theory of cascade showers of cosmic rays; observed slowing of decay rate of high-velocity mesons.

Bianchi, Luigi (1856–1928), Italian mathematician. Contributed to differential geometry and study of noneuclidean geometries; discovered Bianchi identity.

Bichat, Marie François Xavier (1771–1802), French anatomist and physiologist. Founder of animal histology; originated the term "tissues," and distinguished 21 types in his particular scheme.

Bieberbach, Ludwig (1886–1982), German mathematician. Research on complex function theory, differential equations, geometry, and algebra; postulated Bieberbach's conjecture.

Bienaymé, Irénée Jules (1796–1878), French mathematician. Studied calculus of probabilities and its application to financial science; with P. L. Chebyshev, discovered Bienaymé-Chebyshev inequality.

Billet, Felix (1808–1882), French physicist. Invented Billet split lens.

Binet, Alfred (1857–1911), French psychologist. Investigated development and measurement of intelligence.

Binnig, Gerd (1947–), German physicist. With H. Rohrer, developed scanning tunneling microscope; Nobel Prize, 1986.

Biot, Jean Baptiste (1774–1862), French mathematician and physicist. Discovered circular polarization of light; invented a polariscope; with D. Brewster, discovered biaxial crystals; helped formulate Biot-Savart law.

Birkhoff, George David (1884–1944), American mathematician. Investigated differential equations, dynamical systems, ergodic theory, mechanics of fluids, and foundations of relativity and quantum mechanics.

Bishop, John Michael (1936–), American virologist and biochemist. With H. Varmus, researched the genetic basis of human cancers; their work led to the identification of over 50 cellular genes that can become oncogenes; Nobel Prize, 1989.

Bjerknes, Vilhelm Fremann Doren (1862–1951), Norwegian physicist. Research on electric waves; originated the polar-front theory in meteorology.

Black, James (1924–), British pharmacologist. Developed the first beta blocker drug, propranol; also credited with the discovery of another important class of drugs, the H2 antagonists; Nobel Prize, 1988.

Black, Joseph (1728–1799), Scottish physicist and chemist. Established the concepts of latent heat and specific heat, and discovered carbon dioxide.

Blackett, Patrick Maynard Stuart (1897–1974), English physicist. Built an improved cloud chamber used to photograph tracks of a nuclear disintegration and of a cosmic-ray shower; discovered the positron; Nobel Prize, 1948.

Blobel, Günter (1936–), German-born American cell and molecular biologist. Discovered that proteins carry signals that help direct their movement among the organelles of the cell; Nobel Prize, 1999.

Bloch, Felix (1905–1983), Swiss-born American physicist. Discovered a technique for studying magnetism of atomic nuclei in normal matter; Nobel Prize, 1952.

Bloch, Konrad Emil (1912–2000), German-born American biochemist. Traced the transformations of fat and carbohydrate metabolites to cholesterol; Nobel Prize, 1964.

Blodgett, Katharine Burr (1898–1979), American chemical physicist. Studied surface science, and is best know for her work with Irving Langmuir and the development of the Langmuir-Blodgett film.

Bloembergen, Nicholaas (1920–), Netherlands-born American physicist. Contributed to development of maser; made extensive contributions to theoretical and experimental development of nonlinear optics; Nobel Prize, 1981.

Blumberg, Baruch Samuel (1925–), American physician and biologist. Research leading to a test for hepatitis viruses in blood and to an experimental hepatitis vaccine; Nobel Prize, 1976.

Boas, Franz (1858–1942), American anthropologist. Pioneered in physical anthropology; contributed to stratigraphic archeology in Mexico; emphasized importance of linguistic analysis.

Bobillier, Étienne (1798–1840), French mathematician and physicist. Contributed to geometry and statics; discovered Bobillier's law.

Bode, Johann Elert (1747–1826), German astronomer. Prepared a celestial atlas showing about 17,000 stars; formulated Bode's law.

Boerhaave, Hermann (1663–1738), Dutch physician. A great teacher at the University of Leyden; wrote the physiology textbook *Institutiones Medicae.*

Bohm, David (1917–1992), American-born British physicist. Research in quantum theory and new modes of description in physics.

Bohr, Aage (1922–), Danish physicist. With B. R. Mottelson, developed theory which unifies shell and liquid-drop models of the atomic nucleus, and which explains nonspherical nuclei; Nobel Prize, 1975.

Bohr, Niels (1885–1962), Danish physicist. Devised an atomic model; codeveloped the quantum theory, applying it to atomic structure in Bohr's theory; Nobel Prize, 1922.

Boltzmann, Ludwig Eduard (1844–1906), Austrian physicist. An authority on the kinetic theory of gases; demonstrated the Stefan-Boltzmann law of blackbody radiation, Boltzmann's law of energy, and the Boltzmann constant.

Bolyai, János (1802–1860), Hungarian mathematician. Independently of K. F. Gauss and N. I. Lobachevski, originated a system of noneuclidean geometry.

Bolzano, Bernard (1781–1848), Czechoslovakian philosopher, logician, and mathematician. Contributed to theory of real functions; proved Bolzano's theorem and Bolzano-Weierstrass theorem.

Bombieri, Enrico (1940–), Italian mathematician. Made major contributions to the study of prime numbers, partial differential equations and minimal surfaces, univalent functions and the local Bieberbach conjecture, and functions of several complex variables; Fields Medal, 1974.

Bond, George Phillips (1825–1865), American astronomer. Introduced concept of Bond albedo; pioneered astronomical photography.

Boole, George (1815–1864) English mathematician and logician. Developed new system of mathematical logic, which is known as Boolean algebra.

Borcherds, Richard Ewan (1959–), British mathematician. Worked in algebra and geometry; proved the moonshine conjecture, which relates the so-called monster group and elliptical curves, using methods borrowed from string theory in theoretical physics; Fields Medal, 1998.

Borda, Jean Charles (1733–1799), French physicist and mathematician. Introduced Borda mouthpiece; developed instruments for navigation, geodesy, and determination of weights and measures.

Bordet, Jules Jean Baptiste Vincent (1870–1961), Belgian physiologist. Made discoveries in immunology; with O. Gengou, developed the technique of the complement fixation reaction; Nobel Prize, 1919.

Borel, Félix Edouard Émile (1871–1956), French mathematician. Work in infinitesimal calculus and the calculus of probabilities

Born, Max (1882–1970), German-born British theoretical physicist. Pioneered in the development of quantum mechanics; Nobel Prize, 1954.

Bosch, Carl (1874–1940), German chemist. Developed chemical high-pressure methods, and the Haber-Bosch process for ammonia synthesis; Nobel Prize, 1931.

Bose, Jagadis Chandra (1858–1937), Indian plant physiologist and physicist. Founded the Bose Research Institute in Calcutta; investigated photosynthesis, "nervous mechanism" of plants, and other plant subjects.

Bose, Satyendra Nath (1894–1974), Indian physicist. Originated Bose-Einstein statistics to describe photons.

Bothe, Walter (1891–1957), German physicist. Devised the coincidence method for the investigation of nuclear reactions and cosmic radiation; Nobel Prize, 1954.

Bouguer, Pierre (1698–1758), French geodesist, hydrographer, and physicist. Laid foundations of photometry; discovered Bouguer-Lambert law of light intensity.

Bourdon, Eugéne (1808–1884), French inventor. Invented Bourdon pressure gage.

Bourgain, Jean (1954–), Belgian mathematician. Worked in several areas of mathematical analysis, including the geometry of Banach spaces, convexity in high dimensions, harmonic analysis, ergodic theory, and nonlinear partial differential equations; Fields Medal, 1994.

Boussinesq, Joseph Valentin (1842–1929), French mathematical physicist. Research in hydrodynamics; introduced Boussinesq approximation.

Bovet, Daniel (1907–1992), Swiss-born Italian pharmacologist. Research on synthetic compounds that inhibit the action of the vascular system and skeletal muscles; Nobel Prize, 1957.

Bowditch, Nathaniel (1773–1838), American navigator and mathematician. Produced navigation guide; translated and improved P. S. de Laplace's *Mécanique céleste.*

Bowen, Norman Levi (1887–1956), Canadian-born American geologist. Studied physical chemistry of geological processes and phase equilibria of silicates; discovered significance of reaction principle in petrogenesis.

Boyer, Paul D. (1918–) American biochemist. Contributed to understanding how the enzyme ATP synthase catalyses the formation of adenosine triphosphate (ATP); Nobel Prize, 1997.

Boyle, Robert (1627–1691), British physicist and chemist. Conducted experiments on properties of the air pump; Boyle's law concerning gases is named for him; advanced the atomistic theory of matter.

Bradley, James (1693–1762), English astronomer. Discovered aberration of light due to the Earth's motion, and nutation of the Earth's axis; prepared astronomical tables and star catalogs.

Bragg, William Henry (1862–1942), English physicist. Codeveloper, with W. L. Bragg, of the x-ray spectrometer; used x-ray diffraction to determine crystal structure; Nobel Prize, 1915.

Bragg, William Lawrence (1890–1971), British physicist. With W. H. Bragg, developed x-ray analysis of the atomic arrangement in crystalline structures; Nobel Prize, 1915.

Brahe, Tycho or Tyge (1546–1601), Danish astronomer. Made painstaking observations of the planetary system; wrote *Astronomiae Instauratae Progymnasmata.*

Brattain, Walter Houser (1902–1987), American physicist. Investigated properties of semiconductors; research on surface properties of solids; Nobel Prize, 1956.

Braun, Karl Ferdinand (1850–1918), German physicist. Research on cathode rays and wireless telegraphy; Nobel Prize, 1909.

Bravais, Auguste (1811–1863), French physicist. Studied relationship between crystal form and structure; derived Bravais lattices.

Breit, Gregory (1899–1981), Russian-born American physicist. Research on quantum theory, quantum electrodynamics, hyperfine structure, and ionosphere.

Breuer, Josef (1842–1925), Austrian neurologist. Evolved abreaction method for treatment of neuroses.

Brewster, David (1781–1868), Scottish physicist. Formulated Brewster's law on polarization of light; codiscoverer, with J. B. Biot, of biaxial crystals.

Brianchon, Charles Julien (1783–1864), French mathematician. Proved Brianchon's theorem.

Bridgman, Percy Williams (1882–1961), American physicist. Worked in high-pressure physics and thermodynamics of liquids; Nobel Prize, 1946.

Briggs, Henry (1561–1631), English mathematician. Prepared logarithmic tables (later known as common logarithms); devised sophisticated interpolation techniques.

Bright, Richard (1789–1858), English physician. Made biochemical study of disease; researched Bright's disease of the kidneys.

Brillouin, Leon (1889–1969), French physicist. With G. Wentzel and H. A. Kramers, developed Wentzel-Kramers-Brillouin method; originated concept of Brillouin zones.

Brillouin, Louis Marcel (1854–1948), French physicist. Work on crystal structure, viscosity of liquids and gases, radiotelegraphy, and relativity.

Brinell, Johann August (1849–1925), Swedish engineer. Invented the Brinell machine to measure the hardness of alloys and metals in terms of the Brinell number.

Brockhouse, Bertram Neville (1918–), Canadian physicist. Developed the technique of slow neutron spectroscopy to study the dynamics of atoms in solids and liquids; Nobel Prize, 1994.

Broglie, Louis Victor de (1892–1987), French physicist. Worked in nuclear physics; first to link wave and corpuscular theory; Nobel Prize, 1929.

Bromwich, Thomas John I'Anson (1875–1929), English mathematician. Showed how the Heaviside calculus could be developed in a manner acceptable to pure mathematicians through use of contour integrals.

Brönsted, Johannes Nicolaus (1879–1947), Danish chemist. Researched kinetic properties of ions, catalysis, and nitramide; formulated the Brönsted theory of acid-base reactions.

Brown, Herbert Charles (1912–), British-born American chemist. Developed methods for chemical synthesis of diborane and organoboranes; Nobel Prize, 1979.

Brown, Michael S. (1941–), American biochemist and geneticist. With Joseph L. Goldstein, discovered low-density lipoprotein receptors and their function in cholesterol metabolism; Nobel Prize, 1985.

Browning, John Moses (1855–1926), American inventor. Invented the Browning machine gun.

Brun, Viggo (1885–1978), Norwegian mathematician. Worked in number theory, introducing what is now known as the Brun's sieve, which made some progress in the resolution of such problems as Goldbach's conjecture and the twin prime problem.

Brunauer, Stephen (1903–1986), Hungarian-born American chemist. Contributed to surface and colloid chemistry; with P. H. Emmett and E. Teller, developed Brunauer-Emmett-Teller equation for surface area determinations.

Brunel, Isambard Kingdom (1806–1859), English engineer. Constructed great bridges in England; designed important steamships.

Buchner, Eduard (1860–1917), German chemist. Studied alcoholic fermentation of sucrose; Nobel Prize, 1907.

Buckingham, Edgar (1867–1940), American physicist. Worked on thermodynamics and dimensional analysis; derived Buckingham's π theorem.

Buffon, Georges Louis Leclerc, Comte de (1707–1788), French naturalist. Compiled *Histoire Naturelle*, a monumental work on natural history.

Bullen, Keith Edward (1906–1976), Australian applied mathematician. Carried out mathematical studies of earthquake waves; with H. Jeffreys, prepared the Jeffreys-Bullen tables on seismic travel time.

Bunsen, Robert Wilhelm (1811–1899), German chemist. Discovered, with G. R. Kirchhoff, spectrum analysis; invented the Bunsen burner, Bunsen cell, and Bunsen ice calorimeter; formulated law of reciprocity with H. E. Roscoe.

Burbank, Luther (1849–1926), American horticulturist. Experimented on crossing and in-breeding of plant varieties.

Burnet, Frank Macfarlane (1899–1985), Australian immunologist. With P. B. Medawar, studied the body's tolerance of antigenic substances; Nobel Prize, 1960.

Bush, Vannevar (1890–1974) American electrical engineer. Originated the concept of hypertext.

Butenandt, Adolph Friedrich Johann (1903–1995), German chemist. Researched sex hormones; Nobel Prize (declined), 1939.

Buys-Ballot, Christoph Hendrik Didericus (1817–1890), Dutch meteorologist. Devised a system of storm signals; formulated Buys-Ballot's law for determination of wind direction.

Byron, Augusta Ada, Countess of Lovelace (1815–1852), English mathematician. Daughter of the poet Lord Byron; wrote the first program for Charles Babbage's analytical engine; often described as the first computer programmer.

Cailletet, Louis Paul (1832–1913), French chemist. Researched liquefaction of gases; first to obtain liquid oxygen, hydrogen, nitrogen, and air.

Callendar, Hugh Longbourne (1863–1930), English physicist and engineer. Developed platinum resistance thermometer and continuous-flow calorimeter.

Callow, John Michael (1867–1940), English-born American mining engineer and metallurgist. Invented Callow flotation cell and Callow screen.

Calvin, Melvin (1911–1997), American chemist. With J. A. Bassham, traced the path of carbon in photosynthesis; Nobel Prize, 1961.

Cannizzaro, Stanislao (1826–1910), Italian chemist. Promulgated Avogadro's work as related to atomic weights; discovered Cannizzaro's reaction in organic chemistry.

Cantor, Georg (1845–1918), Russian-born German mathematician. Founded set theory; introduced fundamental concepts in topology; worked on the theory and representations of real numbers.

Carathéodory, Constantin (1873–1950), German mathematician. Developed calculus of variations for curves with corners; introduced Carathéodory outer measure; gave mathematical formulation of second law of thermodynamics (Carathéodory's principle).

Cardano, Geronimo, or Jerome Cardan (1501–1576), Italian physician and mathematician. Wrote on algebra, medicine, and astronomy; invented the Cardan shaft.

Carlsson, Arvid (1923–), Swedish pharmacologist. Did research on dopamine, leading to the discovery of its role as a key neurotransmitter in the brain and in the control of movement; the development of L-dopa, a precursor of dopamine, into a drug to treat Parkinson's disease; the elucidation of the mode of action of antipsychotic drugs, which affect synaptic transmission by blocking dopamine receptors; did work (along with that of Paul Greengard and Eric Kandel) leading to the elucidation of the molecular mechanisms involved in slow synaptic transmission in the nervous system; Nobel Prize, 2000.

Carnot, Nicolas Léonard Sadi (1796–1832), French physicist. Formulated Carnot's theorems in thermodynamics.

Carrel, Alexis (1873–1944), French surgeon and biologist. Worked on transplanting organs, suturing blood vessels, treating deep wounds, and prolonging tissue life; Nobel Prize, 1912.

Carrington, Richard Christopher (1826–1875), English astronomer. Investigated motions of sunspots.

Carver, George Washington (1864–1943), American botanist. Did research on industrial uses of the peanut.

Cassegrain, N. (17th century), French physician. Designed the Cassegrain reflecting telescope.

Cassini, Jean Dominique (1625–1712), Italian-born French astronomer. Director of the Paris Observatory; discovered four new satellites of Saturn and the Cassini division in Saturn's ring; conducted pendulum experiments related to the shape of the Earth.

Castigliano, Carlo Alberto (1847–1884), Italian structural engineer. Proved Castigliano's theorem.

Cauchy, Augustin Louis, Baron (1789–1857), French mathematician. Wrote extensively on wave propagation, calculus, and elasticity.

Cavendish, Henry (1731–1810), French physicist and chemist. Determined the density of the Earth and the composition of the atmosphere; studied properties of carbon dioxide and hydrogen.

Cayley, Arthur (1821–1895), English mathematician. Proposed the theory of matrices; developed the theory of invariants and covariants; worked on quantics and the theory of groups.

Cech, Thomas R. (1947–), American chemist. By studying the single-cell organisms *Tetrahymena* and *Thermophila*, discovered that molecules of ribonucleic acid (RNA) have catalytic properties similar to those of enzymes; Nobel Prize, 1989.

Celsius, Anders (1701–1744), Swedish astronomer. Constructed the thermometer using the Celsius (centigrade) scale.

Cerenkov, Pavel Alexeyevich (1904–1990), Soviet physicist. Discovered the Cerenkov effect of radiation; devised the Cerenkov counter for particle detection; Nobel Prize, 1958.

Cerf, Vinton G. (1943–), American computer scientist. Co-invented (with Robert E. Kahn) Transmission Control Protocol/Internet Protocol (TCP/IP).

Cesalpino, Andrea (1519–1603), Italian physician and botanist. First to attempt to classify plants according to characteristics of fruit and seed in his *De Plantis*.

Cesàro, Ernesto (1859–1906), Italian mathematician. Formulated an intrinsic geometry; introduced the Cesàro summation.

Ceva, Giovanni (1647?–1734), Italian mathematician. Formulated a theorem on the concurrency of straight lines passing through the vertices of a triangle.

Chadwick, James (1891–1974), English physicist. Established experimentally the existence of the neutron; Nobel Prize, 1935.

Chain, Ernst Boris (1906–1979), German-born British biochemist. With H. W. Florey, worked on the chemical structure of penicillin and its first clinical trials; Nobel Prize, 1945.

Chamberlain, Owen (1920–), American physicist. With E. G. Segrè, demonstrated the existence of the antiproton; Nobel Prize, 1959.

Chamberlin, Thomas Chrowder (1843–1928), American geologist. Studied the fundamental geology of the solar system.

Chandler, Seth Carlo (1846–1913), American astronomer. Discovered the Chandler wobble.

Chandrasekhar, Subrahmanyan (1910–1995), Indian astrophysicist. Developed a theory of white dwarf stars; Nobel Prize, 1983.

Chaplygin, Sergel Alekseevich (1869–1942), Russian physicist, engineer, and mathematician. Made contributions to fluid mechanics, particularly aerodynamics.

Chapman, Sydney (1888–1970), English mathematician and physicist. Discovered (independently of D. Enskog) gaseous thermal diffusion; studied the daily variations of the geomagnetic field and magnetic storms.

Chaptal, Jean Antoine Claude, Comte de Chanteloup (1756–1832), French chemist. Wrote on technical chemistry; introduced the metric system after the Revolution.

Charcot, Jean Martin (1825–1893), French neurologist. Director of the Salpetrière clinic, where he made systematic clinical studies of chronic nervous disorders, including cerebrospinal disease.

Charles, Jacques Alexandre César (1746–1823), French physicist, chemist, and inventor. Formulated Charles' law, relating gas volume to pressure.

Charpak, Georges (1924–), French physicist. Invented the multiwire proportional chamber, used as a detector in high-energy physics experiments; Nobel Prize, 1992.

Chebyshev, Pafnuti Lvovich (1821–1894), Russian mathematician. Research on convergence of Taylor series, prime numbers, probability theory, quadratic forms, and integral theory.

Chladini, Ernst Florenz Friedrich (1756–1827),

German physicist. Discovered Chladini's figures and used them to study vibrations of solid plates.

Christoffel, Elwin Bruno (1829–1900), Swiss mathematician. Worked in higher analysis, geometry, mathematical physics, and geodesy.

Chu, Ching-Wu (1941–), Chinese-born American physicist. Discovered superconductivity at temperatures over 90 K (−298°F) in yttrium-barium-copper-oxygen compounds.

Chu, Steven (1948–), American physicist. Developed methods to cool and trap atoms with laser light, achieving temperatures equal to the Doppler limit, then thought to be the lowest attainable, and inventing a magnetooptical trap; Nobel Prize, 1997.

Clairaut, Alexis Claude (1713–1765), French mathematician. Studied the shape of the Earth, evolving Clairaut's theorem; made astronomical calculations concerning Halley's Comet.

Claisen, Ludwig (1851–1930), German organic chemist. Developed Claisen condensation; contributed to understanding of tautomerism; worked on rearrangement of allyl aryl ethers into phenols.

Clapeyron, Benoit Paul Émile (1799–1864), French engineer. Developed N. L. S. Carnot's concept of a universal function of temperature.

Clarke, Alexander Ross (1828–1914), British geodesist. Worked on the triangulation of the British Isles; proposed Clarke ellipsoids as geodetic standards.

Claude, Albert (1899–1983), American cytologist, born in Luxembourg. Pioneered in applying electron microscopy to cell studies and in using centrifuge to separate cell components; Nobel Prize, 1974.

Clausius, Rudolf Julius Emmanuel (1822–1888), German physicist. A founder of thermodynamics; worked out the Clausius-Clapeyron equation for the universal temperature function.

Clebsch, Rudolf Friedrich Alfred (1833–1872), German mathematician. Contributions to theory of invariants and algebraic geometry.

Cockcroft, John Douglas (1897–1967), English physicist. With E. T. S. Walton, split nuclei by bombarding them with accelerated protons; Nobel Prize, 1951.

Cohen, Paul Joseph (1934–), American mathematician. Proved that the axiom of choice is independent of the other axioms of set theory, and that the continuum hypothesis is independent of the axiom of choice, a result with profound implications for the foundations of mathematics; Fields Medal, 1966.

Cohen, Stanley (1922–), American biochemist. With R. Levi-Montalcini, made landmark studies of nerve growth factor and its functions; Nobel Prize, 1986.

Cohen-Tannoudji, Claude (1933–), French physicist. Developed methods to cool and trap atoms with laser light, explaining the achievement by W. D. Phillips of temperatures much lower than the Doppler limit, and then developing methods to achieve even lower temperatures that surpass the recoil limit; Nobel Prize, 1997.

Cohn, Ferdinand Julius (1828–1898), German botanist. A founder of bacteriology; did research in plant pathology; first to classify bacteria according to genus and species.

Cole, Kenneth Stewart (1900–1984), American biophysicist. Research on structure and function of living cell membranes and nerve membranes in particular, concentrating on electrical approach; with brother, R. H. Cole, introduced Cole-Cole plot of dielectric behavior.

Cole, Robert Hugh (1914–1990), American chemist and physicist. Research on dielectric properties of matter and intermolecular forces; with brother, K. S. Cole, introduced Cole-Cole plot of dielectric behavior.

Collins, Samuel Cornette (1898–1984), American engineer. Invented Collins helium liquefier.

Compton, Arthur Holly (1892–1962), American physicist. Discovered the Compton effect of x-rays; studied cosmic rays; helped develop the atomic bomb; Nobel Prize, 1927.

Conant, James Bryant (1893–1978), American chemist. Researched free radicals, hemoglobin, and chlorophyll; contributed to atomic energy development.

Condon, Edward Uhler (1902–1974), American physicist. Contributed to the Franck-Condon principle, by extending and giving quantum-mechanical treatment to J. Franck's concept of nuclear motion to molecules in transition from one energy level to another.

Connes, Alain (1947–), French mathematician. Did fundamental work on the theory and application of operator algebras, particularly von Neumann algebras; Fields Medal, 1982.

Coolidge, William David (1873–1975), American physicist. Invented Coolidge tube; discovered method for making tungsten strong and ductile.

Cooper, Leon Neil (1930–), American physicist. Showed that electrons could form Cooper pairs; with J. R. Schrieffer and J. Bardeen, formulated a theory of superconductivity; Nobel Prize, 1972.

Copernicus, Nicolaus (1473–1543), Polish (or Prussian) astronomer. Proposed the Copernican system, with the Sun as the center of planetary orbits.

Corey, Elias James (1928–), American chemist. Developed theories and methods of organic chemical synthesis that have made possible the production of a wide variety of complex biologically active substances and useful chemicals; Nobel Prize, 1990.

Cori, Carl Ferdinand (1896–1984), **and Cori, Gerty Theresa Radnitz** (1896–1957), Czechoslovakian-born American biochemists. Discovered the enzymatic mechanism of glucose-glycogen interconversion and the effects of hormones on this mechanism; Nobel Prize, 1947.

Coriolis, Gaspard Gustave de (1792–1843), French physicist. Contributed to theoretical and applied mechanics; clarified and supplied concepts of work and kinetic energy; derived Coriolis acceleration.

Cormack, Allan MacLeod (1924–1998), American physicist. Contributed to the development of computerized axial tomography; Nobel Prize, 1979.

Cornell, Eric Allin (1961–), American physicist. With C. E. Wieman, succeeded for the first time in producing Bose-Einstein condensates in a dilute gas of alkali (rubidium) atoms, and carried out fundamental studies of their properties, including studies of collective excitations and vortex formation in condensates; Nobel Prize, 2001.

Corner, George Washington (1889–1981), American medical biologist. Contributed to understanding of anatomical details of menstrual cycle and functions of estrogen and progesterone.

Cornforth, John Warcup (1917–), Australian-born British chemist. Investigated stereochemistry of enzyme-catalyzed reactions; Nobel Prize, 1975.

Cornu, Marie Alfred (1841–1902), French physicist. Used Cornu spiral for determination of intensities in interference phenomena.

Coster, Dirk (1889–1950), Dutch physicist. Work in x-ray spectroscopy; with G. von Hevesy, discovered hafnium; with R. Kronig, discovered Coster-Kronig transitions.

Cottrell, Frederick Gardner (1877–1948), American chemist. Invented the Cottrell process for precipitation of particles from gas; researched nitrogen fixation, liquefaction of gases, and recovery of helium.

Coulomb, Charles Augustin de (1736–1806), French physicist. Formulated Coulomb's law of electric charges.

Courant, Richard (1888–1972), German-born American mathematician. Research in geometric function theory, differential equations of mathematical physics, and transition by limiting processes from finite difference equations to differential equations.

Cournand, André Frédéric (1895–1988), French-born American physician. Studied normal and abnormal human cardiovasculatory and pulmonary functions; Nobel Prize, 1956.

Cowan, Clyde Lorrain, Jr. (1919–1974), American physicist. With F. Reines, detected the neutrino.

Crafts, James Mason (1839–1917), American chemist. With C. Friedel, discovered the Friedel-Crafts reaction, wherein anhydrous aluminum chloride acts as a catalyst.

Cram, Donald J. (1919–2001), American chemist. Expanded the field of crown ether chemistry by using crown ethers to synthesize structures that mimic the action of biological molecules; Nobel Prize, 1987.

Cramer, Gabriel (1704–1752), Swiss mathematician. Contributed to Cramer's rule for solving linear equations.

Crick, Francis Harry Compton (1916–), English molecular biologist. With J. D. Watson, proposed a double-helix structure for the deoxyribonucleic acid molecule; Nobel Prize, 1962.

Cronin, James Watson (1931–), American physicist. Collaborated with V. L. Fitch on experiment showing that the principle of time-reversal invariance is violated in the decay of neutral K mesons; Nobel Prize, 1980.

Cronstedt, Axel Fredrik, Baron (1722–1765), Swedish mineralogist. Discovered nickel; developed a chemical classification system for minerals.

Crookes, William (1832–1919), English physicist and chemist. Invented Crookes tube to study electrical discharges in high vacuum, and a radiometer; discovered thallium.

Crutzen, Paul J. (1933–), Dutch-born meteorologist. Contributed to the understanding of how atmospheric ozone forms and decomposes; Nobel Prize, 1995.

Curie, Marie, born **Marya Sklodowska** (1867–1934), Polish physical chemist in France. Explored nature of radioactivity; codiscoverer of radium, and first to separate polonium; Nobel Prize, 1903 and 1911.

Curie, Pierre (1859–1906), French chemist and physicist. Codiscoverer of radium; formulated the Curie point, relating magnetic properties and temperature; discovered the piezoelectric effect; Nobel Prize, 1903.

Curl, Robert F., Jr. (1933–) American chemist. With Harold Kroto and Richard Smalley, discovered fullerenes; Nobel Prize, 1996.

Cushing, Harvey (1869–1930), American surgeon. Innovator in neurosurgical techniques; research on function and diseases of the pituitary gland.

Cuvier, Georges Léopold Chrétien Frédéric Dagobert, Baron (1769–1838), French naturalist. Made a detailed classification of the animal kingdom; wrote on comparative anatomy.

Daguerre, Louis Jacques Mandé (1787–1851), French inventor. Invented the daguerreotype photographic process.

Dale, Henry Hallett (1875–1968), British pharmacologist and physiologist. Isolated acetylcholine and recognized its effect to be similar to that brought about by parasympathetic nerves; Nobel Prize, 1936.

Dalén, Nils Gustaf (1869–1937), Swedish physicist. Invented automatic gas lighting for unsupervised lighthouses and railroad signals; Nobel Prize, 1912.

Dalitz, Richard Henry (1925–), Australian-born British theoretical physicist. Research on properties of mesons and baryons and nuclear interactions of the lambda hyperon; proposed models for elementary particles; introduced the Dalitz plot.

Dalton, John (1766–1844), English chemist and physicist. Proposed the atomic theory of chemical reactions; developed the law of partial pressures of gases; studied color-blindness.

Dam, Carl Peter Henrik (1895–1976), Danish biochemist and nutritionist. Discovered vitamin K and studied its role in human hemorrhagic disease; Nobel Prize, 1943.

Danckwerts, Peter Victor (1916–1984), British chemical engineer. Proposed the surface-renewal model of liquids.

Daniell, John Frederic (1790–1845), English physicist and chemist. Invented the Daniell cell.

Darwin, Charles Robert (1809–1882), English naturalist. Proposed far-reaching theory of evolution of species and theory of natural selection in his *Origin of Species*.

Darwin, George Howard (1845–1912), English mathematician and astronomer. Applied detailed dynamical analysis to cosmological and geological problems.

Dausset, Jean (1916–), French biologist and medical scientist. Studied antigens in human leukocytes and their role in transplant acceptance or rejections; Nobel Prize, 1980.

Davisson, Clinton Joseph (1881–1958), American physicist. Studied magnetism, radiant energy, and electricity; independent of G. P. Thomson, discovered electron diffraction by crystals; Nobel Prize, 1937.

Davy, Humphry (1778–1829), English chemist. Discovered potassium and sodium; invented the Davy safety lamp for use in coal mines; proposed theoretical explanations of electrolysis and voltaic action.

Debye, Peter Joseph William (1884–1966), American physical chemist born in the Netherlands. Worked on dipole moments and the diffraction of x-rays in gases; formulated Debye-Hückel theory on the behavior of strong electrolytes; Nobel Prize, 1936.

Dedekind, Julius Wilhelm Richard (1831–1916), German mathematician. Worked in number theory and analysis, particularly with algebraic integers, algebraic functions, and ideals; defined real numbers by Dedekind cuts.

de Duve, Christian René (1917–), Belgian biochemist and cytologist. Refined centrifuge technique for studying cell components; discovered lysosomes; Nobel Prize, 1974.

De Forest, Lee (1873–1961), American inventor. Pioneer in radio technology; invented audio amplifier and the four-electrode valve; incorporated the grid into the thermionic valve.

de Gennes, Pierre-Gilles (1932–), French physicist. Applied physical principles to the study of complex systems, including liquid crystals and polymers; Nobel Prize, 1991.

de Haas, Wander Johannes (1878–1960), Dutch physicist. Demonstrated the Einstein-de Haas effect; worked on production of extremely low temperatures by adiabatic demagnetization; with P. Van Alphen, discovered de Hass-Van Alphen effect.

Dehmelt, Hans Georg (1922–), German-born American physicist. Developed the Penning trap, which uses magnetic and electric fields to hold ions in a small volume; used the traps to isolate a single electron and carry out extremely accurate measurements of atomic properties; Nobel Prize, 1989.

Deisenhofer, Johann (1943–), German chemist. With R. Huber and M. Hartmut, elucidated the structure of a bacterial protein that performs photosynthesis; Nobel Prize, 1988.

Delbrück, Max (1906–1981), German-born American biologist. Pioneered in molecular biology; research on bacterial viruses; Nobel Prize, 1969.

Deligne, Pierre René (1944–), Belgian mathematician. Solved three conjectures of A. Weil concerning generalizations of the Riemann hypothesis to finite fields—work which brought together algebraic geometry and algebraic number theory; Fields Medal, 1978.

Demoivre, Abraham (1667–1754), French-born English mathematician. Originated two theorems on trigonometrical expansions; proposed methods to approximate functions of large numbers, and the concept of the normal distribution curve.

De Morgan, Augustus (1806–1871), English mathematician. Wrote textbooks and treatises on arithmetic, algebra, and trigonometry; formulated the De Morgan theorem.

Desargues, Gérard (1593–1662), French mathematician. A founder of modern geometry; proposed the theory of involution and transversals.

Descartes, René (1596–1650), French mathematician. Originated Cartesian, or coordinate, geometry.

de Sitter, Willem (1872–1934), Dutch astronomer. Worked on the application of Einstein's theory to the problems of the universe; computed the size of the universe.

De Vries, Hugo (1848–1935), Dutch botanist. Formulated the mutation theory of evolution.

Dewar, James (1842–1923), British chemist. Made pioneering studies of matter at low temperatures; first to liquefy hydrogen; invented the Dewar vacuum flask.

Dick, George Frederick (1881–1967), American physician and bacteriologist. Isolated scarlet fever streptococci, developed scarlet fever streptococcus antitoxin, and developed Dick test.

Dicke, Robert Henry (1916–1997), American physicist. Developed new relativistic theory of gravitation with C. Brans; investigated cosmic blackbody radiation; worked on development of radar.

Diels, Otto Paul Hermann (1876–1954), German chemist. Codiscoverer of the Diels-Alder reaction (diene synthesis); worked on sterol chemistry; discovered carbon suboxide; Nobel Prize, 1950.

Diesel, Rudolf (1858–1913), German inventor. Designed and built the diesel engine.

Diocles (2d century B.C.), Greek mathematician. Contributions to theory of conics and geometry.

Diophantus of Alexandria (3d century), Greek mathematician. Known as the father of algebra; the first to use conventional algebraic notation.

Dirac, Paul Adrien Maurice (1902–1984), English physicist. Worked in quantum mechanics; his theory of negative-energy holes predicted existence of the positron; Nobel Prize, 1933.

Dirichlet, Peter Gustave Lejeuné (1805–1859), German mathematician. Applied higher analysis to the theory of numbers; work on definite integrals.

Djerassi, Carl (1923–) Austrian-born, American chemist. Synthesized the first oral contraceptive.

Dobzhansky, Theodosius (1900–1975), Russian-born American biologist. Elucidated the mechanisms of heredity and variation through studies of *Drosophila*.

Doherty, Peter C. (1940–), Australian immunologist. With Rolf M. Zinkernagel, he discovered how the immune system, and particularly T lymphocytes, recognizes virus-infected cells; Nobel Prize, 1996.

Doisy, Edward Adelbert (1893–1986), American biochemist. Isolated pure crystalline compounds important to human health, such as sex hormones and vitamins; Nobel Prize, 1943.

Domagk, Gerhard (1895–1964), German biochemist. Discovered sulfamidocrysoidin, the first synthetic microbial of broad clinical usefulness. Nobel Prize (declined), 1939.

Donaldson, Simon Kirwan (1957–), British mathematician. Worked on topology of four-manifolds and showed that there exist exotic four-spaces, that is, four-dimensional differentiable manifolds that are topologically but not differentiably equivalent to the standard Euclidean four-space; Fields Medal, 1986.

Donati, Giovanni Battista (1826–1873), Italian astronomer. Studied stellar spectra; discovered six comets.

Donnan, Frederick George (1870–1956), Irish chemist born in Ceylon. Research in chemical kinetics; originated the Donnan theory of membrane equilibrium.

Doppler, Christian Johann (1803–1853), Austrian physicist and mathematician. Formulated Doppler's principle, relating the frequency of wave motion to velocity; described the Doppler effect.

Douglas, Jesse (1897–1965), American mathematician. Solved the Plateau problem (the problem of determining the existence of a minimal surface with a given space curve as its boundary) about the same time as T. Radó and studied generalizations of this problem; Fields Medal, 1936.

Drake, Frank Donald (1930–), American astronomer. Research on solar system, 21-centimeter radio line, search for extraterrestrial intelligence (introduced Drake equation), and radio telescope development.

Draper, Henry (1837–1882), American scientist. Research in spectroscopy; the Draper catalog is named in his honor.

Drinfeld, Vladimir Gershonovich (1954–), Ukrainian mathematician. Worked in algebraic geometry, number theory, and the theory of quantum groups; proved a special case of the Langlands conjecture; Fields Medal, 1990.

Drude, Paul Karl Ludwig (1863–1906), German physicist. Attempted to correlate and account for optical, electrical, thermal, and chemical properties of substances; developed theory of properties of metals based on free electrons treated as a gas.

Duane, William (1872–1935), American physicist and radiologist. Developed treatment of cancer by radioisotopes and x-rays; with F. L. Hunt, discovered Duane-Hunt law of x-rays.

Du Bois-Reymond, Emil (1818–1896), German physiologist. Pioneer work in electrical properties of living tissues, especially nerves.

DuBridge, Lee Alvin (1901–1994), American physicist. Developed Fowler-DuBridge theory of photoelectric emission.

Ducrey, Augusto (1860–1940), Italian dermatologist. Discovered *Hemophilus ducreyi*, the agent of chancroid.

Dufay, Charles François de Cisternay (1698–1739), French chemist. Discovered positive and negative types of electricity.

Dulbecco, Renato (1914–), Italian-born American virologist. Developed techniques for studying animal viruses; investigated interaction between deoxyribonucleic acid tumor viruses and genetic material; Nobel Prize, 1975.

Dulong, Pierre Louis (1785–1838), French chemist and physicist. With A. T. Petit, formulated the law of the constancy of atomic heats; developed the Dulong formula for heat value of fuels.

Dumas, Jean Baptiste André (1800–1884), French chemist. Research on organic compounds; determined many atomic weights.

Dürer, Albrecht (1471–1528), German painter, graphic artist, and mathematician. Work on scientific perspective and mathematical proportion.

du Vigneaud, Vincent (1901–1978), American biochemist. Synthesized a polypeptide hormone, oxytocin; worked on other biologically important sulfur compounds; Nobel Prize, 1955.

Eccles, John Carew (1903–1997), Australian physiologist. Elucidated the action of nerve impulses across zones of close contact between nerve cells; Nobel Prize, 1963.

Eddington, Arthur Stanley (1882–1944), English astronomer and writer. Theoretical research on stellar movements and internal makeup of stars; wrote on theory of relativity.

Edelman, Gerald Maurice (1929–), American biochemist. Worked to determine chemical structure of immunoglobulins; Nobel Prize, 1972.

Edison, Thomas Alva (1847–1931), American electrician and inventor. Invented the gramophone, carbon transmitter for the telephone, incandescent electric lamp, moving pictures, and the diplex method of telegraphy.

Ehrenfest, Paul (1880–1933), Austrian-born Dutch theoretical physicist. Contributed to statistical mechanics and quantum mechanics; developed Ehrenfest's principle; proved Ehrenfest's theorem.

Ehrlich, Paul (1854–1915), German bacteriologist. Research in chemotherapy, notably the discovery of Salvarsan for treatment of syphilis; pioneered in the study of hematology and immunity; Nobel Prize, 1908.

Eigen, Manfred (1927–), German chemist. Devised relaxation techniques to study high-speed chemical reactions; Nobel Prize, 1967.

Eijkman, Christiaan (1858–1930), Dutch physician. Studied dietary deficiency disease, in particular beriberi; Nobel Prize, 1929.

Einstein, Albert (1879–1955), German-born American physicist. Proposed the theory of relativity; extended the application of quantum theory; Nobel Prize, 1921.

Einthoven, Willem (1860–1927), Dutch physiologist born in Java. Used the string galvanometer to record electrical activity of the heart, thereby inventing the electrocardiograph; Nobel Prize, 1924.

Elion, Gertrude Belle (1918–1999), American biochemist. With G. H. Hitchings, pioneered research that led them to the development of drugs for the treatment of leukemia, malaria, gout, herpes, bacterial and fungal infections, and autoimmune diseases and organ-transplant rejection; Nobel Prize, 1988.

Elsasser, Walter Maurice (1904–1991), German-born American geophysicist. Formulated the dynamo theory of the Earth's permanent terrestrial magnetic force.

Elster, Johann Philipp Ludwig Julius (1854–1920), German experimental physicist. With H. F. Geitel, studied atmospheric electricity, radioactivity, and photoelectricity, and invented photocell.

Emmett, Paul Hugh (1900–1985), American chemist. Worked on catalysts for ammonia synthesis and the water-gas conversion reaction; with S. Brunauer and E. Teller, formulated Brunauer-Emmett-Teller equation for surface area determinations.

Encke, Johann Franz (1791–1865), German astronomer. Determined period of revolution of Encke's Comet, discovered by J. L. Pons; measured the distance of the Sun from the Earth.

Enders, John Franklin (1897–1985), American microbiologist. With F. C. Robbins and T. H. Weller, discovered the capacity of poliomyelitis virus to grow in various tissue cultures; Nobel Prize, 1954.

Enskog, David (1884–1947), Swedish physicist. With S. Chapman, developed the Chapman-Enskog theory for solving the Boltzmann transport equation.

Eötvös, Roland, Baron (1848–1919), Hungarian physicist. Research on gravitation and terrestrial magnetism; formulated a law which relates surface tension to temperature of liquids; designed the Eötvös torsion balance.

Erasistratus (3d century B.C.), Greek physician and anatomist. Founder of physiology; distinguished between the cerebrum and cerebellum, and sensory and motor nerves.

Eratosthenes (3d century B.C.), Greek astronomer. Suggested an extra day in the calendar every fourth year; made a determination of the size of the Earth; measured obliquity of the ecliptic.

Erdös, Paul (1913–1996), Hungarian mathematician. Did wide-ranging work in algebra, analysis, combinatorial theory, geometry, topology, number theory, and graph theory; with A. Selberg, gave an elementary proof of the prime number theorem.

Erlanger, Joseph (1874–1965), American physiologist. With H. S. Gasser, created new methods for amplifying and recording electrical impulses in nerves; research on the function of the synapse; Nobel Prize, 1944.

Erlenmeyer, Richard August Carl Emil (1825–1909), German organic chemist. Research on synthesis and constitution of aliphatic compounds; introduced modern structural notation; invented Erlenmeyer flask.

Ernst, Richard R. (1933–), Swiss chemist. Developed methods that transformed nuclear magnetic resonance (NMR) spectroscopy from a tool with a narrow application to a key analytical technique in chemistry as well as many other fields; Nobel Prize, 1991.

Esaki, Leo (1925–), Japanese physicist. Discovered a new negative-resistance characteristic in semiconductor pn junctions, leading to the discovery of the tunnel, or Esaki, diode; Nobel Prize, 1973.

Euclid (ca. 330–ca. 275 B.C.), Greek mathematician. Wrote geometry textbooks; Euclidean geometry is named after him.

Eudoxus of Cnidus (ca. 408–ca. 355 B.C.), Greek astronomer and mathematician. Expounded theory of motion of planets based on homocentric spheres; developed theory of proportions and methods of measuring areas and volumes of geometrical figures.

Euler, Leonhard (1707–1783), Swiss mathematician. Contributed to algebraic series and differential and integral calculus; realized the significance of coefficients (Euler numbers) of certain trigonometrical expansions.

Euler-Chelpin, Hans Karl August Simon von (1873–1964), German-Swedish chemist. Research on enzyme action and fermentation of sugars; Nobel Prize, 1929.

Ewald, Paul Peter (1888–1985), German-born American physicist. Developed dynamic theory of x-ray interference in crystals.

Ewing, William Maurice (1906–1974), American geophysicist. Made fundamental contributions to seismology, geodesy, oceanography, and submarine geology.

Eyring, Henry (1901–1981), Mexican-born American chemist. Pioneered in the application of quantum and statistical mechanics to chemistry; conceived the theory of absolute reaction rates and the significant structures theory of liquids.

Fabricius, Hieronymus, or Girolamo Fabrizio (ca. 1533–1619), Italian anatomist. Made painstaking descriptions of valves in veins; did comparative research in animal embryology.

Fabry, Charles (1867–1945), French physicist. With A. Pérot, invented Fabry-Pérot interferometer; experimentally verified Doppler broadening and Doppler effect.

Fahrenheit, Gabriel Daniel (1686–1736), German physicist. Constructed thermometers; invented the Fahrenheit temperature scale.

Fallopio, Gabriele (1523–1562), Italian anatomist. Discovered Fallopian tubes; gave first clear description of organs of inner and middle ear.

Faltings, Gerd (1954–), German mathematician. Used methods of arithmetic algebraic geometry to prove the Mordell conjecture, which states that there are only finitely many rational points on a curve of genus greater than 1; this was a step toward the later proof of Fermat's last theorem by A. Wiles; Fields Medal, 1986.

Fanning, John Thomas (1837–1911), American civil engineer. Designed water works and water supply systems.

Faraday, Michael (1791–1867), English chemist and physicist. Discovered electromagnetic induction; formulated two laws of electrolysis; invented the dynamo.

Fechner, Gustav Theodor (1801–1887), German psychologist. Founded psychophysics; developed the Fechner law concerning intensity of sensation produced by a stimulus.

Fefferman, Charles Louis (1949–), American mathematician. Worked in Fourier analysis, partial differential equations, and the theory of functions of several complex variables; discovered the dual of the Hardy space H^1; Fields Medal, 1978.

Feit, Walter (1930–), Austrian-born American mathematician. Worked in group theory; with J. G. Thompson, proved that all noncyclic finite simple groups have even order.

Fermat, Pierre de (1601–1665), French mathematician. Founder of the modern theory of numbers; originated Fermat's last theorem, and Fermat's principle in optics.

Fermi, Enrico (1901–1954), Italian-born American physicist. Research on producing radioactive isotopes by neutron bombardment; directed construction of the first atomic pile; Nobel Prize, 1938.

Feynman, Richard Phillips (1918–1988), American physicist. Proposed a theory to eliminate difficulties that had arisen in the study of the interaction of electrons, positrons, and radiation; Nobel Prize, 1965.

Fibiger, Johannes Andreas Grib (1867–1928), Danish pathologist. First to produce cancer experimentally; Nobel Prize, 1926.

Finsen, Niels Ryberg (1860–1904), Danish physician. Originated ultraviolet light therapy for certain diseases; Nobel Prize, 1903.

Fischer, Edmond H. (1920–), American biochemist. With E. G. Krebs, discovered phosphorylation processes that play a critical role in cell-protein regulation; they isolated the first protein kinase, a class of enzymes that transfer phosphate from adenosine triphosphate to proteins; Nobel Prize, 1992.

Fischer, Emil Hermann (1852–1919), German chemist. Synthesized many natural substances, including purines, D-glucose and other sugars, and the first nucleotide; studied polypeptides and proteins; Nobel Prize, 1902.

Fischer, Ernst Otto (1918–), German chemist. Studied how metals and organic molecules combine to form unique molecules with sandwichlike structures; Nobel Prize, 1973.

Fischer, Hans (1881–1945), German organic chemist. Investigated and synthesized pyrrole pigments; studied structure of chlorophylls; Nobel Prize, 1930.

Fisher, Ronald Aylmer (1890–1962), English geneticist and statistician. Developed statistical techniques for analysis of variance, and for use and validation of small samples; developed theory of the evolution of dominance.

Fitch, Val Logsdon (1923–), American physicist. Collaborated with J. W. Cronin on experiment showing that the principle of time-reversal invariance is violated in the decay of neutral K mesons; Nobel Prize, 1980.

FitzGerald, George Francis (1851–1901), Irish physicist. Proposed Lorentz-FitzGerald contraction, relating to a material moving through an electromagnetic field.

Fizeau, Armand Hippolyte Louis (1819–1896), French physicist. First to accurately measure the velocity of light; conducted experiments on the velocity of electricity, use of light wavelength to measure length, and measurement of diameter of stars through the method of interference.

Flamsteed, John (1646–1719), English astronomer. Made a trustworthy catalog of stars; invented conical projection in mapmaking.

Fleming, Alexander (1881–1955), British bacteriologist. Discovered lysozyme and penicillin; Nobel Prize, 1945.

Fleming, John Ambrose (1849–1945), English electrical engineer. Invented the thermionic valve; contributed to widespread application of electric lighting and heating.

Florey, Howard Walter (1898–1968), British pathologist born in Australia. Contributed, with E. B. Chain, to development of penicillin as a chemotherapeutic agent; Nobel Prize, 1945.

Flory, Paul John (1910–1985), American physical chemist. Developed analytic techniques to explore properties and molecular structures of long-chain molecules; Nobel Prize, 1974.

Fock (Fok), Vladimir Alexandrovitch (1898–1974), Soviet theoretical physicist. Contributions to quantum electrodynamics, quantum field theory, electromagnetic diffraction and propagation, and general relativity; developed Hartree-Fock approximation of wave functions.

Forbush, Scott Ellsworth (1904–1984), American geophysicist. Discovered the worldwide decrease in cosmic-ray intensity associated with some magnetic storms.

Forssmann, Werner Theodor Otto (1904–1979), German physician. Developed the technique of cardiac catheterization; Nobel Prize, 1956.

Foucault, Jean Bernard Léon (1819–1868), French physicist. Accurately determined the velocity of light; constructed the Foucault pendulum and the Foucault prism; determined experimentally the rotation of the Earth.

Fourier, Jean Baptiste Joseph, Baron (1768–1830), French geometrician and physicist. Proposed the Fourier series on arbitrary functions; formulated the law of heat propagation.

Fowler, Ralph Howard (1889–1944), English physicist. Applied statistical mechanics to matter at high temperatures and high pressures; explained structure of white dwarf stars; with E. A. Guggenheim, R. F. Peierls, and others, developed Ising model.

Fowler, William Alfred (1911–1995), American physicist. Fundamental contributions to understanding of nuclear reactions that generate the energy of stars and synthesize the elements of the universe; Nobel Prize, 1983.

Franck, James (1882–1964), German physicist. With G. Hertz, studied energy transfer in collisions of molecules; formulated Franck-Condon principle of transition from one energy state to another; Nobel Prize, 1925.

Frank, Ilya Milkhallovich (1908–1990), Soviet physicist. With I. Y. Tamm, proposed a theoretical interpretation of Cerenkov radiation; Nobel Prize, 1958.

Franklin, Benjamin (1706–1790), American physicist, oceanographer, meteorologist, and inventor. Formulated a theory of general electrical "actions"; introduced principle of conservation of charge; showed that lightning is an electrical phenomenon; invented lighting rod.

Fraunhofer, Joseph von (1787–1826), German optician and physicist. First to study the dark lines in the solar spectrum (Fraunhofer lines); invented a heliometer; improved the spectroscope.

Fredholm, Eric Ivar (1866–1927), Swedish mathematician. Developed the theory of integral equations (Fredholm equations).

Freedman, Michael Hartley (1951–), American mathematician. Developed new methods for topological analysis of four-manifolds and applied them to prove the Poincaré conjecture for dimension 4; Fields Medal, 1986.

Frenet, Jean Frédéric (1816–1900), French mathematician. Helped to develop Frenet-Serret formulas.

Frenkel, Yakov Ilyich (1894–1954), Soviet physicist. Pioneered in modern atomic theory of solids; developed quantum-mechanical explanations for electron mean free path in metals, and for paramagnetism and ferromagnetism; postulated excitons, Frenkel excitons, and Frenkel defects.

Fresnel, Augustin Jean (1788–1827), French physicist. Investigated effects (Fresnel's fringes) due to the interference of light; developed a wave theory of light; originated Fresnel's reflection formula.

Freud, Sigmund (1856–1939), Austrian psychoanalyst. Founder of psychoanalysis, with emphasis on dream interpretation and free association; developed a theory of personality involving id, ego, and superego, and stressing importance of the libido.

Friedel, Charles (1832–1899), French chemist and mineralogist. With J. M. Crafts, described the Friedel-Crafts reaction; work on artificial production of minerals; studied crystals, ketones, and aldehydes.

Friedman, Jerome Isaac (1930–), American physicist. Collaborated in experiments that demonstrated that protons, neutrons, and similar particles are made up of quarks; Nobel Prize, 1990.

Frisch, Karl von (1886–1982), Austrian zoologist. Discovered means by which bees communicate information about the distance and direction of food; Nobel Prize, 1973.

Frobenius, Georg Ferdinand (1849–1917), German mathematician. Developed representation theory of finite groups and method for solving linear homogeneous ordinary differential equations.

Froude, William (1810–1879), English engineer. Discovered the Froude law of comparison, concerning the towing of an object in a liquid.

Fubini, Guido (1879–1943), Italian mathematician. Worked in algebra, analysis, and differential projective geometry; proved Fubini's theorem.

Fukui, Kenichi (1918–1998), Japanese chemist. Developed frontier orbital theory, a quantum-mechanical model useful in prediction of the combinative properties of molecules; Nobel Prize, 1981.

Fuller, R. Buckminster (1895–1983), American engineer and architect. Designed geodesic dome; the carbon molecular form C_{60} was named buckminsterfullerene because of its structured resemblance to the geodesic dome, and the name fullerene was given to any closed-cage molecule containing an even number of carbon atoms.

Furchgott, Robert F. (1916–), American pharmacologist. Discovered endothelium-derived relaxing factor (EDRF), a signaling molecule in the cardiovascular system that makes vascular smooth muscle cells relax (Louis J. Ignarro, working independently and with Furchgott, later concluded that EDRF was nitric oxide); Nobel Prize, 1998.

Gabor, Dennis (1900–1979), Hungarian-born British physicist and engineer. Invented holography; Nobel Prize, 1971.

Gajdusek, Daniel Carleton (1923–), American

physician and virologist. Discovered causal virus and transmission mechanism of kuru; Nobel Prize, 1976.

Galen (2d century), Greek physician and medical writer. Wrote treatises long used as textbooks; experimented on animal nervous systems; made anatomical descriptions of structure and functions of body parts.

Galileo Galilei (1564–1642), Italian astronomer. First to use the telescope for observational purposes; made many discoveries related to the planets and the Sun; did theoretical work on classical physics.

Galois, Evariste (1811–1832), French mathematician. Developed Galois theory of polynomials.

Gamow, George (1904–1968), Russian-born American physicist. Made theoretical contributions to nuclear physics, astronomy, and biology; with E. Teller, formulated the selection rule for beta emission; proposed theoretically the genetic code.

Garvey, Gerald Thomas (1935–), American physicist. Research in experimental nuclear physics, particularly nuclear reactions, isobaric spin studies, and weak interactions in nuclear systems.

Gasser, Herbert Spencer (1888–1963), American physiologist. With J. Erlanger, provided a new method for recording electrical impulses of nerves; studied functions of nerve fibers; Nobel Prize, 1944.

Gatterman, Friedrich August Ludwig (1860–1920), German chemist. Originated Gatterman-Koch synthesis of aldehydes; isolated and analyzed nitrogen trichloride; synthesized aromatic carboxylic acids, thionaphthene, and thioanilide.

Gauss, Karl Friedrich (1777–1855), German mathematician, astronomer, and physicist. Formulated the Gauss theorem in the mathematics of electricity; made many contributions to pure and applied mathematics; determined orbits of planets and comets from observational data.

Gay-Lussac, Joseph Louis (1778–1850), French chemist and physicist. Discovered the law of expansion of gases by heat, and the law of combining volumes of gases; studied chemistry of iodine and cyanogen.

Geiger, Hans Wilhelm (1882–1945), German physicist and inventor. Invented the Geiger counter to detect alpha particles; investigated properties of alpha particles, cosmic rays, and artificial radiation.

Geissler, Johann Heinrich Wilhelm (1815–1879), German instrument maker. Developed Geissler pump and Geissler tube.

Geitel, Hans Friedrich (1855–1923), German experimental physicist. With J. Elster, studied atmospheric electricity, radioactivity, and photoelectricity, and invented photocell.

Gell-Mann Murray (1929–), American physicist. Proposed law of conservation of strangeness; used unitary symmetry to classify and explain elementary particles; postulated concept of quarks; Nobel Prize, 1969.

Gesner, Konrad von (1516–1565), Swiss naturalist. Wrote *Historia Animalium*, beginning zoology as a science.

Giaever, Ivar (1929–), Norwegian-born American physicist. Discovered that current-voltage characteristics of an electron tunneling across a thin insulating film separating two metals, one or both of which is in a superconducting state, can be used to obtain electron density of states of superconductors; Nobel Prize, 1973.

Giauque, William Francis (1895–1982), Canadian-born American chemist. Developed adiabatic demagnetization technique for production of extremely low temperatures; collaborated in discovery of isotopes of oxygen; Nobel Prize, 1949.

Gibbs, Josiah Willard (1839–1903), American mathematician and physicist. Made a mathematical treatment of chemical subjects, notably thermodynamics; worked on statistical mechanics, leading to the basis for the phase rule of heterogeneous equilibria.

Gilbert, Walter (1932–), American biochemist. Developed methods for determining nucleotide sequence (independently of F. Sanger), advancing the technology of DNA recombination; Nobel Prize, 1980.

Gilman, Alfred G. (1941–), American pharmacologist. By isolating the G-protein, proved Martin Rodbell's hypothesis asserting the existence of natural signal transducers that assist communication between cells of the body; Nobel Prize, 1994.

Ginzburg, Vitaly Lazarevich (1916–), Soviet physicist. Developed Ginzburg-Landau and Ginzburg-London theories of superconductivity.

Giorgi, Giovanni (1871–1950), Italian electrical engineer, physicist, and mathematician. Developed the meter-kilogram-second-ampere system of units.

Glaser, Donald Arthur (1926–), American physicist. Invented the bubble chamber for detecting the paths of high-energy atomic particles; Nobel Prize, 1960.

Glashow, Sheldon Lee (1932–), American physicist. Contributed to development of theory uniting electromagnetism and weak nuclear interactions; postulated existence of charmed particles; Nobel Prize, 1979.

Glauber, Johann Rudolf (1604–1670), German chemist. Discovered Glauber's salt (sodium sulfate) and hydrochloric acid; conducted experiments on compounds of mercury, arsenic, and antimony.

Gmelin, Leopold (1788–1853), German chemist. Wrote *Handbuch der Chemie*, first systematic treatment of chemical knowledge; devised Gmelin's test for presence of bile pigments; studied cyanides.

Goldbach, Christian (1690–1764), German-born Russian mathematician. Research on number theory and analysis; proposed the Goldbach conjecture.

Goldhaber, Maurice (1911–), Austrian-born American physicist. With J. Chadwick, discovered photodisintegration and disintegration of light elements by slow neutrons; with L. Grodzins and A. W. Sunyar, discovered that the neutrino has left-handed spin.

Goldstein, Joseph L. (1940–), American biochemist and geneticist. With M. S. Brown, discovered low-density lipoprotein receptors and their function in cholesterol metabolism; Nobel Prize, 1985.

Golgi, Camillo (1843–1926), Italian physician. Pioneered in the study of histology of the nervous system; discovered the Golgi bodies; Nobel Prize, 1906.

Gordan, Paul Albert (1837–1912), German mathematician. Worked on the theory of invariants and on solutions of algebraic equations and their groups of substitutions.

Gordon, Walter (1893–1940), German-born Swedish physicist. Contributed to relativistic quantum theory; with O. B. Klein, originated the Klein-Gordon equation.

Goudsmit, Samuel Abraham (1902–1978), Dutch-born American physicist. With G. E. Uhlenbeck, discovered electron spin.

Gowers, William Timothy (1963–), British mathematician. Contributed to functional analysis, in particular, to the theory of Banach spaces, using methods of combinatorial theory; Fields Medal, 1998.

Gram, Hans Christian Joachim (1853–1938), Danish physician. Developed the Gram method for staining and differentiating bacteria.

Gram, Jorgen Pedersen (1850–1916), Danish mathematician. Worked in number theory and analysis; with E. Schmidt, originated Gram-Schmidt process for obtaining orthogonal set of vectors.

Granit, Ragnar Arthur (1900–1991), Finnish-born Swedish physiologist. Research on vision and on motor control by afferent neurons; Nobel Prize, 1967.

Grashof, Franz (1826–1893), German mechanical engineer. Applied mathematics and physics to engineering problems; derived fundamental equations in the theory of elasticity; introduced Grashof number.

Gray, Henry (1825–1861), English anatomist. Wrote *Anatomy of the Human Body*.

Green, George (1793–1841), English mathematician. Worked in analysis; derived Green's theorem and Green's identities; introduced Green's function.

Greengard, Paul (1925–), American neurobiologist. Discovered the mechanism by which dopamine and other chemical neurotransmitters (such as norepinephrine and serotonin) affect the nervous system—work contributing to the elucidation of the molecular mechanisms involved in slow synaptic transmission in the nervous system; Nobel Prize, 2000.

Gregory, James (1638–1675), Scottish geometer. Provided first proof of the theorem of calculus; gave first description of the reflecting telescope; discovered the series from which π can be calculated.

Grignard, François Auguste Victor (1871–1935), French chemist. Discovered organomagnesium compounds, or Grignard reagents, useful in synthesis of organic and organometallic compounds; Nobel Prize, 1912.

Grothendieck, Alexander (1928–), German-born French mathematician. Made fundamental advances in algebraic geometry and related fields, providing unifying themes in geometry, number theory, topology, and complex analysis; introduced the idea of K theory; revolutionized homological algebra; developed theory of schemes, allowing conjectures in number theory to be solved; Fields Medal, 1966.

Gruneisen, Eduard (1877–1949), German physicist. Formulated laws relating specific heat and other properties of solids.

Guillaume, Charles Édouard (1861–1938), Swiss-born French physicist. Studied nickel-steel alloys and invented Invar; Nobel Prize, 1920.

Guillemin, Ernst Adolph (1898–1970), American electrical engineer. Worked on network analysis and synthesis problems; invented Guillemin line; developed network to produce loran pulses.

Guillemin, Roger Charles Louis (1924–), French-born American physiologist. With A. V. Schally, isolated and analyzed peptide hormones secreted in hypothalamic region of brain which control anterior pituitary hormone secretion; Nobel Prize, 1977.

Gullstrand, Alivar (1862–1930), Swedish ophthalmologist. Discovered intracapsular accommodation of the eye lens; improved techniques for studying eye structure; Nobel Prize, 1911.

Gunn, John Battiscombe (1928–), Egyptian-born American physicist. Discovered Gunn effect and used it to develop Gunn oscillator.

Gunter, Edmund (1581–1626), English mathematician and astronomer. Invented Gunter's chain used in surveying, and the logarithmic scale (Gunter's scale) which is the principle of the slide rule.

Gutenberg, Johann (ca. 1397–1468), German inventor. Invented the movable-type printing press.

Haar, Alfred (1885–1933), Hungarian mathematician. Studied orthogonal systems of functions, complex functions, partial differential equations, and calculus of variations; introduced the Haar measure on groups.

Haber, Fritz (1868–1934), German chemist. Developed the Haber-Boch process for synthesis of ammonia; made electrochemical studies; Nobel Prize, 1918.

Hadamard, Jacques (1865–1963), French mathematician. Proved theorem on the asymptotic behavior of the function giving the number of prime numbers less than a given number; introduced concept of "the problem correctly posed" in the solution of partial differential equations.

Haeckel, Ernst Heinrich (1834–1919), German biologist. Studied various invertebrates; classified animals as uni- and multicellular organisms; proposed the theory of recapitulation: ontogeny repeats phylogeny.

Hagen, Carl Ernst Bessel (1851–1923), German physicist. With H. Rubens, conducted experiments confirming Maxwell's electromagnetic theory of light, permitting determination of electrical conductivity of metals by optical measurements alone.

Hagen, Gotthilf Heinrich Ludwig (1797–1884), German hydraulic engineer. Discovered Hagen-Poiseuille law independently of J. L. M. Poiseuille; directed construction of dikes, harbor installations, and dune fortifications.

Hahn, Hans (1879–1934), Austrian mathematician. With S. Banach, proved the Hahn-Banach theorem of linear functionals.

Hahn, Otto (1879–1968), German chemist. With L. Meitner and F. Strassman, discovered that fission of

heavy nuclei was possible by irradiation with neutrons; discovered protactinium with Meitner; Nobel Prize, 1944.

Hahnemann, Christian Friedrich Samuel (1775–1843), German physician. Founder of homeopathy.

Haldane, John Bourdon Sanderson (1892–1964), British geneticist and physiologist. Pioneered in mathematical treatment of population genetics; studied respiration in humans; wrote about enzymes.

Haldane, John Scott (1860–1936), British physiologist. Research on respiration, particularly the effects of high and low atmospheric pressures.

Hale, George Ellery (1868–1938), American astronomer. Established the Mount Wilson and Mount Palomar observatories; proved existence of magnetic fields in sunspots; invented the spectroheliograph.

Hall, Charles Martin (1863–1914), American commercial chemist. Discovered Hall process for extracting aluminum.

Hall, Edwin Herbert (1855–1938), American physicist. Discovered Hall effect and conducted studies of this and other galvanomagnetic and thermomagnetic effects.

Halley, Edmond (1656–1742), English astronomer. Published a southern star catalog; computed orbits of 24 comets; discovered Halley's Comet.

Hamel, Georg Karl Wilhelm (1877–1954), German mathematician. Research on analysis and applied mathematics; introduced Hamel basis of vectors.

Hamilton, William Rowan (1805–1865), Irish mathematician and mathematical physicist. Discovered quaternions; developed mathematical theories encompassing wave and particle optics and mechanics; introduced Hamilton's principle and a form of the Hamilton-Jacobi theory.

Hankel, Hermann (1839–1873), German mathematician. Studied complex and hypercomplex numbers; theory of functions, and Hankel functions; proved that no hypercomplex number system can satisfy all the laws of ordinary arithmetic.

Hansen, Armauer (1841–1912), Norwegian bacteriologist. Discovered the bacillus of leprosy, or Hansen's disease.

Harden, Arthur (1865–1940), English chemist. Research on enzymes and alcoholic fermentation; Nobel Prize, 1929.

Hardy, Godfrey Harold (1877–1947), English mathematician. With S. Ramanujan, discovered formula for number of ways of writing a positive integer as the sum of positive integers; proved that the Riemann zeta function has an infinite number of zeros with real part equal to 1/2.

Harker, David (1906–1991), American crystallographer. Completed development of Patterson-Harker method of x-ray diffraction analysis of crystal structure.

Hartley, Ralph Vinton Lyon (1888–1970), American electrical engineer. Invented Hartley oscillator; developed Hartley principle in information theory.

Hartline, Haldan Keffer (1903–1983), American biophysicist. Elucidated cellular electrical activity in the eye and optic nerve; Nobel Prize, 1967.

Hartmann, Johannes Franz (1865–1936), German astronomer. Derived Hartmann dispersion formula relating index of refraction and wavelengths; devised Hartmann test for telescope mirrors; gave first observational proof of interstellar matter.

Hartree, Douglas Rayner (1897–1958), English mathematician and mathematical physicist. Developed methods of numerical analysis which made it possible to apply Hartree method to calculation of atomic wave functions.

Hartwell, Leland H. (1939–), American geneticist. Discovered more than 100 genes that control the cell cycle and introduced the concept of cell cycle checkpoints, ordered groups of genes and protein which halt progress through the cell cycle if DNA is damaged, allowing time for DNA repair; Nobel Prize, 2001.

Harvey, William (1578–1657), English anatomist

and physician. Described the true circulation of blood and the action of the heart.

Hassel, Odd (1897–1981), Norwegian chemist. Developed concept of conformation by studying three-dimensional structure of cyclohexane molecule, and explaining the orientation of attached atoms or functional groups; Nobel Prize, 1969.

Hauptman, Herbert Aaron (1917–), American chemist. With J. Karle, developed computer-aided mathematical techniques for use in x-ray crystallography to determine three-dimensional structures of molecules; Nobel Prize, 1985.

Hausdorff, Felix (1868–1942), German mathematician. Founded and advanced general topology and the general theory of metric spaces.

Haüy, René Just, Abbé (1743–1822), French mineralogist. Formulated the geometrical law of crystallization; pioneer in the science of crystallography.

Havers, Clopton (1665–1702), English osteologist. Provided the first full discussion of Haversian lamellae and Haversian canals.

Haworth, Walter Norman (1883–1950), English chemist. Synthesized ascorbic acid; studied carbohydrates, including the structure of sugars; Nobel Prize, 1937.

Heaviside, Oliver (1850–1925), English physicist. Proposed the Heaviside layer in the upper atmosphere.

Heeger, Alan J. (1936–) American physicist. Discovered and developed conductive polymers with Alan MacDiarmid and Hideki Shirakawa; Nobel Prize, 2000.

Hefner-Alteneck, Friedrich Franz von (1845–1904), German engineer. Invented Hefner candle as a standard of luminous intensity.

Heine, Heinrich Eduard (1821–1881), German mathematician. Formulated concept of uniform continuity.

Heisenberg, Werner (1901–1976), German physicist. Founder of quantum mechanics; studied structure of the atom and the Zeeman effect; formulated the principle of indeterminancy in nuclear physics; Nobel Prize, 1932.

Heitler, Walter Heinrich (1904–1981), German-born Swiss theoretical physicist. Developed Heitler-London covalence theory of chemical bonding.

Helmholtz, Hermann Ludwig Ferdinand von (1821–1894), German physicist, anatomist, and physiologist. Physiological research on the nervous system and the human eye and ear, and theoretical work on conservation of force in physics; invented the ophthalmoscope.

Hench, Philip Showalter (1896–1965), American physiologist. Discovered that ACTH and cortisone could be used to treat rheumatoid arthritis; Nobel Prize, 1950.

Henle, Friedrich Gustav Jacob (1809–1885), German pathologist and anatomist. Wrote *Handbuch der Rationellen Pathologie*, integrating the study of physiology and pathology; discovered looped portion of the kidney tubules, and the epithelium.

Henry, Joseph (1797–1878), American physicist. Studied electromagnetic induction, solar phenomena, meterology, and acoustics.

Henry, William (1775–1836), English chemist and physician. Formulated Henry's law of solubility of gases in liquid.

Henyey, Louis George (1910–1970), American astronomer. Research on reflection nebulae, interstellar matter, stellar atmospheres and evolution, and optical design.

Hermite, Charles (1822–1901), French mathematician. First to solve a fifth-degree equation; investigated e, the base of natural logarithms.

Hero of Alexandria (3d century or earlier), Greek mathematician. Wrote on the geometry of plane and solid figures, mechanics, and simple machines; showed that the angle of incidence equals the angle of reflection.

Héroult, Paul Louis Toussaint (1863–1914), French metallurgist. Designed the Héroult furnace for electric steel; developed the Héroult process for aluminum extraction.

Herschbach, Dudley Robert (1932–), American

chemist. With Y. T. Lee, developed crossed molecular-beam technique for tracing chemical reactions; Nobel Prize, 1986.

Herschel, John Frederick William (1792–1871), English mathematician, physicist, and astronomer. Discovered many nebulae and clusters; pioneered in celestial photography.

Herschel, William or Friedrich Wilhelm (1738–1822), German-born English, astronomer. Discovered Uranus, two of its satellites, and two satellites of Saturn; discovered the Sun's intrinsic motion; proposed the concept of the form of the Milky Way.

Hershey, Alfred Day (1908–1997), American biologist. With M. Chase, experimented with bacteriophage, confirming earlier indications that the material basis of heredity is contained in nucleic acids; Nobel Prize, 1969.

Hertz, Gustav (1887–1975), German physicist. With J. Franck, studied effects of electron impacts on atoms; Nobel Prize, 1925.

Hertz, Heinrich Rudolph (1857–1894), German physicist. Discovered Hertzian waves in the ether; proved experimentally Maxwell's theories of electricity and magnetism.

Hertzsprung, Ejnar (1873–1967), Danish astronomer. Discovered method of spectroscopic parallax for measuring stellar distances; determined relation between color and luminosity of stars; with H. N. Russell, developed Hertsprung-Russell diagram.

Herzberg, Gerhard (1904–1999), German-born Canadian physicist. Determined electronic structure and geometry of diatomic and polyatomic molecules, particularly free radicals; Nobel Prize, 1971.

Hess, Victor Franz (1883–1964), Austrian physicist. Studied alpha particles from radium; discovered cosmic rays; Nobel Prize, 1936.

Hess, Walter Rudolf (1881–1973), Swiss physiologist. Discovered the organizer function of the middle brain in coordinating activity of internal organs; developed technique of using electrodes to stimulate localized brain areas; Nobel Prize, 1949.

Hevelius or Hewel, Johannes (1611–1687), Danzig-born astronomer. Discovered four comets; charted surface of the Moon; cataloged numerous stars.

Hevesy, George de (1885–1966), Hungarian chemist. Experimented with radioisotope indication, leading to the technique of isotope tracing of biological and chemical processes; Nobel Prize, 1943.

Heyrovský, Jaroslav (1890–1967), Czechoslovakian physical chemist. Developed the technique of polarographic analysis; Nobel Prize, 1959.

Hewish, Antony. (1924–), British astronomer. Pioneered in discovery of pulsars, by means of radio telescopes; Nobel Prize, 1974.

Heymans, Corneille (1892–1968), French-Belgian physiologist. Investigated the carotid sinus in connection with the mechanism of breathing; Nobel Prize, 1938.

Hilbert, David (1862–1943), German mathematician. Contributed to theory of numbers and theory of invariants; applied integral equations to physical problems.

Hill, Archibald Vivian (1886–1977), English physiologist. Worked on heat loss and oxygen consumption in muscle contraction; Nobel Prize, 1922.

Hinshelwood, Cyril Norman (1897–1967), British chemist. Elucidated chain reaction and chain branching mechanisms; Nobel Prize, 1956.

Hippias of Elis (5th century B.C.), Greek philosopher and mathematician. Discovered quadratrix.

Hipparchus of Rhodes (fl. 130 B.C.), Greek astronomer. Calculated the inclination of the ecliptic and the precession of the equinoxes; invented trigonometry; made the first star catalog.

Hippocrates (460?–377 B.C.), Greek physician. Known as the father of medicine; writings attributed to him contain clinical observations of diseases, descriptions of surgical practice, and the Hippocratic doctrine of the four humors.

Hironaka, Heisuke (1931–), Japanese-American mathematician. Worked in algebraic geometry; solved the problem of the resolution of singularities

on an algebraic variety for algebraic varieties of any dimension over a field of characteristic 0, generalizing work of O. Zariski; Fields Medal, 1970.

Hirzebruch, Friedrich Ernst Peter (1927–), German mathematician. Collaborated with M. F. Atiyah in the development of K theory.

Hitchings, George Herbert (1905–1998), American biochemist. With G. B. Elion, pioneered research that led them to the development of drugs for the treatment of leukemia, malaria, gout, herpes, bacterial and fungal infections, and autoimmune diseases and organ-transplant rejection; Nobel Prize, 1988.

Hittorf, Johann Wilhelm (1824–1914), German physicist. Described effects of Hittorf rays in vacuum tubes; studied electrolysis, and electrical discharge in rarefied gases with the Hittorf tube.

Hodgkin, Alan Lloyd (1914–1998), British biophysicist. With A. Huxley, devised a system of mathematical equations describing the nerve impulse; presented evidence for the sodium theory of nervous conduction; Nobel Prize, 1963.

Hodgkin, Dorothy Crowfoot (1910–1994), Egyptian-born British chemist. Determined the structure of the vitamin B_{12} molecule through x-ray crystallographic analysis; Nobel Prize, 1964.

Hodgkin, Thomas (1798–1866), English physician. First to describe Hodgkin's disease, a glandular disorder.

Hoffmann, Roald (1937–), Polish-born American chemist. Developed methods for predicting whether a chemical reaction is possible based on molecular orbital models; Nobel Prize, 1981.

Hofmann, August Wilhelm von (1818–1892), German chemist. Studied reactions of derivatives; developed the Hofmann reaction for preparing primary amines.

Hofmeister, Wilhelm Friedrich Benedict (1824–1877), German botanist. Did fundamental work on plant embryology; explained the alternating life cycles of mosses and ferns.

Hofstadter, Robert (1915–1990), American physicist. Investigated the properties and behavior of the proton and neutron; determined the size and shape of many nuclei; discovered the construction scheme of fundamental atomic nuclei; Nobel Prize, 1961.

Hölder, Otto Ludwig (1859–1937), German mathematician. Contributed to analysis and group theory; introduced Hölder condition; proved Hölder inequality and Jordan-Hölder theorem.

Holley, Robert William (1922–1993), American biochemist. With coworkers, made first determination of a nucleotide sequence of a nucleic acid; Nobel Prize, 1968.

Holmberg, Erik Bertil (1908–2000), Swedish astronomer. Investigations of galaxies, especially photometry of galaxies.

Hooke, Robert (1635–1703), English inventor. Invented the compound microscope, wheel barometer, universal (Hooke's) joint, and the reflecting telescope; formulated theories on light and on the motion of the Earth.

Hooker, Joseph Dalton (1817–1911), English botanist. With G. Bentham, wrote *Genera Plantarum*; prepared works on the flora of New Zealand, Antarctica, and India.

Hopkins, Frederick Gowland (1861–1947), English biochemist. Discovered the amino acid tryptophan and the tripeptide glutathione; did experimental work leading to the discovery of vitamins; Nobel Prize, 1929.

Hopper, Grace (1906–1992), American mathematician and computer scientist. Pioneered in the development of computer software, including the invention of the first compiler. Credited with having discovered the first computer "bug," a moth that literally had to be removed from the wiring of an early computer.

Hörmander, Lars (1931–), Swedish mathematician. Worked on partial differential equations; in particular, made major contributions to the general theory of linear differential operators; Fields Medal, 1962.

Houdry, Eugene Jules (1892–1962), French-born American engineer. Devised catalytic method of producing oil.

Hounsfield, Godfrey Newbold (1919–), British electronics engineer. Invented computerized axial tomography; Nobel Prize, 1979.

Houssay, Bernardo Alberto (1887–1971), Argentine physiologist. Research on the functions and effects of the hypophysis, including its relationship to carbohydrate metabolism; Nobel Prize, 1947.

Howell, William Henry (1860–1945), American physiologist. Discovered heparin; isolated thrombin and thromboplastin; discovered Howell-Jolly bodies; proved that blood platelets are formed in lungs.

Hubble, Edwin Powell (1889–1953), American astronomer. Studied nebulae; formulated Hubble's law of extragalactic nebulae.

Hubel, David Hunter (1926–), American neurobiologist. Contributed to the study of the processing of visual information in the brain; Nobel Prize, 1981.

Huber, Robert (1937–), German chemist. With J Deisenhofer and M. Hartmut, elucidated the structure of a bacterial protein that performs photosynthesis; Nobel Prize, 1988.

Hückel, Erich (1896–1980), German chemist. With P. J. W. Debye, formulated Debye-Hückel theory of strong electrolytes; devised theoretical explanation of electron properties of aromatic hydrocarbons.

Huggins, Charles Brenton (1901–1997), Canadian-born American surgeon and cancer researcher. Developed treatment of cancers using endocrinologic methods; Nobel Prize, 1966.

Hughes, David Edward (1831–1900), English-born American inventor. Invented the Hughes electromagnet, a printing telegraph, microphone, and induction balance.

Hugoniot, Pierre Henry (1851–1887), French physicist. Developed theory of shock waves.

Hulse, Russell Alan (1950–), American astronomer. With J. H. Taylor, discovered a binary pulsar and observed this object to obtain indirect evidence for the existence of gravitational waves, an important confirmation of the general theory of relativity; Nobel Prize, 1993.

Humboldt, Friedrich Heinrich Alexander, Baron von (1769–1859), German naturalist. Founder of physical geography; made scientific explorations of South America and Central Asia.

Hume-Rothery, William (1899–1968), English metallurgist and chemist. Discovered Hume-Rothery rule concerning electron compounds.

Hunt, Franklin Livingston (1883–1973), American physicist. Research on x-ray spectroscopy; with W. Duane, discovered and applied Duane-Hunt law.

Hunt, R. Timothy (1943–), British biologist who discovered cyclins, proteins that regulate cyclin-dependent kinase activity, and found that their periodic degradation is an important general control mechanism of the cell cycle; Nobel Prize, 2001.

Hurwitz, Adolf (1859–1919), German-born Swiss mathematician. Worked on modular functions, number theory, Riemann surfaces, complex function theory, and analytic number theory; formulated condition satisfied by Hurwitz polynomials.

Huxley, Andrew Fielding (1917–), British physiologist. With A. L. Hodgkin, discovered the ionic mechanism involved in excitation in the cell membrane of peripheral nerves; Nobel Prize, 1963.

Huygens, Christiaan (1629–1695), Dutch mathematician, physicist, and astronomer. Discovered Saturn's rings; contributed to dynamics and optics; proposed the wave theory of light.

Hylleraas, Egil Andersen (1898–1965), Norwegian physicist. Applied quantum theory to helium atom, negative hydrogen ion, and other atoms, molecules, and crystals; developed variational method and other methods for mathematical solution of quantum-mechanical problems.

Ignarro, Louis J. (1941–), American pharmacologist. Working independently and with Robert Furchgott, he concluded that endothelium-derived relaxing factor was identical to nitric oxide; Nobel Prize, 1998.

Ingenhousz, Jan (1730–1799), Dutch physician and naturalist. Demonstrated the cycle of photosynthesis in plants.

Ising, Ernest (1900–1998), German-born American physicist. Introduced Ising model of ferromagnetic material; research in solid-state physics and ferromagnetism.

Itō, Kiyosi (1915–), Japanese mathematician. Studied stochastic processes; introduced Itō's integral and Itō's formula.

Jacob, François (1920–), French biologist. Discovered episomes, a class of genetic elements; with J. Monod, proposed the concepts of messenger ribonucleic acid and of the operon; Nobel Prize, 1965.

Jacobi, Karl Gustav Jacob (1804–1851), German mathematician. Worked on elliptic functions and differential equations; developed the theory of determinants.

Jacquard, Joseph Marie (1752–1834), French inventor. Designed and built the Jacquard loom for figured weaving.

Jaeger, Frans Maurits (1877–1945), Dutch crystallographer and physical chemist. Measured physical properties of molten salts and silicates at extremely high temperatures.

James, William (1842–1910), American psychologist and writer. Coformulator of James-Lange theory that emotions are the perception of physiological changes.

Janet, Pierre Marie Félix (1859–1947), French psychologist. Studied hysteria, obsession, and neurosis; wrote a textbook on the theory of hysteria.

Jeans, James Hopwood (1877–1946), English physicist and astronomer. Worked in stellar dynamics; proposed the tidal theory of the origin of planets.

Jenner, Edward (1749–1823), English physician. Discovered vaccination.

Jensen, J. Hans D. (1906–1973), German physicist. With M. G. Mayer, formulated the nuclear shell model; Nobel Prize, 1963.

Jerne, Niels Kaj (1911–1994), Swiss immunologist. Formulated three important theories involving the immune system; how the body produces specific antibodies, how the immune system develops and matures, and how the various interrelated aspects of the immune response are coordinated by the body; Nobel Prize, 1984.

Johnson, Harold Lester (1921–1980), American astronomer. Research in astrophysics, astronomical photometry, infrared astronomy, applications of electronics to astronomy, and Fourier transform spectroscopy; collaborated with W. W. Morgan in developing Johnson-Morgan system of stellar magnitudes.

Joliot-Curie, Irène (1897–1956), French physicist. With M. Curie, discovered projection of atomic nuclei by neutrons; with J. F. Joliot-Curie, discovered artificial radiation; Nobel Prize, 1935.

Joliot-Curie, Jean Frédéric (1900–1958), French physicist. With I. Joliot-Curie, produced an artificial radioactive substance by bombarding boron with fast alpha particles; Nobel Prize, 1935.

Jones, Vaughn Frederick Randal (1952–), New Zealand-American mathematician. Proved an index theorem for von Neumann algebras, discovered a relationship between these algebras and geometric topology, and discovered a new polynomial invariant for knots; this work provided a connecting link for widely separated areas of mathematics and physics; Fields Medal, 1990.

Jordan, Camille (1838–1921), French mathematician. Discovered many fundamental results in group theory; gave a proof of the Jordan curve theorem (later shown to be incorrect).

Jordan, Pascual (1902–1980), German physicist. Contributed to formulation of quantum mechanics; introduced Jordan algebra in an attempt to generalize quantum mechanics.

Josephson, Brian David (1940–), British physicist. Predicted the Josephson effect concerning electron pairs; Nobel Prize, 1973.

Joukowski, Nikolai Jegorowitch (1847–1921), Russian applied mathematician and aerodynamicist. Helped to introduce concept of Kutta-Joukowski airfoil and to prove Kutta-Joukowski theorem.

Joule, James Prescott (1818–1889), English physicist. Formulated a mechanical theory of heat; demonstrated Joule-Thomson effect relating to the fall in temperature of a gas; first to estimate the velocity of a gas molecule.

Jung, Carl Gustav (1875–1961), Swiss psychologist and psychiatrist. Evolved a theory of complexes; founded the analytical school of psychoanalysis and psychotherapy.

Jussieu, Antoine Laurent de (1748–1836), French botanist. Wrote *Genera Plantarum*, the basis of modern natural botanical classification.

Kahn, Robert E. (1938–), American electrical engineer. Co-invented (with Vinton G. Cerf) Transmission Control Protocol/Internet Protocol (TCP/IP).

Kalman, Rudolf Emil (1930–), Hungarian-born American mathematician and electrical engineer. Worked on mathematical theory of control systems; developed the Kalman filter; introduced concepts of controllability and observability.

Kaluza, Theodor Franz Eduard (1885–1954), German mathematical physicist. Developed theory which attempted to unify gravitation and electromagnetism.

Kamerlingh Onnes, Heike (1853–1926), Dutch physicist. Research on cryogenics, critical phenomena, and low temperatures; discovered the phenomenon of superconductivity; Nobel Prize, 1913.

Kandel, Eric R. (1929–), Austrian-born American neurobiologist. Discovered that protein phosphorylation plays an important role in the molecular mechanisms underlying learning and memory formation; his work contributed to the elucidation of the molecular mechanisms involved in slow synaptic transmission in the nervous system; Nobel Prize, 2000.

Kapitza, Pjotr Leonidovich (1894–1984), Russian physicist. Studied magnetism and low temperature; designed hydrogen and helium liquefaction plants; Nobel Prize, 1978.

Kapteyn, Jacobus Cornelius (1851–1922), Dutch astronomer. Studied the proper motion of stars; with P. J. van Rhijn, evolved a theory of the universe.

Karle, Jerome (1918–), American crystallographer. With H. A. Hauptman, developed computer-aided mathematical techniques for use in x-ray crystallography to determine three-dimensional structures of molecules; Nobel Prize, 1985.

Karrer, Paul (1889–1971), Swiss chemist. Pioneering research on vitamins A and B$_2$ and on the flavins and carotenoids; Nobel Prize, 1937.

Kastler, Alfred (1902–1984), French physicist. Developed a double-resonance method to study energy levels of atoms in excited states; Nobel Prize, 1966.

Kater, Henry (1777–1835), English geodesist. Developed Kater's reversible pendulum to obtain accurate values of acceleration of gravity.

Katz, Bernard (1911–), German-born British physiologist. Made discoveries concerning mechanism for release of transmitter substances at nerve-muscle junction; Nobel Prize, 1970.

Keesom, Willem Hendrik (1876–1956), Dutch physicist. Worked in low-temperature physics; first to solidify helium; studied molecular structure of liquids and compressed gases.

Kekulé von Stradonitz, Friedrich August (1829–1896), German chemist. A founder of structural organic chemistry; made theoretical proposal of the structure of benzene.

Kelvin, William Thompson, 1st Baron (1824–1907), British mathematician and physicist. Invented the Kelvin balance; formulated Kelvin's laws concerning electric cables; contributed to thermodynamics.

Kendall, Edward Calvin (1886–1972), American biochemist. Chemical investigation of the adrenal cortex, leading to the isolation of crystalline cortical hormones, especially cortisone; Nobel Prize, 1950.

Kendall, Henry Way (1926–1999), American physicist. Collaborated in experiments that demonstrated that protons, neutrons, and similar particles are made up of quarks; Nobel Prize, 1990.

Kendrew, John Cowdery (1917–1997), British molecular biologist. First to successfully determine the structure of a protein; Nobel Prize, 1962.

Kennelly, Arthur Edwin (1861–1939), American electrical engineer. Discovered the ionized layer in the atmosphere, independently of O. Heaviside; proposed the theory of alternating currents.

Kepler, Johannes (1571–1630), German astronomer. Proposed Kepler's three laws of planetary motion; worked in optics.

Kerr, John (1824–1907), Scottish physicist. Discovered the Kerr magnetooptic effect.

Ketterle, Wolfgang (1957–), German physicist. Produced Bose-Einstein condensates in a dilute gas of sodium atoms (independent of the work of E. A. Cornell and C. E. Wieman) and carried out fundamental studies of their properties, including the production of interference patterns and atom lasers; Nobel Prize, 2001.

Khorana, Har Gobind (1922–), Indian-born American biochemist. Synthesized complicated nucleic acids; proved that genetic code consists of nonoverlapping triplets of bases without gaps between triplets; Nobel Prize, 1968.

Kilby, Jack St. Clair (1923–), American physicist, electronics engineer, and inventor. Participated in the invention of the integrated circuit; Nobel Prize, 2000.

Kirchhoff, Gustav Robert (1824–1887), German physicist. With R. W. Bunsen, discovered method of spectrum analysis; formulated Kirchoff's law of electric currents and electromotive forces in a network.

Kirkwood, Daniel (1814–1895), American astronomer. Discovered Kirkwood gaps; studied nature, origin, and evolution of solar system, particularly role of asteroids, comets, meteors, and meteorites.

Kitasato, Shibasaburo (1852–1931), Japanese bacteriologist. Independently of A. E. J. Yersin, discovered the bacillus of bubonic plague; isolated bacilli of symptomatic anthrax, dysentery, and tetanus.

Klebs, Edwin (1834–1913), German pathologist. Described diphtheria (Klebs-Löffler) bacillus; studied bacteriology of malaria, anthrax, and tuberculosis.

Klein, Christian Felix (1849–1925), German mathematician. Contributed to function theory, noneuclidean geometry, group therapy, and applied mathematics; introduced Klein bottle.

Klein, Oskar Benjamin (1894–1977), Swedish physicist. Codeveloper of Klein-Gordon equation, Klein-Nishina formula, and Klein-Rydberg method; proposed theory of overall structure of the universe.

Kleinrock, Leonard (1934–), American electrical engineer. Developed the basic principles of packet switching for communication networks, the underlying technology of the Internet.

Klitzing, Klaus von (1943–), German physicist. Discovered quantum Hall effect; Nobel Prize, 1985.

Klug, Aaron (1926–), South African-born British biochemist. Developed crystallographic electron microscopy and elucidated biologically important nucleic acid-protein complexes; Nobel Prize, 1982.

Knowles, William S. (1917–), American chemist. Developed chirally catalyzed hydrogenation reactions; Nobel Prize, 2001.

Knudsen, Martin Hans Christian (1871–1949), Danish physicist and hydrographer. Studied flow and diffusion of gases at low pressure; developed Knudsen cell and Knudsen gage; developed methods to measure the properties of seawater.

Koch, Robert (1843–1910), German physician and bacteriologist. Studied cholera, tuberculosis, and bubonic plague; showed a specific bacillus to be the cause of anthrax; discovered the tubercle bacillus; Nobel Prize, 1905.

Kocher, Emil Theodor (1841–1917), Swiss surgeon. Studied the functions and malfunctions of the thyroid gland; Nobel Prize, 1909.

Kodaira, Kunihiko (1915–1997), Japanese mathematician. Did research on harmonic integrals and harmonic forms with applications to Kählerian and more specifically algebraic varieties; demonstrated, by sheaf cohomology, that such varieties are Hodge manifolds; Fields Medal, 1954.

Köhler, Georges Jean Franz (1946–1995), Swiss immunologist. With C. Milstein, discovered a laboratory technique for producing monoclonal antibodies, highly uniform immune bodies that are selective in responding to target substances; Nobel Prize, 1984.

Kohn, Walter (1923–), Austrian-born American physicist. Developed density-functional theory, which solves equations for electron density rather than positions of individual electrons; it is one of the developments that has significantly sped up computational quantum chemistry; Nobel Prize, 1998.

Kolbe, Adolf Wilhelm Hermann (1818–1884), German chemist. Contributed to the synthesis concept of compound formation, doing much to eliminate the division of chemistry into two branches: organic and inorganic.

Kolmogorov, Andrei Nikolaevich (1903–1987), Soviet mathematician. Formulated set-theoretic basis of probability theory.

Kontsevich, Maxim (1964–), Russian mathematician and mathematical physicist. Worked in algebraic geometry, algebraic topology, string theory, and quantum field theory; demonstrated equivalence of two models of quantum gravitation; discovered an invariant for classifying knots; Fields Medal, 1998.

Kornberg, Arthur (1918–), American biochemist. Discovered deoxyribonucleic acid polymerase, providing the first rational enzymatic mechanism for the replication of genetic material of the cell; Nobel Prize, 1959.

Korteweg, Diederik Johannes (1848–1941), Dutch mathematician. Work in applied mathematics, mechanics, and hydrodynamics; with H. de Vries, proposed equation of wave motion with soliton solution.

Kossel, Albrecht (1853–1927), German chemist. Investigated the chemistry of cells and of proteins; Nobel Prize, 1910.

Krafft-Ebing, Richard, Baron von (1840–1902), German neurologist. Authority on psychological disorders and their forensic implications; wrote *Psychopathia Sexualis*, a collection of case histories.

Kramers, Hendrik Anthony (1894–1952), Dutch physicist. Developed quantum theory of dispersion, establishing Kramers-Kronig relation; with G. Wentzel and L. Brillouin, developed Wentzel-Kramers-Brillouin method.

Krebs, Edwin Gerhard (1918–), American biochemist. With E. H. Fischer, discovered phosphorylation processes that play a critical role in cell-protein regulation; they isolated the first protein kinase, a class of enzymes that transfer phosphate from adenosine triphosphate to proteins: Nobel Prize, 1992.

Krebs, Hans Adolf (1900–1981), German-born British biochemist. Elucidated metabolic pathways, including the tricarboxylic and cycle; Nobel Prize, 1953.

Kroemer, Herbert (1928–), German-American physicist and electronics engineer. Developed semiconductor heterostructures used in high-speed and opto-electronics, including fast transistors, laser diodes, and light-emitting diodes; Nobel Prize, 2000.

Krogh, Schack August Steenberg (1874–1949), Danish physiologist. Discovered the regulation of the vasomotor mechanism of capillaries; devised the nitrous oxide method for measuring human circulation; Nobel Prize, 1920.

Kronecker, Leopold (1823–1891), German mathematician. Contributed to theory of elliptical functions, algebra, and number theory, and attempted to unify these disciplines; attempted to base all mathematics on integers and finite processes.

Kroto, Harold W. (1939–), British chemist. With Richard Smalley and Robert Curl, discovered fullerenes; Nobel Prize, 1996.

Kruskal, Martin David (1925–), American mathematician and physicist. Research in plasma physics, asymptotic phenomena, relativity, and minimal surfaces.

Kuhn, Richard (1900–1967), German chemist. Research on the structures and synthesis of vitamins and carotenoids; Nobel Prize, 1938 (declined).

Kundt, August Adolph (1839–1894), German physicist. Used Kundt tube to determine speed of sound in gases; determined ratio of specific heats of monatomic gases; with W. K. Röntgen, demonstrated Faraday effect in gases.

Kurchatov, Igor Vasilievich (1903–1960), Soviet physicist. Discovered nuclear isomers; studied nuclear reactions; developed nuclear weapons and nuclear power.

Kusch, Polykarp (1911–1993), German-born American physicist. Precisely determined the magnetic moment of the electron; Nobel Prize, 1955.

Kutta, Wilhelm Martin (1867–1944), German applied mathematician. Helped introduce concept of Kutta-Joukowski airfoil, prove Kutta-Joukowski theorem, and develop Runge-Kutta method.

Kwolek, Stephanie (1923–), American polymer chemist. Developed Kevlar, poly(*p*-phenylene terephthalamide), a lightweight synthetic fiber that is stronger than steel.

Lacaille, Nicolas Louis de (1713–1762), French astronomer and geodesist. Determined positions of nearly 10,000 stars in southern skies; measured lunar and solar parallax; showed that Earth has equatorial bulge.

Lagrange, Joseph Louis, Count (1736–1813), French geometer and astronomer. Invented the calculus of variations; studied the mathematics of sound; wrote *Mécanique Analytique*, concerning statics and dynamics.

Laguerre, Édmond Nicolas (1834–1886), French mathematician. Discovered Laguerre's differential equations and Laguerre polynomials.

Lamarck, Jean Baptiste Pierre Antoine de Monet, Chevailer de (1744–1829), French naturalist. Proposed theory that changes in animal and plant structure are caused by changes in environment; classified animals into vertebrates and invertebrates.

Lamb, Willis Eugene, Jr. (1913–), American physicist. Made precise atomic measurements leading to a new understanding of the theory of electron interactions and electromagnetic radiation; Nobel Prize, 1955.

Lambert, Johann Heinrich (1728–1777), German physicist. Formulated the Lambert theorem concerning the illumination of a surface.

Lamé, Gabriel (1795–1870), French mathematician, physicist, and engineer. Introduced curvilinear coordinates and applied them to differential equations, elasticity, thermodynamics, and number theory.

Landau, Lev Davydovich (1908–1968), Soviet physicist. Made theoretical explanation of the nature and properties of liquid helium; investigated condensed matter; Nobel Prize, 1962.

Landé, Alfred (1888–1975), German-born American physicist. Introduced Landé g factor; discovered Landé interval rule and Landé Γ-permanence rule.

Landsteiner, Karl (1868–1943), Austrian-born American pathologist. Discovered human blood groups and factors M and N; with A. S. Weiner, discovered the Rh factor; Nobel Prize, 1930.

Lange, Carl Georg (1834–1900), Danish physician and psychologist. With W. James, proposed the James-Lange theory of emotion.

Langerhans, Paul (1847–1888), German pathologist and anatomist. Studied human and animal microscopical anatomy, particularly structures of skin and pancreas; discovered islets of Langerhans.

Langevin, Paul (1872–1946), French physicist. Developed quantitative theories of paramagnetism and diamagnetism; helped to elucidate the theory of relativity; contributed to the development of sonar.

Langley, Samuel Pierpont (1834–1906), American astronomer and airplane pioneer. Studied infrared solar spectrum; constructed in 1896 the first mechanical heavier-than-air machine to fly.

Langmuir, Irving (1881–1957), American chemist. With G. N. Lewis, proposed the Lewis-Langmuir atomic theory; studied surface chemistry and thermionic emission; Nobel Prize, 1932.

Laplace, Pierre Simon, Marquis de (1749–1827), French astronomer and mathematician. Contributed to celestial mechanics, especially to the study of the Moon, Saturn, and Jupiter; formulated the theory of probability; discovered the Laplace differential equation.

Larmor, Joseph (1857–1942), British physicist. Developed electron theory which fused electromagnetic and optical concepts; introduced Larmor precession and derived Larmor formula.

Latimer, Louis Howard (1848–1928), American inventor. A collaborator of Alexander Graham Bell and Thomas Edison, this self-educated son of an escaped slave is best known for key inventions in the field of electric lighting.

Laue, Max Theodor Felix von (1879–1960), German physicist. Proposed the theory of x-ray diffraction by crystals; developed the Laue method of investigating crystal structure; Nobel Prize, 1914.

Laughlin, Robert Betts (1950–), American physicist. Provided a theoretical explanation for the fractional quantum Hall effect by showing how electrons acting together in strong magnetic fields can form new types of quasiparticles with charges that are fractions of electron charges; Nobel Prize, 1998.

Laurent, Pierre Alphonse (1813–1854), French mathematician and physicist. Introduced Laurent series; research on wave theory of light.

Laveran, Charles Louis Alphonse (1845–1922), French physician. Discovered the malaria parasite; researched sleeping sickness; Nobel Prize, 1907.

Lavoisier, Antoine Laurent (1743–1794), French chemist. The founder of modern chemistry; studied combustion and respiration; published a table of the elements.

Lawrence, Ernest Orlando (1901–1958), American physicist. Discovery, development, and use of the cyclotron; Nobel Prize,1939.

Lebesgue, Henry Léon (1875–1941), French mathematician. Developed theory of measure and integration; studied trigonometric series.

Le Chatelier, Henry Louis (1850–1936), French chemist and metallurgist. Research on cement chemistry, gas combustion, blast furnace reactions, chemical equilibria, alloy properties, and chemistry and metallurgy of iron and steel; formulated Le Chatelier's principle.

Leclanché, Georges (1839–1882), French chemist and electrician. Invented the Leclanché galvanic cell.

Lederberg, Joshua (1925–), American geneticist. With E. L. Taturn, discovered genetic recombination in bacteria and organization of genetic material; Nobel Prize, 1958.

Lederman, Leon Max (1922–), American physicist. Collaborated in experiment that demonstrated the existence of two types of neutrino; led an experiment that discovered the upsilon particle; Nobel Prize, 1988.

Lee, David Morris (1931–), American physicist. With D. D. Osheroff and R. C. Richardson, discovered superfluidity in the rare isotope helium-3; Nobel Prize, 1996.

Lee, Tsung-Dao (1926–), Chinese-born American physicist. With C. N. Yang, disproved the parity principle; worked on statistical mechanics, astrophysics, nuclear and subnuclear physics, and field theory; Nobel Prize, 1957.

Lee, Yuan Tseh (1936–), American chemist. With Dudley R. Herschbach, developed crossed molecular-beam technique for tracing chemical reactions; Nobel Prize, 1986.

Legendre, Adrien Marie (1752–1833), French mathematician. Worked on elliptic functions, the theory of numbers, and the method of least squares.

Lehn, Jean-Marie (1939–), French chemist. Studied crown ethers and developed the synthesis of related structures known as cryptands; Nobel Prize, 1987.

Leibniz, Gottfried Wilhelm, Baron von (1646–1716), German mathematician. Contributed to the development of differential calculus.

Leloir, Luis Federico (1906–1987), French-born Argentine biochemist. Discovered sugar nucleotides and their role in carbohydrate biosynthesis; Nobel Prize, 1970.

Lenard, Phillipp Eduard Anton (1862–1947), Hungarian-born German physicist. Studied cathode rays outside the discharge tube; worked on photoelectricity; Nobel Prize, 1905.

L'Enfant, Pierre Charles (1754–1825), French engineer. Designed Washington, DC.

Lennard-Jones, John Edward (1894–1954), English physicist and chemist. Proposed Lennard-Jones potential for interatomic forces; contributed to quantum theory of molecular structure and statistical mechanics of liquids, gases, and surfaces.

Lenz, Heinrich Friedrich Emil (1804–1865), German physicist. Formulated Lenz's law governing induced current.

Leverrier, Urbain Jean Joseph (1811–1877), French astronomer. Studied Mercury, Uranus, and Neptune; credited, with J. C. Adams in England, as discoverer of Neptune.

Levi-Civita, Tullio (1873–1941), Italian mathematician and mathematical physicist. With G. Ricci-Curbastro, developed tensor analysis; introduced concept of parallelism in curved spaces.

Levi-Montalcini, Rita (1909–), Italian biologist. With S. Cohen, made landmark studies of nerve growth factor and its functions; Nobel Prize, 1986.

Lewis, Edward B. (1918–), American developmental biologist. Investigated how genes could control the further development of individual body segments into specialized organs; his work, along with that of American developmental biologist Eric F. Weischaus and German developmental biologist Christiane Nüsslein-Volhard, led to the discovery of important genetic mechanisms which control early embryonic development; Nobel Prize 1995.

Lewis, Gilbert Newton (1875–1946), American chemist. Collaborated in developing the Lewis-Langmuir atomic theory; worked on the electronic theory of valency and chemical thermodynamics.

Leydig, Franz von (1821–1908), German histologist and anatomist. Founder of comparative histology; promoted use of microscope in anatomical study; described cells in testes believed to secrete male hormones.

l'Hospital (l'Hôpital), Guillaume François Antoine de, Marquis de Sainte-Mesme, Comte d'Entremont (1661–1704), French mathematician. Wrote first textbook on differential calculus, which gives l'Hospital's rule.

Libby, Willard Frank (1908–1980), American chemist. Developed the method of radiocarbon dating; Nobel Prize, 1960.

Lie, Marius Sophus (1842–1899), Norwegian mathematician. Originated the theory of tangential transformations.

Liebig, Justus, Baron von (1803–1873), German chemist. Discovered chloroform and chloral; founded agricultural chemistry; invented the Liebig condenser.

Linnaeus, Carolus, real name **Carl von Linné**, (1707–1778), Swedish botanist. Developed the Linnaean system of biological classification.

Lions, Pierre-Louis (1956–), French mathematician. Made important contributions to the theory of nonlinear partial differential equations; Fields Medal, 1994.

Liouville, Joseph (1809–1882), French mathematician. Proved existence of transcendental functions; developed concept of geodesic curvature; originated theory of doubly periodic functions.

Lipmann, Fritz Albert (1899–1986), German-born American biochemist. Formulated general rules for the biotechnology of energy transmission; discovered coenzyme A; Nobel Prize, 1953.

Lippmann, Gabriel (1845–1921), French physicist. Produced the first colored photograph of the light spectrum; invented the Lippmann capillary electrometer; Nobel Prize, 1908.

Lipscomb, William Nunn, Jr. (1919–), American physical chemist. Studied structure and bonding of boranes, providing insight into nature of chemical bonding; Nobel Prize, 1976.

Lissajous, Jules Antoine (1822–1880), French physicist. Invented the vibration microscope, involving Lissajous figures.

Lister, Joseph, 1st Baron (1827–1912), English surgeon. Introduced antiseptics to surgery; pioneered in bacteriology.

Littlewood, John Endensor (1885–1977), British mathematician. Work on diophantine approximation, Tauberian theorems, Fourier series and associated function theory, the zeta function, additive number theory, and inequalities.

Littrow, Joseph Johann von (1781–1840), Austrian astronomer. Studied light refraction; worked on telescope construction.

Lloyd, Humphrey (1800–1881), Irish physicist. Discovered Lloyd's mirror interference; verified W. R. Hamilton's prediction of conical refraction.

Lobachevski, Nikola Ivanovich (1793–1856), Russian mathematician. Originated the first comprehensive system of noneuclidean geometry.

Loewi, Otto (1873–1961), German pharmacologist. Investigated nerve impulses; proved the role of acetylcholine in nerve impulse transmission; Nobel Prize, 1936.

Löffler, Friedrich August Johannes (1852–1915), German bacteriologist. Isolated the diphtheria (Klebs-Löffler) bacillus; developed protective serum against foot-and-mouth disease.

London, Fritz (1900–1954), German-born American physicist. Developed, with W. Heitler, theory of covalent bonding; with H. London, theory of superconductivity; and theory of superfluidity.

London, Heinz (1907–1970), German-born English physicist. Research on electrodynamic and thermodynamic behavior of superconductors and properties of superfluid helium.

Lorentz, Hendrik Antoon (1853–1928), Dutch physicist. Proposed the electron theory to explain electromagnetic properties of materials; proposed the Lorentz-FitzGerald contraction and the Lorentz transformation, contributing to the theory of relativity; studied Zeeman effect; Nobel Prize, 1902.

Lorenz, Konrad Zacharias (1903–1989), Austrian zoologist. Pioneered in study of animal behavior patterns; discovered imprinting in birds; Nobel Prize, 1973.

Loschmidt, Johann Joseph (1821–1895), Austrian physicist and chemist, born in Bohemia. Worked on graphical and structural molecular formulas; attempted to estimate size of air molecules and number of air molecules per unit volume.

Lowry, Thomas Martin (1874–1936), British chemist. Simultaneously with Johannes Brønsted, defined acid-base theory in terms of proton transfer— the Brønsted-Lowry theory (of acids and bases).

Lummer, Otto Richard (1860–1925), German physicist. Codeveloper of Lummer-Brodhun sight box and Lummer-Gehrcke plate; constructed an improved bolometer.

Luria, Salvador Edward (1912–1991), Italian-born American biologist. Devised fluctuation test to demonstrate and study spontaneous mutations in bacteria and viruses; Nobel Prize, 1969.

Lwoff, André Michael (1902–1994), French biologist. Explained the phenomenon of lysogeny in bacteria; Nobel Prize 1965.

Lyapunov, Aleksandr Mikhailovich (1857–1918), Soviet mathematician and physicist. Determined in what cases linear approximations can be used to solve the problem of s ability of a mechanical system with a finite number of degrees of freedom; proved existence of various figures of equilibrium for a rotating liquid.

Lyell, Charles (1797–1875), British geologist. Wrote *Principles of Geology*, refuting catastrophic theory of geological changes.

Lyman, Theodore (1874–1954), American physicist. Observed ultraviolet spectra; clarified nature of Lyman ghosts; discovered Lyman series.

Lynen, Feodor (1911–1979), German biochemist. Research on the formation of the cholesterol molecule; discovered chemistry of biotin; Nobel Prize, 1964.

Lyot, Bernard Ferdinand (1879–1952), French astronomer. Invented the coronagraph and developed monochromatic filters that greatly extended knowledge of the solar corona.

MacDiarmid, Alan G. (1927–), New Zealand-born American chemist. Discovered and developed conductive polymers with Hideki Shirakawa and Alan Heeger; Nobel Prize, 2000.

Mach, Ernst (1838–1916), Austrian physicist. Research on supersonic flight, leading to Mach angle and Mach number; studied airflow over objects at high speeds.

Maclaurin, Colin (1698–1746), Scottish mathematician. Systematized and developed Newton's calculus; introduced Maclaurin series and Maclaurin-Cauchy test.

Macleod, John James Rickard (1876–1935), Scottish physiologist. Shared in discovery of insulin with F. G. Banting and C. H. Best; Nobel Prize, 1923.

Magnus, Heinrich Gustav (1802–1872), German physicist and chemist. Made first quantitative analysis of blood gases; showed that arterial blood has higher oxygen content than venous blood; discovered Magnus effect.

Majorana, Ettore (1906–1938), Italian physicist. Studied properties of elementary particles; postulated Majorana force.

Maksutov, Dmitry Dmitrievich (1896–1964), Soviet physicist and astronomer. Developed general theory of aplanatic optical systems; developed Maksutov system.

Malpighi, Marcello (1628–1694), Italian anatomist. Discovered the capillaries; made microscopic studies in embryology; discovered the Malpighian layer of the epidermis and the Malpighian corpuscles in the kidney.

Malus, Étienne Louis (1775–1812), French engineer and physicist. Formulated Malus' cosine-squared law concerning polarized light and Malus' law of rays.

Mandelstam, Stanley (1928–), American physicist, born in South Africa. Research in theoretical physics of elementary particles; introduced Mandelstam plane and Mandelstam representation.

Marconi, Guglielmo, Marquis (1874–1937), Italian electrician and inventor. Developed commercial wireless telegraphy; Nobel Prize, 1909.

Margulis, Gregori Aleksandrovich (1946–), Russian mathematician. Worked in combinatorics, differential geometry, ergodic theory, dynamical systems, and Lie groups; developed innovative analysis of the structure of Lie groups, describing their discrete subgroups; Fields Medal, 1978.

Mariotte, Edmé (?–1684), French physicist and physiologist. Discovered blind spot; studied circulation of sap in plants, collisions of bodies, properties of air, refraction and color of light, hydrostatics, hydraulics, and meteorology.

Marcus, Rudolph Arthur (1923–), Canadian-born American chemist. Developed the mathematical analysis for electron transfer reactions in chemical systems; Nobel Prize, 1992.

Mark, Herman Francis (1895–1992), Austrian-born American chemist. Elucidated molecular structures of natural and synthetic polymers; developed theory of polymerization; studied relation between structure and properties of macromolecular systems.

Markov, Andrei Andreevich (1856–1922), Russian mathematician. Formulated rigorous proofs of law of large numbers and central-limit theorem; introduced Markov chain.

Martin, Archer John Porter (1910–), English chemist. With R. L. M. Synge, developed partition chromatography; Nobel Prize, 1952.

Mascheroni, Lorenzo (1750–1800), Italian mathematician. Calculated Euler's constant; proved that all plane construction problems that can be solved with a ruler and compass can also be solved with a compass alone.

Mathieu, Emile Leonard (1835–1890), French mathematician and physicist. Worked on solution of partial differential equations; research in celestial and analytical mechanics; studied Mathieu equation and introduced Mathieu functions.

Matthias, Bernd Teo (1919–1980), German-born American physicist. Tested metals and alloys for superconductivity; developed empirical rules to predict new superconducting materials.

Maunder, Edward Walter (1851–1928), British astronomer. Observations of the sun, sunspots and eclipses.

Maupertuis, Pierre Louis Moreau de (1698–1759), French mathematician and astronomer. Discovered the principle of least action; mathematical writings on the properties of curves.

Maxwell, James Clerk (1831–1879), Scottish physicist. Formulated the electromagnetic theory of light and the Maxwell distribution of molecular velocities of gases; invented the Maxwell disk concerning color vision.

Mayall, Nicholas Ulrich (1906–1993), American astronomer. Research on nebulae, globular star clusters, and external galaxies.

Mayer, Julius Robert von (1814–1878), German physicist. Discovered the principle of conservation of energy.

Mayer, Maria Goeppert (1906–1972), German-born American nuclear physicist. With J. H. D. Jensen, discovered nuclear shell structure; Nobel Prize, 1963.

McClintock, Barbara (1902–1992), American geneticist. Discovered mobile genetic elements known as jumping genes; Nobel Prize, 1983.

McCoy, Elijah (1844–1929), Canadian-born American inventor. Educated in Scotland as a mechanical engineer, this son of former slaves is best known for inventing automatic lubricating systems for steam engines and industrial machines, greatly advancing the Industrial Revolution; the popularity of his products is believed to have led to the American expression "the real McCoy."

McMillan, Edwin Mattison (1907–1991), American physicist. Discovered element 93 (neptunium), which led to the creation of element 94 (plutonium); conceived the theory of phase stability; Nobel Prize, 1951.

McMullen, Curtis Tracy (1958–), American mathematician. Worked in hyperbolic geometry and in complex dynamics, also known as chaos theory; Fields Medal, 1998.

Meckel, Johann Friedrich (1781–1833), German anatomist and embryologist. Gave first comprehensive description of birth defects; described Meckel's cartilage; discovered Meckel's diverticulum.

Medawar, Peter Brian (1915–1987), Brazilian-born British biologist and medical scientist. Discovered acquired immunological tolerance; Nobel Prize, 1960.

Meissner, Alexander (1883–1958), Austrian-born German radio engineer. Helped develop improved electrical insulators and continuous-wave transmission; invented Meissner oscillator.

Meissner, Walther (1882–1974), German physicist. Research in low-temperature physics; discovered Meissner effect.

Meitner, Lise (1878–1968), German physicist. With O. Hahn, discovered protactinium; found evidence of four other radioactive elements; with Hahn and F. Strassmann, accomplished fission of uranium.

Mendel, Gregor Johann (1822–1884), Austrian botanist. Formulated Mendel's laws of heredity, the foundation of genetics.

Mendeleev, Dmitri Ivanovich (1834–1907), Russian chemist. Formulated Mendeleev's periodic law and table of the elements; research on interatomic and intermolecular forces.

Menelaus of Alexandria (1st century), Greek mathematician and astronomer. Founded spherical trigonometry.

Mercator, Gerhardus, real name **Gerhard Kremer** (1512–1594), Flemish geographer. Created a chart of the world (Mercator projection); made surveying instruments.

Mercer, John (1791–1866), English chemist. Invented the mercerizing process for cotton.

Merrifield, Robert Bruce (1921–), American biochemist. Developed methods of protein synthesis, including solid-phase peptide synthesis that produces proteins by assembling amino acids sequentially into peptide chains; Nobel Prize, 1984.

Mersenne, Marin (1588–1648), French physicist. Showed that pitch is proportional to frequency and calculated frequencies of musical notes; discovered Mersenne's law for vibrating strings, and similar relations for wind and percussion instruments.

Messier, Charles (1730–1817), French astronomer. Credited with discovering 21 comets; compiled a catalog of nebulae.

Metchnikoff, Elie (1845–1916), Russian-born French zoologist and bacteriologist. Work on cholera and immunology; Nobel Prize, 1908.

Meusnier de la Place, Jean Baptiste Marie Charles (1754–1793), French mathematician, physicist, and chemist. Derived Meusnier's theorem of curvature of surface curve; with A. L. Lavoisier, did research on analysis and synthesis of water.

Meyerhof, Otto Fritz (1884–1951), German physiologist. Studied the glycogen-lactic acid cycle of muscles; Nobel Prize, 1922.

Michaelis, Leonor (1875–1949), German-born American biochemist. Developed theory of kinetics of enzyme-catalyzed reactions.

Michel, Hartmut (1948–), German chemist. With J. Deisenhofer and R. Huber, elucidated the structure of a bacterial protein that performs photosynthesis; Nobel Prize, 1988.

Michelson, Albert Abraham (1852–1931), American physicist. Experimented on the velocity of light with S. Newcomb; invented the Michelson interferometer; performed, with E. W. Morley, an experiment to determine the Earth's motion through the ether; Nobel Prize, 1907.

Mie, Gustav (1868–1957), German physicist. Carried out rigorous electrodynamic calculation of Mie scattering; attempted to formulate theory of matter.

Miller, William Hallowes (1801–1880), British crystallographer and mineralogist. Introduced Miller indices for identifying crystallographic planes.

Millikan, Robert Andrews (1868–1953), American physicist. Determined an accurate value for Planck's constant; originated the "oil drop" experiment to measure electronic charge; work on x-rays and cosmic rays; Nobel Prize, 1923.

Milnor, John Willard (1931–), American mathematician. Proved that a seven-dimensional sphere can have several differential structures, opening up the new field of differential topology; contributions to algebraic K theory, differential geometry, and algebraic topology; Fields Medal, 1962.

Milstein, César (1927–2002), British immunologist. With Georges J. F. Köhler, discovered a laboratory technique for producing monoclonal antibodies, highly uniform immune bodies that are selective in responding to target substances; Nobel Prize, 1984.

Minkowski, Hermann (1864–1909), Russian-born German mathematician. Studied the mathematical basis of relativity, notably the concept of the spacetime continuum.

Minot, George Richards (1885–1950), American physician. With W. P. Murphy, first to recognize the value of liver therapy for pernicious anemia; studied arthritis, cancer, and vitamin B deficiency; Nobel Prize, 1934.

Mitchell, Peter (1920–1992), British chemist. Explained how plant and animal cells store and transfer energy by creating protonic gradients in the oxidative and photosynthetic phosphorylation processes; Nobel Prize, 1978.

Möbius, August Ferdinand (1790–1868), German mathematician and astronomer. Founder of topology; developed the Möbius strip.

Mohl, Hugo von (1805–1872), German botanist. Worked on the anatomy and physiology of higher plant forms; discovered protoplasm.

Mohorovičić, Andrija (1857–1936), Yugoslav meteorologist and seismologist. Discovered Mohorovičić seismic discontinuity.

Mohr, Carl Friedrich (1806–1879), German chemist. Developed titration procedures, including use of Mohr's salt.

Mohr, Christian Otto (1835–1918), German civil engineer. Studied stresses and strains of bodies, and failure of materials; introduced Mohr's stress circle.

Mohs, Friedrich (1773–1839), German mineralogist. Developed Mohs scale of hardness.

Moissan, Ferdinand Frédéric Henri (1852–1907), French chemist. First to isolate fluorine; invented an electric furnace and used it to produce synthetic metal compounds and samples of less common metals; Nobel Prize, 1906.

Molina, Mario J. (1943–), Mexican-born American chemist. Contributed to the understanding of how atmospheric ozone forms and decomposes; Nobel Prize, 1995.

Mollier, Richard (1863–1935), German physicist and engineer. Presented properties of thermodynamic media in form of charts and diagrams; introduced concept of enthalpy and Mollier diagram.

Moniz, Antonio Egas (1874–1955), Portuguese neurosurgeon. Developed cerebral angiography; introduced the prefrontal lobotomy; Nobel Prize, 1949.

Monod, Jacques (1910–1976), French biologist. With F. Jacob, proposed the concepts of messenger ribonucleic acid and of the operon; Nobel Prize, 1965.

Moody, Lewis Ferry (1880–1953), American hydraulic engineer. Made improvements in hydraulic turbines, pumps, and accessories.

Moore, Stanford (1913–1982), American biochemist. With W. H. Stein, developed technique for determining amino acid sequence in proteins, and applied it to ribonuclease; Nobel Prize, 1972.

Mordell, Louis Joel (1888–1972), American-born British mathematician. Worked in number theory; proved the finite basis theorem concerning the finite generation of the group of rational points on an elliptic curve; conjectured that there are only finitely many rational points on a curve of genus greater than 1 (the Mordell conjecture).

Morgagni, Giovanni Battista (1682–1771), Italian anatomist. Founded pathological anatomy; first to describe liver cirrhosis.

Morgan, Thomas Hunt (1866–1945), American geneticist, embryologist, and zoologist. Proposed the chromosome theory of heredity; Nobel Prize, 1933.

Morgan, William Wilson (1906–1994), American astronomer. Developed methods for investigating more precisely the structure of the Milky Way Galaxy and of other galaxies; collaborated in developing the Johnson-Morgan system of stellar magnitudes (with H. L. Johnson) and Bautz-Morgan classification of galaxy clusters (with L. P. Bautz).

Mori, Shigefumi (1951–), Japanese mathematician. Worked in algebraic geometry, particularly on the classification of algebraic varieties of dimension three; Fields Medal, 1990.

Morley, Edward Williams (1838–1923), American chemist and physicist. Associated with A. A. Michelson in an experiment on ether drift; research on variations of atmospheric oxygen content.

Morse, Samuel Finley Breese (1791–1872), American inventor. Invented the receiving and sending instruments for the telegraph, and a code for sending messages.

Moseley, Henry Gwyn Jeffries (1887–1915), English physicist. Discovered Moseley's law for frequency of x-ray spectral lines.

Mössbauer, Rudolf Ludwig (1929–), German physicist. Discovered the property of recoilless resonance absorption, the ability of some nuclei to emit and absorb gamma rays without energy loss; Nobel Prize, 1961.

Mossotti, Ottaviano Fabrizio (1791–1863), Italian physicist. Developed theory of dielectrics, from which he derived the Clausius-Mossotti equation.

Mott, Nevill Francis (1905–1996), British physicist. Applied quantum mechanics to study of charged particle scattering; with R. W. Gurney, developed Gurney-Mott theory of photographic process; introduced fundamental concepts elucidating electronic properties of disordered materials; Nobel Prize, 1977.

Mottelson, Ben Roy (1926–), American-born Danish physicist. With A. Bohr, developed theory which unifies shell and liquid-drop models of atomic nucleus, and which explains nonspherical nuclei; Nobel Prize, 1975.

Muller, Hermann Joseph (1890–1967), American geneticist. Studied genetic mutation rates under natural and artificial conditions; discovered the effect of x-rays on mutation rate; Nobel Prize, 1946.

Müller, Johannes Peter (1801–1858), German physiologist and anatomist. Proposed the principle of specific nerve energies, concerning stimuli to sense organs; discovered the Müllerian duct, an early embryonic structure.

Müller, Karl Alex (1927–), Swiss physicist. With J. G. Bednorz, discovered high-temperature superconductivity in copper oxide ceramic materials; Nobel Prize, 1987.

Müller, Paul Hermann (1899–1965), Swiss chemist. Discovered the insecticidal properties of DDT; Nobel Prize, 1948.

Mulliken, Robert Sanderson (1896–1986), American chemist. Applied principles of quantum mechanics to study of chemical bonding; with F. Hund, systematized electronic states of molecules in terms of molecular orbitals; Nobel Prize, 1966.

Mullis, Kary B. (1994–), American biochemist. Invented the polymerase chain reaction method; Nobel Prize, 1993.

Mumford, David Bryant (1937–), American mathematician. Worked in algebraic geometry, especially on problems of the existence and structure of varieties of moduli and on the theory of algebraic surfaces; Fields Medal, 1974.

Murad, Ferid (1936–), American pharmacologist. Analyzed the action of nitroglycerin and related vasodilating compounds, leading to the discovery that they release nitric oxide, which relaxes smooth muscle cells; his work, along with the research of Robert F. Furchgott and Louis J. Ignarro, led to the discovery of nitric oxide as a signaling molecule in the cardiovascular system; Nobel Prize, 1998.

Murchison, Roderick Impey (1792–1871), British geologist. Studied the order of rock formations in Great Britain; with A. Sedgwick, differentiated the Silurian and Devonian.

Murphy, William Parry (1892–1987), American physician. With G. R. Minot, first to suggest liver diet as a treatment for pernicious anemia; Nobel Prize, 1934.

Murray, Joseph (1919–), American physician. Performed the first successful transplant of a human organ, a kidney; with E. D. Thomas, helped define and then overcome the immunological mechanisms behind organ rejection; Nobel Prize, 1990.

Napier or Neper, John, Laird of Merchiston (1550–1617), Scottish mathematician. Invented the theory of logarithms and developed methods to compute them.

Nathans, Daniel (1928–1999), American biologist. Pioneered in the use of restriction enzymes to study the structure and functions of deoxyribonucleic acid (DNA) molecules; Nobel Prize, 1978.

Natta, Giulio (1903–1979), Italian chemist. Discovered stereospecific polymerization, making possible the production of new classes of macromolecules from inexpensive raw materials; Nobel Prize, 1963.

Navier, Claude Louis Marie Henri (1785–1836), French physicist and engineer. Studied analytical mechanics and its application to strength of materials, machines, and motion of solid and liquid bodies; formulated Navier-Stokes equations.

Néel, Louis Eugène Félix (1904–2000), French physicist. Proposed the theory of behavior of antiferromagnetic and other ferrimagnetic materials in which the crystal lattice is divided into one or more sublattices; Nobel Prize, 1970.

Neher, Erwin (1944–), German biophysicist. With B. Sakmann, using the "patch clamp" technique they developed, showed how individual ion channels control the passage of charged ions into and out of cells; Nobel Prize, 1991.

Nernst, Hermann Walther (1864–1941), German chemist. Proposed the heat theorem (third law of thermodynamics); determined the specific heat of solids at low temperatures; proposed the chain reaction theory in photochemistry; Nobel Prize, 1920.

Neumann, Carl Gottfried (1832–1925), German mathematician. Believed to be founder of logarithmic potentials; developed the potential theory.

Newcomb, Simon (1835–1909), American astronomer. With A. A. Michelson, determined the velocity of light; studied the motions of the Moon and planets.

Newton, Isaac (1642–1727), English mathematician. Proposed a dynamical theory of gravitation; discovered three basic laws of motion which are the foundation of practical mechanics; made discoveries in optics and mathematics.

Neyman, Jerzy (1894–1981), Russian-born American statistician. Developed methodology for making decisions based only on results of experiments or observations subject to chance.

Nicholson, Seth Barnes (1891–1963), American astronomer. Discovered four satellites of Jupiter; with E. Petit, invented a thermocouple to measure surface temperature of planets.

Nicol, William (1763–1851), Scottish physicist. Invented the Nicol prism for investigating the polarization of light.

Nicolle, Charles Jules Henri (1866–1936), French physician. Discovered the louse to be the transmission vector of typhus; Nobel Prize, 1928.

Nicomedes (3d century B.C.), Greek mathematician. Discovered the conchoid.

Nirenberg, Marshall Warren (1927–), American biochemist. Pioneered in deciphering genetic code; Nobel Prize, 1968.

Nishina, Yoshio (1890–1951), Japanese physicist. Pioneer in study of cosmic rays; with O. B. Klein, originated the Klein-Nishina formula.

Nobel, Alfred Bernhard (1833–1896), Swedish chemist and engineer. Invented dynamite and a blasting gelatin containing nitroglycerin; established the annual Nobel prizes.

Noguchi, Hideyo (1876–1928). Japanese bacteriologist. First to produce pure cultures of syphilis spirochetes; discovered the parasite of yellow fever.

Norrish, Ronald George Wreyford (1897–1978), British physical chemist. With G. Porter and colleagues, developed methods of flash photolysis and kinetic spectroscopy for the study of very fast reactions; Nobel Prize, 1967.

Northrop, John Howard (1891–1987), American biochemist. Isolated several enzymes and proved them to be proteins; isolated the first bacterial virus; established the chemical nature of enzymes and viruses; Nobel Prize, 1946.

Novikov, Sergi Petrovich (1938–), Russian mathematician. Worked in algebraic topology; proved the topological invariance of the Pontryagin classes of a differentiable manifold; studied the cohomology and homotopy of Thom spaces; Fields Medal, 1970.

Noyce, Robert Norton (1927–1990), American physicist, electronics engineer, and inventor. Participated in the invention of the integrated circuit.

Noyori, Ryoji (1938–), Japanese chemist. Developed chirally catalyzed hydrogenation reactions; Nobel Prize, 2001.

Nurse, Paul M. (1949–), British biologist. Identified, cloned, and characterized a key regulator of the cell cycle, cyclin-dependent kinase; Nobel Prize, 2001.

Nusselt, Ernst Kraft Wilhelm (1882–1957), German mechanical engineer and physicist. Used dimensional analysis to derive functional form of solutions to equations for heat flux in a flowing fluid.

Nüsslein-Volhard, Christiane (1942–), German developmental biologist. Using *Drosophila*, she and Eric Wieschaus identified and classified a small number of genes that are important in determining the body plan and the formation of body segments; their work, along with that of American development biologist Edward B. Lewis, led to the discovery of important genetic mechanisms which control early embryonic development; Nobel Prize, 1995.

Nyquist, Harry (1889–1976), Swedish-born American physicist and engineer. Discovered conditions necessary to keep feedback control circuits stable; determined Nyquist rate for communications channels.

Ochoa, Severo (1905–1993), Spanish-born American biochemist. Discovered a bacterial enzyme that synthesizes ribonucleic acid from nucleoside diphosphates; first to synthesize a ribonucleic acid; Nobel Prize, 1959.

Ockham, William of (ca. 1284–1347), English philosopher and theologian. Developed nominalist school of thought; postulated Ockham's razor.

Oersted, Hans Christian (1777–1851), Danish physicist, chemist, and electromagnetist. Discovered a fundamental principle of electromagnetism: a magnetic needle turns at right angles to an electric current.

Ohm, Georg Simon (1787–1854), German physicist. Discovered Ohm's law relating electrical resistance to voltage and current.

Olah, George A. (1927–), American chemist. Prepared and studied the first stable carbocations; Nobel Prize, 1994.

Olbers, Heinrich Wilhelm Matthias (1758–1840), German astronomer. Devised new method for computing cometary orbits; proposed Olbers' paradox.

Onsager, Lars (1903–1976), Norwegian-born American chemist. Laid the foundation of irreversible thermodynamics; contributed to theories of dielectrics, electrolytes, and cooperative phenomena; Nobel Prize, 1968.

Oort, Jan Hendrik (1900–1992), Dutch astronomer. Research on the structure and dynamics of the galactic system; investigated the origin of comets.

Oppenheimer, J. Robert (1904–1967), American physicist. Research on nuclear disintegration, quantum theory, cosmic rays, and relativity; directed production of the atomic bomb.

Orr, John Boyd, Baron (1880–1971), Scottish physiologist and nutritionist. Work on animal nutrition; pioneer in science of human nutrition; Nobel Peace Prize, 1949.

Osheroff, Douglas Dean (1945–), American physicist. With D. M. Lee and R. C. Richardson, discovered superfluidity in the rare isotope helium-3; Nobel Prize, 1996.

Ostwald, Friedrich Wilhelm (1853–1932), German chemist born in Latvia. Researches on affinity and mass action; discovered the Ostwald dilution law; worked on the catalytic oxidation of ammonia; Nobel Prize, 1909.

Otto, Nikolaus August (1832–1891), German inventor. Built the first four-stroke internal combustion engine.

Paget, James (1814–1899), English surgeon and pathologist. Studied pathology of tumors and bone and joint diseases; described osteitis deformans (Paget's disease).

Palade, George Emil (1912–), Rumanian-born American cytologist. Applied electron microscope and centrifuge techniques to study of ultrastructure of cells; discovered ribosomes; Nobel Prize, 1974.

Papanicolaou, George Nicholas (1883–1962), Greek-born American cytologist and anatomist. Developed the Papanicolaou test for diagnosis of uterine cervical and endometrial cancers.

Pappus, Alexandrinus (ca. 3d–4th century), Greek mathematician. Wrote *Mathematical Collection*, an account of Greek geometry; formulated Pappus' theorems.

Paracelsus, Philippus Aureolus, real name Theophrastus Bombastus von Hohenheim (1493–1541), Swiss physician. Emphasized use of chemicals in medicine; advocated that diseases are specific and require specific remedies.

Paré, Ambroise (1509–1590), French surgeon. Advocated the treatment of wounds by tying arteries with ligatures rather than by cauterization; proposed improvements in operating methods.

Parkinson, James (1755–1824), English physician and paleontologist. Described parkinsonism.

Parseval des Chenes, Marc Antoine (1755–1836), French mathematician. Introduced an equation from which the theorem now known as Parseval's theorem is derived.

Pascal, Blaise (1623–1662), French mathematician and physicist. Contributed to the geometry of conics; formulated Pascal's law, relating to the pressure of a liquid at rest; applied Pascal's triangle to the calculation of probabilities.

Paschen, Louis Carl Heinrich Friedrich (1865–1947), German physicist. Established Paschen's law; with E. Back, discovered Paschen-Back effect; verified predictions of relativistic fine structure made by Bohr-Sommerfeld theory.

Pasteur, Louis (1822–1895), French biologist. Founder of microbiology; discovered the role of bacteria in fermentation; discovered anaerobic bacteria; developed the pasteurization process; demonstrated the efficacy of vaccination, especially for rabies.

Patterson, Arthur Lindo (1902–1966), New Zealand-born American physicist and crystallographer. Developed Patterson-Harker method of x-ray diffraction analysis of crystal structure.

Paul, Wolfgang (1913–1993), German physicist. Invented the Paul trap, which uses radio-frequency radiation to hold ions in a small volume; Nobel Prize, 1989.

Pauli, Wolfgang (1900–1958), Austrian-born American physicist. Worked on quantum theory; formulated the Pauli exclusion principle; contributed to matrix mechanics; Nobel Prize, 1945.

Pauling, Linus Carl (1901–1994), American chemist. Applied quantum theory to chemistry; research on molecular structure and chemical bonds; contributed to electrochemical theory of valency; Nobel Prize, 1954; Nobel Peace Prize, 1963.

Pavlov, Ivan Petrovich (1849–1936), Russian pathologist. Discovered the nerve fibers affecting heart action and the secretory nerves of the pancreas; research on the physiology of digestive glands; studied conditioned reflexes; Nobel Prize, 1904.

Peano, Giuseppe (1858–1932), Italian mathematician. Pioneer in symbolic logic and foundations of mathematics; promoted axiomatic method in mathematics; formulated postulates for natural numbers.

Pearl, Raymond (1879–1940), American biologist and statistician. Applied statistics to the study of population changes; introduced logistic curve describing population growth.

Pearson, Karl (1857–1936), English applied mathematician, statistician, and biometrician. Pioneered in application of statistics to biology; introduced chi-square test.

Pedersen, Charles J. (1904–1989), American chemist. Developed the synthesis of cyclic polyethers known as crown ethers; Nobel Prize, 1987.

Peierls, Rudolf Ernst (1907–1995), German-born British physicist. Developed theory of heat conduction in nonmetallic crystals; with O. R. Frisch, calculated critical mass of uranium-235.

Peirce, Charles Santiago Sanders (1839–1914), American mathematician, logician, and physicist. Laid foundation for logical analysis of mathematics; contributed to probability theory.

Pelletier, Pierre Joseph (1788–1842), French chemist. Discovered quinine, strychnine, and other alkaloids.

Peltier, Jean Charles Athanase (1785–1845), French physicist. Discovered the Peltier effect in thermoelectricity.

Penrose, Roger (1931–), British mathematician and physicist. Developed twistor theory of spacetime geometry; studied singularities in classical general relativity theory.

Penzias, Arno Allan (1933–), American astrophysicist. With R. W. Wilson, discovered cosmic background radiation, confirming the big bang theory of the origin of the universe; Nobel Prize, 1978.

Perl, Martin Lewis (1927–), American physicist. Discovered the tau lepton, helping to lay the basis for the standard model of elementary particles; Nobel Prize, 1995.

Pérot, Jean Baptiste Gaspard Gustav Alfred (1863–1925), French physicist. With C. Fabry, developed Fabry-Pérot interferometer.

Perrin, Jean Baptiste (1870–1942), French physicist. Research on the particle nature of cathode rays; found values for Avogadro's number, thereby proving the existence of molecules; Nobel Prize, 1926.

Perutz, Max Ferdinand (1914–2002), Austrian-born British crystallographer and molecular biologist.

Worked on the structure of hemoglobin; introduced the method of isomorphous replacement with heavy atoms into protein crystallography; Nobel Prize, 1962.

Petit, Alexis Thérèse (1791–1820), French physicist. With P. L. Dulong, formulated the law of constancy of atomic heats; devised methods for determining thermal expansion and specific heats of solids.

Pettit, Edison (1890–1962), American astronomer. Studied the Sun and formulated laws alleged to govern the movement of prominences; constructed the interference polarizing monochromator; with S. B. Nicholson, devised a sensitive thermocouple to measure the surface temperatures of planets.

Pfaff, Johann Friedrich (1765–1825), German mathematician. Developed theory of Pfaffian differential equations, which is basic to general solution of partial differential equations.

Pfeiffer, Richard Friedrich Johann (1858–1945), German bacteriologist. Discovered Pfeiffer's bacillus in influenza; described Pfeiffer's reaction for determination of cholera.

Phillips, William Daniel (1948–), American physicist. Developed methods to cool and trap atoms with laser light, achieving temperatures much lower than the Doppler limit; Nobel Prize, 1997.

Piaget, Jean (1896–1980), Swiss psychologist. Elucidated development of cognitive functions in the child.

Picard, Charles Émile (1856–1941), French mathematician. Formulated Picard's theorem relating to functions.

Piccard, Auguste (1884–1962), Swiss physicist. Conducted a data-collecting exploration of the stratosphere in an airtight gondola of a balloon; constructed and tested a bathysphere for deep-sea exploration.

Pickering, Edward Charles (1846–1919), American astronomer. Invented the meridian photometer; pioneered in stellar spectroscopy.

Pierce, George Washington (1872–1956), American physicist and electronic engineer. Developed theoretical basis of electrical communications; developed Pierce oscillator; with A. E. Kennelly, discovered concept of motional impedance.

Pitzer, Kenneth Sanborn (1914–1997), American chemist. Pioneered in the development of useful approximations which made possible the calculation of chemical thermodynamic properties of broad classes of chemical substances.

Planck, Max Karl Ernst Ludwig (1858–1947), German physicist. Presented the quantum theory; introduced Planck's constant, or quantum of action; Nobel Prize, 1918.

Planté, Gaston (1834–1889), French physicist. Constructed a storage battery, the first primitive accumulator.

Plateau, Joseph Antoine Ferdinand (1801–1883), Belgian physicist. Experimented with soapy films bounded by wires, noting that the surfaces formed were minimal surfaces; from this he formulated the Plateau problem (the problem of determining the existence of a minimal surface with a given space curve as its boundary).

Podolsky, Boris (1896–1966), Russian-born American physicist. Collaborated in formulation of Einstein-Podolsky-Rosen paradox; research on quantum electrodynamics.

Poggendorff, Johann Christian (1796–1877), German physicist. Introduced the small mirror on a suspended system to magnify small deflections of a light beam; invented the galvanometer.

Poincaré, Jules Henri (1854–1912), French mathematician. Worked on the theory of functions, on differential equations, and on the theory of orbits in astronomy.

Poinsot, Louis (1777–1859), French mathematician. Originated theory of couples.

Poiseuille, Jean Léonard Marie (1797–1869), French physiologist and physicist. Studied physiology of arterial circulation; invented improved methods for measuring blood pressure; discovered Hagen-Poiseuille law independently of G. H. L. Hagen.

Poisson, Siméon Denis (1781–1840), French

mathematician. Worked on mathematical physics; contributed to the wave theory of light; formulated the Poisson ratio concerning the elasticity of materials.

Polanyi, John C. (1929–), German-born Canadian chemist. Studied chemiluminescence, a phenomenon in which the energy states of excited molecules are revealed by their emission of light; Nobel Prize, 1986.

Pomeranchuk, Isaak Yakolevich (1913–1966), Soviet physicist. Showed that energy of cosmic-ray electrons reaching the atmosphere is limited by their radiation in Earth's magnetic field; proved the Pomeranchuk theorem for scattering cross sections.

Pons, Jean Louis (1761–1831), French astronomer. Discovered 37 comets, including Encke's Comet.

Pople, John A. (1925–), British-born chemist. Developed computational methods in quantum chemistry; Nobel Prize, 1998.

Porro, Ignazio (1801–1875), Italian topographer, geodesist, and physicist. Invented optical surveying instruments; Porro prism erecting system, and modern prism binoculars.

Porter, George (1920–), British chemist. With R. G. W. Norrish, developed the technique of flash photolysis to initiate and record very fast chemical reactions; Nobel Prize, 1967.

Porter, Rodney Robert (1917–1985), British biochemist. Research to determine chemical structure of immunoglobulins; Nobel Prize, 1972.

Powell, Cecil Frank (1903–1969), British physicist. Made practical the use of photographic emulsions in nuclear research; with G. P. S. Occhialini and others discovered and investigated production of pions from cosmic radiation in the Earth's atmosphere; Nobel Prize, 1950.

Poynting, John Henry (1852–1914), English physicist. Determined the constant of gravitation and explained why a comet's tail points away from the Sun.

Prandtl, Ludwig (1875–1953), German physicist. Contributed to fluid mechanics, particularly aerodynamics; introduced concept of boundary layer.

Pregl, Fritz (1869–1930), Austrian chemist. Developed microchemical methods of analysis; Nobel Prize, 1923.

Prelog, Vladimir (1906–1998), Yugoslavian-born Swiss chemist. Investigated stereochemistry of organic molecules and reactions; Nobel Prize, 1975.

Prevost, Pierre (1751–1839), Swiss physicist. Developed theory of exchanges, explaining nature of heat.

Priestley, Joseph (1733–1804), English chemist and physicist. Discovered oxygen, ammonia, oxides of nitrogen, hydrochloric acid gas, nitrogen, carbon monoxide, and sulfur dioxide.

Prigogine, Ilya (1917–), Soviet-born Belgian chemist. Contributed to nonequilibrium thermodynamics, particularly the theory of dissipative structures; Nobel Prize, 1977.

Prokhorov, Aleksandr Mikhallovich (1916–2002), Soviet physicist. With N. G. Basov, devised a new method for amplifying electromagnetic radiation; Nobel Prize, 1964.

Prout, William (1785–1850), English physician and chemist. Formulated Prout's hypothesis concerning atomic weights.

Prusiner, Stanley B. (1942–), American neurologist. Discovered and elucidated the mode of action of a new genre of pathogens, "proteinaceous infectious particles" or prions, which are the cause of a group of fatal neurodegenerative disorders (spongiform encephalopathies) that lead to progressive dementia and death in humans and animals; Nobel Prize, 1997.

Ptolemy (2d century), Greco-Egyptian astronomer, geographer, and geometer at Alexandria. Proposed the Ptolemaic system, with the Earth as the center of the universe.

Pupin, Michael (1858–1935), Yugoslavian-born American physicist and electrical engineer. Developed inductance coils for telephone lines; contributed to x-ray fluoroscopy, design of radio transmitters, and network theory.

Purcell, Edward Mills (1912–1997), American

physicist. Developed the method of nuclear resonance absorption; Nobel Prize, 1952.

Purkinje, Johannes Evangelista (1787–1869), Czech physiologist. Discovered the Purkinje effect in eye physiology and Purkinje cells in the cerebral cortex.

Pythagoras (6th century B.C.), Greek mathematician. Originated a system of geometry, including the Pythagorean theorem.

Quételet, Lambert Adolphe Jacques (1796–1874), Belgian statistician. Did pioneer work on statistics; applied the calculus of probabilities to sociological studies.

Quillen, Daniel Gray (1940–), American mathematician. Developed algebraic K-theory, an extension of ideas of A. Grothendieck to commutative rings, which employed geometric and topological methods and ideas to formulate and solve major problems in algebra, particularly ring theory and module theory; Fields Medal, 1978.

Rabi, Isidor Isaac (1898–1988), Austrian-born American physicist. Research on neutrons, magnetism, quantum mechanics, and nuclear physics; Nobel Prize, 1944.

Radó, Tibor (1895–1965), Hungarian-American mathematician. Solved the Plateau problem (the problem of determining the existence of a minimal surface with a given space curve as its boundary) about the same time as J. Douglas.

Radon, Johann (1887–1956), Bohemian-born Austrian mathematician. Work in calculus of variations and integration theory.

Rainwater, Leo James (1917–1986), American physicist. Suggested that shell-model potentials of certain atomic nuclei are not spherical but are deformed into spheroids, and proposed mechanism for this distortion; Nobel Prize, 1975.

Raman, Chandrasekhara Venkata (1888–1970), Indian physicist. Research on diffraction and oscillation; discovered the Raman effect; Nobel Prize, 1930.

Ramón y Cajal, Santiago (1852–1934), Spanish histologist. Isolated the neuron and made discoveries concerning nerve cells in gray matter and the spinal cord; Nobel Prize, 1906.

Ramsay, William (1852–1916), British chemist. With J. W. S. Rayleigh, discovered argon; with M. W. Travers, discovered neon, krypton, and xenon; Nobel Prize, 1904.

Ramsden, Jesse (1735–1800), English mathematical-instrument maker. Invented an eyepiece containing cross-wires as a measuring scale; introduced equatorial mounting for telescopes.

Ramsey, Norman Foster (1915–), American physicist. Invented an accurate method of measuring differences between atomic energy levels that formed the basis for the cesium atomic clock; worked on the hydrogen maser; Nobel Prize, 1989.

Rankine, William John Macquorn (1820–1872), Scottish civil engineer. Contributed to thermodynamics and theories of elasticity and waves; wrote textbooks on the steam engine and civil engineering.

Raoult, François Marie (1830–1901), French chemist. Formulated Raoult's law concerning vapor pressure of a solution.

Rathke, Martin Heinrich (1793–1860), German biologist. Discovered gill slits and gill arches in embryo birds and mammals, and Rathke's pocket in developing vertebrates.

Ray or Wray, John (1627?–1705), English naturalist. Identified the difference between mono- and dicotyledons; arranged plants according to their natural form, the foundation of the natural system of classification.

Rayleigh, John William Strutt, 3d Baron (1842–1919), English physicist. Worked on the theory of sound and on physical optics; with W. Ramsay, discovered argon; Nobel Prize, 1904.

Réaumur, René Antoine Ferchault de (1683–1757), French entomologist. Worked in biology and metallurgy; invented the Réaumur thermometer scale.

Regge, Tullio (1931–), Italian physicist. Played a role in introducing the idea of complex angular momenta into elementary particle physics.

Regiomontanus, real name Johann Müller (1436–1476), German astronomer. Erected the first European observatory, in 1471 in Nürnberg; produced mathematical tables.

Reichstein, Tadeus (1897–1996), Polish-born Swiss organic chemist. Isolated about 30 of the 40 substances produced by the adrenal cortex; synthesized and described the structure and properties of many of these substances; Nobel Prize, 1950.

Reines, Frederick (1918–1998), American physicist. With C. L. Cowan, detected the neutrino, and subsequently did research in neutrino physics; Nobel Prize, 1995.

Reynolds, Osborne (1842–1912), British engineer and physicist. Demonstrated streamline and turbulent flow in pipes, and showed that transition between them occurs at a critical velocity determined by Reynolds' number; introduced Reynolds' analogy.

Riccati, Jacopo Francesco (1676–1754), Italian mathematician. Research on analysis, particularly differential equations, and geometry.

Ricci-Curbastro, Gregorio (1853–1925), Italian mathematician and mathematical physicist. Developed theory of tensor analysis, providing mathematical foundation for general relativity.

Richards, Dickinson Woodruff (1895–1973), American physician. With A. F. Cournand, utilized the technique of cardiac catheterization and proved its value as a diagnostic tool; Nobel Prize, 1956.

Richards, Theodore William (1868–1928), American chemist. Worked on atomic weights; experimentally confirmed the existence of isotopes of lead from uranium and thorium; Nobel Prize, 1914.

Richardson, Owen Willans (1879–1959), English physicist. Studied the emission of electricity from hot bodies and the electron theory of matter; Nobel Prize, 1928.

Richardson, Robert Coleman (1937–), American physicist. With D. M. Lee and D. D. Osheroff, discovered superfluidity in the rare isotope helium-3; Nobel Prize, 1996.

Richet, Charles Robert (1850–1935), French physiologist. Studied serum therapy and discovered anaphylaxis; Nobel Prize, 1913.

Richter, Burton (1931–), American physicist. Independently of S. C. C. Ting, discovered a new heavy elementary particle, which he named the psi particle; Nobel Prize, 1976.

Richter, Jeremias Benjamin (1762–1807), German chemist. Discovered the law of equivalent proportions.

Riemann, Georg Friedrich Bernhard (1826–1866), German mathematician. Originated Riemannian geometry, a noneuclidean system.

Riesz, Frigyes or Frederic (1880–1956), Hungarian mathematician. Did research on abstract and general theories related to mathematical analysis, particularly functional analysis; independently of E. Fischer, discovered Riesz-Fisher theorem.

Righi, Augusto (1850–1920), Italian physicist. Discovered magnetic hysteresis and Righi-Leduc effect, independently of S. A. Leduc; demonstrated that microwaves have all properties characteristic of light waves.

Ritchey, George Wills (1864–1945), American astronomer. Made important astronomical observations, particularly on the Andromeda nebula; with H. Chrétien, developed Ritchey-Chrétien optics.

Ritz, Walter (1878–1909), Swiss-born German physicist. Introduced Ritz combination principle; developed Ritz method for numerical solution of boundary-value problems.

Robbins, Frederick Chapman (1916–), American microbiologist. Discovered that poliomyelitis virus can be grown in various human tissue cultures; Nobel Prize, 1954.

Roberts, Richard J. (1943–), English molecular biologist. He and Phillip A. Sharp independently discovered "split genes," that is, genes in which the coding sequences of deoxyribonucleic acid (exons) are interrupted by noncoding sequences (introns), while studying the genome of the adenovirus; Nobel Prize, 1993.

Robinson, Robert (1886–1975), English chemist. Worked on plant pigments, alkaloids, and phenanthrene derivatives; Nobel Prize, 1947.

Roche, Edouard Adelbert (1820–1883), French physicist, mathematician, and meteorologist. Studied the internal structure and free-surface form of the celestial bodies; applied results to study of cosmogonic hypotheses.

Rodbell, Martin (1925–1998), American biochemist. Discovered the role of G-proteins, later isolated by Alfred G. Gilman, in cellular signal transduction; Nobel Prize, 1994.

Rohrer, Heinrich (1933–), Swiss physicist. With G. Binnig, developed scanning tunneling microscope; Nobel Prize, 1986.

Rolle, Michel (1652–1719), French mathematician. Worked on Diophantine analysis and algebra of equations.

Röntgen, Wilhelm Konrad (1845–1923), German physicist. Discovered x-rays; Nobel Prize, 1901.

Roscoe, Henry Enfield (1833–1915), English chemist. With R. W. Bunsen, evolved the law of reciprocity and invented the actinometer; first to isolate metallic vanadium.

Rosen, Nathan (1909–1995), American-born Israeli physicist. Collaborated in formulation of Einstein-Podolsky-Rosen paradox; research on general relativity and gravitational waves.

Ross, Ronald (1857–1932), British physician. Proved that malaria is transmitted by the female *Anopheles* mosquito; Nobel Prize, 1902.

Rossby, Carl Gustaf Arvid (1898–1957), Swedish-born American meteorologist. Formulated theories of large-scale air movements; derived the Rossby formula, relating speed of propagation of perturbations to airflow and wavelengths of perturbations; devised the Rossby diagram, used to plot air mass properties.

Roth, Klaus Friedrich (1925–), German-born British mathematician. Solved a problem previously studied by A. Thue and C. Siegel concerning the approximation to algebraic numbers by rational numbers; proved that a sequence with no three numbers in arithmetic progression has zero density; Fields Medal, 1958.

Rous, Francis Peyton (1879–1970), American physician and virologist. Produced cancer in chickens by inoculating them with filterable virus procured from tissue of chickens with tumors; Nobel Prize, 1966.

Routh, Edward John (1831–1907), British mathematical physicist. Made contributions to classical mechanics, including procedure for eliminating cyclic coordinates from equations of motion.

Roux, Pierre Paul Emile (1853–1933), French physician and bacteriologist. Helped develop modern serum therapeutics, especially concerning diphtheria.

Rowland, F. Sherwood (1927–), American chemist. Contributed to the understanding of how atmospheric ozone forms and decomposes; Nobel Prize, 1995.

Rowland, Henry Augustus (1848–1901), American physicist. Developed the Rowland grating in spectroscopy; studied electromagnetism and heat.

Rubbia, Carlo (1934–), Italian physicist. Principal architect of experiment that first detected intermediate vector bosons, an important step in confirming theory uniting electromagnetic and weak nuclear interactions; Nobel Prize, 1984.

Rubens, Heinrich (1865–1922), German physicist. With E. B. Hagen, conducted electromagnetic experiments; built new types of galvanometer and bolometer.

Rumford, Benjamin Thompson, Count (1753–1814), British physicist. Carried out research on heat.

Runge, Carl David Tolme (1856–1927), German mathematician and physicist. Research on theoretical and experimental spectroscopy, particularly data reduction and development of series formulas; developed methods for numerical and graphical computation, including Runge-Kutta method.

Ruska, Ernst (1906–1988), German electronic engineer. Developed the electron microscope; Nobel Prize, 1986.

Russell, Bertrand Arthur William (1872–1970), English mathematician and philosopher. With A. N. Whitehead, pioneered in study of mathematical logic.

Russell, Henry Norris (1877–1957), American astronomer and physicist. Analyzed eclipsing binary stars; with E. Hertzsprung, introduced Hertzsprung-Russell diagram; determined abundance of chemical elements in solar atmosphere; with F. A. Saunders, devised theory of Russell-Saunders coupling.

Rutherford, Ernest, 1st Baron (1871–1937), British physicist. Discovered alpha, beta, and gamma rays; suggested the divisible nuclear atom; effected the transmutation of an atom; Nobel Prize, 1908.

Ružička, Leopold (1887–1976), Swiss chemist born in Croatia. Research on many-membered rings and higher terpenes (including male sex hormones); Nobel Prize, 1939.

Rydberg, Johannes Robert (1854–1919), Swedish physicist. Developed a formula for series of spectral lines, involving Rydberg's constant.

Ryle, Martin (1918–1984), British astronomer. Devised aperture synthesis method in radiotelescopy; designed equipment and made observations in radio astronomy; Nobel Prize, 1974.

Sabatier, Paul (1854–1941), French chemist. Discovered, with J. B. Senderens, the process for catalytic hydrogenation of oils to solid fat; Nobel Prize, 1912.

Sabin, Albert Bruce (1906–1993), Polish-born American physician and virologist. Studied nature, mode of transmission, and epidemiology of human poliomyelitis; developed oral polio virus vaccine.

Sabine, Edward (1788–1883), British physicist and astronomer. Headed a magnetic survey of the world which discovered a connection between sunspots and terrestrial magnetic disturbances.

Sabine, Wallace Clement Ware (1868–1919), American physicist. Pioneered in architectural acoustics; discovered law determining reverberation time in acoustics.

Sachs, Julius von (1832–1897), German botanist. Studied the connection between sunlight and chlorophyll; worked on heliotropism and geotropism.

Saha, Meghnad (1894–1956), Indian physicist. Developed theory for degree of ionization of hot gases, a basic component of modern astrophysics.

Sakmann, Bert (1942–), German physiologist. With E. Neher, using the "patch clamp" technique they developed, showed how individual ion channels control the passage of charged ions into and out of cells; Nobel Prize, 1991.

Salam, Abdus (1926–1996), Pakistani physicist. Independently of S. Weinberg, developed theory uniting two of the basic forces of nature, electromagnetism and the weak nuclear interactions; Nobel Prize, 1979.

Salk, Jonas Edward (1914–1995), American physician. Produced killed-virus vaccine effective in preventing poliomyelitis.

Salpeter, Edwin Ernest (1924–), Austrian-born American physicist. Research in quantum theory of atoms, quantum electrodynamics, nuclear theory, energy production of stars, and theoretical astrophysics; with H. A. Bethe, introduced Bethe-Salpeter equation.

Samuelsson, Bengt Ingemar (1934–), Swedish biochemist and medical scientist. Studied prostaglandin metabolism and the formation of prostaglandin from arachidonic acid; Nobel Prize, 1982.

Sanctorius, real name Santorio Santorio (1561–1636), Italian physician. Invented the clinical thermometer; experimented with metabolism.

Sanger, Frederick (1918–), English chemist. Determined the exact order of amino acids in insulin; first to establish amino acid sequence for a protein; developed methods for determining nucleotide sequences (independently of W. Gilbert), advancing the technology of DNA recombination; Nobel Prizes, 1958 and 1980.

Savart, Félix (1791–1841), French physicist. Helped formulate the Biot-Savart law in electromagnetism.

Schally, Andrew Victor (1926–), Polish-born American physiologist. With R. Guillemin, isolated and analyzed peptide hormones secreted in hypothalmic region of brain which control anterior pituitary hormone secretion; Nobel Prize, 1977.

Schawlow, Arthur Leonard (1921–1999), American physicist. Contributed to invention of laser; made numerous contributions to laser spectroscopy, particularly the development of Doppler-free spectroscopy; Nobel Prize, 1981.

Scheele, Karl Wihelm (1742–1786), Swedish chemist. Made many discoveries, including oxygen (independently of J. Priestley), chlorine, and glycerin; synthesized many organic acids.

Schiaparelli, Giovanni Virginio (1835–1910), Italian astronomer. Discovered the connection between comets and meteorites, and the "canals" of Mars.

Schiff, Hugo Josef (1834–1915), German-born Italian organic chemist. Discovered Schiff bases; devised Schiff test; devised an improved nitrometer.

Schmidt, Bernhard Voldemar (1879–1935), Estonian-born German astronomer. Invented Schmidt system for astronomical telescopes.

Schmidt, Erhard (1876–1959), German mathematician. Extended D. Hilbert's work on integral equations; formalized and developed concept of Hilbert space.

Schoenflies, Arthur Moritz (1853–1928), German mathematician and crystallographer. Classified the 230 crystallographic space groups.

Schottky, Walter (1886–1976), Swiss-born German physicist. Discovered Schottky effect; invented screen grid and tetrode; developed Schottky theory of semiconductor-metal junctions.

Schrieffer, John Robert (1931–), American physicist. With J. Bardeen and L. N. Cooper, formulated a theory of superconductivity; Nobel Prize, 1972.

Schrödinger, Erwin (1887–1961), German physicist. Proposed concept of atomic structure based on wave mechanics; contributed to quantum theory and color theory; Nobel Prize, 1933.

Schur, Issai (1875–1941), Russian-born German mathematician. Contributed to representation theory of groups; research on group theory, matrices, algebraic equations, and number theory.

Schwartz, Laurent (1915–), French mathematician. Developed the theory of distributions, which provided an abstract and rigorous mathematical foundation for methods of formal calculation such as the Dirac delta function, and greatly extended their range of application; Fields Medal, 1950.

Schwartz, Melvin (1932–), American physicist. Collaborated in an experiment that demonstrated the existence of two types of neutrino; Nobel Prize, 1988.

Schwarz, Hermann Amandus (1843–1921), German mathematician. Introduced Schwarz reflection principle and Schwarz's lemma while proving the Riemann mapping theorem.

Schwarzschild, Karl (1873–1916), German astronomer. Developed photographic methods for measuring brightness of stars; discovered Schwarzschild solution of equations of general relativity.

Schwarzschild, Martin (1912–1997), German-born American astronomer. Numerical studies of the internal structure and evolution of stars; astronomical observations with balloon-borne telescopes.

Schwinger, Julian Seymour (1918–1994), American physicist. Made fundamental contributions to the quantum theory of radiation; worked out the mathematical formalism of interaction between charged particles and an electromagnetic field; Nobel Prize, 1965.

Seaborg, Glenn Theodore (1912–1999), American chemist. Synthesized and identified eight transuranium elements and over a hundred isotopes; Nobel Prize, 1951.

Secchi, Pietro Angelo (1818–1878), Italian astronomer. Originated the spectroscopic survey of the heavens; made the first classification of stars according to spectral type.

Sedgwick, Adam (1785–1873), English geologist.

With R. I. Murchison, established the Devonian system.

Seebeck, Thomas Johann (1770–1831), German physicist. Investigated thermoelectricity and invented the thermocouple.

Segrè, Emilio Gino (1905–1989), Italian-born American physicist. Codiscovered the elements technetium, astatine, and plutonium, slow neutrons, and the antiproton; Nobel Prize, 1959.

Seidel, Philipp Ludwig von (1821–1896), German astronomer and mathematician. Developed theory of aberrations; made first accurate photometric measurements of stars and planets, and evaluated them with probability theory.

Selberg, Atle (1917–), Norwegian-American mathematician. Worked on generalizations of the sieve methods of V. Brun; proved major results on zeros of the Riemann zeta function; with P. Erdös, gave an elementary proof of the prime number theorem, with a generalization to numbers in an arbitrary arithmetic progression; Fields Medal, 1950.

Semenov, Nikolai Nikolaevich (1896–1986), Soviet chemist. Elucidated the mechanisms of chemical reactions, especially the chain mechanism; Nobel Prize, 1956.

Senderens, Jean Baptiste (1856–1937), French chemist. With P. Sabatier, discovered hydrolysis of oils by catalysis.

Serber, Robert (1909–1997), American physicist. Laid foundations of orbit theory of high-energy particle accelerators; introduced Serber potential to describe nuclear forces.

Serre, Jean-Paul (1926–), French mathematician. Applied spectral sequences to discover fundamental connections between the homology groups and homotopy groups of a space and to prove important results on the homotopy groups of spheres; reformulated and extended some of the main results of complex variable theory in terms of sheaves; Fields Medal, 1954.

Serret, Joseph Alfred (1819–1885), French mathematician and astronomer. Helped develop Frenet-Serret formulas in the theory of space curves.

Servetus, Michael (1511–1553), Spanish physician. Discovered the pulmonary circulation and the purification of the blood by the lungs.

Shannon, Claude Elwood (1916–2001), American mathematician. Developed mathematical theory of communication, making use of analogy between concepts of entropy and information.

Shapley, Harlow (1885–1972), American astronomer. Worked on a theory to explain cepheid variables; made an estimate of the size of the universe; provided a description of the universe's form of construction.

Sharp, Phillip A. (1944–), American molecular biologist. He and Richard J. Roberts independently discovered "split genes," that is, genes in which the coding sequences of deoxyribonucleic acid (exons) are interrupted by noncoding sequences (introns) while studying the genome of the adenovirus; Nobel Prize, 1993.

Sharpless, K. Barry (1941–), American chemist. Developed chirally catalyzed oxidation reactions; Nobel Prize, 2001.

Sherrington, Charles Scott (1861–1952), English physiologist. Studied the neuron and its function and other aspects of the nervous system; Nobel Prize, 1932.

Shirakawa, Hideki (1936–), Japanese polymer scientist. Discovered and developed conductive polymers with Alan Heeger and Alan MacDiarmid; Nobel Prize, 2000.

Shockley, William (1910–1989), English-born American physicist. Discovered the transistor effect for electronic amplification by means of solid-state semiconductors; Nobel Prize, 1956.

Shor, Peter (1959–), American mathematician. Worked in combinatorial analysis and the theory of quantum computing; developed a computational method for factorizing large numbers on quantum computers, which, theoretically, could be used to break many of the coding systems currently employed.

Shubnikov, Aleksei Vasilevich (1887–1970),

Soviet crystallographer. Classified Shubnikov groups; developed techniques for growing crystals, including synthetic rubies used in lasers.

Shull, Clifford Glenwood (1915–2001), American physicist. Developed the neutron diffraction technique to study the atomic structure of solids and liquids; Nobel Prize, 1994.

Siegbahn, Kai Manne Börje (1918–), Swedish physicist. Pioneered the development of high-resolution electron spectroscopy; Nobel Prize, 1981.

Siegbahn, Karl Manne Georg (1886–1978), Swedish physicist. Studied x-ray spectroscopy, in which he discovered the M series; Nobel Prize, 1924.

Siegel, Carl Ludwig (1896–1981), German mathematician. Worked on number theory, functions of one or several complex variables, and differential equations.

Siemens, Ernst Werner von (1816–1892), German engineer and electrician. Developed telegraphy and self-acting dynamo.

Siemens, William or Karl Wilhelm (1823–1883), German inventor in London. Made many inventions, including a differential governor, bathometer, dynamometer, and electric furnace.

Simpson, Thomas (1710–1761), English mathematician. Formulated the Simpson rule for finding the area of a figure, given only a limited number of data.

Singer, Isadore Manual (1924–), American mathematician. Worked in global analysis, especially the theory of elliptic operators and their applications to topology and geometry, and in mathematical physics; collaborated with M. F. Atiyah in proving the index theorem.

Skou, Jens C. (1918–), Danish biophysicist. Discovered the ion-transporting enzyme Na$^+$,K$^+$-ATPase, which maintains the balance of sodium and potassium ions in the living cell; Nobel Prize, 1997.

Slater, John Clarke (1900–1976), American physicist. Introduced Slater determinant describing many-electron systems; developed theory of magnetrons.

Smale, Stephen (1930–), American mathematician. Worked in differential topology, differential equations, and dynamical systems; proved that the sphere can be turned inside out and that the generalized Poincaré conjecture is valid for dimensions greater than 4; discovered strange attractors that lead to chaotic dynamical systems; Fields Medal, 1966.

Smalley, Richard E. (1943–), American chemist. with Robert Curl and Harold Kroto, discovered fullerenes; Nobel Prize, 1996.

Smith, Hamilton Othanel (1931–), American geneticist. Isolated a restriction enzyme that cleaves deoxyribonucleic acid (DNA) molecules at a specific site; Nobel Prize, 1978.

Smith, Michael (1932–2000), British-born Canadian chemist. Made fundamental contributions to the establishment of oligonucleotide-based, site-directed mutagenesis and its development for protein design and studies; Nobel Prize, 1993.

Smith, Robert (1689–1768), English physicist. Developed a particulate theory of light; developed geometric propositions for computing properties of optical systems; derived a special case of the Smith-Helmholtz law.

Snell, George Davis (1903–1996), American immunogeneticist. Demonstrated the x-ray induction of mutational changes in a mammal; contributed to the study of immunological systems and to the development of transplant immunology; Nobel Prize, 1980.

Snell, Willebrod van Roijen (1591–1626), Dutch mathematician. Formulated Snell laws concerning angles of incidence and refraction; conceived the idea of measuring the Earth by triangulation.

Soddy, Frederick (1877–1956), English chemist. With E. Rutherford, developed theory of atomic disintegration of radioactive substances; research on isotopes; Nobel Prize, 1921.

Solvay, Ernest (1838–1922), Belgian industrial chemist. Developed the Solvay process for production of sodium carbonate.

Sommerfeld, Arnold (1868–1951), German physicist. Developed quantum theory, especially in its application to spectral lines and the Bohr atomic model.

Sörensen, Sören Peter Lauritz (1868–1939), Danish biochemist. Did pioneer work on hydrogen ion concentration; invented the symbol pH.

Spemann, Hans (1869–1941), German zoologist. Studied embryonic development and discovered the organizer function of certain tissues; Nobel Prize, 1935.

Sperry, Roger Wolcott (1913–1994), American neuroscientist. Discovered the functional split between the left and right hemispheres of the brain; Nobel Prize, 1981.

Spörer, Gustav Friedrich Wilhelm (1822–1895), German astronomer. Observations of the sun and sunspots.

Stanley, Wendell Meredith (1904–1971), American biochemist. Discovered that a virus is a nucleoprotein and can be crystallized; Nobel Prize, 1946.

Stanton, Thomas Ernest (1865–1931), English engineer. Studied surface friction of fluids; built wind tunnel for wind velocity investigations; studied strength of materials, heat transmission, and lubrication.

Stark, Johannes (1874–1957), German physicist. Studied radiation and atomic theory; discovered the Stark effect on spectrum lines and the Doppler effect in canal rays; Nobel Prize, 1919.

Staudinger, Hermann (1881–1965), German chemist. Conceived and elaborated the explanation of phenomenon of polymerization; Nobel Prize, 1953.

Steenrod, Norman Earl (1910–1971), American mathematician. Worked in topology; introduced Steenrod algebra.

Stefan, Josef (1835–1893), Austrian physicist. Originated Stefan's (or Stefan-Boltzmann) law of blackbody radiation; proposed theory of diffusion of gases; studied gas conductivity.

Stein, William Howard (1911–1980), American biochemist. With S. Moore, developed technique for determining amino acid sequence in proteins, and applied it to ribonuclease; Nobel Prize, 1972.

Steinberger, Jack (1921–), German-born American physicist. Collaborated in an experiment that demonstrated the existence of two types of neutrino; Nobel Prize, 1988.

Steinmetz, Charles Proteus (1865–1923), German-born American electrical engineer. Developed complex number technique for analyzing alternating-current circuits; made numerous electrical inventions; applied mathematical methods to solution of electrical engineering problems.

Stern, Otto (1888–1969), German-born American physicist. Developed the molecular beam method and used it to prove directly the existence of the magnetic moment of atoms and nuclei and to measure their magnitudes; Nobel Prize, 1943.

Stieltjes, Thomas Jan (1856–1894), Dutch-born French mathematician. Developed analytic theory of continued fractions, and Stieltjes integral as a tool for their study.

Stirling, James (1692–1770), British mathematician. Discovered Stirling's formula and Stirling's interpolation formula.

Stokes, George Gabriel (1819–1903), British mathematician and physicist. Originated the idea of determining the chemical composition of the Sun and stars from their spectra; studied double refraction and electromagnetic waves.

Stone, Marshall Harvey (1903–1989), American mathematician. Studied structural aspects of mathematical situations having origins in classic problems of analysis, geometry, and logic.

Störmer, Carl Fredrik Mülertz (1874–1957), Norwegian mathematician and geophysicist. Studied atmospheric phenomena; discovered the Störmer cone concerning cosmic rays.

Störmer, Horst Ludwig (1949–), German-American physicist. With D. C. Tsui, discovered the fractional quantum Hall effect, a manifestation of a new form of quantum fluid with fractionally charged excitations; Nobel Prize, 1998.

Strassman, Fritz (1902–1980), German chemist. With O. Hahn and L. Meitner, discovered nuclear fission; research on uranium and thorium isotopes.

Strömgren, Bengt Georg Daniel (1908–1987), Swedish-born American astronomer. Developed theory of nebulae consisting of hydrogen ionized by hot stars.

Struve, Friedrich Georg Wilhelm von (1793–1864), German-born Russian astronomer. Authority on double stars and nebulae; one of the first to measure a stellar parallax.

Struve, Otto Wilhelm von (1819–1905), Russian astronomer. Discovered some 500 new double stars; calculated the constant of precession.

Sturgeon, William (1783–1850), English electrician and inventor. Constructed the first useful electromagnet and the first moving-coil galvanometer.

Sturm, Jacques Charles François (1803–1855), French mathematician. Formulated the Sturm theorems, concerning real roots of an equation.

Suhl, Harry (1922–), German-born American physicist. Discovered Suhl effect; invented Suhl amplifier; studied resonance in magnetic materials, superconductivity, and general theory of magnetism.

Sumner, James Batcheller (1887–1955), American biochemist. First to isolate an enzyme in pure, crystalline form and characterize it as a protein; Nobel Prize, 1946.

Sutherland, Earl Wilbur, Jr. (1915–1974), American physiologist. Uncovered intermediary role of cyclic adenylic acid in the mechanism of hormone control over human metabolic activities; Nobel Prize, 1971.

Svedberg, Theodor (1884–1971), Swedish chemist. An authority on colloid chemistry (dispersed phase); developed a centrifuge for colloidal particles and protein molecules; Nobel Prize, 1926.

Sydenham, Thomas (1624–1689), English physician. Gave clear descriptions of gout, venereal disease, fevers, hysteria, and Sydenham's chorea.

Synge, Richard Laurence Millington (1914–1994), English chemist. Developed partition chromatography with A. J. P. Martin; Nobel Prize, 1952.

Szent-Györgyi, Albert von Nagyrapolt (1893–1986), Hungarian biochemist. Isolated vitamin C; research on combustion processes in plant and animal tissues, muscular contraction, and cell division; Nobel Prize, 1937.

Talbot, William Henry Fox (1800–1877), English inventor and mathematician. Invented the calotype photographic process.

Tamm, Igor Yevgenevich (1895–1971), Soviet physicist. With I. M. Frank, formulated the mathematical theory explaining the physical origin and properties of Cerenkov radiation; Nobel Prize, 1958.

Tatum, Edward Lawrie (1909–1975), American biochemist and geneticist. Researched the relation of genes to biochemical reactions in bacterial, yeast, and mold cells; with G. W. Beadle, discovered the phenomenon of genetic recombination in bacteria; Nobel Prize, 1958.

Taube, Henry (1915–), American chemist. Elucidated the mechanisms of electron transfer reactions, especially in metal complexes; Nobel Prize, 1983.

Taylor, Brook (1685–1731), English mathematician. Formulated Taylor's theorem and worked on mathematics of physical problems.

Taylor, Geoffrey Ingram (1886–1975), British mathematician. Work in theoretical hydrodynamics, particularly turbulence and effect of rotation on fluid flow.

Taylor, Joseph Hooton, Jr. (1941–), American astronomer and physicist. With R. A. Hulse, discovered a binary pulsar and observed this object to obtain indirect evidence for the existence of gravitational waves, an important confirmation of the general theory of relativity; Nobel Prize, 1993.

Taylor, Richard Edward (1929–), American physicist. Collaborated in experiments that demonstrated that protons, neutrons, and similar particles are made up of quarks; Nobel Prize, 1990.

Taylor, Richard Lawrence (1962–), British mathematician. Assisted A. Wiles in the proof of Fermat's last theorem; collaborated with C. Breuil, B. Conrad, and F. Diamond in extending this work by proving the Taniyama-Shimura conjecture on elliptic curves.

Teisserenc de Bort, Léon Philippe (1855–1913), French meteorologist. Discovered the stratosphere.

Teller, Edward (1908–), Hungarian-born American physicist. With associates, developed the concept which led to the construction of the first hydrogen bomb; with G. Gamow, proposed the Gamow-Teller interaction and Gamow-Teller selection rules.

Temin, Howard Martin (1934–1994), American virologist. Proposed that genetic information is transferred from ribonucleic acid tumor viruses to deoxyribonucleic acid; independently of D. Baltimore, discovered reverse transcriptase; Nobel Prize, 1975.

Tesla, Nikola (1856–1943), American inventor born in Yugoslavia. Invented a high-frequency electric coil; improved design of dynamos, transformers, and electric bulbs.

Thales (ca. 640–ca. 546 B.C.), Greek mathematician and astronomer. First to scientifically predict an eclipse of the Sun; discovered static electricity; credited with formulating several theorems.

Theiler, Max (1899–1972), South African physician and virologist. Developed a vaccine to prevent human yellow fever; Nobel Prize, 1951.

Theorell, Axel Hugo Teodor (1903–1982), Swedish biochemist. Made discoveries concerning the nature and mode of action of oxidative enzymes; Nobel Prize, 1955.

Thiele, F. K. Johannes (1865–1918), German chemist. Research on nitrogen compounds and the theory of unsaturated organic molecules.

Thom, René (1923–), French mathematician. Invented and developed the theory of cobordism in algebraic topology, a classification of manifolds that used homotopy theory in a fundamental way; developed catastrophe theory; Fields Medal, 1958.

Thomas, Edward Donnall (1920–), American physician. Performed the first successful transfer of bone marrow from one individual to another; with J. Murray, helped define and then overcome the immunological mechanisms behind organ rejection; Nobel Prize, 1990.

Thomas, Llewellyn Hilleth (1903–1992), English-born American physicist. Discovered Thomas precession; with E. Fermi, developed Thomas-Fermi atomic model; developed basic theory for Thomas cyclotron.

Thompson, John Griggs (1932–), American-British mathematician. Worked on theory of finite groups; with W. Feit, proved that all noncyclic finite simple groups have even order; determined the finite simple groups whose proper subgroups are solvable (minimal simple groups); Fields Medal, 1970.

Thomson, George Paget (1892–1975), English physicist. Discovered, independently of C. J. Davisson, the diffraction of electrons by crystals; Nobel Prize, 1937.

Thomson, Joseph John (1856–1940), English physicist. Discovered that cathode rays consist of negatively charged particles, or electrons; Nobel Prize, 1906.

't Hooft, Gerardus (1946–), Dutch physicist. With M. J. G. Veltman, elucidated the quantum structure of these interactions and similar particle physics theories on a firmer mathematical foundation by showing how they may be used for precise calculations of physical quantities; Nobel Prize, 1999.

Thouless, David James (1934–), British physicist. Studied many-body problem and its applications to nuclear and condensed matter physics, including phase transitions in superfluid helium films and electrons in disordered systems.

Thue, Axel (1863–1922), Norwegian mathematician. Worked on number theory, especially algebraic numbers and Diophantine equations.

Thurston, William Paul (1946–), American mathematician. Advanced the study of topology in two and three dimensions, showing relationships between analysis, topology, and geometry; suggested that a very large class of closed three-manifolds carry a hyperbolic structure; Fields Medal, 1982.

Tinbergen, Nikolaas (1907–1988), Dutch-born British zoologist. Pioneered in study of social behavior of animals and their responses to complex stimuli; conducted experimental studies of the effects of selection pressures and evolutionary response to them; Nobel Prize, 1973.

Ting, Samuel Chao Chung (1936–), American physicist. Independently of B. Richter, discovered a new heavy elementary particle, which he named the J particle; Nobel Prize, 1976.

Tiselius, Arne Wilhelm Kaurin (1902–1971), Swedish biochemist. Research on electrophoresis and absorption analysis; made discoveries concerning the complex nature of serum proteins; Nobel Prize, 1948.

Todd of Trumpington, Alexander Robertus Todd, Baron (1907–1997), British chemist. Worked on the structure and synthesis of nucleotides, and nucleotide coenzymes, and the related problem of phosphorylation; Nobel Prize, 1957.

Tomlinson, Ray (1941–), American computer engineer. Invented electronic mail, including the use of @ in the address.

Tomonaga, Sin-Itiro (1906–1979), Japanese physicist. Showed the modern theory of quantum electrodynamics to be quantitatively consistent with observed physical phenomena; Nobel Prize, 1965.

Tonegawa, Susumu (1939–), Japanese immunologist. Discovered how a limited number of genes are capable of producing a vast number of diverse antibodies, each designed for a specific invading foreign substance; Nobel Prize, 1987.

Torricelli, Evangelista (1608–1647), Italian physicist. Invented the mercury barometer.

Townes, Charles Hard (1915–), American physicist. Invented the maser; Nobel Prize, 1964.

Townsend, John Sealy Edward (1868–1957), British physicist. Developed collision theory of ionization of gases in an electric field.

Travers, Morris William (1872–1961), English chemist. Discovered, with W. Ramsay, krypton, xenon, and neon; investigated low-temperature phenomena.

Ts'ai Lun (fl. 105), Chinese inventor. Invented paper.

Tsui, Daniel Chee (1939–), Chinese-born American physicist. With H. L. Störmer, discovered the fractional quantum Hall effect, a manifestation of a new form of quantum fluid with fractionally charged excitations; Nobel Prize, 1998.

Turing, Alan Mathison (1912–1954), English mathematician. Developed concept of Turing machine; research on mathematical logic, group theory, and computer technology.

Tychonoff (Tikhonov), Andrei Nikolaevich (1906–1993), Soviet mathematician and geophysicist. Proved Tychonoff theorem in topology and introduced concept of Tychonoff space.

Tyndall, John (1820–1893), British physicist. Studied temperature waves in metals and diathermancy of gases; discovered the effect of atmospheric density on sound transmission.

Uhlenbeck, George Eugene (1900–1988), Javanese-born American physicist. With S. Goudsmit, developed hypothesis of electron spin.

Urey, Harold Clayton (1893–1981), American chemist. Isolated heavy water and thus discovered the heavy isotope of hydrogen; Nobel Prize, 1934.

Urysohn, Pavel Samuilovich (1898–1924), Soviet mathematician. Proved Urysohn's lemma in topology.

Van Allen, James Alfred (1914–), American physicist. Discovered that the Earth is circled by two high-energy radiation belts, leading to major revisions in concepts of the Earth's atmosphere and magnetic field.

Van de Graaff, Robert Jemison (1901–1967), American physicist. Contributed to the development of the direct particle accelerator and invented the electrostatic belt generator.

van der Meer, Simon (1925–), Dutch physicist. Devised method to ensure frequent and efficient collision of accelerated protons and antiprotons in the superproton synchrotron at CERN, contributing to discovery of intermediate vector bosons; Nobel Prize, 1984.

Vandermonde, Alexandre Théophile (1735–1796), French mathematician. Gave first logical exposition of theory of determinants; developed methods to test solvability of algebraic equations.

van der Waals, Johannes Diderik (1837–1923), Dutch physicist. Formulated van der Waals equation; investigated van der Waals forces, concerning intermolecular attraction; Nobel Prize, 1910.

Vane, John Robert (1927–), English pharmacologist. Discovered prostaglandin X (prostacyclin) and the role of aspirinlike drugs as blocking agents in the prostaglandin synthesis; Nobel Prize, 1982.

van Rhijn, Pieter Johannes (1886–1960), Dutch astrophysicist. With J. C. Kapetyn, evolved a theory of the universe.

van't Hoff, Jacobus Hendricus (1852–1911), Dutch chemist. Pioneered in the study of stereochemistry; studied reaction rates, thermodynamics applied to chemistry, and the theory of dilute solutions; Nobel Prize, 1901.

van Vleck, Jan Hasbrouck (1899–1980), American mathematical physicist. Pioneer in the development of the modern quantum-mechanical theory of magnetism; Nobel Prize, 1977.

Varmus, Harold (1939–), American researcher in molecular virology and oncogenesis. With J. M. Bishop, researched the genetic basis of human cancers; their work led to the identification of over 50 cellular genes that can become oncogenes; Nobel Prize, 1989.

Vauquelin, Louis Nicolas (1763–1829), French chemist. Discovered chromium and its compounds and beryllium compounds.

Vega, George, Baron von (1756–1802), Austrian mathematician. Prepared logarithmic tables.

Veltman, Martinus J. G. (1931–), Dutch physicist. With G. 't Hooft, elucidated the quantum structure of the electroweak interactions, placing the theory of these interactions and similar particle physics theories on a firmer mathematical foundation by showing how they may be used for precise calculations of physical quantities; Nobel Prize, 1999.

Verdet, Marcel Émile (1824–1866), French physicist. Determined dependence of Faraday effect on magnetic field strength, wavelength of the light, and index of refraction of the material.

Vernier, Pierre (1580–1637), French technician. Invented the Vernier scale.

Vesalius, Andreas (1514–1564), Belgian anatomist in Italy. Known as the father of modern anatomy; corrected many of Galen's mistaken doctrines.

Viète, François, or Franciscus Vieta (1540–1603), French mathematician. Worked on the solution of equations up to the fourth degree and laid the foundation of modern algebra.

Virtanen, Artturi Ilmari (1895–1973), Finnish biochemist. Research on problems of human nutrition and agriculture; investigated acidity (pH) and biological nitrogen fixation; Nobel Prize, 1945.

Vogel, Hermann Wilhelm (1834–1898), German photochemist. Invented the orthochromatic photographic plate and designed a photometer.

Voigt, Woldemar (1850–1919), German physicist. Introduced transformation equations (later known as Lorentz transformations).

Volta, Alessandro, Count (1745–1827), Italian physicist. Invented the voltaic pile; developed the theory of current electricity.

Volterra, Vito (1860–1940), Italian mathematician. Developed method of solving Volterra equations; pioneered in developing functional analysis.

von Braun, Wernher (1912–1977), German-born American rocket engineer. Directed development of the German V-2 and Wasserfall missiles; instrumental in launch of *Explorer 1*, first American artificial satellite; supervised development of Saturn rockets for the Apollo program.

von Euler, Ulf Svante (1905–1983), Swedish physiologist. Identified norepinephrine as neurotransmitter of sympathetic nervous system; isolated and characterized norepinephrine storage granules in nerves; Nobel Prize, 1970.

von Kármán, Theodore (1881–1963), American aerodynamicist. Theoretical contributions to aerodynamics; formulated von Kármán's theory of vortex streets, an early step in the mathematical treatment of turbulent motion.

von Neumann, John (1903–1957), Hungarian-born American mathematician. Research in logic, theory of quantum mechanics, theory of high-speed computing machines, and mathematical theory of games and strategy.

Wagner von Jauregg (Wagner-Jauregg), Julius (1857–1940), Austrian psychiatrist. Developed use of malarial infection to treat general paresis; Nobel Prize, 1927.

Waksman, Selman Abraham (1888–1973), Russian-born American bacteriologist. Isolated the antibiotic streptomycin; Nobel Prize, 1952.

Wald, George (1906–1997), American biologist and biochemist. Discovered the role of vitamin A in vision; Nobel Prize, 1967.

Walker, John E. (1941–), British molecular biologist. Contributed to the understanding of how the enzyme ATP synthase catalyses the formation of adenosine triphosphate; Nobel Prize, 1997.

Wallace, Alfred Russel (1823–1913), English naturalist. Originated, independently of C. Darwin, theory of natural selection; postulated Wallace's line regarding geographical distribution of animals.

Wallach, Otto (1847–1931), German chemist. Research on essential oils and the terpenes; Nobel Prize, 1910.

Wallis, John (1616–1703), English mathematician. Worked on algebraic curves, interpolation, evaluation of integrals, infinite series, mechanics, and algebra; derived Wallis product and Wallis formulas.

Walton, Ernest Thomas Sinton (1903–1995), British physicist. With J. D. Cockcroft, devised high-voltage apparatus capable of producing fast atomic particles with energies up to 700,000 electronvolts; showed the capability of these particles to disintegrate many light elements; Nobel Prize, 1951.

Wannier, Gregory Hugh (1911–1983), Swiss-born American physicist. Developed harmonization of localized and nonlocalized descriptions of electrons in solids.

Warburg, Otto Heinrich (1883–1970), German physiologist. Worked on chemistry of respiration and on cancer; Nobel Prize, 1931.

Wassermann, August von (1866–1925), German physician. Discovered the Wassermann test for the detection of syphilis.

Watson, James Dewey (1928–), American biochemist. With F. H. C. Crick, determined the double-helix structure of deoxyribonucleic acid; Nobel Prize, 1962.

Watson, John Broadus (1878–1958), American psychologist. Founded the behaviorist school of psychology.

Watt, James (1736–1819), Scottish inventor. Improved the steam engine, making it a commercial success.

Weber, Ernst Heinrich (1795–1878), German anatomist and physiologist. Discovered Weberian apparatus; applied hydrodynamics to study of blood circulation; discovered inhibitory power of vagus nerve; proposed Weber's law of stimuli.

Weber, Heinrich (1842–1913), German mathematician. Introduced Weber differential equation; demonstrated Abel theorem in its most general form; proved that absolute Abelian fields are cyclotomic.

Weber, Wilhelm Eduard (1804–1891), German physicist. Devised instruments for measurement of electrical and magnetic quantities; formulated absolute electrical and magnetic units.

Wegener, Alfred Lothar (1880–1930), German geologist. Presented the idea of continental drift.

Weierstrass, Karl Theodor (1815–1897), German mathematician. Worked on the theory of functions and on the calculus of variations.

Weil, Adolf (1848–1916), German physician. Gave classic description of Weil's disease.

Weil, André (1906–1998), French-born American mathematician. Laid the foundations for abstract algebraic geometry and the modern theory of algebraic varieties, starting a rapid advance in both algebraic geometry and number theory.

Weinberg, Steven (1933–), American physicist. Independently of A. Salam, developed theory uniting two of the basic forces of nature, electromagnetism and the weak nuclear interactions; Nobel Prize, 1979.

Weismann, August (1834–1914), German biologist. Contributed to the theory of heredity, which he attributed to variations in "germ-plasm."

Weiss, Pierre (1865–1940), French physicist. Developed phenomenological theory of ferromagnetism.

Weizsäcker, Carl Friedrich von (1912–), German physicist. Helped develop method for calculating bremsstrahlung in high-energy collisions; developed a theory of origin of solar system.

Weller, Thomas Huckle (1915–), American virologist and parasitologist. Isolated the virus of chickenpox and herpes zoster and proved the common etiology of the two diseases; first to propagate German measles virus; Nobel Prize, 1954.

Wentzel, Gregor (1898–1978), German-born American physicist. Helped develop Wentzel-Kramers-Brillouin method; research on theory of atomic spectra, wave mechanics, quantum electrodynamics, meson field theories, and statistical mechanics of many-body problems, especially superconductivity.

Werner, Alfred (1866–1919), Swiss chemist. Formulated the coordination theory of valency; Nobel Prize, 1913.

Weyl, Hermann (1885–1955), German-born American mathematician and mathematical physicist. Basic research on group representations and Riemann surfaces.

Wheatstone, Charles (1802–1875), English physicist and inventor. Conducted experiments on sound; invented Wheatstone's bridge, an instrument for comparing electrical resistances.

Wheeler, John Archibald (1911–), American physicist. Introduced the concepts of the scattering matrix and resonating group structure into nuclear physics; with N. Bohr, elucidated the mechanism of nuclear fission and predicted the fissibility of plutonium.

Whewell, William (1794–1866), British astronomer. Work on the tides, and on the history and philosophy of science.

Whipple, George Hoyt (1878–1976), American pathologist. Studied anemia and liver treatment; Nobel Prize, 1934.

Whitehead, Alfred North (1861–1947), English mathematician, physicist, and philosopher. With B. Russell, pioneered in mathematical logic and foundations of mathematics.

Whittaker, Edmund Taylor (1873–1956), British mathematician and physicist. Studied special functions of mathematical physics and equations satisfied by them, particularly Whittaker's differential equation; found general integral representation for harmonic functions; made major contributions to analytical dynamics.

Wiedemann, Gustave Heinrich (1826–1899), German physicist and physical chemist. With R. Franz, discovered Wiedemann-Franz law of thermal conductivity of metals; discovered Wiedemann effect.

Wieland, Heinrich (1877–1957), German chemist. Studied bile acids, chlorophyll, and hemoglobin; Nobel Prize, 1927.

Wieman, Carl Edwin (1951–), American physicist. With E. A. Cornell, succeeded in producing Bose-Einstein condensates in a dilute gas of alkali (rubidium) atoms, and carried out fundamental studies of their properties, including studies of collective excitations and vortex formation in condensates; Nobel Prize, 2001.

Wien, Wilhelm (1864–1928), German physicist. Formulated the two Wien laws pertaining to radiation from blackbodies; Nobel Prize, 1911.

Wiener, Norbert (1894–1964), American mathematician. Formulated a mathematical theory of Brownian motion; founded science of cybernetics.

Wieschaus, Eric F. (1947–), American developmental biologist. Using *Drosophila*, he and the German developmental biologist Christiane Nüsslein-Volhard identified and classified a small number of genes that are important in determining the body plan and the formation of body segments; their work, along with that of American development biologist Edward B. Lewis, led to the discovery of important genetic mechanisms which control early embryonic development; Nobel Prize, 1995.

Wiesel, Torsten Nils (1924–), Swedish physiologist. Contributed to the study of the processing of visual information in the brain; Nobel Prize, 1981.

Wigner, Eugene Paul (1902–1995), Hungarian-born American mathematical physicist. With G. Breit, worked out the Breit-Wigner formula for resonant nuclear reactions; proposed the Wigner theorem of conservation of the angular momentum of electron spin; Nobel Prize, 1963.

Wiles, Andrew John (1953–), British mathematician. Proved Fermat's last theorem with assistance of R. Taylor.

Wilkins, Maurice Hugh Frederick (1916–), English biophysicist born in New Zealand. Made x-ray diffraction studies that contributed to the structural determination of deoxyribonucleic acid; Nobel Prize, 1962.

Wilkinson, Geoffrey (1921–1996), British chemist. Research to determine how metals and organic molecules combine to form unique molecules which have sandwichlike structures; Nobel Prize, 1973.

Williamson, William Crawford (1816–1895), English naturalist. Laid the foundation for paleobotany and showed the importance of plant life forms in coal.

Willstätter, Richard (1872–1942), German chemist. Worked on plant pigments; investigated alkaloids and their derivatives; Nobel Prize, 1915.

Wilson, Charles Thomson Rees (1869–1959), British physicist. Worked on ionization; originated the cloud chamber method of studying ionized particles; Nobel Prize, 1927.

Wilson, Kenneth Geddes (1936–), American physicist. Used renormalization group theory to analyze critical phenomena in the behavior of matter at phase transitions; Nobel Prize, 1982.

Wilson, Robert Woodrow (1936–), American astrophysicist. With A. A. Penzias, discovered cosmic background radiation, confirming the big bang theory of the origin of the universe; Nobel Prize, 1978.

Windaus, Adolf (1876–1959), German chemist. Worked on sterols; discovered that ultraviolet light activates ergosterol and gives vitamin D_2; Nobel Prize, 1928.

Witten, Edward (1951–), American mathematical physicist. Applied advanced mathematical tools to theoretical physics, particularly quantum field theory, supersymmetry, and string theory; his physical insights were the basis for major developments in mathematics; Fields Medal, 1990.

Wittig, Georg (1897–1987), German chemist. Work on the linking of carbon and phosphorus (Wittig reaction) made it possible to synthesize new types of compounds, including metal-organic complex compounds; Nobel Prize, 1979.

Wöhler, Friedrich (1800–1882), German chemist. First to synthesize an organic compound, urea.

Wolf, Maximilian Franz Joseph Cornelius (1863–1932), German astronomer. Invented the photographic method of discovering asteroids.

Wollaston, William Hyde (1766–1828), English chemist and physicist. Discovered the lines in the solar spectrum; discovered palladium and rhodium; invented the Wollaston lens.

Woodward, Robert Burns (1917–1979), American chemist. Contributed to the development of total synthesis of complex natural products, and structural determination of several complex natural molecules, later confirmed by total synthesis; Nobel Prize, 1965.

Wright, Almroth Edward (1861–1947), British

physician and pathologist. Studied parasitic disease; introduced inoculation against typhoid.

Wright, Wilbur (1867–1912) **and Orville** (1871–1948), American pioneers in aviation. Built the first successful airplane, which each flew at Kitty Hawk, North Carolina, on Dec. 17, 1903.

Wundt, Wilhelm Max (1832–1920), German physiologist and psychologist. Founded the first laboratory for experimental psychology.

Wurtz, Charles Adolphe (1817–1884), French chemist. Discovered methyl and ethyl amines; evolved the Wurtz reaction for synthesis of hydrocarbons.

Yalow, Rosalyn Sussman (1921–), American medical physicist. Developed a radioimmunoassay technique to detect and measure minute levels of substances such as hormones in the body; Nobel Prize, 1977.

Yang, Chen Ning (1922–), Chinese-born American physicist. With T. Lee, disproved the law of conservation of parity for weak interactions; Nobel Prize, 1957.

Yau, Shing-Tung (1949–), Chinese-born American mathematician. Worked in differential geometry and partial differential equations; solved the Calabi conjecture in algebraic geometry and the positive mass conjecture of general relativity theory; Fields Medal, 1982.

Yersin, Alexandre Émile Jean (1863–1943), Swiss bacteriologist. Discovered the bubonic plague bacillus in Hong Kong, working independently of S. Kitasato, and developed a serum for it.

Yoccoz, Jean-Christophe (1957–), French mathematician. Worked on the theory of dynamical systems; Fields Medal, 1994.

Young, Thomas (1773–1829), English physicist and physician. Discovered the effect of the ciliary muscle on the shape of the eye lens (the mechanism of accommodation).

Yukawa, Hideki (1907–1981), Japanese physicist. Postulated the existence of a new fundamental particle, the meson; Nobel Prize, 1949.

Zariski, Oscar (1899–1986), Russian-born American mathematician. Worked in algebraic geometry, in particular, on local uniformization and reduction of singularities of algebraic varieties.

Zeeman, Pieter (1865–1943), Dutch physicist. Discovered the Zeeman effect in magnetooptics; Nobel Prize, 1902.

Zelmanov, Efim Isaakovich (1955–), Russian mathematician. Contributed to the theory of Jordan algebras and the theory of Lie algebras; solved the restricted Burnside problem, one of the fundamental questions in group theory; Fields Medal, 1994.

Zener, Clarence Melvin (1905–1993), American physicist. Proposed mechanism of Zener breakdown.

Zeno of Elea (ca. 490–425 B.C.), Greek philosophe and mathematician. Formulated a group of paradoxes important for their stimulation of philosophical and mathematical thought, which appear to deny the possibility of motion.

Zernike, Fritz (1888–1966), Dutch physicist. Developed the phase-contrast microscope, making possible the first microscopic examination of the interna structure of living cells; Nobel Prize, 1953.

Zewail, Ahmed H. (1946–), Egyptian-born American chemist. Studied the transition states of chemica reactions using femtosecond spectroscopy; Nobel Prize, 1999.

Ziegler, Karl (1898–1973), German organic chemist Developed a low-pressure process for production o polyethylene; Nobel Prize, 1963.

Zinkernagel, Rolf M. (1944–), Swiss immunologist. Along with Australian immunologist Peter C Doherty, discovered how the immune system, particularly T lymphocytes, recognizes virus-infected cells Nobel Prize, 1996.

Zinsser, Hans (1878–1940), American bacteriologist. Developed methods of immunization against typhus.

Zsigmondy, Richard (1865–1929), German chemist. Studied colloidal solutions; introduced the ultramicroscope; Nobel Prize, 1925.

Zworykin, Vladimir Kosma (1889–1982), Russian-born American physicist. Pioneer in the development of television and the electron microscope.